W9-AFB-203

ENGINEERED MATERIALS HANDBOOK®

Volume 4

CERAMICS AND GLASSES

Prepared under the direction of the
ASM International Handbook Committee

Samuel J. Schneider, Jr. Volume Chairman

Joseph R. Davis, Manager of Handbook Development
Grace M. Davidson, Production Project Manager
Steven R. Lampman, Technical Editor
Mara S. Woods, Technical Editor
Theodore B. Zorc, Technical Editor

Robert C. Uhl, Director of Reference Publications

Editorial Assistance
Heather F. Lampman
Nikki D. Wheaton

**The Materials
Information Society**

First printing, December 1991

Engineered Materials Handbook is a collective effort involving hundreds of technical specialists. It brings together in one book a wealth of information from world-wide sources to help scientists, engineers, and technicians solve current and long-range problems.

Great care is taken in the compilation and production of this Volume, but it should be made clear that no warranties, express or implied, are given in connection with the accuracy or completeness of this publication, and no responsibility can be taken for any claims that may arise.

Nothing contained in the Engineered Materials Handbook shall be construed as a grant of any right of manufacture, sale, use, or reproduction, in connection with any method, process, apparatus, product, composition, or system, whether or not covered by letters patent, copyright, or trademark, and nothing contained in the Engineered Materials Handbook shall be construed as a defense against any alleged infringement of letters patent, copyright, or trademark, or as a defense against liability for such infringement.

Comments, criticisms, and suggestions are invited, and should be forwarded to ASM International.

Library of Congress Cataloging-in Publication Data

ASM International

Engineered materials handbook.
Vol 4-: Samuel J. Schneider, technical chairman.

Includes bibliographies and indexes.
Contents: v. 1. Composites—v. 2. Engineering plastics—
v. 3. Adhesives and sealants—v. 4. Ceramics and glasses

1. Materials—Handbooks, manuals, etc.
I. ASM International. Handbook Committee.
TA403.E497   1987   620.1'1   87-19265
ISBN 0-87170-282-7
SAN 204-7586

Printed in the United States of America

Foreword

The publication of Volume 4, *Ceramics and Glasses* of the *Engineered Materials Handbook* serves as yet another milestone in the history of ASM International. For it is with this Volume that ASM fulfills a major commitment, first made in 1984, to extend its technical scope to all engineered materials. Long known for its Handbooks devoted to metals and metalworking, the past five years have witnessed the expansion of ASM's Handbook library to include composite materials (1987), engineering plastics (1988), electronic packaging materials (1989), adhesives and sealants (1990), and now ceramics and glasses (1991).

Ceramics and Glasses represents the largest, most complete collection of engineering information that has ever been assembled on this subject. While there are other useful texts on ceramic materials, historically their coverage has been somewhat limited, with most emphasizing forming methods or properties. The contents of Volume 4, however, include major sections on every aspect of this important technology: production and synthesis of ceramic powders; forming and sintering processes; final shaping (machining) techniques; glass processing methods—from melting to strengthening (annealing and tempering) of the final product; testing, characterization, and nondestructive evaluation; failure analysis and fractography; design considerations for advanced ceramics; engineering properties, including crystallographic and thermodynamic characteristics; and applications and experience.

Such remarkable coverage would not have been realized without the planning and coordination of Volume Chairman Samuel J. Schneider, Jr., who was the driving force behind the project. On behalf of ASM International, we are pleased to extend our thanks and congratulations to Dr. Schneider and the 20 Section Chairmen and Co-Chairmen for their outstanding work in recruiting an international author list that includes many of the most recognized and respected authorities in the ceramics and glasses field. We also thank the members of the Handbook Committee for their guidance and assistance and the ASM Handbook editorial staff for their tireless efforts. And, of course, we are especially grateful to the more than 400 authors and reviewers who so generously donated their time and efforts to make this Handbook an outstanding source of information.

William P. Koster
President
ASM International

Edward L. Langer
Managing Director
ASM International

Policy on Units of Measure

By a resolution of its Board of Trustees, ASM International has adopted the practice of publishing data in both metric and customary U.S. units of measure. In preparing this Handbook, the editors have attempted to present data primarily in metric units based on Système International d'Unités (SI), with secondary mention of the corresponding values in customary U.S. units. The decision to use SI as the primary system of units was based on the aforementioned resolution of the Board of Trustees, the widespread use of metric units throughout the world, and the expectation that the use of metric units in the United States will increase substantially during the anticipated lifetime of this Handbook.

For the most part, numerical engineering data in the text and in tables are presented in SI-based units with the customary U.S. equivalents in parentheses (text) or adjoining columns (tables). For example, pressure, stress, and strength are shown both in SI units, which are pascals (Pa) with a suitable prefix, and in customary U.S. units, which are pounds per square inch (psi). To save space, large values of psi have been converted to kips per square inch (ksi), where 1 ksi = 1000 psi. Some strictly scientific data are presented in SI units only.

To clarify some illustrations, only SI units are presented on artwork. References in the accompanying text to data in the illustrations are presented in both SI-based and customary U.S. units. On graphs and charts, grids correspond to SI-based units, which appear along the left and bottom axes; where appropriate, corresponding customary U.S. units appear along the top and right axes.

Both nanometers (nm) and angstrom units (Å) are used (1 nm = 10 Å), the former to measure light wavelengths, and the latter as a unit of measure in x-ray crystallography. Data obtained according to standardized test methods for which the standard recommended a particular system of units are presented in the units of that system. Wherever feasible, equivalent units are also presented.

Conversions and rounding have been done in accordance with ASTM Standard E 380, with careful attention to the number of significant digits in the original data. For example, an annealing temperature of 1570 °F contains three significant digits. In this instance, the equivalent temperature would be given as 855 °C; the exact conversion to 854.44 °C would not be appropriate. For an invariant physical phenomenon that occurs at a precise temperature (such as the melting of pure silver), it would be appropriate to report the temperature as 961.93 °C (or 1763.5 °F). In many instances (especially in tables and data compilations), temperature values in °C and °F are alternatives rather than conversions.

The policy on units of measure in this Handbook contains several exceptions to strict conformance to ASTM E 380; in each instance, the exception has been made to improve the clarity of the Handbook. The most notable exception is the use of $MPa\sqrt{m}$ rather than $MN \cdot m^{-3/2}$ or $MPa \cdot m^{0.5}$ as the SI unit of measure for fracture toughness. Other examples of such exceptions are the use of "L" rather than "l" as the abbreviation for liter and the use of g/cm^3 rather than kg/m^3 as the unit of measure for density (mass per unit volume).

SI practice requires that only one virgule (diagonal) appear in units formed by combination of several basic units. Therefore, all of the units preceding the virgule are in the numerator and all units following the virgule are in the denominator of the expression; no parentheses are required to prevent ambiguity.

Preface

Historians often equate the state of human advancement in terms of the level of sophistication of the principal materials needed to perform everyday functions. On this basis descriptive terms like the Stone Age, the Bronze Age, the Iron Age, and the Steel Age came into being. Each title brings to mind a level of technical accomplishment and ceramic materials were the first designated in this era identification scheme. Ceramics were the premier materials of the Stone Age some 10,000 years ago, and have remained a materials class of choice ever since. It is fair to state that ceramics have a leading role in today's world of materials. Overall they have significantly impacted most technological and social sectors of the world's economy—health, agriculture, energy, environment, transportation, information/communication, resource extraction/production, civil construction, and defense.

Simply put, ceramics have a special property character that requires their use in almost every production line, whether it be steel manufacture, textile making, chemicals, or energy production. In one form or another ceramics can be found in products that touch our lives in many ways. Most persons are familiar with the traditional applications of ceramics, such as pottery and china, cement, refractories, construction brick, window glass, floor tiles, and the like. But there are many more, generally less recognized applications. Bioceramics are used for artificial teeth and bone implants. Optical fibers are the new mainstream for telecommunications. Ceramic magnets made portable radios and television possible. Automobiles could not run without the ceramic insulators used in sparkplugs, nor would auto emission control be possible without ceramic sensors and ceramic-based catalyst supports. Hair dryers depend upon ceramic capacitors. Computers use ceramic substrates as chip carriers. Diamonds (yes they are ceramics) are not only gemstones but also are cutting and grinding medium. Even weaponry depends upon ceramics through ceramic armor, radar windows, and a host of other electronic applications. The list of ceramic applications is long and is growing, as for example the high-temperature superconductors being touted for magnetically levitated trains or for new types of automobile engines operating without water-cooling systems.

This Volume emphasizes the technical side of this important materials class. It is intended for use by all engineering technologists whether they have ceramics training or they are uninitiated in the field. The Volume is a "Handbook" and it is a practical engineering guide to ceramics written by experts from the ceramics and glass producer, user, and research community. Authors include those from industry, government, and academia, and thus the Handbook presents an insight to ceramics from different perspectives.

In general terms the Handbook tells what this materials class is, what it is good for, how products are made and used, and what properties are important in engineering production and application. It treats ceramics and glasses as separate technologies, primarily because of different applications, product forms, and processing methods. In the main the demarkation between the two generally separates along the lines of crystalline (ceramics) and non-crystalline (glass). For each of these primary technologies the Handbook is further divided into (1) descriptions of traditional, or conventional applications and (2) advanced, or high technology product lines. Again the distinction between the traditional and advanced product lines is vague at best because the latter category primarily is an evolutionary offshoot of the conventional product type (for example, refractory products used in steelmaking are the forerunners to structural ceramics used in heat engines). Alternate terms sometimes used for traditional product lines and advanced products are commodity products (bulk production) and engineered and technical products (specialized production), respectively.

Within this divisional overlay of ceramics and glass and traditional versus advanced, the Handbook is organized to take the reader through the steps necessary to design, produce, and use ceramics and glasses. The first Section is intended to introduce ceramics and glasses as diverse technologies with many applications. The next Sections address processing stages employed in their production—from powders to product. These are followed by Sections on design, testing and property characterization. The final Sections describe specific applications.

I know of no other text of comparable value to the ceramics field. From a personal standpoint the Handbook represents a text I always wanted to write, but could not because such an undertaking would require expertise far beyond that of one individual. Through ASM International and all the contributing authors from 13 nations this personal goal has been achieved. This Handbook is the result of the efforts of many and each is to be thanked for contributing to the production of this long needed and overdue technical guide to the important field of ceramics and glasses. I commend the Handbook to all students, researchers, and makers and users of ceramics and glasses.

Volume Chairman
Samuel J. Schneider, Jr.
National Institute of Standards and Technology

Officers and Trustees of ASM International

Members of the ASM Handbook Committee (1991–1992)

Previous Chairmen of the ASM Handbook Committee

Authors

James H. Adair
University of Florida
P. Bruce Adams
Precision Analytical
James H. Adams
GTE Valenite Corporation
I.A. Aksay
University of Washington
Kamal E. Amin
3M Company
R.M. Anderson
Rutgers University
Bob Anschuetz
GTE Valenite Corporation
Kenneth J. Anusavice
University of Florida
Patrick F. Auborg
Owens-Corning Fiberglas Corporation
Janet E. Bailey
TechniGlas Consulting
Craig C. Baker
Allied-Signal, Inc.
Donald E. Baker
Allied-Signal, Inc.
U. Balachandran
Argonne National Laboratory
Pronob Bardhan
Corning Incorporated
Roger Bartholomew
Corning Incorporated
Isa Bar-On
Worcester Polytechnic Institute
William C. Bauer
Ponderosa Associates
Edwin K. Beauchamp
Sandia National Laboratories
S.J. Bennison
National Institute of Standards and
Technology
Clifton G. Bergeron
University of Illinois at Urbana-
Champaign
Peter L. Bocko
Corning Incorporated
Dawn A. Bonnell
University of Pennsylvania
Nicholas F. Borrelli
Corning Incorporated
Raymond J. Bratton
Westinghouse Science & Technology
Center
Anthony J. Brown
Cookson Technology Centre
Relva C. Buchanan
University of Illinois at Urbana-
Champaign

S.N. Burchett
Sandia National Laboratories
Alejandro G. Bueno
Glasstech, Inc.
J. Bultitude
Cookson Technology Centre
Ian Burn
E.I. Dupont de Nemours & Co., Inc.
Arthur O. Capp
Laserage Technology Corporation
Roger Cassidy
GE Superabrasives
A.G. Clare
Alfred University, New York State
College of Ceramics
Arthur E. Clark
University of Florida
Stacey Clark-Kerwien
U.S. Army Research
James R. Clifton
National Institute of Standards and
Technology
William S. Coblenz
Defense Advanced Research Projects
Agency
John H. Connelly
Corning Incorporated
George A. Costakis
Advanced Engineering Staff, GMC
Darryl J. Costin
Libbey-Owens-Ford Company
C. Crall
Owens-Corning Fiberglas Corporation
David Cranmer
National Institute of Standards and
Technology
Raymond A. Cutler
Ceramatec, Inc.
Paul S. Danielson
Corning Incorporated
Robert F. Davis
North Carolina State University
Emil W. Deeg
AMP Incorporated
Steven E. DeMartino
Corning Incorporated
John Donohoe
Corning Incorporated
S.E. Dorris
Argonne National Laboratory
Alan L. Dragoo
U.S. Department of Energy
Stephen F. Duffy
Cleveland State University
William H. Dumbaugh
Corning Incorporated

Thomas H. Elmer
Corning Incorporated (Retired)
J.E. Enrique
University of Valencia
Richard A. Eppler
Eppler Associates
Andre Ezis
Cercom Inc.
Pierre Fauchais
University de Limoges
Nancy Faulk
Tennessee Tech University
Matthew K. Ferber
Oak Ridge National Laboratory
Gerald J. Fine
Corning Incorporated
Charles W.P. Finn
Northeastern University
Ross F. Firestone
The Ross Firestone Company
Van Derck Frechette
Alfred University, New York State
College of Ceramics
Stephen W. Freiman
National Institute of Standards and
Technology
Geoffrey Frohnsdorff
National Institute of Standards and
Technology
Patrick K. Gallagher
Ohio State University
M.F. Gazulla
University of Valencia
George E. Gazza
U.S. Army Materials Technology
Laboratory
Randall M. German
Pennsylvania State University
Laurence D. Gill
Mobay Corporation
Alex Goldman
Ferrite Technology Worldwide Inc.
K.C. Goretta
Argonne National Laboratory
Victor A. Greenhut
Rutgers University
Tapan K. Gupta
Alcoa Technical Center
John P. Gyekenyesi
NASA Lewis Research Center
Richard A. Haber
Rutgers University
J. Hadley
Owens-Corning Fiberglas Corporation
Gene H. Haertling
Clemson University

John S. Haggerty
Massachusetts Institute of Technology
John W. Halloran
University of Michigan
William F. Hammetter
Sandia National Laboratories
Stuart Hampshire
University of Limerick
Dale L. Hartsock
Ford Motor Company
Larry L. Hench
University of Florida
Charles D. Hendricks
International Microproducts, Inc.
Thomas P. Herbell
NASA Lewis Research Center
Leif Hermansson
Doxa Certex AB
J. Birch Holt
Lawrence Livermore National Laboratory
Robert L. Holtman
Allison Gas Turbine Division, GMC
Carl J. Hudecek
OI-NEG Technical Products, Inc.
Wolf Hüttner
Schunk Kohlenstofftechnik GmbH
John H. Indge
Peter Wolters of America, Inc.
Alexander G. Jack
Philips Lighting B.V.
Curtis A. Johnson
General Electric Company
Ronald E. Johnson
Corning Incorporated
Ashok V. Joshi
Ceramatec, Inc.
Roy Kamo
Adiabatics, Incorporated
Kevin D. Kannmacher
Allison Gas Turbine Division, GMC
R.D. Kaverman
Owens-Corning Fiberglas Corporation
Fred Kennard
General Motors Corporation
Hyoun-Ee Kim
Oak Ridge National Laboratory
Angus I. Kingon
North Carolina State University
Donald L. Kinser
Vanderbilt University
Lisa C. Klein
Rutgers University
R.N. Kopp
Norton Company
Kunihito Koumoto
University of Tokyo
Leonard P. Krietz
Plibrico Company
Bernard M. Kulwicki
Texas Instruments Incorporated
D.C. Kunerth
Idaho National Engineering Laboratory
Charles C.Y. Kuo
CTS Corporation
Oh-Hun Kwon
Norton Company

W.C. Lacourse
Alfred University, New York State
College of Ceramics
M.T. Lanagan
Argonne National Laboratory
Hans T. Larker
ABB Cerama AB
Dean Larson
Coors Ceramics Company
J.M. Lawson
Cookson Technology Centre
Richard L. Lehman
Rutgers University
H. David Leigh
Clemson University
Stanley R. Levine
NASA Lewis Research Center
Gordon Lewis
Clemson University
Leonard C. Lindgren
Allison Gas Turbine Division, GMC
Meilin Liu
Ceramatec, Inc.
Ronald E. Loehman
Sandia National Laboratories
Donald J. Lopata
Corning Asahi Video Products Company
Layman A. Lott
Idaho National Engineering Laboratory
R.A. Lowden
Oak Ridge National Laboratory
Stanley J. Lukasiewicz
Texas Instruments, Inc.
Subhas G. Malghan
National Institute of Standards and
Technology
J.W. Malmendier
Corning Incorporated
Tiziano Manfredini
University of Modena
Ryuichi Matsuda
Kyocera Corporation
Roger L.K. Matsumoto
Hercules Incorporated
M. John Matthewson
Rutgers University
Brian J. McEntire
Norton/TRW Ceramics
Anna E. McHale
Consultant
Arthur F. McLean
Consultant
Ronald A. McMaster
Glasstech, Inc.
John J. Mecholsky, Jr.
University of Florida
Peter F. Messer
University of Sheffield
Terry A. Michalske
Sandia National Laboratories
D.M. Miller
Owens-Corning Fiberglas Corporation
Herbert A. Miska
Corning Incorproated
Richard E. Mistler
Keramos Industries, Inc.

Masaru Miyayama
University of Tokyo
Howard Mizuhara
GTE Products Corporation
Robert E. Moore
University of Missour—Rolla
A.J. Moorhead
Oak Ridge National Laboratory
Mary A. Moreland
Bullen Ultrasonics, Inc.
Robert M. Morena
Corning Incorporated
Heinrich Mörtel
Institut für Werkstoffwissenschaften Glas
und Keramik
Ronald G. Munro
National Institute of Standards and
Technology
Beebhas C. Mutsuddy
Michigan Technological University
Suzanne R. Nagel
AT&T Bell Laboratories
Noel N. Nemeth
NASA Lewis Reserach Center
E. Ochandio
University of Valencia
Kenneth Okajima
Kyocera Industrial Ceramics Corp.
John C. Oliver
Porcelain Enamel Institute, Inc.
P. Darrell Ownby
University of Missour—Rolla
Toshi Oyama
GTE Products Corporation
Gregory Papatrefon
U.S. Army Research
William P. Parks, Jr.
U.S. Department of Energy
Joseph A. Pask
University of California—Berkeley
Gian Carlo Pellacani
University of Modena
Licio Pennisi
Alfred University, New York State
College of Ceramics
Daniel R. Petrak
Dow Corning Corporation
Jack D. Petty
Allison Gas Turbine Division, GMC
Linda R. Pinckney
Corning Incorporated
Roger B. Poeppel
Argonne National Laboratory
W.R. Prindle
Corning Incorporated
Peter R. Prud'homme van Reine
Philips Lighting B.V.
George D. Quinn
National Institute of Standards and
Technology
Bruce H. Raeder
Corning Incorporated
S. Ramanath
Norton Company
Ernest Ratterman
GE Superabrasives

viii

Reviewers and Contributors

Mufit Akinc
Iowa State University
Ames Laboratory

Terrence Allen
DuPont Company

Ray Ambrogi
Corning Incorporated

I. Fred Anderson
Thermal Technology Inc.

Norman Anderson
Ceradyne, Inc.

W.T. Anderson
Bellcore

Roger J. Austin
Hydro-Lift

Gantam Bandyopadhyay
GTE Laboratories, Inc.

Francis I. Baratta
Army Materials Technical Laboratory

H. Robert Baumgartner
Alcoa Technical Center

Richard W. Beiswenger
Monarch Laboratories

Ram Bhatt
NASA Lewis Research Center

D. Bitko
Fifth Dimension, Inc.

Edward Blanchard
Netzsch Inc.

Thomas A. Bobal
Consulting Metallurgical Engineer

Lewis L. Bognar
Corning Incorporated

J. Gregg Borchelt
Brick Institute of America

Hans Borstell
Grumman Aircraft Systems Division

Animesh Bose
Southwest Research Institute

D. Britko
Fifth Dimension, Inc.

R.J. Brook
University of Oxford (England)

Richard Brow
Sandia National Laboratories

Thomas S. Brown
Babcock & Wilcox
Research and Development Division

Stephen J. Burden
GTE Valenite

V.K. Burriss
ICI Advanced Materials

B. Busovne
Garrett Ceramic Components

T.W. Button
ICI Advanced Materials (United Kingdom)

M. Cable
University of Sheffield (United Kingdom)

Jeff Campbell
Wilbanks, A Coors Ceramics Company

Roger Cannon
Rutgers, The State University of New Jersey

Armand J. Cantafio
Northeast Electronics

Ronald V. Caporali
Consultant

Elis Carlström
Swedish Ceramic Institute (Sweden)

Joseph Carpenter
National Institute of Standards and Technology
Ceramics Division

M. Cellarosi
National Institute of Standards and Technology

Sung R. Choi
NASA Lewis Research Center

W.J. Clegg
ICI Advanced Materials (United Kingdom)

John Coleman
Insaco Inc.

John K. Cook
Chi-Vit Corporation

Richard Cooper
Consultant

Michael Crowley
Consultant

Stephen Danforth
Rutgers, The State University of New Jersey

S.J. Dapkunas
National Institute of Standards and Technology
Ceramics Division

R.F. Davis
North Carolina State University
Department of Materials Science and Engineering

T.C. Dean
TAM Ceramics Inc.

P.M. DiBello
PMD International

Russell J. Diefendorf
Clemson University

Dennis R. Dinger
Clemson University

David Dollimore
University of Toledo

Robert H. Doremus
Rensselaer Polytechnic Institute

Paul Dyer
Air Products and Chemicals, Inc.

Charles Earnest
Berry College

Aicha Elshabini-Riad
Virginia Polytechnic Institute and State University

Kevin G. Ewsuk
Sandia National Laboratories

John Fairbanks
U.S. Department of Energy

Shaji Farooq
IBM

Mark Fishler
Vesuvius International Corporation

William Fredrich
Laser Mechanisms, Inc.

D.W. Freitag
LTV Missiles Division

William D. Friedman
BP America

James Funk
Clemson University

Frank Gac
Los Alamos National Laboratory

A. Gadalla
Texas A&M University

Thomas Gallo
Akzo Chemicals Inc.

W.B. Gardner
AT&T Bell Laboratories

Harmon M. Garfinkel
Engelhard Corporation

Ronald C. Garvie
CSIRO (Australia)
Division of Materials Science & Technology

Michelle M. Gauthier
Raytheon Company

Brian Gold
Superior Technical Ceramics

Steve Gonczy
Allied Signal Research & Technology

Charles Greskovich
General Electric Company

Curt Griffin
Lone Peak Engineering

Toni Grobstein
NASA Lewis Research Center
Lance E. Groseclose
General Motors Corporation
Allison Gas Turbine Operations
T. Grossman
GE Aircraft Engines
C. William Hall
Southwest Research Institute
Wolfgang K. Haller
National Institute of Standards and
Technology
Gregory Hannon
Lanxide Corporation
Kenneth Hatton
Lanxide Corporation
Mark Headinger
E.I. Du Pont de Nemours & Company,
Inc.
Jürgen Heinrich
Hoechst CeramTec AG (Germany)
J.R. Hellmann
Pennsylvania State University
H. Herman
State University of New York
Trevor D. Howes
University of Connecticut
Center for Grinding R&D
Dennis D. Huffman
Timken Company
D.L. Isaac
The American Ceramic Society, Inc.
K. Jakus
University of Massachusetts
E. Mark Jahn
Rohr Industries, Inc.
Sei-Joo Jang
Pennsylvania State University
Materials Research Laboratory
David W. Johnson, Jr.
AT&T Bell Laboratories
Jerry D. Jones
Thomson Consumer Electronics
Charles Kao
CTS Corporation
R.N. Katz
Worcester Polytechnic Institute
W.J. Kelley
Kentucky-Tennessee Clay Company
K. Kendall
ICI Advanced Materials (United
Kingdom)
Paul T. Kerwin
NASA Lewis Research Center
B.L. Keyes
Oak Ridge National Laboratory
Kyekyoon Kim
University of Illinois
Thomas J. Kim
Rhode Island University
Wally Kleem
Southwestern Portland Cement Company
Mitsue Koizumi
Ryukoku University (Japan)
Institute for Science and Technology

Raymond D. Krieg
University of Tennessee
Kevin Krist
Gas Research Institute
John B. Lambert
Fansteel Inc.
Fred F. Lange
University of California—Santa Barbara
Peter W. Lee
Timken Company
James C. Leslie
Advanced Composite Products &
Technology
Paul Lessing
EG&G Idaho, Inc.
Gordon Lewis
Clemson University
Richard P. Lindsay
Norton Company
Stephen Liu
Colorado School of Mines
Robert Locker
Corning Incorporated
John Lucek
Cerbec Ceramic Bearing Company
John W. Lusik
Consultant
S. Malkin
University of Massachusetts
John A. Mangels
Ceradyne, Inc.
S. Manning
Coors Ceramics Company
Alex Marker
Schott Glass Technologies Inc.
Charles G. Marvin
The Refractories Institute
J. Brian Maxwell
East Palestine China Company
Gary Messing
Pennsylvania State University
David Miller
Cooper Power Systems
H.N. Mills
Consultant
B. Mishra
Colorado School of Mines
David Moore
Henny Penny Corporation
John Moore
Colorado School of Mines
A.J. Moulson
University of Leeds (United Kingdom)
Ronald T. Myers
Owens-Brockway
Joseph E. Neely
National Refractories & Minerals
David V. Neff
Metaullics System
R.R. Neurgaonkar
Rockwell International
Robert E. Newnham
Pennsylvania State University
Materials Research Laboratory

David LeRoy Olson
Colorado School of Mines
Dean E. Orr
Orr Metallurgical Consulting Services,
Inc.
Paul Parks, Sr.
PIM Industries
Robert E. Parks
Consultant
Arvid Pasto
GTE Laboratories Inc.
J.L. Pentecost
Georgia Institute of Technology
G.W. Phelps
Rutgers, The State University of New
Jersey (retired)
Donald G. Polensky
Consultant
Lynn Powers
NASA Lewis Research Center
James Price
Corning Incorporated
Carr Lane Quackenbush
Norton Company Advanced Ceramics
Tariq Quadir
W.R. Grace & Company
E.M. Rabinovich
AT&T Bell Laboratories
B. Rand
University of Leeds (United Kingdom)
S.T. Reczek
Ferronics Inc.
Leonard Reed
InTA
Roy W. Rice
W.R. Grace & Company
M. Rosenberg
Philadelphia Decal
Division of Philadelphia Ceramics Inc.
C. Philip Ross, Jr.
Kerr Glass
David Rostoker
Norton Company
Donald W. Roy
Alpha Optical Systems, Inc.
R. Roy
Pennsylvania State University
Edward Ruh
Consultant
Michael L. Santella
Oak Ridge National Laboratory
Vinod Sarin
Boston University
Neil Schattin
American Olean Tile Company
Jane M. Selway
Pfaltzgraff Company
Daniel Shanefield
Rutgers, The State University of New
Jersey
M.C. Shaw
Arizona State University (retired)
G.S. Sheffield
The Edward Orton Jr. Ceramic
Foundation

Samuel S. Shinozaki
Ford Motor Company
Scientific Research Laboratories
Douglas Smith
University of New Mexico
Ferris Engineering Center
Donald M. Smyth
Lehigh University
Joseph Stephens
NASA Lewis Research Center
Harold Stetson
Ceramic Science Association
S. Donald Stookey
Corning Incorporated (retired)
Edmund Storms
Los Alamos National Laboratory
Jeremy C. St. Pierre
Hayes Heat Treating Corporation
Jackson S. Stroud
Schott Glass Technologies Inc.
Kevin Stuffle
Advanced Ceramics Research
E. Lowell Swarts
PPG Industries, Inc.
Glass Research and Development Center

Jenifer A.T. Taylor
Alfred University
New York State College of Ceramics
Darrel Teel
Blount Inc.
John A. Tesk
National Institute of Standards and
Technology
Ceramics Division
Naseem Theilgaard
Danish Technological Institute
(Switzerland)
Wei-Lung Tsai
Toshiba America
H.L. Tuller
Massachusetts Institute of Technology
G.L. Vandillen
Ceramic Magnetics Inc.
A.K. Varshneya
Alfred University
New York State College of Ceramics
Thomas Vasilos
University of Lowell
James G. Vaughan
The University of Mississippi

Brian Wallway
Wilbanks, A Coors Ceramics Company
S.H. Wang
Colorado School of Mines
R.J. Wasowski
Ferro Corporation
Karen E. Weber
Detroit Diesel Corporation
Robin L. Weintraub
American Olean Tile Company
T.J. Whalen
Ford Motor Company
Fred S. Wheeler
PTX-Pentronix
E. Dow Whitney
University of Florida
Douglas Winslow
Purdue University

Contents

Introduction to Ceramics and Glasses

Chairman: Stephen W. Freiman, National Institute of Standards and Technology

Introduction

SINCE THIS VOLUME deals with the properties of ceramics and glasses, our first task is to define these classes of materials in terms of their chemistry and structure. While their structure will be described in some detail later, we can take the broadest possible view and define ceramics as any inorganic, non-metallic, material. Notice that this definition encompasses naturally occurring rocks and minerals which are normally excluded from the category of ceramics, but whose properties are frequently similar to those of technical ceramics. For example, naturally occurring or synthetic gemstones, such as diamond and sapphire, are important for a number of optical applications. In general, however, we will consider ceramics to include only those materials which are man-made.

Inorganic glasses should actually be considered as a subset of ceramics. The difference between the two is that ceramics have a periodic crystal structure, whereas glasses possess a short-range order, but no long-range periodic structure. An example of a naturally occurring glass is obsidian. Some glasses can be converted into ceramics through heat treatments, wherein crystals nucleate and grow in the glass. A classic example of this type of material, known as a glass-ceramic, is Pyroceram, which Corning Incorporated developed for use as cookware.

Most ceramics are composed of oxides, carbides, nitrides, or borides, but other materials qualify as well. For instance, semiconducting materials such as silicon and gallium arsenide have many properties which resemble those of more traditional ceramics. Crystal structures in ceramics can be of many types. The atomic bonding between the atoms in the crystals ranges from purely ionic (for example, magnesium fluoride used for infrared transmitting windows) to purely covalent (for example, silicon carbide used for high-temperature structures such as heat exchangers).

Most ceramics which have technical or engineering applications are used as a polycrystalline form of the material. That is, they consist of individual grains, which are actually single crystals, bonded to each other during processing at elevated temperature. The crystal structures within each grain are usually oriented from one grain to another, leading to potential difficulties in terms of anisotropies in elastic or thermal properties. On the positive side, this misorientation can lead to an increased resistance to the propagation of flaws through the material. The boundaries between the grains can themselves be

sinks for impurities, sources of failure, or in some cases a means of achieving unique overall properties of the material (for example, boundary-layer capacitors).

What we think of as traditional ceramics include bricks and other clay products, whitewares such as porcelain, cements, and even aluminum oxide used in spark plugs and electronic substrates. Traditional glasses include normal window glass, optical glasses, and various glasses used for sealing. While these glass and ceramic materials still represent a sizeable fraction of the ceramics industry, the interest in recent years has focused on what are termed advanced ceramics. Advanced ceramics are materials which have been developed for their particular mechanical, electrical, magnetic, or optical properties. Over the past few years ceramic materials have found increasing uses as both structural and functional components. Why have these materials become such an important factor in new applications? There are a number of motivating factors for the use of advanced glasses and ceramics in many new areas. Their excellent chemical/corrosion resistance in a wide range of environments and temperatures, their optical transparency over a wide range of wavelengths from the ultraviolet to the infrared, their high hardness and resistance to wear, and their unique electrical characteristics all make these materials extremely attractive for a wide range of applications.

As a further example, in the category of advanced ceramics, we include:

- Silicon carbide and silicon nitride used for their strength and oxidation resistance at elevated temperatures
- Boron carbide composited with a metal, used for abrasive machining wheels
- Fiber- or whisker-reinforced composites, which exhibit "graceful" rather than catastrophic failure, for use in various structures
- Complex oxides having unique dielectric or piezoelectric properties (for example, lead zirconate titanate and lead magnesium niobate), which are used in modern multilayer capacitors and transducers
- Recently discovered superconducting ceramics which develop zero resistance at temperatures far above any previously known material

Advanced glasses include high-strength silica fibers which are becoming the primary source of electronic data transmission, non-oxide glasses which transmit radiation well into the infrared, and glasses which act as hosts for lasers.

What are the technical barriers limiting the more widespread use of advanced ceramics? One of the most severe obstacles to the use of ceramics and glasses in structural applications (or in any applications in which relatively high stresses could be encountered) is their relatively low resistance to the growth of small flaws. This small intrinsic fracture resistance is an outgrowth of the fact that at ambient temperatures there are no ductile deformation processes such as yielding which can mitigate the large stresses generated at crack tips. This intrinsic brittleness means that failure-producing flaws are usually only a few micrometers in size, below the detection limit for existing non-destructive evaluation procedures. A second related problem is the difficulty in processing these materials in a "flaw-free" condition. Failure-producing defects can result from pores or inclusions in the bulk of the material, or more commonly, from small cracks generated in the surface during machining to shape or finishing operations. Other problems associated with the processing and use of advanced ceramics and glasses is their sensitivity to small quantities of impurities, the rather high temperatures required to process them to full density, the difficulty in producing identical materials, their slow deformation, that is, creep, at elevated temperatures, the relatively high cost of starting materials, and the expense of machining these hard materials to final shape.

This Section will briefly review the various types of ceramics and glasses and their applications, ranging from bricks to superconductors, windows to data transmission lines. This Section will introduce the reader to the characterization tools available for studying these materials and their key properties. An effort will be made to begin a discussion of processing-structure-property relationships, which are key to the development of improved materials. The knowledge required to design structural elements with these "brittle" materials will be discussed, and concepts of design methodology introduced. Finally, the reader will be guided to other available sources of information and data on ceramic materials.

Overview of Traditional Ceramics

Richard A. Haber and Peter A. Smith, Ceramic Casting Technology Program, Rutgers, The State University of New Jersey

CERAMIC is a term once thought to refer only to the art or technique of producing articles of pottery. Current usage has broadened to include items made of metal oxides, borides, carbides, nitrides, or compounds of such materials. The etymology of the term shows that it derives from the Greek *keramos*, meaning "a potter" or "a pottery." However, the Greek word is related to an older Sanskrit root, meaning "to burn"; as used by the Greeks themselves, its primary meaning was simply "burnt stuff" or "burned earth." The fundamental concept contained in the term was that of a product obtained through the action of fire upon earthy materials.

A traditional ceramic, in the context of this article, refers to the products commonly used as building materials or within the home and industry. Although there is a tendency to equate traditional ceramics with low technology, advanced manufacturing technologies are often used in this industry. Stiff competition among producers has caused the technology to become more efficient and cost effective, by utilizing complex tooling and machinery, coupled with computer-assisted process control.

The oldest ceramic products originated from clay-bearing materials. Early potters found the plastic nature of clay to be useful in forming shapes. Because of its tendency to exhibit a large amount of shrinkage, clay bodies were modified by adding coarse sand and stone, which reduced shrinkage and cracking. In modern clay-based bodies, the typical non-clay additions are silica flour and alkali minerals that are added as fluxes. In traditional ceramic formulations, clay acts as a plasticizer and binder for other constituents.

Traditional Fine Ceramics

Figure 1 classifies traditional ceramics as being either fine (having particles under 0.2 mm, or 8 mils) or coarse (having the largest particle be 8 mm, or 0.3 in.). These, in turn, can be categorized into fired bodies that are either porous (water absorption exceeds 5%) or dense (water absorption does not exceed 5%). Described below are the subordinated categories of traditional ceramics.

Pottery is sometimes used as a generic term for all fired ceramic wares that contain clay, except for technical, structural, and refractory products. Specifically, however, pottery describes the low-temperature fired porous ware that is usually colored. The term is properly applied to the clay products of primitive peoples, or to decorative art products made of unrefined clays by using unsophisticated methods.

Whiteware. The term whiteware was originally applied to white tableware and artware, but has been broadened to include ware that is ivory colored or has a light gray appearance in the fired state. Fine ceramic whitewares are conveniently divided into two classes: formulas consisting primarily of clay minerals, feldspars, and quartz; and nontriaxial bodies made entirely or predominantly of other materials.

Whitewares can further be classified by their degree of porosity. At the extreme ends are porous bodies, which have the capacity to absorb large amounts of water, and vitreous bodies, which are absolutely impervious to water. Between these extremes is a wide variety of bodies having various degrees of porosity (semivitreous) and translucency. In most instances, this variation is directly related to the temperature at which the body is vitrified.

Vitrification is the heat treatment to which the body is subjected. During this process some of the raw materials melt and form a glass. This glass brings about the progressive reduction in body porosity, while acting to bond the remaining body constituents. As vitrification increases, the strength of the body and its resistance to chipping normally increases until a stage of glassiness occurs. At this stage, brittleness begins to develop.

For purposes of this discussion, ceramic whiteware is categorized as earthenware, stoneware, china, porcelain, and technical ceramics. Table 1 lists these classes and their various types, as well as representative products of each type. Table 2 provides typical compositions for these materials.

Earthenware is defined as glazed or unglazed nonvitreous clay-based ceramic ware of medium porosity. Earthenware can be subclassified as:

- Natural clay earthenware, which is made from a single, unbeneficiated clay
- Fine earthenware, which is made from beneficiated clays and nonplastics, and possesses a triaxial body, where triaxial refers to clay-flint-feldspar
- Talc earthenware, which contains a considerable amount of talc
- Semivitreous earthenware, which involves a triaxial body fired to medium porosity

After firing, water absorptions may range from 4 to 9% for semivitreous ware and up to 20% for the high-talc formulas. Colors may range from red, for bodies with a high iron oxide content, to white, for the talc and triaxial formulas. The body is fired at comparatively low temperatures, producing an opaque product that is not as strong as stoneware or china. The product may be glazed or unglazed.

Stoneware is a vitreous or semivitreous ceramic ware of fine texture, made primarily from either nonrefractory fireclay or some combination of clays, fluxes, and silica that matches the forming and fired properties of a natural stoneware. Stoneware can be subclassified as:

- Natural stoneware, which is made from a single, unbeneficiated clay
- Fine stoneware, which is made from beneficiated clays and nonplastics
- Technically vitreous stoneware, which is made from beneficiated materials blended and fired to extremely low porosities
- Jasper stoneware, which is made largely of barium compounds
- Basalt stoneware, which contains high amounts of iron oxide

Stoneware may be made from either a clay or synthesized material. Synthesized stoneware can range from being a highly refined, zero-absorption chemical formula to less-demanding dinnerware and artware formulas. It is typically a low-porosity body (0 to 5% water absorption) that is fired at high temperatures. Although quite durable, it lacks the translucency and whiteness of china, but is resistant to chipping. It differs from porcelain in that it becomes colored as a result of iron and other impurities in the clay.

China is vitreous ware of either zero absorption or low-fired (0 to 5%) absorption used for nontechnical applications. It can be either glazed or unglazed. The expression soft-

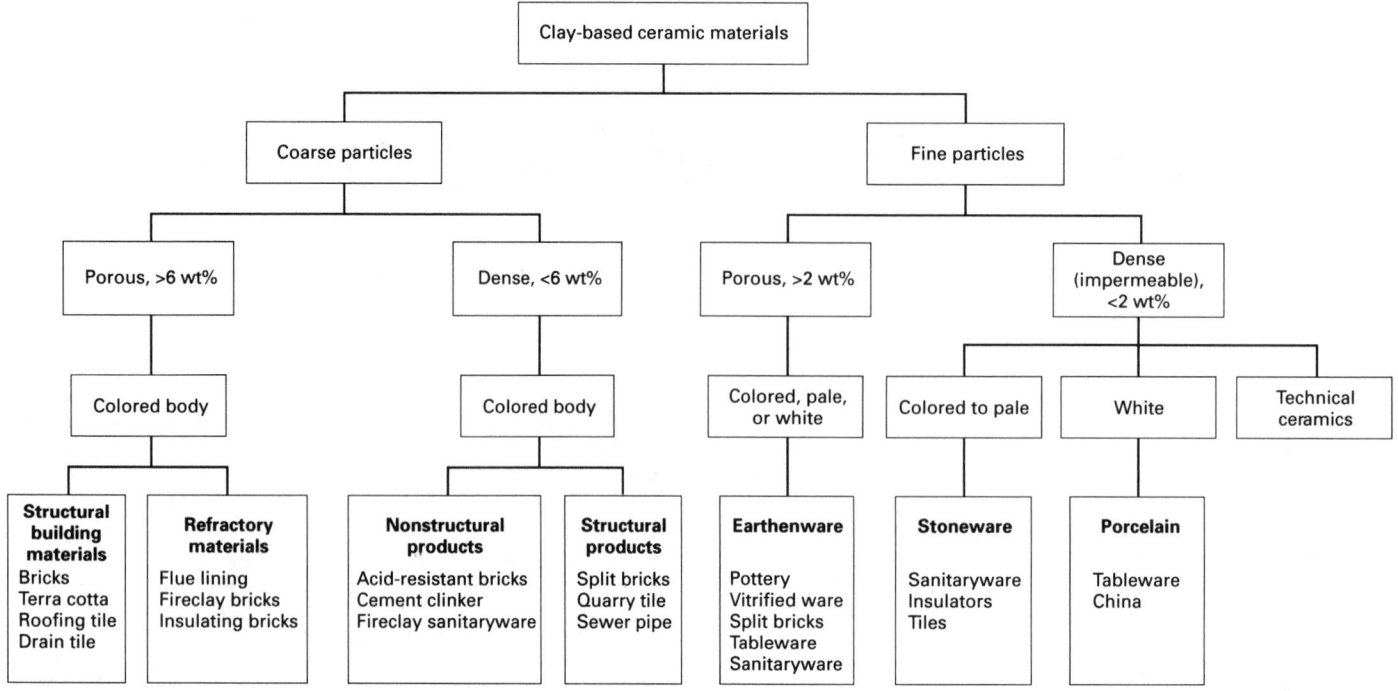

Fig 1 Classification of clay-based ceramics. Weight percent values represent water absorptivity.

Table 1 Ceramic whiteware classes and products

Class and type	Absorption, %	Products
Earthenware		
Natural	+15	Tableware, artware, tiles
Fine	10–15	Tableware, artware, kitchenware, tiles
Talc	10–20	Artware, tiles, ovenware
Semitvitreous	4–9	Tableware, artware
Stoneware		
Natural	0–5	Drain pipe, kitchenware, artware, tiles
Fine	0–5	Cookware, artware, tableware
Technical vitreous	0–0.2	Chemical ware
Jasper	0–1	Artware
Basalt	0–1	Artware
China		
Hotel	0.1–0.3	Tableware
Bone	0.3–2	Tableware, artware
Frit	0–0.5	Tableware, artware
Vitreous plumbing fixtures	0.1–0.3	Lavatories, closet bowls, flush tanks, urinals
Cookware	1–5	Ovenware, stoveware
Porcelain		
Hard	0–0.5	Tableware, artware
Technical vitreous	0–0.2	Chemical ware, ball mill balls and linings
Triaxial electrical	0–0.2	Low-frequency insulators
High-strength electrical	0–0.2	Low-frequency insulators
Dental	0–0.1	Dentures
Technical ceramics		
Steatite	0–0.05	High-frequency insulators, low-loss dielectrics
Electrical porcelains	0–0.2	Low-frequency insulators

paste porcelain and tender porcelain have the same meaning. Formulas can be either simple clay-flux-silica triaxial bodies or bodies containing significant percentages of alumina, bone, ash, frit, or low-expansion cordierite or lithium mineral powders. In general, the composition and properties of china are similar to those of porcelain, except that the body is not translucent and is off-white, although high-quality ware can be white. China can be subclassified further to describe its use or quality: hotel/vitreous china, fine china, and cookware.

Hotel china and vitreous china are bodies specifically engineered for use in commercial operations. Hotel china is used for food containment, whereas vitreous china is most often used for sanitary plumbing fixtures. Both were developed to provide impact resistance, strength, low-water absorption, and glaze durability. Chinaware is fired in a single operation; the body and glaze mature at the same time and at the same temperature. However, a two-firing process may be used for restaurant china where the body and glaze mature at different temperatures. Like fine china, both hotel and vitreous china are vitrified to a water absorption of less than 0.5%.

Fine china refers to a thin, translucent, vitrified body (0.3 to 2.0% absorption). It is generally fired twice: first, at a high temperature to mature the body; and second, to develop the high gloss of the glaze. It is considered the highest quality china. It includes bone china, frit china (or Belleek) and feldspar china.

Cookware is a broad term that is typically applied to china and porcelain, but can include

earthenware. It describes a ceramic having a smooth glaze on the surface where food contacts it. It is strong and resists both chipping and thermal shock. It is produced in a single-firing process.

Porcelain is defined as glazed or unglazed vitreous ceramic ware used primarily for technical purposes. It is frequently used as a synonym for china. Formulations are generally of the triaxial compositions, although some or all of the silica can be replaced by calcined alumina to increase mechanical strength. Compositions may also include low-alkali-containing bodies, such as steatites. The ware can be fired unglazed at high temperature to yield a bisque, or biscuit, with glazing done at low temperature or during single-firing at high temperature.

Technical whiteware ceramics include vitreous nonporous ceramic materials used for such products as dental, electrical insulations, chemical ware, mechanical and structural items, and refractory ware. These are typically similar to restaurant china in composition; substitutions of alumina and zircon for silica are common. Depending on the application, they are glazed in a single-firing operation. The water absorption for these ceramics is again specified to less than 0.5%.

Characterization

The development of clay-based ceramic bodies can be considered empirical, at best. Historically, this entailed adjusting body formulations by manipulating different combinations of raw materials until both the desired green and fired properties were obtained. To

Table 2 Typical ceramic body compositions, %

	Fine whiteware		Earthenware, wall tile	Vitreous floor tile	Chemical stoneware	Tender porcelain	Hard porcelain	Electrical porcelain		Steatite porcelain	Bone china	Hotel china		Cookware	Vitreous china, sanitaryware
	A	B						C	D			E	F		
China clay	30.6	18	22	32	55	30	50.4	20	17	7	37.5	36.8	34.7	62	22.3
Ball clay	14.4	38	25	...	...	...	...	25	35	...	...	6.2	7.1	...	26.7
Feldspar	27.5	32	3	58	30	30	17.1	35	32	6	20	21.7	18.3	10	33.7
Flint	27.5	12	25	8	14	35–37	32.3	...	16	...	5	35.2	27.5	7	17.3
Bone ash [Ca$_3$(PO$_4$)]$_2$	...	...	...	...	...	...	...	...	...	...	37.5	...	...	...	...
Whiting (CaCO$_3$)	...	...	...	...	1	3–5	...	...	...	...	...	...	...	...	...
Alumina	...	...	...	...	...	...	...	20	...	...	...	...	12	...	...
MgCo$_3$	...	...	...	...	...	...	...	...	...	...	...	...	...	21	...
Talc	...	...	10	2	...	...	...	...	...	87	...	...	...	...	...
Pyrophyllite	...	...	15	...	...	...	...	...	...	...	...	...	...	...	...
Firing range (PCE)	8	9	5–6	9	14	8–9	14	10–11	10–11	10	8–10	10	10–12	14	9

A, white body, low plasticity; B, ivory body, high workability; C, high voltage; D, low voltage; E, standard; F, high-strength body

best utilize raw materials and synthetic powders to produce more consistent products, it is important to accurately specify or characterize the material.

Characterization, as it applies to materials, is defined as those features of the composition and structure of a material that are significant for a particular preparation and the study of its properties, and suffice for the reproduction of the material. In relation to ceramic bodies, this entails a correlation between test results and measured properties. This concept has been applied successfully to traditional clay bodies.

Many unfired and fired properties are a consequence of the interaction of two or more key characterizing features. Figure 2 identifies these features for a traditional ceramic and shows their relation to properties that can be evaluated before and after firing. Table 3 represents a complete characterization for two common clay-based ceramics. The values in bold-faced type are the key indicators for the two examples. Any change in a key indicator will, in turn, produce a change in a particular property.

Fired color in a body is controlled by the Fe_2O_3 and TiO_2 content. This is much more important for vitreous china than it is for sanitaryware and must be taken into account. The presence of mica greatly affects the rheological behavior of casting slips and therefore represents an important factor when sanitaryware and vitreous china are slip cast. It is not a key indicator when vitreous china is formed by other techniques. The presence of colloidal organic matter can increase both deflocculant response and green strength. The rheology and both green and fired bulk densities are affected by alterations in particle size distribution. Specific percentages of 15 μm (600 μin.) and 1 μm (40 μin.) size particles are good indicators of any change. The methylene blue index by which surface area is measured correlates with plastic forming properties and green strength of unfired ware.

Mineralogy

A clay-based ceramic body usually consists of one or more clays or clay minerals mixed with nonclay mineral powders such as fluxes (for example, feldspar) and fillers (for example, silica and alumina). Each of these constituents contributes to the plastic forming and fired characteristics of the body.

Clay Minerals. Clay can have various forms that are easy to distinguish as the sticky, plastic component of soil. However, clay can also occur as rock or slate, due to either compression or its geological maturity. Clay is seldom pure. The substances responsible for its characteristic properties are clay minerals, which comprise two primary groups: kaolins and montmorillonites. The kaolins have the empirical formula $Al_2Si_2O_5(OH)_4$, which is 46.53% SiO_2, 39.5% Al_2O_3, and 13.96% H_2O. Montmorillonites are derived from the formula $Al_2Si_4O_{10}(OH)_2$.

The kaolin group includes such minerals as kaolinite, nacrite, dickite, and halloysite. The most important of these is kaolinite, because it is the most abundant constituent in china and ball clays. All kaolin minerals have two layers. One layer has Si—O bonded atoms with the composition $Si_2O_5^{2-}$. This is referred to as the silica, or tetrahedral, layer. It is joined

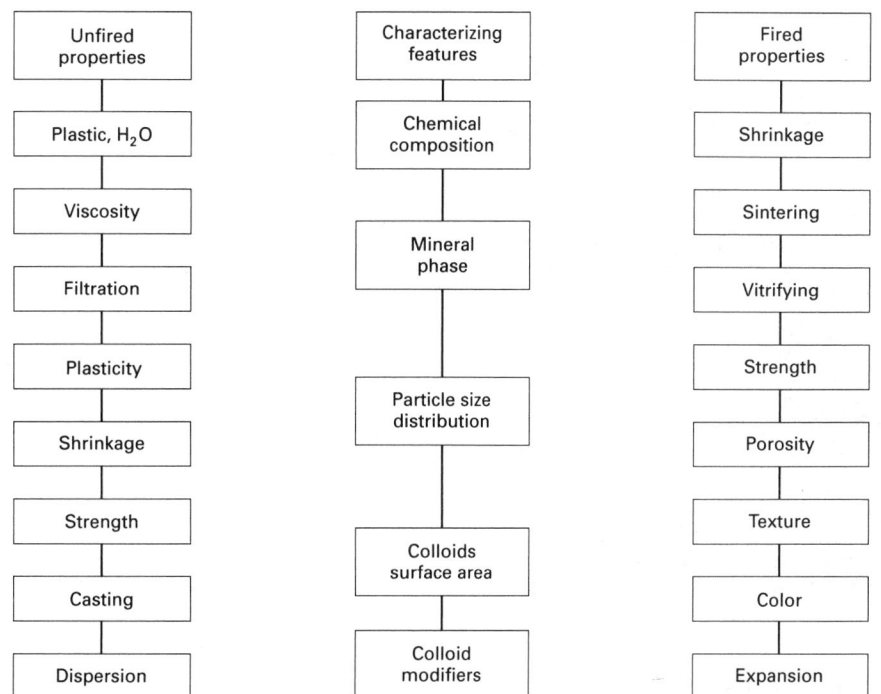

Fig 2 Characteristic features of traditional ceramics and the properties that can be measured in unfired and fired states

Table 3 Characterization of two common clay-based ceramics

Properties	Vitreous sanitaryware	Vitreous china
Compound, wt%		
SiO$_2$	**65.0**	69.4
Al$_2$O$_3$	**23.1**	19.5
Fe$_2$O$_3$	0.44	0.30
TiO$_2$	0.28	**0.14**
CaO	0.33	1.33
MgO	0.13	0.11
K$_2$O	**2.68**	1.45
Na$_2$O	**2.41**	1.14
Ignition loss	5.67	6.46
Mole of flux(a)	**0.0766**	**0.0604**
Minerals, wt%		
Smectite	3.7	3.0
Kaolin group	32.7	33.3
Mica	**8.8**	5.8
Free quartz	**23.7**	**39.6**
Organic	**0.46**	**0.23**
Auxiliary flux	. . .	2.0
Particle size, %		
<20 μm (<800 μin.)	76	76
<5 μm (<200 μin.)	47	45
<2 μm (<80 μin.)	33	36
<1 μm (<40 μin.)	19	21
Surface area, methylene blue index, meq/100 g(b)	**3.3**	**2.7**

(a) Mole of flux is the sum of the percentages of CaO, MgO, K$_2$O, and Na$_2$O, divided by their respective molecular masses. (b) Measured in terms of methylene blue index (MBI); the milliequivalents of methylene blue cation (chlorine salt) absorbed per 100 g of clay

by common oxygen atoms to a similar sheet of Al—O atoms with the composition Al$_2$(OH)$_6$, which is called the gibbsite, or octahedral, layer. The main mineral is tabular in nature, due to the interaction of the gibbsite and silica layers. The kaolinite crystal consists of a large number of two-layer units held together by hydrogen bonds, which act between the OH groups of the gibbsite layer of one unit and the oxygen atoms of the silica layer of adjacent units.

The kaolinite platelets have negatives charges on their face, or basal, planes, due either to occasional Al^{3+} ions missing from the gibbsite layer or to an Si^{4+} missing from the silica layer. When Fe^{2+} and Mg^{2+} replace some of the Al^{3+} in the octahedral layer, the result is a disordered kaolinite. This structure typically has a net negative charge that is balanced by the adsorption of exchangeable cations, usually Ca^{2+}. This results in a distorted octahedral layer, which in turn weakens the bonding between layered units. This ultimately results in smaller-sized kaolin crystals, that is, the type found in secondary kaolins and ball clays.

When the stacking of adjacent layers is slightly displaced, minerals closely associated with kaolinite form. When the stacking of layer units is arranged such that units are stacked one upon another, the mineral nacrite forms. When layers are displaced slightly by a constant amount, dickite and kaolinite form. When the displacement of unit layers is random, the mineral halloysite forms. Because of this random stacking, the

bonding between layers is so weak that water tends to penetrate the layers, forming a hydrated form of kaolinite with the formula Al$_2$Si$_2$O$_5$(OH)$_4$·2H$_2$O. It is this water layer that acts as a lubricant between particles and provides kaolin with its plasticity.

The montmorillonites result from the isomorphous substitution of portions of Al^{3+} and Si^{4+} in the parent mineral, pyrophyllite. With the composition Al$_2$Si$_4$O$_{10}$(OH)$_2$, pyrophyllite is a three-layer mineral formed by the condensation of two silica sheets that sandwich a gibbsite sheet.

When Mg^{2+} replaces some of the Al^{3+} in the octahedral layer, the result is the formation of montmorillonite. However, because magnesium is divalent and only carries two positive charges (vis-à-vis three for Al), there is a charge deficiency. This deficiency has to be satisfied by an external cation, outside of the layered structure. Most commonly, the adsorption of Na$^+$ serves to satisfy this charge, resulting in (Al$_{1.67}$Mg$_{0.33}$)Si$_4$O$_{10}$(OH)$_2$·Na$_{0.33}$. Because the bonding between three-layer units is by weak van der Waals attraction, montmorillonite particles tend to be thin and small.

If one-quarter of the Si^{4+} ions of the tetrahedral layer of pyrophyllite are replaced by Al^{3+}, a charge of sufficient magnitude is produced to bind univalent cations in regular twelve-fold coordination. If the cation is K$^+$, the result is muscovite mica, KAl$_2^+$(Si$_3$Al)$^-$O$_{10}$(OH)$_2$. If the cation is Na$^+$, the result is paragonite mica, Na$^+$Al$_2$(Si$_3$Al)$^-$O$_{10}$(OH)$_2$. If the cation is Ca^{2+}, the result is margarite, Ca^{2+}Al$_2$(Si$_2$Al$_2$)$_2^-$O$_{10}$(OH)$_2$.

When less potassium and more water are found in the lattice of mica, the mineral illite forms. The structure of illite is very close to that of mica. It has been suggested that it is formed by the replacement of the potassium, sodium, or calcium by hydrated hydrogen ions (OH).

When Mg^{2+} replaces all the Al^{3+} in the gibbsite layer, forming a magnesia or brucite layer, the resultant mineral is talc, Mg$_3$Si$_4$O$_{10}$(OH)$_2$.

If alternating layers of talc and brucite are formed, the mineral chlorite, Mg$_5$Al$_2$Si$_3$O$_{10}$(OH)$_8$, is produced. This consists of a substituted talc layer, of the composition Mg$_3$(Si$_3$Al)$^-$O$_{10}$(OH)$_2$, which is similar to that in mica. This layer carries a negative charge, which is satisfied by the substitution of a brucite layer rather than an alkali cation. Commonly, Fe^{2+} is found to replace Mg^{2+} or Al^{3+} in the mineral.

Nonclay minerals that represent fluxes and fillers are described below.

A flux is a substance that, when added to a material, enables it to fuse more readily. In clay-based bodies, fluxes are added to lower the temperature at which liquid or glass forms during firing. This glass, when cooled, binds the grains of the body together. The addition of fluxes allows porcelain to be fired at temperatures from 1000 to 1300 °C (1830 to 2370 °F).

Any material that promotes fusion is generally considered to be a flux. For clay-based bodies, the most common fluxes are based on alkali oxides, that is, K$_2$O, Na$_2$O, and, to a lesser extent, Li$_2$O. Alkaline earth oxides, CaO and MgO, can represent auxiliary fluxes when used in combination with alkali fluxes.

The feldspathic minerals represent the most important source of fluxing oxides in clay-based ceramics. These minerals are all based on a skeletal aluminosilicate. Replacement of Si^{4+} by Al^{3+} results in charge deficits that are balanced by Na$^+$, K$^+$, and Ca^{2+}. This mineral type includes orthoclase, or potash feldspar, KAlSi$_3$O$_8$; albite, or soda feldspar, NaAlSi$_3$O$_8$; anorthite, or lime feldspar, CaAl$_2$Si$_2$O$_8$; and nepheline syenite, or the mixed nepheline [Na(AlSi)O$_4$]-albite-orthoclase feldspar.

Fillers are added to a clay-based body to aid in forming and firing, and to improve the properties of the fired product. The most widely used fillers are silica and, to a lesser extent, alumina.

Silica is considered to be the building block of clay-based ceramics. In the ceramics industry, quartz rock, silica sand, and flint pebbles are the primary sources of silica. These forms consist of small crystals of quartz that may be bound together by water molecules. While fillers are commonly nonreactive, silica (SiO$_2$) is not an inert component of the body. Upon firing, silica can melt into the glassy phase, which alters its properties and/or remain as a particulate component of the microstructure.

Alumina (Al$_2$O$_3$) is not as widely used as silica as a filler. While silica is a naturally occurring raw material, alumina is chemically derived by the digestion of bauxite by caustic to yield Al(OH)$_3$. This is then calcined to produce Al$_2$O$_3$. This leads to significantly higher costs compared to silica. The most common and stable form is alpha-alumina, or corundum. In clay-based ceramics, alumina is used primarily as a particulate reinforcement for improving the strength of a body.

Commercial clays are grouped as being kaolin (china) clay or ball clay. Kaolins are typically primary or residual deposits that remain in the location where they were formed, but may also be located in secondary or sedimentary deposits. Kaolin tends to consist mainly of ordered kaolinite, with some mica and free quartz. Typically, no organic matter is associated with these deposits.

Ball clays are sedimentary deposits that were originally dug out of the ground in "blocks" or "balls." Their particles tend to be very fine grained and are composed of ordered and disordered kaolinite and varying percentages of mica, illite, montmorillonite, free quartz, and organic matter. A number of commercial clays are compared in Table 4, whereas fluxes are compared in Table 5.

Heat Effects. During the firing of clay-based bodies, a number of pyrochemical

Table 4 Characteristics of commercial clays found in the United States

Properties	Coarse kaolin, sedimentary(a)	Fine kaolin, sedimentary(b)	Dark, fine ball clay(c)	Light, coarse ball clay(d)
Compound, wt%				
SiO$_2$	45.7	46.7	50.5	60.4
Al$_2$O$_3$	38.3	38.2	28.7	27.0
Fe$_2$O$_3$	0.41	0.60	0.91	0.93
TiO$_2$	1.55	1.42	1.48	1.62
CaO	0.08	0.12	0.40	0.28
MgO	0.06	0.20	0.30	0.26
K$_2$O	0.06	0.15	0.89	1.70
Na$_2$O	0.14	0.03	0.18	0.50
Ignition loss	13.65	13.79	16.58	7.59
Minerals, wt%				
Montmorillonite	nil	3	8	7
Kaolin group	96	93	58	44
Mica	2	2	10	21
Free quartz	trace	1	14	26
Organic	trace	trace	8	0.5
Particle size, %				
<20 μm (<800 μin.)	95	99	99	98
<5 μm (<200 μin.)	69	88	95	79
<2 μm (<80 μin.)	52	72	82	61
<1 μm (<40 μin.)	35	56	69	43
<0.5 μm (<20 μin.)	28	41	51	29
Surface area, methylene blue index, meq/100 g	1.6	10.5	12.1	5.6

(a) Washington County, Georgia. (b) Wilkinson County, Georgia. (c) Graves County, Kentucky. (d) Weakley County, Tennessee

Table 5 Characteristics of fluxes and fillers

Properties	Feldspathic/feldspathoid fluxes			Flint/quartz fillers		
	Flotation feldspar(a)	Block feldspar(b)	Nepheline syenite(c)	Flint(d)	Quartzite(e)	Silica sand(f)
Compound, wt%						
SiO$_2$	66.8	68.5	60.7	96.6	98.5	99.5
Al$_2$O$_3$	19.6	17.5	23.3	0.2	0.9	0.2
Fe$_2$O$_3$	0.04	0.08	0.07	0.10	0.09	0.06
TiO$_2$	...	...	...	0.01	0.06	...
CaO	1.70	0.30	0.70	0.20	0.02	...
MgO	trace	trace	0.10	0.09	0.03	...
K$_2$O	4.80	10.40	4.60	0.39	0.05	...
Na$_2$O	6.90	3.00	9.80	0.15	0.04	...
Ignition loss	0.20	0.30	0.70	1.58	0.17	0.12
Mole of flux	0.1728	0.1644	0.2220	0.0124	0.0028	...
Minerals, wt%						
Feldspars	92	83	75	...	...	...
Nepheline	...	...	24	...	...	...
Mica	4	7	...	...	...	...
Quartz	4	9	...	96	97	99
Clay	...	...	...	1	1	trace
Organic	...	...	...	trace	...	...
Other	...	...	1	4	2	1
Particle size, %						
<20 μm (<800 μin.)	67	64	68	56	57	58
<5 μm (<200 μin.)	26	26	22	16	18	15
<2 μm (<80 μin.)	11	12	9	5	7	5
<1 μm (<40 μin.)	9	9	3	1	2	1
<0.5 μm (<20 μin.)	trace	6	trace	...	...	...

(a) Mitchell County, North Carolina. (b) Custer County, South Dakota. (c) Ontario, Canada. (d) France. (e) Pennsylvania. (f) California

Table 6 Reactions that can occur when firing clay-based triaxial bodies

Temperature		Reaction
°C	°F	
≤100	≤212	Loss of free moisture
100–200	212–390	Loss of adsorbed water
200–450	390–840	Crystal structure of clay minerals altered by removal of OH groups. Pyrophyllite shows a marked expansion.
400–700	750–1290	Organic matter in the form of lignite is oxidized. Pyrophyllite expands further.
573	1065	Quartz inverts to high-temperature polymorph.
700–950	1290–1740	Pyrophyllite reaches maximum expansion; metakaolin converts to spinel in clay.
950–1100	1740–2010	Mica structure is destroyed. Talc decomposes to protoenstatite and glass. Mullite forms from spinel. Pyrophyllite converts to mullite and glass.
1100–1200	2010–2190	Feldspars melt; clay and cristobalite dissolve. Vitrification begins, porosity decreases, shrinkage increases.
>1200	>2190	Protoenstatite from talc converts to clinoenstatite. Mica breaks down into alumina and glass. Glass content increases, mullite needles grow; only closed porosity remains.

cules, this reaction can occur up to 700 °C (1290 °F)

4. At 573 °C (1065 °F), quartz undergoes an inversion to its high-temperature polymorph

5. At 980 °C (1795 °F), the metakaolin goes through a series of changes as it rearranges itself into a spinel and then quickly into small mullite crystals

6. At 1100 °C (2010 °F), the feldspar begins to melt. Potash feldspars decompose into leucite, KAlSi$_2$O$_6$, and glass, while soda feldspars melt completely. As the glass forms it distributes itself, bringing about the densification of the microstructure

7. As the temperature increases, there is an increase in the glass content as the quartz decreases. The total porosity reaches a minimum. Further increases in temperature lead to a decrease in the viscosity of the glass and deformation of the part due to its own weight

Forming Processes

Traditional ceramic bodies are formed into shapes using many different techniques. The specific forming process is dictated by

reactions can occur, leading to the densification of the green formed part. Whether certain reactions occur depends on the types of raw materials that compose the body. Table 6 summarizes the various reactions.

In a simple clay-flint-feldspar body, the basic processes would be:

1. As the body is heated, the last traces of free water disappear, followed by the removal of water adsorbed on the higher surface area clay particles. This stage is complete at 200 °C (390 °F)

2. The crystal structure of kaolinite is altered by the removal of OH groups. This dehydroxylation of kaolinite to metakaolin, Al$_2$O$_3$·2SiO$_2$, is accompanied by an increase in porosity

3. Organic material in the clay minerals (usually lignite, associated with the ball clay), begins to oxidize. Depending on the molecular weight of the organic mole-

numerous factors, including material characteristics, size and shape of the part, part specifications, production yield, and accepted practices within the geographic region.

Clay-based bodies are heterogeneous mixtures of one or more clays and one or more nonclay powders. Before attaining a final shape, these powders undergo a sequence of unit operations that include raw material preparation, batch preparation, forming, drying, prefire operations, firing, and postfire operations. Figure 3 identifies the possible paths followed by various types of ceramics during their fabrication.

For most traditional bodies, forming techniques can be classified as soft plastic forming, stiff plastic forming, pressing, and casting.

Applied pressure is employed to rearrange and redistribute the raw materials into a better-packed configuration. The rheological behavior of clay-based bodies results from clay mineral interaction with water, which imparts plasticity to the batch. In nonclay bodies, this same type of behavior can be achieved by adding plasticizers. Table 7 shows the superimposed forming pressure and moisture variations within a typical clay-based body for the different forming techniques.

Soft Plastic Forming. The oldest and simplest method of forming a shape is to mold clay or a clay-based body that has a highly plastic rheology. Low-volume production ware may be shaped by hand molding or by throwing on a wheel. Higher-productivity plastic

Table 7 Moisture content and pressure ranges for shaping clay-based bodies

Forming process	Superimposed pressure		Liquid content, %
	MPa	ksi	
Soft plastic	0.1–0.75	0.015–0.109	20–30
Rigid plastic			
Piston	3–15	0.45–2.18	12–16
Auger	20–50	2.9–7.3	15–20
Dry press	100–200	14.5–29	5–15
Dust press	150–250	21.8–36.3	0–2
Isostatic press	200–400	14.5–58.0	0–15
Slip cast	0.1–3	0.015–0.45	20–35

forming is accomplished by jiggering, jolleying, and ram pressing.

Jiggering involves the placement of an extruded slug of a plastic body on a revolving plaster form. A template tool is brought in contact with the slug, which is pressed onto the plaster form while the template tool cuts away excess body. This leaves a shape with one surface that conforms to the shape of the plaster form, while the other conforms to the shape of the template tool. Once the shaping is complete, the part is either allowed to dry on the plaster to a leather-hard state, whereupon it may be removed, or it is "forced" off the plaster by compressed air.

Jolleying is similar to jiggering, except for the use of the template tool. A polished, rotating metal tool presses the slug down on a plaster form. Most often the plaster form is a symmetrical cavity, and the resulting shape is hollow, such as a flower pot or a cup. The metal tool may be heated to assist in the shaping operation. When heated up to 300 °C (570 °F), the tool forms a steam cushion between itself and the body, which prevents sticking.

Ram pressing involves the pressing of an extruded slug of plastic body between two plaster molds to produce a shape. During the operation, water is forced out of the slug by a combination of the pressing force and a vacuum applied to the mold halves. Once the part reaches a leather-hard state, the mold is parted. The upper portion of the mold maintains the vacuum on the part, while pressure is applied to the lower portion of the mold, which acts to separate the part from the bottom. Pressure is then applied to the upper portion and the part is freed from the assembly.

Stiff Plastic Forming. Less-plastic bodies, which result from the reduction in the clay mineral content of the batch or from a reduction in the added water, require greater force to provide flow. Extrusion is employed to form these stiff plastic batches and may be continuous or intermittent, depending on the type of equipment available. During extrusion, a cohesive plastic mass is forced through a rigid die to produce a column of uniform section. The resulting part can then be cut into appropriate lengths as the final product, or it can

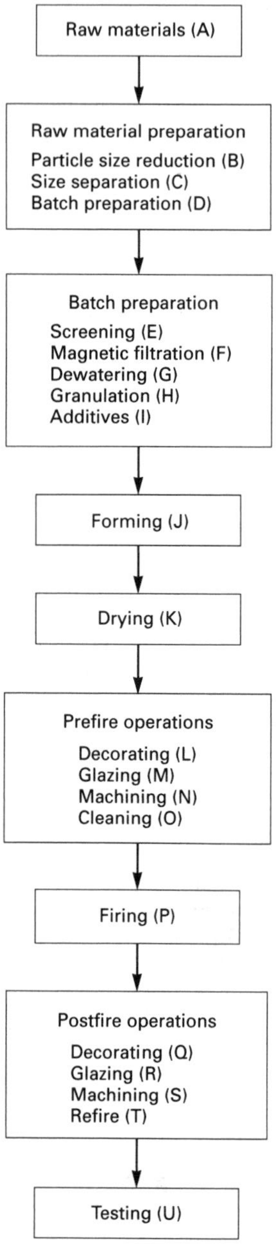

Ceramic type	Sequence of unit operations to produce ceramic type
Cements	A-D/P/L/C/I/U
Structural clay bricks	A-D/J/K/P/U
Structural clay glazed pipe	A-D/J/K/P/R/T/U
Sanitaryware	A/D/E/F/O/M/P/U
Bone china (slip cast)	A/D-F/J/K/O/P/R/T/Q/T/U
Bone china (jiggered)	A/D-G/J/K/O/P/R/T/Q/T/U
Steatite porcelain	A/D-F/J/K/N/M/P/Q/T/U
Refractory, fireclay (extruded)	A-D/J/K/P/U
Refractory, MgO	A-D/J/K/P/U
Wall tile	A/I/D-F/H/E/J/M/K/P/U
Glaze	A/I/D-F/M
Raw materials (washed clay)	A/C-F/K/C/U

Fig 3 Fabrication processes for various ceramic types

be used as a blank or slug for other forming operations.

Piston extrusion is an intermittent operation in which a prepared batch or a de-aired slug is placed in a cylinder and forced through a die. Pressures up to 50 MPa (7.3 ksi) can be achieved. As is the case with all extrusion operations, only parts that are symmetric to the extrusion axis can be produced.

Auger extrusion is a commonly used continuous operation in which high extrusion pressures are not required. The auger extruder consists of a cylinder, feed screw, and die. The feed screw pushes the batch through the die. It is common to have a pug mill in-line with the extruder. A pug mill consists of a trough, fed by a separate de-airing screw that forces the batch through a shredder into a vacuum chamber. The de-aired and shredded batch is then fed by screw into the extrusion cylinder and through the die. Auger extrusion pressures (4 to 15 MPa, or 0.6 to 2.2 ksi) tend to be lower than those used in piston extruders, and a more plastic batch is required.

Pressing is the simultaneous compaction and shaping of a powder or granular mass confined in a rigid die or flexible mold. Pressing can be classified by the direction of pressure application, that is, uniaxial or isostatic.

Uniaxial pressing is what is commonly referred to as die pressing. A die cavity is filled with a metered quantity of powder feed and then is compacted by the action of punches moving along a single axis. Uniaxial pressing can be further classified according to the moisture content of the feed. Dust pressing applies to the compaction of feed containing 5 to 15% moisture at high pressures (140 MPa, or 20.3 ksi), in a steel die. When the moisture content of the feed is reduced to <2%, the compaction is referred to as dry pressing. In dry pressing, higher pressures (200 to 250 MPa, or 29 to 36 ksi) are required than dust pressing. A binder and lubricant are typically employed.

Isostatic pressing involves the application of pressure equally to all surfaces of the feed. An isostatic press consists of a pressure vessel filled with an incompressible fluid. The feed is enclosed in a sealed elastomeric mold, immersed in the fluid, and pressurized. The fluid transfers pressure equally to all sides of the part. Typical pressures are 280 MPa (40.6 ksi).

Slip casting is the process whereby a self-supporting shape is produced by the filtration of liquid from a suspension by the action of a porous mold. In this process a slip, or deflocculated suspension containing 40 to 55 vol% solids, is poured into a plaster mold. After a layer has formed, the excess slip is poured away. The mold continues to draw water away from the cast layer, allowing the layer to assume a rigid plastic state, whereupon it may be removed from the mold. This is referred to as drain casting. If the slip is not drained from the mold and the entire cavity is allowed to solidify, the process is referred to as solid casting.

Structural Clay Products

The use of ceramics in construction dates back more than 4000 years. Urban planners at Mohenjo-Daro in the Indus Valley used fired clay bricks and tile for public buildings, water-supply conduits, and an advanced sewer system. Fired ceramics have been used ever since as structural materials because of their high compressive strength and imperviousness to water penetration. Today, structural clay products constitute over 50% of the entire ceramic industry. Table 8 identifies principal product types and their varied applications. There are many compositions that offer unique

Table 8 Classification of structural clay products

I. Bricks
  A. Facing
    1. Extruded
      a. Solid, smooth, textured, or glazed
      b. Cored, smooth, textured, or glazed
      c. Panel, smooth, textured, or glazed
    2. Sand molded, natural, colored, or glazed
    3. Dry pressed, solid
  B. Paving, solid; extruded
    1. Sidewalk, smooth or textured
    2. Street or highway, smooth or textured
  C. Industrial floor, solid, smooth, or textured; extruded
  D. Chimney, extruded
  E. Industrial chimney lining; extruded
  F. Sewer and manhole; extruded

II. Tiles
  A. Structural for walls; extruded
    1. Nonload-bearing, smooth or textured
    2. Load-bearing, smooth or textured
    3. Facing
      a. Smooth
      b. Textured
      c. Glazed
      d. Acoustical
      e. Lightweight
      f. Through-the-wall
    4. Screen, smooth or textured
  B. Floor
    1. Large-cored, smooth or textured; extruded
    2. Quarry; plastic pressed
    3. Mosaic, unglazed and glazed; dry pressed
  C. Wall, unglazed and glazed; dry pressed
  D. Roofing; extruded and plastic pressed
  E. Flue linings; extruded

III. Pipes, extruded
  A. Sewer and drain, plain or glazed
    1. Bell and spigot end
    2. Plain end
  B. Perforated subdrainage
    1. Bell and spigot end
    2. Plain end
  C. Chemical-resistant, plain or glazed
  D. Conduit, plain or glazed
  E. Drain tile
    1. Plain
    2. Perforated

IV. Liner plates, curved or flat, plain or glazed

V. Filter blocks for trickling filters, plain, or glazed

VI. Chemical-resistant tower packings

VII. Terra cotta specialty shapes

properties specific to their end use. These variations become very large in number if the aesthetics of the product are considered.

The industry is based on improved technologies, rather than improved materials. It can be said that the majority of structural clay products are commodities, rather than specialties. As such, the industry can be considered conservative in that manufacturing issues focus on highly automated processes and on minimizing labor costs.

Materials. With the exception of tile, most structural clay products are coarse-grained ceramics. The primary raw materials are typically located on or near the property of the manufacturing facility. Both clay and argillaceous shale are used with little or no beneficiation, except when the application is tile, for which compositions of beneficiated raw materials are used. Typical bodies contain 35 to 55% clay, 25 to 45% filler, such as silica, and 25 to 55% flux materials. In many cases, the local raw materials contain these constituents in proportions close to that required in the end product. For example, a New Jersey shale deposit contained 67% SiO_2, 18% Al_2O_3, 3% Fe_2O_3, 2% alkaline earth oxides, and 4% alkalies, with a loss on ignition of 4%. This material, by itself, was found to make a satisfactory brick. The various types and colors of clay minerals used in the United States for structural clay products are:

- Crude, residual, kaolinitic clays; white
- Kaolinitic-illitic fireclays; buff
- Illitic-kaolinitic shales; red
- Illitic clays; red
- Illitic shales; red
- Illitic-chloritic shales; red
- Argillaceous-calcareous clays; pink to buff
- Argillaceous-calcareous shales; pink to buff

The properties of structural clay products vary depending on the type of product, the raw materials used, and the firing conditions. The most important properties tend to be compressive strength and water absorption. For example, paving bricks range in compressive strength from 17 to 137 MPa (2.5 to 20 ksi) and 1 to 6% absorption, while facing brick can have a compressive strength ranging from 4 to 137 MPa (0.6 to 20 ksi) and 2 to 25% absorption. The variation is indicative of the different end uses. Table 9 gives the compressive strength and water absorption for four general types of products.

Glazes

A glaze is defined as a continuous adherent layer of glass (or glass and crystals) on the surface of a ceramic body that is hard, non-absorbent, and easily cleaned. The surface may be shiny or matte. A glaze is usually applied as a suspension of glaze-forming ingredients in water. After the glaze layer dries on the surface of the piece, it is fired, whereupon the ingredients melt to form a thin layer of

Table 9 Selected properties of structural clay product types

Product	Compressive strength		Water absorption, %
	MPa	ksi	
Fancy brick	4–137	0.58–20	2–25
Paving brick	17–137	2.50–20	1–6
Structural tile	3.5–69	0.51–10	2–28
	17.5–146	2.53–21	1–8
Drain tile for pipe	10–73	1.45–11	5–16

glass. The glaze may be fired at the same time as the body or in a second firing. Typical glazes can be made to mature from 500 to 1500 °C (930 to 2730 °F), depending on the items to which they are applied.

Glazes are classified as to their optical properties and composition. Glazes tend to be transparent (clear), opaque (enamel), fine interior crystals, or large interior crystals (crystalline). Described below in terms of molar composition are the common types of raw, fritted, and special glazes.

Raw glazes are available in these five compositions: lead-containing, leadless, zinc-containing, porcelain, or slip.

Raw lead glazes are only used on artware, rather than commercial ware, because of the health hazards associated with soluble lead. A typical bright glaze maturing at cone 05 (1031 °C, or 1890 °F) has the composition of 0.5 PbO, 0.2 Al_2O_3, 1.0 SiO_2, 0.3 CaO, and 0.1 Na_2O.

Leadless glazes are designed to lower the maturity of porcelain glazes without the use of lead oxide (1190 °C, or 2175 °F). A typical glaze has the composition of 0.217 K_2O, 0.352 Al_2O_3, 2.77 SiO_2, 0.454 MgO, 0.454 CaO, 0.135 BaO, and 0.82 SnO_2.

Zinc-containing glazes, sometimes referred to as Bristol glazes, are used on stoneware and terra cotta clayware. They are useful when a lower maturing temperature than is possible with a porcelain glaze is desired, and mature at cone 5 (1175 °C, or 2145 °F). A typical glaze has the composition of 0.36 K_2O, 0.5 Al_2O_3, 3.16 SiO_2, 0.40 CaO, and 0.24 ZnO.

Porcelain glazes are designed to mature in the same temperature range as a porcelain body, cones 8 to 10 (1236 to 1285 °C, or 2255 to 2345 °F). A typical bright glaze has the composition of 0.3 K_2O, 0.58 Al_2O_3, 3.75 SiO_2, and 0.7 CaO. A matte porcelain glaze maturing at the same temperature is made by increasing the alumina and decreasing the silica. Its composition is 0.3 K_2O, 0.65 Al_2O_3, 2.25 SiO_2, and 0.7 CaO.

Slip glazes are natural clays used for artware glazing and high-tension electrical porcelain insulators. They have a maturing range of 1200 to 1300 °C (2190 to 2370 °F). Their approximate composition is 0.20 K_2O, 0.60 Al_2O_3, 4.00 SiO_2, 0.45 CaO, 0.08 Fe_2O_3, and 0.35 MgO.

Fritted glazes can either contain lead or be leadless. A frit is a glass ground to a fine powder and used as one of the glaze constituents. The main purpose of fritting is to be able to use water-soluble materials by melting them together to form a relatively insoluble glass. Secondary purposes for fritting are to obtain better working properties of the glaze, to distribute color more uniformly, and to reduce the toxicity of materials such as lead. Fritted glazes, whether lead or leadless, may be totally fritted or partially fritted. The decision to partially frit a glaze, use raw materials with a frit, or to completely frit is usually a cost issue, because frits may increase the cost of the glaze constituents by a factor of ten. Fritted glazes are used for all types of ceramic bodies. A typical leaded frit composition is 0.50 PbO, 0.10 Al_2O_3, 2.70 SiO_2, 0.30 Na_2O, and 0.20 K_2O. A typical leadless fritted glaze containing boric oxide, B_2O_3, which is soluble in water, has the composition of 0.69 CaO, 0.37 Al_2O_3, 2.17 SiO_2, 0.19 Na_2O, 1.16 B_2O_3, and 0.12 K_2O.

Special glazes are available in these four compositions: salt vapor, luster, crystal, or reduction.

Salt glazing is a common method for glazing stoneware and structural clay products. The glaze is formed by throwing common salt, NaCl, into the kiln during the sintering stage of firing. The salt decomposes to form Na_2O and HCl, and the Na_2O combines with Al_2O_3 and SiO_2 on the surface of the product to form a complex silicate glass layer. It is common in firing from cones 5 to 8.

Luster glazes consist of a thin, metallic coating fired on top of a lead-based glaze. The coating is achieved by applying a metallic salt dissolved in an organic resin over a fired glaze. The luster is fired at a low temperature that is high enough to decompose the resin and salt but is lower than the softening point of the glaze. The decomposed resin provides carbon, which acts on the easily reducible oxide to produce the thin, metal film. The metallic salts include bismuth, lead, zinc, cobalt, silver, and gold. The various metal salts produce a variety of reds, yellows, browns, and silvery-white lusters, as well as nacreous and iridescent sheens.

Crystalline glazes are a special type that, under the proper heat treatment, promote crystal growth within the glaze. The most typical glazes are zinc-based, which produce willemite ($ZnO \cdot SiO_2$) crystals. A formulation that is typical of this type of glaze has 0.235 K_2O, 0.162 Al_2O_3, 1.700 SiO_2, 0.087 CaO, 0.202 TiO_2, 0.052 Na_2O, 0.051 BaO, and 0.575 ZnO.

Applications. Glazes may be applied either wet or dry. The process of wet glaze application entails first preparing the glaze suspension. This is typically done by mixing the glaze constituents in a ball mill. During this operation the ingredients are intimately mixed and the particle size is adjusted. Next, the glaze slurry is screened and passed through magnetic filters, thereby removing any mill residue and ferrous contamination that could lead to glaze surface defects. Once the glaze suspension is screened, its rheology is adjusted. The glaze is applied to the ware by either painting, pouring, dipping, or spraying, depending on the type of product and the production volume.

Dry glaze application is done in limited instances. In the case of special tile designs, spray-dried glaze granules are poured over the shape to which an adhesive has been applied, such as a carboxymethyl cellulose gum solution. The glaze granules stick to the shape, which is then dried and fired. The advantages of this method are that unique patterns are possible, glaze waste is minimized, and drying time is reduced.

Portland Cement

Cement is a synthetic mineral mixture that, when ground to a powder and mixed with water, forms a stonelike mass. This mass results from a series of chemical reactions whereby the crystalline constituents hydrate, forming a material of high hardness that is extremely resistant to compressive loading. During hydration, cement forms a noncrystalline paste that has good adhesive properties. When the cement paste has set and hardened, it consists of submicron-sized crystals in a gel-like material that possesses a high surface area value. Cement is characterized by the presence of four main compounds: calcium trisilicate, calcium disilicate, tricalcium aluminate, and tetracalcium aluminoferrite.

The history of cement dates back to the Romans, who found that mixtures of volcanic ash, lime, and clay would harden when wet, and used it extensively to build structures. In 1757, it was found that burned and ground high calcitic clays would harden when placed in water. In 1824, a patent was granted to a British bricklayer who formulated a new type of cement with improved hardness. Because the color of the material after hydration reminded him of the limestone on the Isle of Portland, he named the product portland cement. This cement was made by lightly calcining small batches of lime and clay and grinding the product to fine powder.

The modern manufacturing process is very basic and has not been radically changed since its inception except for the use of computer-controlled equipment, which has greatly improved the consistency of the final product. In addition, the process has now become more energy efficient and there is less waste of material.

The manufacturing process of cement involves four basic operations: 1) quarrying and crushing of raw materials; 2) grinding to high fineness and carefully proportioning the mineral constituents; 3) pyroprocessing the raw materials in a rotary calciner; and 4) cooling and grinding the calcined product, or clinker, to obtain a fine powder.

Because the main ingredient in cement is calcium oxide, production facilities are typically situated near the source of calcareous materials. These can include minerals such as limestone, cement rock, and chalk, or materials such as marl, marine shells, or alkali wastes such as slag or fly ash. In addition, the other constituents of cement include silica, alumina, and iron oxide. The principal sources of these argillaceous materials are clay, shale, slate, cement rock, slag, and ash. Most of these materials contain amounts of iron oxide, but most facilities require additional iron supplied from mill scale, pyrite cinders, or iron ore.

The first step in the manufacturing of cement is the quarrying of the raw materials. The majority of the raw materials originate in the plant quarry. Because it is not uncommon for a large facility to produce 1000 metric tons (2.2×10^6 lb) per day, very large earth-moving equipment is utilized. Mined materials go through a series of primary crushing operations, which can include gyratory or roll crushers. Secondary crushers, such as hammer mills, are used to crush materials to sizes no larger than 10 to 25 mm (0.4 to 1 in.) in diameter. Each material is then stored separately. The various ingredients are then proportioned and finely ground by either wet or dry methods.

The dry process employs large ball mills that further reduce the size of the feed. Continuous milling, coupled with air separators, allows the individual materials to be ground to the desired fineness. The ground material is then transferred to blending silos, where it is blended using compressed air. The blended batch is then transferred to the feed bins of the rotary kiln, or calciner.

The wet process of grinding and blending is predominantly used when one or more of the raw materials contains high amounts of free water, which exists in certain clays. In this process, a proportioned batch is fed into ball or tube mills for grinding. Typically, up to 40% water is added to produce a well homogenized slurry. The slurry is then pumped through filter presses or other dewatering devices to remove excess water from the slurry, which is then shredded into small nodules prior to being fed into the rotary calciner.

The rotary calciner is a refractory-lined steel shell that is 3 to 8 m (10 to 26 ft) in diameter and 50 to 300 m (165 to 985 ft) in length. It rotates at a rate of 1 to 3 rev/min. The length of the calciner depends on whether the wet or dry mixing process is used. In the wet process, large amounts of energy are used to evaporate water. Therefore, long kilns must be used to attain large throughputs; that is, greater than 150 m (490 ft). The dry process lends itself to more efficient use of preheaters and heat exchangers, reducing the overall length of the kiln to about 50 m (165 ft).

The calciner is inclined slightly to the horizontal axis so that material fed into the upper end travels slowly by gravity to the discharge end. The hottest temperatures are developed in a narrow zone at the lower discharge end of the kiln, up to 1400 °C (2550 °F), and become less hot as the upper end is approached. Numerous reactions take place within the calciner. As material enters the hottest regions of the kiln, about 20 to 30% of the mass melts. The reactions that aid in the clinker formation take place primarily in the liquid phase. A clinker is a hard, sintered, marble-sized glass ball that emerges at the discharge end of the kiln. As the clinker emerges, it must be cooled rapidly. This is achieved by passing the material across grate coolers, where it is subjected to cooling by ambient air flowing through the clinker bed. The cooled clinker can either be stored or mixed with 4 to 6% gypsum and directly placed into finish mills. Gypsum ($CaSO_4 \cdot 2H_2O$) acts as a set retarder for the finished cement. Final milling consists of a battery of grinding mills, beginning with roll mills, which reduce the clinker to the fineness of sand, and ending with ball or tube mills, which further reduce the size of the clinker sand to a flour less than 325 mesh (44 μm, or 1.8 mils, in diameter). This powder is then stored in silos and can be packed later in bags or shipped in bulk.

Cement Chemistry. There are four main compounds that compose portland cement: tricalcium silicate, $3CaO \cdot SiO_2$ (C_3S); dicalcium silicate, $2CaO \cdot SiO_2$ (C_2S); tricalcium aluminate, $3CaO \cdot Al_2O_3$ (C_4A); and tetracalcium aluminoferrite, $4CaO \cdot Al_2O_3 \cdot Fe_2O_3$ (C_3AF). Table 10 shows typical compositions that assume ideal stoichiometry.

Tricalcium silicate (C_3S) is an orthosilicate in which four SiO_4^{4-} tetrahedra are bonded to two O^{2-} anions. Five different polymorphic forms of C_3S are found at different temperatures: three triclinic, two monoclinic, and one rhombohedral. The latter is stable above 1050 °C (1920 °F). In general, C_3S is stable and will not convert to C_2S + C between the temperatures of 1250 and 1800 °C (2280 and 3270 °F), but will melt incongruently at 2150 °C (3900 °F). Thus, the cement clinker is fired to approximately 1400 °C (2550 °F) so that the C_3S will convert to the desired rhombohedral form. Solid solutions of the C_3S with impurities enable the high-temperature form to be stable at room temperature. The impure form of C_3S, known as alite, contains ions such as Mg^{2+}, Al^{3+}, and Fe^{3+} in solid solution.

Dicalcium silicate (C_2S) is also an orthosilicate. The polymorphic transformations of C_2S result in large-volume changes. Therefore, it is necessary to develop solid-solution interactions to inhibit polymorphic inversions of the high-temperature forms to the room-temperature forms upon cooling. The form of C_2S that is found in portland cement clinker is known as belite. Belite is most useful in its β and low α forms. However, it is disastrous if the low γ polymorph forms upon cooling. The β to γ inversion is accompanied by a large volume change, which results in the collapse of the crystal structure due to great stresses that form because of the differential volume between the polymorphs. In addition, the γ C_2S is nonhydraulic, meaning it does not react with water and would therefore hinder setting of the cement. As was the case for C_3S, the β to γ C_2S inversion is prevented by the impurity cations, such as Mg^{2+}, Al^{3+}, and Fe^{3+}, and anions such as SO_4^{2-} and PO_4^{3-}.

Tricalcium aluminate (C_3A) does not exhibit polymorphism when pure. However, the addition of alkalies such as Na_2O can induce polymorphic transformations. The cubic structure of C_3A that contains the lowest Na_2O content is the most desirable for cement manufacturing. The C_3A is a necessary component of the portland cement clinker because it reacts well with water and is essential for the hydration process.

The calcium aluminoferrite component (C_4AF) is actually a solid solution of the C_2A and C_2F phases. The C_2F phase is an orthorhombic polymorph that consists of Ca^{2+} ions packed between alternating layers of FeO_6 octahedra and FeO_4 tetrahedra. The C_2F phase is stable at room temperature, while the C_2A phase is only stable at high pressures. Thus, a solid solution of the two phases enhances stability at room temperature. Proper monitoring of the alumina ratio in solid solution of the C_4AF controls its hydraulic setting behavior. It is the ferrite phase that gives cement its color. When MgO is present in the ferrite phase, it becomes blacker and the cement becomes gray. If the MgO content is low, or the ferrite phase contains ferrous iron, the cement becomes browner.

Hydration of Portland Cement. When water is added to portland cement clinker, a series of reactions produce hydrated forms of the C_3S, C_2S, C_2A, and C_4AF phases. The details of the microscopic reactions that occur hold true for all four phases. The hydration products contribute to form the cement paste, which quickly sets and develops the final rigid mass.

The most dominating cement clinker phase that affects the hydration process is trical-

Table 10   Chemical constituents for various types of portland cement

	$3CaO \cdot SiO_2$	$2CaO \cdot SiO_2$	$3CaO \cdot Al_2O_3$	$4CaO \cdot Al_2O_3 \cdot Fe_2O_3$	$CaSO_4$	MgO	Free CaO
Type I	45	27	11	8	3.1	2.9	0.5
Type II	44	31	5	13	2.8	2.5	0.4
Type III	53	19	11	9	4	2	0.7
Type IV	28	49	4	12	3.2	1.8	1.9
Type V	38	43	4	9	2.7	1.9	0.5

cium silicate. Most aspects of the C_3S hydration reaction are similar to the hydration of the other three clinker components. The hydration reaction of C_3S is:

$$2(3CaO \cdot SiO_2) + 7H_2O =$$

$$3CaO \cdot SiO_2 \cdot 4H_2O + 3Ca(OH)_2$$

The $Ca(OH)_2$, or calcium hydroxide, that is produced is commonly known as portlandite, while the $3CaO \cdot 2SiO_2 \cdot 3H_2O$ is known as tobermorite gel. Portlandite is the major product of the hydration reaction, and is also produced to a lesser degree by the other clinker phases. After mixing the C_3S and water, lime and silica ions immediately enter solution, resulting in an increase in lime concentration and pH of the aqueous solution. At this point, the first hydration product begins to form at the surface of the C_3S crystals, and proceeds to cover the crystals. This reaction occurs rapidly. However, as the pH of the aqueous C_3S paste increases to a maximum, and a protective layer of the first hydration layer surrounds the C_3S material, the hydration process slows dramatically. As ions diffuse through this layer, nucleation and growth of the portlandite and tobermorite gel crystals continue.

The reactions involved in the hydration process of dicalcium silicate are similar to the reactions of the C_3S phase. However, the C_2S reactions occur to a somewhat lesser degree.

$$2(2CaO \cdot SiO_2) + 5H_2O =$$

$$3CaO \cdot 2SiO_2 \cdot 4H_2O + Ca(OH)_2$$

The amount of portlandite produced from the C_2S reaction is about one-third the amount produced from the C_3S reaction. There are also marked differences between the hydration process of the C_2S and the hydration process of C_3S. For instance, the maximum rate of hydration for C_2S occurs at about 40 days, when only 20% of the C_2S hydration has developed. Consequently, the C_2S contributes little to the early strength of the cement. However, because its maximum hydration rate occurs after 40 days, C_2S contributes significantly to the late and continued strength of the cement.

Tricalcium aluminate reacts strongly with water to form crystalline hydration products. In fact, the initial reaction occurs so fast that without additions of gypsum, the cement could flash set when mixed with water. The calcium sulfate in the gypsum reacts with the C_3A to form ettringite [$C_3A \cdot CaSO_4 \cdot (31-32)H_2O$]. The retardation of the C_3A hydration is due to the ettringite formation. The hydration of C_4AF is quite similar to the C_3A reaction, with gypsum reacting with the C_4AF to form an iron-substituted ettringite.

Types of Portland Cement. Portland cement is a sintered material that is distinguished from other limes and cements. It is the name of a product made by a given process, not a brand name. The American Society for Testing and Materials (ASTM) defines

portland cement as "a hydraulic cement produced by pulverizing clinker consisting essentially of hydraulic calcium silicates, usually containing one or more of the forms of calcium sulfate as an interground addition." The ASTM specifications designate eight types of portland cement, all of which are made by the same process. They are distinguished by slight variations in the percentages of certain compositional constituents. Aside from the light, ASTM types of portland cement, other special cements are available. Blended cements containing siliceous materials or hydraulic slips mixed with type I are used as replacements to types IV and V. Table 10 identifies the compositions of five types of portland cement, and Table 11 gives typical compressive strengths.

Type I, or general use cement, is utilized for general concrete construction when special properties of the other types are not required. Its applications include pavement and sidewalk, reinforced concrete buildings, bridges, railway structures, tanks, reservoirs, and other structures.

Type II, or moderate heat-of-hardening cement, is used in general concrete construction exposed to moderate sulfate attack, or where moderate heat of hydration is required. It is used in structures of considerable mass, such as piers or heavy retaining walls. It minimizes temperature rise, which is especially important in warm-weather areas.

Type III, or high early strength cement, is used when early high strength of structures is required, such as when the structure must be put into service quickly (usually within a week or less).

Type IV, or low heat cement, is useful when a low heat of hydration is required. This type of cement develops strength at a slower rate than Type I. It is used for massive concrete structures, such as large gravity dams, where the temperature rise resulting from the heat generated during hardening is a critical factor.

Type V, or sulfate-resistant cement, is used when high sulfate resistance is required. It is mainly used where the soil or ground water has a high sulfate level. It gains strength more slowly than Type I.

Types IA, IIA, and IIIA are designated as air-entraining cements by ASTM C175. They correspond in composition to Types I, II, and III, except that small amounts of air-entraining materials are interground with the clinker during manufacture. Air-entraining cements are made by adding a small amount, approximately 0.05%, of an organic agent that causes the entrainment of very fine bubbles in the concrete. Air-entraining portland cement has an air content of 18%, ±3%. This cement helps produce concrete with improved resistance to freeze-thaw action and to scaling caused by chemical attacks.

White portland cement is another ASTM designation of portland cement. It is similar in composition to the other types, but its color is white, rather than gray. The white color is

obtained by limiting its iron content. It is primarily used for architectural purposes.

Portland blast-furnace slag cement is produced by intergrinding blast-furnace slag of selective quality with regular portland cement clinker. An air-entraining additive may also be included. This cement can be used in general concrete construction.

Masonry cement is used for bonding brick in masonry construction. It consists of a mixture of portland cement, limestone, and filler, together with an air-entraining agent or a water-repellent additive.

Oil well cement is used for sealing oil and gas wells. It must be slow setting, resistant to high temperatures (177 °C, or 350 °F) and high pressures (124 MPa, or 18 ksi). The American Petroleum Institute specifications for oil well cement have requirements for six different classes, each covering a certain range of well depths.

Expanding cement is a special type of cement that compensates for the contraction that occurs during the hardening process by expanding an equal or greater amount. It is used for the same applications as regular portland cement, except special provisions must be made to assure an abundance of water during hardening. Expanding cement is made by grinding together 70% portland cement, 20% granulated blast-furnace slag, and 10% calcium sulfoaluminate cement. The latter is prepared by firing a finely pulverized mixture of 50% gypsum, 25% bauxite, and 25% chalk.

Refractory Materials

Refractory materials are traditionally thought of as nonmetallics that resist degradation by corrosive gases, liquids, or solids at elevated temperatures. These materials must withstand thermal shock caused by rapid heating or cooling, failure attributable to thermal stresses, mechanical fatigue due to other material contacting the refractory itself, and chemical attack activated by the high-temperature environment. These materials are required for the manufacture of most ceramic products and are specifically needed in ovens, dryers, furnaces, and high-temperature-bearing engine parts.

The properties that characterize quality refractory materials depend on the nature of the application. The most important aspect of these materials is referred to as "refractoriness." This term refers to the point at which the specimen begins to soften (or melt). Typically, refractories do not have a specific melting point; the phase transition proceeds over a range of temperatures in a phenomenon called softening. This characteristic is often quantified with a pyrometric cone equivalent (PCE), which is a measure of heat content measured by the slumping of a cone during thermal cycling.

A related, and often more useful property, is the temperature of failure under load. Refractories often fail under load at temper-

Table 11 Compressive strength data for typical cements versus time

Cement type	1-day strength		3-day strength		7-day strength		28-day strength		91-day strength	
	MPa	ksi	MPa	ksi	MPa	ksi	MPa	ksi	MPa	ksi
Type I	9.3	1.3	22.5	3.3	32	4.6	42	6.1	50.5	7.3
Type II	14	2.0	27	3.9	36.6	5.3	46.3	6.7	52.5	7.6
Type III	21	3.0	37.5	5.4	44.2	6.4	52.3	7.6	56.0	8.1
Type IV	...	...	9.6	1.4	13.9	2.0	34.3	5.0	...	...
Type V	...	...	22.1	3.2	29.5	4.3	41.3	6.0	...	...
White portland	...	...	26.5	3.8	36.2	5.3	46.5	6.7	...	...
Portland/blast-furnace	8.6	1.2	13.1	1.9	25.3	3.7	45.0	6.5	53.3	7.7

atures much less than the temperature that corresponds to the PCE. In obtaining a value for this parameter, the refractory is subjected to a known load and is subsequently heated. The temperature at which sagging or general deformation occurs is reported. This is of great interest because the value is used to predict mechanical properties during use of the refractory. The load-bearing ability of refractory materials is directly proportional to the amount and viscosity of the glass present.

Another factor that is essential to understanding the performance of a refractory is the dimensional stability. Throughout industrial use, refractory materials are subjected to heating/cooling cycles, which cause the refractory units to either expand or contract. Large changes in the dimensions will reduce stability and may ultimately lead to failure of the refractory-based structure.

A related phenomenon commonly observed with refractory materials is spalling. Spalling is generally considered fracture, splitting, or flaking of the refractory, resulting in the exposure of the inner mass of the material. Spalling is usually brought about by temperature gradients within the material, compression in the structure due to large-volume charges, and variations of the thermal expansion coefficient within the brick. Every effort is made in refractory manufacture to avoid spalling because it reduces the effectiveness of the refractory.

Two related properties that provide significant insight into the performance of the refractory materials are porosity and permeability. In most instances, increasing the porosity will reduce the strength of the material. In arrangements where there is a molten liquid in contact with the refractory, porosity allows penetration and, subsequently, deleterious chemical attack. On the other hand, when heat insulation is desirable, porosity is maximized so that the thermal conductivity is reduced and the heat is maintained within the structure. Permeability is an important characteristic of the refractory because the parameter governs gas transport, which is the route of chemical attack. Gas transport is important for maintaining even temperature controls and to avoid heat loss. The magnitude of the permeability is determined by the size, shape, and connectivity of the porosity.

The capacity to store or transmit thermal energy is crucial to the performance of a

refractory material. This characteristic is reflected in the thermal conductivity of the refractory. The thermal conductivity of refractories varies widely with material and with the temperature of interest. It should be noted that the presence of glass in a refractory material tends to lower its thermal conductivity. However, at elevated temperatures, the thermal conductivity is difficult to predict solely on the basis of the crystalline and glass content of the refractory. In most instances, it is necessary to minimize the thermal conductivity so that heat envelopment is maximized.

Classes of Refractories. There are two classifications of refractory materials: basic and high-duty oxides, the properties of which are summarized in Tables 12 and 13, respectively. The materials within both categories that are based on silica, alumina, or mullite are discussed first, followed by a discussion of the refractories based on magnesite and chrome. Other materials that are not traditionally included are also described.

Silica brick refractories consist almost entirely of SiO_2; a mixture of lime, Fe_2O_3, and Al_2O_3 is usually present due to the nature of beneficiation processes and the need to promote bonding. Because silica refractories are subject to devitrification (a type of phase separation witnessed upon cooling), use temperatures should not exceed 1250 °C (2280 °F). These refractories show a high degree of volume stability and an absence of spalling above 650 °C (1200 °F), compared to fireclay refractories.

Another type of silica is referred to as fused silica. Its (pure SiO_2) composition puts it in the class of high-duty refractory oxides. This material possesses many properties similar to those of ordinary silica refractory but has a much lower use temperature. Additionally, pure SiO_2 refractories are transparent to visible light, which makes the material attractive for use in windows in high-temperature environments.

Fireclay is a basic refractory consisting of primarily hydrated alumino silicates with an SiO_2 content of up to 78% and a content of Al_2O_3 and other minor constituents remaining below 38%. Fireclay materials vary widely in composition and, therefore, in properties. Generally, firebricks based on kaolin show higher refractoriness and load resistance. Increases in the porosity of firebrick reduce

the amount of spalling that occurs. Resistance to chemical attack is reduced with increasing porosity. It is also believed that a higher Al_2O_3 content increases the resistance to attack by molten materials that contact the refractory material.

High alumina refractories typically vary in alumina content, from 80 to 99+%. These refractories fall in the basic class, with the exception of pure alumina oxide, which is considered a high-duty refractory oxide. All alumina refractories possess a greater refractoriness and load-bearing ability, compared to the fireclay refractories. Their chemical resistance is also greater than that of the fireclay refractories. Refractoriness and strength are increased with higher alumina contents. It is also possible to produce a 99+% Al_2O_3 with zero porosity that is highly resistant to chemical attack.

Mullite is a refractory material that is sometimes grouped with the high-alumina refractories. This refractory type is composed primarily of the mineral mullite ($3Al_2O_3 \cdot 2SiO_2$); the remainder of the matrix is usually Al_2O_3. Mullite refractories possess properties similar to those of the high alumina.

Basic Refractories. There are three typical classes of basic refractories whose major raw materials are magnesite ($MgCO_3$) and chrome (Cr_2O_3). The first is simply referred to as magnesite; its composition is dictated by an 80 to 95% MgO content. The name magnesite originates from the mineral that is often used in manufacturing the refractory. Characteristic of the MgO, the magnesite refractories show an extremely high refractoriness. These materials also possess a high thermal conductivity and a good resistance to corrosion by basic materials. Chromite refractories typically contain 30 to 45% Cr_2O_3. The chromite refractories yield excellent resistance to corrosion in a basic environment and a moderate ability to withstand attack in acidic regimes. Chrome-magnesite refractories are composed of greater than 60% MgO with Cr_2O_3. These materials show high refractoriness, high volume stability, and resistance to chemical attack.

Beryllium oxide is a refractory oxide that has limited applications, but is interesting nevertheless. Beryllia has a very high thermal conductivity, high strength, and excellent resistance to spalling. However, it is easily degraded in many acidic environments. Ber-

Table 12 Composition and selected properties of basic refractory materials

Type	Composition	Maximum use temperature in oxygen		Thermal conductivity, kcal/min · °C			Refractoriness under load of 197 kPa (28.5 psi)	
		°C	°F	At 300 °C (570 °F)	At 800 °C (1470 °F)	At 1200 °C (2190 °F)	°C	°F
Silica	93–96% SiO₃	1700	3090	0.8–1.0	1.2–1.4	1.6–1.8	1650–1700	3000–3090
Fireclay	15–45% Al₂O₃ 55–80% SiO₂	1300–1450	2370–2640	0.8–0.9	1.0–1.2	2.5–2.8	1250–1450	2280–2640
Magnesite	80–95% MgO Fe₂O₃, Al₂O₃	1800	3270	3.8–9.7	2.8–4.7	2.5–2.8	1500–1700	2730–3090
Chromite	30–45% Cr₂O₃ 14–19% MgO 10–17% Fe₂O₃ 15–33% Al₂O₃	1700	3090	1.3	1.6	1.8	1400–1450	2550–2640
Chrome magnesite	>60% MgO Fe₂O₃, Al₂O₃	1800	3270	1.9–3.5	1.4–2.5	1.8	1500–1600	2730–2910

Table 13 Composition and selected properties of high-duty refractory oxides

Type	Composition	Melting point		Maximum use temperature in oxygen		Thermal conductivity, kcal/min °C				Refractoriness under load of 196 kPa (28.4 psi)	
		°C	°F	°C	°F	At 100 °C (212 °F)	At 500 °C (930 °F)	At 1000 °C (1830 °F)	At 1500 °C (2730 °F)	°C	°F
Aluminum oxide	100% Al₂O₃	2015	3660	1950	3540	26.0	9.4	5.3	5.0	2000	3630
Beryllium oxide	100% BeO	2550	4620	2400	4350	189.0	56.3	17.5	13.5	2000	3630
Magnesium oxide	100% MgO	2800	5070	2400	4350	31.0	12.0	6.0	5.4	2000	3630
Silicon dioxide	100% SiO₂	...	...	1200	2190	0.8	1.4	1.8	...	...	...
Mullite	72% Al₂O₃ 28% SiO₂	1830(a)	3325(a)	1850	3362	5.3	3.8	3.4	...	...	...

(a) Incongruant

yllia is also subject to volatilization at temperatures of 1000 °C (1830 °F) in the presence of water. Beryllia also has the drawback of being extremely toxic.

Many non-oxide refractory materials exist that are not traditionally included within this subject area, but due to their widespread use, discussion is warranted. These materials are usually separated according to anion and include carbides, nitrides, borides, and silicides. The most common of the carbides is silicon carbide (SiC). In SiC production, bonding of the material is sometimes problematic. Additives are often incorporated to alleviate the difficulty, but the presence of the additives often degrades the mechanical strength and high thermal conductivity of SiC. In some instances, silicon nitride is used to bond the SiC, resulting in an improved refractory because the susceptibility of SiC to undergo deleterious oxidation is reduced.

Boron nitride is a member of the nitride family of refractory materials. This material resists spalling to a large degree. Boron nitride is subject to slow oxidation at temperatures between 700 and 1000 °C (1290 and 1830 °F). It is often used in reducing atmospheres because of its ability to withstand this type of chemical attack.

Applications. Refractory materials are most commonly utilized in the steel and iron manufacturing industries, which account for 63% of refractory consumption in the United

States. Fireclay bricks are used to construct the furnaces used to produce steel. If added protection is required, silica, mullite, or high-alumina refractories may be incorporated as needed. Often, the outside of the refractory is aided in heat exchange by water cooling. As expected, refractory materials in contact with the steel and in the hottest regions of the furnace are subject to the most rapid degradation.

The nonferrous metallurgical industries account for up to 10% of the refractories consumed in the United States. Of the many nonferrous metals, copper and aluminum are most common. Process temperatures for copper can be as high as 1650 °C (3000 °F). Silica brick is the refractory most often used for copper production because it provides the required chemical resistance at the necessary use temperatures. High-alumina refractories are used to construct the furnace used to produce alumina. Typical compositions have 85% alumina with phosphate bonding. The phosphate provides excellent resistance to chemical attack by the molten aluminum metal.

Six percent of all refractories produced in the United States are used by the glass-making industry. The refractories must resist attack by the molten glass contained within the structure. Any leaching of the refractory will alter the purity and quality of the manufactured glass. Corrosion resistance is the most important parameter for refractories in glass

making. The refractories also serve as the container for the glass and control the heat exchange. It is difficult to state the refractory type that is used in glass manufacture, due to the variation in glass composition, manufacturing tolerances, and positional furnace requirements.

There are a wide variety of refractory materials and applications that utilize them. Refractories are necessary in manufacturing environments where high temperatures are present. Application of a particular refractory often depends on use temperature, the nature of the environment, the strength required by the structure, and the nature of the desired heat exchange.

SELECTED REFERENCES

- R.H. Bogue, *The Chemistry of Portland Cement*, Reinhold Publishing, 1947, p 11–52
- W.A. Brown, *The Portland Cement Industry*, Van Nostrand, 1917
- D.B. Butler, *Portland Cements, It's Manufacture, Testing and Use*, Van Nostrand, 1899
- G.C. Bye, *Portland Cement; Composition, Production and Properties*, Pergamon Press, Oxford, 1983
- C150–86, American Society for Testing and Materials, 1989

- W.H. Duda, *Cement Data Book*, 2nd ed., Bauverlag, Wiesbaden, 1977
- *Cement, Encyclopedia of Chemical Technology*, 3rd ed., Pergamon Press, Oxford, 1982
- A.J. Francis, *The Cement Industry 1796–1914: A History*, Newton Abbot, United Kingdom, 1977
- S.N. Ghosh, *Advances in Cement Technology*, Pergamon Press, Oxford, 1983
- Glass: Ceramic Whitewares: Porcelain Enamels, *Annual Book ASTM Standards*, Vol 15.02, 1985
- L.L. Hench and R.W. Gould, *Characterization of Ceramics*, Marcel Dekker, 1971
- H.W. Hennicke, *Ber. Deut. Keram. Gesell.*, Vol 44, 1967, p 209–211
- W.D. Kingery, Ed., *Ceramic Fabrication Processes*, John Wiley & Sons, 1958–1960
- W.D. Kingery, Ed., *Ancient Technology to Modern Science*, Vol 1, *Ceramics and Civilization*, American Ceramic Society, 1984
- W.G. Lawrence and R.R. West, *Ceramic Science for the Potter*, Pergamon Press, Oxford, 1981
- F.H. Norton, *Fine Ceramics: Technology and Applications*, McGraw-Hill, 1970
- F.H. Norton, *Elements of Ceramics*, 2nd ed., Addison-Wesley, 1974
- L.S. O'Bannon, Ed., *Dictionary of Ceramic Science and Engineering*, Plenum Press, 1984
- G.Y. Onoda and L.L. Hench, Ed., *Ceramic Processing before Firing*, John Wiley & Sons, 1978
- G.W. Phelps, *et al.*, *Rheology and Rheometry of Clay Water Systems*, Cyprus Industrial Minerals, 1982
- P. Rado, *An Introduction to the Technology of Pottery*, Pergamon Press, Oxford, 1969
- Refractories, Glass, and Other Ceramic Materials; Terms Relating to Ceramic Whitewares and Related Products; C242–C272.C242–949, Part 13, *Annual Book ASTM Standards*, 1972
- D.W. Richerson, *Modern Ceramic Engineering*, Marcel Dekker, 1982
- H. Salmang, *Ceramics: Physical and Chemical Fundamentals*, 4th ed., Butterworth, London, 1961

Overview of Technical, Engineering, and Advanced Ceramics

David C. Cranmer, National Institute of Standards and Technology

THE TERMS ADVANCED, engineering, or technical ceramics refer to ceramic materials which exhibit superior mechanical properties, corrosion/oxidation resistance, or electrical, optical, and/or magnetic properties. These materials include many monolithic ceramics as well as particulate-, whisker-, and fiber-reinforced glass-, glass-ceramic-, and ceramic-matrix composites. While ceramics have been used for over 3000 years, the materials discussed in this article have generally been developed only in the past 30 to 50 years (Ref 1). These materials are, or have the potential to be, used in a large number of applications where resistance to temperature, stress, and/or environment is required. Specific applications include electronic sensors and devices, optical elements, magnetic devices, tribological components, and structural components. The monolithic materials discussed below are most frequently formed by powder processing techniques followed by firing and consolidation procedures to achieve a dense body. Such densification processes are described in detail in subsequent Sections of this Handbook. Composites can be formed using these same techniques as well as by glass-ceramic fabrication processes (see the Section "Glass Processing" in this Volume).

Development of Advanced Ceramics

Advanced ceramics have been developed using a number of basic principles relating several different levels of structure including atomic, electronic, grain boundary, microstructure, and macrostructure (Fig 1). The interactions of these structural levels result in materials which have properties suitable for specific applications. The successful development of these materials and their successors requires an in-depth knowledge and use of thermodynamics, kinetics, phase equilib-

ria, and crystal structure. An example is the variety of electronic ceramics in use today. Initial dielectric ceramics were based on relatively simple materials such as porcelains, glasses, and steatites. With the discovery of barium titanate ($BaTiO_3$), it was found that much higher dielectric constants could be achieved and controlled over a greater temperature range. Over the years, additives such as strontium have been used which change the temperature response of the dielectric as well as its Curie temperature. Because $BaTiO_3$ is also piezoelectric, it has led to developments in transducer technology, which in turn has driven development of related electrostrictive materials such as lead magnesium niobate (PMN) and lead magnesium titanium niobate (PMTN), respectively.

Continued development of these ceramics will rely on understanding the interactions between processing, microstructure, properties, and performance. While the process of materials improvement has typically been approached empirically, progress in applying statistical experimental design and analysis techniques to materials research for a cutting tool application (Ref 2) has shown that processing windows can be readily established, and correlations made between primary characteristics (processing parameters such as hot pressing temperature, time, and pressure, and matrix powder and reinforcement characteristics, such as whisker type and volume fraction), secondary characteristics (microstructural features such as matrix grain size and chemistry, and whisker distribution), tertiary characteristics (mechanical properties such as hardness, fracture toughness, and strength), and performance characteristics (flank wear, nose wear, notch wear, tool lifetime).

One commonly perceived problem associated with monolithic ceramics is a lack of fracture toughness or damage tolerance. Low fracture toughness (K_{Ic}) can lead to sudden, catastrophic failure of a component. To over-

come this difficulty, the microstructure can be altered to provide toughening sites which can consist of large grains in a fine grain matrix (Ref 3) or a variety of reinforcements (particulate, whisker, and fiber) which increase the toughness can be added or grown in situ. Fiber-reinforced composites are dealt with in Volume 1 of the Engineered Materials Handbook. An additional complication in designing with these materials lies in the fact that a large number of ceramics are susceptible to subcritical crack growth under stress in the presence of environmental contaminants such as water, acids, bases, and some organics (Ref 4). This subcritical crack growth can cause premature failure of a component. However, it is possible to quantify this crack growth behavior and incorporate it into the design of the component, as discussed in more detail in the section "Failure Analysis" in this Volume. In essence, the lifetime of a component can be predicted using a combination of statistics (typically based on Weibull weakest link theory) and fracture mechanics.

Current research areas which may provide engineers with the ability to tailor these materials for specific applications are the measurement of interfacial phenomena including surface forces (Ref 5), which are important in structural (Ref 6) and electronic (Ref 7) ceramics, and modelling and simulation techniques for predicting behavior. Additional advances must be made in crack growth mechanics to relate more fundamental phenomena to macroscopic properties, and in the modelling of these phenomena starting from the properties of the lowest levels of structure (for example, elastic constants) to the highest (for example, stresses and strains as determined from finite element modelling).

In the following sections of this article, the important properties for the applications noted above will be discussed briefly along with some of the specific materials which are currently in use and their range of properties. Additional general information can be found

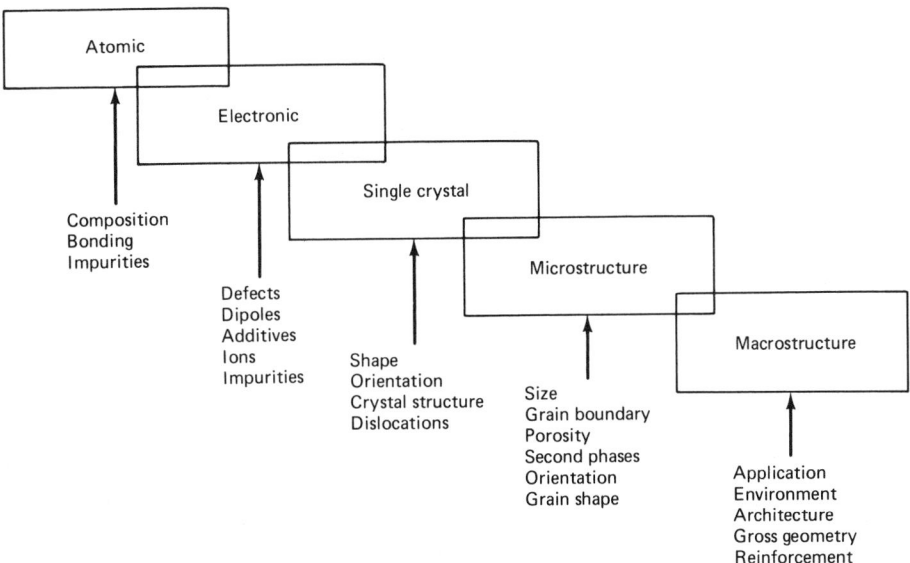

Fig 1 Structural levels and interactions of ceramic materials

in the following Sections and articles of this Volume.

Electronic Ceramics

The electronics industry relies heavily on advanced ceramic materials such as Al_2O_3, Al_2O_3-TiO_2, BeO, and AlN for substrates, barium titanate ($BaTiO_3$), for capacitors; lead zirconate titanate (PZT), lead magnesium niobate (PMN), and lead magnesium titanium niobate (PMTN) for actuators and transducers; ZrO_2-based ceramics for oxygen sensors; ZnO-based materials for varistors; NiO and Fe_2O_3 for temperature sensors; and a variety of glassy materials for a host of devices and packaging. Ceramic oxides which exhibit superconductivity may soon be used in a number of applications (Ref 8). In addition, the ceramics can be combined with polymers or metals to form composites, or hybrids, having unique electrical and structural properties. Ceramics are important in consumer and military electronics, communications systems, power transmission and distribution systems, automotive and other forms of transportation, as well as in computers. Applications for these classes of materials, which currently generate the largest market volume and dollar value of any of the advanced ceramics, are described in the Section "Electrical/Electronic Applications for Advanced Ceramics" in this Volume.

As noted earlier, the successful development of electronic ceramics is the result of detailed knowledge of crystal chemistry, phase equilibria, thermodynamics, kinetics, atomic structure, and electronic structure. These ceramics and hybrids will be the first of the emerging "smart" materials, which can sense their environment and respond to it.

The ceramic properties which are important for electronic applications result from a variety of mechanisms which depend on the bulk material, grain boundary properties, and surface effects. Important properties include dielectric constant, dielectric strength, electrical conductivity, dielectric loss tangent and power loss, Curie temperature, and piezoelectric constant. The dielectric constant is a measure of the amount of charge that a material can hold. The dielectric strength is a measure of the voltage gradient the material can withstand before failure. The critical temperature (T_c) is the temperature at which a material becomes superconductive, that is, has no resistance to the passage of an electrical current. The critical current of a superconductor is the maximum amount of current the superconductor can transmit before reverting to a non-superconductor. For a number of applications, the material must be able to conduct heat away from electronic circuit elements, hence the ceramic must have good thermal conductivity. The physics of electronic ceramics are reviewed in Ref 9.

Dielectrics/Piezoelectrics. Dielectrics, or insulating ceramics, are used generally in the form of capacitor elements and electronic insulators. The dielectric materials can be of either bulk, thick film, or thin film form. Piezoelectrics are materials which deform upon the application of an electric field, or conversely, which develop an electric potential when strained. To be effective, these ceramics require poling, a process which aligns the ferroelectric domains in a common direction in various grains. This process is performed during cooling of the material from above the Curie temperature by exposure to an external electric field. Poling can also result in the introduction of residual stresses due to the volume change in going from one crystal structure to another.

Electrostrictive Ceramics. A related set of materials are the electrostrictive ceramics such as lead magnesium niobate which also develop strains when exposed to an electric field. Both piezoelectrics and electrostrictive materials are commonly used in transducer applications such as microphones, hydrophones, accelerometers, and strain gages. They also find use in complex optical systems, such as actuators to aid in focusing and correcting optical aberrations (Ref 10), and additionally may find application in precision machining and video camera uses.

Pyroelectrics. A variation of these types of materials are the pyroelectrics which develop a voltage in response to an applied temperature difference, or conversely, heat or cool in response to an applied voltage. They include materials such as lithium tantalate ($LiTaO_3$), and are used in applications such as radiometry, Fourier transform infrared spectroscopy, optical pyrometry, blood gas analysis, horizon sensing, proximity fuses, and flame detection.

High-temperature superconductors can be prepared in the same manner as other advanced ceramics, namely via powder processing routes. Their properties depend critically on the chemistry and microstructure. The most common high T_c superconductor is the $YBa_2Cu_3O_{7-x}$ composition. Most of the rare earths can be substituted for Y in this composition (Ref 11). High T_c superconductors find uses in either thin film or bulk forms. Future applications of these superconductors may include superconducting quantum interference devices (SQUIDs), electrical power transport, magnets, motors, and in communication and computer equipment.

Semiconducting ceramics include materials such as silicon carbide (SiC) which have properties similar to silicon, but retain their semiconductive properties at elevated temperatures. These materials are expected to find applications in elevated temperature electronics which might be used in industrial pollution detection and correction systems.

A number of important electronic ceramics are used in the form of thin films. Films can be grown by epitaxial techniques, chemical vapor deposition, or other deposition techniques. This form of electronic ceramics can be used to place devices or components directly on silicon, GaAs, or other semiconductors to fabricate integrated devices such as ferroelectric memories (Ref 12).

Optical Devices. A variety of optical systems require the use of materials which can modify an optical signal by means of an imposed electric field or signal. Materials such as $LiNbO_3$, $LiTaO_3$, $Ca_2Nb_2O_7$, and others are used to accomplish this purpose. Their existence enables the creation of a number of important devices such as oscillators, frequency doublers, and optical modulators of various types (electro-optic phase shifters, beam splitters, beam deflectors, and polarizers) used in laser-based systems for optical communications and will make possible integrated optical devices.

Ionic conductors, such as β''-Al_2O_3, chalcogenides, and halides containing silver and/or copper, find use as fuel cells, batteries, and ion activity probes. These materials rely on their ability to transport ions, rather than electrons, for their conductive properties. The crystal structures which permit this type of ion motion typically have arrangements of anions such that channels or planes are present through which the cations can move. One particular application for β''-Al_2O_3, which is actually an alkali-containing material of composition $NaAl_{11}O_{17}$, is the sodium-sulfur storage battery.

Varistors, typically made from ceramics such as ZnO, sense and limit transient voltage surges. Their behavior can be modified by use of additives such as Bi, Sb, Co, Mn, Ni, Cr, and Si. The varistor takes advantage of the nonlinear current-voltage behavior of ZnO, acting as an insulator up to some voltage (threshold or breakdown voltage), and above that voltage, behaving as a conductor. Additional details can be found in the article "Varistors" in this Volume and in Ref 13.

Electronic packaging is the final application area that needs to be discussed. Packaging consists of the substrate material (Al_2O_3, AlN, BeO, BN) and the coatings applied to the entire finished circuit, or device, to protect it from the environment. A typical device contains many components and interconnections which rely on the structural interactions at all levels of the device. The final device may include the substrate, semiconductor devices, insulators, capacitors, active devices, and protective coatings of glass and/or polymers. Failure to correctly fabricate any one of these parts can lead to overall failure of the device. As the technology advances, typical feature sizes are getting smaller. This size reduction in turn requires advances in fabrication and measurement technologies as well as improved materials which have superior properties. These properties will only be achieved by an increased understanding of the underlying fundamental principles and interactions.

Optical Ceramics

This class of ceramics includes single and polycrystalline ceramics used for optical applications. Glassy materials used for optical applications are discussed in the section "Applications for Traditional and Advanced Glasses" in this Volume. In addition to lenses and windows, other applications for optical ceramics include phosphors for cathode ray tube (CRT) screens, fluorescent lights, and scintillation counters, laser components including lasing crystals, mirrors, and electro-optic and acousto-optic components, windows used in supermarket check-out counters (single-crystal sapphire, Al_2O_3), and radomes. Optical properties are also important in some primarily structural applications such

as dental implants where the appearance of the material is important.

The optical properties of interest include transmission, absorption, color, refractive index, reflectivity, and emissivity. Most of the optical properties are affected by the condition of the surface.

Phosphors. Materials typically used for phosphors include $CaWO_4$ with or without a lead activator (blue), $BaSi_2O_5/Pb$ (UV), Zr_2SiO_4/Mn (green), $CaSiO_3/Pb,Mn$ (yellow to orange), $Cd_2B_2O_5/Mn$ (orange-red), $Ba_2P_4O_7/Ti$ (blue-white), $Sr_2P_4O_7/Sn$ (blue), $Ca_5(PO_4)_3(Cl,F)/Sb,Mn$ (blue to orange and white), $(Sr,Zn)_3(PO_4)_2/Sn$ (orange), $Mg_6As_2O_{11}/Mn$ (red), $3MgO\cdot MgF_2\cdot GeO_2/Mn$ (red), YVO_4/Eu (red), and $MgGa_2O_4/Mn$ (green).

Infrared domes are components which transmit well in the infrared (IR) and are used to protect infrared sensors. In addition to transmission in the IR, these materials must be resistant to erosion from both dust and rain, and have good thermal shock resistance. The materials used can be either single crystal (ZnS, ZnSe, MgF_2, CaF_2, sapphire) or polycrystalline (ZnS, ZnSe, MgF_2, Y_2O_3, ALON, Spinel). The material of choice depends on the wavelength region of interest. For example, ALON (aluminum oxynitride) can be used in the wavelength region from 1 to 4 μm, Y_2O_3 from 1 to 8 μm. The materials can be fabricated by pressureless sintering, hot pressing, hot isostatic pressing, or chemical vapor deposition (CVD).

Lasing Crystals. Another important set of optical ceramics is the lasing crystals (yttrium-aluminum-garnet, or YAG, yttrium-indium-garnet, or YIG, and Alexandrite), and electro-optic components which result in changes to a lightwave as a result of an imposed electric field. A common example of an electro-optic ceramic material is lithium niobate ($LiNb_2O_3$).

Mirrors. Silicon carbide is being considered for application in space- and ground-based mirrors. SiC is attractive because of its higher specific strength and stiffness (compared to glass). Material with good microstructures and uniform properties can readily be made by CVD in sizes up to 0.5 m (1.5 ft).

Magnetic Ceramics

Magnetic ceramics form the basis for numerous devices which rely on soft or hard (permanent) magnets. The soft magnets include materials such as the ferrites and garnets, while the hard magnets include magnetoplumbites and γ-Fe_2O_3. The applications are diverse, from items such as microwave components (for example, magnetrons) to recording tape. They are particularly useful for high-frequency devices and can be found in numerous television and radio applications. In thin films applied to nonmagnetic substrates, these types of ceramics form the basis for magnetic bubble memories for com-

puters. Deposition characteristics, particularly composition, must be precisely controlled as the lattice constant of the film controls the behavior of the magnetic domains. Materials which are suitable for microwave applications include manganese zinc ferrite and a number of garnets. The magnetic properties can also be used to make low-temperature (cryogenic) refrigerators. Such devices rely on materials such as gallium-gadolinium-garnet (GGG) for their success.

As is the case for a number of the electronic ceramics mentioned in the previous section, crystal structure and orientation as well as specific composition (both bulk and grain boundary) are critical to the development of magnetic ceramics. Also, as was the case for the electronic ceramics, the magnetic ceramics can be tailored to the application by careful control of the composition, structure, and orientation. These materials in bulk form are fabricated via the same general process as the electronic and optical ceramics discussed previously. Thin films can be formed via either sol-gel methods or epitaxial growth.

Structural Ceramics

One of the growing uses for advanced ceramics is in the area of structural or load-bearing applications. These uses require materials which have high strength at room temperature and/or retain high strength at elevated temperatures, resist deformation (slow crack growth or creep), are damage tolerant, and resistant to corrosion and oxidation. Ceramics appropriate to these applications offer a significant weight savings over currently used metals. The applications include heat exchangers, automotive engine components such as turbocharger rotors and roller cam followers, power generation components, cutting tools, biomedical implants, and processing equipment used for fabricating a variety of polymer, metal, and ceramic parts. The materials can be either monolithics or composites. A major difficulty still to be overcome before these materials see more widespread use is their cost. Many of the processes used for fabrication are labor-intensive, or have high reject rates, resulting in unacceptably high costs for the final products.

There are a number of factors to consider when attempting to tailor these materials for desired properties. These factors include the trade-offs between a material which has as high a strength as possible so that damage does not initiate; developing a significant flaw tolerance or R-curve (increasing resistance to crack propagation as the crack size increases); or addition or modification of grain boundary phases, or changes in grain size to enhance creep resistance at the expense of strength. Trade-offs in properties are, however, inevitable, and the features which provide exceptional damage tolerance (such as large, anisotropic grains), are not the same features which maximize strength (typically small, equiaxed grains). This is demonstrated in the recent development of a flaw-tolerant Al_2O_3-Al_2TiO_5 composite of 5.8 μm grain size that was compared to two Al_2O_3 materials— one of the same size as the composite, the other of much larger (79.8 μm) grain size (Ref 14). The small grain size Al_2O_3 has a higher strength than the composite, but the composite has a flaw tolerance similar to the large grain size Al_2O_3.

Automotive Components. The materials used in automotive applications must exhibit some combination of good strength, damage tolerance, creep resistance, and oxidation and corrosion resistance at temperatures up to about 1375 °C (2500 °F). Typical automotive engine components include turbocharger rotors, valves, turbine rotors, nozzles, vanes, and piston rings.

Power generation systems include stationary engines, heat-recovery systems, burners and combustors, waste incineration systems, separation and filtration systems, and insulating refractories. The property requirements for power generation systems are similar to those needed for the automotive components. The individual components for these systems require a variety of shapes and sizes including tubes, plates, and mixed thin- and thick-section parts. Maximum application temperatures can range from about 500 °C (900 °F) to approximately 1400 °C (2500 °F).

Materials for tribological applications require resistance to wear in the form of erosion, abrasion, and adhesion. They include bearings, seals, nozzles, brakes, clutches, wear pads, liners, and other rotating or reciprocating parts. These parts can operate at low temperatures such as those required for cam followers or at elevated temperatures such as a high-temperature seal. Coatings of TiN and TiC are commonly deposited on metals to make wear-resistant parts. Another tribological application, machining, is one of the critical unit operations of almost any fabrication process. Typical ceramic components for the machining operation consist of cutting tool inserts and holders, and drill bits. A number of metallic materials used for these applications also rely on ceramic coatings. As a consequence, these combinations of materials must exhibit good adhesion of the coating to the substrate, in addition to hardness and wear resistance.

Processing equipment includes chemical reactors and process equipment used for manufacturing a number of products. Ceramic-based components which go into this type of equipment include reactor vessels, pumps, valves, piping, tanks, heat exchangers, manufacturing or fabrication dies, and storage vessels for toxic and/or corrosive liquids.

Biomedical applications encompass orthopaedic and dental implants and prostheses such as dental restoratives, and knee and hip joint replacements. The key to using advanced ceramics in these applications is the ability to mimic natural bones and teeth. They must have good specific strength and stiffness, be biologically compatible with body fluids, provide a visual appearance which matches the natural system (for teeth), and resist fatigue. The candidate materials for these applications include high-purity Al_2O_3, synthetic hydroxyapatites (based on $Ca_{5-x}(PO_4)_{3-y}(OH)_{1-z}M_xN_yR_z$, where M, N, and R are substitutes for the Ca^{2+}, PO_4, and OH^{1-} ions, respectively), borosilicate glasses, and glass-ceramics.

Primary properties of interest for structural applications are strength (room and elevated temperature), modulus (elastic, shear, and bulk), fracture toughness, Poisson's ratio, creep and creep-rupture behavior, hardness, tribological properties such as abrasion resistance and friction, chemical resistance, and thermal shock. Additional properties such as optical absorption and index of refraction may also be important where visual appearance is a consideration, or electrical properties in a high-temperature sensor application.

Materials Selection. The most important of the bulk monolithic materials for high-temperature structural applications are silicon nitride (Si_3N_4), silicon aluminum oxynitride (Sialon), silicon carbide (SiC), partially stabilized zirconia, ZrO_2, (PSZ), alumina (Al_2O_3), and mullite ($3Al_2O_3 \cdot 2SiO_2$). These materials can exhibit high strengths (>500 MPa, or 70 ksi), moderate to high fracture toughness (4 to 14 MPa$\sqrt{m}$, or 3.6 to 12.7 ksi$\sqrt{in}$.), and low creep rates ($<10^{-9} \cdot s^{-1}$) at 1300 °C (2370 °F). The PSZ is somewhat limited in its application because of its relatively high density (>5 g/cm^3), and the temperature of the phase transformation from tetragonal to monoclinic (~1000 °C, or 1830 °F) on which its favorable properties are based. An important feature of all of the silicon nitrides and Sialons is the incorporation (intentional or otherwise) of glassy grain boundary phases. These glassy phases are present in different compositions and amounts, which result in a range of properties, and a set of materials which can be tailored for different temperature ranges and different applications.

Two-phase ceramics such as siliconized silicon carbides (Si/SiC) can be made by a number of processes, such as a liquid infiltration technique. They are fabricated by first making a green body, or preform, of one material such as SiC, which can be lightly sintered or densified to make a handleable preform. This body is then infiltrated with a liquid such as silicon to fill the open space in the preform. Silicon contents range from about 15 to 50% in these materials.

A second type of two-phase ceramic is the ceramic composite. These materials are designed to have improved damage tolerance or increased toughness through the addition of second-phase reinforcements in the form of particulates, whiskers, or fibers. Particulate reinforcements include several materials including ZrO_2 and SiC. The ZrO_2 reinforcement, when added in the correct amount and particle size, provides a phase transformation-toughening mechanism, which makes the material more flaw tolerant. Whisker reinforcements are typically single crystals of ceramic materials, and are available in several compositions. Whisker-type reinforcements can also be grown *in situ* in some systems such as Si_3N_4. Reinforcing fibers can be single crystals, polycrystalline ceramics, or glass.

Particulate-reinforced ceramics include those to which a second-phase particulate has been deliberately introduced into the composition. The particulates act to deflect cracks from the main propagation path, and can absorb energy by a transformation-toughening mechanism as well. Transformation-toughening relies on the introduction of a material such as ZrO_2, which exhibits a volume-expanding phase transformation from tetragonal to monoclinic. The material is processed in a manner so as to retain the tetragonal phase metastably until it is released by the presence of a crack.

Several whisker-reinforcement compositions including silicon carbide, silicon nitride, and others are available. One perceived problem with whiskers is the potential health hazard associated with them. This perception is based on the size similarity to asbestos, although there is evidence that the similarities do not lead to the same dangers as those related to asbestos (Ref 15). Whisker-reinforced composites such as SiC_w-Al_2O_3 have found use as cutting tools. For alumina, the increase in K_{Ic} is on the order of a factor of 2 versus unreinforced alumina (Ref 16). Similar increases have been observed in mullite- and silicon nitride-matrix composites. The increase in toughness is due to a number of mechanisms including whisker pull-out, crack bridging by unbroken whiskers, and crack deflection around whiskers.

Incorporation of continuous fibers is another method of providing toughening to the ceramic matrix. A number of continuous fibers are available for reinforcement of polymer, metal, glass, glass-ceramic, and ceramic matrices. These include compositions of carbon and graphite, silicon carbide, silicon carboxynitride, silicon nitride, alumina, mullite, and a variety of glasses. Single-crystal fibers of sapphire are also available. Fibers have a range of diameters, elastic moduli, strengths, and other properties, and are used not only in those composites listed above, but also in polymer- and metal-matrix-based composites. The greatest range of properties available currently be found in the carbon and graphite fiber family. These fibers are incorporated into glass, glass-ceramic, and ceramic matrices, as well as a variety of metal and polymeric matrices.

Glass-matrix composites were among the first of the "ceramic" composites to be explored, as reported in 1972 (Ref 17, 18). Their fabrication process relies on the ability to make the glass flow into the fiber preform, thus filling the void space. The use temperature of these composites is limited by the properties of the glass matrix, and is generally restricted to temperatures less than 1000 °C (1830 °F). Composites with glass-ceramic matrices are fabricated initially using the same technology as that used for glass-matrix composites, then undergo heat treatment to crystallize the matrix.

Ceramic-matrix composites are typically fabricated by hot pressing or hot isostatic pressing. A more recent fabrication method is that of chemical vapor infiltration (CVI). This latter method uses a gas flow to permeate a fibrous preform, then a chemical reaction to vapor deposit the matrix on the fibers (Ref 19). The increase in damage tolerance of these composites is the result of fiber debonding and pull-out processes which are controlled by the interfacial properties of the fiber-matrix region. The debonding process relies on the degree of chemical bonding between the fiber and the matrix, while the pull-out process depends on the surface roughness of the fiber and the degree of mechanical clamping of the matrix on the fiber. The clamping stresses applied to the fiber in turn depend on factors such as the thermal expansion mismatch between the reinforcement and the matrix, and the residual stresses present due to the fabrication process (Ref 20).

Carbon-carbon composites are probably the most extensively investigated and engineered of the structural ceramic composite materials (Ref 21). They have very high specific strength and stiffness (strength and stiffness-to-weight ratios) and maintain these properties at elevated temperatures in inert environments. They must, however, be protected when exposed to high-temperature (>600 °C, or 1110 °F) oxidizing atmospheres. This is typically accomplished through the use of coatings or surface sealants (Ref 22). Carbon-carbon composites are used in rocket nozzles, aircraft brakes, and leading edges for vehicles such as the Space Shuttle. The range of properties which can be achieved with these composites is extensive because of the wide range of carbon fibers which are commercially available, the large variations in fiber architecture which can be used, and the degree of graphitization (conversion from carbon to graphite) of fiber and matrix. The degree of graphitization is controlled by heat treatments used during the fabrication process.

The favorable thermal, chemical, and tribological properties of some of the structural ceramics can also be achieved by the use of ceramic coatings on other materials such as metals. A number of these ceramic coatings are in use including ZrO_2, TiN, TiC, and SiC. The coatings are used for thermal barriers and wear-resistant coatings. One recent development in ceramics is the ability to fabricate diamond films from a low-pressure combination of methane and hydrogen. Diamond is the hardest known substance, has excellent thermal conductivity, good transparency in the infrared, and is exceptionally resistant to corrosion. These films are expected to find applications as wear- and corrosion-resistant coatings for structural materials. The ability to make diamond using these methods was first reported in 1911 (Ref 23) but not followed up on until more recent work in the Soviet Union beginning in the 1950s and in Japan in the 1970s and 1980s. A thorough review is given in Ref 24.

Titania (TiO_2) and ZrO_2 also find uses as filters because of their good chemical resistance, high melting points, and good hardness. These oxides, as well as many others, can be produced by a solution-gelation (sol-gel) method. This method involves the initial mixing of organometallics in solution, followed by hydrolyzation and heat treatment to form the desired oxides. The end result can be either a chemically homogeneous, fine particle size powder which can be further processed to form dense bodies, or a monolith of almost any desired density. Variations of the process can result in nitrides, when the hydrolysis step has been replaced by an ammonialysis step, or carbides, if carburization replaces hydrolyzation. A variety of products are possible including the reinforcing fibers mentioned above, optical elements such as windows and lenses (typically but not exclusively glasses), electronic packaging, and protective coatings.

REFERENCES

1. W.D. Kingery, H.K. Bowen, and D.R. Uhlmann, *Introduction to Ceramics*, John Wiley & Sons, 1976
2. E.R. Fuller, Jr., R.N. Kacker, E. Lagergren, R.F. Krause, Jr., J.A. Barta, and G. Shreyer, "Microstructure, Mechanical Properties, and Machining Performance of Alumina-Silicon Carbide Whiskers," NISTIR, 1991
3. S.J. Bennison, J. Rödel, S. Lathabai, P. Chantikul, and B.R. Lawn, "Microstructure, Toughness Curves, and Mechanical Properties of Alumina Ceramics," NATO Advance Research Workshop on Toughening Mechanisms in Quasi-Brittle Materials, 16–20 July 1990, Northwestern University. To be published by Kluwer Academic Publishers, Dordrecht, The Netherlands, 1991
4. S.W. Freiman, Brittle Fracture Behavior of Ceramics, *Am. Ceram. Soc. Bull.*, Vol 67 (No. 2), 1988, p 392–402
5. R.G. Horn, Surface Forces and Their Application to Ceramic Materials, *J. Am. Ceram. Soc.*, Vol 73 (No. 5), 1990, p 1117–1135
6. M.G. Norton and C.B. Carter, Interfaces in Structural Ceramics, *MRS Bull.*, Vol 15 (No. 10), 1990, p 51–59
7. J.B. Goodenough, The Molecular Engineering of Oxides, *MRS Bull.*, Vol 15 (No. 5), 1990, p 23–29
8. J.G. Bednorz and K.A. Müller, Possible High T_c Superconductivity in the Ba-La-Cu-O System, *Z. Phys. B*, Vol 64, p 189–193
9. L.L. Hench and D.B. Dove, Ed., *Physics of Electronic Ceramics*, Marcel Dekker, 1972
10. J. Galvani, Electrostrictive Actuators and Their Use in Optical Applications, *Opt. Eng.*, Vol 29 (No. 11), 1990, p 1389–1391
11. R.J. Cava, Superconductors Beyond 1-2-3, *Sci. Am.*, Vol 263 (No. 2), 1990, p 42–49
12. J.F. Scott and C.A. Paz de Araujo, Ferroelectric Memories, *Science*, Vol 216, 1989, p 1400–1405
13. T.K. Gupta, Application of Zinc Oxide Varistors, *J. Am. Ceram. Soc.*, Vol 73 (No. 7), 1990, p 1817–1840
14. J.L. Runyan and S.J. Bennison, Fabrication of Flaw-Tolerant Aluminum-Titanate-Reinforced Alumina, *J. Mater. Sci.*, 1990
15. S.C. Weaver, Designing Safe Composite Materials, *ASTM Stand. News*, Vol 18 (No. 4), 1990, p 30–32
16. P.F. Becher, C.-H. Hseuh, P. Angelini, and T.N. Tiegs, Toughening Behavior in Whisker-Reinforced Ceramic Matrix Composites, *J. Am. Ceram. Soc.*, Vol 71 (No. 12), 1988, p 1050–1061
17. R.A.J. Sambell, D.H. Bowen, and D.C. Philips, Carbon Fibre Composites with Ceramic and Glass Matrices, Part 1, Discontinuous Fibres, *J. Mater. Sci.*, Vol 7, 1972, p 663–675
18. R.A.J. Sambell, A. Briggs, D.C. Philips, and D.H. Bowen, Carbon Fibre Composites with Ceramic and Glass Matrices, Part 2, Continuous Fibres, *J. Mater. Sci.*, Vol 7, 1972, p 676–681
19. A.J. Caputo and W.J. Lackey, Fabrication of Fiber-Reinforced Composites by Chemical Vapor Infiltration, *Ceram. Eng. Sci. Proc.*, Vol 5 (No. 9–10), 1984, p 654–657
20. C.P. Ostertag, Differential Sintering, *Advances in Sintering*, 1990
21. J.D. Buckley, Carbon-Carbon, An Overview, *Am. Ceram. Soc. Bull.*, Vol 67 (No. 2), 1988, p 364–368
22. J.R. Strife and J.E. Sheehan, Ceramic Coatings for Carbon-Carbon Composites, *Am. Ceram. Soc. Bull.*, Vol 67 (No. 2), 1988, p 369–374
23. R.C. DeVries, Synthesis of Diamond under Metastable Conditions, *Ann. Rev. Mater. Sci.*, Vol 17, 1987, p 161–187
24. K.E. Spear, Diamond-Ceramic Coating of the Future, *J. Am. Ceram. Soc.*, Vol 72 (No. 2), 1989, p 171–191

Overview of Traditional and Advanced Glasses

Donald L. Kinser, Department of Materials Science & Engineering, Vanderbilt University

THE FACTORS that influence the properties of glasses must be understood to ensure the proper selection and application of those glasses to engineering problems. Therefore, a brief history of the development of glasses is presented, along with a description of the evolution of relevant properties as glass technology matured.

Traditional Glasses

Glasses were used in jewelry in Egypt as early as 1480 B.C. because they possessed desirable "gem-like" colors arising from the inclusion of transition metals in the glass composition (Ref 1). Some compositions of the early glasses, described in clay tablets from Assyria dated 1000 B.C., were relatively high in alkali or other species, which reduced melting points but also conferred inferior corrosion resistance on the glasses (Ref 2). Although poor corrosion resistance limited the preservation of examples of this material, the arid climate of Egypt facilitated the preservation that did occur.

The development of vase manufacturing techniques that fused glass on the exterior of a sacrificial clay pattern, allowed the preservation of vases dating back to 1450 B.C. The ability to fuse glass objects larger than "gem size" melts demonstrates an evolutionary step in the developing technology.

The blow pipe, invented circa 250 B.C., necessitated a glass-blowing mechanical aptitude and the modification of glass compositions to include more glass former, generally SiO_2, to attain the requisite viscosity for blowing. Increases in the glass former content meant that higher melting temperatures and longer melting times had to be achieved to melt and homogenize the glass. This technology advancement led to the appearance of glass containers by A.D. 79. Such containers were used only by the extremely wealthy, but numerous examples from this period have been identified.

Further refinements of glass-blowing skills led to the appearance of colored glasses, which were used in church windows by the year 591.

The initial manufacture of such "flat" glass was accomplished by spinning a lump of glass on the end of a handling rod so that a dinnerplate-shaped piece of glass with a thick center and thin edges was produced. Such shapes were either cut to the desired final shape or used in circular, as-manufactured form.

The optical quality and surface finish of early flat glass was poor, by current standards, but had improved by the year 1100 to allow the manufacture of glass mirrors.

Scientific Glass Era

As western Europe emerged from the Dark Ages and scientific studies were begun, glass was used for scientific apparatus. One of the earliest examples was the glass thermometer, which appeared around 1550.

Early work in microscopes and telescopes led to systematic studies of glass optical properties and the development of glass lenses by 1590. Scientists such as Faraday (Ref 3), identified shortcomings in quality and optical properties of the existing optical glasses and conducted extensive studies of new compositions and techniques for improving melting processes. Faraday identified platinum as an excellent crucible material for glass melting and collaborated with his platinum supplier, Wollaston, who developed melting and forming techniques for platinum.

The need for enhanced corrosion resistance of chemical labware, driven by the inception of modern chemistry, led to the development of acceptable corrosion-resistant glasses by 1824. The evolution of glass science prior to 1900 thus consisted primarily of advances in compositional formulation, melting, and forming technology.

Modern Glass Types and Processes

Advances during the 20th century have also focused on formulation, melting, and forming. However, the challenges of property optimization has been extended to include many new physical and chemical properties that had not been examined in previous works.

Silicate Glasses. Early efforts to produce optical-quality flat glass used techniques based on glass blowing, which produced transparent but optically inferior products. Glass blowing-based techniques were supplanted by the Foucalt, or Pittsburgh, forming processes. These processes produced flat glass by drawing molten glass from a melt and subsequently flattening it with rolls. The optical quality of those flat glasses was further improved by the plate glass process, which mechanically polished a glass that had been cast in flat form. The ability to manufacture improved, optical-quality flat glass developed in the early 1900s, but the cost of producing such a material remained relatively high.

Development of the float glass process by Pilkington (Ref 4) permitted production of superior, optical quality flat glass at modest cost. This forming process consists, in simplified form, of pouring molten glass on one end of a pool of molten tin, allowing it to cool as it travels over the molten tin bath, and removing it in a continuous fashion. Pilkington's development represents an especially innovative and efficient technique for forming glass to the desired shape and currently represents the forming technique used for a large fraction of the flat glass being produced.

Solution-gelation, or sol-gel, processes for forming glasses have recently achieved commercial application as processing techniques. The unique advantage of these techniques is that the product may be formed from a gel *without* melting the material. The melting process has always presented difficulties in terms of achieving homogeneity of the glass and avoiding refractory contamination. The energy cost required to fuse glasses is reduced, although the cost of raw materials is increased.

The sol-gel processes permit fabrication of near-net shape glass products by casting the gel forming solution in suitably shaped molds, although considerable shrinkage occurs during subsequent processing. Sol-gel processing techniques also permit the preparation of glass compositions that cannot be prepared utilizing conventional fusion either because a glass component is volatile or because crystallization occurs during cool-down from the fusion temperature.

Sol-gel processes lend themselves to fabrication of composite structures, films, powders, and porous media because the shape may be preserved during processing. This technique is used to fabricate antireflection or color coatings for architectural and other glass products. Films of electrically conductive transparent materials, such as SnO, are prepared by sol-gel techniques and used to deice aircraft windscreens.

Vapor phase processes (Ref 5) used to prepare glasses were initially scientific curiosities, but became mainstream processes to satisfy the demand for higher-purity glasses for long optical paths in fiber optic systems. The optical clarity of such glasses primarily depends on transition metal impurity levels, which are dramatically reduced when vapor phase processes are used. A reaction typical of such vapor phase processes is:

$$SiCl_4 + O_2 \xrightarrow[\text{Heat}]{} SiO_2 + Cl_2 \qquad \text{(Eq 1)}$$

The reactant $SiCl_4$ can be prepurified to produce extremely low levels of transition metal impurities. Thus, the resulting glass has low impurity and low optical absorption.

The reaction in Eq 1 produces a sootlike product that is subsequently sintered to a fully dense glass. The low optical absorption of state-of-the-art fiber optics circa 1990 arises from impurity levels that approach one part per billion for the transition metals. Such processing strategies have permitted development of fiber optic communication systems that use glasses whose optical clarity significantly exceeds that of the air of most urban cities.

Photosensitive glasses (Ref 6), which darken upon exposure to light and clear once the light source is removed, have been prepared by adding silver halide to traditional melted glasses. Photosensitive glasses consist of dispersed spheres of the silver halide in a glassy matrix. As light is incident upon the material, the following reaction occurs:

$$AgCl + h\eta \rightarrow Ag^° + Cl_0 \text{ (g)} \qquad \text{(Eq 2)}$$

where $h\eta$ is the radiation- or light-supplied energy.

When the light source is removed, the silver and chlorine recombine, and the metallic silver responsible for the optical absorption is removed, thus clearing the glass.

This is the same process that occurs in the formation of a latent image in conventional photography, except in that case the chlorine is permitted to escape. In the glass, the chlorine is confined, which causes the reverse reaction. The response speed of the darkening/clearing may be controlled by modifying the particle size of the distributed silver halide.

Nonsilicate Glasses

Chalcogenides. Xerography and laser printing are applications of chalcogenide glasses, which require control of electrical conductivity and photoconductivity. The requisite combination of properties was initially discovered in glasses of the arsenic-selenium family. These glasses were produced using traditional melt fusion techniques, but the production of many xerography drums currently is based on vacuum deposition or reactive sputtering techniques and nontoxic glasses.

Considerable attention was directed to glasses of the chalcogenide family because of their two-state, or switching, behavior. Initial expectations were for the development of computer memory systems based on this unusual electrical behavior. However, as computer speed increased, the transition between the two states proved to be too slow for useful computer memory applications.

Glasses based on arsenic and sulfur have been known for many years, but techniques have only recently been developed to purify these glasses so that their desirable infrared (IR) transmission may be used. Such materials are used in IR lenses, as well as for short lengths of IR fiber optics.

Continuing work in the chalcogenide glass area is focused on the development of solar cells. Although optimism for breakthrough is high, few of the chalcogenide glasses have fulfilled expectations as yet.

Halogen-Containing Glasses. Fluoride and halide glasses were initially examined by Vogel (Ref 7) in Germany during the 1950s, but those glasses, while scientifically fascinating, found few applications. As fiber optic technology matured and interest in IR transmitting glasses increased, Lucas (Ref 8), in France, extended the work to heavy metal fluorides and discovered a number of glasses with improved IR transmission characteristics. These glasses are very challenging to melt and form without introducing impurities that degrade optical properties. Improvements have been made and some of these glasses are or soon will be commercially available.

Hydrogenated amorphous silicon (Ref 9) is an important emerging glass material. It has spawned rapid developments in the amorphous semiconductor area because it is a very versatile material that can be tailored to specified electronic properties. Among its applications are photoreceptors for electrophotographic and laser printing, the switching material in liquid crystal displays, large photovoltaic panels, and other applications requiring semiconductor properties that can be processed into large, curved, or flexible forms.

Metallic glasses are composed of predominantly metallic elements, such as Pd-Si, Ni-P, Nb-Ni, Zr-Fe, Ca-Mg, and Al-La (Ref 10). These glasses generally have metallic luster and electrical conductivities similar to the metals from which they are formed.

Glasses of this type are generally prepared by nonconventional techniques, such as splat or other rapid cooling, vapor deposition, electrochemical deposition, and mechanical amorphization.

Metallic glasses currently represent an area of intense research. Extensive applications of these glasses appear promising, but magnetic applications are, as of the early 1990s, most advanced. Glasses of the Fe-Ni-P-B family have found applications as low loss transformer cores because of the advantages they offer.

Mechanical properties of the metallic glasses are unique in that they are extremely strong, sometimes approaching the theoretical fracture strength, and they also exhibit limited ductility, in contrast to more traditional glasses. The mechanical behavior appears to be based on the fact that dislocations, and therefore the local fracture mechanism, cannot operate. Therefore, rupture of bonds along the plane of maximum resolved shear stress is required for failure.

Solder, or sealing, glasses are formulated to have low melting points and thermal expansion coefficients suitable for joining various combinations of metals, glasses, and glass-ceramics. Glasses of these types find uses in a wide variety of joining applications. They are most common in electronic systems, where they are applied to seal packages and are used for hermetic electrical feedthroughs, as component mounting adhesives, and as components of thick-film resistors, conductors, and dielectrics. Some of these glasses are, in fact, glass-ceramics in that they are intentionally crystallized during processing to enhance properties, as discussed in the section "Glass-Ceramics" in this article.

Oxynitride Glasses. Substitution of nitrogen for oxygen in traditional oxide glasses results in glasses with significant property enhancements over traditional oxide glasses (Ref 11). These glasses are difficult to prepare because of the need for high temperatures and the necessity for low oxygen content in the melting atmosphere. Hybrid sol-gel preparation processes in nitrogen-rich atmospheres hold promise for reducing the problems of preparation, and these glasses may soon reach commercial application. These glasses may be crystallized by a glass-ceramic like process to yield polycrystalline nitride ceramics with exceptional high-temperature properties.

Glass-Ceramics

Reamur conducted several experiments in which he filled wine bottles with sand to prevent slumping and crystallized the bottles to produce large crystals. The mechanical properties of such crystallized materials were so weak that he encountered problems of breakage when removing the sand from the bottles. Until the discovery of controlled crystallization, the crystallization of glass was carefully avoided by glass manufacturers because crystals were normally considered a defect in glasses.

The glass-ceramic process, discovered by Stookey and Armistead (Ref 12), is a technique by which many small crystals are grown in a glass in order to transform the glass article to a fine-grained, fully dense polycrystalline ceramic. Such ceramics generally have mechanical properties that are superior to the glasses from which they were formed and they are much more stable at high temperatures. The advantage of ceramics formed by this process is that they contain no porosity, as is normally encountered in conventionally processed polycrystalline ceramics. The process for manufacturing glass-ceramics consists of melting a glass and forming the desired shape, which is followed by crystallization using a multiple-temperature heat treatment.

Glass-ceramics may be produced in a variety of compositions that permit tailoring physical properties such as thermal expansion. Noteworthy glass-ceramics are the low-expansion types, which can, through compositional and heat-treatment modifications, have near-zero thermal expansion coefficients over limited temperature ranges. Such products have found applications in reflective telescope mirrors, cookware, range tops, heat exchangers, and other products that require dimensional stability with temperature changes.

Modifications of the initial processing technique have led to several other important forms of the basic glass-ceramic (Ref 13). For example, machinable glass-ceramic is tailored to precipitate flakes of mica in the structure, which confer crack arrest sites, resulting in a material that can be machined with conventional high-speed steel tooling. Such a material is useful in low-volume production situations.

Photomachinable glass-ceramics are compositionally modified glass-ceramics that include photoactive species such as silver, which enables the control of crystallization with exposure to light. Those portions of the glassy material that are exposed to light and crystallized by heating may subsequently be removed by chemical dissolution with an etch that preferentially attacks the crystallized portion of the material. Complex shapes may thus be formed using photolithographic like techniques.

Eastern European workers have developed building materials, such as exterior wall cladding and floor tile, using glass-ceramic processes that employ inexpensive raw materials such as slag and other materials. These materials, called "slag sitalls," display remarkable weathering resistance, although their overall high cost has limited applications.

Future Trends

The important elements throughout the historical development of glass have been improvements in compositional formulation, melting, and forming. It is almost certain that important breakthroughs will continue to be classified in these areas, although there may be simultaneous compositional and preparation techniques that produce breakthroughs in the future. This would be a consequence of the fact that many compositions with desirable combinations of properties may be unattainable until a technique is devised for preparing a glass of the composition desired.

Among the research on compositional advances in glass, the work being done on nitride glasses in the early 1990s is likely to produce important improvements in glass properties. The current work being done on halogen-containing glasses may also be expected to lead to glasses with desirable new properties.

Emerging new techniques for preparing and forming glasses include mechanical amorphization, ion implantation, sol-gel-based processes, and reactive sputtering techniques. Many of these preparation techniques are likely to permit production of glasses with new compositions, and, therefore, glasses with desirable property combinations.

REFERENCES

1. H.W.S. De Jong, *Ullmann's Encyclopedia of Industrial Chemistry*, VCH Verlagsgesellschaft mbH, Weinheim, West Germany, 1989, p 365–432
2. A.L. Oppenheim, R.H. Brill, D. Borag, and A. von Saldern, *Glass and Glassmaking in Ancient Mesopotamia*, Corning Museum of Glass, 1970, p 35
3. M. Cable and J.W. Smedly, Michael Faraday—Glassmaker, *Glass Technology*, Vol 30, 1989, p 39–46
4. C.K. Edge, Float Glass, *Adv. Ceram.*, Vol 18, 1986, p 43–49
5. P.C. Schultz, Vapor Phase Materials and Processes for Glass Optical Waveguides, *Fiber Optics*, B. Bendow and S.S. Mitra, Ed., Plenum Press, 1979, p 3–31
6. R.J. Araujo, Photosensitive Glasses, *Commercial Glasses*, Vol 18, *Advances in Ceramics*, D.C. Boyd and J.F. MacDowell, Ed., American Ceramic Society, 1986, p 151–155
7. W. Vogel, *Struktur und Kristallisation der Gläser*, VEB Deutscher Verlag für Grundstoffindustrie, Leipzig, East Germany, 1965
8. J. Lucas and C.T. Moynihan, *Halide Glasses*, Trans Tech Publications, Aedermannsdorf, Sweden, 1986
9. W.E. Spear, The Study of Transport and Related Properties of Amorphous Silicon by Transient Experiments, *J. Non-Crystal. Solids*, Vol 59/60, 1983, p 1–14
10. S.R. Elliott, *Physics of Amorphous Materials*, Longman, London, 1983
11. R.E. Loehman, Oxynitride Glasses, *J. Non-Crystal. Solids*, Vol 42, 1980, p 433–446
12. P.W. McMillan, *Glass-Ceramics*, Academic Press, London, 1964
13. G.H. Beall, Glass-Ceramics, *Commercial Glasses*, Vol 18, *Advances in Ceramics*, American Ceramic Society, 1986, p 157–173

Characterization of Ceramics and Glasses: An Overview

Victor A. Greenhut, Department of Ceramics, Rutgers, The State University of New Jersey

THE CHARACTERIZATION, testing, and nondestructive evaluation of ceramics and glasses are vital to manufacturing control, property improvement, failure prevention, and quality assurance. A fundamental understanding of ceramic and glass materials is requisite, as well. This article provides a broad overview of characterization methods and their relationship to property control, both in the production and use of ceramics and glasses.

Important aspects that are covered in this overview are:

- The means for characterizing ceramics and glasses and the corresponding rationale behind them
- The relationship of chemistry, phases, and microconstituents to engineering properties
- The effects that the *structure* of raw ceramic materials and green (unfired) products and *processing parameters* have on the ultimate structure and properties of the processed piece
- The effects that trace chemistry and processing parameters have on glass properties, which are more significant than the effects of raw glass material microstructure

Those who have only an informal background in this technology should be aware of several sources of potential frustration. First, slight variations in chemistry of the ionic-covalent chemical bond that is typical of ceramics and glasses, and subtle differences in phase distribution in a ceramic are often difficult to observe, but can cause major changes to properties during production or use. Second, experience based on metals and plastics can be misleading because these materials have wider ranges of composition and microstructure, which often lead to indistinguishable properties. Third, preparation procedures associated with metals and plastics are not totally appropriate to the characterization activities associated with ceramics and glasses. For example, the techniques used to polish hard metals prior to microstructural analysis can often lead to artifacts when applied to ceramics, which, in turn, can lead to misinterpretation of key features. Careful sample preparation and application of ana-

lytical techniques are very important to a correct correlation between the structure and properties of ceramics and glasses. Fourth, there are neither comparable standards of preparation or heat treatment, nor compendia of compositions and structures for ceramics and glasses, as there are for metals.

The overviews provided in the preceding articles in this introductory section of the Volume represent a good base of information on the nature of ceramic and glass products. The Section "Testing, Characterization, and NDE" in this Volume has in-depth articles that correspond to each topic presented in this article. Furthermore, the Sections "Ceramic Powders and Processing" and "Failure Analysis" also contain articles that more fully explore specific areas relevant to characterization. Additional sources of information are listed in "Selected References" at the end of this article.

Chemical Analysis

The chemistry of a ceramic or glass is an important starting point in the characterization process because it is almost always vital to specification and property control. Very small quantities of additive can have a major effect on both processing and ultimate properties. Sintering aids are often added at the level of tenths of one percent to promote rapid consolidation of ceramic powders and prevent grain (crystal) growth. A similar level of iron impurity gives a noticeable brown tone to alumina, clay-based whitewares, and silicate glasses. This effect is significantly enhanced by the presence of titanium at trace levels. Small percentages of alkaline elements in alumina can drastically alter its dielectric constant and loss values. Similar levels of alkaline or alkaline earths may significantly deteriorate the chemical durability of alumina.

Bulk Chemistry. The general methods of spectrochemical and wet chemical analysis can be applied to ceramic materials and glasses. Methods of chemical analysis are applicable to both ceramics and glasses, but often char-

acterization techniques are specific to either ceramics or glasses (see the article "Structure and Properties of Glasses" in this Volume). It is generally necessary to dissolve the material because melting and/or vaporization of the ceramic or glass is difficult. This task requires some experience because powerful acids or bases are usually needed to perform the digestion. The whole material must be reduced without leaving either a residue or colloidal material. Interaction effects between species in the ceramic and/or reagent can have major effects on observed responses and sensitivities.

The determination of oxygen, nitrogen, or carbon in oxide, nitride, carbide, and mixed ceramics is frequently calculated by compound balance from the anion determination. This can lead to error when variable valences, complex compounds, or vacancy structures occur. It is important that the analytical facility have experience with ceramics in general and with the specific ceramic and species to be measured. This cannot be stressed strongly enough for the novice who might rely, based on previous good experiences, on an excellent analytical laboratory that is actually inexperienced in ceramic analysis.

Methods. Optical emission spectroscopy (OES) is perhaps the most common method used for major and minor constituent analyses, whereas inductively coupled plasma (ICP) and atomic absorption (AA) are usually chosen for trace element work. The bulk chemistry of solid samples and powders can be determined by x-ray spectroscopy (XRS), down to 0.1% for elements heavier than fluorine, in most cases. Organics in unfired ceramics and important chemical bonds in glass or other transparent/translucent ceramics may be analyzed by infrared spectroscopy (IR) or Fourier transform infrared spectroscopy (FTIR). In some cases, gases given off during drying and firing are analyzed by a gas chromatograph/mass spectrometer (GC/MS). Other chemical analysis methods that yield specific information, are sensitive to light elements, are more suitable for the specific sample, and/or are the most readily available can also be used.

Microchemical Analysis. It is often important to know the local chemical distributions in a ceramic, glass, or composite. This distribution might reveal the homogeneity of raw material powders or the distribution of phases in a fired piece. Such an analysis is most commonly performed down to micrometer levels using x-ray fluorescent analysis or XRS, coupled with a scanning electron microscope (SEM) to perform electron probe microanalysis (EPMA). However, this analysis can also be done at submicrometer levels when coupled with a transmission electron microscope (TEM) or analytical electron microscope (AEM). The electron beam can be localized to an area, line, or spot to perform local qualitative or quantitative chemical analyses. Both TEM/XRS and AEM can be used to analyze cross-sectioned interfaces down to nearly atomic levels. Elemental "maps" can be produced that relate directly to observed microstructure. The emitted intensities of various elements can show the location of phases. Often, heterogeneous microstructure, such as a grain boundary phase, may strongly influence properties.

The surfaces, or interfaces, of a material often have a very different chemistry than the bulk, which can affect processing characteristics such as rheology and sintering rate. The performance of fired ceramics may be dictated by surface resistance to chemical attack and ability to bond to other materials. The SEM/XRS reveals near-surface effects at a depth of about 1 μm (40 μin.). Auger electron spectroscopy (AES), scanning Auger microscopy (SAM), x-ray photoelectron spectroscopy (XPS), secondary ion mass spectroscopy (SIMS), and ion scattering spectroscopy (ISS) can be used to show what is on or within the first few atomic layers of the surface. These techniques yield specific chemical bonding information and identify the elements present and, in contrast to x-ray fluorescent analysis, are sensitive to the lighter elements. Ion sputtering can be used to remove surface layers and, thereby, profile in depth.

Phase Analysis

The specific phases present in a ceramic can strongly influence properties. For example, alumina containing up to about 15 wt% metastable zirconia in an optimum distribution can show a several-fold increase in toughness and strength. The metastable phase in zirconia can be controlled by trace chemical additives, particle size, forming methodology, and thermal treatment. It is increasingly being realized that minor oxide glassy phases, even just a few atoms thick, on grain boundaries of nonoxide and oxide ceramics can promote creep and plastic stress relief at elevated temperature, and affect ambient strength, as well.

High-temperature ceramic superconductors require close control of the formation of specific phases. Many commercial ceramic compositions show a wide multiplicity of phases, depending on thermal history, trace element content, or crystal growth conditions. Zirconia shows three phases that are stable in different temperature domains. Alumina has about five common forms and as many as two dozen minor variants, when trace impurities are considered. Silicon carbide has two major phases and a large set of polytype structures.

Although chemical analysis can provide an important indication of the phases present in a ceramic, when a number of phases are possible, the analysis must be augmented by other techniques. For instance, optical microscopy of thin sections gives optical indices and extinction angles. Crystal shape and color can be used to identify phases in either transmission or reflection modes, providing additional knowledge of the system. An example of a common application is the identification of stones (inclusions) in glass. Either SEM/XRS or EPMA may also be used to infer phases. The coordinated observation of local, microstructural-level chemistry and microstructural features such as grain shape, size, or orientation may help to show a particular phase. Other localized spectral or chemical analysis methods that may reveal specific bonding information useful for indicating phases can also be used.

Diffraction. The most common analytical tool for phase identification is crystal diffraction, which employs x-rays, high-energy electrons, or neutrons that undergo wave interference with the regular arrangement of atoms in a crystalline lattice. Crystal diffraction techniques can be used to give both qualitative and quantitative data on the phases present.

By far the most common technique used is x-ray powder diffraction. The Joint Committee on Powder Diffraction Standards (International Center for Diffraction Data, Swarthmore, PA) publishes a file listing the characteristic patterns of several hundred thousand crystalline phases in terms of interatomic (d) spacings and relative diffraction peak intensities. When an x-ray beam interacts with a material, the diffraction pattern that results is recorded with a spectrometer-like arrangement, called a diffractometer. The d-spacings and relative intensities are used as a "fingerprint" of the crystalline phase(s).

Modern systems are highly computer automated and may analyze as many as 40 samples automatically. Film methods can also be used to record the interference pattern, paradoxically, either when very superior data are required or when a modern system cannot be afforded. Special attachments allow analysis of solid samples at elevated temperature and for films, coatings, liquids, or unconsolidated powders. Some error-causing factors to which ceramic samples can be particularly prone include: composition differences between standards and samples, preferred orientation of plate- or needle-shaped grains, inadequate number of crystals sampled, matrix or microabsorption by some phases in a mixture, interaction with the environment, and sample preparation effects, such as phase segregation during reduction to a fine powder.

Electron diffraction is usually used to identify very fine (<0.1 μm, or 4 μin.) phases and determine their distribution. Because this technique employs a very small sample, care must be taken to ensure that it is typical. Neutron diffraction can enhance information by accentuating differences for phases containing light elements and elements that are close in atomic number.

X-ray diffraction is also the principal method used to determine the atomic arrangement, that is, the crystal structure. Those who have experience with this analytical tool use it to determine preferred crystal orientation (texture), amorphous (glass) content, residual stresses, particle and crystallite size, impurity level, atomic defect structure, and the structure of amorphous material.

Structure and Properties of Glasses

The chemistry of common commercial glasses is the chief feature that requires characterization, because it usually has the greatest correlation with all properties. Often, spectrochemical absorption methods are used to characterize the effects of specific impurities or additives that affect optical properties, such as color. In optical fibers, trace elements must be determined and controlled for species as common as water and iron to either minimize optical transmission losses or to produce an active fiber.

Defects in glasses, such as seeds (small bubbles) and stones (small inclusions), can be analyzed using microstructural analysis and microchemical analysis techniques. Similar methods are used to determine the structure and local chemistry of glasses that either separate into two or more glass phases or partially crystallize to form a glass-ceramic. A polariscope, consisting of crossed polarizers, can be used to determine residual stresses by stress birefringence. It may be important to determine temperature versus viscosity in order to fabricate a glass and allow residual stresses to be either relieved during annealing or incorporated during tempering. Electrical, mechanical, chemical durability, and optical properties can be characterized for specific applications.

Microstructural Analysis

Ceramography, the study of microstructural features at magnifications greater than 50×, is a primary analytical tool that is used to understand the nature of ceramic raw materials, green (unfired) pieces, and fired samples. Important features are the particle size and shape distribution in raw materials, as well as similar features in green and fired ceramics. Other important aspects are the distribution of various phases in a ceramic, the size and spatial distribution of porosity, the nature of surface and volume flaws, and the degree of microstructural uniformity.

Although there is no compendium of microstructures available for ceramics, as there is for metals, someone who is familiar with expected microstructural features can develop a considerable understanding of both structure-property relationships and the types of production changes that may be necessary in order to obtain desired behavior characteristics.

Very fine scale features, such as pores and large grains, can act as critical flaws in a ceramic or glass, because of their brittle nature. A typical desirable ceramic microstructure consists of a minimum volume of small pores and a porosity with fine grain size. Phase heterogeneity and orientation may have a significant effect on all properties. For example, the honeycomb monolithic catalytic converter substrate used in automotive applications must have a carefully controlled porosity to achieve optimum catalytic performance, coupled with controlled grain size and orientation to achieve the desired directional thermal expansion and strength properties. Proper grain orientation is the key to the prevention of thermal shock failure during use.

The preparation of a ceramic sample for microstructural analysis requires great care and attention. Powders can easily separate due to gravity; size and surface chemistry effects can arise during mounting for optical or scanning electron microscopy; and green ceramics can be altered by the drying or embedding procedure used to prepare them for analysis. Although fired samples are extremely hard, polishing cross-sectioned samples can easily cause pull-out of individual grains, leading to a significant error in the interpretation of pore distributions. It is best to polish with diamond using an extremely flat polishing wheel, such as hardwood, with a minimum of polish and a great deal of extender. Light pressure is preferred. This is usually accomplished by polishing sequentially on flat wheels with successively finer polishing compound. At each polishing level, the abrasion scratches of the prior step are removed by polishing normal to the prior direction. Care must be taken to thoroughly remove embedded polishing compound from prior steps. Coarser polishing can introduce pull-out damage and chipping.

Because many ceramics are attacked or altered by common reagents, such as water, preparation artifacts are possible. Etching a ceramic sample to reveal microstructure usually involves the use of strong acids and bases, often at elevated temperatures. Controlling the etch is difficult, and it is complicated by the profound effect that slight differences in chemistry and/or chemical distribution can have. Preparation procedures are not standardized and are frequently undocumented.

Optical microscopy is a primary analytical technique used to examine microstructure. Thick sections and fracture surfaces can be examined with reflected light in order to characterize microstructure. Thin sections can be more revealing and provide a better opportunity for phase analysis, but with increased effort in sample preparation. Frequently, phases and structures that are difficult to show using other microstructural techniques can be distinguished in thin sections. A number of different contrast techniques can reveal structure. Often, slight differences in chemistry can cause a significant difference in color. When using reflected light, care must be taken to focus on the surface of interest; subsurface features may make visualization difficult.

Optical microscopy is often well suited to the characterization of grain structure, glassy bonding phase, and porosity in traditional ceramics. It is also useful for identifying flaws and defects in ceramics and glasses, as well as relatively planar fracture surfaces. However, it is limited in that many ceramic microstructures, particularly those of advanced ceramics, are either at or beneath the scale that an optical microscope can resolve.

Scanning electron microscopy can be a useful tool for looking at finer structures and powders. Secondary electron contrast is provided mainly by topography, atomic number, and conductivity resulting from bombardment with a high-energy electron beam. The most common images, employing secondary electrons, show light and shadow illumination as if the sample were illuminated by an oblique light source. This image shows three-dimensional topographic features. Higher atomic number areas and less electrically conductive microstructural features will appear brighter on polished and etched section samples and, to a lesser extent, topographic samples. Compositional contrast and information can be obtained with a backscatter electron detector and by x-ray fluorescent spectroscopy, which permits analysis of both structure and microchemistry.

The scale of the SEM is particularly suitable for many ceramic parts and raw materials. The depth of focus provided by this instrument makes it very well suited for rough failure surfaces. Indeed, many samples are prepared for SEM by fracturing in order to reveal microstructure. However, caution must be exercised in this case, because failure may follow a specific path in the material that is not typical of the general structure.

Transmission electron microscopy is limited to very detailed, usually scientific, studies at high magnification. It is powerful enough to reveal and identify ultrastructure, submicron chemistry, and very fine crystalline or amorphous phases. Sample preparation is usually difficult for ceramics involving polishing to a very thin 3 mm ($1/8$ in.) disk and dimpling the center. Then a sample is generally ion milled for an extended time to yield a small electron transparent area that may be suitable for TEM study. The transmission electron microscope is particularly useful for elucidating grain boundary phases, interfaces, ultrafine particles, nanoscale microstructures, and crystal defect chemical analysis techniques to give micro-micro chemistry.

Image analysis and feature analysis permit the quantification of morphological aspects of images obtained by optical microscopy, SEM, and TEM. This has become rather convenient to do with the advent of computer processing of images and the application of quantitative stereology to the computer. Major limitations are the contrast that must be obtained for the features of interest and the operator skill that is required to analyze real features. The size and shape distribution of powder particles, grain sizes in a green or fired ceramic, and porosity can all be obtained. Volume fractions of various phases and pores can also be obtained. Because two-dimensional features are analyzed, either a separate normal section must be made or the three-dimensional structure must be inferred.

Ceramic Powder Characterization

The characteristics of a ceramic powder are important to final properties. For example, particle size and surface chemistry affect forming and firing response, as well as ultimate properties. In addition, the degree of powder agglomeration may affect forming and lead to larger-scale porosity.

Particle Size. The microscopic methods described in the section "Microstructural Analysis" in this article are quite suitable for determining particle size. However, more rapid and industrially applicable methods have been developed. Screens of various sizes can be used to measure intervals of particle size in terms of the weight fraction that will pass through a mesh of one size, while remaining on a finer mesh. Settling of a suspension can be used to show particle size distribution in terms of the relative settling rate of different sized particles in a medium. Usually, a simplifying assumption is made that the particles are spherical, a condition infrequently met by ceramics. A particular application is the sedigraph, an instrument in which the turbidity of a constantly falling column is measured by continuously scanning the column. Because these techniques "miss" the fine particles that have a profound effect on forming behavior, newer techniques have been developed to measure submicrometer-sized particles by time-of-flight or interparticle interference.

Surface properties of powders are quite important. Surface chemistry can be obtained using the methods defined in the section "Chemical Analysis" in this article. Surface charge, a property that affects agglomeration, can be determined with a zeta potential meter, whereas the dispersion of a powder can be determined by measuring its rheology. Metal oxide and other ceramics or glasses form an electrical double (Stern) layer in water and other polar liquids. As the extent of the double layer increases, so does the repulsive force between similarly charged particles. As the concentration of surface potential-determining is changed (pH), the layer thickness changes. When the activity of positive and negative sites on the sample is equal, there is no electrical double layer. This zero point of charge gives particle agglomeration and high viscosity. The extent of the double layer can be determined by electrokinetic measurements that give the potential at the shear plane between liquid bound to the particles and free liquid—the zeta potential. The surface charge of the ceramic particles will go through a zero point and particles will agglomerate as a function of additives or pH. When different materials are used, the system must be treated for all components to obtain good forming properties and to heterodisperse the multiple phases for a homogeneous microstructure. Under other conditions, a best dispersion can be developed, which, together with particle size, gives best flow characteristics for forming a ceramic and obtaining best unfired density.

Rheometry is used to measure flow, or viscosity, behavior of a ceramic suspension by shear rate and time. Generally, satisfactory ceramic forming occurs when the ceramic suspension is pseudoplastic and yields and flows readily above a certain shear level. This means that flow will occur when the slurry is forced into a mold, but the material will remain in the mold when the force is removed. For slips, thixotropy is often desired. In this case, agitation thins the material progressively with time so that it will pour into a mold. Upon standing for a time the material "sets" and no longer flows readily.

Porosity can be determined from microstructural analysis, as previously indicated. Alternatively, instruments such as a mercury intrusion porosimeter can be used to determine porosity. Mercury or another nonwetting fluid is forced into open pores under pressure. The pressure required to penetrate with a volume of mercury is measured. This can be related to the diameter of cylindrical capillary pores to yield a pore size distribution based on the assumption of cylindrically shaped pores. Neither closed pores nor small pores (<0.05 μm, or 2 μin.) are determined by this method. Water penetration (boiling) and resulting weight gain can also be used to determine porosity. Ultrasonic attenuation can also furnish some information about porosity, particularly when scanning in frequency.

Density is determined by either pycnometry or the Archimedes experiment (liquid displacement and weight change). Because open porosity may be penetrated, it is not considered in the density measurement. This can be circumvented by coating a sample with wax or lacquer to seal the open pores.

Surface area measurements are frequently made by a technique employing the Brunauer-Emmett-Teller (BET) equation, which is used for high specific surface area materials, such as very fine powders, cata-

lytic substrates, and sol-gel glasses. It determines the adsorption/desorption of a gas in order to calculate the open surface area.

For a simple pore shape, the measurements of pore size, frequency and distribution, density, and surface area should be interrelated, particularly if open and closed porosity is well understood. When discrepancies arise, it is useful to examine whether the microstructure and phase distribution are properly understood or whether a significant experimental error has been made.

Surface roughness can have a significant effect on the mechanical properties of a ceramic or glass. Surface scratches act as flaws that promote failure. The larger and sharper the crack emanating from the bottom of the scratch, the lower the stress at which failure occurs. Chemical exposure, even to atmospheric water, under an applied stress can cause slow crack growth (SCG), historically termed static fatigue and, ultimately, catastrophic failure in a process analogous to stress-corrosion cracking in metals and craze cracking in polymers. The finer the surface finish, the more resistant to failure a ceramic will be. Usually, a fine finish also gives lower friction forces and lower material wear. Surface roughness is often determined by profilometry, optical microscopy, or SEM.

Testing

Strength. A wide variety of mechanical strength tests are used for ceramics, but the most common tests involve three- and four-point loading of bar samples. The typical "dog bone" tensile sample commonly used for metals and plastics is expensive to machine in the case of ceramics, because diamond grinding is required to produce a fine surface finish. Special precautions and fixtures are required to prevent failure in clamping grips.

Recent advances in testing procedures dictate that two lower hardened steel rollers support rectangular bar samples. One central roller on the upper surface transfers load from a universal testing machine in three-point loading. Two symmetrical rollers are used in four-point loading. These are termed three- and four-point modulus of rupture tests. Care must be taken to make the span-to-thickness ratio about 10/1 or more.

Strength is not an innate property of a ceramic. Surface finish has a profound effect on strength. Large flaws will cause brittle failure at a low stress. For this reason, the surface is often machined to a uniform finish (600 grit is common) and edges are chamfered in order to establish a consistency that will show material differences. A sandblasted (damaged) ceramic or glass rod may show a strength value that is several orders of magnitude lower than a polished and lacquered surface. If a typical production surface is to be tested for strength, it should be used as the tensile side of the bar, with carefully chamfered edges.

The results of a three-point test will yield a higher modulus of rupture than a four-point test. This is because the tensile load drops off in the three-point test as one moves away from the center. There is a greater probability that the largest flaw will be in a region of high stress in the four-point case. The two strength values can be related by a statistics of failure argument. The statistical nature of failure also requires a large sample population.

One way of treating strength is to report the probability of failure as a function of stress. Such a procedure gives rise to the so-called "Weibull" plots. This can be viewed as a technique for minimizing flaws by increasing the modulus with refined production methods or surface finish. Failure reliability can be improved by increasing the Weibull modulus (the slope of such a plot), or the strength can be increased by moving the whole probability curve to higher stresses.

Proof testing subjects a part to stresses in order to eliminate parts having large flaws. Understressing uses a relatively low stress to remove most flawed parts and improve the statistics of failure for a moderate load. Overstressing applies loads greater than the design stress to get an additional margin of safety. If the proof test enlarges flaws, it could lessen performance and reliability. Recent studies suggest that proof testing can be used successfully. The article "Strength and Proof Testing" in this Volume provides many additional details.

Toughness. If a flaw of known size and sharpness is put into a test sample and either the tensile stress to cause failure or the work required to fail the material is determined, a fracture toughness value, K_{Ic}, can be attained. This value shows the resistance of the ceramic to catastrophic failure.

Hardness and Wear. A ceramic can be indented with either a Knoop or Vickers shaped diamond indenter and the indentation size measured to give a hardness value. Hardness is related to properties such as wear resistance, which is often measured as the material removal rate that occurs as a pin of similar or differing material is run against a rotating plate. A single-stroke device can also be used. The friction coefficient can also be determined and, along with wear rate, is important for applications such as cutting tools.

Thermophysical Properties. The reversible expansion, as well as the irreversible shrinkage, that occurs during sintering consolidation of a ceramic are tracked by a dilatometer. The irreversible behavior can be used to monitor the rate and extent of consolidation. Resistance to thermal shock is determined by repeatedly heating and cooling a ceramic at a controlled rate and either determining deterioration of strength or onset of failure after a number of repeated cycles. Thermal analysis techniques measure relative or time differential weight change (thermogravimetry), temperature difference relative to a reference (differential thermal analysis) and scanning calorimetry relative energy difference (differential scanning calorimetry). These techniques can explain important phase transformations and reactions that occur during either the firing or use of a ceramic. This information can be used as input to improve firing procedures or to attain a more thermally stable system for use under particular conditions. These systems can be coupled to a GC/MS in order to determine gaseous reaction products. The mechanical behavior of ceramics at elevated temperature in terms of creep resistance, mechanical strength, and toughness is an area of much interest because of the potential of turbine engines and other high-temperature structural applications.

Nondestructive evaluation (NDE) techniques include optical examination, liquid penetrant inspection, ultrasonic testing, acoustic emission, microwave inspection, radiographic inspection, computed tomography, magnetic spin resonance imaging, neutron radiography, thermal inspection, optical holography, and acoustic holography.

As already noted, flaws and cracks in ceramics represent locations where mechanical failure initiates and therefore dictate the mechanical properties. Identifying these flaw sites is necessary in order to eliminate parts with flaws and disqualify parts for which failure is likely. Because flaws and microstructure are frequently of a submicrometer scale for high-performance ceramics, NDE techniques need to progress somewhat in order to be successfully applied to ceramics. Many NDE techniques are appropriate for more conventional materials. The article "NDE Testing and Inspection" in the Section "Testing, Characterization, and NDE" explores the topic in more depth.

Failure Analysis

Fractography is the study of fracture surfaces in order to determine the manner and origin of failure. Based on this determination, a change in design and/or elimination of a class of defects may be undertaken. Ceramic materials fracture in a brittle mode. Several fracture surface features are observable in the optical microscope. Cracks branch as they propagate away from the failure origin, giving rise to a distinctive "river pattern." Tracing these "hackle" and "rib" patterns back to their origins can reveal the imperfection or feature responsible for failure. Initial failure zones that propagate by slow crack growth can often be discerned and their size related to the critical failure load. Products on the surface may give evidence of environmental interaction. The procedures are similar to those employed in the fractography of brittle metals.

The SEM also provides topographic and compositional information from rough fracture surfaces and is therefore used for investigating failure surfaces after optical analysis. The surface analysis techniques can be useful for identifying environmental products that result during slow crack growth. Often, the failure path is important because it can propagate through a weak phase or interface. A modification of such a material to strengthen the weak structure may improve performance.

SELECTED REFERENCES

- L.V. Azaroff, *Elements of X-Ray Crystallography*, McGraw-Hill, 1968
- F.D. Bloss, *An Introduction to the Methods of Optical Crystallography*, Holt Rhinehart and Winston, 1961
- D. Briggs and M.P. Seah, *Practical Surface Analysis*, John Wiley & Sons, 1983
- B.D. Cullity, *Elements of X-Ray Diffraction*, Addison-Wesley, 1978
- J.I. Goldstein, *et al.*, *Scanning Electron Microscopy and X-Ray Microanalysis*, Plenum Press, 1981
- P.B. Hirsch, *et al.*, *Electron Microscopy of Thin Crystals*, Plenum Press, 1965
- *Materials Characterization*, Vol 10, *Metals Handbook*, American Society for Metals, 1986
- *Mechanical Testing*, Vol 8, *Metals Handbook*, American Society for Metals, 1985
- *Metallography and Microstructures*, Vol 9, *Metals Handbook*, American Society for Metals, 1985
- *Metals Handbook Desk Edition*, American Society for Metals, 1985
- L.E. Murr, *Electron Optical Applications in Materials Science*, McGraw-Hill, 1970
- *Nondestructive Evaluation and Quality Control*, Vol 17, *Metals Handbook*, ASM International, 1989
- G.W. Phelps, *et al.*, *Rheology and Rheometry of Clay-Water Systems*, Cyprus Industrial Minerals, 1981
- H.H. Willard, *et al.*, *Instrumental Methods of Analysis*, 6th ed., Wadsworth, 1981

Overview of Ceramic Design and Process Engineering

Richard L. Lehman, Department of Ceramics, Rutgers, The State University of New Jersey

THE SUCCESSFUL APPLICATION of ceramics depends greatly on the ability of the design engineer to develop structures and components in ways that properly utilize the advantageous properties of ceramics and minimize the impact of limiting characteristics. Similarly, the material properties that are achieved for a certain ceramic composition depend almost entirely on the processing of raw materials. Thus, the design and processing of ceramic materials have been the principal research thrusts in ceramics for several decades prior to the 1990s, during which research continues. This article introduces the concepts of ceramic design and processing to familiarize the reader who has no previous experience in these areas. Sections in this Volume that provide greater coverage of these concepts will be cited throughout this article.

Design

Properties of Ceramic Materials. The first step in the design process is to select a material that has the properties required for the application. Historically, performance expectations of traditional ceramics were modest, because hardness, imperviousness, and moderate or low levels of mechanical strength were the only requirements. The classical triaxial porcelain body, based on various mixtures of clay, flint (SiO_2), and feldspar, easily provided the required properties and was engineered principally to facilitate processing rather than to match properties to applications. In the modern era, compositions of ceramic materials that exhibit a wide range of electrical, mechanical, and thermal properties have been developed. To a considerable extent, traditional materials have evolved to meet current performance requirements. However, a range of oxide, nonoxide, and composite ceramic materials that are capable of considerably higher levels of mechanical, thermal, and electrical performance than were exhibited by traditional ceramics have been developed since the 1960s. Overall, this group of high-performance ceramics is called "advanced" ceramics. Those advanced ceramic materials that are noted for

their mechanical performance are referred to as structural, or engineering, ceramics and are the focus of this article.

Ceramic materials are primarily specified by design engineers because of their mechanical strength, thermal stability, hardness, chemical durability, electrical/optical/magnetic properties, and/or thermal conductivity. In particular, structural ceramics are used to achieve mechanical strength at elevated temperatures, usually in the range of 600 to 1600 °C (1110 to 2910 °F). At temperatures below 600 °C (1100 °F), polymers (22 to 300 °C, or 70 to 570 °F) and metals (22 to 600 °C, or 70 to 1110 °F) usually offer more cost-effective thermal performance.

Selection of Materials. Although processing is a key determinant of the ultimate properties of most ceramics, selecting the proper chemistry is necessary, both to achieve the desired overall behavior of the ceramic and to permit fabrication of a cost-effective structure or component. The major types of structural ceramics and their typical properties are identified in Table 1. As a broad generalization, oxide materials are less expensive and are easier to process, whereas the non-oxides offer superior mechanical properties and greater thermal stability. However, each application is different and material selection must be performed on a case-by-case basis.

A very simple design strategy is to consider aluminum oxide as the default material for structural applications and to move to higher or lower performance materials, as required. Aluminum oxide offers very good, but not outstanding, performance with respect to strength, toughness, wear resistance, thermal stability, thermal shock resistance, and other properties. It is excellent in terms of high thermal conductivity, and it is available at relatively low cost. If subsequent design requirements indicate a need for high strength, material selection can move in the direction of silicon carbide or hot pressed silicon nitride. If greater toughness at low temperatures is required, a move to more expensive, but tougher, partially stabilized zirconia may be justified. If catastrophic failure protection is necessary at very high temperature and if high

cost is not an obstacle, a broad range of selection possibilities exists among ceramic-matrix fiber-reinforced composite materials.

Design Methodology. The principal difference between ceramics and other engineering materials, which affects the design process, is the brittle nature of ceramics. Unlike metals, in which a sample population may exhibit a standard deviation of failure stresses of a few percent of the mean, ceramic materials fail according to brittle fracture theory, and thus produce very broad distributions of failure stresses. More importantly, the strength of an individual sample is controlled by flaws induced into the sample, either during processing or during use. This flaw sensitivity of ceramics makes the design process more difficult because a ceramic material cannot be regarded as having a single strength, but, rather, a distribution of strengths determined by processing and the end-use environment. A major component of all research efforts devoted to developing improved ceramic materials addresses the nature of intrinsic flaws in the ceramic microstructure and extrinsic flaws generated either by errors in processing or by machining of the component after firing. Surface flaws generated during use by either scratching or abrasion must also be considered if the ceramic material will be exposed to these adverse conditions.

Three general design techniques are available to the design engineer: empirical, deterministic, and probabilistic design. Empirical design is the simplest approach to design and represents the trial and error approach of matching a material to a particular component and use environment. If the prototype prepared from the selected material fails, a new material is tested. This procedure is repeated until a suitable material is found. Deterministic design relies on measured average properties of materials, combined with a safety factor, to produce successful results.

Materials that have narrow ranges of failure stresses, such as steel, can be successfully designed in a deterministic manner without use of excessive safety factors. The best design approach for brittle materials,

Table 1 Properties of ceramics

Material	Crystal structure	Theoretical density, g/cm³	Knoop or Vickers hardness GPa	10⁶ psi	Transverse rupture strength MPa	ksi	Fracture toughness MPa√m	ksi√in.	Young's modulus GPa	10⁶ psi	Poisson's ratio	Thermal expansion, 10⁻⁶/K	Thermal conductivity, W/m·K
Glass-ceramics	Variable	2.4–5.9	6–7	0.9–1.0	70–350	10–51	2.4	2.2	83–138	12–20	0.24	5–17	2.0–5.4(a) 2.7–3.0(b)
Pyrex glass	Amorphous	2.52	5	0.7	69	10	0.75	0.7	70	10	0.2	4.6	1.3(a) 1.7(c)
TiO₂	Rutile tetragonal	4.25	7–11	1.0–1.6	69–103	10–15	2.5	2.3	283	41	0.28	9.4	8.8(a)
	Anatase tetragonal	3.84	...	...	...	...	...	...	...	...	...	...	3.3(d)
	Brookite orthorhombic	4.17											
Al₂O₃	Hexagonal	3.97	18–23	2.6–3.3	276–1034	40–150	2.7–4.2	2.5–3.8	380	55	0.26	7.2–8.6	27.2(a) 5.8(d)
Cr₂O₃	Hexagonal	5.21	29	4.2	>262	>38	3.9	3.5	>103	>15	...	7.5	10–33(e)
Mullite	Orthorhombic	2.8	...	...	185	27	2.2	2.0	145	21	0.25	5.7	5.2(a) 3.3(d)
Partially stabilized ZrO₂	Cubic, monoclinic, tetragonal	5.70–5.75	10–11	1.5–1.6	600–700	87–102	(f)	(f)	205	30	0.23	8.9–10.6	1.8–2.2
Fully stabilized ZrO₂	Cubic	5.56–6.1	10–15	1.5–2.2	245	36	2.8	2.5	97–207	14–30	0.23–0.32	13.5	1.7(a) 1.9(g)
Plasma-sprayed ZrO₂	Cubic, monoclinic, tetragonal	5.6–5.7	...	...	6–80	0.9–12	1.3–3.2	1.2–2.9	48(h)	7	0.25	7.6–10.5	0.69–2.4
CeO₂	Cubic	7.28	...	...	...	...	...	...	172	25	0.27–0.31	13	9.6(a) 1.2(d)
TiB₂	Hexagonal	4.5–4.54	15–45	1.5–6.5	700–1000	102–145	6–8	5.5–7.3	514–574	75–83	0.09–0.13	8.1	65–120(i) 33–80(j) 54–122(k)
TiC	Cubic	4.92	28–35	4.0–5.1	241–276	35–40	...	...	430	62	0.19	7.4–8.6	33(a) 43(d)
TaC	Cubic	14.4–14.5	16–24	2.3–3.5	97–290	14–42	...	...	285	41	0.24	6.7	32(a) 40(d)
Cr₃C₂	Orthorhombic	6.70	10–18	1.5–2.6	49	7.1	...	...	373	54	...	9.8	19
Cemented carbides	Variable	5.8–15.2	8–20	1.2–2.9	758–3275	110–475	5–18	4.6–16.4	396–654	57–95	0.2–0.29	4.0–8.3	16.3–119
SiC	α, hexagonal	3.21	20–30	2.9–4.4	(l)	(l)	(m)	(m)	207–483	30–70	0.19	4.3–5.6	63–155(a) 21–33(d)
	β, cubic	3.21	...	...	...	...	...	...	...	...	...	...	
SiC (CVD)	β, cubic	3.21	28–44	4.1–6.4	(n)	(n)	5–7	4.6–6.4	415–441	60–64	...	5.5	121(a) 34.6(g)
Si₃N₄	α, hexagonal	3.18	8–19	1.2–2.8	(o)	(o)	(p)	(p)	304	44	0.24	3.0	9–30(a)
	β, hexagonal	3.19	...	...	...	...	...	...	...	...	...	...	
TiN	Cubic	5.43–5.44	16–20	2.3–2.9	...	...	...	...	251	36	...	8.0	24(a) 67.8(q) 56.9(r)

(a) At 400 K. (b) At 1200 K. (c) At 800 K. (d) At 1400 K. (e) At 350 K. (f) 8–9 (7.3–8.2) at 293 K, 6–6.5 (5.5–5.9) At 723 K, and 5 (4.6) at 1073 K, in units of MPa√m (ksi√in.). (g) At 1600 K. (h) 21 (3) at 1373 K, GPa (10⁶ psi). (i) At 300 K. (j) At 1100 K. (k) At 2300 K. (l) Sintered: 96–520 (14–75) at 300 K, and 250 (36) at 1273 K. Hot pressed: 230–825 (33–120) at 300 K, and 398–743 (58–108) at 1273 K, MPa (ksi). (m) Sintered: 4.8 (4.4) at 300 K, and 2.6–5.0 (2.4–4.6) at 1273 K. Hot pressed: 4.8–6.1 (4.4–5.6) at 300 K, and 4.1–5.0 (3.7–4.6) at 1273 K, MPa√m (ksi√in.). (n) 1034–1380 (150–200) at 300 K, and 2060–2400 (300–350) at 1473 K, MPa (ksi). (o) Sintered: 414–650 (60–94). Hot pressed: 700–1000 (100–145). Reaction bonded: 250–345 (36–50), MPa (ksi). (p) Sintered: 5.3 (4.8). Hot pressed: 4.1–6.0 (3.7–5.5). Reaction bonded: 3.6 (3.3), MPa√m (ksi√in.). (q) At 1773 K. (r) At 2573 K

which exhibit a broader range of failure stresses, is probabilistic design. In probabilistic design a complete integration occurs between the mechanical requirements of the application and the statistical failure profile of the material. Consider cutting tools, turbine blades, or other ceramic components that are either complex in shape or experience a diversity of applied stresses over their surfaces as examples. It is immediately apparent that an empirical or deterministic design strategy is not appropriate for such varied applied stresses and such complex components due to the variety of localized conditions.

With probabilistic design, one divides the structure into discrete small volumes, termed finite elements, and assesses the stress exerted within each finite element. This matrix of stress components is then integrated with the known failure behavior of the ceramic to determine the overall performance of the structure. Because the ceramic material has a statistical range of failure stresses that depends on the flaw distribution, merging the finite-element "net" with a statistical description of the strength of the ceramic material becomes a mathematically intricate process for which numerous computer programs have been developed. Nonetheless, the probabilistic design technique is capable of producing excellent results in ceramic design, especially if care is taken in defining the finite-element array and a statistically large number of array elements are analyzed.

Weibull Statistics. Currently, the most popular mathematical means for representing the statistical nature of brittle material failure is a model developed by Weibull (Ref 1) and discussed in greater detail in the Section "Design Considerations" in this Volume. The Weibull equation relates the cumulative probability of failure of all the elements of a sample with n elements to the individual failure stresses of each of the n elements. For the purposes of this introductory article, it is appropriate to skip the derivation of the Weibull equation and to simply state the practical equation:

$$\ln\left[\ln\left(\frac{1}{1-P_f}\right)\right] = m\ln(s_i) \qquad \text{(Eq 1)}$$

where P_f is the cumulative probability of failure of all elements and s_i is the failure stress of the ith element. The constant m is the Weibull modulus and it is a measure of the scatter of strength data. Low *m* values correspond to large variances. High *m* values are desirable because they indicate a narrow range of failure stresses. In practice, a list of strength data for individual elements can be easily displayed on a Weibull graph by ordering the strength values and assigning each value a probability equal to its rank divided by (*n* + 1). Although this probability approximation is not exact, it offers a quick means of displaying data and the error is small (<5% relative) for most populations. The Weibull modulus is determined by a line of best fit to the graphed data.

An example of a Weibull plot of strength data is shown in Fig 1(a) for 20 strength values measured on reaction-bonded silicon nitride. The Weibull modulus is determined by the slope of the line of best fit. It is apparent from these data that two Weibull moduli are required to fully represent the data: one high-modulus value, illustrated by the dashed line, for the high-strength failures and one low-modulus value, denoted by the solid line, for the weak elements. The high modulus indicates nearly single-valued strength behavior for 16 of the 20 samples, which is a very desirable condition. The lower slope reveals a group of highly scattered data at low

strength. Often, the partitioning of Weibull plots into segments such as this reveals groups of samples that failed because of flaws of common origin. These flaw groups then become the subject of subsequent investigations.

After modification of the process to remove the source of these flaws, the entire population should be represented by a single high-modulus value. Thus, Weibull plots can aid in the refinement of ceramic processing to eliminate troublesome flaws. Changes in the Weibull plot that result from process changes are generally of two distinct types, as shown in Fig 1(b). The mean value of the strength and the Weibull modulus are the two parameters of interest. Ideally, it is desirable to increase both parameters, although this is not always possible. Within limits, it is usually more desirable to have a higher modulus and low scatter, than to have a high mean with a low modulus.

Proof testing has become a popular means in some segments of the ceramic industry to assure a certain level of product performance. As the name implies, proof testing consists of testing 100% of production samples at some predetermined proof stress level. Flawed individual samples that do not meet the proof test value fail and are removed from the pop-

ulation. In the case of mechanical strength proof testing, all production samples are exposed to a certain proof stress. Individuals with large flaws fail during the test and are discarded. If individual strengths conform to a normal distribution, the effect of proof testing on the population distribution is as depicted in Fig 1(c). Samples that pass the proof test are assumed to be stronger in end-use than the proof test value. Thus, the low-strength tail of the distribution is completely removed and the average strength of the population is increased.

Two caveats must be mentioned regarding proof testing. First, although it removes flawed samples, the assumption is put forth that the proof testing stress does not grow flaws in passed samples. However, it is possible, and sometimes probable, that flaws (particularly cracks) will grow during the initial proof testing and will cause failure upon repeated loading to the same stress level. Second, one must not apply normal statistics to the proof-tested distribution because this distribution is no longer normal.

Design Flow Chart. Modern ceramic design is a sophisticated process that involves the interaction of design engineers, materials scientists, characterization and testing, and process technology in an integrated, iterative

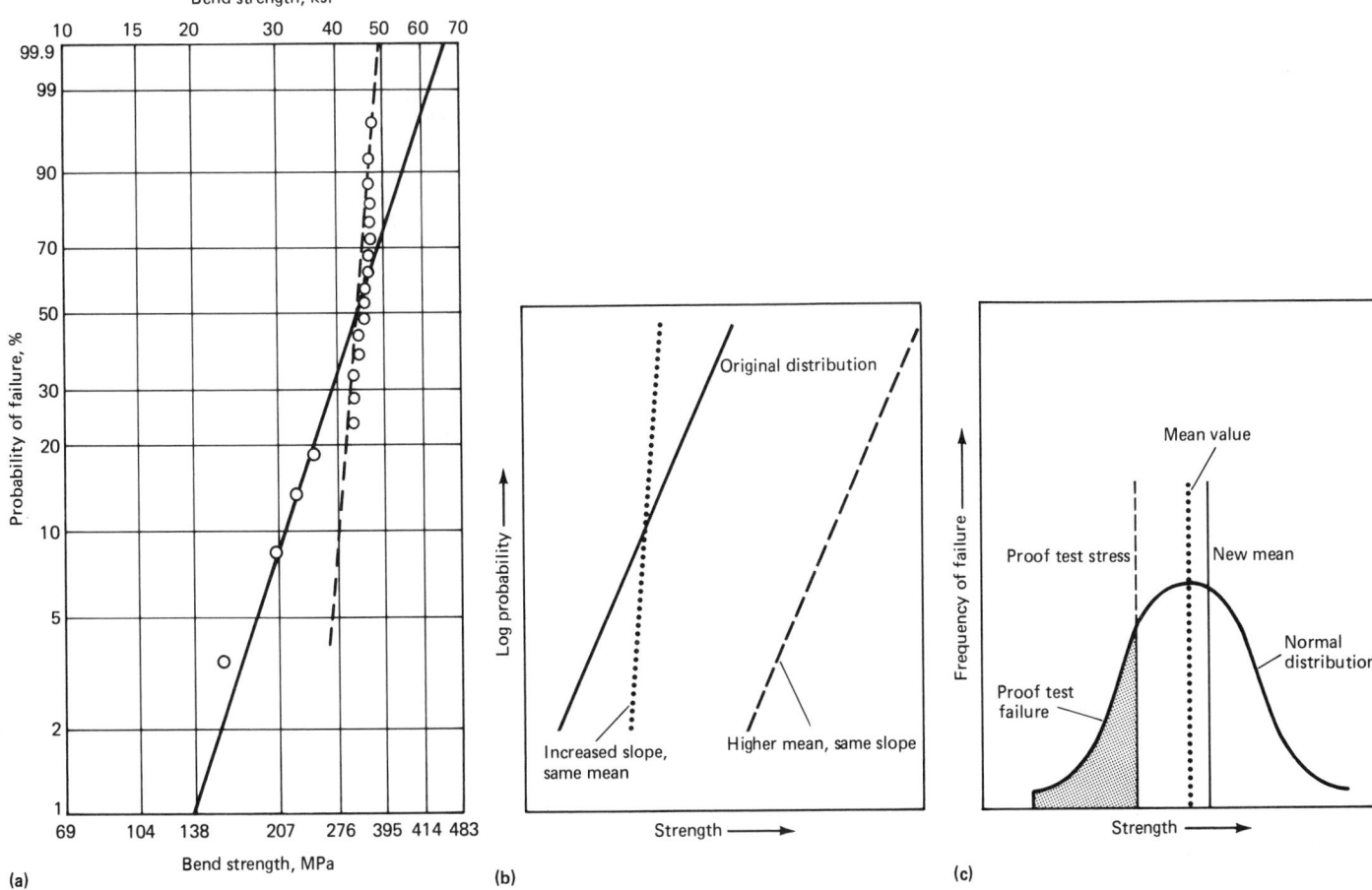

Fig 1 Statistical aspects of material design. (a) Weibull plot for reaction-bonded silicon nitride. (b) Improved Weibull distributions. (c) Modification of normal distribution by proof testing

process. Though it represents a brief introduction, the simplified flow chart in Fig 2 addresses the major functional aspects of the design process. Objectives are stated regarding the dimensions and performance requirements of the structure or component. Based on these objectives, an initial design concept follows. At this point, an iterative materials selection, design, and analysis sequence follows, which leads to the final design concept, fabrication, testing, and evaluation.

Modern testing and evaluation techniques rely increasingly on a combination of traditional test methodologies and nondestructive evaluation (NDE). A detailed discussion of these methods can be found in the Section "Testing, Characterization, and NDE" in this Volume. Two simultaneous activities occur: the evaluations of practical considerations and analytical considerations affecting the design. Practical considerations include material cost and availability, feasibility of the fabrication and manufacturing processes, production scale, and others. Analytical considerations cover a wide array of parameters that must be quantitatively defined or evaluated to assess the limits of the design.

Ceramic Processing Methods

Ceramic processing is a sequential activity that starts with raw materials, proceeds through batch preparation and forming, and concludes with firing and surface finishing, as required by the application. As an overview, Fig 3 illustrates the sequence of steps. Important aspects of each unit operation are discussed below.

Raw Materials

Ceramics are usually prepared from raw materials in the form of powders. However, the raw materials used for traditional ceramics are different from those used for advanced ceramics. The requirements of each ceramic group are defined below.

For traditional ceramics, the raw materials are commonly of mineral origin and are beneficiated to remove impurities and to prepare the raw materials for subsequent ceramic processing. Traditional ceramic raw materials consist of silica, clays, fluxes, and refractory materials. Silica is obtained either from massive quartz rock deposits or from pure quartz sands. The quartz is washed, ground, and classified as to the proper particle size, normally in the range of 200 to 400 mesh.

Clay minerals are enormously diverse, but the most commonly used materials are kaolin ($Al_2O_3 \cdot 2SiO_2 \cdot 2H_2O$) and talc ($3MgO \cdot 4SiO_2 \cdot H_2O$). China clay is predominantly kaolin and is a major component in a wide range of triaxial compositions based on various ratios of clay, silica, and feldspar. Ball clays are finer particle size clays that contain significant amounts of free silica and organic materials. Most clays are used in traditional ceramic compositions because of their ability to generate plasticity in the body when mixed with water. Clays, especially ball clays, contain organic impurities that are important aids to dispersion and to green strength.

Fluxes are minerals, such as feldspars or nepheline syenite, that contain alkali oxides to promote the fusion of silica and alumina to form a glassy phase during firing. The use of these fluxing minerals permits the processing of traditional ceramics at practical temperatures.

Refractory materials are a separate class of ceramics and include oxides, carbides, and other materials that withstand extremely high temperatures. Their principal use is in the preparation of high-temperature forms used as furnace linings to resist the attack of molten glasses and metals.

Advanced Ceramics. Although a broad overlap exists between traditional ceramics and advanced ceramics with regard to raw material sources, raw materials for advanced ceramics are usually chemically prepared powders of high purity. Powders of aluminum oxide, zirconium oxide, silicon carbide, silicon nitride, and aluminum nitride are examples of highly processed raw materials for advanced ceramics. The powder process routes vary greatly, but several common objectives for all processes are high chemical purity, controlled particle size and distribution, high reactivity, and freedom from hard agglomerates, or aggregates, that will not break down during mixing and subsequent processing.

The optimum particle size distribution for ceramic powders is still a controversial issue. Fine, monosized spherical particles of about 100 nm (4 μin.) diameter are best for ordered packing into dense beds by highly dilute settling and are highly reactive. Furthermore, because these particles are of uniform size, they will sinter well with little tendency toward exaggerated grain growth. Unfortunately, even the best packing of monosized spheres, in practice, leaves a 30% interstitial void volume. Hence, a great amount of sintering and shrinkage is required to achieve full density. Alternatively, an extended distribution of particle sizes from approximately 20 μm (800 μin.) to less than 50 nm (2 μin.) allows highly efficient packing with an interstitial pore volume of less than 5%. Thus, much less sintering is required to achieve full density. Unfortunately, the larger particles often grow excessively during sintering, resulting in microstructures that are not ideal. Some intermediate position between these two particle size approaches is likely to produce the best results.

The desire to achieve uniform, fine ceramic powders has generated a wide variety of methods for the preparation of powders. Most methods share the common approach of controlled nucleation from a liquid or gas, followed by a very short and controlled crystal growth period. Such a process for solution precipitation of Al_2O_3 particles is given in Fig 4. The very short nucleation period assures all nuclei of similar age, and the time period in saturation concentration controls the final particle size. Similar approaches are used in vapor atmospheres, where highly localized, high-intensity laser beams control the nucleation and crystal growth periods.

In addition to the raw materials mentioned above, a wide range of specialty raw materials is available. These include materials prepared by chemical vapor deposition, plasma spray, radio-frequency (RF) sputtering, sol-gel, and others, which are not described further in this introductory article.

Raw materials for ceramic composites are the same as those discussed above, except for the addition of reinforcing media, usually whiskers or continuous filament fibers. High-

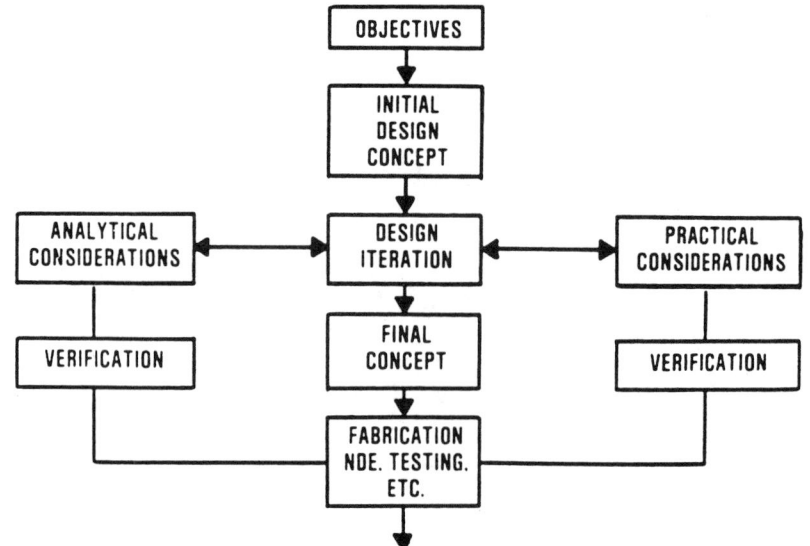

Fig 2 Ceramic design methodology

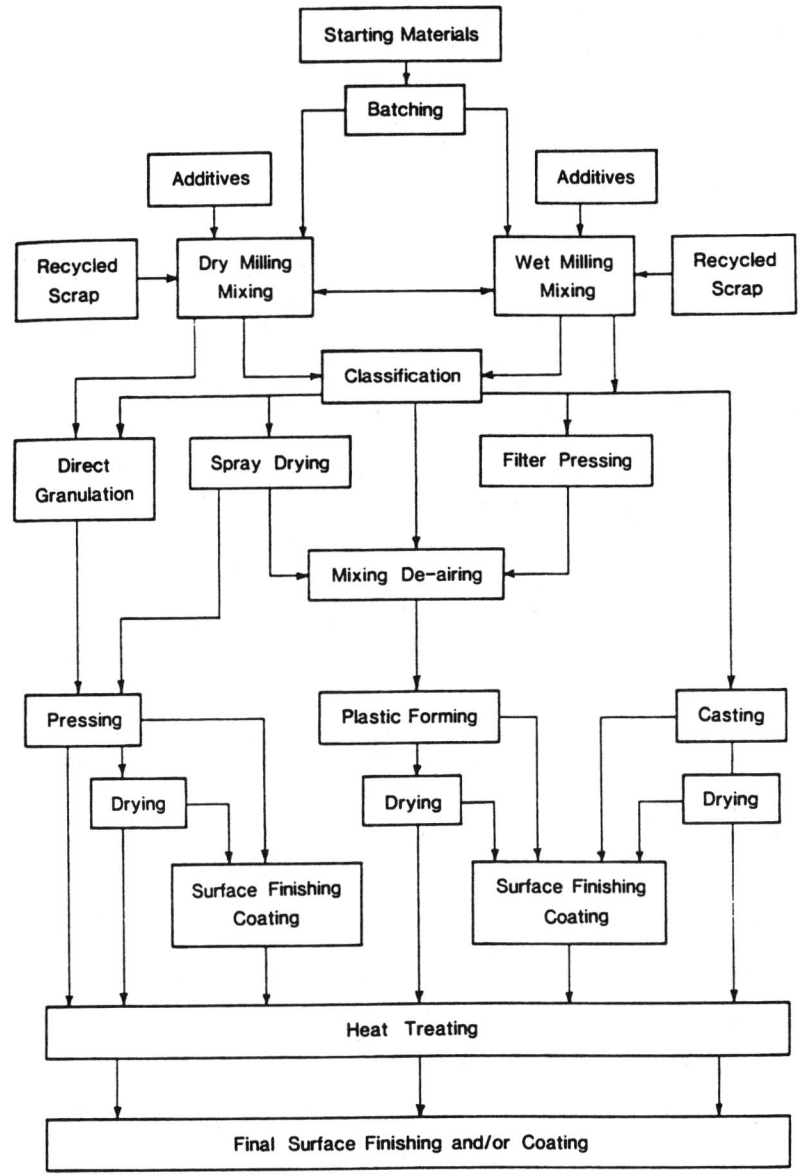

Fig 3 General ceramic processing flow chart

aqueous media to develop a fluid slurry, or "slip." The proper dispersion of ceramic particles requires a thorough understanding of rheology and colloid chemistry. Dispersant type and amount, solids fraction, pH level, and both powder particle size and distribution are important parameters. In the case of aluminum oxide, a satisfactory slip can be prepared by dispersing a submicrometer-sized alumina powder in water to prepare a 75% solids by weight slurry under high shear mixing with the addition of 0.2 to 0.5 wt% (dry powder basis) acrylate dispersant.

Ceramic powders used in either extrusion or injection molding must be prepared into stiff plastic masses with Bingham rheology and sufficient yield points for proper performance in the process. Bingham rheology is the plastic flow behavior in which shear stress is a linear function of shear rate with a non-zero intercept, or yield point. For compositions that do not contain naturally plastic materials, such as clays, it is necessary to add suitable organic lubricants, plasticizers, and binders to achieve the proper plasticity. Mixing of batches for plastic forming is accomplished in high shear devices such as "pug mills" or sigma blade mixers.

Forming Processes

A wide range of forming processes are used; the ceramic engineer can choose whichever process best suits a particular application. Extrusion, dry pressing, and injection molding are three processes that are widely used in polymer, metal, and food technologies, and are therefore familiar to most engineers. Both slip and tape casting of ceramic slurries are processes developed for ceramic materials and are mostly, although not exclusively, used to process them. The proper forming process for a particular component or structure depends on many parameters, the principal ones being the size and shape of the component and the anticipated production volume. Table 2, which defines the relationships between these parameters and the respective forming processes, is intended as a guide to the process engineer. In practice, each forming process has a wide latitude in terms of performance. Although the processes listed in Table 2 are the principal ones used for ceramics, there are many other specialty processes used in the industry, which are not discussed in this article, but are covered in the Section "Forming and Predensification, and Nontraditional Densification Processes" in this Volume.

Dry pressing is a forming process in which nearly dry, free-flowing powders fill a metal die and are compacted under high pressure to the desired shape. Dry pressing is most efficient and economical when large quantities of simple, small shapes are required. In particular, the aspect ratio of the component is important. Shapes with a high aspect ratio, such as a long, cylindrical bar, are not prepared well by dry pressing because die wall

strength fibers made from aluminum oxide, carbon, and silicon carbide are available in filament diameters ranging from about 10 to 150 μm (0.4 to 6 mils). Silicon carbide whiskers are available in diameters of 2 μm (80

μin.) and less, although there is considerable concern over the health effects of these materials.

Preparation for Forming

Before the powder raw materials can be formed, they must be prepared into various states that are compatible with the forming process. Powders used in dry pressing processes are usually spray dried prior to pressing. The spray drying preparation permits the uniform incorporation of additives, such as binders and lubricants, into the powder mix. Because spray drying is a granulation process, it combines many submicrometer-sized powder particles into larger spherical particles that have excellent flow and die-filling characteristics.

Powders used in either slip casting or tape casting must be dispersed in aqueous or non-

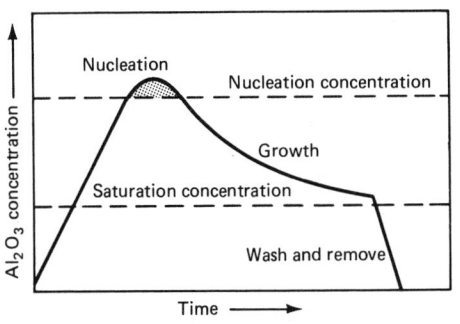

Fig 4 Solution method for preparing monosized ceramic powders

Table 2 Factors influencing ceramic forming process selection

Forming process	Component size	Component shape	Production volume
Slip casting	Large	Complex	Low
Tape casting	Thin sheets	Simple	High
Extrusion	Wide range	Constant cross section	High
Injection molding	Small	Complex	High
Dry pressing	Small to medium	Simple, low aspect ratio	High

Note: Labels in table reflect the most preferred condition for the stated process. In practice most forming processes are used over a range of conditions.

friction precludes the exertion of uniform pressure by the punches. Hence, uniform density is not achieved throughout the component. The development of improved die design that incorporates multisegment punches, floating die cavities, two-directional pressing, and special time/pressure cycles, have reduced density gradients and increased the scope of shapes that can be commercially produced by dry pressing. The use of lubricants, both internal and external to the ceramic powder, reduces friction during the pressing process, facilitating forming, and further reducing undesirable density variations.

When very low density variations are required in the pressed component, or when the component shape is too complex, isostatic pressing is used in place of standard uniaxial dry pressing. Isostatic pressing is the compaction of ceramic powder by hydraulic pressure, which assures uniform pressure in all dimensions during all stages of compaction. Traditionally, ceramic powder is placed in a rubber mold of the proper shape, sealed, and immersed in a hydraulic bath. Pressure is applied to the hydraulic fluid, which transmits the pressure to the powder inside the flexible mold. The powder is compacted by this pressure, yielding a compact with highly uniform density. Currently, it is more common in production to use "dry bag" isostatic pressing for simple shapes such as rods and tubes. In this process, the hydraulic system is completely enclosed and the uniform pressure is transmitted through a flexible membrane to the die cavity.

Plastic Processes. The forming of ceramic shapes by deformation of a plastic mass has historically been popular due to the highly plastic nature of clay-based traditional bodies. In addition to the many traditional ceramic shapes presently made by various plastic forming processes, these processes are also widely used for advanced ceramic compositions that may not contain clay and are not naturally plastic. In aluminum oxide and silicon nitride compositions, for example, organic plasticizer additives are used to impart the required plasticity.

The principal processes used are extrusion, injection molding, plastic pressing, and jiggering. In the plastic pressing process, a preform of plastic raw material is pressed under relatively low pressure to conform to the shape of a die. The die material may be an impervious metal or a porous plaster mold, the latter assisting in water removal from the plastic

mass and increasing the yield point of the molded part. Jiggering is similar to plastic pressing, except that the plastic mass is softly pressed into a plaster die, and the final shape conformance is achieved by rotating the die and shaping the plastic mass with a fixed jiggering tool. Dinner plates and coffee mugs are good examples of jiggered shapes that must have circular cross sections.

Extrusion and injection molding are similar processes that are capable of rapidly forming ceramic shapes by compressing plastic raw feed either through a die aperture of the desired shape (extrusion) or into a mold (injection molding). Equipment for both processes is similar to the extrusion and injection molding machines used in other industries, with some modifications to accommodate the abrasiveness of ceramic raw materials. The articles "Extrusion" and "Injection Molding" in this Volume provide greater detail.

A wide range of component sizes and shapes can be prepared by extrusion, ranging from small thermocouple protection tubes with intricate hole patterns to enormous high-tension insulator blanks. The single requirement, obviously, is that the cross section be constant. Parts that have minor deviations from this requirement can be prepared by green machining after extrusion.

The main advantage of injection molding is that, in principle, large quantities of very intricate parts can be quickly formed, analogous to polymer injection molding. In practice, a large amount of polymer binder/plasticizer is required to generate the required flow behavior. The removal of this binder and the subsequent densification of the component can cause serious problems and is the limiting issue in the application of injection molding to ceramics. Binder burnout is greatly facilitated if the part is small, and recent research into cellulose-based binders offers the possibility of greatly reduced binder levels.

Slurry casting processes, as already noted, include both slip and tape casting, which are described separately below.

Slip casting is perhaps the most versatile ceramic fabrication process. As shown in Fig 5(a), it involves a low-viscosity slurry of ceramic powder that is poured into a porous plaster mold. The porous nature of the plaster mold draws water from the slurry that contacts the mold, thus depositing a layer of solids on the wall of the mold. This process continues as the capillary suction of the mold continues to draw water through an increas-

ing thickness of ceramic solids at the mold wall. When a sufficiently thick layer has been established at the mold wall (ranging from several millimeters for fine ceramic artware to more than 10 mm, or 0.4 in., for heavy sanitaryware), the liquid remaining in the mold is drained. The ceramic shell is permitted to dry for a short period in the mold to develop greater strength. When the cast is sufficiently dry, the mold is removed. Edges and mold seams are trimmed and smoothed with a knife and a wet sponge.

In addition to drain casting, it is possible to fabricate solid shapes by leaving the slip in the mold for a considerably longer period, until the entire cavity is filled with a dewatered ceramic powder. This process, known as solid casting, requires that the mold provide a small reservoir of slip to enable complete filling of the volume, as shown in Fig 5(b).

Several other factors are particularly important in slip casting, in addition to the slip preparation process. The porosity of the plaster mold is controlled by the water content of the plaster during mold fabrication and represents a trade-off between capillary suction (small pores) and water-holding capacity (high porosity). Ideally, a desirable mold has a large volume of small pores. The particle size of the powder should be relatively coarse to permit high levels of vehicle permeability through the cast wall. During casting, the wall thickness increases according to a parabolic time law, thus requiring impractical casting times for very thick wall structures. The casting rate can be substantially increased by the application of external pressure, a process variation known as pressure casting.

Slip casting is widely used commercially because it enables the casting of complex shapes from an almost unlimited variety of ceramic compositions, at a low cost. The most serious disadvantages of the process are low production rates, the relatively short life of the molds, and the labor-intensive nature of mold assembly, casting, demolding, trimming, and other steps.

Tape casting produces thin sheets of flexible ceramic tape. These sheets are prepared from ceramic slurries containing substantial amounts of binder and plasticizer that are cast into thin layers on either a glass plate or an impervious polymer film and allowed to dry. Tape casting is widely used for the production of ceramic substructures in the electronics industry.

There are several important differences between tape and slip casting. No porous mold is used in tape casting. Instead, the vehicle is removed by evaporation. The vehicle used in tape casting is usually a highly volatile organic liquid, compared to the standard aqueous systems used in slip casting. Binders and plasticizers are incorporated into the slurry to impart strength and flexibility to the green tape. Because the tape thickness is small (typically 0.1 mm, or 4 mils) and vehicle

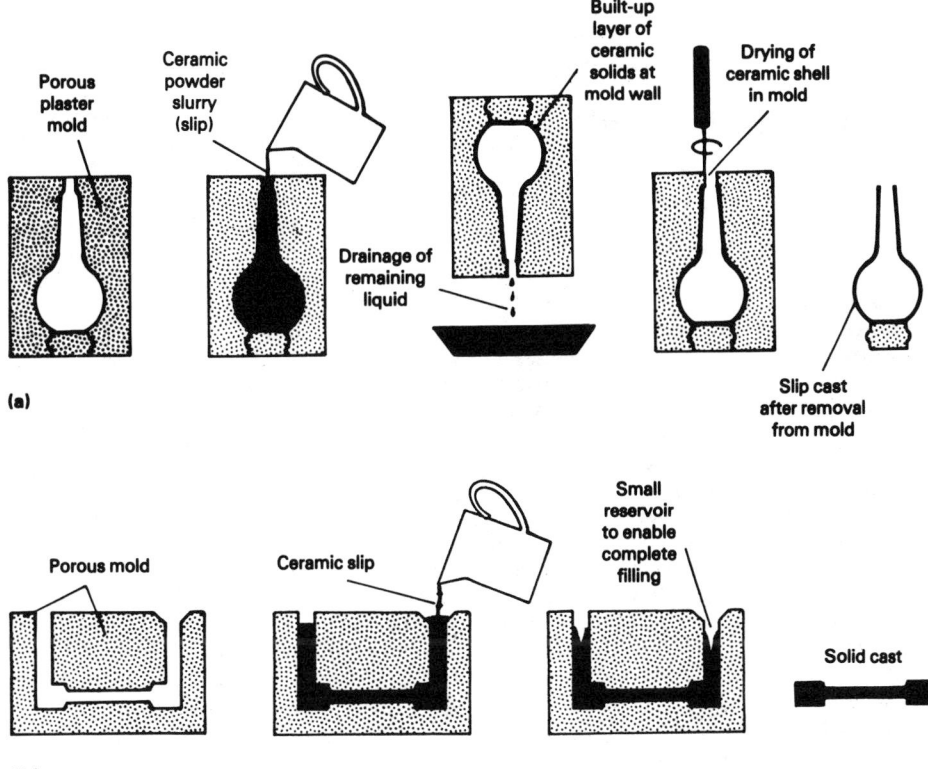

(a)

(b)

Fig 5 Ceramic slip casting process. (a) Drain casting. (b) Solid casting

removal is by evaporation, the particle size of the ceramic can be made finer. Finer particle size slips produce stronger ceramics with improved microstructure.

Composite Fabrication Processes. The challenge in the fabrication of ceramic-matrix fiber-reinforced composites is to uniformly incorporate the matrix phase around the fibers in a way that achieves full density without damaging the fibers. The most common process is slurry infiltration, as shown in Fig 6. This process is an extension of the polymer infiltration process used to make polymer-matrix composites. A slurry of matrix material is infiltrated into the fiber yarn. The slurry must be of sufficiently fine particle size to enable particles to penetrate the voids between the fibers. A binder added to the slurry makes the infiltrated-yarn tapes flexible for subsequent green handling. A similar process uses a sol-gel slurry of matrix material to infiltrate the fibers.

Other processes for forming fiber-reinforced composites include polymer infiltration, followed by polymer pyrolysis, to form a carbide or nitride; and chemical vapor infiltration (CVI) of the matrix phase. CVI composites possess excellent properties due to the high quality of the deposited matrix and the absence of physical damage to the fibers during fabrication. The Lanxide process (Ref 2), which is based on the controlled oxidation of a metal precursor, is a newer process that enables the preparation of well-integrated ceramic-matrix fiber-reinforced composites.

Firing

Exposure to high temperatures results in chemical and physical changes in ceramic powders that lead to the development of strength and other desirable properties. Two

general processes are important in firing ceramics: vitrification and sintering. Vitrification occurs when constituent materials react at elevated temperatures to form a substantial amount of liquid phase, which acts to consolidate the ceramic body through the capillary forces. This liquid forms a glassy phase during cooling. Many traditional ceramics of triaxial composition are densified by this mechanism. Ceramic bodies that do not form substantial amounts of a liquid phase at elevated temperature densify by a solid-state diffusion process known as sintering. Surface-area reduction is the principal driving force for sintering. The sintering process has several stages that involve the following mass transport processes to varying degrees: evaporation/condensation, surface diffusion, and bulk diffusion.

The densification processes discussed above are kinetic processes that are time and temperature dependent. Although temperature is usually considered the most important parameter in vitrification and sintering, because certain minimum temperatures are needed to activate the transport processes, the densification processes are not fast and sufficient time is required to achieve their completion. Thus, a firing schedule is the product of time and temperature, where small trade-offs in these areas can be made without varying the outcome. However, for many materials, this trade-off range is small, and larger variations will result in substantial differences in microstructure and properties.

Atmosphere and pressure are two other factors that strongly affect firing behavior. Oxidizing or reducing atmospheres control chemistry in sensitive oxides and nonoxides,

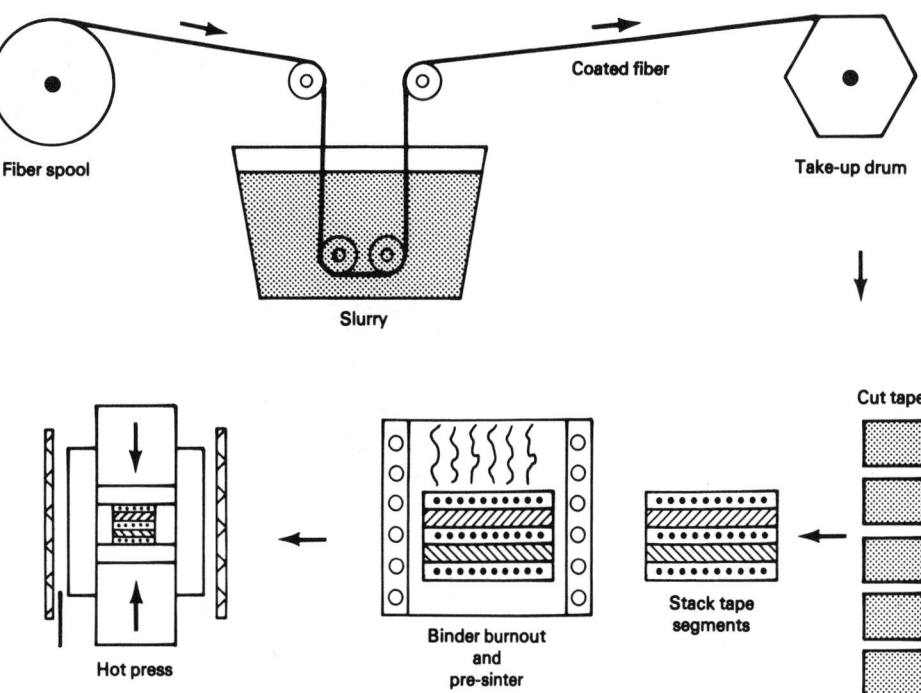

Fig 6 Slurry infiltration process for ceramic-matrix fiber-reinforced composites

and atmospheres of highly mobile gases (H, He) are useful in promoting sintering and pore elimination in systems that have poor sintering behavior. Highly diffusive gases are also useful in developing transparency in crystalline optical ceramics. The simultaneous application of pressure and temperature provides a strong driving force for sintering. The practical application of these two conditions, known as hot pressing (Fig 7) is used to:

- Reduce the required sintering temperatures for sensitive materials
- Densify materials such as carbides and nitrides, which have low diffusivities
- Densify materials that have constrained sintering microstructures, such as ceramic-matrix fiber-reinforced composites

High-temperature processing of ceramics is conducted in gas or electrically fired kilns controlled by programmable logic controllers (PLCs) coupled to the kiln power or fuel supply. These controllers sense temperature in the kilns most often via calibrated thermocouples placed at critical locations or by optical or two-color infrared pyrometers sighted at similar spots in the kiln. Pyrometric cones are a traditional means of measuring time/temperature thermal cycles, and are still one of the most accurate methods for recording the amount of thermal "work" applied to a kiln of ware. Pyrometric cones are small, triangular pyramids prepared from carefully formulated ceramic compounds designed to soften and melt under precise conditions of temperature and time. A series of cones, which are given numbers corresponding to temperatures at which they soften, are placed in the kiln with the ware. The cones are observed during firing to assist in temperature control, and are examined after the fire to indicate and record the degree of firing achieved.

Testing and Evaluation

Ceramic materials and products are used in a wide range of applications, as discussed in the Sections "Applications for Traditional Ceramics" and "Structural Applications for Technical, Engineering, and Advanced Ceramics" in this Volume. The performance of these ceramics in their respective applications is determined by a series of testing and evaluation procedures. Each application will have a specific series of tests and evaluation procedures, many of which are discussed in the Section "Testing, Characterization, and NDE."

A common group of properties tested for nearly all ceramics, particularly structural ceramics, relate to mechanical performance. Strength, the most frequently tested mechanical property, is most easily evaluated by bend testing in a three- or, preferably, four-point flexural fixture. The flexural test applies a mixture of tensile, shear, and compressive loading components to the test specimen. Because ceramics are stronger in compression than in tension, failure almost always occurs along the tensile surface of the test bar. Some fiber-reinforced composites are exceptions, and demonstrate either compressive or shear failure. Generally, shear components can be minimized in flexural tests by proper specimen and loading geometry. Alternatively, pure tensile measurements can be made on ceramics by preparing tapered specimens with a narrow test section over the gage length, as is the practice for other materials. However, due to the high Young's modulus of ceramics, it is important in tensile testing to apply strain gages to the specimen to assure the absence of parasitic bending moments.

Two other important mechanical properties are toughness and creep. Toughness is an important test property for ceramics, given their intrinsic brittle character and the current research emphasis to develop higher toughness in ceramic materials. Toughness is measured either by determination of the critical stress-intensity factor, K_{Ic}, or by a more empirical measure of the work of fracture. Work of fracture is particularly popular for testing fiber-reinforced composites, in which case K_{Ic} has a dubious meaning. The article "Toughness, Hardness, and Wear" provides greater detail. Creep is the slow plastic deformation of a ceramic under load at elevated temperature. It is often important that ceramics used at elevated temperatures strictly retain their dimension in order to perform properly. Turbine blades are a good example. Creep tests are conducted by applying tensile loads to ceramic test specimens at temperatures approximating use conditions.

REFERENCES

1. W. Weibull, A Statistical Theory of the Strength of Materials, *Ing. Ventenstape Akad.*, No. 151, 1939, p 1–45
2. M.S. Newkirk, A.W. Urquhart, and H.R. Zwicker, Formation of Lanxide Ceramic Composite Materials, *J. Mater. Res.*, Vol 1 (No. 1), Jan/Feb 1986, p 81–89

SELECTED REFERENCES

- H.E. Boyer and T.L. Gall, Ed., *Metals Handbook, Desk Edition*, American Society for Metals, 1985, section 32
- W.H. Dukes, *Handbook of Brittle Material Design Technology*, ADS-719–712, NTIS, 1971, p 16–24
- J.E. Funk, Extrusion of Electrical Porcelain Bodies, *Forming*, Vol 9, *Advances in Ceramics*, The American Ceramic Society, 1984, p 184–192
- J.E. Funk, D.R. Dinger, and J.E. Funk, Jr., "Particle Size Distribution Control for Refractories Forming Rheology," presented at the Changes in Refractory Technology Symposium, St. Louis, MO, 1982
- L.G. Johnson, *The Statistical Treatment of Fatigue Experiments*, American Elsevier, 1964, p 46–47
- W.D. Kingery, H.K. Bowen, and D.R. Uhlmann, *Introduction to Ceramics*, 2nd ed., John Wiley & Sons, 1976
- J.A. Mangels and W. Trela, Ceramic Components by Injection Molding, *Forming*, Vol 9, *Advances in Ceramics*, The American Ceramic Society, 1984, p 220–233
- A.F. McLean and D.L. Hartsock, Design

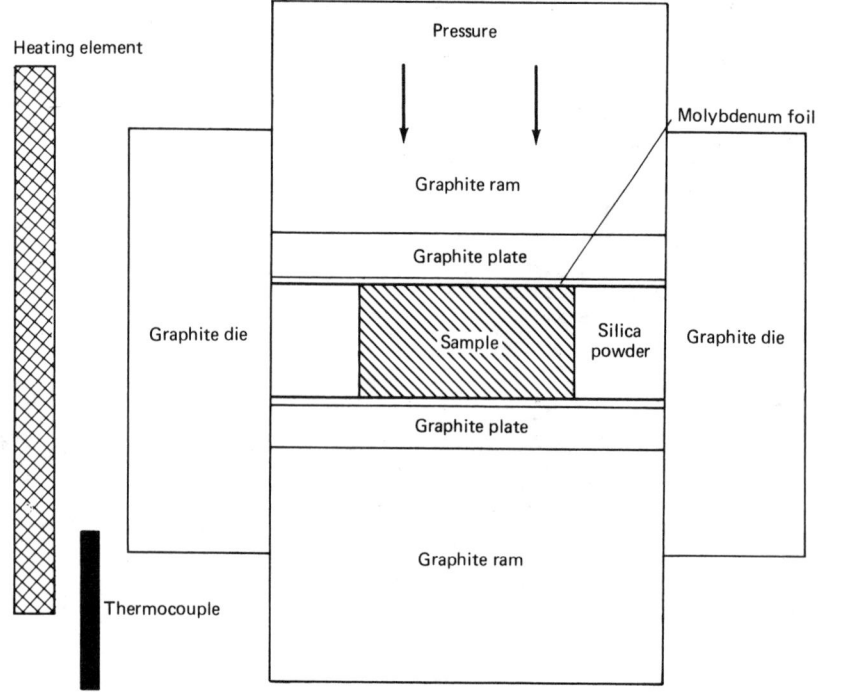

Fig 7 Hot pressing of ceramic materials

with Structural Ceramics, *Structural Ceramics*, J.B. Wachtman, Ed., Academic Press, 1988, p 27–98

- G.L. Messing, J.W. McCauley, K.S. Mazdiyasni, and R.A. Haber, Ed., *Ceramic Powder Science*, The American Ceramic Society, 1987
- F.H. Norton, *Fine Ceramics—Technology and Applications*, 1st ed., McGraw-Hill, 1970
- H. van Olphen, *Clay Colloid Chemistry*, John Wiley & Sons, 1977, p 57–76

- A. Paluszny and W. Wu, Probabilistic Aspects of Designing with Ceramics, *J. Eng. Power*, Vol 99 (No. 4), p 617–630
- G.W. Phelps, S.G. Maguire, W.J. Kelly, and R.K. Wood, *Rheology and Rheometry of Clay Water Systems*, Cyprus Industrial Minerals Co., 1983
- D.B. Quinn, R.E. Bedford, and F.L. Kennard, Dry-Bag Isostatic Pressing and Contour Grinding of Technical Ceramics, *Forming*, Vol 9, *Advances in Ceramics*, The American Ceramic Society, 1984, p 4–15

- J.S. Reed, *Introduction to the Principles of Ceramic Processing*, John Wiley & Sons, 1988, p 3–16
- D.W. Richerson, *Modern Ceramic Engineering*, Marcel Dekker, 1982, p 149–215, 313–321
- M. Velazquez and S.C. Danforth, Casting of Monodisperse Colloidal Silica, *Forming*, Vol 9, *Advances in Ceramics*, The American Ceramic Society, 1984, p 105–114

Guide to Information Sources and Standards

Stacey Clark-Kerwien, Metallic Materials Branch, AED, U.S. Army ARDEC
Gregory Papatrefon, Nuclear Division, FSAC, U.S. Army ARDEC

GENERAL SOURCES of information useful to engineers involved in ceramics and glass technology are described in this article. These resources are presented categorically as trade magazines, research journals, and periodicals; books; information agencies; standards organizations; and data bases provided by professional societies.

When specific problems related to materials and/or equipment are encountered, users are encouraged to seek direct assistance from the vendors involved because vendors not only represent a valuable source of information, but may make available their own published data and handbooks.

Trade Magazines, Research Journals, and Periodicals

The ceramics and glasses literature described in this section have been segregated into three categories:

- General information sources provide overviews on the latest developments and news in the ceramics and glass fields
- Research sources contain in-depth theoretical and experimental research papers about ceramic materials, processing methods, and applications
- Industry-oriented sources provide information and articles pertinent to engineers and managers involved in manufacturing and/or selling ceramics and ceramic-related items

General Information Sources

Advanced Materials and Processes is published by ASM International, Materials Park, OH 44073–9989, (216) 338–5151. It addresses current topics in selecting, designing, processing, and testing advanced materials. Articles provide general overviews as to how new materials and processes can be substituted for current ones in order to solve design or processing problems. Reader service cards facilitate contact with companies that have developed new materials and/or material applications, as well as processing equipment.

The American Ceramic Society Bulletin is published monthly by the American Ceramic Society, 757 Brooksedge Plaza Drive, Westerville, OH 43081–6136, (614) 890–4700. The bulletin contains American Ceramic Society news and information about current ceramic technology, focusing on the fields of ceramic engineering, research, manufacturing, and marketing. Features include a meetings calendar, reader service cards, book reviews, standards activities, and technology updates.

British Ceramics Transactions and Journal is published bimonthly by the Institute for Ceramics, Shelton House, Stoke-on-Trent, England. It contains technical papers, news pertaining to the ceramic industry, and meeting dates and reports.

Ceramics Abstracts, published by the American Ceramic Society, provides abstracts on technical ceramic literature.

Ceramic Forum International is published bimonthly by the Deutsche Deramische Gesellschaft, Bauverlag GmbH, P.O. Box 1460, Wittlesbacherstr. 10, D-6200 Wiesbaden. It contains articles in both English and German, as well as technical papers and news about all aspects of the ceramics industry.

The Glass Industry is published by the Ashlee Publishing Co., 310 Madison Ave., New York, NY 10017. It contains articles about manufacturing and marketing glass and glass products. News pertaining to the glass industry is also included.

Materials Engineering is published by Penton Publishing, 1100 Superior Avenue, Cleveland, OH 44197–8029, 1(800) 321–7003, or (216) 696–7000. It gives engineers an overview of materials fields that may be new or unfamiliar to them and aids in identifying practical solutions to materials, test equipment, and processing problems.

Research Journals

Consulting a research journal to learn about an issue of a very specific nature is often a valuable approach to problem solving. If additional information is desired, the reader can consider contacting the author(s) of the papers directly.

Cement and Concrete Research is published bimonthly by Pergamon Press, Inc., Fairview Park, Elmsford, NY 10523. It deals with research topics such as thermodynamics and kinetics of cement reactions, strength enhancement, and the effects that the environment has on cement and concrete.

The Ceramic Engineering and Science Proceedings, another publication of the American Ceramic Society, is an assemblage of articles originally presented at ceramic technology conferences. Most of the articles share a general theme relating to ceramics, such as glass or refractory materials. Articles are intended to identify, solve, or discuss issues relevant to these general themes.

Ceramics International is published by Elsevier Applied Science Publishers Ltd., Crown House, Linton Road, Barking, Essex IG11 8JU, England. This journal publishes papers that have applicability toward attaining improvements in the quality, reliability, and performance of ceramics. The subjects of the papers range from processing methods to final applications. A broad range of ceramic materials including traditional and advanced ceramics, refractories, electro-optics, and coatings are covered.

Glass and Ceramics is published by Plenum Press, 233 Spring St., New York, NY 10013, (212) 620–8000. It contains Soviet research reports on silicate chemistry, mineralogy, metallurgy, crystal chemistry, solid-state reactions, raw materials, phase equilibria, reaction kinetics, and physiochemical analysis.

The International Journal for High Technology Ceramics, another product of Elsevier Applied Science Publishers, Ltd., is research oriented and focuses on the processing, development, and use of advanced ceramics.

The Journal of the American Ceramic Society, published monthly by the American Ceramic Society, contains ceramic research articles covering a wide range of topics, including, but not limited to new phases of materials, microstructures, development/ fabrication and application of advanced ceramics and composite materials, and research that relates ceramic behavior to structure, processing, and other parameters.

The *Journal of the Ceramic Society of Japan* is published by Marzen Co. Ltd, P.O. Box 5050, 100–31 Tokyo Japan, and contains technical briefs and monographs.

The *Journal of Non-Crystalline Solids* is published by Elsevier Science Publishers BV, North-Holland, Excerpta Medica, P.O. Box 211, 1000 AE Amsterdam, The Netherlands. It contains articles and monographs related to the study of nonoxide glasses, noncrystalline thin films, amorphous metals and semiconductors, glass ceramics, and glass composites.

Materials Science and Engineering is represented by two issues. Issue A, entitled *Structural Materials*, deals with the structure and behavior of a wide range of materials, including ceramics. Issue B, *Solid-State: Materials for Advanced Technology*, concentrates on the processing, characterization, and physical properties of electronic, electro-optic, and magnetic materials. Articles in both issues are research oriented. The publisher is Elsevier Sequoia S.A., P.O. Box 564, 1001 Lausanne 1, Switzerland, (21) 20 73 81.

Physics and Chemistry of Glasses is published by the Society of Glass Technology, Thornton 20, Hallam Gate Rd., Sheffield S10 5BT, England. It contains research and scientific reports dealing with glasses and their structures and properties.

Industry Periodicals

Managers, vendors, manufacturers, and engineers involved with marketing may find the following resources useful because many contain articles on current economic factors and federal laws that may impact the industry.

Ceramic Industry is a monthly publication produced at 5900 Harper Rd., Suite 109, Solon, OH 44139, (216) 498–9214. It focuses on manufacturers in the ceramic industry and vendors that serve the ceramic community. The majority of the articles are about ceramic business trends and forecasts, top ceramic products, and laws being passed in Washington, D.C. that may affect the ceramic industry. Each September, the *Data Book Buyer's Guide* is published and sent to subscribers at no extra charge.

High-Tech Materials Alert is a monthly newsletter produced by Technical Insights, Inc., P.O. Box 1304, Fort Lee, NJ 07024–9967. It features novel materials developments that have major growth potential. The target audience comprises senior executives

and corporate planners and managers who are responsible for long-range planning and new product development.

Industrial Ceramics is published six times per year by Elsevier Applied Science Publishers, Ltd. Articles link applied research with current production practices. Topics include advances in plant design and technology for both the traditional and advanced ceramics industries. Each issue contains news from different areas in the ceramics field such as industry, research, and marketing. Also included is a meetings calendar and useful literature.

Engineering Index, Engineering Information, Inc., 345 E. 47th Street, New York, NY 10017–2387, (800) 221–1044. The index contains bibliographies and abstracts of technical literature in all engineering disciplines.

Industrial Minerals is published by Metal Bulletin Inc., at 220 Fifth Avenue, New York, NY 10001, (212) 213–6202. This journal focuses on all aspects of minerals used for industrial purposes and provides information on processing equipment, industry news, and marketing studies. A different mineral and mineral-producing country is reviewed each month. Details are given as to which companies are mining certain minerals, the production rates and product cost, and consumption characteristics.

Interceram, published by Verlag Schmidt GmbH, Postfach 6605, D-7800 Freiburg 1, Germany, is an international journal focusing on tableware, sanitaryware, tiles, refractories, and technical and advanced ceramics producers and their suppliers.

Materials and Processing Report is published by the Elsevier Science Publishing Co., Inc., 655 Avenue of the Americas, New York, NY 10010. It is an information-packed newsletter that describes the latest in ceramic technological advances and patents. It also contains materials industry analyses and a meetings calendar.

SAMPE Journal is published by SAMPE, P.O. Box 2459, Covine, CA 91722, (818) 331–0616. The journal provides technical information on materials and processes. Also included are industry and membership news and book reviews.

Superconductor Industry is published quarterly by Random Publishing Corp., at 17 South Franklin Turnpike, P.O. Box 555, Ramsey, NJ 07446, (201) 825–2552. The magazine covers high and low T_c materials and their applications, industry news, reports on the federal government's efforts in promoting superconductivity research and, finally, what new developments are taking place in the Far East. Reader service cards and a meetings calendar are also included.

Books

Many excellent books about ceramic science and technology are published each year. Helpful handbooks and books describing fun-

damental principles of ceramic science are listed below. It may also be useful to consult a journal such as the *Ceramic Abstracts*, which is published by the American Ceramic Society, to obtain reviews on some of the latest books or those that are more specific in nature. ASM International also publishes many ceramic materials books and may be contacted at (216) 338–5151 for more information.

Ceramic Raw Materials, by D.J. De Renzo, published by Noyes Data Corporation, Park Ridge, NJ, in 1987, provides a complete listing of properties and compositions for all materials supplied by the major commercial vendors. The listing includes ceramic raw materials and commonly used additives and auxiliary materials used with ceramics.

Encyclopedia of Glass, Ceramic Clay and Cement, is published by John Wiley & Sons, New York, NY. This handbook is categorized into different sections, such as alumina and compounds, borides, carbides, and ceramics. Each section contains information as to how the specific type of material is processed, its physical properties, and its enduse. This book provides general ceramics theory and numerous references, tables, and graphs. It represents a good desk reference.

Fine Ceramics, edited by S. Saito, was published by Elsevier Science Publishing Co., New York, NY, in 1987. In Japan, the term "fine ceramics" is used in place of the term "advanced ceramics," which is traditionally used in the United States. This book contains four chapters devoted to new ceramic processing techniques, characterization, structural ceramics, and electronic ceramics. It provides overviews of some of the latest materials, applications, and challenges facing the advanced ceramics field.

Guide to Materials Characterization and Chemical Analysis, edited by John P. Sibilia, VCH Publishers, Inc., New York, NY, in 1988, discusses various microanalytical techniques, the physical bases behind them, their use, their capabilities, and the information that can be derived from them. The book contains illustrations and references.

Handbook of Glass Data (A,B,C), published by Elsevier Scientific Publications, 52 Vanderbilt Ave., New York, NY, Handbook B, 1985; Handbook C, 1987, lists properties of different types of glasses. Tables and graphs defining the density, coefficient of thermal expansion, specific heat, thermal conductivity, viscosity, transition temperature, strength, chemical durability, and optical and magnetic properties of many different glasses are included.

Handbook of Properties of Technical and Engineering Ceramics, Vol 1, by R. Morrell, HMSO Publications, London, 1985, provides a good general overview of the composite microstructure, density, and porosity, as well as the thermal, mechanical, thermomechanical electrical, optical, and chemical properties of engineered ceramics.

Technical data sheets for various types of engineered ceramics are provided in Vol 2.

Introduction to Ceramics, by W.D. Kingery, published by John Wiley & Sons, Inc., New York, NY, 1976, describes the fundamentals of ceramic science. Chapters are devoted to the characteristics of ceramic solids (crystal structures, imperfections, atom mobility), development of microstructure (phase equilibria diagrams, reactions, grain growth, sintering), and properties of ceramics (thermal, optical, mechanical, and electrical). This book contains references, tables, charts, and illustrations.

Introduction to Phase Equilibria in Ceramics, by C.G. Bergeron and S.H. Risbud, published by the American Ceramic Society, 1984, will help scientists and students develop an understanding of phase equilibria in ceramic systems. It covers thermodynamics, unary, binary, ternary and quaternary systems, phase analysis, and nonequilibrium ceramic reactions. It also contains study problems.

Information Agencies

The American Ceramic Society is a good source for ceramic information. It publishes its own journals and technical books and sponsors Cerabull, an on-line data base, the Technology Update Hotline, and the Ceramics Correspondence Institute. The society is located at 757 Brooksedge Plaza Drive, Westerville, OH 43081–6136, (614) 890–4700.

ASM International is one of the largest materials societies in the world. In addition to sponsoring meetings and short technical courses, it publishes materials-related books and magazines and performs literature searches. Though not solely a ceramics organization, ASM International distributes a wealth of materials information. The society offers seminars and videotape courses on a variety of topics in the realm of engineered materials. Many of these courses give continuing education credits. In addition, the society sponsors or cosponsors expositions and conferences, such as *Materials Week* and the *World Materials Congress*. To ascertain the availability/scheduling of activities related to ceramics and glasses, contact education or conference personnel at 9639 Kinsman Rd, Materials Park, OH 44073–9989, (216) 338–5151.

High Temperature Materials Information Analysis Center (HTMIAC) maintains a data base of the properties of high-temperature materials, dielectric materials, and engineering materials, as well as the mechanical and thermophysical properties of rocks, minerals, and fluids. HTMIAC also produces a newsletter describing their services. The center is located at 2595 Yeager Road, W. Lafayette, IN 47906, (317) 494–9393.

The Metals and Ceramics Information Center (MCIC) provides technical advice and assistance in the areas of metals and ceramics. It also writes summaries of important developments in these fields, prepares handbooks, and has access to reference sources and data bases. The center is located at the Battelle Memorial Institute, 505 King Ave., Columbus, OH 43201–2693, (614) 424–5000.

The National Technical Information Service (NTIS) holds unclassified U.S. government technical reports, foreign technology reports, and reports made by state and local government agencies and contractors on file. For more information contact NTIS at 5285 Port Royal Road, Springfield, VA 22161, (703) 487–4600, and ask for the "Users' Guide to NTIS," PR-786.

The National Institute of Ceramic Engineers (NICE) c/o American Ceramic Society, Westerville, OH 43081–6136, (614) 975–6119. NICE is the professional organization of ceramic engineers.

Standards Organizations

The American Society for Testing and Materials (ASTM) publishes the multivolume *Annual Book of ASTM Standards*. Volume 15 of this publication contains some standards for traditional and advanced ceramics, whereas Volume 10 contains standards for electronic ceramics, and Volume 2, standards for porcelain enamel. ASTM also publishes a monthly journal, *Standardization News*, which provides a calendar for ASTM group meetings and updates on novel standardization methods, as well as special technical publications (STPs).

American National Standards Institute is located at 1430 Broadway, New York, NY 10018.

Department of Defense Index of Standards and Specifications lists specifications and standards of more than 68 organizations, including federal and military organizations. It is updated monthly except in June, July, and August. Contact Commanding Offices, Naval Publications and Forms Center, 5801 Tabor Ave., Philadelphia, PA 19120, (215) 697–3321.

The National Institute of Standards and Technology (NIST), formerly the National Bureau of Standards, publishes two periodicals, the *Journal of Research* and *Dimensions/NBS*, and maintains an information base on international standards. For an overview of NIST laboratories and research activities, contact the Public Information Division, NIST, Gaithersburg, MD 20899, and request a copy of SP-679.

Data Bases

Cerabull, a data base run by Clemson University for the American Ceramic Society, contains on-line ceramic composition calculation programs, a job information bulletin, and much more. The American Ceramic Society or past issues of the *American Ceramic Society Bulletin* can provide more information on this data base, as well as on the phase diagrams for ceramics that it publishes.

Ceramic Abstracts On-Line is a data base that can be used to conduct ceramic abstract searches. Contact STN International, c/o Chemical Abstract Services, Marketing Dept. 40889, P.O. Box 3012, Columbus, OH 43210, (800) 848–6538.

The Materials Information department of ASM International provides extensive coverage of the international literature on all aspects of engineered materials, including composites, plastics, adhesives, and sealants. This information is available monthly from *Engineered Materials Abstracts* (EMA) and its equivalent, the EMA data base. Materials Information also publishes *Polymers/Ceramics/Composites Alert* and its on-line equivalent *Materials Business File*, which emphasizes the business aspects of the materials industry. Materials Information also has access to more than 300 other data bases.

Other computer-searchable data bases are described below.

Standards and Specifications, prepared by the National Standards Association, Inc., is available on line through the Dialog Information Retrieval Service, Palo Alto, CA. It references more than 113,000 U.S. and international documents, including standards from ASTM; American National Standards Institute (ANSI); the federal government; the U.S. Military; the Society of Automotive Engineers, Inc.; and others. In addition to giving information on the standard and its acceptance by ANSI and the U.S. Department of Defense, names of vendors of products and services conforming to the standard are provided. Hard copy of many of the standards is also available from the National Standards Association, Inc.

Standards Search, prepared by ASTM and SAE, is available on line through the Orbit Search Service, McLean, VA. It contains more than 15,000 references (some with abstracts) to the standards included in the *ASTM Book of Standards*, and in the *SAE Handbook*, *Aerospace Index*, and *Index of Aerospace Materials Specifications*.

Military and Federal Specifications and Standards, prepared by Information Handling Services, is available on line through BRS Information Technologies, Latham, NY. This cites more than 80,000 nonclassified U.S. Military and federal standards and specifications, joint Army-Navy specifications, Military Standard Drawings, and Qualified Product Lists.

Combined Industry Standards and Military Specifications is also prepared by Information Handling Services and available on line through BRS Information Technologies, Latham, NY. This source cites more than 150,000 government and industry standards. It covers some 50 U.S. and other national and international standards organizations, including ANSI, the International Standards Organization (ISO), and the National Bureau of Standards Voluntary Engineering Standards data base, which is based on the standards of some 400 societies. It also includes the military and related standards covered by *Military and Federal Specifications and Standards*.

Ceramic Powders and Processing

Chairman: Alan L. Dragoo, Department of Energy

Introduction

THIS SECTION is concerned with the materials and methods used to prepare and characterize powders prior to consolidation, densification, and sintering of ceramic materials. The final outcome of ceramic processing can be influenced to a great extent by the nature and condition of the starting powders and the manner in which they are modified prior to consolidation. Particle size distribution and particle shape can control particle packing and, hence, the amount and size of pores remaining after sintering. Rapid shrinkage of agglomerates during sintering can result in voids and cracks which will limit the strength of a component. Impurities can be exsolved from grains during sintering to form grain-boundary phases which increase high-temperature creep rates or introduce undesirable electrical properties. Although impurities or an undesirable particle size distribution may be compensated for by the addition of a sintering aid or a hotter or more prolonged sintering schedule, such solutions may be at the expense of other desirable properties of the ceramic product. Maximum control of a ceramic process begins with the starting powder and powder preparation. It necessarily requires an accurate and suitable characterization of the materials.

The task of the process engineer is complicated by the fact that there is not a single way in which to fabricate a ceramic material. Traditionally, processing of oxide and silicate ceramics began with starting powders which were close in composition to the final product. These starting materials were prepared from minerals or ores by crushing, classification, purification, and usually a minimal amount of chemical refinement, such as roasting or calcination. However, requirements for greater purity and better density, strength, and electrical properties have led to greater degrees of chemical refinement of the starting materials, so that today these materials may be metals, nitrates, carbonates, chlorides, hydroxides, or organometallics. In the cases of silicon carbide (SiC) and silicon nitride (Si_3N_4), which do not occur naturally, starting materials may be prepared by at least three different chemical processes. Thus, we have attempted to address considerations which are important for the manufacture of modern-day, high-quality engineered ceramic components as well as to bear in mind the continued prevalence and importance of established processes and ceramic materials.

Increasing demands for improved performance of ceramics in load-bearing applications and in functional applications, such as electric circuit elements, have increased the importance of chemical purification and processing of the starting powders. Chemical processing of starting powders is emphasized in the article "Chemical Synthesis." In this article, the physical and synthetic chemistry of thermal decomposition, precipitation, and gas phase reactions are considered.

Methods for physical and chemical characterization of powders are surveyed in the article "Characterization of Ceramic Powders." Although particles in a powder are often conceptualized as spheres for the purpose of defining a dimensional characteristic, they are very complex objects, as indicated by the properties listed in Table 1. Particles generally are not spherical, unless they are formed by carefully controlled precipitation, but are oblong and, especially, angular if the powder was prepared by crushing and grinding. Moreover, the agglomerated particles in a powder may be composed of a few to many thousand grains, depending on how the powder was formed. In addition, such particles may contain more than one solid phase and may contain significant amounts of porosity. Thus, the process engineer may need to employ a battery of physical and chemical methods to characterize a powder adequately.

Dimensional characterization of powder requires both measurement of its mean particle size and of the particle size distribution (PSD). Similarly, classification of a powder may be used to change the mean particle size or to modify the PSD. Indeed, classification alters the mean particle size by changing the PSD; for example, sieving may be used to remove powder fractions with sizes above and below specified upper and lower cut sizes. Methods of powder classification as well as statistical concepts involved in characterizing the PSD are discussed in the article "Particle Sizing." In this article, attention is drawn to the importance and use of control charts for the maintenance of process control.

It is also important to bear in mind that a powder may be significantly altered either intentionally or unintentionally during a series of processing unit operations. To limit characterization to an initial stage out of interest of economy may be to lose sight of the effects of significant physical or chemical operation later in the production

Table 1   Characteristics of ceramic starting powders

Physical characteristics	Chemical characteristics
Grains (primary particles)	Bulk composition
Size	Major elements
Shape	(1 to 100%)
	Minor elements
Agglomerates (secondary particles)	(10 ppm to 1%)
Size	Trace elements
Shape	(<10 ppm)
Porosity	Inorganic species
Amount (see density)	(for example,
Size	sulfates and
Shape	nitrates)
Particle contact	Organic species
Coordination	Water and other
Strength	volatiles
	Phases
Density	
Specific surface area	Surface composition
Permeability	Major elements
Compactibility	Minor elements
Flowability	Trace elements
	Inorganic species
	Organic species
	Water
	Phases

stream. For example, an initially uniform powder may be removed from a hopper in such a way that the fines tend to be removed before the coarse particles. Furthermore, the powder may be altered by further milling and blending. Contamination of the powder may occur during the process due to abrasion or corrosion of a process vessel, pipes, and/or tubing. Accidental debris, such as cigarette ash and rat hairs, may leave voids during sintering, thus introducing strength-limiting flaws in the final product. Thus, further inspection and physical and chemical analysis of the material in the process stream is desirable if the engineer is to have control over the process.

Grinding (comminution) of powders is required to break down agglomerates which may sinter more rapidly than the interparticle regions of a compact. The break up of agglomerates also eliminates pores within the agglomerate which may become trapped during sintering. Comminution may also increase the sinterability of the powder by reducing the mean primary particle size, improving the packing of the powder by modifying the particle size distribution, and increasing the reactivity of particle surfaces. If formation of the ceramic requires solid-state reactions between particles of different compositions, then reduction of particle size will improve the amount of homogeneity attained by reduction of the distances over which reactants may diffuse. Basic concepts, equipment, and control of milling systems are considered in the article "Comminution."

Frequently several chemically different powders are to be combined or sintering aids added so that batching and mixing operations become important to ensure that the ingredients are combined thor-

oughly in the proper amounts and in the proper sequence. The optimum choice of mixing conditions is determined by the rheology of the powder and of the slurry, if wet mixing is used, as well as by equipment requirements and process time. The article "Batching and Mixing" gives special attention to the issue of nonuniformity in mixing. Different mixing methods for different ceramic requirements are also discussed.

Granulation, of which spray-drying is perhaps the most common technique, is employed to obtain a powder with good flowability and packing properties which are required for satisfactory mold filling. Granulation also assures that different chemical components or a range of particle sizes remain intimately mixed throughout the powder during transport between operations and in mold filling. The various methods of granulations, including agitation, pressure granulation, spraying, and atomization, are described in the article "Granulation and Spray Drying." This paper also discusses various processing effects such as hollow granules and contamination.

In the case of oxide ceramics, calcination is often an essential final step in the preparation of the powder prior to compaction and sintering of the ceramic material. Calcination may be used to convert the components from the form of salts or carbonates into oxides. It may also be used to adjust the reactivity of the powder so that a prescribed densification schedule is followed during sintering. The article "Calcination" discusses the basic concepts, including thermodynamics and kinetics, the formation and structure of relics, agglomeration, and effects of the process salt.

Processing of many ceramic materials requires the addition of process aids to control or enable the sintering process, to compensate for impurity effects, or to modify the strength or electrical properties of grain boundaries, such as in the manufacture of varistor materials. Processing aids may also be organic or polymeric compounds which are added as binders, lubricants, and process liquids required for the densification process and to provide strength to a green body prior to sintering. The article "Processing Additives" considers sintering aids and dopants, the use of process aids in forming, solvents, surfactants, wetting agents and dispersants, binders and plasticizers, and lubricants. Although this area of processing has been largely a matter of art and empirical experience, much has been done in recent years to develop scientific concepts underlying the use of processing additives and a number of these important concepts are discussed in this Section.

The reproducibility and efficiency of the subsequent operations of forming and sintering and the reliability of resulting ceramic product begin with, and depend upon, the powder preparation. Preparation of the process powder from starting material until the time it is compacted may involve a number of complex mechanical, thermochemical, and chemical operations which significantly modify the chemical, phase, dimensional, structural, and physical characteristics of the powder. Control of the forming and sintering operations requires a knowledge of these processing operations and how they alter powder characteristics.

Raw Materials

A.J. Brown, J. Bultitude, J.M. Lawson, H.D. Winbow, S. Witek, Cookson Technology Centre, Oxford, England

THE RAW MATERIAL BASE of the ceramics industry is extensive and ranges from mineral products and ores to highly refined chemicals. The uses of these materials include applications in traditional industries such as building products, whitewares, and refractories as well as in newer high-technology industries such as electronics. Materials are selected for these applications mainly on the basis of price and availability while still being consistent with fitness for purpose criteria. These criteria relate to the achievement of market acceptable end product properties through the use of minimum cost high efficiency production processes. There is a clear pattern of national and international trading of a significant number of raw materials to the ceramics manufacturer but there is also a considerable amount of in-house materials sourcing and preliminary materials processing by the producer. The latter is particularly true in the building products sector, where in-house sourcing and processing is the norm, and in high-technology applications where some users favor the production of their ceramic materials from chemical precursors.

In all cases, companies are active in improving and updating their products and processes to match market requirements and in high-technology applications this is complemented by active and vigorous academic research.

This discussion provides an overview of the economically most significant materials in current use and at the same time indicates material technologies that are likely to emerge as commercially viable in the foreseeable future.

Selection of Ceramic Raw Materials Based on Product Applications

The raw materials used as constituents in ceramics all have their primary origin in nature. Some raw materials are incorporated into ceramic products in their natural form while others require treatment and processing prior to use. In general, raw materials for large-tonnage products (such as brick, concrete, refractories, and so on) receive little or no preliminary processing while those for low-tonnage ceramic products (such as ceramics

used for electronic components, ceramic cutting tools, and optical glass) receive extensive beneficiation. This is a consequence of both economic considerations and property requirements. The trend today is toward additional processing of all raw materials, even for large-tonnage applications, because there are more restrictive and narrow specification requirements applied to the properties of ceramic products and because the best natural deposits are being gradually depleted.

Industrial and Residential Building Products (Ref 1—4)

In most countries of the world, naturally occurring clay minerals are processed into bricks, roof tiles, and drainage pipes by plastic forming of the raw material into green-ware that is fired at 900 to 1100 °C (1650 to 2010 °F) in intermittent or tunnel kilns. End product properties are usually defined by local building regulations. In the case of bricks and tiles, there is a need to comply with building style trends and to produce products with an aesthetic appearance in terms of both color and surface texture. The European market for these clay products is very extensive (see Table 1). However, it is difficult to place a monetary value on the raw materials used because in almost every case the end user sources clay from an in-house deposit.

No two clay bodies are the same and indeed there are generally variations in quality, both physical and chemical, throughout the deposit. A typical clay reserve, inasmuch as any can be described as typical, contains an impure kaolinite as a major phase together with a range of other minerals (for example, silica, mica/illite, feldspar, dolomite, calcium sulfate, and calcium carbonate) in variable amounts. In addition, most building clays contain transition elements (for example, iron as pyrites, hematite, and so on) and frequently significant amounts of carbon and other organic matter from the partial degradation of vegetation trapped in the sedimentary clay during its geological formation. This complex mineral array must be processed into a consistent feedstock in order that today's modern high-efficiency automated brick, tile, or pipe plant can meet its production targets in terms of output, product properties, and overall manufacturing costs. The consistency

of the clay feedstock is achieved by the controlled mining of clay from a pit that has been previously characterized chemically, mineralogically, and physically by a forward borehole drilling program. Extracted clay is subsequently stockpiled in layers in quantities that can approach a 6- to 12-month supply of raw material per plant requirements. The stockpile is layered horizontally and the material for operating the plant is then taken vertically to meet the ongoing plant demand.

Historically, most plants were operated with little, if any, product or process additives, although clay deposits were frequently chosen on the basis of the fired color and appearance imparted by some of the impurities. However, as the industry grew, a business in product and process additives slowly emerged. The most significant additives are processing aids such as lignosulfonates, which by acting either as flocculants, deflocculants, or wetting agents, improve clay workability and green strength, and additives such as colorants and antiscum compounds that improve product appearance. There are a significant number of proprietary systems on the market including:

- Body stains and whiteners, based on milled pyrolusite (manganese dioxide), milled hematite (iron oxide), milled chromite (iron chromite), calcium carbonate, zinc oxide, and titania
- Surface colors, which include naturally occurring and synthetic iron oxide pigments and zinc oxide pigments
- Antiscumming compounds, based on barium carbonate that fixes free sulfate in the green brick and minimizes the migration of calcium sulfate to the brick surface during drying

It is projected that the building products additives market will continue to grow as the industry undergoes cost, product design, and environmental pressures. Additives that can increase production rates, improve yields, reduce plant maintenance, meet styling trends set by the consumer and reduce or control sulfur dioxide emissions will be traded on an international basis.

Whiteware (Ref 5—13)

The whiteware sector comprises sanitaryware, wall and floor tiles, and tableware.

Production occurs throughout the world to suit local needs, using mainly locally available raw materials, although high-quality tableware and wall tiles are traded internationally. The raw materials for the industry fall into three distinct categories:

- Body materials
- Glaze materials
- Decoration and color systems

Typical formulations, key components, and physical properties of these three categories are shown in Table 2.

Body materials are chiefly based on white firing kaolinitic clays together with suitable fluxes required to vitrify the body on firing at temperatures in the range of 900 to 1300 °C (1650 to 2370 °F). All the materials are traded on a national and an international basis to major customers and each material is supplied according to agreed specifications that ensure its fitness for each specific application.

Glaze systems are based on glass forming systems, opacifiers (to impart opaqueness), and glaze colors. The glaze systems are carefully matched to conform to customer requirements such as:

- Methods of application
- Physical and chemical match to the body system
- Good durability in use and during cleaning/washing
- Low release of toxic elements into foodstuffs in the case of tableware
- Aesthetic appearance in terms of luster, flatness, and freedom from defects

Decoration and color systems can be applied underglaze, in-glaze, or on-glaze and as such must be chemically and physically compatible with the substrate and the glaze during high-temperature firing and also resistant to normal treatment by end users.

The supply of raw materials to this sector is a multimillion dollar business. The key components for body systems are china clay, ball clay, quartz, nepheline syenite, feldspars, bone ash, and synthetic fluxes. On the other hand, glass frits (prepared from lead, boron, and silica) and milled zircon opacifier are major glaze components. Zirconia is the basis for the zircon colors with synthetic spinels and sphenes adding to the color range.

Marketability of Raw Materials. The market for the above materials is optimistically only increasing slowly in the western world, but third-world countries offer potentially lucrative markets as their standard of living increases. Key factors that will affect the raw materials supplier are the technological changes taking place within the industry. These changes relate to cost, quality, and environmental pressures. The industry is addressing both cost and quality factors by adopting rapid and more automated fabrication techniques (for example, isopressing of

Table 1   European heavy clay products production in 1988

| | Clay brick | | | |
| | Solid and perforated | | Hollow | |
Country	m³	ft³	m³	ft³
Germany	6,177,000	218,100,000	1,639,000	57,880,000
Austria	2,065,000	72,930,000	...	...
Belgium	1,257,000	44,390,000	...	...
Denmark	86,000	3,040,000	...	...
Finland	215,000	7,590,000	...	...
England	2,840,000	100,300,000	...	...
Italy	3,320,000	117,200,000	9,630,000	340,000,000
Netherlands	95,000	3,350,000	...	...
France	685,000	24,200,000	2,281,000	80,550,000
Norway	4,981,000(a)	4,981,000(a)	...	...
Switzerland	1,069,000	37,750,000	...	...
Spain	1,792,000	63,280,000	8,444,000	298,200,000

(a) Pieces. (b) m³. (c) ft³. (d) tonne. (e) ton

spray dried body formulations for tableware and pressure slip casting for sanitaryware) and fast firing technologies while simultaneously attempting to control toxic elements such as lead and cadmium is becoming increasingly important.

Refractories (Ref 14–18)

The refractory business is very dependent on the global iron and steel industry and 50 to 60% of all refractories are supplied to this sector of the market. The remaining portion is used in the production of cement, glass, nonferrous metals, petrochemicals, and ceramic products. In the last two decades, there has been a marked decline in the usage of refractories worldwide by the iron and steel industry due to production cutbacks and dismantling of obsolete facilities and the introduction and consolidation of improved process methods such as basic oxygen steelmaking (BOS), high-powered electric arc furnaces (EAF), and continuous casting. The refractory industry has adapted to these changes by introducing new products with improved properties and longer service life. A clear example of the decline in refractories use is provided by the Japanese refractory industry that has declined from 2.4×10^6 to 1.8×10^6 tonnes (2.6×10^6 to 2.0×10^6 tons) in the period 1981 to 1986, while the Japanese steel output remained approximately constant at 1×10^8 tonnes (1.1×10^8 tons).

Refractory products are supplied to the user industries in the form of as-fired shapes, unfired shapes, and monolithics. There are clear trends toward the use of monolithic products for all applications and toward unfired shapes in the iron and steel industry.

The supply of raw materials to the refractory business has evolved from one that was dependent on locally available raw materials to one that is now dependent predominantly on internationally traded materials. There are some notable exceptions such as silica, dolomite (in Europe and the United States), low-quality clays, and pyrophyllite (in Japan),

while some third-world countries can favor the use of low-quality indigenous materials over the optimum imported product in order to conserve foreign exchange.

The main refractory raw materials are listed in Table 3 in terms of typical properties, applications, and sources of supply. These include the materials that form the body of the product as well as the bond systems (both temporary and permanent) and minor additives. The major tonnage materials include not only standard minerals but also microstructurally designed products such as magnesia and relatively high priced, high-performance materials such as alumina, silicon carbide, and zirconia. Chammotte, bauxite, magnesia, and graphite contribute to a significant percentage of the overall refractory materials business, but the specific consumption of these materials (quoted for example as kg/tonne of steel) will continue to decline as end users demand improved performance and only a strong expansion in customers activity will prevent a downturn in overall supply requirements. It could be inferred that growth opportunities exist for materials such as silicon carbide and zirconia and, while there seems little doubt that these materials will be used more widely, growth may well be limited by astute refractory product design which, by the use of liners and/or coatings, could limit high priced materials to key contact areas. Materials substitution within the industry will thus occur on a value in service basis.

Bonding systems include temporary bonds such as lignosulfates for fired products and permanent bonds such as resins in carbon bonded products and calcium aluminate cements in refractory castables. While body materials have improved in terms of reproducibility and quality, there have been advances in the use of novel bonding systems and this area is forecast to offer opportunities for future development.

The market for refractory insulation materials based on ceramic fiber has increased rapidly in the last two decades as the main

Facing brick		Materials for horizontal structures		Roofing tile	
m³	ft³	m²	ft²	m²	ft²
1,699,000	60,000,000	64,000(b)	2,300,000(c)	25,267,000(b)	892,300,000(c)
...	...	890,000	9,600,000	728,000	7,830,000
666,000	23,500,000	...	...	...	...
382,000	13,500,000	...	...	22,268,000(a)	22,268,000(a)
216,000	7,630,000	...	...	...	...
5,986,000	211,400,000	...	...	...	...
...	...	89,700,000	965,000,000	31,200,000	336,000,000
1,620,000	57,210,000	...	...	45,500,000(a)	45,500,000(a)
...	...	73,000(d)	80,000(e)	2,030,000(d)	2,230,000(e)
19,418,000(a)	19,418,000(a)	...	...	...	...
67,000	2,400,000	...	...	146,053,000(a)	146,053,000(a)
672,000	23,700,000	1,250,000(d)	1,380,000(e)	650,000(d)	720,000(e)

refractory users strive to save energy in their processes and reduce costs. The most used fiber is based on high-quality clay, but fibers based on alumina and zirconia also have widespread applications.

The effect of refractory materials on the environment cannot be overlooked. Products that are hazardous to the health of process workers and customers and that result in disposal problems are already coming under close scrutiny. Three notable examples of hazardous refractory materials are:

- Chrome ore; substitute materials are being actively sought and evaluated
- Baddeleyite, a naturally occurring zirconia, contains radioactive isotopes of zirconium, which make this material and products based on it a registered hazardous material in the United States
- Organic resins, used for making unfired shaped carbon containing refractories, are being closely monitored in terms of product handling and burnout in customer plants

The raw materials suppliers will be under continued pressure, therefore, to control costs, quality, and service. All efforts will be made to reduce the specific energy consumption, to reduce the level of deleterious impurities, and to optimize microstructural properties. Opportunities exist for the development of new materials that can withstand the operating environment and for novel bonding that can improve product quality and/or reduce production costs through improved processing. Key properties for which improvements are being sought are corrosion resistance, oxidation resistance, and thermal shock resistance.

Advanced Ceramics (Ref 19–30)

Whereas the three sectors previously discussed represent traditional ceramics markets that have slow growth or declining markets, the advanced ceramic sector represents new markets for ceramics, some of which have shown phenomenal growth over the last few decades. The major market for state-of-the-art ceramics has been and will continue to be in electronics, but vigorous worldwide research and development programs are continuously searching for new applications and identifying ways of improving ceramic properties such that new markets can be accessed.

Advanced ceramics are produced in Japan, the United States, and Western Europe. The raw materials used in the industry are traded on an international basis, principally as powders, but there is also a significant amount of in-house processing. The estimated United States market for ceramic powders by market and product type for the period 1989 to 1995 is listed in Table 4. While the absolute value of this market is difficult to quantify because of the in-house processing and trends in some markets to use less raw material per component, there is no doubt that both the electronics market and the supply of oxides to this area are dominant. Furthermore, these mar-

Table 2 Composition and physical properties of raw materials used for whiteware applications

Raw material	Composition			Typical properties of fired ware
	Frit materials	Raw glaze materials	Constituents	
Body materials				
Bone china	...	...	Bone ash, china clay, quartz, feldspars	Pure white, translucent, high strength, zero water absorption
Porcelain	...	...	China clay, feldspars, quartz	Pure, bluish, or off white, translucent, high strength, zero water absorption
Hotelware	...	...	Quartz, ball clay, china clay, feldspars	Off white, good physical and chemical durability, zero water absorption
Earthenware	...	...	Quartz/flint, china clay, ball clay, feldspars	Ivory to pure white, lower strength, water absorption 4 to 8%
Sanitaryware	...	...	China clay, ball clay, fire clay, quartz, feldspars, nepheline syenite	...
Wall/floor tiles	...	...	Ball clay, fire clay, quartz, feldspars, limestone	...
Glazes	<1150 °C (2100 °F): Quartz, sodium borate, boric acid, limestone, feldspars, china clay, lead oxides, zircon, zinc oxide, alkali metal, carbonates, and nitrates	...	...	Transparent to opaque, colored or colorless, glossy, matte, vellum, or textured
Pigments	...	>1150 °C (2100 °F): Feldspars, quartz, clays, nepheline syenite, limestone, dolomite, zinc oxide, zircon	...	Transparent to opaque, colored or colorless, glossy, matte, vellum or texured
Zircon-based	...	...	Zirconia, quartz	...
Spinel-based or oxide colorants	...	...	Iron, chromium, zinc, nickel, copper oxides or compounds	...
Sphene-based	...	...	Quartz, limestone, tin oxide	...

Table 3 Composition, applications, and origin of selected body raw materials, bonding materials, and special additives used to produce refractory products

Raw material	Typical composition	Applications	Source of supply
Body materials			
Silica	>95% SiO_2 silica rock	Production of fired silica brick for use in: coke ovens, glass tanks, roofs	Worldwide supplies of suitable material
	Silica sand	Production of sand molds for metal castings	
Chammotte	42–44% Al_2O_3 calcined aluminosilicate Low Fe_2O_3 (typically <1%) Low alkalis Low alkaline earths	Production of fired 42/44 Al_2O_3 aluminosilicate bricks for use in: blast furnaces; cement kilns; production of 42/44 aluminosilicate general purpose castable	Reserve of high-quality clay limited. Main supplies from South Africa, United States, China, and France
Andalusite	58–60% Al_2O_3 aluminosilicate mineral liberated from host rock Low Fe_2O_3 (typically <1.5%) Low alkalis Low alkaline earths	Production of fired 60% Al_2O_3 aluminosilicate bricks for use in: blast furnaces; steel ladles; torpedo ladles; and aluminum anode baking furnace	Reserve of high-quality andalusite limited. Main supplies from South Africa and France. China now entering the market
Bauxite	85–88% Al_2O_3 calcined aluminosilicate Fe_2O_3: <1.5%, TiO_2: <3.5% Low alkalis Low alkaline earths	Production of fired 85–88% Al_2O_3 aluminosilicate bricks for use in: steel ladles, torpedo ladles, and aluminum holding vessels Production of phosphate-bonded brick for aluminum remelt furnaces Production of castables	Reserves of high-quality ore limited to Guyana (gibbsite $Al_2O_3\cdot3H_2O$) and China (diaspore $Al_2O_3\cdot H_2O$)
Doloma	40% MgO, Fe_2O_3: <1.5%, SiO_2: <1%, Al_2O_3: <1%, CaO: ~56–60%	Production of fired doloma brick for use in: cement kiln hot zones; argon oxygen decarburization (AOD) vessels for stainless steel; and steel ladles	Reserves of dolomite for dead burning available in United States, Belgium, Germany, and United Kingdom
Magnesia	High-quality dead burnt magnesia Low porosity (bulk density >3.40 g/cm^3) Controlled $CaO:SiO_2$ ratio: >2.5:1) SiO_2: 0.5%, Fe_2O_3: <0.2%, Al_2O_3: <0.1%, B_2O_3: <0.05%, MgO: <96%	Used in the production of magnesia-carbon brick for basic oxygen steelmaking, electric arc furnaces, and steel ladles Used in the production of fired brick for glass tank regenerators, magnesia-spinel bricks for cement kilns, and slide gate plates High-quality magnesia chrome for use in secondary steel making	High-quality material sources from brine sea water and beneficiated natural magnesite found in Israel, Holland, Ireland, United Kingdom, United States, Greece, and Japan
	MgO: >90%	Used in the production of magnesia chrome, chrome-magnesia fired bricks for cement kilns, nonferrous (copper), and secondary steel making Also as a gumming repair material	Low-quality material sources from sea water and natural magnesite found in Greece, China, Brazil, and Czechoslovakia
Chrome ore	Iron chromite mineral: a mixed spinel of FeO, MgO, Fe_2O_3, Cr_2O_3, and Al_2O_3 Cr_2O_3 levels: 32–56%	Used in the production of fired magnesia-chrome and chrome-magnesia bricks	Supplied after separation from impurities from Phillipines, South Africa, and Zimbabwe
Graphite	Flake graphite mineral is separated from host rock Carbon level 85–95% Remainder aluminosilicate impurities	Used in the production of magnesia-carbon bricks and alumina-carbon, zirconia-carbon continuous casting products	Supplied from China, Norway, Sri Lanka, and Malagasy
Carbon	Electrocalcined or gas calcined anthracite or petroleum coke Carbon level >95% Remainder aluminosilicate impurities	Used in the production of fired carbon for blast furnaces and as the cathode in aluminum reduction cells	Supplied locally to suit market needs in United States, Germany, United Kingdom, and Poland
Alumina	Fused or tabular alumina: >99% Al_2O_3	Used in alumina carbon continuous casting products Fired alumina slide gates High-quality monolithics for the petrochemical industry and for blast furnace application	Produced by fusing or calcining Bayer grade alumina in United States, Holland, United Kingdom, and Japan
Aluminosilicate fiber	42–44% Al_2O_3 fiber produced from high-quality kaolinitic clay	Used as a fiber insulation in kilns, glass tanks, and so on	Produced in United States, Europe, and Japan
Mullite	Sintered or fused 72% Al_2O_3 aluminosilicate, $3Al_2O_3\cdot2SiO_2$	Creep-resistant refractory used as fired shapes in glass tanks, tunnel kilns, continuous casting	Produced in United States, United Kingdom, and Japan
Spinel	Sintered or fused magnesia aluminate, $MgO\cdot Al_2O_3$	Addition to fired magnesia bricks to improve thermal shock resistance for application in cement kilns and glass tanks	Produced in United States, United Kingdom, Germany, and Japan
Olivine	Naturally occurring magnesium silicate	Used as a monolithic coating in tundishes	Scandinavia
Silicon carbide	SiC produced by the Acheson process: >98% SiC Impurities: Si; SiO_2; C; and Fe_2O_3	Used as a nitride Sialon or self-bonded product in blast furnaces and aluminum reduction cells As a silicate-bonded product in incinerators and power plants	Norway and China
Zircon	Naturally occurring zirconium silicate sand, $ZrO_2\cdot SiO_2$, containing small quantities of HfO_2, Al_2O_3, TiO_2, and Fe_2O_3	Used in investment casting, foundries, as a glass contact refractory, an aluminum contact refractory Fluctuating demand as a ladle refractory Raw material for the production of zirconia and zirconia mullite	Australia, South Africa, and United States
Zirconia	Fused zirconia with ZrO_2 > 96% Main impurities: HfO_2; SiO_2 (<0.5%); Al_2O_3 (<0.5%); Fe_2O_3 (<0.1%); TiO_2 (0.2%)	Used in zirconia carbon continuous casting refractories as fired zirconia shapes, nozzles, kiln furniture, and so on, and as an addition to cement kiln doloma refractories to improve thermal shock resistance	Naturally occurring ZrO_2 (baddeleyite) found in South Africa Zirconia derived from zircon in United States, United Kingdom, and Germany
Zirconia mullite	Fused or sintered impurities in zircon diluted by high-purity mullite ($ZrO_2\cdot3Al_2O_3\cdot2SiO_2$)	Used in continuous casting refractories	United Kingdom, United States, and Japan

Table 3 Continued

Raw material	Typical composition	Applications	Source of supply
Bonding materials			
Lignosulfonates	. . .	Temporary green bond in fired refractories	By-product of the paper industry
Plastic clay	. . .	Permanent green bond in fired aluminosilicate refractories and for plastic bond in rammable monolithics	Widespread
Calcium aluminate cements	. . .	Permanent cementitious bond for castables	Japan, United States, France, and United Kingdom
Phosphoric acid	. . .	Permanent bond for low-temperature fired (400 °C, or 750 °F) refractories	Commodity chemical
Sodium silicate	. . .	Air setting bond for mortars	Commodity chemical
Tars/pitches	. . .	Bonding systems for carbon-containing products Materials carbonize on firing in reducing atmosphere Pitches also used to impregnate slide-gate refractories	By-product of coke plants and petrochemical refineries
Phenol formaldehyde resins, and furane resins	. . .	Bonding systems for magnesia graphite, alumina graphite, and zirconia graphite	Commodity chemicals
Special additives			
Volatilized silica	. . .	Used as an addition to low cement castables to improve rheological properties and sintering	By-product of the silicon industry Norway
Calcined alumina	. . .	Used as an addition to castables and high-quality alumina products to improve sintering	Produced from the Bayer process
Chromium oxide	. . .	Used as an addition to improve corrosion resistance in the presence of siliceous slags	Minor use of pigment grade Cr_2O_3
Silicon, aluminum	. . .	Used as powdered additives in carbon-bonded refractories to improve strength and improve oxidation resistance	Commodity materials
Proprietary phosphates	. . .	Used as dispersing aids in monolithic refractories	Commodity materials
Ethyl silicate	. . .	Used as a bond in special precast refractories	Commodity materials

kets are set to expand in real terms over the next few years.

Oxides. The main oxide materials in use today are alumina in spark plugs, substrates and wear applications; zirconia in oxygen sensors, as a component in lead-zirconium-titanate (PZT) piezoelectrics, wear applications, and thermal barrier coatings; titanates in barium titanate capacitors and PZT piezoelectrics; and ferrites in permanent magnets, magnetic recording heads, memory devices, temperature sensors, and electric motor parts.

Carbides and Nitrides. Carbides (mainly silicon carbide and boron carbide) are used in wear applications while nitrides (mainly silicon nitride and Sialon) are used in wear applications and cutting tools. Aluminum nitride with its high thermal conductivity is the primary contending material for part of the electronics substrate market currently

dominated by alumina. In addition, aluminum nitride is a key component in the production of Sialons from silicon nitride:

$$Si_3N_4 + AlN + Al_2O_3 \rightarrow Si_3Al_3O_3N_5 \qquad (Eq\ 1)$$

Mixed Oxide Ceramics. Ceramics research and development efforts are focused on a number of new applications for ceramics that all have enormous potential. Three significant applications are:

- Ceramic superconductors
- Ceramics for solid oxide fuel cells
- Ceramic components for heat engines

Ceramic superconductors are based on a number of mixed oxide systems that include $YBa_2Cu_3O_{7-\delta}$, $Bi_2Sr_2CaCu_2O_8$, and $Bi_2Sr_2Ca_2Cu_3O_{10}$ stabilized with PbO. Solid oxide fuel-cell ceramics are based on ionic conductors in which high-purity stabilized

zirconia is currently the material of choice. Ceramic heat-engine components under investigation are composed of silicon carbide, Sialons, and zirconia either as single-phase ceramics, ceramic-ceramic composites, or metal-matrix composites (MMCs).

Processing Innovations. Research and development activity is generating new technologies for the production of ceramic materials. Precursor derived ceramics are estimated to have a market value of $200,000,000 in 1989, the major part of which is in chemical vapor deposition (CVD) (86% of the total market value). Other segments of this growing market include chemical vapor infiltration (CVI), sol-gel, and polymer pyrolysis. Products that are being successfully produced by these means include continuous ceramic fibers, composites, membranes, and ultra-high-purity/high-activity powders.

Table 4 Estimated market for advanced ceramic powders in the United States from 1989 to 1995 categorized by applications and types of material

Product	1989		1990		1995		1989–1995 growth rate, %/yr
	Percent	Monetary value, 10^6 dollars	Percent	Monetary value, 10^6 dollars	Percent	Monetary value, 10^6 dollars	
Application							
Electronic	81.6	340.5	81.2	374.1	80.8	561.0	8.5
Structures	13.3	55.6	14.0	64.2	15.3	105.9	10.5
Coating	5.1	21.3	4.8	22.2	3.9	27.0	4.0
Total	100.0	417.4	100.0	460.5	100.0	693.9	8.5
Material type							
Oxides	88.2	368.0	87.7	404.0	87.8	608.9	8.5
Carbides	9.8	41.0	10.0	46.0	9.4	65.0	7.1
Nitrides	2.0	8.4	2.3	10.5	2.8	20.0	13.8
Total	100.0	417.4	100.0	460.5	100.0	693.9	8.5

Table 5   Raw materials for advanced structural and magnetic (ferrite) ceramics

End product	Raw materials	Key product properties	Applications
Al_2O_3	Bayer process alumina derived from bauxite	Low permittivity Hardness Wear resistance	Substrates, insulators, spark plugs, wear parts, milling media, thread guides, armor, radomes
ZrO_2	Zirconia derived from zircon by chemical processes	Ionic conductivity Electronic conductivity Wear resistance	Oxygen sensors, fuel cells (potential), high-temperature heater, milling media
CBN	High temperature and pressure transformation of hexagonal form of BN (HBN) (formed by reacting B_2O_3 and urea in nitrogen)	High thermal conductivity High electrical resistivity High hardness	Substrates in electronics Machining of ferrous metals
BeO	Beryllia powder from beryl or bertrandite ores	High thermal conductivity High electrical resistivity	Substrates (heatsinks) in electronics
SiC	Acheson process: $SiO_2 + C \rightarrow SiC + CO$ Pyrolysis of polycarbosilanes	Extreme hardness Resistance to thermal shock	Wear parts As a fiber whisker or particle in MMCs and CMCs
$Al_2O_3 \cdot ZrO_2$	High-quality alumina High-quality zirconia	Improvement in strength and toughness over Al_2O_3	Wear parts
Si_3N_4	Silicon nitride powder derived from silicon and nitrogen	Hardness Resistance to thermal shock	Wear parts
Sialons	Silicon nitride, alumina, 21R polytype (AlN), yttria	Hardness Toughness Resistance to thermal shock	Wear parts Extrusion dies Cutting tools
AlN	AlN powder, prepared by carbothermal reduction of Al_2O_3 in nitrogen, or direct nitridation of Al	Low permittivity High thermal conductivity	Electronic substrates
SnO_2	High-purity tin dissolved in nitric acid and coprecipitated with other oxides	Surface controlled conductivity	Sensors
PZT (lead zirconium titanate)	High-purity oxides Coprecipitated oxides	High piezoelectric coefficients Change of polarization with temperature	Transducers Actuators Pyroelectrics
PMN (lead magnesium niobate)	High-purity oxides Coprecipitated oxides	High permittivity and breakdown voltage Controlled deformation in an applied field	Capacitors Actuators
PLZT (lead lanthanum zirconium titanate)	High-purity oxides Coprecipitated oxides Metal alkoxide-based coatings	Change of birefringence with applied field Controlled deformation in an applied field Change of birefringence with applied field	Electro-optics, head-up displays, flash goggles Actuators
ZTS (zirconium titanium stannate)	High-purity oxides	Stable permittivity at high frequencies over a wide temperature range and very low dielectric and insertion losses	Microwave resonators and filters
PBNT ($PbO \cdot BaO \cdot Nd_2O_3 \cdot TiO_2$)	High-purity oxides	Stable permittivity at high frequencies over a wide temperature range and very low dielectric and insertion losses	Microwave resonators and filters
ZnO	High-purity oxides (derived from metal smelting) plus praeseodymium or bismuth oxides	Change of resistivity with applied field	Varistors
$YBa_2Cu_3O_{7-\delta}$	Barium, strontium, calcium salts (chlorides, carbonates, nitrates, and peroxides)	Superconductivity	Demonstration devices
$Bi_2Sr_2CaCu_2O_8$	Barium, strontium, calcium salts (chlorides, carbonates, nitrates, and peroxides)	Superconductivity Very low insertion losses	Microwave filters
$Bi_2Sr_2Ca_2Cu_3O_{10}$ stabilized with PbO	High-purity oxides Coprecipitation, sol-gel, metal alkoxides, sputtering, chemical vapor deposition	Superconductivity conductor in very high magnetic fields	Nuclear magnetic resonance (NMR) imagers
TiB_2	Powder made by carbothermal reduction of TiO_2 with B_2O_3	Electrical conductivity Resistance to molten aluminum coupled with complete wetting of surface	Potential cathode material in primary aluminum production Evaporator boats (with BN)
B_4C	Carbothermal of reduction B_2O_3	Very hard Abrasion resistant Absorbs thermal neutrons	Shot blast nozzles, bearings, armor, and nuclear energy industry uses
Ferrites			
Hard ferrites			
$SrFe_{12}O_{19}$	$SrCO_3/Fe_2O_3$	High residual flux density High coercive force	Permanent magnets
$BaFe_{12}O_{19}$	$BaCO_3/Fe_2O_3$	High coercive force High residual flux	Motors
Soft ferrites			
$MnZnFe_4O_8$	Mixed oxide, iron oxide derived from thermal hydrolysis of ferric chloride	High initial permeability Low loss	Wide band/pulse, transformers Inductors, telecommunications
$MnNiFe_4O_8$	Mixed oxide, iron oxide derived from thermal hydrolysis of ferric chloride	High initial permeability High saturation flux density	Wide band/pulse, transformers Power transformers, magnetic recording heads
Microwave			
$Y_3Fe_5O_{12}$	Coprecipitation or milling of pure oxides	Narrow line widths and extremely low losses at microwave frequencies	Elements in microwave circuitry
$BaTiO_3$	Barium carbonate and titania Barium chloride and titanium tetrachloride	High permittivity High breakdown voltage Increase of resistance with increase of temperature High piezoelectric coefficients Change of birefringence with field	Capacitors Positive temperature coefficient (PTC) thermistors Transducers, actuators Electro-optics

The main products of the advanced ceramics industry, both existing and potential, are listed in Table 5, which indicates the key end property requirements, the main applications, and the raw material base for the products. The processes used to convert these raw materials to finished products include additional powder processing (for example, milling and spray drying) prior to forming green shapes that are then fired under controlled conditions. The forming processes include die pressing, isostatic pressing, slip casting, tape casting, extrusion, injection molding, hot pressing, hot isostatic pressing (HIP), chemical vapor deposition (CVD), and so on.

Chemical Additives to Aid Ceramic Processing. Each step in the manufacturing process requires careful control so that end product properties are obtained at maximum production efficiency and key effect chemicals are used to optimize powder treatment and green forming. The effect chemicals include milling aids, flocculants and binders, lubricants to effect product release during pressing and minimize wear of die parts, and plasticizers to aid extrusion and injection molding. A list of such chemicals is shown in Table 6. While these materials play an important economic role in production, they are burnt out during firing and play no part in the final product chemistry. The burn out process has to be carefully controlled to avoid residual carbon in the finished products and process research and development is continuously investigating ways of minimizing the levels of effect chemicals used.

In addition to spawning ceramic products and ceramic manufacturing technologies for new applications, the influence of the advanced ceramics industry on the traditional ceramics industry should not be overlooked. It is judged that many high-technology materials and processes will find application in the traditional ceramics industry as the latter strives to reduce manufacturing costs, to improve quality, and to give better value in service to the end user.

Primary Ceramic Raw Materials

From the data presented under the application sections in Tables 3 and 5 it is apparent that there are certain key materials that are either used directly by the ceramics industry or that represent the starting point for the production of added value materials:

- Silica
- Clay
- Alumina
- Magnesia
- Titania
- Iron oxide
- Zircon/zirconia

This discussion will focus on the properties of silica, alumina, and zircon/zirconia.

Silica

Silica, in addition to its use in refractories and whitewares, is also the starting point in the manufacture of elemental silicon, silicon carbide, and silicon tetrachloride. Silicon, in turn, is the starting point for silicon nitride, and silicon tetrachloride is the precursor for a wide range of silicon organics that can be pyrolyzed under controlled conditions to high-quality silicon carbide and silicon nitride. Thus, there is a choice, albeit at a price to obtain silicon nitride and silicon carbide at different states of purity and activity. Currently, silicon carbide (prepared by the Acheson process in which a mix of silica and coke is heated electrically to 2400 °C, or 4350 °F) is used as a refractory in blast furnaces, incinerators, power stations, aluminum reduction cells, and copper refining while high-purity powder (also prepared by the Acheson process) can be used to make engineering wear parts such as seals, pump parts, bearings, and dies. The powder is also used in metal-matrix composites. Silicon carbide has been derived from polycarbosilanes by pyrolysis in both powder and fiber form and, despite its high price, represents a potential for producing ceramic/ceramic composites.

Silicon nitride and its Sialon derivatives and silicon carbide despite their tendency to oxidize, have the potential to meet many of the property targets set by the heat engine market. A feature of silica and the ceramic materials that are derived from silica is that all the elements are readily available in the earth's crust. In this respect, these materials offer the potential of ease of supply in all parts of the world. In practice, however, there is a significant energy input required to produce silicon and silicon carbide. Consequently, manufacture of these materials is by and large limited to countries with cheap and readily available electric power (for example, Scandinavia and North America).

A flowchart relating silica to materials used (or with prospects of use) in the ceramics industry is shown in Fig 1.

Alumina

Alumina is found throughout the earth's crust as a component in aluminosilicate minerals. Economics dictates that alumina be extracted from bauxite using the Bayer process. Bauxite is widespread in the equatorial belt in different states of purity and is divided into two classifications: refractory grade ore and metallurgical ore.

Refractory grade bauxite is supplied by China and Guyana as a high-temperature calcine of the naturally occurring mineral: diaspore ($Al_2O_3 \cdot H_2O$) in China and gibbsite ($Al_2O_3 \cdot 3H_2O$) in Guyana. Both calcines conform to controlled chemical and physical specifications (for example, alumina 80 to 88%, and minimal levels of fluxing impurities) and are used in a wide range of refractory brick and monolithic compositions. During calcination, a complex phase assemblence of corundum (Al_2O_3), mullite, a silica glass, and minor levels of aluminum titanate is formed. The consumption of refractory grade bauxite exceeds 700,000 tonnes/yr (800,000 tons/yr) on a worldwide basis.

Metallurgical grade bauxite is mined in Australia, Jamaica, and West Africa and

Table 6 Selected chemical additives used to optimize powder treatment and green forming of ceramics

Material	Application or function
Polyvinyl alcohol	Binder for advanced ceramics
Polyethylene glycol	Binder for advanced ceramics
Sodium polyacrylate	Deflocculant for slip casting
Tertiary amide polymer	Binder for dry pressing
Starch blended with dry colloidal aluminosilicate	Binder for vacuum forming
Cationic alumina plus organic flocculant	Binder for vacuum forming
Pregelled, cationic corn starch	Flocculant for colloidal silica and alumina binder
High-purity sodium carboxymethylcellulose	Binder
Inorganic colloidal magnesium aluminum silicate	Suspending agent
Medium-viscosity sodium carboxymethylcellulose added to Veegum	Suspending agent, viscosity stabilizer
Ammonium polyelectrolyte	Dispersing agent for casting slips for electronic ceramics
Sodium polyelectrolyte	Dispersing agent binder for spray-dried bodies
Microcrystalline cellulose and sodium carboxymethylcellulose	Thickening agent
Polysilazane	Processing aid, binder, and precursor for advanced ceramics

Source: Ref 30

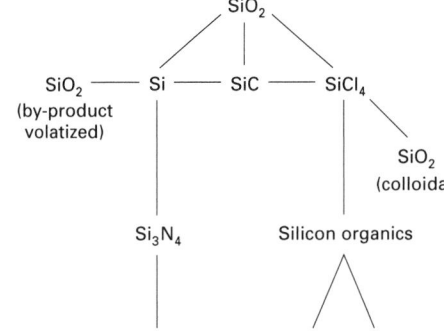

Fig 1 Use of silica as a precursor for ceramics

has a variable alumina level in conjunction with major impurities such as iron oxide and silica. The alumina in the metallurgical ores is extracted from the ore when dissolved by sodium hydroxide, yielding a sodium aluminate solution that is separated from the iron oxide and silica, which are rejected as a waste product in the form of red mud. Essentially, pure aluminum hydroxide is precipitated from the sodium aluminate and then calcined to a number of grades of alumina. In 1989, the world production of alumina was approximately 39×10^6 tonnes (43×10^6 tons), which compares to 37.5×10^6 tonnes (41.3×10^6 tons) in 1988 and 35×10^6 tonnes (39×10^6 tons) in 1987. Approximately 92% of alumina produced by this method is used to produce aluminum by the Hall-Heroult process, with the remainder finding a diverse range of chemical and ceramic applications. Current reserves of metallurgical grade bauxite are more than adequate to meet the demand predicted for the next 200 years.

The high-purity aluminas used in the ceramics industry and derived by the Bayer process are classified as:

- Tabular alumina
- Fused alumina
- Specialty calcined alumina

Tabular alumina is produced by high-temperature ($\sim$2000 °C, or 3630 °F) calcination of low-temperature calcined alumina in large oil-fired rotary kilns. Fused alumina is produced by the electric melting of calcined alumina. Both materials contain >99.5% Al_2O_3 with Na_2O as the major impurity at $\leq$0.3% and have a grain porosity of <5%. Tabular and fused alumina are sold to the refractory industry in crushed and graded form for use in a wide range of high-quality products:

- Continuous casting refractories (for example, single-edge notched or SEN/slide gates)
- Monolithic refractories for application in blast furnaces and the petrochemical industry

Specialty calcined alumina powders are the major raw materials used in the advanced ceramics industry for both electronic and engineering applications. The powders are produced in a wide range of grades against exacting specifications of chemistry, particle size, and crystal type to suit a wide range of end product applications. The normal 0.5% Na_2O level of Bayer process alumina can be reduced to make specialty low soda grades during refining/calcination while physical properties can be adjusted during rotary kiln calcination and subsequent dry grinding. Current trends lean toward improved grinding control to produce materials with median particle sizes up to 0.5 μm (20 μin.) as well as in the supply of spray dried granules that can be fed directly into customer's plant.

There is an established international trade in high-quality aluminas but, with many of the ceramic manufacturers having in-house milling and spray drying facilities, there is clearly a limitation to the growth in the supply of spray dried systems and a continuing need to supply aluminas which match the customer plants so that use of the latter can be optimized at an acceptable price.

Alumina is a significant ceramic material that is available at a high degree of purity. The dominant position of alumina as a ceramic raw material arises because it has desirable properties at a relatively low cost. This cost-effectiveness is attributable to the commodity nature of the business arising from the large demand for alumina by the aluminum industry.

Zircon and Zirconia (Ref 31–34)

The primary source of zirconia is the mineral zircon ($ZrO_2 \cdot SiO_2$), which exists in beach sands principally in Australia, South Africa, and the United States. The production of zircon from 1984 to 1988 is shown in Table 7. In addition, the mineral baddeleyite (ZrO_2) is sourced as a by-product of a phosphate deposit in South Africa.

Zircon extracted from beach sands contains $\sim$2% HfO_2 and traces of Al_2O_3 (<0.5%), Fe_2O_3 (<0.1%) and TiO_2 (<0.1%). In addition, all zircons contain traces of uranium and thorium. The radioactivity inherent in zircon and in zirconia prepared from zircon is generally well within existing hazardous material legislation limits and careful control and monitoring is undertaken to ensure compliance. Baddeleyite also contains similar trace impurities but significantly higher radioactivity and often higher than the 500 ppm limit. The exceeding of the specification limit is making the material increasingly difficult to market in countries such as the United States where legislation is being tightened and regulations strictly enforced.

Zircon is processed by fine grinding to produce a range of milled products of defined particle size. These products have found use in investment casting, foundries, refractory products, and as an opacifier in glazes for whitewares.

Zircon is also the principal source of zirconia. Zircon can be chlorinated in the presence of carbon to give zirconium and silicon tetrachlorides that are then separated by distillation. The zirconium tetrachloride produced can be used to prepare zirconia directly or as a feedstock for other zirconium chemicals. Sintering with alkali or alkaline earth oxides is also used to decompose zircon. Silica is leached from the decomposition products with water, leaving zirconium hydroxide to be further purified by acid dissolution and reprecipitation. Zirconia is then obtained by calcining the hydroxide. Zircon is also converted to zirconia and silica in a plasma at >1800 °C (>3270 °F) with rapid cooling to prevent reassociation. The free silica is removed by dissolution in sodium hydroxide. Fused zirconia is produced in electric arc furnaces from either baddeleyite or zircon/carbon feedstocks. In the latter process, the silica component of zircon is carbothermally reduced to silicon monoxide, which volatilizes prior to the fusion of the residual zirconia.

Zirconia is normally sold as a stabilized or partially stabilized product for use in high-temperature applications so that the effect of the phase transition at $\sim$1100 °C ($\sim$2010 °F) from monolithic to cubic/tetragonal is negated. Stabilization is effected by additives containing Ca^{2+}, Mg^{2+}, Y^{3+}, or Ce^{4+} at the

Table 7 World zircon production from 1984 to 1988

Country	Amount produced annually									
	1984		1985		1986		1987		1988	
	tonne × 10^3	ton × 10^3	tonne × 10^3	ton × 10^3	tonne × 10^3	ton × 10^3	tonne × 10^3	ton × 10^3	tonne × 10^3	ton × 10^3
Australia	458	504	501	551	452	497	457	503	490	540
South Africa	153	168	161	177	140	154	140	154	150	165
United States	113(a)	124(a)	113(a)	124(a)	113(a)	124(a)	113(a)	124(a)	118(b)	130(b)
USSR	80	90	85	94	85	94	85	94	85	94
Brazil	6	7	21	23	15	17	18	20	20	22
Malaysia	8	9	12	13	13	14	18	20	19	21
India	12	13	15	17	16	18	16	18	17	19
China	15	17	15	17	15	17	15	17	15	17
Other countries	4	4.4	5	5.5	6	7	5	5.5	9	10
Total	**849**	**934**	**928**	**1020**	**855**	**940**	**867**	**955.5**	**923**	**1015**

(a) U.S. production capacity estimated by USBM. (U.S. Bureau of Mines). (b) Actual reported to USBM. Source: USBM. Minerals Yearbook, 1988.
Source: Ref 33

precipitation stage or as oxides in the fusion process.

The purity of the zirconias produced varies from process to process. All contain HfO_2, typical fused products contain 0.2 to 0.5% SiO_2, 0.1% Al_2O_3, 0.05 to 0.1% Fe_2O_3, and TiO_2, while high-quality materials via precipitation/plasma routes can be produced with silicon dioxide, aluminum oxide, ferric oxide, and titanium dioxide impurities at the ppm level. The end uses for zirconia include refractories, ceramic colors, thermal barrier coatings, oxygen sensors, additions to lead zirconium titanates, and gemstones. Each application demands different specifications and costs range from $5 to $200/kg ($2.30 to $90/lb). Existing applications all represent growth areas and the potential use of zirconia in solid oxide fuel cells could markedly increase future demand.

Future Outlook

The traditional markets of building products, whitewares, and refractories represent significant businesses with low growth but new industries such as electronics are rapidly becoming as economically important. Significant research and development expenditures are defining new potential for ceramics in terms of materials, methods of production, and applications. It is projected that there will be future growth in ceramics for metal-matrix composites, ceramic-matrix composites, and in advanced processing such as sol-gel, chemical vapor deposition, and polymer pyrolysis.

ACKNOWLEDGMENT

The authors would like to thank Dr. J.S. Campbell, Director of Research, Cookson Group PLC and the Directors of Cookson Ceramics, Cookson Minerals, TAM Ceramics, and Vesuvius International Corporation for permission to publish this article.

REFERENCES

1. J.E. Prentice, The Mineralogy of Brickmaking, *Proceedings of the 3rd Industrial Minerals International Congress,* sponsored by Metal Bulletin, 1978, p 43–47

2. R.E. Grim, Uses of Clays, *Concise Encyclopaedia of Chemical Technology,* Vol 6, 3rd ed., Kirk-Othmer, John Wiley & Sons, 1979, p 207–223

3. European Heavy Clay Products Production in 1988, *Ind. Ceram.,* Vol 10 (No. 4), 1990, p 157

4. R.E. Grim, *Applied Clay Mineralogy,* McGraw-Hill, 1962

5. Annual Minerals Review, *Am. Ceram. Soc. Bull.,* Vol 68 (No. 5), 1989, p 1029–1074

6. Annual Minerals Review, *Am. Ceram. Soc. Bull.,* Vol 69 (No. 5), 1990, p 843–890

7. W.D. Kingery, *Introduction to Ceramics,* John Wiley & Sons, 1960, p 15–321

8. A.C. Bull and J.R. Taylor, *Ceramics Glaze Technology,* Institute of Ceramics, Pergamon Press, 1986, p 13–49

9. J.V. Hamme, Ceramics (Raw Materials), *Concise Encyclopaedia of Chemical Technology,* Vol 5, 3rd ed., Kirk-Othmer, John Wiley & Sons, 1979, p 238–249

10. J.F. Birtles, F. Housley, and W.T. Wilkinson, "Raw Materials for the Pottery Industry: A Survey," British Ceramic Research Association Technical Note No. 234, 1975

11. P. Rado, *An Introduction to the Technology of Pottery,* 2nd ed., Institute of Ceramics, Pergamon Press, 1988

12. W.E. Worrall, *Ceramic Raw Materials,* 2nd ed., Institute of Ceramics, Pergamon Press, 1982

13. W. Ryan and C. Radford, *Whitewares Production, Testing and Quality Control,* Institute of Ceramics, Pergamon Press, 1987

14. F. Kandianis, Raw Materials Impacts on the Refractories Industry, *Am. Ceram. Soc. Bull.,* Vol 67 (No. 7), 1988, p 1158

15. Y. Sakano and H. Takahashi, Outlook for the Refractories Industry in Japan, *Am. Ceram. Soc. Bull.,* Vol 67 (No. 7), 1988, p 1164

16. W.H. McCracken, Refractory Raw Materials in 1992, *Am. Ceram. Soc. Bull.,* Vol 67 (No. 7), 1988, p 1155

17. J.H. Chesters, *Steelplant Refractories,* Lund, Humphries & Co. Ltd, 1963

18. A. Nishikawa, *Technology of Monolithic Refractories,* Plibrico Japan Co. Ltd, 1984

19. Advanced Ceramic Powder Market on Rise: Western World Market Seen to Jump 65%: Study, *Am. Met. Mark.,* 10 Oct 1990, p 4

20. Chem. Systems Study Predicts 13% Advanced Ceramic Growth Rate, *Ceram. Ind.,* Aug 1987, p 12

21. Advanced Ceramic Powders Technology Challenges Conventional Processes, *Am. Ceram. Soc. Bull.,* May 1990, p 768

22. "Ceramic Precursor Technology 1989," Research Studies, Kline & Co., 20 April 1990, p 1–114

23. Aluminum Nitride Powders Predicted to Grow Through 1995, *Ceram. Ind.,* April 1990, p 14

24. JCW Spotlight on Fine Ceramics, *Jpn. Chem. Week.,* 3 Aug 1989, p 6–8

25. Fine Ceramics, *Jpn. Chem. Ann.,* 1986, p 64–66

26. K.H. Jack, Review: Sialons and Related Nitrogen Ceramics, *J. Mater. Sci.,* Vol 11, 1976, p 1135–1158

27. Business Communications Co. Inc., *Am. Ceram. Soc. Bull.,* Vol 69, 1990, p 768

28. Advanced Materials Group, Kline & Co., *Am. Ceram. Soc. Bull.,* Vol 69, 1990, p 1090

29. A.J. Bell, *Fine Chemicals for the Electronics Industry,* P. Bamfield, Ed., Royal Society of Chemistry, London, 1987, p 176–193

30. L.M. Sheppard, The Changing Demand for Ceramic Additives, *Am. Ceram. Soc. Bull.,* Vol 69 (No. 5), 1990, p 802

31. F. Farnworth, S.L. Jones, and I. McAlpine, *Specialty Inorganic Chemicals,* R. Thompson, Ed., Royal Society of Chemistry, 1980, p 248–284

32. A.J.A. Winnubst, W.F.M. Groot Zervert, G.S.A.M. Theunissen, and A.J. Burgraaf, *J. Mater. Sci.,* Vol 25, 1990, p 3449

33. T.E. Garner, Jr., Zircon, *Am. Ceram. Soc. Bull.,* Vol 69 (No. 5), 1990, p 888

34. R. Stevens, *Zirconia and Zirconia Ceramics,* Magnesium Electron Ltd, July 1986

Chemical Synthesis

Patrick K. Gallagher, Departments of Chemistry and Materials Science & Engineering, The Ohio State University

THE GENERAL THERMAL STABILITY of ceramic and metallurgical raw materials has led to their concentration in nature. It is rare, however, that the naturally occurring materials have the precise composition, purity, or morphology necessary to achieve optimum properties. Fortunately, this same thermal stability facilitates the preparation of these materials using a wide variety of synthetic techniques. This article stresses the unconventional, or more chemically oriented methods, as opposed to the conventional methods that typically involve grinding of calcined materials. Because there is an immense amount of literature on this topic, no effort will be made to provide an exhaustive review. Instead, a broad overview with some insight based on experience is provided, along with examples to illustrate the main points.

It is generally necessary to convert chemical precursors into the desired powder by some form of thermal treatment. Because the relevant aspects of these thermal decompositions are common to most methods, this general topic is treated first.

Principles of Thermal Decomposition

Understanding the nature of the thermal decomposition in some detail is vital, because calcination, or prereaction, is a critical multipurpose step in the preparation of ceramic powders. In this step, the precursor compounds are decomposed, the resulting products are reacted when necessary, and the particle size and reactivity are established. Clearly, an appreciation of the nature of these processes and the physical changes associated with them is essential to designing the optimal overall synthesis.

The discussion below focuses on three general areas: the basic stoichiometry of the decomposition, that is, identifying both the chemical steps that take place and the intermediates that are formed as a function of temperature; identifying the actual mechanisms and rates of these decompositions and subsequent reaction steps; and identifying the factors that determine the final particle size and morphology.

Techniques and Stoichiometry. The primary experimental techniques used to follow the stoichiometry during the decomposition of a precursor or the reaction of mixed compounds are thermal analysis and x-ray diffraction. The major thermoanalytical methods are thermogravimetry (TG), differential thermal analysis (DTA) or differential scanning calorimetry (DSC), evolved gas analysis (EGA), and thermomagnetometry (TM). The latter technique can be used to detect the formation of magnetic intermediates or products, as well as the disappearance of magnetic reactants (Ref 1, 2).

Frequently, however, the chemical processes are so complex that it is necessary to utilize supplementary solid-state spectroscopic methods, such as Mössbauer, fluorescence, diffuse reflectance, photoacoustic, solid-state nuclear magnetic resonance (NMR), electron spin resonance, and other techniques. Each of these can offer unique information regarding the microscopic environment associated with key atoms. The major thermoanalytical techniques identified above provide macroscopic information on the overall nature of the reactions.

The thermoanalytical methods are designed to follow the course of the overall reaction as a function of temperature and atmosphere in a dynamic, or scanning mode, which provides data rapidly. However, the supplementary techniques can also be made to run in a dynamic fashion with respect to temperature. The more common approach, though, is to examine samples at room temperature after they have been subjected to the desired thermal treatments. Determining accurate temperatures for various events from measurements taken as the temperature is scanned requires the careful application of standards (Ref 3). The situation becomes even more complex when considering reactions having large values of enthalpy (Ref 4).

The thermal decompositions of two precursors are exemplified below to illustrate the general complexity that may occur and the potential influence of the atmosphere on the course of decompositions. The first example is a freeze-dried mixture of lithium and niobium oxalates, which can serve as a precursor for the preparation of lithium niobate (Ref 5). The second example is the decomposition of a complex cyanide precursor precipitated for the synthesis of highly stoichiometric rare earth orthoferrites (Ref 6).

A comparison of the reaction of conventionally mixed lithium carbonate and niobium oxide with that of a freeze-dried mixture of the hydrated oxalates is indicated by the derivative thermogravimetric (DTG) curves shown in Fig 1 for experiments performed in flowing oxygen. The dehydration and decomposition of the oxalates are responsible for the peaks below 400 °C (750 °F).

Those reactions lead to an intimate mixture of lithium carbonate and niobium oxide, which subsequently reacts around 480 °C (895 °F) to evolve carbon dioxide and form lithium niobate. In contrast, the physical mixture does not react significantly until nearly 700 °C (1290 °F). There is a trace of reaction in the conventional mixed carbonate-oxide material at the reaction point for the oxalate, indicating that even in the coarse physical mixture there is some intimately mixed material.

The intimately mixed nature of these carbonates and oxides resulting from the decomposition of mixed oxalates requires such short diffusion paths for subsequent reactions that the formation of undesirable intermediates is frequently avoided (Ref 7). This better mixing can also be achieved in the traditional manner by increased ball milling time (Ref 8). The problem with this operation is the added contamination that accompanies the longer milling.

Performing the decomposition of the mixed oxalates in an oxidizing atmosphere leads to immediate oxidation of the carbon monoxide that evolved from the decomposition at the surface and in the pores or interstices of the powder. The nature of the decomposition is changed in an inert atmosphere, where the carbon monoxide has the opportunity to disproportionate in accordance with Eq 1.

$$2\,CO = CO_2 + C \qquad \text{(Eq 1)}$$

Although this is thermodynamically favored, it is kinetically inhibited and proceeds to an unpredictable extent, depending on the catalytic nature of the oxides present and on the sample container materials. Clearly, this produces a carbon impurity in the resulting product, which can be readily observed due to the grey-to-black color of the otherwise white product.

The entire synthetic process can be changed by using a reducing atmosphere such as hydrogen, which completely reduces the car-

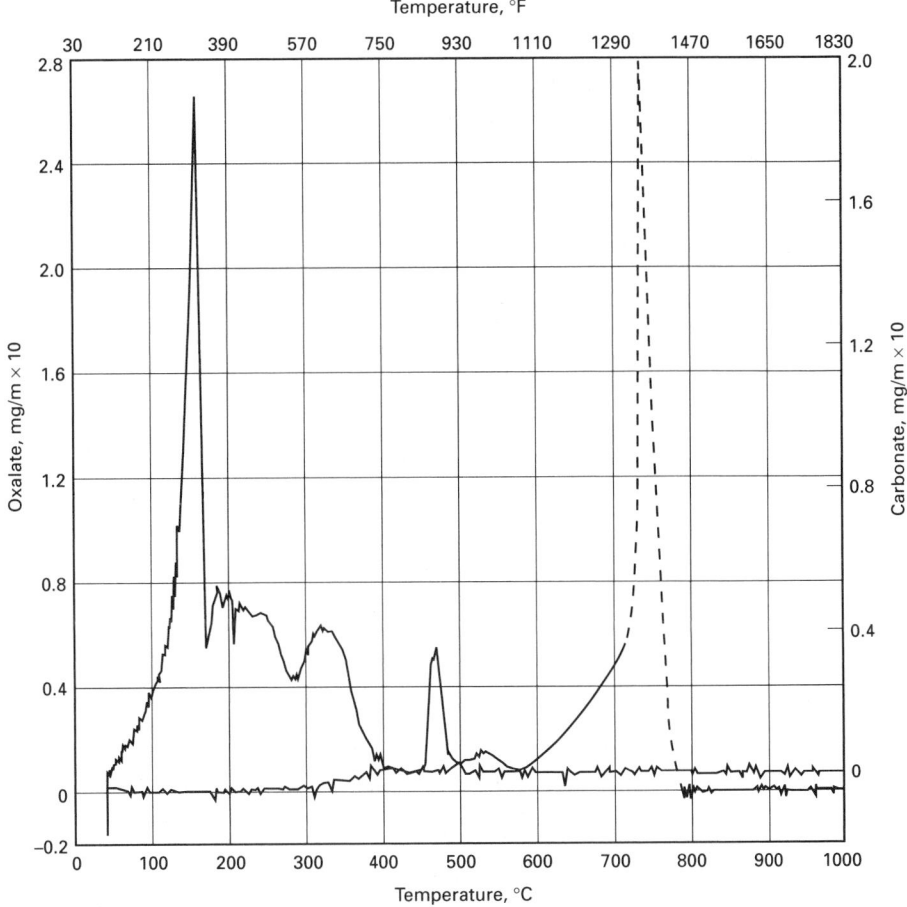

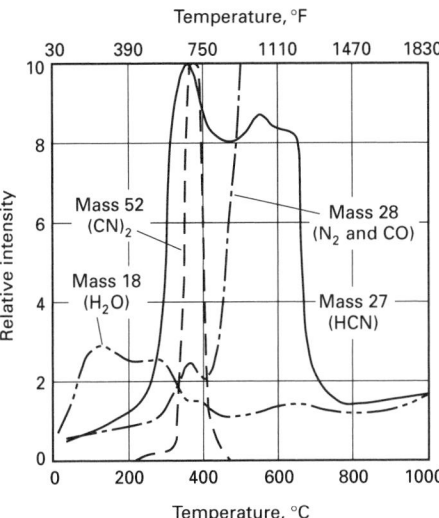

Fig 2 Mass spectroscopic evolved gas analysis of EuFe(CN)$_6$·5H$_2$O

Fig 1 DTG curves in oxygen; Li$_2$C$_2$O$_4$·2HNb(C$_2$O$_4$)$_3$, 2.10 mg (0.03 grain); Li$_2$CO$_3$ + Nb$_2$O$_5$, 9.44 mg (0.14 grain). Peaks represent the periods of rapid weight loss. Source: Ref 5

bon monoxide to elemental carbon. This leads to the incorporation of large amounts of intimately mixed carbon in the initial product that will react at higher temperatures, around 1200 °C (2190 °F), to form the transition metal carbide. It is clear that the nature of the products changes markedly with variations from an oxidizing to an inert to a reducing atmosphere.

The second example concerns the synthesis of europium orthoferrite either from the rare earth, ferricyanide, or the ammonium rare earth, ferrocyanide. Studies based on TG, DTA, x-ray diffraction, and Mössbauer spectroscopy indicated that the decomposition of the precursors was straightforward in an oxidizing atmosphere. At about 300 °C (570 °F), a mixture of Eu$_2$O$_3$ and Fe$_2$O$_3$ was formed, which subsequently reacted around 600 °C (1110 °F) to produce EuFeO$_3$. In an inert atmosphere, however, the decomposition sequence was much more complex. The evidence, primarily Mössbauer spectra, was sufficient to delineate the chain of iron species suggested by the following Eq 2 through 4.

$$2 \text{ EuFe(CN)}_6 \cdot 5 \text{ H}_2\text{O} \rightarrow 2 \text{ Eu(CN)}_3$$
$$+ 2 \text{ Fe(CN)}_2 + 10 \text{ H}_2\text{O} + \text{(CN)}_2 \quad \text{(Eq 2)}$$

$$3 \text{ Fe(CN)}_2 \rightarrow \text{Fe}_3\text{C} + 3 \text{ N}_2 + 5 \text{ C} \quad \text{(Eq 3)}$$

$$\text{Fe}_3\text{C} \rightarrow 3 \text{ Fe} + \text{C} \quad \text{(Eq 4)}$$

The x-ray diffraction patterns and the Mössbauer spectra, however, could not unambiguously establish the sequence of europium-containing products. These were determined with the aid of EGA and infrared (IR) spectroscopy (Ref 9). The mass spectroscopic EGA curves for the decomposition are presented in Fig 2. The results of both the EGA and IR spectra confirmed the series of iron products proposed earlier. The evolution of cyanogen associated with the reduction of iron(III) in Eq 2 is clearly evident in Fig 2. The presence of the large amount of hydrogen cyanide is the key to understanding the decomposition pattern of the europium salt. This indicates a hydrolytic decomposition, as indicated by Eq 5.

$$2 \text{ EuFe(CN)}_6 \cdot 5\text{H}_2\text{O} \rightarrow 2 \text{ EuOOH} + 2\text{Fe(CN)}_2$$
$$+ \text{(CN)}_2 + 6 \text{ H}_2\text{O} + 6 \text{ HCN} \quad \text{(Eq 5)}$$

The IR spectra were consistent with the iron cyanide bonds and the europium hydroxide bonds. The latter persisted to the normal decomposition temperature associated with

europium oxyhydroxide. Numerous other examples exist that illustrate the complex nature of many decompositions and the strong dependence of the decomposition process on the ambient atmosphere. In most cases it is essential to utilize a variety of analytical techniques to establish the sequence of events.

The mechanisms, rates, and reactivity of the products are a fascinating area of research that markedly impacts the success or failure of the syntheses. Generally, the calcination (prereaction) step is at atmospheric pressure in an appropriately oxidizing, inert, or reducing media. However, even under these relatively simple conditions, a variety of effects and reaction paths are possible.

Regardless of the atmosphere initially surrounding or flowing past the sample, it will change during the course of the reaction due to the build up of product gases or the depletion of the reactive gas. For reversible reactions, the equilibrium decomposition temperature is strongly dependent on the partial pressure of the product. Hence, the temperature at the reacting interface must increase during the course of the decomposition as the diffusion of the product gases becomes more difficult. The decomposition of carbonates, hydroxides, hydrates, and oxides are frequently used in ceramic synthesis and they are very likely to be reversible reactions.

This problem of the ill-defined nature of the temperature at the reaction interface is further complicated by the enthalpy of the specific reaction. The amount of energy evolved or absorbed by the reaction is usually very large, compared to the heat capacity of the reactants and products. Consequently, the actual temperature at the reaction's interface may differ substantially from that of the nearby temperature sensor. A good example of this effect is the oxidation of carbon monoxide, which occurs during the decomposition of oxalates in air that was described earlier. The thermocouple in the thermobalance may indicate 300 °C (570 °F) while the actual specimen is glowing a bright red. Conversely, for the more typical endothermic processes, it is essential to supply heat in order to sustain the reaction.

Still another factor that influences the progress of a reaction is the rate, or degree, of structural order that occurs at the reaction interface, particularly on the product side. It must be determined whether the product is ordered immediately or exists initially in an amorphous or microcrystalline state for a significant period of time. If the latter, then the full contribution of the product's lattice energy is not reflected in the apparent activation energy of the process. Such effects are not uncommon (Ref 10). Therefore, the progress of the reaction front for even a virtually perfect material is determined by the complex interactions of the concentration profile of the reactant and/or product gases, the temperature profile and thermal transport properties of the reacting media, and the degree of order present throughout the sample.

The simple decomposition of calcium carbonate

$$CaCO_3 \rightarrow CaO + CO_2 \qquad (Eq\ 6)$$

is a good illustration of these effects. Paulik and Paulik (Ref 11) have demonstrated the dependence of the decomposition temperature on the partial pressure of carbon dioxide by the use of various sample holders. Each of the sample holders shown in Fig 3 allows for a different degree of gaseous exchange with the flowing atmosphere. The reaction temperature increases rapidly as this exchange is impeded and the partial pressure of the car-

bon dioxide is allowed to build. This phenomenon, and the endothermic nature of the reaction, are also shown to have a remarkable influence on the apparent activation energy for this relatively simple reaction as the sample size and heating rate are varied (Ref 12).

Not only can such conditions change the temperature and rate of the reaction, but they can also influence the choice of reaction pathways for energetically similar routes. Consider the somewhat extreme case shown in Fig 4 for the decomposition of aluminum hydroxide, the mineral gibbsite (Ref 13). The arrow at the bottom of the figure indicates how the pathway tends to vary from 3 to 1, with increasing initial particle size, heating rate, total pressure, or partial pressure of water vapor.

The concept of "reactivity" is a difficult one to adequately define. In a study comparing the reactivity of iron oxides prepared from a variety of starting compounds via different processes, it was obvious that the order of reactivity depended on the test applied (Ref 14). The relative order, when determined by solubility in acid at 81 °C (180 °F), or by the extent of the reaction with lithium carbonate at 490 to 840 °C (915 to 1545 °F), or by the ability to sinter to high density at 1200 °C (2190 °F), was different in each case. The reactivity of a powder was a transient phenomenon in that it dissipated itself with increasing temperature. In order to be useful, the event used as a criterion for reactivity had to occur in the unique temperature range associated with the activity of the particular powder being used.

From the standpoint of ceramic synthesis, it is the ability to react with another powder in the intermediate temperature range that is

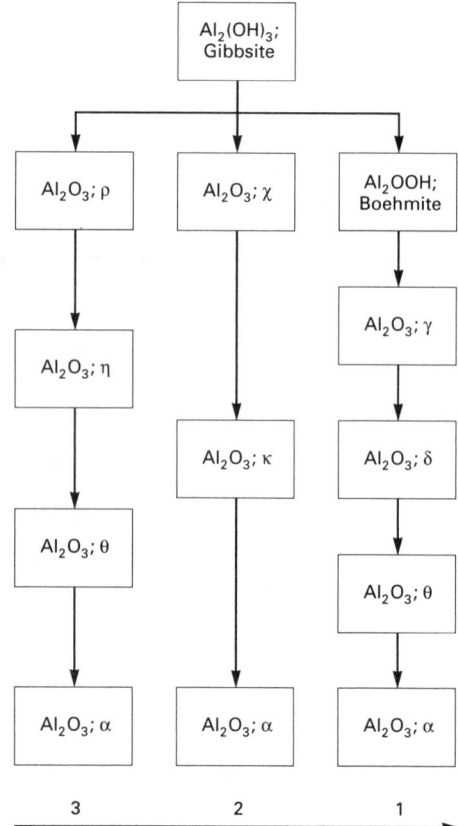

Fig 4 Various routes for the decomposition of gibbsite. Source: Ref 13

most important. Table 1 lists the source of iron oxide and the temperature at which the reaction with lithium carbonate was 85% complete. The temperatures selected for the calcination of each starting compound had been determined by the TG method as the minimum temperature necessary to completely decompose the precursor to iron(III) oxide. All of the oxides affected the lithium carbonate decomposition to some extent as did the presence of niobium oxide described in the "Techniques and Stoichiometry" section of this article. The spread in temperatures, however, revealed a wide range in the relative reactivity among the powders.

Although iron oxide derived from the nitrate was among the most active materials in the two low-temperature tests, it was the least active in the sintering test. Its low-temperature reactivity led to a partial sintering of the powder into hard agglomerates, which were subsequently impossible to sinter to reasonably dense materials. Conversely, the oxides prepared at higher temperatures (for example, from sulfates) sintered well, but were less reactive at temperatures below their formation temperature. That potential low-temperature reactivity had been consumed during formation.

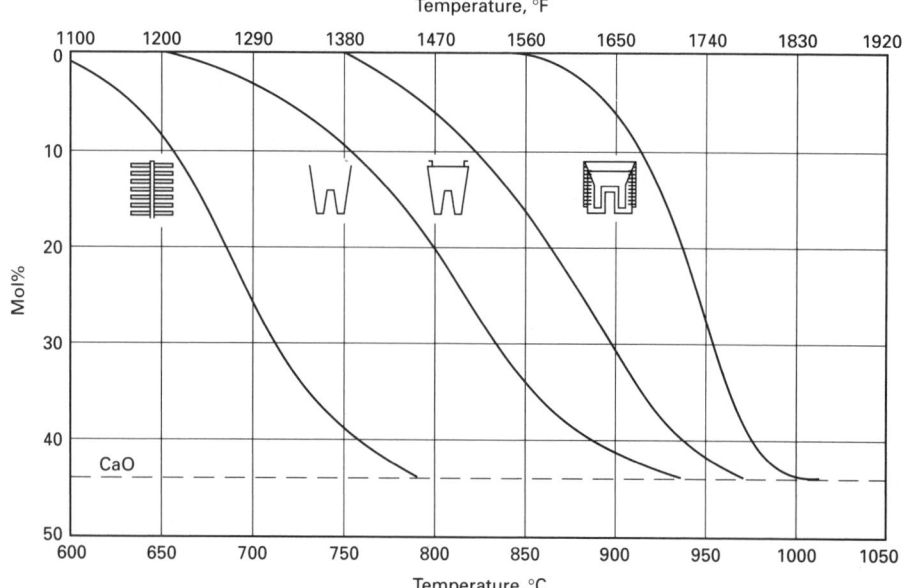

Fig 3 Thermogravimetry curves for the decomposition of calcium carbonate in the indicated containers. Source: Ref 11

Table 1 Decomposition temperatures of iron-lithium mixtures

	Calcined iron salt		85% decomposed Li$_2$CO$_3$	
Iron salt	°C	°F	°C	°F
Iron (III) sulfate	700	1290	535	995
Ammonium iron (III) sulfate	660	1220	495	925
Iron (II) sulfate	720	1330	580	1075
Ammonium iron (II) sulfate	660	1220	490	915
Iron (III) nitrate	390	735	570	1060
Ammonium hexacyanoferrate (III)	670	1240	600	1110
Ammonium hexacyanoferrate (II)	480	895	840	1545
Iron (III) oxalate	325	615	560	1040
Ammonium trisoxalaftoferrate (III)	500	930	560	1040
Ammonium iron (III) citrate	580	1075	800	1470
Magnatite + ethylenediaminetetraacetic acid	525	975	615	1140
Commercial iron (III) oxide	...	...	670	1240
Lithium carbonate	...	...	950	1740

Source: Ref 14

Hedvall (Ref 15) displayed a keen early insight into the reactivity of solids. The major factors associated with reactivity, given in his 1934 review, are:

- Heating
- Deformation
- Impurities in the lattice
- Radiation
- Changes in crystal structure accompanying phase transitions and decompositions

These factors are primarily concerned with the degree of disorder in the solid. Factors that enhance this disorder increase the reactivity of the material. The concept of a brief period of enhanced reactivity associated with a solid$_1$ phase transitioning to a solid$_2$ phase is a somewhat controversial phenomenon referred to as the "Hedvall Effect" (Ref 16, 17). An example of its usefulness in ceramic processing is the higher reactivity achieved when the metastable anatase form, rather than the stable rutile form of titanium dioxide, is used for the preparation of barium titanate (Ref 18).

The qualitative influence of these factors can be deduced from relatively simple experiments, such as those described above. However, quantitative evaluation is usually attained through the measurement of specific rate constants and Arrhenius parameters under the desired sets of conditions. Any technique capable of following the concentration of relevant species as a function of time can yield specific rate constants. If the change of mass or heat can be correlated with concentration, then thermoanalytical techniques are ideally suitable. Meaningful data can be obtained in either an isothermal or programmed temperature mode of operation (Ref 19).

There have been recent studies that suggest that the best approach is to use feedback from the process itself to control the heating rate (Ref 20). For instance, the heating rate is adjusted to give a predetermined rate of weight loss, total pressure, or rate of shrinkage. The latter means of control has been particularly valuable in optimizing the conditions for sintering (Ref 21).

Various rate laws to describe solid-state decompositions, solid-solid reactions, and gas-solid reactions have been developed. Table 2 lists some of the more common expressions as a function of the fraction reacted, α. If the particular model is appropriate, then a plot of the expression versus time should be linear, per Eq 7,

$$f(\alpha) = kt \qquad (Eq 7)$$

and the slope equal to the specific rate constant, k (Ref 12).

Because derivations of the rate laws shown in Table 2 are beyond the scope of this article, several excellent books (Ref 22–24) are recommended for more information. In simple terms, the contracting geometry equations are most applicable for decompositions. They assume a reaction interface moving at a fixed rate and that the overall effective rate is determined by the geometry and crystalline anisotropy of the situation. The reaction initiates at a surface and moves inward in either one, two, or three dimensions. The exact form of the expression depends on the geometrical assumption of the particle shape or preferred direction of reaction. The idealized forms in Table 2 are for a uniform cross section, a circular disk reacting from the edge inward, and a collapsing sphere for the one-, two-, and three-dimensional cases, respectively. Seldom, however, does the particle shape and size range allow for an integral value of n. Because of mass and thermal transport considerations, it may be that a pile, pressed pellet, or other assemblage of powder may react

as if it was a single particle, that is, from the external surface into the center or bottom, depending on transport conditions (Ref 12, 25).

The Erofeev rate laws allow for nucleation and growth under a variety of conditions and considers the case of impinging or overlapping nuclei, as well. Other rate expressions are based on autocatalytic phenomena, or diffusion-controlled rates through product layers, which again are based on certain geometrical assumptions. Clearly, this field of heterogeneous kinetics is difficult, and controversial as well. However, technology often demands that some effort be made to provide a minimum predictive capability and some understanding of the specific mechanisms that occur in these vital processes.

The simplified Arrhenius expression

$$kt = Ae^{-E/RT} \qquad (Eq 8)$$

describes the temperature dependence of the rate constant, where A is the pre-exponential term, E is the apparent activation energy, T is the absolute temperature, and R is the gas constant.

The Arrhenius parameters can also be determined from data taken while undergoing a prescribed temperature program, that is, nonisothermal conditions. Such analyses are based on mathematical combinations of the various rate laws, the Arrhenius equation, and the expression describing the heating rate (usually linear with respect to time). Several numerical analysis methods allow the parameters to be calculated from a single dynamic experiment (Ref 19), while others require a series of experiments performed at different heating rates (Ref 26, 27). Such dynamic experiments are generally faster and more convenient, but are less flexible and informative than the traditional isothermal experimental studies.

Several precautions should be stated. To decide whether mass or thermal transport is the rate-determining step, determine the rate constants or Arrhenius parameters at different sample sizes, heating rates, and flow rates (Ref 12, 25). Do not conclude, based solely on the best-fitting rate law, that the mechanism is necessarily that from which that rate expression was derived (Ref 28). Confirm the mechanism through other techniques, such as microscopy. Keep in mind that because the Arrhenius equation is mathematically ill-conditioned, very precise data are necessary to uniquely establish the values of A and E (Ref

Table 2 Rate law expressions

Power law	α^n, $n = 1, 2, \frac{1}{2}, \frac{1}{3}, \frac{1}{4}$
Contracting geometry	$1 - (1 - \alpha)^{1/n}$, $n = 1, 2, 3$
Erofeev	$[-\ln(1 - \alpha)]^{1/n}$, $n = 1, 1.5, 2, 3, 4$
Diffusion controlled, 2D	$(1 - \alpha)\ln(1 - \alpha) + \alpha$
Diffusion controlled, 3D	$(1 - \frac{2}{3}\alpha) - (1 - \alpha)^{2/3}$
Jander	$[1 - (1 - \alpha)^{1/3}]^2$
Prout-Tompkins	$\ln[\alpha/(1 - \alpha)]$
Second order	$1/(1 - \alpha) - 1$
Exponential	$\ln \alpha$

Source: Ref 12

29). More frequently, ranges of A and E values will fit the data in a manner such that a high value of E will pair with a high value of A in order to predict essentially the same rate. This has been referred to as "the kinetic compensation effect" and is a topic of extensive discussion (Ref 30).

The particle size and distribution, extent of agglomeration, porosity, and morphology are usually established during the decomposition and any subsequent high-temperature reactions. Manipulation of the calcination then offers some degree of control over these factors. Certainly, several characteristics of the starting precursor are also variables in this process. Often, the decomposition may have topochemical aspects that produce oriented agglomerates having the orientation and shape of the initial particle. A final comminution stage may be added to alter the properties from the final thermal treatment. However, one of the goals of chemical synthetic methods is to eliminate, or at least minimize, this requirement.

By lowering the temperature that is necessary to achieve the desired product, it is generally possible to produce the material in a more finely divided stage. This increases the range of particle size that can be achieved, and the time and temperature of calcination can be used to attain some measure of control over the final powder characteristics. At the other extreme, it is possible to prepare amorphous particles or glass-ceramics by preventing or controlling nucleation and growth during cooling from the high-temperature homogeneous melt.

Consider the freeze-dried mixed oxalates of lithium and niobium, used as an example earlier (Ref 5). A plot of surface area as a function of calcination temperature in air is presented in Fig 5. The surface area peaks immediately after the decomposition, as would be expected. The growth of the particles at higher temperatures is indicated by the decrease in surface area and is confirmed by x-ray diffraction. If one assumes uniformly sized spherical particles and a density of 4.6 g/cm^3 for lithium niobate, then it suggests that the "equivalent spherical" particle size could range from about 0.05 to 1.5 µm (2 to 60 µin.) for calcination temperatures below 1000 °C (1830 °F).

The rate of heating can also be a factor. Fisher (Ref 31) has shown that the density and pore structure of calcium oxide formed from the decomposition of calcium carbonate depends not only on the maximum temperature, but also upon the rate. High heating rates tend to heal over the surfaces, inhibiting diffusion and trapping gases. Large, disruptive pores and fissures form under these conditions, while slower heating yields a more dense product that has smaller pores more suitable for subsequent densification. There may be instances, particularly for highly exothermic reactions, where the heating rate is impossible to control, for example, the decomposi-

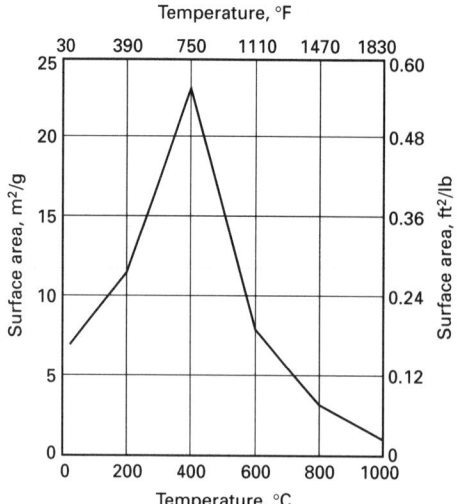

Fig 5 Surface areas for samples of Li$_2$C$_2$O$_4$·2HNb(C$_2$O$_4$)$_3$ calcined in air at the indicated temperatures. Source: Ref 5

tion of oxalates in air or the decomposition of materials with internal oxidizing agents, such as nitrates (Ref 4).

The use of large, deep trays or crucibles, possibly covered, for calcination accentuates all of these problems, forcing different portions of the sample to react at different temperatures and times. Shallow trays in a flowing atmosphere are obviously an improvement, but tumbling and rotating tube furnaces or fluid bed calciners are clearly superior. Mazdiyasni and Brown (Ref 32) prepared rare earth oxides by decomposing the hydroxide with and without tumbling at 800 °C (1470 °F) for 24 h in air. The observed range of particle sizes for the tumbled material was 17.5 to 27.5 nm (0.7 to 1.1 µin.) compared to 25.0 nm (1 µin.) for the untumbled material. Hancock (Ref 33) also discusses the advantages of a rotary calciner for the production of ferric oxide from ferrous sulfate.

Preparation from Solution

There are a number of general considerations concerning the solution process that are applicable to all of the methods. The emphasis of this discussion is primarily on the formation of mixed compounds, as opposed to individual oxides, although much of what is described would be relevant for that purpose, as well. The major concern is on converting the correct ratio of the appropriate cations in solution into the desired ceramic composition that, if not optimal, has at least suitable properties for subsequent ceramic processing.

The principal advantages of starting from solution are homogeneity and improved reactivity. No amount of mixing of solid particles can approach that homogeneity achieved in solution; most of the advantages are a direct consequence of this mixing essentially on an

atomic scale. The challenge in these processes is to retain as much of this chemical homogeneity as possible in the final solid product.

The overall process consists of three major steps: the preparation of the solution, the removal of the solvent, and the conversion of the residue into the desired product. This final stage has already been discussed. Although the first step may seem relatively trivial, there are a few points that deserve consideration, the most obvious of which is solubility. Not only is it desirable that the solubility of the components be high in order to minimize the amount of solvent that must be subsequently removed, but the particular components must also be compatible in solution. For example, iron sulfates could not be combined with barium chloride in order to prepare barium ferrite, because barium sulfate would precipitate immediately from solution. Cost, purity, toxicity, and ease of conversion are also factors to be considered.

A more subtle concern is the choice of the presumably inert anions. As will be discussed later, both the tendencies of these anions to become incorporated into the final product and their subsequent effects on processing and properties can vary markedly. The amount of anion present in the intermediate powder depends largely on the technique utilized for solvent removal. If significant quantities of the anions remain, they must volatilize or decompose into harmless species during the subsequent thermal processing.

It is the second stage, that of solvent removal, that really defines the process. The two basic categories are some form of precipitation and thermal methods, such as evaporation, sublimation, combustion, and others.

The precipitation process may take many forms. One is the precipitation of a unique compound that has the proper stoichiometry corresponding to the desired end product. A second is the precipitation of a solid solution of cations that may not guarantee the proper stoichiometry, but does preserve the mixing on an atomic scale. A third is the simple coprecipitation of a mixture of compounds that has neither of the advantages of the previous two, but which will be more intimately mixed and reactive than could be obtained through conventional processing.

The precipitation process has a large number of variables that influence not only the chemical equilibria involved, but also the purity and physical nature of the precipitate. Some of these variables are:

- Temperature
- Concentration
- Order of mixing
- Time of digestion
- Effect of anions
- pH
- Solvent
- Rate of mixing

- Atmosphere
- Washing

Very seldom have all of these variables been considered for a single system. Certain factors influence the actual equilibrium condition and all influence the extent to which that equilibrium is attained. Although the effects of most of these variables are obvious, some are deserving of further consideration.

The effects of pH are the most complex. Consider the precipitation of aluminum hydroxide from aqueous solution:

$$Al^{3+} + 3(OH)^- = Al(OH)_3 \qquad (Eq\ 9)$$

The solubility product, K_{sp}, is given by

$$K_{sp} = \gamma_{Al^{3+}} \cdot [Al]^{3+} \cdot (\gamma_{OH^-})^3 \cdot [OH^-]^3 \qquad (Eq\ 10)$$

where γ is the activity coefficient of species i and the brackets represent molarity. There are, however, competing equilibria within the solution, for example, Eq 11 and 12.

$$H^+ + (OH)^- = H_2O \qquad (Eq\ 11)$$

$$Al(OH)_3 + (OH)^- = Al(OH)_4^- \qquad (Eq\ 12)$$

Therefore, at low pH, the hydrogen ion successfully competes with the aluminum ion for the hydroxide ion (Eq 11) and aluminum hydroxide does not precipitate. At high pH, the hydroxide ion reacts with the amphoteric aluminum hydroxide (Eq 12) to again prevent precipitation, this time by formation of a soluble complex ion, $Al(OH)_4^-$.

The influence of pH on the precipitation of hydroxides is obvious because the hydroxide ion appears directly in the solubility product. However, it has a more subtle influence on many other precipitation processes. Ammonium oxalate and oxalic acid are commonly used precipitating agents. Figure 6 shows the actual species present in an aqueous solution of oxalate ion as a function of pH. The pH is a major, but indirect, influence on both the solubility products and the extent of complex formation through its effect on the concentration of oxalate ion in solution.

Another indirect effect of pH is indicated by the instability of ferrous ion in alkaline media because the oxidation equilibrium in air is shifted by the nearly complete precipitation of ferric hydroxide from solution.

Ammonium hydroxide is a frequent choice used to adjust the pH because it is inexpensive and the amount that is occluded or absorbed is readily removed during subsequent calcination. There are two potential drawbacks, however, to the use of ammonium hydroxide. First, the solution rapidly becomes buffered by the ammonium salts formed and it is difficult to reach values of pH greater than 9. Second, soluble ammine complexes may form with transition metal ions. There are other bases, for example, tetramethylammonium hydroxide (TMAH) or various amines, that can be substituted at greater cost. However, because they are more dissociated than ammonium hydroxide, lesser quantities are needed to attain a given pH.

The ionic strength is determined by the amount and charge type of the electrolyte remaining in solution. Its effect on the precipitation is through the induced changes in activity coefficients. Generally, increasing the ionic strength lowers the activity coefficients, thereby increasing the solubility of the precipitate. An obvious exception to this is that if the electrolyte has an ion in common with the precipitate, it will lower the solubility in the absence of any complex ion formation. If colloids are a factor, then there is also the influence of the electrolyte on the coagulation of the colloid through its effect on the charged layers surrounding the colloidal particles.

Equilibrium is seldom attained during precipitation processes described herein. Consequently, factors such as pH, solvent, rate of mixing, atmosphere, and washing, are also important. By adding the solution of cations slowly and stirring the solution containing the precipitating agent, there is nearly always a considerable excess of the precipitating agent present, and the solubility products of the various species are most likely to be exceeded simultaneously. This produces a more homogeneous precipitate than if the cations were allowed to precipitate in a stepwise fashion.

In contrast, when dealing with a single cation such as ferrous oxalate or aluminum hydroxide, the order of mixing is less important and a homogeneous precipitation procedure may be more desirable (Ref 34). This technique slowly raises the concentration of the precipitating agent, for example, hydroxide or oxalate ion, by the controlled hydrolysis of urea or ethyl oxalate, respectively, and thereby greatly reduces the degree of supersaturation present.

The degree of supersaturation determines the number of nuclei formed and thus controls the particle size of the precipitate. Consequently, if very finely divided material is desired, the precipitation should be done at lower temperatures with rapid stirring of concentrated solutions. This leads to a relatively impure material that is difficult to filter. In contrast, homogeneous precipitation produces a more pure, coarse powder.

The kinetics of precipitation are quite slow for some materials. Magnesium oxalate, for instance, is notorious for the long induction period that precedes precipitation. Although this would seem to be a serious problem leading to the stepwise or separate precipitation of mixed cations, it is seldom a problem because of the strong tendency to coprecipitate. The nuclei of the more rapidly precipitated compound either catalyzes the formation of nuclei for the other substances or actually serves as a nucleus itself.

The nature of the stirring not only influences the local degree of supersaturation, but can also provide mechanical forces that will influence the degree of agglomeration and even the individual particle size. The application of an ultrasonic field during precipitation can also reduce the average agglomerate size (Ref 35).

Impurities can be included with the precipitate in a variety of ways. Their solubility can be exceeded and they will precipitate as a second phase, or, if their size and charge are appropriate, they may be substitutionally incorporated into the precipitate. Anionic substitutions are likely because of complex ion formation and incomplete replacement by the precipitating agent. This is particularly true when hydroxides are formed. Stoichiometric basic salts can exist, but more often it is a nonstoichiometric, nonequilibrium situation. The order of stability of these basic salts is sulfate > chloride > nitrate > perchlorate (Ref 36).

Impurities can be adsorbed onto the surface. If the precipitation is rapid, those ions can become trapped within the bulk. The occlusion of pockets of solution sometimes occurs. Conditions that favor slow growth, large particle size, and equilibrium will generally produce the cleanest precipitates.

The inclusion of small amounts of impurities from the solution phase can have a surprisingly dramatic effect on subsequent processing (Ref 37). Figures 7 and 8 show the effect of the choice of starting salts and precipitating agents on the densification and the grain size that can occur during otherwise identical precipitation and sintering procedures used to prepare nickel ferrite. Nitrates, for example, led to powders that sintered poorly, and trapped or occluded chloride produced retrograde sintering when volatile chlorides formed after the porosity had closed (see Fig 7). The choice of the specific hydroxide strongly affected the extent of grain growth during sintering as seen from the micrographs in Fig 8. These effects are difficult to predict and are highly specific to the particular system. Clearly, however, they must be given consideration.

There are several possible approaches to precipitating compounds or mixtures from solution without altering their chemical composition by markedly reducing their solubil-

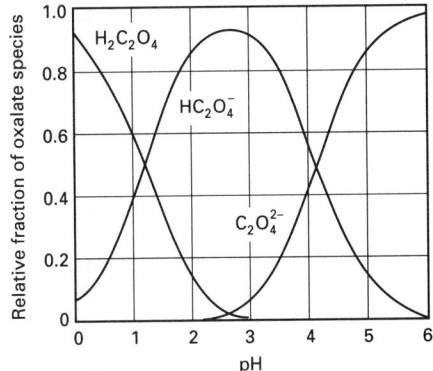

Fig 6 Oxalate species in aqueous solution as a function of pH

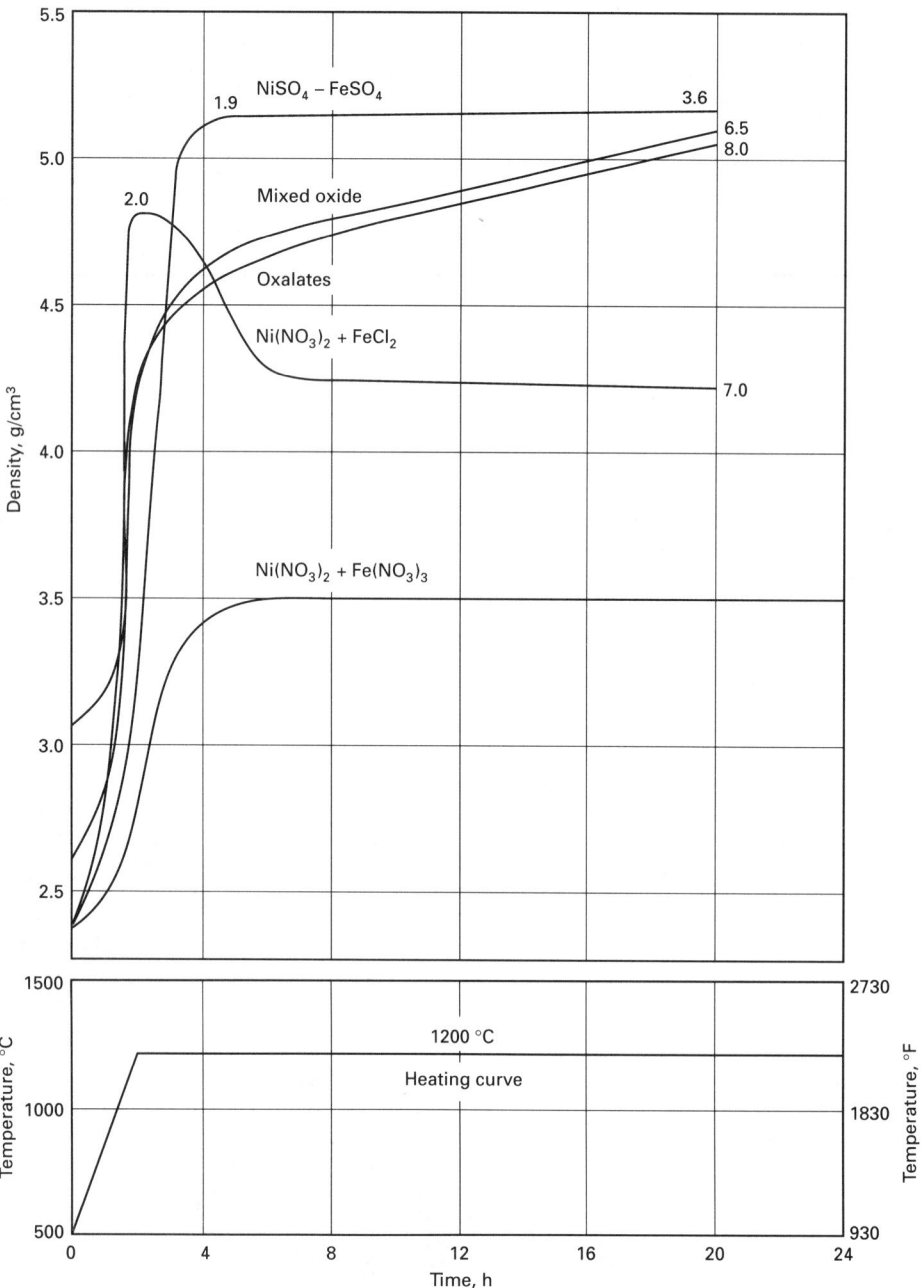

Fig 7 Comparison of sintering behavior of nickel ferrites prepared from a variety of sources. The final grain size in nm is indicated, except for the nitrate sample, which was too porous to measure grain size. Source: Ref 37

ity. Sizable changes in temperature can be used to suddenly reduce the solubility. However, while this may be feasible for a single-component system, it is unlikely that such a process would precipitate a specific mixed composition. A more useful method is to greatly reduce the solubility by mixing the solution with a miscible solvent in which the ions are much less soluble, for example, acetone or alcohols, in the case of aqueous solutions. This latter method has been used to prepare precursors for some ferrites (Ref 38).

In the idealized situation, the ratio of cations in the precipitate will be the same as that in solution. However, it is frequently necessary to compensate for the partial solubility of one or more of the components. This can be done by adding a controlled excess of those components under carefully controlled and reproducible experimental conditions. An alternate approach is to recycle the saturated solution and add only the stoichiometric amounts on subsequent precipitations. Continuous reactors can be designed to facilitate the process. Having considered some of the general aspects of precipitation, a number of representative systems are examined below.

Ferricyanides and Ferrocyanides. In the discussion on thermal decomposition, one of the examples described the ferricyanide and ferrocyanide precipitations as a way of preparing highly stoichiometric rare earth orthoferrites. This is also applicable to the use of cobalt or manganese hexacyanno complexes (Ref 6). Other specific compositions could be achieved such as Ba_2FeO_x using barium ferrocyanide or $Sr_3Co_2O_x$ from strontium cobalticyanide. The unique stoichiometry of these compounds provides a fixed cation ratio in the decomposed product. The excess cation, or cyanide complex, that may have been inadvertently placed in solution remains behind in the filtrate.

The oxalate ion has been particularly useful for chemical synthesis of oxides. The ability to precipitate most divalent ions directly as solid solutions has been used in the synthesis of ferrites for many years (Ref 39–41). More interesting is the ability to form complex anions with higher valent cations. Unique compounds similar to those described above for the cyanides are possible. The most important of these has been the $BaTiO(C_2O_4)_2 \cdot 4H_2O$ precursor for barium titanate developed by Claybaugh et al. (Ref 42). When rare earth doped barium titanates became important for their anomalous positive temperature coefficient of electrical resistivity, this method was exploited to provide doped materials directly (Ref 43–45). Substantially less lanthanum, for instance, was necessary using this technique than was required by conventional additions (Ref 18).

The anomaly of the sharp positive temperature coefficient occurs in the region of the tetragonal-to-cubic phase transformation. This temperature is around 120 °C (248 °F) for pure barium titanate, but it can be raised or lowered by the substitution of lead or strontium for barium, respectively. Modifying the oxalate precipitation process requires adjustment of the pH, which also alters the nature of the precipitate. In producing the pure strontium compound, it is necessary to raise the pH to a value where there is partial substitution of hydroxide ion for oxalate ion, that is, $SrTiO(OH)_2(C_2O_4)$. This change in composition can be predicted by examination of the pH dependence of the species in solution, as shown in Fig 9 (Ref 46).

More recently, there has been a wave of interest in the chemical preparation of the high-temperature superconductors. Oxalate coprecipitation was among the first methods tried (Ref 47–49). The formation of barium carbonate and the copper and yttrium oxides during the decomposition of the mixed oxalates produced the same mixture (Ref 50) as was used in conventional processing (Ref 51), but in a better-mixed, more reactive state.

The oxalate coprecipitation method can be modified to avoid the presence of extraneous ions that might cause difficulties as described earlier. If carbonates, hydroxides, or active oxides are used as the starting compounds, they can be reacted with oxalic acid to form the mixed oxalate precipitate with only carbon dioxide and water as the products. The

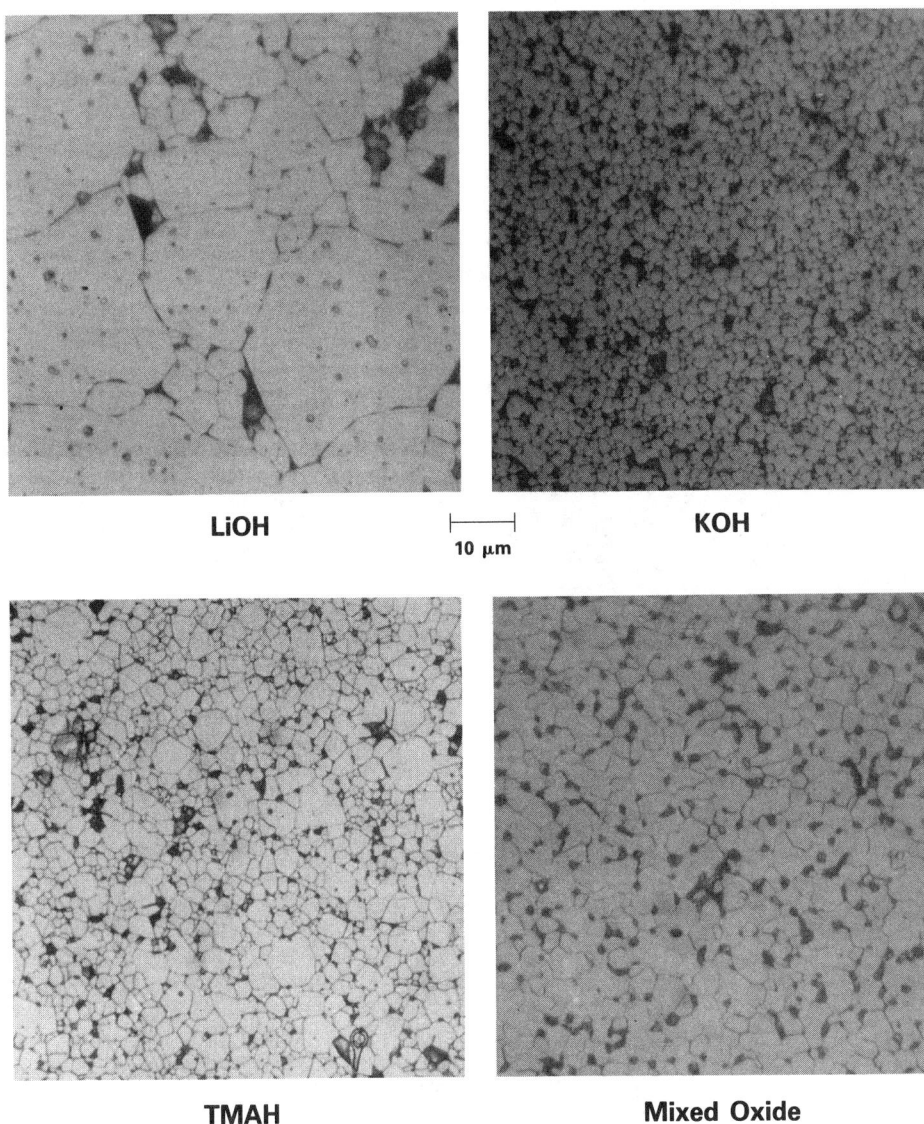

Fig 8 Comparison of microstructures of nickel ferrites precipitated by various hydroxides at 1250 °C (2280 °F) for 4 h in nitrogen. Source: Ref 37

LiOH

KOH

10 μm

TMAH

Mixed Oxide

resulting solution is essentially neutral with zero ionic strength. Both of these factors markedly decrease the compensation required for partial solubility of the oxalate (Ref 40). This method has been successfully used to prepare ferrites (Ref 41) and superconductors (Ref 52).

Hydroxides. The precipitation of hydroxides, or hydrated oxides, is undoubtedly the most widely used approach to the chemical synthesis of ceramic materials. There are many other approaches, such as by directly adding hydroxide ions or by using a slower hydrolysis step. These processes can be performed at room temperature, moderately elevated temperature, or under hydrothermal conditions. Each of these methods can yield distinctly different products.

Direct precipitation at near room temperature usually yields a polyphase that is well mixed and finely divided if it is precipitated simultaneously. This advantage has permitted substitutions that are not possible in the conventional mixed oxide approach, for example, silver and lanthanum substitutions into hexagonal barium ferrite (Ref 53).

There are innumerable publications on the preparation of ferrites by the coprecipitation of hydroxides and hydrated oxides. The solubilities are usually low, and nearly quantitative yields can be obtained with compositions that closely approach those in the starting solution (Ref 54). Low-temperature calcination will form a finely divided spinel phase. The aggregates tend to be more chainlike than those formed from oxalates.

If the precipitation and prolonged digestion are done at high values of pH and near the boiling point, a spinel phase can form directly (Ref 55). The ease of dehydration of the hydroxide is an important factor in determining whether the spinel forms in solution. Pure magnesium ferrite does not form under these conditions, but manganese ferrite

does. However, the presence of a more easily dehydrated hydroxide, for example, manganese or zinc, will induce the formation of a ternary magnesium spinel phase. In evaluating the precipitate by x-ray diffraction, it is important to keep in mind the strong similarities in the patterns of the ferrite, magnetite, and γ-iron(III) oxide. When magnetic precipitates are formed, an external magnetic field can increase the efficiency during washing and filtration (Ref 56). Orthoferrites and garnets can also be prepared in this manner (Ref 57).

Carrying out the precipitation at still higher temperature, that is, hydrothermally, produces a highly crystallized spinel directly from hydroxides having even higher dehydration temperatures (Ref 58–61). Figure 10 shows the improvement in crystallinity with increasing temperature (Ref 59). X-ray diffraction indicates that the spinel first appeared at 150 °C (300 °F).

There are many examples of hydroxide precipitation outside of the ferrite area. Magnesium aluminate can be formed at 400 °C (750 °F) by coprecipitation, but it does not form until above 1200 °C (2190 °F), depending on particle size, when using conventional mixed oxide techniques (Ref 62).

The morphology of the particles can be altered by various modifications of the precipitation process. There are dramatic differences in the zinc oxide powders from hydroxide based precipitations. Spheres, flowerlike crystals, and thin needles are among the shapes of zinc oxide crystals that can be produced by chemical synthesis (Ref 63). Freezing the solution of cations and subsequent treatment of the resulting solid with a base has been used to reduce the degree of segregation that occurs (Ref 64).

Ferroelectric and piezoelectric compositions have been precipitated by hydroxides at elevated temperatures. Barium titanate has been precipitated directly from an aqueous solution of barium hydroxide and titanium esters (Ref 65), and single crystals have been grown hydrothermally (Ref 66). Similarly, ultrafine powders of zirconates (Ref 67),

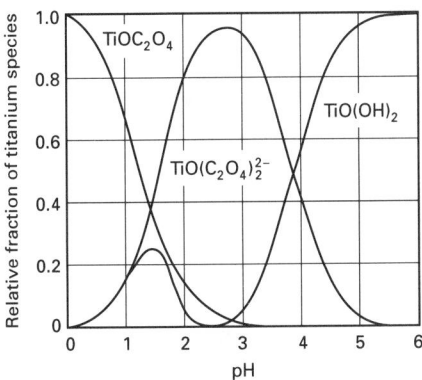

Fig 9 Titanium species in an aqueous oxalate solution as a function of pH. Source: Ref 46

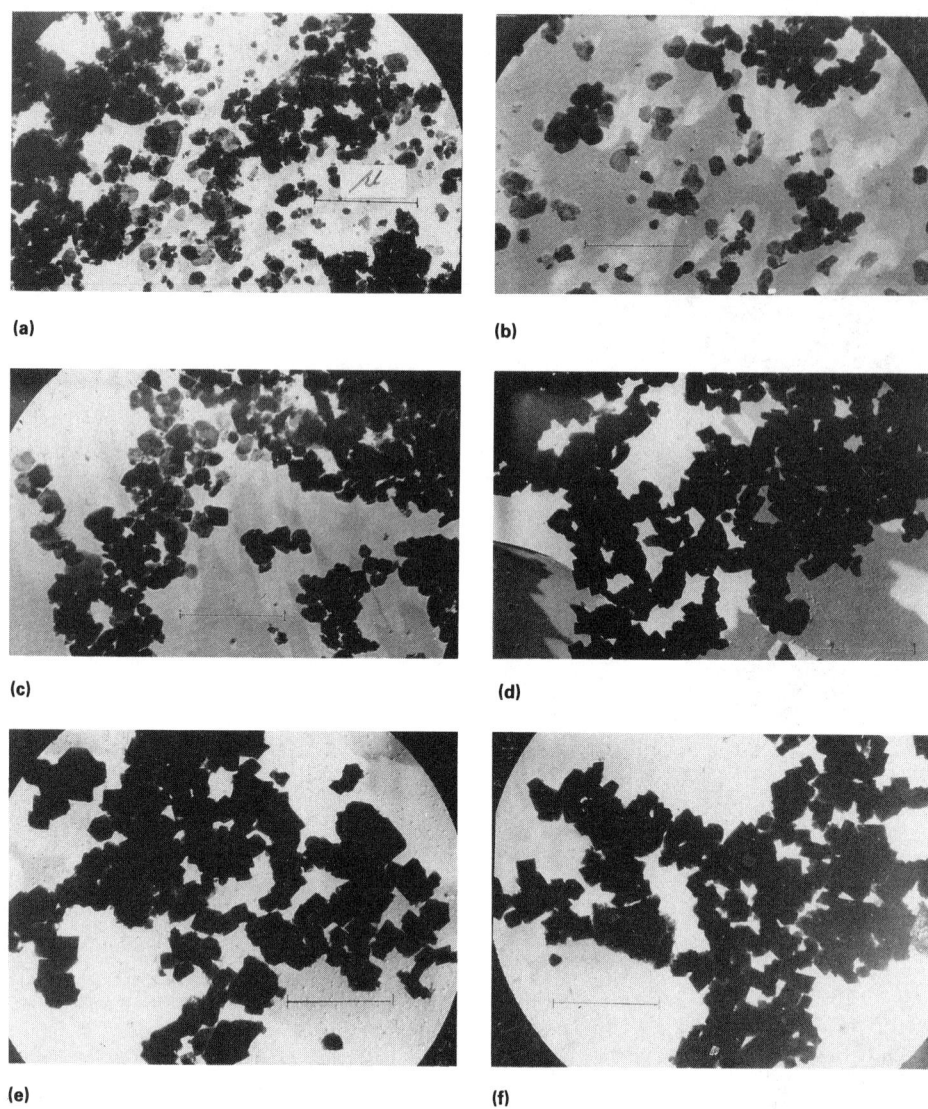

Fig 10 Micrographs of manganese zinc ferrites prepared in autoclave for 4 h under various hydrothermal conditions. (a) Control. (b) At 150 °C (300 °F). (c) At 200 °C (390 °F). (d) At 250 °C (480 °F). (e) At 300 °C (570 °F). (f) At 350 °C (660 °F). Source: Ref 59

stannates (Ref 68), and strontium titanates (Ref 69) have also been prepared.

Metal-organics. Mazdiyasni and co-workers have successfully pioneered metal-organics for the preparation of a variety of materials, for example, barium titanate (Ref 70) and yttria-stabilized zirconia (Ref 71). High-purity alkoxides are synthesized, dissolved in a suitable solvent, and mixed with high-purity water. The latter step leads to hydrolysis and the formation of very pure and finely divided hydrated oxides. These can be dehydrated by controlled calcinations to produce the desired particle size.

This process has been extensively modified and extended in recent times to even produce single crystals of a double alkoxide having a reported composition of $BaTi(OC_3H_7)_6$ (Ref 72). Polytitanates can also be prepared with a variety of results as regards the success in achieving a particular single-phase material (Ref 73–76).

Mixed Reagents. The discussion so far has been limited to a single precipitating agent. However, there are situations in which a single agent either will not suffice, or, for other reasons, is not desirable. Magnesium (Ref 62), calcium (Ref 77), and lanthanum (Ref 78) aluminates are examples for which oxalate-hydroxide mixtures were used. The pH was kept low enough to not redissolve the aluminum hydroxide, and the oxalate of the remaining cation precipitated simultaneously. An oxalate, citrate, and hydroxide combination has been used to precipitate zirconates (Ref 79), a stearate-hydroxide combination for lithium ferrite (Ref 80), and a carbonate-hydroxide mixture for a high-temperature superconductor (Ref 81).

Coated Colloids or Slurries. Precipitation can also be done in a slurry of suspended particles. Thoria-doped nickel is an excellent example in which the suspended particles are the minor component (Ref 82). A solution of

nickel and a colloidal suspension of thoria are mixed simultaneously in basic solution. Nickel hydroxide forms a thick coating on each particle of thoria. After filtering and drying, this is converted to nickel metal by heating in hydrogen. The resulting powder is sintered to produce a metal that has been significantly hardened by the dispersion of the unreduced particles of thoria.

Examples in which the solid in suspension represents the major phase occur when the solution phase is used either to dope the solid in a well-dispersed manner (lanthanum to barium titanate) (Ref 45) or to make minor corrections to adjust stoichiometry (iron or yttrium to yttrium-iron-garnet) (Ref 83).

Miscellaneous Techniques. There are other relatively minor techniques that have also been used successfully to prepare precursor precipitates. A novel approach involves electrochemically generating the proper concentrations and types of cations in a solution of an electrolyte, such as sodium sulfate, by controlling the amount of current through appropriate metallic anodes (Ref 84). The hydrated oxides are coprecipitated by bubbling oxygen through the solution surrounding the cathode and thereby preventing undesirable precipitation on the anodes. Ferromagnetic Fe_3S_4 has also been prepared electrochemically (Ref 85).

Aqueous solutions of citrates and formates have been sprayed into alcohol, yielding mixed fine powders for subsequent calcination to form various ferroelectric and piezoelectric materials (Ref 86). Precipitation of precursors from molten salts is also practical, provided there is a suitable solvent to dissolve the flux and separate it from the precursor. A high degree of supersaturation is necessary to obtain fine powders.

Since 1980, considerable interest has developed in the formation of discrete, mono-sized, spherical particles in solution and the methods that might pack these into arrays suitable for relatively low-temperature sintering. The pioneering efforts of Matjevic and his group in the formation of well-dispersed colloidal solutions by a variety of techniques has been largely responsible (Ref 87). Other investigators have extended the work and formed multi-cation materials, such as aluminum titanate, by precipitation of a hydrated oxide of one cation on a colloidal particle of the other cation by carefully controlled hydrolysis reactions (Ref 88). An alternative route for the formation of relatively homogeneous, discrete, multi-cation particles is to limit the ability to segregate by confining the solution to tiny droplets by emulsification (Ref 89).

Thermal methods of solvent removal utilize a great variety of specific techniques. Simple evaporation is obviously the most direct approach. However, in the case of mixed cations, there will be segregation unless this evaporation is very rapid. Magnesium spinel and yttrium-iron-garnet have been prepared by mixing the solid nitrates, melting them, carefully dehydrating them at low temperature in vacuum, decomposing the nitrates in air at an intermediate temperature, and, finally, firing the oxides at 1000 °C (1830 °F) in air (Ref 90). Rapid dehydration has been used to form amorphous masses of citrates, tartrates, and other materials that can be calcined to form the desired material. Several perovskites, spinels, and garnets have been made in this manner (Ref 91).

Spray Drying. Because segregation is confined within the individual droplets, aerosols and fine sprays that dry rapidly can produce reasonably homogeneous materials. Starting with aqueous solutions of sulfates, deLau has produced powders that yield ferrites or magnesium alumina spinel upon calcination (Ref 92). Titanates have been prepared from nitrate solutions (Ref 93). Adding colloidal carbon to the solution prior to spray drying produces a material that can be processed to yield metallic powders. Subsequent reoxidation of this powder forms finely divided oxides (Ref 94). If the solution contains both oxidizing and reducing anions, the dried powders will react strongly to form the oxide (Ref 4, 95).

Emulsion Methods. The segregation can also be restricted to a fine droplet size by using emulsion methods such as ultrasonically dispersing an aqueous solution in an immiscible liquid, such as kerosene. The emulsion is heated in a vacuum dryer to give a suspension of the dehydrated salt particles. A deflocculant is added, and the powders are filtered and calcined (Ref 96).

Spray Roasting. If the temperature in the spray dryer or furnace is raised to a point that not only dehydrates but also decomposes the salts, then the process can yield the final product directly without subsequent calcination. This process has been called spray roasting, and it is used to prepare ferrites (Ref 97, 98). Other investigators have sprayed into tube furnaces rather than using spray dryers to produce many ceramics of interest (Ref 99). A particularly useful modification of this latter approach is to use fluid bed conditions. Controlling the atomization of the solution, the gaseous fluidizing flow, and the furnace temperature allows optimization of the process and continuous operation (Ref 100). Uranium and aluminum oxides have been prepared in this manner (Ref 101).

Combustion of an organic solvent can be used both to remove the solvent and simultaneously perform the calcination. Hydrolysis of aluminum, silicon, and titanium chlorides in a flame at around 800 to 1400 °C (1470 to 2550 °F) has long been used to prepare these oxides in a finely divided state. A number of ferrites have similarly been made by dissolving the nitrates in alcohol, atomizing the resulting solution with oxygen, burning the alcohol, and, finally, collecting the powder (Ref 102). At the flame temperatures, the oxides react to form the spinel phase. Aggregates on

the order of tenths to several microns in size are normally obtained and the average crystallite size is tens of nanometers.

Collection of Powder. One of the major problems associated with this process is the collection of the powder. Some devices that have been used are a cyclone separator (Ref 102), water curtain (Ref 103), and an electrostatic precipitator (Ref 104). There can also be problems associated with the volatility of some components, such as zinc (Ref 101), or with the effects of oxygen partial pressure on the nature of the product (Ref 103). Many materials other than ferrites have been made by this flame spray process (Ref 105).

Thermal Evaporation. In the types of thermal evaporations described in the preceding sections, the resulting particle is generally a hollow shell with a hole that formed by escape of the vapor phase. Similarly, if a porous support is dipped into a solution of catalyst precursor or suspension of finely divided catalyst, capillary forces will again cause the solution or fine particles to migrate to the surface during drying. This migration markedly decreases the homogeneity or distribution of the powder.

Freeze drying, on the other hand, immobilizes the solids during the drying process, producing a uniform distribution within the individual particle or porous support. Figure 11 shows how well the distribution of the cobalt nitrate is maintained on the porous support when freeze dried (Ref 106). In contrast, the oven drying has brought most of the cobalt to the surface.

Bulk freeze dried powder also retains the size and shape of the droplet of frozen solution from which it was derived. This produces low-density, readily friable, spherical powders having good flow properties. The flow properties can be destroyed if the material is hygroscopic at this stage. However, the ideal shape is generally preserved through the calcination stage, producing spherical aggregates of the final material that have a crystallite size established by the conditions of calcination. The general texture of particles within the spherical aggregates, both before

and after calcination, is chains of crystallites arranged in a radial pattern. This results from the structure of ice established during the rapid freezing process.

The freeze drying process was initially applied to the production of ceramics by Schnettler *et al.* (Ref 107). It is best described using the phase diagram for a typical aqueous solution (Fig 12) (Ref 108). A solution of the appropriate salts, frequently sulfates, acetates, and oxalates, is rapidly frozen (from point 1 to 2 in Fig 12) in order to minimize any segregation. A common method is to spray the solution as small droplets into a chilled hexane bath, but continuous processes have been developed that produce droplets in the range of 0.01 to 5 mm (0.4 to 200 mils) (Ref 109).

The frozen droplets are separated by filtration or skimming and transferred in the frozen state to a freeze dryer. The material is kept frozen and the total pressure reduced to below the triple point, from point 2 to 3 in Fig 12. A simple mechanical pump is adequate for this purpose. Heat is supplied to the sample in a sufficiently controlled manner so as to sublime the water without formation of a liquid phase. The water sublimes isothermally at point 4 in Fig 12 until the material is dehydrated. It is essential that heat not be supplied so rapidly that water vapor is evolved at a rate beyond the capability of the pumping and condensing system.

A number of oxides have been made in this manner, for example, ferrites (Ref 107), lithium niobate (Ref 5), alumina (Ref 110), iron oxide (Ref 14, 111), beryllia (Ref 112), as well as perovskite catalysts (Ref 113), and superconductors (Ref 114).

The sol-gel process is an interesting one that has undergone a remarkable renaissance in popularity. In many respects, it is a hybrid between the simple precipitation and evaporation categories presented herein. The method was originally used by Ferguson *et al.* (Ref 115) to prepare small dense spheres of nuclear fuel. Fletcher and Hardy (Ref 116) discussed the general process and expanded it to other oxides of industrial importance. In that pro-

Fig 11 Comparison of distribution of cobalt resulting from oven drying or freeze drying porous ceramic saturated with a cobalt nitrate solution. Source: Ref 106

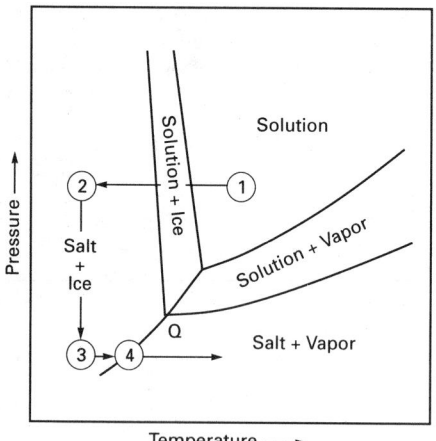

Fig 12 Phase diagram of an aqueous solution showing the steps (1 to 4) in the freeze drying process. Source: Ref 108

cess, a gel was formed from a colloidal solution by the extraction of water or, alternatively, removing the solubilizing species, such as hydrogen ion, by exchange with a second liquid phase, such as an amine. In that fashion, the suspended liquid drops were converted to solid spheres of gel. A surface active agent was generally present to prevent coalescence of the droplets.

The spheres were collected and sintered at 1200 to 1475 °C (2190 to 2690 °F) to form dense spheres with diameters of approximately 50 to 150 μm (2.6 to 6 mils), with an average crystallite size of 60 to 70 nm (2.4 to 2.8 μin.) and very high strength. This density was only achievable with conventional processing by sintering at 1800 °C (3270 °F), producing a much larger crystallite size and reduced strength. If colloidal carbon was added to the solution and incorporated in the gel, then carbides could be prepared by firing in either vacuum or inert atmospheres (Ref 117).

These earlier techniques have been extensively modified and expanded during the recent decade to produce a wide range of products in many forms, from thin films to large preforms for drawing optical fibers. The gels can even be cast into shapes, although large shrinkages will occur during drying and sintering. The literature is voluminous and the interested reader is directed to recent review articles, such as Ref 63, 75, and 118; the Materials Research Society's series, "Better Ceramics Through Chemistry" (Ref 75); or a recent book on the subject (Ref 119).

The two basic approaches to gel formation are to either start with a controlled hydrolysis of metal-organics, such as alkoxides, to form colloidal solutions or start directly with a colloidal suspension of at least one of the components and then adding the remaining components. Either of these systems are then induced to gel in the desired shape, and, depending on the application, very carefully dried and sintered to the desired porosity and density.

Much chemistry is involved in controlling the rates of hydrolysis or gelation to achieve the desired porosity and microstructure within the gel. The choice of catalyst for the hydrolysis step, for instance, can greatly influence the nature of the cross-linking and the amount of water incorporated into the gel. This will markedly influence the subsequent drying and sintering properties as well as the nature of the final product (Ref 120). Although a well-sintered material is the usual goal, there are times when a large degree of fine, open porosity is desirable to form aerogels. Aerogels have interesting and valuable mechanical, thermal, optical, and acoustic properties (Ref 118).

Examples that represent applications of the sol-gel process to the synthesis of ceramic materials include products such as high-temperature superconductors (Ref 121, 122), dielectrics (Ref 123, 124), ionic conductors (Ref 125), and glass (Ref 126, 127).

Gas Phase Reactions

Ceramic powders have been prepared in various ranges of particle size by evaporation of parent oxides and subsequent controlled condensation. These techniques generally use high-intensity arcs, plasma jets, or lasers to vaporize fabricated electrodes or targets. Many oxides, such as alumina, silica, and magnesia, and some carbides have been produced in this way (Ref 128). These methods can produce very fine powders that must then be separated from the carrier gas using electrostatic precipitators, filters, and other devices.

If a plasma is used to form molten droplets that are not significantly vaporized, then the particle size is large and the shape spherical (Ref 129, 130). Spherical particles can also be made by melting irregular-shaped particles in an oxygen-hydrogen flame, which is also commonly used to hydrolytically decompose volatile chlorides and metal-organics. Large quantities are made of very finely divided silica and alumina (Ref 131, 132). Special burner designs have been developed to prevent clogging of the burner orifices (Ref 133).

Other examples of vapor phase reactions are the partial combustion of natural gas used to prepare enormous quantities of commercially important carbon black (Ref 134) and the production of finely divided metals by the decomposition of gaseous metal carbonyls (Ref 135). Because of the pyrophoric nature of these finely divided metals, the formation of oxide particles requires a very carefully controlled oxidation process. However, the burning of bulk metals is used to prepare magnesia or zinc oxide powders (Ref 101). This is economically feasible for high-purity oxides when the metal is the more easily purified material. A novel approach to metallic combustion uses an electric arc in an insulating liquid to simultaneously fracture and oxidize the metal (Ref 136).

Organometallics and silicon-based polymers are being developed that will thermally degrade to form oxides under oxidizing conditions, for example, silica (Ref 137) or carbides, borides, nitrides (Ref 138), and silicides when pyrolysis in an inert atmosphere is performed.

Currently, the ultimate tool for the preparation of fine powders from the gas phase is the laser. Homogeneous gas phase reactions can be induced in a containerless environment. Tremendous process control is achieved by manipulating flow rates, optical path lengths, laser intensity, partial pressures of reactants, and other parameters. Unagglomerated, very finely divided powders of various oxides, borides, carbides, and silicides, as well as small metallic clusters with potential interest as catalysts, have already been prepared by this promising method (Ref 139).

The previously discussed gas-solid reaction methods have relied almost exclusively on the external supply of energy as their driving energy. However, there is a class of self-sustaining high-temperature synthesis reactions that offer the potential for energy savings and simultaneous formation and densification. Munir (Ref 140) has recently reviewed the method in some detail. A criterion for the self-sustaining aspect is that the enthalpy of the reaction, compared to the heat capacity of the product, must be adequate to provide a condition analogous to a thermal runaway situation. Under these conditions, the reaction front will proceed rapidly through the material once the initial reaction or ignition temperature has been reached, provided that adequate contact is maintained between the reactants.

The nitridation of titanium is an example. In this case, the product must be sufficiently porous to allow for the continuous and rapid supply of nitrogen to the reaction zone. The pressure of the reactant gas and its solubility in both the solid product and reactant phases are important (Ref 141).

REFERENCES

1. S. StJ. Warne and P.K. Gallagher, *Thermochim. Acta*, Vol 110, 1987, p 269–279
2. P.K. Gallagher and S. StJ. Warne, *Thermochim. Acta*, Vol 43, 1981, p 253–267
3. Z. Zhong and P.K. Gallagher, *Thermochim. Acta*, 1991, to be published
4. K. Kourtakis, M. Robbins, and P.K. Gallagher, *J. Solid State Chem.*, Vol 82, 1989, p 290–297
5. P.K. Gallagher and F. Schrey, *Thermochim. Acta*, Vol 1, 1970, p 465–476
6. P.K. Gallagher and F. Schrey, *Thermal Analysis*, Vol 2, R.F. Shwenker and P.D. Garn, Ed., Academic Press, 1989, p 929–952

7. P.K. Gallagher and D.W. Johnson, *Thermochim. Acta*, Vol 4, 1972, p 283–289

8. G.S. Grader, P.K. Gallagher, and D.A. Fleming, *Chem. Mater.*, Vol 1, 1989, p 665–670

9. P.K. Gallagher and B. Prescott, *Inorganic Chem.*, Vol 9, 1970, p 2510–2512

10. O. Chaix-Pluchery and J.C. Niepce, *Reactiv. Solids*, Vol 8, 1990, p 323–325

11. F. Paulik and J. Paulik, *J. Therm. Anal.*, Vol 5, 1973, p 253–270

12. P.K. Gallagher and D.W. Johnson, *Thermochim. Acta*, Vol 6, 1973, p 67–83

13. W.H. Gitzen, Ed., *Alumina as a Ceramic Material*, The American Ceramic Society, 1970, p 14–20

14. P.K. Gallagher, D.W. Johnson, F. Schrey, and D.J. Nitti, *Am. Ceram. Soc. Bull.*, Vol 52, 1973, p 842–849

15. J.A. Hedvall, *Chem. Rev.*, Vol 15, 1934, p 139–169

16. P.D. Garn and T.S. Habash, *J. Phys. Chem.*, Vol 83, 1979, p 229–232

17. P.K. Gallagher and D.W. Johnson, *J. Phys. Chem.*, Vol 86, 1982, p 295–297

18. H.A. Sauer and J.R. Fisher, *J. Am. Ceram. Soc.*, Vol 43, 1970, p 297–302

19. D.W. Johnson and P.K. Gallagher, *J. Phys. Chem.*, Vol 76, 1972, p 1474–1479

20. J. Rouquerol, *Thermochim. Acta*, Vol 144, 1989, p 209–224

21. H. Palmour and D.R. Johnson, *Sintering and Related Phenomena*, G.Y. Kuczynski *et al.*, Ed., Gordon & Breach, 1967, p 779–785

22. D.A. Young, *Decomposition of Solids*, Pergamon Press, 1966

23. W.E. Brown, D. Dollimore, and A.K. Galwey, *Comprehensive Chemical Kinetics*, Vol 22, Elsevier Scientific, Amsterdam, 1980

24. H. Sschmalzried, *Solid State Reactions*, Verlag Chemie, Basel, 1981

25. J. Huang and P.K. Gallagher, *Thermochim. Acta*, 1991, to be published

26. T. Ozawa, *Bull. Chem. Soc. Jpn.*, Vol 38, 1965, p 1881–1891

27. H.E. Kissinger, *Anal. Chem.*, Vol 29, 1957, p 1702–1710

28. P.K. Gallagher, E.M. Gyorgy, and H.E. Bair, *J. Chem. Phys.*, Vol 71, 1979, p 830–835

29. M. Arnold, G.E. Veress, J. Paulik, and F. Paulik, *Thermal Analysis*, Vol 1, H.G. Wiedemann, Ed., Birkhauser Verlag, Basel, 1980, p 69–73

30. P.K. Gallagher and D.W. Johnson, *Thermochim. Acta*, Vol 14, 1976, p 255–261

31. H.C. Fisher, *J. Am. Ceram. Soc.*, Vol 38, 1955, p 245–250

32. K.S. Mazdiyasni and L.M. Brown, *J. Am. Ceram. Soc.*, Vol 54, 1971, p 479–485

33. K.R. Hancock, *Ultrafine Grain Ceramics*, J.J. Burke *et al.*, Ed., Syracuse University Press, 1970, p 43–48

34. L. Gordon, M.L. Salutsky, and H.H. Willard, *Precipitation from Homogeneous Solution*, John Wiley & Sons, 1959

35. F.K. Kukoz and E.M. Feigina, *Zh. Prikl. Khim.*, Vol 42, 1969, p 1978–1982

36. J.F. Coetzee, *Treatise on Analytical Chemistry*, Vol 1, I.M. Koltoff and P.J. Elving, Ed., Interscience, 1959, p 803

37. H.M. O'Bryan, P.K. Gallagher, F.R. Monforte, and F. Schrey, *Am. Ceram. Soc. Bull.*, Vol 48, 1969, p 203–208

38. R.E. Jager and T.J. Miller, *Am. Ceram. Soc. Bull.*, Vol 50, 1971, p 755–763

39. D.G. Wickman, "Synthesis of Ferrites and Preparation of Cobalt-Ferrite Single Crystals," Tech. Report 89 ONR Contracts, Massachusetts Institute of Technology, Oct 1954

40. P.K. Gallagher and F. Schrey, *J. Am. Ceram. Soc.*, Vol 47, 1964, p 434–437

41. P.K. Gallagher, H.M. O'Bryan, F. Schrey, and F.R. Monforte, *Am. Ceram. Soc. Bull.*, Vol 48, 1969, p 1053–1059

42. W.S. Claybaugh, E.M. Swiggard, and R. Gilchrist, *J. Res. Natl. Bur. Stand.*, Vol 56, 1956, p 289–295

43. P.K. Gallagher, F. Schrey, and F.V. DiMarcello, *J. Am. Ceram. Soc.*, Vol 46, 1963 p 359–365

44. P.K. Gallagher and F. Schrey, *J. Am. Ceram. Soc.*, Vol 46, 1963, p 567–573

45. W.R. Northover, *J. Am. Ceram. Soc.*, Vol 48, 1965, p 173–177

46. F. Schrey, *J. Am. Ceram. Soc.*, Vol 48, 1965, p 359–365

47. K. Kaneko, H. Ihara, M. Hirabayashi, N. Terada, and K. Senezaki, *Jpn. J. Appl. Phys.*, Vol 26, 1987, p L734-L736

48. T. Ozawa, A. Negishi, Y. Takahashi, R. Sakamoto, and H. Ihara, *Thermochim. Acta*, Vol 124, 1988, p 147–156

49. G.R. Paz-Pujalt, M.K. Mehrotra, S.A. Ferranti, and J.A. Agostinelli, *Solid State Ionics*, Vol 32/33, 1989, p 1179–1185

50. A. Nehishi, T. Takahashi, R. Sakamoto, M. Kamimoto, and T. Ozawa, *Thermochim. Acta*, Vol 132, 1988, p 15–23

51. M.F. Yan, H.C. Ling, H.M. O'Bryan, P.K. Gallagher, and W.W. Rhodes, *IEEE Trans. Componen., Hybrids, Manuf. Technol.*, Vol 11, 1988, p 401–406

52. P.K. Gallagher and D.A. Fleming, *Chem. Mater.*, Vol 1, 1989, p 659–664

53. K.K. Laroia and K.K. Vadhera, *Indian J. Technol.*, Vol 6, 1968, p 61–66

54. P.K. Gallagher and F. Schrey, *J. Am. Ceram. Soc.*, Vol 47, 1964, p 434–437

55. V.P. Chalyi and E.N. Lukachina, *Isv. Akad. Nauk SSSR*, Vol 7, 1971, p 652–655

56. F.G.W. Strickland, British patent No. 914,773, Jan 1963

57. M. Robbins, G.K. Wertheim, A.R. Storm, and D.N.E. Buchanan, *Mat. Res. Bull.*, Vol 7, 1972, p 233–237

58. V.P. Chalyi, E.N. Lukachina, and K.N. Potomkin, *Isv. Akad. Nauk SSSR, Neorg. Mater.*, Vol 5, 1969, p 2136–2140

59. F. Schrey, *Am. Ceram. Soc. Bull.*, Vol 44, 1967, p 788–792

60. P.E.D. Morgan, *J. Am. Ceram. Soc.*, Vol 57, 1974, p 499–500

61. M. Kiyama, *Bull. Chem. Soc. Jpn.*, Vol 47, 1974, p 1646–1650

62. R.J. Bratton, *Am. Ceram. Soc. Bull.*, Vol 48, 1969, p 759–763

63. D.R. Ulrich, *Chem. Eng. News*, Vol 68 (No. 1), 1990, p 28–40

64. G.E. Sunbury and D.R. Meisser, U.S. patent No. 3,681,011, Aug 1972

65. S.S. Flashen, *J. Am. Ceram. Soc.*, Vol 77, 1955, p 6194–6196

66. A.N. Christiansen, *Acta Chem. Scand.*, Vol 24, 1970, p 2447–2450

67. T.R.N. Kutty, R. Vivekanadan, and S. Philip, *J. Mater. Res.*, Vol 5, 1990, p 3649–3658

68. T.R.N. Kutty and R. Vivekanadan, *Mater. Res. Bull.*, Vol 22, 1987, p 1457–1465

69. M. Avudaithai and T.R.N. Kutty, *Mater. Res. Bull.*, Vol 22, 1987, p 641–650

70. K.S. Mazdiyasni, R.T. Dolloff, and J.S. Smith, *J. Am. Ceram. Soc.*, Vol 52, 1969, p 523–526

71. K.S. Mazdiyasni, C.T. Lynch, and J.S. Smith, *J. Am. Ceram. Soc.*, Vol 50, 1967, p 532–537

72. K.W. Kirby, *Mater. Res. Bull.*, Vol 23, 1988, p 881–886

73. J.J. Ritter, R.S. Roth, and J.E. Blendell, *J. Am. Ceram. Soc.*, Vol 69, 1986, p 155–160

74. T. Kikuchi and T. Saito, Japanese patent No. JP6121916A2 [86/21916], Jan 1986

75. P.P. Rhule and S.H. Risbud, *Better Ceramics Through Chemistry*, Vol 3, 1988, p 275–281

76. H. Hirano and S. Naka, *Advances in Ceramics*, Vol 19, 1986, p 139–146

77. F. Schrey, personal communication

78. A.M. Golub, T.N. Maidukova, and T.F. Limar, *Isv. Akad. Nauk, SSSR, Neorg. Mater.*, Vol 2, 1966, p 111–115

79. G. Paris, G. Szabo, and R.A. Paris, *Compt. Rend. C*, Vol 266, 1968, p 554–562

80. A.L. Micheli, *IEEE Trans. Magnet.*, Vol 6, 1970, p 606–611

81. B.C. Bunker and J. Voight, *Better Ceramics Through Chemistry*, Vol 3, 1988, p 373–384

82. G.B. Alexander and P.C. Yates, U.S. patent No. 3,082,084, Mar 1963

83. P.K. Gallagher, F.R. Monforte, and F.

Schrey, *Am. Ceram. Soc. Bull.*, Vol 42, 1965, p 912–915

84. H.B. Beer and G.V. Planer, *Br. Comm. Electron.*, Vol 9, 1958, p 939–945

85. S. Yamaguchi and T. Moori, *J. Electrochem. Soc.*, Vol 119, 1972, p 1062–1068

86. B.J. Mulder, *Am. Ceram. Soc. Bull.*, Vol 49, 1970, p 990–996

87. E. Matejevic, *Acc. Chem. Res.*, Vol 14, 1981, p 22–35

88. H. Okamura, E.A. Barringer, and H.K. Bowen, *J. Am. Ceram. Soc.*, Vol 69, 1986, p C22-C24

89. M. Aktinc and K. Richardson, in *Better Ceramics Through Chemistry*, Vol 2, 1986, p 99–109

90. D.R. Messier and G.E. Gaza, *Am. Ceram. Soc. Bull.*, Vol 51, 1972, p 692–698

91. C. Marcilly, P. Courtney, and B. Delmon, *J. Am. Ceram. Soc.*, Vol 53, 1970, p 56–63

92. J.G.M. deLau, *Am. Ceram. Soc. Bull.*, Vol 49, 1970, p 572–580

93. I.A. Kudrenko and T.F. Limar, *Isv. Akad. Nauk SSSR, Neorg. Mater.*, Vol 6, 1970, p 1998–2002

94. J.E. Hardy and M.E. Jordan, U.S. patent No. 3,406,228, Oct 1968

95. K. Kourtakis, M. Robbins, and P.K. Gallagher, *J. Solid State Chem.*, Vol 83, 1989, p 230–236

96. A.L. Stuiijts, *Proc. Int. Conf. on Ferrites, Japan 1970*, Y. Hosimo, S. Iida, and M. Sugamoto, Ed., University Park Press, 1971, p 108–113

97. T. Akashi, T. Titsuji, and Y. Onada, *Sintering and Related Phenomena*, G.C. Kuczynski, N.A. Hooton, and C.F. Gibbon, Ed., Gordon & Breach Science Publishers, 1967, p 747–752

98. M.J. Ruthner, H.G. Richter, and I.L. Steiner, *Proc. Int. Conf. Ferrites*, Y. Hoshino, S. Ieda, and M. Sugimoto, Ed., University Park Press, 1971, p 75–80

99. R. Roy, *Int. J. Powder Metall.*, Vol 6, 1974, p 25–28

100. O.M. Todes, V.A. Seballo, Y.Y. Kaganovitch, A.P. Goltsikin, S.P. Nalimov, and O.M. Rozanov, *Khim. Prom.*, Vol 46, 1970, p 612–615

101. A.A. Jonke, E.J. Petkus, J.W. Loeding, and S. Lawroski, *Nucl. Sci. Eng.*, Vol 2, 1957, p 303–319

102. J.F. Wenkus and W.Z. Leavitt, *Papers Conf. Magnetism Magnetic Mater.*, AIEE Pub. No. T-91, 1957, p 526–535

103. W.W. Malinofsky and R.W. Babbit, *J. Appl. Phys.*, Vol 32, 1961, p 2375–2380

104. R.M. Glaister, N.A. Allen, and N.J. Hellicar, *Proc. Brit. Ceram. Soc.*, Vol 3, 1965, p 67–77

105. M.L. Nielsen, P.M. Hamilton, and R.J. Walsh, in *Ultrafine Particles*, W.E. Kuhn, Ed., John Wiley & Sons, 1963, p 181–195

106. D.W. Johnson, P.K. Gallagher, F.J. Schnettler, and E.M. Vogel, *Am. Ceram. Soc. Bull.*, Vol 56, 1977, p 786–792

107. F.J. Schnettler, F.R. Monforte, and W.W. Rhodes, *Science of Ceramics*, Vol 4, G.H. Stewart, Ed., The British Ceramic Society, 1968, p 79–85

108. M.D. Rigterink, *Am. Ceram. Soc. Bull.*, Vol 51, 1972, p 158–165

109. H.A. Sauer and J.A. Lewis, *AIChE, Journal*, Vol 18, 1972, p 435–445

110. D.W. Johnson and F.J. Schnettler, *J. Am. Ceram. Soc.*, Vol 53, 1970, p 440–446

111. Y.S. Kim and F.R. Monforte, *Am. Ceram. Soc. Bull.*, Vol 50, 1971, p 532–535

112. D.W. Johnson and P.K. Gallagher, *J. Am. Ceram. Soc.*, Vol 55, 1972, p 232–233

113. D.W. Johnson, P.K. Gallagher, F. Schrey, and W.W. Rhodes, *Am. Ceram. Soc. Bull.*, Vol 55, 1976, p 520–523

114. K.H. Song, H.K. Liu, S.X. Dou, X. Shi, and C.C. Sorrell, *J. Am. Ceram. Soc.*, Vol 73, 1990, p 1771–1773

115. D.E. Ferguson, O.C. Dean, and D.A. Douglas, *Int. Conf. Peaceful Uses of Atomic Energy, A/Conf., 28/P 237, Geneva 1964 Proceed.*, Vol 10, 1965, p 307–317

116. J.M. Fletcher and C.J. Hardy, *Chem. Ind.*, Vol 1968, 1968, p 48–58

117. R.G. Wymer, *Proceed. of Panel on Sol-Gel Processing for Ceramic Nuclear Fuels*, International Atomic Energy Agency, Vienna, 1968, p 131–137

118. J.D. LeMay, R.W. Hopper, L.W. Hrubesh, and R.W. Pekala, *MRS Bull.*, Vol 15, 1990, p 30–36

119. G.W. Scherer and C.J. Brinker, *Sol-Gel Science: The Physics and Chemistry of Sol-Gel Processing*, Harcourt, Brace, Jovanovich, 1990

120. K. Nassau, E.M. Rabinovitch, A.E. Miller, and P.K. Gallagher, *J. Noncryst. Solids*, Vol 82, 1986, p 78–85

121. J.M. Tarascon, P.B. Barboux, B.G. Bagley, L.H. Greene, and G.W. Hull, *Mater. Sci. Eng.*, Vol B1, 1988, p 29–36

122. G. Kordas, *J. Noncryst. Solids*, Vol 121, 1990, p 436–442

123. J.M. Hayes, T.R. Gururaja, G.L., Geoffroy, and L.E. Cross, *Mater. Lett.*, Vol 5, 1987, p 396–402

124. J.B. Blum and S.R. Gurkovitch, *J. Mater. Sci.*, Vol 20, 1985, p 4479–4483

125. L.C. Klein, *Solid State Ionics*, Vol 32/33, 1989, p 639–645

126. D.W. Johnson, *Am. Ceram. Soc. Bull.*, Vol 64, 1985, p 1597–1602

127. J.B. Mac Chesney, D.W. Johnson, D.A. Fleming, and F.W. Walz, *Electron. Mater.*, Vol 23, 1987, p 1005–1007

128. J.D. Holmgren, J.O. Gibson, and C. Sheer, *J. Electrochem. Soc.*, Vol 111, 1964, p 362–366

129. H.J. Hedger and A.R. Hall, *Powder Metall.*, Vol 8, 1961, p 65–71

130. F. Walten, *Tech. Mitt.*, Vol 56, 1963, p 179–185

131. G.J. Duffy, *Ind. Eng. Chem.*, Vol 51, 1959, p 232–238

132. K.A. Loftman, *Ultrafine Particles*, W.E. Kuhn, Ed., John Wiley & Sons, 1963, p 196–205

133. M. Formenti, F. Juillet, P. Meriaudeau, S.J. Teichner, and P. Vergnon, *J. Colloid Interface Sci.*, Vol 39, 1972, p 79–89

134. E.J. Mezy, *Vapor Deposition*, C.F. Powell, J.H. Oxley, and J.M. Blocher, Ed., John Wiley & Sons, 1966, p 423–430

135. J.H. Oxley, *Vapor Deposition*, C.F. Powell, J.H. Oxley, and J.M. Blocher, Ed., John Wiley & Sons, 1966, p 452–457

136. Trade literature describing the Iwatani Process, Iwatani and Co. Ltd., Osaha, Japan

137. B.G. Bagley, P.K. Gallagher, W.E. Quinn, and L.J. Amos, *Proc. Mater. Res. Soc.*, Vol 32, 1984, p 287–292

138. L.V. Interrante, L.E. Carpenter, C. Whitmarsh, W. Lee, M. Garbauskas, and G.A. Slack, *Better Ceramics Through Chemistry*, Vol 2, 1986, p 373–382

139. R.L. Woodin, D.S. Bomse, and G.W. Rice, *Chem. Eng. News*, Vol 68 (No. 51), 1990, p 20–31

140. Z.A. Munir, *Am. Ceram. Soc. Bull.*, Vol 67, 1988, p 342–349

141. S. Deevi and Z.A, Munir, *J. Mater. Res.*, Vol 5, 1990, p 2177–2183

Characterization of Ceramic Powders

S.G. Malghan, Ceramics Division, National Institute of Standards and Technology
A.L. Dragoo, U.S. Department of Energy

THE PHYSICAL AND CHEMICAL characteristics of ceramic powders strongly influence their behavior during processing and, thus, impact the microstructure and performance of the ceramic component. For high-technology components in particular, detailed information on powder characteristics is required to achieve adequate control of large-scale production processes. The intended application of the ceramic components partly governs the choice of powder characteristics on the basis of measurement and control. For example, monitoring the silica content in a powder may be required to produce either an electronic oxide ceramic, in which silica would cause a grain-boundary phase with a high resistivity, or a structural silicon nitride ceramic, in which silicates might result in undesirable creep properties. Regardless of which properties are monitored, consistent, reproducible, and calibrated laboratory practice is required to obtain meaningful data.

Because the sintering process is driven by the reduction of particle surface area, measurements of particle size and surface area are important. However, powder particles are complex structures and usually consist of many submicrometer-sized primary particles that have agglomerated into larger, secondary particles. Primary particles comprise one or more grains. Particles in the agglomerates are either bound weakly (soft agglomerates) or strongly (hard agglomerates). A soft agglomerate deforms under compression. Therefore, the identities of the agglomerates within a compact are destroyed and a uniform density results. Hard agglomerates maintain their identities under compaction. Therefore, a discontinuity in structure and density exists between the interior of the agglomerates and the interagglomerate regions. Differences in density and structure between these two regions are very likely to result in different densification rates during sintering.

The characterization of a powder is complicated by the complex shapes of particles, the range of particle sizes, variations in composition from particle to particle, and the presence of more than one phase, either within or between particles. Although the assumption of spherical particles or cylindrical pores may be required to obtain a particle diameter or a pore radius from a measurement, the real nature of the particles should not be forgotten. Because a range of particle sizes is present, it is important to measure not only the mean size, but the particle size distribution as well.

Macroscopic physical and chemical methods treat a powder sample en masse without consideration of the differences between or even within individual particles. Analytical microscopic methods are required to determine such variations.

Because the particles utilized in advanced ceramics are submicrometer-sized, their surface chemistry is a major factor in deciding their interactions. Surface impurities and phases are significant in their effects on processing and the subsequent microstructure.

Thus, the preceding discussion demonstrates that a number of properties are important in the characterization of a given powder. Table 1 summarizes the major physical and chemical characteristics of ceramic powders, many of which are discussed in this article. This article also provides a general description of the major methods of powder characterization, and brief descriptions of their principles of measurement, range of application, and deficiencies. References to more detailed sources are also provided.

Particle Size and Size Distribution

Particle size distribution is affected by the starting powder characteristics, as well as subsequent size reduction unit operations, such as milling and deagglomeration. Therefore, it is often necessary to measure the size distribution at various stages of powder processing. Currently, several techniques are available to measure size distribution. These techniques utilize a variety of measurement principles. Often, instrument inadequacy can be secondary to the sample preparation steps.

Steps in Size Analysis

The major steps in the measurement of particle size distribution are shown in Fig 1. To obtain an accurate and reproducible measurement, each step should be followed meticulously.

Sampling from Powder Lot. Procedures that are applicable to the collection of representative samples are different than those used with dry powders and slurries. However, in either case, probabilistic sampling is the preferred method. According to Gy (Ref 1), sampling is distinguished by the selection process:

- *Probabilistic*. All elements of the lot are submitted, along with a probability of being selected
- *Nonprobabilistic*. Rather than being based on probability, the selection can be deterministic, for example, as in grab sampling

Both sampling and analysis are error-generating steps, with the consequence that the overall estimation error (OE) is the sum of the total sampling error (SE) and analytical error (AE). Sampling and analysis involve several preparation stages, alternated with selection stages, all of which can potentially generate errors. Therefore, total overall estimation error is represented by:

$$OE = \sum_n (PE_n + SE_n) + AE_n \qquad (Eq\ 1)$$

where PE_n and SE_n are preparation and selection errors, respectively, of each stage n ($n = 1,2,...$). Various errors described by Allen and Davies (Ref 2) are shown in Fig 2. Relative particle sizes and errors depend on specific powders. For submicrometer-sized ceramic powders, dispersion-related errors appear to dominate. An international round robin study (Ref 3), in which five ceramic powders were examined by various techniques for particle size distribution measurement, presents the same conclusion.

Wetting and Dispersion. A chosen solvent should completely wet the particle surface. The powder dispersion includes

Table 1 Characteristics of ceramic powders

Physical
Grains (primary particles)
    Size and shape
Agglomerates (secondary particles)
    Size and shape
    Porosity
        Amount
        Size
        Shape
    Particle contact
        Coordination
        Strength
    Density
    Specific surface area
    Permeability
    Compatibility
    Flowability
Chemical
Bulk composition
    Major elements (1 to 100%)
    Minor elements (10 ppm to 1%)
    Trace elements (<10 ppm)
    Inorganic species
    Organic species
    Water and other volatiles
Phases
    Crystallinity
    Amorphous
Surface composition
    Major elements
    Minor elements
    Trace elements
    Inorganic species
    Organic species
    Water
    Phases

deagglomeration and formation of a stable suspension. Of the several methods used for deagglomeration (ultrasonic probe/bath, stirring, tumbling), ultrasonication with a probe is the most effective in achieving the separation of particles held together by weak forces. An optimum level of power input to

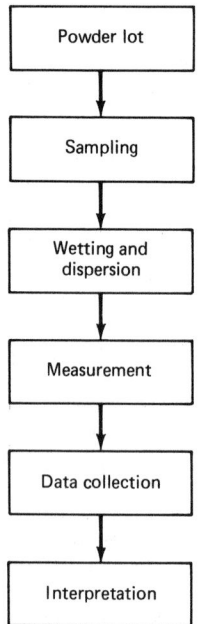

Fig 1 Steps in size distribution analysis

Powder lot

Sampling

Wetting and dispersion

Measurement

Data collection

Interpretation

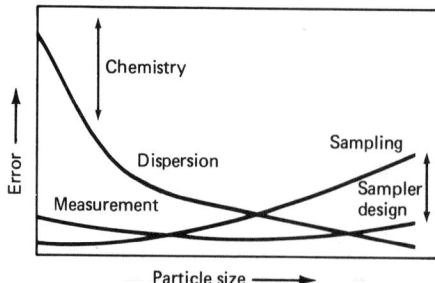

Fig 2 Dispersion error related to particle size

the suspension is required to achieve enhanced data reproducibility. The probe diameter, volume of suspension, power input, time of ultrasonication, and rate of power input are some of the parameters that affect deagglomeration.

The surface chemistry of the solvent-powder interface also controls the preparation of a stable suspension. A repulsive force of sufficient strength to overcome attractive forces between the particles in order to disperse them is required. The major attractive force that hinders dispersion is the van der Waals type. Electrical double layer forces are the counterforces that can effectively provide repulsive forces. The optimum range of electrical double layer forces in polar liquids is provided by high surface charge, moderate electrolyte concentration, and an adsorbed layer of polyelectrolyte of an appropriate type and concentration. A combination of electrical double layer repulsion and steric stabilization has been found to be highly effective for dispersing powders in polar solvents. In nonpolar solvents, steric stabilization alone is often sufficiently effective (Ref 4), and represents the only choice available.

Measurement, Data Collection and Interpretation. There are two basic methods of defining particle size. The first method is to inspect the particles and make actual measurements of their dimensions. Microscopic techniques, in principle, measure many dimensional parameters. The second method utilizes the relationship between particle behavior and its size. This often implies an assumption of an equivalent spherical size. Techniques have been developed using suitable relationships. For example, sedimentation is used to measure size according to Stokes' law, and light scattering is used to measure the size according to the Fraunhofer-Mie theories. Because the physical parameter measured is determined by the relationship used, the particle size range to which a technique is sensitive is also determined by the applicability of the physical law.

Methods of Analysis

Within the general principles of measurement, data collection, and interpretation, as described in this article, hundreds of particle

sizing methods have been developed and reported (Ref 2, 5, 6). Table 2 summarizes methods that are commonly employed, many of which are described below.

Sieving measures the minimum square aperture through which the particles can pass. A variety of sieve apertures are currently used, the most popular of which are the ASTM standard E 11–61 and the American Tyler Series. The laboratory sieving operation can be carried out wet or dry, either for a fixed period of time or until less than 0.2% of the sample passes through the mesh in any 5-min sieving period. Electroformed micromesh sieves have extended the size range to 5 μm (200 μin.). Several methods of wet sieving using micromesh sieves are available, such as rinsing, vibration, reciprocating action, vacuum, and/or ultrasonics (Ref 7). Periodic calibration of the sieves is carried out to determine the difference between the true and the nominal aperture size, and thereby promote accurate sieving (for example, ASTM E 11–87 method by microscopic examination).

Field Scanning Methods. Optical and electron microscopy are the most widely used methods in which the individual particles are directly observed and measured.

Optical Microscopy. Although optical microscopy is used to examine particles in the range of 150 to 1 μm (6000 to 40 μin.), its limitation is a small depth of focus, which is 10 μm (400 μin.) at 100× and about 0.5 μm (20 μin.) at 1000× (Ref 5). An acceptable procedure consists of placing a small, well-dispersed, representative sample on a slide and measuring the diameters of particles using a calibrated filar micrometer eyepiece. Manual counting methods are tedious. Automatic counting or quantitative image analysis techniques have advanced significantly (Ref 2, 8). The image of a particle seen in a microscope is two dimensional. From this image, an estimate of the particle size has to be made.

The image analyzers accept samples in a variety of forms: photographs, electron micrographs, and direct viewing. The particles are scanned by a video camera, displayed on a console, and electrical information that describes the sample is passed to a detector. Examples of the parameters measured are particle count, size distribution either by area or by some statistical diameter; projected length and intersect count; and chord-size distribution.

Transmission Electron Microscopy (TEM) (Ref 8, 9). The TEM produces an image on either a fluorescent screen or a photographic plate by means of an electron beam for particles in the range of 5 to 0.001 μm (200 to 0.04 μin.) at magnifications up to 100,000×. When preparing samples for TEM examination, the fact that all the matter is opaque to electrons must be considered. Hence, particles are examined after being either deposited on or incorporated into a very thin membrane. The accuracy of size measurement using TEM is within ±5%. The size distribution

Table 2   Methods of particle size analysis and nominal size ranges

Method	Nominal particle size		Measurement diameter
	μm	mil	
Sieving			
Dry	>10	>0.4	Geometrical
Wet	>2	>0.08	
Field scanning			
Optical microscopy	0.5–1000	0.02–40	Image
Electron microscopy	0.01–10	4×10^{-4}–0.4	
Stream scanning			
Resistivity	0.05–500	2×10^{-3}–20	Dynamic/Stokes
Optical	1–500	0.04–20	
Ultrasonic attenuation	100–1000	4–40	
Column hydrodynamic chromatography	0.1–1.0	4×10^{-3}–0.04	
Sedimentation field flow fractionation	0.01–1.0	4×10^{-4}–0.04	
Laser Doppler velocimetry	0.01–3.0	4×10^{-4}–0.12	
Gravity sedimentation			
Pipette	1–100	0.04–4	Stokes
Photoextinction	0.5–100	0.02–4	
X-ray absorption	0.1–130	4×10^{-3}–52	
Centrifugal sedimentation			
Photoextinction	0.05–100	2×10^{-3}–4	Dynamic/Stokes
Mass accumulation	0.05–25	2×10^{-3}–1	
X-ray absorption	0.1–5	4×10^{-3}–0.2	
Other			
Gas absorption	0.005–50	2×10^{-4}–2	Equivalent spherical
Mercury intrusion	0.01–200	4×10^{-4}–8	
Gas permeability	0.1–40	4×10^{-3}–16	
Cascade impaction	0.05–30	2×10^{-3}–12	
Brownian motion	0.01–3	4×10^{-4}–0.12	Geometrical

obtained by TEM may differ from that obtained by optical microscopy, due to the higher resolution of the TEM.

Scanning Electron Microscopy (SEM) (Ref 9). In SEM, a fine beam of electrons of medium energy (5 to 50 keV) scans across the sample in a series of parallel tracks. These electrons interact with the sample, producing secondary electrons, back-scattered electrons, cathodoluminescence, and x-rays. Each of these signals can be detected and displayed on the screen of a cathode-ray tube or recorded on photographic films. Samples as large as 25 by 25 mm (1 by 1 in.) can be accommodated, and parts can be viewed at magnifications from 20 to 100,000×, with resolutions of details finer than 20 nm (800 μin.). The depth of focus is 300 times that of an optical microscope. Several current methods of particle size analysis can be adapted to the measurement of images in SEM photographic records. There is also an active interest to develop automated analysis techniques that use three-dimensional image presentation.

Electrical sensing, a technique based on the Coulter principle, determines the number and size of particles suspended in an electrolyte by causing them to pass through a small orifice, on either side of which an electrode is immersed. The changes in resistance as particles pass through the orifice generate voltage pulses with amplitudes proportional to particle volumes. In this method, the size limits may be extended from 400 to 0.6 μm (16 to 0.024 mils), using a multiple tube technique. An advantage of the method is its independence from particle density, which allows powder mixtures to be sized.

Radiation scattering methods utilize the interaction between a radiation beam and an assembly of particles suspended in a fluid medium. When a beam of radiation strikes an assembly of particles, some of it is scattered (diffracted), some is absorbed, and some is transmitted. In a typical laser diffraction based instrument (Ref 8–10), a He-Ne laser emits a beam of 632.8 nm (25.3 μin.), the flux of which is enlarged by a beam expander and radiated upon the particles suspended in the liquid. After being diffracted and dispersed by the particles, the laser beam passes through the condenser lens, and its image is formed at the photocell detector located at the focal point of the lens (Fig 3). The diffraction pattern that appears at the detector is a pattern of light and dark concentric rings that corresponds to the particle size distribution. By using both the Fraunhofer diffraction and Mie scattering theories, the intensity values are then used to calculate the particle size distribution of the sample. The scattered radiation includes the diffracted, refracted, and reflected parts. The absorbed radiation is transmitted at a longer wavelength and is not picked up by the detecting device. Several different devices

(Ref 11) are available that use these basic principles.

Microtrac (Ref 12) is one of the highly successful instruments that uses the low-angle, forward-scattering optical system. The size distribution is determined by flowing particles through a sample cell that is illuminated by a continuous laser beam. The suspended particles scatter the light, which is collected by a series of lenses and a rotary optical filter. The Fraunhofer plane diffraction pattern for the particles is interpreted with the aid of the specially designed optical filter and a digital microprocessor to yield a complete size distribution. Particle concentrations must be sufficiently high to obtain good counting statistics, but sufficiently low to avoid multiple scattering.

Ultrasonic Attenuation. When ultrasonic energy in the frequency range of 0.1 to 10 MHz is transmitted through a slurry, losses occur due to viscous, scattering, and diffraction effects (Ref 13). The absorption coefficient is a function of frequency, particle size distribution, and volume concentration. In the PSM System-100 (Ref 14), which is based on this principle, careful choice of operating frequencies enables the evaluation of both particle size and solids concentrations.

Sedimentation Methods. The Stokes' equation is a relationship between settling velocity, U_{st}, and particle diameter, D:

$$U_{st} = \frac{(\rho_s - \rho_f)gD^2}{18\eta} \qquad (Eq\ 2)$$

where ρ_s is solid density, ρ_f is fluid density of viscosity η, and g is gravitational acceleration (Ref 6). The particle size distribution may be determined by examining a sedimenting suspension of the powder. Changes with time in the concentration or density of the suspension at known depths are determined, and size distribution is determined from these data (Ref 5).

The Andreasen's Pipette. In the pipette method of size analysis, the concentration changes that occur within a settling suspension are followed by using a pipette to draw off definite volumes. Stokes' law is used to calculate the size distribution.

In photosedimentation, a narrow horizontal beam of parallel light is projected through a suspension at a depth, h, onto a photocell. In a homogeneous suspension, if the particles are allowed to settle, the number of particles that leave the light beam will be

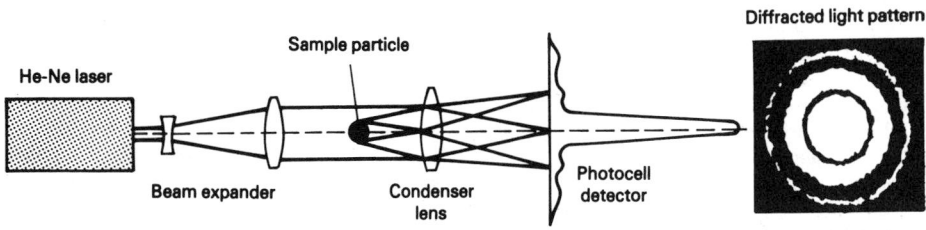

Fig 3   Principle of laser diffraction based particle size distribution

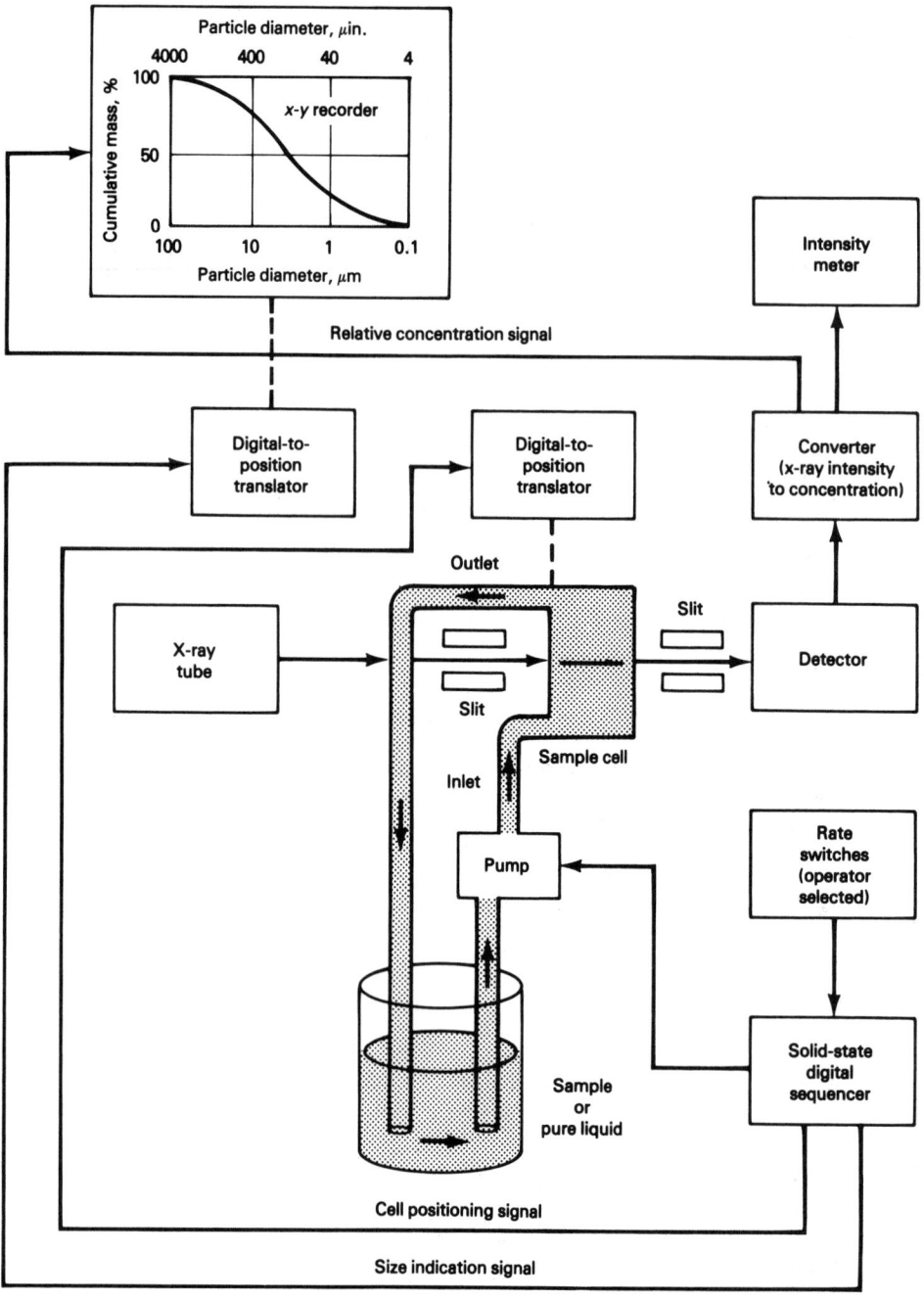

Fig 4 Sedigraph schematic

The Sedigraph (Ref 14) is a well-known instrument based on the gravitational settling and x-ray absorption analysis. This instrument automatically presents the cumulative percentage frequency, and the sedimentation cell is driven in such a way that the concentration is recorded directly as a function of the Stokes' diameter. The x-rays from a 22.6 keV source are collimated into a narrow beam and pass through a fixed thickness of suspension. The sedimentation cell (Fig 4) is filled and emptied with the suspension by a built-in pump. The transmitted radiation is detected as pulses by a xenon-filled scintillation detector. The counting electronics give a voltage that is proportional to the powder concentration. Particle size distribution is calculated by the application of Stokes' law to these data.

Specific Gravity Balance. The changes in density within a settling suspension are followed by using a specific gravity balance. This instrument comprises a bob on each arm of a beam balance, one bob being immersed in a clear fluid and the other in the suspension. The depth of immersion of bobs can be varied, and the change in buoyancy counterbalanced by means of solenoids that are connected to a pen-recorder. From the trace on the pen-recorder, the particle size distribution can be determined (Ref 5).

Centrifugal Techniques. Because the centrifuging process speeds up the rate of settling of particles, it overcomes one of the serious disadvantages of the gravitational sedimentation techniques, which take an unduly long time for particles below 5 μm (200 μin.). In addition, most sedimentation devices suffer from the effects of convection, diffusion, and Brownian motion. Centrifugal sedimentation may be carried out by either a two-layer or homogeneous suspension technique, either cumulatively or incrementally. The two-layer techniques can give rise to streaming, a phenomenon most likely to affect cumulative analyses, in which the fraction sedimented against time is determined (Ref 5).

Photoncorrelation Spectroscopy (PCS). Instruments are available that utilize laser light scattering and digital electronics to measure particle size (Ref 15). This spectroscopy method is used to extract information on the dynamics of a system of scatterers by the calculation of the correlation function of the scattered radiation. The correlation function of a scattered light signal with itself, that is, the autocorrelation function, is the Fourier transform of the power frequency spectrum of the input signal. This information is analyzed to quantify the dynamics of the scattering process that generated the signal. Some of these instruments have the capability to measure zeta potentials of the particles, by the application of laser Doppler velocimetry.

Hydrodynamic Chromatography. In this method (Ref 16), particle separation occurs due to hydrodynamic interaction between the particles and velocity profile near solid sur-

balanced by the number entering it from above. However, after the largest particle, D_m, in the suspension has fallen from the surface to the measurement zone, the emergent light flux will begin to increase, because no more particles of this size will enter the measurement zone from above. Hence, the concentration of particles in the light beam at any time, t, will be the concentration of particles smaller than D_{st}, where D_{st} is given by the Stokes' equation (Ref 2). The particle size distribution is calculated from the relationship between the attenuation of the light beam and the projected surface area of the particles.

X-Ray Sedimentation. Similar to the use of a light beam, x-rays can be used in combination with the gravity settling of particles. In this case, the x-ray density is proportional to the weight of powder in the beam:

$$I = Io \exp(-BC) \qquad \text{(Eq 3)}$$

where B is a constant, and C is the concentration of powder in the beam. The x-ray density, X_d, is defined as:

$$X_d = \log_{10}\left(\frac{I}{Io}\right) \qquad \text{(Eq 4)}$$

faces. Two basic techniques of hydrodynamic chromatography are utilized: capillary tube and packed column. The separation of particles in dilute colloidal suspensions is carried out by passing the particles through a column containing a long capillary tube or packed uniform spheres. As the particles exit the separation zone, their concentration and size are measured by a detector. A significant advantage of this method is its ability to separate and size the particles simultaneously.

Sedimentation field flow fractionation (Ref 8, 16) combines the principles of chromatography and centrifugation to separate and characterize particles in the range of 10 nm to 1 µm (0.4 to 40 µin.) (Ref 17). The particles in a suspension are separated in a flat, ribbon-like channel of elongated hexagonal shape, which is contained in a continuous flow centrifuge rotor. Under equilibrium conditions, the separated particles are eluted. Small, lightly retained particles elute first, followed by large, more tenaciously held particles. Eluted particles are counted by an ultraviolet (UV) detector.

Morphological Analysis

The whole ensemble of properties, that is, size, shape, and texture, is often collectively termed "morphological analysis." The shape of a particle is defined as the pattern of all of the points of the surface or profile of the particle image. The particle texture is defined as the pattern of all of the points of the image of the particle. These definitions lead to the development of mathematical descriptors (Ref 18–21). The relationships between the particle size, shape, and textural descriptors and the fundamental physical and chemical properties of powder, their mode of formation, and their behavior are the focus of morphological analysis.

Most shape analysis is performed using microscopic methods, either optical or electron, in which the image is recorded with a video detector and processed with a computer. Sophisticated software enables a variety of shape factors to be derived, compiled, and statistically analyzed. If the microscope, video recording, and computer analysis are automated, then it is possible for a statistically significant number of samples and particles to be processed with a reasonable amount of operator time and effort.

It must be borne in mind that regardless of the microscopic method and its sophistication, all of these methods survey the particles lying on a surface. Consequently, only two dimensions of a particle are viewed, with respect to its most stable position on the surface. The shortest dimension is not observed. However, by directionally coating samples on a substrate at a low angle with respect to the horizon of the surface, it is possible to shadow the particles in a manner that reveals the shortest dimension when the particles are viewed from above.

A qualitative visual analysis of particle shape is by far the quickest approach and may be satisfactory for nonexacting work. However, qualitative descriptions are highly subjective and usually have little statistical validity because an insufficient number of particles and samples are examined. Definitions of particle shape are presented in Table 3.

There are three basic approaches to quantitative shape description: shape factors, Fourier analysis, or other special functions (Ref 18), and fractal analysis. There are many schemes within these three approaches.

Beddow and his collaborators have developed the Fourier and other special functions to morphological analysis (Ref 18–20). The methods of the calculus of variations were applied to the one-dimensional, two-dimensional, and three-dimensional problems of morphological analysis. The result of these efforts is the derivation of Luerkens equations (Ref 18), shown in Table 4. These equations indicate the derivation of the boundary functions of the problems.

Beddow (Ref 18) has shown that the solution using the variational principle for a one-dimensional case is a Fourier equation, which is useful for shape analysis. For the two-dimensional case in Table 4, the gray level is defined in terms of the radius vector, r, and angle, θ. It is a combination of two functions, a Fourier and a Bessel. This relationship is employed in the surface analysis and texture morphology analysis. The three-dimensional solution in Table 4 defines R in terms of the two angles, θ and φ, by means of a Fourier series and associated Legendre functions, $P_j^n(\phi)$. This equation is applicable to analysis of particle behavior problems.

Kaye (Ref 21) has investigated fractal analysis as a means of characterizing the projected area and perimeters of particles. This analysis yields a noninteger dimension (Ref 8) for typical particles, which may be related to the process by which the particles were formed.

Specific Surface Area

The average surface area, $\langle S \rangle$, of a particle in a powder composed of deagglomerated particles that have regular shapes and are in a narrow size distribution is closely related to the mean particle size. If the particles are idealized as spheres, then the specific surface area, S_w (the ratio of the total surface area, S_t, and the mass, M, of the powder) provides an estimate of the equivalent spherical diameter, D, because:

$$S_w = \frac{S_t}{M} = \frac{N\langle S \rangle}{N\langle V \rangle} = \frac{6}{\rho D} \qquad \text{(Eq 5)}$$

where N is the number of particles, $\langle V \rangle$ is the average volume of a particle, $\langle S \rangle$ is the average surface area of a particle, and ρ is the density of a particle. Thus, the equivalent spherical particle diameter of a powder with a specific surface area of 10 m^2/g (50 × 10^3 ft^2/lb) and a density of 3 g/cm^3 is 0.2 µm (8 µin.).

However, the particle diameter calculated from a surface area measurement can exceed that obtained by either sedimentation or light scattering by a factor of 10 or more because these methods measure the size of agglomerates, whereas surface area methods measure total accessible surface. Some particle surface areas of tightly bound clusters of primary particles may be excluded from detection. Therefore, the equivalent spherical diameter calculated using Eq 5 will be less than the size of agglomerates, but greater than the size of the primary particles.

Two common approaches to the determination of specific surface area are to determine gas adsorption according to the theory of Brunauer, Emmett, and Teller (Ref 22) and to conduct permeametry. Two other methods, the relative Harkins-Jura method and the absolute Harkins-Jura method, are included briefly in the discussion below. More detail on these methods can be found in books by Lowell and Shields (Ref 23) and Allen (Ref 5).

Gas Adsorption. The specific surface area of a powder (the adsorbent) can be obtained from a determination of the weight, W, of vapor (the adsorbate) adsorbed as a function

Table 3 Particle shapes

Acicular	Needle-shaped
Angular	Sharp-edged or with a roughly polyhedral shape
Crystalline	Freely developed in a fluid medium to have a geometric shape
Dendritic	Branched crystalline shape
Fibrous	Regularly or irregularly thread-like
Flaky	Plate-like
Granular	Equidimensional, approximately irregular shape
Irregular	Lacking any symmetry
Modular	Rounded, irregular shape
Spherical	Global shape

Note: Definitions based on British Standard 2955

Table 4 Luerkens equations

Dimension	Equation	Application
1-D	$R(\theta) = a_0 + \Sigma a_n \cos n\theta + b_n \sin n\theta$	Shape analysis
2-D	$G(r,\theta) = \sum_n \sum_j (a_{nj} \cos n\theta + b_{nj} \sin n\theta) J_j(r)$	Surfaces, texture
3-D	$R(\theta,\phi) = \sum_n \sum_j (a_{nj} \cos n\theta + b_{nj} \sin n\theta) P_j^n(\phi)$	Particle behavior

of vapor pressure, P, of the adsorbate vapor at constant temperature. The curve that represents this relationship is called an adsorption isotherm, an example of which is shown in Fig 5, representing the adsorption of N_2 by silicon powder (Ref 24). As an alternative to the determination of W, the volume, V, of gas adsorbed can be determined, for example, with a calibrated hot-wire thermal conductivity gage. Brunauer *et al.* (Ref 25) discussed five typical adsorption isotherms that have been experimentally observed. Determination of the specific surface area requires a theoretical model of the adsorption process, such as the BET model.

Brunauer, Emmett, and Teller (Ref 22) modeled the equilibrium between a vapor and adsorbate in which multiple layers of vapor molecules are adsorbed without complete adsorption of underlying layers. Second and higher layers were assumed to have properties of the liquid state. The resulting so-called BET equation can be expressed as:

$$\frac{1}{W(P_0/P - 1)} = \frac{1}{W_m C} + \frac{C - 1}{W_m C}\left(\frac{P}{P_0}\right) \qquad \text{(Eq 6)}$$

where W_m is the weight of one monolayer of adsorbed vapor and C is the BET constant, which based on N_2 as the adsorbate, is in the range of 50 to 300 for many adsorbents (Ref 23). The total surface area is given by:

$$S = \frac{W_m \cdot N_A \cdot A}{M} \qquad \text{(Eq 7)}$$

where N_A is Avogadro's number, A is the cross-sectional area of an adsorbate molecule, and M is its molecular weight. For nitrogen, which is often used as a standard adsorbate, a cross-sectional area of 0.162 nm^2 (2.6×10^{-10}) at its boiling point is a commonly accepted value.

The most accurate and reproducible surface area measurements are obtained on clean, dense powders that have low porosity and with adsorbates that are weakly physically adsorbed. Nitrogen and the inert gases are often suitable adsorbates. Because adsorption of a vapor will occur on any exposed surface of a powder, the total surface area of a powder will be greatly enhanced if many pores open on the outside of the particles. Other adsorbed vapors, such as water or carbon

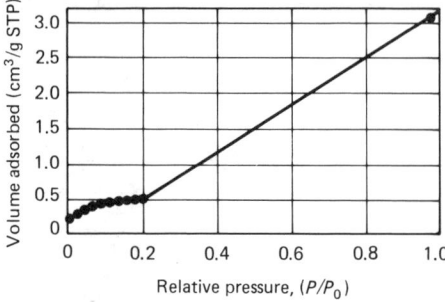

Fig 5 BET multipoint data plot of volume of gas adsorbed versus relative pressure

dioxide, will modify the adsorptive properties of the surface or may desorb nonuniformly during the measurements, leading to nonreproducible and inaccurate results.

Typical samples sizes are 1 to 5 g (0.04 to 0.18 oz), although samples as small as 0.1 to 0.5 g (0.004 to 0.018 oz) can be used with certain instruments or sample tubes. Degassing of the sample is usually carried out in a vacuum oven or a special stage of the instrument at temperatures greater than 100 °C (212 °F) for times up to 1 h or longer. The degassing conditions must be established by practice.

BET surface area measurements are accomplished by either the singlepoint method or the multipoint method. For reasonably high values of the BET constant, C, the intercept of Eq 6, is small compared to the slope. Therefore, the total surface area is given approximately by:

$$S = W\left(1 - \frac{P}{P_0}\right)\frac{N_A}{M} A \qquad \text{(Eq 8)}$$

A singlepoint measurement is carried out at a relative pressure greater than that required to produce a monolayer coverage of the surface. For example, $P/P_0 = 0.3$ is often adequate to obtain good agreement with multipoint measurements and to avoid condensation in pores for all but microporous samples.

Multipoint BET measurements are usually carried out at three or more relative pressures in the range of $0.05 < P/P_0 < 0.35$. Linear behavior is usually observed in this range. An example of a BET plot is shown in Fig 6 (Ref 24) for the data given in Fig 5. Deviation from linear behavior can indicate condensation in pores, chemisorption, or nonhomogeneity of the surface. Measurements can be performed under static or dynamic gas flow conditions. For dynamic flow conditions, N_2/He gas mixtures are frequently used.

Permeametry. Permeation of a fluid through a packed bed of a powder is described by Darcy's law (Ref 26), which states that the average velocity, u, measured over the whole area of the bed is directly proportional to the pressure difference, ΔP, and inversely proportional to the thickness, L, of the bed.

If the viscous flow is imagined to occur along tortuous channels through many small capillaries, then permeation can be described by the Carmen-Kozeny equation (Ref 27, 28):

$$u = \epsilon\left(\frac{L}{L_e}\right)\frac{\epsilon^2}{(1 - \epsilon)^2}\frac{V_s^2}{2\eta S_W^2}\frac{\Delta P}{L_e} \qquad \text{(Eq 9)}$$

where L_e is the average length of a capillary through the bed, ϵ is the porosity (volume fraction of voids), V_s is the volume of solids, and η is the viscosity of the fluid. The value (L/L_e) is replaced by a tortuosity factor, $\sqrt{k_1}$, which can be theoretically estimated or experimentally determined for ideal cases.

A determination of the specific surface area requires the following input values:

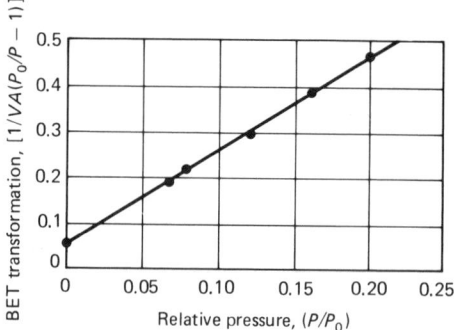

Fig 6 BET transformation of data in Fig 5

- ΔP from measurement of the input pressure while venting to atmospheric pressure
- ρ, the true powder density
- u, the volumetric flow rate divided by the cross-sectional area of the powder bed
- ϵ, (volume of powder bed minus the volume of powder) divided by volume of the powder bed

Two basic designs of permeameters are the constant pressure and constant volume types. Perhaps the best known of the constant pressure type instrument is the Fisher sub-sieve sizer. This instrument is designed to yield the surface-volume mean diameter:

$$D_{sv} = \frac{6V}{S} \qquad \text{(Eq 10)}$$

The Blaine apparatus is an example of a constant volume type instrument.

Because the Carmen-Kozeny equation is based on the assumption that the pores resemble a bundle of capillaries, it is subject to the criticism that this model of the pore structure is greatly oversimplified. In addition, a significant number of small pores may not be detected if the pore size distribution is large. Reproducible results may be obtained if the bed porosity is in the range of 0.4 to 0.8 and if the average pore diameter is within a factor of 2 of the largest diameter.

The Harkins-Jura Methods. Two methods were proposed by Harkins and Jura. Their relative method (Ref 29) is based on the adsorption theory of Gibbs (Ref 30) and relates the vapor pressure of liquid film on a surface to the weight of the adsorbed film. Emmett (Ref 31) showed that within the BET range of pressures, linear Harkins-Jura plots are obtained if the BET C-value is between 50 and 250. For the case of nitrogen, Emmett found that if a cross-sectional area of 18.62 $\mathring{A}^2$ is used rather than 13.62 $\mathring{A}^2$, then the Harkins-Jura method with $K = 4.06$ yields the same surface area as the BET method.

The absolute Harkins-Jura method (Ref 32) is based on calorimetric determination of the heat of immersion, H_i, of the solid in the vapor and is independent of the adsorption isotherm. For powder samples, the heat of immersion is proportional to the surface area.

Values of the heat of immersion per unit area have been tabulated for many solids by Gregg and Sing (Ref 33). Determination of surface area by this method may be subject to great errors if the heat of immersion per unit area is not precisely known, or if the purity of the liquid and sample are not carefully controlled. Although this has questionable utility as a routine method, if it is carried out rigorously and with great care, it may provide absolute confirmation of measurements by other procedures.

Density

Theoretical density can be calculated based on the overall atomic composition and the lattice parameters, determined by x-ray diffraction, for each crystalline phase. If more than one crystalline phase is present, then the relative volume percent of each phase must be determined by quantitative x-ray diffraction. If a significant amount of a glassy phase is present, its amount, composition, and density must be estimated.

The bulk density of powders is a measure of the true, or skeletal, density plus closed internal porosity. This measurement is required for application in several particle size analysis methods and for the determination of specific surface area by permeametry.

Measurement of the powder volume is a key to the determination of the powder density. The volume is most often measured with a helium pycnometer because helium will penetrate all open pores. Measurement of the volume of gas displaced provides the determination of the volume of the powder. The largest sample possible should be used to maximize the volume of gas displaced. Frequent calibration of the pycnometer is necessary to provide close control of the instrument. The ASTM D 854–83 procedure, "Standard Test Method for the Specific Gravity of Soils," may be consulted for guidance.

Tap density is a relative measure of the fill density. The tap density is obtained from the volume that a weighed powder occupies after a set number of taps have been applied to the bottom of its container. As such, the tap density provides a measure of the compactibility of a powder. Highly flocculated or agglomerated powders can be expected to retain a large amount of void space and, thus, to show a low tap density.

Measurement of the tap density is carried out with a prescribed weight of powder, often 100 g (3.5 oz), in a graduated cylinder that is positioned on a tapping device for which the number of taps are preset. The measured tap density is strongly dependent on the manner in which the powder is loaded, the number of taps, the force imparted with each tap and the wall effects of the graduated cylinder. For guidance in conducting these measurements, consult ASTM D 3347–86, "Flow Rate and Tap Density of Electrical Grade Magnesium Oxide for Use in Sheathed-Type Electric Heating Elements."

Porosity

Two principal methods used for porosity measurements, mercury porosimetry and gas adsorption (Ref 5, 23) are described below. Two additional methods, nuclear magnetic resonance and small angle neutron scattering, are also discussed briefly.

Mercury Porosimetry. When a nonwetting liquid, such as mercury, is forced into a cylindrical capillary, the interfacial tension results in a force:

$$F_{out} = 2\pi r\gamma \cos \theta \qquad \text{(Eq 11)}$$

opposing the intrusion of mercury into the capillary. The force driving the mercury into the pores is:

$$-F_{in} = P\pi r^2 \qquad \text{(Eq 12)}$$

where P is the applied pressure. Equating these two forces yields the Washburn equation (Ref 34):

$$Pr = -2\gamma \cos \theta \qquad \text{(Eq 13)}$$

The contact angle, θ, for mercury on many materials is between 135° and 142°, with 140° taken as the customary value (Ref 23). The surface tension of mercury is 0.480 J/m^2 (0.032 ft · lbf/ft^2), at room temperature.

Mercury porosimetry data are usually displayed as a plot, or porogram, of the volume of mercury penetrated into the solid, on one axis, versus the applied pressure, on the other axis. The curves are plotted both for the intrusion and extrusion of mercury, for one or more intrusion-extrusion cycles.

Porosimetry measurements are typically made between ambient pressure (about 14 Pa, or 2 psia) and 415 Pa (60 ksia). This corresponds to a range of pore radii from >2000 μm (80 mils) down to about 2 nm (0.08 μin.).

Lowell and Shield (Ref 23) identified five common characteristics of porosimetry curves. These are:

1. For porous powder samples, during intrusion, mercury penetrates the interparticle voids at low pressure and pores within the particles at high pressure
2. All porosimetry curves exhibit hysteresis, that is, at a given applied pressure a greater volume of mercury is present in the sample during extrusion than during intrusion, especially for the first cycle
3. Upon completion of the first intrusion-extrusion cycle, some mercury is retained by the sample so that the porogram curve does not return to the origin, but intersects the volume axis at a finite value
4. Successive intrusion-extrusion cycles will show less hysteresis, so that eventually the intrusion-extrusion loop will close, showing that entrapment of mercury has ceased
5. Each step on the intrusion curve has a corresponding step on the extrusion curve, but at a lower pressure

Hysteresis results from entrapment of mercury in the pores when the applied pressure is reduced. Lowell and Shield (Ref 23) postulated that the entrapment is due to a larger pore potential in narrow pores. The hysteresis also was explained by an "ink-bottle" effect (Ref 35), in which narrow necks of pores physically restrain the extrusion of mercury. Androutsopoulos and Mann (Ref 36) proposed a model in which smaller pores delay the filling of larger pores, which yields a result resembling the ink-bottle effect.

A porogram also yields an estimate of the pore surface area, S_p. This estimate depends on the applied pressure and the intruded volume and is not dependent on a specific model for the pore shape. Data are plotted to show the distribution of a number of pores of a given size versus pore size. The distribution may be expressed as the number of pores per unit pore radius, $D_v(r)$; the number of pores per unit logarithm of the pore radius, $D_v(\ln r)$; or the pore surface area per unit pore radius $D_s(r)$. Modern computer-controlled porosimeters provide automatic analysis of the porogram to yield these various distributions, as well as mean pore radius, total pore volume, and surface area.

Gas Adsorption. In the range of 1.5 to 100 nm (0.06 to 4 μin.), gas adsorption can be used to determine pore size. Adsorption is assumed to be favored in small capillaries due to the overlapping surface potentials, resulting in condensation of vapor up to a capillary radius, r. At a radius, r, condensation and evaporation are equal, and r is assumed to be related to the vapor pressure, P, in the capillary by the Kelvin equation:

$$\ln\left(\frac{P}{P_0}\right) = -\frac{2\gamma\bar{V}}{rRT}\cos \theta \qquad \text{(Eq 14)}$$

where P_0 is the equilibrium vapor pressure of the same liquid over a plane surface, τ is the surface tension of the liquid, $\bar{V}$ is the molar volume of the liquid, and θ is the contact angle between the liquid and the wall. The contact angle is usually taken to be 0° (cos θ = 1), resulting in only a small error in the estimate of r.

Adsorption-desorption isotherms can exhibit hysteresis due to differences in the rates of adsorption and desorption throughout a system of pores of various sizes. Because it can be shown that the desorption isotherm corresponds to a more stable adsorbate condition, it is generally used to determine the pore size distribution, except in the case of bottleneck pores.

In the microporosity range (less than a 1.5 nm, or 0.06 μin., pore radius), the pore dimensions are in the range of molecular dimensions. Adsorption in this range often shows a Langmuir-type isotherm, a characteristic of monolayer adsorption. Adsorption in micropores is discussed by Lowell and Shield (Ref 23).

Research Methods. Small-angle neutron scattering (SANS) and nuclear magnetic resonance (NMR) have been used for pore size measurement. These methods require either a nuclear reactor with a SANS beamline or an NMR spectrometer, respectively, which are generally not available to many analysts.

SANS techniques have been used to measure pore sizes in the range of 1 to 10^4 nm (0.04 to 400 μin.) and volume fractions of 1 to 50% (Ref 37). The NMR technique was used by Smith *et al*. (Ref 38) to measure porosity by observing the shift in the spin-lattice relaxation for water adsorbed on pore surfaces relative to bulk water in the pores. The method yields an estimate of the pore volume to surface area ratio.

Chemical Composition

Although the importance of measuring and controlling the chemical composition of starting powders is widely recognized, accurate determination of chemical species remains problematic. Chemical analyses requiring decomposition/dissolution of the material are complicated by the general chemical refractiveness of the compounds. Therefore, failure to recover all of the component as a reaction product and contamination by the apparatus influence the results.

In the analysis of ceramic powders, there is major interest in the determination of bulk composition and impurity elements (Ref 39, 40). A generalized listing of the methods available for powders analysis is shown in Table 5. These methods are described below with respect to the range of applicability, limitations, and special requirements.

Table 5 Methods for bulk chemical analysis of ceramic powders

Major chemical components (10–50 wt%)
X-ray fluorescence spectroscopy
Atomic absorption spectroscopy
Inductively coupled plasma emission spectroscopy
Direct current plasma emission spectroscopy
Arc emission spectroscopy
Gravimetry
Combustion
Kjeldahl
Electrochemical
   Coulometry
   Selective-ion potentiometry
   Potentiometric titration
   Argentometric
Neutron activation analysis
Mass spectrometry
Nonmetallic impurities (0.01–1.0 wt%)
Combustion
Electrochemical
Fast neutron activation analysis
Ion chromatography
Mass spectrometry
Metallic impurities (0.01–1.0 wt%)
Electrochemical
Atomic absorption spectroscopy
Inductively coupled plasma
Direct current plasma
Optical emission spectroscopy

Source: Ref 3, 24, 40–44

X-ray Fluorescence (XRF) Spectroscopy (Ref 41). In the XRF method, the characteristic x-ray fluorescence of elements in a sample is excited, and the resulting x-ray intensities are analyzed. The XRF spectroscopy method is primarily used for both semiquantitative and quantitative determination of major elemental constituents and impurities with atomic numbers greater than 9. Rapid, multielement determinations on a single sample can be carried out. Reference standards are required for the conversion of x-ray intensities to absolute concentrations. Though XRF can be carried out nondestructively, it is preferred that either sample dissolution or fine powders pressed into pellets are used.

Atomic Absorption Spectroscopy (AAS) (Ref 41). In AAS, radiation from an external light source, emitting the spectral lines that correspond to the energy required for an electronic transition from the ground state to an excited state, is passed through a flame, into which the analyte has been atomized. The absorption of the radiation from the light source depends on the population of the ground state, which is proportional to the solution concentration sprayed into the flame. Absorption is measured by the difference in transmitted signal in the presence and absence of the test element. The AAS analyzes at least 67 metallic elements.

Inductively Coupled Plasma (ICP) (Ref 42). The ICP uses the excitations of atoms in a plasma to obtain their characteristic emissions. This is one of the most widely used techniques for quantitative and semiquantitative elemental analysis. The powder is dissolved, and the resulting solution is injected as an aerosol into the axial channel of a toroidal argon plasma. Powder dissolution and solution preparation are critical steps.

Direct Current Plasma (DCP) (Ref 42). The DCP emission spectroscopy utilizes a dc plasma jet as the excitation source. An analyte solution is nebulized, similar to that in ICP, or a finely ground powder is dispersed in a gas stream and carried into the plasma jet. In both ICP and DCP, the emission spectra are analyzed by photodetection.

Atomic Emission Spectroscopy (AES) (Ref 43). Common methods of excitation use an ac arc, a dc arc, a flame, and an ac spark. The function of the excitation source is to introduce the sample into the source in a vaporized form and to excite electrons in the vaporized atoms to higher energy levels. These excited electrons tend to return to the ground state, and, in so doing, emit the extra energy as a photon of radiant energy. The resulting spectra are analyzed for quantitative determination.

Gravimetry (Ref 44). This method refers to the determination of a component by isolating it in a solid form that can be weighed, thus providing an independent determination of that component. Estimating the amount of a component by subtracting the amounts of all other components from the total weight of the sample does not constitute an independent measurement, although such an estimate may be useful for the purpose of comparison. Much depends on the procedure required to separate the component and to precipitate or collect it in such a manner as to exclude other chemical species.

Combustion (Ref 44) or fusion is used when the chemical component can be liberated in a gaseous form from the sample by a decomposition reaction. Usually, a reactive flux is used. The quantity of gaseous product obtained is determined from a property of the gas. This method is widely used for the analysis of C, O, N, and S in fine powders.

Kjeldahl (Ref 24) is a special method used for the determination of N by neutralization titration. Nitrogen in the powder is liberated as ammonia, which is recovered in an acid or a base. The concentration of ammonia is determined by neutralization-titration with either a standard strong acid or a base.

Electrochemical (Ref 44). The major techniques that constitute electrochemical reactions for chemical analysis are coulometry, ion-selective potentiometry, potentiometric titration, and argentometric titration. Coulometric titrations determine the quantity of electricity passed to an electrode by a chemical reaction involving the ion of interest. This method is particularly suited for the milligram to microgram concentrations of trace and volatile impurities (Cl, Br, I, and others).

In ion-selective potentiometry, the activity of a specific ion (thermodynamically effective free-ion concentration, not the total concentration) is obtained by direct measurement of an electrode potential. Selective-ion electrodes are utilized for this determination. Interference by other ions in the solution is a problem.

In potentiometric titration, the solution containing the unknown constituent is titrated against a standard solution until the equivalence point is reached. Changes in the electromotive force of a cell with the addition of a titrant can be used to locate the equivalence point of a reaction. The concentration of the unknown is calculated from the volume of the standard solution added.

Argentometric titrations are a specific type of potentiometric titration in which silver-related equilibria are employed. Two specific equilibria are associated with the solubility of AgCl and with the formation of an $Ag(CN)_2$ complex. Ion-selective electrodes can be used in conjunction with the titration process to determine the Ag^+ or an anion concentration.

Neutron Activation Analysis (Ref 44). Most elements, when irradiated by thermal neutrons, give rise to a radioactive species that has the same atomic number but is one mass unit greater than its progenitor through an (n, γ) reaction. Immediately after neutron capture, a gamma ray is emitted, the energy of which is equal to the neutron binding energy plus the kinetic energy of the neutron. The chemical identification is carried out by analyzing the gamma-ray spectrum. This is a highly sensitive method for impurities as low as 10^{-3} to 10^{-7} µg/g.

The mass spectrometer (Ref 44) produces charged particles consisting of the parent ion and ionic fragments of the original molecule, and sorts these ions according to their mass/charge ratio. The mass spectrum is a record of the numbers of different kinds of ions that are characteristic of a given compound, and the relative numbers of each. Sample (solids and liquids) weight requirements are from a few milligrams to submicrograms.

Phase Composition

X-ray powder diffraction (XRD) is the principal method used to identify the phase composition of crystalline materials. Recent developments in this field are computer-automated diffractometers and processing of x-ray diffraction spectra. Automation has provided improved diffractometer performance, as well as a reduction in operator time, and computer processing of x-ray data has greatly facilitated phase identification and quantitative analysis of x-ray data. In addition, crystallite size and residual strain can be determined from the XRD data.

The XRD method is applied to solid materials in the form of powders and densified ceramics. However, a well-mixed powder sample that is prepared in a manner to minimize preferred orientation of grains is necessary for enhanced accuracy in quantitative analysis. Different approaches for the treatment of data have been devised.

In an x-ray diffractometer that has the Bragg-Brentano geometry (Fig 7), an x-ray beam from a source at point 1 irradiates a specimen at point 2 with a divergent, collimated, and sometimes monochromated x-ray beam. These x-rays are diffracted by the crystalline planes of the sample and are received by a detector that is placed beyond the slits, point 3. The angles through which the x-rays are scattered are characteristic of the crystal structure, and the intensity of the scattered radiation is characteristic of both the atomic composition and the atomic packing of the diffracting planes of the atoms.

For quantitative x-ray analysis, sample preparation is enhanced by minimizing the effects of preferred orientation, texture, particle size broadening, and other material effects. Preferred orientation of grains causes erroneously enhanced or reduced intensities with respect to different crystallographic orientations. The presence of coarse particles (>10 µm, or 400 µin.), can reduce or distort intensities due to the detection of a statistically insufficient number of diffracting grains. For grains that are under 0.1 µm (4 µin.), broadening of diffraction lines occurs. Spray drying of powders eliminates many of these problems.

Quantitative analysis should be carried out by calculating integrated line intensities, which is preferable to the use of peak heights. However, integrated intensities often are not used due to the additional experimental difficulty of performing the integration, a task that is often encumbered by overlap of significant peaks. Instead, relative concentrations of two or more phases are estimated from heights of one or more isolated peaks of each of the phases present.

Automated data processing can now be carried out. Rietveld refinement methods are also available and yield relative concentrations of minor crystalline phases, in addition to material parameters (strain).

Amorphous phase content cannot be determined by routine XRD techniques. Recently, some success has been claimed by the application of Rietveld refinements. Yet another technique for the determination of amorphous phase content is NMR spectroscopy. Magic angle spinning is employed to quantitatively determine the amorphous phase content.

Surface Composition

The surface composition of ceramic powders is determined by a number of techniques, including x-ray photoelectron spectroscopy, Fourier transform infrared spectroscopy, and micro-Raman spectroscopy. The primary purpose of these methods is to determine quantitative and qualitative data on surface species of powders. These data are instrumental in determining surface interactions in dispersion and powder processing. In

addition, when the powder is in contact with a liquid, the electrokinetic techniques are used to determine interface properties.

X-ray photoelectron spectroscopy (XPS) is also known as electron spectroscopy for chemical analysis (ESCA) (Ref 44). Powder samples are irradiated by monoenergetic x-rays, and the emitted electrons are analyzed by passing through an electron spectrometer. The ESCA method is primarily suitable for surface analysis because the ejected electrons are easily stopped by even a minute thickness of solid. All elements except hydrogen can be identified, and the oxidation state and bonding of the element is usually determined.

Fourier Transform Infrared (FTIR) Spectroscopy (Ref 44). IR spectroscopy (wavelengths from 0.7 to 500 µm, or 0.028 to 20 mils) involves the twisting, bending, rotating, and vibrational motions of atoms in a molecule. Upon interaction with IR radiation, portions of the incident radiation are absorbed at particular wavelengths. The multiplicity of vibrations that occur simultaneously produces a highly complex absorption spectrum, which is uniquely characteristic of the functional groups that compose the molecule, and the overall configuration of the atoms as well. The FTIR spectrometer differs from the interferometer in its basic configuration, and rather than dispersing polychromatic radiation, the FTIR spectrometer performs a frequency transformation.

Raman Spectroscopy (RS) (Ref 44). A unique feature of this method is that each line has a characteristic polarization, and polarization data provide additional information related to molecular structure. The Raman effect is important in the elucidation of molecular structure, and for locating various functional groups or chemical bonds in molecules, as well as for the quantitative analysis of complex mixtures, particularly for major components. Raman spectra provide complementary information to the FTIR. Vibrations that are active in Raman may be inactive in IR, and vice versa.

The NMR spectroscopy method (Ref 44) involves radio-frequency-induced transitions between quantized energy states of magnetic nuclei that have been oriented by magnetic fields. Significant information is obtained by NMR because of the effects of inter- and intramolecular interactions on the values of the magnetic field strength at the nuclear sites in molecules. Questions on specific nuclei, such as its identity, its location in the molecule, its quantity, its relation to neighbors, and the identification and location of neighbors are all answered by the spectra.

Electron spin resonance (ESR) (Ref 44) is a form of absorption spectroscopy, in which radiation of microwave frequency induces transitions between magnetic energy levels of electrons of unpaired spins. The magnetic energy splitting is created by a static magnetic field. Unpaired electrons are present in odd molecules, free radicals, triplet elec-

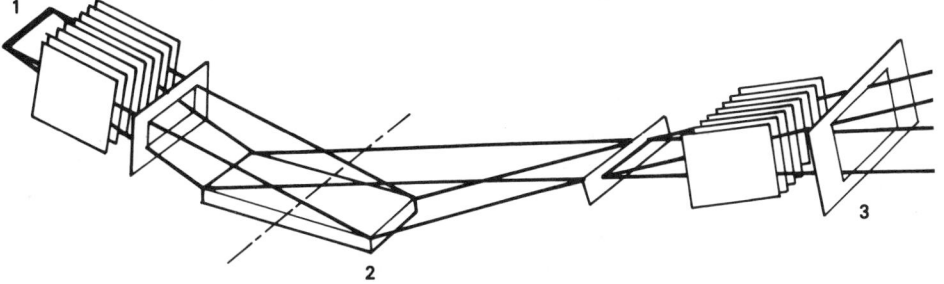

Fig 7 Geometry of Bragg-Brentano x-ray diffractometer. Source: Ref 45

tronic states, and transition metal and rare earth ions. There has been much interest in the unpaired electrons of free radicals.

Electrokinetic (Ref 4) properties of powders are an important set of parameters that describe electrical double layer phenomena. There are four methods of determining electrokinetics: electrophoresis, electro-osmosis, streaming potential, and sedimentation potential. The most commonly used method is electrophoresis, in which an electric field is applied to a dispersion and the velocity of the particles is measured. Based on this mobility data, zeta potential (potential at the Stern plane of the electrical double layer surrounding the particle) is calculated. The measurement of zeta potential at several pH values allows the determination of an isoelectric point (a pH at which the particles carry a net zero charge). Parameters such as zeta potential and isoelectric point are very widely used in the study of colloidal-sized ceramic particles. A limitation of these measurements is the low concentration of particles in the suspension. Acoustophoresis-based techniques have been developed to study dense suspensions. In these systems, the suspension is subjected to sound or electric vibrations (~ 1 MHz). The particles are then set in motion at the same frequency as the input impulse. The amplitude of vibration of the particles is considered to be proportional to the zeta potential.

REFERENCES

1. P.M. Gy, Chapt. 4, *Sampling of Particulate Materials—Theory and Practice*, Elsevier Scientific Publishing, 1979
2. T. Allen and R. Davies, Modern Aspects of Particle Size Analysis, *Advances in Ceramics*, Vol 21, *Ceramic Powder Science*, G.L. Messing, *et al.*, American Ceramic Society, 1987, p 721–746
3. S.G. Malghan, S.M. Hsu, A.L. Dragoo, *et al.*, Physical and Chemical Characterization of Ceramic Powders in an International Interlaboratory Comparison Program, *Proceedings of CIMTEC World Ceramic Congress*, P. Vincenzini, *et al.*, Ed., June 1990
4. G.D. Parfitt, Chapt. 1–3, *Dispersion of Powders in Liquids*, 3rd ed., Applied Science Publishers, London, 1981
5. T. Allen, *Particle Size Measurements*, 3rd ed., Chapman and Hall, London, 1981
6. L. Svarovsky, *Solid-Liquid Separation*, Butterworths, 1977, p 15–23
7. M.W. Daescher, *Powder Technol.*, Vol 2 (No. 6), 1969, p 349–355
8. H.G. Barth, Ed., *Modern Methods of Particle Size Analysis*, John Wiley & Sons, 1984
9. D.H. Kay, *Techniques for Electron Microscopy*, 2nd ed., Blackwell Scientific Publications, Oxford, 1965
10. "Horiba Laser Diffraction Particle Size Distribution Analyzer," LA-500, Bulletin HRE-3610B, 1990
11. E.L. Weiss and H.N. Frock, Rapid Analysis of Particle Size Distributions by Laser Light Scattering, *Proceedings of the 7th Annual Fine Particle Society Conference*, Philadelphia, Aug 1975
12. R. Davies, Recent Progress in Rapid-Response and On-Line Methods for Particle-Size Analysis, *Amer. Lab.*, Apr 1978, p 97–106
13. B.F. Osborne, A Complete System for On-stream Particle Size Analyzers, *Can. Min. Metall. Bull.*, Sept 1972, p 1–11
14. Micromeritics Instrument Corporation product literature, Norcross, Georgia
15. D.G. Dalgleish, Measurement of Electrophoretic Mobilities and Zeta Potentials of Particles from Milk Using Laser Doppler Electrophoresis, *J. Dairy Res.*, Vol 51, 1984, p 425–438
16. T. Provder, Ed., *Particle Size Distribution*, American Chemical Society, 1987
17. "SF³-Sedimentation Field Flow Fractionator," Brochure E-84626, E.I. Du Pont de Nemours & Company, Inc., 1988
18. J.K. Beddow, Morphological Analysis, *Advances in Ceramics*, Vol 21, *Ceramic Powder Science*, G.L. Messing, *et al.*, Ed., The American Ceramic Society, 1987
19. J.K. Beddow and T.P. Meloy, Ed., *Advanced Particle Technology*, CRC Press, 1977
20. J.K. Beddow, Ed., *Particle Characterization Technology*, Vol I and II, CRC Press, 1984
21. B.H. Kaye, *Direct Characterization of Fineparticles*, John Wiley & Sons, 1981
22. S. Brunauer, P.H. Emmett, and E. Teller, Adsorption of Gases in Multimolecular Layers, *J. Am. Chem. Soc.*, Vol 60, 1938, p 309–319
23. S. Lowell and J.E. Shields, *Powder Surface Area and Porosity*, 2nd ed., Chapman and Hall, London, 1984
24. S.M. Hsu *et al.*, "Characterization of Ceramic Powders: Data and Analysis-Final Report," Ceramics Division, National Institute of Standards and Technology, March 1988
25. S. Brunauer, L.S. Deming, W.W. Deming, and E. Teller, On a Theory of the van der Waals Adsorption of Gases, *J. Am. Chem. Soc.*, Vol 62, 1940, p 1723–1732
26. H.P.G. Darcy, *Les Fontaines de la Ville de Dijon*, Victor Dalamont, 1856
27. J. Kozeny, *Ber. Wien. Akad.*, Vol 136A, 1927, p 271
28. P.C. Carmen, *Flow of Gases Through Porous Media*, Butterworths, 1956
29. W.H. Harkins and G. Jura, Surfaces of Solids. X. Extension of the Attractive Energy of a Solid into the Adjacent Liquid Film, the Decrease of Energy, and the Thickness of Films, *J. Am. Chem. Soc.*, Vol 66, 1944, p 919–927
30. J.W. Gibbs, *Collected Works of J.W. Gibbs*, Vol 1, Longmans and Green, 1931
31. P.H. Emmett, Multilayer Adsorption Equations, *J. Am. Chem. Soc.*, Vol 68, 1946, p 1784–1789
32. W.H. Harkins and G. Jura, Surfaces of Solids. XII. An Absolute Method for the Determination of the Area of Finely Divided Crystalline Solids, *J. Am. Chem. Soc.*, Vol 66, 1944, p 1362–1366
33. S.J. Gregg and K.S.W. Sing, *Adsorption Surface Area and Porosity*, Academic Press, 1967
34. E.W. Washburn, *Phys. Rev.*, Vol 17, 1921, p 273
35. G.P. Androutsopoulos and R. Mann, Evaluation of Mercury Porosimeter Experiments Using a Network of Pore Structure Model, *Chem. Eng. Sci.*, Vol 34, 1979, p 1203
36. C. Orr, Jr., *Powder Technol.*, Vol 25, 1970, p 37
37. K.A. Hardman-Rhyne, Pore Morphology Analysis using Small-Angle Neutron Scattering Techniques, *Advances in Ceramics*, Vol 21, *Ceramic Powder Science*, G.L. Messing and J.W. McCauley, Ed., American Ceramic Society, 1987, p 767–778
38. D.M. Smith, T.E. Holt, D.P. Gallegos, and D.L. Sternier, Pore Structure Analysis via NMR, Mercury Porosimetry, and Dynamic Methods, *Advances in Ceramics*, Vol 21, *Ceramic Powder Science*, G.L. Messing and J.W. McCauley, Ed., American Ceramic Society, 1987, p 779–791
39. W.F. Davis and E.J. Merkle, Dissolution of Bulk Samples of Silicon Nitride, *Anal. Chem.*, Vol 53, 1981, p 1139
40. G. Czupryna and S. Natansohn, Analysis of Silicon Nitride, *Advances in Materials Characterization*, Vol 15, *Materials Science Research*, D.R. Rossington, R.A. Condrate and R.L. Snyder, Ed., Plenum Press, 1983
41. T.J. Vickers, Atomic Fluorescence and Atomic Absorption Spectroscopy, *Physical Methods in Modern Chemical Analysis*, Vol 1, Academic Press, 1972, p 189–254
42. T.J. Vickers and J.D. Winefordner, Flame Spectrometry, Chapt. 7, *Analytical Emission Spectroscopy*, Part II, E.L. Grove, Ed., Marcel Dekker, 1972
43. P.W.J.M. Boumans, Excitation of Spectra, *Analytical Emission Spectroscopy*, Part II, E.L. Grove, Ed., Marcel Dekker, 1972, p 192–204
44. H.H. Willard, L.L. Merritt, and J.A. Dean, X-ray Methods, *Instrument Methods of Analysis*, D. Van Nostrand Company, 1974, p 258
45. R. Jenkins, "X-Ray Technology," Norelco Reporter, No. 31 (2XR), 1984, p 18–26

Comminution

S.G. Malghan, Ceramics Division, National Institute of Standards and Technology

COMMINUTION is frequently described in the technical literature as milling, grinding, and size reduction. All these terms are used interchangeably for a process in which small particles are produced by reducing the size of large ones. Invariably, dry ceramic powders are received from powder manufacturers in agglomerated form. The ceramic component manufacturers utilize these powders to prepare the required size distributions via comminution techniques. The wet-milling technique predominates because dry milling has limitations in the production of particle agglomerates that are finer than a few micrometers. Although there are other applications in which dry milling is used to produce powders of varying size distributions, the final powders still have agglomerate size distributions of the order of a few microns. Therefore, this article does not describe the crushing and autogenous milling of coarse materials. Rather, it focuses on the milling process as applicable to fine powder production.

Objectives of Comminution

Current technology for processing ceramic powders requires that the particles in the powder meet certain criteria that depend on the specific material and its applications. Comminution is an essential step in almost all conventional powder preparation processes, because it achieves these important objectives:

- Deagglomerates by separating particles from clusters
- Decreases the size of powder to eliminate unwanted coarse particles above a certain size
- Increases specific surface area by producing a large quantity of very fine size particles
- Provides surface activation without causing large-scale size reduction
- Homogenizes powder-solvent mixtures
- Carries out surface chemical and bulk chemical reactions

The constituents and requirements of the milling process are shown in Fig 1. It should be noted that there are instances in which the goal of comminution is just to thoroughly mix constituents. In many current structural ceramic applications that use silicon nitride, as well as electronics applications using aluminum nitride, the size reduction of raw material powders may not be required. Because the as-received powders only require deagglomeration, mixing represents the critical aspect of formability and final properties.

A disadvantage of comminution is that it sometimes cannot be utilized for a single purpose. This is because mixing, surface area, particle size distribution, and chemistry are generally interrelated.

Breakage Phenomena. Particle breakage is considered to be the elementary process of comminution. Particles are stressed by contact forces that deform and cause stress fields that produce cracks when the stress is large enough. The number of cracks and their directions determine the size and shape of daughter fragments. A schematic of particle breakage events that describe the interdependence of these phenomena is shown in Fig 2.

Crystalline materials tend to exhibit characteristic mechanical properties that arise, in part, from their crystal structure and associated parameters, such as crystallite size, and the arrangement of the crystallites within the macro crystal or powder particle. In the size reduction of homogeneous materials theoretically, the applied stress will be uniform throughout the particle if the crystal lattice were to be perfect. Therefore, when the stress reaches a level equal to that required for failure, the crystal structure breaks down to produce particles of the same order of size as the primary crystallites. The fracture of a particle involves the propagation of cracks that are either already present or initiated in the particle. The stress required for fracture is given by the Griffith relationship (Ref 2):

$$\sigma = \sqrt{\frac{2E \cdot \gamma}{L}} \qquad \text{(Eq 1)}$$

where E is Young's modulus, γ is fracture energy, and L is crack length. For brittle materials, to which most powders belong, the fracture energy is 1 to 10 $\mu J/m^2$ (10^3 to 10^4 ergs/cm^2).

In an ideal size reduction process, the particles from a homogeneous material would be of the same strength, regardless of size. Therefore, a plot of fineness attained versus work required to obtain that fineness would be a straight line (Ref 3). Since no crystal is perfect in a real material because of defects (lattice defects, flaws, cracks, and presence of a mosaic structure, that is, elementary blocks having a size of the order of 100 lattice dimensions), its behavior is expected to be nonideal. For such materials, the relationship between the fineness attained and work required to achieve that fineness is expected to contain several straight lines, each with a different slope.

In practice, two phenomena are observed:

1. When a particle is repeatedly fractured, each new particle tends to be stronger
2. As fine particles are produced, an increasing degree of particle aggregation takes place

During fracture of particles, the larger cracks propagate first, leaving behind smaller cracks. The probability of finding a flaw decreases for a given minimum fracture stress. As fragmentation proceeds, the required fracture stress may increase so much that some plastic deformation occurs. Subsequently, the particle cannot be further reduced in size. A limit of fineness in milling may exist, depending on the actual process being used. As milling proceeds, there is an increasing degree of particle aggregation, which makes it extremely difficult to separate the primary particles from the aggregates. One underlying reason for this effect is probably the increase in the magnitude of surface forces in inverse proportion to the particle size (Ref 4).

Comminution Process

Comminution is an inefficient and energy-intensive process. Typically, only 7 to 13% of the input energy is utilized for size reduction during ball milling, while the remaining energy dissipates mostly as heat (Ref 5). The energy provided to the milling process is distributed among different subprocesses within a mill, which involve (Ref 3–7):

- Increased surface energy of powder
- Plastic deformation of the particles
- Elastic deformation of the particles
- Lattice rearrangements (gliding, slipping, twinning) within a particle

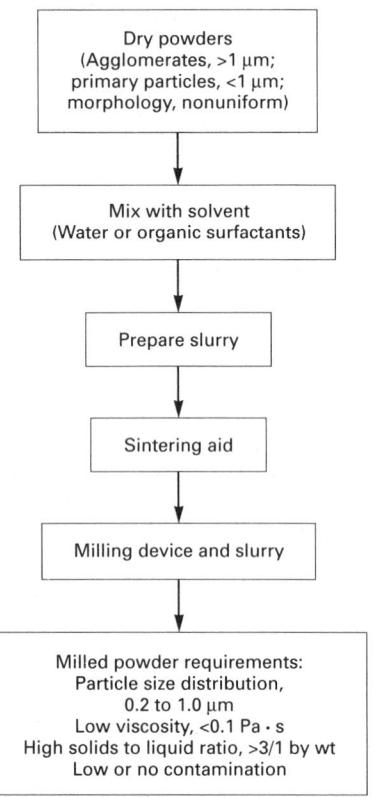

Fig 1 Process flow for silicon nitride powder milling

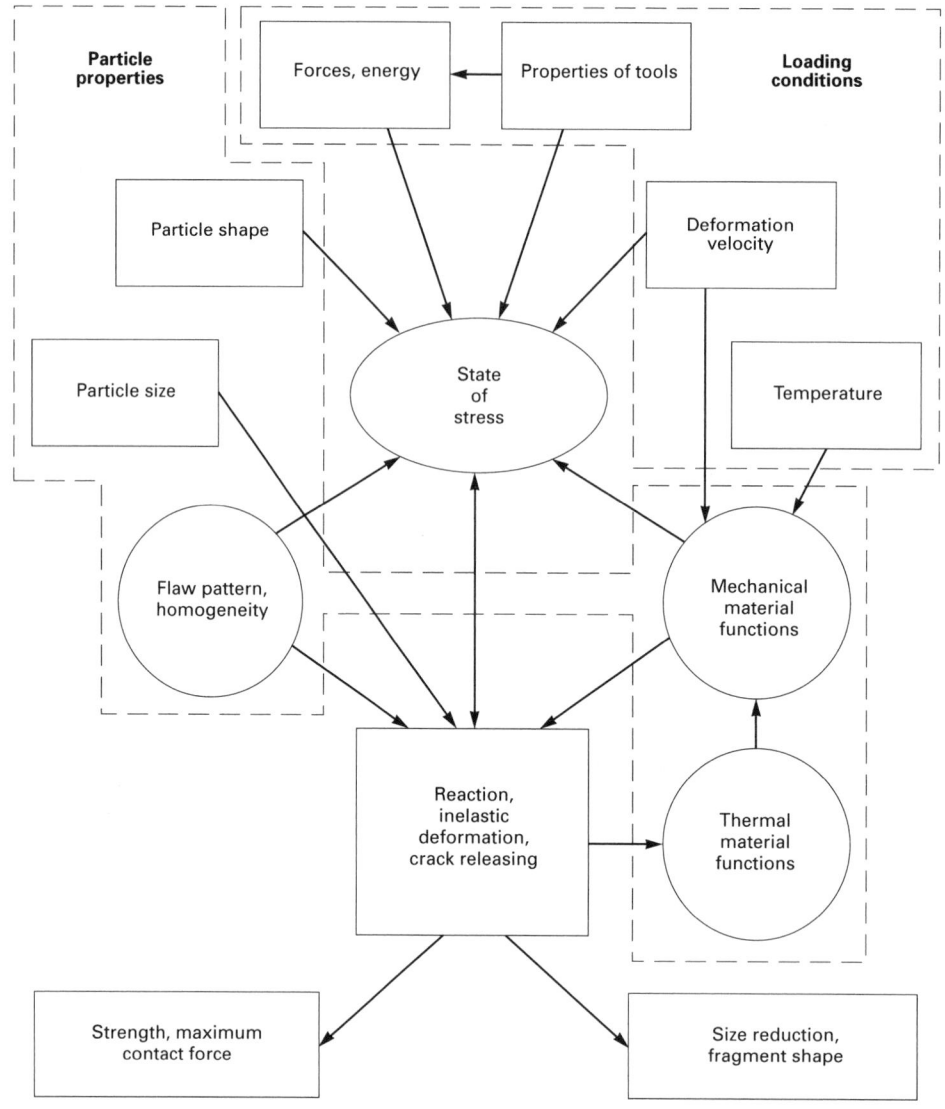

Fig 2 Particle breakage schematic. Source: Ref 1

The size reduction that occurs in a mill is based on a combination of the following mechanisms (Ref 8): impact with media, abrasion with media, and attrition with particles and media.

An understanding of the subprocesses, mechanisms, and their impact on particle breakage is important to the development of a size reduction knowledge base. These subprocesses are affected by the fluid medium in the mill (Ref 7). In dry comminution, air or an inert gas is used to keep the particles in suspension. The tendency of fine particles to agglomerate in response to van der Waals attractive forces limits the capabilities of dry processes. Because wet milling can control the reagglomeration tendency of fine particles, it is used for comminution to submicrometer-sized particles. Water is the most commonly used liquid, although alcohols or other organics are used where oxidation by water is undesirable.

The technical literature (Ref 9–11) shows that the actual differences between wet and dry comminution are often very large. For example, the power to drive a wet ball mill is as much as 30% less than that required by a dry mill. This difference in energy requirements can be substantially decreased by the appropriate choice of equipment and operating conditions. Another important difference is that wet milling allows the introduction of surfactants, sintering aids, binders, and other additives, and results in a better mixing than in dry milling.

Comminution Models. Theoretical models of comminution presented by Rittinger, Kick, and Bond to describe the comminution process have been unsuccessful (Ref 3, 9). This is because each of the models describes a particular aspect of this complex process. The comminution process can be better understood if it is analyzed by mechanistic models (Ref 10). The quantitative investigation of the phenomena involved has been analyzed using a first order rate equation, for the general size interval, i. A size and mass balance for batch grinding gives (Ref 10):

$$\frac{dW_i(t)}{dt} = -S_i t + \sum_{\substack{j=1 \\ i>1}}^{i-1} b_{i,j} S_j W_j(t), \quad n \geq i \geq j \geq 1$$

(Eq 2)

where $b_{i,j}$ is the breakage function, which is a fraction of the material broken from a larger size, j, which reports to a smaller size, i, on primary fracture; S_i is the breakage rate function of size interval, i; $W_i(t)$ is the mass frac-

tion of total charge, W, which is of size i at time, t, of milling; and n is the sink size interval defined by the lowest measurable size.

The mechanistic models have been applied to the analysis of milling a number of materials to micrometer sizes. In addition, these models are used for the scale-up, simulation, and process control of laboratory, pilot-scale, and commercial tumbling mills. The basic concepts of these models, including the parameters and their determination, are shown in Fig 3.

Mechanochemical Effects in Comminution

Because comminution is an energy-inefficient process, various attempts have been made to improve its energy utilization. In one such attempt, Rahbinder proposed that the presence of adsorbed surface active agents decreased the surface energy and hardness of

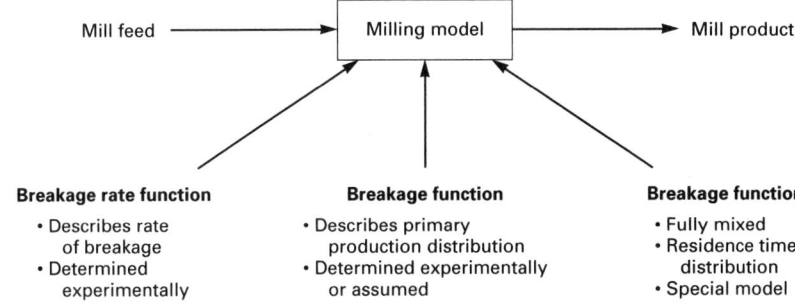

Fig 3 Mechanistic model concepts

Mill feed → Milling model → Mill product

Breakage rate function
- Describes rate of breakage
- Determined experimentally

Breakage function
- Describes primary production distribution
- Determined experimentally or assumed

Breakage function
- Fully mixed
- Residence time distribution
- Special model

powders being milled (Ref 11). In the presence of adsorbed surfactant, the cohesive forces that bond the surface centers of crystal lattices at the surface of the particle are lowered by an amount necessary for the adsorption to take place (in accordance with Gibb's equation). Thus, the microscopic cracks and fissures that are present in brittle materials are more easily widened because of the adsorption process. Based on this hypothesis, Rahbinder arrived at two conclusions:

1. The strength of all sizes of particles will be affected by modifying the conditions at the crack tip, which results in a higher stress concentration, and
2. The dispersive effect will be markedly increased

Numerous researchers (Ref 12–14) have attempted to explain this phenomenon, called the mechanochemical effect, in which mechanical aspect of milling is influenced by the addition of chemical additives. Uncertainty remains as to whether the chemical additives can influence milling kinetics. Furthermore, the mechanism by which the chemical additives act is not very clear; several mechanisms actually exist. Rahbinder proposed that surface energy reduction takes place in the presence of chemical additives. Westwood (Ref 15) has proposed the decrease of surface hardness. However, the size reduction process in a ball mill, for example, is similar to striking a small particle with a large hammer. Fracture occurs irrespective of the presence of surface flaws. It is possible that the stress waves produced by massive high-speed impact can activate flaws throughout the solid, and a number of fractures propagate at a speed comparable to the speed of sound. Because surfactant molecules cannot diffuse down the cracks at the speed of sound, they may not affect the energy at the propagating tip of the crack. Hence, the mechanochemical effect may not explain the influence of milling aids, and the explanation of the observed changes should be sought in other effects of additives.

As already noted, comminution is a complex process in which several simultaneous and interrelated physical and chemical processes can occur. No single mechanism can explain the influence of all types of chemical additives. Recent work suggests that the individual and cumulative effects of additives on various system properties (such as pulp fluidity, flocculation/dispersion state of the slurry in the mill, and charge characteristics) and on crack initiation and propagation have to be taken into account when studying the mechanisms involved (Ref 16). Although there is sufficient experimental verification of the advantageous effect of milling additives, no sound scientific explanation has yet been offered that explains or predicts the general behavior of additives.

Alteration of Powder Properties

Although the primary objective of milling is to produce fine particles for further processing, the physical and chemical characteristics of the powders frequently undergo significant changes. There is sufficient evidence in the literature that not only do the desired changes in physical properties occur, such as powder specific surface area and particle morphology, but changes in chemical properties, such as catalytic activity, can take place during prolonged milling (Ref 13, 14).

During milling, bonds between the atoms of particles undergo rupture, resulting in surfaces with unsatisfied valencies. In addition, fine particles possess high surface energy, which favors physico-chemical reactions between the solid and surrounding medium. Weyl (Ref 17) reported that quartz, upon prolonged milling in a dilute solution of silver nitrate, actually reacted with the silver nitrate to form a monomolecular film of yellow silver nitrate. Additionally, after milling, the fresh surface acquired hydrophilic properties and adsorbed relatively large quantities of water vapor. In the milling of silicon nitride in dry and aqueous environments, Imamura et al. (Ref 18) have demonstrated the formation of amorphous surface layers on the particles. Wet-milled powders contained amorphous layers of oxynitrides up to 1.0 nm (0.04 μin.) thick, whereas the dry-milled powders showed oxynitrides of a 10.0 nm (0.4 μin.) thickness.

In addition, the zeta potentials and isoelectric points of the two types of powders were found to be different. The dry-milled silicon nitride powder surface behaved more like that of amorphous silica. In a study of high-energy agitation ball milling of silicon nitride powders, the isoelectric points (pH_{iep}) of aqueous milled powders were found to be different by varying the milling conditions (Ref 19). The increased abrasion of powder surface was suspected to shift pH_{iep} to a lower value. The explanation behind this effect was based on the increased exposure of freshly created surfaces and their oxidation. Enhanced oxidation of silicon nitride appears to convert the surface to that of silica (Ref 19, 20). Hence, a highly oxidized surface had a lower pH_{iep} than that of a less oxidized surface.

Oxygen is one of the major impurities picked up by silicon nitride powder during milling, but other impurities include carbon, iron, aluminum, and other elements from milling fluid and mill hardware (Ref 20). The primary source of oxygen in surface oxide formation is the result of Si-O bond formation at the sites of freshly broken Si-N bonds. The oxygen content of milled powder was found to increase as a function of milling time (Ref 20, 21).

Control of Milling Systems

The general purpose of a milling system is to produce powders of a specific product size distribution. Consequently, control of the milling circuits involves maintaining optimum size distribution without producing too many particles that fall outside of the desired size distribution. In addition, maintaining or maximizing the mill throughput is often required. Therefore, the common control objective is to operate the mill at the maximum throughput that will produce the desired particle size distribution. Based on operator experience, these objectives can be achieved by controlling process variables such as feed size distribution, mill power or speed, ratio of solid to liquid, feed rate, and flow rate. This type of control does not work when the operating conditions deviate from the norm. Thus, some form of on-line measurement of particle size is required to achieve reproducibility in powder milling, so that compensating changes can be made based on the deviations.

Recently, laser-based light scattering techniques have been developed (Ref 22) for on-line measurement of particle size distribution from slurries. Slurry viscosity, pH, and conductivity are some of the additional parameters that can be measured with relative ease. Automation of the milling circuits has become a necessity because of technical and economic factors, such as escalation of capital and operating costs, increased need for product quality enhancement, and availability of reliable sensing instruments. The automation and control of powder milling circuits is probably one of the first steps required in the intelligent processing of advanced materials.

Equipment

Comminution equipment can be classified based on the range of particle size distribution that can be produced (Table 1). Therefore, primary constraints are the particle size distribution of both the feed and the desired product, as well as the dry or wet state of each (Ref 23, 24).

Dry-Milling Methods. Four basic mechanisms are employed in milling feed particles of several hundred micrometers to product sizes of 1 to 10 μm (40 to 400 μin.). An important requirement of these mills is the continuous removal of fines that have reached the desired product size distribution. Otherwise, process efficiency decreases drastically. Because of technology limitations and the interaction of surface forces (reagglomeration of dry particles begins to dominate), the classification process becomes extremely inefficient as the particle size distribution approaches approximately 10 μm (400 μin.). Ball and rod mills utilize a combination of milling mechanisms that incorporate mechanical impact, pressure crushing, and, to a lesser extent, particle-particle attrition. In dry milling, a fluid energy mill is the only type that utilizes particle-particle impact and attrition to carry out size reduction.

Wet-Milling Methods (Ref 3, 24). Wet ball milling, which uses balls as the media to break particles and aggregates, is the dominant method. The term ball mill refers to a mill with a length and diameter of approximately the same dimensions, in which the media of spherical form function by applying sufficient force to break particles that are suspended in a liquid medium. Common types of wet-milling devices are categorized by the method they use to impart motion to the media:

- Tumbling (balls or rods)
- Vibratory (balls)
- Planetary (balls)
- Tube mills (balls)
- Agitation (balls or beads)

These mills can be operated in batch, open-circuit continuous, or closed-circuit continuous modes. In the latter, a classifier is used to separate the finished particles so that milling efficiency can be increased. A rod mill, normally considered useful for particles larger than 2000 μm (80 mils), may find special applications involving selective breakage of

either larger particles or agglomerates. When the rods roll, the fine particles (under 2000 μm, or 80 mils), stay unaffected in the interstices, while the coarse aggregates get crushed by the rods. Tube mills are an extension of ball mills, in that the mill length exceeds its diameter by a factor of 2. Tube mills are used only where a long residence time is required in a single pass. In planetary ball mills, spherical media are swirled around at speeds that exceed the critical speed of the oval- or cylindrical-shaped mills.

The milling efficiency of some wet milling devices used for barium ferrite size reduction is shown in Fig 4. Of the four types of mills examined, the ball mill shows the lowest rate of milling. This rate reflects the poor utilization of energy for size reduction. Most of the energy used in ball milling is expended in lifting the media to the top of the mill so that they can carry out size reduction (Ref 24).

The size of the media used in a ball mill is an important parameter because it determines the product size distribution (Fig 5). Widening the size distribution using spherical and irregular-shaped media is especially dominant in vibratory and agitation ball mills. The maximum particle size in the feed determines the lower limit on media size required for adequate force application to break these particles. The smaller the size of the media, the larger the number of media that are required per unit volume. This leads to more efficient milling if the media size is sufficiently large to break the particles.

The wear of milling media not only affects milling efficiency, but also introduces con-

tamination. Media charged to a mill that consists of a distribution of sizes will undergo wear at different rates, depending on their specific surface area. The smaller media will wear faster than the larger media.

The selection of a milling media is largely a function of composition, hardness, toughness, bulk density, and shape (Ref 3, 5, 26). Because the bulk density of the media has an effect on the milling rate, the media should be selected based on laboratory evaluation. However, selection is often based on the media composition that can decrease contamination. Media are commercially available in a variety of compositions, sizes, and shapes. Some spherical media are made from alumina (density of 3.6 to 3.9 g/cm^3), zirconia (5.4 to 6.2 g/cm^3), and silicon nitride (3.1 to 3.2 g/cm^3).

Tumbling ball mills have a wide range of applications because they come in various sizes (1 to 10^5 L, or 0.26 to 26,000 gal) to fulfill different types of industrial requirements. These mills can be categorized according to the feed method and final product removal method (Ref 3, 24):

- Batch mills
- Trunnion overflow mills (Fig 6a)
- Grate or diaphragm discharge mills (Fig 6b)

In a batch mill, the powder is loaded with the desired quantity of a liquid. After completion of a milling period, the entire material is removed in one batch. Such a method of operation is limited to small quantities of powders. The capacity of batch mills can be

Table 1 Types of dry- and wet-milling equipment and basic mechanisms of comminution

Dry mills	Wet mills	Basic mechanism
Rollers, rings	...	Pressure
Hammer	...	Mechanical impact
Rod, pebble, ball	Ball	Pressure and mechanical impact
Pulverizers	...	High-velocity impact
Fluid energy	...	High-velocity impact and attrition
Agitation	Agitation	Pressure and attrition

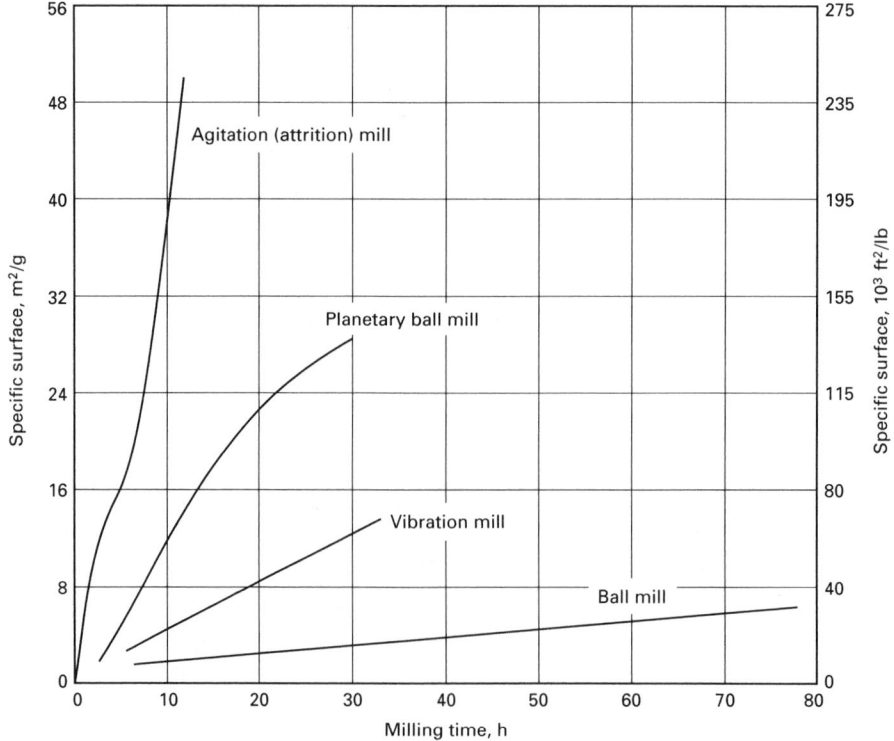

Fig 4 Milling kinetics of barium ferrite in different mill types. Source: Ref 25

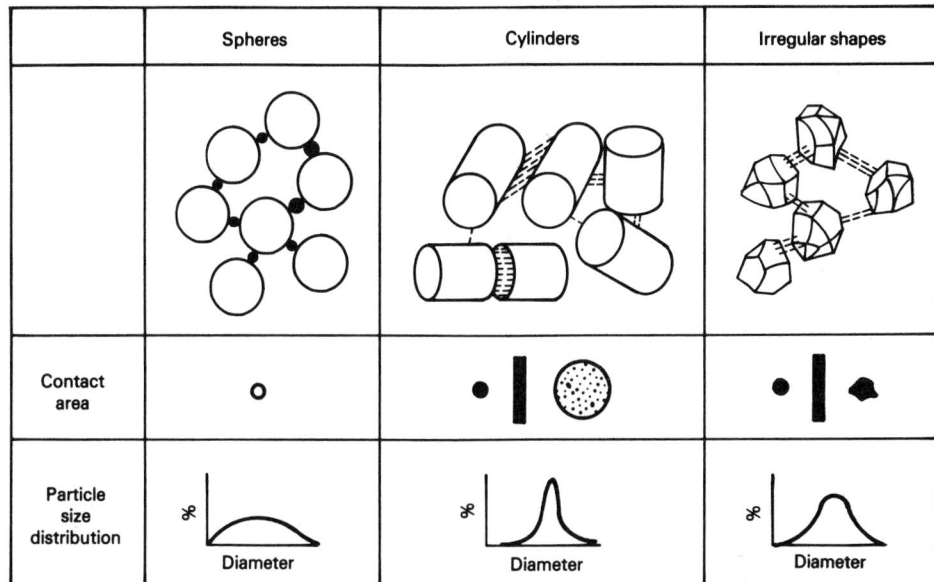

	Spheres	Cylinders	Irregular shapes
Contact area	o	•	○	•	•	◆
Particle size distribution	(curve) % Diameter	(curve) % Diameter	(curve) % Diameter

Fig 5 Expected particle size distribution from various shapes of milling media. Source: Ref 26

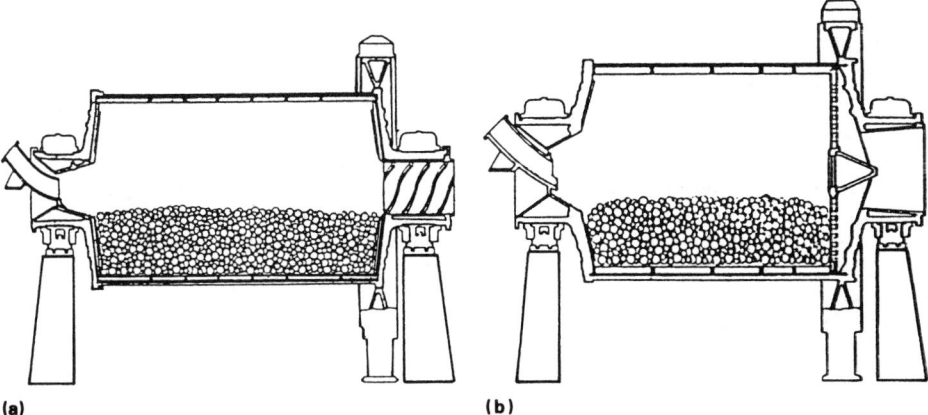

(a)   (b)

Fig 6 Conventional tumbling mills. (a) Overflow discharge. (b) Grate or diaphragm discharge. Source: Ref 24

increased by recirculating the slurry through an external holding tank.

Tumbling ball mills are normally used for size reduction of feed as coarse as 350 μm (14 mils) to produce fine product as small as 1 to 10 μm (40 to 400 μin.) (Ref 3). Multiple-compartment mills are frequently used to produce very high reduction ratios (product/feed size) in a single mill. The mill is divided into two or more compartments so that the ball charge for each compartment can be sized to achieve the most efficient milling based on the size of the powder entering the compartment.

Tumbling ball mills are extremely flexible in both geometry and the speed with which they can be operated (Ref 27, 28). Mills with lengths between three and five times their diameter are selected for applications where the specific surface area of the product is critical and high recycle load is not desirable. On the other hand, mills with lengths between one and two times their diameter are selected for applications where the narrow size distribution of the product is required.

The speed of the tumbling mill is calculated as a function of its critical speed, N_c, by using (Ref 3):

$$N_c = \frac{423}{\sqrt{D - d}} \qquad \text{(Eq 3)}$$

where the critical speed is the lowest rev/min value that causes infinitely small particles on the mill shell liner to centrifuge, D is the internal diameter of the mill in centimeters, and d is the diameter of the media in centimeters. The motion of charge (media, powder, and liquid) inside the mill is largely a function of mill speed, and, to a lesser extent, of media filling, particle and media size, and powder charge. The mill speed determines the trajectories followed by the balls in a mill (Fig 7), where the purpose is to make the descend-

ing balls fall on the toe of the charge and not on the mill liner. The impact of the balls upon the shell can lead to unduly rapid wear of the shell, high maintenance costs, and contamination. For a media filling of 40 to 50% of mill volume, the highest rate of milling exists at speeds in the range of 65 to 75% of the critical speed in conventional tumbling ball mills.

Vibratory ball mills (Ref 29–31) are being used increasingly because they provide a higher production rate at a lower capital cost, finer and more uniform product size distribution, and lower power consumption. These mills are often called close-packed media mills because the media occupy more than 90% of the total mill volume. If the grinding chamber is vibrated three-dimensionally by applying a circular displacement in the horizontal plane with a superimposed vertical oscillation, it is possible to cause the contained media to pack closely. However, the amplitude should be just sufficient to produce the necessary alignment because excessive amplitude will disrupt the close packing. An example of a vibratory ball mill is shown in Fig 8.

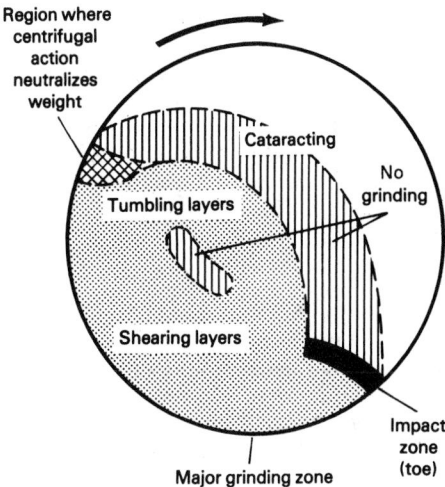

Fig 7 Motion of milling media in a tumbling ball mill. Source: Ref 24

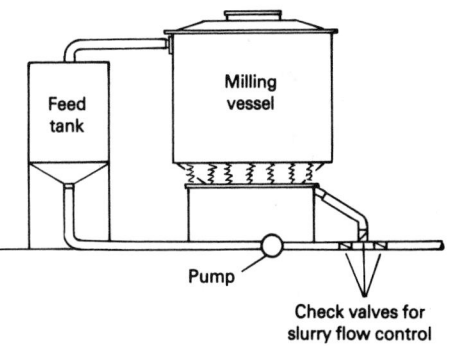

Fig 8 Vibratory ball mill with external tank for slurry recirculation. Source: Ref 32

The three-dimensional characteristics of the forces applied to the milling chamber induce complex movements in the media and charge. The packed mass of media slowly gyrates in the horizontal plane, while it rises in the vertical plane in proximity to the outer wall, and descends as it approaches the inner wall of the annulus. This type of motion assists in distributing the charge; in wet milling, it serves to maintain the suspension of solids. The degree of circulation is controlled by varying the angle of lead between the out-of-balance weights on the motor shaft. Small auxiliary weights can be added to adjust the power input in relation to the rated horsepower (Ref 31). The range of application of the vibratory ball mills depends on (Ref 29):

● The feed size of the materials, in which the maximum size depends on the amplitude of vibration, size of milling media, and degree of close packing of the media
● Viscosity of the slurry, which should not unduly impede movement of the media
● Energy developed in the mill, which should be used for milling, rather than for ancillary processes, such as mixing

The rate of milling is a function of the shape of the media. Although there is no general agreement on the effects of shape of the milling media, cylindrical media are preferred. Because they spin on an axis, they produce small shear forces that are of great importance in maintaining highly polished surfaces of the media. However, the same polishing of the media by shear forces results in more media wear. Therefore, if one is interested in decreasing the wear rate of the media, cylindrical shapes should be avoided.

In addition, the rate of milling to submicrometer sizes has been found to be a function of media size, feed size, and powder filling (Ref 29). There is no advantage in using small balls beyond a media/particle diameter ratio of 500 to 1000. The ratio of the weight of powder present to that required to fill the interstitial voids between the balls strongly affects the rate of milling. The rate of milling increases rapidly with increasing quantities of powder, but is fairly constant over a considerable range where interstitial filling equals 100%. A scale-up and milling rate analysis of the vibratory mills has been attempted using dimensional analysis, in which functional relationships are developed for individual applications (Ref 30).

Fluid energy mills (Ref 33–35) represent a rapid means of size reduction to less than 10 μm (400 μin.). A sharply defined particle size distribution can be achieved by using a classifier as an integral part of the mill. The operation of a fluid energy mill consists of the interaction of one or more high-velocity streams of fluid, bearing the material to be milled, with another stream. Milling is effected by collision between particles in the working fluid, rather than by collision with the walls or by impact with balls. This type of mill offers

totally contamination-free milling by the special design of the milling chamber. High throughputs of a wide range of feed materials can be achieved in a relatively compact, continuously operating mill.

In a typical fluid energy mill, the milled product leaves the mill in the emergent fluid stream and is collected in a cyclone chamber outside the mill. A small fraction of the product may not be collected in the cyclone and is therefore removed by a bag filter connected to the fluid discharge from the cyclone. The mean particle size and size distribution of the product are a function of several parameters, including the physical dimensions of the mill, its classifying chamber, feed size and distribution, rate of energy input to the fluid streams, and the hardness and elasticity of the feed particles (Ref 33).

Multiple collisions between the particles enhance the size reduction process. Therefore, multiple jet arrangements are normally incorporated in the mill design. The elastic fluid is normally air, at pressures up to 100 psig (6.9 bars) or steam, at up to 300 psig (20.7 bars). Inert compressed gases, such as nitrogen or carbon dioxide, are used under certain circumstances involving volatile or oxidizable materials. The fluid is introduced into the milling zone through specially designed nozzles that convert the pressure energy into kinetic energy at its associated velocity. Feed material that is simultaneously introduced into this turbulent zone accelerates with the expanding gas stream, such that multiple particle-particle collisions are induced at sonic or supersonic velocities. The velocity of the driving fluid exiting the nozzles is directly proportional to the square root of the absolute temperature entering the nozzles. Thus, to obtain maximum energy from a given mass of fluid, it is desirable to elevate the incoming temperature to the maximum tolerance temperature of the feed material.

Because it is customary to further comminute after fluid energy milling, it is necessary to classify any oversize material. Centrifugal energy already present in the powder stream is used to effect classification. A typical fluid energy mill is shown in Fig 9.

A simple criteria for particle separation can be considered as the situation in which viscous drag force on a single particle exceeds centrifugal force. Although this is an oversimplification, it has previously been applied to air classifiers with some success. On this basis, it can be shown that the maximum diameter, d, of the particle able to leave the chamber is given by (Ref 33):

$$d = \sqrt{\frac{18 U_r \mu}{\rho r \omega^2}} \qquad \text{(Eq 4)}$$

where μ is the viscosity of the fluid, ω is the angular velocity of the particle, ρ is the density difference between the particle and fluid, and U_r is the radial velocity determined by

the fluid entry pressure and velocity. The angular velocity is determined by the volumetric flow rate of high-pressure fluid supplied to the jets.

Mori (Ref 35) has shown that in a circulating type of fluid energy mill, the size reduction takes place in the jet stream in the vicinity of the nozzle. In addition, data are given on the effects of feed rate, nozzle pressure, and nozzle area on the product size distribution.

Agitation Ball Mills (Ref 37–40). The agitation ball mill (ABM) is also referred to as an attrition, or stirred mill. The ABM has made a significant stride into powder production technology because of the following advantages:

● High energy efficiency
● Ability to handle high solids loading (up to 85 wt%) and high viscosities (up to 750 Pa · s)
● Narrow size distribution of milled product
● Ability to produce highly homogeneous slurries of desired size distribution in either single or multiple passes
● Advanced engineering and design, utilizing components that are wear resistant or made from the same material as that being milled

Recent designs of the high-speed agitated mills are compact, heavy-duty, and totally enclosed to contain fumes and dust. Moreover, these mills can be easily equipped with process control and automation devices for improved reliability and cost-effective operation.

The modern ABM (Fig 10) has an enclosed grinding chamber with an agitator that rotates at high speeds in either a vertical or horizontal direction and contains fine-sized media (Ref 37). The primary mechanism of size reduction in these mills is imparted by a combination of compression and shear forces. The high efficiency of milling is due to high power density. For example, a high-speed ABM is operated at a power density in excess of 1 kW/L, whereas a typical 0.3 by 0.4 m (1.0 by 1.3 ft) tumbling ball mill is operated at a power density of 0.025 kW/L (0.09 kW/gal), and a 28 L (7.5 gal) vibratory ball mill utilizes a power density of 0.06 kW/L (0.23 kW/gal). The agitator speeds are higher than the critical speeds, and centripetal accelerations are in excess of 50 times the gravitational acceleration (Ref 23).

Essentially, an agitator is located in the ABM at the center of a size reduction chamber filled with fine-sized milling media that occupy 60 to 95% of void volume in the mill. The agitator rotates at a high peripheral speed of 4 to 20 m/s (13 to 65 ft/s). The suspension that contains solids to be milled is fed in a continuous mode and discharged at the other end.

The availability of grinding media in a variety of wear-resistant materials in fine sizes and a range of densities has made significant

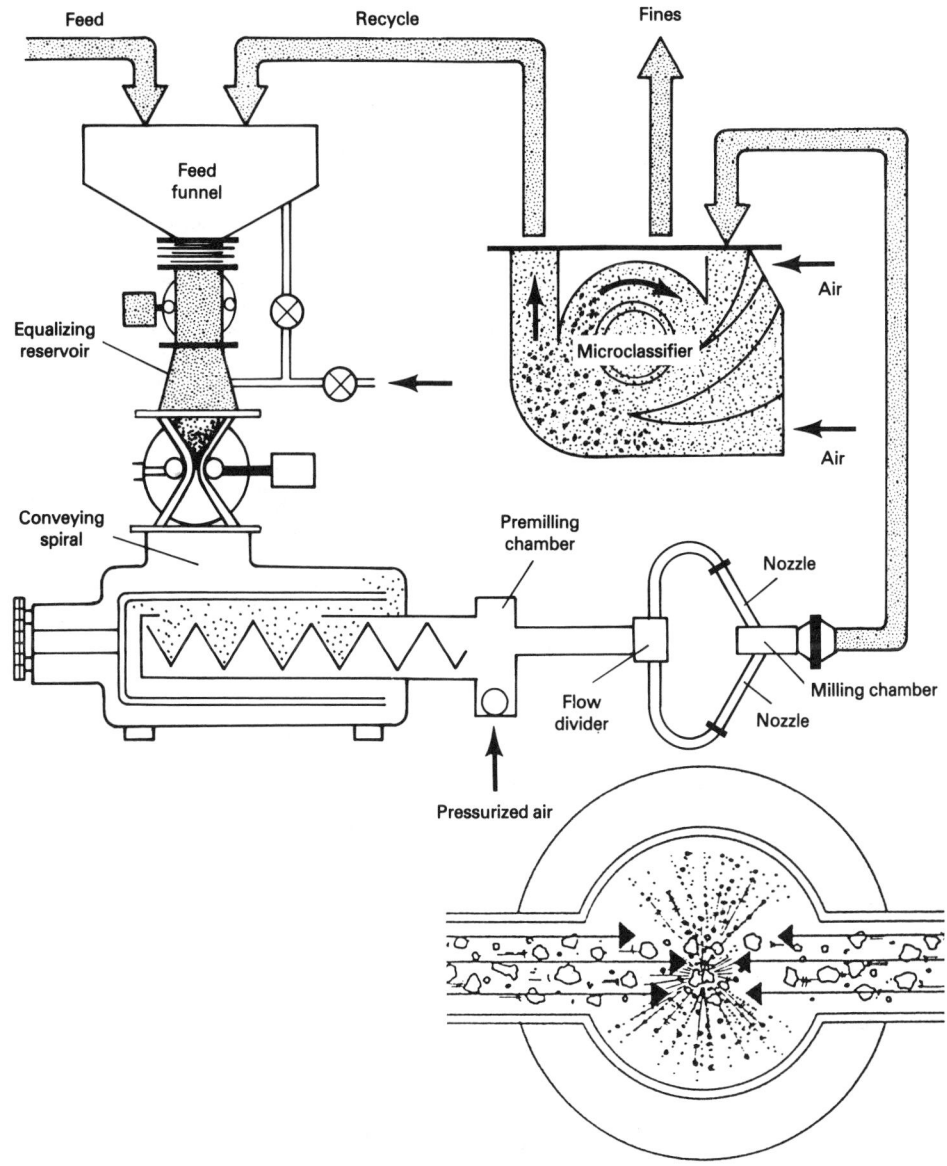

Fig 9 Fluid energy mill. Source: Ref 36

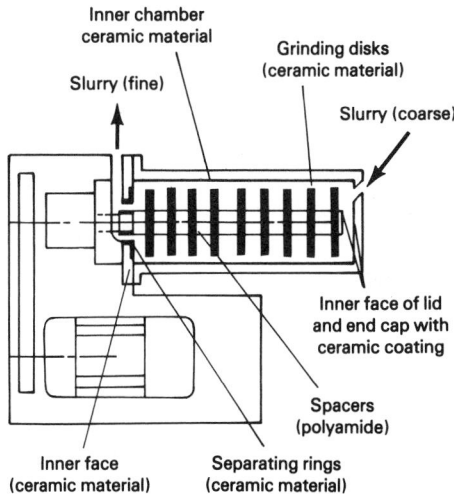

Fig 10 High-energy agitation ball mill especially designed for milling ceramic powders. Source: Ref 40

contribution to the success of the ABM. However, the availability of high-purity media for ultrapure ceramic powders production is a concern. The media sizes required in the ABM range from 0.2 to 15 mm (8 to 600 mil) diameter spheres (Ref 23). A media separator is an essential component of the ABM. It allows discharge of suspension only, while retaining the media inside the mills. Several designs being used are a frictional gap separator and both conventional and hydrodynamically shaped screen cartridges (Ref 37).

The major variables of this type of mill are (Ref 38–40): agitator speed, suspension flow rate, residence time, slurry viscosity, solids in feed (wt% or vol%), media size, density and charge, and product fineness. A number of these variables interact, which leads to a nonlinear response of milling performance as a function of a given variable. Power, P,

drawn by the agitator shaft is found to be a strong function of (Ref 38) agitator speed, N, agitator chamber volume, V, media diameter, d_m, and media density, ρ_m:

$$P = 2.45 \times 10^{-4} \times V^{1.75} \times N^{1.37} \times d_m^{0.48} \times \rho_m^{1.09}$$

This empirical model was found to be valid over a range of values of N (1.6 to 8.0 m/s, or 5.2 to 26.3 ft/s), ρ_m (2.74 to 8.0 g/cm^3), and d_m (2.38 to 6.35 mm, or 0.10 to 0.25 in.). As the speed of the agitator increases, power input to the media increases drastically, which produces enhanced size reduction.

The ABM is designed to provide the required size reduction in a single pass with a mean residence time in the range of 4 to 10 min. If the required size reduction is not achieved in a single pass, multiple passes are used. Short residence time milling allows size

reduction of mostly coarse particles in the feed, and a minimum size reduction of the fine particles, which are already in the product size range. To improve the efficiency of the ABM, high solids loading is necessary. However, the upper limit of the solids content is dictated by slurry rheology inside the mill. Particulates concentration in the suspension should be as high as possible without creating excessive flow-resistance or too much hydraulic cushioning at the media contact points (Ref 23). Energy requirements in the ABM are lower than those for other mills. However, powder hardness has a strong influence. The selection of process variables, agitator speed, flow rate, media size, and rheology plays a major role in achieving high efficiency (Ref 41).

The high efficiency of the ABM is illustrated by examining sintering aids dispersion and milling kinetics. The sintering aids disperson by the attrition milling of silicon nitride powder was found to be higher, compared to that in a ball mill (Ref 12). The dispersion of Y_2O_3 was evaluated by backscattered imaging on the scanning electron microscope. In addition, attrition milling for only 15 min using silicon nitride media was equivalent to 24 h of ball milling.

REFERENCES

1. K. Schonert, Advances in the Physical Fundamentals of Comminution, *Advances in Mineral Processing*, P. Somasundaran, Ed., Society of Mining Engineers/AIME, 1986, p 19–32

2. A.A. Griffith, The Phenomena of Rupture and Flow in Solids, *Philos. Trans. R. Soc. (London), A*, Vol 221, 1920–1921, p 163–168

3. C.C. Harris, On the Limit of Comminution, *Trans. AIME*, Vol 238, 1967, p 17–30

4. H.E. Rose and R.M.E. Sullivan, *A Treatise on the Internal Mechanics of Ball, Tube and Rod Mills*, Chemical Publishing, 1958

5. C.J. Straimond, The Energy Efficiency of Milling Processes, *Dechema-Monographien*, NR 1549–1579, Verlag Chemie, GMBH, Weinheim/Bergstrasse, 1976, p 1–18

6. B. Beke, *Principles of Communition*, Akademi Kiado, Budapest, 1964

7. S.J. Gregg, Surface Chemical Study of Comminuted and Compacted Solids, *Chem. Ind.*, 11 May 1968, p 611–617

8. D.D. Crabtree, *et al.*, Mechanisms of Size Reduction in Comminution Systems, 1, Impact, Abrasion and Chipping Grinding, *Trans. AIME*, Vol 229, 1964, p 201–206

9. H. Rumpf, Physical Aspects of Comminution and New Formulation of a Law of Comminution, *Powder Technol.*, Vol 7, 1973, p 145–149

10. L.G. Austin, A Review—Introduction to the Mathematical Description of Grinding as a Rate Process, *Powder Technol.*, Vol 5, 1971/72, p 1–17

11. P.A. Rahbinder, Hardness Reduction Through Adsorption of Surface Active Agents, *J. Tech. Phys. (USSR)*, Vol 2, 1932, p 726–755 (English translation)

12. L.C. Zarman and R.A.L. Drew, Dispersion Methods of Yttria in Silicon Nitride by Comminution, *Comm. Am. Ceram. Soc.*, Vol 72 (No. 3), March 1989, p 495–498

13. P. Somasundaran, Theories of Grinding, *Ceramic Processing Before Firing*, G. Onoda and L. Hench, Ed., John Wiley & Sons, 1978, p 105–121

14. Y.J. Burton, Change in the State of Solids Due to Milling Processes, *Trans. Inst. Chem. Eng.*, Vol 44, 1966, p 37–41

15. A.R.C. Westwood, Environment Sensitive Mechanical Behavior, Status, and Problems, *Environment Sensitive Mechanical Behavior*, A.R.C. Westwood and N.S. Stoloff, Ed., Gordon and Breach, 1966, p 1–65

16. I.J. Lin and P. Somasundaran, Alterations in Properties of Samples During Their Preparation by Grinding, *Powder Technol.*, Vol 6, 1972, p 171–179

17. W.A. Weyl, Effect of Environment Upon the Properties of Solid, *Adv. Chem. Ser.*, Vol 33, 1961, p 72–85

18. Y. Imamura, K. Ishibashi, and S. Shimodaira, "Characterization of the Surface of Si_3N_4 Particles," Proceedings of the 10th Annual Conference on Composites and Advanced Ceramic Materials, American Ceramic Society, Jan 1989

19. S.G. Malghan and L.-S.H. Lum, Kinetics of Si_3N_4 Powder Milling in a High Energy Agitation Ball Mill, *Powder Technol.*, Vol 66, April-June 1991

20. T.P. Herbell, T.K. Glasgow, and N.W. Orth, Demonstration of a Silicon Nitride Attrition Mill for Production of Fine Pure Silicon and Silicon Nitride Powders, *Ceram. Bull.*, Vol 63 (No. 9), 1984, p 1176–1178

21. M.R. Freedman, J.D. Kiser, and T.P. Herbell, Factors Influencing the Ball Milling of Si_3N_4, *Ceram. Eng. Sci. Proc.*, Vol 6 (No. 7–8), 1985, p 1124–1134

22. D.F. Nicolin, J.S. Wu, and Y.J. Chang, On-Line Submicron Particle Sizing, *Am. Lab.*, Oct 1990, p 70–79

23. D.G. Bossi, Ultrafine Grinding Equipment Types, Capabilities and Choices, *Ultrafine Grinding and Separation of Industrial Minerals*, S.G. Malghan, Ed., Society of Mining Engineers, 1983, p 3–8

24. E.G. Kelly and D.J. Spottiswood, *Introduction to Mineral Processing*, John Wiley & Sons, 1982, p 127–165

25. M.H. Stanczyk and I.L. Feld, "Comminution by the Attrition Grinding Process," Bulletin 670, U.S. Department of the Interior, U.S. Bureau of Mines, 1980

26. R.F. Conley, Attrition Milling of Industrial Minerals, *Ultrafine Grinding and Separation of Industrial Minerals*, S.G. Malghan, Ed., Society of Mining Engineers, 1983, p 39–48

27. C. Greskovich, Milling, *Treatise on Materials Science and Technology*, Vol 9, *Ceramic Fabrication Processes*, F.F.Y. Yang, Ed., Academic Press, 1976, p 15–33

28. J.L. Pentecost, Powder Preparation Processes, *Treatise on Materials Science and Technology*, Vol 9, *Ceramic Fabrication Processes*, F.F.Y. Yang, Ed., Academic Press, 1976, p 1–14

29. H.L. Podmore and E.S.G. Beasley, Vibration Grinding in Close Packed Media Systems, *Chem. Ind.*, 26 Aug 1967, p 1443–1450

30. H.E. Rose, Some Observations on the Application of Vibration Mills, *Chem. Ind.*, 19 Aug 1967, p 1383–1390

31. E.A. Smith, Some Special Functions of the Vibrational Ball Mill, *Chem. Ind.*, 26 Aug 1967, p 1436–1442

32. "Sweco Vibro-Energy Grinding Mill," Model M-18, Sweco Inc., Los Angeles

33. B. Dobson and E. Rothwell, Particle Size Reduction in a Fluid Energy Mill, *Powder Technol.*, Vol 3, 1969/70, p 213–217

34. V.C. Grimshaw and J.F. Albus, Fluid Energy Grinding and Drying for Fine Powders Production, *Ultrafine Grinding and Separation of Industrial Minerals*, S.G. Malghan, Ed., Society of Mining Engineers, 1983

35. Y. Mori, Fluid Energy Milling, *Proceedings of the 2nd European Symposium on Size Reduction*, *Dechema Monograph*, Vol 57, Amsterdam, 1966, p 695–698

36. LAROX FP Jet Mill, Larox, Inc., Columbia, MD

37. N. Stehr, "Recent Development in Stirred Ball Milling," Engineering Foundation Conference: Recent Developments in Comminution, Hawaii, 8–13 Dec 1985

38. J.A. Herbst and J.L. Sepulveda, "Fundamentals of Fine and Ultrafine Grinding in a Stirred Ball Mill," Proceedings of the Powder and Bulk Solids Handling Conference, Chicago, 1978

39. N. Stehr, Residence Time Distribution in a Stirred Ball Mill and Their Effect on Comminution, *Chem. Eng. Proc.*, Vol 18, 1984, p 73–83

40. Bulletin, "The Bead Mill," Netzsch, Inc.

41. D.E. Wittmer, "Use of Bureau of Mines Turbomill to Produce High Purity Ultrafine Nonoxide Ceramic Powders," No. 8854, U.S. Department of the Interior, Bureau of Mines, 1983, p 1–12

Particle Sizing

Ben C. Wood, Eastman Kodak Company

A QUANTITATIVE APPROACH to particle sizing is necessary in a modern competitive market. High-quality products require a better understanding of both statistical methods and particle-size measurement in order to maintain competitive processes and develop new ones. Both measurement, in general, and particle size, in particular, are often felt to be simple when in fact they require a degree of science to achieve the desired quality level.

Particle size impacts sintering, density, magnetic susceptibility, and other properties (Ref 1–5). For high-temperature ceramics, size affects castability of the suspension, rheological properties, casting rate constant, porosity, and texture of the green bodies (Ref 6). Size even has a major impact on how manufacturing processes affect the environment (Ref 7). Thus, a deeper understanding of particle size, particle-size measurement, and particle-size analysis methods is required to meet the high-quality, high-tech needs of ceramics today (Ref 8).

Sampling is the first step in the proper analysis of a powder or slurry. It is essential that the sample used by the laboratory is representative of the bulk. Allen's "golden rule" of powder sampling requires taking a sample from a moving stream and catching all of the stream for a short time (Ref 9). Subsampling of the sample obtained in this manner is best accomplished by riffling, especially spin riffling (Ref 10, 11) (Fig 1).

Sampling procedures are described in Ref 12 to 14, whereas the latest guidelines used when sampling powders for both chemical and physical analyses are described in terms of particle sizing in Ref 15 and 16, and shape analysis in Ref 17. In addition, chemical and statistical considerations must be recognized (Ref 18). Because all the statistics, expensive instruments, and analysts cannot overcome a nonrepresentative sample, the referenced guidelines should be followed.

General Characteristics of Instruments

Before a sizing method can be chosen for a specific ceramic powder, some general characteristics of the method should be understood. Usually, size distribution output is plotted as a differential plot, with the x axis

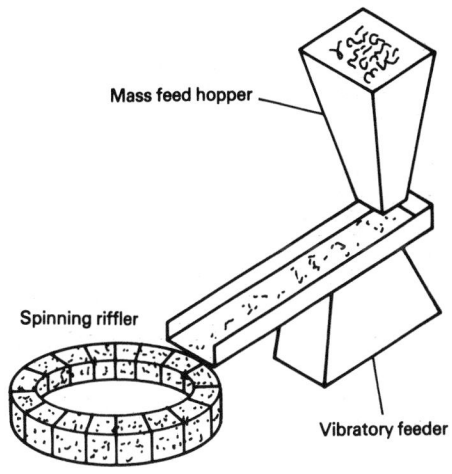

Fig 1 Spinning riffler

representing size and the y axis representing some weighted response relative to the number of particles. Many other forms that use a variety of notations and presentations are also possible, including cumulative distributions (some are on specialized papers, such as log-probability or Rosin-Rammler). Excellent background reading is provided in Ref 9 and 19 to 24.

Size Range. The size of a particle is most often described by its equivalent spherical diameter, that is, the diameter of a sphere that would match the property of the particle, as measured by the instrument being used. For example, the diameter of a sphere that would just pass through a sieve mesh would be known as a sieve diameter or, if it sediments at the same rate as the particles under study, it would have a Stokes' diameter. For convenience, this linear "diameter" can be plotted on a log scale along the x axis. Because size, or equivalent spherical diameter, is being converted to a linear dimension, it becomes easier to discuss the size range of the instruments that measure using various principles.

Figure 2 depicts the general classes of sizing instruments and their size ranges, expressed in terms of linear dimension, or equivalent spherical diameter. Note that this convention is also applied to specific surface area devices. The article "Characterization of Ceramic Powders" in this Section of the Vol-

ume provides details on this topic, as does Ref 25.

It should be noted that the y axis often confounds many analysts. Europeans (for example, Ref 26) generally plot the amount in each of several equal *linear* size intervals on the x axis, whereas Americans and many instrument manufacturers typically display the amount in each of several equal *log* size intervals. Knowledge of both schemes is desirable, because powders are an international commodity and ISO standards will be emphasized by the mid-1990s. Furthermore, it is possible to mathematically transform either method to the other.

Figure 3 displays histograms of a sieve analysis, reflecting both schemes. In both cases, the x axis uses a logarithmic scale, but the y axis in one case is based on the weight per unit linear size, whereas the other is based on the weight per unit log size. This produces the apparent shift in distribution weighting, because it is the area under the curve that matters, rather than the height of the histogram bar.

Both differential curves may be integrated over the size range to give a cumulative weight distribution. In one case, the integration is over linear diameter, and the other, over log diameter. The cumulative plots in both cases are *identical*. However, because both differential curves are often called weight distributions, but are *not* identical, there has been confusion.

To prevent confusion, the notation favored by Aitchison and Brown (Ref 27) can be used when equal log intervals are assumed, as is often done on many instruments. It can then be stated that bar height (BH) is

$$BH \propto ND^J \qquad \text{(Eq 1)}$$

where the bar height times the equal log intervals gives equal areas under the histogram bar, N is the number of particles in the interval, and D is the geometric midpoint of the interval. As used by Aitchison and Brown, J is then the weighting factor, which will be unambiguous. Although both schemes in Fig 3 are considered mathematically correct, a J notation is necessary to avoid the ambiguity in naming.

As a further example, Fig 4 shows how another histogram display might look. Because the x axis is composed of equal log intervals, making the histogram bar height on the left

Analytical principle and method

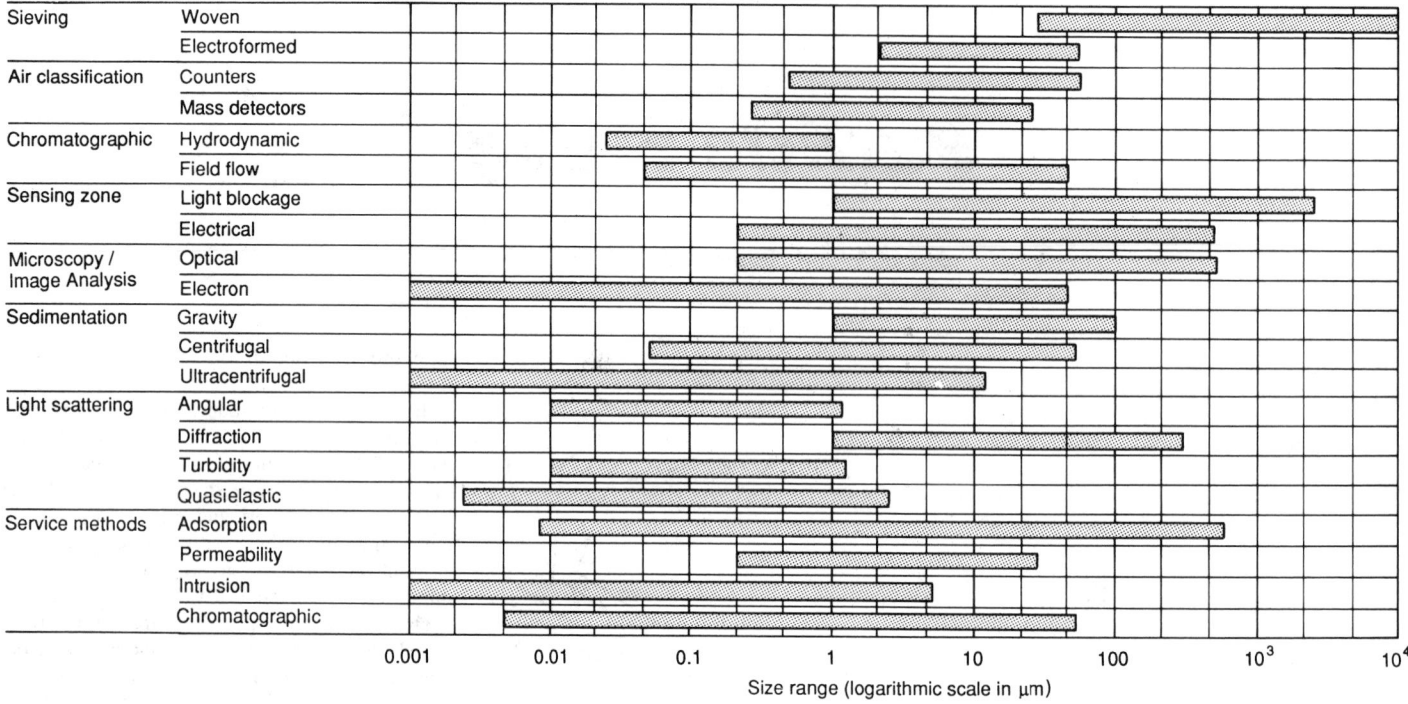

Fig 2 Analytical principle versus size range. Source: Ref 79

($J = 0$) proportional to the number in the interval will give an American-style population, or number distribution. In the European parlance, this would be a length-weighted distribution. Similarly, the one on the right would be an American volume, or weight distribution, which the European nomenclature would term a "moment" distribution.

J Concept for Means and Distribution. There is another reason for using the J notation. If one considers the many "means" of particle size (Ref 9), then one must realize that they are moment estimates of means (Ref 26 or 27). A common model for particle-size distributions is the lognormal distribution, for which the geometric mean is the parameter of central tendency analogous to the arithmetic mean for the normal distribution (Ref 28–31). If one computes the appropriate geometric mean according to Ref 24, 26, 32, and 33, then a plot of the size distribution in the American fashion causes the geometric mean to fall at the mode of the differential distribution curve (see Fig 3 and 4) or at the median of the cumulative distribution curve. This reinforces the relationship between the "means" and the distribution curves, which are not merely moment estimates subject to truncation and weighting errors, but true means estimated by nonlinear regression techniques such as maximum likelihood (Ref 28).

Given this background, the applicability of Table 1 in selecting an analysis method should be obvious, since the distribution weighting factor is critical. Distributions with high J values are more heavily weighted to large particles, and those with low J values, to small particles. In other words, if there is a series of powders that produces different properties in a usage test, a high J value sizing method may be needed if the property depends on a few large particles in the distribution. A low J value may be needed in order to see an effect that is due to fines and to get the property to correlate to particle size as measured. Thus, the ranking of the powders by size may depend upon J; it may not even be possible to convert but will require a method that directly deter-

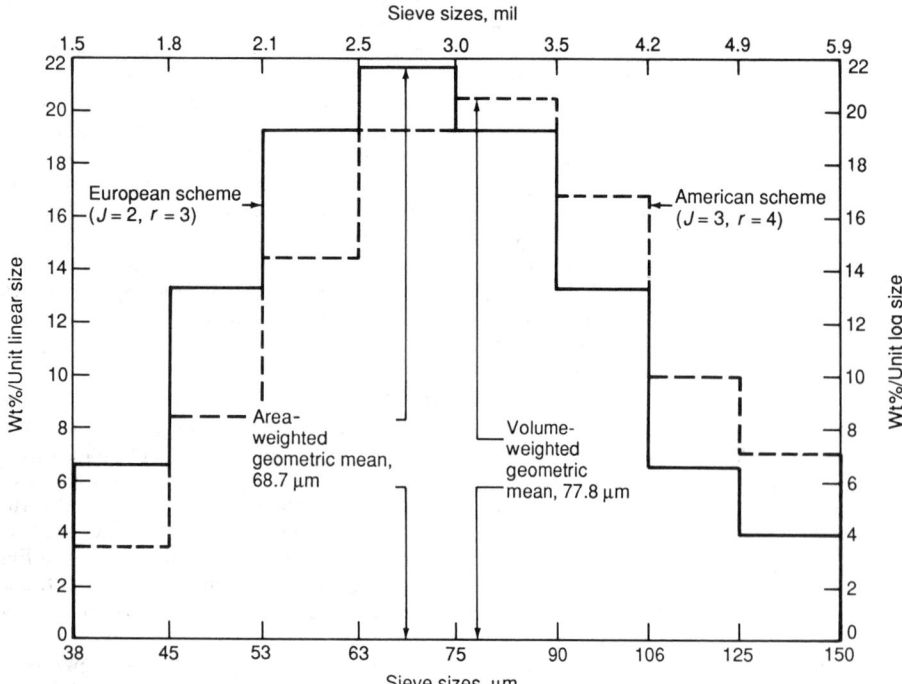

Fig 3 European versus American histograms of sieve data. The r value is the European analog of J for equal linear intervals.

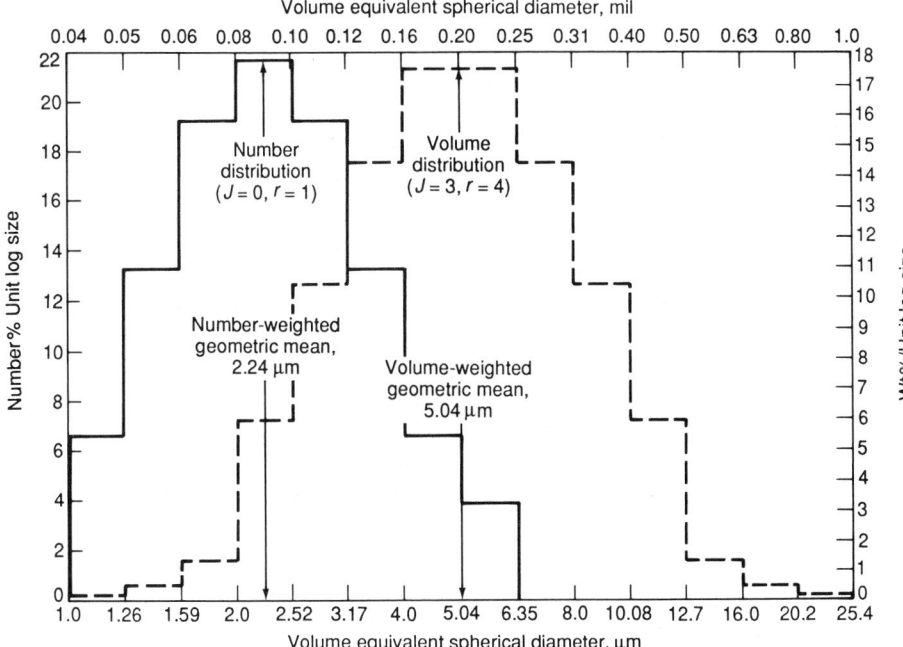

Fig 4 Coulter counter/electrozone histograms for $J = 0$ and $J = 3$

mines a high or low value of J as its primary measurement.

The primary J value (Table 1) of an instrument is determined by its x axis intervals, whether equal linear or equal log intervals, and by its method of particle detection. In addition, there may be a secondary value of J, which the distribution may be converted to without a shape assumption (for example, from $J = 0$ to 3 for a Coulter counter, as in Fig 4, where the volume equivalent spherical diameter on the x axis allows conversion from 0 to 3 because no shape assumption is required

to get from the number to the volume distribution).

By using only equivalent spherical diameter on the x axis, one can both make sense of the y axis using J and organize the means in a systematic way. Reference 28 lists the various means, all defined in terms of the J values that they would be assigned. They can all be treated as moment estimates of geometric means, which are an invariant feature of the lognormal model, as shown in Ref 23 to 31. These can easily be extended for other values of I and J, and there is no mathemat-

ical reason for them to be restricted to integral or half-integral values. The value I is related to the size, or x axis, somewhat like J is related to the y axis depending on whether the initial size measurement is derived from the area, volume, or other size property of the particle. In fact, one can take the limit of such a summation as I goes to zero to obtain the common computational equation for a geometric mean. This is just e to the power of the arithmetic mean of the natural logarithms of the values for which one wishes to find the geometric mean (see Appendix D of Ref 28).

Statistical Uses of Particle-Size Measurements

When plots such as those described above have been created, a qualitative notion of the size distribution of the powder of interest usually exists, and is applied to these plots for comparison purposes. For quantitative comparisons, however, parameters of these curves (such as means) must be compared. After taking a representative sample, measuring it on the appropriate instrument with regard to size range and J value, and computing an appropriate mean size to represent the obtained distribution, the numbers obtained should be put to good use. Four statistical uses of the data are:

- Statistical process control (SPC)
- Specification testing/acceptance sampling
- Designed experiment response variable
- Computations

Although the old concept of significant figures has been much used in the past, the statistical approach is much more modern (Ref 34–36). Often, the difficulty is in obtaining enough significant figures in the data, given the old-fashioned bias against uncertainty and modern digital devices, such as A to D Convertors (Ref 37).

Statistical Process Control. A particle-size measurement used in a statistical manner must meet certain criteria, one of which is that the measurement process itself must be under SPC, with known precision and bias (Ref 38–40). Although many references describe the techniques for preparing Shewhart or control charts (Ref 38, 41–43), they often fail to describe the three types of control charts related to measurement.

A Type I control chart on the measurement process itself is shown in Fig 5. As with all control charts, subgrouping is key. Here, the subgroup is replicate samples from a large quantity of a control material chosen as being representative of the samples to be measured and riffled into unit sample containers. Bias can be computed using a traceable control sample. The S or R chart (see Shewhart control chart descriptions in Ref 38, 41–43) of an in-control chart is the best estimate of the method precision. It is preferable to the common practice in which one sample is run 10

Table 1 Analytical principle versus weighting factor

Analytical principle and method		Weighting factor, J		
		Primary	Secondary	Variable(a)
Sieving	Woven	3		
	Electroformed	3		
Air classification	Counters	0		
	Mass detectors	3		
Chromatographic	Hydrodynamic			2–5
	Field flow			2–5
Sensing zone	Light blockage	0	2	
	Electrical	0	3	
Microscopy/image analysis	Optical	0	2	
	Electron	0	2	
Sedimentation	Gravity			2–3
	Centrifugal			2–5
	Ultracentrifugal			2–5
Light scattering	Angular	−1		
	Diffraction	3		
	Turbidity	4.5		
	Quasielastic			2–6
Surface methods	Adsorption	2.5		
	Permeability	2.5		
	Intrusion	2.5		
	Chromatographic	2.5		

(a) Correction factor for light scattering relationships. Source: Ref 79

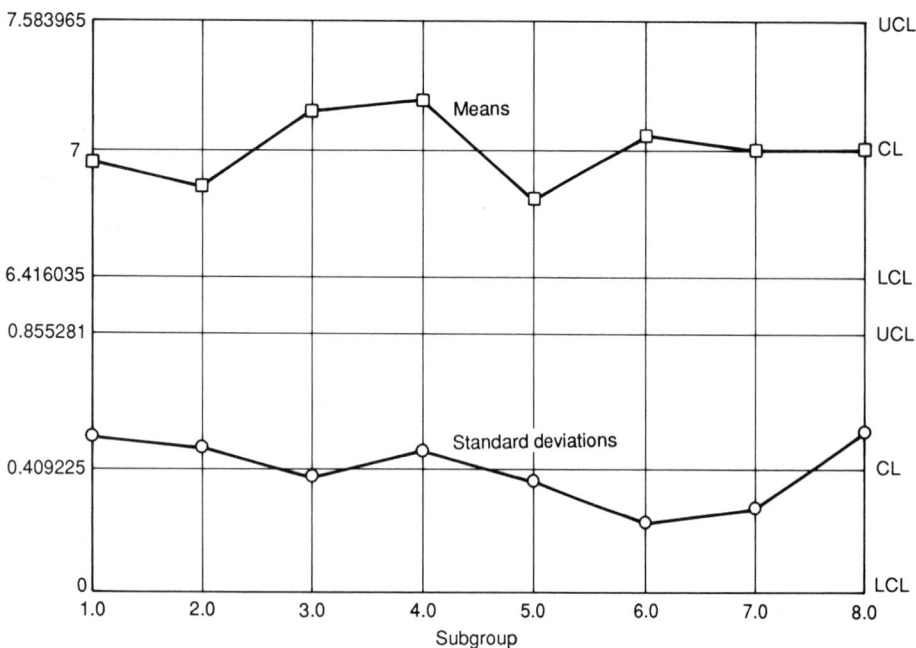

Fig 5 Type I control chart (measurement process in control). UCL, upper control limit; CL, control limit; LCL, lower control limit

times by one operator on one instrument in one day.

Figure 6, a Type II control chart, is the least understood of the three types. The subgroup is now replicate measurements on production batches as they are produced. This is called an error of measurement study in Ref 43, and is detailed in Ref 38 as well. As the name implies, the replicate measurement variability must be small relative to the process batch-to-batch variation. The S or R chart should be in control as before but now the X-bar chart *must be* out of control! Unless this

is true, the measurement is not adequate for detecting batch-to-batch differences; it can be quantified as the discrimination ratio (DR), as defined by Wheeler and Lyday (Ref 38), which is

$$DR = \left\{ \left[\frac{2\sigma(total)^2}{\sigma^2(testing)} \right] - 1 \right\}^{0.5} \quad (Eq\ 2)$$

This ratio should be at least 4 and preferably 8 for a good test. The measurement replicate must include replicate samples, because this is often greater than the variability seen in the

Type I chart. If the S-bar on the Type II chart is much larger than that on the Type I chart, then sampling may be the most significant contributor to measurement variability. This is the true variability seen in the measurements used, not just that from the Type I.

The Type III control chart is shown in Fig 7, and is most important to production personnel. Here, the subgroup is on production batches and should be in control. This is true even if more than one measurement of particle size or other test is necessary to achieve the desired DR, because multiple numbers are averaged together to describe a single batch.

Table 2 summarizes the requirements a measurement should meet before it is used for statistical purposes. Given these requirements, in addition to the standard requirement for robustness, turnaround time, and cost, one can then rely on the interpretation of the Type III control chart for computing the process capability index (Ref 44, 45) and for process improvement (Ref 46).

Specification testing, another statistical use of data, requires the consideration of other parameters, which are incorporated in Eq 3 and enable sample size to be determined:

$$n \approx \frac{(Z_{\alpha/2} + Z_\beta)^2 \sigma^2}{\delta^2} \quad (Eq\ 3)$$

where n is the number of measurement replicates necessary to test a given batch of material to determine whether it meets specifications; δ is the difference between the aim, μ_0, and the specification limit, μ_1; σ is the best estimate of test precision including sampling (from a Type II control chart, if possible); α is the so-called producer's risk, or Type I error, which is the chance of rejecting a batch that meets specification just because of measurement variability; β is the consumer's risk or Type II error, which is the chance of accepting a batch that does not meet specification just because of measurement variability; and Z is the standard normal deviation found by looking in a table of Gaussian values or the normal distribution corresponding to the appropriate values of α and β. This process is also known as sampling inspection by variables (Ref 47, 48). Additional information on the consumer risk value (β) is provided in Ref 49 to 52, whereas Ref 53 explains how to derive the formula when the test precision value (σ) is only estimated from a small sample.

To fully appreciate these concepts, both the number of correct decisions such a scheme produces (Ref 54) and the costs associated with incorrect decisions (Ref 55) need to be considered. The total testing cost per batch can be stated as:

Total testing cost/batch

= n(measurement cost)

+ α(cost of rejecting good batch)

+ β(cost of accepting bad batch) (Eq 4)

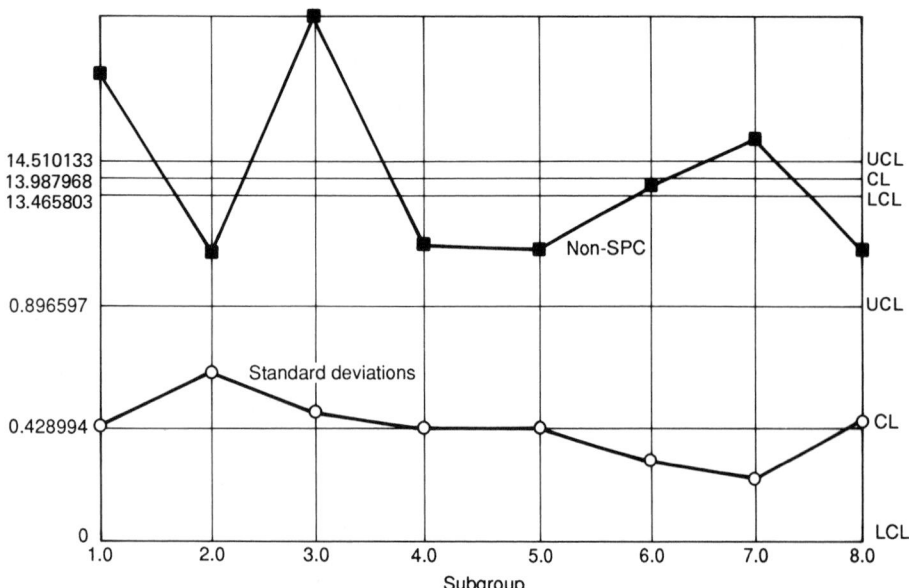

Fig 6 Type II control chart (measurement tracks production variability). UCL, upper control limit; CL, control limit; LCL, lower control limit

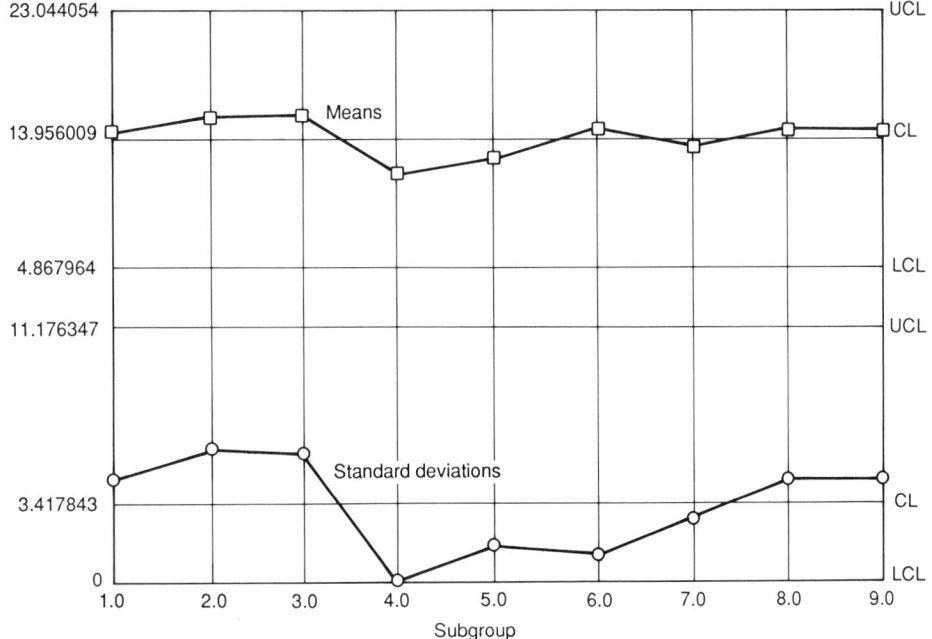

Fig 7 Type III control chart (production process in control). UCL, upper control limit; CL, control limit; LCL, lower control limit

ances plus two times the covariance, and the variance of a difference is the sum of the variances minus two times the covariance. If the computational formula is more complex, one must use error propagation (Ref 60) or a Taylor series expansion (Ref 61). The simplest way to handle even very complex situations is to use Taguchi's method (Ref 62) derived from Evans' work (Ref 63, 64). This may be easily extended and programmed on a small personal computer for even the worst cases.

Application of Measurement and SPC to Ceramics

Both particle-size measurement and SPC are increasingly being used. Frantz *et al.* (Ref 65) and Sheppard (Ref 66) have recently reviewed some of the latest developments in instrumentation. These reports should all be read while acknowledging the above requirements, in addition to the usual concern for size range, that is, that Table 1 can be as important as Fig 2. Other recent reports describe sieving techniques (Ref 67, 68), small-angle x-ray scattering and dynamic light scattering (Ref 69, 70), transmission electron microscopy (Ref 71), sedimentation (Ref 72), and even small-angle neutron scattering (Ref 73), as well as the application of these techniques to real-world problems (Ref 74).

Statistical process control used in the ceramic field should increasingly be based on the principles described above with regard to particle sizing, and is gradually being used more frequently (Ref 75–77) in these areas. However, the future belongs to what Tucker *et al.* are calling algorithmic SPC, a potent blend of SPC and process-control algorithms (Ref 78). These workers are applying in industry what Box and MacGregor *et al.* (as discussed in Ref 78) have developed.

When the costs of testing are known, rejection of a good batch and acceptance of a bad batch for each situation allows a computation (using Eq 4) of the optimum scheme for holding down testing costs. If the product is inexpensive, it may be most economical to not perform testing and instead depend on statistical control of the process to prevent the use of bad batches. If the product is expensive and testing is relatively inexpensive, then the optimum may be to take a large *n* giving low values of both α and β. Ideally, the specifications are well known, making this scheme easy to apply.

However, when working with particle size, there is often difficulty in obtaining "real" specifications. When no real specifications are available, the relationship between a Type III control chart for the property of interest and the particle size can be used as a measure of the maximum sensitivity model. Because the data are random within the chart, one cannot simply draw a regression line between the particle-size data and the variable of interest, but instead, must use the control limits. This is a conservative scheme and ensures that no problems occur if these specifications are met.

Table 2 Six requirements of a good measurement

Characteristic	Demonstrated by
Stable	Type I control chart
Precise	Sigma value from SPC
Accurate	Traceable standards
Representative	Type II control chart, discrimination ratio >4
Site comparable	Control charts, audits
Effective	Statistical measures

However, there may be no relationship at all between the two variables, that is, the correlation may be zero; this should be examined by a statistically designed experiment.

Designed-Experiment Response Variable. Measurement of particle size or other variables can also be used as a designed-experiment response variable. To determine whether a measurement is adequate when the measured value is to be used in this manner, one should apply the approach described above and in Eq 3 to determine sample size (as described in Ref 56–59). If this is not possible, then the discrimination ratio (Ref 38) should be applied, using the mean square error of the ANOVA as the process variance. This approach is best when using experimental design to study the measurement process, such as when a Type I control chart is out of control and it is desirable to make the analytical procedure more robust.

When experimental design is being used for production processes, then specifications (if only from the maximum sensitivity model) should be available and used. When experimental design is used for research or development with pilot processes, then the mean square error of the pilot processes must match that of production, even if this requires replication of the experiment. Specifications can be based on product characteristics to begin with, but may be replaced with a particle size or other measurement when the pilot processes are shown to be scalable.

Computations represent yet another use of measurements. Simple computations, such as addition and subtraction, can utilize the formulas in Appendix 3A of Ref 56, where the variance of a sum is the sum of the vari-

REFERENCES

1. I.-K. Jeong, D.-Y. Ki, Z.G. Khim, and S.J. Kwon, Exaggerated Grain Growth During the Sintering of Y—Ba—Cu—O Superconducting Ceramics, *Mater. Lett.*, Vol 8 (No. 3–4), May 1989, p 91–94

2. F. Mahloojchi, F.R. Sale, N.J. Shah, and J.W. Ross, Production and Sintering of Oxide Superconductors Via the Citrate Gel Process, *Br. Ceram. Proc.*, No. 40, March 1988, p 1–14

3. T.T. Kodas, E.M. Engler, and V.Y. Lee, Aerosol Flow Reactor Production of Fine $Y_1Ba_2Cu_3O_7$ Powder: Fabrication of Superconducting Ceramics, *Appl. Phys. Lett.*, Vol 52 (No. 19), 9 May 1988, p 1622–1624

4. Y. Yoshisato, T. Yokoo, T. Ikemachi, and Y. Kuwano, Magnetic and Electric Properties of YBCO Ceramics Prepared by Coprecipitation Method, *J. Ceram. Soc. Jpn.*, Vol 96 (No. 4), 1988, p 459–462

5. M.S. Multani, P. Guptasarma, V.R. Palkar, P. Ayyub, and A.V. Gurjar, Size Effects in Microparticle High-Temperature Superconducting Ceramics, *Phys. Lett. (Netherlands)*, Vol 142 (No. 4–5), 11 Dec 1989, p 293–299

6. A. Bellosi, C. Galassi, R. Lapasin, and E. Lucchini, Casting Conditions and Rheological Properties of Si_3N_4 Based Materials, *Br. Ceram. Proc.*, Vol 38, Dec 1986, p 179–196

7. *U.S./U.S.S.R. Symposium on Particulate Control (3rd) Held at Suzdal, U.S.S.R. on September 10–12, 1979*, Report No. EPA/600/9–85/011, Environmental Protection Agency, Apr 1985

8. A.L. Dragoo, C.R. Robbins, and S.M. Hsu, A Critical Assessment of Requirements for Ceramic Powder Characterization, *NIST Special Publication 766*, National Institute of Standards and Technology, 1989, p 1–18

9. T. Allen, Chapt. 1, *Particle Size Measurement*, Chapman and Hall, 3rd ed., 1981

10. T. Allen and A.A. Khan, *Chem. Eng.*, Vol 238, 1970, CE108–CE112

11. A.E. Hawkins, Demonstrating the Sampling of Heterogeneous, Free-Flowing, Segregating Powders, *Particle, Particle System Characterization*, Vol 5, 1988, p 23–28

12. J. Visman, A General Sampling Theory, *Mater. Res. Stand.*, Vol 9 (No. 11), Nov 1969, p 8–13, 51–66

13. A.J. Duncan, Comments on 'A General Theory of Sampling,' *Mater. Res. Stand.*, Vol 11 (No. 1), Jan 1971, p 25

14. J. Visman, Further Discussion: 'A General Theory of Sampling,' *Mater. Res. Stand.*, Vol 11 (No. 8), Aug 1971, p 32–37

15. D. Wallace and B. Kratochvil, Visman Equations in the Design of Sampling Plans for Chemical Analysis of Segregated Bulk Materials, *Anal. Chem.*, Vol 59 (No. 2), Jan 1987, p 226–232

16. G.H. Fricke, P.G. Mischler, F.P. Staffieri, and C.L. Housmyer, Sample Weight as a Function of Particle Sizes in Two-Component Mixtures, *Anal. Chem.*, Vol 59 (No. 8), Apr 1987, p 1213–1217

17. A.E. Hawkins and K.W. Davies, Sampling Powders for Shape Description, *Part. Char.*, Vol 1 (No. 1), July 1984, p 22–27

18. "Quality Assurance for the Chemical and Process Industries, A Manual of Good Practices," American Society for Quality Control, Chemical and Process Industries Division, 1987

19. J.D. Stockham and E.G. Fochtman, *Particle Size Analysis*, Ann Arbor Science, 1977

20. B.H. Kaye, *Direct Characterization of Fineparticles*, Vol 61, Chemical Analysis Monograph Series, Wiley-Interscience, 1981

21. H.G. Barth, Ed., *Modern Methods of Particle Size Analysis*, Vol 73, Chemical Analysis Monograph Series, Wiley-Interscience, 1984

22. T. Provder, Ed., *Particle Size Distribution, Assessment and Characterization*, ACS Symposium Series, Vol 332, American Chemical Society, 1987

23. Deutsche Normen, "Darstellung von Korn-(Teilchen-)grossemverteilungen, Grundlagen," DK 539.215.2. (083/084), DIN 66141, Feb 1974

24. I.C. Edmundson, Particle Size Analysis, *Advances in Pharmaceutical Sciences*, Bean, Beckett and Carless, Ed., Vol 2, Academic Press, 1967

25. S. Lowell and J.E. Shields, *Powder Surface Area and Porosity*, 2nd ed., Powder Technology Series, Chapman and Hall, 1984

26. K. Leschonski, Representation and Evaluation of Particle Size Analysis Data, *Part. Char.*, Vol 1 (No. 3), Nov 1984, p 89–95

27. J. Aitchison and J.A.C. Brown, *The Lognormal Distribution*, Cambridge at the University Press, 1957

28. B.C. Wood, "A Coherent and Unified Particle Size Measurement System," Ph.D. thesis, University of Bradford, 1988

29. T. Hatch and S.P. Choate, Statistical Description of the Size Properties of Non-Uniform Particulate Substances, *J. Franklin Inst.*, Vol 205, 1929, p 369–387

30. M. Hartmann and H. Dautzenberg, Die Verallgemeingerte logarithmische Verteilung, *Acta Polym.*, Vol 34 (Heft 1), 1983

31. E.L. Crow and K. Shimizu, Ed., *Lognormal Distributions, Theory and Applications*, Marcel Dekker, 1988

32. P. Besancon, J. Chastang, A. Lafaye, and J.-M. Lafaye, Conversion Dimensionnelle des Distributions Granulometriques, *Powder Technol.*, Vol 60, 1990, p 205–214

33. M. Alderiesten, Mean Particle Diameters Parts I and II, preprint, *Part. Char.*

34. D.B. DeLury, Computations with Approximate Numbers, reprint, *Precision Measurements and Calibration, Statistical Concepts and Procedures*, H.H. Ku, Ed., Vol 1, *Special Publication 300*, National Bureau of Standards, 1969

35. R.L. Anderson, Practical Statistics for Analytical Chemists, Van Nostrand Reinhold, 1987

36. C. Eisenhart, H.H. Ku, and R. Colle, *Expression of the Uncertainties of Final Measurement Results: Reprints*, *NBS Special Publication 644*, National Bureau of Standards, 1983

37. W.E. Harris, "Sampling, Manipulative, Observational, and Evaluative Errors," *Am. Lab.*, Jan 1978, p 31–39

38. D.J. Wheeler and R.W. Lyday, *Evaluating the Measurement Process*, SPC Press, 1989

39. G. Wernimont, Statistical Control of Measurement Processes, *Validation of the Measurement Process*, J.R. DeVoe, Ed., Symposium Series 63, American Chemical Society, 1977

40. J.K. Taylor, *Quality Assurance of Chemical Measurements*, Lewis Publishers, 1988

41. ASTM Committee E-11 on Quality and Statistics, *Manual on Presentation of Data and Control Chart Analysis: 6th Edition*, Manual Series: MNL 7, Revision of STP 15 D, American Society of Testing and Materials, 1990

42. *Glossary and Tables for Statistical Quality Control*, 2nd ed., American Society for Quality Control, Statistics Division, 1983

43. *Statistical Quality Control Handbook (AT&T)*, Western Electric Co., May 1984

44. V.E. Kane, Process Capability Indices, *J. Qual. Tech.*, Vol 18 (No. 1), Jan 1986, p 41–52

45. M. Marcucci, "Some Statistical Properties of Capability Indices," General Foods Corp., Technical Center, 1988

46. A.L. Bopp and R.P. Grant, Statistical Process Control Based on Function Analysis, *Tappi J.*, Apr 1989, p 77–79

47. A.H. Bowker and H.P. Good, *Sampling Inspection by Variables*, 1st ed., McGraw-Hill, 1952

48. T.W. Calvin, *How and When to Perform Baynesian Acceptance Sampling*, Vol 7, How-To series, American Society for Quality Control, 1990

49. A.E. Mace, *Sample Size Determination*, Reinhold Publishing, 1964

50. K.A. Brownlee, *Statistical Theory and Methodology in Science and Engineering*, 2nd ed., John Wiley & Sons, 1965

51. W.J. Dixon and F.J. Massey, Jr., *Introduction to Statistical Analysis*, McGraw-Hill, 1957

52. J.M. Juran, F.M. Gryna, Jr., and R.S. Bingham, Jr., *Quality Control Handbook*, McGraw-Hill, 3rd ed., 1979

53. G.E.P. Box, L.R. Connor, W.R. Cousins, O.L. Davies, F.R. Himsworth, and G.P. Sillitto, *The Design and Analysis of Industrial Experiments*, O.L. Davies, Ed., 2nd ed., Longman Group Limited, London, 1978

54. K.F. Speitel, "Making Your Measurement Effort Pay Off," 31st Annual Technical Conference of American Society for Quality Control, Philadelphia, 16 May 1977

55. J.R. Rutherford, A Logic Structure for Experimental Development Programs, *Chem. Tech.*, Mar 1971, p 159–164

56. G.E.P. Box, W.G. Hunter, and J.S. Hunter, *An Introduction to Design, Data Analysis, and Model Building*, John Wiley & Sons, 1978, p 87

57. C.R. Hicks, *Fundamental Concepts in the Design of Experiments,* 2nd ed., Holt, Rinehart & Wilson, 1973

58. C.K. Bayne and I.B. Rubin, *Practical Experimental Designs and Optimization Methods for Chemists,* VCH Publishers, 1986

59. G.D. Stocker, Reducing Variability—Key to Continuous Quality Improvement, *Manuf. Sys.,* Mar 1990, p 32–36

60. C.W. Sill, *Propagation of Errors,* Report from Health Services Laboratory, Analytical Chemistry Branch, U.S. Energy Research and Development Administration

61. ANSI Standard on Measurement Uncertainty, ANSI/ASME PTC 19.1, 1965, p 36–39

62. G. Taguchi, Performance Analysis Design, *Int. J. Prod. Res.,* Vol 16, 1978, p 521–530

63. D.H. Evans, An Application of Numerical Integration Techniques to Statistical Tolerancing, *Technometrics,* Vol 9, 1967, p 441–456

64. D.H. Evans, Statistical Tolerancing: The State of the Art: Part I, *J. Qual. Technol.,* Vol 6, 1975, p 188–195, Part II, Vol 7, 1975, p 1–12

65. F. Frantz, I. Burn, and J. Chase, Comprehensive Ceramic Particle-Size Measurements, *Advances in Ceramics,* Vol 11, 1984, p 131–141

66. L.M. Sheppard, Automation of Particle Analysis, *Am. Ceram. Soc. Bull.,* Vol 67 (No. 5), May 1988, p 878–889

67. C.E. Goss, Correlating Particle-Size Distribution of Clay to Water Absorption, *Am. Ceram. Soc. Bull.,* Vol 67 (No. 5), Apr 1988, p 888–890

68. D.A. Okongwe and V. Paranthaman, Effect of Particle Size of Coal Additive on the Burnt Properties of Clay Brick, *Am. Ceram. Soc. Bull.,* Vol 68 (No. 11), Nov 1989

69. W. Blau, T. Gerber, W. Goecke, B. Himmel, and R. Kranold, Measurements of Particle-Size Distributions and Agglomerate Structures by Small-Angle X-ray Scattering and Light Scattering, *Freiberg. Forschungsh. A,* Vol 778, 1988, p 168–180

70. D.C. Agrawal, Characterization of Dilute Ceramic Suspensions by Photon Correlation Spectroscopy, *Ind. Ceram. Soc.,* Vol 47 (No. 4), July-Aug 1988, p 97–102

71. M.G. Kim, M.J. Cima, and A.W. Searcy, Comparison of Sizes of Magnesia Crystals Calculated from X-ray Diffraction Line Broadening and TEM Observations, *J. Am. Ceram. Soc.,* Vol 72 (No. 8), 1989, p 1514–1516

72. J. Taylor, Submicron Particle Sizing by Light Scattering and Sedimentation, *Materials Research Society Symposia Proceedings,* Vol 24, 1984, p 327–334

73. N.F. Berk and K.A. Hardman-Rhyne, Phase Shift and Multiple Scattering in Small-Angle Neutron Scattering: Application to Beam Broadening from Ceramics, *Optics,* Vol 136 (No. 1–3), Jan-Feb 1986, p 218–222

74. J.S. Chappell, J.D. Birchall, and T.A. Ring, The Origin of Defects Arising in Colloidal Processing of Sub-Micron, Monosize Powders, *Br. Ceram. Proc.,* Vol 38, Dec 1986, p 49–57

75. D.J. Martins, E.A. Matney, and B.L. Leach, Statistical Process Control in the Raw Materials Industry, *Am. Ceram. Soc. Bull.,* Vol 67 (No. 5), May 1988, p 866–871

76. E.J. Schwarz, J.J. Kanet, and H.D. Leigh III, Quality Control Practices in the American Ceramic Industry, *Am. Ceram. Soc. Bull.,* Vol 68 (No. 3), Mar 1989, p 530–544

77. D. Holmes, SPC Techniques for Low-Volume Batch Processes, *Am. Ceram. Soc. Bull.,* Vol 69 (No. 5), May 1990, p 818–821

78. F.W. Faltin, G.J. Hahn, W.T. Tucker, and S.A. Vander Wiel, "Algorithmic Statistical Process Control," General Electric Corporate R&D, Schenectady, NY, 1990

79. B.C. Wood, Selection Criteria for Particle Size Analysis Equipment, *Powder and Bulk Solids,* Apr 1986, p 35–39

Filtration and Washing

James H. Adair, Materials Science and Engineering, University of Florida

MANY SINTERABLE GRADES of ceramic powders are now produced via chemical synthesis techniques because superior physical and chemical properties are more readily available with these powders (Ref 1–3). However, the recovery of chemically derived powders from solution and the removal of extraneous ions associated with the precursor compounds have presented additional processing issues not found with powders produced by conventional preparation routes such as calcination and dry milling (Ref 4, 5). The general goal of powder recovery is to concentrate the powders into a form amenable to green forming of the ceramic. Removal of the extraneous ions is required because evaporation of these species during green forming or sintering can disrupt the ceramic body by producing cracks or voids. Residual species can also compromise the properties of the final ceramic product.

The procedures to recover and wash ceramic powders have become more demanding with the synthesis and use of particles in the size range from 0.1 to 10 μm. Powders in this size range are more difficult to recover than coarser powders because the finer particles tend to occlude and foul filtration media more readily, thus reducing filtration efficiency. Yet, fine particles, particularly over the size range from 0.1 to 1 μm, are required for the ultimate consolidation of the powders via sintering. An additional prerequisite for the powder recovery system is that the formation of irreversibly formed aggregates must be avoided to ensure that green bodies formed from the powders have a uniform pore structure to minimize grain growth during sintering.

The objective of this article is to present a general outline of powder recovery techniques based on microfiltration procedures and to discuss strategies involved in washing extraneous ions from ceramic powders. Included in the discussion are methods to avoid the formation of irreversible aggregates via the use of organic flocculants that can be used to enhance recovery and washing without compromising the attractive physical chemistry of chemically derived powders.

Powder Recovery Techniques

A discussion of the vast number of powder recovery techniques is beyond the scope of this article. More detailed information can be found in a number of comprehensive reviews (Ref 6–9). Figure 1 provides a summary of solid-liquid separation techniques for given particle size ranges.

There are essentially two general techniques to recover or concentrate fine powders—centrifugation and filtration. Centrifugation is based on the sedimentation of powders in a gravitational or imposed centripetal field. Filtration, the primary focus herein, is based on the physical collection of particles on a porous filter or membrane. Filtration is the primary focus of this article because it is more amenable to industrial processes as well as being easier to prevent irreversible aggregation of the fine particles. Thus, centrifugation will be briefly reviewed followed by a more extensive discussion of filtration techniques.

Centrifugation

The height, H, at which a particle of diameter, a, will settle in time, t, is given by the Stokes equation where (Ref 10):

$$H = \frac{a^2(\rho_P - \rho_L)gt}{18\,\eta} \qquad \text{(Eq 1)}$$

where η is the solution viscosity, ρ_P and ρ_L are the particle and solution densities, respectively, and g is the acceleration due to gravity or centrifugation. Thus, the height to which a particle of a specific size settles is directly proportional to both time and the acceleration due to gravity or the applied centrifugation.

Ceramic suspensions may be centrifuged at different speeds depending on the particle size distribution to concentrate the particles from suspension. Unfortunately, the force acting on the particles to sediment them within reasonable times is usually great enough to bring the particles in close proximity and, thus, promote irreversible aggregation. The close proximity of the particles is not a problem if centripetal casting is to be used as the green body (Ref 11). However, the aggregated particles will become an issue if redispersion of the particles is desired or the dried powder is to be used in a green-forming operation such as uniaxial compaction. Polymeric flocculants, which enhance powder recovery by centrifugation or gravitational sedimentation and minimize irreversible aggregates, are discussed later in this article.

Microfiltration

The recovery of sinterable ceramic powders in the size range from 0.1 to 1 μm via filtration is more properly described as microfiltration. As shown in Fig 2, the recovery of ceramic powders can range from filtration (that is, particles greater than 10 μm) to ultrafiltration with particles recovered in the range from 0.001 to 0.1 μm. While the trend in ceramic particle sizes is toward finer sizes, processing considerations associated with dispersion and packing of powders currently limits most ceramic powders to greater than 0.1 μm. Thus, microfiltration techniques are of the current greatest interest although ultrafiltration techniques will continue to take on increasing importance.

The historical development of microfiltration is summarized in Table 1 with the earliest application of microfiltration in 1748 based on osmosis through animal bladders (Ref 12). Most current filter materials are based on copolymers of cellulose nitrate and cellulose acetate. The former component was the basis for the first synthetic filter in 1855. After World War II, many synthetic filter materials including polypropylene and polytetrafluoroethylene were developed. Although not indicated in Table 1, there is a growing market for ceramic filters for applications in which the higher strength combined with their chemical resistance are favored over polymeric membranes (Ref 13). The higher strength has an important functional significance because greater filtration rates can be achieved at higher pressures across the filtration membrane as discussed below.

Non-elastic Filtration. Adcock and McDowell were among the first to describe the filtration rate of fine particles on a filter membrane (Ref 14). In fact, their theory was motivated by the desire to better understand the slip casting kinetics in ceramic systems. Filtration of nonelastic particle compacts is similar to slip casting because the fine pore structure produced by the accumulation of the fine, filtered particles at the membrane surface limits the filtration rate. The fundamental kinetic relationship for filtration rate as developed by Adcock and McDowell is

$$L^2 = \frac{2\,PgE^3}{5S_P^2\eta\,(y-1)\,(1-E)^2}\,t \qquad \text{(Eq 2)}$$

where L is the filtrate thickness, P is the pres-

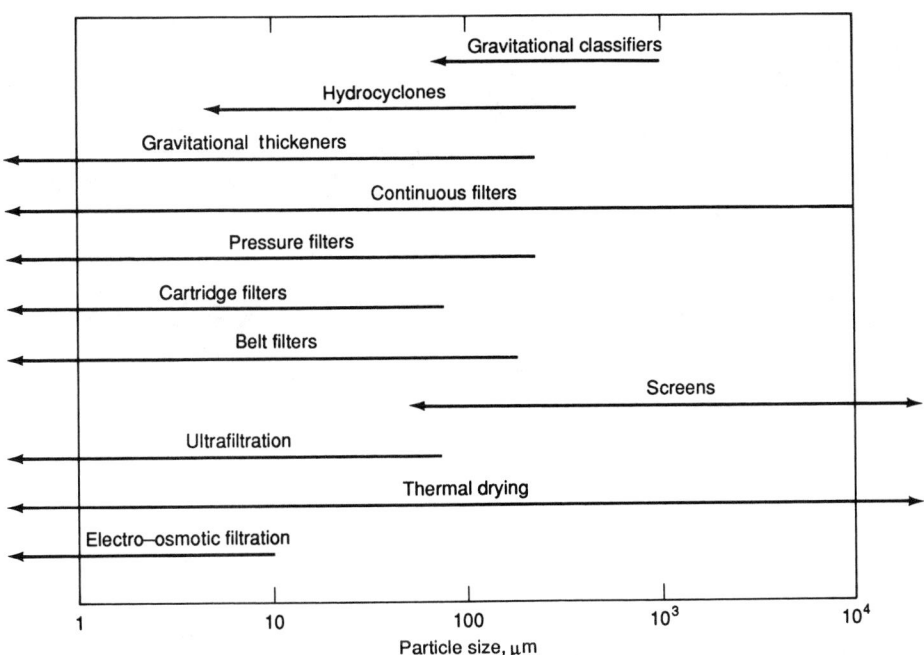

Fig 1 Particle size range over which various solid-liquid separation methods operate. Source: Ref 7

Table 1 Historical development of microfiltration

Year	Filter development
1748	Friar Abbe Nollet: • Osmosis through animal bladders
1855	Fick: • First synthetic membrane (nitrocellulose)
1918	Zsigmondy and Bachmann: • Development of commercial process
1927	Sartorious Werkes: • Began commercial production
1954	Millipore Corp.: • First membrane company formed
1964	Gelman Sciences Inc.: • Fabric-reinforced membranes
1970	Celanese Corp.: • Polypropylene membrane
1970	W.L. Gore and Assoc.: • Expanded PTFE
1984	Gelman Sciences Inc.: • Sunbeam process
1984	Pall Corp.: • Affinity membranes

Source: Ref 12

sure differential across the filtrate and filter membrane, g is acceleration due to gravity, E is the void fraction of particles in the filtrate, S_p is the surface area of particles in the filtrate, η is the fluid viscosity, y is the volume of suspension containing $(1 - E)$ of solids, and t is filtration time.

The validity of Eq 2 is demonstrated in Fig 3 which shows a linear relationship for L^2 as a function of time. Figure 3 also demonstrates that significantly faster filtration rates can be obtained if a greater pressure differential is applied. Thus, the filtration rate is

significantly greater at 415 kPa (60 psi) applied pressure than without any applied pressure as shown in Fig 3. Part of the motivation for the development of ceramic and metal membranes is their superior strength over polymeric membranes and, therefore, the greater pressure differentials that can be used.

The filtration-rate data in Fig 3 is somewhat deceptive as presented. As the fine particles accumulate in a relatively close-packed condition at the membrane surface, the filtration rate decreases as a function of $t^{-1/2}$. However, even with this limitation, the fil-

tration rate at a high-pressure differential is still significantly larger than the more "passive" filtration without applied pressure. This limitation in filtration rate created by the close packing of particles at the filtration surface may be reduced with polymeric flocculating agents to create larger filtration units that have larger pores and therefore less inherent resistance to fluid flow through the pore channels.

Filtration of Elastic Systems. Adcock and McDowell's analysis summarized in Eq 2 is based on the assumption that the filter cake is not elastic and thus does not become more compact as a function of filtration time. However, non-elastic filter beds are generally not encountered in most practical cases because of the low inherent filtration rate associated with non-elastic systems relative to elastic filter beds. Elasticity of the filter bed can be controlled by control over the state of flocculation. In the present context, flocculation

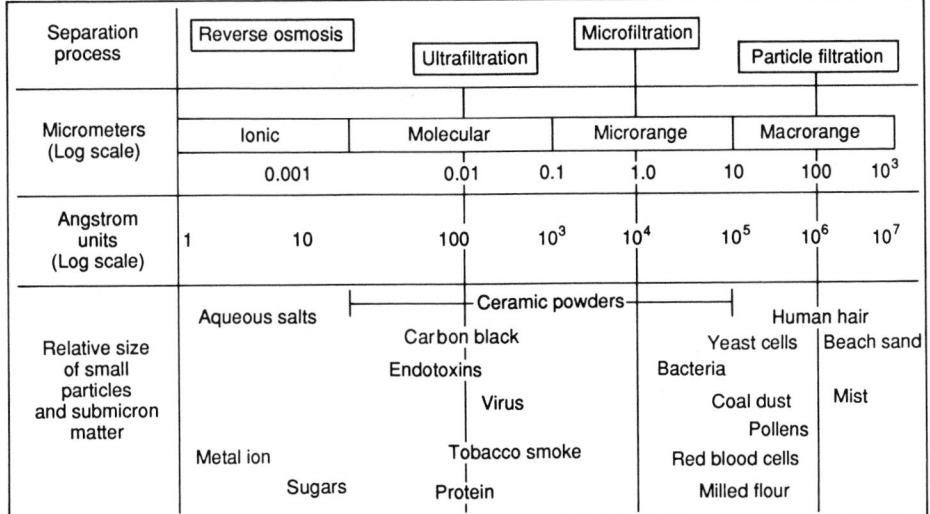

Fig 2 Range of particle sizes for which four filtration processes are capable of operating. Microfiltration ranges between 0.1 and 10 µm. Source: Ref 12

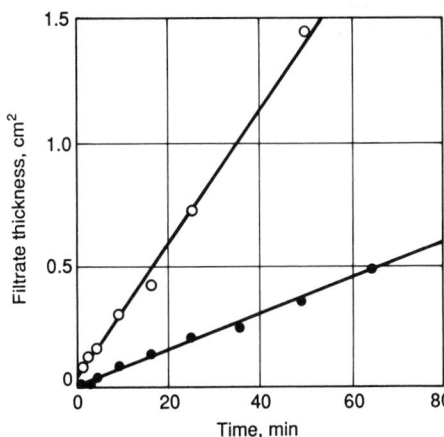

Fig 3 The square of the depth of filtered material as a function of filtration time. Closed circles are for filtration with no pressure differential. Open circles are data for filtration with a pressure differential equal to 415 kPa (60 psi). Source: Ref 14

is the preferred term over aggregation because of the irreversible nature of the agglomeration process implied by the latter term, a condition that is to be avoided for subsequent green-forming operations.

A quantitative description of the filtration kinetics for elastic systems is beyond the scope of this article. The interested reader is referred to the relevant literature (Ref 6–10). Elastic filtration requires the formation of flocs of the particles to be filtered. Floc formation is usually achieved by the intentional addition of polymers that promote bridging in and among the ceramic particles. The net effect is to increase the particle size which in turn leads to increased pore sizes in the filter bed and increased permeability. However, the pressure differential across the filter bed may be great enough to deform the relatively weak flocs within the filter bed leading to a decreased filtration rate although still significantly larger than filtration rates obtained for non-elastic systems.

The Role of Flocculation

There are a variety of ways that particles may agglomerate in suspension that will enhance filtration rates. Some of these are summarized in Fig 4. The favored way to promote agglomeration to enhance particle recovery without forming irreversible aggregates is via polymer bridge formation (Ref 10, 15). The major advantage of polymer flocculation is that after particle concentration via filtration, introduction of further polymer will redisperse the ceramic particles as demonstrated by Lange and coworkers (Ref 16). Thus, irreversible aggregation is relatively easily avoided and yet filtration rates can be greatly enhanced. Healy and LaMer demonstrated that maximum flocculation is achieved when one-half of the maximum concentration of polymer required to fully coat the particle surfaces is present (Ref 17). In practice, the optimum polymer dosage to achieve maximum flocculation is most easily determined by evaluating the state of flocculation as a function of the polymer concentration, that is, the degree of sedimentation in a series of suspensions as a function of polymer dosage and time (Ref 15, 18).

Moudgil and Shah have provided selection criteria for flocculants and flocs for enhanced

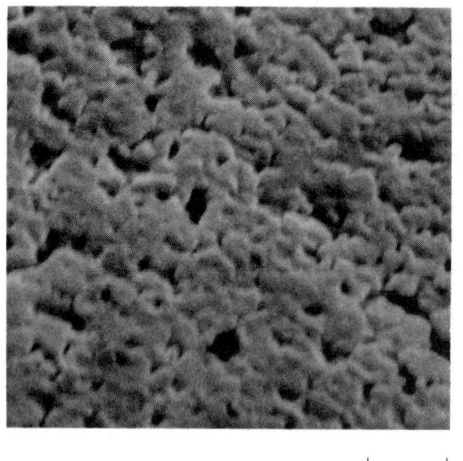
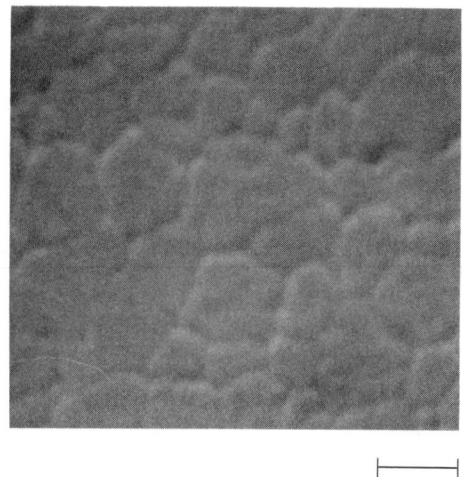

(a)   1 μm   **(b)**   0.4 μm

Fig 5   Sintered microstructures of ZrO_2 containing 3 mol% Y_2O_3 before and after removal of chloride. (a) 0.75 wt% Cl^- present. (b) Chloride reduced by successive washings to parts per million levels

filtration rate (Ref 15). Flocs should be porous, strong, and permeable. Some of the polymers used as flocculants are summarized in Ref 17. Some of these include:

- *Nonionic polymers* such as polyacrylamides and polyethyleneoxide
- *Anionic polymers* such as acrylamide copolymer and polyacrylics
- *Cationic polymers* such as polyamines and acrylamide co-polymers

In general, nonionic flocculants are preferred for ceramic systems because subsequent washing to remove co-ions (for example, Na^+ associated with polyacrylate or Cl^- associated with cationic polymers) is not required. Thus, polyethylene oxide and polyacrylamide are often used to promote concentration of particles for ceramic systems. A much wider selection of polymer types and commercial sources of polymer flocculants is available in the literature (Ref 19).

Powder Washing

Powder washing or laundering is required to ensure that salts that may affect either downstream processing or ultimate properties of the ceramic are removed after synthesis. For example, most commercial zirconia (ZrO_2) powders are precipitated from aqueous $ZrOCl_2$ solutions. The sintered microstructures for the 3 mol% Y_2O_3-ZrO_2 materials shown in Fig 5 dramatically demonstrate the influence that residual chloride has on the density of this ceramic. The material with 0.75 wt% residual Cl has a density of only 91% of theoretical while the washed powder has a density exceeding 99%. Scott and Reed (Ref 20, 21) have demonstrated that residual chloride has a deleterious effect on several ZrO_2 powders while Mistler and Shanefield have shown that residual sodium resulting from the Bayer process to recover alumina from bauxite compromises electrical insulating properties of

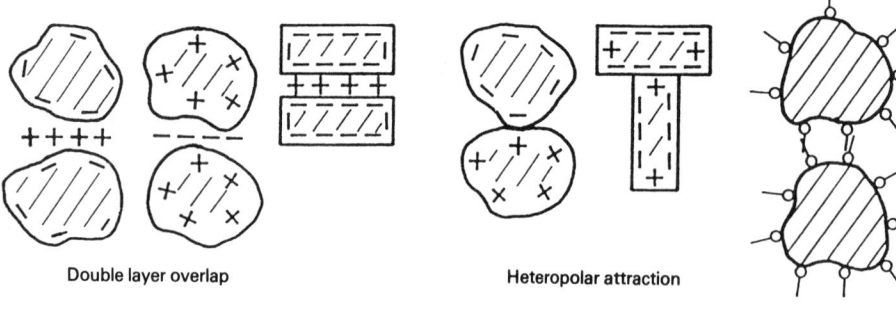

Double layer overlap   Heteropolar attraction

Coagulation   Hydrophobic

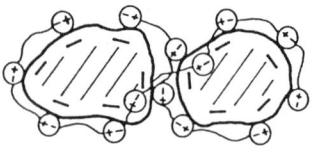
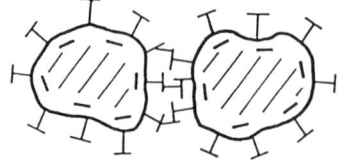

Polymer bridge   Heteropolar colloid bridge

Flocculation

Fig 4   Models of different ways that particles can coagulate or flocculate. Source: Ref 10

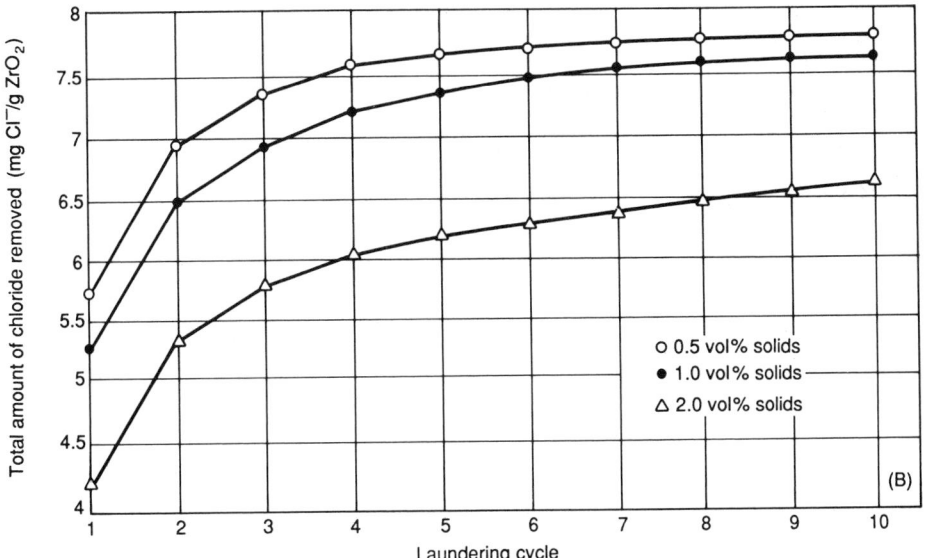

Fig 6 Amount of chloride removed after successive washes for submicron ZrO_2 powder. Source: Ref 20

sintered alumina (Ref 22, 23). Thus, washing of powders to remove residual, deleterious substances is an important aspect of powder recovery from solution (Ref 24).

The basis for washing ceramic powders is relatively straightforward. Since the residual ions to be removed are usually present on the surfaces of particles (Ref 20, 23), washing with deionized water promotes an ion exchange between the undesirable substance on the powder surface and water molecules via an equilibrium reaction such as:

$$Cl^- \text{ (surface)} + H_2O \text{ (bulk solution)}$$

$$= H_2O \text{ (surface)} + Cl^- \text{ (bulk solution)} \quad (Eq\ 3)$$

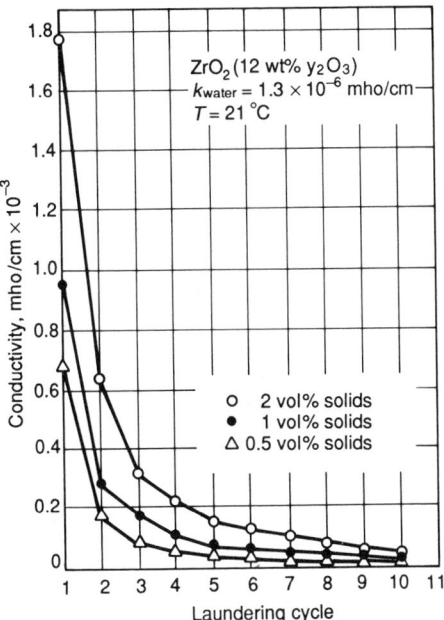

Fig 7 Specific conductivity of the supernatant of ZrO_2 powder containing 12 wt% Y_2O_3 after successive washings. Source: Ref 7

Successive washings are required to remove the undesired substance because of the nature of the chemical equilibrium as shown in Fig 6 for ZrO_2 powder.

The extent of washing may be followed by chemical analysis of the washed powder or, more easily, via specific conductivity measurements of the supernatant after each washing iteration. As shown in Fig 7, the specific conductivity of the supernatant decreases rapidly with each washing iteration depending on the solids content being washed. The speed at which residual ions can be removed also depends on the surface area of the powder. Powders with higher surface areas and smaller particle sizes will require more washing iterations to reduce the level of residual contaminant than coarser powders.

If a conductivity bridge is unavailable, the formation of insoluble precipitates in the supernatant can be used to monitor the extent of washing. For example, if halides (that is, Cl^-, Br^-, or I^-) are the residual ions to be removed, a 0.01 mole/L silver nitrate solution can be used to monitor the level of halide ion in the supernatant. Chloride concentrations less than 10^{-6} moles/L in the wash water can be detected by visual observation of the AgCl precipitates formed by the addition of the silver nitrate solution.

REFERENCES

1. D.W. Johnson, Jr., Nonconventional Powder Preparation Methods, *Am. Ceram. Soc. Bull.*, Vol 60 (No. 6), 1981, p 221–224, 243
2. B. Fegley, Jr., P. White, and H.K. Bowen, Processing and Characterization of ZrO_2 and Y-Doped ZrO_2 Powders, *Am. Ceram. Soc. Bull.*, Vol 64 (No. 8), 1985, p 1115–1120
3. K.S. Mazdiyasni, Powder Synthesis from Metal-Organic Precursors, *Ceram. Int.*, Vol 8 (No. 2), 1982, p 42–56
4. D.L. Hankey and J.V. Biggers, Solid State Reactions in the System PbO-TiO_2-ZrO_2, *J. Am. Ceram. Soc.*, Vol 64, 1981, p C172-C173
5. B.V. Hiremath, A.I. Kingdon, and J.V. Biggers, Reaction Sequence in the Formation of Lead Zirconate—Lead Titanate Solid Solution, *J. Am. Ceram. Soc.*, Vol 66, 1982, p 790–793
6. F.M. Tiller and C.S. Yeh, Introduction to Solid-Liquid Separation: Principles and Theoretical Aspects, *Advances in Solid-Liquid Separation*, H.S. Muralidhara, Ed., Battelle Columbus Laboratories, 1986
7. D.A. Dahlstrom, Selection of Solid-Liquid Separation Equipment, *Advances in Solid-Liquid Separation*, H.S. Muralidhara, Ed., Battelle Columbus Laboratories, 1986
8. L. Svarovsky, Ed., *Solid-Liquid Separation*, 2nd ed., Butterworths, 1981
9. *Solid Liquid Separation Practice 3*, The Institute of Chemical Engineers, Symposium Series No. 113, The Chameleon Press, Ltd., London, 1989
10. J.S. Reed, *Introduction to the Principles of Ceramic Processing*, John Wiley & Sons, 1988
11. I.A. Aksay, F.F. Lange, and B.I. Davis, Uniformity of Al_2O_3-ZrO_2 Composites by Colloidal Filtration, *J. Am. Ceram. Soc.*, Vol 66 (No. 10), 1983, p C190-C192
12. E.C. Gregor, The State of the Art of Microfiltration: Current and Future Applications, *Tappi J.*, April 1988, p 123–128
13. L.M. Sheppard, Corrosion Resistant Ceramics for Severe Environments, *Bull. Am. Ceram. Soc.*, Vol 7 (No. 7), 1991, p 1146–1158
14. D.S. Adcock and I.C. McDowell, The Mechanism of Filter Pressing and Slip Casting, *J. Am. Ceram. Soc.*, Vol 40 (No. 10), 1957, p 355–362
15. B.M. Moudgil and B.D. Shah, Selection of Flocculants for Solid-Liquid Separation Processes, *Advances in Solid-Liquid Separation*, H.S. Muralidhara, Ed., Battelle Columbus Laboratories, 1986
16. F.F. Lange, B.I. Davis, and E. Wright, Processing-Related Fracture Origins: IV, Elimination of Voids Produced by Organic Inclusions, *J. Am. Ceram. Soc.*, Vol 69 (No. 15), 1986, p 66–69
17. T.W. Healy and V.K. La Mer, *J. Colloid. Sci.*, Vol 19 (No. 4), 1964, p 323–332
18. M.A. Hughes, Coagulation and Flocculation, *Solid-Liquid Separation*, L. Svarovsky, Ed., 2nd ed., Butterworths, 1981
19. *McCutcheon's Functional Materials*, 1990, North America ed., McCutcheon Division, MC Publishing Co.
20. C.E. Scott and J.S. Reed, Analysis of Cl^- Ions Laundered from Submicron

Zirconia Powders, *Bull. Am. Ceram. Soc.*, Vol 57 (No. 8), 1978, p 741–743

21. C.E. Scott and J.S. Reed, Effect of Laundering and Milling on the Sintering Behavior of Stabilized Zirconia Powders, *Bull. Am. Ceram. Soc.*, Vol 58 (No. 6), p 587–590

22. R.E. Mistler and D.J. Shanefield, Washing Ceramic Powders to Remove Sodium Salts, *Bull. Am. Ceram. Soc.*, Vol 57 (No. 7), 1978, p 689

23. N.Z. Misak and H.F. Ghoneimy, Ion Exchange Properties of Hydrous Zirconia in Mixed Solvents: Thermodynamics of Alkali Cation Exchange in Methanolic Solutions, *Coll. Surf.*, Vol 7, 1983, p 89–104

24. D.A. White and S. Bashir, Counter Current Washing of Ion Exchange Flocs—Theory and Optimisation, *Solid Liquid Separation Practice 3*, The Institute of Chemical Engineers, Symposium Series No. 113, The Chameleon Press, London, 1989

Batching and Mixing

P.F. Messer, School of Materials, University of Sheffield, United Kingdom

THE OBJECTIVES OF BATCHING are to combine the required proportions of the components of a system in a safe, reproducible manner that suits the subsequent processing operations. This involves weighing out the solid components, metering the volumes of liquids, or, in some continuous processes, controlling the feed rates of the components.

The objectives of mixing are to produce a mixture with a composition and/or particle size distribution that approaches the uniform state on a scale of size appropriate to the use of the mixture, in a manner that suits the subsequent processing operations. Common operations in the ceramics industry are to mix different types of particulate solids with each other, and/or with liquids that may contain components in solution. The machine used to mix the components must be capable of causing the individual particles to move relative to one another. The type of machine selected depends on the rheological properties of the mixture.

Batching

A batch can range from several tens of grams, representing the small laboratory scale, to thousands of tonnes, as in the heavy clay industries. Care must be taken when handling dry powders because of the associated dust hazard (Ref 1). Such operations should be carried out under dust hoods or in suitably designed enclosures to prevent inhalation of particles by the operator.

Allowances must be made for the loss or pickup of material that occurs during processing, which can alter the composition of the final product. For example, powders, particularly those with high specific surface areas (total surface area per unit mass), can pick up moisture from the atmosphere in amounts that vary with humidity and may require drying to ensure accurate and reproducible batching of powder mixtures. The solubility of components in water can result in losses when water is removed by either filter pressing or decanting. Pickup occurs in milling. For example, when batching ferrites (ceramic magnets), the iron that is picked up from attrition of the steel balls and the mill lining must be considered.

In the production of whitewares, batching is carried out wet, because of the variable moisture contents in the clays, and because it is a safer handling procedure. Suspensions of the clays, flux, and filler are prepared, and their densities are measured to determine the amount of solids they contain. The required quantities of the suspensions, measured by either weight or volume, are then blended together. The suspensions must be continuously stirred to prevent sedimentation, and air bubbles must be minimized because they affect the accuracy of batching.

When preparing a batch that includes a minor component below the 1% level, it is common practice to add the component in a water-soluble form so that it can be adequately dispersed. Alternatively, the technique of masterbatching can be used, in which a mixture is prepared with a higher concentration of the minor component than is required in the final product. The batch is then prepared using the appropriate amounts of this more concentrated mixture and the other components. This approach ensures that the initial disposition of the minor component in the mixer is favorable for rapid mixing and avoids the possibility of it becoming localized in regions of the mixer where little or no mixing occurs. This approach can be further improved by calcining and milling the more concentrated mixture, provided that all of the minor component can go into solid solution with either the major component or the compound that forms from the major components upon calcination.

Mixing

Mixing operations have been studied more extensively by chemical engineers than by ceramists, but often only for very simple model systems. Very few publications (Ref 2–12) deal specifically with ceramic systems, which is surprising considering that mixing is so commonplace and that demixing can occur in many operations, such as slip casting and spray drying.

In some particulate systems, an ordered mixture can develop because of either physical or chemical differences between the particulate components. If ordering and demixing do not occur, the best mixture that can be obtained is a random homogeneous mixture, in which chance alone determines the positions of the particles. Because systems are generally nonuniform on the scale of size of interest, it is more appropriate to address the nonuniformity of composition and particle size.

Assessment of Nonuniformity

The nonuniformity of the composition or particle size on a particular scale of size can be assessed using either a direct or indirect approach. In the former, nonuniformity is examined after the mixing or demixing process, or after a later processing operation that does not significantly alter nonuniformity. In the indirect approach, nonuniformity is examined after a later processing stage by means of an obvious effect on either the intermediate or final product. In both approaches, the assessment can be either quantitative or qualitative. Qualitative methods may be suitable for quality control.

The direct approach requires the selection of the scale of size appropriate to the use of the mixture and the methods to sample, analyze, and describe the nonuniformity of the mixture. It can be difficult and time consuming to assess nonuniformity via a direct, quantitative examination with the very small scales of size encountered in ceramics. The initial data must then be related to the characteristic or behavioral properties of either a subsequent intermediate product or the final product. With the indirect approach, one or more of these properties is examined and used as the means of assessment.

Scale of Size. The nonuniformity of the characteristic properties of a given system depends on the scale of size being considered. This scale is known either as the scale of scrutiny or the characteristic volume. The latter term is used in this article. It is defined as the amount of material at any location throughout the system in which the positions of the individual particles or system components are considered unimportant. This parameter can be specified in a number of ways, such as bulk volume, mass, or number of particles. Thus, the characteristic volume is the scale of size on which information about the nonuniformity of the system is required.

The nonuniformity of a system can be determined directly on a quantitative basis by analyzing spot samples for the amounts of each component that they contain. These samples are either withdrawn from the system or analyzed *in situ*. The size of the spot sample must be the same as that of the appropriate characteristic volume when the direct approach is employed. However, because the characteristic volumes are generally very small fractions of the size of a given ceramic system, the analysis of appropriately sized spot samples may be impractical. Either a direct, qualitative approach or an indirect approach may have to be employed.

Demixing can sometimes occur during mixing or in subsequent operations. Nonuniformity can then be detected by analyzing spot samples that are fairly large fractions of the size of the system. In this case, the change in the characteristic properties, from one of these large characteristic volumes to the next, is usually systematic, and the variation throughout the system possesses a degree of symmetry. In the early stages of mixing, the nonuniformity of the characteristic properties can also be detected on a similar scale of size, although the change from one large characteristic volume to the next is usually not systematic. In both cases, the systems can be described as being macroscopically nonuniform.

If a system is found to be nonuniform on a certain scale of size, it is consequently nonuniform on all smaller scales. A system in which nonuniformity can only be detected by analyzing spot samples that are a very small fraction of the system is essentially uniform on the macroscale, but nonuniform on the microscale. In this case, the change in the characteristic properties, from one very small characteristic volume to the next, is not systematic. The system will exhibit both macroscale and microscale nonuniformities either in the early stages of mixing or when demixing occurs.

The Direct Approach. When an overall measure of system nonuniformity on a certain scale of size is required, spot samples of the appropriate size must be selected at random to obtain a representative set. The nonuniformity of particle size could be treated in a similar manner to that of composition by denoting particles in selected size ranges as different components of the system. The fractional amounts of a chosen component in the spot samples will be distributed about the mean value for that component in the system. Either the standard deviation or variance can be used as a measure of nonuniformity for the chosen component. Each component of the system can be chosen in turn and its nonuniformity assessed.

Alternatively, a mixing index can be constructed. Many indices have been devised (Ref 13). In some, the standard deviation or vari-

ance of the random homogeneous mixture (RHM) is introduced so that the approach to uniformity can be assessed relative to the most uniform attainable state, assuming that ordering and demixing do not occur. An ordered system would be more uniform than the random system. Variance is preferred to standard deviation because it has additive properties. An example of a mixing index is (Ref 14):

$$M = \frac{S^2 - \sigma_R^2}{\sigma_0^2 - \sigma_R^2} \qquad \text{(Eq 1)}$$

where S^2 is the variance for a chosen component, σ_R^2 is the variance for the RHM (Ref 14, 15), and σ_0^2 is that for the unmixed state (Ref 14).

For a low level of uncertainty, a large number (~100) of spot samples should be analyzed (Ref 15). If the index is to be followed throughout mixing, fewer spot samples (20 to 30) for each mixed state may be sufficient, because all the data will be used to establish how the index varies as a function of mixing time.

Most sampling and analysis techniques are not suitable for studying microscale nonuniformities. For example, sampling with a thief probe (Ref 16), which can withdraw a powder spot sample from a chosen location, is limited to spot samples that are moderate fractions of a gram. A technique that may be suitable when the components are of different compositions is quantitative microscopy using image analysis equipment. The spot sample will be the field of view that is analyzed to determine the volume fraction of each component present. Appropriately sized spot samples can be selected using variable magnification. Using optical microscopy, the image might be obtained by direct observation of a polished section of a resin-impregnated mixture (Ref 9). If the different components cannot be distinguished optically, micrographs obtained from electron probe microanalysis can be examined, provided that there is sufficient contrast between the components.

Qualitative assessment might be suitable in some circumstances. Consider, for example, the distribution of a temporary, organic binder in a green (unfired) ceramic. To perform its function and avoid pore formation upon its removal, the binder should coat the individual particles. Hence, the appropriate characteristic volume is related to the average particle size. Examination of fracture surfaces of the green ceramic using scanning electron microscopy (SEM) (Ref 11) may reveal localized concentrations of the binder.

When a system that might exhibit demixing requires examination, moderately large spot samples can be analyzed. The compositions or average particle sizes for these spot samples, together with information on their locations, can be used to construct a map that shows the variation of composition or average particle size throughout the system.

The indirect approach, in principle, can be applied to all cases. Inadequate mixing or the occurrence of demixing both mean that the behavioral properties of the fired product, such as tensile strength or electrical resistivity, will be adversely affected. Thus, the change of relevant properties of either an intermediate or final product could be followed as a function of either mixing time or another processing parameter that affects uniformity. Two examples are described below to illustrate the approach.

Consider the dispersal of a low volume fraction of a liquid-forming component into a powder that, subsequent to forming, will be liquid-phase sintered. The appropriate characteristic volume would be a small group of the particles that mainly remain solid during sintering. The characteristic volume is larger than one of these particles because, during sintering, the wetting liquid phase will spread to a limited extent over the particle surfaces. The distribution of the liquid-forming component can readily be investigated indirectly, whereas it would not be practical to use the direct, quantitative approach. In the regions of the green ceramic where the liquid-forming component is either deficient or excessive, the local densification and grain growth that occur during sintering will be either retarded or advanced. An examination of the microstructure of the fired ceramic will reveal the nonuniformity of porosity and grain size, which reflect the nonuniform distribution of the liquid-forming component. Such observations could be used as a qualitative assessment.

Alternatively, these observations could be coupled with the measurement of a relevant behavioral property to allow the effectiveness of the mixing operation to be quantitatively assessed. A common reason for inadequate dispersal is that the particle size of the liquid-forming component is too coarse. Progressive size reduction, coupled with a quantitative assessment, would allow the most suitable particle size to be selected.

Consider the mixing of oxides and carbonates prior to calcination to form a desired compound. The appropriate characteristic volume is determined by the distances that ions can diffuse during calcination. It depends on particle size, the contact between particles, and the severity of the heat treatment. Hence, the characteristic volume can be changed by varying the heat treatment. It may contain up to a few hundred particles.

The effect of mechanical mixing can be followed by measuring the amount of the desired compound that is formed by a given heat treatment. By determining the proportion of the compound in a large spot sample withdrawn from the calcined product, the average proportion for a vast number of characteristic volumes is determined. This corresponds to a mixing index, D_A, proposed by Hixson and Tenney (Ref 17). The compound formation depends on both mechanical mixing and thermal mixing by diffusion, the effects of which cannot be separated. However, by using a heat treatment of low severity, the effect of mechanical mixing can be followed for a long period. This index is plotted in Fig 1 for the wet mixing of barium carbonate and iron oxide in a ball mill, followed by calcination to form barium hexaferrite ($BaO \cdot 6Fe_2O_3$) at either 1140 °C (2085 °F) with no soaking period or 1280 °C (2335 °F) for 6 h (Ref 10). The index D_A is plotted against the log of mixing time in seconds. The decrease in D_A at long mixing times was not caused by demixing, but by the pickup of iron from the steel balls.

It has been shown theoretically, for model mixtures, and experimentally, for the mixing of barium carbonate and iron oxide (Ref 10), that the values of D_A for mixtures with minor components can be increased if an excess of the minor component is used. In the latter case, more barium hexaferrite is formed when excess barium carbonate is used. The excess required to maximize D_A decreases as the characteristic volume increases, and as the required proportion of the minor component in the mixture increases.

Mixture Behavior

Rheological behavior, as well as the mechanisms of mixing and demixing, are described below.

The rheological behavior encountered when mixing ceramic systems, such as dry powders, suspensions, pastes, or plastic bodies, varies widely. Machines of different design and power are required to effect the relative particle movement necessary for the composition and/or particle size to approach the uniform state on the required scale of size.

Noncohesive particles or granules exhibit free-flowing behavior. This results from the effect of gravity on individual particles or granules, which overcomes the forces of attraction between them when they make contact.

Suspensions of particles in liquids exhibit a wide range of behaviors that may be time dependent (thixotropic) and will be temperature dependent, because the viscosities of liquids generally decrease with increasing temperature. Generally, behavior depends on the viscosity of the liquid, the solids concentration, the particle size distribution, the particle shapes, and whether the particles attract or repel one another (Ref 18).

With nonattracting particles in suspension at low solids concentration, the behavior will be time-independent and may be either Newtonian or pseudoplastic. The apparent viscosity is independent of shear rate in Newtonian behavior, but decreases with increasing shear rate in pseudoplastic behavior. This decrease may result from the flow-induced orientation of anisometric particles. The apparent viscosity of suspensions increases strongly as the solids content increases. With a higher solids content of nonattracting particles, an increase in apparent viscosity with an increase in shear rate may be encountered as a result of particle interaction. This is dilatant behavior.

When particles in suspension attract one another, the shear stress must exceed a certain value, that is, the yield point, before the particle structures are disrupted and flow can occur. The behavior observed then might be Bingham, pseudoplastic with a yield point, or dilatant with a yield point. In addition, the flow behavior may be time dependent (thixotropic), with the apparent viscosity decreasing with the extent of working. At high solids content, a system might be described as a paste or plastic body. The presence of clays and water in mixtures can confer plasticity, that is, the ability of the material to be deformed significantly without rupture and to remain deformed when the applied forces are removed. Bodies without clay can be plasticized by the addition of polymeric materials.

Mixing and Demixing Mechanisms. Lacey (Ref 14) proposed three mechanisms by which particulate solids mix. These are convective, shear, and diffusive mixing.

Convective mixing involves the interchange of parts of the mixture as circulation patterns are set up in the mixer. Breakup into smaller and smaller parts occurs as mixing proceeds and the interfacial area between the different components increases. This mechanism is effective in the early stages of mixing. Convective mixing can occur in mixtures with any type of rheological behavior.

Shear mixing involves relative movement within the mixture, that is, drawing out and making thinner regions that contain different components to improve the uniformity. Shear mixing can occur in mixtures of all types.

Diffusive mixing, which is analogous to molecular diffusion in fluids, can occur only in mixtures consisting of free-flowing particles or granules or in suspensions of nonattracting particles. This mechanism involves individual particles moving under the action of the mixer and gravity. The particles or granules are scattered randomly as they collide with one another. Although the mechanism can cause the composition to approach that of the uniform state if particles are closely similar in size, very small differences in size have been shown (Ref 15) to cause demixing.

The three primary ways (Ref 15) in which demixing occurs because of size difference are:

- Trajectory segregation, which occurs when powders are thrown about in the mixer, with the largest particles traveling the farthest
- Percolation of fines through the spaces between coarse particles
- The rise of coarse particles upon vibration, whereupon fines move into the vacated spaces. Hence, coarse particles tend to be segregated at the top of dry mixtures. The reverse occurs for suspensions, in which coarse particles settle more rapidly than fines

Although other physical differences can lead to demixing, size difference is the most important in the dry mixing process. Density differences usually have only a small effect on demixing in dry mixing but in suspensions, settling occurs faster for particles of a given size as density increases. Demixing that

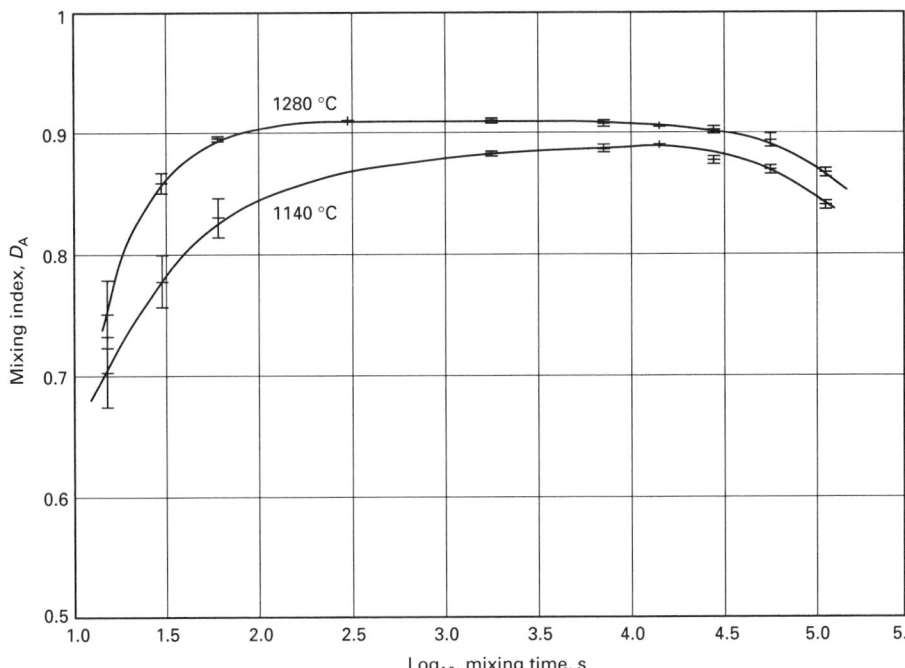

Fig 1 Effects of mixing time and heat treatment on the mixing index of a 1:6 barium carbonate and iron oxide powder mixture. Bars indicate a range of ±1 σ.

results from size and density differences can be prevented if the powders are mixed as a paste or if the particles are reduced in size so that they become cohesive.

If one of the components in a mixture is coarse and another is very fine, the fines may stick to the surfaces of the coarse particles to form an ordered mixture. This type of mixture is prepared in the manufacture of vitreous-bonded grinding wheels. The coarse abrasive grit is made sticky by first mixing it with a solution of dextrose in water. The fine components, which will form the glassy bond, are then added so that they can be picked up on the surfaces of the coarse particles. The mixing continues for a time that is sufficient for the fines to coat the grit, but not long enough for the solution to migrate through the coating. In this state, the mixture can be poured into the die cavities.

Demixing can occur in the slip casting process when coarser or denser particles settle preferentially. This is avoided by only partially deflocculating the slip, retaining sufficient particle-particle structure to prevent the coarser or denser particles from settling.

Demixing can also occur in the spray drying process. Fine particles tend to migrate to the surfaces of the granules during the drying process (Ref 11). In addition, components in solution may become concentrated at the granule surfaces.

Mixing Practice

Both mixer selection and mixing operations are described below.

Mixer Selection. Many different designs of particle mixers exist to handle different types of mixtures and quantities of material to be processed. Mixers are designed for either batch or continuous operation. Most have been developed to process materials that are not as hard or abrasive as ceramics, but these may be suitable if the mixer parts that contact the mixture are either made from wear-resistant materials or coated with a tough polymeric material.

Selecting the most suitable mixer is a difficult task that involves trials with the particular mixture and assessment of the quality of the mixture produced.

The mixing action results from the complex interplay of the movements of the mixing tools and/or body of the mixer and the rheological behavior of the mixture, which tends to change as mixing proceeds. Consequently, the mixing action of a given mixer varies with the material being mixed. For example, turbulent flow, which can disperse agglomerates, may be induced in suspensions of low apparent viscosity, but not in those with a high value.

Mixers usually must be capable of breaking down agglomerates so that the individual particles can move relative to one another. However, granules must not be broken down when mixing a powdered lubricant, such as zinc stearate, into a spray-dried powder. In this case, the gentle tumbling action of a double cone or twin shell mixer would be useful.

In some cases, a reduction in particle size may be required so that uniformity can be approached on a certain scale of size. In other cases, particles must not be reduced, such as when mixing either the batch for grinding wheels or the components to make porous filters with a specified pore size.

When selecting a mixer that must mix different batches, an important requirement is the ease with which it can be cleaned. If several mixers fulfill the technical requirements, selection would then be based on the costs involved.

Mixing Operations. A manufacturing scheme can include one or more mixing operations. However, it must be realized that the uniformity of composition may improve in other processing operations. In the mixed-oxide solid-state reaction route, for example, the approach to uniformity is affected not only by the particulate mixing operation and thermal diffusion during calcination, but also by the particulate mixing and size reduction that occurs during milling of the reacted oxides and diffusion that occurs in subsequent sintering.

The material may also be made less uniform, as already discussed, by the demixing of particulate components of dissimilar size or density in operations in which relative particle movement can occur. In addition, large-scale nonuniformities of composition can arise during firing because of the loss of volatile components from the surfaces of ceramics, as well as incomplete oxidation.

Examples of mixing operations that are drawn from the wide range encountered in ceramics are given in Table 1. These examples include the mixing of dry powders, pastes, plastic bodies, and suspensions.

Table 1 Examples of mixing operations

Product/stage of mixing	State of mixture	Components of mixture	Type of mixer	Effective mixing mechanisms
1. High-alumina ceramics/1st stage	Dry	Alumina powder, fluxes, grain growth inhibitor and milling aid	Ball mill	Convection and shear
2. High-alumina ceramics/3rd stage	Dry	Spray-dried granules and zinc stearate	Double cone tumbler	Convection and diffusion
3. Reaction-bonded silicon carbide/1st stage	Dry	Graphite and silicon carbide powders	Y-cone tumbler	Convection and diffusion
4. Vitreous bonded grinding wheels. (a) 1st stage. (b) 2nd stage	Paste	(a) Coarse abrasive grit and dextrose solution. (b) Fine components for vitreous bond and sticky coarse grit	Planetary mixer	(a) Shear, convection, and wetting by liquid. (b) Shear and convection
5. Porous filters. (a) 1st stage. (b) 2nd stage	Paste	(a) Fired clay grog and sodium silicate solution. (b) Fine glass powder and sticky grog	Planetary mixer	(a) Shear, convection and wetting by liquid. (b) Shear and convection
6. Refractories	Paste	Graded refractory components and sulfite lye	Pan mixer	Shear, convection, and wetting by liquid
7. Heavy clay products/1st stage	Plastic	Natural clays and water	Muller mixer	Shear, convection, and wetting by liquid
8. Heavy clay products/2nd stage	Plastic	Plastic body	De-airing pug mill	Shear and convection of plastic mass and diffusion of water
9. Whitewares/2nd stage (homogenization of moisture content)	Plastic	Filter cake	De-airing pug mill	Shear and convection of plastic mass and diffusion of water
10. Whitewares/1st stage	Wet	Clays, flux, filler, and water, plus deflocculant	Blunger	Wetting of solids by water, convection, shear, and diffusion
11. High-alumina ceramics/2nd stage	Wet	High alumina mixture, water, and deflocculant	High-speed blunger	Wetting of solids by water
12. Reaction-bonded silicon carbide/2nd stage	Wet	Graphite and silicon carbide powder mixture, polymers for plastic forming, and solvent to produce slurry	Y-cone tumbler	Wetting of solids by solvent, dissolution of binder, and its diffusion and convection
13. Electrical and magnetic ceramics (mixed oxide route)	Wet	Oxides, carbonates plus additives (water-soluble dopants, liquid-forming additions or grain-growth inhibitors), water, and deflocculants	Ball mill or high-speed blunger	Wetting of solids by water, convection, shear, and diffusion

Dry mixing can be used as a preliminary mixing stage before the addition of a liquid, as in examples 1, 3, 11, and 12 in Table 1. Tumbling, which can be carried out in a variety of rotating containers with different shapes, granulates powders that have any tendency to agglomerate. This can be countered by fitting the mixer with an intensive mixing bar, which can be rotated rapidly to break up agglomerates.

Dry mixing in a ball mill, as in example 1, can only be used in cases where the powders do not form compacted layers on the milling media or the lining of the mill. The use of a surfactant, such as ethylene glycol, at the fractional percentage level can reduce this problem. Milling reduces the size of the particles, which should lead to a more uniform mixture on the microscale.

The equipment used to mix pastes, such as planetary and pan mixers, must be capable of shearing the mixtures, which will have high to very high apparent viscosities, so that the agglomerates are broken down and the liquid component contacts all the individual particles. In a planetary mixer, the mixing tools move in a complex pattern, stirring and plowing the mixture in a cylindrical pan, which may either be stationary or may rotate. In pan mixers, the mixing is carried out by the stirring action of blades and by plows that move through the mixture in either a stationary or rotating pan.

Stiff pastes or plastic bodies can be mixed in high-shear sigma blenders. Heated mixtures of ceramic powders and polymers used for injection molding are commonly prepared using this mixer, but it has been found (Ref 19) that mixing with a twin-screw extruder causes less oxidative degradation of the polymer.

Pastes used in screen printing are often mixed using triple-roll mills. The paste adheres to the rolls and is sheared between them because of their differential rotational speed. This action disperses the agglomerates. It is also possible to mix such pastes in static in-line mixers, where the paste is pumped at high pressure through a tube containing a set of fixed mixing elements, which shear the mixture.

The preliminary stage of mixing heavy clay products is carried out in a muller mixer. The clay, contained in a cylindrical pan with water, is sheared and crushed between the heavy muller wheels and the surface of the pan. Plows direct the mixture under the wheels. Further mixing and de-airing is performed in a pug mill. The mixture is sheared by rotating knife blades, which push it through a multi-orifice plate into a chamber that is partially evacuated. The shredded, moist clay falls onto an auger, which pushes the plastic mass out of the pug mill through an extrusion die. During this stage, the plastic mass is consolidated.

A de-airing pug mill is also employed to homogenize the moisture content of whiteware bodies produced by filter pressing. Before filter pressing, these bodies are mixed as a suspension in a blunger, for which various designs exist. The suspension is stirred either by a slowly rotating propeller in an octagonal tank or, in more recent designs, by a rapidly rotating impeller.

High-speed blungers are used to mix other suspensions, employing turbulence and cavitation to disperse the agglomerates (example 11). However, the most effective way to disperse agglomerates is to wet mill a suspension of powder that contains a deflocculant, by using a ball, attritor, or vibroenergy mill. This method reduces particle size to improve uniformity on the microscale and is commonly employed for the "mixed oxide" approach (example 13).

Detailed descriptions of the mixers used in the ceramics industry are provided in the chemical engineering literature (Ref 16, 20), and in literature from the manufacturers.

REFERENCES

1. E. King, Health Risks of Fine Powders, Chapt. 12, *Principles of Powder Technology*, M.J. Rhodes, Ed., John Wiley & Sons, 1990

2. R.C. Rossi, R.M. Fulrath, and D.W. Fuerstenau, Quantitative Analysis of the Mixing of Fine Powders, *Am. Ceram. Soc. Bull.*, Vol 49 (No. 3), 1970, p 289–292

3. H. Ries, The Importance of Mixing Technology for the Preparation of Ceramic Bodies, *Interceram*, Vol 3, 1971, p 224–228

4. J.S. Reed, R.J. Williams, and T.J. Robert, Microstratified Mixing, *Am. Ceram. Soc. Bull.*, Vol 51 (No. 12), 1972, p 908–909

5. J.M. Pope and K.C. Radford, Blending of UO_2-ThO_2 Powders, *Am. Ceram. Soc. Bull.*, Vol 53 (No. 8), 1974, p 574–578

6. T.D. Croft, J.S. Reed, and R.L. Snyder, Microstratified Mixing of Ceramic Systems: Viscosity Dependence, *Am. Ceram. Soc. Bull.*, Vol 57 (No. 12), 1978, p 1111–1115

7. R. Hogg, Grinding and Mixing of Nonmetallic Powders, *Am. Ceram. Soc. Bull.*, Vol 60 (No. 2), 1981, p 206–211, 220

8. P.F. Messer, Uniformity in Processing, *Brit. Ceram. Soc. Trans. J.*, Vol 82 (No. 5), 1983, p 156–162

9. Y. Lin, R.L. Bowland, and P.E. Messer, Assessment of the Uniformity of Composition, *Processing of Advanced Ceramics*, J.S. Moya and S. de Aza, Ed., Soc. Esp. Ceram. Vidr., Madrid, 1986

10. Y. Lin and P.F. Messer, Mixing Prior to Calcination, *Brit. Ceram. Soc. Trans. J.*, Vol 86 (No. 3), 1987, p 85–90

11. J.S. Reed, Batching and Mixing, Chapt. 17, *Introduction to the Principles of Ceramic Processing*, John Wiley & Sons, 1988

12. E. Bouteloup, C. Michel, J.M. Haussone, P. Boudois, and O. Regreny, Mixing of Powders: Main Parameter of Synthesis of a Material by Calcining, *Sci. Ceram.*, Vol 14, 1988, p 163–168

13. L.T. Fan, S.J. Chen, and C.A. Watson, Solids Mixing, *Ind. Eng. Ceram.*, Vol 62 (No. 7), 1970, p 53–69

14. P.M.C. Lacey, Developments in the Theory of Particle Mixing, *J. Appl. Chem.*, Vol 4 (No. 5), 1954, p 257–268

15. J.C. Williams, Mixing of Particulate Solids, Chapt. 16, *Mixing: Theory and Practice*, Vol III, V.W. Uhl and J.B. Gray, Ed., Academic Press, 1986

16. C.W. Clump, Mixing of Solids, Chapt. 10, *Mixing: Theory and Practice*, Vol II, V.W. Uhl and J.B. Gray, Ed., Academic Press, 1967

17. A.W. Hixson and A.H. Tenney, Quantitative Evaluation of Mixing as the Result of Agitation in Liquid-Solid Systems, *Trans. Am. Inst. Chem. Eng.*, Vol 31, 1935, p 113–128

18. J.S. Reed, Rheological Behaviour of Slurries and Pastes, Chapt. 15, *Introduction to the Principles of Ceramic Processing*, John Wiley & Sons, 1988

19. K.N. Hunt, J.R.G. Evans, and J. Woodthorpe, The Influence of Mixing Route on the Properties of Ceramic Injection Moulding Blends, *Brit. Ceram. Soc. Trans. J.*, Vol 87 (No. 1), 1988, p 17–21

20. D.C.H. Cheng and S.H.R. Zaidi, "A General Review of the Problem of Mixing Equipment Selection and a Survey of Commercially Available Industrial Mixing Equipment," Report from Warren Spring Laboratory, Stevenage, Herts, United Kingdom, 1970

Granulation and Spray Drying

Stanley J. Lukasiewicz, Texas Instruments

GRANULATION is the intentional agglomeration of fine particles into larger clusters in order to improve certain powder properties. For example, bulk powders typically have a low bulk density, do not readily flow, are dusty, and have low thermal conductivity. When properly granulated, the same powder pours easily, exhibits a high and uniform bulk density, does not experience dusting losses, and more efficiently transfers thermal energy.

Granulated powders used in the ceramics industry are intermediaries. They are typically used as feedstock for either a shape-forming technique, such as powder pressing, injection molding, or extrusion, or a thermal treatment operation, such as calcining or melting. The intended use of the granulated powder dictates the properties that are most important, and, therefore, the optimum granulation technique.

For example, in the case of a powder used as a feed material for high-speed presses, the granules should typically be greater than 50 μm (2 mils), but less than 1000 μm (40 mils) in diameter, have a spherical shape, and should not be strong enough to retain their identity in the compacted part. The most commonly used techniques are spray drying and spray granulation. However, when powder is granulated prior to use in a calcining or melting operation, the granule size can be larger and a spherical shape is not necessary. In this case, tumbling and pressure granulation methods are more commonly used.

Granulation Techniques

The granulation methods used by the ceramic industry can be categorized as agitation, pressure, or spray techniques. Granulation by agitation involves bringing moist particles into contact by mixing or tumbling so that bonding forces can cause agglomeration. Pressure granulation is accomplished either by compacting powder into briquettes or pellets, or by extruding powders that are in a plastic state through a perforated plate or orifice. Granules are formed by spray techniques either by atomizing a powder-liquid suspension into a hot, dry gas or by spraying a liquid over a bed of fluidized powder. Granules formed by these methods tend to have a spherical geometry.

Bonding Mechanisms

Although van der Waals forces, electrostatic forces, or particle interlocking can contribute to the strength of granulated ceramic powders, most adhesion is achieved by using mobile liquids or binders.

Mobile, low-viscosity liquids bind particles by capillary suction. When the liquid occupies less than approximately 30 vol% (Ref 1) of the void space in a granule, discrete lens-shaped rings form at particle-particle contacts. This is known as the pendular state (Fig 1a). Increased liquid levels cause the rings to coalesce and form a continuous liquid network with interspersed air, known as the funicular state (Fig 1b). The capillary state is achieved when more than 80% of the pore volume is occupied by the liquid (Fig 1c). The tensile strength of a granule in the capillary state is approximately three times that achieved in the pendular state (Ref 1). Granules in the funicular state are intermediate in strength.

Binders agglomerate by forming bridges among the particles within a granule (Fig 1d). The strength of the granules is controlled by the physical properties of these bridges and is generally higher than that achieved with mobile liquids. The binders can be either inorganic or polymeric in nature.

Granulation by Agitation

Tumbling Methods. A drum granulator consists of a rotating cylinder that is inclined up to 10° from the horizontal position to assist material transport through the drum (Ref 2). The tumbling motion imparted to the particles tends to form spherically shaped granules. The drum may contain a dam ring at the exit to increase powder residence time and scrapers may be used to prevent internal build-up. The drum must be long enough to allow sufficient retention time to form granules of the necessary size. Powder can be introduced to the drum in a dry or moist condition. If it is introduced dry, a liquid (typically water or a binder solution) is sprayed onto the powder near the entrance. Because there is no inherent size classification mechanism in a drum granulator, a screening operation may be necessary to remove excessively large or small granules from the final product.

A second type of tumbling granulator, an inclined disk (Ref 3, 4), is shown in Fig 2. It consists of a tilted rotating disk with an attached rim to retain the powder charge. The feed is centrally loaded onto the disk and either water or a binder solution is added to promote agglomeration. The powder repeatedly cascades down the disk and is discharged as spherical granules over the retaining rim. Scrapers are used to prevent caking and build-up. One advantage of a disk granulator is its ability to produce uniformly sized granules. Further size classification is not generally necessary.

When using an inclined disk, the angle of inclination, rim height, rotational speed, and amount of admixed liquid will control the size of the granules formed. The granule shape is influenced by the amount of admixed liquid, the particle size of the feed powder, and the growth mechanism of the granules (Ref 4). Conditions that cause rapid agglomeration, such as high liquid content and very fine powders, will result in irregular shapes. Spherical granules result when granule growth occurs by a layering process.

Blenders and mixers, such as v-blenders and continuous rotating drums, can also be used to granulate powders. Agglomeration occurs in the presence of a small quantity of liquid or binder solution, and the resulting granules are often weakly bonded. Granules are typically several millimeters or larger in size, and require screening if a specific size distribution is desired.

Mixing Methods. Powders can be granulated using paddle mixers, high-speed shaft mixers, such as peg or pin granulators, or horizontal pan mixers. They exert force on the powder mass by means of multiple agitators, which combine mixing with size enlargement. The kneading action of the paddle mixers and the high-speed shaft mixers can form strong, dense, irregularly shaped granules. The mixers are capable of handling materials in a plastic condition.

Paddle mixers are single- or double-shaft horizontal trough vessels; each trough contains a mixing shaft (Ref 2). In a double-shaft

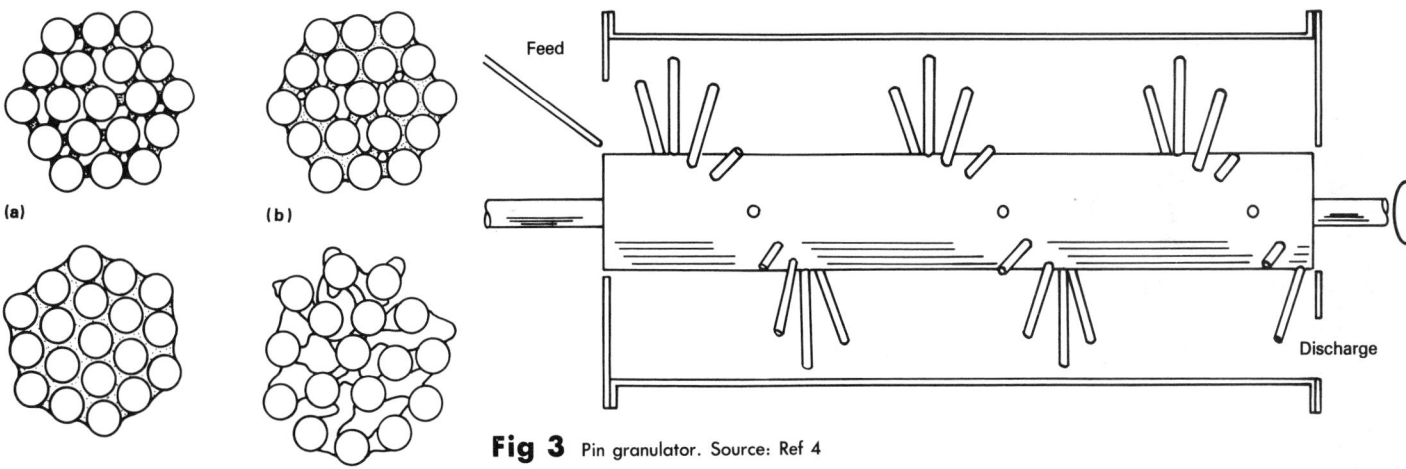

Fig 3 Pin granulator. Source: Ref 4

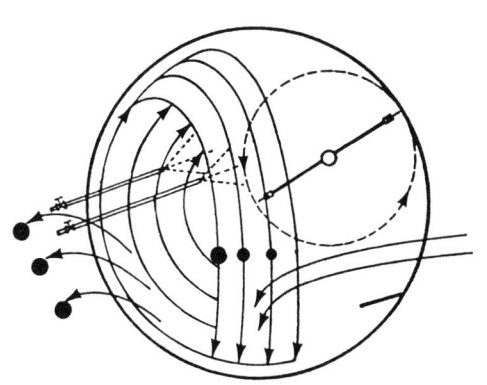

Fig 1 Bonding mechanisms in granule formation. (a) Pendular state. (b) Funicular state. (c) Capillary state. (d) Polymeric binder

arrangement, the shafts rotate in opposing directions and move the powder toward the center. Multiple paddles of various design are attached to the shaft along its length. Feedstock can be added at different points along the mixer to fully utilize mixing capacity.

High-speed mixers are typically single-shaft devices and attain more intensive granulation action because of their higher rotational velocities (Ref 5). Either pegs or pins are generally used, instead of paddles. A pin mixer is depicted in Fig 3. Because agglomeration occurs quickly in high-speed mixers, required residence times are short.

A *horizontal pan mixer* comprises a rotating mixing pan with eccentrically mounted mixing blades that rotate in a direction opposite to the pan rotation (Ref 4). The addition of a small amount of liquid forms small granules that can be used as feedstock for a subsequent granulation method. Horizontal pan mixers are operated in a batch mode.

The agitation granulation techniques, both tumbling and mixing, are typically used to prepare feedstock for calcining or melting operations. Bulky powders are granulated prior to these operations if the improvements in heat transfer, flow properties, and bulk density can be justified on either a cost or technical basis.

Granule Size Limitations. Granules formed by agitation techniques are subjected to destructive forces, within the moving powder charge, that oppose those forces causing agglomeration. The balance between these forces determines the maximum granule size possible under given process conditions.

The tensile strength, S, of a granule formed from monosized spherical particles is given by (Ref 6):

$$S = \frac{9}{8} \frac{NF(1-E)}{\pi D^2} \qquad \text{(Eq 1)}$$

where E is the volume fraction of void space in the granule, D is the diameter of the monosized particles, N is the coordination number of the particles, and F is the bonding force per particle contact point. It can be seen from this equation that decreasing the particle size of the powder charge will increase granule tensile strength and, therefore, will increase the maximum attainable granule size. Also, when using a mobile liquid as the bonding agent, either increasing the amount present or adding a binder will increase the bonding force per particle contact, further increasing granule strength and the maximum possible granule size. Conversely, increasing the intensity of agitation will increase the destructive forces acting in a system and will decrease the maximum attainable granule size.

Granulation by Pressure

Size enlargement by pressure granulation is accomplished by compressing a powder while it is held in a confined space. Pellets formed are usually of a regular, but nonspherical, geometry. Common techniques include roll briquetting, pelletizing, and extrusion.

In roll briquetting, powder is compacted in the gap between two rolls rotating at the same speed (Ref 3). Granules produced by this method are commonly in the form of a "pillow" shape as the powder is pressed between matching indentations in the rolls. Pressure is applied to the rolls hydraulically. A safety valve prevents excessive pressure build-up in the event that a foreign object reaches the rolls.

Pellet mills form cylindrical shapes by compressing and forcing moist powder through an orifice (Ref 2). One of the more popular designs is shown in Fig 4. A binder can be added to the powder to promote adhesion, increase pellet strength, and act as a lubricant. Pellet sizes range from approximately 2 to 30 mm (0.08 to 1.2 in.).

Extrusion is accomplished by forcing a plastic powder mass through either an orifice or a perforated plate. The extruder can be either the piston or auger type. Water and binder solutions must be added to materials that are not already in a plastic condition. Pellet strength and density are increased by de-airing the feed prior to formation. The granules formed tend to be several millimeters or larger in size, and their shape is defined by the geometry of the orifice or plate.

Pressure granulation techniques typically produce large granules and are used in the

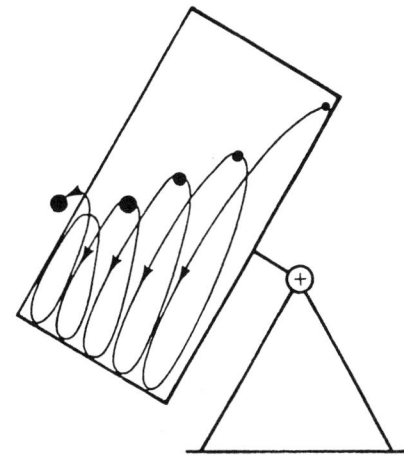

Fig 2 Inclined disk granulator

Fig 4 Pellet mill. 1, Loose material feeds into pelleting chamber; 2, Rotation of die and pressure of roller forces material through die, compressing it into pellets; 3, Adjustable knives cut pellets to desired length

ceramics industry to form a feed material for calcining or melting operations. They can also be used to prepare the feedstock for injection-molding processes. They are not used to produce press-powders for automatic presses.

Granulation by Spraying

Spray techniques are often utilized to prepare granulated powder for automatic presses because they tend to yield dust-free, highly flowable powders of constant bulk density. The most commonly used methods are spray drying and spray granulation. The feed for a spray dryer is a solid-liquid suspension. For a spray granulator, it is usually a dry or slightly moist bulky powder.

Spray granulation (Ref 7) forms granules by atomizing a liquid or a binder solution into a fluidized powder bed (Fig 5). Fluidization is achieved by directing a heated gas, which is usually air, through a distributor at the bottom of the powder bed. The gas imparts a vigorous motion to the particles, which prevents the formation of large lumps. The binding liquid is usually sprayed into the powder bed with a two-fluid nozzle. Spray granulators can be designed to operate in either a batch or a continuous mode.

The formation of granules in a spray granulator occurs through the random nucleation of small seed agglomerates, followed by the growth of these seeds to the desired size. Growth occurs either by the layering of powder onto the seeds or by the agglomeration of seeds to form larger granules. Granule growth by seed agglomeration forms irregular shapes.

Granule size increases as the fraction of the bed exposed to the binding liquid is reduced and as the spray nozzle is adjusted to give coarser droplets (Ref 4). As in the agitation methods, increasing the intensity of agitation of the bed by increasing the fluidizing gas velocity will decrease the size of the granules. There is an upper limit on granule size

because of the tendency of the powder bed to de-fluidize. However, spray granulation can form larger granules than is usually possible by spray drying because of longer residence times.

Spray granulation is commonly used in the pharmaceutical industry to prepare feedstock for tablet presses. It is used less frequently in the ceramics industry for producing press-powder because many ceramic powders are wet-milled prior to granulation and would have to be dried before being spray granulated.

Spray Drying

Spray drying is the most common granulation technique for producing ceramic press-powders. It is a continuous operation that produces a free-flowing granulated powder with uniform and repeatable properties.

The feed for a spray dryer is usually a water-based suspension, called a slurry, and the drying medium is heated air. The slurry is pumped to an atomizer located in the drying chamber, where it is broken into a large number of droplets. These droplets quickly achieve a spherical shape because of surface tension effects, and their large surface area-to-volume ratio allows high evaporation rates. The dried and granulated powder is collected at the base of the unit, while the fines are separated from the exhaust air by either a cyclone or a bag collector. Figure 6 represents a typical spray dryer utilizing a disk atomizer with co-current air flow. Examples of commercially spray-dried powders are shown in Fig 7.

Because spray drying is the most common method of producing ceramic press-powders, and because the character of the granulated

spray-dried powder can strongly influence the properties of the pressed articles, a thorough discussion of the important parameters of this process follows.

Slurry Preparation. Because most spray-dried ceramic powders are produced from water-based slurries, this discussion is limited to such systems. If the slurry is formulated with a flammable or explosive liquid, closed-cycle dryers that use nitrogen as the drying medium are required. Although the mechanisms of granule formation are the same in both cases, the reader is referred to Masters (Ref 8) for a discussion of spray drying using closed-cycle dryers.

The slurry is prepared by uniformly dispersing the ceramic powder in water. Any aggregates present should be eliminated through a milling or screening procedure. A high weight percent solids slurry is usually desired because it results in a higher powder output rate from a dryer with a given evaporative capacity. However, as solids content increases, slurry viscosity also increases and mixing becomes difficult.

Slurry stability and viscosity are important, and the methods by which they can be controlled are described in the article "Processing Additives" in this Section of the Volume, as well as in Ref 9 to 12. The roles of dispersants and binders are also described in the above-cited article. Additional information on dispersants is provided in Ref 13 to 20, whereas binders are described in detail in Ref 21 to 25.

Spray-dried powders typically contain between 0.5 to 4.0 wt% binder, based on the dry weight of the ceramic powder. An excessive amount can produce spray-dried granules that resist deformation during compaction

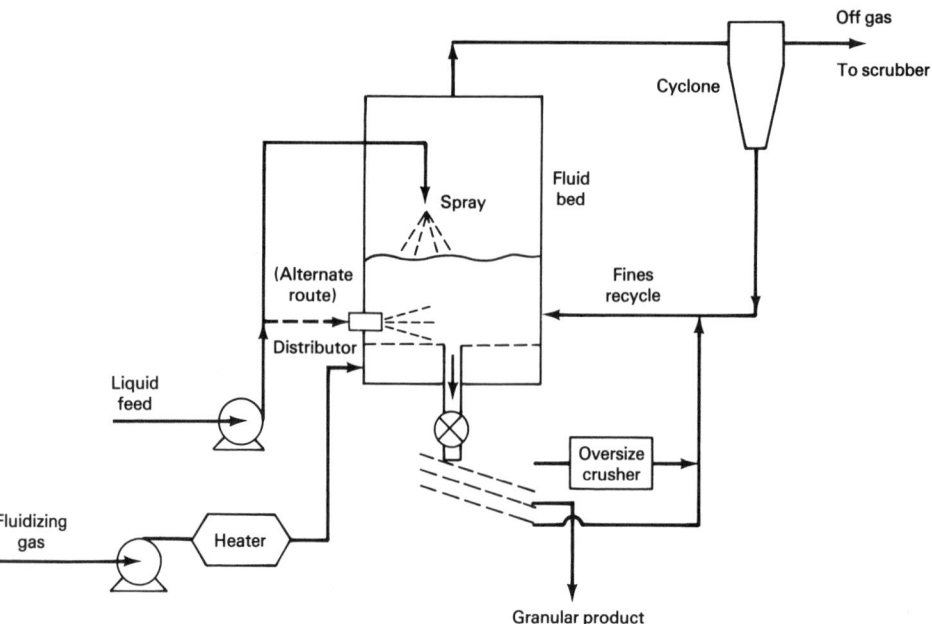

Fig 5 Fluidized bed spray granulator. Source: Ref 4

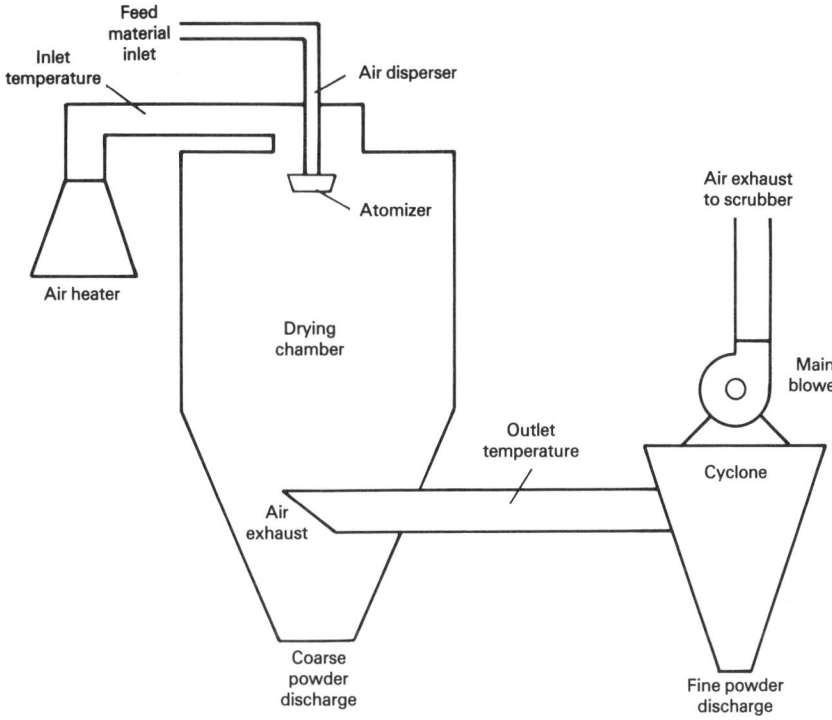

Fig 6 Typical spray dryer arrangement

(Ref 22) and may generate excessive gaseous pressure within the compacted article during burnout. An insufficient amount can result in low green strength.

The most frequently used binder system is probably polyvinyl alcohol (PVA). Numerous grades that differ in percent hydrolysis of the acetate groups and degree of polymerization are available. These variations affect the chemical and physical properties of the resin and, therefore, can also affect the spray-drying operation. Because PVA is considered a hard binder, it is usually softened by the addition of between 10 and 50 wt% plasticizer, based on the weight of PVA.

The presence of organic agglomerates in a binder solution can cause defects in pressed articles (Ref 26, 27), which may not be elim-inated during sintering (Ref 28). Therefore, it is desirable to add binder to the slurry as a prepared solution, because this allows preliminary screening to remove undissolved or undispersed organic material (Ref 21).

Slurries that contain binder solutions have a tendency to foam when subjected to high shear strain rates during mixing. The resulting trapped air is often incorporated in the granules and lowers their density. The addition of a defoaming agent (Ref 20) can be helpful when foaming cannot be prevented.

When selecting a binder, it should be recognized that its impurities will remain in the ceramic powder after spray drying. A measure of the metallic ion contamination in the organic is its ash content. Many commercially available binders have an ash content of between 0.5 and 2%. This results in an increased metallic ion contamination of approximately 100 to 500 ppm in the ceramic powder (Ref 24).

Another consideration in binder selection is spray dryer clean-up. Binders that are cold-water soluble allow easy removal of wall deposits from the drying chamber. Hot-water soluble binders result in deposits that are more difficult to remove. Emulsions tend to produce deposits that are the most difficult to remove. In addition, excessive drying temperatures can further harden the binder in these wall deposits. Extended dryer down times for cleaning can result in increased manufacturing costs.

Once the slurry has been prepared, it should be filtered prior to transferring to either a storage tank or the spray dryer. The smallest reasonable mesh size should be selected to remove large agglomerates that may be present. Use of a vibrating screen will take advantage of the pseudoplastic behavior of

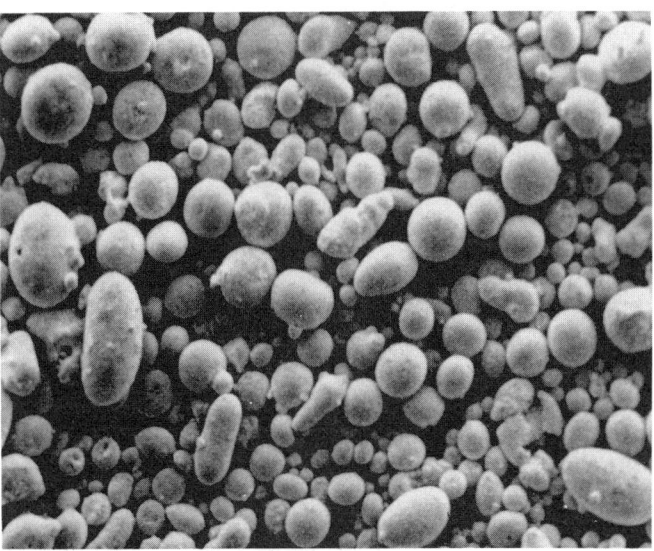

(a)

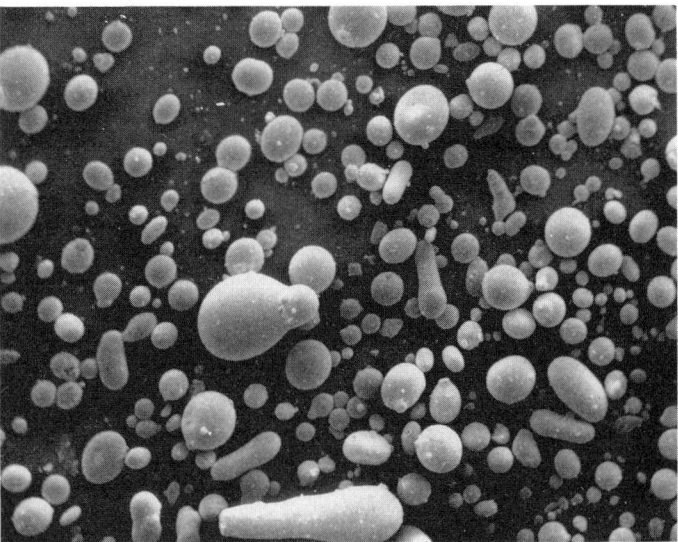

(b)

Fig 7 Micrographs of commercially spray-dried granules. (a) Ferrite, 75×. (b) Zirconia, 30×

most slurries that contain an organic binder (Ref 24). During storage, a low-speed mixer should be used to maintain slurry homogeneity and prevent particle settling.

Atomization of the ceramic slurry generates a large number of small droplets from the bulk fluid and results in a large increase in the surface area-to-volume ratio. The cooling effect from the resulting high evaporation rates allows high temperatures to be used in the dryer without product degradation.

Atomizers can be classified according to the manner in which energy is supplied to produce these droplets (Ref 8). The most common types are rotary atomizers, which use centrifugal energy; pressure nozzles, which use pressure energy, and pneumatic nozzles, which use kinetic energy. A fourth, less common type is an ultrasonic nozzle, which uses high-frequency vibration to generate droplets. Selection of an atomizer is based on the required drying capacity, the slurry character, and the desired granule size distribution. Typical granule size distributions from a pressure nozzle and a rotary (disk) atomizer are given in Table 1.

A rotary atomizer consists of a spinning wheel or disk centrally located in the upper portion of the drying chamber. The slurry, which is pumped to the atomizing device at pressures less than 690 kPa (100 psi), is accelerated outward into the chamber and disintegrates into droplets because of frictional forces between the airborne liquid and the drying air (Ref 8). Various wheel and disk designs have been developed for producing different droplet size distributions (Ref 8, 30).

Because the slurry is transferred to the atomizer under low pressures, inexpensive pumps are utilized in the feed system. Rotary atomizers do not contain small internal passages and are not prone to blockage. Rotary atomizers are expensive to purchase, cannot properly handle high-viscosity slurries (Ref 30), and require large-diameter drying chambers. However, they do provide operational flexibility in that feed rate and wheel rotational speed can be independently controlled. Granule size increases with increasing slurry feed rate, increasing slurry viscosity, and with decreasing wheel/disk rotational speed and diameter (Ref 8).

In pressure atomization, pressure nozzles atomize the slurry by accelerating it through a large pressure differential and injecting it into the drying chamber at a high velocity (Ref 8). Flow rate of the slurry is limited by the size of the metering passages within the nozzle and is proportional to the pump pressure. Pressure nozzles are subject to wear from highly abrasive slurries and will become plugged if large particles or agglomerates are allowed to reach the nozzle.

Pressure in excess of several hundred psi is required for proper operation of the nozzle, and the pumps are relatively expensive and difficult to maintain. The nozzles are inexpensive, but even slight wear can cause significant variations in droplet spray patterns. Because the droplets exit the nozzle with a high vertical velocity, the drying chambers are characteristically long. Granule size increases with increasing slurry viscosity, increasing orifice diameter, and decreasing pump pressure (Ref 31).

The spray patterns produced by pressure nozzles are classified as either hollow cone or solid cone. When using a hollow-cone nozzle, the majority of the droplets are concentrated near the periphery of the spray cloud. With a solid-cone nozzle, the droplets are more uniformly distributed throughout the spray, with a majority being near the center.

Pneumatic atomization occurs when the slurry is impacted by a stream of high-velocity air from a pneumatic nozzle (Ref 8). This can occur within the nozzle (internal mixing) or after the slurry is ejected from the nozzle orifice (external mixing). Internal mixing nozzles have greater energy efficiency and generate a more uniform droplet size, but they do not have the versatility of external mix units, where both the feed rate and air flow can be varied over a wide range.

Pneumatic nozzles operate well at low feed rates and can handle both low- and high-viscosity slurries. Granule size increases with increasing slurry viscosity, decreasing relative velocity between the atomizing air and the slurry, and a decreasing mass ratio of atomizing air to feed.

Ultrasonic nozzles atomize by passing the slurry over a metal surface that is vibrating at high frequency (Ref 32). A two-dimensional wave pattern develops, which becomes unstable and collapses into droplets. Droplet size is inversely proportional to vibration frequency, and a mean droplet size of 19 μm

(760 μin.) is reported to be achievable for a nozzle spraying water and operating at 120 kHz (Ref 33).

Ultrasonic nozzles are low-pressure devices that do not have small internal metering passages that can become blocked by large particles. The maximum slurry flow rate for the lowest frequency nozzles is in the range of 32 L/h (8.5 gal/h) (Ref 33). Nozzle capacity decreases with increasing atomizing frequency. Minimum flow rates at any frequency approach zero. Spray ejection velocity is very low, in the range of 0.4 m/s (1.3 ft/s), compared to 10 m/s (33 ft/s) for pressure and pneumatic devices (Ref 32). These low-ejection velocities allow higher chamber residence times and lower drying temperatures. Ultrasonic atomizers are useful for producing very small granules and for spray drying limited quantities of slurry.

Slurry Transfer Pumps. Transferring the slurry to the atomizer is usually accomplished with either a progressing cavity or a diaphragm pump. For pressures up to several hundred psi, the progressing cavity can be used; higher pressures may require a diaphragm type. With the progressing cavity, output is proportional to operating speed, while wear is exponentially dependent on the operating speed (Ref 34). Therefore, it is important for highly abrasive slurries that the pump be properly sized to maintain the required output at low rotor speeds (that is, less than 100 rpm). Diaphragm pumps must be specially designed to properly handle abrasive, high-solids content slurries. Consultation with the manufacturer and evaluation using a typical slurry formulation are recommended prior to purchasing and installing a pump of this type.

Droplet-Air Mixing. The air flow pattern in the drying chamber controls the completeness of moisture removal from the droplet, the maximum temperature that the granules will experience, and the formation of wall deposits. Droplet-air mixing in spray dryers is determined by the location of the air disperser and the atomizing device, and is classified according to the relative direction between the droplets and the air. These classifications are termed co-current, counter-current, and mixed-flow mixing.

Co-current conditions exist when the atomizing device is located near the air disperser at the top of the dryer, as in Fig 8(a). The droplets are exposed to the hottest air immediately after formation, but high evaporation rates maintain low product temperatures. As the moisture content of the droplets decrease, they come into contact with cooler air and surface temperature does not increase appreciably. Co-current air flow is common in dryers equipped with rotary atomizers.

Counter-current conditions occur when the atomizing device is placed at the top of the chamber and the air disperser is located at the bottom, as in Fig 8(b). Immediately after formation, the droplets contact cool, humid air.

Table 1 Granule size distributions as obtained by sieve analysis using different atomizers

	Wt% of granules larger than:					
Atomizer type	420 μm (17 mils)	250 μm (10 mils)	177 μm (7 mils)	120 μm (5 mils)	60 μm (2.5 mils)	40 μm (1.6 mils)
Disk	0	6	27	55	84	94
Nozzle	6	60	76	86	96	98
Disk and nozzle	6	32	52	73	94	96

Source: Ref 29

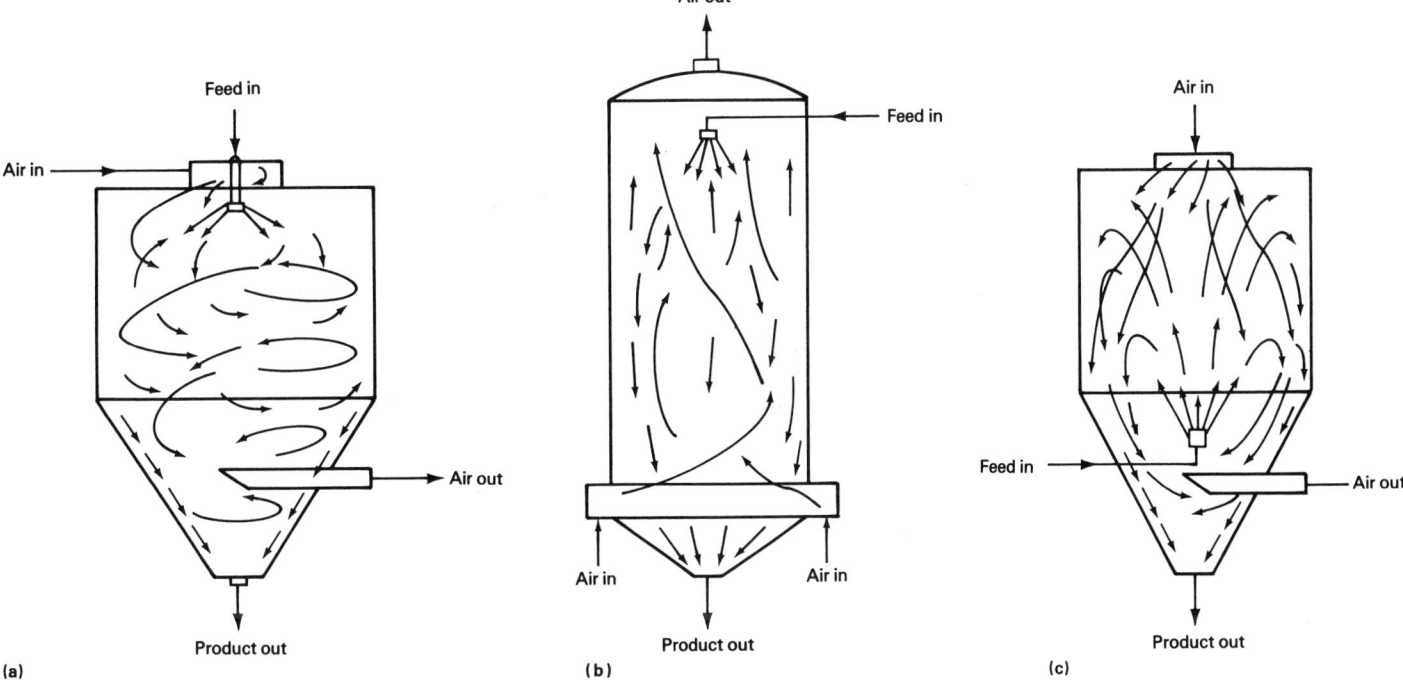

Fig 8 Types of droplet-air mixing in spray dryers. (a) Co-current. (b) Counter-current. (c) Mixed-flow. Source: Ref 8

However, as their moisture content decreases, they are exposed to increasingly hotter air. Because high internal temperatures can be realized in the granules, the organic binders should not be heat sensitive.

Mixed-flow conditions are a combination of co-current and counter-current air flow. This type is commonly utilized in fountain-type dryers, where a nozzle atomizer is located at the base of the drying chamber and the air disperser is placed at the top of the chamber, as in Fig 8(c). Mixed-flow conditions are frequently used in combination with pneumatic nozzles in small laboratory dryers because the fountain-like spray pattern increases droplet trajectory and provides sufficient airborne time to dry large droplets in small chambers.

Wall deposits are possible wherever atomized droplets or dry granules contact the chamber walls. Deposit formation is undesirable, because continued exposure of the powder to the high temperatures within the chamber can cause either thermal decomposition or partial evaporation of the admixed organics, which will alter compaction behavior. The three types of wall deposits and their causes are defined in Table 2. Careful examination of wall deposits after completing a spray-drying run is helpful in determining the formation mechanism. Analyses of wall deposits are described in detail in Ref 8.

Droplet Drying. Moisture evaporation from a freshly atomized droplet in a spray dryer is illustrated in Fig 9. Upon entering the chamber, the droplet achieves its maximum evaporation rate (curve A-B). This maximum rate, called the constant-rate drying period (curve

B-C), continues as long as moisture can migrate from the droplet interior rapidly enough to maintain entire surface saturation (Ref 35). The droplet temperature remains low and constant. The length of the constant-rate drying period depends on the droplet moisture content, the viscosity of the migrating fluid, and the temperature and humidity of the drying air. At point C, the critical moisture content, the entire droplet surface can no longer be maintained saturated by moisture migration, and the falling-rate drying period begins (curve C-E). The evaporation rate decreases, and droplet temperature begins to increase. At point D, the entire surface is dry, and the evaporation rate continues to decrease as the plane of evaporation moves inside the droplet. Further removal of moisture is a function of the moisture permeability of this dry surface crust.

Table 2 Wall deposit types and causes

Type	Cause
Semi-wet droplets that hit the chamber before their surfaces dry	Large droplet size, incomplete atomization, improper air flow, and low-solids-content slurry
Granules that have a sticky surface	Function of organic additives
Surface dusting by dry granules	Wall geometry and cleanliness, local air flow conditions, electrostatic forces, and low-density granules

Source: Ref 8

Hollow granules generated during spray drying will yield low bulk density press-powders. This is undesirable for press-powders because large compaction ratios and, therefore, density gradients within pressed articles, can result. Four possible mechanisms of hollow granule formation are (Ref 8):

- A low-permeability surface layer forms on the droplet. Moisture evaporates within the droplet interior and causes the surface to "balloon"
- Moisture migrates to the droplet surface, where evaporation and salt crystallization occur (soluble salts). Evaporation exceeds the diffusion of salt back into the droplet interior and internal voids are formed
- Moisture migration to the droplet surface carries along particles by capillary action (insoluble solids). The liquid evaporates

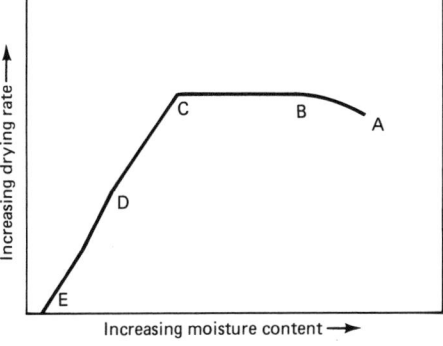

Fig 9 Drying behavior of atomized droplet in spray dryer; points along curve are described in text. Source: Ref 37

and internal voids are created within the granule

- Air trapped in the feed slurry (that is, foam) is incorporated within the atomized droplets and retained in the dry granules

In the absence of film formation and ballooning, high-solids-content slurries tend to result in high-density spray-dried granules. Because moisture migration can only occur if there is a continuous liquid network within the droplet pore channels, a high-solids-content slurry has a short constant-rate drying period. Internal moisture is removed through the vapor phase, particle migration is reduced, and void volume is minimized.

Binder Migration. Organic binders, especially those that are water soluble, will follow moisture flow (Ref 36) during droplet drying. This binder migration results in a nonuniform distribution of organics within the spray-dried granule. If a large amount of interior moisture flows to the droplet surface, the accompanying binder migration causes an increased binder concentration on the granule surface and a decreased concentration in the interior. This can result in a granule having a tough surface layer that resists fracture and deformation during compaction. However, if most of the moisture evaporates within the droplet interior, corresponding to a short constant-rate drying period, binder migration to the surface is minimized.

The simplest model for moisture migration during droplet drying is that of a Newtonian fluid flowing through a narrow, cylindrical tube. If the driving force for liquid flow is assumed to be from capillary forces, the linear flow rate of the liquid through the tube can be given by (Ref 37):

$$\frac{dl}{dt} = \frac{r\gamma\cos\theta}{4\eta L} \qquad \text{(Eq 2)}$$

where dl/dt is the linear flow rate of the liquid, r is the radius of the tube, $\cos\theta$ is the liquid contact angle, η is the viscosity of the liquid, γ is the surface tension of the liquid, and L is the length of the tube.

Although the interparticle passageways within a droplet are neither cylindrical nor of constant radius, this relation does illustrate the parameters that control macroscopic moisture flow and binder migration during droplet drying. The flow rate increases with increasing liquid surface tension, contact angle, and passageway radius. It decreases with increasing fluid viscosity. The passageway radii (that is, the radii of the pore channels), which are controlled by the particle size of the powder, and the surface tension and contact angle do not vary greatly in a given manufacturing process. It is the liquid viscosity within a spray-drying run and between one run and another that will tend to vary. These variations occur because of inconsistent slurry formulation, different binder lev-

els or types, or simply because of water additions during the spray-drying run.

A change in the fluid viscosity of the slurry will affect binder migration, and, possibly, compaction of the spray-dried powder. It is important to note that it is the viscosity of the migrating liquid (that is, water plus soluble organics) that is rate controlling, not the viscosity of the slurry. Low-viscosity-grade binders, which are beneficial for achieving a high-solids-content slurry, may increase the amount of binder migration during droplet drying.

Powder Collection. After the atomized droplets have been dried, the granules must be separated from the exhaust air and collected. Either all of the powder can be removed from the base of the drying chamber or the coarse and fine fractions can be separated and collected individually. When spray drying ceramic press-powders, size segregation is preferred, because it gives more control over powder properties such as flowability and bulk density.

Coarse granules are typically collected from the base of the drying chamber. Because spray dryers operate at pressures other than atmospheric, chamber integrity must be maintained during product discharge. Coarse granules that exit at the base of the dryer must pass through either a flap or a rotary valve. Caution must be exercised when using rotary valves. The discharge should be guarded to prevent an operator from inadvertently placing a finger or hand inside the device.

After the coarse fraction has been separated, the drying air and entrained fines leave the chamber. A wide variety of duct designs is available, and each design has a specific advantage (Ref 8). The relative proportion of powder carried with the exhaust air and labeled as "fines" is controlled by the size and density of the granules, the design of the exhaust duct, and the air movement near the duct. Excessive air flow rates can result in a high percentage of granules being pulled from the population falling toward the chamber discharge valve.

The two most common techniques of removing fines from the exhaust air use either a bag collector or a cyclone. Bag collectors are more difficult to clean when changing powder type. They are also more expensive to purchase and maintain (Ref 8). However, if they are chosen because of processing requirements, singed bags should be used. These are bags that have been flame treated to remove surface fibers (Ref 38), which are a possible contamination source.

Dryer Operation. Spray dryers operate according to this energy balance: Energy input equals energy consumed minus the amount of energy lost. The input energy is determined by the quantity and temperature of the air entering the drying chamber. Because the volume of air flowing through the system is usually constant, the temperature of the incoming air, called the inlet temperature, is

a measure of input energy. The energy loss from the dryer consists of conduction through the walls to the external environment and the hot, humid air leaving through the exhaust duct. The energy consumed is that required for droplet drying.

Spray dryers are operated under conditions of constant energy input (that is, constant inlet temperature). The optimum inlet temperature, usually between 250 and 400 °C (480 and 750 °F) for ceramic powders, is determined empirically. A low temperature decreases the dryer evaporative capacity, whereas too high a temperature can result in lower granule density (Ref 39) and either evaporation or thermal decomposition of the organics. The dryer is controlled by adjusting the slurry feed rate to maintain a given outlet temperature. Higher rates result in increased water evaporation per unit time and, therefore, a lower outlet temperature. Lower rates cause increased outlet temperatures. High outlet temperatures are detrimental to compaction behavior when powders containing heat-sensitive binders are dried under co-current or mixed-flow conditions. Feed rate also determines the final moisture content of the granules, because lower outlet temperatures correspond to higher moisture levels.

There are many problems that can occur while spray drying, which result in reduced press-powder quality, decreased yields, or even disruption of the dryer operation. Normally, the sensors that monitor the system detect a problem and alert the operator. However, there are occasions when a process problem will not cause a noticeable sensor deviation, and the dryer continues to operate. Although it is not feasible to list every possible event that can occur, several representative situations that can aid in detecting problems that result in nonoptimal performance are described below.

Most ceramic spray dryers operate at less than atmospheric pressure. As a result, leaky seals around fixtures such as the air disperser, chamber door, atomizer, temperature sensors, or duct system will allow cool external air to enter. A large air leak results in a lower than normal outlet temperature. A small leak will not noticeably lower the outlet temperature, but can adversely affect air movement within the chamber. Unusual or localized wall deposits can provide evidence of air leaks in the dryer.

The temperature sensors (typically thermocouples) that monitor the inlet and outlet air temperatures can cause operational problems. A drift in calibration is always a possibility. A small air leak near the sensor can cause the indicated temperature to be lower than the actual temperature. Lack of a rapid response in the outlet temperature to variations in the slurry feed rate should be questioned. Because the air retention time in the drying chamber is usually measured in seconds, any change in the amount of water evaporated per unit time should result in a rapid change in the outlet temperature. A sluggish response can result from either the temperature sensor not being fully inserted into the air stream or a coating of powder covering the sensor. This coating forms a low thermal conductivity layer between the thermocouple and the air. A separate thermocouple, along with a portable read-out device, can be used to confirm that the temperature sensors are functioning properly.

Nozzle atomizers are susceptible to blockage when large particles are present in the slurry. Pressure and pneumatic nozzles have gages to measure feed pressure. Pneumatic nozzle systems also contain gages to monitor the atomizing air pressure and flow rate. A blockage of the metering passages in a nozzle will result in a rapid increase in the feed pressure and the outlet temperature. A slow build-up of powder at the nozzle orifice will translate into an upward drift in the feed pressure and the outlet temperature. This type of build-up can occur when the nozzle is exposed to high temperatures, such as when operating with high outlet temperatures under countercurrent or mixed-flow conditions. A powder build-up in the air-metering orifice of a pneumatic nozzle results in decreased air flow and incomplete atomization. The outlet air temperature will increase noticeably if the blockage is sufficient.

Powder Contamination. Care should be taken to avoid contamination of the granulated powder during spray drying. Slurries that flow through pumps and feed lines at high velocities can cause abrasion. Feed lines should be sized to reduce flow velocity without causing particle settling. When using a two-fluid nozzle atomizer, the pressurized air source should be clean. If plant compressed air is being used, it should be adequately filtered to remove oil and particulates. Also, because the chamber drying air is drawn from the plant environment, most manufacturers design their dryers to allow filtering of ambient dust. Filters should be changed as necessary.

Characterization of the feed slurry and the spray-dried powder is important for maintaining a consistent process. Characterization is also required in order to solve the occasional problems that can occur. The slurry properties that should be measured are density, solids content, weight percent of organics, and viscosity.

The slurry density, which is easily measured, is used to determine the amount of foam present. A convenient amount of the slurry should be poured into a graduated cylinder with care to avoid splashing any material on the sides. The net weight of the slurry divided by the volume it occupies is the density. The amount of foam can be calculated by comparing this measured density to the calculated theoretical slurry density (Ref 37).

The weight percent of solids in the feed is determined by heating a small quantity at a temperature sufficient to eliminate the water and organics. Comparison of this measured solids content with the theoretical value, which is determined from the slurry composition, reveals errors that may have occurred during formulation.

Selective elimination of low-boiling-point organics during the drying cycle results in an altered compaction response in the spray-dried powder. For example, partial evaporation of polyethylene glycol, a common plasticizer for polyvinyl alcohol, results in relatively hard granules. Selective evaporation can be detected by comparing the level of organics in the feed material with the amount present in the spray-dried powder. This is accomplished by subjecting measured amounts of the slurry and the spray-dried powder to appropriate heat treatments. Preliminary low-temperature evaporation of water from the slurry is advisable to prevent it from masking the results. Thermogravimetric analysis (TGA) can be utilized on a nonroutine basis (Ref 40) to study the problem.

One of the slurry properties that determines the size of the droplets formed during atomization is feed viscosity. However, it is the viscosity of the slurry at the shear strain rate of the atomizing device that is the controlling factor, rather than the viscosity at the strain rate of the measuring viscometer. The pseudoplastic behavior exhibited by most feed materials that contain an organic binder is usually not a problem. If a suspension exhibits dilatancy, the high shear strain rate of the atomizing device can increase the viscosity to a magnitude sufficient to prevent proper droplet formation. It is important to measure the feed viscosity over a range of strain rates (for example, zero to several thousand $(seconds)^{-1}$).

Characterization of the spray-dried powder includes granule size analysis, powder flow rate and bulk density, percent residual moisture, percent organics, compaction behavior, and visual inspection of the granules. If complete testing of each powder lot is not possible, only those properties that previous experience has shown to be critical need to be measured. A small quantity of the powder (several hundred grams) should be saved for a specified period of time for possible future analysis.

Powder bulk density, flowability, and size distribution are measured using the procedures outlined in ASTM standards B212, B213, and B214, respectively (Ref 41). The percent residual moisture and percent organics are determined by either appropriate low-temperature heat treatment of the powder or thermogravimetric analysis. Because of the plasticizing effect of water on many soluble binders, such as PVA, it is important to monitor residual moisture closely. However, it should be noted that unsealed powders tend to achieve equilibrium with ambient humidity conditions (Ref 42).

A compaction curve of the spray-dried powder (Ref 43) not only indicates its pressing behavior, but also detects any temperature degradation or selective organic evaporation that may have occurred. A more rapid, but less thorough, method of measuring powder behavior is to determine the density that can be achieved by compacting the spray-dried powder at a given pressure according to ASTM B331 (Ref 41).

Finally, the granules should be examined visually with a stereo microscope. Nonspherical, hollow, or donut-shaped granules may indicate improper atomization, excessive drying temperatures, or low-solids-content slurries. The internal structure can be viewed by either carefully impregnating the granules with epoxy and polishing or using scanning electron microscopy to view a fracture surface (Ref 43) of a partially sintered compact pressed from spray-dried powder.

REFERENCES

1. H. Rumpf and H. Schubert, Adhesion Forces in Agglomeration Processes, *Ceramic Processing Before Firing*, G.Y. Onoda Jr. and L.L. Hench, Ed., John Wiley & Sons, 1978, p 357–376
2. Size Enlargement, *Encyclopedia of Chemical Technology*, Vol 21, 3rd ed., John Wiley & Sons, 1983, p 77–105
3. Size Enlargement, *Chemical Engineers Handbook*, R.H. Perry and C.H. Chilton, Ed., McGraw-Hill, 1973, p 8–57 to 8–65
4. C.E. Capes, Particle Size Enlargement, *Handbook of Powder Technology*, Vol 1, J.C. Williams and T. Allen, Ed., Elsevier Scientific, 1980
5. C.E. Capes and A.E. Fouda, Agitation Methods, Chapt. 7, Part 5, *Handbook of Powder Science and Technology*, M.E. Fayed and L. Otten, Ed., Van Nostrand Reinhold, 1984, p 286–294
6. H. Rumpf, The Strength of Granules and Agglomerates, *Agglomeration*, W.A. Knepper, Ed., Interscience, 1962, p 379–418
7. J.S. Reed, *Introduction to the Principles of Ceramic Processing*, John Wiley & Sons, 1988
8. K. Masters, *Spray Drying*, 4th ed., John Wiley & Sons, 1985
9. B.V. Dejaguin and L. Landau, *Acta Physicochim*, Vol 14, 1941, p 633
10. E.J. Verwey and J.Th.G. Overbeek, *Theory of the Stability of Pyrophobic Colloids*, Elsevier, 1948
11. P. Sennett and J.P. Oliver, Colloidal Dispersions, Electrokinetic Effects and the Concept of Zeta Potential, *Ind. Eng. Chem.*, Vol 57 (No. 8), 1975, p 32–50
12. W. Black, *Dispersions of Powders in Liquids*, G.D. Parfitt, Ed., Halsted Press, 1973, p 132–174
13. R.E. Kirk and D.F. Othmer, Ed., Dispersants, *Encyclopedia of Chemical Technology*, Vol 7, 3rd ed., John Wiley & Sons, 1979, p 833–848

14. *McCutcheon's Functional Materials*, 1980 Annual, MC Publishing
15. M. Ash, *Encyclopedia of Surfactants*, Vol 1–3, Chemical Publishing, 1980
16. J.P. Sisley and P.J. Wood, *Encyclopedia of Surface-Active Agents*, Vol 1, 1952, and Vol 2, 1964, Chemical Publishing
17. *McCutcheon's Emulsifiers and Detergents*, North American ed., MC Publishing, 1981
18. M.J. Rosen, *Surfactants and Interfacial Phenomena*, John Wiley & Sons, 1978
19. E.M. Vogel, Dispersants for Ferrite Slurries, *Am. Ceram. Soc. Bull.*, Vol 58 (No. 4), 1979, p 453–454, 458
20. T. Morse, *Handbook of Organic Additives in Ceramic Body Formulations*, Montana Energy and MHD Research and Development Institute, 1979
21. E.R. Hoffman, Importance of Binders in Spray Dried Pressbodies, *Am. Ceram. Soc. Bull.*, Vol 51 (No. 3), 1972, p 240–242
22. J.W. Harvey and D.W. Johnson, Binder Systems in Ferrites, *Am. Ceram. Soc. Bull.*, Vol 59 (No. 6), 1980, p 637–639, 645
23. S.L. Levine, Organic (Temporary) Binders for Ceramic Systems, *Ceramic Age*, Jan 1960, p 39–42, Feb 1960, p 25–28, 32–36
24. G.Y. Onoda, The Rheology of Organic Binder Solutions, *Ceramic Processing Before Firing*, G.Y. Onoda and L.L. Hench, Ed., John Wiley & Sons, 1978, p 235–250
25. A.G. Pincus and L.E. Shipley, The Role of Organic Binders in Ceramic Processing, *Ceramic Industry*, Vol 92, 1969, p 106–109, 146
26. G.Y. Onoda, Green Body Characteristics and Their Relationship to Finished Microstructure, *Ceramic Microstructures '76*, R.M. Fulrath and J.A. Pask, Ed., Boulder, CO, Westview Press, 1977, p 163–183
27. W. Malinowski and A. Withop, Organic Sources of Voids in Ceramics, *Am. Ceram. Soc. Bull.*, Vol 57 (No. 5), 1978, p 523–524
28. T.J. Carbone and J.S. Reed, Dependence of Sintering Response with a Constant Rate of Heating on Processing-Related Pore Distributions, *Am. Ceram. Soc. Bull.*, Vol 57 (No. 8), 1978, p 748–750, 755
29. H.J. Helsing, Advanced Processing of Oxide-Ceramics by Spray Drying, *Powder Metall. Int.*, Vol 1 (No. 2), 1969, p 62–65
30. W.R. Marshall, Atomization and Spray Drying, *Chemical Engineering Progress Monograph Series*, Vol 50 (No. 2), 1954
31. "Delavan Industrial Nozzles and Accessories," Technical bulletin 1622A-1077, Delavan Corporation, West Des Moines, Iowa
32. H.L. Berger, Ultrasonic Nozzles Take Pressure Out of Atomizing Processes, *Res. Dev.*, Sept 1984
33. H.L. Berger, Ultrasonic Nozzles Atomize Without Air, *Mach. Des.*, 21 July 1988, p 58–62
34. "Moyno Pumps," Technical bulletin 110, Moyno Pump Division, Robbins and Myers, Inc., Springfield, OH
35. R.H. Perry and C.H. Chilton, Ed., Solids-Drying Fundamentals, *Chemical Engineers Handbook*, 5th ed., McGraw-Hill, 1973, p 20–4 to 20–16
36. J.E. Comeforo, Migration Characteristic of Organic Binders, *Ceram. Age*, April 1945, p 132–136
37. S.J. Lukasiewicz, Spray-Drying Ceramic Powders, *J. Am. Ceram. Soc.*, Vol 72 (No. 4), 1989, p 617–624
38. M.A. Maxwell, *Chem. Eng.*, Vol 88 (No. 26), 28 Dec 1981, p 54–56
39. J.A. Duffee and W.R. Marshall, Jr., *Chem. Eng. Progr.*, Vol 49 (No. 9), 1953, p 480–486
40. S.L. Levine, A Quantitative Determination of Binder and Lubricant in Ceramic Materials by TGA, *Am. Ceram. Soc. Bull.*, Vol 48 (No. 2), 1969, p 230–231
41. *Annual Book of ASTM Standards*, Vol 2.05, American Society for Testing and Materials, Philadelphia, 1986
42. J.A. Brewer, R.H. Moore, and J.S. Reed, Effect of Relative Humidity on the Compaction of Barium Titanate and Manganese Zinc Ferrite Agglomerates Containing Polyvinyl Alcohol, *Am. Ceram. Soc. Bull.*, Vol 60 (No. 2), 1981, p 212–215, 220
43. S.J. Lukasiewicz and J.S. Reed, Character and Compaction Response of Spray Dried Agglomerates, *Am. Ceram. Soc. Bull.*, Vol 57 (No. 9), 1978, p 798–801, 805

Calcination

John W. Halloran, University of Michigan

CALCINATION processes are endothermic decomposition reactions in which an oxysalt, such as a carbonate or hydroxide, decomposes, leaving an oxide as a solid product and liberating a gas. This class of reactions has been the subject of several extensive reviews (Ref 1, 2). However, these reviews do not directly pertain to ceramics. The purpose of this article is to review calcination as a method to produce "active" ceramic powders with excellent sinterability.

Producing this type of powder is convenient and economical because the as-calcined oxide is extremely finely divided. It is easy to obtain a particle size much finer than 1 μm (40 μin.) and a specific surface area in the range of 100 m^2/g (488×10^3 ft^2/lb). The surface area and particle size are easily modified. Low calcination temperatures produce very fine, high surface area powders. Heating at temperatures well above the decomposition temperature reduces the surface area (and increases the particle size) to any degree that is desired, even to the point of making very coarse and inactive "dead burnt" lime, dolomite, and periclase for use in refractories.

Calcination is often the final step in the production of high-purity ceramic powders. Soluble salts are commonly used as the ceramic precursor to make an easily purified solution. After purification, a calcinable salt is precipitated from solution by adding hydroxide, carbonate, sulfate, and similar materials. The salt is then calcined to yield the pure oxide. It is important to note that the "high purity" designation may only pertain to metallic impurities. The "pure" oxide may have significant residual carbon, sulfur, or hydrogen if it has been obtained from a carbonate, sulfate, or hydroxide, particularly when being prepared as a high surface area active powder.

Calcination reactions often play a role in the synthesis of multicomponent ceramic compounds. A classical solid-state synthesis uses mixtures of both oxide raw materials and calcinable salts. For example, the production of barium titanate typically occurs via:

$$TiO_2 + BaCO_3 \rightarrow BaTiO_3 + CO_2$$

The solid-state reaction to form the titanate is accelerated by the highly reactive barium oxide created by the carbonate decomposition. Synthesis of ceramic powders by "wet chemical" techniques also usually involves calcination. For example, a homogeneous mixture of several elements can be made by co-precipitation of solutes using hydroxides, carbonates, or oxalates, which are later converted to oxide compounds by calcination.

In this article, calcination is reviewed from the perspective of preparing sinterable ceramic powders. The physical chemistry of the process is examined to determine how the reaction depends on temperature and atmosphere. The mechanisms of the process are also examined to explain how the powder particles are formed. Two examples, MgO and Al_2O_3, are described in some detail. Finally, the optimization of calcination in order to produce sinterable powder is addressed.

Thermodynamics and Kinetics of Calcination

It is useful to briefly examine the thermodynamics and kinetics of calcination reactions to understand the important factors. Consider, as an example, the decomposition of magnesium carbonate:

$$MgCO_3 = MgO + CO_2$$

The heat of the reaction is determined from the standard enthalpies of formation (ΔH_F°) by:

$$\Delta H_R^\circ = \Delta H_F^\circ(CO_2) + \Delta H_F^\circ(MgO) - \Delta H_F^\circ(MgCO_3)$$

At 1000 K, these values are:

$$\Delta H_R^\circ = (-94.05) + (-143.7) - (-265.7)$$

$$= 27.95 \text{ kcal/mol}$$

Therefore, to decompose a mole of magnesium carbonate (84.3 g, or 2.97 oz) 117 kJ (27.95 kcal) must be supplied to sustain the calcination reaction. This is a significant amount of heat, representing about half the enthalpy needed to heat MgO from room temperature and melt it. It is easy to see why heat transfer considerations become very important in the control of the process.

The equilibrium conditions can be examined by calculating the Gibbs free energy change $\equiv (\Delta G_R^\circ)$ for the reaction:

$$\Delta G_R^\circ = \Delta G_F^\circ(CO_2) + \Delta G_F^\circ(MgO) - \Delta G_F^\circ(MgCO_3)$$

which, at 1000 K, has the values:

$$\Delta G_R^\circ = (-94.6) + (-117.8) - (-199.7)$$

$$= -12.7 \text{ kcal/mol (or } -53.2 \text{ kcal/mol)}$$

The standard Gibbs free energy change is negative, indicating that the decomposition of magnesium carbonate will proceed at 1000 K under *standard* conditions, that is, when the solids are pure and the carbon dioxide partial pressure is 0.1 MPa (1 atm). However, there are conditions where the magnesium carbonate is stable. These can be found by evaluating the equilibrium constant, K^{eq}:

$$K^{eq} = \exp\left(\frac{-\Delta G_R^\circ}{RT}\right) = \frac{a_{MgO}P_{CO_2}}{a_{MgCO_3}}$$

which relates the activities of the products and reactants at equilibrium and where R is the gas constant and T is the absolute temperature. The magnesium carbonate and oxide are nearly pure solids; therefore, to a good approximation, $a_{MgO} = a_{MgCO_3} = 1$. The equilibrium constant is essentially equal to the equilibrium partial pressure of the carbon dioxide: $K^{eq} = P_{CO_2}^{eq}$. At 1000 K, P^{eq} is 61 MPa (60 atm). Therefore, $MgCO_3$ is stable only at this high carbon dioxide pressure; MgO is stable at all lower pressures.

At every temperature, there is an equilibrium CO_2 partial pressure. This can be expressed by:

$$K^{eq} = \exp\left(\frac{-\Delta G_R^\circ}{RT}\right) = P_{CO_2}^{eq}$$

$$= \exp\left(\frac{\Delta S_R^\circ}{R}\right)\exp\left(\frac{-\Delta H_R^\circ}{RT}\right)$$

The equilibrium partial pressure is 0.1 MPa (1 atm) at 672 K, the "decomposition temperature" of magnesium carbonate in pure CO_2 at atmospheric pressure. At reduced CO_2 partial pressure, decomposition occurs at a lower temperature. In ordinary air (30 Pa, or 0.0003 atm CO_2), the equilibrium decomposition temperature is 480 K.

The fact that magnesium carbonate does not decompose at 480 K in air brings us to the issue of the kinetics of the reaction. Significant superheating is required to achieve observably fast decomposition. Typically, superheating at about 50 K is required to initiate the reaction, and considerably higher superheat is required to sustain the reaction at a practical rate.

There is no general theory that is valid for the full range of kinetic phenomena displayed by calcination reactions (Ref 3). The situation was summarized by Bertrand in 1978, who wrote "the kinetics of decomposition reactions still remains a complex subject where theories and experimental results are numerous but not in accordance" (Ref 4). The decomposition rate can be controlled by several factors, depending on the driving force, as expressed alternately as either superheat above the decomposition temperature (for a given ambient partial pressure) or reduction of the ambient partial pressure below P^{eq} (for a given temperature).

When the reaction is proceeding vigorously (at large driving force), it will be limited by heat or mass transport. Heat must be supplied to the material by convection from the gas stream to the particles, and by transport through the product oxide layer to the reaction interface. Inward heat transfer is coupled to outward mass transport, because the product gas must be transported through the same porous oxide layer and away into the gas stream. Heat and mass transfer from the gas stream to the reacting particle can be treated as a boundary layer transport problem (Ref 5), which can describe the reaction kinetics for cases of industrial importance. The transport kinetics may also be limited by the oxide layer.

The oxide layer usually grows on the salt particle, gradually converting it to a "relict," that is, a porous agglomerate of oxide particles with the same size and sample as its parent salt particle. The structure of the relict— a body with ultrafine porosity—makes it an excellent thermal insulator and a good barrier to gas diffusion. Models exist for transport through these layers (Ref 6), but require detailed information on the pore size and porosity, both of which can change during calcination as the product oxide sinters.

When the driving force is low, decomposition is slow, and reaction kinetics are often controlled by some interface reaction. The kinetics in the interface-controlled regime are difficult to generalize because of the many different processes that could be rate limiting. Large salt particles, such as macroscopic calcite crystals, can be limited by nucleation. However, nucleation is often not rate limiting in powders. The reaction rate in powders is usually proportional to the total surface area, making the kinetics very sensitive to the particle size of the salt. A number of extrinsic factors, such as grinding and impurities, can accelerate the decomposition in ways that are generally difficult to predict. Beyond the stage of nucleation control, there are still a variety of chemical mechanisms that could control the kinetics of the interface reaction. Searcy

and Beruto (Ref 7) considered four reaction steps that could be rate controlling for the case of steady-state decompositions in vacuum. From their general solution, they derived six rate-limiting equations, each of which involved three or four different kinetic constants.

Large-scale industrial calcination operations are usually conducted with rotary kilns in which the charge is lifted and agitated in countercurrents of hot air. This provides excellent heat and mass transfer, assuring that the temperature and atmosphere are uniform throughout the batch. Because rotary kilns are not practical for small-scale calcination, laboratory or batch calcinations are frequently conducted in pans or crucibles. In this case, the bed thickness is very important. The relatively poor heat and mass transport through the bed will create variation in product quality. Material near the outer surface decomposes early in the heating cycle, and during the remainder of the process coarsens by sintering.

Meanwhile, the material in the interior decomposes slower because of the lower temperatures and high local pressures of the product gas. In fact, when the reaction is fast enough to be practical for producing powder, that atmosphere deep inside the bed can be 100% product gas. It is a better practice to flow air through the bed to sweep away the product gas and provide more uniform heat and mass transfer. This can be accomplished in a shaft kiln. Fluidized bed reactors would also provide good transport, but they do not seem to be used much in the ceramics industry.

Sweeping away the product gas can be crucial when the decomposition is difficult. One example is in the synthesis of $YBa_2Cu_3O_{7-x}$ from Y_2O_3, CuO, and $BaCO_3$. One must decompose the barium carbonate and react the compound below about 900 °C (1650 °F), where melting begins. However, because barium carbonate is quite stable at this temperature (ΔG_R° is about 80 kJ/mol, or 19 kcal/mol), the equilibrium carbon dioxide pressure is just barely above the CO_2 content of air. The barium carbonate cannot be successfully decomposed in a thick bed in static air, but it can be easily decomposed by flowing air through the bed to sweep away the CO_2.

Production of Fine Particles

Very fine particle size is the major virtue of calcined powders. It is worthwhile to briefly examine the mechanisms that create these fine particles. Upon decomposition, a relatively coarse, low surface area salt becomes an oxide with very high surface area. This is remarkable, because it is impossible to achieve such a high surface area by the mechanical grinding of oxides. It occurs conveniently with calcination because part of the free energy change that drives the decomposition creates surface free energy. For example, consider decomposing $MgCO_3$ at 900 K to produce

MgO with a specific surface area of 100 m^2/g (488 × 10^3 ft^2/lb). This is a large surface area, and the MgO has an excess surface free energy of about 100 J/g (680 cal/oz). However, because the total free energy change for the decomposition chemical reaction was −1 kJ/g (−6.8 kcal/oz) of MgO, only a small fraction of the total free energy was used to create the surface.

Several mechanisms seem to be at play in accomplishing this subdivision. Frequently, subdivision occurs in transformations, and fine particle size can be a simple result of transformation to the product phase. In cases where nucleation is involved, each nucleation event creates an oxide particle; the particle size therefore depends on the nucleation rate. Growth kinetics may also be involved. These decomposition reactions occur at temperatures where diffusion is very slow in the solid state. Counterdiffusion of the fugitive species (CO_3^{2-} or OH^-) with other species is required for growth. To propagate the reaction at a faster rate, the system is subdivided into small particles to minimize diffusion distances. Another factor, possibly the most important, is strain from coherency.

The product oxide always bears a topotactic orientation relationship with the parent salt. A ubiquitous feature of calcination is that the oxide lattice attempts to maintain coherence with the salt along preferred orientations. [This is common with phase transformations in solids, because energy can be minimized by preserving atomic planes and directions between the parent and the product (Ref 8).] Lattice coherence exacts a high cost in strain energy, due to the very large difference in molar volume and lattice spacings between the salt and the oxide.

Coherency strains cause subdivision in several ways. For decompositions that are heterogeneous (that is, where the salt and the oxide have a distinct interface), maintaining coherence creates strains in both the oxide and the underlying salt, which can fracture both phases. Another mechanism is proposed for cases in which the underlying salt becomes depleted in the fugitive species below the interface. Here, the fugitive species diffuses out of the salt along specific crystallographic directions, leaving a lacunar structure behind. The lacunar structure later reorganizes into the oxide structure (perhaps via a diffusionless shear transformation), and the resulting transformation strain fractures the product into a state of fine subdivision (Ref 9).

Magnesium Oxide and Aluminum Oxide

The production of magnesium oxide and the production of aluminum oxide powders from their respective hydroxides are exemplified below. Because the reactions in each process are quite different, these two examples illustrate behavior extremes.

The decomposition of Mg(OH)₂ and the decomposition of MgCO₃ occur as a heterogeneous reaction that forms MgO directly, with no intermediate phases. Therefore, these are good examples of discrete heterogeneous decomposition reactions.

The case of brucite, Mg(OH)₂, has been widely studied. In this reaction, the hexagonal brucite decomposes to yield cubic periclase, MgO, and water vapor:

$$Mg(OH)_2 = MgO + H_2O$$

The lattice parameters of partially decomposed Mg(OH)₂ and freshly formed MgO are the same as in macroscopic crystals. Upon decomposition, the brucite loses half of its oxygen and all of its hydrogen, resulting in a 25% reduction in mass. The molar volume is reduced by 35%. This volume change is accommodated by subdividing the MgO into tiny crystallites separated by pore spaces. Calcination of brucite can produce exceedingly fine material, particularly when decomposition is conducted at a very low temperature. Moodie and Warble (Ref 10) decomposed Mg(OH)₂ in water-free air at 300 °C (570 °F) to produce MgO crystals in the form of little cubes that were only two to three unit cells long in edge length. Figure 1, a high-resolution micrograph of MgO particles produced similarly by Kim *et al*. (Ref 11), shows the very small size, cubic shape, and orientation relationship between adjacent MgO particles.

The topotactic relationship between the brucite and periclase is well known. The basal plane of brucite $(0001)_b$ is parallel to the $(111)_p$ of periclase, and the $[11\underline{2}0]_b$ direction of brucite is parallel to the $[10\underline{1}]_p$ of MgO. Recently,

Kim *et al*. (Ref 12) rationalized these relationships by pointing out that the oxygen octahedra are preserved in the brucite and periclase and that the orientation relationships of the decomposition correlate with the orientation of these octahedra.

The calcined powder is agglomerated, with the largest agglomerate structures often being pseudomorphs of the starting brucite particles. Sufficiently thin brucite crystals become porous relicts, consisting of ultrafine (2 to 10 nm, or 0.08 to 0.4 μin.) MgO crystals separated by pores in the 1 to 5 nm (0.04 to 0.2 μin.) size range. Within the relict, the relative orientation of the MgO crystallites is perfect enough to give single crystal-like electron diffraction patterns. Larger brucite crystals are fractured by the coherency strains (Ref 13), forming micrometer-sized brucite particles. These later become porous relicts.

Calcining at higher temperatures causes sintering of the MgO crystallites within the porous relicts. This causes a gradual decrease in surface area. An example appears in Fig 2, which shows data from Phillips *et al*. (Ref 14) on the surface area of the MgO obtained from a variety of Mg(OH)₂ precipitates after calcining 30 min in air. The steady reduction in surface area as temperature increases is a common feature of calcination. However, notice the difference in the surface area reduction and the dramatic difference in the 700 °C (1290 °F) surface area obtained from the various samples. This is a typical observation: Different precursors—even Mg(OH)₂ from different precipitation conditions—can behave differently, for reasons that are not well understood.

The surface area reduction is very sensitive to the atmosphere, as is particle growth. Very small quantities of water vapor can have a dramatic effect with MgO. Anderson and

Morgan (Ref 15) found a 1000-fold increase in the rate of surface area reduction in 665 Pa (5 torr) water vapor, compared to vacuum. They attributed this to enhanced surface diffusion, but recent work suggests that water vapor catalyzes the rearrangement and coalescence of particles (Ref 16).

Aluminum oxide is quite a different case. Rather than directly transforming to the stable hexagonal corundum structure (α-alumina), most aluminum salts decompose to one or more "transition aluminas." These transition aluminas are metastable phases, usually observed only as poorly crystallized ultrafine powders. There is quite a variety of transition aluminas (Ref 17). The transition alumina from aluminum trihydrate, Al(OH)₃, and monohydrate, AlOOH, have been thoroughly studied, because they are the basis of the Bayer aluminum process, and are widely used as catalysts (Ref 18, 19). Sequences of transition phases have also been reported for nitrates and sulfates (Ref 20) and ammonium alum, NH₄Al(SO₄)₂·12H₂O (Ref 21, 22).

These phase sequences are complicated. Figure 3 illustrates the sequence of Al(OH)₃, both as the mineral gibbsite and in the form of bayerite (the precipitate from the Bayer process), as well as the sequence for AlOOH, both as boehmite and diaspore. This diagram (Ref 23) shows the temperature range of occurrence of the various transition aluminas under different decomposition conditions. Notice that the relatively rare diaspore is the only one that decomposes directly to α-alumina (Ref 24). (Because the monohydrate diaspore, AlOOH, already has hexagonal close-packed oxygen, it decomposes directly to corundum.) Under hydrothermal conditions, gibbsite and bayerite first decompose to the boehmite form of AlOOH (path A, Fig 3), which then transforms to γ, δ, and then θ forms of alumina before finally converting to α-alumina. Hydrothermal-like conditions prevail whenever the ambient water vapor pressure is high, or when diffusion of water away from the interface is difficult. The latter situation occurs with large (>100 μm, or 34 mils) particles of Al(OH)₃. Boehmite does not form under dry conditions, or with small Al(OH)₃ particles. Rather, the decomposition follows path B (Fig 3), which means that gibbsite forms the χ-alumina phase, which later transforms to κ-alumina, and then α-alumina.

The metastable transition aluminas are ultrafine. Dehydration around 250 °C (480 °F) can create specific surface areas as large as 500 m²/g (2440 × 10³ ft²/lb). As the decomposition proceeds, the Al(OH)₃ particles retain their shape, becoming porous relicts, with pores as fine as 2 nm (0.08 μin.). Upon further heating, the pores coarsen and the surface area is reduced, dropping to about 100 m²/g (488 × 10³ ft²/lb) at 500 °C (930 °F).

Wilson has studied the dehydration mechanism of boehmite in detail (Ref 25, 26), showing that the reaction occurs by counter-

Fig 1 High-resolution electron micrograph of 2 nm (0.08 μin.) cubic MgO particles formed on the surface of Mg(OH)₂, imaged under conditions that resolve the 0.21 nm (0.008 μin.) {200} lattice spacings. Note the small particle size, shape, and alignment. Source: Ref 12

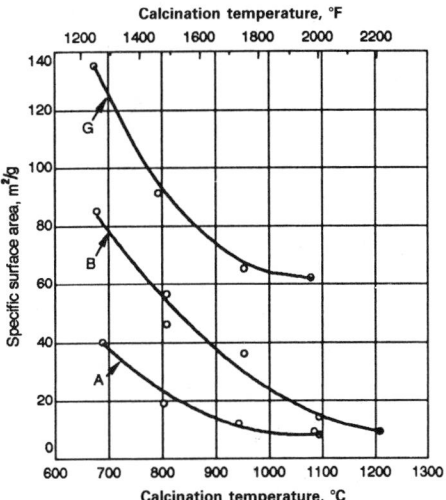

Fig 2 Specific surface area versus calcination temperature for Mg(OH)₂ precipitates calcined 30 min. A, B, and G refer to precipitation conditions in Ref 14.

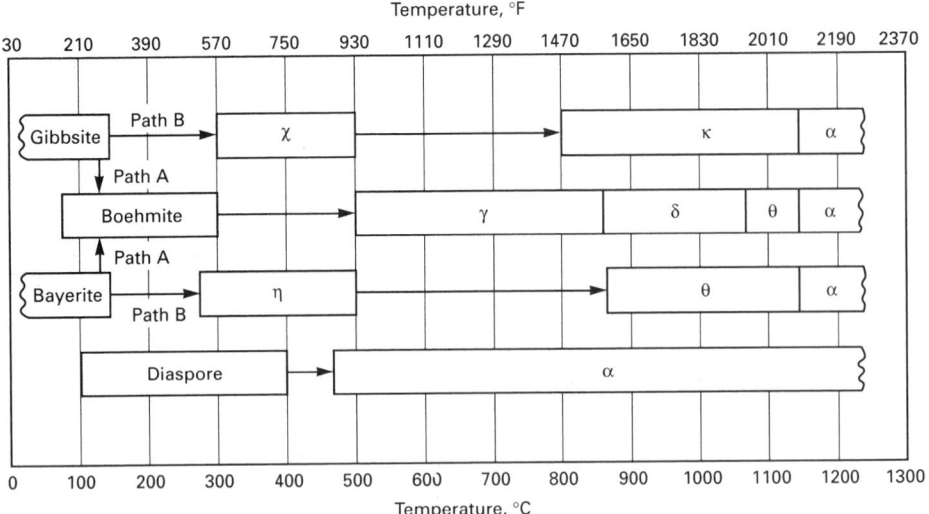

Fig 3 Phase transformation sequence of aluminum hydroxides. Source: Ref 23

diffusion of protons and aluminum ions along the directions of the hydrogen-bonded chains in the AlOOH structure. This preferred direction of diffusion leads to complete topotaxy of the γ-alumina with the boehmite. Micropores are created to accommodate the change in volume. The result is a porous skeleton of γ-alumina that is coherent with the boehmite lattice, with sheets of regularly spaced pores on the γ-alumina (001) plane. These pores are exceedingly small. Wilson observed 0.8 nm (0.03 μin.) pores spaced 3.5 to 4 nm (0.14 to 0.16 μin.) apart. In slowly heated boehmite, the pores are uniformly spaced in an ordered lamellar structure. Similar microvoids are formed by dehydration of other layered hydroxides. The void formation is an integral part of the dehydration process.

The transition phases can be an end product themselves, as with active alumina catalysts, but the stable phase is usually desired for most ceramic powders. The characteristics of the α-alumina are not directly determined by the dehydration step (that is, boehmite to γ-alumina), but, rather, by the subsequent phase transformations (such as θ-alumina to α-alumina). The nature of the transformation to the α phase has been controversial, but recent work has made it clear that the transformation occurs via nucleation and growth (Ref 27, 28). The transformation rate fits nucleation and growth kinetic models and microstructural evidence consistent with nucleation and growth.

Figure 4 is an electron micrograph of γ-alumina that has been partially transformed to α-alumina. The α-alumina appears as the coarser phase in colonies with a wormy (or vermicular) particle morphology. Similar vermicular morphologies have been observed in aluminas obtained from aluminum salts (Ref 29–32). The transformation to an α form is accompanied by a large decrease in surface area. The γ-alumina particles in Fig 4 are about

20 nm (0.8 μin.) in diameter, while the cross-sectional diameter of the vermicular α-alumina grains ranges from 75 to 200 nm (3 to 8 μin.). Because each colony is the result of a separate nucleation event, the individual "grains" within the particle are all part of a porous single crystal.

Agglomeration

Calcined powders are invariably agglomerated, often with a hierarchy of agglomerates. The largest agglomerates are usually

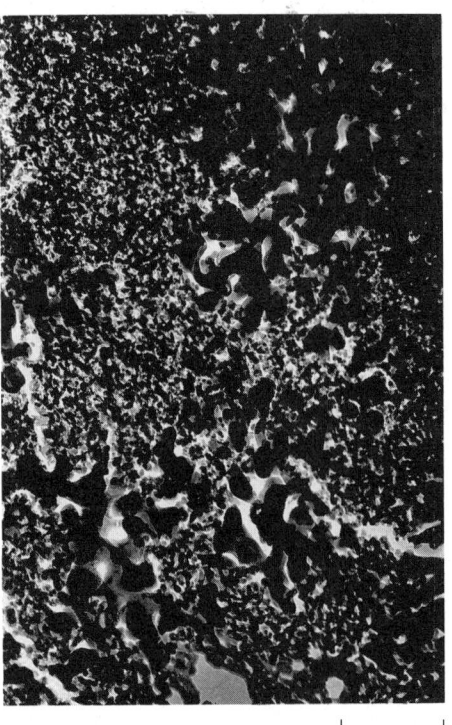

Fig 4 Growth of vermicular α-alumina in γ-alumina

0.7 μm

porous relicts that are pseudomorphs of the salt particles. The origin of the porosity is the large difference in molar volume between the parent salt and the oxide. For example, magnesium carbonate occupies 0.0285 m^3/kg (0.45 ft^3/lb) and decomposes to create magnesium oxide at 0.0113 m^3/kg (0.18 ft^3/lb). This difference can create 60 vol% porosity if the relict is the same size as the parent salt. Sintering will reduce the porosity, but at the cost of creating a stronger agglomerate.

Rhodes and Weunch (Ref 33) showed that the agglomerates resulting from the relicts retard the sintering of MgO. They compare magnesium hydroxide, carbonate, and oxalate, and noted that the hydroxide and oxalate-derived powder suffered exaggerated grain growth during hot pressing, whereas the carbonate-derived powder did not. The difference was related to the greater tendency to form large oriented relicts in the hydroxide, which can form the nuclei for exaggerated grain growth.

In aluminas produced by calcination, extensive agglomeration is present at several scales. At the level of the particles, hard agglomerates are formed by the nucleation and growth of α-alumina in the transition aluminas, as discussed above. On a coarser scale are the pseudomorphic relicts of the salt particles, as in the case of bayerite.

Powders produced by calcining ammonium aluminum sulfate (alum) present an interesting case. Alum is used to prepare powders with very high purity (with respect to metallic impurities, "99.5% pure" alumina can actually contain a significant amount of sulfur). Upon heating, the alum melts to form a syrupy liquid. The molten salt froths as the H_2O vaporizes, leaving a foamy solid of anhydrous ammonium aluminum sulfate. Ammonia is lost at about 300 °C (570 °F), followed by sulfur trioxide at about 800 °C (1470 °F). The result at 1000 °C (1830 °F) is a foamy cake of low bulk density (few vol% solids), consisting of a transition alumina with a surface area of about 30 m^2/g (145 $\times$ 10^3 ft^2/lb), and a crystallite size of about 50 nm (2 μin.).

This foamy structure is preserved throughout the process and leaves its relicts in the ultimate agglomerates of the alumina. Within the cells of this foam, the sulfate undergoes decomposition to create flaky, porous relicts consisting of the transition alumina. Alpha-alumina forms by nucleation and growth in the porous relicts, creating the vermicular structure discussed above (Ref 34). A typical 1 μm (40 μin.) size alumina powder made by this process consists of these agglomerates, as shown in Fig 5. These agglomerates are quite hard and easily survive compaction processes, reducing green density and retarding sintering (Ref 35, 36).

Calcined oxides can be milled to "deagglomerate" the powder. The milling process does not change the surface area or ultimate particle size, which is too small for further

Fig 5 Agglomerates in a commercial 1 μm (40 μin.) alumina powder derived from calcined alum. Source: Ref 35

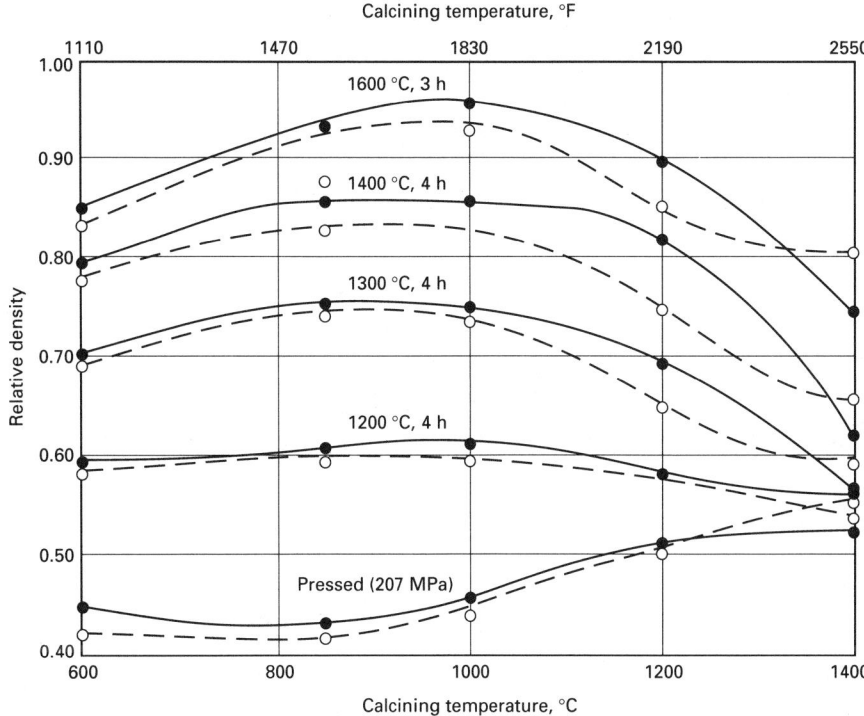

Fig 6 Effect of calcining treatment of reactive spinel on pressed and sintered densities. Solid curves, 2 h calcine; dashed curves, 6 h calcine. Source: Ref 39

comminution. Rather, it crushes the large relicts and improves the state of agglomeration. Such a treatment can dramatically improve the quality of the powder, improving bulk and green density, and enhancing sintering.

Effect of Process Variables

Nature of the Precursor. Salt precursors commonly used in ceramics include the carbonates, hydroxides, acetates, oxalates, sulfates, and nitrates. Carbonates and hydroxides are by far the most common. Although sulfates are also frequently used, they have a few disadvantages, such as emission of SO_3, an obnoxious pollutant. As a class, the sulfates tend to be more stable at high temperature. Therefore, oxides prepared from them can be contaminated with residual sulfur. Oxalates are frequently used as precursor salts, especially when co-precipitations are involved. Acetates and nitrates are less common, but are used when the salt precursor is obtained by desiccation (for example, freeze drying), rather than precipitation. Many nitrates melt before decomposition, which is inconvenient and restricts their use.

In any particular case, it is observed that certain precursor salts can be calcined to highly sinterable powders, whereas other salts are inferior. The oxides may differ in surface area, particle morphology, or state of agglomeration. They may also differ in purity, particularly anion impurity, such as residual carbonate or hydroxyl.

In a typical study (Ref 37), magnesia from acetates was found to be more sinterable than magnesia from oxalates or tannate, at least under certain conditions. The behavior can also be different for the same salt, prepared in a different manner. Recall the data in Fig 2 for brucite precipitates. In another example, Kato *et al*. (Ref 38) reported on the character of α-aluminas prepared by decomposition of hydrated aluminum sulfate, $Al_2(SO_4)_3 \cdot 18H_2O$. This salt was precipitated from an aqueous solution under conditions in which the precipitate was either thin flakes or small cubes. The samples differed in decomposition rate from the sulfate to the transition phase η-alumina. They also differed in the kinetics of the η- to α-alumina transformation and had quite different α-alumina crystallite size. Moreover, the character of the aggregates were spectacularly different: the α-alumina agglomerates from one solution were porous flakes with a two-dimensional vermicular morphology, whereas the particles from another precipitate were skeletal cubes, with a three-dimensional vermicular structure.

Temperature and Time. For given processing conditions, an optimal calcining temperature, time, and atmosphere can be determined empirically. However, the "optimal" conditions depend entirely on what one intends to optimize. The conditions that give the more active *powder* tend to be at lower temperatures where surface reactions are rate limiting, and the reaction rate is rather slow. Hence, these conditions may not optimize the *process*, in terms of efficiency or throughput.

Here the focus is on optimizing the quality of the powder. One must recognize that quality is only meaningful in relation to particular properties and particular processing parameters. For example, sintered density can be optimized for a particular compaction pres-

sure and sintering temperature. Generally, sintering behavior improves with increasing surface area. However, forming techniques become more difficult at high surface area levels, because characteristics such as compaction behavior, slip viscosity, and moldability are degraded. This often means that green density is lower for higher surface area powders, as illustrated in Fig 6, which gives

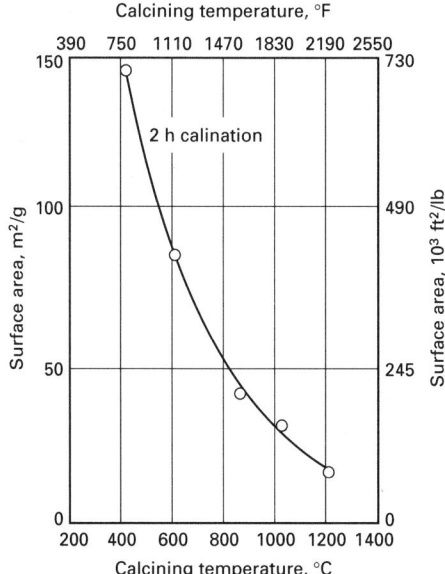

Fig 7 Effect of calcining temperature on specific surface area of reactive spinel. Source: Ref 39

data for very fine spinel powders (Ref 39). The sintered density (at 1600 °C, or 2910 °F), increases with calcination temperatures from 600 to 1000 °C (1110 to 1830 °F), but then decreases with higher calcination temperatures.

This can be rationalized by noting the improvement in the pressed green density (Fig 6) with increased calcination temperature (improving sintering) and the decrease in surface area with calcination temperature (Fig 7), which degrades sinterability. This trade-off leads to an "optimum" temperature for calcination. In a similar way, calcination time, atmosphere, or precursor salt characteristics can be optimized. Of course, there is nothing fundamental about this sort of optimal temperature, because any change in conditions (forming pressure, green density, deagglomeration) will result in a different optimal calcination treatment.

REFERENCES

1. D.A. Young, Endothermic Decompositions, Chapt. 3, *Decomposition of Solids*, Pergamon Press, 1966
2. C.H. Bamford and C.F.H. Tipper, Ed., Reactions in the Solid State, *Comprehensive Chemical Kinetics*, Vol 22, Elsevier, Amsterdam, 1980
3. H. Schmalzried, Chapt. 8, *Reactivity of Solids*, Academic Press, 1974
4. G. Bertrand, Comments on Kinetics of Endothermic Reactions. 2. Effect of the Solid and Gaseous Products, *J. Phys. Chem.*, Vol 82 (No. 23), 1978, p 2536–2537
5. W.D. Kingery, H.K. Bowen, and D.R. Uhlmann, Chapt. 9, *Introduction to Ceramics*, 2nd ed., John Wiley & Sons, 1976
6. A.W. Searcy and D. Beruto, Kinetics of Endothermic Decomposition Reactions. 2. Effect of the Solid and Gaseous Products, *J. Phys. Chem.*, Vol 82 (No. 2), 1978, p 163–167
7. A.W. Searcy and D. Beruto, Kinetics of Endothermic Decomposition Reactions. 1. Steady-State Chemical Steps, *J. Phys. Chem.*, Vol 80 (No. 4), 1976, p 425–429
8. A.G. Khachaturyan, *Theory of Structural Transformations in Solids*, Wiley, 1983
9. J.C. Niepce, J.C. Mutin, and G. Watalle, *Reactivity of Solids*, Plenum Press, 1977, p 131
10. A.F. Moodie and C.E. Warble, MgO Morphology and the Thermal Transformation of Mg(OH)$_2$, *J. Cryst. Growth*, Vol 74 (No. 1), 1986, p 89–100
11. M.G. Kim, U. Dahmen, and A.W. Searcy, Shape and Size of Crystalline MgO Particles Formed by the Decomposition of Mg(OH)$_2$, *J. Am. Ceram. Soc.*, Vol 71 (No. 8), 1988, p C373-C375
12. M.G. Kim, U. Dahmen, and A.W. Searcy, Structural Transformations in the Decomposition of Mg(OH)$_2$ and MgCO$_3$, *J. Am. Ceram. Soc.*, Vol 70 (No. 3), 1987, p 146–154
13. R. Gordon and W.D. Kingery, Thermal Decomposition of Brucite: I, *J. Am. Ceram. Soc.*, Vol 49 (No. 12), 1966, p 6544–6560
14. V.A. Phillips, H. Opperhauser, and J.L. Kolbe, Relations among Particle Size, Shape, and Surface Area of Mg(OH)$_2$ and Its Calcination Product, *J. Am. Ceram. Soc.*, Vol 61 (No. 1–2), 1978, p 75–81
15. P.J. Anderson and P.L. Morgan, Effects of Water Vapor on Sintering of MgO, *Trans. Faraday Soc.*, Vol 60 (No. 5), 1964, p 930–937
16. D. Beruto, R. Botter, and A.W. Searcy, H$_2$O-Catalyzed Sintering of 2-nm Cross-Section Particles of MgO, *J. Am. Ceram. Soc.*, Vol 70 (No. 3), 1987, p 155–159
17. A.J. Leonard, F. Van Cauwelaert, and J.J. Fripiat, Structure and Properties of Amorphous Silicon Aluminas: III, Hydrated Aluminas and Transition Aluminas, *J. Phys. Chem.*, Vol 71 (No. 3), 1967, p 695–707
18. H.C. Stumpf, A.S. Russel, J.W. Newsome, and C.M. Tucker, Thermal Transformations of Aluminas and Alumina Hydrates, *Ind. Eng. Chem.*, Vol 42 (No. 7), 1950, p 1398–1403
19. S.C. Carniglia, Thermochemistry of the Aluminas and Aluminum Trihalides, *J. Am. Ceram. Soc.*, Vol 66 (No. 7), 1983, p 495–500
20. A.M. Kalinina, Polymorphism and the Course of Thermal Transformation of Aluminum Oxide, *Zh. Neorg. Khim.*, Vol 4 (No. 6), 1959, p 1260–1269
21. J.L. Henry and H.J. Kelley, Preparation and Properties of Ultrafine High Purity Alumina, *J. Am. Ceram. Soc.*, Vol 48 (No. 4), 1965, p 217–218
22. A.N. Ryobov, I.I. Kozhina, and J.L. Kozlov, Influence of the Conditions of Preparation of Alpha Alumina on its Polymorphic Transformations, *Russ. J. Inorganic Chem.*, Vol 15, 1970, p 311–313
23. K. Wefers and G.M. Bell, "Oxides and Hydroxides of Alumina," Technical Paper No. 19, Alcoa Research Laboratories, Pittsburgh, 1972
24. G. Ervin, Jr., Structural Interpretation of the Diaspore-Corundum and Boehmite-Gamma-Alumina Transitions, *Acta Crystallogr.*, Vol 5 (No. 1), 1952, p 103–108
25. S.J. Wilson, The Dehydration of Boehmite, Gamma-AlO(OH) to Gamma-Al$_2$O$_3$, *J. Solid State Chem.*, Vol 30, 1980, p 247
26. S.J. Wilson, J.D.C. McConnell, and M.H. Stacey, Energetics of Formation of Lamella or Porous Microstructures in Gamma-Al$_2$O$_3$, *J. Mater. Sci.*, Vol 5, 1980, p 3081
27. S.J. Wilson and J.D.C. McConnell, A Kinetic Study of the System Gamma-AlOOH/Al$_2$O$_3$, *J. Solid State Chem.*, Vol 34, 1980, p 315
28. F.W. Dynys and J.W. Halloran, Alpha Alumina Formation in Alum-Derived Gamma-Alumina, *J. Am. Ceram. Soc.*, Vol 65 (No. 2), 1982, p 442–448
29. S. Kato, T. Iga, S. Sano, and E. Ishii, Formation Process of Alpha Alumina by Thermal Decomposition of Aluminum Sulfate, *Yogyo-Kyokai-Shi*, Vol 77 (No. 2), 1969, p 32
30. P.A. Badkar and J.E. Bailey, The Mechanism of Simultaneous Sintering and Phase Transformation in Alumina, *J. Mater. Sci.*, Vol 11, 1976, p 1794
31. F.W. Dynys, M. Ljungberg, and J.W. Halloran, Microstructural Transformations in Alumina Gels, *Better Ceramics through Chemistry*, Materials Research Society Symposium Proceedings, Vol 32, C.J. Brinker, D.E. Clarke, and D.R. Ulrich, Ed., Elsevier, 1984, p 312
32. D.W. Johnson and F.J. Schnettler, Characterization of Freeze-Dried Al$_2$O$_3$ and Fe$_2$O$_3$, *J. Am. Ceram. Soc.*, Vol 53 (No. 8), 1970, p 440
33. W.H. Rhodes and B.J. Weunch, Relation between Precursor and Microstructure in MgO, *J. Am. Ceram. Soc.*, Vol 56 (No. 9), 1973, p 495–496
34. F.W. Dynys and J.W. Halloran, Alpha Alumina Formation in Alum-Derived Gamma Alumina, *J. Am. Ceram. Soc.*, Vol 65 (No. 9), 1982, p 442–448
35. F.W. Dynys and J.W. Halloran, Compaction of an Aggregated Powder, *J. Am. Ceram. Soc.*, Vol 66 (No. 9), 1983, p 677–680
36. F.W. Dynys and J.W. Halloran, Influence of Aggregates on Sintering, *J. Am. Ceram. Soc.*, Vol 67 (No. 9), 1984, p 596–601
37. C.M. Pathak and V.K. Moorthy, I: Influence of Calcination Treatments on the Development of Morphology in MgO Powders, *Trans. Indian Ceram. Soc.*, Vol 35 (No. 5), 1976, p 89–96
38. E. Kato, K. Diamon, and M. Nambu, *J. Am. Ceram. Soc.*, Vol 64 (No. 8), 1981, p 436–443
39. R.J. Bratton, Characterization and Sintering of Reactive MgAl$_2$O$_4$ Spinel, *Ceram. Bull.*, Vol 48 (No. 8), 1969, p 1069–1076

Processing Additives

Anna E. McHale, Consultant

THE PROCESSING OF CERAMICS into useful products requires the transformation of powdered raw materials to a dense, uniform body through the application of consolidation techniques and subsequent thermal processing or sintering. During the sintering process, densification, grain growth, and either phase formation or transformation can all take place simultaneously. These processes require diffusional transport along and across interparticle interfaces, as well as within individual grains or crystallites.

The characteristics of the final ceramic are determined by the nature and relationship of the phases in the final microstructure. Any inhomogeneity of local composition, grain size, interface structure, porosity, and other aspects can lead to unpredictable behavior and unsatisfactory performance. Inhomogeneity in either the initial consolidated mass or green (dried) compact can lead to inhomogeneity in the final product. Flaws introduced because of inadequate elastic response to shear stresses in either forming or postforming processes or handling cannot usually be reversed or healed in later thermal processing stages.

Ceramics processing relies on process additives to make the manufacture of new and traditional ceramics products practical, reproducible, and forgiving to process and production variables. The goal is to achieve uniformity under a reasonable range of conditions at each stage of processing, without incorporating defects or inhomogeneities that will affect either further processing stages or final properties. Each product requires a "system" of additives, the selection and function of which are dependent on the means by which the ceramic product will be formulated from its raw materials, shaped before densification, incorporated with other production processes, and densified.

New application areas within the ceramics industry for traditional plastics-based processing methods, such as injection molding, require that processing requirements, as well as ceramic materials properties, are well understood. New preparative techniques for both oxide and non-oxide raw materials are constantly under development, and each new application further enhances the understanding of even the most traditional forming methods.

A ceramic body comprises a number of essential ingredients. Primarily, the permanent chemical constituents (oxides or non-oxides with sintering aids and dopants) determine the final chemistry and phase assemblage, thus determining the ultimate properties attainable. The means by which these properties are consistently attained are the function of the transient additives: solvents, dispersants, binders, and the like. Selection of transient additives is dictated by the forming method and processing details (including cost). The overriding requirement is that the additives leave no residue, or ash, or other chemical contamination that will interfere with the sintering process or adversely affect ultimate properties. Additives, such as carbon-based organics and water, are usually transient, but they may be permanent parts of the chemical make-up of the product. In certain processes, particularly in non-oxide ceramics processing, both functions may be fulfilled.

In a general processing scheme, the ceramic raw materials, including sintering aids, must be dispersed and deagglomerated for effective milling and later homogenization with binders. The viscosity and viscoelastic properties are then adjusted after dispersion to be appropriate to the forming method through the use of a binder and/or plasticizer. A preservative may be necessary to ensure that microbial action does not degrade binder properties during holding periods. The solvent used with all additives must be compatible with the process after forming (that is, drying and sintering), or with subsequent stages in processing which may include storage before further process steps. Stability against environmental and biological degradation are practical concerns, as are toxicity and waste handling.

Sintering Aids and Dopants

Sintering aids and dopants are intentional modifications of the base chemistry of the ceramic. They can be categorized according to their concentration and function, both during and after sintering. Sintering aids are combinations of one or more oxides that are added primarily to control grain growth, and thus enhance densification. They are generally added in the range of several weight percent in total, exceeding the solubility limit of these impurities in the base ceramic. They can be effective in the modification of grain growth by several means, depending on whether or not a liquid phase is formed during sintering and whether or not the sintering aid is soluble in that liquid. If the sintering aid is reasonably insoluble, or if no liquid forms, the additive may act simply as a dispersed second phase, inhibiting grain growth through interference with grain-boundary motion.

In the case of either liquid-phase sintering or transient (metastable) liquid-phase sintering, there can be many effects, including:

- Modification of surface energy, altering growth habit to minimize exaggerated grain growth due to growth anisotropy
- Increased dissolution/reprecipitation rate to enhance reactive sintering interaction and growth of preexistent nuclei, as from calcination steps
- Increased nucleation from the liquid of either the desired product phase or precursor phase, thus effectively limiting maximum grain size due to growth impingement of many crystallites

Control of these many facets is complex and often requires detailed study of microstructure and relevant phase equilibria to determine the optimal control mechanism.

Because of their relatively large and important contribution to overall chemistry, sintering aids must be considered as chemical components in a thermodynamic sense. Although transient additives can contribute greater mass or volume to the ceramic body before firing, their function is to promote the particle-particle contacts and pore structure of the oxide constituents to allow reaction, rather than to alter the course or outcome of that reaction toward the equilibria phases.

Dopants are also oxide chemical additives that, through their interaction with surface chemistry and subsequent alteration of surface energetics, can affect grain growth and densification. Dopants are usually oxides of metals that have valences different from the primary oxide. These oxides are soluble in the primary oxide and are added primarily for their effects on the electronic and ionic transport properties of the primary oxide (as used after sintering), once they are incorporated into the oxide crystal structure. During densification, the "doped" ceramic can exhibit altered diffusional properties, which can enable either more rapid densification or solid-state reaction. If the dopant is incorporated in the primary oxide prior to compounding with transient additives, as via coprecipitation or other means, alteration of the surface interaction with solvents and dispersants can occur because of change in either the surface charge or surface structure, which is often dependent on ionic lattice defect concentrations. Modification of the properties of a ceramic via doping is an intensively studied topic, especially in the electronics and solid-state battery industries.

There is often some small solubility of oxide "sintering aids," as well, and there can be a dopant effect. Such an effect is observed most critically in dielectric properties and high-temperature mechanical properties, such as creep deformation. Because of the relatively large concentration of sintering aids used, a dopant effect can be difficult to analyze. The analysis and control of dopants requires that they be operative primarily in the "dilute solution" concentration regime. The temperature dependence of the saturation concentration of the additive oxide in the primary oxide or phase will dominate observed effects if high concentrations of oxide additives are used.

Maximum uniformity of dispersion of sintering aids and dopants is critical for reproducibility of their effectiveness. For this reason, dopants are often incorporated during either powder synthesis or the initial milling and dispersion stage of processing. Dopants expected to have strong surface activity can be incorporated in solution during milling if a suitable solvent can be found, that is, it must be either compatible with further processing or removable without residue during a preprocessing calcination step.

It must also be noted that any residues from transient additives or solvents can also function as dopants or interfere with sintering. Particular problems can result from residual sulfate, chloride, or unoxidized carbon, resulting from decomposition of organics. Metal contaminants that act as "unintentional" dopants can be introduced from these organics, either intrinsic to the organic solvent or arising from later processing and/or filtration steps. Ashing tests and chemical analysis of residual material are recommended in most processes that use additives, especially where chemical purity is essential to functional, or device, and properties.

Forming

For thousands of years, ceramics have been formed from fluid (slip) or plastic bodies based on naturally occurring clays. Only recently has there been an effort to apply traditional ceramic-forming methods to nonclays and nonoxide ceramics. Utilization of nontraditional forming methods adapted from other materials industries is also becoming more common.

In the process of this revolution, an understanding of the rheology of "loaded" systems, the meaning of dispersion and dispersion stability, the interaction of organic and inorganic species with surfaces and solvents, and the mechanics of plastic and pseudoplastic materials under stress have all evolved. The publications of the American Ceramic Society, *The American Ceramic Society Bulletin*, and *Journal of the American Ceramic Society*, and *The Journal of Materials Science* are primary sources of valuable information on process development, as well as ideas and information on new methods, systems, and applications.

Within each process type, general requirements of viscoelastic behavior (viscosity and shear strength) must be met, along with the overall requirement of chemical homogeneity and "green" density. Further requirements for strength and handling characteristics, such as flexibility and machinability, can also be imposed. Although not discussed here, the selection of appropriate raw materials characteristics, particularly particle size distribution, also affects the properties and performance of the consolidated powder after forming.

It is unsatisfactory to classify forming process by a single characteristic or product type. However, a broad classification of forming processes can be based on the viscoelastic requirements of the process. Subclassification, as shown in Table 1, is based on the handling characteristics of the ceramic before forming, that is, in dry, semidry (plastic), or wet (slurry or slip-like) states. A single ceramic product can involve several interim forming steps. A ceramic forming process, such as dry pressing or injection molding, can involve significant wet forming during slurry production prior to granulation and final-shape forming.

The selection of a forming process is dictated by the complexity of the shape to be formed, the nature of the raw materials, the surface finish required, and other aspects of device manufacture. Additives are used at various stages during each process to ensure practical and reproducible processing.

Solvents

Solvents are necessary both to impart fluidity and as a vehicle for the dissolution and uniform distribution of all additives. Solvents are either polar or nonpolar and can be either aqueous or nonaqueous. Generally, polar molecules are soluble in polar solvents, and nonpolar molecules are soluble in nonpolar solvents. Polar solvents interact strongly with ionic (oxide) surfaces.

Water is a common polar solvent, as are simple alcohols. Polyhedric alcohols of low molecular weight (glycols) are also water soluble. Organic polar solvents, such as the fatty

Table 1 Viscoelastic properties required by various ceramic-forming processes

Process	Degree of saturation	$y = \dfrac{V_L}{V_{void}}$	Forming pressure		Flow property	Viscosity range, Pa·s	Shear stress rate, 1/s	Comment
			MPa	ksi				
Dry(a)								
Dry pressing	0.01–0.05	~0.3	20–200	2.9–29	...	...	$0–10^3$(b)	Stability of binder against
Isostatic pressing	...	...	...	...	Newtonian or elastic	...	$0–10^3$(b)	hydration and degradation essential during storage
Semi-wet plastic forming								
Extrusion, pressing, and jiggering	>0.9	~1	1–20	0.15–2.9	Newtonian	100–1000	$10^2–10^3$	Lubricants used
Injection molding	>0.9	~1	1–20	0.15–2.9	Pseudoplastic	<1000	$10^2–10^3$	Values measured at 200 °C (390 °F); both soluble and solid lubricants used
Wet								
Slip casting	>1	1.5–2.0	~1	~0.15	Pseudoplastic	0.1	$1–10^3$(c)	Thixotropic behavior sometimes desirable
Tape casting	>1	1.5–2.0	~1	~0.15	Pseudoplastic	0.1	$>10^3$	Flexibility after solvent removal required
Screen or thick-film printing	>1	1.5–2.0	~1	~0.15	Pseudoplastic	0.1	$>10^4$(d)	Thixotropy must be carefully controlled
Spray-dry granulation	>1	1.5–2.0	>1	>0.15	Newtonian	0.1	$1–10^3$(c)	Values measured at ~115 °C (240 °F)

(a) Based on granulated feedstock. (b) 0 represents static storage of feedstock. (c) Highest during mixing. (d) During some application methods

acids or amines, can also have an acid or base character, respectively. Nonpolar solvents include benzene and toluene. Common solvents are identified in Table 2.

Although organic-derived solvents from industrial sources are usually of known quality and purity, local water supplies are often used. As a highly effective polar solvent, water can contain high levels of dissolved impurities, such as metals (Ca^{2+}, Na^+, Mg^{2+}) and anions (sulfate or carbonate). Halogens, such as Cl^- or F^-, are often present in city water supplies. General water quality can be estimated by measuring the specific conductance, which is related to the total dissolved solids content (TDS) via the relationship:

$$TDS(ppm) = \frac{C(\mu S/cm) - 0.055}{2.5}$$

where the factor 2.5 is a weighted average of the specific conductances of the most common ionized contaminants in water supplies, and S units are ohm^{-1}.

It is useful to standardize water quality using filtration, distillation, and/or deionization, which should be checked frequently using chemical analysis. Chelating agents may be necessary to remove heavy metal contaminants. If they are not removed, the dissolved substances in water could be the greatest single source of irreproducible processing results. Dissolved metals can adsorb on the primary oxide surface, changing its ionic nature and altering the effects of surfactants. Some metals react with the surfactant molecules, forming insoluble compounds and lessening surfactant effectiveness.

A classic example is the interaction between hard water and soap. Not only does the Ca^{2+} interfere with the surfactant action of the soap, but a hard, insoluble precipitate that can be damaging to process equipment is formed.

Metallic contaminants can act as dopants, which are especially deleterious in the processing of high-quality dielectric ceramics, or can be concentrated at flaws in the final microstructure and contribute to unpredictable dielectric and mechanical breakdown.

The solvating action of a polar solvent such as alcohol is often altered by solution in another polar organic solvent. The solvent may be "activated," that is, acquire true solvent properties, in such a solution.

The solubility of the primary oxide and sintering aids can also be a factor for consideration in the use of polar solvents. Metal contamination from this source can interfere with reliable processing, just as an initially contaminated water or solvent source would. Naturally occurring clay minerals are ranked by their characteristic cation exchange capacity (CEC), which is the ability to exchange surface cations with water, and is usually stated in milliequivalents per 100 g. Knowledge of the nature of charged species and contaminants arising from these sources is essential in controlling the dispersant action when using clays. Colloidal clays, particularly those containing high fractions of montmorillonite, have a high capacity for attracting dissolved ions and are ranked using the methyl blue index, or MBI. The combined CEC and MBI values of all clays in the product formulation is also a factor in the later selection of dispersants and flocculants.

In addition to solvent properties, the volatility of the solvent is a necessary factor in the solvent or solvent system selection process. Solvent removal is facilitated by high volatility, preferably at temperatures below binder burnout, so that clean, uniform pyrolysis is promoted by the presence of open-pore channels. The wall or part thickness of the product governs the selection of appropriate solvent volatility. In thick-walled pieces, very high volatility can lead to deleterious internal pressure build-up and excessive drying stress if porosity is very fine or tortuous. A high boiling point is desirable to avoid blistering or bubble formation during removal stages.

The solvent can also function as a plasticizer for binders or effectively modify binder properties such as characteristic glass transition temperature. Although pure organic liquids can act as dispersants for particular oxide powders, water-based systems and most organic solvents will require the use of either a dispersant or electrolyte to maintain stability over time.

Surfactants

Additives other than binders, plasticizers, and some preservatives are generally categorized as surface active agents, or surfactants. Extensive literature exists as to the nature and characterization of surfactants as applied in the broad spectrum of industrial chemical processes. A surfactant acts only as part of a system with three or more components. This system includes one or more solvents, a dispersed phase or phases, and the surfactant itself, which acts to modify the interfacial characteristics between the dispersed phase(s) and solvent(s). The surfactant macromolecule is depicted in Fig 1. To be effective, the head, or hydrophobic group, must represent the appropriate type to be attractive to the dispersed phase, and the tail, or hydrophilic group, must be soluble in the chosen solvent system, often water. For water solubility, chain length in the tail is generally from 8 to 16 carbons. The lowest molecular weight polar molecules are the most soluble.

There are four types of surfactant molecules, based on the nature of the functional "head" group:

- Anionic, which are most common in ceramics processes and include common soaps. Many examples of this type are acidic in nature and may either be corrosive or interfere with binder performance
- Cationic, which are not commonly used and are often toxic, but may have good germicidal effects. Generally basic, this type acts to neutralize acidic additives or impurities, and thus is used as a corrosion

Table 2 Solvents commonly used in ceramics processing (room temperature)

Solvent	Dielectric constant	Surface tension, mN/m	Viscosity, Pa · s
Polar, aqueous			
Water	80	73	1.0
Polar, organic			
Alcohols			
Methyl alcohol (CH_3OH)	33	23	0.6
Isopropyl alcohol (C_3H_7OH)	18	22	2.4
n-octyl alcohol ($C_8H_{17}OH$)	10	28	10.6
Benzyl alcohol (C_7H_7OH)	13.1	35.5	5.8
Ethylene glycol ($C_2H_6O_2$)	37	48	20
Furfuryl alcohol [$2-(CH_4H_3O)CH_2OH$]	...	38	4.6
Acids			
Propionic acid ($CH_3CH_2CO_2H$)	3.3	26.7	1.1
Octonoic acid [$CH_3(CH_2)_6CO_2H$]	2.45	28	4.6
Aldehydes			
Octanal ($C_8H_{17}OH$)	10	28	10.6
Benzaldehyde (C_6H_5CHO)	16.8	38.5	1.32
Esters			
Ethyl acetate ($C_4H_8O_2$)	6.15	23.52	0.42
n-butyl n-butyrate	5	23.72	0.98
Ethers			
Isopentyl ether	2.82	22.8	0.4
Tetrahydrofuran (C_4H_8O)	7.85	27.4	0.53
Ketones			
Acetone (C_3H_6O)	20.7	25.1	0.32
Heptatone [$CH_3(CH_2)_4COCH_3$]	9.8	26.7	0.76
Methyl ethyl ketone (C_4H_8O)	18	25	0.4
Halogenated hydrocarbons			
Trichloroethylene (C_2HCl_3)	3	...	...
Nonpolar, organic			
Benzene (C_6H_6)	2.3	30.22	0.65
Toluene ($C_6H_5CH_3$)	2.4	28.5	0.6
n-hexane (C_6H_{14})	1.89	18.4	0.33

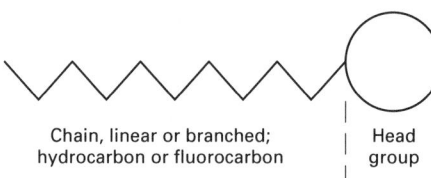

Fig 1 Surface-active molecules

Chain, linear or branched; hydrocarbon or fluorocarbon | Head group

inhibitor. This type is generally not compatible with binder systems

- Zwitterionic, which have two functional groups and include many naturally occurring substances, such as lecithin and triglycerides. Not generally considered water soluble
- Non-ionic, which are not numerically common. These molecules tend not to interfere with binder systems, which makes them valuable in processing. They may exhibit ionic character in certain pH ranges

The most commonly used surfactants are of the water-soluble anionic type. They may be used in combination with nonionic and/or zwitterionic surfactants. Surfactant effect is cancelled, either totally or partially, by mixing cationic and anionic types.

Surfactant interaction with ceramic powders is dominated by the surface charge distribution typical of crystalline or amorphous ionic materials, as modified by the dielectric properties of the solvent system in polar solvents.

Surfactants have several functions in the additive system, which may be inseparable. As wetting agents, they enable effective total wetting of the solids by the solvent. As dispersants, they promote deflocculation and stability. As antifoaming agents, they minimize bubble-type pores that result from mixing or forming processes and which may entrain air. In addition, they may alter the rheological properties of the solvent-solids system and act as plasticizers and/or lubricants.

Wetting Agents and Dispersants

Powdered raw materials tend to agglomerate because of surface-adsorbed moisture and simple cohesive forces between fine particles, such as the van der Waals attraction. The function of a dispersant/solvent system is, first, to wet the oxide surfaces so that these attractive forces can be overcome, and, second, to modify the surface properties so that the particles will not recombine, or flocculate. In certain processes, it may be desirable to reverse the state of dispersion of the system later in the process. Therefore, in cases such as slip casting, a dispersant that is sensitive to pH or temperature may be useful.

To be effective, the interaction of any surface active agent with the powder surface must be unhindered by other adsorbed species, such as water or other solvent molecules remain-

ing from earlier processing steps. The availability of soluble cations on clays is ranked by the cation exchange capacity (CEC). Dissolved cations may remain in loose association with the clay particles in a solvated gel layer, affecting interaction with other surfactants. In addition to interference between dispersant and surface because of saturation or alteration of active sites, surface contaminants may be displaced and either contaminate the solvent or interfere with binder effectiveness. Oxide powders are usually dried before processing with solvents and surfactants to increase reproducibility.

Wetting is defined in Fig 2 as a simple force balance at the junction between solid, liquid, and vapor. The tendency of the liquid to spread on the solid surface is generally described by the wetting angle, θ. Measurement of wetting angle is subjective and quite difficult in practice, because the characteristics of the wetted surface tend to change over time as chemical interaction between the surface and fluid occurs. The wetting angle is often high for the fresh interaction of a proceeding junction, but is generally lower for the receding junction of the same system. Often, both angles are reported.

Wetting agents are surfactants that are primarily soluble in the liquid phase. They are also attracted to a particle surface, lowering its surface energy, while altering the interfacial characteristics of the liquid so that it spontaneously wets the particle surface. Perfect wetting involves the displacement of all previously adsorbed water, air, and contaminant species by a new liquid film that penetrates all surfaces of the particle (including surface porosity). A perfectly wetted powder will also be dispersed. Practical dispersion requires that the wetted particles remain separated in the liquid over time, without a tendency to recombine. Dispersive surfactants can also be wetting agents, but their interaction with the powder surface is more important than their effect on the surface tension of the liquid phase.

The work of dispersion of a dense, non-agglomerated powder in a fluid is equal to the sum of: the work necessary for adhesion of the powder and liquid (W_a), the work of immersion of the particle (W_i), and the work of spreading the liquid over the surfaces of the particle (W_s). Thus,

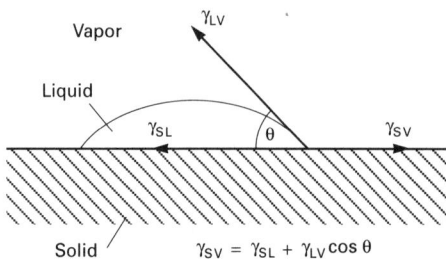

Fig 2 Definition of wetting angle and surface tension force balance

$$\gamma_{SV} = \gamma_{SL} + \gamma_{LV} \cos \theta$$

$$W_d = W_a + W_i + W_s$$

With reference to force balance of Fig 2, it can be shown that

$$W_d = -6 \gamma_{LV} \cos \theta$$

where θ is the solid-liquid contact (wetting) angle. Most commercial wetting agents effectively reduce the wetting angle to near zero and also effectively reduce γ_{LV} and enhance wetting.

Because ceramic powders generally agglomerate, an effective wetting agent or dispersant must penetrate intra-agglomerate surfaces. It can be shown (Ref 1) that effectiveness in dispersing agglomerated powders requires that the difference ($\gamma_{SV} - \gamma_{SL}$) be maximized, with γ_{LV} being as small as possible, regardless of the magnitude of the wetting angle. This added complexity often means that the best wetting/dispersing agent or solvent/dispersant system for an individual powder is found by trial and error.

Nonvolatile surfactant molecules can serve as wetting agents, provided that they are soluble in the fluid medium. Common types are anionics with medium chain (C_{12} to C_{17}) lengths in the general families of alkali sulfonates, sulfates, lignosulfates, carboxylates, and phosphates. Effectiveness in terms of wetting a particular powder can be estimated using a flotation test, in which the time to "sink" all particles is inversely proportional to wetting efficiency and spreading kinetics. A concentration dependence of time to full immersion is usually observed, and a plot of $\log(t)$ versus $\log(c)$ is usually linear for a given wetting agent.

Surfactants used as wetting agents are most effective at or near the critical micelle concentration (CMC). Typical values of CMC are listed in Table 3 for examples of each surfactant type, generally decreasing within type with increasing molecular weight. This concentration represents the location of a thermodynamic phase boundary between a single-phase monomer solution and a two-phase solution consisting of a monomer solution plus crystal or micelle. It is the activity of the monomer in solution that determines surfactant behavior. Concentrations above the CMC will not increase effectiveness.

Two critical temperatures have been defined. One is the Krafft point, or the temperature at which surfactant solubility becomes exactly equal to the CMC. Below this temperature the surfactant solution is metastable. The other critical temperature is the cloud point, or the minimum temperature above which liquid-liquid immiscibility is observed. Practical application of wetting agents and other surfactants requires that all process temperatures be between these points or at least below the cloud point temperature. Operation below the Krafft temperature, in the case of a supersaturated solution, could ensure a fixed activity of monomer and highly reproducible behavior if temperature excursions can be

Table 3   Typical surfactants and critical micelle concentration values at 25 °C (77 °F)

	Critical micelle concentration	
Surfactant type	mol/L	mol/gal
Anionic (sodium stearate and sodium disopropylnaphthalene sulfonate)	0.001–0.1	0.004–0.4
Cationic (dodecyltrimethylammonium chloride)	~0.01	~0.04
Non-ionic (ethoxylated nonylphenol and ethoxylated tridecyl alcohol)	0.0001–0.01	0.0004–0.04

avoided. In practice, wetting agents are used at concentration levels from 0.01 to 0.2 wt%, near the expected CMC at room temperature (Ref 2).

Another important function of surfactants that is related to wetting is the destabilization of foams and bubbles. Surfactants act to weaken the elastic properties of the interface between liquid and vapor, hastening destabilization. Foams can be created through mechanical action, as in high shear blending, or can result from a sudden supersaturation of a volatile phase that is due to a change in temperature or pressure. Bubbles formed by the latter process can be particularly deleterious because they may nucleate at any point in the ceramic thickness and are difficult to dissipate through viscous binder systems. Internal bubbles may coalesce into large internal pores if they are not destabilized. Finely distributed pores can be eliminated through diffusion at sintering temperatures and can actually be somewhat effective in pinning grain boundaries.

Although a perfectly wetted powder is also dispersed, particular surfactants and electrolytes will act to stabilize the dispersion of particles against re-agglomeration, or flocculation. Thus, dispersants are also called deflocculants (Table 4). The distinction between the two terms is not clear, although dispersant is the term generally given to organic macromolecules, whereas deflocculant refers most often to simpler, nonorganic polyelectrolytes. Flocculation and deflocculation are reversible through alteration of the electrolyte balance and pH of the system. Macromolecular dispersion is not easily reversible and is often valued for this stability.

Polyelectrolytes act to alter the surface charge "double-layer" characteristic of a charged particle in a polar solvent so that the electrostatic repulsive forces will overcome any van der Waals-type attraction. This interaction is described by the combined theories of several authors (Ref 3, 4) and is referred to as the DVLO theory. It is shown in Fig 3.

Table 4   Common dispersant/ deflocculant polyelectrolytes

Inorganic	Organic
Sodium carbonate	Sodium polyacrylate
Sodium silicate	Ammonium polyacrylate
Sodium borate	Sodium citrate
Tetrasodium pyrophosphate	Sodium succinate
	Sodium tartrate
	Sodium polysulfonate
	Ammonium citrate

The existence of the secondary minimum (Fig 3) is a function of the surface charge density and the dielectric constant of the solvent, and only exists in those systems where the boundary layer thickness is smaller than the dispersed particle size. The extent of the double-layer repulsion is dependent on the dielectric properties of the solvent, as described by the Gouy-Chapmann theory, and may be diminished either by the action of a counterion or a system that is too highly concentrated in electrolyte. If a system finds equilibrium in the secondary minimum separation, it is said to be coagulated, or loosely flocculated. Coagulation is seldom desirable, whereas flocculation resulting from the reversal of dispersion using counter-ions after homogenization has been achieved, is extremely useful.

A properly flocculated slip has advantageous properties for casting or filter pressing because of the efficient removal of solvent. In cases of full deflocculation, it is still possible that random Brownian motion can supply sufficient energy to overcome the electrostatic potential barrier and cause particle approach within the primary minimum distance, resulting in adhesion. A well-dispersed suspension settles over time (days to weeks) into a dense layer, whereas a coagulated system with the same settling time has a greater sediment volume, as indicated in Fig 4.

The action and effectiveness of macromolecular dispersants, particularly in nonpolar solvents with low dielectric constant, is often said to be due to "stearic hindrance" to particle approach within the primary minimum, because of the bulk of the adsorbed molecule. In fact, fully effective dispersion requires judicious selection of the organic dispersant/solvent system. The "head" group characteristic must be selected for appropriate charge character to ensure efficient absorption on the ionic surface. This may require consideration of the relative size of the functional group with respect to the charge site distribution on the surface, in order to obtain optimal absorbed concentration. The hydrophilic group, or tail, must be soluble in the solvent system, without a tendency to react between the solvated ends of adjacent mole-

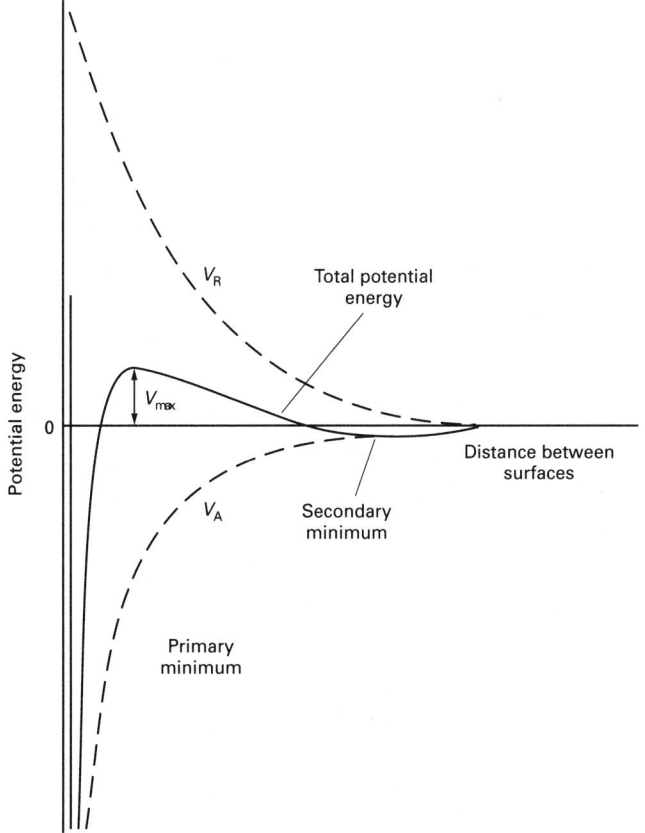

Fig 3 Energetics of the interaction of charged particles in a polar solvent, as described by DVLO theory. V_R, repulsive potential energy; V_A, attractive potential energy; V_{max}, maximum total potential energy ($V_R + V_A$). Source: Ref 5

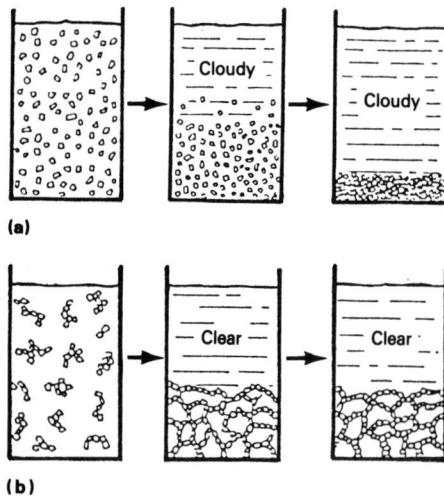

Fig 4 Sedimentation behavior. (a) Deflocculated suspension. (b) Coagulated suspension. Source: Ref 2

cules. Stearic hindrance is then, in fact, a function of the entropy of the system. Too close an approach demands that the solvated tails interact and partially align, decreasing the system entropy and thus increasing the total free energy. Any tendency toward exothermic reaction between solvated molecules would mediate the effect and could result in either coagulation over time or the formation of a gel structure.

When evaluating a dispersant, primary consideration is given to compatibility with the binder/solvent system. As discussed below, binder selection is dictated by the forming method and finishing requirements. One protocol for selection that has been suggested is (Ref 5):

- Consider only those dispersants that are readily soluble in the binder/solvent system
- Determine surface charge of the powder in the solvent. The appropriate dispersant type (anionic, cationic or non-ionic) should maximize surface adsorption. Determine surface charge through measurement of zeta-potential or by observing "drift" toward an immersed electrode
- Determine which dispersant candidates yield minimum viscosity at high solids loading, about 60 to 70 vol% solids. Test only the solvent/dispersant system because binder viscosity characteristics could mask dispersant properties
- Observe which candidate has minimum settling volume, either gravitational or centrifugal. Solids loading for settling experiments can be as low as 4 vol%

The best candidates determined on the basis of this selection protocol will need to undergo further testing for ultimate compatibility with the binder. The dispersant can either have deleterious effects on the binder rheology or it can act beneficially, as a minor binder or

plasticizer. Dispersants used in lower-viscosity process systems can also function as a binder, imparting some strength to the compact after solvent removal.

Binders and Plasticizers

Binder selection is based on: the viscoelastic properties required for the selected fabrication process, the type of casting surface (porous or nonporous), cost considerations, process temperatures, and stability of properties with respect to long-term storage. The function of the binder is to impart sufficient strength and appropriate elastic properties for handling and shaping during the postforming stage.

The viscosity requirements of common forming processes can be categorized as being low (10^{-2} to 10^{-1} Pa/s), medium ($\sim$1 Pa/s) and high (10 to 1000 Pa/s). The binder concentration typical for each process is about 3 wt% in dry processes, 3 to 17 wt% in wet processes, and about 7 to 20 wt% in plastic forming (corresponding to 8 to 15 vol% after solvent removal). Binders are categorized by their viscosity grade at concentration in the relationship:

$$\log \eta = kc$$

where η is viscosity in mPa/s, k is a constant, and c is the concentration. Values of k have these ranges:

- Very low, $0 < k < 0.133$
- Low, $0.133 < k < 0.333$
- Medium, $0.333 < k < 1.0$
- High, $1.0 < k < 3.0$
- Very high, $3.0 < k <$ infinity

Naturally occurring colloidal clays, such as bentonite, can also be effective as binders and are commonly used to impart useful rheological properties while maintaining dispersion in slurries, inks, and paints.

Common water-soluble and nonwater-soluble binders are identified in Table 5. Bind-

ers are generally used with a plasticizer, which modifies the properties of the binder to increase its suitability for the process and associated processing conditions. A thermoplastic binder used in injection molding may need to have its viscosity at temperature either raised or lowered to enable good fill properties. Organic binders have characteristic glass transition behavior. It is important for the glass transition temperature (T_g) to be below the temperature of plastic forming processes and postforming handling and machining. The judicious choice of solvent, as well as plasticizer, can enable easy forming and toughness-in-handling properties as the T_g of the binder/plasticizer changes during solvent removal (drying) stages.

Conversely, solvent characteristics can be dictated either by the nature of the primary oxide or non-oxide or by the solvent removal characteristics required by the product shape. Highest volatility is suitable for the thinnest castings, whereas lower volatility is required for thicker parts to minimize drying stresses. The solvent can then dictate the choice of binder system, with secondary consideration being given to other factors.

The nature of the rheological behavior under shear stress is very important in binder selection. The behavior of the binder system can be normal, or Newtonian, offering constant resistance to flow (often after a threshold stress), in which case the viscosity/concentration relationship is log-normal, as described above. Or, the behavior can be dilatant or pseudoplastic, either of which can exhibit yield point characteristics, as well. A dilatant system exhibits increasing resistance to flow with increasing shear rate, whereas a pseudoplastic system becomes more fluid as shear rate increases. A pseudoplastic system can either return instantly to the high-viscosity state when shear stress is removed or be thixotropic, retaining low-viscosity characteristics for a significant time after shear stress is removed.

Table 5   Binders for aqueous and nonaqueous systems

Binder	Viscosity grade	Electrochemical type
Aqueous systems		
Colloidal types		
Cellulose	Low	...
Clays		
Carbohydrate-derived organics		
Methyl cellulose, other cellulose ethers	Medium-high	Non-ionic
Starches and dextrins	Low-medium	Non-ionic
Sodium alginate, ammonium alginate, natural	Medium-high	Anionic
gums	High	Anionic
	High-very high	Any type
Noncarbohydrate-derived organics		
Polyvinyl alcohol	Low-medium	Non-ionic
Polyvinyl butyral	...	
Acrylic resins	High	Non-ionic
Polyethylene glycol	Low-high	Non-ionic
Waxes	(Temperature dependent)	Non-ionic
Polyethyleneimine	Low-medium	Cationic
Nonaqueous (organic solvent) systems		
Polyvinyl butyral	...	...
Polyvinyl formal	...	...
Polymethylmethacrylate	...	...

Dilatant behavior is seldom desired. However, pseudoplastic behavior is required in some wet processes, such as screen printing or tape casting. Controlled thixotropic behavior can be beneficial for the rapid removal of mechanical foams and bubbles after remixing operations and for the self-leveling of drain surfaces in slip-casting operations.

A gel structure can also form, and because it does not flow, it will tear under shear stress in excess of its yield point. Gelation may be either thermal, in which case it is usually reversible, or chemical, in which case it is due to irreversible interaction of the molecular species present. Gel behavior is noted in unacceptably slow drying rates, because solvent removal is limited by the minute gel pore structure.

Several processes require the consideration of high-temperature rheological properties, such as the preparation of slurries for spray-dry granulation (~115 °C, or 240 °F) or the compounding of thermoplastic binders and filler for ceramic powder (~200 °C, or 390 °F).

During the preparation of a spray-dried granulated product for subsequent compaction in dry or isostatically pressed bodies, the binder must impart significant strength to the granule to avoid deformation or crushing during storage. Resistance to water absorption is considered important for good die-filling properties. Plastic flow properties and lubricity during pressing are also required.

Special binders that have high-viscosity, pseudoplastic behavior at ~200 °C (390 °F) are suitable for injection molding of high-performance ceramics. These include polypropylene microcrystalline wax, low-density polyethylene, polypropylene glycol, polyethylene wax, silicone oil, vegetable oil, or dioctyl phthalate, used as minor binders or plasticizers, as well as ethyl cellulose, polystyrene, polyethylene, pre-ceramics (Ref 6), borodiphenyl siloxane, polycarbosilane (*n*-hexane), and polysilazane (toluene).

Lubricants

High shear rate processes, such as injection molding and extrusion, can require the use of lubricants to minimize friction with mold or die surfaces, as well as between particles. Surfactants often function as adequate lubricants to reduce sliding friction between particles. Nonsoluble or solid lubricants, such as graphite, may be necessary to ensure flow in some cases, particularly when forming at higher temperatures. Commonly used lubricants are listed in Table 6.

Postforming Operations

The successful use of additives requires that they are removed from the ceramic without causing flaws that are due either to drying stresses or the byproducts of pyrolysis of organic compounds. The removal of additives is usually sequential.

Table 6   Common lubricants

Paraffin wax
Aluminum stearate
Butyl stearate
Lithium stearate
Magnesium stearate
Sodium stearate
Steric acid
Zinc stearate
Oleic acid
Polyglycols
Talc (platey form)
Graphite, boron nitride

Initially, solvent removal occurs through evaporation at moderate temperature, well below the pyrolysis temperatures of any organics present, which opens up the internal surfaces of the green, or dry, ceramic to the atmosphere. If a continuous network of distributed pores has not developed prior to the pyrolysis temperatures of the organics, significant ashing may occur under anaerobic conditions.

While the product is in a green state, further surface finishing can be undertaken. The binder must impart sufficient strength and plasticity to withstand significant handling. Upon removal of the solvent, it is normal and desirable to find that the T_g of the binder has increased above room temperature, and that the piece is no longer plastically deformable. If plasticity is desired for postforming operations, alternate plasticizers or co-solvents may be required to tailor the mechanical properties.

Carbon residues must be fully oxidized prior to densification. For carbon-based organics, onset of "burn-out" is around 300 °C (570 °F) in air. High rates of removal occur between 350 and 600 °C (660 and 1100 °F). A residual of 0.5 to 2% ash is common. If the ash is concentrated because of poor binder distribution, it contributes to the total impurity content, which is deleterious in high-performance ceramics. Non-ionic organics contain no metals but contribute significant nitrogen, which is believed to affect sintering behavior.

The residues of some organics, such as the nitrosilanes, are considered "pre-ceramic" in the non-oxide ceramic formulations, SiC and Si_3N_4 (Ref 6). Inorganic binders, such as clays, are also pre-ceramic in the sense that they contribute to the behavior of the ceramic during densification and in application.

REFERENCES

1. Th.F. Thadros, Ed., *Surfactants*, Academic Press, 1984
2. J.S. Reed, *Introduction to the Principles of Ceramic Processing*, Wiley Interscience, 1988
3. B. Derjaguin and L. Landau, *Acta Physicochim.*, Vol 14, 1941, p 633
4. E. Verwey and J.Th.G. Overbeck, *Theory of the Stability of Lyophobic Colloids*, Elsevier, Amsterdam, 1948
5. K. Mikeska and W.R. Cannon, Dispersants for Tape Casting Pure Barium Titanate, *Advances in Ceramics*, Vol 9, *Forming of Ceramics*, J.A. Mangels, Ed., American Ceramic Society, 1984, p 164–183
6. B. Schwartz and D.J. Rowcliffe, Modelling Density Contributions in Preceramic Polymer/Ceramic Powder Systems, *J. Am. Ceram. Soc.*, Vol 69 (No. 5), 1986, p c106-c108

SELECTED REFERENCES

- G.A. Ackley and J.S. Reed, Body Parameters Affecting Extrusion, *Advances in Ceramics*, Vol 9, *Forming of Ceramics*, J.A. Mangels, Ed., American Ceramic Society, 1984, p 193–200
- M.J. Edrisinghe and J.R.G. Evans, Properties of Injection Moulding Formulations, Part 1, Melt Rheology, *J. Mater. Sci.*, Vol 22 (No. 1), 1987, p 269–277
- J.E. Funk, Slip Casting and Casters, *Advances in Ceramics*, Vol 9, *Forming of Ceramics*, J.A. Mangels, Ed., American Ceramic Society, 1984, p 76–84
- E.R. Hoffman, Importance of Binders in Spray Dried Press Bodies, *Am. Ceram. Soc. Bull.*, Vol 51 (No. 3), 1972, p 240–243
- I. Mellan, *Industrial Solvents Handbook*, 2nd ed., Noyes Data Corp., 1977
- R.E. Mistler, D.J. Shanefield, and R.B. Runk, Tape Casting of Ceramics, *Ceramic Processing before Firing*, G.Y. Onoda and L.L. Hench, Ed., Wiley Interscience, 1978, p 411
- G.Y. Onoda, The Rheology of Organic Binder Solutions, *Ceramic Processing before Firing*, G.Y. Onoda and L.L. Hench, Ed., Wiley Interscience, 1978, p 235–251
- M.V. Parish, R.R. Garcia, and H.K. Bowen, Dispersions of Oxide Powders in Organic Liquids, *J. Mater. Sci.*, Vol 20, 1985, p 996–1008
- V.K. Pujari, Effect of Powder Characteristics on Compounding and Green Microstructures in Injection Molding Processes, *J. Am. Ceram. Soc.*, Vol 72 (No. 10), 1989, p 1981–1984
- M.J. Rosen, *Surfactants and Interfacial Phenomena*, Wiley Interscience, 1978
- V.N. Shukla and D.C. Hill, Binder Evolution from Powder Compacts: Thermal Profile for Injection Molded Articles, *J. Am. Ceram. Soc.*, Vol 72 (No. 10), 1989, p 1797–1803
- E.S. Tormey, R.L. Pober, H.K. Bowen, and P.D. Calvert, Tape Casting—Future Developments, *Advances in Ceramics*, Vol 9, *Forming of Ceramics*, J.A. Mangels, Ed., American Ceramic Society, 1984, p 140–149
- Wright, J.R.G. Evans, and M.J. Edrisinghe, Degradation of Polyolefin Blends Used for Ceramic Injection Molding, *J. Am. Ceram. Soc.*, Vol 72 (No. 10), 1989, p 1822–1828

Forming and Predensification, and Nontraditional Densification Processes

Chairman: David W. Richerson, Ceramatec, Inc.

Introduction

THIS SECTION addresses (1) fundamentals of compaction and removal of compaction aids (fluids and organics), (2) techniques of forming a shaped compact of a ceramic powder, and (3) techniques which achieve compaction of a powder plus densification and techniques which start with non-powder precursors.

Fundamentals of Forming. Most ceramic-forming processes start with a powder and consist of compaction of the powder into a porous shape. The objective is usually to achieve the greatest degree of particle packing and a high degree of homogeneity. Close packing of the particles reduces the amount of porosity which must be removed during densification (sintering) and thus reduces the total shrinkage of the part during densification. Close packing also enhances the kinetics of sintering. Homogeneous powder packing decreases the likelihood of distortion or microstructural inhomogeneity during sintering.

A variety of parameters must be controlled to achieve optimum particle packing. These include particle size, particle size distribution, degree of agglomeration or dispersion, mode of flow of powder into tooling, condition of the tooling, cycle time, and removal of the compacted part from the tooling. The first article, "Mechanical Consolidation," discusses the critical parameters, the interrelationships of the forming process with the starting powder and the final properties, and the role of additives in achieving optimum compaction.

The ceramic compact generally consists of the ceramic powder plus a fluid and organic compounds. The fluid and organics are trapped in the pores. These must be removed before the ceramic powder can be heated to high temperature to achieve sintering. Too-rapid removal can cause cracking, bloating, distortion, or fracture. The second article, "Drying," reviews the critical parameters, equipment, and other aspects of removing fluids from the particulate compact. The article

"Organics Removal" addresses the critical parameters, equipment, and procedures for successful removal of organic additives.

Powder Compaction Processes. The primary processes for forming a shaped compact of a ceramic powder include dry pressing, slip casting, tape casting, extrusion, injection molding, and green machining.

Pressing involves mechanical compaction of a relatively dry powder. Two types of pressing are utilized. The most common and widespread is uniaxial dry pressing. The powder is mixed with a minimum amount of fluid and with organic additives and compacted between two punches in a die cavity. This process can be easily automated and can produce hundreds of relatively simple parts per minute. The variations and capabilities of uniaxial pressing are presented in the article "Dry Pressing."

The second type of pressing is referred to as cold isostatic pressing. The ceramic powder with suitable organic additives is sealed into a flexible shaped bag of an elastometer such as polyurethane or rubber. In one process (wet bag), the sealed bag is immersed in water in a high-pressure container. Pressure is applied to the fluid which deforms the elastomer die and produces a shaped powder compact. In an alternate process (dry bag), pressurized fluid is forced through channels in the elastomer die which transmits pressure to form a shaped compact. Isostatic pressing can form parts with complex contours, high aspect ratios, and thin walls which cannot be formed by uniaxial pressing. The article "Cold Isostatic Pressing" describes the types of equipment and tooling, the key parameters, and some of the applications.

A third shape-forming process is slip casting. Slip casting involves a fluid suspension of ceramic particles in a liquid. Typically, the pH is controlled or organic compositions are added to achieve maximum dispersion of the ceramic particles. This can result in a flowable suspension that contains over 50 vol% ceramic particles. To accomplish slip casting, the suspension is poured into a porous shaped mold. The pores are small enough that the particles cannot enter, but large enough that the liquid can be absorbed by capillary pressure. As the liquid is extracted through the pores, the ceramic particles deposit on the mold walls and progressively form a uniform compact inside the mold. The article "Slip Casting" describes slip casting approaches and applications and discusses rheology of slips and key processing parameters.

A fourth forming process is tape casting. Tape casting utilizes a suspension of ceramic particles in a liquid which is similar in flowability to a slip casting suspension, but which contains a higher content of an organic binder. The slip is poured or spread onto a flat, smooth surface to form a thin film typically under 2 mm (0.075 in.) thick. The fluid is removed by heat or flowing air to leave a thin tape consisting of moderately close-packed ceramic particles bonded together by the organic binder. The tape is flexible and can be cut, punched, metallized or other operations prior to binder removal and densification. The article "Tape Casting" describes procedures for tape casting, critical process parameters, and important applications.

Pressing utilizes a relatively dry powder and casting utilizes a fluid suspension. Other forming techniques utilize an intermediate mixture of powder and liquid that has a stiff dough-like consistency and requires the application of pressure to form a shape. These techniques are referred to as plastic forming techniques. Two important plastic forming techniques are extrusion and injection molding.

Extrusion of ceramic powders is a room-temperature operation. The powder is mixed with a liquid (usually water) plus either an inorganic or organic binder. The function of the binder is to form a uniform, stiff mixture which can be forced under high pressure through a shaped orifice and retain the cross-section shape imposed by the orifice. The article "Extrusion" describes the equipment, procedures, typical formulations, and applications for extrusion.

Injection molding is conducted at elevated temperature. The ceramic powder is mixed at temperatures with typically a thermoplastic polymer. The mixture is injected into a shaped mold. The mold is cooler than the injected powder-binder mixture and causes the thermoplastic polymer to harden. This results in a rigid compact of powder bonded with the polymer which can be removed from the mold with a rapid cycle time. The binder can then be extracted and the ceramic powder compact densified at high temperature. The article "Injection Molding" describes the equipment, procedures, rheology, typical formulations, and applications for injection molding.

An objective of ceramic-forming processes is to achieve the desired shape as closely as possible. Often a combination of a forming process such as pressing or casting plus machining is required. The machining can be conducted on the compacted powder before densification, at a partially-densified bisque stage or after final densification. Machining a densified ceramic requires diamond tooling and can be very expensive. Machining an unfired ceramic can usually be done less expensively and without diamond tooling. This is referred to as "green machining." The article "Green Machining" describes the tooling, techniques, and key procedures.

Nontraditional/Densification Processes. Some forming techniques combine consolidation or shape forming with densification. Examples include hot pressing, hot isostatic pressing, molten particle deposition, sol-gel, vapor deposition, and reaction forming.

Hot pressing involves compaction of a ceramic powder under uniaxial pressure in a die at a high enough temperature that the powder can densify. Because pressure and temperature are applied simultaneously, the powder can typically sinter more quickly and at a lower temperature than if only temperature is applied. This results in several benefits: (1) usually less porosity in the final part, which correlates with higher strength and hardness; (2) decreased grain growth due to lower temperature and shorter time at temperature, which also yields

higher strength; and (3) capability to densify some ceramic powders with less additives, which results in improved properties. The article "Hot Pressing" describes the equipment options, the typical procedures, the limitations, some typical properties of hot pressed materials, and some applications.

Hot pressing also has limitations. The major limitation is shape capability. Shapes other than simple blocks or short cylinders are difficult and require expensive diamond grinding. Hot isostatic pressing (HIP) has been developed to extend the capability of hot pressing to complex shapes. Hot isostatic pressing is analogous to cold isostatic pressing, except that high-pressure argon is used instead of water. Rather than using rigid tooling, the powder to be compacted and densified is sealed into flexible tooling. However, devising flexible tooling which can operate at temperatures in the range of 1200 to 2000 °C (2190 to 3630 °F) is a major challenge. The greatest success has been to use glass encapsulation. The article "Hot Isostatic Pressing" describes the equipment, procedures, properties of the resulting materials, shape capabilities, and applications.

All of the processes discussed so far have involved compaction of a powder. A different type of process melts ceramic particles and deposits them on a surface. This is referred to as the molten particle deposition approach and is accomplished with several different types of equipment. This equipment and the resulting materials, properties, and applications are reviewed in the article "Molten Particle Deposition." These processes are primarily used to form coatings, especially for thermal barrier and wear resistance applications.

All of the aforementioned processes use powders as the precursor. To obtain optimum properties, the powders must be carefully sized and the purity and homogeneity controlled. This is often difficult to achieve reproducibly in practice. An alternate processing approach referred to as "sol-gel," utilizes solutions or colloidal suspensions as precursors. The article "Sol-Gel Process" describes the theories, techniques, advantages, disadvantages, and applications.

Ceramics can also be achieved from polymer precursors. The ceramic composition results as the polymer is thermally decomposed (pyrolyzed). This occurs at substantially lower temperature than normally used to densify ceramics and can yield a very fine-grained high-strength microstructure. An important example is the synthesis of high-strength SiC-based fibers from polymer precursors. Fibers with strengths greater than 2500 MPa (360 ksi) are commercially available. The article "Polymer-Derived Ceramics" reviews the state of technology of processing ceramics from polymer precursors.

Ceramics can also be achieved via vapor-phase processes. One type of vapor-phase process involves chemical reaction of vapor species on a heated substrate. This is referred to as chemical vapor deposition (CVD). It has been used to produce very high purity, dense coatings of a wide variety of oxide, carbide, nitride, boride, and other compositions. It has also been used to infiltrate fibrous or particulate preforms to achieve composite structures. Another vapor deposition approach is referred to as physical vapor deposition (PVD). It involves vapor-phase removal of a material from a target and deposition on a substrate. Examples include sputtering and electron beam deposition. Both CVD and PVD processes are discussed in the article "Vapor Deposition."

Ceramics can also be formed by solid-solid, vapor-liquid, and liquid-solid reactions. One type of solid-solid reaction capable of forming ceramic compacts is described in the article "Self-Propagating, High-Temperature Synthesis." This approach involves the use of exothermic reactions to provide the heat to synthesize a composition and potentially simultaneously achieve consolidation. Several vapor-liquid and liquid-solid reaction approaches are described in the article "Reaction-Forming Processes." Another process of substantial current interest is the directed metal oxidation process. A molten metal is allowed to react at the surface with oxygen to grow ceramic matrices around pre-placed filler or reinforcement materials. The processing, microstructures, and applications of such materials are described in the article "Directed Metal Oxidation."

Mechanical Consolidation

Roger L.K. Matsumoto, Hercules Advanced Materials and Systems Company, Hercules Incorporated

MECHANICAL CONSOLIDATION is an operation performed on prepared ceramic powders in order to form a shape that can be sintered. Many methods have been developed to accomplish this, including uniaxial dry pressing, cold isostatic pressing (using either wet or dry bag techniques), hot uniaxial pressing, slip casting, tape casting, extrusion, injection molding, and papermaking, as well as some newer methods. Although each of these techniques has different powder requirements, they share one goal—uniform density in the consolidated (green) part. This goal is often compromised by the specific requirements of a particular method. These compromises are discussed in this article, as well as in subsequent articles on specific processes in this Section of the Volume.

The necessity for uniform density in the green part is due to the nature of the sintering process. In this process, the total free energy of the powder system is minimized by the reduction of surface area. For most systems, the result is densification; an ideal compact sinters isotropically to yield a fully dense part. Although two compacts with different green densities can both fully densify, the shrinkage experienced by each is different. The part with the lower green density shrinks more than the part with the higher green density in order to attain the same final density. If these two green compacts were joined as one piece, the region of higher green density would shrink less to reach full density, whereas the adjacent region of lower green density would be constrained from shrinking further and would not reach full sintered density. This phenomenon, known as differential sintering, is depicted schematically in Fig 1. Thus, the goal of any powder consolidation technique is to obtain a uniform density throughout the green part. Obstacles to this goal are due to either the powder system or the consolidation technique. Often, the two are inseparable. The powder system must be carefully prepared and characterized, and the consolidation method must be both selected and conducted properly.

The topics discussed in this article include particle packing, the effects of the starting powders on consolidation, the effects of the particle compact on the sintered microstructure, the role of the additives used in the many compaction methods, and test methods for powder consolidation systems.

Powder Packing

The way in which the packing of powder occurs in the green compact influences subsequent processing steps, especially sintering. The green microstructure must be uniform on the local, as well as the overall, scale. Any inhomogeneities will adversely affect the final sintered part (Ref 1, 2).

The simplest way to approach the packing of powders is to consider monosized spheres. Calculations show that the densest packing of these spheres results in about 75% of the volume being filled. Smaller spheres of a size that just fits the voids between the larger spheres can be used to increase the percentage of the volume filled, thereby increasing the packed density. This sequence can continue, at least theoretically, with the addition of ever-smaller spheres of discrete sizes until the entire volume is filled. Unfortunately, most powders in real systems are not spherical, nor are they available in precisely determined particle size distributions. Real powders comprise blocky, fibrous, platy, or other nonspherically shaped particles with a varying size distribution. Even when agglomerated into spherical shapes, such as by spray drying, the particle size distribution of the agglomerates is continuous and does not contain the discrete sizes required for maximum packing. In addition, packing in real systems does not proceed in the layer-by-layer process that would be necessary to obtain maximum packing from a powder with the required size distribution. Fortunately, it is only required that consolidated particles be uniformly packed, rather than packed to ultimate density, although it is desirable to maximize packing density.

Characteristics of Starting Powders

It is necessary to know the particle shape, the average particle size, and the particle size distribution before a powder system can be designed for compaction. Particles that closely approximate spheres, such as relatively equiaxed shapes like cubes, produce higher packed densities. Shapes such as large aspect ratio flakes or whiskers result in low packed densities. Special consideration must be given to those systems that include both types of particle shapes, such as whisker-reinforced ceramic systems.

The average particle size is important when filling molds and dies or when dispersing powders in a fluid. Gravitational effects are stronger with large particles, whereas surface charge effects dominate smaller particles. It is desirable to have large particles when filling molds and dies, because more uniform fill densities result. Small particles tend to collect electrostatic charges and resist packing. On the other hand, large particles can settle out of suspensions used for both slip and tape casting, resulting in nonuniform products.

For most advanced ceramic systems in which densification proceeds via a solid-state sintering process, powders that have high surface free energy are required. These highly reactive particles, either formed by milling or by means such as chemical precipitation, are generally very fine powders, and thus have a high surface area. In order to handle these powders in compaction techniques such as uniaxial or isostatic pressing, the powders are usually agglomerated by spray drying or pelletizing. Large uniform agglomerates are advantageous in uniaxial pressing because they are flowable and can pack uniformly.

As described in the previous section "Powder Packing" in this article, the particle size distribution is important in determining the ultimate green density that can be achieved with a powder. Although there have been advocates for monosized powders and agglomerates (Ref 3), practical systems comprise a range of particle sizes. In a classic paper, Farris (Ref 4) predicted the resulting viscosity of a filled liquid when various populations of particle size distributions were used. While it may not appear applicable in a general sense, the packing of powders must be accomplished by movement of particles en masse. Thus, the "viscosity" of the powder blend, composed of particles either alone or suspended in a liquid, influences the ultimate packing potential of the system. It was predicted that a broad particle size distribution results in the lowest relative viscosity (hence, the highest packed

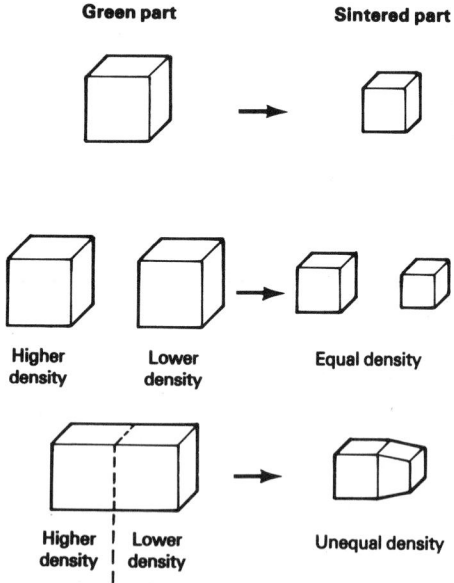

Green part   **Sintered part**

Higher density — Lower density — Equal density

Higher density | Lower density — Unequal density

Fig 1 Shrinkages due to sintering of compacted green parts and effects of green density variations

density), which has been demonstrated in numerous real systems. Recalling that a population of discrete particle sizes is required for maximum packing, a broad particle size distribution can be viewed as the superposition of discrete populations of many starting sizes of spheres.

Particle Compact

The stated goal of any compaction method—to yield a part with uniform green density—is sometimes not realized. Most of the problems encountered in sintering can be traced back to nonuniform green densities. In addition, the microstructure and chemical composition of the sintered ceramic are influenced by the powder compact. These relationships are described below.

Although ceramics are normally considered to be homogeneous, most materials are quite heterogeneous on the microscopic level. This is due to the way most ceramics are fabricated, that is, by mixing powders of different compositions and expecting compositional equilibrium to be reached during sintering. These powders often have different densities and shapes, which can result in either settling (in a suspension) or demixing (in dry powders). In addition, the surface charges of powders in a liquid can be different. This also affects the dispersion. In extreme cases, one constituent may agglomerate, while other constituents are well dispersed. These effects result in chemical inhomogeneities that cause differential sintering to occur.

Sintered ceramics usually contain two types of porosity. The first type is represented by the space between grains, normally at triple points, where several grains meet. In powder systems that densify via a liquid phase, these spaces may be filled with glass. When present, these pores do not present much of a problem because they are smaller than the grains. The second type is detrimental to the material, and is represented by larger pores that have various origins. In wet consolidation methods, air bubbles are frequently trapped or not properly removed. These bubbles become incorporated into the sintered body. Large pores can also result when agglomerated structures are not eliminated during the consolidation process (Ref 5). In dry consolidation methods, this results when the pressing pressure is not sufficient to break down spray-dried agglomerates. In wet consolidation methods, especially slip casting, these pores develop when agglomerated structures are created during the consolidation process (Ref 6). The interagglomerate porosity will survive in the sintered part. Unlike bubbles, these pores have a crack-like shape and are often the sites of fracture initiation.

Consolidation Methods and Use of Additives

Many additives are employed in powder systems to maximize packing in the green body. These additives are included for specific purposes, for example, to control viscosity of a slip, help the flowability of a powder agglomerate, or lubricate powders during compaction. Other additives, such as binders, are not necessarily used to enhance packing. Although some binders do help, most actually hinder packing. For instance, a binder that forms a rigid structure at low pressures prevents pressure transmission to the unconsolidated powders, resulting in a very nonuniform green body. Thus, the effects of all components in a powder system should be known in order to develop the proper formulation.

Consolidation methods can be classified by two broad categories: dry and wet. Dry techniques are those in which the inorganic powder portion predominates. They include uniaxial pressing, wet and dry bag isostatic pressing, and hot pressing. Wet techniques are those in which the inorganic powders are in an aqueous, polymeric, or organic solvent carrier. They include slip casting, tape casting, extrusion, and injection molding. Each consolidation method has unique requirements and is therefore described below separately, along with the role of each additive. A summary is provided in Table 1.

Dry Consolidation Methods

The majority of ceramic parts are made by dry consolidation techniques, principally uniaxial pressing. The powders that are used must be agglomerated in order to evenly fill the dies and molds using automated die-loading equipment. Even filling is necessary to obtain predictable compression ratios and fill densities, which are especially important parameters for automated uniaxial pressing and dry bag isostatic pressing, because it is difficult to externally assist the packing of the dies prior to compression. Wet bag isostatic pressing molds and hot press dies are normally filled manually, independent from the machines in which they will be used. Thus, it is possible to help the die-filling procedure through tapping or vibration, which leads to a more uniform fill when using powders that may not be fluid enough for uniaxial pressing. Dry consolidation methods produce simple shapes that must normally be green or final machined to size.

Uniaxial pressing uses two rams to press the powders inside a die body. The geometry of pressed parts is either cylindrical or some variation thereof, such as a hollow cylinder.

Binders used in uniaxial dry pressing must be soft enough to allow the powder agglomerates to yield below the final compaction pressure. When practical, plasticizers are used to reduce the glass transition temperature (T_g) of the binder phase. It is often helpful if the T_g is either at or just above the pressing temperature. Thus, the agglomerates will be hard enough to withstand handling, but the binder will soften from the heat generated during compaction. At the same time, pressure must be transmitted uniformly through the entire compact. Lubricants can aid the pressure transmission. The use of both binders and lubricants must yield a compact that is strong enough to sustain the stress due to springback upon ejection from the mold. A separate lubricant is often used on the die walls to reduce die wall friction. Additives must not cause the compact to stick to any die parts.

Table 1 Methods of consolidation and constituents of powder formulation

Consolidation method	Powder form	Binder	Lubricant	Dispersant	Others
Uniaxial dry pressing	Dry, agglomerated	Yes	Yes	No	Plasticizers
Cold isostatic pressing	Dry, agglomerated	Yes	Yes	No	Plasticizers
Hot uniaxial pressing	Dry	No	No	No	No
Slip casting	Dispersed in fluid	Yes	No	Yes	Viscosity modifiers
Tape casting	Dispersed in fluid	Yes	Yes	Yes	Plasticizers
Extrusion	Plastic mix	Yes	Yes	Yes, or flocculant	Viscosity modifiers
Injection molding	Plastic mix	Yes	Yes	Yes	Viscosity modifiers
Papermaking	Dispersed in fluid	Yes	No	Flocculant	No

Dry bag isostatic pressing uses a fixed elastomeric mold that is pressurized externally with a fluid. Because the mold is normally attached to nonyielding parts, the pressurization occurs inward in a radial direction. This results in a quasi-isostatic condition, and parts produced in this method are usually cylindrical. Dry bag pressing is typically an automated operation.

Many of the same requirements for uniaxial pressing apply to dry bag isostatic pressing. An additional requirement is that the binder not adhere to the elastomeric bag. Because the pressure transmission is quasi-isostatic, a binder and lubricant that are different from those used in uniaxial pressing may be needed. This is because vaulting may occur, although it is more likely to occur when wet bag isostatic pressing is used. The phenomenon is described below.

Wet bag isostatic pressing differs from the dry bag method in terms of the nature of the mold. The wet bag method uses a free-standing elastomeric mold that is placed into a fluid-filled pressure chamber. Many molds can be placed in the chamber, depending on size. Because the mold is pressurized from all directions, this method can provide true isostatic pressing, depending on mold design. Wet bag isostatic pressing is typically a manual batch operation.

Powder systems used in this method have the same general requirements as those used in the dry bag method. However, because the size of the agglomerate is less critical, fine agglomerates (from the cyclone of a spray dryer, for example) can be used. In cases where mold design results in true isostatic pressurization, the binder must be chosen to prevent vaulting, which occurs when a strong shell forms on the outside of a compact at a low pressure, preventing pressure transmission into its center. The resulting green body will invariably crack during densification. Both the binder and lubricant system must be able to prevent this from occurring.

Hot Pressing. In a mechanical sense, hot pressing is identical to uniaxial pressing. The difference is that the die is heated, which sinters the powders during the pressing operation. This results in densification at lower temperatures, compared to pressureless sintering.

Hot pressing is unique among the mechanical consolidation techniques in that additives are not normally used. Because this method leads directly from the powder to a densified body, a temporary binder is not required. In fact, any organic additive would generally leave an undesirable carbon residue under typical hot press conditions.

Wet Consolidation Methods

Wet consolidation methods are usually specialty techniques that are used to fabricate complex structures. Parts are often used with minimal machining after sintering. In these methods, the inorganic powder can be considered as a filler in the organic or aqueous carrier. Thus, the properties of the system are determined mainly by the fluid phase. The powders for all methods are normally required to have a wide particle size distribution, in order to maximize the solids loading for a given viscosity.

Slip casting is one of the oldest techniques for forming ceramic articles. In this technique, a colloidal suspension (usually in water, but can be in other solvents) is poured into a porous mold (typically, plaster of paris). As the liquid enters the mold walls, a layer of powder consolidates at the mold surface. This layer continues to grow as more liquid enters the mold walls. The article takes the shape of the mold interior and can be either hollow or solid, depending on whether the suspension is removed from the mold or not. In nonclay systems, binders must be added to the suspension to give the molded article sufficient strength for handling. All systems contain dispersants, and perhaps, other additives to control viscosity.

The powders used in slip casting must have a small average particle size with a wide particle size distribution. Coarse powders will preferentially settle, resulting in a nonuniform cast article. A narrow size distribution will yield a suspension with a low solids loading, resulting in nonoptimum packing. Such an article will experience a large drying shrinkage, which can more easily produce cracks. Very fine particles will close the pores in the mold, resulting in the necessity of very long times to cast a given thickness.

Tape casting, in its simplest form, involves spreading a colloidal suspension over an impermeable substrate with a blade set to yield a certain film thickness. The cast tape can dry from only the top surface. In practice, the impermeable substrate, or carrier, is a polymer film that is rolled up as the tape is cast onto its surface. Because the tape must be "dry" before the carrier can be rolled, the use of volatile solvents is common.

The suspension used in this method is similar to that required for slip casting, with a few important differences. Both require high solids loadings at low slip viscosities. However, because tape casting does not use a porous mold, the solvent is normally a more volatile organic liquid. The other major difference is that the cast tape is flexible, unlike the rigid slip cast part. Thus, the organic polymer binder system is more complicated and often requires additives to plasticize the binder. Because the tape is cast onto a nonporous substrate, one-sided drying can result in powder or organic content variation through the thickness of the tape. Thicker tapes are more susceptible to this variation. A typical response to this variation occurs during sintering, when differential shrinkage causes the tape to curl.

Extrusion of a plastic mix makes possible the formation of very complex two-dimensional geometries. The main limitation of this method is that the third dimension does not change. Rods, tubes, and honeycomb structures are commonly extruded. The plastic mix comprises inorganic powders, a binder, and a liquid, such as water. This formulation is designed so that a sharp decrease in viscosity is experienced at high shear rates, with a corresponding sharp increase in viscosity at low shear rates. By adjusting the mix so that this transition occurs at the shear rate experienced at the die, it becomes possible to extrude rapidly while the extrudate maintains the molded shape. Binders are needed in nonclay systems to promote green handling strength. Lubricants are helpful in reducing die wall friction. Because the mix is never a low-viscosity fluid, much larger particles are tolerated, although very large particles present problems during densification.

Injection molding is one of the newest methods developed to form ceramics; the earliest reference to it was during the 1940s (Ref 7). Although it has great potential for making extremely intricate parts, it is often the last choice of all forming methods. In theory, the method is simple: Mix powders into a molten organic thermoplastic polymer, inject into a cool mold to set the part, and then remove the finished green part from the mold. In practice, the method has developed slowly, with the greatest difficulty being the removal of the polymer from the part during firing. A variation to this method uses thermoset polymers and a heated mold to set the polymer. A host of additives are required, including the binder (which is actually the carrier for the powder), dispersants to keep the viscosity of the mix low, other viscosity modifiers, and lubricants to reduce friction in the gate and mold.

Papermaking is a method of making wide, flat sheets of green ceramics. This method is not widely practiced and literature is scarce. In paraphrased form, a typical patent (Ref 8) states that the method comprises making a slurry consisting essentially of water; 100 parts of feltable fiber, consisting essentially of mineral fiber, of which at least 83.3% by weight is synthetic mineral fiber; and from 10 to 250 parts of water-dispersible particulate ceramic material and particulate organic heat activatable binder. Then, the slurry should be flocculated by adding a surface-active flocculating agent, thereby forming flocs suspended in substantially clear water. The amount of the surface-active agent should be in the range from 0.013% to 0.073% of the dry weight of the suspended content of the slurry. The resulting suspension of flocs should be dewatered, thereby forming a wet mat of solids consisting essentially of felted network of fibers and particulate materials. The binder should be activated by heat. A rigid, handleable, bonded, porous dry mat should be formed by drying the wet mat. The dry mat should be fired at a temperature sufficient to burn out organic matter and to achieve ceramic bonding throughout the piece.

Test Methods

Powder systems used with wet consolidation methods can be tested by determining the viscosity of the mix. Because the viscosity values are different for each system, the instrumentation required is different. Viscometers are available with an array of probes that can measure the thin slips needed for slip casting or the plastic mix used in extrusion.

The powder systems used with dry consolidation methods are best tested by dynamic response techniques. These include the compaction response diagram (Ref 9–11), as well as the more recently developed compaction rate diagram (Ref 12). A compaction response diagram relates the density of a compact as a function of applied pressure (Fig 2). This diagram can be generated by compacting parts at different pressures and measuring the green densities obtained.

An alternative method is to make one part and measure its dimensions during pressing. This method is rapid and, if the compliance of the test equipment is accounted for (Ref 13), accurate. A compaction response diagram shows the highest density achievable with a given powder.

Much more information is obtained from a compaction rate diagram. The compaction rate diagram is derived from the compaction response diagram by taking the first derivative of the curve. This diagram shows the rate at which compaction occurs at a given pressure. High values on this diagram show when large changes in density result from small

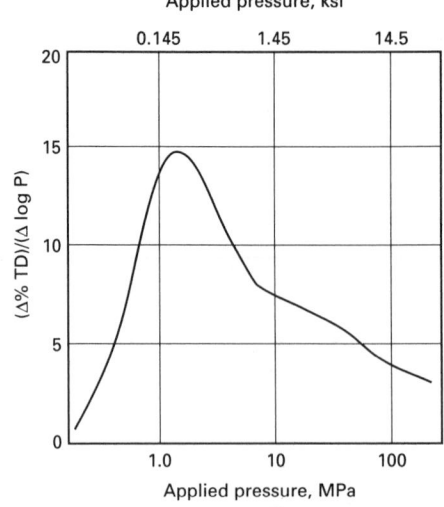

Fig 3 Compaction rate diagram of powder agglomerates with a soft binder; agglomerates have low density. TD, theoretical density; P, pressure

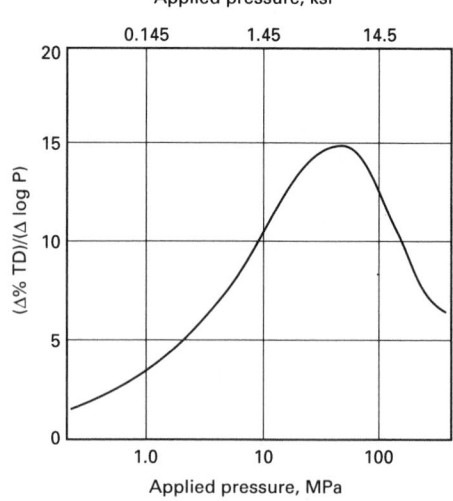

Fig 4 Compaction rate diagram of powder agglomerates with a hard binder; agglomerates have low density

changes in pressure. Low values are obtained when the same small change in pressure results in a small change in density. Large changes in density result when an open structure is being crushed. In the compaction rate diagram, agglomerated powders peak at the pressure at which the agglomerate structure collapses. The location of this peak is a function of the binder used. Harder binders can retain the structure to higher pressures. Binders with glass transition temperatures have hardnesses that are a function of temperature. In those cases, plasticizers can be used to change the T_g.

Spray drying often results in hollow agglomerates. When these are crushed, a large compaction rate occurs. Figure 3 shows the compaction rate of such an agglomerate with a soft binder, which is crushed at a relatively low pressure. A similar structure, but with a hard binder, requires a much higher pressure to crush, as shown in Fig 4. A solid agglomerate will not crush, but, rather, will deform. The pressure at which this deformation occurs is a function of the binder. The compaction rate of a solid agglomerate with a soft binder phase is shown in Fig 5, whereas the compaction rate of an unagglomerated powder is shown in Fig 6. Because there is minimal structure in this material, only particulate rearrangements are evident in the diagram.

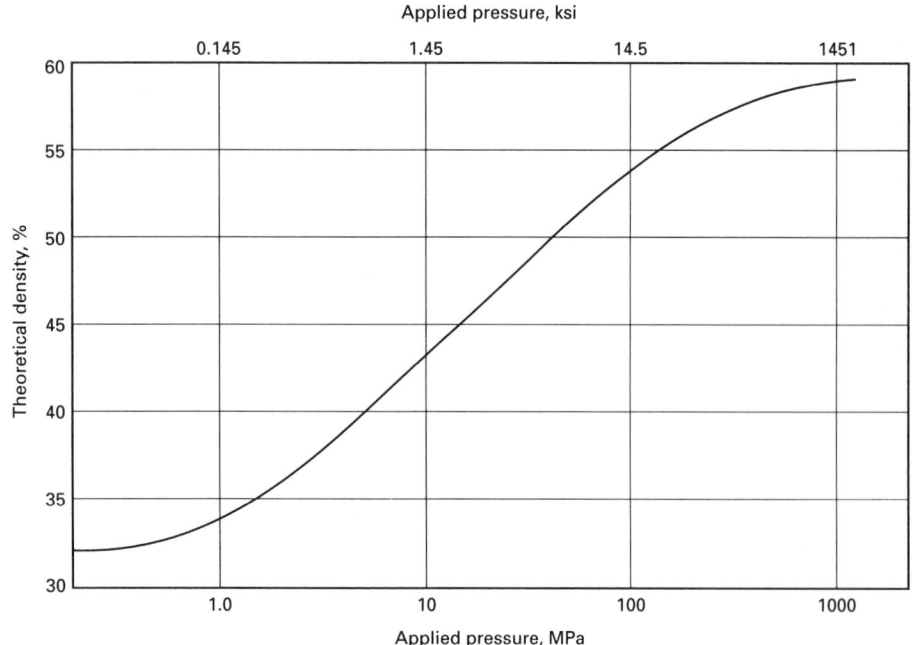

Fig 2 Typical compaction response diagram

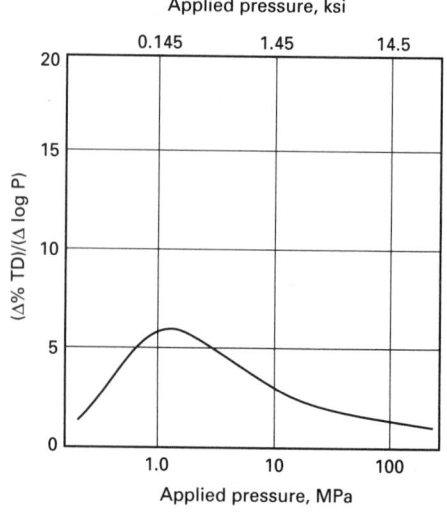

Fig 5 Compaction rate diagram of powder agglomerates with a soft binder; agglomerates have high density

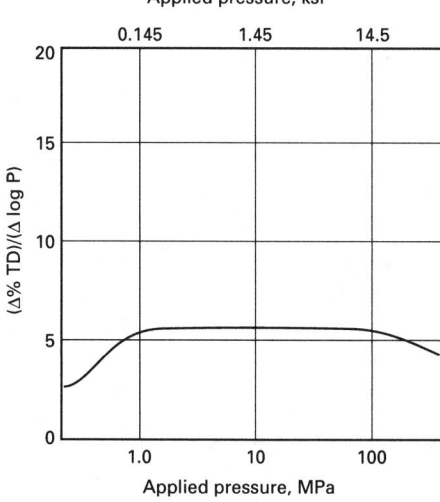

Fig 6 Compaction rate diagram of an unagglomerated powder

Thus, a low, constant value of the compaction rate results.

In general, it is desirable to have as small a change in green density with a change in pressure as possible. Large changes mean that there are regions in the pressed part that have large local density differences. Upon sintering, these areas will have different shrinkage rates and the part will warp or crack. It is especially critical that these large variations do not occur either near or at the maximum pressing pressure.

If vaulting is a problem in isostatic pressing, a compaction rate diagram will indicate the pressure range in which the vaulting is occurring. It will then be possible to alter the binder, or to select another binder to eliminate the vaulting.

REFERENCES

1. A.G. Evans, Considerations of Inhomogeneity Effects in Sintering, *J. Am. Ceram. Soc.*, Vol 65 (No. 10), 1982, p 497–501
2. B. Kollett and F.F. Lange, Stresses Induced by Differential Sintering in Powder Compacts, *J. Am. Ceram. Soc.*, Vol 67 (No. 5), 1984, p 369–371
3. E.A. Barringer and H.K. Bowen, Formation, Packing, and Sintering of Monodisperse TiO_2 Powders, *J. Am. Ceram. Soc.*, Vol 65 (No. 12), 1982, p C199–C201
4. R.J. Farris, Prediction of the Viscosity of Multimodal Suspensions from Unimodal Viscosity Data, *Trans. Soc. Rheology*, Vol 12 (No. 2), 1968, p 281–301
5. M. Ciftcioglu, M. Akinc, and L. Burkhart, Effect of Agglomerate Strength on Sintered Density for Yttria Powders Containing Agglomerates of Monosized Spheres, *J. Am. Ceram. Soc.*, Vol 70 (No. 11), 1987, p C329–C334
6. I.A. Aksay, Microstructure Control Through Colloidal Consolidation, *Advances in Ceramics*, Vol 9, *Forming of Ceramics*, American Ceramic Society, 1984
7. K. Schwartzwalder, Injection Molding of Ceramic Materials, *Am. Ceram. Soc. Bull.*, Vol 28 (No. 11), 1949, p 459–461
8. J.E. Cadotte, U.S. patent 3,510,394, May 1970
9. S.J. Lukasiewicz and J.S. Reed, Character and Compaction Response of Spray-Dried Agglomerates, *Am. Ceram. Soc. Bull.*, Vol 57 (No. 9), 1978, p 798–801
10. J.A. Brewer, R.H. Moore, and J.S. Reed, Effect of Relative Humidity on the Compaction of Barium Titanate and Manganese Zinc Ferrite Agglomerates Containing Polyvinyl Alcohol, *Am. Ceram. Soc. Bull.*, Vol 60 (No. 2), 1981, p 212–220
11. G.L. Messing, C.J. Markoff, and L.G. McCoy, Characterization of Ceramic Powder Compaction, *Am. Ceram. Soc. Bull.*, Vol 61 (No. 8), 1982, p 857–860
12. R.L.K. Matsumoto, Generation of Powder Compaction Response Diagrams, *J. Am. Ceram. Soc.*, Vol 69 (No. 10), 1986, p C246–C247
13. R.L.K. Matsumoto, Analysis of Powder Compaction Using a Compaction Rate Diagram, *J. Am. Ceram. Soc.*, Vol 73 (No. 2), 1990, p 465–468

Drying*

James S. Reed, New York State College of Ceramics at Alfred University

DRYING is the removal of liquid from a porous material by means of its transport and evaporation into a surrounding unsaturated gas or, in some cases, a desiccating liquid. It is an important operation prior to firing in processing bulk raw materials, products shaped by plastic forming and casting, and decorations and coatings on surfaces.

The evaporation of processing liquids is relatively energy-intensive, and drying efficiency is always an important consideration. Drying must be carefully controlled, because stresses produced by differential shrinkage or gas pressure may cause defects in the product. In this article, drying practices, mechanisms of drying, and causes of defects will be considered.

Drying Systems

Drying costs are a significant factor in the selling price of industrial minerals, and the natural drying action of the sun and wind is considered in mining and storing these materials. Wet-processed raw materials are commonly dried in large rotary dryers in which the material is tumbled, on belts in continuous tunnel dryers, by spray-drying, or supported on trays in a chamber dryer. The flow of a warm dry gas through a permeable material, called fluidized bed drying, is used when a granular material is very temperature-sensitive. Material that has been partially dewatered mechanically and is in the form of a ribbon, pellets, granules, and so on, is usually dried more uniformly.

Freshly cast plaster molds and very large wet-processed shapes are commonly dried slowly by "open-air drying." Shaped ceramic products requiring drying and working molds are usually dried in a controlled manner in fabricated metal dryers. Chamber dryers are used for drying large, free-standing shapes and smaller products supported on shelves or suspended from the structural framework on dryer cars. Continuous dryers may convey the ware up and down through a baffled chamber by means of shelves supported on continuous chains in a mangle dryer, or on rack dryer cars in a tunnel dryer. Air circulation is maintained and controlled by means of fans. Heat

sources include direct fired air heaters, steam coils, waste warm air from kilns and furnaces, and infrared or microwave radiation.

Damp material may be heated by the mechanism of convection, conduction, and radiation. When using convective heating, which is most common, the product temperature during drying is lower than the temperature of the hot circulating gas. Conductive heating in which heat passes to the product from a heated surface supporting the product is used in drying some thin substrates and slurries in belt and rotary drum dryers. Infrared radiation that is of a long wavelength does not penetrate deeply into wet ceramics but may be absorbed by the liquid and transported into the interior by conduction. It is more often used for drying thin substrates, films, and coatings. Heating is produced by the coupling between the infrared radiation of 4 to 8 μm and the molecular vibration of O—H groups. Radiation of a very long wavelength in the microwave range penetrates deeply into most ceramics; energy dissipation from the polarization of the water molecules absorbs the radiation and heats the liquid. Dielectric and microwave drying are used for the drying of liquid saturated products where the drying must be relatively rapid, the maximum temperature of solids must be relatively low, or the product integrity is sensitive to liquid concentration gradients and capillary stresses. Large cast and extruded products containing water have also been partially dried by inserting electrodes and passing a high direct current (dc) at low voltage through the piece.

The design and capacities of industrial dryers vary considerably and are a function of the size, shape, and drying behavior of the product; the dryer loading; setting configuration; the production rate; the temperature, humidity, and velocity of the drying air; and the mode of heating. The temperature and humidity of inlet air and the circulation of air within the dryer must be monitored to control the performance of an industrial dryer.

Mechanisms in Drying

Drying involves the transport of energy into the product; liquid is transported through pores

to the meniscus, where evaporation occurs, and by vapor transport through pores. In a drying system, heat energy must be brought to the surface of the product, and vapors must be carried away. The mechanisms of evaporation and mass and thermal transport must be considered before discussing the drying process.

Liquid placed in a closed container will evaporate to establish a pressure of vapor above the liquid. This pressure will increase with the temperature of the liquid and is dependent on the radius of curvature of the meniscus. The constant vapor pressure at a particular temperature is referred to as the saturation vapor pressure. Because evaporation involves the loss of molecules having the highest kinetic energy, evaporation is a cooling process, and heat must be supplied to maintain an isothermal condition. The difference between the temperature of thermometers with dry and wet bulbs in flowing air indicates the cooling due to evaporation. Heats of vaporization for several processing liquids are listed in Table 1. The boiling point of a liquid is the temperature at which its vapor pressure becomes equal to the external pressure acting on the surface of the liquid, which for water is 100 °C (212 °F) at 10^5 Pa (760 mm Hg, or 1 atm). The critical temperature and pressure of a liquid are the conditions for which the physical properties of the liquid and vapor become identical (no meniscus); for water these are 374 °C (705 °F) and 22×10^3 kPa (220 atm).

The evaporation rate of water into air below its boiling point varies directly with the temperature and surface area of the liquid and inversely with the concentration of that liquid in the air. Above the boiling point, the evaporation rate is dependent on the rate at which heat is supplied and independent of the concentration of liquid vapor in the air. The humidity is defined as the mass of liquid/mass of air, and the relative humidity as the humidity/maximum possible humidity at that temperature. The dew point is the temperature at which air of a certain humidity becomes saturated and precipitates as liquid droplets; it is lower when the content of liquid in the air is lower. During evaporation into moving

*Reprinted from *Introduction to the Principles of Ceramic Processing*, John Wiley & Sons, 1988, p 411–425. With permission

Table 1 Latent heat of vaporization of several processing liquids

Temperature, °C (°F)	Heat of vaporization, kJ/kg		
	H_2O	CH_3OH	C_2H_5OH
20 (70)	2.45	1.17	0.91
40 (105)	2.40	1.14	0.90
60 (140)	2.36	1.11	0.88
80 (175)	2.31	1.06	0.85
100 (212)	2.26	1.01	0.81

air, a static boundary layer of air and vapor exists between the mobile air and the stationary product. Diffusion through this boundary layer controls the liquid transport between the air and the product. The apparent evaporation rate at a particular temperature varies indirectly with the salt content of the liquid and the forces bonding liquid on the surfaces of solids.

Thermal transport to the product may occur by convection, conduction, and radiation. Convective heating of the product is limited by the heat transfer coefficient (h_b) for the static boundary layer between the moving air and the static liquid on the surface. Mobile air of higher velocity produced by forced convection reduces the thickness of boundary layer and increases thermal transport and the apparent evaporation rate. If this heating is the only source of energy for evaporation, the rate of evaporation (R_E) is:

$$R_E = \frac{h_b(T_A - T_L)}{L_E} \qquad \text{(Eq 1)}$$

where $T_A - T_L$ is the difference in temperature across the boundary layer and L_E is the latent heat of evaporation. In contrast, radiation transport to the product is relatively independent of the static boundary layer and the airflow conditions.

Within the product, thermal transport occurs by conduction and radiation. Conduction through the solid particles occurs by phonon propagation, and in the liquid between particles and within pores by molecular vibrations. The thermal conductivity of the ceramic particles is significantly higher than the conductivity of water. Infrared radiation does not commonly penetrate deeply into the product, because the waves are scattered by the particle system; microwaves of much longer wavelengths may penetrate deeply into an electrically insulating ceramic material but are absorbed by water.

Liquid in a porous solid may be chemically or physically adsorbed on solid surfaces, exist as bulk liquid in pores of microscopic size, and be distributed as an external surface film. The migration of liquid to the surface may occur by means of capillary flow, chemical diffusion, and thermal diffusion. Capillary migration is motivated by capillary forces and is very dependent on the radius of the pores and the surface tension and viscosity of the liquid. Liquid and vapor may diffuse along a

gradient in concentration. Thermal diffusion is the migration of liquid or vapor along a thermal gradient and is important when using conduction heating and in dielectric and microwave drying.

The Drying Process

Drying is often regarded as occurring in three stages corresponding to the ranges of liquid content for which the drying rate is increasing, constant, and decreasing, as shown in Fig 1.

In a saturated material, liquid is initially removed by evaporation from the external surface (Fig 2a). The drying rate, expressed as a weight loss per unit time, increases on heating when the relative humidity is less than 100%. The drying rate is strictly constant when the evaporation rate and evaporation surface area are constant. In this stage, the product temperature is normally equal to the wet-bulb temperature of the environment. The mass of water evaporating per unit area (area·time) (R_E) is:

$$R_E = K_E(P_w - P_0) \qquad \text{(Eq 2)}$$

where K_E is the evaporation constant that is dependent on air flow conditions, P_w is the vapor pressure of the liquid at the evaporation temperature, and P_0 is the partial pressure of liquid in the surrounding atmosphere. Evaporated liquid may be replenished by interparticle liquid transported to the external surface by fast diffusion and capillary flow. The loss of interparticle liquid coupled with external capillary stress may cause dimensional shrinkage (Fig 2b) and an increase in the plastic shear strength. Adsorbed binders, binders of higher molecular weight, and gelled binder will resist migration with the liquid, but dissolved salts and dispersed salts and dispersed colloidal particles or molecules may migrate to the surface. Shrinkage and chemical migration may cause a perceptible lowering of the drying rate. For the constant rate (CR) period, the practical drying rate (R_{CR}) is given by the equation:

$$R_{CR} = \frac{W_1 - W_2}{At} \qquad \text{(Eq 3)}$$

where W_1 and W_2 are the liquid contents before and after drying for a time t, and A is the surface area for evaporation. Dynamic changes in the temperature of the product or absorptive capacity of the atmosphere can reduce or preclude the appearance of a constant rate period.

Evaporation from the menisci of liquid in pores (Fig 2c) begins to occur when the rate of internal liquid transport is lower than the evaporation rate and when the pores in the body become unsaturated; the concomitant decrease in the drying rate is called the falling rate period (Fig 1). The temperature of the surface of the product may rise rapidly to the dry-bulb temperature in those areas where the evaporation surface recedes into the pores.

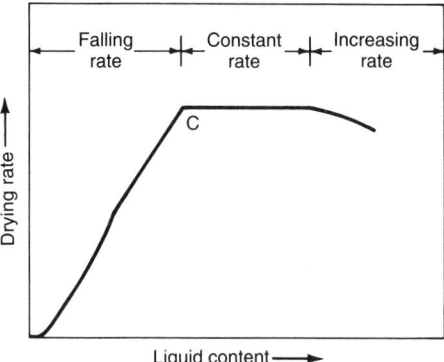

Fig 1 Changes in apparent drying rate when drying a product under moderate conditions. The critical point C corresponds to the termination of shrinkage in all regions only when the liquid is distributed uniformly throughout the volume.

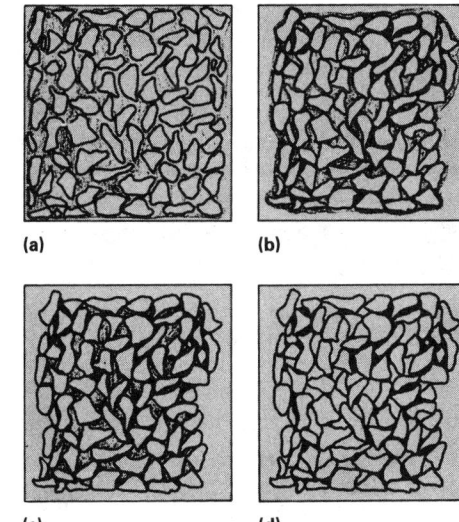

Fig 2 Apparent volume and distribution of liquid among particles during slow drying. (a) As cast with interparticle contacts. (b) Just preceding the end of the constant rate period. (c) On entering the falling rate. (d) Near the end of the falling rate period (liquid in capillaries with a sharp meniscus and liquid physically adsorbed on surfaces remains)

Thermal diffusion into the body is retarded by heat consumed in supplying the latent heat of vaporization. Transport by vapor diffusion through pores increases when the liquid menisci recede into the body. The termination of capillary migration may cause a noticeable change in slope in the falling rate period. Large pores and large accessible interstices are emptied first (Fig 2d) in an isothermal body; an increasing amount of energy is needed to remove water from smaller interstices and interparticle contacts where the radius of curvature of the liquid meniscus is relatively small. Dissolved salts in residual liquid may also reduce the vapor pressure slightly.

A larger capillary stress is produced as the menisci recede. The shape of the drying curve during the falling rate period is strongly influenced by the driving force for liquid evaporation and the pore structure of the product which moderates liquid and vapor transport. Spatially nonuniform drying may occur in bodies with a nonuniform microstructure.

Adsorbed liquid remains after the bulk liquid has been removed (Fig 2d). Heating above the boiling point of the liquid or long exposure to a dessicating environment can remove physically adsorbed liquid and may sometimes remove chemically combined liquids such as the water of crystallization in some inorganic salts and gypsum molds.

During drying, heating increases both the vapor pressure of the liquid and the adsorptive capacity of the drying air. The forced convection of hot air of lower relative humidity maintains the product temperature and flushes away air of higher humidity. Impinging air thins the boundary layer and increases the evaporation rate, particularly during the constant rate period. Setting patterns and the circulation of air into cavities in products must be considered to improve the uniformity of drying as well as the efficiency. Heating becomes more important than air flow during the falling rate period.

Drying Shrinkage and Defects

Shrinkage occurs during drying as the liquid between the particles is removed and the interparticle separation decreases. When the shrinkage is isotropic, the volume shrinkage ($\Delta V/V_0$) is related to the linear shrinkage ($\Delta L/L_0$) by the equation:

$$\frac{\Delta V}{V_0} = 1 - \left(1 - \frac{\Delta L}{L_0}\right)^3 \qquad \text{(Eq 4)}$$

The linear shrinkage is proportional to the mean reduction in interparticle spacing $\Delta \bar{l}$ and the mean number of interparticle liquid films per unit length $\bar{N}_1$ when particle sliding and rearrangement do not contribute significantly to the shrinkage:

$$\frac{\Delta L}{L_0} = \bar{N}_1 \Delta \bar{l} \qquad \text{(Eq 5)}$$

A variation of $\bar{N}_1$ or $\Delta \bar{l}$ with direction due to particle orientation or liquid gradients may cause the linear shrinkage to be anisotropic; variations with position produce differential shrinkage.

Shrinkage during drying can be reduced by forming the product at a lower liquid content to reduce $\Delta \bar{l}$ and increasing the mean particle size to decrease $\bar{N}_1$. Bonds that develop a chemical set may reduce or eliminate $\Delta \bar{l}$ and the shrinkage on drying.

For common extruded and cast products, the linear drying shrinkage is usually in the range of 1.5 to 4%. The dependence of the volume of the piece on the liquid content for homogeneous drying where the liquid content is spatially uniform is shown in Fig 3; shrink-

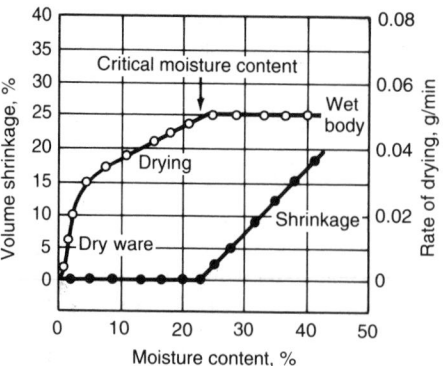

Fig 3 Change in bulk volume on drying a ceramic body. Source: Ref 1

age ceases at a particular liquid content, which is called the leatherhard liquid content.

During drying, a liquid concentration gradient at the beginning of the falling rate period may cause differential shrinkage and a tensile stress in the surface. The transition from a plastic body to an elastic body occurs with a loss of liquid content of about 5 to 15 vol% (Fig 4). Unfired ceramics are normally quite weak and can sustain stresses of only a few 100 kPa (14.5 psi) without deformation or local fracture. Surface cracks may form when material near the surface becomes brittle and the differential shrinkage produces a stress that exceeds the tensile strength. Regions of low strength such as laminations are especially susceptible. Stresses produced during drying may be increased by the pressure of vapor in pores. Small cracks called checks are produced by differential shrinkage in a small region. Parameters that increase the dry strength may increase the stress required to form cracks. Case hardening occurs when shrinkage or surface material has ended, but internal material with a higher liquid content continues to shrink; tensile stresses may produce internal cracks. Differential shrinkage may be produced by gradients in particle size and at junctures between differentially oriented particles of high aspect ratio.

Differential shrinkage due to a differential liquid content when the product was formed or a differential drying rate across the surface

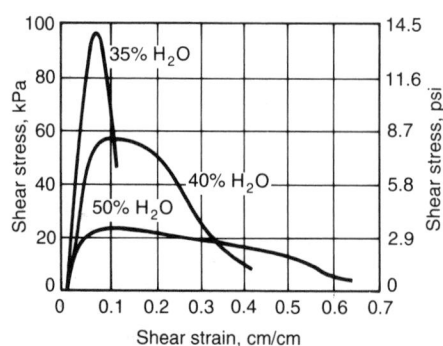

Fig 4 Stress-strain diagram for an uncompressed plastic clay with different liquid contents (liquid concentration in wt%). Source: Ref 2

of the product may also produce warping. Warping is caused by stresses accompanying nonsymmetrical shrinkage, which produces plastic elongation in regions with a lower shrinkage rate. Warping is reduced by increasing the uniformity of drying and reducing the average drying shrinkage of the body. The tendency for warping may be increased by nonuniform external films or coatings, particle orientation, or binder migration, which produce a nonuniform surface permeability, and by nonuniformities in the circulation or temperature of drying air due to setting patterns and supporting hardware. Ideally, drying should occur symmetrically in an isotropic material, and all shrinkage should end before entering the falling rate period. An insufficient rate of transport to the surface relative to the evaporation rate will reduce the constant rate period and increase the transition range during which the deleterious differential shrinkage can occur.

The mechanical restraint of shrinkage may also produce stress and cracks. Contact friction between a rigid support and the shrinking product may cause a crack, especially when the product is heavy and the surface is rough. Restrained shrinkage is also produced between sections drying at different rates, as in a nonuniform cross section, and when material adheres to a porous mold. When drying shapes with a deep cavity, the forced convection of drying air into the cavity may be required to prevent the formation of a crack due to differential drying rates or the restraint of material adhering to the mold.

Other sources of defects are the migration of colloids to the surface producing a skin having different properties and internal gas pressure produced by too rapid volatilization of liquids and insufficient permeability of the pores.

Modes of Drying

Conduction and convection drying are commonly used for drying ceramic products. Floors or shelves supporting the product may be heated by waste heat or steam. Convection drying is used to heat the product and remove vapors, and the air circulation may be designed to facilitate the drying of a particular size, setting, or shape. Infrared drying can be used to reduce the drying time and for the drying of coatings such as decorations and glazes. Vacuum-assisted drying reduces the partial pressure of vapors. Other modes of drying deserving special comment are discussed below.

Controlled Humidity Drying. For every product with a finite drying shrinkage, there is some critical drying rate that will cause defects in the product. When drying thick products or a product of very low permeability (K_P), drying must be conducted in such a way that the liquid concentration gradients are not large. The isothermal capillary flow rate to the drying surface of a slab is proportional to K_P/η_L. The safe drying rate may be increased by reformulating the body to increase K_P or by using controlled humidity drying. Heating the product in humid air arrests the surface evaporation and reduces the viscosity of the liquid (η_L) prior to drying. On reducing the humidity of the drying air, drying can occur at a greater rate without an increase in the liquid concentration gradient.

Microwave Drying. Microwaves are generally reflected by electrical conductors, transmitted by electrical insulators, and adsorbed by dielectrics. Liquid water behaves like a dielectric, because the molecule is polar and the direction of polarization cycles when subjected to a microwave field. Microwave adsorption causes heating in proportion to the field strength and the product of the frequency and dielectric loss factor. At a particular field strength, penetration of the dielectric varies inversely with power adsorption. Microwave energy may be used to heat and evaporate liquid in large cross sections relatively rapidly, independently of the thermal conductivity of the solid. When drying ceramic insulators, the microwaves are preferentially adsorbed by the water, and the product temperature during drying may never exceed 50 °C (120 °F); that is, the high surface temperatures in conventional drying are avoided. The apparent penetration of microwaves increases as water is vaporized and diffuses as a gas to the surface. Potential uses for microwave drying are in the processing of temperature-sensitive products, more rapid drying, drying products of large cross sections and large gypsum molds, and drying products containing colloidal materials such as gels, pigments, and clay of extremely low liquid permeability.

Slurry Drying. Slurried raw materials supported on a metal belt and tape-cast films are dried continuously and relatively rapidly. In drying a mineral slurry on a metal belt, heat is supplied by conduction through the belt and by the hot drying air, which may exceed 400 °C (750 °F). Turbulence produced by boiling may retard the formation of a low-permeability surface layer, as shown in Fig 5. The drying rate is dependent on the solid's concentration of the slurry and the viscosity and permeability of the partially dried material.

In tape casting, the thin slurry film cannot be dried too rapidly, because the drying is one-directional and the liquid permeability of the tape is very low. The concentration of liquid in the air at the beginning of drying is relatively high, as in controlled humidity drying. The maximum liquid temperature must be well below the boiling point of the liquid during the constant rate period and is commonly less than 50 °C (120 °F) for nonaqueous systems. Filtered air enters the exit end and is solvent-loaded on exiting above the entering tape. The air flow and temperature are carefully controlled to maximize the constant rate period and minimize liquid concentration gradients until shrinkage has ceased

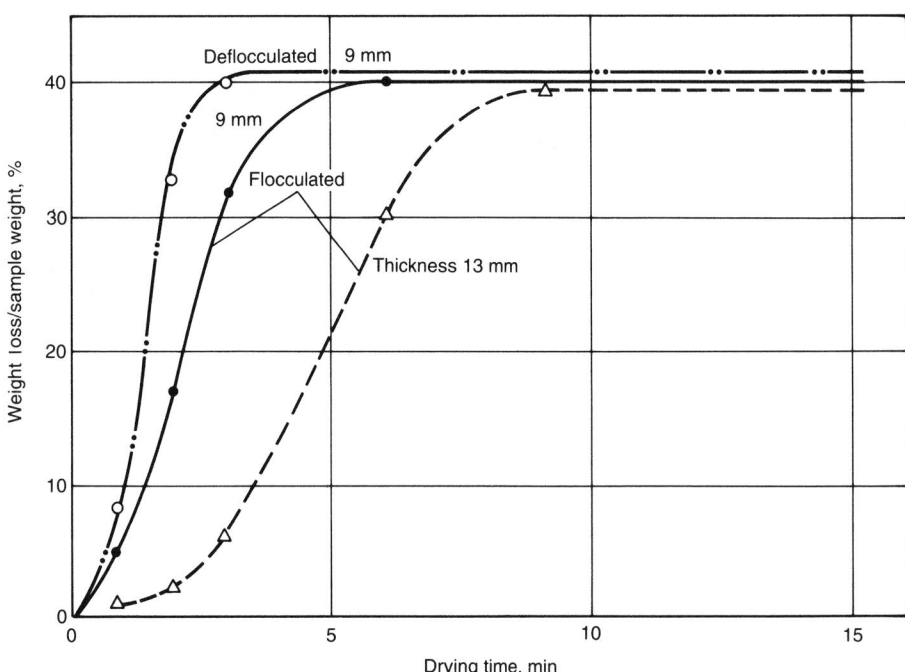

Fig 5 Drying behavior on rapidly heating to 500 °C (930 °F) for a slurry that develops a low-permeability skin

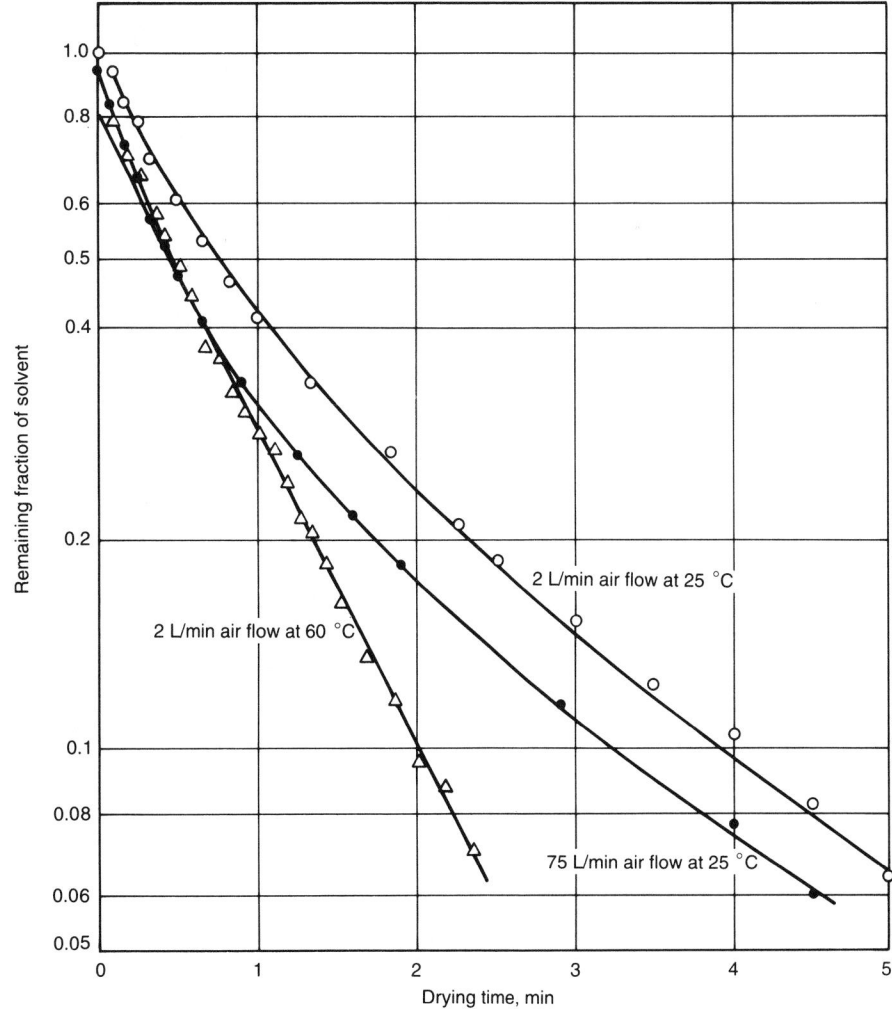

Fig 6 Drying behavior for a tape-cast slurry. Source: Ref 3

(Fig 6). Nonaqueous solvent vapors are collected and reclaimed.

Supercritical Drying and Freeze-Drying. Supercritical drying may be used to minimize effects of the surface tension of the liquid during drying. In supercritical drying, the product is heated in an autoclave until the liquid becomes a supercritical fluid. Supercritical drying occurs during isothermal depressurization. Freeze-drying is used for drying products where a liquid formation and product heating must be avoided.

Spray-drying is relatively efficient compared to the convection drying of formed products, because the material is well dispersed in the drying medium, the diffusion path is shorter, and the high specific surface area contributes to a higher rate of evaporation per unit mass of product.

Summary

Drying is an important process in producing ceramic raw materials and shaped products ready for firing. During drying, heat is transported to the liquid in the body, and evaporated liquid is transported into the surrounding atmosphere. The drying rate depends on the temperature of the liquid in the body and the temperature, humidity, and flow rate of the drying air. Radiation may be used to augment conduction and convective heating or as the primary heating source. After initial heating, the product dries at a constant rate during which shrinkage commonly occurs. Transition to a decreasing drying rate occurs when the external surface of the product is incompletely covered with liquid. When the drying rate is very fast, or nonuniform, the constant rate period is relatively short, and the differential shrinkage can cause cracks. Warping is produced by nonuniform drying when the body is shrinking and can deform plastically. Dried products are commonly hygroscopic and may readsorb moisture in proportion to the relative humidity of the atmosphere.

REFERENCES

1. W.D. Kingery, *Introduction to Ceramics*, Wiley-Interscience, 1960
2. F.H. Norton, *Elements of Ceramics*, Addison Wesley Press, 1952
3. R.E. Mistler, *et al.*, *Ceramic Processing Before Firing*, G.Y. Onoda and L.L. Hench, Ed., Wiley-Interscience, 1978

SELECTED REFERENCES

- W.E. Brownell, *Structural Clay Products*, Springer-Verlag, 1976
- A.R. Cooper, Quantitative Theory of Cracking and Warping During the Drying of Clay Bodies, *Ceramic Processing Before Firing*, G.Y. Onoda and L.L. Hench, Ed., Wiley-Interscience, 1978
- R.W. Ford, *Institute of Ceramics Textbook Series*, Vol 3, *Drying*, Maclaren and Sons, London, 1964
- J.E. Funk, Simultaneous Weight Loss and Shrinkage of Clays, *Am. Ceram. Soc. Bull.*, Vol 53, 1974, p 450–452
- R.W. Grimshaw, *The Chemistry and Physics of Clays*, Wiley-Interscience, 1971
- P. Hrma, Effect of Surface Saturation on Transfer of Water in a Saturated Body, *J. Am. Ceram. Soc.*, Vol 66 (No. 5), 1983
- R.B. Keey, *Introduction to Industrial Drying Operations*, Pergamon Press, 1978
- W.O. Williamson, Dimensional Changes and Microstructures of Unfired Clay Products, *Interceram*, Vol 31, 1968, p 199–204

Organics Removal

Leif Hermansson, DOXA CERTEX AB, Sweden

ORGANICS are used extensively in ceramic processing to improve homogeneity and facilitate both powder compaction and the shaping of components. However, organic processing aids (dispersants, binders, plasticizers, and lubricants) must be removed completely prior to sintering. The removal process is a crucial step in the production of ceramics, particularly when components are large, complex-shaped, or thick, and when forming methods require large amounts of organics. Problems related to the removal process increase in complexity, from powder pressing to tape casting to injection molding. The removal process associated with the injection molding method is well documented in the literature.

Because the removal of organics is a very time-consuming step, it is of vital economic interest to reduce the removal cycle time, which is reported as being several days to weeks for injection molded components (Ref 1). Extremely slow debinding is necessary to avoid harmful defects, such as swelling, skinning , cracking, blistering, delaminating, and warping. Defects caused by improper organics removal are impossible to heal by subsequent sintering, except in the case of encapsulated hot isostatic pressing, in which external pressure squeezes the powder compact and thereby heals larger internal void defects during densification (Ref 2).

This article describes the fundamentals of the removal process, as well as the principles and techniques used to optimize it, such as how to find the right compromise between increased removal rates and minimal structural rearrangements. It is appropriate to mention that neither definitive understanding nor treatment of the binder removal process exists. However, some guidelines can be drawn from both the vast experimental and recent theoretical work. Examples of organics removal from ceramic bodies with different levels of binders are presented, as are techniques for studying the removal process. In addition to cited sources of information, various patented methods are identified in the section "Selected References" in this article.

Removal Process Overview

The complexity of the organics removal process is due to several factors:

- The characteristics of the ceramic powder (solids content, composition, particle size, and particle size distribution)
- The binder system (composition of the blend and related chemical stability and rheology)
- The homogeneity of the consolidated green body and the component geometry
- The process milieu (temperature, pressure, and chemical environment and gradients)

There are both chemical and physical aspects of the removal process that must be considered. The chemical aspects include decomposition, depolymerization, recombination of volatile species, formation of residues, oxidation, and solvent extraction. The physical aspects include heat and mass transport mechanisms (diffusion, gas permeation, micromigration, capillary flow, and supercritical extraction) and related stress development. Debinding techniques with a chemical basis are immersion extraction, supercritical extraction, and thermally activated chemical reactions, whereas techniques with a thermal basis are diffusion, permeation, and wicking.

Thermal degradation plays an important role in almost all practical removal techniques. Thermally activated removal processes are frequently discussed in terms of three stages, each of which is discussed below.

Stage 1 is the initial heating of the organic blend up to its softening point. Only a minor portion of the organics is removed in this stage, but the corresponding problems can include shrinkage, the release of residual stresses, and the explosion of easily volatile species or entrapped gases. Heating at a rate that is too slow (less than 10 °C/h, or 18 °F/h), as well as too rapid (more than 100 °C/h, or 180 °F) can cause cracking, which, in the case of slow heating, is due to entrapped gases having enough time to expand upon heating, and in the case of rapid heating, is due to a thermoshock condition. Initial heating rates of approximately 50 to 60 °C/h (90 to 108 °F) have been reported to be appropriate (Ref 3).

This information should provide reasons not to disregard the critical initial stage. Although the amount of organics removed in this stage is limited, it contributes to the main shrinkage associated with the overall debinding process. The small quantities of organics removed in this stage means that it is impossible to control the process by using weight loss techniques. Rather, pure temperature control must be used (Ref 4).

Extremely small quantities of organics in the form of volatile species (<1 μg, or 3.5×10^{-8} oz) may theoretically induce an internal pressure of 15 MPa (2.2 ksi), which is enough to open up the structure (Ref 3). However, high internal pressure caused by volatile species has been questioned due to the possibility of gas diffusion or dissolution in the polymer (Ref 4).

In this stage, organics are often removed via a combination of heating and capillary action (wicking). Wicking is accomplished by embedding the component in a powder bed of fine-grained alumina or graphite (Ref 5–7). The particle size of the bed material should be finer than that of the ceramic body to promote the flow of fluid into the powder bed, or wick. In addition to its capillary function, the powder bed: reduces the temperature gradients to a minimum and helps avoid surface radiant heating; may reduce the gas partial pressure gradients at the surface; and functions as a support to prevent sagging (Ref 8, 9). Stress release is favored by micromigration induced at increased temperature (Ref 10).

In a system with either slight or no thermal degradation, the analogies between removal by wicking and drying of green bodies are apparent. During the constant rate drying, the liquid (water or low molecular weight polymers) is transported to the surface by capillary forces. Resistance is caused by the viscosity of the liquid. Wicking is thus favored by elevated temperatures.

Stage 2 is the stage at which most of the now-fluid organic blend is heated above its softening point and subsequently removed. This stage involves degradation and evaporation of the organic system, in addition to capillary flow. The mechanisms of degradation depend on the specific organic system and the removal atmosphere (vacuum, inert, or oxidative). The polymers decompose into volatile species within a narrow temperature range. There is increased internal pressure, as well as the risk of swelling and cracking. Because temperature control is very essential in this narrow range, extremely slow heating is used (as low as 1 to 2 °C/h, or 2 to 4 °F/h for injection-molded components) (Ref 11). To spread the decomposition over a wider range of temperatures, the organic system is often based on a multicomponent mixture of polymers with different melting characteristics. During this stage, the maximum weight loss occurs, and the maximum tangential tensile stresses develop (Ref 12).

Stage 3, the final removal of organics at high temperatures, involves decomposition and evaporation in an open structure with a substantial void volume, which facilitates organics extraction and reduces the tensile stresses in the body. During the final stage of the removal cycle, the heating rate is normally increased (Ref 13). The improper choice of a removal atmosphere yields either inorganic residues or residues of high molecular weight species by recombining the free radicals originally formed by oxidative degradation, which may adversely affect sintering (Ref 14). Heating to high temperatures in oxidative atmospheres can result in the unwanted oxidation of carbide and nitride ceramic powders, such as formation of silica in silicon-based ceramics.

Organics Degradation and Removal Principles

The degradation of polymers has been studied intensively and reported in review articles and books (Ref 15–18). Thermal degradation is the principal mechanism involved in the removal of organics from consolidated ceramics. Combinations of thermal and chemical degradation (oxidation, solvent extraction, and others) as well as pressure-induced degradation (evaporation and supercritical extraction) are reported in the literature as well (Ref 18). However, it is worth noting that a ceramic powder can influence the degradation of the polymer by catalysis and that additives used to reduce the viscosity during injection molding can modify the degradation behavior of the polymer.

Thermal Degradation. Polymers are not stable at elevated temperatures, which is why the removal of organic processing aids can be accomplished simply by heating in an inert (vacuum or nitrogen) atmosphere. Decomposition occurs via two principal mechanisms. For polymers such as polyolefins, degradation occurs by scission at random locations on the backbone carbon chain, which reduces the viscosity of the system. Continued degradation reduces the chain lengths further, and these shorter segments can then evaporate. The monomer yield is extremely small. For polymers such as polymethyl methacrylate (PMMA, used in tape casting) and polytetrafluoroethylene (PTFE), the bond breaks between the terminal group and the rest of the chain. This unzipping reaction (depolymerization) yields an increasing amount of volatile monomers. Polyvinyl butyrate (PVB) and polystyrene (PS) have intermediate behavior. At temperatures above 400 °C (750 °F) the degradation is rapidly completed and the pyrolysis products are entirely volatile.

Oxidative Degradation. Heating in an oxidative atmosphere changes the decomposition mechanisms. This is due to oxidation reactions, in addition to thermal reactions (Ref 19).

Polymers are vulnerable to oxidation. Unstabilized polymers undergo slow aging in air, even at room temperature. The oxidation, which is heavily pronounced at elevated temperatures, involves free radical chain reactions. This is true even for quite inert polymers such as polyethylene.

The oxidation of polymers (especially in pure oxygen) often initiates decomposition at a somewhat lower temperature and speeds up the removal rate, compared to thermal treatment in inert atmospheres (Ref 20). However, data suggest that a fraction of the degradation products may recombine to form large molecular weight components and, probably, cross-linked compounds. Such compounds often require pyrolysis in oxygen (or water vapor) at high temperatures for complete removal (Ref 11, 21). Oxidation of low molecular weight polyethylene (waxes) in air is not completed until 500 °C (930 °F), and generates more diverse and higher molecular weight species than does thermal degradation. Whereas thermal degradation mainly yields alkanes and alkenes, oxidative degradation favors the formation of aldehydes and ketones (Ref 14).

Theoretical Aspects. A limited number of articles deal with the theoretical aspects of the organics removal process (Ref 10, 12, 22, 23). In these articles, efforts are made to relate the degradation process not just to the stability of the polymers but to the physical interaction between the powder and the binder system. Equations for the mechanisms of diffusion and permeation (transport of volatile species), as well as capillary flow (the wicking, or extraction, of the liquid phase to the surface) are given in Ref 22. The influence of several parameters on these mechanisms is shown in Table 1.

The most comprehensive theoretical evaluation as of 1990 (Ref 12) relates heat and mass transport to the stresses that develop with these mechanisms. The model that is presented takes into account capillary action, gas transport (diffusion and permeation), phase changes by chemical reactions, and evaporation. The relative rates of capillary liquid flow toward the sample surface and of convective mass transfer from the surface to the external medium are shown to significantly

influence the magnitude of internal tensile stresses. Profiles of temperature, gas and liquid hold-up, gas phase composition, and stresses as a function of removal progress (time) and position within the body are predicted. All these considerations are important to the design of the ceramic removal process.

From the theoretical work, it is evident that the size and spatial arrangement of the particles and the pore structure are key factors in the removal process. It can be stated that:

- A too-fine powder should be avoided
- The powder bed material should be finer than the ceramic powder
- Packing that is too dense and clusters of particles should be avoided
- Even (broad) particle size distribution should be used, and bimodal particle size distribution should be avoided (Ref 24)
- Creating porosity early in the removal process is essential for increased diffusion and optimized removal
- Control of the tangential stresses that develop is more important than control of either the heating rate or weight loss rate
- Large cross sections should be avoided (Ref 22, 23)
- Reduced binder content is favorable

Removal of Binders from Powder Compacts and Tape-Cast Films

The factors that govern organics removal from injection-molded bodies are also applicable to organics removal from ceramic greenware produced by other methods. For powder compacts and tape-cast films, reduced levels of binders and reduced cross-section thickness facilitate the debinding process and speed up the removal process. In powder compacts, the content of pressing aids is 2 to 5 vol%, whereas the binder content in tape-cast films is approximately 30 vol%.

Powder Compacts. Significantly increased heating rates are associated with powder compacts. For materials sintered in an oxidative atmosphere, burnout is included as a first step in the sintering cycle. A typical cycle for the removal of polyethyleneglycol (PEG) from silicon nitride compacts in H_2 atmosphere is a heating rate of 10 to 20 °C/min (18 to 36 °F/min) up to 200 to 300 °C (390 to 570 °F), followed by 0.5 to 1 °C/min (0.9 to 1.8 °F/min) up to 400 °C (750 °F) and a hold time of 0.5 to 1 h before the actual sintering cycle (Ref 25).

Table 1 Effect of parameter on debinding time of three mechanisms

Parameter	Diffusion	Permeation	Wicking
Particle diameter, D	D^{-1}	D^{-2}	$D_w/D_c(D_c - D_w)$
Porosity, E	E^{-1}	$(1 - E)^2/E^3$	$(1 - E_c)^2/E_c^2$
Section thickness, H	H^2	H^2	H^2
Viscosity, G	...	G	G
Pressure difference, P	$(P - P_0)^{-1}$	$P/(P - P_0)^2$	...

Note: The subscript w refers to the wick material, and c refers to the ceramic material.

The choice of specific additives in powder pressing is not primarily related to the removal process, but is based on the optimization aspects of several functions, including moldability, strength of granules, and green strength of the compact, in addition to removal-related behavior. The ease of removing organic materials used as processing aids can be ranked from easiest to least easy in this order: polyacrylates, methylcellulose, polyvinyl alcohol, carboxymethylcellulose, hydroxyethylcellulose, polysaccharides, and alginates (Ref 26).

Tape-Cast Films. The removal of organics from tape-cast films, or sheets, is discussed in recent articles (Ref 10, 21, 23, 27–34). An improved removal situation for tape-cast structures is basically due to heavily reduced cross sections, the presence of pores, and the "unzipping" binder system (Ref 22, 23, 27). Heating rates used in the removal process of tape-cast films are in the same order of magnitude as those used with injection-molded bodies. The burnout and the sintering cycle of a ceramic multilayer capacitor is shown in Fig 1. However, increased heating rates, from approximately 2 °C/h (3.6 °F/h) in conventional burnout to corresponding values of approximately 5 to 30 °C/h (9 to 54 °F/h) have been demonstrated by reaction-controlled binder burnout (Fig 2). This is explained further in the section "Weight-Loss Control" in this article.

Binders frequently used in tape casting are PMMA and PVB. Their depolymerization mechanisms facilitate the removal process. However, changes in polymer degradation have been reported, depending on the preparation history of the polymer, the removal atmosphere, and the interaction between polymers and "ionic" ceramic surfaces (Ref 28, 32). Somewhat higher temperatures are required for complete degassing of mixtures of alumina-PMMA mixtures, compared to pure PMMA. The removal rate can also be influenced by the addition of small amounts of compounds such as manganese acetate. The influence of processing parameters (atmosphere, gas flow, heating rate) on the removal process is described in Ref 30.

Polymer Removal Techniques

The parameters used in different techniques to control degradation and the removal rate of organic processing agents are temperature, time, weight-loss rate, and gas partial pressure. Although the chemical environment and capillary forces (internal or external) influence removal-related behavior, they are not used as control parameters.

Temperature-Time Control. An approximate temperature-time (heating rate) curve for a specified component geometry can be generated based on general information related to the specific ceramic-organic system, such as its physical properties, heat and mass transport equations (Ref 12), tangential stress equations (Ref 12), dynamic thermogravimetry (Ref 10, 19, 27), differential thermal analysis (Ref 10, 28), gas permeability (Ref 22, 23), gas composition analyses (Ref 18, 28), mass spectrometry (Ref 14), Fourier transform infrared spectroscopy (Ref 10, 21), hot-stage microscopy (Ref 35), mercury porosimetry (Ref 10), scanning electron microscopy (Ref 27), and other considerations (Ref 36). This conceptual approach is discussed in Ref 19, and the conceptual binder-removal profile is presented in Fig 3.

To ensure the safe removal of organics from the ceramic body, a conservatively slow heating is often used within a major portion of the temperature interval necessary for complete removal. Using pure heating rate control, the geometry of components is limited to a thickness of approximately 10 to 15 mm (0.4 to 0.6 in.) (Ref 3).

Weight-Loss Control. A technique called rate-controlled (thermal) extraction has been proposed to improve the control parameters during heating. The vital feature of this system is regulation of the temperature by direct coupling to the weight loss of the sample during polymer removal. Time, temperature, and component weight data are monitored continuously by a microcomputer.

The temperature follows a preprogrammed curve (similar to the profile shown in Fig 3), as long as the weight-loss rate does not exceed a preset value. When this set critical value is reached, the temperature is immediately reduced to moderate the reaction. When the weight-loss rate is stabilized, the programmed heating is reapplied until the value is reached again. In this way, the degradation rate is kept at a high, but controlled, level with, hopefully, no harmful effects on the green compact. The system is thus self-adjustable and offers documentation of the extraction process history (Ref 37).

Injection-molded thick-walled (>20 mm, or 0.8 in.) geometries (Ref 38) and complex-shaped components (turbocharger rotors) have been successfully treated by the rate-controlled thermal method based on an improved

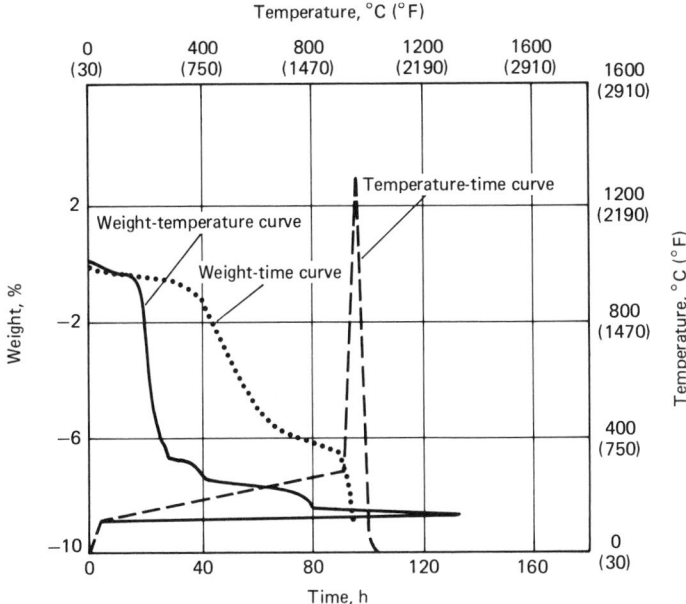
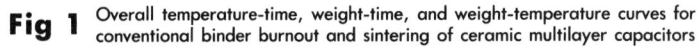

Fig 1 Overall temperature-time, weight-time, and weight-temperature curves for conventional binder burnout and sintering of ceramic multilayer capacitors

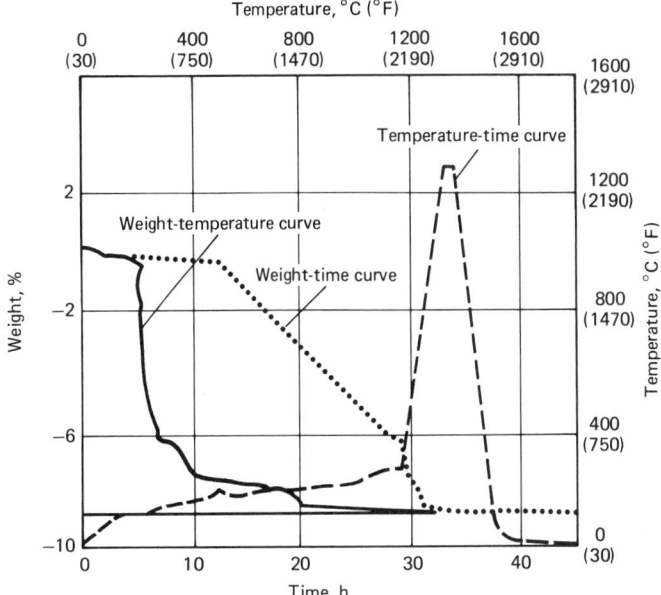

Fig 2 Overall temperature-time, weight-time, and weight-temperature curves for reaction-controlled binder burnout and sintering of ceramic multilayer capacitors

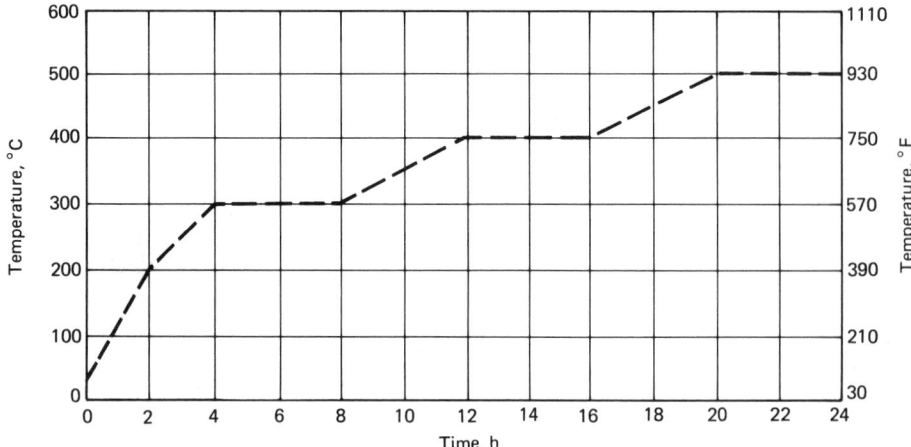

Fig 3 Conceptual binder-removal profile

regulation technique (Ref 39) and heavily reduced removal times (Fig 4). The reaction-controlled binder burnout technique (Fig 2) is also suggested for ceramic multilayer capacitors (Ref 34).

Pressure-Temperature Control. Successful removal of low molecular weight compounds has been demonstrated through the use of pressure-temperature control. The molded components are placed in a furnace that enables control of the partial gas pressure of the organic species and temperature. Mixing polymers and using pressure to control the evaporation rate reduce the risk of damaging the molded body. However, this measure of safety is gained by sacrificing the speed of the removal process. As described in Ref 40, the organics were recondensed and reused.

Other techniques include solvent extraction, supercritical extraction, and capillary action.

Fast extraction of organics from ceramic green compacts can be accomplished by either immersing the body in or washing it with boiling solvents. A two-component vehicle has been used wherein one component was soluble in the extraction solvent. Combinations of thermoplastic waxes and thermosetting resins have been employed. The solvents proposed in Ref 41 can be distilled and reused. Use of this method may be limited due to modern safety regulations (Ref 3).

Supercritical extraction, or high-pressure solvent treatments might represent an alternative approach to fast removal (Ref 22, 42, 43). The molded body and the solvent are processed above the critical point of the solvent (CO_2). However, the process may be too costly to implement in practice (Ref 22).

Capillary action, or wicking, in the removal process has already been discussed in general terms. A capillary extraction process for low-

viscosity waxes, employed in low-pressure molding of ceramics (Ref 44), uses the temperature range of 50 to 120 °C (120 to 250 °F) and evaporation of the remaining binder up to 300 °C (570 °F). As already stated, the powder bed material should be somewhat finer than that of the molded article. Powder bed techniques are frequently used as a first process step, preceding either thermal or chemical degradation techniques (Ref 9, 44).

Acknowledgment

The author expresses his gratitude to Dr. Elis Carlström for valuable suggestions during the preparation of this article.

REFERENCES

1. C.L. Quackenbush, K. French, and J.T. Neil, Fabrication of Sinterable Silicon Nitride by Injection-Molding, *Ceram. Eng. Sci. Proc.*, Vol 3, 1982, p 20–34
2. L. Hermansson, T. Holmström, and H.T. Larker, Some Aspects of Flexibility, Consistency and Cost Efficiency of Glass Encapsulated HIP, *3rd International Conference on Isostatic Pressing, Metal Powder Report*, Vol 2, 1986, p 17.1–17.9
3. J.G. Zhang, M.J. Edirisinghe, and J.R.G. Evans, Initial Heating Rate for Binder Removal from Ceramic Mouldings, *Mater. Lett.*, Vol 7, 1988, p 15–18
4. E. Carlström, Ph.D. thesis, "Defect Minimization in Silicon Carbide, Silicon Nitride and Alumina Ceramics," Chalmers University of Technology, Gothenburg, 1989, p 33
5. D.L. Mann, "Injection-Molding of Sinterable Silicon Base Non-Oxide Ceramics," Technical Report AFML-TR-78–200, Air Force Materials Laboratory, Dec 1978
6. V.M. Sleptsov, O.D. Shcherbina, and G.V. Trunov, Removal of Binder from Silicon Nitride Specimens, *Sov. Powder Metall. Met. Ceram.*, Vol 14, 1975, p 596–598 (in Russian)
7. J. Ichiharu and Y. Imamura, "Binder Removing Materials for Injection Molded Ceramic Products," Japanese patent 88 45165 (63 45165)
8. M.J. Edirisinghe and J.R.G. Evans, Review: Fabrication of Engineering Ceramics by Injection Moulding II. Techniques, *Int. J. High Technol. Ceram.*, Vol 2, 1986, p 249–278
9. B. Mattson and J. Nilsson, ABB Cerama AB, private communication
10. M.J. Cima, J.A. Levis, and A.D. Devoe, Binder Distribution in Ceramic Greenware During Thermolysis, *J. Am. Ceram. Soc.*, Vol 72, 1989, p 1192–1199
11. N.R. Lonkar, R.O. Loutfy, and C.V. Cox, "Process for Removing Carbon and Carbon Compounds Comprised in Binder from Non-Oxide Ceramics and Sintera-

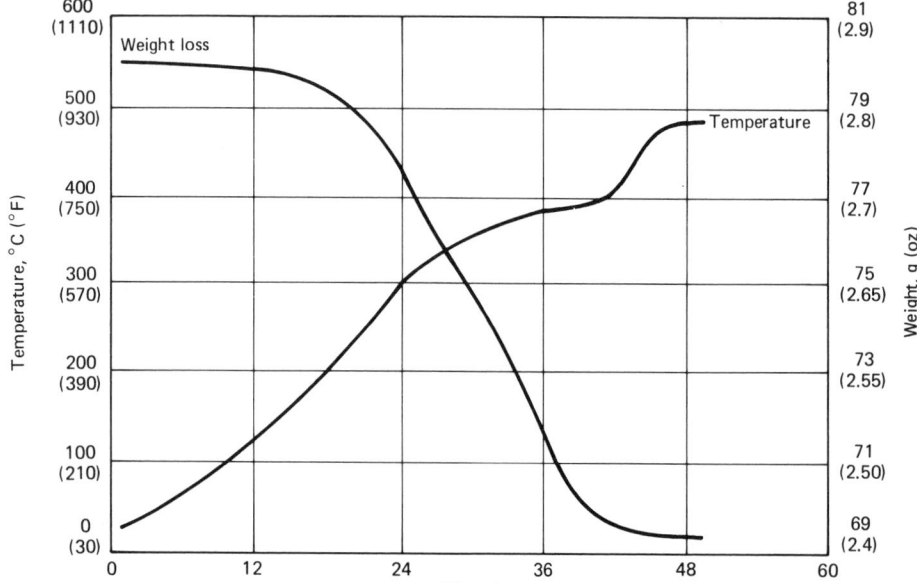

Fig 4 Temperature and weight-loss curves for proportional control. Set point is 10 mg/min (3.5×10^{-4} oz/min)

ble Bodies obtained Therefrom," European patent 0276146, 1988

12. G.C. Strangle and I.A. Aksay, Simultaneous Momentum, Heat and Mass Transfer with Chemical Reactions in a Disordered Porous Medium: Application to Binder Removal from a Ceramic Green Body, *Chem. Eng. Soc.*, Vol 45, 1990, p 1719–1731

13. S. Wada and Y. Oyama, Thermal Extraction of Binder Components from Injection Molded Bodies, *Second International Symposium on Ceramic Materials and Components for Engines*, W. Bunk and H. Hausner, Ed., Deutsche Keramische Gesellschaft, 1986, p 225

14. G.C. Strangle, D.J. Rhee, I.A. Aksay, J.C. Seferis, and R.B. Prime, The Relative Importance of Thermal Cracking and Reforming during Binder Removal from Ceramic/Polymer Composite, *Proceeding of ANTEC 89*, Society of Plastics, 1989, p 1066–1069

15. N. Grassie, Degradation, *Polymer Science*, Vol II, A.D. Jenkins, Ed., North-Holland, Amsterdam, 1972

16. R.T. Couley, *Thermal Stability of Polymers*, Marcel Dekker, 1970

17. C. Hall, *Polymer Materials*, 2nd ed., MacMillan Education, London, 1989

18. *Engineering Plastics*, Vol 2, *Engineered Materials Handbook*, ASM International, 1988

19. B.C. Mutsuddy, Oxidative Removal of Organic Binders from Injection-Molded Ceramics, *Non-Oxide Tech. Engineering Ceramics*, S. Hampshire, Ed., Elsevier, London, 1986, p 397–408

20. R. Gilissen and A. Smolders, Binder Removal from Injection-Molded Ceramic Bodies, *High Tech. Ceramics*, P. Vincenzini, Ed., 1987, p 591–594

21. M.J.Cima and J.A. Levis, Firing-Atmosphere Effect on Char Content from Alumina Polyvinyl Butyral Films, *Ceram. Trans.*, Vol 1, p 567–574

22. R. German, Theory of Thermal Debinding, *Int. J. Powder Metall.*, Vol 23, 1987, p 237–245

23. P. Calvert and M. Cima, Theoretical Models for Binder Burnout, *J. Am. Ceram. Soc.*, Vol 73, 1990, p 575–579

24. N. Theilgaard, "Injection Moulding of Technical Ceramics, Technical Report," Plastics Technology, Danish Technological Institute, 1989, ISBN 87-7511-983-8

25. M. Hatcher, Permascand AB, private communication

26. R. Bast, Organische Additive für Trockenpressverfahren, *Keram. Z.*, Vol 42, 1990, p 155–156

27. D.W. Sporson and G.L. Messing, Organic Removal Processes in Closed Pore Powder Binder Systems, *Ceram. Trans.*, Vol 1, 1988, p 528–537

28. Y.N. Sun, M.D. Sacks, and J.W. Williams, Pyrolysis Behavior of Acrylic Polymers and Acrylic Polymer/Ceramic Mixtures, *Ceram. Trans.*, Vol 1, 1988, p 538–548

29. W.K. Shih, M.D. Sacks, G.W. Scheiffle, Y.K. Sun, and J.W. Williams, *Ceram. Trans.*, Vol 1, 1988, p 549–558

30. G.W. Scheiffle and M.D. Sacks, *Ceram. Trans.*, Vol 1, 1988, p 559–566

31. N. Sarkar and G. Greminger, Methylcellulose Polymers as Multifunctional Processing Aids in Ceramics, *Ceram. Bull.*, Vol 62, 1983, p 1280–1284, 1288

32. W. Farneth, R. Staley, and T. Budzichovski, Reaction Mechanisms in Organic Binder Removal during Ceramic Processing: PMMA/Coordierite as a Prototype System, *MRS Symposium Proceedings*, Vol 108, 1987, p 95–99

33. E.S. Tormey, R.L. Pober, H.K. Bowen, and P.D. Calvert, Tape Casting—Future Developments, *Forming of Ceramics*, Vol 9, *Advances in Ceramics*, J.A. Mangels and G.L. Messing, Ed., American Ceramic Society, 1984, p 140–149

34. H. Verweij and W.H.M. Bruggink, Reaction-Controlled Binder Burnout of Ceramic Multilayer Capacitors, *J. Am. Ceram. Soc.*, Vol 73, 1990, p 226–231

35. C. Dang and H.K. Bowen, Hot Stage Study of Bubble Formation During Binder Burnout, *J. Am. Ceram. Soc.*, Vol 72, 1989, p 1682–1687

36. M.R. Barone, J.C. Ulicny, R.R. Hengst, and J.P. Pollinger, Removal of Organic Binders Ceramic Powder Compacts, *Ceramic Powder Science II*, G.L. Messing, E.R. Fuller, and H. Hausner, Ed., American Ceramic Society, 1988, p 575–583

37. A. Johnsson, E. Carlström, L. Hermansson, and R. Carlsson, Rate-Controlled Extraction Unit for Removal of Organic Binder from Injection-Moulded Ceramics, Materials Science Monograph 16, *Ceramic Powders*, P. Vincenzini, Ed., Elsevier, Amsterdam, 1983, p 767–772

38. A. Johnsson, E. Carlström, L. Hermansson, and R. Carlsson, Rate Controlled Thermal Extraction of Organic Binders from Injection-Molded Bodies, *Advances in Ceramics*, Vol 9, J.A. Mangels and G.C. Messing, Ed., 1983, p 241–245

39. E. Carlström, M. Sjöstedt, B. Mattson, and L. Hermansson, Binder Removal from Injection-Molded Ceramic Turbocharger Rotors, *Sci. Ceram.*, Vol 14, 1987, p 199–204

40. R.E. Wiech, Jr., "Method and Means for Removing Binder from Green Body," U.S. patent 4,395,756, 1981

41. M.A. Strivens, "Improvements in or Relating to the Formation of Moulded Articles from Sinterable Materials," U.K. patent 779242, 1957, and U.K. patent 808583, 1959

42. S. Yokawa and N. Nakashima, Supercritical Fluid Extraction Technology for Ceramic Binder Removal, *Kagaku Kogaku*, Vol 52, 1988, p 513–515

43. N. Enoshima, "Readily Removable Binders for Ceramic Production," Japanese patent 8864977 (6364977)

44. I. Peltsman and M. Peltsman, Low Pressure Moulding of Ceramic Materials, *Interceram*, Vol 4, 1984, p 56

SELECTED REFERENCES

- N.A. Bakhmet'eva, A.K. Kozlov, and E.A. Mikutskaya, Kinetics of Removal of a Thermoplastic Binder from Ceramic Parts, *Steklo Keramika*, 1987, p 23–24

- G. Bandyopadhyay and K. French, "Method for Injection Molding and Removing Binder from Large Cross Section Ceramic Shapes," U.S. patent 4,704,242, 1987

- E. Carlström, A. Johnsson, and L. Hermansson, "Method of Removal of Organic Binding Agents from Molded Bodies," U.S. patent 4,534,936, 1985

- M. Hashimoto, "Removal of Binder from Molded Ceramic Materials," Japanese patent 8741769 (6241769)

- Y. Hirano, "Temperature Control of Pressurized Furnace for Removing of Binders from Injection-Molded Ceramic Bodies," Japanese patent 861974740 (61197470)

- M. Inoe, T. Sakai, and Y. Kihara, "Binder Removal from Extruded Articles of Ceramic and Metal Powder," Japanese patent 86201673 (61201673)

- S. Kamiya, "Binder Removal from Ceramic Green Body," Japanese patent 87113773 (62113773)

- S. Kato, "Removal of Organic Binders from Formed Ceramics," Japanese patent 8653147 (6153147)

- N. Kurata, "Removal of Binder from Ceramic Injection-Molded Body," Japanese patent 87148376 (62148376)

- K. Maeda, "Method for Removing Binder from Ceramic Green Body," Japanese patent 87297271 (62297271)

- K. Maeda, "Method of Removing Organic Binder from Ceramic Greenware," Japanese patent 88297275 (63297275)

- R.E. Mistler, Tape Casting: The Basic Process for Meeting the Needs of the Electronics Industry, *Ceram. Bull.*, 1990, p 1022–1026

- S. Mizuno, "Binder Removal from Ceramic Injection-Molded Bodies," Japanese patent 86117166 (61117166)

- S. Mizuno, "Binder Removal from Injection-Molded Ceramic Greenware," Japanese patent 8787464 (6287464)

- S. Mizuno, "Removal of Binder from Injection-Molded Ceramic Green Body," Japanese patent 8772572 (6272572)

- S. Mizuno, "Removal of Binder from Ceramic Injection-Molded Body," Japanese patent 87148375 (62148375)

- M. Okumura and A. Nakazawa, "Binder Removal from Ceramic Greenware," Japanese patent 87191474 (62191474)
- T. Sakai, M. Inoe, Y. Kihara, and Y. Kawabata, "Binder Removal from Ceramic Molded Body," Japanese patent 85215578 (60215578)
- T. Sakai, M. Inoe, Y. Kihara, and Y. Kawabata, "Binder Removal from Ceramic Molded Bodies," Japanese patent 85215577 (60215577)
- V.N. Shukla and D.C. Hill, Binder Evolution from Powder Compacts: Thermal Profile for Injection-Molded Articles, *J. Am. Ceram. Soc.*, Vol 72, 1989, p 1797–1803
- Sumitomo Electric Industries Ltd, "Ceramic Molding Composition and Binder Removal," Japanese patent 84199570 (59199570)
- Toyota Motors Co, "Removal of Binder from Injection-Molded Ceramic Articles," Japanese patent 85118675 (60118675)
- S. Wada, H. Masathi, and T. Honma, "Removal of Binder from Ceramic Molded Body," Japanese patent 8746969 (6246969)
- S. Watabiki, "Method for Binder Removal in Manufacture of Ceramics," Japanese patent 89219070 (01219070)
- S. Yasuhara and T. Sasaki, "Binder Removal from Ceramic Greens by Supercritical Fluid," Japanese patent 88134574 (63134574)
- S. Yoshida, "Binder Removal from and Sintering of Molded Ceramic Body," Japanese patent 86205673 (61205673)

Dry Pressing

B.J. McEntire, Norton/TRW Ceramics

DRY PRESSING is the process by which ceramic powders are consolidated inside a cavity into a predetermined shape through the use of applied pressure acting in a uniaxial direction. The process is sometimes referred to as dust-pressing, die-pressing, or uniaxial-compaction. The compaction force is generally applied only in the vertical direction by counteracting mechanical or hydraulic rams. Complex shapes possessing multiple holes, levels, and diameters can be easily formed by the appropriate use of engineered compaction tools. Components produced by dry pressing include diverse products such as porcelain tile, cutting tools, grinding wheels, chip carriers, refractories, insulators, seal rings, and nozzles.

Production rates vary depending on part geometry and equipment type. Complex or large components such as refractories and grinding wheels can be produced at rates of 1 to 15 parts per minute. Simpler or smaller parts (that is, seal rings and nozzles) can be produced at rates up to several hundred per minute. Small flat parts like insulators, chip carriers, or cutting tools can be produced at rates up to several thousand per minute. Pressing pressures in the range of 20 to 300 MPa (3 to 45 ksi) are used. Low pressures are common for clay-based materials, whereas higher pressures are required for fine technical ceramics. Because of its flexibility and low-cost, dry pressing is the most widely used high-volume forming process for ceramics.

The general process for forming of a component by dry pressing is common to all forms of equipment and production rates. There are three general stages:

- Die fill
- Compaction
- Ejection

For the most common type of dry press, known as a double-acting press, the stages of the pressing operation are shown schematically in Fig 1. Die fill begins when the powder is pulled into the die cavity from the powder shoe by the downward movement of the lower punch (step 1). The powder shoe retracts across the die face leaving a known volumetric quantity of powder within the cavity (step 2). The upper-punch enters the cavity and begins the work of compaction (step 3). As pressing proceeds, both the upper and lower punches move toward each other in a synchronized manner (step 4). After compaction is complete, the part has residual compressive stresses that lock it into the die cavity (step 5). These are overcome by the movement of the bottom-punch which pushes the part toward the upper die face (step 6). During ejection, the part springs back to relieve residual stress (steps 7 and 8). The powder shoe then enters the process again to push the component off of the die face and to restart the cycle (step 9).

Types of Equipment

The details of particular pressing cycles vary with equipment type. Dry presses can be one of three types:

- Single action
- Double action
- Floating die

Tool motion for each of these types is described in Table 1.

Single-action presses are commonly known as anvil presses. They are mechanically actuated and generally follow a master cam. They have only one moving punch—generally moving from the bottom. The powder is loaded via a powder shoe from the top. The die cavity is closed by a fixed anvil. The bottom punch then compacts the powder against the anvil. After compaction, the anvil is removed and the bottom punch ejects the part. Timing of the movement of the powder shoe, anvil, and punch is provided by the master cam. Such presses are used for producing simple flat shapes at high production rates.

Double-Action Presses. At the next level of complexity, double-action presses can have fixed cavities and movable upper and lower punches. Mechanical rotary presses belong to this equipment type. Also, some hydraulic presses are fixed die, particularly those used in the production of large symmetrically flat components such as grinding wheels or refractories.

Floating-Die Presses. However, the most common equipment, particularly for technical ceramics, is the floating-die press. In this type of press, both of the punches, the die set, and even the core rods can be timed to move during the press cycle. Typical operation of a floating-die press is shown in Fig 2. Multiple synchronized movement of all aspects of the tooling are often required to define complex features, including multiple levels, holes, slots, or tapers. Combined motion of the entire tool also aids in minimizing density gradients within the pressed part.

State-of-the-Art Press. More recently, hydraulic and mechanical press technology has advanced due to the incorporation of computer numerical control (CNC) systems into the equipment. Automation of pressing cycles using computers allows controlled pressure and timing previously unavailable on cam-driven equipment. It is anticipated that this trend will continue and the level of equipment sophistication will increase dramatically over the next decade.

Punches and dies are generally produced from high-strength steels. Brazed or shrink-fit carbide inserts are attached to these tools in areas of high pressure or wear. Control over dimensional tolerances and surface finish during the construction of tools is critical. Dies and punches must fit together with clearances generally ≤ 50 μm (≤ 0.002 in.) and finishes of ≤ 0.2 μm (≤ 8 μin.). By complying with these requirements, entrainment of ceramic particles between critical tooling components is eliminated; and sticking of powders to die and punch surfaces is minimized. The construction of the dies, punches, and entire insertable press tools is a well developed art. Companies dedicated to the design and construction of dies and punches for general or specialized presses are available for the novice and expert alike.

Mechanics of Dry Pressing

While the production of a powder compact can be carried out in any number of similar types of presses, the basic process of compaction remains the same regardless of the equipment type. All presses rely on the application of pressure in a controlled manner to define component geometry and density. A knowledge of the relationship between pressure and green density is essential. Pressure, more than any other single factor, controls compact porosity, strength, and green density

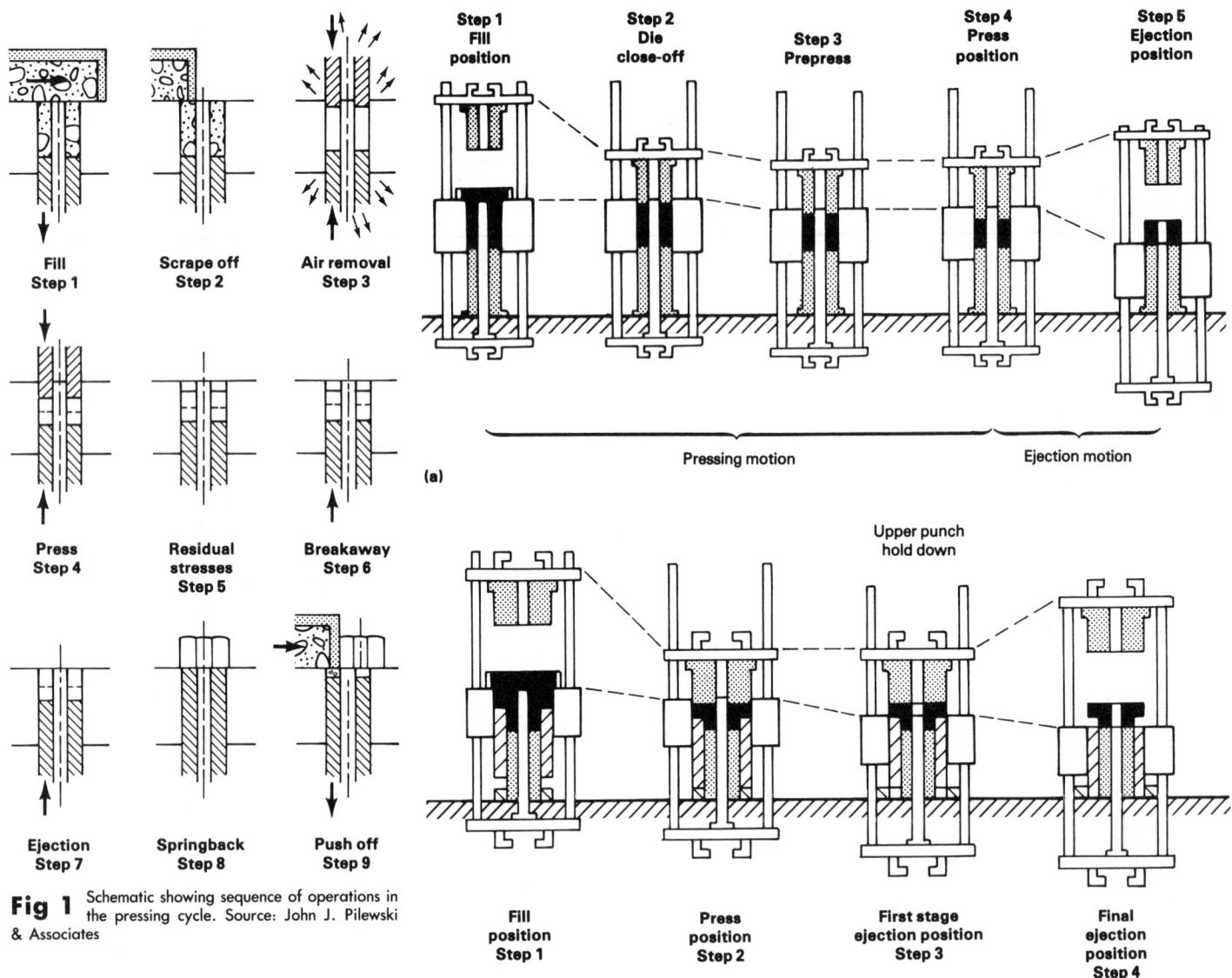

Fig 1 Schematic showing sequence of operations in the pressing cycle. Source: John J. Pilewski & Associates

Fig 2 Pressing cycle for a floating die press used to form workpieces of increasing complexity. (a) Simple one-level part. (b) Two-level part. Dashed lines indicate motion of press components.

uniformity. In turn, these factors play major roles in achieving adequate component properties. Consequently, a clear understanding of the mechanics of compaction will lead to improvements in final properties.

Basic Compaction Stages

There are five overlapping stages of compaction:

- Die fill
- Transitional restacking
- Low flow and fragmentation
- Bulk compression
- Springback

Die Fill. For any reliable pressing operation, powders must flow readily to produce a reproducible volumetric fill for each press cycle. Spherical particles are preferred and are generally produced from fine powders by spray drying or build-up granulation. For high levels of green density, broad size distributions are preferred. Generally, for fine technical ceramics, agglomerates in the range of 40 to 200 μm (0.0016 to 0.008 in.) are desirable. Conversely, for sanitaryware, refractories, or grinding wheels, agglomerates ranging in size from approximately 100 μm to 0.1 mm (0.004 to 0.040 in.) are commonly used. The presence of excessive amounts of fines in the powder severely inhibits flow and leads to poor uniformity in die fill. In addi-

tion, fines can become trapped between the punches and die wall, leading to increased frictional forces, excessive tool wear, sticking, and fracture of the compacts. Consequently, agglomerated powders are essential for any modern press operation. Upon completion of an ideal die fill, the prepressed component typically has a fill density in the range of 20 to 35%. There is considerable retained air within the powder that will be removed during the next stage of compaction.

Transitional Restacking. As the punch enters the die cavity, the rate of volume change is high. Almost half of the tool movement in the cycle occurs during this stage of compaction. Pressures are generally less than 10 MPa (1.5 ksi). Particles move to fill interstices and maximize packing. Air is driven out between the punch and the die walls. The effect in this early stage of compaction is that

Table 1   Tool motion for dry pressing of powder compacts

Press type	Motion		
	Die	Top punch	Bottom punch
Single action	Stationary	Stationary	Movable
Double action	Stationary	Movable	Movable
Floating die	Movable	Movable	Stationary or movable

Source: Ref 1

the material increases in density by particle sliding and rearrangement according to the forces exerted on them. No significant particle deformation occurs during this stage, although some fragmentation is likely during the rearrangement process.

Local Flow and Fragmentation. As the punches continue their movement, considerable resistance is generated by the compact. Deformation, compaction, extensive crushing, and fragmentation of the granulated powders occur. Interstices between granules still persist, but significant knitting and fragmentation lead to the filling of these void spaces. The primary particles comprising the larger agglomerates begin to play a role at this stage. A network of particle-particle contacts is set up throughout the compact. Generally, based on their morphology, hardness, chemical nature, and the presence or absence of compaction aids, the particles will either continue to reorient, fragment and flow, or limit continued movement of the punches. Punch movement during this stage of compaction comprises almost the remaining 50% of the total. Compaction pressures in this stage reach upwards of 50 to 100 MPa (7 to 15 ksi).

Bulk Compression. Following the stage when agglomerates are no longer able to reorient or fragment because of few remaining cavities, the predominant remaining compaction process is bulk compression. Fragmentation of primary particles follows leading to further, although minor improvements, in density. Compaction pressures reach their peak in this stage and can range from 50 to 200 MPa (7 to 30 ksi), depending upon material and press type. With increasing pressure, the apparent density will gradually approach theoretical packing levels. Compact stresses are high at this stage; and green density uniformity is either at its best (for engineered powders) or worst (for poorly prepared powders).

Springback. As the punches retract from the compact, the residual stress within the part is translated to a relaxation of dimensions. Upon ejection from the die cavity, dimensional growth occurs. The extent of this growth is highly dependent upon the material itself, the additives used, and the compaction pressure. Springback is often thought of as a constant, but this is not empirically correct because higher compaction pressures lead to larger springback. Springback also creates some common problems such as end-capping, fracture, and delamination of the compact. Most modern presses possess adjustable punch holddown to minimize springback until the compact is ejected from the die. Under punch holddown conditions, axial pressure is applied to the compact during ejection. This pressure prevents springback in the axial direction and therefore aids in eliminating ejection defects.

Compaction Models

Stemming from the need to predict green density from compaction pressure, a number of both theoretical and practical formulas have been developed and are in general use. These range from the empirical models developed from soil science to rigorous complex relationships that closely simulate actual die pressing cycles. However, all of them attempt to account for the basic pressure density relationship shown in Fig 3.

Empirical based models emphasize this observed relationship by equating porosity (ϕ), green density (D_g), or compact volume (V_g) to an exponential or power function of applied pressure (P_a). For dry pressing, the more basic of these formulas were developed by Seelig (Ref 2) concurrently with Smith (Ref 3). Their equations, given below, are still in wide use today (Ref 2, 3):

$$\phi/\phi_0 = \exp[-(P_a/R)] \tag{Eq 1}$$

$$D_g - D_0 = S \cdot P_a^2 \tag{Eq 2}$$

$$D_g - D_0 = T \cdot P_a^{1/3} \tag{Eq 3}$$

where R, S and T are empirical constants that vary based on the material hardness or plasticity, the particle geometry, and the testing method; and D_0 is the initial density.

Even extensive models developed to simulate detailed aspects of compaction still rely on these three basic equations (Ref 4). Unfortunately, they are not universally applicable. The constants are experimentally derived, and fundamental relationships between powder character and compaction behavior are lacking.

Semiquantitative Based Models. Others, most notably Kawakita (Ref 5), Niesz (Ref 6), and Lukasiewicz (Ref 7), developed models that were more universal and easier to apply. Of these, Lukasiewicz's model has been given considerable attention. It equates relative changes in green density to relative changes in applied pressure. Because of its ease of use, it has gained a reputation as a quality control tool for monitoring the compaction process of agglomerated powders. The equation is simple in its form and application (Ref 7):

$$D_g = D_f + m \cdot \ln(P_a/P_y) \tag{Eq 4}$$

where D_g is the compact density at applied pressure; D_f is the fill density of powder; m is the compaction constant, dependent on packing and deformability of powders; and P_y is the apparent granule yield pressure.

Its graphical interpretation is shown in Fig 4, in which a comparison is given of the compaction behavior of three types of powders:

- Bulk nonagglomerated material
- Solid spherical agglomerates
- Mixed solid and hollow spherical agglomerates

Significant differences are indicated in Fig 4 for these three types of powders. Upon fill, agglomerated powders provide higher fill densities than bulk powders due to the good flow characteristics of the material. In the earlier stages of compaction, the bulk powder compacts extensively to achieve a break point

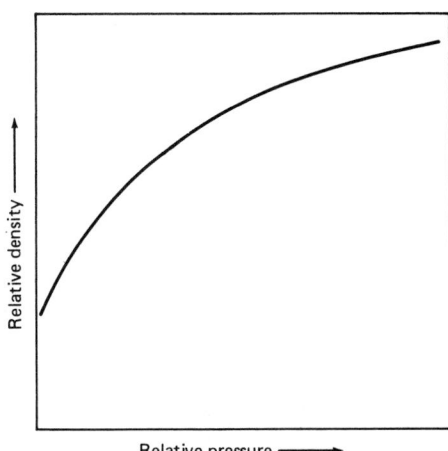

Fig 3 Compressibility curve showing typical pressure-density relationship for a powder compact

at point A in the curve at relatively low pressure, P_A. This break point represents fundamental changes in the compaction process. For the bulk powder, the break point indicates the point at which rearrangement and sliding cease, interstices are filled, and further compaction occurs via particle fracture. For agglomerated powders, work is required to rearrange and initially deform the agglomerates. A break point in this process is reached when interstices between agglomerates are filled and the compact density is equivalent to the bulk density of the agglomerates (points $B_{case\ 1}$ and $B_{case\ 2}$). If the agglomerates are hard and resist further compaction, then ultimate density values are limited to less than that of the bulk powder. Conversely, provided the agglomerates are engineered correctly, they will continue to be crushed and achieve densities equivalent to the bulk powder. The pressure at which green density equality is achieved occurs at the joining pressure (P_C). At this point (points $C_{case\ 1}$ and $C_{case\ 2}$), intraagglomerate and interagglomerate porosity are equivalent. Microstructural inhomogeneities resulting from agglomerates are therefore eliminated. While this model has been explored and applied extensively by several investigators (Ref 8–15), it also fails to fully explain fundamental processes occurring during compaction and has been shown to be extremely sensitive to powder preparation and testing procedures (Ref 16).

Quantitative Models. Predating the work of Lukasiewicz, a quantitative model for compaction was proposed by Cooper and Eaton (Ref 17). Although more difficult to apply than the model of Lukasiewicz, it correctly accounts for the various stages of compaction and is broadly applicable to a wide range of ceramic materials. The model equates the reduction in volume of a compact to two statistically distributed and pressure-activated processes:

- Filling of large voids
- Filling of small voids

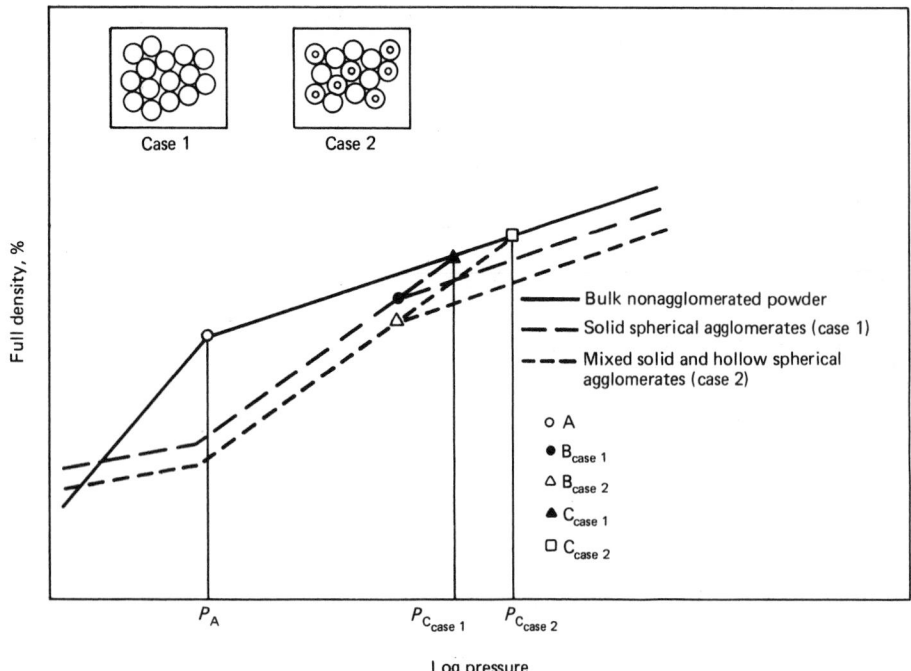

Fig 4 Compaction models for bulk nonagglomerated and spherical agglomerated ceramic powders. P_A, yield pressure; P_C, joining pressure. Source: Ref 7

The compaction model is as follows:

$$V^* = (V_0 - V)/(V_0 - V_\infty)$$

$$= A_1 \cdot \exp(-k_1/P) + A_2 \cdot \exp(-k_2/P) \quad \text{(Eq 5)}$$

where V^* is the fractional volume compaction; V_0 is the initial volume of the compact; V is the volume of compact at pressure, P; V_∞ is the volume of compact at theoretical density; and A_1, A_2, k_1, and k_2 are constants associated with the mechanics of compaction.

When interpreted, the model shows that reduction in volume of the compact is related exponentially to the inverse of pressure and is controlled by pressure activation constants, k_1 and k_2. These experimentally derived constants determine when compaction is dominated by either agglomerates or primary particles. Unlike the model by Lukasiewicz, Cooper's derivation demonstrates that compaction is a continuous process and not separate distinct mechanisms. No break points in curves are noted, although the mechanism change from compaction of agglomerates to primary particles is correctly accounted for. A schematic diagram showing selected stages of compaction is presented in Fig 5. Evidence for experimental verification of the model is presented in Fig 6. As depicted in Fig 5, agglomerates or granules of powder first fill in a uniform packing array. Prior to the application of pressure, they contain both intragranular and intergranular porosity. The compact has a broad bimodal pore size distribution. During early stages of compaction, the collapse of intergranular voids dominates but not entirely at the exclusion of a reduction in intragranular porosity. Compaction

results in an overall reduction in porosity, but a bimodal distribution persists. Finally, in the later stages of compaction, a unimodal distribution is observed as individual granules are meshed with their neighbors. At this point, soft agglomerates will completely lose their character and intergranular porosity will be eliminated. However, the bimodal distribution will persist if granules are too hard. If this is the case the activation pressure constant for agglomerates will be large. Appropriate use of organic additives and proper processing techniques will minimize these activation constants.

While Cooper's model supplies physical meaning to the compaction process, it too is lacking in certain respects. The model does not account for the interactive effects between punches, die walls, and powders. For many materials, frictional effects at the die walls dominate pure compaction. In fact, friction

between the powder and die wall is the leading problem in dry pressing. Poor or nonuniform green density, die sticking, lamination, and end-capping all occur because of these frictional effects. Consequently, many researchers have attempted to model and measure die wall interaction during compaction (Ref 18–20). Frictional effects leading to nonuniform green densities have been known for nearly a century and were first modeled by Seelig (Ref 2). A schematic of this effect is presented in Fig 7 for single-acting punch and dies. Figure 7 shows that high-density wedges generally form just below the die face in the areas designated by the letter A. Material in this region is under a high degree of shear associated with the moving punch and fixed walls. As pressure increases, more material is compacted into these zones. Resulting shear stresses are transmitted towards the center of the compact along the lines denoted by a_1 and a_2. This in turn produces a high level of density toward the center of the part surrounded by lower values. Simultaneously, nonuniform radial stresses are generated in the component due to natural springback tendencies of the material. Upon ejection from the die, end-capping or lamination can often occur, or warpage and cracking may occur during densification.

In an attempt to explain the effects and problems of die interactions, Thompson (Ref 21) developed a rigorous theoretical model and then verified it using existing data. While the complexity of his derivation is beyond the scope of this summary article, Thompson used principles first presented by Smith (Ref 3) to identify five factors controlling green density and compact integrity. The effect of each of these factors is given in Table 2.

Powder Fluidity Index. The first of these, the powder fluidity index, α, was defined as the ability of the material to redistribute stress under applied pressure. Higher values lead to more uniform stress and density. In the ideal case, the material behaves as a Newtonian fluid and therefore supports no stress or density gradients. Unfortunately, higher values also result in larger radial stresses and therefore a greater likelihood of lamination or cracking upon ejection. For the most part, powder flu-

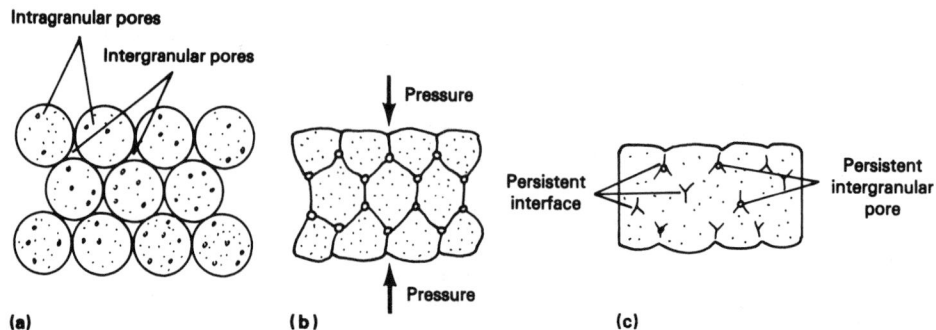

Fig 5 Three stages of compaction for agglomerated powders. (a) Packed spherical granules. (b) Deformed, packed granules. (c) Pressed piece. Source: Ref 2

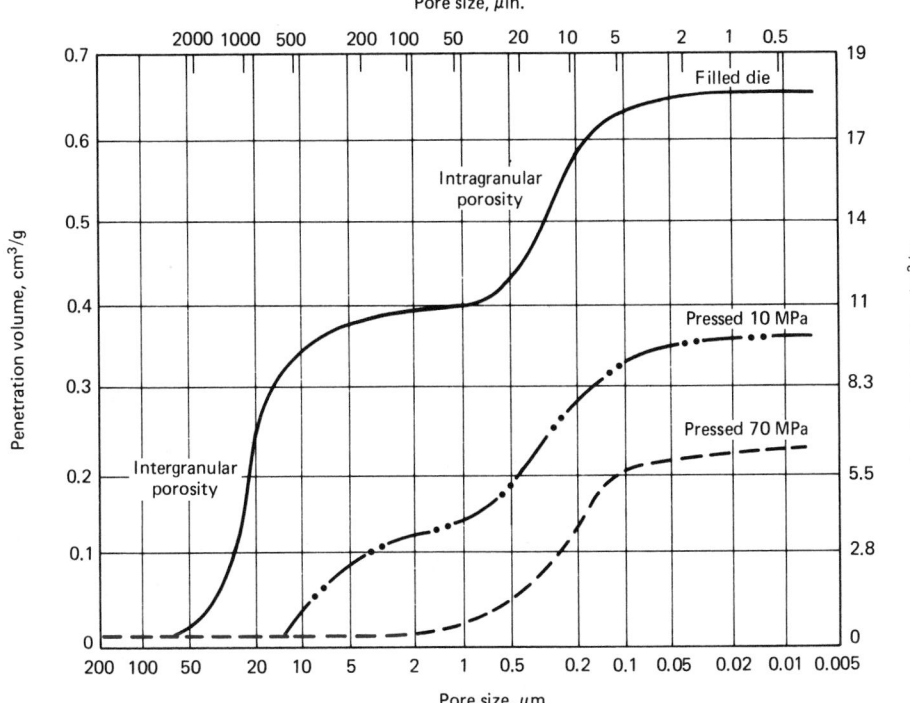

Fig 6 Plot of cumulative pore size distribution versus penetration volume as a function of compaction pressure. Source: Ref 2

idity is a fundamental property of a given material and is a function of its internal friction coefficient, in accordance with the Mohr-Coulomb failure law.

The Mohr-Coulomb failure law, developed from soil mechanics, predicts the amount of shear stress a compact will support before rearrangement or flow of the particles ensues. The law:

$$\tau_f = C + \sigma_f \cdot \tan \phi \qquad \text{(Eq 6)}$$

states that shear stress at failure (or flow), τ_f, is equal to the sum of a constant, C, and the applied axial load, σ_f, multiplied by the tangent of the angle of internal friction, ϕ. The internal friction angle is a fundamental material property that is dependent upon physical, mechanical, morphological, and surface chemical properties of the material in question. It can generally be reduced through the use of lubricants.

Certain ceramics, particularly clay-based materials, possess low values of internal friction and will shear under applied pressure to

redistribute stress. Other materials such as alumina, silicon carbide, or silicon nitride have higher values and will support high stress levels without shear redistribution.

Powder Die Wall Coefficient of Friction. The second factor (that is, μ, the powder die wall coefficient of friction) was found to be detrimental for both green density gradients and lamination or end-capping. Low friction coefficients are always preferred. To minimize this effect, lubricants are generally added to the agglomerated powders, although Thompson argued that placing the lubricant directly on the die wall was preferable to additions to the powder.

Length-to-Diameter Ratio. The third factor, L/D, length-to-diameter ratio, suggests that there is a fundamental limit to uniaxial pressing. High values (long narrow cylinders) lead to poor green densities and lamination. Dry pressing is therefore in practice only applicable to L/D ratios of ≤ 4.

Upper Punch Hold-Down. The fourth factor discussed by Thompson was upper punch hold-down pressure P_H. This is a built-in feature of most modern presses. It allows continuous axial pressure on the component during the ejection cycle by the upper punch. This axial pressure minimizes shear stresses within the compact and therefore prevents lamination. However, because it is only applicable to fully compacted components, upper punch hold-down pressure does not significantly affect green density.

Green Tensile Strength. Lastly, improvements in green tensile strength, σ, via the addition of binders or additives aids in the prevention of end-capping and lamination. Provided the binder is selected for its lubrication abilities, it can also markedly improve green strength. Thompson was rigorous in his quantitative determination of these five factors. However, when collectively understood

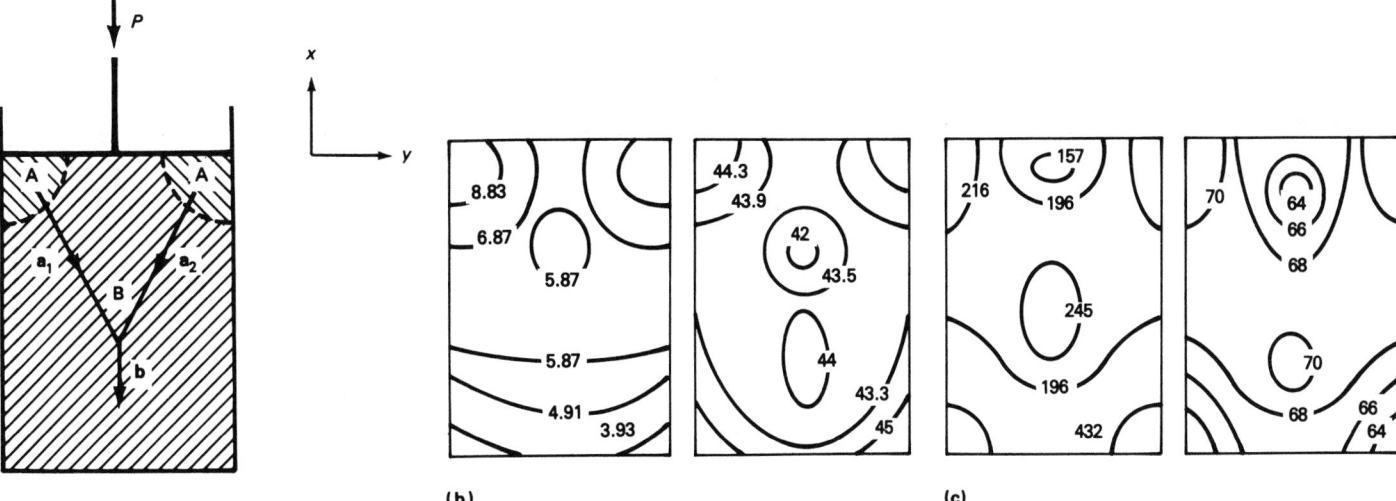

Fig 7 Diagrams showing stress effects present in uniaxial pressing of a magnesium carbonate (MgCO₃) powder compact. (a) Typical shear stresses are transmitted from regions A located just below die face via lines a_1 and a_2 into center region of compact (region B) to concentrate shear stresses along line b. (b) Pressure contour (left) and density contour (right) of compact at an applied pressure of 8.83 MPa (1.28 ksi). (c) Pressure contour (left) and density contour (right) of compact at an applied pressure of 200.1 MPa (29.01 ksi). Pressure contour data given in MPa; density contour data given in %

Table 2 Effect of increasing process parameters on the green density and end-capping (lamination defects) of a powder compact

Parameter	Symbol	Effect of increase in parameter value	
		On green density	On end-capping or lamination
Powder fluidity index	α	Advantageous	Detrimental
Powder die-wall coefficient of friction	μ	Detrimental	Detrimental
Length-to-diameter ratio	L/D	Detrimental	Detrimental
Upper punch hold-down pressure	P_H	None	Advantageous
Green tensile strength	σ	Advantageous	Advantageous

Source: Ref 4

and correctly practiced, they can eliminate a host of daily problems faced in dry pressing.

Powder Specifications Required for Dry Pressing

Uniform mold filling is perhaps the most crucial powder parameter governing dry pressing. Because of this fact, most dry-press operations require a consistent agglomerated powder. Agglomeration of the material through such practices as either build-up granulation or spray drying produces a material that exhibits excellent flow behavior. Coarser granules flow more readily than finer ones. Consequently a size distribution of between 20 μm and 200 μm (0.0008 to 0.008 in.) is often preferred. If particles less than about 20 μm (800 μin.) are present in quantities above about 10%, significant impediments to flow are encountered, along with sticking of the upper and lower punches.

Of all agglomeration techniques available, spray drying is generally preferred. In order to prepare spray-dried material of adequate quality, additives are included with the batch. In general, the following additives are used in batch preparation:

- *Dispersant.* Used to increase batch solids-loading prior to spray drying and reduce batch viscosity. High solids contents are preferred because they lead to high bulk-density powders
- *Binders.* Used to improve green strength
- *Plasticizer.* Used to aid softening of the binder and to reduce the moisture sensitivity of the powder-binder mixture
- *Lubricants.* Used to reduce the internal friction coefficient of the powder
- *Wetting agent.* Used to aid in controlling rheology, the behavior of the slurry during batch preparation, or actual spray drying
- *Defoamer.* Used to aid in controlling behavior of slurry during batch preparation or actual spray drying

A list of common additives for spray drying has been discussed by several authors, including Onoda (Ref 22) and Reed (Ref 1). By far the most common binder system in use today is polyvinyl alcohol (PVA) plasticized by polyglycols. This system has received intense review and discussion by a number of researchers (Ref 14–16). Other common binder systems include microcrystalline wax emulsions or acrylic wax binders. Lubricants are generally selected from the stearate family of materials, waxes, talcs, or clays. Dispersants are as varied as are their individual formulations, but among the most commonly used are complex anionic carboxylic acids.

The engineering of these agglomerates is no trivial matter. Recent articles have emphasized the importance of proper attention to the agglomeration process (Ref 9–16). The nature of these agglomerates determines to a large extent the final properties of the product. If they are produced incorrectly (that is, low density, excessively strong, without lubricants or with too little or excessive residual moisture) they will fail to crush completely away during the latter stages of the press cycle. Excessive residual porosity, poor strength, and poor surface features are often the result.

REFERENCES

1. J.S. Reed, *Introduction to the Principles of Ceramic Processing*, John Wiley & Sons, 1988, p 331–333
2. R.P. Seelig, Introduction to Seminar—Review of Literature on Pressing Metal Powders, *Trans. Inst. Min. Metall.*, Vol 171, 1947, p 506–517
3. G.B. Smith, Compressibility Factor, *Met. Ind.*, Vol 72, 1948, p 427
4. R.A. Thompson, Mechanics of Powder Pressing I-III, *Am. Ceram. Soc. Bull.*, Vol 60 (No. 2), 1981, p 237–245
5. K. Kawakita, Characteristic Constants in Kawakita's Powder Compression Equation, *J. Powder Bulk Sol. Technol.*, Vol 1 (No. 2), 1977, p 3–8
6. D.E. Niesz *et al.*, Strength Characterization of Powder Aggregates, *Am. Ceram. Soc. Bull.*, Vol 51 (No. 9), 1972, p 677
7. S.J. Lukasiewicz and J.S. Reed, Character and Compaction Response of Spray-Dried Agglomerates, *Am. Ceram. Soc. Bull.*, Vol 57 (No. 9), 1978, p 798–801, 805
8. J.W. Halloran, Role of Powder Agglomerates in Ceramic Processing, *Advances in Ceramics*, Vol 9, J.A. Mangels, Ed., American Ceramic Society, 1984, p 67–74
9. G.L. Messing, C.J. Markhoff, and L.G. McCoy, Characterization of Ceramic Powder Compaction, *Am. Ceram. Soc. Bull.*, Vol 61 (No. 8), 1982, p 857–860
10. R.G. Frey and J.W. Halloran, Compaction Behavior of Spray-Dried Alumina, *J. Am. Ceram. Soc.*, Vol 67 (No. 3), 1984, p 199–203
11. F.W. Dynys and J.W. Halloran, Compaction of Aggregated Alumina Powder, *J. Am. Ceram. Soc.*, Vol 66 (No. 9), 1983, p 655–659
12. R.A. Youshaw and J.W. Halloran, Compaction of Spray-Dried Powders, *Am. Ceram. Soc. Bull.*, Vol 61 (No. 2), 1982, p 227–230
13. J.A. Brewer, R.H. Moore, and J.S. Reed, Effect of Relative Humidity on the Compaction of Barium Titanate and Manganese Zinc Ferrite Agglomerates Containing Polyvinyl Alcohol, *Am. Ceram. Soc. Bull.*, Vol 60 (No. 2), 1981, p 212–220
14. R.A. DiMilia and J.S. Reed, Effect of Humidity on the Pressing Characteristics of Spray-Dried Alumina, *Advances in Ceramics*, Vol 9, J.A. Mangels, Ed., American Ceramic Society, 1984, p 38–46
15. R.A. DiMilia and J.S. Reed, Dependence of Compaction on the Glass Transition Temperature of the Binder Phase, *Bull. Am. Ceram. Soc.*, Vol 62 (No. 4), 1983, p 484–488
16. R.L.K. Matsumoto, Generation of Powder Compaction Response Diagrams, *J. Am. Ceram. Soc. Commun.*, Vol 69 (No. 10), 1986, p C246-C247
17. A.R. Cooper, Jr. and L.E. Eaton, Compaction Behavior of Several Ceramic Powders, *J. Am. Ceram. Soc.*, Vol 45 (No. 3), 1962, p 97–101
18. S. Strijbos, Phenomena at the Powder-Wall Boundary during Die Compaction of a Fine Oxide Powder, *Ceramurgia Int.*, Vol 6 (No. 4), 1980, p 119–122
19. S. Strijbos, Powder-Wall Friction: The Effects of Orientation of Wall Grooves and Wall Lubricants, *Powder Technol.*, Vol 18, 1977, p 209–214
20. H.H. MacLeod and K. Marshall, The Determination of Density Distributions in Ceramic Compacts Using Autoradiography, *Powder Technol.*, Vol 16, 1977, p 107–122
21. R.A. Thompson, Mechanics of Powder Pressing I-III, *Bull. Am. Ceram. Soc.*, Vol 60 (No. 2), 1981, p 237–245
22. G.Y. Onoda, Jr. The Rheology of Organic Binder Solutions, *Ceramic Processing before Firing*, G.Y. Onoda, Jr. and L.L. Hench, Ed., John Wiley & Sons, 1978, p 235–252

Cold Isostatic Pressing

Dr. Fred Kennard, AC Rochester Division, General Motors Corporation

ISOSTATIC PRESSING is a process used to form ceramic components from a dry powder by uniform pressing from all directions. It is also used with many other materials, such as metals, plastics, graphite, and carbon. The process is accomplished by enclosing the powder in a deformable mold and then collapsing that mold using a fluid medium to apply hydrostatic pressure.

This article is limited to a review of cold isostatic pressing (CIP), which is carried out at or near room temperature, as opposed to hot isostatic pressing (HIP), which takes place at elevated temperatures. Cold isostatic pressing is used extensively on both laboratory and production scales. In production, it is used to form a diverse array of parts, including spark plug insulators, oxygen sensors, many large refractory components, and dinnerware.

Probably the earliest documented use of isostatic pressing was described in 1913 in a patent by Madden (Ref 1). In this instance, the process was used to form refractory metal billets for subsequent processing into lamp filaments. In the 1930s, additional patents were granted for using isostatic pressing to form various ceramic articles (Ref 2–4). Since that time, this forming method has come to be extensively used by the ceramic industry. The highest-volume example of a component formed by cold isostatic pressing is probably spark plug insulators, with a worldwide daily volume of five to six million (Ref 5).

The general advantages of cold isostatic pressing are:

- Very few size or dimensional limitations, because the uniform application of pressure associated with this process obviates the size limitations of many other processes, particularly the length-to-diameter problems of dry pressing
- Very uniform pressed compacts, due to the uniform application of pressure, which leads to very consistent densities and shrinkage yielding a reproducible process
- Generally moderate tooling costs, particularly in cases of prototype and low-volume production, where the tooling can be quite simple
- Short overall process times, which do not require long binder burnout or drying times

The general limitations of this forming technology include:

- Relatively poor shape and dimensional control. Generally, green machining (machining of the pressed compact before sintering) is required, although the dimensions defined by a metal mandrel can be controlled very well
- Only fairly simple shapes are possible. The overall complexity of the parts to be formed are only limited by the ability to make tooling that allows the part to be removed after pressing. However, only the simplest of parts have been produced to net shape and green machining is almost always required

There are two basic types of cold isostatic pressing, wet-bag and dry-bag. Each has its own advantages and disadvantages, other than those listed above, and is discussed below, separately.

Wet-Bag Isostatic Pressing

Wet-bag isostatic pressing was the process developed first and is the most popular type of isostatic pressing. It is frequently used for the production of experimental or prototype parts because parts can be formed to a rough shape and then green machined to their final shape more quickly and less expensively than any other forming method. It is also used for low-volume production of many parts, particularly very large parts or those with length-to-diameter ratios that are difficult to form by other methods.

In this process, the elastomeric mold is filled with powder and sealed. Then, it is placed into the pressure vessel, which is filled with a fluid (normally, a soluble oil/water mixture), and hydrostatically pressed. Pressures typically vary between 21 and 690 MPa (3 and 100 ksi) (Ref 6, 7). After pressing, the mold is removed from the pressure vessel, and the pressed component is removed from the mold. This process is shown schematically in Fig 1.

Advantages of the wet-bag isostatic pressing process are:

- Very few size or dimensional limitations
- Very uniform density of pressed compacts
- Low tooling costs, because elastomeric molds are generally inexpensive and can be made in-house, if desired

Disadvantages of the process are:

- Poor shape and dimensional control. Although fairly complex shapes can be formed by wet-bag pressing, the dimensions formed by the elastomeric bag are not well controlled. It is difficult to achieve tolerances better than ±3% (Ref 7, 8)
- Low production rates. Although cycle times vary considerably depending on the equipment, tooling, part design, and many other factors, typical cycles would be between 5 and 60 min (Ref 9)
- Difficult to automate; generally very labor intensive

The significant advantages of this process have made it the method of choice for many parts. Wet-bag pressing is also used when very high pressures, >275 to 1380 MPa (>40 to 200 ksi), are required. The disadvantage of poor shape and dimensional control can be overcome by producing oversized parts and then green machining. However, this can be expensive and can introduce defects into the parts. Because of the low achievable production rates and the difficulty (and cost) in automating this process, it is generally not used for high-volume production.

Dry-Bag Isostatic Pressing

In dry-bag isostatic pressing, the elastomeric mold is an integral part of the press itself. This mold is contained within a second elastomeric bag which then contains the hydraulic fluid. As shown in Fig 2, the mold is filled, the powder pressed, and the pressed compact removed without disturbing the mold itself. Typical pressing pressures are between 21 and 275 MPa (3 and 40 ksi) (Ref 7).

Advantages of the dry-bag process are:

- Very few size or dimensional limitations; however, more limited than wet-bag pressing because a straight withdrawal of the part from the mold is required. Parts up to 150 mm (6 in.) in diameter by 610

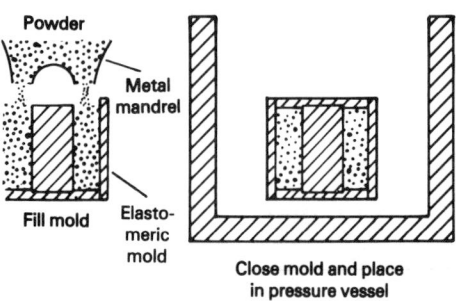
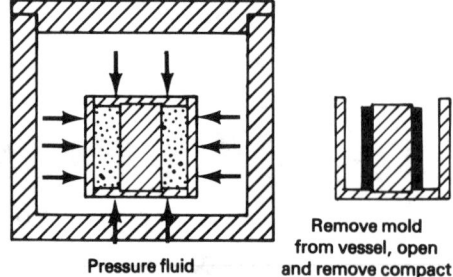

Fig 1 Wet-bag isostatic pressing

mm (24 in.) long have been reported (Ref 7)

- Very uniform density of pressed compacts, but not quite as good as wet-bag pressing because some restriction occurs at the end of the mold, where the part is withdrawn. Good enough to be generally accepted for high-volume production systems
- High production rates; up to 60 parts/min have been reported (Ref 6)
- Moderate tooling costs; although somewhat higher than wet-bag tooling due to increased complexity, cost is still low compared to many other forming methods, especially considering the high production rates that can be achieved
- Automation; Fully automated units are available from a number of vendors and have been operating in high-volume production more than 15 years

Disadvantages of this process are:

- Only simple shapes possible. Because the mold stays in the press and the part must be withdrawn from it, dry-bag pressing cannot produce as complex a part as wet-bag pressing (Ref 10)
- Poor shape and dimensional control, because the dimensions formed by the elastomeric mold cannot be precisely controlled. However, compacts with thin walls and small total bag movement (such as lamp tubes) have been made in production with dimensional controls of about ±0.05 mm (±2 mils)

Dry-bag pressing is a frequent choice for production when the volume is high enough to justify its somewhat higher machine and tooling costs, versus wet-bag pressing. It is rarely used for prototype or experimental work where wet-bag pressing is easier and less costly.

The advantages and disadvantages of this process are generally similar to those for wet-bag pressing, notably in the excellent properties of the green and sintered compact. The major differences are that dry-bag pressing is limited to smaller fairly simple shapes, but is highly automatable with very high production rates.

Materials

As with all forming processes, cold isostatic pressing has its own unique material requirements. However, the requirements are generally less stringent than they are for many other processes. Important basic material parameters include the particle size distribution, surface area, binder properties, and amount of each constituent, particularly the binder(s) (Ref 5, 11). These primary material parameters control the green (and fired) body properties. The green body properties that are most important to the isostatic pressing process are the green density and green strength of the pressed compact. In addition, the flowability of the powder to be pressed is often very important. Chemical pressing aids are frequently added to the powder to improve green density.

Good flowability is required for fast and uniform mold filling. This is most critical in automatic dry-bag pressing, where a uniform charge is required and where it helps to control dimensional accuracy, surface finish, and filling speed (Ref 7). Powders with good flowability are most often produced by spray drying (Ref 12, 13), although other granulation techniques are sometimes used. The wet processing that is generally done as part of a spray drying process also helps to ensure uniform properties. Vibration of the mold has also been shown to help, particularly with less-flowable powders, although segregation can occur if vibration is overdone (Ref 14).

The green strength of the pressed blank must be sufficient to survive subsequent handling. In many cases, this would include green machining to form the final dimensions of the part. Appropriate green strength is normally obtained by adding a small amount (approximately 1 to 2%) of binder, such as polyvinyl alcohol, polyethylene glycol, and various waxes to the powder prior to pressing. A small amount of a lubricant added at the same time as the binder is frequently helpful in maximizing the green density. Because any binder or other organic additive must be removed during the firing cycle, no more than is necessary to obtain satisfactory green properties should be used. Binders may also create problems during green machining, particularly during contour grinding (Ref 5). In this case, both the binder selection and the ceramic powder properties are based as much on the requirements of the operation as on any other requirements. Fairly hard waxes are desirable for contour grinding (Ref 5).

Care must also be taken when selecting the binder type and amounts to assure that the granules in the flowable powder collapse and deform when pressed at the desired pressure (Ref 15, 16). This property is not only a function of the amount and type of binder, but of the properties of the ceramic powder and the processing conditions as well. The powder properties that are the most significant are particle size distribution and surface area. Processing conditions, particularly spray drying parameters such as temperature and

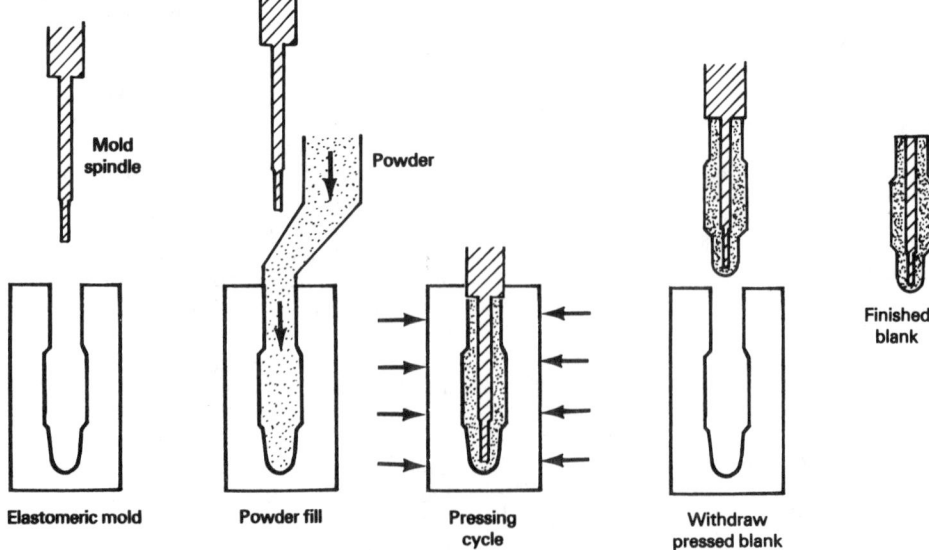

Fig 2 Dry-bag isostatic pressing

final moisture content, are also important in controlling the properties of the spray-dried granule.

The green density of the pressed compact is important in controlling the fired density and fired shrinkage. The ceramic raw materials, binders and pressing aids, powder processing conditions, and pressing pressure should be controlled as much as possible to maximize the green density and consequently to maximize the fired density and minimize shrinkage. However, in some circumstances, this may not be desirable. Frequently, powder properties are controlled to achieve some other end result. Green machining processes, such as contour grinding, may impose their own requirements on the system. Pressed green densities do not typically exceed 60% of the theoretical density (Ref 17).

The overall system of ceramic powder, organic additives, and processing conditions must also be optimized so as to minimize the adherence of the pressed compact to any tooling used in the isostatic pressing process. This is rarely a problem with low-volume, low-speed processes, but may be a problem with a high-speed, automated process, such as the process used for spark plug insulators (Ref 5, 18). Low binder contents, harder binders, and lubricants all can play a role in controlling this problem. The surface nature of the tooling itself is also very important (Ref 5).

Tooling

Proper tooling selection and design are also very important facets of an isostatic pressing operation. This is particularly true when producing anything beyond the most elementary of shapes. In its simplest form, a tooling set consists of a flexible, elastomeric mold (or bag) and, usually, a rigid metal mandrel. Generally, the mold defines the external dimensions and shape of the component, while the mandrel forms the internal dimensions and shape. However, this can be reversed through the use of a dilatory mold. The method of choice depends on the dimensional requirements of the exterior versus the interior and whether green machining is desirable. The dimensions and shape defined by the rigid metal mandrel can be controlled quite precisely, and, generally, the component needs no further finishing.

Elastomeric molds have these requirements:

- Physical and chemical compatibility with the powder to be pressed, which generally does not represent a problem
- Chemical compatibility with the pressing fluid, because many elastomers are degraded by the oil used either for or in the pressing fluid. In dry-bag pressing, the mold is not exposed to the pressing fluid, and only the secondary mold, or bag, is in contact with the fluid and needs to be resistant to it
- Resiliency, which must allow the mold to

return to its original size and shape when pressure is released. This is critical for high-volume applications where a long life is necessary, as well as for applications where the part is pressed to a net shape
- Durability, in terms of sufficient strength, tear resistance, hardness, and other properties, to survive the required number of cycles

A variety of elastomeric mold materials is available, including polyurethanes, polyvinyl chloride, natural rubber, neoprene rubber, nitrile rubber, polysulfides, latex, and silicones (Ref 11, 19–21). Each has its own advantages and limitations. Polyurethane molds are extensively used because of their favorable properties (Ref 11). This includes the availability of a wide variety of hardnesses, good tear and wear resistance, oil resistance, and compatibility with most powders. Polyvinyl chloride materials are easy to form by pouring or dripping, and can be produced in a wide variety of hardnesses and thicknesses. They also have the advantage of not shrinking when formed. However, they are not as oil resistant as polyurethane. Natural rubbers have very good physical properties, such as tensile and tear strength, abrasion resistance, and resilience, but are not resistant to oils. Neoprene rubbers have physical properties similar to but not quite as good as the polyurethanes. They are not as oil resistant as the polyurethanes, either. Nitrile rubbers have very good oil resistance, but only fair physical properties. Silicones generally have poor physical properties but are very heat resistant. Latex molds have a number of drawbacks including relatively poor corrosion resistance and interaction with the powders being produced. Latex molds are also generally more difficult and costly to produce when compared to many of the other materials. Polysulfides are extremely corrosion resistant but have poor mechanical properties.

The material used for the mandrel is also important. For low-volume and/or low-pressure applications, steel or even aluminum may be used. However, for most applications, these materials would not have sufficient durability and it would be necessary to use a harder material, such as a hardened tool steel, tungsten carbide, or even a ceramic for the mandrel (Ref 20). The mandrel surface finish is also important. A rough finish may cause sticking of the pressed part and either green cracks or other defects when pulled free. Material build-up on the mandrel can cause similar problems (Ref 22), which can be avoided by proper binder selection, low moisture content in the powder, periodic cleaning of the mandrel, and/or by surface treatment of the mandrel with a lubricant or permanent casting.

The design of satisfactory tooling depends on many factors, such as the type of press (wet or dry bag), the properties of the material to be pressed, the tooling materials, vol-

ume of parts required, and the size, shape, and dimensional tolerances of the desired part. Generally, tooling designs are developed in an empirical manner (Ref 20, 21).

The first step would be to determine the properties of the powder to be pressed. The specific material parameters that must be considered when designing the tooling are the compression ratio of the powder (typically 2–2.5/1), the amount of springback that occurs when the pressure is released (typically 0.1 to 0.3% of the nominal diameter, per Ref 11), and the firing shrinkage. These properties can be determined by isostatically pressing simple shapes and making the appropriate measurements.

The second step is to use the data defined in the first step to design an appropriate tooling set. If the part to be pressed is simple, or if there is an experience base associated with similar parts, then this may be all that is required. However, if the appropriate experience level does not exist, if the part is more complex than normal, or if the final tooling set is very expensive, then the second step will be to design an inexpensive, easy to make tooling set that closely approximates the required set. This allows a better determination of the final design parameters, which can then be used for the final tooling set.

Other design issues that need to be addressed include:

- Mold collapse in different directions, due to variations in the wall thickness of the mold, interaction between the mold and mandrel at their junction, and other factors. This may lead to unequal material compaction and shrinkage variations
- Abrupt changes in the cross section of the part being pressed, which may lead to powder movement during pressing, and cause necking or lamination (Ref 11)
- Mold size and shape, because molds that have thin walls or are unusually large may require special provisions for their support (Ref 22), as do long, thin mandrels
- Mold hardness, because a low hardness will give a poor as-pressed surface finish. Harder molds improve the surface finish, but may cause damage to the compact due to unequal rates of recovery when the pressing pressure is released (Ref 11). A good starting point would be 55 to 60% of Shore A hardness (Ref 9)

Equipment

A wide variety of isostatic pressing equipment is available from vendors throughout the world. This includes equipment for one-at-a-time laboratory use to very high volume units. Equipment can also be built to order. There are three main factors in choosing a piece of equipment.

The first is size and shape of the component to be pressed, because its size determines the cavity size. However, a single wet-bag cavity can hold more than one tool set, and many dry-bag presses have more than one cavity. The shape of the part will determine whether dry-bag pressing is an option.

The second factor is the amount of required pressure. Although most parts are pressed at pressures below 205 MPa (30 ksi), higher pressures are needed occasionally. Dry-bag presses rarely exceed 275 MPa (40 ksi) (Ref 10), and highly automated, multiple cavity presses usually have a maximum pressure capability of 70 MPa (10 ksi) or less. Wet-bag presses capable of 690 MPa (100 ksi) and higher are available.

The third factor is the required production rate. This determines the degree of automation needed and whether multiple tool sets in a wet-bag press or multiple cavities in a dry-bag press are required.

A typical cold isostatic press consists of the following subsystems (Ref 23): pressure vessel, closure with the opening and closing mechanism, reservoir(s) with the attached filtering system, high-pressure generator, depressurization system, fluid transfer system, control system, and tooling. Each equipment vendor has an individual approach to designing a complete cold isostatic pressing system. Most vendors can either offer standard units or design a unit to suit specific needs. Safety considerations are a primary design parameter and are carefully taken into account by the equipment manufacturer. However, the user has the responsibility to follow all the recommendations of the manufacturer as to proper use and maintenance.

The smallest, least automated units are typically used for laboratory development work and low-volume production of small parts. Figure 3 shows a small wet-bag press with an internal diameter of 150 mm (6 in.) and a depth of 460 mm (18 in.).

Figure 4 shows a semiautomatic dry-bag press capable of producing components approximately 75 mm (3 in.) in diameter by 430 mm (17 in.) long at pressures up to 240 MPa (35 ksi). Cycle times are 2 to 5 min, and a tool change takes 2 to 3 min (Ref 24).

Large wet-bag presses are available with cavity diameters of over 1.8 m (6 ft) and lengths up to 3.7 m (12 ft). Pressures of up to 1380 MPa (200 ksi) are also available. This type of unit is available in varying degrees of automation. A schematic of an automated wet-bag setup is depicted in Fig 5.

A highly automated dry-bag press is shown in Fig 6. Pressures up to about 275 MPa (40 ksi) are available in dry-bag presses. Production rates up to 3600 parts/h can be achieved with a six-cavity press.

In addition to these standard types of units, specialized units that utilize the basic principles of isostatic pressing to produce very specific shapes are available. One example is the continuous pressing of tubes in the manner of an extruder (Ref 25). Another example is the isostatic pressing of tableware (Ref 17).

Another variant of isostatic pressing is the use of temperatures ranging from 100 to 300 °C (212 to 570 °F) (Ref 26, 27). This "warm isostatic pressing" is useful for polymers and

Fig 4 Semiautomatic dry-bag isostatic press. Courtesy of Simac Limited

similar materials, but may also find applications for ceramics.

Applications

Cold isostatic pressing is used commercially to form a diversity of components. It is generally used when one of its unique attributes represents an advantage. Components as different as spark plug insulators (Ref 18, 22, 28), very large refractory blocks (Ref 29–32), dinnerware (Ref 17, 33, 34), and radomes (Ref 17) are produced by this process. A number of examples are described below to illustrate the benefits of isostatic pressing, when used appropriately.

As already noted, spark plug insulators are probably the highest-volume application of isostatic pressing. Dry-bag isostatic pressing is a proven, low-cost, and high-volume process for forming insulator blanks, which are then green machined to their final shape by contour grinding. Production rates for one multicavity press are reported to be as high as 3,000 parts/h (Ref 10).

A similar pressing setup has been used to produce grinding balls at rates of up to 10,000 balls/h depending on their size (Ref 28).

The production of translucent alumina tubes for high-pressure sodium vapor lamps is another example of the use of isostatic pressing, although it should be noted that one major manufacturer uses an extrusion process to form essentially the same product (Ref 5). This is

Fig 3 Small laboratory- or pilot-scale wet-bag isostatic press

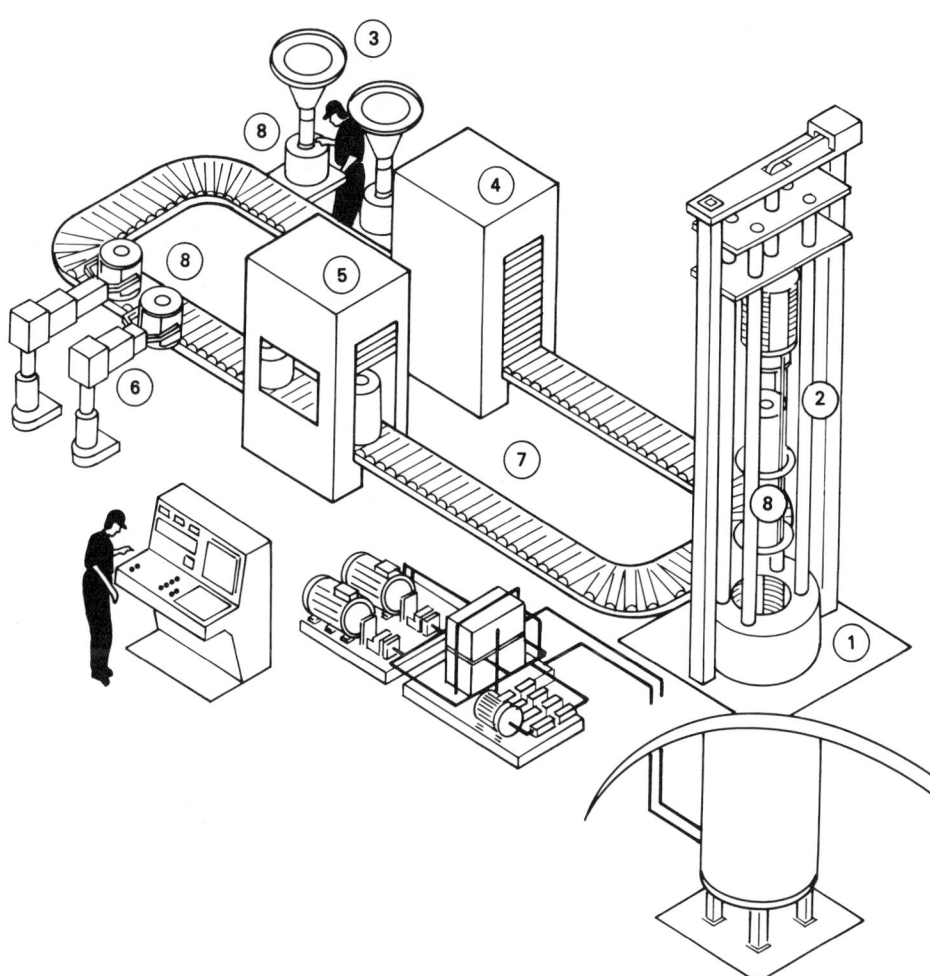

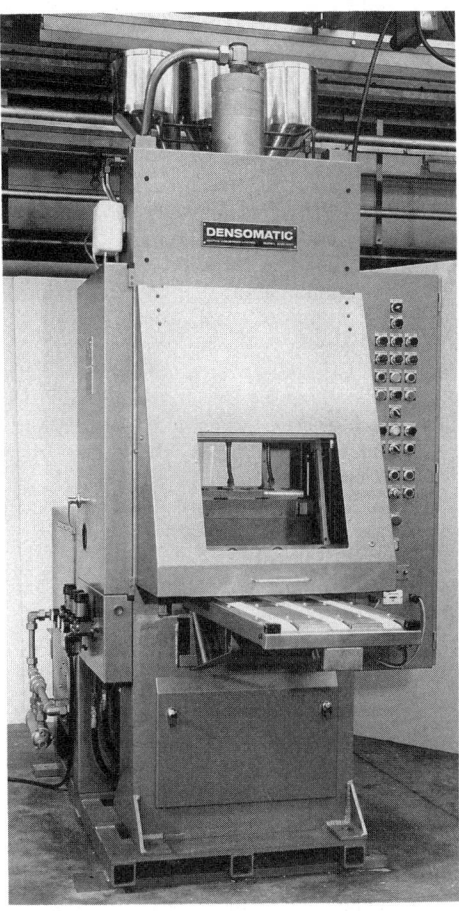

Fig 6 Automated dry-bag isostatic press. Courtesy of Simac Limited

Fig 5 Automated wet-bag isostatic pressing line. 1, automated wet-bag press. 2, Equipment for loading mold into high-pressure vessel. 3, Powder dosing and mold-filling equipment. 4, Washing booth. 5, Drying booth. 6, Compact removal unit. 7, Tooling conveyor. 8, Tooling

REFERENCES

1. H.D. Madden, U.S. patent No. 1,081,618, 1913
2. B.A. Jeffery, U.S. patent No. 1,863,854, 1932
3. B.A. Jeffery, U.S. patent No. 1,864,365, 1932
4. H.W. Daubenmeyer, U.S. patent No. 1,983,602, 1932
5. F.L. Kennard, Ceramic Component Fabrication, *Ceram. Eng. Sci. Proc.*, Vol 7, 1986, p 1095–1111
6. F.S. Wheeler, Comparison of Wet-Bag versus Dry-Bag Isostatic Pressing, *Ceram. Eng. Sci. Proc.*, Vol 7, 1986, p 1242–1244
7. P.W. Kaveney and J.R. Schultz, "Dry-Bag Isostatic Pressing," presented at First International Conference on Isostatic Pressing, Sept 1978
8. P.J. James, Principles of Isostatic Pressing, Chapt. 1, *Isostatic Pressing Technology*, Applied Science Publishers Ltd., London, 1983
9. K.J. Morris, Tooling for Isostatic Pressing, Chapt. 4, *Isostatic Pressing Technology*, Applied Science Publishers Ltd., London, 1983
10. P.W. Kaveney, "Dry-Bag Isostatic Press

also one of the few examples, at least one for which the volume is relatively high, where the component is essentially net shape as pressed. This is possible because of its simple shape, which is generally a hollow cylinder.

Of particular importance in successfully using isostatic pressing for this application is the use of spray drying to produce powders that have uniform additive and binder distribution and good flowability to uniformly fill the mold (Ref 13, 35, 36). The use of vibration during mold filling is also reported to be beneficial (Ref 35). Although both wet-bag and dry-bag pressing are used for this application, dry-bag pressing has evolved as the main production technique. This is due to its higher production rate capability and to its better dimensional control ability.

For very large parts, such as refractory blocks and power line insulators, isostatic pressing is a method that avoids the size limitation of dry pressing, as well as the long drying times and shrinkage problems of wet processes such as extrusion, slip casting, ramming, and others. For large insulators, cold isostatic pressing has been reported to have reduced the overall manufacturing time by 75%. Refractory blocks as large as 1 Mg (2200 lb) have been isostatically pressed (Ref 30). Other reported advantages of isostatic pressing are lower capital and tooling costs (Ref 19) although this appears to be true only in the case of relatively high volumes (Ref 30).

Isostatic pressing has also been developed as a process for forming tableware. The claimed advantages of this process, versus a more traditional process, include high production rates, high quality, low tooling costs, and even savings in refractories and firing due to closer loading. The major drawback is the high capital investment required, both for the isostatic pressing equipment, and the powder processing equipment, such as a spray dryer.

Equipment," presented at First International Conference on Isostatic Pressing, Sept 1978

11. B.J. McEntire, Tooling Design for Wet-Bag Isostatic Pressing, *Ceram. Eng. Sci. Proc.*, Vol 9, 1984, p 16–31

12. W.D. Kingery, Hydrostatic Molding, Chapt. 8, *Ceramic Fabrication Processes*, John Wiley & Sons, 1958

13. G.A. Fryburg and F.B. Makar, Improved Fabrication of Al$_2$O$_3$ Tubing for Sodium Vapor Lamps, *Ceram. Eng. Sci. Proc.*, Vol 9, 1984, p 32–37

14. R.M. Gill and J. Byrne, Application of Isostatic Pressing Techniques to the Production of Dense Ceramic Bodies, Chapt. 7, *Science of Ceramics*, Vol 4, The British Ceramic Society, 1968

15. D.W. Richerson, Shape Forming Processes, Chapt. 6, *Modern Ceramic Engineering*, Marcel Dekker, 1982

16. R.J. Street and J.H. Duncan, "Thin Walled Beta-Alumina Electrolyte Tubes," presented at First International Conference on Isostatic Pressing, Sept 1978

17. R. Swindells and G. Deplace, Cold Isostatic Pressing Applications, Chapt. 6, *Isostatic Pressing Technology*, Applied Science Publishers Ltd., London, 1983

18. D.M. Baron and J.R. Perry, The Fabrication of Spark Plugs, *Fabrication Science 3, Proc. Brit. Ceram. Society*, No. 33, 1983, p 79–87

19. P. Popper, Tooling, Chapt. 4, *Isostatic Pressing*, Heyden Press, 1976, p 33–43

20. C. Deplace and F. Cornil, "Production Application for Isostatic Pressing and Related Mold Technology," presented at First International Conference on Isostatic Pressing, Sept 1978

21. P. Popper, I.K. Bloor, R.D. Brett, and D.E. Lloyd, Pressatura Isostatica con Particolare Riferimento Agli Stampi, *Ceramurgia*, Vol 9, 1979, p 113–124

22. D.B. Quinn, R.E. Bedford, and F.L. Kennard, Dry-Bag Isostatic Pressing and Contour Grinding of Technical Ceramics, *Ceram. Eng. Sci. Proc.*, Vol 9, p 198

23. E.L.J. Papen, Cold Isostatic Pressing—Processes and Equipment, Chapt. 3, *Isostatic Pressing Technology*, Applied Science Publishers Ltd., London, 1983

24. K.J.A. Brookes, SIMAC Bangs the Drum for Dry-Bag Pressing, *Met. Powder Rep.*, Vol 42, 1987, p 619–622

25. H. Ittner, Cold Isostatic Pressing with the RTS System, *Ceram. Eng. Sci. Proc.*, Vol 7, 1986, p 1391–1405

26. T. Nishimoto, Y. Kishi, and T. Naoi, Recent Trends in Cold and Warm Isostatic Pressing Equipment, *Met. Powder Rep.*, Vol 42, 1987, p 614–616

27. T. Naoi and Y. Narukawa, Recent Trends in CIP and HIP Equipment, *Kobelco Tech. Review*, Vol 2, 1987, p 23–27

28. E.L.J. Papen, Automated Isostatic Pressing, Chapt. 6, *Special Ceramics 4*, The British Ceramic Research Association, 1968, p 69–81

29. A. Traff and P. Skotte, Isostatic Pressing of Ceramic Materials, Methods and Trends, *Powder Metall. Int.*, Vol 8, 1976, p 65–68

30. T.M. Wehrenberg, E.R. Begley, and R.F. Patrick, Isostatic Pressing Large Refractory Blocks, *Ceram. Bull.*, Vol 47, 1968, p 642–645

31. E.F. Cooper, D.T. Dorrit, and B.G. Newland, "Low Cost Approach to Isostatic Pressing of Refractories and Industrial Ceramics," presented at Second International Conference on Isostatic Pressing, Sept 1982

32. A. Traff, "Isostatic Press Equipment and Isostatic Pressing of Metallic and Non-Metallic Materials," presented at First International Conference on Isostatic Pressing, Sept 1978

33. P. Popper, Applications, Chapt. 5, *Isostatic Pressing*, Heyden Press, 1976, p 44–52

34. The Future of Forming: What to Expect in the '90s, *Ceram. Ind.*, May 1990, p 19–21

35. P. Hing, "Use of Cold Isostatic Pressing for the Fabrication of Some Lamp Ceramic Materials," presented at Second International Conference on Isostatic Pressing, Sept 1982

36. G.A. Fryburg and F.B. Makar, Productivity Effect of Industrial Processing Techniques for Al$_2$O$_3$ Tube Manufacturing, *Advances in Ceramics*, Vol 11, The American Ceramic Society, 1983, p 11–18

Slip Casting

C.H. Schilling, Pacific Northwest Laboratory*
I.A. Aksay, Department of Materials Science and Engineering; and Advanced Materials Technology Center, Washington Technology Centers, University of Washington

SLIP CASTING is an economical process that is widely used to produce complex shapes from a broad range of ceramic-based materials (Ref 1–8). Applications are numerous, including artware, chinaware, sanitaryware, crucibles, filter media, structural tubing, bone implants, and heat engine components.

A slip is a suspension of colloidal powders in an immiscible liquid (usually water). Slip casting entails pumping or pouring the slip into a permeable mold (usually gypsum). Capillary suction of the mold causes the liquid to be filtered from the suspending medium and a densely packed layer of particles to be deposited against the mold wall. This process is shown in Fig 1.

Several slip casting variations are used, depending on the application requirements. Drain casting, the most common process, involves casting for a limited time period to deposit a layer of the desired thickness. The remaining slip is subsequently drained from the mold.

Another common process is solid casting, which is identical to drain casting except that slip is continually added until a solid cast is made. Pressure filtration entails pressurizing the slip to increase the casting rate.

A less common approach is vacuum casting, in which a vacuum is applied to the outside of a mold. This process can be used to increase casting rates with either the drain or solid casting approach.

Centrifugal casting is a recently developed process in which the spinning of a mold produces large forces that serve to increase consolidation rates (Ref 9, 10). These molds are not necessarily porous and can be fabricated from various metals and plastics.

The production of highly complex shapes can be accomplished by slip casting, using insoluble mandrels attached to a mold. One example is fugitive wax slip casting, which is based on the much older technology of investment casting (Ref 5, 11).

Surface Chemistry and Rheology

Particle surface chemistry is a dominant factor in the slip casting process because it regulates interparticle attraction and repulsion forces, which, in turn, have significant effects on slip rheology, casting rate, and microstructure evolution (Ref 12–16). The control of surface chemistry while preparing low-viscosity suspensions that are highly concentrated with powders is a major requirement of slip casting. Low viscosities (<1 Pa · s) are needed for pourability, and high solids concentrations (up to 60 vol% solids) are needed to maximize the casting rate and "green" density. Traditional methods for achieving low-viscosity, concentrated suspensions entail the use of polydisperse particle size systems and repulsive interparticle forces that arise from surface-adsorbed ions or polymers (Ref 13). Recently, it has been shown that low-viscosity, highly concentrated suspensions can also be formed by the use of attractive forces between particles that are coated with lubricating polymers (Ref 13, 17).

Regulating surface forces by altering the slip chemistry is also fundamental to eliminating cast layer heterogeneities such as large, isolated pores; size-segregation of particles; and spatial variations of packing density. Each of these defects can create fracture origins in fully sintered ceramics and can cause cracks to form during drying and sintering because of nonuniform shrinkage stresses. The basic relationships that exist in surface chemistry, surface forces, and slip rheology are introduced in this Section. The effects of process parameters on the casting rate are described in the section "Process Considerations" in this article.

Surface Forces and Rheology. A prevailing trend in ceramic processing is the development of smaller and smaller powders (≤1 μm) in order both to enhance sintering rates and to reduce the size scale for mixing uniformity in powder blends. For example, slip casting of advanced composites using very fine particles can produce unique properties resulting from the submicrometer-scale architecture of the component phases (Ref 6, 18–21). A concern with the use of dry, submicrometer-sized powders is spontaneous agglomeration resulting from the coupling of electronic oscillations of the atoms in adjacent particles (van der Waals-London attraction). Ceramic particles generally contain atoms that are easily polarizable, thus creating a strong interparticle attraction. Agglomerates defeat the purpose of reducing the scale of mixing. They also can produce spatial variations of packing density that lead to cracking and impaired mechanical properties.

A preferred approach for eliminating agglomerates is to establish interparticle repulsion forces by mixing powders in a liquid with dissolved ions or polymers, which then adsorb to the particle surfaces (Ref 3–7, 16–26). The former technique (electrostatic stabilization) is most commonly used in slip casting and entails spontaneous formation of a "diffuse double layer" of dissolved ions surrounding each particle due to the presence of oppositely charged surface sites. The latter method involves adding either neutral (steric stabilization) or charged (electrosteric stabilization) polymers to a suspension.

A broad range of rheological properties and slip cast microstructures can be produced from a given powder using the above stabilization techniques. In general, interparticle repulsion results in dispersed slurries that exhibit Newtonian rheology (viscosity is independent of shear rate) at low particle concentrations. At high solids concentrations, the slurries become dilatant (viscosity increases with shear rate), because the apparent volume must increase upon shearing to allow closely spaced particles to slip past one another. Dispersed

*Pacific Northwest Laboratory is operated by Battelle Memorial Institute for the U.S. Department of Energy under Contract DE-AC06–76RLO 1830

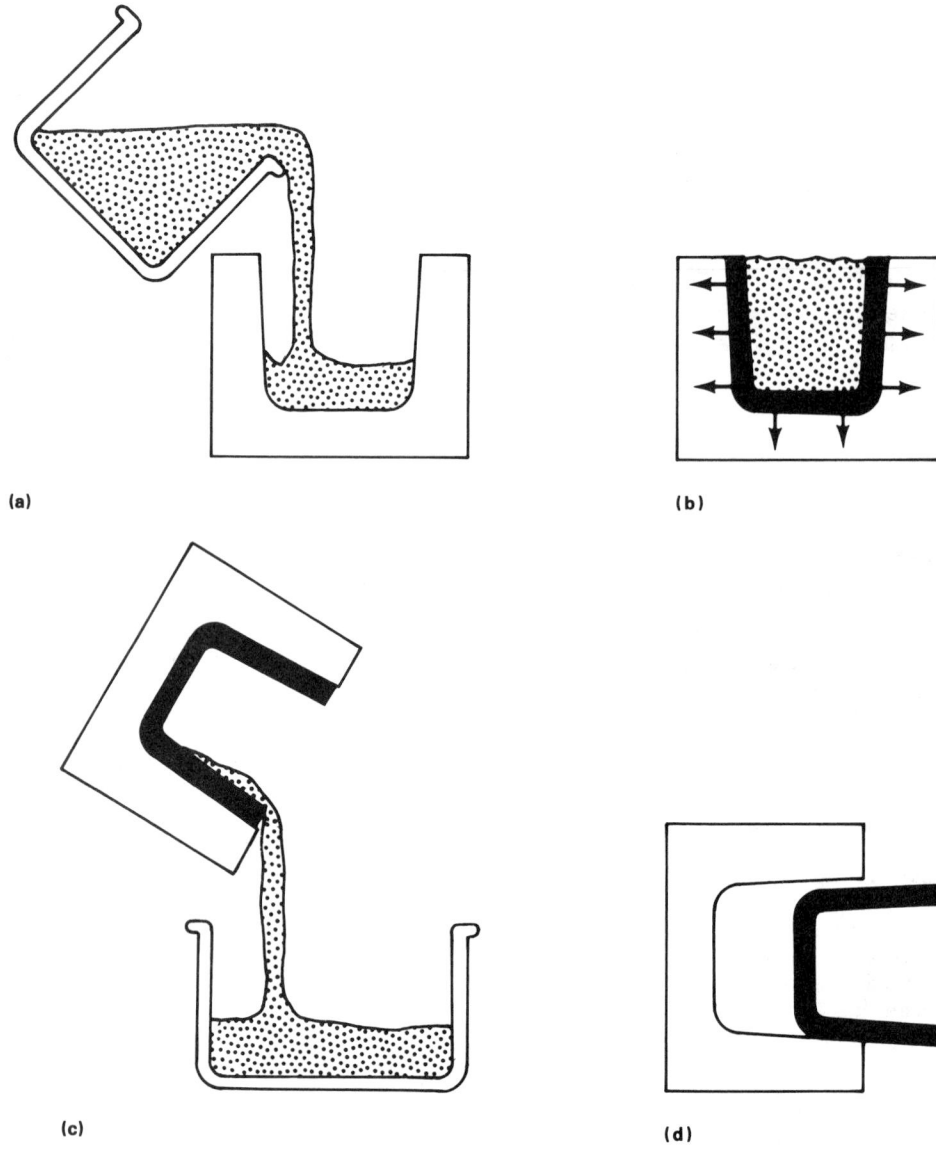

Fig 1 Drain casting illustration. (a) Permeable mold is filled with slip. (b) Liquid is extracted from the mold, while forming compacts along mold walls. (c) Excess slip is drained. (d) Casting is removed after partial drying. Source: Ref 5

slips generally result in the slow casting of densely packed microstructures that have low permeability. In contrast, attractive interparticle forces result in flocculated slips that become exceedingly viscous above 10 to 20 vol% solids unless lubricating adlayers are present (Ref 17). In general, flocculated slips exhibit pseudoplastic, thixotropic rheology (viscosity decreases with increasing shear rate). Casting rates are usually rapid, because the microstructures are typically porous and permeable. After stirring, these slips may exhibit increased viscosity as a result of time-dependent agglomeration (rheopexy).

Electrostatic Stabilization. When suspended in aqueous media, particles carry a surface charge that is balanced by counterions in solution. With most clays, the surface charge arises from a valence deficiency resulting from isomorphous substitution in a crystal lattice (Ref 8). With nonclay materials, the surface charge results from the ionization of surface moieties. For example, oxides will possess surface charges that are determined by the following surface hydroxyl reactions:

$$M—OH^0 + H^+ = M—OH_2^+ \qquad \text{(Eq 1)}$$

$$M—OH^0 = M—O^- + H^+ \qquad \text{(Eq 2)}$$

where M denotes a surface cation, such as Al or Si (Ref 27–29). The net surface charge will depend on the solution pH, which describes the relative concentrations of potential-determining ions (H^+ and OH^-) in solution. Based on the above reactions, a net positive or negative charge occurs at low pH and high pH, respectively; the isoelectric point is the pH at which the net surface charge is approximately zero. Potentiometric titration is used to mea-

sure the concentration of potential-determining ions adsorbed to particle surfaces (Ref 30).

At a given pH, the net surface charge also depends on the valence and population density of the surface cations that are bonded to the hydroxyls. Based on this consideration, Parks and de Bruyn (Ref 27) reported an empirical relationship for estimating the isoelectric point of oxide ceramics. This relationship states that decreases in the isoelectric point are linearly proportional to the ratio Z/R, where Z and R equal the cation valence and radius, respectively (in general, the smaller the cation radius, the smaller the number of surrounding oxygen atoms and the greater the number of cation sites per unit area). Extensive tabulations of isoelectric points are reported by Parks and de Bruyn (Ref 27) and James (Ref 31).

There is no good quantitative theory to describe the dispersion and rheology of concentrated suspensions; however, there are well-developed theories on the surface forces between two particles (Ref 23, 25, 28–33). The double-layer theory describes the electrostatic repulsion between two particles that results from overlapping their electrical double layers. This overlap causes a local increase in the electrolyte concentration, giving rise to an osmotic pressure that prevents the particles from being brought together by the van der Waals force. The theory states that the net potential energy (pair potential) between two spheres of the same diameter is approximately equal to the algebraic sum of the van der Waals attractive potential, U_A, and the electrostatic repulsive potential, U_R. The U_A depends on the particle radius, a, and the separation distance between particle centers, r, and the Hamaker constant, A, which is an intrinsic measure of the dipole attraction for the particular solid-liquid system (Ref 23). The U_R depends on additional factors, including the dielectric constant of the liquid, ϵ_r, the surface potential, ψ_0, and the double-layer thickness, κ^{-1} (that is, the distance from the particle surface at which $U_R \approx \psi_0/2.7183$):

$$U_A = \frac{-Aa}{12(r - 2a)} \qquad \text{(Eq 3)}$$

$$U_R = 2\,\epsilon_r a\psi_0^2 \ln[1 + \exp - \{\kappa(r - 2a)\}] \qquad \text{(Eq 4)}$$

$$\kappa^{-1} = \frac{\epsilon_r\epsilon_0 kT}{F^2\Sigma(N_i Z_i^2)} \qquad \text{(Eq 5)}$$

In these expressions, T is temperature, F is the Faraday constant, k is the Boltzmann constant, ϵ_0 is the permittivity of vacuum, and Z_i and N_i are equal, respectively, to the valence and concentration of the dissolved ions in solution.

An important consequence of the double-layer theory is that interparticle repulsion is increased by increasing κ^{-1}, which in turn increases ϵ_r and decreases both Z_i and N_i. Another significant aspect of this theory is that the repulsion energy increases rapidly as a function of ψ_0^2. Although it is impossible to directly measure ψ_0, it is approximately equal to the zeta potential, ζ, which in turn is equal to the potential at the liquid shear plane and is determined by electrophoretic mobility measurements (Ref 28, 29). The zeta potential depends on the net surface charge and hence the chemical equilibria expressed above. As with the surface charge, ζ is zero at the isoelectric point and reaches maximum and minimum values under more acidic and basic conditions, respectively.

An important tool for interpreting rheological behavior is the double-layer interaction diagram (Fig 2), which plots U_A and U_R as a function of the separation between spherical particles. This diagram shows that for a given particle-fluid system there is a critical zeta potential and range of double-layer thicknesses for which U_R exceeds U_A, producing an energy barrier to agglomeration. The kinetic energy of colloidal particles from Brownian motion is of the order of 10 kT, and at 20 °C (68 °F), an energy barrier corresponding to a zeta potential of approximately 25 mV is needed to minimize agglomeration (Ref 7). A larger barrier is needed to retard agglomeration during pouring and mixing processes, because they produce a greater kinetic energy. A secondary minimum in the pair potential may also occur for some systems in which the separation is of the order of the particle size. The primary minimum shown in Fig 2 is typical of all systems and occurs when the separation approaches molecular dimensions.

Polymeric Stabilization. Adsorption of nonionic surfactants or polymers may be used to produce repulsive interparticle forces. When particles approach each other closely, the adlayers overlap, resulting in an osmotic pressure that pushes the particles apart (steric hindrance). Charged macromolecules, that is, polyelectrolytes, can produce both repulsive electrostatic and steric forces. Theoretical treatments describing the pair potential as a function of adlayer properties and interparti-

cle separation distance have been reported by several authors (Ref 23, 26, 28, 29, 32, 33). In general, these models predict steep increases in repulsion as the adlayers of two particles begin to contact one another. Although the science of interparticle forces has a strong theoretical base verified through direct surface force measurements (Ref 25), the choice of the best polymer-solvent system is still a matter of trial and error for most ceramic systems. Typical dispersants include polyethylene oxides, polyacrylic acids, polymethacrylic acids, polyvinyl alcohols, polyvinyl butyrals, polyacrylamides, alginates, and cellulose-based polymers.

An important requirement for dispersing particles and achieving high packing densities is to maximize the quantity of surface-adsorbed polymer. With low adsorption densities, steric repulsion may be insufficient to prevent close approach of particles and subsequent agglomeration. In addition, upon close approach of particles, polymer segments extending from one particle surface may attach to available surface sites on another particle, causing bridging flocculation and the formation of low-density casts (Ref 34). Polymer and solvent molecules can also compete for surface adsorption sites; therefore, the polymer adsorption densities will depend on the particular chemical equilibria for a given system. In oxide powders, polymer adsorption densities may also depend on the amount of surface hydroxyl groups, which in turn varies with powder heat treatment (Ref 34). Adsorption densities are experimentally determined by photon absorption techniques (adsorption isotherms) (Ref 23). A simple means to analyze polymer adsorption effects on compaction entails the measurement of sediment densities as a function of polymer chemistry and concentration.

Stabilization is also promoted when surface-adsorbed polymers are well solvated, causing the "loop" and "tail" segments of a given polymer to extend from the particle surface into the surrounding liquid. In poor solvents, polymer segments are self-attracting, which leads to less-extended configurations and bridging flocculation. A simple way to assess solvent quality is through viscosity measurements. In good-quality solvents, polymer chains are extended and the larger radii of gyration result in higher intrinsic viscosities. The opposite trend is observed with poorer solvents, because the polymer chains tend to coil up. Poor solvation or insufficient surface adsorption can lead to premature flocculation as the solids concentration is increased during slip casting. In turn, low green densities may result (Ref 34).

An important aspect of electrosteric repulsion concerns the dissociation behavior of polyelectrolytes and the resulting effects on suspension viscosity (Ref 35). As an example, polymethacrylic acid and polyacrylic acid are commonly used to produce aqueous slips with high solids loading and low viscosity.

These linear polymers consist of carboxylic acid pendants (that is, COOH groups) that are attached to a C—C backbone. Depending on the solvent condition (that is, pH and ionic strength) the fraction of these pendants that is dissociated (that is, COO^-) will vary. At high pH levels, the fraction dissociated increases, and the polymers assume an extended configuration due to the electrostatic repulsion between COO^- groups; at low pH levels, the fraction dissociated decreases and the chains coil up (Ref 35).

An important consequence of this result is that a minimum viscosity occurs at a pH where the polyelectrolyte is fully dissociated (Ref 35). In addition, as the pH is raised, the amount of surface adsorption required to prevent flocculation (that is, the saturation adsorption) significantly decreases because of the greater amount of charge per molecule. The slip viscosity is also influenced by the quantity of dissolved ions in the solvent. It is important to minimize the ionic strength in order to raise the zeta potential and hence the interparticle repulsion. Finally, it is important to minimize solvent viscosity in highly concentrated suspensions in which the solvent volume is relatively low. Solvent viscosity increases with the amount of nonadsorbed polyelectrolyte (which depends on pH) and its molecular weight (Ref 35).

Colloidal Phase Equilibria. Equilibrium phase diagrams for colloidal suspensions that are also useful for interpreting rheological and slip casting behavior in terms of surface forces have recently been developed (Fig 3) (Ref 12–15, 36, 37). These diagrams are similar to the phase diagrams of atomic systems, in which the equilibria between gas, solid, and liquid states are regulated by principles of free-energy minimization. For example, in a one-component atomic system, the free energy is minimized by varying the temperature and the number of atoms per unit volume. Phase equilibria for colloidal systems can be analyzed in an analogous manner, although these phase transitions take place isothermally and free energy is determined by varying any other parameter that changes the pair potential (such as zeta potential or particle size).

For example, in the diagram of Fig 3, the stability regions of dispersed and flocculated states in a one-component colloidal system (that is, monosize, single-phase particulates) are calculated as a function of the volume fraction of solids, c, and a reduced temperature, kT/E, where k is the Boltzmann constant, T is the temperature, and E is the binding energy between two particles (particle repulsion increases with kT/E). Here, the colloidal fluid refers to the state of a dispersed slip, which has a repulsive barrier to agglomeration and thus generally displays a low viscosity. The colloidal solid refers to the cast state, which displays a significantly higher viscosity and which is subsequently transformed into a denser component through sintering.

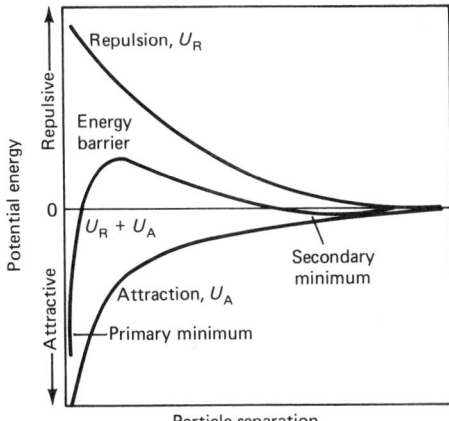

Fig 2 Interparticle potential energy as a function of separation distance resulting from electrostatic stabilization. Source: Ref 7

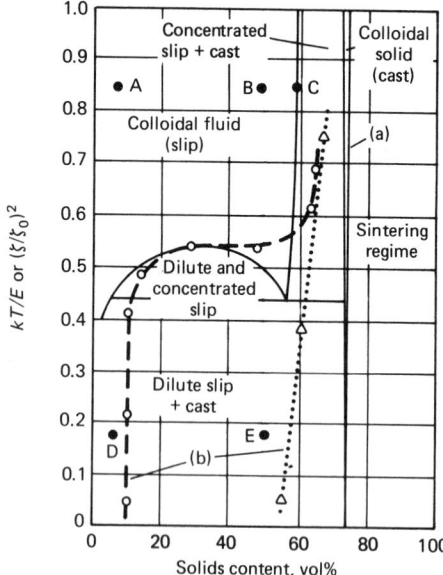

Fig 3 (a) Phase diagram (solid lines) for a colloidal system of only one type of particle. Points A through E are defined in text. In electrostatically interacting systems, the reduced temperature scale is approximately equal to $(\zeta/\zeta_0)^2$, where ζ is the zeta potential and ζ_0 is a normalization constant. Source: Ref 36. (b) Experiments with Al_2O_3 powders indicate that sedimentation (dashed line) and slip cast (dotted line) densities are lower than the predicted dense packing value of 74% as a result of hierarchically sized pores in the samples. Source: Ref 15

An important aspect of this diagram is that two distinct regions are displayed at potentials above and below $kT/E \approx 0.5$. Above $kT/E \approx 0.5$, particles repel one another, and a slip shows Newtonian behavior at low solids concentrations (point A). This Newtonian behavior resembles the ideal gas-like behavior of atomic systems. As the particle concentration increases (for example, during slip casting), a continuous transition from Newtonian (point A) to thixotropic (point B) behavior is experimentally observed. This continuous transition can be interpreted as being analogous to the gas-to-liquid transitions of atomic systems. Finally, beyond a critical particle concentration (point C), dense-packed agglomerates are nucleated from single particles. If these particles are spherical, these agglomerates display crystal-like structures and are referred to as colloidal crystals. Because these liquid-like to crystal-like transitions are first order, a two-phase coexistence of the colloidal liquid and crystal structures is expected. This behavior has been observed in sedimentation experiments in which the boundary between the sediment and the suspension corresponds to this phase transition boundary (Ref 13, 36). Similar transitions are believed to occur during slip casting of highly dispersed suspensions.

In contrast, below $kT/E \approx 0.5$, interparticle attraction predominates, and fluid-like states transform to condensed states through first-order transitions at the calculated phase boundary. In systems where E is sufficiently

low, the energy barrier in Fig 2 diminishes altogether and particles form aggregated networks that exhibit viscoelastic properties with a well-defined yield point (that is, Binghamplastic flow behavior). In this case, slip casting consolidation from point D to point E in Fig 3 entails plastic collapse of these aggregated networks.

Mechanics of Slip Casting

The pressure at the gypsum-cast boundary is approximately equal to the capillary suction pressure, P, and is approximated by the Laplace equation: $P = S \sigma \cos \gamma$, where S is the specific surface area of gypsum, σ is the surface tension of water, and γ is the contact angle ($\cos \gamma = 1$ as gypsum is completely wetted by water). Capillary suction for water in gypsum varies between 0.03 and 0.1 MPa (4.5 and 14.5 psi) (Ref 38, 39).

Slip casting is generally limited to the production of thin-walled articles (≈ 15 mm, or 0.6 in., maximum thickness) because of casting rate limitations imposed by the hydraulic resistances of the cast and mold. The effects of process parameters on casting rates are revealed using a filtration kinetics model, which is based on an incompressible cake having a uniform cross-sectional area as shown in Fig 4 (Ref 40–42). The model predicts that thickening rates diminish as a parabolic function of time:

$$\xi_c = \frac{\xi_s^0 - \xi_s}{n}$$

$$= \frac{-\epsilon_m \xi_m}{n} = \left[\frac{2Pt}{n\eta\epsilon_m(n\alpha_m + \alpha_c)} \right]^{1/2} \quad (Eq\ 6)$$

$$n = \frac{1 - c - \epsilon_c}{c} \quad (Eq\ 7)$$

In these expressions, ξ_c is the cast layer thickness, t is time, P is the total pressure effecting filtration, ξ_s^0 is the position of the suspension-air interface at $t = 0$, ξ_s is the position of the suspension-air interface during filtration, ξ_m is the position of the filtrate-air interface in the mold, η is viscosity of the suspending medium, c is the volume fraction of solids in the slip, and n is a mass balance factor. The values ϵ_m and ϵ_c are the void ratios of the mold and cast, respectively, whereas α_m and α_c are the specific hydraulic resistances (reciprocal permeabilities) of mold and cast, respectively. Equations 6 and 7 reveal that increases in casting rate result from decreasing α_m and α_c as well as increasing P, c, ϵ_m, and ϵ_c. Gypsum molds typically have large porosities, rendering small values of α_m. The total flow resistance is usually dominated by much larger values of α_c, which result from the high packing densities of typical casts. Tiller and Tsai (Ref 43) extended the above model for compressible cakes, and Hampton et al. (Ref 44) modified the approach to consider transportation of fine particles to the bottom of a cake.

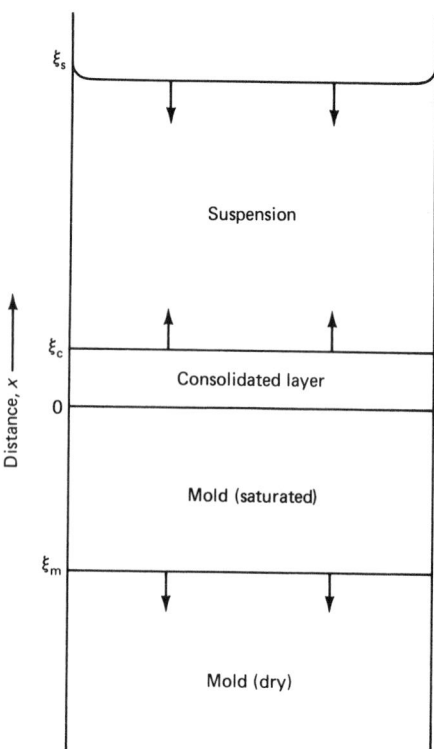

Fig 4 Model slip casting. Source: Ref 42

Useful information on the homogeneity of a cast layer can be obtained by correlating Eq 6 with laboratory measurements of filtration kinetics (Ref 14, 40, 42, 45). Typical procedures entail casting a suspension of known c in an apparatus having a constant cross-sectional area; measurements of ξ_c or $(\xi_s^0 - \xi_s)$ are plotted as a function of $t^{1/2}$. A strictly linear plot is desirable, because it indicates constancy of the microstructure-related parameters in Eq 6 and 7 and thereby suggests uniformity of the cast-layer microstructure. Possible causes of deviations from linearity and thus homogeneity are associated with either the presence of large flow units (that is, hard agglomerates) that could settle to the bottom layer as sediment or with changes in α_c that may result from time-dependent flocculation of a suspension. In addition, packing densities may vary across a given cast layer due to spatial variations of the effective stress (that is, the stress supported by the particle network) (Ref 43, 46). Effective stresses are typically lowest at the top of a cast layer and increase to maximum values at the bottom of the cast layer as a result of cumulative drag forces. For flocculated slips, this can result in the localized compaction of a high-density, low-permeability layer at the mold-cast layer interface (Ref 43, 46).

Process Considerations

Slip Control. Constancy of slip and mold properties is essential for achieving property reliability (Ref 8). Critical parameters include rheological properties, settling rate, casting rate, drain properties, drying shrinkage, mold-release properties, and green strength. The slip must be easily reproduced and should be insensitive to slight variations in solids content, chemical composition, mixing conditions, and storage time. The viscosity must be low enough to allow complete filling of the mold, but the solids content must be high enough to achieve a reasonable casting rate. Empirical methods are traditionally used to determine optimum conditions. The slip must also be free of air bubbles (or chemical reactions producing bubbles) that may become critical defects in the sintered part. Slips are usually aged to keep properties consistent, and then final viscosity measurements are performed before deairing and casting.

Organic polymers are common slip ingredients that serve in a number of functions, such as dispersants, viscosity modifiers, plasticizers, and binders (in reference to their utility for improving green-body strength). Aqueous-suspending media are typically used to dissolve these polymers, although various organic solvents can also be used (such as alkanes or alcohols).

Binders are usually 10,000 to 30,000 molecular weight polymers such as acrylics (for example, polyacrylic acid or polymethacrylic acid), vinyls (for example, polyvinyl alcohol), alginates, and various cellulose-based polymers. Plasticizers are typically small- to medium-sized molecules that decrease cross-linking among binder molecules and make castings more pliable (for example, polyethylene glycol or butyl benzyl phthalate). Extensive reviews of organic polymers and their effects on slurry rheology are presented by several authors (Ref 4, 7, 8).

A concern with the use of organic polymers is that they are transported into the capillaries of the molds and may be difficult to remove. This can change the mold permeability, suction behavior, and shorten the life of the mold. The use of organic chemicals may also pose health hazards. There are also concerns associated with the burn-out of organics, such as the production of toxic gases and the formation of carbonaceous residues in sintered microstructures (Ref 6).

It is important to control the particle size distribution in order to maximize the packing density of a cast layer and thereby minimize dimensional shrinkages during drying and sintering (Ref 4, 7, 8). Several packing models are available to aid in formulating mixtures of different-sized particles to achieve maximum packing efficiency (Ref 7, 8). Various size fractionation methods can be used to alter the particle size distribution, such as sieving, air classification, sedimentation, and centrifugation. In addition, comminution techniques (for example, milling or crushing) are used to pulverize agglomerates and reduce the average particle size. Particle size distribution measurements are easily measured by a variety of methods, including sieving, optical

and electron microscopy, x-ray absorption, light scattering, light intensity fluctuation, and electrical sensing (Ref 7, 8).

Another reason for controlling particle size distribution is to eliminate microstructure heterogeneities, such as large particles and hard agglomerates. For example, the sedimentation of heavy particles or hard agglomerates during casting can result in thickness and density variations, which may lead to defects in greenware and cracking during drying or sintering. It is desirable to prepare slips with repulsive interparticle forces to break apart weak agglomerates, fractionate inclusions greater than a given size, and mix different fractionated powders. Once fractionated and mixed, dispersed suspensions can be subsequently cast or the interparticle forces can be made attractive in order to form a low-density, deformable network that prevents mass segregation. Another approach to minimize segregation of heavy particles or agglomerates is to increase the solids content of a dispersed slip, which increases the casting rate, as discussed in the section "Mechanics of Slip Casting" in this article.

Proper control and characterization of slip rheology is perhaps the most important parameter in slip casting (Ref 8). Rheological properties are not only dependent on interparticle forces, but also depend strongly on the particle size and concentration. In general, the viscosity increases with smaller particles and higher solids concentrations because of increased particle-particle interactions. For the same reason, reduced viscosities generally result at a given solids concentration when spherical particles are used instead of platelets or rods.

Characterization of slips generally entails measurements of shear stress and viscosity as a function of shear rate over a range of shear rates of 1 to 10 s^{-1}. An important aspect of dense-packed colloidal structures in slip-cast layers is the ability to store and dissipate energy upon deformation (viscoelasticity). A full characterization of dense-packed networks can be obtained using experimental procedures such as stress growth, forced oscillation, stress relaxation, or creep compliance. There are several excellent references on the rheology of ceramic suspensions (Ref 1, 3, 4, 7, 8, 16, 22, 33, 34, 47).

Once casting is completed, a given part begins to dry and shrink away from the mold. This shrinkage is necessary to release the part from the mold. If sticking occurs, the part can be damaged and rejected. Mold release can be aided by coating the walls of the mold with a release agent such as silicone oil, olive oil, or a dilute aqueous solution (5 wt%) of sodium alginate. However, these coatings may alter the casting rate.

The gypsum mold is a very important element in slip casting, because the capillary suction of this porous material is responsible for slip dewatering. Advantages of gypsum as a mold material are the ease of manufac-

turing into a wide variety of shapes, low cost, good dimensional accuracy, and repeatable water absorption properties. There are several excellent reviews of gypsum molds for slip casting (Ref 8, 38, 39, 48–50); key points from these reviews are mentioned below.

Processing of gypsum molds begins with mixing a plaster-water slurry and pouring it into a case mold having the same shape as the final greenware product. Case molds are typically made of gypsum, epoxy resin, or a combination of these materials. After filling the case mold, the slurry sets into a rigid shape via the chemical hydration of plaster ($CaSO_4 \cdot 1/2H_2O$) to gypsum ($CaSO_2 \cdot 2H_2O$). This process entails the nucleation and growth of needle-shaped gypsum crystals which entangle and become tightly interlocked with random orientation. The final size of the gypsum crystals is approximately 2 to 5 μm (80 to 200 $\mu in.$) in diameter by up to 20 μm (800 $\mu in.$) in length. The pore spaces among these crystals interconnect in all directions to produce irregularly shaped, continuous capillaries (≈ 0.1 μm, or 4 $\mu in.$, diameter) that are filled with excess water. The freshly set mold is separated from the case mold and the excess water is subsequently evaporated.

Theoretically, only 8.4 kg (18.6 lb) of water are required for the hydration of 45 kg (100 lb) of plaster. However, in actual practice more water is added to impart fluidity to a given slurry. Most gypsum molds are made using 30 to 36 kg (64 to 80 lb) of water per 45 kg (100 lb) of plaster, depending on the particular application. Increases in the consistency (water-to-plaster ratio) increase the pore volume of the hydrated gypsum, which in turn increases water absorption and decreases the mechanical strength. The slurry volume can expand by as much as 0.3% during setting; this expansion increases with decreases in the consistency.

Metal ion impurities in the mixing water and slips should be kept at minimum levels in order to avoid chemical reactions with gypsum that produce soluble salts (for example, sodium sulfate or magnesium sulfate). These salts migrate to the surface of the mold during drying (efflorescence), creating hard spots on the mold surface that might cause sticking of the greenware or other problems with the mold. As a general guideline, any compound having a greater solubility than gypsum (≈ 2 kg/m^3, or 0.13 lb/ft^3) can produce efflorescence (Ref 50).

Careful control of mixing procedures is needed to minimize trapped bubbles and to maintain consistent strength and water absorption. The solubility of gypsum in water has a maximum temperature range between 21 and 38 °C (70 and 100 °F). It is recommended that mixing take place in this temperature range to maintain consistent properties of molds. In general, increases in the mixing water temperature reduce the setting time and increase the compressive strength. Generally, the greater the mixing energy input (speed and duration), the greater the strength of the mold and the lesser the water absorption. For example, Whiteside (Ref 50) reported an increase in dry compressive strength from 9.7 to 14 MPa (1.4 to 2 ksi) for samples that were mixed for approximately 2.3 and 9 min, respectively. He also reported data for 80 consistency molds, where the amount of water absorption doubled after increasing the mixing time from approximately 3.5 to 7 min.

The size of a batch should also allow it to be completely poured into the molds no later than 5 min after mixing. Large batches are difficult to handle promptly and may reduce the performance of a mold. Careful pouring of the slurry is necessary to both prevent air entrapment and provide smooth surfaces. Agitation of the filled case mold can prevent air bubbles at or near the mold surface. If a mold has cracks or subsurface bubbles, air can collect in these regions, causing nonuniform suction and pinholes in the ware.

Case molds are precoated with a soap film that seals the surfaces, promotes mold release, and prevents surface erosion. Insufficient soaping of gypsum case molds can cause absorption of water from the plaster slurry. During subsequent slip casting, this region of the working mold will absorb less water than the surrounding surfaces because it has had less water to create void volume during setting. Similarly, too much soap on the case mold will cause a deposit of insoluble material on the face of the working mold, and this spot may absorb less water than the surrounding mold. Neutral or potassium soaps are more desirable than highly alkaline compounds, which are sources of sodium. The best soaps are those that react with the gypsum case mold to produce an insoluble film such as calcium stearate.

Prior to slip casting, the excess water is evaporated from the molds using drying ovens or a drying room. Optimum conditions entail the use of uniform temperatures and uniform air circulation. Molds should not be heated above 50 °C (120 °F), because gypsum will degrade from calcination. Slow cooling to room temperature after drying is also recommended to avoid thermal shock. Evaporation rates are typically rapid ($\approx$0.1 kg/m², or 0.02 lb/ft²) during the first few hours of drying and thereafter become much slower ($\approx$0.0014 kg/m², or 0.0003 lb/ft²). The compressive strength of gypsum also sharply increases during the final stages of drying (Ref 50).

It is usually necessary to bring a new mold to a proper water content in order to control the rate of water absorption. This "conditioning" process is typically performed by completely drying the mold, and then rewetting to a proper moisture content that past experience has deemed appropriate (typically 5 to 15% water). If the slip is dewatered too rapidly, regions of the cast layer that are adjacent to the mold can shrink excessively, which consequently reduces the casting rate and may result in surface cracks or premature release of the ware. If absorption is too slow, the cast layer does not build up at a satisfactory rate and is likely to be mucky and stick to the mold. An intermediate condition is required and is usually achieved by empirical methods. The amount of free water in the mold has significant effects on the rate of absorption. For example, when a dry mold is exposed to water, absorption rates are initially rapid ($\approx$4.5 kg/m², or 0.90 lb/ft²) and then parabolically decrease to lower values ($\approx$0.2 kg/m², or 0.04 lb/ft²) (Ref 39). The absorption rate of new molds increases appreciably during the first 3 cycles of immersion and drying due to the dissolution of gypsum from the pore walls (Ref 39).

Pressure Casting

Pressure casting is a process that resembles slip casting, except that the slip is pressurized in order to increase the filtration rate (Ref 1, 3, 16, 51–53). A wide range of intricate shapes can be formed with generally higher production rates than conventional slip casting. The following review is largely derived from Norton (Ref 3) and Blanchard (Ref 51).

Rapid improvements in casting rate are produced by increasing the slip pressure (Eq 6). For example, Blanchard (Ref 51) reported experiments in which the time required to cast a clay body with a thickness of 8.5 mm (0.33 in.) could be reduced from 80 to 21 min by increasing the pressure from 0.025 to 0.4 MPa (0.004 to 0.06 ksi). Increases in casting rate are also made possible by using flocculants that reduce the packing density of the cast layer and increase its permeability.

There are two main types of pressure casting equipment: medium-pressure (0.3 to 0.4 MPa, or 0.04 to 0.06 ksi) and high-pressure (up to 4 MPa, or 0.6 ksi) systems. The materials of construction and tolerances of mold fit become more expensive as the pressures increase. Medium-pressure systems are used on large, high-production articles such as sanitaryware. In this case, the molds consist of two mating halves that are made of gypsum plaster, porous plastic, or a combination of these materials. A series of up to 15 molds are then stacked together, and slip is simultaneously pumped into each mold. After the casting process is completed, the saturated, empty molds are subsequently purged with compressed air to remove the liquid occupying the pores. The entire production cycle takes approximately 75 to 90 min with a 15-mold bench.

Higher-pressure casting can become more advantageous for smaller shapes, such as dinnerware. Although the operating principles are similar, production levels may be lower than with medium-pressure systems, because the equipment typically operates with a single mold. High-strength plastic molds are needed for high-pressure casting because plaster-based molds tend to fracture. The molds are as large as 500 × 600 mm (20 by 24 in.) and can be built with either single or multiple cavities.

Pressure casting has several economic advantages relative to slip casting. For example, a major problem with slip casting is the slow filtration rate, which requires a large number of plaster molds for commercial production. Another problem that is eliminated is the drying of plaster molds, a slow process involving large amounts of floor space. Pressure casting has a shorter production cycle, which requires fewer molds and consequently less floor space and labor. Fluctuations in the quality of greenware are also reduced with pressure casting, due to machine automation.

Unlike plaster, plastic molds are not as prone to surface wear; this improves the degree of fit with mold parts, decreases the surface roughness of greenware, and increases the consistency of edges and sharp borders. Plastic molds are also less prone to chemical attack from metal ions contained in slips. Mold conditioning is also avoided, because, unlike plaster, the slip pressure is not significantly affected by variations in mold saturation. Another advantage of pressure casting over slip casting is that the greater pressures result in greater green densities and less moisture. Hence, drying and sintering shrinkages are reduced, and dimensional tolerances for near-net-shape processing are improved. One disadvantage, however, is that plastic molds are more expensive than plaster molds.

Acknowledgment

This work was supported by the U.S. Department of Energy Office of Basic Energy Sciences through a subcontract by Pacific Northwest Laboratory under Contract No. 063961-A-F1.

REFERENCES

1. W.D. Kingery, *Ceramic Fabrication Processes*, MIT Press, 1963
2. R.R. Rowlands, A Review of the Slip Casting Process, *Bull. Am. Ceram. Soc.*, Vol 45 (No. 1), 1966, p 16–19
3. F.H. Norton, *Fine Ceramics, Technology and Applications*, Krieger Publishing, 1970
4. G.Y. Onoda and L.L. Hench, *Ceramic Processing before Firing*, John Wiley & Sons, 1978
5. D.W. Richerson, *Modern Ceramic Engineering*, Marcel Dekker, 1982
6. G.L. Messing, E.R. Fuller, and H. Hausner, Ed., *Ceramic Transactions*, Vol 1, *Ceramic Powder Science II*, A and B, American Ceramic Society, 1988
7. J.S. Reed, *Introduction to the Principles of Ceramic Processing*, John Wiley & Sons, 1988
8. G.W. Phelps, S.G. Maguire, W.J. Kelly, and R.K. Wood, *Rheology and Rheometry of Clay-Water Systems*, Kentucky-

Tennessee Clay Company (Cyprus Industrial Minerals), 1989

9. F.F. Lange, "Forming a Ceramic by Flocculation and Centrifugal Casting," U.S. patent 4,624,808, Nov 1986

10. P.K. Bachmann, P. Geitner, E. Krafczyk, H. Lydtin, and G. Romanowski, Shape-Forming of Synthetic Silica Tubes by Layerwise Centrifugal Particle Deposition, *Am. Ceram. Soc. Bull.*, Vol 68 (No. 10), 1989, p 1826–1831

11. A. Ezis and J.T. Neil, Fabrication and Properties of Fugitive Mold Slip-Cast Si_3N_4, *Bull. Am. Ceram. Soc.*, Vol 58 (No. 9), 1979, p 883

12. I.A. Aksay, Microstructure Control Through Colloidal Consolidation, *Advances in Ceramics*, Vol 9, *Forming of Ceramics*, J.A. Mangels and G.L. Messing, Ed., American Ceramic Society, 1984, p 94–104

13. I.A. Aksay, Principles of Ceramic Shape-Forming with Powder Systems, *Ceramic Transactions*, Vol 1, *Ceramic Powder Science II, B*, G.L. Messing, E.R. Fuller, and H. Hausner, Ed., American Ceramic Society, 1988, p 663–673

14. I.A. Aksay and C.H. Schilling, Colloidal Filtration Route to Uniform Microstructures, *Ultrastructure Processing of Ceramics, Glasses, and Composites*, John Wiley & Sons, 1984, p 439–447

15. I.A. Aksay, W.Y. Shih, and M. Sarikaya, Colloidal Processing of Ceramics with Ultrafine Particles, *Ultrastructure Processing of Advanced Ceramics*, J.D. MacKenzie and D.R. Ulrich, John Wiley & Sons, 1988, p 393–406

16. F.F. Lange, Powder Processing Science and Technology for Increased Reliability, *J. Am. Ceram. Soc.*, Vol 72 (No. 1), 1989, p 3–15

17. T.K. Yin, I.A. Aksay, and B.E. Eichinger, Lubricating Polymers for Powder Compaction, *Ceramic Transactions*, Vol 1, *Ceramic Powder Science II, B*, G.L. Messing, E.R. Fuller, and H. Hausner, Ed., American Ceramic Society, 1988, p 654–662

18. L.L. Hench and D.R. Ulrich, Ed., *Ultrastructure Processing of Ceramics, Glasses, and Composites*, John Wiley & Sons, 1984

19. L.L. Hench and D.R. Ulrich, Ed., *Science of Ceramic Chemical Processing*, John Wiley & Sons, 1986

20. G.L. Messing, K.S. Mazdiyasni, J.W. McCauley, and R.A. Haber, Ed., *Advances in Ceramics*, Vol 21, *Ceramic Powder Processing Science*, American Ceramic Society, 1987

21. J.D. MacKenzie and D.R. Ulrich, Ed., *Ultrastructure Processing of Advanced Ceramics*, John Wiley & Sons, 1988

22. J.A. Mangels and G.L. Messing, *Advances in Ceramics*, Vol 9, *Forming of Ceramics*, American Ceramic Society, 1984

23. P.C. Hiemenz, *Principles of Colloid and Surface Chemistry*, 2nd ed., Marcel Dekker, 1986

24. B.V. Velamakanni, J.C. Chang, F.F. Lange, and D.S. Pearson, New Method for Efficient Colloidal Particle Packing via Modulation of Repulsive Lubricating Hydration Forces, *Langmuir*, Vol 6, 1990, p 1323–1325

25. R.G. Horn, Surface Forces and Their Action in Ceramic Materials, *J. Am. Ceram. Soc.*, Vol 73 (No. 5), 1990, p 1117–1135

26. D.H. Napper, Steric Stabilization, *J. Colloid Interface Sci.*, Vol 58 (No. 2), 1977, p 390–407

27. G.A. Parks and P.L. de Bruyn, The Zero Point of Charge of Oxides, *J. Phys. Chem.*, Vol 66, 1962, p 967–973

28. R.J. Hunter, *Zeta Potential in Colloid Science, Principles and Applications*, Academic Press, 1981

29. R.J. Hunter, *Foundations of Colloid Science*, Oxford University Press, 1986

30. R.O. James and G.A. Parks, Characterization of Aqueous Colloids by Their Electrical Double-Layer and Intrinsic Surface Chemical Properties, *Surface and Colloid Science*, Vol 12, E. Matijevic, Ed., Plenum Press, 1982, p 119–216

31. R.O. James, Characterization of Colloids in Aqueous Systems, *Advances in Ceramics*, Vol 21, *Ceramic Powder Processing Science*, G.L. Messing, K.S. Mazdiyasni, J.W. McCauley, and R.A. Haber, Ed., American Ceramic Society, 1987, p 349–410

32. R.E. Johnson, Jr. and W.H. Morrison, Jr., Ceramic Powder Dispersion in Nonaqueous Systems, *Advances in Ceramics*, Vol 21, *Ceramic Powder Processing Science*, G.L. Messing, K.S. Mazdiyasni, J.W. McCauley, and R.A. Haber, Ed., American Ceramic Society, 1987, p 323–348

33. J.W. Goodwin, Rheology of Ceramic Materials, *Bull. Am. Ceram. Soc.*, Vol 69 (No. 10), 1990, p 1694–1698

34. M.D. Sacks, C.S. Khadilkar, G.W. Scheiffele, A.V. Shenoy, J.H. Dow, and R.S. Sheu, Dispersion and Rheology in Ceramic Processing, *Advances in Ceramics*, Vol 21, *Ceramic Powder Processing Science*, G.L. Messing, K.S. Mazdiyasni, J.W. McCauley, and R.A. Haber, Ed., American Ceramic Society, 1987, p 495–515

35. J. Cesarano III and I.A. Aksay, Processing of Highly Concentrated Aqueous α-Al_2O_3 Suspensions with Polyelectrolytes, *J. Am. Ceram. Soc.*, Vol 71 (No. 4), 1988, p 1062–1067

36. I.A. Aksay and R. Kikuchi, Structures of Colloidal Solids, *Science of Ceramic Chemical Processing*, L.L. Hench and D.R. Ulrich, Ed., John Wiley & Sons, 1986, p 513–521

37. W.Y. Shih, I.A. Aksay, and R. Kikuchi, Phase Diagrams of Charged Colloidal Particles, *J. Chem. Phys.*, Vol 86 (No. 9), 1987, p 5127

38. P.H. Dal and W.H.H. Berden, The Capillary Action of Plaster Molds, *Science of Ceramics*, Vol 4, G. Stewart, Ed., The British Ceramic Society, 1968, p 113–131

39. B.W. Nies and C.M. Lambe, Movement of Water in Plaster Molds, *Am. Ceram. Soc. Bull.*, Vol 35 (No. 8), 1956, p 319–324

40. D.S. Adcock and I.C. MacDowall, The Mechanism of Filter Pressing and Slip Casting, *J. Am. Ceram. Soc.*, Vol 40, 1957, p 355–362

41. P.H. Dal and W. Deen, Die Scherbenbildung beim Keramischen Giessverfahren, *Proceedings of the Sixth International Ceramic Congress*, Wiesbaden, Deutsch Keramische Gesellschaft, E.V., 1958, p 219–242

42. I.A. Aksay and C.H. Schilling, Mechanics of Colloidal Filtration, *Advances in Ceramics*, Vol 9, *Forming of Ceramics*, J.A. Mangels and G.L. Messing, Ed., American Ceramic Society, 1984, p 483–491

43. F.M. Tiller and C. Tsai, Theory of Filtration of Ceramics: I, Slip Casting, *J. Am. Ceram. Soc.*, Vol 69 (No. 12), 1986, p 882–887

44. J.H.D. Hampton, S.B. Savage, and R.A.L. Drew, The Clogging Effects in Slip Casting, *Ceramic Transactions*, Vol 1, *Ceramic Powder Science II, B*, G.L. Messing, E.R. Fuller, and H. Hausner, Ed., American Ceramic Society, 1988, p 749–757

45. C.H. Schilling, "Microstructure Development by Colloidal Filtration," M.Sc. thesis, University of California at Los Angeles, 1983

46. W.-H. Shih, S.I. Kim, W.Y. Shih, C.H. Schilling, and I.A. Aksay, Consolidation of Colloidal Suspensions, *Better Ceramics Through Chemistry IV*, MRS Symposium Proceedings, Vol 180, B.J. Zelinski, C.H. Brinker, D.E. Clark, and D.R. Ulrich, Ed., Materials Research Society, 1990, p 167–172

47. F. Moore, *Rheology of Ceramic Systems*, Mac Laren & Sons, London, 1965

48. C.M. Lambe, Plaster Molds for Mechanized Forming of Clay Ware, *Am. Ceram. Soc. Bull.*, Vol 34 (No. 8), 1955, p 251–255

49. C.M. Lambe, Preparation and Use of Plaster Molds, *Ceramic Fabrication Processes*, W.D. Kingery, Ed., MIT Press, 1963, p 31–40

50. E.L. Whiteside, Quality Control in the Plaster Mold Shop, *Am. Ceram. Soc. Bull.*, Vol 45 (No. 11), 1966, p 1022–1026

51. E.G. Blanchard, Pressure Casting Improves Reliability, *Am. Ceram. Soc.*

Bull., Vol 67 (No. 10), 1988, p 1680–1683

52. D.V. Miller and J.S. Reed, Packing Density of Alumina Blends for Substrates Evaluated From Pressure Filtered-Pressed Compacts, *Ceramic Transactions*, Vol 1, *Ceramic Powder Science II, B,* G.L. Messing, E.R. Fuller, and H. Hausner, Ed., American Ceramic Society, 1988, p 733–740

53. F.F. Lange and K.T. Miller, Pressure Filtration: Consolidation Kinetics and Mechanics, *Am. Ceram. Soc. Bull.*, Vol 66 (No. 10), 1987, p 1498–1504

Tape Casting

Richard E. Mistler, Keramos Industries, Inc.

TAPE CASTING is a very specialized ceramic fabrication technique. It is used to form sheets that have large surface areas with very thin cross sections (Ref 1). These ceramic sheets are essentially two dimensional in nature and are used as the basic building blocks for many electronic substrates and packages.

In the ceramics industry, tape casting is a process that is most similar to slip casting, in that they both utilize a fluid suspension of ceramic particles. However, there are subtle differences. Tape casting is usually based on nonaqueous liquid, because the drying process is evaporative in nature, whereas slip casting is an absorptive process that utilizes a porous plaster of paris mold. The article "Slip Casting" in this Section of the Volume fully describes that process.

Tape casting was originally developed during the 1940s as a method of forming thin sheets of piezoelectric materials and capacitors (Ref 2). The patent (Ref 3), issued in 1952, focuses on the use of aqueous and nonaqueous slurries applied to moving plaster batts by a doctor blade device. This technology was improved in a patent issued in 1961 (Ref 4), in which the plaster batts were replaced by a moving organic carrier such as Mylar, and the process became continuous in nature. The improvements made since that time have been predominantly in the formulation phase of the process. Nearly all of the patents issued since then have dealt with this type of improvement.

This article first describes the fabrication process, its variables, and material parameters associated with tape casting. It then describes the fabrication steps that transform tape-cast sheets into multilayered ceramics, a major application of tape-cast products. Alternate thin-sheet forming methods are also described and compared to the tape-casting method.

Processing Variables and Material Parameters

A ceramic substrate produced by the tape-casting process is shown in Fig 1. It exemplifies one of the advantages of the tape process, that is, the ability to punch holes of various sizes and shapes in the "green," or unfired, tape prior to sintering. The ability to punch a variety of external shapes is another distinct advantage of the tape-casting process.

The process itself and the variables associated with it are described below, followed by a discussion on basic formulations and critical parameters associated with the materials used in the process.

The Process

A typical tape-casting process is depicted in Fig 2.

Inorganic powders are added to a ball mill (or other comminution device), along with a deflocculant/dispersant that usually has been predissolved in a solvent or a combination of solvents. The materials are milled until the proper dispersion of solid particles in the solvent is accomplished.

The ball milling procedure may be as simple as breaking down agglomerates into individual particles or as complex as breaking large particles into smaller ones. The specific milling procedure depends on the raw materials selected to undergo the complete process. Usually, the inorganic materials are selected for specific properties that they can impart to the final product. However, the simplification of other processing steps, such as lowering sintering temperatures, can factor into the material selection process. Milling for 12 to 24 h is usually sufficient to accomplish the comminution of the powders.

The production of a well-dispersed suspension of the particles created during comminution represents a critical step in the tape-casting process. In order for a tape to reach its highest green bulk density, the particles must stay in suspension during the evaporative drying process. This allows the distribution of particles of many sizes to be uniform throughout the thickness of the tape and prevents warping and distortion during sintering. Many studies conducted in the 1980s were aimed at the deflocculation/dispersion process associated with tape casting (Ref 5–7).

Mixing. The next process step is to mix the plasticizer and binders into the well-dispersed slurry produced during milling. The plasticizer(s) are added to the mill first, because the binder is usually more soluble in plasticizers than it is in the solvent-based slurry. Some manufacturers prefer to predissolve the binders and plasticizers in a portion of the solvent. This shortens the mixing step

considerably. The time required to complete the homogenization of the binder/plasticizer into the slurry usually ranges from 12 to 24 h. Because the viscosity of the slip increases considerably during this phase (usually >2000 mPa · s), the milling media are kept in the mill to help with the homogenization.

De-Airing. Upon completion of the milling and mixing steps, a considerable amount of either entrapped air or solvent vapor bubbles is usually present. De-airing, usually by vacuum, is necessary because bubbles in cast tapes act as initiation sites for cracks. A simple method is to evacuate the slurry in a vacuum chamber, such as a desiccator. Typically, a roughing pump type of vacuum (approximately 635 mm, or 25 in., of mercury) is sufficient to remove most of the bubbles. The time required for de-airing depends on the volume of slip. A good rule of thumb is 8 to 10 min/10 L (2.6 gal) of slip. Other techniques are also used to de-air slurries. Either slow stirring or slow rotation on a set of rollers for about 24 h will also remove the entrapped bubbles.

Filtering. For the most critical applications, such as very thin tapes used as multilayers or as very smooth thin-film substrates, a slip filtration step is usually incorporated just prior to the casting step. A simple nylon mesh screen can be used for most applications. The mesh size is selected on the basis of the raw materials used, the slurry viscosity, and the debris to be removed.

A simple system has been described in the literature (Ref 8). In this system, the slip is pumped or pressure-fed up through a set of two screens in series. The screen openings are described as 37 μm (1.5 mils) and 10 μm (0.4 mil), respectively. Other processes use screens with openings as large as 50 μm (2 mils) to remove very large pieces of undissolved binder, pieces of the mill, or large agglomerates. Using a closed system, whether it is a "finger-pump" or a gas pressurized reservoir, prevents the introduction of air during the slip transfer step. The filtration step also removes any bubbles that may remain subsequent to de-airing.

The actual tape forming is done by a doctor blade that is either stationary, used with a moving casting surface, or moves along a stationary casting surface. In either case, a thin film of slip is cast onto the carrier sub-

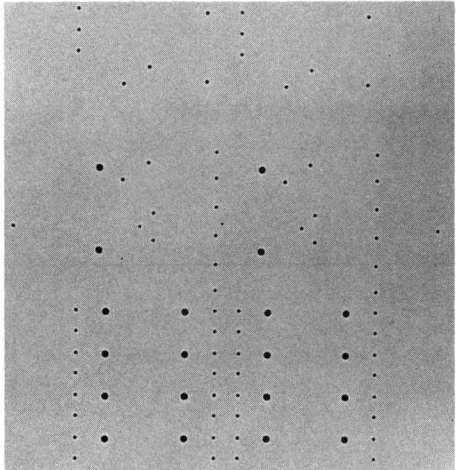

Fig 1 Tape-cast ceramic substrate with punched holes

Fig 3 Typical doctor blade assembly

strate. The thickness of the cast layer is metered by adjusting the gap between the doctor blade and the carrier. This gap is usually set by means of screw micrometers on a spring-loaded doctor blade in a sturdy frame-reservoir combination. A typical doctor blade assembly is shown in Fig 3.

Blades as wide as 1.25 m (4.1 ft) have been manufactured to produce wide tapes. The use of dual doctor blades (two in series) has also been cited (Ref 9). These blades provide very precise thickness control of the final tapes. For most slips, a 2/1 ratio is used to set the blade gap for casting. For example, to produce a 0.76 mm (30 mils) thick dried tape, the blade gap setting would be about 1.52 mm (60 mils). The resulting tape, when sintered, would yield a fired thickness of about 0.63 mm (25 mils).

The final thickness of the tape is also dependent on other factors, such as slip viscosity, casting speed, percent solids in the slurry, and depth of the slip in the reservoir behind the doctor blade. All of these parameters must be controlled precisely to yield the proper thickness of dried tape. Casting speed is determined by the size of the casting machine, the desired production rate, and the thickness of the tape being produced.

For example, thin tapes can be produced in small machines at very rapid rates. Tapes in the range of 0.025 to 0.125 mm (1 to 5 mils) are sometimes produced at rates approaching 15 to 18 m/min (50 to 60 ft/min). Most thicker tapes (>0.50 mm, or 20 mils) are produced at rates between 120 mm (4.8 in.)/min and 500 mm (20 in.)/min.

Tape-Casting Machines. Doctor blades are used in concert with tape-casting machines, which are either stationary blade/moving carrier machines or moving blade/stationary carrier machines. Stationary blade machines are by far the most popular, and are

the only type of machine capable of producing from hundreds to thousands of feet of tape product on a continuous basis. Moving blade machines are predominantly used in small-scale laboratory operations.

A typical continuous production machine that utilizes the stationary blade concept is shown in Fig 4. The basic elements of any tape-casting machine are a sturdy frame with a solid, level plate, on which the doctor blade is positioned during casting and an enclosed drying chamber, where the solvents can be removed under controlled conditions as the tape moves continuously toward the exit end and take-up spool.

The casting surface carrier, or substrate, can be any smooth material that is compatible with the slip being cast. The most important property of a casting surface is its release characteristics. A surface to which the dried tape sticks is not a good carrier. Materials used as moving substrates include stainless steel, blued steel, various polymeric films, and release film coated steel, paper, and glass. In many cases, cost is also an important factor in the selection of a casting surface. Millions of meters of capacitor tapes are cast on a coated "butcher-type" paper. In most cases, the carrier can only be used once.

The machine shown in Fig 4 has a Mylar carrier, which is fed at the entrance end and is rolled up with the tape product at the exit end. The tape can also be stripped at the exit end and can be spooled separately. The speed of casting is controlled by means of a ratio-control Variac and is continuously variable from zero to the maximum specification of the take-up drive system (usually 1 m/min, or 3.3 ft/min, or greater).

Tape-casting machines range in size from 2 to more than 35 m (6.5 to 115 ft) in length and from about 100 mm (4 in.) to more than 1.25 m (4 ft) in width. The length of the machine partly determines the rate at which tapes can be produced on a continuous basis.

Most modern machines include precision-oriented features that are incorporated on the lab-scale model shown in Fig 4. For example, there is a self-contained air dryer with a continuously variable speed control to regulate the volume and velocity of air flowing counter-current to the movement of the cast tape in the drying chamber. This counter-current flow of air is essential in tape casting to produce a gradient of solvent vapors in the drying chamber, which prevents rapid skin formation, and, therefore, promotes slower overall drying throughout the machine. Air velocity in the drying chamber should be in the laminar flow regime (usually less than 3 m³/h, or 106 ft³/h). This machine also has an in-line air heater, whereby a portion of the flowing air can be heated to accelerate drying of the tape. The drying process has been described in detail (Ref 1). Glass plates can be used as covers for the drying chamber so that the cast product can be observed during the casting and drying process.

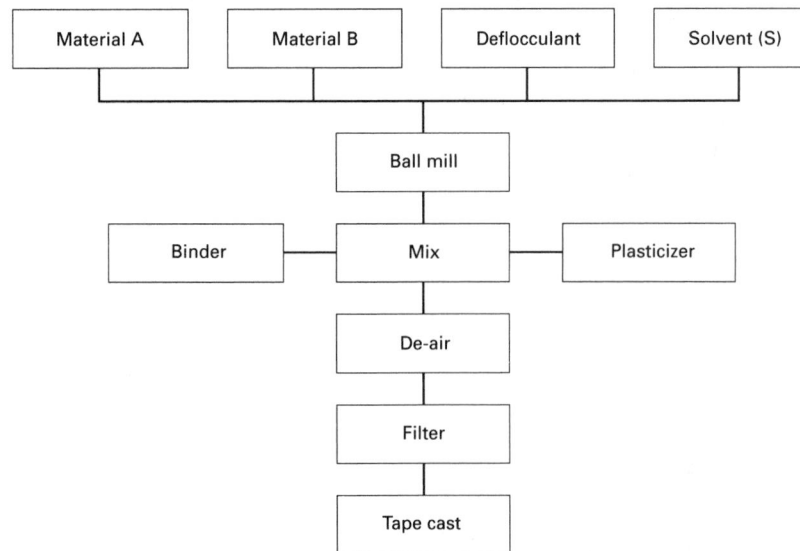

Fig 2 Typical tape-casting process

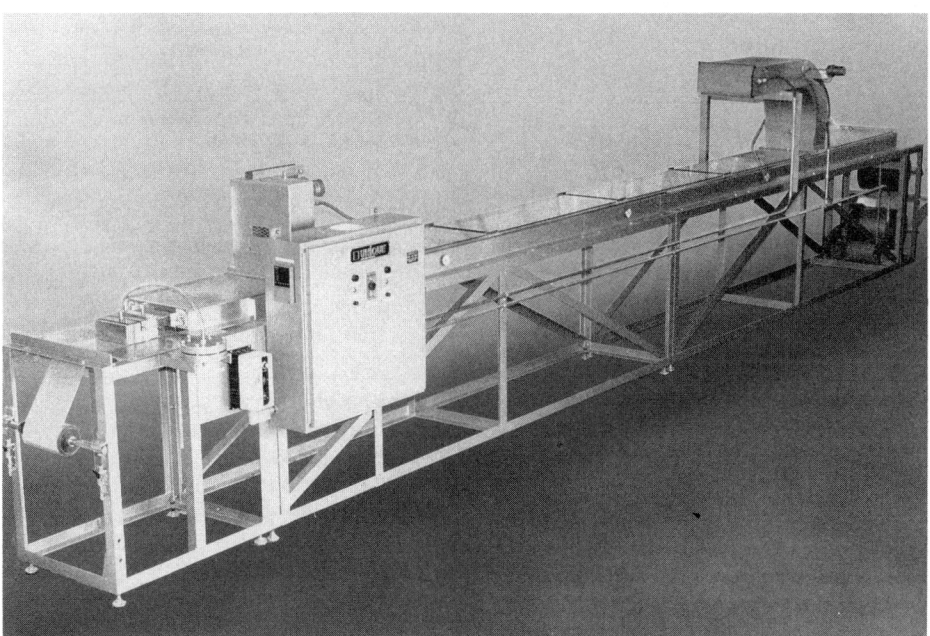

Fig 4 Continuous tape-casting machine. Courtesy of Unique/Pereny Furnaces, Kilns, and Dryers

Table 2 Typical nonaqueous system formulation used in nonoxidizing sintering atmospheres (Ref 11)

Formulation component	Batch weight	
	g	oz
Menhaden fish oil (dispersant)	1.0	0.035
Methyl ethyl ketone (solvent)	10.0	0.35
Ethyl alcohol (solvent)	10.0	0.35
Barium titanate (ceramic powder)	100.0	3.5
Butyl benzyl phthalate (plasticizer)	4.0	0.14
Polyethylene glycol (plasticizer)	4.0	0.14
Cyclohexanone (homogenizer)	0.7	0.02
Acrylic in methyl ethyl ketone (30% solution) (binder)	13.32	0.47

Formulations and Critical Materials Parameters

After selecting the deflocculant/dispersant, the next most important decision related to organic additives is the selection of a binder/solvent/plasticizer formulation for the tape system. The binder is an essential ingredient because it forms the backbone that holds the inorganic materials in place during the forming and shaping process. In the case of tape casting, the binder must:

- Form a tough, flexible film (a long-chain polymer is usually selected)
- Be soluble in a relatively inexpensive, volatile solvent system
- Be easily removed during the binder removal phase of the sintering process; in a nitrogen atmosphere, it should "unzip" and evaporate, whereas in an oxidizing atmosphere, it should convert into its gaseous components

Other criteria that must be considered when selecting a binder are the thickness of the tape to be produced, and, therefore, the viscosity of the slip, as well as the casting surface to be used, and whether or not the binder will stick to it.

Plasticizers are added to binders to extend their flexibility by changing the glass transition temperature of the system. Plasticizers for specific binders are usually recommended by the binder manufacturers. A detailed list of binder/plasticizer references is provided in Ref 10.

Tables 1 and 2 give representative formulations of tape systems for oxidizing and nonoxidizing binder removal situations, respectively. The quantity of solvent in these formulations should be adjusted to allow for differences in powder surface area and particle size distribution. Complex organic systems such as those shown in Tables 1 and 2 often are rooted in the paint and coatings literature. It is highly recommended that anyone interested in tape-casting formulations become familiar with this information source.

Tape Applications

Tape casting was originally developed to produce thin layers of dielectric materials that could be electroded, stacked, and sintered into monolithic multilayered capacitors (Ref 2). The basis for modern multilayered ceramics and capacitors is the ability to fire, or sinter, the ceramic tape and metallization simultaneously. The history of this technology is

Table 1 Typical nonaqueous system formulation used in oxidizing sintering atmospheres

Formulation component	Batch weight	
	g	oz
Menhaden fish oil (dispersant)	1.8	0.06
Xylene (solvent)	21.0	0.74
Anhydrous ethyl alcohol (solvent)	13.7	0.48
Aluminum oxide (ceramic powder)	100.0	3.5
Mixed phthalates (plasticizer)	3.6	0.13
Polyalkylene glycol (plasticizer)	4.3	0.15
Polyvinyl butyral (binder)	4.0	0.14

reviewed in Ref 12, and the metallized via concept used in all multilayer ceramic interconnection packages is described in Ref 13. The remainder of this article examines the use of tape-cast ceramics in multilayered structures.

The Multilayer Process. A schematic of the post-tape fabrication steps involved in the production of multilayered ceramics is shown in Fig 5. Because there are usually no holes to punch or metallize, the production process for multilayered capacitors is initially slightly different.

Blanking and Hole Fabrication. The initial step in multilayer processing is to punch out "cards" that are used throughout the rest of the process. Simultaneously, several sets of alignment holes are punched around the periphery of the piece of tape. The alignment holes are used to ensure that registration from step to step and from piece to piece is maintained. A new set of alignment holes is used for each subsequent step in the multilayer process. This is essential when processing "green" tape, because the holes can stretch

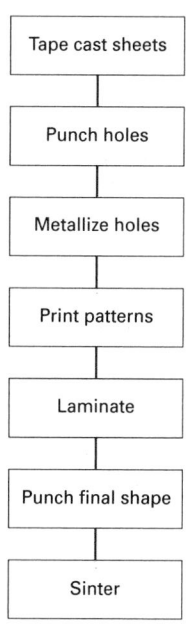

Fig 5 Typical multilayered ceramic process

and/or distort during any of the process steps. Usually, the alignment holes are from 3 to 6 mm (0.12 to 0.24 in.) in diameter and are spaced about 10 mm (0.4 in.) from one another.

If the number of via, or interconnection, holes is small, the holes can be punched concurrent with the blanking/alignment hole operation. However, this is usually not the case, because hundreds to thousands of holes are normally required. Clusters, or groups, of holes are usually punched by high-speed programmable machines. The hole size for vias can be 80 μm (3 mils) or larger.

Punch and die design is critical for tape-cast ceramics. Some important criteria unique to tape punching are:

- Use hard punches and softer die material. Tungsten carbide punches are recommended for diameters of 0.254 mm (0.01 in.) and larger. Borided tool steel should be used for smaller diameter punches. This will provide excellent wear and some flexibility. Die material, in this case, can be tool steel hardened to a Rockwell 60 value
- Design the punch so that it extends through the die into a larger chamber when the tool is closed. This will push the tape "slugs" completely through the die
- Apply a partial vacuum to the chamber under the die to assist "slug" removal. An air jet that produces a Bernoulli effect is sufficient
- Punch-to-die clearance should be 0.038 mm (1.5 mils) on all sides, for example, a 0.305 mm (12 mil) diam punch should be used in a 0.380 mm (15 mil) diam die
- Use a stripper plate to hold the tape down during punch withdrawal and to prevent hole chipping

Metallization. It is beyond the scope of this paper to discuss the process of metallization in any detail. Generally, patterns are applied by screen printing and, in some cases, via holes are filled simultaneously. Some details specific to the metallization of "green" tape are that it is essential to use a vacuum hold-down fixture and that pastes, or "inks," have to be formulated to match the shrinkage of the tape.

One major advantage of "green" sheet metallization is that the conductor width and spacing shrink with the substrate to yield very fine lines that are close together. Another advantage is that line spreading is virtually eliminated due to solvent penetration into the substrate, which increases the paste viscosity, locking the printed shape.

Lamination. After the various screen-printed layers dry, they are stacked and laminated using the alignment holes to provide accurate registration of the patterns and metallized interconnection holes. The lamination process was originally conceived and described in a patent issued in 1965 (Ref 14). Innovations have been introduced into the technology in

recent years, such as using a closed die set and 90° rotation of alternate layers of tape to minimize differential shrinkage of the laminated package (Ref 15).

Heat and pressure are applied to laminate the layers into a monolithic "green" package. Temperatures from 20 to 110 °C (70 to 230 °F) and pressures ranging from 1.38 to 138 MPa (0.2 to 20 ksi) are commonly used. The time at temperature is determined by the number of layers, or overall thickness, of the package.

The final shape of the package, once lamination is completed, is formed by punching or cutting with a tool or blade.

Sintering is then performed to produce the multilayered ceramic package. The sintering process is thoroughly discussed in the Section "Firing/Sintering: Densification" in this Volume.

Alternate Thin-Sheet Forming Methods

In addition to tape casting, two other methods are used to form thin sheets of ceramic materials: roll compaction and extrusion. The article "Extrusion" in this Section of the Volume provides complete details on that process, whereas roll compaction is briefly described below in terms of the basic process and its advantages and disadvantages when compared to tape casting.

The roll compaction process is related to dry pressing in that it utilizes spray-dried powders as feedstock to produce the continuous strip of tape. Its major difference, compared to dry pressing, is the binder/plasticizer content of the powders, which is considerably higher in order to produce a flexible, continuous roll of tape.

Basically, the process involves a powder-feed system, which continuously replenishes a reservoir above a pair of polished steel rollers that rotate in opposing directions. The powder is restrained by lips on one of the rollers, which meshes with the opposing roller and squeezes, or compacts, the powder into a continuous strip. This strip is then edge slit and collected on a take-up roller. More details on the process are available in the patent literature (Ref 16, 17).

The advantages of roll compaction, compared to tape casting, are:

- It utilizes powders prepared from water-based slurries
- It can be incorporated in a manufacturing facility that already has a spray dryer set up for dry pressing
- It is more akin to conventional ceramic processes than is tape casting
- It produces cost-effective, lower-quality substrates, such as 96% alumina, for thick-film processing
- It produces tapes that sinter flatter, with more uniform shrinkage

Its disadvantages, compared to tape casting, are:

- A spray dryer is necessary if the process is not located in a conventional ceramic manufacturing facility
- Very thin tapes (<0.254 mm, or 10 mils) cannot be produced easily, which precludes the production of multilayered ceramic tapes
- The tapes cannot be laminated due to the low binder concentration
- Surface quality of a rolled product is not as good as that of a tape-cast product

In terms of applications, roll compaction is used by most U.S. producers of thick-film substrates. Tape casting is used by all manufacturers of thin-film substrates and multilayered ceramics and capacitors.

REFERENCES

1. R.E. Mistler, R.B. Runk, and D.J. Shanefield, *Ceramic Processing Before Firing*, G.Y. Onoda and L.L. Hench, Ed., John Wiley & Sons, 1978, p 411–448
2. G.N. Howatt, R.G. Breckenridge, and J.M. Brownlow, Fabrication of Thin Ceramic Sheets for Capacitors, *J. Am. Ceram. Soc.*, Vol 30, 1947, p 237–242
3. G.N. Howatt, "Method of Producing High-Dielectric High-Insulation Ceramic Plates," U.S. patent 2,582,993, 1952
4. J.L. Parks, Jr., "Manufacture of Ceramics," U.S. patent 2,966,719, 1961
5. E.S. Tormey, "The Absorption of Glyceryl Esters at the Alumina/Toluene Interface," Ph.D. thesis, Massachusetts Institute of Technology, 1982
6. K. Mikeska and W.R. Cannon, Dispersants for Tape Casting Pure Barium Titanate, *Advances in Ceramics*, Vol 9, *Forming of Ceramics*, J.A. Mangels and G.L. Messing, Ed., American Ceramic Society, 1984, p 164–183
7. T. Chartier, E. Streicher, and P. Boch, Phosphate Esters as Dispersants for the Tape Casting of Alumina, *Am. Ceram. Soc. Bull.*, Vol 66 (No. 11), 1987, p 1653–1655
8. D.J. Shanefield and R.E. Mistler, Filter for Ceramic Slips, *Am. Ceram. Soc. Bull.*, Vol 55 (No. 2), 1976, p 213
9. R.B. Runk and M.J. Andrejco, A Precision Tape Casting Machine for Fabricating Thin, Organically Suspended Ceramic Tapes, *Am. Ceram. Soc. Bull.*, Vol 54 (No. 2), 1975, p 199–200
10. R.E. Mistler, Tape Casting: The Basic Process for Meeting the Needs of the Electronics Industry, *Am. Ceram. Soc. Bull.*, Vol 69 (No. 6), 1990, p 1022–1026
11. R.J. MacKinnon and J.B. Blum, Particle Size Distribution Effects on Tape Casting Barium Titanate, *Advances in Ceramics*, Vol 9, *Forming of Ceramics*, J.A.

Mangels and G.L. Messing, Ed., American Ceramic Society, 1984, p 150–157

12. H.W. Stetson, Multilayer Ceramic Technology, *Ceramics and Civilization*, Vol III, *High Technology Ceramics—Past, Present, and Future*, W.D. Kingery, Ed., American Ceramic Society, 1987, p 307–322

13. H.W. Stetson, "Method of Making Multilayer Circuits," U.S. patent 3,189,978, 1965

14. W.J. Gyurk, "Methods for Manufacturing Multilayered Monolithic Ceramic Bodies," U.S. patent 3,192,086, 1965

15. J.I. Steinberg, S.J. Horowitz, and R.J. Bacher, Low Temperature Co-fired Tape Dielectric Materials Systems for Multilayer Interconnections, *Proceedings of the Fifth European Hybrid Microelectronics Conference*, Stressa, Italy, 1985, p 302–316

16. R.C. Ragan, "Method for Continuous Manufacture of Ceramic Sheets," U.S. patent 3,007,222, 1961

17. R.C. Ragan, "Method for Continuous Manufacture of Ceramic Sheets," U.S. patent 3,097,929, 1963

Extrusion

Irving Ruppel, The Carborundum Company

EXTRUSION can be a very effective and efficient method of forming material continuously or semi-continuously using relatively simple equipment. The advantages of extrusion as a forming and consolidation process have been recognized and utilized by manufacturers of nearly all materials. If a material can be melted, softened, or mixed into a plastic state so that it can be forced through a die, then it can be and probably has been extruded.

In terms of material and energy conservation, the net shape and continuous forming capabilities of the extrusion process are very attractive. Extrusion has been used for many years in the clay/porcelain industries. More recently, it has been used with fine, technical ceramics such as silicon carbide, silicon nitride, and oxide materials. Shape capability has also expanded greatly, from simple rods and tubes to complex profiles, sheets/films, and honeycombs.

Extrusion has limitations and cannot be used to make all products. It is best suited to fabricate shapes that are of a constant cross section and can be linearly formed. Typical products formed by extrusion are:

- Tubes or pipes, with either open or closed ends
- Profiles of numerous shapes
- Rods
- Honeycombs
- Plates; solid, hollow, or ribbed
- Films

The disadvantages of extrusion exist in the drying, binder removal, and firing operations. The binder and solvent that have been added to the mix must be either totally or partially removed in the drying process. This operation may be performed over a wide temperature range, depending on the characteristics necessary in the green (unfired) product to enhance green machining or firing. Generally, however, solvent/binder removal can result in the formation of cracks, laminations, and other flaws, and relatively large amounts of shrinkage can occur, disrupting dimensional stability and causing warpage. Drying problems can usually be solved by changes in processing and/or materials.

Firing problems also can cause defects that parallel those found in drying. Parts may need to be fixtured to maintain straightness, roundness, and cope with large dimensional shrinkage. This especially applies to large or long parts, where shrinkage can result in retained stresses, flaws, and difficulty in maintaining targeted tolerances. Again, most problems can be solved by following the proper process path and instituting process controls so that process predictability narrows the possible material/process choices that cause problems.

The choice of materials, equipment, and processing methods to produce a specific part are the keys to ensuring that the green extruded ware brought to the furnace is of the highest quality and has the necessary attributes to reduce risk of loss. In many fine ceramic processes, firing costs represent over 50% of the total cost to produce a part. It therefore makes sense to do everything possible to make the right green part, and thus greatly increase the chances of making the right fired part.

Equipment and Tooling

There are three basic extruder categories: ram or piston, pug mill-auger, and screw-fed plasticator. Each type has distinct benefits and drawbacks, which are described below. The function of the die that serves all three types is also discussed.

Piston. The advantage of a piston extruder is its capability to reach very high pressures. It is usually powered by hydraulic pumps, which are easy to service and maintain (Fig 1). Because of its simple plunger design, maintenance on this type of unit is usually minimal and wear from abrasive materials is limited to the wall of the mix cylinder and the tooling. A piston extruder is usually capable of mix cylinder evacuation. Temperature control of the mix and equipment is simplified because the mix is transported by low shear plunger action.

A disadvantage of a piston extruder is that it is a batch machine that holds only a limited amount of material. This can cause problems if low production cost is a significant driving factor and if extrusion represents a large percentage of total cost. Another disadvantage is that incremental reloading (layering of mix) in the mix cylinder can cause disruptions in the flow pattern in the cylinder/die, as well as trapped air, which can result in changes in extrusion behavior and flaws in the green parts. Careful control of mix rheology and mix cylinder loading methods can greatly minimize these potential pitfalls.

A pug mill can be thought of as a mixer and extruder in one unit (Fig 2). Material usually undergoes a preblend/mixing step after which it is fed into the mixing section of the machine. Large paddles, or augers, chop and mix the material while transporting it to a shredder. At the shredder, the material is cut or torn into small pieces and dropped into a vacuum chamber, where de-airing takes place. The de-aired mix is then transported and consolidated by an auger screw, which forces the material through a tapered section and through the die opening.

The advantage of this type of unit is that continuous mixing, de-airing, consolidation, and extrusion all take place in one piece of equipment. However, turning augers and shredders can wear out, which not only adds to the maintenance cost, but can introduce unwanted contaminants into the mix and/or extruded part. In addition, pug mills require a material that will not "slip" and can be moved by the screw action and consolidation. The angle of the screw flights is important, and a process compromise may be necessary. For example, the steeper the angle on the flights, the faster material can be transported. However, pressure capability is reduced by steeper angles. In this case, mix properties

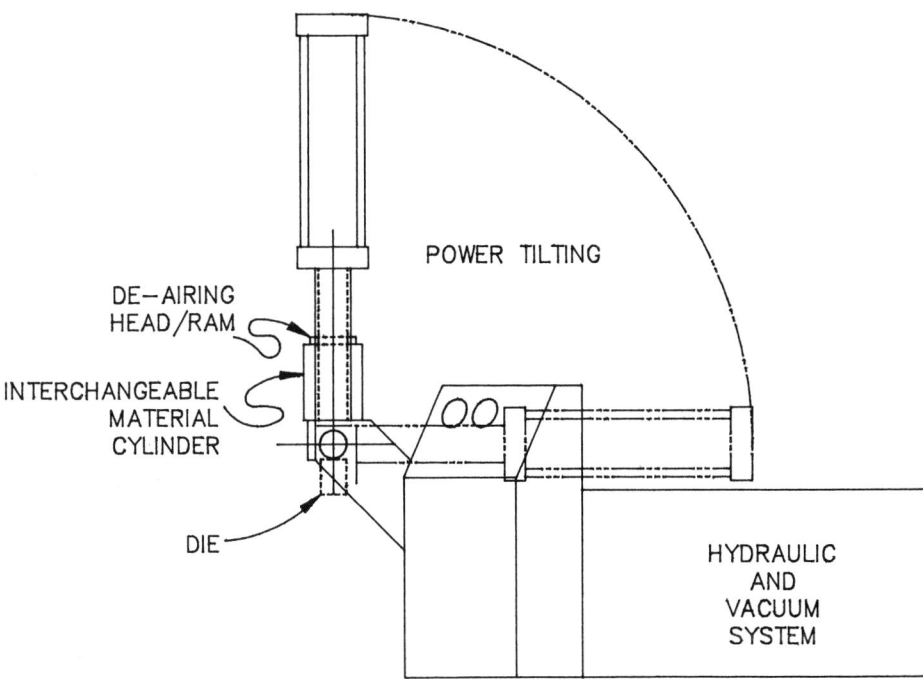

Fig 1 Hydraulic (RAM) piston extruder

can be customized so that a reasonable process compromise is reached.

It should be noted that Japanese and German equipment manufacturers have made great strides in improving the design and performance of screw-type nonplasticating extrusion equipment for fine ceramic materials. These machines are purported to offer many of the advantages of a screw machine, with few of the disadvantages. Unlike a pug mill (and similar to plasticating equipment), these machines are single-screw extruders with large length/diameter (L/D) ratios, and are designed for ceramic (solvent) extrusion.

Screw-fed plasticating extruders were developed to extrude plastics. There is a large variety of sizes and designs. The basic machines are either a single or twin screw design. Their barrels are generally heated externally, and dies can be heated and/or cooled. The screw is inside the barrel and can be turned at variable rev/min. The barrel/screw assembly can be broken down into three sections, as shown in Fig 3.

A pelletized material, such as ceramic powder and plastic binder, is fed into the first section and conveyed to the melting section, where frictional and externally applied heat soften and melt the plastic binder. It is then moved into the mixing/pumping section, where high shear and a reduction in viscosity result in further mixing and the formation of a relatively fluid mixture. The mixture is then transported into the pumping section of the machine, where the highest pressure is developed to deliver material with the desired rheology to the die. In simple terms, the plas-

ticating extruder can be thought of as a screw-type pump.

Like the pug mill, this equipment offers continuous operation and improved mix homogeneity. However, it also can produce high maintenance costs and contamination problems that are due to screw and barrel wear. These problems can be somewhat modified by the choice of plastic binder and material used to construct the screw and barrel.

The die, which forms the plastic mass being passed through it into a required shape, is a key piece of extrusion equipment. Some general basic guidelines that apply to all die designs are described below. First, tooling should be simple. All interior (mix) flow paths should be streamlined. Turbulence, temper-

ature gradients, hard mix pockets, and hard-to-predict flow patterns can result from steep tooling angles, flat tooling areas, and large tooling defects, such as gouges and dents.

Second, if a spider is used (Fig 4), as much open flow area should be provided as possible, and the spider vanes should be thin and streamlined. A "bullet-point" to separate the mix and lead it to the spider opening should be provided at the back of the spider.

The generic die design shown in Fig 4, which produces a tube, is used in all three types of equipment described earlier. The diverter section is usually used for plasticating-type extrusion so that mix pressure can be maintained at the back of the die and at the pumping section of the extruder, before the mix flows into the spider and forming sections of the die. The mix must travel a smooth, relatively unrestricted path through the die in order to ensure predictable extrudate behavior and consistently high-quality extruded components.

Third, entrance angles should usually be 25° or less. Fourth, L/D ratios in the range of 2:1 to 4:1 work well with most materials and equipment, but this rule of thumb can sometimes be violated without detriment. If a spider is used in the tooling, a longer finishing section (larger ratio) may allow the formed mix to knit well, and the spider "parting lines" to heal.

Fifth, the economics of tooling should be considered. Dies can be very costly, but if their use is limited to only a few parts, additional investment may not be required for hardened or specially treated metal. On the other hand, accelerated die wear can result from ceramic extrusion. Therefore, if many parts are to be extruded, the cost of special tool materials can be justified. In any case, die wear should be monitored so that die dimensions and die cost per part can be determined and the fabrication of parts that do not meet specifications can be avoided.

Last, uniform cross sections are an important aspect of tool and/or part design. Extruded parts should be designed with uniform walls

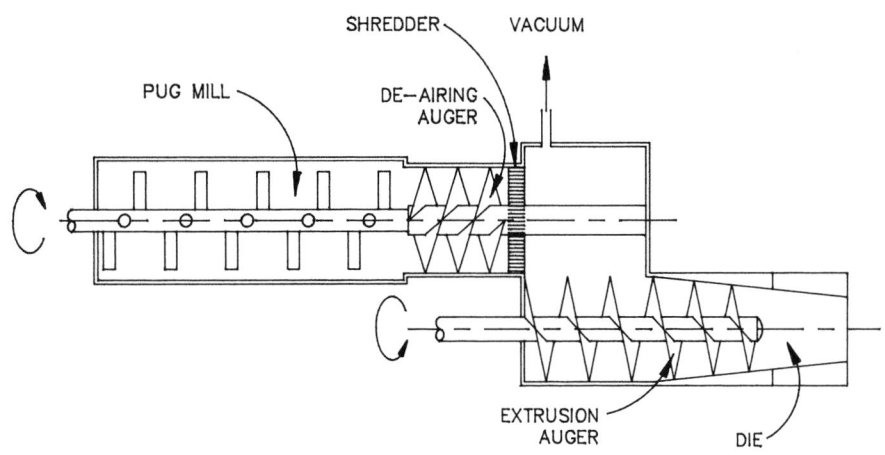

Fig 2 Pug mill with de-airing chamber and extrusion auger

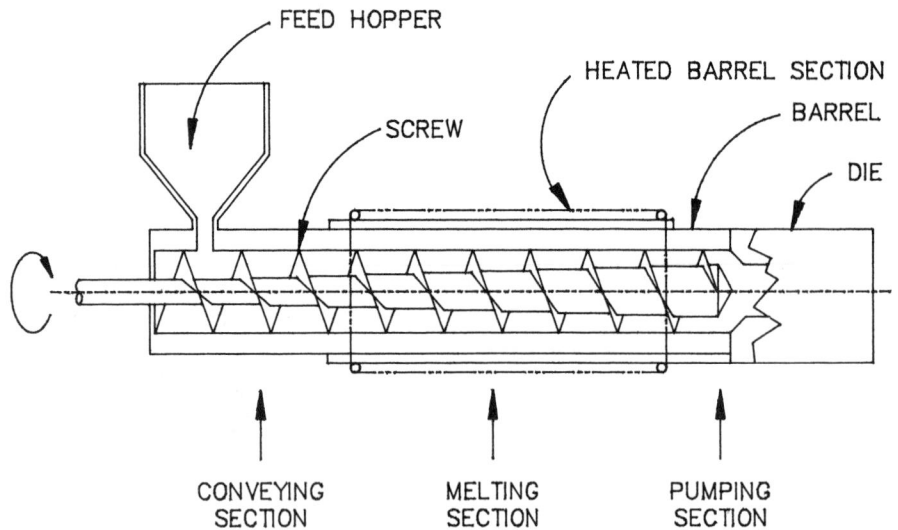

Fig 3 Single-screw plasticating extruder

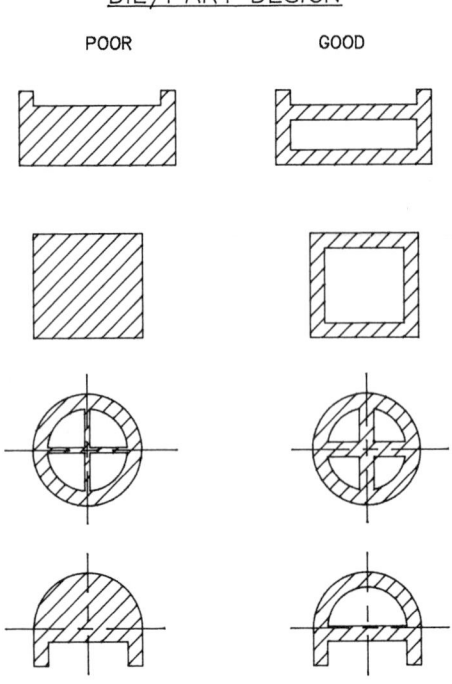

Fig 5 Die/part design

and no sharp edges or corners. Nonuniform cross sections can cause flow rate differentials that will affect extrudate behavior, as well as produce density variations that will affect flaw generation, shrinkage, and dimensional integrity. Sharp interior edges and corners tend to be highly stressed regions and susceptible to cracking, whereas sharp external edges tend to easily chip. Both poor and good part designs are shown in Fig 5.

Extrusion Process Flow

Typical process flowcharts for conducting both room-temperature and plasticating ceramic extrusion are shown in Fig 6 and 7, respectively. Each process step may require several substeps to produce the desired result, depending on factors such as type of ceramic filler, type and amount of binder, and any special requirements. The choice of which process to use to form a particular ceramic part is important, and may be governed by factors such as available equipment, type and cost of ceramic powder, particle size and shape, type of binder, cost of including the binder and then removing it, and cost considerations relative to development and projected part fabrication/processing.

As a general guideline, the desired result of the extrusion process and the limiting factors that influence the choices should be analyzed. For example, if the ceramic material being formed reacts with water, then the choice of water-soluble binder is probably not wise; or if the average particle size of the ceramic filler is very coarse, the use of plasticating extrusion equipment may be inappropriate because of high equipment wear and maintenance costs.

The process choice can also be narrowed simply on the basis of capacity; that is, piston extruders are batch-type machines appropriate for small runs, although they can also be supplied to large extrusion processes. On the other hand, plasticating extruders are usually used for large continuous runs of one particular shape, and are not economical for short runs because of the time required to achieve extrusion stability/equilibrium. The rule to follow is to know the process limitations and targeted goals, and then select the process that gives the simplest, most cost-effective result.

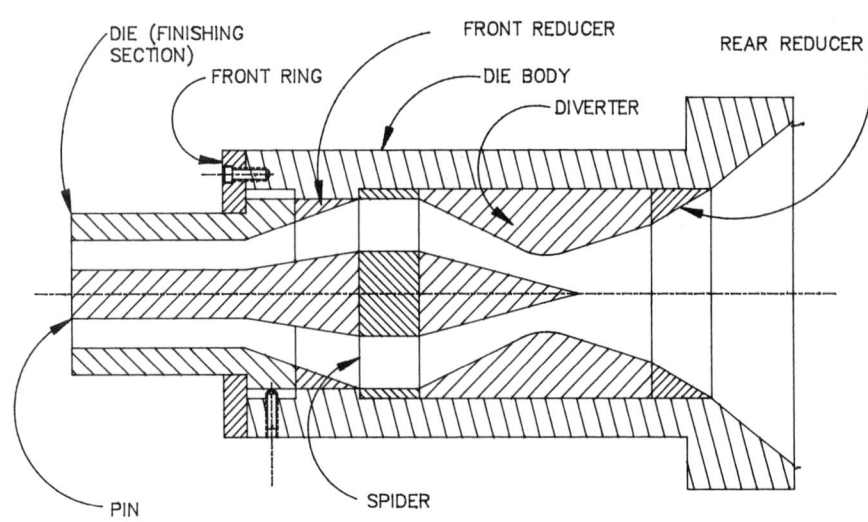

Fig 4 Tube extrusion die

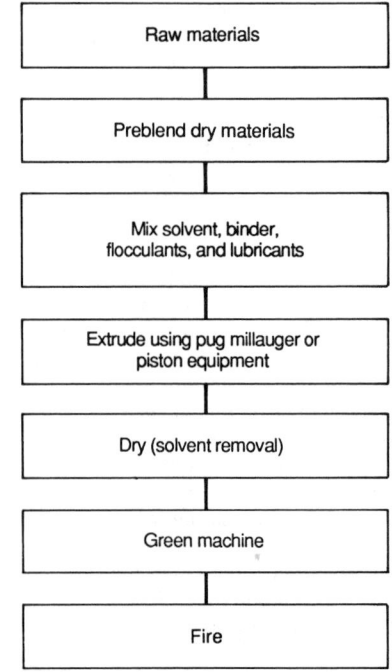

Fig 6 Generic solvent extrusion process

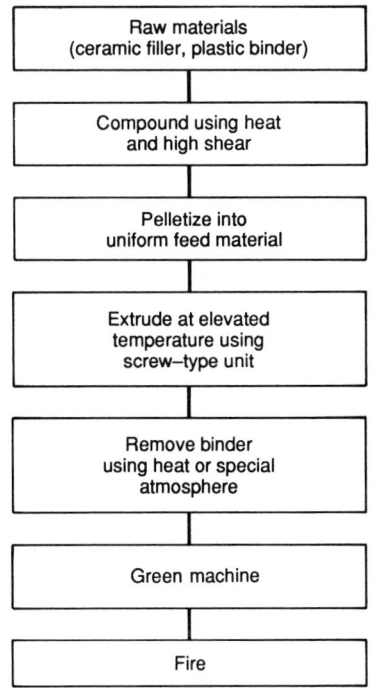

Fig 7 Generic plasticating extrusion process

perature extrusion in either a piston or pug mill extruder, binder/extrusion enhancement choices need to be made in these general categories:

- Liquid/solvent
- Binder
- Wetting agent (surfactant)
- Deflocculant
- Coagulant
- Lubricant
- Plasticizer
- Preservative

A good rule of thumb is to use the proper materials in the minimum amount needed to "do the job." Some extrusions perform well using a dual-system binder, that is, a solvent plus a binder, whereas others require from three to five or more binder components to produce the desired results.

For example, if bodies of alumina or silica are being extruded, clay is often used in an aqueous system with one or more organic binders. On the other hand, if silicon carbide is being extruded, clay may result in disruptions during the furnace operations because of its chemistry. In the processing of fine ceramics, organic polymers such as cellulose ethers are often used as primary binders. They are preferred because of their water solubility, lubricity, good green strength, availability of ranges of molecular weight, clean burnoff, and purity.

Methyl cellulose binders are capable of thermal gelling, which means that their viscosity and rheological behavior change with temperature (Fig 8). This property can be used advantageously during mixing and processing. For example, to achieve maximum homogeneity, mixing should be carried out when the binder is at its lowest viscosity. Generally, the methyl cellulose should be a high-viscosity type in order to increase elastic-compressive behavior (material will exhibit pseudoplastic behavior under pressure) and still allow for high green strength of the hydrated, extruded body.

The cellulose binder can be added to the ceramic powder as either a dry "flour" or a liquid. The choice depends on the type of cellulose chosen and the concentration (binder/solvent) required. In a typical mix preparation, the dry ceramic powder and methyl cellulose binder are blended in high-speed, high-intensity mixers (Fig 9).

The mix is then transferred to a high-shear mixer, where water is added and the material is mixed to a paste or semipaste consistency. Temperature control at this stage of mixing is usually beneficial. Depending on the mix constituents, type of mixer, and desired homogeneity, this mixing step can take from 15 min to 2 h. A torque rheometer is often used to check the mix during the mixing cycle, as well as to measure the "completeness" of mixing at the end of the cycle. However, because of the thermal gelling characteristics of methyl cellulose, care must be taken to

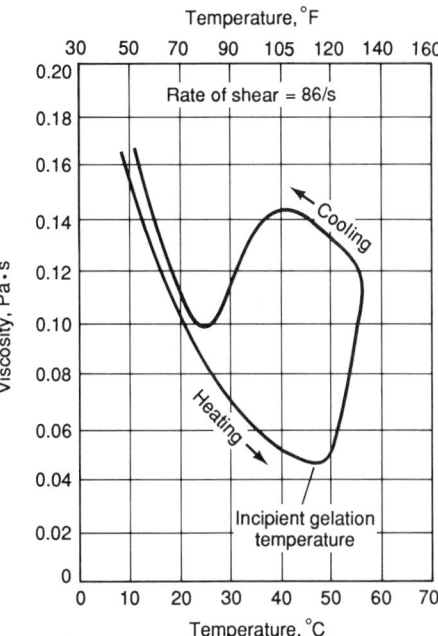

Fig 8 Temperature-viscosity behavior of methyl cellulose (Methocel). Source: Dow Chemical Company

control mix temperature in order to produce valid readings on this tool. A mix should be stored in an airtight container prior to extrusion. Refrigeration is usually unnecessary, but exposure of the mix to temperature extremes during storage is usually undesirable.

A typical plastic binder/ceramic extrusion mix requires a science and process different in many respects from an aqueous (solvent) extrusion mix. Although the mixing of plastic materials is usually simple, it is intensive. The ceramic powder and polymer binder are dry blended at room temperature, compounded using heat, plasticizers, and lubricants, and then pelletized. The dry materials can be blended in a V-type or conical blender using conventional procedures. In the compounding step, the material is run through a high-shear screw-type compounder, where it is mixed, melted, remixed under high-shear conditions, and pumped through a die to form a strand. The strands are usually cut at the die face to form a pellet, and either recompounded or used as "feed" (extrusion mix) to extrude components. The highly filled plastic extrusion compound can rapidly wear extrusion screws, barrels, and tooling. The cost and possible contamination effects should be considered.

A wide choice of binder materials are available, and selection is based on the type of part, cost considerations, polymer characteristics, and other factors, just as in creating room-temperature aqueous ceramic mixes. Developing a plastic extrusion compound can be a complex and expensive task. Several companies have spent years and millions of dollars to develop compositions that

---

As with all choices, the decision is less risky and less costly if the proper background information is obtained and analyzed.

The Extrusion Mix

The goal in any ceramics extrusion process is to produce a uniform, homogeneous mix that can be forced under pressure through a die to form the desired part shape, with targeted physical properties. Usually, the characteristics of the final fired component dictate the choice of raw ceramic material. Particle shape and size and particle size distribution profoundly affect packing. Depending on end-product characteristics, it may be desirable to maximize particle packing or control it to some degree, that is, to produce a high-density green product or one with porosity. The choice of particle size and particle size distribution strongly impacts the amount of densification achieved. Other factors can also control particle packing, or density, such as the mechanics and interactions between the ceramic filler and the binder/solvent.

Solvent-based systems, in which the primary binder(s) are organic-type polymers and/or clay-type materials, are often chosen for ceramics extrusion. Clay can act as a very good plasticizer and binder, but its chemistry can make it unsuitable for certain ceramic materials because of contamination. Generally, these solvent-based room-temperature systems are relatively simple and low in cost.

Because ceramic particles generally exhibit poor flow properties, processing additives are needed in order to extrude or consolidate/shape these particles. To conduct room-tem-

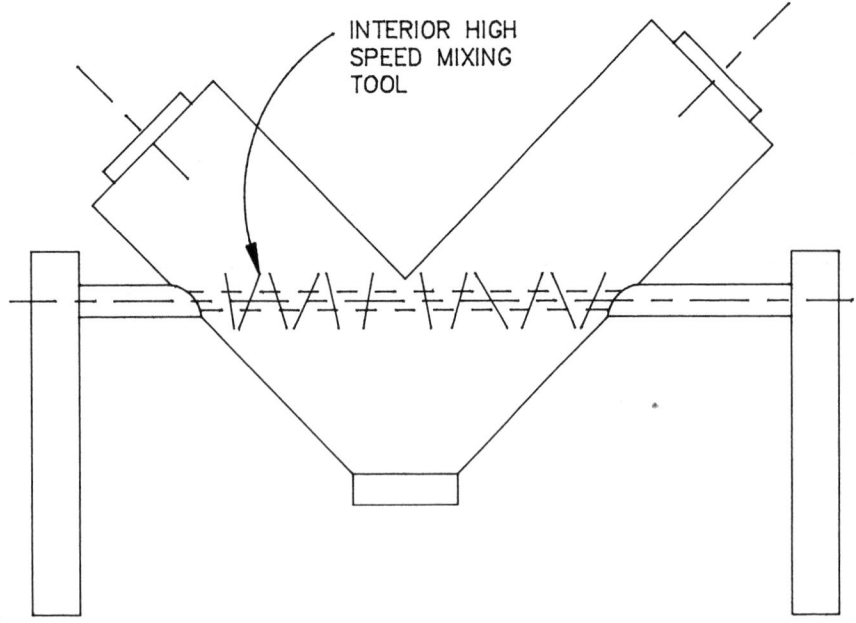

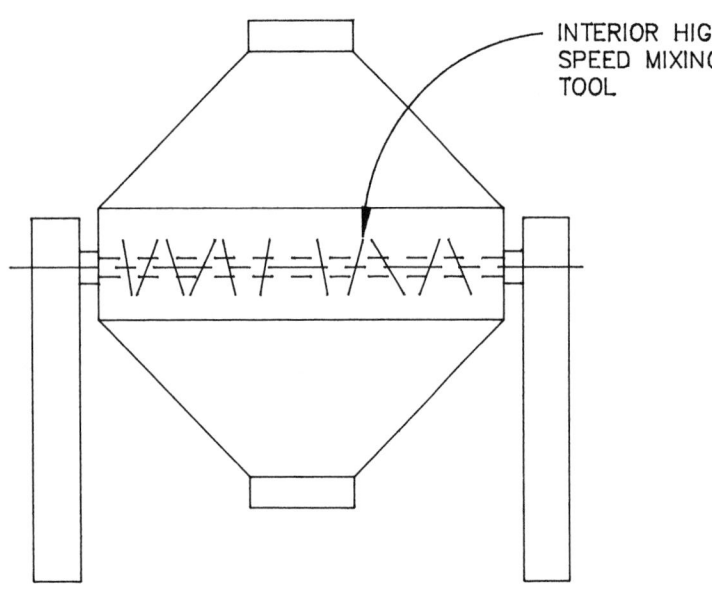

Fig 9 V-type (a) and conical-type (b) blenders

it leaves the high-pressure, controlled environment of the die may also affect postextrusion handling. How these potential problems are handled depends on the type of extrusion process, equipment type and orientation (that is, horizontal or vertical extrusion), the geometry and dimensions/tolerances of the extruded part, and green extrudate properties, such as strength, stiffness, porosity, and solvent content.

Solvent Extrusion. In the solvent (room-temperature) extrusion process, the extruder is oriented either vertically or horizontally. Table 1 lists the major advantages and disadvantages of each.

In cases where extrudate green strength can withstand the stress of increased mass as it extends from the die, vertical extrusion can offer advantages. This stress can stretch the parts and change diameter/wall thickness as length increases. When stress exceeds green strength, it may simply break the part. Supports under the part can minimize this stress, but they must be designed to provide *constant* upward force as the part increases in length. This can be a difficult engineering task. Parts can also be extruded into a high-density medium, such as a thick, high-viscosity oil, but this can become messy, and the medium

Table 1   Advantages and disadvantages of vertical and horizontal extrusion

Orientation	Advantages	Disadvantages
Vertical		
	Assistance of gravity to take up material; no friction on extrudate	Gravity can stress material as extrudate mass approaches or exceeds green strength; dimensional changes/flaws can result
	Reduction of surface flaws related to take-up fixtures	
	Tilting unnecessary	Extrudate length limited, such as from die to bottom of pit or floor
	Minimal floor space required	
	Safer if die assembly fails at high pressure	Handling of extruded part after being cut at die
	Multiple (turret-type) mix cylinders possible	
Horizontal		
	Extrudate length virtually unlimited	Reloading mix cylinder on piston extruders may require cessation of extrusion and tilting to reload
	Take-up equipment/fixtures can help maintain size/tolerances	Equipment/fixtures may be required to take up extruded part to maintain size and reduce effects of gravity and friction

perform in a desired way. Therefore, the literature is presently lean on information that can lead to a plastic-ceramic composition. Basically, it is safe to assume that plastic compositions are a complex mixture of polymer, plasticizer, and lubricant, which requires extensive mixing/compounding at polymer melt temperatures in order to produce a homogeneous extrusion mix. The complexity and difficulty of the task intensifies because it is desirable to minimize plastic binder content and maximize ceramic content. This represents a tribulation of sorts; that is, it takes a lot of trouble and energy to put the plastic in, and it takes a lot of trouble and energy to

take it out again after extrusion. However, this aspect may be offset by the continuous, precise (dimensions and tolerances), broad size capability, wide range of shape capability, and green strength/toughness afforded by plastic extrusion.

Postextrusion Processing

After an extruded part exits the die, it needs to be handled so that it maintains physical and dimensional integrity. In simple terms, the part must be "taken up" from the die so that frictional and/or gravitational forces can be dealt with. The properties of the green extrudate as

may interact with the binder/solvent and cause other problems.

Vertically oriented extrusions can benefit from the force of gravity and lack of take-up fixtures. For example, surface flaws that are due to movement of the extrudate over guides/fixtures would be eliminated. The effects of gravitational sagging on walls of tubes and hollow beams are also minimized. Other than length considerations, the main detraction of a vertical orientation is that the part is released (cut) from the die at some point and placed in a fixture, rack, or mandrel to dry. This procedure can require the part to be flexed or bent. This source of stress may cause other flaws, such as cracks, or deformations, which may have to be trimmed from the part and scrapped. Determining the best removal and handling methods may require detailed engineering, as well as good operator training.

Most extrusion equipment is horizontally oriented. The take up and handling of parts may require substantial engineering and fixture design. Gravity, friction, and the increasing part mass as part length increases can cause a decline in dimensional integrity, as well as introduce flaws. Typical problems are part "buckling" at the die, surface flaws generated as the part is extruded onto a fixture, sagging of hollow parts, and cracks caused by surfaces that dry too fast. Rheological changes made in the mix so that the extrudate can be "manufacturing toughened" and better adapted to postextrusion processing can solve some of these problems. However, the green horizontally extruded parts often require design/engineering intervention to produce desired results.

Intervention may involve extruding the part onto a take-up fixture for the reasons mentioned above. Coatings may be added, either permanently or temporarily, on take-up fixtures to reduce drag and friction. Extruding onto fluid films is another way to reduce drag and friction. Moving belts, rollers, or bearings can also be used. The movement can either be supplied by an external source, such

as a motor, or transmitted to the take-up mechanism by the moving extrudate. Extrusion into a controlled environment chamber is often done to either accelerate drying or slow it. Rapid heating of the part as soon as it exits the die often has a positive effect on dimensional control; for example, out-of-roundness of tubes. In specific cases, the thermal gelling properties of methyl cellulose binders can be utilized. A final intervention method is to extrude into a controlled tensioning mechanism (Fig 10). It is important to control take-up speed when using these machines, which can be a difficult task in the case of ceramics extrusion. In addition, the green extrudate must be able to withstand the compressive forces of the pulling equipment. These machines are well suited for continuous extrusion processes, but can be used for periodic batch extrusions. The simpler the postextrusion take-up method, the fewer problems there will be.

To remove solvent (usually water) and/or set the binder before firing, the green as-extruded part can be heated in a forced-air (convection) oven while residing on a fixture, such as the one it was extruded upon, or one made especially for the drying/curing operation. Although it is usually best not to move the part from fixture to fixture until it is cured, this can be done with proper care. Green body characteristics and requirements for the fired components are considerations. The cycle in the drying process is very important. Rapid heating up to a temperature to set the binder sometimes works fine. Usually, heat-up cycles that allow gradual solvent removal without causing laminations, cracks, warpage and/or spalling of the part are required. Very porous green parts can withstand rapid cycles whereas dense parts may not, because volatiles cannot escape fast enough to avoid catastrophic pressure buildup in the pore spaces. Although simple convection heating is probably the most commonly used drying method, others may be more suitable for a particular process, such as drying in a multiple controlled environment or continuous drying, either immediately as the part exits the extruder or as a batch operation, separate from extrusion. Various manual methods to control drying rate involve part encapsulation using paper or plastic to wrap or bag the part in order to slow or control the release of solvents. This method often works well where temperature/humidity control are desired, but humidity-controlled drying equipment is not available.

Less conventional heating/drying methods that may fit specific process or material needs include microwave, induction, dielectric, and freeze-drying. Manufacturers of these types of equipment generally have laboratories where tests can be run.

Ceramic extruded bodies usually shrink during drying. The shrinkage may range from near zero to several percent. During drying, high-shrinkage materials present special problems that relate to generation of flaws in

the material, stresses retained in the dried green part, and fixturing the part to prevent the frictional forces associated with shrinkage from pulling apart the part. In addition, differential (nonuniform) drying can cause cracks, warpage, laminations, and other problems. Shrinkage can be minimized by:

- Changes in the mix and green part density
- Reducing friction on drying fixtures
- Hanging the part vertically in the dryer
- Moving/rotating the part during drying
- Using longer, slower drying cycles

Plasticating Extrusion. Thermal plastic and ceramic-filled extrudate is very soft and putty-like as it exits the die. One advantage of this property is that the material can be vacuum formed while still hot and pliable. Vacuum equipment is used to keep tube and pipe shapes round and maintain accurate dimensions for puller take-up processing, as well as meet part specifications. A second advantage is that a puller can be used to put the hot extrudate under tension or compression to help control dimensions. This allows very close tolerances and precise parts. Final (fired) product properties, such as shrinkage, can be affected by stressing the deformable part during extrusion. Third, features can be stamped, punched, or rolled into the hot extrudate, and fourth, water or gas cooling can be used to "freeze" the desired shape and dimensions after they are achieved.

The major disadvantage is that the very fluid part that exits the die has very little green strength and must be handled and taken up in a manner that maintains part integrity. The proper balance between mix rheology, extrusion conditions, and cooling/take-up procedures must be established.

Depending on the type of part produced, other take-up methods that are modifications of puller mechanisms can be used. Sheets, plates, filaments, profiles, and other forms can be effectively extruded using plasticating technology and equipment to obtain precise, high-quality greenware.

Standard extruded greenware is dried or cured using routine methods. Plastic extruded ceramic greenware does not generally require secondary processing to cure the binder and/or remove solvents. Processing this material prior to firing requires the partial or total removal of the plastic binder. This commonly involves heating the part to high temperature in order to volatilize the plastic to drive it from the part. This process requires strict control of temperature and atmosphere to assure optimum binder removal rates and the cleaning of the atmosphere in the bake furnace chamber. Process and environmental changes can strongly influence the choice of methods for binder removal. Economic limitations may also be factors.

Fixturing during binder removal is also important because high shrinkage can be encountered, as well as a substantial loss of green strength as the plastic binder is depleted.

Fig 10 Tensioning/take-up equipment used for plasticating extrusion; note small-diameter tube being extruded

Debonded ware can be difficult to handle and fire, depending on the amount of intergranular bonding. The plastic binder and/or ceramic powders are often enhanced by adding a high-temperature binder material that survives bake-out operations and increases green strength, but does not detrimentally affect firing properties. One way to circumvent some of the debonded strength issues is to partially bake out the binders, that is, some portion of the plastic binder is left in the green body to minimize handling and firing problems. The drawback of this method is that controlled removal of residual binder in the final firing operation often creates tricky problems involving heat-up rate, as well as environmental and contamination issues. There is no absolute guideline for postprocessing plastic extruded parts. It is up to the engineer to balance requirements with all process and cost factors.

REFERENCES

- J.J.Benbow, T.A. Lawson, E.W. Oxley, and J. Bridgwater, Prediction of Paste Extrusion Pressure, *Am. Ceram. Soc. Bull.*, Oct 1989, p 1821–1824
- K.E. Burnfield and D. Schweiger, "Use of Cellulose Ethers for Rheological Control of Advanced Ceramics," Technical paper, Society of Manufacturing Engineers, 1989
- J.M. McKelvey, *Extrusion Processing of Polymers*, Washington University
- J. Reed, *Principles of Ceramic Processing*, New York State College of Ceramics, Alfred University
- J. Reed, *Extrusion of Ceramics*, Short Course, Alfred University, 9 July 1987

Injection Molding

Beebhas C. Mutsuddy, Institute of Materials Processing, Michigan Technological University

INJECTION MOLDING is a cyclic process in which a granular ceramic-binder mix is heated until softened, and then forced into a mold cavity, where it cools and resolidifies to produce a part of the desired shape.

Injection molding is not a direct analog of the plastic molding process. Successful ceramic injection molding depends on several process steps. These steps involve:

- Tailoring the ceramic powder
- Developing organic binder formulations
- Producing a homogeneous ceramic and organic binder mix (pelletizing)
- Forming parts by injection molding (and recycling scrap for pelletizing)
- Processing the parts to remove the organic binder
- Densifying the parts

The simplicity of the injection molding process must not lead to the assumption that fully dense parts of predictable properties and dimensional control are easily producible. A number of process-related problems include binder removal, dimensional stability, reproducibility, and tooling costs, all of which must be solved before wide-scale commercial applications of injection molding are feasible.

Ceramic products that have successfully been produced by injection molding include spark plugs, ceramic cores, thread guides, electronic parts, welding nozzles, and dental braces. Despite these examples of successful applications, the market for injection molding ceramics has remained both relatively small and fragmented. In recent years, injection molding has received special attention as a method for forming either high-volume and low-value ceramic parts or high-value and low-volume complex ceramic parts, both advanced and traditional in nature.

Powder Characteristics

The concentration of ceramic powder that can be incorporated into a continuous binder matrix to provide a high green density and close dimensional accuracy in the "as fired" state depends on the particle packing characteristics of the powder.

Although particle packing is a common aspect of all ceramic processes, it has received little attention. Thus, there is a limited theoretical base from which to predict packing behavior. However, empirical research on a variety of particle types, generally coarse and spherical, has led to a basic understanding.

The current trend in ceramic industries is to use powder with high reactivity and fine particle size (less than 1 μm, or 40 μin.) to achieve better end-product properties and to reduce sintering temperature and time. The use of such powder presents a particle packing problem due to particle agglomeration. As particle size decreases, the surface-area-to-volume ratio increases, and surface effects begin to influence particle behavior. Electrostatic forces, van der Waals forces, and hydrogen bonding from adsorbed moisture lead to the formation of small particle agglomerates that resist being broken down into individual particles. The high porosity in these agglomerates results in a low packing density. For random distributions that contain agglomerates comprising a larger number of spherical particles, the maximum packing density decreases. Further, particle packing and powder loading in injection molding mixes are major factors in attaining close dimensional tolerances in sintered parts. Both these factors influence green density and shrinkage, which in turn control the dimensional stability.

Additives are usually required to achieve adequate dispersion of particles and thereby obtain optimum flow and packing characteristics. Dispersion is usually achieved in three stages: wetting the ceramic particles, breaking up the agglomerates to form smaller particles, and preventing coalescence of the individual particles. Each of these three areas is equally important.

Binder Formulation

The selection of a binder formulation that gives the correct flow properties to the ceramic-binder combination is of primary importance. Correct flow properties are essential to achieving a void- and stress-free, uniformly dense, molded shape.

Ideally, the mix that leaves the gate of the mold cavity should spread like a liquid, sweeping the air before it and forcing that air to escape through mold clearances, or vents.

The absence of a yield stress will cause the molded part to deform during the dewaxing stage. On the other hand, if the mix has too high a yield stress, it will extrude into the mold cavity much like a thread of ribbon that coils upon itself (jetting), tending to trap air and induce mechanical stresses.

To obtain the best results, a binder formulation must have a yield point consistent with the shape, size, and dimensional stability of the molded part. In other words, the ceramic-polymer/wax formulation should have either Bingham or pseudoplastic flow characteristics, which are explained in the section "Rheological Behavior" in this article. Furthermore, the binder should be easily degradable or extractable in a subsequent dewaxing process. Low-molecular-weight polyethylene, polystyrene, polypropylene, ethylene vinyl acetate, ethylene ethyl acrylate, and both paraffin and microcrystalline waxes that contain small amounts of other organic modifiers all have been successfully used as binders for ceramic injection molding.

Water-Soluble Binder. Methylcellulose and hydroxypropyl cellulose have also been successfully used as binders for injection molding (Ref 1). The physical properties of an aqueous methylcellulose solution that modify the rheology of a ceramic mix include molecular weight, molecular weight distribution, degree of hydration, extent of intramolecular and intermolecular interaction, temperature, concentration, and the presence of additives. When dissolved in water and heated, these polymers form gels that revert to liquid upon cooling. The gelation temperature decreases with increasing polymer concentration. It also decreases in the presence of additives, which may compete with the polymer for water.

Gelation of the methylcellulose or hydroxypropyl methylcellulose solution is primarily caused by the hydrophobic interaction between molecules containing methoxyl substitutions. In a solution at lower temperatures, molecules are hydrated and there is little polymer-polymer interaction, other than simple entanglement. As the temperature is increased, the molecules gradually lose their water of hydration. Eventually, when a sufficient but not complete dehydration of the polymer occurs, a polymer-polymer association takes place, and the system approaches an infinite network structure.

Rheological Behavior

Rheology is the study of flow and deformation of matter. Shear, temperature-dependent viscosity is the most accepted parameter used to define the rheological behavior of injection molding mixes. The types of viscous flow that injection molding mixes can exhibit are shown in Fig 1 and defined as:

- St. Venant, in which the viscosity is inversely proportional to shear rate
- Bingham plastic, which does not show any flow unless the shear stress reaches a critical value. The nonlinearity has been attributed to changes in the network structure and alignment within the network structure of the flowing material as the stresses change. The network structure can support a stress before any material flow can be initiated. This finite stress is known as the yield stress
- Pseudoplastic, in which the viscosity decreases continually with shear rate
- Dilatant, in which the viscosity increases with shear rate

Measurements. Several rheological parameters can be investigated in a variety of experimental modes. Capillary rheometry provides apparent viscosity versus shear rate data, which can be used in injection flow analyses. In capillary flow, non-Newtonian material flows through a tube. The velocity gradient near the wall is high, and the shear stress is measured at the wall. Thus, the condition of the material at the wall controls the recorded viscosity, which is the ratio of shear stress to shear rate. In many filled polymeric systems, the particles migrate to areas of low shear stress when subjected to steady shear. In the capillary rheometer, this means that the material near the wall may have a lower concentration of filler than the bulk material. This effect is more pronounced with large filler particles than with small ones, and is also more

noticeable at the highest shear rates. Therefore, it is not surprising that viscosity versus shear rate curves for filled polymers tend to converge at high shear rates for a given polymer.

The difficulties encountered with capillary measurements can be overcome by investigating the low shear rate behavior of the filled polymers. The effect of particle migration can be minimized by performing the experiments in a dynamic oscillatory mode rather than in steady shear. The experimental facility consists of parallel plates, one of which oscillates sinusoidally at frequencies between 0.001 and 100/s. The instrument measures dynamic viscosity and modulus as a function of oscillation frequency at the temperature of interest.

In measuring rheological properties with the oscillating plate method, the percent strain must be selected carefully, because the strain is proportional to the total angle to which the plate is turning. The larger the angle through which the plate oscillates, the greater the strain. If the angle is too large, the material being measured may undergo internal cracking or segregation. In either case, the results would not be reproducible and the shear modulus (G') of the material would decrease.

Rheology of Mix Formulations

Rheological behavior is explained below using two Al6-SG alumina formulations as examples and capillary rheometer and oscillating plate techniques at 0.1% strain.

Viscosity. Al6-SG alumina was mixed in the atactic polypropylene at loadings of 32, 48, and 57 vol%. Figure 2 shows the dynamic viscosity of the different mixes. It is obvious that the viscosity increases by one order of magnitude (at high shear rates) with the addition of alumina powder (56.7 vol%). Because this particular grade of atactic polypropylene is highly plastic, even in the presence of a large volume fraction of ceramic, it was necessary to reduce its plasticity by blending with

BASF-A wax. The replacement of atactic polypropylene with BASF-A wax does not result in much change in viscosity (Fig 3), compared to the mix containing atactic polypropylene alone as a binder. However, physical examination of the mix containing BASF-A shows some reduction in plasticity.

The addition of processing aids (for example, oleic acid) to a mix containing simply Al6-SG alumina and atactic polypropylene reduces the viscosity markedly. For example, at a frequency of 10 rad/s, the viscosity of the mixes with and without processing aids is 0.75×10^3 and 0.5×10^4 Pa · s, respectively. Furthermore, the addition of processing aids in the mix containing both atactic polypropylene and BASF-A wax brings the viscosity of the mix close to the viscosity of pure atactic polypropylene.

The viscosity profile in Fig 4, for an alumina and ethylene-vinyl acetate (EVA) mix, is significantly different from alumina-atactic polypropylene, particularly at shear rates approaching 100/s, where the viscosity of the alumina-EVA is close to that of pure EVA. The addition of butyl stearate as a plasticizer does not cause any noticeable change in viscosity, whereas the addition of a processing aid to the mix reduces the viscosity at higher shear rates to a level close to the viscosity of pure EVA.

The viscosity behavior of these systems at low shear rates indicates the presence of a network structure within the melt, which could result from the interaction of filler particles, possibly through active sites on their surfaces. The effect of this structure on the viscosity of these systems is dependent on the dynamic equilibrium between the breakup and re-formation of the filler particle structure. As the shear rate is increased, the equilibrium between breakup and re-formation shifts toward a decreasingly network-like structure, until at relatively high shear rates (greater than 1.0/s), breakup of the network is essentially complete. During this process, it is also likely that some of the polymeric binder may tend to migrate and cause an abrupt change in viscosity.

The role of polyethylene-derived wax (BASF-A) to atactic polypropylene and the use of butyl stearate as a plasticizer for EVA are fairly complex, and a clear explanation has not yet been developed. The observations can only be based on assumptions. For example, a plasticizer does not, as a rule, react with the polymer; rather, it forms a mechanical mixture wherein plasticizer molecules are interspersed between polymer molecules. This causes a change in the intermolecular energy and brings about a stiffness in the structure, as in the case of the atactic polypropylene and BASF-A mixture, whereas in the case of EVA, a small dilution might have occurred in the presence of butyl stearate. On the other hand, the presence of processing aids has a pronounced influence on the viscosity and uniform dispersion of ceramic powders in these

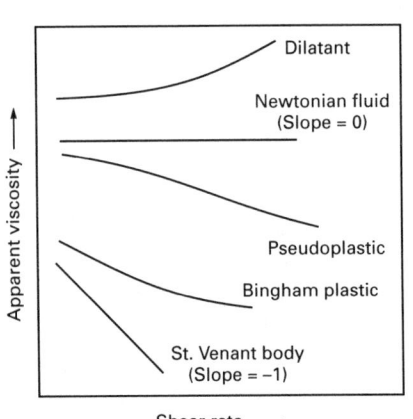

Fig 1 Types of flow behavior

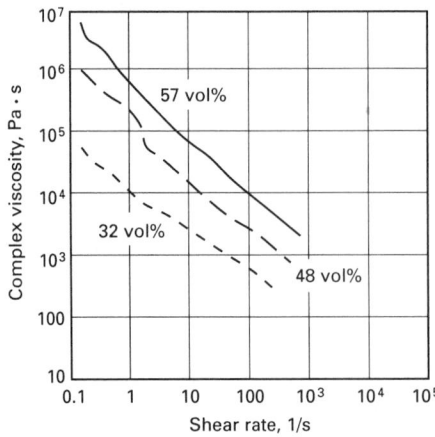

Fig 2 Viscosity curves for atactic polypropylene at various Al6-SG alumina filler levels

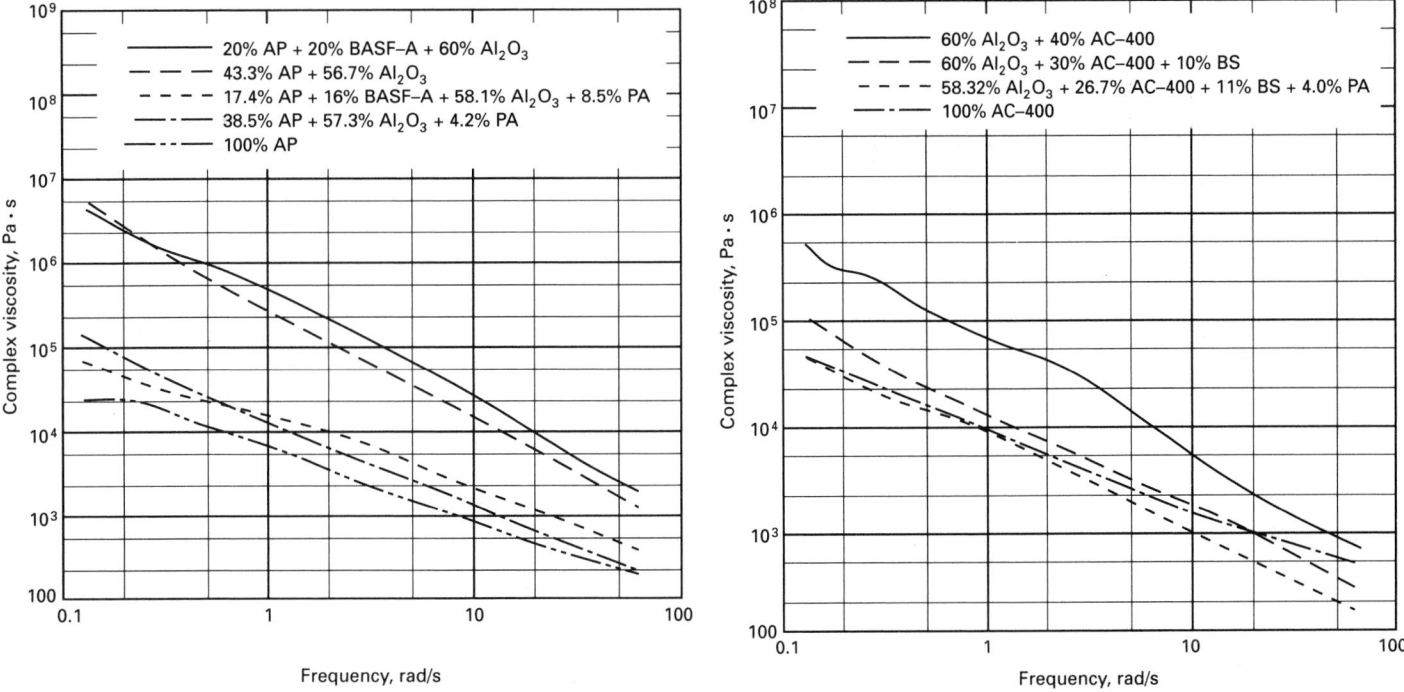

Fig 3 Viscosity of Al6-SG Al$_2$O$_3$ and atactic polypropylene (AP) mixes. All percentages given in vol%. PA, processing aids

Fig 5 Shear modulus of Al6-SG Al$_2$O$_3$ with atactic polypropylene (AP) mixes. All percentages given in vol%. PA, processing aids

mixes. Functionally, processing aids can provide a multitude of effects, including modification of cohesive forces between the polymer and ceramic powder. These materials combine both polar and nonpolar properties (usually, a small polar head and long, nonpolar chain). The polar head is adsorbed on the ceramic powder surface, allowing the long tail to become soluble with the polymer. At the same time, a macromolecule adsorbed on the particle surface can maintain a uniform dispersion in a concentrated mix by counteracting the van der Waals forces, thereby minimizing particle-particle interactions.

Dynamic Shear Modulus. The dynamic shear modulus versus frequency curves for these mixes are shown in Fig 5 and 6. With both polymers, the irregularities (for example, nonlinear) in shear modulus are pronounced with mixes that do not contain either plasticizer and/or processing aids. These

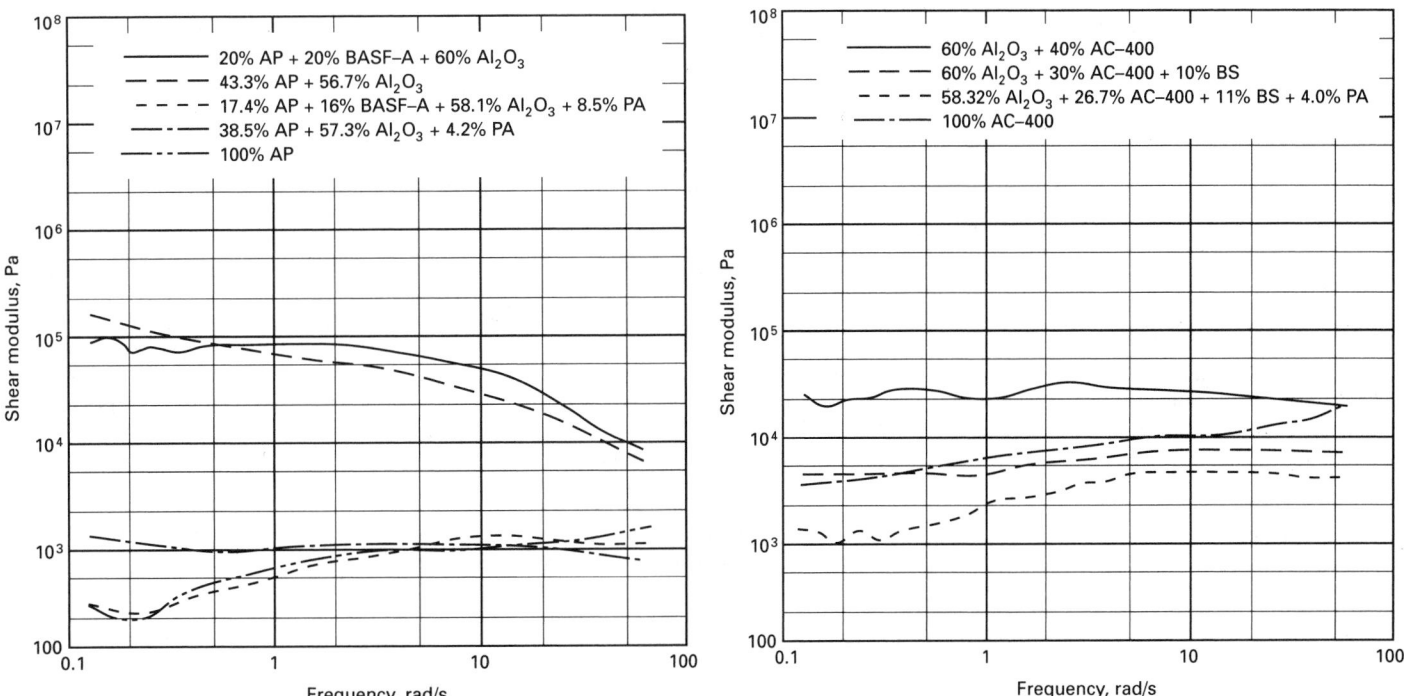

Fig 4 Viscosity of Al$_2$O$_3$ with ethylene-vinyl acetate (AC-400). All percentages given in vol%. BS, butyl stearate; PA, processing aids

Fig 6 Shear modulus of Al6-SG Al$_2$O$_3$ with ethylene-vinyl acetate (AC-400) mixes. All percentages given in vol%. BS, butyl stearate; PA, processing aids

irregularities indicate that the mix does not respond as a continuum, but that some form of internal rearrangement occurs. It is likely that weakly bonded agglomerates start to break down at oscillatory frequencies above 2/s. With the addition of BASF-A wax and processing aids to atactic polypropylene mixes, the overall shear modulus drops significantly, whereas in EVA mixes, the drop in shear modulus is relatively small. In addition, the fact that the dynamic shear modulus drops or reduces in slope indicates that the maximum packing fraction has apparently increased, signifying that agglomerates have been broken down by the shear force imparted by the oscillating plate.

Temperature has a significant effect on the flow behavior of liquids and slurries, including polymer melts, solutions, and filled compounds. Viscosity decreases both for Newtonian and non-Newtonian fluids, primarily because of a decrease in intermolecular forces. The increased flexibility of a polymer backbone chain also contributes to reduced viscosity. In fact, temperature and shear rate are superimposable with respect to the viscoelastic behavior of polymers. For a filled compound, the effect of temperature on particle-particle interaction must be considered, in addition to the effect of temperature on inter- and intramolecular interaction in polymers. This temperature effect depends on the level of loading. If the temperature is sufficiently high to allow for migration of binder from particle-particle interfaces, then the compound may exhibit dilatancy in place of pseudoplasticity.

The effect of temperature is similar to that of shear rate on polymer rheology, but its effect on particle-particle interaction is unpredictable. Both high temperature and high shear rate weaken polymer-particle interaction. However, it is important to remember that the extent of the effect (viscosity and interaction reduction) for a particular change in shear rate or temperature is not predictable. Both factors, exerting their effects at different rates, contribute to the complexity of the system.

Mix Preparation

Mixing is a major factor in achieving a homogeneous injection molding mix that is free from agglomerates, has optimum ceramic-binder content, and still maintains sufficient fluidity for molding. In the mixing process, one attempts to uniformly disperse a relatively small fraction of binder components in a large portion of fine ceramic powder, while maintaining the ultimate binder constituents without intrinsic changes. When selecting a mixing system, factors that should be considered are: optimizing the best achievable mixing conditions, determining the factors that affect the performance of a mixing device, identifying the causes of segregation in mixtures, and knowing and defining when a homogeneous mix is attained.

Mixing operations involve material transport to produce the desired spatial arrangement of the individual components. Therefore, transport mechanisms are important, and can be categorized as dispersive, diffusive, or laminar.

Among the more important functions of dispersive mixing is the incorporation of ceramic powder into molten organic binder. The high shear stresses in dispersive mixing promote dispersion. With any given system, there is some critical stress below which no dispersion can occur. When the shear stress is only slightly greater than the critical stress, only those agglomerates that are favorably oriented initially will be dispersed. If the flow in the mixer is completely unidirectional, only those agglomerates that have a favorable initial orientation will be ruptured. The other agglomerates will simply be aligned with the flow, and continued mixing will result in no further dispersion. However, if the mixer produces a flow pattern that continually changes direction, continued mixing should result in further dispersion. Given time, each agglomerate will eventually be caught by the shear stresses in a position of favorable orientation and rupture may occur.

Three types of mixers are considered suitable for preparing injection molding mixes. Dispersive mixing, which is usually a batch operation, is conducted primarily in internal mixers and roll mills. The internal mixers are batch machines that disperse ceramic powder uniformly within the binder. The typical internal mixer consists of a figure-eight-shaped chamber that fits over two sigmoid, counter-rotating blades. The mixing process is scrutinized by monitoring the torque required to produce a uniform mix. Mixing torque is proportional to the viscosity of the mix and is therefore an indication of the work required to mix the ingredients. Typically, the torque exhibits a rapid rise after binder is added to the ceramic powder. Once the torque reaches a steady state, no additional mixing occurs, and it reflects a uniformity in mix viscosity. Mixing time is determined by the time required for a mix to reach a steady state.

The results obtained with different mixing heads indicate that the breakdown of the agglomerates formed during batch-type mixing depends largely on the geometry of the mixing head, rather than on speed, time, or temperature. A roll mill mixing head gives the best result. However, if mixing is continued briefly at a temperature at which the entire mix is fluid, followed by mixing at a reduced temperature at which the shear action becomes more effective, the number of agglomerates in the mix is markedly reduced. This improvement in deagglomeration is independent of the mixing-head geometry.

It should also be noted that agglomerates can be substantially reduced, depending on the sequence in which various binder ingredients are introduced to the mix. The surface characteristics of the powder and the compatibility of different binder ingredients dictate the sequence of addition. Surface charges and compatibility can be determined from oil absorption, contact angle, and electrophoretic mobility data.

Birchall et al. (Ref 2) achieved a very homogeneous mixture of microdefect-free cement at high solids loading using roll mill mixing. This type of mixing involves passing a mixture of ceramic powder and binder through a relatively wide nip (1 to 10 mm, or 0.04 to 0.4 in.) to disperse and heat the mixture. Slabs of material are then produced on a batch basis for subsequent molding. The system geometry comprises a parallel pair of rigid, rotating cylindrical rolls, with a larger length than diameter and a minimum separation that is very small. The confined flow in the nip can be treated according to unidirectional lubrication theory, but will not be discussed here.

Both single- and twin-screw extruders are also used for mixing, with the latter being more common. An obvious characteristic of screw extruders is their remarkable variety and variation of channel geometry. Such geometry involves interruptions to the screw flight, reversals of helicity, barriers in the main channel, grooves in the barrel, and outside clearances. Although the flow characteristics of such extruders have not been well-defined, useful information is provided in Ref 3 to 5.

Mechanical mixing processes are not usually very effective in producing a high degree of fine-scale homogeneity, because of the dominance of adhesive forces in fine powders. These problems can be alleviated by mixing in a large amount of fluid/solvent/binder and applying turbulence, such as milling. However, the solvent must be removed from these suspensions prior to molding.

Molding Equipment

Injection molding machines must be able to hold a supply of material and provide a method of forcing the material into a mold, a method of opening and closing the mold, and methods for controlling the times, temperatures, and pressures associated with the process. Two basic types of molding machines, the screw type and the plunger type, are shown in Fig 7. These two types differ in terms of material conveying and plasticizing. The plunger machine heats the material by conduction and convection only, whereas in a screw machine, a major portion of the heat comes from the frictional forces between screw, material, and cylinder. In a plunger machine, the pressure loss in the cylinder could be significant, and the pressure at the nozzle would be affected. In a screw machine, the ram pressure and nozzle pressure are almost the same.

The selection of a specific type of machine is influenced by the requirements of the

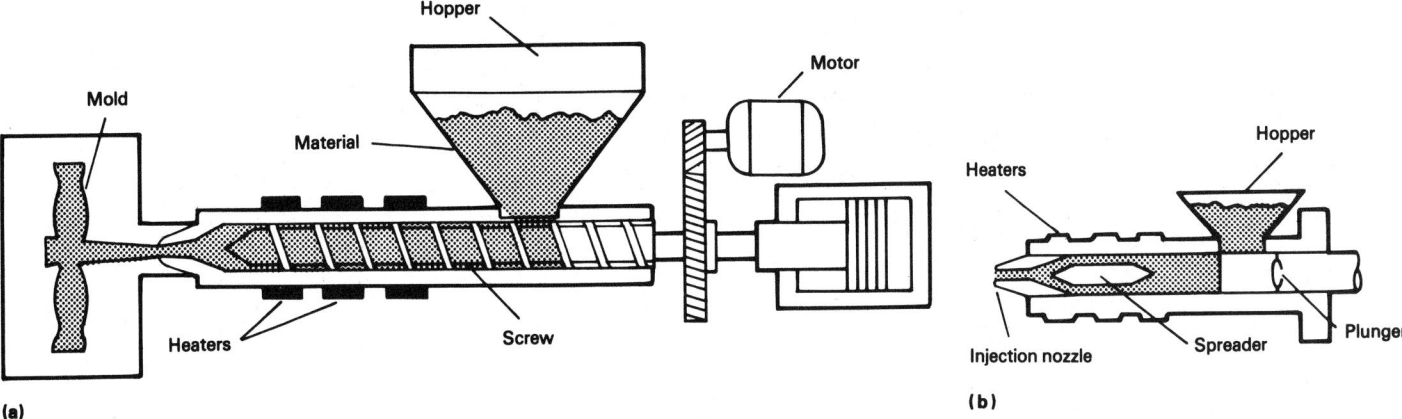

Fig 7 Cross-sectional side view of (a) screw-type machine and (b) plunger-type machine

molded product and the degree of versatility desired. Regardless of type, the machines must have the features shown in Fig 8.

Molding Machines for Ceramics

Until recently, plastic molding machines have been used extensively, without modifications. However, a screw-type machine that is specifically designed for ceramics is now being marketed (Ref 6). This machine is designed to cope with highly abrasive ceramics based on "super-hard" materials.

Another manufacturer (Ref 7) has designed a type of molding machine that uses a wax-type binder and is particularly suitable for low-viscosity mixes. Operation of this machine involves injecting the material through a well-controlled, hydraulic-operated system in which a piston drives a column of the wax-based molding mix. Hydraulic control exists on both sides of the piston. The molding mix is pulled down from the heated reservoir, where it is agitated and placed into a chamber in which hydraulic pressure is applied. After a valve shift, it is injected into the cavity of the cold die, where it solidifies. The machine is available with both vertical and/or horizontal clamping options. Further, the size of the machine can vary from 0.12 to 2.9 MN (12 to 300 kgf) in clamping force, depending on the size and number of parts to be molded in one cycle.

Another molding machine operates at low pressure, typically less than 34.5 MPa (5 ksi). It is claimed that low pressure and low temperature extend the machine life and provide greater operator safety. Still another low-pressure (0.2 to 0.6 MPa, or 0.03 to 0.09 ksi) molding machine developed in Russia for making electronic components is now commercially manufactured in the United States. The machine consists of a tank connected to a feeder pipe, each having independent temperature-control systems. A planetary mixer located in the tank is used for blending the ceramic and binder mixture. The tank is also connected to a vacuum pump that de-airs the mix during preparation. The die is clamped to the opening of the feeder pipe with an air cylinder. Molding takes place by applying air pressure to the surface of the ceramic mix in the tank. The molding cycle takes place automatically. A third low-pressure molding machine has been developed in Japan (Ref 8).

A special machine designed to control the internal stress in large, injection-molded parts has also been described (Ref 9). The pressure on the material is made to fluctuate over a wide range, causing extended heat flux in the sprue that prolongs solidification of the material in this section of the mold. Oscillating pressure is applied by a high-pressure valve assembly that can be maintained at high temperatures, and is mounted between the nozzle and the mold. The machine nozzle sits on one end of the valve block. The other end of the

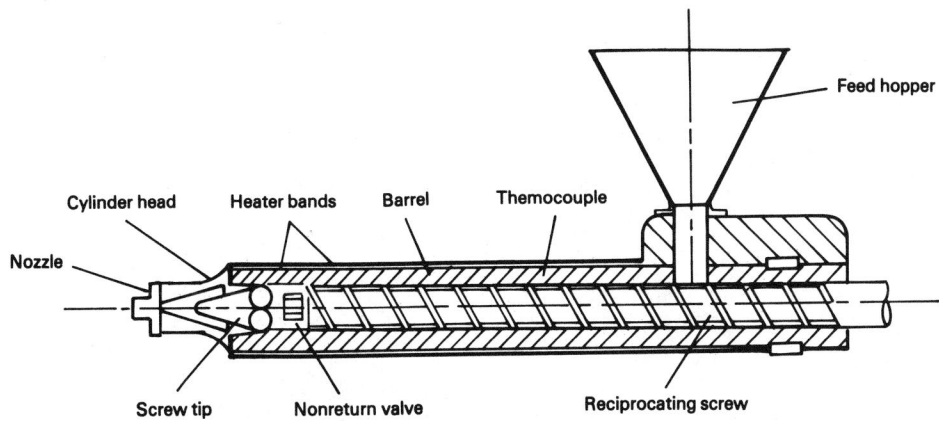

(a)

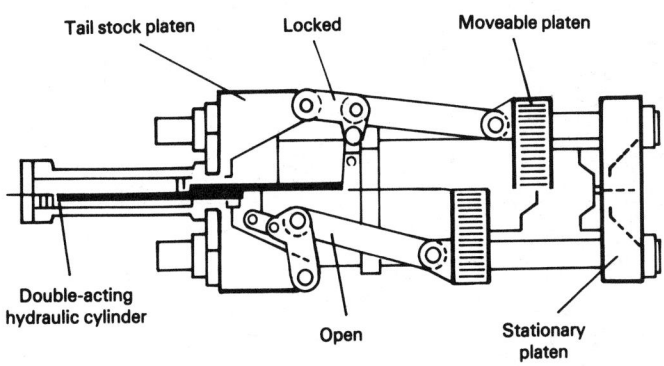

(b)

Fig 8 Components of a reciprocating screw machine. (a) Injection unit. (b) Clamping unit

block contains a nozzle that meets the sprue bush during injection. A passage exists through the block so that material can enter the mold via the pressure unit. It also contains a cylindrical hole inclined to the horizontal axis in which a hydraulically driven piston can alternately compress and decompress the fluid in the cavity by closing the inlet port at the beginning of each compression stroke.

Binder Removal

The binder must be completely removed from injection-molded parts prior to densification. Binder removal, a key operation in successful injection molding, is not well understood. The binder systems may comprise more than one organic ingredient with different melting points and decomposition ranges. In such systems, attempts are made to remove the ingredients slowly and to gradually increase passages (porosity) within the part with increasing temperature and time. Thus, the major binder ingredient can escape at higher temperatures without causing any failure.

Thermal degradation is the principal method for binder removal (in oxidizing atmospheres). It is a common practice to use standard commercial ovens to remove the binders at atmospheric pressure either in air or in a nonoxidizing gas atmosphere. Use of a bed material is often considered to provide additional support, better thermal uniformity, and wicking out of the binder through capillary action.

German (Ref 10) proposed three mechanisms involved in thermal binder removal: diffusion control, permeation control, and fluid wicking. German favors the use of a partial vacuum for diffusion- or permeation-controlled binder removal.

Some furnace manufacturers have offered high-temperature equipment that incorporates a binder removal system as an integral unit. One such unit (Ref 11) uses a carrier, or sweep, gas that enters the sides of the chamber through a series of control systems, providing safe operation in a closed vessel. The sweep gas passes into the hot zone of the furnace, where it entrains the binder vapor being released by the parts, and then passes out of the furnace through an internal manifold to a burner tower, where the gas and vapor are oxidized. Often, a portion of the binder is removed by evaporation under vacuum and is condensed in special condensers outside the furnace. Subsequently, the sweep gas can be used to remove the remaining binder at a pressure of a few torr. Figure 9 depicts a binder removal cycle as a relation between time, temperature, and pressure.

Solvent Extraction of Binder. The potential value of solvent extraction for the removal of the binder derives from the high level of solubility of the binder in solvents, which in many ways are gas-like. Furthermore, viscosity is low, diffusion is high, and

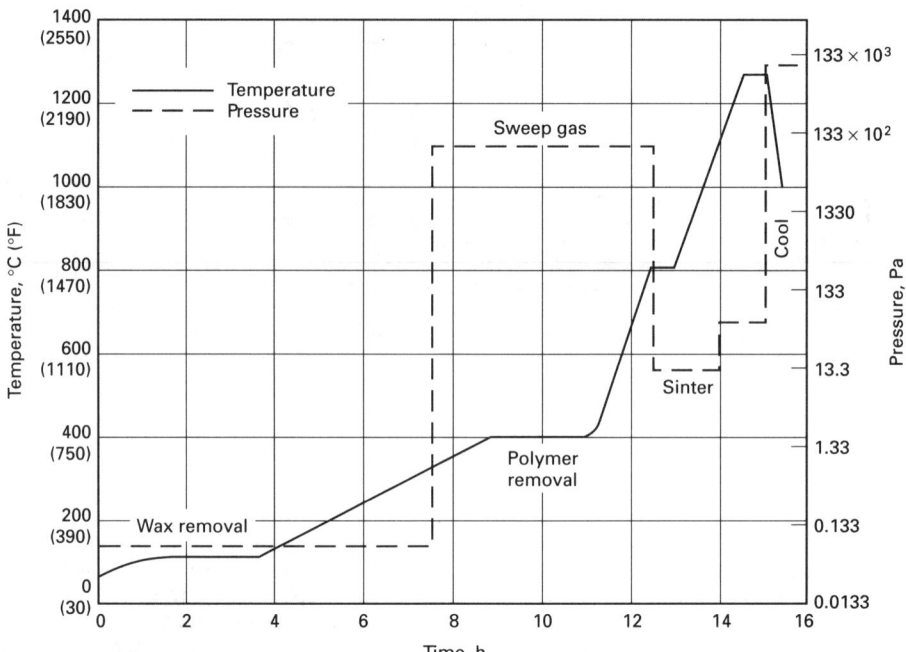

Fig 9 Pressure-induced binder removal cycle. Source: Ref 11

surface tension is minimal. With the proper choice of solvent conditions, it should be possible to remove the binder without part disruption.

If the ceramic part is placed in a solvent, the only force acting on the binder to remove it is the composition gradient. This results in a very slow removal rate, but it is not dis-

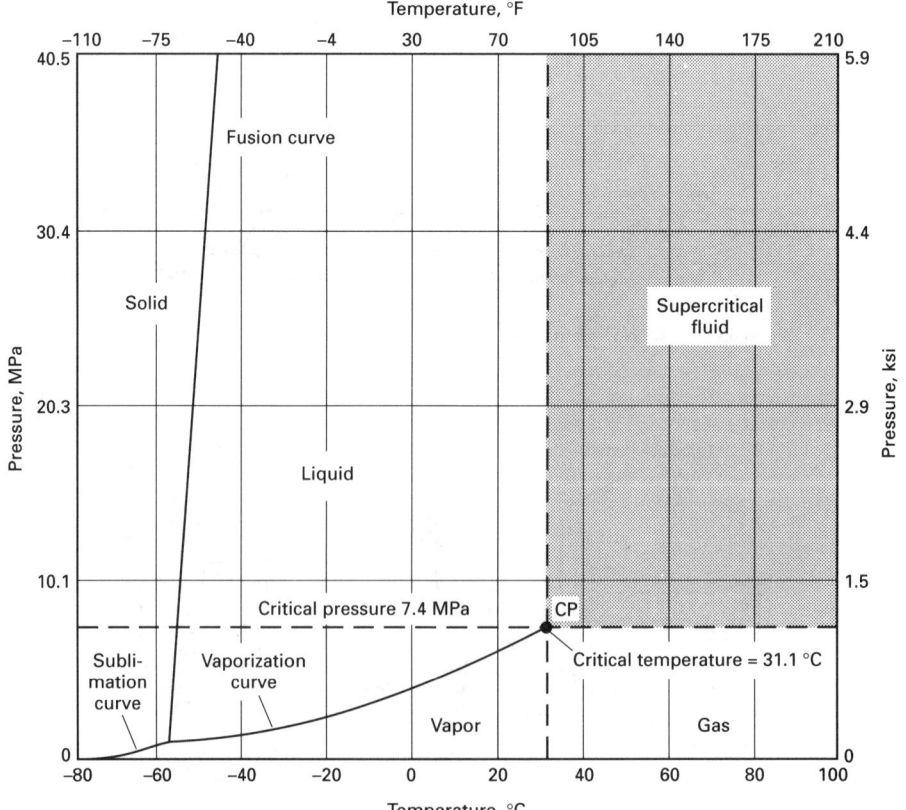

Fig 10 Phase diagram of carbon dioxide. CP, critical point

ruptive. Combinations of waxes and thermosetting resins are used. Extraction has been effected by immersion in, or washing with, boiling solvent prior to drying. The process can be automated by using a conveyor belt that passes through solvent washing and drying zones. A wide range of solvents has been used; many require expensive equipment for accurate control. Solvents can be distilled and reused.

The advantage claimed for solvent extraction (Ref 8) was faster removal of the organic vehicle, and it was also claimed to be more economical, requiring a less expensive plant.

Supercritical Extraction. Figure 10 (Ref 8) shows a phase diagram of carbon dioxide, which enters the supercritical condition at the critical point (CP). The supercritical fluid has a diffusion coefficient similar to that of the gas, and a solubility similar to that of the liquid, but has no surface tension. Therefore, the supercritical fluid is likely to avoid the occurrence of capillary attraction, which causes an irregular rearrangement of packed grains, and either microdefects or density variation in a molded component.

In addition, it is possible to greatly increase the binder removal rate, because there is no generation of large quantities of gas within the weakly bound molded body, as in the case of binder removal using a thermal process. Because there are few limitations on part size, this method is an excellent means of binder removal. Figure 11 is a supercritical extraction unit schematic. Various solvents are applicable as extraction media, but carbon dioxide is excellent in terms of safety and cost performance.

Pyrolysis. Many of the non-oxide ceramics cannot be exposed to oxygen (air) during binder removal. The organometallic polymers that are being used as binders for injection molding of some non-oxide ceramics also cannot be exposed to an oxidizing environment. Therefore, the thermal treatment of these materials is being carried out either in inert or reducing atmospheres. Current understanding of the pyrolysis of such binders is very limited.

Defects in Dewaxed Parts

Four major defects that are frequently encountered in dewaxed parts are skin formation, knit lines, cracks, and delamination.

Skin Formation. The main effect of skin formation is primarily associated with the chemical compatibility of different binder ingredients and the ceramic powder. A cause-and-effect analysis is by no means comprehensive. A better understanding of surface chemistry, the chemical interaction of various binder ingredients, and binder compatibility with ceramic is necessary.

Skin can also form during the molding process. This can be due to either excessive restriction in the mold or to excessive cavity pressure with too low or too high a mold temperature.

Knit lines, also known as weld or flow lines, are flaws in a molded part that occur when two flow fronts meet during the molding operation. Improper gate design, poor flow characteristics of the mix, and/or improper molding conditions cause this defect.

Cracks represent the most difficult defect to remedy. They become progressively worse with an increasing complexity in part shape. These cracks are generally believed to be caused by localized stress concentrations resulting from uneven packing during molding, the part geometry, or nonuniform shrinkage during the cooling period and dewaxing. Shrinkage or expansion could be minimized by increasing the filler loading in the mix and maintaining a mold temperature that allows the stress to be relieved. Stress cracks that originate from adhesion of the part to the mold can be reduced by applying a mold-release agent before every molding cycle.

Another major source of cracks stems from the density variation in a part, which may be related to a number of factors. Density variation is thought to be responsible for the microcracks that are often seen in injection molded parts. These cracks can be identified in dewaxed parts by nondestructive evaluation.

Delamination presents the most challenging problem. However, success is still largely based on trial and error.

Other flaws also can occasionally be seen in dewaxed parts. These flaws include blisters, pinholes, and crazing, which are associated with inhomogeneous mixing and uncontrolled heating during the dewaxing process. Speckles are normally caused by foreign inclusions. Slumping is primarily a function of excess binder, uncontrolled heating rate, and, most likely, the particle shape of the ceramic powder.

REFERENCES

1. N. Sarkan and G.K. Gremminger, Jr., Methylcellulose Polymers as Multifunction Processing Aids in Ceramics, *Am. Ceram. Soc. Bull.*, Vol 62 (No. 11), 1983, p 1280–1284, 1288
2. J.D. Birchall, A.J. Howard, and K. Kendall, A Concrete Approach to the Energy Crisis, *Met. Mater. Technol.*, Vol 15, 1983, p 35–38
3. Z. Tadmor and C.G. Gogos, *Principles of Polymer Processing*, John Wiley & Sons, 1979, p 434
4. V.W. Uhl and J.B. Gray, Ed., Chapt. 1 & 2, *Mixing: Theory and Practice*, Academic Press, 1967
5. H.F. Irving and K.L. Saxton, Chapt. 1 & 2, Mixing of High Viscosity Materials, *Mixing: Theory and Practice*, V.W. Uhl and J.B. Gray, Ed., Academic Press, 1967
6. M. Inoue, Y. Kihara, and Y. Arakida, Injection Molding Machine for High-Performance Ceramics, *Interceram*, Vol 38 (No. 2), 1989, p 53–57
7. J.R. Peshek, Machinery for Injection Molding of Ceramic Shapes, *Advances in Ceramics*, Vol 9, *Forming of Ceramics*, J. Mangels and G.L. Messing, Ed., American Ceramic Society, 1984

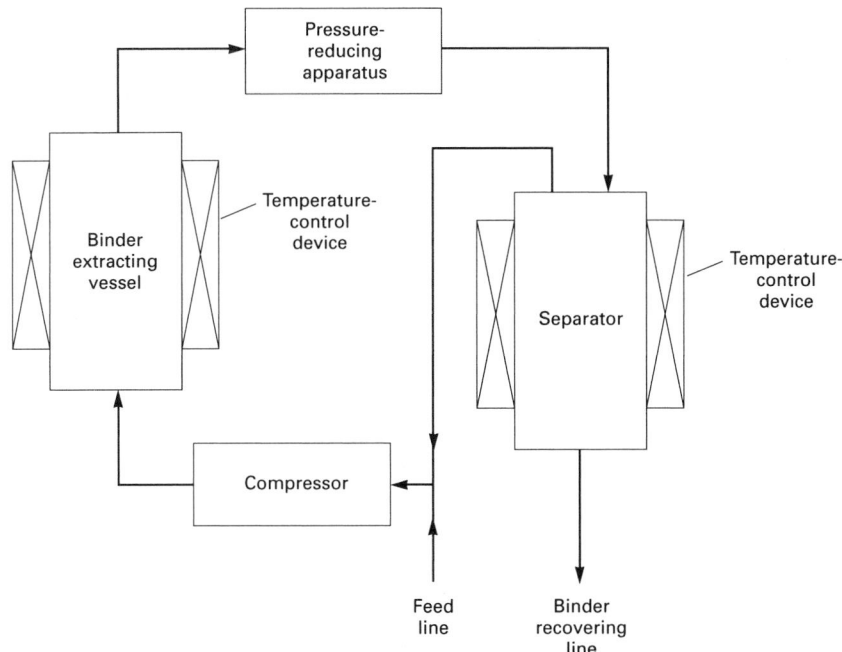

Fig 11 Schematic of apparatus for supercritical extraction. Source: Ref 12

8. A. Miyamoto, H. Nishio, and T. Miyashita, "Slip Injection Molding and Supercritical Extraction-SIMSE," presented at AUSTCERAM '90, Perth, Australia, 26–31 August 1990

9. M.J. Edirisinghe and J.R.G. Evans, Systematic Development of the Ceramic Injection Molding Process, *Mater. Sci. Eng.*, Vol A109, 1989, p 17–26

10. R.M. German, Theory of Thermal Debinding, *Int. J. Powder Metall.*, Vol 23 (No. 4), 1987, p 237–245

11. S.W. Kennedy, Advanced High Temperature Equipment Used for New, Structural Ceramic Processing, *Ind. Heat*, Vol 53 (No. 4), 1986, p 28–33

12. N. Nakajima *et al.*, U.S. patent 4,731,208, 1988

Green Machining

Dean Larson, Coors Ceramics—Structural Division

THE OPERATIONS classified as green machining are any of those that occur following the initial compaction of the ceramic body yet prior to the part being fully sintered. While green machining is normally considered feasible after such processes as isostatic pressing, where the final shape is not yet completed, it can also be very important in making available the ultimate capabilities of other processes such as dry pressing, casting, and injection molding. Some shapes simply cannot be formed without these secondary green machining operations.

Not only is material removed with these processes, but there is also the option of adding material at this stage via slip bonding. This allows the component designer additional flexibility in meeting both configuration and property specifications of the as-fired part. Additional information is available in the article "Slip Casting" in this Volume.

Because of limitations imposed by the binder system used in some ceramic formulations, it may be necessary to partially fire the part prior to using the green machining method. This process is called bisque firing and because the machining process required to be used in that condition is more like green machining than machining on fully fired parts, it has been included in this article.

As with any other part production, not only should the short-term apparent effects of the machining operations be considered in defining requirements of the manufacturing processes, but the long-term effects further along the manufacturing stream through to the ultimate part use must also be evaluated. Recommendations will be made regarding the subsequent effects of various techniques as well as the inspection techniques that are available and necessary to produce a quality ceramic product.

Objectives and Advantages of Green Machining

The purpose of all green forming operations is to produce a component that comes out of the firing process as close to the final shape as possible. The benefits of these additional manufacturing steps are gained in the areas of economics, capability, and quality.

Cost-Effectiveness of Process. The closer the as-fired part is to final configuration specifications, the less time will be devoted to the costly grinding operation. In addition, less grinding tooling is consumed, which can be a considerable savings if the cutting edge consists of diamond abrasives. The equivalent amount of material will require up to $10\times$ more machining time to remove after firing than in the green state and could cost $20\times$ more in tool wear expenditures.

With the application of computer-aided machining (CAM) technology to the green forming process, numerous other economic benefits are provided. Virtually instantaneous programming of computer numerical control (CNC) tools makes it feasible to use a single set of pressing tools to make numerous shapes from the same pressed blank. This not only saves on the initial tooling cost but also reduces leadtime for the fabrication of the required tools. Easy production of design variations also encourages empirical development of the optimum shape for firing thus resulting in the best possible overall production yields.

Process Capability. The most obvious increase in manufacturing capability through the use of green machining is the ability to send completed parts to the kilns that would be impossible to form directly via any pressing or casting process. Some processes such as jiggering or casting leave features such as seams or feather edges that have to be removed before firing. Other examples of features only possible through green machining include zero-draft inside diameters, undercuts (with good green pressed density in all remaining areas), or shapes unfillable due to excessive length (Fig 1). Less obvious are the increases in the ability to treat the surface. For example, it can produce smoother surfaces via sanding or buffing.

In addition, the slip-bonding process joins multiple parts, producing shapes that would be impossible to form because of their complexity or prohibitively expensive to press and form.

Product Quality Control. Every step in the manufacturing cycle should be evaluated on the basis of how the process can improve the final quality of the product. Green machining can be beneficial here as well.

By judiciously selecting the areas to machine in the green state, the yield of the firing process can be significantly improved. Cutting away areas of low density makes for a more uniform shrinkage. Chamfering or radiusing sharp outside corners or relieving inside fillets will reduce the incidence of stress raisers throughout the firing process as well as in handling of the component afterwards (Fig 2).

By the application of appropriate surface treatments, the residual stresses generated in the green compact because of nonuniform pressing can be relieved to reduce the chance of fired cracks from burnout degassing or thermal shock.

Green machining can also smooth out defects due to variations in the raw materials or other process steps up to this point. Common differences in green bulk density, compaction ratios, and shrinkage are caused by variations in the compaction pressure. With the advent of CAM techniques, programs used in CNC machine tools can be modified virtually instantaneously and adapted to any idiosyncrasies of the green powder while either maintaining pressure values or leaving compaction die setups intact.

Equipment and Tooling

Machining Specifications. Virtually all conventional or CNC machine tools can be adapted for use in green ceramic machining. Vertical and horizontal lathes (either mounted between centers or in chuckers), mills, drill presses, and sanders are all needed in a production operation. Machining specifications are considerably different for green machining than the specifications required for metal cutting.

Power Requirements. While tolerance requirements and the mass of the workpieces are comparable to that of metal cutting, the power requirements of green machining are significantly reduced. This results in specifying stiff and substantial frames with disproportionately low power requirements. Green machining power requirements for a milling or turning operation may be as low as one-fifth the requirement for comparable metal-cutting operations.

Surface speed requirements will vary somewhat from typical metal-cutting operations and from one ceramic formulation to another. Speeds to consider as a starting point for fully-dense alumina bodies are shown in Table 1.

Reduction of Dust Generated by Machining. Machining ceramic green bodies, by nature, generates more dust than chips. This creates the need for different types of waste removal methods as well as dust protection for the machinery. Dust protection is crucial both in the area of machine way and ball-screw protection and needs to be taken into consideration for electronic control isolation and protection as well. Machine ways that may last a decade in metalworking applications will degrade significantly in only six months while in steady green or bisqued ceramic machining if the wear areas are left unprotected. The dust produced is virtually a diamond lapping compound and can move tolerance capabilities from ±0.0125 to ±0.125 mm (±0.0005 to ±0.005 in.) very quickly as well as making sophisticated CNC controls unreliable or completely inoperable.

Successful methods of dust isolation include capturing as much dust as possible in the air behind the tool before it has a chance to settle, compound boots at all areas of entry to the machine works, positive air pressure within the machine works and very slow weep lubrication of the exposed machine ways. High cooling load and sensitivity of CNC controls makes the adequate volume and filtration of incoming cooling air an often overlooked priority in otherwise well-designed machine tools.

Liquid coolants cannot be depended upon as dust reducers, not only because of their detrimental effect on the part being machined but also because of the reduction in value of the potentially reclaimable excess ceramic material.

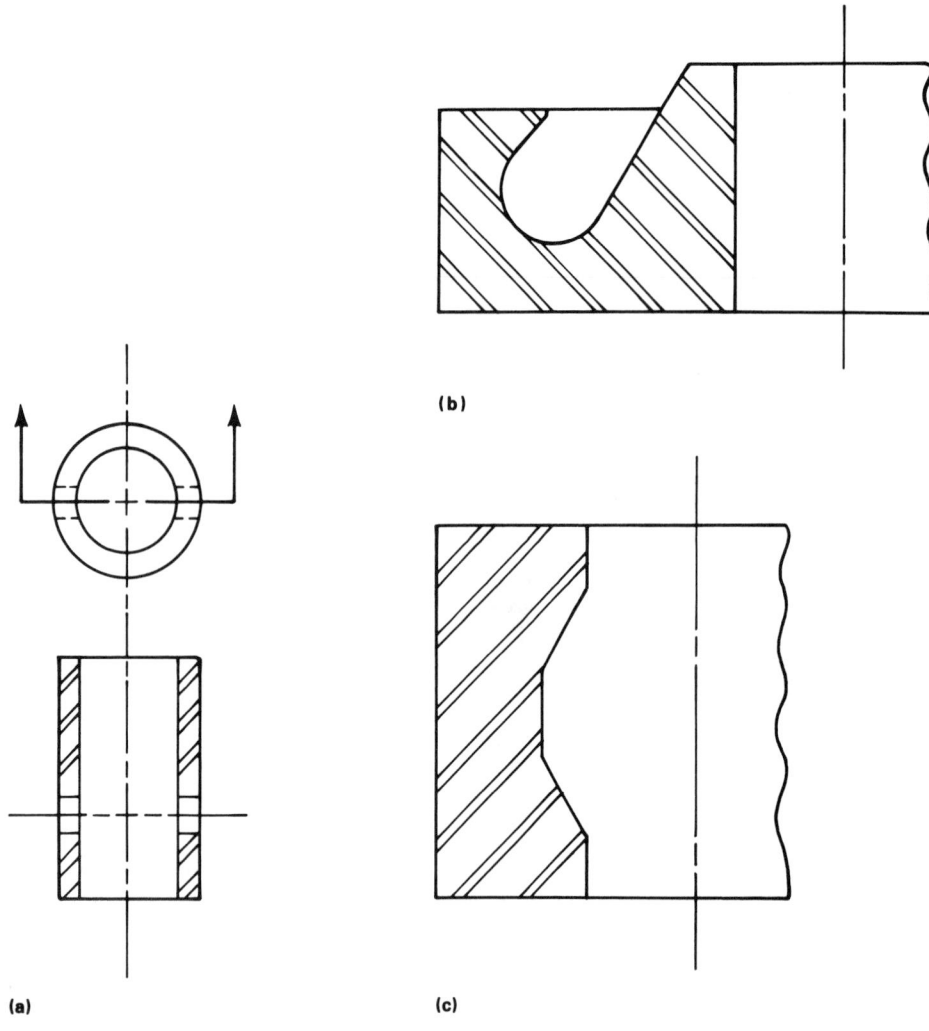

Fig 1 Cross sections of unpressable ceramic shapes that can be produced by green machining techniques. (a) Cylindrical pipe with through holes perpendicular to walls and centerline of pipe. (b) Wheel hub with a blind relief. (c) Pipe wall having varying wall thickness due to changing inside diameters

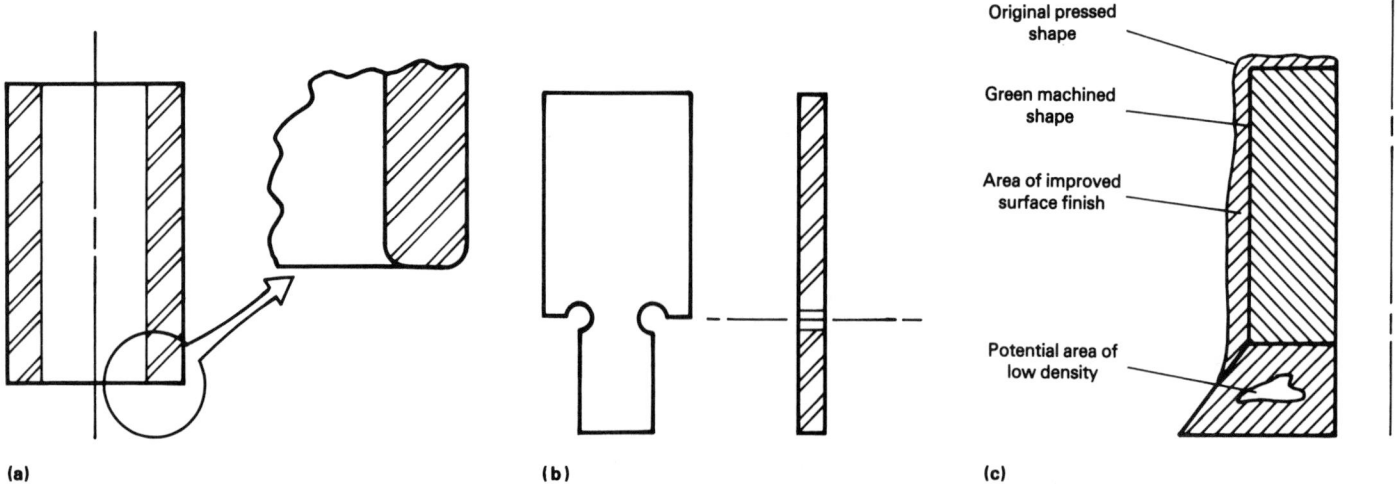

Fig 2 Green machining operations that improve the product quality of ceramics prior to firing in a kiln. (a) Radiusing sharp corners to obtain chip reduction. (b) Removal of areas of low density to improve uniformity of density and surface quality. (c) Relieving inside fillets to reduce stress risers

Table 1 Speeds and feeds for green machining of fully dense alumina ceramic bodies

Depth of cut		Feed		Surface speed	
mm	in.	mm/rev	in./rev	m/min	sfm
0.13	0.005	0.05	0.002	76	250
0.38	0.015	0.13	0.005	46	150
0.76	0.030	0.25	0.010	30	100
1.5	0.060	0.38	0.015	23	75

Environmental Effects. The high degree of dust generation also requires assessment of the environmental impact upon the people who work in the area. The very fine particle size, the variety of binders, and the possible toxicity of some components make this a serious consideration that can be solved with engineered solutions much more readily at the initial stage of production machine specification rather than when the machining operations are already in progress.

Computerized Numerical Control. The addition and sophistication level of CNC control options should be strongly considered in the machine specifications because of the potential advantages of real-time control in cutting programs. These can account for variations in the raw ceramic materials and make a more consistent final product.

Machining Processes. Because of the unique characteristics of pressed green ceramic, not only can the traditional tooling or workholders associated with metalworking applications be used but also toolholders normally only considered for use in woodworking applications. Most pressed green ceramics or bisqued ceramics have the strength characteristics of blackboard chalk and appropriate fixtures must be used for machining operations.

Single-Point Tooling. For the purpose of maximum setup flexibility or maximum material removal, single-point tooling is used. Although the geometry of the point may not be as critical as the geometry required for metalcutting, the sharpness of the tool is still a crucial factor in minimizing the amount of surface damage incurred by the workpiece and residual stresses. In addition, because the power requirements for green ceramic applications are less than the requirements for metal-working applications, the cutting tool support does not need to be as rigid as the support used in metal-cutting applications. This increases the number of options available when selecting fixturing shapes. For example, the fixture design may specify a mechanism having simultaneous multiple-tool engagement or extended reaches.

Form cutting or grinding is an extension of the single-point cutting edge. The entire contour of a turned part can be produced on a carbide blade or formed grinding wheel and moved into a turning part in one motion. If the volume of parts requiring green machin-ing is sufficient, this can be a highly automated and efficient way of reducing pressed stock. If contour cutting is done with a carbide blade, care must be taken to monitor the rate of wear on the blade over a span of time to ensure that dimensionally correct parts are produced prior to sintering in the kilns. If the quantity of parts and production speed requirements are great enough, it may be necessary to contour the blades with custom diamond grinding wheels to ensure consistent dimensional accuracy.

Coining and Stamping. The malleable nature of green ceramic (but not bisqued parts) allows machining operations such as coining or stamping to be applied to the workpiece. Typically, this would be used to produce two-dimensional shapes composed of thin stock such as tapecast parts, but it could also be used to add a third dimension to thicker parts. Caution must be exercised to minimize the effect of density changes caused by third-dimensional stamping or by residual stresses that can lead to failure of the workpiece during firing or subsequent machining operations.

Laser-Beam Machining. Just as the machining capabilities of the industrial laser are commonly used for cutting fired ceramics, laser beam machining (LBM) can also be used in green machining applications. Possibilities include scribing eventual break lines, adding z-axis features (typically, ≤1.5 mm, or 0.060 in., deep), or cutting very small and detailed edge configurations. Laser production tends to be considerably more expensive than other cutting techniques, but it may be the only available option for low production quantities or very detailed work.

Surface Finishing. A family of operations not often included in the field of machining includes surface finishing operations such as sanding, sponging, or polishing. These operations can result not only in a smoother as-fired surface, but they can also eliminate stress risers on surfaces that have been cut too rapidly with a single-point tool or to break up sharp external corners, which could promote the formation of chips and cracks. Any sharp edges can create potential for stress risers that propagate through the part after firing and should be eliminated by the various kinds of surface finishing.

Cutting Edge Specifications. Selection of the proper cutting edge for green machining applications is based primarily on economics as it is in the case of other materials. In the case of ceramics, however, the decision is easier because the tool life of the high-speed steels cutting ceramics is so short that it would be difficult to find circumstances under which they would be preferable in green machining applications. The alternative cutting tool materials are carbides and diamond. The long tool life of diamond makes it preferable for single-point cutting, while the flexibility of shape fabrication for carbide makes it preferable for form turning. The use of grinding operations in green machining is generally optimized with resin-bonded diamond wheels providing the cutting edge.

The specific type and grade of polycrystalline diamond or other cutting edge used will also vary with the specific type of green ceramic being cut (Table 2). The tendency of the green ceramic binders to build up on the grinding wheel makes the size of the grit and the pattern of diamond deposition on the wheel surface important factors in the selection of the cutting edge specifications.

Fixturing. The unique physical characteristics of green or bisqued ceramic require careful attention to the method of holding the blank while it is being machined. The demand for stiff, secure retention of the workpiece to allow maximum energy input at the tool edge must be weighed against the demand for even, modulated contact that does not damage or deflect the soft ceramic blank.

Although properties vary widely, green ceramic properties such as a green modulus of elasticity (MOE) and an ultimate tensile strength that are a small percentage of fired ceramic properties (or most industrial metals) need to be taken into account in process design. The obvious failure of cracking in the machining operation is not the only concern because internal, undetectable failures can be produced yet detected only after firing—or even worse, in end use. Furthermore, even if the ultimate strength is never exceeded, the relatively low MOE value in combination with inappropriate fixturing can cause deflection during machining that results in parts which are bowed or oval even before firing.

While numerous types of clamps can hold green ceramic, the concerns listed in the previous paragraph tend to narrow the choice to clamps that maximize the surface area of contact. When combined with careful attention to the elimination of sharp edges (for example, burrs), the maximum point loading on the part to be machined will be minimized while the overall holding effectiveness will be maximized. Careful monitoring of these variables will minimize the potential of green cracks in the workpiece.

The use of either air pressure or vacuum chucks should be considered because of their ability to maximize the area of contact as well as their ability to control the air pressure to the maximum force level that still does not produce excessive distortion. Furthermore, this allows a positive grip on the workpiece with small dimensional deviations. Figures 3 and 4 show the application of a pressure chuck (Fig 3) and vacuum chucks (Fig 4) in green machining.

Feed Rates. The determination of the optimum feed and speed rates for machining green ceramic is based on the rate at which the material is removed versus how much power can be consumed in machining of the workpiece before the physical strength is exceeded and damage occurs. Extreme speeds, even if the parts do not fly apart due to centrifugal force, may cause either internal cracks

Table 2   Comparison of cutting speed versus tool life for selected abrasives showing cost-effectiveness of green machining a green compact of pure silicon prior to nitriding operation to obtain reaction-bonded silicon nitride product

| | Cutting rate | | | Tool life | | |
| | Surface speed | | | Workpiece material removed | | |
Tool material	m/min	sfm	Cost factor	cm³	in.³	Cost factor
Cubic boron nitride	30	100	x	8.2	0.50	x
Titanium-coated carbide	30	100	x	13.1	0.80	x
Cobalt-bonded tungsten carbide	30	100	x	19.7	1.20	x
Diamond	90	300	10x	8500	520	10x

Source: Ref 1

or may damage the surface with resulting failure when sintered. Surface damage may occur on either the workpiece being machined or the workpiece surface being held by the chuck should there be slippage. Tool wear as a function of the material removal rate (MRR) is not significantly influenced by the rate of feed.

Surface Finish. While virtually any surface finish may be obtained with careful control of the cutting tool, care must be exercised to ensure that neither the surface to be fired is damaged nor is it covered with minute but sharp stress risers that can cause outright failure during the manufacturing process or fatigue failure in eventual application in the field.

Material Requirements

The traditional design of ceramic parts optimizes the material formulation for final functional requirements or processing limitations such as raw material processing, compaction, or firing. While those steps are important, the green machining process requirements need to be considered as well to optimize the overall manufacturing process.

The primary concern in the selection of the best binder formulation for green machining is the green strength of the pressed part itself. The material has to be strong enough not only to handle the stresses of machining but also to survive part handling and movement to the next operation. On the other hand, if the body is too strong in the green state, machining can become very difficult due to increased power requirements, chatter, deflection of the holding fixtures, and, most importantly, faster tool wear.

The optimum strength for the machining operation can vary with the design of the part. Thick workpiece sections can stand the high forces resulting from aggressive cutting of tough body formulations while thin workpiece sections need additional strength to overcome collateral damage incurred during handling elsewhere in the manufacturing process.

The ceramic in the pressed green condition is relatively fragile in terms of its resistance to environmental conditions. Both shelf life and long-term creep characteristics need to be taken into consideration. While the temperature and humidity of the process rooms need to be monitored for the optimum control of the overall manufacturing process, the best assurance of absolute uniformity in workpiece properties and dimensions require binder formulations that provide good resistance to corrosive atmospheric attack as well as having predictable springback and creep characteristics.

Although the machine tools that will be used for green machining applications need to be adapted to the dust generation of this material both for maintenance and health reasons, the most efficient method to reduce dust is to control the amount of dust generated at the source or tool point itself. Through careful selection of the binder chemistry and loading, dust generation can be minimized and excess ceramic will end up in the reclaim area rather than on the surface of the machine ways or in the lungs of the employees.

Processing parameters that can be varied to maximize green strength include increasing the compaction pressure, increasing or changing the type or loading of the binder, and varying the distribution of ceramic particles. Dust generation is controlled by selecting alternative binders.

Capabilities and Limitations

Size Requirements. It is the restrictions of section thickness and overall size imposed by the sintering operation and not the restrictions imposed by the actual green machining operation that determine the size limits of the green machining process. Nevertheless, size is a key factor because of the inherent limited structural strength in either green ceramic or bisqued ceramic. In making a preliminary evaluation of the green machining specifications of a specific ceramic, it is advantageous to note that the green strength of most ceramics is comparable to that of blackboard chalk. Bisqued ceramic is not much stronger and in certain circumstances the green strength may be even weaker. The application of improper handling devices can damage the workpiece surfaces with concentrated loading, can cause impact with other parts if movement is not constrained or, worse yet, can cause internal cracking that is not detected until the workpiece is fired.

Tolerance Requirements. If the proper machine tool has been selected for green machining operations, the tolerance control in the machining process will be tighter than that capable of being held in subsequent firing operations. The organic binders present in the ceramic do cause some differences in machining specifications compared to typical metal-cutting operations. Depending on the

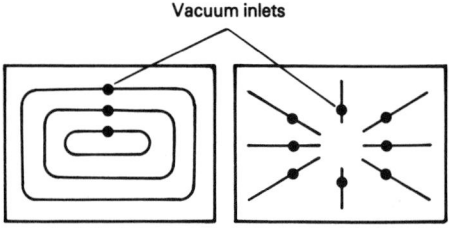

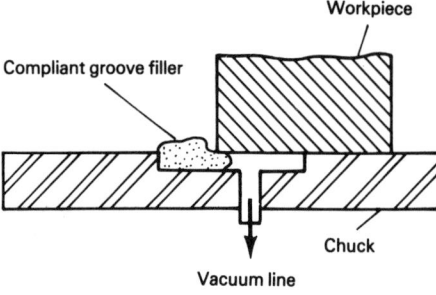

Fig 3 Typical flat vacuum chuck system used in green machining operations. (a) Plan view of chuck surface incorporating either circular path grooves (left) or straight path grooves (right) to provide sites for vacuum inlets. (b) Sectional elevation showing use of compliant filler to provide seal between the chuck surface and the workpiece surface

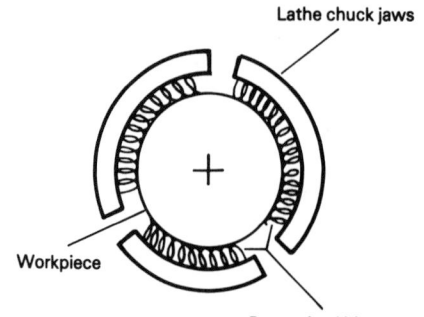

Fig 4 Schematic showing use of pressurized bladders to hold a ceramic workpiece in the jaws of a chucking lathe in green machining operations

elapsed time and the environmental conditions under which the parts are machined or stored prior to firing, a degree of swelling (caused by moisture absorption or springback from pressing) or warpage (caused by relief of internal stresses during green machining operations) may occur. This may amount to ≤0.5% in linear dimensional change and needs to be either controlled or compensated for when making parts whose as-fired tolerances need to be tight or where grinding stock needs to be minimized. These and other factors for making compacts with uniform characteristics prior to firing in a kiln are shown in Table 3. These effects are not encountered on machined, bisqued parts that maintain their stability indefinitely.

Surface Finish Capabilities. Surfaces can be very smooth in the as-fired condition if the proper green processing steps are followed. These steps include final skimming cuts taken with properly sharpened tools and then sanding with progressively finer papers or sponges. A 1.6 µm (64 µin.) R_a as-fired surface finish is typical with only standard machining practices and it is possible to get a 0.40 µm (16 µin.) R_a surface finish with some care when using a fully dense ceramic.

Quality Control Inspection Techniques. Because of the relatively fragile nature of the ceramic binder used in green machined ceramics, inspection techniques are somewhat limited, thus making it imperative that processing of the ceramic be diligently monitored through every step of its production. Although it is impossible to detect all possible defects, careful and predictable processing can reduce the possibility of their formation in the first place.

Density Measurements. The density of the part can be checked to ensure proper compaction and resulting shrinkage. This can be done either by weighing the part directly or by measuring and subsequently weighing test slugs pressed at the same time. This technique is restricted to an aggregate estimation

Table 3 Green machining tolerance allowances and processing steps required to obtain from ±0.5% to ±5% in as-fired tolerance levels

Tolerance, %	Procedure
±1.0	Characterize each ceramic body batch by making and firing test blocks to predict shrinkage
	Calibrate and monitor pressing pressure ±700 kPa (±100 psi) or about 1% of pressing pressure
±0.1	Characterize each ceramic body batch by making and firing test blocks to predict shrinkage
	Calibrate and monitor pressing pressure ±700 kPa (±100 psi) or about 1% of pressing pressure
	Characterize each pressed block
	Control temperature and relative humidity for all green ceramic areas
	Use one-half the feed rate normally used to obtain ±1.0% tolerance
	Calibrate and monitor pressing pressure to ±350 kPa (±50 psi) or about 0.5% of pressing pressure
	Control the queue time from final green machining to firing to, at best, ensure all residual stress relief or creep is over or, at least, to ensure a uniform amount of creep

of part density. More precise analysis of a particular region in a part can be done by x-ray analysis. However, it is limited to a relatively small cross section of ≤12.7 mm (0.500 in.), but this dimension can be crucial for predicting shrinkage uniformity in dry-pressed parts because different regions in a part may receive different levels of compaction.

Dimensional inspection of green machined parts can be accomplished by techniques applied to any other materials, including the use of either contact or noncontact probes using state-of-the-art robotic coordinate measuring machines. The same concerns specified in the section "Material Requirements"

in this article related to flexibility and creep need to be evaluated when parts are held for inspection.

Surface cracks can be detected via the application of selected penetrant dyes. The vendor supplying the dye has to be consulted to determine specific dyes to be used for absorbent materials and to recommend a minimum amount of material that should be used to limit the possible detrimental effects to the binder system during burnout. Careful dye selection and proper testing procedures can provide effective nondestructive testing techniques.

There are radiographic and thermographic methods under development to detect internal cracks, porosity, and variations in density. However, these techniques are only in the early stages of development and can often give ambiguous outputs. The most effective method to ensure a uniform internal structure is careful control and monitoring of all steps from raw material preparation through firing in the kiln.

ACKNOWLEDGMENT

The author would like to thank Tom Riley, John Wilson, Binh Nguyen, and Steve Manning for their help with this article.

REFERENCE

1. D.W. Richerson and M.W. Robare, Turbine Component Machining Development, *The Science of Ceramic Machining and Surface Finishing*, N.B.S. Special Publication No. 562, National Bureau of Standards, 1979

SELECTED REFERENCES

● F.H. Norton, *Elements of Ceramics*, 2nd ed., Addison-Wesley, 1974
● D.W. Richerson, *Modern Ceramic Engineering*, Marcel Dekker, 1982

Hot Pressing

Andre Ezis, Cercom Inc.
Jack A. Rubin, Ceramic Consultant

HOT PRESSING, or pressure sintering, is a commonly used fabrication method that couples both thermal and mechanical energy to effect densification (Ref 1). Externally introduced shear and compression, the mechanical energy components, strongly influence the ceramic densification process (Ref 2–4). In general, the process is initiated by placing a powder mixture or a precompacted form into a pressing cavity and then sintering by applying an appropriate time-temperature-pressure profile. The superimposed uniaxial load provides the mechanical energy and is usually applied simultaneously with heat during processing. The resultant increase in particle mobility and contact-stress within the powder or preform rapidly accelerates the kinetics for densification. Calculations show that the densification rate can be increased by as much as twentyfold by the addition of mechanical energy during sintering (Ref 5).

The advantages of the hot pressing process relate directly to the additional energy source and the accompanying increased densification rate. Fully dense, fine-grained ceramic bodies can be fabricated at lower temperatures and at shorter cycle times than those required by conventional sintering techniques, using equivalent starting powder parameters.

Microstructures can be engineered with respect to grain size and, if required, pore size and distribution. Large cross-sectional bodies can be manufactured and materials that normally are "not sinterable" can be densified by hot pressing. Starting powders generally do not require large surface areas, and minimal amounts of densification (sintering) aids are needed.

The disadvantages of the hot pressing process are also related to producing the mechanical energy component. External pressure must be generated and, by definition, contained during processing. Therefore, furnaces and associated equipment are costly when compared to traditional ceramic fabrication systems. Tooling is an added expense in cases where die and punch sets have a limited life. The process is also limiting in shape capability and is generally performed on a batch basis.

Hot Pressing Process

In simple form, the process requirements are:

- Constructing a die or mold assembly that consists of a die body (structural component) and inserts, which together define the desired pressing cavity
- Assembling pistons or punches that transmit an externally generated pressure to the die cavity
- Introducing a powder or preform into the die cavity and placing the assembly within an appropriate furnace chamber
- Engaging the pistons with a pressure source
- Establishing a desired furnace environment
- Applying pressure as temperature is increased
- Maintaining temperature and pressure until densification is achieved
- Cooling to allow for disassembly, part removal, and inspection

The resultant part is generally in a billet or flat plate form and requires diamond grinding as a finishing operation.

Uniaxial Hot Pressing

Uniaxial hot pressing is the most commonly used fabrication technique for both commercial and laboratory applications. It is also the most restrictive technique. Because pressure is applied from one or two directions (unidirectional or bidirectional), component shaping is normally limited to right-angle stock in the form of plates, disks, and cylinders. Innovative tooling concepts can broaden the shape-making capability, and, with appropriate die body design, the unidirectional system can be made to simulate a bidirectional condition. Additional limitations are placed on component thickness and/or the number of components that can be hot pressed simultaneously. Die wall friction and pressure distribution limit most cross-sectional heights, L, to approximately one-quarter of the nominal component diameter, D. When the L/D ratio is less than 2, multiple components can be processed by placing rigid spacers between each unit. However, the cumulative L/D ratio is still limited to approximately 4.

The hot pressing furnace can have a variety of configurations, but is generally classified as being either an ambient (0.1 MPa, or 1 atm) or vacuum setup. Most ambient furnaces are "box" type in construction and generate a nonoxidizing environment during operation (Fig 1). These furnaces are heated inductively or resistively and use carbon black and carbon felt as the insulating medium. The carbon black is not only an effective insulating material, but also serves as a "getter" for oxygen, thereby protecting the die assembly and the powder charge. Ambient furnaces are relatively inexpensive to construct, but their use is limited because of inherent safety concerns.

The furnace of choice for most applications is a vacuum unit with "back-filling" capability (Fig 2). Furnaces can range in size from small laboratory models to units that can produce components with diameters up to 915 mm (36 in.). Most units are constructed with water-cooled induction coils that "couple" into a graphite susceptor, which then serves as the heating element. In some designs, the die body also doubles as the susceptor. Resistance heating can also be used in vacuum furnace construction but is generally considered operationally restrictive.

Hot pressing may require temperatures up to 2500 °C (4530 °F) or higher. Furnaces and die assemblies constructed from graphite are capable of achieving these elevated temperatures. However, temperatures between 1400 and 2000 °C (2550 and 3630 °F) are used for most applications.

Furnace environments are dictated by the die assembly materials, the material being pressed, and the processing requirements. Typical environments are a vacuum, nitrogen, argon, or helium. In practice, usually more than one environment is used during the hot press cycle. If an oxidizing atmosphere such as an air or O_2 is required, the furnace will typically have either a platinum group metal heating element wound on an alumina

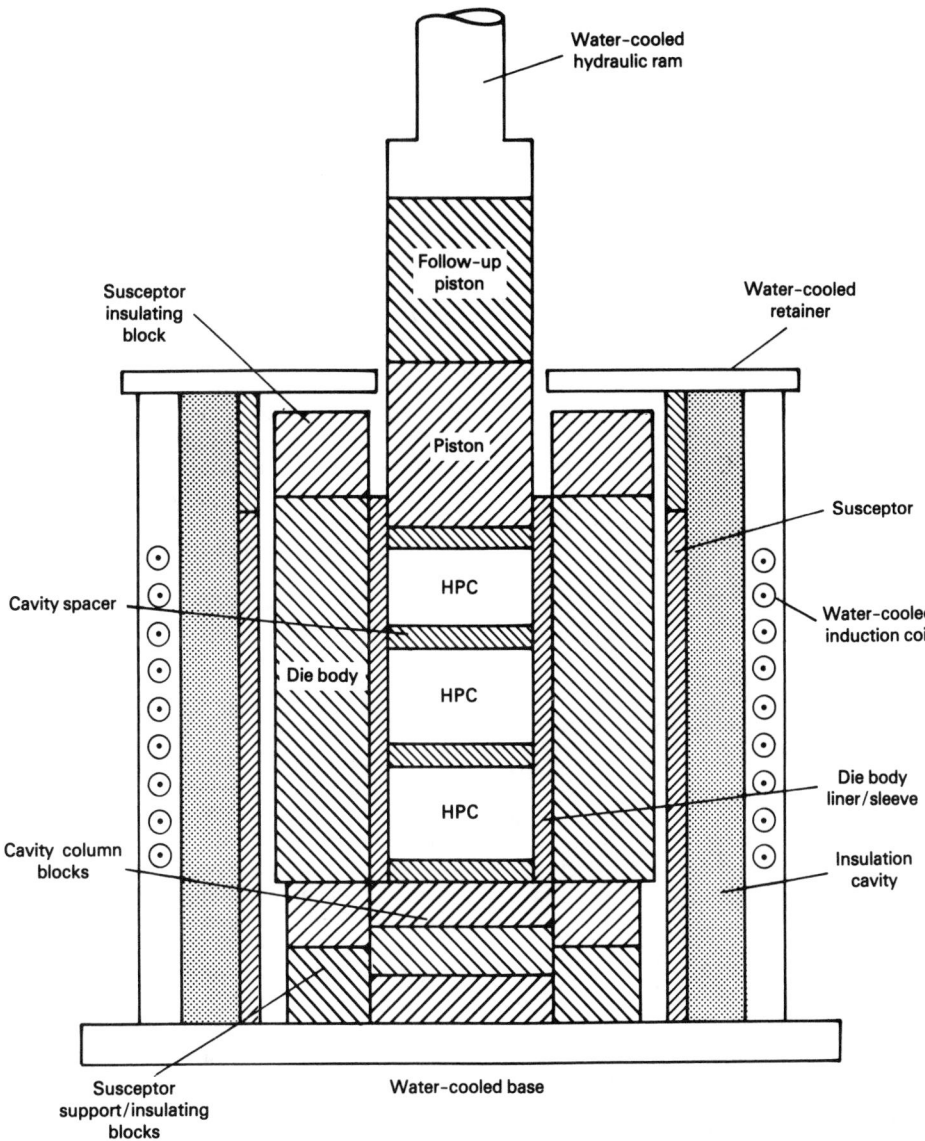

Fig 1 Typical ambient hot press (uniaxial, unidirectional). HPC, hot press cavity

refractory or silicon carbide heating elements (Ref 6).

The sources of pressure for both ambient and vacuum furnaces are hydraulic cylinders mounted in press frames modified to accept the desired furnace design. Press capacity is determined by the required densification pressure and by the component size and cross section. Most uniaxial hot pressing is achieved in the range of 6.9 to 69 MPa (1 to 10 ksi). However, some investigators have used pressures above 138 MPa (20 ksi) (Ref 7).

One often overlooked aspect of a hydraulic system is ram stroke, or travel, which must accommodate the compaction ratio of the total powder or preform charge. The compaction ratio is an indicator of the bulk density of the powder or preform and represents the amount of ram travel required to achieve full density; that is, the ratio compares the height of the initial powder or preform to the height

of the theoretically dense material. Most powders have compaction ratios between 2–3/1, whereas others can have compaction ratios as high as 8/1.

The die assembly consists of a die body, sleeve or insert set, spacers, and pistons. Material selection and assembly design are critical for reliable performance and for maintaining the chemical integrity of the component to be fabricated. Graphite is the most commonly used material for constructing the die assembly. It is easily machined, relatively inexpensive, has moderate room-temperature strength properties that increase with temperature, and has good creep resistance up to 2500 °C (4530 °F). Other properties include excellent thermal conductivity and a range of relatively low coefficients of thermal expansion (CTE). The latter property is an important consideration when designing a die assembly that is compatible with the hot pressed

part because the part must be released from the die body upon cooldown.

A variety of graphite grades are commercially available and are classified according to fabrication history. The isostatically pressed grades are generally dense, fine-grained, high-strength materials. They are considered the structural grades and are used as die bodies and pistons. The extruded, pressed, or molded graphites are weaker, coarser grained, and anisotropic with respect to properties. These grades are suitable for die liners or inserts and spacers.

The major disadvantage of graphite is its lack of oxidation resistance, which means it can only be used in nitrogen, inert, or vacuum environments. An additional concern is the reactivity of graphite to many ceramic materials. Liners or washes are used to preclude direct contact between the graphite and ceramic. Tantalum, molybdenum, and tungsten foils are frequently used as lining materials and boron nitride is commonly used as a wash. Due to the reaction of carbon with residual air, the atmosphere in a graphite-based hot press will always contain a partial pressure of carbon monoxide. Therefore, some ceramics, particularly oxides, will experience compositional and stoichiometric changes during fabrication. In cases where chemical integrity is of prime importance, alternate die assembly materials can be used. Potential materials and their respective properties are identified in Ref 9.

Ceramic dies of fully dense aluminum oxide (Al_2O_3) and silicon carbide (SiC) are frequently used in oxidizing atmospheres. Both materials have excellent oxidation resistance, good to excellent high-temperature strength, and good wear resistance. Aluminum oxide has acceptable utility to 1200 °C (2190 °F) and silicon carbide can be used to 1400 °C (2550 °F). Crushed, fused magnesia (MgO) or zirconia (ZrO_2) are used to separate the powder from the die wall to prevent reactions or welding (Ref 5).

Powder selection is generally based on average-particle size and distribution, surface area, chemistry, and purity. Additional considerations include crystallite size, particle shape/morphology, maximum particle size, and bulk density. Many powders require additives in the form of densification aids, grain-growth inhibitors, and/or special second-phase materials, as in the case of composites. Because several additives may be functional for any one ceramic system, additive selection is based on the desired property development within the densified end-product. Typical additives for single-phase polycrystalline oxides, carbides, nitrides, and borides are provided in Table 1.

After the basic "purity" has been selected, particle size and its associated surface area are generally the overriding considerations. Normally, the finer the powder, the more rapidly densification develops. In addition, likelihood for achieving a fully dense state

Fig 2 Vacuum hot press. Courtesy of Vacuum Industries

Table 1 Typical additives for hot pressed ceramics

Ceramic	Additive	
	Sintering aid	Grain-growth inhibitor
Al_2O_3	CaO, MgO, SiO_2, Y_2O_3	MgO, NiO, Y_2O_3, SiC, ZrO_2
BeO	Li_2O, MgO, CaO, Y_2O_3	SiC, C
MgO	LiF, NaF	SiC, C
Y_2O_3	La_2O_3, ThO_2, SrO	. . .
Si_3N_4	Y_2O_3, SiO_2, MgO, Al_2O_3, CeO_2	. . .
AlN	Y_2O_3, CaO, MgO, BaO, SrO	. . .
BN	B_2O_3, $Ca_3(PO_4)_2$, SiO_2	. . .
B_4C	Al, Cu, B_2O_3	. . .
SiC	Al, B, Be, Al_2O_3, BeO, B_4C, AlN	. . .
TiB_2	Ni, Cr, Fe, Co, ZrO_2	ZrC, SiC

(theoretical density) is improved. The powder in all instances should be less than 325 mesh, that is, the maximum particle size should not exceed 44 μm (1.7 mils). Ideally, the average particle size is between 0.1 and 1.5 μm (4 and 60 μin.). A relatively broad distribution around the average is desirable so that a high green density, upon cold compaction, is achieved.

Hot pressing requirements for powders are less stringent because of the additional energy source available for densification. Generally, powders that are two to four times larger in average particle size than their sinterable counterparts can be hot pressed. For example, with all other powder variables being equal, it will only take a hot pressing powder of 1 to 2 μm (40 to 80 μin.) to produce an equivalent sintered product from a 0.5 μm (20 μin.) powder.

This ability to hot press relatively coarse powders is a primary advantage for both commercial and laboratory users. On the laboratory level, hot pressing is a convenient way for evaluating or screening potential new ceramic compounds using commercially available powders. If resulting properties show promise, then further effort can be directed toward optimizing the powder and eventually developing a pressureless sinter process. From the commercial perspective, average particle size reflects costs. Costs rapidly increase as average particle size decreases. These increased costs are related to comminution and powder manufacturing techniques. Comminution procedures such as jet, turbo, or attrition milling, and so forth, add expense and sometimes, unfortunately, contamination by way of media or mill component wear. The ultrafine powders (0.5 μm, or 20 μin., or less) are either chemically or plasma synthesized and are costly because they generally represent pilot plant or other expensive manufacturing efforts. Furthermore, these powders become difficult to process because of their low bulk density and their extremely high compaction ratios.

Powder Loading. After powder selection, uniform placement into the hot pressing cavity is required. The most common placement technique, particularly on a laboratory scale, is to load the powder directly into the cavity, mechanically de-air it, and cold press it to improve green density. If there is room, separate, additional layers can be produced by introducing a spacer and then repeating the process. The advantages to this technique are ease and simplicity, but the problems include potential contamination, lack of uniformity, and poor green density. The green density directly relates to compaction ratio, and, therefore, to the hot pressing production efficiency, that is, the volume of product produced per hot press cycle.

Higher efficiencies can be achieved by using an alternate procedure in which the powder is green-formed prior to loading into the die. Useful green-forming techniques include: die pressing at elevated pressure, cold isostatic pressing (CIP), and injection molding. Each of these processes will generally produce preforms with relatively high green densities. For example, an equiaxed, silicon nitride powder with an average particle size of 0.4 μm (16 μin.) and a specific surface area of 6 m²/g (1845 ft²/oz) can be compressed in a graphite die assembly at 10 MPa (1.5 ksi) to a green density of only 35% of theoretical. The same powder, when injection molded, can achieve a preform density of 61%. The higher density represents an increase in hot pressing efficiency of almost 50%, a significant achievement since hot pressing is considered a batch process.

Additional hot pressing economic advantages can be realized by using preforms or powders that have a small compaction ratio. These economies are related to the number of spacers (spacer volume) necessary to stabilize the moving column of material during hot pressing. The frictional effects between the die wall and the compressing material, known as side-wall drag, lead to pressure gradients across the pressing cavity. These pressure gradients can cause material flow within the preforms, resulting in density variations and plate distortion. This type of distortion becomes more pronounced with powders or preforms with high compaction ratios. Placing rigid spacers between pre-compacts redistributes the load within the column and maintains flatness. However, the use of spacers reduces the volume of material that can be processed. An approach that minimizes the required spacer volume and is an economically useful technique for powders or preforms that have large L/D ratios and compaction ratios less than 2/1 is possible; see Ref 8 for more details.

Hot press densification is a process characterized by three distinct stages (Ref 10). The first stage involves particle rearrangement, the second stage includes viscous and plastic flow, and the third stage is the result of diffusional transport. Two basic types of theoretical models have been proposed in an attempt to understand and quantify the observed kinetics. The oldest type is derived from pressureless-sintering models (Ref 11) and the most recent type is based on creep theory (Ref 12). These models continue to be topics of debate and are beyond the scope of this analysis, however, excellent reviews are given elsewhere (Ref 13, 14).

The primary variables of concern for hot press densification are time, temperature, and pressure; temperature is usually addressed first. When the hot press temperature of a material is not known, two-thirds of the absolute melting point is usually a good approximation. However, for strongly covalently bonded materials, such as nitrides, borides, or carbides, the hot press temperature can be somewhat higher. If pressureless-sintering conditions are known, then hot pressing can be accomplished at the sintering temperature, less approximately 200 °C (360 °F) (Ref 8).

Additionally, there are several systematic methods used to determine optimum hot press conditions. The isothermal technique (IT) involves conducting a series of hot pressings under a set of different isothermal temperatures (Ref 15). Material is slowly heated to a hot pressing temperature (the isotherm); pressure is applied and then maintained until densification is complete.

The densification rate is indirectly determined by monitoring relative piston movement with respect to time. Axial piston movement can be measured by attaching dial indicators, strain gages, or linear transducers. Data from a series of these experiments are plotted in a form shown in Fig 3. Each data point represents the time increment required to reach a specific density at a given temperature. When plotted, two lines at different slopes develop. The intersection of these two lines gives an optimum hot pressing temperature that will yield a fine-grained, dense product. Experimental hot pressings of this type can be repeated at different pressure levels, providing a composite relationship between time, temperature, and pressure. Although the isothermal technique is time consuming because of the many required pressings, an optimum hot pressing temperature can be determined.

An alternate technique, climbing temperature program (CTP), can provide the same information with fewer experimental pressings (Ref 16). This technique requires applying full pressure at room temperature and then increasing temperature at a slow rate until densification is achieved. Full density is assumed when linear piston movement stops for a predetermined increment of time. Compaction data (density versus time at constant pressure) is then plotted in the manner shown in Fig 4, which clearly illustrates the three stages of densification. The second stage of densification is approximately linear and can be fitted using regression techniques. The resultant linear equation is used to find the isothermal pressing temperature.

Hot pressings conducted below the CTP-derived optimum temperature are often sluggish and inefficient. Those conducted above that temperature densify relatively quickly but yield materials with inferior mechanical properties and coarsened microstructures.

Although these techniques can define effective hot press temperatures and times, they do not provide an optimum procedure for applying pressure during hot pressing. Application of full pressure at room temperature (isothermal technique) may place graphite die assemblies at risk and may prevent the removal of entrapped and adsorbed gases, whereas application of full pressure once temperature is reached (CTP technique) may actually retard the densification kinetics by allowing localized particle necking. The pressure-temperature profile is predominantly dependent on the powder characteristics, and, at best, is determined empirically for each system. However, Leipold does recommend an initial starting procedure that should be considered (Ref 14): Apply one-third the maximum pressure at approximately one-half the hot pressing temperature, then increase pressure linearly, achieving maximum load before maximum temperature.

Hot Pressed Product Properties. Hot pressing as a forming technique can yield ceramic products with uniform, fine-grained microstructures at or near theoretical density. Materials can be densified with or without minimal use of sintering aids, and fine-grained microstructures may be developed without the use of grain-growth inhibitors. These processing advantages allow the development of optimum room- and high-temperature mechanical properties, such as strength, fracture toughness, hardness, thermal conductivity, and creep resistance. Furthermore, certain unique properties, such as transparency, can only be developed in some materials by hot pressing. Additionally, the fine-grained nature of hot-pressed materials allows them to be processed to exceptional surface finishes, which is an important criterion for many wear-related applications.

Electrical properties, particularly in the ferroelectric and ferrite ceramics, are strongly influenced by density and by certain microstructural features, such as grain size. For example, the dielectric constant, remanet polarization, planar coupling, and squareness ratio all increase as the density of a ferroelectric ceramic approaches the theoretical value. In contrast, the coercive field and the mechanical Q are little affected by density, but are strongly influenced by grain size (Ref 6). Additionally, theoretically dense, hot

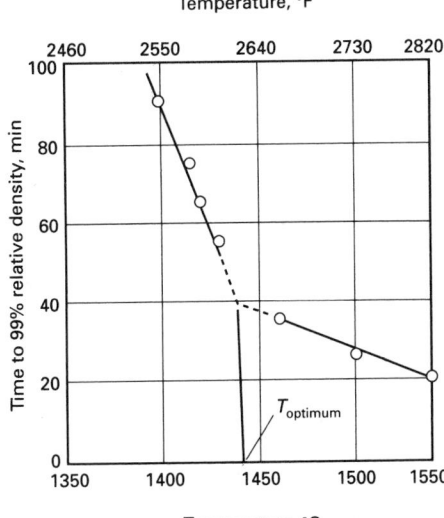

Fig 3 Optimum hot pressing temperature derived from isothermal data for a commercial Al_2O_3 powder

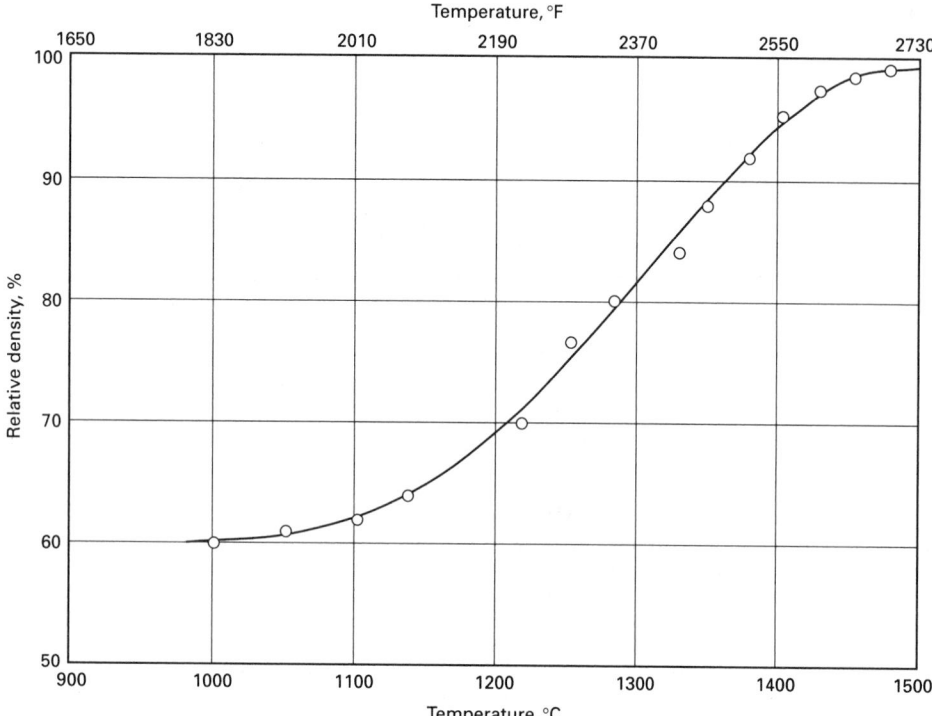

Fig 4 CTP behavior for a commercial Al_2O_3 powder

pressed ferrites have an optimum saturation and magnetic permeability and can be polished to a fine surface finish (Ref 17), which is an essential requirement for magnetic recording head applications.

Sintering aids generally remain in the final product as grain boundary phases (secondary phase), which in turn influence the performances of the material, particularly at elevated temperatures. By eliminating or minimizing the second phase, the use temperature of the material can be extended and certain application ranges can be widened. For example, the corrosion and wear properties of many ceramics can be significantly improved by controlling the content and volume of the second phase.

Hot pressing is an engineering tool capable of imparting or improving certain material characteristics and properties not available by conventional sintering techniques. However, unlike sintering, hot pressing can produce directional properties in some materials by a grain orientation mechanism. This occurs in powder systems that have grains shaped like rods, needles, or plates. During densification the grains with large aspect ratios will align themselves with their long axis perpendicular to the hot pressing direction. Therefore, the resultant microstructure becomes preferentially oriented and will produce properties that vary according to direction within the ceramic. These effects are clearly evident in hot pressed whisker composites such as Al_2O_3-SiC_w and Si_3N_4-SiC_w. For example, the fracture toughness for 30 vol% SiC whisker-reinforced Al_2O_3 is approximately 30% higher in the plane par-

allel to the hot pressing direction than that measured in the perpendicular plane. Other properties, such as flexural strength, sonic velocity, and volume resistivity, also show similar directional behavior.

Preferred orientation can also occur in systems that undergo a type of liquid phase sintering known as solution-recrystallization (for example, silicon nitride). During recrystallization, silicon nitride predominantly devel-

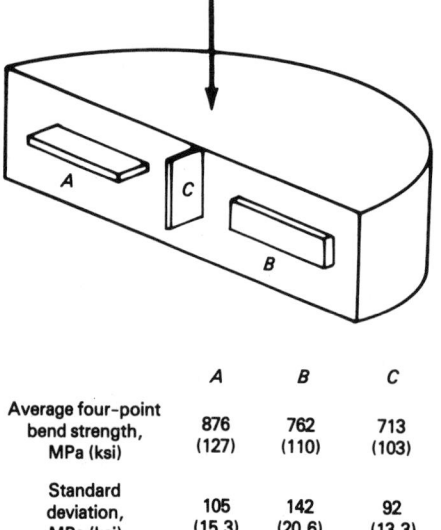

	A	B	C
Average four-point bend strength, MPa (ksi)	876 (127)	762 (110)	713 (103)
Standard deviation, MPa (ksi)	105 (15.3)	142 (20.6)	92 (13.3)

Fig 5 Variations in strength of Si_3N_4 as a function of hot press direction (after D.W. Richerson, Ref 5)

ops into hexagonal-shaped rods of varying aspect ratios. The large aspect ratio rods orient themselves randomly in a plane perpendicular to the hot press axis, thereby producing directional properties. The anisotropic nature of hot pressed silicon nitride is illustrated in Fig 5 (Ref 5). As shown, strength is a directional property and is greatest in the plane perpendicular to the hot pressing direction.

Virtually all hot pressed ceramics exhibit some degree of microstructural anisotropy and therefore directional properties. For comprehensive design studies, the anisotropic materials require complete directional characterization. Unfortunately, such a characterization is costly, and therefore most property data, with the exception of strength, are given as "typical" values. In contrast, flexural strength is generally measured from test specimens cut from the plane perpendicular to the direction of hot pressing (strong direction). Here again, the data are not true representations of material strength, but appear to be sufficient for most design efforts.

Table 2 lists "typical" mechanical and thermal properties for selected hot pressed ceramics. The flexural strength was measured from the strong direction.

Applications

Although there is always pressure to convert hot pressed parts into components produced by more "conventional" means, because of the real or perceived expense, many commercial products such as those described below are produced by the hot press method.

Traditionally Hot Pressed Ceramics. Two materials, boron carbide (B_4C) and hexagonal boron nitride (BN) are still produced only by the hot press method (Ref 1). Besides diamond and cubic boron nitride, B_4C is among the hardest of all substances. It is therefore used in wear-resistant applications such as sand blast nozzles and wear plates. Hot-pressed B_4C has gained wide acceptance as a lightweight ceramic armor, which will be discussed below.

Hexagonal boron nitride is a "soft" ceramic material with excellent lubricity. It has high thermal conductivity, a relatively low dielectric constant, and a high electrical resistivity, which allow its use in electronic applications. This material is easily machined, and components can be made from hot pressed blocks and cylinders by conventional machining techniques. "Powdering" hot pressed BN produces a ceramic grain that can be used as a filler for resins and as a lubricant ("white" graphite).

Optical Windows. Because hot pressing is one of the few ceramic fabrication methods that can produce parts of true theoretical density, it is employed to make materials for the transmission of various wavelengths of light. In single-phase polycrystalline materials at theoretical density, there are neither pores nor grain boundary phases to scatter the light and

Table 2 Typical properties of hot pressed ceramics

Ceramic	Density, g/cm³	Mechanical							Poisson's ratio			Thermal			
		Modulus of rupture(a)		Modulus of elasticity		Fracture toughness				Knoop hardness		Coefficient of thermal expansion, 10^{-6}/K	Conductivity		Thermal shock parameter, ΔT_c
		MPa	ksi	GPa	10^6 psi	MPa$\sqrt{m}$	ksi$\sqrt{in.}$			GPa	10^6 psi		W/m·K	(Btu·in./h·ft²·°F)	
Si₃N₄	3.30	850	125	310	45	6.1	5.6	0.27		14.7	2.13	3.4	29.5	205	700
Al₂O₃-SiC_w	3.64	640	95	390	56	7.7	7.0	0.23		17.6	2.55	5.9	35	245	900
Al₂O₃-TiC_p	4.26	760	110	395	57	4.5	4.1	0.22		18.1	2.63	8.0	40	275	300
SiC	3.18	610	90	430	62	5.2	4.7	0.17		24.5	3.55	4.5	110	760	440
TiB₂	4.43	345	50	540	78	5.0	4.6	0.11		26.5	3.84	8.2	75	520	200
B₄C	2.48	350	51	470	68	3.5	3.2	0.17		28.4	4.12	5.6	16	110	200
Al₂O₃	3.98	550	80	390	56	4.5	4.1	0.23		15.2	2.20	8.1	29	200	150
AlN	3.25	300	45	318	46	4.5	4.1	0.25		13.5	1.96	5.6	180	1245	550

(a) Measured in strong direction

reduce transmissivity. A wide application of this technology has been to produce the "eye," or window, in the tip of infrared-seeking missiles. This class of material is typified by hot pressed magnesium fluoride (MgF_2), zinc sulfide (ZnS), and zinc selenide (ZnSe).

Light scattering can also take place in nonisotropic materials such as theoretically dense alumina (Al_2O_3) because of the difference in the velocity of light along different crystallographic axes in this hexagonal material. This problem can be overcome by using an isotropic (cubic) material as the window. Solid solutions of yttrium oxide (Y_2O_3) and spinel ($MgAl_2O_4$) have been hot pressed to produce large-diameter, thin windows used in the visible light range (Ref 18).

Ceramic armor has revolutionized the way in which military threats are "defeated" (Ref 19). The best application was the development of large, thin plates of hot pressed B_4C used to stop armor-piercing projectiles. Because of its low theoretical density (2.51), this material became known as lightweight ceramic armor and is used in airborne applications, especially on helicopters. Other hot pressed materials, such as titanium diboride (TiB_2) and SiC, are successful armor materials. Even Al_2O_3, most commonly made by pressureless sintering, performs better as armor when hot pressed (especially in large, thin sheets). Hot pressed aluminum nitride (AlN) and particulate composites of Al_2O_3 are also being considered as armor materials.

Cutting tools have represented an outstanding commercial success for hot pressed ceramics (Ref 20). One of the first commercially successful materials was a ceramic particulate composite of alumina and titanium carbide (Ref 21). Other hot pressed materials that make excellent cutting tools are Si_3N_4 (Ref 22, 23) and Si_3N_4-AlN-Al_2O_3 solid solutions. The silicon carbide whisker-reinforced alumina matrix composite is now finding wide application as a cutting tool (Ref 24).

Tooling. Strong, tough, wear-resistant, fine-grained, hot pressed ceramics make useful molds and dies, especially in areas where abrasion, erosion, and corrosion occur. For example, hot pressed Al_2O_3-SiC_w composites are useful in making dies for forming two-piece aluminum cans (Ref 24). This same material can be diamond ground into "twist" drills for use in drilling abrasive materials such as fiberglass-reinforced epoxies. Additionally, hot-pressed SiC and Al_2O_3 have been used as die bodies for hot pressing electro-optic and piezoelectric materials in oxidizing atmospheres (Ref 6).

Sputtering Targets. A large-diameter, thin disk of hot pressed material is frequently used as a source for the vacuum deposition of materials used in the semiconductor industry (Ref 25). These disks, known as sputtering targets, must be dense, strong, and have a high degree of mechanical integrity. Chromium-silicon monoxide (Cr-SiO) is one of the hot pressed targets used for making thin-film resistors. Other target materials include TiN, Si_3N_4, B_4C, and Al_2O_3.

Heat engine components used in various propulsion applications, such as gas turbines and diesel engines, require high-performance materials to extend their operating temperature range to achieve improved Carnot efficiencies. Special materials with high melting points; excellent thermal shock resistance; high strength; good erosion, corrosion, and abrasion resistance; outstanding oxidation resistance; and good creep resistance are needed to build components for heat engines operating under these extreme conditions. Hot pressed Si_3N_4 and SiC (Ref 26) have been used as turbine vanes, shrouds, piston caps, valves, and bushings in this application (Ref 27, 28).

Nuclear Industry Components. Cetain ceramic particulate composites used in nuclear reactors cannot be sintered readily, and, therefore, are made by hot pressing. Three types of materials produced by hot pressing are: burnable "poisons" such as boron carbide-alumina; burnable fuels, such as europium oxide-urania; and fuel-moderator combinations, such as urania-beryllia (Ref 29).

Ceramic bearings that can run with little or no lubricants in harsh or severe environments are made of hot pressed, fully dense ceramics (Ref 30). The leading material for this application is Si_3N_4. Its low coefficient of rolling friction, low thermal expansion, high strength, moderate elastic modulus, good surface finish, light weight, and resistance to oxidation and various chemical environments makes it an ideal candidate for the fabrication of roller and ball bearings.

Microwave absorbers are those materials that are used in the vacuum environment inside of high-powered microwave tubes, such as traveling wave tubes (TWT), to absorb unwanted energy "spikes." Traditionally, porous SiC pieces used as "whetstones" were employed, but were found to cause outgassing and could only handle small amounts of power due to their poor thermal conduction properties. Hot pressed magnesia-silicon carbide particulate composites were developed for this application (Ref 31), and are still used for normal requirements, but hot pressed beryllia-silicon carbide is now used for the most demanding microwave-absorbing requirements, while alumina-silicon carbide is used for the less-demanding applications.

Varistors are materials that can protect electrical circuits from large surges in voltage. These materials undergo an extremely large, rapid decrease in resistance, which causes the current to be sidetracked away from the device, to afford protection when voltage spikes occur. Zinc oxide (ZnO) ceramics containing special oxide additives are used for high-performance varistors. The highest breakdown voltage (field) handling capabilities are found in these ZnO varistors fabricated by hot pressing (Ref 32).

Electro-optic materials are those that change their direction of polarization upon the application of an electric field. It was discovered that a polycrystalline ceramic, lead lanthanum zirconate titanate (PLZT), could be hot pressed to transparency to act as an electro-optic element (Ref 6). This hot pressed material has found wide application, especially in the manufacture of nuclear blast protection goggles and windows. In this application, a sensor picks up the sudden change in light intensity and causes the plane of polarization to immediately rotate in one of two joined sheets of PLZT. This causes the material to immediately "blacken" because the light cannot be transmitted through the "crossed" polarizers. This protects the observer from the blinding flash of the nuclear detonation.

Ferrites are magnetic ceramics based on the mineral magnetite (Fe_3O_4) or the spinel phase (M_3O_4 or AB_2O_4). These materials are normally produced by dry pressing and sintering. However, when high performance is necessary or a tightly controlled microstructure is required, as in magnetic recording heads, hot pressed ferrites are used (Ref 17). In the recording head, the ferrites must be capable of acquiring a fine, mirror-like finish. This is best accomplished by lapping and polishing hot pressed ferrite ceramics (Ref 33).

Titanates. In recording heads, where top-quality ferrites are employed, a stable, compatible nonmagnetic holder must be used. A ceramic dielectric, such as conventionally sintered Al_2O_3, had been traditionally used. However, this proved unsatisfactory in two areas: the thermal expansion coefficient of the Al_2O_3 was not close enough to that of the ferrite, and the sintered Al_2O_3 could not take a high surface finish comparable to the ferrite, both of which had to glide smoothly over the magnetic tape. Ceramic titanates were found to have an expansion coefficient that was very close to that of the ferrites. It was further observed that if these titanates were hot pressed, they could be lapped and polished to the same high degree of surface finish as that achieved with the ferrites (Ref 34). Two materials, barium titanate ($BaTiO_3$) and calcium titanate ($CaTiO_3$), are primarily used for this application, either alone or in combination.

Microelectronic ceramic "packages" are traditionally made by sintering laminated green sheets of metallized ceramics in a process known as cofiring, which simply means that the ceramic and metal conductors are fired at the same time. With some advanced, high-performance materials, such as AlN, it is difficult to conventionally fire large, flat, laminated "packages" together. In this instance, cofiring of tungsten-metallized multilayer AlN ceramic has been successfully accomplished by hot pressing (Ref 35). This fabrication technique produces a low thermal expansion, high thermal conductivity, mechanically strong, multilayered, cofired, microelectronic ceramic package.

Resistors. In special applications where high-power, thermal shock resistant resistors are required, ceramic composites made of unusual materials are the only possible candidates. Because of the special nature of the ceramic constituents, these composites must be hot pressed. Typical combinations that are being developed are a Si_3N_4 matrix with particulates (Ref 36) of either TiB_2, TiC, TiN, or SiC as the dispersed conductive phase, as well as SiC with TiB_2 (Ref 37).

Other Materials. When special high performance is required from complex oxide- or intermetallic-based ceramics, hot pressing is normally the fabrication method of choice. Typical examples of these applications and the typical ceramic used in them are:

- Magnets (samarium cobalt, $SmCo_5$)
- Superconductors (niobium-tin, Nb_3Sn)
- Thermoelectrics (lead telluride, PbTe)
- Piezoelectric materials (lead zirconate, titanate, PZT) (Ref 6)

REFERENCES

1. J.A. Rubin, "Sintering under Pressure: Aspects of Processing (Sintering Symposium)," presented at the American Ceramic Society Pacific Coast Regional Meeting, Anaheim, CA, Oct 1971
2. K.R. Venkatachari and R. Raj, Shear Deformation and Densification of Powder Compacts, *J. Am. Ceram. Soc.*, Vol 69 (No. 6), 1986, p 499–506
3. M.N. Rahaman *et al.*, Densification and Shear Deformation in $YBa_2Cu_3O_{6+\delta}$ Powder Compacts, *Adv. Ceram. Mater.*, Vol 3 (No. 4), 1988, p 393–397
4. M.N. Rahaman *et al.*, Effect of Shear Stress on Sintering, *J. Am. Ceram. Soc.*, Vol 69 (No. 1), 1986, p 53–60
5. D.W. Richerson, *Modern Ceramic Engineering: Properties, Processing, and Use in Design*, Marcel Dekker, 1982, p 399
6. G.H. Haertling, Piezoelectric and Electrooptic Ceramics, *Ceramic Materials for Electronics*, R.C. Buchanan, Ed., Marcel Dekker, 1986, p 170–173
7. R.M. Spriggs, L.A. Brissette, M. Rosetti, and T. Vasilos, Hot-Pressing Ceramics in Alumina Dies, *Am. Ceram. Soc. Bull.*, Vol 42 (No. 9), 1963, p 477–480
8. A. Ezis, E.C. Beckwith, and W.B. Copple, "Method of Fabrication Hot Pressed Silicon Nitride Billets," U.S. patent 4,632,793, Ford Motor Company, Dec 1986
9. M.H. Leipold, Hot Pressing, *Treatise on Materials Science and Technology*, Vol 9, *Ceramic Fabrication Processes*, F.F.Y. Wang, Ed., Academic Press, 1976, p 95–134
10. A. Mocellin, Stress Assisted Hot Formation of Ceramics, *Progress in Nitrogen Ceramics*, F.L. Riley, Ed., Martinus Nijhoff, the Hagae, Netherlands, 1983, p 253–273
11. P. Murray, E.P. Rodgers, and A.E. Williams, Practical and Theoretical Aspects of Hot-Pressing of Refractory Oxides, *Trans. Brit. Ceram. Soc.*, Vol 53 (No. 8), 1954, p 474–510
12. R.C. Gifkins, Transitions in Creep Behaviour, *J. Mater. Sci.*, Vol 5, 1970, p 156–165
13. J.M. Vieira and R.J. Brook, Kinetics of Hot Pressing: The Semilogarithmic Law, *J. Am. Ceram. Soc.*, Vol 67 (No. 4), 1984, p 245–249
14. P. Barnier, C. Brodhag, and F. Thevenot, Hot Pressing Kinetics of Zirconium Carbide, *J. Mater. Sci.*, Vol 21, 1986, p 2547–2552
15. P.H. Crayton and J.J. Price, Prediction of Effective Isothermal Hot Pressing Temperature, *Am. Ceram. Soc. Bull.*, Vol 63 (No. 5), 1984, p 715–717
16. G.J. Leshkivich and P.H. Crayton, Identification of Hot-Pressing Conditions Producing Optimum Strength and Microstructure in Alumina, *Am. Ceram. Soc. Bull.*, Vol 64 (No. 5), 1985, p 684–686
17. T.G. Reynolds III and R.C. Buchanan, Ferrite (Magnetic) Ceramics, *Ceramic Materials for Electronics*, R.C. Buchanan, Ed., Marcel Dekker, 1986, p 255
18. D.W. Roy and J.L. Hustert, Polycrystalline $MgAl_2O_4$ Spinel for High Temperature Windows, *Ceramic Engineering and Science Proceedings*, Vol 4 (No. 7–8), 1983, p 502–509
19. S.K. Chung, Fracture Characterization of Armor Ceramics, *Am. Ceram. Soc. Bull.*, Vol 69 (No. 3), 1990, p 358–366
20. W.W. Gruss, Ceramic Tools Improve Cutting Performance, *Am. Ceram. Soc. Bull.*, Vol 67 (No. 6), 1988, p 993–996
21. S.J. Burden, J. Hong, J.W. Rue, and C.L. Stromsborg, Comparison of Hot-Isostatically Pressed and Uniaxially Hot-Pressed Alumina-Titanium-Carbide Cutting Tools, *Am. Ceram. Soc. Bull.*, Vol 67 (No. 6), 1988, p 1003–1005
22. A. Ezis, "Method of Making Silicon Nitride Based Cutting Tools—I," U.S. patent 4,264,548, Ford Motor Company, April 1981
23. A. Ezis, "Method of Making Silicon Nitride Based Cutting Tools—II," U.S. patent 4,264,550, Ford Motor Company, April 1981
24. G.C. Wei and P.F. Becher, Development of SiC-Whisker-Reinforced Ceramics, *Am. Ceram. Soc. Bull.*, Vol 64 (No. 2), 1985, p 298–304
25. Technical brochure, CERAC Corporation, 1984
26. M. Shimada, Structural Nonoxide Ceramics by Hot Pressing, *Am. Ceram. Soc. Bull.*, Vol 65 (No. 8), 1986, p 1153–1155
27. E.M. Lenoe and J.L. Meglen, International Perspective on Ceramic Heat Engines, *Am. Ceram. Soc. Bull.*, Vol 64 (No. 2), 1985, p 271–275
28. D.W. Richerson, Evolution on the U.S. & Ceramic Technology for Turbine Engines, *Am. Ceram. Soc. Bull.*, Vol 64 (No. 2), 1985, p 282–286
29. J.A. Rubin, "Method of Producing an Oxidation-Resistant UO_2BeO Fuel Body," U.S. patent 32,867,489, U.S. Atomic Energy Commission, Feb 1975 (classified version allowed 1962)
30. C.F. Bersch, Overview of Ceramic Bearing Technology, *Ceramics for High Performance Applications—II*, J.J. Burke, E.N. Lenoe, and R.W. Katz, Ed., Brook Hill Publishing, 1978, p 397–405
31. L.E. Gates Jr. and W.E. Lent, "Method for Providing Improved Lossy Dielectric Structure for Dissipating Electrical Microwave Energy," U.S. patent 3,634,566,

Hughes Aircraft Company, Jan 1972

32. E. Sonder, T.C. Quinby, and D.L. Kiner, ZnO Varistors Made from Powders Produced Using a Urea Process, *Am. Ceram. Soc. Bull.*, Vol 65 (No. 4), 1986, p 665–668

33. K. Miyoshi and D.H. Buckley, Surface Chemistry, Friction, and Wear of Ni-Zn and Mn-Zn Ferrites in Contact with Metals, *Ceram. Eng. Sci. Proc.*, Vol 4 (No. 7–8), 1983, p 674–681

34. Technical brochure, Kyocera Inc., 1989

35. J.H. Enloe, J.W. Lau, J. Stephan, and M.R. Ehlert, Concepts in Advanced Aluminum Nitride Packaging, *Proceedings of the 1990 NEPCON West Technical Program*, Feb-Mar 1990, p 1143–1152

36. D.W. Richerson, "Hot Pressed Silicon Nitride Containing Finely Dispersed Silicon Carbide or Silicon Aluminum Oxynitride," U.S. patent 3,890,250, Norton Company, 1975

37. C.H. McMurtry, W.D.G. Boecker, S.G. Seshadri, J.S. Zanghi, and J.E. Garnier, Microstructure and Material Properties of SiC-TiB$_2$ Particulate Composites, *Am. Ceram. Soc. Bull.*, Vol 66 (No. 2), 1987, p 325–329

Hot Isostatic Pressing

Hans T. Larker, ABB Cerama AB, Robertsfors, Sweden

VERY STRONG INTERATOMIC BONDS characterize engineered ceramic materials. It is these bonds that provide ceramics with desirable properties such as high hardness, stiffness, and strength in applications requiring elevated temperatures for prolonged periods of time. However, a consequence of the strong bond characteristic is that the ceramic components are often very difficult to fully densify and to machine to the required final shape.

Hot isostatic pressing (HIP) has in many cases been found to provide unique solutions for problems related to densification and the forming of ceramics. The process utilizes uniform and omnidirectional pressure at elevated temperature to enhance densification and interparticle bonding (Fig 1). An inert gas (usually argon because of cost and availability factors) is used to transmit pressure. The gas must be prevented from penetrating into any voids in the processed body that are to be sealed. Metals, ceramics or, most commonly, glasses may be used for such a gas impervious encapsulation. If the body is brought to a state with no surface-connected porosity by preliminary processing steps such as pressureless sintering, the HIP process can alternatively be carried out without separate encapsulation using a process called cladless HIP or sinter-HIP.

No rigid tools with limited strength (for example, graphite tools in uniaxial hot pressing) are needed to transmit the pressure to the body, which enables high pressure levels (typically 100 to 320 MPa, or 15 to 50 ksi) to be used even above 2000 °C (3630 °F). In isostatic pressing, the gas pressure is applied perpendicularly to and at the same magnitude on all accessible surfaces of the body. Under ideal conditions, no change of shape (just change of size) of the body occurs. The absence of shear forces in the body (on a scale substantially larger than particle size) and the absence of any die friction, combined with the 5 to 10 times higher pressure level than in uniaxial hot pressing, makes hot isostatic pressing an even more powerful densification method for ceramics than hot pressing.

Even more important from a commercial standpoint is the inherent ability of the HIP process to produce components with a complex and well-defined shape. The high gas pressure transmitted to the porous body via the encapsulation makes the body extremely rigid and insensitive to external influence, which is particularly important in the initial stages of densification when bonds between particles are still very weak. This inherent characteristic of the encapsulated HIP process can actually result in better shape precision from a uniform green powder body than any other sintering process for dense ceramics of complex shape.

Hot isostatic pressing is thus a very versatile process with many advantages. In general, voids, particularly large ones (Ref 1), in most materials can be efficiently reduced in size and frequency. Through the use of encapsulation, ceramics can be densified at relatively low temperatures. Even extremely difficult to sinter ceramics can be fully densified (for example, high-purity silicon-nitride powders can be sintered without any additives at all; the same condition applies to ceramic/ceramic composites with a high loading of particles, whiskers, or long fibers). The reduced sintering temperature means that grain growth and undesirable reactions can be controlled or avoided. A very high uniformity in properties (for example, density) can be obtained.

The encapsulated HIP process also provides chemical isolation of the material from the processing gas. The chemical activity of trace substances in the gas increases with the gas pressure and might, if not prevented by the encapsulation, attack and deteriorate the properties of the powder body surface. On the other hand, possible gaseous products formed in the powder body cannot escape. That is a great advantage for materials like nitrides that tend to decompose. The gaseous decomposition products are contained by the encapsulation that is supported by the high-pressure gas. As soon as equilibrium pressure is reached, decomposition stops. For materials that require removal of gaseous reaction products before final densification, adequate preprocessing must, however, be adopted (for example, necessary removal of carbon oxides from ceramics containing oxides and carbonaceous material must be carried out before the encapsulation is sealed).

Alternatively, porous preforms can also be saturated with controlled amounts of gases for reactive sintering of encapsulated parts. Cladless HIP, on the other hand, allows gases to escape from the powder body until the pores are disconnected from the surface. Chemically active pressure gases (for example, nitrogen for the limitation of decomposition of nitrogen ceramics or oxygen/argon mixtures for zirconium oxide materials or oxide superconductors) offer possibilities to control processing chemistry.

HIP Equipment

Pressure-Vessel Furnace Construction. Equipment for hot isostatic pressing must be able to simultaneously maintain a high, isostatically acting pressure and a high temperature in the powder body to be processed for the desired length of time under tightly monitored conditions. Today, practically all equipment used is of the cold pressure-vessel wall, internal furnace, type (Fig 2). It resembles a much more common type of processing equipment: the cold-wall, high-temperature, high-vacuum furnace. In both pieces of equipment, the pressure difference to ambient is taken by the cooled wall of the equipment. The heat in the workspace is generated by electrical furnaces usually having resistance heating elements and a thermal insulation barrier tailored to the requirements. At the low pressure level in high vacuum furnaces, typically ten orders of magnitude below ambient, only heat transport by radiation needs to be considered. At the pressure required in HIP equipment, typically 3 to $3\frac{1}{2}$ orders of magnitude over ambient or 100 to 320 MPa (15 to 45 ksi), the gases used to produce the pressure (usually argon, but sometimes nitrogen is also used), become very dense. In areas near ambient temperature (for example, just inside the pressure-vessel wall), the density of the gas may actually exceed the density of water but still not be transformed into a liquid. High gas density gradients in combination with low gas viscosity at high temperatures (being less than a factor of two higher at 200 MPa, or 30 ksi, than at atmospheric pressure) makes the tendency to convection extremely strong. Control of gas convection is consequently crucial but based on principles of H. Larker (Ref 2) and C. Boyer (Ref 3), heat losses by

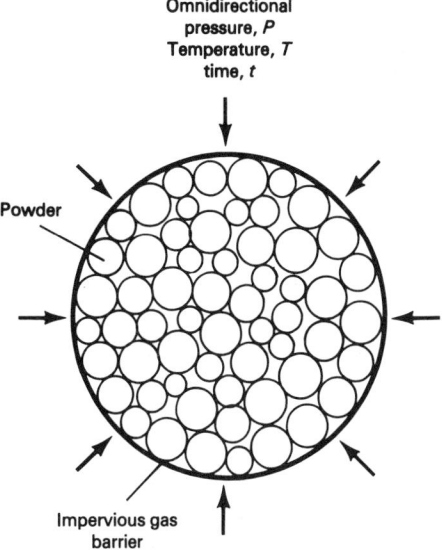

Fig 1 Schematic showing key components of the HIP process. A porous material confined by a barrier or container is subjected to a pressure, *P*, at a temperature, *T*, for a time, *t*. The compressive forces generated by the pressure of the gas are directed radially inward perpendicular to the circumference of the gas barrier that encloses the powder compact.

convection (present even in large industrial units) can be controlled. Furthermore, a relatively thin heat insulation wall that provides a large available work space within a given pressure vessel size can be effective in con-

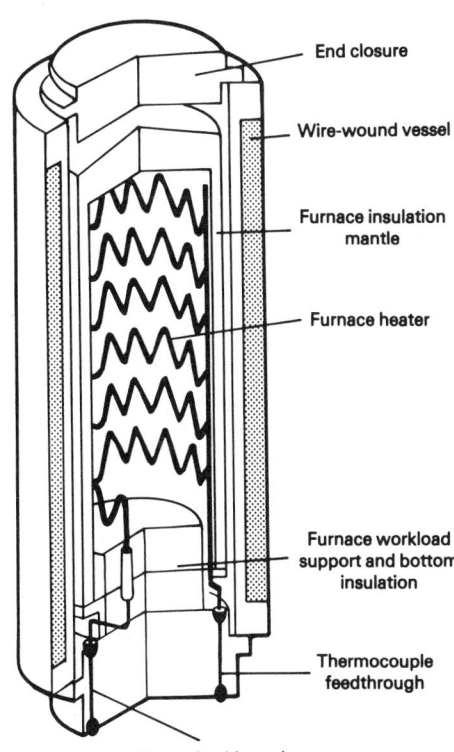

Fig 2 Typical construction of a cold pressure-vessel, internal-type HIP furnace. Courtesy of Asea Brown Boveri

trolling convection heat losses. Heat conduction can become the dominant heat treat mechanism and can determine the minimum thickness of the heat insulation wall.

Heating elements as well as gas-tight sheaths for convection control in the heat insulation are typically made of refractory metals such as molybdenum for uses up to 1400 °C (2550 °F) and by graphite, graphite paper, and graphite/graphite composites at temperatures exceeding 1400 °C (2550 °F). Platinum is used in some research and development units that can operate with partial oxygen atmospheres such as 20% O_2 and 80% Ar.

Temperature Measurement Devices. Temperature is usually measured with platinum-rhodium thermocouples (Pt-6Rh or Pt-30Rh) up to 1750 °C (3180 °F) and with tungsten-rhenium thermocouples up to 2000 °C (3630 °F). Thermocouple life is, however, shortened at the higher temperatures (>1600 °C, or 2910 °F, for platinum-rhodium and >1800 °C, or 3270 °F, for tungsten-rhenium) and the readout may be uncertain, especially in nitrogen gas. Particularly important for thermocouple life is the quality of thermocouple insulators. Hot isostatically pressed, high-purity boron nitride that can be made with very low boron oxide content appears to be a good choice of insulator material. Boron-carbide/carbon thermocouples can be used up to 2400 °C (4350 °F) (longer term up to 2200 °C, or 3990 °F) (Ref 4) and an all-graphite system based on mechanical principles can be used up to 3000 °C (5430 °F) (longer term, 2700 °C, or 4890 °F) (Ref 5).

Pressure-Vessel Safety Guidelines. The furnace system with temperature control devices is perhaps the most critical part of a HIP installation from a technical standpoint. From a standpoint of safety, however, the pressure vessel and high-pressure gas installation is the most significant component.

The safety codes as prescribed by authorities in different countries must be strictly followed as the stored energy in HIP vessels under pressure is very high. The pressure vessel may be made with thread-type closures for small-sized units or those installations requiring lower pressure applications. For large-volume or high-pressure applications, yoke closures are preferred (Fig 3) and vessels with wire-wound cylinders and yokes can incorporate leak-before-break designs in their construction (Ref 6).

Workpiece Fixtures. Typical tooling and fixtures required to support workpieces being processed by hot isostatic pressing are made of graphite. Cylindrical trays are often used for small furnaces to encompass the full workspace diameter of the unit. For larger diameter HIP units, seven trays in each plane are typically used as fixtures. The graphite may react on surfaces that come in contact with the ceramic and is therefore covered by a boron-nitride wash, sometimes in combination with molybdenum foils.

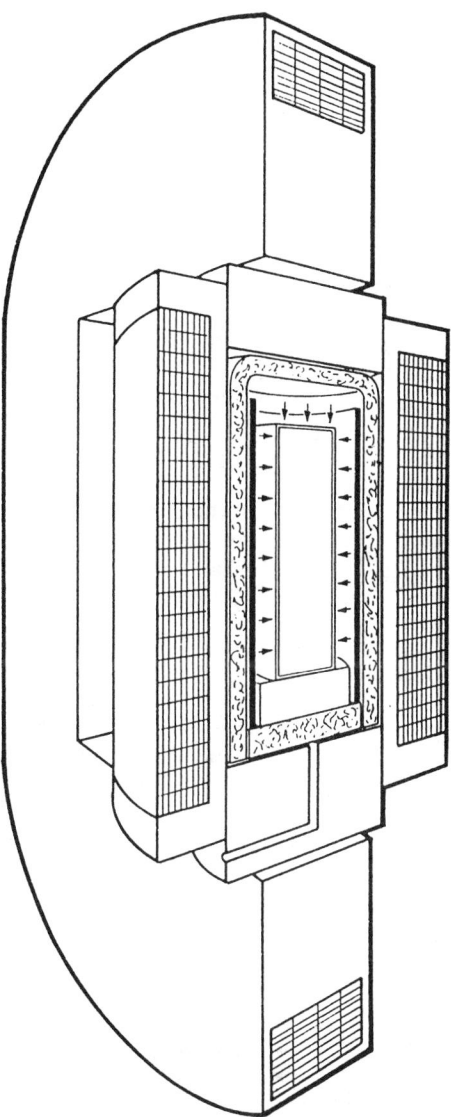

Fig 3 HIP pressure vessel incorporating a yoke closure. The pressure vessel is of leak-before-break design. Courtesy of Asea Brown Boveri

HIP Technology Applied to Ceramics

There are several processing options available to produce shaped and dense ceramic parts using hot isostatic pressing (Fig 4). The HIP method chosen has, however, a great effect on the results that can be achieved for many high-temperature ceramic materials.

Filling into a Shaped Capsule. The method listed at the extreme left in Fig 4 (hot isostatic pressing with a shaped container) is the most common for near-net-shape manufacture of powder metallurgy (P/M) parts. Sheet-metal or glass containers are used. These containers have a shape similar to the workpiece being produced but are enlarged to compensate for the difference between the fill density and the final density of the powder. The removal of the encapsulation after hot isostatic pressing for such materials can often be combined with finish machining.

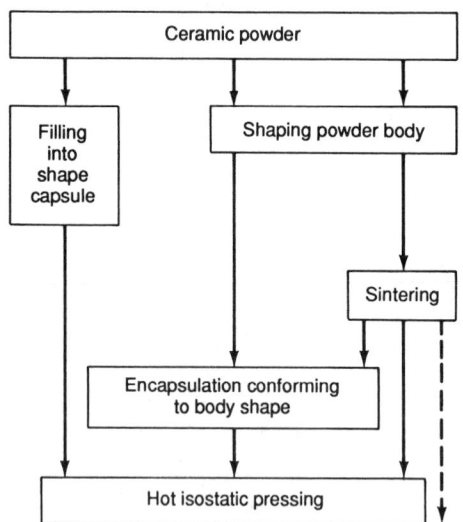

Fig 4 Alternative methods to manufacture shaped and dense parts from a ceramic powder

This HIP method was used for large α-alumina canisters in a technical feasibility study for making long-term resistant containment for spent nuclear fuel (Ref 7). About 1600 kg (3500 lb) of 99.8% purity Al_2O_3 powder was carefully packed in a low-carbon steel container that was 3000 mm (120 in.) in length and 600 mm (24 in.) in diameter. A fully dense alumina canister with a diameter of 500 mm (20 in.) having a tolerance of 1 to 2 mm (0.04 to 0.08 in.) was produced (Fig 5). Such a tolerance was acceptable in this case but could

only be reached through the use of an efficient system to uniformly pack the alumina powder into the annular steel container and because of the relatively simple geometry of the workpiece. For several reasons, it is difficult to meet close tolerances, to find suitable capsule material, and to control and limit reactions with the part produced for ceramics that require high processing temperatures. These findings have all contributed to make the interest in this HIP method very limited for high-performance ceramic products with a defined shape. Tantalum cans have, however, been used with good results for making material samples of silicon nitride (Ref 8) and boron (Ref 9).

Cladless Hot Isostatic Pressing. The alternative to the right-hand side of Fig 4, cladless HIP (often called sinter-HIP or sinter plus HIP), was invented as early as 1938 by Johan Romp (Ref 10). It was used for one of the first commercial applications of hot isostatic pressing about 20 years ago. Large cemented carbide products like rolls for rolling mills were post-densified. As the parts after liquid-phase sintering already had reached over 99% of theoretical density, the remaining porosity was well separated from the surface. The high-pressure gas (argon) could therefore

be allowed to act directly on the surface of the part and the positive effect of closure on internal pores was utilized. The HIP process provided a typical reduction in the frequency of a certain size of pore by two orders of magnitude when compared to the pore size prior to hot isostatic pressing.

Cladless HIP is a widely used method to hot isostatically press ceramics. It works particularly well with oxide ceramics, and the production of alumina-based cutting tools by sinter-HIP is an example of a commercial application. A necessary prerequisite operation prior to the HIP-stage is to sinter the ceramic powder components to such a density, typically 92 to 96% of theoretical, that the pores cease to be interconnected to the surface of the part. Hot isostatic pressing is then carried out with gas pressure acting directly on the surface of the part. The gas composition is important for many materials. When processing silicon nitride, for example, nitrogen gas is commonly used to reduce the dissociation of the material.

The necessary presintering to about 95% density limits the composition to materials that can be pressureless sintered to a high density without unacceptable deterioration of structure, shape, and other properties. Many

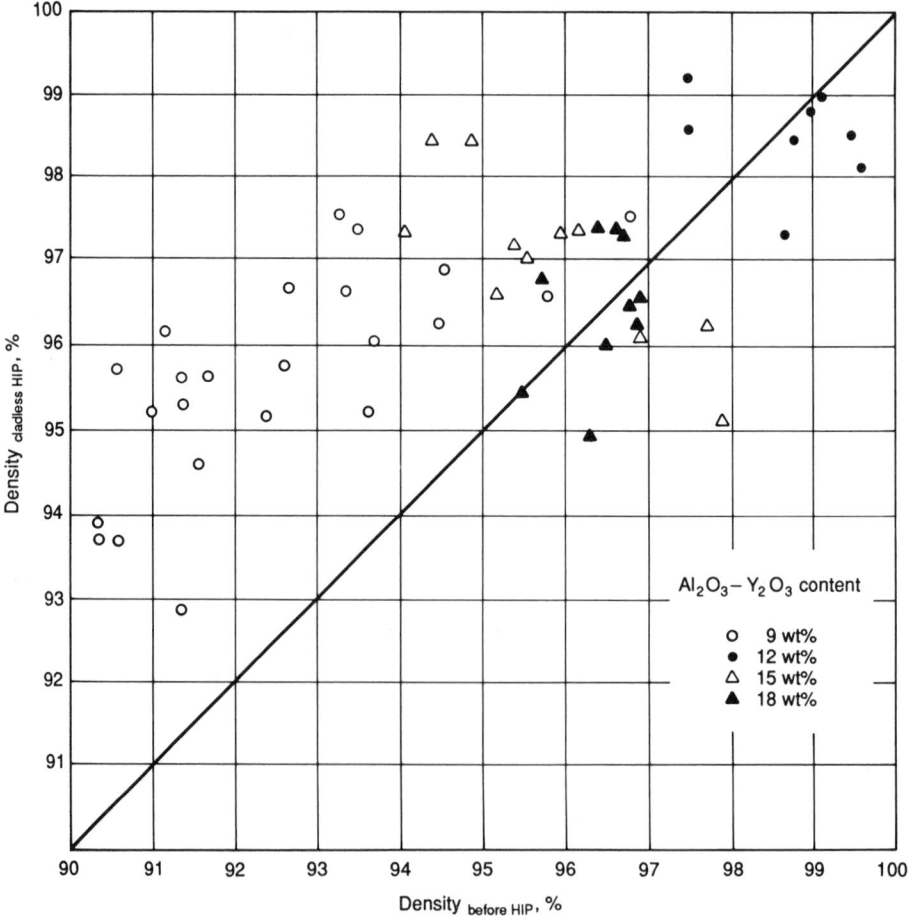

Fig 5 Large high-purity α-alumina canister produced by the HIP process using a shaped mild-steel container. The 500 mm (20 in.) canister with a 100 mm (4 in.) wall thickness is being assembled to join a similarly made lid using HIP bonding.

Fig 6 Plot of the density of cladless HIP process versus the density before the HIP process for selected concentrations of yttria-alumina doped silicon nitride. Source: Ref 12

ceramic materials with high strength at high temperature are prone to other limitations. In the presintering of silicon nitride, for example, elongated β-Si₃N₄ grains are formed. These new grains often form a very rigid interlocking structure. With the yttria-doped systems, several investigators such as Ziegler and Woetting (Ref 11) and Kim and Baik (Ref 12) have shown that typically only a small percent density increase is obtained regardless of the initial density of the ceramic body (Fig 6). Considering the high strength and low creep rates of the pure β-Si₃N₄ grains, this may not be surprising. For magnesia-doped silicon nitride, the densification obtained is more effective and close to full density.

If there is a sufficiently high-volume fraction of glassy boundary phase present, the densification is often better in both alloying systems. Kim and Baik, however, reported some cases where the gas appeared to penetrate into the glassy phase to such an extent that the density was actually reduced by the HIP treatment. Generally, the densification is positively affected by long processing times (typically 6 h or more) (Ref 13). In addition, in silicon-carbide materials the grains formed during the presintering stage often resist further densification. If grain growth in the presintering stage can be restricted and a fine-grain material is used, good densification during hot isostatic pressing can be obtained (Ref 14).

Generally, an inert gas (usually argon) is used for the cladless HIP method. The use of a chemically active pressure gas is, however, preferred in some cases. Some oxide materials tend to be slightly reduced in an inert atmosphere, causing discoloration and deterioration of material properties. An oxygen partial pressure can be used for white hot isostatic pressing of oxides such as zirconia (Ref 15). Furthermore, oxide superconductors are hot isostatically pressed under similar conditions. For materials that tend to decompose or dissociate at sintering temperature, a partial pressure of a gaseous decomposition product will reduce or eliminate the decomposition. Nitrogen gas is therefore used for hot isostatic pressing of nitride ceramics, usually as pure nitrogen gas. In all cases where gases that are not fully inert are used, suitable inert furnace materials must be used. In spite of these precautions, there is often a reduction in the service life of furnace components such as thermocouples. In cladless hot isostatic pressing with any type of gas, it is important to be aware that small amounts of impurities in the pressure gas can become chemically much more active in proportion to the gas pressure and may react with furnace components or processed material.

Encapsulation Conforming to Body Configuration. In the HIP method in the center portion of Fig 4 (encapsulated HIP of shaped powder body), the green powder body is encapsulated before any significant shrinkage of the body occurs. The method is also sometimes designated direct-HIP in order to separate it from sinter-HIP. The positive effect of pressure with this method can be fully utilized throughout the shrinkage and sintering stage. In the first published realization of this principle for silicon nitride (Ref 16, 17), a conformable high-silica glass container was used (Fig 7). That method is still well-suited for material samples. With Vycor glass, which yields better results than quartz glass, the experimental conditions can be particularly well-defined. Hot evacuation can be made at relatively high temperatures (~1000 °C, or 1830 °F). Shapes like turbine blades have been processed (Ref 18) by the use of a boron-nitride powder bed filling the space between the green powder body and glass container. The powder bed should have about the same relative density as the green powder body in order to reduce shape distortion. The boron nitride also provides a chemical separation to the glass.

Containers of lower softening point glasses like Pyrex have been used (Ref 19) to surround the sample packed in a boron-nitride bed and wrapped with molybdenum foil (for example, in silicon-nitride/silicon-nitride joining experiments).

Encapsulation techniques using glass particles (Fig 8) are the most commonly used methods in industry. They have also found extensive use in the manufacture of prototype silicon-nitride components (for example, turbine rotors or stators) for heat engines.

Different applications of particle-type encapsulation have been published in the patent literature. There are single- or multiple-layer techniques (Fig 9a) and there are techniques using an open vessel to contain the glass (Fig 9b). In the open vessel method, low-viscosity glasses can be used. Separation layers,

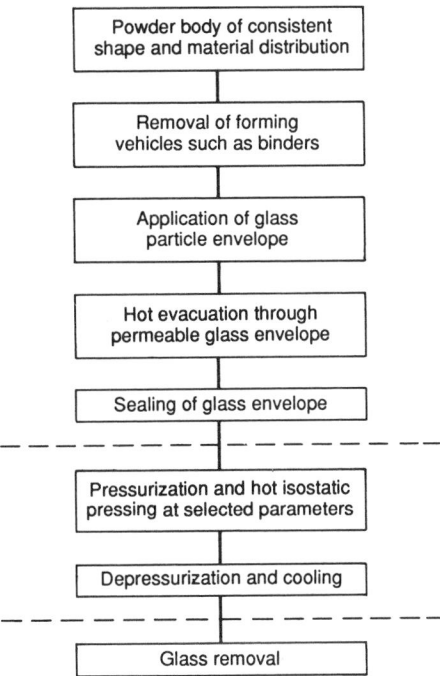

Powder body of consistent shape and material distribution

↓

Removal of forming vehicles such as binders

↓

Application of glass particle envelope

↓

Hot evacuation through permeable glass envelope

↓

Sealing of glass envelope

- - - - - - - - - - - -

Pressurization and hot isostatic pressing at selected parameters

↓

Depressurization and cooling

- - - - - - - - - - - -

Glass removal

Fig 8 Flowchart showing processing sequence of glass particle encapsulation and the HIP process. Source: Ref 20

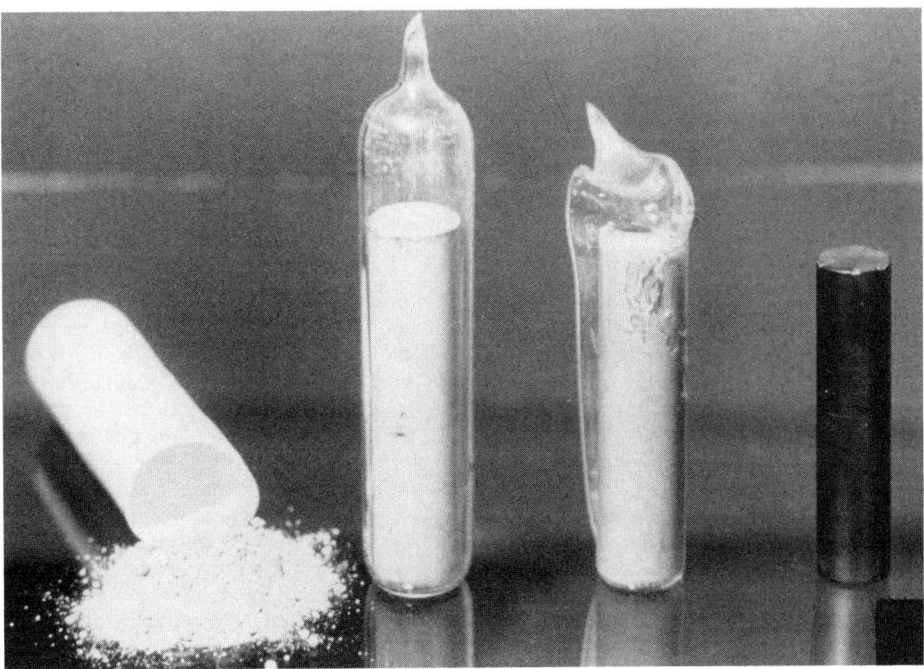

Fig 7 Steps required to use encapsulated HIP process for a shaped powder body. Encapsulation of a cold isostatically pressed (CIP) cylinder of silicon nitride powder with 1 wt% yttria additive in a Vycor high-silica glass container (left). Evacuated and sealed container ready for HIP (center left). Container halfway through the HIP cycle after heating to softening of the glass (about 1300 °C, or 2370 °F) and pressurizing the glass to conform to the shaped powder body (center right). Fully dense part after HIP process at 1750 °C (3180 °F) and 200 MPa (30 ksi) for 1 h and finishing by sand blasting (right)

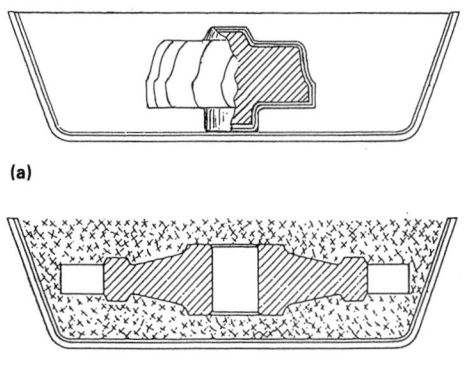

Fig 9 Two methods used for particle-type encapsulation: (a) single/multiple-layer technique and (b) technique using an open vessel to contain the glass. Source: Ref 21-24

usually to some part containing boron nitride powder, may often be a part of the encapsulation system in order to control possible reaction and penetration phenomena. The encapsulation may be mechanically removed after hot isostatic pressing (for example, by sand blasting), a method also used for the ceramic-ceramic multiple layers (Ref 23).

The process route (right of center in Fig 4) leading from sintering to encapsulation and hot isostatic pressing is of interest for full densification of reaction-sintered but porous materials such as reaction-bonded silicon nitride. An advantage of this process is the low linear shrinkage during full densification: 6 to 9% compared with typically 14 to 18% from a green powder body.

Temperature Selection. In cladless hot isostatic pressing (sinter-HIP), it is advisable to limit the temperature to just below incipient melting for materials that have been liquid-phase sintered to closed porosity. A similar consideration should be made for materials containing low-viscosity glasses as second-phase materials. Otherwise, the liquid phase or low viscous glass may be squeezed into the remaining pores and a desired uniform densification of the body would not be obtained.

In encapsulated HIP, it may often be possible to obtain fully dense high-temperature ceramic products at temperatures 300 to 400 °C (540 to 720 °F) lower than those required in pressureless sintering (for example, fully dense, 99.99% pure Al_2O_3 at 1200 °C, or 2190 °F, at 1 h, 200 MPa, or 30 ksi) (Ref 24). Existing conventional hot pressing methods can be used as a guideline. A lowering of the temperature by ≥100 °C (≥180 °F) in relation to hot pressing temperatures may be possible, however, particularly for materials with very high strength levels at high temperature (for example, pure TiB_2 at 1600 °C, or 2910 °F, and 200 MPa, or 30 ksi, for 1 h) (Ref 14).

Great progress has recently been made in the modeling of HIP processes (Ref 26, 27).

Through the use of computer software, the complex results can readily be available to material engineers.

Because material properties are so variable, it must be recognized that a tuning process for the results of real HIP runs will be necessary.

Full densification by hot isostatic pressing can thus generally be achieved at comparatively low temperatures. The development of desired material phases and microstructures can, however, often depend less on the pressure than on the temperature and time and these two variables may consequently be chosen for the optimization of material properties.

Removal of Adsorbed Substances. In encapsulated hot isostatic pressing, the powder body becomes isolated from its surroundings typically <1000 °C (<1830 °F) for metal encapsulation and often from 1000 to 1300 °C (1830 to 2370 °F) for glass encapsulation. Adequate selection and handling of the powder and the powder body is therefore needed in order to avoid having those deleterious substances becoming trapped in the material. Fluorine (from hydrofluoric acid treatments of powder used to remove metal impurities) is an example of a very tenacious substance (Ref 28) that may adhere to powder surfaces at typical temperatures for the sealing of the encapsulation. Humidity, which is always present and adheres to powder surfaces, will, if not removed, also reduce the high-temperature properties of materials such as silicon nitride.

High-Precision Processing. An inherent feature of hot isostatic pressing is to produce parts with very accurate shape. In order to obtain highly precise dimensions in the final product, directional and spatial uniformity of the green powder body is needed as well as precision in the size and in the shape of its boundary. In addition to several factors that cause a lack of uniformity in the green body, distortion can also be caused by too rigid an encapsulation or the force of gravity itself. Metal containers are usually rather rigid, but models have been proposed (Ref 29) to better understand and to compensate for such deformation. If a density front in the powder body is formed by too rapid heating of the part while the part is under pressure, similar effects appear and large changes of shape may result. An analysis (Ref 30) proposes that this may happen when the densification rate is greater than the characteristic heat flow rate. A second effect in this case, as well as in the hot isostatic pressing of green bodies with large spatial variation of density, is the high residual stresses that may be built into the densified part. Stress relieving by high-temperature annealing appears to be one solution, but tailored heating rates and improved green body forming methods are recommended. In parts with asymmetric spatial density variations, it has been observed that most of the shape distortion occurs during the final portion of the densification process as the part approaches

the fully dense condition. If the remaining low-density regions are squeezed dense later than the rest of the part by the high HIP pressure, it is obvious that such shape changes can occur.

With methods such as injection molding used to form the green powder body (which can provide very high uniformity and precision) and with glass powder encapsulation (which is flexible enough not to restrain the shrinkage of the ceramic powder body), it appears that almost ideal conditions can be achieved (Ref 31). The linear size change from green body to final shape under ideal conditions is given by the ratio F:

$$F = \left(\frac{D_0}{D_f}\right)^{1/3} \qquad \text{(Eq 1)}$$

where D_0 and D_f are the relative green density and relative final density, respectively. On a 59.7 mm (2.35 in.) long curved component of silicon nitride (top section of Fig 10), the standard variation of the length was found to be only 0.042 mm (0.0017 in.), or 0.07%, in a production batch of 768 parts.

A standard deviation limited to only a few tenths of a percent in initial density of the green powder body would in fact leave room for no other shape-distorting effect whatsoever during sintering, if a low dimensional variation such as 0.07% is to be attained. This is obvious from the equation when knowing that the standard deviation of the final density is typically some hundredths of a percent.

Fig 10 Series-produced high-precision silicon-nitride components for advanced textile machinery applications manufactured to net shape by injection molding, glass-encapsulated HIP process, and with vibratory tumbling as the finishing operation

An explanation for this exceptional dimensional consistency was presented by H. Larker (Ref 32). The gas pressure, P_{gas}, that acts isostatically via the glass encapsulation on the boundary of the porous powder body has a very large effect on the effective pressure between particle contacts (Fig 11). As a result, gravity-induced tensile-stress components, which could cause shape deformation, will become virtually negligible (dotted line in Fig 11). The powder body shape in a typical glass-powder encapsulated HIP envelope is kept intact by the gas pressure acting on its boundary throughout the entire shrinkage and sintering process. The rigidity of the porous powder body may be visualized by the common vacuum-packed plastic package of ground coffee. The ceramic powder, however, is much harder and the pressure difference over its boundary more than a thousand times higher! In contrast, when sintering without encapsulation and consequently also in sinter-HIP or gas-pressure sintering (a process that may be categorized as sintering and low-pressure HIP in one sequence), such an effect cannot be taken advantage of because the major part of the shrinkage and densification process has to occur before the force exerted by the gas pressure becomes a factor. It should be noted

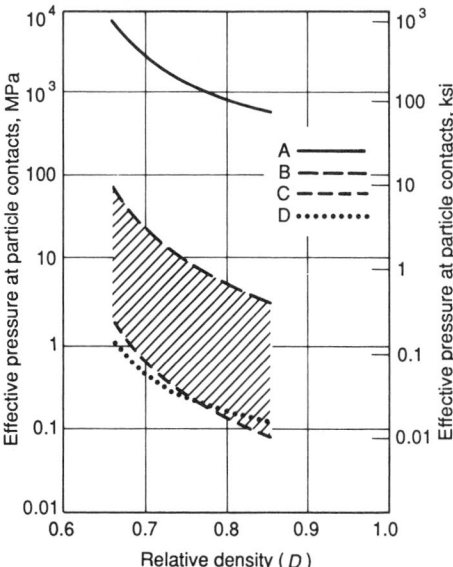

Fig 11 Driving force for densification (effective pressure at particle contacts) caused by external gas pressure (curve A) and surface tension, γ (curves B and C). Curve A values are typically 1000× greater than comparable values of curve D, which are the gravity-induced tensile stress components. Contribution due to external gas pressure: A, 200 MPa (30 ksi) pressure that is independent of the average particle radius, R. Contribution due to surface tension; B, γ is typically 1 J · m^{-2} (9 × 10^{-5} Btu · ft^{-2}) for crystalline materials (Al$_2$O$_3$, SiC) and R is 1 μm (40 μin.) for Al$_2$O$_3$-SiC powder; C, γ is typically 0.3 J · m^{-2} (3 × 10^{-5} Btu · ft^{-2}) for glasses and R is set as 10 μm (400 μin.) to illustrate the effect of occasional large voids in the powder body. Contribution due to tensile stress component; D, data for horizontal beam composed of Si$_3$N$_4$ powder having a length-to-height ratio of 10. Source: Ref 32

that there is no longer any such effect in the encapsulated HIP process when a part is fully densified and a uniform pressure is already established throughout the part.

Special Processing Options. H. Seino, K. Ishizaki, and M. Takata (Ref 33) pointed to an interesting possibility to control the oxygen chemistry in glass-encapsulated hot isostatic pressing of high-T_c ceramic superconductors where T_c is the critical temperature. By using the PbO/PbO$_2$-system, this group of researchers was able to produce highly dense material having the specified oxygen balance to give very good superconducting properties without any additional treatment in oxidizing atmospheres that previously was considered necessary.

M. Koizumi (Ref 34) has applied high-pressure self-combustion sintering (HPCS) to process TiB$_2$. A powder mixture of titanium metal and amorphous boron was formed into a rod by cold isostatic pressing (CIP). This TiB$_2$ powder was sealed under vacuum in a glass capsule (of the type described in Ref 19) together with a tungsten heating wire. The glass was heated to the softening point, 100 MPa (15 ksi) argon gas pressure was applied to the capsule assembly, and ignition was made electronically with a tungsten wire. The best densification was obtained using an excess amount of titanium metal.

Synthesis with as much as 50% of the volume liquid at a processing temperature of 1900 to 1950 °C (3450 to 3540 °F) (some 300 °C, or 540 °F, above the eutectic temperature) and with good part integrity obtained when using glass-encapsulated hot isostatic pressing has been reported by R. Larker and B. Loberg (Ref 35). A block formed by cold isostatic pressing of an equimolar mixture of silicon nitride and silica powders devoid of other oxides was hot isostatically pressed at a temperature of 1950 °C (3540 °F) and a pressure of 200 MPa (30 ksi) for 2 h with better than 90% of the composition converted to silicon oxynitride, a material that dissociates at atmospheric pressure above a temperature of 1700 °C (3090 °F).

Quality Control

Controlling the product quality of ceramic parts produced by hot isostatic pressing must not only be a check-before-delivery control but also requires that quality assurance be incorporated into all the processing steps. The objective of process optimization techniques is to make the process more efficient, easier to control, and more reliable. Knowledge of the process technology is the key factor for controlling the parameters and thus assuring product quality. Whenever possible, the processing parameters should be set to correspond to a specific area that the literature designates as the processing map or response surface in which the effect of parameter variations or uncontrolled variations in the process on product properties is minimized.

The basic rule in quality control, which is to check instruments and sensors against calibrated masters at predetermined intervals, must be adhered to. The high-temperature measurement sensors in the HIP equipment must be closely monitored.

Typical HIP Applications

Any of the selected HIP methods are, in general, effective in reducing or even eliminating porosity in materials that have been brought to a state of closed porosity without compromising the desired composition, microstructure, or other essential properties of the final product.

Cladless hot isostatic pressing is successfully used for the production of tool bits composed of oxide ceramics (for example, alumina base). In manganese-zinc and nickel-zinc ferrites, freedom from pores results in high permeability, high saturation induction, and improved wear resistance. The uniform fine-grain size that can be obtained gives better high-frequency characteristics and low ferrite noise. In piezoelectric ceramics, which are used in applications such as surface acoustic wave filters and oscillators, the uniform fine-grain size permits the processing of disks as thin as 0.05 mm (0.002 in.). The absence of pores is important in order to improve strength and to obtain a surface without defects.

Encapsulated hot isostatic pressing allows greater selection in the choice of composition of materials to take advantage of the high-temperature properties of powders and composites.

Glass-Encapsulated HIP Components. Important silicon nitride materials in the current U.S. heat engine programs such as NT 154 (Norton-TRW), GN10 (Garrett Ceramic Components), and PY6 (GTE Laboratories) are all processed by glass-encapsulated hot isostatic pressing. An extreme case is high-purity silicon nitride powder (UBE SNE-10), containing <0.01 wt% metal impurities, which without any sintering additives was fully densified at a temperature of 1950 °C (3540 °F) and a pressure of 250 MPa (35 ksi) for 2 h (Ref 36). The four-point bend strength at 1350 °C (2460 °F) for this alloy appears to be significantly higher than that obtained at room temperature. The fracture toughness is low, however, due to a microstructure having equiaxed grains. Tanaka et al. (Ref 37) found a very high resistance to slow crack growth at temperatures up to 1400 °C (2550 °F) in undoped high-purity silicon nitride in low yttria-doped high-purity silicon nitride. A similar material (3.5% Y$_2$O$_3$) made by the glass-encapsulated HIP process exhibited nearly the same bend strength (630 MPa, or 90 ksi) at 1400 °C (2550 °F) as that obtained at room temperature (680 MPa, or 100 ksi) (Ref 38). Furthermore, the material also showed very good oxidation resistance after 100 h at 1400 °C (2550 °F).

Silicon-nitride or silicon-nitride-based composites are utilized for high-precision parts by the textile industry (Fig 10); silicon-nitride or boron-carbide nozzles (Ref 39) are used in flue-gas desulfurization plants by the utility industry; silicon-nitride ball- and roller-bearing components (Ref 40); and alumina-based cutting tools are examples of products being commercially produced by the glass-encapsulation HIP process.

Ceramic-Ceramic Composites. Encapsulated HIP is generally very effective in densifying any type of ceramic-ceramic composite (particulate, whisker, or fiber-reinforced). If whiskers or fibers are of random orientation, they are aligned in a plane or are unidirectional to heavily affect the mode of shrinkage during densification. With whiskers of random orientation (15 to 35% silicon-carbide whiskers in Si_3N_4) (Ref 41), the most striking result was the decreased scatter in the strength data. An *m*-value (Weibull modulus or distribution parameter) over 30 at a three-point bend strength above 1000 MPa (150 ksi) was observed for a material with a 15% SiC whisker content in Si_3N_4 doped with 2.5 wt% Y_2O_3. This is, however, consistent with observations by Becher *et al.* (Ref 42). The anticipated breakage of long fibers does not seem to occur with carbon-fiber bundles in a silicon-nitride matrix (Ref 43).

REFERENCES

1. A.G. Evans and C.H. Hsueh, Behavior of Large Pores during Sintering and Hot Isostatic Pressing, *J. Am. Ceram. Soc.*, Vol 69 (No. 6), 1986, p 444–448
2. H.T. Larker, "Vertical Tube Electric Furnace," U.S. patent 3,470,303, Oct 1966
3. C.B. Boyer and F.D. Orcutt, "High-Pressure Furnace," U.S. patent 3,427,011, Nov 1967
4. K. Hunold, A Thermocouple for High Temperatures, *Adv. Mater. Proc.*, No. 9, 1986, p 5–6
5. A. Traeff, "QUINTUS Hot Isostatic Presses," ISO4 Conference, 5–7 Nov 1990, MPR Publishers, in press
6. G. Haerkegaard, J. Liljeblad, and L. Ohlsson, "A Comparative Study of Various Design Concepts for High-Pressure Vessels Subject to Cyclic Operation," No. 84-PVP-116, American Society of Mechanical Engineers, 1984
7. H.T. Larker, Hot Isostatic Pressing—Characteristics and Prospects in Industrial Use, *High Pressure Science and Technology*, B. Vidar and Ph. Marteau, Ed., Pergamon Press, 1980, p 571–582
8. D.W. Richerson and J.M. Wimmer, Properties of Hot-Pressed Silicon Nitride, *Commun. Am. Ceram. Soc.*, 1983, p C173-C179
9. C. Hoenig, R. Otto, and W. Stutler, "Densification Studies of Refractory Materials using Hot Isostatic Pressing (HIP) and Tantalum Containment," Proceedings of the 7th CIMTEC—World Ceramics Congress and Satellite Symposia, Montecatini Terme, Italy, 24–30 June 1990
10. Johan Romp, "Method of Making Sintered Hard Metal Alloys," U.S. patent 2,263,520, Nov 1938
11. G. Ziegler and G. Woetting, Post-Treatment of Presintered Silicon Nitride by Hot Isostatic Pressing, *Int. J. High Technol. Ceram.*, Vol 1 (No. 1), 1985, p 31–58
12. S.S. Kim and S. Baik, Hot Isostatic Pressing of Sintered Silicon Nitride, *J. Am. Ceram. Soc.*, Vol 74 (No. 7), 1991
13. K. Hirota, T. Ichizaki, Y. Hasegawa, and H. Suzuki, Densification of Si_3N_4 by Hot Isostatic Press in N_2 Atmosphere, *Proceedings of the 1st International Symposium on Ceramic Components for Engines*, Hakone, 17–19 Oct 1983, S. Somiya, E. Kanai, and K.-I. Ando, Ed., KTK Scientific Publishers, 1984, p 434–441
14. K. Hunold, Hot Isostatic Pressing of High Temperature Ceramics, *Interceram*, (No. 2), 1985, p 38–43
15. N. Claussen, K.H. Heussner, R. Janssen, E. Lutz, and N.A. Travitsky, White HIPing of Oxide Ceramics, *Hot Isostatic Pressing—Theories and Applications*, T. Garvare, Ed., CENTEK Publishers, 1988, p 395–397
16. H.T. Larker, J. Adlerborn, and H. Bohman, "Fabricating of Dense Silicon Nitride Parts by Hot Isostatic Pressing," Technical Paper No. 770335, Society of Automotive Engineers, 1977
17. J. Adlerborn and H.T. Larker, "Method of Manufacturing Bodies of Silicon Nitride," U.S. patent 4,455,274, Nov 1974
18. M. Boehmer and J. Heinrich, German patent 30,37,237, 1980
19. M. Shimada, K. Tanihata, T. Kaba, and M. Koizumi, Diffusion Bonding of Al_2O_3 and Si_3N_4 Ceramics by HIPing, *Emergent Process Methods for High-Technology Ceramics*, R.F. Davis, H. Palmour III, and R.L. Porter, Ed., Plenum Press, 1984, p 591–596
20. H.T. Larker, HIP Silicon Nitride, *AGARD*, CP-276, 1979, p 18/1–4
21. S.-E. Isaksson and H.T. Larker, "Method of Manufacturing a Sintered Powder Body," U.S. patent 4,339,271, March 1971
22. J. Adlerborn and H.T. Larker, "Method of Manufacturing an Object of Silicon Nitride," U.S. patent 4,112,143, Jan 1976
23. J. Heinrich and M. Boehmer, "Process by Compacting a Porous Structural Member for Hot Isostatic Pressing," U.S. patent 4,812,272, March 1989
24. J. Adlerborn, H.T. Larker, B. Mattsson, and J. Nilsson, "Method of Manufacturing an Object of Metallic or Ceramic Material," German patent 2,950,158 and U.S. patent 4,478,789, Dec 1978
25. A.K. Tjernlund and L. Hermansson, "Hot Isostatic Pressing of Some Oxides at Low Temperatures," presented at the 89th Annual Meeting of the American Ceramic Society, Pittsburgh, 26–30 April 1987
26. M.F. Ashby, The Modelling of Hot Isostatic Pressing, *Hot Isostatic Pressing—Theories and Applications*, T. Garvare, Ed., CENTEK Publishers, 1988, p 29–40
27. A.S. Helle, K.E. Easterling, and M.F. Ashby, Hot Isostatic Pressing Diagrams: New Developments, *Acta Metall.*, Vol 33, 1985, p 2163–2174
28. L.A.G. Hermansson, M. Burstroem, T. Johansson, and M. Hatcher, High-Temperature Behavior of Chemically Treated Silicon Nitride Powders, *Commun. Am. Ceram. Soc.*, April 1988, p C183-C184
29. J. Besson and M. Abouaf, Behaviour of Cylindrical HIP Containers, to be published in *Int. J. Solids Struct.*
30. W.B. Li, M.F. Ashby, and K.E. Easterling, On Densification and Shape Change during Hot Isostatic Pressing, *Acta Metall.*, Vol 35, 1987, p 2831–2842
31. H.T. Larker and E. Karlsson, "Net Shape Series Production of Ceramics by Glass Encapsulated HIP," ISO4 Conference, 5–7 Nov 1990, MPR Publishers, in press
32. H.T. Larker, Recent Advances in Hot Isostatic Pressing Processes for High Performance Ceramics, *Mater. Sci. Eng.*, Vol 71, 1985, p 329–332
33. H. Seino, K. Ishizaki, and M. Takata, HIPed High Density Bi-(Pb)-Sr-Ca-Cu-O Superconductors Produced without any Additional Treatment, *Jpn. Appl. Phys.*, Vol 28, 1989, p L78-L81
34. M. Koizumi, R & D Today, Applications Tomorrow, *Hot Isostatic Pressing—Theories and Applications*, T. Garvare, Ed., CENTEK Publishers, 1988, p 287–296
35. R. Larker and B. Loberg, Diffusion Reactions between Silicon Oxynitride and Silicon Nitride during HIP-Synthesis, *J. Physique*, NC-5, Vol 49, 1988, p 219–225
36. J. Adlerborn, M. Burstroem, L. Hermansson, and H.T. Larker, Development of High Temperature High Strength Silicon Nitride by Glass Encapsulated Hot Isostatic Pressing, *Mater. Des.*, Vol 8 (No. 4), 1987, p 229–232
37. I. Tanaka, G. Pezzotti, J. Zeng, T. Okamoto, and Y. Miyamoto, "Highly Slow-Crack-Growth Resistant Si_3N_4 at Elevated Temperatures and Its Composites," Proceedings of the 7th CIMTEC—World Ceramics Congress and Satellite Symposia, Montecatini Terme, Italy, 24–30 June 1990
38. I.P. Tuersley, G. Leng-Ward, and M.H. Lewis, Silicon-Nitride-Based Ceramics for Gas Turbine Applications, *Third*

International Symposium on Ceramic Materials and Components for Engines, 27–30 Nov 1988, V.J. Tennery, Ed., American Ceramic Society, 1989, p 856–870

39. H.T. Larker, "Forming and Full Densification of Advanced Ceramics by Encapsulated Hot Isostatic Pressing," SME Paper EM87–117, 1987

40. Silicon Nitride for High Performance Bearings, Editorial (based on information by J.F. Chudecki), *Ceram. Bull.*, Vol 69, 1990, p 1113–1115

41. H.T. Larker and J. Adlerborn, Undoped and Low-Doped Silicon Nitride with Whisker Reinforcement Made by Glass Encapsulated HIP, *Third International Symposium on Ceramic Materials and Components for Engines*, 27–30 Nov 1988, V.J. Tennery, Ed., American Ceramic Society, 1989, p 227–236

42. P.F. Becher and T.N. Tiegs, Whisker-Reinforced Ceramics, *Composites*, Vol 1, *Engineered Materials Handbook*, ASM International, 1987, p 941–944

43. R. Lundberg, R. Pompe, and R. Carlsson, HIPed Carbon Fiber Reinforced Silicon Nitride Composites, *Ceram. Eng. Sci. Proc.*, Vol 9 (No. 7–8), 1988, p 901–906

Molten Particle Deposition

P. Fauchais, University of Limoges, France

MOLTEN PARTICLE DEPOSITION represents a group of processes in which finely divided surfacing materials are deposited in a molten or semi-molten state on a prepared substrate to form a spray deposit (Ref 1). The following factors determine the feasibility of using the molten particle deposition process in an application:

- Any material (plastic, metallic, ceramic, or cermet) that has a melting temperature at least 300 K (540 °F) lower than its vaporization or decomposition temperature can be sprayed (see Table 1)
- The heat source accelerates and heats the particulates being separated from the substrate. The surface temperature of the substrate and the coating can be kept at <150 °C (<300 °F) by cooling the coating surface during spraying with air (or gas) jets or even maintained at lower temperature by using a liquid gas (cryogenic cooling) or carbonic snow jets (Ref 2)
- The particulates in flight interfere in some way chemically and physically with the environment during the flight. As the flames or the plasma jets pump their surrounding atmosphere down very rapidly, spraying in air increases the oxidation drastically with the increase in the temperature of the particulates. However, plasma spraying in a controlled atmosphere or under a soft vacuum to avoid oxidation requires higher investments (in a ratio of 3 to 10 compared to ambient atmospheric spraying)
- The accelerated molten or semi-molten particulates strike the surface, flatten, and form thin platelets (splats) that conform and adhere to the irregularities of the prepared surface and to each other. After impact, the splats freeze very rapidly (generally in a few tens of a microsecond), with the next particulate impinging on an already solidified splat. Thus, the formed coating has a lamellar structure (see Fig 1) and the deposited layer generally has properties and a chemistry different from that of the same bulk material. The properties of the coatings depend primarily on the temperature and to a lesser extent on the velocity of the impinging particulates (thus, on the source required

to accelerate and heat them). Another factor affecting coating properties are the preparation and nature of the substrate, its preheating, heating or cooling of substrate and coating during spraying, and the cooling of the substrate and coating after spraying. These heating and cooling problems are obviously more critical for ceramics with low thermal conductivities

- The substrates onto which thermal sprayed coatings are applied include metals, ceramics, glass, most plastics, composites, and wood. Substrate preparation is mandatory and comprises two steps: (1) cleaning the surface to eliminate contamination and (2) roughening the surface to provide minute asperities or irregularities to enhance coating adhesion and to provide a greater effective surface area (for example, by sand blasting) for the substrates that must be sprayed shortly after preparation (typically within 30 min)

Deposition Methods

Transfer of Heat and Momentum. Whatever the heat source, the flame (combustion or plasma) has a limited length, L, and the critical factor is the type of gun used to melt the largest size powder of diameter d in a flight restricted to the distance L. The selection of the gun type is complex because it requires the solution (assuming the flame velocity and temperature distributions are known) of the equations of motion of a single particulate with a given injection velocity to determine its trajectory velocity and then calculate the heat transfer along its trajectory (Ref 3, 4). Moreover, such calculations have to be repeated to take into account the size and injection velocity distributions (Ref 5). However, simplified calculations assuming uniform temperature, T_∞, and velocity of the flow, U_∞, can help to define the guidelines. For example, the simplified calculation of Savkar et al. (Ref 6) shows that the particulate heating is governed by the ratio, $L/\sqrt{U_\infty \cdot d^2}$. This suggests that the lower the plasma speed, the larger the powder diameter that can be melted, assuming L is held constant. Using similar arguments, Engelke (Ref 7) has shown that a quantity, A, related to the properties of the

flame must be $>Q_L^2 d^2/16 \cdot \rho$ where Q_L is the particulate heat content per unit volume of the liquid at its melting point and ρ is its mean density. Thus, the ratio $Q_L \cdot d/\sqrt{\rho}$ may be considered a measure of the difficulty of melting a particular powder. The values obtained from this expression are comparable with experimental observations. It is also interesting to consider the parameter A to compare the different guns:

$$A = L(\bar{K})^2/(U_\infty \cdot \mu) \qquad \text{(Eq 1)}$$

where $\bar{K}$ is the integrated thermal conductivity of the gas

$$\bar{K} = \int_{T_s}^{T_\infty} K(s)ds \qquad \text{(Eq 2)}$$

with T_s being the particulate surface temperature and μ the mean viscosity of the plasma. For spraying plasma jets (in most cases a mixture of Ar-20 vol% H_2 or Ar-60 vol% He) (Ref 8), the addition of H_2, drastically increasing $\bar{K}$ for $T > 4000$ K (Ref 9), helps in melting the most refractory materials especially with T_∞ of the order of 10,000 K (17,500 °F) (at the maximum 14,000 K, or 24,700 °F). For argon-helium mixtures where $\bar{K}$ is higher than that of the argon-hydrogen mixtures only for $T > 9000$ K (15,700 °F), it is more difficult to melt the same refractory material because μ is three times higher than that of the argon-hydrogen mixture.

Equation 1 shows the advantage of plasmas compared to combustion flames where the maximum temperature is about 3500 K (5840 °F) and where $\bar{K}$ is lower than in plasmas. However, for plasma jets even at atmospheric pressure, the drag and Nusselt coefficients used in the equations of motion and energy may be significantly reduced by particulate evaporation and noncontinuum effects (Ref 3, 5). Fast evaporation is obtained in plasmas with high values of the mean integrated thermal conductivity, $\bar{K}$, where:

$$K = \bar{K}/(T_\infty - T_s) \qquad \text{(Eq 3)}$$

(which is the case in argon-hydrogen mixtures) compared to that of the particulates, K_p. As soon as $K/K_p > 0.03$ (Ref 9) heat propagation phenomenon occurs and the particulate surface starts to evaporate long before its center has reached the melting temperature.

Table 1   Properties and applications of ceramic and cermet coatings commonly used in molten particle deposition

Classification	Plasma spray material	Maximum service temperature(a)		Applications	Properties
		°C	°F		
Cermets	WC-W$_2$C with Co or Ni (7, 12, 17, 20 wt%)	450	840	Used for wear resistance	Metal matrix lowers wear resistance but increases resistance to mechanical or thermal shocks Erosion (cavitation) or abrasion depends on porosity, carbon content, and mean diameter of WC grains Sliding wear resistance generally poor except if WC decomposed
	Cr$_3$C$_2$-NiCr (80–20 wt%)	≤800	≤1470	Same use as WC-Co but at higher temperatures (mostly in aeronautics)	...
Carbides	TiC-Cr$_3$C$_2$	...	...	...	Good oxidation resistance up to 800 to 900 °C (1470 to 1650 °F)
	NbC-TaC	...	...	...	High corrosion resistance at high temperature High tenacity (oxidation resistance limited to 600 to 700 °C, or 1110 to 1290 °F)
Borides	TiB$_2$	...	...	Wear resistance	Low reactivity with molten metals High neutron-capture cross section
	CrB$_2$	...	...	Wear resistance	Low reactivity with molten metals High neutron-capture cross section
	HfB$_2$	...	...	Wear resistance	Low reactivity with molten metals High neutron-capture cross section
	TaB$_2$	...	...	Wear resistance	Low reactivity with molten metals High neutron-capture cross section
Oxides	ZrO$_2$ stabilized with CaO or MgO	<1000	<1830	Thermal barriers	Good wear resistance Good mechanical properties
	ZrO$_2$ stabilized with Y$_2$O$_3$ or CeO$_2$	<1350	<2460	Thermal barriers	Good wear resistance Good mechanical properties
	Al$_2$O$_3$	≤950	≤1740	...	Good wear resistance Dielectric High reactivity with molten salts Poor resistance to thermal shocks
	Al$_2$O$_3$ + xTiO$_2$ (x = 3, 13, 40 wt%)	≤800	≤1470	...	Addition of TiO$_2$: Lowers Al$_2$O$_3$ porosity, improves adhesion, gives an excellent surface finish, and good wear resistance
	Cr$_2$O$_3$	1200	2190	...	Excellent wear resistance Low porosity Excellent finish Used on sliding surfaces
	ZrSiO$_4$	1100	2010	...	Good resistance to liquid glass and combustion gases Not wetted by liquid metals: casting
	TiO$_2$	≤800	≤1470	...	Excellent surface finish Different colors (from white to black) due to oxygen losses

(a) Related to the coating detached from its substrate and assuming no severe thermal shock upon heating

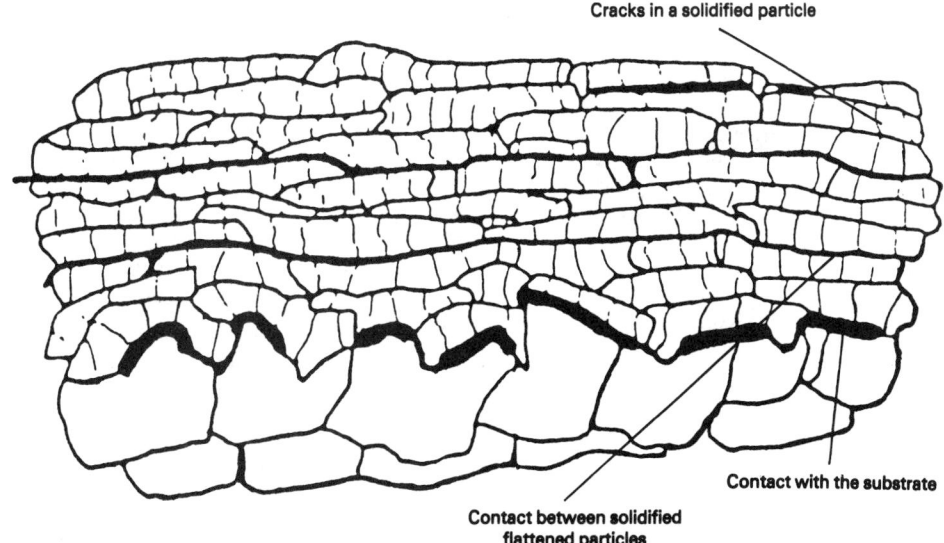

Fig 1 Schematic showing cross-sectional view of the lamellar structure of the coating obtained with molten particle deposition. Source: Ref 1

Noncontinuum effects are encountered when the Knudsen number $N_k = l/d$ is >0.01. The mean free path, l, is proportional to $k \cdot T / \sigma \cdot P \cdot \sqrt{2}$ where σ is the total collision cross section of the plasma particles (typically, $\sigma \approx 10^{-19}$ m^2), P is the pressure in the jet, and k is the Boltzmann constant (1.38×10^{-23} J/molecule · K). Thus, at 10,000 K and 10^5 Pa, $l \sim 5$ μm and for soft vacuum spraying at 5000 K and 10^4 Pa, $l \sim 25$ μm. Finally, it must be remembered that according to the length, L, of the flames or plasma jets used in spraying the particulate velocity, v_p, even small particles (a few μm in diameter) never reach the flame or approach plasma jet velocity (with hypersonic flames $v_p < 500$ m/s, or 1600 ft/s).

Typical Spraying Processes for Ceramics. Among the devices used to process ceramics are the detonation gun, hypersonic flame spraying, and the direct current (dc) plasma gun.

Detonation Gun. The detonation gun or D-gun (Ref 10, 11) (Fig 2) consists of a water-cooled barrel about 1 m (3 ft) in length with an inside diameter of approximately 25 mm (1 in.) and associated gas and powder metering equipment. Initially, a mixture of oxygen and acetylene is fed into the barrel along with a charge of powder ($d < 45$ μm, or 0.0018 in.). The gas is then ignited, and the gas accelerates the powder to a velocity of ~800 m/s (~2600 ft/s). The maximum free burning temperature of oxygen/acetylene mixtures occurs with 45 vol% C_2H_2 and is ~3140 °C (~5680 °F), and it is estimated that under detonation conditions the temperature reaches 4000 °C (7230 °F). Even higher maximum free burning temperatures are obtained in the super D-gun.

The great success of this process has been in decomposition sensitive materials such as cermets that require the metal matrix to be slightly melted to avoid carbide dilution. The distance that the powder is entrained is much larger than that obtained in a plasma device. After the powder has exited the barrel, a pulse of nitrogen purges the barrel and the cycle is repeated about four to eight times a second. Each pulse results in the deposition of a coating within a circular region approximately 25 mm in diameter to a thickness of ≤20 μm (≤800 μin.). The total coating is in turn produced by a gridwork of numerous overlapping coated regions that are circular in configuration. The pattern of overlapping, as encountered in other spraying techniques, has to be carefully controlled to minimize substrate heating and residual stresses (see the section "Coating Formation" in this article). An adhesion pressure of 69 MPa (1 ksi) has been measured for cobalt tungsten-carbide coatings.

Hypersonic Flame Spraying. In recent years, the field of thermal spraying has seen the introduction of hypersonic combustion flame spray (HCFS) guns. In the Jet-Kote (JK) method (the first gun of this type was commercially available in 1983), a high-velocity combustion process (chemical energy) accelerates and melts (convective heat transfer from the hot flame) the particulates. The system consists basically of three parts: a main console, a pressurized powder feeder, and a portable torch (see Fig 3). An internal combustion flame is produced in a water-cooled combustion chamber by utilizing a continuous flow of oxygen and fuel gases. This flame makes a right angle bend and passes through four vortices that focus the hot gases into a narrow beam, which is accelerated through an extended length water-cooled nozzle. The combustion process creates a high back pressure, thus requiring a pressurized powder feeder to propel particulates axially from the rear of the nozzle into the stream of hot gases. According to the values reported by Wagner (Ref 12), at a distance of 300 mm (12 in.) from the nozzle tip, the mean velocity of the gas (oxygen to propylene of 7:1) was found

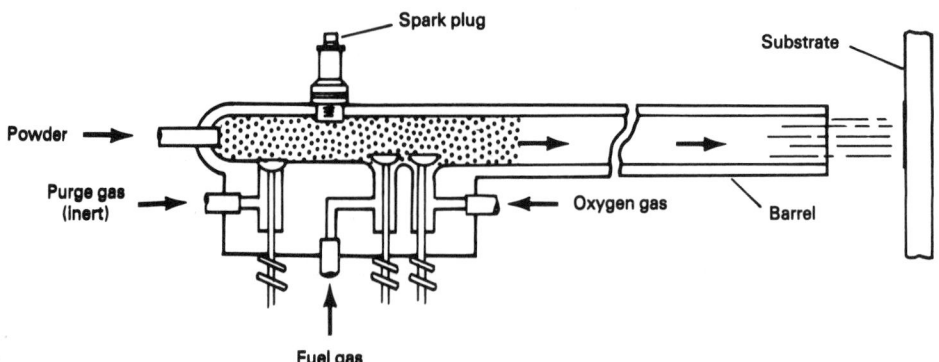

Fig 2 Schematic showing key components of a detonation gun. Source: Ref 2

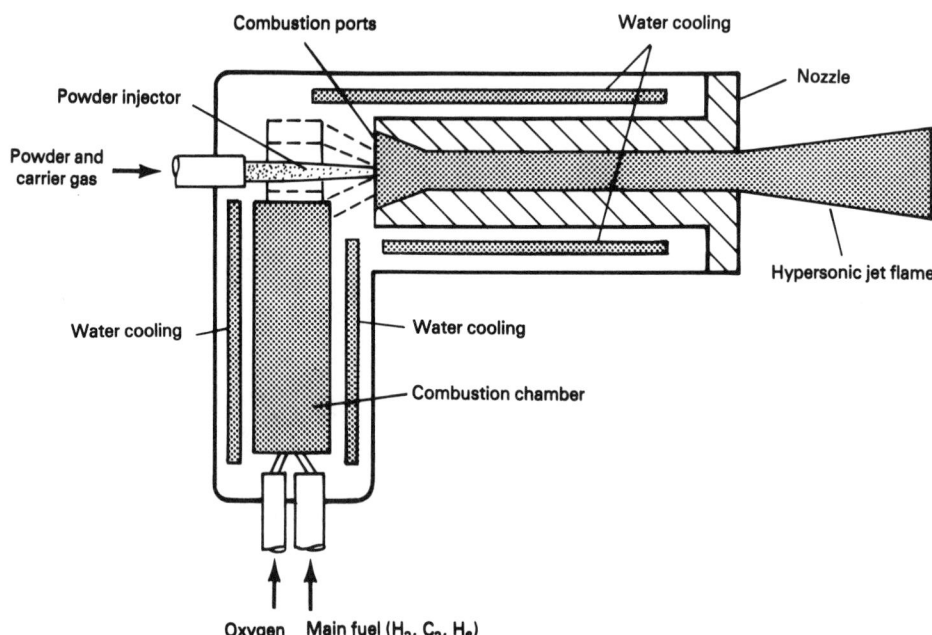

Fig 3 Schematic showing key components of a hypersonic flame spraying torch assembly. Courtesy of Jet Kote

to be 1770 m/s (5810 ft/s) and the temperature of the flame estimated to be about 2900 °C (5250 °F). Typically, for tungsten-carbide cobalt particulates in the size range −45 to 5 μm, the particulates velocity was found to be in the range 500 to 100 m/s (1600 to 330 ft/s) with an average value of 315 m/s (1030 ft/s). Gausert and Herman (Ref 12) have compared 60% Al_2O_3-40% TiO_2 coatings sprayed by Jet-Kote and by air plasma spraying (APS). Their conclusions were as follows:

- Particulates must have a diameter <25 μm (<1000 μin.) and only composite powders could be JK sprayed (sintered powders are not successfully sprayed)
- X-ray diffraction shows that the JK method produces coatings which are essentially a composite structure of alumina embedded within the relatively soft titania matrix

whereas the APS coatings consist of β-Al_2TiO_5 and α-Al_2O_3
- Bond strength of JK method coatings was ~40 MPa (6 ksi) versus 33 MPa (5 ksi) for the APS coating
- The hardnesses were comparable
- Wear tests were inconclusive because they depended on the testing method and not the thermal spray process used

When considering WC-Co coatings, it can be seen that ceramics coated with Jet-Kote are comparable to those obtained with vacuum plasma spraying (VPS), in which the carbide grains are not decomposed or diluted as they are with the APS method. This results in hardnesses up to 1300 HV 0.5 with the JK method versus 700 to 900 HV 0.5 with the APS method. However, oxidation cannot be avoided with the JK method due to the excess of oxygen. Such results may explain the recent interest in expanding the number of HCFS gun installations.

Dc Plasma Gun. Figure 4 shows a plasma spraying torch designed to work at atmospheric pressure (Ref 2). The arc strikes between a stick-type thoriated tungsten cathode with a conical tip and an anode nozzle, which in the APS method is generally a convergent followed by a cylindrical tube. The plasma is created at the cathode tip by thermoionic emission and part of the plasma gas injected along the cathode or in a vortex is pumped into the expanding plasma column. This column is stabilized by the cooling of its fringes both by the cold part of the plasma gas and by the water-cooled copper walls of the anode. When the cold boundary layer, close to the wall, is sufficiently heated by the hot boundary layer of the plasma column, this column becomes turbulent and the arc strikes at the anode wall. To cross the cold gas layer surrounding the plasma column, the arc constricts itself in one or a few tiny plasma columns that fluctuate under the action of the drag force and the electromagnetic forces (Ref 13). These fluctuations of the arc root on a rather large distance ($\sim 2D$, where D is the diameter of the cylindrical tube) help to preserve the integrity of the anode wall because the thermal fluxes at the arc root approach 10^{11} W/m^2 (3×10^{11} Btu/ft$^2 \cdot$ h). After the formation of the arc root, the extinguishing plasma can be maintained inside the nozzle and outside the nozzle (over distances ≤60 mm, or $2^3/_8$ in., in air) because of the heat released by the recombination of the ions and electrons as well as that of the atoms (for diatomic gases). Because of the hot tungsten cathode tip, no oxidizing gases can be used and the main plasma gases are Ar-H$_2$ and Ar-He and, for some cases, Ar, N$_2$, and N$_2$-H$_2$. With pure argon, the viscosity of the plasma increases (about 6 times that of cold gas) up to 10,000 K (17,500 °F) and then decreases for higher temperatures (due to the presence of ionized species). This viscosity increases with the addition of He and decreases with the addition of H$_2$. It should also be noted that the density of the plasma is $^1/_{10}$ to $^1/_{30}$ that of the cold gas. The plasma column has a temperature ≤20,000 K (35,500 °F) close to the cathode tip and ≥7000 K (12,100 °F) in its fringes (below that temperature the electrical conductivity is too low for a self-sustained arc) (Ref 8). As the plasma jet exits the nozzle, it encounters a steep laminar shear. The large velocity difference causes the flow around the nozzle exit to roll up into a ring vortex that is pulled downstream with the flow to allow the process to repeat itself again at the nozzle exit. Adjacent vortex rings have the tendency to coalesce and form larger vortices. It is the entanglement process of adjacent unstable vortices that results in the first large-scale engulfment or entrainment of external air. This phenomenon depends on the gas flow rate and viscosity, on the arc current, and on the nozzle design. When air is

entrained, its dissociation at 3500 K (5840 °F) cools down the plasma jet rapidly. A typical temperature distribution (Ref 14) is presented in Fig 5 for an Ar-H$_2$ plasma jet at 27 kW (36 hp). Velocities of 1000 m/s (3300 ft/s) at the nozzle exit have been recently measured under these conditions (the sound velocity at 14,000 K, or 24,700 °F, is close to 1800 m/s, or 5900 ft/s). It is worth noting that the design of the arc chamber and gas injector are of primary importance, the length of the plasma jet being increased by a factor up to 1.5 when this design is optimized (Ref 14). Typical power levels of APS torches range from 15 to 60 kW (20 to 80 hp). Because the surrounding atmosphere pumping increases with the arc current for currents higher than 500 to 700 A (depending on the gas nature, flow rate, and nozzle design), the length of the jet does not increase any more while the gas flow velocity increases and the heat transfer to particulates decreases. Very high flow rates (150 to 200 sfm, or 45 to 60 m/min), with power levels ≤80 kW (≤110 hp)

are thus used for the production of supersonic flows in air mainly to spray carbide cermets and to limit their decomposition.

If the surrounding atmosphere is argon (inert plasma spraying, or IPS, in a controlled atmosphere chamber, or CAC), the plasma jets used are longer (up to 1.4 times) and broader (up to 1.6 times) than those used with the APS method, and thus, besides the drastic reduction of oxidation, the heat-transfer properties are improved as a result of a longer plasma jet but similar particulate velocity. When the pressure in the controlled atmosphere chamber is progressively reduced (with soft vacuum plasma spraying, the controlled atmosphere chamber has to be double walled and water cooled or have a big volume of 20 to 30 m^3, or 700 to 1100 ft^3), the plasma jet length increases progressively due to the reduction of the turbulences in its fringes and can reach up to 0.6 m (2.0 ft) at 8 kPa (1 psi). The plasma gun nozzle obviously has to be adapted (Ref 2) to keep the pressure close to 1 atm (100 kPa) in the arc chamber and

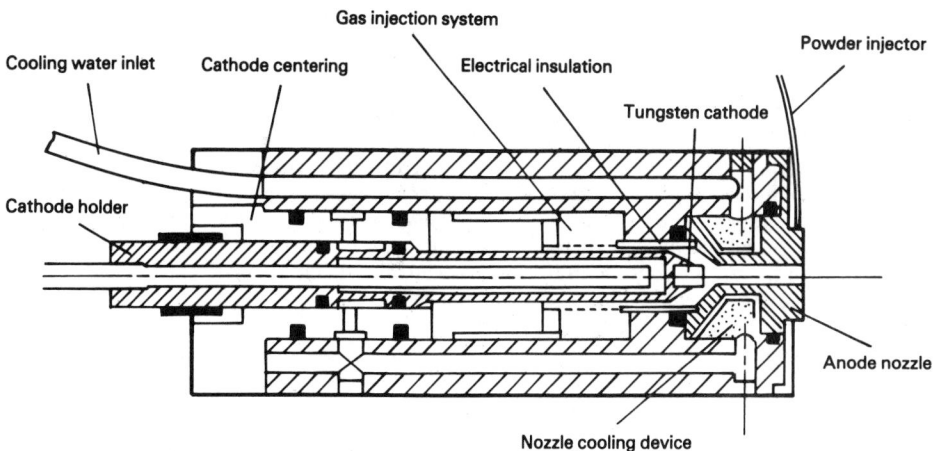

Fig 4 Schematic showing typical construction of a spraying plasma torch designed to operate at atmospheric pressure. Source: Ref 2

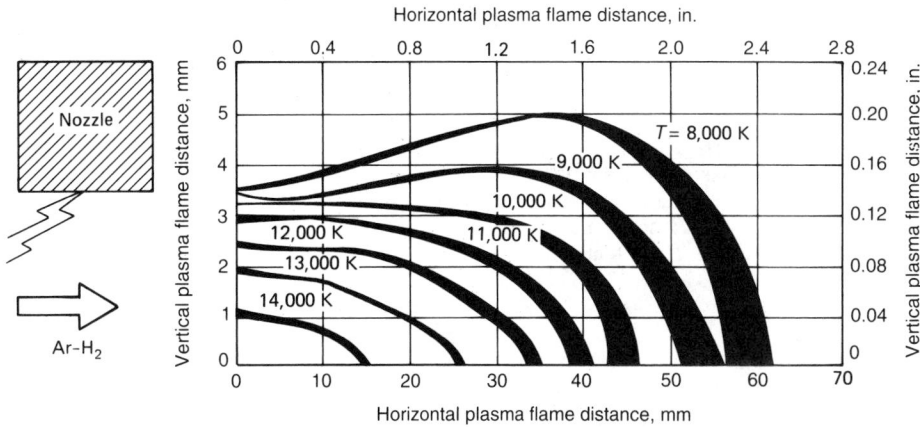

Fig 5 Typical temperature distribution of an Ar-H$_2$ plasma jet (40 sLm, or 1.4 scfm Ar flow rate and 8 sLm, or 0.30 scfm, H$_2$ flow) with a power rating of 27 kW (36 hp) (current, *I*, is 475 A; voltage, *V*, is 57 V); and a nozzle diameter of 8 mm ($^5/_{16}$ in.). Isotherm data is from Ar(I) line at absolute intensity where wavelength, λ, is 727.2 nm (7272 Å). Temperature of plasma, *T*, is in K. Source: Ref 14

the nozzle terminated with a Laval nozzle to account for the fast expansion of the jet (very often below 15 kPa, or 2 psi, diamonds appear and the plasma is said to be at Mach 1, 2 according to the number of diamonds). However, with the pressure reduction the temperature of the jet is drastically reduced (Ref 2, 15) (down to 4000 to 5000 K, or 6700 to 8500 °F, at 10 kPa) and as the mean free path of the plasma increases, N_K also increases and the heat transfer to the particulates decreases (Ref 3). Thus, at <30 kPa (<4 psi), the VPS method is not adaptable for ceramic plasma spraying, and its main function is to spray dense alloy bond coats (no oxidation, cleaning of the metallic substrate by a reverse transferred arc, or TA, and densification of the coating by heating of the substrate and the coating up to 1000 °C, or 1830 °F, with a direct transferred arc during spraying). Typical power levels range from 40 to 80 kW (55 to 110 hp). With the VPS method, the reduction of heat transfer is also quite useful to spray carbide cermets because only the metal matrix is melted, and, with the absence of oxygen, almost no carbide is decomposed. However, because of the capital investments required for vacuum plasma spraying and the quality of cermet coatings obtained with the D gun or the HCFS gun, VPS and IPS techniques are mostly used for spraying cermets.

In contrast to the D gun and the HCFS gun, the particulates in plasma spraying are not introduced along the axis of the barrel or the nozzle but orthogonally to the plasma jet axis either outside the nozzle or inside the nozzle but downstream of the arc root. With internal injection in the VPS method, the user must avoid having the injector located in the path of a shock wave. Typical powder flow rates for plasma spraying are in the range of 1 to 3 kg/h (2 to 7 lb/h) with a deposition efficiency ranging from 40 to 60%. In both flame spraying and plasma spraying, however, it must be kept in mind that particulates have size distribution and injection velocity distribution and that the trajectory distribution depends on both of these factors. Thus, the particulates will have different heat treatment before impacting the substrate.

Coating Generation

Effects of Particulate Flattening.

Madejski (Ref 16) has carried out a theoretical treatment on the impact of a molten droplet with a solid substrate, expressing the ratio of the lamella diameter, D, to that of the initial particulate size, d, as:

$$D/d = 1.29(\rho \cdot v_d/\mu)^{0.2} \qquad \text{(Eq 4)}$$

where ρ and μ are the density and viscosity of the particulate, respectively, and v_d is the droplet impact velocity. Equation 4 immediately shows the advantage of having molten particulates upon impact because μ is drastically reduced. The surface temperatures of ceramic particulates ($d < 45$ μm, or 0.0018

in.) in plasma jets have been measured up to 4000 K (6740 °F) with velocities up to 250 m/s (820 ft/s) versus <3000 K (<4940 °F) in HCFS guns with velocities in the range of 100 to 500 m/s (330 to 1640 ft/s) for WC-Co particulates. Typical values of D/d with ceramic particulates in plasmas are in the range of 4 to 7, which is in good agreement with Eq 4 (Ref 2) versus D/d values of 1.5 to 3 with alumina-titania particulates in an HCFS gun. The melting of the particulates obviously depends on the one hand on their size (for example, 40 μm, or 0.0016 in., diam zirconia particulates are completely molten in a 40 kW, or 55 hp, Ar-H$_2$ plasma jet while 60 μm, or 0.0024 in., particulates are not completely molten) (Ref 17), and on the other hand on the technique used to produce the particulate. This last point is especially crucial for ceramics having poor thermal conductivity properties and those subjected to conditions enhancing any heat propagation phenomenon. Thus, when using agglomerated particulates (which are rather porous) instead of fused and crushed particulates, K_p is lowered, and the particulates impact onto the substrate with a completely molten shell separated from a solid core (Ref 17). This results in completely different splats and forms a coating with lower mechanical and thermal properties. Moreover, Houben (Ref 18) has shown that shock waves propagated by the impacting particulates results in very different types of splats depending on the particulate diameter d, velocity v, and temperature T upon impact. Pancake types, for which d and v are relatively small, seem to have a very good adherence over the whole contact surface with microcracks generated to release stresses as the particulates freeze. Flower types have a central zone that adheres well to the substrate and the lateral free-standing material exhibits no cracks. Finally, exploded particulates correspond to overheated particulates. In turn, the contacts between the lamellae control the mechanical and thermal properties of the coating (Ref 2).

The nature of coating formation by impact and solidification of separate droplets results in some porosity that generally lies in the 5 to 20% range and depends both on spraying conditions and the properties of the sprayed material. All of the particulates have the same mean direction when impacting the substrate or the previously deposited layers and a shadow effect occurs which causes porosity. The liquefied and ductile particulates have little opportunity to accommodate completely the irregularities of the surface (that is, narrow holes cannot be filled out due to the fast freezing, a valley in the surface can include some air or gas, and the particulates may burst upon impact). A material-dependent effect is observed. For example, with TiO$_2$, 50% Al$_2$O$_3$-50% TiO$_2$, and Cr$_2$O$_3$, porosities below 4% can be achieved while with mullite, ZrO$_2$, and Al$_2$O$_3$, porosity values are in the range of 4 to 10%. Porosity also depends on the

particulate size distribution and on the incomplete molten particulates embedded in the coating (this factor will tend to increase the porosity).

Heat-transfer calculations as well as recent two-color pyrometry measurements (Ref 19) have shown that the freezing of particulates occurs in a few microseconds with a temperature drop of 1000 to 2000 K (1800 to 3600 °F). Thus, as the arrival of the next particulate on the same spot takes about 10^2 to 10^4 times the freezing time, the heat content removal from one particulate is not severely affected by the new incoming particulate. With freezing times at such low magnitudes, almost no diffusion phenomenon can take place and the adhesion for ceramics is almost purely mechanical (Ref 20). Grit blasting provides an ideal surface topology for interlocking and bond strength increases with increasing surface roughness because the lamella follow the peak contour and the contact is greatly improved upon cooling due to the lamella contraction. When grit blasting is not feasible (for example, composite materials), the surface can be treated with corrosive reagents. Metallic parts can also be sputtered with a reverse polarity transferred arc (using substrate as a cathode) utilizing the VPS method at low pressure (about 5 kPa, or 0.7 psi). When the substrate is too thin to be sand blasted, the solution is to spray (using VPS method) a bond coat of superalloy in which the last layer is sprayed with coarse particulates (60 to 80 μm, or 0.0024 to 0.0032 in., in diameter). In general, better adhesion is achieved if the particulates have (Ref 20):

- Good plasticity or low viscosity when they are liquid
- Good wettability
- Sufficient impact velocity
- A high coefficient of expansion

The roughness average, R_a, of the substrate has to be adapted to the size of the particulates (a poor adhesion is obtained with an R_a ~ 15 μm, or ~600 μin., and particulates of d ~ 5 μm, or ~200 μin.). Adhesion of ≤60 MPa (≤9 ksi) can be obtained for ceramic coatings.

The crystalline structure is linked to this fast cooling that may result in the formation of metastable phases (completely molten alumina particulates crystallize in γ phase) or even suppress crystallization (for example, in the first layers of alumina close to the substrate).

Lastly, chemical reactions may occur in particulates (mainly at their surface because it is a very slow diffusion process, requiring ~1 s, compared to the flight time, ~1 ms, of the particulates) by reaction with the plasma gas or the entrained surrounding atmosphere. Oxidation is detrimental to all of the non-oxide ceramics and either the IPS method or the VPS method is recommended or even mandatory in certain cases. Moreover, to limit or to avoid the decomposition of carbides, gases such as methane must be added to the plasma gas (nitrogen must be present in the plasma gas in order to spray nitrides). For example, pure carbide coatings such as tantalum carbide have been sprayed using the IPS method modified with cryogenic cooling by the Commissariat à l'Energie Atomique (CEA) at Bruyères-Le-Châtel in France.

Coating Formation. Substrate and torch have to be moved with well-defined relative velocities and patterns (for example, by a computer-controlled robot for complex shapes). For parts having cylindrical symmetry, a modified lathe setup with the torch mounted on what would normally be a tool port with the parts to be coated rotated in the headstock can be used. Predetermined torch-to-part surface speeds and overlap can be maintained by varying the rotation of the cylinder and traverse torch speeds. The sprayed bead thickness [in flight the particulates number flux distribution has roughly a Gaussian shape (Ref 4), resulting in a deposited Gaussian bead] and the bead overlap, which determines the thickness of the passes, have to be carefully monitored to limit the temperature gradients within the coatings (especially for sprayed ceramic materials with low thermal conductivities) (Ref 21). Besides the movement of the torch relative to the substrate, the coating temperature during spraying has to be limited by gas jets (in most cases, air is used as the medium) blown at the coating surface orthogonally to the plasma jet plume. With no preheating of the substrate, it results in a temperature gradient in the first tenths of mm with a final stationary temperature ranging from 150 to 700 °C (300 to 1290 °F). When preheating the substrate, the temperature gradient can be avoided. The final temperature can be drastically reduced (<0 °C, or <32 °F) when cooling the coating surface with liquid argon droplets during spraying. The temperature gradients and the stationary temperature of the coating are closely linked to residual stresses in coatings that can give rise to deformation of coated workpieces, spallation and cracking of the coating, modifications of adhesion, strength resistance to thermal shocks, and corrosion resistance. The following sources of stress generation emerge during the spraying process (Ref 22).

One source is stress due to the construction of individually sprayed splats as they rapidly cool to the substrate temperature. Quenching stress is always tensile and partly relaxed by edge relaxation, through-thickness yielding, interfacial sliding, microcracking, and creep. Very little is known about these sources of stress, and it appears that in ceramics they are insignificant due to their relaxation by microcracks.

Another source is differential thermal contraction stress due to temperature gradients and expansion mismatch between the coating and the substrate. They can be limited or controlled by carefully monitoring substrate and coating temperatures before, during, and after plasma spraying. Such stresses, depending on the differences between thermal expansivities and temperature gradients, can be either tensile or compressive forces or both in the coating. For example, extremely high temperature gradients within the coating (even if the expansion mismatch with the substrate is low) can limit its thickness very rapidly (to a few tenths of a mm) while cryogenic cooling allows the ceramic coating to obtain a thickness of at least a few centimeters. To limit the mismatch caused by heat expansion, bond coats can be used with a coefficient of expansion somewhere between that of the substrate and the coating (Ref 1). They are also selected to provide protection for the substrate as well. Bond coat thickness usually lies between 0.08 and 0.18 mm (0.0032 and 0.0071 in.) and when corrosion resistance is the main goal, they have to be sprayed and densified by using VPS techniques. Grooving is also employed to reduce internal stresses by dividing the stresses into smaller randomly distributed components and cancelling each other out. Such grooving is recommended for coatings that have edges.

The last source is volume change associated with any solid-state phase transformation. For example, in sprayed alumina coatings crystallized in γ-phase, the transformation of γ- to α-phase, which occurs when the coatings are heated above 1373 K (2012 °F), results in a change in density from 3.60 to 4.00 g/cm^3 (0.130 to 0.144 lb/in.^{3}) that causes spallation of the coating.

Coating Heat Treatments. Ceramic coatings, especially free-standing bodies, can be densified by HIP methods. However, the use of hot isostatic pressing for ceramic coating applications is rather expensive. The coatings can be heat treated by laser to generate cracks orthogonal to the substrate (controlled segmentation) resulting in a much better resistance to thermal shock but also yielding a material more prone to corrosion. Surface grooving is also used in coating laser-grooved printer rolls.

Surface finishing of the coatings is mainly achieved by grinding. However, the proper choice of the grinding wheel and wheel pressure is critical for coatings in which ductility and thermal conductivity are low (Ref 1). Typically, an R_a <0.4 μm (<16 μin.) can be obtained.

Applications

The technologies of plasma spraying for ceramics and D-gun, or hypersonic, flame spraying for cermets are essentially limited to materials engineering applications and have great potential in connection with:

- Improvement in the performance of parts and machine components by the pairing of optimum base material and surface coating properties to obtain a combination of characteristics that would not be possible with homogeneous materials

- Optimum use of resources through limited use of materials that are expensive, rare, or otherwise difficult to obtain
- Technical products innovation through the introduction of new ceramic materials whose properties and characteristics make the ceramics applicable to the manufacture of composite powders via spray deposition techniques. Typically, these products are metals or alloys with dispersed ceramic grains whose particle size and quantity can be varied easily, mixed ceramic powders incorporating ceramic grains with fibers to reinforce the strength of the coatings, and ceramics combined with plastics to obtain very porous coatings. However, because the methods used to produce the ceramic and cermet coatings introduce many variables into the processing (particulates with different size and trajectory distributions within the jet, temperature monitoring of the substrate prior to spraying, and temperature variations in the thickness of the substrate and coating during and after spraying), the coatings contain many flaws in terms of random size, configuration, and orientation and the strength of the coating varies from part to part. Thus, the statistics of these parameters must be taken into account to successfully predict failure (for example, by using the Weibull modulus, which is a parameter as significant as the material strength itself)

Because the aircraft and aerospace industries have provided an ideal proving ground for the testing and integrating of new coating concepts, the spraying technology used for ceramics and cermets has advanced steadily to the point that it has increased the credibility and the reliability of these coatings. The success of the coatings in aircraft component manufacturing applications (still the largest market but shrinking in terms of market share due to growth of ceramics in other fields) has led to applications in markets such as the paper, printing, steel, metal processing, textile, synthetic fibers, chemical, oil and gas, automotive, plastic, glass, pumps, pneumatic and hydraulic systems, mechanical engineering, shaped part, turbine, nuclear, electronics, and electrical industries.

The most common applications today for such coatings are discussed below (Ref 1, 2, 23–27).

Thermal barrier coatings (TBCs). Used on diesel engine power assemblies (piston domes, heads, and valves) and on components in automotive combustion chambers, the coatings provide the following benefits:

- 5 to 10% increase in horsepower
- Elimination of carbon build up
- Twice the service life (lower running tests)
- Increased combustion efficiency
- Extended intervals between routine maintenance

When used on gas turbines and aircraft engines (for example, combustion chamber, vane, and flame holder applications), a two- to fourfold increase in component life results. A recently developed Pratt and Whitney engine consisted of 59 TBC engine parts.

These coatings are generally two-component systems consisting of a metallic layer (a superalloy) overcoated with stabilized zirconia. The bond coat (sprayed using the VPS method and fully dense) protects against oxidation and corrosion. Thermal barrier coatings protect mandrels and dies used in the fabrication of high-temperature extrusions. These parts can be easily recoated after multiple extrusions.

Protection against Antifretting and Wear. Wear is caused by fretting, sliding, impact, abrasion, erosion, and other service conditions. Wear-resistant coatings have to be hard and exhibit resistance to heat and chemical attack. Typical applications include:

- Al_2O_3 and Al_2O_3-TiO_2 coatings used in the textile industry
- Cr_2O_3 coatings used in pumps and printing presses (in general, Cr_2O_3 is the most common ceramic coating material used to significantly reduce mechanical metal-to-metal wear)
- Cr_2O_3-SiO_2 coating used because silica functions as a damping medium that absorbs energy from impact loading
- WC-Co and Cr_3C_2-NiCr coatings are used in aviation applications (for example, landing gear components and compressor blades) and by the turbine industries (a Pratt and Whitney engine contained 253 parts that were sprayed to improve component wear resistance)
- Thermal spraying of ceramics or cermets on plastic or brick clay extruders, valves used in oil drilling equipment, Yankee dryers in paper machines (usually coated on-site), large rollers in sheet steel production, grinding wheel mold rings, and cam followers

Corrosion Protection. All plasma ceramic coatings have varying degrees of interconnected porosity and either the protection is obtained with a dense bond coat (as in thermal barrier coatings) or by a sealant such as an asbestos-filled Teflon or epoxy heated to 150 °C (300 °F), or copper and zinc deposited by electrolysis for higher temperatures. Sintering hot isostatic pressing (HIP) is also used for self-standing bodies.

Surfaces with Special Electrical Properties. Electrically insulating (very often Al_2O_3) or conducting surfaces (typically, resistors made of carbon and ceramics or with superalloys and alumina particulates).

Self-lubricating coatings that contain free carbon.

Free-Standing Ceramic Bodies. Production of near-net shapes that are costly to process using conventional processing methods.

Composite parts such as carbon fiber-reinforced graphite tubes internally coated with ceramics that can withstand internal pressures up to 100 MPa (1000 atm). A cermet coated composite crankshaft has been used in cars produced by a European automobile manufacturer for the past year.

Medical implants coated with hydroxylapatite [$Ca_5(PO_4)_3OH$] for bone reimplantation and Al_2O_3 or Cr_2O_3 for friction parts (for example, in hip joints).

Color deposits include red brick, yellow, brown, green, blue, violet, gray, and black.

Future Trends (Ref 28)

Over the last five years, sales of thermal sprayed ceramic and cermet coatings for commercial applications have steadily increased at an average annual rate of 15%. This is due not only to the advances in equipment, but also due to a better understanding of the physical and chemical phenomena encountered during coating generation. Currently, major emphasis has been directed toward improving gun efficiency and greatly increasing the deposition rate (especially for free-standing bodies required for large parts produced at rates approaching 90 to 230 kg/h (200 to 500 lb/h).

REFERENCES

1. *Thermal Spraying, Practice, Theory, and Application*, American Welding Society, 1985
2. P. Fauchais, A. Grimaud, A. Vardelle, and M. Vardelle, *Ann. Phys. Fr.*, Vol 14, 1989, p 261–310
3. E. Pfender, *Plasma Chem. Plasma Process.*, Vol 9 (No. 1), 1989, p 1675
4. P. Fauchais, M. Vardelle, A. Vardelle, and J.F. Coudert, *Metall. Trans. B*, Vol 20B, 1989, p 263
5. A. Vardelle, Ph.D. thesis, University of Limoges, France, July 1987
6. S.D. Savkar and P.A. Siemers, Some Recent Developments in Rapid Solidification Technology, Workshop on Industrial Applications, *ISPC 9*, M.I. Boulos, Ed., University of Sherbrooke, 1989, p 80
7. J.L. Engelke, "Heat Transfer to Particles in Plasma Flame," A.I.Ch.E. meeting, Los Angeles, 5 Feb 1962
8. B. Pateyron, M.F. Elchinger, G. Delluc, and P. Fauchais, Thermodynamic and Transport Properties of Ar-H_2 and Ar-He Plasma Gases used for Spraying at Atmospheric Pressure Part 1: Properties of the Mixtures, submitted to *Plasma Chem. Plasma Process.*
9. E. Bourdin, P. Fauchais, and M. Boulos, *Int. J. Heat Mass Transf.*, Vol 26, 1983, p 567
10. B. Gill, J.M. Quets, T.A. Taylor, and R.C. Tucker, ASME paper 82-GT-266, American Society of Mechanical Engineers
11. Union Carbide Coating Service France, Industrie CFE, Vol 4, 1990, p 34
12. D.J. Gausert and H. Herman, *Thermal Spray Technology, New Ideas and Processes*, ASM International, 1989, p 139
13. P. Fauchais, J.F. Coudert, A. Vardelle, M. Vardelle, A. Grimaud, and P. Roumilhac, *Thermal Spray: Advances in Coatings Technology*, ASM International, 1988, p 11
14. P. Fauchais, P. Roumilhac, and J.F. Coudert, "Influence of the Arc Chamber Design and of the Surrounding Atmosphere on the Characteristics and Temperature Distributions of Ar-H_2 and Ar-He Spraying Plasma Jets," Proceedings of Materials Research Society, Spring Meeting, 1990
15. S. Weissman, M. Smith, and W. Chambers, *Mater. Res. Symp. Proc.*, Vol 98, 1987, p 197
16. J. Madejski, *Int. J. Heat Mass Transf.*, Vol 19, 1976, p 1003
17. P. Fauchais, M. Vardelle, A. Vardelle, and A. Denoirjean, "Heat Treatment of Ceramic Powders with Different Morphology under Thermal Plasma Conditions," Proceedings of Materials Research Society, Spring Meeting, 1990
18. J.M. Houben, Ph.D. thesis, University of Eindhoven, Dec 1989
19. C. Moreau, P. Cielo, M. Lamontagne, S. Dallaire, and M. Vardelle, *Meas. Sci. Technol.*, Vol 1, 1990, p 807
20. J.M. Zaat, *Ann. Rev. Mater. Sci.*, Vol 13, 1983, p 9
21. D. Bernard and P. Fauchais, Influence of the Torch Substrate Relative Movements and Cooling of the Coatings during Spraying on the Mechanical Properties of TBCs, Proceedings of the National Thermal Spray Conference 90, Long Beach, CA, ASM International, to be published
22. S. Kuroda and T.W. Clyne, The Quenching Stress in Thermally Sprayed Coatings, submitted to *Thin Solid Films*, Aug 1990
23. E. Lugsheider, *Thermal Spray: Advances in Coating Technology*, ASM International, p 105
24. C. D'Angelo and H. El Joundi, *Adv. Mater. Process.*, Vol 12, 1988, p 41
25. A. Benett, "Rolls-Royce Experience with TBCs," RR Ltd, Great Britain, 1984
26. E. Muehlberger, *1st Plasma Technik Symposium*, Vol 3, Plasma Technik, 1988, p 105
27. S.G. Lee and S. Safai, *Advances in Thermal Spraying*, Pergamon Press, 1986, p 197
28. R.W. Smith, The State of the Art—and Future—from a Surface to a Materials Processing Technology, *2nd Plasma-Technik Symposium*, Vol 1, 1991, p 17

Sol-Gel Process

Lisa C. Klein, Ceramics Department, Rutgers, The State University of New Jersey

SOL-GEL PROCESSING is a chemical synthesis of oxides involving hydrolyzable alkoxides that undergo a sol-gel transition. This definition is often interpreted to mean any system that undergoes a sol-gel transition, including colloidal sols and soluble salts. Nevertheless, the majority of systems investigated in the last ten years fit the narrow definition that the sol-gel transition represents a linking of nanometer-sized units into an oxide network of infinite molecular weight. The physics and chemistry of this process have been treated in a comprehensive way by Brinker and Scherer (Ref 1).

Sol-gel processing has been used to prepare glasses, glass-ceramics, and ceramics (Ref 2–6). Either amorphous or crystalline materials can result depending on the composition, the precursors, the handling, and heat treatments. These factors lead to either a powder process or a powder-free process as indicated in Fig 1. A powder process refers to any process that uses discrete particles, which in this case are generated from a sol-gel process. A powder-free process refers to any process that does not involve an aggregation of discrete particles. Starting from hydrolyzable alkoxides, a powder-free process would generate a porous preform of a glass or ceramic in the desired geometry by linking molecular species directly. The desired geometry may be a thin film, a fiber, or a bulk shape called a monolith.

Whether sol-gel processing is carried out as a powder process or as a powder-free process, it is growing in importance among ceramic-forming and predensification processes. One of the reasons for this growth is the purity of the starting materials. Other factors should become clearer as the various stages of processing are described.

Solutions versus Sols

When it comes to the sol-gel process, the initial step is always the formulation of the sol or solution. Although the designation sol-gel process implies that the starting point is a sol, to be precise it should be specified when sol is an abbreviation for solution and when sol is truly a colloidal dispersion with some degree of stability (Ref 7).

One-Component Solutions and Sols. Beginning with a one-component solution (for example, a silica system), there are many silicon alkoxides that are readily hydrolyzed and go through a sol-gel transition (Ref 8). Of the available silicon alkoxides, tetraethylorthosilicate (TEOS) is commonly used. This alkoxide reacts more slowly with water than tetramethylorthosilicate (TMOS) and comes to equilibrium as a complex silanol. The clear TEOS liquid is the product of the reaction of $SiCl_4$ or Si with ethanol. The reaction produces HCl or H_2 along with the ester $Si(OC_2H_5)_4$. This colorless liquid has a density of about 0.9 g/cm^3 (0.03 lb/in.^{3}) and is easy to handle. After multiple distillation, it is extremely pure.

Tetraethylorthosilicate is insoluble in water. In order to initiate the hydrolysis reaction, the TEOS alkoxide and water must be introduced into a mutual solvent. Often, the mutual solvent is ethanol. A typical formulation is 43 vol% TEOS, 43 vol% ethanol, and 14 vol% water. An electrolyte such as HCl or NH_4OH may be used. The apparatus needed for laboratory scale processing is relatively simple, consisting of a three-necked flask, a mechanical stirrer, a reflux condenser, and a temperature probe controlling a constant temperature bath. The neck of the flask that is filled by the temperature probe may also be used for electrolyte additions or sampling.

The chemical reactions that occur when water and TEOS alkoxide are dissolved in ethanol are hydrolyzation and condensation polymerization. The solution is activated once the water reacts with alkoxy groups on the silicon to form hydroxyl groups and alcohol. This hydrolyzation produces silanols and ethanol but this never goes to completion; that is, the water is not used up to form silicic acid. Instead, condensation polymerization takes partially hydrolyzed units and makes larger units with bridging oxygens. While hydrolyzation uses water as a reactant, polymerization regenerates water as a product. The kinetics of this process are very complex. Indeed, the mechanisms for reactions catalyzed by acids are different from those catalyzed by bases. Some of the names for the species present in the solution during these reactions are indicated in Table 1.

During the first step, all components must be mixed to form a clear solution. Cloudiness or precipitation indicates a segregation of components. Once all of the components are mixed, the water and alkoxides react to generate the structure leading to the sol-gel transition. While being continuously stirred, the fluid solution will become increasingly more viscous. At a definite point, the viscous solution becomes an elastic gel. At this point, for example, bubbles cease rising. One way to picture the gel is as a saturated sponge now filling the volume that was once filled by the sol. It is important to note that the molecular structure at the sol-gel transition determines the probability that a gel will dry in one piece or whether it will break into fragments. It is convenient to think of this transition as the formation of the last bond needed to create an infinitely long molecule.

While tetraethylorthosilicate is a good example of a one-component solution, a good example of a one-component sol is the commercial product Ludox (Ref 9). This is an aqueous colloidal dispersion. The particles are discrete, dense spheres classified according to size (typically 10 nm, or 100 Å) with the stabilizing ion usually either sodium or ammonia. The specific gravity is about 1.30. Gelling occurs when the pH is changed. The sol is said to go through a sol-gel transition. When hydroxyl groups on the surface of the particles condense to form siloxane bonds, the particles coalesce in an irreversible sol-gel transition. When coalescence occurs without forming siloxane bonds, the sol-gel transition may be reversible.

Multicomponent solutions and sols can be prepared by mixing the appropriate precursors for the various oxides. However, problems arise because the precursors hydrolyze at different rates and this can create inhomogeneities. Some strategies for forming the compositions in Table 2 include partially hydrolyzing the less reactive alkoxide (TEOS) before adding the more reactive alkoxide such as titanium isopropoxide or aluminum sec-butoxide. Other components such as lithium can be added as a salt (lithium nitrate).

The Sol-Gel Transition. The phenomenon of the sol-gel transition for the solution

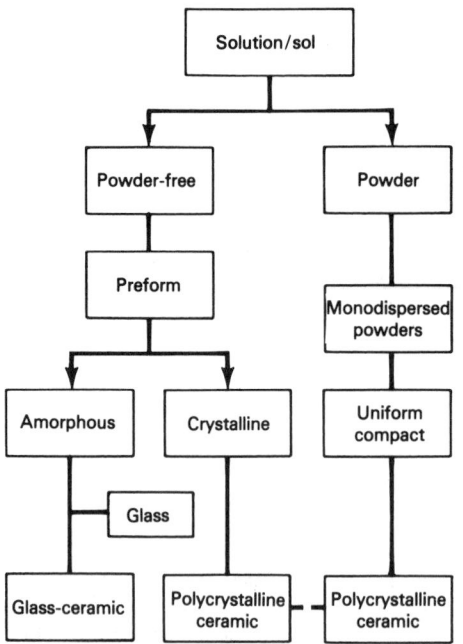

Fig 1 Preparation of glasses, glass-ceramics, and ceramics by the sol-gel process

and the sol is shown in Fig 2. The difference between the sol-gel transition for the hydrolyzing alkoxide solution and for the coalescing sol is that in the solution the transition is irreversible, whereas in the sol the transition may be reversible or irreversible. The transition in the solution occurs with no change in volume (that is, the alcogel, which is the oxide network when its molecular weight becomes infinite, fills the same volume that the solution filled). On the other hand, the volume of the hydrogel, which is the coalescence of the colloidal particles in the aqueous medium, may be smaller than the volume the sol once filled.

In practice the sol-gel transition is determined by inspection when an abrupt increase in viscosity occurs. The emphasis in many empirical studies is to find the optimum combination of variables that gives a structure at the sol-gel transition which can be processed to variously form fibers, beads, frits, microballoons, shapes, seals, coatings, or membranes.

Alkoxide-Derived Gels

Having taken the solution or sol through the sol-gel transition, the system now consists of two phases: an oxide skeleton and a solvent phase in the pores. The solvent phase has to be removed. The product after drying is either an aerogel or xerogel in most cases. The special syntheses leading to sonogels, cryogels, and vapogels will be mentioned. Other products include ceramers and ormocers that have an organic component. Some composite schemes create diphasic gels or gels that can be infiltrated after preparation. Also,

postdrying treatments in gases can give oxyfluoride and oxynitride gels.

The Problem of Drying. In the next section (see "Ceramic Preforms" in this article), the various preforms will be discussed, but before considering films and fibers in which drying is not a problem, the problem of drying has to be dealt with in bulk samples known as monoliths.

The major problem of making bulk samples by the sol-gel process is the large volume shrinkage involved and the usual cracking of the piece (Ref 10). The problem arises when the surface energy for solid-liquid is less than the surface energy for solid-vapor. In the wet region of a gel, the gel contracts to lower the solid-liquid surface area. When the gas phase enters the gel in the dry regions, the solvent tries to cover the solid-vapor interface. Meanwhile, the dry region wants to contract to lower the solid-vapor surface area.

Experimentally, it is found that the plot of linear shrinkage versus time in a drying gel monolith gives an S-shaped curve. Initially the shrinkage is slow when there is a high concentration of water in the gel. This is when there is excess water in the capillaries and syneresis is operating. The viscosity of the network is low in the beginning, then the shrinkage accelerates to a linear rate for a period of time. This portion of the shrinkage curve is not unlike the drying seen in conventional slip cast bodies, where the evaporation rate at the surface is set equal to the diffusive flux to the surface. As the drying continues, the water content decreases and the

viscosity of the network increases. Eventually the viscosity of the solid phase will become so high that the evaporation rate will be greater than the shrinkage rate. The meniscus enters the capillary. As the gel approaches dryness, the shrinkage again is slow.

As the network shrinks further, more hydroxyls are brought close together so that more condensation occurs. Eventually, the dry fraction of the gel is larger than the wet fraction. At that point, the question is whether or not strains in the materials are great enough to cause fracture. The final period is the critical time for cracks.

The pore size in alkoxide-derived gels is small. Similarly, the thickness of gel material between pores or the width of the wall separating two capillaries is small. In many cases, the size of the pore is nearly the same as the thickness of the wall. As such, the gel structure is somewhat stronger than expected in the early stages of drying.

The drying model by Scherer predicts the behavior in alkoxide gels rather closely (Ref 11). It accounts for the elastic response and the viscous response. While the model cannot be used to design a drying schedule as yet, it helps to explain why monoliths are hardest to dry crack-free and thin films and fibers are much easier. The various types of dried gels are now defined.

Aerogels. The first method of drying is to place a sol or solution in an autoclave and remove the fluid by hypercritical evacuation. These gels are called aerogels (Ref 12).

Table 1 Stages in the hydrolysis and polymerization of tetraethylorthosilicate (TEOS)

Reactants	Intermediates	Products
Alkoxide (TEOS): $Si(OC_2H_5)_4$	Ethoxysilanol: $-Si-OH$ $-Si-OC_2H_5$	Alcogel
H_2O		H_2O
	Polysiloxanes: $-Si-O-Si$	C_2H_5OH
One phase	One phase	One phase → Two phases

Table 2 Typical formulations for single-component and multicomponent

Precursor	Molecular weight, g	Density at 20 °C (70 °F)		One component: 100 mol% SiO_2			
				Solution concentration		Oxide content	
		g/cm³	lb/in.³	wt%/100 g	vol%/100 mL	mol	g/100 mL
$Si(OC_2H_5)_4$	208	0.936	0.0338	45	43	...	...
$Ti(OC_3H_7)_4$	284	0.955	0.0345	...	...	...	...
$Al(OC_4H_9)_3$	246	0.967	0.0349	...	...	...	...
$LiNO_3$	69	2.380	0.0860	...	...	...	...
C_2H_5OH	46	0.789	0.0285	40	43	4	...
H_2O	18	1.000	0.0361	16	14	4	...
Oxide (Si + Ti + Al + Li)	...	...	...	...	...	1	11.3

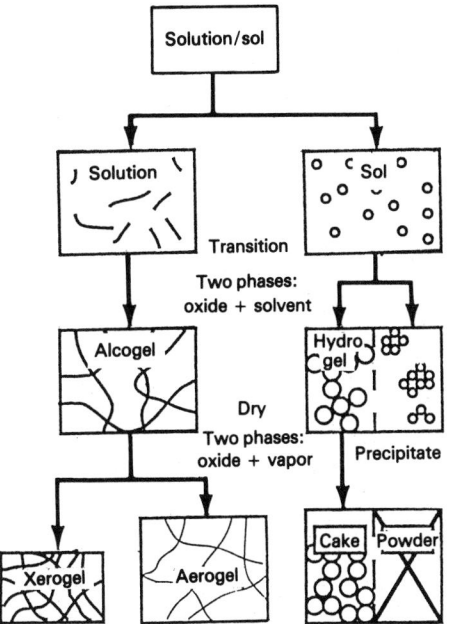

Fig 2 Flowchart comparing gel formation from solutions versus gel formation from sols

Aerogels are processed by increasing the temperature and pressure above the critical point to by-pass the liquid-vapor interface and then to vent out the vapor leaving a network of about 95% porosity. The pore size distribution may be bimodal. This network must be de-aired before the pores close during firing.

Xerogels. The second method of drying is by natural evaporation of the solvent and water to the atmosphere. While the liquid is evaporating, the gel structure is collapsing on itself. Gels obtained this way are called xerogels (Ref 13). Xerogels may have pores as small as 2 nm (20 Å). Large stresses develop when liquid is removed from the capillaries which leads to shrinkage. The conditions used for drying xerogels are more or less controlled by humidity conditions. At first, the drying is slow to permit the gel itself to become stronger by further polymerization. Subsequently, the samples may be heated to lower solvent viscosity. If the temperature is raised too high too soon, the solvent escapes rapidly and breaks the gel network. A compromise in

temperature is required to accelerate drying without disrupting the gel. Ultimately, the dried gel has about 40 to 60% of the fired density. The interest in xerogels, despite long drying times, is that xerogels are more dense than aerogels and are simpler to prepare.

It is also possible to obtain xerogels by adding a drying control chemical additive (DCCA) (Ref 14). When this chemical additive (for example, formamide) is mixed in with the solution, the amine groups interact with the inorganic polymer chains. The resultant network has a uniform pore size, about 4 nm (40 Å), and remains intact through all stages of drying. However, these organics may be difficult to remove during firing.

Sonogels. Another method for preparing gels before drying them with autoclave treatments is to subject the solution to ultrasound in the 20 kHz range. Gels obtained this way are called sonogels (Ref 15). The process involves cavitation. The end result of this treatment in a system such as P_2O_5-SiO_2 is to ensure better retention of P_2O_5 in the gel after autoclaving.

Cryogels. A method, similar to the formation of aerogels in that the liquid phase is absent, is to prepare aqueous hydrogels that are dried in a freeze-drying apparatus by subliming the water at a pressure below the triple point so only ice and water vapor are present. Normally, a pressure of about 130 mPa (10^{-3} torr) is used, and drying time may exceed 24 h. Gels obtained this way are called cryogels (Ref 16). Because of the expansion of ice upon freezing, the process produces a finely divided porous powder and not a monolithic piece.

Vapogels. A method for rapid formation of a gel involves injecting a steady stream of liquid $SiCl_4$ or $SiCl_4$ vapor into acidified water, along with other organometallics such as $GeCl_4$, while stirring. The gelling is almost instantaneous. Gels obtained this way are called vapogels (Ref 17). This method allows a higher incorporation of GeO_2 into a SiO_2 gel. The gel is then dried to a xerogel.

Additional Gels. A number of methods have been developed to prepare hybrid materials that are both organic and inorganic. Precursors can be synthesized where all of the alkoxy groups are not the same. Some of the bonds are metal-oxygen bonds and the rest of

the bonds are metal-carbon bonds. The hybrid gels derived from these precursors are called ceramers or ormocers (Ref 18, 19). The hybrid may have organic groups linked to a stiff inorganic backbone or the organic and inorganic may be interpenetrating networks.

Several systems have been used in which a gel host is formed with a precipitated second phase uniformly distributed in the pores. These nanoscale composites are called diphasic gels (Ref 20). These gels can be compositionally diphasic or structurally diphasic.

Finally, oxyfluoride and oxynitride gels can be prepared. In the case of fluorine-doped silica, the process required either SiF_4 or H_2SiF_6 along with TEOS (Ref 21). In the case of nitrogen-doped silica, the process required a treatment in ammonia (Ref 22). In both cases, the incorporation of fluoride or nitrogen exceeded the incorporation from more traditional methods.

Ceramic Preforms

When talking about the sol-gel process, it is difficult to separate the processing from the applications (see Table 3). The many applications and their technology have been collected in an edited volume by Klein (Ref 23). The suitability of the sol-gel process for a film versus a fiber versus a monolith is determined at the time the sol or solution is formulated. The solutions selected for thin films are typically those with fast gelling and rapid drying. The solutions for fibers must show gradual increases in viscosity rather than abrupt increases. Lastly, solutions for monoliths must have an excessive amount of water in the alkoxide solutions. This ensures that the cross-linking is sufficient to increase the strength of the gel so that the gel can withstand capillary forces during solvent removal.

Films and Coatings. The most common ceramic preforms obtained from the sol-gel process are thin films and coatings (Ref 24). At this time, the majority of sol-gel coatings are applied by dipping (Ref 25). This is a simple approach that allows the properties of the solution to control the deposition. A substrate is lowered into a vessel containing the solution. A meniscus develops at the contact of the liquid and the substrate. As the substrate is withdrawn, the meniscus, according to the viscosity and surface tension of the solution, generates a continuous film on the substrate. In general, the film thickness increases with increasing withdrawal rate. The film thickness, for a given withdrawal speed, increases with an increase in oxide content. An increase in oxide content generally means a higher viscosity and shorter time-to-gel, so it is possible to select the proper film thickness. The process is frequently a batch process, although it has been scaled up as an industrial process. When the substrate is flexible, such as a fiber, a continuous process can be designed.

alkoxide solutions

Two components: 94 mol% SiO_2 + 6 mol% TiO_2				Three components: 15 mol% Li_2O + 3 mol% Al_2O_3 + 82 mol% SiO_2			
Solution concentration		Oxide content		Solution concentration		Oxide content	
wt%/100 g	vol%/100 mL	mol	g/100 mL	wt%/100 g	vol%/100 mL	mol	g/100 mL
11	11	...	...	35	34	...	...
1	1	...	...	...	...	...	...
...	...	...	...	1	1	...	...
...	...	...	...	3	1	...	...
36	41	14	...	29	34	4	...
52	47	50	...	33	30	8	...
...	...	1	3.15	...	...	1	10.6

Table 3   Typical compositions, characteristics, and applications of ceramic preforms produced by the sol-gel process

Preform	Composition	Sol-gel process advantage	Typical product application
Thin film	Titania-silica	Simplicity	Interference filter
Membrane	Alumina	Porous	Ultrafilter
Fiber	Alumina-zirconia-silica	Low temperature	Reinforcement
Bulk	Silica	Purity	Lens
Aerogel	Silica	Low density	Thermal insulation

A variation on dip coating is drain coating. In this case, the substrate remains stationary and the liquid is drained from the vessel. For one side coating, a technique that is often used is spin coating. In this case, the substrate (usually a silicon wafer) is placed on a spinner, rotated at perhaps 200 rev/min while the solution is dripped on the center of the substrate.

When it comes to coatings, 50 to 500 nm (500 to 5000 Å) coatings are easy to make but thicker coatings are more difficult to obtain. Repeated dipping builds up a thicker film. The film needs to be dried at this point. Because of the one thin dimension, films generally do not crack on drying. The films are more or less porous and require further treatments to promote epitaxial crystal growth or densification.

Fibers. After coatings, the next common ceramic preform for the sol-gel process is fibers. Basically, two major approaches are used for making ceramic fibers by a sol-gel process.

Controlled Hydrolysis of Metal Alkoxides. The first approach involves the controlled hydrolysis of metal alkoxides for direct fiber drawing from solution (Ref 26). Solutions which are suitable for this process show a gradual increase in viscosity with time. The cross section of these fibers is usually elliptical.

Dispersion of Colloidal Particles. The second approach involves the dispersion of colloidal particles, either in aqueous medium or nonaqueous medium (Ref 27). For the dispersion technique, the colloidal particles are generally mixed with a binder and extruded. An exploratory approach to making fibers from a sol-gel process involves coating a shell or concentric shells of sol-gel glass on a fiber of plastic (Ref 28). The organic host is sacrificed, and the shell is collapsed to a fiber. This approach is similar to film technology and can result in hollow fibers.

Bulk Shapes and Monoliths. At this time, the most problematical of the ceramic preforms obtained from the sol-gel process are three-dimensional objects where no one dimension is thin. The problem with bulk shapes is drying (see the section "The Problem of Drying" in this article).

In general, the approach to monoliths is to formulate a sol or solution and cast it into a mold of the desired shape (Ref 29). The final shape is intended to be a miniature replica of the mold. The sol or solution is allowed to gel in the mold. After the shape has gone through the sol-gel transition, its two phases are oxide skeleton and solvent in the pores. The pores are interconnected. The solvent is removed from the pores by one of the drying schemes. Hypercritical evacuation results in aerogels and natural evaporation results in xerogels.

In the case of xerogels, it is recommended that the oxide skeleton have sufficient time to age to be able to withstand drying stresses. These stresses are absent in hypercritical evacuation but the gels resulting in this case have low density and the same dimensions as the original mold. In aerogels, the pore size is on the scale of 10 to 50 nm (100 to 500 Å). In xerogels, the pore size is on the scale of 2 to 5 nm (20 to 50 Å).

In porous materials, typical methods to avoid cracks are to avoid small pores, to keep pores consistently the same size, and to use low surface tension solvents. These suggestions are helpful but not always practical in xerogels. Nevertheless, crack-free xerogels are achieved due in part to the uniformity of the pores despite the small size. Because xerogels have textures that are required for applications such as adsorbents and catalyst supports, further work on xerogels is needed.

The dry condition of a gel is difficult to define. When there is no further weight loss with time is one definition. When the sample shows no adverse effects from ambient conditions is another. Most gels will show a readsorption of water after heating to 150 °C (300 °F). No matter which way drying is accomplished, the gel is not truly dry until after some stabilizing heat treatment.

There has been a lot of emphasis in the sol-gel processing of monoliths on the duplication of the physical properties of the dense oxide by conventional methods. The justification for the sol-gel process over a conventional process is working in uncontaminated conditions and at lower temperatures than the melting point.

Before crystals form, the mechanism for densification is viscous sintering. Along with viscous flow, volume relaxation and condensation reactions contribute to densification. For example, silica has been densified using constant rate heating, isothermal heating, and step-heat treatments (Ref 30). In each case, as the temperature is increased the viscosity decreases making viscous flow easier. At the same time, the viscosity is tending to increase due to the decomposition of silanols. This decomposition also increases the surface ten-sion which, in turn, increases the driving force for sintering. Because of all these effects, the sol-gel process is a combined consolidation and densification process.

For a silica xerogel from a powder-free process, the pore volume and surface area initially increase during densification. The pore fraction increases up to 400 °C (750 °F) and then decreases continuously for higher temperatures. The surface area increases to a maximum at 400 °C (750 °F), then levels off before decreasing. The increase in pore volume and surface area up to 400 °C (750 °F) results from the removal of water and alcohol trapped in the dried gel. The decrease in pore volume and surface area above 400 °C (750 °F) indicates an increase in the degree of polymerization and eventually the removal of pores. For a gel from a powder process, the surface area decreases continuously as the structure ripens.

In densifying gels, what results is an optimum heating rate. Fast heating rates may give rapid densification at first but the samples may bloat. Slower heating rates may give densification without bloating but may not achieve full density. The rate should be fast enough to provide shrinkage yet slow enough to allow the elimination of gases before trapping the gases in isolated pores.

Processing Diagnostics and Quality Control Procedures

While there are a number of commercial applications for the sol-gel process (notably coatings, fibers, and microballoons), there is not much information on how quality control (QC) procedures are implemented during processing. Some of the obvious diagnostics are indicated on the flow chart given in Table 4.

Formulation of the Solution or Sol. The first step is formulation of the solution or sol. A suitable diagnostic for this step is a measure of light scattering. A true solution should show no scattering or immiscibility. A sol, depending on the particle size, may show some scattering.

Table 4   Processing diagnostics used to monitor quality of sol-gel products at selected stages of production

Processing steps	Processing diagnostics
Formulation of solution/sol	Light scattering
Gelation to porous preform (thin films; fibers; and monoliths)	Rotational viscosity
Drying to xerogel or aerogel	Thermal weight loss
	Density
	Texture
Heat treatment	Thermal analysis
Densification and properties evaluation	X-ray diffraction analysis
	Dilatometry

Gelation. The second step is gelation. Typically gelation is monitored by measuring the viscosity. Rotational viscometry is one approach where the spindle experiences drag from the medium as the viscosity increases and finally the spindle cores the medium once it has gelled. Similarly, tilting a container with the solution will show flow until the solution gels and shows no flow.

Gel Drying. The third step is drying of the gel. The condition of dryness is often defined as the state where there is no further weight loss. Therefore, weight loss is a convenient diagnostic for this step. Another important feature to note during this step is the occurrence of cracks. Especially for monoliths, the idea is to make crack-free shapes.

The characterization of the dried gel at this point is concentrated on microstructural features. The density is one important property. The other is the porosity, generally referred to as texture (Ref 31). Since the texture of gels is usually difficult to resolve with the electron microscope, the texture of gels is probed with nitrogen sorption. Nitrogen is used for microporosity (<1.5 nm, or <15 Å) and mesoporosity (<50 nm, or <500 Å). Mercury porosimetry is used for macroporosity (>50 nm, or >500 Å). The adsorption isotherm gives the surface area according to the Brunauer-Emmet-Teller (BET) equation (see the article "Porosity, Density, and Surface Area Measurements" in this Volume). The desorption gives the pore size distribution from the Kelvin equation. Hysteresis in these curves is interpreted according to previously characterized materials.

Aerogels have very low bulk density and xerogels have higher bulk density. Xerogels from acid-catalyzed solutions are characterized by extremely high bulk densities. Xerogels from colloidal gels behave like agglomerations of compact spheres.

Stabilizing Heat Treatment. The fourth step is a stabilizing heat treatment. While sintering is not expected until temperatures approach two-thirds of the melting temperature, some densification may precede sintering in these materials. Thermal analysis of the samples is useful in this step to register any phase transformations. Depending on how the gel has been handled up to this point, the product may be a porous glass, a glass ceramic, or a polycrystalline ceramic. From this point on, the diagnostics for the process would be identical to those for a conventional ceramic process.

Advantages and Limitations of Sol-Gels

The advantages of the sol-gel process in general are high purity, homogeneity, and low temperature. Because the precursors can be multiple distilled and filtered, the products can be relatively free of impurities. Because the

mixing is accomplished in solution, components mix on the nanometer scale in relatively short times. If the intention is to use this process to make a dense material, the porous dried gel can be compacted at temperatures one-third to one-half lower than the melt temperature in Kelvin.

For a lower temperature process, there is a reduced loss of volatile components. In fact, some compositions that cannot be made by conventional means because of phase separation or devitrification can be made by this process. In some systems, the sol-gel process extends the glass-forming range.

One of the primary advantages of the sol-gel process, especially in the form of films, is that the material use is efficient. Any excess material is recovered and can be used again. The only disadvantages at this time are the raw material costs. This tends to limit the use of the process for bulk ceramics but is a minor factor for specialty applications or those cases where conventional technology fails.

REFERENCES

1. C.J. Brinker and G.W. Scherer, Sol-Gel Science, *The Physics and Chemistry of Sol-Gel Processing*, Academic Press, 1990
2. B.J.J. Zelinski and D.R. Uhlmann, Gel Technology in Ceramics, *J Phys. Chem. Sol.*, Vol 45, 1984, p 1069–1090
3. D.R. Ulrich, Chemical Sciences Impact on Future Glass Research, *Am. Ceram. Soc. Bull.*, Vol 64, 1985, p 1444–1448
4. S. Sakka, Sol-Gel Synthesis of Glasses: Present and Future, *Am. Ceram. Soc. Bull.*, Vol 64, 1985, p 1463–1466
5. D.W. Johnson, Sol-Gel Processing of Ceramics and Glass, *Am. Ceram. Soc. Bull.*, Vol 64, 1985, p 1597–1602
6. P. Colomban, Gel Technology in Ceramics, Glass-Ceramics and Ceramic-Ceramic Composites, *Ceram. Int.*, Vol 15, 1989, p 23–50
7. R.K. Iler, *The Chemistry of Silica*, Wiley Interscience, 1979
8. L.C. Klein, Sol-Gel Processing of Silicates, *Ann. Rev. Mater. Sci.*, Vol 15, 1985, p 227–248
9. R. Roy, Ceramics by the Solution-Sol-Gel Route, *Science*, Vol 238, 1987, p 1664–1669
10. J. Zarzycki, M. Prassas, and J. Phalippou, Synthesis of Glasses from Gels: the Problem of Monolithic Gels, *J. Mater. Sci.*, Vol 17, 1982, p 3371–3379
11. G.W. Scherer, Stress and Fracture during Drying of Gels, *J. Non-Cryst. Solids*, Vol 121, 1990, p 104–109
12. J. Fricke, Ed., *Aerogels*, Springer-Verlag, Berlin, 1986
13. J. Zarzycki, Monolithic Zero- and Aerogels for Gel-Glass Processes, *Ultrastructure Processing of Ceramics, Glasses and Composites*, L.L. Hench and D.R. Ulrich, Ed., Wiley Interscience, 1984, p 27–42
14. L.L. Hench, Use of Drying Control Chemical Additives (DCCAs) Controlling Sol-Gel Processing, *Science of Ceramic Chemical Processing*, L.L. Hench and D.R. Ulrich, Ed., Wiley Interscience, 1986, p 52–64
15. L. Esquivias and J. Zarzycki, Sonogels: An Alternative Method in Sol-Gel Processing, *Ultrastructure Processing of Advanced Ceramics*, J.D. Mackenzie and D.R. Ulrich, Ed., Wiley Interscience, 1988, p 255–270
16. G.M. Pajouk, M. Repellin-Lacroix, S. Abouarnadasse, J. Chaouki, and D. Klvana, From Sol-Gel to Aerogels and Cryogels, *J. Non-Cryst. Solids*, Vol 121, 1990, p 66–67
17. J.W. Fleming and S.A. Fleming, Vapogel—A New Glass Formation Technique, *Better Ceramics through Chemistry III*, Vol 121, C.J. Brinker, D.E. Clark, and D.R. Ulrich, Ed., Materials Research Society, 1988, p 673–678
18. H.H. Huang, B. Orler, and G.L. Wilkes, Ceramers: Hybrid Materials Incorporating Polymeric/Oligomeric Species with Inorganic Glasses by a Sol-Gel Process, *Polym. Bull.*, Vol 14, 1985, p 557–564
19. H. Schmidt and H. Wolter, Organically Modified Ceramics and Their Applications, *J. Non-Cryst. Solids*, Vol 121, 1990, p 428–435
20. R. Roy, Y. Suwa, and S. Komarneni, Nucleation and Epitaxial Growth in Diphasic (Crystalline and Amorphous) Gels, *Science of Ceramic Chemical Processing*, L.L. Hench and D.R. Ulrich, Ed., Wiley Interscience, 1986, p 247–258
21. T. Tsukada, M. Shinmei, and T. Yokokawa, Synthesis of Fluorine Doped Silica Gel and its Properties at High Temperature, *J. Non-Cryst. Solids*, Vol 100, 1988, p 435–439
22. C.J. Brinker and D.M. Haaland, Oxynitride Glass Formation from Gels, *J. Am. Ceram. Soc.*, Vol 66, 1983, p 758–765
23. L.C. Klein, Ed., *Sol-Gel Technology for Thin Films, Fibers, Preforms, Electronics, and Specialty Shapes*, Noyes Publications, 1988
24. H. Dislich, Glassy and Crystalline Systems from Gels: Chemical Basis and Technical Application, *J. Non-Cryst. Solids*, Vol 57, 1983, p 371–380
25. C.J. Brinker, A.J. Hurd, G.C. Frye, K.J. Ward, and C.S. Ashley, Sol-Gel Thin Film Formation, *J. Non-Cryst. Solids*, Vol 121, 1990, p 294–302
26. S. Sakka and K. Kamiya, The Sol-Gel Transition in the Hydrolysis of Metal Alkoxides in Relation to the Formation of Glass Fibers and Films, *J. Non-Cryst. Solids*, Vol 48, 1982, p 47–64
27. H.G. Sowman, A New Era in Ceramic Fibers via Sol-Gel Technology, *Am.*

Ceram. Soc. Bull., Vol 67, 1988, p 1911–1916

28. D.C. Cranmer, A Perspective on Fiber Coating Technology, *Ceram. Sci. Eng. Proc.*, Vol 9 (No. 9–10), 1988, p 1121–1124

29. L.L. Hench and W. Vasconcelos, Gel-Silica Science, *Ann. Rev. Mater. Sci.*, Vol 20, 1990, p 269–298

30. L.C. Klein and T.A. Gallo, Densification of Sol-Gel Silica: Constant Rate Heating, Isothermal and Step Heat Treatments, *J. Non-Cryst. Solids*, Vol 121, 1990, p 119–123

31. A.J. Lecloux, Texture of Catalysts, *Catalysis—Science and Technology*, Vol 2, J.R. Anderson and M. Boudart, Ed., Springer-Verlag, Berlin, 1981, p 171–230

Vapor Deposition

D.P. Stinton, T.M. Besmann, R.A. Lowden, and B.W. Sheldon,
Metals and Ceramics Division, Oak Ridge National Laboratory

FABRICATION OF CERAMIC MATE-RIALS by chemical vapor deposition (CVD) and physical vapor deposition (PVD) techniques offer many potential advantages over conventional processes. It is often more economical to apply a thin coating with a desired set of properties to an inexpensive substrate than to fabricate the entire component from a more costly material. Fabrication costs are further reduced for refractory ceramic coatings because the materials are difficult or impossible to densify by conventional fabrication processes. Other advantages of vapor deposition techniques include ease of application to nonuniform shapes, fabricability to desired shape with minimum or no machining, elimination of ceramic-metal joining requirements, tailoring of properties by control of coating process variables, and use of less material, which is an important consideration if expensive or strategic materials are involved.

Chemical vapor deposition also offers great potential for preparation of composite materials that overcome the traditional brittleness of refractory materials. Ceramic-matrix composites, which consist of ceramic fibers that reinforce a ceramic matrix, exhibit excellent toughness, maintain high-temperature strength, and do not fail catastrophically. Using CVD to produce the matrix of these fiber-reinforced ceramic composites is referred to as chemical vapor infiltration (CVI).

Both chemical and physical vapor deposition techniques have represented active areas of research for many years. The techniques have grown in acceptance as commercial methods for the fabrication of advanced ceramic components. This article describes chemical and physical vapor deposition techniques, as well as the advantages and disadvantages of each. Issues such as adherence and morphology, which are critical to the deposition of coatings by both techniques, will be summarized.

Chemical Vapor Deposition

Processes for the chemical vapor deposition of SiC, Si_3N_4, SiO_xN_y, and BN have been used to fabricate microelectronic devices for over 20 years (Ref 1). The success achieved by these materials led to the rapid expansion of CVD technology into many other areas of ceramic processing. The technique is used to produce advanced ceramics, such as high-performance tool bits, fiber-reinforced composites, high-efficiency solar cells, and high-temperature oxidation protection coatings. Reviews are available that summarize the deposition of an extensive list of oxides, carbides, nitrides, and borides for a variety of applications (Ref 2–5). Table 1 lists the precursors used, the deposition temperatures, and the techniques used to apply a variety of coatings.

Conventional CVD Process. In this process, a solid material is deposited onto a heated substrate surface as a result of chemical reactions in the gas phase. Chemical vapor deposition systems thus have several components in common: reactant supply systems, the CVD reactor (a furnace or similar system for heating the substrate), and an effluent gas handling system for gases that are often corrosive, toxic, or both (Fig 1). At room temperature, the reactants, which are usually metal halides, may be gases, liquids, or solids. Liquid and solid reactants are usually warmed sufficiently to raise their vapor pressure to a level at which a carrier gas (normally Ar or H_2) flowing through a vessel that contains the reactants can transport the vapors into the reactor at the desired rate. Some metals, however, do not form high vapor pressure halides, and therefore a number of metal-organic vapor sources have been developed, opening a new area of research. In the reactor the substrate can be suspended by a wire, lie on a platform (often a radiofrequency-heated susceptor), be immersed in a fluidized bed of particles, or (if the substrate is small particulates) form the fluidized bed. The effluent from the reactor usually contains the carrier gas, partially decomposed reactants, and HCl or HF vapors that must be removed by a caustic scrubber or cold trap.

The deposition reaction can be of several types: reduction, thermal decomposition, or displacement. Hydrogen reduction of a metal chloride or fluoride is by far the most common type of process, and an example of such a reaction is:

$$TiCl_4(g) + 2BCl_3(g) + 5H_2(g)$$
$$\rightarrow TiB_2(s) + 10HCl(g) \qquad (Eq 1)$$

Thermal decomposition reactions involve precursors such as metal carbonyls, hydrides, halides, or organometallics that decompose in the absence of hydrogen. An example of such a reaction involving a carbonyl is

$$Ni(CO)_4(g) \rightarrow Ni(s) + 4CO(g) \qquad (Eq 2)$$

An example of a displacement reaction is

$$SiCl_4(g) + CH_4(g) \rightarrow SiC(s) + 4HCl(g) \qquad (Eq 3)$$

CVD reactors are classified as either hot-wall or cold-wall (Ref 1, 6). Hot-wall reactors are heated chambers (furnaces) in which the substrate is placed for coating. As expected, the interior of the furnace also becomes coated, resulting in both maintenance problems and lower efficiency. In cold-wall reactors, only the substrate is heated, either inductively, radiatively, or resistively. Although these reactors are more complex, they allow greater control over the deposition process.

Considerable progress has been made in the past 15 years in understanding and predicting the thermodynamic (Ref 7, 8) and kinetic (Ref 9, 10) behavior of the CVD process. For example, thermodynamic calculations based on minimization of the Gibbs free energy of the gas-solid system are useful in predicting the influence of process variables, such as temperature and inlet gas concentrations on the phase assemblage and stoichiometry of the deposit, the equilibrium deposition efficiency, and the concentration and molecular species in the exhaust gas. Similarly, kinetic considerations can guide process variables so that the deposition process is either limited by gas-phase diffusion to the substrate surface or by reaction at the substrate. Achieving this level of control over the process is valuable because, for example, geometric surface irregularities (grooves or corners) in the substrate are coated uniformly in a kinetically controlled process, but in a process controlled by diffusion, a protrusion receives a thicker coating while a depression is thinly coated (Ref 10). Process conditions that lead to kinetic control are low temperature, low pressure, low concentration of reactants, and high gas flow rates.

Table 1 Ceramic materials produced by chemical vapor deposition

Coating	Chemical mixture	Deposition temperature °C	Deposition temperature °F	Method	Application
Carbides					
TiC	$TiCl_4$-CH_4-H_2	900–1000	1650–1830	CVD	W
	$TiCl_4$-$CH_4(C_2H_2)$-H_2	400–600	750–1110	PACVD	E
HfC	$HfCl_x$-CH_4-H_2	900–1000	1650–1830	CVD	W, C/O
ZrC	$ZrCl_4$-CH_4-H_2	900–1000	1650–1830	CVD	W, C/O
	$ZrBr_4$-CH_4-H_2	>900	>1650	CVD	W, C/O
SiC	CH_3SiCl_3-H_2	1000–1400	1830–2550	CVD	W, C/O
	SiH_4-C_xH_y	200–500	390–930	PACVD	E, C
B_4C	BCl_3-CH_4-H_2	1200–1400	2190–2550	CVD	W
B_xC	B_2H_6-CH_4	400	750	PACVD	W, E, C
W_3C	WF_6-CH_3OCH_3-H_2	350–500	660–930	CVD	W
W_2C	WF_6-CH_4-H_2	400–700	750–1290	CVD	W
Cr_7C_3	$CrCl_2$-CH_4-H_2	1000–1200	1830–2190	CVD	W
Cr_3C_2	$Cr(CO)_6$-CH_4-H_2	1000–1200	1830–2190	CVD	W
TaC	$TaCl_5$-CH_4-H_2	1000–1200	1830–2190	CVD	W, E
VC	VCl_2-CH_4-H_2	1000–1200	1830–2190	CVD	W
NbC	$NbCl_5$-CCl_4-H_2	1500–1900	2730–3450	CVD	W
Nitrides					
TiN	$TiCl_4$-N_2-H_2	900–1000	1650–1830	CVD	W
	$TiCl_4$-N_2-H_2	250–1000	480–1830	PACVD	E
HfN	$HfCl_x$-N_2-H_2	900–1000	1650–1830	CVD	W, C/O
	HfI_4-NH_3-H_2	>800	>1470	CVD	W, C/O
Si_3N_4	$SiCl_4$-NH_3-H_2	1000–1400	1830–2550	CVD	W, C/O
	SiH_4-NH_3-H_2	250–500	480–930	PACVD	E, C/O
	SiH_4-N_2-H_2	300–400	570–750	PACVD	E
BN	BCl_3-NH_3-H_2	1000–1400	1830–2550	CVD	W
	BCl_3-NH_3-H_2	25–1000	77–1830	PACVD	E
	$BH_3N(C_2H_5)_3$-Ar	25–1000	77–1830	PACVD	E
	$B_3N_3H_6$-Ar	400–700	750–1290	CVD	E, W
	BF_3-NH_3-H_2	1000–1300	1830–2370	CVD	W
	B_2H_6-NH_3-H_2	400–700	750–1290	PACVD	E
ZrN	$ZrCl_4$-N_2-H_2	1100–1200	2010–2190	CVD	W, C/O
	$ZrBr_4$-NH_3-H_2	>800	>1470	CVD	W, C/O
TaN	$TaCl_5$-N_2-H_2	800–1500	1470–2730	CVD	W
AlN	$AlCl_3$-NH_3-H_2	800–1200	1470–2190	CVD	W
	$AlBr_3$-NH_3-H_2	800–1200	1470–2190	CVD	W
	$AlBr_3$-NH_3-H_2	200–800	390–1470	PACVD	E, W
	$Al(CH_3)_3$-NH_3-H_2	900–1100	1650–2010	CVD	E, W
VN	VCl_4-N_2-H_2	900–1200	1650–2190	CVD	W
NbN	$NbCl_5$-N_2-H_2	900–1300	1650–2370	CVD	W, E
Oxides					
Al_2O_3	$AlCl_3$-CO_2-H_2	900–1100	1650–2010	CVD	W, C/O
	$Al(CH_3)_3$-O_2	300–500	570–930	CVD	E, C
	$Al[OCH(CH_3)_2]_3$-O_2	300–500	570–930	CVD	E, C
	$Al(OC_2H_5)_3$-O_2	300–500	570–930	CVD	E, C
SiO_2	SiH_4-CO_2-H_2	200–600	390–1110	PACVD	E, C
	SiH_4-N_2O	200–600	390–1110	PACVD	E
SiO_2	$Si(OEt)_4$-O_2	650–750	1200–1380	CVD	E
SiO_2	SiH_2Et_2-O_2	350–450	660–840	CVD	E
TiO_2	$TiCl_4$-H_2O	800–1000	1470–1830	CVD	W, C
	$TiCl_4$-O_2	25–700	77–1290	PACVD	E
	$Ti[OCH(CH_3)_2]_4$-O_2	25–700	77–1290	PACVD	E
ZrO_2	$ZrCl_4$-CO_2-H_2	900–1200	1650–2190	CVD	W, C/O
Ta_2O_5	$TaCl_5$-O_2-H_2	600–1000	1110–1830	CVD	W, C, E
Cr_2O_3	$Cr(CO)_6$-O_2	400–600	750–1110	CVD	W
Borides					
TiB_2	$TiCl_4$-BCl_3-H_2	800–1000	1470–1830	CVD	W, C, E
MoB	$MoCl_5$-BBr_3	1400–1600	2550–2910	CVD	W, C
WB	WCl_6-BBr_3-H_2	1400–1600	2550–2910	CVD	W, C
NbB_2	$NbCl_5$-BCl_3-H_2	900–1200	1650–2190	CVD	W, C
TaB_2	$TaBr_5$-BBr_3	1200–1600	2190–2910	CVD	W, C
ZrB_2	$ZrCl_4$-BCl_3-H_2	1000–1500	1830–2730	CVD	W, C, E
HfB_2	$HfCl_x$-BCl_3-H_2	1000–1600	1830–2910	CVD	W, C

CVD, conventional CVD; PACVD, plasma-assisted CVD; W, wear-resistant coatings; E, electronics; C, corrosion-resistant coatings; O, oxidation-resistant coatings

The most important inherent advantage of CVD over other coating techniques is its high deposition rate, typically greater than tens of micrometers/h, which is exceeded only by plasma spraying. Because the technique does not require line-of-sight with the vapor source, the coatings can be uniformly deposited over substantial contours. It is also a process that allows the preparation of refractory materials at temperatures much lower than the usual fabrication temperature. Indeed, the moderate temperature, together with the ease of purifying the reactants, allow the deposition of exceptionally high-purity materials.

Another advantage of CVD processes is the ability to tailor the stoichiometry, morphology, and even crystal structure and orientation of the coating by controlling the deposition parameters (Fig 2) (Ref 6, 11–14). Microstructural control is, of course, necessary to optimize the mechanical properties of a material (Ref 15). Fibrous deposits result from conditions of low reactant concentrations and sufficiently rapid surface diffusion. Deposits consisting of randomly oriented fine grains near the substrate surface grow into large columnar grains of preferred orientation at higher reactant concentrations and more limited diffusion (Ref 16). The most desirable coatings for mechanical strength and fracture toughness consist of very fine, equiaxed grains. This microstructure is typically produced at still higher concentrations and conditions under which surface diffusion is limited. The supply of reactants must be sufficient to rapidly renucleate new grains while the movement of atoms to preferred crystal lattice sites is restrained. Addition of trace quantities of impurities or catalysts may promote renucleation.

A recently developed method to tailor the microstructure or properties of a coating for a specific application is the simultaneous deposition of more than one phase (Ref 17–20). The second phase can be selected to modify the thermal conductivity, electrical conductivity, fracture toughness, mechanical strength, and other properties of the composite coating. For example, Si_3N_4 has been codeposited with TiN, BN, C, and SiC to improve the mechanical and physical properties of the coatings (Ref 17, 18).

The principal disadvantage of conventional CVD is that the temperature required for deposition frequently exceeds the maximum use temperature of metallic substrates. Other disadvantages are the need to use or generate corrosive, toxic, and flammable reactants or by-products; the relatively high cost of the precursors; and expensive process equipment that is necessary to perform CVD and handle process streams. The reactants or gaseous products may also adversely affect the substrate materials at the temperatures required to promote the CVD reaction.

Applications. A major market for coatings is the cutting tool industry, where nearly 50% of the cobalt-bonded tungsten carbide tool inserts are coated with TiC, TiN, or titanium carbonitride (TiC_xN_{1-x}) (Ref 2, 3, 21–27). Coatings extend the life of these cutting inserts by a factor of five because of their increased hardness and wear resistance and because they prevent the reaction of the cobalt binder with the metal workpiece at the elevated temperatures of machining. The complementary properties of the carbide and nitride reduce sliding and crater wear and further improve their performance. Unfortunately, titanium carbide and nitride are susceptible to oxidation at temperatures above approximately 800 °C (1470 °F). Thus, for high-temperature, high-speed operations, tool bits with very hard Al_2O_3 coatings on the outer surface

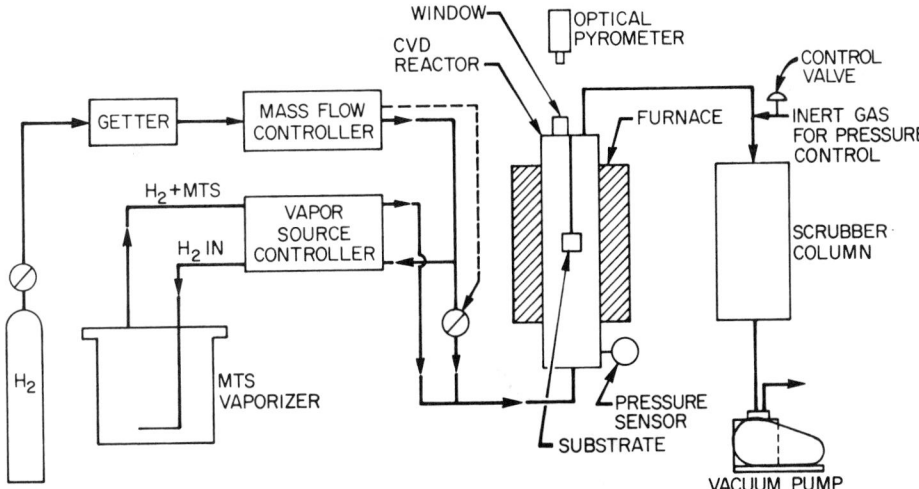

Fig 1 Simple chemical vapor deposition system. MTS, methyltrichlorosilane, or CH₃SiCl₃

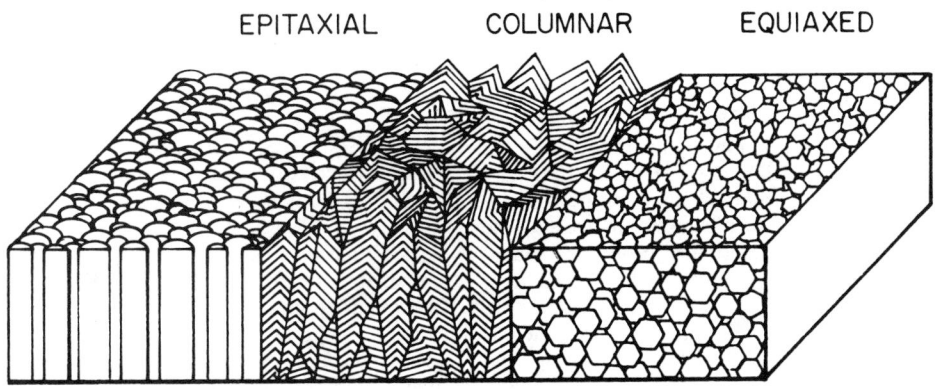

Fig 2 Polycrystalline structures deposited by CVD and PVD. Left, columnar grains with domed tops; middle, faceted, columnar grains; right, equiaxed grains

are used. At temperatures greater than 800 °C (1470 °F), Al₂O₃-coated inserts have a lifetime that is twice as long as comparable TiC-coated inserts. Multiple layers of different materials are routinely utilized to combine the best properties of each layer. For example, TiC interlayers are frequently used between the substrate and either TiN or Al₂O₃ coatings to improve adherence (Ref 28).

Adherence of compounds such as TiC, TiN, and TiC$_x$N$_{1-x}$ on cobalt-bonded tungsten carbide and steel cutting tools is due to the interdiffusion of elements at the relatively high processing temperatures, which tightly bonds the coating to the substrate. Unfortunately, the high temperatures which benefit adherence soften the substrates and necessitate subsequent heat treatments to reharden the inserts. Some distortion of the tool occurs, however, and this may be unacceptable for some applications.

Hard coatings have also found application in erosion protection in gas turbine engines, rocket nozzles, and letdown valves in fossil energy systems (Ref 29). Erosion of metallic blades by abrasive particles in gas turbine engines significantly shortens the life of the

blades and reduces efficiency. Titanium carbonitride coatings nearly eliminate erosive wear in this application and extend the lifetime by a factor of 70 or 80 (Ref 30).

The high-temperature mechanical and physical properties of carbon fibers and carbon-carbon composites have led to their widespread use as refractory materials in aerospace heat shield and structural components, and, more recently, in turbine propulsion systems (Ref 31–33). These and other applications have been restricted by the low oxidation resistance of carbonaceous materials at elevated temperatures. Efforts have expanded to develop protection systems for carbon materials that can withstand long-term exposure to high-temperature oxidizing environments. CVD processes, particularly for SiC, have been adopted to produce theoretically dense, single- and multi-layered coatings that fulfill oxidation protection requirements.

Free-standing shapes can also be fabricated by CVD techniques. Components are produced by depositing a relatively thick coating onto a mandrel and subsequently removing the mandrel. The dimensions of the

component are, therefore, controlled by the shape of the mandrel and the thickness of the applied coating. An application of this technique is the production of optically transparent windows for missile guidance systems. Chemically vapor deposited polycrystalline ZnS or CdS, with a thickness ranging from 8 to 12 μm (320 to 480 μin.), which is necessary to perform in this application, have the requisite hardness, fracture toughness, erosion resistance, and, most importantly, are transparent to electromagnetic radiation (Ref 34, 35). The microstructure, mechanical properties, and optical properties of these materials are optimized by control of the deposition temperature, reactant gas concentration, and growth rate. Free-standing BN shapes, such as crucibles and cylinders, are also fabricated using CVD techniques. Boron nitride is typically applied to graphite shapes, and, if properly designed, the graphite mandrel is easily removed (Ref 36).

There are numerous electronic industry applications for advanced ceramic materials attained by CVD. Oxides such as Al₂O₃ and SiO₂ have been employed as dielectrics in metal-oxide semiconductor transistors and capacitors, as composite insulator structures in similar applications, and as diffusion barriers and protective coatings in a number of micro- and optoelectronic systems. Refractory metal nitrides, amorphous Si₃N₄, AlN, BN, and transition metal carbides and borides find application as insulators, dielectrics, and semiconductors, and are extensively used in the fabrication of solid-state devices (Ref 37–42).

Plasma-Assisted CVD. A variation of the conventional CVD technique that has been widely employed in the fabrication of electronic devices is plasma-enhanced or plasma-assisted CVD (PACVD) (Ref 43–49). In the PACVD technique, the reactants are introduced into a high-energy radiofrequency (rf) or microwave glow discharge, where they decompose and subsequently deposit condensable species as a thin film. Plasma deposition processes produce coatings very efficiently at low substrate temperatures. The technique has found application in the deposition of films for semiconductor devices and in the manufacture of more efficient photovoltaic cells.

Plasma-activated chemical vapor deposition has also been used extensively to prepare hard, diamond-like carbon coatings (Ref 50–52) and, more recently, polycrystalline diamond films (Ref 53–59) from hydrocarbon plus hydrogen precursors. Diamond-like films are hard, electrically insulating, thermally conducting, chemically inert, and optically transparent, although they lack long-range order and contain large amounts of hydrogen or inert gas. Because of their chemical inertness, diamond-like carbon coatings function as protective coatings against corrosion and as a protective or insulating coating for electronic components. This unique combination of properties also makes diamond-like carbon an ideal candidate for optical coatings, particularly because the refractive index can be controlled by deposition conditions.

The recent phenomenal success and explosion of activity in the superconductor area has prompted the development of deposition techniques to apply these materials to microelectronic components. The previous class of superconductors, such as NbN systems, were typically prepared by CVD (Ref 60, 61). Researchers have demonstrated the ability of CVD-produced thin films of the new high-temperature superconductors ($YBa_2Cu_3O_7$) to carry high currents; and thus, sizable efforts aimed at developing CVD processing of these materials are underway (Ref 62).

Physical Vapor Deposition

Physical vapor deposition is a process in which vapor created by evaporation or sputtering is transported by line-of-sight (normally in a vacuum) from the source to the substrate, where a solid film or coating is to be deposited. Different PVD processes are distinguished by the way the vapor phase is generated. PVD is normally used to deposit metals or alloys but can be modified for the deposition of ceramics. Three basic PVD processes are evaporation, ion plating, and sputtering, each of which is described below.

Evaporation is an established, versatile deposition technique in which the material to be deposited is heated in a vacuum until a sufficiently high vapor pressure is achieved to cause material to condense on a nearby substrate at a suitable deposition rate (Ref 63, 64). Several methods can be used to heat the evaporant, including electrical resistance, radiation, induction, arcs, laser ablation, and electron beams. Most modern systems use high-vacuum electron-beam technology, which allows high evaporation rates, freedom from contamination, good control, and high thermal efficiency. One or more beams can be deflected through 270° (Fig 3) onto one or more evaporants, which are usually held in water-cooled copper crucibles. The ability to deflect the electron beam allows the gun to be kept clean because it can be positioned away from the vapor.

Reactive evaporation is a modification of electron-beam evaporation, which is particularly applicable to ceramic coatings (Ref 65). With this method, either a gas, such as oxygen or nitrogen, or a hydrocarbon is metered into the vacuum chamber, where it reacts with the vapor from a metallic evaporant to form a metal oxide, nitride, or carbide coating. An extension of this modification obtained by using a plasma is termed "activated reactive evaporation."

In the ion-plating process, a vapor phase is created by evaporation, as described previously, and transported through a gaseous glow discharge prior to arrival at the sub-

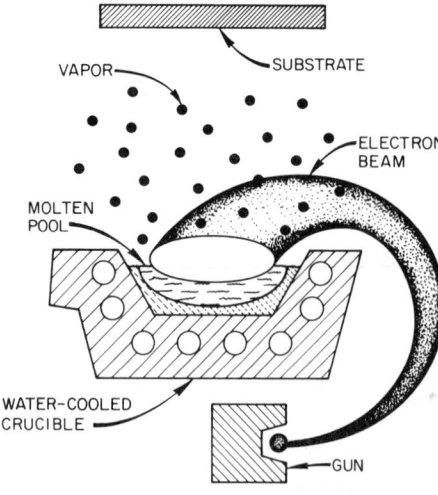

Fig 3 PVD process for application of a coating by electron-beam evaporation. Modifications of this process include reactive evaporation where oxygen and nitrogen are present to form oxides or nitrides and activated reactive evaporation where a plasma is also present.

strate. The glow discharge is normally created by negatively biasing the substrate in the presence of a small argon pressure. Exposure to the glow discharge ionizes some fraction of the vapor, which causes changes to the substrate surface and film properties (adhesion, morphology, density, and residual stress) when compared to conventional evaporation. Variations of the ion-plating process, such as vacuum ion plating, reactive ion plating, and chemical ion plating, have been developed in recent years. Reactive ion plating is particularly applicable to ceramic coatings because metallic species interact with reactive gases to form oxides, nitrides, and carbides.

The primary advantage of ion plating is the increased adherence that is obtained when compared to evaporated or sputtered coatings (Ref 66). Ion bombardment of the substrate surface sputters away material, which roughens the surface and enhances the mechanical bond. Ion bombardment also increases the substrate surface temperature without radically changing the bulk temperature. The increased surface temperature enhances diffusion and chemical reactions at the interface for better adhesion. The morphology of the growing coating is also modified because of better mixing and higher defect concentrations.

Sputtering involves bombarding the source of the material to be deposited with energetic ions instead of depending on heat to vaporize the material, as in evaporation (Ref 63, 67, 68). The ion source can be an ion gun, or more typically, positive ions from a plasma, that are attracted to the negatively charged target (Fig 4). These ions rapidly transfer their energy through knock-on collisions to atoms of the target, which form a secondary-atom collision cascade. When this cascade intersects the surface of the target, material is

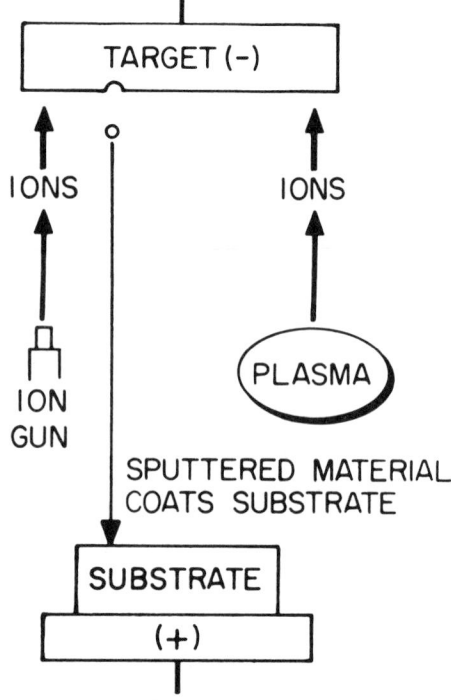

Fig 4 PVD process used to apply coatings by sputtering

ejected (that is, sputtering occurs). The sputtered material then coats the nearby substrate. Use of an rf power supply prevents a sustained build-up of positive charge on the target, which tends to occur with electrical insulators, and allows deposition of insulating ceramics. Deposition rates of 1 μm/min (40 μin./min) are obtained with modern systems (Ref 69). Thus, one former major disadvantage of sputtering, low deposition rates, has been lessened.

An advantage of sputtering over other processes, such as CVD, is that the substrate can be held at low temperatures, which allows deposition onto nonrefractory substrates and minimizes reactions with the substrate. The lower substrate temperature also minimizes room-temperature stresses induced by thermal expansion mismatch between the coating and substrate because the temperature change, upon cooling from the deposition temperature, is reduced. Perhaps the major disadvantage of sputtering is that atoms of the gas used to sustain the plasma are incorporated into the coating. Other disadvantages of the use of low deposition temperatures are that the coating contains residual stresses and it frequently has a fibrous structure, which is less stable upon heating to high temperatures, as compared to stress-free material of equiaxed grain size.

The morphology of PVD coatings deposited by any of these techniques is influenced by many variables; however, the characteristic columnar microstructure is always present (Ref 64, 70). Columnar microstructures result because PVD is a line-of-sight process (vapor species coming from only one direction) with relatively little atom movement because of the low temperatures as compared to CVD. Combining the low mobility with a limited number of nuclei results in long, narrow (fibrous) crystallites with rounded tops and considerable porosity between columns (Fig 2). Deposition of coatings at higher temperatures with increased surface mobility produces coarser columns with faceted tops and less porosity between columns. Coatings deposited at still higher temperatures, where bulk diffusion dominates, contain very large, densely packed grains that maintain the columnar nature. Ion bombardment, as present in some PVD processes, increases the surface diffusion and results in microstructures characteristic of higher-temperature deposits.

The applications of physically vapor deposited coatings cover several different areas. Because of its characteristic gold color, decorative coatings such as TiN are applied to wristwatch bezels and bands, and similar items. Inexpensive metals or even plastics can be coated with ceramic materials, because the coating process takes place at room temperature or only tens of degrees above room temperature. There are numerous applications for wear- and erosion-resistant coatings. With the increased adherence offered by ion plating and variations of other PVD techniques, more cutting tools are being coated with TiC, TiN, and TiC$_x$N$_{1-x}$ using this process. PVD technology is particularly useful for steel cutting tools because the low processing temperatures prevent thermal damage to the substrates that frequently occurs in CVD techniques.

Materials deposited by PVD are often used in the fabrication of electronic devices. Thin metal, alloy, or cermet films with varying resistivities are used in the fabrication of resistors. Cermet films such as Cr-SiO, Au-SiO, and Au-Ti$_2$O$_5$ have higher resistivities (10^2 to 10^4 $\mu\Omega \cdot$ cm) than alloy or metal films and are stable up to 500 °C (930 °F). Capacitors are fabricated from multiple layers of conductors and insulators. Silicon monoxide films are used as dielectrics, insulation layers, and protective coatings. Silicon dioxide films are used as diffusion masks, insulating layers, and protective coatings. Silicon nitride films are used as diffusion masks and insulating layers in the fabrication of integrated circuits.

Adherence

Coating adherence is a critical problem that must be addressed in every coating application, regardless of the deposition technique selected. The coating's coefficient of thermal expansion must be closely matched with that of the substrate to prevent cracking or spalling when high-temperature processing or applications are involved. Intermediate coatings with thermal expansion values between those of the substrate and surface layer are often necessary to minimize the residual stresses created at the interface. More recently, coat-

ings have been produced in which the composition of the interlayer gradually varies from that of the substrate to that of the outer coating. This variation not only reduces the residual stresses but also avoids interface effects. Coating compositions must also be selected that avoid significant reaction at the coating-substrate interface. Reaction layers frequently create large residual stresses that result in cracking or spalling.

Fiber-Reinforced Composites

Chemical vapor infiltration (CVI) is an attractive process for fabricating fiber-reinforced composites because it is one of only a few processes capable of incorporating continuous SiC or alumino-silicate fibers in a ceramic matrix (often SiC) without chemically, thermally, or mechanically damaging the relatively fragile reinforcing fibers (Ref 71–74). The high strength ($\approx$400 MPa, or 58 ksi) and exceptional fracture toughness ($>$20 MPa$\sqrt{m}$, or 18 ksi$\sqrt{in}$.) of these composites, combined with their refractoriness and resistance to erosion, corrosion, and wear, make them ideal candidates for numerous advanced high-temperature structural applications (Ref 75–77).

Most CVI composites are fabricated commercially by an isothermal, isobaric process

in which reactant gases diffuse into a freestanding preform as they flow through the furnace (Ref 78–81). Matrix formation is inherently very slow because diffusion is required to carry the reactants into the center of the preform, while the gaseous reaction products are carried back to the preform surface. To enhance uniform infiltration, the process must be slowed by combinations of low temperatures, low-reactant concentrations, and low pressures to avoid coating and sealing the outer surface of the preform and depleting the reactants before they reach the inner volume.

In isothermal CVI, more matrix material is deposited on fibers near the surface of the preform, as compared to fibers near the center. The density gradient depends on process parameters: temperature, pressure, and reactant composition. Deposition, of course, eventually closes the porosity near the surface and prevents gases from fully densifying the center of the preform. For thick preforms, the process is interrupted to allow intermediate diamond machining operations to remove material from the surface and open passageways to the interior of the preform. Commercial production of composites is economically viable because a large furnace can simultaneously densify hundreds of preforms. However, the total production time, particularly for high-density composites, can be very long (months).

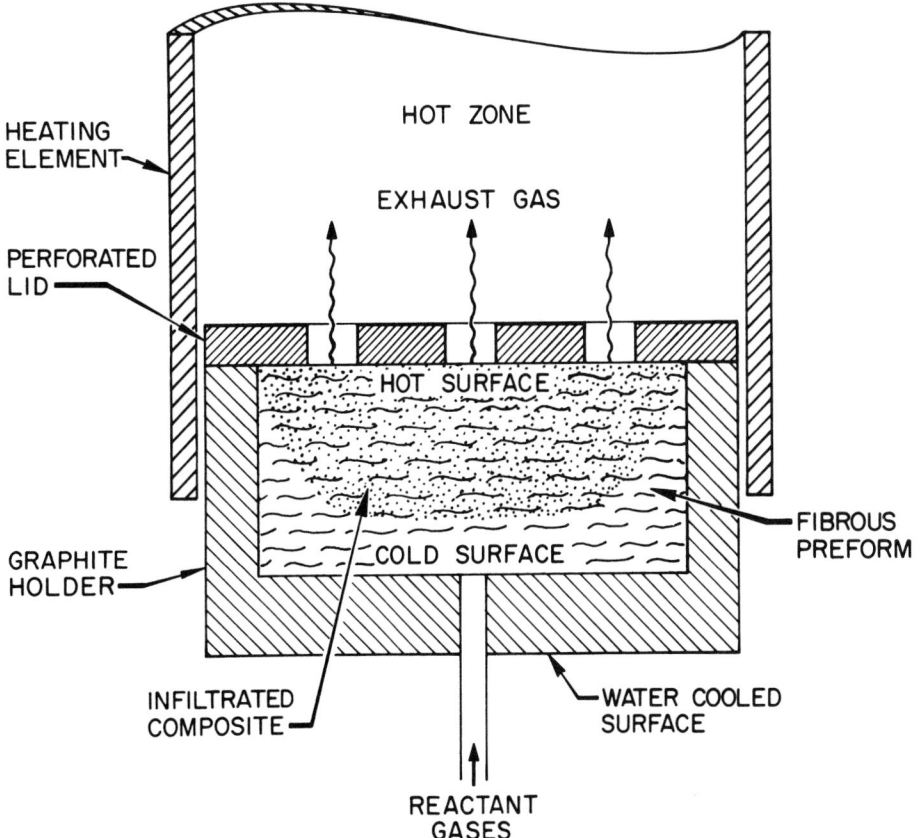

Fig 5 CVI process utilizing thermal and pressure gradients for the fabrication of fiber-reinforced composites

A different process for the fabrication of fiber-reinforced ceramic composites has been under development at Oak Ridge National Laboratory (ORNL) for several years (Ref 82, 83). This process simultaneously utilizes thermal and pressure gradients to reduce the infiltration time from weeks to less than 24 h (Fig 5). Fibrous preforms are retained within a graphite holder that contacts a water-cooled metal gas distributor, thus cooling the gas inlet and side surfaces of the substrate. The opposite end of the fibrous preform is exposed to the hot zone of the furnace, creating a steep temperature gradient across the preform. The reactant gases are forced into the cooled side of the fibrous preform but initially do not react because of the low temperature. The gases continue from the cooled portion of the preform into the hot portion, where the matrix begins to deposit on the fibers. Deposition of matrix material within the hot region of the preform increases the density, and therefore the thermal conductivity of the preform. Thus, the deposition zone moves progressively from the hotter regions toward the cooler regions. The process continues until reduced permeability of the densified composite prevents sufficient flow of reactant gases into the preform (a backpressure of ≈0.2 MPa, or 0.03 ksi). Disk shapes (75 mm, or 3 in. diam, by >20 mm, or 0.8 in., thick) and thick-walled (10 mm, or 0.4 in.) tubes have been routinely fabricated at ORNL by forced CVI. Disk shapes greater than 200 mm (8 in.) in diameter have been fabricated in industry.

Acknowledgment

The authors acknowledge the information obtained from the U.S. Department of Energy's Fossil Energy (AR&TD) Materials Program; the Office of Transportation Technologies, as part of the Ceramic Technology for Advanced Heat Engines Project of the Advanced Materials Development Program; and the Assistant Secretary for Conservation and Renewable Energy, Office of Industrial Technologies, Advanced Industrial Concepts Division, Advanced Industrial Materials Program under contract DE-AC05-84OR21400 with Martin Marietta Energy Systems, Inc. We also appreciate the assistance of J.I. Kelly for preparation of the manuscript and H.R. Livesey for the graphics.

REFERENCES

1. K.E. Bean, Chemical Vapor Deposition Applications in Microelectronics Processing, *Thin Solid Films*, Vol 83, 1981, p 173–186
2. K.K. Yee, Protective Coatings for Metals by Chemical Vapor Deposition, *Int. Met. Rev.*, Vol 23 (No. 1), 1978, p 19–42
3. H.E. Hinterman, Tribological and Protective Coatings by Chemical Vapor Deposition, *Thin Solid Films*, Vol 84, 1981, p 215–243
4. D.P. Stinton, T.M. Besmann, and R.A. Lowden, Advanced Ceramics by Chemical Vapor Deposition Techniques, *Am. Ceram. Soc. Bull.*, Vol 67 (No. 2), 1988, p 350–355
5. W.J. Lackey, D.P. Stinton, G.A. Cerny, A.C. Schaffhauser, and L.L. Fehrenbacher, Ceramic Coatings for Advanced Heat Engines—A Review and Projection, *Adv. Ceram. Mater.*, Vol 2 (No. 1), 1987, p 24–30
6. H.O. Pierson, Ed., *Chemically Vapor Deposited Coatings*, American Ceramic Society, 1981
7. K.E. Spear, Applications of Phase Diagrams and Thermodynamics to CVD, *Proceedings of Seventh International Conference on Chemical Vapor Deposition*, T.O. Sedgwick and H. Lydtin, Ed., Electrochemical Society, 1979, p 1–16
8. T.M. Besmann and K.E. Spear, Analysis of the Chemical Vapor Deposition of Titanium Diboride, I. Equilibrium Thermodynamic Analysis, *J. Electrochem. Soc.*, Vol 124, 1977, p 786–790
9. T.M. Besmann and K.E. Spear, Analysis of the Chemical Vapor Deposition of Titanium Diboride, II. Modeling the Kinetics of Deposition, *J. Electrochem. Soc.*, Vol 124, 1977, p 790–797
10. C.H.J. Van der Brekel, Characterization of Chemical Vapor Deposition Processes, *Philips Res. Rep.*, Vol 32, 1977, p 118–133
11. D.M. Mattox, Ion Plating Technology, *Deposition Technologies for Films and Coatings*, R.F. Bunshah *et al.*, Ed., Noyes Publications, 1982, p 244–287
12. R.F. Bunshah and D.M. Mattox, Applications of Metallurgical Coatings, *Physics Today*, May 1980, p 50–55
13. J.A. Thornton, High Rate Thick Film Growth, *Ann. Rev. Mater. Sci.*, No. 7, 1977, p 239–260
14. W.A. Bryant, Review: The Fundamentals of Chemical Vapor Deposition, *J. Mater. Sci.*, Vol 12, 1977, p 1285–1306
15. V.K. Sarin, Structure/Property Relationship of CVD-TiC Coatings, *Proceedings of First International Conference on the Science of Hard Materials*, Plenum Press, 1983
16. V.K. Sarin, "TEM Studies of Wear-Resistant TiC Coatings," Proceedings of the Seventh CVD Conference, Los Angeles, 1979
17. K. Niihara and T. Hirai, *Proceedings of Seventh International Conference on Vacuum Metallurgy—Metallurgical Coatings*, The Iron and Steel Institute of Japan, Tokyo, p 180–186
18. T. Hirai, Emergent Process Methods for High Technology Ceramics, *The Nineteenth University Conference on Ceramic Science*, North Carolina State University, Raleigh, 8–10 Nov 1982, R.F. Davis, H. Palmour III, and R.L. Porter, Ed., Plenum, 1984, p 329–345
19. D.P. Stinton, W.J. Lackey, R.J. Lauf, and T.M. Besmann, Fabrication of Ceramic-Ceramic Composites by Chemical Vapor Deposition, *Ceram. Eng. Sci. Proc.*, Vol 5 (No. 7–8), 1984, p 668–676
20. D.P. Stinton and W.J. Lackey, Simultaneous Chemical Vapor Deposition of SiC-Dispersed Phase Composites, *Ceram. Eng. Sci. Proc.*, Vol 6 (No. 7–8), 1985, p 707–713
21. P.J.M. van der Straten and G. Verspui, Chemical Vapor Deposition of Wear-Resistant Coatings on Tool Steel, *Phillips Tech. Rev.*, Vol 40 (No. 7), 1982, p 204–210
22. R. Bonetti, Hard Coatings For Improved Tool Life, *Met. Prog.*, June 1981
23. K. Akamatsu, M. Ikenaga, and K. Kamei, Adherence of Chemical Vapor Deposition To Some Tool Steels, *Technol. Rep. Kansai Univ.*, No. 25, March 1984
24. N. Archer, Hard and Wear-Resistant Coatings by Chemical Vapour Deposition, *Ceramic Surfaces and Surface Treatments*, R. Morrell and M.G. Nicholas, Ed., British Ceramic Society, Staffs, United Kingdom, 1984, p 187–194
25. H.E. Hintermann, Tribological and Protective Coatings by Chemical Vapor Deposition, *Thin Solid Films*, Vol 84, 1981, p 215–243
26. S. Ramalingam, New Coating Technologies for Tribological Applications, *Wear Control Handbook*, M.B. Peterson and W.O. Winer, Ed., American Society of Mechanical Engineers, 1980, p 385–411
27. H.E. Hintermann, Tribological and Protective Coatings by Chemical Vapour Deposition, *Thin Solid Films*, Vol 84, 1981, p 215–243
28. V.K. Sarin, Cemented Carbide Cutting Tools, *Advances in Powder Technology*, American Society for Metals, 1982
29. A.J. Caputo, W.J. Lackey, and I.G. Wright, Chemical Vapor Deposition of Erosion-Resistant TiB$_2$ Coatings, *Proceedings of the Ninth International Conference on Chemical Vapor Deposition*, McD. Robinson, C.H.J. van den Brekel, G.W. Cullen, J.M. Blocher, Jr., and P. Rai-Choudhury, Ed., Electrochemical Society, 1984, p 782–799
30. C.L. Yawsend and G.F. Wakefield, Scale-up Process for Erosion Resistant Titanium Carbonitride Coating, *Proceedings of the Fourth International Conference on Chemical Vapor Deposition*, G.G. Wakefield and J.M. Blocher, Jr., Ed., Electrochemical Society, 1973, p 577–587
31. H.W. Chang and R.M. Rusnak, Oxidation Behavior of Carbon/Carbon Com-

posites, *Carbon*, Vol 17, 1979, p 407–410

32. P. Ehrburger, L. Lahaye, and C. Bourgeois, Characterization of Carbon-Carbon Composites-II: Oxidation Behavior, *Carbon*, Vol 19, 1981, p 7–10

33. E. Fitzer, The Future of Carbon-Carbon Composites, *Carbon*, Vol 25 (No. 2), 1987, p 163–190

34. K.L. Lewis, A.M. Pitt, J.A. Savage, J.E. Field, and D. Townsend, The Mechanical Properties of CVD-Grown Zinc Sulphide and Their Dependence on the Conditions of Growth, *Proceedings of the Ninth International Conference on Chemical Vapor Deposition*, McD. Robinson, C.H.J. van den Brekel, G.W. Cullen, J.M. Blocher, Jr., and P. Rai-Choudhury, Ed., Electrochemical Society, 1984, p 530–539

35. C. Hayman, Ceramic Coatings by CVD, *Ceramic Surfaces and Surface Treatments*, R. Morrell and M.G. Nicholas, Ed., British Ceramic Society, Staffs, United Kingdom, 1984, p 175–186

36. N. J. Archer, The Preparation and Properties of Pyrolytic Boron Nitride, *High Temperature Chemistry in Inorganic and Ceramic Materials*, F.P. Glasser and P.E. Potter, Ed., The Chemical Society, London, 1977, p 167–180

37. P.B. Ghate, Deposition Technologies and Microelectronic Applications, *Deposition Technologies for Films and Coatings*, R.F. Bunshah *et al.*, Ed., Noyes Publications, 1982, p 514–547

38. R.D. Dupuis, Chemical Vapor Deposition for III-V Compounds Semiconductor Devices, *Chemical Vapor Deposition, Second International Conference*, J.M. Blocher, Jr. and J.C. Withers, Ed., Electrochemical Society, 1970, p 503–515

39. J.L. Vossen and W. Kern, Thin-Film Formation, *Physics Today*, May 1980, p 26–33

40. M.L. Green and R.A. Levy, Chemical Vapor Deposition of Metals for Integrated Circuit Applications, *J. Met.*, June 1985, p 63–69

41. J.A. Amick and W. Kern, Chemical Vapor Deposition Techniques for the Fabrication of Semiconductor Devices, *Chemical Vapor Deposition, Second International Conference*, J.M. Blocher, Jr. and J.C. Withers, Ed., Electrochemical Society, 1970, p 551–570

42. A.K. Sinha, Refractory Metal Silicides for VLSI Applications, *J. Vac. Sci. Technol.*, Vol 19 (No. 3), 1981, p 778–785

43. R. Reif, Plasma Enhanced Chemical Vapor Deposition of Thin Crystalline Semiconductor and Conductor Films, *J. Vac. Sci. Technol. A*, Vol 2 (No. 2), 1984, p 429–435

44. B. Gorowitz, T.B. Gorczyca, and R.J. Saia, Applications of Plasma Enhanced Chemical Vapor Deposition in VLSI, *Solid State Technol.*, June 1985, p 197–203

45. H. Itoh, M. Kato, and K. Sugiyama, Plasma-Enhanced Chemical Vapour Deposition of AlN Coatings on Graphite Substrates, *Thin Solid Films*, Vol 146, 1987, p 255–264

46. T.D. Bonifield, Plasma Assisted Chemical Vapor Deposition, *Chemical Vapor Deposition, Second International Conference*, J.M. Blocher, Jr. and J.C. Withers, Ed., Electrochemical Society, 1970, p 365–383

47. S. Veprek, Plasma-Induced and Plasma-Assisted Chemical Vapor Deposition, *Thin Solid Films*, Vol 130, 1985, p 135–154

48. A. Sherman, Plasma-Assisted Chemical Vapor Deposition Processes and Their Semiconductor Applications, *Thin Solid Films*, Vol 113, 1984, p 134–149

49. J. Mort and F. Jansen, *Plasma Deposited Thin Films*, CRC Press, 1986

50. D. Kuppers, Recent Developments in Plasma Activated Chemical Vapor Deposition, *Proceedings of the Seventh International Conference on Chemical Vapor Deposition*, T.O. Sedgwich and H. Lydtin, Ed., Electrochemical Society, 1979, p 159–175

51. T.D. Bonifield, Plasma Assisted Chemical Vapor Deposition, *Deposition Technologies for Films and Coatings*, R.F. Bunshah *et al.*, Noyes Publications, 1982, p 365–384

52. J.C. Angus, P. Koidl, and S. Domitz, Carbon Thin Films, *Plasma Deposited Thin Films*, J. Mort and F. Jansen, Ed., CRC Press, 1986, p 89–128

53. K.E. Spear, personal communication, 1986

54. S. Matsumoto, Y. Sato, M. Kamo, and N. Setaka, Vapor Deposition of Diamond Particles from Methane, *Jpn. J. Appl. Phys. Lett.*, Vol 21 (No. 3), 1982, p L183-L185

55. S. Matsumoto, Y. Sato, M. Tsutsumi, and N. Setaka, Growth of Diamond Particles from Methane-Hydrogen Gas, *J. Mater. Sci. Jpn.*, Vol 17, 1982, p 3106–3112

56. A. Sawabe and T. Inuzuka, Growth of Diamond Thin Films by Electron Assisted Chemical Vapor Deposition, *Appl. Phys. Lett.*, Vol 46 (No. 2), 1985, p 146–147

57. Y. Saito, S. Matsuda, and S. Nogita, Synthesis of Diamond by Decomposition of Methane in Microwave Plasma, *J. Mater. Sci. Lett.*, Vol 5 (No. 5), 1986, p 565–568

58. K. Kitahama, K. Hirata, H. Nakamatsu, and S. Kawai, Synthesis of Diamond by Laser-Induced Chemical Vapor Deposition, *Appl. Phys. Lett.*, Vol 49 (No. 11), 1986, p 634–635

59. K.E. Spear, personal communication, 1986

60. A.C. Warren *et al.*, Fabrication of Sub-100 nm Linewidth Periodic Structure for Study of Quantum Effects from Interference and Confinement in Si Inversion Layers, *J. Vac. Sci. Technol. B*, Vol 4 (No. 1), 1986, p 365–368

61. T.H. Geballe and J.M. Rowell, Vapor-Deposited Metastable Superconductors, *Thin Solid Films*, Vol 91, 1982, p 33–43

62. C. Vahlas and T.M. Besmann, "Thermodynamic Approach to the CVD of the $YBa_2Cu_3O_{7-x}$ Phase Deposited from Organometallics," Eleventh International Conference on Chemical Vapor Deposition, Seattle, 14–19 Oct 1990

63. *Physical Vapor Deposition*, Research Development Department of Airco Temescal, 1976

64. R.F. Bunshah, Evaporation, *Deposition Technologies for Films and Coatings—Developments and Applications*, Noyes Publications, 1982, p 83–169

65. R.F. Bunshah, Reactive Evaporation, *Science and Technology of Surface Coating*, B.N. Chapman and J.C. Anderson, Ed., Academic Press, 1974, p 361–368

66. D.M. Maddox, Ion Plating Technology, *Deposition Technologies for Films and Coatings—Developments and Applications*, Noyes Publications, 1982, p 244–268

67. L. Holland, The Basic Principles of Sputter Deposition, *Science and Technology of Surface Coating*, B.N. Chapman and J.C. Anderson, Ed., Academic Press, 1974, p 369–385

68. J.A. Thornton, Coating Deposition by Sputtering, *Deposition Technologies for Films and Coatings—Developments and Applications*, Noyes Publications, 1982, p 170–243

69. S.D. Dahlgren, Vapor Quenching Techniques, *Proceedings of International Conference on Rapidly Quenched Metals III*, Vol 2, B. Cantor, Ed., The Metals Society, London, 1978, p 36–47

70. J.A. Thornton, High Rate Thick Film Growth, *Ann. Rev. Mater. Sci.*, Vol 7, 1977, p 239–260

71. E. Fitzer, D. Hegen, and H. Strohmeier, Chemical Vapor Deposition of Silicon Carbide and Silicon Nitride and Its Application for Preparation of Improved Silicon Ceramics, *Proceedings of the Seventh International Conference on Chemical Vapor Deposition*, T.D. Sedwick and H. Lydtin, Ed., Electrochemical Society, 1979, p 525–535

72. W.H. Pfeifer *et al.*, Consolidation of Composite Structures by CVD, *Second International Conference on Chemical Vapor Deposition*, J.M. Blocher, Jr. and J.C. Withers, Ed., Electrochemical Society, 1970, p 463–483

73. J.C. Theis, Jr., The Process Development and Mechanical Testing of a Car-

bon/Carbon Composite Fabricated by Chemical Vapor Infiltration of a Filament-Wound Substrate, *Third International Conference on Chemical Vapor Deposition*, F.A. Glaski, Ed., American Nuclear Society, 1972, p 561–573

74. F. Christin, R. Naslain, and C. Bernard, A Thermodynamic and Experimental Approach of Silicon Carbide-CVD Application to the CVD-Infiltration of Porous Carbon-Carbon Composites, *Proceedings of the Seventh International Conference on Chemical Vapor Deposition*, T.O. Sedgwick, Ed., Electrochemical Society, 1974, p 499–514

75. J. Aveston and A. Kelly, Theory of Multiple Fracture of Fibrous Composites, *J. Mater. Sci.*, Vol 8, 1973, p 352–362

76. D.B. Marshal and A.G. Evans, Failure Mechanisms in Ceramic-Fiber/Ceramic Matrix Composites, *J. Am. Ceram. Soc.*, Vol 68, 1985, p 225–231

77. J.W. Warren, Fiber and Grain-Reinforced Chemical Vapor Infiltration (CVI) Silicon Carbide Matrix Composites, *Ceram. Eng. Sci. Proc.*, Vol 6 (No. 7–8), 1985, p 684–693

78. R.E. Fisher, C.V. Burkland, and W.E. Bustamante, Ceramic Composites Based on Chemical Vapor Infiltration, *Proceedings of the Metal and Ceramic Matrix Composite Processing Conference*, Battelle Columbus Laboratories, 14 Nov 1984

79. E. Fitzer and R. Gadow, Fiber Reinforced Silicon Carbide, *Am. Ceram. Soc. Bull.*, Vol 65 (No. 2), 1986, p 326–335

80. P.J. Lamicq, G.A. Bernhart, M.M. Dauchier, and J.G. Mace, SiC/SiC Composite Ceramics, *Am. Ceram. Soc. Bull.*, Vol 65 (No. 2), 1986, p 336–338

81. R.W. Rice, Mechanisms of Toughening in Ceramic Matrix Composites, *Ceram. Eng. Sci. Proc.*, Vol 2 (No. 7–8), 1981, p 661–701

82. D.P. Stinton, A.J. Caputo, and R.A. Lowden, Synthesis of Fiber-Reinforced SiC Composites by Chemical Vapor Infiltration, *Am. Ceram. Soc. Bull.*, Vol 65 (No. 2), 1986, p 347–350

83. D.P. Stinton, Ceramic Composites by Chemical Vapor Infiltration, *Proceedings of the Tenth International Conference on Chemical Vapor Deposition*, G.W. Cullen, Ed., Electrochemical Society, 1987, p 1028–1040

Polymer-Derived Ceramics

Daniel R. Petrak, Dow Corning Corporation

POLYMER-DERIVED CERAMICS offer a unique approach to fabricating ceramic materials. This route to processing non-oxide chemistries has been applied to ceramics such as SiC, Si_3N_4, AlN, BN, and TiN (Ref 1). Perhaps the most widely recognized ceramic products produced from polymers are fibers. However, other reported uses of polymer-derived ceramics include coatings for oxidation protection, matrices for ceramic fiber-reinforced composites and nonfugitive binders for ceramics prepared from powders. While a wide variety of ceramic compositions can be derived from polymer precursors, the predominant chemistries of greatest current interest are those that produce silicon carbide and silicon nitride. This discussion will focus on these two silicon compounds.

Ceramics from Organosilicon Polymers

Early work by Verbeek and Winter (Ref 2), then Yajima *et al.* (Ref 3, 4) demonstrated that the pyrolysis of polycarbosilane would produce a char containing silicon carbide. Since then, a number of polymers such as polysilastyrene (Ref 5), vinylic polysilane (Ref 6), and various polysilazanes (Ref 7–9) have been shown to produce Si-C or Si-N-C chars useful in the preparation of silicon-base ceramics. Details of the synthesis and structure of many of these polymers have been reviewed by Baney and Chandra (Ref 10).

Processing preceramic polymers often involves heating to a melt or softening temperature to form a controlled shape. This behavior is advantageous in forming fibers and coatings, and in processes for molding fine powders. A cure or cross-linking of the polymer must be introduced in order to maintain the shape during subsequent pyrolysis. Polycarbosilane, for example, can be cross-linked by heating to 200 °C (390 °F) in air (Ref 11). Conversion of the polymer to a stable ceramic phase (that is, pyrolysis), is done by heating in a nonreactive atmosphere such as argon, helium, or nitrogen to 1000 °C (1830 °F) or higher. One important exception is the ammonia pyrolysis of various polymers to form Si_2N_2O or Si_3N_4 powders or fibers (Ref 12, 13). Pyrolysis is accompanied by substantial shrinkage caused by weight loss and densification. Densities of preceramic polymers are typically 1.1 to 1.2 g/cm^3 (0.039 to 0.043 $lb/in.^3$), while polymer chars show densities of 2.3 to 2.6 g/cm^3 (0.083 to 0.094 $lb/in.^3$). Weight-based char yields range from 60 to over 90%, while volumetric yields are more typically 25 to 35%. Shrinkage in fiber processing is of only minor consequence, but polymer shrinkage in the preparation of coatings can be a major concern.

Figure 1 shows a thermogravimetric analysis (TGA) curve for a cross-linked polysiloxane preceramic polymer run in helium. Weight loss begins at approximately 400 °C (750 °F) and is substantially complete by 800 °C (1470 °F). Many other polymers show similar weight change behavior. Actual char yield of a given polymer may depend upon the extent of cross-linking, polymer molecular weight, atmospheric pressure, and polymer composition.

The chemistry and structure of polymer chars is described in the sections that follow. Four types of polymer-derived ceramics will be reviewed:

- Fibers
- Ceramic matrix composites (CMC)
- Coatings
- Nonfugitive binders

Processing methods for these materials are shown in Fig 2.

Fibers. Only two polymer-derived ceramic fibers, Nicalon (Fig 3) and Tyranno, are commercially available, although a third, HPZ fiber (Fig 4), is also available in the United States in pilot scale quantities. The Nicalon fiber discussed here is a ceramic grade (CG) Nicalon, although other grades such as high volume resistivity (HVR) and low volume resistivity (LVR) are available. An earlier standard grade (SG) has been discontinued. Only one grade of Tyranno fiber is currently available but reportedly a low oxygen ($\approx 10\%$) version of Tyranno will soon be marketed. All three fibers have diameters less than 15 μm (600 μin.) and are considered textile-grade fibers (that is, they are weaveable).

Table 1 shows the chemical analysis for these fibers. As determined by x-ray defraction, the structures of the fibers are primarily amorphous. Nicalon does exhibit a broad band peak for β-SiC. Lipowitz *et al.* (Ref 14), reported that β-SiC crystallites in Nicalon are 1.7 nm (17 Å) and constitute 35 vol% of the structure. Those same investigators conducted infrared spectroscopy, Raman spectroscopy, and NMR studies on Nicalon and HPZ fibers. They reported evidence for Si-C, Si-N, Si-O, and C-C bonds in HPZ fiber and all but the Si-N bond for Nicalon. They suggest that while bonding approaches random, there is some memory of the bonding present in the original preceramic polymer. Nicalon is derived from polycarbosilane and HPZ fiber is made from hydridopolysilazane polymer (Ref 15). Lipowitz *et al.*, also point to lower densities in the fibers than would be expected for amorphous materials, which suggests the presence of extremely fine porosity (Ref 14).

Tyranno fiber is prepared from polytitanocarbosilane (Ref 16). Titanium is present reportedly to limit crystallization of SiC. As-made Tyranno fiber is amorphous per x-ray diffraction analysis, but after aging at 1300 °C (2370 °F) in argon for 75 h, the titanium precipitates as TiC and further aging to 1400 °C (2550 °F) in argon for 170 h promotes crystallization of SiC and SiO_2 (cristobalite) (Ref 17).

The tensile strength and elastic modulus values for the three fibers discussed above are shown in Table 1. These room-temperature strengths are 50 to 100% higher than other available ceramic fibers (for example, Nextel mullite fiber from 3M Corporation and Al_2O_3 Fiber FP from DuPont). Stability of strength of the nonoxide fibers may be limited to a temperature range of 1000 to 1300 °C (1830 to 2370 °F) depending on local atmospheric conditions. Single-fiber tensile strength determined by Pysher *et al.* (Ref 18) showed Nicalon and Tyranno fibers to degrade rapidly at 1200 °C (2190 °F) in air. Aging studies for 100 h in moist air at 1000 °C (1830 °F) showed similar degradation, that is, reduction of 20 to 40% in original strength (Ref 19) for Nicalon and Tyranno but virtually no change in strength for HPZ fiber.

Crystallization of these amorphous structures at temperatures $\geq$1300 °C ($\geq$2370 °F)

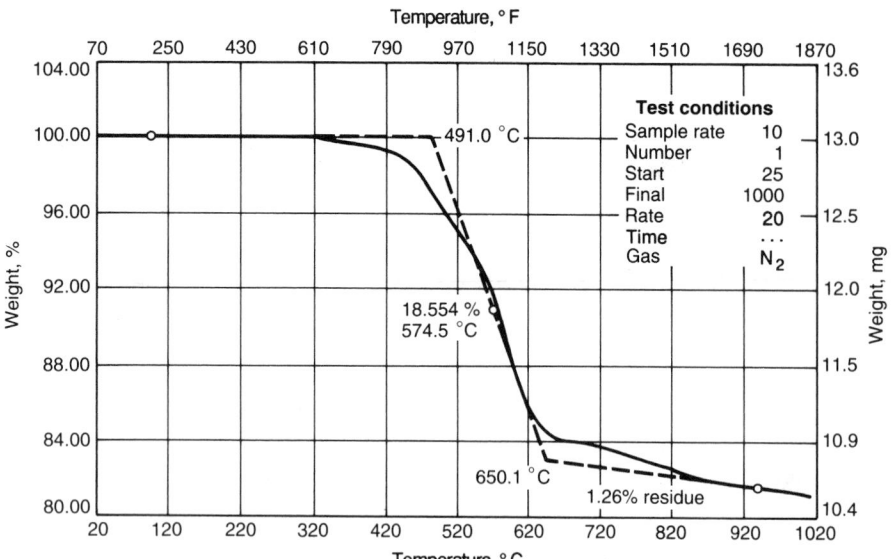

Fig 1 Thermogravimetric analysis of cured siloxane polymer

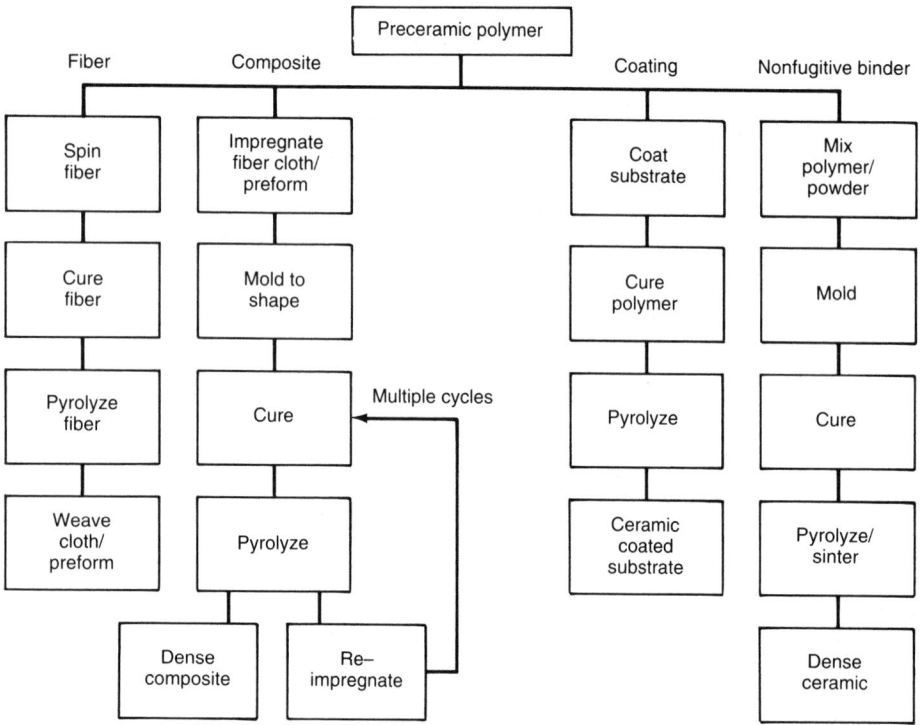

Fig 2 Major processing steps necessary to prepare the four types of polymer-derived ceramics

in vacuum, argon, or helium, accompanied by weight loss from the evolution of gases such as carbon monoxide, silicon monoxide, and nitrogen, may also be a source of strength degradation (Ref 20).

While these fibers are the only polymer-derived fibers that are commercially available, other fibers are under development. They include a Si-N-O fiber from Tao Nenryo Kogyo K.K. (Tonen), a Si-N-C fiber from Rhone-Poulenc, and a crystalline silicon-car-bide fiber being developed by Dow Corning Corporation, with sponsorship by NASA Lewis Research Center.

Ceramic Matrix Composites. Ceramic fiber-reinforced ceramic matrix composites have been under development for many years. Methods of fabrication often used include hot pressing of glass-ceramic matrices and chemical vapor infiltration (CVI). It is also possible to prepare ceramic matrix composites by a polymer-impregnation-pyrolysis tech-

nique. As shown in Fig 2, this involves preparation of a prepregged fiber tape, cloth, or a woven fiber preform. Subsequent molding, similar to the method used to make organic matrix composites, permits a wide range of shapes and large sizes. These methods include compression molding, autoclave processing, resin transfer molding, and filament winding. Once a shape is formed, the preceramic polymer is cured and the part is pyrolyzed. Because the polymer char shrinks during pyrolysis, substantial porosity is produced in the composite. Multiple impregnation-pyrolysis cycles are required for adequate densification to produce a high-strength part. The number of cycles required may be four to 10 depending on the fiber concentration and char yield of the polymer.

Chi and Stark (Ref 21, 22) and others (Ref 23) reported composites using this process. Composites prepared using a modified version of the Chi and Stark patent have been prepared and characterized. The reinforcing fiber was CG Nicalon in the form of eight-harness satin woven cloth and matrix was derived from a polysiloxane preceramic polymer. The polymer produces an amorphous char with a yield of 80% by weight. The nominal chemistry of the char is 35Si-45C-20O after pyrolysis to 1200 °C (2190 °F). After six pyrolysis cycles, the composite open porosity was reduced to 6%, and the composite density was 2.15 g/cm^3 (0.0777 lb/in.^{3}).

Table 2 lists the room-temperature mechanical properties for two-dimensional reinforced composites prepared with Nicalon reinforcement and processed by the multiple impregnation pyrolysis of polysiloxane. Figure 5 shows a scanning electron microscopy (SEM) micrograph of a fractured specimen.

This particular matrix has not shown stable high-temperature properties. However, research is continuing to solve these problems through polymer development and fiber-matrix interface studies. The feasibility of producing high strength-tough composites using polymer-derived matrices has been demonstrated.

Coatings. Polymer-derived coatings have been reported to be effective in protecting carbon and metal substrates against oxidation and wear. Niebylski (Ref 24) blended boroxines with polycarbosilane to prepare a coating formulation. Both polymers were dissolved in a suitable organic solvent at a total concentration of 40 to 60% and applied to the substrate by spraying or brushing. After evaporating the solvent and heating at 200 to 250 °C (390 to 480 °F) to ensure that all the solvent was removed, the thickness of the coating should be 10 to 250 μm (0.0004 to 0.01 in.). The coatings were then pyrolyzed at 600 to 900 °C (1110 to 1650 °F). Thicker coatings can be applied, but they should be built up in thin layers. The char chemistry of the polymer was not described. Results of oxidation tests of the coating on graphite substrates are shown in Table 3.

Similar results were reported for a silicon-

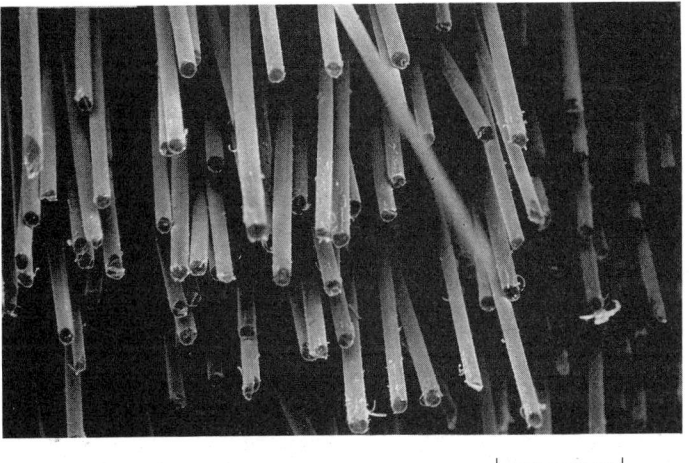

100 μm

Fig 3 SEM photomicrograph of Nicalon polymer-derived ceramic fiber. 164×

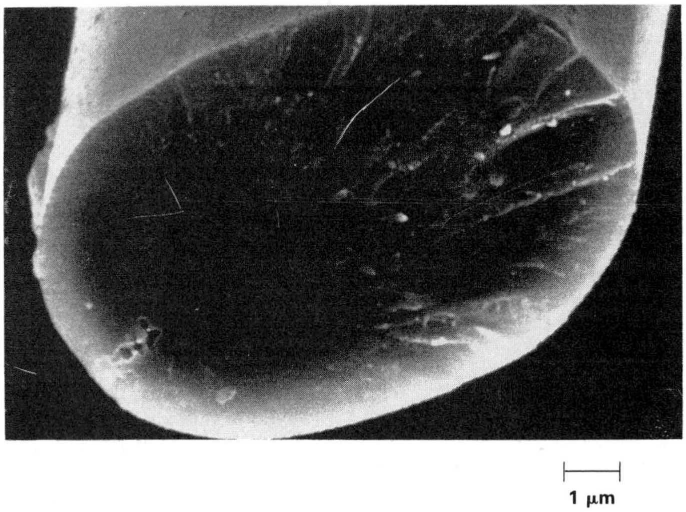

1 μm

Fig 4 SEM photomicrograph of a HPZ fiber showing oval cross section of this polymer-derived ceramic. 6150×

1000 μm

Fig 5 SEM photomicrograph showing fracture surface of a composite flexural strength test bar. Fibrous fracture indicates good toughness properties. 35×

deposition (CVD) or plasma spraying, has as yet not been reported.

Nonfugitive Binders. The use of a preceramic polymer as a binder or lubricant in green part forming can promote high green strength and reduce volumetric shrinkage during sintering. Both parameters have important practical significance in the production of sintered ceramics. Higher as-molded strength can reduce chipping and breakage during handling and green machining, and a reduction in total shrinkage should allow finer control of as-sintered dimensions.

Schwartz and Rowcliffe (Ref 26) have discussed the modeling of the density contributions in preceramic polymer-ceramic powder systems. They derived the following equation:

$$\rho_g = V_p\rho_p + V_s\rho_s \qquad (\text{Eq 1})$$

where ρ_g is the green density of a molded blend of polymer and ceramic powders, V_p and ρ_p are the volume fraction and the density, respectively, of the polymer, and V_s and ρ_s are the volume fraction and the density, respectively, of the solid powder. Because the polymer has a lower density than the ceramic powder, the maximum green density of a molded part is controlled by the maximum particle packing of the powder. Most sinterable ceramic powders are limited to a volume fraction of ≈0.63. For example, a molded part of silicon-carbide powder and a preceramic

carbide powder-filled polycarbosilane-derived coating referred to as Nicalocoat from Nippon Carbon (Ref 25). Careful surface preparation prior to coating application must be considered. Control of coating thickness to less than 60 μm (0.0024 in.) was recommended to avoid cracking and peeling when thermal cycling is anticipated. One advantage of polymer-derived coatings should be their ease of application. A comparison of performance and relative economics with other coating methods, such as chemical vapor

Table 1 Composition and room-temperature mechanical properties of commercially available polymer-derived monofilament ceramic fibers

| | Composition, wt% | | | | | Mechanical properties | | | | | |
| | | | | | | Tensile strength | | Elastic modulus | | Density | |
Fiber	Si	N	C	O	Ti	GPa	ksi	GPa	psi × 10⁶	g/cm³	lb/in.³
CG Nicalon	58.4	0.1	31.2	10.1	. . .	2.97	430	193	28	2.55	0.0921
Tyranno	49.3	0.3	27.9	20.2	1.2	2.80	406	193	28	2.40	0.0867
HPZ	59.4	28.9	10.1	3.1	. . .	2.60	377	193	28	2.45	0.0885

Table 2 Room-temperature properties of two-dimensional CG Nicalon fiber-reinforced composites with polysiloxane-derived matrices

Property	Value
Bulk density, g/cm³ (lb/in.³)	2.15–2.20 (0.0777–0.0795)
Four-point flexural strength, MPa (ksi)	345–415 MPa (50–60)
In-plane tensile strength, MPa (ksi)	300 (44)
Tensile modulus, GPa (psi × 10⁶)	15 (2.2)
Double notch shear strength, MPa (ksi)	14 (2.0)
Coefficient of thermal expansion, 10⁻⁶/°C (10⁻⁶/°F)	4.3 (7.8)
Fiber volume, %	47

Table 3 Time-dependent weight loss of boroxine-polycarbosilane-derived coatings on graphite at 650 °C (1200 °F) in air

Coating thickness		Weight loss at selected times, %		
μm	in.	3 h	8 h	24 h
0	0	...	32	100
80–100	0.0032–0.004	0.1	0.3	4.2
160–200	0.0064–0.008	0.1	0.2	1.9

Source: Ref 24

polymer with a density of 1.2 g/cm³ (0.043 lb/in.³) would therefore be limited to a density of 2.46 g/cm³ (0.0888 lb/in.³) if the part had no porosity. The green part would have 18 wt% polymer.

The contribution of the polymer to the final density of the sintered part will be dependent on the char yield of the polymer. If the polymer had a char yield of 50 wt% and no other losses occurred during pyrolysis or sintering, the polymer would contribute 9% weight to the finished part. This would be equivalent to molding a 70% dense silicon-carbide part.

In practice, there may be deviations from the example cited above. It may be difficult to achieve a volume fraction of 0.63 particle packing. It may be desirable to incorporate some porosity by reducing polymer content to permit pyrolysis gas to be released without disrupting the part. On the other hand, increasing polymer content may promote ease of part forming by techniques such as in injection molding.

S.R. Su (Ref 27) reported using 10 to 40% levels of polysilanes or polysilazane polymers to densify Si_3N_4 powders. The compo-

sitions of Si_3N_4 included Al_2O_3 and Y_2O_3 sintering additives. Molded parts were pyrolyzed in N_2 to 900 °C (1650 °F) and then sintered at temperatures ranging from 1700 to 1900 °C (3090 to 3450 °F) with an overpressure of nitrogen. The densities of sintered parts ranged from 2.94 to 3.30 g/cm³ (0.106 to 0.119 lb/in.³) and flexural strengths were over 700 MPa (100 ksi) at room temperature. It was reported that this method enhanced green strength, reduced shrinkage, and produced parts free of the internal cracking problem that was prevalent when using other methods.

Atwell, Burns and Saha (Ref 28, 29) reported the use of organopolysiloxane to form silicon-carbide parts. In this case, the char yield and the chemistry was first determined for a given formulation of polymer. The amount of excess carbon in the silicon carbide plus carbon char was used to determine the level of polymer to be added to the powder formulation. Enough polymer was added to the silicon-carbide powder plus sintering aid blend to produce from 0.2 to 3% free carbon in the blend. Molding methods used were pressing of powders and transfer molding with higher levels of polymer. Readily moldable green parts were produced. Sintering was done at temperatures of 1900 to 2200 °C (3450 to 3990 °F) to produce specimens with densities ranging from 2.4 to >3.1 g/cm³ (0.087 to >0.11 lb/in.³).

Future Outlook

The development of silicon-base ceramics from polymer precursors provides unique approaches to processing ceramics. As progress is made in understanding the stability of amorphous phases produced from polymer chars and their transition to more stable crystalline phases, polymer-derived ceramics will have an increasing impact on the ceramic industry. Investigations of silicon-base systems should serve as a guide to developing many other non-oxide ceramic chemistries with unique structures and properties.

REFERENCES

1. M. Peuckert, T. Vaaks, and M. Bruck, Ceramics from Organometallic Polymers, *Adv. Mater.*, Vol 2 (No. 9), 1990, p 398–404
2. W. Verbeek and G. Winter, German patent 2,236,078, 1974
3. S. Yajima, J. Hayashi, and M. Omori, *Chem. Lett.*, 1975, p 931
4. S. Yajima, T. Iwai, T. Okamura, and Y. Hasegawa, *J. Mater. Sci.*, Vol 16, 1981, p 1349
5. R. West, L. Hench, and D. Ulrich, Ed.,
Ultrastructure Processing of Ceramics, Glasses and Composites, Wiley Interscience, 1984, p 235
6. C. Schilling, Polymeric Routes to Silicon Carbide, *Br. Polym. J.*, Vol 18 (No. 6), 1986, p 355–358
7. W. Verbeek, U.S. patent 3,853,566, 1974
8. J. Cannady, U.S. patent 4,535,007, 1985
9. D. Seyferth, G. Wiseman, L. Hench, and D. Ulrich, Ed., *Ultrastructure Processing of Ceramics, Glasses and Composites*, Wiley Interscience, 1984, p 265
10. R. Baney and G. Chandra, *Encyclopedia of Polymer Science and Engineering*, Vol 13, 2nd ed., 1988, John Wiley & Sons, p 312–344
11. S. Yajima, *Philos. Trans. R. Soc. (London) A*, Vol 294, 1980, p 419–426
12. K. Okamura, M. Sata, and Y. Hasegawa, Silicon Nitride and Silicon Oxynitride Fibers Obtained by the Nitridation of Polycarbosilane, *Ceram. Int.*, Vol 13, 1987, p 55–61
13. J. Rabe and C. Bujalski, U.S. patent 4,761,389
14. J. Lipowitz, H. Freeman, R. Chen, and E. Prack, *Adv. Ceram. Mater.*, Vol 2 (No. 2), 1987
15. G. LeGrow, T. Lim, J. Lipowitz, and R. Reaoch, *Am. Ceram. Soc. Bull.*, Vol 66, 1987, p 363
16. S. Yajima, *Am. Ceram. Soc. Bull.*, Vol 62, 1983, p 893
17. V. Murthy, M. Lewis, M. Smith, and R. Dupree, *Mater. Lett.*, Vol 8 (No. 8), 1989, p 263
18. D. Pysher, K. Goretta, R. Hodder, and R. Tressler, *J. Am. Ceram. Soc.*, Vol 72, 1989, p 284
19. R. Jones and J. Rabe, Dow Corning Corp., private communication
20. T. Mah, N. Hecht, D. McCullum, J. Hoenigmen, H. Kim, A. Katz, and H. Lipsitt, *J. Mater. Sci.*, Vol 19, 1984, p 1191
21. F. Chi and G. Stark, U.S. patent 4,460,639
22. F. Chi and G. Stark, U.S. patent 4,460,640
23. L. Haluska, U.S. patent 4,460,638
24. L.M. Niebylski, U.S. patent 4,873,353
25. Product literature for Nicalocoat, Dow Corning Corp., Midland, MI
26. K. Schwartz and D. Rowcliffe, *J. Am. Ceram. Soc.*, Vol 69 (No. 5), 1986, p C106–C108
27. S.R. Su, European patent application 0,305,759
28. W. Atwell, G. Burns, and C. Saha, U.S. patent 4,888,376
29. W. Atwell, G. Burns, and C. Saha, U.S. patent 4,929,573

Self-Propagating, High-Temperature Synthesis*

J.B. Holt, University of California, Lawrence Livermore National Laboratory

SELF-PROPAGATING, HIGH-TEMPERATURE SYNTHESIS (SHS) is potentially an energy-efficient process to synthesize many inorganic materials, including intermetallics, ceramics, and ceramic composites. In the production of some ceramic materials by the SHS process, the synthesis and densification steps may be combined into one operation. The process utilizes the energy of chemical reactions in a unique way. To understand the basic concept, we may consider a simple solid-state combustion experiment. Assume that titanium and carbon powders have been thoroughly mixed in stoichiometric proportions to form titanium carbide:

$$Ti + C = TiC \qquad \text{(Eq 1)}$$

The powdered mixture is cold pressed into a cylinder or poured into a refractory crucible. Next, the powder is placed into a chamber that may be evacuated or filled with an inert gas to prevent reaction with air. The top surface of the sample is ignited with an appropriate hot thermal source, such as a heated coil, electric match, or laser beam. From the point of ignition, a combustion wave rapidly self-propagates through the reactants and converts them into the titanium-carbide product (Fig 1). The calculated adiabatic temperature or maximum theoretical temperature that the product can reach for this reaction is 2930 °C (5305 °F). Another variation of solid combustion is thermal explosion (Ref 1) in which the entire sample is rapidly heated to the ignition temperature where it spontaneously reacts throughout the volume of the powder mixture without movement of a combustion front.

Process Characteristics

Characteristics of the process in the combustion-wave mode are self-generated high temperatures (800 to 3500 °C, or 1470 to 6330 °F), relatively rapid propagating combustion fronts (0.1 to 10 cm · s^{-1}, or 0.04 to 4 in. · s^{-1}), high rates of heating (up to 10^6 deg · s^{-1}) and thermal gradients (up to 10^7 deg · cm^{-1}) at the combustion front. These distinctive features exhibit the nonequilibrium conditions of the SHS process. The exact values of temperature, wave velocity, thermal gradients, and rate of heating are a function of the particular chemical system and experimental parameters. Both solid-solid and gas-solid combustion reactions are used to produce a variety of advanced technological materials.

Process Advantages

When compared to other methods of synthesis, the purported advantages of the SHS process are demonstrated by the Ti + C combustion reaction. High temperatures are reached without the use of external furnaces. The high temperatures cause near complete conversion of the reactants and volatilize many impurities at the combustion front in a manner similar to zone refining. Therefore, the SHS product is usually completely reacted and contains less impurities than the initial reactants. SHS reactions have been utilized to produce pure homogeneous preforms subsequent to their processing by other methods into the final product. For example, lithium niobate (LiNbO$_3$) is synthesized by the SHS process before its use as a seed material in the growth of single crystals (Ref 2). Another advantage of the SHS process is the capability to produce complex multicomponent products that are difficult or impossible to synthesize with ordinary methods. One may refer to the synthesis of titanium carbohydride (TiC$_x$H$_y$) as typical of this type of reaction (Ref 3). In certain chemical reactions the nonequilibrium conditions at the combustion front favor the formation of metastable compounds. An example of the synthesis of a metastable compound is the formation of cubic tantalum nitride (TaN) (Ref 4) which normally requires high nitrogen pressures to produce.

In addition, the nonequilibrium conditions are claimed to produce materials with high concentrations of defects that should affect mechanical properties. One such claim is the improved abrasive nature of SHS-formed TiC powders compared with those prepared by conventional methods (Ref 5). Another important advantage is the relative time of reaction. With SHS combustion, the reaction times are typically seconds, compared to minutes or hours with other methods of material synthesis. As energy costs for the processing of ceramic materials continue to rise, SHS processing should be considered as an alternative method of production. In making an economic assessment, however, the cost of producing the raw materials should be considered because energy has already been expended in producing the elemental reactants.

Historical Development of SHS Process

Various metals and alloys have been produced for decades by metallothermically reducing oxides (Ref 6, 7). In the late 1950s and early 1960s, Krapf (Ref 8), Abramovici (Ref 9), and Walton (Ref 10) published papers describing the use of exothermic reactions to form metal-ceramic composites. In 1967, Merzhanov and colleagues (Ref 11) at the Institute of Chemical Physics in the Soviet Union were searching for a gasless combustion reaction. They selected, for a trial experiment, the reaction of titanium and boron to form titanium diboride:

$$Ti + 2B = TiB_2 \qquad \text{(Eq 2)}$$

Upon ignition, a high-temperature combustion wave travels through the mixed powders producing titanium boride. Based on the results of that initial combustion reaction, experimental procedures were developed to synthesize many inorganic materials utilizing the chemical energy of the reactants. The process was designated self-propagating, high-temperature synthesis (SHS) by the Soviet scientists. Today, more than 300 compounds have been synthesized by SHS processing. The center of SHS Research in the Soviet Union

*Work was performed under the auspices of the U.S. Department of Energy by the Lawrence Livermore National Laboratory under Contract No. W-7405-ENG-48.

Fig 1 Combustion wave generated by the ignition of a cold-pressed mixture of titanium powder and carbon powder to form a titanium-carbon refractory compound (top portion of cylinder above combustion front). The mixture was ignited at the top portion of the cylinder and the wave front has proceeded approximately one-half the distance down the cylinder.

is the relatively new Institute of Structural Macrokinetics in Chernogolovka. At least eight other institutes are involved in SHS research.

Because of space limitations, we will refer only to a few of their numerous theoretical and experimental research reports. The Proceedings of the 1st International Symposium on Combustion and Plasma Synthesis of High Temperature Materials (Ref 12), two recent review articles (Ref 13, 14), and one compilation of Russian references (Ref 15) will assist those interested in the Russian approach and will also provide an assessment of their current activities.

In the early 1980s small research efforts were started in the United States, Japan, and Australia. Since then, research and development work has been initiated in South Korea, China, Germany, Brazil, Italy, and India. Some of the current research in the forementioned countries is reported in Ref 12. To avoid confusion, we should point out that in some places the use of exothermic reactions in the synthesis of materials is called combustion synthesis (CS) or reaction hot pressing (RHP), which are either synonymous with SHS processing or refer to a specific application. The number of patents dealing with the synthesis of ceramic products is rapidly growing and one interested in pursuing product development should consult the patent literature. Commercial products produced by the SHS process will be described in the section "Potential Applications" in this article.

Theory of Combustion

Solid-Solid Combustion. The fundamental principles of gaseous combustion were used in the initial efforts to model SHS reactions. Both systems are characterized by the movement of a combustion front. The general heat balance expression (Eq 3) is used to derive a formula describing a self-propagating reaction.

$$C_p \rho \cdot \partial T/\partial t = k \cdot \partial^2 T/\partial x^2 + Q\rho \cdot \Phi(T,\eta) \qquad \text{(Eq 3)}$$

where Φ and η are defined by:

$$\Phi(T,\eta) = \partial\eta/\partial t \qquad \text{(Eq 4)}$$

and C_p is the specific heat of the product; ρ is the density of the product; η is the fraction reacted; T is the temperature; t is the time; k is the thermal conductivity of the product; x is the axial distance; Q is the heat generated in the reaction; and Φ is the reaction rate (kinetic function). Equation 3 assumes no heat losses, that is, the sample has an essentially infinite diameter and the radial temperature distribution is uniform so that the combustion front is planar. Assuming a homogeneous reaction, then the kinetic function $\partial\eta/\partial t$ may be expressed as:

$$\partial\eta/\partial t = u \cdot \partial\eta/\partial x$$
$$= k_0 \cdot \exp(-E/RT) \cdot (1 - \eta)^n \qquad \text{(Eq 5)}$$

where k_0 is a preexponential constant; R is the universal gas constant; and n is the order of reaction. The homogeneous assumption is supported by the experimental observation that at least one of the reactants melts at the combustion front and by capillary action covers the particles of the other reactant. Therefore, there is a large area of contact between the reactants and thus a more homogeneous reaction.

By substitution of Eq 4 and Eq 5 into Eq 3 and assuming a narrow zone of reaction,

Khaikin and Merzhanov (Ref 16) derived the solution:

$$u^2 = \frac{[f(n)^a k_0 C_p RT^2]}{[QE \cdot \exp(-E/RT)]} \qquad \text{(Eq 6)}$$

where $f(n)$ is a function that depends upon the order of reaction, n, u is the velocity of the combustion front, and a is the thermal diffusivity of the product. We see that u is a function of the temperature T. The temperature T may be increased by raising the initial temperature T_0, or decreased by adding selected amounts of inert diluents to the reactants before combustion. By plotting the log of u/T versus $1/T$ the apparent activation energy, E, is calculated from the slope of the Arrhenius plot. The activation energy is compared to values of other processes known to occur in chemical reactions in order to deduce the controlling mechanism. The results of several studies (Ref 17, 18) suggest that a solution-precipitation mechanism is the rate-controlling step in many combustion reactions rather than a diffusion mechanism through a solid product layer.

The assumption of a narrow reaction zone applies only in a limited number of cases. Most combustion reactions occur with a broad reaction zone or afterburn region as shown in Fig 2. The afterburn region is the result of chemical reactions still proceeding after the passage of the combustion front. Because the concept of afterburn is not a part of gaseous combustion theory, then a new approach was required to model combustion of solid reactants. The SHS process is a combination of combustion with a restructuring of the products including nucleation of grains, grain growth, sintering, and so on under extreme nonequilibrium conditions. Taking the afterburn region into account, a formula was derived for rates of combustion:

$$u^* = A(T^*\eta^*) \cdot \exp(-E/RT) \qquad \text{(Eq 7)}$$

where A is a constant; T^* is a value of temperature below the adiabatic temperature, T_{ad}; and η^*, the fraction reacted, is less than 1.

Gas-Solid Combustion. Combustion of solids in active gases such as hydrogen and nitrogen to form hydrides and nitrides, respectively, is an entirely different type of reaction from that discussed in the section "Solid-Solid Combustion." At 0.1 MPa (15 psi) pressure, there is not enough gas within the pore space of a powder compact (40 to 60% porosity) for complete conversion. Therefore, a pressure drop exists from the combustion front to the outside surface of the powder sample. The permeation or filtration of gas through the pore space is a critical experimental parameter.

One method to circumvent the problems of permeation is the use of high gas pressures. For some nitrides, the gas pressure needed for full conversion is on the order of 100 MPa (15 ksi). However, high-pressure gas brings its own problems of high heat loss and difficult ignition. An important factor in gas-solid

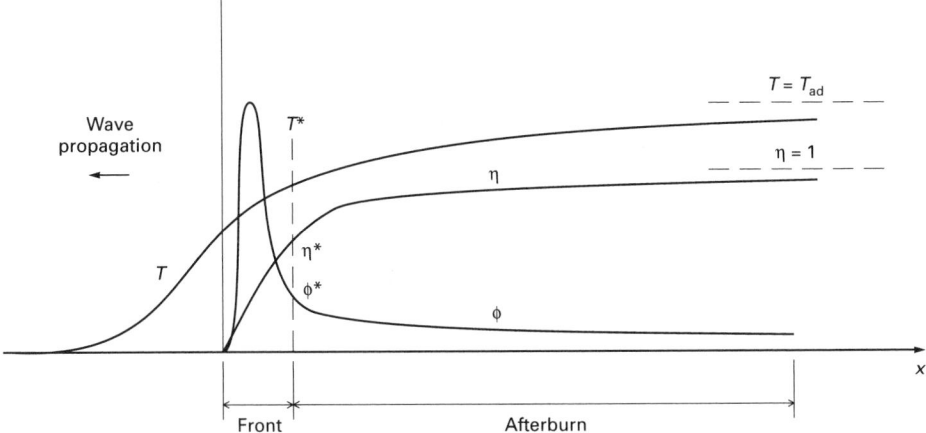

Fig 2 Temperature profile of a combustion front moving from right to left. T is the temperature; ϕ, the rate of heating; T_{ad} is the adiabatic temperature; and η is the fraction reacted. Note that ϕ does not decrease to 0, but extends a long distance after the front; $\eta < 1$; and T is below the adiabatic temperature.

combustion is the dissociation pressure of the product phase as a function of temperature. If the dissociation pressure of the product at the combustion temperature is higher than the ambient pressure, then the solid reactant will not ignite.

The modeling of filtration combustion is a new scientific endeavor and entails greater complexity than the treatment of solid-solid combustion. Filtration combustion is also accompanied by afterburn that contributes to the difficulty in the development of models. Soviet scientists have been the most active in the study of filtration combustion (Ref 19, 20).

Combustion Modes. Up to this point, only stable combustion, in which a planar front moves with a constant velocity, has been considered. However, it should be noted that there are combustion reactions which occur with unstable modes such as oscillatory and spin. The criterion for delineation of combustion modes, α, is given by:

$$\alpha = \left(\frac{9.1 C_p R T^2}{QE}\right)\left(1 - 0.27Q}{RT}\right) \qquad \text{(Eq 8)}$$

where α is the thermal diffusivity of the product. When $\alpha \geq 1$, then the reaction proceeds in a stable mode. When $\alpha < 1$, oscillatory, spin, or other unstable modes predominate. As α becomes more negative, the oscillations or spin are more intense.

The mode of combustion affects the structure of the product. Generally, unstable combustion reactions cause incomplete conversion and affect the macrostructure of the product. For example, in the oscillatory mode, the product is composed of layers whose thickness corresponds to the frequency of oscillation. If possible, stable combustion reactions are the preferred mode for the synthesis of high-quality products.

Experimental Procedures

Thermodynamic Considerations. High calorific content of a chemical system is nec-

essary for the occurrence of a self-propagating combustion reaction. An exact threshold value of calorific content or exothermicity has not been defined. However, calculated adiabatic temperatures serve as a general guide for the probability of a particular chemical reaction occurring in the combustion regime.

For refractory products such as borides, carbides, and nitrides, it has been suggested (Ref 21) that the adiabatic temperature must be at least 1800 to 2000 °C (3270 to 3630 °F). With the appropriate thermodynamic data, the calculation of an adiabatic temperature for any reaction is a relatively simple operation. Novikov *et al.* (Ref 22) and Holt and Munir (Ref 23) explain the method of calculation and Ref 22 lists the thermodynamic data for many important reactions. Computer programs (Ref 24) are available to simplify the calculations.

Table 1 is a list of common SHS reactions, ceramic products, and their adiabatic temperatures. When considering synthesis of a ceramic product by the SHS process, the first step should be the calculation of the adiabatic temperature. It should be pointed out, however, that lower melting compounds have also

been synthesized by the SHS process. Therefore, we suggest that the likelihood of a self-propagating reaction should be determined by experiment.

Facilities. The synthesis of ceramic products does not require elaborate equipment and facilities. In solid-solid combustion reactions, the thorough blending of the reactant powders is necessary for high-quality products. Dry ball-milling is not recommended as a method of mixing because sufficient heat may be generated to cause an explosion. Adsorbed gases on the as-received reactant powders will evolve during combustion. It may be necessary to remove these gases through either thermal or chemical means before reaction. The mixed powders are placed into a refractory crucible whose design and construction depend on the maximum temperature and amount of heat generated. The mixture may be ignited by pyrofuse, an electric match, a laser beam, or any other suitable source of intense energy. If an oxide contaminant is not a problem, then combustion may take place in air. Otherwise, it may be ignited in either a vacuum or an inert gas.

The gas-solid combustion reactions take place in combustion chambers. If high gas pressures are required, commercial units are available where the maximum pressure is $\geq$100 MPa (15 ksi). Some nitrides, such as Si_3N_4, have been synthesized at pressures of 10 to 20 MPa (1.5 to 3 ksi) with the aid of inert diluents (Ref 25). Titanium nitride, hafnium nitride, and zirconium nitride have been formed at 0.1 MPa (15 psi) using solid sources of nitrogen, such as NaN_3 (Ref 26). The sodium is volatilized during reaction and is not part of the nitride product. Overall, the synthesis of nitrides and hydrides requires more extensive facilities than those used for the solid-solid combustion reactions.

Experimental Parameters. Self-propagating, high-temperature synthesis reactions may be controlled and the microstructure of the product affected by experimental parameters such as sample porosity, average grain size of the reactant powders, grain size dis-

Table 1 Ceramic products obtained from SHS process reactions and the calculated adiabatic temperatures

Reaction components	Product(s)	Adiabatic temperature	
		°C	°F
Ti + C	TiC	2930	5305
4B + C	B$_4$C	730(b)	1345(b)
Ta + C	TaC	2430	4405
Ti + 2B	TiB$_2$	2920	5290
Zr + 2B	ZrB$_2$	3040	5505
Al + $^1/_2$N$_2$	AlN	2630	4765
3Si + 2N$_2$	Si$_3$N$_4$	4030(c)	7285(c)
B + $^1/_2$N$_2$	BN	3430	6205
Mo + 2Si	MoSi$_2$	1630	2965
SiO$_2$ + 2Mg + C	SiC + 2MgO	2340	4245
WO$_3$ + 2Al + C	WC + Al$_2$O$_3$	3800	6870
Zn + H$_2$	ZnH$_2$	950	1740
0.5Y$_2$O$_3$ + 2BaO$_2$ + 3Cu + xO$_2$	YBa$_2$Cu$_3$O$_{5.5+x}$(a)	...	...

(a) Superconducting oxide. (b) Will not support combustion. (c) Actual combustion temperature is much lower.

tribution, temperature, chemical composition, and additives. A careful study of the combustion reaction as a function of these parameters should reveal the best method of synthesizing the powders. Two drawbacks of the SHS process are that the use of expensive metal powders in some reactions and the exothermicity of certain reactions is not high enough to support combustion (for example, the formation of B_4C from the elements is not possible in a combustion regime). However, B_4C may be synthesized using the thermite-type reaction:

$$2B_2O_3 + 6Mg + C = B_4C + 6MgO \qquad \text{(Eq 9)}$$

The magnesium oxide must be removed by leaching with HCl or other acids, which is an additional step that will increase the cost of production. The B_4C synthesized by the SHS process is a high-quality submicron powder.

Properties and Microstructures

Cast Parts. If the adiabatic temperature for a SHS reaction is higher than the melting temperature of the product(s), then the product(s) will be molten during reaction. Soviet researchers have carried out this type of combustion in a centrifuge to produce a dense cast product (Ref 27). For example, the reaction to synthesize Mo_2C is:

$$2MoO_3 + 4Al + C = Mo_2C + 2Al_2O_3 \qquad \text{(Eq 10)}$$

The calculated adiabatic temperature for this reaction is 4927 °C (8901 °F), which is higher than the melting temperatures of both product phases. The centrifugal force causes the molten products to segregate and on cool down they can be mechanically separated from each other. The Mo_2C product had a microhardness of 20 to 25 GPa (2000 to 2500 kgf/mm^2) and contained up to 0.05 wt% O_2. Odawara (Ref 28) used a similar thermite-type process to internally coat iron pipes with a layer of aluminum oxide.

Powders, or loosely bonded porous (50 to 60%) solids are the product of SHS reactions when the adiabatic temperature is below the melting point of the product(s). The porous solids are caused by differences in the molar volume of reactants compared to that of the products, the evolution of adsorbed gases, and porosity in the initial reactant compact. In some cases, the powder product can be used as abrasive powders or in sintering applications without any further modification. Usually, the powder must be ground to attain the desired particle size. Whether or not the thermal history of the combustion process generally produces powders with enhanced physical properties due to high concentrations of defects is debatable.

Dense Products. A more energy-efficient process would combine the synthesis and product consolidation steps into one operation (Ref 23). To achieve consolidation of the combustion product, pressure is applied either during synthesis or immediately after reaction while the product is still at a high temperature. If pressure is applied during synthesis, the evolution of adsorbed gases causes problems of containment. When pressure is applied after reaction, the lag time is a critical parameter. Different techniques have been used to apply pressure. Explosive compaction has been used to densify titanium carbide and titanium boride (Ref 29). A relatively low pressure of 10 MPa (1.5 ksi) in uniaxial dies densified TiC-NiAl composites (Ref 30). Gas-pressure sintering is a process that is used to densify titanium carbide, titanium carbide plus nickel, and titanium boride (Ref 31). Dense AlN (92%) was produced at 100 MPa (15 ksi) N_2 (Ref 32). Isostatic pressure of an inert gas was used to densify a plasma-formed Ti-2B reactant compact (Ref 33). Near-net shape specialty items may be the initial product of the simultaneous synthesis and consolidation process. Further research is necessary to develop procedures for the production of commercial components.

Potential Applications

Products of the SHS process are expected to have a wide variety of applications. Powders of TiC, TiB_2, B_4C, AlN, Si_3N_4, hexagonal BN, and TiN are commercially available as abrasives and for hot pressing or sintering. Recently, Soviet scientists announced the synthesis of superconducting oxides, including $YBa_2Cu_3O_{7-x}$ (Ref 34). Additional thermal treatment was not required for formation of the orthorhombic phase. Independent measurements have determined that the superconducting powder is a high-quality material with a sharp transition temperature (Ref 35).

Functionally graded materials (FGM) are an intriguing class of products that can be synthesized by SHS reactions. Sata (Ref 36) was one of the first to use the SHS process in the preparation of a TiC + Ni FGM. A FGM composite varies continuously in composition from 100% ceramic at one surface to a metal at the opposite surface. Because of the gradient structure, the physical and mechanical properties can be significantly enhanced compared to a typical metal-ceramic composite.

Large-area (300 mm, or 12 in., diameter) hard-alloy plates, composed of either TiC or TiB_2 and a metal binder phase, have been densified in a uniaxial press during combustion (Ref 37). The production of such large-area components is a relatively easy task with the SHS process. Another interesting application of the SHS process is the production of $MoSi_2$ furnace rods that are extruded immediately after the combustion reaction. Research is being carried out to develop methods for the SHS process to weld ceramics or metals together (Ref 37) and the formation of thin ceramic coatings on refractory materials (Ref 37).

This discussion is a brief introduction to applications of the SHS process in the synthesis of ceramic materials. It is not intended to be a complete survey. When the SHS process is considered as a possible method for the synthesis of ceramic materials, then the cost and quality of products should be judged on the basis of the specific application. Certain SHS reactions may offer real advantages for specific applications, while other applications may be less effective. Therefore each individual SHS reaction should be considered on its own merits.

REFERENCES

1. V.I. Itin, A.D. Bratchikov, and L.N. Postnikova, Use of Combustion and Thermal Explosion for the Synthesis of Intermetallic Compounds and Their Alloys, *Sov. Powd. Metall. Met. Ceram.*, Vol 19 (No. 5), 1980, p 315–317
2. V.V. Boldyrev, N.P. Novikov, V.V. Aleksandrov, and V.I. Smirnov, Self-Propagating High-Temperature Synthesis Processes with Participation of Oxygen-Containing Compounds, *Fiz. Khim. Okislov. Met.*, 1981, p 115–126
3. N.A. Martirosyan, S.K. Dolukhanyan, and A.G. Merzhanov, Critical Phenomena in Combustion of Mixtures of the Type $A_s + B_s + C_g$ (Example of the Titanium-Carbon-Hydrogen System), *Comb. Explos. Shock Waves*, Vol 167 (No. 4), 1981, p 369–374
4. I.P. Borovinskaya, A.G. Merzhanov, N.P. Novikov, and A.K. Filonenko, Self-Propagating High Temperature Synthesis of Tantalum Nitrides, *Comb. Process in Chem. Tech. and Metall.*, Chernogolovka, 1975, p 141–146
5. A.G. Merzhanov, G.G. Karynk, I.P. Borovinskaya, S.Y. Sharivker, E.I. Moshkovskii, V.K. Prokudina, and E.G. Dyadko, Titanium Carbide Produced by Self-Propagating High-Temperature Synthesis: Valuable Abrasive Material, *Sov. Powd. Metall. Met. Ceram.*, Vol 20 (No. 10), 1981, p 709–713
6. R.A. Cutler, A.V. Virkar, and J.B. Holt, Synthesis and Densification of Oxide-Carbide Composites, *Ceram. Eng. Sci. Proc.*, Vol 6 (No. 7–8), 1985, p 715–728
7. V. Hlavacek, A Combustion Synthesis: A Historical Perspective, *Bull. Am. Ceram. Soc.*, Vol 70 (No. 2), 1991, p 240–243
8. S. Krapf, Process Using Heat of Reaction, *Ber. Deut. Keram. Ges.*, Vol 31 (No. 1), 1954, p 18–21
9. R. Abramovici and M. Enache, Alumino-Thermal Production and Study of Mixed Ceramic Bodies in the System Al_2O_3-$MoSi_2$, *Keram. Zeits.*, Vol 19 (No. 11), 1967, p 699–702

10. J.D. Walton, Jr. and N.E. Poulos, Cermets from Thermite Reactions, *J. Am. Ceram. Soc.*, Vol 42 (No. 1), 1959, p 40–49

11. A.G. Merzhanov and I.P. Borovinskaya, Self-Propagated High Temperature, *Dokl. Akad. Nauk. SSSR*, Vol 204 (No. 2), 1972, p 429–432

12. *Combustion and Plasma Synthesis of High Temperature Materials*, Z.A. Munir and J.B. Holt, Ed., VCH Publishers, 1990

13. Z.A. Munir, Synthesis of High Temperature Materials by Self-Propagating Combustion Methods, *Bull. Am. Ceram. Soc.*, Vol 67 (No. 2), 1988, p 342–349

14. J.B. Holt, The Use of Exothermic Reactions in the Synthesis and Densification of Ceramic Materials, *Mat. Res. Soc. Bull.*, Vol 12, 1987, p 60–64

15. W.L. Frankhauser, M.G. Kieszek, and K.W. Brendley, *Gasless Combustion of Refractory Compounds*, Noyes Publications, 1985, p 152

16. B.I. Khaikin and A.G. Merzhanov, Theory of Thermal Propagating of a Chemical Reaction Front, *Comb. Explos. Shock Wave*, Vol 2 (No. 3), 1966, p 22–27

17. V.V. Aleksandrov and M.A. Korchagin, Mechanism and Macrokinetics of Reactions Accompanying the Combustion of SHS System, *Comb. Explos. Shock Wave.*, Vol 23 (No. 5), 1987, p 557–564

18. A.N. Tabchenko, T.A. Panteleeva, and V.I. Itin, Interaction of Titanium and Carbon in the Presence of a Solution-Melt, *Comb. Explos. Shock Wave.*, Vol 20 (No. 4), 1984, p 387–391

19. A.P. Aldushin, B.S. Seplyarskii, and K.G. Shkandinskii, Theory of Filtration Combustion, *Comb. Explos. Shock Wave*, Vol 16 (No. 1), 1980, p 33–40

20. S.D. Dunmead, Z.A. Munir, and J.B. Holt, Gas-Solid Reactions under a Self-Propagating Combustion Mode, *Solid State Ionics*, Vol 32/33, 1989, p 474–481

21. A.G. Merzhanov, Theory of Gasless Combustion, *Arch. Procesow Splania*, Vol 5 (No. 1), 1974, p 17–39

22. N.P. Novikov, I.P. Borovinskaya, and A.G. Merzhanov, Thermodynamic Analysis of Self-Propagating High-Temperature Synthesis Reactions, *Comb. Processes Chem. Eng. and Metall.*, Chernogolovka, 1975, p 174–188

23. J.B. Holt and Z.A. Munir, Combustion Synthesis of Titanium Carbide: Theory and Experiment, *J. Mater. Sci.*, Vol 21, 1986, p 251–259

24. B. Rapp, personal communication, Lawrence Livermore National Laboratory, Dec 1990

25. K. Hirao, Y. Miyamoto, and M. Koizumi, Synthesis of Silicon Nitride by a Combustion Reaction under High Nitrogen Pressure, *J. Am. Ceram. Soc.*, Vol 69 (No. 4), 1986, p C60-C61

26. J.B. Holt, A New Process for the Synthesis of Refractory Nitride Materials, *J. Ind. Res. Dev.*, April 1983, p 86–98

27. A.G. Merzhanov, V.I. Yukhuid, and I.P. Borovinskaya, Self-Propagating High-Temperature Synthesis of Cast Inorganic Refractory Compounds, *Dokl. Acad. Sci. SSSR Chem.*, Vol 255 (No. 1), 1981, p 503–506

28. O. Odawara and J. Ikeuchi, Ceramic Composite Pipes Produced by Centrifugal-Exothermic Process, *J. Am. Ceram. Soc.*, Vol 69 (No. 4), 1986, p C80-C81

29. A. Niiler, L.J. Kecskes, and T. Kottke, Shock Consolidation of Combustion-Synthesized Ceramics, *Combustion and Plasma Synthesis of High Temperature Materials*, Z.A. Munir and J.B. Holt, Ed., VCH Publishers, 1990, p 309–315

30. S.D. Dunmead, Z.A. Munir, J.B. Holt, and D.D. Kingman, Combustion Synthesis in the Ti-C-Ni-Al System, *Combustion and Plasma Synthesis of High Temperature Materials*, Z.A. Munir and J.B. Holt, Ed., VCH Publishers, 1990, p 229–237

31. Y. Miyamoto and M. Koizumi, Simultaneous Synthesis and Densification of Ceramic Components under Gas Pressure by SHS, *Combustion and Plasma Synthesis of High Temperature Materials*, Z.A. Munir and J.B. Holt, Ed., VCH Publishers, 1990, p 163–170

32. S.D. Dunmead, J.B. Holt, and D.D. Kingman, Simultaneous Combustion Synthesis and Densification of AlN, *Combustion and Plasma Synthesis of High Temperature Materials*, Z.A. Munir and J.B. Holt, Ed., VCH Publishers, 1990, p 186–194

33. J.B. Holt and M.D. Kelly, "Processes for Forming Exoergic Structures with the Use of a Plasma and for Producing Dense Refractory Bodies of Arbitrary Shape Therefrom," U.S. patent 4,923,241, June 1990

34. A.G. Merzhanov, Self-Propagating High-Temperature Synthesis of Ceramic (Oxide) Superconductors, *Ceram. Trans.*, Vol 13, 1990, p 519

35. H.B. Radousky, personal communication, Lawrence Livermore National Laboratory, Jan 1991

36. N. Sata, N. Sanada, T. Hirano, and M. Niino, Fabrication of a Functionally Gradient Material by Using a Self-Propagating Reaction Process, *Combustion and Plasma Synthesis of High Temperature Materials*, Z.A. Munir and J.B. Holt, Ed., VCH Publishers, 1990, p 195–204

37. A.G. Merzhanov, Self-Propagating High-Temperature Synthesis: Twenty Years of Search and Findings, *Combustion and Plasma Synthesis of High Temperature Materials*, Z.A. Munir and J.B. Holt, Ed., VCH Publishers, 1990, p 1–54

Directed Metal Oxidation

A.W. Urquhart, Lanxide Corporation

DIRECTED METAL OXIDATION technologies for ceramic composites will be reviewed in this article. A key feature of these technologies is the flexibility to engineer the properties of the composites to meet the needs of a wide range of applications. Thus, for example, the reinforcement, the matrix, and the microstructural features of the composite can be varied quite widely and controlled to provide the needed performance characteristics. The processing methods allow the economical production of finished components over a wide range of sizes in complex configurations and to net or near-net shape.

Initial work in this area began in 1983. First sales of products of these technologies occurred in 1989 primarily for armor and wear part applications. Sales of components to be used or to be evaluated for use in gas turbine engines, rocket engines, piston engines, heat exchangers, high-temperature furnaces, and other applications have also begun.

What follows is a brief description of the processing methods, examples of some of the composites produced and their properties, and an indication of applications being addressed. These technologies are still in an early stage of development and new composite systems, as well as further improvements in the properties of existing composites can be anticipated.

Composite Processing

The composite formation process uses accelerated oxidation reactions of molten metals to grow ceramic matrices around preplaced filler or reinforcement materials (Ref 1, 2). The simplified schematic of Fig 1 illustrates the process steps.

Typically, the reinforcement materials are shaped into a preform of the size and shape desired in the final component (Ref 3). For a particulate reinforcement, this step might utilize any of the ceramic green body forming processes (for example, pressing, isostatic pressing, slip casting, extrusion, injection molding, and so forth). Alternatively, cloth layup, braiding, weaving, or filament winding might be used for a fiber reinforcement. A growth barrier material is placed on the preform surfaces to stop the matrix growth process and preserve the preform shape in the final composite body (Ref 3).

The preform and the parent metal alloy are then heated to the growth temperature where a rapid oxidation process occurs outward from the metal surface and into the preform, such that the oxidation reaction product becomes the matrix surrounding the reinforcing material in the preform. After the growth process is complete, the setup is cooled, and the composite article is recovered. Although Fig 1 shows matrix growth occurring unidirectionally, other geometric arrangements are used to make a wide variety of complex shapes.

Growth of the matrix into a preform typically involves little or no change in dimensions. Thus, problems associated with densification shrinkage in traditional ceramic processing are avoided. As a result, it becomes possible to make very large components quite readily, as well as to maintain improved control of component dimensions.

Because of the relatively inexpensive starting materials and modest processing condi-

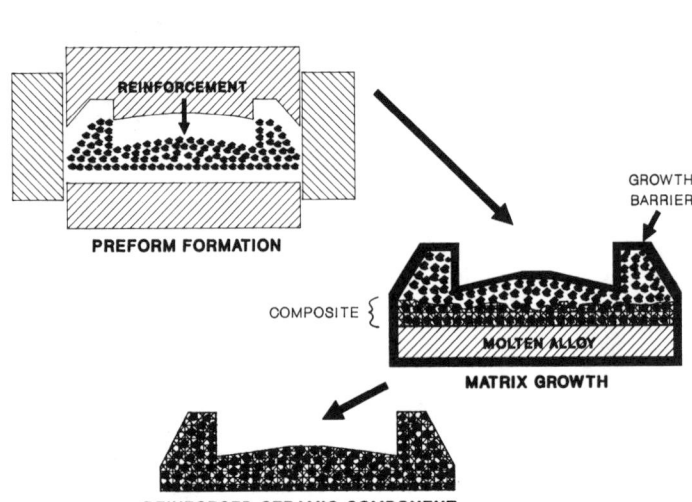

Fig 1 Schematic of DIMOX™ directed metal oxidation process for making a shaped ceramic-matrix composite component. Courtesy of Lanxide Corporation

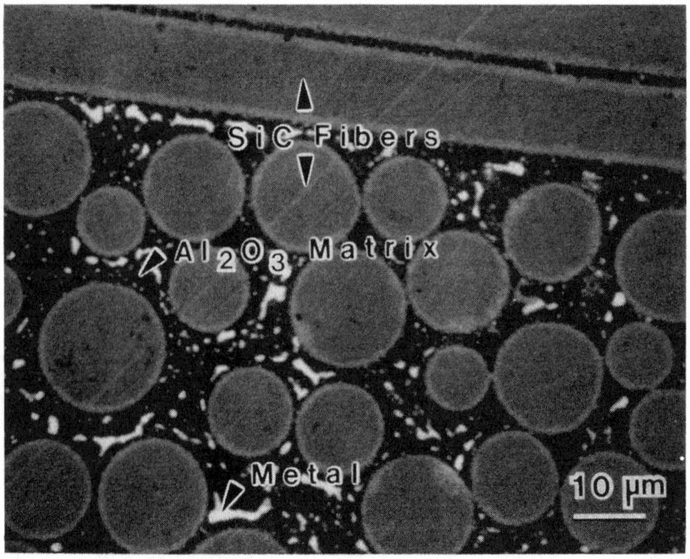

Fig 2 Optical micrograph of an Al₂O₃/Al matrix system reinforced with woven Nicalon™ fibers. Impurities in the uncoated fibers tend to react with the molten alloy during matrix growth, which leads to the reaction layer at the fiber surfaces. Source: Ref 2

Table 1 Examples of directed metal oxidation ceramic-matrix systems

Parent metal	Reaction product
Aluminum	Oxide, nitride, boride, titanate
Silicon	Nitride, boride, carbide
Titanium	Nitride, boride, carbide
Zirconium	Nitride, boride, carbide
Hafnium	Boride, carbide
Tin	Oxide
Lanthanum	Boride

Source: Ref 2, 3

tions used (especially for aluminum-based systems, as described in Ref 2), particle-reinforced composites can be produced quite economically. Fiber-reinforced composites tend to be more expensive because of the cost of the reinforcement.

The directed metal oxidation technology has been used to produce a large number of composite matrices, as listed in Table 1 (Ref 2, 3). Typically, the reaction involves a molten metal and a gaseous oxidant; however, the directed reaction of molten zirconium with solid B_4C has been used to make a zirconium diboride platelet-reinforced zirconium carbide composite, as described below.

Composite Examples and Applications

SiC Fiber-Reinforced Al_2O_3 Matrix Composites. Fiber-reinforced Al_2O_3 matrix composites are being evaluated for applications requiring excellent mechanical properties at high temperatures. As described in Ref 4, such composites have been made with a two-dimensional reinforcement by stacking layers of woven Nicalon™ fiber cloth to form the preform and growing the Al_2O_3 matrix

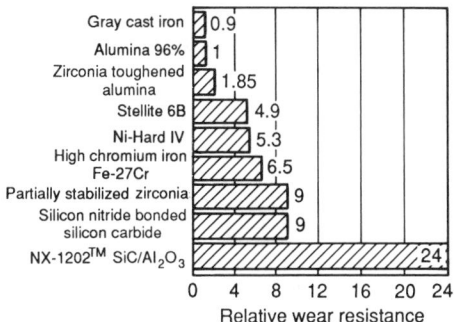

Fig 3 Relative slurry erosion wear performance of several materials. The NX-1202™ SiC/Al_2O_3 material was made using the directed metal oxidation process. This evaluation compares measured material losses in a slurry pot test in which 14 mm (0.54 in.) diam by 60 mm (2.4 in.) long test pins were rotated at 10 m/s (33 ft/s) for 20 h in a slurry of 40 wt% 300 to 600 μm SiO_2 particles in neutral water. Courtesy of Lanxide Corporation

Fig 4 Examples of components produced for use in applications requiring resistance to slurry erosion. The composite contains SiC particles in an Al_2O_3 matrix. The pump housing is approximately 640 mm (25 in.) in diameter. Large wear rings are currently being produced up to 1.2 m (48 in.) in diameter. Courtesy of Alanx Products

around the fibers. Nicalon fiber consists of ultrafine SiC crystals in an amorphous Si—C—O matrix.

The microstructure of such a composite with uncoated fibers (Fig 2) demonstrates that the matrix grows well into the interstices of the fiber bundles (Ref 2). Figure 2 also illustrates a characteristic feature of these ceramic composites, namely that the matrix typically contains some metal, which is associated with the wicking of the molten parent metal through the matrix to sustain the oxidation reaction. For high-temperature applications, the residual metal may either be removed to leave behind a small amount of porosity or treated to prevent further oxidation during use (Ref 4).

In practice, the Nicalon fibers are given a proprietary coating prior to matrix formation. The coating protects the fibers during the oxidation reaction and provides for debonding and fiber pullout during fracture for increased toughness (resistance to crack propagation). The mechanical properties of one such composite are summarized in Table 2. Good

strength and excellent toughness values are obtained over a wide range of temperatures through 1400 °C (2550 °F) (Ref 4, 5). The observed high-toughness values result from extensive fiber pullout that yields a "graceful" (non-catastrophic) failure mode.

Preliminary results indicate that these composites show good resistance to mechanical properties degradation during prolonged elevated-temperature exposure. For example, if the composite is held in air at the 1200 °C (2200 °F) test temperature prior to testing, its strength decreases to 400 MPa (58 ksi) but apparently stabilizes at this value, since there is no further loss in strength on increasing the hold time from 24 to 100 h. Longer time and higher temperature exposure studies are continuing.

The thermal shock resistance of these fiber-reinforced composite materials is particularly impressive (Ref 4). For example, specimens heated to 1000 and 1200 °C (1830 and 2200 °F) and rapidly quenched in room-temperature water suffered only minor damage (as measured by strength losses of only 7 and

Table 2 Mechanical properties of a two-dimensional Nicalon™ fiber-reinforced Al_2O_3 matrix composite

Test temperature		4-point flexural strength			Chevron notch toughness	
°C	°F	MPa	ksi	Weibull Modulus	MPa√m	ksi√in.
21	70	461 ± 28	67	15.5	27.8	25.3
1200	2190	488 ± 22	71	20.6	23.3	21.2
1300	2370	400 ± 12	58		19.2	17.5
1400	2550	340 ± 11	49		15.6	14.2

Source: Ref 4, 5

16%, respectively). Even after the quench cycle was repeated five times from 1000 °C (1830 °F) the strength loss was only 17%. By contrast, monolithic ceramics like Al_2O_3 and SiC are typically badly damaged by quenching from temperatures as low as 200 to 400 °C (400 to 750 °F).

This SiC fiber-reinforced Al_2O_3 composite is currently being evaluated for component applications in gas turbine engines. Examples of such components include flameholders, turbine shrouds, exit vanes, and combustor parts.

SiC Particle-Reinforced Al_2O_3 Matrix Composites. An Al_2O_3 matrix composite reinforced with coarse SiC particles has been optimized for applications requiring slurry erosion resistance (Ref 6, 7). Processing in this case is similar to that of the fiber-reinforced composite except that ceramic pow-

der-forming methods are used to obtain a shaped preform prior to matrix growth. Comparative wear test results are given in Fig 3. Materials of this type are being used successfully in a range of applications including slurry pump components, hydrocyclone liners, chute liners, and material handling systems. Figure 4 shows several examples of actual wear components that are available commercially. A similar composite has been commercialized for armor applications.

Another composite also using SiC particles in an Al_2O_3 matrix has been developed for high-temperature applications, such as furnace and heat exchanger components (Ref 8). In this case, the SiC particles are much finer and the growth alloy and processing conditions are selected to optimize high-temperature stability and mechanical characteristics. Elevated-temperature flexural strengths for

three such composites are given in Fig 5. Tests on similar composites have shown that these high-temperature strengths are unaffected by holding at temperatures at least up to 1500 °C (2730 °F) for a thousand hours or more. Prototype heat exchanger tubes and furnace components have been tested successfully, including a rig test of a burner tube involving temperature cycling to 1530 °C (2790 °F). Figure 6 shows examples of heat exchanger and furnace components made of this composite.

Another variant of a SiC particle filled Al_2O_3 matrix composite has been optimized for resistance to sliding and rolling contact wear (Ref 9). This composite uses fine SiC particles with modifications to the residual metal in the Al_2O_3 ceramic matrix. Successful prototype test results have been obtained for piston engine cam follower rollers made of this material.

ZrB_2 Platelet Reinforced ZrC Composites. To form this composite, molten zirconium is directionally reacted with B_4C powder in a graphite mold at temperatures of about 1850 to 2000 °C (3350 to 3630 °F) (Ref 10). The B_4C is completely reacted to form the two ceramic phases ZrB_2 and ZrC. The ZrB_2 takes the form of hexagonal platelets in a ZrC matrix containing controlled amounts of free zirconium, as illustrated by the microstructure shown in Fig 7.

Typical room-temperature mechanical properties for composites in the 3 to 12 vol% Zr range include flexural strengths of 800 to 900 MPa (116 to 130 ksi), Weibull moduli of 20 to 30, and fracture toughnesses (chevron notch method) of 12 to 15 MPa$\sqrt{m}$ (11 to 14 ksi$\sqrt{in.}$) (Ref 11). The fracture toughness values depend on the metal content. Toughness values of 10 to 12 MPa$\sqrt{m}$ (9 to 11 ksi$\sqrt{in.}$) at very low metal contents result from platelet toughening mechanisms (that is, crack deflection and crack bridging/platelet pullout). As the metal content increases, the toughness increases due to an additional contribution from ductile rupturing of metal ligaments during crack propagation.

This composite has been successfully tested in very high temperature (>2700 °C, or 4900 °F), short time applications in rocket engines. These results derive from the highly refractory nature of the ZrB_2 and ZrC phases and the excellent thermal shock resistance of the composite. Longer term structural applications at high temperatures may not be possible because of the limited resistance of these materials to high-temperature oxidation and creep.

This composite is also being evaluated for applications in prosthetic devices based on its strength, fracture toughness, wear resistance, and biocompatibility. Mechanical seals and other special wear parts are additional promising application areas for these materials.

Other ceramic composite systems based on the directed metal oxidation technology are

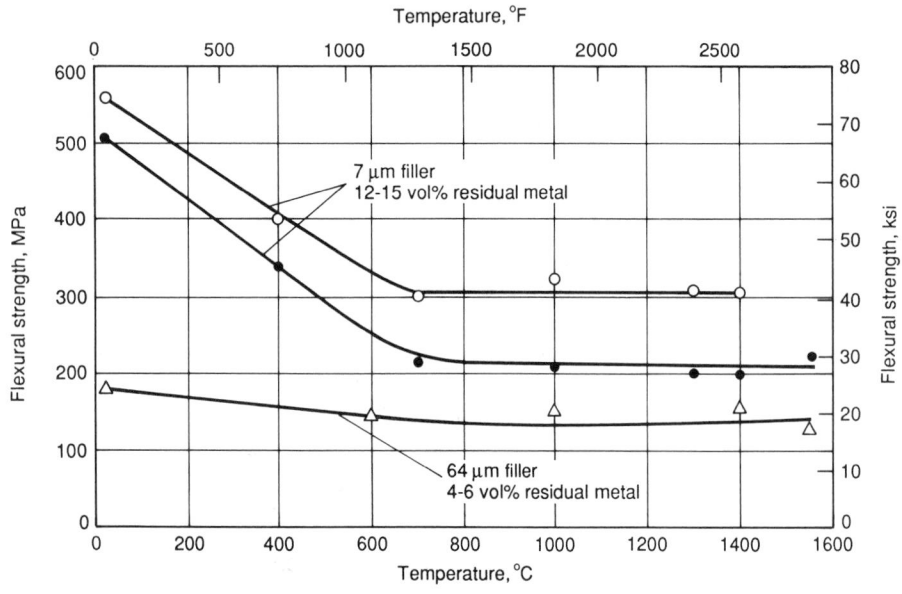

Fig 5 Four-point flexural strengths of three SiC particle-filled Al_2O_3 matrix composites over a range of temperatures. The upper two curves represent process differences using the same nominal filler particle size. Source: Ref 8

Fig 6 Heat exchanger and furnace components made from a SiC particle-filled Al_2O_3 matrix composite. Courtesy of Du Pont Lanxide Composites Inc.

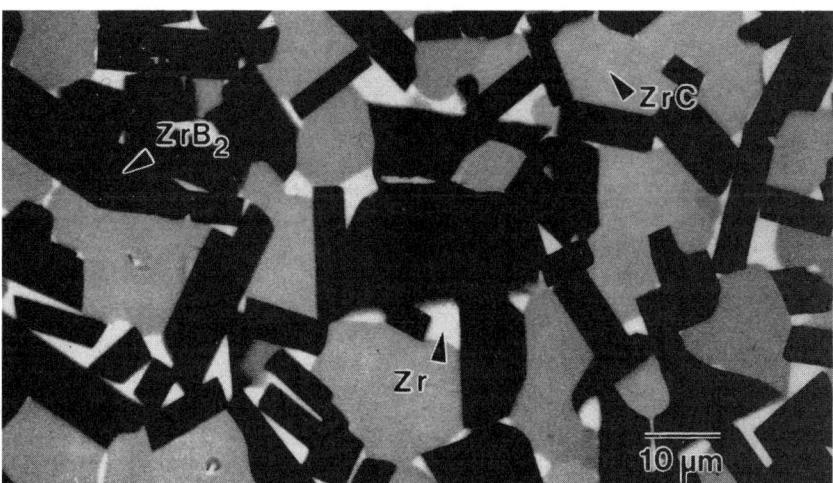

Fig 7 Microstructure of a ZrB₂ platelet-reinforced composite with a ZrC matrix that also contains controlled amounts of zirconium metal. Courtesy of Lanxide Corporation

in various stages of development. Additional examples of composites being commercialized include a particle-filled AlN matrix composite for ballistic applications, an Al_2O_3 particle-filled Al_2TiO_5 composite for applications involving thermal insulation and thermal shock resistance, and a carbide-based coating system for graphite for a wide range of possible uses. Laboratory development work is underway on a Si_3N_4-matrix system. Methods of combining more than one composite in a single component are also being developed, with a particular focus on combining ceramic-matrix composites with metal-matrix composites.

ACKNOWLEDGMENT

Research and development work leading to the technologies described in this article were partially supported by the U.S. Defense Advanced Research Projects Agency through the Office of Naval Research. Portions of this article were previously published in the journal *Materials Science and Engineering* and are reprinted by permission.

REFERENCES

1. M.S. Newkirk, A.W. Urquhart, and H.R. Zwicker, Formation of Lanxide™ Ceramic Composite Materials, *J. Mater. Res.*, Vol 1 (No. 1), 1986, p 81–89
2. M.S. Newkirk, H.D. Lesher, D.R. White, C.R. Kennedy, A.W. Urquhart, and T.D. Claar, Preparation of Lanxide™ Ceramic Matrix Composites: Matrix Formation by the Directed Oxidation of Molten Metals, *Ceram. Eng. Sci. Proc.*, Vol 8 (No. 7–8), 1987, p 879–885
3. Lanxide Corporation, Newark, DE, unpublished work, 1985
4. A.S. Fareed, B. Sonuparlak, C.T. Lee, A.J. Fortini, and G.H. Schiroky, Mechanical Properties of 2-D Nicalon™ Fiber Reinforced Lanxide™ Aluminum Oxide and Aluminum Nitride Matrix Composites, *Ceram. Eng. Sci. Proc.*, Vol 11 (No. 7–8), 1990, p 782–794
5. A.S. Fareed, B. Sonuparlak, and A.J. Fortini, Lanxide Corporation, Newark, DE, unpublished work, 1990
6. J. Weinstein and B.R. Rossing, Application of a New Ceramic/Metal Composite Technology to Form Net Shape Wear Resistant Composites, *Proceedings of the TMS 6th Northeast Regional Symposium on High Performance Composites for the 1990's*, TMS/AIME, 1990, p 339–360
7. B.R. Rossing and M.A. Rocazella, Slurry Erosion of Silicon Carbide Particulate Reinforced Alumina Composites, *Proceedings of the 4th Berkeley Conference on Corrosion-Erosion-Wear of Materials at Elevated Temperatures*, National Association of Corrosion Engineers, to be published
8. D.J. Landini and K.S. Hatton, Lanxide Corporation, Newark, DE, unpublished work, 1989
9. R.K. Dwivedi and J.F. Ramsay, Lanxide Corporation, Newark, DE, unpublished work, 1988
10. W.B. Johnson, T.D. Claar, and G.H. Schiroky, Preparation and Processing of Platelet Reinforced Ceramics by the Directed Reaction of Zirconium with Boron Carbide, *Ceram. Eng. Sci. Proc.*, Vol 10 (No. 7–8), 1989, p 588–598
11. T.D. Claar, W.B. Johnson, C.A. Andersson, and G.H. Schiroky, Microstructure and Properties of Platelet Reinforced Ceramics Formed by the Directed Reaction of Zirconium with Boron Carbide, *Ceram. Eng. Sci. Proc.*, Vol 10 (No. 7–8), 1989, p 599–609

Reaction-Forming Processes*

REACTION-FORMED CERAMICS are those ceramics which increase in mass during firing by reaction with either a gas or a liquid phase. These bodies differ from sintered and pressure-densified ceramics which are made with powders that, after compacting, will bond together at elevated temperatures by mechanisms of mass transport between particles to form bond necks. Frequently, such bodies will shrink as grains grow together, and porosity will decrease. The sintering process may involve new phase formation such as in porcelain or in the reaction of zircon with alumina to form mullite-zirconia bodies. The distinction as compared to reaction-formed ceramics is the lack of a mass change for the sintered body as a whole. Hot pressed bodies are made by subjecting the ceramic grains and powders to pressure at elevated temperatures usually in a graphite mold, again with no change in mass.

Reaction forming utilizes a mass transport mechanism with a vapor or a liquid phase in which the initial reactants combine to bond the grains together, whereas in sintered ceramics, bonding is through grain contacts which grow to become grain boundaries. Mass transport in reaction-formed materials occurs over distances which correspond to the part or compact size in contrast to sintered or pressure-densified bodies in which mass transport is restricted to the scale of the particle size.

Modeling of reaction bonding combines microscopic chemical kinetics with macroscopic mass and heat transfer. The overall reaction rate is initially controlled by the chemical reaction rate while, for large components, the final reaction rate is nearly always determined by heat or mass transport, thus restricting the size of bodies which may be made by the process.

Reaction forming may be combined with normal sintering to produce shrinkage either during the reaction-forming process as in the formation of some silicon oxynitride bonded systems described later in this article or as a post-sintering step such as is used for sintered reaction-bonded silicon nitride (SRBSN) (Ref 1).

Several ceramics can be made by reaction forming. The most well known is reaction-bonded silicon nitride (RBSN); this has been well documented in the literature (Ref 2). An excellent comprehensive review on the formation and properties of RBSN has been written by Moulson (Ref 3). Silicon oxynitride has also been made by a reaction-bonding technique (Ref 4).

Reaction-formed bodies in which a liquid and a solid are the precursors are impervious sintered silicon carbide bodies. In these cases the liquid is silicon, and the solid is a body of silicon carbide containing carbon. The reaction to form silicon carbide in the pores of the body together with infiltration of liquid silicon may be controlled to achieve a body with no porosity.

Since the field of reaction-formed ceramics is so broad, only a few pertinent examples are selected in this article.

Reactions with Gas Phases

For those reaction-formed bodies in which one of the reactants is gaseous, porosity is inherent. These bodies generally do not exhibit shrinkage during formation. In this case, the gas reacts with a compact of solid reactant or reactants when the temperature is raised above a certain level. The gas diffuses from the outer surfaces of the compact toward the center of the mass. Slow, controlled increases in temperature can result in reacted material filling the pores of the body internally with no shrinkage of overall body. This sequence of gas diffusion, reaction, and pore filling with no shrinkage, is found in both reaction-bonded Si_3N_4 (Ref 5) and reaction-bonded Si_2ON_2 (Ref 4).

In order for the reaction to proceed to completion, a continuous pore structure must be maintained for diffusion of reactant gases. If the number of pores and pore channels is decreasing, then the depth to which the reactions may proceed becomes a limitation. The original density, particle size of the initial reactants, and the thickness of the compact become important factors. The greater the density, the less the depth to which the reaction may proceed. Likewise, the coarser the particle, the more difficult the nitridation. Figure 1 shows these effects for silicon nitride bodies of different thicknesses, grain size, and density. Coarse-grain compacts showed lower weight gains than fine-grain compacts because of slower gas diffusion in the larger grains and fewer number of pore channels. Fine-grain compacts with lower density showed higher weight gains and more nitridation for thicker compacts because of more reactive surface area and an increased number of pore channels through which the gases may diffuse. In Fig 1 weight gains for coarse-grain compacts with particles ≤60 μm (−250 mesh) fall above 13 mm (0.5 in.) thick. For fine-grain, lower-density compacts with particles ≤20 μm, weight gains fall off above 19 mm (0.75 in.).

Reaction-Bonded Silicon Nitride

In the case of nitrogen reacting with elemental silicon to form silicon nitride, after the physical dimensions of a compact of silicon metal powder have been established and during firing, internal pores decrease significantly in size and number. The density of silicon is 2.33 g/cm^3, and the density of silicon nitride is 3.18 g/cm^3. From these values and a weight gain of 66.6% for the reaction, it can be calculated that a particle of silicon expands 22% in volume upon reacting to form Si_3N_4. Pore sizes decrease two or three orders of magnitude during the nitriding process (Ref 6). New mass being added to the body must expand into surrounding pores. As the pores decrease in number, the pore channels close off and the diffusion of nitrogen stops. In low-density bodies the reaction may proceed to completion before the channels close off. A suggested sequence for the nitridation is that N_2 gas molecules are chemisorbed onto a surface of silicon and nuclei of Si_3N_4 form and grow laterally, eventually approaching each other (Ref 7). As silicon from the surrounding decreasing area is depressed by the removal of silicon for the reaction, the depressions become deeper and undercut the nitride. As

*Adapted from "Reaction-Formed Ceramics," M.E. Washburn and W.S. Coblenz, *Ceramic Bulletin*, American Ceramic Society, Vol 67 (No. 2), 1988, p 356–363. With permission

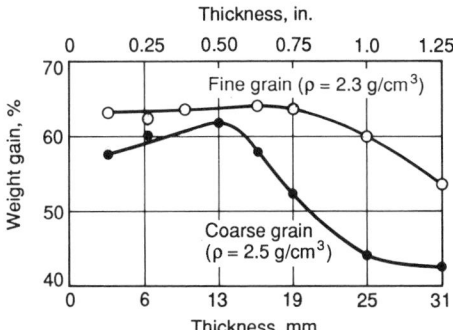

Fig 1 Effect of thickness of compact of RBSN

the channels through which silicon is supplied to the outer surface and nitrogen to the inner face close off, the reaction rate decreases, and with closure of all the channels, the reaction effectively stops. In high-density compacts, channel closure may occur before the reaction has gone to completion. Fully dense, fully reacted bodies are therefore not possible. Achieving densities >90% is impossible because of the limited number of open pore channels in the initial compact. Under highly controlled conditions with thin cross sections, densities of 85% have been reported for injection-molded turbine blades (Ref 8, 9).

Two forms of Si_3N_4, alpha and beta, are both hexagonal. Either may form, depending upon the temperature, nature of the silicon, impurities, and nature of the gaseous phase. Oxygen partial pressure and the amount of combined oxygen found on the particles of silicon result in the formation of Si_2ON_2 if levels are high enough. Frequently all three nitrides can be found together.

Alpha Si_3N_4 forms from a gaseous state at temperatures below the melting point of silicon, as a white matte of fibers. Densification proceeds as the pores of the compact become filled with needles of Si_3N_4 which grow together to form a matte. Near and above the melting point of silicon, the matte becomes denser and crystallites of β-Si_3N_4 grow into the molten silicon. The resultant structure consists of discrete grains of β-Si_3N_4 in a matte of α-Si_3N_4 (Ref 10). The ratio of α- to β-Si_3N_4 depends upon several factors: temperature, impurities, and atmosphere. Liquid phases such as molten silicon or silicon alloys at higher temperatures may result in greater formations of β-Si_3N_4 and in lower α/β ratios (Ref 11, 12). The addition of hydrogen to the reactant nitrogen appears to result in higher α/β ratios (Ref 13). Hydrogen can assist the reaction by keeping oxygen partial pressure low, reducing an inherent SiO_2 film on the silicon and improving the diffusion of gases into and out of the compact. Iron additions may disrupt the SiO_2 layer and initiate whiskers of α-Si_3N_4, resulting in a denser, more completely reacted matte formation. Temperature gradients in the compact due to the exothermic reaction do not result in gradients

in the α/β ratio (Ref 14). Other factors that can influence the α/β ratio include time, pressure, gas flow, and surface area. Table 1 shows some effects on the product.

Beta Si_3N_4 can grow at high temperatures where liquid phases may result. The major growth of β-Si_3N_4 occurs in a liquid phase and, to a minor extent, as the result of the reaction between solid silicon and nitrogen (Ref 3). Iron additions will favor the formation of liquid phase, but β-Si_3N_4 formation as well as a conversion of α- to β-Si_3N_4 is favored by the liquid phase rather than by the presence of iron. Above the melting point of silicon or when the melting point is depressed by the addition of iron-forming FeSi, β-Si_3N_4 spikes have been found in the silicon (Ref 16). Large β-Si_3N_4 grains have been observed to grow in liquid iron silicide, whereas smaller α-Si_3N_4 grains grow nearby from vapors. A mixture of α- and β-Si_3N_4, including iron silicide, can form between the two zones (Ref 17).

Beta Si_3N_4 also forms when high nitrogen pressures are used during the nitridation (Ref 18). Nitriding silicon compacts at pressures as high as 50 MPa (7.25 ksi) increases the Si_3N_4 nuclei that develop, resulting in a uniform fine distribution of pores and mostly β-Si_3N_4. The bodies have improved fracture strength and toughness.

When high-purity silicon free of an oxide layer is used, and under firing conditions in which small items are rapidly brought to >1380 °C (>2515 °F) and held for a longer than normal period of time, bodies can be formed that are high in β-Si_3N_4 with α/β < 1 (Ref 19). Bonding is less developed with a microstructure that is particulate in nature. These bodies may be gray tan colored and can be easily machined to various shapes. Table 2 shows properties of high-purity Si_3N_4. Fracture surface energies are quite high and are higher than for predominately α-Si_3N_4 bodies (Ref 20).

Formation of Si_3N_4 Shapes. For Si_3N_4 shapes, several factors, that is, particle size and distribution, atmosphere nature and control, time, and temperature of firing, are important. Complete nitriding is difficult if the initial particle size is >70 μm (200 mesh).

Coarse particles of silicon during nitridation may be covered with a thin film or a shell of Si_3N_4. The nitride layer cracks, and nitrogen can penetrate through the microcracks (Ref 21). Frequently, ≤60 μm (250 mesh) is used for normal ceramic shapes. For high-performance ceramics, however, particle size control is essential and silicon must be <20 μm with an average size of 2 μm. Even particles as small as 10 μm can result in inclusions in fired pieces which can act as flaws and crack initiators (Ref 22).

Compacts of silicon may be made by several techniques including hydraulic pressing, isostatic pressing, slip casting, and injection molding. Slip casting requires control of the liquid slip with deflocculants that will not react with the silicon to form gas bubbles in the slip. Most deflocculants such as sodium silicate or methyl ethyl amine are chemically basic and will react with silicon to form hydrogen gas. Frequently, slips are allowed to sit until the silicon is covered with a passive coating. Silicon made by the arc furnace method can also contain calcium carbide which can react with water and result in gas bubble formation.

Compaction of the silicon powder mixed with binders such as polyvinyl acetate (PVA) or polyethylene glycol by isostatic pressing is frequently used to form either a finished shape or a billet that can be machined to a finished shape. Machining may be done on a green compact or on a compact that has been prefired to provide sufficient strength for machining. Prefiring has been successfully done in nitrogen or in argon. Prefiring in argon has advantages in that a network of sintered silicon grains is established which, with a fine uniform distribution of pores from the fine starting powders, nitrides to form a uniform, high-strength ceramic body (Ref 23). Table 2 shows strength and properties of a high-performance Si_3N_4 made using an argon pre-sintering technique.

Figure 2 shows flexural strengths for fine-grain, high-performance bodies up to 1375 °C (2505 °F). It should be noted that the high strengths at 1375 °C (2505 °F) compared to low-temperature strength are usually not found with hot pressed or sintered Si_3N_4. Grain

Table 1   Influence of reaction variables on RBSN products

Reaction variable	Influence on product
Time	Early reaction at low temperature produces high α:β
Low temperature (<1350 °C, or 2460 °F)	High α:β Si_3N_4; fine structure
Medium temperature (1350–1410 °C, or 2460–2570 °F)	Texture becomes coarser with increasing temperature; pore size increases
High temperature (>1410 °C, or 2570 °F)	Coarsens structure; β-Si_3N_4 encouraged
High surface area	α:β Si_3N_4 increases
High pressure	Texture becomes finer; β-Si_3N_4 increases
High oxygen	Increases α:β Si_3N_4
Hydrogen	Increases α:β Si_3N_4; texture becomes finer
Iron	Increases β-Si_3N_4 at high temperature
Flowing nitrogen atmosphere	Encourages clean porosity
Heating rate	Various rates appear to alter α:β Si_3N_4

Source: Ref 15

Table 2 Properties of RBSN

Type of RBSN/form of silicon compact	Predominant phase	Density, g/cm³	Modulus of elasticity, GPa	Modulus of rupture at 2.2 g/cm³, MPa (ksi)	at 2.35 g/cm³, MPa (ksi)	at 2.4 g/cm³, MPa (ksi)	at 2.55 g/cm³, MPa (ksi)	Fracture surface energy at 25 °C, W/m²	at 2000 °C, W/m²	Cryolite resistance (loss rate) at air/bath notch 2.29 g/cm³, mm/yr (in./yr)	2.46 g/cm³, mm/yr (in./yr)
High purity/isostatically pressed high-purity silicon	β	2.2	55 (8)	96 (14)	...	...	...	23	20	...	...
Slip cast/slip cast technical-grade silicon	α	2.2–2.4	145 (21)	165 (24)	...	186 (27)	...	20	14	168 (6.6)	137 (5.4)
High-performance/ isostatically pressed 2 μm average silicon	α	2.35–2.55	172 (25)	...	248 (36)	...	362 (52.5)	...	...	...	109 (4.3)

boundaries in either hot pressed or sintered Si_3N_4 frequently contain a glassy phase which can soften at the higher temperature resulting in a decrease of strength.

The reaction of nitrogen with silicon is highly exothermic, with a heat of reaction of ≈733 kJ/mol. The adiabatic reaction temperature is high enough to result in silicon melting which must be avoided to achieve a complete and nondisruptive reaction. Three basic approaches are used to control the reaction exotherm. First, the overall reaction rate is controlled by the flow rate of nitrogen into the furnace (Ref 24). This puts an upper limit on the reaction rate and heat generation within the furnace. Second, the addition of an inert gas such as argon or helium either in the nitrogen stream or as a cyclic purge, can control the formation of local hot spots within the compact. A third approach is to add a nonreactive material such as prereacted Si_3N_4, SiC, or BN. The heat generation within the body is thereby reduced, and the added material acts as thermal ballast.

Additions of inert gas such as argon or helium to the nitrogen gas flow have several important implications in the nitridation of silicon. Diffusivity of the gas, which depends upon its molecular weight, results in the enrichment of the heavier gas in the hottest part of the furnace (Ref 25). Thus, if argon

is used, the faster diffusing nitrogen will decrease in concentration in the hot zone of the furnace in the absence of convection. Additions of helium or hydrogen, on the other hand, will increase the nitrogen concentration in the hot zone of the furnace. Viscosity of the gas phase controls the flow rate of nitrogen into and out of the compact driven by changes in pressure resulting from consumption of nitrogen from the reaction. Argon and helium increase the viscosity, and hydrogen decreases it. Hydrogen and helium also greatly increase the thermal conductivity of the gas phase. This will reduce thermal gradients in the compact created by the reaction.

Si_3N_4 Products. The properties of reaction-bonded Si_3N_4 make it an attractive material for many uses. Because very little dimensional change occurs on firing, precision machined into the unfired or green body will be maintained in the fired body. Experimental rotors and stators for automotive gas turbines have been fabricated (Ref 26). Injection molding has been used successfully to shape turbine blades. Various RBSN stators have been tested for 30,000 thermal cycles to 1200 °C (2190 °F) and 9000 cycles to temperatures of 1370 °C (2500 °F) (Ref 27).

Oxidation of RBSN occurs in two stages and can be both internal, in the open pores of the structure (stage I) and external, on the surfaces (stage II) (Ref 28). At 1000 °C (1830 °F) the initial stage I oxidation may take 3 or 4 days. At 1200 or 1400 °C (2190 or 2550 °F), however, it is essentially complete in ≤15 min. Stage II oxidation at 1000 °C (1830 °F) is not generally evident but presumably occurs over longer periods of time (>4 days). At ≥1200 °C (≥2190 °F), however, stage II oxidation is faster than at 1000 °C (1830 °F) although much slower than for stage I. At 1000 °C (1830 °F) in still air, a 2.53 g/cm³ sample showed a weight gain >5% after several days. At 1200 °C (2190 °F) a similar sample gained only 2%. The 1000 °C (1830 °F) sample oxidized internally, whereas the 1200 °C (2190 °F) sample oxidized on the surface, sealing the internal pores (Ref 29). Surface oxidation can be influenced by the surface condition. The as-fired surface frequently is light gray and dusty. X-ray diffraction analysis has

shown this dust to be finely divided whiskers of α-Si_3N_4, having formed by vapor reactions outside the physical parameters of the compact. When this surface is ground off, oxidation at 1375 °C (2505 °F) decreased from 1.3 to 0.6% (Ref 23). The α-Si_3N_4 dust oxidized to form a coating of silica on the sample. Flexural strength can decrease significantly with crack initiation occurring in cracks in the silica coat. When the coat or film is ground off, flexural strength can return to its original value. A decrease in flexural strength of 51% was found on an as-fired surface test bar, but upon grinding the oxidized surface off, it returned to its original strength value of 276 MPa (40 ksi) (Ref 23).

Reaction-bonded silicon nitride has good corrosion resistance to many severely corrosive environments. An interesting example of this is its good resistance to attack by molten cryolite. Cryolite, $3NaF \cdot AlF_3$, in its molten state dissolves virtually all oxides. Because it will dissolve Al_2O_3-forming ionic species, it is used in aluminum cells for the electrolytic deposition of aluminum metal. In a program to find a material to act as a thermocouple protection tube for the aluminum cell, RBSN was found to be one of the best materials to resist attack by the molten cryolite bath (Ref 30). Corrosion in the bath occurs at the air-bath interface where dissolved oxygen at the surface of the bath can oxidize Si_3N_4 to SiO_2, which in turn can dissolve in the bath. Relative corrosion resistance was determined by measuring the depth of the notch that developed at the air-bath interface over a period of time. Loss rates were compared to a standard test bar of known loss rate. The test was made on 150 by 10 by 5 mm (6 by 0.4 by 0.2 in.) bars partially immersed in molten cryolite and aluminum at 800 °C (1470 °F) for 100 h. The RBSN samples showed loss rates of 100 to 150 mm/yr (4 to 6 in./yr), with higher density samples with less porosity showing lower loss rates than lower density samples. By comparison, samples of dense Al_2O_3 completely disappeared, and samples of both sintered and hot pressed AlN showed loss rates of 3560 and 3380 mm/yr (140 and 133 in./yr). Reaction-bonded SiC showed loss rates of 150 to 200 mm/yr (6 to 8 in./yr). One

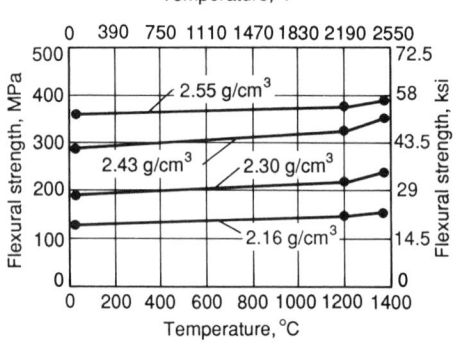

Fig 2 Flexural strength for high-performance reaction-bonded RBSN of different densities. Measured on 6 by 3 mm (0.24 by 0.12 in.) ground test bars in three-point loading on a 25 mm (1 in.) span

difficulty with RBSN, however, is that molten cryolite can pass freely through the pores, presenting problems with containment.

Silicon Oxynitride

It is believed that orthorhombic crystals of Si_2ON_2 grow from a liquid phase which has been saturated with dissolved nitrogen (Ref 4). A glassformer such as CaO or MgO establishes a liquid phase with SiO_2. The solution of nitrogen in these liquid phases is similar to that observed in iron slags (Ref 31). Nitrogen dissolves in the molten glass and crystals of Si_2ON_2 precipitate out as depicted in Fig 3. In the diagram, the glass is depicted as a thin film coating a particle of silicon moving from the SiO_2 to the silicon. Nitrogen combines to form crystals of Si_2ON_2 and is replenished from the nitrogen atmosphere in a process that continues until all the SiO_2 and silicon have been consumed. The final structure is a network of Si_2ON_2 crystals in which pores are partially filled with a glass containing nitrogen (Ref 4).

Properties of Si_2ON_2 ceramic bodies are influenced by the nature of the nitrogen glass. As shown in Fig 4, flexural strength values have a tendency to fall off at >1200 °C (>2190 °F). This is believed to be caused by softening of the glass. Density of the compact has a significant effect on flexural strength (Fig 5). Theoretical density of Si_2ON_2 is 2.8 g/cm³.

Reaction-formed Si_2ON_2 can be made with densities >90% of theoretical (Ref 32). By using reactive precursors that are very finely divided, with 2 μm silicon, SiO_2 fume, and CaF_2 as the glassformer, a body can be made that undergoes simultaneous weight gain and shrinkage upon firing. These bodies completely react to form a network of Si_2ON_2 crystals with glass in the pores. Table 3 shows some properties of reaction-bonded Si_2ON_2 of both normal density, having 71 to 78% of theoretical density, and high density, with 89 to 94% of theoretical density.

Reaction-bonded Si_2ON_2 shapes have been made by hydraulic pressing, isostatic pressing, and slipcasting (Ref 33). Shapes have been fired in large furnaces controlled with argon injection (Ref 34). Thick cross sec-

$$3Si + (1 + x) SiO_2 + xCaO \rightarrow 2Si_2O \cdot xCaO + xSiO_2$$

$$2Si_2O \cdot xCaO + xSiO_2 + 2N_2 \rightarrow 2Si_2ON_2 + xCaO \cdot SiO_2$$

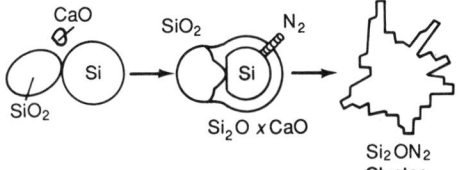

Fig 3 Schematic diagram of the formation of crystal clusters of Si_2ON_2 showing a thin liquid film around a particle of silicon

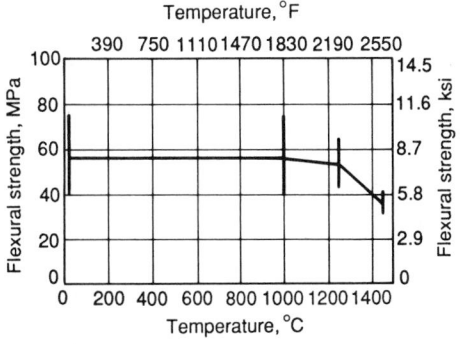

Fig 4 Flexural strength of normal density, 2.2 g/cm³, reaction-bonded Si_2ON_2 showing strength fall off at >1200 °C (>2190 °F)

tions to 75 mm (3 in.) have been nitrided successfully.

Silicon oxynitride ceramics show good resistance to attack by molten chlorides and chlorine gas. The material has been successfully used as a lining for cells used for the electrolysis of $AlCl_3$ to aluminum. Silicon oxynitride plates showed little deterioration after prolonged usage in a cell operating at 700 °C (1290 °F) in which molten $AlCl_3$, NaCl and KCl, aluminum, and chlorine gas were abundant.

Silicon oxynitride has also been successfully used as a lining in a cell for the electrolysis of zinc chloride to zinc metal. After 45 days at 500 °C (930 °F), in a $ZnCl_2$, KCl, and NaCl bath with chlorine gas and molten zinc, no significant deterioration or surface attack was found (Ref 4).

Although its resistance to molten cryolite is not as good as Si_3N_4 as seen in Tables 2 and 3, Si_2ON_2 can act as a container for the molten salt. Cryolite does not pass through the pores as it does for Si_3N_4 and bath levels in crucibles will remain fairly constant for periods >100 days at 800 °C (1470 °F). Reactions occurring between Si_2ON_2 and cryolite appear to result in a material with a viscous nature that plugs the pores of the crucible (Ref 35). A comparative loss rate for a sample of sialon was 580 mm/yr (23 in./yr).

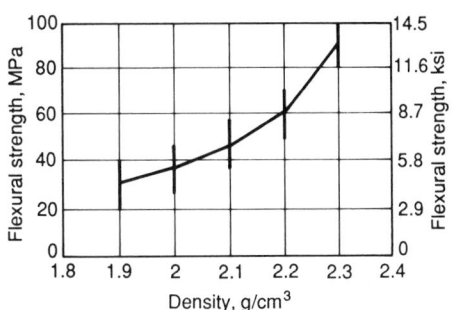

Fig 5 Flexural strength of normal density Si_2ON_2 at room temperature measured on 229 by 57 by 19 mm (9.0 by 2.2 by 0.75 in.) test bars broken in three-point loading on a 200 mm (7.9 in.) span

Comparison of these values with the lower values of 100 to 150 mm/yr (4 to 6 in./yr) found for Si_3N_4, shows that molten cryolite will react with an oxygen atom in a ceramic body even though it is combined in a nitride.

Reaction with a Liquid Phase

Reaction-bonded SiC is manufactured by infiltration of SiC-carbon shaped bodies with liquid silicon. The ratio of SiC to carbon varies greatly in practice. On one extreme, carbon fiber, cloth, or felt has been infiltrated with liquid silicon (Ref 36), while at the other extreme, an impervious sintered silicon carbide body is manufactured with a small amount of carbon (Ref 37). Most RBSC is made with formulations which contain an organic polymer with carbon and α-SiC grain (Ref 38). Such mixes are well suited to near-net-shape forming by extrusion or injection molding. The density of carbon contained in SiC is equal to the density of SiC times the atomic weight of carbon divided by the molecular weight of SiC. This density is 0.963 g/cm³ as compared to graphite, which has a theoretical density of 2.25 g/cm³ (Ref 38). The density of carbon surrounding the α-SiC starting grain should be <43% of the theoretical density of 0.963 g/cm³ to prevent the formation of a skin of fully dense SiC which would impede infiltration of silicon during the reaction-bonding process.

Capillarity provides the driving force for the silicon infiltration. Characterizing the green pore structure as an array of parallel tubes of radius r (Ref 40), the equilibrium height, h, for a column of liquid would be given by:

$$h = \frac{2\gamma \cos \theta}{\rho g r} \qquad \text{(Eq 1)}$$

where γ is the surface tension (1 N/m), ρ is the density of liquid (2×10^3 kg/m³), g is the acceleration of gravity (9.8067 m/s²), and θ is the wetting angle (0, for complete wetting).

For silicon, assuming a pore radius of 1 μm, a surface tension of 1 N/m, and a wetting angle of 0, the maximum height which may be infiltrated would be ≈100 m. Sufficient capillary driving force is therefore unlikely to be limiting for any shapes of interest, provided that the surfaces are wet by liquid silicon. Oxide and nitride surfaces are less wetting than carbon and pose problems for infiltration.

The kinetics of infiltration are controlled by the viscous flow of liquid silicon through the pores of the body as determined by Poiseuille's law (Ref 41). The rate of climb of silicon, dh/dt, is given by (Ref 39):

$$\frac{dh}{dt} = \frac{r \gamma \cos \theta}{4 h \eta} \qquad \text{(Eq 2)}$$

and

Table 3   Properties of reaction-bonded Si₂ON₂

Material density	Density		Porosity		Modulus of rupture			Modulus of elasticity, GPa (psi × 10⁶)	Oxidation in air(b)		Crystolite resistance (loss rate) at air/bath notch, mm/yr (in./yr)
	g/cm³	Percent of theoretical(a)	Open, %	Closed, %	at 20 °C, MPa (ksi)	at 1000 °C, MPa (ksi)	at 1450 °C, MPa (ksi)		at 24 h, mg/cm²	at 100 h, mg/cm²	
Normal	2.0–2.2	71–78	26–19	3	55 (8)	55 (8)	34 (5)	83 (12)	4.9	6.4	890 (35)
High	2.5–2.6	89–94	7–1	...	290 (42)	214 (31)	46 (6.7)	221 (31.7)	0.8	1.2	710 (28)

(a) Theoretical density = 2.8 g/cm³. (b) At 1100 °C (2010 °F)

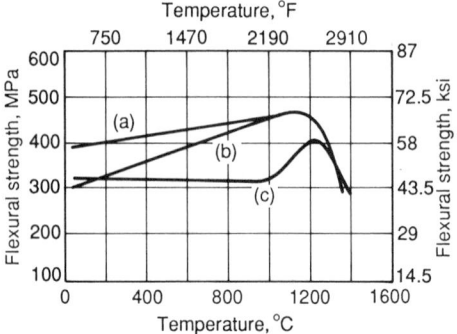

Fig 6 Flexural strength of RBSC. (a) Norton Crystar. (b) Coors Si/SiC, SC-2. (c) UKAEA/BNF Refel. Source: Ref 44

$$h^2 = \frac{r\gamma\cos\theta}{2\eta} \times t \qquad \text{(Eq 3)}$$

where η is the viscosity coefficient.

The parabolic kinetics showed that 1 h was required to infiltrate 50 mm (2 in.), while 36 h were required to infiltrate 300 mm (12 in.). The infiltration rate is proportional to the pore size, and so components made with larger particle sizes and resultant pore sizes are more readily infiltrated. Strength, however, is adversely affected by large grain sizes. A large dropoff of strength is found with grain sizes >100 µm. Between 12 and 100 µm, for example, flexural strength was found to vary between 410 and 275 MPa (59 and 40 ksi).

An interesting method of increasing pore size is to convert the carbon to SiC by reaction with SiO prior to infiltration. Two-phase mixtures of SiC and SiO₂ may be used to generate high vapor pressures of SiO vapor by the reaction (Ref 42):

$$SiC + 2SiO_2 \rightleftarrows 3SiO(g) + CO(g) \qquad \text{(Eq 4)}$$

The equilibrium SiO vapor pressure between 1800 and 1900 K varies between 3.3 and 11.2 kPa (1.6 psi). The reaction of SiO vapor with carbon to form SiC involves little, if any, volume change since the carbon atom is removed for each silicon atom added to the body. Siliciding with SiO vapor therefore does not close off pores. In fact, a substantial increase in the silicon infiltration rate is found following the reaction with SiO (Ref 39).

The high-temperature strength of reaction-bonded SiC is reduced by the silicon phase. Figure 6 shows three typical strength curves which fall off in strength at >1200 °C (2190 °F). Strength increases between 500 and 1200 °C (930 and 1290 °F) are attributed to crack healing by the SiO₂ oxidation product and to increases in fracture toughness associated with plasticity in the residual silicon phase. An increase in the effective surface energy with temperature for crack initiation from ≈35 J/m² at 1400 °C (2550 °F) has been found (Ref 43). The strength-limiting flaws in several commercial siliconized SiC materials have been summarized (Ref 44). At room temperature, failures may initiate at pores, large grains, or machining flaws. At 1000 to 1200 °C (1830 to 2190 °F), oxidation pits, pores, and large grains were noted as fracture origins with plastic deformation noted in several materials.

Shapes of Reaction-Bonded SiC. Reaction-bonded SiC is widely used for diffusion furnace components for semiconductor processing (Ref 45) and for engine applications (Ref 46). Slipcasting is commonly used to form shapes. Table 4 shows some properties of three different types of reaction-bonded SiCs.

REFERENCES

1. A. Giachello and P. Popper, Post-Sintering of Reaction Bonded Silicon Nitride, *Ceram. Int.*, Vol 5 (No. 3), 1979, p 110–114

2. D.R. Messier and M.M. Murphy, "An Annotated Bibliography on Silicon Nitride for Structural Application," MCIC-79–41, Battelle Columbus Laboratories, July 1979

3. A. Moulson, Review: Reaction-Bonded Silicon Nitride: Its Formation and Properties, *J. Mater. Sci.*, Vol 14, 1979, p 1017–1051

4. M.E. Washburn, "Silicon Oxynitride Refractory Materials," 85th National Meeting of AICHE, microfiche No. 27, 7 June 1978, Norton Co., Advanced Ceramics Div.

5. H.M. Jennings and M.H. Richman, Structure, Formation Mechanism, and Kinetics of Reaction-Bonded Silicon Nitride, *J. Mater. Sci.*, Vol 11, 1976, p 2087–2098

6. A. Atkinson, P.J. Leatt, and A.J. Moulson, The Nitriding of Silicon Powder Compacts, *J. Mater. Sci.*, Vol 7, 1972, p 482–484

7. A. Atkinson, A.J. Moulson, and E.W. Roberts, Nitridation of High-Purity Silicon, *J. Am. Ceram. Soc.*, Vol 54 (No. 7–8), 1976, p 285–289

8. J.A. Mangels, "Development of Moldable High Density RBSN," 7th Quarterly Progress Rept. No. 21, Contract No. DEN 3–20, NASA-Lewis Research Center, Jan 1980

9. J.A. Mangels and G.J. Tennenhouse, Densification of Reaction Bonded Silicon Nitride, *Am. Ceram. Soc. Bull.*, Vol 59 (No. 12), 1980, p 1216–1218, 1222

10. D.R. Messier and W.J. Croft, "Silicon Nitride," AMMRC TR-82–42, Army Materials and Mechanics Research Center, June 1982

11. P.E.D. Morgan, The Alpha/Beta Question, *J. Mater. Sci. Lett.*, Vol 15, 1980, p 791–793

12. D. Campos-Loriz and F.L. Riley, The Alpha/Beta Question, *J. Mater. Sci. Lett.*, Vol 15, 1980, p 2385–2386

13. N.J. Shaw, The Combined Effects of Fe and H₂ on the Nitridation of Silicon, *J. Mater. Sci. Lett.*, Vol 1, 1982, p 337–340

14. A.A. Atkinson and A.D. Evans, Temperature Gradients in Nitriding Silicon Powder Compacts, *Trans. Br. Ceram. Soc.*, Vol 73, 1974, p 43–46

15. H.M. Jennings, Review on Reactions Between Silicon and Nitrogen, Part 1

Table 4   Properties of RBSC

Material	Bulk density, g/cm³	Porosity, vol%	Flexural strength four-point (RT)		Young's modulus	
			MPa	ksi	GPa	psi × 10⁶
Norton NC-435	2.96	2.4	395	57	343	49.7
UKAEA Refel 1	3.10	0.9	309	45	363	52.6
Coors Si/SiC, SC-2	3.10	0	306	44	400	58

Source: Ref 44

Mechanisms, *J. Mater. Sci.*, Vol 18, 1983, p 951–967

16. H.M. Jennings, S.C. Danforth, and M.H. Richman, Microstructure Analysis of RBSN, *Metallogr.*, Vol 9, 1976, p 427–446

17. S. Shinozaki and M.E. Milberg, Electron Microscopy Study of the Effect of Iron in Reaction-Sintered Silicon Nitride, *J. Am Ceram. Soc.*, Vol 64 (No. 7), 1981, p 382–385

18. J. Heinrich, Nitridation of Silicon under High Pressure, *Adv. Ceram. Mater.*, Vol 2 (No. 3A), 1987, p 239–242

19. Personal communication, Norton Co. Advanced Ceramics Laboratory

20. J.A. Coppola, R.C. Brandt, D.W. Richerson, and R.A. Alliegro, Fracture Energy of Silicon Nitrides, *Am. Ceram. Soc. Bull.*, Vol 51 (No. 11), 1972, p 847–851

21. S.N. Ruddlesden, Electron Microscopy of Ceramics, *Special Ceramics 1962*, P. Popper, Ed., The British Ceramic Research Association, 1963, p 341–354

22. M.E. Washburn, "Reaction Bonded Silicon Nitride," U.S. patent 4,127,630, Nov 1978

23. M.E. Washburn and H.R. Baumgartner, High Temperature Properties of RBSN, *Ceramics for High Performance Applications*, J.J. Burke, A.E. Gorum, and R.N. Katz, Ed., Brook Hill, 1975, p 479–492

24. J.A. Mangels, Effect of Rate-Controlled Nitriding and Nitriding Atmospheres on the Formation of Reaction-Bonded Si_3N_4, *Am. Ceram. Soc. Bull.*, Vol 60 (No. 6), 1981, p 613–617

25. H. Kim and C.H. Kim, The Influence of the Various Transport Properties of the Nitriding Atmosphere on the Formation of Reaction Bonded Si_3N_4, Part 1, Molecular Diffusivity and Viscosity, *J. Mater. Sci.*, Vol 20, 1985, p 141–148

26. M.L. Torti, R.A. Alliegro, D.W. Richerson, M.E. Washburn, and G.Q. Weaver, High-Temperature Properties of Silicon Carbide and Silicon Nitride, *High-Temperature Materials in Gas Turbines*, Elsevier, Amsterdam, 1974

27. A.F. McLean, Ceramic Technology for Automotive Turbines, *Am. Ceram. Soc. Bull.*, Vol 61 (No. 8), 1982, p 861–865, 871

28. R.W. Davidge, A.G. Evans, D. Gilling and P.R. Wilyman, Oxidation of Reaction-Sintered Silicon Nitride and Effects on Strength, *Special Ceramics 5*, P. Popper, Ed., British Ceramic Research Association, 1972, p 329–343

29. "Engineering Property Data on Selected Ceramics, Vol 1, Nitrides," MCIC-HB-07, Battelle Columbus Laboratories, Mar 1976

30. Personal communication, Norton Co. Advanced Ceramics Laboratory

31. E.A. Dancy and D. Janssen, The Dissolution of Nitrogen in Metallurgical Slags, *Can. Metall. Q.*, Vol 15 (No. 2), 1976, p 103–110

32. M.E. Washburn, "High Density Silicon Oxynitride," U.S. patent 4,331,771, May 1982

33. M.E. Washburn, Silicon Oxynitride Refractories, *Am. Ceram. Soc. Bull.*, Vol 46 (No. 7), 1967, p 667–671

34. M.E. Washburn, "Control System in an Apparatus for Reacting Silicon with Nitrogen," U.S. patent 3,854,882, Dec 1974

35. M.E. Washburn, "Behavior of Silicon Oxynitride in Contact with Molten Cryolite," Norton Co. Advanced Ceramics Div., 1969

36. W.B. Hillig, Tailoring of Si/SiC Composites for Turbine Applications, *Ceramics for High Performance Applications II*, J.J. Burke, E.N. Lenoe, and R.N. Katz, Ed., Brook Hill, 1978

37. M.G. Rogers, High Pressure Slip Casting of Silicon Carbide, *Ceramics for High Performance Applications*, J.J. Burke, A.E. Gorum, and R.N. Katz, Ed., Brook Hill, 1975, p 87–97

38. T.J. Whalen, J.E. Noakes, and L.L. Terner, Progress on Injection Molded, Reaction Bonded SiC, *Ceramics for High Performance Applications II*, J.J. Burke, E.N. Lenoe, and R.N. Katz, Ed., Brook Hill, 1978, p 179–189

39. C.W. Forrest, P. Kennedy, and J.V. Sherman, The Fabrication and Properties of Self-Bonded Silicon Carbide Bodies, *Ceramics for High Performance Applications*, J.J. Burke, A.E. Gorum, and R.N. Katz, Ed., Brook Hill, 1975, p 99–123

40. F. Daniels and R.A. Alberty, *Physical Chemistry*, 3rd ed., Wiley, 1967, p 278–279

41. N.K. Adams, *The Physics and Chemistry of Surfaces*, Oxford University Press, 1941, p 191

42. W.S. Coblenz and D. Lewis, "SiC Coating of Carbon-Carbon Composites for High Temperature Oxidation Resistance," unpublished research, U.S. Naval Research Laboratory, 1983

43. J.R. McLaren, G. Tappin, and R.W. Davidge, "The Relationship Between Temperature and Environment, Texture, and Strength of Silicon Bonded SiC," AERE-R 6705, U.K.A.E.A. Research Group Rept., Harwell, Berkshire, U.K., 1971

44. D.C. Larsen and W.J. Adams, "Property Screening and Evaluation of Ceramic Turbine Materials, Final Technical Report," AFWAL-TR-83-4141, Air Force Wright Aeronautical Laboratories, April 1984

45. B.D. Foster and R.E. Tressler, Silicon Processing with Silicon Carbide Furnace Components, *Solid State Technol.*, Vol 27 (No. 10), 1984, p 143–146

46. M.L. Torti, J.W. Lucek, and G.Q. Weaver, "Densified Silicon Carbide An Interesting Material for Diesel Applications," Eng. Tech. Paper Series No. 780071, Society of Automotive Engineers, 1978

Firing/Sintering: Densification

Chairman: William S. Coblenz, Defense Advanced Research Projects Agency

Introduction

THIS SECTION describes the firing (or sintering) of ceramic powder compacts. The firing process is operationally quite simple even though the mechanisms and processes which determine the final microstructure and properties may be quite complex.

The first article in this Section, "Furnaces and Related Equipment," describes the choices of furnace types and controls available to fire ceramic articles. Furnaces may be run in a periodic or continuous mode. Gas-fired and electrically heated (resistive or inductive) furnaces are in common usage, while microwave sintering has received a great deal of attention from the research community as of late. Special atmosphere control may be required to prevent oxidation of non-oxides or to achieve optically dense oxides where gas entrapment can dominate final stage removal of porosity. Temperature control and uniformity are critical to produce consistency and uniformity.

Materials development is often done with simple pressed pellets or tiles on which sintering temperatures and material properties are determined. The manufacture of components usually requires modification of heating rates and hold times to accommodate heat and mass transfer. The article "The Firing Process" describes some of the reactions that occur in a ceramic body during firing. Methods for determining the optimum firing curve for various ceramics are also introduced. The latter is particularly critical because an improper firing schedule may introduce flaws such as blistering or cracking due to binder burnout or the decomposition of hydrated materials.

The article entitled "Fundamentals of Sintering" is intended to give the reader an understanding of the processes which give rise to the final microstructure and thus control properties and performance of the final product. Reaction-bonded ceramics are characterized by their method of densification which involves an increase in mass rather than shrinkage. For example, reaction-bonded SiC is formed by silicon infiltration of SiC and/or carbon bodies to form a dense bonded product. Reaction-bonded Si_3N_4 is formed by the nitriding of silicon metal powder compacts. Shrinkage of glass powder compacts occurs via viscous flow driven by surface tension. Mass transport resulting in particle contact formation, with or without shrinkage, can occur by surface, volume, or grain-boundary diffusion in addition to vapor-phase or liquid-phase transport. Lattice diffusion, grain-boundary diffusion and liquid-phase transport transfer material from grain contacts

to pore surfaces and result in shrinkage. The relative contributions of each mechanism depend upon temperature, particle size, and local geometry as the sintered structure evolves. Much of the literature on the sintering of specific materials is concerned with the use of sintering aids which alter the mass transport coefficient and therefore the sintering kinetics and microstructural evolution. Shrinkage measurements in isothermal, constant heating rates are often used to evaluate the sintering process. Changes in surface area can be used to assess the structural evolution in the absence of shrinkage. Analysis of fractured or polished and etched surfaces can be a powerful tool for monitoring and understanding the sintering process.

The articles entitled "Solid-State Sintering," "Liquid-Phase Sintering," and "Reaction Sintering" are intended to build on the previous article ("Fundamentals of Sintering"). They give practical examples of products sintered under conditions where these specific mechanisms predominate. The use of sintering aids is discussed both in regards to their influence on the densification mechanisms and their effects on the final product properties.

Pressureless densification requires the greatest of care in formation of the green body to avoid inter-agglomerate porosity or the crescent shaped pores characteristic of hard aggregates. The article "Pressure Densification" describes the advantages of the application of pressure during densification which results in near theoretical density and may change the strength-limiting flaws from bulk to surface. The three major aspects of the pressure densification mechanisms are:

- Enhanced driving force for shrinkage mechanisms
- Creep densification to eliminate large pores
- Conversion to a closed or semi-closed system which conserves volatile liquid phases or other densification aids

The availability of equipment for hot isostatic pressing (HIP) combines the advantages of pressure densification with near-net-shape forming which greatly reduces machining costs.

The final article in this Section is entitled "Grain Growth." Normal grain growth is usually slow below about 85% of theoretical density since porosity acts to pin grain boundaries according to the Zener criterion. Below 85% theoretical density, grain growth may proceed by surface diffusion or vapor-phase transport as is observed in SiC or Cr_2O_3 respectively (without sintering aids). At high densities, breakaway grain growth is commonly observed in ceramic systems. Pores trapped within large grains as a result of breakaway grain growth are exceedingly slow to shrink away. Phase transformations during densification may also promote grain growth. The α to β transformation in Si_3N_4 for example, results in elongated grain shapes which are critical in obtaining high fracture toughness. Isolated large grains within a fine-grained matrix result from abnormal grain growth. They are important for structural ceramics since they may constitute the strength-limiting flaw population. This article describes the role of sintering aids in controlling grain growth and trouble-shooting methods for grain growth related problems.

Furnaces and Related Equipment

C.W.P. Finn, Materials Science Division, Department of Mechanical Engineering, Northeastern University

FURNACES can be divided into categories based on their method of heating (combustion or electric), their method of applying heat (direct or indirect), and their method of operation (periodic or continuous). The type of furnace chosen for specific ceramic firing and sintering applications is based on factors such as the number, size, value, and physiochemical nature of the ceramic object treated. Small indirect electrically heated batch furnaces are often used for laboratory studies while large continuous combustion furnaces are often used for whiteware.

Furnace parameters that must be controlled include temperature, pressure, heating and cooling rates, and atmosphere composition. It is a primary principle of control theory that one can only control a parameter if that parameter can be measured.

Choosing the correct furnace for processing materials is a complex procedure involving economic, technical, and safety considerations. One might choose a small batch furnace incorporating interchangeable heating elements during the research and development phase, a medium-size batch furnace during the scale-up or pilot-plant phase, and a large continuous furnace during the production phase of a project. One might choose an electrically heated furnace with elaborate computer controls for processing high-value added materials in an area where labor costs are high and the infrastructure exists for technical support; and a fuel-fired furnace with manual controls for processing low-value materials in a less developed area. Regardless of the furnace chosen, safety should be the prime consideration in all cases.

While the obvious safety hazard of high temperature is well known, there are several other safety issues encountered during furnace operation. The possibility of electrocution must be addressed in electrically heated furnaces, while the possibility of poisonous gases is an issue with both combustion and electrically heated furnaces. The dangers of binder breakdown products are often overlooked, while the possible interactions between the ceramic parts being treated and the furnace environment must also be considered.

Types of Ceramic Firing and Sintering Furnaces

As mentioned in the introduction, furnaces used to fire and sinter ceramic components can be grouped into four categories:

- Combustion furnaces
- Electrically heated furnaces
- Periodic or batch furnaces
- Continuous furnaces

Combustion Furnaces

The use of combustion predates recorded history and is shrouded in ancient myths such as Vulcan's forge. The abundant archaeological evidence of the use of fire in the smelting of metals and the firing of clay artifacts dates from well before the beginning of the Christian era. Clearly, at the time of the earliest writings on materials processing, the use of fire was a well-established technology based on the writings of Vannoccio Biringuccio in 1540 (Ref 1). Indeed, Biringuccio's description of the manufacture of crucibles is remarkably similar to current manufacturing practices.

Combustion heating takes advantage of the thermodynamics of the reactions:

$$H_2 + \tfrac{1}{2}O_2 \rightarrow H_2O$$

$$\Delta H_{298K} = -242 \text{ kJ/mol} \qquad (Eq\ 1)$$

and

$$C + O_2 \rightarrow CO_2$$

$$\Delta H_{298K} = -394 \text{ kJ/mol} \qquad (Eq\ 2)$$

From the negative values of the enthalpy change at 25 °C (75 °F) ($\Delta H_{298\ K}$) for these reactions (Ref 2), it can be seen that large amounts of energy are liberated during combustion. This energy can be used to heat a furnace and the parts therein. An excellent review of combustion furnaces was done by Dryden (Ref 3) in 1975.

Fuel for combustion furnaces can be solid, liquid, or gaseous hydrocarbons that are mixed with air or oxygen-enriched air. In ceramic processing, gaseous fuel is preferred for several reasons, including measurement, con-

trol, and purity. Thus, nongaseous fuels will be mentioned only in passing and the reader is referred to Dryden for further details.

Solid fuels include coal and its refined by-products (metallurgical char and coke) as well as petroleum coke, which is a by-product of petroleum refining. The major disadvantages of solid fuels are impurity content, especially in the form of inorganic ash and sulfur compounds, and the necessity for grinding in order to increase the rate of combustion. Pulverized solid fuels are used in furnaces for primary treatment of minerals such as the calcining of limestone and cement in rotary kilns. The major advantage of solid fuel is cost. Per unit of contained energy, solid fuels are usually the least expensive form of fuel.

Liquid fuels are easier to measure and to control than solids, but low molecular weight fractions are generally not used due to cost considerations. Higher molecular weight hydrocarbons are very viscous at normal temperature, requiring preheating prior to combustion. Complex burners, including atomizers and air/fuel mixers, must be used in order to ensure the necessary rates of combustion. Finally, high molecular weight liquid hydrocarbons available to the materials processing industry contain high sulfur content.

Gaseous Hydrocarbons. Thus, gaseous hydrocarbons are the most common fuels used. Measurement and control of the required fuel/air ratio is simpler with gases than with other forms of fuel. The maximum temperature that can be achieved in a gas-fired furnace is controlled by thermodynamics and heat transfer. The first is quantified by the adiabatic flame temperature that is calculated by assuming all the energy provided by the combustion reaction(s) is transferred to the products of combustion.

As a simplified example, consider the combustion of methane (CH_4) with stoichiometric air. The reaction is:

$$CH_4 + 2O_2 \rightarrow CO_2 + 2H_2O \qquad (Eq\ 3)$$

and its enthalpy change at room temperature can be calculated from:

$$\Delta H_{298K} = \Sigma \Delta H_{Products} - \Sigma \Delta H_{Reactants} \qquad (Eq\ 4)$$

by using data from Barin as:

$$\Delta H_{298K} = \Delta H_{CO_2} + 2\Delta H_{H_2O} - \Sigma\Delta H_{CH_4}$$

$$- 2\Delta H_{O_2} = -803 \text{ kJ/mol} \qquad (\text{Eq 5})$$

Thus, each mole of methane burned supplies 803 kJ and requires 2 mol O_2 for combustion. The air supplying this oxygen contains 79% N_2. The products of combustion of methane in air are:

Compound	Quantity, mol/mol CH_4
H_2O	2
CO_2	1
N_2	7.52

The enthalpy balance can be written:

$$\Delta H_{298K} + \int_{298K}^{T} \Sigma C_p dT = 0 \qquad (\text{Eq 6})$$

where C_p is the heat capacity at constant pressure.

Solving this enthalpy balance, one can calculate the maximum possible temperature as 2325 K (3725 °F). The actual temperature is limited due to endothermic decomposition of the reaction products, such as:

$$H_2O \rightarrow H + OH \qquad \Delta H = 485 \text{ kJ/mol} \qquad (\text{Eq 7})$$

$$N_2 \rightarrow 2N \qquad \Delta H = 940 \text{ kJ/mol} \qquad (\text{Eq 8})$$

$$CO_2 \rightarrow CO + O \qquad \Delta H = 527 \text{ kJ/mol} \qquad (\text{Eq 9})$$

These reactions occur at high temperatures. These corrections, which are beyond the scope of this article, result in the significantly lower adiabatic flame temperatures (Table 1).

Heat transfer considerations limit the ultimate temperature of a fuel-fired furnace to well below the adiabatic flame temperature. The products of combustion can only transfer energy to the furnace if they are hotter than the furnace interior. They must leave the heated zone at high temperature carrying with them a significant portion of their enthalpy.

Direct and Indirect Heating Methods. There are two distinct methods of heating a fuel-fired furnace: direct and indirect (Fig 1). In direct heating, the products of combustion are introduced into the same part of the furnace as the work so that the processed parts are exposed to the flame. In indirect heating, the products of combustion are separated from the work by a wall so that the processed parts can be maintained in an atmosphere with a different chemical composition. Direct heating is by far the most efficient from a heat transfer point of view while indirect heating is more flexible from a process point of view.

A major limitation of fuel-fired furnaces is the necessity of using refractories that can tolerate the corrosive properties of the products of combustion while the major advantage of fuel-fired furnaces is economy.

Electrically Heated Furnaces

With the availability of mass produced electricity during the late 1800s, electrically heated furnace technology developed rapidly. The earliest electric furnaces, similar to the earliest electric lights, used direct current (dc) arcs between carbon electrodes for heating. Later, with the development of alternating current (ac) power, resistance and inductive heated furnaces were developed along with ac arc furnaces. More recently, microwave heated furnaces have been developed. All forms of electric heating use the principle that a voltage applied across a resistance results in the flow of current via Ohm's law:

$$V = IR \qquad (\text{Eq 10})$$

where V is the potential drop (in volts); I is the current (in amperes); and R is the resistance (in ohms).

As the current flows through the resistance, power is dissipated.

$$P = VI = I^2R \qquad (\text{Eq 11})$$

where P is the power (in watts).

In an arc, the resistance is formed by the ionized gas in the arc. An inductive circuit uses the principle that an alternating electric field forms an alternating magnetic field at right angles. This magnetic field can induce a magnetic field in a body in the vicinity if that body has the appropriate property. This magnetic field in turn causes an electric field that dissipates power. Microwave heating is an application of induction heating using extremely high frequency electric power.

The main advantages of electric power for heating are ease of measurement and control as well as cleanliness. An atmosphere completely independent of the heating source can be maintained. The main disadvantage is the inefficiency of production of electricity that results in higher net cost per energy unit. Most electricity is generated by the combustion of hydrocarbon fuels to produce the steam to run the turbines. The second law of thermodynamics limits the maximum conversion efficiency to:

$$E = \frac{T_H - T_C}{T_H} \qquad (\text{Eq 12})$$

where E is the maximum thermodynamic

Table 1 Adiabatic flame temperatures of various gas mixtures

	Adiabatic flame temperature			
	Fuels with air		Fuels with O_2	
Gas	K	°F	K	°F
H_2	2450	3950	3395	5650
CH_4	2276	3635	3849	6470
C_2H_2	2657	4325	3737	6265

Source: Ref 4

efficiency and is dimensionless; T_H is the steam inlet temperature (in K); and T_L is the steam outlet temperature (in K).

Practical conversion efficiencies in the neighborhood of 10 to 15% are common.

In addition to the descriptions of electric furnaces provided by Dryden, an excellent general overview is provided by Orfeuil (Ref 5) and specific details related to induction heating by Zinn and Semiatin (Ref 6).

Arc furnaces are generally used for smelting and melting operations. Carbon electrodes carry the high-current/low-voltage power to the work to be heated. They are rarely used for sintering operations.

Induction furnaces are generally used for melting and surface hardening operations. They are sometimes used in sintering operations in conjunction with a hot pressing operation. Microwave furnaces currently are used for research and small-scale production because the available power sources are limited in size.

The majority of electric heated sintering furnaces use resistance heating. They can be conveniently classified based on the type of material used as the resistance heating element:

- Graphite
- Refractory metal
- Non-oxide ceramics
- Oxide ceramics

Graphite Heating Elements. Graphite is an excellent material for resistance heating elements because it has an extremely low vapor pressure (Fig 2). Even at 4000 K (6740 °F), the vapor pressure of the three gaseous species of carbon (C, C_2, and C_3) is less than atmospheric pressure. Graphite remains solid over this entire temperature range. The major disadvantage of graphite is its tendency to react with ceramics, particularly oxide ceramics. This can be seen from the Gibb's free energy of formation for the general reaction between graphite and a metal oxide:

$$M_xO_y + y C \rightarrow x M + y CO \qquad (\text{Eq 13})$$

As the temperature increases, the free energy of formation of carbon monoxide becomes more negative than all the metal oxides (Fig 3). Hence, in principle, Eq 13 will take place for the most refractory oxides if the temperature is high enough. In practice, the necessity for protecting graphite heating elements from oxidation by the atmosphere requires the use of an inert gas or a vacuum. This, coupled with the low vapor pressure of graphite, reduces the rate at which graphite can react with oxides provided that direct contact between the materials is avoided.

A typical graphite resistance furnace is shown in Fig 4. The entire furnace is contained within a water jacketed steel shell containing input and output ports for electric power cables, temperature and pressure monitoring sensors, and vacuum pumping apparatus. The furnace is insulated with several layers of graphite felt.

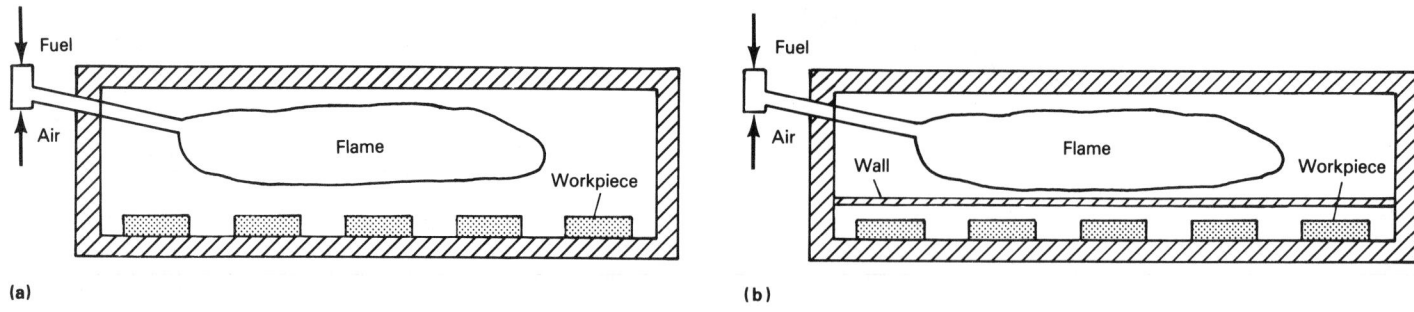

Fig 1 Relation of flame to workpiece in two types of fluid-fired combustion furnaces. (a) Direct heating. (b) Indirect heating

The resistivity of graphite, ρ_C, which is $10 \times 10^{-6} \, \Omega \cdot m$ at 0 °C, is low. The resistance is related to resistivity by the relationship:

$$R = \rho_C \frac{L}{A} \qquad \text{(Eq 14)}$$

where R is the resistance (in ohms); ρ_C is the resistivity (in $\Omega \cdot m$); L is the length (in meters); A is the cross-sectional area (in m^2).

A typical heating element in a small furnace is 300 mm (12 in.) long, 100 mm (4 in.) high, and 8 mm ($^5/_{16}$ in.) thick. Solving for R in Eq 14, its resistance is:

$$R = 10 \times 10^{-6} \, \Omega \cdot m \frac{0.3 \, m}{0.1 \, m \times 0.008 \, m} \qquad \text{(Eq 15)}$$

$$= 3.75 \times 10^{-3} \, \Omega$$

which results in typical power inputs of the order of magnitude of 10 to 100 volts and kiloamperes of current generating kilowatts of power. From Eq 10, for example, with a voltage of 15 V the current will be 4000 A, resulting in a power input (from Eq 11) of 60 kW. The heating elements typically consist of rectangular plates having alternating thin grooves machined from each side of the plate to increase the resistance (because grooves transform heating elements into equivalent of wire with greater length, L) by an order of magnitude.

The resistance of graphite is a function of temperature, initially decreasing with a minimum around 500 °C (930 °F) then rapidly increasing (Fig 5). This requires that the power supply be capable of varying voltage over a wide range in order to maintain the required power as the temperature changes. This is especially true at low temperature where the resistance decreases with increasing temperature.

As noted above, graphite reacts readily with oxygen to form carbon monoxide, making it imperative that the furnace atmosphere exclude oxygen. If an inert gas is used, it must be very dry because water vapor reacts with graphite:

$$H_2O + C \rightarrow CO + H_2 \qquad \text{(Eq 16)}$$

This reaction can result in the destruction of the heating elements and insulation and can also affect the parts being processed by providing a transport mechanism for carbon.

Hydrogen and forming gas (a mixture of hydrogen and nitrogen) are often used in atmosphere sintering furnaces as a protective blanket to exclude oxygen. In a graphite furnace these gases should be avoided. Dry hydrogen is commonly used in graphite furnaces (for example, debindering applications). Hydrogen reacts with graphite to form methane via the reaction:

$$2H_2 + C \rightarrow CH_4 \qquad \text{(Eq 17)}$$

Using data from Barin, the methane pressure in equilibrium with graphite and hydrogen can be calculated. It depends both on the temperature and the total pressure. Figure 6 shows the methane pressure at 1 atm (100 kPa) total pressure in equilibrium with pure hydrogen gas. Up to 1000 K (1340 °F), methane comprises more than 1% of the atmosphere. This methane formation consumes graphite and can severely affect the chemistry of the parts being processed. Even when the furnace interior is very hot, the insulation remains at temperatures low enough to allow methane formation. Thus, it is recommended that hydrogen be excluded from graphite furnaces.

Graphite can react with nitrogen at high temperature to produce the toxic gas cyanogen (CN). Much more serious is the reaction among graphite, water vapor, and nitrogen at moderate temperatures to produce hydrogen cyanide (HCN), which is also toxic. These reactions are discussed in the section "Chemical Safety" in this article.

A final consideration concerning inert gas is the possibility of plasma formation when using inert gases. The neon in the common neon light used in advertising signs and the argon used in common fluorescent lights are examples of inert gas plasmas. The voltage-pressure-temperature relationship for the initiation and the maintenance of an argon plasma is beyond the scope of this article. However, qualitatively, plasmas are favored by high temperature, high voltage, and intermediate pressures. Typically, at temperatures in excess of 1400 °C (2550 °F) and argon pressures in the 0.5 to 10 torr (0.65 to 1.3 kPa), voltages should be kept below 40 V to avoid plasma formation.

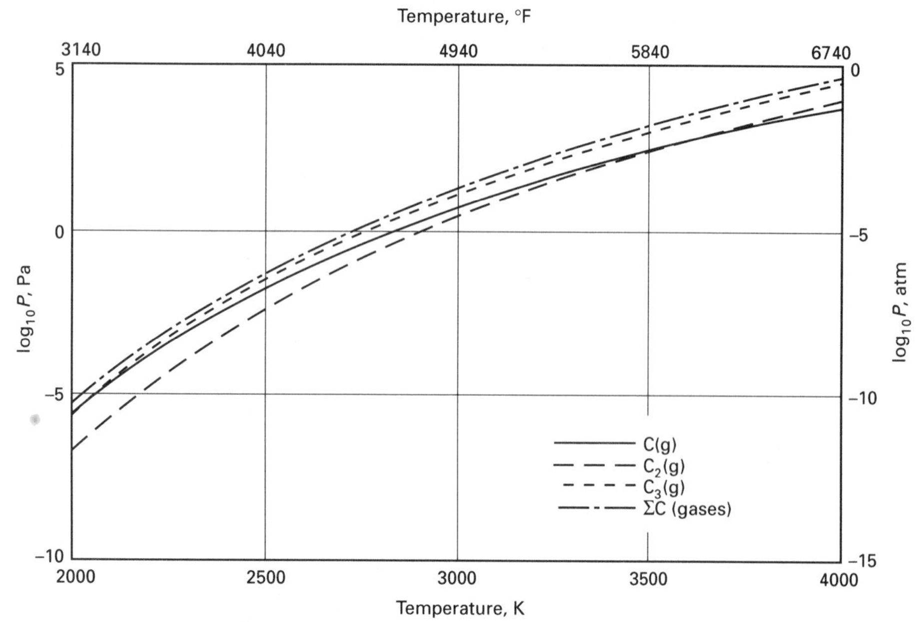

Temperature, °F

Fig 2 Vapor pressure of graphite from 2000 to 4000 K (3140 to 6740 °F)

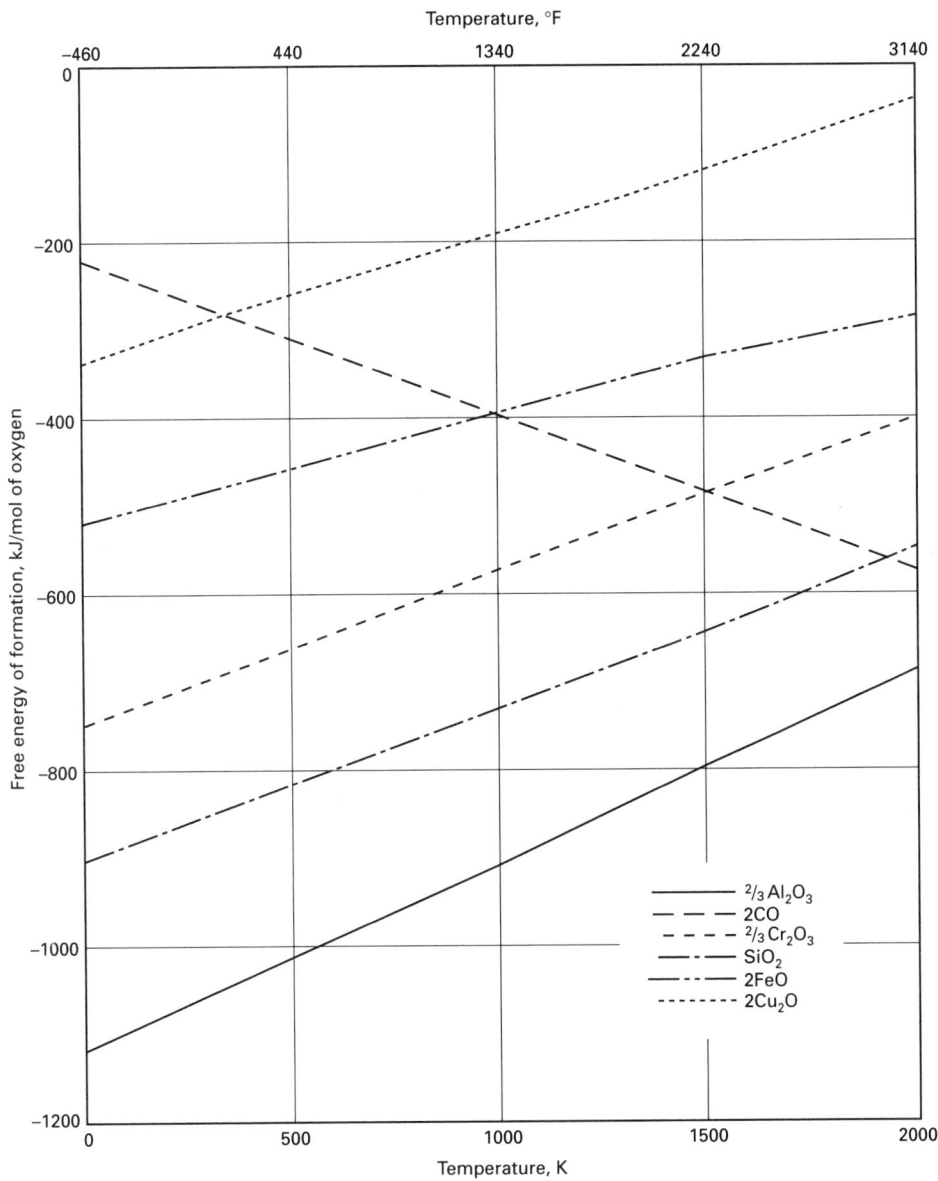

Fig 3 Free energy of formation for selected oxides

Legend:
- —— $^2/_3 Al_2O_3$
- – – – $2CO$
- – – $^2/_3 Cr_2O_3$
- –·– SiO_2
- –··– $2FeO$
- ····· $2Cu_2O$

(y-axis: Free energy of formation, kJ/mol of oxygen; x-axis: Temperature, K)

Refractory Metal Heating Elements. The refractory metals (molybdenum, niobium, rhenium, tantalum, and tungsten) can be used as resistance heating elements. Because they are metals, they can be fabricated into rods, foils, and wires. However, niobium, rhenium, and tantalum are seldom used as heating element materials because of the high cost of these metals. Rhenium has the additional drawback of being very brittle and difficult to work. Collectively, the major disadvantages of refractory metals are their reaction with oxygen, brittleness of the metals, and cost. They all have very low vapor pressures (Fig 7). A typical refractory metal furnace is shown in Fig 8(a). The insulation consists of layers of metal foil that act as radiation shields. The heating elements are formed from 5 mm ($^3/_{16}$ in.) diameter rods bent into a hairpin shape (Fig 8b).

Molybdenum Heating Elements. The most common refractory metal used is molybdenum. It is relatively easy to work and less expensive than the other refractory metals. It can be used in the form of plates, wires, or rods. The room-temperature resistivity of molybdenum ($9 \times 10^{-8} \Omega \cdot m$) is two orders of magnitude lower than graphite, but it increases rapidly with temperature. At 1600 °C (2910 °F), the resistance of molybdenum is nearly an order of magnitude higher than it is at room temperature.

The resistance of the heating element shown in Fig 8(b) can be calculated from its length ($6 \times 0.1 = 0.6$ m) and radius (0.0025 m) using Eq 14:

$$R = 9 \times 10^{-8} \, \Omega \cdot m \, \frac{0.6 \, m}{\pi(0.0025 \, m)^2}$$

$$= 2.75 \times 10^{-3} \, \Omega \qquad \text{(Eq 18)}$$

This resistance is of the same order of magnitude as a graphite heating element. Thus, a good furnace producer can design a furnace with interchangeable graphite and molybdenum hot zones using the same power supply. Such a furnace is especially useful in a research and development laboratory.

Molybdenum reacts with oxygen forming a volatile oxide (molybdenum trioxide, MoO_3) at temperatures in excess of 700 °C (1290 °F) so that the heating elements and radiation shields must be protected. Hydrogen, forming gas, or a vacuum are all used for this purpose. Moisture will also react with molybdenum but very high H_2O/H_2 ratios can be tolerated at high temperature. Cofiring (sintering in a wet hydrogen atmosphere) of ceramic substrates with metals takes advantage of this phenomenon. Sufficient moisture is used to react with any carbonaceous binder in the substrate while sufficient hydrogen is present to keep the metal in its reduced form.

The upper limit temperature for a molybdenum heated furnace, as determined by mechanical properties (especially creep), is approximately 1500 °C (2730 °F).

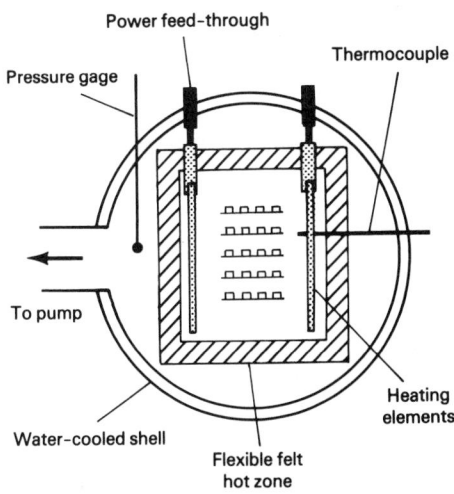

Fig 4 Schematic showing cross section of a typical graphite resistance furnace

(labels: Power feed-through, Thermocouple, Pressure gage, To pump, Water-cooled shell, Flexible felt hot zone, Heating elements)

Operation in a vacuum is the preferred method for processing ceramics in a graphite furnace. The level of vacuum is limited by the vapor pressure of the graphite and furnace cleanliness. Sublimation of graphite, which occurs at 2200 to 2300 °C (3990 to 4170 °F), is also a limiting factor when considering use of a vacuum. Small traces of moisture can lead to prolonged pumping times when attempting to reach the ultimate vacuum. Typical vacuum levels reached in a furnace are given in Table 2.

Graphite resistance heated furnaces can be used for the sintering and densification of ceramics. However, care must be taken to avoid reactions between the graphite of the furnace and the ceramic workpieces. Oxide ceramics are particularly prone to reaction with graphite while nitrides and carbides are less sensitive to graphite.

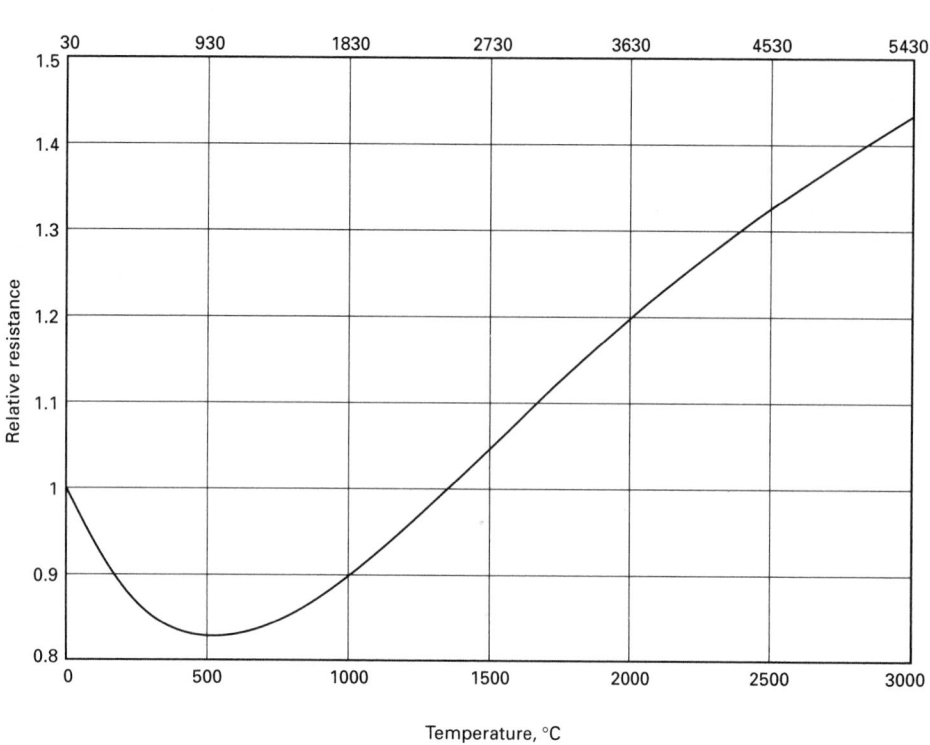

Fig 5 Relative resistance of graphite as a function of temperature from 0 to 3000 °C (30 to 5430 °F)

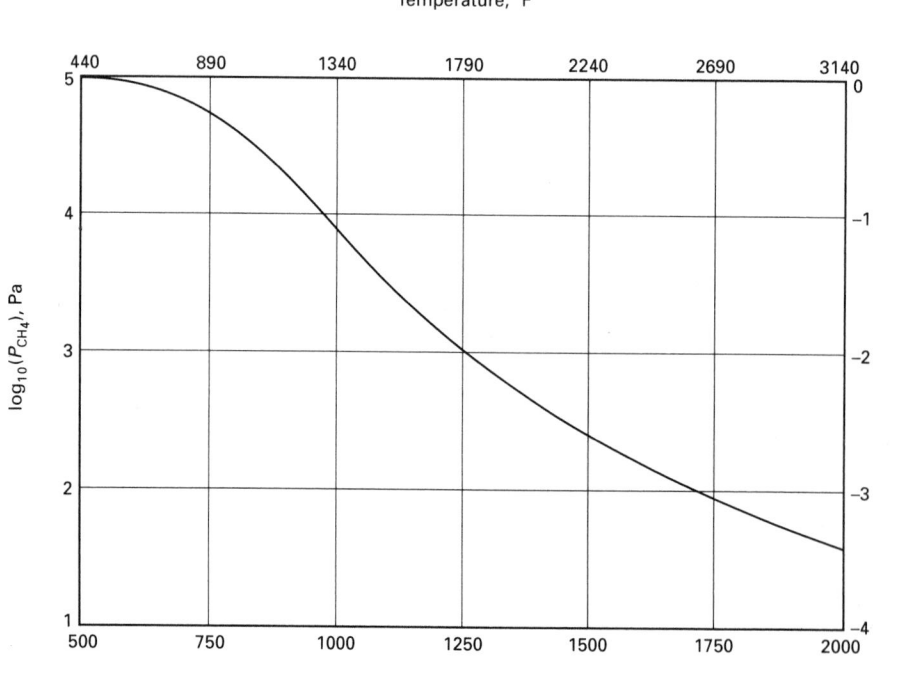

Fig 6 Methane formation from pure hydrogen gas at a total pressure of 100 kPa (1 atm)

Table 2 Typical ultimate vacuum levels

Maximum temperature		Ultimate pressure	
°C	°F	Pa	torr
1200	2190	10^{-5}	10^{-7}
1600	2910	10^{-3}	10^{-5}
2200	3990	0.1	10^{-3}
2500	4530	10	0.1
3000	5430	10^4	100

Tungsten and Tantalum Heating Elements. High temperature can be achieved using tungsten heating elements that can be fabricated as hairpins like molybdenum or as a mesh element for ultrahigh-temperature applications. The temperature-resistance phenomena for tungsten is very similar to that of molybdenum so interchangeable elements are possible with proper design. Tungsten furnaces are available for temperatures up to 3000 °C (5430 °F).

Tantalum is much more ductile, dense, and expensive than either tungsten or molybdenum. It is rarely used for heating element applications.

Non-oxide Ceramic Heating Elements. Silicon carbide (SiC) and molybdenum disilicide (MoSi₂) are classified as nonoxide ceramic heating elements.

Silicon carbide is the principal non-oxide ceramic used for furnace heating elements that is commercially available in numerous shapes. Silicon carbide has a negative temperature coefficient of resistance at temperatures ranging from 500 to 750 °C (930 to 1380 °F). Resistance drops by a factor of two, then increases in a near linear fashion to twice its room-temperature value at 1500 °C (2730 °F). This presents a challenge in terms of control and power supply. Silicon carbide is extremely brittle; therefore, care must be used to avoid breakage of the heating elements.

Because the resistivity of silicon carbide is two orders of magnitude higher than graphite, higher voltages and lower currents are used. Commercial silicon carbide elements are designed with low resistance connections at the ends to reduce overheating at the connection points. The upper temperature limit is set by the decomposition of silicon carbide by the equation:

$$SiC\ (s) + O_2\ (g) \rightarrow SiO\ (g) + CO\ (g) \qquad (Eq\ 19)$$

This reaction manifests itself in the form of a large white plume of silica smoke followed shortly by heating element failure. The exact temperature depends on design factors, especially power density, but is typically in the vicinity of 1500 to 1600 °C (2730 to 2910 °F). The elements undergo an aging phenomenon in which their resistance slowly increases with time. This is caused by intergranular oxidation. Silicon carbide heating elements are usually installed vertically and wired in series (Fig 9). If one element fails it is recom-

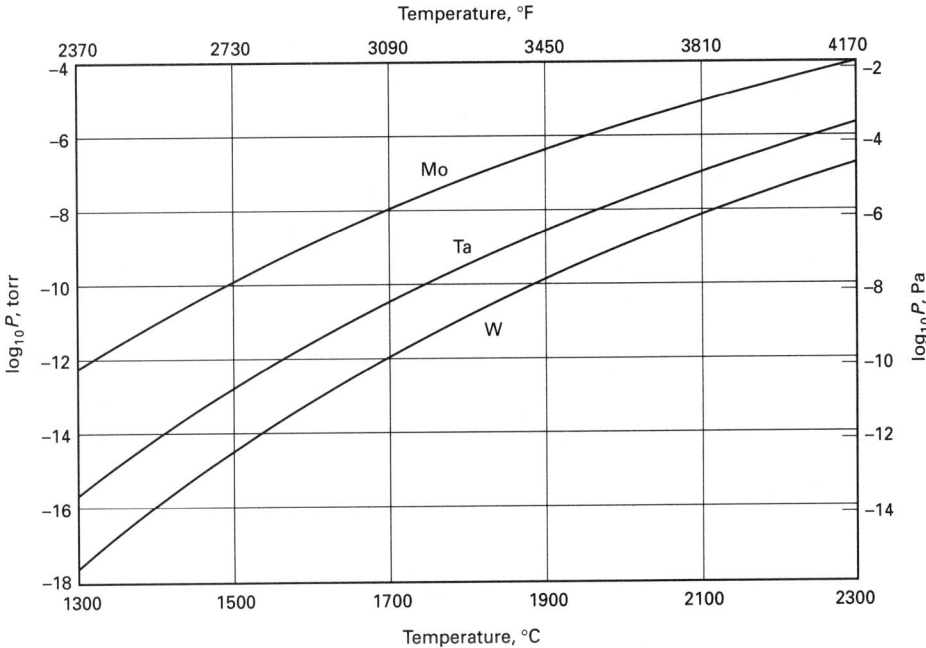

Fig 7 Vapor pressure of selected refractory metals from 1300 to 2300 °C (2370 to 4170 °F)

mended that all silicon carbide elements be replaced because the new elements will have a lower resistance. If only the failed element is replaced it will run cold due to its lower resistance in the series circuit. Conversely, if the elements are connected in parallel, the new element will run hot and age prematurely.

Molybdenum Disilicide. The other common non-oxide ceramic heating element is molybdenum disilicide, which is produced by pressing and sintering. These heating elements are available in the form of hairpins with a thick cross section at the ends to reduce connection temperature. The resistivity of molybdenum disilicide is $3.5 \times 10^{-7} \, \Omega \cdot m$

at room temperature and increases nearly linearly to $4 \times 10^{-6} \, \Omega \cdot m$ at 1700 °C (3090 °F).

Molybdenum disilicide elements form a surface coating of silicon dioxide (SiO_2) in air that prevents the oxidation of molybdenum disilicide. The breakdown of this coating limits the maximum operating temperature, which is highest in air and lowest in reducing atmospheres (Table 3).

Molybdenum disilicide elements are not usually used in vacuum furnaces. Their max-

imum operating temperatures are 1200 °C (2190 °F) at 0.13 Pa (10^{-3} torr) and 1450 °C (2640 °F) at 130 Pa (1 torr).

Molybdenum disilicide is very brittle at room temperature, but it becomes quite plastic at elevated temperatures. Because of this phenomenon, the elements are generally hung in a vertical position.

Oxide Ceramic Heating Elements. Two oxide ceramic heating elements are commercially available: lanthanum chromite ($LaCr_2O_4$) and stabilized zirconia (ZrO_2).

Lanthanum chromite offers a rare combination of properties for an oxide being both extremely refractory (with a melting point near 2500 °C, or 4530 °F) and an electrical conductor. Its resistivity decreases from 1×10^{-8} $\Omega \cdot m$ at room temperature to $1 \times 10^{-11} \, \Omega \cdot$ m at 1800 °C (3270 °F). Lanthanum chromite is stable in air up to 1850 °C (3360 °F) but decomposes in reducing atmospheres. The very high cost of lanthanum chromite limits its use to laboratory furnaces or small pilot plant furnaces.

Stabilized zirconia is an extremely refractory oxide that conducts electricity via the transport of oxygen ions. It has to be preheated to 1000 to 1200 °C (1830 to 2190 °F) using an auxiliary furnace (often using $MoSi_2$ or $LaCr_2O_4$ heating elements) before its resistivity becomes low enough for significant power generation. The auxiliary furnace then has to be removed prior to increasing the temperature to the maximum operating temperature of 2200 °C (3990 °F) in air. Zirconia is susceptible to damage caused by physical and thermal shock. These difficulties limit the use of ZrO_2 furnaces to laboratory applications where extremely high temperatures are required in an oxidizing atmosphere.

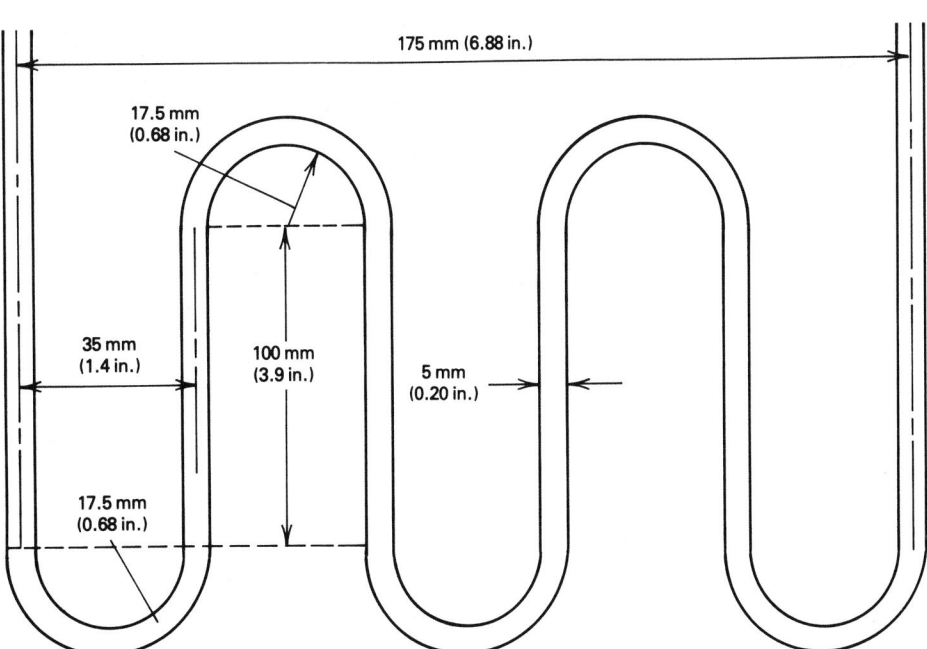

Fig 8(a) Schematic showing cross section of a typical refractory metal resistance heated furnace

Fig 8(b) Typical U-shaped configuration and dimensions of a refractory metal heating element used for the furnace shown in Fig 8(a)

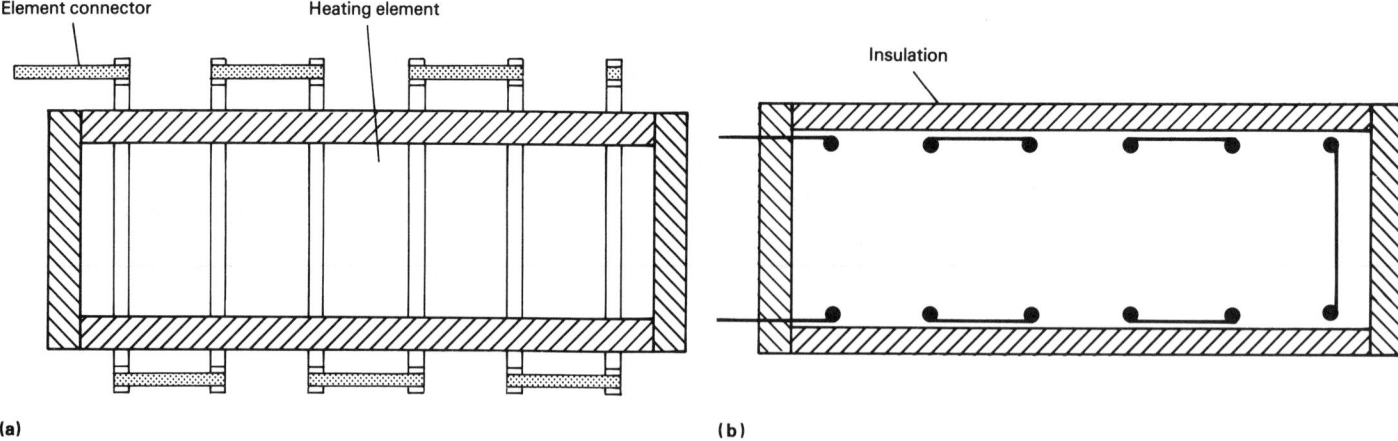

Fig 9 Schematic showing two views of a furnace using silicon-carbide heating elements connected in series. (a) Elevation view. (b) Plan view

Periodic or Batch Furnaces

A periodic furnace processes the ceramic material by placing the material in the furnace at (or near) room temperature, heating the ceramic to operating conditions, and then cooling the ceramic to room temperature. The principle advantages of this type of furnace, which is generally used for process development and evaluation, are simplicity and flexibility. The principle disadvantage of this type of furnace concerns scale-up. As the size increases, the surface-to-volume ratio generally decreases. If one considers a cylindrical chamber of length, L, and diameter, D, increasing the diameter by a factor of two increases the volume by a factor of four while increasing the surface by a factor of two. Thus, four times the material can be processed but the parts in the interior are twice as far from the heating elements. The major challenge to the furnace designer is to ensure that all the parts see the same temperature-time conditions. If one increases the length, the surface-to-volume ratio remains constant but the

Table 3 Recommended maximum operating temperatures for molybdenum disilicide heating elements in selected atmospheres

Atmosphere	Maximum operating temperature	
	°C	°F
Air	1800	3270
Oxygen	1700	3090
Nitrogen, carbon dioxide	1600	2910
Inert gases	1550	2820
Carbon monoxide	1500	2730
Hydrogen, 15 °C (60 °F) dew point	1460	2660
Dry hydrogen, methane	1350	2460

Source: Ref 7

introduction and removal of parts becomes more difficult. It is a tribute to commercial furnace designers that this problem has been overcome in most cases. A bottom loaded elevator furnace is one design typically used in batch furnaces.

Continuous Furnaces

In a continuous furnace, the temperature at any location is constant with time. The parts are moved through the temperature-distance profile at a velocity such that the desired process occurs at the desired rate. Continuous furnaces are best for mass production where large quantities of material are subjected to the same conditions. The major disadvantages of continuous furnaces include energy consumption because the furnace must be maintained at temperature throughout the process and the lack of flexibility because the only significant process variable is velocity. If a different temperature profile is required, a new furnace (or a redesign of an existing furnace) is required. At temperatures up to 1200 °C (2190 °F), the materials required for transporting parts through a furnace are cost-effective and readily available. As the temperature increases beyond 1200 °C (2190 °F), materials problems increase approximately exponentially.

Furnace Measurement and Control

The prime tenet of measurement and control is that a parameter must be measured if it is to be controlled. Fortunately, modern measurement techniques are very advanced. Sensors are available to generate electrical and/or pneumatic outputs that are directly related to the parameter measured. The principle parameters in furnace operation are temperature, pressure, and atmosphere composition. The general principles and practices of engineering measurements are given in detail by Holman (Ref 8).

Temperature Measurement

Specific information on temperature measurement is available from McGee (Ref 9). The concept of temperature as the hotness and coldness of things is a common misconception. If a wooden door and its brass doorknob are both at the same temperature, the doorknob feels colder than the door. This is related to the difference in heat transfer due to the difference in thermal conductivity of brass versus wood.

Historical Background. The earliest record of temperature measurement comes from Philo of Byzantium in a manuscript from the end of the 2nd century B.C. Philo described a form of a gas thermoscope that reacted to being placed alternately in the sun and in the shade. The first instrument to actually measure temperature was developed by Galileo early in the 17th century. The thermoscope proved to be inadequate because its readings varied with air pressure as well as temperature.

The liquid-in-glass thermometer was developed in the middle of the 17th century. Each of these thermometers was individually made and calibration was arbitrary. At the start of the 18th century, Fahrenheit introduced the concept of two fixed points to calibrate thermometers. He chose the lowest melting mixture of ice, water, and salt as the zero point, and the temperature of the human body was designated as 100°. With refinements, this results in the awkward Fahrenheit temperature scale still in everyday use in the United States. In 1742, Anders Celsius of Uppsala, Sweden inverted the centigrade scale with 100 divisions between the normal freezing and boiling points of water. Rather strangely, Celsius set the boiling point at 0 and the freezing point at 100! Independently, in 1743, Jean Pierre Christin in Lyons, France produced the "right-side-up" scale used today. In 1948, the term Celsius was officially adopted for this 0 to 100 scale. Later, the triple point (solid, liquid, vapor in equilibrium) of water (0.006 °C, or 32.0 °F) was adopted as the primary calibration point and there are no longer 100 degrees between the normal melting point (0 °C) and the normal boiling point (99.97 °C) of water. However, for all practical purposes, the Celsius scale of 0 to 100 °C is the common temperature scale used worldwide.

Experiments by Robert Boyle in 1660 showed that, at constant temperature, gases followed the rule:

$$PV = \text{constant} \qquad \text{(Eq 20)}$$

In 1787, Jacques-Alexandre Charles investigated the behavior of the volume of a gas as a function of temperature with the pressure held constant. In 1802, Joseph Gay-Lussac extended the work of Charles and showed that all gases expanded at the same rate when heated at constant pressure from the freezing point of water to the boiling point of water. This is expressed in terms of the isobaric coefficient of expansion, $\bar{\alpha}$, as:

$$V_{100°C} = V_{0°C}(1 + 100\bar{\alpha}) \qquad \text{(Eq 21)}$$

which can be rearranged to:

$$\bar{\alpha} = \frac{V_{100°C} - V_{0°C}}{100\,V_{0°C}} = \frac{1}{V_{0°C}}\left(\frac{\Delta V}{\Delta T}\right) = \frac{1}{267} \qquad \text{(Eq 22)}$$

In 1847, Victor Regnault obtained a more accurate value of $\bar{\alpha} = 1/273$.

Later experiments conducted with better precision showed that all gases had different behaviors, but that at low pressures the gases approached a common value of $\bar{\alpha}$:

$$\bar{\alpha} = \frac{1}{V_0}\left(\frac{\partial V}{\partial T}\right)_P = \frac{1}{273.15} \qquad \text{(Eq 23)}$$

From Eq 22, it is apparent that the volume will approach zero as the temperature approaches −273.15 °C. This established the concept that absolute zero as a negative volume is nonsensical.

In 1834, Emile Clapeyron combined Boyle's law and Gay-Lussac's law to obtain the equation of state for a mole of a perfect gas as:

$$PV = RT \qquad \text{(Eq 24)}$$

where R is the gas constant (in 8.3145 J/mol·K).

With the realization that temperature could be determined on an absolute scale using the expansion of a gas, the first gas thermometer was set up in 1851. The gas thermometer is now used by standards laboratories worldwide. Because it is inconvenient, alternate temperature measurement instruments are compared to the gas thermometer to develop a practical temperature scale. The normal melting point of several elements is used as a calibration point. One interesting feature is the variation in the melting point throughout history. As an example, Table 4 shows the variation for silver.

Table 4   Variation in the experimentally determined melting point of silver from the year 1828 to the present

Year	Melting point	
	°C	°F
1828	999	1830
1836	1000	1830
1863	960	1760
1879	954	1750
1892	985	1805
1892	971	1780
1898	962	1765
1900	961.5	1763
IPTS-68	961.93	1763.5

Temperature is now a defined property. The International Practical Temperature Scale (IPTS) was introduced in 1948 and redefined several times. Currently IPTS-68 (with some revisions accepted in 1976) is used. A new temperature scale, renamed International Temperature Scale (ITS-90), has been proposed and awaits international acceptance by the various national standards institutes. The major changes affect ultralow-temperature measurements near absolute zero and the dropping of the thermocouple as a primary standard.

ITS-90 defines temperature over several ranges:

- 0.5 to 4.221 K (−458.8 to −452.1 °F) based on the vapor pressure of helium
- 4.2 to 24.6 K (−452 to −415 °F) measured by gas thermometry
- 13.8 to 963 K (−435 to 1275 °F) measured by platinum resistance thermometry
- Above 965.08 K (1277.5 °F) measured by optical pyrometry

All secondary measurements are referenced to the primary techniques by comparison.

Practical Temperature Measurement. In theory, any physical property that has a unique value at a given temperature can be used. However, at the elevated temperatures used in sintering, only two techniques are in common use:

- Thermoelectric response of dissimilar metals (thermocouples)
- Electromagnetic radiation emission (optical pyrometers)

Thermocouples. In 1821, Thomas Seebeck discovered that when two wires of different composition were connected at their ends only to form a closed circuit, an electric current would flow if one of the connections was heated. The measurement of the potential (emf) causing this current is the basis of temperature measurement with thermocouples.

In 1885, Henri LeChatelier experimented with platinum and its alloys as a thermocouple system. He proposed the pure platinum and 90% Pt-10% Rh thermocouple that was used extensively until IPTS-68 replaced it as a primary standard for temperatures up to the melting point of silver.

There are several requirements for a thermocouple metal to be effective and practical in temperature measurement applications:

- Unique emf-temperature response
- Large change in emf per unit change in temperature (called the Seebeck coefficient)
- High resistance to change with time
- Reasonable price/performance ratio
- Ease of fabrication and use

Many types of thermocouples are available to cover temperatures ranging from −273 to 2000 °C (−459 to 3630 °F). For sintering applications, three types of thermocouples are in common usage. Their Instrument Society of America (ISA) designations are type K, type R, and type C. The actual alloy compositions used are determined by the manufacturer and the ISA designation refers to the standard emf-temperature response. The current thermocouple reference tables are available in National Bureau of Standards (NBS) Monograph 125 (Ref 10).

Type K, type R, and type C thermocouples are recommended for the following applications:

- Type K is based on nickel-chrome versus nickel-manganese-silicon-aluminum alloys and is used for temperatures up to 1200 °C (2190 °F). The thermocouple wires are protected by a thin layer of oxide on the surface. At temperatures in excess of 1000 °C (1830 °F), diffusion through the oxide layer increases rapidly and thermocouple deterioration gradually increases
- Type R is based on platinum-rhodium alloys and is used for temperatures up to 1600 °C (2910 °F). Platinum-base thermocouples have limited life in the presence of reducing conditions (especially carbonaceous gas). Deterioration by diffusion at the hot junction increases as the upper limit of operation is approached
- Type C is based on tungsten-rhenium alloys and is used for temperatures up to 2000 °C (3630 °F). These thermocouples are limited to operation in reducing atmospheres, high vacuum, and inert gases due to their tendency toward oxidation

Optical Pyrometry. At any temperature above absolute zero, all objects emit electromagnetic radiation that is proportional to T^4. This is expressed as the Stefan-Boltzmann law:

$$W = \sigma T^4$$
$$= 5.6718 \times 10^{-8}(\text{J/K}^4 \cdot \text{m}^2 \cdot \text{s})\,T^4(\text{K}) \qquad \text{(Eq 25)}$$

for an ideal black body where W is the total radiation, T is the absolute temperature of the body, and σ is the Stefan-Boltzmann constant. IPTS-68 defines temperatures in excess of the melting point of gold (1064.43 °C, or 1947.97 °F) using this effect while ITS-90 begins the definition of temperature via radiation at the melting point of silver (691.93 °C, or 1277.47 °F). There is no theoretical upper temperature limit to the technique. An ideal black body, which emits all the radiation possible, is a theoretical concept but can be closely approximated by a long narrow isothermal cavity. Practical optical pyrometers can be calibrated against the NBS black body standard. Corrections for the nonideal behavior of real objects is complex, but by the use of multiple frequency pyrometers (generally at two wavelengths), these corrections can be achieved in industrial equipment. One major limitation of optical pyrometry, the necessity of line-of-sight access to the object, can be easily overcome using optical fiber technology.

Pressure Measurement

Pressure is a measure of the force per unit area exerted on a surface by the random motion of gas molecules. Because the earth's atmosphere exerts a constant pressure (approximately), it is common to express this quantity in terms of gage or relative pressure. A positive gage pressure represents a pressure greater than that of the atmosphere while a negative gage pressure represents a vacuum. The common units of pressure in use in the U.S. are pound force per square inch gage (psig) and the pound force per square inch absolute (psia). However, the SI unit of pressure is the newton per square meter (N/m^2) which is named the pascal (Pa). The standard atmospheric pressure is defined as 14.696 psia, which equals 101,325 Pa. Early experiments by Torricelli in the 1600s showed that the atmosphere would support a column of mercury 760 mm (30 in.) high at sea level, which is the basis of another common unit of pressure, the torr or 1 mm Hg (130 Pa).

An unknown pressure can be measured by comparing the unknown pressure to a known pressure using a U tube filled with a fluid (Fig 10a); absolute pressure can be measured with a closed end evacuated tube containing a fluid, as shown in Figure 10(b).

$$\Delta P = \rho g h \qquad (Eq\ 26)$$

where ΔP is the pressure difference (in pascals); ρ is the fluid density (in kg/m^3); g is the acceleration due to gravity (9.806 m/s^2); and h is the height (in meters).

This technique has several disadvantages. Because it does not produce an electrical signal for process control applications and data logging, any tube device requires the use of a fluid that may affect the material processed and is insensitive to small changes in pressure.

An alternate method is the capacitance manometer that is the electrical analog of a diaphragm gage. A thin metallic diaphragm separates the two pressures and its displacement is measured by a change in capacitance. These gages can be designed to measure pressures from 100 Pa (0.015 psi) to hundreds of MPa.

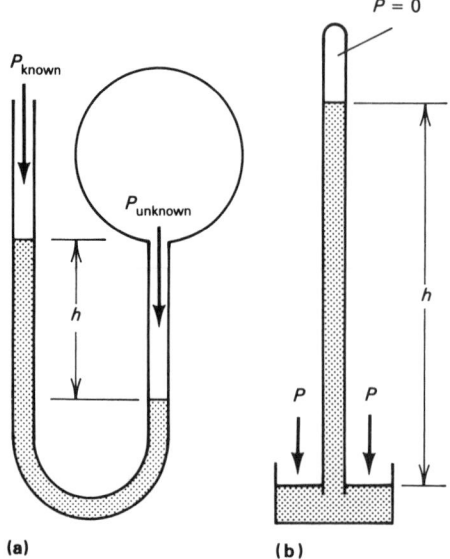

Fig 10 Pressure measurement using fluid levels. (a) Gage pressure. (b) Absolute pressure

The measurement of subatmospheric pressures is reviewed by Van Atta (Ref 11).

At lower pressures, thermocouple gages are used. The heat transferred from a hot wire to a fluid is a function of pressure; hence, the temperature of a wire with a current flowing can be related to pressure. This technique allows pressures to be measured from 1 to 1000 Pa (1.5×10^{-4} to 0.15 psi). At lower pressures, radiation exceeds convection while at higher pressures heat transfer is approximately constant.

At very low pressures, ionization of a gas occurs when a high dc potential is applied. Measurement of the current from anode to cathode can be related to the pressure at pressures from 10^{-7} to 0.1 Pa (1.5×10^{-11} to 1.5 $\times 10^{-5}$ psi). At lower pressures, the current becomes more difficult to measure; at higher pressures, very little ionization occurs.

Measurement of Atmosphere Composition

Methods of measuring gas composition include *in situ* techniques that use in-line sensors and external techniques that require taking a representative sample for eventual transfer to the analytical apparatus.

In Situ Techniques. Measurement of atmosphere composition *in situ* is difficult at elevated temperatures, but significant progress has been made in the last several decades. Two techniques show special promise: electrochemical sensors and infrared analysis.

Electrochemical Sensors. Stabilized zirconia (Ref 12) is a specific anion conductor for O_2^- ions over a wide range of oxygen partial pressures at high temperatures. The voltage from such a sensor is related to the oxygen partial pressure across the electrolyte by the Nernst equation:

$$E = \frac{RT}{n\mathscr{F}} \ln \left[\frac{(P_{O_2})_{\text{reference}}}{(P_{O_2})_{\text{unknown}}} \right] \qquad (Eq\ 27)$$

where E is the cell emf (in volts); R is the gas constant (8.314 J/mol·K); T is the temperature (in K); n is the number of equivalents transferred (in equivalents/mol); $\mathscr{F}$ is the Faraday constant (96,500 J/V equivalent); $(P_{O_2})_{\text{reference}}$ is the oxygen partial pressure at the reference electrode; and $(P_{O_2})_{\text{unknown}}$ is the oxygen partial pressure at the unknown electrode.

Zirconia sensors are used routinely in automotive emission control systems. Sensors for other gases such as hydrogen, hydrocarbons, and water vapor are currently available. Sensors for additional gases are under development (Ref 13).

Infrared Analysis. The infrared (IR) absorption of gases such as CO and CO_2 at characteristic wavelengths allows their analysis if the process gas stream can be directed through an optically clear detector. These IR analyzers must be calibrated with standard reference gases. The reference gas is usually supplied in a sealed container that should also contain the same matrix of gases as the system to be analyzed. The principle of IR analysis is very straightforward but in practice is complicated by the presence of particulate matter such as dust, smoke, and so on.

External Techniques. The two principle external methods used are gas chromatography and mass spectroscopy.

Gas Chromatography. Gases passing through a column containing an appropriate granular material travel at differing velocities depending on their molecular size. In gas chromatography, a sample of process gas is injected into a stream of inert carrier gas (usually helium) and passed through a long (typically 1 m, or 0.3 ft) tube where the gases separate. Each gas has a characteristic retention time and is detected at the exit of the tube. This method is useful for controlling processes that change slowly with time as a typical analysis takes 10 to 30 min.

Mass Spectroscopy. In a mass spectrometer, gas atoms are ionized and accelerated by a high-voltage dc field through a curved magnetic field. The radius of curvature in the field is given by:

$$r = 1.12 \times 10^{12} \left(\frac{mV}{z} \right)^{1/2} \frac{1}{B} \qquad (Eq\ 28)$$

where r is the radius of curvature (in meters); m is the mass of the ion (in grams); V is the applied potential (in volts); z is the charge on the ion; and B is the magnetic flux density (in Gauss).

Thus, by measuring the flux of ions at a given radius while varying the applied potential, gaseous ions can be detected based on their mass-to-charge ratio. The resulting spectra, while quite complex, can be used as both a qualitative and (with calibration) a quantitative technique. More detailed information can be found in Van Atta (Ref 11).

Furnace Control

Process control can be as simple as lighting a fire and standing back to a completely computerized feedback system controlled from a central location. Numerous manuals and texts on process control exist, but they all follow the central theme that any process can be controlled if it is understood. Temperature is controlled in a constant pressure process by the enthalpy balance:

$$\Delta H_{in} - \Delta H_{out} = \left(\frac{\partial \Delta H}{\partial t}\right) = C_p\left(\frac{\partial T}{\partial t}\right) \qquad \text{(Eq 29)}$$

If the energy input equals the energy output of a process, the temperature remains constant with time. Most temperature control techniques consist of measuring the temperature and adjusting the rate of power input. The simplest control technique is an off-on control: When the temperature is below the desired level, turn the power on; when the temperature is above the desired level, turn the power off. This results in a coarse form of control (Fig 11).

More elaborate control strategies involve so-called proportional integral differential (PID) control. Proportional action is the control mode by which a control signal is generated that is proportional to the difference between the measured and the set point temperature. As the set point is approached, the power input is reduced. Proportional control alone often results in a steady but offset temperature. Integral action corrects for offset. Differential control corrects for overshoot and undershoot of the temperature by adjusting the power input based on the rate of approach to the set point.

Furnace Safety

When working with furnaces used to process ceramics, the primary hazards focus on the electric power source and the toxic materials generated during the processing.

Electrical Safety

The dangers associated with electricity are well known but are often ignored. They are related to excessive voltage, current, and frequency.

High Voltage. The dangers of high voltage are well known. All potentials in excess of a few tens of volts must be properly insulated and physically isolated before any maintenance. Proper labelling and lock out procedures should be standard procedure. Fortunately, most industrial furnaces are well designed and labelled. Also, low-voltage designs are common because it reduces the danger of electrocution due to the current passing through a human body to ground.

High Current. The low voltages used lead to extremely high currents in order to produce the high power requirements. It is not unusual for a 20-V furnace to generate 200 kW of power. This requires a current of 10 kA. High current *per se* is not the danger. Because the resistance of a human being is very high, 20 V above ground causes very little current flow. The danger is two fold. First, if a person is wearing jewelry (such as a ring or metal watch band) and causes a short circuit, the resultant high current can cause severe local burns. Second, high current leads to high induced fields that surround the furnace bus work. This electric field can affect electronic devices, such as watches and pacemakers. High-current bus work should be shielded to protect both the operators and the equipment.

High Frequency. Induction furnaces often have high frequency power supplies. These power supplies can cause severe burns if an operator places a conductive object in the field. All induction coils should be shielded to protect the operators.

Chemical Safety

When materials are heated to high temperature, numerous chemical reactions occur. Many result in the formation of undesirable and potentially lethal compounds.

Carbon Monoxide. The hazards associated with carbon monoxide (CO) are well known. However, CO formation can occur by the reaction of moisture trapped in furnace insulation and any carbon present. If the insulation is graphite felt, special precautions are required to avoid carbon monoxide fumes. The simplest is the maintenance of a dry furnace. Any insulation exposed to humid atmospheres should be heated and evacuated at low temperature. It is simple common sense that the discharge from the furnace and the vacuum system (if any) be safely vented outside the immediate work area. Simple CO alarms are a wise investment against the dangers of carbon monoxide, the silent killer.

Another danger in furnaces using nitrogen gas at high temperatures is the formation of cyanide complexes such as cyanogen (CN) and hydrogen cyanide (HCN).

Cyanogen forms via the reaction:

$$C + {}^1/_2 N_2 \rightarrow CN \qquad \text{(Eq 30)}$$

This reaction occurs at high temperatures in graphite furnaces containing a nitrogen atmosphere (Fig 12). Above temperatures of 2200 °C (3990 °F), special precautions must be implemented because the CN concentration exceeds the lethal concentration guidelines. Fortunately, cyanogen is unstable and can be destroyed by passing the exit gases through an oxidizing flame.

Hydrogen cyanide forms via the reaction:

$${}^1/_2 H_2 + C + {}^1/_2 N_2 \rightarrow HCN \qquad \text{(Eq 31)}$$

This reaction can occur readily in graphite furnaces containing H_2/N_2 (commonly called forming gas) atmospheres that are used in metal sintering furnaces. Unknown to many, water vapor can react with graphite and nitrogen via:

$$H_2O + 2C + N_2 \rightarrow 2HCN + CO \qquad \text{(Eq 32)}$$

thus giving a double dose of lethal gas. Special precautions must be taken to maintain dry conditions as the thermodynamics of these reactions are highly favorable at all furnace temperatures. Proper venting is a must.

Chemical Vapor Deposition. Operations involving chemical vapor deposition (CVD) use extremely dangerous gases such as silane, arsine, phosphine, titanium tetrachlo-

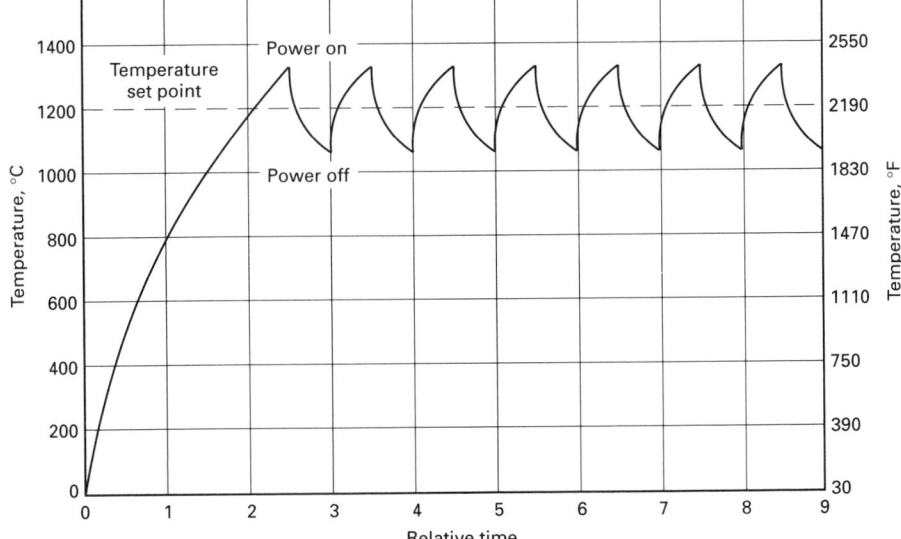

Fig 11 Alternate on-off cycling of a furnace operated by an off-on power control about the temperature set point

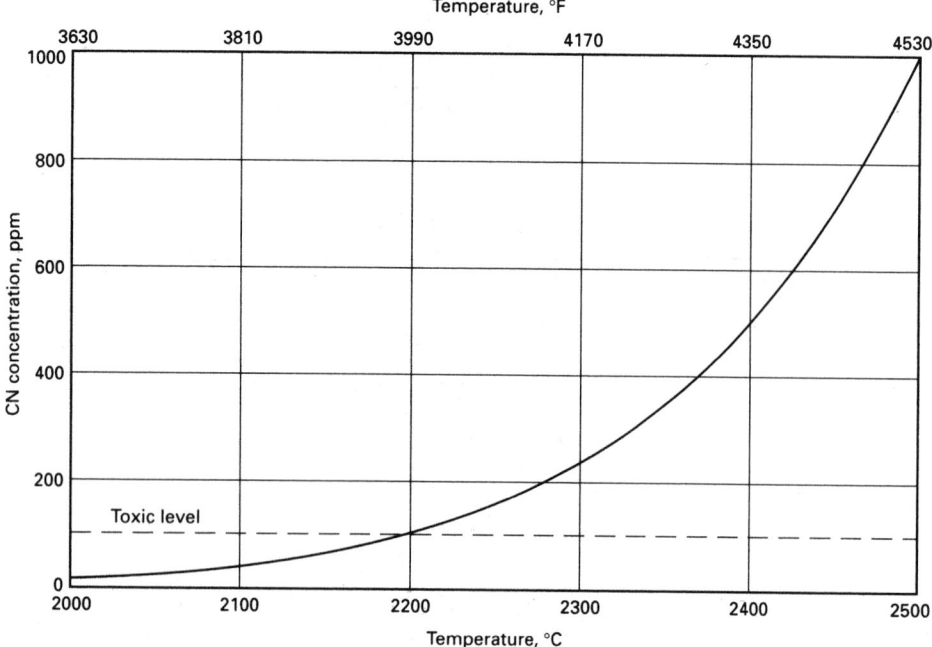

Fig 12 Cyanogen formation at 100 kPa (1 atm) nitrogen pressure in a graphite hot zone

ride, and so on. Fortunately, these gases are so dangerous that elaborate safety features are built into the system by the furnace vendor, but common sense is still required.

All furnace discharges should be assumed to be potentially dangerous and require proper ventilation.

Hydrogen atmospheres create a potential explosive hazard. Being lighter than air, hydrogen is relatively safe *unless* it is trapped in a closed area. If the concentration of hydrogen is low or high, it will not burn. The explosive limits of hydrogen in air are 6 to 75% so that it is easy to develop an explosive mixture. Proper venting and burn-off of exit gases is the only truly safe method of handling hydrogen. *Do not tamper with any safety interlocks.*

The organic chemicals used as binders and lubricants can decompose during binder removal; this is commonly called binder burnout in the ceramics industry. Often, the organics are literally burned to be transformed into carbon dioxide and water, which are benign. However, subatmospheric binder removal can result in depolymerization reactions that generate dangerous gaseous and liquid chemicals. The need for proper handling of binder breakdown products is essential.

REFERENCES

1. V. Biringuccio, *Pirotechnia*, Venice, 1540. Reprinted by M.I.T. Press, 1966
2. I. Barin, *Thermochemical Data for Pure Substances*, VCH Verlagsgesellschaft mbH, Weinheim, Germany, 1989
3. I.G.C. Dryden, *The Efficient Use of Energy*, Chapt. 8, IPC Business Press Ltd., Surrey, England, 1975
4. R.L. Brown, D.A. Everest, J.D. Lewis, and A. Williams, High-Temperature Processes with Special Reference to Flames and Plasmas, *J. Inst. Fuels*, Vol 41, 1988, p 433
5. M. Orfeuil, *Electric Process Heating, Technologies/Equipment/Applications*, Battelle Memorial Institute, Trans. from French, 1987
6. S. Zinn and S.L. Semiatin, *Elements of Induction Heating*, Electric Power Research Institute, 1988
7. *The Kanthal Handbook*, AG Kanthal, Hallstahammar, Sweden, 1985
8. J.P. Holman, *Experimental Methods for Engineers*, 3rd ed., McGraw-Hill, 1978
9. T.D. McGee, *Principles and Methods of Temperature Measurement*, Wiley-Interscience, 1988
10. Thermocouple reference tables, NBS Monograph 125, U.S. Department of Commerce, 1974
11. C.M. Van Atta, *Vacuum Science and Technology*, McGraw-Hill, 1965
12. E.C. Subbarao, *Solid Electrolytes and Their Applications*, Plenum Press, 1980
13. Proceedings of the International Meeting on Chemical Sensors, Fukuoka, Japan, T. Seiyama, K. Fueki, J. Shiokawa, and S. Suzuki, Ed., Elsevier, Amsterdam, 1983

The Firing Process

Licio Pennisi, New York State College of Ceramics, Alfred University and
Dipartimento di Chemica, University of Modena, Italy

THE FIRING PROCESS is usually the final stage in ceramic manufacturing. It is the stage at which a green (unfired), mechanically weak, newly formed workpiece is transformed into a strong, durable product. It is also the processing operation at which fabrication flaws or errors that may not be noticeable during manufacturing will usually manifest themselves. An improper firing schedule may also introduce flaws or even cause failure to ceramic components that were fabricated defect free.

There are several distinct regions in a typical firing curve that are critical and should be considered for the optimization of the firing process. The composition of the ceramic workpiece will usually determine which region of the firing curve is the most critical. For compositions having clay as the major constituent, the reactions that occur during firing are well understood (Ref 1). Some of the reactions that must be evaluated for developing an efficient firing curve are the dehydration of mechanical water and dehydroxylation of chemical water. Chemical water is bonded in the clay's matrix (for example, $xSiO_2 \cdot yAl_2O_3 \cdot zH_2O$), while mechanical water is a component that is added during manufacturing. Other common reactions include the volatilization of binders and lubricants introduced during the manufacturing process, the oxidation of organic materials that are found in many clays (especially in sedimentary clays), and reversible solid-state reactions during the firing cycle such as phase transformations and glass transition.

Table 1 shows some of the common and specific reactions that occur in a typical ceramic body along with the temperature ranges commonly associated with these reactions. Figure 1 shows a typical firing curve for a specific ceramic composition with selected reaction areas designated on the curve. It should be noted, however, that a firing curve designed for a particular ceramic composition will vary somewhat from this specific curve shown in Fig 1 and will depend on the final sintering temperature required and the actual reactions that occur.

Determination of Reactions during Firing

The determination of reactions during firing can best be accomplished by using techniques that measure chemical and physical changes as the sample being tested is fired:

- Thermogravimetric analysis (TGA)
- Differential thermal analysis (DTA)
- Dilatometry

Thermogravimetric analysis is a technique in which the weight (mass) of a sample is continuously measured as the sample is subjected to a selected firing profile that is usually linear. Reactions that occur during heating (for example, loss of absorbed water, loss of chemically bound water, and oxidations and reductions which are accompanied by a change in weight) can be identified within the temperature interval in which they occur.

Differential thermal analysis employs two thermocouples connected differentially so that their electromotive force (emf) outputs are in opposition. One of the thermocouple junctions is imbedded in the sample being tested while the other thermocouple junction is imbedded in a reference standard. The reference standard material, usually alumina (Al_2O_3), undergoes no reactions during heating. As the sample and standard are heated at a controlled linear rate, reactions occurring in the sample are recorded as peaks (indicating exothermic reactions) or valleys (indicating endothermic reactions).

The same reactions that result in a weight loss for TGA monitoring have an associated thermal effect which can also be detected using differential thermal analysis. There are solid-state reactions such as crystallization, however, that are not accompanied by a weight change. Figure 2 shows a typical curve and a typical DTA TGA curve plotted versus their common temperature axis for a sample of Bandy Black raw powder.

Dilatometry is the continuous measurement of the length of a sample as the specimen is subjected to a controlled linear heating rate. There are two common dilatometric tests that can be done on a sample: reversible dilatometry (TE) and irreversible dilatometry (ITE).

Reversible dilatometry is so named because the test can be repeated on the same sample (usually with similar results) and the test is used mainly to determine the thermal expansion of a fired sample. This test is used to evaluate a sample usually fired well beyond the testing temperature and thus further heating during the test has no effect on the thermal expansion coefficient of the sample. This test is also useful in identifying glass transi-

Table 1 Common and specific reactions that occur when clay minerals are fired at selected temperature ranges

Reaction		Temperature	
Type	Mechanism	°C	°F
Common	Loss of mechanical water	250–350	480–660
	Oxidation of organic matter	250–450	480–840
	Dehydroxylation	450–670	840–1240
	α-β quartz inversion of free silica	700–850	1290–1560
	Decomposition of carbonates	790–870	1455–1600
	Crystallization, liquid development, densification	880–960	1615–1760
Specific	Gibbsite dissociation	RT–200	RT–390
	Dehydroxylation of illite	170–700	340–1290
	Dehydroxylation of kaolinite	250–900	480–1650
	Dehydroxylation of montmorillonite	575	1065
	Dissociation of dolomite	800–950	1470–1740
	Dissociation of calcite	820–1020	1510–1870

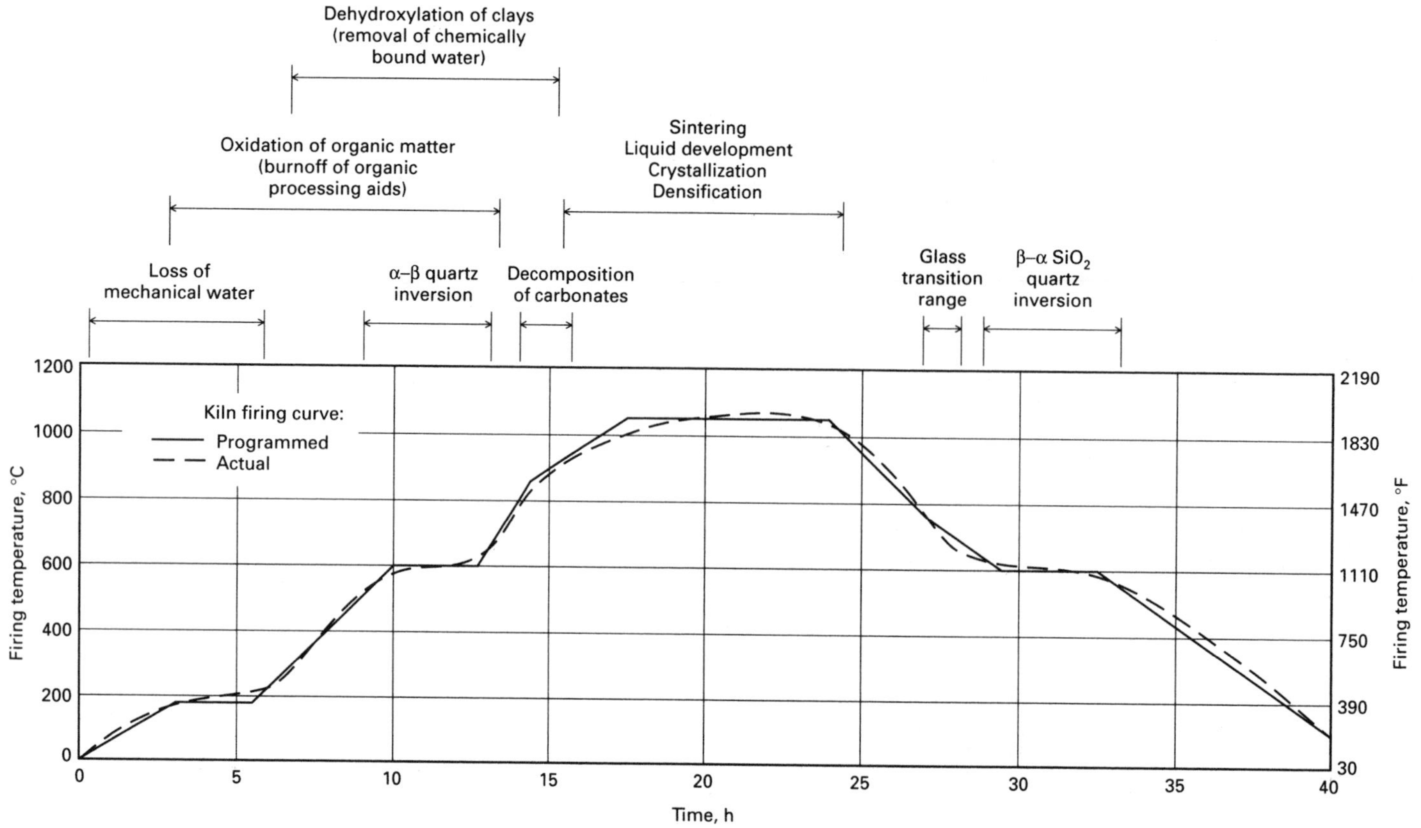

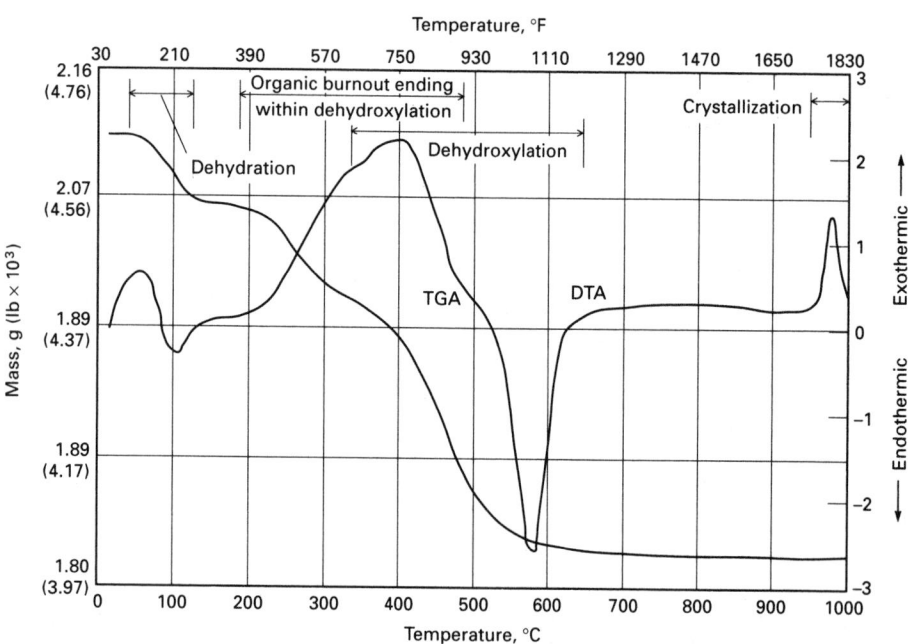

Dehydroxylation of clays (removal of chemically bound water)
Oxidation of organic matter (burnoff of organic processing aids)
Sintering
Liquid development
Crystallization
Densification
Loss of mechanical water
α–β quartz inversion
Decomposition of carbonates
Glass transition range
β–α SiO2 quartz inversion
Kiln firing curve:
Programmed
Actual
Firing temperature, °C
Firing temperature, °F
Time, h

Fig 1 Typical firing curves with corresponding reactions obtained for clay-bearing ceramic compositions

Temperature, °F
Organic burnout ending within dehydroxylation
Crystallization
Dehydroxylation
Dehydration
TGA
DTA
Mass, g (lb × 10³)
Exothermic
Endothermic
Temperature, °C

Fig 2 Thermogravimetric analysis and differential thermal analysis of Bandy Black raw powder fired at 10 °C/h (18 °F/h)

tion ranges and phase transformations such as the α-β quartz inversion that can be critical during the cooling of ceramic ware after sintering.

Irreversible dilatometry is run on a green (unfired) sample to obtain data that differ from the data obtained with reversible dilatometry. The data obtained includes the firing shrinkage of a composition and more importantly, the onset and completion of the sintering cycle. This test is irreversible because once the test is run, the sample has then been transformed from a green unfired sample to a dense sintered ceramic.

Sintering is a time-dependent and a temperature-dependent phenomena. Firing a composition at the lower end of the sintering range requires extended firing time while firing the composition at the higher end of the sintering zone requires less firing time to obtain the same fired density.

Note that the heating rate is a critical factor in all three of the above-mentioned thermal analysis tests. Reactions occur over shorter times and broader temperature ranges with increased heating rates. Many are detected at later times, and hence, at higher temperatures (Fig 3). In order to obtain consistent and repeatable results, heating rates must be identical from test to test. Note that although reaction temperatures for many transformations in materials can be found in the literature (Ref 1–3), numerous authors disagree on the actual temperatures at which the reactions are initiated and sometimes disagree on the temperature ranges over which the reactions occur. This disagreement is due to differences in the heating rate, sample holder geometry, furnace design, sample particle size, sample density, and ambient atmosphere dur-

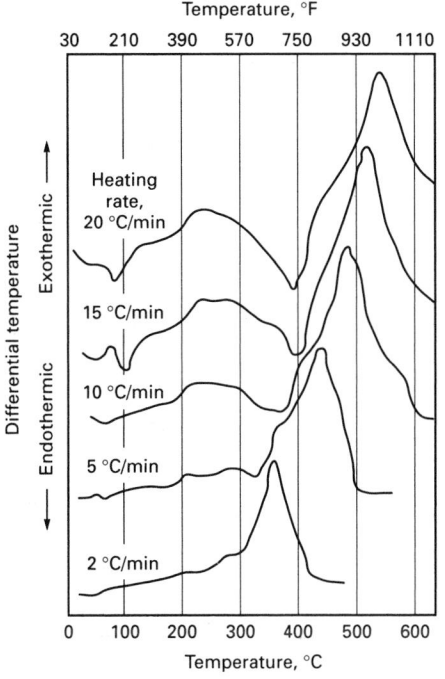

Fig 3 Differential thermal analysis of unfired dielectric material bound with wax and acrylic binder heated at selected heating rates: 2 °C/min (3.6 °F/min), 5 °C/min (9 °F/min), 10 °C/min (18 °F/min), 15 °C/min (27 °F/min), 20 °C/min (36 °F/min)

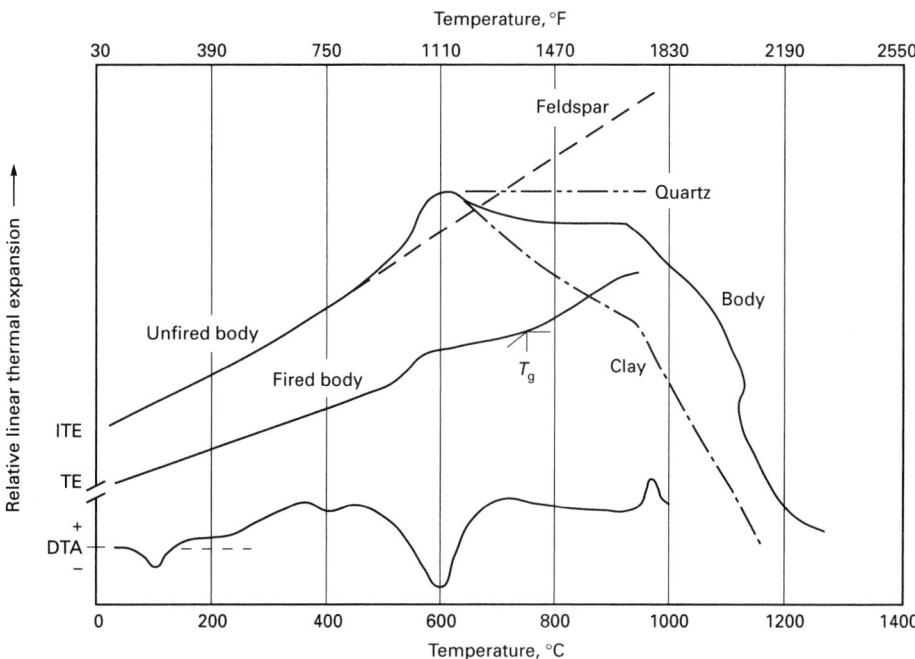

Fig 4 Thermal analyses used to determine firing curve: ITE, irreversible thermal dilatometry of green body (approximate thermal dilatometry of body components is also shown); TE, thermal expansion of fired ceramic body; DTA, differential thermal analysis of green body; T_g, glass transition temperature

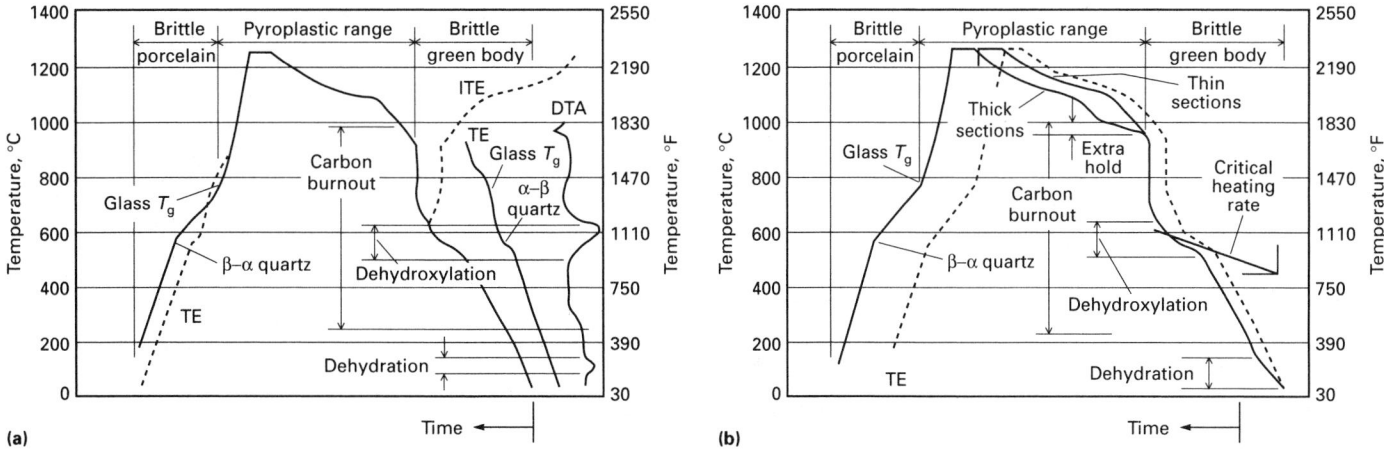

Fig 5 First stage of theoretical firing curve for ceramics in the porcelain family. (a) Initial curve plotted from ITE and TE curves. (b) Modified curve obtained when effects measured by differential thermal analysis of green body are included in plot

ing testing. By maintaining consistency in the aforementioned variables, comparable results can be obtained during thermal analysis.

Determination of the Firing Curve

In 1982, Funk (Ref 4) published a method for designing the optimum firing curve for ceramics in the porcelain family. By using reversible and irreversible thermal expansion curves (Fig 4), Funk showed that it is possible to obtain the first stage of a firing curve (Fig 5a). Modifications were then carried out on this curve by imposing the effects obtained by DTA analysis (Fig 5b). Once the theoretical firing curve shape is complete, there remains the task of determining the critical heating rate through dehydroxylation that would in effect establish the time axis for the firing curve. Funk described how this critical rate can be determined empirically by carefully monitoring the temperatures of the surface, center, and other locations of an actual ceramic product for which the initial firing curve has been designed. In essence, when applying this technique at various heating rates, the failure of the sample tested will be detected when the temperature sensors at different locations within the sample begin to approach a uniform temperature due to their exposure to the kiln atmosphere after the sample has failed. The heating rate just below the critical heating rate that caused sample failure is the fastest heating rate that is allowed through the zone of maximum thermal expansion rate, which usually occurs at the α-β transition, and the clay mineral dehydroxylation. By establishing the heating rate through this critical zone, the total time axis for the firing curve can be interpolated.

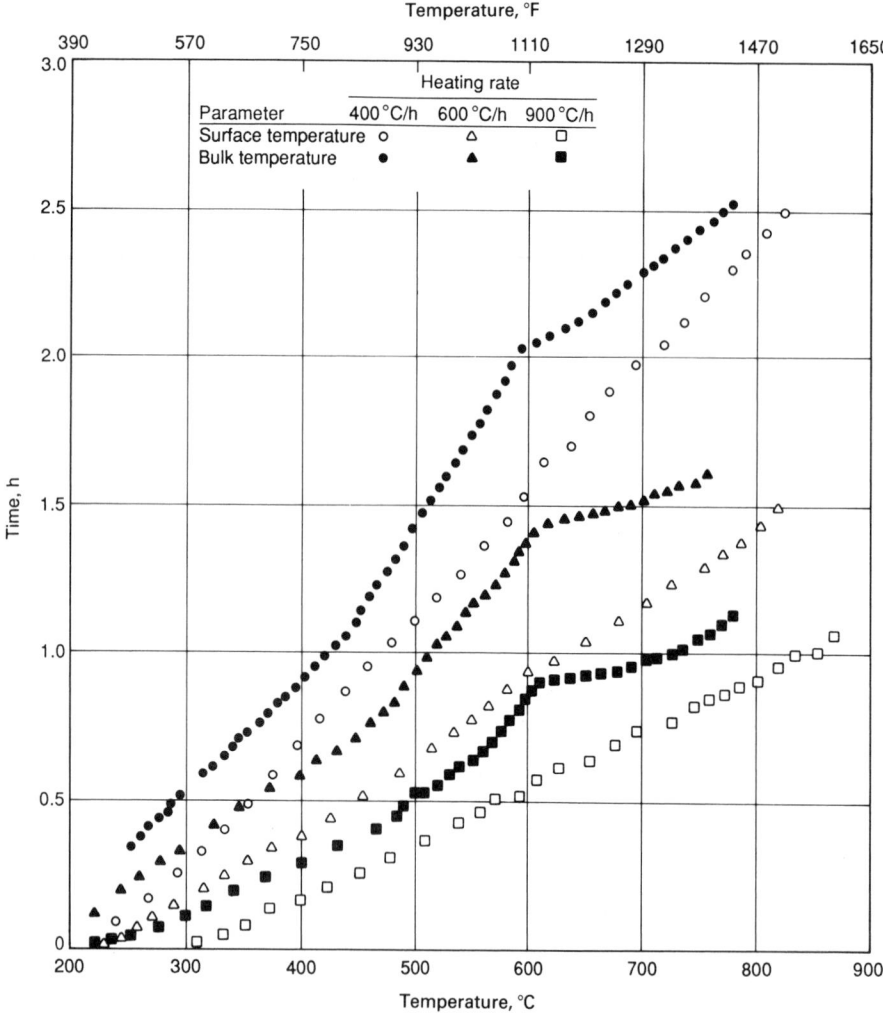

Fig 6 Temperature profiles of the surface and the center of typical 25 mm (1 in.) cube whiteware bodies fired at heating rates of 400, 600, and 900 °C/h (720, 1080, and 1620 °F/h)

DTA and TGA analyses will indicate if further modification is necessary because of excessive amounts of carbonaceous material in the composition. Funk also described how, by doing draw trials on samples, the time extension to fully allow for total oxidation of this material can be determined. This is another critical phase in firing because the incomplete burnout of carbon-containing material will lead to black coring or blue coring and other defects in glazed ware after the surface is sealed by sintering. Black coring is defined by American Society for Testing and Materials (ASTM) as a black or gray course in the interior of a brick, usually associated with carbonaceous clays that have had insufficient oxidation treatment (Ref 5). Blue coring is the color caused by residual reduction of any iron- or titanium-bearing materials in the body, after the organics have been burned out. These minerals must be reoxidized to eliminate the blue core.

Once all modifications have been included and the best firing curve has been determined, there remains one final step when transferring laboratory results to production scale. Figure 6 shows the temperatures of the surface and the center of a 25 mm (1 in.) cube fired at various linear heating rates. The composition of the cube is a typical whiteware body. Figure 7 shows the temperature difference between the surface and bulk as a function of the surface temperature while Fig 8 shows a differential thermal analysis carried out at 10 °C/min (18 °F/min) on the same composition.

It is important to note that the DTA dehydroxylation peak occurs at approximately 595 °C (1100 °F) (Fig 8). The same reaction determined by the actual measurement of the surface and bulk of a production size sample

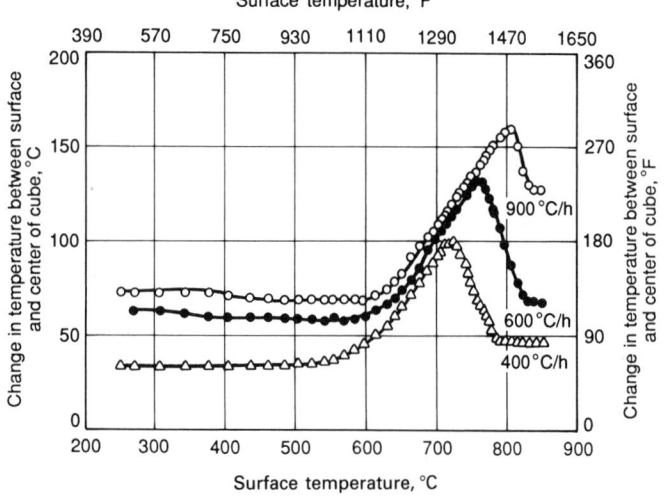

Fig 7 Plot of the temperature difference between the surface and the center of a 25 mm (1 in.) whiteware body cube versus the surface temperature of the cube at three selected heating rates

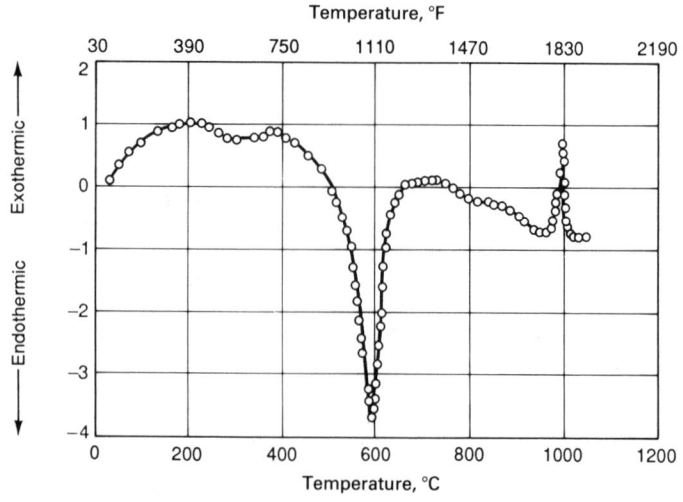

Fig 8 Differential thermal analysis of the whiteware body used in obtaining Fig 6 and 7 fired at 600 °C/h (1080 °F/h)

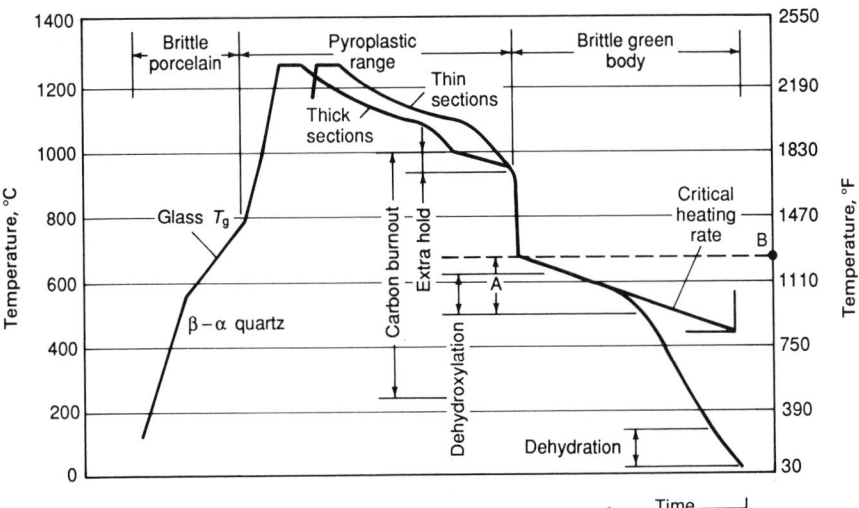

Fig 9 Final modification of firing curve in the dehydroxylation zone due to product cross-sectional area: A, reduction zone extension, proportional to the thickness of the product; B, maximum change in temperature between surface and center of product determined by testing (see Fig 7)

Future Outlook

Long firing schedules in the sintering of ceramic products have the potential to significantly reduce deleterious effects of the reactions occurring during firing and to allow the safe firing of virtually all ceramic compositions. However, with long firing times, one also incurs higher fuel costs and lower production output. The optimization of a firing profile for a ceramic product will improve productivity by increasing production output, reducing fuel costs, improving product quality, and reducing potential product liability.

ACKNOWLEDGMENT

The author wishes to acknowledge the collaboration of Prof. James Funk and his encouragement in the writing of this article.

REFERENCES

1. R.W. Grimshaw, *The Chemistry and Physics of Clays and Other Ceramic Materials*, John Wiley & Sons, 1971
2. L. Padoa, *La Cottura dei Prodotti Ceramici*, Faenza Editrice S.p.A., 3rd ed., 1982
3. R.C. Mackenzie, Ed., *Differential Thermal Analysis, Vol I: Fundamental Aspects, Vol II: Applications*, Academic Press, 1970
4. J.E. Funk, Designing the Optimum Firing Curve for Porcelains, *Ceram. Bull.*, Vol 62 (No. 6), 1982, p 632–635
5. E.C. Van Schoik, Ed., *Ceramic Glossary*, The American Ceramic Society, 1963

SELECTED REFERENCES

- M.J. Pope and M.D. Judd, *Differential Thermal Analysis*, Heyden, 1977
- W.J. Smothers and Y. Chiang, *Handbook of Differential Thermal Analysis*, Chemical Publishing, 1966
- R. West, Analysis and Tests for Ceramic Raw Materials, *Ceramics III Laboratory Handbook*, University Press, 1976

shows the maximum temperature change corresponding to dehydroxylation to be ≈710, 750, and 790 °C (1310, 1380, and 1455 °F) for heating rates of 6.6, 10, and 15 °C/min (12, 18, and 27 °F/min), respectively (see Fig 7). Although there is some peak shift in laboratory samples with respect to heating rate, the shift due to actual sample size is much more dramatic and significant. Similar data were obtained for a typical shale composition and for a typical porcelain composition.

Funk modified his theoretical firing curve by inserting an extra hold to allow for carbon burnout of thick sections (Fig 5b). However, a final modification must be made to allow for reaction lags in actual production. This lag can be determined by monitoring the internal and surface temperatures of the sample with similar dimensions as the production ware. By analyzing the temperature differences, the actual kiln temperature at which the critical dehydroxylation reaction has ceased in the interior of the sample may be determined. Funk's critical heating rate must then be continued to this temperature. Figure 9 shows an example of this shift. In essence, the theoretical firing curve has to be shifted to allow for sample dimensions. This shift increase should be proportional to the product's cross-sectional area for similar compositions. A similar analysis may be obtained for the cooling cycle to allow for temperature lags in the glass transition and phase inversion zones. However, this area is not as critical because of the increased strength of the product after sintering. Funk's modifications in this area are quite sufficient to allow for optimum safe cooldown.

Fundamentals of Sintering

Randall M. German, Engineering Science and Mechanics Department, Pennsylvania State University

SINTERING provides the interparticle bonding that generates the attractive forces needed to hold together the otherwise loose ceramic powder mass. Instead of setting a sintering schedule by trial and error, a better approach is to understand the fundamentals of interparticle bonding. The effects of both intrinsic material characteristics and extrinsic powder and atmosphere characteristics are understood. To study sintering, one monitors the density, surface area, shrinkage, or some other easily measured parameter that correlates favorably with the sintered properties.

Sintering is the result of atomic motion stimulated by a high temperature. Diffusive processes generally are dominant. Several variables influence the rate of sintering. These include the initial density, material, particle size, sintering atmosphere, temperature, time, and heating rate. The geometric progression associated with sintering can be divided into three stages: initial, intermediate, and final. During the initial stage, particles form bonds at the particle contacts. As densification proceeds, new contacts will form; hence, there will be variations in the degree of sintering from point-to-point in the microstructure due to the delayed contact formation. With prolonged sintering, the pore structure becomes smoothed, leading to the intermediate stage, which corresponds to the open, continuous pore structure that exists between densities of approximately 70 and 92% of theoretical. In the intermediate stage, the sintering rate is continuously decreasing and is very sensitive to the pore-grain boundary morphology. Grain growth occurs in the later portion of sintering, during which the pores become spherical and isolated. Elimination of isolated pores is difficult at this point. The final stage of sintering corresponds to closed, spherical pores that shrink slowly by vacancy diffusion to grain boundaries. Such densification is sensitive to the relative grain size and the attachment of pores to grain boundaries. Any atmosphere trapped in the pores will inhibit densification.

Several potential mass transport paths can be active during sintering. The characteristic distinction is between surface and bulk processes. Bulk transport provides densification, while surface transport contributes to interparticle bonding without densification. Be-

cause of complexities in the particle-pore geometry, mass transport processes, and stages of sintering, it is typical to resort to computer simulations to quantify the sintering models (Ref 1, 2, 3–5). These simulations have accurately predicted the behavior of several systems. One major benefit is that they allow various processing options to be evaluated without the need for repetitive and painstaking laboratory measurements.

Additive phases that improve diffusion rates during sintering are used in many ceramic materials. These phases can be used to stabilize desirable crystal structures or, more typically, to form a liquid phase to increase the rate of sintering. Liquid-phase sintering is an attractive option for many of the high-performance materials because of the rapid processing cycle, high final density, and excellent final properties.

The atomic motion in sintering contributes to the formation of weld bonds between the particles and the elimination of pores. The sintered material exhibits improved strength and other properties. To enhance the overall process, external force (pressure) or chemical additives can be used.

Sintering Fundamentals

Sintering is the bonding together of particles when heated to high temperatures. On a microstructural scale, this bonding occurs as cohesive necks (weld bonds) grow at the points of contact between particles. Figure 1 shows scanning electron micrographs of neck formation between sintering spheres. Such neck growth causes all of the important property changes associated with sintering. There are several laws that guide the application of sintering fundamentals to practical situations (Ref 6–10). It is the objective of this section to highlight the fundamentals of sintering.

Particles sinter together by atomic motions that act to eliminate the high surface energy associated with an unsintered powder. The surface energy per unit volume is inversely proportional to the particle diameter. Thus, smaller particles have more energy and sinter more rapidly than larger particles. However, not all of the surface energy is necessarily available as a driving force for sintering. For a crystalline solid, nearly every particle con-

tact will evolve a grain boundary with an associated grain boundary energy. These grain boundaries prove important to atomic motion because the boundaries are defective regions with high atomic mobility.

During sintering, neck growth by the movement of mass to the neck is desirable because it reduces the surface energy by decreasing the total surface area. The structural changes associated with neck growth depend on several possible transport mechanisms, most of which are diffusion processes. Diffusion is thermally activated, meaning that there is minimum energy necessary for atomic or ionic movement and there must be available sites. This motion depends on the atoms or ions attaining an energy equal to or above the activation energy necessary to break free from their present sites and moving into other available sites. The population of available sites and the number of atoms with sufficient energy to move into these sites both vary with the Arrhenius temperature relation,

$$N/N_0 = \exp(-E/kT) \qquad \text{(Eq 1)}$$

where N/N_0 is the ratio of available sites or activated atoms to total atoms, E is the appropriate activation energy, k is the Boltzmann constant, and T is the absolute temperature. Thus, sintering is faster at higher temperatures because of the increased number of active atoms and available sites.

Because the elimination of surface energy is the main objective of sintering, one obvious gage in achieving this goal is the surface area. The surface area declines rapidly from an initial value and provides a gage of the degree of sintering. The surface area can be measured using microscopic analysis, gas adsorption, or gas permeability techniques (Ref 7).

Another measure of sintering is the relative neck size ratio, X/D, defined as the neck diameter, X, divided by the particle diameter, D (Fig 2). In addition to undergoing neck growth, a sintered compact usually shrinks, densifies, and increases in strength. The shrinkage that accompanies neck growth is illustrated in Fig 3 for two temperatures. The shrinkage, $\Delta L/L_0$, is the change in compact length from the initial dimension, ΔL, divided by the initial dimension, L_0. Because of shrinkage, the compact densifies from the

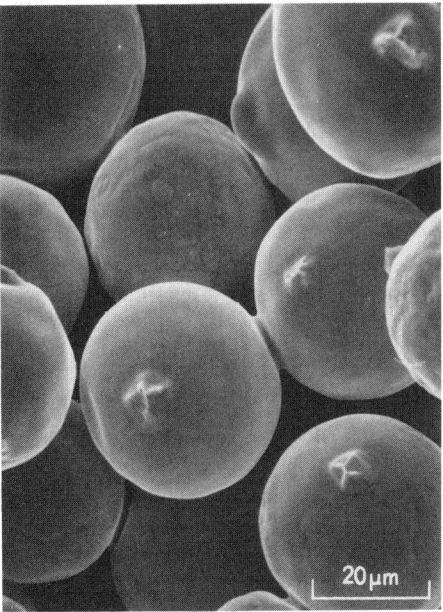 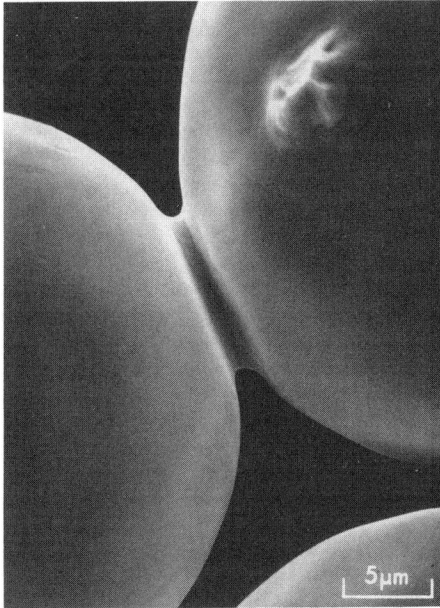

Fig 1 Scanning electron micrograph of neck formation between monosized spherical particles induced by sintering

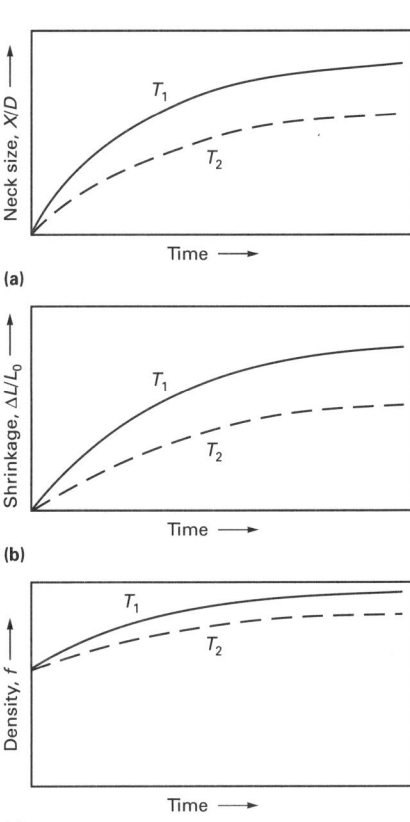

Fig 3 Effect of sintering temperature and sintering time on several factors: (a) neck growth; (b) shrinkage; (c) density. Data are shown at temperatures $(T_1 > T_2)$ with more rapid sintering at the higher temperature.

fractional green density f_g to the fractional sintered density f_s (Fig 3) according to the relation

$$f_s = f_g(1 - \Delta L/L_0)^{-3} \qquad \text{(Eq 2)}$$

The densification, Ψ, is the change in density due to sintering divided by the change needed to attain a pore-free solid:

$$\Psi = (f_s - f_g)/(1 - f_g) \qquad \text{(Eq 3)}$$

Densification, final density, neck size, surface area, and shrinkage are related measures of the pore elimination process during sintering. At high temperatures, sintering is more rapid, consequently shorter times are needed to attain an equivalent degree of sintering.

Kinetics of Sintering

Consider two spherical particles in contact with each other (Fig 2). In actual powder compacts, there are many such contacts on each particle (Fig 4a). As the bond between the particles grows, the microstructure changes as described in Fig 4. The bonds between contacting particles enlarge and merge. At each point of contact, a grain boundary grows to replace the solid-vapor interface. The initial stage of sintering occurs while the neck size ratio, X/D, is less than 0.3 (Fig 4b). During this stage, the kinetics are dominated by the curvature gradients near the interparticle neck. The pore structure is open and fully interconnected, although the pore shape is not very smooth.

In the intermediate stage, the pore structure is much smoother and has an interconnected, cylindrical structure (Fig 4c). The compact properties are developed predominantly in this stage. It is common for considerable grain growth to occur in the latter portion of the intermediate stage of sintering. This is accompanied by possible pore isolation. A small grain size is very important to maintaining a high sintering rate. Grain growth and pore separation from the grain boundaries are unfavorable events with respect to densification of a powder compact.

The open pore network becomes geometrically unstable when the porosity has shrunk to approximately 8% (92% of theoretical density). At this point, the cylindrical pores collapse into spherical pores, which are not as effective in slowing grain growth (Fig 4d). The appearance of isolated pores signals the beginning of the final stage of sintering and slow densification. Gas in the pores at this point will limit the end-point density; accordingly, vacuum sintering can be beneficial for achieving high final densities as long as the compound does not decompose or evaporate.

There are no clear distinctions between the sintering stages. The initial stage generally corresponds to a microstructure with large curvature gradients. The neck size ratio and shrinkage are both small and the grain size is no larger than the initial particle size. In the intermediate stage, the pores are smoother and the density is between 70 and 92% of theoretical. Grain growth occurs late in the intermediate stage; thus, the grain size is larger than the initial particle size. By the final stage of sintering, the pores are spherical and closed, and grain growth is evident. The four micrographs shown in Fig 5 provide a perspective on actual sintering behavior. These photographs show the changes in density, grain size, and pore structure that are characteristic of sintering. The scanning electron micrograph in Fig 5(d) shows fracture along the grain boundaries of a compact sintered to the final stage. The spherical pores are present on the fractured grain boundaries, a condition that is desirable for continued final stage densification.

Transport mechanisms are the methods by which mass flow occurs in response to the driving forces. The two classes of transport mechanisms are surface transport and bulk transport. These are composed of the actual

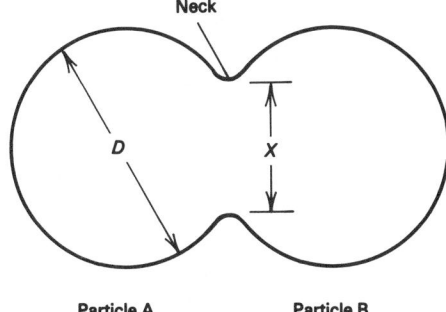

Fig 2 Schematic showing neck formation and growth in a two-particle model. Neck size ratio, X/D, is defined as the diameter of the interparticle neck, X, divided by the particle diameter, D.

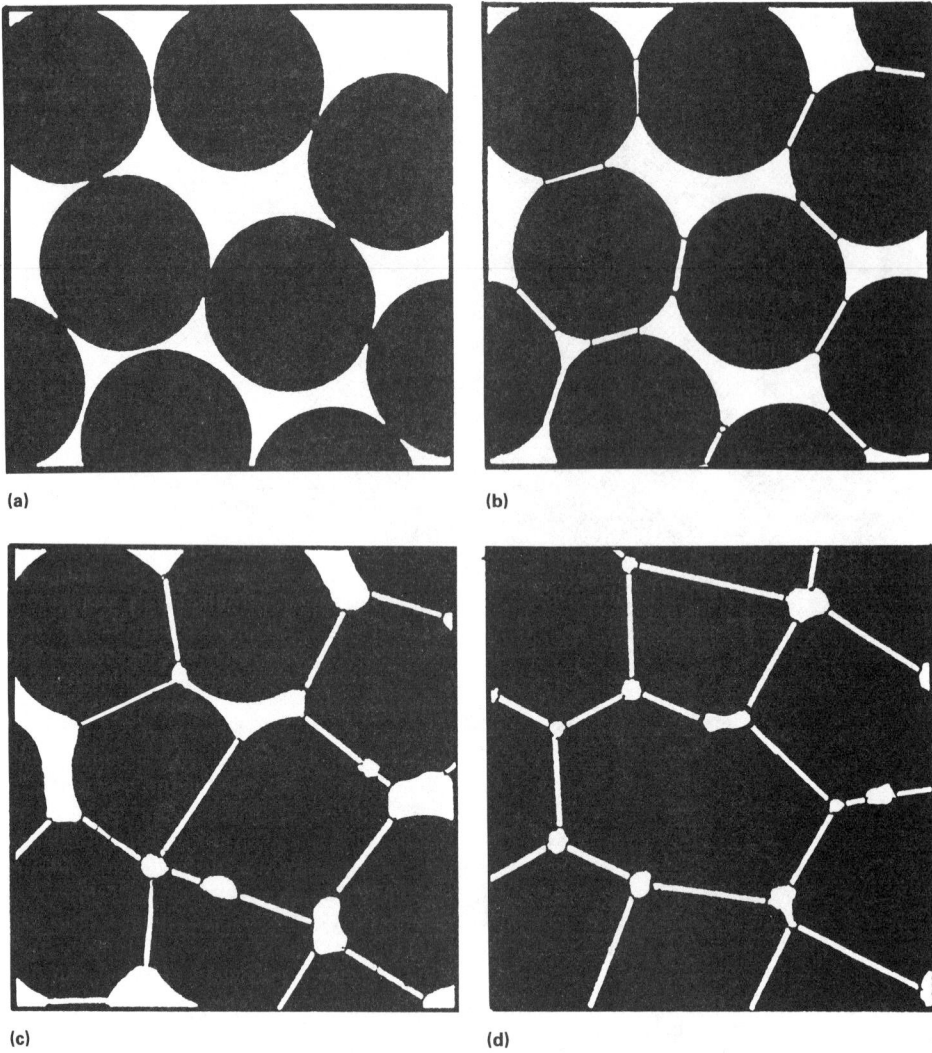

(a)

(b)

(c)

(d)

Fig 4 Development of the interparticle bond as the ceramic microstructure is transformed during the sintering process. (a) Loose powder (start of bond growth). (b) Initial stage (the pore volume shrinks). (c) Intermediate stage (grain boundaries form at the contacts). (d) Final stage (pores become smoother)

atomistic mechanisms contributing to mass flow (Fig 6).

Surface transport involves neck growth without a change in particle spacing (no densification) due to mass flow originating and terminating at the particle surface. Surface diffusion and evaporation-condensation are the two most important contributors during surface transport-controlled sintering. Surface diffusion dominates the sintering of many covalent solids (such as pure boron and silicon carbide). Evaporation-condensation is not as important, but is observed to dominate the sintering of low-stability ceramics such as sodium chloride.

Bulk Transport. In contrast, bulk transport-controlled sintering results in shrinkage. The mass originates at the particle interior with deposition at the neck. Bulk transport mechanisms include volume diffusion, grain boundary diffusion, plastic flow, and viscous flow. Plastic flow is usually not important for crystalline ceramics, because any dislocation

motion would produce identical ionic neighbors, a condition that is undesirable. Also, the stresses due to surface tension encountered during sintering are generally insufficient to generate dislocations. However, some transmission electron microscopy (TEM) observations indicate dislocation motion in the neck region due to the thermal stresses associated with rapid heating. In contrast, amorphous materials (such as glasses and plastics) sinter by viscous flow, where the particles coalesce at a rate that depends on the particle size and material viscosity. Viscous flow is also possible for crystalline ceramics containing glass phases on the grain boundaries. Grain boundary diffusion is fairly important to densification for most crystalline materials and appears to dominate the densification of many common ceramics, including alumina (Al_2O_3). Volume diffusion is often restricted by the defect structure in the ceramic. In ionic substances, the mobility of the slower ion species dictates the sintering rate.

While both surface and bulk transport processes give neck growth, the main difference is in density (or shrinkage) during sintering. Generally, bulk transport processes are more active at higher temperatures. The actual behavior is dependent on the material, particle size, sintering stage, temperature, and several other process variables.

Isothermal Neck Growth Determined by Initial Stage Sintering Model. The simple models of sintering all inherently focus on isothermal neck growth as measured by the neck size ratio, X/D:

$$(X/D)^n = Bt/D^m \qquad \text{(Eq 4)}$$

where X is the neck diameter, D is the particle diameter, t is the isothermal sintering time, and B is a collection of material and geometric constants. The values of n, m, and B depend on the mechanism of mass transport (Table 1). Generally, the model represented by Eq 4 is valid for $X/D < 0.3$.

Although not very accurate, Eq 4 does illustrate some key processing factors:

- High sensitivity to the inverse particle size, smaller particle sizes giving more rapid sintering
- In all cases, temperature appears in an exponential term, meaning that small temperature changes can have a large effect
- Time has a relatively small effect in comparison to temperature and particle size

A similar kinetic form is applicable to the surface area loss during initial stage sintering (Ref 7). The surface area provides an index of the degree of sintering by measuring the cumulative pore changes throughout the compact. However, surface area should not be used to monitor the later stages of sintering because pore closure will hinder accurate assessment of the surface area.

Shrinkage during initial stage sintering follows a similar kinetic law:

$$(\Delta L/L_0)^{n/2} = Bt/(2^n D^m) \qquad \text{(Eq 5)}$$

where $n/2$ is typically between 2.5 and 3, D is the particle diameter, and t is the isothermal time. The parameter B is exponentially dependent on temperature:

$$B = B_0 \exp(-Q/kT) \qquad \text{(Eq 6)}$$

where k is the Boltzmann constant, T is the absolute temperature, and B_0 is a collection of constants that depends on surface energies, atomic size, atomic vibration frequencies, and system geometry. The activation energy Q is a measure of the energy needed to stimulate atomic motion. Shrinkage is easily measured using samples heated to various temperatures for various times. Alternatively, dilatometers or direct imaging techniques can be used to continuously record shrinkage during heating. A common technique with dilatometry is to scan over all possible sintering temperatures by heating the compact at a constant rate

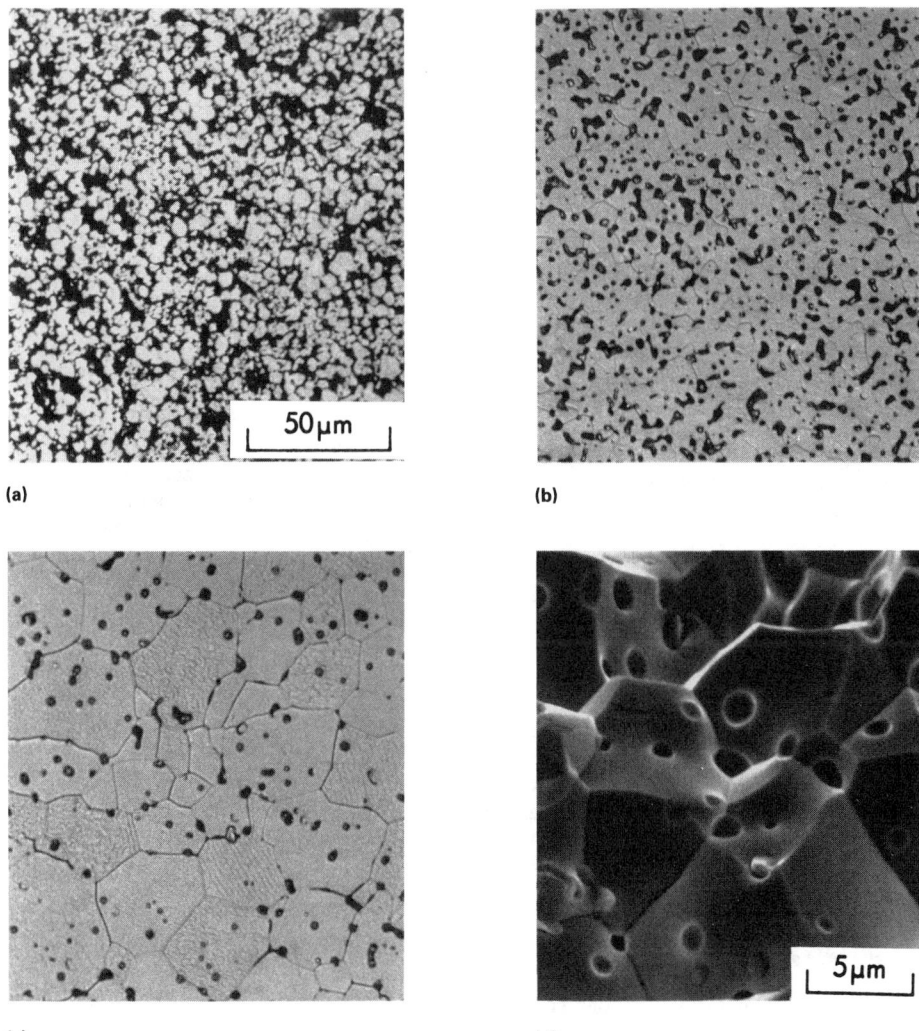

(a)

(b)

(c)

(d)

Fig 5 Progressive densification and grain growth at several stages of sintering: (a) initial stage; (b) intermediate stage; (c) final stage; (d) fracture surface. The fracture surface micrograph shows the desirable placement of spherical pores on grain boundaries in the final stage of sintering.

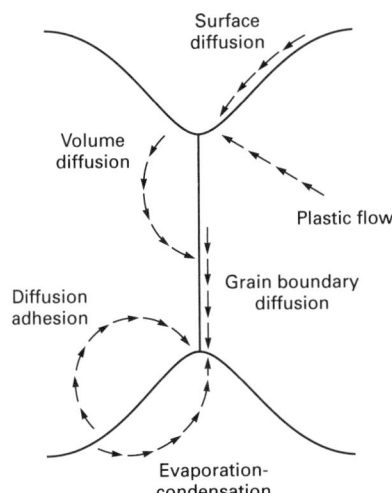

Fig 6 Mass flow paths sketched with respect to the neck region between two sintering particles. Surface diffusion and evaporation-condensation are both surface transport mechanisms. Volume diffusion, diffusion adhesion, grain boundary diffusion, and plastic flow are all classified as bulk transport mechanisms.

of temperature increase, such as 5 or 10 K/ min (9 or 18 °F/min). The specimen length or shrinkage versus temperature is recorded. This approach uses a single experiment to identify the temperatures where significant densification occurs. This is a considerable improvement over isothermal experiments. Various adaptations have evolved using closed-loop feedback control, where density versus time is programmed and the control computer determines the necessary temperature increase needed to obtain the desired density. This leads to nonlinear heating schedules that reflect sensitivities to grain growth, pore coarsening, densification, and other thermally driven microstructural changes.

Sintering is not a single mechanism process. Many materials sinter by a combination of actions involving multiple modes of mass flow. With a change in particle size, temperature, or time, it is possible to shift the dominant sintering mechanism because of the differing sensitivities to process parameters.

Viscous flow dominates for amorphous materials. Volume diffusion is prevalent with very high temperature sintering of large particles. Surface diffusion and grain boundary diffusion are usually dominant with smaller particle sizes in crystalline materials. Furthermore, small particles exhibit faster neck growth and need less sintering time or a lower sintering temperature to achieve an equiva-

lent degree of sintering. Generally, large particles will sinter slower and require higher sintering temperatures or longer sintering times to attain an equivalent degree of densification. The improved sintering densification with decreasing particle size is a major reason for using small particles.

Another demonstration of the effect of small particles on sintering is given in Fig 7. Figure 7 plots the fractional sintered density for bimodal mixtures of alumina powders with sizes of approximately 0.5 and 5 μm (20 to 200 μin.). The sintered density increased with the fraction of small particles, demonstrating the potent role of small particles. Model experiments with continuous particle size distributions indicate maximum sintering densification with distributions of intermediate to large widths (Ref 12).

During sintering, bulk transport will create a change in the interparticle spacing as neck growth takes place. The result is shrinkage, increased density, and higher strength for the entire powder compact. Figure 8 plots the sintered density of monosized titania (TiO$_2$) powder sintered at various temperatures and times. Longer times, higher temperatures, and

Table 1 Nomenclature for the initial stage sintering equation: $(X/D)^n = Bt/D^m$

Mechanism	n	m	B(a)
Viscous flow	2	1	3 $\gamma/(2\eta)$
Plastic flow	2	1	9 $\pi\gamma\mathbf{b}D_v/(kT)$
Evaporation-condensation	3	1	3 $P\gamma/\rho^2[\pi M/(2kT)]^{1/2}$
Lattice diffusion	5	3	8 $D_v\gamma\Omega/(kT)$
Grain boundary diffusion	6	4	20 $\delta D_b\gamma\Omega/(kT)$
Surface diffusion	7	4	56 $D_s\gamma\Omega^{4/3}/(kT)$

(a) γ, surface energy; η, viscosity; $\mathbf{b}$, Burgers vector; D_v, volume diffusivity; k, Boltzmann's constant; T, absolute temperature; P, vapor pressure; ρ, theoretical density; M, molecular weight; Ω, atomic volume; δ, grain boundary width; D_b, grain boundary diffusivity; and D_s, surface diffusivity

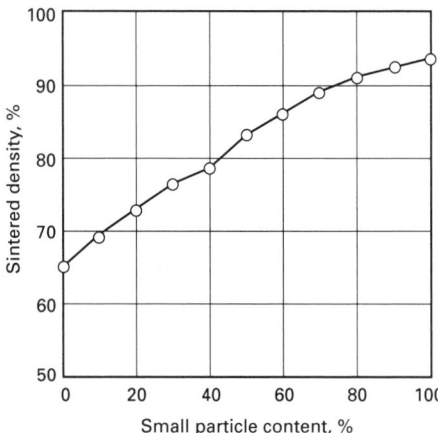

Fig 7 Plot of fractional sintered density versus composition of bimodal alumina powders to show that the highest density occurs with the greatest content of small particles. Bimodal alumina, which was sintered at 1600 °C (2910 °F) for 60 min, consisted of particle diameters of 0.5 and 5 μm (20 and 200 μin.). Source: Ref 11

higher initial packing densities all contribute to a higher sintered density. Note that the rate of densification continuously decreases during sintering. This is the consequence of a relaxation in the driving force and the need for larger diffusion distances as the neck enlarges.

Shrinkage mandates an oversizing of the tooling to bring the final sintered part into an acceptable size. Density gradients in the compact translate into differential shrinkages during sintering. Consequently, warpage and nonuniform shapes result from inhomogeneities in the initial packing density of the powder.

The initial stage sintering model for shrinkage represented by Eq 5 is only valid for approximately the first 3% of shrinkage. After this point, the driving force for sinter-

ing changes and an intermediate stage model is appropriate:

$$f_s = f_i + B_i \ln(t/t_i) \tag{Eq 7}$$

where f_s is the sintered fractional density, f_i is the fractional density at the beginning of the second stage, B_i follows Eq 6, t_i is the time corresponding to the onset of the intermediate stage, and t is the time (where $t > t_i$). Typically, B_i will vary inversely with the cube of the grain size, reflecting the strong role played by grain boundaries in the densification of crystalline ceramics.

The intermediate stage is important in determining the properties of the sintered compact. This second stage is characterized by densification coupled with grain growth. The mean grain size G increases with time t according to:

$$G^3 = G_0^3 + Kt \tag{Eq 8}$$

where G_0 is the initial grain size, and K is a thermally activated parameter that is similar to the factor B in Eq 4 and 5. During the intermediate stage of sintering the pore structure becomes smooth but remains interconnected until the final stage. The solid-vapor surface area of the compact is now less than 50% of the initial surface area of the powder. Indeed, the rate of surface area loss can be used to follow sintering well into the intermediate stage.

Effect of Sintering on Pore Structures

The geometry of the grain boundary and the pore during intermediate stage sintering controls the sintering rate. At the beginning of the intermediate stage, the pore geometry is highly convoluted and the pores are located at grain boundary intersections. With contin-

ued sintering, the pore geometry approaches a cylindrical shape in which densification occurs by decreasing of the pore radius (Ref 14). During sintering, the interaction between pores and grain boundaries can have three forms:

- Pores can retard grain growth
- Pores can be dragged by the moving grain boundaries during any grain growth
- Grain boundaries can break away from the pores, leaving pores isolated in the grain interior

At the temperatures typical of sintering, most materials exhibit moderate to high grain growth rates. Any differences in initial grain sizes generate forces on the grain boundaries that cause grain growth. As the temperature is increased, the rate of grain boundary motion increases. Breakaway of the boundaries from the pores occurs because the pores are slower moving than the grain boundaries (Ref 15). Under the tension of a moving grain boundary, pores can move by volume or surface diffusion or even by evaporation-condensation across the pore, but this requires close control of the heating rate, using a process termed rate-controlled sintering (Ref 16).

Consider the two possible pore-boundary configurations illustrated in Fig 9. Pores can occupy sites located at the grain edges or in the interior of the grains. The system energy is lower for the pore occupying the grain edge because the pore reduces the total grain boundary area (and energy). Should the pore and boundary become separated, the system energy is increased in proportion to the amount of newly created interfacial area. As a consequence, the pore has a binding energy in relation to the grain boundary that increases as the porosity increases. Hence, at the beginning of the intermediate stage, little separation of boundaries from pores is expected.

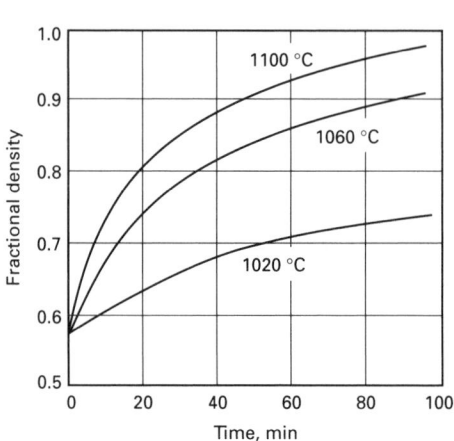

Fig 8 Sintered density of 0.34 μm (14 μin.) titania monosize particles as a function of the sintering time for selected temperatures. Source: Ref 13

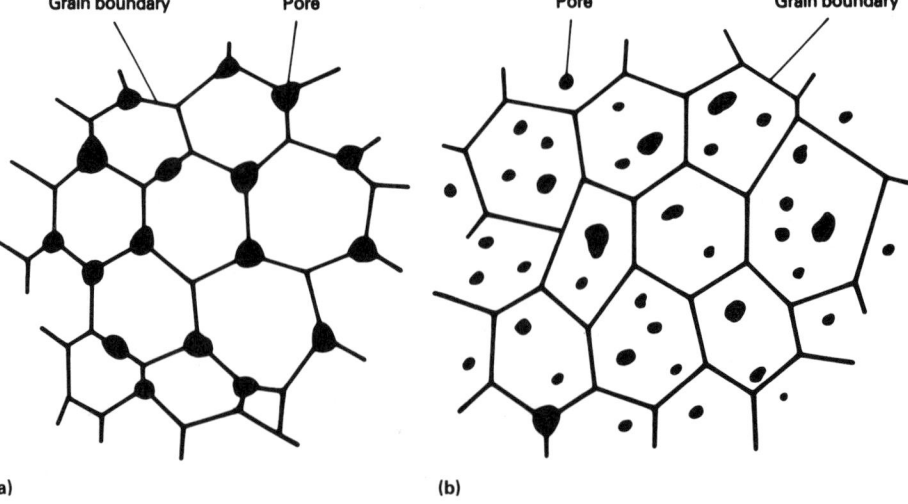

Fig 9 Two possible pore-grain boundary configurations in the intermediate stage of sintering. (a) With densification. (b) Without densification. The pores located on grain boundaries provide densification while the isolated pores do not densify.

Densification. As densification proceeds, the lower mobility of the pores coupled to the diminishing pinning force allows breakaway. Separation of the pores from the boundaries limits the potential final density. Consequently, it is important to minimize breakaway by careful processing control. A combination of large pore size and grain size leads to breakaway during grain growth (Fig 10). Ideally, the large pores are immobile in the early stages of sintering and are pinned against the grain boundaries to maintain a small grain size. In the later stages of sintering, the pores become fewer in number and diminish in size due to shrinkage. Although the grains are relatively large, the pores are sufficiently mobile to migrate with the boundaries. Densification is dependent on the rate of pore shrinkage while this situation persists and a high grain boundary diffusivity is helpful. As the pore size decreases, there is less inhibition to grain growth and pore mobility is a greater concern. For the pores to remain on the moving grain boundaries, it is necessary to increase their mobility, for example, by increasing surface diffusivity. Rapid grain growth is to be avoided because invariably the densification rate is low (Ref 17).

The mechanisms whereby the breakaway event can be avoided are of particular importance to ceramic materials that exhibit a high sensitivity to residual porosity. Although the rate of densification is dependent on factors such as grain size, density, temperature, and time, the dominant factor is the effect of temperature. If bulk diffusion is inactive at the end of the initial stage of sintering, there will be no densification, although pore growth and possibly grain growth might be active. Pore rounding occurs simultaneously with densification.

The final stage of sintering occurs when the pores spheroidize into a closed structure, at approximately 8% porosity. Most materials are sintered to densities over 92% of the-

oretical and pass into the final stage. Surface transport is active during intermediate stage sintering. Its effect is to smooth the pore structure and to allow pore migration with grain boundaries during grain growth. Surface transport does not contribute to densification or shrinkage. The specific sintering events are dependent on the microstructure (grain size, pore size, and pore spacing). Furthermore, since the microstructure is continually changing, the influence of temperature can be quite pronounced. The densification rate is enhanced by a high diffusivity and a small grain size (Ref 18). Pores, dispersoids, and second phase inclusions slow grain growth and can be used to improve densification.

Final stage sintering is a slow process wherein spherical pores shrink by a diffusion mechanism. If the pores have a trapped gas, then solubility of the gas in the matrix will influence the rate of pore elimination. For this reason, it is preferable to sinter in vacuum or to use an atmosphere that is soluble in the sintering material. When a gas is sealed in the pores, the densification rate is controlled by the internal gas pressure. Pores begin to pinch closed at approximately 80% of theoretical density, as illustrated by the urania (UO_2) data in Fig 11. At approximately 92% density the pores make a rapid shift to a closed structure. If the closed pores are mobile enough to stay coupled to the grain structure, then shrinkage will continue into the final stage. A homogeneous grain size and sintering in a vacuum aid densification in the final stage.

In most materials, the distributions in particle size and packing create a pore size distribution in the final stage. With long sintering times, the number of pores decreases and pore size coarsens while the total porosity may even increase. Differences in pore curvature will lead to growth of the larger pores at the expense of the smaller, less stable pores. This process is known as Ostwald ripening. Attaining 100% density by sintering requires precise manipulation of the initial powder mi-

crostructure and heating cycle, because several factors can inhibit final pore elimination.

Sintering Diagrams

The sintering diagram is a useful concept for representing sintering behavior (Ref 1, 2). It shows the density or neck size versus temperature for several isothermal sintering times. These diagrams combine the several mechanisms of sintering with the changing geometry to show the effects of the main process variables. Because of the complexity of such an analysis, sintering diagrams are generated from a materials data base using computer simulations. A sintering diagram for 0.4 μm (16 μin.) alumina is given in Fig 12. Densification is dominated by grain boundary diffusion, although surface diffusion makes a considerable contribution to the overall mass flow. This plot shows the relative density versus isothermal sintering temperature for four selected times. Because grain growth during sintering reduces the densification rate, actual sintering behavior tends to give lower densities. From such plots, an assessment of the interactions between the key processing variables of particle size, sintering temperature, and sintering time can be made. A change in particle size has a large effect on each diagram. At a given temperature, a higher density is attained from small powders.

The sintering diagram provides a computer generated first estimate of the sintering behavior for a given combination of parameters. One important aspect of sintering diagrams is the inclusion of multiple mechanisms in the calculations. Most materials do not sinter by a single mechanism. Rather, several mass transport processes usually are active during sintering. The effect of these simultaneous mechanisms is to increase the overall rate of sintering.

Effects of Sintering on Compact Properties

The main reason for sintering is to improve such compact properties as hardness, strength, transparency, toughness, electrical conductivity, thermal expansion, magnetic saturation, and corrosion resistance. The sensitivity of each of these properties to the degree of sintering can be quite different, but in general it improves with the degree of densification. Various quantitative models of the effects of porosity on properties are listed in Table 2. In this table, each property is given as a function of the bulk material property (subscript "0"), fractional density f, and various adjustable parameters that depend on the pore size, shape, and spacing. The dynamic properties (such as impact strength) prove to be the properties that are most sensitive to sintering.

Table 3 contrasts the advantages and disadvantages of some of the adjustable processing variables in sintering. From a sinter-

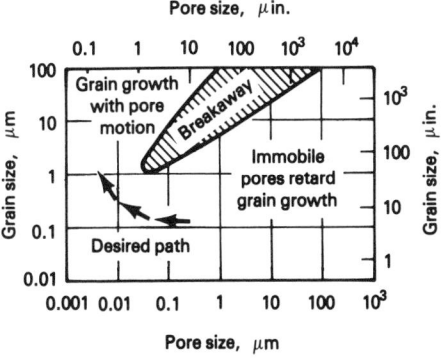

Fig 10 Plot of grain size versus pore size during sintering to show the conditions under which grain growth can cause a breakaway from the pores, eventually leading to inhibited densification. Desirable sintering pathways avoid the breakaway condition by sustaining pore shrinkage at low temperatures where the grain growth is inhibited.

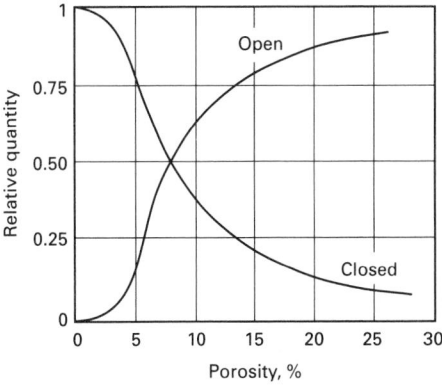

Fig 11 The variation in the open and closed pore fractions versus the total porosity for sintering urania at 1400 °C (2550 °F). Source: Ref 19

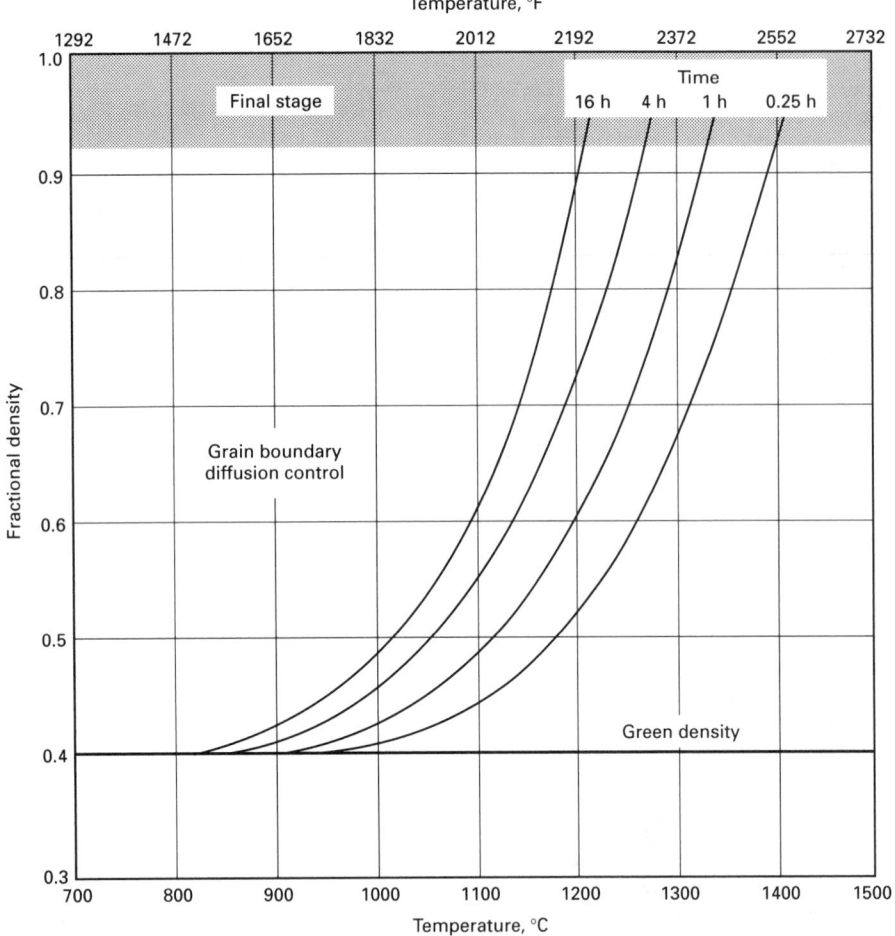

Fig 12 Sintering diagram for 0.4 μm (16 μin.) particle size and 0.1 μm (4 μin.) grain size alumina with the sintered density expressed as a function of the sintering temperature and the time. Regions of dominance for the various mechanisms are indicated on the diagram, but in each case the minor mechanisms still contribute to mass flow.

ing standpoint, a decrease in particle size is favorable, because smaller particles lead to more rapid sintering rates and higher strengths. Figure 13 illustrates such an effect by plotting the sintered strength of zirconia (ZrO_2) versus the particle size for a constant processing cycle. A long sintering time will typically improve the degree of sintering but will lead to more expense in processing. The sintering temperature is one of the most influential of all sintering variables. Higher sin-

tering temperatures give more rapid sintering. This is not always favorable, because of the greater expense in furnace design, the higher energy costs, and the possible grain boundary separation from the pores. In the same manner, compositional changes can improve sintered properties. By adjusting the composition it is possible to considerably strengthen the material and aid sintering densification. Thus, another goal of sintering is to attain desirable diffusional homogenization responses

during the high-temperature portion of the sintering cycle.

Mixed-Powder Sintering

Mixed powders provide flexibility in sintering because it is easy to change composition and because some unique composite microstructures are possible. Two types of sintered structures are possible from mixed powders:

- Solid solutions, which involve the homogenization of the powder mixture by diffusional processes
- Composites, which involve cosintering two distinct phases

Homogenization during sintering is an alternative to forming compacts from chemically prehomogenized powders. Sintering of multiple component powder systems is more complex than sintering single component compacts because of the simultaneous phase reactions and homogenization effects. Diffusional leveling of the chemical concentration gradients is dominant early in the sintering cycle.

Mixed-phase sintering has some recognized problems. The compositional gradients require control of both time and temperature to ensure homogenization. The degree of homogenization H is defined as the point-to-point chemistry variation, and it varies with diffusion rate and particle size as follows (Ref 24):

$$H \approx D_v t / Y^2 \qquad \text{(Eq 9)}$$

where Y is the scale of the microstructural segregation, D_v is the volume diffusivity, and t is the time. The scale of segregation as measured by Y depends on the particle size and concentration of the powders. Equation 9 shows that homogenization is faster with small particles where the diffusion distances are small. If the diffusion rates of the two components are very different, then swelling may occur. The phase diagram is important to understanding the possible phase reactions during mixed phase sintering (Ref 25, 26).

Composites with two interlaced phases can be formed if a condition of cosintering is attained (Ref 27). Success in cosintering requires specific particle size ratios and uses of compositions that exhibit similar shrinkage versus temperature behaviors.

Enhanced Solid-State Sintering

To provide the maximum properties, it is often necessary to enhance densification. The common approaches are (Ref 7):

- Phase stabilization
- Activated sintering
- Reactive sintering
- Liquid-phase sintering

Table 2 Effect of porosity changes induced by sintering on selected properties of compacts

Compact property	Symbol	Equation incorporating effect of porosity on compact property(a)
Electrical conductivity	Ω	$\Omega = \Omega_0 A f^2$
Magnetic saturation	B	$B = B_0(A_1 + A_2 f)$
Strength	σ	$\sigma = \sigma_0 A_1 [1 - A_2(1 - f)^{2/3}]$
Elastic modulus	E	$E = E_0 f^{3.4}$
Shear modulus	G	$G = G_0 f^3$
Poisson ratio	ν	$\nu = 0.068 \exp(1.37 f)$

(a) Property calculated as a function of bulk material property (indicated by subscript zero); f, fractional density; and A_1 and A_2, adjustable constants.
Source: Ref 6, 7, 20–22

Table 3 Results of modifications in processing designed to aid the sintering process

Result of processing change	Decrease in particle size	Increase in time	Increase in temperature	Increase in packing density	Increase in alloying level	Use of sintering aids
Thermal decomposition	...	...	X	...	...	...
Slower debinding	...	...	...	X	...	...
Reduced productivity	...	X	...	...	...	...
Pore coarsening	...	...	X	...	...	...
More rapid densification	...	...	...	...	...	X
Lower toughness	...	...	...	...	...	X
Lower sintering temperature	...	...	...	...	...	X
Less shrinkage	...	...	...	X	...	...
Less binder	...	...	...	X	...	...
Increased health hazards	X	...	...	...	...	...
Homogeneity concerns	...	...	...	...	X	...
Higher strength	...	...	...	...	X	...
Higher impurity level	X	...	...	...	...	...
Higher final density	...	...	...	...	...	X
Higher density	...	...	X	...	...	...
Greater expense	X	X	X	...	...	...
Grain growth	...	X	X	...	...	X
Furnace limitations	...	...	X	...	...	...
Faster sintering	X	...	X	...	...	...
Distortion	...	...	...	...	...	X
Degraded creep strength	...	...	...	...	...	X

Phase Stabilization. Diffusivity of a material is determined by several factors, including the temperature, crystal structure, and defect configuration. The degree of sintering densification depends on the phases present at the sintering temperature. Additionally, mixed phase microstructures resist grain growth during sintering, especially in the intermediate stage. It is possible to adjust the composition or stoichiometry of a powder system to stabilize a phase with a high sintering rate. This can involve use of a second phase to pin grain growth during sintering (Ref 17).

Activated sintering allows for either a lower sintering temperature, a shorter sintering time, or improvement in properties from a chemical addition to the powder. One of the most dramatic examples of activated sintering occurs when tungsten or molybdenum are treated with certain transition metals (Ref 28, 29). The successful additive acts to improve the rate of bulk transport during sintering by providing a fast diffusion path. To perform this role, the activator segregates to the interparticle contacts where it provides a high diffusivity short-circuit path by forming a low melting temperature phase. The lower melting point ensures a lower activation energy for diffusion. Typically, an activator that decreases the liquidus and solidus of the base material is most successful. Figure 14 shows an ideal phase diagram for activated sintering systems. At temperatures slightly above the activated sintering range, a liquid phase forms. Forming of a liquid phase is another means of enhancing sintering (see the section "Liquid-Phase Sintering" in this article).

Reactive Sintering. Additives are also used to prevent stoichiometry loss during sintering. An excess of the higher stability species might be needed to avoid decomposition of the compound during sintering. Also, surface contaminants may attack one of the constituents, leading to preferential attack or depletion. For this reason, it is common practice to add excess carbon to carbides (tungsten carbide or silicon carbide) during sintering. Indeed, silicon carbide can be formed by the use of free carbon and free silicon to produce a silicon carbide reaction product simultaneously with sintering. Likewise, there are special sintering situations where mass can be gained by the compact by reaction with the sintering atmosphere. The most common form of such processing involves the use of excess silicon and a nitrogen atmosphere. In sintering, the silicon forms silicon nitride and the reaction product swells to fill the pore space. These latter techniques rely on chemical reactions during sintering, and the many variants of these approaches are generally known as reactive sintering.

Liquid-Phase Sintering. In two-phase systems involving mixed powders, liquid formation is possible because of differing melting ranges for the components or the formation of a low melting phase (including a glass phase). In such a system, the liquid may provide for rapid transport and therefore rapid sintering if certain criteria are met (Ref 6, 25, 30). The liquid must form a film surrounding the solid phase. Thus, wetting is the first requirement. Secondly, the liquid must have a solubility for the solid. Finally, the diffusive transport for the solid atoms dissolved in the liquid should be high enough to ensure rapid sintering. The formation of a liquid film provides the benefit of a surface tension force acting to aid densification and pore elimination. In this sense, the liquid phase acts like a low-pressure external stress. An example of systems involving liquid phase formation during sintering are WC-Co and Si_3N_4-Y_2O_3. In liquid-phase sintering, the densification rate is much faster than in solid-state sintering, and times as short as 15 min at the maximum temperature can be successful in producing full density.

When a mixture of powders is heated, liquid forms and then flows to wet the particles. Figure 15 provides a diagram of the densification stages of liquid-phase sintering. The combination of wetting, liquid flow, and particle rearrangement all contribute to densification. With continued heating, the solid phase dissolves into the liquid and the amount of liquid grows until is it saturated with the solid component. The liquid phase then becomes a carrier for the solid-phase atoms in a process termed solution-reprecipitation, wherein the small particles dissolve and reprecipitate on the large particles. This sequence of events does not significantly change the amount of liquid or solid. It does provide for densification and coarsening of the solid phase. Ac-

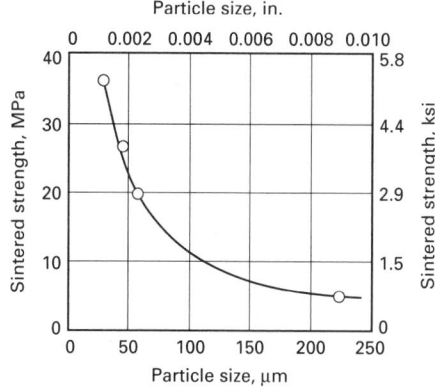

Fig 13 Plot of sintered strength versus the particle size of zirconia sintered for 4 h at 2200 °C (3990 °F). The more active sintering of the smaller particles leads to improved strength. Source: Ref 23

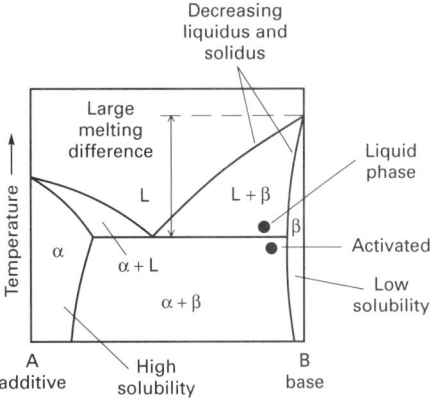

Fig 14 Idealized phase diagram for an additive that aids in the sintering densification of a base material. The additive acts to enhance the mass transport rate during sintering. Source: Ref 25

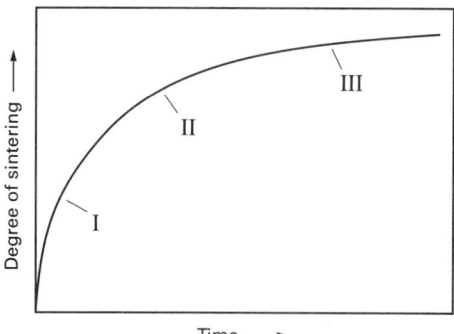

Fig 15 Overlapping of the three processing stages of densification for a powder compact during liquid-phase sintering: stage I, rearrangement (melt flow and penetration repacking, rapid densification or swelling, and particle sliding); stage II, solution-reprecipitation (diffusion control densification, shape accommodation, grain growth, and neck formation); stage III, solid state (rigid structure neck growth, grain growth, coalescence, and pore coarsening). Source: Ref 30, 31

10 μm

Fig 16 Scanning electron micrograph of a liquid phase sintered WC-Co composition containing angular carbide grains dispersed in the solidified liquid. The angular grain shape indicates anisotropic solid-liquid surface energy and reaction-controlled grain growth.

cordingly, the solid phase can achieve a higher packing density through particle shape accommodation which allows better packing (Ref 32). Consequently, the sintered microstructure consists of contacting solid grains with an interspersed (solidified) liquid matrix. The solid grains exhibit shape accommodation, where flat faces occur on neighboring grains, allowing them to fit into close proximity to one another. This releases liquid to fill the pore space.

Small particles aid densification in short sintering times. In general, coarsening is to be avoided because it degrades mechanical properties. Densification is also aided by large amounts of liquid and high solid solubilities in the liquid. At liquid contents of approximately 35 vol%, full density is achieved in the rearrangement stage and considerable compact slumping occurs. Consequently, large amounts of liquid phase cannot be used because the compact lacks sufficient rigidity during sintering to maintain its shape.

During liquid-phase sintering, the solid phase will coarsen at a rate such that the average grain size, G, enlarges as a function of $t^{1/2}$ or $t^{1/3}$ (Ref 33). Typically, grains with rounded shapes exhibit solution-reprecipitation kinetics at a rate of $t^{1/3}$. In contrast, angular grains, such as the carbides in WC-Co (Fig 16), have a coarsening rate limited by the availability of surface reaction sites and exhibit coarsening that approximates $t^{1/2}$. In solution-reprecipitation controlled coarsening, the grain size enlarges and the coarsening rate increases as the amount of liquid decreases, due to the smaller diffusion distances for grain growth (the liquid layers are thinner). The activation energy for coarsening corresponds to either that for diffusive transport through the liquid or reaction at the particle surface. The effect of temperature is to raise the transport rate at an exponential rate thus giving faster grain growth.

Liquid-Phase Sintering versus Activated Sintering. Liquid-phase sintering and activated sintering are similar densification enhancement techniques. Both involve the use of a second phase at the sintering temperature to provide rapid mass transport of an effective short-circuit path. As shown in Fig 14, the phase diagram characteristics applicable to both of these enhanced sintering techniques are similar. A major difference between liquid-phase and activated sintering is in the amount of second phase present at the sintering temperature. Liquid phase systems typically have several times more second phase. This is beneficial, because the second phase is the continuous phase in the sintered product. Consequently, liquid phase sintered products often are more capable of sustaining strain and deformation, which leads to improved mechanical properties. A new and potentially useful variant is transient liquid-phase sintering, where liquid forms from mixed powders but persists only for a short time. This allows the liquid to induce densification without causing grain coarsening. Swelling can occur in some cases, even though the compact undergoes substantial strengthening.

Supersolidus liquid-phase sintering is an alternative processing approach that relies on the use of prealloyed (PA) powders (Ref 30). This technique will be applicable to rapidly solidified and atomized ceramic powders. The powders are heated to a temperature at which a liquid phase nucleates within each particle. The approach is attractive because of the high densification possible with large particle sizes. In supersolidus liquid-phase sintering, the temperature is held intermediate between the liquidus and solidus for the composition. As a consequence, the individual particles become mushy, and they can densify by capillary-induced rearrangement with subsequent solution-reprecipitation. The overall process is similar to viscous flow controlled sintering except that the liquid is formed within the particles rather than between the particles. Temperature and composition are the most important process variables because these variables dictate the volume fraction of liquid. The process can be applied to systems having large melting ranges. The principles of supersolidus liquid-phase sintering have sufficiently advanced that materials can be designed for the process. A first application in ceramics is for the densification of alumina-glass powders formed using plasma atomization.

REFERENCES

1. M.F. Ashby, A First Report on Sintering Diagrams, *Acta Metall.*, Vol 22, 1974, p 275–289
2. F.B. Swinkels and M.F. Ashby, A Second Report on Sintering Diagrams, *Acta Metall.*, Vol 29, 1981, p 259–281
3. C. Sierra and D. Lee, Modeling of Shrinkage during Sintering of Injection Molded Metal Powder Compacts, *Powder Metall. Int.*, Vol 20 (No. 5), 1988, p 28–33
4. K.S. Hwang and R.M. German, Analysis of Initial Stage Sintering by Computer Simulation, *Sintering and Homogeneous Catalysis*, G.C. Kuczynski, A.E. Miller, and G.A. Sargent, Ed., Plenum Press, 1984, p 35–47
5. L. Berrin and D.L. Johnson, Precise Diffusion Sintering Models for Initial Shrinkage and Neck Growth, *Sintering and Related Phenomena*, G.C. Kuczynski, N.A. Hooton, and C.F. Gibbon, Ed., Gordon and Breach, 1967, p 369–392
6. F.V. Lenel, *Powder Metallurgy Principles and Applications*, Metal Powder Industries Federation, 1980, p 211–267
7. R.M. German, *Powder Metallurgy Science*, Metal Powder Industries Federation, 1984, p 145–200
8. M.B. Waldron and B.L. Daniell, *Sintering*, Heyden and Sons, London, 1978
9. F. Thummler and W. Thomma, The Sintering Process, *Metall. Rev.*, Vol 12, 1967, p 69–108
10. H.E. Exner, Principles of Single Phase Sintering, *Rev. Powder Metall. Phys. Ceram.*, Vol 1, 1979, p 1–251
11. J.P. Smith and G.L. Messing, Sintering of Bimodally Distributed Alumina Powders, *J. Am. Ceram. Soc.*, Vol 67, 1984, p 238–242
12. B.R. Patterson and J.A. Griffin, Effect of Particle Size Distribution on Sintering of Tungsten, *Modern Developments in Powder Metallurgy*, Vol 15, E.N. Aqua and C.I. Whitman, Ed., Metal Powder Industries Federation, 1985, p 279–288

13. E.A. Barringer, R. Brook, and H.K. Bowen, The Sintering of Monodisperse Titania, *Sintering and Heterogeneous Catalysis*, G.C. Kuczynski, A.E. Miller, and G.A. Sargent, Ed., Plenum Press, 1984, p 1–21

14. R.L. Coble, Sintering Crystalline Solids. 1. Intermediate and Final State Diffusion Models, *J. Appl. Phys.*, Vol 32, 1961, p 787–792

15. R.J. Brook, Pore-Grain Boundary Interactions and Grain Growth, *J. Am. Ceram. Soc.*, Vol 52, 1969, p 339–340

16. H. Palmour and T.M. Hare, Rate Controlled Sintering Revisited, *Sintering '85*, G.C. Kuczynski, D.P. Uskokovic, H. Palmour, and M.M. Ristic, Ed., Plenum Press, 1987, p 17–34

17. M.F. Yan, Sintering of Ceramics and Metals, *Advances in Powder Technology*, G.Y. Chin, Ed., American Society for Metals, 1982, p 99–133

18. M.F. Yan, R.M. Cannon, and K.H. Bowen, Grain Boundary Migration in Ceramics, *Ceramic Microstructures '76*, R.M. Fulrath and J.A. Pask, Ed., Westview Press, 1977, p 276–307

19. S.C. Coleman and W.B. Beere, The Sintering of Open and Closed Porosity in Urania, *Philos. Mag.*, Vol 31, 1975, p 1403–1413

20. R. Haynes, The Mechanical Behaviour of Sintered Metals, *Reviews on the Deformation Behaviour of Materials*, Vol 3, 1981, p 1–101

21. G.F. Bocchini, The Influence of Porosity on the Characteristics of Sintered Materials, *Int. J. Powder Metall.*, Vol 22, 1986, p 185–202

22. J.C. Wang, Young's Modulus of Porous Materials. Part 1. Theoretical Derivation of Modulus-Porosity Correlation, *J. Mater. Sci.*, Vol 19, 1984, p 801–808

23. Y.L. Krasulin, V.N. Timofeev, S.M. Barinov, and A.B. Ivanov, Strength and Fracture of Porous Ceramic Sintered from Spherical Particles, *J. Mater. Sci.*, Vol 15, 1980, p 1402–1406

24. R.W. Heckel, R.D. Lanam, and R.A. Tanzilli, Techniques for the Study of Homogenization in Compacts of Blended Powders, *Advanced Experimental Techniques in Powder Metallurgy*, J.S. Hirschhorn and K.H. Roll, Ed., Plenum Press, 1970, p 139–188

25. R.M. German and B.H. Rabin, Enhanced Sintering through Second Phase Additions, *Powder Metall.*, Vol 28, 1985, p 7–12

26. D.L. Johnson and I.B. Cutler, The Use of Phase Diagrams in Sintering of Ceramics and Metals, *Phase Diagrams*, Vol 2, A.M. Alper, Ed., Academic Press, 1970, p 265–291

27. L.S. Klein and R.M. German, Controlled Thermal Expansion Metal-Ceramic Composites by Co-Sintering, *Int. J. Powder Metall.*, Vol 24, 1988, p 39–46

28. R.M. German and Z.A. Munir, Activated Sintering of Refractory Metals by Transition Metal Additions, *Rev. Powder Metall. Phys. Ceram.*, Vol 2, 1982, p 9–43

29. G. Petzow, W.A. Kaysser, and M. Amtenbrink, Liquid Phase and Activated Sintering, *Sintering—Theory and Practice*, D. Kolar, S. Pejovnk, and M.M. Ristic, Ed., Elsevier Scientific, Amsterdam, 1982, p 27–36

30. R.M. German, *Liquid Phase Sintering*, Plenum Press, 1985

31. F.V. Lenel, Sintering in the Presence of a Liquid Phase, *Trans. AIME*, Vol 175, 1948, p 878–896

32. W.J. Huppmann, Sintering in the Presence of a Liquid Phase, *Sintering and Catalysis*, G.C. Kuczynski, Ed., Plenum Press, 1975, p 359–378

33. Y. Masuda and R. Watanabe, Ostwald Ripening Processes in the Sintering of Metal Powders, *Sintering Processes*, G.C. Kuczynski, Ed., Plenum Press, 1980, p 3–21

Solid-State Sintering*

Man F. Yan, AT&T Bell Laboratories

SINTERING is the single most important step during powder processing because it is at this stage that a powder compact is exposed to the maximum temperature. Densification is an important objective of sintering. More importantly, many electrical, magnetic, optical, and mechanical properties are determined by the physical and chemical changes during sintering at high temperatures. Once these properties are acquired during sintering, they cannot be easily modified by low-temperature processes. Powder processing prior to sintering can determine the success of sintering. Conversely, a critical examination of the sintering process and the sintered products can lead to useful inference about the way in which the pre-firing processing must be modified to obtain satisfactory products.

While both sintering and hot pressing have been used to prepare ceramic compacts, sintering is the more preferred process because of economical reasons and its suitability to fabricate parts with complex shapes. In this article, the discussion will be restricted to solid-state sintering. However, many principles discussed here are also applicable to hot pressing.

Theoretical studies of sintering models have been reviewed by Swinkels and Ashby (Ref 1). They have constructed sintering diagrams for several metals to predict the sintering density as a function of time and temperature. In systems with simple geometries, for example, wires and spheres, the predicted sintering behaviors are in good agreement with the experimental data. However, when sintering data from powder aggregates with irregular particle shapes are compared with the predicted sintering diagrams, the agreement is less satisfactory because most practical compacts consist of $>10^{10}$ particles with rather irregular size, shape, and spatial arrangement.

In this article, the physical, chemical, and kinetic factors on sintering of ceramic powder compacts will be discussed. In particular, the effects of physical characteristics of starting powders and microstructures of green compact on sintering will be evaluated. It has been shown that powder preparation and compaction processes have a significant effect on these physical characteristics. Chemical reactions, among different constituents in a powder compact and between the sintering atmosphere and the specimen, take place simultaneously with the sintering process. *In situ* physical and chemical changes during sintering also have important influences on the resultant density and microstructures. Furthermore, it has been demonstrated that a quantitative measurement of the densification kinetics can provide a useful data base to fine tune the firing schedule for its optimization. These important factors will be illustrated by reviewing recent data and analyses in the literature. From this review, there emerge some general guidelines which are useful to promote densification during sintering of ceramics.

Physical Characteristics of Powders

Characteristics of the starting powder, for example, particle size, size distribution, particle shape, particle aggregates, and degree of agglomeration, have a profound influence on densification and microstructural development. An ideal powder should have the attributes of small particle size, nonagglomeration, equiaxed particle shapes, narrow size distribution, and high purity or controlled dopant content.

Particle Size. It is generally agreed that a smaller particle size in the starting powder leads to a faster densification rate. The theoretical basis of this argument is due to Herring's scaling law (Ref 2) which states that there are simple laws governing the times required to produce, by sintering at a given temperature, geometrically similar changes in two or more systems of solid particles that are identical except for a difference in particle dimensions. For example, if time, t_1, is required to sinter an array of spherical particles with radius, r_1, at a given temperature, then the time, t_2, required to sinter another geometrically identical array with particle radius, r_2, is given by

$$t_2 = \left(\frac{r_2}{r_1}\right)^n t_1 \qquad \text{(Eq 1)}$$

where the value of n depends on the mass transport mechanism for sintering. Typical values of n range from 3 to 4 for lattice and grain-boundary diffusion. Equation 1 shows that when the particle size decreases from 1 to 0.01 μm, the sintering time decreases by a factor of 10^6 to 10^8. Consequently, those interested in high sintered density or reduced sintering temperatures and times strive for fine starting powders. However, the success in sintering of the fine powder relies on the removal of agglomerates and aggregates.

Agglomerates and Aggregates. Agglomerate is a term describing a small mass of primary particles bonded together by surface forces, while the term aggregate describes a coarse constituent of strongly bonded and/or reacted particles. Both agglomerates and aggregates are bonded by the surface force between constituent particles. The surface force per unit weight is inversely proportional to the particle size. Thus, the problems of agglomeration and aggregation are most pronounced in powder with a submicron particle size. The voids between aggregates and agglomerates are much larger than those between the constituent particles, and the larger voids obviously require a much longer sintering time. Furthermore, densification of the individual agglomerates or aggregates leads to their shrinkage from each other and the voids between them become even larger.

Agglomerates and aggregates are formed during several stages of powder processing. The attractive and repulsive forces between particles in liquid and solid media are shown schematically in Fig 1. During the mixing and milling steps, a ceramic powder is usually suspended in a liquid medium. Brownian motion leads to very frequent collisions between solid particles, which then adhere together to form agglomerates if there exists a net attractive force between particles. The van der Waals force is the major attractive force leading to the formation of agglomerates in a liquid medium (Ref 3). The van der Waals force originates from the interaction

*Adapted from *Advances in Ceramics*, Vol 21, *Ceramic Powder Science*, American Ceramic Society, 1987, p 635–669. With permission

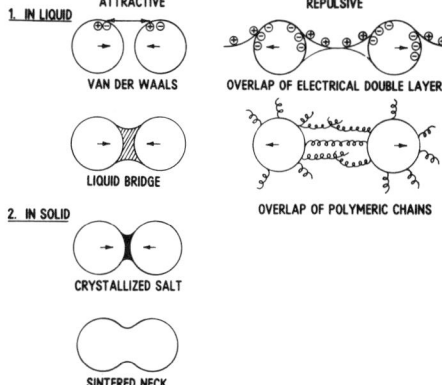

Fig 1 Schematic diagram showing the attractive and repulsive forces between the fine particles in liquid and solid environments to form agglomerates and aggregates

between a pair of isolated molecules; the force itself is rather short range, that is, it has a r^{-6} dependence where r is the distance between the two interacting molecules. The macroscopic interaction between particles is due to the summation of the pairwise interactions of the constituent molecules in the individual particles. Hamaker has shown that the interparticle force due to the summation of these pairwise interactions becomes rather long range and has a R^{-3} dependence, where R is the interparticle distance (Ref 4).

The van der Waals attractive force is usually balanced by other repulsive forces. A major repulsive force is due to the overlapping of electrical double layers associated with the particle surfaces. It has been shown that oxide surfaces usually contain ionizing sites from which H^+ (acid dissociation) or OH^- (basic dissociation) may be released, giving rise to a net surface charge and a potential difference (zeta potential) between the oxide surface and the liquid medium far from the oxide surface. The electric charge on the particle surfaces gives rise to an adjacent diffuse charge layer in the liquid in order to maintain the charge neutrality. When two particles are brought together, their associated diffuse charge layers become overlapped. Since the diffuse charge layers on both particles carry the same electrical charge, their overlap leads to an electrostatic repulsive force. The state of surface ionization depends on the nature of the oxide and on the pH value of the aqueous solution with which it is in equilibrium. It has been demonstrated that the degree of dispersion of several oxide powders is controlled by the pH value of the aqueous solution (Ref 5). Furthermore, the electrolytes dissolved in the aqueous solution can lead to flocculation of the powder dispersion (Ref 3, 6). The effectiveness of the electrolyte as a flocculation agent increases with the valence of the ions that form diffuse layers in the electrolyte. Thus, the finite solubility of the ceramic powder itself or the unreacted

precursor salts in the aqueous solution can lead to flocculation.

Polymeric solutions have been used extensively to stabilize colloidal dispersions (Ref 7, 8). Before a polymeric dispersant becomes effective, the adsorption of polymer chains onto the solid surfaces is required. When polymer-coated particles approach each other the overlap between the adsorbed polymer layers provides a repulsive energy. The repulsive energy is caused by an increase in the free energy of mixing between the polymer and the solvent due to a loss of configurational freedom of the adsorbed polymer molecules (Ref 7).

Experimental techniques have been developed to quantitatively measure the forces between two oxide surfaces maintained in a controlled spacing of 0 to 100 nm and in various electrolytes (Ref 9, 10) and polymer-containing solutions (Ref 11–13). These measurements have provided quantitative data of the efficacy of macromolecules and electrolytes in the stabilization of colloidal dispersion. Experimental measurements of the surface forces among ceramic powders in different processing media can provide a useful knowledge base for ceramic processing.

During the drying process, a residual liquid forms a liquid bridge between the particle necks. The capillary pressure in the liquid bridge provides an attractive force between particles. It has been shown that the attractive force due to the liquid bridges is about the same order of magnitude when compared to the van der Waals force, that is, 10^{-2} to 10^{-3} dyne between particles of 0.1 μm diameter (Ref 14).

Solid bridges may be formed during drying in crystallizing salts. The adhesive strength of agglomerates bonded by these salt bridges depends on the strength of these salt bridges and the bonding between the crystallizing salt and the solid particles. The average neck size of these salt bridges is determined by the concentration of the salt solution in the powder slurry. Thus, in principle, the salt bridges can be eliminated by rinsing or chemical treatment to precipitate the salts prior to drying.

Solid bridges during calcination essentially result from partial sintering or neck growth between solid particles. When the particles have already formed loose agglomerates during the powder preparation processes, heat treatment during calcination will change these loose agglomerates into much harder aggregates. Since the size of the sintered neck increases with the calcination temperature, the bonding strength of aggregates also increases with temperature. Thus, mechanical force is often used to break down the aggregates during the milling process.

The effect of agglomerates on densification can be illustrated by the sintering study of β-Al_2O_3 (Ref 15). The starting powders were prepared by freeze-drying and the powder composition was designed such that the sintered body has a β instead of a β'' phase.

Microstructures of a powder prepared by freeze-drying of a sulfate solution are shown in Fig 2(a). The powder had a very fine particle size of about 0.1 μm. However, a significant amount of agglomeration was observed in this powder. During calcination, these agglomerates formed plate-shaped aggregates, as illustrated by the fractograph of a sintered sample shown in Fig 2(b). Densities of sintered compacts from this highly aggregated powder never exceeded 90% of the theoretical value.

Particle size and extent of aggregation in Al_2O_3 and ferrite powders have been related to densification kinetics (Ref 16, 17). It was observed that Fe_2O_3 powder derived from a nitrate salt had a large surface area and a rapid low-temperature densification. However, further densification became difficult because of

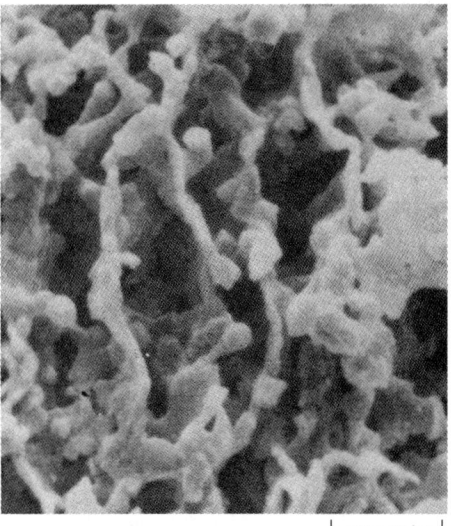

(a)   15 μm

(b)   3 μm

Fig 2 Micrograph (a) of freeze-dried Ni-doped aluminum- and sodium-sulfate powders, and fractograph (b) of sintered sample from same powder. Source: Ref 15

a high incidence of powder aggregation (Ref 17). In hydrothermally synthesized Al_2O_3 powder, it has been shown that small particle and agglomerate sizes and a low agglomeration factor, which is related to the number of particles in the agglomerates, are required for densification to full density (Ref 18). It has been reported that in some cases, $BaTiO_3$ powder having a smaller particle size actually ended up with a lower sintered density due to an inhomogeneous agglomeration in the fine powder (Ref 19).

These studies showed that data derived from the Brunauer-Emmet-Teller (BET) surface area analysis may not bear any meaningful correlation with the densification behavior of powders having different morphologies. Powders prepared by chemical methods usually have a large surface area, but the extent of agglomeration can have a more important impact on their sinterability.

Deagglomeration treatments increase the sinterability of ceramic powder. The most remarkable success has been obtained during sintering of Y_2O_3 stabilized ZrO_2 submicron powder (Ref 20). Fine powder was prepared by organometallic methods and the low temperature calcined powder has a particle size of about 0.01 μm. The powder was suspended in an aqueous solution of HCl with pH = 1.2 and allowed to settle. The small crystallites which remain in suspension after 72 hours were cast by centrifuge to form compacts with 74% green density. These compacts sintered to 99.5% after 1 h at 1000 °C (2010 °F), which is 300 °C (540 °F) lower than necessary for agglomerated powder of the same particle size.

Recently TiO_2 powder has been prepared from hydrolysis of $Ti(OC_3H_7)_4$ precursor to yield a fine particle size (diameter) of 0.04 μm in the calcined powder (Ref 21, 22). Deagglomeration was achieved by ball milling in an aqueous solution of HCl with a pH value of 2.1, and a sedimentation process was used to remove coarse calcined aggregates. The centrifuge-compacted TiO_2 samples have a high densification rate at low temperatures, as indicated in Fig 3. The data show that the prepared powder sintered to greater than 99% of the theoretical density at 840 °C (1545 °F), whereas the commercial powder required sintering above 1230 °C (2245 °F) to achieve only 96% density.

Particle Shape. Powder with equiaxed shapes is desirable for enhanced densification. This is demonstrated by sintering β-Al_2O_3 (Ref 15). Figure 4 shows the microstructure of β-Al_2O_3 powder prepared by decomposition of gels from citrate solution. In process 1, the powder was calcined at 1200 °C (2190 °F) and a β″ phase was formed. The powder had a needle shape with an aspect ratio of nearly 20. The poor alignment among these needle-shaped particles prevented densification, and samples less than 85% dense were obtained. However, good densification was observed in local regions where the elongated

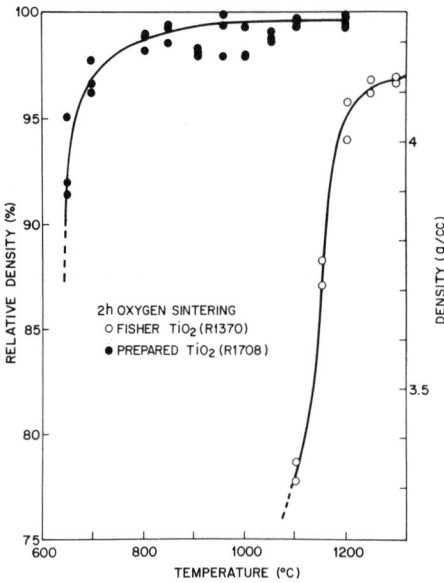

Fig 3 TiO_2 sintered densities versus temperature after 2 h sintering cycle in O_2. In commercial TiO_2, sintering temperatures of 1230 °C (2245 °F) or above are required to densify to 96% of the theoretical density. TiO_2 prepared by hydrolysis of $Ti(OC_3H_9)_4$ reached the same density at 680 °C (1255 °F), which is 550 °C (990 °F) lower than that for commercial powders. Source: Ref 21

particles had good alignment. Thus, it is desirable to inhibit the growth of particles with a high aspect ratio. For example, in process 2 of Fig 4 when the power prepared from gel was calcined at a lower temperature (about 900 °C, or 1650 °F), powder having λ phase, equiaxed particle shapes and 0.01 μm size was obtained. The λ powder was ground, centrifugally classified, and compacted. *In situ* transformation to the β phase occurred during sintering. The particle packing evidently restricted the lateral growth into plate-shaped grains and allowed sintering to ~97% density as shown in process 2 in Fig 4.

During liquid-state sintering the particle shape can have a significant effect on the densification rate and the required volume fraction of liquid. Cahn and Heady have shown that there exist significant differences in the interparticle capillary forces between wetted spherical and irregularly shaped particles (Ref 23, 24). Figure 5 plots the interparticle force versus liquid volume in particles with different shapes. When a small volume of liquid is placed in the contact region of spherical particles, the normal force is very large; the force decreases slowly as the liquid volume increases. For irregularly shaped particles with pointed contacts, the normal force increases rapidly from zero as the liquid volume

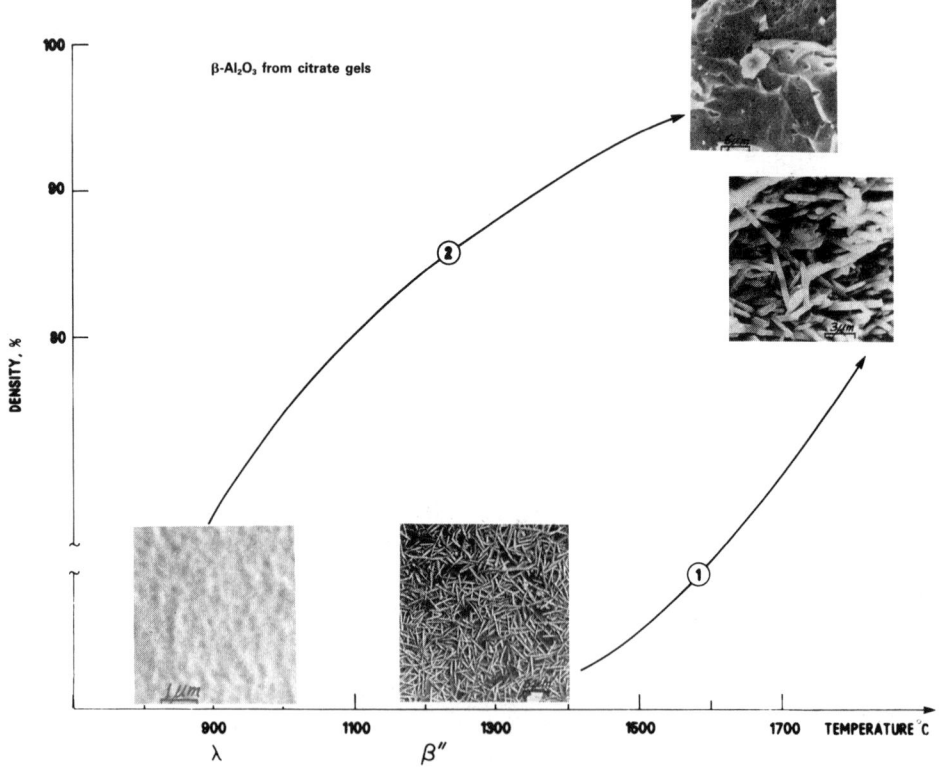

Fig 4 Microstructures of β-Al_2O_3 powder from calcination of citrate gels. In process 1, poor alignment along needle-shaped β″ phase powders from 1200 °C (2190 °F) calcination gives a low sintered density. In process 2, an equiaxed particle-shaped λ-phase powder from 900 °C (1650 °F) calcination yields a significant increase in the sintered density. Source: Ref 15

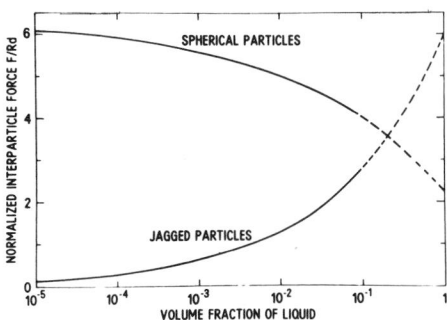

Fig 5 Interparticle forces versus the liquid volume for spherical and jagged particles

increases. Metal powders usually have round-shaped particles because metals tend to have isotropic surface energies. Thus, in wetted metal particles, a large interparticle force exists even when only a very small liquid volume is present. However, the surface energies of ceramics are generally more anisotropic, and their particle shapes are often angular. Thus, a large quantity of liquid is required to provide a cohesive force to ceramic powders with angular particle shapes. This is consistent with the general observation that only a very small liquid volume (less than 1%) is used in liquid-phase sintering of metal powders, whereas a large liquid volume (up to ~30%) is usually required during liquid-phase sintering of ceramic powders (Ref 24).

In addition, Cahn and Heady have shown that in powder with angular particle shapes, there exist torque and shear components in the interparticle force (Ref 24). These force components contribute greatly to particle rearrangement, which is an important factor during the initial stage sintering.

Particle-Size Distribution. The effect of particle-size distribution on the final achievable density can be studied by analyzing the force balance between the pore drag and the driving force for grain growth in compacts with different grain-size distributions (Ref 25, 26). It is reasonable to expect that the grain-size distribution at the final stage of sintering is similar to the particle-size distribution of the starting powder if a significant particle coarsening does not occur. The analysis shows that a narrow grain-size distribution is imperative for obtaining a high sintered density. The conditions for pore separation from grain boundaries during sintering were analyzed; densification usually stops under such conditions. It is shown analytically that a minimum solute drag force is required to prevent pore separation. In microstructures with a uniform grain-size distribution, the dopant concentration needed to reach the minimum solute drag force is typically 14 times smaller than that in microstructures with a nonuniform size distribution. If the drag force is below the required minimum, pore separation will occur after the compact sinters to a critical density. This critical density is 99.3% in compacts with a narrow grain-size distribu-

tion, but only 90.6% in compacts with a nonuniform size distribution.

Techniques for synthesizing monodisperse oxide powders [for example, $Cr(OH)_3$ (Ref 27), AlOOH (Ref 28), TiO_2 (Ref 29), and SiO_2 (Ref 30)] through controlled nucleation and growth of particles in dilute solution have been established by colloid chemists. However, the sintering behavior of the synthesized monodisperse powders has not been studied systematically until the recent work of Barringer and Bowen (Ref 31). They have prepared monodisperse spherical TiO_2 powder with a particle size of 0.3 to 0.6 μm by a controlled hydrolysis reaction between 0.1 to 0.2 M $Ti(OC_2H_5)_4$ solution in ethanol and 0.3 to 0.5 M water in ethanol. The synthesized powder was dispersed in an NH_4OH solution with a pH of 9.5 and compacted by gravitational or centrifugal settling. Figure 6(a) shows a SEM micrograph of the top surface of a uniform green TiO_2 compact with 65% density, and Fig 6(b) shows the same sample sintered to

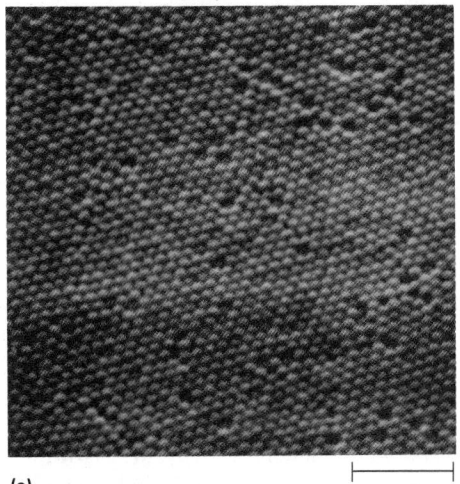

(a)   2 μm

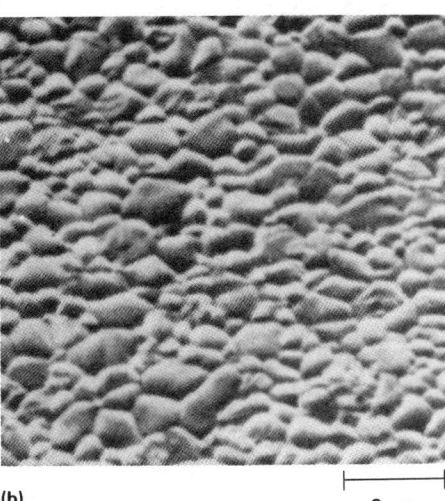

(b)   2 μm

Fig 6 Scanning electron micrograph of (a) top surface of dense, uniform green compact of TiO_2 and (b) microstructure sintered to >99% of the theoretical density in 90 min at 1050 °C (1920 °F). Source: Ref 31

>99% of the theoretical density in 90 min at 1050 °C (1920 °F) (Ref 31).

Microstructural Changes. During powder processing before firing, the powder characteristics can be considered as a set of constant parameters. However, significant microstructural changes have been observed prior to or concurrent with the densification process during sintering. Particle coarsening at the initial stage of sintering has been reported in Al_2O_3 (Ref 32) and SiC (Ref 33). Since particle coarsening is detrimental to densification, certain dopants have been selected to reduce coarsening and thus to give the desired densification, for example, B dopant in SiC (Ref 33).

In the intermediate and final stages of sintering, grain growth during densification is a commonly observed phenomenon and has been reported in Cu, Al_2O_3, BeO, ZnO (Ref 34), and Ag (Ref 35). Figure 7 shows grain growth during sintering in TiO_2 by plotting the density and grain size versus temperature (Ref 36). At temperatures below 1300 °C (2370 °F), the compact densified readily up to about 85% of the theoretical density with little grain growth and the grain size remaining at ~1 to 2 μm. At higher temperatures, the densification rate was much reduced while significant grain coarsening occurred, and the grain size increased to 9 μm.

Recently, a theory of grain growth during densification has been proposed (Ref 37, 38). In this theory, it is assumed that densification results from material transport between pores and grain boundaries by lattice or boundary diffusion, and grain growth is controlled by the pore mobility which is limited by either surface, lattice, or vapor transport. Furthermore, two competing processes can lead to changes in pore size. Pore shrinkage occurs by vacancy diffusion from pores to grain boundaries and dislocations. Pore growth is due to either the coalescence of several pores

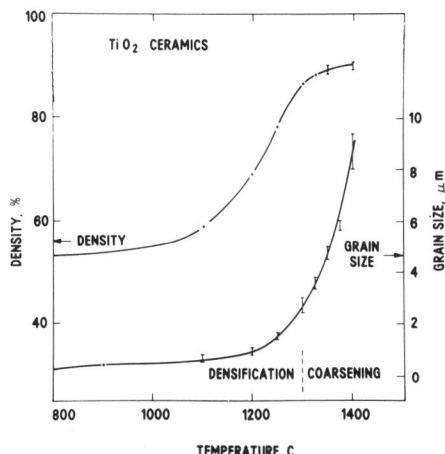

Fig 7 Density and grain size of TiO_2 versus sintering temperatures to show rapid densification with little grain growth at low temperatures and rapid grain growth with little densification at high temperatures

dragged by the migrating grain boundaries (Ref 39) or Ostwald ripening of pores by a concurrent diffusion of vacancies from smaller pores to larger ones (Ref 40). This theory predicts that the extent of grain coarsening and the achievable final density depend on the rate-controlling mechanisms in coarsening and densification kinetics. In particular, a small starting particle size and high green density can lead to high sintered density with minimal grain growth as a result of enhanced grain-boundary diffusivity at the expense of surface diffusivity and vapor transport. Thus, this study provides a theoretical framework for the selection of dopants and sintering atmosphere for microstructural control.

Shape changes in grains are often observed during sintering of ceramics. The most notable examples are found in recent investigation in ceramic superconductors. Anisotropic grain growth in $YBa_2Cu_3O_x$ superconductors usually leads to grain faceting in sintered samples. Since the basal (001) plane is the slowest growing surface in $YBa_2Cu_3O_x$, the basal plane is frequently observed as the boundary plane (Ref 41). It has been estimated that roughly 75% of the total grain boundary area is occupied by the (001)-faced grain boundaries. This observation is consistent with the type of facets developed on large millimeter-sized single crystals in which the (001) plane is also the preferential growth facet (Ref 42, 43). The anisotropic grain growth is used to develop texture in $YBa_2Cu_3O_x$ superconductors such that the basal planes are oriented preferentially parallel to the specimen surface (Ref 44). It has been shown that a substantially higher current density can be transported in the direction parallel to the basal plane than in the normal direction (Ref 45).

In $Bi_2Sr_2CaCu_2O_8$ superconductors, the anisotropic grain growth has been shown to cause retrograde densification in the temperature range 850 to 890 °C (1560 to 1635 °F) during which the material becomes less dense as the sintering temperature is raised (Ref 46). Within this temperature range a significant grain growth occurs. $Bi_2Sr_2CaCu_2O_8$ has a crystal growth habit yielding platelike crystallites that have an aspect ratio as large as 50 (Ref 47). The basal plane is the preferential growth facet in this superconducting compound. It is believed that during sintering the growing platelike crystallites push each other apart, leading to a reduction in density. At higher temperatures (900 to 905 °C, or 1650 to 1660 °F), a liquid phase forms with a composition close to that of the starting powder, and specimen densities increase by a liquid-phase sintering process (Ref 46).

Physical Characteristics of the Powder Compact

Compaction of a ceramic powder to give a variety of shapes is an important processing step. Green microstructures, distribution profiles of stress, strain, and density in green powder compacts affect the subsequent densification and microstructural development processes. In this Section, the compaction behaviors of ceramic powders will be described with particular emphasis on the effects of compact inhomogeneities pertinent to the subsequent sintering process.

Compaction behavior of Al_2O_3, SiO_2, MgO, and $CaCO_3$ has been studied by Cooper and Eaton (Ref 48). They have proposed two broad classes of consolidation processes during powder compaction. In the first process, holes similar in size to the original particles are filled by an elastic deformation of particles to slide past each other. In the second process, voids with a much smaller size than the original particles are filled by plastic flow and fragmentation of the particles. Probability functions have been assigned to determine the likelihood of hole fillings by these two processes at a given pressure. The compaction data of the above oxides agree reasonably well with the model proposed by Cooper and Eaton (Ref 48). Their data showed that filling of the smaller holes requires considerably higher pressures of ~10 to 20 times greater than those needed to fill the larger ones. Furthermore, it was found that softer powders tend to yield a greater fractional volume compaction at a given pressure. This may be explained by the difference in the predominant process for pore filling. Harder particles depend primarily on fracture to fill the small void spaces with fragments. The requirements for the size, shape, and location of a fragment that will effectively contribute to densification become more demanding as the pores become smaller. However, for the softer powders, plastic flow occurring at lower pressures can contribute significantly to powder compaction (Ref 48).

In most nonplastic ceramic powders, organic binders are introduced to increase the cohesive strength of the powder compact. In the absence of any frictional force, the weakest links in an unfired powder compact are the organic bonds between particles (Ref 49). The tensile strength of the powder compact depends on the cohesive strength of the binder, the distribution of binder among the particles, and the porosity of the powder compact. Low molecular-weight binders, such as waxes, tend to have low cohesive strengths of ~0.7 to 3.5 MPa (100 to 500 psi) (Ref 49). However, high molecular-weight resins, for example, polyvinyl alcohol, can have high strengths of ~21 to 70 MPa (3000 to 10,000 psi) (Ref 49). Furthermore, the distribution of binder among the solid particles has a significant effect on the tensile strength of the compact. For example, a nonwetting binder liquid does not contribute to the tensile strength, because the binder does not coat the neck regions between particles (Ref 49). The porosity of the powder compact also affects the tensile strength, because the number of touching particles around each particle increases with a decrease in porosity.

Thompson has studied the mechanics of powder pressing (Ref 50). According to the earlier experimental results of Unkel (Ref 51), on the top and bottom surfaces of a cylindrical die, the axial pressures have a parabolic radial distribution and are symmetrical about the central axis of the die. Based on these experimental results, Thompson has calculated the stress and density distributions in the powder compacts formed in a single-acting press and a symmetrical double-acting press (Ref 50). The gradients in compact forming pressure and green density are due to the friction between the die wall and the powder. Thus, lubrication at die walls and reducing the coefficient of friction should minimize the pressure and density gradients in the powder compact. Butyl stearate, graphite powder, metallic aluminum powder, oleic acid, and neofat are typical lubricants used in ceramic processing (Ref 8).

The frictional force between the die wall and the powder is directly proportional to the radial stress at the wall. During a uniaxial pressing, the applied stress is in the axial direction and is parallel to the die wall. For a given axial stress, the resultant radial stress depends on the fluidity of the powder under compaction. For example, both the radial and axial stresses are identical when a liquid is compacted. However, when a nonelastic and incompressible solid is under an axial compaction, there should not be any radial stress. Thus, it is desirable to decrease the powder fluidity such that the radial and frictional stresses are minimized. Consequently, powder lubrication should be avoided in order to minimize the density and stress gradients in the powder compact. Furthermore, irregularly shaped, fine, and sticky powders are preferred to the spherical, coarse, and free-flowing powders in order to obtain a uniform density distribution (Ref 50). However, a decrease in the powder fluidity also reduces flowability, which is important for a uniform and consistent filling of die cavities, especially during an automatic pressing process.

Onoda and Janney have reviewed the basic concepts in soil mechanics with emphasis on their applicabilities to ceramic processing (Ref 52). They have shown that the principles in soil mechanics provide a sound framework in calculating the stresses and strains in ceramic bodies during processing, and that the concepts of pore liquid pressure, capillarity, and effective stress are pertinent to systems containing liquids. In particular, they have proposed the critical state theory of Schofield and Wroth (Ref 53–55) as the most promising in providing a unified semiquantitative explanation of the material behavior during processing.

Effects of Compact Inhomogeneities. It is very important to obtain defect-free green compacts with uniform microstructures prior to sintering. Flaws, usually in the form of physical and chemical inhomogeneities, may be introduced during the preparation of green compacts. They originate from lamination during die pressing, hollow agglomerates in some spray dried powders, binder accumulation and incomplete burnout, and inclusion of foreign particles (Ref 56). The firing process usually amplifies rather than corrects these defects (Ref 57).

Evans has analyzed the inhomogeneity effects in sintering (Ref 58). It is reasonable to expect that compact inhomogeneities can induce nonuniform shrinkage rates, which in turn, create transient stresses at the interface between regions with different shrinkage rates. Evans has used the Eshelby procedure (Ref 59) to calculate these stress states. His analysis showed that the peak stresses due to the inhomogeneous shrinkage are about the same magnitude as the effective sintering stress within the region of unconstrained shrinkage. Thus, these stresses can lead to defect formation in a powder compact. During the intermediate stage sintering, pores have a cylindrical shape and they are usually intersected by three grain boundaries. In the absence of any stress, the cylindrical pores assume an equilibrium shape with a cross section shown in the dashed line in Fig 8. However, transient stress, induced by the inhomogeneous shrinkage rates, may impose a tensional force on the grain boundaries intersecting the cylindrical pores. Equilibrium-shaped pores grow at locations where the tensional stress exceeds the local sintering stress. Furthermore, the tensional stress on the grain boundary will cause material transport from the neck region, at the intersection between the pore and grain boundary, to the grain-boundary region via boundary diffusion. Thus, the pore will change from the equilibrium to a cracklike morphology. Evans has shown that the transitions to cracklike pores usually occur along the circumferential boundaries near the inhomogeneity interface, where the transient tensions are the largest (Ref 58). Furthermore, a larger circumferential shrinkage crack may occur by coalescence of the cracklike cavities and these shrinkage cracks may be converted into a large void by further shrinkage.

One can estimate the limiting size of defects formed by the inhomogeneity effects proposed by Evans. If it is assumed that all porosity is collected at the circumference between the two shrinkage zones, the limiting void spacing, Δr, after densification is given as:

$$\frac{\Delta r}{R} = (\rho_0^{1/3} - \rho^{1/3}) \tag{Eq 2}$$

where R is the radius of the fine-grained, low-density zone, ρ_0 and ρ are the relative density of the matrix and the fine-grained, low-density zones, respectively, as shown in Fig 9. In Eq 2, it was assumed that, except on the circumference region, both zones are fully densified. Figure 10 shows the calculations of $\Delta r/R$ for different ρ_0 and ρ. Calculations show that for a typical matrix density of 0.5, a void spacing about 6% of the size of the inhomogeneity zone can result if the latter zone has a density of 0.4, which is about 20% lower than the matrix zone. Calculations also show that for a typical inhomogeneity zone of ~100 μm, which is roughly the size of an agglomerate, a void spacing of about several microns, which is about the same magnitude of a typical grain size, may result. Conversely, this author's calculations show that in order to limit the resulting crack size to less than 1 μm, the unfired powder compact should be uniform with $\Delta\rho/\rho \leq 0.1$ within a zone of 50 to 100 μm.

Since submicron dry powders show a strong tendency to form agglomerates due to van der Waals forces, it may be difficult to remove agglomerates and the subsequent flaws using the dry powder consolidation routes. Aksay *et al.* have demonstrated the potential of colloidal filtration to avoid the problem of agglomeration by introducing either an electrostatically or a sterically repulsive force between particles (Ref 60). In fact, these authors have prepared uniform Al_2O_3-ZrO_2 composites by colloidal filtration, and sintered compacts have a flexural strength 1.6

times that of bodies consolidated by isostatic pressing of dry powders.

In the near-net shape forming processes, such as injection molding, high concentrations of organic binders are needed to enhance greenware moldability. In these processes, binder removal must be carefully controlled to avoid damaging the green compacts. Barone and Ulicny have analyzed the basic stages during the binder removal process and calculated the resultant stresses on the ceramic compacts (Ref 61). In that study, the binder system investigated was a mixture of paraffin wax (melting point at 56 °C, or 133 °F) and amorphous polyethylene. During the initial heatup above the binder melting point, the liquid binder expands substantially to fill the void space within the ceramic skeleton, and the liquid binder imposes a hydraulic stress on the green compact. The effective hydraulic pressure is proportional to the product of the heating rate, binder viscosity, binder expansion coefficient, nonceramic-containing void fraction, and the square of the specimen dimension (Ref 61). Calculations based on the binder properties show that for 50 mm (2 in.) specimens, a slow heating rate of ~0.5 °C/min (~0.9 °F/min) is required such that the hydraulic stress induced by the liquid binder will not exceed 5×10^5 Pa (5 atm), which is the fracture strength of a typical unfired ceramic body after binder removal. A lower hydraulic stress on ceramic compacts also results when a binder system having a lower viscosity and reduced thermal expansion coefficient is chosen.

Chemical Effects of Dopant Addition

During powder processing, dopants are usually added to improve densification, and some dopants have been discovered such that the near-theoretical density can be achieved. Magnesia in aluminum is a classical example (Ref 62). In general, the effect of a dopant

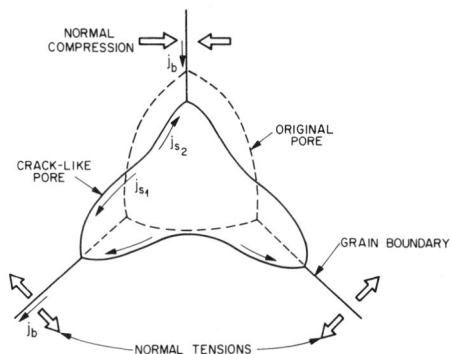

Fig 8 Schematic illustrating transition from equilibrium to cracklike morphology for normal tension at the lower two grain boundaries. See text for details. Source: Ref 58

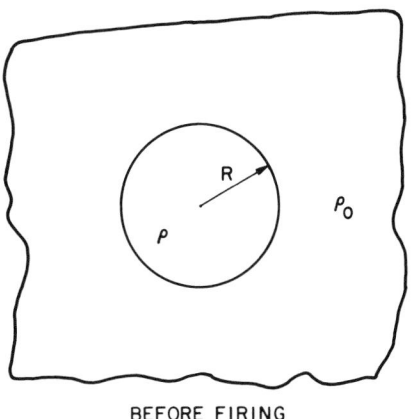 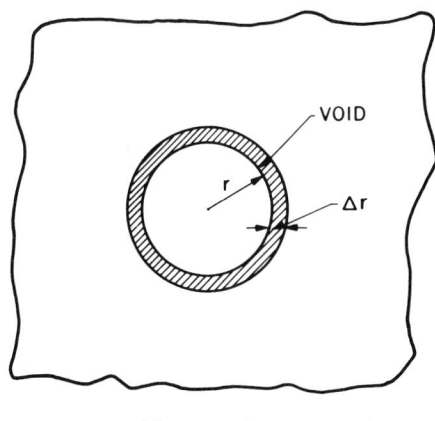

Fig 9 Schematic showing an inhomogeneity zone within a matrix with densities ρ and ρ_0, respectively, and with $\rho < \rho_0$. Both the homogeneity zone and the matrix are assumed to be fully densified, leaving a void spacing, Δr.

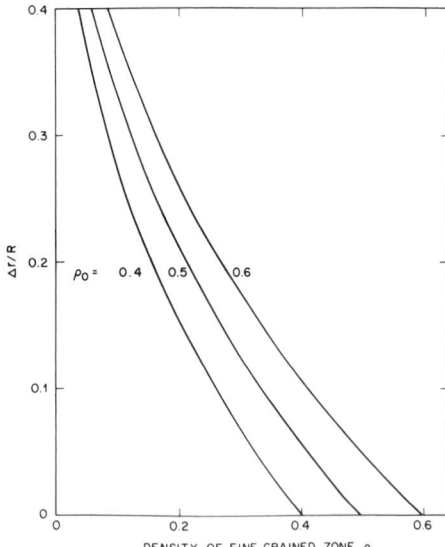

Fig 10 Void space, Δr, between an inhomogeneity zone and the matrix, normalized with respect to the size of the inhomogeneity zone, and calculated as a function of densities in both zones

depends on whether it is in solid solution with the host at the sintering temperature. Some possible beneficial influences of dopants on densification are in preserving the ideal characteristic of the starting powder and in increasing the packing density of powder compacts. Dopants are often used to enhance the material transport rate during sintering by increasing the lattice diffusivity and by providing high diffusivity paths due to a preferential dopant segregation at grain boundary regions. Dopants can also be used to prevent exaggerated grain growth and to permit a complete densification with a minimal porosity entrapment. Some of these effects will be discussed in the following sections.

Preservation of Ideal Power Characteristics. In the Section of this article on physical characteristics of powders, it was shown that significant particle coarsening prior to densification has been observed during sintering. Certain dopants can prevent particle coarsening and thus preserve the ideal particle characteristics for densification. For example, it has been shown that boron is essential to densification of SiC because it impedes surface diffusion in SiC and, thus, inhibits particle coarsening in the powder compact at temperatures <1500 °C (<2730 °F) (Ref 33). Since the rate of particle coarsening increases rapidly with a decrease in particle size, dopants are particularly important in suppressing particle coarsening during sintering of ultrafine particles.

In ceramic processing, the precursor powder used for the green-body fabrication may go through a phase transformation during high-temperature sintering. The phase transformation process can have a significant effect on the microstructural development prior to densification (Ref 63). Thus, a properly con-

trolled transformation process can result in a uniform microstructure and thus enhanced densification. Kumagai and Messing have shown that seeding the boehmite (AlOOH) sol-gel with α-Al_2O_3 particles can control the phase transformation and can substantially improve the sinterability (Ref 64, 65). In particular, boehmite sol-gel seeded by 2 wt% α-Al_2O_3 was sintered to 98% of the theoretical density, whereas unseeded gel had to be sintered at 1600 °C (2910 °F) to reach 94% of the theoretical density (Ref 64, 65). It is believed that an increased number of nucleation sites limits the resultant particle size of the transformed α-Al_2O_3 grains. Thus, a uniform, fine-grained α-Al_2O_3 microstructure is developed upon transformation instead of an extensive vermicular microstructure as observed in unseeded gels.

Enhancement of Lattice Diffusion. The densification rate can be significantly affected by the type and concentration of lattice defects which are important for material transport. In most ceramic materials, the type and concentration of lattice defects are determined by the trace impurities and dopants because the ionic nature of the material requires a charge balance between the vacancies and the solutes. Furthermore, the energy of vacancy formation in many oxide ceramics is very high, such that the intrinsic vacancy concentration is very small (about 10^{-8} vacancy/lattice site even at temperatures close to the melting points of most ceramics). This intrinsic vacancy concentration is much smaller than the trace impurity concentration, which is $\geq 10^{-5}$ ions/lattice site in most ceramics. In metals, the intrinsic vacancy concentration can be much larger than the impurity concentration because the energy of vacancy formation in metals is rather low, such that the vacancy concentration can be $\geq 10^{-4}$ vacancy/lattice site.

In most oxides, the oxygen ions are usually less mobile than cations; thus a dopant favoring an increase in the oxygen vacancy concentration may promote densification (Ref 66–68). Effects of stoichiometry have been demonstrated in the sintering studies of nickel ferrite ($NiFe_2O_4$), and nickel aluminate ($NiAl_2O_4$). It has been shown that densification is very difficult in samples with excess Fe and Al, from the lack of oxygen vacancies in the ferrite and aluminate (Ref 68). Figure 11 plots the sintered densities versus composition, x, in $(1 - x)$ NiO + xAl$_2$O$_3$ (nickel aluminate). The data show that near-theoretical densities can be achieved in the aluminates with excess Ni compositions because oxygen vacancies are abundant in these compositions. However, the aluminates become very porous when they are Ni-deficient.

Prevention of Exaggerated Grain Growth. It is well known that exaggerated grain growth during sintering usually leads to densification well below the theoretical density. An extensive review on grain-boundary migration has been presented (Ref 69, 70). In general, the grain-boundary mobilities

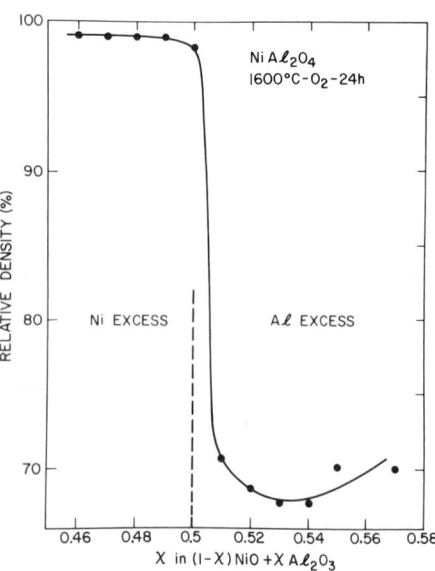

Fig 11 Sintered density of nickel aluminates versus the composition. Near-threshold density was achieved in compositions with excess nickel. Source: Ref 68

reported in the literature are much less than the upper bound estimate of the intrinsic mobility based on the Turnbull model (Ref 71). The lower velocity measured experimentally may be caused by impurity and second-phase drag. Theories of impurity and second-phase drag on grain-boundary migration have been proposed and the theories agree qualitatively with the experimental data (Ref 72).

Densification to near-theoretical density can be achieved if certain dopants are used to prevent discontinuous grain growth such that pores remain on grain boundaries until they are eliminated. In the case of sintering Al_2O_3 to near-theoretical density with the aid of MgO dopant, it is generally agreed that the success is due to the suppression of exaggerated grain growth in Al_2O_3 by the dopant. Figure 12 shows that evaporation of MgO from the surface region of sintered Al_2O_3 leads to the growth of exaggerated grains (Ref 73). Burke et al. have estimated that the grain-boundary mobility in undoped Al_2O_3 is about 100 times faster than in MgO-doped Al_2O_3 (Ref 73). However, there is a lack of agreement on the mechanism by which MgO suppresses the exaggerated grain growth in Al_2O_3. For example, Johnson and Coble have demonstrated that MgO segregation at grain boundaries was not detected by Auger analysis and that $MgAl_2O_4$ second-phase pinning is not necessary to inhibit grain growth (Ref 74). Based on the microstructural evidences, Heuer (Ref 75) and Bannister (Ref 76) have proposed that the MgO addition enhances the surface diffusivity and thus increases the pore mobility in Al_2O_3. With an increase in the pore mobility, pores remain on grain boundaries until they are eliminated. However, if MgO can indeed increase the surface diffusivity of

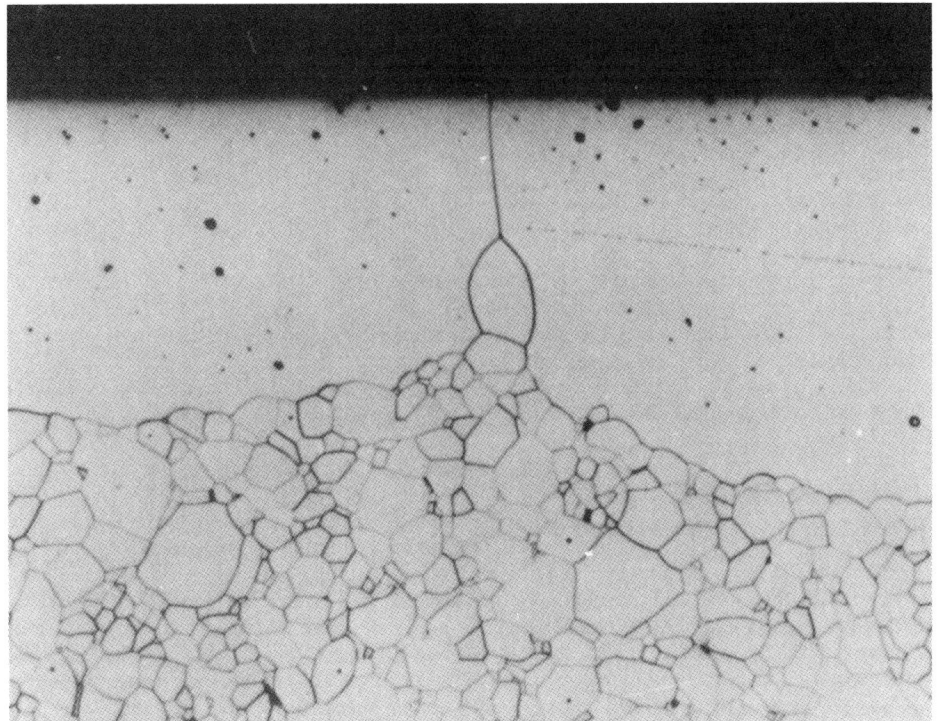

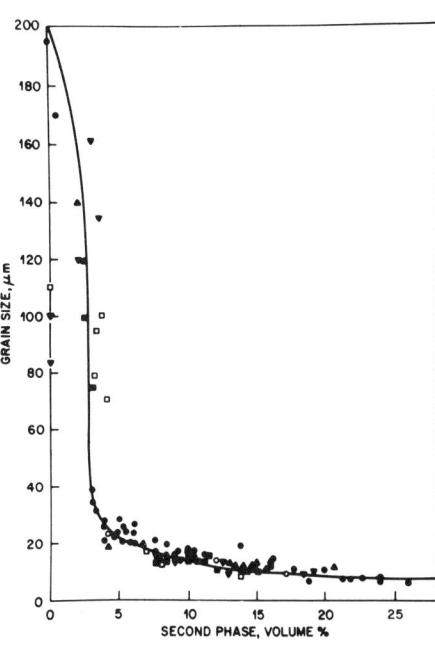

Fig 13 Grain size in Ti-doped MnZn ferrites is plotted versus the volume fraction of the hematite second phase

Fig 12 Evaporation of MgO from the surface region of sintered MgO-doped Al₂O₃ that led to exaggerated grain growth. Source: Ref 73

Al₂O₃, it may also increase the particle coarsening of MgO-doped Al₂O₃ powder. However, the surface area data of Burke *et al.* do not support this prediction (Ref 73). Thus, this problem is still very much open to further research.

When the dopant concentration is in excess of the solubility limit, a second phase forms in the specimen. In commercial high-alumina (96% Al₂O₃) (Ref 77) and zirconia ceramics (Ref 78), discrete and continuous second phases are found extensively. Continuous glassy phases are found wetting most grain boundaries of 96% Al₂O₃. The glassy phases can wet grain boundaries provided that the alumina-glass interfacial energy is less than half of the Al₂O₃ grain-boundary energy (Ref 77). While these glassy phases in commercial alumina have a wide range of compositions, 50 to 80% SiO₂, 12 to 21% Al₂O₃, 7 to 30% MgO + CaO, and trace amounts of other constituents, the compositional variations within the range reported do not have significant effects on the glass-alumina interfacial energy and thus the wettability of glasses on alumina grain boundaries. However, low-energy grain boundaries, such as low-angle boundaries as well as basal and rhombohedral twin boundaries, are not wetted by the glassy phases (Ref 77). Faceted interfaces are often observed in alumina because the surface free energy is a strong function of surface orientation. Anisotropic grain growth is observed in both the undoped alumina (Ref 79, 80) as well as commercial 96% Al₂O₃ (Ref 77).

In MnZn ferrite (Ref 81) and Y₂O₃ (Ref 82), second-phase particles have been used to inhibit exaggerated grain growth leading to near-theoretical densities. It has been observed that in an oxidizing atmosphere, TiO₂ stabilizes the hematite second phase among the spinel major phase in MnZn ferrite. The hematite second phase inhibits grain growth in ferrite as illustrated in Fig 13, which plots grain size versus volume fraction of the hematite second phase. Figure 14 shows the microstructures of undoped and 2% Ti-doped MnZn ferrite. Densification to near-theoretical density is achieved in Ti-doped samples because the hematite second phase inhibits the exaggerated grain growth. The hematite second phase is not stable in an atmosphere with low-oxygen partial pressure. Figure 15 shows the phase boundary between the spinel single phase and hematite-spinel two-phase region as a function of temperature and *p*O₂, the oxygen partial pressure. Thus, after the near-theoretical density is achieved during sintering, single-phase ferrite with the spinel structure can be obtained by nitrogen annealing (Ref 81).

Phase relations in a number of different rare earth oxide combinations, for example, La₂O₃-Y₂O₃ as shown in Fig 16, show a two-phase field at high temperatures and a single phase of the same composition at a lower temperature; both phase fields are at temperatures ≳80% of the melting point. Rhodes made use of these unique phase relations to sinter La-doped Y₂O₃ to full density (Ref 82). At high

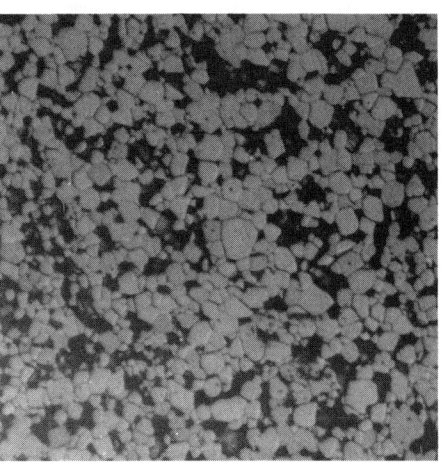

(a)

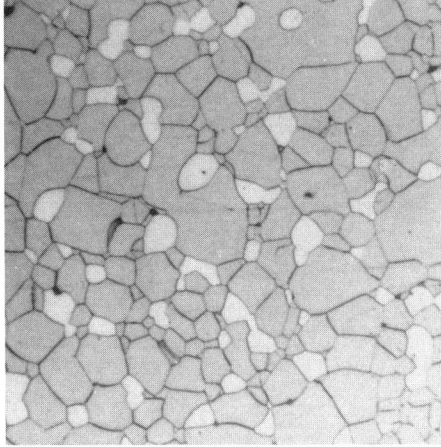

(b)

Fig 14 Microstructure of undoped (a) and 2.4 wt% MnTiO₃-doped (b) MnZn

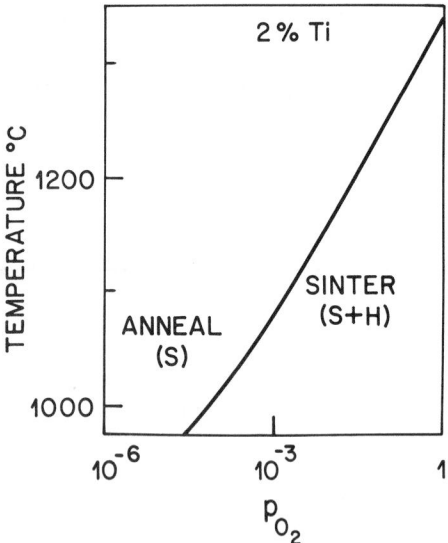

Fig 15 Phase diagram of 2% Ti-doped MnZn ferrite showing the phase boundary between the spinel single-phase and the hematite-spinel two-phase regions. S and H refer to spinel and hematite phases, respectively.

temperatures, the hexagonal phase is used to retard grain growth so that pores remain at grain boundaries. The dense specimen is then annealed in the single-phase region at a lower temperature, and the structural transition yields a pore-free single-phase body.

As indicated in the above discussion, only in very few systems can a quantitative prediction be made about the effect of a given dopant on densification and microstructural control. There exist data on experimental techniques to measure the effect of a dopant on the defect chemistry, lattice diffusivities, and the phase relations in most ceramic mate-

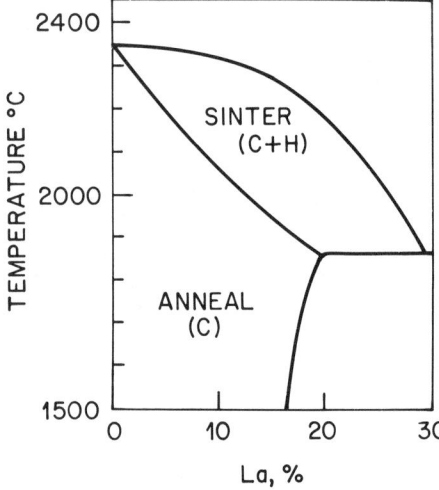

Fig 16 Phase diagram of La₂O₃-Y₂O₃ showing a two-phase field at high temperatures and a single phase at a lower temperature. C and H refer to cubic and hexagonal phases, respectively. Source: Ref 88

rials. Analytical techniques to measure the grain-boundary chemistry, for example, solute or second-phase segregation, are gaining acceptance. Analytical measurements in several ceramic systems have been used to correlate the sintering data (Ref 74, 83). However, significant progress is required to understand the role of dopants in vapor transport, in grain-boundary and surface diffusivities, and in surface and grain-boundary energies of ceramic and metal systems. These kinetic and thermodynamic parameters of the internal and external surfaces have a significant influence on densification and microstructural development. Progress in these areas can bring technological breakthroughs in ceramic powder processing.

Chemical Effects of Sintering Atmosphere

Effects of the sintering atmosphere on densification and microstructural development are related to the gas solubility, the reactions with dopants, and the powder itself. These factors will be discussed in the following sections.

Gas Solubility. First, the influence of sintering atmosphere is related to the gas solubility in ceramics. For example, MgO-doped Al₂O₃ can be sintered to full density in O₂ or H₂. However, the theoretical density cannot be achieved in air, N₂, He, and Ar, which have a limited solubility in Al₂O₃. These insoluble gas species become trapped within closed pores and prevent further densification (Ref 84). Similarly, when Ar at high pressure (>5 MPa, or 725 psi) is used in sintering SiC, the gas trapped within the pores prevents densification above 90% density since Ar does not readily diffuse through SiC. The use of vacuum in sintering of SiC yields the highest density. However, vacuum sintering also leads to significant thermal decomposition of SiC. The best compromise is probably the use of inert gas at atmospheric pressure (0.1 MPa, or 14.7 psi) (Ref 85).

Reactions with Dopants. The sintering atmosphere may react selectively with certain dopants which are vital to the densification process. For example, it has been shown that excess boron is necessary to enhance densification of SiC. While the precise role of boron in SiC activated sintering is not known, it is believed that boron inhibits particle coarsening and provides a high diffusivity path along grain boundaries. Certain sintering atmospheres, such as N₂ and CO, react with and remove the excess boron from SiC particle surfaces and the densification process is inhibited (Ref 85). In the case of the nitrogen sintering atmosphere, N₂ reacts with the excess boron to form BN. Since BN is highly soluble in SiC as a solid solution, boron remains in the SiC lattice as BN. Thus, the presence of excess boron in the SiC particle surfaces can be reestablished and densification can be restored by decreasing the nitrogen atmosphere below the equilibrium pressure of BN.

In the case of CO sintering atmosphere, boron is oxidized by CO to form BO which has a relatively high vapor pressure of about 10^2 Pa (10^{-3} atm). Thus, boron is irreversibly removed from SiC when a CO sintering atmosphere is used.

Reactions with Powder. Sintering atmospheres have a significant effect on the surface energies of many ceramic systems. For example, Barsoum and Ownby have reported that at 1430 °C (2605 °F) the surface energies of SiC, AlN, and Si₃N₄ decrease by 4.5 to 6% for every tenfold increase in pO_2 between 10^{-16} and 10^{-14} Pa (10^{-21} and 10^{-19} atm) (Ref 86). Consequently, the contact angles between silicon and SiC, AlN, and Si₃N₄ ceramics were found to decrease with decreasing pO_2 due to the increase in the surface energies of these solids (Ref 86). Furthermore, Chaklader, Gill, and Mehrotra have developed a theoretical model to predict the pO_2 dependence of the interfacial energies between molten metals (Cu, Ag, Fe, and Ni) and Al₂O₃ substrates, and they have obtained a reasonably good agreement with the experimental data (Ref 87).

In a more extreme case, the sintering atmosphere affects the oxidation states of cations, for example, Cr, in oxide ceramics and has a significant effect on the densification kinetics of Cr-containing oxides (Ref 88). Figure 17 plots the porosity of several chromites versus the pO_2 of the sintering atmosphere. At temperatures ranging from 1600 to 1740 °C (2910 to 3165 °F), the chromites do not densify in a pO_2 higher than 10^{-3} Pa (10^{-8} atm). However, 99% dense chromites can be obtained by sintering in an atmosphere with a lower pO_2 between 10^{-6} and 10^{-5} Pa (10^{-11} and 10^{-10} atm). In an oxidizing atmosphere, Cr₂O₃ becomes unstable and vaporizes as CrO₂ or CrO₃. A high vapor pressure enhances evaporation and condensation processes, increases the neck growth, and coarsens particles but does not contribute to densification (Ref 89-91).

Magnesia dopant in Cr₂O₃ leads to a further increase in density by inhibiting grain

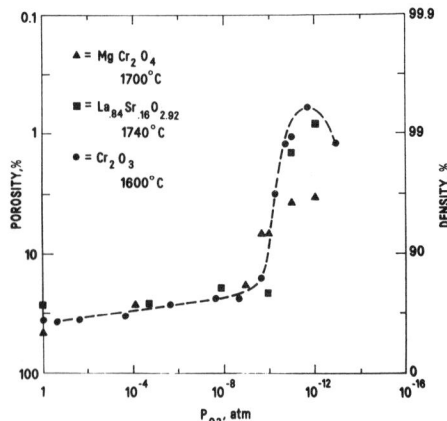

Fig 17 Dependence of density or porosity of sintered chromites on the partial pressure in the sintering atmosphere. Source: Ref 88

growth with MgCrO₄ second-phase particles nucleated along with the grain boundaries (Ref 89). However, MgO dopant is effective only if the pO_2 of the sintering atmosphere is sufficiently reduced to maintain the Cr_2O_3 phase and the undoped powder itself is sinterable to high density. Recently, TiO_2 dopants have been used such that a higher pO_2 (10^4 instead of 10^{-3} Pa when undoped) in the sintering atmosphere and a lower temperature (1300 °C, or 2370 °F, instead of >1600 °C, or 2910 °F, when undoped) are permitted during sintering of Cr_2O_3 (Ref 92). It has been proposed that Ti^{+4} ions replace the Cr ions in a valence state higher than 3. Thus, the Ti dopant prevents particle coarsening of Cr_2O_3 by suppressing the formation of volatile compounds such as CrO_2 and CrO_3. Furthermore, this dopant creates Cr vacancies and thus enhances the densification kinetics (Ref 92).

In transition metal oxides, stoichiometry and lattice defects can be readily controlled by the oxygen partial pressure in the sintering atmosphere. Thus, it is expected that densification and microstructural development in these oxides may depend on pO_2. Figures 18 and 19 show the densification data of MnO (Ref 93) and Fe_3O_4 (Ref 94), respectively, versus pO_2. Both the type and concentration of lattice defects are expected to vary over this wide range of pO_2. In both cases the sintered densities are relatively insensitive to the atmosphere. However, pO_2 has a pronounced influence on the microstructural development of Fe_3O_4 (Ref 94). In an oxidizing atmosphere, a small grain size is obtained, whereas the grain size becomes much larger in a reducing atmosphere.

Several studies have been reported on the effect of water vapor on sintering of MgO (Ref 95–97). Among these studies, it is generally agreed that the presence of water vapor increases the rates of densification and particle coarsening. For example, Whittemore and Varela reported that both the densification and coarsening rates are proportional to water vapor pressures to the 1.5 power during the initial stage sintering of MgO at 1100 °C (2010 °F) (Ref 97). It has been proposed that water vapor increases surface diffusivity of MgO and thus enhances the kinetics of particle coarsening and densification (Ref 97).

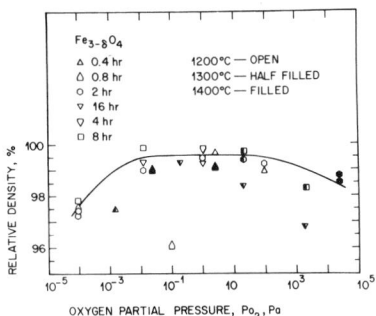

Fig 19 Densities of sintered Fe_3O_4 versus the oxygen partial pressure in the sintering atmosphere

Kinetic Effects of Firing Schedule on Preparation and Sintering Processes

Kinetics of heat treatment during powder preparation can have a significant effect on the powder characteristics and thus its densification behavior. The firing schedule can also have a significant effect on densification and microstructural development during sintering.

Kinetics of Powder Preparation. During powder preparation, the kinetics of heat treatment can have a significant effect on powder characteristics and thus densification behavior. For example, a glassy phase of cordierite ($2MgO·2Al_2O_3·5SiO_2$) can be retained by a rapid quench from the melt. Fine powder can be prepared by granulation of the quenched glass, and the resultant powder compact can be fully densified by a viscous flow sintering mechanism at a temperature between the glass-transition temperature (810 °C, or 1490 °F) and the softening point (860 °C, or 1580 °F) (Ref 98). Furthermore, crystalline cordierite phases can be obtained by annealing the densified glass compact at higher temperatures. However, if a crystalline powder of mullite and cordierite phases was first formed by slow cooling from the molten state, the crystallized powder cannot be densified to ≥75% of the theoretical density even up to 1400 °C (2550 °F) (Ref 99). Thus, this example illustrates a useful method to achieve a complete densification by sintering a glassy phase precursor and a subsequent conversion to the desired crystalline phases. A similar approach can be applied to electronic ceramic materials, such as $BaTiO_3$, as inferred from work of Herczog (Ref 100, 101).

Effect of Heating Rate on Densification. The kinetics of particle coarsening, densification, and grain growth can have different temperature dependencies. For example, several stages of grain growth as a function of temperature have been observed in both ceramic and metallic systems, and the activation energies are significantly different in different temperature regimes (Ref 102–104). During sintering of ceramic systems, several competing processes, for example, grain growth and densification, have been observed. In many cases, densification and grain-growth processes operate at different temperature regimes. Since a high densification rate without grain growth usually occurs at a sufficiently low temperature, a firing schedule can be designed to prolong the dwell time in the densification regime at intermediate temperatures.

A rate-controlled firing schedule has been developed such that a compact is densified following a controlled densification profile (Ref 105). Rate-controlled and conventional profiles in sintering MgO-doped Al_2O_3 are shown in Fig 20 along with the resultant microstructures from these firing schedules. It has been reported that the rate-controlled profile minimizes the gas entrapment and prevents excessive grain growth. Both the rate-controlled and conventional firing schedules give a density of about 99%, but a much smaller grain size has been obtained by rate-controlled sintering.

A new fast-firing process has been developed for β-Al_2O_3 (Ref 106), α-Al_2O_3 (Ref 107), and ferrites (Ref 108). In this process, powder compacts undergo rapid high-temperature firing by being passed through a short hot-zone. The short dwell time at high temperatures results in dense, small-grained ceramics with microstructures similar to those prepared by hot pressing (Ref 106, 107). A rapid increase ($\sim10^5$ °C/h, or 1.8×10^5 °F/h) in temperature favors the kinetic processes with a high activation energy; thus, particle coarsening prior to densification is greatly reduced. Furthermore, very little grain growth occurs within a short dwell time at high temperatures.

Thus, high density without grain growth can be achieved in two different ways: long dwell time at intermediate temperatures in the case of rate-controlled sintering and short dwell time at high temperatures in the case of fast sintering. This can be understood from a simplified model by assuming three competing processes: grain growth, densification, and particle coarsening with activation energies of Q_g, Q_d, and Q_c, respectively. For grain growth limited by the impurity drag, Q_g is the same as the activation energy of the impurity lattice diffusion. It is further assumed that densification occurs by grain-boundary diffusion, and particle coarsening is controlled by surface diffusion. Thus, it is expected that $Q_g > Q_d > Q_c$. Figure 21 shows a schematic diagram of these three kinetic processes versus the reciprocal temperature. In rate-controlled sintering, particle coarsening is avoided by heating rapidly to an intermediate temperature, and grain growth is prevented by a limited exposure to high temperatures. The specimen sinters at the intermediate temper-

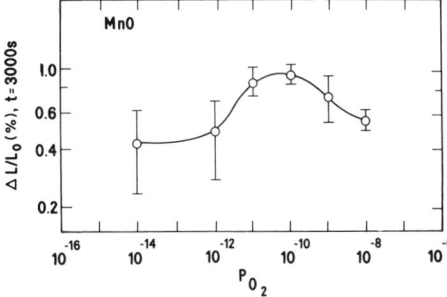

Fig 18 Shrinkage in MnO at 1000 °C (1830 °F) versus the oxygen partial pressure in the sintering atmosphere. Source: Ref 93

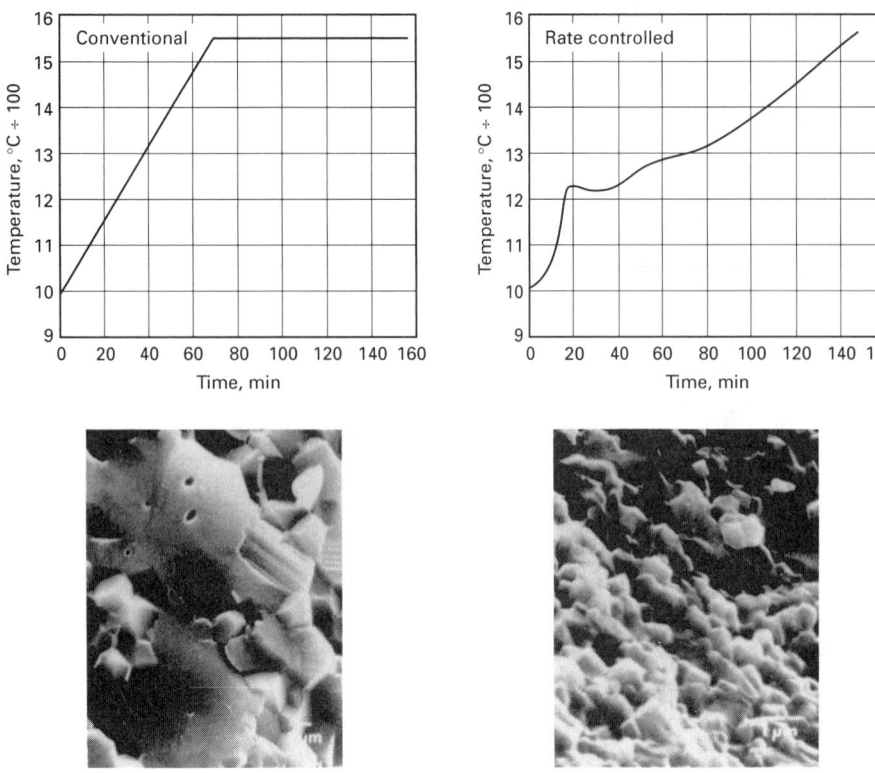

Fig 20 Temperature profiles and resultant microstructures from conventional and rate-controlled sintering process. Source: Ref 105

exists a temperature range within which fast firing can lead to densification without grain growth.

Effects of Cooling Rate on Second-Phase Evolution. Heterogeneous phase distributions observed in many sintered ceramic samples are due to the cooling rate during the firing process. For example, second-phase precipitation has been observed on surfaces of slowly cooled Sr- and Ba-doped TiO_2 specimens (Ref 109, 110). In Ba-doped TiO_2, researchers have observed the precipitation of the $Ba_2Ti_9O_{20}$ phase with a blocklike and equiaxed prism morphology and the $BaTi_4O_9$ phase with an acicular prism and large transparent grain-like morphologies (Ref 109). Microstructures of such materials are shown in Fig 22. In Sr-doped TiO_2, $SrTiO_3$ is the only second phase precipitated on the specimen surfaces. However, the morphologies of the $SrTiO_3$ second phase on TiO_2 ceramic surfaces depend on the annealing temperatures. Between 1250 and 1400 °C (2280 and 2550 °F), Sr segregates to grain boundaries to form a continuous second phase. Below 1200 °C (2190 °F), this continuous second phase breaks up into discrete precipitates due to shape instability (Ref 110), as shown in Fig 23. The second-phase precipitation in both Ba- and Sr-doped TiO_2 ceramics is probably due to segregation of dopants that have a large size misfit with the

atures for a long time, and densification without grain growth is achieved. Since grain growth is usually limited by the residual porosity in the specimen, it occurs only after a relatively high density is reached. As indicated by the dashed line in Fig 21, grain growth in a porous compact occurs at a higher temperature than in a dense body. Thus, there

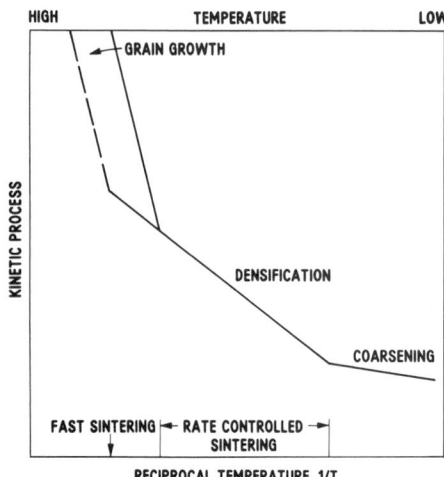

Fig 21 Schematic diagram showing grain growth, densification, and coarsening kinetics versus reciprocal temperature. Temperature ranges suitable for rate-controlled sintering and fast firing are indicated.

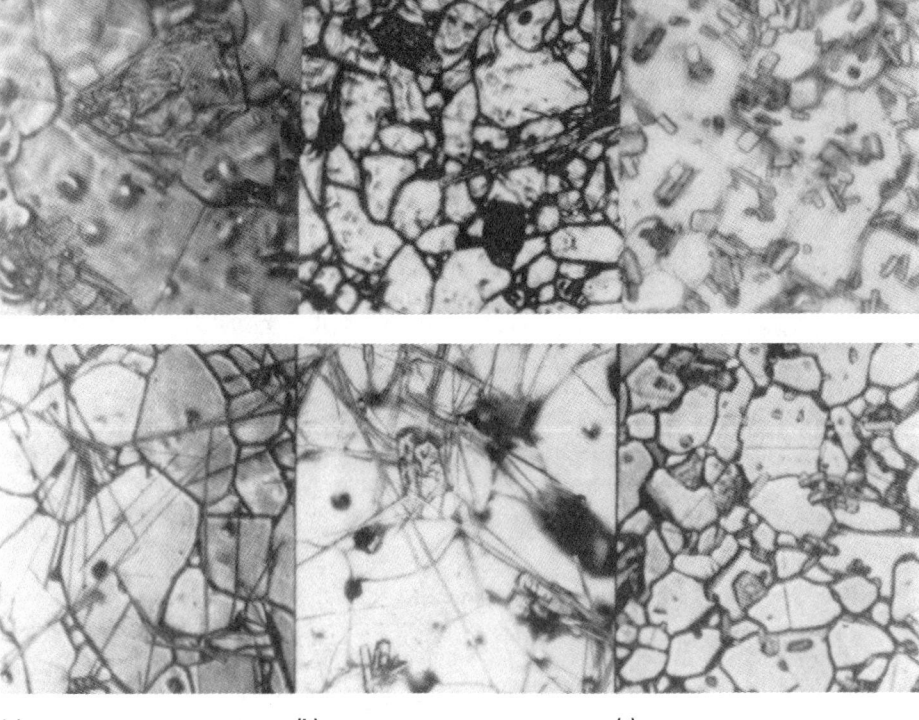

(a)　　　(b)　　　(c)

Fig 22 Microstructures of sintered surfaces of (a) TiO_2 + 0.2 mol% BaO, (b) TiO_2 + 0.4 mol% BaO, and (c) TiO_2 + 2 mol% BaO. The upper micrographs show samples cooled at 100 °C/h (180 °F/h) from 1400 °C (2550 °F). The lower micrographs show 1250 °C (2280 °F) annealed samples.

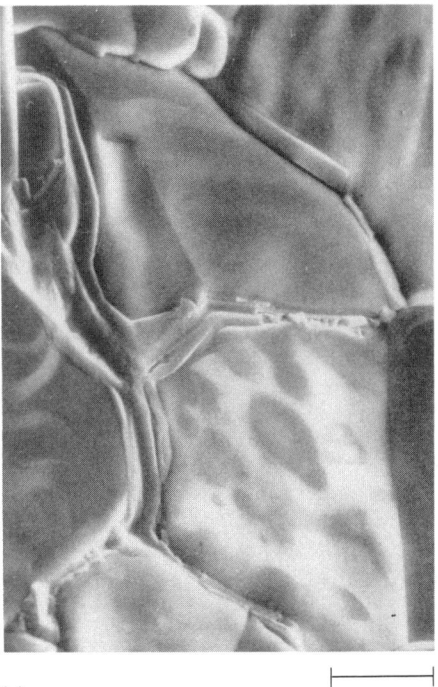
(a)
10 μm

(b)
10 μm

(c)
10 μm

Fig 23 Different second-phase morphologies in TiO_2 + 0.5 mol% SrO. Large transparent particles, sintered at 1400 °C (2550 °F) for 2 h and annealed at 1200 °C (2190 °F) for 24 h. Discrete grain-boundary particles sintered at 1400 °C (2550 °F) for 2 h, then quenched to room temperature followed by reheating to 1200 °C (2190 °F) for 2 h. Continuous grain-boundary second phase sintered at 1400 °C (2550 °F) for 2 h and annealed at 1250 °C (2280 °F) for 2 h

matrix. Furthermore, heterogeneous phase distributions resulting from a temperature-dependent phase boundary and slow reaction kinetics have been observed in $Ba_2Ti_9O_{20}$ ceramic microwave resonators (Ref 111). Microstructural development during cooling or post-sintering annealing of these ceramics has significant effects on their electrical properties and device reliability.

The $YBa_2Cu_3O_x$ perovskite compound is stable in either the tetragonal or orthorhombic phase depending on its oxygen content (Ref 112–114). The tetragonal phase has a lower oxygen content ($x < 6.5$ to 6.6) and it is stable at high temperatures. The tetragonal phase is a semiconductor, and it does not exhibit superconductivity. However, the orthorhombic phase contains more oxygen ($x \geq 6.5$ to 6.6), and it becomes superconducting at or below 90 to 95 K (-183 to -178 °C, or -297 to -288 °F). The tetragonal to orthorhombic phase-transition temperature depends on the ambient oxygen partial pressure, pO_2, and this transition temperature decreases from 686 °C (1267 °F) at $pO_2 = 1.0$ atm to 587 °C (1089 °F) at $pO_2 = 0.02$ atm. Thus, the $YBa_2Cu_3O_x$ ceramics have a tetragonal structure when sintered at temperatures above 700 °C (1290 °F) in oxygen.

The semiconducting phase, which is stable at the sintering temperature, must be converted to the orthorhombic phase for superconducting device applications. The phase conversion can be achieved by annealing or slow cooling in an oxidizing atmosphere at temperatures ranging from 400 to 600 °C (750 to 1110 °F). The required oxidation time depends on the specimen size, microstructure, and density. In very dense ($\geq 98\%$ of the theoretical density) samples with dimensions ≥ 10 mm (≥ 0.4 in.), the oxidation anneal may take several days at 500 °C (930 °F) (Ref 115). The oxidation kinetics can be increased by raising the annealing temperature to near the orthorhombic to tetragonal transition temperature. However, a higher annealing temperature also results in a lower equilibrium oxygen content in $YBa_2Cu_3O_x$. The phase conversion can be readily monitored by x-ray diffraction analyses. The birefringence of the orthorhombic phase can also be used to delineate the microscopically oxidized region near the specimen surface (Ref 115).

Conclusions

In this article, the physical, chemical, and kinetic factors during sintering of ceramics have been reviewed with the emphasis being placed on solid-state sintering. In particular, the influences of powder characteristics, powder compaction, dopant addition, sintering atmosphere, and firing schedules on densification and microstructural development during sintering have been examined. The following general guidelines appear to be useful to optimize the densification process and to control the microstructural development:

- To achieve a dense fine-grained microstructure, it is desirable to have a starting powder with small particle size and a narrow size distribution, nonagglomerated particles with equiaxed shapes, and high purity or controlled dopant content

- Particle coarsening must be avoided or minimized if a dense ceramic is desirable. This can be achieved by dopant addition or atmosphere control to suppress vapor or surface transport

- Uniform and consistent microstructures in sintered ceramics can be obtained only from defect-free green compacts with uniform microstructures. Any existing defects in green compacts are usually amplified during sintering

- Since an increase in the defect concentration of the slower moving ionic species can promote densification, a study of defect chemistry can guide the choices of an appropriate dopant and sintering atmosphere

- Certain dopant layers on powder surfaces and grain boundaries may provide high diffusivity paths for material transport during densification. A critical examination of diffusivities in these dopant layers may lead to a proper choice of dopants to enhance densification

- Exaggerated grain growth is usually detrimental to densification, and certain dopants have been chosen based on some empirical rules, for example, the misfits in ionic sizes and valences between the dopants and the matrix, to reduce grain growth

- Some unique features in the phase relations of certain systems can be exploited to provide transient second phases to inhibit grain growth and to increase the sintered density
- A proper choice of an appropriate sintering atmosphere is determined by the gas solubility, the reactions between the atmosphere and the dopants, and the powder compact. Gases with limited solubility in the sintered compact lead to residual porosity. The sintering atmosphere may react and remove the desired dopants. Interfacial energies of many ceramic systems are also affected by the ambient atmosphere. More importantly, the stoichiometry, valence, and oxidation states of a powder compact may be affected by the sintering atmosphere
- Enhanced densification can be achieved by sintering a glassy phase precursor through the viscous flow mechanism with subsequent conversion to the desired crystalline phases
- The heating schedule can affect the relative rates of the densification and grain-growth kinetics. Both rate-controlled sintering and fast-firing processes have been developed to provide high-density and fine-grain materials. Certain heterogeneous phase distributions in ceramic parts can be controlled by a suitable cooling and annealing process after sintering

REFERENCES

1. M.F. Ashby, A First Report on Sintering Diagram, *Acta Metall.*, Vol 22, 1974, p 275; F.B. Swinkels and M.F. Ashby, A Second Report on Sintering Diagram, *Acta Metall.*, Vol 29, 1981, p 259–281
2. C. Herring, Effect of Change of Scale on Sintering Phenomena, *J. Appl. Phys.*, Vol 21 (No. 4), 1950, p 31–303
3. P.C. Hiemenz, Chapt. 10, Principles of Colloid and Surface Chemistry, Marcel Dekker, 1977
4. H.C. Hamaker, The London-van der Waals Attraction Between Spherical Particles, *Physica IV*, No. 10, 1937, p 1058–1072
5. P.J. Anderson and P. Murray, Zeta Potentials in Relation to Rheological Properties of Oxide Slips, *J. Am. Ceram. Soc.*, Vol 42 (No. 2), 1959, p 70–74
6. J.Th.G. Overbeek, Stability of Hydrophobic Colloids and Emulsions, *Colloid Sci.*, H.R. Kruyt, Ed., Elsevier, p 302–341
7. G.D. Parfitt and J. Peacock, Stability of Colloidal Dispersions in Nonaqueous Media, *Surface and Colloid Sci.*, Vol 10, E. Matijevic, Ed., Plenum Press, 1978, p 163–226
8. T. Morse, "Handbook of Organic Additive for Use in Ceramic Body Formulation," Montana Energy and MHD R&D Institute, Butte, Montana, 1979
9. J.N. Israelachvili and G.E. Adams, Measurement of Forces Between Two Mica Surfaces in Aqueous Electrolyte Solutions in the Range 0–100 nm, *J. Chem. Soc. Faraday Trans.*, Vol 74, 1978, p 975–1001
10. S. Patel and M. Tirrell, Measurement of Forces Between Surfaces in Polymer Fluids, *Ann. Rev. Phys. Chem.*, Vol 40, 1989, p 597–635
11. G. Hadziioannou, S. Patel, S. Granick, and M. Tirrell, Forces Between Surfaces of Block Copolymers Adsorbed on Mica, *J. Am. Chem. Soc.*, Vol 108, 1986, p 2869–2876
12. S. Granick, S. Patel, and M. Tirrell, Direct Measurement of Intermolecular Forces Between a Polymer Layer and Mica, *J. Chem. Phys.*, Vol 85 (No. 9), 1986, p 5370–5371
13. S. Patel and M. Tirrell, A Simple Model for Forces Between Surfaces Bearing Grafted Polymers Applied to Data on Adsorbed Block Copolymers, *Coll. Surf.*, Vol 31, 1988, p 157–179
14. H. Rumpf and H. Schubert, Adhesion Forces in Agglomeration Process, *Ceramic Processing Before Firing*, G.Y. Onoda, Jr. and L.L. Hench, Ed., John Wiley & Sons, 1978, p 357–376
15. U. Chowdhry and R.M. Cannon, Microstructural Evolution During the Process of Sodium β-Alumina, *Materials Science Research*, Vol 11, H. Palmour III, R.F. Davis, and T.M. Hare, Ed., Plenum Press, 1978, p 443–455
16. D.W. Johnson, Jr., D.J. Nitti, and L. Berrin, High Purity Reactive Alumina Powders: II Particle Size and Agglomeration Study, *Am. Ceram. Soc. Bull.*, Vol 51 (No. 12), 1972, p 896–900
17. P. Gallagher, D.W. Johnson, Jr., and F. Schrey, Some Effects of the Source and Calcination of Iron Oxide on Its Sintering Behavior, *Am. Ceram. Soc. Bull.*, Vol 55 (No. 6), 1976, p 589–593
18. R.T. Tremper and R.S. Gordon, Agglomeration Effects on the Sintering of Alumina Powders Prepared by Autoclaving Aluminum Metal, *Ceramic Processing Before Firing*, G.Y. Onoda and L.L. Hench, Ed., John Wiley & Sons, 1978, p 153–176
19. T.J. Carbone and J.S. Reed, Microstructure Development in Barium Titanate: Effects of Physical and Chemical Inhomogeneities, *Am. Ceram. Soc. Bull.*, Vol 58 (No. 5), 1979, p 512–515
20. W.H. Rhodes, Agglomerate and Particle Size Effects on Sintering Yttria Stabilized Zirconia, *J. Am. Ceram. Soc.*, Vol 64 (No. 1–2), 1981, p 19–22
21. M.F. Yan and W.W. Rhodes, Low Temperature Sintering of TiO_2, *Mater. Sci. Eng.*, Vol 61 (No. 1), 1983, p 59–66
22. L. Springer and M.F. Yan, Sintering of TiO_2 from Organometallic Precursors, *Ultrastructure Processing of Ceramics, Glasses and Composites*, L.L. Hench and D.R. Ulrich, Ed., John Wiley & Sons, 1984, p 464–475
23. R.B. Heady and J.W. Cahn, Analysis of Capillary Forces in Liquid-Phase Sintering of Spherical Particles, *Metall. Trans.*, Vol 1 (No. 1), 1970, p 185–189
24. J.W. Cahn and R.B. Heady, Analysis of Capillary Forces in Liquid-Phase Sintering of Jagged Particles, *J. Am. Ceram. Soc.*, Vol 53 (No. 7), 1970, p 406–409
25. M.F. Yan, R.M. Cannon, U. Chowdhry, and H.K. Bowen, Effects of Impurities and Pores in Grain Boundary Mobility, *Am. Ceram. Soc. Bull.*, Vol 56 (No. 3), 1977, p 291
26. M.F. Yan, R.M. Cannon, U. Chowdhry, and H.K. Bowen, Effect of Grain Size Distribution on Sintered Density, *Mater. Sci. Eng.*, Vol 60, 1983, p 275–281
27. D. Demchak and E. Matijevic, Preparation and Particle Size Analysis of Chromium Hydroxide Hydrosols of Narrow Size Distribution, *J. Coll. Interface Sci.*, Vol 31 (No. 2), 1969, p 257–262
28. R. Brace and E. Matijevic, Aluminum Hydrous Oxide Sols, I, *J. Inorg. Nucl. Chem.*, Vol 35 (No. 11), 1973, p 3691–3905
29. E. Matijevic, M. Budnick, and L. Meites, Preparation and Mechanisms of Formation of Titanium Dioxide Hydrosols of Narrow Size Distribution, *J. Coll. Interface Sci.*, Vol 61 (No. 2), 1977, p 302–311
30. W. Stoerber, A. Fink, and E. Bohn, Controlled Growth of Monodisperse Silica Spheres in the Micron Size Range, *J. Coll. Interface Sci.*, Vol 26, 1968, p 62–69
31. E.A. Barringer and H.K. Bowen, Formation, Packing and Sintering of Monodisperse TiO_2 Powders, *Comm. Am. Ceram. Soc.*, 1982, p C199-C201
32. C. Greskovich and K.W. Lay, Grain Growth in Very Porous Al_2O_3 Compacts, *J. Am. Ceram. Soc.*, Vol 55 (No. 3), 1972, p 142–146
33. C. Greskovich and J.H. Rosolowski, Sintering of Covalent Solids, *J. Am. Ceram. Soc.*, Vol 59 (No. 7–8), 1976, p 336–343
34. T.K. Gupta, Possible Correlation Between Density and Grain Size during Sintering, *J. Am. Ceram. Soc.*, Vol 55 (No. 5), 1972, p 276–277
35. S.C. Samanta and R.L. Coble, Correlation of Grain Size and Density During Intermediate-Stage Sintering of Ag, *J. Am. Ceram. Soc.*, Vol 55 (No. 11), 1972, p 583

36. M.F. Yan and W.W. Rhodes, Bell Laboratories, unpublished data, 1979

37. M.F. Yan, U. Chowdhry, and R.M. Cannon, Conditions for Discontinuous Grain Growth with Pore Entrapment or Pore Coarsening, *Am. Ceram. Soc. Bull.*, Vol 57 (No. 3), 1978, p 316

38. M.F. Yan, R.M. Cannon, and U. Chowdhry, "Theory of Grain Growth During Densification," Bell Laboratories, unpublished report, 1980

39. W.D. Kingery and B. Francois, Grain Growth in Porous Compacts, *J. Am. Ceram. Soc.*, Vol 48 (No. 10), 1965, p 546–547

40. T.K. Gupta and R.L. Coble, Sintering of ZnO: II Density Decrease and Pore Growth During the Final Stage of the Process, *J. Am. Ceram. Soc.*, Vol 51 (No. 9), 1968, p 525–528

41. S. Nakahara, G.J. Fisanick, M.F. Yan, R.B. van Dover, and T. Boone, On the Defect Structure of Grain Boundaries in $Ba_2YCu_3O_{7-x}$, *J. Cryst. Growth*, Vol 85, 1987, p 639–651

42. L.F. Schneemeyer, J.V. Waszczak, T. Siegrist, R.B. van Dover, L.W. Rupp, B. Batlogg, R.J. Cava, and D.W. Murphy, Superconductivity in $YBa_2Cu_3O_7$ Single Crystals, *Nature*, Vol 328, 1987, p 601–602

43. D.L. Kaiser, F. Holtzberg, M.F. Chisholm, and T.K. Worthington, Growth and Microstructure of Superconducting $YBa_2Cu_3O_x$ Single Crystals, *J. Cryst. Growth*, Vol 85, 1987, p 593–598

44. I.-W. Chen, X. Wu, S.J. Keating, C.Y. Keating, P.A. Johnson, and T.Y. Tien, Texture Development in $YBa_2Cu_3O_x$ by Hot Extrusion and Hot-Pressing, *Comm. Am. Ceram. Soc.*, 1987, p C388-C390

45. P. Chaudhari, R.H. Koch, R.B. Laibowitz, T.R. McGuire, and R.J. Gambino, Critical Current Measurements in Epitaxial Films of $YBa_2Cu_3O_{7-x}$, *Phys. Rev. Lett.*, Vol 58, 1987, p 2684

46. D.W. Johnson, Jr. and W.W. Rhodes, Retrograde Densification in $Bi_2Sr_2CaCu_2O_8$ Superconductors, *J. Am. Ceram. Soc.*, Vol 72 (No. 12), 1989, p 2346–2350

47. S.A. Sunshine, T. Siegrist, L.F. Schneemeyer, D.W. Murphy, R.J. Cava, B. Batlogg, R.B. van Dover, R.M. Fleming, S.H. Glarum, S. Nakahara, R. Farrow, J.J. Krajewski, S.M. Zahurak, J.V. Waszczak, and W.F. Peck, Structure and Physical Properties of Single Crystals of the 84K Superconductor $Bi_{2.2}Sr_2Ca_{0.8}Cu_2O_{8+\delta}$, *Phys. Rev. B: Condens. Matter*, Vol 38, 1988, p 893–896

48. A.R. Cooper, Jr. and L.E. Eaton, Compaction Behavior of Several Ceramic Powders, *J. Am. Ceram. Soc.*, Vol 45 (No. 3), 1962, p 97–101

49. G.Y. Onoda, Jr., Theoretical Strength of Dried Green Bodies with Organic Binders, *J. Am. Ceram. Soc.*, Vol 59 (No. 5–6), 1976, p 236–239

50. R.A. Thompson, Mechanics of Powder Processing I, II, and III, *Am. Ceram. Soc. Bull.*, Vol 60 (No. 2), 1981, p 237–251

51. K. Unkel, The Mechanism of the Pressing of Metal Powders, *Arch. Eisenhüttenwes.*, Vol 18, 1945, p 161–167

52. G.Y. Onoda and M.A. Janney, Application of Soil Mechanics Concepts to Ceramic Particulate Processing, *Advances in Powder Technology*, G.Y. Chin, Ed., American Society for Metals, 1982, p 53–74

53. A. Schofield and C.P. Wroth, *Critical State Soil Mechanics*, McGraw-Hill, London, 1968

54. J.H. Atkinson and P.L. Bransby, *The Mechanics of Soils*, McGraw-Hill, London, 1978

55. C.P. Wroth, Applications of Plasticity Theory to the Processing of Ceramics, Glasses, and Composites, *Proceedings of Ultrastructure Conference*, Gainesville, FL, 13–17 Feb 1983

56. R.W. Rice, Processing Induced Sources of Mechanical Failure in Ceramics, *Materials Science Research*, Vol 11, H. Palmour III, R.F. Davis, and T.M. Hare, Ed., Plenum Press, 1978, p 303–319

57. W.D. Kingery, Firing—The Proof Test for Ceramic Processing, *Ceramic Processing Before Firing*, G.Y. Onoda, Jr. and L.L. Hench, Ed., John Wiley & Sons, 1978, p 291–305

58. A.G. Evans, Considerations of Inhomogeneity Effects in Sintering, *J. Am. Ceram. Soc.*, Vol 65 (No. 10), 1982, p 497–501

59. J.D. Eshelby, Determination of the Elastic Field of an Ellipsoidal Inclusion and Related Problems, *Proc. R. Soc. (London)*, Vol 241A, 1957, p 376–396

60. I.A. Aksay, F.F. Lange, and B.I. Davis, Uniformity of Al_2O_3-ZrO_2 Composites by Colloidal Filtration, *Comm. Am. Ceram. Soc.*, Vol 10, 1983, p C190-C192

61. M.R. Barone and J.C. Ulicny, Liquid-Phase Transport During Removal of Organic Binders in Injection-Molded Ceramics, *J. Am. Ceram. Soc.*, Vol 73 (No. 11), 1990, p 3323–3333

62. R.J. Coble, Sintering of Crystalline Solids: II, *J. Appl. Phys.*, Vol 32 (No. 5), 1971, p 793–799

63. G.L. Messing, J.L. McArdle, and R.A. Shelleman, The Need for Controlled Heterogeneous Nucleation in Ceramic Processing, *Mater. Res. Soc. Symp. Proc.*, Vol 73, 1986, p 471–480

64. M. Kumagai and G.L. Messing, Enhanced Densification of Boehmite Sol-Gels by α-Alumina Seeding, *Comm. Am. Ceram. Soc.*, Vol 11, 1984, p C230-C231

65. M. Kumagai and G.L. Messing, Controlled Transformation and Sintering of a Boehmite Sol-Gel by α-Alumina Seeding, *J. Am. Ceram. Soc.*, Vol 68 (No. 9), 1985, p 500–505

66. D.W. Readey, Mass Transport and Sintering in Impure Ionic Solids, *J. Am. Ceram. Soc.*, Vol 49 (No. 7), 1966, p 366–369

67. D.W. Readey, Equilibria, Interfacial Effects and Transport in Multiphase Oxides, *Advances in Ceramics*, Vol 1, L.M. Levinson, Ed., American Ceramic Society, 1981, p 453–468

68. P. Reijnen, Non-Stoichiometry and Sintering of Ionic Solids, *Reactivity of Solids*, J.W. Mitchell, R.C. DeVries, R.W. Roberts, and P. Cannon, Ed., John Wiley & Sons, 1969, p 99–114

69. M.F. Yan, R.M. Cannon, and H.K. Bowen, Grain Boundary Migration in Ceramics, *Ceramic Microstructures '76*, R.M. Fulrath and J.A. Pask, Ed., Westview Press, 1977, p 276–307

70. M.F. Yan, Microstructural Control in the Processing of Electronic Ceramics, *Mat. Sci. Eng.*, Vol 48, 1981, p 53–72

71. D. Turnbull, Theory of Grain Boundary Migration Rates, *Trans. AIME*, Vol 191, 1951, p 661

72. J.W. Cahn, The Impurity-Drag Effect in Grain Boundary Motion, *Acta Metall.*, Vol 10 (No. 9), 1962, p 789–798

73. J.E. Burke, K.W. Lay, and S. Prochazka, The Effect of MgO on the Mobility of Grain Boundaries and Pores in Aluminum Oxides, *Materials Science Research*, Vol 13, G.C. Kuczynski, Ed., Plenum Press, 1980, p 417–425

74. W.C. Johnson and R.L. Coble, A Test of the Second-Phase and Impurity Segregation Models for MgO-Enhanced Densification of Sintered Alumina, *J. Am. Ceram. Soc.*, Vol 61 (No. 3–4), 1978, p 110–114

75. A.H. Heuer, The Role of MgO in the Sintering of Alumina, *J. Am. Ceram. Soc.*, Vol 62 (No. 5–6), 1979, p 317–318

76. M.J. Bannister, Comment on the Role of MgO in the Sintering of Alumina, *J. Am. Ceram. Soc.*, Vol 63 (No. 3–4), 1980, p 229–230

77. C.A. Powell-Dogan and A.H. Heuer, Microstructure of 96% Alumina Ceramics, *J. Am. Ceram. Soc.*, Vol 73 (No. 12), 1990, p 3670–3691

78. M. Ruhle, N. Claussen, and A.H. Heuer, Microstructural Studies of Y_2O_3-containing Tetragonal ZrO_2 Polycrystals (Y-TZP), *Science and Technology of Zirconia II*, N. Claussen, M. Ruhle, and A.H. Heuer, Vol 12, *Advances in Ceramics*, American Ceramic Society, 1984, p 352–370

79. W.A. Kaysser, M. Sprissler, C.A. Handwerker, and J.E. Blendell, Effect of a Liquid Phase on Morphology of Grain Growth in Alumina, *J. Am. Ceram. Soc.*, Vol 70, 1987, p 339–343

80. J. Rodel and A. Glaeser, Anisotropy of Grain Growth in Alumina, *J. Am. Ceram. Soc.*, Vol 73 (No. 11), 1990, p 3292–3301

81. M.F. Yan and D.W. Johnson, Jr., Sintering of High Density Ferrites, *Materials Science Research*, Vol 11, H. Palmour III, R.F. Davis, and T.M. Hare, Ed., Plenum Press, 1978, p 393–402

82. W.H. Rhodes, Controlled Transient Solid Second-Phase Sintering of Yttria, *J. Am. Ceram. Soc.*, Vol 64 (No. 1), 1981, p 13–19

83. D.R. Clarke, High Spatial Resolution Analysis of Grain Boundaries: Techniques and Applications, Vol 1, L.M. Levinson, Ed., American Ceramic Society, 1981, p 67–90

84. R.L. Coble, Sintering of Alumina: Effect of Atmospheres, *J. Am. Ceram. Soc.*, Vol 45 (No. 3), 1962, p 123–127

85. S. Prochazka, C.A. Johnson, and R.A. Giddings, Atmosphere Effects in Sintering of Silicon Carbide, in *Proceedings of the International Symposium Factors in Densification and Sintering of Oxide and Non-Oxide Ceramics*, S. Somiya and S. Saito, Ed., Association for Science Documents Information, Tokyo Institute of Technology, Ookayama, Meguro, Japan, 1978, p 366–381

86. M.W. Barsoum and P.D. Ownby, The Effect of Oxygen Partial Pressure on the Wetting of SiC, AlN, and Si_3N_4 by Si and a Method for Calculating the Surface Energies Involved, *Surfaces and Interfaces in Ceramic and Ceramic-Metal Systems*, J. Pask and A. Evans, Ed., Plenum Press, 1981, p 457–466

87. A.C.D. Chaklader, W.W. Gill, and S.P. Mehrotra, Predictive Model for Interfacial Phenomena Between Molten Metals and Sapphire in Varying Oxygen Partial Pressures, *Surfaces and Interfaces in Ceramic and Ceramic-Metal Systems*, J. Pask and A. Evans, Ed., Plenum Press, 1981, p 421–432

88. H.U. Anderson, Fabrication and Property Control of $LaCrO_3$ Based Oxides, *Materials Science Research*, Vol 11, H. Palmour III, R.F. Davis, and T.M. Hare, Ed., Plenum Press, p 469–477

89. P.D. Ownby and G.E. Jungquist, Final Stage Sintering of Cr_2O_3, *J. Am. Ceram. Soc.*, Vol 55 (No. 9), 1972, p 433–436

90. H.U. Anderson, Influence of Oxygen Activity on the Sintering of $MgCr_2O_4$, *J. Am. Ceram. Soc.*, Vol 55 (No. 9), 1974, p 34–37

91. P.W. Ownby, Oxidation State Control of Volatile Species in Sintering, *Materials Science Research*, Vol 6, G.C. Kuczynski, Ed., Plenum Press, 1973, p 431–437

92. W.D. Callister, M.L. Johnson, I.B. Cutler, and R.W. Ure, Jr., Sintering Chromium Oxide with the Aid of TiO_2, *J. Am. Ceram. Soc.*, Vol 62 (No. 3–4), 1979, p 208–211

93. R.L. Porter, P.S. Nicholson, and W.W. Smeltzer, Sintering of Non-Stoichiometric MnO, *Materials Science Research*, Vol 13, G.C. Kuczynski, Ed., Plenum Press, 1980, p 405–415

94. M.F. Yan, Grain Growth in Fe_3O_4, *J. Am. Ceram. Soc.*, Vol 63 (No. 7–8), 1980, p 443–447

95. P.F. Eastman and I.B. Cutler, Effect of Water Vapor on Initial Sintering of Magnesia, *J. Am. Ceram. Soc.*, Vol 49 (No. 10), 1966, p 526–530

96. J.A. Varela and O.J. Whittemore, Structural Rearrangement During the Sintering of MgO, *J. Am. Ceram. Soc.*, Vol 66 (No. 1), 1983, p 77–82

97. O.J. Whittemore and J.A. Varela, Initial Sintering of MgO in Several Water Vapor Pressures, *Structure and Properties of MgO and Al_2O_3 Ceramics*, W. D. Kingery, Ed., Vol 10, *Advances in Ceramics*, American Ceramic Society, 1984

98. E.A. Geiss, J.P. Fletcher, and L.W. Herron, Isothermal Sintering of Cordierite-Type Glass Powders, *J. Am. Ceram. Soc.*, Vol 67 (No. 8), 1984, p 549–552

99. B.H. Mussler and M.W. Shafer, Preparation and Properties of Mullite-Cordierite Composites, *Bull. Am. Ceram. Soc.*, Vol 63 (No. 5), 1984, p 705–711

100. A. Herczog, Barrier Layers in Semiconducting Barium Titanate Glass-Ceramics, *J. Am. Ceram. Soc.*, Vol 67 (No. 7), 1984, p 484–490

101. A. Herczog, Microcrystalline $BaTiO_3$ by Crystallization from Glass, *J. Am. Ceram. Soc.*, Vol 47 (No. 3), 1964, p 107–115

102. M.F. Yan, R.M. Cannon, H.K. Bowen, and R.L. Coble, Grain Boundaries and Grain Boundary Mobility in Hot-Forged Alkali Halides, *Deformation of Ceramic Materials*, R.C. Bradt and R.E. Tressler, Ed., Plenum Press, 1975, p 549–569

103. C.J. Simpson, K.T. Aust, and W.C. Winegard, The Four Stages of Grain Growth, *Metall. Trans.*, Vol 2 (No. 4), 1971, p 987–991

104. C.J. Simpson, K.T. Aust, and W.C. Winegard, Activation Energies for Normal Grain Growth In Lead and Cadmium Base Alloy, *Metall. Trans.*, Vol 2 (No. 4), 1971, p 993–997

105. M.L. Huckabee and H. Palmour III, Rate Controlled Sintering of Fine Grained Al_2O_3, *Am. Ceram. Soc. Bull.*, Vol 51 (No. 7), 1972, p 574–576

106. I.W. Jones and L.J. Miles, Production of β-Al_2O_3 Electrolyte, *Proc. Brit Ceram. Soc.*, Vol 19, 1971, p 161–178

107. M. Harmer, E.W. Roberts, and R.J. Brook, Rapid Sintering of Pure and Doped α-Al_2O_3, *Trans. J. Br. Ceram. Soc.*, Vol 78 (No. 1), 1979, p 22–25

108. A. Morell and A. Hermosin, Fast Sintering of Soft Mn-Zn and Ni-Zn Ferrite Pot Cores, *Am. Ceram. Soc. Bull.*, Vol 59 (No. 6), 1980, p 626–629

109. H.M. O'Bryan and M.F. Yan, Second-Phase Development in Ba-Doped Rutile, *J. Am. Ceram. Soc.*, Vol 65 (No. 12), 1982, p 615–619

110. H.C. Ling and M.F. Yan, Second Phase Development in Sr-Doped TiO_2, *J. Mater. Sci.*, Vol 18, 1983, p 2688–2696

111. H.M. O'Bryan and J. Thomson, $Ba_2Ti_9O_{20}$ Phase Equilibria, *J. Am. Ceram. Soc.*, Vol 66 (No. 1), 1983, p 66–68

112. P.K. Gallagher, H.M. O'Bryan, S.A. Sunshine, and D.W. Murphy, Oxygen Stoichiometry in $Ba_2YCu_3O_x$, *Mater. Res. Bull.*, Vol 22, 1987, p 995

113. P.K. Gallagher, Characterization of $Ba_2YCu_3O_x$ as a Function of Oxygen Partial Pressure: Part I: Thermoanalytical Measurements, *Adv. Ceram. Mater.*, Vol 2 (No. 3B), 1987, p 632–639

114. H.M. O'Bryan and P.K. Gallagher, Characterization of $Ba_2YCu_3O_x$ as a Function of Oxygen Partial Pressure: Part II: Dependence of the O-T Transition on Oxygen Content, *Adv. Ceram. Mater.*, Vol 2 (No. 3B), 1987, p 640–648

115. H.M. O'Bryan and P.K. Gallagher, Kinetic of the Oxidation of $Ba_2YCu_3O_x$ Ceramics, *J. Mater. Res.*, Vol 3 (No. 4), 1988, p 619–625

Liquid-Phase Sintering

Oh-Hun Kwon, Advanced Ceramics, Norton Company

LIQUID-PHASE SINTERING (LPS) is an important means of manufacturing dense ceramic components from powder compacts. Various technically important ceramics are manufactured by liquid-phase sintering, including alumina substrates, mechanical seals, silicon-carbide glow plugs, silicon-nitride structural parts, zinc-oxide varistors, $BaTiO_3$ capacitors, lead-lanthanum-zirconate-titanate (PLZT) piezoelectric components, and composites. Two major advantages of LPS are:

- Enhanced sintering kinetics
- Tailorable properties

Some disadvantages of LPS are that ceramics densified by LPS have a susceptibility to shape distortion and it may be difficult to control the sintering parameters due to additional complications from the liquid phase (for example, temperature-dependent dissolution and crystallization).

There are three general requirements for LPS (Ref 1, 2):

- A liquid must be present at sintering temperature
- There must be good wetting of a liquid on solid (that is, low contact angle)
- There must be appreciable solubility of solid in liquid

Liquid-phase sintering of metals has been extensively studied and there are comprehensive monographs on the subject (Ref 3, 4). In contrast to metal systems, ceramic systems are characterized by viscous grain-boundary phase, limited mutual solubility, slow reactions between constituents, and anisotropic material properties.

Liquid-phase sintering initially requires that at least two solid powders are homogeneously mixed by dry and wet mixing techniques. The mixed powder is usually formed into a green compact of 50 to 65% relative density by various forming methods [for example, uniaxial die pressing, cold isostatic pressing (CIP), slip casting, injection molding]. Accordingly, the structure of a ceramic green compact is best described as a random loose or dense packing. A liquid phase, typically 1 to 20 vol%, is formed upon heating the mixed powder compact. While the mixture of solid particles and liquid sinters together, porosity of a powder compact gradually diminishes to form a dense ceramic part.

Three Stages of Liquid-Phase Sintering

Densification during LPS is divided into three distinct stages as defined by three different rate-controlling mechanisms (Ref 1, 2). With increasing density, the densification mechanism progressively changes from (1) rearrangement to (2) solution-precipitation to (3) final pore (or vapor-phase) removal stages (Fig 1). However, there exists significant overlapping between connecting stages in actual powder compacts. In general, the densification rate significantly decreases as the sintering progresses (typical values range from 10^{-3}/s to 10^{-6}/s).

To quantitatively describe three regimes of sintering, Kwon and Messing (Ref 5) have developed a ternary LPS diagram (Fig 2). Stages of LPS and dominant sintering mechanisms are mapped as a function of relative volume fractions of solid (V_s), liquid (V_l), and pores (V_p). Changes in the relative volume fractions during densification for solid-state sintering (SSS), LPS, viscous composite sintering (VCS), and viscous glass sintering (VGS) are shown as densification loci. For LPS, a porous powder compact at O is expected to densify by traversing three regions of consecutive mechanisms (regions I, II and III in Fig 2), along the arrow to a dense compact, Q. Two boundaries between three regions can be determined by geometrical analyses of compact structures. The rearrangement of particles will cease at $V_s = 0.74$ by achieving the close packing. The boundaries for densification by solution-precipitation are conservatively determined as a triangular region (DEF in Fig 2):

$$0.74 < V_s < 0.92$$
$$0 < V_l < 0.20 \qquad (Eq\ 1)$$
$$0.08 < V_p < 0.26$$

Final stage pore removal starts after pore closure (that is, $\rho > 0.92$, where ρ is the density) at a later stage of solution precipitation.

Driving Force for Sintering and Densification Mechanisms

Figure 3 illustrates simplified two-grain contacts (so called two-sphere model) depicting a solid-liquid-vapor assemblage. Densification during LPS is driven by the thermodynamic driving force to minimize interfacial free energy of the system. In general, the change in free energy ΔG going from one configuration to another (see Fig 1a) in a solid-liquid-vapor system is given by:

$$\Delta G = \Delta A_{sv}\gamma_{sv} + \Delta A_{ss}\gamma_{ss} + \Delta A_{sl}\gamma_{sl} + \Delta A_{lv}\gamma_{lv}$$
$$(Eq\ 2)$$

where ΔA_{sv}, ΔA_{ss}, ΔA_{sl}, and ΔA_{lv} are changes in the various interfacial areas and γ_{sv}, γ_{ss}, γ_{sl}, and γ_{lv} are their corresponding interface energies (subscripts s, l, and v represent solid, liquid, and vapor, respectively). If good wetting of the solid by a liquid as assumed, the values ΔA_{sv} and ΔA_{ss} are unimportant. Also, when there is no grain growth, ΔA_{sl} is negligible. Therefore, ΔA_{lv} is the primary and the most important variable in determining the driving force for LPS.

Rearrangement

In the initial LPS stage, a number of consecutive and simultaneous processes may occur including melting, wetting, spreading, and liquid redistribution (Ref 4). Both solid and liquid are subject to rearrangement (Ref 6) because of unbalanced capillary forces around solid particles as dictated by particle contact and liquid meniscus geometries that result in shearing and rotational movements of particles. Liquid films between particles act as a lubricant. The rearrangement of particles proceeds in the direction of reducing porosity. As density increases, particles experience increasing resistance to further rearrangement due to crowding by neighboring particles until the formation of a closed packing structure.

Earlier models based on axial symmetry (Fig 3) do not adequately explain the driving force for rearrangement. The driving force for rearrangement arises because of an imbalance in capillary pressure as a result of:

- Particle size distribution
- Irregular particle shape
- Local density fluctuation in the powder compact
- Anisotropic material properties

If the geometry of the particle contact is known, the driving force for rearrangement can be calculated for various particle shapes and contact geometries (Ref 7, 8).

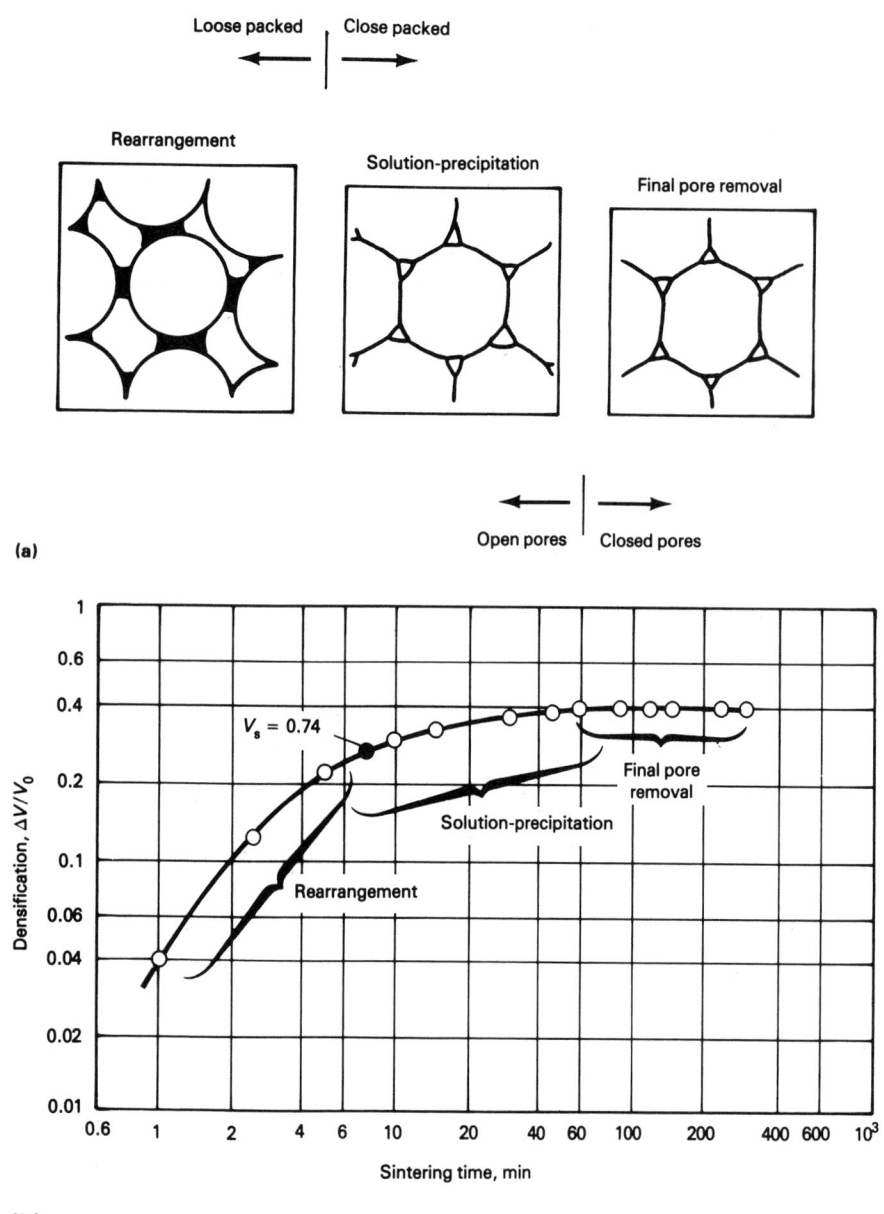

Fig 1 Role of densification during LPS as a function of rearrangement, solution-precipitation, and final pore removal. (a) Schematic of typical microstructure and pore size for three stages of liquid-phase sintering. (b) Plot of densification versus sintering time for Al$_2$O$_3$-MAS glass sintered at 1600 °C (2910 °F) with 3.6 μm (145 μin.) particle size alumina. Initial liquid volume fraction, V_l^0 was 5 vol%; MAS, magnesium-alumino-silicate

by dissolution of the solid at grain contacts, thus resulting in the center-to-center approach of particles. The solubility is proportional to the normal traction at grain contacts arising from the capillary forces (Laplacian forces) that draw the solid particles together.

The dissolved solute transfers to the uncompressed part of the grain structure by diffusion through a liquid phase followed by reprecipitation on uncompressed solid surface for a multicomponent system (Fig 4). This mass transfer results in contact-point flattening and corresponding linear shrinkage in the powder compact. The dissolution rate of the solid decreases as the contact area increases due to simultaneous reduction of effective stress at the contact area. Accordingly, the densification (shrinkage) rate decreases as the density of the powder compact increases. At the later stage of solution precipitation, the interconnected pore structures pinch off to form isolated (closed) pores.

With appropriate geometrical models for grain, liquid, and pore structures, parametric relationships for densification rate can be derived. Assuming pores are located at the edges or corners of a tetrakaidecahedron (Ref 11, 12), the driving force can be determined by geometries of a solid-liquid-vapor assemblage.

In general, there are two rate-limiting processes for solution-precipitation. When material transport is limited by diffusion through the liquid phase, the densification rate is (Ref 5):

$$\frac{d(\Delta\rho/\rho_0)}{dt} = B(g)\frac{\delta D_b c_1 \gamma_{lv}\Omega}{kT} r_s^{-4} \qquad \text{(Eq 4)}$$

where $B(g)$ is the geometrical constant depending on V_s, V_1, ρ, and the dihedral angle; δ is the thickness of liquid boundary; D_b is the grain-boundary diffusion constant of the solute, c_1 is the solubility of solute, Ω is the molecular volume of solute; k is the Boltzmann constant; and T is the absolute temperature. If material transport is interface reaction controlled, the densification rate is:

$$\frac{d(\Delta\rho/\rho_0)}{dt} = C(g)\frac{Kc_1\gamma_{lv}\Omega}{kT} r_s^{-2} \qquad \text{(Eq 5)}$$

where $C(g)$ is the geometrical constant and K is the interface reaction constant. It is important to note that both Eq 4 and 5 show that densification strongly depends on r_s. Further analysis (Ref 5) predicts that the interface reaction control is more likely with small particles, which is consistent with a simple geometrical analysis, in that a larger grain requires a longer diffusion distance from the grain contact to pore sites for densification. If grain growth occurs rapidly, then the rate-controlling mechanism may shift from interface reaction to diffusion controlled.

There exist unresolved issues concerning the structure of the thin liquid intergrain boundaries in LPS systems and its load-bearing capacity. Island structure (Ref 13), semi-

This model is based on the theory that the viscous flow of a liquid sandwiched between solid particles limits the rearrangement process (Ref 9). Assuming a Newtonian liquid, the deformation rate is proportional to shearing stress. Accordingly, the densification rate is given by:

$$\frac{d(\Delta\rho/\rho_0)}{dt} = A(g)\frac{\gamma_{lv}}{\eta r_s} \qquad \text{(Eq 3)}$$

where ρ is the relative density, t is the time, $A(g)$ is the geometrical constant, γ_{lv} is the surface tension, η is the viscosity of liquid, and r_s is the radius of the solid particle. The value of $A(g)$ decreases with increasing vol-

ume fraction of solid and relative density. The V_1 is an important factor in determining the extent of rearrangement. At approximately 30 to 35 vol% liquid, full densification can be achieved by rearrangement alone (Ref 1, 2). Sintering behavior of powder compacts with excessive amount of liquid (Ref 10) is quite different from LPS. In contrast, rearrangement in solid-state sintering can be limited by the absence of liquid capillarity and lubricating films between particles.

Solution-Precipitation

When rearrangement becomes insignificant, additional densification can be achieved

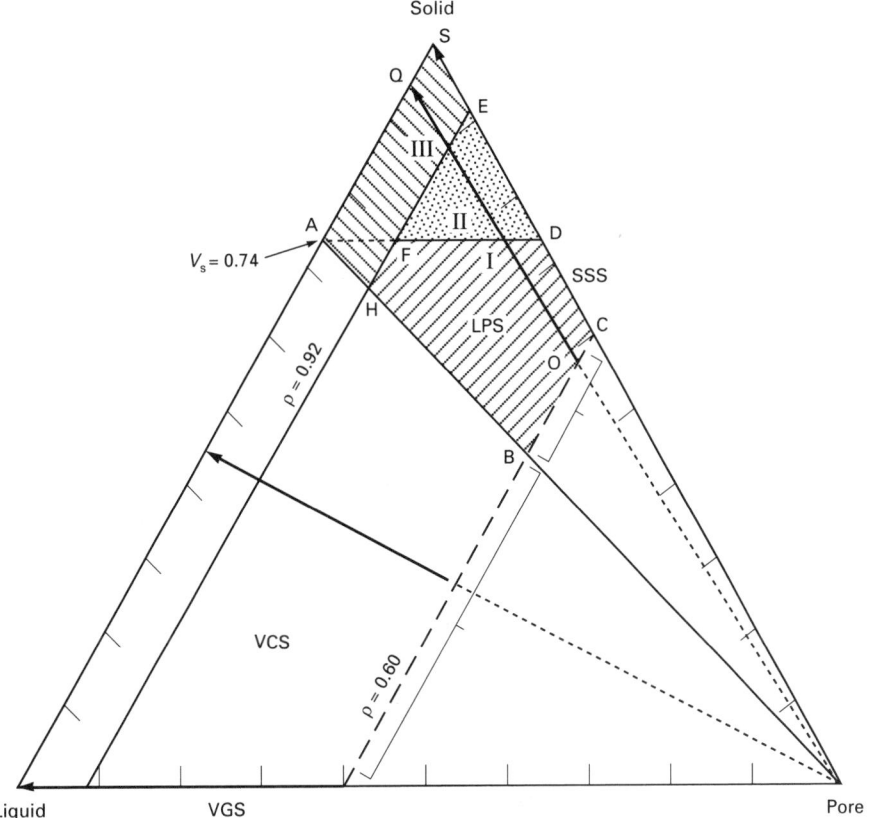

Fig 2 Ternary diagram showing volumetric phase relationships during densification by SSS, LPS, viscous composite sintering (VCS), and viscous glass sintering (VGS). Arrows represent loci of volumetric change of phases with initial compact density of 60%. In the LPS region (ABCS), subdivisions for dominant mechanisms are also displayed: (I) rearrangement; (II) solution-precipitation; (III) pore removal. Source: Ref 5

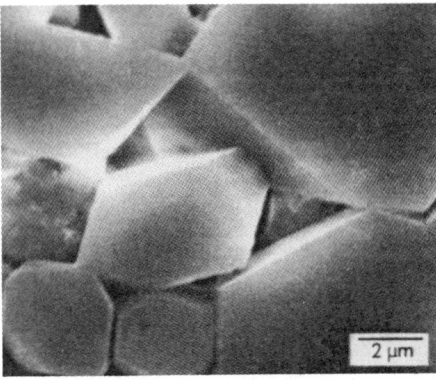

(a)

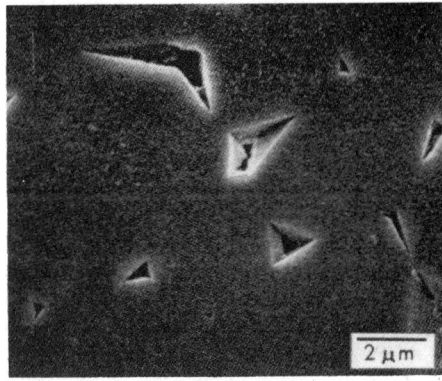

(b)

Fig 5 Typical grain structure of heavily etched alumina-MAS glass specimens showing (a) initial stage and (b) later stage of solution-precipitation. Note that the grain contacts are significantly flattened with densification. Source: Ref 17

crystalline interface (Ref 14, 15), and hydrodynamic squeeze film (Ref 5, 16) models have been proposed to date. The existence of a continuous liquid intergrain boundary, typically ~2 nm (~20 Å) thick, has been documented in liquid-phase sintered ceramics by high-resolution transmission electron microscopy (TEM) (Ref 15).

Kwon and Messing (Ref 17) analyzed solution-precipitation kinetics of an alumina-glass system densified by isothermal liquid-phase sintering. The controlling mechanism was successfully tested by evaluating the particle size dependency of densification rate based on the above models. Figure 5 shows

the grain structure of an LPS alumina. It is clear that initial point contacts between particles (Fig 5a) were significantly flattened (Fig 5b) by solution precipitation controlled densification.

Pore Removal

During intermediate stage sintering, interconnected pore channels pinch off to form closed pores in the density range from 0.9 to 0.95 depending on the material. The final stage

of LPS starts after pore closure. The closed pores usually contain gaseous species from sintering atmosphere, vaporized liquid, and decomposed solid. After pore closure, the driving stress for densification is:

$$D.S. = \frac{2\gamma_{lv}}{r_p} - \sigma_p \qquad \text{(Eq 6)}$$

where σ_p is the vapor pressure inside pore. If the pore radius, r_p, and σ_p remain small (that is, $D.S. > 0$), the densification will proceed. As contacts between solid particles flatten, the densification rate by solution precipitation decreases. However, if r_p increases due to growth and coalescence of pores, and σ_p increases by any internal gas evolution, the driving force will decrease resulting in dedensification (that is, pore growth and bloating) in some cases.

Several processes can occur simultaneously during final stage LPS (including growth and coalescence of grains and pores, dissolution of liquid into solid, phase transformations, and formation of reaction products between liquid and solid) (Ref 4). Lack of critical experiments and models for these simultaneous processes adversely impacts the predictability of final stage densification in LPS (that is, final density and microstructure).

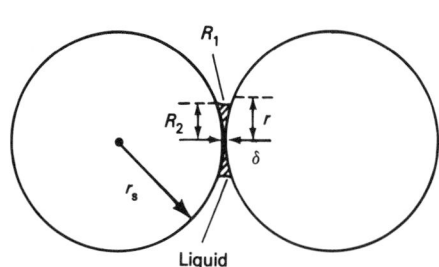

Fig 3 Classical two-sphere model for liquid-phase sintering. The Laplacian force develops at a grain contact due to the concave liquid lens, resulting in a compressive normal force at the contact.

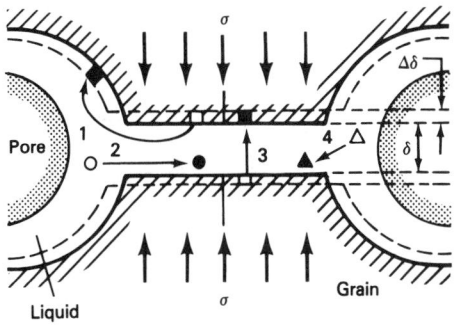

Fig 4 Schematic of two-grain contact during LPS by solution-precipitation, showing three paths for mass transfer: 1, out-diffusion of solute; 2 and 4, influx of solution components into the grain contact region; 3, dissolution-reprecipitation of solute within the contact region. Source: Ref 5

Pressure-assisted sintering techniques [for example, hot pressing and hot isostatic pressing (HIP)] may be used to lower sintering temperature, to achieve higher final density, and to produce a more homogeneous microstructure. Pressure-assisted sintering is often employed to prepare structurally reliable or optical quality components. The applied stress (or pressure) enhances the driving force for densification in all three LPS stages. During final stage of LPS, the driving stress with an applied stress is given by:

$$D.S. = \frac{2\gamma_{lv}}{r_p} + \sigma_a - \sigma_p \qquad \text{(Eq 7)}$$

where σ_a is the applied stress. Consequently, the upper boundary of solution-precipitation can move to a higher density with an applied stress. For instance, the isodensity line EF in Fig 2 can move toward the line SA.

Grain Growth and Liquid-Film Migration

While the dissolution of solid in a liquid contributes to densification, the differential dissolution between different shapes and sizes of particles results in grain growth by Ostwald ripening (Ref 18). The dissolved solute from small or sharp corners of particles tends to reprecipitate on coarser particles. Thus, coarse particles grow as fine particles are eliminated. The rate of grain growth is either diffusion or interface reaction controlled, depending on the materials and sintering conditions.

The amount of liquid can be a dominant variable for grain growth. A small concentration of additives in the liquid can strongly influence the kinetics and morphology of grain growth. For example, in comparison to MgO, when CaO is added as a sintering additive with SiO_2 in alumina, samples with CaO addition result in faster grain growth and more faceting in alumina (Ref 19).

In recent investigations, it has been demonstrated that the liquid film can migrate by changing the chemical composition of ceramic alloys. In the 8% Y_2O_3-ZrO_2 system with a small amount of silica present as an impurity, annealing at 700 to 1400 °C (1290 to 2550 °F) resulted in enrichment of the Y_2O_3 content of the cubic-ZrO_2 grains, assisted by the liquid-film migration (LFM) involving the ubiquitous silicate grain boundary phase (Ref 20). The resulting microstructure may result in a significant property modification.

Real Powder Compacts

In real powder compacts with different particle shapes, sizes, and phase distributions, a number of complex processes may occur in addition to the foregoing idealized processes. Liquid penetration into aggregates and agglomerates of polycrystalline particles can limit densification (Ref 21). Liquid pools may appear in a sintered microstructure due

to heterogeneous liquid redistribution and liquid pore-filling processes (Ref 22, 23). Also, additional rearrangement may occur in liquid-rich regions having a loose compact structure as observed by the particle-liquid mixture flow in different material systems (Ref 24, 25).

Sintering atmosphere may influence final stage densification. If gaseous species are entrapped in closed pores, the diffusion of gases out of the pores to the surface of the powder compact can control densification as well as the size and distribution of pores. It has been demonstrated that a change of sintering atmosphere in the final stage is effective in achieving a higher final density due to the increased driving force for transfer of entrapped gases from pores (Ref 24, 25).

Figure 6 illustrates a typical microstructural heterogeneity that can be created during liquid-phase sintering. Large pores may result from burnout of the organic inclusions or the melting of a liquid-forming particle as shown in Fig 6(a), acting as critical flaws for the structural reliability. The coarse pores may be filled with liquid or particle-liquid mixture flow (Fig 6b) by extended sintering or by pressure-assisted densification (Ref 26).

Use of Phase Diagrams

Information gained from phase equilibria is critically important in understanding liquid-phase sintering and designing LPS systems because the amount of liquid increases as temperature increases above the eutectic temperature. In addition, the viscosity of liquids is exponentially related to temperature, and diffusion in a liquid is much faster than it is in a solid. As a result, liquid-phase sintering takes place rapidly because both the quantity of liquid and diffusion increases (or viscosity decreases) with increasing temperature (Ref 27). Compositions for liquid-phase sintering should be chosen away from eutectics such that the volume fraction of liquid increases slowly with increasing temperature. This will minimize warpage and migration of liquid phase associated with temperature gradients during heating.

Figure 7 is a ternary phase diagram of MgO-Al_2O_3-SiO_2 (Ref 28) displaying many technically important ceramics. Alumina substrates are manufactured by mixing a glass (G_1) and alumina. Upon heating, the alumina content of the glass will be increased to G_2 at 1600 °C (2910 °F), if an equilibrium is reached. The diagram indicates that corundum (α-Al_2O_3) is the only solid phase in equilibrium with a liquid, G_2. The optimum sintering temperature can be estimated using phase equilibria and viscosity data (Ref 29). The amount of liquid in the final microstructure can be estimated from solubility data and the density of a liquid. The phase diagram information can also be used to optimize cooling or heat-treatment cycles after sintering to promote or inhibit crystallization of grain

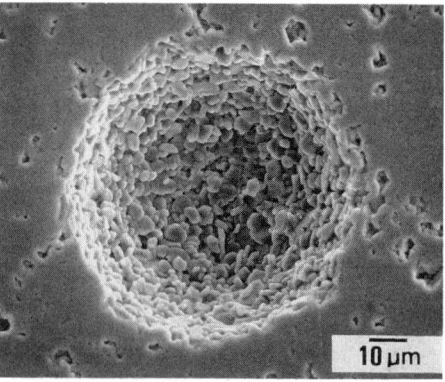

(a)

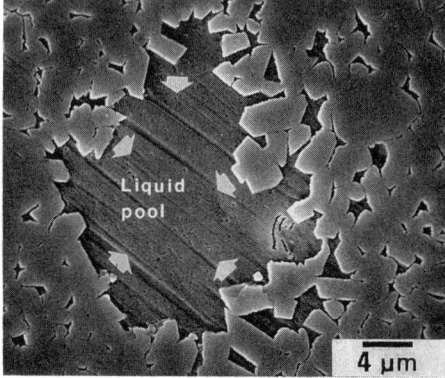

(b)

Fig 6 (a) Macropore formed by burning out of an organic inclusion, and (b) liquid pool formed when the liquid flows into a macropore. Source: Ref 23

boundary phases (for example, mullite, spinel, sapphrine, and cordierite) (Ref 30).

Reactive and Transient Liquid-Phase Sintering

There are many ceramics in which a chemical reaction takes place during sintering. These are comprised of mixtures of powders of different phases, or coated particles, or of a phase that undergoes a phase transformation during heat treatment. It has been realized that the chemical driving force due to various reactions can be much larger than the driving force from interfacial energies in some multiphase LPS systems (Ref 31).

Liquid-phase sintering is often used to densify Si_3N_4 due to its refractoriness and susceptibility to decomposition. A small amount of liquid-forming additives such as MgO, Y_2O_3, Al_2O_3 + Y_2O_3, $BeSiN_2$ and other rare earth oxides are known as effective sintering aids. During sintering, a silicate liquid phase promotes the α-Si_3N_4 to β-Si_3N_4 transformation at approximately 1800 °C (3270 °F), providing a rapid mass transfer path. In order to achieve full density the temperature-time cycle must be optimized for the component size to allow for gas transport from pores (Ref

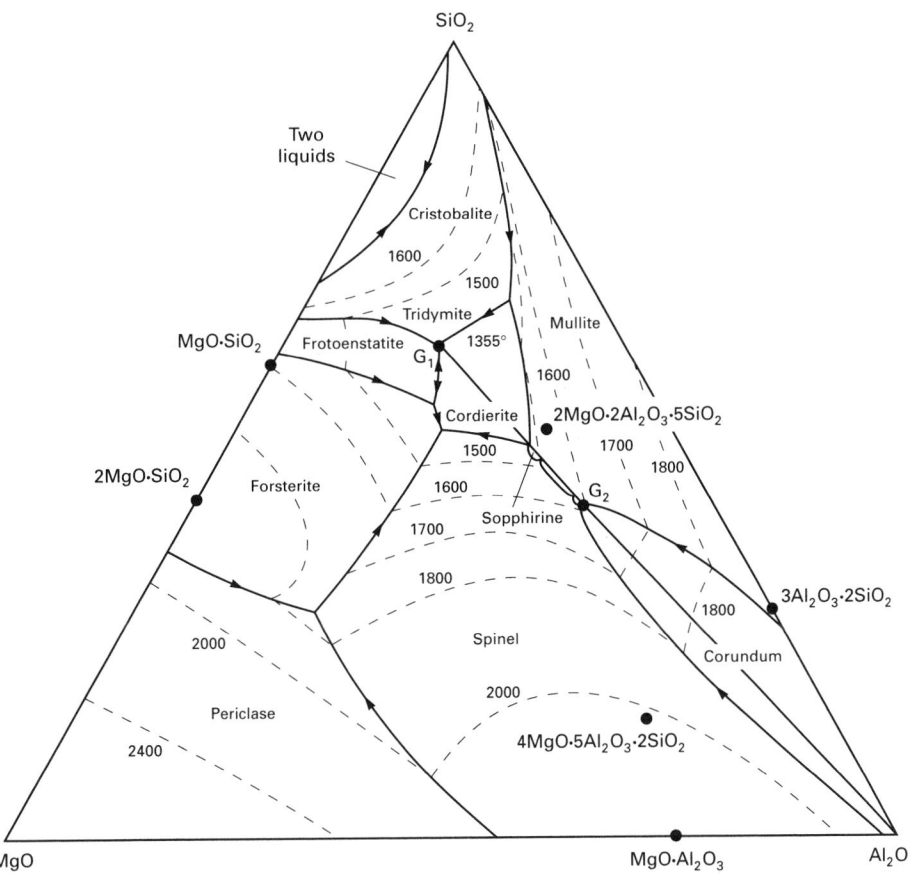

Fig 7 Ternary phase diagram of the system MgO-Al$_2$O$_3$-SiO$_2$. Dashed lines show isotherms in degrees centigrade for the presence of liquid phase. Source: Ref 28

32). Complex chemical reactions between the sintering atmosphere and the powder compact can also occur at high temperatures and are partially responsible for incomplete densification during liquid-phase sintering.

A transient liquid phase has been used to enhance densification kinetics in some special systems. The transient liquid may be removed by evaporation or remain in the compact by forming solid solution or crystalline reaction products with solids. For example, MgO specimens hot pressed with a small amount of LiF (Ref 33) and annealed at 1300 °C (2370 °F) for 3 h in air approached the theoretical density of MgO, had a lattice parameter equal to that of MgO, and became colorless and transparent. Development of transparency started in the center of specimens as the transient liquid was evaporated from the surface. Likewise, the transient liquid-phase sintering can be an alternative route to fabricating fully dense polycrystalline ceramics if the direct solid-state sintering is difficult.

Future Outlook

Liquid-phase sintering is a commercially important process to fabricate a variety of dense ceramic components for both structural and electronic applications. The process can be divided into three different stages on the basis of distinctive mechanisms. Liquid-phase sintering is not fully understood scientifically in its atomistic processes. Therefore, further fundamental understanding will open up more opportunities in the future (for example, uniquely tailored composites with improved performances).

REFERENCES

1. W.D. Kingery, Densification during Sintering in the Presence of a Liquid Phase. I. Theory, *J. Appl. Phys.*, Vol 30 (No. 1), 1950, p 301–306
2. W.D. Kingery and M.D. Narasimhan, Densification during Sintering in the Presence of a Liquid Phase. II. Experimental, *J. Appl. Phys.*, Vol 30 (No. 1), 1950, p 307–310
3. V.N. Eremenko, Yu.V. Naidich, and I.A. Lavrinko, *Liquid Phase Sintering*, Consultant Bureau, 1970
4. R.M. German, *Liquid Phase Sintering*, Plenum Press, 1985
5. O.-H. Kwon and G.L. Messing, A Theoretical Analysis of Solution-Precipitation Controlled Densification During Liquid Phase Sintering, *Acta Metall.*, Vol 39 (No. 9), 1991, p 2059–2068

6. T.M. Shaw, Liquid Redistribution during Liquid-Phase Sintering, *J. Am. Ceram. Soc.*, Vol 69 (No. 1), 1986, p 27–34
7. R.B. Heady and J.W. Cahn, An Analysis of the Capillary Forces in Liquid-Phase Sintering of Spherical Particles, *Metall. Trans.*, Vol 1 (No. 1), 1970, p 185–189
8. J.W. Cahn and R.B. Heady, Analysis of Capillary Forces in Liquid Phase Sintering of Jagged Particles, *J. Am. Ceram. Soc.*, Vol 53 (No. 7), 1970, p 406–409
9. O.-H. Kwon, "Liquid-Phase Sintering and Hot Isostatic Pressing of an Alumina Glass Composite: Modeling and Experiments," Ph.D. thesis, Pennsylvania State University, 1986
10. K.G. Ewsuk and L.W. Harrison, Densification of Glass-Filled Alumina Composites, *Sintering of Advanced Ceramics*, C.A. Handwerker, J.E. Blendell, and W.A. Kaysser, Ed., The American Ceramic Society, 1990, p 436–451
11. D.W. Budworth, Theory of Pore Closure During Sintering, *Trans. Brit. Ceram. Soc.*, Vol 69 (No. 1), 1970, p 29–31
12. P.J. Wray, The Geometry of Two-Phase Aggregates in Which the Shape of the Second Phase Is Determined by Its Dihedral Angle, *Acta Metall.*, Vol 24 (No. 2), 1976, p 125–135
13. R. Raj and C.K. Chyung, Solution Precipitation Creep in Glass Ceramics, *Acta Metall.*, Vol 29 (No. 1), 1981, p 159–166
14. J.E. Marion, C.H. Hsueh, and A.G. Evans, Liquid-Phase Sintering of Ceramics, *J. Am. Ceram. Soc.*, Vol 70 (No. 10), 1987, p 708–713
15. D.R. Clark, On the Equilibrium Thickness of Intergranular Glass Phases in Ceramic Materials, *J. Am. Ceram. Soc.*, Vol 70 (No. 1), 1987, p 15–22
16. F.F. Lange, Liquid-Phase Sintering: Are Liquids Squeezed Out from Between Compressed Particles, *J. Am. Ceram. Soc.*, Vol 65 (No. 2), 1982, p C-23
17. O.-H. Kwon and G.L. Messing, Kinetic Analysis of Solution-Precipitation During Liquid-Phase Sintering of Alumina, *J. Am. Ceram. Soc.*, Vol 73 (No. 2), 1989, p 275–281
18. D.S. Buist, B. Jackson, I.M. Stephenson, W.F. Ford, and J. White, The Kinetics of Grain Growth in Two-Phase (Solid-Liquid) Systems, *Trans. Brit. Ceram. Soc.*, Vol 64, 1965, p 173–209
19. W.A. Kaysser, M. Sprissler, C.A. Handwerker, and J.E. Blendell, Effect of a Liquid Phase on the Morphology of Grain Growth in Alumina, *J. Am. Ceram. Soc.*, Vol 70 (No. 5), 1987, p 339–343
20. R. Chaim, A.H. Heuer, and D.G. Brandon, Phase Equilibration in ZrO$_2$-Y$_2$O$_3$ Alloys by Liquid-Film Migration, *J. Am. Ceram. Soc.*, Vol 69 (No. 3), 1986, p

243–248

21. G. Petzow and W.A. Kaysser, Liquid Phase Sintering, *Science of Ceramics,* Vol 10, H. Hausner, Ed., Deut. Keram. Gesellschaft, 1980, p 269–278

22. O.-H. Kwon and D.N. Yoon, The Liquid Phase Sintering of W-Ni, *Sintering Processes,* G.C. Kuczynski, Ed., Plenum Press, 1980, p 203–218

23. O.-H. Kwon and G.L. Messing, Macropore Filling during Hot Isostatic Pressing of Liquid Phase Sintered Ceramics, *Sintering '85,* G.C. Kuczynski, D.P. Uskokovic, Hayne Palmour III, and M.M. Ristic, Ed., Plenum Press, 1987, p 165–171

24. J.-J. Kim, B.-K. Kim, B.-M. Song, D.-Y. Kim, and D.N. Yoon, Effect of Sintering Atmosphere on Isolated Pores During the Liquid-Phase Sintering of MgO-CaMgSiO$_4$, *J. Am. Ceram. Soc.,*

Vol 70 (No. 10), 1987, p 734–737

25. S.-J. Kang, P. Greil, M. Mitomo, and J.-H. Moon, Elimination of Large Pores During Gas-Pressure Sintering of β'-Sialon, *J. Am. Ceram. Soc.,* Vol 72 (No. 7), 1989, p 1166–1169

26. S.-J.L. Kang and K.J. Yoon, Densification of Ceramics Containing Entrapped Gases, *J. European Ceram. Soc.,* Vol 5, 1989, p 135–139

27. D.L. Johnson and I.B. Cutler, The Use of Phase Diagram in the Sintering of Ceramics and Metals, *Phase Diagrams: Materials Science and Technology,* A.M. Alper, Ed., Academic Press, 1970, p 265–291

28. E.M. Levin, C.R. Robins, and H.F. McMurdie, *Phase Diagrams for Ceramists,* 2nd ed., The American Ceramic Society, 1964, p 246

29. E.F. Riebling, Structure of Magnesium

Aluminosilicate Liquids at 1700 °C, *Can. J. Chem.,* Vol 42 (No. 12), 1964, p 2811–2821

30. W.A. Zdaniewski and H.P. Kirchner, Toughening of a Sintered Alumina by Crystallization of the Grain Boundary Phase, *Adv. Ceram. Mater.,* Vol 1 (No. 1), 1986, p 99–103

31. R.L. Coble, Reactive Sintering, *Sintering—Theory and Practice,* S. Pejovnik and M.M. Kristic, Ed., Elsevier Scientific, 1982, p 145–151

32. C. Greskovich, Preparation of High-Density Si$_3$N$_4$ by a Gas-Pressure Sintering Process, *J. Am. Ceram. Soc.,* Vol 64 (No. 12), 1981, p 725–730

33. P.E. Hart, R.B. Atkin, and J.A. Pask, Densification Mechanisms in Hot-Pressing of Magnesia with a Fugitive Liquid, *J. Am. Ceram. Soc.,* Vol 53 (No. 2), 1970, p 83–86

Reaction Sintering

John S. Haggerty, Materials Processing Center, Massachusetts Institute of Technology

MAKING CERAMIC PARTS by reaction forming mechanisms represents an important alternative to the sintering processes previously discussed in the articles "Fundamentals of Sintering," "Solid-State Sintering," and "Liquid-Phase Sintering" in this Volume. For polycrystalline ceramics made by reaction forming processes, consolidation between constituent particles occurs by chemical reactions rather than by neck-growth mechanisms induced by relatively weak surface energy forces. The boundaries between reaction sintering, reaction bonding, and combustion synthesis are increasingly diffuse. The materials and processes constituting the reaction formed ceramic topic area are extremely varied, as is shown by the following alphabetic listing of the most active subjects:

- Chemical vapor deposition (CVD) and chemical vapor infiltration (CVI) derived monoliths and ceramic matrix composites
- Exchange reaction formed monoliths and ceramic matrix composites
- Direct oxidation and nitridation Dimox type monolithic and composite materials
- Polymer-derived monoliths and ceramic matrix composites
- Reaction-bonded silicon-nitride (RBSN) monoliths and ceramic matrix composites
- Reaction-bonded silicon-carbide (RBSC) monoliths and ceramic matrix composites
- Self-propagating high-temperature synthesis (SHS) types of monolithic and composite materials

As a group, these processes have the advantage of synthesizing materials in potentially complex and large shapes with near-net shape capability. Reducing, or even eliminating, post-consolidation machining has an obvious direct cost benefit; minimizing, if not eliminating, the uncontrolled flaw population that results from grinding operations adds considerable value to these processes. The absence of shrinkage during consolidation is particularly important for fabricating ceramic matrix composites, because useful matrix densities and reinforcement contents are achieved without applied pressure and liquid-phase sintering aids.

A number of reaction forming processes yield materials with higher purities than typical polycrystalline ceramics containing sintering aids. This improved purity can permit achievement of high-temperature mechanical, corrosion, and electrical property levels that are generally inaccessible to conventional ceramics unless they are consolidated with high applied pressures (see the article "Pressure Densification" in this Volume).

Processing temperatures are frequently much lower and processing times are sometimes much shorter than for conventional sintering. One exceptional result is that service-temperature limits can substantially exceed processing temperatures. Another factor is that nonequilibrium phase chemistries or combinations of phases that are inaccessible to conventional processing routes can be achieved. These options are particularly important for the composites, which will almost certainly employ complex marginally stable layers and interfaces between matrices and reinforcements.

Because many of the reaction formed ceramics employ gaseous reactants or generate gaseous products, these materials tend to have much higher levels of frequently interconnected, residual porosity than would be considered acceptable in conventional brittle materials. Although it has been shown that porosity levels of 25 to 30% can be accommodated in RBSN material without degradation of strength or oxidation resistance, residual porosity remains an important issue. The most serious problem is that there is little to guide the microstructural optimization of these porous materials. Interconnected porosity can be expected to be troublesome for oxidation phenomena that do not form protective coatings, for cases where hermeticity is required, and for liquid-soaked parts that are heated rapidly to temperatures above the boiling points of the liquids. Thus, protective coatings may be necessary for these materials. However, precise control of the distribution of residual porosity offers the opportunity to substantially improve strength-to-density ratios. The reduced modulus that intrinsically accompanies uniformly distributed porosity facilitates load transfer to reinforcements in properly designed composites and also yields proportionally higher strains to failure.

The reactions used to form ceramics can be difficult to control due to inherent instabilities. Heat management is a significant issue in those reactions that are exothermic because overheating causes loss of microstructural control and phase chemistry. Achieving single-phase bodies with highly stoichiometric materials is a particularly difficult problem. Even though reactions are normally carried out at relatively low temperatures, the combination of long reaction times and the reactivity of constituents (that is, composite reinforcement) can pose serious detrimental interactions. Understanding and modeling the reactions remains an area of great need.

This review of reaction forming processes will be limited to discussing fundamental details and citations of important results for reaction-bonded silicon-nitride and silicon-carbide materials as well as brief descriptions of the other listed processes. The reader is referred to other more extensive reviews (Ref 1–3) for more complete discussion of each process as well as the primary references. References not included in these manuscripts are included in this text.

Reaction-Bonded Silicon Nitride

Reaction-bonded silicon nitride, in which a green body made from silicon powder is nitrided by N_2 gas, is the most well developed of the processes based on gas-phase reactions. Consolidation in the net-shape fabrication process occurs by filling the void space between silicon particles with the Si_3N_4 reaction product. Two important advantages of this type of process for crystalline ceramic matrix composites (CMCs) are the absence both of shrinkage during reaction bonding and of liquid-phase sintering aids.

Interest in RBSN material has been limited despite its potential advantages. Low strength (typical modulus of rupture, or MOR, ≈ 100 to 200 MPa, or ≈ 15 to 30 ksi) and only moderate resistance to internal oxidation have been the principal sources of hesitancy about using this material either as a monolith or as a matrix for ceramic matrix composites. Fortunately, sufficient important progress has been achieved with RBSN monoliths, RBSN/SiC ceramic matrix composites, and related hot pressed and sintered Si_3N_4/SiC ceramic matrix composites to show the potential of these materials.

Conventional RBSN Processes. Conventionally, RBSN material is made from commercial grades of silicon powder having the characteristics summarized in Table 1. The silicon is first shaped into either a billet or the shape of the final part. Shaping is done by a variety of processes including isostatic pressing, slip-casting, extrusion, injection-molding, die-pressing, and flame-spraying. Frequently, the silicon part is either presintered in argon at approximately 1200 °C (2190 °F) or is partially nitrided to produce a part that can be machined to final dimensions by conventional tooling. After shaping, the part is nitrided to completion, usually in atmospheric pressure N_2 at temperatures in the 1250 to 1450 °C (2280 to 2640 °F) range. Many reported nitriding schedules (see Table 2) employ temperatures in excess of $\approx$1410 °C (2570 °F) (the silicon melting temperature) and times up to 150 h. Residual porosity is typically in the range of 20 vol%, it is usually fully interconnected, and 80% of the pores are smaller than 0.1 μm (4 μin.).

The mechanisms involved in the formation of RBSN material are extremely complex and are still subject to active research. For conventional purity materials and process conditions, the dominant mechanism is the formation of α-Si_3N_4 via a CVD reaction between N_2 and vaporized silicon species. The β-Si_3N_4 forms via a vapor feed gases-liquid catalyst-solid crystalline whisker growth (VLS) type of mechanism with N_2 in solution in liquid alloys formed between silicon and either impurities or intentionally introduced additives such as iron, nickel, and cobalt. Minor

amounts of β-Si_3N_4 form by direct nucleation on the solid silicon particles with growth occurring by surface diffusion of silicon to the reaction site.

High-Purity Silicon Powder RBSN Processes. At Massachusetts Institute of Technology (MIT), very high quality silicon powder has been synthesized from silane (SiH_4) gas. Powder handling, dispersion, shaping, and drying techniques have been developed by which defect-free, RBSN parts can be made rapidly at low temperatures without introducing contaminants. If as-synthesized purities are maintained until the nitriding step, these silicon parts will react to completion under unusually mild combinations of time and temperature, such as 4 h at 1150 °C (2100 °F) or 10 min at 1250 °C (2280 °F) in isothermal cycles, or in <2 h at 1100 °C (2012 °F) with the use of a pre-nitriding nucleation step (see Table 2). Nitriding studies with high-purity silicon/silicon carbide mixtures indicate that these high reaction rates can be maintained at relatively low temperatures and that RBSN/SiC composite parts can be nitrided to completion without damaging VLS-SiC whiskers. These results demonstrate the effect impurities have on reaction kinetics because silicon carbide is generally thermodynamically unstable since the right side of Eq 1 is favored under most nitriding conditions.

$$3SiC + 2N_2 \rightleftarrows Si_3N_4 + 3C \qquad \text{(Eq 1)}$$

For these high-purity silicon powders, the reaction mechanism differs from those observed with conventional processes; nucle-

ation occurs directly on the solid silicon particles and growth occurs via vapor transport of silicon to the reaction sites. The relative formation rates of equiaxed α-Si_3N_4 and β-Si_3N_4 grains are determined by heteroepitaxial nucleation events on a limited number of crystallographic planes whose rates are extremely sensitive to trace impurity levels.

Silane-originating RBSN parts provide clear evidence of the potential advantages that are accessible to even porous reaction formed ceramics if defect structures are controlled. Lapped samples exhibit average strengths up to 592 MPa (85.8 ksi) (maximum strength is 870 MPa, or 125 ksi), a level that is 2.5 to 5.0 times higher than other reported values. With the low specific gravities ($\approx$2.4) that result from the intrinsic porosity, RBSNs specific strengths (up to 3.6×10^5 m^2/s^2) equal or exceed those of the best metallic, intermetallic, and ceramic alternatives while having substantially higher temperature capability. The combination of high strength and reduced modulus gives substantially improved strain failure values of $\approx$0.3%. Exposure to 1000 and 1400 °C (1830 and 2550 °F) air for 1 and 50 h had no measurable effect on the room-temperature strengths of etched and unetched samples. The SiH_4-originating RBSN material exhibits over an order of magnitude lower oxidation weight gain than the more dense RBSN material, and up to an order of magnitude less than hot pressed silicon nitride (HPSN) material. Protective films, less than 1.5 μm (60 μin.) thick, avoided normal cracking problems in oxidized RBSN material and the strength-limiting surface defects in HPSN material. The plane-strain fracture toughness (K_{Ic}) of the SiH_4-derived RBSN is also somewhat higher than optimized commercial RBSN material. These property improvements result directly from the highly controlled pore structure and the high purity of the material, characteristics that should be achievable in other porous reaction formed ceramics.

Recently, there has been some interest in sintered or hot isostatically pressed RBSN matrices in which a sintering aid for Si_3N_4 is added to the silicon plus reinforcement mixture. After the silicon powder is nitrided, parts are densified with normal Si_3N_4 densification heat treatments. It is probable that improved densities will be gained at the expense of high-temperature properties and other characteristics (for example, creep and corrosion resistance) that are dominated by grain-boundary phases.

The type of gas-phase reaction used for RBSN material can be applied advantageously to other ceramic materials. For instance, reaction bonding of silicon carbide via the vapor phase is an established process and recently reaction-bonded aluminum oxide has been made in air (Ref 4). Other reaction formed borides, carbides, nitrides, oxides, and silicides can be contemplated. Also, novel two-material parts such as the RBSN-TiN high

Table 1 Typical impurity levels and particle dimensions of commercial grade and MIT-experimental grade of silicon powders used to produce RBSN

	Impurity concentration, wt%					Mean particle size		Surface area	
Grade	Fe	Al	Ca	Ti	O	μm	μin.	m^2/g	in.^{2}/lb
Commercial	0.9	0.2	0.3	0.1	1.0	15.0	600	1.5	1.2×10^6
MIT-experimental	...	...	...	...	<0.1	0.2	8	10.0	7.8×10^6

Table 2 Comparison of nitrided RBSN material produced from silicon powder using conventional methods versus high-quality silicon powder synthesized from silane (SiH_4) gas source

Producer	Material preparation	Sample no.	Heating time, h	Nitriding		
				Time, h	°C	°F
Ford Motor Company	Conventional	1	...	60	1250–1460	2280–2660
	Conventional	2	$\approx$10	150	1000–1400	1830–2550
Army Materials and Mechanics Research Center (AMMRC)	Conventional	1	4	68	1300–1400	2370–2550
	Conventional	2	4	75	1350–1460	2460–2660
	Conventional	3	4	54	1250–1460	2280–2660
	Conventional	4	4	110	1150–1390	2100–2535
Brown University	Conventional	1	20	84	1100–1400	2010–2550
Massachusetts Institute of Technology	Dry pressing and sintering	1	1.3	120	1100–1410	2010–2570
	Dispersion and pressing	2	20	7	1200–1400	2190–2550
	Dry pressing	3	0.1	1	1150	2100
	Dry pressing	4	0.1	0.2	1250	2280
	Prenucleated	5	0.1	1	1050	1920

power density electric motor armatures can be made by nitriding both phases simultaneously (Ref 5).

Reaction-Bonded Silicon Carbide

Reaction-bonded silicon carbide represents the most important example of the processes based on reactions between a porous solid and an infiltrating liquid phase. These processes share the near-net shape and near-net dimension capabilities of gas-phase reaction bonding, as well as the reduced processing temperatures relative to solid-state sintering. Also, they may have significant advantages in the processing rates and material densities that are achievable.

Reaction-bonded silicon carbide was first developed in the 1950's by the United Kingdom Atomic Energy Authority (UKAEA, Springfield) as a means of bonding silicon carbide grain into a refractory body. A class of fibrous silicon/silicon-carbide composites was later developed by General Electric Company using carbon fiber preforms. Other carbides have also been synthesized by the liquid metal infiltration of carbonaceous preforms and, in principal, compounds such as borides can be synthesized as well, although much less work has been done in the area. Recently, boride-carbide-metal composites based on reactions between molten zirconium, titanium, and hafnium have been reported (NX-3400).

Conventionally, RBSC material is made by infiltrating a porous carbon shape with liquid silicon and then allowing the two elements to react to form silicon carbide at temperatures somewhat above the melting point of silicon ($\approx$1410 °C, or $\approx$2570 °F). This compares very favorably with hot pressing that is carried out at temperatures in excess of 1800 °C (3270 °F) and sintering at $T \geq 2000$ °C ($T \geq 3630$ °F). Generally, infiltration and reaction processes occur simultaneously and involve kinetics that are extremely complex in detail. The process can be extremely fast, due to:

- Rapid infiltration kinetics that result from good wetting of carbon by liquid silicon and low silicon melt viscosities
- Rapid reaction kinetics aided by a large exothermic heat of reaction

In fact, processing problems such as thermal stress cracking more often arise from too fast a reaction rate rather than too slow a reaction rate, as is also the case in SHS processes.

Controlling the residual phase content is a particularly important issue for RBSC material because silicon carbide exists essentially as a line compound (compound having negligible range of stoichiometry) so that either excess carbon or silicon will appear in the fired part. Free silicon is particularly troublesome because the molten silicon expands upon freezing, frequently cracking the part. The kinetics of simultaneous reaction and infiltra-

tion are of particular concern when fine-scale microstructures and low free-silicon fractions are desired, because these kinetics determine practical issues such as the ultimate processable dimensions, the processing rates, and the spatial uniformity of resulting materials. When alloyed-melts are used, there is an additional level of control over the reaction rate as well as the second-phase content. For example, boron additives enhance the silicon-carbide reaction rate, molybdenum has a negligible effect, and aluminum inhibits the rate.

The mechanical properties achievable in RBSC material are a function of microstructure and phase content. Commercially available materials vary in room-temperature flexural strengths from 200 to 600 MPa (30 to 85 ksi), whereas a laboratory material prepared by Hucke (using microporous carbon preforms derived from organic precursors to obtain finer-grained microstructures) has demonstrated strengths up to $\approx$700 MPa ($\approx$100 ksi). At high temperatures, the principal deficiency is a marked degradation in strength and creep resistance upon exceeding the melting point of the residual silicon ($\approx$1410 °C, or $\approx$2570 °F). Furthermore, a high free-silicon content can be detrimental to mechanical properties; failure flaws apparently initiate at and propagate along the weak silicon/silicon-carbide interface at low temperatures. At elevated temperatures near the melting point of silicon (1200 °C < T < 1410 °C, or 2190 °F < T < 2570 °F), the silicon phase is ductile and has mixed effects. While the presence of the ductile phase can improve fracture toughness and strength by modest amounts, it can also be the initiation source for cavities that lead to creep rupture at long elapsed times.

An approach for eliminating the residual silicon phase that substitutes alloyed silicon infiltrants for pure silicon has recently been investigated at Massachusetts Institute of Technology. The alloying constituent(s) are selected on the basis of their ability to form refractory silicides that are stable with respect to silicon carbide at high temperatures. Upon reactive formation of silicon carbide, the solutes are rejected into the remaining melt, enriching it until the refractory silicide forms. Using the silicon-molybdenum melt system as a model, the reaction is:

$$(x + 2y)Si + yMo + xC \leftrightarrow xSiC + yMoSi_2$$

$$(Eq\ 2)$$

This approach demonstrates that materials free of residual silicon phase that retain the important processing advantages of liquid-phase reaction bonding are possible.

With respect to the processing of fiber-reinforced composites, it is worth noting that the introduction of an inert particulate phase into preforms has been widely used since the initial development of RBSC material. Substitution of fibers and whiskers for the particulate phase should present no particular difficulties except perhaps in the preparation

of preforms. The major issue of concern is the high reactivity of the liquid phase and the severe thermal transients that accompany infiltration and reaction. These are likely to present problems in the control of matrix-fiber interfacial properties. Fitzer and Gadow have prepared a variety of silicon carbide and carbon fiber-reinforced RBSC composites and obtained the best results (inelastic deformation with peak flexural strengths in excess of 300 MPa (45 ksi) and strains to failure in excess of 0.3%) for carbon fiber composites when the pore volume is too low for significant attack of the fibers by the infiltrating silicon melt. The development of appropriate fiber coatings is thus another obvious future direction for these composites.

A recent development in liquid-solid reaction bonding is a class of multiphase composites consisting of a carbide, a boride, and a residual metal that is processed by the reaction between B_4C and a molten metal. For the zirconium system, the reaction is:

$$(2.2 + x)Zr + 0.6B_4C \rightarrow ZrB_2 + ZrC_{0.6} + xZr$$

$$(Eq\ 3)$$

The reaction of Eq 3 is necessarily carried out at temperatures 1850 to 2000 °C (3330 to 3600 °F) above the melting point of zirconium. This is again an exothermic reaction, with processing times on the order of 1 to 2 h. The resulting cermet microstructure contains hexagonal platelets of the diboride in a matrix of the carbide and metal; the residual metal distribution is either discrete or interconnected depending on its volume fraction. The reported properties of this material are excellent and include values for room-temperature flexural strengths of 800 to 900 MPa (120 to 170 ksi) and fracture toughness of 16 to 18 MPa $\sqrt{m}$ (15 to 16 ksi $\sqrt{in}$.).

Additional Reaction-Bonding Processes

Chemical vapor infiltration represents one of the most promising, yet intrinsically difficult, reaction forming processes for making ceramic matrix composites. CVI processes are used to chemically vapor deposit the matrix phase(s) within a porous preform, typically made from the reinforcing fiber phase. Conventional CVI processes were first applied successfully in Europe by researchers at Société Européenne de Propulsion (SEP), Laboratoire de Chimie du Solide du CNRS (French Atomic Energy Commission), and the Institut für Chemische Technik der Universität Karlsruhe, primarily in composites with silicon carbide fibers and matrices. Recent advancements have made important progress with respect to long-standing obstacles.

The potential advantages of the CVI process include:

- Good throwing power (achieving uniform coverage)
- Ability to manipulate the stoichiometry,

crystal structure, crystal orientation, and morphology of the deposit
- Maintenance of high purity levels
- Processing at relatively low temperatures

Especially important features for ceramic matrix composites are the absence of both matrix shrinkage around dimensionally stable reinforcements and of applied pressure in addition to the ability to operate with complex preforms in nominally isothermal conditions.

The problems that are inherent to the process are also significant. These include issues that are typical of most CVD processes (for example, they have expensive, corrosive, toxic, and explosive reactants and low deposition rates). However, the most serious problems are related to achieving uniform penetration throughout a preform without either closing off the surfaces or damaging the reinforcements and their coatings. The penetration requirement has forced most researchers to deposit material at very slow rates under low temperature, low partial pressure, and low total pressure conditions. Using conditions optimized for uniform penetration, even thin parts require 2 weeks to several months to fully infiltrate. Even under these process conditions, it is claimed that density gradients become an issue for thicknesses >3 mm (>$^1/_8$ in.). In addition, a part typically must be diamond ground on external surfaces several times during the deposition process to open blocked channels.

Researchers at the Oak Ridge National Laboratory (ORNL) have improved upon the conventional CVI process by conducting the infiltration-deposition process with a gaseous reactant stream flowing up an imposed temperature gradient in the preform. Uniform depositions can thus be achieved for parts up to 25 mm (1 in.) thick, in time scales on the order of 24 h, rather than weeks to months, without intermediate surface grinding.

Solid-Solid Exchange Reactions. The use of exchange or displacement reactions between two or more condensed phases to obtain a thermodynamically preferred phase assemblage are interesting alternatives for monolithic and composite ceramics. Exchange reactions are generally less exothermic than those of the SHS variety, and are sometimes endothermic (for example, when the reaction entropy is largely due to the generation of a gaseous product phase). Moderately high firing temperatures (1300 to 1600 °C, or 2370 to 2910 °F) are required in the following examples of solid-solid exchange reactions:

$$Al_2O_3 + ZrSiO_4 \text{ (zircon)} \leftrightarrow 3Al_2O_3 \cdot 2SiO_2$$

$$\text{(mullite)} + ZrO_2 \qquad \text{(Eq 4)}$$

$$2MgO + ZrSiO_4 \leftrightarrow Mg_2SiO_4 + ZrO_2 \qquad \text{(Eq 5)}$$

$$3ZrO_2 + 4AlN \leftrightarrow 3ZrN + 2Al_2O_3 + 0.5N_2 \qquad \text{(Eq 6)}$$

$$3TiO_2 + 4AlN \leftrightarrow 3TiN + 2Al_2O_3 + 0.5N_2 \qquad \text{(Eq 7)}$$

The principal advantage envisioned for these processes is the ability to form *in situ* multiphase composites with fine and uniform microstructures using inexpensive reactants. In the widely studied zircon-based reactions (Eq 4 and 5), the generation of transformation-toughened zirconium oxide composites has been one obvious objective. The zirconium nitride and titanium nitride based composites are materials of interest for their high hardnesses and electrical properties.

While these processes are free of the thermal runaway difficulties experienced with more highly exothermic SHS reactions, the need for simultaneous reaction and sintering often renders control of the process sensitive to starting material characteristics and thermal history. Dense materials sintered without applied pressure can be obtained in the zircon systems when sintering additives such as excess silicon dioxide (present in nonstoichiometric zircon starting materials), magnesium oxide, or calcium oxide are used to form a liquid sintering phase. Otherwise, hot pressing seems necessary for high densities. Reactions that yield a gaseous product also clearly require gas transport outward for the reaction to proceed. More fundamentally, the processing of fiber-reinforced composites based on these systems is ultimately limited by the same sintering constraints as for nonreactive ceramic matrices.

Directed Oxidation and Nitridation Processes. One of the recent developments in reaction processing of monolithic and composite ceramics is based on the directional oxidation or nitridation of molten metals (Dimox process). The reaction between the liquid metal and the gas forms a ceramic phase with continuous pore channels through which, with appropriate additives, wetting and percolation of the liquid alloy metal occurs, thereby allowing continued oxidation or nitridation at the gas-composite interface. Principal attractive features of the technology are:

- Relatively low processing temperatures involved (900 to 1100 °C, or 1650 to 2010 °F)
- Capability for forming shaped components by carrying out the process within the confinement of a mold
- Ease of incorporating particulate and fibrous reinforcements
- Variety of systems (including Al_2O_3/Al, ZrN/Zr, TiN/Ti) to which the process can be adapted

Inherent to the process is the formation of an interconnected residual metal network (typically 5 to 30 vol%) that improves fracture toughness. It remains somewhat questionable whether or not the limitations a relatively low-melting temperature metal phase places on high-temperature mechanical properties can be overcome.

Self-Propagating High-Temperature Synthesis. At the very fast extreme of reaction forming processes are the self-propa-

gating high-temperature synthesis (SHS) reactions, in which a large exothermic heat of reaction is utilized to sustain a combustion wave through the reactants. This method can be considered generally applicable: over 200 compounds, up to and including cuprate superconductors, have been synthesized by the SHS process to date. A partial listing appears in Table 3. Principal advantages in the structural ceramics area are the very rapid reaction rate and low energy input required, even for highly refractory compounds. Consequently, most work has focused on carbides, borides, nitrides, and aluminides as well as multiphase systems based on these compounds that are difficult to synthesize by conventional techniques.

The high (sometimes near-explosive) rate of many SHS reactions is also a principal liability if the process is considered for component manufacture rather than for powder synthesis alone. Even more so than for liquid-solid RBSC reactions, an uncontrolled reaction rate leads to internal thermal stresses, violent volatilization of impurities and reactants, and a porous final product. Control of the reaction rate is thus an area of obvious interest, as are approaches to simultaneous reaction and densification. The latter include the use of liquid-phase sintering aids (for example, TiC-10% Ni and TiB_2-10% Ni) or the application of forming pressures during reaction. Reasonably large-scale components of titanium boride (for example, armor plates several inches in thickness) have been fabricated by simultaneous hot pressing and the SHS process. Additional information is available in the article "Self-Propagating, High-Temperature Synthesis" in this Volume.

Polymer-Derived Ceramics. Synthesis of ceramic materials via pyrolysis of polymeric precursors is a relatively new processing option that is receiving increasing attention in the ceramics community. Using a wide range of polymers, researchers have successfully produced ceramic materials, such as silicon carbide, silicon nitride, boron carbide, boron nitride, and aluminum nitride and others, in fibrous, film, and bulk forms. Many of the processes have been based on simple pyrolysis technologies. However, researchers are increasingly employing reactive atmos-

Table 3 Adiabatic temperature rise obtained when product material is produced by SHS reaction

Reactants	Product(s)	Adiabatic temperature (T_{ad})	
		K	°F
Ti + 2B	TiB_2	3190	5280
Zr + 2B	ZrB_2	3310	5500
Ti + C	TiC	3200	5300
Al + 1/2N_2	AlN	2900	4760
SiO_2 + 2Mg + C	SiC + 2MgO	2570	4165
$3TiO_2$ + 4Al + 3C	$3TiC + 2Al_2O_3$	2320	3715
$3TiO_2$ + 4Al + 6B	$3TiB_2 + 2Al_2O_3$	2900	4760

spheres, mixtures of polymers, and solid reactants.

The polymer pyrolysis route presents a difficult set of processing issues that must be resolved. Virtually all of the polymers emit troublesome quantities of gaseous products from the beginning of the pyrolysis reactions through the final crystallization steps at very high temperatures. Also, the loss of a large mass fraction causes reduced yields from generally expensive reactants and large amounts of shrinkage if the parts consolidate to eliminate the resulting porosity. Retaining the as-formed shape throughout the pyrolysis steps is often difficult because the polymers become too fluid to hold their shapes over the time scales needed for pyrolysis; shape stabilization has been achieved either by producing oxide skins or by cross-linking the polymeric parts before pyrolysis is initiated. Identifying individual polymers and pyrolysis schedules that will leave constituent elements in precisely stoichiometric ratios is unlikely. Thus, most of the single-phase pyrolysis products tend to be amorphous and the crystalline products tend to be poly-phase.

Despite these difficult problems, the polymer pyrolysis route has been used successfully for some ceramic materials and applications. The most successful have been fibers made by Nippon Carbon, Dow Corning Corporation, Ube, Rhone-Poulenc, Atochem Incorporated, and Tonen (Toa Nenryo Kogyo K.K.). Ethyl Corporation has announced commercially available polymer-derived ceramic coatings for carbon, graphite, and metal substrates. The Ethyl Corporation and Nippon Steel Corporation have reported making monolithic parts by using polymeric precursors to bind ceramic powders. Despite high porosity levels ($\approx$30%), parts exhibited strengths up to 420 MPa (61 ksi). Polysilazane infiltrants are being used to partially fill the voids in RBSN material, thereby increasing its density. Also, Massachusetts Institute of Technology is investigating the use of polysilazane binders for silicon powders to make reaction-bonded silicon nitride. Detailed information is available in the article "Polymer-Derived Ceramics" in this Volume.

Future Outlook

Though not presently as advanced as conventional processes for making monolithic and composite ceramics, the reaction-based processing methods offer particular promise in two regimes. For applications at all service temperatures, many of these processes have important advantages for making large, complex, net-shape and net-dimension parts that are difficult or expensive with conventional densification-based ceramic processes. Additionally, for high-temperature applications, many of these processes appear uniquely capable of fabricating parts with useful properties because of the extremely high purity levels and unusual phase chemistries that can be achieved.

REFERENCES

1. J.S. Haggerty, *Ceramic-Ceramic Composites with Reaction Bonded Matrices*, *Mat. Sci. Eng.*, Vol A107, 1989, p 117–125

2. Y.-M. Chiang, J.S. Haggerty, R.P. Messner, and C. Demetry, Reaction-Based Processing Methods for Ceramic-Matrix Composites, *Am. Ceram. Soc. Bull.*, Vol 68 (No. 2), 1989, p 420–428

3. J.S. Haggerty and Y.-M. Chiang, Reaction Based Processing Methods for Ceramics and Composites, 14th Annual Conference on Composites and Advanced Ceramics, American Ceramic Society, Cocoa Beach, FL, Jan 1990, *Ceram. Eng. Sci. Proc.*, Vol 11 (No. 7–8), 1990, p 757–781

4. N. Claussen, N.A. Travitzky, and S. Wu, Tailoring of Reaction-Bonded Al_2O_3 (RBAO), 14th Annual Conference on Composites and Advanced Ceramics, American Ceramic Society, Cocoa Beach, FL, Jan 1990, *Ceram. Eng. Sci. Proc.*, Vol 11 (No. 7–8), 1990, p 806–820

5. Y. Yasutomi and M. Sobue, Development of Reaction-Bonded Electro-Conductive TiN-Si_3N_4 and Resistive Al_2O_3-Si_3N_4 Composites, 14th Annual Conference on Composites and Advanced Ceramics, American Ceramic Society, Cocoa Beach, FL, Jan 1990, *Ceram. Eng. Sci. Proc.*, Vol 11 (No. 7–8), 1990, p 857–867

Pressure Densification

George E. Gazza, U.S. Army Materials Technology Laboratory, Ceramics Research Branch

DENSIFICATION OF CERAMICS can be achieved using a variety of techniques that are based on the application of external force. These techniques include uniaxial hot pressing, hot isostatic pressing, gas pressure sintering, and various modifications. A general overview of the issues, approaches, and parameters used in pressure-assisted densification processes is provided in this article. First, pressure densification theory is described in terms that are not intended to oversimplify complex processes but to simply outline some models and parameters used for the analysis and interpretation of densification behavior. Second, the various densification techniques are described using approaches and results from densification studies. These approaches continue to be examined and modified in order to provide further understanding of the densification processes and to produce ceramic component materials of suitable quality for emerging and expanding markets.

Pressure Densification Theory

Densification occurs when the rate of densification exceeds competitive processes, such as grain and pore growth (coarsening of the microstructure) or thermal decomposition of the material. Applied pressure is particularly useful in promoting an effective densification rate while maintaining microstructural control. This is accomplished by increasing the transport of matter via deformation or diffusional processes, which permits the use of lower processing temperatures to restrict grain growth, increases particle contact and packing densities, and suppresses material volatility or dissociation reactions.

Microstructural control is an important means for preventing pore growth and keeping pores from becoming entrapped within the grains. Although the principal parameters that influence pressureless sintering (temperature, time, atmosphere, composition, and powder characteristics) continue to affect the densification process when pressure is applied, the addition of the pressure parameter alters the driving forces and mechanisms responsible for densification.

Models developed for pressure densification have been divided into initial, interme-

diate, and final stages. In the final stage, the pores become isolated in the matrix. Both the reduction in interfacial and surface energies and the applied pressure serve as driving forces. However, when pressures are sufficiently high, as in hot pressing or hot isostatic pressing, the surface energy reduction as a driving force becomes sufficiently small in the latter stages of densification and is usually neglected for purposes of the models.

In the initial stage, where the starting bulk density of the compacted powder is typically 50 to 60% of the theoretical density, the effective pressure at particle-to-particle contacts is much higher than in the latter stages of densification, where contact surface areas increase appreciably. The stress parameter, for purposes of the models, is recalculated as an effective stress using equations that involve the actual applied stress, relative porosity, and pore/particle geometry (Ref 1). Several mechanisms (Ref 2, 3) that are suggested to contribute to densification at the initial stage are particle fragmentation, grain boundary sliding, plastic flow, and diffusion. Grain boundary sliding effectively breaks down powder agglomerates and promotes grain rearrangement into a denser configuration. Consequently, grain rearrangement and/or change in grain shape are the principal modes of densification. Although both the applied pressure and reduction of surface energy may be driving forces at this stage, the dominant mechanism depends on the magnitude of the pressure, temperature, and starting powder characteristics.

Early densification models derived by MacKenzie and Shuttleworth (Ref 4), and later, by Murray et al. (Ref 2), were based on densification by viscous plastic flow, that is, materials with Newtonian behavior. The models predicted an end-point density for a given temperature and stress. Such behavior was verified for the hot pressing of fused silica (Ref 5) and conformed to the rate equation derived for the model. However, when applied to crystalline oxide materials, it was determined (Ref 6) that plastic flow was unlikely to be a major contributing mechanism to densification under conventional hot pressing conditions.

Subsequently, magnesia (MgO) and alumina (Al_2O_3) were hot pressed to full density

(Ref 7, 8) over a range of temperatures, supporting the concept of a diffusion-controlled process for densification in the intermediate and final stages. Models proposed to define the diffusional-creep mechanisms are described as Nabarro-Herring diffusional creep (Ref 9, 10), Coble diffusional creep (Ref 11), and power-law dislocation creep (Ref 12). The Nabarro-Herring model is based on the diffusion of vacancies (opposite diffusion of atoms) through the lattice (bulk diffusion) in a stress-directed flow, from boundaries in tension to those in compression. It is generally described by an equation in which the strain rate or densification rate is proportional to the applied stress and the lattice diffusion coefficient, and inversely proportional to $1/G^2$, where G is the grain size. In a porous body, the stress term would be calculated as the effective stress.

Alternatively, the creep rate may be controlled by diffusion along the grain boundaries and is known as Coble creep (Ref 11). This model is described by an equation where the densification rate is proportional to the grain boundary width, the boundary diffusion coefficient, and effective stress. It is inversely proportional to $1/G^3$.

In both models, it is evident that the size of the grains has a significant influence on the densification and strain rates. When high stresses and temperatures are used, where dislocation climb and glide are active, power-law creep (dislocation creep) becomes a significant contributory mechanism to densification (Ref 12). It is particularly important in the intermediate stage and early in the final stage of densification. An applicable equation for the process was derived, in which the strain rate is proportional to the Burgers vector, the shear modulus, and the effective stress, where the values of stress exponents in the equation are related to the particular mechanism of deformation. Dislocation creep is dependent on the chemistry and microstructure of the material, as well.

To assist the densification process in ceramic compositional systems where the diffusion coefficients are low and/or where lower processing temperatures are desired, compositional modifications can be made to produce some liquid phase at particle boundaries. The required characteristics of the liquid are that it wets and penetrates the particle interfaces and that the solid has some solubility in the liquid. The effect of pressure on densification in the presence of a liquid phase was reported (Ref 13, 14) to increase both the rate of initial stage densification by particle rearrangement and the rate of solution into the liquid at particle contacts. The enhanced particle rearrangement process may decrease substantially with changing particle morphology, for materials such as silicon nitride (Si_3N_4), as the phase transformation occurs from the alpha-to-beta forms (Ref 14). Transport rates may also increase by accelerated diffusion through the liquid phase and by the increase in the effective grain boundary width. Identification of the controlling mechanism is generally related to either exponential values of time (Ref 13) or a grain-size parameter (Ref 15), where densification is proportional to $1/d$ for an interface reaction-controlled process and $1/d^3$ if the process is diffusion controlled.

The use of the described models to identify densification mechanisms and predict a strategy for modifying the approach to enhance the densification behavior of ceramic materials has been recognized as being more qualitative than quantitative. This is due to the limitations of the models for strict interpretation of data. The need for further modifications and refinements based on an expanded data base of densification behavior correlated with microstructural analyses has been discussed in research-oriented papers and studies (Ref 16–18). The integration of models for competitive processes, such as coarsening of grains and pores, into densification behavior analysis was suggested to be an improved approach (Ref 18). The competitive models were based on nondensifying mechanisms, such as surface diffusion or vapor transport. Models that use diffusional creep or other deformation-flow models that produce densification rate values could be compared with competitive process rates under given process conditions to determine a densification/coarsening ratio. The effect of processing modifications on this ratio could then be evaluated.

A difficulty with identifying responsible transport mechanisms that are operative during densification is the probability of more than one mechanism being active at a given point in the process. The relative contribution of each mechanism changes as densification proceeds and would be influenced by the pressure-temperature parameters. A greater contribution from plastic deformation would be expected at high stresses and temperatures and for materials with lower yield stresses (weaker bonding) and multiple active slip planes. For example, a greater contribution from deformation would be favored for ionically bonded ceramics with higher symmetry, but diffusive processes would predominate for high-strength, covalently bonded ceramics.

With various mechanisms operating in different ways, depending on the processing

conditions, Ashby (Ref 19) devised a method of graphic representation (deformation maps) to identify the dominant mechanisms of densification under given process conditions. They are constructed by combining densification models and determining relative contributions from each mechanism for different stress levels as a function of temperature.

As shown in Fig 1, field boundaries are developed on a stress-temperature plot, indicating a continuous range of stress-temperature values, over which a particular mechanism is dominant. Another parameter, the densification time or strain rate ($\dot{e}$), is also included and allows prediction of the third parameter if two are known. Initially developed to graphically illustrate mechanical behavior at high temperatures, the technique was readily adaptable to define dominant densification mechanisms under the various stress-temperature conditions found in uniaxial hot pressing (Ref 20–22) and was later applied to hot isostatic pressing (Ref 23).

Pressure Densification Techniques

Pressure densification techniques have certain advantages and limitations that are either inherent in the technique itself or are restricted by the characteristics and process parameter requirements of the material to be densified. The process parameters commonly associated

with each technique, along with their advantages and disadvantages, are shown in Table 1.

Uniaxial hot pressing is the most widely used approach to pressure densification. The application of pressure to assist sintering allows reductions in the time-temperature parameters and suppresses coarsening of the microstructure. It is often used as a "screening tool" to densify new ceramic compositions for evaluation before more difficult densification techniques using sintering can be developed. The hot pressing method is best suited to densifying simple geometric shapes, although some shape curvature can be produced.

The most common approach to hot pressing involves the use of graphite tooling, plungers, and dies to attain pressures typically up to 35 MPa (5.1 ksi) (moderately priced commercial graphite) and temperatures over 2000 °C (3630 °F). Pressure can be applied as either a single-acting system, in which only the upper plunger moves into the die, or a double-acting system, in which both upper and lower plungers are movable. The die system is usually contained within a cold wall chamber that can be evacuated for hot pressing in a vacuum environment or evacuated and refilled with a gas to provide high-quality environmental control.

Temperature measurement and control are performed either optically or with thermocouples, such as Pt-Pt/10–13%Rh or W/5%Re-W/26%Re, depending on processing

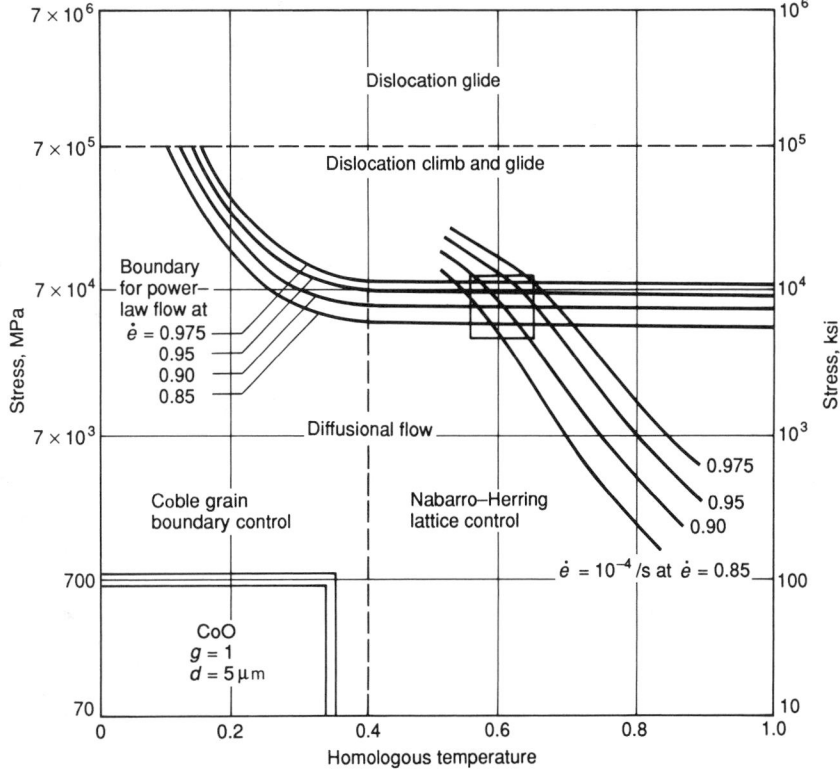

Fig 1 Map demonstrating motion of mechanism boundaries and contour of constant densification rate as a function of density. Source: Ref 22

Table 1 Comparison of parameters and conditions in various pressure-densification processes

	Uniaxial hot pressing, die/plunger	Gas pressure sintering, cladless	Hot isostatic pressing, cladded
Pressure, MPa (ksi)	10–70 (1.5–10)	0.1–10 (0.015–1.5)	70–200 (10–30)
Usual temperature range, °C (°F)			
Oxides	900–1700 (1650–3090)	1100–1600 (2010–2910)	900–1300 (1650–2370)
Nitrides	1600–1800 (2910–3270)	1700–2100 (3090–3810)	1500–1800 (2730–3270)
Carbides	1900–2200 (3450–3990)	1800–2200 (3270–3990)	1700–2100 (3090–3810)
Pressure cycle	Single	Dual	Single
Shape capability	Simple	Complex	Complex

temperature and environment. Uniaxial pressure up to 100 MPa (14.5 ksi) has been achieved by using various other die/plunger materials (Ref 7, 24). Although the use of alumina, titanium diboride, or TZM molybdenum alloy as tooling permits higher hot pressing pressures, temperatures are restricted to <1200 °C (2190 °F). With the development of high-strength graphite and carbon composite materials, high pressures can be attained at temperatures of 1600 °C (2910 °F) (Ref 25) or higher.

Die design is influenced by the temperature-pressure requirements for densification and the thermal and chemical characteristics of the ceramic material being hot pressed. If the thermal expansion coefficient of the material being hot pressed to high density is significantly lower than the die material, which is typically graphite, a compression fit of the die onto the hot pressed specimen will result upon cooling. The specimen can be extracted safely from the die if appropriate die designs are used, such as those that include inner sleeves. Chemical reactions between hot pressed material and die or plunger faces can be minimized by using graphitized paper coated with a reaction barrier powder, such as boron nitride, between the die surfaces and the powder to be hot pressed.

In order to generate densification data during hot pressing, the temperature, applied stress, and shrinkage, or densification, rate must be monitored. For the latter measurement, either a dial gage or, preferably, a dilatometer is used to continuously record ram travel as the powder being hot pressed densifies and shrinks. The densification rate can then be determined as a function of temperature and stress level as the variable parameters. Before powder is hot pressed, a trial run should be made using a complete load train system, but without powder, to generate data showing changes in ram movement attributed to thermal expansion of the system during heating.

The data can subsequently be used for correction purposes when the actual hot pressing of powder is conducted using similar temperature-pressure parameters. Changes in specimen density during a hot pressing experiment can be determined from shrinkage measurements and the final density calculation. Because the specimen mass, M, and the area, A, are fixed quantities, the change in specimen thickness, l, can be related directly

to the change in specimen density, ρ, as:

$$\frac{d\rho}{dt} = \frac{[(M/Al)/\rho_{max}]}{dt} = \rho\left(\frac{-dl}{l\,dt}\right) = \dot{e} \qquad \text{(Eq 1)}$$

After hot pressing, the final density of the specimen can be determined from standard density measurement techniques. Changes in density with time during the hot pressing run can then be calculated from the shrinkage measurements. The strain rate, $\dot{e}$, can be calculated from $(1/\rho)(d\rho/dt)$ where ρ is the relative density and $d\rho/dt$ is the rate of densification. Strain rate can be related to temperature and stress by an Arrhenius-type equation of the form:

$$\dot{e} = K(\sigma)^n e^{-(Q/RT)} \qquad \text{(Eq 2)}$$

where $\dot{e}$ is the strain rate, K is a constant, σ is the applied stress, n is the stress exponent, Q is the activation energy, R is the gas constant, and T is the absolute temperature.

Plots of strain rate versus $1/T$, derived from relative density versus time plots for various temperatures at a fixed stress, can be prepared for given relative density values. The slopes of these plots yield activation energy values for densification. Similarly, plots of strain rate versus stress can be established from relative density versus time curves for various stress levels and at a fixed temperature. Slopes of these plots represent the stress exponent values for given relative densities. Values of the activation energy and stress exponent, supplemented by microstructural data, can be used to interpret densification behavior in order to suggest densification mechanisms and identify rate-controlling species.

Studies of hot pressing (Ref 6, 7, 8, 20, 26) initially focused on oxide ceramics because of the greater availability of powder, lower pressure-temperature requirements, and their suitability for modeling. Some difficulties encountered with producing dense, uniform microstructures in the hot pressed products were attributed to impurity segregation and particle agglomeration (Ref 24). In addition, effects such as variations in light transmission were observed in hot pressed alumina because of density differences produced by agglomerates. The fabrication of transparent oxide ceramics, as reviewed by Rhodes (Ref 27), was shown to be largely dependent on the quality of the starting powder and on controlling grain growth to maintain active pore removal mechanisms during densification.

In covalently bonded ceramics, self-diffusion coefficients are low, compared to those in ionically bonded ceramics. In order to densify these materials, high temperatures (generally higher than 1900 °C, or 3450 °F), high pressures, and/or additives to assist densification are required. Covalent ceramics that have high-temperature stability, such as B_4C, TiC, SiB_6, and TiB_2, can be densified by using high sintering temperatures and/or high pressures to promote diffusion and/or activate deformation processes, such as grain boundary sliding or power-law creep. Ceramics with high-temperature instability, such as silicon nitride or aluminum nitride (AlN) may require densification aids (usually small amounts of oxides) to produce a liquid phase at the grain boundaries.

It has been demonstrated that in order to fabricate high-strength, hot pressed Si_3N_4, it is preferable to start with a powder that contains a high percentage of α-Si_3N_4. During hot pressing at temperatures above 1500 °C (2730 °F), the α-phase transforms to the β-modification via a solid/liquid/solid mechanism, that is, solution/reprecipitation. The morphology of the beta grains are elongated, or prismatic, in shape. The aspect ratio is dependent on the starting materials, the processing conditions (pressure, temperature, and time), and the amount and viscosity of the liquid phase.

The parameters and mechanisms known to influence the microstructure of densified Si_3N_4 are shown in the process sequence in Fig 2. The pressure-densified regime is divided into three regions: the formation of a liquid phase; the α to β transformation, where the principal thermodynamic driving force for the formation and growth of β is assumed to be the difference in free energy of the α and β structures at temperature; and the Ostwald ripening, or growth, of the β grains.

Transformation and densification were suggested to be interface reaction controlled (Ref 25) from the analysis of densification behavior, using the pressure, temperature, and the α/β phase ratio of the starting powders as the principal variable parameters. Further studies (Ref 28) to identify the mechanisms of densification during the hot pressing of Si_3N_4 with MgO additive also concluded that the densification rate was controlled by reaction kinetics based on the dependence of the densification rate on grain size. Other additives (Y_2O_3, CeO_2, ZrO_2, Al_2O_3, and other

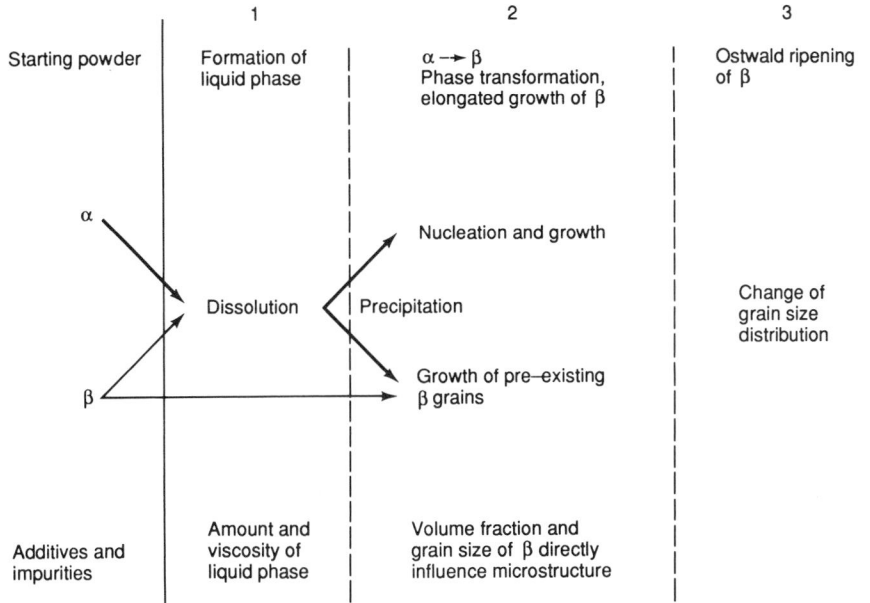

Fig 2 Parameters and mechanisms influencing microstructure during hot pressing

rare-earth oxides) have also been used successfully to densify Si_3N_4, principally by pressure-assisted processes.

Extending the pressures to a range from 500 to 5000 MPa (72.5 to 725 ksi) (ultrahigh-pressure densification/hot pressing) has been used to densify carbides and borides (Ref 29–33). Special die/plunger materials or anvil-type pressure systems were required to produce densified products with relatively small dimensions in these studies. Deformation by boundary sliding and deformation by grain fragmentation were considered the dominant densification processes. Other mechanisms, such as dislocation creep and diffusional processes, also played a significant role, particularly in the final stages of densification.

Hot isostatic pressing was first explored in the mid-1950s and was initially known as gas-pressure bonding. It is now used on a worldwide basis for a variety of applications, including tooling, wear parts, engine ceramics, and electronic materials. The hot isostatic pressing (HIP) process involves densifying a powder compact or other porous green (unfired) body with a high gas pressure, typically ranging from 70 to 200 MPa (10 to 30 ksi). An inert gas, usually argon, is used as the pressurizing medium. More recently, gases such as nitrogen or oxygen have been used in a functional role, as well. Nitrogen is useful in suppressing the thermal decomposition of nitrides at elevated temperatures, whereas oxygen is beneficial in maintaining the stoichiometry of oxides during high-temperature densification.

Because high gas pressure is ineffective in densifying compacts with open porosity, a cladding around the compact is used to prevent gas penetration. Mild steel sheet, refractory metal sheet, or glass are used as the

principal cladding materials. The materials selected should have sufficient strength and formability (as well as viscosity, for glass) at HIP densification temperatures to transfer the isostatic stress to the specimen without cracking or permitting gas penetration through the cladding. For glasses, crystallization should be avoided at processing temperatures. Additionally, any reaction between the cladding material and the specimen should be minimal. The use of an effective reaction barrier material between the specimen and cladding may be required. Hot isostatic pressing can be used for near-net-shape fabrication and has the advantageous capability to densify complex shapes.

Methods for encapsulating the specimen to be densified by HIP vary with specimen type and composition, specimen shape, and densification temperature. Glass encapsulation methods usually involve placing the specimen in a glass tube or capsule, evacuating the container, and then sealing the glass to maintain the specimen in a vacuum environment (usually required when encapsulating a specimen). Other approaches include immersing the specimen in a glass bath, use of a glass powder coating applied to the specimen, and pressure forming a glass powder around a specimen in a mold. After specimen densification, removal of the cladding is usually accomplished by mechanical or chemical means.

Hot isostatic pressing can be used either directly from the unfired body-encapsulated stage or as a post-treatment to reduce or eliminate residual defects in sintered ceramics having closed porosity. The use of HIP with different approaches has been shown (Ref 34–38) to improve the properties and reliability of the material, versus those in the as-sin-

tered condition. However, it has been demonstrated that some large pores may not be eliminated by post-HIP processing, resulting in end-point densities (Ref 39). In systems where a glassy phase is present, the pressure may force glassy material into the pore, reducing its potential detrimental effect on properties (Ref 40). For materials that normally require very high processing temperatures, as well as additives to promote densification, HIP can be used to reduce processing temperatures and produce materials with finer grain size and isotropic properties. The amount of additives required for densification can be eliminated or reduced, resulting in improved high-temperature properties (if required) and, in general, a more uniform microstructure.

Gas pressure sintering, a process developed in the early 1970s, is presently receiving considerable attention. A major advantage of this technique is that it does not require the cladding and decladding of specimens. The method generally uses a pressurized gas to promote high-temperature densification of specimens from their green-body form. Early studies focused principally on oxide materials, for example, Al_2O_3 (Ref 41), $Pb(Zr,Ti)O_3$, $BaTiO_3$, and $SrTiO_3$ (Ref 42).

The technique usually consists of two steps (Fig 3). The first step involves sintering the specimen to the closed pore stage, either under vacuum or at a relatively low pressure using a gas with high diffusivity through the material being densified. In this first step, oxygen can be used for the sintering of oxide ceramics. In the second step, the external gas pressure is raised to a high level to promote pore elimination (densification). Pressure levels are usually up to 10 MPa (1.5 ksi) during gas pressure sintering, and 70 to 200 MPa (10 to 30 ksi) during the sinter-HIP process. The pressurizing gas used in the second step should not have an appreciable diffusivity in the material being densified.

If pores become isolated and contain a gas that has low solubility and diffusivity through the ceramic being densified, the pressure within the pore resists pore closure. This is given by:

$$P_{int} = \frac{2\gamma}{r} + P_{ext} \qquad (Eq\ 3)$$

where P_{int} is the pressure inside the pore, γ

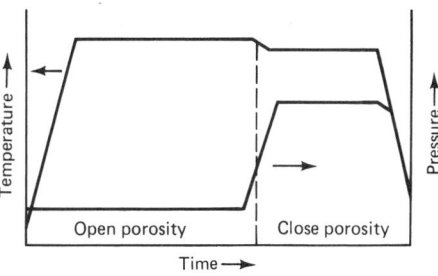

Fig 3 Gas pressure sintering process

is the solid-vapor surface energy, r is the pore radius, and P_{ext} is the external gas pressure used in the process. Complete elimination of porosity does not occur, resulting in end-point densities. The maximum attainable density was found (Ref 43, 44) to be a function of initial pore size and the magnitude of the external pressure used for densification. If the pores contain a soluble gas that has high diffusivity through the ceramic, reduction in pore size occurs as a function of gas solubility and diffusion kinetics, with resulting high densities.

As in other pressure-densification methods, the temperatures used should restrict significant grain growth (before complete densification) to avoid either pore entrapment within the grains or pore growth. However, higher temperatures may be required to reach the closed-pore stage in the cladless approach. Using a relatively low pressure sintering step to reach the closed-pore stage may result in a somewhat coarser microstructure after the second-step HIP process than that obtained when conventional HIP with cladded specimens is used entirely for densification.

A comparison of microstructures for gas pressure sintered and hot isostatically pressed Si_3N_4 is observed in Fig 4. The gas pressure sintered material exhibits a coarser structure with high aspect ratio grains, whereas the pressed material has a finer, more equiaxed structure. The densification temperature for the pressed material was 200 °C (360 °F) lower than for the sintered material. In addition, surface chemistry changes can sometimes occur either by reaction with or volatilization into the sinter-HIP environment. This is opposed to the problem of surface reaction with cladding materials in the conventional HIP approach.

The use of a high-pressure gas can serve a functional purpose, as well as to simply act as a stress-related driving force parameter. For example, early studies (Ref 45, 46) on the densification of Si_3N_4 by gas pressure sintering were conducted using nitrogen gas pressures of up to 2 MPa (0.3 ksi). The principal role of the nitrogen gas was to suppress the thermal decomposition of Si_3N_4 and allow higher sintering temperatures to be used for densification. Subsequently, a two-step process was used to densify Si_3N_4 (Ref 44, 47–50) and other ceramics (Ref 50).

To determine the amount of nitrogen gas pressure required to stabilize the Si_3N_4 at a given temperature, a diagram (Fig 5) defining the region of sinterability was thermodynamically determined from the reaction in which the nitrogen pressure and silicon vapor pressure are in equilibrium with Si_3N_4. This diagram is presented and discussed in Ref 51.

In addition to suppressing the decomposition of Si_3N_4 into Si and N_2, high N_2 gas pressure will limit oxygen removal from the specimen by suppressing the dissociation reactions producing SiO and N_2. The use of a cover powder in which the specimen is embedded is also effective for this purpose. The application of higher gas pressures (70 to 200 MPa, or 10 to 30 ksi) in the second step of the process combines gas pressure sintering with hot isostatic pressing (cladless) and is termed "sinter-HIP." Its effectiveness is related to the increase in the pressure difference between the external pressure and the pressure within the isolated pores resulting in a smaller equilibrium pore size. Higher pressures may also contribute to densification by increasing the solubility of entrapped gas into the liquid boundary phase.

Modified Techniques. The pressure-densification approach has been combined with other unconventional techniques to produce dense ceramics. These techniques include pressure calcintering, reactive hot pressing, and press forging, as well as superplasticity.

In the calcintering process, the starting materials are precursors of the final densified material, which undergoes decomposition during the densification process. For example, $Mg(OH)_2$, $Al(OH)_3$, and $BaTiO(C_2O_4)_2 \cdot 4H_2O$ have been pressure calcintered to high density to form MgO, Al_2O_3, and $BaTiO_3$, respectively (Ref 52, 53). The Al_2O_3 was evaluated for use as tooling, and the $BaTiO_3$, for use in electronic applications. Compaction kinetics of fireclay, china clay, and boehmite (Ref 54, 55) have also been studied using this technique.

Reactive hot pressing has been used for many years to form a dense product from the mixed compositional components of the material. These components react under pressure to form a high-density body with the desired compositional stoichiometry. Examples of materials that have been hot pressed using this approach include $MgAl_2O_4$ from Al_2O_3 and MgO (Ref 56) and SiAlON from Si_3N_4, SiO_2, and AlN (Ref 57).

Hot forging (thermomechanical forming) has been successfully applied to sintered ceramics (Ref 58, 59) to produce dense materials with preferred crystallographic orientation (texturing). The method involves applying pressure to a partially densified compact, where the material may flow without lateral constraint. The strain rate is adjusted to produce a selected deformation rate to avoid extensive cracking or cavity formation. Kellett and Lange (Ref 59) demonstrated that

(a)

(b)

Fig 4 Microstructures of Si_3N_4 containing Y_2O_3 sintering additive. (a) Gas pressure sintered. (b) Hot isostatically pressed. 2000×

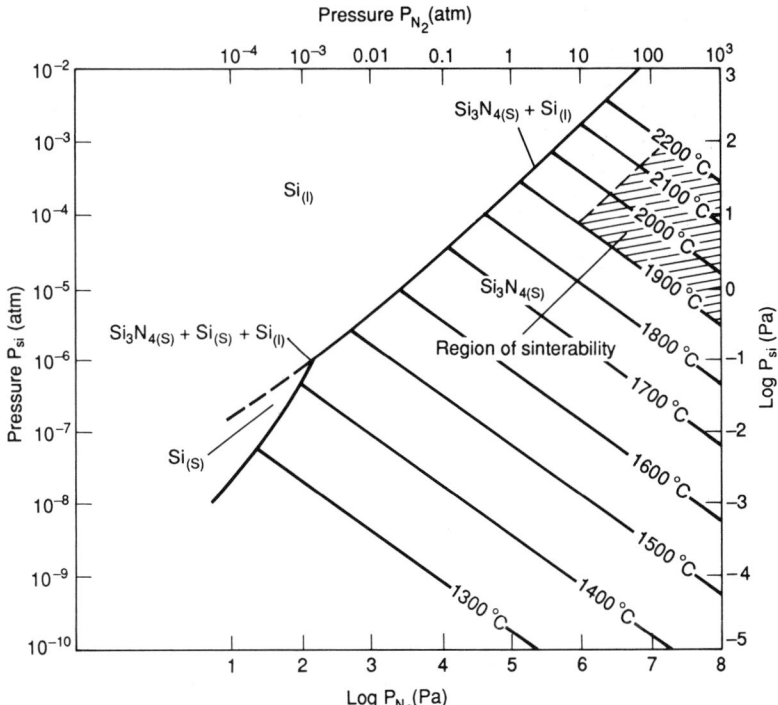

Fig 5 Stability diagram for Si₃N₄, where Si vapor is in equilibrium with Si₃N₄ as a function of N₂ pressure and temperature. Source: Ref 51

forging can be more effective in eliminating pores than post-HIP treatments.

Superplastic behavior of ceramics is a relatively new phenomenon that demonstrates the ability to produce large amounts of strain or deformation in a ceramic body without cracking. Characteristics of a ceramic that exhibits superplastic behavior are the presence of an ultrafine grain size, usually in the nanometer range, and sufficient grain boundary strength and cohesiveness to avoid the formation of cavities at grain interfaces. The suggested models that explain this phenomenon are based on a diffusional mechanism, where strain rates are inversely related to the cube of the grain size. As such, nanometer-sized particles have exceptional impact on the strain rate and rates of diffusion. Therefore, a significant problem arises if dynamic grain growth occurs during processing. Some ceramics that have successfully demonstrated superplastic behavior are zirconia, alumina, and silicon nitride. Extensive reviews of this technical area have been reported in the lit-

erature (Ref 60, 61). The materials that have been densified by the techniques described in this article are identified in Table 2.

REFERENCES

1. R.L. Coble, Diffusion Models for Hot Pressing with Surface Energy and Pressure Effects as Driving Forces, *J. Appl. Phys.*, Vol 41 (No. 12), 1970, p 4798–4807
2. P. Murray, D.T. Livey, and J. Williams, Hot Pressing of Ceramics, *Ceramic Fabrication Processes*, W.D. Kingery, Ed., Technology Press of Massachusetts Institute of Technology and John Wiley & Sons, 1958, p 147–171
3. J.D. McClelland, A Plastic Flow Model of Hot Pressing, *J. Am. Ceram. Soc.*, Vol 44 (No. 10), 1961, p 526
4. J.K. MacKenzie and R. Shuttleworth, Phenomenological Theory of Sintering, *Proc. Phys. Soc. (London)*, Vol 62 (No. 360B), 1949, p 833–852
5. T. Vasilos, Hot Pressing of Fused Silica, *J. Am. Ceram. Soc.*, Vol 43 (No. 10), 1960, p 517–519
6. R.L. Coble and J.S. Ellis, Hot Pressing Alumina-Mechanisms of Material Transport, *J. Am. Ceram. Soc.*, Vol 46 (No. 9), 1963, p 438–441
7. T. Vasilos and R.M. Spriggs, Pressure Sintering: Mechanisms and Microstructures for Alumina and Magnesia, *J. Am. Ceram. Soc.*, Vol 46 (No. 10), 1963, p 493–496
8. R.C. Rossi and R.M. Fulrath, Final Stage Densification in Vacuum Hot-Pressing of Alumina, *J. Am. Ceram. Soc.*, Vol 48 (No. 11), 1965, p 558–564
9. F.R. Nabarro, Deformation of Crystals by Motion of Single Ions, *Report of a Conference on the Strength of Solids*, University of Bristol, July 1947, p 75–90
10. C. Herring, Diffusional Viscosity of Polycrystalline Solids, *J. Appl. Phys.*, Vol 21 (No. 5), 1950, p 437–445
11. R.L. Coble, Sintering of Crystalline Solids, *J. Appl. Phys.*, Vol 32 (No. 5), 1961, p 787–792
12. D.S. Wilkinson and M.F. Ashby, Pressure Sintering by Power Law Creep, *Acta Metall.*, Vol 23, 1975, p 1277–1285
13. W.D. Kingery, J.M. Woulbroun, and F.R. Charvat, Effects of Applied Pressure on Densification During Sintering in the Presence of a Liquid Phase, *J. Am. Ceram. Soc.*, Vol 46 (No. 8), 1963, p 391–395
14. W.A. Keysser and G. Petzow, Liquid Phase Sintering of Ceramics, *Emergent Process Methods for High Technology Ceramics*, Vol 17, *Materials Science Research*, R.F. Davis, H. Palmour III, and R.L. Porter, Ed., Plenum Press, 1984, p 225–231
15. R. Raj and C.K. Chyung, Solution-Pre-

Table 2 Pressure densification techniques and parameters for various ceramic materials

Technique and material	Pressure		Temperature		Ref
	MPa	ksi	°C	°F	
Uniaxial hot pressing					
Al₂O₃	12–40	1.7–5.8	1150–1350	2100–2460	8
SiO₂	7–17	1.0–2.5	1100–1200	2010–2190	5
NiO	4–130	0.6–18.8	800–1100	1470–2010	26
CoO	75	11	950–1100	1740–2010	20
MgO	5–30	0.7–4.4	1440–1790	2625–3255	63
YBa₂Cu₃Oₓ	10–40	1.5–5.8	850–950	1560–1740	62
Hot isostatic pressing					
TiN	100–200	15–30	1000–1600	1830–2910	64
TaC; WC	70–100	10–15	1500–1760	2730–3200	34
SiC	138	20	1850–2100	3360–3810	65
Gas pressure sintering(a)					
Si₃N₄ (Be₃N₂)	2/7	0.3/1.0	2050/1950	3720/3540	44
Al₂O₃	Vacuum/100	Vacuum/15	1650	3000	41
Si₃N₄ (Y₂O₃/Al₂O₃)	0.1/2	0.015/0.3	1780	3235	47
PLZT; BaTiO₃	0.1/20	0.015/3	1170	2140	42

(a) Values separated by virgule indicate first step/second step.

cipitation Creep in Glass Ceramics, *Acta Metall.*, Vol 29 (No. 1), 1981, p 159–166

16. R.L. Coble and R.M. Cannon, Current Paradigms in Powder Processing, *Processing of Crystalline Ceramics*, Vol 11, *Materials Science Research*, H. Palmour III, R.F. Davis, and T.M. Hare, Ed., Plenum Press, 1978, p 151–170

17. J.M. Vieira and R.J. Brook, Kinetics of Hot Pressing: The Semilogarithmic Law, *J. Am. Ceram. Soc.*, Vol 67 (No. 4), 1984, p 245–249

18. R.J. Brook, Fabrication Principles for the Production of Ceramics with Superior Mechanical Properties, *Proc. Brit. Ceram. Soc.*, Vol 32 (No. 7), 1982

19. M.F. Ashby, A First Report on Deformation Mechanism Maps, *Acta Metall.*, Vol 20 (No. 7), 1972, p 887–896

20. P. Urick and M.R. Notis, Final Stage Densification During Pressure Sintering of CoO, *J. Am. Ceram. Soc.*, Vol 56 (No. 11), 1973, p 570–574

21. T.G. Langdon and F.A. Mohamed, Deformation Mechanism Maps for Ceramics, *J. Mater. Sci.*, Vol 11, 1976, p 312–327

22. M.R. Notis, Advances in Ceramic Hot Forming and Pressing: Theory and Practice, *Ceramurgia Int.*, Vol 3, 1977

23. A.S. Helle, K.E. Easterling, and M.F. Ashby, Hot Isostatic Pressing Diagrams: New Developments, *Acta Metall.*, Vol 33 (No. 12), 1985, p 2163–2174

24. T. Vasilos and W. Rhodes, Fine Particulates to Ultrafine-Grain Ceramics, *Ultrafine Grain Ceramics*, 15th Sagamore Army Materials Research Conference, J.J. Burke, N.L. Reed, and V. Weiss, Ed., Syracuse University Press, 1970, p 137–172

25. H. Knoch and G.E. Gazza, On the α to β Phase Transformation and Grain Growth During Hot-Pressing of Si₃N₄ Containing MgO, *Ceramurgia Int.*, Vol 6 (No. 2), Apr-June 1980, p 51–56

26. R.M. Spriggs, L.A. Brissette, and T. Vasilos, Pressure-Sintered Nickel Oxide, *Am. Ceram. Soc. Bull.*, Vol 43 (No. 8), 1964, p 572–577

27. W.H. Rhodes, Polycrystalline Oxides for Optical Applications, *Ceramics: Today and Tomorrow*, S. Naka, N. Soga, and S. Kume, Ed., The Ceramic Society of Japan, Kanto Dentsu Printing, Tokyo, 1986, p 149–164

28. R.R. Wills, M.C. Brockway, and G.K. Bansal, Relationship Between Densification and High Temperature Mechanical Properties of HIPed Silicon Nitride, *Emergent Process Methods for High Technology Ceramics*, Vol 17, *Materials Science Research*, R.F. Davis, H. Palmour III, and R.L. Porter, Ed., Plenum Press, 1984, p 603–605

29. R. Chang and C.G. Rhodes, High Pressure Hot Pressing of Uranium Carbide Powders and Mechanism of Sintering of Refractory Bodies, *J. Am. Ceram. Soc.*, Vol 45 (No. 8), 1962, p 379–382

30. M. Shimada, N. Ogawa, and M. Koizumi, High Pressure Sintering of Non-Oxide Materials, *J. Jpn. Soc. Powder Powder Metall.*, Vol 25 (No. 8), 1978, p 27–28

31. D. Kalish and E.V. Clougherty, Densification Mechanisms in High Pressure Hot-Pressing of HfB₂, *J. Am. Ceram. Soc.*, Vol 52 (No. 1), 1969, p 26–29

32. S. Prochazka and W.A. Rocco, High Pressure Hot-Pressing of Silicon Nitride Powders, *Processing of Crystalline Ceramics*, Vol 11, *Materials Science Research*, H. Palmour III, R.F. Davis, and T.M. Hare, Ed., Plenum Press, 1978, p 615–625

33. E.D. Whitney, High Pressure Processing of High Technology Ceramics, *Emergent Process Methods for High Technology Ceramics*, Vol 17, *Materials Science Research*, R.F. Davis, H. Palmour III, and R.L. Porter, Ed., Plenum Press, 1984, p 765–781

34. E.S. Hodge, Elevated Temperature Compaction of Metals and Ceramics by Gas Pressures, *Powder Metall.*, Vol 7 (No. 14), 1964, p 169–201

35. G.K. Watson and J.J. Moore, "Hot Isostatic Pressing of Structural Ceramics at NASA," Proceedings of Department of Energy Contractor's Coord. Meeting, Dearborn, MI, Oct 1981

36. R.R. Wills, M.C. Brockway, and L.G. McCoy, Hot Isostatic Pressing of Ceramic Materials, *Emergent Process Methods for High Technology Ceramics*, Vol 17, *Materials Science Research*, R.F. Davis, H. Palmour III, and R.L. Porter, Ed., Plenum Press, 1984, p 559–570

37. H. Larker, J. Adlerborn, and H. Bohman, "Fabricating of Dense Silicon Nitride Parts By Hot Isostatic Pressing," Proceedings of the International Automotive Engineering Congress and Exposition, Detroit, Society of Automotive Engineers, 28 Feb-4 Mar 1977

38. E.M. Lenoe, Survey of Hot Isostatically Pressed Ceramics, *Ofc. Naval Res. Far East Sci. Bull.*, Vol 13 (No. 2), 1988, p 33–58

39. K.G. Ewsuk and G.L. Messing, Microstructural Changes During Hot Isostatic Pressing of Sintered Lead Zirconate Titanate, *Emergent Process Methods for High Technology Ceramics*, Vol 17, *Materials Science Research*, R.F. Davis, H. Palmour III, and R.L. Porter, Ed., Plenum Press, 1984, p 609–619

40. O, -H. Kwon and G.L. Messing, Microstructural Changes in Hot Isostatically Pressed Alumina-Glass Composites, *J. Am. Ceram. Soc. Comm.*, Vol 67 (No. 3), 1984, p C43-C45

41. E.A. Bush, "Corning Glass Works, High Temperature Gas Isostatic Pressing of Crystalline Bodies Having Impermeable Surfaces," U.S. patent 3,562,371, Feb 1971

42. K.H. Hardtl, Gas Isostatic Hot Pressing Without Molds, *Am. Ceram. Soc. Bull.*, Vol 54 (No. 2), 1975, p 201–207

43. K.J. Yoon and S.L. Kang, Densification of Ceramics Containing Entrapped Gases During Pressure Sintering, *J. Europ. Ceram. Soc.*, Vol 6, 1990, p 201–202

44. C. Greskovich, Preparation of High Density Si₃N₄ by a Gas Pressure Sintering Process, *J. Am. Ceram. Soc.*, Vol 64 (No. 12), 1981, p 725–730

45. M. Mitomo, Pressure Sintering of Si₃N₄, *J. Mater. Sci.*, Vol 11, 1976, p 1103–1107

46. H.F. Priest, G.L. Priest, and G.E. Gazza, Sintering of Si₃N₄ Under High N₂ Pressure, *J. Am. Ceram. Soc.*, Vol 60 (No. 1–2), Jan-Feb 1977

47. G.E. Gazza, R.N. Katz, and H.F. Priest, Densification of Si₃N₄-Y₂O₃/Al₂O₃ by a Dual N₂ Pressure Process, *J. Am. Ceram. Soc.*, Vol 64 (No. 11), 1981, p C161-C162

48. E. Tani, S. Umebayashi, K. Kishi, K. Kobayashi, and N. Nishijima, Gas Pressure Sintering of Si₃N₄ With Concurrent Addition of Al₂O₃ and 5% Rare Earth Oxide: High Fracture Toughness Si₃N₄ With Fiber-Like Structure, *Am. Ceram. Soc. Bull.*, Vol 65 (No. 9), 1986, p 1311–1315

49. I. Iturriza, J. Echeberria, I. Gutierrez, and F. Castro, Densification of Silicon Nitride Ceramics Under Sinter-HIP Conditions, *J. Mater. Sci.*, Vol 25, 1990, p 2539–2548

50. H.U. Kessel and W.P. Engel, Gas Pressure Sintering With Controlled Densification, *Powder Metall. Int.*, No. 5/6, 1989

51. C.D. Greskovich and S. Prochazka, Stability of Si₃N₄ and Liquid Phase(s) During Sintering, *J. Am. Ceram. Soc.*, Vol 64 (No. 7), 1981, p C96-C97

52. P.E.D. Morgan and E. Scala, The Formation of Fully Dense Oxides by Pressure Calcintering of Hydroxides, *Proceedings of the International Conference on Sintering and Related Phenomena*, June 1965, Gordon & Breach, 1967, p 861–894

53. E. Scala, S.K. Dutta, and G.E. Gazza, "Fine Grained Ceramics: The Role of Pressure Calcintering," Proceedings AIAA/ASME 10th Structures, Structural Dynamics, and Materials Conference, New Orleans, Apr 1969

54. A.C.D. Chaklader and L.G. McKenzie, Reactive Hot Pressing of Clays, *Am. Ceram. Soc. Bull.*, Vol 43, 1964, p 892

55. A.C.D. Chaklader and R.C. Cook, Kinetics of Reactive Hot Pressing of Clays and Hydroxides, *Am. Ceram. Soc. Bull.*, Vol 47 (No. 8), 1968, p 712

56. K. Hamano and S. Kanzaki, Fabrication

of Transparent Spinel Ceramics by Reactive Hot Pressing, *Yogyo-Kyokai-Shi*, Vol 85 (No. 5), 1977, p 225–230 (in Japanese)

57. M.N. Rahaman, F.L. Riley, and R.J. Brook, Mechanisms of Densification During Reaction Hot Pressing in the System Si-Al-O-N, *J. Am. Ceram. Soc.*, Vol 63 (No. 11–12), 1980, p 648–653

58. R.W. Rice, Hot Forming of Ceramics, *Ultrafine Grain Ceramics*, 15th Sagamore Army Materials Research Conference, J.J. Burke, N.L. Reed, and V. Weiss, Ed., Syracuse University Press, 1970, p 203–250

59. B.J. Kellett and F.F. Lange, Experiments on Pore Closure during Hot Isostatic Pressing and Forging, *J. Am. Ceram. Soc.*, Vol 71 (No. 1), 1988, p 7–12

60. I.W. Chen and L.A. Xue, Development of Superplastic Structural Ceramics, *J. Am. Ceram. Soc.*, Vol 73 (No. 9), 1990, p 2585–2609

61. Y. Maehara and T.G. Langdon, Review: Superplasticity in Ceramics, *J. Mater. Sci.*, Vol 25, 1990, p 2275–2286

62. W. Chen, X.W. Wu, S.J. Keating, C.Y. Keating, P.A. Johnson, and T.Y. Tien, Texture Development in $YBa_2Cu_3O_x$ by Hot Extrusion and Hot Pressing, *J. Am. Ceram. Soc.*, Dec 1987, p C388-C390

63. M. Vieira and R.J. Brook, Hot Pressing of High Purity Magnesium Oxide, *J. Am. Ceram. Soc.*, Vol 67 (No. 7), 1984, p 450–454

64. T. Yamada, M. Shimada, and M. Koizumi, Fabrication of TiN by Hot Isostatic Pressing, *J. Ceram. Soc. Jpn.*, Vol 89 (No. 11), 1981, p 621–625

65. S. Dutta, High-Strength Silicon Carbides by Hot Isostatic Pressing, *Proceedings of the 3rd Symposium on Ceramic Materials and Components for Engines*, Las Vegas, V.J. Tennery, Ed., The American Society Inc., 1989, p 683–695

SELECTED REFERENCES

• D.A. Buckner, H.C. Hafner, and N.J. Kreidl, Hot Pressing Magnesium Fluoride, *J. Am. Ceram. Soc.*, Vol 45 (No. 9), 1962, p 435–438

• C. Greskovich, Hot Pressed β-Si_3N_4 Containing Small Amounts of Be and O in Solid Solution, *J. Mater. Sci.*, Vol 14 (No. 12), 1979, p 2427–2438

• A.B. Harker, P.E.D. Morgan, and J.F. Flintoff, Hot Isostatic Pressing of Nuclear Waste Glasses, *J. Am. Ceram. Soc. Comm.*, Vol 67 (No. 2), 1984, p C26-C28

• T. Hattori, M. Yoshimura, and S. Somiya, High Pressure Hot Isostatic Pressing of Synthetic Mica, *J. Am. Ceram. Soc. Comm.*, Vol 69 (No. 8), 1986, p C182-C183

• M.Y. Kim, S.J. Park, H. Haneda, A. Wanatabe, J. Tanaka, and S. Shirasaki, Hot Pressing of Perovskite $LaNiO_3$, *J. Ceram. Soc. Jpn.*, Vol 97 (No. 10), 1989, p 1129–1133

• T. Sakai and N. Hirosaki, Hot-Pressing of SiC With Additions of BaO and C, *J. Am. Ceram. Soc. Comm.*, Vol 68 (No. 8), 1985, p C191-C193

• A.A. Solomon, K.M. Cochran, and J.A. Habermeyer, Isostatic Hot Pressing of UO_2, *J. Am. Ceram. Soc.*, Vol 64 (No. 8), 1981, p 472–479

Grain Growth

S.J. Bennison*, Ceramics Division, National Institute of Standards & Technology

GRAIN GROWTH occurs during the high-temperature fabrication of ceramics in response to the excess energy associated with interfaces or boundaries between neighboring grains (Ref 1–4). The control of grain growth is a major goal of the ceramics processor for two reasons. Firstly, the size and morphology of grains generally play important roles in the resulting material properties, as, for example, in strength (Ref 5) or electrical behavior (Ref 4, 6). Secondly, the attainment of high density requires that coarsening processes be suppressed during firing (Ref 7–9). Associated with both these motivations is the prevention of abnormal grain growth in which a few grains grow rapidly to sizes many times the matrix grain size (Ref 1, 10).

In this article, the fundamental grain growth mechanisms of relevance to ceramics are reviewed. Particular attention is devoted to the interactions between densification and coarsening together with the problem of abnormal grain growth. Recommendations for processing strategies aimed at fabricating high-density ceramics with controlled microstructures are offered.

Mechanisms and Kinetics of Grain Growth

Driving Force. The change in volume of an individual grain is determined by migration of its component boundaries under the influence of chemical potential gradients arising from grain boundary curvature. Curvature develops in order to balance surface tension forces acting between intersecting boundaries under the constraint of filling space analogous to curved interfaces in soap films (Ref 11). The resulting gradient induces a flux of ions from the concave side to the convex side of the boundary, causing it to migrate against the flow of matter. The driving force for boundary motion, F_b', expressed as the chemical potential gradient, $d\mu/dx$, per ion across the interface, is related to the boundary curvature, κ, by:

$$F_b' = 2\kappa\gamma_b\Omega/\delta_b \equiv d\mu/dx \qquad \text{(Eq 1)}$$

where Ω is the volume of matter transported along with the rate limiting ion; δ_b is the interface thickness; and γ_b is the interfacial energy (Ref 12).

The specific grain shape determines the direction of boundary migration relative to its center, and therefore, whether shrinkage or growth occurs. Figure 1 is a micrograph of a normal grain structure in an MgO-doped Al_2O_3 ceramic along with a two-dimensional model representation. Inspection of the hypothetical microstructure reveals that grains with more than six sides grow and grains with less than six sides shrink. The sign and magnitude of local boundary curvature is related approximately to the size of the grain containing the boundary, ℓ', and the size of the stable six-sided grain, ℓ^*, by Hillert's expression (Ref 13):

$$\kappa = \xi \cdot (1/\ell^* - 1/\ell') \qquad \text{(Eq 2)}$$

with ξ a constant ($3/2 \geq \xi \geq 1$) (Ref 3). Substituting for κ in Eq 1 yields:

$$F_b' = 2\xi\gamma_b\Omega(1/\ell^* - 1/\ell')/\delta_b \qquad \text{(Eq 3)}$$

F_b is positive (with respect to motion of the boundary) for $\ell' \geq \ell^*$ (that is, growth) and is negative for $\ell' \leq \ell^*$ (shrinkage). During normal grain growth the microstructure scales in a continuous self-similar manner with the following characteristics:

- $\ell_{max} \leq 2\ell^*$
- Average grain size $\ell = \psi\ell^*$, where ψ is a geometrical constant ($\psi \simeq 8/9$ for three dimensions) (Ref 13)

The average driving force for boundary motion, F_b, derived from Eq 3 is (Ref 13):

$$F_b = 2\xi\gamma_b\Omega/\delta_b\ell \qquad \text{(Eq 4)}$$

The driving force scales linearly with inverse grain size. More detailed analyses of the geometrical changes accompanying grain growth have yielded alternative values for the constant in Eq 4 with no change in its form.

General Kinetic Formulation. The average rate of grain growth, $d\ell/dt$, is related directly to the average velocity, v_b, of the migrating boundaries through the differential equation:

$$d\ell/dt = \beta v_b \qquad \text{(Eq 5)}$$

where β is a constant ($\simeq 2$). The velocity can in turn be represented as a product of the average grain boundary mobility, M_b, and the average driving force, F_b, namely:

$$v_b = M_b \cdot F_b \equiv (1/\beta)d\ell/dt \qquad \text{(Eq 6)}$$

A specific mechanism governing boundary migration is treated via the mobility term and kinetic laws describing the grain growth are derived by integrating the complete form of Eq 6 with respect to time (Ref 2, 3).

Intrinsic Mechanism. The intrinsic motion of a grain boundary in a pure system is determined by the diffusional transfer of matter from the contracting grain to the expanding grain. The mobility of an ion crossing the boundary, M_0, is given by the relationship:

$$M_0 = D_b^*/kT \qquad \text{(Eq 7)}$$

where D_b^* is the diffusion coefficient of the rate limiting ion crossing the boundary by the easiest route, k and T are the Boltzmann constant and the absolute temperature, respectively (Ref 14). Combining Eq 4 to 7 yields:

$$d\ell/dt = 2\beta\xi D_b^* \Omega\gamma_b/(kT\delta_b\ell) \qquad \text{(Eq 8a)}$$

Therefore,

$$\ell^2 - \ell_0^2 = [2\beta\xi D_b^* \Omega\gamma_b/(kT\delta_b\ell)] \cdot t \qquad \text{(Eq 8b)}$$

where ℓ_0 is the initial grain size at $t = 0$. Note that the main assumption made in deriving the parabolic law is that all properties are considered to be isotropic and independent of the detailed atomic structure of the boundary structure.

Observations of the parabolic rate law are rare in single-phase ceramic systems. Deviations from the parabolic rate law in which the sensitivity to grain size is decreased (for example, linear kinetics) are usually associated with the presence of soluble pinning precipitates or the initiation of abnormal grain growth. An increased sensitivity to grain size (for example, cubic or higher order kinetics) generally results from the impeding influence of attached species such as solutes and second phases.

Solute Segregation. Impurities present in solid solution tend to segregate to or away from the grain boundaries in response to an elastic driving force resulting from the lattice parameter mismatch and/or electrostatic driving force(s) resulting from valency differences and space charge effects (Ref 15). During migration, the impurities must diffuse

---

*Guest Scientist on leave from the Department of Materials Science and Engineering, Lehigh University

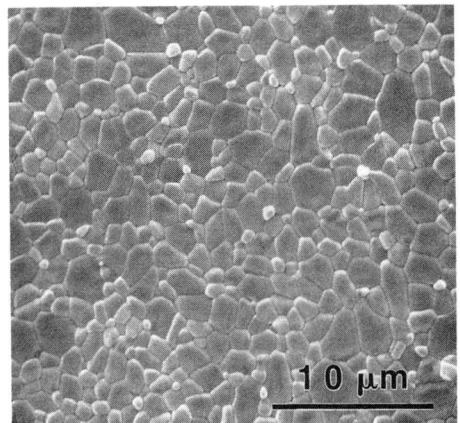

(a)

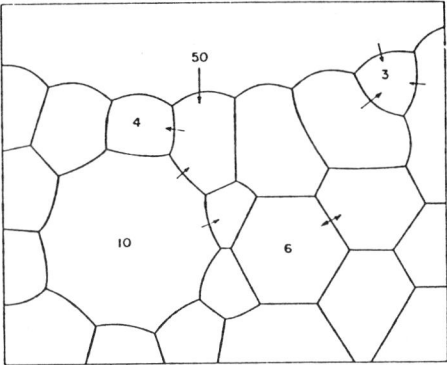

(b)

Fig 1 (a) Photomicrograph of typical grain structure of MgO-doped Al_2O_3 ceramic. (b) Schematic representation of normal microstructure assuming isotropic boundary energies. Numbers indicate number of sides. With the six-sided grain, size ℓ^* is predicted to be stable in this configuration whereas grains with greater than six sides grow and grains with less than six sides shrink. Source: Ref 1

along to maintain the energetically favorable spatial configuration.

For the case of normal grain growth under low driving force and strong segregation the velocity of the boundary may be expressed as:

$$v = F_b M_0 \cdot \{1/(1 + 4\Phi\delta_I Q C_0 \Omega^{2/3})\} \quad \text{(Eq 9)}$$

where Φ is the volume concentration of impurity ions, δ_I is the width of a zone over which the impurities interact with the boundary, Q is the partition coefficient for the impurity between the boundary and bulk crystal, and C_0 is the bulk impurity concentration (Ref 16). The velocity is linearly related to the inverse of the impurity concentration and therefore significant retardation arises from strong segregation.

A kinetic law for the case of impurity drag can be obtained by recognizing that the bulk impurity concentration, C_0, is related to the total impurity concentration, C, and the grain size via (Ref 17):

$$C_0 = C \cdot [1 + f\{(\delta_I/\ell)(Q - 1)\}] \quad \text{(Eq 10)}$$

where $f\{(\delta_I/\ell)(Q - 1)\}$ is a linear function of grain size, ℓ. Substituting for F_b, M_0, and C_0

in Eq 9 and expressing the equation in terms of the grain size dependencies alone:

$$v \equiv d\ell/dt = Y \cdot (1/\ell)^2 \quad \text{(Eq 11a)}$$

Therefore,

$$\ell^3 - \ell_0^3 = Y \cdot t \quad \text{(Eq 11b)}$$

where Y is a constant. Cubic kinetics are therefore expected if a strong solute drag mechanism controls the boundary motion. Note that several assumptions have again been made in the derivation of Eq 11b; if these assumptions do not hold, then segregation may produce more complex kinetic behavior.

Observations of cubic kinetics have been made for several single-phase systems containing either unintentional impurities (Ref 18–20) or deliberate solute additions (Ref 21–23). It may be argued that the majority of ceramics are sufficiently impure such that background impurities always control boundary migration in single-phase systems.

Grain growth suppression through a solid-solution mechanism is used to great effect in the fabrication of translucent Al_2O_3 (Lucalox) (Ref 18, 24). In this case, trace magnesium oxide additions reduce boundary migration rates by up to an order of magnitude (Ref 21), suppressing abnormal grain growth (Ref 1, 10) and leading to high fired density.

Liquid Films. Many commercial ceramics are manufactured via liquid-phase sintering or vitrification (Ref 25, 26) in which the presence of a wetting liquid during firing aids densification of a powder compact. Migration of the grain/liquid/grain interface is then controlled by one of two processes:

- Diffusion of matter through the liquid film
- Dissolution/reprecipitation (interface) reactions

Diffusion of Matter through Liquid Film. The first case has been treated in a similar fashion to Ostwald ripening of spherical particles of average diameter, ℓ, in a liquid in which the solubility of matter is controlled by the curvature of the interface. The velocity of the film, v_f, is given by (Ref 27):

$$v_f = \theta D_f C_f \Omega/[kT\ell \cdot \delta_f(\ell)] \quad \text{(Eq 12)}$$

where the subscript f denotes liquid film quantities; C_f is the grain solubility in the liquid; and θ is a geometrical constant. For a fixed volume fraction of liquid the film thickness, $\delta_f(\ell)$, scales linearly with grain size. Integrating Eq 12 results, therefore, in a cubic growth law.

Dissolution/Reprecipitation Reactions. For the second case of interface controlled migration, the kinetics are controlled by an exponentially temperature-dependent reaction constant, Γ. The velocity relationship becomes (Ref 27–29):

$$v_f = \eta\Gamma\Omega\gamma_f/(kT\ell) \quad \text{(Eq 13)}$$

where η is a constant. Integration of Eq 13 leads to familiar parabolic growth kinetics.

Anisotropic Grain Growth. Interpretation of grain growth data for liquid-phase sintering has tended toward control by diffusion

through the liquid film (Ref 26). In many cases, however, the development of highly anisotropic grain structures has prevented clear understanding of the grain growth mechanism. Anisotropic grain growth is considered in detail in the section "Abnormal Grain Growth" in this article.

Isolated Second Phases. Second phases are generally present in ceramic microstructures and take the form of either intentionally added compounds to adjust properties (Ref 30, 31) or porosity resulting from incomplete densification. Such phases invariably play a controlling role during grain growth because of their tendency to be attached to grain boundaries. The attachment force arises from the reduction of grain-boundary energy associated with the removal of boundary area equal to the particle cross-sectional area; this force, F_p, is given by:

$$F_p = \pi r \gamma_b \quad \text{(Eq 14)}$$

for a spherical particle of radius r (Ref 32). The modes through which second phases may influence grain growth are classified into two groups according to mobility criterion.

Immobile second phases act as effective boundary pinning agents if the dragging force equals the driving force for migration. Initially the microstructure coarsens until a limiting matrix grain size, ℓ_+ is obtained. For a uniform distribution of particles of radius, r, and the volume fraction, v_f, the limiting grain size is given by:

$$\ell_+ = \chi r/v_f \quad \text{(Eq 15)}$$

where χ is a constant ($\approx 8/3$) (Ref 33). Normal grain growth can only proceed if the second-phase particles coarsen or change in volume fraction. Ostwald ripening by either lattice or grain-boundary diffusion leads to particle coarsening kinetics following $r^3 \simeq t$ and $r^4 \simeq t$, respectively (Ref 27, 34). Assuming a constant v_f grain growth kinetics therefore follow:

$$\ell_+^3 \simeq \omega \cdot t \quad \text{(lattice diffusion)} \quad \text{(Eq 16a)}$$

$$\ell_+^4 \simeq \omega' \cdot t \quad \text{(boundary diffusion)} \quad \text{(Eq 16b)}$$

where ω is a constant. If the second phase is soluble in the matrix, the particle radius decreases at a constant rate for a given number density of particles and linear growth kinetics ensue.

Boundary pinning has proved to be an effective method of controlling grain growth in ceramics (Ref 35). Microstructures composed of equal volume fractions of two insoluble phases are observed to be extremely resistant to coarsening and show promise for applications requiring microstructural stability during prolonged high-temperature exposure (Ref 36).

In many instances however, a second class of behavior is observed where second-phase inclusions tear away from the grain boundaries and lead to abnormal grain growth. This occurs under microstructural conditions where the driving force is always greater than the pinning force. Breakaway phenomena are treated in the section "Abnormal Grain Growth" in this article.

Mobile Second Phases. Under conditions where the second phase is mobile, three different classes of behavior may result depending on the relative mobilities of the boundary and second phase and the relative driving forces for migration. For particles that are attached to grain boundaries moving at a velocity, v_p, the following relation has been derived:

$$v_p = v_b = F_b[M_bM_p/(NM_b + M_p)] \qquad \text{(Eq 17)}$$

where M_p is the particle mobility and N is the density of particles at the grain boundary (Ref 37).

The first class of behavior is observed for larger, less mobile particles acted on by low driving force. In terms of Eq 17, the boundary velocity $v_b \simeq F_bM_p/N$ and migration is therefore controlled by the motion of attached particle. The second class of behavior arises for smaller, more mobile particles and higher driving force for migration. The boundary velocity, $v_b \simeq F_bM_p$, and therefore the boundary mobility (intrinsic or with attached solute) controls migration. The third case concerns breakaway of the particle from the boundary.

Derivation of growth kinetics requires knowledge of the dependence of particle mobility on scale and the limiting migration behavior. A general relation between particle mobility, M_p, and particle radius, r, is:

$$M_p = AD\Omega/(\pi kTr^n) \qquad \text{(Eq 18)}$$

where A is a constant, and n and D are the growth exponent and the diffusion coefficient, respectively, pertaining to a specific transport mechanism. Expressions and values for these are listed in Table 1.

Consider, for example, the case of particle-controlled boundary migration where an attached pore is moving by surface diffusion. Assuming the areal pore density can be expressed in terms of a grain size dependence, $N \sim 1/\ell^2$, and that the pore radius scales with grain size, $r \sim \ell$, then:

$$d\ell/dt \equiv \beta v_b \simeq \beta(FM_p/N) \qquad \text{(Eq 19a)}$$

$$d\ell/dt \simeq 2\xi\beta\gamma(\Omega/\delta_b\ell)\{\delta_sD_s\Omega/\pi kT\omega\ell^4\}/\{\epsilon\Omega^{2/3}/\ell^2\} \qquad \text{(Eq 19b)}$$

Therefore,

$$\ell^4 - \ell_0^4 = K \cdot t \qquad \text{(Eq 20)}$$

Note Eq 20 is derived assuming the system remains at constant density and changes in pore size due to coalescence are accompanied by opposite changes in area density.

A summary of the possible kinetic growth laws discussed is presented in Table 2. It should be noted that observation of a particular growth exponent is not proof alone of a boundary control by any one mechanism (Ref 2, 19–21). Supporting microstructural and

chemical evidence is generally needed to unequivocally identify a specific mechanism.

Abnormal Grain Growth

During abnormal grain growth, a fraction of the grain population grows at a rapid rate, generally leading to a broad or even bimodal grain size distribution (Fig 2). Abnormal grain growth during fabrication should be avoided for two reasons. Firstly, a degradation in material property may result from the ensuing microstructural heterogeneities; for example, large grains are effective nucleation sites for large creep cavities that limit high-temperature component life (Ref 38). Secondly, abnormal grain growth is often associated with the separation of second phases from grain boundaries; if the second phase is porosity densification, processes are effectively terminated (Ref 1).

Initiation of abnormal grain growth is often linked to the occurrence of local fluctuations in boundary or particle velocities. Common mechanisms are characterized in terms of either an intrinsic or extrinsic origin and whether the velocity fluctuation arises from variations in mobility or the driving force.

Intrinsic Condition. Experimental and theoretical evidence is available to show

anisotropy in both intrinsic grain-boundary energies (Ref 39) and mobilities (Ref 40, 41). Computer simulations of grain growth, in which an orientation dependence of either the driving force (energy) (Ref 4) or the mobility (Ref 42, 43), show abnormal structures quickly develop when such anisotropies exist. Boundary anisotropies may be enhanced or diminished by the presence of impurities either relieving or exacerbating abnormal grain growth.

When impurities exist in sufficient quantities to form a liquid phase during firing a characteristic platelike grain morphology may develop. Figure 2(b) is an example of such grain morphology in an Al_2O_3 material containing trace (≈ 0.2 wt%) impurities of Na_2O, SiO_2, and CaO. High resolution transmission electron microscopy reveals that the liquid wets the long basal {0001} type facets but not the facets making up the grain ends. This anisotropic wetting is believed to lead to strong variability in grain interface mobilities. The shapes that evolve in the Al_2O_3 material shown therefore result from a combined anisotropy in surface energies and film/boundary mobilities (Ref 44).

Extrinsic Condition. Extrinsic influences embody a variety of microstructural phenomena and are the most common cause of abnormal grain growth.

A wide distribution in particle size of the starting powder tends to promote abnormal grain growth. Grains larger than twice the critical size may develop abnormally due to the large driving force from highly curved boundaries between the large grain and the matrix (see the section "Driving Force" in this article) (Ref 13). The driving force for matrix grain growth is considerably lower (5 to 10 times) being dictated by the lesser curvatures in the matrix. This disparity in driving force leads to linear growth kinetics of the abnormal grain. In many cases, after a pronounced growth stage, abnormal grains that initiate via this mechanism impinge on each other and their migration rate diminishes as they now effectively become matrix grains. The smaller grains continue growing and reach the size of the larger matrix grains leading to a normal but coarser microstructure. Bimodal microstructures may in some instances be transient in nature if initiated by a wide distribution in size of the starting powder.

Distributions in other microstructural characteristics may also trigger abnormal grain

Table 1   Parameters in the mobility expression of particles: $M_p = AD\Omega/\pi kTr^n$. Transport through the vapor phase applies to pores only

Controlling mechanism	D	n	A
Evaporation/condensation	α'(a)	2	$3\Omega P_V/4(2m\pi kT)^{1/2}$(e)(f)
Vapor diffusion	D_G(b)	3	$3_\Omega P_V/4kT$(e)
Lattice diffusion	D_L(c)	3	3/2
Surface/interface diffusion	$\delta_s D_s$(d)	4	3/2

(a) Gas sticking coefficient. (b) Vapor diffusion coefficient. (c) Lattice diffusion coefficient. (d) Surface diffusion zone thickness/surface diffusion coefficient product. (e) P_V, vapor pressure. (f) m, molecular weight of gas specie

Table 2   Summary of grain growth kinetic laws for various controlling mechanisms

Grain growth mechanism	Exponent, q(a)
Boundary control	
Pure system	2
Impure system	
Impurity drag (low velocity)	3
Impurity drag (high velocity)	2
Solution of solid second phase	1
Coalescence of solid second phase (lattice diffusion)	3
Coalescence of solid second phase (boundary diffusion)	4
Continuous liquid phase (interface control)	2
Continuous liquid phase (diffusion through film)	3
Pore control	
Vapor transport (interface control)	2
Vapor transport (diffusion control)	3
Lattice diffusion	3
Surface diffusion	4

(a) For a law of the form $\ell^q - \ell_0^q = K \cdot t$ where K is a rate constant and t is the time

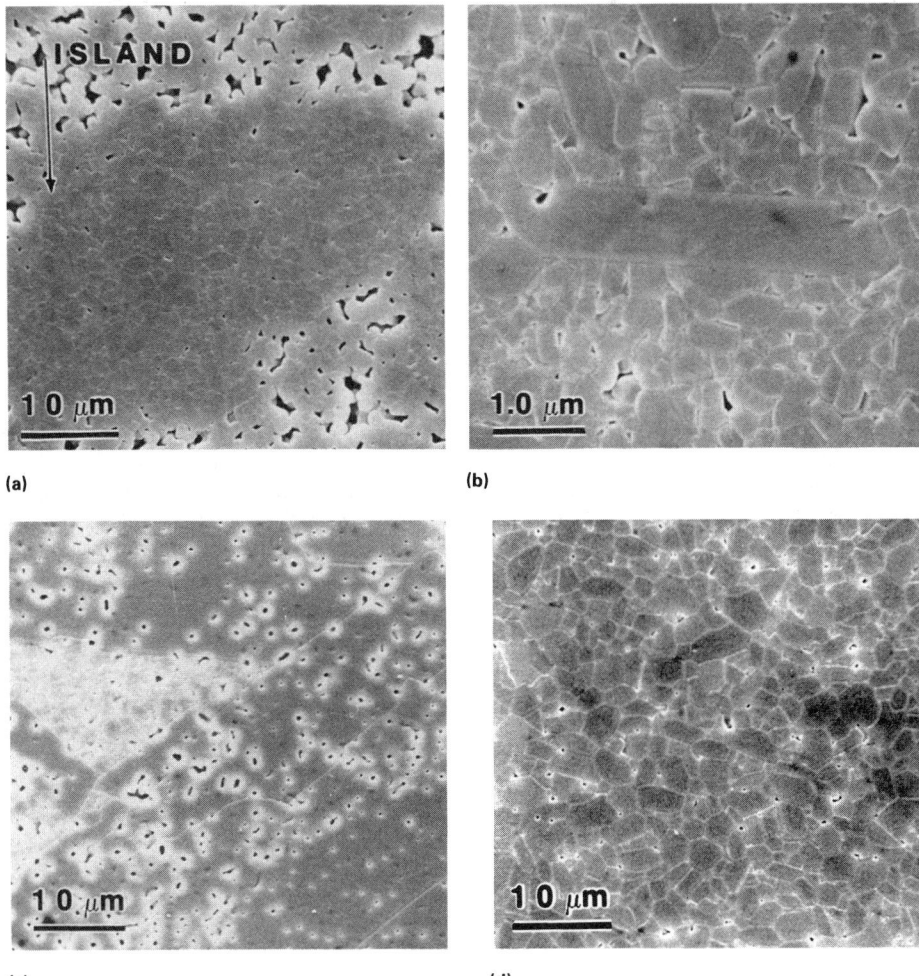

Fig 2 Sequence of abnormal grain growth development in Al₂O₃. (a) Dense islands form during intermediate stages of sintering. (b) Initiation of lathlike grains in dense island. (c) Breakout of lathlike grains into porous matrix. (d) Same material processed with 0.25 wt% MgO added showing equiaxed grain structure. Source: Ref 10

growth. Porosity gradients, for example, lead to variable growth rates when pore drag controls boundary motion. Such gradients may be introduced during green fabrication, for example, from nonuniform powder compaction during pressure forming. In addition, agglomeration of the starting powder often results in nonuniform packing and shrinkage during firing (Ref 45). Poor distribution of a dopant or contaminant may also affect local densification and grain growth indirectly or the dopant may influence grain growth directly through solute or second-phase drag mechanisms (Ref 46). The common link between these mechanisms is that spatial variations in the boundary velocities occur leading to the relatively rapid growth of a few grains.

A further cause of grain-boundary velocity fluctuations is the occurrence of the breakaway phenomena. Analysis of the solute drag mechanism indicates that a boundary may exhibit two distinct velocity characteristics depending on the relative values of the driving force and drag effect (Ref 16). The so-called low velocity limit is discussed in the section "Intrinsic Mechanism" in this article

in connection with normal grain growth. However, if impurity diffusion is slow compared to intrinsic ion motion across the interface, the majority of segregated solute will be shed from the boundary allowing it to assume a high near-intrinsic velocity (Ref 3). Variations in driving force and solute content from boundary to boundary may result in velocity transitions that can initiate abnormal grain growth. Fluctuations along a boundary may also occur leading to unstable velocity perturbations and abnormal grains with a distinctive fingerlike morphology (Ref 47). In general, solute breakaway is believed to be uncommon because as grain growth proceeds the driving force decreases and solute accumulates, which tends to push migration behavior toward the low velocity control regime.

A more common form of breakaway phenomenon that leads to abnormal grain growth involves second phases. The most important breakaway event concerns the separation of pores from grain boundaries that results in a kinetic barrier to densification. The separation criterion is calculated for microstructural

conditions under which the maximum driving force for grain growth, F_b^{max}, is greater than the maximum drag force exerted by the particle or pore. From Eq 14 and 17, this is given by:

$$F_b^{max} > \pi r \gamma (N + M_p/M_b) \qquad (Eq\ 21)$$

The breakaway conditions can be calculated as a function of grain size and pore size for a specific mechanism controlling boundary migration and the mechanism controlling pore motion. For surface diffusion controlled pore drag and intrinsic boundary motion, separation is given by solution to the following quadratic (Ref 48):

$$[(\delta_s D_s \Omega)/(kTM_b r^3)] \cdot \ell^2 - 2\eta(1 - \ell/\ell^*) \cdot \ell$$
$$+ 24r = 0 \qquad (Eq\ 22)$$

The criteria for pore separation are mapped as a function of microstructural conditions in Fig 3. Such a map is useful in describing qualitatively the effects of changing mobility values (for example, through the use of additives). A modified presentation of this map in which breakaway is calculated as a function of grain size and density is used later to demonstrate the principles of sintering to high density.

Grain Growth and Densification

Two criteria must be met in order to sinter a powder to a density near the single crystal value. The first is thermodynamic in nature and requires that the end point state of the reaction is favorable; the second is based on kinetics and requires that full density is attainable within a reasonable time (typically a few hours at temperature). Grain growth affects both these criteria by modifying pore shapes and increasing diffusion distances. These effects are considered in the discussion that follows in connection with final stage sintering.

Thermodynamic Considerations

Densification is driven by local chemical potential gradients between the grain boundary and pore (free) surface. In order to promote densification, matter should be transported from the grain boundary to the neck formed at the intersection of the pore and boundary (Ref 7, 49). For this to occur, the radius of curvature of the pore should be positive with respect to the pore center (that is, concave).

The curvature is determined by the pore coordination number and the force balance between tensions associated with the boundary and pore (Ref 50, 51). The force balance is characterized by the dihedral angle, ϕ, where:

$$\phi = 2 \cdot \cos^{-1}(\gamma_b/2\gamma_s) \qquad (Eq\ 23)$$

and γ_s is the surface energy. The criterion for pore shrinkage as a function of the pore size to grain size ratio (reflecting the coordination

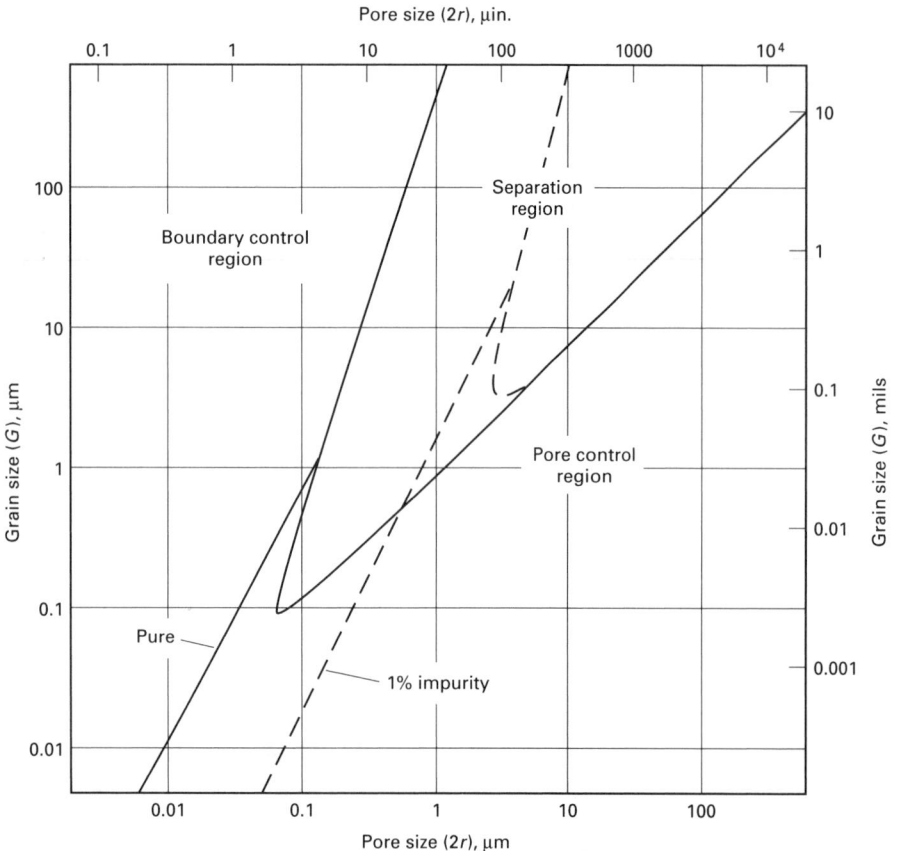

Fig 3 Map describing pore-grain boundary interactions for surface-diffusion controlled pore migration. The microstructural conditions for separation can be adjusted through reductions in the boundary mobility. Source: Ref 37

number) and the dihedral angle in two dimensions (Ref 5) is shown in Fig 4. For a given dihedral angle (~120° for alumina), the pore size should be kept below the grain size in order to satisfy the thermodynamic criterion for densification. For conditions where the pore curvature is convex (that is, when the pore is several times larger than the surrounding grains), shrinkage is predicted to cease and further densification can only occur if grain growth causes a reduction in coordination number with an attendant change to a concave curvature. From a thermodynamic viewpoint, it appears that coarsening is necessary under certain microstructural conditions to remove large pores (Ref 51, 52).

Kinetic Considerations

The kinetic criterion for densification within reasonable times depends on the specific diffusion mechanism(s) controlling densification and the scale over which matter is to be transported. Analysis of the diffusion processes for final stage sintering in which there is no thermodynamic barrier to densification

yields the following general equation relating density to time:

$$\rho - \rho_0 = (BD\gamma_s\Omega/kT\ell^m) \cdot t \qquad \text{(Eq 24)}$$

where B is a constant, D is the pertinent diffusion coefficient, and m is an exponent (Ref 18); expressions and values for these parameters are given in Table 3. Note that the time required to densify is strongly dependent on microstructural scale and that grain growth will lead to a rapidly diminishing densification rate. The effect of grain growth on densification can be calculated by either solving a differential equation incorporating both densification and coarsening dependencies on scale (Ref 4) or by taking the separate expressions for a specific densification process and a coarsening process and employing a numerical method to simulate microstructure evolution (Ref 7).

Consider, for example, the scenario of a microstructure densifying via lattice diffusion and coarsening via a surface diffusion controlled pore drag mechanism. The respective equations describing densification at constant grain size and coarsening at constant density during final stage sintering are:

$$\rho - \rho_0 = [(287D_L\gamma_s\Omega)/(kT\ell^3)] \cdot t \qquad \text{(Eq 25a)}$$

$$\ell^4 - \ell_0^4 = [(440\delta_s D_s\gamma_B\Omega)/(kT(1-\rho)^{4/3})] \cdot t \qquad \text{(Eq 25b)}$$

Microstructure evolution is then calculated using a time/step iteration and the resulting grain size-density trajectory presented as a map such as the one shown in Fig 5. The trajectory is a measure of the competition between coarsening and densification and is an extremely useful guide as to how changes in the relative diffusivities controlling each process affect microstructure evolution.

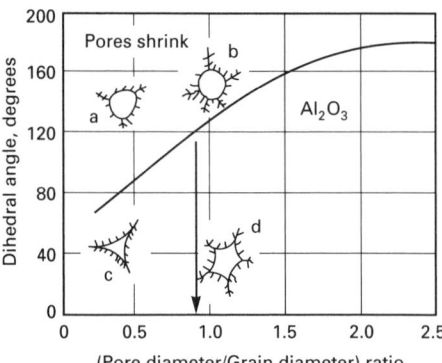

Fig 4 Thermodynamic criterion for pore shrinkage as a function of the dihedral angle and the ratio of pore size to grain size (coordination number) (Ref 50) in a two-dimensional microstructure. The hypothetical pores in cases (a) and (b) will shrink and disappear whereas the pores in (c) and (d) will be stable and require grain growth and recoordination in order to shrink further.

Table 3 Expressions and values for the parameters in the expression for densification during sintering: $\rho - \rho_0 = (BD\gamma_s\Omega/kT\ell^m) \cdot t$

Controlling mechanism	D	m	B
Lattice diffusion	D_L	3	287
Grain-boundary diffusion	$\delta_b D_b$(a)	4	xxx

(a) Boundary diffusion zone thickness/boundary diffusion coefficient product

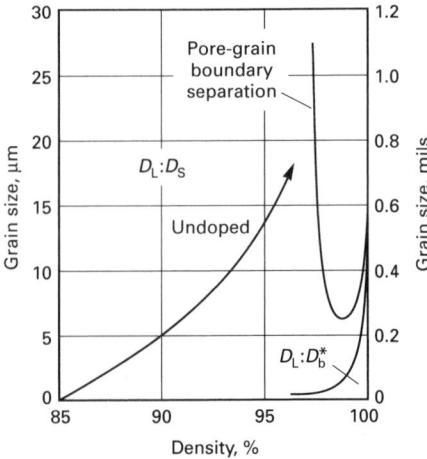

Fig 5 Microstructure development map for Al_2O_3. The grain size-density trajectory was calculated for densification as determined by lattice-diffusion control and grain growth determined by surface-diffusion-controlled pore motion. The pore-grain boundary separation region is in the top right of the diagram. Source: Ref 8

Also plotted in Fig 5 is the pore separation region presented in Fig 3 as a map of grain size versus pore size. [The density and pore size are related by $\rho = 1 - 48(r/\ell)^3$ for a tetrakaidecahedral grain with pore at four-grain junctions (Ref 18)]. Note that the separation region is also dependent on the same geometrical characteristics of the microstructure and the diffusivities controlling grain boundary and pore motion. If abnormal grain growth and pore separation occurs (that is, pores are trapped within grains, away from grain boundaries), diffusion distances become prohibitively large leading to a kinetic barrier to densification. Note that matter must be transported in the stoichiometric ratio during diffusion. For attached pores, anion transport occurs with relative ease in the boundary core and cation diffusion either through the grain boundary or crystal lattice is rate limiting. After breakaway, the anion must diffuse in the lattice at a rate that is orders of magnitude slower than the anion (Ref 53). The resulting microstructure development map indicates whether a given microstructure will densify completely, the aim being to keep the trajectory as flat as possible and avoid the separation region.

Attainment of High Density

The thermodynamic and kinetic criteria lead, therefore, to competing effects of coarsening in sintering: grain growth leads to a higher fraction of pore shapes that are unstable and shrink, but grain growth also leads to kinetic barriers to densification of these pores. Naturally the question arises as to whether grain growth should be suppressed or enhanced during sintering.

Consider the effect of enhanced coarsening, for example, by an additive that increases the surface diffusivity on microstructure evolution. The grain size-density trajectory will be steepened leading to a kinetic barrier to removal of the thermodynamically unstable pores due simply to scaling effects. Figure 6 is a microstructure development map which displays the effect of enhanced coarsening on the trajectory. Interestingly, the diagram also indicates that pore separation region is moved to a higher grain size which generally is a favorable consequence of doping. However, the greatly steepened trajectory suggests the chance of pore separation is enhanced in the present case. Thus, the argument to enhance coarsening in order to favorably recoordinate large pores will lead to diminished densification of the finer scale matrix pores. The analysis further suggests that even if larger pores are eventually recoordinated correctly, densification will be prohibitively slow (Ref 54). The recommendation, therefore, is to achieve good packing density during green processing in order to prevent the formation of large stable pores and to increase the den-

sification rate/coarsening rate ratio in order to readily remove finer scale porosity and prevent pore-grain boundary separation. Various strategies may be employed to increase the densification rate/coarsening rate ratio.

The simplest route to increasing the ratio is by adjusting the firing temperature. The diffusion processes controlling densification processes and coarsening processes are thermally activated but generally have different activation energies. In the case of alumina the activation energy for densification via lattice diffusion is approximately twice that for coarsening controlled by surface diffusion (Ref 7). Increased firing temperatures, therefore, lead to in increase in the relative densification rate/coarsening rate ratio (Ref 55, 56). This effect is exploited in the practice of fast-firing where alumina (Ref 55) and other ceramics [for example, barium titanate (Ref 57), β-alumina (Ref 58)] are sintered to full density within a few minutes at high homologous temperatures ($>0.9\,T_\mathrm{m}$). The choice of firing temperature requires foreknowledge of the relative activation energies for the material of interest. Such data is often absent. Therefore, in practice, it is recommended that extremes in firing temperature are tested.

An alternative approach to increasing the ratio is to enhance densification directly using the technique of hot pressing or hot isostatic pressing (HIP). In both cases, the application of pressure to the powder compact increases the driving force for densification without affecting the driving force for grain growth (Ref 59). Pressure is applied uniaxially via a die and punch arrangement in hot pressing, which limits the technique to the fabrication of simple shapes (Ref 60). Particle rearrangement and a shearing effect that result from constrained shape changes during hot press-

ing also aid densification processes. For the case of hot isostatic pressing, pressure is applied hydrostatically using a gas phase and as such is well suited to the fabrication of complex shapes (Ref 61). Parts are either presintered into the closed pore state or encase (canned) with an impervious deformable material. The application of pressure is an expensive option for the fabrication of many ceramics. However, where high reliability components are required (for example, structural applications), pressure assistance is often the only option.

The third approach to improving densification employs the use of additives such as solutes, solid second phases, and controlled atmospheres during firing.

The most celebrated example of an additive effect is the fabrication of translucent polycrystalline Al_2O_3 (Lucalox) with the aid of MgO solute additions and a hydrogen firing atmosphere (Ref 18, 24). The potency of this host additive combination lies with the multiple effects MgO has on the parameters controlling sintering. Independent measurements have revealed that at 1600 °C (2910 °F), MgO solute:

- Increases lattice diffusion by a factor of three through modification of point defect concentrations (Ref 62)
- Increases surface diffusion by a factor of four (Ref 63)
- Reduces the grain-boundary mobility by a factor of five to fifty depending on powder purity (Ref 20, 21)

The combined effect is displayed on the microstructure development map presented in Fig 7. The trajectory is flattened appreciably and the position of the separation region is

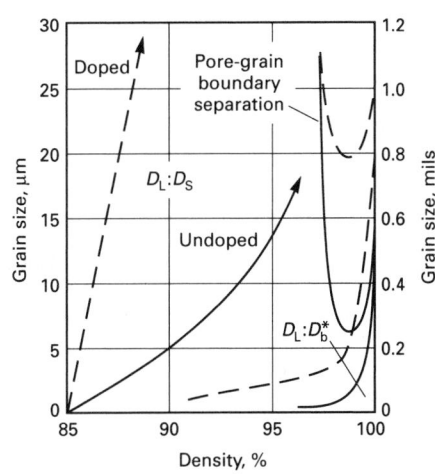

Fig 6 Microstructure development map showing the calculated effect of enhanced coarsening on the grain size-density trajectory and pore-boundary separation region. A kinetic limit to densification is predicted along with an increased likelihood of pore separation. Calculated for alumina under same conditions as Fig 5. Source: Ref 8

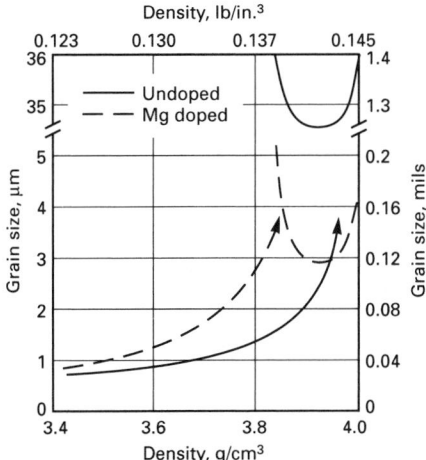

Fig 7 Microstructure development map displaying the effect of doping with MgO solute on the sintering behavior of Al_2O_3. The additive has successfully increased densification processes over coarsening (flattened trajectory) and decreased the chance of pore separation.

shifted to larger grain size. The effect of the additive is to enhance the densification rate/coarsening rate ratio, to prevent pore separation, and to stabilize the microstructure against abnormal grain growth.

Figure 2(a to c) is a sequence of micrographs illustrating the development of abnormal grains in an impure alumina ceramic (Ref 10). The first stage (Fig 2a) involves the formation of dense islands in the compact due to either fluctuations in green density or impurity contamination. Platelike grains nucleate in the dense islands (Fig 2b) due to mobility differences caused by anisotropic wetting by a liquid film; the lack of porosity encourages rapid grain growth in these islands. The platelike grains grow unchecked in the islands and burst out into the surrounding porous matrix (Fig 2c) leading to a final microstructure consisting of a large distribution in sizes and shapes, entrapped pores, and associated low end-point density. Figure 2(d) shows the final microstructure of an MgO-doped alumina prepared from the same powder following an identical processing sequence; the additive has successfully suppressed abnormal grain growth (Ref 10) and led to a high density ceramic.

The primary role of MgO has, therefore, been identified as a grain growth inhibitor (Ref 64). The additive operates successfully in aluminas of varying purity containing solid and liquid second phases. The detailed atomic mechanisms through which MgO realizes this role remain uncertain and make *a priori* additive selection for other host systems difficult. Nevertheless, a general guide to solid-solution additive selection can be pieced together by combining the various following suggestions:

- Choose a compound which forms a restricted solid solution (<5 mol%) and dope at the solubility limit for maximum effect
- Choose a compound with a combination of cation valence and size misfit to promote strong grain-boundary segregation. If the cation valence is one unit different from the host, point defects are created that affect segregation behavior and may be beneficial in enhancing the diffusivities controlling densification
- Choose a compound with high volatility to promote even distribution during doping (Ref 65)

Note that these requirements may run counter to one another in some cases (for example, requirement of some solid solubility and strong segregation). Therefore, a compromise will often be required. The suggestions presented should be only taken as a rough guide and experimentation is usually necessary to find the best additive. It should be mentioned that the MgO/Al$_2$O$_3$ system is somewhat exceptional and that multiple additives, in which different chemical species affect specific

components of the sintering reaction, may prove effective in some cases (Ref 66).

The function of atmospheres may be similar to that of additives if redox reactions affect point defect concentrations. Also, adsorption effects can adjust surface energies and modify the driving force for densification (Ref 67). A strict requirement for an atmosphere is that it be soluble at least in the grain-boundary region so it may be removed from closed porosity during the final stages of sintering (Ref 68).

Finally, the densification rate/coarsening rate ratio may be enhanced indirectly through the pinning effect of second-phase additives on grain-boundary motion. The criterion for such additives are:

- Compound forms an equilibrium two-phase microstructure when added to the host
- Pinning phase is immobile (low interface diffusivity)
- Two phases display little mutual solid solubility in order to diminish the mobility and to prevent coarsening by an Ostwald ripening mechanism (see the section "Immobile Second Phases" in this article)
- Second-phase pinning additive should be present at significant volume (≥15 vol%)

Second-phase pinning has been used successfully in the fabrication of several ceramic systems including Y$_2$O$_3$(La$_2$O$_3$) (Ref 35) and magnetic ferrites (Ref 69).

Future Trends

A rich variety of phenomena affect grain growth in ceramic systems. Manipulation of coarsening behavior by exploitation of the various mechanisms shows potential for the design of microstructure with specific properties that are scale sensitive. The attainment of high densities requires that coarsening is severely restricted at least until the desired density is realized. The *a priori* selection of additives that restrict coarsening behavior remains a major challenge to the ceramics processor.

ACKNOWLEDGMENT

The author thanks J. Rödel and E.P. Butler for their useful discussions and contributions to this article. Financial support from the United States Air Force Office of Scientific Research and the E.I. DuPont de Nemours & Co. Inc. is gratefully acknowledged.

REFERENCES

1. R.L. Coble and J.E. Burke, Sintering in Ceramics, *Prog. Ceram. Sci.*, Vol 3, 1963, p 197–251
2. R.J. Brook, Controlled Grain Growth, *Ceramic Fabrication Processes*, Vol 9, *Treatise in Materials Science and Tech-*

nology, F.F.Y. Wang, Ed., Academic Press, 1976, p 331–364
3. M.F. Yan, R.M. Cannon, and H.K. Bowen, Grain Boundary Migration in Ceramics, *Ceramic Microstructures '76*, R.M. Fulrath and J.A. Pask, Ed., Westview Press, 1977, p 276–307
4. M.F. Yan, Microstructural Control in the Processing of Electronic Ceramics, *Mater. Sci. Eng.*, Vol 48, 1981, p 53–72
5. P. Chantikul, S.J. Bennison, and B.R. Lawn, Role of Grain Size in the Strength and R-Curve Properties of Alumina, *J. Am. Ceram. Soc.*, Vol 73 (No. 8), 1990, p 2419–2427
6. H.C. Graham, N.M. Tallen, and K.S. Mazdiyasni, Electrical Properties of High-Purity Polycrystalline Barium Titanate, *J. Am. Ceram. Soc.*, Vol 54 (No. 11), 1971, p 548–553
7. R.J. Brook, Fabrication Principles for the Production of Ceramics with Superior Mechanical Properties, *Proc. Brit. Ceram. Soc.*, Vol 32, 1982, p 7–24
8. M.P. Harmer, Use of Solid-Solution Additives in Ceramic Processing, *Structure and Properties of MgO and Al$_2$O$_3$ Ceramics*, Vol 10, *Advances in Ceramics*, W.D. Kingery, Ed., The American Ceramic Society, 1984, p 679–696
9. M.P. Harmer, Science of Sintering as Related to Ceramic Powder Processing, *Ceramic Powder Science*, Vol 1B, *Ceramic Transactions*, G.L. Messing, E.R. Fuller, Jr., and H. Hausner, Ed., The American Ceramic Society, 1988, p 824–839
10. S.J. Bennison and M.P. Harmer, Microstructural Characterization of Abnormal Grain Growth Development in Al$_2$O$_3$, *Advances in Materials Characterization*, Vol 15, *Materials Science Research*, D.A. Rossington, R.A. Condrate, and R.L. Snyder, Ed., Plenum Press, 1983, p 309–320
11. C.S. Smith, Grains, Phases and Interfaces: An Interpretation of Microstructure, *Trans. AIME*, Vol 175 (No. 1), 1948, p 15–51
12. J.E. Burke and D. Turnbull, Recrystallization and Grain Growth in Metals, *Prog. Metall. Phys.*, Vol 3, 1952, p 220–292
13. M. Hillert, Theory of Normal and Abnormal Grain Growth, *Acta Metall.*, Vol 13 (No. 3), 1965, p 227–238
14. P.G. Shewmon, Chapt. 4–3, *Diffusion in Solids*, McGraw-Hill, 1963
15. W.D. Kingery, Plausible Concepts Necessary and Sufficient for Interpretation of Ceramic Grain Boundary Phenomena: II. Solute Segregation, Grain Boundary Diffusion and General Discussion, *J. Am. Ceram. Soc.*, Vol 57 (No. 2), 1974, p 74–83
16. J.W. Cahn, The Impurity-Drag Effect in Grain Boundary Motion, *Acta Metall.*, Vol 10 (No. 9), 1962, p 789–798

17. R.J. Brook, The Impurity-Drag Effect and Grain Growth Kinetics, *Scripta Metall.*, Vol 2 (No. 7), 1968, p 375–378

18. R.L. Coble, Sintering of Crystalline Solids—II. Experimental Test of Diffusion Models in Porous Compacts, *J. Appl. Phys.*, Vol 32 (No. 5), 1961, p 793–799

19. A. Mocellin and W.D. Kingery, Microstructural Changes During Heat Treatment of Sintered Alumina, *J. Am. Ceram. Soc.*, Vol 56 (No. 6), 1973, p 309–314

20. S.J. Bennison and M.P. Harmer, Effect of MgO Solute on the Kinetics of Grain Growth in Al$_2$O$_3$, *J. Am. Ceram. Soc.*, Vol 66 (No. 5), 1983, p C90–C92

21. S.J. Bennison and M.P. Harmer, Grain Growth Kinetics for Alumina in the Absence of a Liquid Phase, *J. Am. Ceram. Soc.*, Vol 68 (No. 1), 1985, p C22–C24

22. W.A. Kaysser, M. Sprissler, C.A. Handwerker, and J.E. Blendell, Effect of Liquid Phase on the Morphology of Grain Growth in Alumina, *J. Am. Ceram. Soc.*, Vol 70 (No. 5), 1987, p 339–343

23. J. Rödel and A.M. Glaeser, Anisotropy of Grain Growth in Alumina, *J. Am. Ceram. Soc.*, Vol 73 (No. 11), 1990, p 3292–3301

24. R.L. Coble, "Transparent Alumina and Method of Preparation," U.S. patent 3,026,210, Mar 1962

25. W.D. Kingery, Densification During Sintering in the Presence of a Liquid Phase. I. Theory, *J. Appl. Phys.*, Vol 30 (No. 3), 1959, p 301–306

26. D.S. Buist, B. Jackson, I.M. Stephenson, W.F. Ford, and J. White, The Kinetics of Grain Growth in Two-Phase (Solid-Liquid) Systems, *Trans. Brit. Ceram. Soc.*, Vol 64, 1965, p 173–209

27. G.W. Greenwood, Growth of Dispersed Precipitates in Solutions, *Acta Metall.*, Vol 4 (No. 4), 1956, p 243–248

28. C. Wagner, Theory of Precipitate Change by Redissolution, *Z. Electrochemie*, Vol 65 (No. 7–8), 1961, p 581–591

29. K.W. Lay, Grain Growth in UO$_2$-Al$_2$O$_3$ in the Presence of a Liquid Phase, *J. Am. Ceram. Soc.*, Vol 51 (No. 7), 1968, p 373–376

30. E.H. Lutz, N. Claussen, and M.V. Swain, K^R-Curve Behavior of Duplex Ceramics, *J. Am. Ceram. Soc.*, Vol 74 (No. 1), 1991, p 11–18

31. J.L Runyan and S.J. Bennison, Flaw-Tolerant Aluminum Titanate-Reinforced Alumina, *J. Eur. Ceram. Soc.*, 1991, in press

32. D.A. Porter and K.E. Easterling, Chapt. 3.3.5, *Phase Transformations in Metals and Alloys*, Van Nostrand Reinhold, Wokingham, England, 1984

33. C. Zener, private communication to C.S. Smith, 1948, see Ref 11

34. M.V. Speight, Growth Kinetics of Grain-Boundary Precipitates, *Acta Metall.*, Vol 16 (No. 1), 1968, p 133–135

35. W.H. Rhodes, Controlled Transient Solid Second-Phase Sintering of Yttria, *J. Am. Ceram. Soc.*, Vol 64 (No. 1), 1981, p 13–19

36. J.D. French, M.P. Harmer, H.M. Chan, and G.A. Miller, Coarsening-Resistant Dual-Phase Interpenetrating Microstructures, *J. Am. Ceram. Soc.*, Vol 73 (No. 8), 1990, p 2508–2510

37. R.J. Brook, Pore-Grain Boundary Interactions, *J. Am. Ceram. Soc.*, Vol 52 (No. 1), 1969, p 56–67

38. A.G. Evans, Structural Reliability: A Processing Dependent Phenomenon, *J. Am. Ceram. Soc.*, Vol 65 (No. 3), 1982, p 127–137

39. J.F. Shackleford and W.D. Scott, Relative Energies of {1100} Tilt Boundaries in Aluminum Oxide, *J. Am. Ceram. Soc.*, Vol 51 (No. 12), 1968, p 688–692

40. K.T. Aust and J.W. Rutter, Grain Boundary Migration in High-Purity Lead and Dilute Lead-Tin Alloys, *Trans. AIME*, Vol 215, 1959, p 119–127

41. H. Glieter, Theory of Grain Boundary Migration Rate, *Acta Metall.*, Vol 17 (No. 7), 1969, p 853–862

42. M.P. Anderson, D.J. Srolovitz, G.S. Grest, and P.S. Sahni, Computer Simulation of Grain Growth—I. Kinetics, *Acta Metall.*, Vol 32 (No. 5), 1984, p 783–791

43. D.J. Srolovitz, M.P. Anderson, P.S. Sahni, and G.S. Grest, Computer Simulation of Grain Growth—II. Grain Size Distribution, Topology and Local Dynamics, *Acta Metall.*, Vol 32 (No. 5), 1984, p 793–802

44. C.A. Bateman, S.J. Bennison, and M.P. Harmer, A Mechanism for the Role of MgO in the Sintering of Al$_2$O$_3$ Containing Small Amounts of a Liquid Phase, *J. Am. Ceram. Soc.*, Vol 72 (No. 7), 1989, p 1241–1244

45. R.E. Mistler and R.L. Coble, Microstructural Variation due to Fabrication, *J. Am. Ceram. Soc.*, Vol 51 (No. 4), 1968, p 237

46. T.J. Carbone and J.S. Reed, Microstructure Development in Barium Titanate: Effects of Physical and Chemical Inhomogeneities, *Am. Ceram. Soc. Bull.*, Vol 58 (No. 5), 1979, p 512–517

47. A.M. Glaeser, H.K. Bowen, and R.M. Cannon, Grain Boundary Migration in LiF: II, Microstructural Characteristics, *J. Am. Ceram. Soc.*, Vol 69 (No. 4), 1986, p 299–309

48. K. Uematsu, R.M. Cannon, R.D. Bagley, M.F. Yan, U. Chowdhry, and H.K. Bowen, Microstructure Evolution Controlled by Dopants and Pores at Grain Boundaries, *Proceedings of the International Symposium of Factors in Densification and Sintering of Oxide and Non-Oxide Ceramics*, S. Somiya and S. Saito, Ed., Gakujutsu Bunken Fukyu-Kai, Tokyo, 1979, p 190–205

49. M.F. Ashby, A First Report on Sintering Diagrams, *Acta Metall.*, Vol 22 (No. 3), 1974, p 275–289

50. W.D. Kingery and B. Francois, Sintering of Crystalline Solids: I—Interactions Between Grain Boundaries and Pores, *Sintering and Related Phenomena*, G.C. Kuczynski, N.A. Hooton, and G.F. Gibbon, Ed., Gordon Breach, 1967, p 471–498

51. F.F. Lange, Sinterability of Agglomerated Powders, *J. Am. Ceram. Soc.*, Vol 67 (No. 2), 1984, p 83–89

52. R.M. Cannon and W.C. Carter

53. R.M. Cannon and R.L. Coble, Review of Diffusional Creep of Al$_2$O$_3$, *Deformation of Ceramic Materials*, R.C. Bradt and R.E. Tressler, Ed., Plenum Press, 1975, p 61–100

54. J. Zhao, "Effects of Pore Distribution on Microstructure Development," Ph.D. Dissertation, Lehigh University, Bethlehem, PA, 1990

55. M.P. Harmer, E.W. Roberts, and R.J. Brook, Rapid Sintering of Pure and Doped α-Al$_2$O$_3$, *Trans. Brit. Ceram. Soc.*, Vol 78 (No. 1), 1979, p 22–25

56. M.P. Harmer and R.J. Brook, Fast Firing: Microstructural Benefits, *Trans. J. Brit. Ceram. Soc.*, Vol 80 (No. 5), 1981, p 147–148

57. H. Mostagachi and R.J. Brook, Fast-Firing of Non-Stoichiometric BaTiO$_3$, *Trans. J. Brit. Ceram. Soc.*, Vol 80 (No. 5), 1981, p 148–149

58. I.W. Jones and L.J. Miles, Production of ß-Al$_2$O$_3$ Electrolyte, *Proc. Brit. Ceram. Soc.*, Vol 19 (No. 3), 1971, p 161–178

59. R.L. Coble, Diffusion Models for Hot Pressing with Surface Energy and Pressure Effects as Driving Forces, *J. Appl. Phys.*, Vol 41 (No. 12), 1970, p 4798–4807

60. M.H. Leipold, Hot-Pressing, *Ceramic Fabrication Processes*, Vol 9, *Treatise in Materials Science and Technology*, F.F.Y. Wang, Ed., Academic Press, 1976, p 95

61. HIP review

62. M.P. Harmer and R.J. Brook, The Effect of MgO Additions on the Kinetics of Hot-Pressing in Al$_2$O$_3$, *J. Mater. Sci.*, Vol 15 (No. 12), 1980, p 3017–3024

63. S.J. Bennison and M.P. Harmer, Effect of Magnesia Solute on Surface Diffusion in Sapphire and the Role of Magnesia in the Sintering of Alumina, *J. Am. Ceram. Soc.*, Vol 73 (No. 4), 1990, p 833–837

64. S.J. Bennison and M.P. Harmer, A History of the Role of MgO in the Sintering of α-Al$_2$O$_3$, *Sintering of Advanced Ceramics*, Vol 7, *Ceramic Transactions*, C.A. Handwerker, J.E. Blendell, and W.A. Kaysser, Ed., The American Ceramic Society, 1990, p 13–49

65. M.O. Warman and D.W. Budworth,

Criteria for the Selection of Additives to Enable the Sintering of Alumina to Proceed to Theoretical Density, *Trans. Brit. Ceram. Soc.*, Vol 6 (No. 6), 1967, p 253–264

66. S. Wu, E. Gilbart, and R.J. Brook, Solid-State Sintering: The Attainment of High Density, *Structure and Properties of MgO and Al₂O₃ Ceramics*, Vol 10, *Advances in Ceramics*, W.D. Kingery, Ed., The American Ceramic Society, 1984, p 574–582

67. S. Prochazka and R.M. Scanlan, Effect of Boron and Carbon on Sintering of SiC, *J. Am. Ceram. Soc.*, Vol 58 (No. 1–2), 1975, p 72

68. R.L. Coble, Sintering of Alumina: Effect of Atmospheres, *J. Am. Ceram. Soc.*, Vol 45 (No. 3), 1962, p 123–127

69. M.F. Yan and D.W. Johnson, Jr., *Mater. Sci. Res.*, Vol 11, 1978, p 393

Final Shaping and Surface Finishing

Co-Chairmen: K. Subramanian, Norton Company
Ross F. Firestone, The Ross Firestone Company

Introduction

THIS SECTION covers the final step in the production of ceramic parts before they are put to use. Ceramics are finished before final use to meet shape, size, finish or surface, and quality requirements. The extent of finishing depends upon the application. For example, bricks are finished when the setting sand is brushed off as they are removed from the kiln while ceramic turbine blades require extensive shaping which may cost as much or more than all the previous processing steps combined. Most ceramics fall in between these extremes. Traditional ceramics are trimmed and fettled after drying to remove seams, fins, and spares and sometimes glazed after firing to produce a smooth surface and increase their strength. Engineering ceramics are often shaped after forming and again after firing; they are also sometimes glazed and/or metallized. Advanced ceramics are rarely glazed but they often require extensive shaping when they are fully hard and dense since most are produced by processes (such as hot pressing) which combine forming and firing to give these ceramics maximum density and strength.

Machining is the most common method of shaping. It is the application of energy to the part to remove stock and create new surfaces which define a shape. Such energy transfer is usually accomplished through the use of abrasives. Machining processes utilizing abrasives are referred to as abrasive machining methods. Methods which result in such energy transfer without the use of abrasives are called non-abrasive machining methods.

Abrasive machining is performed on green (unfired), white (partially fired), and hard (fully dense) ceramic components. Abrasive machining and surface finishing methods include grinding, lapping, honing, polishing, abrasive fluid jet cutting, and ultrasonic machining.

Grinding methods used on ceramic components are the most versatile. They employ diamond abrasives held fixed in a grinding wheel and applied against the work surface in a variety of configurations.

Abrasives can also be applied to the work surface without being rigidly held in a grinding wheel. These methods are called free-abrasive machining. Some of these methods are:

- *Lapping*. Loose or bonded abrasives in a low-pressure, low-speed operation achieve high geometric accuracy, correct minor shape errors, improve surface finish, or provide tight fits between mating surfaces
- *Honing*. Resembles lapping in principle, but usually is reserved for finishing internal surfaces of cylindrical spherical parts
- *Polishing*. Chosen primarily to improve surface finish; employs loose abrasives of fine particle size and preselected hardness
- *Ultrasonic machining*. Selective erosion of component produced by imparting mechanical energy to abrasive particles suspended in liquid medium and excited to vibrate at ultrasonic frequencies
- *Water jet/abrasive water jet machining*. When a high-velocity liquid stream carrying abrasive particles erodes cavities of required shape or profile, the process is referred to as abrasive water jet, or abrasive fluid jet machining. Utilizing a high-pressure fluid stream without abrasive particles to create surfaces is called fluid jet machining

Diamond is the abrasive most widely used for ceramics. Conventional abrasives, however, are generally selected for lapping operations, with diamond powders and lapping slurries reserved for special applications. Also, green or white-state components are occasionally machined using conventional abrasives, carbide saws, or single-point cutting tools made of carbide, ceramic, or diamond.

Non-abrasive machining methods utilize a variety of means to transfer energy to the ceramic material surface. These include laser beam machining, electrical discharge machining, ion-beam machining, electron-beam machining, chemical machining, and electrochemical machining.

Both abrasive and non-abrasive machining methods are covered in this Section. Major emphasis is placed on abrasive machining methods since these are the most common processes. The Section begins with a discussion of the principles of abrasive machining, which provides critical insight into the mechanics of material removal associated with abrasive machining methods. Following this, abrasives are discussed. While diamond is the most common abrasive as well as the most cost effective, other abrasives are also considered. Production grinding methods and techniques are reviewed next with particular reference to ceramic materials and components. This is followed by articles on free abrasive machining, ultrasonic machining, and abrasive fluid jet machining.

The non-abrasive methods most widely used for machining are covered last. These include laser beam machining and electrical discharge machining. Other surface finishing techniques such as glazing, ceramic coatings, and metallizing are described elsewhere in this Handbook.

Principles of Abrasive Machining

K. Subramanian and S. Ramanath, Norton Company

SUCCESSFUL GRINDING OF CERAMICS requires that the initiation and propagation of brittle fracture be minimized. In addition, if the grinding operation occurs predominantly in the plastic mode, high strength, good surface finish, and complex geometries can be generated by the grinding of ceramics. This requires a careful attention to parameters that can be grouped into four categories:

- Machine tool
- Work material
- Wheel selection
- Operational factors

This integrated method to optimize the grinding process results is known as the systems approach. This approach can generate ceramic parts/surfaces considered impossible to machine until recently (for example, ceramic springs, mirror finished ceramics, thin wall tubing). These guidelines can also be applied in a variety of components ranging from electronic devices to traditional ceramics to advanced/structural ceramics. Recognition and application of these guidelines should lead to an orderly transition for engineers familiar with the grinding of metals to produce a quality part when grinding ceramics.

Grindability of Ceramics Versus Metals

Table 1 compares typical properties of selected ceramic materials with those of metals and a plastic.

The strength of ceramic materials varies widely, depending on the material chosen. Even for a given material, like silicon nitride, for example, the strength depends on the sintering aids used and the sintering methods (that is, pressureless sintering, hot pressing, hot isostatic pressing (HIP), and so on) applied. In this regard, the ceramic materials are analogous to metals in which the composition, microstructure, and strength influence their grindability.

In general, ceramic materials have higher stiffnesses (Young's modulus, E) than metals. This would imply that elastic deformation of ceramic materials during grinding would be less than the metals for the same applied grinding forces. This is indeed true, and as a result, ceramic materials can be machined to closer tolerances, more precise geometry, and better flatnesses and parallel-

ism than is possible for metals. This principle is the basis for choosing ceramic materials over metals in precision instrument parts, machine tool beds or ways, as well as for gage blocks.

Ceramics are generally more chemically stable than metals. Hence the burn sometimes observed on metals during grinding is rarely observed on ceramic materials. Furthermore, the hot hardness and recovery hardness of metals are much lower than those of ceramics. In other words, any thermal softening that aids in the grinding of metals can rarely be counted on in the grinding of ceramics.

Thermal conductivity of ceramic materials varies widely. This property determines the ability to conduct heat away from the grinding zone. Poor thermal conductivity in metals such as titanium and Inconel can lead to high temperatures and great difficulty in grinding. Similarly, it is more difficult to grind poorly conducting ceramics and the problem becomes worse when they are also poor in thermal shock resistance.

One characteristic that significantly distinguishes ceramics from metals is their low fracture toughness or the resistance to propagation of a preexisting crack or notch. With all the above similarities between metals and ceramics in mind, it would appear that it is possible to achieve successful grinding of ceramics if the generation and propagation of cracks during the grinding process can be minimized.

Ceramics include a wide range of materials with their thermophysical properties dependent on the type, composition, microstructure, and the processing methods. In this regard, ceramics are a family of materials, just as we recognize metals as a family of materials. Once we recognize this commonality, many of the principles applicable to abrasive finishing of metals—an industry with over 100 years of history—become applicable to the abrasive machining of ceramics. However, for the same properties considered (thermal conductivity, Young's modulus, high-temperature resistance, fracture toughness, and so on), metals and ceramics fall into distinctly different ranges. If we recognize these differences, then the principles of abrasive machining of metals can be modified for successful abrasive machining of ceramics.

Many competing technologies are being evaluated for finish machining of ceramic materials, including abrasive and nonabra-

sive machining processes. Grinding with diamond abrasives is a well-established abrasive machining process, extensively used for the finishing of electronic ceramics, such as magnetic heads, microchips, substrates, and so on (Ref 1, 2). These processes are also used effectively in the fabrication of ceramic cutting tools, ball bearings, engine components, and so on. A variety of other well-established ceramic applications such as quartz tubing, arc magnets, refractories, and so on also utilize diamond grinding processes extensively (Ref 3–5). However, precision grinding of ceramics may cause damage leading to strength degradation. This degradation in strength is associated with the mechanism of material removal dominated by the brittle fracture of ceramics (Ref 6). Recent developments indicate that this may not be the only mechanism of material removal in ceramics. However, it is a mechanism that must be and can be minimized to achieve high-strength ceramics of required surface quality and geometry (Ref 7).

Figure 1 shows a comparison of the hardnesses of work materials and the abrasives used to machine them. It is observed that ceramic work materials are of comparable hardness to conventional abrasives such as alumina or silicon carbide. Diamond abrasives, because of their hardness and wear resistance, are the abrasive of choice for the surface finishing of ceramics. Table 2 summarizes the properties of selected abrasives. Additional information is available in the article "Abrasives" in this Volume.

Interactions at the Grinding Zone

The interactions at the grinding zone may be characterized as shown in Fig 2. As the grinding wheel is applied to the work material at a given wheel speed, work speed, and feedrate, there are generally four types of surface interactions taking place at the grinding zone where the interference between the grinding wheel and the workpiece occurs. The abrasive grain interferes with the work material at the stress, strain, and strain rate conditions dependent on the grinding process variables, abrasive geometry, and work material properties. Simultaneously, there are three frictional interactions due to the rubbing of:

Table 1 Physical and mechanical properties of selected ceramics, metals, and a polymer

| | Physical properties | | | | | | Mechanical properties | | | | | |
| | Density | | Melting/ decomposition temperature | | Thermal conductivity, k | | Coefficient of thermal expansion, 10^{-6}/K | Tensile strength | | Modulus of elasticity, E | | Fracture toughness | |
Material	g/cm³	lb/in.³	°C	°F	W/m·K	Btu·in./ft²·h·°F		MPa	ksi	GPa	10^6 psi	MPa$\sqrt{m}$	ksi$\sqrt{in.}$
Ceramics													
Aluminum oxide	3.4–4.0	0.12–0.14	2050	3720	27	190	8	205–550	30–80	385	55	2.0–3.0	1.8–2.7
Silicon carbide	3.0–3.2	0.11–0.12	2500	4530	63–155	435–1075	4–5	415–550	60–80	420	60	2.5–3.5	2.3–3.2
Silicon nitride	3.2–3.5	0.12–0.13	1900	3450	9–30	60–210	3	620–1100	90–160	305	44	3.5–5.0	3.2–4.6
Zirconium oxide	5.8	0.21	2500	4530	2	14	9–10	965–1380	140–200	140	20	6.0–8.0	5.5–7.3
Ferrite	5.0–6.0	0.18–0.22	2000	3630	8	55	7.5	140–170	20–25	205	29	1.0–1.3	0.9–1.2
Fused quartz	2.2	0.079	1670	3040	1.4	9.7	0.55	110	16	77	11	1.0	0.9
Metals													
Inconel 718	8.1	0.29	1400	2550	12.4	86.0	7	1240–1405	180–204	215	31	24	22
Tool steel	7.84	0.283	1500	2730	52	360	7	1405–2000	180–290	210	30	98	89
52100 bearing steel	7.85	0.284	1500	2730	30–40	210–280	11	415–550	60–80	210	30	56	51
Aluminum	2.2	0.079	660	1220	204	1420	13	275–550	40–80	70	10	36	33
Polymer													
Polyethylene	1.1	0.040	120	250	0.32	2.2	167	7–20	1–3	3.5	0.5	0.25	0.23

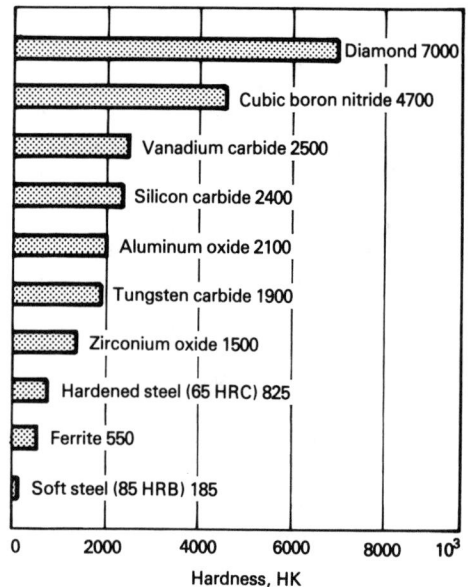

Fig 1 Typical hardness values of abrasive materials and some work materials

- Chips produced, against the bond matrix of the wheel
- Bond matrix in the wheel against the work material
- Chips (entrained in the wheel) against the work material

The influence of coolants should also be included in these interactions. The grinding process differs from the cutting process in two key areas:

- Chip-bond interaction and the bond-work interaction are absent in single or multiple-point cutting processes such as turning, milling, and so on
- Cutting geometry of the abrasive constantly changes in the grinding processes

Table 2 Typical physical and thermal properties of selected abrasives

| | | Density | | Relative thermal conductivity | Coefficient of thermal expansion, 10^{-6}/K | Threshold temperature for degradation | |
Abrasive	Chemical composition	g/cm³	lb/in.³			°C	°F
Diamond	C	3.52	0.127	100–350	4.8	800	1470
Aluminum oxide	Al₂O₃	3.92	0.142	1	8.6	1750	3180
Silicon carbide	SiC	3.21	0.116	10	4.5	1500	2730

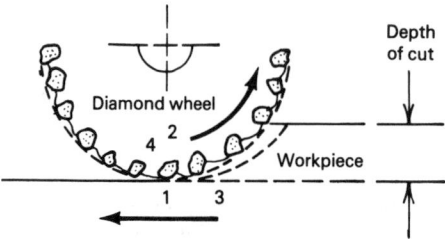

Fig 2 Schematic illustrating interactions in the grinding zone of a grinding wheel/workpiece interface. 1, abrasive/work interface; 2, chip/bond interface; 3, chip/work interface; 4, bond/work interface. Source: Ref 8

An efficient grinding process attempts to maximize the abrasive/work interaction, and to minimize the other three frictional interactions (chip/bond, chip/work, and bond/ work interfaces). The ability to preferentially control the abrasive/work interaction depends to a large degree on our understanding of the principles of grinding ceramics.

Effect of Grinding Direction on Ceramic Strength. Figure 3 shows the strength (modulus of resilience, MOR) of a hot pressed silicon nitride (HPSN) after it has been precision ground (Ref 9, 10). It is observed that the grinding direction has a significant influence on the strength of the material (Ref 11). As the abrasive grain size decreases, this anisotropy observed in the strength gradually

diminishes. Thus, it appears beneficial to grind with finer abrasive particles in order to obtain high retained strength as well as strength which is independent of the grinding direction.

Surface Finish and the Retained Strength. Figure 4 shows that the strength improvement with fine abrasive particles is also associated with an improvement in surface finish (Ref 12). However, improvement in surface finish alone is not adequate to improve the strength. There appears to be no improvement in strength with surface finish using coarser abrasive grits. This would suggest that the damage caused by coarser abrasive grains may not be removed simply by further burnishing of the work surface to achieve better surface finish. The finer abrasive grits, on the other hand, may not cause the damage to begin with, thus assuring high strength. The damage referred to here is predominantly crack generation and propagation through brittle fracture.

Effect of Abrasive Grain Size on Retained Strength. Figures 3 and 4 indicate that the finer the abrasive grain size, the better the strength of the ground ceramic and the better the surface finish. Table 3 shows the variation of surface finish as a function of grit size and other grinding parameters. Tests have shown that surface finish for alumina ceramic is generally unchanged by the depth of cut or the table speed when coarse abrasive grain is

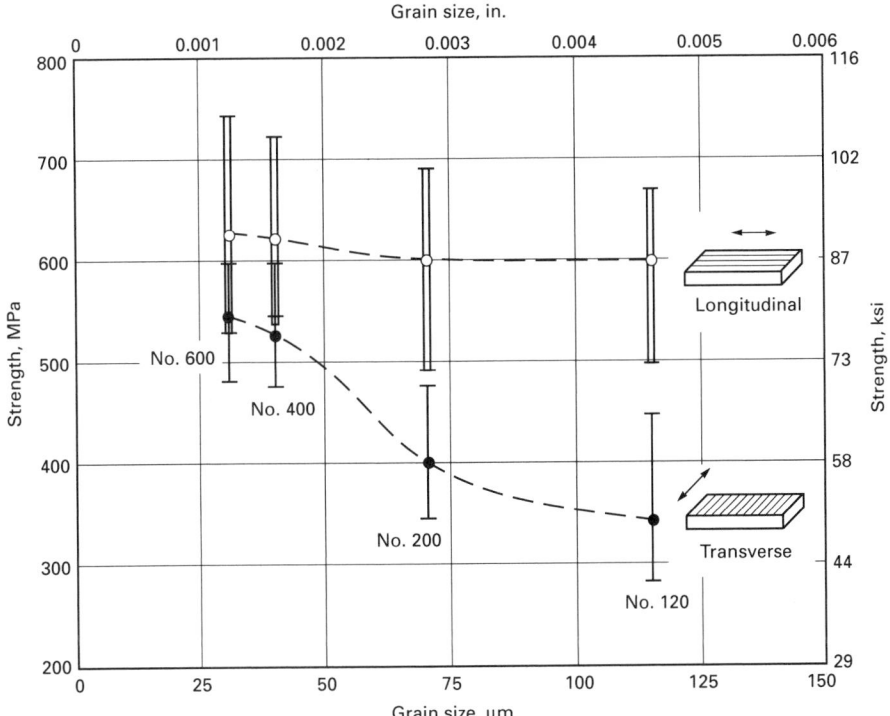

Fig 3 Plot of strength versus grain size of ground ceramic material as a function of grinding direction. Source: Ref 9

used. However, the finish improves gradually as the grit size is made finer.

Figure 5 shows the specific energy (energy required per unit volume of material removed)

required in the grinding of Sialon material as a function of the abrasive grain size. It is observed that as the grain size decreases, the specific energy increases rather dramatically.

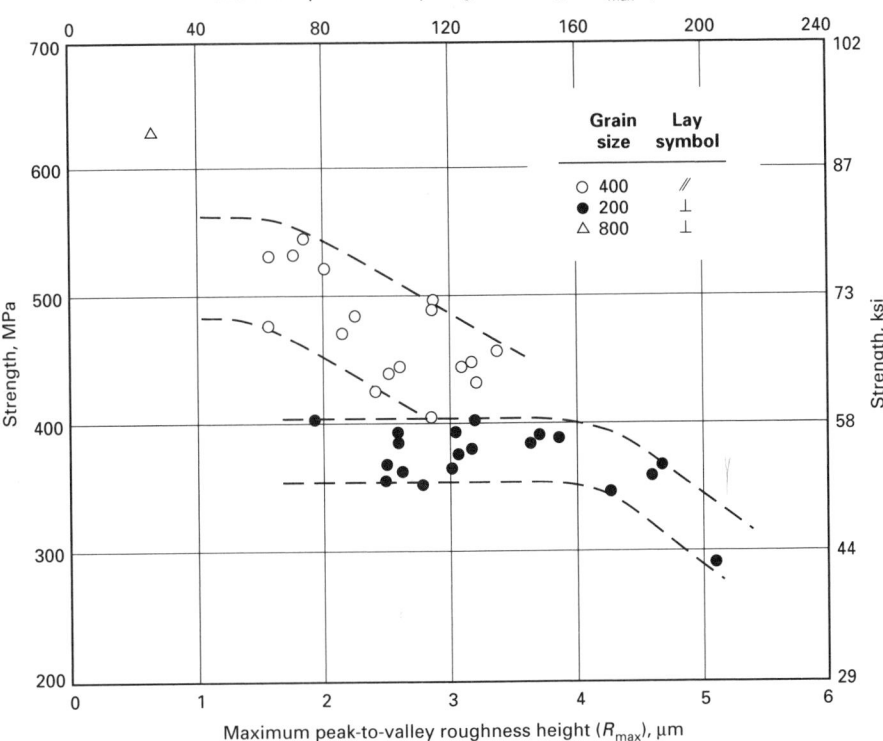

Fig 4 Plot of strength versus surface finish of ground ceramic material as a function of abrasive grain size. Source: Ref 9

Figure 6 shows a similar effect of abrasive grain size on the total grinding forces.

Material Removal Mechanism in the Grinding of Ceramics

Indentation Fracture Mechanism. A commonly used model to describe the process of surface generation in ceramics is referred to as the indentation fracture model (Fig 7) (Ref 14–16). In this model, it is assumed that the grain acts like an indenter, which under a normal load (f_n) initiates a large median crack. Associated with this are lateral or vent cracks, which when propagated back to the surface, remove or lift a piece of material off the work surface. During such a brittle fracture-dominated process, the surface finish can be conceived to be independent of grinding process parameters (Ref 17–19).

This model serves to explain the beneficial effect of fine abrasive particles to enhance the strength of the ground ceramic material. Figure 8 shows the calculated normal force/ abrasive grain as a function of abrasive grain size. It is observed that the normal force/grain decreases significantly with grit size. This could account for the decrease in the size of the median cracks, thus leading to the higher strength of the ceramic material. While this model seems to simplify a rather complicated problem, there are several limitations of this model (Ref 21, 22).

Ductile Regime Grinding Model. It was noted (see the section "Indentation Fracture Mechanism" in this article) that the force/grain decreases with decrease in abrasive grain size. However, if this force is compared to the volume of material removed, there appears to be a disproportionate increase in their value. Similarly, the grinding energy or power required per unit volume of material removed also appears to increase significantly as the abrasive particle size or chip volume removed per abrasive grain decreases. These increases in force and energy with the decrease in grit size, depth of cut, and material removal rate are remarkable and more than required for a purely brittle fracture process (that is, the force is higher than that which would cause lateral cracks).

In an attempt to explain the higher grinding forces and energies associated with a smaller depth of cut, Miyashita (Ref 23) has designated this process as ductile regime grinding. In this process, the abrasive grinding wheel is trued to a high degree of accuracy. This wheel, when mounted on a very stiff and high-precision spindle of a machine tool, can feed at very small increments, resulting in very small chip thickness. Under such conditions, the ceramic material is proposed to be deformed at highly plastic conditions, leading to ductile regime grinding. This is a good approach because it recognizes

Table 3 Surface finish (roughness average, R_a) as a function of grinding parameters

Coolant	Wheel	Table speed mm/min	in./min	0.0025 mm (0.0001 in.) downfeed μm	μin.	0.0050 mm (0.0002 in.) downfeed μm	μin.	0.013 mm (0.0005 in.) downfeed μm	μin.	0.025 mm (0.001 in.) downfeed μm	μin.	0.050 mm (0.002 in.) downfeed μm	μin.
100% oil	D150-N100	355	14	...	...	1.3	52	1.08	43	1.00	40	...	...
		710	28	...	...	0.95	38	1.03	41	1.08	43	0.95	38
		1270	50	...	...	1.3	52	1.03	41	1.03	41	1.00	40
2.5% oil in water	SD320-R100B95	1145	45	...	...	0.86	34.5	...	...	...	...	...	...
	SD320-R100B95 (after spark out)	1145	45	...	...	0.81	32.5	...	...	...	...	...	...
	SD20/30μ-R100B69	1145	45	...	...	0.45	18	...	...	...	...	...	...
	SD20/30μ-R100B69 (after spark out)	1145	45	...	...	0.40	16	...	...	...	...	...	...
	SD4/8μ-R100B69	1145	45	0.24	9.5	...	...	...	...	...	...	...	...
	SD4/8μ-R100B69 (after spark out)	1145	45	0.18	7	...	...	...	...	...	...	...	...

the possibility of plastic deformation in the grinding of ceramic materials. The limitation of this approach is the primary dependence on the depth of cut as the controlling factor for plastic deformation during grinding. Even so, the coarse abrasive grains could leave a pattern of microcracks on an otherwise plastically deformed surface due to their dull cutting action and excessive friction during grinding.

Chip Formation Model for Precision Grinding of Ceramics (Ref 20). It is possible to treat hard ceramic materials like any other material subjected to machining or grinding processes. Under such processes, shear deformation at the cutting or grinding zone produces the chip (Fig 9). A sequential removal of chips leads to the generation of the machined surface. The nature of the chips produced (that is, continuous or discontinuous) with varying degrees of plastic deformation, will depend on the cutting tool geometry, depth of cut, cutting velocity, work material properties, and so on. This is the classical cutting model proposed by Merchant, *et al.* (Ref 24), and successfully used in metal grinding (Ref 25–28).

Such a model would suggest that materials of high strength and fracture toughness, such as ZrO_2 and hot pressed silicon nitride, would exhibit greater plastic deformation during grinding. On the other hand, low strength and low fracture toughness materials (for example, ferrite), could produce a large degree of discontinuous brittle fractured chips. It would also be reasonable to state that grinding conditions that remove a large volume of material per grain (such as with a large depth of cut and coarse abrasive grains) are likely to result in brittle fracture due to the increased probability for the presence of a large number of defects or brittle fracture sites in the volume removed. Conversely, dense ceramics with low flaw population are likely to resist brittle fracture, particularly as the size of the chip decreases. In the absence of brittle fracture, these hard ceramics are likely to offer large resistance to deformation, possibly accounting for their higher grinding forces and the energy required as the grit size or depth of cut decreases.

In addition to the work material properties, the degree of plastic deformation in a cutting or chip formation model will also depend on the geometry of the cutting tool, abrasive grain size, and the depth of cut per grain or chip thickness. There are several literature references that discuss potential localized deformation in small sections of brittle work materials caused by small indenter forces and dependent on the indenter geometry. It has been suggested that it is impossible to fracture brittle materials at small sizes because they cannot store enough elastic energy to propel a crack through the particle before plastic flow takes place (Ref 29). The exis-

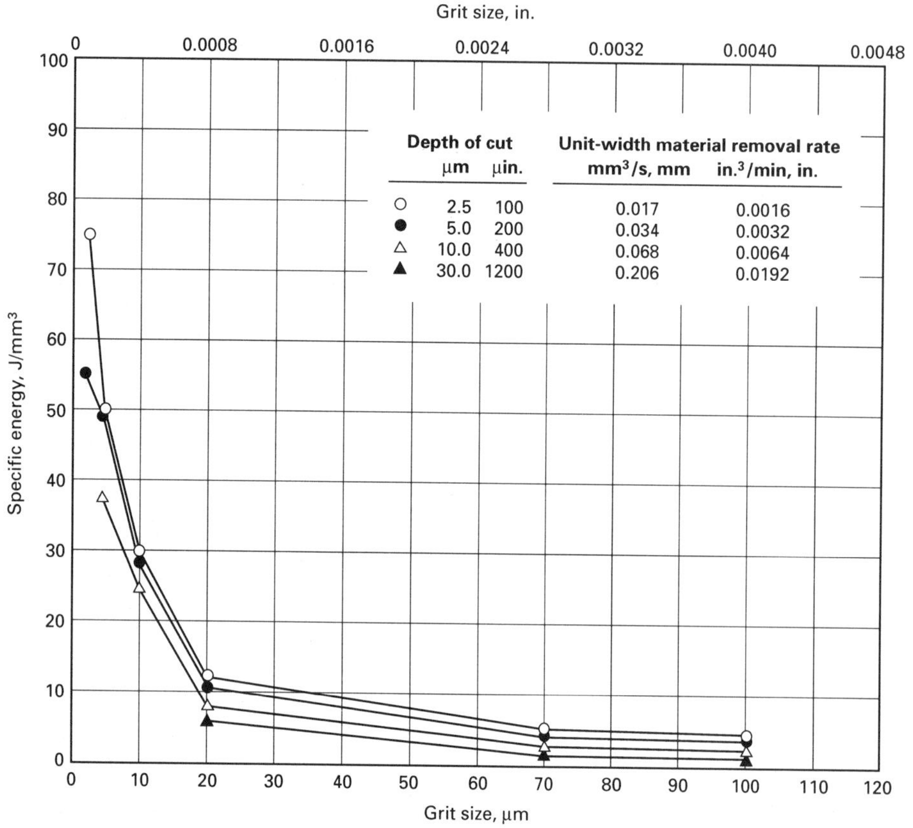

	Depth of cut μm	μin.	Unit-width material removal rate mm³/s, mm	in.³/min, in.
○	2.5	100	0.017	0.0016
●	5.0	200	0.034	0.0032
△	10.0	400	0.068	0.0064
▲	30.0	1200	0.206	0.0192

Fig 5 Plot of specific energy versus grit size of ground Sialon ceramic as a function of depth of cut and unit-width material removal rate. Source: Ref 13

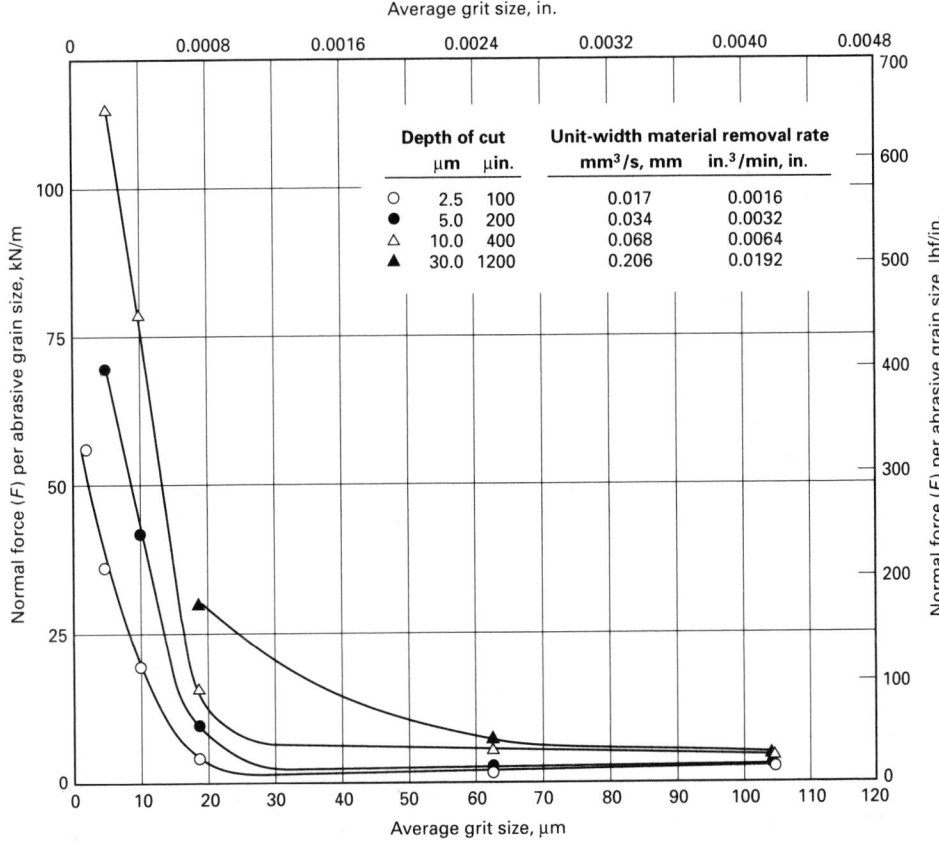

Fig 6 Plot of (normal force/abrasive grain size) versus average grit size of ground Sialon ceramic as a function of depth of cut and unit-width material removal rate. Source: Ref 13

	Depth of cut		Unit-width material removal rate	
	μm	μin.	mm³/s, mm	in.³/min, in.
○	2.5	100	0.017	0.0016
●	5.0	200	0.034	0.0032
△	10.0	400	0.068	0.0064
▲	30.0	1200	0.206	0.0192

cutting/chip formation model, there is always a degree of plastic deformation associated with the grinding of brittle or ceramic materials. The nature and extent of this plastic deformation may depend on several grinding process input variables. A systematic understanding of these interactions will be the key to achieving high retained strength in ceramic components of complex geometry to the desired level of tolerances and surface quality.

Chip Formation and Surface Generation. Figure 11 shows a low magnification

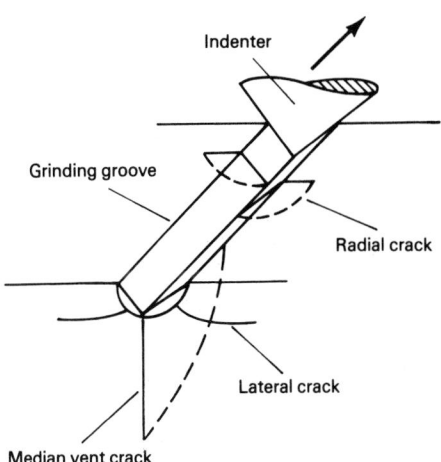

Fig 7 Indentation fracture model of the ceramic grinding process. Source: Ref 16

tence of a critical depth and a critical load per grain, below which plastic deformation occurs in preference to elastic loading and brittle manufacture, has also been observed (Ref 30).

The modes of localized deformation and fracture due to indentation are shown in Fig 10 (Ref 30). Under lightly loaded conditions and a small radius of the indenter, plastic deformation would be promoted in the brittle material subjected to the indentation force. Under a large radius of the indenter, the so-called cone crack is initiated. Thus, even in situations where ductile regime grinding (see the section "Ductile Regime Grinding Model" in this article) may be promoted by a precisely trued wheel, using an ultrasmall depth of grinding, it is possible that such deformation may be accompanied by surface cracks caused by the large apparent radius of the indenter (abrasive grain). However, it may be possible to obtain plastically deformed surfaces without initiation of surface cracks in the high-density high-strength ceramics if a small indenter radius and small force/grit conditions are applied during grinding.

Studies of residual stresses after grinding indicate the presence of compressive residual stress for a variety of ceramic materials (Ref 31, 32). Such residual stress is present even in the grinding with coarse abrasive grains (180 grit, 91 μm, or 0.0036 in.), and is dependent on the work material. Consistent with the

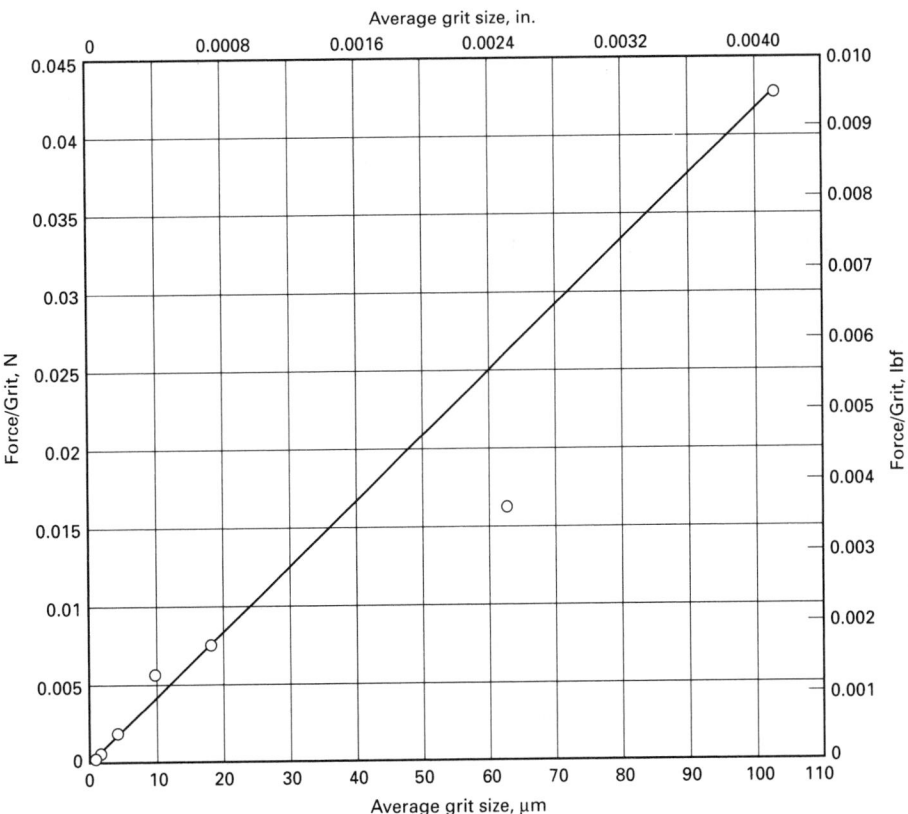

Fig 8 Plot of force/grit versus average grit size for the plunge grinding of Sialon ceramic using a 200 × 5 × 32 mm (8 × 0.2 × 1.25 in.) resin bonded grinding wheel with a 150 concentration. Depth of cut was 0.005 mm (0.0002 in.). Source: Ref 20

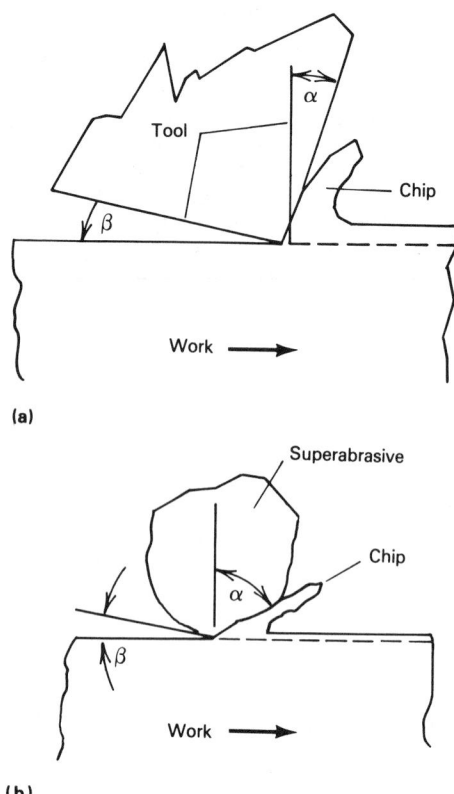

(a)

(b)

Fig 9 Schematic of the cutting or chip formation model generated by: (a) single-point cutting tool; (b) superabrasive. Clearance angle, β; rake angle, α

picture of the chips collected in the grinding of ceramics and steel. The average chip size for ceramics is of the order of 1 to 10 μm (40 to 400 μin.). This is substantially smaller than the chips produced in grinding steel, which may be of the order of 100 μm (0.004 in.) in size.

Figures 12(a), 12(b), and 12(c) compare the chips produced in ferrite, alumina, and zirconia, respectively. As the fracture toughness increases, the nature of the chips produced changes from purely brittle fracture to a combination of brittle fracture and plastic deformation. The extent of plastic deformation would seem to increase with the fracture toughness of the material.

A similar conclusion may also be drawn from Fig 13, which compares the surface finish achieved in these four materials. The extent of plastic deformation is further enhanced by using finer grain sizes as noted in Fig 14.

Systems Approach for Abrasive Machining

Input variables to the grinding process are rather large in number and may be grouped into four broad categories:

- Machine tool
- Wheel specification
- Work material properties
- Operational factors

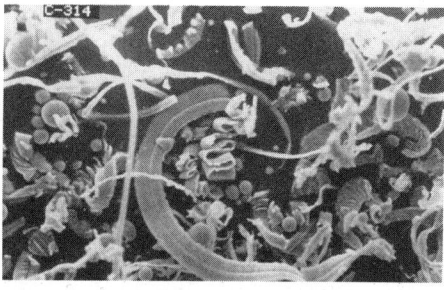

(a)

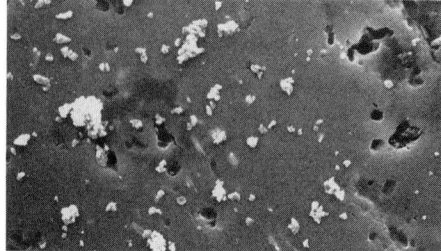

(b)

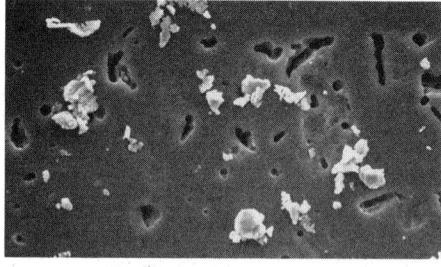

(c)

Fig 11 Comparison of typical chips produced by the grinding of metal and ceramics. (a) Steel. 435×. (b) Si_3N_4 ceramic. 580×. (c) ZrO_2 ceramic. 580×

The influence of these input variables, through the process variables of grinding forces and energy, result in the output of the grinding process (that is, part geometry, tolerances, retained strength, surface quality, and so on). This input/output model and its use to optimize the grinding process is called the systems approach (Ref 7, 27, 33).

A flowchart of the systems approach for precision production grinding is shown in Fig 15. This readily shows that the precision ground ceramic part is the output of a large number of variables. It is not simply a process of an abrasive grain scratching on the work material surface, as modeled in some instances (Ref 34). Figure 16 summarizes the typical grinding system inputs.

A detailed discussion of all the four input variables and each of the factors involved in the grinding system will be rather extensive. Hence, a few key factors and their relevance for ceramic grinding will be reviewed in the remainder of this article. Information that is specific to a given machine tool or production configuration is covered in the article

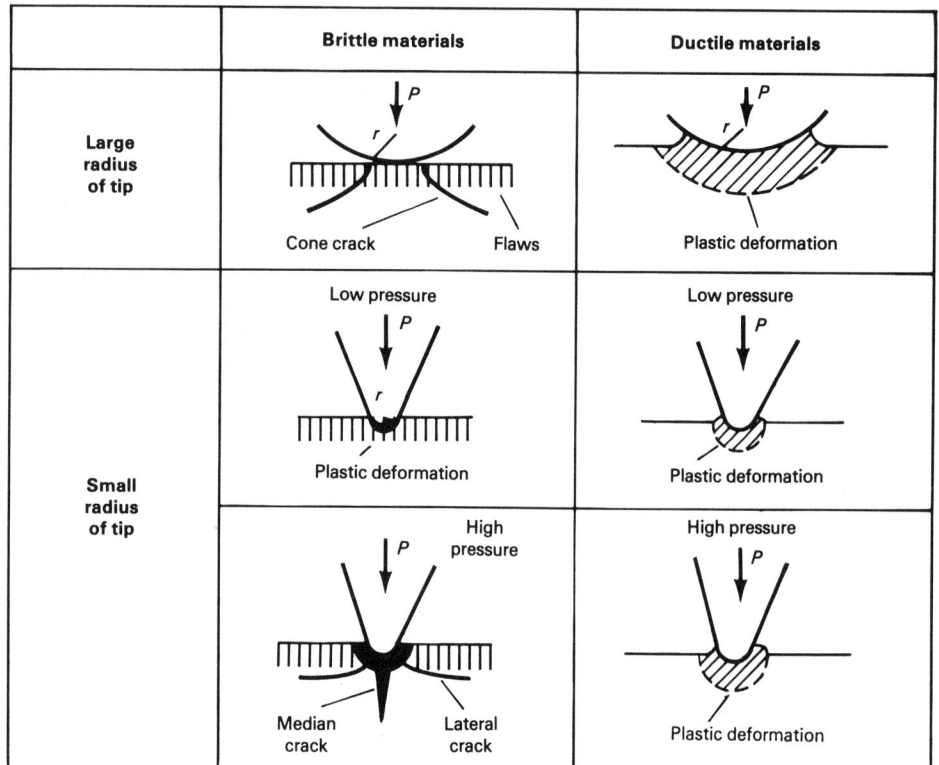

Fig 10 Localized deformation and fracture generated by an indenter in both brittle and ductile materials under varying conditions of indenter radii, *r*, and pressure, *P*. Source: Ref 30

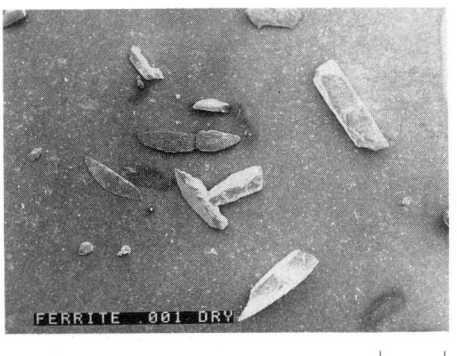

Fig 12(a) Photomicrographs of brittle chips produced with the dry grinding of ferrite ceramic; (left) low magnification; (right) high magnification. Source: Ref 20

Fig 12(c) Photomicrograph of plastically deformed chips produced when sintered ZrO_2 is wet ground with 180 grit abrasive. 5400×

Fig 12(b) Photomicrographs of chips produced when grinding 96% Al_2O_3 ceramic. Chips produced by predominately brittle fracture of material (left). Plastically deformed chips produced by grinding: (center) dry ground; (right) wet ground. 5400×

"Production Grinding Methods and Techniques" in this Volume.

Machine Tool Parameters (Ref 35)

The following machine tool parameters and operational factors predominantly determine the product quality of precision ground ceramic components:

- Rigidity/stiffness
- Vibration level
- Coolant systems
- Creep-feed grinding
- Precision movements and positioning
- On-machine dynamic balancing
- Truing and dressing systems
- Multiaxis CNC capability
- Materials handling systems

Rigidity/Stiffness. In general, grinding forces for dense ceramics are higher than those encountered in the grinding of carbides. Figure 17 shows a comparison of normal grinding forces for a few selected ceramic materials. Figure 17 shows that the grinding forces required for hot pressed silicon nitride are significantly higher than those encountered in

tungsten carbide grinding. This implies that in order for the grinding wheel to produce required straightness, flatness, or similar surface requirements, we need a spindle assembly of greater resistance to deflection. This higher stiffness is required in the spindle/wheel/work fixture/table assembly. In addition, this stiffness property is required at operating speeds (that is, dynamic stiffness) rather than static stiffness alone.

Vibration Level. In some respects, vibration level is a measure of stiffness. In addition, it is a measure of the damping characteristics in the spindle itself and the machine tool assembly as a unit. Low levels of vibration are critical to minimizing chippage for grinding of ceramics with thin sections or precision forms, especially for low-strength ceramics such as ferrite and/or low fracture toughness materials.

Coolant Systems. The direction, pressure, and flow of coolant applied are very critical in the grinding of ceramics. Contrary to commonly held views, oil in some instances may be preferable to water combined with rust inhibitor as a coolant such as normally used for ceramic grinding. This selection will

depend heavily on the work material properties as well as the grinding process conditions. The state-of-the-art coolant systems employed for production grinding of conventional materials will be equally valuable for production grinding of ceramics.

Figure 11 compares chips produced in grinding steel versus advanced ceramic materials. In general, the ceramic chips are of the order of 1 to 10 μm (40 to 400 μin.) compared to the 100 μm (0.004 in.) size of steel chips. These fine particles of ceramic chips with their low density tend to float rather easily in a coolant stream or a coolant tank. Hence, finer filters and novel methods of filtration such as centrifuging, staging, and so on will be necessary for the production grinding of ceramics.

Also, because of their fine size and flotation, ceramic chips may get carried into the ways and guides of the machine tool more easily than steel chips. Such entrainment could accelerate the wear of machine tool parts that are not adequately sealed.

Creep-Feed Grinding. Ceramic grinding processes will be driven towards lower grinding forces per abrasive grain to obtain max-

Fig 13 Photomicrographs of surfaces obtained after grinding selected ceramics with coarse abrasive grits. (a) ZrO_2. (b) Si_3N_4. (c) Al_2O_3. (d) Ferrite. All 2225×

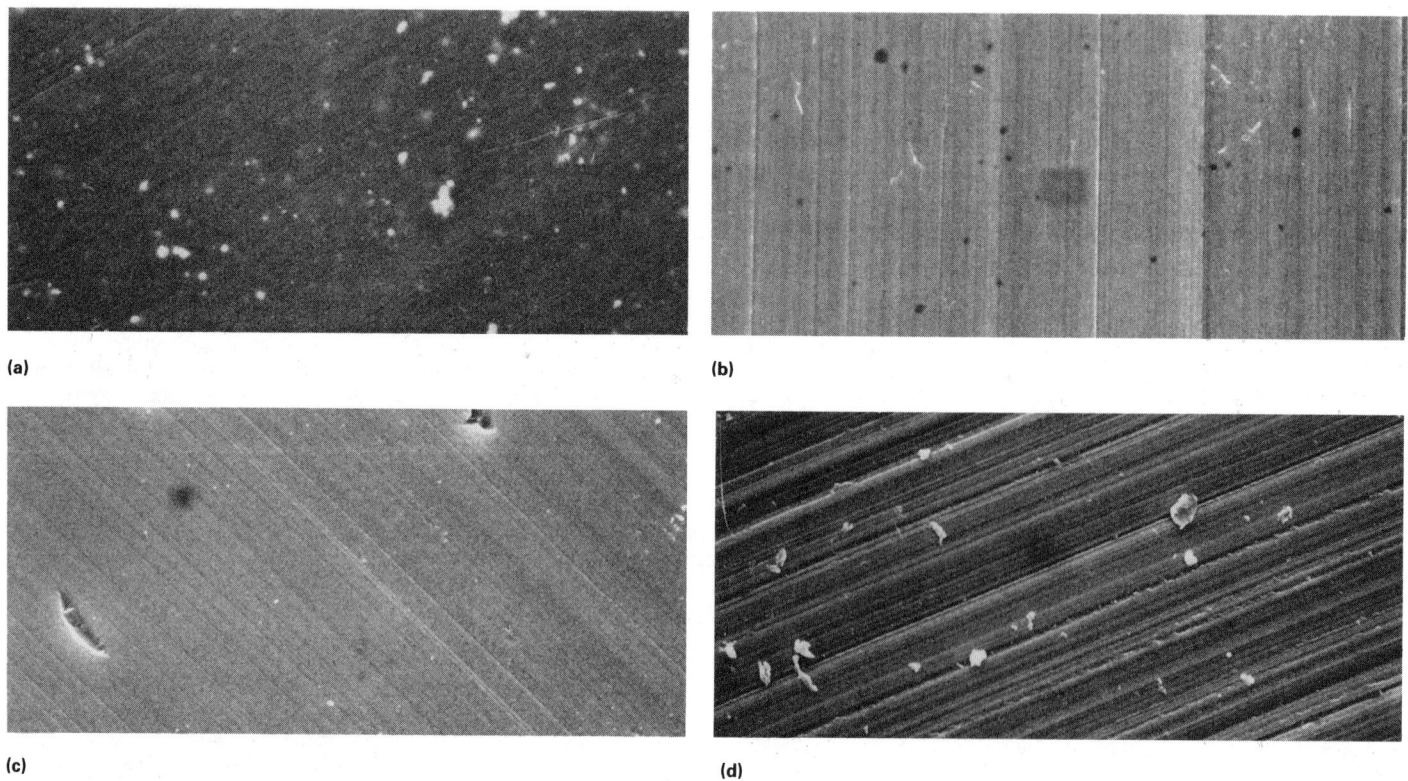

Fig 14 Photomicrographs of surfaces obtained after grinding selected ceramics with fine abrasive grits. (a) HPSN. (b) ZrO_2. (c) Al_2O_3. (d) Ferrite. All 2225×

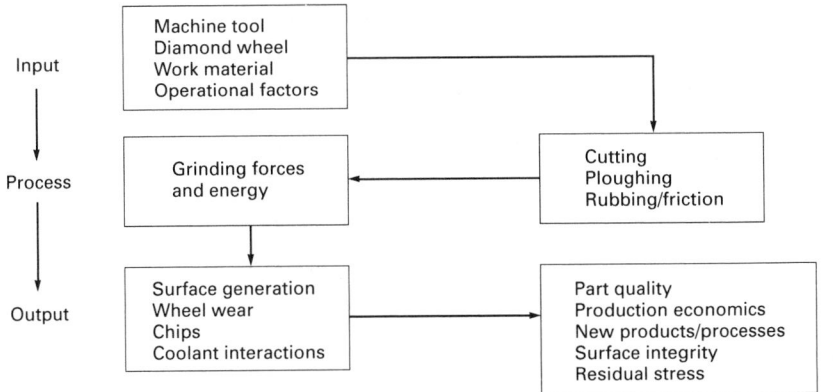

Input
Process
Output

Fig 15 Flowchart showing key parameters that comprise the systems approach to precision production grinding of ceramics

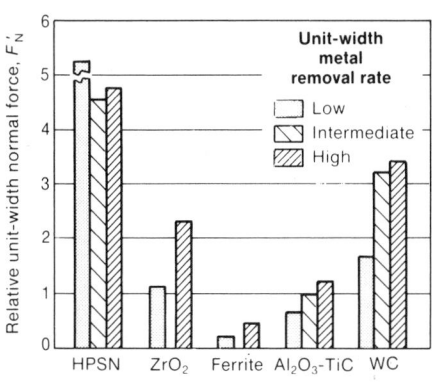

Fig 17 Relative unit-width normal force required to machine various structural and electronic ceramics. Unit-width material removal rates classified as low (2 mm³/s, mm; or 0.2 in.³/min, in.), medium (5 mm³/s, mm; or 0.5 in.³/min, in.), and high (10 mm³/s, mm; or 1.0 in.³/min, in.). Source: Ref 36

imum retained strength. A comparison between creep-feed and surface grinding of hot pressed silicon nitride, for example, reveals that while creep-feed process requires high total forces and power, the intensities of contact stress and power at the grinding zone are significantly lower (Ref 37). Creep-feed grinding processes generally utilize large depths of cut and very low work speeds. Machines capable of such creep-feed grinding in a variety of situations would find preferential use (if machining allowance is large enough) in the production grinding of ceramics.

Precision Movements and Positioning. Ceramic materials, because of their superior hardness and thermal stability, find applications where tolerances are much tighter than tolerances for metal parts. The surface finishes required in such ceramic parts will also be smoother compared to their metallic counterparts. Such precision tolerances and finish, in turn, will require machine tools capable of precision movements, with a high degree of repeatability, stability, and positioning accuracy (Ref 38, 39). These features of the machine tools will be enhanced by the proper selection and application of diamond wheels.

On-Machine Dynamic Balancing. As mentioned in the section "Rigidity/Stiffness" of this article, high static and dynamic stiffness as well as accurate vibration monitoring systems are critical for machine tools used in ceramic grinding applications. The grinding wheel is the final dynamic element of the machine tool system that contacts the work material. Hence, it is critical that grinding wheels operate at low levels of vibration. In many instances, this will require on-machine balancing systems that correct for any degree of imbalance while the grinding wheel is rotating at operating speeds (Ref 40, 41).

This on-machine balancing operation should be applied as part of the truing process prior to the use of the grinding wheel. In addition, such on-machine balancing is required periodically during the use of the grinding wheel in the grinding operation. Several such systems are in commercially available grind-

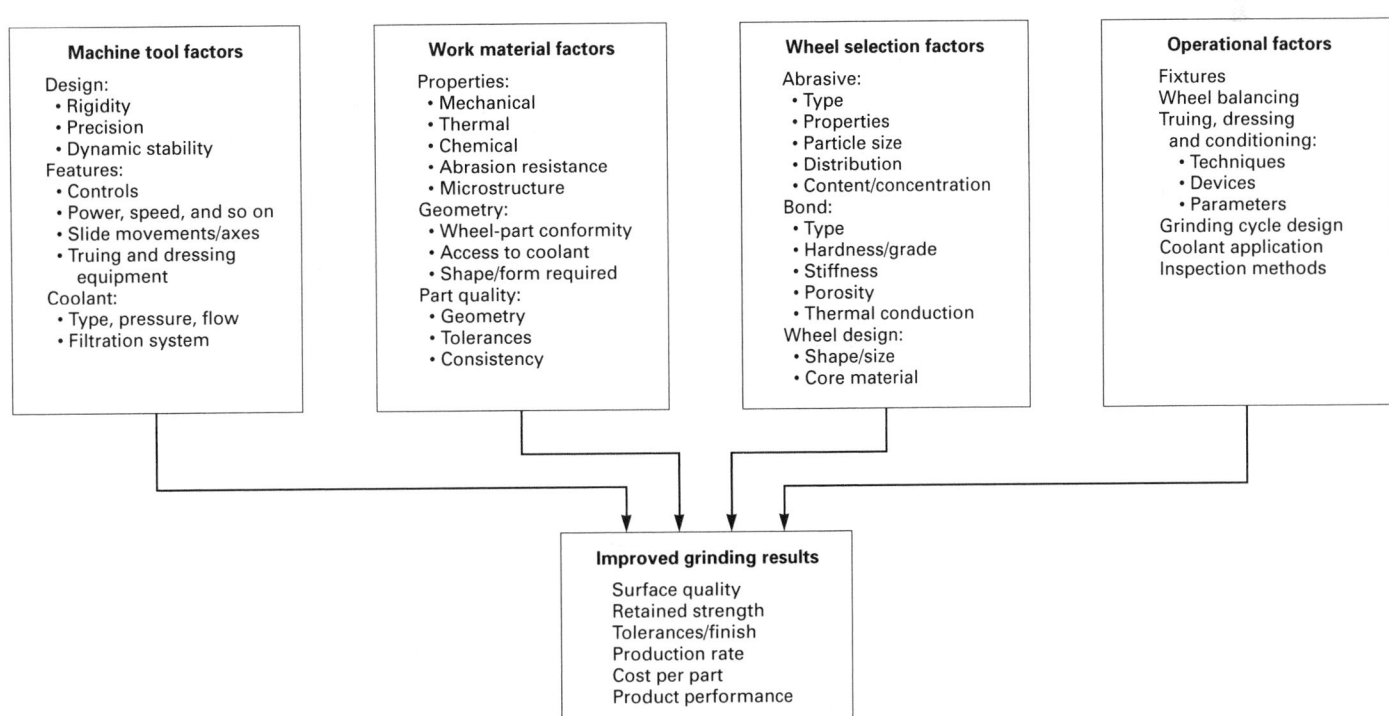

Fig 16 Selected parameters that affect the superabrasive grinding system

ing equipment used for electronic ceramics finishing operations.

Truing and Dressing Systems. Truing is the process of generating a concentric wheel face with accurate form or straightness as required (Fig 18a). Dressing is the process of exposing the diamond abrasives above the bond matrix for efficient grinding operation (Fig 18b).

Truing and dressing methods applied to diamond grinding wheels differ significantly from the methods used for conventional abrasive wheels. There are at least six important reasons for this difference:

- Diamond abrasives (being the hardest material known on earth) are difficult to cut or shape as required by the truing process
- The tools used for truing diamond wheels undergo rapid wear unless the truing system is properly designed and implemented. This rapid wear of the truing tool has serious implications in setting up automated production grinding cycles
- The amount of diamond wheel removed during the truing process should be minimized to make the production process cost effective. Conventional truing methods with silicon carbide wheels erode the bond matrix of the diamond wheel, thereby removing large wheel volume
- Diamond abrasive can be damaged if the truing process is harsh and carried out with high forces. Similarly, the resin bond matrix can be thermally damaged if the truing and dressing processes are not controlled

- The precision required in truing diamond wheels for ceramic grinding will be significantly tighter than the precision required in current practice for job shop uses of diamond wheels or for applications such as production grinding of glass or carbide parts
- Current production methods for steel grinding normally use vitrified bonded wheels. There will be situations where diamond wheels of resin or metal bond are likely to be used in addition to vitrified bonded wheels for production grinding of ceramics. This will be analogous to the cubic boron nitride (CBN) wheel bond types used for high-production grinding of steel parts. Such situations require dressing methods and equipment distinctly different from the truing equipment. Automation and consistency of such a dressing process will be key requirements

Truing and dressing equipment or devices that meet the above requirements will be a key requirement for the precision production grinding of ceramics. Integration of such equipment with the machine tool will be necessary for production viable grinding of ceramics.

Multiaxis CNC Capability. One of the key requirements of production grinding economics is the decrease in the total cost of grinding. Set-up time, machine-to-machine movement time, as well as in-process inventory costs contribute heavily to the total cost of fabrication. These costs can be significantly decreased if the part can be fabricated on one machine using a single set up. This

approach is being used successfully via the application of multiaxis CNC grinding systems specifically developed for use with CBN wheels to grind steel parts. Similar concepts and cycle design strategies are likely to find many uses in the precision production grinding of ceramics.

Materials Handling Systems. While ceramic materials can meet a wide range of applications, it is imperative that their potential for chippage be recognized throughout the production process until their eventual installation into a finished assembly. This necessitates a sequence of materials handling systems compatible with the ceramic material properties. In many respects, these operations are similar to the materials handling systems developed for the processing of electronic ceramic parts.

New Machine Tools Versus Retrofitting of Current Machines. A frequently asked question is: Can the machine tools used for ceramic grinding be adapted from existing machines or will a new generation of machine tools be required? Initially there will probably be a great deal of effort to adapt existing equipment. However, the level of machine tool developments or modifications required are rather involved. Hence, machine tools developed specifically for ceramics grinding may find greater success and acceptance for precision production grinding of ceramics.

Wheel Selection Criteria

The size, configuration, and features of the diamond wheel bond types are described in detail in the article "Production Grinding Methods and Techniques" in this Volume. Wheel manufacturers should be consulted concerning details of each specification and their effect on grinding operations. There are also general guidelines for factors that are applicable to all grinding operations.

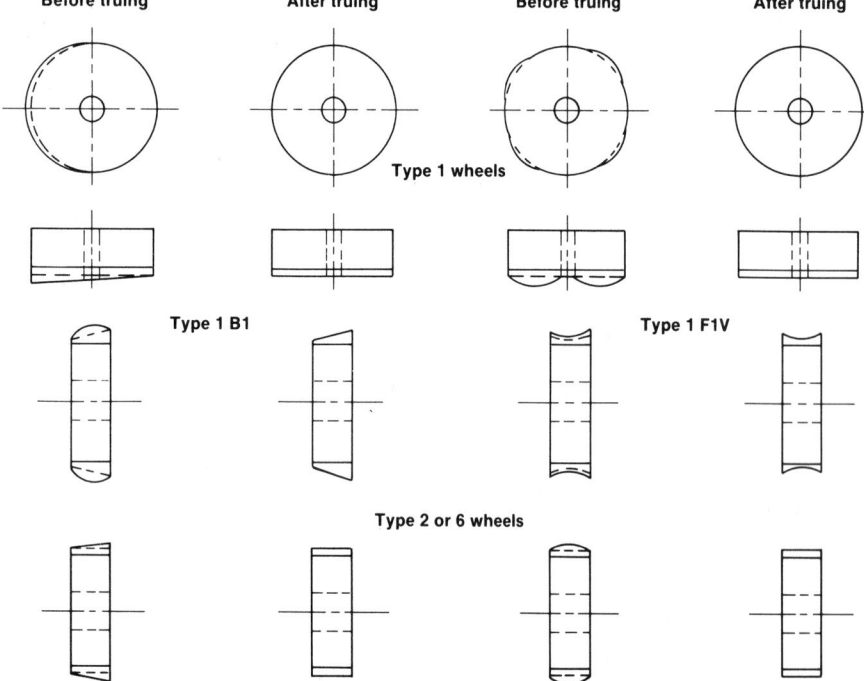

Before truing   After truing   Before truing   After truing

Type 1 wheels

Type 1 B1   Type 1 F1V

Type 2 or 6 wheels

Fig 18(a) Typical examples of conditions that require truing

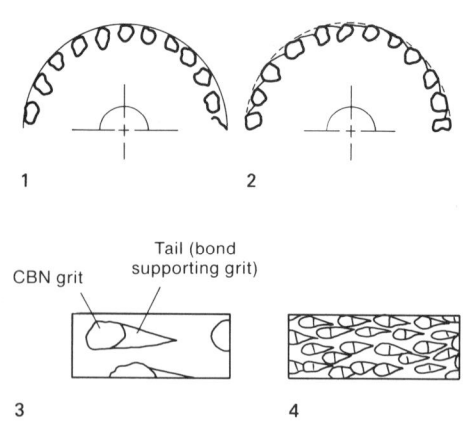

1   2

CBN grit   Tail (bond supporting grit)

3   4

Fig 18(b) Schematic of a wheel that has been trued and dressed. 1, after truing, wheel face is smooth and closed; 2, after dressing, wheel face is open with grits exposed and ready for efficient grinding; 3, after dressing, bond supports the grit; 4, after dressing. Note path connecting the tails for efficient cooling and chip flow.

Table 4   Advantages and limitations of diamond abrasive bond types

Resin bond	Vitrified bond	Metal bond	Layered products
Readily available	Free cutting	Very durable	Single abrasive layer plated on a premachined steel preform
Easy to true and dress	Easy to true	Excellent for thin slot, groove, cutoff, simple form, or slot grinding	Extremely free cutting
Moderate freeness of cut	Does not need dressing (if selected and trued properly)	High stiffness	High unit-width material removal rates
Applicable for a wide range of operations	Controlled porosity to enable coolant flow to the grinding zone and chip removal	Good form holding	Form wheels, easily produced
First selection for determining the optimum use of diamond wheels	Intricate forms can be crush formed on the wheels	Good thermal conductivity	Form accuracy dependent on preform and plating accuracy
	Suitable for creep-feed or deep grinding, ID grinding, or high-conformity grinding	Potential for high-speed operation	High abrasive density
	Potential for longer wheel life than resin bond	Generally requires high grinding forces and power	Generally not truable
	Excellent with oil as coolant	Difficult to true and dress	Generally poorer surface finish than bonded abrasive wheels

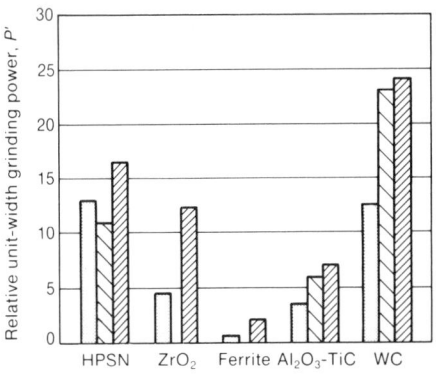

Fig 21 Relative unit-width grinding power required to machine various structural and electronic ceramics. Unit-width material removal rates classified as low (2 mm^3/s, mm; or 0.2 $in.^3$/min, in.), medium (5 mm^3/s, mm; or 0.5 $in.^3$/min, in.), and high (10 mm^3/s, mm; or 1.0 $in.^3$/min, in.). Source: Ref 36

Diamond grinding wheels used for ceramic grinding are generally of four bond types: resin, metal, vitrified, and single-layer achieved by electroplating the diamond onto a steel preform. The advantages and limitations of each bond type are listed in Table 4. Figure 19 shows a typical specification for a resin bonded wheel.

Abrasive grit plays a critical role in the proper grinding of ceramics. Along with Table 3, Fig 3 and 4 show the effect of grit size on strength and surface finish obtained by the proper selection of grit size. Figure 20 shows the diamond grain size as a function of application.

With the wide range of operations and grinding configurations, it is difficult to set guidelines for the selection of bond type. However, flexibility, ease of use, and resilience are the most common factors in favor of resin bonded diamond wheels. Vitrified bonded diamond wheels have several advantages for production grinding, including: form holding, higher stiffness, tighter tolerances,

and lighter weight. Metal bonded diamond wheels are normally chosen when durability or long life is the primary objective (for example, large contact area grinding, slot grinding). In some instances, metal bonded grinding wheels have been used in a machining center to grind complex profiles using small diamond wheels (Ref 42). In general, higher grinding forces and power are the most frequent limiting factors in the use of metal bonded wheels. Many traditional ceramics are cut, finished, and sawed using metal bonded diamond wheels under dry grinding conditions.

Effect of Workpiece Material Properties

Grinding force requirements vary depending on the ceramic material chosen (Fig 17).

Grinding power requirements also vary with the workpiece material (Fig 21). Table 5 compares the properties of hot pressed silicon nitride and tungsten carbide. Creep-feed grinding of these two materials indicates that the HPSN material with its higher hardness (and hence resistance to penetration) requires higher normal grinding forces than the tungsten carbide material (Fig 21). However, the tungsten carbide material (with higher strength) requires higher grinding power than the HPSN material.

Porosity, grain size, and microstructure could have a major effect on surface finish and surface quality. Surface finish can be controlled in ceramic workpiece materials by proper selection of grinding parameters (for example, unit-width material removal rate,

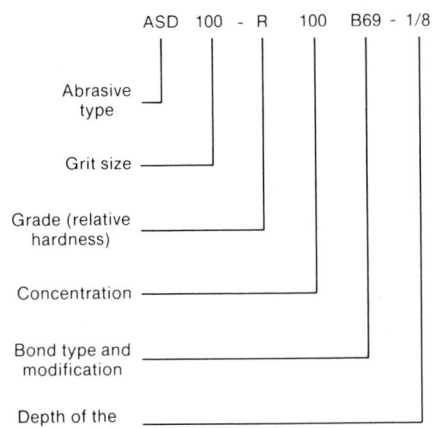

Fig 19 Typical specifications used for superabrasive wheels

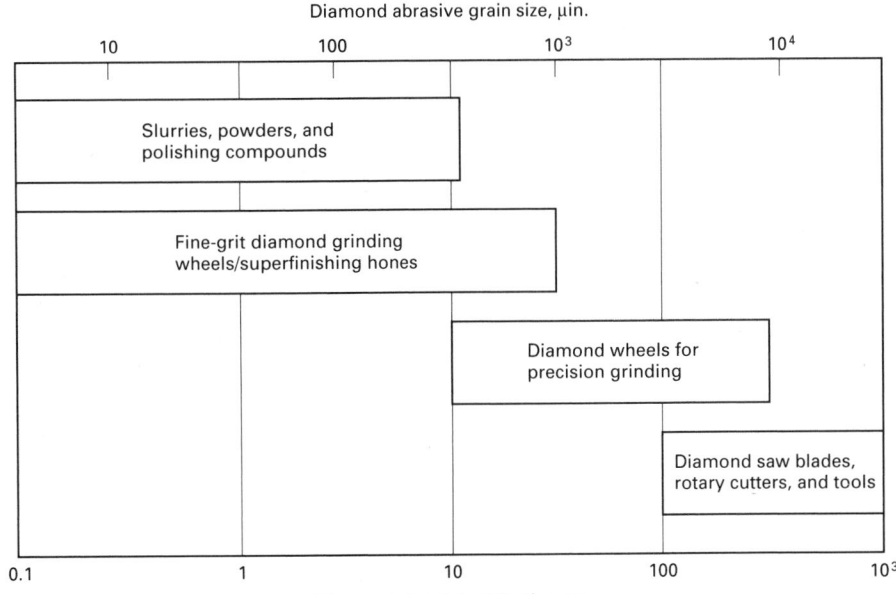

Fig 20 Diamond grinding application range as a function of diamond abrasive grain size

Table 5 Comparison of properties for tungsten carbide and hot pressed silicon nitride ceramics

Material	Density		Hardness, HK	Strength, MOR						Relative thermal shock resistance(a)			Thermal conductivity at 20 °C (70 °F), W/m·K
	g/cm³	lb/in.³		RT		800 °C (1470 °F)		1000 °C (2010 °F)		RT	800 °C (1470 °F)	1000 °C (2010 °F)	
				MPa	ksi	MPa	ksi	MPa	ksi				
WC	12.5–15.1	0.452–0.546	1700	1550	225	1000	145	414	60.03	0.33	0.23	0.07	121
HPSN	3.2–3.25	0.116–0.117	2000	793	115	793	115	758	110	1.0	0.65	0.65	32

(a) Obtained by normalizing the values with reference to the property of HPSN at room temperature

grit size, and depth of cut) as shown in Table 3. If the ceramic grinding is primarily dominated by brittle fracture, generally there is poor control of the surface finish. However, by proper selection of the abrasive grains, a rigid grinding system, proper truing, dressing, and balancing, extremely fine finishes of the order of ≤0.025 μm (≤1 μin.) can be readily obtained. With mirror finish grinding technology, finishes of the order of 1 nm (0.040 μin.) have been produced (Ref 21).

The surface finish obtained in workpiece materials is also a function of the type of workpiece material. Figure 22 shows the surface finish for three different work materials. For low-fracture toughness materials such as alumina, the surface finish obtained is poor when compared to HPSN or ZrO₂ materials with the same grinding conditions. Surface finish is also related to the thermal shock resistance of the workpiece material and its effect via the coolant application. Poor thermal shock resisting materials such as alumina are prone to thermal cracking during the grinding process. Pullout of grains can occur, particularly in poorly bonded workpiece materials with coarse grain structure, especially when fine abrasive grained wheels are used to obtain ultrasmooth surfaces.

Residual stress and surface damage are two factors that have attracted significant attention in the grinding of ceramics (Ref 31, 32, 43). Figure 23 shows the factors that govern the residual stress caused by the grinding pro-

cess in the workpiece material. While the research in this area will continue for years to come, proper attention to optimizing the grinding process (that is, to minimize grinding forces, power, contact stress, and heat input to the work material) will result in significant near-term and practical end results.

Operational Factors

While the machine tool, workpiece material, and the grinding wheel are the discrete inputs into the grinding system, the actual grinding process occurs when these three inputs interact through the operational factors such as grinding parameters, type of grinding, coolant interactions, truing and dressing, and so on.

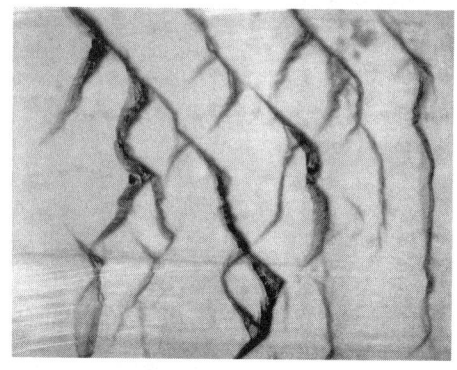

(a)

(b)

Grinding Cycle. The proper designs of grinding cycles for ceramic materials are in their evolutionary stage. Many of the principles applicable to metal grinding (Ref 44–50) are applicable to ceramic materials as well. It is clear that grinding of ceramics can occur in one of three modes:

- Brittle fracture
- Plastically deformed surface
- Combination of brittle fracture and plastic deformation

There are well-defined parameters that delineate this process. For example, small depth of cut, small abrasive grains, and small depth of cut per abrasive grain as well as low indentation force per grain are all conditions that promote plastic deformation. The objectives for production grinding of ceramics will be to obtain all of the above conditions while simultaneously maintaining the desired cycle time. When it is implemented properly, it is possible to obtain complex geometries in ceramic parts having cycle times of the order of two to three times those for comparable metal parts.

Self-Excited Vibrations (Chatter). The consequences of inadequate understanding of the principles of grinding ceramics can be dramatic. For example, Fig 24(a) shows a smooth plastically deformed surface that is required to obtain optimum surface finish and strength. The same material ground with the same wheel in the same machine tool can exhibit poor surface conditions due to brittle fracture of the surface (Fig 24b). In general, chatter in the machining of metallic materials leads to poor surface finish or waviness. When machining ceramic materials, chatter usually translates into brittle fracture leading to very poor surface finish, lower strength, or chippage.

Chippage is one of the predominant limitations in the grinding of ceramics. This is accelerated when the ceramic is of low theoretical density or poor strength and when the grinding forces are large or widely variable in magnitude. Chippage is invariably due to a brittle fracture process. Any effort to minimize brittle fracture will also minimize chippage. Great care is taken when grinding electronic ceramics to ensure uniform and vibration-free conditions that minimize chippage. Figure 25 shows some common causes of chippage in the machining of ceramics.

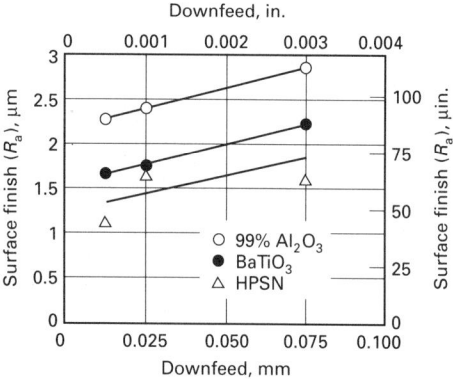

Fig 22 Plot of surface finish (roughness average, R_a) versus downfeed (depth of cut) for three ceramics that were surface ground with a resin bonded diamond wheel. Source: Ref 33

Fig 23 Comparison of the surface finish of sapphire (Al_2O_3) ground by two different production techniques. (a) Surface grinding, which produced cracks visible on surface. (b) Creep-feed grinding, which yielded no detectable cracks.

(a)

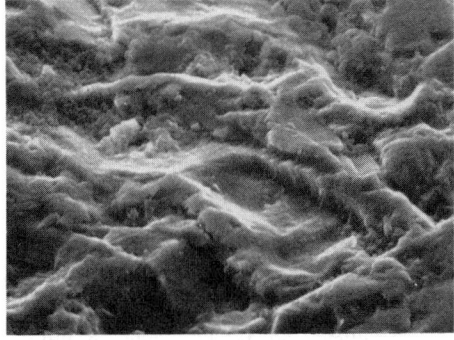

(b)

Fig 24 Photomicrographs comparing surface quality of Al_2O_3 ceramic. (a) Proper grinding procedure was followed to yield a smooth plastically deformed surface that met surface finish specifications. (b) Incorrect grinding techniques generated chatter that led to subsequent brittle fracture to yield poor surface finish.

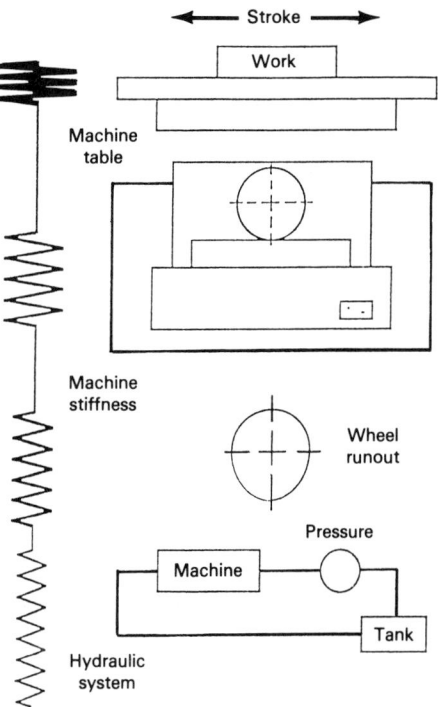

Fig 25 Schematic showing varying degrees of chippage caused by selected components in an abrasive grinding installation

Vibrations induced by machine reversals are of low frequency and can produce large chips. The machine spindle/workpiece/holder assembly constitutes a dynamic system with its own natural frequency of vibration. Resonance at the natural frequency of vibration of the system can lead to chippage. Wheel runout is a key factor for uneven grinding that leads to chippage. Vibrations induced by the hydraulic pump impeller can induce a high-frequency vibration leading to fine chippage.

These numerous sources of vibration can be analyzed by the simulation of springs of various stiffness or natural frequency. Springs of higher stiffness and smaller natural frequency produce large-scale chippage or fracture.

Cracking of Ceramics. Mechanically or thermally induced cracks limit the grinding of ceramics. Conditions that produce cracking in ceramics include high grinding forces, thermal shock, poor removal of heat from the grinding zone, and contact stresses at the grinding zone. Figure 23 compares the surfaces produced with surface grinding versus creep-feed grinding of ceramics. For the same work material, the choice of the grinding process can make a significant difference in the number of surface cracks generated.

REFERENCES

1. H.K. Tonshoff, W. Scheiden, I. Inasaki, and A. Spur, Abrasive Machining of Silicon, *Ann. CIRP*, Vol 2, 1990, p 230–243

2. K. Subramanian and R.N. Kopp, Grinding Advanced Ceramics: A Forecast for 1990's, *Am. Ceram. Soc. Bull.*, Vol 67 (No. 12), 1988, p 1892–1902

3. J.A. Dudley, Precision Finishing and Slicing of Ceramic Materials with Diamond Abrasives, *Fourth International Grinding Conference*, 1990, p 550

4. K. Subramanian, Precision Finishing of Ceramic Components with Diamond Abrasives, *Am. Ceram. Soc. Bull.*, Vol 67 (No. 6), June 1988, p 1026–1029

5. P. Daniel, Ceramic Bearings for Volume Applications, *Ind. Diamond Rev.*, Vol 50 (No. 538), Mar 1990, p 122–124

6. D.W. Richerson, *Modern Ceramic Engineering*, Marcel Dekker, 1982, p 260–273

7. K. Subramanian, S. Ramanath, and Y.O. Matsuda, Precision Production Grinding of Fine Ceramics, *Proceedings of the First International Conference on New Manufacturing Technology*, Chiba, Japan, 1990, p 309–315

8. K. Subramanian, Superabrasives for Precision Production Grinding—A Case for Interdisciplinary Effort, *Proceedings of the Symposium on Interdisciplinary Issues in Materials Processing and Manufacturing*, Vol 2, 1987, p 665–676

9. M. Ohta, K. Miyahara, and K. Matsuo, Effect of Grinding Parameters on the Strength of Ceramics, *Jpn. Soc. Prec. Eng.*, 1987, p 753–754

10. C.A. Anderson and R.J. Bratton, Effect of Surface Finish on the Strength of Hot Pressed Silicon Nitride, *The Science of Ceramic Machining and Surface Finishing II*, Special Publication 562, National Bureau of Standards, p 351–378

11. Y. Matsuo, *et al.*, Analytical and Experimental Study on the Bending Strength of SiC Specimens with Ground Surfaces, *Intersociety Symposium on Machining of Advanced Ceramic Materials and Components*, Apr 1987, American Ceramic Society, p 235–248

12. M. Ota and K. Miyahara, The Influence of Grinding on the Flexural Strength of Ceramics, *Proceedings of the Fourth International Grinding Conference*, 1990, p 538

13. Y. Ichida, Mirror Finish Grinding of B-Sialon with Fine Grained Diamond Wheels, *Yogyo Kyohoi-Shi*, Vol 94 (No. 1), 1986

14. B.R. Lawn, A.A. Evans, and D.B. Marshall, Elastic/Plastic Indentation Damage in Ceramics: The Median/Radial Crack System, *J. Am. Ceram. Soc.*, Vol 63 (No. 9–10), 1980, p 574–581

15. J.C. Conway, Jr. and H.P. Kirchner, The Mechanics of Crack Initiation and Propagation Beneath a Moving Sharp Indentor, *J. Mater. Sci.*, Vol 15, 1980, p 2879–2883

16. R.W. Rice and J.J. Mecholsky, The Nature of Strength Controlling Machining Flaws in Ceramics, *The Science of Ceramic Machining and Surface Finishing II*, Special Publication 562, National Bureau of Standards, 1976, p 351–378

17. I. Inasaki, Grinding of Hard and Brittle Materials, *Ann. CIRP*, Vol 36 (No. 2), 1987, p 463–471

18. G. Spür, C. Stark, and T.H. Tio, Grinding of Non-Oxide Ceramics Using Diamond Grinding Wheels, *Machining of Ceramic Materials and Components*, PED-Vol 17, American Society of Mechanical Engineers, 1985, p 33

19. L.G. Pukaite and K. Subramanian, Creep Feed Grinding of Silicon Nitride Tool Material, *Proceedings of the Society of Carbide and Tool Materials Annual Meeting*, Phoenix, Feb 1987

20. K. Subramanian and S. Ramanath, Mechanism of Material Removal in the Precision Grinding of Ceramics, to be published

21. J.C. Conway, Jr. and H.P. Kirchner, Crack Branching as a Mechanism of Crushing During Grinding, *J. Mater. Sci.*, Vol 69 (No. 8), Aug 1988, p 603–607

22. S. Malkin and J.E. Ritter, Grinding Mechanisms and Strength Degradation

for Ceramics, *Intersociety Symposium on Machining of Advanced Ceramic Materials and Components*, American Society of Mechanical Engineers, 1988, p 57–72

23. M. Miyashita, Ultraprecision Centerless Grinding of Brittle Materials, *First Annual Precision Engineering Conference*, North Carolina State University, Raleigh, 1985

24. H. Ernst and M.E. Merchant, Surface Treatment of Metals, *American Society of Metals*, 1941, p 299–378

25. R.P. Lindsay, Principles of Grinding, *Metals Handbook*, Vol 17, *Machining*, ASM International, 1989, p 421–429

26. R.S. Hahn and R.I. King, *Handbook of Modern Grinding Technology*, Chapman and Hall, 1986

27. K. Subramanian and R.P. Lindsay, A Systems Approach for the Use of Vitrified Bonded Superabrasive Wheels for Precision Production Grinding, *Proceedings of the Symposium on Grinding Technology*, Winter Annual Meeting of the American Society of Mechanical Engineers, Nov 1989

28. S. Malkin, Grinding of Metals: Theory and Application, *J. Appl. Met. Work.*, Vol 3 (No. 2), 1984, p 95

29. Z. Feng and J.E. Field, Dynamic Strengths of Diamond Grits, *Ind. Diamond Rev.*, Vol 3, 1989, p 104

30. I. Nakajima, Y. Uno, and T. Fujiwara, Cutting Mechanism of Fine Ceramics with a Single Point Diamond, *Prec. Eng.*, Vol 11 (No. 1), Jan 1989, p 19–26

31. R. Samuel, *et al.*, Effect of Residual Stress on the Fracture of Ground Ceramics, *J. Am. Ceram. Soc.*, Vol 72 (No. 10), 1989, p 1960–1966

32. S. Chandrasekar, *et al.*, Morphology of Ground and Lapped Surfaces of Ferrite and Metal, *Proceedings of the ASME Winter Annual Meeting*, PED-17, American Society of Mechanical Engineers, 1985

33. K. Subramanian, Advanced Ceramic Components: Current Methods and Future Needs for Generation of Surfaces, *Intersociety Symposium on Machining of Advanced Ceramic Materials and Components*, Apr 1987, American Ceramic Society, p 10–32

34. H.P. Kirchner and J.C. Conway, Jr., Mechanisms of Material Removal and Damage Penetration During Single Point Grinding of Ceramics, *Intersociety Symposium on Machining of Advanced Ceramic Materials and Components*, PED-17, American Society of Mechanical Engineers, 1985, p 53

35. K. Subramanian and S. Ramanath, Machine Tool Developments Required for the Precision Production Grinding of Ceramics, *Proceedings of the Third Biennial International Manufacturing Technology Research Forum*, 1989

36. K. Subramanian and P.P. Keat, Parametric Study on Grindability of Structural and Electronic Ceramics—Part I, *Proceedings of the Symposium on Machining of Ceramic Materials and Components*, Winter Annual Meeting, American Society of Mechanical Engineers, 1985

37. L.G. Pukaite and K. Subramanian, Creep Feed Grinding of Silicon Nitride Tool Material, *Proceedings of The Society of Carbide and Tool Materials Annual Meeting*, Phoenix, Feb 1987

38. J. Yoshioka, F. Hashimoto, M. Miyashita, and M. Daito, Ultraprecision Grinding Technology for Brittle Materials: Application to Surface and Centerless Grinding Processes, *Milton C. Shaw Grinding Symposium*, Vol 16, Production Engineering Division, American Society of Mechanical Engineers, p 209–227

39. T.A. Bifano, T.A. Dow, P. Blehe, and R.O. Scattergood, Precision Machining of Ceramic Materials, *Proceedings of the Intersociety Symposium on Machining of Advanced Ceramics*, PED-17, American Society of Mechanical Engineers, 1985, p 99–120

40. M.H. Layne and W.C. Heck, Grinding Wheel Balancing: Sources and Solutions, *Am. Eng. Soc.*, Fall 1989, p 12–17

41. J. Yoshioka, F. Hashimoto, and M. Miyashita, Application of Grinding Wheel to Ultraprecision Machining: Machining for Precise Surface Generation on Grinding Wheel, *Intersociety Symposium on Machining of Advanced Ceramic Components*, Apr 1987, American Ceramic Society, p 50–69

42. T. Nakagawa, K. Suzuki, and T. Uematsu, Three Dimensional Creepfeed Grinding of Ceramics by Machining Center, *Proceedings of the ASME Winter Annual Meeting*, PED-17, American Society of Mechanical Engineers, 1985, p 1–8

43. L.R. Clarke, *et al.*, Ultrasonic Measurement of Machining Damage Surface Defects in Ceramics, *Proceedings of the ASME Winter Annual Meeting*, PED-17, American Society of Mechanical Engineers, 1985

44. R.P. Lindsay, CBN Grinding Principles, *Proceedings of the Abrasives Engineering Society Annual Meeting*, Apr 1986, Cincinnati

45. R.P. Lindsay, "Grinding Cycles," Paper No. 87–818, Society of Manufacturing Engineers, 1987

46. R.P. Lindsay, Laboratory Investigations in Support of CBN Production Grinding, *ASME Proceedings*, PED-16, American Society of Mechanical Engineers, 1985

47. K. Subramanian, Vitrified Bonded CBN Wheels—Current Applications and Future Directions, *Proceedings of the Industrial Diamond Association—Japan Symposium*, Oct 1988

48. B.A. Cooley and H.O. Juchem, Vitrified Bonds—No Longer a Synonym for Conventional Abrasive Tools, *Superabrasives '85*, Society of Manufacturing Engineers, 1985, p 2–14 to 2–56

49. A. Kishimoto, Success of Vitrified Bond CBN Wheels for Grinding of Precision Parts in Production Lines, *Superabrasives '85*, Society of Manufacturing Engineers, p 9–1 to 9–20

50. S. Malkin, Current Trends in CBN Grinding Technology, *Ann. CIRP*, Vol 34 (No. 2), 1985

Abrasives

Ernest Ratterman and Roger Cassidy, GE Superabrasives

THE GRINDING OF CERAMICS AND GLASSES presents certain unique problems not found in the grinding of other engineered materials. Ceramics and glasses are the largest single category of materials that undergo predominantly microbrittle fracture in the process of chip formation in grinding. In addition, most of the important technical ceramics have hardness values much greater than any of the other engineered materials. Aluminum oxide, silicon carbide, silicon nitride, and titanium nitride are widely used in the form of abrasive grains or cutting tool tips to grind and machine other engineered materials such as steels, aluminums, superalloys, tool steels, cast alloys, and so on. All of these engineered metals usually undergo a highly plastic chip-forming process when being ground or machined.

In order to rationally identify the optimum abrasives for ceramics and glasses, a system of comparing physical properties has been developed. The key properties of ceramics, glasses, abrasives, and other engineered materials can be readily compared and it is shown that diamond abrasives are most effective. Guidelines to selecting the optimum diamond abrasives for most glass and ceramic materials are provided along with supporting examples.

Physical Properties of Engineered Materials

The relative ease or difficulty of grinding any brittle material may be principally (but not solely) determined by the combined properties of hardness and modulus of resilience (MOR). The acronym MOR is broadly understood to mean modulus of rupture. However, for the information presented in this article, MOR will be used as the acronym for modulus of resilience. MOR is defined as:

$$\text{MOR} = \frac{\sigma^2}{2E} \qquad \text{(Eq 1)}$$

where σ is the ultimate tensile strength of the material, and E is its elastic modulus. This calculation is simply the area under the classic stress strain diagram for the given material. As such, the calculation yields a value of specific energy in units of J/mm^3, which is an approximation of the specific energy

required to precipitate fracture or failure of the given material. Ratterman (Ref 1) has developed and expanded the hardness/MOR map shown in Fig 1. The MOR calculation for ductile metals may underestimate the true energy required to produce chips, but the calculation may be used to approximately position these materials on the map itself. Because grinding ceramics and glasses is the subject of this article, a detailed summary of the generic physical properties of selected ceramics and glasses is shown in Table 1.

Before proceeding to position abrasive grains themselves on the map in Fig 1, a few observations about the positions of major families is useful. Naturally occurring granites, marbles, slates, sandstones, limestones, and engineered concrete materials are among the easiest of all commercially important materials to process with abrasives. They can all be ground or sawed at very high speeds compared with all other materials on the map. Situated across the very center of the map are much tougher and harder families of workpieces. These include the refractories, the alumina ceramics, all of the glasses, and the modern technical ceramics.

For sheer toughness, there are no materials superior to the cemented carbides, tool steels, and especially the high-speed tool steels. The hardness and toughness properties severely limit the rate at which these materials can be ground as well as the life of the grinding tools.

There are factors that may significantly influence the relative ease or difficulty of grinding or machining a given workpiece other than the hardness and the modulus of resilience:

- The degree to which there may be chemical interaction between the workpiece and the abrasive grain in the grinding process may severely limit tool life. A few outstanding examples are diamond and silicon carbide with iron and aluminum oxide with glass
- The degree to which a workpiece itself may be classified as highly abrasive may significantly influence the performance and selection of suitable grinding wheels

No attempt will be made to assess the influence of these two factors on the hardness/toughness map.

Position of Abrasives on the Hardness/MOR Map

There are two major categories of industrial abrasives available for the grinding of the spectrum of materials shown on the map in Fig 1: conventional abrasives, (a), and superabrasives, (b). The two major conventional abrasives are aluminum oxide and silicon carbide. The two forms of superabrasives are diamond and cubic boron nitride (CBN). A summary chart of the range of important physical properties of these four types of abrasives is shown in Table 2.

The modulus of resilience of abrasives is extremely important as abrasive grains are brought in contact with the workpiece under high velocity conditions. These will typically be 25 to 30 m/s (82 to 98 ft/s), but in certain very high efficiency operations they may reach the range of 50 to 120 m/s (165 to 395 ft/s). Thus, the fracture characteristic of the abrasive grain, while not as important as hardness, is nonetheless a key element in the performance of any grinding system. It must always be taken into account in selecting wheels with appropriate abrasives to grind a specific workpiece.

The determination of modulus of resilience depends on established values for the elastic modulus and tensile strength of the material in question. These are readily determined by standard test procedures on relatively large bodies of work materials cited above. In the case of abrasive grains, however, such procedures cannot be applied simply due to the basic physical form of abrasives. This information is simply not available and must be estimated from the knowledge of such properties of these materials as they appear in other forms.

Given the very significant difference in compressive strengths between superabrasives and conventional abrasives, one can infer that the tensile strengths of conventional abrasives will be but a fraction of the tensile strength of superabrasives. In combination with the lower elastic moduli, the modulus of resilience of conventional abrasives must be less than that of superabrasives.

In the case of diamond, Field (Ref 4) provides data on elastic modulus and theoretical tensile strength of single-crystal diamond. A

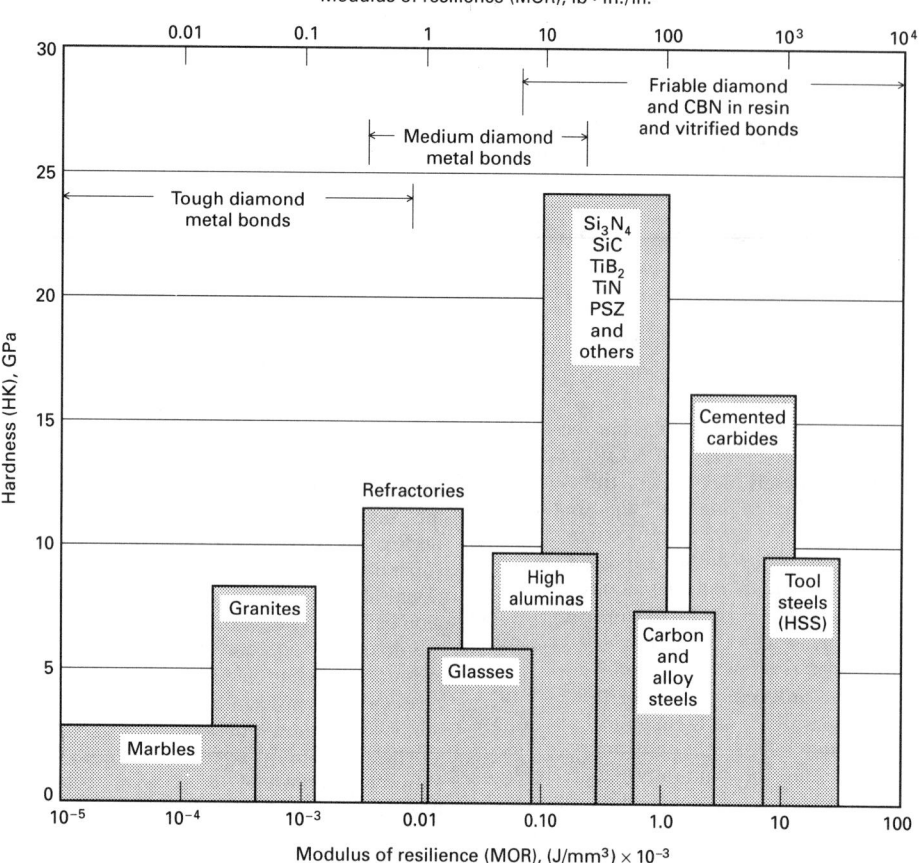

Fig 1 Plot of hardness versus modulus of resilience of selected ceramics and glasses to determine relative difficulty of grinding these brittle materials. Also shown is the effective application range for metal, resin, and vitrified wheel bonding systems.

described in the section "Controlling Abrasive Properties" in this article to estimate the modulus of resilience of CBN abrasives. Such tests show that the most friable forms of cubic boron nitride are somewhat tougher than the weakest diamond. The toughest forms of cubic boron nitride, however, do not nearly approach the toughness of the toughest diamond.

All values of the strength and elastic moduli of both aluminum oxide and silicon carbide are significantly less than diamond and cubic boron nitride. They are an integral part of the ceramics family. Thus, they must fall below diamond and cubic boron nitride on the map.

The hardness/MOR map may now be completed by locating diamond, cubic boron nitride, aluminum oxide, and silicon carbide on the map (Fig 2). Some immediate observations can be made by inspection of the full map:

- *Diamond* is by far the hardest and strongest of all abrasives available. As such, it should be the superior abrasive of choice for grinding all types of materials on this map. In fact, diamond is used effectively on all materials—ceramics, glasses, concretes, natural stones, cemented carbides, other nonferrous metals, and other nonmetallic materials. However, because diamond is an allotrope of carbon, it inherently reacts with ferrous metals at the typical temperatures encountered in the grinding process. The resulting rapid wear of diamond abrasives make them generally uneconomical in grinding ferrous metals, except in certain low-speed honing applications
- *Cubic boron nitride* is a manufactured superabrasive, second in hardness only to diamond. Cubic boron nitride has properties specifically suited to grinding a broad range of hard ferrous metals, superalloys, and cast iron. Based on its position on the map, cubic boron nitride appears to offer some potential in the effective grinding of technical ceramics. The inability of CBN

calculation of the modulus of resilience of diamond using these data places diamond well into the MOR range of cemented carbides. Cemented carbide rolls and spheres are commonly used to reduce raw diamond abrasive for sizing purposes. It therefore seems unreasonable that the MOR values for diamond in the range of 2.0 to 3.5 J/mm^3 × 10^{-3} (280 to 510 lbf · in./in.^{3}) are representative.

Gigl (Ref 5) provides experimental data on the elastic modulus and transverse rupture strength of polycrystalline diamond. Use of these values yields a modulus of resilience of 1.0 J/mm^3 × 10^{-3} (145 lbf · in./in.^{3}). This value has been used as an arbitrary upper limit of the modulus of resilience of diamond.

In the case of cubic boron nitride, one can only rely on the abrasive toughness tests

Table 1   Mechanical properties of selected ceramics and glasses

Material	Hardness (HK), GPa	Transverse rupture strength MPa	Transverse rupture strength ksi	Young's modulus GPa	Young's modulus psi × 10^6	Fracture toughness, K_{Ic} MPa√m	Fracture toughness, K_{Ic} ksi√in.	Modulus of resilience (J/mm^3) × 10^3	Modulus of resilience lbf·in./in.^{3}
Al$_2$O$_3$	10–20	276–700	40–102	380	55	2.7–4.2	2.5–3.8	0.1–0.6	15–87
Reaction-bonded Si$_3$N$_4$	8–19	250–345	36–50	304	44	3.6	3.3	0.1–0.2	15–29
Hot pressed Si$_3$N$_4$	8–19	700–1000	100–150	304	44	4.1–6.0	3.7–5.5	0.8–1.6	116–233
Sintered Si$_3$N$_4$	8–19	414–650	60–94	304	44	5.3	4.8	0.03–0.68	4.4–99
Partially stabilized ZrO$_2$ (PSZ)	10–11	600–700	87–102	205	30	8–9	7.3–8.2	0.87–1.19	127–173
Fully stabilized ZrO$_2$	10–15	245	35.5	97–207	14–30	2–8	1.8–7.3	0.2	29
Hot pressed SiC	20–30	230–825	33–120	207–483	30–70	4.8–6.1	4.4–5.6	0.08–0.98	12–143
Sintered SiC	20–30	96–520	14–75	207–483	30–70	4.8	4.4	0.013–0.39	1.9–57
TiB$_2$	15–36	700–1000	102–145	514–574	74.5–83.2	6–8	5.5–7.3	0.45–0.92	65–134
Pyrex glass	5	60	8.7	70	10	0.75	0.68	0.026	3.8
Glass ceramics	6–7	70–350	10–51	83–138	12–20	2.4	2.2	0.22–0.55	32–80
Quartz	8–10	70	10	74	11	...	...	0.033	4.8
Cemented carbide (16% Co)	18	1860(a)	270(a)	524	76	...	...	3.3	480

(a) Ultimate tensile strength. Source: Ref 2

Table 2 Mechanical and thermal properties of selected conventional abrasives and superabrasives

Material	Hardness (HK), GPa	Young's modulus		Comprehensive strength		Thermal conductivity	
		GPa	psi × 10^6	MPa	ksi	W/m·K	Btu/ft·h·°F
Superabrasive							
Diamond	70–100	1,075(a)	156(a)	10,444	1,515	2,092	1,209
Cubic boron nitride	45	662(a)	96(a)	7,061	1,024	1,300	750
Conventional							
SiC	27	207	30	569	82.5	41.5	24.0
Al$_2$O$_3$	25	308	45	2,942	426.6	33.5	19.4

(a) Internal General Electric Company sonic evaluation. Source: Ref 3

Table 3 Conventional abrasives and superabrasives applied to the machining of selected ceramic and glass components

Material	Bond type	Applications
Conventional		
Silicon carbide	Vitreous	Grind soft ceramics
		Flame spray ceramic coatings
		Glass beveling
	Rubber	Cutoff glass tubing
	Coated	Seaming plate glass
		Finishing picture tubes
		Scientific stemware
Aluminum oxide	Vitreous	Glass beveling
Zirconia alumina	Coated	Seaming plate glass
Ceramic alumina	...	...
Superabrasives		
Diamond	Metal	Precision pencil edging and beveling of tempered plate and safety glass
		Optical lens production
		Glass/quartz tube cutoff
		General precision grinding of all types of glass
		Production grinding of aluminas, quartz, and refractories.
	Resin	All production and finish grinding of silicon carbide, titanium nitride, silicon nitride, and partially stabilized zirconia, and other technical ceramics
Cubic boron nitride	...	...

abrasive to grind ceramics and glasses will be described in the section "Ceramic Machining Guidelines" in this article

- *Conventional abrasives.* Both aluminum oxide and silicon carbide abrasives have properties that make them an integral part of the entire family of ceramic materials. As such, their utility is limited in comparison with that of diamond. The hardness of conventional abrasives may be below, equal to, or marginally higher than the ceramic material, leading to inefficient grinding

Abrasives for Grinding Ceramics and Glasses

An overview of the major uses of both conventional abrasives and superabrasives in grinding ceramics and glass is found in Table 3. All high-production grinding that also demands a significant level of precision and control over all aspects of the ground surface are carried out today with diamond abrasives.

The use of conventional abrasives is limited to roughing, finishing, and cutting off relatively easy to grind materials where precision and finish are not major requirements.

Matching Abrasive and Bond Properties

It is obvious that the hardness of abrasives is not matched by their modulus of resilience or toughness. All abrasives must be used to grind materials of much higher MOR values. In order to take advantage of abrasive grain hardness in the grinding process, it is nec-

essary to design an appropriate bonding system. The bond system must have either friability or flexibility characteristics that can effectively absorb the impact of grinding. By absorbing the impact of grinding, the hardness and strength of the abrasive itself can then be exploited for chip formation. Thus, vitreous and resin bonds have been developed for both conventional abrasives and superabrasives for the grinding of medium and high MOR workpiece materials.

For workpieces with medium to low MOR characteristics, the modulus of resilience of the abrasive ceases to be an issue, and hard, high-strength metal bonds may be employed with diamond abrasives. The properties of conventional abrasives do not lend themselves for use in metal bonds.

The ranges of applications for the most effective wheel bonding systems are identified in Fig 1.

Controlling Abrasive Properties

The high-pressure, high-temperature processes for manufacturing superabrasives are quite different from the electric furnace and sintering processes used in the manufacture of aluminum oxide and silicon carbide. Yet

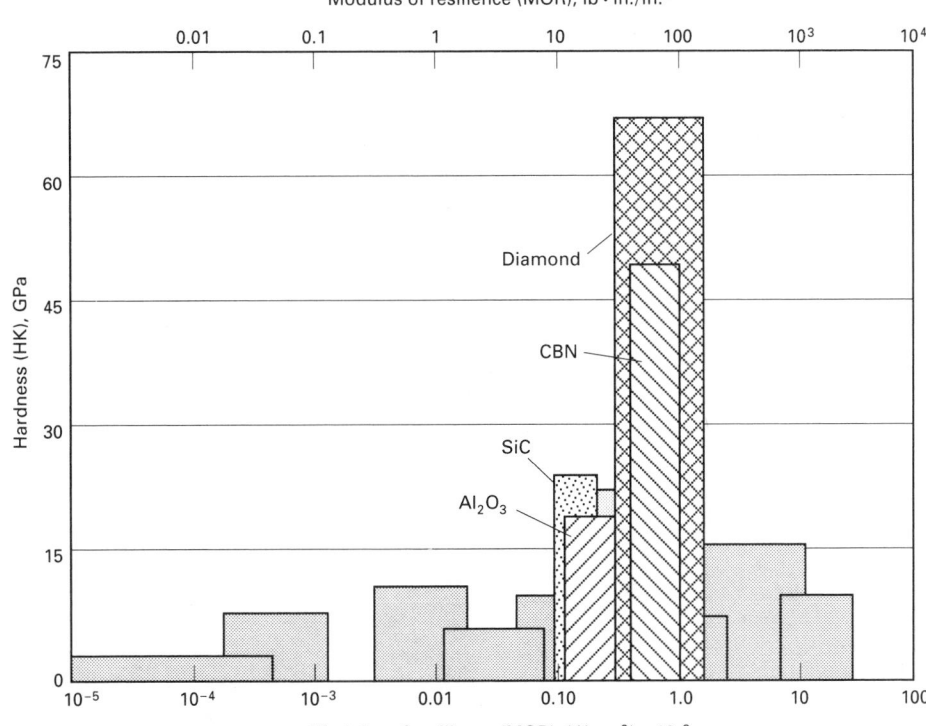

Fig 2 Plot of hardness versus modulus of resilience to compare grinding properties of selected conventional abrasives and superabrasives

all are highly controlled processes that can be designed and regulated to produce discrete products having a broad range of properties and characteristics. Most important in the manufacture of abrasive grains is control of the fracture toughness or friability of the abrasive. Thus, diamond, cubic boron nitride, aluminum oxide, and silicon carbide are each available in a range of friabilities or toughnesses that tailor these abrasives to specific areas of application. This range is reflected in Fig 2 by the range of modulus of resilience for each abrasive type. Aluminum oxide, for example, can be alloyed with zirconium oxide and titanium dioxide to create a range of tough alumina abrasives. The white α aluminas have relatively low fracture strengths and this fills out the spectrum of applications for aluminum oxide abrasives.

Friability. The friability of diamond abrasives is controlled through selection of the time, pressure, and temperature parameters of the crystal growth cycle. This allows for the controlled introduction of defects and second phases that alter the fracture properties of the resulting products. This tailors the abrasive to specific bonds and workpiece properties. In addition, the size of the abrasive grains, their morphology, and surface characteristics are specifically determined in this growth process.

Bondability. A critical element in the development of an abrasive bond system for a specific area of application is the bondability of the surfaces of the abrasives. Both aluminum oxide and silicon carbide readily react with the glasses used to manufacture conventional grinding wheels. This factor provides a high degree of flexibility in combination with other controllable factors in the manufacturing process to create grinding wheels adapted to a broad range of uses.

Superabrasives (diamond and cubic boron nitride) have relatively inert surfaces and present a different set of problems in the manufacture of grinding wheels. With the exception of vitreous-bonded CBN wheels, superabrasives are retained in the grinding wheel almost solely by the mechanical strength of the bond system. It has therefore been important to develop diamond and CBN abrasives having a very broad range of toughness properties for optimum performance on the total spectrum of workpiece MOR values.

A very broad range of products with sharply differentiated fracture, shape, and surface properties are available to diamond tool manufacturers.

Toughness Index. The fracture characteristics of diamond abrasives can be measured using a device that subjects a measured sample of abrasive to a series of controlled impacts. This is effected by oscillating the sample of diamond in a small capsule with a hard metal ball. After a precise number of impacts, the degree to which the diamond abrasive sample has been reduced in size is then determined by a sieving process. The

percent of original sample by weight that has not been broken down (reduced in size) by the controlled impact test is called the toughness index. Thus, a diamond abrasive with a high toughness index has a high resistance to fracture, and one with a low toughness index has a low resistance to fracture and is quite friable.

Because the applications for diamond demand a very broad range of fracture characteristics, several different toughness test conditions are required in order to obtain the desired precision of measurement for each type of product. However, in order to illustrate the range of toughness characteristics of manufactured diamond abrasives, one set of conditions has been selected for the evaluation of several selected types of diamond. Discrete mesh size samples of several selected types of diamond and one type of CBN abrasive were subjected to the toughness test. The results of such tests are shown in Fig 3, where the variation of abrasive toughness or fracture strength is graphed as a function of grit size or mesh size of the abrasive. This graph reveals

the classic effects of flaw density with respect to size of abrasive for all types of abrasives. Finer grain sizes exhibit greater resistance to fracturing and breakdown; the coarser abrasive sizes exhibit the opposite characteristics

The most applicable bond system and scope of modulus of resilience for each of three ranges of diamond toughness are also found in Fig 3.

Figures 4 to 6 illustrate diamond abrasives from the three major categories shown in Fig 3.

Coatings to Improve Bond Properties

Relatively weak, resilient resin bonds are required for grinding high MOR workpieces such as cemented carbides. Because of the low-strength bond, even irregularly shaped, friable diamond abrasives are lost from the surface of the wheel after undergoing only slight wear. While such a system offers high grinding rates, wheel life is relatively short.

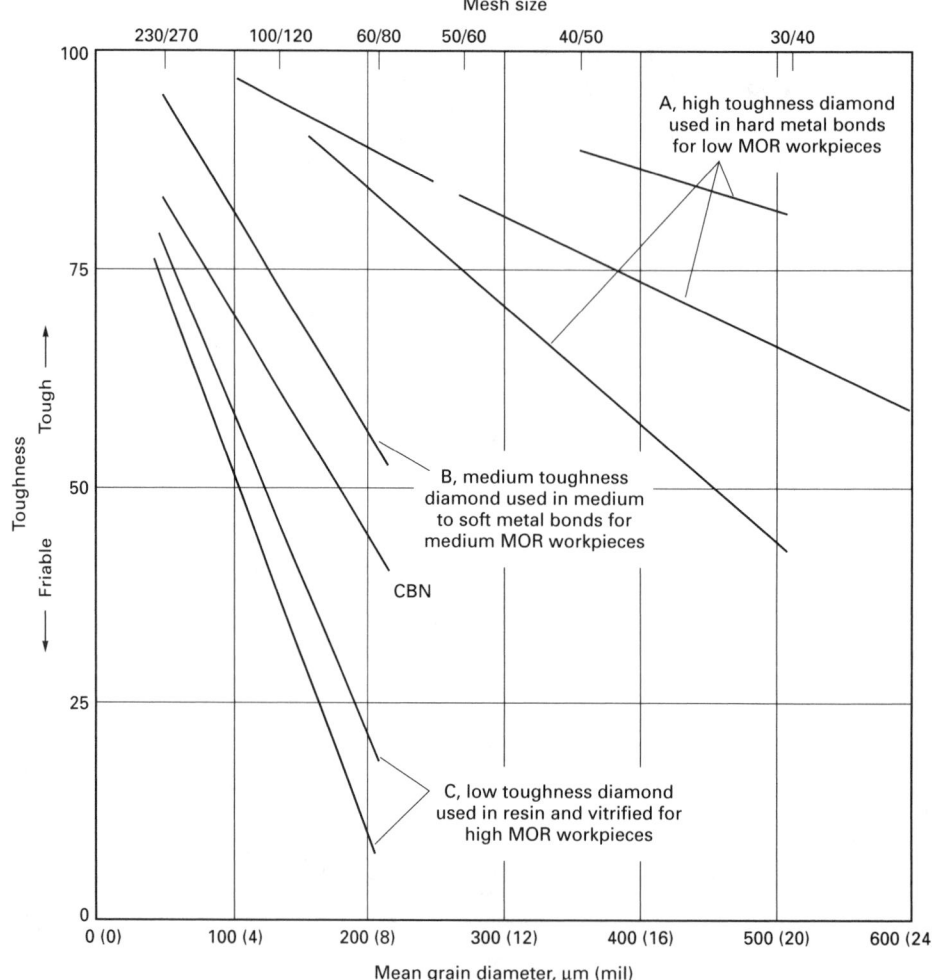

Fig 3 Toughness of superabrasives shown as a function of abrasive grain size. In addition, the most effective bond system for each of the three ranges of diamond toughness—A (high toughness), B (medium toughness), and C (low toughness)—is described in terms of the MOR value.

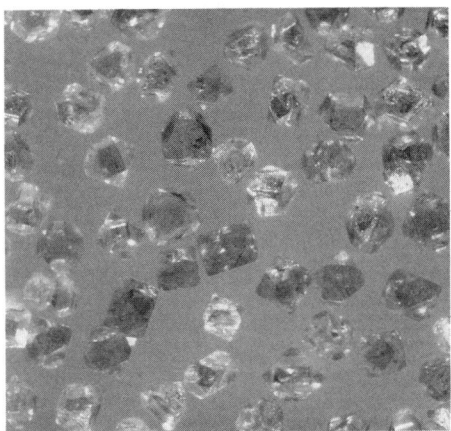

Fig 4 Typical high toughness (A range) diamond abrasive for metal bonds used in the grinding and sawing of low MOR workpieces (for example, natural stones and concrete)

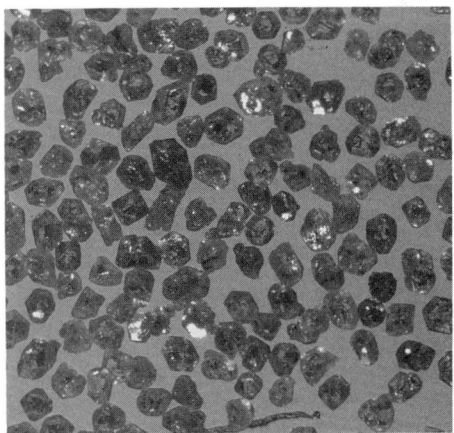

Fig 5 Typical medium toughness (B range) diamond abrasive for soft to medium metal bonds for use in the grinding of high-alumina ceramics, glasses, and other medium to low MOR materials

Fig 6 Low toughness (C range, high friable) diamond abrasive with a heavy metal coating for use in resin bonds. Such a system is specifically applied to the grinding of medium to high MOR workpieces (for example, silicon carbide, silicon nitride, zirconias, and the cemented carbides).

Nickel Metal Coatings. In order to alleviate this problem, thick nickel metal coatings for friable diamond were introduced in 1965 (Ref 6). There are several important advantages that thick metal coatings bring to the abrasive/resin bond system:

- High strength reinforces retention of abrasive
- Relatively low thermal conductivity metal buffers degradable resin from grinding heat
- High surface area and superior wettability increase bonding strength over uncoated abrasive

These advantages translate into grinding wheel life two to three times that of wheels with uncoated diamond.

Metal coated diamond is now used almost exclusively in the grinding of high MOR workpieces. This same coating process is used with CBN abrasives intended for use in resin bonds on hardened steels and superalloys.

Diamond Grinding Applications

In order to support the guidelines for selecting diamond for the grinding media for ceramics and glasses presented in this article, additional terminology and some specific examples with further guidelines are required.

Nomenclature

A glossary of grinding terminology is helpful in the interpretation of grinding performance data:

Specific grinding rate (Q'). This rate quantifies the volumetric rate of material removal for a specified set of grinding parameters. It is expressed in terms of volume of material ground away per unit of grinding wheel width: $mm^3/mm \cdot s$.

Grinding ratio (G). The volumetric efficiency of the grinding wheel is the ratio of the volume of workpiece removed to the volume of grinding wheel worn away per unit time at constant Q', and is dimensionless: $(cm^3/s)/(cm^3/s)$. Large values indicate high efficiency (long wheel life) while low values indicate low efficiency (short wheel life).

Specific grinding energy. The energy required to grind away a unit volume of workpiece per unit time at a specified grinding condition: $W \cdot h/cm^3$.

Grit size. This number in the grinding wheel specification denotes the approximate number of openings per lineal inch in the final screen size used in sizing abrasive grain. Typical coarse grits are 46, 60, and 80. Typical fine grits are 150, 220, and 300.

Ceramic Machining Guidelines

The following recommendations are guidelines to provide optimum results when grinding and finishing ceramics.

Grinding of High-Alumina Ceramics. Referring again to Fig 1, high-alumina ceramics have a medium to low modulus of resilience. Consequently, such materials can be effectively ground by diamond in a metal bond. Surface grinding studies to determine the performance of both a medium- and a high-toughness diamond abrasive in grinding high-alumina ceramics are available in the literature (Ref 7). The metal bond wheels contained 120 grit diamond abrasive. The ceramics selected were rectangular bodies of an 85% fine-grained Al_2O_3, and a 99.5% coarse-grained Al_2O_3. The grinding conditions and grinding ratio results are shown in Fig 7. These results show that on the higher MOR 85% Al_2O_3 (MOR of 0.22 J/mm^3, or 32 lbf · in./in.^{3}), the medium toughness diamond provided the most effective results in terms of wheel life. On the lower MOR 99.5% Al_2O_3 (MOR of 0.14 J/mm^3, or 20 lbf · in./in.^{3}), the toughest diamond provided performance equal to the medium toughness diamond. These results clearly demonstrate the sensitivity of diamond abrasive toughness or fracture properties to the modulus of resilience of the workpiece material.

Grinding of Glass. All commercial glasses fall within the medium to low range of MOR values and can be readily ground with metal bonded diamond wheels. But again the relationship between the diamond toughness and workpiece modulus of resilience can be illustrated in an actual grinding test. In this case, a white crown optical lens grade of glass has been surface ground with both medium- and high-toughness diamond of 100 grit in a metal bond (Ref 8). By surface grinding at two different material removal rates, the change in wear characteristics in the two types of diamond can be illustrated. The results of this work are shown in Fig 8. The wheel with medium-toughness diamond shows somewhat higher grinding ratio than the wheel with tougher diamond at the lighter grinding rate ($Q' = 279$ $mm^3/mm \cdot s$). But at the heavier grinding rate ($Q' = 373$ $mm^3/mm \cdot s$), the results reverse, and the wheel with the tougher diamond exhibits higher grinding ratio. Thus, the selection of optimum diamond properties

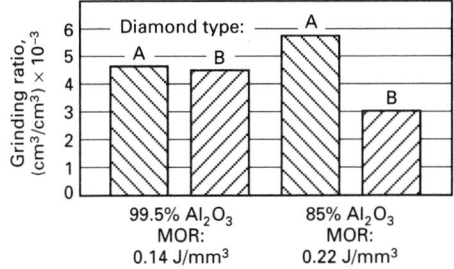

Fig 7 Rating the performance of both a tough and a medium diamond abrasive wheel in the grinding of two different alumina ceramics (99.5 Al_2O_3 and 85% Al_2O_3) with Q' of 157 $mm^3/mm \cdot s$

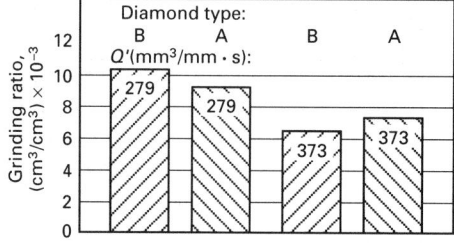

(a)

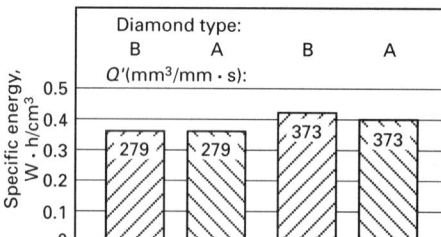

(b)

Fig 8 Comparing the ratings of two types of diamond abrasives at two specific grinding rates when grinding optical glass. (a) Grinding ratio. (b) Specific energy

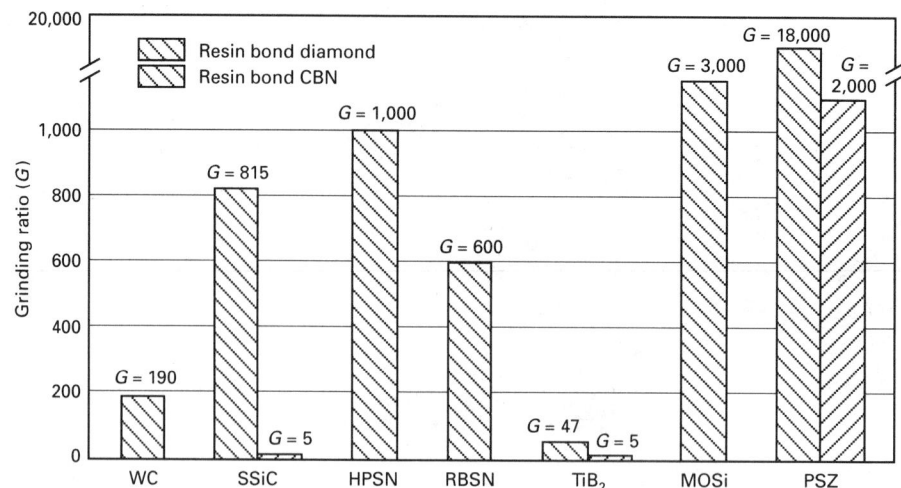

(a)

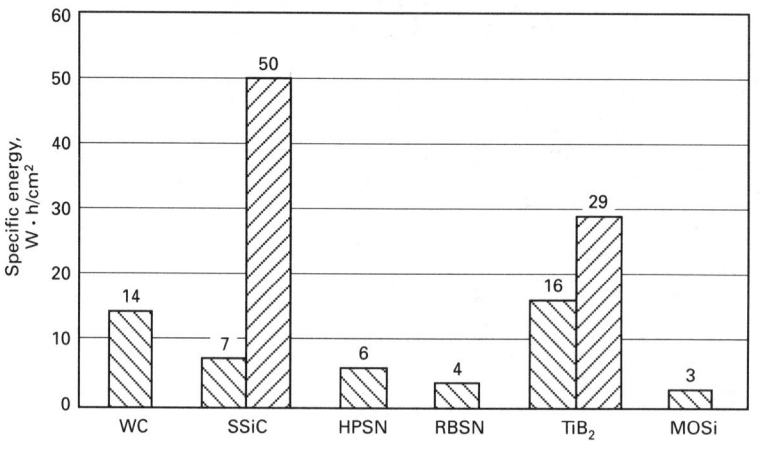

(b)

Fig 9 Grindability survey comparing resin-bonded wheels containing 150-grit metal coated abrasives to evaluate capabilities of both friable diamond and CBN abrasives in grinding technical ceramics. (a) Grinding ratio. (b) Specific energy. The specific grinding rate, Q', is 2.55 mm³/mm · s in all cases with the exception of partially stabilized zirconia (PSZ), where Q' is 1.90 mm³/mm · s. Source: Ref 9

is also shown to be a function of material removal rate or feed rate, as well as the modulus of resilience of the workpiece. This aspect of diamond selection becomes even more critical on workpiece materials of very low MOR value at the far left of Fig 1.

Grinding of Technical Ceramics. Prophesies and expectations of a significant growth in the uses of technical ceramics in industrial and consumer products have spawned intense research aimed at solving the many problems surrounding the processing and finishing of these ceramics. Many approaches are being investigated. Diamond grinding, however, remains the most economically and technically effective method for the precision shaping and finishing of technical ceramics. This discussion will only cite results obtained in surface grinding several different types of technical ceramics with the most appropriate type of diamond grinding wheel. The results of a ceramic grindability survey are presented in Fig 9. This work was conducted using resin bonded wheels containing 150-grit metal-coated friable diamond and 150-grit metal-coated cubic boron nitride. The objective of this work was only to evaluate the relative ease or difficulty of grinding a range of ceramic materials. No attempt was made to correlate these results with the remnant strength following the grinding process. The bibliography, however, references several studies that do correlate specific grinding conditions with surface integrity. The quantitative results are the grinding ratio and the average specific energy

required in watt · h/cm³. Each of these materials are discussed below.

Cemented Carbide (tungsten carbide, MOR of 3.3 J/mm³, or 480 lbf · in./in.³). Die grade of 16% cemented carbide that has been ground simply as a reference point from which to judge the grindability, or ease of grinding, of the other materials evaluated.

Sintered Silicon Carbide (SSiC, MOR of 0.01 to 0.04 J/mm³, or 1.4 to 6 lbf · in./in.³). This sample was ground with both a diamond and a CBN wheel to verify the superiority of diamond on the broad family of ceramics. Sintered silicon carbide is relatively easy to grind with a resin-bonded diamond wheel. A CBN wheel, by contrast, wore extremely rapidly and required an extraordinary amount of power to grind. This particular workpiece could be successfully ground

at $Q' = 7.5$ mm³/mm · s before grinding operations began to damage workpiece surface through chipping and cracking.

Hot-Pressed Silicon Nitride (HPSN, MOR of 0.8 to 1.6 J/mm³, or 115 to 230 lbf · in./in.³). This is a relatively easy to grind workpiece. This material can be ground at $Q' = 10$ mm³/mm · s. However, as in the case of sintered silicon carbide, final grinding may have to be carried out at $Q' = 0.5$ mm³/mm · s in order to heal inherent surface damage.

Reaction-Bonded Silicon Nitride (RBSN, MOR of 0.1 to 0.2 J/mm³, or 15 to 30 lbf · in./in.³). A relatively easy workpiece material to grind. This material can be ground at $Q' = 7.5$ mm³/mm · s but finish grinding at very low Q' values of ≤0.5 mm³/mm · s must be used for final grinding.

Titanium Diboride (TiB₂, MOR of 0.4 to

0.9 J/mm^3, or 60 to 130 lbf · in./in.^{3}). This is a far more difficult material to grind, as evidenced by the low grinding ratio and the much higher power required. The difficulty in grinding TiB$_2$ cannot be explained by its physical properties alone. Both diamond and CBN demonstrate extremely rapid wear rates on this material. This strongly suggests that this is another example of solution wear. This wear may be the result of the interaction of TiB$_2$ with both diamond and CBN at the temperatures generated in this process.

Monocrystalline Silicon (MOSi). An extremely brittle but readily ground material. Selecting grinding conditions and fixturing will be important in controlling edge chipping and fracture during grinding.

Partially Stabilized Zirconia (PSZ, MOR of 0.8 to 1.2 J/mm^3, or 115 to 170 lbf · in./in.^{3}). This material was creep feed ground at $Q' = 1.9 \text{ mm}^3/\text{mm} \cdot \text{s}$. It is clearly an extremely easy material to grind, with an extremely high grinding ratio under this relatively low grinding condition. Even the CBN wheel could grind this workpiece fairly readily under these conditions. However, the CBN wheel is totally uneconomical in comparison with diamond on this material.

REFERENCES

1. E. Ratterman, Grinding Structural Ceramics—A Systems Approach, *Ceram. Eng. Proc.*, 1985
2. W.J. Lackey, D.P. Stinton, G.A. Cerny, A.C. Schaffhauser, and L.L. Fehrenbacher, Ceramic Coatings for Advanced Heat Engines—A Review and Projection, *Adv. Ceram. Mater.*, Vol 2 (No. 1), 1987, p 24–30
3. *Metals Handbook*, Vol 2, 10th ed., Oct 1990
4. J.E. Field, *The Properties of Diamond*, Academic Press, 1979
5. P.D. Gigl, "The Strength of Polycrystalline Diamond Compacts," Airapt Conference, July 1977
6. W.F. Matthewson, E. Ratterman, and K.H. Gillis, An Analysis of the Coated Diamond/Bond System, *Grinding and Finishing*, 1965
7. Ceramic Finishing with Diamond, Part II—The Abrasive and the Bond System, *Ceram. Ind.*, Mar 1968
8. Unpublished internal report, EB 83–39, GE Superabrasives, 1983
9. Unpublished internal report, EB 83–32, GE Superabrasives, 1983

SELECTED REFERENCES

- A. Antenen and R. Ohnsorg, "Superabrasives for Ceramic Grinding and Finishing," Advanced Ceramics '89, 1989
- R.W. McEachron, "Grinding of Structural Ceramics with Diamond Abrasives," Superabrasives '85
- R.W. McEachron and S.C. Lorence, Superabrasives and Structural Ceramics in Creep-Feed Grinding, *Am. Ceram. Soc. Bull.*, Vol 67 (No. 6), 1988
- K. Subramanian, "Parametric Study on Grindability of Structural and Electronic Ceramics—Part 1," The Production Engineering Division, American Society of Mechanical Engineers, Nov 1985
- J.J. Tuzzeo, "Correlation of Diamond Properties and Performance in Surfacing of Ophthalmic Glass Lenses," Diamond Wheel Manufacturers Institute, 1975
- K. Unno and T. Imai, "Performance of Diamond Wheel in Grinding Ceramics," 6th International Conference on Production Engineering, Osaka, 1987
- G. Willmann, Finish Machining & the Strength of Ceramic Parts, *Ind. Diamond Rev.*, 1985

Production Grinding Methods and Techniques

K. Subramanian and R.N. Kopp, Norton Company

PRODUCTION GRINDING SPECIFICATIONS AND APPLICATIONS associated with the machining of ceramics are discussed in this article as a supplement to the general principles of the grinding of ceramics covered in a previous article "Principles of Abrasive Machining" in this Volume.

Ceramic Grinding Applications and Methods

Electronic Ceramics Grinding. As the name implies, these applications of ceramics are based on their electronic properties. These applications are found in computer components such as microchips, magnetic heads, substrates, and so on.

Electronic ceramic components require extensive grinding. Figure 1 shows the sequence of operations used in the fabrication of semiconductor chips. The single-crystal silicon log is cropped at the ends, cylindrical ground, and a reference flat machined on it using diamond grinding wheels. These operations are followed by slicing, using annual saw blades. These are thin flat sheet steel stretched closed to the yield point to achieve a stiff membrane. The inner surface of this annular membrane carries a layer of abrasives, which cause the cutting off of the silicon logs into thin slices or wafers. These are then lapped and polished to desired thickness. To prevent chippage during handling, these wafers are ground along the edges to form a radius or other profile. After the circuitry is deposited, these wafers are ground on the backside. Then they are cut off into individual chips using dicing blades. These blades are of the order of 0.02 to 0.05 mm (0.0008 to 0.002 in.) in thickness.

Figure 2 shows the operation of a magnetic head on a hard disk. The head literally flies on the top of the disk at the constant spacing or height. This spacing or gap width is being constantly reduced to achieve enhanced performance (Fig 3). The reliability and consistency of this gap is achieved by precision finishing of the surfaces on the magnetic head. The head is made of ferrite composites (that

is, a barium or calcium titanate body with a ferrite core) or an alumina/titanium carbide ceramic.

Figure 4 shows the typical sequence of operations required in the grinding of magnetic heads. As shown, many of these operations require thin precision grinding wheels operating with a high degree of accuracy. Figure 5 shows a typical profile ground in a ferrite ceramic for a magnetic head application. Thin precision wheels assembled in a precision gang are often used in magnetic head fabrication.

In general, electronic ceramics grinding requires extreme precision and tolerances. Understanding the residual stresses created by grinding and their influence on part geometry is very critical in these components. The tolerances on straightness and parallelism are held typically 0.005 mm (0.0002 in.). This is in addition to the smaller thickness of the wheels used primarily to reduce the kerf width or the material removed during the grinding. Minimizing the kerf loss helps to increase the number of components made per wafer. Micron-sized diamond grains are extensively used in these operations. The grain size uniformity, quality, and distribution heavily impact the grinding performance. In addition, the consistency, tolerances, homogeneity, and balance limits of the grinding wheels also influence the grinding results. Several specially built grinding machines are used for various operations in electronic ceramic grinding. A variety of conventional abrasives, as well as diamond abrasives, are used in the form of powder, slurry, or compounds for the lapping and polishing of electronic ceramic components.

Technical ceramics designate applications where the tolerances required are critical. However, they are not generally as small or as precise as those required for electronic ceramics. These applications include quartz tubing used for lighting, semiconductor component casings, orthodontics, nuclear, biomedical, fiber optics, and so on (Fig 6).

Technical ceramics grinding often involves cutting a large part or tube into subcomponents having minimal chippage. Frequently,

finishing of flat surfaces must be scratch free or a mirror finish is also required. Figure 7 shows typical arrangements used in the cutting of tubes using both single wheels and gang wheels.

A wide range of bond types and grit sizes are used in these applications. The typical products used for grinding technical ceramics are cut-off blades and Blanchard or disk wheels. Grinding process optimization, the tolerances of grinding wheels, and the work material features, are among the factors that heavily influence the success of these grinding operations. Both resinoid and metal bonded grinding wheels are used for these applications. Loose abrasives (both conventional and diamond) are also extensively used in the finishing of technical ceramics for lapping, polishing, and so on.

Industrial ceramics are ceramics classified as traditional ceramics of low density and low strength. These products are extensively used for applications requiring the following properties:

- Thermal resistance
- Corrosion resistance
- Electrical insulation

Typical examples of industrial ceramics are:

- Refractories
- Furnace liners
- Electrical parts
- Coated ceramics

The grinding operations for industrial ceramics vary widely in shape and configuration. Metal bonded grinding wheels are mostly used for these operations to cut off parts or to generate large flat surfaces. Metal bonded and electroplated diamond wheels are extensively used for these applications to achieve long durability for wheels. The chippage criteria and tolerances for these applications are generally rather broad relative to other applications. Hence, they can tolerate the higher grinding power and forces generated by the metal bonded wheels or the poor finish and damage produced by the electroplated wheels. However, applications requiring tighter tolerances, the choice of metal bond, and its

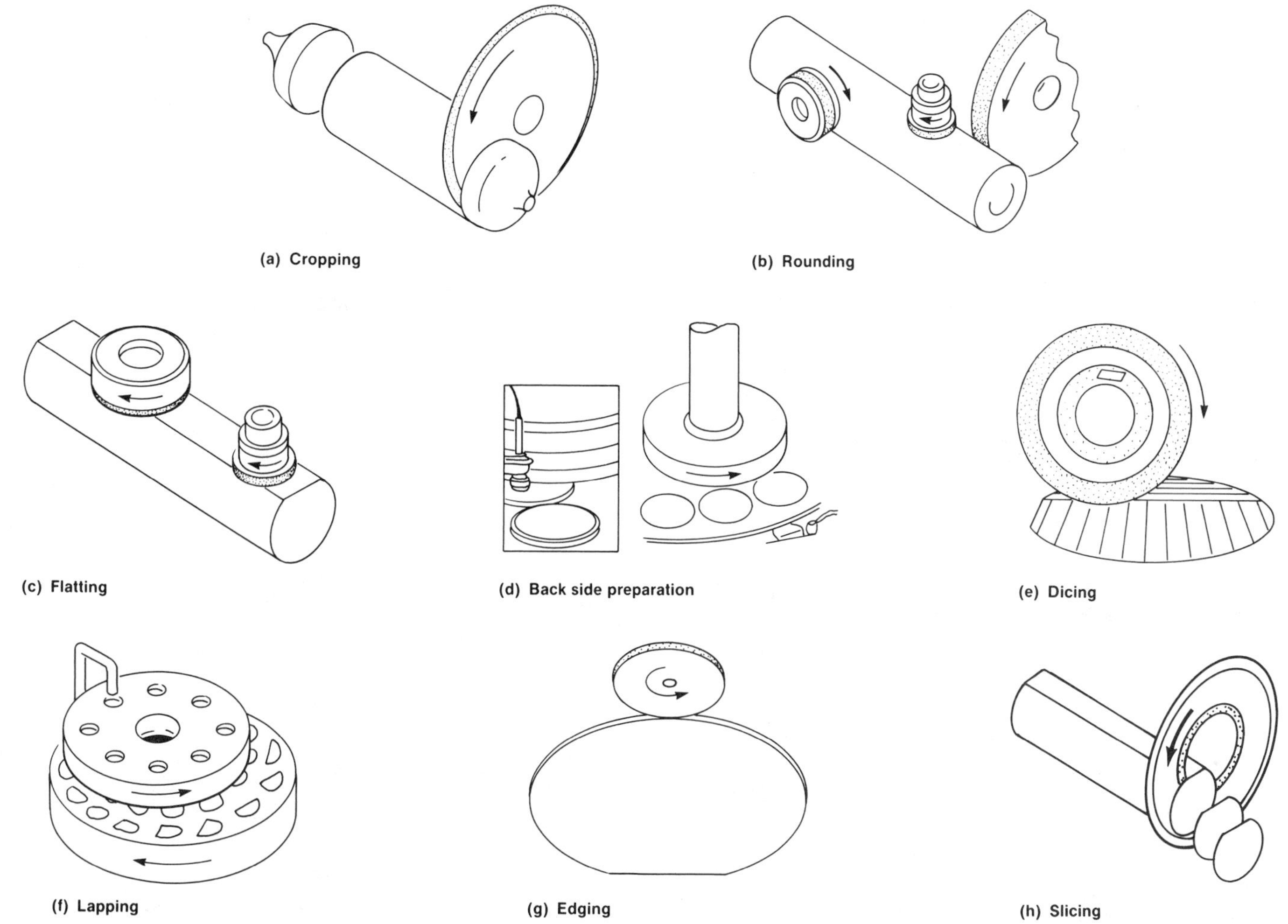

Fig 1 Diamond abrasive machining operations used to fabricate semiconductor components

(a) Cropping

(b) Rounding

(c) Flatting

(d) Back side preparation

(e) Dicing

(f) Lapping

(g) Edging

(h) Slicing

properties, as well as the dressing practices used, will require careful consideration.

Advanced ceramics are the high-strength, high-density (low-porosity) ceramics being

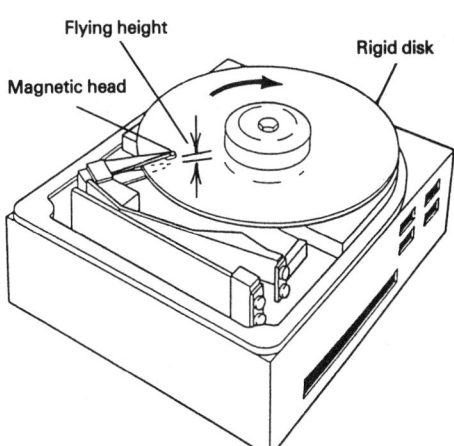

Fig 2 Magnetic head used for read-write functions in a disk drive. Flying height is the space between the head and the disk when the head is operational.

Flying height

Rigid disk

Magnetic head

evaluated for mechanical or structural applications. These materials are the most difficult to grind due to their high hardness and strength. Figure 8 shows a range of advanced ceramic parts after grinding. Typical requirements for these components are high retained strength after grinding and also production viable grinding methods (that is, short cycle time, economic grinding process, as well as consistent part quality (Ref 3).

The range of grinding operations required for advanced ceramics parallel the grinding operations used for metal and carbide components (Ref 4). Figure 9 shows the typical range of diamond grinding processes available for advanced ceramic components grinding. Figure 10 shows the typical range of diamond abrasive products used in the grinding of advanced ceramics.

The principles described in the article "Principles of Abrasive Machining" in this Volume should be closely followed to achieve successful grinding of ceramics. In the absence of such a fundamental and systematic approach, severe damage can be caused during the grinding process. Through the use of suitable grinding methods, principles, and procedures, advanced ceramics can be ground

to the required geometry and tolerances at competitive cycle times (Table 1).

Diamond Grinding Wheel Construction

The diamond grinding wheels used for ceramic materials grinding can range widely in their shape, size, and configuration. The successful application of grinding wheels will depend both on a thorough understanding of these aspects of diamond wheels and the principles outlined in the article "Principles of Abrasive Machining" in this Volume.

Diamond grinding wheels are available in a wide range of shapes as shown in Fig 9 and 10. These shapes and their selection will depend on the grinding process or configuration. The most common shape designations are shown in Fig 11.

Diamond abrasives are very expensive compared to conventional abrasive grains. Hence, it is not economical to make the entire wheel with abrasives, as is common for conventional abrasive wheels. Hence, diamond wheels are often designed with a core section that does not carry the diamond abrasive. The diamond abrasive section forms a layer on the

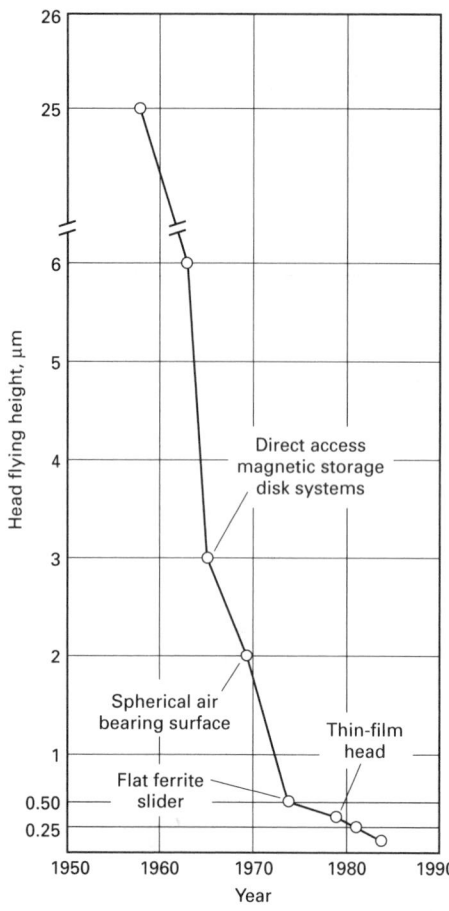

Fig 3 Reduction in gap width between magnetic head and hard disk of commercial storage systems from the years 1960 to 1990. Source: Ref 1

outer section of the core (Fig 12). This abrasive section consists of diamond abrasives and a bond matrix to retain the abrasive grains.

The features of the bond heavily influence the grinding performance. Grinding wheel manufacturers should be consulted for the

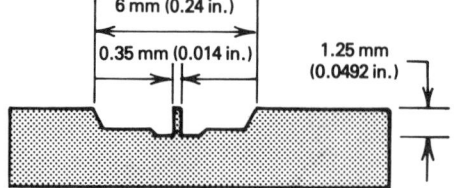

Fig 5 Schematic showing typical cross-section obtained by grinding of a ferrite magnetic head

proper selection and application of the various bond types available. There are four families of bonds used for diamond grinding wheels:

- *Resin bond* refers primarily to a polymer binder holding the abrasive grains

- *Metal bond* refers to metallic alloys used to retain the abrasive grains
- *Vitrified bond* refers to glass matrix containing the abrasive grains
- *Single-layer bond* refers to electroplating or brazing a layer of abrasive grains onto a premachined preform

The key characteristic of any bond material is to retain the abrasive grains efficiently. This implies strong retention of the abrasives during the grinding process, yet preferential breakdown to shed the worn abrasives and expose fresh abrasives for cutting action. In addition, the bond should contribute minimally to the frictional interactions during grinding. On occasion, the bond is also designed to contain porosity, which minimizes the frictional interactions, enhances coolant entrainment, and allows for a con-

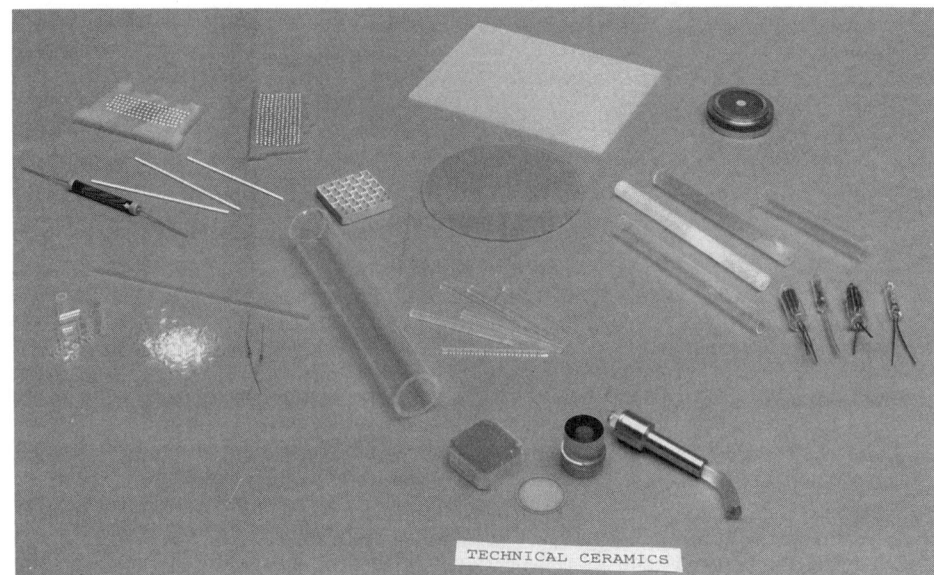

Fig 6 Typical technical ceramic components produced by grinding processes

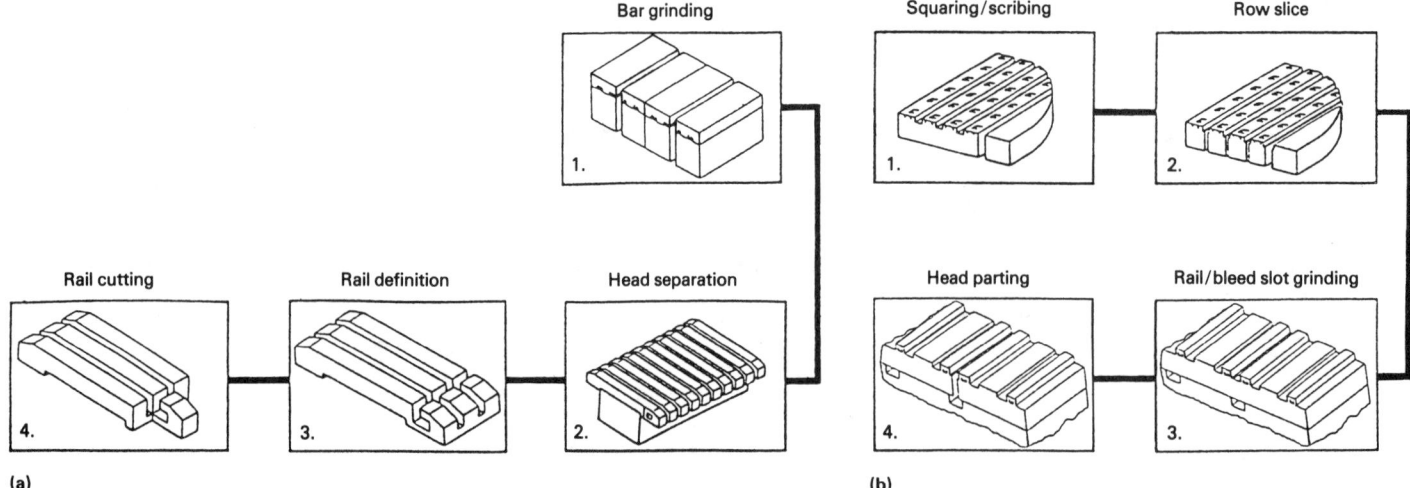

Fig 4 Typical grinding operations required to fabricate two types of magnetic heads. (a) Composite heads. (b) Thin-film heads

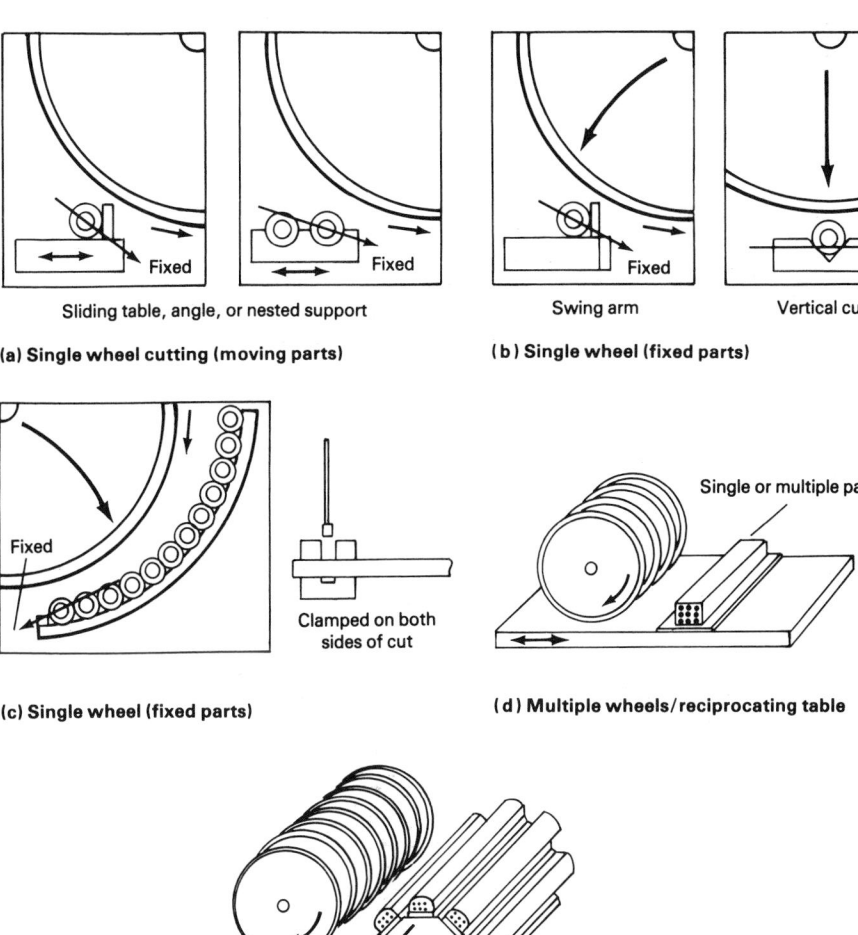

(a) Single wheel cutting (moving parts)

Sliding table, angle, or nested support

(b) Single wheel (fixed parts)

Swing arm · Vertical cut

(c) Single wheel (fixed parts)

Clamped on both sides of cut

(d) Multiple wheels/reciprocating table

Single or multiple parts

(e) Gang wheels/multiple fixtured

Parts totally fixed

Fig 7 Schematic configurations available for slicing technical ceramic tubes. Source: Ref 2

Fig 8 Selected advanced ceramic workpieces finished by production grinding techniques

trolled breakdown of the bond (Ref 5, 6). Other factors that affect wheel selection are shown in Table 2.

The abrasive types and their applications are discussed in the article "Abrasives" in this Volume. Each application for ceramic grinding will require a carefully selected grit size range. However, depending on the application, the chosen grit size may fall in a wider band than is common for conventional abrasives. For instance, micron grit sizes ranging from 0.1 to 1 μm (4 to 8 μin.) may be used for ultrasmooth finishing of electronic ceramics. However, specific application of this type may require 0.1 to 0.2 μm (4 to 8 μin.) or 0.5 to 1 μm (20 to 40 μin.) range, and so on (Fig 13).

Ceramic grinding requires precision movements and uniform, vibration-free operation to achieve these requirements (Ref 7, 8). The tolerances required for diamond grinding wheels also become tighter. For instance, the thickness tolerances used in precision slicing of ceramics is on the order of ± 0.005 mm (± 0.0002 in.) or better. Other dimensional tolerances that must be closely monitored are shown in Fig 14. This aspect of close tolerances in diamond grinding wheels should not be ignored. Otherwise, chipping and cracking will result, leading to poor yield and high grinding costs.

Practical Aspects of Grinding Machines and Process

As shown in Fig 1, 4, 7, and 9, there are a variety of grinding methods or configurations to choose from, depending on the machine tool available, work material geometry, and so on (Ref 9–11).

Creep Feed Versus Reciprocating Surface Grinding. The surface generation in the grinding process may be achieved by rapid traversals of the grinding wheel, with a small depth of material removed per traversal. This is called surface grinding. On the other hand, the work material can be traversed at a very low table speed, while a large depth of the material is removed per traversal. This is called creep-feed grinding. These two methods are compared in Fig 15. The key aspects that distinguish creep-feed grinding are the low work speed and the large depth of cut. Generally, creep-feed grinding also involves the contact between the work and the entire thickness or profile of the wheel (Fig 15). The area of contact between the wheel and the work is substantially larger in creep-feed grinding in comparison to surface grinding. This generally results in a larger total force and power required for creep-feed grinding when compared to surface grinding (Table 3).

These process variables can be normalized with reference to the contact area to determine the contact stresses and the energy dissipated per unit area per unit time (power intensity).

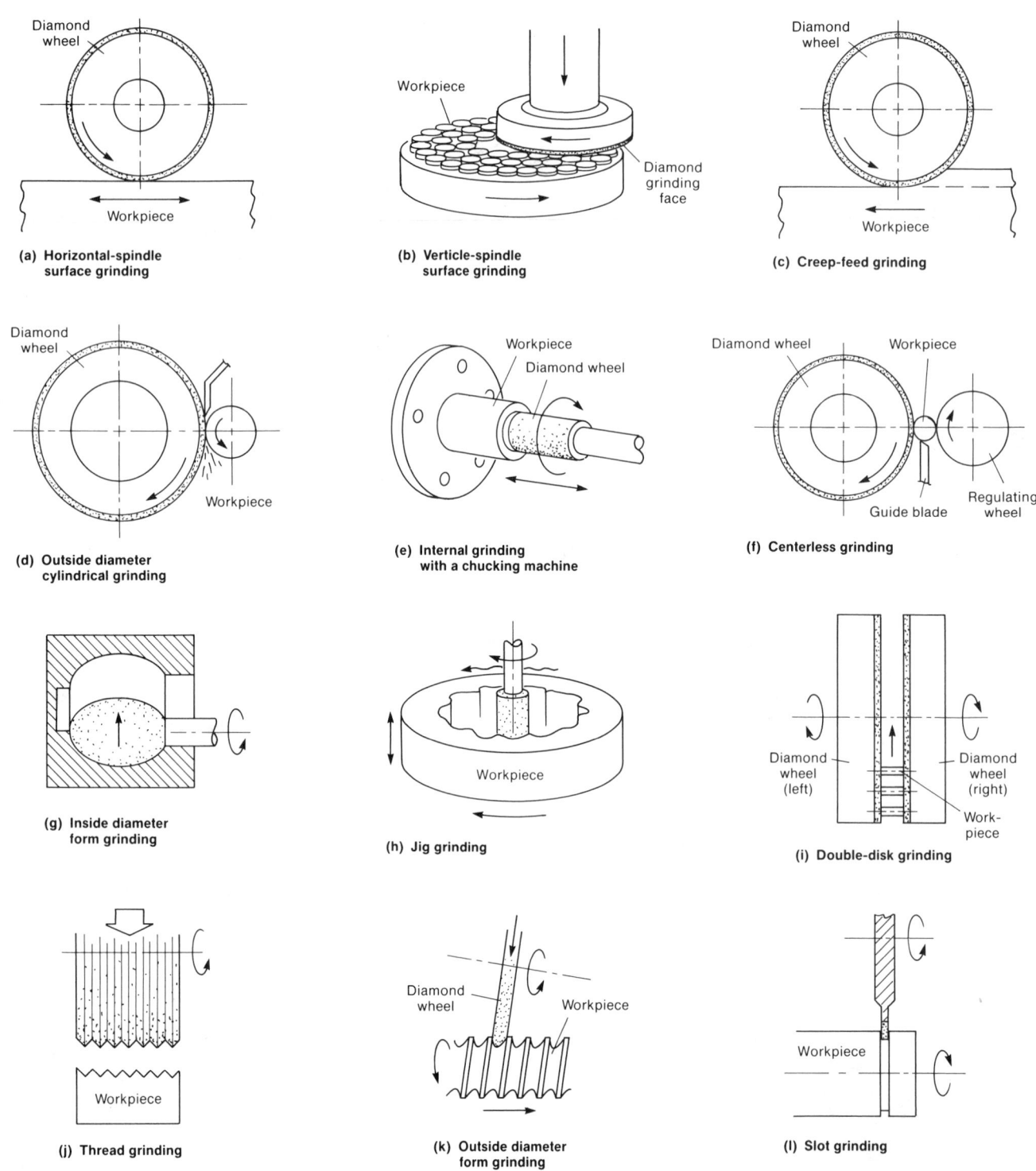

Fig 9 Production grinding applications of diamond grinding wheels

It is observed that creep-feed grinding leads to lower grinding stresses and power intensities when compared to surface grinding. These are very beneficial in ceramics grinding. Lower grinding stresses could imply high strength after grinding. Lower power intensities generally result in small thermal gradients and less thermal cracking of ceramics.

Creep-feed grinding is also beneficial to the grinding wheels. The lowered stresses and power intensities help to retain the form longer and to achieve closer tolerances in the geometry of the workpiece. However, the wheel design should accommodate the higher total grinding forces and power requirements. The machine tool design also is very critical for the successful creep-feed grinding of ceramics.

Form Grinding. Very often contours or forms are required to be machined on work materials. This requirement will be a growing need, particularly for advanced ceramic components that will replace carbide or metal parts. A typical example of form grinding is shown in Fig 16. The form grinding may be achieved by a profiled grinding wheel, which translates its form on to a work material (Fig 16a).

Fig 10 Typical diamond abrasive products required for grinding advanced ceramics

The form may also be generated by a diamond grinding wheel by computer numerical control (CNC) path generation (Fig 16b).

Form grinding can be accomplished in a variety of machines. The relative motions between the wheel will determine the nature of the machine chosen. For example, cylindrical forms may be plunge ground using standard OD or ID grinding machines and suitably profiled wheels. Creep-feed grinding, a common method used for form grinding, has the following key requirements:

● Rigidity
● Precision movements
● Truing and dressing methods/devices for form generation

Form generation using CNC motion is a developing technology. The advantages of this method are flexibility and lower stiffness required in the machine tool. Single wheels of simple shapes can be used for a variety of forms, depending on the CNC motions. The truing of a simple shaped wheel is also relatively easy and less expensive. Because the unit-width material removal rate (MRR') is small, the total forces are usually low. Hence, the high rigidity of the machine may not be required.

Form Grinding Versus CNC Generation. Choosing between form grinding and CNC generation depends on the production needs. Production grinding of large volumes may preferably be accomplished by form grinding. When the lot size is small and constant variation in part geometry is the norm, CNC generation of forms may be preferable. The benefit of both these approaches may be combined using multiaxis CNC grinders built with high rigidity. A CNC machine and its multiaxis CNC motion capabilities are illustrated in Fig 17.

Cylindrical grinding is defined as the grinding of parts to achieve concentric surfaces. This may be an OD or ID grinding operation. Several methods of holding the workpiece are used in cylindrical grinding (Fig 18). Many of these methods are often modified to accommodate ceramic material properties. For instance, the end support used in metal grinding rests on a countersunk hole in the work material. Because these are difficult to machine in ceramics, other means are used, such as attaching a metal end with a center hole to the ceramic material, which in turn is used to rest the end support.

The collets or chucks used for ceramics need to be modified to ensure uniform pressure around the periphery and to avoid stress concentration. The same principles also apply for the face plate and mandrels used for holding ceramic parts. Magnetic chucks are often replaced with vacuum chucks to directly hold ceramic parts. Sometimes a metal adaptor may be attached to the ceramic. The adaptor is in turn held by the magnetic chuck.

CNC grinding is also used to generate multiple surfaces in a single set up in cylindrical grinding operations. A CNC cylindrical grinder and the surfaces it can machine are shown in Fig 19.

Centerless Grinding. This is the method where the work material is ground on the OD, without being held on by its centers. In centerless grinding, the part is supported by the work rest blade (Fig 20a) and a regulating or feed wheel (Fig 20b). The rotation and translation of the part to achieve work speed is obtained by a slight incline of the regulating wheel relative to the grinding wheel. Centerless grinding may be one of four types depending on the process configuration (Fig 21).

Centerless grinding of ceramics is one of the methods where significant research has been conducted (Ref 12). The results indicate that a high degree of concentricity and good surface finish can be achieved in the centerless grinding of ceramics.

Wear of the work rest blade is one of the key concerns in centerless grinding of ceramics. In addition, the regulating wheel should be both wear resistant and deformation resistant. However, the regulating wheel should also be compliant to prevent chatter during grinding and to prevent cracking and poor surface finish in the ceramic materials.

Resin bonded diamond wheels are commonly used for centerless grinding of advanced ceramics. Vitrified bonded diamond wheels have been successfully evaluated to achieve good part quality while retaining long wheel life. Metal bonded grinding wheels are used in centerless grinding of low-density or low-strength ceramics where tolerances and finish are not critical but extreme durability of the wheel is desired.

Disk grinding refers to operations where the grinding wheel, in the shape of a large flat surface, is used to grind the ends of a

Table 1   Current and potential production capabilities for the grinding of advanced ceramics of three widely varying cycle times

	Parameter						
	Unit-width material removal rate (MRR')		Surface finish		Cycle time		
Process status	mm²/s, mm	in.³/min, in.	μm	μin.	h	min	s
Current practice	0.10–1.0	0.01–0.10	0.75–1.25	30–50	6–10	45–120	x(a)
Obtainable today with optimized grinding practice	1.0–30	0.1–3.0	0.05–0.10	2–4	<1	5–30	y(b)
Future	. . .	. . .	0.0003–0.001	0.012–0.040	. . .	. . .	. . .

(a) Not available, but on the order of minutes. (b) Not available, but on the order of several seconds

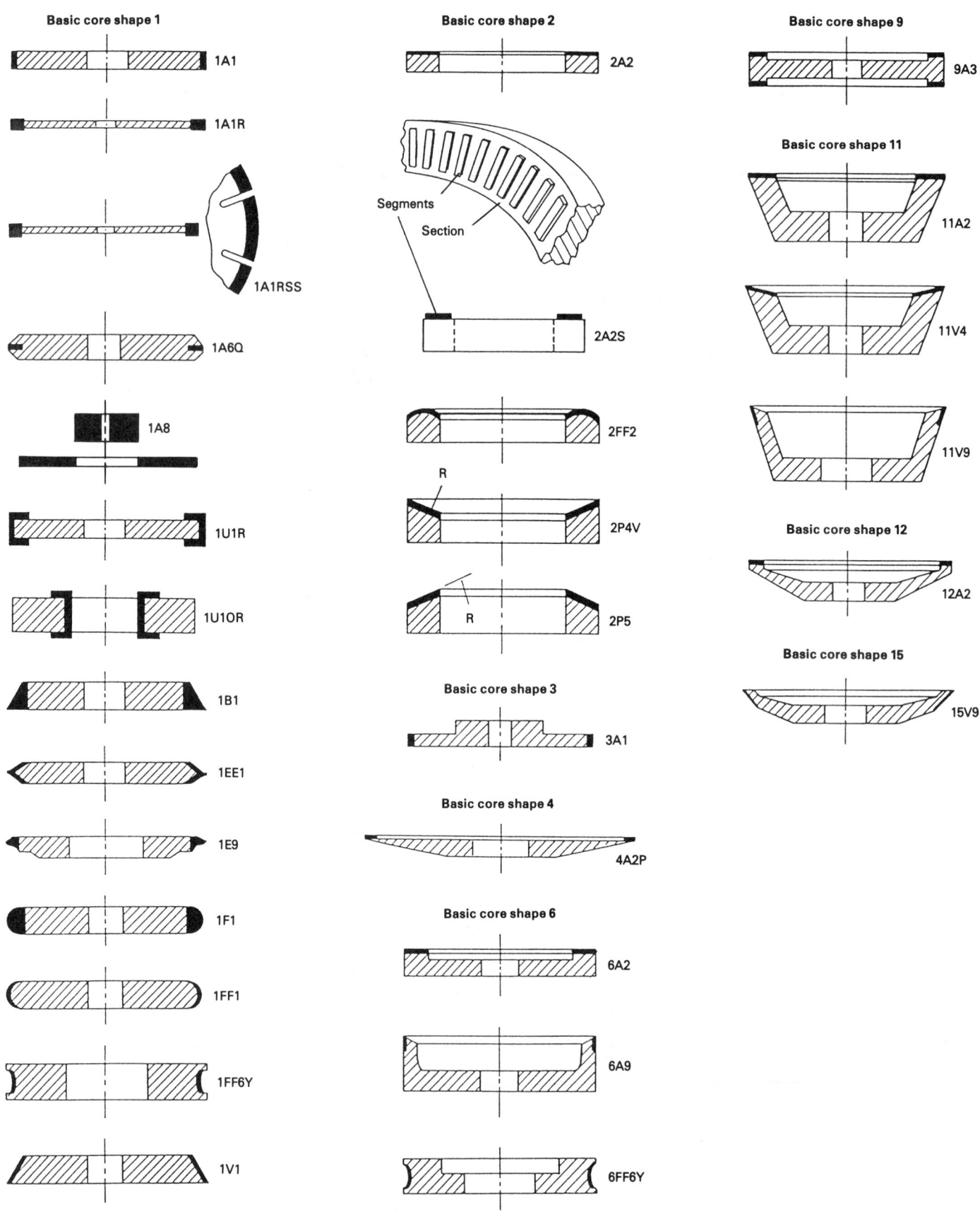

Fig 11 Superabrasive wheel configurations and their designations

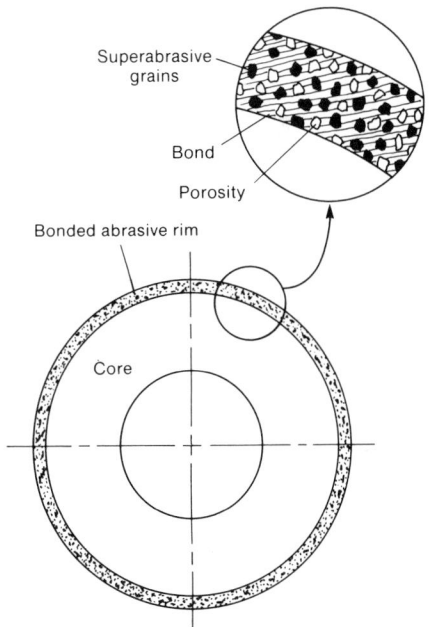

Fig 12 Construction of a typical superabrasive wheel

component to achieve the desired length or parallelism. These operations may be carried out using a single wheel or two wheels grinding in parallel. These are called single or double disk grinders. Figure 22 shows typical double disk grinding configurations.

Disk grinding is very often observed in the finishing of industrial ceramics. This method is used in the finishing of magnets, insulators, refractories, thermal parts, and so on. These operations are most often carried out dry. To accommodate this and ensure sufficient wheel life, metal bonded disk wheels are commonly used. Occasionally, electroplated disk wheels are also used for these applications.

Disk grinding will also be used for the finishing of advanced ceramics as they reach production quantities. In such situations, resin bonded disk wheels will be the first choice. As the process gets established and the volume of parts increases or the control limits are brought closer, vitrified bonded diamond disk wheels will find an increasing number of applications.

The setting up of the disk grinding machines is very critical and almost viewed as an art. In situations that seem impossible, fine tuning of machine settings, coolant applications, and work support fixturing often help to achieve the desired grinding results.

Precision slicing and slotting is one of the most common grinding methods observed in ceramics grinding. The ceramic materials are often manufactured as flat sheets, disks, rods, tubes, and so on. These materials are then cut or sliced into several smaller components for further use. Achieving this subdivision with a high degree of precision and consistency often requires diamond wheel cut

Table 2 Parameters affecting diamond abrasive and wheel bond type selection for machining ceramics

Parameter	Diamond abrasive designation	Effect on grinding process
Type	Natural or synthetic	Abrasive strength, consistency, and sharpness of abrasive grit
	Monocrystalline or polycrystalline	Self-sharpening characteristics
	Coated or uncoated	Grit retention strength
Particle size	Coarse	High MRR′, poor finish, poor strength
	Fine (including micron powders)	Low MRR′, improved finish and good strength
Particle size distribution	Inconsistent	Nonuniform grinding results
	Uniform/consistent	Reliable grinding at low chippage
Content concentration	Low	Free-cutting action, low life
	High	Long life, higher grinding forces, and power consumption
Bond properties	Hardness/grade	Freeness of cut
	Stiff/resilient	Dampening characteristics, deflection and chippage, slot or form accuracy
	Porosity	Freeness of cut, coolant entrainment
	Thermal conductivity	Heat removal from grinding zone, thermal gradients

off operations (Ref 13). The tolerance required in such operations varies widely depending on the applications. Magnetic heads and microchips are sliced with no visible chippage at 100× magnification. Industrial ceramics of low density are sliced to avoid gross fracture. While the tolerance level changes depending on the application, there are certain common principles that apply to all ceramic slicing/slotting operations.

The wheels must be flat, parallel, and concentric. The tolerance for these factors depends very much on the application. In general, the closer the tolerances of the wheel, the more homogeneous is the grinding process and the less chippage. Care must be exercised in the mounting of thin wheels to prevent distortion. Unequal flanges or excessive torsion on the flanges could distort the wheels, leading to excessive side runout (Fig 23).

Proper work support and minimizing of the vibration at the entry and the exit are also critical. Figure 24(a) shows concoidal fracture that can occur due to material defects propagated by the vibration as the wheel exits the cuts. Figure 24(b) shows fracture of a ceramic tube caused at the exit side when there are no material defects.

Work material properties also affect the quality of the cut. Figure 25 shows the comparison of the quality of the two different materials under similar grinding conditions.

Row bow is a key criteria in the quality of ceramic material slicing. This is the deformation or distortion of a thinly sliced part caused by residual stresses. Coarse abrasive grains reduce the row bow due to the brittle fracture during grinding and the absence of residual stresses. However, the chippage and cracks will increase substantially. Finer abrasive grits reduce chippage and cracking, but they also lead to deformation of surface layers of material leading to residual stresses. Hence, a key balance has to be achieved

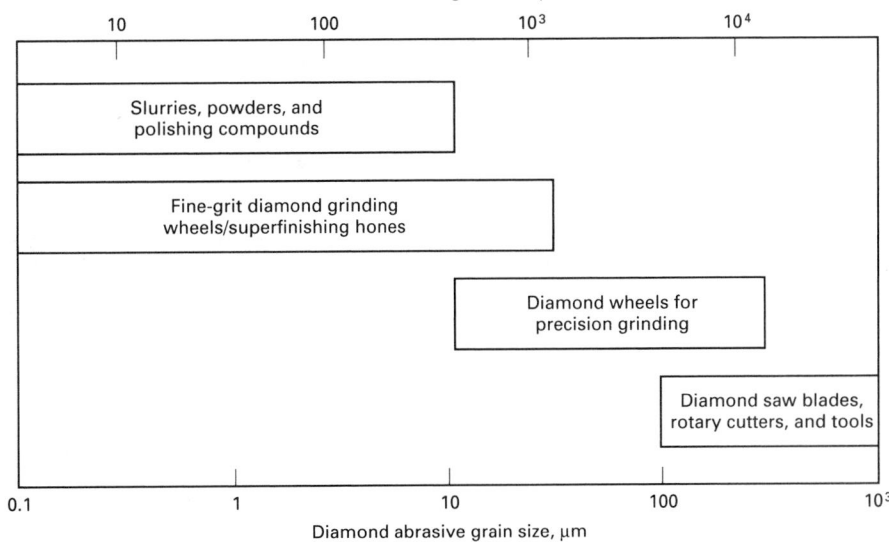

Fig 13 Selected diamond grinding methods and applications based on abrasive grain size

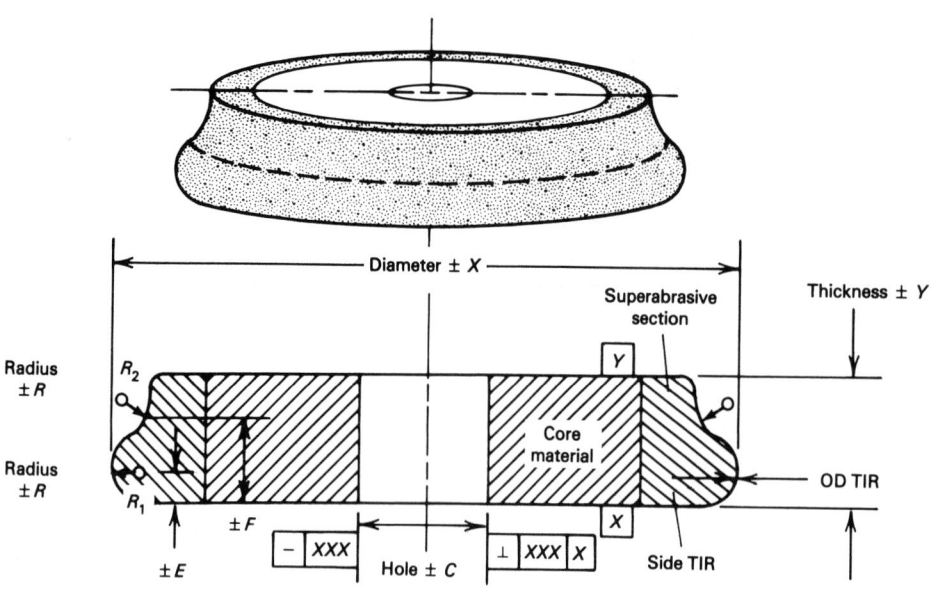

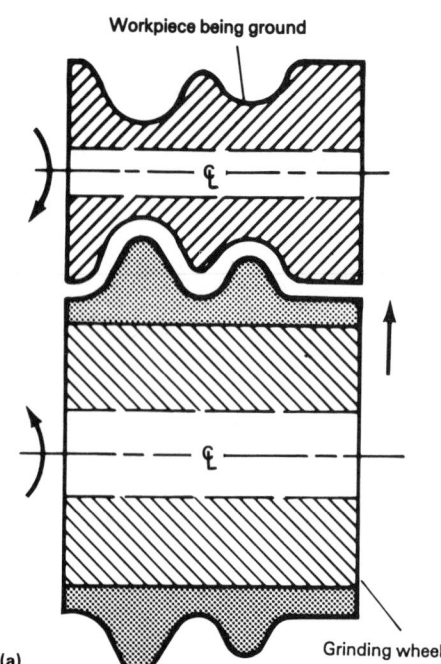

Fig 14 Tolerance specifications for diamond grinding wheels. X, Y, C, E, F, R are typical dimensional tolerances that must be controlled and monitored for precision grinding. TIR, total indicated runout; ⊥, perpendicularity; and —, straightness of hole

(a)

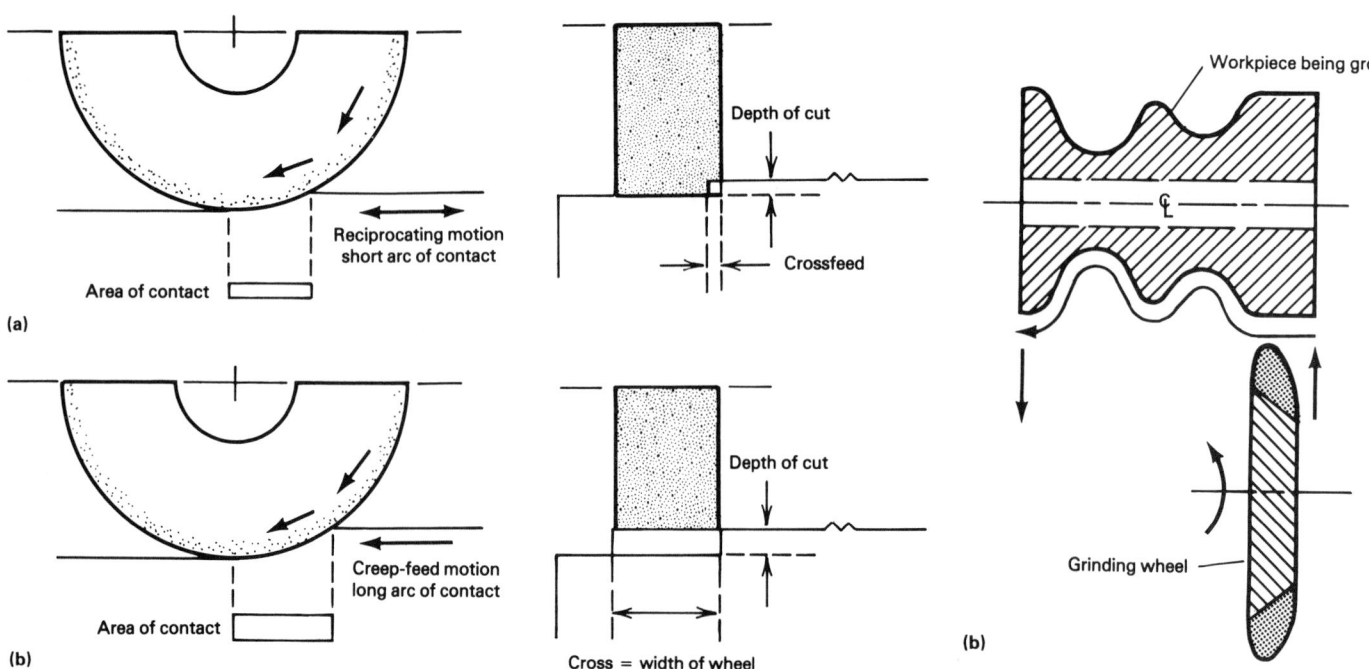

Fig 15 Schematics showing front view and side view of contact arc lengths. (a) Conventional surface grinding. (b) Creep-feed grinding

Fig 16 Two form grinding techniques. (a) Plunge grinding with a profiled grinding wheel. (b) Contour grinding by CNC path generation

Table 3 Recommended machining parameters for creep-feed grinding and reciprocating surface grinding of hot-pressed silicon nitride (HPSN) ceramic when using a SD180R100B69 wheel at a wheel speed of 1700 m/min (5500 ft/min)

										Parameter								
	Table speed		Downspeed		Crossfeed		MRR'		Normal force		Power		Estimated average normal stress at the grinding zone		Estimated average power intensity at the grinding zone			
Grinding process	mm/min	in./min	mm	in.	mm	in.	mm³/s, mm	in.³/min, in.	MPa	psi	hp/mm	hp/in.	MPa	psi	hp/mm²	hp/in.²
Creep feed	255	10	2.54	0.100	4.76	0.1875	10.0	1.00	3.5	500	0.68	17.0	4.5	650	0.0340	21.9
Reciprocating surface	9140	360	0.05	0.002	1.3	0.050	7.2	0.72	0.93	135	0.12	3.0	8.46	1227	0.0423	27.3

Fig 17 Complex components and machine tools that can be produced by the multiaxis CNC grinding wheels shown alongside the workpieces

between the two extremes to minimize the row bow. This can be further aided by reducing the friction during the grinding process. Several wheel design/configurations and excellent coolant applications help to reduce the friction and thus produce small row bow.

Optimizing Precision Grinding Parameters

Precision grinding is a systems approach. This involves the optimization of the grinding process by controlling four key input aspects:

- Machine tool
- Work material
- Grinding wheel
- Operational factors

Each of these four aspects is addressed in detail in the article "Principles of Abrasive Machining" in this Volume. The section here addresses the practical aspects of truing, dressing, conditioning, and vibration control.

Balancing of the superabrasive wheels prior to mounting and subsequent on-machine balancing are critical to achieving high wheel performance, microsmooth surface finishes, and high precision geometry of the surfaces produced with minimal chatter or chipping.

It is generally thought that diamond wheels are hard, that they cannot be trued and dressed, or that special machines are required. When properly selected, diamond wheels can be trued, dressed, and made to operate in current production operations. However, a systematic understanding of the difference between conventional wheel truing and diamond wheel truing is required. For the successful use of grinding wheels, the wheel surface must be concentric, must be free of lobes and straight across the thickness of the wheel, and it must have the correct profile for form wheels. The process to achieve these requirements is called truing. Resin and metal bond wheels must be trued and dressed. Vitrified bond wheels generally require truing only and do not require dressing. Electroplated wheels do not require either truing or dressing, except in special situations.

Truing Methods for Production Grinding. A variety of methods are available for

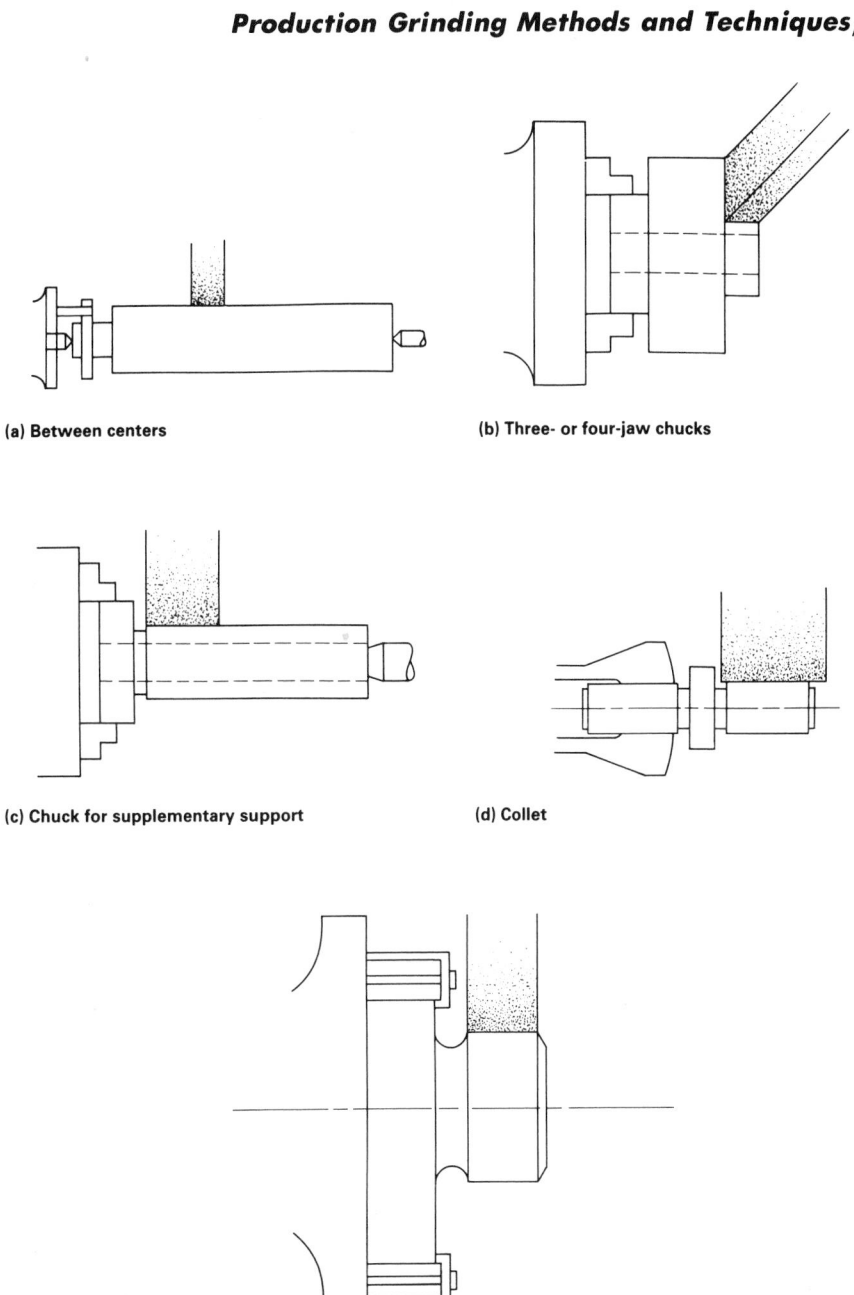

(a) Between centers

(b) Three- or four-jaw chucks

(c) Chuck for supplementary support

(d) Collet

(e) Face plate

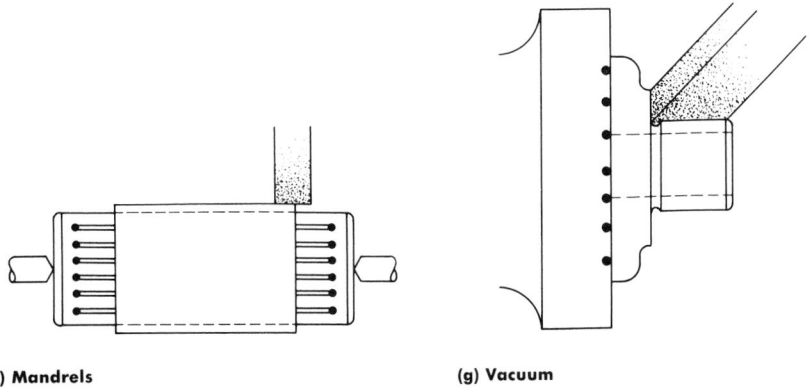

(f) Mandrels

(g) Vacuum

Fig 18 Workholding methods and devices utilized in the cylindrical grinding of ceramics

Fig 19 CNC cylindrical grinder adjacent to components produced to show its motion capabilities

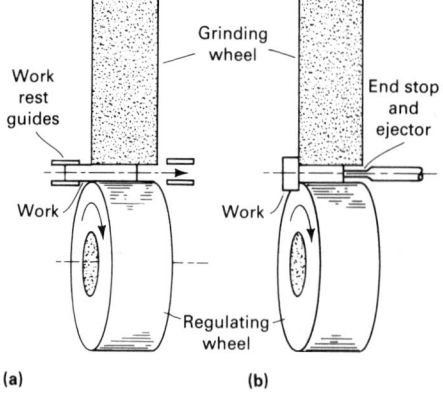

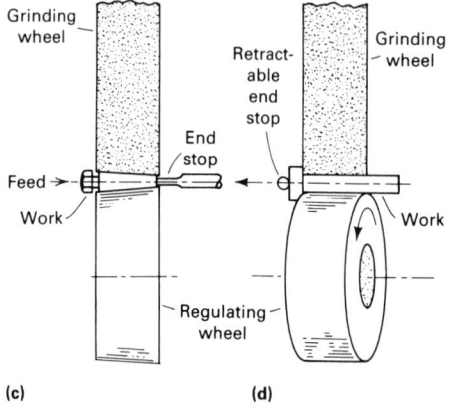

Fig 21 Four basic types of centerless grinding operations. (a) Throughfeed. (b) Infeed. (c) Endfeed. (d) Combination infeed/throughfeed. An inclined regulating wheel (always used for throughfeed as well as endfeed grinding and in some cases for infeed grinding) imparts a light axial force to the workpiece in all four cases.

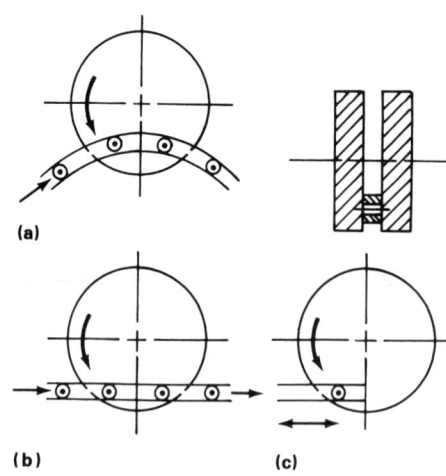

Fig 22 Double-disk grinding equipment for ceramic applications. (a) Carousel type. (b) Throughfeed type. (c) Reciprocating type

truing in production grinding operations. These methods can be broadly divided into three categories:

- Stationary tool truing
- Powered truing
- Form truing

The truing of diamond wheels in production grinding is still evolving.

Stationary tool truing involves passing a nonrotating truing tool across the face of a rotating wheel. The tool may consist of a single diamond point, a cluster tool containing three to five points, or a nib containing many diamond particles. Stationary tool truing can be used for very small vitrified wheels or for small scale infrequent production. However, the diamond point of the tool wears rapidly, requires frequent turning, and demands considerable operator attention. As a result, the stationary tool truing of diamond wheels is generally not recommended. Nibs containing small diamond particles (Fig 26) work well for micron-sized diamond wheels.

Powered Truing Methods. Commercially available rotary powered truing devices are versatile to use. They also generate low truing forces, offer consistent results, and are easily automated (Fig 26b). Wheels with a straight face or simple forms can be trued with a traversing diamond rotary cutter. Plunge truing with a diamond roll can also be used, but it is still an evolving method.

Form Truing. Complex form wheels can be trued by plunge truing with diamond rolls or crush truing methods (Fig 26c). Another method of truing form wheels is the use of a thin rotary cutter (considerably thinner than the wheel width). This rotary cutter can be traversed across the wheel profile with a mechanical arrangement that generates the wheel profile; for example, using a form bar or a CNC-controlled form generator. Dia-

mond form wheels have been successfully trued with this approach in order to generate simple radius or complex profiles.

Truing Methods for Batch Production. For operations in which the wheel is used for short production runs with each run containing different parts (job shop grinding), powered rotary truing methods are preferred and should be used whenever possible. However, for batch production the truing operations can also be carried out using abrasive wheels.

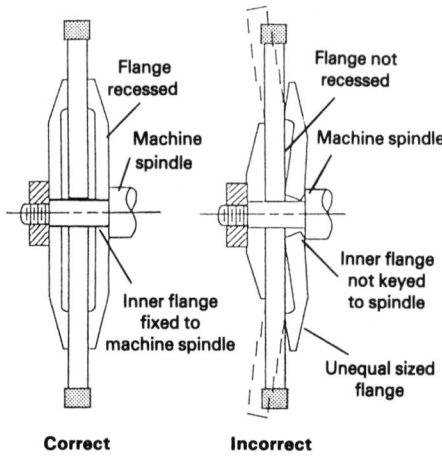

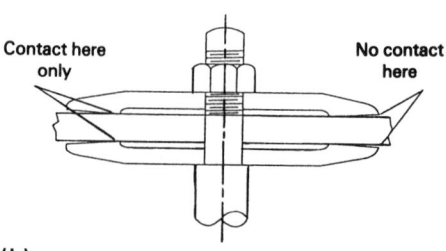

Fig 23 Parameters present in the mounting of thin precision grinding wheels. (a) Correct (left) wheel mounting and incorrect (right) wheel mounting due to unequal flange system. (b) Incorrect wheel mounting because of sprung flanges caused by excessive flange tightening

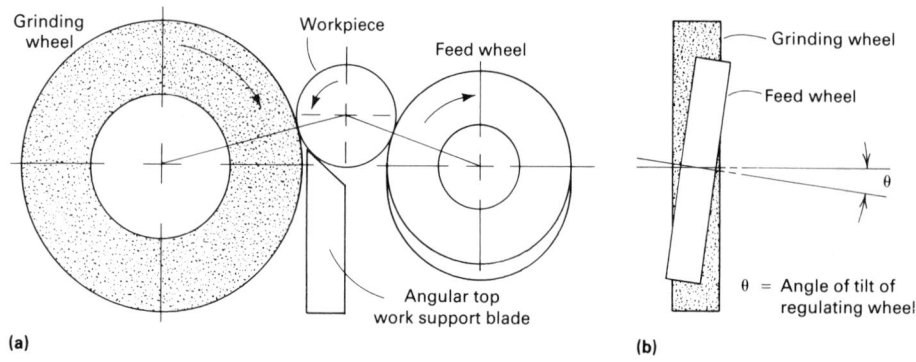

Fig 20 Basic elements of the centerless grinding operation showing the workpiece located between the grinding wheel and the feed wheel (a) and the inclination angle (θ) of the regulating or feed wheel (b)

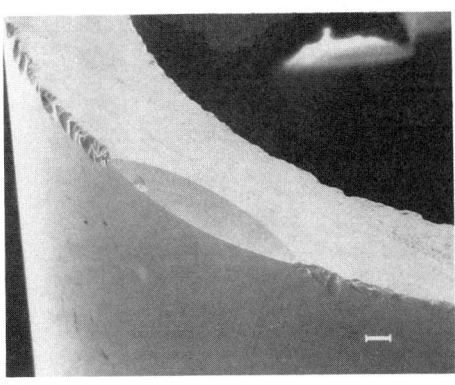

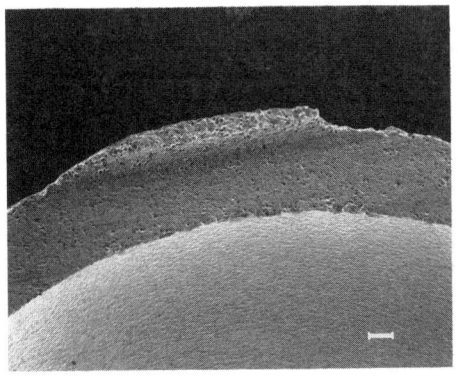

(a)   (b)

Fig 24 Types of fractures encountered in the cutoff of quartz tubes. (a) Concoidal fracture. (b) Tube fracture caused by grinding wheel exit after cut is complete. The mark in photo indicates 1 mm.

Truing with Abrasive Wheels. Brake-controlled and powered rotary truing devices use conventional abrasive wheels (Fig 27). The precision with which a trued wheel surface is generated by these methods is lower than in the methods described earlier. In general, a relatively harder wheel with silicon carbide abrasive will true the superabrasive wheel faster at lower levels of abrasive consumption. However, an open wheel surface is obtained with a softer wheel. Occasionally, the diamond wheels can be finished to the required form or geometry by using a grinder with an optical attachment.

Recommendations for Successful Truing. In a properly managed truing effort, it is necessary to:

- Minimize truing forces
- Minimize truing time
- Control the exposure and extent of flatness on diamond grits to obtain the desired surface finish
- Minimize loss of abrasive grain
- Automate the truing process to achieve consistent wheel surface condition

Quantitative Understanding of Truing Parameters. The following variables apply to any truing operation:

(a)   (b)

Fig 25 Effect of type of material on chippage and quality of cut. (left) Quartz. (right) Sapphire

- Truing method used: stationary or powered rotary truing
- Type of truing tool: stationary (single point or nib); rotary (disk, cup, diamond roll, bonded or plated cutter, reverse plated cutter); conventional abrasive; wheel; and so on
- Specification of truing tool: abrasive type, grit size, concentration, and geometry
- Diamond wheels to be trued: bond type, grit size, concentration, and wheel size
- Truing process variables: speed, infeed increments, speed ratio, overlap ratio, relative motion, crossfeed, coolant flow, and machine rigidity
- Output variables: truing forces, power, chatter or lobing, total indicated run-out (TIR), accuracy of form, wheel surface quality, truing time, and grind quality after truing

Practical Dressing Methods. Several methods are available for dressing diamond wheels. These include:

- Abrasive stick dressing, manually or via the use of mechanical devices
- Abrasive jet dressing

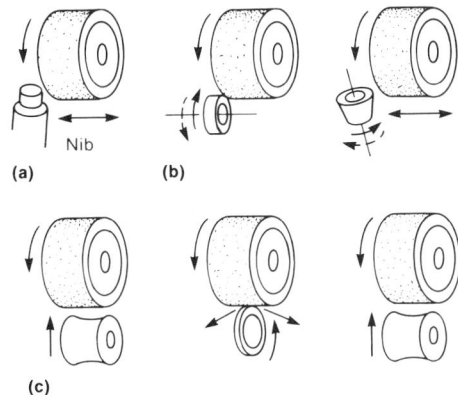

(a)   (b)

(c)

Fig 26 Production truing methods used for diamond wheels. (a) Stationary tool truing with a nib. (b) Powered truing using a rotary cutter (left) or a rotary cup (right). (c) Form truing using a diamond roll (left), tracing with rotary cutters (center), and a crush roll (right)

- Slurry dressing
- High-pressure waterjet dressing

Abrasive stick dressing using mechanical devices is the most frequently recommended practice and is suitable for a wide variety of applications. This method consists of pushing an abrasive stick (called a conditioning stick) onto the wheel face with a constant force or a constant infeed rate. Several variations of the stick dressing arrangement are used in industry.

Dressing devices are available that can be adapted to grinding machines. In this method, the abrasive stick can be pushed into the wheel face at a constant infeed or a constant force level. As the stick is pushed onto an as-trued wheel face at a constant infeed rate, the forces increase rapidly and then reach a lower steady-state level. For a given wheel and stick specification, the steady-state force will reach a constant value that is determined only by the infeed rate and is independent of the abrasive volume consumed, after a minimum quantity. In general, if it is determined that the wheel is consuming the stick excessively, but is not grinding satisfactorily after dressing, the operator should consider a coarser grit stick or a higher infeed rate.

Manual stick dressing is used in a number of batch production operations. However, from the standpoints of safety and consistency, manual stick dressing should be discouraged. The stick dressing can be mechanized whenever possible by mounting the stick on the machine table and rapidly feeding it onto the wheel with machine controls. Proper guarding is essential in such cases. The conditioning sticks used for dressing the wheels should be distinguished from the soft rubber sticks used for cleaning the wheel face.

The dressing block can be used as a work material in surface grinding, and the wheel can be dressed by grinding such blocks. Care should be taken to prevent excessive wear of the diamond wheel or abusive conditions that can lead to uneven wear.

Diamond wheels, particularly resin or vitrified bond wheels, can be dressed by erosion of the bond during grinding. This invariably results in a loss of wheel material during dressing. In addition, depending on the grinding conditions, the wheel may not be adequately dressed at any time.

Abrasive jet dressing consists of impacting the wheel face with a jet of fine grit abrasive propelled by high-pressure air. This method is also suitable for dressing a contour or form on the wheel face. The system is portable so that it can be transferred from machine to machine within a plant.

High-pressure waterjet dressing consists of a high-pressure waterjet directed at the wheel face to erode the bond in a controlled manner. Preliminary test results show that these methods generally required machining the wheel face with a waterjet entrained with abrasives rather than a waterjet alone.

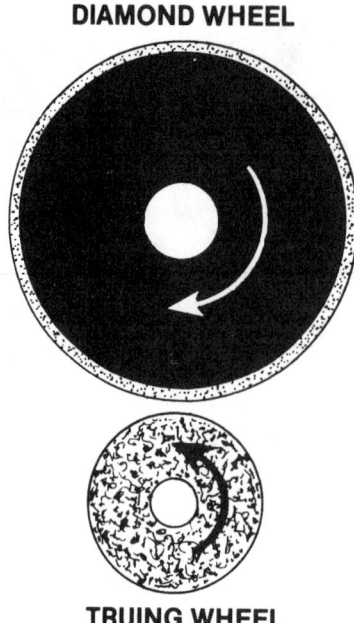

DIAMOND WHEEL

TRUING WHEEL

Fig 27 (a) Photograph of brake-controlled powered rotary truing device. (b) Schematic showing motion of both diamond wheel and truing wheel

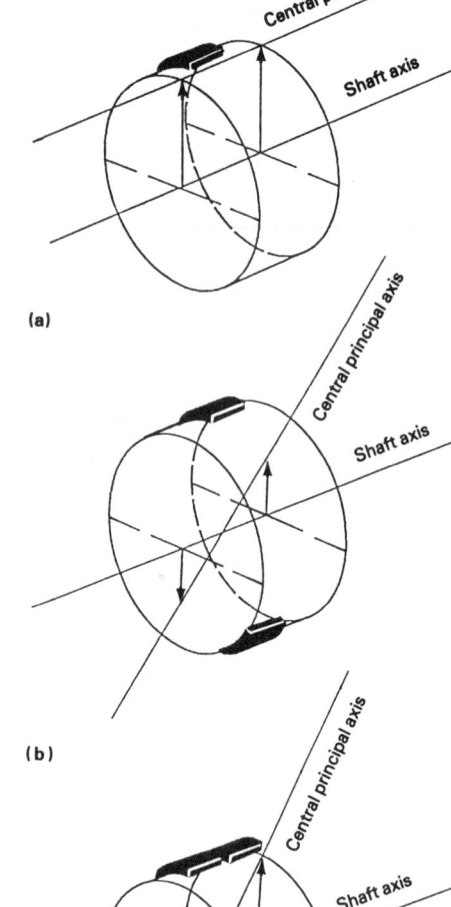

(a)

(b)

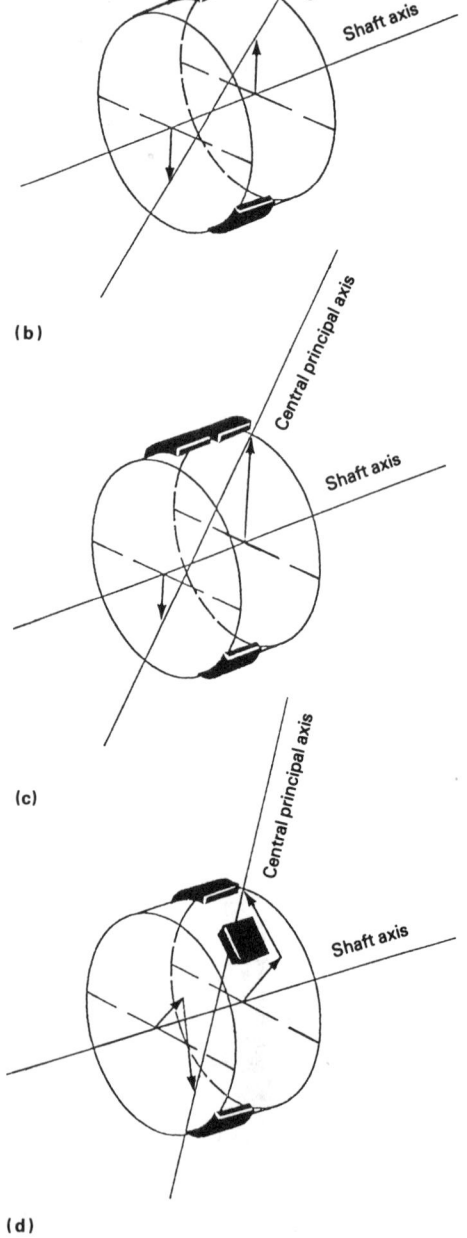

(c)

(d)

Fig 28 Modes of imbalance in a grinding wheel and spindle assembly. (a) Static unbalance. (b) Couple unbalance. (c) Quasistatic unbalance. (d) Dynamic unbalance

The cost of such units for slurry dressing, waterjet dressing, and the specialized nature of the equipment involved do not justify these methods for a wide variety of applications. However, these methods may become useful under special machine and production conditions.

Conditioning is a periodic correction to the wheel face to restore its geometry or to expose the diamond abrasive grains. Frequent conditioning of the wheels may render the grinding wheel application uneconomical.

A properly trued and dressed diamond wheel grinds with a minimum of grinding power that remains constant as good-quality parts are produced from the beginning of the production run. The wheel wears gradually until the point when surface finish changes, chippage increases, or changes in geometry occur. Any one of these conditions could lead to an increase in grinding power or grinding forces. Therefore, it is necessary to correct the wheel face at that point. This may involve precise truing of the wheel face or slight dressing. The precision involved and the extent of truing and dressing greatly depend on the application (that is, part size, tolerances required, and wheel specifications).

The repetitive and predictable nature of the conditioning interval is a key aspect of abrasive applications in production grinding. This strongly depends on wheel selection, the initial truing and dressing scheme, and on the grinding process parameters chosen.

From this discussion, it is observed that conditioning of the wheel is a high-precision operation. The increments or infeed for retruing or dressing are much smaller in the conditioning operation than in the initial truing and dressing of the superabrasive wheel. It is imperative that grinding machines have such precision capabilities and mechanical means of dressing.

Vibration Control and Balancing. The importance of vibration control to achieve good ceramic grinding results was discussed in the article "Principles of Abrasive Machining." There are several commercially available devices for monitoring vibration. Manufacturers should be contacted for details on the methods available to control vibration and achieve proper balancing.

It is important to recognize conditions that lead to vibration and distinguish them from conditions of unbalance. There are several problems that can reveal vibration characteristics similar to unbalance (for example, misalignment, resonance, eccentricity, and so on). These problems cannot be totally corrected by balancing. Therefore, a thorough analysis of the source of the vibration must be carried out.

Unbalance in a rotating system can be one of four types: static, couple, quasistatic, and dynamic (Ref 14).

Static unbalance is the simplest form of unbalance and occurs when the central principle axis is displaced parallel to the shaft axis (Fig 28a). Static unbalance is the only type of unbalance that can be totally solved by making weight corrections in a single reference plane.

Couple unbalance is the condition that occurs when the central principal axis intersects the shaft axis at the center of gravity of the rotating member (Fig 28b). Couple unbalance becomes apparent only when the wheel is rotated.

Quasistatic unbalance is caused when the central principal axis intersects the rotating center line but not at the center of gravity of the rotating member (Fig 28c).

Dynamic unbalance is a random combination of static and couple unbalances (Fig 28d). Dynamic unbalance problems can be solved completely by making weight corrections in a minimum of two separated reference planes.

Balancing Techniques. The nature and degree of unbalance may be entirely dependent on the grinding system and the output requirements. As tolerance and surface finish requirements become critical, the degree of balance required will become highly critical.

There are several practical methods used for balancing. Static balancing is usually achieved by locating the grinding wheel assembly between two bearing surfaces that allow for free rotation of the assembly. Balance weights are added or removed until the assembly can come to rest without preference to any rotational angle. Dynamic balance is complex and must be performed on the grinding machine for the best results. Sometimes dynamic balance may be achieved using special adapters with movable weights. There are many other automatic balancing systems built in the machine tool that are used for precision grinding of metals. Many of these systems will be adapted with the growth in precision production grinding of ceramics.

Cost-Effective Production Grinding of Ceramics

Like any other manufacturing process, production grinding of ceramics must be amenable for competitive manufacturing costs. Grinding costs as high as 50 to 70% of component costs are quoted for ceramics. In comparison, a well-established metal component may require only 5 to 25% as the cost of grinding. The difference between the two can become obvious as one looks at the total cost equation:

Total cost/Part

$$= \text{Raw material (powder cost)/Part}$$

$$+ \text{Processing cost/Part}$$

$$+ \text{Finishing cost/Part} \qquad \text{(Eq 1)}$$

Finishing (grinding) cost = Abrasive cost/Part

$$+ \text{Machine tool cost/Part}$$

$$+ \text{Overhead cost/Part}$$

$$+ \text{Labor cost/Part} \qquad \text{(Eq 2)}$$

Abrasive cost/part = Wheel cost/Number

$$\text{of parts per wheel} \qquad \text{(Eq 3)}$$

Machine tool cost = Machine rate × Cycle time

$$\text{(Eq 4)}$$

Overhead cost = Burden rate × Cycle time

$$\text{(Eq 5)}$$

Labor cost = Labor rate × Cycle time $\qquad$ (Eq 6)

Yield = Good quality parts/Total number

$$\text{of parts machined} \qquad \text{(Eq 7)}$$

Number of parts/Wheel = Total number

$$\text{of parts machined}$$

$$× \text{Yield} \qquad \text{(Eq 8)}$$

T = Total time consumed

$$= \text{Set up time}$$

$$+ \text{Time for truing and dressing}$$

$$+ \text{Grinding time}$$

$$+ \text{Inspection time} \qquad \text{(Eq 9)}$$

Cycle time = Total time consumed/Number of

$$\text{parts produced in time } T \qquad \text{(Eq 10)}$$

It can be shown through the use of Eq 1 to 10 that the cost per part can be reduced by decreasing the powder cost, the processing cost, and the total time consumed.

Total time consumed can be further decreased by looking at Eq 9 (that is, set up time, truing and dressing time, and inspection time) in addition to the actual grinding cycle time.

The total cost per part is also influenced by the rates for machine, burden, and labor. The total cost can also be decreased by increasing wheel life (number of parts per wheel) and the yield. Further improvements in total cost can be achieved by lowering the wheel cost.

Thus wheel cost, labor rate, and machine rate are only three among the possible eleven variables that control the total cost/part. This is a significant point because it identifies the several possible methods to reduce the total cost of ceramic parts and to render them cost effective. Thus, expensive wheels, machine tools, or operators should not be discarded. On the other hand, by using such means and optimized grinding systems, it is feasible to reduce cycle time, to increase yield or the number of parts per wheel, each of which could contribute to obtaining the desired economic result. For instance, it has been shown that by using state-of-the-art grinding machines and diamond wheels, complex ceramic parts can be machined from simple solid geometry. The total cost/piece of this process is lower than the near-net shape forming and finishing methods in some instances.

Emerging Ceramic Machining Technologies

As the ceramic materials find a growing number of applications, creative approaches

will be pursued to achieve production economics, to provide unique surface features, or to obtain enhanced quality. Progress in each of these areas can be observed both in industrial practice and in research activities.

Mirror finish grinding of ceramics is the term used to designate reflective surfaces on ceramic materials. These are applications where the surface finishes obtained are of the order of a few microinches or better. Finishes of the order of 0.5 to 1.0 nm (0.02 to 0.04 μin.) have been reported. These results are achieved using extremely fine micron-sized abrasives, ultra-precision machinery and fine-grain ceramic materials.

Flat Honing. Because extreme fine finishes can be achieved by grinding, it is possible to replace lapping processes by methods using fixed abrasives or diamond grinding wheels. These methods are called flat honing, and they are in their evolutionary stage.

Grinding from Solid. When total cost considerations are taken into account, some of the traditional approaches of near-net shaping and minimum grinding can be challenged. When efficient and economical grinding practices and methods are in place, it may be more economical (lower total cost/ part) to grind ceramic parts from a simple solid shape instead of near-net shape fabrication. There are several approaches in place to achieve such results. These will parallel the advancements in grinding of metals in which solid rods are through-hardened and finish ground to rotary tools without resorting to prefinishing operations.

Diamond Belt Machining. Coated diamond abrasive products in fine abrasive sizes are being developed and evaluated. These applications focus on finishing of ceramic parts to obtain the desired finish without a serious attempt to alter the component geometry.

CNC Machining. Machining of multiple surfaces in a single set up offers enormous benefits by reducing set-up costs and improving geometric tolerance on related surfaces. This concept is practiced by using machining centers. However, machining centers generally have low rev/min spindles that reduce the grinding efficiency. To overcome this, machining centers are being retrofitted with high-speed spindles. There are also many CNC grinding machines that produce similar results for the grinding of ceramic parts.

REFERENCES

1. P.A. McKewon, High Precision Manufacturing, *Ann. CIRP*, Vol 37, Feb 1987
2. J.A. Dudley, "Precision Finishing and Slicing of Ceramic Materials with Diamond Abrasives," Fourth International Grinding Conference, Society of Automotive Engineers, 1990, p 550
3. K. Subramanian, Advanced Ceramic Components: Current Methods and Future Needs for Generation of Surfaces, *Proceedings of the Intersociety Symposium on Machining of Advanced Ceramic*

Components, The American Ceramic Society, Apr 1987, p 10–32

4. W.N. Ault, Grinding Equipment and Processes, *Machining*, Vol 16, *Metals Handbook*, ASM International, 1989, p 430–452
5. K. Subramanian, Superabrasives, *Machining*, Vol 16, *Metals Handbook*, 1989, p 453–471
6. K. Subramanian, Precision Finishing of Ceramic Components with Diamond Grinding Wheels, *Am. Ceram. Soc. Bull.*, Vol 67 (No. 6), June 1988, p 1026–1029
7. A.H. Slocum, Kinematic Design Principles for Precision Machines, Proceedings of the 3rd Annual Precision Engineering Conference, American Society for Precision Engineering, Atlanta, 1988
8. N. Taniguchi, Current Status in, and Future Trends of Ultra Precision Machine and Ultra Fine Materials Processing, *Ann. CIRP*, Vol 32 (No. 2), 1983
9. S.F. Krar and E. Ratterman, *Superabrasives Grinding and Machining with CBN and Diamond*, McGraw-Hill, 1989
10. S. Ohishi and Y. Furukawa, Accuracy of Work Pieces Produced by One Pass Creepfeed Grinding, *Prec. Eng.*, Vol 8 (No. 3), July 1986, p 144–150
11. C. Bateja and R.P. Lindsay, *Grinding, Theory Techniques and Troubleshooting*, Society of Manufacturing Engineers, 1982
12. J. Yoshioka, F. Hashimoto, M. Miyashita, and M. Daito, Ultra-precision Grinding Technology for Brittle Materials: Application to Surface and Centerless Grinding Processes, *Milton C. Shaw Grinding Symposium*, Vol 16, American Society of Mechanical Engineers, p 209–227
13. H.K. Tonshoff *et al.*, Abrasive Machining of Silicon, *Ann. CIRP*, Vol 2, 1990, p 230–243
14. "A Practical Guide to In-Plane Balancing," Technical Paper No. 116, IRD Mechanalysis, Inc.

Lapping, Honing, and Polishing

John H. Indge, Peter Wolters of America, Inc.

MOST CERAMIC MATERIALS are machined with an abrasive, and this article will focus on the technology of using loose rather than bonded abrasives as the cutting tool. The loose abrasive may roll, slide, or be restricted in its movement. However, it is not bonded to a substrate as are grinding wheels, honing tools, or coated abrasives. The processes described in this article apply to the precision machining of flat and cylindrical ceramic components by lapping.

Abrasive Cutting Mechanism in Lapping

When loose abrasive is used to machine ceramic, it may slide, roll, or imbed, and more than likely do all three depending on the shape of the abrasive grain and the composition of the back-up surface. There are so many different situations for flat ceramic machining that it is not realistic to assume only a few principles will apply to all kinds of applications. In this article, however, three basic theories will be described in simplified terms:

● Single-side flat lapping
● Double-side flat lapping
● Cylindrical lapping outside diameters of cylindrical shaped parts

Abrasives come in a wide variety of forms: soft to hard, strong to brittle, coarse to fine, uniform to irregular. The most common abrasive for machining softer materials is calcined alumina and garnet. Harder materials can be cut with silicon carbide, boron carbide, and diamond.

The hardnesses of selected abrasives are listed in Table 1.

The size and shape of abrasive grains have an effect on the lapping action. A broad size distribution (Fig 1a) may cause scratches and be slower cutting than an abrasive grain with a tight size distribution (Fig 1b).

A weak-shaped blocklike abrasive (Fig 2a) may cut a softer material faster than a strong shape. However, for lapping hard material, the strong shape (Fig 2b) has the advantage of withstanding more force before breaking down.

Abrasives such as calcined alumina have a flat or platelike configuration (Fig 3) and tend not to roll when used for lapping. Boron carbide is more sliver-shaped and very hard. When it breaks, new sharp edges are produced so that it can last longer than silicon carbide, which tends to dull the longer it is used. Diamond and cubic boron carbide, the hardest of all abrasives, are normally blocklike in shape (Fig 4).

Table 1 Hardness of selected abrasives used in the lapping, honing, and polishing of ceramics

Abrasive	Hardness	
	Mohs	Knoop
Zirconia	8	1160
Garnet	8	1360
Calcined alumina	9	2100
Silicon carbide	9–10	2480
Boron carbide	9–10	2750
Cubic boron nitride	10	4500
Diamond	10	7000

Rolling Abrasives. Lapping takes place when abrasive grain entrained in a liquid vehicle, often known as slurry, is guided across the surface to be lapped and backed up by a lapping plate. Because the abrasive grains used for lapping have sharp, irregular shapes and each grain is backed up by a lapping plate, when a relative motion is induced and pressure applied, the sharp edges of the grains are forced into the workpiece material to be lapped and either make an indentation or cause the material to chip away microscopic particles. Even though the abrasive grains are irregular in size and shape, they are used in large quantities. Thus, the cutting

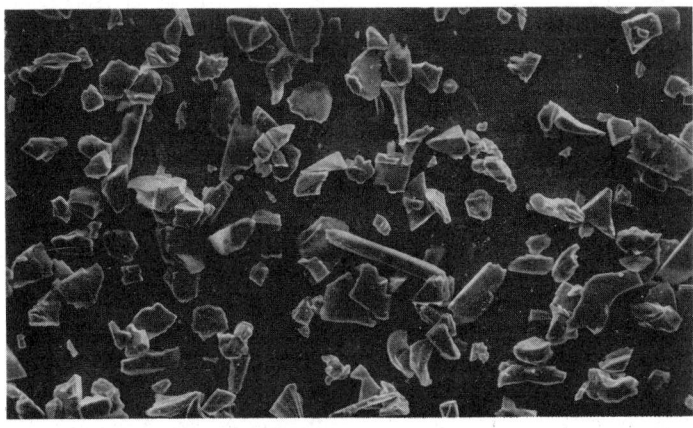

(a)

⊢──────────────┤
100 µm

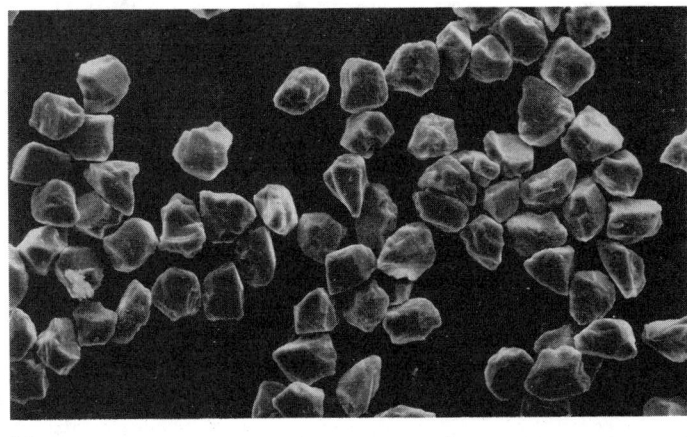

(b)

⊢──────────────┤
50 µm

Fig 1 Photomicrographs showing abrasive grain size distribution of silicon carbide used for lapping. (a) Broad size distribution. (b) Tight size distribution. Courtesy of Norton Company

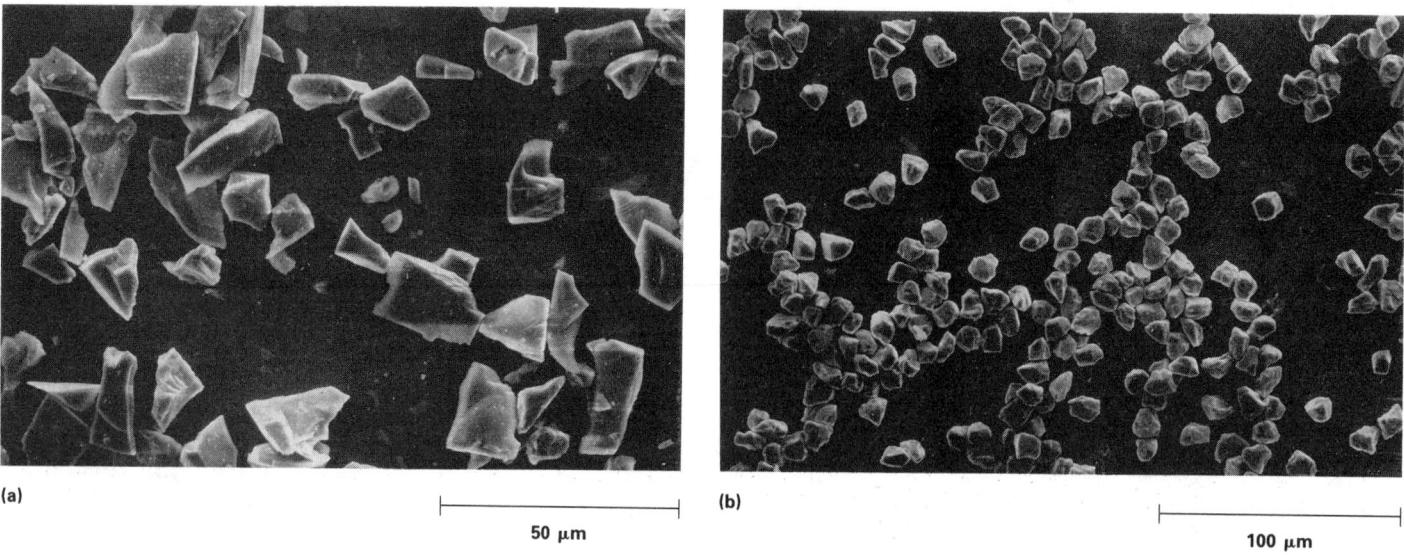

(a)

(b)

50 μm

100 μm

Fig 2 Photomicrographs showing two types of blocklike silicon carbide abrasives used for lapping. (a) Weak-shaped abrasive used for soft materials. (b) Strong-shaped abrasive used for hard materials. Courtesy of Norton Company

action takes place continuously over the entire surface of the workpieces that come in contact with an abrasive slurry backed by a lapping plate.

Fig 3 Platelike calcined alumina abrasive used for lapping applications. Courtesy of Norton Company

1 μm

Fig 4 Blocklike diamond abrasive used for lapping applications. Courtesy of Norton Company

With regard to the movement of the abrasive grains, special importance has to be assigned to the liquid vehicle. The abrasive grains are rarely evenly round in shape, while slates and slivers are much more common. Simplified, one can visualize the process within the slurry as shown in Fig 5. As the workpiece moves at velocity, *v*, the adherent liquid moves with the workpiece. However, the velocity of the liquid at the lapping plate is zero. Ideally, a distribution of velocity with a gradual transition would develop and be disturbed by the abrasive grains contained in the slurry. Vortices which develop in the liquid pick up and upright even grains that are lying flat. These grains are thereby also forced to work as shown in Fig 5.

Sliding Abrasives. The initial conditions for lapping with sliding abrasives are similar to that of rolling abrasives; however, because the abrasive grains are more flat or platelike in configuration, the cutting action simulates tiny scrapers. The exact movement of the abrasive grains has not really been established; however, the platelike abrasive grains are believed to stack on top of each other similar to tipped-over dominos, thus providing many cutting edges to machine away the ceramic being lapped (Fig 6).

The direction of cut is constantly shifting by the constantly changing direction of movement of the workpieces against the back-up lapping plates, thus utilizing all edges of the abrasive grains.

With both rolling and sliding abrasive processes, the abrasive grit mixed with a water vehicle is carefully metered and fed automatically to the work area. This motion makes it ideal for production lapping by providing a consistent and uniform cutting action.

Charged Plate Abrasives. An early form of lapping, which is still used today for some applications, utilizes the embedding of abrasive in the lapping plate or tool. The abrasive grains that are doing most of the work become embedded and act as microscopic scraping tools. These abrasive grains eventually dull or break and are replaced by fresh grains that are added periodically by hand or from an abrasive paste which works up from slots in the lap plate.

As indicated in Fig 7, the larger abrasive grains that become embedded provide the most aggressive lapping action when a relative motion takes place between the workpiece and the lapping plate. As these larger grains are worn down or break down, the smaller grains start to embed and also to work. Most lapping plates used for this kind of processing have grooves in a waffle pattern and use a paste form of the lapping compound. The paste may be used as purchased or it can be mixed with an oil to the consistency of a heavy cream. The grooves hold the lapping compound that works its way out to keep the lap plate lubricated and provides a recharging effect on the lapping surface.

This type of lapping is more labor intense than the two previous systems and requires the operator to manually apply the proper amount of compound. The cutting action changes as the abrasive grains are used up, thus requiring more operator attention.

The cutting action of a charged plate abrasive is actually a combination of rolling, sliding, and embedded abrasive processes.

Lap Plate Materials

Cast Iron. The most commonly used material for making laps is cast iron, which has a special close-grained microstructure that has no porosity or other defects to effect the lapping action. Cast iron is very versatile and can be used for rolling abrasive, sliding abrasive, and embedded abrasive applications, and for polishing with hard abrasives.

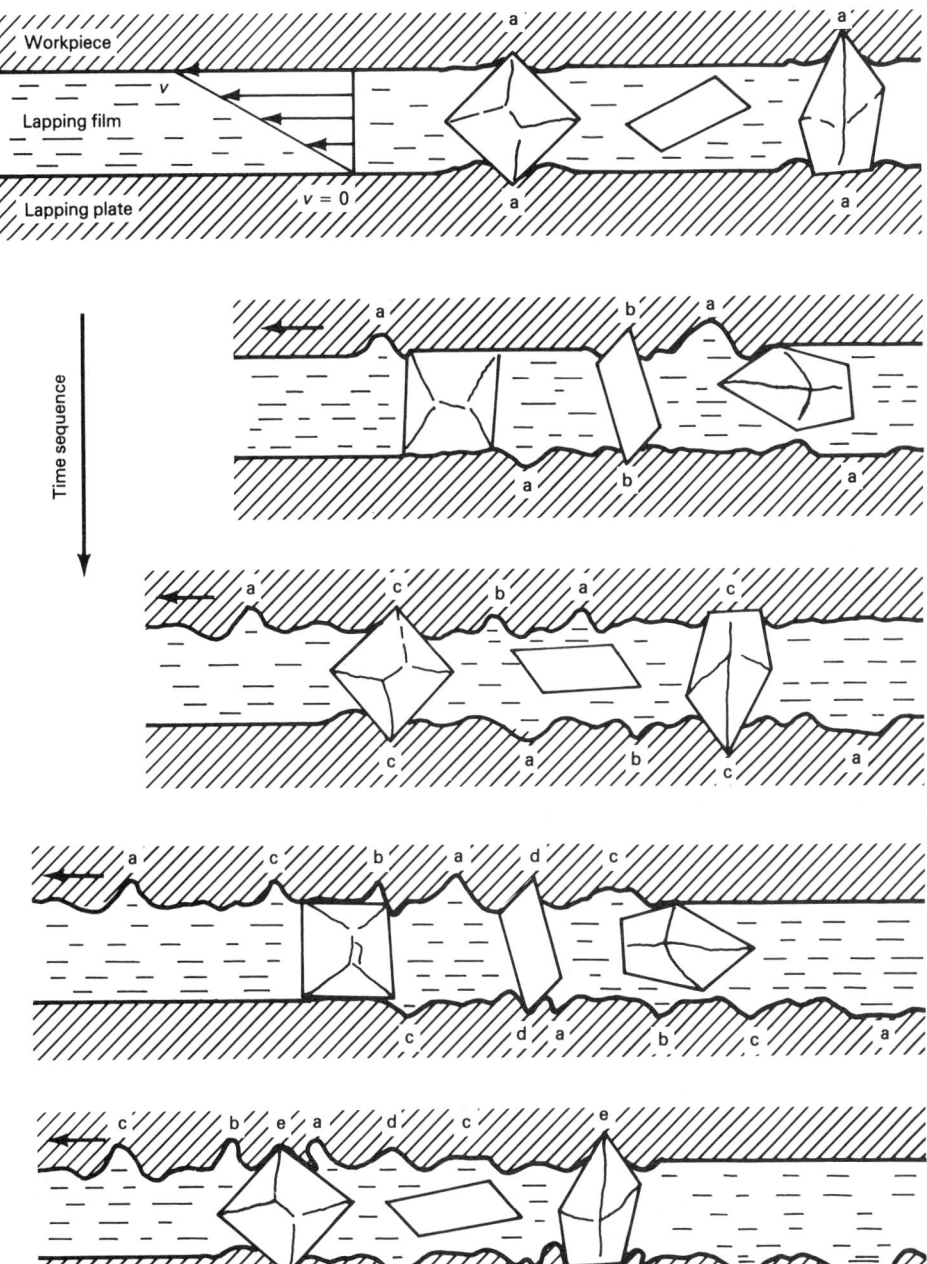

Fig 5 Schematic showing effect of workpiece motion at velocity *v* on the abrasive grains rolling in a lapping film. a to e are identifications of lapping impressions in chronological sequence.

Fig 6 Schematic showing cutting action of platelike abrasive grains as the grains slide and scrape the region between the workpiece and the lapping plate

The close-grained iron has the unique capability to grip the abrasive grains and to make them roll or shave effectively (as in the case of the platelike aluminum oxide), giving them a foothold to work against the workpiece efficiently.

Steel Alloys Hardened to 60 HRC. Another material sometimes used for the lapping wheel (plate) for one wheel single-side flat lapping machines is made of a steel alloy hardened to 60 HRC. This plate is reported by the manufacturer to resist embedding of the rolling abrasive grains with the purpose of obtaining a deeper indentation of each working abrasive grain into the workpiece. A high pressure is applied to the workpiece, and a generous amount of abrasive slurry is fed to the lapping area to prevent the workpieces from making extremely close contact with the lapping plate.

Additional Plate Materials. Other materials, used less frequently, are nodular iron, aluminum, and nonmetallics such as granite.

Lapping Vehicle Fluids

Water-based vehicles are commonly used for lapping although oil vehicles can and are being used. Oil is a good vehicle; however, it is often objectionable due to its tendency to penetrate into the pores of some materials. In addition, oil is expensive to purchase and its disposal is difficult and expensive because it is now considered to be classified as a hazardous waste by the government.

The purpose of any vehicle is to carry the abrasive grains and to position them to work most efficiently. The vehicle also lubricates the surfaces and carries away the abraded material removed from the workpieces. To prevent rust to the machine components, inhibitors are often added to water vehicles. Depending on the equipment, the vehicle is sometimes also depended upon to carry away heat generated by the lapping process.

Suspension agents are occasionally added in order to prevent settling of the abrasive. These suspension agents do not necessarily improve the actual lapping action. There are a variety of methods that the abrasive can be kept in suspension with minimal negative effect at the area where the machining takes place.

Lapping Processes and Equipment

The machining methods used in lapping operations can be classified as:

- Single-side flat lapping
- Double-side flat lapping
- Cylindrical lapping between flat laps

Single-Side Flat Lapping

Many ceramic workpieces (for example, seals) require one flat surface. Lapping is the most frequently used machining process for

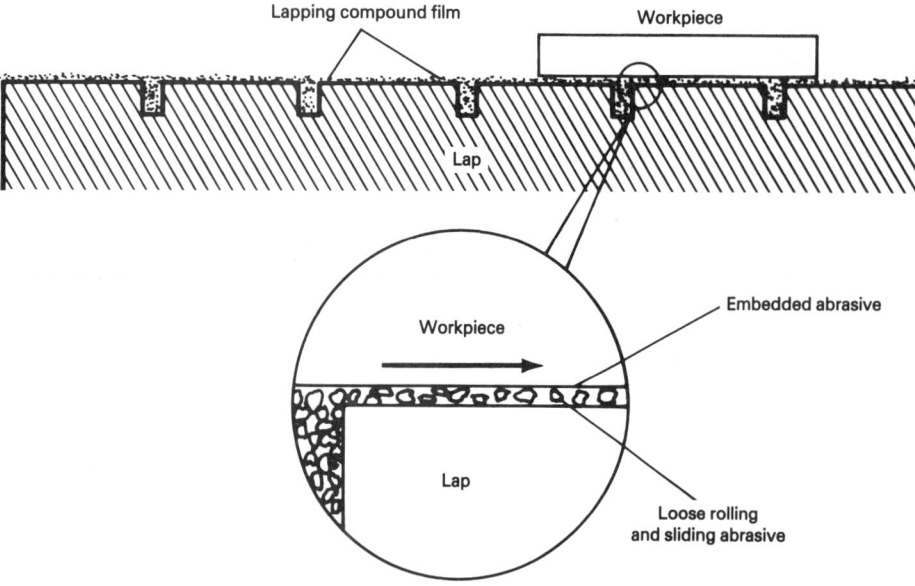

Fig 7 Schematic showing cutting action of abrasive grains for a charged lapping plate

- Keeping the wheel flat
- Applying a uniform and predictable pressure
- Applying and maintaining a uniform and consistent flow of abrasive

Because of these three variables, lapping has often been considered an art and such processes relied a great deal on operator experience and judgment. Considerable strides have been made in today's modern single-face lapping machines to reduce these variables and to provide more control.

Control of Process Flatness Parameters. There are two major causes of wheel out-of-flatness condition: deflection and uneven wear.

Deflection in the lap wheel due to pressure applied on top of the workpieces must be kept to a minimum by providing heavy and sturdy support of the rotating wheel. Thermal expansion due to heat can cause an out-of-flatness condition; therefore, a temperature controlled lapping wheel is an essential requirement. Modern production machines have devices to carry away heat caused by lapping. One method is to provide a wheel mount that incorporates a labyrinth where chilled water can be circulated to carry away the heat generated by the lapping process. Ideal systems are designed to keep hot spots and cold spots to a minimum.

Uneven Wheel Wear. When lapping on a single-wheel machine, the lapping action on the workpieces tends to wear the wheel away where the workpieces contact the wheel, causing an out-of-flatness condition to develop. In most instances, conditioning rings are

producing such flat surfaces (Fig 8). The advantages of lapping are that many workpieces can be machined at one time, workholding is simple, cut rates are consistent, and close accuracies are inherent with the process.

Processing Equipment. Single-side flat lapping is perhaps one of the best known and most widely used processes. From the most simple hand lap arrangement to today's sophisticated lapping machines, there have been and still are a wide variety of systems available. Basically, single-side machines have

a rotating, annular-shaped wheel (also known as the lapping plate) and the workpieces are applied to the flat rotating wheel (Fig 9). Numerous devices are employed to keep the wheel flat and also to guide the workpieces while pressure is applied. In addition, the abrasive slurry is then added, often automatically to the wheel surface.

Designers have strived to overcome some of the variables in single-side flat processing. The three most difficult procedures to control have been:

Fig 8 Typical ceramic components produced by single-side flat lapping process

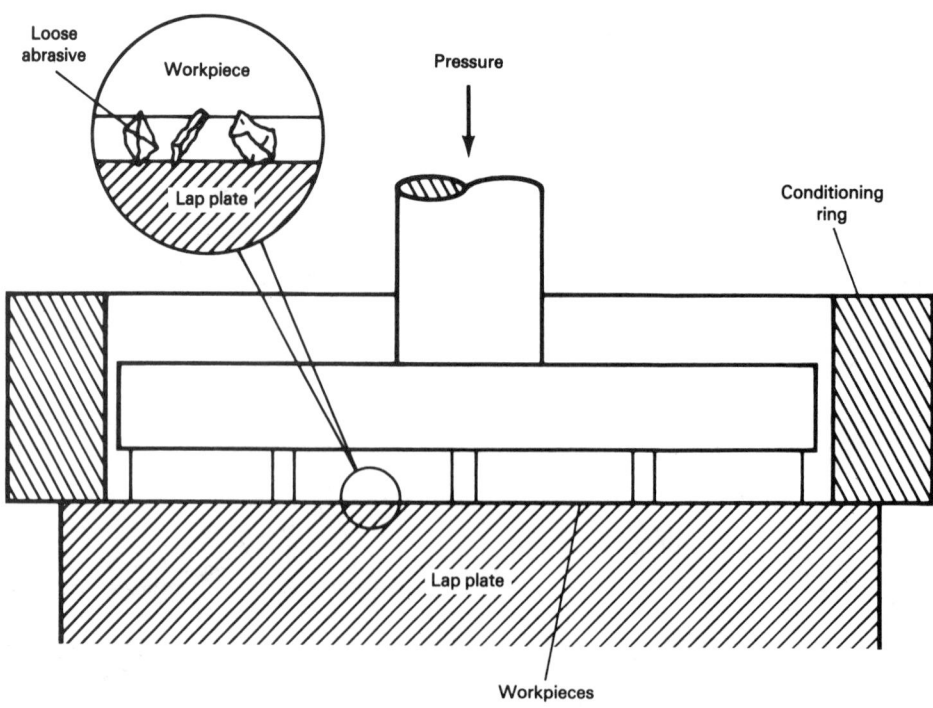

Fig 9 Schematic showing key components of a single-side flat lapping process

applied to wear away the higher points on the wheel that do not contact the workpieces, thus keeping the wheel face flat. These conditioning rings are rotated either by the force of friction from the rotating lap wheel face or by some drive device in the center of the wheel. To provide control of lap flatness and uniformity of lapping action, the conditioning rings should be rigidly supported and always power driven in one direction only. In addition, positive means should be provided to correct for either a concave or a convex lap wheel condition.

Old-fashioned means to correct for concave or convex wear conditions such as positioning the rings more towards the center or more towards the periphery present problems for pressure cylinder positioning and can also cause small parts to fall over the edge of the lap. The other method of reversing the direction of the conditioning ring with a gear drive in the center will cause a different lapping action, thus causing variations in cycle time, and so on. Modern methods to maintain lap flatness without changing the lapping action or to make setups difficult have made this type of processing more predictable, thus requiring less reliance on operator judgment.

Maintaining Uniform Pressure. Single-face lapping normally requires pressure on top of the workpieces. Production-type processing requires reliable pressure systems. Normally, pneumatic cylinders are used and positioned over the center of the conditioning rings. The cylinders need to be arranged with a three-stage pressure control system: preload, mainload, and postload. A predictable and repeatable pressure needs to be applied to each station. Therefore, the cylinder-piston design should have minimum slipstick. The lapping forces should be carried by the guide rollers for the conditioning rings to prevent a strain being placed on the cylinder piston rod that could cause more slippage.

Uniform Abrasive Slurry Flow. Reliable and consistent results with single-face lapping will normally require a once-through type abrasive feed system. The amount and distribution of the abrasive slurry should be monitored at all times. If used abrasive is recirculated, the cutting action will change and variables will be introduced that will require the judgment of the operator to solve. Once-through feed systems can be very efficient, and those systems that utilize a rich slurry of fairly thick consistency, keeping it agitated and moving at all times, will get the most use from the abrasive before it exits over the edge of the lapping wheel. The feeding system should distribute the abrasive slurry evenly over the working area of the wheel utilizing forces such as gravity, centrifugal force, and the rotating action of the wheel and conditioning rings.

Double-Side Flat Lapping

The most accurate method in terms of parallelism and uniformity of size is abrasive machining of both sides of the workpiece simultaneously using batch mode-type processing. The same abrading action obtained when processing on a single-side machine is thus applied to both sides of the workpiece. With double-side processing, there is no chance that foreign particles can be introduced into the process to settle between the workpiece and the device applying the pressure because of the relative motion between the wheels. When the lapping or wheel surfaces are flat, the pressure is applied to the thicker portion of each workpiece, thereby reducing its size at these specific points (Fig 10).

Double-side processing increases lapping efficiency, because the abrasive machining of two surfaces simultaneously takes approximately the same time it takes to machine one surface alone, and it does so all in one handling.

Operational costs by double-side lapping are reduced because:

● Conditioning rings are not wearing away the lap and themselves
● More efficient use of the abrasive slurry takes place
● Cost of consumables is less per workpiece

Typically, double-side lapping done on a batch mode unit utilizes rotating work holders that guide the workpieces uniformly between the rotating lapping wheels.

Variable Pressure during Lapping Cycle Increases Machine Efficiency. To lap away taper, bow, and size variation, it is necessary to accurately control lapping pressure. Normally, when a load of workpieces is placed in the machine and the top wheel is lowered to contact the top of the workpieces, the top lap actually is only contacting a few high points. Old-style machines with deadweight top wheels produced excessive pressure initially and not enough pressure later in the process cycle to remove stock. Today's lapping machines have provisions to start the cycle with the proper amount of reduced pressure and gradually increase the pressure as the high points are machined away. Once the high points are removed, the pressure should be maintained uniformly and at a level sufficient for the type of material being processed. When the workpieces are machined almost to finish size, means should be available to automatically reduce the pressure in order to provide a light cutting action and to allow sufficient time to remove any imperfections caused by the higher pressure.

Consistent and dependable results by lapping on today's sophisticated double-side lapping machines are obtained because approach (soft touch) and pressure can be under full control. Systems are now available that measure the weight of the top wheel, monitor each cycle, and automatically compensate for the effect of weight change incurred due to wheel wear. In addition, the very light starting pressure needs to be monitored at the beginning of each cycle, and wheel startup should take place only after the top wheel is actually making contact on top of the parts, a very important control for high yield when processing thin and delicate parts. The rate of pressure change, along with setting of the mainload pressure and postload pressure, should be programmable and repeatable.

Abrasive Slurry Flow Rate. A continuous flow of abrasive slurry applied evenly to the lapping area results in accuracy in size, flatness, surface finish, and uniform production. The abrasive slurry must be consistent in its concentration and the flow rate uniform in order to maintain the desired uniform cutting action.

Processing Equipment. A typical machine set up for flat double-side lapping is shown in Fig 11. Six sprocket-type workholder carriers are laying flat on the lower wheel. The workpieces are nested, one in each aperture, seven per carrier. The carriers are driven by a center inner pin ring driving device to rotate the carriers in an epicyclic motion within the confines of the nonrotating outer pin ring. The carriers are always thinner than the finish thickness of the workpieces and carry the workpieces in an everchanging pattern across the lap faces.

A typical process cycle consists of the following sequences:

● Loading of the workpieces, in which the top wheel swings from the load position on the left to center over the lower wheel
● Top wheel is lowered gently to rest lightly on top of the workpieces
● Rotation of the lower wheel, upper wheel, and carrier drive units starts slowly either

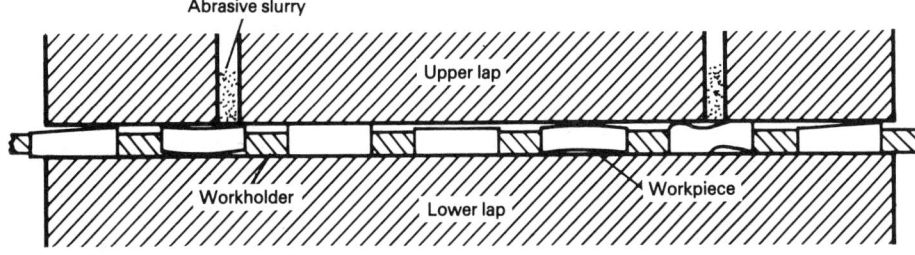

Fig 10 Schematic showing errors in parallelism and flatness being corrected with double-side lapping process

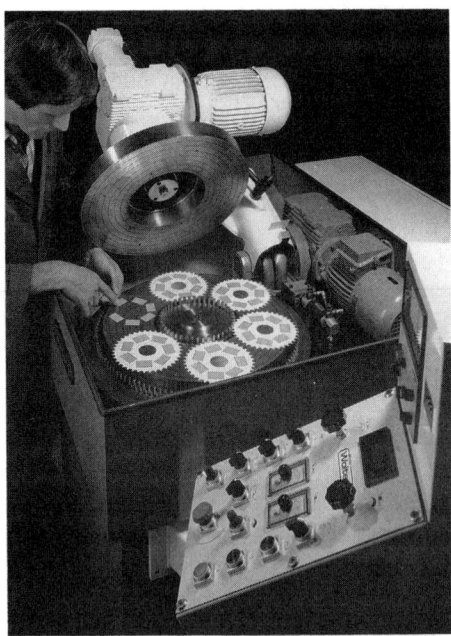

Fig 11 Flat double-side lapping machine that incorporates six sprocket-type workholder carriers on the lower wheel

automatically, or it can be controlled by the operator

- Abrasive for lapping is automatically fed through the top wheel during the cycle to provide the necessary abrasive action on both sides of the complete load of workpieces
- Pressure of the top wheel against the workpieces is initially light and as the high points are machined away, the pressure is gradually increased automatically until the optimum pressure, best suited for the job, is reached
- Cycle continues until the desired size is obtained, at which time the rotation stops, the top wheel lifts and swings to the left for unloading the finished workpieces

Process Advantages. Double-side flat lapping offers the following advantages:

- Ability to machine two sides of a workpiece in the same time as required to machine one side
- Process produces a large number of workpieces simultaneously
- Ability to hold nonmagnetic material
- Capability of machining any kind of stable material from plastic to diamond
- Best available method to obtain close tolerances for flatness, parallelism, and size
- Removing stock from both sides of a workpiece simultaneously helps to relieve internal stress of the workpiece, thus making it easier to achieve flatness
- Simple workholder design with no need to clamp or rigidly hold the workpiece eliminates stresses in the workpiece, thus improving tolerances for flatness, parallelism, and size

- Accuracy with double-face lapping is achieved by using flat lap faces and a free-floating top wheel. No critical machine alignment, precision high-speed spindle bearing, or accurately machined sliding ways are involved
- Workpiece exposed to minimal stress and surface damage because lapping generates no heat
- Cut rate is uniform and repeatable. No dulling of the abrasive takes place because fresh sharp abrasive particles are gradually fed to the lapping area continuously during the processing cycle
- Lower operating costs often result because of less handling, higher efficiency, and the feasibility of combining machining operations

Cylindrical Lapping between Flat Laps

The outside diameter of cylindrical-shaped parts can be lapped between flat laps. Machines for cylindrical lapping utilize two annular laps, each mounted on a vertical spindle. One or both of the laps rotate, depending on the type of machine. Parts are placed in a workholder that guides them between the lap faces to produce an abrading action. The workholder is generally disk-shaped and thinner than the diameter of the workpieces. The workholder is guided in the center by a rotating pin that can be adjusted to move eccentric to the center of the lower lap. The cylindrical workpieces are placed in slots, the centerline of which are tangent to a circle in the center of the workholder. The rolling action of the parts causes the workholder to rotate. Controlled lapping occurs as the parts slide and slip during rolling, caused by the nonradial position of the workholder slots (Fig 12).

Process Capability. Accuracy of size and straightness to <0.125 μm (<5 μin.) are obtainable by this process on selected parts. Lap flatness and operational procedure are the two major factors that determine the machine capability to lap to extremely close tolerances.

Surface Finish. Finishes of 0.025 μm (1 μin.) have been readily obtained on hard,

dense ceramics as a result of optimum abrasive and lubrication selection. This finish is unique to this form of lapping and cannot be duplicated by any other process. Although measuring instruments such as profilometers may record an identical or a comparable reading on parts lapped between flat laps and those lapped or polished on centerless and hone-type superfinishers, there is a marked difference. Perhaps the best why to describe the difference is to say that the finish pattern is more multidirectional on parts lapped between flat laps. A finish produced by centerless lapping or hone-type superfinishers has a distinct directional pattern. The multidirectional finish provides long wear life and a uniform surface contact between the lapped part and its mating surface.

Roundness. Parts that have been carefully ground between dead centers prior to cylindrical lapping can be lapped round to <0.125 μm (<5 μin.).

Parts that have been previously ground by centerless grinding have a certain degree of multilobe out-of-roundness that is not appreciably improved or corrected by lapping between flat laps. If there is a roundness requirement on parts that have been centerless ground or centerless lapped, the required roundness would have to be obtained prior to lapping, or by ring lapping each individual part.

Flat Honing

Sometimes called fine grinding, flat honing finds the abrasive grain mixed into a bonded wheel or a disk that also acts as the reference surface. The abrasive grains are oriented in a random way that exposes sharp edges for the surface of a part to pass over. A relative motion and uniform pressure applied between the wheel and the part surface cause the protruding abrasive edges to penetrate the workpiece surface and scratch away material. This cutting action takes place continuously over the entire surface of the workpiece that comes in contact with the flat abrasive surface.

Flat honing requires an abrasive wheel(s) that will produce the required surface finish and accuracies of flatness and parallelism. To

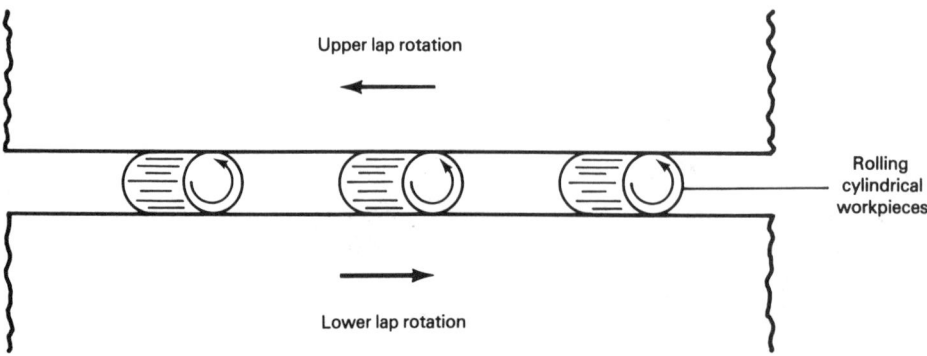

Fig 12 Schematic showing motion of laps and cylindrical workpieces when cylindrical lapping between flat laps

be efficient, the wheel(s) must continue to cut, load after load, without requiring frequent truing. The accuracy obtained is dependent on the flatness of the wheel surfaces. If the faces wear, then they must wear away evenly in order to keep the wheel surfaces flat.

Presently, the use of diamond or cubic boron nitride (CBN) wheels for flat honing is relatively new. Research is presently being done to flat hone hard ceramic materials with metal bonded fine grit diamond wheels. Anticipated benefits of diamond wheel honing over conventional grinding are cost-effectiveness, less surface damage, better accuracies and finishes, less chipping, and easier fixturing.

The coolant normally used in the flat honing process provides a wetting of the small particles generated during the machining process and also carries the particles away. In addition, the coolant serves to cool the workpiece and to provide lubrication for the flat honing process. Mineral oil or mineral seal oil are traditional fluids that meet these requirements and still have a high enough flash point and low odor, making both oils feasible for use based on property and safety requirements. Water-based coolants have specific applications and some workpieces composed of carbon and other ceramics are often processed using water.

Double-Side Flat Honing Process. Similar to double-side lapping, flat honing is capable of producing highly accurate products in terms of parallelism and uniformity of size when abrasive machining is used on both sides of a workpiece simultaneously in a batch-type mode. Although considered to be the more widely used mode, flat honing is also available as a through-feed or single-side process.

The batch mode uses two rotating flat wheels mounted on vertical spindles with the upper wheel able to be raised and lowered (Fig 13). The lower wheel has heavy support for thrust.

Fig 14 Typical workholder setup in a double-side flat honing machine used for batch mode processing

A number of workpieces are nested in workholders called carriers. The workpieces are normally spaced evenly over the area of the hone surface (Fig 14).

The top hone wheel is lowered onto the workpieces coming to rest on the highest points provided by the workpieces (Fig 15).

The honing takes place as the hone wheels are rotated and pressure is applied to the top wheel. To provide uniform hone wheel wear and most accuracy to the workpieces, the carriers are also rotated during the hone cycle.

Because of the uneven surface of the plane created by a batch of workpieces, the flat hone surfaces make contact initially at only a few locations in the batch (Fig 15).

As the high points are honed away, more and more of the flat surface of each workpiece comes in contact with the flat abrasive surfaces of the upper and lower hones. The

best possible flatness and plane parallelism can be obtained by this method for a number of reasons. Because the hone wheel faces are maintained in precision flatness, the surfaces being generated by honing actually mirror this flatness (Fig 16).

If there are any internal stresses in the material, removing stock from both sides simultaneously helps to normalize the part and enables the best possible flatness on both sides to be obtained. It is known that if workpieces have any internal stresses, machining one side at a time often will cause the workpiece to bow as the stresses are being relieved at different times. When working to flatness tolerances of <1 μm (<40 μin.), such stresses are a critical factor and should be kept to a minimum.

Batch mode processing enables a quantity of workpieces to be honed simultaneously, and each workpiece in the load will normally be honed to the same thickness. Basically, the greater the number of workpieces in a load, the more uniform the thickness of each part in the load will be.

Applications. Flat honing is ideally suited for processing the faces of a wide variety of components. Traditionally, the parts that lend themselves to this process require precision geometric tolerances with a surface finish free from directional grinding lines. Typical applications include components with dimensional or wear surfaces such as bearings, pumps, seals, and disk drives.

The finish tolerances obtained by flat honing will vary depending upon the basic characteristics of the component being processed. The most important are material type, hard-

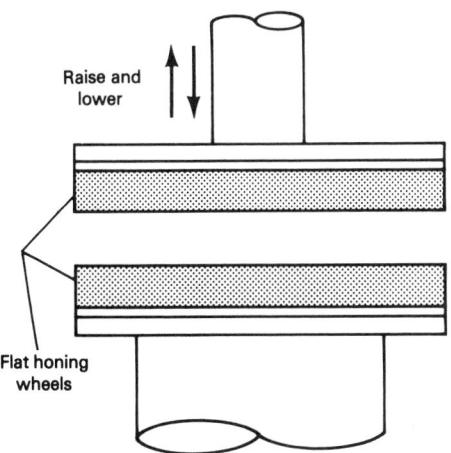

Fig 13 Schematic showing motion of both upper and lower honing wheels in a double-side flat honing machine

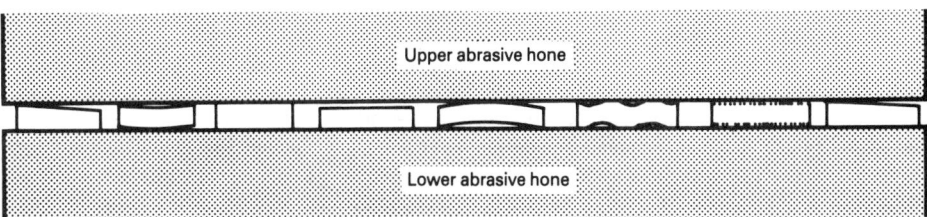

Fig 15 Correction of typical machining defects in workpieces through the use of the flat honing process

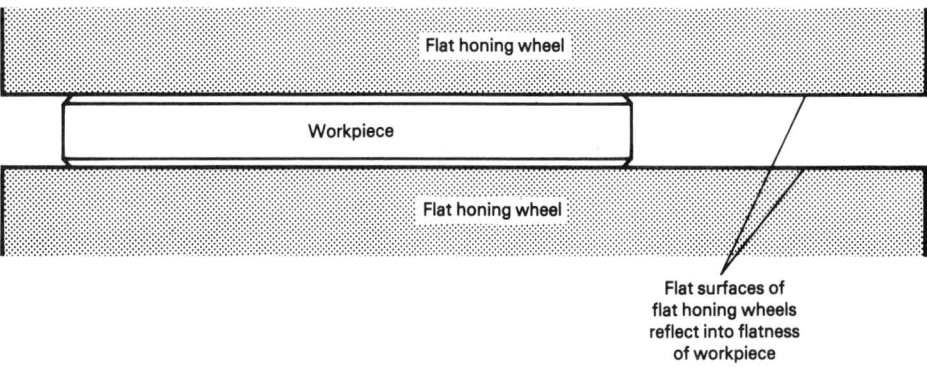

Fig 16 Schematic showing optimum flatness and plane parallelism obtained by machining with a double-side flat honing machine

ness, configuration (area), and tolerances from prior operation.

Precision Flat Polishing

Traditionally, this type of polishing is applied to previously lapped surfaces to improve flatness, to reduce surface roughness, to remove damage to the surface caused by lapping or grinding, to provide a reflective wear surface, and to supply a clear finish for transparent materials such as glass optics.

Briefly, there are two common polishing processes: mechanical polishing and a combination of both chemical and mechanical polishing. Mechanical polishing is basically a process that uses a fine abrasive to rub against the surface and thus remove small minute particles from the workpiece surface. The chemical/mechanical polishing process often uses a fluid mixed with an extremely fine abrasive. The type of fluid used causes a chemical action to take place on the surface of the part, thus etching the surface of the workpiece. The fine abrasive mixed with the fluid mechanically removes the etched material from the surface to produce a polished surface.

The most widely used mechanical polishing processes utilize a pad as the polishing tool, and the abrasive mixed with a water-based solution is applied with a continuous feed to the tool. The pad can be firm or soft, thick or thin. Because of the flexibility of the pad, the flatness of the surface being polished is usually not improved, and in most cases, the flatness actually deteriorates the longer the polishing operation takes. The softness of the pad follows the out-of-flatness of the workpiece and, in addition, tends to round off (wash off) the edges (Fig 17).

Mechanical Polishing. In all cases, the abrasive used for mechanical polishing is harder than the material being polished. The abrasive is normally embedded into the polishing tool or at least is applied to a surface that restricts the rolling movement of the abrasive grains, causing them to stroke the surface. For most accurate flat polishing applications, the polishing tool is of a soft metal composition and the abrasive, usually a fine diamond powder, is mechanically embedded into the soft metal to hold the abrasive particles (Fig 18).

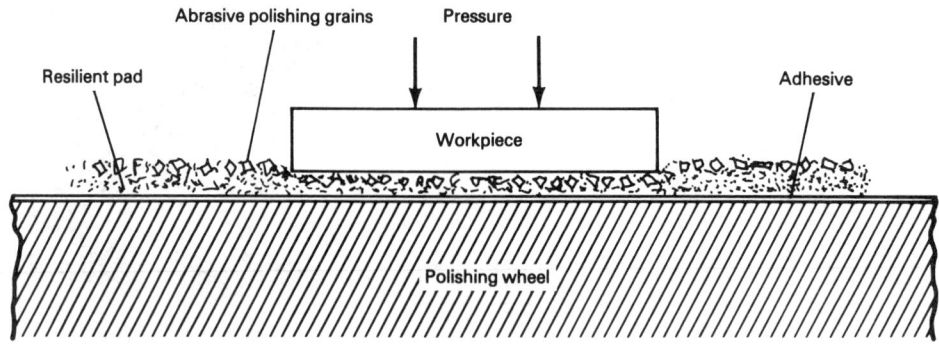

Fig 17 Schematic showing abrasive grains held in a soft pad for a mechanical polishing application

Chemical/Mechanical Polishing. Most chemical/mechanical polishing processes use pads. The chemical etching action is usually supplied by either an acid or a caustic solution. Therefore, machines used with this type of processing often need to be constructed of corrosion-resistant materials in the area that comes in contact with the polishing media. Because of the chemical reaction, heat is sometimes developed during polishing and often serves to improve or optimize the polishing process. Most polishing compounds used with the chemical/mechanical process consist of extremely fine abrasives such as submicron silica particles having a mean particle size ranging from 50 to 70 nm (500 to 700 Å). The abrasive used is of the softer type and is used primarily to abrade away the etched workpiece material.

Flat Polishing Equipment. Today, both single-side and double-side polishing are commercially available for production applications. The equipment is basically the same as that used for lapping (see the section "Abrasive Cutting Mechanism for Lapping" in this article). In some cases, however, more power and higher pressures are required. In addition, the recommended polishing compound feed systems are custom designed for each application. Along with suitable rinse systems (described in the section "Chemical/Mechanical Polishing" in this article), chemical/mechanical polishing necessitates the use of corrosion-resistant components in the work area.

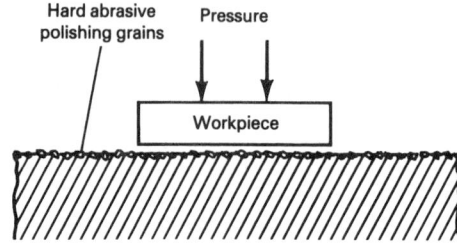

Fig 18 Schematic showing hard abrasive grains embedded in a firm polishing wheel used in mechanical polishing applications

SELECTED REFERENCES

- A.M. Corsini, "Precise Abrasive Machining Flat Surfaces in Production," SME Technical Paper MR 90–399, Society of Manufacturing Engineers, 1990
- J.H. Indge, "Double-Side Precision Abrasive Machining Ceramics," SME Technical Paper EM 90–139, Society of Manufacturing Engineers, 1989
- J.H. Indge, "Flat Precision Machining Ceramic Materials," SME Technical Paper MR 88–132, Society of Manufacturing Engineers, 1988
- D. Rostoker, J.H. Indge, and G.T. Emond, "Lapping and Polishing of Ceramic Materials," American Ceramic Society, SME, Abrasive Engineering Society Symposium on Machining Ceramic Materials and Components, April 1987

Ultrasonic Machining

Mary A. Moreland, Bullen Ultrasonics, Inc.

ULTRASONIC MACHINING (USM; also known as ultrasonic abrasive machining, UAM; or ultrasonic impact grinding) is a nontraditional machining method used in the machining of hard, brittle materials. This mechanical grinding process provides the capability of producing a limitless variety of configurations with the virtual absence of stress.

In ultrasonic machining, electrical energy is converted into mechanical motion. The linear motion is typically only 0.025 mm (0.001 in.), but at a rate of 20 kHz. The low amplitude vibration is acoustically transmitted to a toolholder and specially designed tool. When combined with an abrasive slurry, the workpiece material is microscopically ground away and the machined area becomes a mirror or counterpart of the vibrating tool.

Ultrasonic machining is nonthermal, nonchemical, and nonelectrical. Therefore, no change takes place in the metallurgical, chemical, or physical properties of the workpiece. Commonly machined materials include:

- Aluminum oxides
- Silicon
- Silicon carbide
- Silicon nitride
- Glass
- Quartz
- Sapphire
- Ferrite
- Fiber optics
- Germanium
- Zirconium
- Graphite
- Beryllium oxide
- Ceramic composites

While it can cut any material, ultrasonic machining is most effective on materials harder than 40 HRC.

The stroke or amplitude of the stroke is produced by the combination of an electronic generator coupled to a transducer package (magnetostrictive or piezoelectric). The generator converts typical line voltage into the voltage and frequency required to energize the transducer coupled to the generator.

The transducer is connected to a transmitting connecting body. The physical size is designed to resonate naturally at the same frequency as the electric current. A threaded stud is used on the end of the connecting body to connect the toolholder. Figure 1 shows an ultrasonic transducer and toolholder assembly.

Tooling Requirements

The metal toolholder and tool combination must be designed to transmit the acoustic energy properly and resonate within the bandwidth of the specific transducer used. The resonant length partially depends on the material used. Monel or 304 stainless steel are typically used because their soldering and brazing characteristics meet requirements to attach the cutting tool.

A nonamplifying toolholder usually ranges from 25 mm (1 in.) to 102 mm (4 in.) in diameter and reproduces the stroke coming from the connecting body. An amplifying toolholder generally ranges from 6.35 mm to 88.9 mm (0.250 to 3.50 in.) in diameter. The maximum stroke usually required for grinding applications is 0.064 mm (0.0025 in.). Toolholders made of high-strength materials can have an output stroke of 0.20 mm (0.008 in.). Figure 2 shows several silicon carbide

Fig 2 Silicon carbide engine components machined using toolholder/tool combination shown with the workpieces

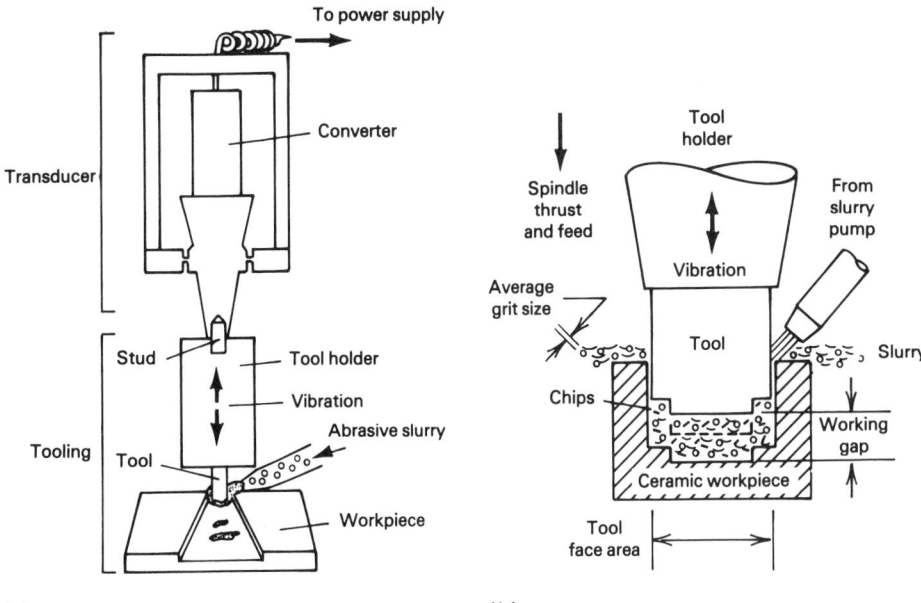

(a)

(b)

Fig 1 Schematic showing key components of a typical USM installation. (a) Transducer assembly coupled to tooling assembly of unit. (b) Close-up view of tooling assembly being used to machine a ceramic

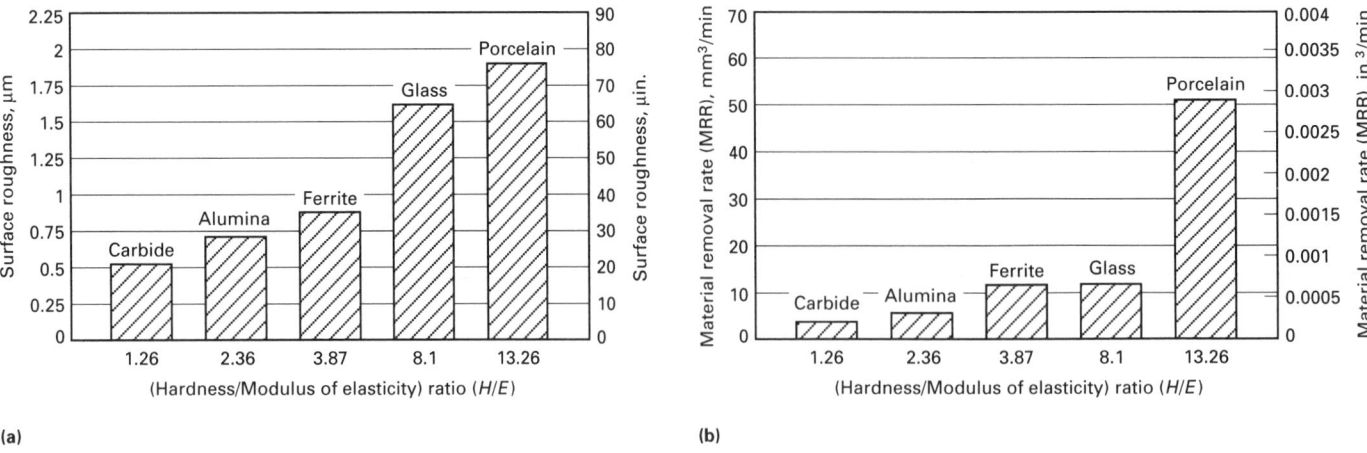

Fig 3 Comparison of the machining properties of selected ceramics processed under identical USM conditions. (a) Surface roughness. (b) Material removal rate

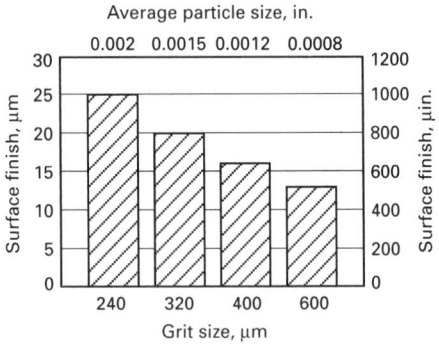

Fig 4 Effect of abrasive grit size on the surface finish of USM machined ceramics

engine components with the toolholder/tool combination used in their machining.

The attachment of the tool to the toolholder can be made by several different methods. The actual application, physical size, and mass of the tool determine the method of choice. Silver brazing forms a strong joint and is probably the most commonly used method. It provides a bond that allows the tool to have a stroke of 0.064 mm (0.0025 in.). Soft soldering of pins for multiple hole drilling is widely used, but the maximum recommended stroke is 0.038 mm (0.0015 in.). Both of these methods provide a lower tooling cost because the toolholder can be used repeatedly after the tool has worn out.

The cutting tool can be machined directly into the end of the toolholder. This forms the strongest tool/toolholder unit because the cutting tool is held by the strength of the parent material.

High-strength adhesives can be used if the stroke is kept at <0.013 mm (<0.0005 in.). Finer abrasives are required at this point, but they provide better finishes and tolerance capability. In addition, tools are easily exchanged and tooling costs decreased.

Some high-volume applications use mechanical connections. They provide for simpler tool changing procedures and allow the tool designer to use different materials for the tool and toolholder (for example, a titanium toolholder and a stainless steel tool). The maximum stroke recommended for this application is 0.064 mm (0.0025 in.).

Selection of Abrasives

An abrasive is the most important element in ultrasonic machining. The specific abrasive used effects the cutting speed, surface finish, and tolerance capability of the entire process.

Boron carbide and silicon carbide are the most commonly used abrasives. Boron carbide is recommended for machining advanced ceramics (silicon carbide, silicon nitride, aluminum oxide, zirconium oxide, and so on) and metal parts. Silicon carbide is more frequently used on glass, quartz, and single-crystal silicon.

The carbide abrasive is suspended in water at a 20 to 50% concentration and constantly recirculated to the workpiece. The high volume of slurry cools the tool and the workpiece, supplies fresh abrasive to the cutting site, and removes the particles abraded. The rapid motion of the tool impacts the sharp grain edges into the workpiece material and microscopically machines it away.

Surface roughness or finish and material removal rates are dependent upon the abra-

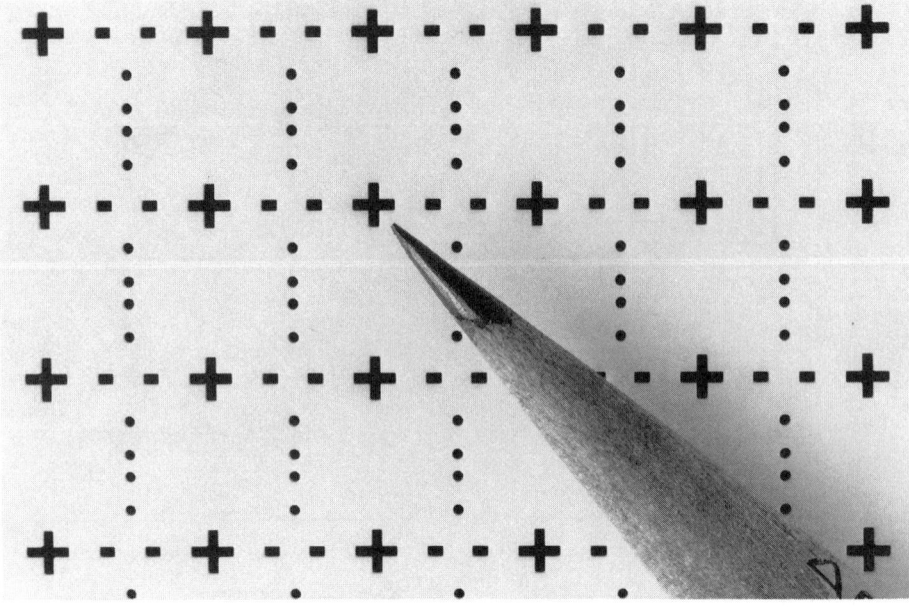

Fig 5 Fine holes and intricate slots produced in a 99.6% Al_2O_3 substrate with ultrasonic machining

sive chosen. The mechanical properties and fracture behavior of the workpiece materials also play a large role. Figure 3(a) shows the surface roughness values of various materials under like machining parameters. As a ratio of hardness (H) to the modulus of elasticity (E) increases, the workpiece roughness values also increase. In Fig 3(b), it is interesting to note that the same is true in material removal rates (MRR). Higher material removal rates are evident in higher H/E ratios. Thus, a low fracture toughness will result in a higher material removal rate and an increase in workpiece roughness values. Figure 4 shows the effect of grit size on finish for some of the more commonly used abrasive sizes.

Process Capabilities

Ultrasonic machining provides the capability to machine a limitless number of different cuts and shapes. The high-speed vibrations of the tool edge against the sharp abrasive grains reproduce the shape of the tool face into the workpiece. Round and odd-shaped holes (oval, elliptical, triangular, square, diamond, star, and so on), slots, blind cuts and cavities, outside diameter (OD) and inside diameter (ID) work, and decorative engraving are a few of the more common applications. Figures 5 and 6 demonstrate these cuts in advanced ceramic applications.

Tool wear is controlled by dressing off the bottom of the tool face. This procedure may also be required to reduce taper. Dressing is used to maintain the wear surface of the tool in order to propel the abrasive grains directly into the intended cut and not outward toward the walls of the workpiece material. Failure to keep the abrasive grains directed at the hole and simple abrasive rubbing or chafing are two mechanical problems that generate the taper. The use of trepanning methods greatly reduces material removal and is preferable to solid drilling.

Holes as small as 0.076 mm (0.003 in.) have been drilled. However, the practical limitation for production applications is 0.25 mm (0.010 in.).

Advantages of Ultrasonic Machining

Ultrasonic machining is frequently used to provide faster deliveries and to lower tooling costs for the machining of ceramics. In the early design stages when changes are required, alternatives can be made quickly and economically through the application of impact grinding.

Frequently, ultrasonic machining provides the only method that is capable of machining odd-shaped cuts or shapes in hard, brittle material. Sharper radii and tighter tolerance capability are also process advantages in some applications. Because of its nonthermal characteristics, ultrasonic machining produces virtually stress-free machined surfaces. Fig-

ure 7 compares a typical hole obtained with ultrasonic machining methods to a hole of the same size obtained using a laser machining process. The surface of an ultrasonic hole is compared to an as-fired surface and is readily metallized. The stress-free characteristics of ultrasonic machining make it the method of choice in many high-reliability applications because it preserves the strength and integrity of the workpiece material better than any other material removal method used to machine ceramics.

Equipment

Due to the wide diversity of ultrasonic machining applications, impact grinders are built to serve a great variety of requirements ranging from laboratory use to high-volume production. In their simplest forms, the machines have manually operated axis control movement. The x and y axes are typically used to position the workpiece under the cutting tool, and the z axis is normally used to

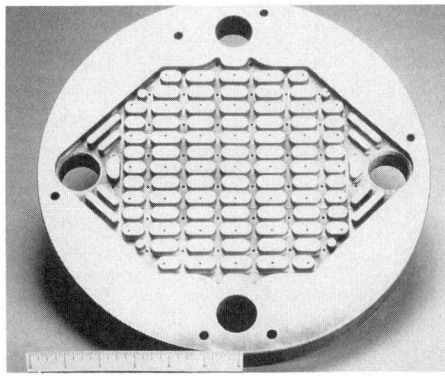

Fig 6 Channels and holes machined ultrasonically in polycrystalline silicon

control the movement of the cutting tool into the workpiece.

The cutting tool can be operated to cut at its own rate in many applications. This is true when:

- Amplitude of the cutting tool is greater than the abrasive particle size being used
- Physical strength of the cutting tool is enough to handle the ultrasonic vibration as well as the tool pressure used to feed it into the workpiece
- Depth of cut is less than 6.4 mm (0.25 in.)

The cutting-tool feed rate can also be manually controlled. This may be necessary when using very sensitive and miniature cutting tools. Feed rates as low as 4 μm/s (160 μin.) may be required for machining ceramics. Periodic manual reversing of direction also allows the abrasive to flush underneath the cutting tool, a procedure that may be necessary when the cut is deep enough to create abrasive starvation. High aspect ratio holes and cavities often require removal of the cutting tool from the workpiece periodically to ensure uninterrupted abrasive slurry flow to the cutting zone.

Processing equipment is also available with computer numerical control (CNC) of all axes that has a position resolution of 0.25 μm (10 μin.). The feed can be programmed for specific feed rates and the depth of cut desired. Reversing of the tool during the program permits abrasive flushing when a preset cutting force is exceeded.

Preset cutting force limits can be programmed to actuate the momentary extraction of the tool from the cut. This procedure ensures the maintenance of proper abrasive conditions. Cutting forces increase when abrasive slurry conditions depart from optimum settings. Automatic tool-wear compen-

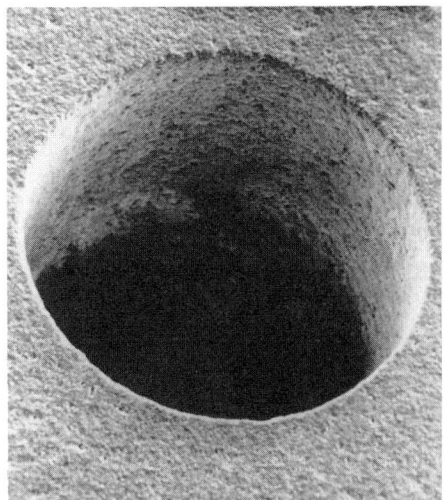

(a)

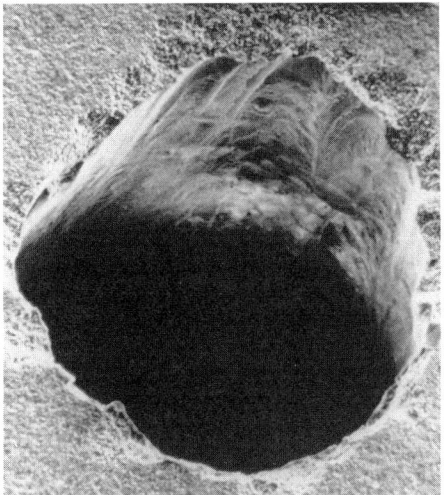

(b)

Fig 7 Scanning electron microscopy (SEM) photomicrographs of 0.64 mm (0.025 in.) holes drilled into alumina by two different methods. (a) Ultrasonic machining. (b) Laser beam machining

sation can be programmed in to the CNC unit. The versatility of computer numerical control is especially beneficial in production applications.

Ultrasonic impact grinders are classified on the basis of power ratings of the ultrasonic transducer. The size of the transducer governs the size of the toolholder that can be used. Impact grinders which range from table-top models rated at 500 W (0.7 hp) to standing floor models rated at 2000 W (3 hp) are available with solid-state power supplies. Automatic tracking to the toolholder's resonant frequency compensates for variations in the temperature of the toolholder and for varia-

tions in the length of the cutting tool caused by wear.

Rotary Ultrasonic Machining (RUM). An ultrasonic assist added to a rotating tool enhances material removal rates, finish capabilities, and overall drilling efficiency. Rotary ultrasonic machining uses a diamond-plated drill and is water cooled because the diamond-impregnated abrasive provides the cutting edge.

Rotary ultrasonic machining can be used for milling, drilling, threading, and grinding applications. However, one drawback of rotary ultrasonic machining is that its use is limited to round tool configurations.

SELECTED REFERENCES

- M. Komaraiah, M.A. Manan, P. Narisimha Reddy, and S. Victor, Investigation of Surface Roughness and Accuracy in Ultrasonic Machining, *Prec. Eng.*, Vol 10, 1988, p 59–65
- M.A. Moreland and D. Moore, Versatile Performance of Ultrasonic Machining, *Ceram. Bull.*, Vol 67 (No. 6), 1988, p 1045–1047
- *Machining Data Handbook*, Metcut Research Associates Inc., 3rd ed., Vol 2, p 10–43 to 10–63

Abrasive Fluid Jet Machining

Pawan J. Singh, Ingersoll-Rand Waterjet Cutting Systems

THE USE OF FLUID-TRANSPORTED ABRASIVE for material removal had been limited to deburring and descaling applications until the recent advent of abrasive waterjet cutting. Since 1985, the use of abrasive-entrained ultrahigh-pressure (>240 MPa, or 35 ksi) waterjets to cut a wide variety of metallic and nonmetallic materials has rapidly expanded, creating a whole new industry and an ever-expanding roster of new applications. The technique known as abrasive waterjet cutting (AWJ) offers several distinct advantages over alternative processes and is, in fact, one of the available cost-effective techniques for cutting several composites. Besides cutting, the technique is under development for turning, milling, drilling, and trepanning, but the primary application at this time is limited to cutting.

Principles of Abrasive Jet Machining

Conventional abrasive jet machining refers to the use of abrasive-gas jets to propel particles against a surface and to remove the material from that surface via particle impact and shear. Typically, abrasives in quantities up to 0.05 kg/min (0.1 lb/min) are propelled through a nozzle at inlet pressures ranging from 170 to 860 kPa (25 to 125 psi) in a gas (nitrogen or air) medium. Outlet velocities are generally supersonic and the volume of material removed varies as particle velocity raised to a power between 2 and 3. Sand blasting and deburring are the two most common applications for abrasive jet machining. Material removal rates (MRR) can be enhanced by increasing the inlet pressure (experiments up to 12 MPa, or 1.8 ksi have been carried out) and by optimizing nozzle design. However, difficulties in controlling shock-laden jet structure and particle distribution within the jet make this process impractical for precision machining. Extremely high noise level (>85 dB) is another problem.

Fundamentals of Abrasive Waterjet Cutting

In abrasive waterjet cutting, high-velocity water, rather than gas, is used to entrain and accelerate abrasive particles. Unlike gas jets, waterjets can be confined to a small diameter without spreading and they can accelerate particles to desired velocities in a relatively short distance. In addition, relatively high abrasive mass flow rates (1.4 to 1.8 kg/min, or 3 to 4 lb/min) can be easily accommodated. The result is an extremely high-energy jet that can easily cut through a solid steel plate more than 102 mm (4 in.) thick.

System Components

A typical abrasive waterjet system is made up of the following components:

- Water intensifier to produce discharge pressures of up to 415 MPa (60 ksi)
- Cutting head (including waterjet orifice and mixing nozzle) to convert high-pressure water to a high-velocity (up to 760 m/s, or 2500 ft/s) abrasive jet
- Abrasive delivery system to transfer and meter abrasive from a storage tank to the cutting head
- Manipulator to move the cutting head (or the material) according to the desired cut profile
- Catcher tank to entrain spent abrasive, to remove material, and to provide some means of abrasive disposal

Figures 1 and 2 show a schematic diagram and a flow diagram, respectively, of abrasive waterjet system components. The intensifier (Fig 3) is a hydraulically driven reciprocating intensification pump, typically 19 to 75 kW (25 to 100 hp), with 4.9 L/min (1.3 gal/min) flow capacity at pressures selectable up to 380 MPa (55 ksi). Because of high pressures and extremely high velocities involved in the process, inlet water needs to be free of all impurities up to low-micron levels. It is recommended that water undergo deionization or filtration through a reverse osmosis membrane prior to entry into the intensifier. Multiple intensifiers, sometimes mounted on a single frame, are used to increase the flow rate.

High-pressure water is passed through sapphire orifices to create a high-velocity jet that, due to the jet-pump effect, induces vacuum in a tiny mixing chamber past the orifice (Fig 4). This chamber is connected to a tube that brings precisely metered abrasive flow from an abrasive source to the mixing chamber. Most of the real mixing between water and abrasive takes place in a 50 to 75 mm (2 to 3 in.) long ultrahard tungsten-carbide tube aligned concentrically with the sapphire orifice. The stand-off distance between the jet exiting the carbide tube and material to be cut is typically less than 6.4 mm (0.25 in.). A fast-response, high-pressure on-off pneumatic valve, mounted upstream and close to the sapphire orifice controls the jet and limits the water-hammer effect to a very small region.

Spent abrasive and removed kerf material are captured in a catcher tank whose grated top typically supports the material to be cut. The catcher tank can be a simple large metal box with an open top or a well-designed system with baffle plates, fluid recirculation, and automatic abrasive removal and filtration arrangement. Catchers also come in various styles: slit catcher, point catcher, and follower catcher. The correct style depends on the relative movement of the abrasive jet and the material to be cut. For example, when material moves along both x and y axes in a plane and the abrasive jet remains stationary, a point catcher—generally a piece of pipe containing energy-absorbing material such as spherical balls—is adequate.

For abrasive waterjet applications, two- or three-axis gantry manipulators are most common. These manipulators are driven by dedicated or programmable controller (PC) based controllers that make waterjet equipment operate similar to any other state-of-the-art machine tool.

Process Capabilities

The abrasive particle energy density impacting the material is so high that virtually any material can be cut, including common metals such as aluminum and various grades of steel, aerospace materials such as titanium and high-nickel alloys, and brittle materials such as glass and marble, green and reinforced composites, honeycomb and sandwiched materials, and selected ceramics.

Advantages. Besides its wide applicability, abrasive waterjet offers several other unique advantages for machining ceramics:

- Cut surface is hardly affected by thermal distortion or oxidation, or any other

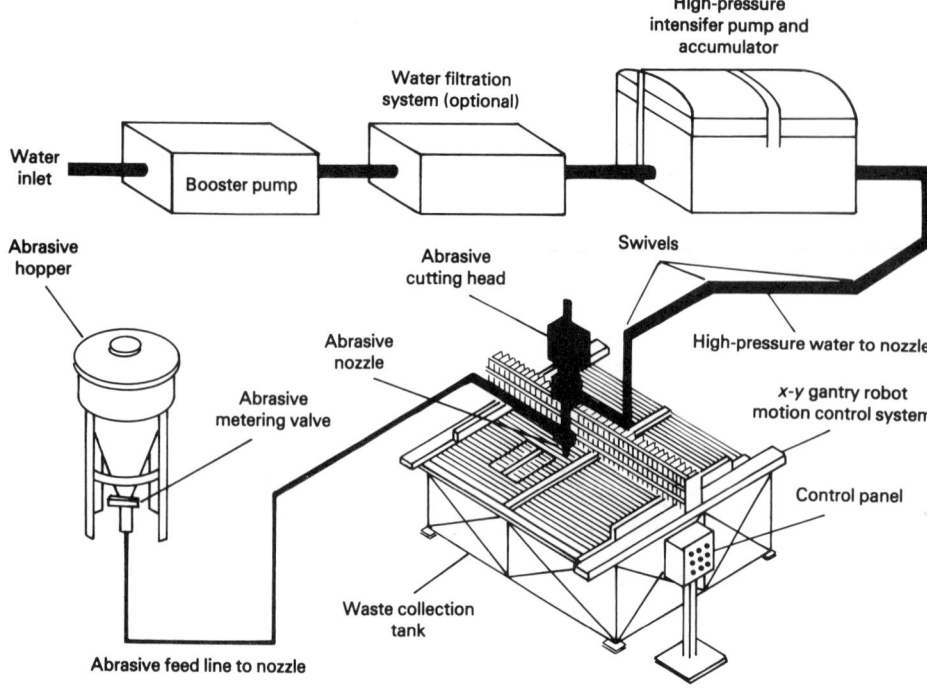

Fig 1 Schematic showing key components of a state-of-the-art abrasive waterjet installation incorporating an x-y gantry robotic motion system to control movement of the cutting head

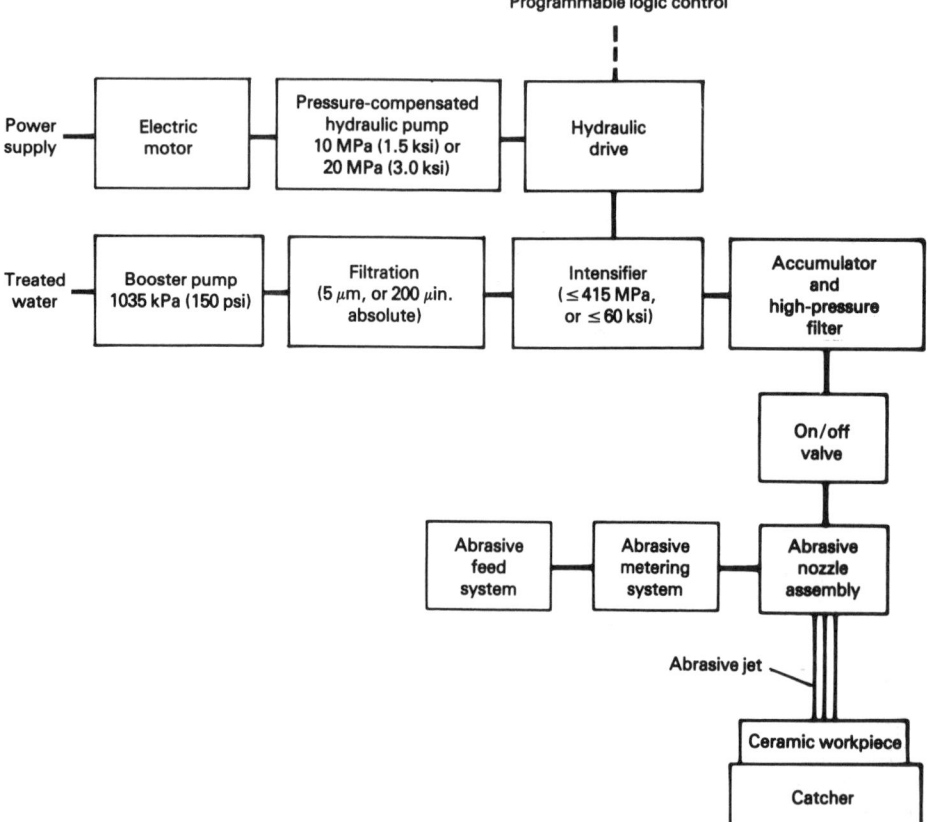

Fig 2 Flowchart showing key components and operational parameters of selected components in an abrasive waterjet cutting system

localized structural change such as coldworking

• The process is omnidirectional, allowing cutting of complex contours in one continuous operation
• Because of low nozzle weight and inertia, the abrasive waterjet can be easily integrated with computers, robots, and controllers. Much of the popularity of AWJ systems can be attributed to this flexibility
• Airborne dust is eliminated, making the process safer and environmentally less hazardous for many materials that generate dust with traditional cutting techniques
• The process is much safer in hazardous atmospheres relative to other processes. Because there is no open flame or heat buildup, fire hazards are minimized
• Same tooling and system can be used to handle a wide range of materials

Limitations. The AWJ technique also has some limitations:

• Initial capital investment cost is high (>$150,000) and operating costs ($25 to $40/h) are also relatively high. The process is thus not competitive for machining of low-cost materials, except in limited cases
• Noise levels are high (85 to 95 dBa) but can be reduced to <80 dBa level by enclosing the cutting area
• Abrasive disposal and effluent filtration can be cumbersome and sometimes costly
• Some materials, such as thick sections of some ceramics, cannot be effectively cut

Effect of Cutting Parameters on Surface Quality

The abrasive waterjet cutting process is characterized by many parameters and the operator has, at least in theory, a broad set of choices to cut at the desired speed. Manufacturers recommend certain conditions, although their recommendations vary widely.

The choice of parameters includes: water pressure, sapphire orifice diameter, carbide tube diameter, abrasive flow rate, type and size of abrasive, and cutting speed. The surface quality (roughness and kerf taper) also depends on these parameters although cutting speed and surface quality are directly related. In general, the slower the cutting speed, the better the surface quality. The following trends can be clearly noticed:

• Cut depth increases with water pressure beyond a critical pressure value below which cutting stops. The rate of increase is mostly linear
• Increase in abrasive flow rate increases cut depth up to a certain point and then reduces it. The reduction is caused by oversaturation of abrasive resulting in

Fig 3 Rear view (with covers removed) of an intensifier pump used to generate the maximum 415 MPa (60 ksi) fluid pressure required to cut ceramics with an abrasive waterjet

reduced particle velocities and excess loss of energy in particle collisions
• Water flow rate and power consumption increase with an increase in sapphire orifice diameter. Cut depth depends primarily on abrasive flow rate and somewhat on water flow rate. However, with increased water flow, abrasive loading can proportionally increase before saturation prevails. There also exists an optimum combination of sapphire orifice and carbide tube diameters
• Cut depth is higher with coarser abrasive
• Best improvement in surface quality is obtained by decreasing cutting speed. Finer abrasives also produce smoother cut surfaces

The following parameter combinations are typically used in AWJ industrial applications:

Operating pressure, MPa (ksi)	205–345 (30–50)
Sapphire orifice diameter, mm (in.)	0.25–0.46 (0.010–0.018)
Carbide tube diameter, mm (in.)	0.76–1.57 (0.030–0.062)
Carbide tube length, mm (in.)	50–102 (2–4)
Type of abrasive	garnet, 60–120 mesh
Abrasive flow rate, kg/min (lb/min)	0.23–0.90 (0.5–2)
Water flow rate, L/min (gal/min)	1.9–3.8 (0.5–1)
Intensifier power, kW (hp)	19–75 (25–100)

Besides garnet, silica sand and olivine is sometimes used for cutting glass and softer materials because of its lower cost. Very hard abrasives such as aluminum and silicon carbide are needed to cut some ceramics.

Table 1 shows the range of cutting speeds recommended for various materials. The wide ranges in cutting speed result from great flexibility in the selection of operation parameters.

Cut Surface Properties. An abrasive waterjet cut surface finish depends greatly on operation parameters, with cutting speed having the most significant effect. At slow speeds, the resulting surface is relatively smooth with surface roughness as low as $R_q = 16$, where R_q is the root mean square (rms) value. At higher speeds, the bottom part of the surface is striated due to bending of the slowing jet and material removal mechanism changing from cutting through wear to cutting through deformation. With even higher speeds, striations become longer and coarser and eventually the jet cannot cut through at all. The material removal rate also slows down after penetration of the jet to a certain distance. Thus the kerf is narrower at the bottom than at the top of the material, leading to kerf taper. Like the surface roughness, kerf taper depends mostly on cutting speed. In summary, surface quality can vary from excellent to poor, depending on what the primary objective is: high throughput or tight tolerances. For comparison purposes, even under worst cases, surface quality is better than plasma or oxy-flame cut surfaces.

Abrasive Slurry Jets

In abrasive slurry jets, a thin slurry mixture of abrasive and water is directly pumped to a high pressure and then passed through a nozzle to form the jet. A slurry jet is more efficient because energy losses due to mixing and particle collision are much reduced. In addition, high abrasive loading can be readily achieved. However, other problems easily outweigh these advantages. Pump and component life is greatly reduced and uniform mixing of the slurry is a problem. A British company has introduced a continuous-batch lower pressure (20 to 105 MPa, or 3 to 15 ksi) system for primarily portable applications.

Machining of Ceramics with Abrasive Waterjets

The following applications are representative of the use of abrasive waterjet machining to process ceramics:

• Aircraft manufacturers (green and cured composites)
• Automotive (window glass and dashboards)
• Job shops (shape cutting of a wide variety of materials)
• Heavy equipment manufacturers (glass and deburring nonmetallics)
• Glass manufacturers (architectural glass and glass shape cutting)
• Marble and architectural stone shape cutting

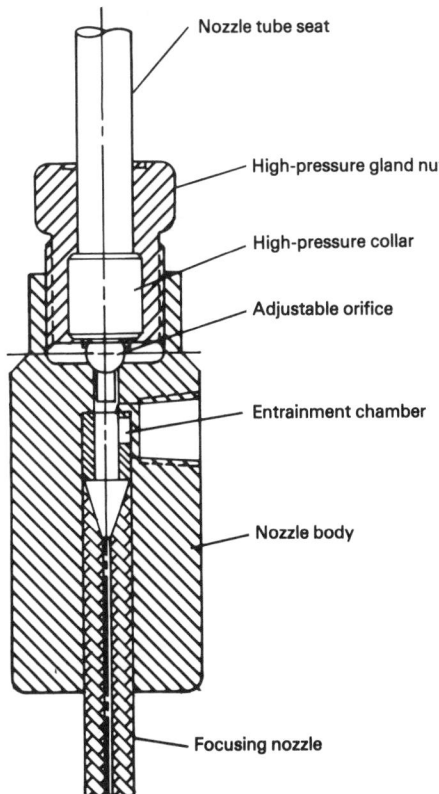

Nozzle tube seat

High-pressure gland nut

High-pressure collar

Adjustable orifice

Entrainment chamber

Nozzle body

Focusing nozzle

Fig 4 Cross-sectional view of abrasive nozzle assembly used in an abrasive waterjet

Table 1   Recommended abrasive waterjet cutting speeds for selected ceramics. Data are for 80-mesh garnet abrasive with 35 kW (50 hp) maximum intensifier

Material	Thickness		Cutting speed	
	mm	in.	mm/min	in./min
Glass	4.75	0.188	1020–3050	40–120
Laminated glass	6.35	0.250	510–2030	20–80
Optical glass	82.55	3.250	2.5–13	0.1–0.5
Glass filters	2.0	0.080	255–1020	10–40
Safety glass	6.35	0.250	760–2540	30–100
Stained glass	3.18	0.125	1525–6350	60–250
Glass (television tube)	12.7	0.500	380–1270	15–50
Ceramic (96% Al)	153	6.027	50–205	2–8
Machinable ceramic	6.10	0.240	510–2540	20–100
Ceramic wall tile	6.35	0.250	510–2540	20–100

Abrasive Waterjet Applications

The abrasive waterjet has gained a strong foothold in glass-cutting applications involving complex cut shapes. With the waterjet, the cut is initiated by drilling a small hole at the starting point in the material. To avoid shattering of brittle glass, this hole is drilled at a pressure (typically, 105 to 125 MPa, or 15 to 18 ksi) lower than operating pressure (275 to 310 MPa, or 40 to 45 ksi). Once the hole is drilled through, the pressure is raised and the cutting starts. Because this pressure cycling can undermine the fatigue life of high-pressure components, manufacturers have devised ways to ramp the pressure up or down gradually. To avoid damage around the cut from stray abrasive particles, glass is often cut either under a thin layer of water or sprayed with fresh water in the cut region. Like other materials, the surface finish depends on cutting speed, abrasive type, and loading. At slow speeds, excellent surface finish can be obtained. With enough cutting power, glass blocks as thick as 0.3 m (1 ft) have been cut.

Ceramic applications are usually difficult, and therefore ceramic cutting with a waterjet is in its early development phase. Green ceramics are cut easily, most requiring no abrasive at all. Harsher abrasives such as aluminum oxide or silicon carbide are often required to cut ceramics and cermets. These abrasives are also tough on the carbide mixing tube, shortening its life to an impractical level of less than an hour. High-pressure waterjet, often without abrasives, is proving very beneficial for the removal of ceramic coatings without affecting the base material. These coatings are very difficult to remove by other means. Similarly, ceramic-coated metals can be easily cut with an abrasive waterjet. Architectural materials such as marble and stone are also cut economically with a waterjet.

Future Outlook

Abrasive waterjet technology is rapidly evolving. Recent advances in abrasive transfer system, carbide tube life, and other system components have increased the technology's acceptance in basic industries such as automotive and metal fabrication. Although cutting is the primary application now, the focus of new research is to expand the range of applications to other machining functions such as milling, drilling, and turning, and to make the process more user-friendly.

Laser Beam Machining

Arthur O. Capp, Jr., Laserage Technology Corporation

THE USE OF THE LASER to machine production quantities of ceramic substrates for the fledgling hybrid microelectronic industry began in the late 1960s. Because both of these emerging technologies were new at that time, they merged in a unique way to complement and to facilitate the growth of each other.

Ceramic substrates [principally, alumina (Al_2O_3)] were the foundation upon which the entire hybrid circuitry technology program was built. The thin-film metallization segment of the industry focused on 99+% Al_2O_3 while thick-film metallization focused on 96% Al_2O_3 (balance glass frit binder). Technical reasons for these preferences are beyond the scope of this article, but suffice it to say that the laser proved very useful to both segments of the electronics market.

The substrate material property requirements included rigidity, stability, an appropriate thermal coefficient of expansion, and a good dielectric constant. Unfortunately, with these desirable properties came other less desirable properties such as hardness and brittleness that caused real production fabrication challenges. Conventional fabrication methods such as mechanical cutting, milling, and drilling were totally inadequate.

Because it is a ceramic, an alumina substrate is produced by conventional refractory techniques. This typically consists of blending a slurry of very small alumina particles, mechanically shaping the slurry, and then kiln firing the alumina workpiece. To obtain the number of comparatively thin substrates required by both thick-film and thin-film circuitry applications, tape-casting and subsequent green punching of sizes and shapes was the best method available. Dimensional precision was not a process capability. The product made with this method shrinks ≈20% after being fired in the kiln. The dimensional variations caused by this shrinkage prohibited the serious automation of the process to make it a high-volume production method. Conventional shrinkage tolerance at the time was ±1%. Even today, shrinkage tolerances of ±0.5% command a premium price and are produced only by careful batch selection. For example, on a ceramic substrate board 125 mm (5 in.) long, the edge location could vary ±1.3 mm (±0.050 in.). Similarly, green-punched hole locations at the extremities of such a board could often drift further than the diameter of the hole.

The first multiple-part arrays (duplicate parts repeated in one substrate) were scored in the green state with a razor-sharp metal cutter. This blade partially penetrated the top surface of the substrate material during punching impact. Two significant problems were associated with this method:

- Wide dimensional variations due to the shrinkage
- Score-line snap variations resulting from the abrasive nature of the alumina particles wearing away the sharp edge of the blade

This wear reduced the score penetration depth that caused the subsequent breakage characteristics to vary beyond reasonable production limits.

Laser Beam Machining Equipment

It was into this setting that the laser moved rapidly out of the research laboratory onto the manufacturing floor as a viable production tool. Laser beam machining (LBM) the substrate after firing the ceramic in a kiln eliminated the above-mentioned shrinkage tolerance problem.

Wavelength Specifications. The CO_2 laser emits a wavelength of 10.6 μm (41.7 μin.). This wavelength was determined to optimally couple with the absorption characteristics of the fired alumina surface. This particular type of laser delivers enormous thermal energy with pinpoint precision. The intense heat of the focused laser beam causes fired alumina, which is so hard that conventional diamond drills and saw blades dull or wear out when machining of the alumina is attempted, to vaporize very rapidly. Furthermore, the pinpoint precision offered dimensional control necessary for volume production requirements that were previously unavailable. The speed with which the laser beam could be pulsed electronically resulted in practical production scribing and/or machining rates that could meet high-volume quantity demands at affordable costs.

Power Requirements. An industrial CO_2 laser system is an expensive tool. A complete system costs roughly $45,000 per 100 W (140 hp) of laser power. In the early days a 100 to 150 W (140 to 200 hp) laser system was the maximum size laser that was commercially available for this kind of work. More powerful lasers were developed and soon became commercially available. Today, a laser system of practical size for machining ceramic is rated 500 W (680 hp). The cost of this unit, which includes a numerical controller and a custom workstation complete with optics and a micropositioning table, is approximately $225,000. This wattage rating provides adequate power to machine ceramic thicknesses up to 3.18 mm (0.125 in.) or, by optically splitting the beam, to machine two substrates at a time that are <0.76 mm (<0.030 in.) thick at optimum rates. Units with enough power to cut more than two substrates at a time can cost up to $450,000 per system. Laser machining of ceramic is a capital-intensive technology. Therefore, it must produce a precision product day in and day out with a minimum of maintenance or downtime and accomplish this for many years to justify the high initial cost. Actually, fifteen to twenty year old lasers are still operational today in volume production applications producing quality products as they first did over a decade ago. These units often perform better due to improvements in optics, fixturing, and electronics.

Gaussian Mode Waves. The laser machining of ceramic requires a laser beam with a Gaussian or transverse electromagnetic ($TEM_{0,0}$) mode (Fig 1a). This mode, as opposed to the top hat or donut mode, $TEM_{0,1}$, causes rapid vaporization of the ceramic at the center of the focal point. Laser energy is absorbed in the process of vaporization, thus minimizing thermal shock to the surrounding area. A poor $TEM_{0,0}$ beam mode is shown in Fig 1(b). It has secondary hot spots located around the periphery of the center hot spot that can cause excessive heating of the bulk material and inefficient vaporization. This results in the formation of irregular molten material puddling, thermal cracks, glassy edges, and excessive debris when the ceramic is machined.

With the CO_2 laser, beam mode is to a great extent dependent on laser resonator design. Axial gas flow resonators, slow or fast, gen-

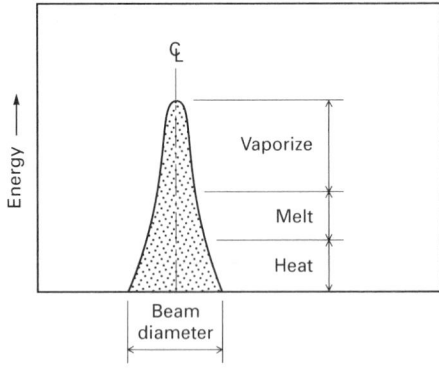

(a)

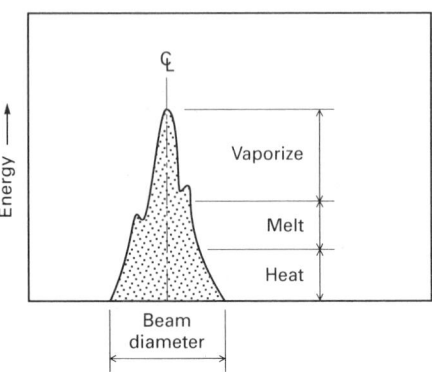

(b)

Fig 1 Plot of energy versus beam diameter of a Gaussian mode laser beam utilized in the laser-beam machining of ceramics. (a) Recommended envelope to minimize thermal shock to workpiece. (b) Unacceptable envelope consisting of hot spots along periphery of the curve resulting in excessive heating of workpiece

erally provide a Gaussian ($\text{TEM}_{0,0}$) mode. Other mechanical means to improve the mode include resonator apertures, reflective mirror curvature design, output window shape, and spatial filters/beam expanders. Many times,

the solution to optimum mode quality is a combination of several of these parameters.

Laser power is the second most significant factor in laser machining of ceramic. Increased power is axiomatic when machining thicker materials. However, there is a threshold power below which insufficient vaporization occurs and causes excessive liquid material to accumulate resulting in thermal shock, glassy edges, and excessive debris. This threshold power varies with the thickness of the ceramic material being machined, the rate of machining, and the width of kerf.

Focusing Lens Specifications. Ideally, the focused spot size for machining of a ceramic should be as small as possible. This is limited by the optics of the laser system. This suggests the use of the shortest focal length lens available. However, radiant heat and molten debris, both of which damage the lens surface, generally dictate using an $f2.5$ in. lens. A longer focal length lens may be needed for certain specific applications, but an $f5$ in. lens on a 500 W (680 hp) CO_2 laser generally produces a spot size too large for good vaporization characteristics on alumina.

Laser Machining Processes

Scribing. The most extensive use of the CO_2 laser in ceramic processing is for laser scribing. Prior to the advent of the laser, a ceramic was sectioned by scoring the material with a very hard sharp tool that scratched the substrate surface. This created a stress line along which the material would break uniformly. The laser system made an entirely new technique available. Partial perforation caused by a rapidly pulsing laser beam produced a snap line of extremely accurate dimensional precision. Centerline-to-centerline tolerances of ±0.025 mm (±0.001 in.) became a production reality that was quickly specified by

industry. A typical pulsed laser setup uses a cross air jet to blow the hot plume and molten debris away from the beam path.

A typical scribe is defined as a 0.13 mm (0.005 in.) diameter entry hole spaced 0.15 mm (0.006 in.) apart and penetrating conically to approximately one-third to one-half of the thickness of the substrate material. Figure 2(a) shows a section of a good scribe line. The contrast in the photomicrograph is achieved by coloring the scribe line with a black or other dark-colored felt-tip marker before snapping the material along the scribe line. Figure 2(b) shows a section of a poor scribe line generated by an irregular laser beam. This particular condition is usually corrected by means of circular polarizing optics. These optics are available as an off-the-shelf accessory or the device may be specified on new machines. Recommended scribe dimensions and routine, commercially available tolerances for various substrate thicknesses are listed in Table 1. Scribe breaking characteristics may be altered by increasing or decreasing either scribe spacing or the depth individually or collectively.

Scribe breaking or snapping characteristics may also vary dramatically due to set-up variations. Ideally, the focal point of the laser beam should be exactly at the top surface of the substrate. The surface plane of the micropositioning table should be parallel to the plane of motion. Camber of the ceramic substrate must be watched carefully because the focal point will change relative to the top surface of a moving warped substrate. Commercially standard camber variation of 0.003 mm/mm (0.003 in./in.) is acceptable on smaller substrates measuring 50×50 mm (2×2 in.). However, for larger substrates measuring 115×115 mm ($4\frac{1}{2} \times 4\frac{1}{2}$ in.) or 100×150 mm (4×6 in.), camber runout can accumulate to as much as 0.46 mm (0.018 in.). A premium camber variation of 0.002 mm/mm

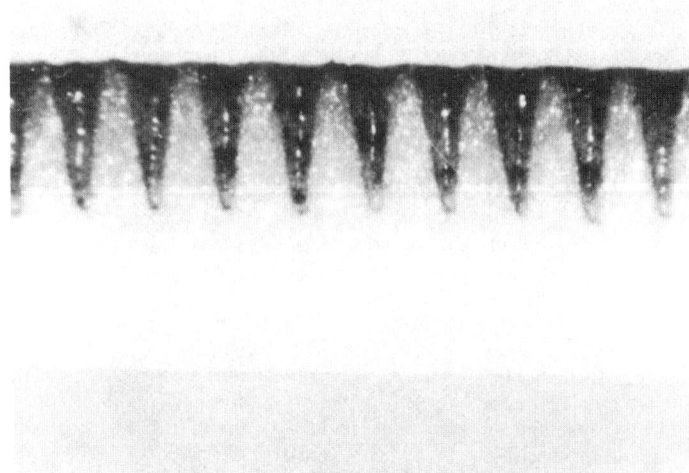

(a)

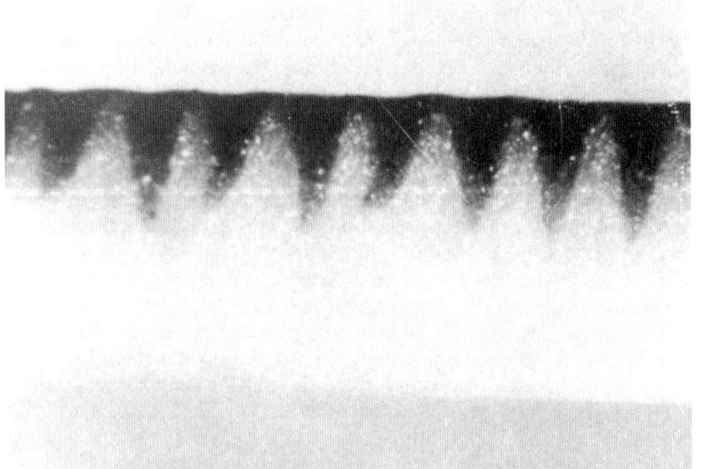

(b)

Fig 2 Cross sections of laser scribe lines in 0.64 mm (0.025 in.) thick aluminum substrate. (a) Good quality lines obtained because of uniform beam power. (b) Poor quality line obtained because of nonuniform beam power

Table 1 Recommended depth and spacing dimensions and commercially available tolerances for scribing selected ceramic workpiece thicknesses in LBM operations

Workpiece thickness		Scribe parameters					
		Depth				Spacing(a)	
		Dimension		Tolerance			
mm	in.	mm	in.	mm	in.	mm	in.
0.25	0.010	0.10	0.004	±0.038	±0.0015	0.10	0.004
0.38	0.015	0.15	0.006	±0.038	±0.0015	0.13	0.005
0.51	0.020	0.20	0.008	±0.05	±0.002	0.15	0.006
0.64	0.025	0.25	0.010	±0.05	±0.002	0.15	0.006
0.76	0.030	0.36	0.014	±0.064	±0.0025	0.15	0.006
0.89	0.035	0.41	0.016	±0.08	±0.003	0.15	0.006
1.02	0.040	0.43	0.017	±0.08	±0.003	0.15	0.006
1.27	0.050	0.51	0.020	±0.10	±0.004	0.18	0.007
1.52	0.060	0.64	0.025	±0.10	±0.004	0.20	0.008

(a) Tolerance: ±0.013 mm (±0.0005 in.)

were made one at a time utilizing the single beam of the laser. Sample parts, prototype quantities, and small-volume production runs became common. It then became evident that the cost to laser machine certain small configurations having multiple parts in a single array pattern (Fig 3) could be more than offset by subsequent savings in manufacturing costs. Decreased handling and improved yields resulted from the precision of the laser. By the mid-1980s, this was carried a step further with the design of multiple-beam lasers capable of machining several multiple-part arrays at one time. This resulted in an additional reduction in laser machining costs per unit part. Through-hole metallization, which facilitated the use of both sides of the ceramic substrate for circuitry applications, began to develop as the cost of precision hole drilling decreased.

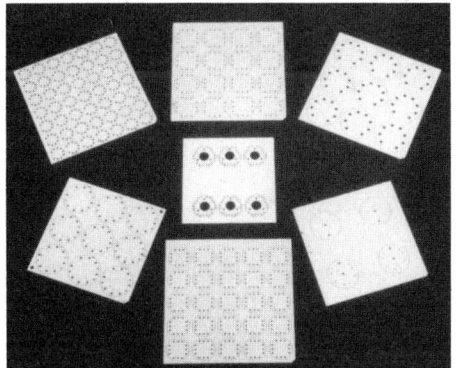

Fig 3 Typical multiple-part arrays of duplicate components on a single substrate for cost-effective processing of thick-film ceramics

(0.002 in./in.) should be considered for all large substrate sizes.

Drilling and Contour Machining. Initially, the laser drilling of holes and the machining of shapes in alumina was relatively slow and expensive. However, the cost was easily justified by the fact that with a laser a new part shape could be made in hours without the previous cost or time delay of very expensive hard tooling which had been required for every new part punched from green ceramic tape. In addition, it was soon learned that there were definite manufacturing advantages to parts laser machined to exact finished dimensions after the ceramic had been kiln fired.

Precision Machining. With early LBM equipment, precision laser machined parts

Laser Lens Protection and Debris Removal

Laser scribing, drilling, and machining all produce a violent eruption of both gaseous emissions and liquidus debris from the point at which the focused laser beam strikes the surface of the ceramic. The almost instantaneous vaporization of the ceramic causes rapid expansion and subsequent eruption of the gases from the immediate area that carries with it molten particles that must be controlled both in the air above the ceramic and on the surface of the ceramic being machined. The lens must also be protected. A conical aperture or a strong jet of air individually, or in tandem, may be used to prevent particulates from depositing on the lens. Such deposits cause

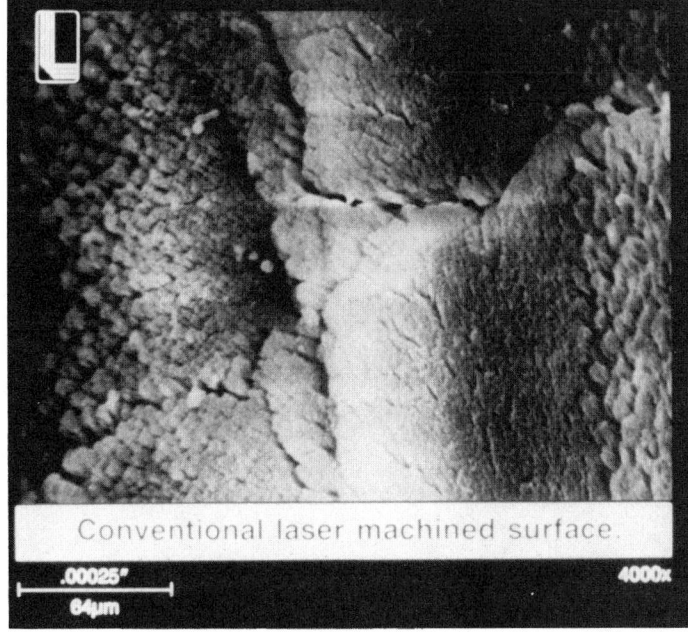

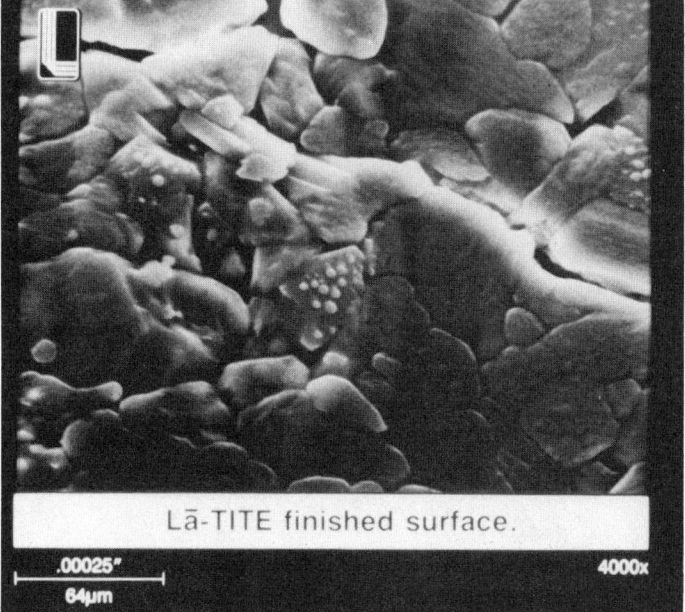

(a)

(b)

Fig 4 Surface finish of LBM ceramic edge. (a) Conventional as-machined material. (b) Post-treated machined material. 4000×

the lens to develop hot spots because of insufficient light transmission and lead to premature failure of the lens. A periodic maintenance schedule for cleaning the lens is essential to obtain peak efficiency of the laser beam.

Molten debris falling on the surface of the ceramic resolidifies and sticks to the surface. This can cause printing and adhesion problems with metallization during subsequent processing. Deposits on the surface of the substrate may be minimized by the application of a water-soluble coating prior to laser beam machining. These water-soluble coatings are commercially available. Rinsing in either tap water or distilled water after laser beam machining restores the original surface condition.

The use of air jets to control the vaporous plume and flying debris is also important. For laser scribing, only a stream of cross air is necessary. For laser drilling and machining, an axial air stream produced with a pressurized cone mounted under the lens with an aperture just large enough to allow the laser beam through should be used. The air flow,

parallel to the beam, forces most of the debris down under the part.

However, most applications require all three functions to operate during the same set-up. Carefully directed off-axis air jets are often a compromise solution. Whichever method is employed, the air jet nozzles must be regularly monitored for debris buildup in the aperture that can adversely affect the airflow.

Effect of Laser Beam Machining on Ceramic Adhesion Properties

In the mid-1980s, reports from the hybrid microelectronics industry indicated that certain developing metallization processes encountered adhesion problems on the laser-machined edges of the ceramic. While most thick-film metallization materials were fritted with glass particles that adhered regardless of the affected edge condition, the problem warranted investigation. Microscopic investigation results published in 1985 (Ref 1) and 1988 (Ref 2)

disclosed an interesting phenomenon. The laser beam caused a very fragile surface on the hole wall (Fig 4a). Post heat treating restored the surface (Fig 4b), correcting the adhesion problem.

Today the CO_2 laser is an essential tool in the machining of ceramics and in helping to sustain the steady growth of the hybrid electronic industry into the 21st century.

REFERENCES

1. M.L. Capp and R.R. Luther, "Analysis of the Effect of Laser Machining on 96% Alumina Ceramic Substrates and the Advantages of New Látite Finish," Proceedings of the 1985 International Symposium on Microelectronics

2. M.L. Capp and C. Missele, "An Empirical Adhesion Study of a Reactive Bonded Silver Thick Film Ink to: (1) As-Laser-Machined, (2) Thermally (Látite) Treated, (3) Chemically Treated Surfaces of Alumina Substrates," Proceedings of the 1988 International Symposium on Microelectronics

Electrical Discharge Machining

Nancy Faulk, Chemical Engineering Department, Tennessee Tech University

ELECTRICAL DISCHARGE MACHIN-ING (EDM) is derived from conventional diesinking, first developed as a practical metalworking method by Lazarenko and Lazarenko (Ref 1) after World War II. Much theoretical and experimental work has been conducted since that time in order to identify the basic processes involved. Recent research has started to extend this metal-finishing technique to processing ceramics and composites (Ref 2–3). These studies have shown that EDM can be successfully applied to ceramics, including single phases, cermets, and ceramic-matrix composites (CMCs) if the electrical resistivity is below 300 $\Omega \cdot$ cm (Ref 4).

Electrical discharge machining has been developed into an outstanding technique for precision machining. Diesinking and wire EDM machines allow intricate cutting and shaping of appropriate solid materials to nearly any three-dimensional size or shape. Tolerances below 1 μm (40 μin.) have been achieved (Ref 4, 5), and EDM can produce a mirror finish, even on ceramics, giving a surface roughness of less than 0.3 μm (12 μin.) (Ref 6, 7). This technique is especially important for machining hard, brittle, and high-melting-point materials, which cannot be machined by conventional techniques such as grinding.

According to Lee and Feick (Ref 8) and Weimer (Ref 9), EDM has become a principal method for shaping hard metals and composites containing carbides, nitrides, or abrasive oxides. They report that operators have cut holes in superalloys, refractory metals, and tungsten carbides in metallic matrices and can shape hardened steels and alloys that are difficult to work by other conventional methods. The rate of material removal is influenced mainly by the electrical and thermal properties of the workpiece (the metal or other material), but not at all by the hardness of the material. This is an advantage for ceramics, which are often limited by their hardness and/or brittleness. In EDM, the workpiece and the shaping tool are the electrodes separated by a liquid dielectric. Because there is no direct physical contact between the electrodes, no mechanical stress is placed on the workpiece.

Types of EDM Machines

Depending on the application, there are two types of machines used for EDM operation. The diesinking machine, also known as the ram-type, plunge, or vertical erosion machine (Fig 1), can be used for tapping, cutting holes, and helical machining, and is capable of cutting or sinking very complicated shapes into a workpiece. The workpiece is the cathode and the shaping tool (die) is the anode in this type of machine. It is desirable to remove material from the workpiece, but not from the tool, in order to use the die repeatedly for the same application. Die materials range from brass, steel, and other improved alloys, to specially made EDM graphite. These machines have begun employing automatic toolchangers and have extended the electrode life because current and voltage peaks are matched precisely to the distance between electrodes, the workpiece material, and the cutting conditions. Heavy hydrocarbons and kerosene are most commonly used in this conventional form of EDM.

Wire-cutting machines (Fig 2) evolved from research on diesinking machines (Ref 5). Instead of using a shaped electrode as the tool, wire-cutting machines use a thin metallic wire under tension to cut like a jigsaw. Some erosion of the wire is allowable because the wire is continuously being wound between spools as it travels through the material, but excessive wire erosion is the major cause of wire breakage. This breakage is the dominant practical and economic problem of wire-cut EDM. In this type of machining, water is the usual dielectric, and brass-, steel-, or molybdenum-based wire is used for the electrode. The polarity is switched for wire-cutting machines: The workpiece is the anode, and the wire is the cathode. Since the early 1980s, most of the research and development work in the EDM area has focused on wire-cutting EDM. Refinements include increased cutting speeds, extended taper angles, and independent programming of the upper and lower wire guides so that operators can cut shapes that have different geometries at the top than at the bottom. Also in very limited use as an attachment to the wire machine is a rotary spinning or indexing fixture. This process, also referred to as rotary EDM, allows for better flushing and for cutting circular parts that would otherwise be impossible to cut.

The most dramatic advance in EDM technology came with adoption of computer numeric control (CNC) in the late 1970s and early 1980s. Using CNC, operators can make changes in machine parameters, cutting speeds, or cutting directions without having to stop and reprogram. Most significant, from an industrial point of view, is the tremendous increase in the length of time a machine can run unattended. Some wire-cutting machines can now provide up to 120 h of untended cutting; 200 h are predicted sometime in the early 1990s (Ref 6). This autonomy allows a single operator to run a number of machines. The CNC capability has helped EDM to become a practical and repetitive science.

As already noted, the shaping tool places no mechanical stress on the workpiece. This can be especially important in the machining of materials that cannot be machined by conventional techniques, such as hard, brittle, and high-melting-point materials. Much work has recently centered on cutting nonmetallic materials and composites (for example, conducting ceramics) by wire EDM, and it has been found that many erode at rates comparable to and sometimes better than rates for many steels (Ref 10).

The Erosion Process

Electrical discharge machining is based on electric discharge erosion. Figure 3 illustrates how this discharge process works, with discussion following the work of Lee and Feick (Ref 8), Foster (Ref 5, 11), and Eubank and coworkers (Ref 12, 13). The three necessary components, that is, the two electrodes and the dielectric fluid flowing in the gap between them, can be seen in Fig 3. Note that the dielectric field contains ultrafine solid impurities. An electric field generated by applying approximately 200 volts dc across the gap, with an average width of 40 μm (1.6 mils), generates a magnetic field between the workpiece and the electrode. At first, the two pieces are insulated by the dielectric, so no current flows across the gap. The resulting electric field, however, causes the micron-size par-

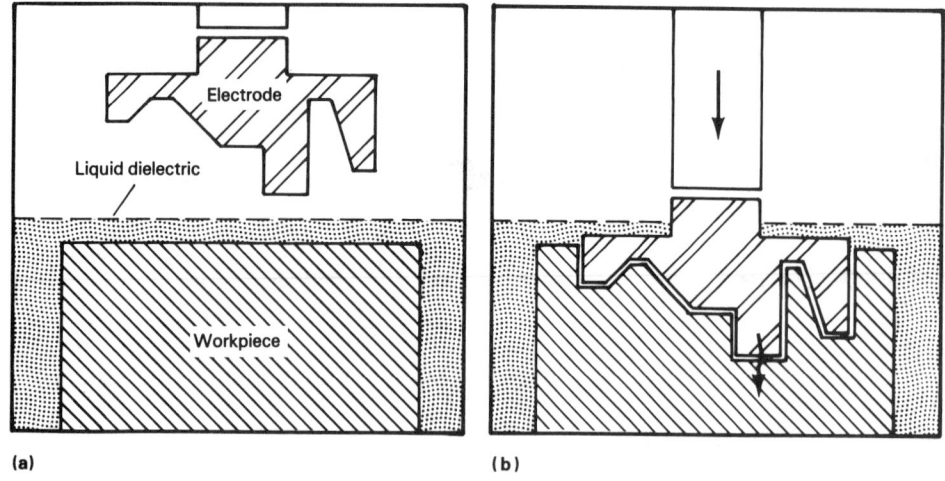

(a) **(b)**

Fig 1 Diesinking machine, schematic. (a) Before EDM. (b) After EDM

ticles to be suspended and to form a bridge. This causes the breakdown, or deionization, of the dielectric. The voltage (*V*) falls to about 25 V, and the current (*I*) rises to a constant value set by the operator. The plasma channel, seen forming in panel 3 of Fig 3, grows during the pulse time (*t*), or "on-time." The intense heat produced causes a vapor bubble to form around this channel. The surrounding liquid dielectric restricts plasma growth, thus concentrating the input energy, *VIt*, into a very small volume. Energy densities up to 3 J/mm^3 (50 Btu/in.^{3}) result, which cause the plasma temperature to reach nearly 40,000 K (72,000 °F), and the plasma pressure to rise as high as 300 MPa (2700 atm) due to inertial and viscosity effects (Ref 12).

During this on-time, the high-energy plasma affects both electrodes in different ways, depending on the properties of the materials involved. For most applications, the plasma melts both electrodes (Ref 11), as explained below. As the current flow halts, the bubble implodes, drawing the molten material into the gap. The dielectric fluid flushes and solidifies this molten material and then carries it away, while also cooling the electrode surfaces. The cycle is then repeated.

The pulse duration is one operating condition that greatly influences the material removal rate (MRR). The optimum pulse duration depends on which type of EDM machine is being used. For diesinking machines, optimum times of 10 to 100 μs are common, whereas for wire-cutting machines, times are less than 2 μs (Ref 3). The difference is explained by the following discussion. Because of bombardment of fast-moving electrons at the start of the pulse, the anode melts rapidly first, but then begins to resolidify after a few microseconds. Recall that for the diesinking machine the anode is the

shaping tool. This resolidification is believed to be due to the rapid expansion of the plasma radius at the anode, which causes a decrease in the local heat flux at the anode surface. This expansion is referred to in the literature as channel widening (Ref 14) or expanding circle heat source (Ref 13). Melting of the cathode occurs one to two orders of magnitude later because of the lower mobility of the positive ions. Thus, for the diesinking machines, time must be allowed for the resolidification of the die, but for the wire-cutting machine, much lower times may be used to allow high erosion rates of the anodic workpiece.

Effect of Operating Conditions on MRR and Surface Quality

Material removal, as previously discussed, depends on the length of time of the pulse, that is, it depends on starting and stopping the electrical spark. The amount of material removed is affected by the energy that is released during the spark, which in turn is determined by the voltage between the electrodes (gap voltage), the current, and the on-time of the spark. Older EDM equipment used a bank of capacitors in the power supply to store energy for the spark. This design is known as a capacitant discharge power supply. These capacitors store electrical energy until the equipment senses the correct gap voltage, and then they form the spark and release their energy. After the capacitors dump their charge, the spark is extinguished and the capacitors begin to recharge for the next spark.

The main problem with this system is that the spark is initiated solely on the basis of what gap voltage will deionize the dielectric. The capacitors may be fully charged for one spark, releasing the maximum amount of energy and removing a large amount of material. However, for another spark, the gap voltage may reach the proper value before the capacitors reach full charge, and the ensuing spark will remove less material. Thus, material removal across the surface of the workpiece may be quite uneven.

Another power supply developed more recently is the pulse-type power supply, and it addresses the problem described above (Ref 15). In an effort to control the spark more effectively, this power supply releases each spark with the same amount of energy. No new spark is released until the energy from the previous spark has been used. The sparks may now occur more randomly, but because the same amount of energy is released and nearly the same amount of material removed, the surface is of a much higher quality. Each eroded cavity has roughly the same diameter and depth. The amount of material recast back into the cavity is also greatly minimized. If this recast layer must be removed in subsequent finishing steps, the amount of material to be removed is also minimized.

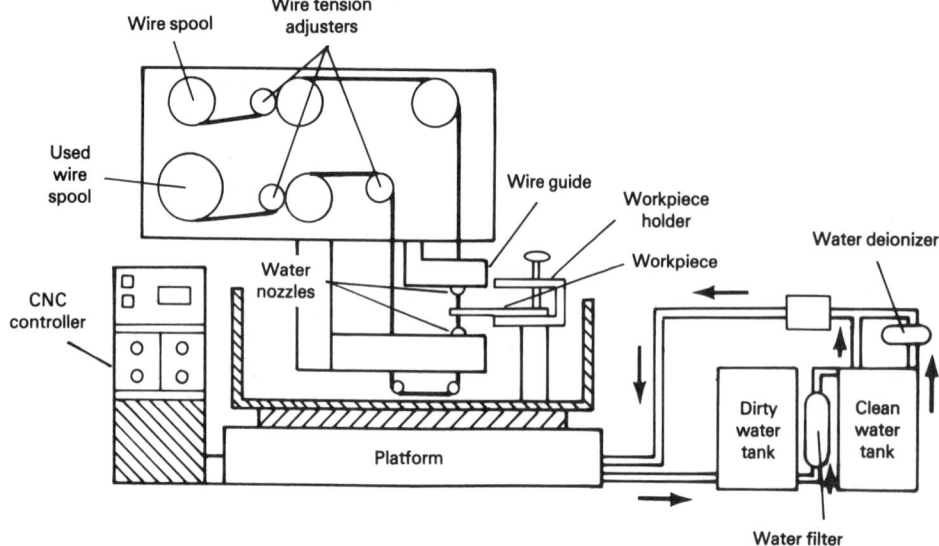

Fig 2 Wire-cutting machine, schematic

Other properties of the dielectric also play an important role. The dielectric should have a sufficiently low viscosity for effective circulation, because the liquid must carry away the debris. In addition, the fluid acts as a bulk heat-transfer medium, cooling the two electrodes in the off-time between pulses. It is necessary for the fluid to be nonflammable and chemically inert to the electrode materials. It must not emit toxic vapors, and, to be practical, it must be inexpensive and readily available.

Kerosene was the most commonly used dielectric for diesinking machines in the past. Experimental results have shown that small additions of fine graphite powder (4 μm, or 160 μin., size in concentration of 4 g/L, or 0.5 oz/gal) can increase the MRR and the tool wear rate (Ref 16). This effect may be attributed to decrease in the voltage necessary for breakdown of the dielectric. However, the high risks associated with kerosene and the increased scarcity of petroleum products have caused many operators to choose the heavier hydrocarbon oils. An ideal dielectric oil should have a low viscosity (2×10^{-6} m^2/s, or 22 $\times 10^{-6}$ ft^2/s, at 40 °C, or 103 °F), a flashpoint of 80 °C (175 °F) or higher, and an aromatic content below 1% (Ref 17). Many of the heavier oils with flashpoints around 130 °C (265 °F) also have high viscosities, which of course reduces the flushing action. This has led some researchers to consider other mediums and perform comparisons. Water is a natural choice, because it is plentiful, cheap, and has a very high surface tension, but workers at first hesitated to use it because it has a limited resistivity and electrochemical action.

The use of a conducting liquid causes more material to be eroded, which can prevent sharp, precise edges from being formed and can produce a lower-quality surface finish. Ordinary (tap) water can always be used for rough machining, provided that the electrochemical reactions do not damage the workpiece. In addition, water can be used for wire-cutting machines, and indeed it is the dielectric most often used in wire-cutting applications, because the pulse times are so much smaller than for die-sinking machines due to the reversed polarity discussed previously. Thus, use of a conducting liquid such as water is desirable.

Filtration System. It is necessary to filter the dielectric fluid to remove the eroded particles that it flushes from between the two electrodes. Three main types of filtration systems are being used to clean the fluid: the disposable cartridge, or paper, filter; the precoated filter; and the edge filter. The type employed depends on the application. Paper filters are inexpensive, easy to use, and have a low capital cost. However, if they are not changed frequently enough, they can cause dirty working conditions, which in turn can prevent a fine finish from being achieved. The precoated filter provides better filtration, to

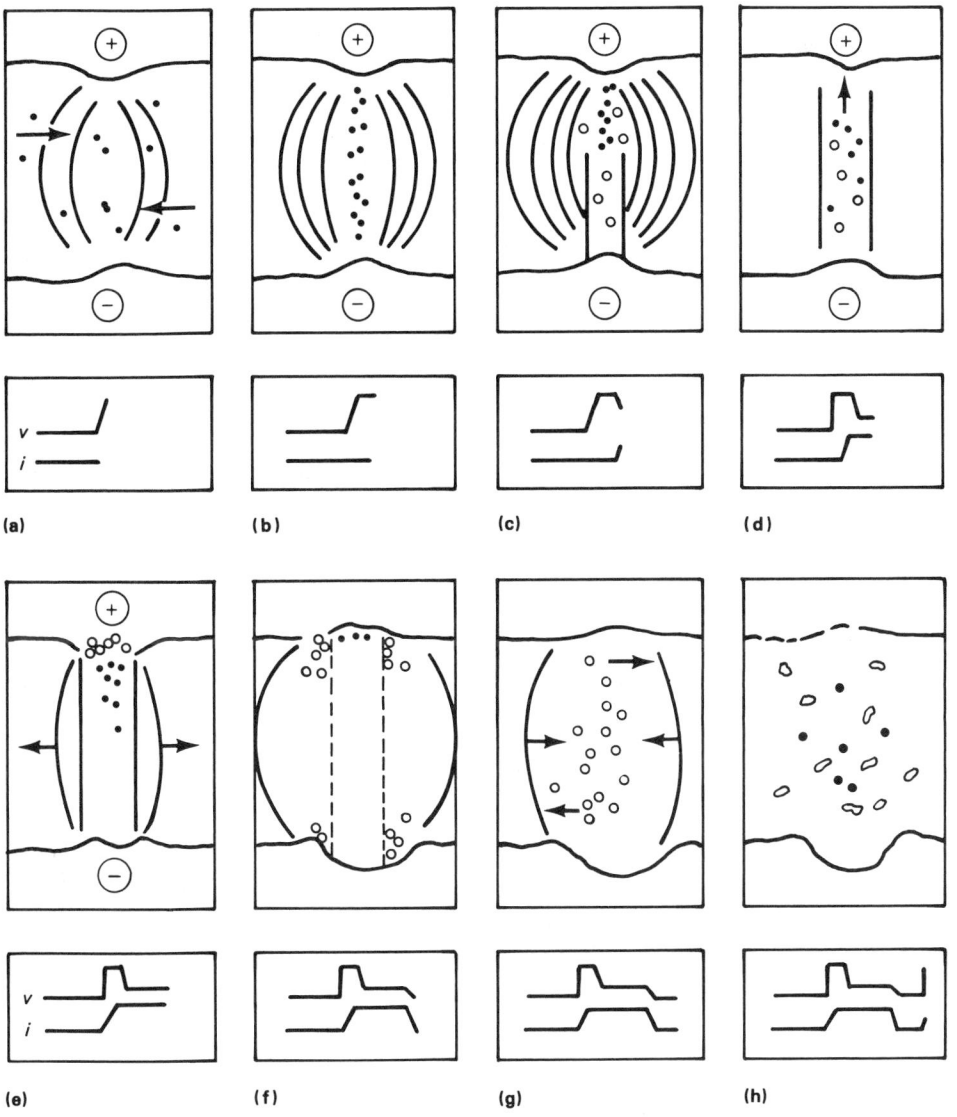

Fig 3 Sequential EDM process steps showing how the dielectric fluid breaks down in the gap between the electrodes. (a) Buildup of an electric field. (b) Formation of a bridge by conductive particles. (c) Beginning of discharge due to emission of negative particles. (d) Flow of current by means of negative and positive particles. (e) Development of discharge channel due to a rise in temperature and pressure; formation of vapor bubble. (f) Reduced heat input after drop in current; explosion-like removal of material. (g) Collapse of vapor bubble. (h) Residues: material particles, carbon, and gas

Diesinking machines have always produced higher surface qualities than have the wire-cutting machines. Surface quality is a very important consideration when ram-type technology is utilized for making injection molds, pressure casting dies, and diamond draw dies, the surface quality of which must range from a smooth, matte finish to a shiny, polished finish. Part of the reason for the higher quality obtained from a ram-type EDM is the size of the electrode involved in each process. The electrode here has a relatively large cross-sectional area, and the spark characteristics can be controlled much more easily than can happen when the electrode is a thin (for example, 150 μm, or 6 mil) wire, as is used for the wire-cutting application.

The dielectric fluid that is used also has an effect on both the MRR and the surface quality obtained. Initially (before the discharge occurs), the fluid must act as an insulator. Then, as discussed previously, the electric field generated by the applied voltage begins to break down the dielectric. A plasma channel is then formed between the two electrodes and is surrounded by a vapor bubble. The dielectric fluid must surround and restrict this plasma growth, which implies one important characteristic for a successful dielectric application: surface tension. High surface tension provides a low plasma-spreading velocity, which is the rate at which the radius of the expanding plasma channel moves. A low plasma-spreading velocity concentrates the heat source to melt the workpiece. The dielectric fluid must deionize quickly after the spark charge is complete in order to act as an insulator again.

the 1 μm (40 μin.) level, but they are labor intensive, the process can be expensive, and carcinogenic powders are frequently used. Edge filters provide filtration to the submicron level, but are also labor intensive and have a high capital cost.

Various machine parameters that the operator can control also have an effect on both the MRR and the surface quality. In many cases, an increase in one parameter occurs at the expense of the other. A good example of this for the wire-cutting machine is the wire feed rate. With a higher feed rate, the regulating systems react more quickly to disturbances, thus giving a higher average MRR. However, the cutting process itself is rougher, resulting in larger contour errors.

The pulse time has already been discussed to some extent. A longer pulse has the advantage of reducing the risk of wire breakage. The work of Tsai (Ref 18) on tungsten carbide-cobalt (WC-Co) composites indicates that for wire-cutting machines (low pulse times) the cutting rate increases with increasing pulse duration before levelling off. Longer durations also gave better surfaces with low roughness, less segregation of the phases, and no surface cracks. When the pulse time is increased, the pulse frequency should be reduced in roughly the same proportion. A higher discharge frequency produces a higher discharge power, which may break the wire. The cited work showed that high frequencies increased the energy consumed without a net increase in cutting rate. The amounts of microcracks and surface deterioration increased with increasing frequency.

Increasing the wire thickness may prevent breakage, as well as creating a thicker workpiece, distributing the number of discharges over a greater cutting area. Increasing current through the wire is the third main factor in wire breakage. The working current controls the frequency of the pulse discharges. The higher the frequency, the higher the cutting rate will be. It is therefore desirable to have the current set as high as possible without thermally overloading the wire and causing breakage.

Mechanisms Involved in EDM

Two mechanisms for ceramic erosion have been reported: melting and evaporation, and thermal spalling. Each mechanism is discussed below, along with examples of the materials that each erodes.

Melting and evaporation of the electrodes is the most common method for erosion and the one traditionally explained and modelled (Ref 11–14, 19, 20). For metals and most ceramics, the plasma melts both electrodes. As the current flow ceases, a violent collapse of the plasma channel and vapor bubble causes superheated, molten liquid on the surfaces of both electrodes to explode into the gap, where the liquid dielectric resolidi-

fies the molten material. Analysis of the debris was carried out on a WC-Co system by Gadalla and Tsai (Ref 22–23), and they concluded that, in some cases, the plasma temperature causes evaporation of material from the electrode. Their studies clearly show that material is removed in both the liquid and gaseous state; perfectly spherical particles are apparently resolidified particles from the gaseous state, whereas irregularly shaped particles are solidified from the liquid state (Ref 24).

Not all the molten material can be removed after the spark due to surface tension, tensile strength, and bonding forces between liquid and solid. Estimates of the amount resolidified, also referred to as the recast layer, vary. Erden (Ref 20) suggests 10 to 20% is reasonable, but Pandit and Rajurkar (Ref 19) use 60% in their calculations. The pulse-type power supply, developed to replace the capacitance power supply, helps to minimize the depth of the recast layer because most of the material is removed by vaporization, not melting. It is the molten material that is able to be recast into the crater. Different machines may therefore have different percentages of resolidified material, which may explain the variation in the work cited above. Crosby reports that for carbides cut with a combination of diesinking and wire-cutting EDM, the recast layer has a depth of about 5 μm (0.2 mil) (Ref 25).

Pure ceramics eroded by melting include B_4C, SiC, and graphite; CMCs such as Al_2O_3-TiC (70/30%, respectively), Sialon-TiN (80/20%), Si_3N_4-TiN (75/25%) and SiC-TiB_2 (80/20%), and cermets, including WC-Co (94/6%) and Si-SiC (5/95%) (Ref 10, 22). Many of the CMCs listed have a conductive phase added to decrease the resistivity (Ref 2). For Si_3N_4-TiN, the resistivity dropped from $>10^{13}$ to 1.5×10^{-3} $\Omega \cdot$ cm to allow for EDM processing (Ref 26). Its polished surface can be seen in Fig 4(a), where the light phase is the TiN. The EDM surface with resolidified melt-formation droplets is shown in Fig 4(b). Surface roughness measurements show the roughness to be <4 μm (<16 mils) (Ref 2). This reflects, in part, the small grain size of the material. This is further discussed in the section "Application of EDM to Advanced Ceramics" in this article.

Thermal spalling is usually defined as the mechanical failure of a material, without melting, due to the internal stresses created by heating and/or cooling, which overcome the bond strength. This occurs as a material expands and contracts during the sudden temperature changes associated with EDM, resulting in tension and compression sufficient to cause tensile or compressive failure, respectively.

To calculate the spalling temperature, equations that show the thermal stress, σ, created due to a temperature gradient of ΔT can be used (Ref 27). During heating, the specimen is exposed to high temperatures in

(a)

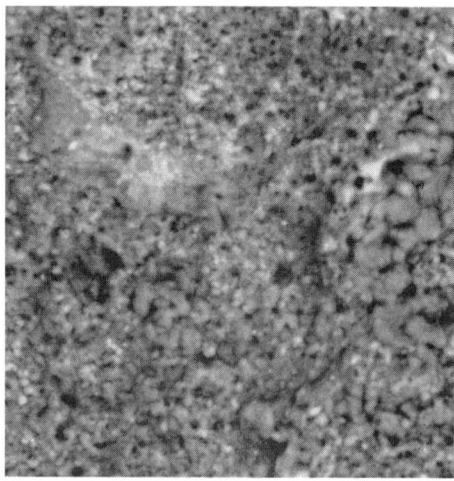

(b)

Fig 4 (a) Polished, uncut surface of Si_3N_4-TiN. (b) Electrical discharge machined surface of Si_3N_4-TiN, showing resolidified melt-formation droplets

the plasma; fluxes on the order of 10^{10} to 10^{11} W/m² (3.2×10^9 to 3.2×10^{10} Btu/h/ft²) are reported for EDM (Ref 28). Due to heat transfer from the plasma to the workpiece, a compressive force perpendicular to the surface of the workpiece is created. Moreover, because the surface attempts to expand, but cannot, due to the constraint from the bulk of the specimen, compressive stress is created parallel to the surface. When the stress exceeds the ultimate compressive strength, σ_c, of the material, failure by thermal stress is assumed, and can be calculated by the following equation, assuming a brittle fracture at the end of the elastic limit:

$$\sigma_c = \frac{E\alpha\Delta T}{1 - \nu}$$

where E is the elastic modulus, α is the coefficient of thermal expansion, ν is Poisson's ratio, and ΔT is the temperature difference causing fracture parallel to the surface if the planes of maximum atomic density (slip planes) are parallel to the machined surface. During cooldown (flushing), the reverse occurs; quench rates of 10^5 to 10^9 K/s are observed (Ref 28). The compressive forces created by the plasma disappear and tensile forces are created in the specimen as the surface attempts to contract. Failure occurs if the stress exceeds the ultimate tensile strength, σ_t, of the material, using the above equation. The $(1 - \nu)$ term included is associated with the biaxial nature of the stresses; it arises from Hooke's Law. It is thus possible to calculate the minimum temperature rise (or decrease) necessary under these rapid conditions for spalling to occur.

Materials with high thermal expansions and low thermal conductivity are subjected to higher stresses due to more severe gradients. If a material has a low ultimate strength or Young's modulus, as well, the tendency to spall is enhanced. When the sparks are localized in the same spot, the material may also break by fatigue after a limited number of cycles. An additional property that plays a role in whether or not a material spalls is the melting point. If the melting point is low enough, the material simply melts or evaporates, as discussed above, and no stresses build up to cause fracture. Petrofes (Ref 10) reports that materials with melting points below 2800 °C (5100 °F) are cut by the melting mechanism.

An example of a ceramic that erodes by spalling is pure TiB$_2$. Figure 5 shows the polished, uncut surface of this material; note the absence of melt droplets or a recast layer as was seen in Fig 4. Instead, fracture of crystals and flaking are observed in this high-melting-point ceramic. The liquid droplets seen on the surface are not from the surface melting but rather from the brass wire (Ref 2). Energy-dispersive x-ray spectrometry (EDS) proves that these droplets contain copper and zinc only. Other than these droplets, there is no recast layer on the machined surface. The small holes present in the grains themselves are due to dislocations initially present in the crystals, which the thermal process of EDM revealed by thermal etching.

Application of EDM to Advanced Ceramics

Studies have been performed to answer several questions that are important from an industrial point of view, such as the degree of contamination of the surface; the effects of EDM on the hardness, surface stresses, and strength; the MRR attainable; and the effect of the grain size of the material. These are considered separately below.

Contamination of the Surface. Gadalla and Petrofes studied the surfaces created by applying EDM to several samples of advanced ceramics (Ref 29). Elemental mapping was performed on many of the samples in order to determine the extent of contamination on the surface by the material or impurities in the dielectric, which in the cited case was water. The brass wire that was used (63/37% copper/zinc) caused some contamination. Copper alloy resolidified droplets with a lower concentration of zinc and a fine dispersion of zinc exist over the entire surface. The zinc dispersion is expected due to its higher vapor pressure. Viewed by EDS, the surface sub-jected to EDM showed traces of sodium, potassium, and chlorine from the tap water added to the deionized water (the dielectric fluid used) to adjust its conductivity.

Changes in Hardness. The Vicker's microhardness of surfaces prepared by EDM was compared to that of diamond sawn and polished surfaces (Ref 10). It was concluded that for those materials that were machined by the melt mechanism, EDM produced surfaces with lower hardnesses than those obtained with a diamond saw. This conclusion implies that the internal stresses created by EDM on quenching the liquid phase are less than those created by mechanical erosion. On the other hand, the hardness obtained for spalled materials by EDM was greater than that obtained by a diamond saw. This indicates that the hardness obtained by the spalling mechanism is comparable to mechanical erosion. In some cases, regardless of the erosion mechanism, polished surfaces gave the highest hardness due to the high mechanical stresses created, capture of debris by a softer matrix material, and/or possible redistribution of phases.

Surface residual stresses were determined by Gadalla and Tsai for surfaces of WC-Co composites subjected to EDM and were found to be anisotropic (Ref 22). In another study (Ref 10), it was found that flaking due to spalling causes residual compressive stresses, but the contracting due to solidification of melted surfaces causes residual tensile stresses.

Strength. The decrease in strength of a ceramic machined by EDM is its biggest drawback. Saito (Ref 30) reports that even

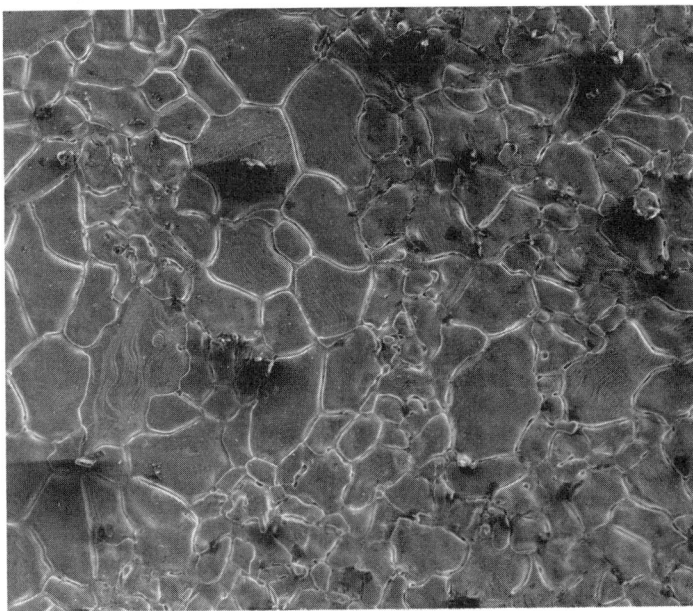

(a)

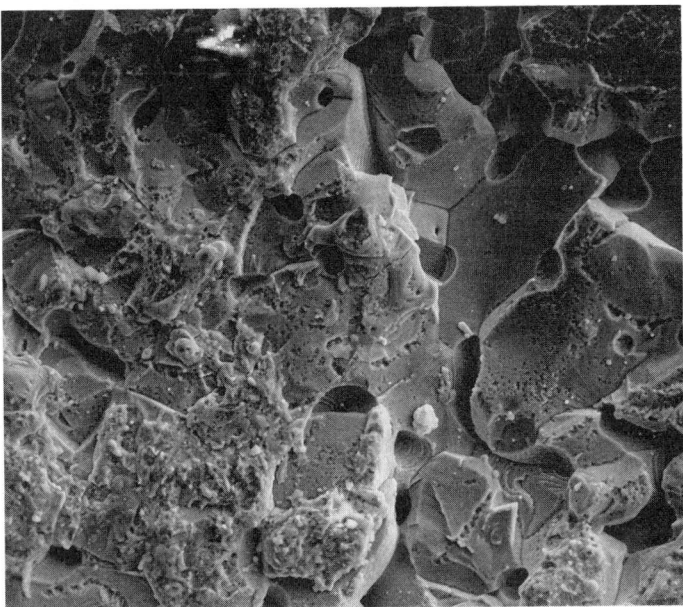

(b)

Fig 5 (a) Polished, uncut surface of TiB$_2$. (b) Electrical discharge machined surface of TiB$_2$, showing fracture of the grains. The droplets seen are resolidified brass from the wire.

when no cracks are observed on the machined surface, the value of the bending strength is between 40 to 50% of its initial value for several ceramics and CMCs. Nakamura does report, however, that when no cracks are observed on the surface, the bending strength can be recovered by removing entirely the roughness with a diamond fine grinding wheel of No. 400 or more (Ref 31).

Material Removal Rates Achieved. For a wide variety of advanced ceramics, including pure ceramics, cermets, and CMCs, Petrofes reports cutting rates of 10 to 50 mm²/min (0.02 to 0.08 in.²/min), but does not claim that these numbers are optimum (Ref 10). Saito reports rates of 100 to 200 mm²/min (0.2 to 0.4 in.²/min) for conductive Sialon (contains TiN) and Si_3N_4 using optimum cutting conditions (Ref 29).

Grain Size Effects. Gadalla and Petrofes report that larger grain sizes of the nonconductive phase (for composites) cause higher roughness and a higher removal rate because the loss of a single grain leaves a large crater and removes more material, compared to a small grain (Ref 29). They also report the case of an Al_2O_3—TiC composite of grain sizes 20 and 30 μm (0.8 and 1.2 mils), respectively, in which the x-ray diffraction pattern of the machined surface showed much broader peaks than did its polished surface. This indicated that the melted material produced by EDM precipitated finer crystals with high residual stresses on quenching.

Expanding Use of EDM for Ceramics

Advanced materials that are difficult or impossible to machine by conventional abrasive methods can be machined into complex shapes using EDM, if they have low enough electrical resistivity (<300 $\Omega\cdot$cm). Because the tool and the workpiece are separated by a dielectric field, extremely hard and brittle materials can be cut. The EDM process takes place by melting, evaporating, or fracturing the surface, depending on the melting point of the material.

Several materials have been successfully cut, and as awareness of the requirements for EDM grows, more ceramics manufacturers are adding conductive phases to their products to make them machinable. Use of this technique is expected to grow as operators begin to see ways to incorporate EDM into production.

REFERENCES

1. B.R. Lazarenko and N.I. Lazarenko, Machining by Erosion, *Am. Mach.*, Vol 91 (No. 26), 1947, p 120–121
2. N.F. Petrofes and A.M. Gadalla, Electrical Discharge Machining of Advanced Ceramics, *Am. Ceram. Soc. Bull.*, Vol 67 (No. 6), p 1048–1052
3. N.F. Petrofes and A.M. Gadalla, Processing Aspects of Shaping Advanced Materials by Electrical Discharge Machining, *Adv. Mater. Manuf. Proc.*, Vol 3 (No. 1), 1988, p 127–153
4. R.F. Firestone, Abrasionless Machining Methods for Ceramics; The Science of Machining and Surface Finishing II, *Natl. Bur. Stand. Spec. Publ.*, No. 562, B.J. Hockey and R.W. Rice, Ed., 1979, p 261–281
5. R.F. Foster, Evolution of Wire EDM, *Tool. Prod.*, Vol 47 (No. 12), 1982, p 90–93
6. J.C. Quinlan, What's New in EDM, *Tool. Prod.*, Vol 52 (No. 12), 1987, p 40–49
7. L. Houman, EDM Surfaces: The Persistent Problem of Measurement, *Carbide Tool J.*, Vol 14 (No. 4), 1982, p 24–26
8. D.W. Lee and G. Feick, The Techniques and Mechanisms of Chemical, Electrochemical, and Electrical Discharge Machining of Ceramic Materials; The Science of Machining and Surface Finishing, *Natl. Bur. Stand. Spec. Publ.*, No. 348, 1972, p 197–211
9. G.A. Weimar, EDM/ECM: The Exotic Becomes the Routine, *Iron Age*, 1983, p 75–81
10. N.F. Petrofes, "Shaping Advanced Ceramics with Electrical Discharge Machining," Ph.D. dissertation, Texas A&M University, 1989
11. R.F. Foster, Advanced Wire Cutting EDM, *Carbide Tool J.*, Vol 14 (No. 4), 1982, p 22–23
12. D. DiBitonto, M.R. Patel, and P.T. Eubank, Theoretical Models of the Electrical Discharge Machining Process. Part I: A Simple Cathode Erosion Model, *J. Appl. Phys.*, Vol 66 (No. 9), 1989, p 4095–4103
13. M.R. Patel, M.A. Barrufet, P.T. Eubank, and D.D. DiBitonto, Theoretical Models of the Electrical Discharge Machining Process. Part II: The Anode Erosion Model, *J. Appl. Phys.*, Vol 66 (No. 9), 1989, p 4104–4111
14. C. van Osenbruggen, High-Precision Spark Machining, *Philips Tech. Rev.*, Vol 30 (No. 6, 7), 1969, p 195–208
15. R.L. Bormann, Cut Surface Quality—The Pluses and Minuses of EDM, *Tool. Prod.*, Vol 50 (No. 10), 1985, p 37–50
16. M.L. Jeswani, Effect of the Addition of Graphite Powder to Kerosene Used as the Dielectric Fluid in Electrical Machining, *Wear*, Vol 70, 1981, p 133–139
17. R. Mann, How to Improve EDM Filtration Systems, *Mach. Prod. Eng.*, Vol 16 (No. 23), 1981, p 42–43
18. W. Tsai, "The Effect of WC Grain Size and Cobalt Content on Properties and Electrical Discharge Machining of WC-Co Composites," Master's thesis, Texas A&M University, 1988
19. S.M. Pandit and K.P. Rajurkar, "A New Approach to Thermal Modelling Applied to Electro Discharge Machining," No. 81-WA/HT-3, American Society of Mechanical Engineers, 1981
20. A. Erden, Effect of Materials on the Mechanism of Electric Discharge Machining (E.D.M.), *Trans. ASME*, Vol 105 (No. 2), 1983, p 132–138
21. S.T. Jilani and P.C. Pandey, An Analysis of Surface Erosion in Electrical Discharge Machining, *Wear*, Vol 84 (No. 3), 1983, p 275–284
22. A.M. Gadalla and W.-L. Tsai, Machining of WC-Co Composites, *Adv. Mater. Manuf. Proc.*, Vol 4 (No. 3), 1989, p 411–423
23. A.M. Gadalla and W.-L. Tsai, Electrical Discharge Machining of Tungsten Carbide-Cobalt Composites, *J. Am. Ceram. Soc.*, Vol 72 (No. 8), 1989, p 1396–1401
24. A. Erden and B. Kaftanoglu, Investigations on Breakdown Phase of Sparks in Electric Discharge Machining, *Middle East Tech. Univ. J. Pure Appl. Sci.*, Vol 13 (No. 1), 1980, p 45–74
25. H. Crosby, Evolution in EDM, *Machine and Tool Blue Book*, Vol 80 (No. 2), 1985, p 48–52
26. E. Kamijo, M. Honda, M. Higuchi, H. Takeuchi, and T. Tanimura, Electrical Discharge Machinable Si_3N_4 Ceramics, *Sumitomo Electr. Tech. Rev.*, Vol 24, 1985, p 183–190
27. R.W. Davidge, *Mechanical Behavior of Ceramics*, Cambridge University Press, 1979, p 118–122
28. A.E. Berkowitz and J.L. Walter, Spark Erosion, *J. Mater. Res.*, Vol 2 (No. 2), 1987, p 277–288
29. N. Saito, A Review of Special Machining of Fine Ceramics, *Sumitomo Electr. Tech. Rev.*, Vol 25, 1986, p 1–19
30. M. Nakamura, K. Kubo, and S. Hirail, *Electr. Mach. Tech.*, Vol 8 (No. 22), 1984, p 1–12
31. A.M. Gadalla and N. Petrofes, Surfaces of Advanced Ceramic Composites Formed by Electrical Discharge Machining, *Mater. Manuf. Proc.*, Vol 5 (No. 2), 1990, p 253–271

SECTION 6

Glass Processing

Co-Chairmen: W.R. Prindle, P.S. Danielson, and J.W. Malmendier, Corning Incorporated

Introduction

THE ARTICLES in this Section are in three general groupings. First, there are three articles that describe the basic ways in which a very large majority of the commercially important glasses are processed. These begin with considerations relating to raw materials selection, their treatment and combination into "glass batches" and the feeding of these mixtures into glass melting furnaces (see the article "Raw Materials/Batching").

The next article in the first group provides an overview of the various types of melting units employed, the form and function of their respective components, the general attributes of their molten products and the special utility of each of the described glass melting/fining systems (see the article "Melting/Fining"). Machine forming processes, and the ancient hand forming processes that they emulate, are described in the third article along with more recent innovations for producing like products, such as the float process (see the article "Forming"). The high volume of inexpensive, more common glass products produced by these processes are usually made from soda-lime silicate glass compositions.

In these three articles the authors have attempted to convey the relative ranges of possibilities, with respect to product quality and cost, for each of the processes described. Their purpose was to provide the reader with a rudimentary appreciation of the concerns of glass makers and of the cost/benefit impact of alternative approaches to the manufacture of glass products by these processes.

Special processes to form glass and/or glass-ceramic products are described in the next eight articles. The included processes are "special" in the sense that they were developed for unique products and are not a mechanized version of hand forming operations(s). These processes are therefore more limited in their applicability than are the forming processes described in the article "Forming" in this Section. The article on "Laminated Glass" is a good example. While hand lamination is still done, primarily for artware, today's commercially important laminated glass products are produced by a process that only remotely corresponds to the ancient hand forming method.

Except for the products covered under "Optical Fibers" and those made by, and described in the article "Sol-Gel Processes," most all of the commercially important products resulting from the other "special processes" are now formed primarily from molten glass. The products within this latter group that are formed from other than molten glass are few in number; hollow glass spheres being one example.

High-silica glass optical fibers have been formed from molten glass by the double crucible method; however, this process is presently of negligible commercial importance. The chemical vapor deposition processes, by which most optical fibers are formed, and the sol-gel processes were developed primarily as low-energy methods to produce high-silica glass products. Such glasses are characterized by high melting temperatures and viscosities that make it extremely difficult to form them with conventional techniques. Products made via chemical vapor deposition and sol-gel processes can be made with much lower contaminant levels and/or better controlled dopant levels.

Glass and glass-ceramic products discussed in the articles "Porous and Reconstructed Glasses," "Phase-Separated Glasses and Glass-Ceramics," and "Photosensitive Glasses and Glass-Ceramics" are for the most part initially formed by the processes reviewed in the first group of articles in this Section. They were categorized in the "Special Processes" subsection due to the additional thermal and/or radiative treatments that are needed to impart their special properties. Also, glass composition ranges for the achievement of these unique characteristics are limited and very process-related. In all these cases, as a result of the special post-forming process(es), the original glass undergoes a substantial phase separation and/or structural transformation.

Secondary Processing. The final group of four articles discusses secondary processes that are performed on glass or glass-ceramic products that have previously been shaped into their basic useful form. In most instances, the processes described in the articles "Annealed and Tempered Glass," "Ion-Exchange," and "Grind/Polish" would most often be performed by the manufacturer of the previously formed glass item. However, tempering and ion-exchange strengthening of ophthalmic lenses are high-volume processes normally done by lens manufacturers. Similarly, most grinding and polishing of ophthalmic and optical glass products are done by the manufacturers of the final products. For processes covered in the article "Decorating" in this Section as well as sealing and bonding processes described elsewhere in this Volume, the glass manufacturer's product is almost always used directly by a customer, or the customer of a distributor of the product.

In all of the articles in this Section, the authors' major objectives are to convey to the reader that both the glass and glass-ceramic materials families have a multitude of members, that these offer many unique combinations of properties, and that they can be obtained in virtually any shape. Many of the possible products have not yet become commercially available only because they have not been asked for. Additional information on glass products can be found in the section on "Applications for Glasses" in this Handbook.

Raw Materials/Batching

William C. Bauer, Ponderosa Associates
Janet E. Bailey, TechniGlas Consulting

FOUR MAJOR CATEGORIES of manufactured glass are: flat glass, containers/tableware, fiberglass, and specialty glass. The first two categories utilize soda-lime-silica glass, and constitute the highest volume of products. Fiberglass is often based on borosilicate glass and is discussed in detail in the article "Fiberglass" in this Section of the Volume. Specialty glasses sometimes, but not always, use materials that are common to the other glasses produced, although nearly every specialty glass product has a unique composition. The materials discussed in this article are those used in most commercially available glasses. The four glass product categories and their various applications are summarized in Table 1.

Important Minerals

Silica sand is the major "glass former" used in commercial glasses. Silica (SiO_2), one of the most abundant natural minerals, occurs in several forms. The form of most significance to the glass industry is quartz. The important silica sand deposits in the United States include the Oriskany quartzite found in Pennsylvania, West Virginia, and Virginia, and the St. Peter sandstone found in deposits stretching from Illinois through Missouri to Arkansas. The Oil Creek sandstone found in Oklahoma is a high-purity sandstone with a finer particle size distribution than is found in any of the other deposits.

Sandstone deposits of lower quality are found in New Jersey, North Carolina, Florida, and Tennessee. Some of these have been successfully beneficiated and upgraded to glass-quality silica sands. The West Coast has only two deposits that have been upgraded to this level. The remainder of the west coast sand is a mixture of quartz and feldspar (feldspathic). Although this type has potential for use in glass, some of the deposits vary too much in feldspar content.

Sandstone deposits are mined using conventional techniques in open-pit or near-surface mines. Generally, sandstones are loosely consolidated and require minimal crushing to separate the individual quartz grains, which are quite rounded and abraded in appearance. St. Peter and Oil Creek sandstones require only

washing to remove clay impurities. Sand grains are quite angular. After the sand has been dried, sizing is generally accomplished by passing the production stream over screens that either separate out "base grades" for reblending to customer specifications or simply remove an unwanted coarse fraction and leave the remaining distribution unchanged. Oriskany quartzite is a hard rock that requires crushing.

Those deposits requiring additional beneficiation to upgrade them to glass-quality sand generally require treatment to remove refractory heavy mineral (RHM) impurities. These impurities have densities greater than that of sand and are more refractory than sand (that is, they melt at a higher temperature). Froth flotation is the most frequently used technique to remove RHM impurities. Sand from the deposit is washed to remove fines and clays and then conditioned with chemicals that cause the sand to sink and the denser RHMs to rise in special agitated cells. The RHMs overflow to waste, and the sand is recovered, dried, and sized as required.

Although almost all glass batch is wet before melting, the use of wet sand does not prevail in U.S. sand plants, which are designed and built to size, handle, store, load, and ship sand after it is dry. Much of the sand moves by rail in cold climates, where dry sand has the advantage of not freezing in transit. U.S. glass plants are also built to unload, handle, and store dry sand. Thus, the infrastructure in place in the United States simply is not designed for wet sand. In Europe and many other regions of the world, however, wet sand is the prevalent product.

Most glass-quality sand is at least 99% SiO_2. A large proportion is 99.5% SiO_2, minimum. For most white glass, the iron content (Fe_2O_3) must be less than 0.03%. Float glass, colored container glass, and fiberglass can tolerate a higher iron content. Other impurities, including RHMs, chrome, and nickel are particularly detrimental. Rather than melting, they form discrete defects, which necessitates the scrapping of any glass containing them.

Chemical analysis of glass-quality sand generally involves quantitative spectrographic analysis (QSA) combined with atomic absorption spectroscopy (AAS) and/or wet

chemical methods. Heavy mineral analysis utilizes tetrabromoethane (density of 2.98 g/cm^3) as the separation medium. The sand floats, and all minerals with a density greater than 2.98 sink. These minerals are either recovered and identified petrographically or seeded into laboratory melts to determine their propensity for forming defects in the glass. The retention of heavy mineral particles coarser than 60 mesh is either limited to a percentage of the sand or a counted number of particles in a sample of given size.

The preferred size distribution of sand contains no more than 5 to 10% of particles retained on 40 mesh, and no more than 5 to 10% can pass 140 mesh. Particle size distribution analysis for all the major batch materials generally utilizes a nest of sieves on a shaker. Because of the segregation that occurs with any handling or transfer of material, the greatest challenge in a particle size analysis is to obtain a representative sample. Segregation is discussed at length in the section "Segregation of Materials in Shipping, Storing, and Mixing" in this article.

Size distribution is frequently compromised to obtain a cost savings, often without the data for a reliable cost versus benefit comparison. The fine particle size distribution of the Oklahoma sandstone increases dusting and carryover in regenerative furnaces, which must be controlled by special material handling techniques. Wet batch methods must be used, and a caustic wet batch is usually preferred. There is a cost trade-off between bringing in a coarser sand from a greater distance and the cost of caustic wetting, increased carryover, reduced furnace life, and special handling techniques. The disadvantages are balanced, to some extent, by the faster melting rate of the finer sand.

Another typical cost/quality trade-off involves the iron content of the sand. Generally, the higher the iron content, the lower the cost. The iron contained in flint glass used for containers contributes a residual color, which is undesirable for many applications. However, this coloration can be masked by the use of certain expensive additives, such as selenium and cobalt. Thus, a glassmaker may opt for a cheaper high-iron sand, but must compensate for this with higher decolorizing

Table 1 Glass product types and applications

Glass product type	Application
Flat glass	Automotive: cars and trucks Architectural: commercial buildings, storefronts Residential: windows, doors, sunrooms, skylights Patterned glass: shower doors, privacy glass
Containers/tableware	Beverage Liquor, beer, wine Food Pharmaceutical, drugs Glasses, plates, cups, bowls, serving dishes
Fiberglass	Wool: insulation, filters Textile: plastic or rubber tire reinforcements, fabrics, roof shingle and roll goods reinforcement
Specialty glass	Artware, stained glass, lead and lead crystal, lighting, TV picture tubes, ovenware and stovetop, ophthalmic, aviation, tubing, foamed glass, marbles

costs to the extent required to meet customer specifications.

Limestone is either a high calcium ($CaCO_3$) or dolomitic [$CaMg(CO_3)_2$] product. Technically, dolomitic limestones contain less than the theoretical 46% $MgCO_3$; those that contain the theoretical 46% $MgCO_3$ and 54% $CaCO_3$ are called dolomite. In reality, nearly all U.S. dolomitic limestones are close to the theoretical composition, and dolomite is used almost exclusively to refer to any calcium-magnesium-carbonate. Most of the glass industry uses these materials in their carbonate form, although there is still some limited use of their oxides, known as lime, quicklime, burned lime, or dolomitic quicklime.

Oxides are formed by heating the stone under controlled conditions to drive off carbon dioxide (CO_2). When the carbonate form is used in glass batches, the CO_2 is driven off during melting. This reaction "stirs" the glass melt as the CO_2 rises to the surface and escapes. The CaO and MgO act as fluxes to hasten the melting of silica. The substitution of MgO for some of the CaO in soda-lime-silica glass increases meltability, reduces the tendency for devitrification, and extends the working range (the time it takes for the glass to become rigid). This is one reason for limiting the MgO content in many container glasses, where high-speed forming techniques are used.

Two mineral forms of $CaCO_3$ are used: calcite (limestone) and aragonite. Limestone of suitable quality is found in numerous locations in the world. The aragonite used in the United States is mined in the Caribbean and shipped to ports along the Atlantic and Gulf Coasts. Glass-grade dolomite deposits are found in Ohio, Oklahoma, California, and Connecticut.

Conventional mining techniques are used in open-pit or near-surface mines. The rock is crushed through primary and secondary crushers and separated into several size fractions. Coarser rock is used to produce burned products (oxides and hydrates). The fraction that is finer than about 6.4 mm ($\frac{1}{4}$ in.) is dried, ground, air separated to remove fines, and screened to produce the grades used by the glass industry.

Limestone should be 96% $CaCO_3$ minimum, with $MgCO_3$ and silica as the major impurities. Dolomite generally can range in content from 52 to 54% $CaCO_3$ and 40 to 46% $MgCO_3$. However, for any one raw material, the maximum variation in CaO and MgO cannot be greater than 1%. There must be no silica present in a particle size coarser than the sand being used. An iron content of less than 0.10% Fe_2O_3 is desirable. Chemical analysis generally utilizes wet chemical methods or AAS. Acid insoluble fraction is determined by solution of the carbonates in hydrochloric acid (HCl). This test concentrates the silica and other impurities, including RHM, so that they can be analyzed for size and defect potential. A quantitative spectrographic analysis can determine the potentially detrimental minerals that may be present by looking for elements such as zirconium, chromium, nickel, manganese, and others.

Another important characteristic of limestone and dolomite is decrepitation, the term used when these carbonates give up their CO_2 (or chemically combined H_2O), often quite explosively, upon heating. The resulting oxide particles are very fine and dusty. Decrepitation may begin as low as 400 °C (750 °F) in some dolomite, creating dusty conditions around the furnace doghouse (where batch is fed into the furnace) and increased carryover within the furnace. Decrepitation varies with particle size. The coarse fractions (+40 mesh) in limestone generally decrepitate more severely than finer fractions. The opposite is generally true of dolomite. The coarser fractions are less prone to decrepitation.

Ideally, the particle size distribution of all batch materials should be similar. This reduces the segregation and inhomogeneity problems discussed later in this article. Matching particle size distributions provide the proper blend of fluxes and accelerators for each sand grain size and allow more uniform melting, with less homogenization required after melting. However, the particle size distribution of the limestones and dolomite are often chosen to minimize decrepitation, rather than to match the sand size distribution. Hence, there is no standard particle size distribution for these materials. Each producer tends to make what works best for that particular deposit and plant, with input from customers. The Ohio dolomite producers make at least three grades for the glass industry.

Feldspar and nepheline syenite are the primary U.S. sources of alumina. Feldspathic sand is used in some areas where it is both available and of adequate quality. Alumina used in glass reduces the coefficient of expansion, increases tensile strength, and makes the glass more resistant to abrasion, weathering, and attack by acids. The various feldspar products and nepheline syenite are composed of several minerals and contain silica and alumina with varying amounts of calcia, magnesia, soda, potassia, and iron oxide.

The most important source of nepheline syenite is in Ontario, Canada. It is mined by conventional techniques, and minimal beneficiation is required. This material is now available in a grade with a low +40 mesh content. Feldspars are widely found, and important deposits reside in Connecticut, North Carolina, and Georgia. Feldspars generally require more beneficiation than nepheline and still contain a significant portion of +30 mesh material (up to 10%). Aplite, found in Virginia, is lower in +30 mesh (about 2%).

All of these materials contain 60 to 70% SiO_2, 18 to 23% Al_2O_3, and varying amounts of CaO, MgO, Na_2O, and K_2O, depending on which minerals are present. Iron oxide is generally limited to 0.10%. More important than the exact amount of any oxide in a given material is its variation. The batch is calculated based on an analysis of the oxides present. The amount of each oxide must then remain constant for the material in use. As with other materials, analysis is usually done by AAS and/or wet chemical methods, with QSA used to determine the presence of other potentially detrimental elements.

The alumina material is usually relatively expensive, compared to other batch materials, and its choice depends on price, availability, and meltability. Melting problems result when feldspars coarser than the sand are used. In the late 1960s and early 1970s, cost and meltability were the driving forces for removing alumina from float glass. As of the early 1990s, there is some indication that alumina may again be added to flat glass to increase weatherability. The alumina content of container glass varies from 1 to 3%, but is virtually always present to increase durability (resistance to attack by the product being contained).

Soda ash, or sodium carbonate (Na_2CO_3), is the third major constituent of soda-lime-silica glasses and is the main source of Na_2O in any glass that contains soda. It acts as a flux, reducing the temperature required to melt the silica.

Soda ash is still made by the Solvay process (which reacts salt, or NaCl, and limestone to form Na_2CO_3, with $CaCl_2$ as a by-product) in some parts of the world. This material always contains NaCl as an impurity, up to 1% in some plants, but usually under 0.5%. The NaCl is volatile when glass batch is melted. It becomes part of the carryover in furnaces, reacting with and attacking furnace and checker refractories. There is one small Solvay process plant in Canada and one in Mexico.

The major source of soda ash for the United States and much of the rest of the world is the trona deposit in Wyoming. Trona, a hydrated sodium carbonate sodium bicarbonate ore ($Na_2CO_3 \cdot NaHCO_3 \cdot 2H_2O$), is found in "beds" that range from less than 1 m (a few feet) to 6 m (20 ft) in thickness 180 to 670 m (600 to 2200 ft) below the surface in southwest Wyoming. Five producers now operate there, with plant capacities from 0.9 to 1.8 Mg (1 to 2×10^6 tons) each, annually. Deep mining techniques with special equipment are used to mine the ore and transport it to the surface for further processing.

Solution mining has been commercialized by one producer. Once on the surface, the ore is calcined and dissolved, insolubles are removed, and it is filtered and dried to obtain the carbonate. The point in the process at which the calcining step occurs determines whether the Na_2CO_3 crystal will be rhombohedral in shape or cylindrical (needle shaped). Most of the soda ash produced is the denser, rhombohedral-shaped crystal referred to as the "monohydrate." The only other U.S. producer is located in California and produces approximately 0.9 Mg (1×10^6 tons) of soda ash annually from a complex brine obtained from Searles Lake.

These "natural" soda ashes are very pure and contain a minimum of 99.5% Na_2CO_3; several parts per million iron; less than 0.05% NaCl; and less than 0.25% Na_2SO_4. The California material also contains small amounts of magnesium and boron compounds. Chemical analysis can utilize AAS or wet chemical methods. QSA can be used to determine minor contaminants. A water insolubles test can also be run to isolate and concentrate impurities. A typical particle size distribution will have less than 5 to 10% retained on 30 mesh and less than 10% will pass 140 mesh. Soda ash is friable, and additional fines will be created during handling systems that involve numerous transfers, drops, and vibration.

The soda ash property that is the most problematic for glass manufacturers is the propensity to form hydrates. The monohydrate ($Na_2CO_3 \cdot H_2O$) forms at temperatures below about 110 °C (230 °F). Below 35 °C (95 °F), heptahydrates ($7H_2O$) and decahydrates ($10H_2O$) form. The serious problems that result when this reaction is ignored are discussed later in this article.

Salt cake generally refers to an impure form of anhydrous sodium sulfate (Na_2SO_4). In the glass industry, salt cake is used to refer to both pure and impure forms. This material is a melting accelerator. Under the proper redox conditions, the sodium sulfate and soda ash form a eutectic, which melts at a lower temperature than either material alone and forms a fluid liquid that attacks and dissolves the silica grains.

Salt cake sources include natural deposits in Texas and western Canada, as well as the by-products of rayon, hydrochloric acid, and chrome manufacture. The Texas material is of quite high purity and is sized more closely to sand and soda ash, with most of the fines removed. It is recovered as a brine, and the sodium sulfate is separated out by a differential chilling process. The resulting material is calcined to the anhydrous form and sized using screens and air separation. The Canadian material is less pure, being allowed to crystallize naturally in evaporation ponds. It is then calcined to the anhydrous form. The by-product materials can be high in purity or contain different impurities, depending on the process and primary product. Most of these materials contain as much as 25% -200 mesh material. Salt cake should be at least 99% Na_2SO_4, with no detrimental impurities. Analysis is accomplished by AAS, wet chemical, QSA, and water insolubles techniques.

Gypsum ($CaSO_4 \cdot 2H_2O$) is sometimes used in place of salt cake. It reacts with soda ash in the batch to form Na_2SO_4 and $CaCO_3$, performing the same functions as salt cake. Its use reduces the use of limestone as the CaO source. It does not cake in storage, as salt cake tends to do. There are a number of deposits of sufficient purity in the United States, but few producers make a glass grade. Most gypsum is ground for wallboard and plasters. Glass-grade material should have a size distribution between 20 and 200 mesh, with as little fine material as possible. Impurities include chlorides, MgO, SiO_2, Al_2O_3, and iron. The choice between salt cake and gypsum is based on availability and batch cost.

Sodium nitrate is an oxidizing agent. Its primary function in glass is to stabilize/maintain color. One U.S. producer manufactures sodium nitrate from chemical reaction. Much of the material in use is imported from natural deposits in Chile. The sodium nitrate is formed into beads from a melt and is very coarse, compared to the other batch materials. However, it is the lowest melting point material used. It is also very hygroscopic, cakes readily as a result, and cannot be stored for any appreciable length of time without special precautions to keep it dry. The U.S. material is of very high purity, but the Chilean material is also quite pure, containing only minor amounts of impurities.

Borate materials are used extensively in fiberglass, as well as in borosilicate specialty glasses used for oven and stovetop cookware, laboratory and pharmaceutical ware, and ophthalmic lenses. Borate materials lower the expansion, decrease viscosity, and shorten the working range, allowing increased fabrication speeds. A number of sources of B_2O_3 are used. Most are produced from borax mined in California, or from a California brine extraction process. Colemanite from Turkey is used where low soda is required. The borate materials available are 5 mol borax ($Na_2O \cdot 2B_2O_3 \cdot 5H_2O$), anhydrous borax ($Na_2O \cdot 2B_2O_3$), boric acid ($B_2O_3 \cdot 3H_2O$), and colemanite ($Ca_2B_6O_{11} \cdot 5H_2O$). Ulexite ($Na_2O \cdot 2CaO \cdot 5B_2O_3 \cdot 8H_2O$) and probertite ($Na_2O \cdot 2CaO \cdot 5B_2O_3 \cdot 5H_2O$) use is increasing in fiberglass. The article "Fiberglass" in this

Volume further discusses the use of borates in fiberglass.

Lead oxides and silicates are used in the production of specialty glass for optical glass, electrical glass, and lead-crystal tableware. In cases that require lead oxide in the glass composition, lead silicates are replacing oxides as the preferred raw material in order to reduce exposure to harmful lead dusts.

Minor Materials

Melting accelerators include the sulfates, lithia, and fluorspar (CaF_2). Of these, sodium sulfate (salt cake) is the most widely used. The redox state of the glass batch is important in order to optimize the action of the sulfate. This is discussed further in the section "Batch Formulation" in this article. Lithia may be supplied by lepidolite ($LiF \cdot KF \cdot Al_2O_3 \cdot 3SiO_2$), spodumene ($LiO_2 \cdot Al_2O_3 \cdot 4SiO_2$), or lithium carbonate (Li_2CO_3). Very small additions (0.1 to 0.2% LiO_2 in the glass) of lithia are reported to reduce operating temperatures and fuel consumption and increase pull and furnace life (Ref 1–3), but its use is not widespread. Fluorspar is no longer widely used because of the emission of fluorine.

Fining agents are used to minimize seeds, blisters, and bubbles. These agents include sulfates, arsenic, antimony, fluorides, phosphates, and chlorides. Fining is a very complex process that is dependent on viscosity, composition, raw materials, and redox conditions in the batch and melter, to name just a few variables. The combination of sodium sulfate and a reducing agent represents the most common fining system used for soda-lime-silica glasses. Calumite slag (a calcium-aluminum-silicate by-product of steel manufacture) is an effective reducing agent because of its sulfide and carbon content. It also supplies other glass-forming oxides. Its main drawback is its heavy mineral content. The use of arsenic and antimony is decreasing for environmental reasons. Fluorspar is an effective fining agent for glasses that contain more than 1.5% Al_2O_3. Its use is also limited by environmental concerns. Sodium nitrate and sodium chloride are sometimes used, but are not as effective as the sulfate system, in most glasses. Sulfates are not effective fining agents for reduced glasses, because they are oxidizing in nature; chlorides, bromides, and iodides have been used instead. Calumite is claimed to be effective in both oxidized and reduced glasses.

Colorants and decolorizers include selenium, cobalt oxide, iron oxides or salts, manganese dioxide, rare-earth compounds, nickel, copper, cadmium, and chrome compounds. The most common decolorizing system involves the use of selenium and cobalt to mask the blue and yellow (green) color imparted to glass by iron impurities in the raw materials. The selenium pink complements the iron yellow, and the cobalt (blue) produces a neutral gray to the eye. Generally, sodium nitrate is also added as an oxidizing agent to stabilize the color. Used in different proportions, these same materials produce the bronze, gray, and blue architectural glasses used in many buildings. The rare earths (lanthanum, cerium, praseodymium, and neodymium) are being used in specialty glasses because of their ability to provide glass with a high refractive index and low dispersion, or to absorb ultraviolet wavelengths, or to impart color to the glass.

Table 2 lists the various materials used in glass manufacture.

Cullet

Cullet is the term used in the glass industry for waste or broken glass, as well as glass that is deformed, or is unsuitable for sale as product. Virtually every glass operation generates some amount of in-house cullet, which is reused as a batch material. This cullet is usually of the same composition as the glass being produced, and no special batch adjustments are required in order to utilize it. Foreign cullet (from sources other than the manufacturing plant) generally requires careful analysis and batch adjustments, like any other batch material, in order to accommodate the compositional differences. Foreign cullet in "post-consumer" form is increasing in importance to the container manufacturers as recycling expands. This cullet must be carefully processed to remove metal, ceramic, and other detrimental contaminants before it is used by the glass manufacturer (Ref 4, 5).

Although cullet has been used in ratios ranging from 0 to 100% of the glass, 30 to 60% cullet is the most effective range. The amount used is normally determined by the amount of cullet available.

Cullet does have beneficial effects on glass batch melting. Because it is already a glass, there is no need for a particle solution step, as there is with batch raw materials. Cullet size is important, and a size that is coarser than the raw materials seems to work best. Cullet can be incorporated in a number of ways. After the batch has been mixed, it can be added to the mixer, or added at the furnace doghouse (under, over, or layered with the batch), or added almost any other way that is practical. However, the batch and cullet should be mixed somewhat before they are fed to the furnace. Melt stability cannot be attained if the furnace is fed a slug of batch, followed by a slug of cullet.

Batch Formulation

Batch formulation defines the proportions in which various raw materials are used to yield a glass of the desired composition and properties. The properties are generally defined by customer requirements and specifications. A composition or range of compositions is chosen to meet these requirements. Data on available raw materials, such as chemical and particle size analyses, material cost, and transportation cost are gathered. A batch formulation is then devised to provide the desired composition. The cost of this batch can be calculated and compared with alternatives to provide a glass that meets end-use requirements at the lowest cost.

Inherent in this process is a consideration of how the proposed batch will melt, fine, and homogenize. It is important to consider the particle size distribution of each available material and its match or mismatch with other available materials, as well as the addition of minor ingredients to promote rapid melting and fining. In soda-lime-silica compositions, this usually means the addition of sulfate to the batch, as either salt cake or gypsum, and the inclusion of a reducing agent (a form of carbon or calumite is generally used) to attain the proper balance of oxidizing and reducing forces (Ref 6, 7).

Because the state of oxidation (redox) of the molten glass in the furnace has a powerful influence on both the rates of melting and fining, it is now common practice to control the redox of the batch during the formulation step. A calculation procedure based on both theoretical and empirical information has been developed for this purpose (Ref 8). Each component in the batch has been assigned a value that represents its relative influence on the overall redox of the glass. A redox number is determined from these values and the composition of the batch. This number can then be altered to obtain an optimum redox state for each given glass and the melting

Table 2 Materials used in glass manufacture

Material	Purpose	Material	Purpose	Material	Purpose
Antimony oxide (Sb_2O_3)	Decolorizing and fining agent	Dolomite/dolomitic limestone [$CaMg(CO_3)_2$]	Source of calcium and magnesium	Manganese dioxide/pyrolusite (MnO_2)	Colorant
Aplite (K, Na, Ca, Mg, alumina silicate)	Source of alumina	Feldspars (Ca, Mg, Na, K, alumina silicates)	Sources of alumina	Nepheline syenite	Source of alumina, made up of nepheline and feldspars (approximately 60% SiO_2, 23% Al_2O_3, 10% Na_2O, 5% K_2O, <1% CaO, MgO, Fe_2O_3)
Aragonite ($CaCO_3$)	Source of calcium oxide	Albite ($Na_2O \cdot Al_2O_3 \cdot 6SiO_2$)			
Arsenic oxide (As_2O_3)	Fining and decolorizing agent	Anorthite ($CaO \cdot Al_2O_3 \cdot 2SiO_2$)			
Barite/barytes ($BaSO_4$)	Flux and fining agent, source of barium	Microcline ($K_2O \cdot Al_2O_3 \cdot 6SiO_2$)		Potash (K_2O) and potassium carbonate (K_2CO_3)	Sources of K_2O
Barium carbonate ($BaCO_3 \cdot$	Source of barium for specialty glasses	Mixtures of these, plus Fe_2O_3 and free quartz; see also nepheline syenite		Potassium dichromate ($K_2Cr_2O_7$)	Colorant in some artware
Borate materials	Sources of B_2O_3			Pyrite (FeS_2)	Colorant in amber glass
Sodium tetraborate ($Na_2O \cdot 2B_2O_3 \cdot 10H_2O$)		Fluorspar (CaF_2)	Used with feldspar as an opacifier in opal glasses	Salt cake/anhydrous sodium sulfate (Na_2SO_4)	Melting and fining aid
5 mol borax ($Na_2O \cdot 2B_2O_3 \cdot 5H_2O$)				Selenium (Se)	Aids decolorizing, and is used in colored glasses
Anhydrous borax ($Na_2O \cdot 2B_2O_3$)		Gypsum ($CaSO_4 \cdot 2H_2O$)	Flux and fining agent		
Boric acid ($B_2O_3 \cdot 3H_2O$)		Iron oxides/rouge (FeO, Fe_2O_3, Fe_3O_4)	Colorants	Silica sand/quartz (SiO_2)	Glass former
Calcium-aluminum-silicate/calumite slag	Reducing agent, fining aid	Lead oxides	Source of PbO for lead glasses	Feldspathic sand (mix of feldspar and quartz, source of SiO_2 and Al_2O_3)	
Caustic soda/sodium hydroxide (NaOH)	Solution (50%) is used for batch wetting	Litharge (PbO)			
Cerium oxide (CeO)	Ultraviolet absorber used in specialty glasses	Red lead (Pb_3O_4)		Soda ash (Na_2CO_3)	Major flux used in all soda-lime-silica glass
Chromite ($FeO \cdot Cr_2O_3$)	Colorant for green bottles	Lead silicates ($2PbO \cdot SiO_2$, $PbO \cdot SiO_2$, $4PbO \cdot SiO_2$)		Sodium antimonate ($2Na_2O \cdot 2Sb_2O_5 \cdot H_2O$)	Fining and decolorizing agent
Cobalt oxide ($Co_2O_3 \cdot CoO$)	Strong colorant (blue)			Sodium nitrate ($NaNO_3$)	Oxidizing and fining agent
Colemanite ($Ca_2B_6O_{11} \cdot 5H_2O$)	Source of B_2O_3	Limestone/calcite ($CaCO_3$)	Source of CaO		
Cryolite (Na_3AlF_6)	Flux and opacifier in opal glasses	Lithia minerals	Source of LiO_2, flux, melting accelerator	Tin oxides (SnO and SnO_2)	Colorants used in artware
Cullet	Crushed or powdered glass, may be internal or foreign	Lepidolite ($LiF \cdot KF \cdot Al_2O_3 \cdot 3SiO_2$) Spodumene ($LiO_2 \cdot Al_2O_3 \cdot 4SiO_2$)			

conditions. Minor additions of salt cake and/or carbon are usually made to the batch composition to obtain the desired situation. Utilizing this procedure, it is possible to control such factors as the amount of SO_3 that will dissolve in the glass and the ratio of ferrous-to-ferric iron.

Typical composition ranges for both container and flat glasses are given in Table 3. Using typical raw material analyses, an example batch has been calculated to yield a glass within this composition range (Table 4).

Raw Materials Handling

In addition to raw material selection and batch formulation, attention must be given to handling of the raw materials during shipping, storage, and mixing. Because of the strict chemical and physical requirements for acceptable glass products, the proper handling of the raw materials during shipment and storage is essential. In terms of both the chemical and physical homogeneity of the molten glass issuing from the furnace, the melt should ideally be a perfect solution on a random molecular basis. Only this condition can yield glass products with the desired maximum strength and aesthetic appearance. Obtaining this state is not a simple matter.

The process of making glass from its raw materials usually involves two mixers in series. The first of these is in the "batch house," where the materials are blended at ambient temperature in some sort of a rotary mechanical device. The second mixer is the furnace itself, where the blending action occurs by diffusion and convection within a very viscous medium at a very high temperature. It is essential that the mixed batch charged to the furnace be as perfectly blended as possible to reduce the mixing load imposed on the furnace. Various characteristics of the raw materials themselves combine to make a perfect random mixture difficult, if not impossible, to attain.

Segregation of Materials during Shipping, Storage, and Mixing

Free-flowing streams of particles are encountered during the initial production of each individual raw material, the loading and unloading of the hoppers required for their transportation, the filling and discharging of the storage silos at the glass manufacturing plant, and, finally, the filling and discharge

Table 3 Typical glass compositions

Oxide	Container glass, %	Float glass (U.S.), %
SiO_2	71.0–73.0	72.8–73.2
Al_2O_3	1.0–3.0	0.1–0.2
Fe_2O_3	0.04–0.06	0.10–0.14
CaO	10.0–11.0	8.55–8.85
MgO	0.1–1.0	3.85–4.00
Na_2O	13.0–14.0	13.65–13.85
K_2O	0.1–1.0	0.01–0.04
SO_3	0.2	0.25–0.30

Table 4 Typical batches

Material	For 100 units glass	Based on 455 kg (1000 lb) sand(a)	For 100 units glass	Based on 455 kg (1000 lb) sand(a)
Sand	65.3	1000	73.0	1000
Limestone	18.6	284.8	5.6	76.7
Dolomite	...	...	18.6	254.8
Feldspar	10.9	166.9	...	...
Soda ash	22.3	341.5	22.6	309.6
Salt cake	0.6	9.2	1.3	17.8
Rouge	...	...	0.07	1.0
Carbon	0.04	0.6	0.08	1.1
Total	**117.74**	**1803.0**	**121.25**	**1661.0**
Glass yield, %	85		82.4	

(a) Calculating batches based on 455 kg (1000 lb) sand is a glass industry convention.

of the mixed batch storage bin at the furnace charger. At each of these steps, size segregation occurs. Because these steps are in series, the overall effect is generally a cumulative and additive one. In many cases, variations in product quality have been related directly to such variations in sizing.

The principal bulk raw materials required for glass making are fine, granular solids that vary in size from about 10 to 14 mesh, at the coarse end, to a small quantity finer than 200 mesh. These materials flow freely when allowed, for example, to tumble by gravity down the slope of the typical cone that forms in a storage bin. Unfortunately, when this free-flowing action occurs, the various factors that come into play quickly create major demixing and segregation within the bed of solids.

Because of the necessity for charging a well-mixed, homogeneous batch to the furnace, it is very important that the factors that lead to demixing and segregation of free-flowing particles be understood and controlled as much as possible. The characteristics of solids that affect solid-solid mixing are:

- Grain size distribution
- Nominal grain size, *per se*
- Effective particle density or specific gravity
- Particle shape
- Surface characteristics, lubricity, and others
- Surface conductivity, or tendency to hold a static charge
- Bulk density
- Flowability
- Angle of repose
- Resistance to agglomeration

Generally, in the case of glass-making raw materials, only the first three characteristics have major influence in causing segregation among the particles. Of these, grain size distribution is by far the most important, perhaps accounting for more than 90% of the overall segregation encountered in any stream of free-flowing particles.

The powerful, controlling influence of grain size distribution in the demixing and segregation of moving particles has been described in a number of sources (Ref 9–15). In somewhat exaggerated form for emphasis, Fig 1 shows the segregation conditions that develop

during the filling and emptying of a typical storage bin. In Fig 1(a), the largest particles in the material are able to surmount the smaller sizes and thus roll down the cone of solids (representing the angle of repose of the material) to the outside wall of the bin. The smallest particles rapidly sift down through the larger ones and immediately come to rest at the center of the bed. Thus, the filled bin contains a core of the finest particles throughout the entire height of the solid bed. The largest particles are concentrated at the outside wall of the bin.

During the discharge cycle of the bin, the first material is primarily the finest fraction from the center core of the bin, as shown in Fig 1(b). This condition is maintained until the inverted angle of repose develops, at which point the coarsest particles at the wall tumble down the cone and flow out with the finer fractions. An example of the sizing variation that results as a bin is discharged from full to empty is shown in Fig 2. In this case, the bin

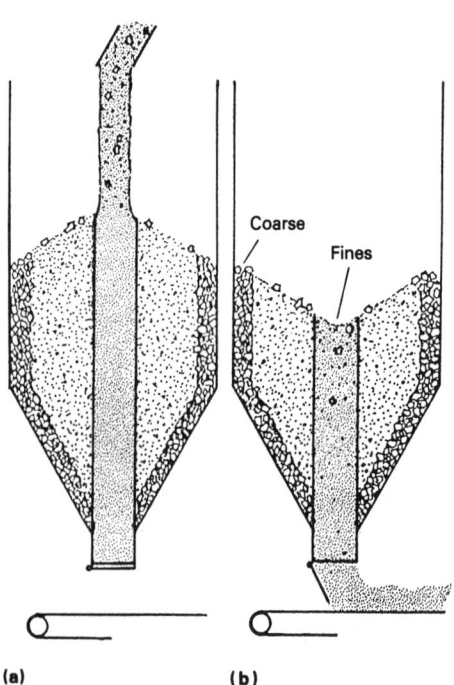

Fig 1 Patterns of bin segregation. (a) During filling. (b) During discharge

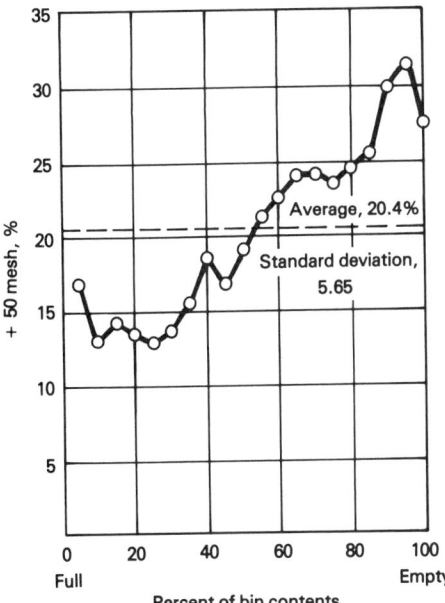

Fig 2 Segregation of mixed batch; variation in screen sizing during discharge

Fig 3 Segregation pattern formed during mixing; dark bands represent the +35 mesh fraction that has been dyed green

contained a mixed batch ready for charging to the furnace. Twenty samples were taken during the discharge, and these were subsequently screened in the laboratory. The amount of each sample larger than 50 mesh is plotted against the percent of bin contents that had passed from the bin at the time of sampling. A typical segregation pattern is indicated. During approximately the first half of the cycle, the batch was finer than the true average. Correspondingly, the last half of the flow was above average. Note that the standard deviation (a good index for defining the degree of segregation) was 5.65.

When a batch segregated to this extent is charged, the furnace must correct for the demixing. This requires that the furnace be operated at a reduced throughput rate with higher than necessary temperatures and, thus, increased energy use and air-polluting emissions.

Segregation can also occur in the mixer (Fig 3). In this case, a typical glass grade of limestone was carefully screened to remove the 35% of particles coarser than 35 mesh. This fraction was dyed a dark green and added back to the sample. The mixture was then rotated in the plastic tube shown in Fig 3. Within perhaps one minute, a segregation pattern developed, which was not altered by further rotation, indicating that a steady-state condition had occurred.

Two observations illustrate the powerful influence exerted by various factors that promote demixing in moving beds of free-flowing granular particles. First, a steady-state condition was attained, and this occurred relatively rapidly. Second, the sharp line of demarcation between the light and dark particles illustrates how completely the demixing occurred. In this case, a separation occurred between the largest of the white particles (35 mesh) and the smallest of the dark particles (roughly, at 34 mesh).

The necessity of both measuring the amount of segregation being generated in any given batching system and subsequently minimizing this condition is obvious. An extensive amount of literature on this topic is available, and representative contributions include Ref 10 to 15, as well as those sources cited in their bibliographies.

The purpose of the preceding discussion has been primarily to emphasize how important it is that the glassmaker recognize the seriousness of the size segregation problem. The major effects that occur when a stream of particles is allowed to flow freely under the force of gravity can be summarized as:

- The forces of demixing rapidly create a steady-state pattern of segregation
- Under the same flow conditions, this pattern is reproducible
- Size difference is the principal factor contributing to the segregation, but with glass-making raw materials, density difference is also a factor to a lesser extent
- A size difference ratio between particles that is as little as 1.05 will lead to measurable demixing
- The forces that create the demixed state in moving beds of particles operate whenever moving streams of particles are present, such as in all types of rotary mixers, the filling and discharge of bins, and wherever else the coning of solids occurs

Wetting of Glass Batch

It is now obvious that the forces of demixing in streams of flowing solids result from the free-flowing action of individual particles. Reducing this effect during the unloading and storage of batch materials is a matter of recognizing that segregation occurs whenever materials are handled and, therefore, designing the unloading and storage facilities to minimize the effect.

Minimizing the effect once the materials are weighed and conveyed to the mixer is a matter of suppressing their free-flowing characteristics. One method of suppression which has become a common practice, is to wet the materials with sufficient liquid during the mixing operation to permit a thin film to coat each particle. Generally, either water (with or without an added wetting agent) or a 50% solution of caustic soda (NaOH) is used. In addition to reducing segregation, this practice has been very beneficial in reducing the dusting that can occur within the furnace; an equivalent reduction in the rate of regenerator plugging and in the amount of particulate stack emission also occurs. Properly carried out, wet batching can also lead to increased melting rates and lower furnace temperatures, which in turn reduce the amount of air pollution per ton of glass melted.

Although batch wetting has many advantages, the procedure is not without its problems. Because the function of wetting is to make the batch tacky to impede the normal free-flowing properties, this condition also complicates handling of the batch during subsequent conveying and storage. In many cases, modifications to the batch handling system are too costly to properly implement. Therefore, the tackiness of the batch is reduced by decreasing the degree of wetness. This may or may not solve the handling problem, and may in fact create new problems that negate the advantages of wetting. When the batch is properly wetted at the mixer, it is possible to charge a batch to the furnace with excellent homogeneity in terms of the chemical composition. However, wetting the batch does not solve the size segregation that occurs in raw material handling and storage prior to mixing. Thus, even a chemically homogeneous batch may vary widely in sizing from point to point and from time to time.

The controlling step in the melting of glass batch is the rate of dissolution of the sand particles. This rate is controlled by the size of the individual silica particles. When this size varies over a period of time, the overall rate of melting will also vary accordingly, and maintenance of a smooth, steady-state furnace operation becomes difficult.

Size variation with time also complicates the wetting procedure itself. The desired tacky state requires that each particle be covered with a thin film of the wetting liquid. The exact amount of liquid required is thus a function of the total surface area of the batch particles, which in turn is a function of the particle diameters. When significant size variation occurs as the result of segregation in the raw materials storage system, continuous adjustments in the rate of liquid addition are required to prevent batch condition variation, from too wet and sticky to too dry and dusty.

Batch wetting with water has been the most common procedure. A number of requirements must be met to ensure a successful application. The desired tacky state of a wetted batch occurs when sufficient liquid has been added to coat the surface of every particle. In the case of water, allowance must be made for its chemical reaction with the soda ash (sodium carbonate) to form various hydrates, that is, the deca-, hepta-, or monohydrates. The temperature of the batch determines which hydrate will form. The actual surface wetting of all of the particles cannot occur until the soda ash particles are completely hydrated. The monohydrate form is, therefore, preferred to minimize the total amount of water required for the wetting action.

In order for the monohydrate form to be stable, a temperature in excess of 35 °C (95 °F) must prevail after the addition of the water. Although some heat is evolved by the hydration of the soda ash, it may be necessary to preheat the raw materials and/or the water to ensure that the necessary final temperature is obtained. Typically, the additional water required will amount to about 3.5 to 4.0% of the final batch weight. Once the batch is wetted, the temperature must be maintained above the 35 °C (95 °F) level up to and through the furnace charger; otherwise, the monohydrate will convert to the heptahydrate, effectively drying up the batch and cementing it into a solid block.

Batch wetting with caustic soda solution results in a simplified wetting system. When a 50% solution of sodium hydroxide (NaOH) is used, the soda ash will not hydrate (because of the phase relationship in the sodium carbonate-sodium hydroxide-water system), and thus remains in the anhydrous form. For this reason, only about 1.5% of water equivalent needs to be added as the caustic solution, in order to attain the desired tacky state. Because the caustic solution will react with limestone and dolomite fines, the amount of caustic required will vary with the amount of fines in these materials. This reaction is accelerated with heat and will also cause the batch to dry and cake. Although the addition of the caustic replaces an equivalent amount of soda ash, caustic solution is generally a more costly source of sodium, and its use is normally limited to market conditions that put caustic into a surplus position (it is generally a co-product with the production of chlorine).

Collecting, Weighing, Mixing, and Filling

Present-day glass manufacturing operations require utilization of the latest technology in the batch preparation. To remain competitive with alternative products, the industry has found it necessary to take advantage of the best-available equipment and instrumentation to improve product quality and reduce rejects to a minimum.

Collecting. In recent years, the industry trend has been toward a completely auto-mated batching system. For each batch, the required quantity of the individual raw material is measured in a weigh hopper at the bottom of each storage bin. These quantities are then transported to the mixer by conveyor belts. Although these systems require continuous maintenance and calibration, they represent an accurate and efficient means for preparing a batch with the precise amount of each material required for the desired final glass composition.

Weighing. Impressive improvements have been made in electronic circuitry and microprocessors. These have been utilized, along with the application of process loop techniques, to bring a high degree of sophistication to glass batch preparation. Electronic weighing systems with a sensitivity of about one part in 4000 allow adequate precision for glass batch preparation. The use of these latest developments, including such details as digital readout and computer systems, has made possible essentially foolproof and versatile raw materials handling installations.

Mixing. A number of suitable mixers are available. Two general classes are the rotary type (similar to a cement mixer) and the pan type, which has a stationary flat bed, with the mixing action provided by paddles that rotate on a vertical axis. The actual choice of mixer is dictated by such factors as the variations anticipated in production rates and whether or not wet batch will be used. The mixer should be located near the furnace, preferably directly over the mixed-batch storage bin. Such an arrangement allows for the best utilization of the wet batching procedure. The problems of conveying the wet batch are eliminated, and there is minimum time for the batch to cool and hydrate. This is not always possible because of retrofit and space limitations.

Filling. A number of significant advances in glass technology have enabled major increases in glass furnace pull rates. High melting temperatures are necessary in order to transfer the required massive inputs of energy to the melting batch. The charging of the batch to the furnace is a key factor in this energy transfer process, because the cold batch is involved in establishing the convection currents within the melt that are essential for mixing and homogenizing. Uniform charging of the batch is required to maintain a smooth, steady-state flow of glass. The use of automatic chargers is now common. These are regulated by sophisticated glass-level controllers, which increase or decrease the rate of batch charging as required for glass level regulation in the feeders to the glass-forming operations.

The chargers must be properly sized to allow development of the desired configuration of the batch blanket required for the massive energy input. Poor charging patterns interfere with the steady furnace operation needed for suitable glass quality. The use of wet batch must also be considered in the selection of the charging units.

Cost/Quality Trade-offs

Cost/quality trade-offs occur throughout the process of material selection, material handling, batch mixing, wetting, conveying, charging, and melting. Examples of some of these trade-offs have already been described. Size distribution and chemistry of the raw materials are often compromised to attain material at lower cost. Silos are still mostly single-entry, single-exit designs, although a multiple-discharge design has been shown to reduce segregation and resulting problems. Mixers are more frequently found in the batch house than over the furnace. One reason for this is the expense of redesigning existing plants.

Melting and fining are carried to the point of acceptable product quality, not necessarily to the level that could be attained under ideal conditions. Melting is generally carried to completion because discrete defects (stones from unmelted batch) weaken the glass product and make it unsuitable for its intended use. The presence of stones will cause the product to be scrapped (and probably returned to the process as cullet).

Fining is more difficult to carry to completion. A common plant control point involves the number of residual bubbles (commonly called seeds, or blisters) in the final product. Some glass customers will accept product that contains 7 seeds/g (200 seeds/oz) of glass. Others prefer a glass with essentially no visible bubbles. Although there are several variables that control the number of residual bubbles in the glass, an important one is the pull, or production, rate on the furnace, that is, the residence time the molten glass remains in the furnace. Generally, the lower the production rate at a given furnace temperature or energy input rate, the lower the seed count. However, it is also true that the lower the production rate, the higher the unit cost of a ton of glass thus produced. The goal of every cost/quality trade-off is to produce a product of acceptable quality at a price the customer is willing to pay.

REFERENCES

1. T.M. Mike, Chemical Boosting of a Glass Melting Furnace Using Spodumene, *Collected Papers from the 32nd Annual Conference on Glass Problems*, University of Illinois Urbana-Champaign, 1971
2. D.J. Kingsnorth, Adding Lithium Minerals Can Reduce Melting Costs, *Glass Ind.*, Mar 1988, p 18–25
3. C.E. Larson, How Lithium Benefits Production Glasses, *Glass Ind.*, Dec 1986, p 14–19, 39
4. F. DeNapoli and W. Kilpatrick, Effective Beneficiation of Recycled Cullet for Use in Glass Container Furnaces, *Ceram. Bull.*, Vol 67 (No. 11), 1988, p 1798–1801
5. D. Schendel, How To Make Good Qual-

ity Cullet, *Glass Ind.*, Feb 1990, p 10–13, 22

6. A.R. Conroy, W.H. Manring, and W.C. Bauer, The Role of Sulfate in the Melting and Fining of Glass Batch, *Glass Ind.*, Feb-Mar 1966

7. W.H. Manring and R.E. Davis, Controlling Redox Conditions in Glass Melting, *Glass Ind.*, Vol 59 (No. 5), 1978, p 13–16, 23, 24, 30

8. P.M. DiBello, Controlling the Oxidation State of a Glass as a Means of Optimizing Sulfate Usage in Melting and Refining, *Glass Technol.*, Vol 30 (No. 5), Oct 1989

9. F.V. Tooley, *Handbook of Glass Manufacture*, Ashlee Publishing, 1984

10. J.F. Van Denburg and W.C. Bauer, Segregation of Granular Materials in Storage Bins, *Chem. Eng.*, 28 Sept 1964, p 135

11. W.H. Manring and W.C. Bauer, Influence of the Batch Preparation Process on the Melting and Fining of Glass, *Glass Ind.*, July-Aug 1964

12. J.F. Van Denburg and W.F. Manring, How Batch Segregation Leads to Glass Defects, *Ceram. Ind.*, June-July 1965

13. W.H. Manring and W.C. Bauer, Solid-Solid Mixing Efficiencies, *Glass Ind.*, July 1962

14. H. Campbell and W.C. Bauer, Cause and Cure of Demixing in Solid-Solid Mixers, *Chem. Eng.*, 12 Sept 1966, p 179–184

15. W.H. Manring and W.C. Bauer, Batch Homogeneity Testing by Screen Analysis, *Collected Papers from the 36th Annual Conference on Glass Problems*, University of Illinois Urbana-Champaign, 1975

Melting/Fining

Frank E. Woolley, Corning Incorporated

THE MELTING PROCESS converts the granular or powdered raw materials, prepared as described in the article, "Raw Materials/Batching" in this Volume into a homogeneous liquid, free of visible defects, ready for forming by the processes discussed in the article "Forming" in this Volume. The product requirements of each application determine the quality of raw materials, the type of melter, and its operating conditions. This article describes:

- Chemical and physical processes that occur within glass melters
- Product defects that result from these processes
- Major types of melters
- Selection of a specific type of melter for each type of glass product
- Limitations on product quality imposed by present melters
- Trends in melter development

The melting process should minimize the internal defects that limit the function and appearance of glass products:

- Undissolved or precipitated solids, called stones
- Gas bubbles, called seeds or blisters
- Streaks of off-composition glass, called cord or stria

In addition, the melter should deliver to the forming process a melt that is uniform in viscosity, and hence in composition and temperature, on a scale which is much smaller than the product dimensions.

Several excellent books discuss the melting process (Ref 1–4) and the melting and forming processes for specific applications (Ref 5–7).

Fundamentals of Melting

Glass melting has three basic steps:

- Batch melting, the conversion of the batch (powdered raw materials plus crushed scrap glass) into a liquid essentially free of batch stones (undissolved crystalline inclusions)
- Fining, the removal of the bubbles remaining after batch melting

- Homogenizing, the elimination of chemical and thermal variations in the molten glass

In a continuous glass melter, these three steps take place simultaneously as the material moves through the furnace. A final step, conditioning, is needed to bring the glass to a uniform lower temperature for forming. Cable (Ref 8) presents an excellent discussion of the basic processes of glass melting.

Heat Transfer to Batch Materials. Batch melting begins when the cold batch materials reach a temperature at which reactions are rapid. Because the batch pile is a poor thermal conductor, heat transfer limits the overall batch melting rate. Batch piles float on the molten glass as they melt because the bulk density of the batch is usually about one-half that of the glass melt. Typically, 65 to 75% of the melting occurs on the top surface of the batch pile, in a fuel-fired melter. The balance occurs on the bottom. Radiation is the dominant mechanism for heat transfer, both from the flames above and from the translucent melt below the batch. At the surfaces of the batch pile, a sharp temperature gradient exists. The batch reactions take place in a thin layer on the surfaces while the interior remains cold. Measured thicknesses and temperatures for a soda-lime batch (Ref 9) are shown in Fig 1. The surface layer contains enough liquid to allow it to flow off the batch pile, thus exposing underlying cold batch to the heating process.

Batch melting, the elimination of crystalline phases, occurs in four steps that overlap in terms of time and temperature:

- Drying and dehydrating the powdered raw materials
- Melting the salts and other low-melting ingredients (fluxes)
- Decomposing some of the fluxes, producing gases
- Dissolving the silica sand and other refractory materials in the molten fluxes

Reactions between solid particles are extremely slow. The practical melting process begins when some of the fluxes melt, creating a liquid phase. Most glass batches contain carbonates, and sometimes also sulfates, nitrates, and hydroxides of the alkalies

and alkaline earths. These compounds melt around 400 to 800 °C (700 to 1500 °F), producing a molten salt of very low viscosity (1 to 4 mPa · s, or 1 to 4 cP). As the refractory ingredients begin to dissolve, the molten salt becomes enriched in silica and alumina. This causes some of the anions in the liquid to begin to decompose, producing large volumes of gases (CO_2, H_2O, N_2, O_2, and SO_2). The total volume of gases may be more than 100× greater than the volume of batch. These gases escape by bubbling through the liquid. This stirs the liquid, accelerating the dissolution of refractory particles.

The addition of silica to the molten salt will typically cause the liquid viscosity to increase by four orders of magnitude, reaching 10 to 40 Pa · s (100 to 400 P) in the final melt. Higher viscosities slow the dissolution of refractory particles and promote foaming. Thus, the dissolution of silica too early in the process reduces the melting rate. Eliminating very fine particles of sand and cullet in the batch avoids this problem. The addition of so-called melting accelerators, such as sodium sulfate and various borates and fluorides, lowers the liquid viscosity and increases the melting rate. Various methods of batch consolidation, such as batch wetting, pelletizing, granulating, and compacting also increase the melting rate (Ref 10).

Fining. Bubbles in the melt are generated by decomposition reactions during batch melting and by the reaction of the melt with the refractories and metals used as container and electrodes (see the section "Reboil" in this article). Bubbles are removed in three steps:

- Gases produced by chemical reactions in the melt enlarge existing bubbles
- Enlarged bubbles rise to the surface and break
- Bubbles that do not reach the surface dissolve during cooling

Removal of the last few bubbles is usually the slowest step in the glass-melting process. This step often limits the pull rate (throughput) achieved by a melter of a given size.

The enlargement of bubbles is by far the most important step, since the rate of rise is proportional to the square of the bubble

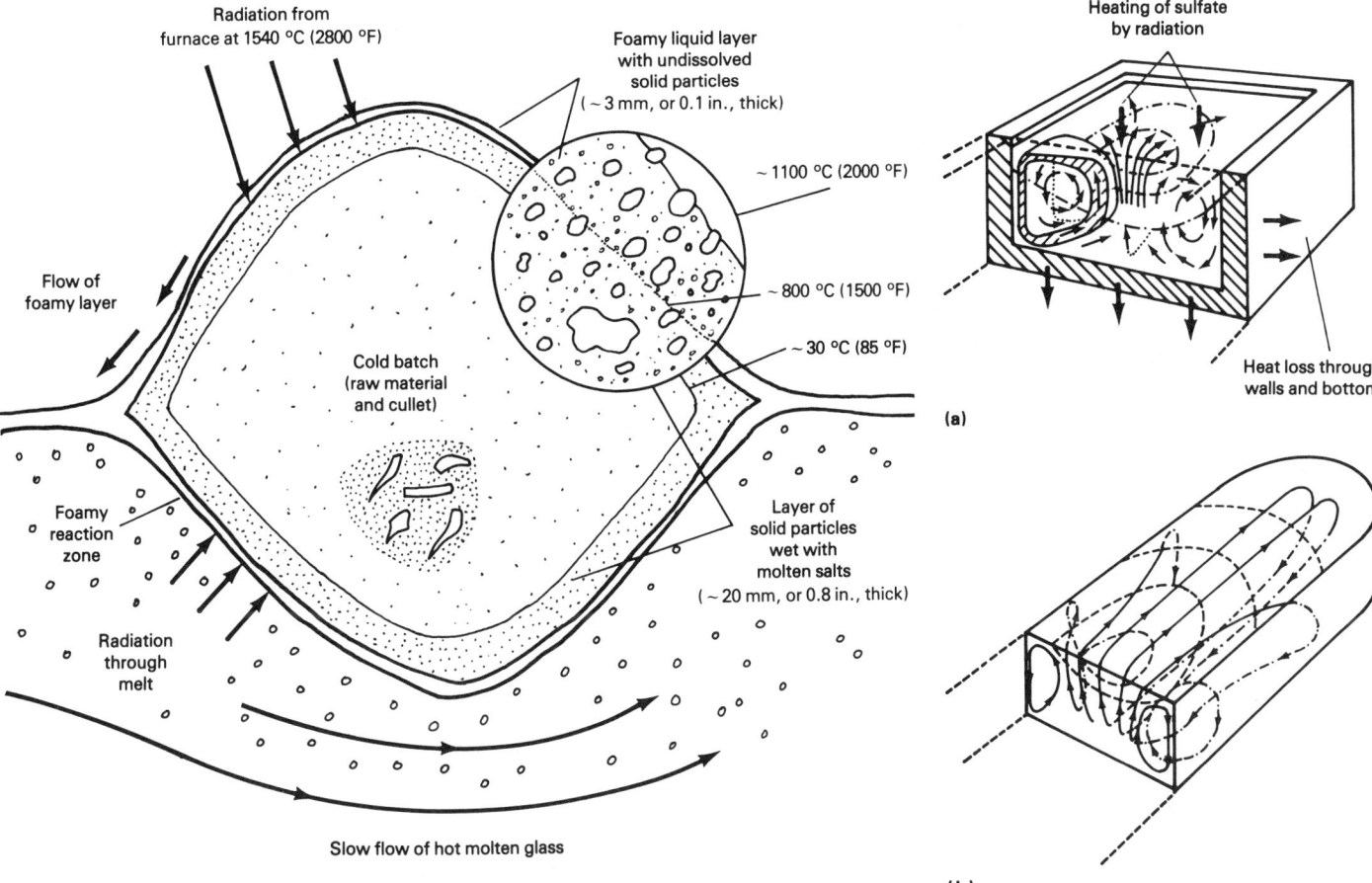

Fig 1 Cross section of batch pile floating on molten glass. Source: Ref 9, modified

Fig 2 Three-dimensional free convection flow in continuous glass melters. (a) Short wide furnace. (b) Long furnace. Source: Ref 12

diameter. Fining agents enlarge bubbles by producing gases as the temperature of the melt increases, either by a redox (reduction-oxidation) reaction or by volatilization. The most common redox fining agent is sulfate, dissolved during batch melting and reduced at high temperature to SO_2 gas. An increase in temperature reduces oxides of metals such as arsenic and antimony to their lower oxidation states. This produces oxygen that enlarges existing bubbles. Volatilizing fining agents are effective because their vapor pressures over the melt increase rapidly as the temperature increases. Specialty glasses are fined with chlorides, fluorides, and sometimes other halides. All halides present air pollution problems.

Bubbles that do not have enough time to reach the surface and break may dissolve during cooling. Bubbles generated at a refractory surface after the melt passes the fining zone in the furnace may also dissolve as the melt cools. The redox and volatilization reactions are reversible, so O_2, H_2O, and halide vapors usually go back into solution in the melt. However, SO_2, CO_2, N_2, and argon redissolve slowly and usually remain behind as bubble defects.

Because the major fining process occurs when the melt temperature increases, melting furnaces are designed to bring the melt to its highest temperature after batch melting is complete. The time-temperature history in the melter determines the fining capability of the furnace. Cable (Ref 8) and Swarts (Ref 11) present excellent discussions of fining.

Homogenizing. Compositional variations, called cord, result in permanent defects in the product. Temperature variations in the melt lead to losses in forming. Diffusion reduces both types of variations. Thermal diffusion in most melts is rapid, so thermal homogenization depends primarily on avoiding excessive cooling or heating close to the point at which the melt is delivered to forming. In contrast, mass diffusion is very slow in silicate melts because of their high viscosity and strong atomic bonding. Therefore, elimination of cord is more difficult.

Cord results from:

- Melting segregation during batch melting
- Volatilization from the melt surface
- Reaction of the melt with the refractories

Cord is reduced by a four-step strategy. First and most important, the sources are minimized. Next, skimmers, overflows, and bottom taps separate and discard the cordy portion of the melt. Third, orientation of cord reduces its effect on product properties. For example,

cord oriented parallel to the surface of flat glass has no visual effect in normal use and cord located near the surface of cast bars of optical glass is removed during machining of the final product. Finally, stirring helps eliminate remaining cord by reducing diffusion distances.

Three stirring mechanisms occur in melters. First, bubble stirring by bubbles during batch reactions removes most of the chemical variations present in the batch materials. Next, convection flows stir the melt in the furnace. Finally, mechanical stirrers remove remaining cord during conditioning (see the section "Conditioning and Delivery Systems" in this article) where the product requirements justify the additional cost.

Convection in Melters. Flow in glass melters results primarily from free convection. Heating the melt from above by flames and from within by electricity creates temperature gradients in the melt. The cooling effect of incoming batch and heat losses through the furnace walls add to the gradients. These temperature gradients produce density differences that drive three-dimensional convective flows in the melt (Fig 2). Maximum velocities are low, typically 1 to

10 m/h (3 to 30 ft/h), because of the high viscosities. Flow produced by the continuous filling of batch and the withdrawal of molten glass for forming is usually an order of magnitude slower than free convection flows.

Free convection serves three critical functions in melters:

- Separates the steps by preventing incompletely melted material from mixing forward into the fining zone, and incompletely fined material from entering the cooling zone
- Increases the melting rate by bringing hot melt under the batch blanket and removing cooler partially melted material
- Aids homogenization by shearing and stretching the melt

The control of the flow patterns is the primary aim of the design and operation of a glass furnace. These flow patterns are extremely complicated and difficult to determine, and much of the art of glass melting relates to the control of convection. Tank geometry and the amount of fuel and electric power introduced into each part of the melter determine temperature gradients. Bubblers and powered electrodes create local upward currents. Major thermal flow patterns extend over the full dimensions of the melters (Fig 3). Local flows near bubblers, electrodes, and mechanical stirrers are on a scale of tens of centimeters.

Competing Processes

Both the necessary processes discussed in the section "Fundamentals of Melting" in this article and a variety of undesirable competing processes result from the same conditions. These competing processes create defects in the glass and increase the cost of glass melting. The most important of these are listed below:

- Refractory corrosion
- Electrode corrosion
- Reboil
- Volatilization
- Devitrification

Refractory corrosion is important for three reasons:

- Produces stones and cord in the product
- Limits the temperature and hence the melting rate of the furnace
- Forces the periodic replacement of the furnace, an expensive operation that removes the furnace from production for several weeks or more and requires a large capital investment

Refractories consist of some of the same materials found in the batch (principally silica, alumina, and zirconia) and dissolve by the same diffusion-controlled process. The rate of corrosion typically doubles for every 50 to 100 °C (100 to 200 °F) increase in temperature.

Many refractories consist of more than one crystalline phase. Stones result when the weakest phase dissolves more rapidly, leaving particles of the more resistant phases free to enter the melt.

Vapors and batch dust attack those refractories not in direct contact with the molten glass, limiting the furnace life. This attack also may produce stones and cord in the glass as the stone-containing liquid formed on the refractory surfaces drips or runs down into the melt.

Electrode corrosion contaminates the glass with corrosion products and limits the amount of electrical power that can be introduced through electrodes of a given design. Electrodes are made of metals, graphite, and semiconducting tin oxide. Molybdenum is the most widely used material. Graphite is the least expensive but is limited to uses where the

constant release of CO bubbles is acceptable. Platinum and tin oxide are used where their chemical inertness justifies their high cost.

Metal electrodes are corroded by simple oxidation. Molybdenum and graphite usually cannot be used in glasses with high levels of easily reducible oxides such as PbO, SO_3, As_2O_3, and Sb_2O_3. These components are lost from the melt and the electrodes are rapidly oxidized. Electrodes are also corroded by irreversible electrochemical reactions, usually generating oxygen bubbles or causing alloying. Electrochemical corrosion of platinum electrodes often produces small platinum inclusions in the glass.

Reboil refers to the nucleation and growth of bubbles in a previously bubble-free melt. Bubbles nucleate on dissolving batch particles and on refractory and metal surfaces throughout the melter. Bubble growth is driven by a decrease in solubility of dissolved gases, by shifts in equilibrium in redox reactions, and by electrochemical reactions.

Large amounts of H_2O, CO_2, and SO_2 dissolve in molten silicates during batch melting (typically 10^2 to 10^4 wt ppm, equivalent to 1 to 100 volumes of gas at 1500 °C (2700 °F) per volume of melt). The less reactive gases N_2 and argon also dissolve but to a very limited extent (<1 wt ppm or <1 vol%). Oxygen solubility is linked to the amount and oxidation state of polyvalent cations (see the section "Fining" in this article). The amount of oxygen liberated by a temperature increase may be as much as 0.1 wt%, or 12 volumes/volume of melt. The solubilities of the most important dissolved gases decrease as the temperature or the concentration of silica and other refractory oxides increase. Thus, bubbles are generated as the melt reaches its highest temperature and as batch particles and refractories go into solution.

Molten glass is an ionic conductor, as are refractories at high temperatures. Electrochemical cells are created wherever electronic conductors contact the melt in two locations. Powered electrodes and metallic thermocouple sheaths, bubblers, and melt level detectors make direct contact with the melt. Support steel, cooling devices, and cladding to limit corrosion make contact through hot refractories. The cells created by these combinations of materials and strong temperature gradients can discharge, often producing O_2 bubbles or destructive alloying of the metals.

Volatilization. Loss of material by evaporation from the batch pile or melt surface is important for three reasons:

- Makes it impractical to melt glasses containing high levels of halides, sulfur, selenium, and some of the heavy metals in ordinary fuel-fired or air-atmosphere melters
- Results in cord and stone defects when depleted surface layers are not completely remixed and homogenized in the melt

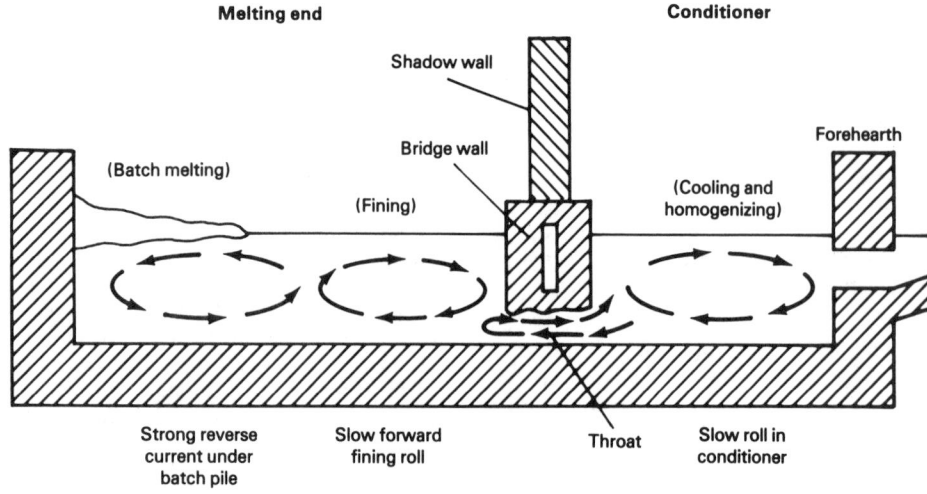

Fig 3 Major convection flows in a fuel-fired tank furnace (longitudinal vertical section on tank centerline)

- Produces vapors and condensate that reduce furnace life by attacking the superstructure refractories and require expensive gas-cleaning equipment to prevent air pollution

The highest volatilization losses are with alkalies (sodium and potassium), boron, heavy metals (lead, zinc, arsenic, antimony, cadmium, bismuth, and so on), halides (fluorine, chlorine, and bromine), sulfur, selenium, and tellurium. Volatilization losses are minimized by reducing melt surface temperatures, lowering gas velocities near hot surfaces, and reducing the melt-free surface area (see the section "Melting Furnaces" in this article).

Devitrification is the nucleation and growth of crystals in a molten glass as it cools. It is an important source of stones and knots. Buildup of crystalline material in flow channels and orifices can interfere with continuous forming by restricting flow. Devitrification normally occurs in the coldest part of the melting system, usually the forehearth or delivery orifice. It limits the methods of forming that can be used with a given glass because it prevents delivering the melt at a higher viscosity to the forming process.

The liquidus of a glass is the highest temperature at which any crystalline material can remain in equilibrium with the melt. It is a fundamental property of that composition. Crystallization does not occur in glass-forming melts if they are cooled quickly from above their liquidus to a low temperature. However, crystals will form and grow in all glass-forming melts if they are held for a long enough time at temperatures somewhat below their liquidus. In addition to the primary phase that forms below the liquidus, other (secondary) crystalline phases can form at lower temperatures. The liquidus and the devitrification rate of a melt often vary strongly with composition. Persistent devitrification problems are solved by modifying the glass composition.

Devitrification often occurs in a surface layer altered in composition by dissolution of refractories or by volatilization from the melt surface. In some borosilicates, volatilization produces a silica-rich surface layer in the melter and devitrification creates silica stones in this layer. Subsequent convection and stirring can convert these into silica-rich knots and cord.

Melting Defects

Solid inclusions reduce the mechanical strength of glassware, because they create regions of high stress. They are classified by their mechanism of formation as batch, refractory, or devitrification stones.

Batch stones are incompletely reacted material intentionally or accidentally included in the batch. They are usually large grains or agglomerates of the most refractory ingredients of the batch (zirconia, alumina, and silica) or slowly soluble contaminants (chromite, zircon, and rutile). They often are rounded and surrounded by a solution sac (cordy glass of altered composition).

Refractory stones include undissolved crystals released by corrosion from the interface (primary stones) and crystals formed by devitrification of glass enriched with dissolved refractory (secondary stones). Refractory stones can come from superstructure refractories as well as from glass-contact refractories (see the section "Refractory Corrosion" in this article).

Stones usually dissolve as they pass through the melter from their point of origin. If they dissolve completely, adequate time may still not remain for complete chemical homogenization to occur and a defect called a knot may remain that is high in silica, alumina, and so on. Convection or stirring can result in strings of stones, knots, and cord. The origin and characteristics of solid inclusions are discussed in the literature (Ref 13–14).

Bubbles in glass are called blisters if they are large (typically greater than 0.5 mm, or 0.02 in.) and seeds if they are smaller. They generally affect the appearance of glassware more than its function. Bubbles arise from the sources discussed in the sections "Fining" and "Reboil" in this article. The source of bubbles may be deduced from their chemical composition and sometimes from their orientation in the ware. The composition of the gases in bubbles as small as 0.1 mm (0.004 in.) can be reliably analyzed by carefully breaking them into a mass spectrometer or gas chromatograph.

Chemical Inhomogeneities. Various terms are used, sometimes interchangeably, to describe the distortions of optical paths through glass. Usually, cord means a one-dimensional vein, while stria is a two-dimensional layer, and ream is a three-dimensional array of cord. The effects of inhomogeneities on the product depend very much on the application, their intensity (chemical difference or gradient), their scale or size, and their orientation in the product. The gradient of composition (and hence of properties) is often more important than absolute differences. Stress produced by thermal expansion gradients can cause delayed breakage in glassware. Detection methods are usually optical, and produce only qualitative measures of inhomogeneity.

Melting Furnaces

Movement of glass through the melter can be periodic or continuous. Pot furnaces and day tanks are filled with a fixed amount of material. The temperature is then varied to melt, fine, homogenize, and cool the melt. Continuous melters are filled steadily to maintain a constant melt level as glass is withdrawn continuously for forming. The temperature at each point remains constant, while the material flows from the batch filling region through the furnace to the delivery area.

The melt can be contained in a leakproof crucible heated externally or in a basin of refractory blocks heated internally. Pot furnaces use single-piece refractory pots or metal crucibles, and are operated periodically. Tank furnaces are assembled from blocks. The melt does not leak through the joints because of the low temperature maintained on the outside surface of the furnace.

Glass melters are heated by the combustion of fuels and by direct or indirect electrical heating. Electrically boosted fuel-fired melters and all-electric cold-top melters use direct electrical heating. Small pot furnaces and optical glass melters sometimes are indirectly heated with resistance elements.

Pot furnaces are the oldest type of glass melter (Ref 15). They are still used for some intermittent or low-volume production, because each pot can contain a different composition. Up to 16 refractory pots are arranged in a fuel-fired furnace. Fireclay pots hold up to 0.4 m^3 (11 ft^3) of glass and last about eight weeks before having to be replaced. Platinum crucibles eliminate contamination by refractories and allow higher temperatures.

Fuel-fired tank furnaces are reverberatory furnaces in which flames heat a shallow bath of molten glass from above. Natural gas is the preferred fuel. Oil is used when gas is unavailable. Heat transfers by radiation directly from the flame and indirectly by reflection and reradiation from the hot refractory superstructure (crown and walls). The heat transfer rate and the energy efficiency of the furnace both increase as the temperature increases. For this reason, maximum air preheat and minimum excess air are most efficient. The temperature in the furnace is limited by crown deformation or collapse, which depends on the crown span and material.

A typical design for melting container glass is shown in Fig 4. Batch is fed continuously through the doghouse into the back end of the tank where free convection flows confine it until the batch is melted. Fining takes place as the melt moves forward before it flows through a throat to a separate chamber for cooling and homogenizing. Fuel and air enter through ports along the sides, or in the back end of smaller furnaces. In furnaces larger than 15 to 20 m^2 (150 to 200 ft^2) melting area, air is preheated by regenerators or recuperators that are beside or below the tank. Day tanks are periodic furnaces of the same design as small continuous tank furnaces.

All melters must limit forward mixing of material undergoing the batch melting, fining, and homogenizing steps of melting. Partially processed material must be separated from that which is more completely prepared. Most designs depend on convection flows for primary separation. Mechanical barriers provide more positive control of flow (Fig 3). The dimensions of the tank and distribution of fuel to the side ports control the convec-

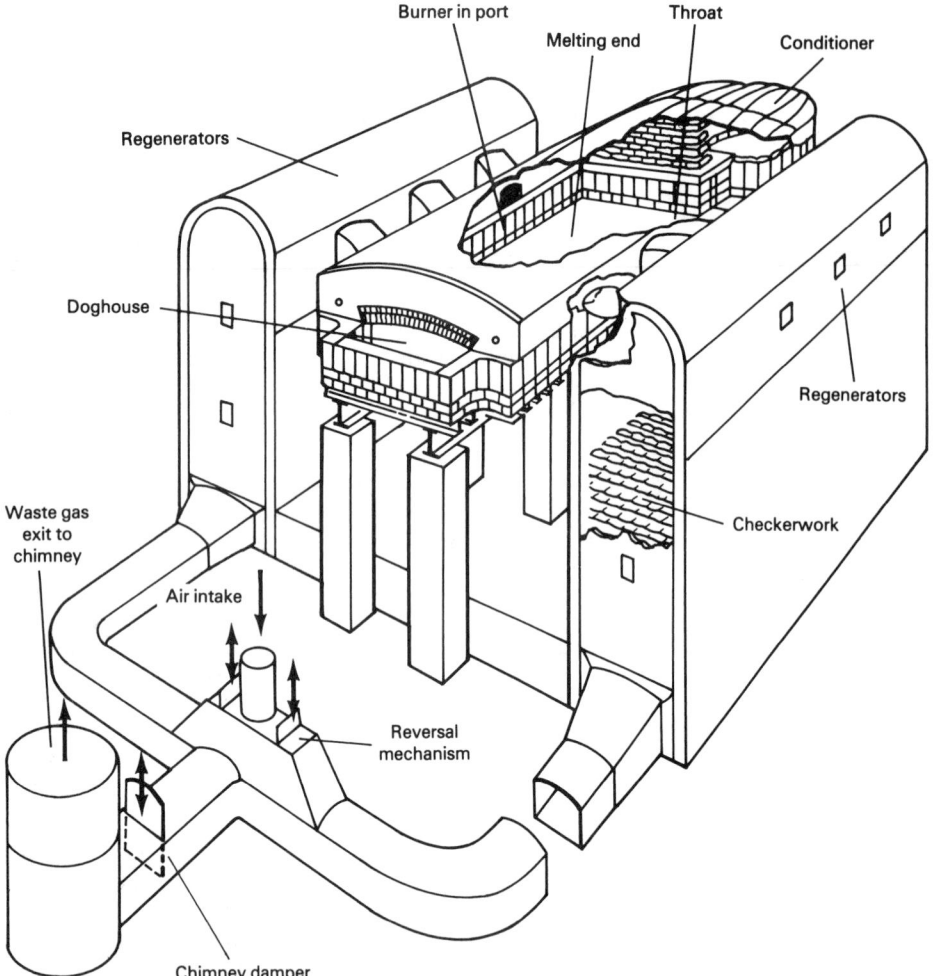

Fig 4 Tank furnace construction typical of side fuel-fired container glass melters. Source: Ref 4

tion flows. The melting chamber is separated from the conditioning chamber by an internally cooled bridge wall. Molten glass flows through the throat or under a skimmer. Material undergoing batch melting is sometimes prevented from entering the fining zone with rows of bubblers or electrodes, or with a second wall or a submerged barrier. The melter design is a compromise. The need to prevent forward mixing of material leads to complex multichamber melters. Long life and low maintenance require simplicity.

Electrically boosted fuel-fired tank furnaces get a part (typically 10 to 30%) of the total energy input from electrodes immersed in the melting end. Direct electrical heating has increased rapidly since the development in the early 1950s of practical molybdenum electrodes. Electric boosting provides better control than flames alone of the location and amount of local heating. Thus, boosting gives better control of convection currents. This often increases the furnace output (hence the term boosting) at relatively low cost. It may also permit the melt surface temperature and

gas flow rate to be lowered, reducing emissions.

Energy saving methods for fuel-fired furnaces include regenerators, recuperators, and use of oxygen for combustion.

Regenerators and recuperators transfer heat from the exhaust gases to the combustion air, thus reducing overall energy consumption by 50 to 70%. Regenerators employ two large refractory chambers in which packings of bricks or special shapes are stacked to provide open vertical passages for gas flow. The flame direction in the furnace is reversed every 20 to 30 min. The hot exhaust gases and the cold combustion air pass alternately through each chamber and the packings cycle over a range of 200 to 400 °C (400 to 800 °F). Recuperators exchange heat continuously through a ceramic or metal wall separating the exhaust and air streams. There are no reversals so that air preheat temperature is stable. Both regenerators and recuperators are severely attacked by dust and vapors in the exhaust gas. This causes furnace performance to deteriorate throughout the furnace

cycle and frequently limits the life of the furnace.

Direct-fired melters exhaust the combustion gases without heat recovery. They are more economical below 10 to 20 m² (100 to 200 ft²) melting area because of the high capital cost and heat losses from regenerators and recuperators. A typical design is a channel with many burners along each side.

Oxygen Used to Promote Combustion. Replacing combustion air with oxygen reduces the volume of exhaust gases by about 70%. This reduces the energy carried by the hot exhaust gas sufficiently to make direct firing economical in much larger furnaces. Elimination of nitrogen from the furnace atmosphere also prevents formation of NO_x in the flame. Oxygen is also injected below burners in fuel-air furnaces late in the life of the furnace to compensate for deterioration of the regenerators and to maintain the furnace pull rate.

All-electric cold-top melters are heated entirely by passing alternating current through the melt between immersed electrodes. The melt surface is completely covered by a floating blanket of batch that is typically 50 to 150 mm (2 to 6 in.) thick. Batch melting occurs in a thin layer at the bottom and gases generated by decomposition escape through the unmelted batch above. Volatilized glass components such as alkalies, boron, heavy metals, and halides condense in the cold upper part of the batch blanket and practically eliminate volatilization losses. Fining occurs by bubble rise immediately below the batch layer and by dissolution of gases in the cooler lower part of the tank.

The primary reason for the installation of all-electric melters is the reduction of volatilization losses that cause air pollution and worker hazards. Electric melters are also sometimes justified by their lower energy consumption because the thermal efficiency of large electric melters is two to three times higher than that of fuel-fired melters with regenerators. In most locations, however, this does not completely offset the considerably higher cost of units of electrical energy compared with energy sources such as natural gas or oil. Small electric melters have a significant energy cost advantage over direct-fired fuel-air melters.

The principle disadvantages of all-electric melters are a shorter furnace life because of refractory corrosion near electrodes and the limited flexibility for the pull rate and the glass composition. Electric melters will not operate at low pull rates because continuous cooling by incoming batch is required to maintain the cold batch layer. Without the insulating batch layer, heat losses are so high that melting temperature cannot be maintained. Electrode corrosion (see the section "Competing Processes" in this article) limits the compositions that can be melted. In addition, some glass compositions are impractical to melt electrically because of the instability produced by

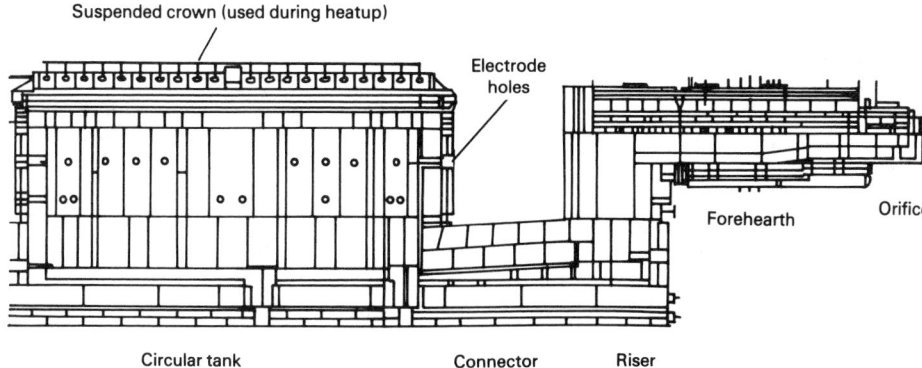

Fig 5 Vertical section view of a circular all-electric tank furnace. Source: Ref 16

interlocking parameters such as temperature, viscosity, electrical resistivity, and thermal conductivity.

All-electric melters are often deeper than fuel-fired melters, and may be circular or rectangular. A typical circular design for a large furnace is shown in Fig 5. Electrodes are inserted through the walls at one or several levels. Electrodes may also enter through the batch layer. The largest all-electric melters produce about 250 tonne/day (280 U.S. ton/day), considerably smaller than large regenerative melters. The maximum pull rate is directly proportional to the melting surface area for a given glass and is limited by heat transfer to the batch. In operation, the crown that covers the tank during heatup with burners is removed and $x-y$ conveyers or rotating chutes distribute batch over the entire surface. Power distribution between the batch and the upper and lower electrodes controls the convection currents in the tank. Stanek (Ref 12) gives a complete discussion of the principles and practices of electric melting.

Conditioning and Delivery Systems. After melting and the removal of defects, the melt must be cooled and made uniform in temperature before it can be formed. This conditioning begins when the melt passes the hottest spot in the tank and ends when it is delivered to the forming process. Most of it occurs in the forehearth, a channel in which the melt cools while energy is applied to prevent large temperature gradients. Typically, the energy required by the cooling portions of the furnace is about a third of the total energy input to the furnace. There are usually several forehearths on a single melter.

Gas-fired forehearths use burners to keep the crown temperature hot enough to reduce the rate of cooling of the open surface of the melt.

In electric forehearths, power is introduced through electrodes and heat is lost through the refractory walls. With no burners, refractory cover blocks can be below the glass level. This eliminates the free surface and thus prevents volatilization cord or stones.

Mechanical stirring is used to homogenize the melt chemically if the relatively high cap-

ital and maintenance costs of stirrers are justified. Stirring is usually done before the melt has cooled to forming temperature to give more opportunity for diffusion after stirring. Stirrer shapes are kept simple to improve their operating life. Refractory stirrers often take the form of a helix while a molybdenum rod with crossarms may be used in more corrosive melts. Platinum-clad stirrers are used where cord quality requirements are highest.

Refractories for melters are selected primarily to reduce defects in the glass; secondly, to increase furnace life, and finally to reduce initial cost. The corrosion resistance of all refractories results from the chemical inertness at high temperature of the oxides of silicon, aluminum, magnesium, and especially zirconium and chromium. They are produced by two processes: sintering and fusion casting. Sintered silica and various aluminosilicates are used in superstructures. Sintered zirconium silicate, alumina, chromia, chromia-alumina, and zirconia-aluminosilicates are used in glass contact.

Fusion-cast refractories are produced by melting the raw materials in direct arc electric melters. Simple shapes are cast in molds of sand, graphite, or water-cooled metal. The most widely used glass contact refractory is

fusion-cast AZS. It consists of alumina and zirconia crystals in an aluminosilicate glass matrix. It has excellent resistance to corrosion and thermal stress. Fusion-cast alumina, chromia-AZS, and zirconia are also used in critical applications.

Control Systems. Complete closed-loop automatic control of glass melters is not possible at present because the relationships between the many process inputs and outputs are not well enough understood. Individual inputs are frequently held at manually set levels by automatic controllers, maintaining outputs such as temperatures and glass level on target. Large computer systems are often installed to monitor and to record hundreds of sensor outputs. They may also apply automatic control to limited parts of the furnace. Fully automatic control is practiced in forehearths where process dynamics are better understood.

Furnaces for Specific Applications

The types of melters employed and their operating conditions vary considerably depending on the glass composition and on the glass quality and flow rate needed for forming. About 90% of United States glass production is melted in regenerative furnaces that are usually fired with natural gas and often electrically boosted. Typical size ranges are given in Table 1.

The production breakdown based on product lines of the roughly 500 glass melters in the United States yields the following data (Ref 17):

- 33% produce soda-lime glass containers
- 8% produce soda-lime flat glass, primarily by the float process
- 22% are used for fiberglass
- 37% make a wide variety of specialty glasses

Container glass melters are designed for low energy cost and for a low cord level under

Table 1 Typical melting rates of glass-melting furnaces

| | Melting Parameters | | | |
| | Area | | Rate | |
Type of furnace	m²	ft²	tonne/day · m²	ft²/ton/day(a)
Fuel-Fired Melters				
Container	55–185	600–2000	2–3	3–5
Float	230–510	2500–5500	1–2	5–10
Soda-lime tableware	14–46	150–500	1	10
Television panel	46–280	500–3000	1–1.5	6–10
Hard borosilicate	28–7	300–700	0.8–1.5	6–12
Day tanks	2–9	20–100	0.4–0.8	10–20
All-Electric Cold Top Melters				
Clear container	>9	>100	2.5	4
Colored container	>9	>100	1.6	6
Borosilicate	0.9–28	10–300	2–3	3–5
Radioactive waste	0.6–0.8	6–8	0.5–1.5	6–20
Pot furnaces (pot area)	0.4–0.8	4–8	0.3–0.8	10–30

(a) For U.S. melting rate in ft²/ton/day, divide 9.77 by the metric specific melting rate.

a range of pull rates. The most common type is a cross-fired regenerative tank furnace (Fig 4). Container tanks typically have melting areas of 60 to 180 m² (650 to 1900 ft²) and deliver 100 to 500 tonne/day (100 to 500 U.S. ton/day). AZS glass-contact refractories are used, the maximum melt temperatures are around 1600 °C (2900 °F), and melter life is typically 8 to 10 years.

Float glass melters must produce glass of very low cord and inclusion levels. Cross-fired regenerative tank furnaces are used. An open-surface section of reduced width (called a waist) separates the conditioning zone from the melting zone. The bridgewall with throat found in container tanks is not used because it would disrupt the smooth flow needed to keep cord oriented parallel to the sheet surface. Forehearths are not used for the same reason and cooling occurs in a prolongation of the melter. Float tanks are the largest glass melters, typically having five to seven ports on each side, melting areas of 200 to 500 m² (2100 to 5400 ft²), and delivering 300 to 700 tonne/day (300 to 800 U.S. ton/day). AZS glass-contact refractories are used in the melting zone with fused alumina used in the conditioning zones. Maximum melt temperatures are typically 1600 °C (2900 °F) and major tank repairs are made every 10 to 12 years.

Fiberglass melters are smaller gas-fired recuperative or all-electric melters. These are described in detail in the article "Fiberglass" in this Volume.

Specialty Glass Melters. Lead crystal glasses have been melted in smaller electrically boosted fuel-fired melters. However, the need to control lead emissions has made the all-electric melter much more common. AZS glass-contact refractories are usually used.

Hard borosilicate glasses, such as those used for chemical ware, are increasingly melted in all-electric furnaces or in very heavily boosted regenerative melters to reduce emissions. The difficulty of dissolving sand in these glasses requires melt temperatures >1600 °C (>2900 °F). Fusion-cast AZS and zirconia and sintered zirconium silicate are usually used in glass contact.

Aluminosilicate glass-ceramics are melted in regenerative fuel-fired melters. The difficulty of removing bubbles from the viscous melts requires melt temperatures approaching 1700 °C (3100 °F).

Optical glasses were formerly melted in pots but continuous melters have become common. These are small fuel-fired or electrically heated tank furnaces. Fining and conditioning are done in platinum tubes to eliminate cord and bubbles due to refractory contact. The extremely high capital cost of these melters is somewhat offset by their flexibility.

Alkali silicate and phosphate frits are melted to obtain a vitreous material in which defect quality is not critical. Open tanks, without a bridgewall or forehearths, are normally used.

Melters have been developed to incorporate high-level radioactive wastes in durable borosilicate glasses for long-term storage. These are all-electric cold-top melters in which the melt is contained in a refractory-lined metal shell. Forming is completed by intermittent casting into storage containers.

Furnace Parameters

Melting Rate. The maximum pull rate at which a melter can produce satisfactory glass is roughly proportional to the melting area. Melting area is defined as the surface area from the doghouse to the throat, or to the waist for open tanks, or the cold top area of an all-electric melter. Melters of various types and sizes can be compared by considering their specific melting rates which is the pull rate per unit melting area. (In the United States, melting rates are usually expressed in square feet per US ton per day). Typical rates given in Table 1 are for production of commercially acceptable product quality.

Inclusion level depends primarily on the application, because the design and operation of the melter can be adapted to produce whatever glass quality is required in the market. Typical inclusion levels required to meet product specifications economically are given in Table 2. Inclusions usually have a broad size distribution. The size range included in the count should be considered when comparing inclusion levels. Both solid and gaseous inclusions decrease rapidly in number per volume of glass as the size of the inclusion increases. Thus the difficulty of meeting an inclusion specification is usually more strongly influenced by the size limits than by the number limits.

Melting (glass preparation) cost is the cost of converting raw materials into hot glass

ready for forming. It varies widely with different compositions and applications but is typically 10 to 30% of the total manufacturing cost. Roughly half the melting cost may be attributed to energy, another fourth to the repair of the melter, and the balance to the direct costs of batch preparation and furnace operation.

Cost of losses due to defects in glass is high in products that require low defect levels:

- Float glass
- Textile fiberglass
- Many specialty glasses

In the specialty industry, glass losses (stones, knots, bubbles, and cord) are typically 3 to 30% of total production. In these glasses, the cost of melting-related defects may equal the sum of the energy, repair, and operations costs.

Like other chemical processes, both construction and operating costs per unit output drop rapidly as capacity increases but increase rapidly if the process operates below full capacity. Thus, fewer but larger melters reduce costs but are less flexible. A volatile demand for a product or the use of oversized production facilities lead to frequent production changes. The cost of changing the composition or size of a continuously formed glass product is high. The obvious costs are the labor and materials used and the value of production lost while making the change. Much larger costs often result from abnormally high losses before process stability is reestablished. This can take many days or even weeks because of the complexity of the melting process. Thus, frequent changes result in very inefficient production of glass products.

Environmental Impact. The escape of hazardous chemicals from the process is the primary issue in melting because it contrib-

Table 2 Inclusion levels in commercially available glass products

Application	Defects			
	Size		Quantity	
	mm	in.	No./dm³(a)	No./lb
Bubbles				
Best art lead crystal	>0.03	>0.001	0.3	0.04
Color TV panel	>0.25	>0.010	0.06	0.01
	>0.05	>0.002	30	5
Float automotive	>0.5	>0.020	0.06	0.01
Float architectural	>1.0	>0.040	0.02	0.003
	>0.1	>0.004	0.8	0.15
Soda-lime tableware	>0.03	>0.001	400	60
Glass-ceramic ovenware	>0.2	>0.008	2000	400
Containers	>0.05	>0.002	3000	500
	>0.8	>0.031	20	3
Wool fiberglass	>0.2	>0.008	10⁴	2000
Textile fiberglass	>0.007	>0.0003	0–900	0–160
Solid Inclusions				
Float architectural	...	...	0.001	0.0002
Containers	>0.03	>0.001	0.06	0.01
Glass-ceramic ovenware	>0.5	>0.020	1	0.2

(a) dm, decimeter

utes to the pollution of the environment and because it affects the health and safety of the workers at the plant. Environmental concerns strongly influence not only melter designs and operating conditions but also factory location and capacity. Worker health issues impose limits on the glass compositions and raw materials that can be used.

Most materials escape as dusts or gases. The principal sources are:

- Batch handling
- Volatilization from batch and melt surfaces
- Products of combustion of fuel

The primary control strategy is to reduce losses at the source, with various devices serving to capture unavoidable losses. Dusting losses are reduced by wetting batches, by avoiding fine materials, and by lowering gas velocities near the batch. Volatilization losses decrease with lower surface temperatures and gas velocities. Many techniques are used to reduce NO_x formation in flames. Sulfur gases are controlled by limiting the amount in the batch and the fuel.

Capturing emissions outside the tank requires costly equipment. It usually consumes energy and produces hazardous powders or contaminated water. Scrubbers, electrostatic precipitators, and cloth filters remove vapors and particulates from exhaust gases. Disposal of captured dust is expensive; often it is more economical to recycle it into the batch.

Emerging Processes

Environmental concerns have been the driving force behind the most sweeping changes in glass-melting processes and the determining factor in choosing materials during the past several decades. All-electric cold-top melters and fuel/oxygen combustion are becoming more widely used to reduce emissions of NO_x and volatile glass components.

Increased process flexibility is required to provide customers with more choices and faster response while at the same time reducing inventories. Smaller melters are more flexible because the glass composition can be changed more quickly. Compartmentalized melters that depend less on free convection to prevent foward mixing are also more flexible. Melters can also be made smaller if the melting, fining, and homogenizing processes are accelerated. An attempt to accomplish this objective with a mechanically stirred melter and centrifugal finer (Ref 18) has not yet been commercially applied but it shows the direction that the glass industry is headed toward in the future.

REFERENCES

1. P.J. Doyle, *Glassmaking Today*, Portcullis Press, Redhill, England, 1979
2. L.D. Pye, H.J. Stevens, and W.C. LaCourse, *Introduction to Glass Science*, Plenum Press, 1972
3. F.V. Tooley, *The Handbook of Glass Manufacture*, 3rd ed., Ashlee Publishing, 1984
4. W. Trier, *Glass Furnaces (Design, Construction and Operation)*, K.L. Loewenstein, Trans., Society of Glass Technology, England, 1987
5. D.C. Boyd and J.F. MacDowell, Ed., *Commercial Glasses*, Vol 18, *Advances in Ceramics*, American Ceramic Society, 1986
6. G.W. McLellan and E.B. Shand, *Glass Engineering Handbook*, 3rd ed., McGraw-Hill, 1984
7. D.R. Uhlmann and N.J. Kreidl, Ed., *Glass: Science and Technology*, Vol 2, *Processing I*, Academic Press, 1984
8. M. Cable, Principles of Glass Melting, *Glass: Science and Technology*, Vol 2, *Processing I*, Academic Press, 1984, p 1–44
9. J.J. Hammel, Some Aspects of Tank Melting, *Commercial Glasses*, Vol 18, *Advances in Ceramics*, American Ceramic Society, 1986, p 177–185
10. W.C. Bauer and F.V. Tooley, Advanced Methods of Batch Preparation, *The Handbook of Glass Manufacture*, 3rd ed., Ashlee Publishing, 1984, p 1139–1157
11. E.L. Swarts, Gases in Glass, 46th Conference on Glass Problems, Urbana, IL, 12–13 Nov 1985, *Ceram. Eng. Sci. Proc.*, Vol 7 (No. 3–4), 1986, p 390–403
12. J. Stanek, *Electric Melting of Glass*, Elsevier Scientific, Amsterdam, 1977
13. E.R. Begley, *Guide to Refractory and Glass Reactions*, Cahners, 1970
14. C. Clark-Monks and J.M. Parker, *Stones and Cord in Glass*, Society of Glass Technology, Sheffield, 1980
15. R.W. Douglas and S. Frank, *A History of Glassmaking*, G.T. Foulis & Co., Ltd., Henley-on-Thames, Oxfordshire, 1972
16. J. Woltz, Operating Experience with Electrically Heated Tanks for Production of C-Glass, *Glastech. Ber.*, Vol 55 (No. 5), 1982, p 88–95
17. J.T. Brown, 100% Oxygen-Fuel Combustion for Glass Furnaces, 51st Conference on Glass Problems, Columbus, OH, 1 Nov 1990
18. R.S. Richards, Rapid Glass Melting and Refining System, *Advances in the Fusion of Glass*, American Ceramic Society, 1988, p 50.1–50.11

Forming

Harrie J. Stevens, New York State College of Ceramics, Alfred University

WITH THE FORMATION of the first glass article, man began the race to complete its shape before the viscosity of the glass rose to a value that prevented further flow. This race still occurs in most forming operations, although times have been reduced as in most competitive sports.

The challenge of the glass-forming engineer has been to design processes that capitalize on the unique glass properties while minimizing adverse effects. Most current glass-forming operations have evolved in rather small steps. When tracing a process back to its origin, it is generally found that the first stage of evolution was to mimic a hand operation. In a few and very successful cases, such as the ribbon machine and the float process, the forming processes represented radical departures from in-place procedures. These two inventions completely changed the light bulb and flat glass industries and permitted their substantial growth.

This article focuses on common glass-forming operations, their evolution, and the glass property relationships that influenced their design and, in many cases, limit their capability. The production techniques covered include hand forming, hot glass delivery, container manufacture, pressed ware, sheet glass, tubing and cane production, casting, extrusion, and glass marbles/spheres.

Viscosity. The glass-forming process generally begins with glasses that have viscosities ranging from 100 to 1000 Pa · s. In a soda-lime-silica glass, this typically occurs between 1200 and 1000 °C (2190 and 1830 °F) (Fig 1). If this glass is then sheared or formed against a metal mold, the surface temperature immediately drops to around 550 °C (1020 °F) and a subsequent viscosity of 10^{12} Pa · s. Cooling rate measurements have indicated that this occurs in around 20 ms. At this point, the glass surface essentially won't flow, and the interior that is still fluid has a temperature near 1000 °C (1830 °F).

The forming operation then changes from dealing with a somewhat uniform sample (in terms of temperature and viscosity) to a material that has a wide range of flow properties. Subsequent operations will cause selective heat losses as the object is formed and a redistribution of the internal heat to permit the selective flow and/or healing of

the glass. In some cases, such as hand forming, additional heat may be applied to permit longer forming times.

The automation of glass forming produced a more mechanically stable process. However, because of the viscosity changes involved, it was also necessary to produce a process that was thermally stable. Even today, many process engineers end up addressing thermal problems as often as mechanical problems.

Glass Transition. As illustrated in Fig 2, glasses also change their specific volume as a function of cooling rate. This occurs when the viscosity becomes sufficiently high to prevent the glass from rearranging into its equilibrium structure. The consequence is that a higher-temperature structure is frozen in the glass. Because this rearrangement is a dynamic process, the faster the cooling rate, the less time for rearrangement, and the faster the structure becomes frozen in the glass.

Because different regions of the glass are cooled at different rates during the forming operation, stresses develop that generally need to be relieved after the part is formed. This is done by annealing, during which the glass must flow sufficiently to permit structural rearrangement without allowing the shape of the object to distort. Normal annealing procedures have been developed to accomplish this on standard items, but recent product developments that require much closer dimensional tolerances have brought glass transition range behavior to the forefront of processing concerns.

Hand Operations

The basic elements of glass forming were developed nearly 2000 years ago, with workers blowing, pressing, shaping, and drawing glass articles by hand. Many current processes are mechanical adaptations of initial hand operations.

Handmade Containers. The basic tool of the glass handworker is the blowpipe, generally a hollow, high-temperature steel rod about 1.5 m (5 ft) long, which is tapered at the mouth, or blowing, end and enlarged to a conical shape at the gathering end. The glass blower places the pipe above the molten glass in the furnace. When the pipe is heated, it is

dipped into the molten glass and rotated as a sufficient amount of glass gathers on its end. While the rotation continues, the pipe is withdrawn and the surface of the glass is permitted to cool, thus forming a skin of higher-viscosity glass.

With experience, the worker learns when to puff a small amount of air into the pipe and merely slip his finger over the end of the pipe, which permits the introduced air to heat and expand, thereby creating a bubble inside the glass. During most of the forming process, the glass is rotated to maintain its symmetry. Frequently, the glass is reheated to permit further blowing, additional glass is added for handles or bases, and the piece is shaped with various tools.

The blocks and paddles used to shape the glass are frequently made of fruit wood that has been charred and soaked in water. This treatment is probably due to the fact that fruit woods are hard and have a dense cellulose structure. Consequently, upon charring, they form a dense but somewhat porous carbon working surface. The absorbed water then produces a layer of steam that the glass is actually worked against. Not only is this steam layer nonabrasive to the glass, but it probably lowers the surface viscosity of the glass to enable the formation of a smoother surface.

In some cases, it is necessary to transfer the article to a solid rod, called a punty rod, which enables the worker to cut and trim the lip of a glass or vase. Like the blow pipe, the punty rod is first heated and a small amount of glass is gathered on its end. After shaping and possibly reheating, the end of the punty rod is attached to the previously blown object. The worker then scores the blown object at the end of the blow pipe and, with the addition of a little water and a tap on the rod, the piece is transferred to the punty rod.

At this point the piece may be reheated and further shaped or cut. Cutting can be accomplished with regular metal shears. It is interesting to ponder how molten glass can be cut. If glass is considered as a perfect viscous material, then application of the shearing stress of the shears should just stretch it, not cut it, because the viscosity should not change with shear stress. Because it does not appear that the glass is fractured, the ability to cut it implies that at high shear stress, glass is not

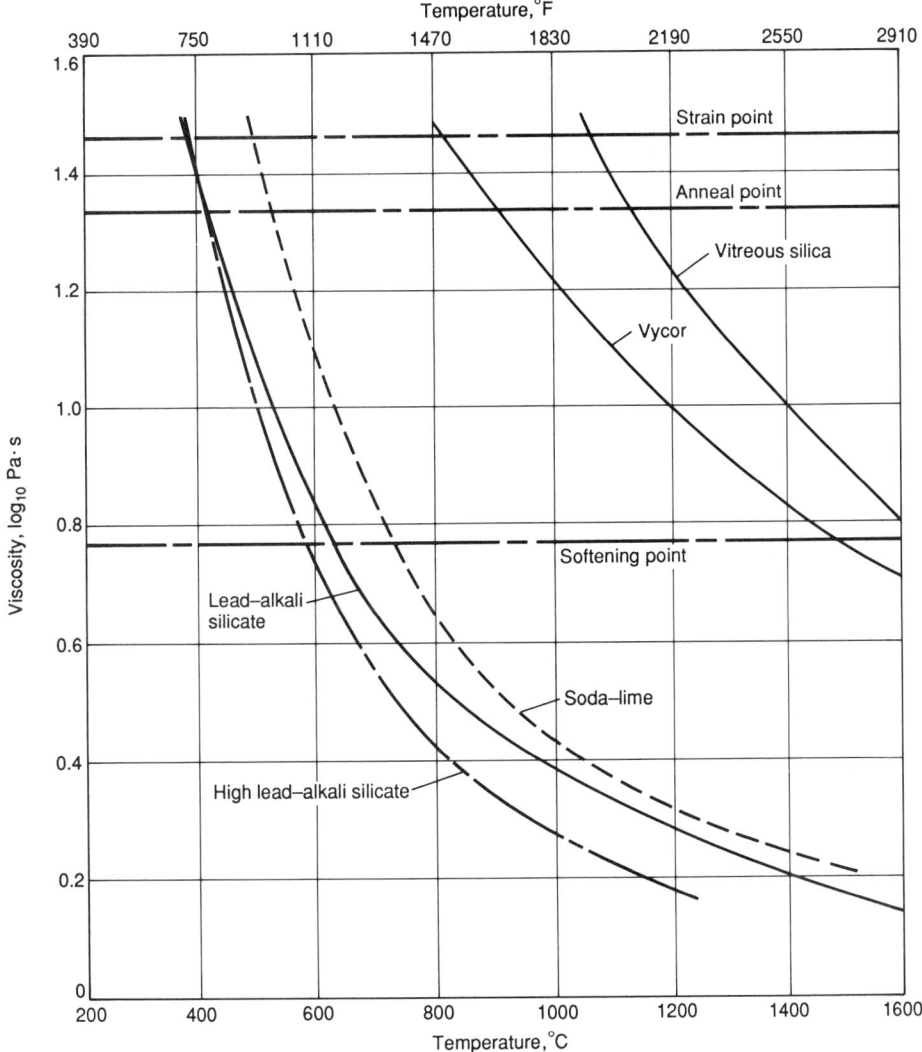

Fig 1 Viscosity of commercial glasses

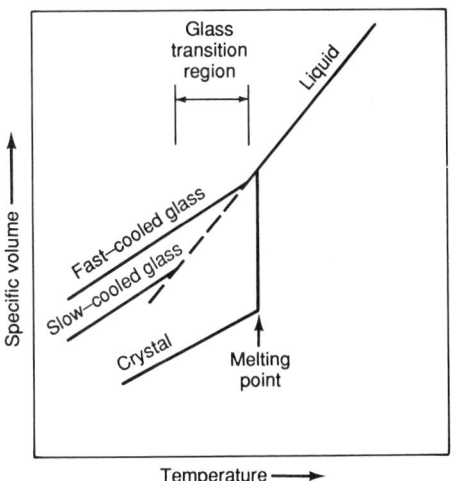

Fig 2 Glass transition behavior

a perfect viscous material, since its viscosity changes nonlinearly with extreme shearing and permits the cutting action. Although workers have been cutting molten glasses for hundreds of years, only recently have glass scientists begun to investigate this nonlinear viscosity.

Cane and Tubing. The process for drawing cane (rod) and tubing begins in the same manner as blown containers, except that the gathering rod does not need to be hollow to produce cane. After the glass has been gathered on the end of the rod, it is rolled against a flat stone or plate (marvered) into the shape of a long, truncated cone. A second, smaller gather is also made on a punty rod, marvered to a smaller truncated cone, and attached to the end of the first gather.

Cane drawing is accomplished with one person standing still and slowly rotating the first gather while a second worker pulls and rotates the second attached punty rod while walking away from the first. Sometimes, for greater control over the diameter, a third worker cools the drawn cane with a hand fan as it is formed. The freshly drawn rod is then laid down on wooden slats along the drawing floor.

The drawing differs from cane in that a bubble of air is introduced into the first gather and maintained by blowing as the tubing is drawn. This process has a 25% yield and was used until 1917 to draw all glass tubing.

Sheet made by the crown glass process is begun by gathering the glass with a blow pipe. A large, hollow, cone-shaped piece of glass is then blown and transferred to a punty rod after being cracked off the blow pipe. This open-ended cone is then reheated and spun in the furnace (glory hole) until it suddenly flattens out (flashes) by the centrifugal force. Upon annealing and cooling on a flat refractory surface, glass panes are cut from all but the center of the circular sheet. This process has been used since Roman times to produce small, fire-polished panes. Panes formed in this manner are generally thicker on the edge

that is closer to the center. If this pane was then installed with the thicker edge down, it would explain why some ancient cathedral glass panes have been found to be thicker at the bottom. The center-most portion of the spun sheet, the bullseye, was often used in decorative leaded glass windows and is highly prized by collectors today.

Other workers found that if they blew a tall cylinder instead of a sphere, they could cut the cylinder in half (lengthwise) and flatten each section by reheating. Glass panes formed in this manner (so-called broad glass) were larger than panes produced by the crown method, but lacked the fired, polished surface because of the reheating and flattening process.

Automatic Hot Glass Delivery

Robot-like feeders have been developed to mimic the hand gathering operation. These computer-controlled gathering devices reach into a forehearth, gather the molten glass on the end of a rod, withdraw it from the furnace while rotating, flow it into the forming mold, and shear it automatically. Such machines can feed multistationed presses, blow machines, glass spinners, and stem presses. Because such devices do not see into the tank, the gathered quality is as good as the glass at the specific gathering location.

Gob feeders are devices that precisely deliver preshaped glass gobs of a specific weight to individual molds for pressing or blowing operations. The initial feeders were mechanical devices that were run by cam drives. In the early 1980s, ballscrew servo-controlled machines were developed and, more recently, screw or auger, feeders have been introduced.

The feeder, which is attached to the end of the forehearth, traditionally consists of four components: plunger, rotating tube, orifice

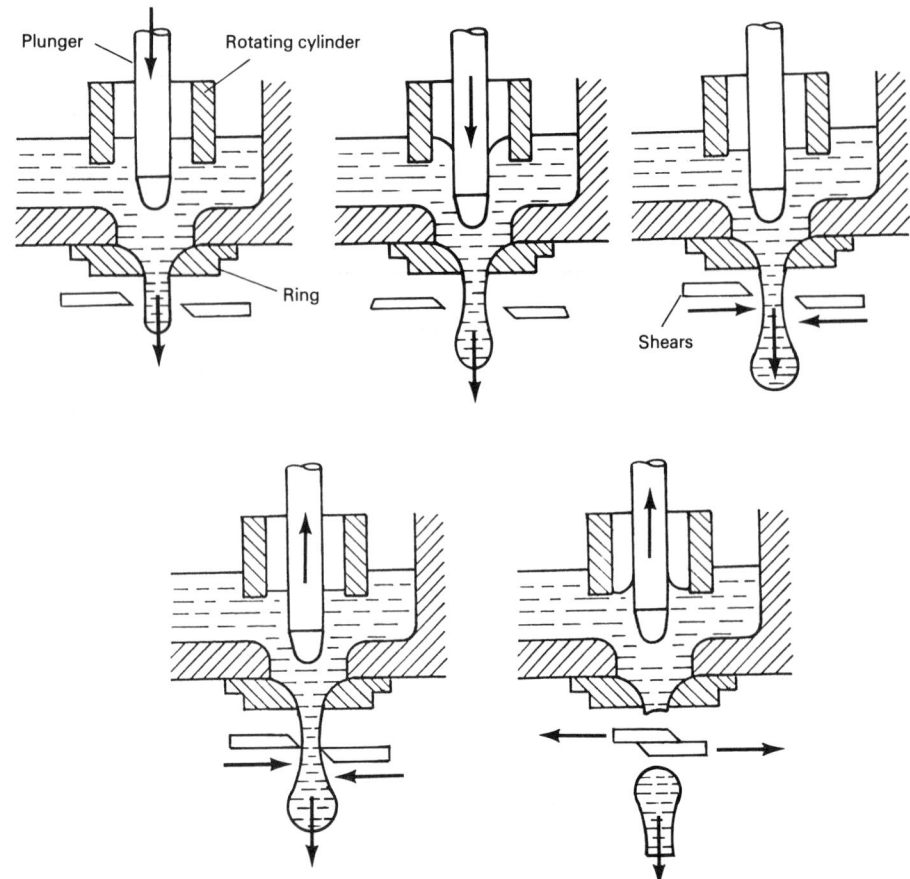

Fig 3 Gob feeder process

plate, and shears (Fig 3). The orifice plate limits the glass flow by virtue of its diameter. Further flow control is achieved by adjusting a screw mechanism that controls the height of the tube. Some feeders rotate the tube to increase the thermal and physical homogeneity of the glass.

The cam-driven plunger cycles up and down on the order of a few inches. It first extrudes the glass out through the orifice plate. Then, on the upstroke, it draws the glass back into the orifice which is important for two reasons. First, it lifts the glass off the shears so that the hot glass-shear contact is minimized. Second, it permits the freshly sheared glass, which has been chilled by the shearing action, to be lifted up into the orifice and reheated, thus removing the majority, if not all, of the shear mark. This drawing back of the glass is only possible if the stroke of the plunger is timed properly and the viscosity of the glass is suitable.

Ballscrew-driven feeders were developed to avoid cam tracking problems when more plunger force was required to feed the glass. Because the cam follower was held against the cam just by the force of gravity, the follower would not track properly if the plunger was forced down too hard. To overcome this, and to eliminate some mechanical variability

in cam-driven systems, a ballscrew servo-driven system was developed.

The screw feeder is a relatively new design, where the plunger is replaced by a screw, or auger. The glass can first be extruded through the orifice by rotating the screw down. Then, by reversing the rotation, the glass is drawn back up. This feeder allows the delivery of lower-viscosity glass in a controlled manner.

Rotating Shears. Bridging the gap between gob feeders and continuous feeding is the so-called flying, or rotating, shearing mechanism. This feeder starts with a continuous flowing stream of glass that is severed by a pair of counter-rotating shears. Gob length and size can be controlled by adjusting the flow of the glass and the rotating speed of the shears. Although this method delivers glass at a rapid speed, a disadvantage is the substantial shear marks that result at the top and bottom of the gob. To overcome this, the shear marks are frequently placed outside the useful processing area and are removed with the cutoff after the piece is manufactured.

Continuous Glass Flow. Certain operations, such as those to make drawn, rolled, or floated sheet glass, tubing, cane, incandescent bulb envelopes, and fiberglass, require a continuous flow of molten glass to the forming operation. In these cases, flow is

controlled by the depth of the glass, the size of the orifice (if any), and the height of either a refractory tube or a rectangular dam under which the molten glass flows.

Automatic Container Manufacturing

Blow and Blow. Narrow-necked glass containers are generally made on an individual section (IS) machine, that is, each bottle is completely manufactured at a specific location, or section, of the machine. A series of these individually controlled sections (6, 8, 12, or 16) are assembled side by side to create the forming machine.

This method of manufacturing glass articles can be contrasted to earlier container machines that used one or two rotating tables to move the container around. A specific forming operation was performed at each position, or station, of the machine, corresponding to each step of the handworker at a specific position. The IS machine, introduced in the 1920s, greatly increased the forming rate of glass containers. The first machines were controlled pneumatically via a timing drum at the rear of each section, whereas modern machines are electronically controlled.

Although several variations have been developed, the basic concept of the original IS machine is still being used (Fig 4). The gob is delivered into the first split mold, and is formed in an inverted position into the initial shape, called the parison. The forming operation actually starts with the shaping and cutting of the gob as previously described. The gob is then dropped into the parison mold via a chute. A blow head seals the top of the mold and provides a settle blow, or puff of air, and then pushes the gob down into the bottom of the mold, where, in this inverted position, the finish (top) of the container is formed. Then, from the bottom, a counter blow is introduced into the center of the gob, which creates the inside of the container.

The parison mold then opens and an arm swings the parison over 180° and places it into a second split (finish) mold for the completion of the container. Because the exterior of the parison was chilled during its contact with the first mold, a brief pause is introduced into the forming process to allow the glass surface to reheat. Upon reheating, a second puff of air is introduced to blow the final container shape. Finally, a pick-up mechanism lifts the container out of the mold and sets it on a conveyer. Although the speed with which a container can be made depends on several factors, it generally depends on the weight of the article being produced. A 0.33 L (0.09 gal), 200 g (7 oz) bottle generally can be made in 5 to 6 s.

Variations on the basic process continue to be introduced to increase container quantity and quality. The first modification, in the mid-1960s, had feeders with two orifices and plungers and delivered two gobs simulta-

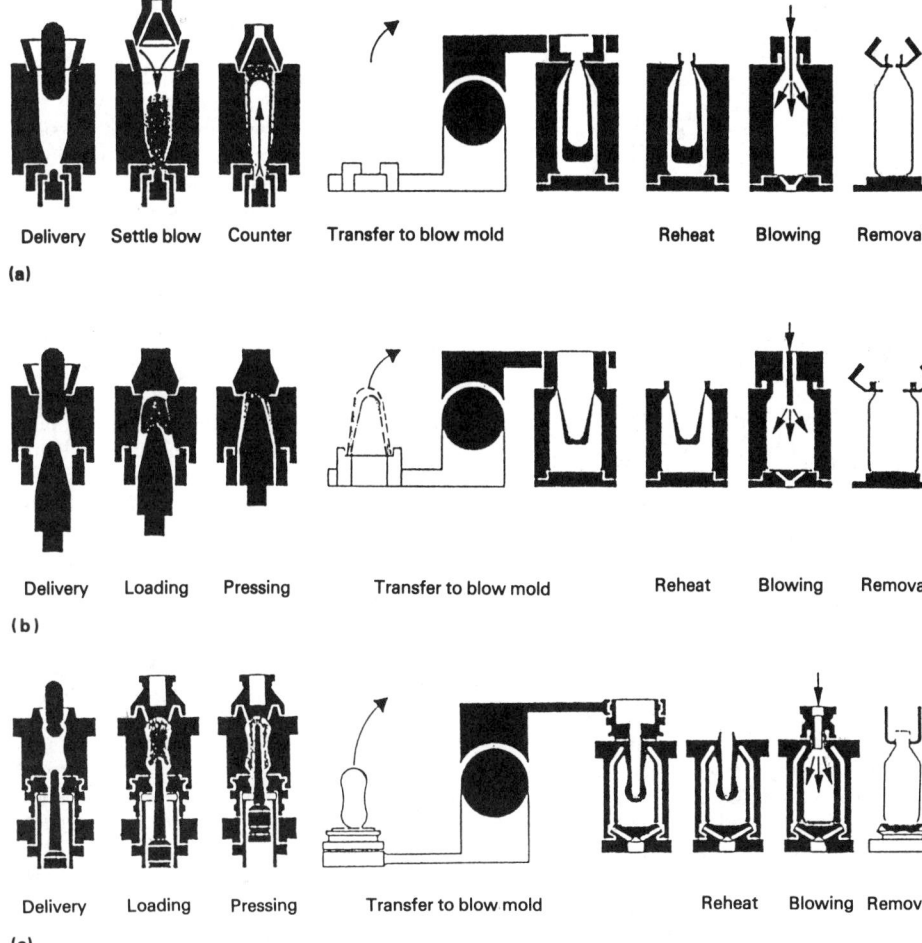

Fig 4 Automatic container manufacturing variations. (a) Blow and blow. (b) Wide-mouth press and blow. (c) Narrow-necked press and blow

neously to a pair of molds in the same section, thus almost doubling the capacity of the machine. This concept has been expanded to triple and even quadruple gob machines, the latter of which are generally limited to small-sized containers, but are capable of producing over 600 bottles/min.

It was also found that increased production rates could be achieved, especially in the newer lightweight containers, by using two finish, or final, molds with each parison. The 1:2 process permits a longer parison reheat, which allows the production of lighter weight containers at higher production speeds. This method has received limited acceptance because the industry has a tremendous capital investment in existing machines and a marginal improvement cannot be justified.

A press and blow operation can be used to produce large-mouth containers and some of the newer, light-weight beverage containers. This process modification, which yields greater control over the uniformity of the container wall, uses a plunger, rather than an air blow, to press the parison. After the parison is formed, it is flipped over, and the final shape is then produced using the standard blow process.

It has been found that the glass contact with the plunger frequently results in the creation of defects on the glass surface, which decreases the strength of the container. Work is underway to introduce a porous plunger that will overcome this problem by pressing the glass with an air or steam layer situated between the molten glass and the plunger.

Paste-mold processing is generally used on articles such as glass tumblers and light bulb envelopes to avoid the seam mark that is introduced by the split molds of IS machines.

Paste molds are metal molds, usually cast iron, whose interior surface has been cleaned, heated to 60 °C (140 °F), coated with a thin coat of varnish, and then dusted with either a cork or wood powder. This "paste" is then fired to around 300 °C (570 °F) to carbonize the coating. Small holes are also drilled through the mold surface. During the process, these molds are dipped into or sprayed with water before they contact the molten glass. The heat from the glass causes this wet, porous carbon coating to produce a layer of steam, against which the glass is formed. The drilled holes permit the release of the heated steam, so that there is not an excessive pres-

sure buildup that deforms the piece. The various forming machines rotate either the mold or the glass during forming, thus eliminating seam marks and producing a high-luster surface.

The Hartford H-28 is a continuous rotary machine with either 12 or 18 sections. After gob feeding, the object is formed with a press and blow operation as it rotates around the table. The 12-section machine is capable of producing 60 to 80 tumblers/min and the 18-section machine can produce 45 to 115 tumblers/min.

Olivotto has developed a new electronic blowing machine in which each section has its own electronic timing controller. Consequently, different articles from the same size gob can be made side by side. The gob is delivered and pressed into a biscuit that is then placed at the top of the split paste mold. The biscuit is rotated while air blows the glass into its final shape in the stationary paste mold below. By varying the size of both the table and the working heads, machines with 9 to 30 heads are assembled that can produce 6 to 90 pieces/min with a weight of 30 to 700 g (1 to 25 oz).

After paste-molded articles are formed, the top, or neck, portion must be removed from the machine. The final rim is either fire-polished, which leaves a slight bead, or ground, which is more expensive and is generally reserved for finer articles where a bead is not desired. The removal of this excess material (the moil) means that only a portion of the glass melt is found in the final product. The waste can range from 45 to 140% of the weight of the finished product. Many regard this poor glass utilization as the chief drawback of this process.

The ribbon machine, the most highly mechanized paste-mold machine, is used to produce nearly all light bulb envelopes, as well as most glass ornaments used as holiday decorations. The ribbon machine consists of three long, moving chains that merge to produce each light bulb envelope (Fig 5). The glass is delivered in a stream to two rollers, one that is flat and one that has recesses or pockets cut into it. As the molten glass passes between them, it is rolled into a ribbon, and biscuits are pressed into the surface. This ribbon is then laid down on top of a moving chain such that the glass biscuits fall directly over holes in the chain.

As the chain begins to move the length of the machine, the glass sags down through its holes. Next, a second chain drive with blow heads appears above the first, each blow head matches up with each biscuit, and air pressure blows the envelope. From below, a third chain drive with presoaked split paste molds comes up, closes around each glass article, and begins to rotate as it travels along with the glass ribbon and the blow heads. This combination of glass, blow head, and rotating mold continues until the article is formed. Then, the mold opens and drops below to be

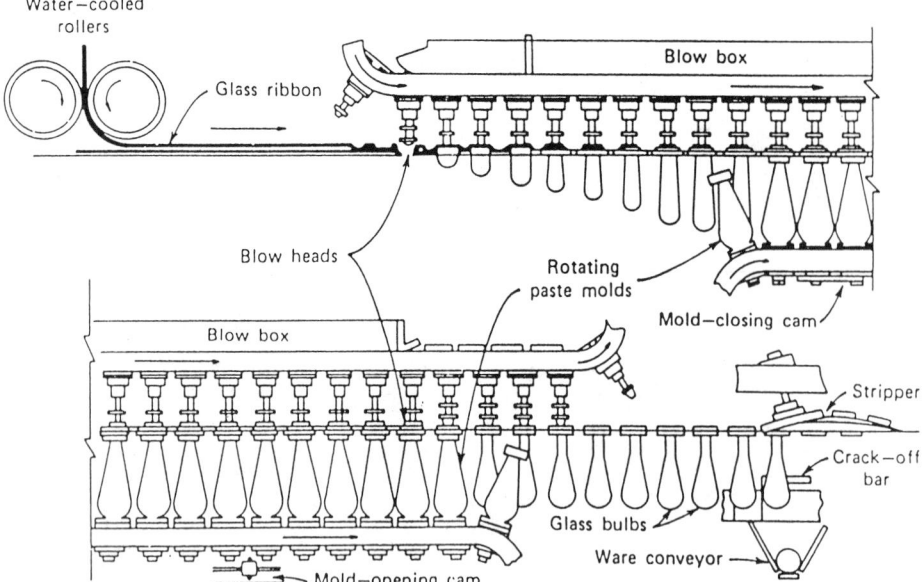

Fig 5 Ribbon machine. Courtesy of Corning Inc.

cooled, soaked, and returned to meet another glass article. The blow heads lift off and travel above the glass ribbon to the front of the machine.

Subsequently, the glass ribbon and its formed articles travel to the end of the machine, where a rotating hammer breaks the envelopes off onto a conveyer. The remaining glass ribbon is generally returned to the furnace as cullet, but in some cases, it is used to make other glass products. Finally, the ribbon-transporting chain turns under the machine and returns to pick up some new glass. Although the output of this machine depends on the size of the bulb being blown, production speeds commonly range from 1200 to 2000 envelopes/min.

The Turret chain paste-mold machine is also used to mass manufacture items such as coffee pots, bulb envelopes, and tumblers. It is somewhat of a combination of the ribbon and Olivotto machines in that a traveling chain carries the glass around an oval track, but pressed glass biscuits are used instead of a glass ribbon. Blow heads are positioned above and paste molds below as the article moves in a circular path at one end of the machine. Because a biscuit is used instead of a ribbon, there is less glass waste than with the ribbon machine, but the production speeds are an order of magnitude less.

Pressed Ware

Although conceptually simple, pressing molten glass economically into an exact shape with reasonable dimensional tolerances is an extremely difficult task. Not only does one have to distribute the glass properly throughout the mold, but it has to be done in a few seconds, while encountering temperature variations of 600 °C (1080 °F) and five or six orders of magnitude in viscosity.

Free Pressing. If the object to be made does not have a fixed edge or outer dimension, it can be pressed between the mold and plunger and the edge allowed to vary somewhat. The shape is controlled by the loading of the mold and glass distribution. Examples of such free-pressed pieces are custard cups and eyeglass blanks, which can tolerate slight variations in glass flow.

Ring Pressing. Most pressed ware requires that a spring-loaded ring be added to the molding operation. This ring forms the top edge of the piece and fits over the bottom mold, forming a collar that the plunger presses through. During the process, some glass flows up to meet the ring and presses against it while the remainder of the glass is being pressed into the mold.

Although the operation can be adequately performed with one station or mold, a circular rotating table generally is set up with a maximum of 24 molds or sections per table. Articles that range from 0.005 to 16 kg (0.01 to 35 lb) are formed with pressing pressures of 515 to 690 kPa (75 to 100 psi). As the table is rotated, the glass gob (at 200 to 400 Pa · s) is fed into the base mold at the first position. The table is then rotated so that the loaded mold is at its second position, where the plunger and ring press the glass. The pressed article is then rotated to the remaining positions, where it is cooled and removed (10^5 to 10^8 Pa · s).

With accurate loading, automatically pressed ware can be made with variations of 0.5% in bottom thickness and ±0.05% in diameter (Ref 1). The removal of the ware usually occurs several rotation positions from the pressing position. The remaining rotation positions provide the necessary time to cool the molds before they receive the next gob of glass. This cooling stage is important to maintaining correct thermal balance and preventing overheating of the molds. With traditional cast iron or high-temperature steel molds, surface temperature must be about 450 °C (840 °F) when the glass is dropped into the mold, and less than 500 °C (930 °F) when the glass is removed. If mold temperature is below this level, the glass is chilled too rapidly and small cracks, or checks, are created. If the mold exceeds 500 °C (930 °F), the glass tends to stick to its surface.

Tooling represents approximately half the cost of operating a glass press. Consequently, extensive work has focused on developing designs and utilizing materials that efficiently cool the glass without imparting defects such as chill marks and checks. The material must have sufficient thermal conductivity to permit the removal of heat from the glass, and should resist oxidation, corrosion, deformation, and grain growth. In addition, the material should be machinable, accept and hold a high degree of polish, and be economical.

Depending on the length of the production run, three materials that meet these requirements are generally used. Fine-grained cast iron molds, which are the least expensive, are used for short runs. Stainless steel is commonly used for runs of intermediate length, and high-temperature metals, such as Inconel, are used for long runs. Although these materials are very costly to machine and finish, the significant increase in their useful life justifies the increased finishing costs.

Tooling designers frequently build cooling fins into the interior of the mold to increase the heat flow in selected areas. With the addition of cooling air or water, heat can be selectively removed from specific mold regions, thus tailoring the mold temperatures to enhance glass flow and/or cooling. Sophisticated computer design programs are used to achieve the correct thermal balance before the mold is cast.

Sheet Glass

In an effort to mechanize the hand cylinder method described in the section "Hand Operations" in this article, the Lubber's process was developed by The American Window Glass Company in 1903 (Ref 2). A blow pipe with a large flanged metal disk, or "bait," was lowered into a pool of molten glass. As the pipe and bait were raised, air was introduced into the blow pipe and a cylinder of glass was blown. When the final length was reached, the cylinder was broken off at the bottom, turned sideways, split in half, and flattened. Diameters of 0.6 m (2 ft) and lengths of just over 9 m (30 ft) were eventually possible. Although this method produced a superior finish and replaced the hand methods, it was, in turn, replaced by mechanized sheet-drawing operations.

Sheet Drawing. One method of drawing a sheet from a glass melt is known as the Fourcault process, which was put into production in 1913. A slotted refractory bar, known as a debiteuse, is forced into the glass melt, and hydrostatic pressure is used to force the glass out through the slot while water-cooled tubes in proximity cooled the newly formed sheet. The sheet is then grabbed from above and drawn up until it was cool enough to be cut.

In another sheet-forming method, the Colburn process, glass is pulled directly from the melt, gripped at the outside edges by knurled rollers to prevent it from contracting, reheated slightly after rising a few feet, and turned 90° over rollers. This process eliminates the marks that were left on the glass from the slot in the debiteuse, but creates additional marks from the rollers as the sheet is turned.

In a method that combines the Fourcault and Colburn processes, the glass is drawn directly from the melt and continues vertically. Just above the glass surface, the edges of the sheet are gripped by water-cooled, knurled rollers to maintain sheet dimensions. Submerging a refractory "draw bar" just below the surface of the glass assists in dimensional control. As the glass flows over this bar, it is thermally conditioned and the temperature/viscosity variations are minimized.

By the mid-1900s all three processes were responsible for the world's production of sheet glass, with 72% being made by the Fourcault process, 20% by the Colburn process, and 8% by the combined process.

Plate Glass. Because each of the above methods produce glass with surface defects, applications such as mirrors or storefront windows require surface grinding and polishing. Rolled sheet can be used, because the surface is ground away. In the continuous grinding and polishing process, each pair of adjacent heads rotate in the opposite direction to equalize the torque on the continuous sheet. This process is costly and produces much waste.

Float Glass. In a process commercialized in 1959, molten glass is floated on top of a bath of molten tin. The bottom glass surface is supported only by molten tin, whereas the top surface is exposed only to the atmosphere above the bath (Fig 6). Gravity allows the glass to flow out, while its surface tension holds it back to an equilibrium thickness of 7.1 mm (0.281 in.). Sheet glass with the surface quality of plate glass can be made using this process. Thicknesses from 1.5 to 25 mm (0.06 to 1.0 in.) can be made by either stretching the glass with knurled rollers or compressing the glass with graphite paddles.

A single float glass line may reach 215 m (700 ft) in total length and produce 23 Mg (50,000 lb) of finished glass per hour. The tin bath can be 45 m (150 ft) long and of sufficient width to produce a glass sheet 4 m (13 ft, 4 in.) wide that is later finished to a 3.7 m (12 ft) width. The atmosphere inside

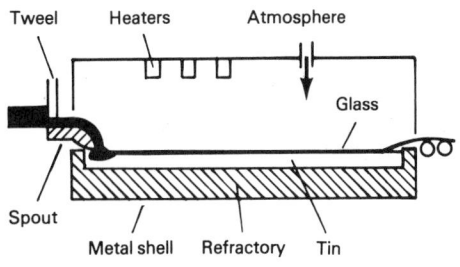

(a)

Tweel · Heaters · Atmosphere · Glass · Spout · Metal shell · Refractory · Tin

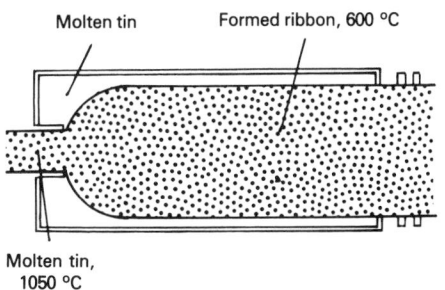

(b)

Molten tin · Formed ribbon, 600 °C · Molten tin, 1050 °C

Fig 6 Float glass process. (a) Molten tin bath. (b) Equilibrium mode of operation

the bath is kept slightly reducing by a mixture of hydrogen and nitrogen (forming gas). The glass enters the bath at approximately 1150 °C (2100 °F) and 1000 Pa · s and exits it near 650 °C (1200 °F). Because of both its temperature limitations and the reducing atmosphere above the bath, this process has been limited to the production of soda-lime-silica sheet glass.

By the early 1970s, the float glass process had eliminated all plate glass polishing operations in the United States and all but one sheet-drawing process that is used to make special glasses.

Using the fusion process, a wide variety of glasses, ranging from those with softening points of 660 °C (1220 °F) to hard glasses with softening points of 865 °C (1590 °F), can be formed into sheet product with the same surface quality as float glass. Although somewhat slower than the float process, fusion is the only direct-melt process capable of producing such a wide range of precision sheet glass.

In this process, molten glass flows from the melter into either a trough or overflow pipe at about 20,000 Pa · s. The glass divides and overflows each side of the trough and rejoins below the refractory fishtail or root (Fig 7). Thus, a sheet of glass is formed with two untouched surfaces.

The thickness is independent of flow rate, and it is controlled by the pulling velocity. Glass as thin as 0.5 to 1 mm (0.02 to 0.04 in.) can routinely be made, and thicknesses up to 15 mm (0.6 in.) have been produced. Production widths are half the size produced by the float process.

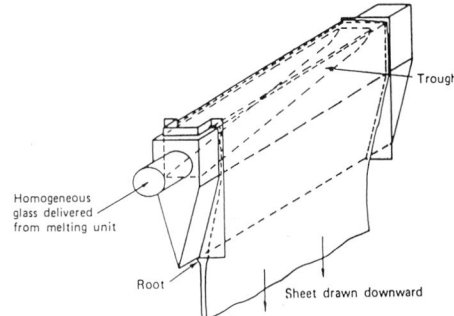

Trough · Homogeneous glass delivered from melting unit · Root · Sheet drawn downward

Fig 7 Overflow pipe used in fusion process. Courtesy of Corning Inc.

Redrawn sheet, as a process, has had limited success. The process involves forming hard glass of normal thickness by conventional means, and then reheating and redrawing it into a thin sheet.

Microsheet drawing can be used to produce very thin sheet (0.075 to 0.15 mm, or 3 to 6 mils). Glass is drawn through a small slit in a platinum plate, and upon emerging from the slit, is passed through a cooler, or annealing region before being gripped between two rollers, which pull the sheet. The process is primarily suitable for very thin thicknesses. Drawing thicker glass by this method would be difficult, because the glass weight would exceed the pulling force.

Tubing

The Danner process is a hand-drawing method that was first automated in 1917. In this process, molten glass is allowed to flow out of the tank onto a rotating refractory mandrel (Fig 8). The glass coats the mandrel and flows down toward its end, while air is blown through the center of the mandrel. The glass is then pulled off the end and out of the furnace, as tubing. With the center-blown air shut off, the result is rod or cane. This machine can replace 90 workers drawing glass by hand.

The refractory mandrel is inclined between 15 and 20° from the horizontal, and is usually made from a refractory material, although high-temperature steels have been used. Special care is taken in the design and construction of the bearing system that supports and rotates the mandrel, because any variation or wobble in the rotation adversely affects tubing quality. Sizes are controlled by adjusting temperature, blowing air pressure, and drawing speed. Tubing diameter of up to 50 mm (2 in.) with a maximum wall thickness of 2.5 mm (0.1 in.) can be made. More advanced systems use a closed-loop control system for air-pressure control. Typical manufacturing tolerances are ±0.15 mm (6 mils) on 10.75 mm (0.43 in.) outside diameter tubing and ±0.03 mm (12 mils) on a 0.5 mm (0.02 in.) wall thickness. At high production speeds, this process tends to produce side-to-side variations in wall thickness, which can be mini-

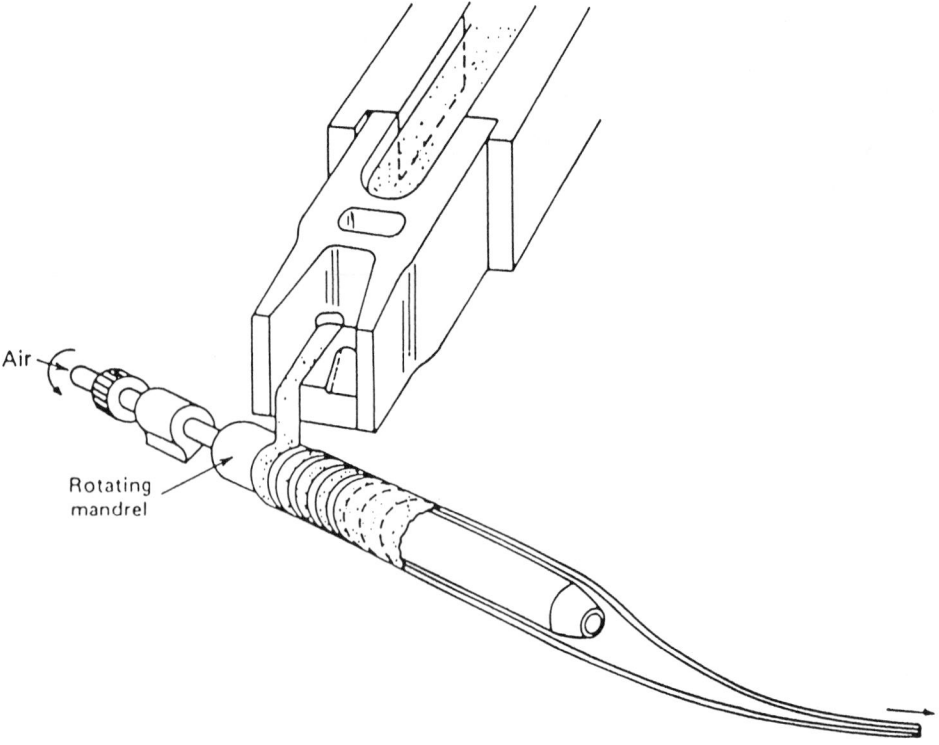

Fig 8 Rotating mandrel used in Danner process

Updraw. It is difficult to produce large-diameter tubing, especially thick-walled tubing, using either the Danner or Vello process because the weight of the drawn tubing pulls too much on the molten glass. Consequently, a slow but effective updraw process is used to produce large-diameter tubing.

In the updraw process, glass is pulled from the surface of a refractory cup that has a bottom-mounted air nozzle in its center. Cooling rings located above the glass surface chill the skin of the drawn glass, which is further attenuated as it is drawn vertically up several feet to the cutting station. A wide range of sizes can be manufactured at pull rates between 2.7 and 3.6 Mg (3 and 4 tons) per day. Because the drawing forces are better controlled and the geometry is better suited, it is possible to produce very small tubing (2 to 3 mm, or 0.08 to 0.12 in.) using this process.

Casting

Gravity Casting. A limited number of pieces with unusual shapes or close dimensional tolerances can be made by pouring molten glass into refractory molds. The largest piece produced by this gravity casting process was a 6.6 m (22 ft) borosilicate telescope mirror that weighed 33 Mg (36 tons). The process is extremely slow, and is only suited for making small quantities of an article. Care must be taken to avoid trapped air or inclusions from the mold material. Machined graphite is frequently used, for laboratory-scale molds, whereas standard refractory materials are used for larger commercial castings. The metal mold can then be heated to keep the glass hot and prevent it

mized by carefully controlling line cooling and the drawing machine.

The Vello or downdraw processes are frequently used when very large quantities of standard tubing are required. Either version of the process can draw 59 Mg (65 tons) per day from a single line, compared to the Danner process, which produces just under 45 Mg (50 tons) per day. These larger quantities offset the large capital costs of the Vello/downdraw facility.

In both setups, the melted glass flows into a refractory bowl area that has an orifice plate attached to its bottom. The Vello and downdraw setups differ mainly in the shape of the refractory bowl. Frequently, a rotating refractory cylinder stirs the glass above the orifice to ensure uniform temperature and viscosity. Inserted into or through the center of the orifice is a hollow refractory or platinum "bell," which forms the interior diameter of the tubing when air is blown into the tube. Below the orifice, which is sized for a specific range of tubing, is a temperature-controlled muffle that maintains a constant cooling rate and contributes significantly to the uniformity of the resultant tubing. Upon emerging from the muffle, the tubing is turned to the horizontal position and pulled down the draw by the pulling or tractor machine located from 45 to 60 m (150 to 200 ft) away.

Ovality problems can occur when the tubing is turned from the vertical to the horizontal, and compensatory adjustments in the bell can be made. Because any given bell and

orifice setup produces a limited range of sizes, conversion to other sizes requires substantial setup time. Consequently, the Danner process is preferred for short production runs, and the Vello process is preferred for longer runs.

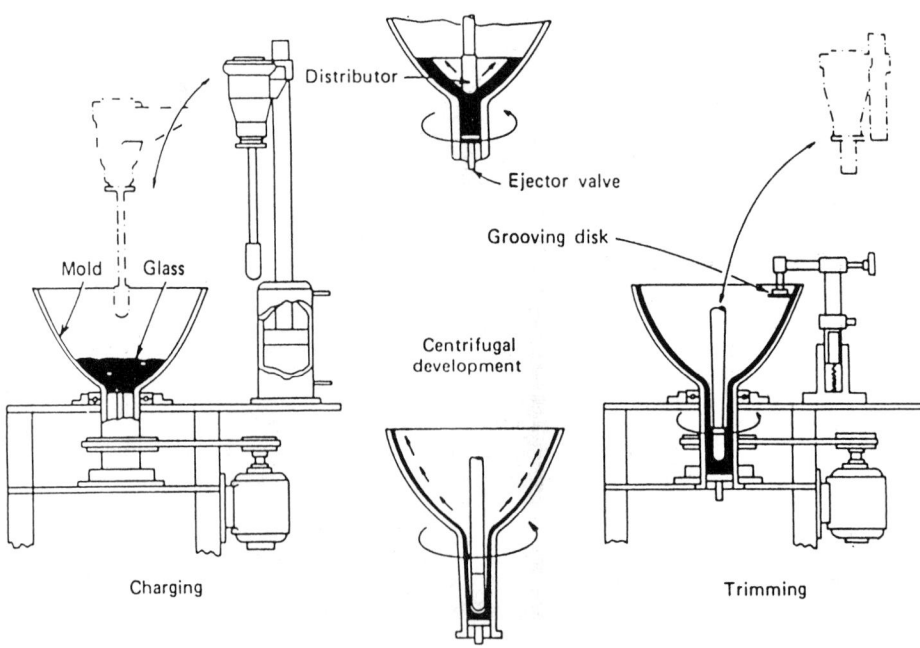

Fig 9 Centrifugal casting. Courtesy of Corning Inc.

from setting before the mold is completely filled. The entire piece and mold can also be slowly annealed in a furnace to avoid residual stresses.

Centrifugal casting can be used to form conical-shaped ware, such as missile radomes and television tube funnels. Molten glass is poured into a rotating mold and centrifugal force pushes it to the outside walls. By gradually increasing the rotation speed, the glass can be made to "walk" up the sides of the mold, forming a piece with relatively uniform wall thickness (Fig 9). An interior profiling tool is sometimes used to improve the uniformity of the wall thickness, especially for pieces with a large height/diameter ratio.

Extrusion

Glass billets at relatively high viscosities (10^5 to 10^7 Pa · s) can be extruded through a die to produce very unique shapes with close tolerances and sharp corners. Ram pressures of the order of 100 MPa (15 ksi) are used.

One of the most critical aspects of hot glass extrusion is the selection of the die material which must be able to withstand the temperatures and pressures of the process and have a low coefficient of friction with the glass. If the friction forces are large, there will be a substantial shear gradient across the extruded cross section. Upon emerging from the die, the glass will exhibit extrudate swelling, which is typical of a viscoelastic material. Materials such as graphite or boron nitride can be used as die materials to minimize this swelling. The most effective metal dies seem to be alloys with a high nickel content.

Marbles and Spheres

The production of glass marbles or large spheres begins with the shearing of a stream of molten glass into small gobs. In the case of "cat's eye" marbles, a second thin stream of colored glass is blended with the major stream prior to the shearing operation.

Each sheared gob is directed to the top of two inclined, parallel counterrotating screws. Each screw has a thread depth equal to approximately one-half the marble diameter. As the screws rotate, the glass gob is rolled into a sphere. Upon reaching the end of the screw, the finished marble drops off.

These screws can be from 2.4 to 3.0 m (8 to 10 ft) in length with an incline that is 10 to 20° from the horizontal. Screw diameters are generally a few inches. The glass used is frequently cullet from another operation.

Smaller glass spheres of about 6.4 mm ($^1/_4$ in.) can be made by depositing very small gobs of glass in a circular hole cut into a graphite plate. When this plate slides over a graphite surface while the glass is still soft, the rolling action creates a sphere from the small gob. A series of multihole plates can then be strung together in a conveyer-type system, allowing the coproduction of hundreds of spheres.

The very small glass spheres used in reflective highway paints and signs are formed by dropping a high-index glass through a tall, vertical furnace/drop tower. Presized glass or agglomerated raw materials are introduced at the top, and will melt and become spheres (through surface tension) as they fall through the hot zone.

REFERENCES

1. D.C. Boyd and D.A. Thompson, *Encyclopedia of Chemical Technology*, 1980
2. R.W. Douglas and S. Frank, *A History of Glass Making*, 1972, p 150

SELECTED REFERENCES

- W.C. Hynd, Flat Glass Manufacturing Processes, *Glass: Science and Technology*, Vol 2, Academic Press, 1984
- W. Trier, *Glass Machines*, Springer-Verlag, 1969

Fiberglass

Fay V. Tooley, Industrial Consultant and Professor Emeritus of Glass Technology, University of Illinois

A SIMPLE EXPERIMENT can be conducted by any curious person who wonders about the processes involved in converting earthen materials to the vast number of glass wool and textile products that exist. This disarmingly simple experiment, which greatly facilitates understanding, requires an individual to recall the experience of painting a house, wall, ceiling, or the like. After dipping the brush, wiggling it a little, and withdrawing it to a level several inches above the surface of the paint in a can, a "fiber" of paint flows off the edge of the brush. If this fiber were to be cooled rapidly to a solid immediately after its formation, it could be wound up as a continuous fiber on a revolving drum.

Also consider, much more realistically, the formation of carnival cotton candy, whereby molten sugar syrup falls in a continuous stream upon a rapidly revolving disk, from which it is thrown off the edge as a liquid fiber that cools immediately into a solid fiber mass. It is indeed remarkable that when, with inventive genius, hot molten glass was substituted for the paint or the molten sugar syrup, and the fiber-forming process was greatly refined, the result was a multimillion dollar U.S. industry.

Glasses utilized in glass fiber manufacture are closely related, in terms of raw materials, composition, and melting procedures, to most of the glasses used in the container, flat glass, technical, and specialty segments of the industry. The glass used in these fields is defined as an inorganic product of fusion that has cooled to a rigid condition without crystallizing (Ref 1). Solid glass objects are formed upon cooling molten glass rapidly; sensible heat is extracted at such a high rate that equilibrium structural changes cannot keep pace with the demands of falling temperature. Taking into account the increase in viscosity with falling temperature, the consequence is a structural state that is not characteristic of true thermodynamic equilibrium. Instead, a disordered higher-temperature state is "frozen in" and persists at room temperature. A simple structure of SiO_2-Na_2O glass is depicted in Fig 1 to illustrate its structurally disordered nature (Ref 2). To promote an understanding of current glass-fiber technology, a brief history is provided below.

Historical Review. The knowledge to create glass in fiber form existed in ancient Egypt, where glass makers trailed glass threads

of brilliant colors up and down around a cooling glass vessel to form wavy patterns. During the 1800s, mineral wool was produced in Germany and Wales (1840s) (Ref 3), and the steam jet method of manufacturing mineral wool insulation from molten slag was patented in 1870 (Ref 4). The earliest production of slag wool in the U.S. reportedly occurred in New Jersey in 1875 (Ref 5, 6). In 1880, a cloth made of coarse-spun fibers gathered in bundles and woven together with silk threads was patented (Ref 7). A dress was made from this cloth in 1893 (Ref 8), and was modelled at the Columbia Exposition in Chicago. The successful manufacture of rock wool in 1897 (Ref 9) resulted from an investigation by the manager of a steel plant as to the fiber-forming possibilities of a local rock used for fluxing purposes in the plant. Fiber forming was effected by shattering a falling stream of molten rock or slag with a steam jet. By 1929, seven companies operated eight U.S. plants and numerous melting facilities, all producing insulating wool from rock or slag raw materials.

At the turn of the century, a prominent European method for manufacturing glass wool involved drawing fibers from rods horizontally to a revolving drum, a process later modified by vertical, downward drawing from glass rods (Ref 10). A variation that came to be known as the Gossler process replaced the use of rods with streams of molten glass issuing from holes at the bottom of a melting facility (Ref 11). The resulting product was cut from the drum, carded, and "loosened up" to provide the insulating medium widely used. This product was placed on subfoundation paper, burlap, or wire mesh and processed by sewing machines into a mattress, which in turn was cut into bands of desired widths.

Glass fibers were also produced by a method resembling cotton candy spinning, later called the Hager process. In this process, which was patented in 1929 (Ref 12), molten glass was dropped onto a horizontal revolving disk, from which fibers were thrown off by centrifugal force.

In the late 1920s, the Owens-Illinois Glass Company and Corning Glass Works, each searching independently for new markets for glass, mounted research and development activities aimed at the industrial production of glass fibers. These two companies merged in 1938 and became "the undisputed leader

in development, marketing, and technology of this industry" (Ref 13).

Modern fiberglass technology is characterized by the product phases in the 1935–1940 period. A dichotomy of research and development interests existed, represented by current and prospective wool, as differentiated from textile glass products. Wool, following an earlier history of rock and slag wool products, was viewed as a fluffy mass of discontinuous fiber used for domestic and industrial insulation purposes. Textile fibers were primarily viewed in terms of their prospects as continuous fibers adaptable to conversion into woven or mat products, such as electrical insulating tapes, filter cloths, decorative fabrics, and battery separator plates. Product avenues have changed considerably over the years. Although the insulating function has continued to be the dominant field for wool glass batts, greater market areas in the field of tire cord and composite reinforcement fibers have developed in the textile sector.

This article describes fiberglass composition and raw materials, furnace types, the preparation process, and forming technology. Additional information is available in Ref 14 and 15.

Glass Compositions

When considering the glass compositions that are currently used to produce glass fibers, it is useful to realize the determining factors behind their adoption. Glass composition choices have been based on their adequacy with respect to (1) final product performance; (2) glass preparation characteristics (melting, fining, homogenizing); (3) fiber-forming characteristics; (4) environmental considerations; and (5) raw materials, fuel, and processing costs.

Table 1 gives the average glass compositions of the most used glass types in modern technology, that is, Types T, E, C, and S. Type A, a rather common early container composition, is included merely to indicate the first glass considered for wool production. It was soon replaced by T_1 compositions in order to achieve increased chemical durability and fiber-forming characteristics. Glasses of the T_2-type replaced T_1 during the period when modern rotary-spinner technology replaced the method based on shattering molten glass streams that issued vertically from

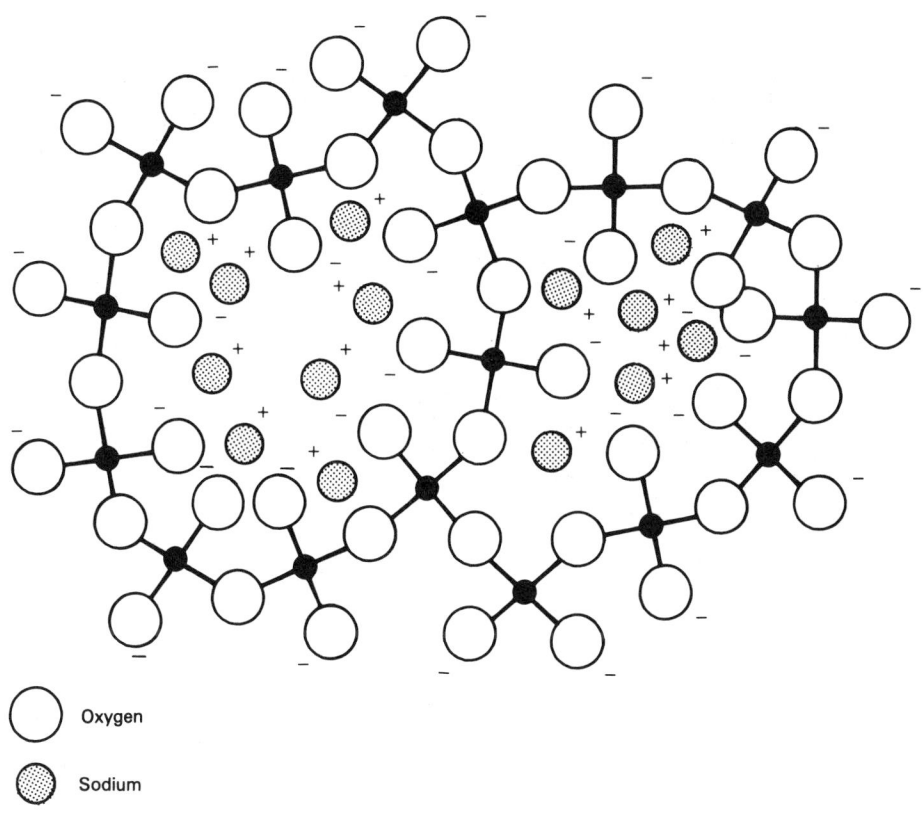

○ Oxygen

◉ Sodium

● Silicon

Fig 1 Reaction between SiO_2 (crystal or glass) and Na_2O to give Na_2O-SiO_2 glass structure

holes in metallic bushings positioned in fore-hearth extensions of the melting furnace.

Type E (electrical) was originally used in the manufacture of electrical insulation tape. It became the favored glass for most of the first textile applications, as well as tire cord and plastic reinforcement products. Over the years, composition modifications to improve product performance and fiber-forming characteristics have chiefly involved moderate changes in the content of B_2O_3, CaO-MgO,

F_2, and minor oxides. Type C (chemical) textile glass has superior acid resistance. Type S (strength) exhibits superior strength and elastic modulus, compared to E glass, and is used in highly specialized applications where structural properties are important.

These compositions are sometimes modified, either by impurities in the raw materials, or by materials that are purposely added to optimize fiber production or to modify final product characteristics.

Raw Materials. Table 2 categorizes the raw materials used in glass compositions. The major class includes materials that, in some combination, contribute from 97 to 99 wt% of the final glass composition. The remaining 1 to 3 wt% is represented by minor constituents. These materials can be either manufactured chemicals or natural minerals. Soda ash, borax, and boric acid are chemically or semichemically manufactured. Blast furnace slag, a by-product of steel manufacture, is a hybrid that does not really fit either classification. The rest of the materials in Table 2 are naturally occurring minerals, some of which are purchased in calcined form.

All raw materials are generally purchased on the basis of chemical quality and uniformity, grain size distribution, predictability of shipping schedules, and cost. These factors underlie the ability of the fiber manufacturer to produce a consistently uniform product.

Furnace Types

Furnaces that are used to melt glass for fiber production do not differ, in principle, from those commonly used to produce container, flat, technical, or specialty glasses. Operationally, there are some moderate differences. Because of fiber-forming process requirements, the molten glass is delivered to the forming equipment at a much higher temperature than in the case of container, other holloware, and flat glass operations. For the same tonnage production, glass fiber technology employs a large number of smaller furnaces than is the case in other traditional glass-manufacturing operations. Also, fiber-forming furnaces have forehearth, or channel, structures to transfer the molten glass to the forming point(s). These structures are not characteristic of traditional furnaces.

Regenerative. There has been a great decrease in the use of regenerative furnaces for fiber production. They are included here briefly because of their historical significance in the development of the industry in the U.S.

Table 1 Compositions of glass used for wool and textile products

Raw material	Wool			Textile		
	Containers, Type A	Thermal insulation		Electrical, Type E	Chemical resistant, Type C	Strength/stiffness, Type S
		T_1	T_2			
SiO_2	72–72.5	63	58.6	52–56	64–68	64
Al_2O_3	0–2.0	5	3.2	12–16	3.5	25
MgO	2.5–4.0	3	4.2	0–6	2–4	10
CaO	5.5–10.0	14	8.0	16–25	11–15	...
$Na_2O + K_2O$	10–16	10	15.1	0–2	7–10	...
B_2O_3	...	5	10.1	5–10	4.6	...
BaO	...	...	...	...	0–1	...
TiO_2	...	...	...	0–1.5	...	...
F_2	...	...	...	0–1	...	...
Fe_2O_3	...	...	...	...	0.8	...
FeO	...	...	...	0.8	...	...

Table 2 Basic raw materials used in glass fiber production

Major constituents	Minor constituents
Silica sand	Salt cake
Calcined alumina	Niter
Feldspar	Carbon
Nepheline syenite	Gypsum
Limestone	Iron oxide
Dolomite	Fluorspar
Burned dolomite	Rutile
Magnesite	Zinc oxide
Soda ash	Manganese
Borax	oxide
Boric acid	Potassium
Colemanite	carbonate
Ulexite	
Kaolin clay	
Beneficiated blast furnace slag	
Cullet	

They were the primary melting facility for wool glass production from the mid-1930s to the early 1970s, before being replaced by the electric melter.

Operating on either a gas or fuel oil basis, this type of furnace utilizes a heat-exchange technique aimed at the economic use of fuel. Regenerative chambers, in which heat exchange takes place, are located on both sides of a melting tank. These chambers contain many ceramic refractory elements, strategically selected and placed as to composition, shape, and pattern with respect to heat absorption, heat release, and necessary exhaust velocity. Overall heat exchange is effected by periodic reversals of precombustion air traveling through one of the previously heated chambers and uniting with gas or oil at the ports. This provides both the necessary energy for glass melting and conditioning for the forming process. The process is reversed on a time basis of 20 to 30 min, depending on furnace size, type, and operational protocol.

Electric. Wool-type fiberglass was first melted in an all-electric furnace in 1969. The prime reason for melting wool glass electrically is that the cold-top melter has a natural emission control capability (Fig 2). The batch blanket, or crust, provides a medium in which condensation of the off-gases takes place, with predominantly H_2O and CO_2 escaping to the atmosphere. The batch blanket is also a highly effective batch preheater. It enables the cold-top electric melter to operate at high efficiency (70 to 80%), depending on melter capacity.

The electrical resistivity of wool insulation glass is comparable to the relatively low resistance of soda-lime glass. As a result, the wool insulation glass is ideal for electric melting. The predominant type of melter has been the two-phase, single, or double electrode furnace. The single square unit is used up to about 118 Mg/day (130 tons/day), and the double square unit up to 250 Mg/day (275 tons/day), with maximum capacity believed to be about 270 Mg/day (300 tons/day).

Compared to fossil fuel melters, cold-top units are low in capital cost, because no emission control equipment or heat recovery devices are required. In spite of good efficiency, energy cost is relatively high, because of the cost of electricity. Electric melting is usually the most cost-effective choice, when all factors are considered.

The melting of textile fiberglass (E glass) is more complicated. There is only one proven method for melting E-glass all electrically: the Pochet (DeBussy) pot melter (Fig 3). This simple melter can be designed to operate on single- or three-phase power. Maximum melting capacities are about 18 Mg/day (20 tons/day).

Some fossil-fuel furnaces for melting E glass are electrically boosted. A considerable majority uses vertical rod electrodes, as shown in Fig 2. These electrodes are placed strategically in the bottom of the furnace and

arranged for two- or three-phase power application. Generally, these boosts are of a low level (20% or less) and are employed to either increase production or reduce emissions. However, boost capacities greater than 20% are possible using current technology.

Recuperative. Like the regenerative furnace, a typical recuperative furnace involves a heat-exchange technique aimed at the eco-

nomic use of fuel (Fig 4). The ability to increase the temperature of incoming combustion air by heat recovery from the process exhaust gas represents an economic advantage. Heat exchange is effected by heating combustion air in a double conduit (or conduits) immediately above the stack port. Exhaust gases from the melting tank enter the outer segment of the conduit and combustion

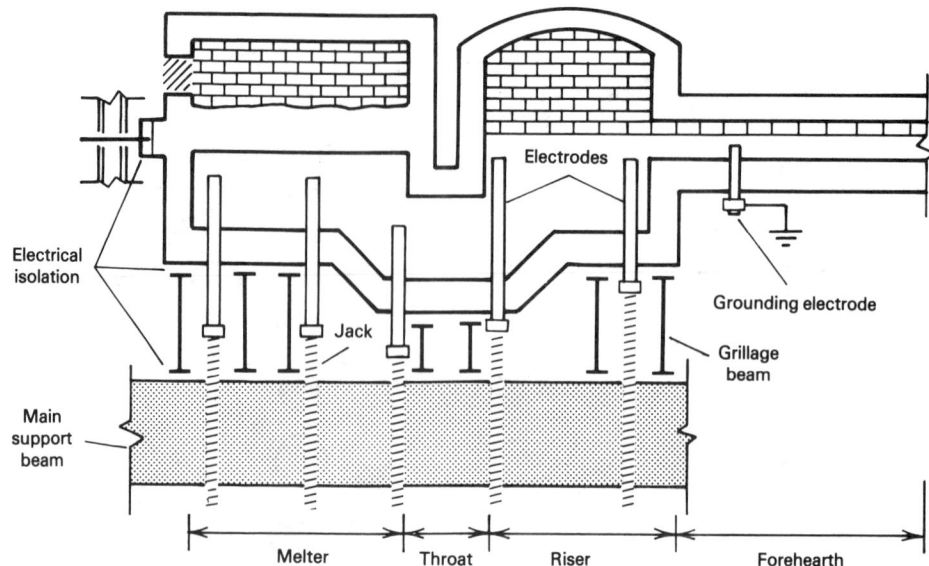

Fig 2 Electric melter using vertical electrodes

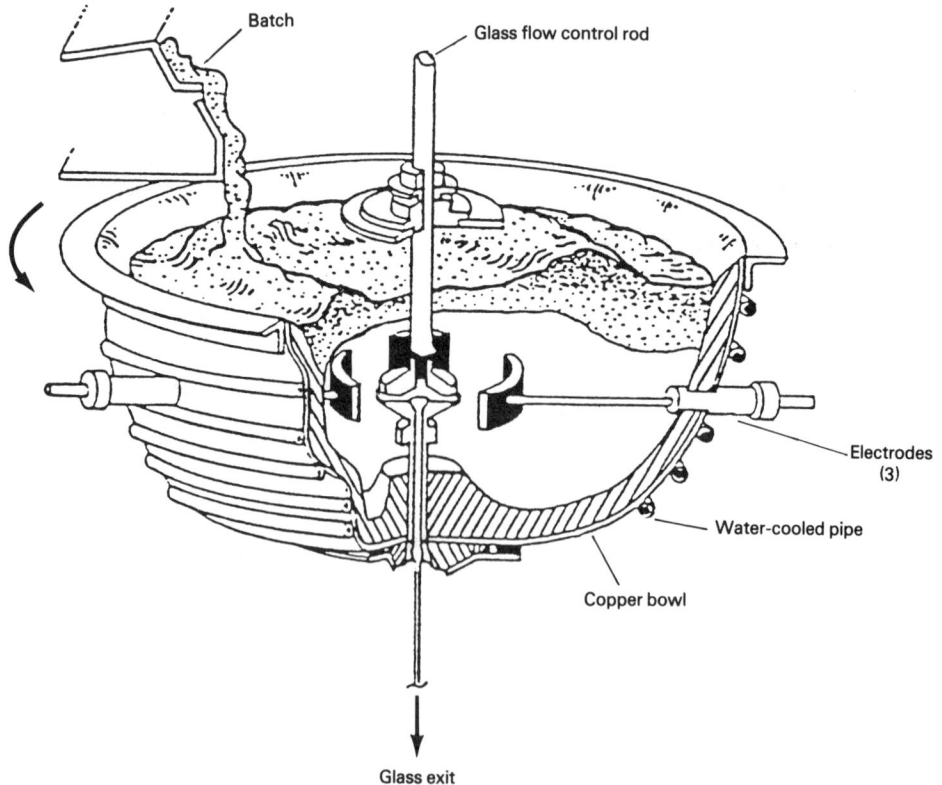

Fig 3 Pochet melter

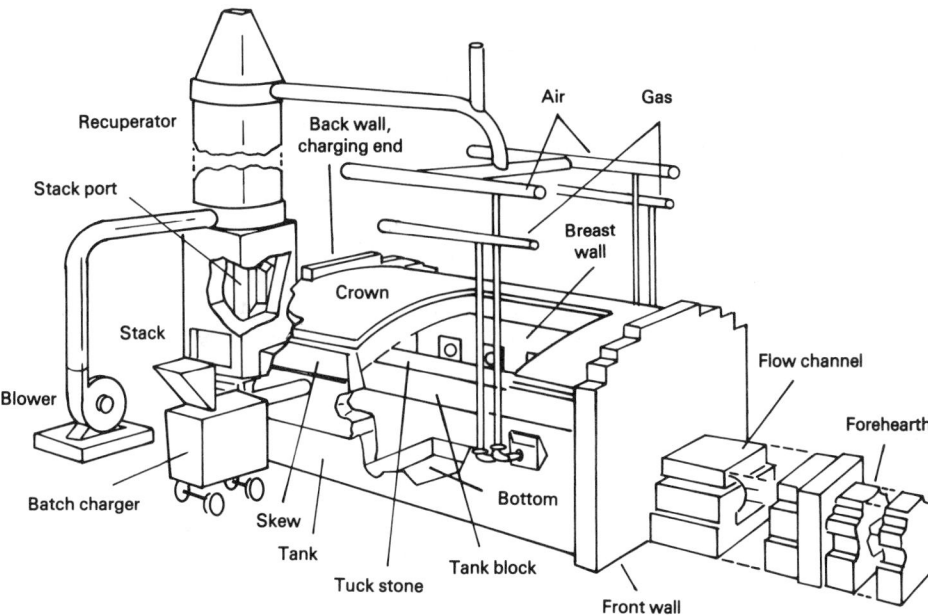

Fig 4 Recuperative fired furnace

air is blown upward in the inner chamber, thus absorbing heat prior to its delivery to the burners.

Recuperative furnaces (as well as direct-fired furnaces) supply a more consistent operating temperature than do the regenerative type because of slight, but nevertheless important, temperature fluctuations caused by reversals of firing direction. This factor is not of serious consequence in many traditional forming operations that involve relatively large masses of the article being formed. However, it is particularly important in textile fiber production, where slight changes in glass-processing conditions (temperature-viscosity-flow patterns) can be detrimental to uniformity of the product.

Direct-fired furnaces are widely used in glass melting for fiber production. A common modification of the example shown in Fig 5 is the use of a single, rather than a double, stack. They are furnaces used to prepare glass for marbles used in some textile production or other relatively small-tonnage production of special products. Based on furnace size and operating parameters, the value of the heat recovered can be compared to the expense of furnace rebuild plus the heat recovery system. When justified, the recuperative principle is usually adopted to lower furnace operating costs.

Mixed Melters. Most furnaces, whether regenerative, recuperative, or electric, can be pushed beyond their normally rated production or quality by judiciously employing electric, oxygen, or gas "boosting." This term connotes that the production of any large-tonnage furnace can be increased by the use of gas or gas-oxygen burners or electrodes at strategic points in the tank walls. In some instances, the alternate heating system is larger than the "boosting" category. Mixed melting systems, in which substantial portions of the total melting and conditioning energy are supplied by both electricity and gas or oil, have been utilized. Table 3 lists the sites of U.S. glass fiber manufacturing facilities and the types of furnaces employed.

Glass Preparation Process

Regardless of the type of furnace used to prepare glass for fiber manufacture, the input material must undergo four basic stages during its transformation into a glass adequate for the fiber-forming process. As shown in Fig 6, these stages are melting, fining, homogenizing, and heat conditioning. Melting connotes continuous conversion of the solid input batch to a noncrystalline liquid. In the early stage of melting, the resultant liquid is a mixture of many glasses of different composition, depending on the chance proximity of batch particles in a given-temperature environment.

Concurrent with initial melting, the fining process commences. The fining process eliminates gas bubbles in the molten glass mass. These vapors originate from the gases emitted from the batch materials and from entrapped air and furnace gases as melting occurs. Also concurrent with or shortly following the start of the melting process, homogenization is initiated to produce uniformity of composition throughout the glass mass. Beyond the melting phase, fining and homogenizing processes continue until a chemically uniform, reasonably void-free mass of molten glass, suitable for the fiber-forming process and end-product performance, is achieved.

At some intermediate stage of fining and homogenizing, the fourth phase of glass preparation begins. In this heat-conditioning phase, the glass mass is brought to a uniform temperature to prepare it for a satisfactory forming process.

Although three of these four stages begin simultaneously, it should be noted that there is a definitive point at which the melting stage is complete. Beyond this point, the completion of fining, homogenizing, and heat conditioning depend on the suitability of the glass mass for the forming process.

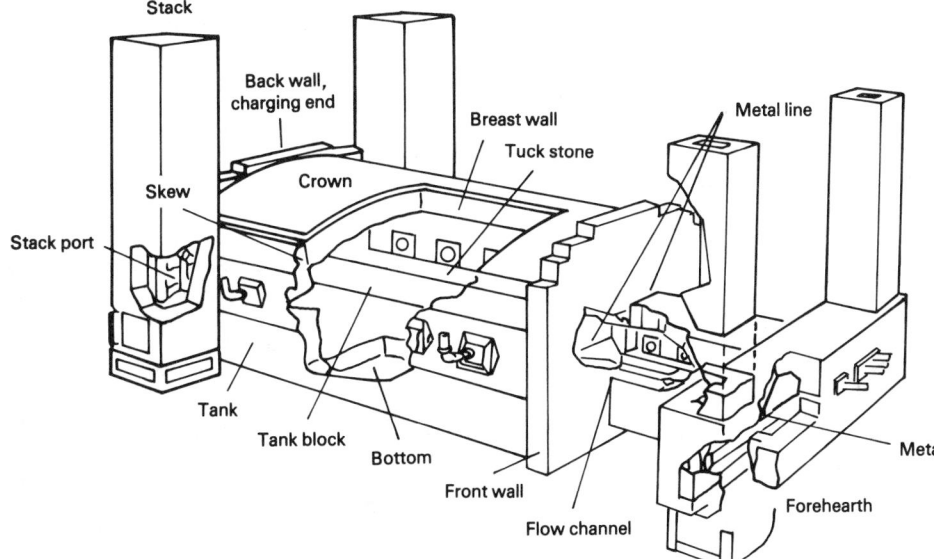

Fig 5 Direct-fired furnace

Table 3 U.S. glass fiber manufacturers

Location	No. of furnaces	Furnace type(a)	Boost		Fiber	
			Electric	Gas/O2	Wool	Textile
Alabama						
LaNette	1	Recuperative			X	
Arizona						
Eloy	1	All electric				X
California						
Chowchilla	1	Recuperative	X	X	X	
Santa Clara	2	All electric			X	
Willows	5	All electric			X	
Georgia						
Athens	1	All electric			X	
	1	Recuperative			X	
Fairburn	3	All electric			X	
Winder	1	All electric			X	
Indiana						
Richmond	2	Unit melter			X	
Shelbyville	3	All electric			X	
	1	Direct-melt gas			X	
Kansas						
Kansas City	1	Recuperative	X		X	
	1	Recuperative			X	
	2	Gas-electric			X	
	1	All electric				
McPherson	3	All electric				
Michigan						
Albion	5	All electric			X	
Mississippi						
Mineral Wells	3	All electric			X	
N. Carolina						
Lexington	7	Recuperative				X
Shelby	10	Recuperative				X
	1	Recuperative	X			X
New Jersey						
Berlin	1	Recuperative			X	
Penbryn	3	All electric	X	X	X	
New York						
Delmar	2	All electric			X	
Ohio						
Bremen	24(b)	Gas				X
Defiance	2	Gas			X	
Newark	2	All electric			X	
	2	Electric-gas			X	
	3	Recuperative	X		X	
Shawnee	20(b)	Gas				X
Waterville	2	Recuperative				X
Oregon						
Corvallis	2	All electric			X	
Pennsylvania						
Huntingdon		Paramelt				
Mountain Top	2	Recuperative		X	X	
S. Carolina						
Aiken	8	Recuperative				X
	3	Recuperative	X			X
	6	Recuperative		X		X
Anderson	1	Recuperative	X			X
	2	Recuperative	X			X
Tennessee						
Etowah	1	All electric			X	
	1	Recuperative				X
Texas						
Amarillo	2	Recuperative	X			X
Cleburne	2	All electric			X	
	1	Recuperative			X	
Waxahachie	1	Gas-electric			X	
	2	All electric			X	
Wichita Falls	4	Recuperative	X			X
Utah						
Salt Lake City	1	All electric			X	
	1	All electric				X
	10(b)	Gas				X
West Virginia						
Parkersburg	1	Gas		X	X	

(a) Production range, collectively, for wool furnaces is from 45 to 135 Mg/day (50 to 150 tons/day) and for textile furnaces, from 14 to 250 Mg/day (15 to 275 tons/day). (b) Modigliani units (small refractory gas-fired), capable of 204 kg/day (450 lb/day)

The issue of achieving glass homogeneity that is adequate for forming and product performance is an object of considerable study in the industry, both in the laboratory and on the furnace line. The early stage of melting usually produces a glass that lacks adequate homogeneity. Improvement is effected by establishing a glass flow pattern in which portions of glass that differ slightly in composition are induced into flow tangentially against each other with a differential velocity. At the interface, the compositional difference between two such glasses is a strong driving force in the direction of achieving homogeneity by ion diffusion. However, diffusion alone is a slow process, particularly in a liquid of high viscosity. When a high degree of shearing action is induced throughout the glass during melting and conditioning, the number of interfaces between glasses that differ in composition are increased. Then, diffusion can be more effective in achieving compositional uniformity, because it only needs to take place over relatively short distances. The ultimate concern is which method is the most efficient means of achieving an adequate degree of shearing action in the glass mass.

There are two principal means of accomplishing this objective, one of which is mechanical, and the other, thermal. The mechanical approach involves either devices that mechanically stir the glass or bubblers that are built into the tank bottom and inject air under pressure into the glass mass. The large bubbles that rapidly rise through the glass induce a stirring motion.

The thermal means of achieving homogeneity initially involves the natural convection characteristic of the particular furnace being used. "Cold" glass masses that are adjacent to furnace walls sink, while hotter glass toward the middle of the tank rises. This pattern induces a circular motion within the glass mass, which is quite effective in achieving adequate homogeneity. Strategic positioning of electrodes, bubblers, auxiliary burners, cooling air sources, and, in some cases, shadow walls are all useful in influencing the desired convection flow of the glass streams within the mass.

Fiber Forming

Wool Process. The dominant method for producing high-tonnage insulation glass wool is the rotary, or spinner, process (Fig 7). In this process, molten glass from the furnace flows into a cylindrical container fitted with small holes in its wall. The rapid rotation of the vessel produces multiple horizontal glass streams by centrifugal force. Immediately after their issue, downward-directed blowers eject a continuous high-velocity blast of combustion gases (air or steam), which converts the molten glass streams into fibers. During their descent from the spinner to an underlying traveling belt, the fibers are sprayed with a binder designed to minimize fiber-to-fiber

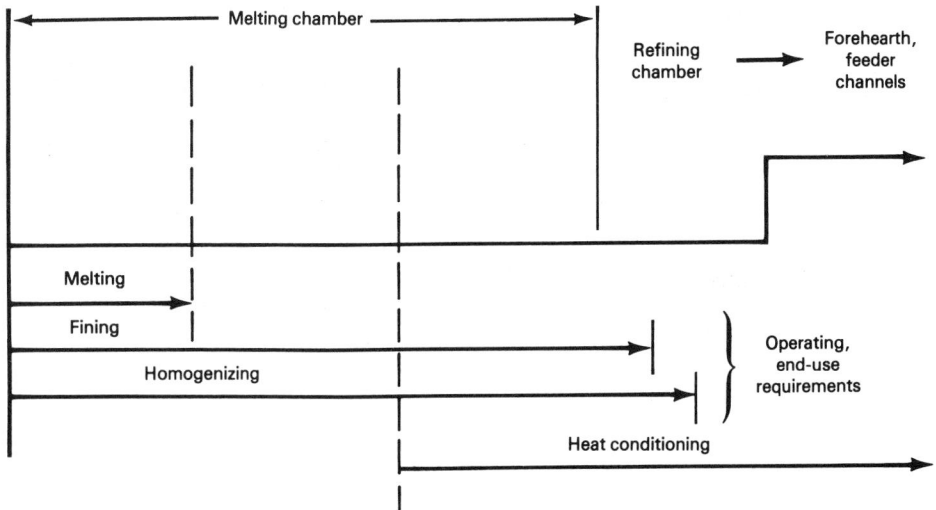

Fig 6 Relation between melting, fining, homogenizing, and heat-conditioning stages

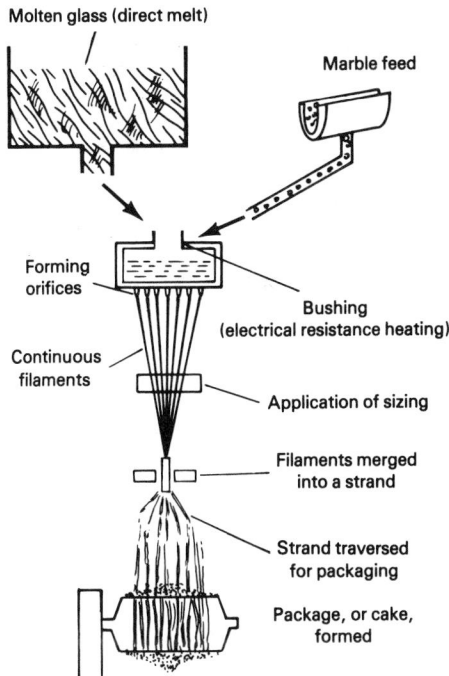

Fig 8 Fiber forming using either direct melt or marble process

abrasion while providing the chemical basis for bonding the wool into uniform products of desired density. Bonding is accomplished by curing the resinous binder while the wool bed is compressed to a constant desired thickness. The products of this process are insulation wool in either batt or rolled form.

Textile fibers are formed by either of two processes, direct melt or marble melt. In the direct melt process, molten glass from the melting furnace is fed directly to the forming apparatus. In the marble melt process, solid glass balls (marbles) are fed to the forming device, where they are both melted and formed into fiber. The fiber-forming process for either case is shown in Fig 8. Glass fed to a dedicated bushing is either in direct melt or marble form. The forming bushing is constructed from a platinum-rhodium alloy. The necessary melting or holding temperature is established by electrical resistance heating. Bushings are fitted with from 204 to 3000 or more orifices, depending on the intended basic product.

The molten glass streams pass quickly through the orifice and are rapidly cooled to form solid, continuous fibers. A short distance below the forming point a size is applied. As in the case of wool fibers, the size is intended to minimize fiber-to-fiber abrasion and supply chemical coatings basic to the particular product function.

Individual fibers are then gathered into a strand, the elemental building block of yarns, rovings, and fabrics, by largely traditional textile processes. To produce a suitable package for these operations, the strand is traversed to facilitate a convenient form for further processing.

On a relatively small production scale, textile fiber can be produced by air-shattering streams of glass as they issue from a bushing into discontinuous fibers, which are collected on a vacuum drum. The tangled mass is drawn from the drum through a merging orifice and packaged as a continuous staple sliver.

Chopped Fiber. An important final fiber-forming operation is that of chopping the fiber into short lengths. Chopped fiber is produced either on-line or fed from packages of yarn mounted in a creel. Either wet or dry chopping is practiced. Fiber length and diameter are chosen according to suitability for the final product. Continuous strands, yarns, or rovings are usually preferable in cases where maximum directional strength is required in the final product, whereas chopped fiber is preferable when uniform multidirectional strength is required in reinforced plastic products. Chopped fiber mats or fiber-resin mixtures are well suited to high-rate production processes, such as press and injection molding, pultrusion, and resin injection molding.

REFERENCES

1. Standards: Glass and Glass Products, Definition of Glass Terms, C-162, American Society for Testing and Materials
2. *The Handbook of Glass Manufacture*, Vol

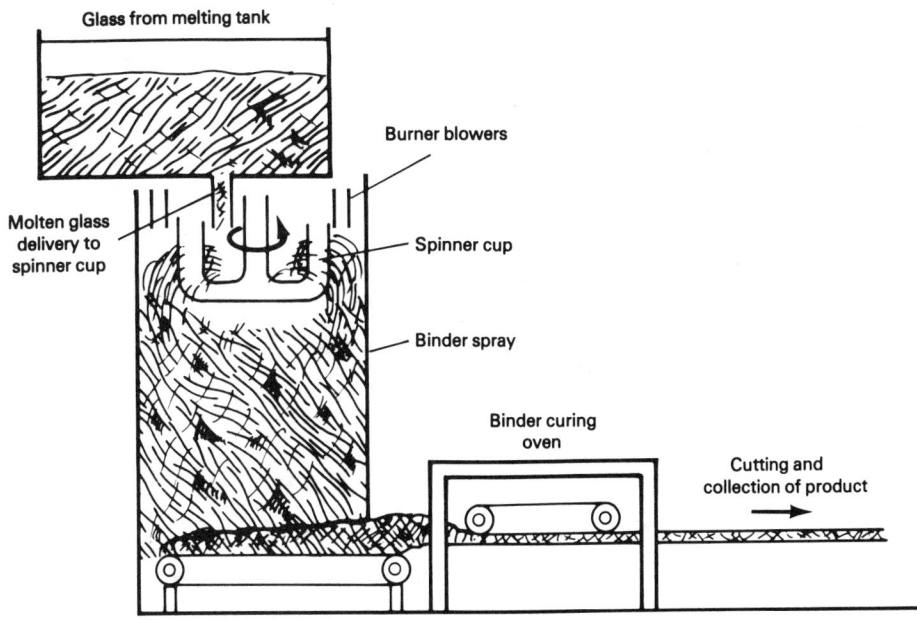

Fig 7 Steps for forming insulating wool; rotary or spinner process

II, 3rd ed., F.V. Tooley, Ed., Ashlee Publishing, 1984, p 1001–1020

3. The Origin of Rock Wool, *Stone,* Vol 57 (No. 12), 1936

4. J. Player, U.S. patent 103,650, May 1870

5. H. Lang, Designing and Operating a Slag Wool Plant, *Chem. Metall. Eng.,* 27 Aug 1933, p 365

6. J.M. VanVoorhis, "The Mineral Wool Industry in New Jersey," Bulletin 56, Department of Conservation and Development, State of New Jersey, 1942

7. H. Hammesfahr, U.S. patent 232,122, Sept 1880

8. K. Field, "The Drama of Glass," brochure of Libbey Glass Company, 1893

9. J.R. Thoenen, "Mineral Wool," Information Circular 6142, U.S. Bureau of Mines, June 1929

10. A.D. Saborsky, Glass Wool Heat Insulation in Europe, *J. Am. Ceram. Soc.,* Vol 6, May 1923, p 674–684

11. O. Gossler, British patent 354,763, May 1930

12. Hager process, U.S. patent 2,234,087, Mar 1941

13. K.L. Lowenstein, *The Manufacturing Technology of Continuous Glass Fibers,* 2nd ed., American Elsevier, 1983, p 1

14. F.V. Tooley, Ed., *The Handbook of Glass Manufacture,* Vol II, 3rd ed., 1984, p 717

15. J.C. Watson and N. Raghupathi, Glass Fibers, in *Composites,* Vol 1, *Engineered Materials Handbook,* ASM International, 1987, p 107–111

Optical Fibers

John F. Stroman, Corning Inc.

THE EMERGENCE of fiber optics is an important event in the evolution of communication technology. Information density, speed, and transmission distance are improved significantly with fiber-optic systems, compared to systems based on metallic wire conductors. Fiber-optic systems offer immunity to electromagnetic interference (EMI) and provide easy detection of efforts to violate the security of transmitted information. An increasing number of sensor functions are uniquely performed optically. This overview of optical fiber principles includes discussions on the physics of light propagation in optical fiber, fiber performance and design, and the various processes used to manufacture optical fiber.

Basic Physics of Optical Fibers

An optical fiber is a medium through which information may be transmitted. In a typical system, an electrical signal is converted into light by a laser diode (LD) or light-emitting diode (LED). The optical signal is then launched into the fiber and propagates through it until it is received by an optical detector, which converts the optical signal back into an electrical signal. Signals typically are generated between the 780 and 1600 nm wavelengths, which are bordered by the visible and far infrared regions.

Signal propagation in optical fibers is made possible by total internal reflection, which occurs at the boundary between two solid regions within a fiber, the core and the cladding. Because of total internal reflection, optical signals are largely contained within the core region, as illustrated in Fig 1. In all fiber designs, the core glass has a higher refractive index than the cladding glass. The refractive index is a dimensionless parameter that is defined by the ratio of the speed of light in a vacuum to the speed of light in a given material. The refractive index varies with wavelength. The dependence of the speed of light on refractive index accounts for the ability of a prism to separate white light into its component colors. The difference in refractive index between the core and cladding plays a critical role in the performance and manufacture of fiber.

Some basic optical properties of fiber can be described by a simplified representation of light within a fiber as rays. To describe these properties, it is necessary to consider the mechanisms of internal reflection. Total internal reflection occurs when a light ray in a medium of some refractive index n_1 is directed toward a surface of a medium with an index n_2, where n_1 is greater than n_2; and the ray is incident at some angle greater than what is known as the critical angle. As shown in Fig 2, the critical angle is measured relative to a normal drawn to the surface at the point of incidence of the light ray. The critical angle is calculated readily from Snell's law:

$$n_1 \sin(\theta_1) = n_2 \sin(\theta_2) \qquad \text{(Eq 1)}$$

In Fig 2, the light is directed toward a dielectric surface at some incident angle θ_1, measured with respect to a normal, drawn from the surface. The exit angle, θ_2, is determined by the ratio of refractive indices for the two media, such that:

$$\theta_2 = \sin^{-1}\left(\frac{n_1 \sin \theta_1}{n_2}\right) \qquad \text{(Eq 2)}$$

From Snell's law, the minimum angle that an incident ray must take for reflection to take place can also be calculated. At some critical angle for a given refractive index ratio, θ_2 equals 90°; therefore, $\sin \theta_2$ equals unity. For angles of incidence greater than the critical angle, all of the incident light is totally reflected, there is no refracted ray, and θ_1 equals θ_2.

$$\sin(\theta_2)_{max} = 1 \qquad \text{(Eq 3)}$$

Substituting for $\sin \theta_2$ in Eq 1, the critical angle of incidence is therefore written as:

$$n_1 \sin(\theta_{critical}) = n_2(1) \qquad \text{(Eq 4a)}$$

$$\sin \theta_{critical} = \frac{n_2}{n_1} \qquad \text{(Eq 4b)}$$

$$\theta_{critical} = \sin^{-1}\left[\frac{n_2}{n_1}\right] \qquad \text{(Eq 4c)}$$

Why is the critical angle important to optical fiber function? Total internal reflection is a requirement to propagate light signals within the fiber core, and results only when the incident angle equals or exceeds the critical angle. In Fig 3, light has been launched into a fiber and has reached the cladding surface. For incident angles less than the critical angle, the light signal is refracted out of the core. Once it is out of the core, the light is absorbed by the cladding glass or the outer protective coating layers, which is discussed later. In some cases, the signal may continue to propagate in the cladding. In either circumstance, however, the signal cannot be recovered.

The maximum launch angle that results in internal reflection of a light ray is the acceptance angle. The sine of the acceptance angle is called the numerical aperture. Numerical aperture is important in determining the ability of a fiber to capture light from nonideal signal sources, where some of the light is received at relatively wide angles. Numerical aperture (NA) is calculated as:

$$NA = \sqrt{n_1^2 - n_2^2} \qquad \text{(Eq 5)}$$

where n_1 is the core refractive index and n_2 is the cladding refractive index. Through substitution of values into Eq 5, it is found that to guide light within a fiber, large differences in refractive indices between the core and cladding are not required.

A typical example of a single-mode step index fiber design is based on only a 0.4% relative difference in indices. Describing other properties that affect fiber performance, design, and manufacture requires that the simple model of light as rays be abandoned. In fact, a rigorous understanding of light propagation within fiber necessitates analysis of wave solutions to Maxwell's equations. It may be recalled from classical electromagnetic theory that four equations bring together various unconnected analytical descriptions of electric and magnetic field behavior. Light propagation within fiber is governed by these statements. Two constraints are imposed:

- Solutions to Maxwell's equations in the core and cladding must be independent of each other
- The electric and magnetic fields associated with the light signal are continuous at the core/cladding boundary

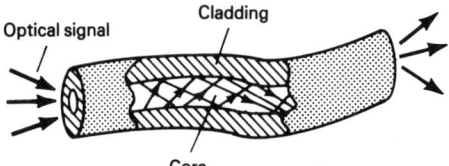

Fig 1 Regions within an optical fiber

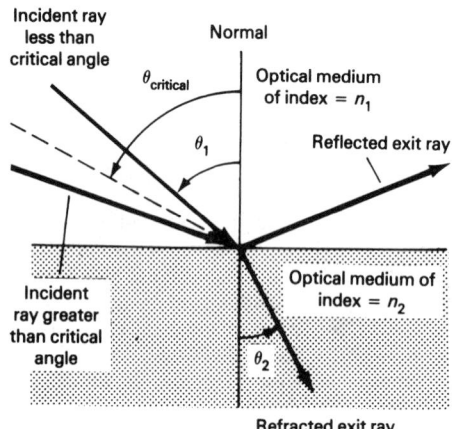

Fig 2 Mechanics of internal reflection

Solution of the equations results in an exact description of light propagation in an optical fiber, for which the following observations are made.

Modes. First, note that many solutions to the four Maxwell equations are possible by describing different spatial distributions of light. These distributions, or patterns, are referred to as modes. This result contradicts the intuitive notion of the ray mode: that light launched into a fiber bounces its way down the fiber in a random path. One of the most important features of a fiber design is whether the fiber supports the propagation of many modes or only a single mode. Accordingly, fibers are referred to as multimode or single mode. The following expression is used to calculate the total number of modes that can propagate in a fiber:

$$N = \frac{\alpha}{\alpha + 2} \, n_1^2 k^2 a^2 \delta \qquad \text{(Eq 6)}$$

where N is the number of modes, a is the core diameter, k is $2\pi/\lambda$ (where λ is the signal wavelength), n_1 is the core index, n_2 is the cladding index, α is the index profile constant, and δ equals the relative refractive index difference, which equals $(n_1^2 - n_2^2)/2n_1^2$.

The alpha term is a dimensionless constant, called the fiber index profile constant, which is based on the variation of fiber refractive index with radial position within the fiber radius. Typical index profiles are shown in Fig 4. Step index profiles feature constant index of refraction from the core center out to the cladding boundary, where a step decrease down to the cladding index value occurs. Typical graded index profiles exhibit parabolic decreases in index with increasing radial position. The shape of the index profile in the core is described by an expression that gives the value of index at any radial position, r, up to the cladding:

$$n(r) = n_1 \left[1 - 2\delta\left(\frac{v}{a}\right)^{\alpha} \right]^{0.5}, \text{ for } r \le a \qquad \text{(Eq 7a)}$$

$$n(r) = n_1(1 - 2\delta)^{0.5}, \text{ for } r \ge a \qquad \text{(Eq 7b)}$$

where n_1 is the core index, δ equals the relative refractive index difference, which equals $(n_{\text{core}}^2 - n_{\text{cladding}}^2)/2n_{\text{core}}^2$, a is the core radius, α is the refractive index profile constant, and r is the radius ($0 \le r \le a =$ core, and $r > a$ = cladding).

Figure 4 also illustrates the relationship between the profile shape and the value of the profile constant, where other parameters in the above expression are held constant. The number of modes supported is inversely proportional to the square of the signal wavelength. When other parameters in the expression are held constant, as wavelength increases, the number of modes decreases. The cutoff wavelength is the wavelength beyond which only one mode, the fundamental, can propagate. Systems that are designed for single-mode operation must be operated above the cutoff wavelength to ensure single-mode performance.

Scattering is an important mechanism, along with absorption and others, which results in attenuation in fiber. Attenuation is the loss of signal power, and is undesirable in any system. Scattering is described as the absorption of light by charges, which are caused to oscillate. Where the oscillation of charges causes secondary radiation at the incident light wavelength in different directions, scattering results. There are small refractive index fluctuations in all fiber materials, caused by inhomogeneities in material density or construction. These fluctuations are small, compared to the wavelength of light that propagates and results in Rayleigh scattering. The optical attenuation, experienced as a result of Rayleigh scattering, varies inversely with the fourth power of the wavelength, as shown in Fig 5.

Absorption is loss of the light signal, which is due to either conversion of the electromagnetic signal to molecular motion realized as heat or reradiation at a wavelength that is different from the incident wavelength. Absorption is present in fiber in two forms. The first, intrinsic absorption, occurs because of the inherent nature of glass itself, and is therefore not lessened by efforts to either refine the purity of the glass material or modify the manner in which the fiber is deployed. Intrinsic absorption of ultraviolet (UV) light occurs because of the presence of strong transition bonds in silica glass material, which causes light in the UV range (160 to 400 nm) to induce electrons in the material to jump energy bands. Reradiation of the light signal at unusable wavelengths results when these electrons return to lower energy states. Ultraviolet absorption causes high attenuation at shorter wavelengths, as shown in Fig 6. Infrared (IR) absorption, which is also important, is experienced because silica glass molecules are held together by bonds that are thermally excitable, or resonant, in the IR frequency range. A light signal of an IR wavelength causes atoms in silica glass to vibrate, thereby exchanging light energy for heat. Infrared absorption, as shown in Fig 6, increases significantly above 1600 nm and, in fact, limits the use of optical communications systems above this wavelength.

The second type of absorption of interest is due to impurities, particularly transition

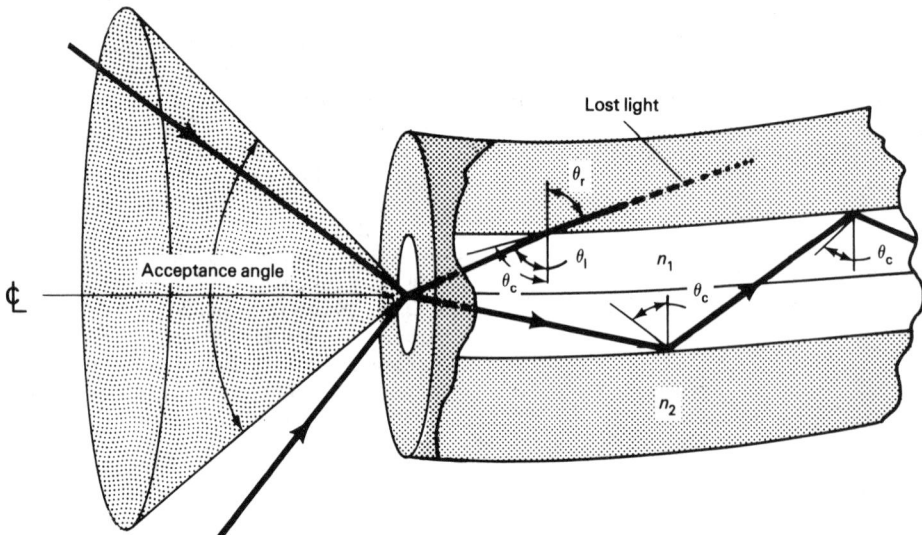

Fig 3 Light incident at angle greater than numerical aperture (acceptance angle) is not guided because critical angle is not exceeded. Refraction occurs at fiber-air boundary, according to Snell's law (θ_c not to scale).

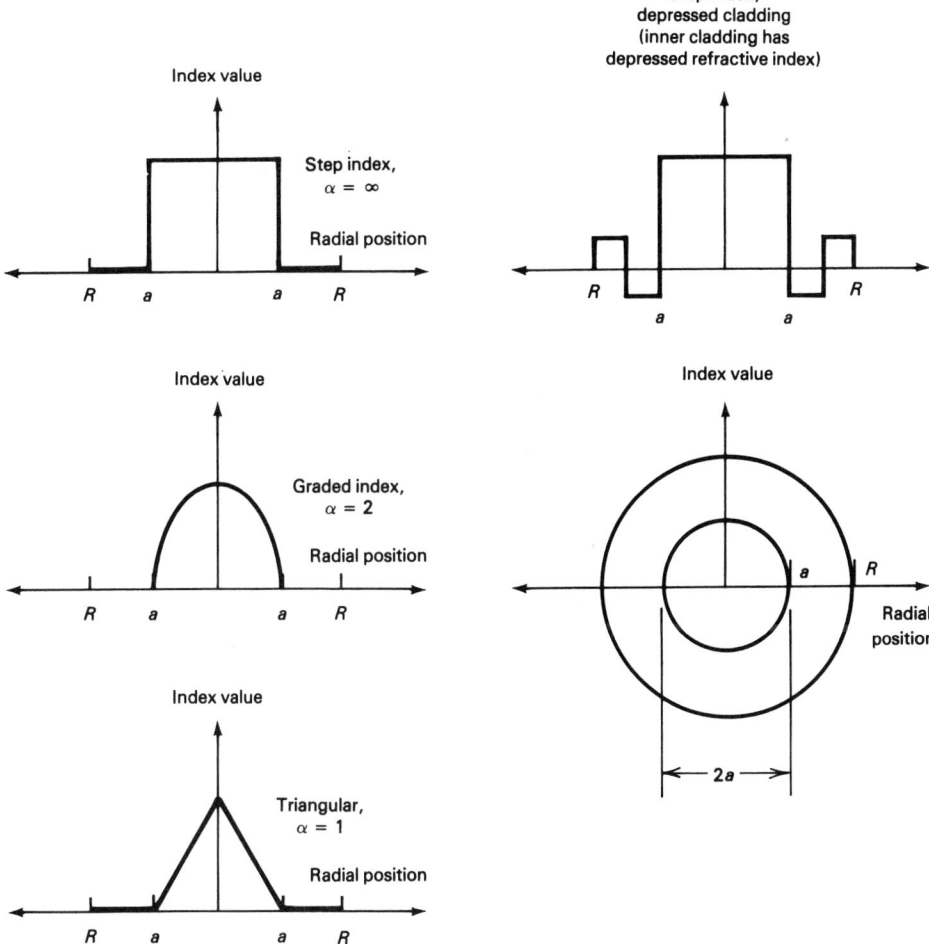

Fig 4 Typical fiber index profiles, which plot refractive index with radial position

metal ions and OH^- ions. Absorption in transition metals, such as iron (Fe), chromium (Cr), and copper (Cu), occurs because of the induced movement of electrons through energy states, as in UV absorption. The absorption of OH^-, which is pernicious to optical communication, is induced by vibration of OH^- ions in the wavelength of optical signals around 950, 1280, and 1400 nm. Optical systems are designed to operate in windows, or wavelength ranges, that avoid these OH^- ion absorption peaks.

A second observation relative to the Maxwell equation solutions is that the presence of light within the cladding is predicted. The energy present in the cladding is called the evanescent field. Associated with a light wave are sinusoidally varying electromagnetic field vectors that are perpendicular to the direction of propagation. As such, light is commonly referred to as a transverse wave. The polarization state of a light signal is defined as the spatial and temporal, or time-based, orientation of these two fields. In an unpolarized state, the field vectors are uniformly distributed in all directions, and the phase relationship between the two is random. Polarized light is characterized by electric and magnetic field vectors that have a fixed-phase relationship and either a uniform or constantly varying spatial orientation. Both reflection and refraction have their origin in the scattering process. When light is reflected on a dielectric surface, it undergoes a phase change. The

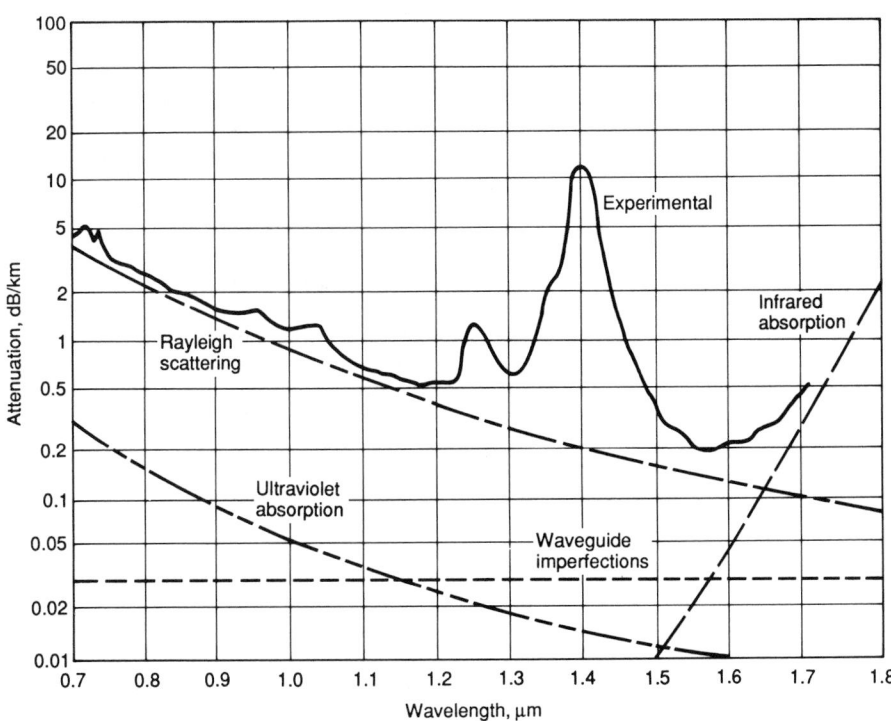

Fig 5 Optical loss behavior of fiber

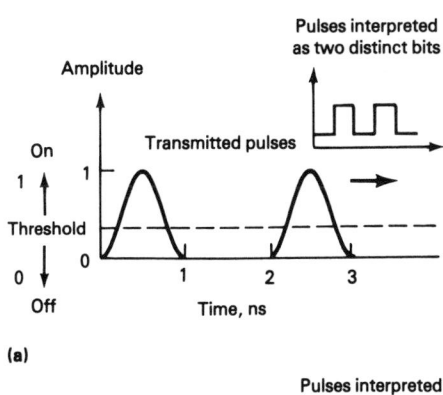

(a)

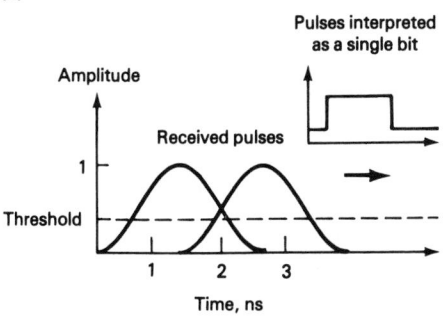

(b)

Fig 6 (a) Input pulses, which undergo pulse spreading. (b) Output pulses, which are interpreted as one pulse because the received power does not fall below the on-off bit threshold

magnitude of the change is determined by the state of polarization of the incoming light and the angle of incidence. It is the difference in phase of the incident and reflected waves that accounts for the presence of the evanescent field. Although the evanescent field exists in all fiber, its effect is much more pronounced in single-mode fiber.

Mode Field Diameter. The presence of an evanescent field is also important for determining how to characterize the light-carrying ability of fiber. Because of the prominence of the evanescent field in single-mode fiber structures, a parameter called the mode field diameter is defined. This parameter is a measure of the actual diameter of light, or spot size, that propagates in a a fiber. In a single-mode fiber, the mode field diameter is typically greater than the physical glass core diameter, whereas in a multimode fiber, the mode field diameter is roughly equal to the physical glass core diameter. Hence, multimode fibers are characterized by core diameter and numerical aperture, and single-mode fibers are described by mode field diameter. The mode field diameter depends on the core radius, wavelength of light, and indices of refraction of the core and cladding. An empirical expression for mode field diameter, w_o, is:

$$w_o = 2a(0.65 + 1.619V^{-1.5} + 2.879V^{-6}) \quad (Eq 8)$$

where a is the fiber core radius and V is the normalized frequency. In Eq 8, the core diameter and normalized frequency determine the mode field diameter. The normalized frequency is a very useful mathematical tool for characterizing the optical behavior of fiber, and therefore appears in many mathematical representations of the behavior of light in fiber. The expression for normalized frequency groups the fiber index, core size, and the signal wavelength in one expression, where

$$V = \frac{2\pi a}{\lambda \sqrt{n_1^2 - n_2^2}} \quad (Eq 9)$$

It was previously noted that a fiber operating above the cutoff wavelength permits only single-mode light propagation. Above the cutoff wavelength, the normalized frequency value, V, is less than or equal to 2.405.

Dispersion

Fiber-optic systems have the ability to support a very high information transmission rate. These rates can be so high in fiber-optic systems that they cannot even be obtained in copper-based media. Notwithstanding, there are limitations on information transmission rate, and the communications industry has instituted measures that make it possible to draw relative comparisons between different fiber designs with respect to rate capacity. To understand the mechanisms that limit transmission rate, it is necessary to separately discuss transmission rate in single-mode and

multimode fibers because the mechanisms that limit it are different. Several concepts that are germane to either fiber type are first discussed below.

A digital signal is a stream of light pulses of some amplitude measured in dBm, the $\log_{10}$ of the optical power measured relative to a 1 mW reference signal. A typical fiber-optic system is designed to transmit at a bit rate that typically exceeds 10^6 bits/s, or 1 Mbps. Analog systems are based on either amplitude- or frequency-modulated sinusoidal carrier waves and normally operate well in excess of 1 MHz.

The mechanism that limits the transmission rate is dispersion. The net result of dispersion is straightforward. In a digital system, the ability to distinguish an 'on' bit from an 'off' bit is lost because the threshold power of the leading edge of one pulse cannot be discerned by the detector from the trailing end of an adjoining pulse. In an analog amplitude-modulated system, when a detector cannot discern a given light amplitude from some other value, the result is distortion of the signal carried in the fiber or conducting cable. In Fig 6, the problem is illustrated in a digital system where two pulses are launched. Ideally, these are sharp step index pulses. In practice, pulses are slightly rounded as launched into the fiber, as shown in Fig 6. After propagating through the fiber for some distance, the pulses spread in the time domain. The threshold between an on and off state is difficult to discern because of overlap of the received pulses. This phenomenon of pulse broadening is called dispersion.

The types of dispersion found in fiber, with consideration for whether the fiber is single-mode or multimode, are summarized in Table 1. In multimode fiber, two types of dispersion are present: intramodal dispersion and intermodal dispersion.

Intermodal dispersion is unique to multimode propagation. Its existence is a consequence of the presence of more than one mode propagating in a fiber. To describe it, the ray model of light propagation developed earlier needs to be considered. In Fig 7(a), two rays are shown. One ray, which travels down the center of the fiber core, is characterized as a lower-order mode. It concentrates energy along the center region of the core. The second ray, which takes much steeper angles relative to the centerline, is a higher-order mode. It tends to concentrate energy toward the outer region of the core. Because a higher-order mode takes steeper angles, it has a longer distance to travel to reach the end of the fiber. In a multimode fiber, a pulse may consist of hundreds of modes distributed among high- and low-order modes. Consequently, the time of flight of the high- and low-order modes differs because of the path length differences. The result is that the low-order modes reach the light detector first, whereas the high-order modes arrive later. This is intermodal dispersion. An elegant remedy

Types of dispersion found in single-mode and multimode fibers

Mode	Dispersion
Multimode	Chromatic(a)
	Waveguide(c)
	Material
	Intermodal(b)
Single-mode	Chromatic(a)
	Waveguide
	Material

(a) Also known as intramodal dispersion. (b) Also known as modal dispersion. (c) Negligible

for this delay distortion is the graded index profile fiber.

Graded index fibers feature progressively lower indices of refraction as the distance from the fiber center increases. Refractive indices are highest in the core center and diminish parabolically (for α = 2) to the lowest value at the core/cladding boundary. As discussed before, light speed in a material varies inversely with the index of the material. The result is that light propagating in the core center will travel slower than light in the outer core region. Path length differences between modes ideally are compensated by introduction of flight speed differences through control of the index profile. A practical example is considered below.

For a 20 Mbps digital bit stream, pulses are separated by about 50 ns. An expression that gives the maximum difference in arrival times between modes in step index fiber is:

$$\delta t = \frac{L n_{core}}{c} \frac{n_{core} - n_{cladding}}{n_{core}} \quad (Eq 10)$$

where n_{core} is 1.50, n_{clad} is 1.485, c is the speed of light (2.988×10^8 m/s, or 9.836×10^8 ft/s), and L is the fiber length (1 km, or 3.3×10^3 ft). Substitution of the values provides a delay of 50 ns. After a 1 km (3.3×10^3 ft) flight in this fiber, the leading edge of a pulse may be indistinguishable from the trailing edge of a neighboring pulse. The signal cannot be demodulated accurately at this rate, and the system experiences a high bit error rate. On the other hand, a graded index fiber design is capable of 0.5 ns/km (0.15 ns/10^3 ft) delays, enabling this system to run 100 times faster, or at the same speed more than 100 times the distance.

Intramodal, or chromatic, dispersion dominates in single-mode fiber, although it occurs in multimode fiber, as well. Single-mode transmission rates are much faster because this type of dispersion is typically much less pronounced than intermodal dispersion. Chromatic dispersion consists of two phenomena, material and waveguide dispersion.

Material dispersion, like all forms of dispersion, causes a pulse to broaden. Broadening occurs either when some component of a light pulse travels faster or slower than other components, or when the optical path lengths

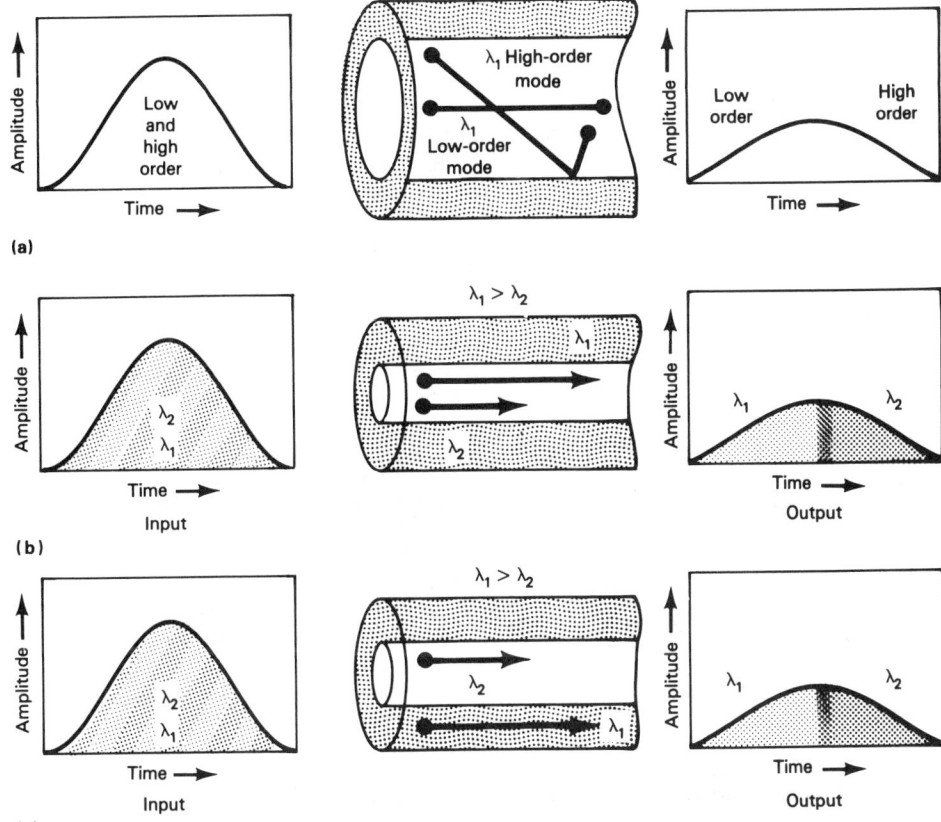

Fig 7 Intermodal dispersion (a) and chromatic (intramodal) dispersion in two forms, material (b) and waveguide (c)

are different for different components. In the preceding discussion of multimode intermodal dispersion, the components described were modes. A single-mode fiber supports the propagation of only one mode, the fundamental mode, and thereby eliminates intermodal dispersion.

In single-mode fiber, dispersion originates because the multiple wavelengths that are present in a pulse travel at different speeds. Although light sources are designed to supply light at a single wavelength, such as 1300 nm, they emit over a range of wavelengths, in practice. The actual output from a laser designed for 1300 nm operation may normally occur over a 4 to 5 nm spread around 1300 nm, such as 1298 to 1303 nm. Less expensive LEDs emit over a wider spectral width. This means that the light source is not strictly monochromatic; many wavelengths are present in the same mode within the fiber. Material dispersion exists because different wavelengths in a pulse propagate at different speeds, which causes their arrival times to the detector to vary. The existence of speed differences follows from the dependence of the index of a material on the wavelength, and the dependence of the speed of light on the index of the medium through which it travels.

The speed of a light component in a material is dependent on the wavelength of the light component. Generally, higher wavelengths

go faster than lower wavelengths, as is depicted in Fig 7(b). It also is true that by building a more narrow spectral width source, pulse broadening that is due to material dispersion is lessened.

Waveguide dispersion, shown in Fig 7(c), has its basis in the speed difference between light in the cladding, where the index is usually lowest, and the core light, where the index is higher. It has been concluded that for a given wavelength, light that propagates in a low-index material travels faster than that moving in a high-index material. The speed difference between core and cladding light is the cause of waveguide dispersion. It follows that a fiber that could restrict nearly all light to within the core (as in multimode fiber) has nearly zero waveguide dispersion.

Material and waveguide dispersion curves for a typical single-mode fiber are shown in Fig 8. At the zero dispersion wavelength, the sum of the material and waveguide dispersions equals zero. The zero dispersion wavelength occurs at about 1310 nm. This is the wavelength at which many long-distance communication systems are designed to operate. Also, from the 1/wavelength[4] dependence of attenuation that is due to Rayleigh scattering, it is desirable to operate a long-distance communication link at a higher wavelength, where lower optical signal attenuation is experienced. By shifting the zero

dispersion wavelength to 1550 nm, high transmission rates are possible over longer distances. A fiber design that accomplishes this does so with an innovative index profile design, and is called a dispersion-shifted fiber.

Bandwidth. In the discussion of dispersion, optical power was represented as some signal amplitude changing with time, as is the case when describing a digital pulse. The dispersion performance of a fiber can be described in the frequency domain; that is, in terms of amplitude versus frequency of pulses. A Fourier transform is a mathematical tool that is used to convert a description of a signal from the time domain into the frequency domain. Bandwidth is a parameter that is typically used to quantify the transmission rate capacity for multimode fiber. By convention, bandwidth is defined as the frequency at which the amplitude in the frequency domain falls to one half, or 3 dB, of its value at dc or zero frequency.

Optical Fiber Glass Materials

Silica, the most prominent constituent of optical fiber, is also one of the most abundantly found materials on earth. As beach sand, silica (SiO_2) is a crystalline solid. Crystalline materials are produced when liquids are cooled at a rate that provides enough time for rearrangement of molecular structure into an ordered and symmetric state. However, optical fiber, which is also glass, is amorphous rather than crystalline. Amorphous materials, which are somewhat random in structure, are produced by cooling at a rate that prevents formation of an ordered structure.

Glass, including optical fiber silica, is a brittle material. Up to the point of failure under load, glass exhibits elastic behavior. Although the stress-strain curve for glass is not strictly linear, ideally, it can be said that when a linearly increasing stress is applied to glass, the strain that is experienced is proportional to the applied stress up to the point of failure. This behavior is unlike that of ductile materials, which, when subjected to such loading, undergo linearly increasing strain up to the yield point, where plastic, nonlinear strain with permanent deformation occurs. It is the presence of surface flaws that causes glass to break under load. The theoretical modulus of optical fiber glass, without flaws, is about 21 GPa (3×10^6 psi), which is many times that of steel. However, this theoretical strength is not realized practically. A typical glass fiber can withstand a 3.4 to 4.8 GPa (0.5 to 0.7×10^6 psi) proof stress, compared to 0.34 to 1.4 GPa (0.05 to 0.2×10^6 psi) for hardened steels. Glass fiber is commonly screened during manufacture to meet a minimum 0.34 to 0.69 GPa (0.05 to 0.1×10^6 psi) proof stress level.

The requirement for low absorption drove early fiber material research. Optical systems were considered impractical because early designs exhibited very high optical attenuation. The high attenuation values observed were largely the result of an inability to remove

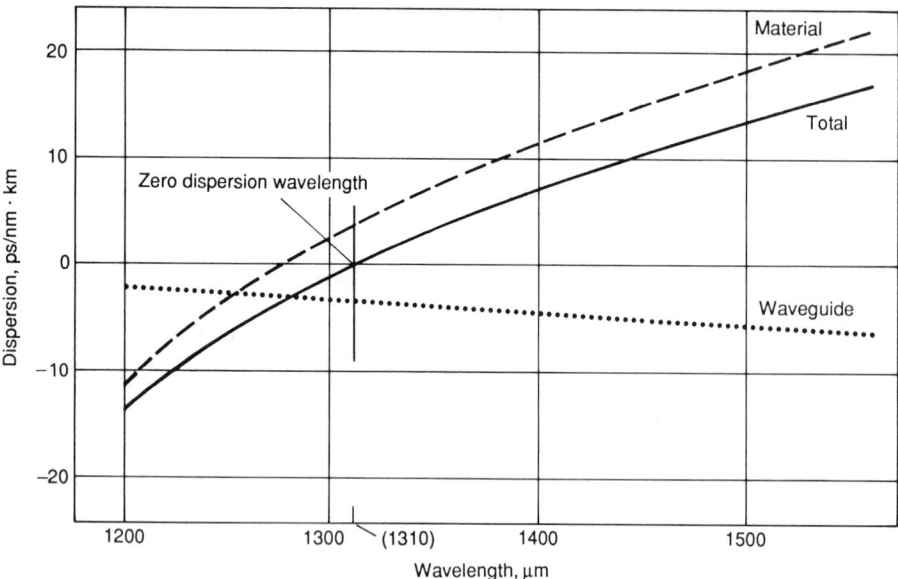

Fig 8 Chromatic dispersion curves for single-mode fiber. At 1310 nm, sum of material and waveguide dispersion equals zero, which is where system information rates are highest.

contaminants such as iron and chromium, which are key contributors to absorption. Of the many types of glasses in existence, fused silica is used in nearly all fiber designs because it can be manufactured to very high levels of purity, with the contaminant level reduced to the part per billion level.

Fiber Manufacture

The production of a hair-thin strand of glass capable of low-loss signal transmission over miles requires extremely pure raw materials and a process that allows the controlled introduction of dopant materials as a function of radial position within the fiber, while preventing the introduction of contaminants. Conventional glass manufacturing methods, which involve large-scale melting of materials in refractory-lined furnaces, were not appropriate for fiber manufacture because the presence of impurities caused high optical absorption losses and low-strength behavior. Therefore, the method used to deposit semiconductor materials, also called sputtering, became the precursor to current fiber-making techniques. Transistor operability is made possible by a precisely controlled deposition of organic metallics (metal halides) in thicknesses of the order of 10^{-10} m (4×10^{-9} in.). In optical fiber manufacture, this concept is applied to produce silica glass preforms that are thousands of times larger than a transistor. A large mass of glass of the required purity can then be drawn to produce fiber.

Outside vapor deposition (OVD) and modified chemical vapor deposition (MCVD) are the two principal methods of fiber manufacture. The four processes involved in each method are laydown, sometimes called preform deposition; consolidation; draw; and

measurement. In some cases, these processes can be combined.

OVD Laydown and Consolidation. The goal in laydown is to deposit glassy material onto a surface, layer by layer, eventually building up a mass that can be drawn into fiber, after other operations are performed. In the reactive sputtering process, which is heterogeneous, reaction of the halides occurs on the surface or substratum.

In the OVD process, the reaction of oxygen with metal halides (such as germanium tetrachloride and silicone tetrachloride) occurs in the air while the materials being deposited are still in the vapor phase. This is a homogeneous reaction that defines the first part of the laydown process, soot deposition. A soot particle is tiny, typically less than 0.2 μm (8 μin.) in diameter.

To form soot, a vapor stream of $SiCl_4$ and O_2 is first passed through a methane-oxygen flame. The halide reactants are introduced to the flame by a transport process. The most widely used transport system used to be the bubbler system, where a carrier gas is passed through containers of reactants with high equilibrium vapor pressures. Already possessing a strong tendency toward vaporization, the reactants are swept along with the carrier gas, which flows into the burner. The loose pickup or bonding of reactant material by the carrier gas is called entrainment. Now, a direct vaporization system is used because it allows more precision in the control of dopant material.

The flame is directed toward a rod known as the target, or bait rod, which is inspected dimensionally to ensure minimal runout when rotated during laydown. Although a number of reactions occur in the flame, the fundamental homogeneous reaction is:

$$SiCl_4 + O_2 \rightarrow SiO_2 + 2Cl_2$$

As is characteristic of hydrocarbon reactions, one by-product is water. Because the presence of water, in the form of OH ions, is detrimental to the attenuation performance of fiber, the goal of a later process step is to eliminate it entirely. Water combines with the chloride formed by the reaction of silica tetrachloride and oxygen, producing hydrochloric acid and oxygen:

$$2H_2O + 2Cl_2 \rightarrow 4HCl + O_2$$

The preform is built up layer by layer. Soot deposition is aided by thermophoresis, which is the tendency of a gas particle suspended in a thermally gradient environment to move from a high temperature to a lower temperature. To understand this, an idealized situation is considered, in which a gas particle is bordered on one side by high-temperature molecules and on the other, by low-temperature molecules. Higher temperatures are characterized by more robust molecular oscillatory motion. One correctly expects that a net force in the direction of the lower-energy molecules would ensue from dynamic interaction of the electric fields of the molecules contained within the particles. During laydown, one observes that because of thermophoresis, soot particles are deposited about the circumference of the target rod, including cooler regions not directly in the flame path. Deposition of soot particles results in a porous, translucent mass that requires further processing.

Before it can be drawn into optical fiber, the deposited material must be fully sintered. Sintering is a high-temperature process that results in the consolidation of soot particles into a large, void-free bulk mass. Soot particles that strike and bond to the target rod are only partially sintered. During sintering, bonds between soot particles are developed either through production of a liquid phase or by solid diffusion. Two surfaces lie between two silica particles. These surfaces represent high-energy regions because atoms at the boundaries of the surfaces have neighbors only on one side, toward the interior of the particle. Application of the thermodynamic second law would establish a preferential state, where a single interface or grain boundary, rather than two surfaces, is produced, which is the result sought through sintering. The target rod rotates while the flame stream is swept back and forth to achieve a uniform thickness. It is porous because of those unconnected, or unsintered, surfaces. Index profiles are implemented by varying the concentration of dopant materials as the flame traverses the rotating rod. After a sufficient mass of soot is deposited, the target rod is removed and the next OVD process step, consolidation, commences. Figure 9 depicts both the OVD transport and laydown processes.

The major function of consolidation is to dry the mass of deposited soot particles, called a preform, and complete the sintering process. Densification of the preform and elimination of trapped gas bubbles occurs during consolidation. Sintering is completed by surface tension forces that act to minimize the surface area, and by diffusion of molecules to accommodate the process. The preform undergoes a reduction in volume because of the combination of surfaces between soot particles. Sintering is conducted in a controlled atmosphere at temperatures from 1000 to 1600 °C (1830 to 2910 °F), and results in a solid glass blank. The blank is placed in an electrically heated furnace, where Cl_2 is introduced to the center hole of the preform, created by removal of the target rod. The Cl_2 is introduced to remove residual water in the form of OH^- ions and H_2O. This forms hydrochloric acid, which is a vapor at consolidation temperatures, and is swept away in the exhaust gas flow.

MCVD Laydown and Consolidation. Like the OVD process, the MCVD process allows the production of various multimode and single-mode fiber designs at one laydown lathe station by introducing changes in the reactant flow composition and rate. The process chemistry, similar to that of the OVD process, features a homogeneous reaction of vapor-entrained materials. However, MCVD fiber laydown does have unique features, which are highlighted below.

A fused silica tube is obtained, inspected dimensionally, cleaned in acid, and mounted into a lathe. As in the OVD process, the lathe installation is checked to ensure that minimal runout occurs during rotation of the tube during laydown. A leak-free fluid connection is made to the inlet side of the tube. This helps establish low contaminant levels during the reaction and deposition. Next, a fire polish is performed to eliminate air bubbles and smooth the surface on which deposition will occur. The reactant vapor flow is then started.

Reactants are introduced to the tube using the bubbler system, where entrainment of $SiCl_4$ and dopant halides onto oxygen or a noble gas, such as argon or helium, occurs. Unlike the OVD process, reactant vapor flow and reaction do not occur within the flame. This leads to some interesting differences in MCVD process chemistry.

In the OVD process, reactant vapors are sprayed through nozzles which themselves are surrounded by nozzles that supply the methane and oxygen required for the flame. The reactants that produce the flame also join the reaction that produces soot particles. In the MCVD process, there is no such intermingling of flame and soot production materials. As shown in Fig 10, the flame in MCVD is directed toward the outside of the rotating tube, while reactant flow is in the inside. Thermophoresis also occurs during MCVD laydown. The flame impacts the outside surface of the tube. In the transverse plane of flame contact, the tube interior diameter surface and air temperatures are highest. Further down the tube, both the surface and gas temperatures are lowest. Particles undergo a homogeneous reaction and, through thermophoresis, deposit downstream on these cool regions. Figure 10 shows the reactant flow during laydown. Not all the soot particles reside on stream lines

that reach the surface. Those that do not are swept through the exhaust end of the tube, where processing to recover expensive dopant materials is initiated.

In both the OVD and MCVD processes, GeO_2 is used most often as the dopant. An important consideration in the selection of dopants is the influence that dopant composition has on the sintering temperature of the reactants. For instance, phosphorus may be used to decrease the required temperatures. Dopant selection also influences the optical loss behavior of the fiber. Other dopants, such as fluorine, are used to reduce the core or cladding index below that of silica. Profile designs are implemented by controlling the introduction of dopants into the reactant flow stream.

Consolidation in MCVD occurs continuously during the laydown step. As soot is deposited, the flame traverses the outside diameter of the tube and sinters the deposited particles. After a sufficient mass is agglomerated, collapse of the tube is initiated by increasing the flame temperature. Traversal of the rotating tube is continued until the rod is collapsed into a nonporous, clear solid ready for draw. Here again, elimination of hydroxyl, OH^-, is accomplished using chlorine, which reacts with the unwanted water to produce hydrochloric acid.

In the MCVD process, the starting silica tube forms the outer cladding after the draw. Therefore, deposition begins with the inner cladding, which has a much greater purity than the outer silica tube. Deposition of high-purity clad material helps prevent the entrance of OH ions that cause high optical attenuation through absorption near the 1300 nm wavelength. Deposition of successive layers proceeds inward, ultimately forming the core of the fiber, while index profiles are constructed by controlling dopant material flow rates into the carrier gas stream.

Vapor axial deposition (VAD), which is a derivative of OVD, is another fiber manufacturing method. In OVD laydown, deposition of layers occurs radially outward, by an end-to-end traversal of a rotating target rod by a reactant flame flow. In VAD laydown, burners are positioned at the end of a vertically oriented target rod, so that buildup of the preform occurs axially, not radially. The raw material transport mechanism and process chemistry are otherwise conceptually alike. In VAD, GeO_2 is also used as the primary dopant material.

Deposition starts on a rotating silica target. The burners are fixed. The chemical reaction, as in the OVD process, is homogeneous and may include combustion gases. As deposition of soot particles proceeds, the preform grows axially. To maintain the correct distance between the flame and the preform, the preform is lifted as it grows, in accordance with a feedback system based on optical position sensing of the preform end. Index profile control in VAD fiber manufacture is accom-

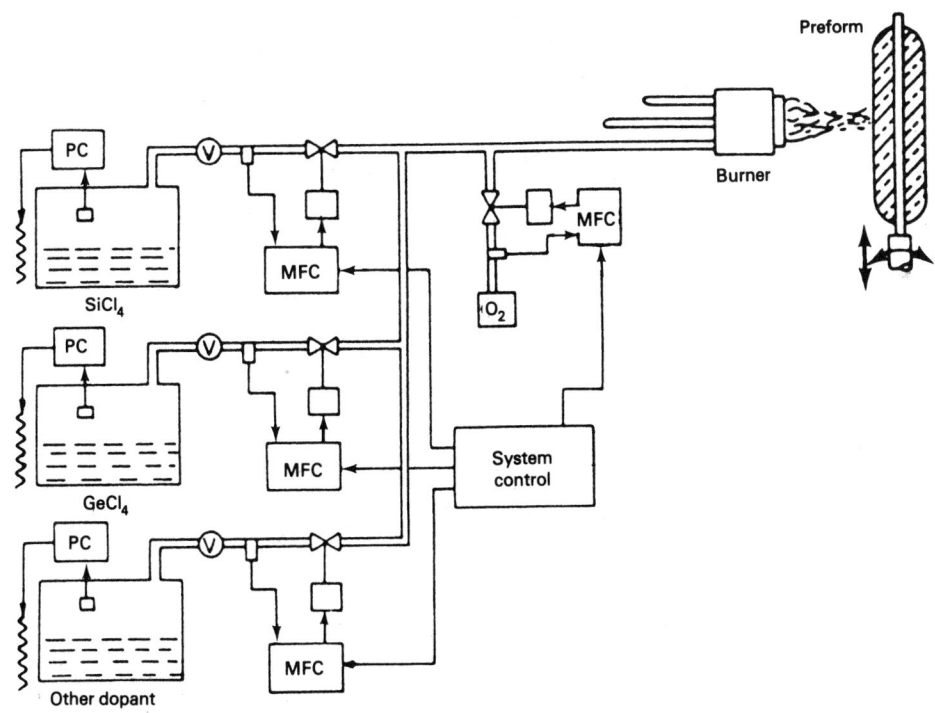

Fig 9 Outside vapor deposition transport and laydown. PC, pressure controller; V, valve; MFC, mass flow controller

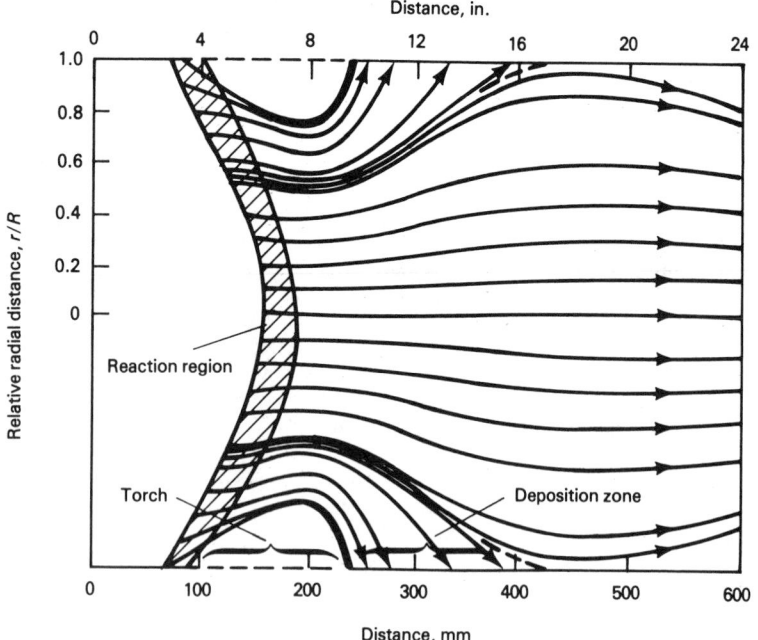

(a)

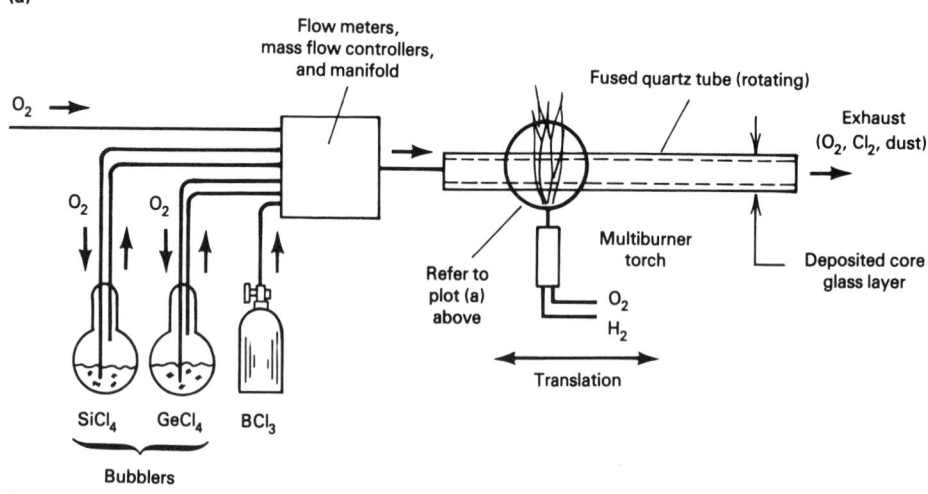

(b)

Fig 10 (a) Particle trajectories resulting from temperature field. (b) MCVD glass preform fabrication

draw tower. A draw furnace produces temperatures greater than 1800 °C (3270 °F), causing the tip of the blank to flow. An operator grabs the melted tip and pulls it into a tractor reel assembly, which is manually controlled to continue the initial draw until the fiber strand is loaded through a diameter measurement sensor, which is also part of the tractor system.

The function of the tractor system is very simple, conceptually. The sensor provides an instantaneous measurement of fiber diameter. The diameter of the fiber is inversely proportional to the speed at which it is drawn by the tractor pulleys. A control system governs the speed of the draw using a fiber diameter feedback loop. Such systems are capable of a very precise fiber diameter control (less than ±2 μm, or 80 μin.) of variation. The fiber is fed through the diameter feedback control system until its diameter falls within the specification tolerance. At this point, the startup length of fiber is broken off, and the intolerance follow-on fiber is fed into tractor pulleys, which establish the draw at a steady rate. Downstream, the fiber is directed through

plished through two mechanisms: preform temperature distribution and raw material mixing. Temperature distribution is controlled by varying the fuel and oxidizer delivery to the flame. The flame is directed at the preform tip, which is rounded. Experiments have demonstrated that for a correctly sized flame, parabolic temperature distributions result at the tip of the preform. The shape of the distribution is varied by the flame temperature and flow conditions. These conditions include flame incident angle and flow speed. The ability to influence the temperature distribution at the preform tip can result in an ability to control the index profile as described below. The intensity of the reaction between silica tetrachloride ($SiCl_4$) and germanium tetrachloride ($GeCl_4$) (the dopant raw material) is influenced by the surface tem-

perature. If the surface temperature is increased, a higher index profile constant is achieved, and an opposite result is obtained where the surface temperature is lowered. Therefore, at the tip where temperatures are highest, the index is highest, because of greater inclusion of dopant material in the silica matrix. Raw material mixing is also required to achieve accurate index profile control. As described earlier, the torch features multiple concentric jet rings. Oxygen and hydrogen flow from the outermost ring. Dopant vapors are added one ring inward. The center jet may add silica tetrachloride. By changing the ratio of dopant to silica flow, the index profile can be changed.

Fiber draw is basically the same for all fiber manufacturing processes. A fiber blank is placed in a furnace located at the top of a

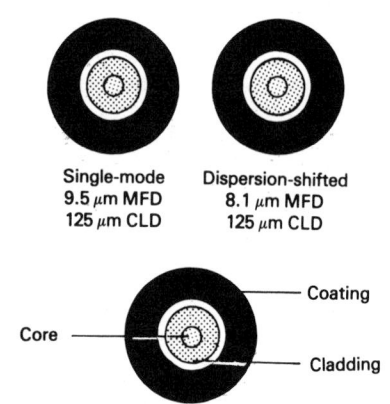

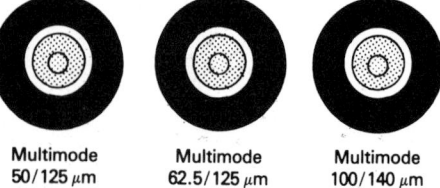

Core
• Transmits the light signal
• Germania doped silica glass

Cladding
• Keeps the light signal within the core
• Pure silica glass

Coating
• Protects optical fiber from abrasion and external pressures
• UV cured acrylate

Fig 11 Typical fiber sizes. MFD, mode field diameter; CLD, cladding diameter; NA, numerical aperture

a coating application system. In most systems, one or two coats are applied, although one manufacturer uses two acrylate urethane coatings. The coatings are cured by exposure to UV radiation supplied by lamps downstream from the coating applicator nozzles. A variety of fiber sizes are currently produced to support telephony, private premise wiring, data communication, and sensor applications. Figure 11 depicts typical fiber sizes.

Quality control measurements are very important to the manufacture of fiber for two reasons. First, comprehensive measurements help ensure that a product meets customer requirements. Second, measurements provide feedback to the manufacturing process, and therefore serve as tools to identify areas for further process refinement and trends that indicate future problems.

Fiber performance measurements for multimode fibers usually include:

- Attenuation at 850 and 1300 nm
- Bandwidth at 850 and 1300 nm
- Numerical aperture

- Core diameter
- Continuous strength proof testing
- Glass geometry
- Coating geometry

Measurements for single-mode fibers include:

- Attenuation at 1300 and 1550 nm
- Chromatic dispersion at 1310 and 1550 nm
- Cutoff wavelength
- Mode field diameter
- Continuous strength proof testing
- Glass geometry
- Coating geometry

ACKNOWLEDGMENT

The author wishes to thank Mike Blankenship, Anne Bussard, Margot Botecho, Doug Wolfe, John Ritter, Mark Robinson, Melody Setter, and Dr. Kumar Vethanayagum for providing valuable technical and stylistic comments on this article.

SELECTED REFERENCES

- F.C. Allard, *Fiber Optics Handbook for Engineers and Scientists*, McGraw-Hill, 1990, p 1.1–1.44
- P.J. Dean, *Optical Fiber Communications, Principles and Practice*, Prentice Hall International, 1985, p 68
- P. Gossing and G. Mahlke, *Fiber Optic Cables*, John Wiley & Sons, 1987, p 14, 26
- J. Hecht, *Understanding Fiber Optics*, Howard W. Sams, 1988, p 50–80
- T. Li, *Optical Fiber Communications*, Vol 1, Academic Press, 1985, p 3, 8, 74, 184
- S.E. Miller and I.P. Kaminow, *Optical Fiber Telecommunications II*, Academic Press, 1988, p 121–163, 175
- H. Murata, *Handbook of Optical Fibers and Cables*, Marcel Dekker, 1988, p 43–49
- J.C. Palais, *Fiber Optic Communications*, Prentice Hall, 1984, p 123
- S.D. Personick, *Fiber Optics, Technology and Applications*, Plenum Press, 1985, p 25

Glass Spheres

Charles D. Hendricks, International Microproducts, Inc.

GLASS SPHERE production methods are critically dependent upon the size and character of the spheres. The production methods used to manufacture a wide variety of spheres have been thoroughly monitored to examine in great detail the parameters of the processes.

The types of glass spheres can be divided into three general classes: hollow, porous, and solid. Hollow and porous spheres can be subdivided according to shell thickness and pore size and whether the pores are connected through channels or are separate and distinct closed cells. It may also be important from an application standpoint if the outer surface of a porous sphere is a smooth gas-enclosing layer or the pores connect to the surrounding space.

Another obvious division of the general class of glass spheres is on the basis of size. However, aside from reference to particular applications, there are no specific size divisions other than small, medium, and large. Industrially useful glass spheres range in size from under 10 nm (100 Å) to a few centimeters in diameter. Decorative spheres and spheres used for some optical applications and other special purposes may be as large as 0.5 to 1.0 m (1.6 to 3.3 ft) in diameter. It should be noted that some techniques lend themselves to the production of particular sizes of spheres and are totally inappropriate for the generation of spheres of other sizes.

The spheres have applications in the fields of recreation, optics, medicine, industry, architecture, and research and development. Among the many uses and applications for glass spheres of various sizes are:

- Bead blaster cleaners
- Filters made of sintered glass beads
- Reflecting particles in paints
- Spacers in liquid crystal displays
- Chemical boiling balls
- Marbles
- Fine powder reference spheres
- Fission reactor fuel carriers
- Laser fusion targets
- Shot peening of metals
- Distillation columns
- Reinforcement filler for plastic furniture products
- Optical elements
- Decorative elements

- Crystal balls for fortune tellers
- Medical treatment carriers
- Catalyst carriers

The list consists of typical applications and is not a complete tabulation of all the uses and applications of glass spheres of various types.

Glass spheres can also be classified on the basis of composition. Glass includes a very wide variety of materials:

- Single compounds such as glasses of a single-element oxide (for example, silicon dioxide or quartz)
- Oxides other than those of silicon that are glass formers but are usually found in combination with other compounds to form glasses with particularly desirable properties (for example, quartz, borates, lead glasses, beryllium and phosphate glasses, and soda-lime glasses and some special-purpose glasses such as those compounded from the oxides of tantalum, tungsten, yttrium, and zirconium)

The line between glass, glassy spheres, and glasses that have a definitive crystal structure is finely drawn by some, but the focus in this article will be on the manufacturing or the fabrication technique and the process technology. The science of glass formulation and the control of chemical and physical properties of glasses by composition are separate fields by themselves and both subjects are only mentioned because of specific applications for spheres having unique glass compositions.

The formation of a glass sphere is a fairly simple process. Because glass can be melted and resulting liquids in free fall will be transformed into spheres by surface tension forces, a very logical way to make spheres is to melt a lump of glass and let it cool while it falls through the air. If the molten glass has a low viscosity and high surface tension and if the fall time is long enough, the glass will spheroidize and cool before it hits the bottom of the chamber. Small spheres are often produced using this simple and obvious technique.

Solid Glass Spheres

Several interesting and useful processes have been developed for the industrial production of small solid glass spheres. The production of glass spheres used in reflecting paint requires a molten column of a glass of appropriate composition having a high index of refraction that emanates vertically downward from a furnace (Ref 1, 2). A high-temperature, high-velocity flame is caused to impinge on the column of molten glass perpendicular to the side of the column and blows the glass into small particles that remain fluid long enough for surface tension to transform them into perfect or nearly perfect spheres. This process unavoidably generates a broad distribution of sizes that may require sieving to provide the correct size for a particular application. Nevertheless, the process is in use commercially to generate many tons of small glass spheres annually.

Glass sphere applications range from esoteric applications such as spheres used as an *in situ* medical treatment for patients suffering from liver cancer to the production of glass spheres used to manufacture toy marbles.

Marble Production. Some of the individual marbles are truly works of art that may have been individually handmade from sections of various colored glass canes which were melted and twisted into exotic patterns, cut into sections, fused, and finally lapped into nearly perfect spheres. Indeed, some of the spheres were made from naturally occurring glasses and other minerals such as obsidian, flint, malachite, or natural marble or similar materials. However, the high cost of such works of art usually prevented their use on the playground. A cost-effective mass production technique was needed to produce large quantities of glass marbles that were affordable.

A simple system is used to make glass spheres for marbles, which are approximately 13 mm ($1/2$ in.) in diameter. The glass materials are raised to a working temperature in the furnace and combined in the desired amounts to form patterns, swirls, and so on. A soft, workable stream of glass emerges from the furnace and is cut mechanically into sections having the correct mass to form spheres of the desired size. The soft pieces of glass fall into the space between two counter-rotating coaxial cylinders in which there are opposing grooves. The hot soft glass pieces move along the grooves and are simultaneously rolled into spheres and cooled to a nonworkable temperature. By the time the glass pieces have moved through the entire length of the two cylinders, they are relatively cold and have been transformed into marbles.

Glass Beads for State-of-the-Art Medical Treatment. From the production of nostalgic glass marbles, we turn to the production of spheres of glass that have a unique composition, are much smaller than marbles, and are used for more sophisticated applications.

There are niches in today's high-technology world for specialty products, and this is also true in glass sphere technology. As an example of a special use for glass spheres, consider the application of uniquely formulated glass spheres to combat liver cancer.

The treatment of liver cancer is difficult because of the characteristics of the organ and its location in the body. It is desirable to introduce radiation treatment for liver tumors directly to the tumor from the inside of the liver in order to increase the efficiency of the treatment and to avoid damage to other surrounding organs. To irradiate tumors inside the liver, spheres of a unique formulation of glass are first activated in a nuclear reactor and then injected into the liver through its blood supply. This unique glass includes as a major component an oxide of yttrium that was formulated by Dr. Delbert Day of the University of Missouri at Rolla and MOSCI, Inc. When it is neutron activated, the $^{90}_{39}Y$ in the glass becomes a β-emitter (2.273 MeV) with a 64-h half-life. The oxides and other basic glass-forming components are mixed, melted, cooled, ground, sieved, and remelted into spheres 20 μm (800 μin.) in diameter. Spheres of this size can be injected into the blood stream at a point that leads to the liver so the spheres can find their way into the capillaries and become lodged directly inside the liver. There they emit β-particles (high-energy negative electrons) that irradiate the liver tumor from within. The half-life range of the β-particles is short enough that subsidiary damage to healthy tissue is significantly reduced.

The production of the medicinal spheres illustrates some critical aspects of glass sphere production:

- There was a need for a specialty glass that Dr. Day was able to formulate and produce
- Spheres were required to have a specific diameter to permit injection and flow in the blood stream
- Spheres needed to lodge in the capillaries of the liver blood supply and radioactively decay to irradiate the cancerous tumor

This is an example of a need for a particular composition and size of glass spheres, the success in formulating the chemical composition of the glass, and developing and using the techniques for producing the spheres.

Hollow Glass Spheres

Glass spheres need not always be solid. In fact, many applications for glass spheres require them to be hollow with very thin walls. If the fly ash from such large coal-fired power plants such as the one at Four Corners (the geographical point where Utah, Arizona, New Mexico, and Colorado intersect) is examined carefully, it is found that most of the small particles are hollow glassy spherical shells. The shell walls are generally filled with impurity particles and tiny bubbles and have widely variable thicknesses. There may be holes in the walls and the shells may not be perfect spheres but they are hollow glassy shells.

The shells in furnace fly-ash are formed when small coal particles containing silica, sulfates, and other inorganic materials are raised to high temperatures in the firebox of the furnace. When the coal particles are injected into the high-temperature region of the firebox, the outer layers of the particle are flash heated and the carbon is burned away. The remaining materials melt and become a molten glassy material that is probably still permeable to oxygen and other gases. As the inner materials burn and the sulfur-containing compounds decompose and the outer layers become more impermeable, the gaseous products such as carbon dioxide and sulfur oxides blow the glassy material into a tiny bubble. Because of the irregularities in the particle size, shape, and composition, the resulting shells cannot be expected to be high-quality uniform-wall spheres. However, the shell layer consists mostly of glass and is gas impermeable. The size distribution of the shell is remarkably narrow considering the irregularities of the entire process. The smallest shells are probably a few micrometers in diameter, and the largest shells are a few hundred microns in diameter.

Although there are some applications for the fly-ash glass shells, probably the primary importance to the engineer is the possibility of making higher-quality hollow glass spheres by some well-controlled process in a hot flame or a furnace.

Process for Forming Glass Particles in Combustible Gas Using a Burner Head. A process developed in the 1930s for the production of glass shells was different enough from the fly-ash process that it was patentable, but it did have some similarities. The glass-forming chemicals such as sodium silicate, calcium carbonate, potassium compounds, borates, and so on, were mixed into a slurry with water and then dried. The resultant cake was pulverized into small particles that could be sieved for size and introduced into a high-temperature flame or furnace (Ref 3). In the high-temperature region, the particles melted to form a glass and simultaneously were expanded into hollow shells by the gases produced by the decomposition of the glass-forming components. Because surface tension acting on a liquid body tends to make the body spherical, the hot glass shells became spheres. When the product was examined under a microscope, it was found that the shells were imperfectly shaped, though to a lesser extent than were the glass shells in the fly-ash generated by power plant furnaces. These glass shells contained wall imperfections, bubbles, irregular wall thicknesses, and non-spherical shells. The general bulk applications of the shells did not require perfect spheres and hence, minimal further processing of the shells was necessary.

Controlled Composition Process Yields Shells Having Uniform Walls. A more sophisticated process for producing higher quality and more controlled composition glass spherical shells was developed in the period from 1960 to 1970. In this process, soluble glass-forming components are dissolved into an aqueous solution that allows ideal mixing of the components. In addition, a specific blowing agent such as urea is added to ensure adequate and controlled blowing of the spheres during the glass formation process. The solution is dispersed through a spray nozzle into a heated chamber. The liquid droplets in the spray then dry into solid but irregular particles in the hot chamber and are scooped up, bagged, and stored for future use. To form glass spheres, the dry particles are introduced into either the gas supply or the air/oxygen supply of a burner. As the dry particles travel through the burner flame, the particles melt to form a glassy sphere and the blowing agent decomposes to form gases that blow the glassy sphere into a glassy shell. The trajectory of the particle carries the particle through both the higher and the lower temperature regions of the flame. In the high-temperature environment, the particle becomes more fluid and also expands because the gas inside is heated. In the lower temperature regions, the thin portions of the shell cool more rapidly and become less fluid than the thick portions. There is still an overpressure in the sphere as the decomposition products continue to evolve and the less viscous, thick portions are further expanded and thinned. As the particle finds itself again in a hotter zone of the flame, the entire particle is heated and continues to expand. Under the influence of surface tension forces, of internal pressure fluctuations, and of the heating and cooling of the walls, the glass shell is ultimately formed into a spherical shell with relatively uniform walls.

Glass shells formed in this manner are of relatively high quality and have applications as a filler in the plastic used for bottle-type containers, boat hulls, microwave lenses, and so on. These applications require relatively uniform hollow spheres but not the perfect surface finishes nor the perfectly uniform walls that are required for some other purposes. However, the shells are satisfactory and can be produced in large lots with a specified composition and are exceptionally useful for many applications. Spherical shells made by these techniques have been marketed by Emerson-Cuming Corporation under the tradename Eccospheres. The distribution of shell diameters is from 20 μm (20 μin.) to

300 μm (0.0008 to 0.012 in.). Wall thicknesses vary from 0.5 μm to a few micrometers. Typical commercial applications for small hollow glass shells include:

- Polymer filler (for boat hulls, furniture, microwave lenses, building materials, plastic containers, and reinforced fiberglass-epoxy)
- Mixer and filler for solid and slurried explosives
- Paint filler and extender
- Paper coating materials
- Filler for artificial rubber products
- Inertial fusion fuel containers (noncommercial at present)
- Filler for low-density foams
- Floating layer to prevent petroleum product evaporation in storage
- Hydrogen storage containers for hydrogen-burning engines

Controlled Composition Process for Solid Glass Beads. A similar technique has been used to form solid glass beads (Ref 3). Particles of glass bead-forming materials entrained in a gas stream that have both oxidizer and combustible components are projected from a burner head, and the gas is burned. One component of the combustible gas mixture with the entrained particles is introduced along one pathway to the burner head. The second gas component is introduced transversely into the first pathway, and is then thoroughly mixed with the first gas component and the entrained particles. The mixture is introduced into the burner head and with the resulting flame, the particles are melted to form glass beads.

Inertial Confinement Fusion Research Program. Hollow glass spheres with very specific quality characteristics were needed as containers for deuterium (D) and tritium (T) in the inertial confinement fusion (ICF) research program (Ref 4–6). The glass shells were to have the following characteristics:

Diameter, μm (in.)	20–1000 (0.0008–0.0400)
Wall thickness, μm (in.)	0.5–50 (0.00002–0.002)
Surface finish	R_y < 10 nm, where R_y is the peak-to-valley roughness height. Fewer than 5 peaks of 100 nm altitude with base width equal to wall thickness
Sphericity	>1%
Composition	As required by the ICF experiments and which is permeable to D_2, T_2, and DT at ~400 °C (~750 °F) and will contain DT gas up to 10 MPa (100 atm) at room temperature

Other glass shell criteria were the availability of reasonable quantities in predetermined diameters and wall thicknesses. This package of characteristics could not be met by commercially available glass shells at that time (1974). It was decided to go forward with a research and development program based

on work previously performed by Hendricks and his colleagues at the University of Illinois. Extensive research had been done at Illinois on liquid drop formation, shells of organic materials, metals, some inorganics, and liquid cryogens such as hydrogen, oxygen, nitrogen, and argon. Because of this previous work, the techniques were applied to the production of thin-wall glass shells with exceptional and reproducible quality and geometric parameters.

The basic process for the production of high-quality hollow glass shells begins with a very careful preparation of an aqueous solution of glass-forming components. The solution is introduced into a droplet generator that disperses the liquid into a single in-line stream of uniform drops, each of which contains just the correct quantity of solute to form one glass shell of the correct predetermined diameter and wall thickness. The drops are passed vertically downward through a warm column in which the drops dry to form small particles, each containing all the solute from the drop in which it started. Further downward, the particles pass into a region that is at a high enough temperature (~1500 °C, or 2730 °F) to melt the particles to form glass. Gaseous products generated in the glass-forming process from decomposition of some of the components produce enough pressure inside the molten particle to blow the particle into a bubble or shell that is refined as it undergoes the temperature variations in the high-temperature column. The shells are collected at the bottom of the high-temperature column. The fractional standard deviation in size has been made <0.1 for sizes of interest and >99% of the shells in a batch will meet or be better than the specified quality criteria. Figure 1 shows the basic process for making the glass shells. Figure 2 is an artist's rendition of the equipment and the technique.

The filling of the glass shells with the gases necessary to run the ICF experiments is accomplished by a variety of methods. As mentioned earlier in this section, deuterium and tritium can be permeated into the shells at high temperatures (T > 350 °C, or T > 660 °F) and kept in the shells by reducing the temperature while the external gas pressure is maintained at the desired internal pressure. After the temperature is reduced to normal ambient values, the external deuterium and tritium gases may be pumped away and the shells brought to normal atmospheric pressure for actual use or for storage.

Gases to which the glass is impermeable can be introduced into the shells during the formation process by filling the vertical column with the gases (for example, xenon, argon, krypton, and so on). During the glass sphere formation process, the shells are permeable to all gases, and as the shell is transformed into molten glass and becomes gas impermeable, gas at the column pressure is trapped in the shell and retained when the shell is finally collected at the bottom of the column.

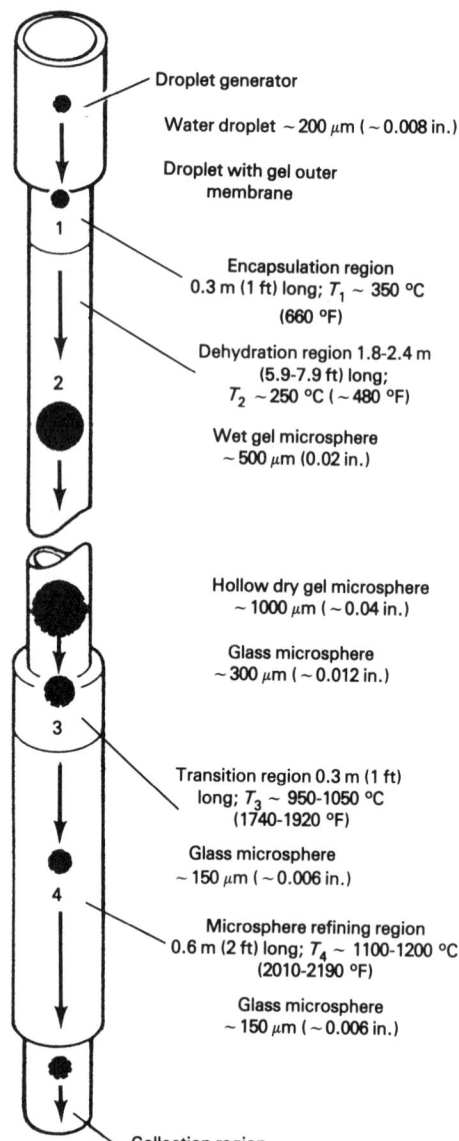

Fig 1 Schematic of a droplet generator used to produce glass shells for the ICF program

It has also been demonstrated that small holes (1 to 2 μm, or 40 to 80 μin. diameter) can be drilled through the shells, gases introduced into the shell, and the holes filled and sealed to retain the fill gases. To achieve the smoothness required in the sealing process is difficult, but it is not impossible.

Rotating Disk Process. Another method of generating glass spheres also lends itself to the production of fibers used in making glass wool insulation (Ref 7). A stream of molten glass impinges on the surface of a rotating metal disk. Depending on the particular process, the disk may have a serrated edge or spines around the edge or it may have a smooth edge. The stream of glass may fall on the disk at the center of rotation or it may be incident on the disk very near the edge. If the glass temperature is only slightly into the working region, the flow will be highly viscous and

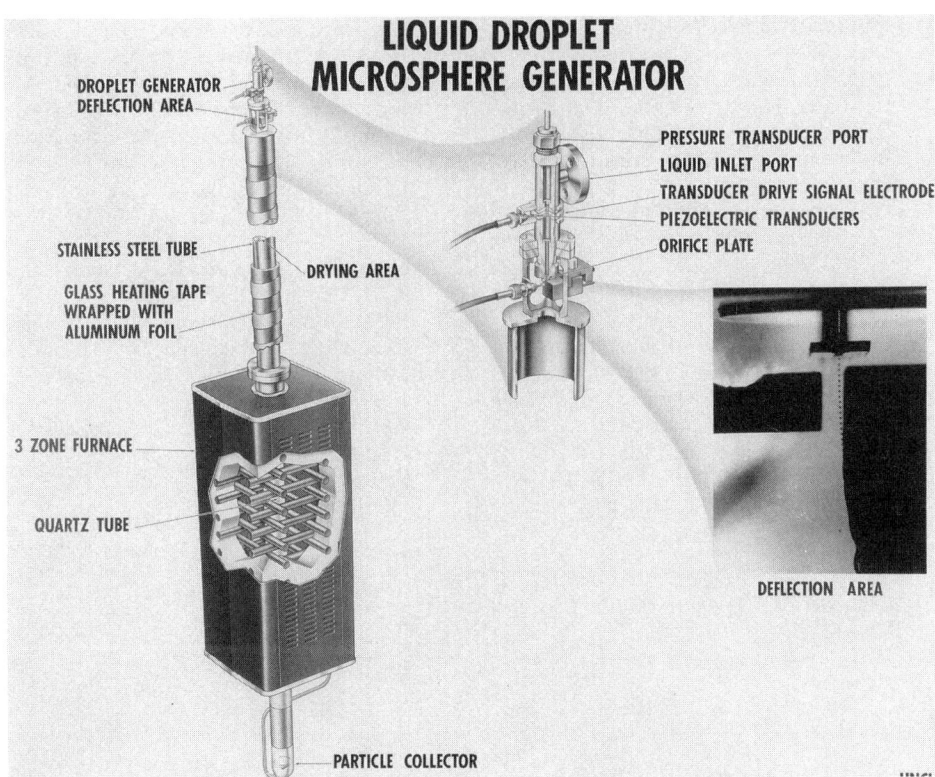

LIQUID DROPLET MICROSPHERE GENERATOR

DROPLET GENERATOR DEFLECTION AREA

STAINLESS STEEL TUBE

DRYING AREA

GLASS HEATING TAPE WRAPPED WITH ALUMINUM FOIL

3 ZONE FURNACE

QUARTZ TUBE

PARTICLE COLLECTOR

PRESSURE TRANSDUCER PORT
LIQUID INLET PORT
TRANSDUCER DRIVE SIGNAL ELECTRODE
PIEZOELECTRIC TRANSDUCERS
ORIFICE PLATE

DEFLECTION AREA

Fig 2 Key components of a liquid droplet microsphere generator used to produce hollow spheres

as the glass is thrown off the edge of the disk, the glass will form into long fibers that will cool down enough that there will be no breakup of the fibers into separate particles. Consequently, long small-diameter fibers will be produced. On the other hand, if the disk is hot, the glass remains hot, and if the rotational velocity is set appropriately, the filaments that form at the periphery of the disk will separate into short segments which are sufficiently fluid to form into spheres before they cool. To enhance the filament disintegration, hot gas jets can be used to break up the filaments. By using the rotating disk process for making glass spheres, very high rates of production can be achieved, a fact which makes the process attractive to industry.

Porous Glass Beads

Thixotropy. Exceptionally small spheres of silica have a property which is very useful in many fluid applications. If a small percentage by weight of 10 to 20 nm (100 to 200 Å) silica spheres is put into any of a large number of liquids, the liquids will gel. That is, they become almost like gelatin that is dissolved in hot water and allowed to cool. However, if the liquid that has gelled because of the addition of silica spheres is stirred, it is transformed into a flowing liquid (that is, it has become thixotropic). The material is virtually a solid unless it is subjected to shearing forces strong enough to make it move, at which point it becomes a low-viscosity liq-

uid. Paints that claim they will not run down the brush handle and down your arm when you are painting above your head are thixotropic. When the paint brush is dipped into the paint, the paint forms a jellylike blob on the bristles but when the brush and paint are stroked on the surface to be painted the paint flows as any paint should flow. The small spheres of silica are only a few tens of nanometers diameter and are marketed under several trade names (for example, Cab-O-Sil and Thixo) and are made by burning silane (SiH_4) in oxygen to form water and silica (SiO_2) spheres as products of the reaction. Another use for such spheres is to produce very smooth and creamy drink mixes (such as canned or bottled Tom Collins or daiquiri mixes). If silica or silicon dioxide are listed as an ingredient on the label, there is a strong possibility that you will consume small silica spheres in your mixed drink.

Hydrolysis of Silicon and Zirconium Alkoxides. A production method for porous glass beads was reported in the patent literature in 1987 (Ref 8) wherein the glass beads were made in a water solution. An initial solution of silicon and zirconium alkoxides in a water-soluble solvent (for example, a lower aliphatic alcohol) is stirred in a liquid dispersant to form droplets of uniform size that gel into hardened beads of condensed silicon and zirconium hydroxides. The beads are separated from the liquid phases and dried at a high enough temperature to produce the desired mixed-oxide structure for the beads.

Glass beads of titanium, aluminum, and other metal oxides may also be formed by using this technique.

Droplet Generation Method and Sol-Gel Processing. Professor K. Kim at the University of Illinois has developed an interesting method of preparing hollow glass shells that may have either porous or gas-impermeable walls (Ref 9). In this process, a solution of organic silicon compounds is formed into uniform liquid shells around a gas bubble. The shells pass through a suitable atmosphere (for example, a gas containing ammonia) and are converted into a rigid gel material. Subsequent drying converts the shells into silicon dioxide aerogel. A wide range of characteristics is obtained by using this process, which is currently being utilized to produce the shells for ICF research.

Rotating Wheel Process. A unique method of dispersing a stream of hot molten glass into small particles involves the interaction of an electric current in the glass stream and an externally applied magnetic field (Ref 10). The transverse force (measured in newtons) on the stream of glass owing to the interaction of the electric and magnetic fields is $\mathbf{I} \times \mathbf{B}$, where $\mathbf{I}$ is the magnitude of the electric current in amperes and $\mathbf{B}$ is the magnitude of the magnetic flux density in tesla. If either the electric current or the magnetic field are time alternating functions, the stream of glass will undergo an alternating transverse force which, if made large enough, will cause the stream to disintegrate into spherules of glass that spheroidize because of the surface tension before cooling.

Production of Large Spheres

While the generation of spheres as small as <5 to 10 mm (<0.20 to 0.4 in.) in diameter is typical, it should be pointed out that there are also techniques for making large spheres. One method is to start with a multisided, fairly regular lump of glass that is roughly spherical. This large lump of glass is then put into a lapping machine that produces a sphere which has very good characteristics as far as the desired radius and overall surface finish and sphericity are concerned. The process is rather expensive when compared to the cost of producing small spheres in ton lots, but if large spheres are needed, the expenditure per sphere may be cost effective. Other large spheres are used for decorative and specialized optical purposes, and such spheres may first be cast into molds and then lapped to achieve the desired characteristics.

Future Outlook

The ancient art of glass working combined with a large number of modern and highly imaginative and creative techniques have made it possible to obtain inexpensive glass spheres of excellent quality in sizes ranging from a few tens of nanometers to many centimeters in diameter. There will be continual advances and developments in the art, science, and technology of making glass spheres in the future.

REFERENCES

1. C.C. Bland, "Apparatus for Production of Glass Beads by Dispersion of Molten Glass," U.S. patent 3,252,780, May 1966
2. C.C. Bland, "Method and Apparatus for Production of Glass Beads by Dispersion of Molten Glass," U.S. patent 3,243,273, Mar 1966
3. H. Neusy, "Method and Apparatus for Manufacturing Rounded Vitreous Beads; Entrainment of Glass Forming Particles in Combustible Gas and Projection from a Burner Head," U.S. patent 4,487,620, Dec 1984
4. C.D. Hendricks, Fabrication of Targets for Laser Fusion, UCRL-76380, Lawrence Livermore National Laboratory, *Bull. Am. Phys. Soc.*, Series II, Vol 20, Oct 1975, p 1230
5. C.D. Hendricks and J. Dressler, Production of Glass Balloons for Laser Targets, UCRL-78481–1, Lawrence Livermore National Laboratory, *Bull. Am. Phys. Soc.*, Oct 1976, p 1137
6. C.D. Hendricks, A. Rosencwaig, R.L. Woerner, J.C. Koo, and J. Dressler, "Fabrication of Glass Sphere Laser Fusion Targets," UCRL-81415 (Rev 1), Lawrence Livermore National Laboratory, *Bull. Am. Phys. Soc.*, 11 May 1979
7. E.M. Guyer and J.E. Nitsche, "Method and Apparatus for Manufacturing Glass Beads," U.S. patent 3,313,608, Apr 1967
8. O. DePous, C.J.G. Oliver, and M. Schneider, "Porous Spherical Glass Filtrating Beads and Method for the Manufacturing Thereof by Hydrolysis, the Condensation of Alkoxide of Silicon and Zirconium," U.S. patent 4,713,338, Dec 1987
9. N.K. Kim, K. Kim, D.A. Payne, and R.S. Upadhye, Fabrication of Hollow Silica Aerogel Spheres by a Droplet Generation Method and Sol-Gel Processing, *J. Vac. Sci. Technol.*, Vol A7, 1989, p 1181
10. P.D. Law, "Method of and Apparatus for Production of Glass Beads by Use of a Rotating Wheel," U.S. patent 3,310,391, Mar 1967

Laminated Glass

William H. Dumbaugh, Corning, Incorporated

GLASS LAMINATION involves fusing two or more glasses to produce a material with properties that would have been either difficult or impossible to obtain in a single glass. There are many multi-step lamination processes. Usually, a glass is formed during one process and is subsequently laminated with another glass during a separate process. Probably the earliest example of glass lamination is the decoration of one glass with another to impart an attractive appearance. A brief summary of this decorative process is provided in Ref 1. The discussion below only describes lamination processes that occur simultaneous to the forming of the glass article, resulting in strong composites.

Thermal tempering has historically been the standard method used for strengthening (Ref 2). It is accomplished by chilling the surface of a glass article as it cools from an elevated temperature. The body of the article continues to shrink after the surface has become rigid, thus subjecting the surface to compression. The article will not break until this compressive stress has been overcome. If a glass with a specific thermal expansion is clad during the forming process with a glass that has a lower expansion, strengthening occurs, and characteristics different from those associated with thermal tempering are attained.

The stress profiles that result from strengthening methods are shown in Fig 1. Much higher ratios of maximum compression to maximum tension can be obtained in laminated ware than in tempered ware. The violence of breakage of tempered ware is therefore greater for the same maximum compressive stress because tempered ware has a much higher internal tensile stress. In addition, tempered glass must be relatively thick in order to achieve effective strengthening, whereas laminated glass need not be.

Lower expansion cladding glasses have been used to achieve strength in composite systems with bodies of optical glasses, glass-ceramics, ceramic insulators, and china. The process of applying the overlay is discontinuous, that is, two or more process steps are required. However, the lamination of glasses during the forming process not only eliminates process steps, but also results in defect-free fused glass surfaces.

Process

The basic process comprises melting the two glasses in separate furnaces, and then passing them through a platinum delivery system such as the one shown in Fig 2. The core glass passes through an inner chamber and the cladding glass passes through smaller chambers on both sides of the inner chamber. The glasses meet at slotted orifices at the bottom of the delivery chambers. The total thickness and the body-to-cladding thickness ratio are determined by the size and relative position of the orifices. Figure 2 shows the delivery of a flat sheet through rollers that adjust dimensions, and then deliver the sheet to another forming machine. In this case, the forming machine is a rotating wheel with molds placed around its circumference. As the wheel rotates, the sheet overlays the molds and a portion sags into a shape, such as a dinner plate. A trimmer then cuts the plate from the sheet, partially wrapping the cladding glass around the rim of the core glass. As the wheel rotates, a vacuum pick-up removes the plate from the mold and places it on a conveyer belt while the rest of the sheet drops into a cullet (glass discard) bin.

This hot-lamination process can be adapted to other shapes, such as rods (Ref 4) or tubing (Ref 5), through proper design of the delivery system. It is also possible to manufacture composites with more than three layers by building a delivery system with the appropriate number of chambers and slots.

Applications. Two laminated glass products that are commercially available are dinnerware (Ref 3, 6) and C3 resistor substrates (Ref 4) produced by Corning, Inc. The process to manufacture the dinnerware is essentially that shown in Fig 2, using opal as the core glass. In the case of resistor substrates, a laminated rod of about 1.5 mm (0.06 in.) in diameter is used. This rod is coated with tin oxide, and the resistor is manufactured from this composite.

Glass Requirements

Glasses that are used to form strong composites have special requirements beyond those usually required when manufacturing a single glass.

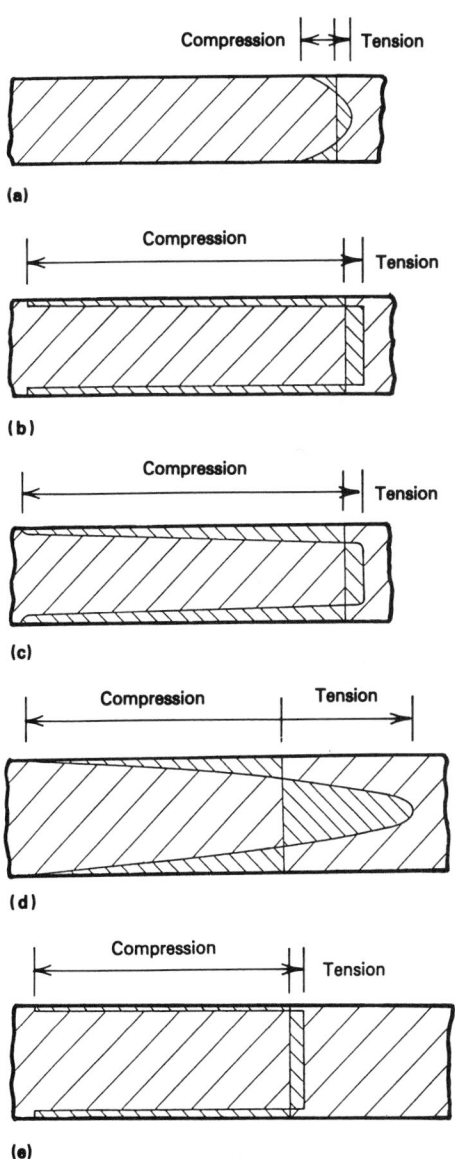

Fig 1 Stress profiles of strengthened glass. (a) Tempered. (b) Laminated, having same maximum internal stress as in (a). (c) Tempering superimposed on laminated system. (d) Tempered. (e) Laminated, having same maximum compressive stress as (d). Source: Ref 3, 4

The viscosity of the cladding, at the temperature where the glasses join, should preferably be the same as or lower (by up to a factor of 4) than that of the body. The dense, white opal body of the patented-process dinnerware presents special problems because the opacity is due to phase separation that occurs during the forming process. The glass must phase-separate very rapidly during the forming process, but cannot cause an inordinately large and sudden viscosity increase, thereby preventing further shaping of the composite.

Composition Compatibility. The core and cladding should not differ radically in their chemical nature for two reasons. First, there is a certain amount of interdiffusion of the mobile ions at the interface at elevated temperatures, and it is possible to generate undesirable reactions that produce bubbles or devitrification. If conductive orifice materials are used, an electrolytic cell may be set up, causing gaseous evolution at a metal-glass interface.

Second, when shapes are formed from a larger sheet, the resulting cullet must be remelted for use as raw material for the core glass. Therefore, the cladding glass must not contain elements that are not present in the core glass. Compositions that meet this requirement can be generated using:

$$X = \frac{GY(1 - F) + Z(1 - G)}{1 - GF} \qquad \text{(Eq 1)}$$

where X is the percent of a given oxide in the core glass, Y is the percent of a given oxide in the cladding glass, Z is the percent of a given oxide in the batch fill, F is the fraction of laminate that is cladding, G is the fraction of fill that is cullet. This equation is based only on the oxide content of the batch and does not factor in the volatile losses from such batch materials as carbonates.

Thermal Expansion. The strength of the composite depends primarily on the relative thermal expansion of the two glasses. Because Young's modulus and Poisson's ratio do not vary from composition to composition nearly as much as expansion values do, they are of secondary importance. The cladding must have a lower expansion than the core glass to induce surface compression.

Chemical durability and weathering characteristics are controlled solely by the cladding. The unique feature of composites is that body durability is not important. Therefore, many glasses that have poor durability can be useful as the core in a composite.

Cladding and Body Glasses

Two examples of composite glass pairs, R_b and R_C versus C_b and C_c, where the subscripts represent body and cladding, respectively, are shown in Table 1. The R pair represents the first pair of glasses that was

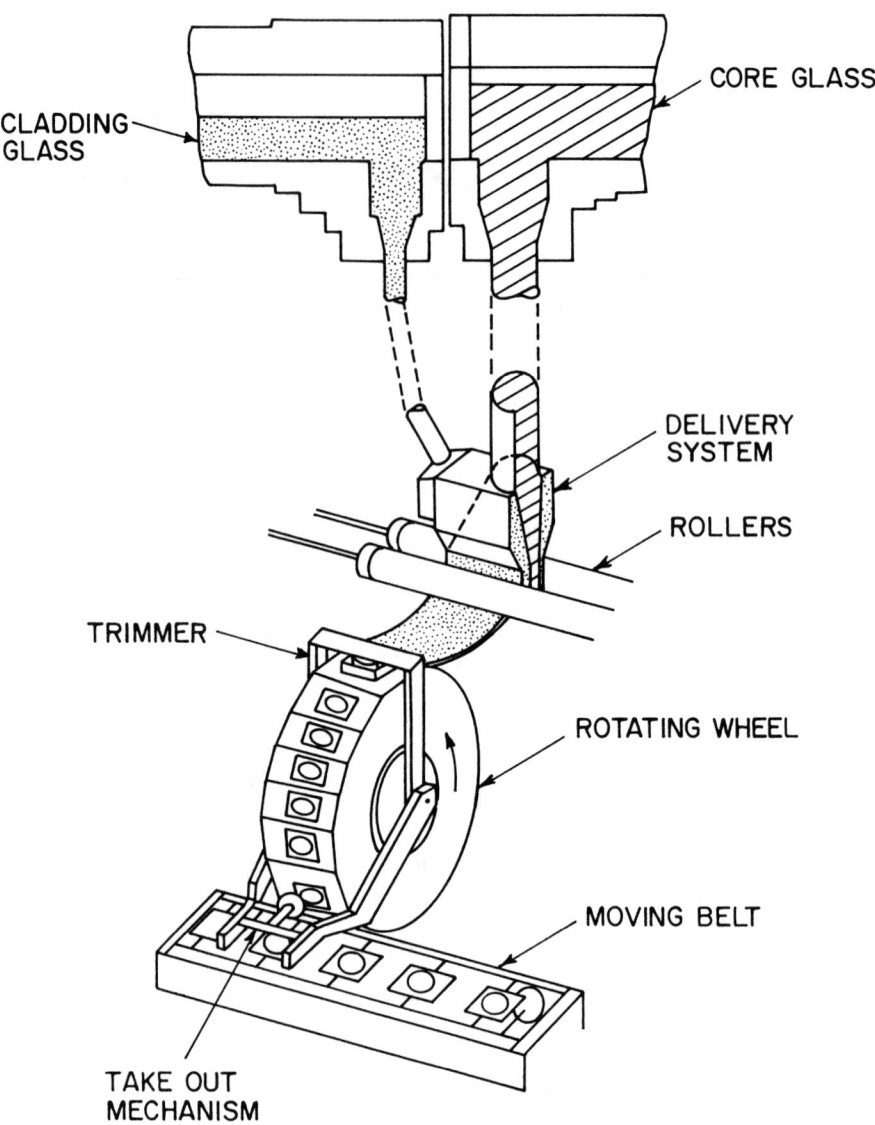

Fig 2 Process for forming laminated sheet glass

Table 1 Glass compositions and properties

	Body clear, R_b	Cladding clear, R_c	Body opal, C_b	Cladding clear, C_c
Composition, wt%				
SiO$_2$	58	58	64	58
Al$_2$O$_3$	20	15	6	15
B$_2$O$_3$	...	4	5	6
Na$_2$O	13	...	3	...
K$_2$O	4	...	3	...
MgO	2	7	1	6
CaO	3	10	15	15
BaO	...	6	...	...
F	...	...	3	...
Softening point, °C (°F)	863 (1585)	910 (1670)	...	890 (1635)
Annealing point, °C (°F)	633 (1171)	712 (1314)	610 (1130)	710 (1310)
Strain point, °C (°F)	588 (1090)	665 (1230)	563 (1045)	670 (1240)
Coefficient of thermal expansion from 0–300 °C (32–570 °F), 10^{-7}/K	92 (198)	46 (115)	71 (160)	48 (118)
Density, g/cm^3	2.48	2.63	2.47	2.57
Liquidus temperature, °C (°F)	1058 (1935)	1114 (2037)	...	1089 (1990)
Young's modulus, GPa (10^6 psi)	74.5 (10.8)	86.2 (12.5)	75.1 (10.9)	85.5 (12.4)
Poisson's ratio	0.22	0.24	0.22	0.25
Stress in cladding, MPa (ksi)(a)	344 (50)		207 (30)	
Stress in cladding, from tempering, MPa (ksi)(a)	414 (60)		241 (35)	

(a) Laminate property

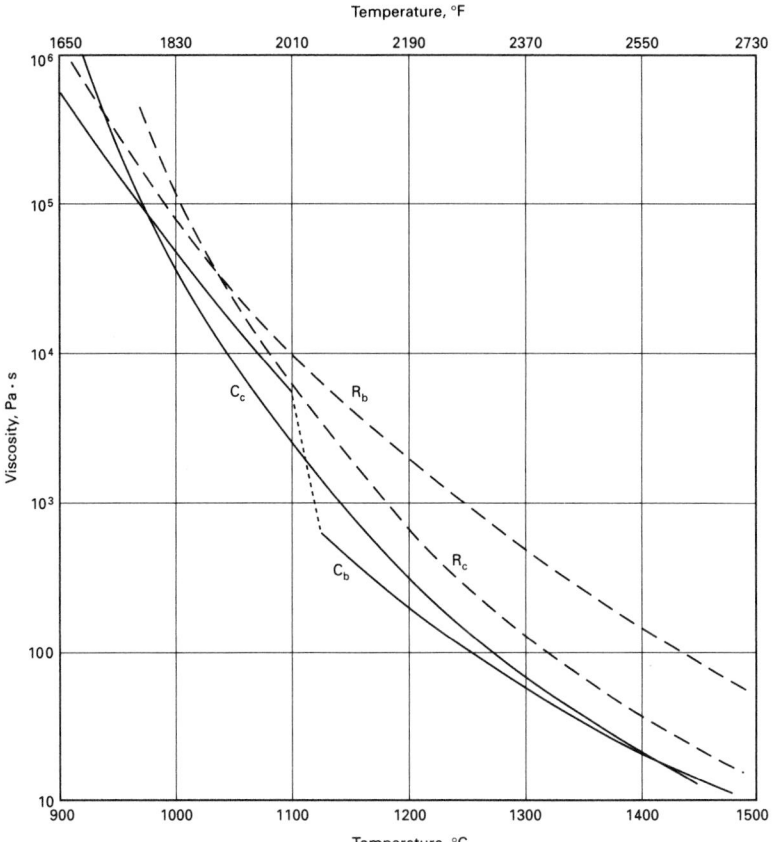

Fig 3 Viscosity curves for R_b, R_c, C_b, and C_c glasses

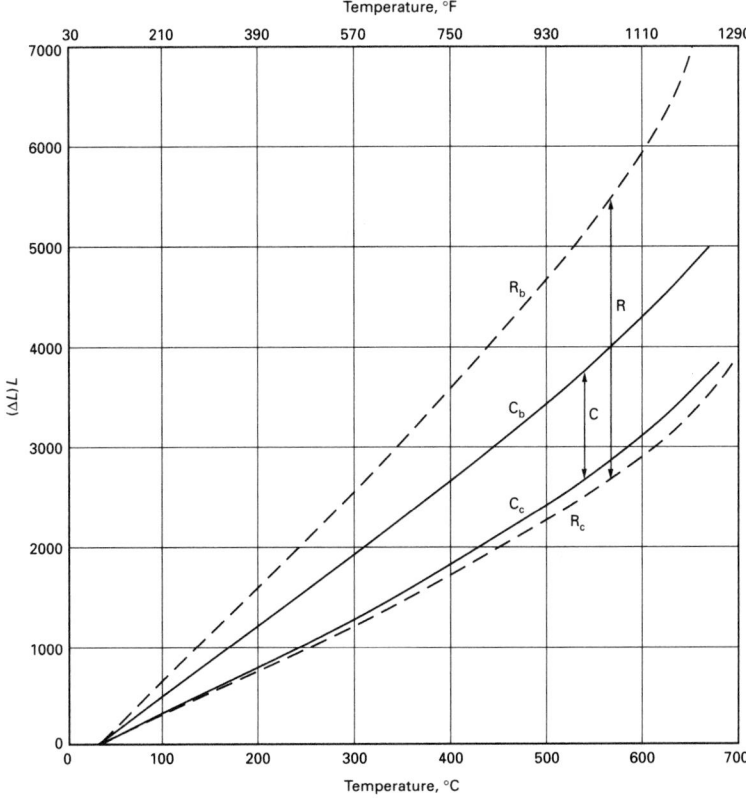

Fig 4 Thermal expansion curves for R_b, R_c, C_b, and C_c glasses. Expansion mismatch at the setting point of the lower-viscosity glass is indicated by R and C.

successfully laminated using the continuous process previously described. However, the expansion mismatch between cladding and body (see Table 1) is a bit too high, and they fail the chemical compatibility requirement. Glass R_b is not designed to accept composite cullet because it does not contain barium. Under certain conditions (using a platinum delivery system for cane), an electrolytic cell was established, although this phenomenon was not encountered with a sheet geometry. Viscosity curves for each glass are shown in Fig 3, whereas expansion curves are shown in Fig 4.

The C pair exemplifies the effective use of all glass requirements described above. Their viscosities are close enough (in the forming region) to easily form laminated sheet in a controlled manner. The body glass separates into two liquid phases as it leaves the delivery orifice. This separation is sufficiently large to give dense white opacity but the material is noncrystalline and remains fluid at forming temperatures causing no problem in further shaping of the sheet. Glass C_b contains every element that exists in glass C_c; more than 60% cullet can be incorporated in the raw material used for glass C_b.

Composite Stresses

Using Eq 2, the magnitude of surface compression can be predicted, and glasses can be designed for desired strength characteristics. This equation is appropriate for laminated sheet or rod; similar equations can be derived for other geometries.

$$S_c = \frac{A_b E_c E_b [(\Delta L)/L]_t}{A_b E_b (1 - \sigma_c) + A_c E_c (1 - \sigma_b)} \qquad (\text{Eq 2})$$

where S_c is the stress in cladding, A_b is the cross-sectional area of the body, A_c is the cross-sectional area of the cladding, E_b is the Young's modulus of the body, E_c is the Young's modulus of cladding, σ_b is the Poisson's ratio of body, σ_c is the Poisson's ratio of the cladding, and $[(\Delta L/L)]_t$ is the total strain determined from expansion mismatch between body and cladding at the setting point of softer glass. (The setting point is defined as strain point $+5\ °C$, or $9\ °F$.)

The residual stresses in a laminated system are the result of stored energy much of which may be released when a piece is fractured. If internal tension is too great, the piece breaks into small pieces via dicing. The R pair of glasses shown in Table 1 has a tendency to display this characteristic because of the high central tension generated by the large expansion mismatch. Internal tension can be reduced by decreasing the expansion mismatch or increasing the body-to-cladding ratio. Surface compression varies only slightly between a 10 and 30 body-to-cladding ratio.

Thermal tempering can be combined with lamination to give the profile shown in Fig

1(c) (Ref 7). This is done because of bruise check-type damage that can result from glass-to-glass impact. The compressive layer may be shallow enough to permit propagation of the bruise check into the tensile layer over time, causing a phenomenon called delayed breakage. Thermal tempering superimposes a deeper compression layer on the stress produced by the differential expansion of the two glasses. As indicated in Fig 1(c) and Table 1, the overall stress does not increase significantly when tempering and lamination are combined, but the compressive stress becomes sufficiently deep to inhibit crack propagation into the tensile layer.

REFERENCES

1. J.A. Stewart, Glass Decorating and Finishing, Section 15, *The Handbook of Glass Manufacture*, Vol II, 3rd ed., F.V. Tooley, Ed., Ashlee Publishing Co., 1984, p 833–852.3

2. H.A. McMaster, Annealing and Tempering, Section 14, *The Handbook of Glass Manufacture*, Vol II, 3rd ed., F.V. Tooley, Ed., Ashlee Publishing Co., 1984, p 799–832.13

3. J.W. Giffen, D.A. Duke, W.H. Dumbaugh, J.E. Flannery, J.F. MacDowell, and J.E. Megles, "Method for Continuously Hot Forming Strong Laminated Bodies," U.S. patent 3,849,097, Nov 1974

4. J. Spiegler, "High Strength Resistor," U.S. patent 3,437,974, April 1969

5. J.E. Megles and T.W. Richardson, "Apparatus and Method for Producing Composite Glass Tubing," U.S. patent 4,023,953, May 1977

6. W.H. Dumbaugh, J.E. Flannery, J.E. Megles, and J.A. Smith, "Method for Making Multi-Layer Laminated Bodies," U.S. patent 3,737,294, June 1973

7. J.E. Megles, "Strengthened Laminated Glass Bodies," U.S. patent 3,649,440, March 1972

Porous and Reconstructed Glasses

Thomas H. Elmer, Corning, Incorporated (Retired)

GLASSES WITH HIGHER HEAT SHOCK RESISTANCE AND GREATER RESISTANCE TO DEFORMATION than that found in ordinary commercial glasses have been sought for decades by glass technologists. Fused quartz is an ideal glass in many ways, but fused quartz is difficult to produce in clear bubble-free form. Even at high temperatures, the glass is viscous, and the release of bubbles is slow. After melting, the glass presents difficulties in forming shapes other than tubing or sheet because the temperature required for blowing or pressing is beyond the range of known materials used for molds and plungers.

Porous and Reconstructed Glass Processing

Glasses having the desirable properties of fused silica have been on the market for nearly a half century under the trademark Vycor. These glasses are almost pure vitreous silica. They are prepared by a unique process discovered by Hood and Nordberg (Ref 1) that circumvents the need for high temperatures in melting and forming. A relatively soft alkali-borosilicate glass is melted in a conventional manner and is then pressed, drawn, or blown into the desired but oversized shape by standard processes used in glass production. The resultant workpiece, which occasionally is given additional finishing operations, is subjected to a heat treatment above the annealing point but below the temperature that would produce deformation. During this heat treatment, two continuous closely intermingled glassy phases are produced. One phase is rich in alkali and boric oxide and is readily soluble in acids. The other phase is rich in silica and is insoluble.

After heat treatment, the workpiece is immersed in a hot dilute acid solution. The soluble phase is slowly dissolved, leaving behind a porous high-silica skeleton. The resulting porous article is commonly known as thirsty or porous glass.

In the final step of the process, the workpiece is slowly heated to >1200 °C (>2190 °F) whereby the porous structure is consolidated into a clear impervious glass known as Vycor brand 96% SiO_2 glass or reconstructed glass. A flow chart of the process for making reconstructed glass is shown in Fig 1.

Composition of Leachable Alkali-Borosilicate Glasses

The starting glasses used for the production of reconstructed glasses generally contain small amounts of alumina. A portion of the alumina is retained in the porous high-silica glass that is obtained after leaching, rendering the final sintered glass more stable against devitrification and deformation than alumina-free glass. A typical composition of a starting glass is Glass A shown in Table 1 (Ref 2). The optimum starting glasses contain up to 4% Al_2O_3 for the quaternary system SiO_2-B_2O_3-Al_2O_3-Na_2O.

Glass compositions in the ternary system SiO_2-B_2O_3-Na_2O also have been evaluated for making shaped glass components. However, these glasses generally contain a substantially higher percentage of B_2O_3 and Na_2O after leaching than glasses prepared from quaternary compositions. A typical composition that has been used extensively by numerous workers in connection with phase separation studies and the preparation and characterization of porous glass is Glass B in Table 1.

Phase Separation

Since the discovery by Hood and Nordberg that 96% SiO_2 glasses can be made from heat-treated alkali-borosilicate glasses, considerable attention has been given to the study of glasses that can be phase separated. Skatulla, Vogel, and Wessel (Ref 4) used electron microscopy to show that phase separation in the Na_2O-B_2O_3 system is caused by a boron anomaly that is also responsible for phase separation in the Na_2O-B_2O_3-SiO_2 system and that leachable glasses are not homogeneous (this can be deduced from light scattering measurements and the observation of opalescence of glasses that had been subjected to heat treatments). Growth of microphases with temperature was shown or discussed by Watanabe, Noake, and Aiba (Ref 5), Kühne and Skatulla (Ref 6), and others (Ref 7–10) in electron micrographs obtained by replica and transmission methods.

Some investigators proposed that the inhomogeneity in leachable glasses was caused by the formation of clusters of ions or of molecular groups similar to chemical compounds. Nordberg (Ref 11) suggested that the mutual compatibility of oxide constituents for certain glasses, particularly for alkali-borosilicate glasses, decreases on cooling from the melt and thereby leads to submicroscopic phase separation. Roy and Ruiz-Menacho (Ref 12) suggested that the phase separation in such glasses is related to the well-known phenomena of a liquid mixture separating into immiscible phases, but because of the high viscosities of glass melts which prevail at subliquidus temperatures, the liquid phases remain finely dispersed causing no light scattering. However, on subsequent heating of these glasses the submicroscopic phases start to grow, increasing in size with temperature and time (Fig 2).

Leaching

Leaching involves the removal of the soluble constituents from the heat-treated glasses. The glasses are totally immersed in hot dilute acid solutions containing HCl, HNO_3, or H_2SO_4. These solutions are characterized by the presence of H^+ or H_3O^+ ions that react with the alkali-rich phase. In glasses prepared in the SiO_2-B_2O_3-Na_2O system, the sodium oxide is converted to NaCl, $NaNO_3$, or Na_2SO_4, depending on the type of acid used, and the boric oxide is hydrolyzed to H_3BO_3. These reaction products diffuse into the leach solution. Electroneutrality is retained even in the smallest spaces in the glass structure as the pores are filled with acid. The interaction of the acid solution with the glass leads to the gradual removal of the soluble constituents,

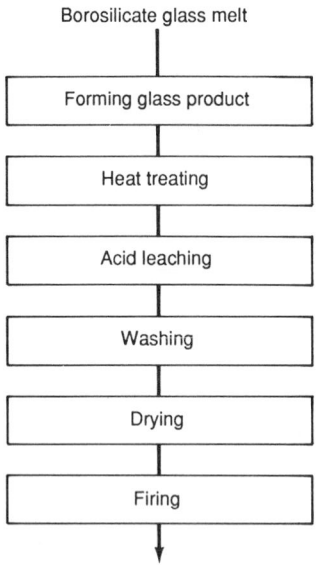

Borosilicate glass melt

→ Forming glass product

→ Heat treating

→ Acid leaching

→ Washing

→ Drying

→ Firing

High–silica reconstructed glass

Fig 1 Flow chart showing production process for reconstructed glass

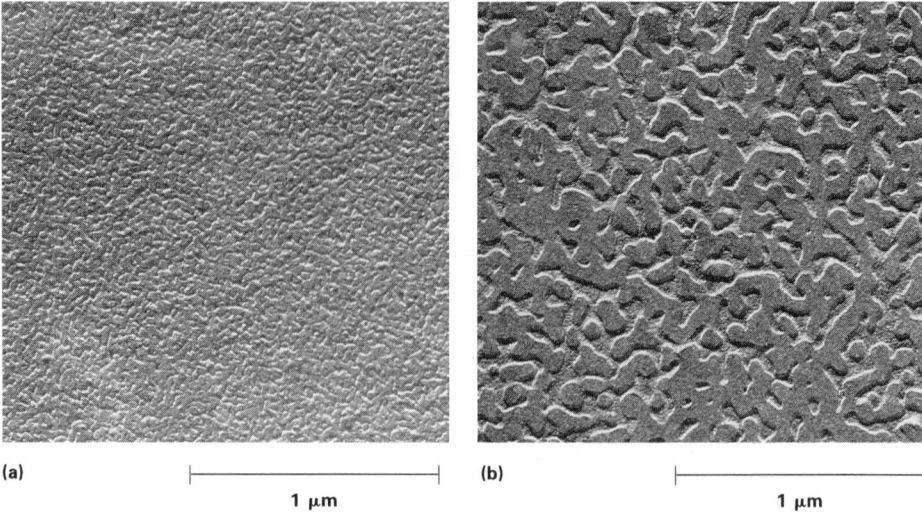

(a)   1 μm   (b)   1 μm

Fig 2 Photomicrographs of phase separation in 67.4 SiO_2-25.7 B_2O_3-6.9 Na_2O alkali-borosilicate glass. (a) Sufficiently resolved microstructure indicates phase separation has occurred after being heat treated at 580 °C (1075 °F) for 3 h. (b) Raised and smooth surface, which indicates silica-rich phase, has markedly coarsened after being heat treated at 650 °C (1200 °F) for 3 h. Coarsening is attributed to further coalescence of silica-rich phase into an interconnecting network caused by heating.

leaving behind a porous high-silica structure commonly known as porous glass. This glass is generally slightly opalescent.

Strain develops in the leached layer as leaching progresses and it can be observed under a polarizing microscope. Swelling of the leached layer develops tensile stresses in the yet undisturbed glass while shrinkage develops tensile stresses in the porous leached layer. Excessive strain can lead to fracture following leaching. However, the strain can be controlled by the proper selection of the starting glass and by heat treatment of this glass. The patents by Hood and Nordberg that were mentioned in the section "Composition of Leachable Alkali-Borosilicate Glasses" in this article focus on this leaching problem.

The diffusion of salts and acid through the porous glass determines the rate of leaching. The rate of leaching decreases with increasing thickness and is essentially a function of the square root of the leaching time. The process is therefore best adapted to relatively thin ware with a maximum thickness of 10 mm (0.4 in.) being recommended for general purposes. However, plates ≈25 mm (≈1 in.) in thickness have been prepared, demonstrating that it is possible to make heavier ware for

certain applications when the additional cost of slow leaching can be justified.

After leaching, the porous glass is washed in dilute acid and distilled water to remove the leachant in the pores. This washing procedure can take from an hour in duration to as long as overnight depending on the thickness of the porous glass being leached.

The washed porous glass can be safely dried at room temperature. During the initial stages of drying, which involves the removal of capillary water, the glass turns briefly opaque (white) and eventually becomes slightly opalescent as drying continues. The white opaque appearance is due to the temporary formation of a myriad of H_2O-menisci inside the partially-dried porous glass. These menisci serve as centers for the scattering of light that is responsible for the opaque appearance of the glass.

Properties of Porous Glass

Commercially available porous glass, the properties of which are given in Table 2, is an intermediate glass prepared by heat treating and leaching a special alkali-borosilicate glass. It has a surface area of 150 to 200 m^2/g as determined by the Brunauer-Emmett-Teller (BET) method (Ref 13) using nitrogen as the adsorbate and an internal pore volume of 28%. Its pore size distribution is generally very narrow (Fig 3), with about 96% of the pores in the glass being ±0.3 nm from the average radius that for commercial porous glass is about 4 to 6 nm (40 to 60 Å).

Enlarging Pores in Porous Glass

In making porous glass articles in which the pores are subsequently closed, the pore

size is generally unimportant. However, for certain applications it is desirable to have larger pore diameters. This can be accomplished by impregnating porous glass with an aqueous solution of a weakly reactive fluorine-containing compound such as 10% aqueous solution of ammonium bifluoride ($NH_4F \cdot HF$), reacting the compound *in situ* with a mineral acid to release hydrofluoric acid at temperatures sufficient to dissolve a portion of the glass body, and washing the treated glass to remove the soluble constituents (Ref 14). Discs and tubing with ≈20 nm (200 Å) pore size have been made from commercial glass by using such treatment or multiple treatments.

Table 1 Typical compositions for leachable alkali-borosilicate glasses

Material	System	Composition, wt%			
		SiO_2	B_2O_3	Na_2O	Al_2O_3
Glass A	Quaternary	62.7	26.9	6.6	3.5
Glass B	Ternary	65.0	26.0	9.0	...

Source: Ref 2, 3

Table 2 Composition and properties of commercially available porous glass

Composition on basis of its ignited weight:	
SiO_2	96.3
B_2O_3	2.95
Na_2O	0.04
$R_2O_3 + RO_2$	0.72(a)
Appearance	Opalescent
Refractive index	1.33(b)
Apparent density (dry), g/cm^3 ($lb/in.^3$)	1.5 (0.054)(c)
Internal pore volume, %	28
Average pore diameter, nm (Å)	5 (50)
Internal surface area, m^2/g	200
Water adsorption at saturation, %	25
Modulus of rupture, MPa (ksi)	42 (6.0)(d)
Young's modulus at 22 °C (72 °F), GPa (psi)	17.6 (2.5×10^6)
Loss tangent at 22 °C (72 °F), 100 Hz	0.007(e)
Dielectric constant at 22 °C (72 °F), 100 Hz	3.1(e)

(a) Chiefly $Al_2O_3 + ZrO_2$. (b) Depends on amount of moisture in pores. (c) Depends on relative humidity. (d) Abraded 6.4 mm ($^1/_4$ in.) rods at 22 °C (72 °F). (e) Appreciably affected by moisture; values are for specimens activated at 400 °C (750 °F), cooled in a desiccator, and then immediately tested

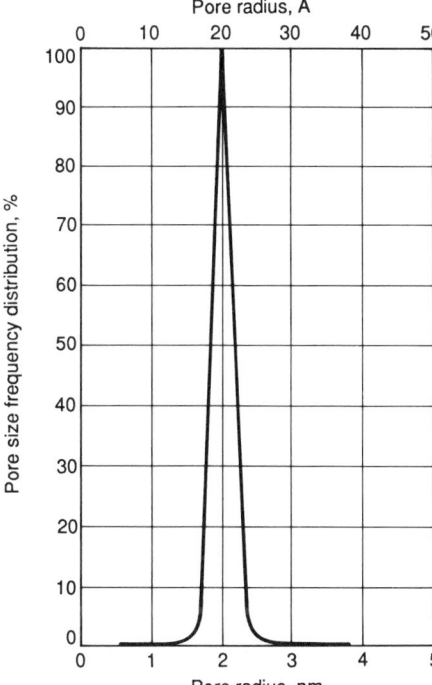

Fig 3 Pore size distribution of a porous glass having an average pore diameter of 4.1 nm (41 Å)

Still larger pores can be obtained by starting with glass compositions that have been especially modified to promote the formation of coarse rather than fine interconnecting phases with heat treatment (Ref 15).

Adsorption of Water Vapor by Porous Glass

Porous glass has been referred to as "thirsty glass" because of its affinity for moisture. Its primary advantage over other adsorbents is that it can be fabricated into relatively strong, nondusting shapes including tubing, rods (cane), disks, squares, various hollow shapes, fibers, and granules. In such forms, it lends itself particularly well for use as a nondusting drying agent or moisture getter in scientific instruments. Small disks and thin rods of this material activated at 180 °C (360 °F) in an electric oven will adsorb appreciable moisture (Fig 4).

The adsorption capacity for water at high relative humidities is not impaired by heating porous glass from room temperature to 800 °C (1470 °F). Such heating results in partial loss of hydroxyl groups [water of constitution in the form of silanols ($\equiv$SiOH) and boranols ($=$BOH)] but no appreciable loss in internal surface area of the glass (Ref 16, 17). However, to ensure maximum adsorption of water vapor at low relative humidities, it is important to heat the porous glass to <200 °C (<390 °F), a temperature at which the loss of OH$^-$ groups and the loss in surface area of colloidal deposits in the pores of leached glass

(Ref 16), resulting from acid leaching of the heat-treated starting glass, are minimal.

Removal of OH⁻ Groups from Porous Glass

The thermal removal of hydroxyl groups from porous glass is a function of both temperature and time. The temperature for maximum removal of water decreases as the hold time increases. The optimum temperatures result from a balance in the rates of water diffusion and glass sintering that retards diffusion. A thorough discussion of thermal dehydration of porous glass is given in a paper by Elmer and Nordberg (Ref 18).

It should be pointed out that if the porous glass is rapidly heated, the capillary water (molecular water) is expelled violently causing it to break and shatter. This problem is easily avoided by equilibrating the porous glass in room air preferably at <65% relative humidity and then heating it slowly in an electric oven.

The chemistry of silanol groups in porous glass can be studied in both aqueous and non-aqueous media and in gaseous atmospheres. The hydroxyl groups in porous glass can be replaced with chloride and amide groups by treating the glass with thionyl chloride ($SiOCl_2$) or exposing it at elevated temperatures to atmospheres containing chlorine, hydrogen chloride, or ammonia diluted with nitrogen as shown in Eq 1 through 3:

$$\equiv SiOH + SOCl_2 \rightarrow \equiv SiCl + SO_2 + HCl \quad (Eq\ 1)$$

$$\equiv SiOH + HCl \rightarrow \equiv SiCl + H_2O \quad (Eq\ 2)$$

$$\begin{array}{ccc} =Si-OH & & =Si \\ | & & | \diagdown \\ O & + NH_3 \rightarrow & O \quad NH + 2H_2O \\ | & & | \diagup \\ =Si-OH & & =Si \end{array}$$

$$(Eq\ 3)$$

Impregnation of porous glass with aqueous ammonium fluoride solutions followed by drying and gradual heating to 800 °C (1470 °F) also results in the removal of hydroxyl groups in such glass. This is due to a replacement of OH$^-$ groups by fluoride ions:

$$\equiv Si-OH + NH_4F \rightarrow \equiv Si-F + NH_3 + H_2O$$

$$(Eq\ 4)$$

There is a reduction by about a factor of 10 in the amount of water adsorbed at room temperature and low relative humidities by ammonium fluoride-treated porous glass compared to untreated porous glass under the same water vapor pressure. The substitution of fluoride ions for hydroxyl groups changes the interfacial forces between glass and water making the porous glass, which is normally hydrophilic, more hydrophobic.

Cleaning of Porous Glass

Because porous glass adsorbs not only moisture but also organic molecules from the atmosphere, it has also been used as an organic

getter. Organic impurities in the atmosphere cause it to turn yellow or brown with time and generally necessitate the glass to be cleaned prior to use. The adsorbed organic molecules can be removed by immersing the glass in an oxidizing acid such as nitric acid to which has been added a small amount of sodium or potassium chlorate, or by treating the glass in 30% H_2O_2 solution preferably at about 90 °C (195 °F). After subsequent washing in distilled water, the glass can be kept in a desiccator or stored in distilled water prior to use. Instead of wet oxidation, the organic contamination in porous glass can also be removed by gradually heating it in air or oxygen from room temperature to about 500 °C (930 °F) and holding at this temperature until all of the contamination is removed by oxidation.

Precleaned porous glass has found use in liquid chromatographic and electrophoretic separations (Ref 19).

Commercially available porous glass generally has a mean pore diameter of 4 to 6 nm (40 to 60 Å) that is much smaller than the mean free path (10 to 100 nm, or 100 to 1000 Å) of most gases at atmospheric pressure and temperature. This also makes porous glass a suitable material for porous membranes to be used at elevated pressures for the separation of gases.

By using special leaching, washing, and various post treatments, it also is possible to prepare semipermeable porous glass membranes with superfine structure capable of purifying water that has been contaminated with mineral salts and of desalinating sea water (Ref 20–22).

The above information, which is by no means complete, shows that porous glass is a versatile inorganic material. It has been and continues to be of interest to workers in diverse scientific fields chiefly because of its unique properties (Table 2). The mechanical strength of porous glass is discussed in a recent paper by Elmer, Helfinstine, and Seward III (Ref 23).

Controlled-Pore Glass

Haller (Ref 24, 25) developed porous glasses with very sharp pore distribution that can be tailor-made for having mean pore diameters over a range from 7.5 to 300 nm (75 to 3000 Å). These glasses have become known as controlled pore glass (CPG). These glasses have found wide application for size-exclusion and adsorptive chromatography of proteins, nucleic acids, viruses, and high polymers. They have been covalently derivatized and are used as such for affinity chromatography, as solid support for bioactive molecules in bioreactors and diagnostic kits, and as substrates for sequential analysis and synthesis of deoxyribonucleic acid (DNA) and proteins.

Controlled-pore glass is generally made in the form of granules and is now offered for sale in various pore sizes by a company called

CPG, Inc. of Fairfield, New Jersey. The primary application is in permeation chromatography, where the chemical resistance and mechanical stability of porous glass significantly exceeds that of polymer materials used for the same purpose. Because of the organophilic nature of porous glass, its internal surface can be modified with organo-functional silane coupling agents. It can also be modified by using the methods of classic biochemistry. The surface-treated controlled-pore glasses have found uses for the preparation of specialized diagnostic products and for the immobilization of enzymes in fixed-bed reactors.

Reconstructed Glass Products Obtained from Consolidation of Treated and Impregnated Porous Glass

A whole family of reconstructed glasses can be made by starting with

- Alkali-borosilicate glass and varying the conditions of heat treatment, leaching, and firing
- Porous glass and subjecting the leached glass to additional processing steps (for

example, impregnation with various coloring oxides prior to consolidation into an impervious glass)

Consolidation of the leached glass (porous glass) generally involves slowly heating it to >1200 °C (>2190 °F) so that the porous structure is consolidated (sintered) into an impervious clear high-silica glass. The end product undergoes shrinkage of 35% in volume or about 14% in linear dimensions. For this reason, the starting glass that is melted in a conventional manner and is blown, drawn, or pressed must be formed into oversize shapes.

Products made from consolidated reconstructed 96% SiO_2 glass are comparable in performance and properties to those of fused quartz and silica. The end products have excellent heat shock resistance, allowing them to be heated repeatedly to red heat and then plunged into ice water without breakage. Special firing and outgassing *in vacuo* significantly reduce the hydroxyl content in such glass to ensure greater heat resistance. The lower water content (hydroxyl groups) in the glass workpiece reduces any tendency of the glass to deform at high operating temperatures and allows the glass to be used continuously at 900 °C (1650 °F) and intermittently at >1200 °C (>2190 °F). The consolidated glasses are mechanically strong and their strength increases at higher temperatures. Selected properties of reconstructed glass are given in Table 3.

Transmittance Properties of Reconstructed Glasses

The ultraviolet transmittance of 96% SiO_2 glass can be engineered to produce glasses with cutoffs at nearly any desired wavelength in the ultraviolet spectrum. This is accomplished by varying the leaching and firing conditions and by impregnating the porous glass with appropriate ultraviolet absorbing inorganic salts that upon firing supply network forming or modifying ions (for example, Ce^{4+}, Sn^{4+}, Ti^{4+}, V^{4+}, V^{3+}, and Fe^{3+} ions).

Reconstructed 96% SiO_2 glasses effectively transmit infrared radiation. Degassing of the porous glass in a vacuum prior to its consolidation substantially increases the infrared transmittance of the glass but does not completely eliminate the absorption band at 2.73 µm (109 µin.) (Fig 5), that is attributed to the presence of OH^- groups in the structure of the glass. Impregnation of the porous glass with aqueous fluoride-containing solutions, followed by drying and firing in air or in a vacuum eliminates hydroxyl groups, thereby substantially improving the usefulness of such a glass as an infrared window (Ref 26).

The OH^- groups in porous glass can also be eliminated by subjecting the porous glass to a stream of gas containing combined or free fluorine plus a carrier gas (for example,

Table 3 Properties of consolidated 96% silica glass

Refractive index	1.458
Average thermal expansion coefficient (0 to 300 °C, or 32 to 572 °F), °C^{-1} (°F^{-1})	7.5×10^{-7} (4.2×10^{-7})
Density, g/cm^3 (lb/in.^{3})	2.18 (0.0788)
Annealing point, °C (°F)	1020 (1870)(a)
Specific heat, cal/g · °C	0.18
Thermal diffusivity, cm^3/s	0.009
Thermal conductivity, W/m · K (cal/cm · s · °C)	1.38 (0.0033)
Total normal emissivity, 100 °C (212 °F)	0.87
Young's modulus at 22 °C (72 °F), GPa (psi)	68 (9.8×10^6)
Shear modulus at 22 °C (72 °F), GPa (psi)	28 (4.0×10^6)
Poisson's ratio	0.19
Modulus of rupture at 22 °C (72 °F), abraded surface MPa (ksi)	48 (7.0)
Hardness, HK$_{100}$ kgf/mm^2	487
Dielectric constant at 22 °C (72 °F):	
1 MHz	3.8
8.6 GHz	3.8
Electrical resistivity, Ω · cm:	
Log$_{10}$ ρ at 250 °C (480 °F)	9.7(b)
Log$_{10}$ ρ at 350 °C (660 °F)	8.1(c)
Chemical durability, weight loss, mg/cm^2	0.0005(d)(e), 0.07(f), 0.90(g)

(a) 1080 °C (1975 °F) for special processed ware. (b) 11.4 for special processed ware. (c) 9.7 for special processed ware. (d) 5% HCl at 100 °C (212 °F) for 24 h. (e)Reconstructed Pyrex glass is ten times as durable in 5% HCl as Pyrex No. 7740 glass (which itself has excellent acid resistance). (f) N/50 Na$_2$CO$_3$ at 100 °C (212 °F) for 6 h. (g) 5% NaOH at 100 °C (212 °F) for 6 h

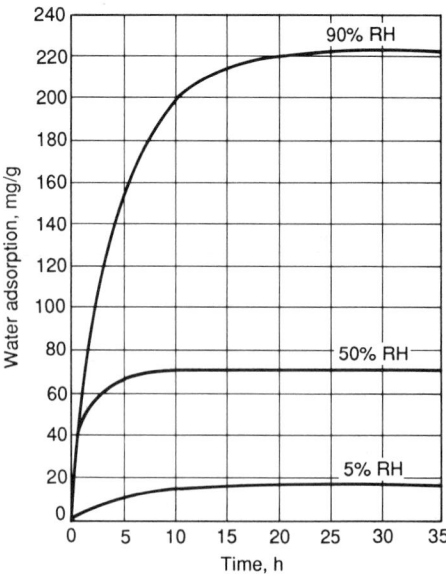

Fig 4 Water adsorption at room temperature in mg/g of activated porous glass at relative humidity (RH) levels of 5, 50, and 90%. The glass was activated at 180 °C (360 °F).

nitrogen). Complete removal of OH^- groups is accomplished by exposing the porous glass to a chlorine-containing atmosphere at elevated temperature prior to consolidation in a vacuum (Ref 27). The annealing point of the final glass product is markedly increased by the above treatments that result in the elimination of the OH^- band in the infrared spectrum by both thermal dehydroxylation and substitution of F^- or Cl^- ions at OH^- sites in the porous glass.

The infrared transmittance can be further increased by subjecting the porous glass to a stream of ammonia and nitrogen at elevated temperature (Ref 28, 29) and after cooling,

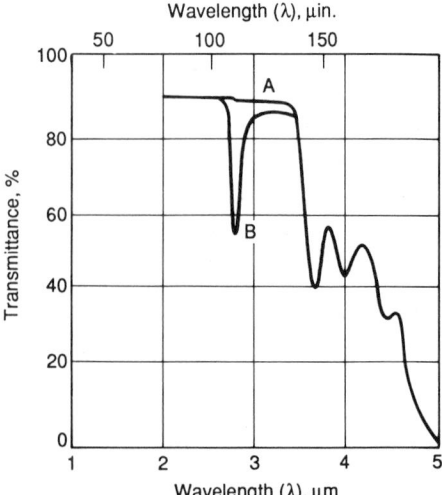

Fig 5 Infrared spectra of 96% SiO_2 glass of 1.5 mm (0.06 in.) thickness: A, treated with ammonium fluoride; B, untreated

releaching the nitrided porous structure in hot dilute acid, drying, and then consolidating the chemically treated porous glass in a vacuum at 1350 °C (2460 °F). These glasses have annealing points in excess of 1150 °C (2100 °F), indicating that they are substantially harder than reconstructed glass prepared by conventional processing methods which have annealing points around 1025 °C (1880 °F).

Colored Reconstructed Glasses

Porous glass is well suited for producing a wide variety of reconstructed colored glasses, including white, opaque, and black glasses. This is accomplished by simply impregnating the porous glass in aqueous solutions containing the desired coloring ions, which are generally selected from nitrate or chloride salts. The impregnated glass is rinsed in distilled water to remove excess salt solution from its surface. The impregnated glass is then dried and consolidated in air or oxygen at 1225 °C (2240 °F). In contrast to porous glasses that are colored by dipping in organic coloring solutions and then used in the air-dried state, the color of consolidated glasses is not degraded by heat.

Red reconstructed glass is made by impregnating porous glass in an aqueous solution containing nitrate salts (that is, of Fe, Ni, and Al), drying, and subsequently slowly heating in a stream of air to 1225 °C (2240 °F). X-ray diffraction studies (Ref 30) show that spinel crystallites, which are present in colloidal sizes, are responsible for the red color. The glass absorbs most of the visible light from a tungsten filament heated to 2425 °C (4395 °F) but effectively transmits infrared radiation and consequently is used in the form of tubing and sheet in heating applications.

Black Reconstructed Glass. By impregnating porous glass with furfuryl alcohol ($C_5H_6O_2$) solutions, polymerizing the alcohol in the glass to a resin, pyrolyzing, and then firing to consolidate the porous structure to an impervious glass, it is possible to make a black glass. By varying the concentration of furfuryl alcohol in the impregnating solution, it also is possible to prepare glasses with resistivities ($\log_{10} \rho$) at 22 °C (72 °F) ranging from 0.1 to 8.5 (Ref 31). The carbon phase in the fired glasses is amorphous. Carbon is an excellent dehydroxylating agent, removing silanols and boranols (OH^- groups) in porous glass that, if left in the structure, soften the end-product glass. Consequently, the annealing point of reconstructed glass can be markedly increased by incorporating carbon in porous glass and firing in a dry nonoxidizing atmosphere (Ref 32). The carbon-containing glasses are not only black but totally opaque to radiation in the ultraviolet, visible, and part of the infrared spectra. Furthermore, the carbon-containing glasses have excellent chemical durability and superior devitrification resistance at high temperatures compared to that of carbon-free glass, which is comparable to that of fused quartz and silica glasses.

Specialized Applications. For special heating lamp applications, porous glass tubing may be unilaterally impregnated by introducing the solution containing the inorganic glass-coloring agents in the porous glass from the external surface only. By controlling the depth of penetration of the impregnating solution, it is possible to retain a layer of clear glass along the inner wall of the tubing, indicating no colorant additives. Consolidated glass tubes prepared by this method avoid the possibility of the coloring additives reacting with other lamp components (for example, tungsten from the filament in an infrared heating lamp that could produce an adverse effect on lamp operation and/or light transmission) (Ref 33).

Graded Seals

Thermal seals with expansion matched to fused quartz or silica and glass of higher expansion can be made by starting with short pieces of porous glass tubing (Ref 34). One end of each tube to be converted to a graded seal is impregnated with an aqueous solution composed of alkali borates and then washed to remove a part of the alkali borates from the pores near the surface of the impregnated glass. The tube is then consolidated to an impervious clear glass to prepare it for use in sealing glasses of different expansion characteristics.

Product Applications

The above information gives some insight into how reconstructed high-silica glass products are prepared. Porous glass is not only a unique material with interesting properties but a marvelous starting material for making a whole family of reconstructed glasses that have many applications in both research laboratory and commercial industrial applications.

Reconstructed glass has favorable optical characteristics, high electrical resistivity, and excellent chemical durability and heat resistance properties that make it applicable for a wide range of applications:

- Laboratory ware
- Lamps (germicidal, photochemical, halogen-cycle, and radiant types)
- Heat sheath tubing
- Thermocouple protection tubing
- Welding torch tips
- High-temperature trays and jars
- Sight glasses
- Photomultiplier tubes
- Camera and space windows
- Exhaust tubing
- Radiant heaters
- Precision pore tubing
- Fritted glassware
- Special coatings for space vehicles

REFERENCES

1. H.P. Hood and M.E. Nordberg, "Treated Borosilicate Glass," U.S. patent 2,106,744, Feb 1938
2. H.P. Hood and M.E. Nordberg, "Method of Treating Borosilicate Glasses," U.S. patent 2,286,275, June 1942
3. H.P. Hood and M.E. Nordberg, "Borosilicate Glass," U.S. patent 2,221,709, Nov 1940
4. W. Skatulla, W. Vogel, and H. Wessel, Phase Separation and Boron Anomaly in Simple Sodium Borate and Technical Alkali-Borosilicate Glasses, *Silikat Technol.*, Vol 9 (No. 2), 1958, p 51–62
5. M. Watanabe, H. Noaka, and T. Aiba, Electron Micrographs of Some Borosilicate Glasses and Their Internal Structure, *J. Am. Ceram. Soc.*, Vol 42 (No. 12), 1959, p 593–599
6. K. Kühne and W. Skatulla, Physical and Chemical Investigations of Glasses in the Ternary System SiO_2-B_2O_3-Na_2O in the Region of VYCOR Type Glasses, *Silikat Technol.*, Vol 10 (No. 3), 1959, p 105–119
7. F. Oberlies, Phase Separation in Glass, *Naturwissenschaften*, Vol 43 (No. 10), 1956, p 224
8. W. Vogel, *Structure and Crystallization of Glasses*, Pergamon Press, 1971
9. D.I. Levin, S.P. Zhdanov, and E.A. Porai-Koshits, Structure and Opalescence of Sodium Borosilicate Glasses, *Izv. Akad. Nauk SSSR*, 1955, p 31–39
10. W. Haller, Rearrangement Kinetics of the Liquid-Liquid Immiscible Microphases in Alkali Borosilicate Melts, *J. Chem. Phys.*, Vol 42 (No. 2), 1965, p 686–693
11. M.E. Nordberg, unpublished work from 1940
12. R. Roy and C. Ruiz-Menacho, Subsolidus Nucleation in the System Li_2O-Al_2O_3-SiO_2, *Am. Ceram. Soc. Bull.*, Vol 38 (No. 4), 1959, p 229
13. S. Brunauer, P.H. Emmett, and E. Teller, Adsorption of Gases in Multimolecular Layers, *J. Am. Chem. Soc.*, Vol 60 (No. 2), 1938, p 309–319
14. I.D. Chapman and T.H. Elmer, "Porous High Silica Glass," U.S. patent 3,485,687, Dec 1969
15. T.H. Elmer, M.E. Nordberg, G.B. Carrier, and E.J. Korda, Phase Separation in Borosilicate Glasses as Seen by Electron Microscopy and Scanning Electron Microscopy, *J. Am. Ceram. Soc.*, Vol 53 (No. 4), 1970, p 171–175
16. T.H. Elmer, I.D. Chapman, and M.E. Nordberg, Changes in Length and Infrared Transmittance During Thermal Dehydration of Porous Glass at Temperatures up to 1200 °C, *J. Phys. Chem.*, Vol 66, 1962, p 1517–1519
17. T.H. Elmer, Sintering of Porous Glass, *Am. Ceram. Soc. Bull.*, Vol 62 (No. 4), 1983, p 513–516

18. T.H. Elmer and M.E. Nordberg, Thermal Dehydration of Porous Glass, *Glastech. Ber.*, Vol 61 (No. 5), 1988, p 140–142

19. H.L. MacDonell, J.M. Noonan, and J.P. Williams, Porous Glass as an Adsorption Medium for Chromatography, *Anal. Chem.*, Vol 35, 1963, p 1253–1255

20. E.V. Ballou, M.I. Leban, and T. Wydeven, Stabilization of Porous Glass Hyperfiltration Membranes by Aluminum Chloride, *Nature (London), Phys. Sci.*, Vol 229 (No. 4), 1971, p 123–124

21. P.W. McMillan and C.E. Matthews, Microporous Glasses for Reverse Osmosis, *J. Mater. Sci.*, Vol 11 (No. 7), 1976, p 1187–1199

22. T.H. Elmer, Evaluation of Porous Glass as Desalination Membrane, *Am. Ceram. Soc. Bull.*, Vol 57 (No. 11), 1978, p 1051–1054

23. T.H. Elmer, J.D. Helfinstine, and T.P. Seward III, Strength of Porous Glass in Water, *Glastech. Ber.*, Vol 61 (No. 8), 1988, p 214–217

24. W. Haller, Chromatography on Glass of Controlled Pore Size, *Nature*, Vol 206, 1965, p 693–696

25. W. Haller, Application of Controlled Pore Glass in Solid Phase Biochemistry, Chapt. 11, *Solid Phase Biochemistry*, W.H. Scouten, Ed., John Wiley & Sons, 1983, p 535–597

26. T.H. Elmer, "Method of Removing Dissolved Water from 96% SiO_2 Silica Glass," U.S. patent 2,982,053, May 1961

27. T.H. Elmer, Dehydroxylation of Porous Glass by Means of Chlorine, *J. Am. Ceram. Soc.*, Vol 64 (No. 3), 1981, p 151–154

28. T.H. Elmer and M.E. Nordberg, Nitrided Glasses, *VIIth Congrès International du Verre*, Bruxelles 1965, C.R. Chaleroi, Ed,. 1965, p I.3.1 to I.3.11

29. T.H. Elmer, Chlorine Treatment of Nitrided Porous Glass, *Glastech. Ber.*, Vol 61 (No. 1), 1988, p 24–28

30. T.H. Elmer and H.J. Holland, Solid-State Reactions of Fe, Ni, and Al in 96% SiO_2 Glass, *Am. Ceram. Soc. Bull.*, Vol 66 (No. 8), 1987, p 1265–1269

31. T.H. Elmer, Electrical Properties of Carbon-Containing Reconstructed Silica Glasses, *Am. Ceram. Soc. Bull.*, Vol 55 (No. 11), 1976, p 999–1003

32. T.H. Elmer and H.E. Meissner, Increasing of Annealing Point of 96% SiO_2 Glass on Incorporation of Carbon, *J. Am. Ceram. Soc.*, Vol 58 (No. 5–6), 1976, p 206–209

33. T.H. Elmer and M.E. Nordberg, "Lamp Having a Colored Bulb," U.S. patent 3,258,631, June 1966

34. R.V. Lukes, "Method of Making Sealing Glasses," U.S. patent 2,522,524, Sept 1950

Phase-Separated Glasses and Glass-Ceramics

Linda R. Pinckney, Corning, Incorporated

MANY SPECIALTY GLASSES (for example, opals, photosensitive glasses, and the large family of glass-ceramic materials), share a common origin in controlled phase separation. Phase separation is the phenomenon whereby an amorphous homogeneous phase is divided into two immiscible phases of different composition. In glasses, this separation may occur spontaneously on cooling from the molten state, as is evident in many opal glasses, or upon reheating. The resulting microstructure generally consists of finely dispersed droplets of glass or colloidal crystals in a matrix of the host glass. Given an appropriate glass composition and the nature of the phase separation, this assemblage can then serve as a template for controlled crystallization to a glass-ceramic.

Phase-Separated Glasses

Phase-separated glasses derive their unique optical and microstructural characteristics from the presence of a small amount of a finely dispersed amorphous or crystalline phase. Generally, this phase consists of uniformly distributed discrete droplets or particles. It is also possible, given certain glass compositions that are subjected to carefully controlled conditions, to obtain a structure comprising two interconnected glassy phases on a very fine scale. This microstructure, in which one of the interconnected phases is selectively removed by leaching, is the basis of the commercial porous and reconstructed high-silica glasses such as Vycor (see the article "Porous and Reconstructed Glasses" in this Volume). The photosensitive glasses and glass-ceramics (see the article "Photosensitive Glasses and Glass-Ceramics" in this Volume) also entail phase separation.

Opal glasses commonly consist of a fine dispersion of CaF_2 or NaF crystals (3 to 10% by volume) in a host glass (Ref 1). Borate and phosphate opals, which are less common, are typically based on liquid-liquid phase separation. The optical properties of opal glasses result from light refraction and scattering between the separated phases. The degree of opacity or translucency is controlled by the degree of phase separation, the size and quantity of the separated particles, and the difference in the refractive indices of the two phases. Most commercial opal glasses are based on CaF_2 crystals and phase-separate spontaneously on cooling from the melt. Commercial opal glass products include lighting components, bottles, and tableware such as Corelle dinnerware.

Other commercial phase-separated glasses include colored glasses that are derived from the controlled nucleation and crystallization of absorbing particles, particularly metal colloids. The decorative ruby glasses, for example, are based on gold colloids in lead glass (Ref 2). Colored filter glasses (for example, Corning 2403 sharp-cut red filter glass) can be made by precipitating colored sulfoselenide crystals (Ref 3). The first glass-ceramics were actually derived using silver colloids as nucleating agents (Ref 4).

Glass-Ceramics

Glass-ceramics are polycrystalline materials formed by the controlled crystallization of special glasses. They are differentiated from phase-separated glasses by their high crystallinity. Glass-ceramics are by definition $\geq 50\%$ crystalline by volume and generally are $>90\%$ crystalline. The range of glass-ceramic compositions is extremely broad, requiring only the ability to form a glass and control its crystallization.

Glass-Ceramic Design

Glass-ceramics are true engineered materials: the properties of the parent glass can be tailored for the ease of manufacture while simultaneously tailoring those of the glass-ceramic for a particular application. The three key variables in the design of a glass-ceramic (Ref 5) are:

- Glass composition
- Glass-ceramic phase assemblage
- Nature of the crystalline microstructure

The glass composition controls much of the workability of the material, including glass viscosity as well as the effectiveness of nucleation and rapidity of crystallization. A glass-ceramic base glass must meet the seemingly contradictory requirements of resistance to uncontrolled devitrification during melting and forming while simultaneously meeting specifications of high nucleation density and homogeneous crystallization during subsequent heat treatment.

The glass-ceramic phase assemblage (that is, the types of crystals and the proportion of crystals to glass) is responsible for many of the physical and chemical properties (including thermal and electrical characteristics, chemical durability, and hardness). Finally, the nature of the crystalline microstructure (crystal size and morphology and the textural relationship among the crystals and glass) is the key to many mechanical and optical properties, including transparency/opacity, strength, fracture toughness, and machinability.

Glass-Ceramic Processing

Glass-ceramic components can be fabricated by means of either bulk or powder processing methods.

Bulk Glass-Ceramic Processing. In this most common method of glass-ceramic man-

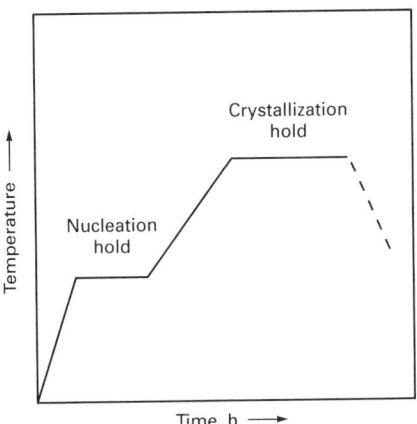

Fig 1 Typical heat-treatment cycle for a glass-ceramic material

ufacture, components are melted and fabricated to shape while in the glass state. Most forming methods may be employed, including rolling, pressing, spinning, casting, and blowing. The article is then crystallized using a heat treatment designed for that material. This process is known as ceramming. Because crystallization occurs at high viscosity, article shapes are typically preserved with little or no shrinkage (1 to 3%) or deformation during the ceramming.

Nucleation. Heterogeneous nucleation, in which the primary crystalline phases grow epitaxially upon nuclei of an earlier-formed phase (internal nucleation) or from a surface defect, is most common in glass-ceramic systems. To effect internal nucleation, certain nucleating agents generally are added to the batch. These melt homogeneously into the glass but promote very fine-scale phase separation on reheating. The dispersed phase, typically a metal, titanate, zirconate, or fluoride, is structurally incompatible with the host glass and is normally highly unstable as a glass. It therefore precipitates tiny crystalline nuclei on heating at temperatures near the annealing point of the host glass. These crystals serve as the sites for subsequent nucleation of the primary crystalline phases.

Crystallization. Nucleation is followed by one or more higher temperature treatments to promote crystallization. During this stage of the heat treatment, the primary crystalline phases nucleate and grow upon these nuclei until they impinge on neighboring crystals. Given a high nucleation density, the resulting microstructure is highly uniform, consisting of fine-grained randomly oriented crystals in a matrix of minor residual glass. Depending on composition and heat treatment, crystal size ranges from <0.i μm (<4 μin.) to >10 μm (>400 μin.).

A typical heat-treatment cycle (Fig 1) comprises both nucleation and crystallization temperature holds, but some glass-ceramics are designed to nucleate and crystallize during the ramp itself, thus eliminating the need for multiple holds.

Powder Glass-Ceramic Processing. The manufacture of glass-ceramics from powdered glass using conventional ceramic processes (for example, extrusion, spraying, or slip-casting), extends the realm of useful glass-

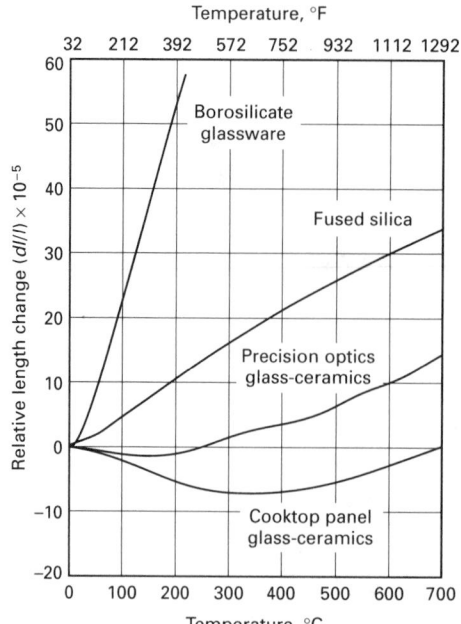

Fig 2 Plot of relative change in length versus temperature of two β-quartz glass-ceramics designed for precision optics and cooktop panels. Curves for fused silica glass and borosilicate laboratory glassware are given for comparison. Source: Ref 7

ceramic compositions by taking advantage of surface crystallization. In these materials, the crystal phases nucleate and grow from sites on the surfaces of glass grains. The glass undergoes viscous sintering just before the crystallization process is completed. The final crystalline microstructure is essentially the same as that produced from the bulk process. Glass-ceramic sealing frits, for example, are applied like glass frits and then crystallized during the soldering process.

Secondary Processing (Strengthening). Because their polycrystalline microstructures provide resistance to crack propagation, glass-ceramics possess mechanical strength inherently superior to that of glass. This strength may be augmented by a number of techniques that impart a thin (generally several mils thick) surface compressive stress to the body. These techniques induce a differential surface volume or expansion mismatch by means of ion-exchange (see the article "Ion-

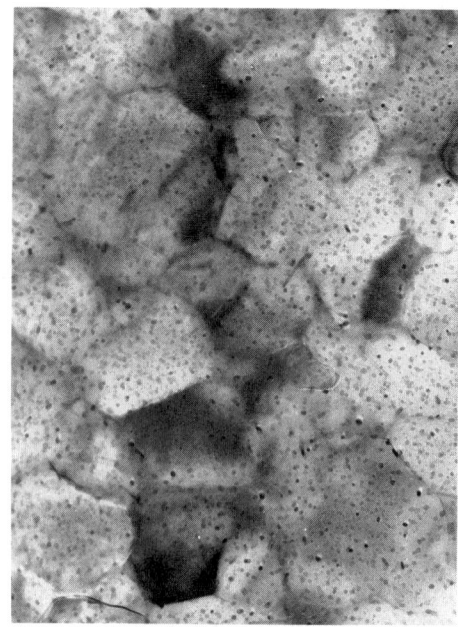

Fig 3 TEM photomicrograph of the microstructure of transparent β-quartz solid solution glass-ceramic

Exchange" in this Volume), differential densification during crystallization (Ref 6), or by employing a lower-expansion surface glaze.

Advantages of Glass-Ceramics. Glass-ceramics can provide significant advantages over conventional glass or ceramic materials by combining the ease and flexibility of forming and inspection of glass with improved and often unique physical properties in the glass-ceramic. Unlike conventional ceramic materials, glass-ceramics are fully densified with zero porosity and their uniform microstructure guarantees that the physical properties of glass-ceramics are highly reproducible. Complex shapes are readily produced. Glass-ceramics are stronger than their parent glasses and can be tailored to meet a wide range of thermal expansion requirements (from negative to high positive thermal expansion coefficients that can match those of metals). They also can be designed to provide excellent thermal shock resistance,

Table 1 Base compositions and applications of transparent glass-ceramics based on β-quartz solid solutions

| Material | Composition, wt% | | | | | | | | | | | | | | Commercial applications |
	SiO_2	Al_2O_3	MgO	Na_2O	K_2O	ZnO	Fe_2O_3	Li_2O	BaO	P_2O_5	F	TiO_2	ZrO_2	As_2O_3	
Vision(a)	68.8	19.2	1.8	0.2	0.1	1.0	0.1	2.7	0.8	...	...	2.7	1.8	0.8	Transparent cookware
Zerodur(b)	55.5	25.3	1.0	0.5	...	1.4	0.03	3.7	...	7.9	...	2.3	1.9	0.5	Telescope mirrors
Ceran(b)	63.4	22.7	(d)	0.7	(d)	1.3	(d)	3.3	2.2	(d)	(d)	2.7	1.5	(d)	Black infrared transmission cooktop
Narumi(c)	65.1	22.6	0.5	0.6	0.3	...	0.03	4.2	...	1.2	0.1	2.0	2.3	1.1	Rangetops; stove windows

(a) Manufactured by Corning Glass Works. (b) Manufactured by Schott. (c) Manufactured by Nippon Electric. (d) No data available

Table 2 Compositions, properties, and applications of glass-ceramics based on β-spodumene (keatite) solid solutions

| | Composition, wt% | | | | | | | | | | | Mechanical properties | | | | | Thermal properties | | |
| | | | | | | | | | | | | Modulus of rupture | | Young's modulus | | | Coefficient of thermal expansion, $K^{-1} \times 10^{-6}$ | | |
Material	SiO$_2$	Al$_2$O$_3$	MgO	Na$_2$O	K$_2$O	ZnO	Fe$_2$O$_3$	Li$_2$O	TiO$_2$	ZrO$_2$	As$_2$O$_3$	MPa	ksi	GPa	10^6 psi	Hardness, HK$_{100}$	0–500 °C (32–930 °F)	0–1000 °C (32–1830 °F)	Commercial applications
Corningware(a) (code 9608)	69.7	17.8	2.6	0.4	0.2	1.0	0.1	2.8	4.7	0.1	0.6	100	15	81	12.8	660	1.2	...	Cookware; hot plate tops
Cercor(a) (code 9455)	72.5	22.5	...	...	...	...	...	5.0	...	...	...	...	...	...	...	...	...	0.5	Heat exchangers; regenerators

(a) Manufactured by Corning Glass Works

dielectric properties, chemical durability, or high strength and toughness.

Limitations of Glass Ceramics. Glass-ceramics manufacture is subject to several limitations. Compositions are restricted to those constituents that can form glasses and whose crystallization can be controlled. In addition, as also required in glass manufacture, sufficient product volume is necessary to justify the costs of both glass melting and forming. Such costs, however, are often more than balanced by the process economy inherent in a high-speed continuous operation.

Commercial Glass-Ceramic Systems

Worldwide annual sales of glass-ceramic products exceeded $500,000,000 in 1990. These products range from transparent, zero-expansion materials with excellent optical

Table 3 Composition, properties, and applications of a cordierite-based glass-ceramic (code 9606)

Composition, wt%(a)	
SiO$_2$	56.1
Al$_2$O$_3$	19.8
MgO	14.7
CaO	0.1
TiO$_2$	8.9
As$_2$O$_3$	0.3
Fe$_2$O$_3$	0.1

Properties:	
Crystalline phases	Cordierite
	Cristobalite
	Rutile
	Mg-dititanate
Coefficient of thermal expansion (0–700 °C, or 32–1290 °F), $K^{-1} \times 10^{-6}$	4.5
Modulus of rupture, MPa (ksi)	250 (36)
Fracture toughness, MPa$\sqrt{m}$ (ksi$\sqrt{in.}$)	2.2 (2.0)
Young's modulus, GPa (10^6 psi)	120 (17.4)
Thermal conductivity, W/m·K (cal/cm·s·°C)	38 (0.09)
Hardness, HK$_{100}$	700
Dielectric constant at 8.6 GHz	5.5
Loss tangent at 8.6 GHz	0.0003
Softening temperature, °C (°F)	>1300 (>2370)

Commercial applications

Radomes

(a) Corning Glass Works specifications

properties and thermal shock resistance to jadelike highly crystalline materials with excellent strength and toughness. The highest volume is in cookware and tableware consumer items, architectural cladding, and stovetops and stove windows.

Aluminosilicate Glass-Ceramics. These glass-ceramics are based on framework structures of aluminosilicate tetrahedra and have widespread commercial applications. Because these crystal structures have very low bulk thermal expansions, their glass-ceramics also possess low thermal expansion with the consequent benefits of exceptional thermal stability and thermal shock resistance. Commercial glass-ceramics based on framework structures comprise compositions from the Li$_2$O-Al$_2$O$_3$-SiO$_2$ (LAS) and the MgO-Al$_2$O$_3$-SiO$_2$ (MAS) systems.

Glass-ceramics in the LAS system have great commercial value because of their very low thermal expansion and excellent chemical durability (Ref 7). These glass-ceramics are based on essentially monophase assemblages of either β-quartz or β-spodumene (keatite) solid solution with only minor residual glass or accessory phases. Partial substitution of MgO and ZnO for Li$_2$O improves the working characteristics of the glass while lowering the material cost. Figure 2 illustrates the range of thermal expansion coefficients of some β-quartz glass-ceramics compared with those of fused silica and borosilicate glassware.

Glass-ceramics containing β-quartz or β-spodumene can be made from glass of the same composition by modifying its heat treatment:

- β-quartz is formed by ceramming at ≤900 °C (≤1650 °F)
- β-spodumene by ceramming at >1000 °C (>1830 °F)

β-Quartz Solid Solution. A mixture of ZrO$_2$ and TiO$_2$ produces highly efficient nucleation of β-quartz, resulting in very small (<100 nm, or 1000 Å) crystals. This fine crystal size, coupled with low birefringence in the β-quartz phase and closely matched refractive indices in the crystals and residual glass, result in a transparent yet highly crystalline body (Fig 3a). Near-zero thermal expansion behavior

combined with transparency, optical polishability, excellent chemical durability, and strength greater than that of glass has made these glass-ceramics highly suitable for use as telescope mirror blanks, thermally stable platforms, ring laser gyroscopes, infrared-transmitting rangetops, woodstove and other heat-resistant windows, and cookware. Representative compositions and applications of β-quartz based glass-ceramics are given in Table 1.

β-Spodumene Solid Solution. Opaque low-expansion glass-ceramics are obtained by ceramming these LAS materials at temperatures >1000 °C (>1830 °F). The β-quartz to β-spodumene transformation takes place between 900 and 1000 °C (1650 and 1830 °F) and is accompanied by a five- to ten-fold increase in grain size (to between 1 to 2 μm, or 40 to 80 μin.). When TiO$_2$ is used as the nucleating agent, rutile development accompanies the silicate phase transformation. The combination of larger grain size and the high refractive index and birefringence of rutile gives the glass-ceramic a high degree of opacity. Secondary grain growth is sluggish, giving these materials excellent high-temperature dimensional stability. Because of their larger grain size, these glass-ceramics are stronger than those based on β-quartz. They also can be strengthened by means of standard surface strengthening techniques.

β-spodumene glass-ceramics have found wide application as cookware, architectural sheet, benchtops, hot-plate tops, valve parts, ball bearings, sealing rings, heat exchangers, and matrices for fiber-reinforced composite materials (Ref 8). Table 2 gives the compositions of two commercial β-spodumene glass-ceramics. One material is internally nucleated with TiO$_2$ and is used for cookware, while the other material, produced as a honeycomb product for a turbine engine heat exchanger, is manufactured using powder processing methods and requires no additional nucleating agent.

Cordierite Glass-Ceramics. These glass-ceramics in the MAS system combine high strength, good thermal stability, and thermal shock resistance with excellent dielectric properties at microwave frequencies. Corning produces a silica-rich cordierite-based glass-ceramic with a complex phase assemblage that includes the high-expansion cristobalite phase. While this phase raises the overall thermal expansion, it provides for a unique post-ceramming strengthening process. The surface of the material is treated with hot alkali solution to selectively leach the cristobalite phase, leaving a porous skin about 0.38 mm (0.015 in.) thick. The skin serves as an abrasion-resistant layer that inhibits flaw initiation and increases the strength from 120 to 240 MPa (17 to 35 ksi). This material, with its high transparency to radar, is the standard glass-ceramic used for missile radomes. Its composition and properties are given in Table 3.

Sheet Silicate Glass-Ceramics. Machinable glass-ceramics based upon sheet silicates of the fluorine-mica family have unique microstructures composed of interlocked platelike mica crystals. Because micas can be easily delaminated along their cleavage planes, fractures propagate readily along these planes but not along other crystallographic planes. The random intersections of the crystals in the glass-ceramic, therefore, cause crack branching, deflection, and blunting, arresting crack growth. As a result, these glass-ceramics possess higher intrinsic mechanical strength and toughness than that obtained with framework silicates. In addition, the combination of ease of fracture initiation with almost immediate fracture arrest enables these glass-ceramics to be readily machined. Compositions and properties of two glass-ceramics based on fluorine mica are given in Table 4.

Macor. The commercial glass-ceramic Macor, based on a fluorophlogopite ($KMg_3AlSi_3O_{10}F_2$) mica, is capable of being machined to high tolerance (±0.01 mm, or 0.0004 in.) by conventional high-speed metalworking tools. By suitably tailoring its composition and nucleation temperature, relatively large mica crystals with high two-dimensional aspect ratios are produced, thus enhancing the inherent machinability of the material. This house-of-cards microstructure is illustrated in Fig 4. In addition to precision machinability, Macor has high dielectric strength, very low helium permeation rates, and is unaffected by radiation or oxyacetylene flame. This glass-ceramic has been employed in a wide variety of applications including high-vacuum components and hermetic joints, precision dielectric insulators and components, seismograph bobbins, sample holders for field ion microscopes, boundary retainers for the space shuttle, and γ-ray telescope frames.

Dicor. Recently, another machinable glass-ceramic, Dicor, has been developed for use in dental restorations (Ref 9). Based on a tetrasilicic fluormica ($KMg_{2.5}Si_4O_{10}F_2$), this material has higher strength (150 MPa, or 21 ksi) and improved chemical durability over that of Macor. Dicor glass-ceramic has a hardness, radiographic density, and translucency closely matching natural tooth enamel. The desired translucency is achieved by maintaining a fine-grained (≈1 μm, or 40 μin.) crystal size and by roughly matching the refractive indices of the crystals and glass. Ceria (CeO_2) is added to simulate the fluorescent character of natural teeth. The glass-ceramic may be accurately cast using a lost wax technique and conventional dental laboratory molds. The high strength, low thermal conductivity, and transparency to x-rays of Dicor provide working advantages over conventional metal-ceramic systems.

Chain Silicate Glass-Ceramics. Naturally occurring, massive aggregates of chain silicate amphibole crystals (for example, nephrite jade) are well known for their durability and high resistance to impact and abrasion. Such aggregates consist of compact intergrowths of randomly oriented acicular or fibrous crystals. This crystalline microstructure is the key to their strength and toughness because in order for a fracture to propagate

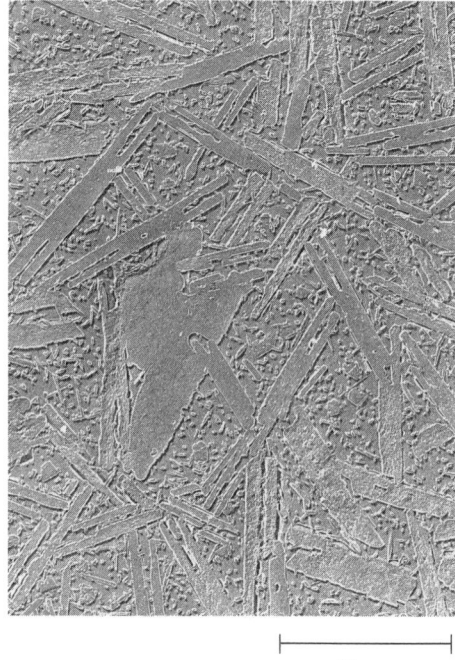

Fig 4 Replica micrograph showing house-of-cards microstructure in a machinable fluormica glass-ceramic

through the material, it generally will be deflected and blunted as it follows a tortuous path around each crystal. Hence, interlocking bladelike or rodlike crystals can serve as important strengthening or toughening agents, much as fiberglass is used to reinforce polymer matrices. Glass-ceramics based on chain silicate crystals have the highest toughness and body strength of any materials in the glass-ceramic family. Compositions, properties, and applications for a number of chain silicate glass-ceramics are given in Table 5.

Slagsitall. Blast furnace slags, with the addition of sand and clay, have been used in

Table 4 Compositions, properties, and applications of sheet silicate (fluormica) glass-ceramics

	Composition, wt%								Modulus of rupture		Young's modulus		Hardness, HK$_{100}$	Fracture energy, J·m^2
											Mechanical properties			
Material	SiO$_2$	Al$_2$O$_3$	MgO	K$_2$O	F	ZrO$_2$	B$_2$O$_3$	CeO$_2$	MPa	ksi	GPa	10^6 psi		
Macor(a) (9658)	47.2	16.7	14.5	9.5	6.3	...	8.5	...	100	15	65	9.4	250	8.2
Dicor(b)	56–64	0–2	15–20	12–18	4–9	0–5	...	0.05	152	22	...	...	362	...

	Thermal properties			Maximum use temperature (unstressed)		Electrical properties	Optical properties	
Material	Coefficient of thermal expansion at 25–600 °C (77–1110 °F), K^{-1} × 10^{-6}	Thermal conductivity		°C	°F	Dielectric strength (dc, 25 °C, or 80 °F, 0.25 mm, or 0.010 in. thick), KV·cm^{-1}	Refractive index	Commercial applications
		W/m·K	cal/cm·s·°C					
Macor(a) (9658)	12.9	1.3–1.7	0.0030–0.0040	900	1650	1200	...	Machinable components
Dicor(b)	7.2	1.7	0.0040	...	...	...	1.52	Dental restorations

(a) Manufactured by Corning Glass Works. (b) Manufactured by Dentsply

Table 5 Compositions, properties, and applications of glass-ceramics based on chain silicates

Base	Composition, wt%													
	SiO_2	Al_2O_3	MgO	CaO	Na_2O	K_2O	ZnO	MnO	Fe_2O_3	S	Li_2O	BaO	P_2O_5	F
Slagsitall														
White (USSR)	55.5	8.3	2.2	24.8	5.4	0.6	1.4	0.9	0.3	0.4	...	...	...	...
Gray (Hungary)	60.9	14.2	5.7	9.0	3.2	1.9	...	2.0	2.5	0.6	...	...	...	...
K-richterite(a)	67.3	1.8	14.3	4.7	3.0	4.8	...	...	...	...	0.8	0.3	1.0	3.5
Canasite(a)	54–62	1–4	0–2	17–25	6–10	6–12	0–2	...	...	...	...	...	...	4–8

Base	Crystalline phases	Mechanical properties						Hardness, HK$_{100}$	Thermal properties		Electrical properties	Commercial applications
		Modulus of rupture		Fracture toughness		Young's modulus			Coefficient of thermal expansion, $K^{-1} \times 10^{-6}$		Dielectric breakdown strength (50 Hz, 20 °C, or 70 °F), $kV \cdot mm^{-1}$	
		MPa	ksi	MPa√m	ksi√in.	GPa	10^6 psi		0–300 °C (32–570 °F)	0–500 °C (32–930 °F)		
Slagsitall												
White (USSR)	Wallastonite	65–100	9.4–15	...	...	75–88	11–13	590–790	...	9.1–9.5	40–50	Cladding, industrial products
Gray (Hungary)	Diopside	80–120	12–17	...	...	88–108	13–16	640–740	...	7.2–7.6	40–50	Cladding, industrial products
K-richterite(a)	F-K-richterite, Cristobalite, M mica	150–200	22–29	3.2	2.9	95	14	...	11.5	...	...	Tableware
Canasite(a)	Canasite, CaF_2	250–300	36–44	4–5	3.6–4.6	80	12	500	12.5	...	...	Rigid disk substrates

(a) Manufactured by Corning

Eastern Europe for over twenty years (Ref 10) to manufacture inexpensive nonalkaline glass-ceramics. The primary crystalline phases are wollastonite ($CaSiO_3$) and diopside ($CaMgSi_2O_6$) in a matrix of aluminosilicate glass. Metal sulfide particles serve as nucleating agents. The relatively high residual glass levels of these materials (typically >30%), coupled with comparatively equiaxial (less rod-shaped) crystals, confers only moderately high mechanical strengths of ≈100 MPa (≈15 ksi). High hardness, good to excellent wear and corrosion resistance, and low cost are the chief attributes of these materials.

Slagsitall materials have found wide use in the construction, chemical, and petrochemical industries. Applications include abrasion- and chemical-resistant floor and wall tiles, industrial machinery parts, chimneys, plungers, parts for chemical pumps and reactors, grinding media, and coatings for electrolysis baths. These materials presently constitute the largest volume applications for crystallized glass.

Potassium Fluorrichterite. More recently, two other families of chain silicate glass-ceramics have found commercial application. Glass-ceramics based on the amphibole potassium fluorrichterite (KNaCaMg$_5$Si$_8$O$_{22}$F$_2$) have a microstructure consisting of tightly interlocked fine-grained rod-like amphibole crystals in a matrix of minor cristobalite, mica, and residual glass. The flexural strength of these materials can be further enhanced (150 to 200 MPa, or 22 to 29 ksi) by employing a compressive glaze. Richterite

glass-ceramics have good chemical durability and are usable in microwave ovens. When glazed, they resemble bone china in their gloss and translucency. These glass-ceramics are currently utilized for high-performance institutional tableware and as mugs and cups for the Corelle dinnerware line.

Fluorcanasite. An even stronger and tougher microstructure of interpenetrating acicular crystals (Fig 5) is obtained in glass-ceramics based on fluorcanasite (K$_{2-3}$Na$_{4-3}$Ca$_5$Si$_{12}$O$_{30}$F$_4$) crystals. Cleavage splintering and high thermal expansion anisotropy augment the intrinsic high fracture toughness of the chain silicate microstructure. The inexpensive batch materials and low melting and ceramming temperatures make these materials particularly attractive for a number of applications. Fluorcanasite glass-ceramics are being developed for use as magnetic memory disk substrates. Other potential applications include architectural materials and thin consumer ware.

Commercial Sintered Glass-Ceramic Products. These products take advantage of the complete densification, thermal stability, and wide range of physical properties possible in glass-ceramic systems. Glass-ceramic frits provide a number of advantages over conventional glass frits. Products incorporating glass-ceramic seals, for example, can be reheated to their soldering temperature without deformation.

Neopariés is a sintered glass-ceramic-based material with about 40% crystallinity. Its primary crystalline phase is wollastonite, a chain silicate. This material can be manufactured in

flat or bent shapes by molding during heat treatment. It has a texture similar to that of marble but with greater strength and durability than granite or marble. Neopariés is used

Fig 5 Fracture surface (replica micrograph) of fluorcanasite glass-ceramic showing interlocking blade-shaped crystals and effects of cleavage splintering

1 μm

as a construction material for flooring and exterior and interior cladding.

Frits, Coatings, and Matrices. Other commercial products include sealing frits for glasses, ceramics, and metals, high-temperature coatings, and matrices for fiber-reinforced composite materials. Glass-ceramic frits also have found numerous applications in the electronics and television industries (for example, vacuum seals, insulation layers in integrated circuits, heat-resistant cements, corrosion-resistant coatings for metals, and honeycomb structures in heat regenerators). Manufacturers of these components include Owens-Illinois, Corning, and Schott Glaswerke.

Glass-ceramic theory and systems are described more fully in the books listed in the Selected References, and in the review articles in Ref 5 and 6. Commercial glass-ceramic compositions and properties, as well as comparisons with glass and ceramic materials, are also covered in depth in the books listed in the Selected References.

REFERENCES

1. J.E. Flannery and D.R. Wexell, Opal Glasses, *Commercial Glasses*, Vol 18, *Advances in Ceramics*, D.C. Boyd and J.F. MacDowell, Ed., 1986, p 141–150
2. W.A. Weyl, *Coloured Glasses*, Society of Glass Technology, Sheffield, 1951
3. D.C. Boyd and D.A. Thompson, Glass, *Kirk-Othmer: Encyclopedia of Chemical Technology*, Vol 11, 3rd ed., John Wiley & Sons, 1980, p 807–880
4. S.D. Stookey, Catalyzed Crystallization of Glass in Theory and Practice, *Ind. Eng. Chem.*, Vol 51, 1959, p 805
5. G.H. Beall, Design of Glass-Ceramics, *Reviews of Solid State Science*, Vol 3, 1989, p 333–354
6. G.H. Beall and D.A. Duke, Glass-Ceramic Technology, *Glass: Science and Technology*, Vol 1, Uhlmann/Kreidl, Ed., Academic Press, 1983
7. H. Scheidler and E. Rodek, Li_2O-Al_2O_3-SiO_2, Glass-Ceramics, *Am. Ceram. Soc. Bull.*, Vol 68, 1989, p 1926–1930
8. K.M. Prewo, J.J. Brennan, and G.K. Layden, Fiber Reinforced Glasses and Glass-Ceramics for High Performance Applications, *Am. Ceram. Soc. Bull.*, Vol 65, 1986, p 305–313
9. K.A. Malament and D.G. Grossman, *J. Prosthetic Dentistry*, Vol 57, 1987, p 62
10. A.I. Berezhnoi, *Glass-Ceramics and Photo-Sitalls*, Plenum Press, 1970

SELECTED REFERENCES

- P.W. McMillan, *Glass-Ceramics*, 2nd ed., Academic Press, 1979
- R. Morrell, Part 1: An Introduction for the Engineer and Designer, *Handbook of Properties of Technical and Engineering Ceramics*, National Physical Laboratory, London, 1985
- Z. Strnad, *Glass-Ceramic Materials*, Glass Science and Technology, Vol 8, Elsevier, 1986
- W. Vogel, *Chemistry of Glass*, The American Ceramic Society, 1985

Photosensitive Glasses and Glass-Ceramics

N.F. Borrelli and T.P. Seward, III, RD&E Division, Corning Incorporated

PHOTOSENSITIVE GLASSES are sensitive to light or other electromagnetic radiation in the sense that some directly or indirectly measurable change is produced in the glass by exposure to the radiation. In some cases the effect is readily observable immediately upon irradiation. In other cases, a thermal treatment is required to bring about the observed change.

The darkening of window glass after prolonged exposure to sunlight is well documented in glass history (Ref 1). However, this photosensitive behavior, known as solarization, is generally unwanted. The modern glass manufacturing process includes precautions to prevent it or measures to disguise its effects (Ref 2). This article concentrates on more useful, controlled photosensitive phenomena.

Modern photosensitive glass technology can perhaps be said to have originated in the 1930s and 1940s with the work of Dalton (Ref 3, 4), Stookey (Ref 5–8), and others (Ref 9, 10), who studied the photosensitivity of colors developed in copper-, silver-, and gold-containing glasses. This technology is actively evolving today, as evidenced by the recent work that is referenced in this article.

Photosensitive Mechanism. Optical absorption induced in glasses by exposure to light or other electromagnetic radiation results from either one or the other of these two mechanisms:

- Color centers that result from the trapping of electrons and holes produced by ionizing radiation
- Photothermal reduction of metal ions by electrons produced by ionizing radiation and subsequent aggregation of the reduced metal ions to form submicroscopic, light-absorbing metal particles

Both mechanisms require a source of photoelectrons. There are basically two ways to produce electrons in glass via irradiation. The first method utilizes short-wavelength ionizing radiation, such as x-rays or γ-rays, to produce free electrons from energy levels within the valence band of the glass. The second method utilizes a photoionizable impurity ion in the glass, such as Ce^{3+}. For photoexcitation of an electron from Ce^{3+}, the wavelength lies in the near-ultraviolet region, ~310 nm. Here, the radiation promotes the electron from the Ce^{3+} to an excited state, which is strongly coupled to conduction states of the glass. In either method, an electron is created in the conduction band and, equally importantly, a hole is created in the valence band. Other ions that may provide photoelectrons are Eu^{2+} and Cu^{+}. The subsequent deep trapping of the electron and hole is the next necessary step.

A further distinction as to the kind of photosensitive effect to be considered depends on the other constituents of the glass and subsequent thermal treatment, if any. Photosensitive glass can be categorized by three types:

- Direct photosensitive, in which absorption develops directly from the light exposure. This involves the trapping of both the photoelectron and the hole. The absorption results from energy transitions of these trapped species
- Developed photosensitive, in which a thermal step follows the exposure with the intention of freeing the trapped photoelectron, as well as providing mobility of a metal ion, so that reduction of the metal ion will be favored. A metal particle results from this reduction and is the source of the absorption
- Photosensitive nucleated, in which the same mechanism as that described for the developed type occurs, but the thermal treatment effects a nucleation and growth of a separate phase within the glass through the nucleating action of the reduced metal particle formed

Each type is described in more detail below, followed by a description of its processing. Photochromic glass types and processing are then examined.

Photosensitive Glass Types

Direct Photosensitive. In general, most glasses are photosensitive to some degree when exposed to high-energy photons. The electrons and holes produced by the energetic photons, $hv > E_{gap}$, find traps either within the basic network structure of the glass (for example, nonbridging oxygens) or polyvalent impurity ions, such as the transition metal ions. These trapped species can often be observed through the use of electron spin resonance spectroscopy (Ref 11–15). The optical absorption is thus associated with the optical transitions of the trapped species.

For x-ray "browning," as the optical absorption induced by x-ray exposure is called (Ref 16), the induced absorption band is broad, as shown in Fig 1. The absorption spectrum is displayed for different x-ray dosages. The depth of the traps is apparently shallow, because heating the glass to a temperature of 100 °C (212 °F) bleaches the absorption. The extent of this coloration is an important consideration for the glass used for TV bulbs where x-rays are generated as a by-product of the electron beam. Methods used to alleviate this problem usually provide competing trapping ions within the glass of such a nature that the resulting absorption of the trapped electrons, after exposure, lies outside, rather than inside, the visible portion of the spectrum (Ref 17, 18).

The role of glass composition in determining the extent of induced absorption is relatively minor, with the exception of the necessary inclusion of photoelectron source ions, such as Ce^{3+} and trapping site impurities (Ref 5, 6, 19–21), and the exclusion of competitive strong absorbers, such as Pb^{2+} and Ti^{4+}. There is also a dependence on the number of nonbridging oxygens, which is discussed below. There is little, if any, commercial market for this direct photosensitive effect, as of 1991.

Developed Photosensitive. The mechanism for this glass type involves the photoreduction of metal ions as the ultimate step in the mechanism described above (Ref 5–8, 19–21). The metal ions are usually noble metal ions such as Au^{+} or Ag^{+}, but can also be ions such as Cu^{+} or V^{3+}. After exposure, the glass is heated through a thermal schedule, usually in the range of 500 to 600 °C (930 to 1110 °F) for about 1 h. As a consequence of the heat

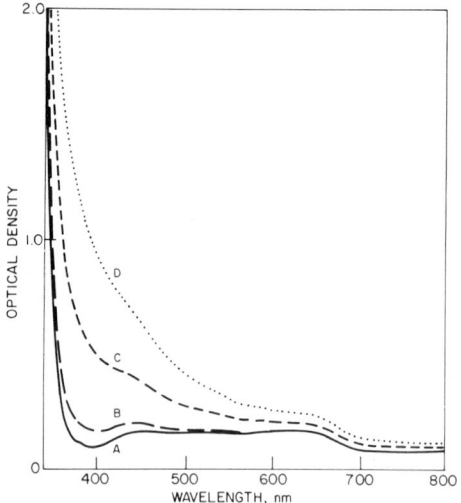

Fig 1 Optical absorption versus wavelength of a glass exposed to x-rays. Curve A, no exposure; Curve B, 10^4 r; Curve C, 10^5 r; and Curve D, 3×10^5 r

treatment, the photoelectrons are thermally freed, while the noble metal ions are very mobile in the glass structure. The reduction of the metal proceeds through stages of nucleation and growth, attaining particle diameters of the order of 50 nm (2 μin.). The optical pattern developed by this photosensitive mechanism is due to the optical absorption associated with the metal particles. The interaction of light with small particles can be understood through scattering theory (Ref 22–25). The spectral position of the peak absorption is a function of the optical constants, size and shape of the metallic particle, whereas the breadth of the absorption band is determined primarily by the particle size (Ref 26–28). For particle sizes ≤ 20 nm (0.8 μin.), the spectral position of the peak absorption is essentially independent of particle size. The spectra of the developed absorption for Ag, Au, and Au-Pd particles in glass are shown in Fig 2.

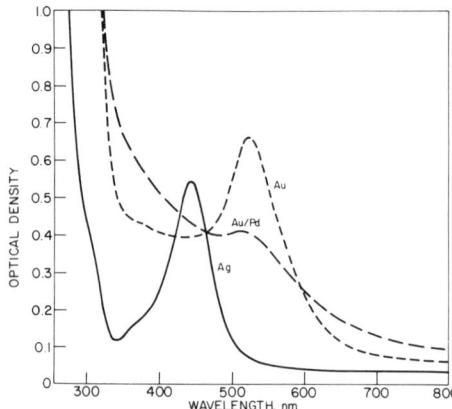

Fig 2 Optical absorption spectra of developed photosensitive glasses containing (a) silver, (b) gold, and (c) gold-palladium particles

The processes discussed above can thus be summarized by the following set of reactions. With an ionizing radiation greater than 5 eV,

$$h\nu = e + h \qquad \text{(Eq 1a)}$$

$$e + T = T_e \qquad \text{(Eq 1b)}$$

$$h + T' = T'_h \qquad \text{(Eq 1c)}$$

Or, with an impurity excitation less than 4 eV and where $\lambda = 300$ to 350 nm,

$$Ce^{3+} + h\nu = Ce^{4+} + e \qquad \text{(Eq 2a)}$$

$$e + T = T_e \qquad \text{(Eq 2b)}$$

$$Ce^{4+} + T' = T_h + Ce^{3+} \qquad \text{(Eq 2c)}$$

followed by particle growth by the successive addition of atoms,

$$T_e + M^+ + \text{heat} = M^0 + T \qquad \text{(Eq 3a)}$$

$$M^0 + e + M^+ = (M^0)_2 \qquad \text{(Eq 3b)}$$

$$(M^0)_n + e + M^+ = (M^0)_{n+1} \qquad \text{(Eq 3c)}$$

where e and h are electrons and holes, T and T' represent electron and hole traps, respectively, T_e and T'_h represent trapped electrons and holes, M represents a metal ion, and $h\nu$ represents a photon.

The optical sensitivity of the development of an image in the photosensitive glass, that is, the amount of energy per unit area required to produce a visually observable contrast, depends primarily on the metal ion content, through Eq 3(a). Typical values of the sensitivity, utilizing the impurity excitation mechanism, are in the range of 10 to 100 mJ/cm² (0.048 to 4.75 ft · lbf/in.²), at wavelengths in the range of 300 to 350 nm (Ref 29).

The glass composition, insofar as it determines the glass structure, governs whether the photosensitive mechanism is operative. It has been found that a sufficient amount of singly bonded oxygens, called "nonbridging oxygens," must be present in the glass before it can exhibit this type of photosensitivity (Ref 30). The proposed explanation is that the nonbridging oxygen is the deep hole trap represented in Eq 1(c) and/or 2(c). Without it, the hole would be mobile in the valence band and eventually recombine with the trapped electron. The composition of a typical photosensitive glass is shown in Table 1. It should be noted that the growth of metal particles also includes diffusion of the metal ion, which is taken to be included in Eq 3(b) and (c).

Table 1 Typical composition of photosensitive glass

Glass component	Wt%
SiO₂	71
Na₂O	17
Al₂O₃	7
ZnO	5
Sb₂O₃	0.2
Ag	0.01–0.1
CeO₂	0.05

Thermochemical reductions of metal ions also can be important in some compositions.

Photosensitive nucleated glass is a type in which the formation of the small metal particle (5 to 10 nm, or 0.2 to 0.4 μin.), through the preceding mechanism, nucleates a microcrystalline phase during the thermal treatment (Ref 31, 32). The composition of the glass determines the phase that will be formed. Examples of the phases that have been produced are lithium metasilicate and sodium fluoride. In lithium metasilicate, the photonucleated phase is contiguous and constitutes over 20% of the glass. In sodium fluoride, the crystal-phase content is about 1%. Photoelectron-micrographs of the developed phases are shown in Fig 3.

The commercial utility of the lithium metasilicate material stems from the very different physical and chemical properties of the optically developed crystal-glass composite, compared to those of the unexposed glass (Ref 33). In particular, this material is chemically machinable in photosensitively controlled patterns, such as the Fotoform products of Corning Inc. Recently, this material has been applied in the area of microoptics, namely spherical microintegrated lens arrays. The formation of the composite phase leads to a material with a higher density than that of the unexposed glass. At the development temperature, which is above the softening point of the parent glass, the densification exerts a squeezing effect that causes an extrusion of the unexposed glass beyond the surface, ultimately yielding a spherical profile suitable for a lens (Ref 34–36).

The photosensitive sodium fluoride material is marketed as two products, one of which has been used to form opal-like patterns in clear glass (Fotalite by Corning Inc.), and the other, a range of clear colors (Polychromatic by Corning Inc.) (Ref 37, 38). The transparent color production involves another step beyond the thermal development of the sodium fluoride. The mechanism proposed is that the initial sodium fluoride crystallite morphology produces a tip-like (pyramidal) protrusion from the cubic face that can be reduced to metallic silver particles of controlled anisotropic shapes by further exposure and thermal treatment. The initial development exposure time ultimately controls the particle shape and the resulting color. The transmission versus wavelength curves for this glass are shown in Fig 4 as a function of the initial exposure time.

Photosensitive Glass Processing

For most of the examples of photosensitive glasses given above, the physical processing is the same. The glass is melted and formed in a conventional manner, usually as a rolled sheet. Careful control over composition and oxidation state, as discussed below for photochromic glasses, is generally required. The glass is ground and polished before exposure

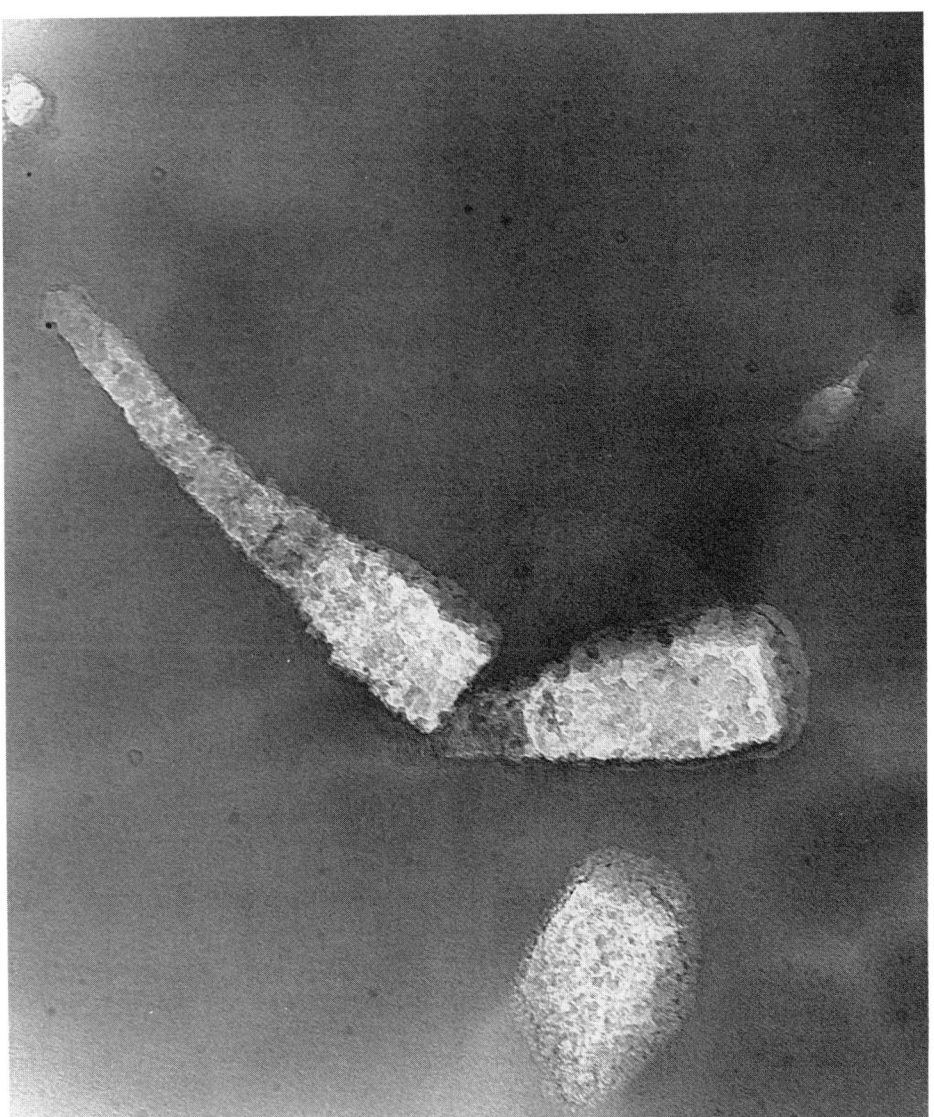

Fig 3 Photoelectron micrograph of photosensitive-developed NaF crystal

to the activating ultraviolet light, if high resolution is required. The exposure light source is often a Hg or Hg-Xe arc lamp commonly used for photolithography. A variety of masking techniques can be used. Exposure times vary according to the light intensity, but are typically in the range of seconds to minutes for 1 kW lamps. The thermal development schedule depends on the specific photosensitive glass. However, the nucleating silver speck development occurs above 500 °C (930 °F), whereas the subsequent development of the second phase, that is, sodium fluoride or lithium metasilicate, usually requires temperatures above 550 °C (1020 °F).

Photochromic Glasses

Photochromic glasses comprise a class of photosensitive glasses whose transmittance changes dramatically upon exposure to electromagnetic radiation (light). Particularly useful glasses are ones whose visible light transmittance decreases (darkens) upon exposure to ultraviolet and blue light, and then reversibly increases (fades) upon cessation of the irradiation.

There are many types of glasses that show photochromism (Ref 39–44), but most commercial photochromic glasses owe their unusual performance to a uniform dispersion of very small (approximately 10 nm, or 0.4 μin. diameter) copper-doped silver halide particles within the glass. The darkening and fading is ascribed to a process very similar to latent image formation in a photographic emulsion (Ref 22). Tiny specks of silver metal are generated within the silver halide, or at the silver halide-glass interface, under the influence of the ultraviolet light. The photosensitive process is similar to that described in the section "Photosensitive Mechanism" in this article except here Cu^+ dopant ions within the silver halide act as the hole traps, whereas silver ions, located at appropriate silver halide crystal defects, act as the electron traps. The reduced silver atoms, which result when the electrons are trapped, aggregate at ambient temperature to form light-absorbing silver metal specks. The process reverses when irradiation ceases.

The degree of darkening, the rates of darkening and fading, the temperature dependence of those rates, and even the color of the darkened glass, are all controlled by the chemical compositions of the host glass and the silver halide particles, as well as by the size and shape of the silver halide particles (Ref 45).

Fading is a diffusion-controlled process (Ref 46). During darkening the electrons and holes are trapped at finite distances apart (less than the diameter of the silver halide particle). Fading is believed to involve a diffusion of Cu^{2+} ions (trapped holes) to within efficient electron tunneling distances from the silver metal specks, at which time a trapped hole recombines with an electron from the silver speck. To maintain charge neutrality, a silver ion diffuses away from the silver speck. This process continues until most, if not all, of the silver ions are redispersed in the silver halide. This mechanism explains why glasses with large particles fade more slowly than do glasses with small ones and why there is a dependence of both equilibrium darkening and the rates of darkening and fading on the Cu^+ concentration in the crystal (Ref 46).

Photochromic Glass Processing. Photochromic glasses are generally made by combining all the necessary ingredients in the batch, melting the batch ingredients, forming a glass article, and heat treating that article in order to grow the needed dispersion of silver halide particles in a controlled manner. Glass compositions are chosen in part for their ability to precipitate the silver halide in a well-controlled manner, for example, by providing a significant difference in solubility of silver halide between the batch melting temperature and the heat-treatment temperature (generally between the softening point and the annealing point of the glass) (Ref 47) and in part for the physical and chemical properties required of the final glass product. Borosilicates are the most common bases for commercial photochromic glasses.

Heat-treatment temperatures between 500 and 650 °C (930 and 1200 °F) are common for the borosilicate photochromic glasses. Sometimes two-step treatments are used to first nucleate and then grow the silver halide crystals (Ref 48); the first treatment is in the range of 500 to 600 °C (930 to 1110 °F) and the second between 550 and 650 °C (1020 and 1200 °F).

Some care must be exercised in melting photochromic glasses, particularly with regard to maintaining precise control of the copper, silver, and halogen concentrations in the glass. These elements are generally present in amounts much less than 1% by weight,

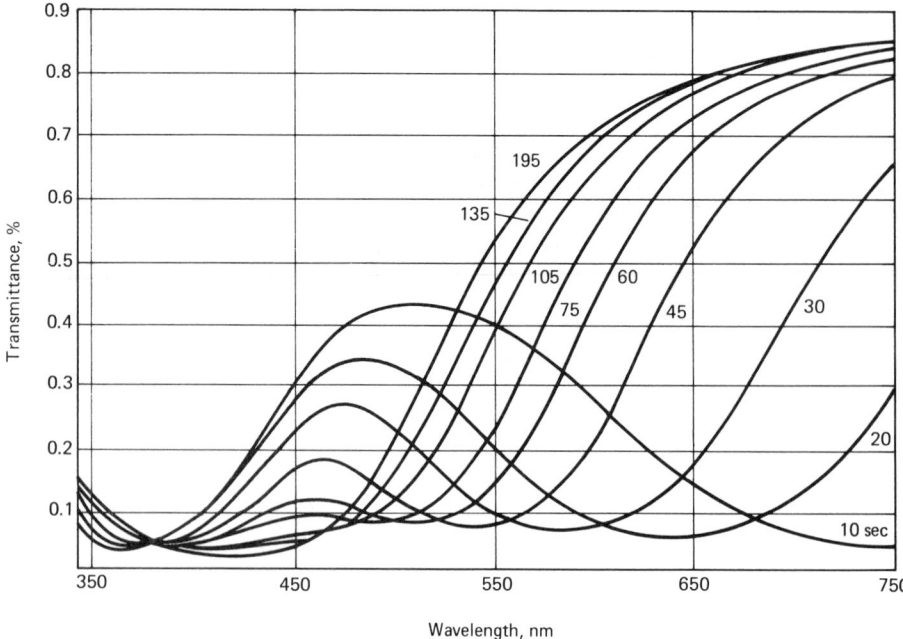

Fig 4 Transmission versus wavelength of developed Polychromatic glass as a function of duration of initial exposure in seconds. Source: Ref 37

sometimes less than tenths of a percent. In addition to careful batching, careful control of the melting conditions must be maintained to keep the volatilization and oxidation state of the dopants constant.

Photochromic glass has found its widest commercial application as eyeglass lenses. For this purpose the glass is generally press-formed as lens blanks, which are subsequently heat treated to develop the desired photochromic behavior, and then ground and polished into ophthalmic lenses. Photochromic flat glass can also be produced, but care must be taken during heat treatment to avoid damage to the glass surface. For sunglass applications, compositions have been developed that allow drawing from the melt a glass sheet that is capable of being simultaneously heat treated and sagged to the appropriate lens curvature without touching the mold (Ref 49, 50). Large sheets of thin photochromic glass can also be made by suitably supporting the glass, such as on an air hearth, during heat treatment. Examples of photochromic ophthalmic and sheet glass compositions are shown in Table 2. Darkening/fading and spectral transmittance curves for a commercial photochromic ophthalmic lens glass are shown in Fig 5. Corresponding data are provided in Table 3.

Photochromic glasses are generally clear (colorless and highly transmitting in the faded state), but can be made with a fixed tint superimposed on the photochromic behavior, either by including coloring ions in the batch composition or by chemically reducing the silver or other metals, below the glass surface, such as by treating in a reducing gas atmosphere at temperatures of a few hundred degrees Celsius (Ref 51).

Photochromic glass objects can be physically strengthened by air tempering and chemically strengthened by ion-exchange. The article "Ion-Exchange" in this Volume provides additional information. However, sometimes the desired photochromic properties are altered by the further heat treatment inherent to the strengthening processing (Ref 48). Glasses of both compositions listed in Table 2 can be chemically strengthened, using sodium-for-lithium ion-exchange, without adversely affecting the photochromic properties.

Variations to be described below include bistable, optically bleached, and polarizing photochromic glasses. As already noted, the darkening and fading rates of silver halide-based photochromic glasses are temperature

Table 2 Typical compositions of photochromic ophthalmic and flat glass

Component	Ophthalmic glass (a),wt%	Flat glass (drawn sheet)(b),wt%
SiO$_2$	56.46	60.4
B$_2$O$_3$	18.15	17.7
Al$_2$O$_3$	6.19	11.8
Li$_2$O	1.81	2.1
Na$_2$O	4.08	5.9
K$_2$O	5.72	1.6
PbO	. . .	0.28
ZrO$_2$	4.99	. . .
TiO$_2$	2.07	. . .
Ag	0.207	0.16
Cl	0.166	0.48
Br	0.137	0.10
F	. . .	0.22
CuO	0.006	0.007

(a) Ref 45. (b) Ref 50

dependent. Compositions have been found for which the fading at room temperature is so slow that the glass is essentially bistable (Ref 52). To fade the glass, its temperature must be significantly raised. Some photochromic glasses are optically bleachable in the sense that the darkened state can be faded by exposure to intense light of visible or near-infrared wavelengths (Ref 53, 54). Effective optical bleaching at 150 mJ/cm^2 (0.71 ft · lbf/in.^{2}) has been reported (Ref 53). Optically bleachable bistable glasses can be used for information storage and/or image storage.

Photochromic glasses can be made to polarize light by at least two processes. In the first, the heat-treated photochromic glass is mechanically deformed at temperatures near, but below, its softening point, under high stress, so as to elongate the silver halide particles along a common axis (Ref 55, 56). Extrusion and redraw are two techniques used to suitably deform the glass. In this condition, the glass is polarizing in the darkened state, but, of course, is not polarizing in the faded state. A permanent polarization can be added to these glasses by chemically reducing the elongated silver halide particles, below the surface of the glass, to give elongated silver metal particles. This can be accomplished by heat treatment in a reducing atmosphere (Ref 57, 58). This treatment must be done at

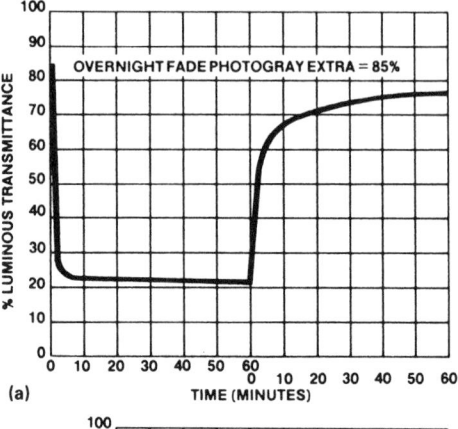

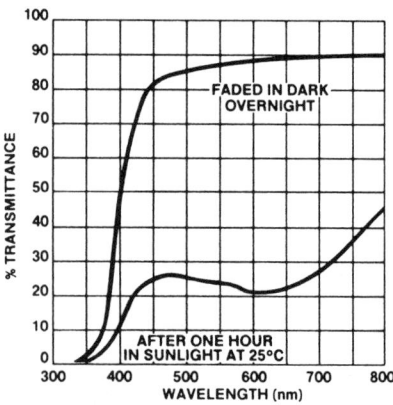

Fig 5 Luminous transmittance versus time for a commercial photochromic ophthalmic lens during darkening and fading (top) and spectral transmittance for the same lens in the darkened and faded states (bottom). Source: Corning Inc. product literature

Table 3 Darkened transmittance after 1 h in direct sunlight

	For 2 mm (0.08 in.) thick lens			For 4 mm (0.16 in.) thick lens		
Tempering state	At 0 °C (32 °F)	At 25 °C (77 °F)	At 40 °C (104 °F)	At 0 °C (32 °F)	At 25 °C (77 °F)	At 40 °C (104 °F)
Annealed (untempered)	20%	22%	38%	6%	11%	30%
Chemtempered	19%	22%	39%	5%	11%	31%
Heat-tempered	23%	24%	30%	7%	9%	20%

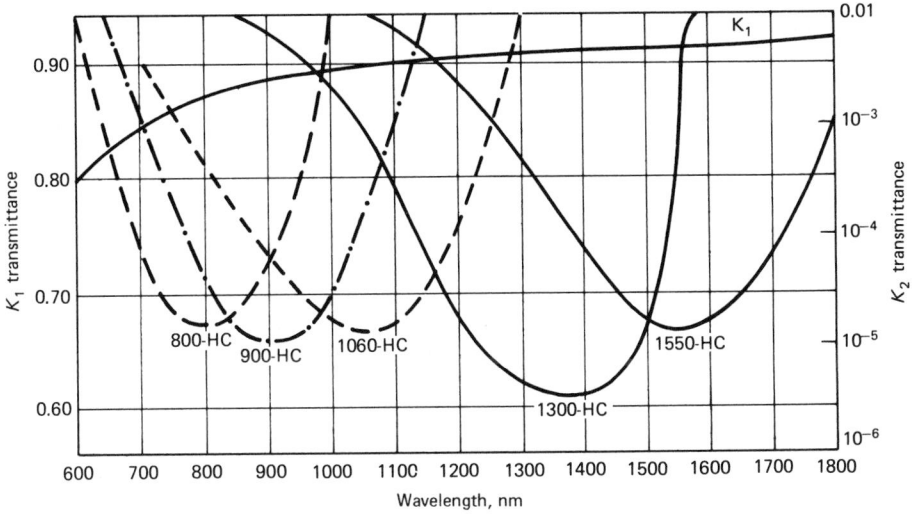

Fig 6 Transmittance versus wavelength for commercial near-infrared polarizing glass filters for polarized light oriented with its electric field parallel (K_2) and perpendicular (K_1) to the elongation axis of the glass. Note that the K_1 curve applies to all the reported filters, whereas the five K_2 curves vary, depending on the filter. HC, high contrast. Source: Corning Inc. product literature

temperatures below the lesser of the annealing point of the glass or the melting point of the silver halide; generally below 450 °C (840 °F). The polarizing properties so generated are permanent. Figure 6 shows transmittance versus wavelength curves for commercial near-infrared polarizing glass filters, with two different orientations of polarized light. High extinction as well as high transmittance are provided by the high-contrast polarizers.

Alternatively, some photochromic glasses can be rendered polarizing by optically bleaching them from the darkened state using intense polarized light (Ref 54). Unfortunately, unlike the mechanically deformed polarizing glasses, those made polarizing by this technique tend to return to their nonpolarizing condition upon repeated darkening-fading cycling.

REFERENCES

1. M. Faraday, *Ann. Chim. Phys.*, Vol 25, 1825, p 99
2. W.A. Weyl, Chapt. XXXI, *Colored Glasses*, The Society of Glass Technology, Sheffield, 1951 and 1986
3. R.H. Dalton, "Glass Article and Method of Making It," U.S. patent 2,326,012, Aug 1943
4. R.H. Dalton, "Glass Article," U.S. patent 2,422,472, June 1947
5. S.D. Stookey, "Photosensitive Gold Glass and Method of Making It," U.S. patent 2,515,937, July 1950
6. S.D. Stookey, "Photosensitive Copper Glass and Method of Making It," U.S. patent 2,515,938, July 1950
7. S.D. Stookey, Photosensitive Glass, a New Photographic Medium, *Ind. Eng. Chem.*, Vol 41, 1949, p 856
8. S.D. Stookey, Colorization of Glass by Gold, Silver, and Copper, *J. Am. Ceram. Soc.*, Vol 32, 1949, p 246
9. A.E. Badger and F.A. Hummel, Effect of Ultraviolet Light on Glass-Containing Silver, *Phys. Rev.*, Vol 68, 1945, p 231
10. W.H. Armistead, "Silver-Containing Photosensitive Glass," U.S. patent 2,515,936, July 1950
11. H. Imagawa, ESR Studied on ns^1 Centers in Glass, *J. Non-Cryst. Solids*, Vol 1, 1969, p 335–338
12. Yu.N. Alenko, R.A. Zhitnikov, V.K. Krasikov, and D.P. Peregud, ESR Investigations of Silver Centers in Photochromic Glass, *Sov. Phys. Solid State*, Vol 18 (No. 6), 1976, p 902–904
13. R. Yokota and H. Imagawa, ESR Studies of Radiophotoluminesce Centers in Ag-Activated Phosphate Glasses, *J. Phys. Soc. Jpn.*, Vol 20, 1965, p 1537–1538
14. R. Yokota, Formation of Ag Centers and Mechanism of Ag Formation in Photosensitive Glasses, *J. Ceram. Soc. Jpn.*, Vol 78 (No. 8), 1970, p 39–40
15. J.S. Stroud, Color Centers in a Ce-Containing Silicate Glass, *J. Chem. Phys.*, Vol 37, 1962, p 836–841
16. J.L. Lineweaver, Oxygen Outgassing Caused by Electron Bombardment of Glass, *J. Appl. Phys.*, Vol 34 (No. 6), 1963, p 1786–1791
17. J. de Gier, U.S. patent 3,725,710, April 1973
18. J.J. Van der Geer *et al.*, U.S. patent 4,337,410, June 1982
19. S.D. Stookey, "Photosensitive Glass," U.S. patent 2,515,275, July 1950
20. S.D. Stookey, "Photosensitive Glass Containing Palladium," U.S. patent 2,515,942, July 1950
21. S.D. Stookey, "Photosensitive Glass Article and Composition and Method of Making It," U.S. patent 2,515,943, July 1950
22. T.P. Seward III, Coloration and Optical Anisotropy in Silver-Containing Glasses, *J. Non-Cryst. Solids*, Vol 40, 1980, p 499–513
23. U. Kreibig and C.V. Fragstein, The Limitation of Electron Mean Free Path in Small Silver Particles, *Z. Phys.*, Vol 224, 1969, p 307–323
24. H.C. van de Hulst, *Light Scattering by Small Particles*, John Wiley & Sons, 1957
25. C.F. Bohren and D.R. Huffman, *Absorption and Scattering of Light by Small Particles*, John Wiley & Sons, 1983
26. R.H. Doremus, Optical Properties of Small Gold Particles, *J. Chem. Phys.*, Vol 40 (No. 8), 1964, p 2389–2396
27. U. Kreibig, Kramers Kronig Analysis of the Optical Properties of Small Silver Particles, *Z. Phys.*, Vol 234, 1970, p 307–318
28. U. Kreibig and P. Zacharias, Surface Plasma Resonances in Small Spherical Silver and Gold Particles, *Z. Phys.*, Vol 231, 1970, p 128–143
29. N.F. Borrelli, Corning Inc., private data, 1989
30. R.J. Araujo, Corning Inc., private communication, 1990
31. S.D. Stookey, Chemical Machining of Photosensitive Glass, *Ind. Eng. Chem.*, Vol 45, 1953, p 115–118
32. S.D. Stookey, Catalyzed Crystallization of Glass, *Progress in Ceramic Sciences*, Vol 2, J.E. Burke, Ed., Pergamon, 1962
33. S.D. Stookey, Controlled Nucleation and Crystallization Lead to Versatile New Glass-Ceramics, *Chem. Eng. News*, Vol 39 (No. 25), 1961, p 116–125
34. N.F. Borrelli, D.L. Morse, R.H. Bellman, and W.L. Morgan, Photolytic Technique for Producing Microlenses in Photosensitive Glass, *Appl. Opt.*, Vol 24, 1985, p 2520–2525
35. N.F. Borrelli and D.L. Morse, Micro-

lens Arrays Produced by a Photolytic Technique, *Appl. Opt.*, Vol 27, 1988, p 476–479

36. D.M. Baranowski, R.H. Bellman, L.G. Mann, and N.F. Borrelli, Glass Microlens Array, *Laser Focus World*, Nov 1989, p 139–143

37. S.D. Stookey, G.H. Beall, and J.E. Pierson, Full Color Photosensitive Glass, *J. Appl. Phys.*, Vol 49 (No. 10), 1978, p 5114–5123

38. N.F. Borrelli, J.B. Chodak, D.A. Nolan, and T.P. Seward III, Interpretation of Induced Color in Polychromatic Glasses, *J. Opt. Soc. Am.*, Vol 69 (No. 11), 1979, p 1514–1519

39. R.J. Araujo, Photochromic Glass, *Treatise on Materials Science and Technology*, Vol 12, R.H. Doremus and M. Tomozawa, Ed., Academic Press, 1977, p 91

40. R.J. Araujo, "Photochromic Glass Article and Method of Making It," U.S. patent 3,325,299, June 1967

41. R.J. Araujo and P.A. Tick, "Refractive Index Corrected Copper Cadmium Halide Photochromic Glasses," U.S. patent 4,166,745, Sept 1979

42. D.M. Trotter, J.W.H. Schreurs, and P.A. Tick, CuCd Halide Photochromic Glasses, *J. Appl. Phys.*, Vol 53 (No. 7), 1982, p 4657–4672

43. D.L. Morse and T.P. Seward III, "Optically Clear Copper Halide Photochromic Glass Articles," U.S. patent 4,222,781, Sept 1980

44. D.L. Morse, Copper Halide Containing Photochromic Glasses, *Inorg. Chem.*, Vol 20, 1981, p 777

45. G.B. Hares, D.L. Morse, T.P. Seward III, and D.W. Smith, "Photochromic Glass," U.S. patent 4,190,451, Feb 1980

46. R.J. Araujo, N.F. Borrelli, and D.A. Nolan, The Influence of Electron-Hole Separation on the Recombination Probability in Photochromic Glasses, *Philos. Mag. B.*, Vol 40, 1979, p 279

47. R.J. Araujo and D.W. Smith, The Effect of Boron Coordination on Halogen Solubility, Nickel Colour, and Photochromium in Potassium Aluminoborosilicate Glasses, *Phys. Chem. Glasses*, Vol 21, 1980, p 114

48. G.B. Hares and T.P. Seward III, Thermo-Chemical Effects in Photochromic Glasses, *J. Non-Cryst. Solids*, Vol 38 and 39, 1980, p 205

49. D.J. Kerko and T.P. Seward III, "Photochromic Sheet Glass Compositions and Articles," U.S. patent 4,018,965, April 1977

50. J.P. Mazeau and T.P. Seward III, "Photochromic Glasses Suitable for Simultaneous Heat Treatment and Shaping," U.S. patent 4,130,437, Dec 1978

51. N.F. Borrelli and B. Wedding, Optical Properties of Chemically Reduced Photochromic Glasses, *J. Appl. Phys.*, Vol 63, 1988, p 2756

52. R.J. Araujo, Properties and Applications of Photochromic Glasses, *Recent Advances in Display Media*, National Aeronautics and Space Administration, 1968, p 63

53. T.P. Seward III, Thermally Darkenable Photochromic Glasses, *J. Appl. Phys.*, Vol 46, 1975, p 689

54. N.F. Borrelli, J.B. Chodak, and G.B. Hares, Optically Induced Anisotropy in Photochromic Glasses, *J. Appl. Phys.*, Vol 50, 1979, p 5978–5987

55. R.J. Araujo, W.H. Cramer, and S.D. Stookey, "Photochromic Polarizing Glasses," U.S. patent 3,540,793, Nov 1970

56. R.J. Araujo, W.H. Cramer, and S.D. Stookey, "Method of Forming Photochromic Polarizing Glasses," U.S. patent 3,653,863, April 1972

57. N.F. Borrelli and K.K. Lo, "Method for Making Polarizing Glasses by Extrusion," U.S. patent 4,304,584, Dec 1981

58. N.F. Borrelli, F. Coppola, D.L. Morse, D.A. Nolan, and T.P. Seward III, "Infrared Polarizing Glasses," U.S. patent 4,479,819, Oct 1984

59. T.P. Seward III, Glass Polarizers Containing Silver, *Proc. SPIE*, Vol 464, 1984, p 96

Sol-Gel Processes

Robert D. Shoup, Corning Incorporated

SOL-GEL PROCESSING, whereby solution or sol becomes dense glass, consists of the steps shown in Fig 1, which is a simplified version of the process. Several additional steps may be involved in preparing the precursor solution or sol, especially in multicomponent systems. In any sol-gel process, the chemistry of the precursor material is controlled to induce particle-particle interactions or polymer chain interactions that result in the structuring of colloidal solids to encapsulate solvent(s). Without the addition of dispersed solids, the open networks typically undergo at least 50% linear shrinkage during solvent removal and thermal consolidation to form dense glass.

Fused silica represents the most explored area of bulk glass formation by sol-gel processing. The reasons for this are that silica forms a single-component oxide glass, and there is an increasing demand for high-quality (and high-purity) fused silica optical components with high transmission in ultraviolet (UV) and infrared (IR) wavelengths. Near-net-shape casting of gels provides an alternative to expensive finishing.

Different sol-gel processing techniques are described in this article, primarily in terms of silica-based developments. Two types of sol formulations are included: gels produced by the destabilization of colloidal sols and polymeric gels produced by the hydrolysis of metal-organic compounds. Because it is necessary to limit the scope of this review, several general texts (Ref 1–8) are recommended as sources of additional details on the chemistry and physics of sol-gel processes.

The advantages that sol-gel processing has over conventional melting can be enumerated:

1. Colloidal particle compacts have very high surface energies, allowing sintering well below their melting temperatures. Lower sintering temperatures not only mean lower energy costs, but also translate into higher glass purity because less metal oxide contaminants are likely to be released from refractory-lined furnaces. This is particularly true for fused silica, which requires melting/forming temperatures of about 2000 °C (3630 °F)
2. High-purity raw materials, such as tetraethylorthosilicate (TEOS), that contain less than 100 ppb total metals do exist. If care is exercised, these levels can be retained or reduced in the glass product
3. In theory, improved homogeneity of multicomponent species can be obtained by blending a variety of metal alkoxides, colloidal dispersions, or easily diffused soluble salts. The primary concerns are whether the reactivity of the various species can be controlled to produce the desired level of homogeneity, and whether that distribution can be retained throughout the remaining processing steps
4. Low-temperature sintering of near-molecularly dispersed components could lead to new noncrystalline compositions that might otherwise phase separate or crystallize if produced by the conventional melting approach
5. Gelation permits the molding of near-net shapes in applications where machining of those shapes is very costly. Although shrinkage factors must be considered, the shape and surface configuration will be retained, despite the large dimensional change
6. Special materials or configurations have been developed, such as films that are easily applied in thicknesses under 1 μm

(40 μin.) controlled-size spherical powders, fibers, and others
7. Environmental issues are minor, compared to corrosive by-products of the $SiCl_4$ chemical vapor deposition (CVD) process for fused silica

Disadvantages of sol-gel processing include:

1. The colloidal gel monoliths have very small pore structures and relatively low densities. Removal of the solvents from these open networks and the overall shrinkage in processing require special techniques to avoid cracking. In addition, thermal processing must take into account high surface water and carbonaceous residues that can lead to bloating, residual bubbles, or crystal formation if not properly removed. These barriers tend to limit the size of monolithic glass parts that can be produced by direct casting procedures
2. The high-purity alkoxides are relatively expensive raw materials
3. Multi-step processing adds time and expense. It is unlikely that high-volume glass products, such as containers, cooking ware, flat glass, and common fiber, would be produced economically by sol-gel technology, which has found its niche in the specialty products area

Sol Formulations

In this discussion, the distinction between polymer gels (hydrolyzed alkoxides) and gels produced from particulate sols is based on the classical definition of colloidal sols. The sols are considered to contain colloidal particles in the range of 1 to 1000 nm (0.04 to 40 μin.), which gel by aggregation and condensation into a three-dimensional network encompassing the liquid phase. On the other hand, the polymer gels are considered to be suspensions of highly branched macromolecules in which there are no detectable particles. Low solubility of the silica in acidic aqueous/alcohol media apparently does not permit breakage of siloxane bonds to allow rearrangement of chains into particles. Therefore, the polymer gels are considered to follow the polymerization behavior of polyfunctional organics.

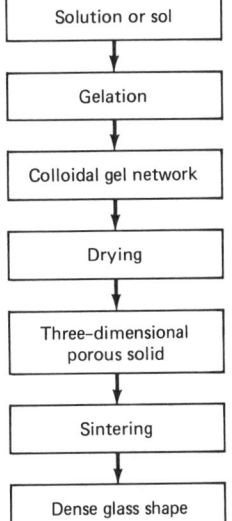

Fig 1 Simplified sol-gel process

- Solution or sol
- Gelation
- Colloidal gel network
- Drying
- Three-dimensional porous solid
- Sintering
- Dense glass shape

Both gel structures have very high surface energies. Small pore (<10 nm, or 0.4 μin.) silica gels can be sintered to dense glass at relatively low temperatures, below about 1200 °C (2190 °F). As particle size increases (above 50 nm, or 2 μin.) and pore diameters exceed about 0.1 μm (0.004 μin.), temperatures of about 1400 °C (2550 °F) are required for complete consolidation of silica matrices.

Gels Produced by Sol Destabilization, Aggregation, and Aging

To provide an understanding of the problems associated with particulate gel processing, the formation of silica sol as outlined by Iler (Ref 9) is described below.

Silica Sol Formation. Freshly prepared diluted (1 to 3%) silicic acid sols (pH 2 to 3) contain about 1 nm (0.04 μin.) sized particles, which undergo rapid growth to 2 to 4 nm (0.08 to 0.16 μin.). Above a pH level of 7, silica solubility increases significantly, so that the growth of particles to 4 to 6 μm (160 to 240 μin.) can take place by the coalescence of particles (Fig 2). The presence of monomeric silicic acid or soluble silica plays a role in cementing the particles together after the initial formation of Si–O–Si bonds. Through a series of processing steps that involve concentration, the addition of monomeric silica, and the control of pH (9 to 10), silica sols that contain up to 50% solids and have particle sizes between 10 and 25 nm (0.4 and 1 μin.) have become available commercially. The stability of these silica sols depends on the maintenance of a repulsive double layer (alkali counterions) that serves to prevent particle collisions.

Gelation of aqueous silica sols can be induced by several techniques. Reducing the pH level to between 5 and 7 causes rapid gelation of these sols. At a given pH, the addition of soluble salts can cause the repulsive double layer to collapse, thereby allowing particle aggregation and gelation. Multivalent cations are more effective than monovalent species. The addition of organics can result in ionic attraction, hydrogen bonding (low pH), reduction in surface tension, or other bridging mechanisms that can lead to flocculation or gelation.

In general, these gel structures are composed of loosely bound, closely stacked spherical particles that produce very small pores (<10 nm, or 0.4 μin., diam). The high capillary forces involved in removing water from very small pores create stresses in the shrinking network that induce cracking. Because of this strong tendency for gel cracking, early investigators used silica sols to bond aggregates, such as fused silica and quartz (Ref 10–12). The high reactivity of colloidal particles (small size and surface hydroxyls) induced densification at lower temperatures than previously observed. Colloidal silica was also combined with salts of glass-former

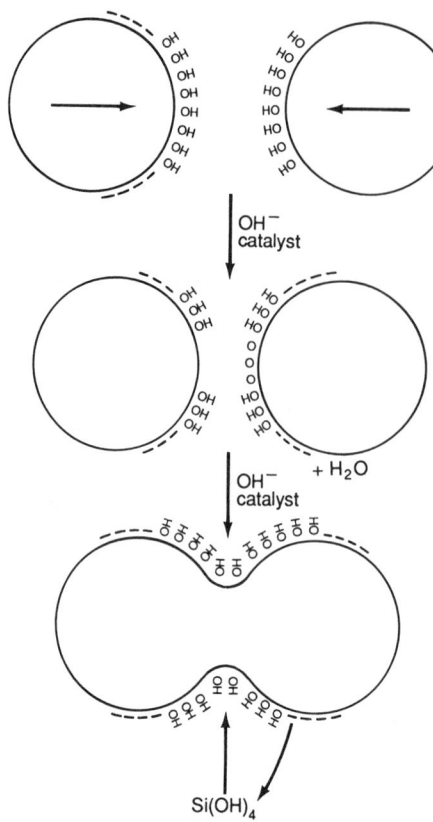

Fig 2 Bond formation between silica particles. Source: Ref 9

components to form glass powders, which were used to make either glass compacts or coatings for either glass or ceramics (Ref 13–17). None of these approaches produced monolithic glass shapes or high-quality bulk glasses.

Monolithic silica gels were prepared at one time by gelation of polysilicic acid sols, followed by long-term (weeks to months) drying at constant humidity. Even then, only the most persistent workers produced an uncracked gel body. A technique known as supercritical drying was demonstrated on polymer gels (Ref 18, 19) to overcome the destructive capillary forces involved in drying very small pore gel structures.

Other approaches have been used to improve the strength and crack resistance of these colloidal gel structures. Iler (Ref 20) discusses the use of a mixture of particle sizes to produce more densely packed silica gel structures. Monosized particle gels gave either very soft films or cracked (mud-flat) layers that would easily spall off the surface. Combining three additional small-size-range sols with 100 nm (4 μin.) colloidal particles produced a hard film (with a packing density of 80.4%) that shrank very little and did not crack.

An idealized experiment for the mechanical packing of spherical particles was performed by McGeary (Ref 21) using metal shot and vibratory compaction techniques. He

achieved 95.1% of theoretical density with four ideally sized sphere lots. This approach is applicable to both coatings and monoliths, but its difficulty lies in producing the appropriate size ranges and achieving the maximum packing arrangement (note that Iler obtained only 80.4% density). Binders, such as silicic acid sols, were shown to develop interparticle bonds and form stronger gels (Ref 22). Reinforcement of 100 nm (4 μin.) particle gels would create two positive effects. First, large particles create large pores, which have lower capillary forces than small pores. Second, reinforcement strengthens the gel network to counter drying stresses.

Shoup (Ref 23, 24) discovered a system that incorporates both principles: the growth of large particles to form large pore gel structures with strongly reinforced bonds between the aggregates. The precursor gel mixture comprises potassium silicate, a 40% silica sol (14 nm, or 0.6 μin., particles), and formamide. The latter is a water-miscible gel reagent that hydrolyzes to induce gelation by lowering the pH. By controlling the ratio of colloidal silica to soluble silicate, strong monolithic gel networks are produced in the preferred pore size range of 100 to 250 nm (4 to 10 μin.). The microstructure shown in Fig 3 has an open silica network. The reinforced particle chains are able to survive the forces involved in rapid (minutes to hours) drying in microwave ovens.

This gel technology led to the formation of monolithic near-net shapes in large sizes (600 mm, or 24 in., plate) that could survive rinsing, drying, and sintering to fully dense fused silica despite more than 50% linear shrinkage. Further development led to the formation of titania/silica (TiO$_2$/SiO$_2$) gel compositions (Ref 25). Basic titania sols (stabilized with quaternary ammonium counterions) were substituted for silica sols that were used in nucleation and growth of the 100% SiO$_2$ gels. The result is a gel network con-

1 μm

Fig 3 Scanning electron micrograph of potassium silicate gel network with mean pore diameter of 240 nm (10 μin.) and porosity of about 80%

taining homogeneously dispersed TiO_2 that can be sintered to produce ultralow-expansion glass shapes.

Purity can become an issue in the formation of glass (for example, fused silica) from the aqueous particulate sols. The commercial silica sols are produced from sodium silicate solutions, which are products derived from melting impure raw materials. Consequently, there are high levels of oxides of aluminum, titanium, iron, and others in the sols, which are retained in the gel matrix. In fused silica, these trace metals cause absorption in the UV wavelengths. Purer silica sols can be produced from high-purity raw materials, but costs increase significantly.

Multicomponent glass compositions from particulate sols face a number of barriers in the production of homogeneous monolithic gel structures. In an aqueous system, it is preferable to deposit the additive in the silica gel matrix as an insoluble species. This reduces migration and the formation of concentration gradients during drying. One possible approach is to mix various metal oxide sols with the base SiO_2 sol prior to gelation. Matijevic (Ref 26, 27) reviewed a broad range of materials prepared as monodisperse sols in spherical and polyhedral shapes (Fig 4). These materials include oxides of Ce, Si, Ti, Cr, Cd, Fe, Al, Zn, V, and others.

A difficulty with mixing sols is that incompatible sols will spontaneously flocculate to reduce homogeneity. There are chemical means that can be used to develop compatible sol mixtures. One approach involves charge reversal, such as that described for alumina-modified silica sols (Ref 28). However, it may be very difficult to carry out the chemical modifications beyond two or three additives. Dispersed colloidal particles prepared from aerosols, CVD deposition, and other methods are alternative measures.

The addition of metal salts or complexes to silica sol may also result in sol instability because of increased cation concentrations. These soluble species have an added disadvantage. Unless they can be made to precipitate, migration during drying will create concentration gradients. Limited success has been reported on the development of multicomponent glass compositions by gelation of mixed sol monoliths (Ref 25), but multicomponent particles and frits have been developed as potential coating materials (Ref 15–17, 29).

Dispersion processing techniques have been developed to overcome the difficulties encountered in processing small-pore monolithic gels. Most of these processes are combinations of ceramic processing and gel casting of colloidal silica particles. Their common feature is the aggregation of colloidal particles that, when cast, produce pore diameters between about 0.05 and 0.5 μm (2 and 20 $\mu in.$). The result is a porous casting that dries faster, shrinks less, and has less of a tendency to crack. Despite slightly higher sin-

tering temperatures, this combined processing permits more flexibility in casting larger, more intricate glass shapes. Rabinovich, *et al.* (Ref 30, 31) developed "double" processing of fumed silica to cast large shapes that shrank little upon drying. The initial dispersed colloidal silica was dried, heated to 900 °C (1650 °F) and redispersed in water. The second dispersion contained significantly larger aggregates that produced larger pore castings. These bodies were sintered to full density at about 1300 °C (2370 °F).

Clasen (Ref 32, 33) used a larger-diameter fumed silica (Degussa OX-50) to cast green bodies that were 40 to 50% dense. The keys to this concept were high solids suspensions,

fluoride reactivity, and particle reinforcement upon gelation.

Schermerhorn, *et al.* (Ref 34) started with an alkoxide gel, which was dried rapidly to form particles of high surface area. These particulates were heated and milled in water to form slurries containing 70 to 80% solids. Castings that were sintered to full density produced high-quality fused silica disks as large as 370 mm (15 in.) in diameter and 70 mm (3 in.) thick.

Nonaqueous solvents have been shown to be effective media for dispersion casting. Scherer, *et al.* (Ref 35) demonstrated the formation of 30% solids dispersions of fumed silica (Degussa OX-50) in chloroform. Steric

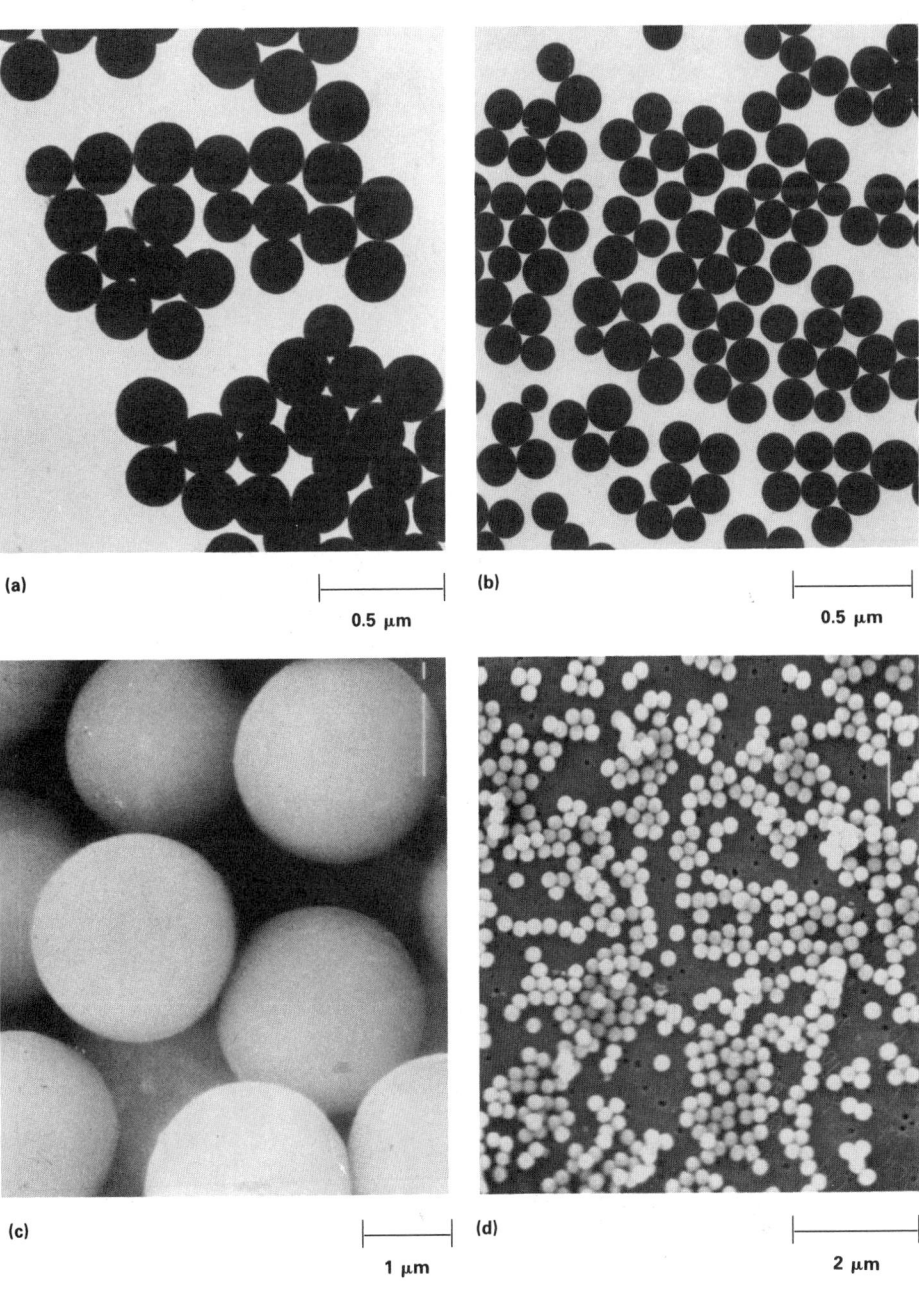

(a)   0.5 μm

(b)   0.5 μm

(c)   1 μm

(d)   2 μm

Fig 4 Transmission electron micrographs of chromium hydroxide (a) and terbium basic carbonate (b). Scanning electron micrographs of zinc sulfide (c) and cerium oxide (d). Source: Ref 26

stabilization of the particles with moderate length chain alcohols could be upset by exposure to ammonia vapors. The resulting gelation could be used to cast flat-plate, as well as layered, cylindrical shapes. Both single-mode and gradient index optical fiber were fabricated. Larger pore structures and a low surface tension solvent combine to reduce capillary forces involved in the drying step.

Gels Produced by Hydrolysis and Polycondensation of Metal Alkoxides

Gel technology based on hydrolysis of metal-organics has experienced a strong growth since the 1970s. This is due to two key properties: a very uniform microstructure composed of very small particles (pores) that replicate a mold surface and maintain that shape and surface despite more than 50% linear shrinkage, and the greater potential for composition manipulation through cross-linking of a wide selection of commercially available metal alkoxides.

The alkoxide sol-gel reactions can be represented by these equations:

$$-\mathrm{Si-OR} + H_2O \quad -\mathrm{Si-OH} + R\mathrm{OH} \quad (Eq\,1)$$

$$-\mathrm{Si-OR} + \mathrm{HO-Si-}$$

$$-\mathrm{Si-O-Si-} + R\mathrm{OH} \quad (Eq\,2)$$

$$-\mathrm{Si-OH} + \mathrm{HO-Si-}$$

$$-\mathrm{Si-O-Si-} + H_2O \quad (Eq\,3)$$

Hydrolysis (Eq 1) is the attack of water (OH) upon the reactive alkoxide group (OR) where R is the alkyl group, C_xH_{2x+1}. Condensation reactions (Eq 2 and 3) produce the siloxane bonds that lead to longer chains and branches. More detailed explanations of these reactions are provided in Ref 36.

The type of gel structure that can be obtained from an alkoxide such as TEOS depends on the pH level and amount of water used in hydrolysis. At a low pH, particularly below 3, complete hydrolysis (four moles of water per mole of TEOS) produces a highly branched polymeric fractal structure, as depicted in Fig 5. There are no discernible particles and the pores are very small (<1 nm, or 0.04 μin.). As the pH of gelation increases towards 7, there are more dissolution and condensation reactions and the gel structure should coarsen to some extent.

Above a pH of 7, there is a maximum particle growth because increased silica solubility results in depolymerization of siloxane

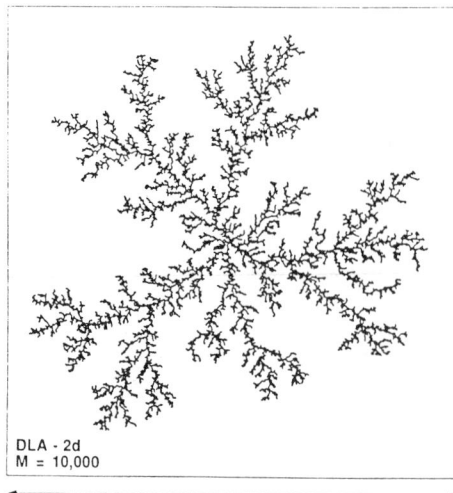

DLA - 2d
M = 10,000

350 DIAMETERS

Fig 5 Fractal polymer made by branching of poly-functional monomer with $f > 2$ (computer simulation of two-dimensional aggregation). Source: Ref 37

bonds and provides monomeric silica for the aging process (not unlike Ostwald ripening). Aggregation/growth is depicted in Fig 6. Stober, et al. (Ref 38) formed monosized spherical particles in a similar manner. Other investigators report the formation of 0.3 to 1.0 μm (12 to 40 μin.) particles in fairly concentrated solutions. Acidic gels typically have specific areas of about 600 m²/g (2.9×10^6 ft²/lb) whereas 0.3 μm (12 μin.) particles formed in basic conditions gave surface areas as low as 5 m²/g (24×10^3 ft²/lb) (not sintered).

Monolithic Gels. Interest in producing monolithic gel glasses from acid-hydrolyzed alkoxide gels was rekindled by the work of Yoldas (Ref 40) on forming transparent alumina and by the developments of Yamane, et al. (Ref 41) on silica. Both investigations utilized long, tedious drying techniques to preserve the integrity of the gel.

Efforts were directed at finding better ways to dry these very small pore gels. Prassas, et al. (Ref 19) successfully used supercritical drying to remove the solvent before sintering the gel to glass. Henning and Svensson (Ref 42) extended this operation to large auto-

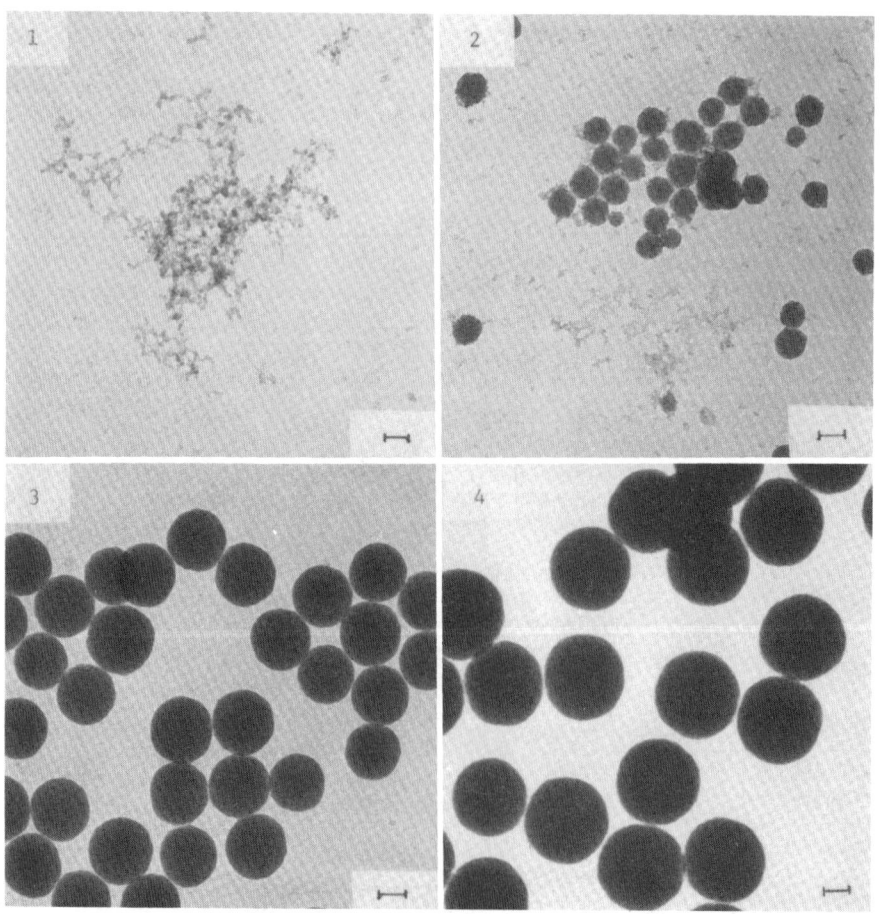

Fig 6 Silica particles precipitated from solution of 0.17 M tetraethylorthosilicate, 1.3 M ammonia, and 2.0 M water in ethanol at 25 °C (77 °F). Grids taken 2 min, 8 min, 30 min, and 20 h after initiation of reaction. Aggregation of small particles probably occurs as they are concentrated on the grid by drying. Bar = 100 nm (4 μin.). Source: Ref 39

claves for the production of large (transparent) aerogels.

In the mid-1980s, Hench (Ref 43) announced that drying control chemical additives (DCCA) in neutral or acidic alkoxide gels was responsible for improving drying. It was found that the addition of formamide to a tetramethylorthosilicate (TMOS) sol produced a sol-gel network with uniformly larger pores than the gels made with methanol alone. An acidic DCCA, oxalic acid, also produced a narrow distribution of pores, but they were somewhat smaller in size. The DCCAs were described as affecting the rates of hydrolysis and polycondensation. The uniform growth of the network led to greater reinforcement (strength) to resist drying stresses. A similar result was recorded for dimethyl formamide, which produced even larger pores after aging at ~150 °C (~300 °F) (Ref 44). Although these DCCAs decreased drying times, the subsequent thermal treatment required careful control to remove carbonaceous materials and water before sintering at about 1050 to 1150 °C (1920 to 2100 °F).

Toki, et al. (Ref 45) were successful in generating larger pore structures by dispersing colloidal silica, either Degussa OX-50 or base-catalyzed TEOS gel particles, in a TEOS sol before hydrolysis. The resultant gel was opaque, which demonstrated the greater permeability of the network. This open network dried substantially faster than gels without a dispersed phase, because capillary drying forces were decreased. The coarser gels required higher sintering temperatures between 1300 and 1500 °C (2370 and 2730 °F), but the yield of larger glass structures and faster processing more than offset this requirement.

Multicomponent gels have been generated by various approaches. The mixed metal species can be added in the form of soluble salts, soluble complexes, dispersed colloidal particles, or metal alkoxides. The preferable method is to produce a totally miscible alkoxide system, so that intimate polymerization of all the metal species will yield the highest homogeneity. However, because of the varying reactivities of different metal alkoxides toward nucleophilic reagents such as water, it is unlikely that simple mixing of the metal alkoxides will produce the desired homogeneity. The more reactive species would hydrolyze to form clusters before the less reactive alkoxide started to polymerize.

Two general schemes were devised by Dislich (Ref 46) and Thomas (Ref 47) for dealing with these differences in reactivity. Thomas used a sequential addition process, in which the least reactive alkoxide was partially hydrolyzed, whereupon the more reactive alkoxide was added, followed by additional partial hydrolysis, and so on. This allows the newly added unhydrolyzed alkoxides to react and condense with the hydrolyzed sites of the previously reacted species. This approach has the potential to distribute the metals randomly along the polymer chain, rather than in random clusters. More detail on the reactivity of various metal alkoxides is provided in the review by Livage, et al. (Ref 48). Some of the glass compositions that have been developed are described in the section "Multicomponent Glasses" in this article.

Drying and Densification

Drying of gels results in network shrinkage in response to the capillary forces that actively transport the liquid to the surface for evaporation. The very small pore (1 to 5 nm, or 0.04 to 0.20 μin.) polymeric gels exhibit a higher degree of shrinkage than do the larger pore (100 to 250 nm, or 4 to 10 μin.) silicate gels (Ref 23). This is due to at least two factors. The smaller pores produce significantly higher capillary forces, and the silicate gels have more SiO_2 reinforcement at the necks of the aggregated particles. Even with this increased matrix strength, combined with large pores/low stress, it was found that uniform drying techniques, as provided by microwave processing, were necessary.

It is obvious that the highly branched polymer gels are extremely flexible in the early stages of shrinkage. However, as the network stiffens, it must survive the stress that accompanies the recession of the air-liquid interface into the pore structure. It is surprising that the DCCA gels (Ref 43), with only a slight increase in pore diameter (from about 1 to 5 nm, or 0.04 to 0.20 μin.), can survive the vigors of drying better than the gels that contain none of these reagents. The uniformity of the pore distribution, with a slight assist from increased condensation (strength), must contribute significantly to reduction in cracking.

The dispersion gels seem to be gaining in popularity, because their open networks can be dried more rapidly with less cracking. Nonaqueous dispersion gels have a reduced tendency to crack, despite low bond strength. This is primarily due to the lower stress, imparted by lower surface tension, of organic solvents, compared to that of water.

Supercritical drying techniques are primarily directed at producing large aerogel structures used in specialized applications. There is no known effort to develop this technique for glass processing. The key to the process is to raise the temperature of the gel in an autoclave above its critical point (T_c, P_c), at which time the densities of the liquid and the vapor phases are the same. The vapor is vented off the sample at constant temperature with no concern for capillary forces because there is no liquid-vapor interface. By using liquid carbon dioxide, it is possible to work at lower temperatures (T_c of 31.1 °C, or 88 °F), but solvent exchange procedures may be required because of immiscibility with certain solvents.

Densification. The sintering of silica gels is driven by the interfacial energy that a particular gel possesses. Because small-particle (pore) gels possess high surface areas, they should sinter at lower temperatures than large-pore/lower surface area gels. The polymer gels with surface areas of 600 m^2/g (2.9 × 10^6 ft^2/lb) show complete densification at about 1100 °C (2010 °F). Large-pore silicate gels with surface areas of 100 m^2/g (490 × 10^3 ft^2/lb) or less require temperatures of about 1400 °C (2550 °F) to completely sinter. Of course, that is not the complete situation. High surface area indicates high surface hydroxyls, which contribute to condensation reactions and further aid in sintering.

There are other considerations that have been shown to inhibit densification of the polymer gels. Along with water, residual organics must be removed before pore closure, or else they can contribute to reboiling and bloating at higher temperatures. The smaller the pore, the greater the difficulty in removal, which is a slow process that requires careful consideration of the atmosphere. There is a strong potential for buildup of reaction products in 1 to 5 nm (0.04 to 0.20 μm) pores, which could shift the equilibrium toward incomplete oxidation of residual carbon. This event, combined with fast heating rates, produces gels with encapsulated carbon. There are some organic reagents (for example, acetylacetonate complexes) that seem to resist oxidation, even in 100% O_2 atmospheres.

Hydroxyl removal is a two-step process. First, thermal processing removes the molecular H_2O and most of the silanol groups. The final OH must be chemically reacted with chlorine (or fluorine) at elevated temperatures. Chlorine is the most commonly used reactant. It requires temperatures above 800 °C (1470 °F), and preferably above 1000 °C (1830 °F) for faster reaction rates. This could conceivably cause a problem when dealing with a very small pore gel that starts to sinter at temperatures below 1000 °C (1830 °F). The structure could collapse to trap H_2O and chlorine (or Cl^-), either in pores or in the matrix. Both could reboil at higher temperatures.

Certain advantages can be attributed to the lower sintering temperatures exhibited by polymer gels. Higher temperatures can subject the structure to higher levels of contaminants (from refractories), and increased crystallization could occur. However, most of the large-pore gels being sintered at 1400 °C (2550 °F) or above are being consolidated in vacuum furnaces. This is a very clean environment, and at 1750 °C (3180 °F), any surface-nucleated cristobalite is remelted (not to reappear upon cooling).

The dispersion processing technique described earlier resulted in larger mean pore diameters that required sintering temperatures higher than 1100 °C (2010 °F). This is partly due to surface activity, but particle size effects usually play a role. Many of these dispersions are less than ideal. Either the particle size distribution is difficult to control or the particles tend to aggregate and, possibly, settle. Settling can be controlled by adding surface active reagents that can provide surface charge or steric stabilization. However, if purity is an issue, then what is being added and whether impurities can be removed in processing are cause for concern.

Forms, Compositions, and Properties

Thin films were the first, and are now the largest, application of sol-gel processing. Schroeder, Dislich, and others (Ref 49–51) have reviewed the development of thin films as optical coatings. The films are applied by either dip or spin coating techniques using the polymeric alkoxide sols. By controlling precursor chemistry and deposition conditions, one can control film pore size, porosity, surface area, and refractive index. A major development was the formation of multiple transparent layers that produce a gradual change in refractive index from the air interface to that of the substrate. One manufacturer has developed the technology to the extent that millions of square meters of glass are coated yearly. One product is a coated architectural glass in which titania controls the reflectivity and palladium provides absorption. Another product is an antireflection coating based on a triple application of titania/silica, titania, and silica. It is recommended for store-front windows to allow light transmission, but reduce glare.

Antireflection coatings are also used in solar cells and laser optics. A heads-up display is a unique application of these coatings. By controlling the ratio of reflectance to transmission, the speed of an automobile can be displayed at eye level on the windshield, without distorting the field of view. The driver no longer needs to shift his eyes to the instrument panel. Other sol-gel coating applications include films for planarizing surfaces up to several micrometers in surface irregularity, protective films for preventing the corrosion and abrasion of mirror surfaces, and electronic passivation coatings.

Fibers have been drawn directly from polymer sols by controlling the viscosity of acid-catalyzed alkoxide gels (Ref 52, 53). High viscosity stabilized the fibers against spheroidization during the draw operation. A key to preventing premature gelation is to maintain water at a lower level. Both silica and titania fibers have been formed, but without sintering, they remain porous and contain some unreacted alkoxide. It was reported that Asahi Glass Co. of Japan produces and sells silica glass fibers made by this technique.

A second application of fibers produced by sol-gel processing is optical waveguides. The dispersion-casting technique of Scherer, et al. (Ref 35) produced silica fiber with optical losses of about 3.5 dB/km. Several other investigators (Ref 30, 33, 45, 54, 55) have used variations of the dispersion-casting technique to produce waveguide preforms that

were sintered to high-quality fused silica. In some cases, the casting was directed toward forming the less-sensitive cladding for the core fiber. At this point in time, none of these approaches is known to be used commercially.

Monoliths. One manufacturer has commercialized the DCCA gel process described by Hench, et al. (Ref 56), and offers two versions of fused silica. One version, a fully densified glass, is available in almost any shape or lens curvature (Fig 7). Overall size can be as large as 65 mm (2.6 in.) in diameter and 10 mm (0.4 in.) thick, but miniature lenses as small as 1.5 mm (0.06 in.) are also available. Because of high purity and low water content, these glasses have broad-band transmission in the UV and IR wavelengths. A key advantage is the ability to gel cast shapes and surfaces that require no finishing. The surface specifications are a function of the mold quality. The second version is a porous fused silica that has not been completely consolidated. It possesses surface hydroxyls, but light transmission exceeds 80% over most of the spectra. The pores are capable of being infiltrated with a variety of materials for specialty applications.

Potassium silicate gels (Ref 24, 57) are capable of near-net-shape casting, and shapes can be machined from solid gel blocks as large as 560 mm (22 in.) long by 460 mm (18 in.) wide by 200 mm (8 in.) thick. The top half of Fig 8 shows a silica shape (before and after sintering) that was machined in the porous gel state by a water-jet technique to simulate a lightweight mirror core segment. The bottom

half of Fig 8 shows a hemispherical silica shape (before and after sintering) that was cast in a mold of that configuration.

The structures retained their shapes despite 50% linear shrinkage throughout the processing and sintering to 1750 °C (3180 °F), which is above the softening point of fused silica. The open pore network can be reacted with chlorine to remove surface OH to less than 0.1 ppm in the final glass. Total residual trace metals are of the order of 10 ppm, because the silica source has high levels of Fe, Al, Ti, and other elements. Therefore, the UV wavelength cutoff is about 230 nm. The other fused silica properties are comparable to CVD-processed fused silica. This process has strong potential for producing large reflective optics. TiO_2/SiO_2 compositions have been produced in plate form, and zero expansion was obtained over the temperature region of 0 to 300 °C (32 to 570 °F).

One corporation has developed a monolithic process that combines dispersed silica colloids with polymeric gel casting (Ref 45). At last report, high-quality fused silica can be produced in plate form (300 mm, or 12 in., diam by 5 to 10 mm, or 0.2 to 0.4 in., thick) and cylindrical forms up to 1 m (40 in.) long. Applications include photomask substrate for integrated circuit processing and preforms or cladding for optical fiber production.

Gradient index lenses also have been prepared by sol-gel processing. Yamane, et al. (Ref 58) first produced gradient composition gel structures by ion exchange and interdiffusion of ions. Later, a technique of

Fig 7 Monolithic glass shapes produced from DCCA-fortified tetramethylorthosilicate polymer gels. Courtesy of GelTech, Inc.

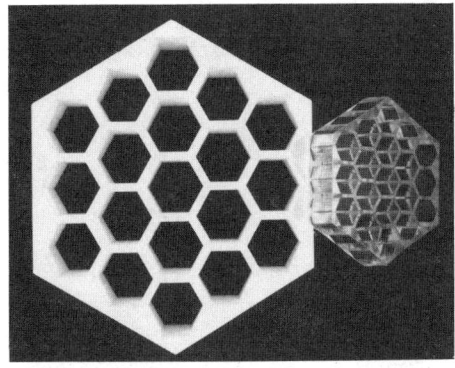

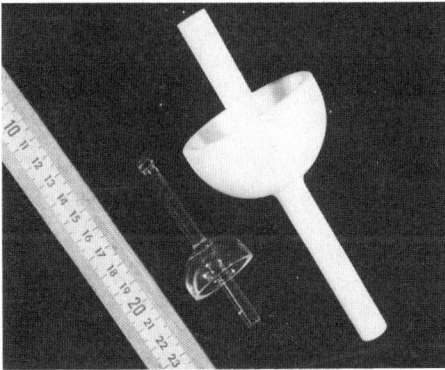

Fig 8 Machined 300 mm (12 in.) porous silica preform and sintered 150 mm (6 in.) mirror core segment (top). Before and after sintering of hemispherical silica monolith (bottom)

selectively leaching titania produced a compositional gradient that was retained in the sintered glass lens.

Multicomponent Glasses. Mixed-oxide sol-gel processing is evolving slowly because of the difficulty in developing homogeneous compositions from precursors that vary greatly in reactivity. Some multicomponent glasses that have been reported include:

- TiO_2/SiO_2 systems for very low expansion glasses
- SiO_2/ZrO_2 systems for alkali-resistant glasses, films, fibers, and photoresponsive polymers
- SiO_2/Fe_2O_3 made from TEOS and $Fe(OEt)_3$
- SiO_2/Y_2O_3 for high-temperature glasses [$Y(NO_3)_3$ salt was used]
- $SiO_2/TiO_2/ZrO_2$ glasses and films prepared by dip-coating
- $SiO_2/ZrO_2/Al_2O_3/Na_2O$ for alkali-resistant glasses
- $SiO_2/TiO_2/Al_2O_3/Li_2O$ for low-expansion glass-ceramics
- $Na_2O/B_2O_3/V_2O_5/SiO_2$ colored gel glasses, as well as $Co(OAc)_2$ and $Ni(OAc)_2$

Additional information and references are provided in Ref 48.

REFERENCES

1. R.K. Iler, *The Chemistry of Silica*, Wiley Interscience, 1979
2. C.Q. Brinker and G.W. Scherer, *Sol-Gel Science*, Academic Press, 1990
3. L.L. Hench and D.R. Ulrich, Ed., Ultrastructure Processing of Ceramics, Glasses, and Composites, Wiley Interscience, 1984
4. L.L. Hench and D.R. Ulrich, Ed., *Science of Ceramic Chemical Processing*, Wiley Interscience, 1986
5. J.D. Mackenzie and D.R. Ulrich, Ed., *Ultrastructure Processing of Advanced Ceramics*, Wiley Interscience, 1988
6. C.J. Brinker, D.E. Clark, and D.R. Ulrich, Ed., *Better Ceramics Through Chemistry*, North-Holland, 1984
7. C.J. Brinker, D.E. Clark, and D.R. Ulrich, Ed., *Better Ceramics Through Chemistry II*, Materials Research Society, 1986
8. C.J. Brinker, D.E. Clark, and D.R. Ulrich, Ed., *Better Ceramics Through Chemistry III*, Materials Research Society, 1988
9. R.K. Iler, *The Chemistry of Silica*, Wiley Interscience, 1979, p 172–311
10. R.J. Walsh, U.S. patent 3,161,468, 1964
11. C. Potter and J.W. Linderthal, U.S. patent 3,177,057, 1965
12. E.C. Broge and R.K. Iler, U.S. patent 2,680,677, 1954
13. R.K. Iler, U.S. patent 3,761,936, 1973
14. R. Roy, *J. Am. Ceram. Soc.*, Vol 39 (No. 4), 1956, p 145–146
15. R. Roy, *J. Am. Ceram. Soc.*, Vol 52 (No. 6), 1969, p 344
16. G.G. McCarthy, R. Roy, and J.M. McKay, *J. Am. Ceram. Soc.*, Vol 54, 1971, p 637
17. R.F. Caroselli, U.S. patents 2,754,221–2,754,224, 1956
18. S.S. Kistler, *J. Phys. Chem.*, Vol 36, 1932, p 52–64
19. M. Prassas, J. Phalippou, and J. Zarzycki, *J. Mater. Sci.*, Vol 19, 1984, p 1656–1665
20. R.K. Iler, *The Chemistry of Silica*, Wiley Interscience, 1979, p 370–372
21. R.K. McGeary, *J. Am. Ceram. Soc.*, Vol 44 (No. 10), 1961, p 513–522
22. P.C. Yates, Canadian patent 878,444, 1971
23. R.D. Shoup, *Colloid and Interface Science*, Vol 3, Academic Press, 1976, p 63–69
24. R.D. Shoup, *Ultrastructure Processing of Advanced Ceramics*, J.D. Mackenzie and D.R. Ulrich, Ed., Wiley Interscience, 1988, p 347–354
25. R.D. Shoup, U.S. patent 4,786,618, 1988
26. E. Matijevic, *Chemtech*, Mar 1991, p 176–181
27. E. Matijevic, *Science of Ceramic Chemical Processing*, L.L. Hench and D.R. Ulrich, Ed., Wiley Interscience, 1986, p 463–481
28. R.K. Iler, *The Chemistry of Silica*, Wiley Interscience, 1979, p 408–411
29. R.D. Shoup, U.S. patent 3,678,144, 1972
30. E.M. Rabinovich, D.W. Johnson, Jr., J.B. MacChesney, and E.M. Vogel, *J. Am. Ceram. Soc.*, Vol 66 (No. 10), 1983, p 683–688
31. E.M. Rabinovich, *Sol-Gel Technology for Thin Films, Fibers, Preforms, Electronics, and Specialty Shapes*, L.C. Klein, Ed., Noyes, 1988, p 260–294
32. R. Clasen, *Glastech. Ber.*, Vol 60, 1987, p 125–132
33. R. Clasen, U.S. patent 4,726,828, 1988
34. P.M. Schermerhorn, M.P. Teter, and R.V. Vandewoestine, U.S. patent 4,789,389, 1987
35. G.W. Scherer and J.C. Luong, *J. Non-Cryst. Solids*, Vol 63, 1984, p 163–172
36. C.J. Brinker and G.W. Scherer, *Sol-Gel Science*, Academic Press, 1990, p 108–216
37. P. Meakin, *Ann. Rev. Phys. Chem.*, Vol 39, 1988, p 237–267
38. W. Stobler, A. Fink, and E. Bohn, *J. Colloid Interface Sci.*, Vol 26, 1968, p 62–69
39. G.H. Bogush and C.F. Zukoski, *Proceedings of the 44th Annual Meeting of the Electron Microscopy Society of America*, G.W. Bailey, Ed., San Francisco Press, 1986, p 846–847
40. B.E. Yoldas, *J. Mater. Sci.*, Vol 12, 1977, p 1203–1208
41. M. Yamane, A. Shinji, and T. Sakaino, *J. Mater. Sci.*, Vol 13, 1978, p 865–870
42. S. Henning and L. Svensson, *Phys. Sci.*, Vol 23, 1981, p 697–702
43. L.L. Hench, *Science of Ceramic Chemical Processing*, L.L. Hench and D.R. Ulrich, Ed., Wiley Interscience, 1986, p 52–64
44. T. Adachi and S. Sakka, *J. Non-Cryst. Solids*, Vol 99, 1988, p 118–128
45. M. Toki, S. Miyashita, T. Takeuchi, S. Kanbe, and A. Kochi, *J. Non-Cryst. Solids*, Vol 100, 1988, p 479–482
46. H. Dislich, *Angew. Chem., Int. Ed., Engl*, Vol 10, 1971, p 363
47. I.M. Thomas, U.S. patent 3,791,808, 1974
48. J. Livage, M. Henry, and C. Sanchez, *Prog. Solid State Chem.*, Vol 18 (No. 4), 1988, p 259–341
49. H. Schroeder, *Physics of Thin Films*, Vol 5, G. Hass, Ed., Academic Press, 1969, p 87–141
50. H. Dislich, *Sol-Gel Technology for Thin Films, Fibers, Preforms, Electronics, and Specialty Shapes*, L.C. Klein, Ed., Noyes, 1988, p 50
51. H. Dislich and E. Hussmann, Thin Solid Films, Vol 77, 1981, p 129
52. M. Sacks and R. Sheu, *J. Non-Cryst. Solids*, Vol 92, 1987, p 383–396
53. S. Sakka, *Better Ceramics Through Chemistry*, C.J. Brinker, D.E. Clark, and D.R. Ulrich, Ed., North-Holland, 1984, p 91
54. J.W. Fleming, D.W. Johnson, Jr., J.B. MacChesney, and S.A. Pardenek, U.S. patent 4,872,895, 1989

55. D.W. Johnson, J.B. MacChesney, and E.M. Rabinovich, British patent 2,103,202B, 1985
56. L.L. Hench, S.H. Wang, and J.L. Nogues, *Multifunctional Materials,* Society of Photo-Optical Instrumentation Engineers, Vol 878, p 76–85
57. J.A. Bohlayer, G.F. Foster, and R.D. Shoup, U.S. patent 4,940,675, 1990
58. M. Yamane, J.B. Caldwell, and D.T. Moore, *Better Ceramics Through Chemistry II*, C.J. Brinker, D.E. Clark, and D.R. Ulrich, Ed., Materials Research Society, 1986, p 765

Annealed and Tempered Glass

Ronald A. McMaster, Donivan M. Shetterly, and Alejandro G. Bueno, Glasstech, Inc.

THE GLASS COMPONENTS used in architectural, automotive, furniture, and appliance glazing applications exist in one of three states:

- Annealed
- Laminated
- Tempered

Annealed glass is the product of the float glass method of sheet glass manufacturing introduced by Pilkington Brothers, Ltd. in the late 1950s by which practically all sheet glass for glazing applications is made today. The glass produced by this method is commonly called float glass. The float glass production demand in the United States was about 4.45 $\times$ 10^9 ft^2 in 1989. Of this total figure, about 2.02 $\times$ 10^9 ft^2 or 45% was safety glass, 8.4 $\times$ 10^8 ft^2 was laminated glass, and 1.18 $\times$ 10^9 ft^2 was tempered glass.

Float glass is produced in a virtually stress-free state known as annealed glass. Annealed glass can be cut, ground, drilled, and beveled, as needed. However, glass in the annealed state is not very strong and can easily be broken by thermal gradients, wind loads, impact, and so on. Moreover, when annealed glass breaks (Fig 1a), it typically forms dagger shaped shards radiating from the origin of failure. Concern over personal injuries and even deaths resulting from accidents involving annealed glass has led to the legislation

of codes requiring safety glass in buildings and vehicles in the United States and elsewhere.

Safety glass exists largely in two forms: laminated glass and tempered glass. Laminated glass consists of two pieces of annealed glass sandwiching a central layer of polyvinyl butyral (PVB) plastic. The main difference between laminated and annealed glass is that when laminated glass breaks the shards adhere to the tough PVB interlayer instead of falling free (Fig 1b).

Tempered glass is produced from annealed glass. As a result of a thermal treatment, tempered glass is about five times stronger than annealed glass when loaded in a bending

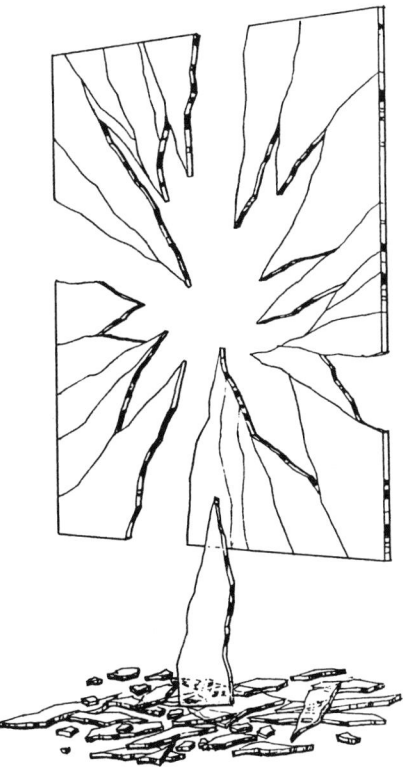

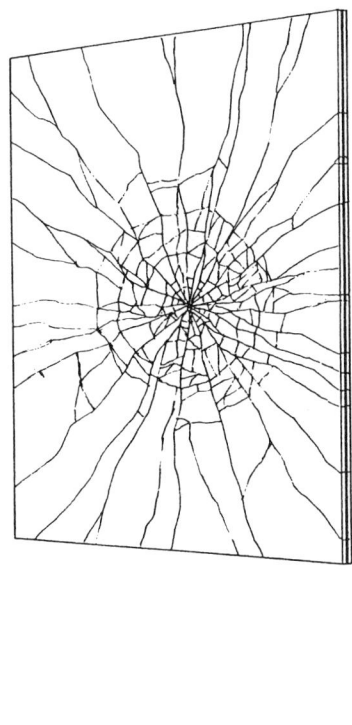

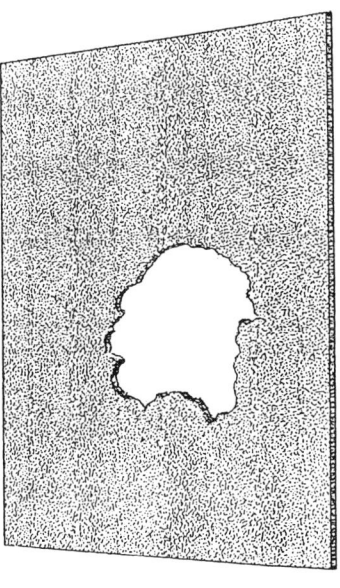

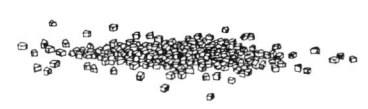

(a)   (b)   (c)

Fig 1 Break pattern of three states of glass used in commercial and consumer glazing applications. (a) Annealed. (b) Laminated. (c) Tempered

mode. When broken (Fig 1c), tempered glass breaks into small rocksalt-like particles, incapable of causing major injuries because of their small size, blunt shape, and dull edges.

The shape of annealed, laminated, and tempered glass parts may be flat, like the float glass from which they are made, or curved, depending on the application. The latter requires that the original flat glass article be bent or formed to the desired shape. This is accomplished at an elevated temperature. The cooling rate thereafter determines whether the shaped glass will be in the annealed or tempered state. Slow cooling rates are necessary to produce the annealed state, while fast cooling rates produce the tempered state.

Types of Tempering Systems

A typical, continuous tempering system for flat architectural glass parts consists of loading, heating, quenching, cooling, and unloading sections in an in-line arrangement (Fig 2a). The furnace is typically 37 m (120 ft) long, the quench is 4.9 m (16 ft), and the cooler 15 m (50 ft). The glass parts are conveyed through the system in a horizontal orientation by a roller conveyor. The quench establishes the temper and the cooler reduces the glass part to a comfortable handling temperature. A 1520 mm (60 in.) wide system is capable of tempering more than 14 tonnes (15 tons) of glass per day. Smaller capacity systems are available that utilize shorter furnaces and quenches which are batch loaded.

A typical continuous forming and tempering system for shaped automotive glass parts is shown in Fig 2(b). The system is very similar in configuration to the architectural tempering system (Fig 2a), with the addition of a forming station at the end of the furnace before the quench station. In the forming station, the glass part is taken from the furnace conveyor, formed between upper and lower forming tools, and transferred to a lower quench tool that carries it into the quench to complete the tempering process. The tempered glass part is transferred to the cooler where it is cooled to a comfortable handling temperature.

Tempering Process

During the tempering process, the glass is first heated and then rapidly cooled or quenched (Fig 3a). The heating portion (Fig 3b) of the process usually takes about 40 s/mm of glass thickness in a modern furnace where the primary mode of heating is radiant energy exchange between electric heating elements and the glass. When the glass temperature has been raised into the tempering range ($\approx$620 to 640 °C, or 1150 to 1185 °F) the glass leaves the furnace and enters the quench where it is cooled rapidly (Fig 3c) by means of radiation exchange with the envi-

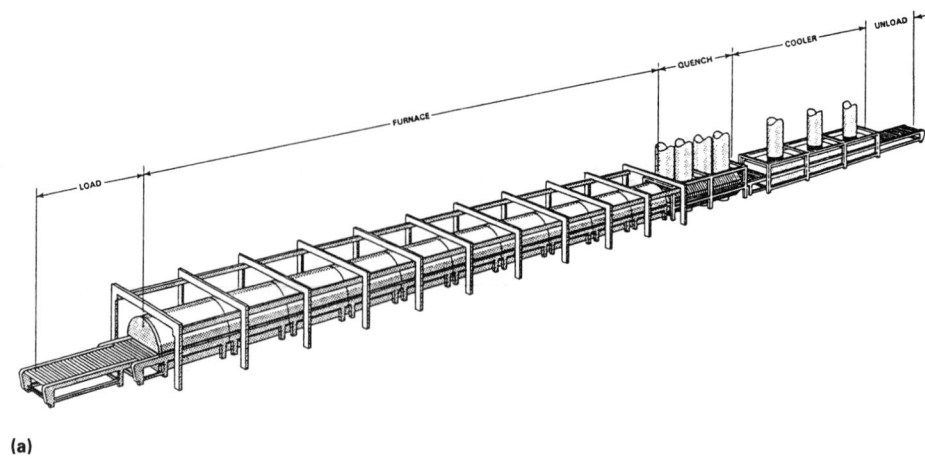

(a)

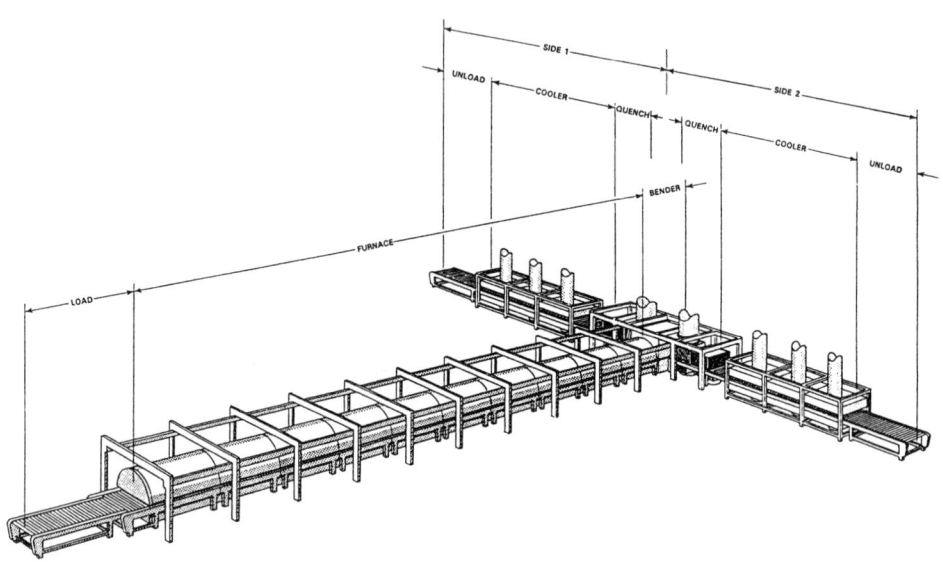

(b)

Fig 2 Two types of continuous tempering systems. (a) Flat glass system for production of architectural glass. (b) Quick sag system for production of formed automotive glass components

ronment and forced convection by an array of impinging air jets.

In the quenching operation, most of the cooling is achieved by forced convection. The heat transfer coefficient required to fully temper glass (midplane stress, σ_m, = 60 MPa, or 9 ksi) depends on the initial temperature of the glass entering the quench and the thickness, t, of the glass (Fig 4). Colder glass needs a higher heat transfer coefficient to achieve a full temper, as does thinner glass. However, there is an upper limit on the heat transfer coefficient that can be applied to a glass of a given thickness and initial temperature without causing failure. During the quenching operation, transient stresses are set up in the glass as described in the section "Physics of the Tempering Process" in this article. If the transient tensile surface stresses exceed about 30 MPa (4 ksi), failure is highly probable. Hotter glass needs a lower heat transfer coefficient to achieve a full temperature. However, if the glass temperature is much more than about 640 °C (1185 °F), the glass will

be so soft that objectionable distortion results. Consequently, there is a narrow range of temperatures that result in commercially acceptable yields and distortion levels. Within this range, the normal practice is to run thin glass hotter than thick glass. Interestingly, the heat transfer coefficient varies approximately as $t^{-1.05}$. The heat transfer coefficient is a function of the nozzle arrangement, type, size, and air pressure used to cause the air to flow. The air power (Fig 5a) needed to fully temper glass is more than inversely proportional to the glass thickness; consequently, much more power is required to temper thin glass than to temper thick glass.

The quenching time (Fig 5b) needed to achieve full temper depends on the glass thickness, with thick glass requiring a longer quenching time than thin glass. If the quenching time is too short, the residual stresses have a chance to relax (as described in the section "Physics of the Tempering Process"), and full temper is not achieved. Usually, the quench is continued at the same or

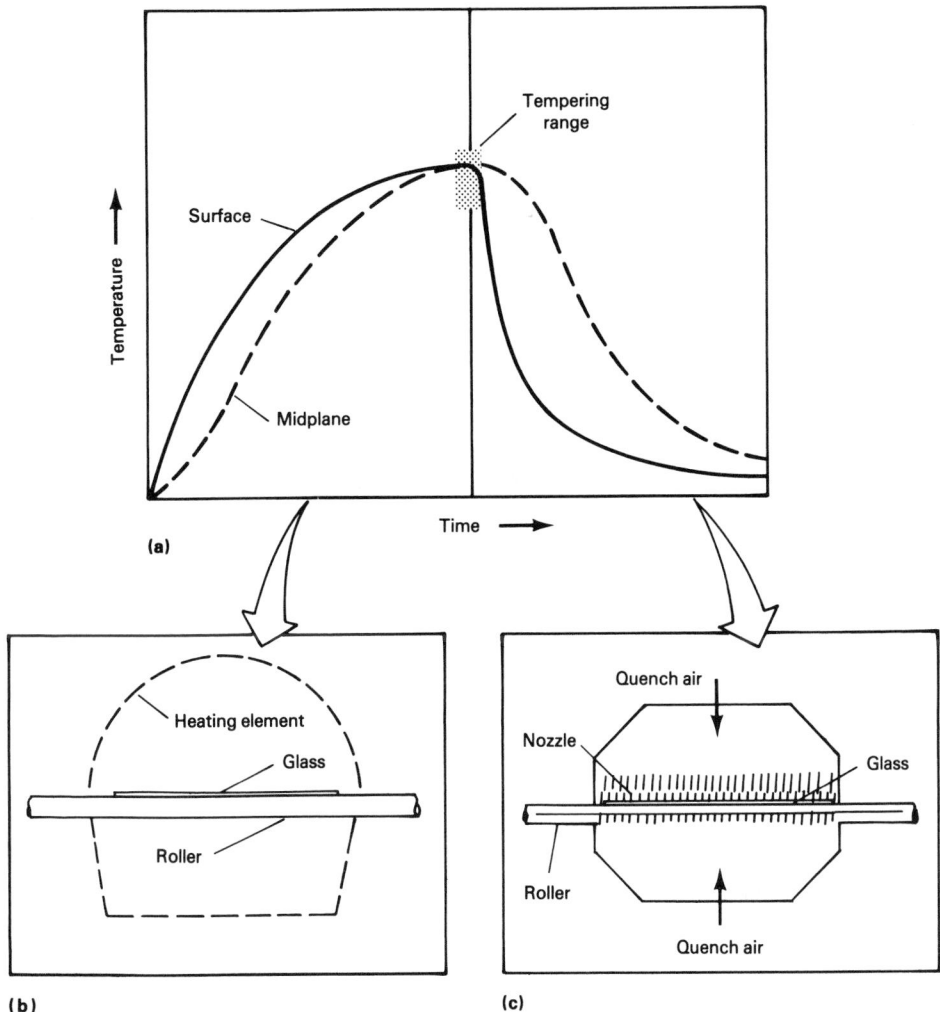

Fig 3 Processing and equipment used to temper glass. (a) Plot of surface and midplane glass temperatures versus time to show narrow tempering range. (b) Cross-section of heating portion of a tempering furnace. (c) Cross-section of quenching portion of a tempering furnace

a lower level to cool the glass to a convenient handling temperature.

The tempering process is usually carried out with the glass in a horizontal or vertical orientation. The thickness of glass which is tempered commercially ranges from 2.8 to 19 mm (0.11 to 0.75 in.). During the tempering process, it is important that the heating and quenching of the glass be similar on both sides of the glass so that the glass stays flat and distortion is minimized.

Physics of the Tempering Process

Although glass at room temperature exhibits characteristics typical of a solid crystalline material, it is in reality an amorphous (that is, noncrystalline) material somewhat like an undercooled liquid. Its behavior with increasing temperature has consequently been defined in terms of viscosity (Fig 6) and stress relaxation time (Fig 7). Up to the strain point ($10^{14.5}$ P, or $10^{13.5}$ Pa · s, at 520 °C, or 970 °F) any stress in the glass will be removed in a matter

of an hour or more. At the annealing point (10^{13} P, or 10^{12} Pa · s, at 540 °C, or 1005 °F) any stress in the glass will relax in a matter of several minutes. Between the lower and upper temperature limits of the tempering range ($10^{10.2}$ P, or $10^{9.2}$ Pa · s, at 620 °C, or 1150 °F; $10^{9.5}$ P, or $10^{8.5}$ Pa · s, at 640 °C, or 1185 °F) any stresses in the glass will be removed in a matter of several seconds. Consequently, by the time the glass has been heated to a temperature within the tempering range, it is largely stress free, even though a temperature gradient may still exist across its thickness.

During the quenching portion (Fig 8) of the tempering process, the surface is cooled faster than the interior and a temperature gradient of parabolic shape is established across the thickness of the glass. As the temperature gradient is established, a contraction gradient is also established because of the thermal expansion property of glass. As long as the glass temperature is above the strain point, stresses resulting from the thermal contraction gradient tend to relax. By the time the

surface temperature reaches the strain point (~1.5 s), the thermal gradient has largely been established. As the strain point temperature front progresses from the surfaces to the midplane, two events are taking place: the portion of the glass toward the interior is relaxing stresses, and the portion toward the exterior is developing stresses in response to the strains generated by the thermal gradient. The latter results in the surface of the glass being put into a state of compression, the magnitude of which steadily increases until all of the glass is below the strain point temperature. When the strain point temperature front reaches the midplane (~12 s), the glass is in an elastic state with a stress distribution and a thermal gradient resulting from the temperature history thus far. As the glass is cooled further, the temperature gradient is converted into a strain gradient that adds to the stress gradient already established. When the glass reaches room temperature throughout (∞), the temperature gradient has been completely transformed into the residual stress distribution. The residual stress distribution is parabolic in shape; consequently, the magnitude of the surface compression is twice that of the midplane tension, because the sum of the residual tensile and compressive stresses must be zero to be in equilibrium.

A peculiarity of glass is that its thermal expansion (Fig 9) in the solid state is dependent upon how fast it was cooled from the liquid state. In the liquid state, the atoms of which glass is comprised have enough mobility that atomic rearrangements can occur, resulting in more efficient atomic packing arrangements as the temperature is reduced. Consequently, the relationship between length (or volume) and temperature is a property (line B-B). As the temperature of the glass is reduced across the transformation range towards the strain point, however, the mobility of the atoms becomes restrained to that of thermal vibration only as the atoms become locked into the molecular structure of the glass typical of the solid state. The intersection of the glass and liquid expansion lines defines the transition or fictive temperature. The increasing viscosity hinders the formation of the most efficient molecular packing, resulting in a length expansion to temperature relationship dependent on the cooling rate. As the cooling rate increases so does the length expansion at a given temperature, although the slope (that is, coefficient of expansion) is unchanged (lines A-A, A'-A', A''-A''). Thus, there is a density gradient in tempered glass in which the surface is less dense than the interior that is responsible for a portion of the residual stresses. The residual stresses in tempered glass, then, are partly of a structural origin as well as being of a viscomechanical origin.

Another aspect of the thermal expansion characteristic of cooling glass is that the transition from the liquid state to the glass state is not an instantaneous event at the fictive

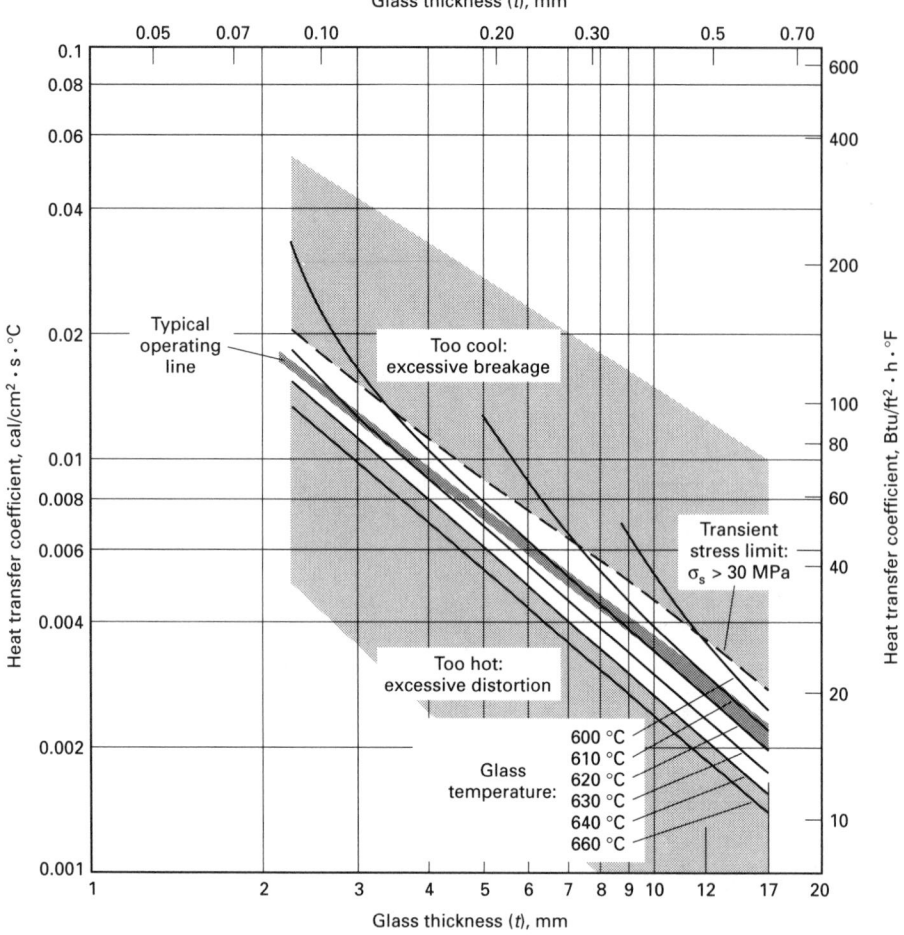

Fig 4 Plot of heat transfer coefficient required for full temper (σ_m is 60 MPa, or 9 ksi) versus glass thickness, t, as a function of glass temperatures ranging from 600 to 660 °C (1110 to 1220 °F). Source: Glasstech computer models of glass

temperature but takes place gradually over the transformation range. This results in a relaxation of the structural and viscomechanical stresses until the strain point is reached.

Strength of Glass

Glass is a brittle material. When it fails, it fails instantaneously and completely. The ultimate strength of glass is not so much a function of its chemistry as it is of its surface condition. Pristine glass fibers (that is, those without flaws) have exhibited tensile strengths of about 14 GPa (2030 ksi). However, ordinary everyday glass is never in pristine condition. The surfaces of flat glass in particular are cumulatively damaged on a microscopic scale during the initial float glass manufacturing process as well as subsequent operations (for example, handling, cutting, grinding, and tempering). Hence, there is a wide range of results when glass specimens are subjected to strength tests. The lower limit of the tensile strength of annealed float glass can be taken to be about 30 MPa (4 ksi).

The strength increase of tempered glass is due to the surface flaws being closed up by the surface compression of the tempered glass. The surface compression of fully tempered thin glass is about 120 MPa (17 ksi). Thus, the tensile strength of tempered glass is ≈150 MPa (that is, 30 MPa + 120 MPa), or ≈22 ksi. The strength of tempered glass, then, has been increased over that of annealed glass by a factor of ≈150/30 = 5.

The surface compression of tempered glass also increases the capability of the glass surface to resist damage from stone impact.

Break Pattern of Tempered Glass

The break pattern that results when tempered glass is broken is a result of the virtually instantaneous running and bifurcation of cracks initiated at the point of failure. The energy needed to propagate the break pattern comes from the strain energy associated with the residual tensile stresses of the interior. It has been established that the magnitude of the midplane tension must be above a threshold value (Fig 10) for a break pattern to be propagated. As the magnitude of the midplane stress is raised above the threshold value, the break pattern becomes finer.

Measurement of Stress in Glass

All techniques for the quantitative measurement of stress in glass rely on the fact that glass which is nonuniformly cooled develops birefringence. Birefringence is a property of a material whereby light is transmitted through the material at different velocities depending on the direction the light travels through the material. As applied to glass, the fundamental law of photoelasticity states that the difference in velocity of the propagation of light in different planes of the material is directly proportional to the difference in stress in those planes. The constant of proportion-

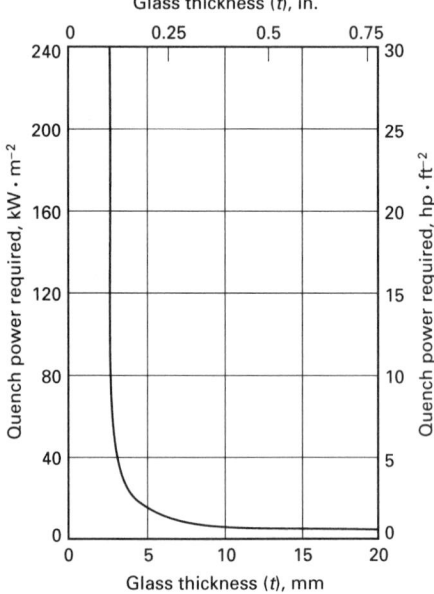

(a)

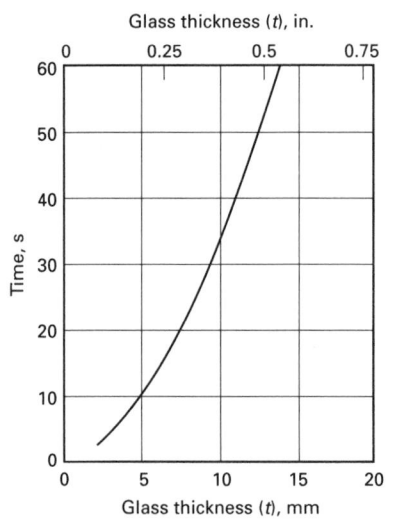

(b)

Fig 5 Effect of glass thickness on (a) air power and (b) quenching time required for full temper. Exit temperature, 620 °C (1150 °F). Source: Glasstech computer models of glass

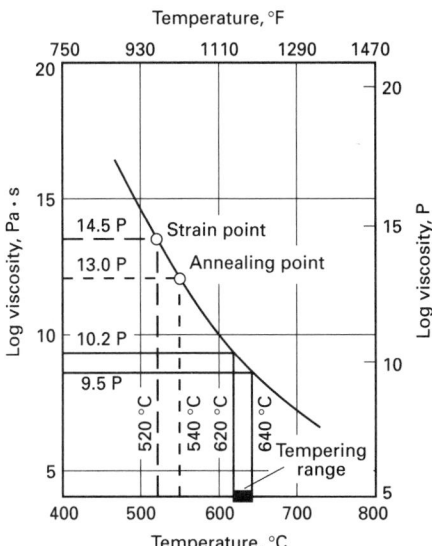

Fig 6 Plot of log viscosity versus temperature to obtain optimum tempering range. After data from Institut National DuVerre, Belgium

ality between stress and birefringence is called the stress optical coefficient or Brewster constant.

For glass, the amount of stress and hence the amount of birefringence present is generally proportional to the magnitude of the temperature gradients developed as it cools through its strain point. Thus, glass cooled at any finite rate, however slow, develops stress and birefringence.

As stated earlier, temperature gradients throughout the glass develop in proportion to the rate of cooling applied to its surfaces. In general, all glass surfaces develop compressive stresses while the interior of glass is typically in a state of tensile stress (see the section "Physics of the Tempering Process" in this article). The distribution of stress from surface to center is approximately parabolic. Mathematically, this dictates that the magnitude of the surface compression will be about twice the magnitude of the center tension. Thus, if either the surface compression or the center tension is measured, the other value can usually be approximated.

Manufacturers and users of glass are often interested in both surface compression and center tension levels because:

- Only glass that has very low levels of stress (annealed glass) can be successfully cut
- In the case of heat-strengthened or tempered glass, the surface compression determines the resistance of the glass to breakage from flexure or impact (see the section "Strength of Glass" in this article) while the center tension determines the size of fragments that result when the glass breaks (see the section "Break Pattern of Tempered Glass" in this article)

By far the most common method of inferring stress levels in tempered glass is by breaking it. A sharp impact to the glass surface will result in its breaking into fragments. The size of the fragments is a good qualitative measure of the center tension present in the glass. Although this method is generally qualitative, data exists for correlating the number or size of fragments to the level of stress.

Polarimeter. The most common instrument for qualitatively and sometimes quantitatively measuring stress is the polarimeter. The glass sample is placed between crossed polarizing sheets in front of an ordinary light source. The pattern observed reveals average stresses through the glass. Areas where there is no average stress are dark while areas of stress show up as light. Refinements of this technique using so-called wave plates reveal stress levels as brightly colored fringes whose color can be used to determine stress. The limitation of polarimeter techniques is that they only reveal average stresses through the thickness of the glass.

Differential Surface Refractometer. To quantify surface compression, an instrument called a differential surface refractometer (DSR) is commonly used. This instrument directs light along and slightly below the surface of the glass. The surface ray travels at a different velocity if stress is present in the surface. An optical telescope is used to accurately read the stress.

Babinet Compensator. To measure center tension, a quartz wedge or Babinet Compensator is used. This equipment requires that the sample be viewed edge on and only small samples or fragments can be used. The quartz wedge has a variable known birefringence along its length. The point at which its birefringence offsets that of the sample can be used to read the stress in the glass sample.

Scattered Light Polarimeter. A light scattering technique can also be used to measure center tension. In this technique, polarized light in perpendicular planes is passed through the center of the sample. Because of the birefringence, the two polarized components go in and out of phase, which results in alternating areas of constructive and destructive interference. The observer sees a dashed line of scattered light. The pitch of the dashes is proportional to the level of stress.

Automotive Glass Forming

Flat sheets of glass are formed into complex shaped surfaces (for example, automotive glass) by heating them to the viscoelastic temperature range and pressing them into the desired shape. At these temperatures the stresses, induced under the forming loads, relax in a timely manner and allow the strains to remain even after the loads are removed. The relaxation of these stresses are closely described by a generalized Maxwell model (Fig 11) and the equation:

$$G(t) = G_\infty + \sum_{i=1}^{n} G_i \exp\left(-t/\tau_i\right) \quad \text{(Eq 1)}$$

where t is the time; $G(t)$ is the shear relaxation modulus; G_∞ is the long term shear modulus; G_i are the short term moduli; μ_i are the viscosity components; τ_i, the ith relaxation time component, is equal to G_i/μ_i; and σ is the transient stress.

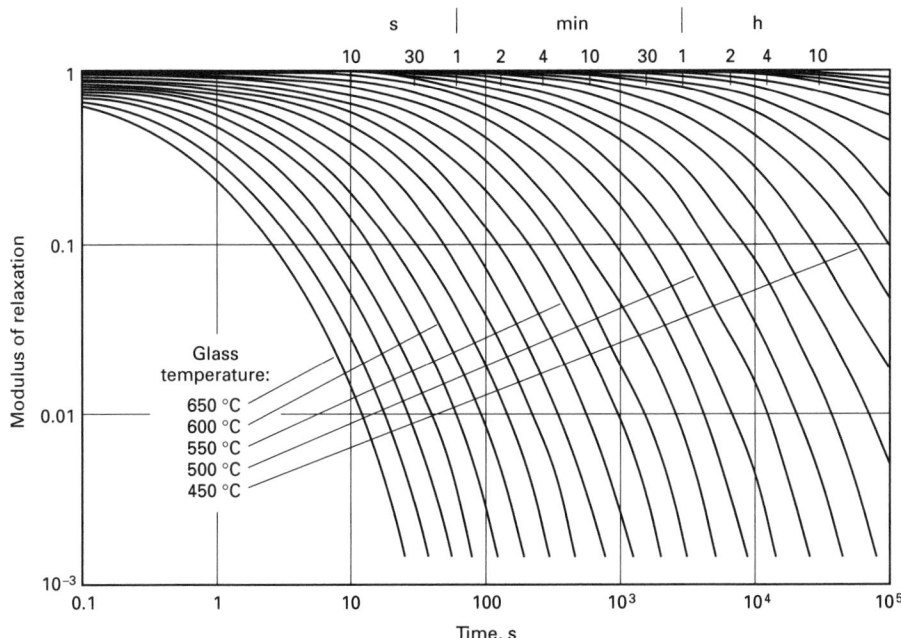

Fig 7 Plot of modulus of relaxation versus stress relaxation time as a function of temperature. Source: Glasstech computer model of glass

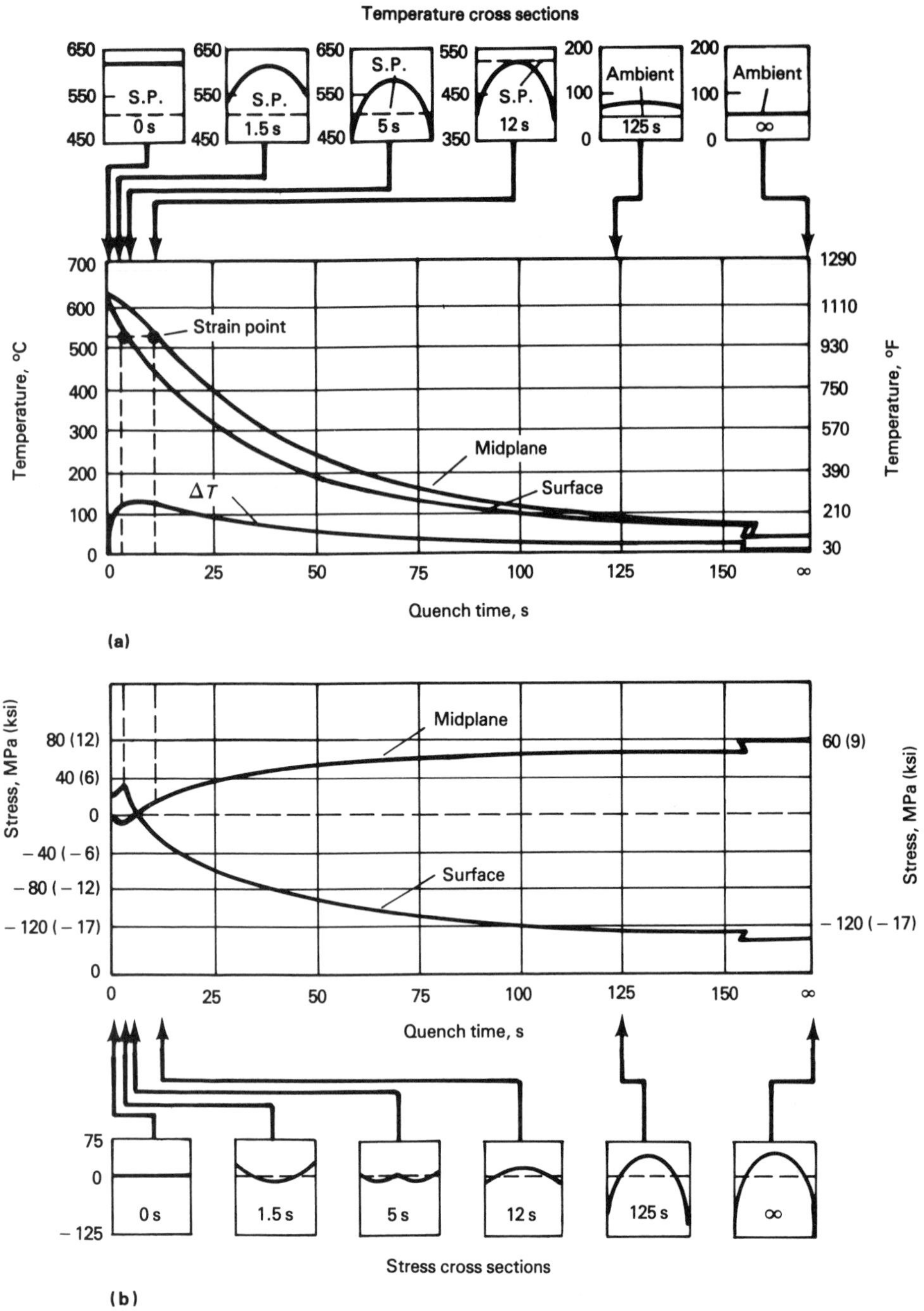

Temperature cross sections

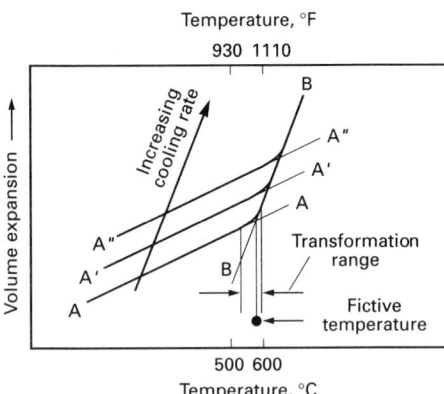

Fig 9 Plot of volume thermal expansion versus temperature as a function of cooling rate. After data from Institut National DuVerre, Belgium

(a)

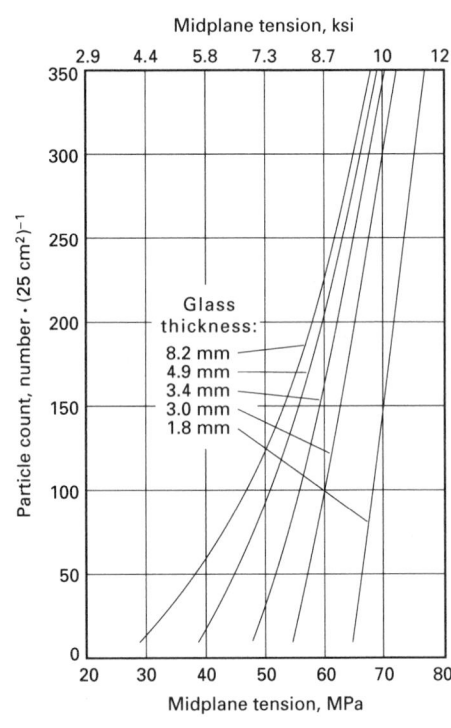

Fig 10 Dependency of break pattern on midplane stress. Exit temperature, 660 °C (1220 °F). After data shown in K. Akeyoshi *et al.*; see Selected References

Stress cross sections

(b)

Fig 8 Cross-sections of 6.1 mm (0.24 in.) thick glass showing (a) temperature distribution and (b) stress distribution as functions of quenching time. SP, strain point. Quench parameters: heat transfer coefficient, 0.0050 cal/cm^2 · s · °C (37 Btu/ft^2 · h · °F); exit temperature, 620 °C (1150 °F); and ambient temperature, 30 °C (85 °F). Source: Glasstech computer model of glass

This viscoelastic property is highly temperature dependent. Fortunately, for the purpose of computational analysis, the glass behaves as a thermorheologically simple material where the pattern of the relaxation function remains constant but its position shifts along the time scale as the temperature varies (Fig 12).

Ideally, the forming is done at the cold end of the viscoelastic range in order to minimize surface damage due to localized loads (that is,

points of contact of the glass with the forming tool). Simultaneously, the glass needs to be sufficiently resilient to the forming forces acting on it. In practice, the operational range of the forming tools is very narrow, typically between 620 °C (1150 °F) and 640 °C (1185 °F) for tempered glass. The optimum temperature depends on the complexity of the surface and the required degree of temper.

In the process of changing from a flat, two-

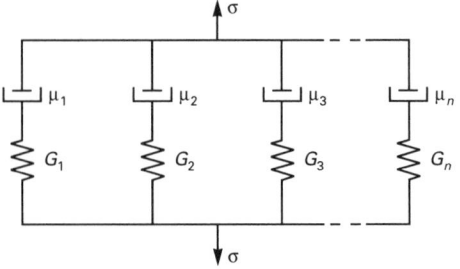

Fig 11 Generalized Maxwell model to describe stresses in a flat glass sheet. The mechanical-viscous model is represented by the four spring dashpots (μ_1, μ_2, μ_3, and μ_4).

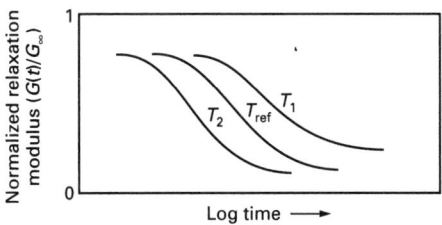

Fig 12 Plot of relaxation modulus versus time showing shift of relaxation modulus along time scale for $T_1 < T_2$ where T is the temperature

dimensional sheet to a three-dimensional surface with multiple curvatures, the glass must yield primarily by stretching proportionally to the local complexity of the desired surface. The rate of change of strain as a function of position should be minimized to reduce the possibility of creating optical distortion of the transmitted light. This is difficult to achieve in some recent model designs where the back light of an automobile may have areas where the curvature must change from positive to negative (that is, from convex to concave). The transition region from one curvature sign to the opposite will naturally tend to produce areas of high strain rates that can cause distortions which result in poor optical quality. These distortions can be minimized by optimizing the sequence in which the loading forces are applied to the glass.

SELECTED REFERENCES

- K. Akeyoshi *et al.*, *Rept. Res. Lab.*, Asahi Glass Company, Ltd., Vol 17 (No. 1), 1967, p 23–36
- R. Gardon, *Ceram. Bull.*, Vol 66 (No. 11), 1987, p 1594–1599
- J.E. Gordon, *The Science of Structures and Materials*, Scientific American Books, 1988
- G.W. McLellan and E.B. Shand, *Glass Engineering Handbook*, 3rd ed., McGraw-Hill, 1984
- L.H. Van Vlack, *Elements of Materials Science*, 2nd ed., Addison-Wesley, 1964

Ion-Exchange

Roger F. Bartholomew, Corning Inc.

AN ION EXCHANGER is an insoluble material that carries exchangeable cations or anions. The first materials identified as showing ion-exchange properties were naturally occurring silicates that were based on clays and zeolites. Later, synthetic organic resins were demonstrated to exhibit the property of ion-exchange (Ref 1). Today, zeolites and organic resins are used in countless ways, such as water softeners in detergents, treatment of industrial wastes, and use in preparative chemistry.

Schulze (Ref 2) was the first to demonstrate that cations in glasses could be exchanged when he immersed a soda-lime glass into a bath of molten silver nitrate. In the intervening years, the ion-exchange properties of glass have been used to explain the functioning of the pH glass electrode and the durability of glass. Ion-exchange is used in numerous applications, such as chemical strengthening of glass, gradient index (GRIN) lenses, as well as planar waveguides.

A generalized ion-exchange reaction,

$$\bar{A} \text{ (glass)} + B \text{ (salt)} = \bar{B} \text{ (glass)} + A \text{ (salt)} \quad \text{(Eq 1)}$$

can be written where $\bar{A}$ and $\bar{B}$ are the counterions in the exchanger phase, and A and B are the counterions in the liquid phase.

Kinetics of Ion-Exchange

It has been clearly established that the rate at which ion-exchange occurs is controlled by diffusion of the ions in the silicate matrix. Extensive studies of the diffusion processes, both self-diffusion and interdiffusion, have been made in glass, as well as in glass-ceramics (Ref 3, 4). The binary ion-exchange reaction shown in Eq 1 involves two diffusing species, ions A and B, which carry charges and generally tend to diffuse at different rates. One ion moves faster than the other, leading to a build-up of electrical charge. There is also a gradient in electrical potential which acts to slow down the faster ion and speed up the slower ion. This leads to the fluxes of the two ions being equal and opposite, thus preserving electrical neutrality.

The driving force for transport is the negative of the gradient in total chemical potential, that is, the sum of the gradient in activity and in electrical potential. This assumes that no other gradients, such as thermal or pressure, exist. With the additional assumption that the mobilities of the ions are the same in self-diffusion and interdiffusion, the flux of diffusing species, J_A, in the x-direction is:

$$J_A = -D_A \left[\left(\frac{\delta \bar{c}_A}{\delta x} \right) \left(\frac{\delta \ln \bar{a}_A}{\delta \ln \bar{c}_A} \right) + \bar{c}_A \left(\frac{FE}{RT} \right) \right]$$

$$\text{(Eq 2)}$$

where D_A is the self-diffusion coefficient, $\bar{c}_A$ is the ionic concentration, $\bar{a}_A$ is the activity of species A, and E is the potential gradient or electric field. A similar equation exists for the flux J_B. When univalent-for-univalent exchange is under consideration, this set of equations is referred to as the Nernst-Planck equations. Electroneutrality requires that $J_A = -J_B$. It is easily shown that:

$$J_A = \left(\frac{D_A D_B}{\bar{N}_A D_A + \bar{N}_B D_B} \right) \left(\frac{\delta \ln \bar{a}_A}{\delta \ln \bar{c}_A} \right) \quad \text{(Eq 3)}$$

where $\bar{N}$ is the cation fraction in the exchanger and $\bar{N}_A + \bar{N}_B = 1$. A similar expression can be written for J_B. The activity is related to the concentration by:

$$a_A = \bar{c}_A^n \quad \text{(Eq 4)}$$

The interdiffusion coefficient, D, is related to the self-diffusion coefficients of the diffusion ions by:

$$D = \left(\frac{D_A D_B}{\bar{N}_A D_A + \bar{N}_B D_B} \right) (n) \quad \text{(Eq 5)}$$

The temperature dependence of diffusion, both interdiffusion and self-diffusion, follows the Arrhenius equation:

$$D = D_0 \exp\left(\frac{-\Delta E}{RT} \right) \quad \text{(Eq 6)}$$

where ΔE is the activation energy. Two other equations should be noted. To describe the variation of concentration, c, with distance, x, D independent of concentration, and a constant source of concentration, c_0, then,

$$c = c_0 \, \text{erfc} \, x/2(Dt)^{1/2} \quad \text{(Eq 7)}$$

To determine M_t, the amount taken up at time, t, under the same conditions,

$$M_t = 2c_0 \left(\frac{Dt}{\pi} \right)^{1/2} \quad \text{(Eq 8)}$$

For further reading on the application of the diffusion equations under different boundary conditions, see Ref 3.

Experimental methods for measuring both self-diffusion and interdiffusion are well developed (Ref 4, 5). Radioactive-labeled ions are the main source for measuring self-diffusion, although nuclear magnetic resonance (NMR) data have also been used. Electron microprobe data, NMR data, weight change calculations, and sectioning by etching, followed by chemical analysis of the removed layer (or, if radioactive tracers are used, then counting the etchate or remaining sample) have all been used to measure interdiffusion.

Both self-diffusion and interdiffusion measurements have been made on many alkali-oxide-containing glasses (Ref 5, 6). Table 1 shows self-diffusion data for alkali ions in binary alkali silicate glasses. Considerable scatter has been observed in results among different laboratories, even for identical glass compositions. The most likely explanation is that the thermal history of the glasses varies. In general, the larger the cation is, the lower the self-diffusion coefficient is at a given temperature in a glass of comparable molar composition. Self-diffusion rates for alkaline earth ions are two orders of magnitude less than for alkali ions in comparable glasses, while ΔE values are approximately double or more for alkaline earth ions compared with alkali ions.

The applicability of Eq 5 relating interdiffusion and self-diffusion coefficients was verified for the interdiffusion of Ag^+ and Na^+ in a soda-lime-silica glass, with $n = 1$ (Ref 10). Garfinkel (Ref 11) fitted concentration profiles from interdiffusion for a diffusion coefficient that varied with concentration according to $D = D_0/(1 - \alpha C)$, where D_0 is the coefficient at $C = 0$, and α is a constant. Agreement was good for the Na^+ and Li^+ system in an 11.4 Li_2O/16.5 Al_2O_3/71.5 SiO_2 (mol%) glass, with $D_{Li}^+/D_{Na}^+ = 1.3$ and $n = 1.9$.

Thermodynamics

An equilibrium constant, or selectivity coefficient, K_{AB}, can be defined from Eq 1, such that:

$$K_{AB} = \frac{\bar{a}_B a_A}{\bar{a}_A a_B} \quad \text{(Eq 9)}$$

Table 1 Self-diffusion coefficients of alkali ions in alkali silicate glasses

Glass composition	Alkali oxide, mol%	D_o		E		D at 400 °C (750 °F) (a)		Ref
		cm²/s	in.²/s	MJ/kmol	kcal/mol	cm²/s	in.²/s	
Na₂O-SiO₂	10	7.9×10^{-4}	1.22×10^{-4}	73.2	17.5	1.64×10^{-9}	0.25×10^{-9}	7
	20	1.3×10^{-3}	0.20×10^{-3}	73.5	17.4	2.91×10^{-9}	0.45×10^{-9}	7
	30	2.0×10^{-3}	0.31×10^{-3}	67.4	16.1	1.18×10^{-8}	0.18×10^{-8}	7
K₂O-SiO₂	10	2.0×10^{-3}	0.31×10^{-3}	81.2	19.4	1.00×10^{-10}	0.16×10^{-10}	7
	20	3.8×10^{-4}	0.59×10^{-4}	77.0	18.4	4.03×10^{-10}	0.62×10^{-10}	7
	33.3	9.33×10^{-4}	1.45×10^{-4}	66.9	16.0	5.94×10^{-9}	0.86×10^{-9}	7
Rb₂O-SiO₂	20	3.39×10^{-4}	0.53×10^{-4}	75.7	18.1	4.5×10^{-10}	0.70×10^{-10}	7
Cs₂O-SiO₂	16.7	8.7×10^{-3}	1.35×10^{-3}	103	24.5	9.64×10^{-11}	1.49×10^{-11}	8
	16.7	1.3×10^{-3}	0.20×10^{-3}	78.2	18.7	1.1×10^{-11}	0.17×10^{-11}	9

(a) Calculated using Eq 6

The $\bar{a}_i$ values in Eq 9 are the respective activities in the exchanger phase and the liquid phase. The value of K_{AB} depends on the reference state chosen to define activities. For the salt, it is convenient to use the pure material as the reference state, so in the limit as N_A tends to 1, the activity coefficient equals 1. The reference state for the solid exchanger is that in which all of the exchangeable cations are of the ion in question.

Ion-exchange equilibria are characterized by ion-exchange isotherms. The ratio of the activities of the ions in the exchanger is given by:

$$\frac{\bar{a}_B}{\bar{a}_A} = \left(\frac{\bar{N}_B}{\bar{N}_A}\right)^n \qquad \text{(Eq 10)}$$

where $\bar{N}$ is the cation fraction. This is the so-called n-type behavior. Most molten nitrate salts can be characterized as regular solutions. Thus, the equation

$$\log \frac{N_B}{N_A} - \frac{A(1 - 2N_A)}{2.30 \, RT} = n\log \frac{\bar{N}_B}{\bar{N}_A} - \log K_{AB}$$

$$\text{(Eq 11)}$$

can be derived. Therefore, a plot of $\log(a_B/a_A)$ for the salt versus $\log(\bar{N}_B/\bar{N}_A)$ for the exchanger should be linear with a slope equal to n.

Two simple experimental techniques have been used to test these equations. The first is to powder the exchanger, then equilibrate in the salt bath for several hundred hours. The powder is then analyzed for the cation concentration. This is done for different salt bath concentrations. Another technique is to determine the profile of an exchanged rod and extrapolate back to the surface. A typical ion-exchange isotherm is shown in Fig 1. From these data, using Eq 11, an n value of 2.4 was determined (Ref 12). Note how insensitive the equilibrium isotherm is to temperature.

Practical Aspects

To exchange ions in glass in an acceptable time frame requires elevated temperatures. Temperatures generally range from above 200 °C (390 °F) to above the annealing point of the glass—in some instances, as high as

800 to 900 °C (1470 to 1650 °F). Most practical ion-exchange is done in molten nitrate melts. Sodium nitrate (melting point of 310 °C, or 590 °F) and potassium nitrate (melting point of 337 °C, or 640 °F) can be used up to temperatures of 450 and 525 °C (840 and 975 °F), respectively, before the onset of the decomposition of the nitrate to the nitrite and oxygen. The upper use temperature of both nitrates can be extended another 100 °C (180 °F) by additions of sulfate.

For higher temperatures (>600 °C, or 1110 °F), combinations of salts have been used, including Na₂SO₄, Li₂SO₄, K₂SO₄, NaCl, and KCl. This is by no means an exhaustive list. The choice is dictated by kinetic and thermodynamic properties. Lists of melting points of single salts and mixtures are readily available in the literature (Ref 13). Care has to be taken in the selection of the container material. Most salts, particularly the nitrates, can be contained in stainless steel (309 is recommended). Because melts containing chlorides tend to be corrosive to stainless steel, they are best melted in Vycor vessels.

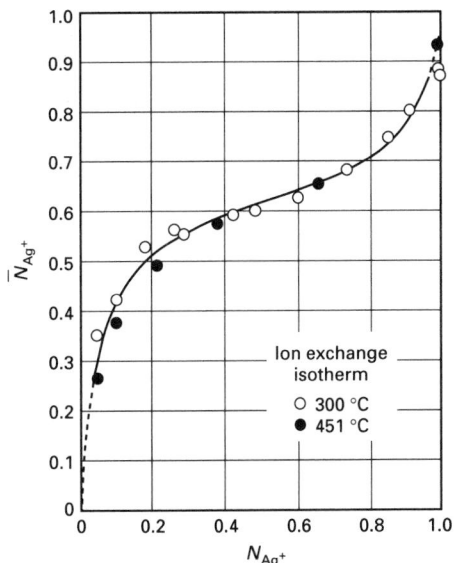

Fig 1 Ag⁺-Li⁺ ion exchange in 11.4 mol% Li₂O, 16.5 mol% Al₂O₃, 71.5 mol% SiO₂ glass

Alternatives exist for high-temperature ion-exchange, one of which is to use pastes that contain the required cation, mixed with an inert carrier in a vehicle. The resulting slurry can be adjusted to the required viscosity to allow for either spray or dip coating. The most often used carrier is ochre, a mixture of iron oxide and clay; the vehicle can be water.

The second alternative is the use of vapor phase methods. These can be divided into two distinctly different types: dealkalization and use of volatile salts. Dealkalization occurs when glass is exposed at high temperature to acidic gases, of which the best known are the oxides of sulfur. Water is necessary to complete the exchange reaction. The result of such a treatment is the formation of a "bloom" of alkali sulfate on the glass surface. Ion-exchange from the vapor phase has been tried using cuprous halides (Ref 5).

Applications

Chemical Strengthening. Glass is intrinsically a very strong material. Theoretically, its strength should approach 13.8 GPa (2×10^6 psi). However, due to the presence of surface flaws, called Griffith flaws, the measured strength is in the region of 55 to 70 MPa (8 to 10 ksi). One of the most interesting applications of ion-exchange in glass is that of chemical strengthening, or "stuffing." Glass fails in tension without exception, due to the surface flaws mentioned above. A surface compressive layer can be attained by either thermal or chemical tempering, that is, high-temperature ion-exchange.

There are two different mechanisms to strengthen by ion-exchange. One mechanism produces a physical property difference at the surface, such as expansion, whereas the other represents the "stuffing" principle, based on a chemical change. Typically, a larger ion is stuffed into the same site in the silicate network as a smaller ion below the glass transition temperature, T_g. This is the temperature below which the glass network cannot relax in a short time. The further below the T_g that the exchange occurs, the less the amount of stress relaxation that occurs, but the slower the exchange rate is. There are many examples in the literature (both open and patent) of this approach (Ref 4). The ion-exchange pairs most used are the Na^+ for Li^+, or the K^+ for Na^+.

The two process variables of most importance for chemical strengthening are temperature and treatment time. The measure of strength is the modulus of rupture, either abraded or unabraded. Flexibility in tailoring the final properties is possible, leading to a range of strength (modulus of rupture), central tension and hence dicing, and the depth of compression layer. For each temperature there is a time at which the modulus of rupture goes through a maximum. The higher the temperature, the shorter the time required to attain the maximum strength. However, the maximum strength is generally lower as temperature increases. This effect can be explained by considering that the rate of build-up of the integral stress is proportional to the rate of ion-exchange minus the loss in stress due to relaxation of the glass. This is illustrated in Fig 2 (Ref 14). Another mechanism for the loss of strength is thermal decay during sustained use at elevated temperature. By heating the ion-exchanged glass in the absence of a source of ions, a loss of strength results (Ref 4).

In addition to the kinetic effects (time and temperature), the bath composition (thermodynamic) influences the strength of the treated glass. Contamination of the bath by the counterion reduces the surface concentration at the glass-molten salt interface; hence, the concentration profile and gradient for diffusion. The extent of reduction depends very much on the equilibrium described in Eq 6. Because the stress profile can be related to the concentration profile (Ref 15), any change caused by contamination of the bath alters the strength.

The formation of a deep exchange layer has been demonstrated by the electrical assist method (Ref 16), which is carried out by applying an electric field across the sample, usually in the order of tens of volts per centimeter field strength. Control of the concentration profile, and, hence, the stress profile, is possible using a finite concentration of exchanging ion in the salt bath.

The simple replacement of one ion for another can be augmented by multistep methods. To exemplify a two-step technique, the first step may involve a small ion that is exchanged for a large ion above T_g, followed by a second step that can consist of exchanging a large ion for a small ion below T_g, or reheating in air below T_g, or exchanging a small ion for a large ion above T_g. An example of a three-step technique would be exchanging a small ion for a large ion above T_g, followed by the exchange of a small ion for a large ion below T_g, and then a large ion for a small ion below T_g. Other examples are given in Ref 4.

The area of interest in these various treatments is not the maximum strength obtained, but the tailor-made stress profile that results. Profiles can be introduced into the glass to determine whether breakage is violent or nonviolent and whether dicing into fine particles or no dicing at all occurs, and to meet the need to maintain strength at elevated temperatures. Dicing characteristics are a function of the tensile strain energy and are the product of the central tensile stress and the thickness of the glass.

Another major factor influencing the increase in strength of a glass after ion-exchange is the composition of the glass. Not all glasses can be strengthened by ion-exchange. For example, because borosilicate glasses do not contain enough alkali, they cannot build up enough stress. The common soda-lime glasses tend to show high rates of stress relaxation. The compositions that give the best strength results are the alumina-silica compositions.

Figure 3 shows the dependence on strength of ion-exchanged silica-alumina glasses on the (Li or Na)/Al ratio (Ref 17). Note the maximum at the ratio equal to 1. It is at this ratio that the glass contains no nonbridging ions and where the self-diffusion and interdiffusion properties show a maximum and relaxation a minimum.

Ophthalmics and Optics. In 1971 the Federal Drug Administration published regulations defining minimum impact conditions for all eyeglasses sold in the U.S. As a result, most glass prescription eyeware that is sold today is given an overnight treatment to exchange sodium ions in the glass for potassium ions from the bath. In the case of photochromic glasses, the treatment is carried out at 400 °C (750 °F) in a mixed sodium nitrate-potassium nitrate bath. For white crown compositions, the bath is potassium nitrate, and the temperature is 460 °C (860 °F). Lenses have to pass a test using a stainless steel ball (16 mm, or $^5/_8$ in. diam) dropped from a height of 1.3 m (50 in.). In 1989, approximately ten million pairs of both white crown and photochromic lenses were chemically strengthened.

In the field of optics, the first use of ion-exchange to make a commercially feasible product was the GRIN (gradient refractive index) lens, which is used to make arrays with a one-to-one magnification for application in copiers and fax machines. Reduction in the object-to-image distance to as little as a few inches is possible with these arrays, resulting in major size reductions in the product. Refractive index gradients in GRIN lenses are obtained by ion-exchange of a cylinder of glass. The applications call for an increase of index on the outside of the cylinder, or rod, relative to the interior.

This property is also required for another application in optics that utilizes ion-exchange of glass to produce channel waveguides. Light can be guided in a region where the refractive index in the guiding medium is higher than the surrounding matrix. Various

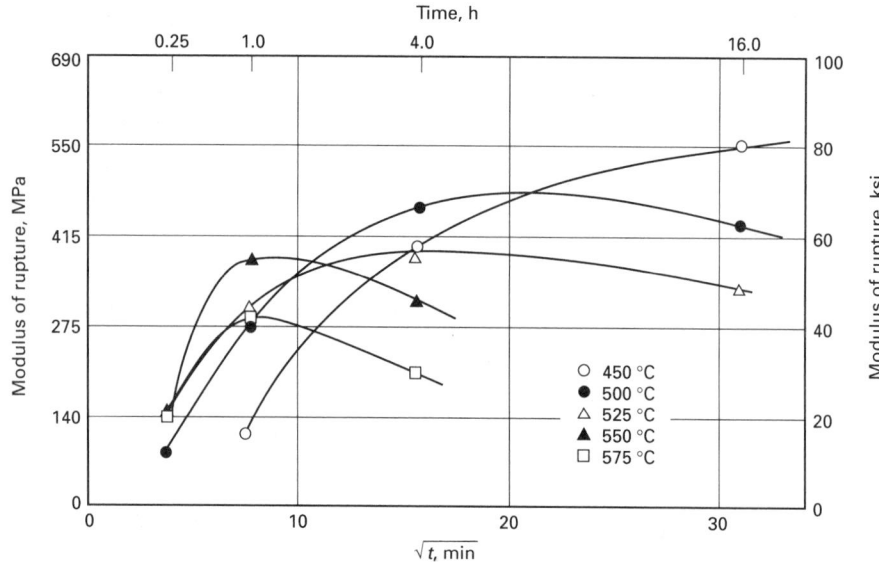

Fig 2 Effect of temperature on the strength of potassium-exchanged $1.1Na_2O-1Al_2O_3-4SiO_2$ glass

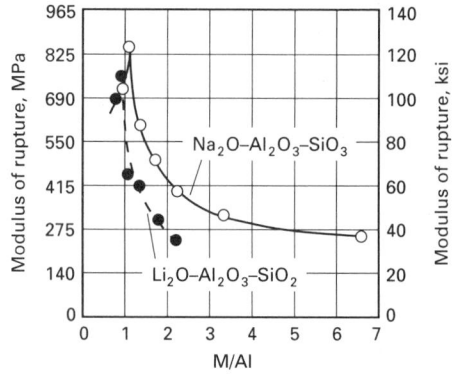

Fig 3 Dependence of the strength of ion-exchanged Al_2O_3 glasses on the M(Li or Na)/Al ratio

ions have been used to increase the index of a base glass for both GRIN and waveguide uses. Most of the exchange couples involve sodium in the glass initially. Therefore, those ions of interest that would increase the refractive index would be K^+, Rb^+, Cs^+, Ag^+, and Tl^+. There are some disadvantages to using each of the listed ions. Thallium is toxic; Ag^+ can be reduced, causing color in the glass; Cs^+ is the slowest to diffuse; Rb^+ is not much better; and K^+ does not have a very large refractive index difference. An additional constraint is the need to avoid stress following the ion-exchange reaction.

The basic process for making a channel waveguide requires at least seven steps, only one of which is the ion-exchange (Ref 18). The initial steps involve cleaning the glass surface and using photolithography to form the mask before the ion-exchange process begins. For this application the electrical-assist technique has been extensively developed, and many ingenious techniques proposed. Buried waveguide structures are possible by performing a second ion-exchange, using the ion originally present in the glass, following the first ion-exchange. Integrated optical waveguides are now available as coupler products, such as splitters and wavelength division multiplexers. Other applications for ion-exchanged glass include windows for airplane cockpits and space vehicles, laboratory pipettes, consumer ware, and glass computer tape reels.

With the judicious use of ion-exchange time, temperature, and thickness, many useful property combinations can be obtained (Ref 19). These are illustrated in Table 2. A low central tension means that the article dices like an air-tempered material, whereas a high central tension means that it dices into small pieces. A larger depth of compression guards against damage due to abrasion. The glass codes 0313, 0315, and 0319 are derived from the same base soda-alumina-silica composition, whereas code 8361 is a potash-soda-lime-zinc silicate and code 8111 is an alkali alumina-borosilicate. Both base glasses have T_g's in the region of 540 °C (1000 °F); the strengthening is accomplished by potassium-for-sodium exchange.

Table 2   Properties and use of five chemically strengthened glasses

Corning code	Thickness		Abraded modulus of rupture		Center tension		Depth of compression		Applications
	mm	mils	MPa	ksi	MPa	ksi	mm	mils	
0313	1.270	50	310	45	55	8.0	0.1778	7	Cladding for aircraft windshields
	2.159	85	310	45	31	4.5	0.1778	7	Tape reels
0315	2.159	85	448	65	62	9.0	0.2286	9	Spacecraft windows
	5.080	200	517	75	24	3.6	0.2286	9	High-strength lens systems
0319	1.270	50	227	33	89	13.0	0.3048	12	Cladding for aircraft windshields
	2.159	85	276	40	55	8.0	0.3048	12	Cladding for aircraft windshields
8111	2.00	79	220	32(a)	2.8	0.42	NA		Photochromic eyeglass lenses
8361	2.00	79	380	55(a)	NA		NA		White crown eyeglass lenses

NA, Not available. (a) Unabraded modulus of rupture

REFERENCES

1. F. Helfferich, *Ion Exchange*, McGraw-Hill, 1962
2. G. Schulze, Experiments Relating to the Diffusion of Silver into Glass, *Ann. Phys.*, Vol 40, 1913, p 335–367
3. J. Crank, *Mathematics of Diffusion*, Oxford University Press, London, 1956
4. R.F. Bartholomew and H.M. Garfinkel, Chemical Strengthening of Glass, *Glass Science and Technology*, Vol 5, D.R. Uhlmann and N.J. Kreidl, Ed., Academic Press, 1980
5. G.H. Frischat, Ionic Diffusion in Oxide Glasses, *Trans Tech Publications*, Aedermannsdorf, Switzerland, 1975
6. R. Terai and R. Hayami, Ionic Diffusion in Glasses, *J. Non-Cryst. Solids*, Vol 18, 1975, p 217–264
7. K.K. Evstrop'ev and V.K. Pavlovskii, Ion Diffusion and Electrical Conductivity of Single Alkali Glasses, *Inorg. Mater.*, Vol 3, 1967, p 592–596
8. R. Terai, Self-Diffusion Coefficient of Cesium Ions in Cesium Silicate Glasses, *J. Ceram. Soc. Jpn.*, Vol 77, 1969, p 38–40
9. R. Hayami and R. Terai, Diffusion of Alkali Ions in Na_2O-Cs_2O-SiO_2 Glasses, *Phys. Chem. Glasses*, Vol 13, 1973, p 102–106
10. R.H. Doremus, Exchange and Diffusion of Ions in Glass, *J. Phys. Chem.*, Vol 68, 1964, p 2212–2218
11. H.M. Garfinkel, Cation-Exchange Properties of Dry Silicate Membranes, *Membranes*, Vol 1, *Macroscopic Systems and Models*, G. Eisenman, Ed., Marcel Dekker, 1972
12. H.M. Garfinkel, Ion-Exchange Equilibria between Glass and Molten Salt, *J. Phys. Chem.*, Vol 72, 1968, p 4175
13. G.J. Janz, *Molten Salt Handbook*, Academic Press, 1967
14. H.M. Garfinkel, Strengthening Glass by Ion-Exchange, *Glass Ind.*, Vol 50, p 28–31 and 74–76
15. H.M. Garfinkel and C.B. King, Ion Concentration and Stress in a Chemically Tempered Glass, *J. Am. Ceram. Soc.*, Vol 53, 1970, p 686–691
16. H. Ohata and M. Hara, Ion-Exchange in Sheet Glass by Electrolysis, *J. Ceram. Soc. Jpn.*, Vol 78, 1970, p 158–164
17. M.E. Nordberg, E.L. Mochel, H.M. Garfinkel, and J.S. Olcott, Strengthening by Ion-Exchange, *J. Am. Ceram. Soc.*, Vol 47, 1964, p 215–219
18. L. Ross, Integrated Optical Components in Substrate Glasses, *Glastech. Ber.*, Vol 62 (No. 8), 1989, p 285–297
19. H.A. Miska, Understanding the Basics of Chemically Strengthened Glass, *Mater. Eng.*, 1976, p 6–76

Grind/Polish

Bruce H. Raeder and H. Gordon Shafer, Jr., Corning Incorporated

OPTIMUM FINISH MACHINING of glass and glass-ceramics requires a thorough analysis of the numerous parameters that affect the machinability of the workpiece material (Fig 1). Although the differences between the finishing operations of glass versus glass ceramics may appear to be subtle, even minor modifications in processing can significantly affect the successful machining of the workpiece. These machining and inspection guidelines become even more narrow and demanding as the product specifications become progressively more exacting.

The product specifications should be, and usually are, a reflection of specifications required to ensure the product will perform as designed. The polish and dimensional specifications range from a commercial polish and features such as flatness measured in thousandths of an inch, all the way to the extremely precise where the surface finish is given in angstroms peak-to-valley (PV) and the figure or shape is measured in fractions of a wavelength. Some products and their approximate positions in the precision range are shown in Fig 2. When the finishing processes for all of these products are examined, there are many more similarities than differences. The major differences are in the accuracy and sophistication of the equipment, the sophistication of the metrology, and the skill of the practitioners. The flow diagram shown in Fig 1 is an attempt to show the parameters that must be considered as part of the grinding and polishing process.

Glass Preparation

The initial finishing step to consider is establishing the blank shape. In the case of lenses, most optical and ophthalmic parts are supplied in a near-net shape from the hot glass forming process and no preparatory work is required. However, for small volume lenses or specialty substrates such as photomasks, a preshaping is required. In addition, special features such as holes, reference flats, or depth countersinks are added at this time. The most commonly used processes using commercial equipment and tooling are, in order of convenience and cost:

- Scoring (for example, with a diamond tip or friction driven rotating scribe wheel)
- Sawing and edge grinding (for example, diamond blades and diamond wheels)
- Core drilling (diamond, loose abrasive, or ultrasonic methods)
- Water-jet cutting (using abrasive entrained in water at a pressure of 380 MPa, or 55 ksi)
- Laser cutting

Once the part is machined or formed to dimensions and configuration suitable for grinding, the next step is to determine how the part will be held in the first grinding operation. The holding device can be attached to the part temporarily or it can be attached to the part permanently throughout the polishing process.

In the case of optical and many ophthalmic operations, the lenses are cemented in place with a wax, a low-temperature metal, or a temporary bond contact cement to a special block, either one at a time or nested together in multiples on a spot block (Ref 1).

Caution must be used in selecting the block fit to the part, the depth of wax, and the type of wax because movement and flexing during finishing will affect the final figure and lead to irregularity in the finished part, especially for thin parts or parts with a large thickness variation from the center to the edge of the workpiece.

When blocking the part to finish the second side, extra care must be taken to avoid damaging the finished surface that will now be against the block. Many people use a spray or strippable coating (for example, Universal Strip Cote) to protect the polish and in some cases the spot for blocking can be slightly undercut so the lens contacts the block only at the very edge. Care must again be taken as the undercutting may result in distortion if the lens is thin or if the part gets excessively warm during polishing.

Other methods of holding the workpiece, especially for grinding, include vacuum chucks, metal fencing around the parts for a magnetic chuck, or mechanical-type collet chucks.

Grinding Operations

The purpose of the grinding operation is to establish the shape of the part quickly and at low cost. In many cases, the grinding step will produce an accuracy that is nearly as good as the finish requirements, but the surface is very rough and opaque.

The surface finish and geometry obtained on the workpiece and the rate of machining (cost) will depend on the machine tool, the diamond wheel design, the operational process, and finally the material composition of the workpiece being ground.

Processing Equipment

Figure 3 shows three different systems that describe most glass grinding equipment.

The first system is a vertical spindle surface grinder and the diamond tool is referred to as a ring tool. Equipment of this design can be obtained with table sizes ranging from a few inches up to well over several feet. Equipment of this type is commercially available with continuous rim wheels having a ≤510 mm (≤20 in.) diameter. For wheels >510 mm (>20 in.), the wheels are usually constructed from segments or buttons.

The second system is very similar except that the angle of the spindle is adjustable to produce spherical parts. This system is commercially available from vendors for use as ophthalmic and optical generators. This system also makes it possible to add a traverse feature to the spindle to generate nonspherical curves. For large optics, Campbell grinders that incorporate this feature have been manufactured with table sizes up to several feet.

The third design is the horizontal spindle grinder that can be used either as a conventional or a creep-feed grinder. In general, the vertical spindle and curve generation system surface grinders are more efficient, while the third produces less total load on the part but at the sacrifice of efficiency.

In addition to these machines, there are many special-purpose grinders such as toric and cylinder generators for ophthalmic lenses, center edge and bevel machines for edge grinding of lenses, cam machines for edging, numerical control (NC) machines for aspheric lenses, and several variations of belt-type grinders.

Processing and Tooling

After the equipment to produce the correct shape has been chosen and the workpiece is affixed to the table, the type of tooling and processing that will scratch, gouge, and chip the part to the desired shape with an acceptable finish must be determined.

For most glass and glass-ceramic grinding, the tooling used contains a diamond grit held in place by a metal or resin bond. In the

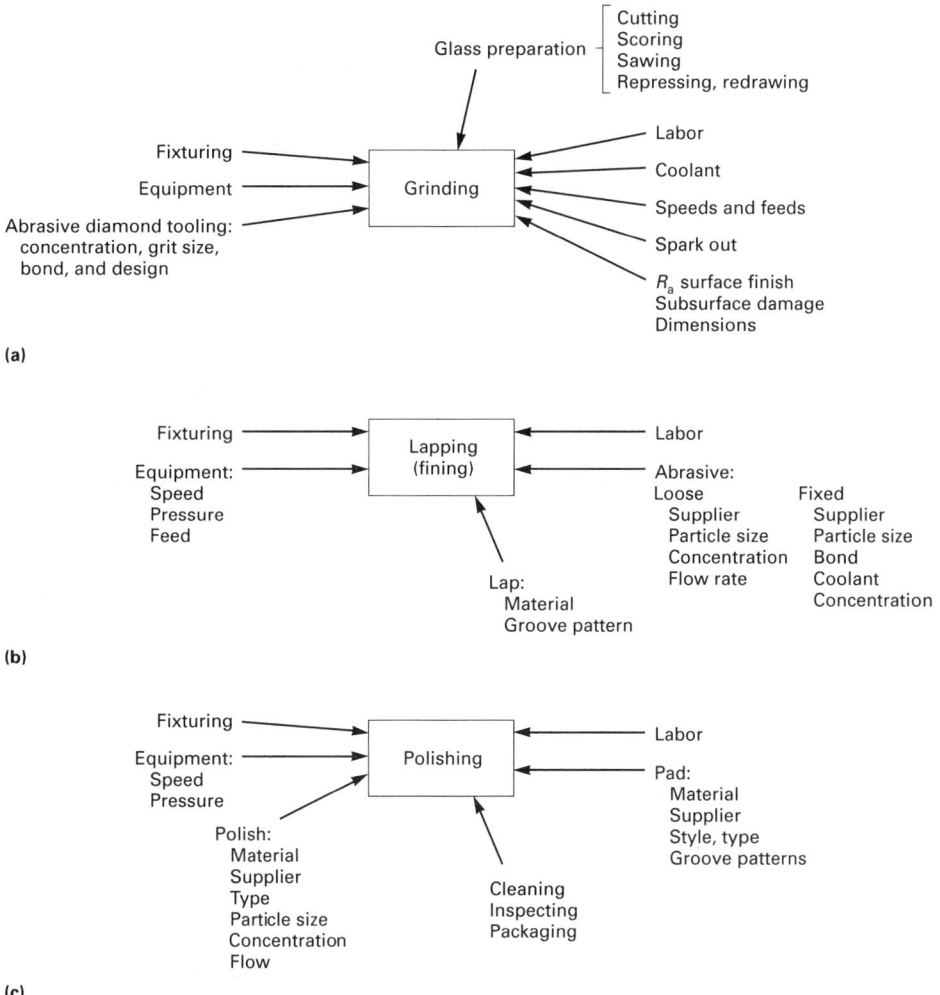

Fig 1 Flow diagrams showing key parameters and machining operations that affect both glass and glass-ceramic finishing operations. (a) Grinding. (b) Lapping. (c) Polishing

Fig 3 Three basic setups used to grind glass and glass-ceramics. (a) Vertical-spindle surface grinder. (b) Two-spindle curve generator. (c) Horizontal surface grinder

grinding process, each exposed diamond grit creates a conchoidal-type fracture comparable to a single-point indenter in a Rockwell hardness tester (Ref 2). Because the indenter is in motion in the grinding operation, a long ragged scratch is produced. The bulk of the spalling occurs after the diamond grit passes by as the surface goes from compression to tension and the lateral vent cracks create a spall. The size of the scratches produced will be primarily dependent on the effective size of the diamond grit and the load (feed rate) with lesser dependence on parameters such as the machine and wheel run-out, wheel design, coolant, rotational speed, and workpiece material.

The rough surface created by grinding is usually characterized by stylus measurement and then referred to by R_a (arithmetic roughness average) or R_t (total peak-to-valley roughness height). In addition to roughness, the grinding process also creates a microcracked region or subsurface damage. Figure 4 shows these features in addition to the waviness that must be considered when characterizing a ground surface. Waviness is caused by machine inaccuracy (runouts) or chatter (Ref 3).

In the ideal grinding case, the process will operate continuously piece-to-piece without altering the process. This means that as diamonds are worn down or fractured, the glass

spalling mentioned earlier in this section will wear away the bond at just the right rate, thereby exposing new diamonds to continue the cutting process. The wheel is designated as being self-dressing, self-sharpening, or free cutting. The objective of the engineer is to design the wheel to be free cutting at the right feed rate to produce the desired roughness.

There are a number of guidelines or formulas that relate grit size in a free cutting mode to surface roughness and subsurface damage (Ref 4, 5). As a first estimate, assume that the R_t value is equal to about $^1/_6$ to $^1/_8$ of the particle size as a good approximation. The subsurface damage is then about two to four times the roughness. The factor is two for the average subsurface damage while a factor of

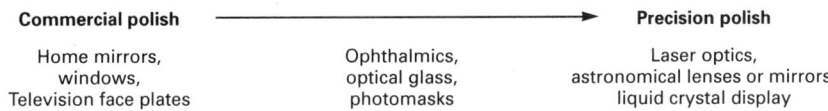

Commercial polish	$\longrightarrow$	Precision polish
Home mirrors, windows, Television face plates	Ophthalmics, optical glass, photomasks	Laser optics, astronomical lenses or mirrors, liquid crystal display

Fig 2 Typical products produced by commercial polish and precision polish processing

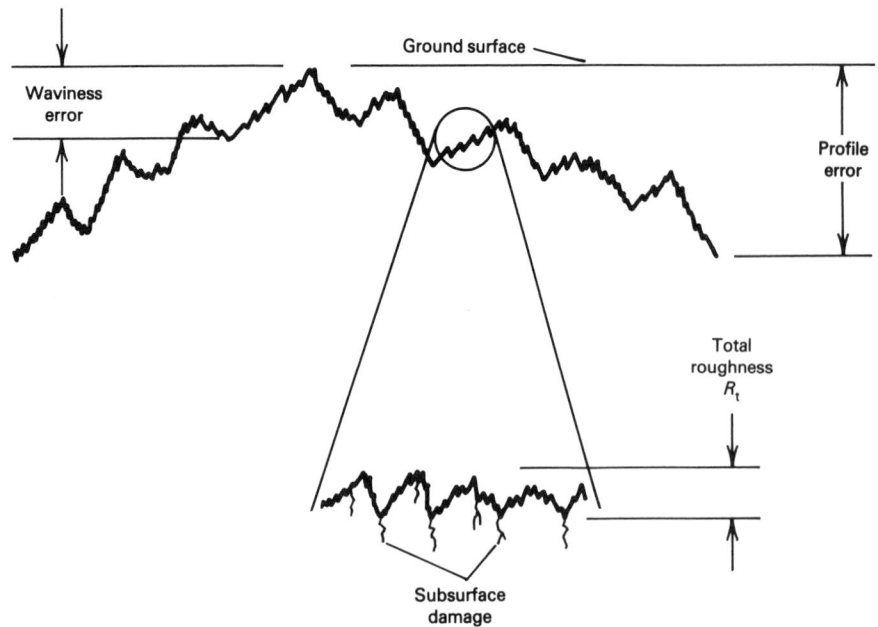

Fig 4 Schematic showing parameters used to evaluate surface finish roughness generated by grinding process. Also shown is subsurface damage caused by grinding.

four may be required to remove the deepest check (Ref 4).

Some data presented by Ardamatskii (Ref 4) with reference to Lepitova support the above approximation. In summary, about 0.5 of the grain protrudes from the bond for a free cutting wheel while the depth of cut is 12 to 15% of the particle diameter.

In selecting a grinding wheel, the above exercise will get the user in the correct operating region. We then need to consider the other aspects of the wheel and how aggressively we can or want to run the process. To specify a diamond wheel, the following items must be determined and specified:

- Diamond type D (synthetic or natural, tough, friable, and so on)
- Grit size
- Bond grade letter (J to T is normal, soft to hard)
- Concentration (17 to 100+)
- Type of bond (resin or metal)

Any special modifications are then usually identified by a letter and then the depth of abrasive section. A typical designation (Fig 5) for a wheel according to the above parameters is MD-100 N-25-M-¹/₄.

In order for the wheel to stay free cutting, the bond must wear at just the right rate to expose new diamonds. The desired feed rate is dictated by the cost or time available to grind (that is, from a few seconds for a small lens to days and even weeks for a large mirror). The key variables to be controlled in order to match the wear rate and the cutting rate are the wheel design, concentration, and bond type. Each of these parameters has a significant impact on the cutting process.

Wheel design determines the contact area, the total load on the part and, therefore, how much the wheel must be pushed to remain free cutting. A very narrow wheel will be more free cutting than a wide wheel. A typical wheel width for a Blanchard grinder is 9.5 mm (³/₈ in.) at 510 mm (20 in.) diameter, while a generator ring of 75 to 100 mm (3 to 4 in.) diameter would require a 3 to 9.5 mm (¹/₈ to ³/₈ in.) wheel width. The contact area for a Blanchard grinder may be the full wheel, while the contact area for a generator ring is usually less than the full wheel. A wide face wheel of perhaps 25 mm (1 in.) will probably never be free cutting. Another wheel design that allows the contact area to be reduced is a button construction which can work well for shaped wheels or larger diameter wheels (Ref 6, 7).

Concentration. The relationship between the concentration, the particle size, and the number of particles per carat is shown in Table 1. An accepted convention in the industry is that 100 concentration is equal to 72 carats/in.³ and is the maximum particle packing most vendors prefer to use. Given that a grit size has some distribution, concentrations to about 150 in a single grit size and to higher than 200 in mixed grits can be obtained. The effect of increasing the concentration is to increase the number of cutting edges, thereby increasing the wear resistance of the grinding wheel. Large contact area wheels such as Blanchard wheels will generally use low concentration (17 to 35), while very aggressive optical generator rings with a smaller contact area can use concentration values up to 100 or higher.

Bond Type and Hardness. The two types of bonds used for glass grinding are metal bond and resin bond. The metal bond is usually used for heavy cutting and is more durable than the resin bond, which tends to be used for lighter cutting where finer finishing is required.

For large contact area grinding (for example, Blanchard grinders), an N or M bond will be satisfactory while for more aggressive optical grinding an R or S will be required.

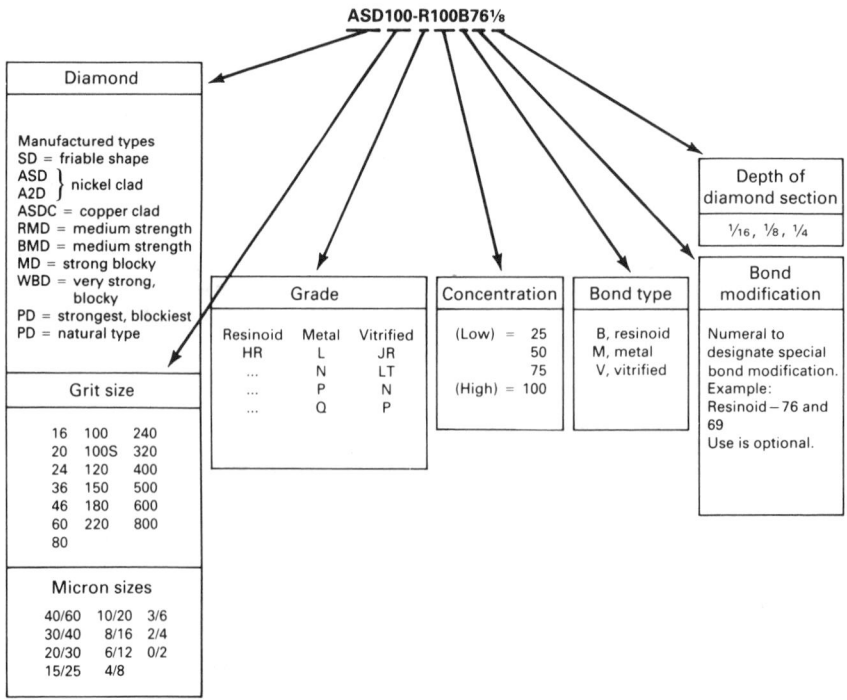

Fig 5 Standard marking system for diamond superabrasive grinding wheels

Table 1 Relationship between diamond particle size and particles per carat

Size			Particles per carat(a)
US mesh	μm	in.	
$3\frac{1}{2}$	5660	0.223	0.32
4	4760	0.187	0.52
5	4000	0.157	0.88
6	3360	0.123	1.4
7	2830	0.111	2.8
8	2380	0.094	4.2
10	2000	0.079	7
12	1680	0.066	12
14	1410	0.055	20
16	1190	0.047	33
18	1000	0.039	57
20	841	0.033	97
25	707	0.028	160
30	595	0.023	282
35	500	0.019	460
40	420	0.016	770
45	354	0.014	1334
50	297	0.012	2080
60	250	0.010	3240
70	210	0.008	6140
80	177	0.007	1.04×10^4
100	149	0.006	1.714×10^4
120	125	0.005	2.092×10^4
140	105	0.004	4.94×10^4
170	88	0.0035	8.34×10^4
200	74	0.0029	1.4×10^5
230	63	0.0025	2.52×10^5
240	60	0.0024	2.64×10^5
270	53	0.0021	3.84×10^5
325	45	0.0018	6.2×10^5
325	44	0.0017	6.6×10^5
400	37	0.0015	1.12×10^6
600	30	0.0012	2.046×10^6
1200	15	0.0006	1.696×10^7
1800	9	0.00035	7.86×10^7
3000	6	0.00024	2.62×10^8
8000	3	0.00012	2.05×10^9
14000	1	0.00004	6.2×10^{10}

(a) Particle counts shown are only approximate because the actual number will vary depending on the particle shape.

These are usually used with grit sizes from 60 up to about 320 for roughing operations. When a finer finish is required, the less wear-resistant resin bonds are used with grit size from 320 down to a few microns in particle size.

Care must be exercised in two areas: bond and hardness designations of wheel manufacturers differ. Furthermore, hardness is probably not the right term. The proprietary bonds of a vendor differ in hardness, toughness, and abrasion resistance. Some add fillers of other abrasives or salts to alter the porosity and the wear resistance of the workpiece (Ref 4).

Additional Factors. The optimum wheel rotational speeds for diamond grinding are in the range of 20 to 30 m/min (4000 to 6000 sfm) (Ref 4, 5). Without changing the feed rate, slower speeds produce excessive depth of cut and diamond pull out while higher speeds result in excessive diamond wear without wearing of the bond and termination of the cutting action.

Coolant Flow Parameters. The last consideration is the coolant and the method of application. For glass grinding, synthetic oils are usually used in water at a rate of about 1 part oil per 50 parts water. The coolant should be directed at the wheel/workpiece interface. Flow rates up to 150 to 300 L/min (40 to 80 gal/min) are typical for creep grinding and flow rates of 75 to 150 L/min (20 to 40 gal/min) for Blanchard grinding.

Many people have analyzed the relationship of glass properties to resultant feed rates and surface finish and to date no entirely satisfactory connection among these parameters has been established. The glass properties normally studied are hardness, elastic modulus, and the softening point. People may refer to a hard glass relative to grinding to indicate that it must be ground at a slower feed rate than a soft glass. As a general guideline, harder glasses (which have higher hardness, elastic modulus, and softening point than soft glasses) will machine slower, self-sharpen the wheel at lower feed rates, and have a smoother surface after grinding. The lapping hardness data supplied by most glass producers can be used as a guideline for the grinding hardness of glass components (Ref 8).

Lapping (Fine Grinding, Smoothing, or Fining)

After the workpiece configuration has been generated by the grinding process, the next step is lapping, the goal of which is to refine the shape and to reduce the surface roughness and subsurface damage prior to the polishing operation. There is a trade-off between cycle time (stock removal rate) and surface/subsurface damage for both the grinding and lapping process. Furthermore, each subsequent process removes stock at a slower rate with grinding rates ranging from 0.025 to 2.5 mm/min (0.001 to 0.1 in./min), lapping rates at 0.0025 to 0.025 mm/min (0.0001 to 0.001 in./min), and polishing rates ranging from 0.00025 to 0.0025 mm/min (0.00001 to 0.0001 in./min). This is about an order of magnitude drop for each step.

The fining of the surface is done with either fixed abrasives or a loose abrasive slurry with any one of several types of processing equipment. The majority of conventional machines use a metal plate on a vertical spindle. The glass is held against the plate by either positioning the workpiece in a ring that is held in place with roller bearings around its outer diameter or by another vertical spindle positioned above the workpiece that is held in place and oscillated by a ball and socket into the workpiece backer plate. For large optics, the plate and workpiece are reversed (that is, the workpiece is affixed to the lower vertical spindle and the smaller lap is controlled from above). A variation on this concept is the spherical lap for which the upper spindle is angled to conform to the radius of curvature. For some high-speed optical lapping, a machine similar to a generator is used but with a tool shaped to the workpiece contour rather than a ring tool. A more sophisticated version for flat lapping is the dual-face equipment in which the workpieces are held between two plates with a rotating gear arrangement.

Tooling for high-speed diamond lapping is usually fabricated from diamond buttons arranged in a pattern from the inside diameter to the outside diameter to ensure even wear of the workpiece. Computer models have been generated based on velocity vectors to predict wear. Another form of fixed abrasive that is used, especially for ophthalmics, is abrasive silicon carbide, alumina oxide, or diamond paper film with pressure-sensitive backing attached to a backer plate that has the desired concave or convex shape.

Most of the lapping is done using loose abrasives such as garnet, silicon carbide, or fused alumina while the lap plates are fabricated from cast iron, brass, glass, steel, or ceramic material. In addition to these abrasives, there is also a calcined alumina that has a platelike shape, rather than the normal blocky grain.

The lap plates usually have a groove pattern generated in the surface to help distribute slurry and to give any oversize abrasive particle a place to hide and to prevent the generation of scratches. Common patterns are checkerboards, spirals, radial, arc radial, and special grooves. The removal rates are fastest using cast iron with brass being about 50 to 60% of the cast iron removal rate and the removal rate with glass or ceramic lap plates at ≤50%.

The primary variables that must be controlled in the lapping process are the type and concentration of grit, grit size, lap speed, and pressure.

The blocky-type grains cut by rolling and digging into the surface cause conchoidal or lateral-type fractures (Ref 5) and lead to subsurface damage, as discussed earlier. The platelike particles cut more by digging at the surface like a razor blade or chisel as opposed to the pickax cut of the blocky grain. The surface finish, subsurface damage, and removal rate are proportional to the grain size and it is this relationship that will be used to determine what grit size or grit sizes to use for the lapping operation. The total surface roughness, R_t, plus the subsurface damage are about equal to the particle size, with a few deep checks going to 1.5 to 1.8 diameters. A typical plot showing this relationship is shown in Fig 6. The abrasion resistance of the glass will alter the damage about ±20% with harder glasses such as fused silica incurring less damage.

For most high-speed operations, a single step lap is used. Again, the trade-off involves a large abrasive that cuts fast with a rougher surface (longer polishing time is required) and a finer abrasive that cuts slower but leaves a smooth and less damaged surface. For large optics and very precise finishing, several steps of lapping with different grit sizes may be used, starting at ≈30 μm (≈1200 μin.) and finishing at ≈5 μm (≈200 μin.). From Fig 4, the abrasive used for lapping should remove

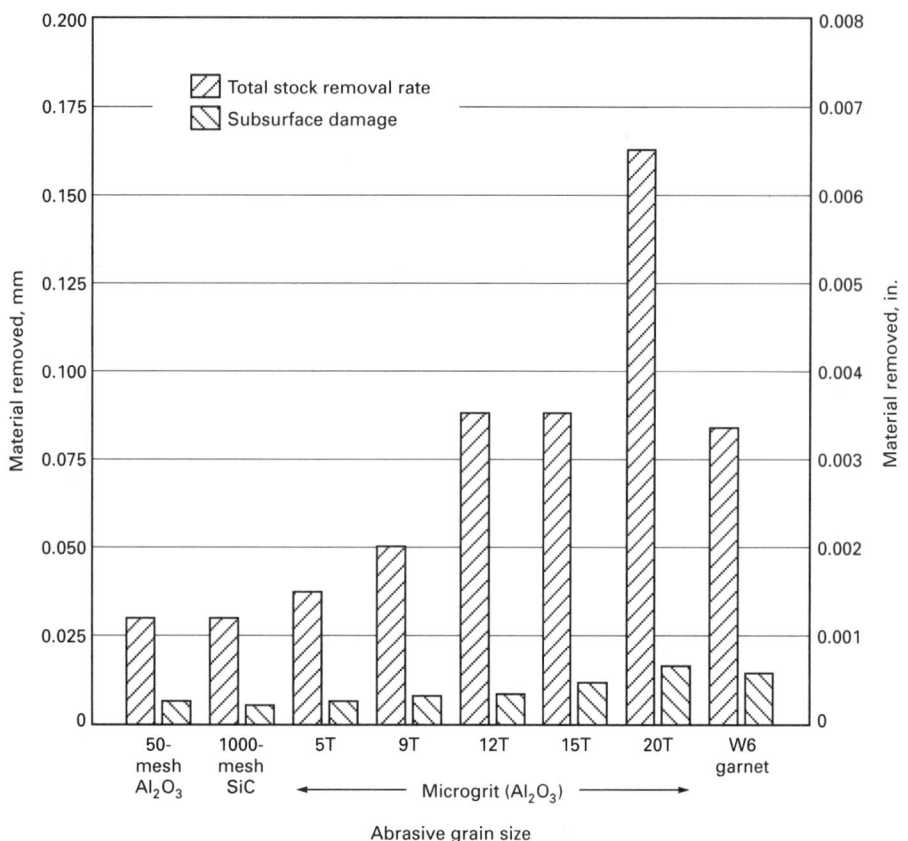

Fig 6 Effect of abrasive grain size on material removal rate and subsurface damage. Workpieces were lapped with an abrasive that was 25% solid by weight at a 10 to 14 kPa (1.5 to 2 psi) pressure for 10 min.

all of the damage except for the loose abrasive damage. In the ideal case, the bottom of the grinding subsurface damage will be at the same depth as the lapping damage and no unnecessary lapping has been done. Hence, another guideline that can be used to determine the subsurface damage is to measure the R_t value and then multiply by four.

The set of machine variables that can be manipulated to enhance the rate of cut of a given machine configuration includes the pressure applied and the relative speed of the lap to the workpiece. The Preston equation developed during work on plate glass polishing during the 1920s states that the removal rate is proportional to the pressure and velocity (Ref 5, 9, 10):

$$H/t = C \cdot (L/A) \cdot (S/t) \qquad \text{(Eq 1)}$$

where H is change in height (thickness), t is lap time, L is the load, A is the area, S is the distance traveled, and C is the proportionality constant. At the beginning of the lapping process, the thickness change will be high because only the high points of the ground surface are being lapped (high pressure), and the rate will gradually decrease to a steady state where the surface roughness is defined by the lap grit. Factors such as grit size, lap material, groove pattern, and grit type will change the Preston coefficient. Of more interest is what is the effect of workpiece material.

Relative to lapping and polishing, practitioners refer to hard glasses or soft glasses, two designations that indicate the ease or rate at which the materials can be machined. One supplier catalogs their materials relative to lapping hardness with a ratio of ≈4:1 from softest to hardest. Therefore, the Preston coefficient will be altered by about a factor of four dependent only on the workpiece

material. At least a portion of this factor can be attributed to elastic by the modulus (Fig 7). Studies to relate other glass factors (for example, hardness and softening point) have found more exceptions than rules; however, in most cases, these properties tend to track the elastic modulus (Ref 11). Such factors as water durability may help to predict these exceptions (see the section "Cleaning and Inspection of Polished Components" in this article).

Polishing

The purpose of the polishing operation is to bring the workpiece surface finish within tolerance and produce a transparent part or highly reflective surface. The equipment used is generally the same as used for lapping. However, the process materials are changed. What really takes place during the polishing process has spawned all kinds of theories and many scholarly papers have been written on the subject. The most accepted theory is that the polishing process is a combination of chemical action with the water-based slurry and mechanical action of the polishing pad and the particles in the slurry (Ref 9, 12–14). The choice of the polishing compound and the pad determines the degree of chemical and mechanical action.

Polishing compounds used for glass are usually cerium oxide (30 to 99% CeO_2), zirconium oxide, iron oxide, and, in some special cases, fine or colloidal silica, or diamond. The relative rate of cut for some polishing materials is shown in Fig 8. The oxides are used in a water slurry of ≈1050 to 1250 g/L (≈8.8 to 10.4 lb/gal) final density. For high-speed polishing that is required for commercial or semiprecision finishing, the compound is fed at high volumes (typically 4 to 8 L/min, or 1 to 2 gal/min for a 660 mm, or 26 in. diameter machine operating at 50 to 60 rev/min). This contrasts with pitch pol-

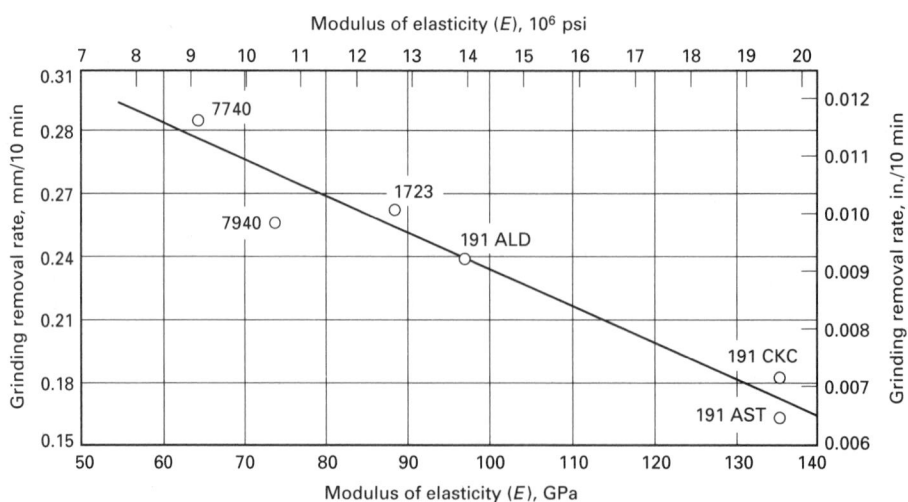

Fig 7 Plot of grinding removal rate versus modulus of elasticity for selected polished glasses

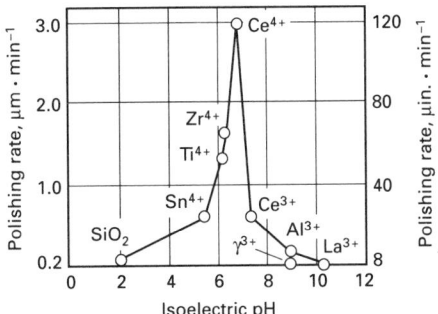

Fig 8 Plot of polishing rate versus isoelectric pH of selected polishing compound cations. Source: Ref 13

ishing in which the surface is barely kept wet or the slurry is charged or painted onto the lap. In some specialized pitch polishing operations, the entire pitch plate and work are kept in a slurry bath maintained at a very uniform temperature.

Most of the polishing compounds have a mean particle size ranging from 1 to 3 μm (40 to 120 μin.); however, some may have individual particles over 10 μm (400 μin.). For high precision polishing, specialty wet milled products can be obtained with average size near 1 to 2 μm (40 to 80 μin.) and oversize held to $\approx$3 μm ($\approx$120 μin.). Again, these precision slurries are milled wet and are usually supplied as a wet concentrate to eliminate agglomerates.

When mixing slurries from dry compounds, a number of things such as premixing, rolling, or decanting can help to improve the surface uniformity and to reduce scratches and sleeks.

As mentioned previously in this section, pitch plates are used for the most exacting polish. Pitches of varying viscosities are mixed to obtain the desired viscosity. Fillers such as felt or wood flour are sometimes added to extend the operating range of the pitch. The groove pattern and depth are critical parameters to help control the rate of change and also control the shape of the workpiece. In general, these materials are run at very slow speeds (a few rev/min), and polishing time is measured in hours or even days (Ref 1, 8, 14).

Pad Materials. For high-speed polishing, a large number of pad materials are available. The urethanes, either plain or filled with cerium or zirconia, are at the hard end of the range (Shore A, 70 to 97) followed by the polyester or wool felt materials that are urethane impregnated (Shore A, 64 to 75), and finally the poromerics and straight felts are used for commercial polishing or as a last step in the polish process to remove light sleeks or scratches.

Polishing Process Parameters. During the polishing process, the polishing media and water react with the glass to form a silica gel on the surface that is continuously abraded away by the mechanical action of the process

(Ref 13). As in lapping and grinding, the process follows the Preston equation in that the stock removal rate is directly proportional to the velocity and pressure. Again, the rate of thickness reduction is high at the beginning (high pressure on the lapping peaks) and is reduced to the steady state as the subsequent damage is removed. For some high-speed optical polishing, speeds to 3000 rev/min and pressures of 14 to 70 kPa (2 to 10 psi) are used. The final finish can be obtained in an elapsed time of 30 s to 1 to 2 min. Also, for many dual-face polishing processes, the polishing is accomplished within 10 to 20 min.

For many polishing operations, a secondary operation using a soft pad is used after the polyurethane step to remove the light sleeks or scratches. This secondary operation is a light buff and in general only removes $\leq$1 μm ($\leq$40 μin.).

Polishing of the fine grain glass-ceramics such as Zerodur and 9600 glass-ceramic (see the article "Phase-Separated Glasses and Glass-Ceramics" in this Volume) is not different from glass, except that a little more time may be required because these materials, like fused silica, are not very chemically reactive. Polishing a coarser-grained or irregular grained glass-ceramic such as canasite or 9608 glass-ceramic is more difficult. With these materials, the ratio R_a/R_t is $\approx$10 rather than a value of near 5 typically expected for glass, thus indicating grain pull-out during lapping. These materials behave more like metals and require a harder pad and a less reactive slurry to obtain a good finish. In most cases, a finish equivalent to glass cannot be obtained. Figure 9 is

a Wyko trace of a canasite (chain-silicate) glass-ceramic disk polished with 482 Transelco cerium and a polyurethane pad. The 24.8 nm peak-to-valley finish is rougher than would be expected for a glass surface with an R_a of 2.59 nm. This condition is explained by the theory that the chemical activity is greater with the glass interfaces of the grains and that some grain pull-out or break-off has occurred.

Cleaning and Inspection of Polished Components

The drawings and specifications usually refer in some way to the following items that must be certified in the final part by 100% inspection or some acceptable quality level (AQL) or statistical process control (SPC) guideline:

- Shape; figure; power (irregularity)
- Surface finish (actual measurement, percent transmission, or reflection)
- Defects [military specifications (MIL), Deutsche Institut für Normung (DIN), Japan Industrial Standard (JIS), and so on] referring to scratches, digs, chips, stains, and so on
- Coatings (anti-reflective, reflective, and protective)
- Cleaning (optional)

After polishing, it is very critical that the polishing media be removed before the liquid vehicle is allowed to dry. Once the liquid containing the media is dried, it may be impossible to remove the material even with a secondary buff polish or vigorous scrubbing. If the workpiece was blocked, addi-

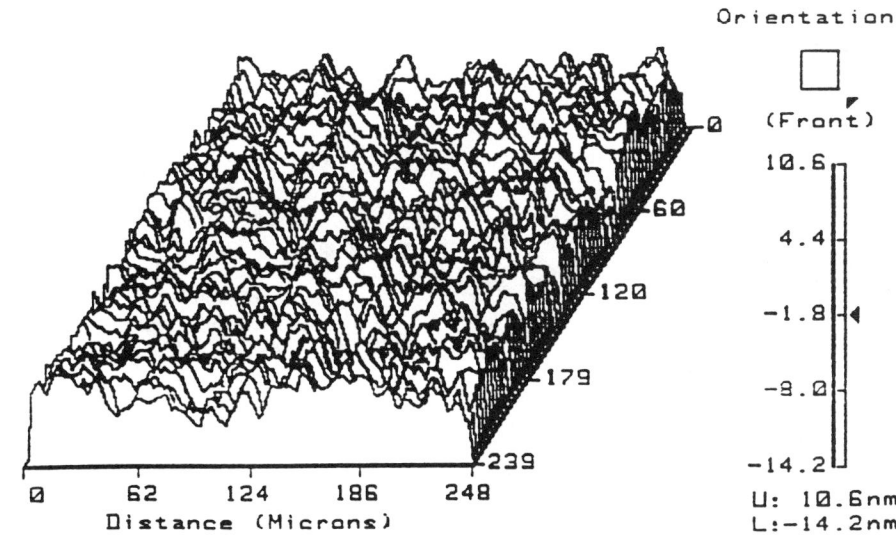

Fig 9 Wyko plot of a canasite glass-ceramic disk polished with cerium and a polyurethane pad

tional solvents are required to ensure removal of the blocking compound.

A recommended process for cleaning of glass after polish is:

- Soak the workpiece in an ultrasonic bath with a detergent for 2 to 5 min
- Remove the workpiece from the ultrasonic bath and scrub surfaces with a clean room sponge
- Spray rinse in deionized water (DI)
- Deionized water rinse in overflow cascade tank
- Dry by Freon/Genesolv, hot DI water pull, air knife, or spin dry

Depending on the shape and size of the part, there are commercial systems available from numerous manufacturers of equipment that will accomplish the above steps.

The shape and irregularity of the finished part can be measured in many ways with commercially available equipment. The process used will depend on the part specifications and the budget available to purchase equipment. Listed below are some of these techniques and a relative figure of merit for their precision:

Contact measurement systems:

- Flat plate and shim stock about 0.025 mm (0.001 in.)
- Template with 1 to 5.0 μm (~50 to 200 μin.) of power
- Spherometer or variation thereof, 1.25 μm (50 μin.) power
- Test plate, 0.050 to 0.075 μm (2 to 3 μin.)
- Stylus, 0.10 to 0.25 μm (4 to 10 μin.)

Noncontact measurement systems:

- Interferometer, 0.050 to 0.075 μm (2 to 3 μin.)
- Functional test (transmission, projection), not direct measure but could measure acceptability

Surface Finish. The specification for surface finish can go all the way from commercial polish [that is, you can see through it well to very exacting specs such as R_a, R_t, or rms (root-mean-square) roughness (R_q) in microinches, microns, angstroms, or nanometers]. It is believed that for some laser optics, finishes are required down to an R_a value of 1 to 2 Å and measurements at that level can be made. Again, there are a number of techniques that can be used to directly measure or infer the roughness, including those listed below:

- Stylus: measures from 0.005 μm (0.2 μin.) to as low as a few angstroms

- Optical stylus (Mireau interferometer), Wyko plot: device measures to within a few angstroms
- Microscope (especially with differential interference): estimate to angstroms
- Scatter light: measures to a few nanometers
- X-ray analysis: measures to a few angstroms

Subsurface damage was discussed earlier in the section "Lapping (Fine Grinding, Smoothing, or Fining)" in this article. If the lapping process removes the grinding damage and the polishing process takes off sufficient material ($\approx$ one to 1.8 times the lapping particle size or four times the R_t value after lapping), then subsurface damage should not be a concern. As a process control tool, an occasional lens can be examined by dipping one-half the lens in aqueous ammonium bifluoride ($NH_4F \cdot HF$) for 10 to 15 s. Subsurface damage from grinding, if present, will be visible as a series of scratches in the shape of the ring tool pattern or in a pattern that looks like a series of short arcs in a line (the edge of the conchoidal fractures). Subsurface damage from lapping usually does not have a pattern but appears as small digs or bruise checks. The reason for concern about subsurface damage is that checks present therein will increase light scattering or absorption for lenses and mirrors and, for products where strength is important, the checks may propagate with time and cause part failure.

Surface Defects. JIS, DIN, and United States military specifications, MIL, specify the light sources and inspection methods to be used depending on the size and number of defects allowed. The techniques go from very low level light sources (when only obvious scratches are rejectable), to very high levels (around 30 K lx, or 2800 lm/ft^2) for small defects (microns).

Advanced Grinding Technologies. Some of the new areas of glass finishing are not discussed in this article; however, listed below are several literature references and human resources from which more information can be obtained:

- Creep-feed grinding: lower surface damage, full depth cutting, Trever Howes, University of Connecticut
- Plastic regime grinding: with the right equipment (rigid, precision, expensive) the grinding process can produce a polished surface. University of Rochester, Lawrence Livermore National Laboratory (LLNL), Cranfield Unit for Precision

Engineering (CUPE) in the United Kingdom, Kodak, Japan
- Aspheric finishing: Computer-controlled grinding and polishing machines. LLNL, Kodak, CUPE, Japan
- Ion polishing (deterministic shape correction—no surface finish improvement) (Kodak)

REFERENCES

1. F. Twyman, *Prism and Lens Making*, A. Hilger Ltd., London, 1942
2. B. Lawn and R. Wilshaw, Review Indentation Fracture: Principles & Applications, *J. Mater. Sci.*, Vol 10, 1975, p 1049–1081
3. W. Rupp, Surface Structure of Fine Ground Surfaces, *Opt. Eng.*, Vol 15 (No. 5), 1976, p 392–397
4. A.L. Ardamatskii, Principles of Diamond Tool Operations, *Sov. J. Opt. Technol.*, Vol 47 (No. 10), 1980, p 613–622
5. N.J. Brown, "Precision Optical Fabrication," SPIE's 33rd Annual International Technical Symposium on Optical & Optoelectronic Applied Science & Engineering, Aug 1989
6. D. Ketelsen and R.E. Parks, Spherical Grinding Wheel Use & Fabrication, *Sci. Opt. Fin.*, Vol 9, June 1990, p 95–98
7. J. Taeyaerts, Guidelines to Determine Diamond Concentration Required to Grind Glass, *Ind. Diamond Rev.*, Feb 1976, p 48–52
8. N.J. Brown, Some Speculation on the Mechanisms of Abrasive Grinding and Polishing," Fourth International Precision Engineering Seminar, May 1987
9. F.W. Preston, The Theory & Design of Plate Glass Polishing, *J. Soc. Glass Technol.*, 1927, p 214–256
10. G.S. Khodakov, Abrasive Grinding Mechanism of Glass, *Sov. J. Opt. Technol.*, Vol 52 (No. 5), May 1985, p 287–291
11. K.G. Kumanin, *Generation of Optical Surfaces*, The Focal Library, 1962
12. J.R. Hutchins, III and R.V. Harrington, Glass, Reprinted from *Encyclopedia of Chemical Technology*, Vol 10, 1966, p 533–604
13. L.M. Cook, Chemical Processes in Glass Polishing, *J. Non-Cryst. Solids*, Vol 120, 1990, p 152–171
14. W.J. Rupp, Conventional Optical Polishing Techniques, *Optica Acta*, Vol 18, June 1971, p 1–16

Decorating

Ronald E. Johnson and John Donohoe, Corning Incorporated

THE ART OF DECORATING glass and ceramic (pottery) is almost as old as the materials themselves. Since about 15,000 B.C., metal oxides have been used for coloring, but for many centuries just a few colors were available because the oxides of iron and manganese provided only yellows, browns, and black. Cobalt blue was not introduced until the ninth century A.D.

Throughout many centuries, patterned decorations were almost exclusively applied by hand. Hand painting, assisted sometimes with a simple stencil, was almost the sole process used for intricate-colored designs until the introduction of copper-plate printing onto transfer tissue in the 1750s (Ref 1, 2). However, it was not until the 1840s that the use of this printing method for applying decorations on pottery became widespread. By 1937, the decorating options had increased to include spraying, silk screening, machine banding, rubber stamping, and lithographic-printed decals on duplex paper (Ref 3). Since then, decorating process options have greatly expanded. Even the older options have evolved into more sophisticated processes that have greater design and process capabilities.

The trend has been toward automated application and more colors. Although the color trend is now somewhat threatened by growing demands for stiffer regulations on lead and cadmium release, color manufacturers are responding by developing improved lead-free and cadmium-free colors (Ref 4–6). Decorated products have also expanded from the traditional tableware and container industries to many new markets.

Decorating Processes

The consumer demand for more decorative ware and continually changing decorations continues to fuel the need for new, economical mechanical decorating systems, as well as more creative ways to utilize the older process options. Meeting these decorating challenges requires careful attention to all the factors involved in producing decorations. A well-matched system between the substrate, inorganic materials, organic media, process, and machine is essential to produce quality decorated products economically.

Decorating process options are classified on the basis of where the design is initially created. Direct decorating methods, where the ultimate design is created directly on the ware, include hand painting, crayoning, spraying, banding, and ground laying. In semidirect methods, the pattern is initially created in a permanent form, which is used to apply the enamel or color directly to the ware. This type of method includes stamping, stenciling, and screen printing. Indirect methods are those in which the design and the enamels are produced on a separate temporary carrier, such as paper or silicone, from which it is transferred to the ware. This includes all forms of decals and offset printing. The decorating techniques associated with multicolor processes are sometimes classified simply as direct or indirect, depending on whether the full multicolor pattern is created in register prior to application to the ware, or is created by applying each color to the ware in consecutive steps. Decals and offset collector processes are the only two indirect methods, based on this definition.

Many of the processes named above have been detailed in the literature and references are provided in this article. The preparation and physical properties of the materials used in these processes are critical, whether for single color or multicolor designs, and are unique to the process. The organic media required have been customized to optimize the manufacturing performance in terms of design appearance, difficulty of operation, and cost.

Direct Methods

Hand painting, brushing, crayoning, and ground laying techniques are the older, labor-intensive processes that have been well described in the literature (Ref 7, 8). These techniques are still used in cases where skilled labor is available and product cost is favorable.

The spraying process is widely used to produce uniform colors on glass and ceramics whenever the area coverage is substantial, particularly if it is all over. In the conventional air spray system, compressed air is supplied to both the spray gun and coating container. At the spray gun, the air and coating mix, propelling the atomized coating toward the ware. Much of the volatile portion of the coating composition evaporates before the coating reaches the ware, but a sufficient amount must remain to enable the coating to flow out to a uniform layer.

Conventional spraying has improved continually since the 1960s, but the basic principles have changed very little. Recent developments include airless spray systems, electrostatic spray systems, and low-pressure, high-volume spray systems. All spray processes are often used with ware masks either to protect against overspray contamination or to limit area coverage. The principles of the various spraying processes are explained by Schneberger (Ref 9), who also discusses the factors influencing the application of enamel, as well as the equipment and problems associated with the process.

Equipment varies in complexity and capability, from a simple one-station spray booth with a turnable and a hand-held spray gun to a fully automated, high-speed production line. An automated line is generally based on a chain-on-edge conveyor that carries rotating ware chucks and transports the ware through spray booths with a series of spray guns. Ware preheaters and dryers for the sprayed enamel may also be necessary. The chain-on-edge conveyors can be continuous motion or indexing types. Each configuration is selected to customize the actual process application.

Sprayed coatings can range from high gloss to matte. Design effects include single-color and multicolor uniform coatings, shaded (light-to-dark) coatings, and stencil designs, as well as speckled and other two-color or two-tone effects. One two-color technique imparts a stoneware or clayware appearance, whereas others can simulate marble or stone agate.

Curtain and Roll Coating. Spraying does have disadvantages, even when applying an allover color. In particular, air-borne lead-based frits are a major health and safety concern. Consequently, curtain coating, which is an alternative process for applying a uniform coating to a glass surface, has been used extensively in flat glass applications (Ref 10). Roll coating, which has also been used for flat-glass decorating (Ref 11), can apply both uniform colors and textured effects.

Banding is a low-cost form of decorating that has been used for many years. In the earliest version, bands were applied by hand with a brush, and utilized a simple turntable for the ware. This simple process is still used for both enamel and precious metal designs, but automated banding using a roller or wheel is more common.

Products with an axis of rotation can be automatically decorated with ease. Many forms of semiautomatic and fully automated equipment are available in single-color and multicolor versions. The key factors in machine banding are that the application roller, or wheel, must:

- Contact the ware on the centerline and be perpendicular to the ware at the point of contact
- Rotate in the same direction as the ware and at the same surface speed at the point of contact

Hoover (Ref 12) discusses these principles and is an excellent source of banding information.

Standard tungsten carbide banding wheels can apply enamel lines from less than 1 mm (0.04 in.), up to 6 mm (0.24 in.) wide in one pass. Edge or rim enamel bands are applied with a roller, usually made of hardened steel. Precious metals can be applied using either liquid material with a lining wheel or paste formulations with neoprene or urethane rollers in a special applicator unit.

The banding process is a simple and reliable process. It not only is used as a primary design, but as a low-cost, secondary decoration for cups, saucers, or bowls in dinnerware sets to complement designs applied by other more expensive processes, such as offset gravure printing, screen printing, or decals.

Chemical frosting, or acid etching, is well known as a means to provide decorating effects. The frosting can be either all over or patterned using acid-resist masking or screenable frosting pastes. Etchants were historically hydrofluoric acid-based, but hydrochloric acid-based etchants have been developed recently (Ref 13). The use of an acid-resist bright gold provides brilliant accents to frosted glass (Ref 14). Other techniques for decorating glass by removing part of the glass surface include sand blasting, or sand carving (Ref 15, 16), water-jet cutting (Ref 17), and laser etching (Ref 18).

Semi-direct Methods

These methods involve rubber stamping, stenciling, and screen printing, each of which is described below.

Rubber stamping, or flexography, produces designs from a raised profile that has been created in an elastomeric stamp cast from a mold. A plate or roll, usually metal, is coated with an even layer of enamel, and the stamp is first pressed against the coated surface and subsequently against the ware to transfer a thin layer of enamel. The stamp usually has a soft backer to contour to the ware. Over the years, various levels of automation have been developed. Although machines have been built to apply large decorations, including multicolor (Ref 19), to a variety of ware shapes, the print quality has been limited in terms of definition, resolution, registration, and toning capabilities. However, the process is low cost and is widely used in trademarking and date coding. Additional details on the rubber stamping process are provided in Ref 20.

Stenciling. Stencils are simply patterns that have been cut in an impervious material, through which ink is applied by a roll, brush, blade, or other means. The art of decorating by stenciling is described in Ref 21. Stenciling is a forerunner to screen printing, and screens are sometimes referred to as screen stencils (Ref 2). True stenciling is not often employed now, although stencils are often used as spray masks.

In screen printing, the screen is simply a fine-mesh material stretched taut on a rigid frame. Some holes are open for ink passage, while others are blocked, thereby creating a patterned design. The printing color is forced through the open mesh areas with a flexible squeegee, thereby reproducing the design on the substrate. The screen holes can be blocked either by adhering a stencil to the surface or by applying a screen resist. A light-sensitive emulsion is the most commonly used screen resist. The screen is coated with the emulsion, dried, exposed to light through a photopositive, and then the unexposed emulsion is washed away (Ref 22).

Screen printing has been used for ceramics since the 1920s, but it continues to develop, both in terms of equipment and materials, to provide an ever-improving process for decorating glass and ceramics. Advances have been made in understanding screening from a theoretical basis (Ref 23), as well as in both close-tolerance printing (Ref 24) and improved resolution (Ref 25–28). Equipment and processes have been developed to print on various dinnerware shapes, including flatware (both center and rim) and holloware. Screen sizes range from small trademark screens to screens 1.2 m (4 ft) wide or more for flat glass. Screens can also be either planar or rotary.

The key to direct screen decorating is ensuring that the screen mesh can be held parallel to the ware during contact with the squeegee. The squeegee will then produce a well-defined print at the point of contact. This requirement limits screening from being used on highly contoured surfaces.

Materials for screen printing, such as the squeegee, screen, and emulsion, are all readily available from a wide range of manufacturers (Ref 29). The squeegee is typically made of urethane. The screen is usually from 125 to 305 mesh and can be made from a variety of materials, including nylon, polyester, or stainless steel, which are listed in the order of stretch and inverse order of registration capabilities.

The most commonly used emulsion for screen making is the liquid type of diazo-sensitized polyvinyl alcohol (PVA). The screen is scoop-coated with the light-sensitive emulsion and dried. The emulsion is exposed to a light source of suitable wavelength, while the required photopositive is in contact with the screen. The exposed screen is washed out with water, leaving the pattern area free of emulsion. This method gives a durable, long-lasting screen, especially for printing on glass and ceramics.

Gaining in popularity is the use of capillary sheet emulsion. These preformed sheets of diazo-sensitized PVA have uniform thickness and deliver excellent print thickness control. Screens are manufactured by placing the capillary sheet in close contact with the water-saturated mesh, whereupon the emulsion softens and is drawn by capillary forces into the mesh. The screen and emulsion are subsequently dried and readied for exposure and development. Resin-based photopolymer emulsions other than PVA used for screen resists are a recent development and offer promises of higher resolution, increased durability, and longer life.

Multicolor screening directly onto ware became a practical proposition with the development of thermoplastic, or hot-melt, media. These materials operate with heated screens. Each color "freezes" on the ware as it is deposited, so that it does not interfere with the subsequent colors. Equipment for multicolor screening is generally either semi- or fully automatic, and the equipment manufacturers offer trade-offs between cost and performance, speed, registration, decorating area, and ware-handling automation.

Indirect Methods

The indirect processes can generally be classified as either offset printing or decals. The offset methods include screening, gravure, flexographic, and even lithographic processes for printing onto an intermediate (offset) surface. The processes most often used are screening and gravure. The offset surface is almost always an elastomer, because of its ability to conform to irregular ware surfaces. The offset member can be a cylinder, cone, membrane, or pad, depending on the ware shape. Large pads or membranes are most often used for dinnerware shapes. Pad geometry must be carefully selected based on specific ware geometries, because a rolling motion must be maintained during pad engagement and disengagement from the ware. Even when membranes are employed, they are generally pressed against the ware with a configured pad. Pad durometer hardness can range from 20 to 80 on a Shore 00 scale.

Silicone Pad Printers. The use of silicone in printing was developed in the 1960s for printing onto plastics and other materials. In the mid-1970s, the British Ceramic Research Association (BCRA) adapted silicone pad printing to ceramic applications as an improvement over the gelatin pads then in use (Ref 30–33). BCRA also identified the need for specialized printing enamels and developed these with the local pigment manufacturers in Stoke-on-Trent, United Kingdom (Ref 34).

Because problems were encountered in maintaining complete transfer with oil- and solvent-based mediums, thermoplastic enamels became preferred, particularly for use with on-glaze ceramics and glass. A British firm was one of the first companies to build single-color and multicolor ceramic and glass decorating machines specifically designed for silicone pad printing using thermoplastic enamels. Other firms quickly followed with similar machines. These hydraulic silicone pad printers were capable of printing dinnerware with a diameter up to 305 mm (12 in.) at a rate of typically 8 to 12 pieces/min.

The original printing pad type of decorating machines, like the Murray-Curvex (Ref 31) and Dekram (Ref 34) used handmade, chrome-plated, engraved copper plates to create the pattern and gelatin pads to effect transfer. The silicone pad printers, on the other hand, use etched hardened steel plates. The pattern is produced by coating the steel plate with a photoresist, exposing it through a photopositive, washing away the unexposed resist, and etching. Etch depths range from 30 to 75 μm (1.2 to 3 mils), with 50 μm (2 mils) being typical. Photopolymer plates are sometimes used for short production runs. Laser-etched ceramic plates have been utilized for long production runs.

Multicolor patterns use one etched plate per color. The normally heated etched plate is flooded with its own colored enamel, and the excess enamel is cleaned from the etch surface with a doctor blade mechanism, which leaves enamel in the etched recesses of the steel plate. The silicone pad is compressed onto the etched plate, picks up the enamel, and then transfers it to ware upon being pressed against the ware surface. Properly formulated ink will completely transfer to the ware surface without leaving a shadow (residue) on the pad. The ware then indexes through the machine and is reprinted with each color until the design is completed.

Etched-plate printing processes are excellent for printing dots, which can be round, square, or elliptical in the photopositive. The preferred shape is elliptical because it gives the smoothest transition of density in the midtone (40 to 60%) range. For glass enamels used with silicone transfer methods, whether pads or rolls, the practical capability of the etch process is 4 lines/mm (100 lines/in.), although a higher level has been demonstrated.

Tone range is an inverse function of etch depth, and excellent designs with tones ranging from 10 to 90% are easily attained at etch depths of 40 μm (1.6 mils). Solid-color areas are difficult to achieve, and generally are printed in densities from 70 to 90%. The remaining pillars of unetched steel support the doctor blade, thereby avoiding scooping of the ink by the blade, which would result in lighter-colored areas within the solid color. Inks must exhibit sufficient flow either during printing or upon subsequent firing to achieve a 100% dense print. Solid, straight lines or large-diameter curves in the design should not be placed with open spans parallel to the doctor blade, because the result is usually either rapid wear of the etched plate or damage to the doctor blade.

Decalcomania (decals) have been used for many years to decorate ceramic and glass articles (Ref 35). Originally, ceramic decals comprised a gravure-printed design on thin tissue paper, which was pressed against the ware, and the paper soaked off with water. This pottery-tissue decal gradually evolved into the duplex paper type, and later, the water-slide-off type decals (Ref 36).

The printing process employed for most of the decals prior to the 1940s was lithography, in which the printing of varnishes onto the paper alternated with dusting of the ceramic colors onto the varnishes. Although this process produced high-quality artistic decals, color layers were thin, thereby necessitating high pigment concentrations. Consequently, for over-glaze ceramics and glass, durability was poor and toxic metal release was relatively high.

Screening began to be used for decals in the 1940s, sacrificing the fine detail and toning capabilities in order to achieve improved durability and toxic metal release by using heavier deposits with lower pigment concentrations. With improvements in design capabilities, including halftones, screening eventually became the principal process used for ceramic decal manufacture. Application to ware remained labor intensive, spurring the development of decals suitable for automatic application.

In the mid-1950s, a heat-release decal composed of ceramic colors screened onto a special wax paper was developed. Although the heat-release decal was ideally suited for automatic application, fully successful application equipment was not available until the early 1970s, when the system known as Roll-Mark was introduced. Other decal types suitable for automatic application were developed, including a cold-release system that used silicone release papers, but none achieved commercial success. Today, water-slide-off and heat-release are the two most commercially significant types of ceramic decals. Water-slide-off decals represent the most flexible manufacturing option.

Gravure and offset gravure processes have also been used to print heat-release decals for glass (Ref 37, 38), but the use of gravure-printed heat-release decals has been relatively limited. Screening is the most widely used process for ceramic decal manufacture.

There have been considerable advances in terms of design capabilities and application methods related to decals (Ref 39). In one of the most promising application methods, heat-release decals are first transferred to a silicone pad, which subsequently transfers the design to the ware. This technique overcomes the shape limitations of conventional heat-release decals. Stretchable decals have also been proposed as a means to expand decal shape capabilities (Ref 39, 40). Decal design capabilities include textures, metallic lusters, iridescent colors, etched effects, and sand-carved effects.

Improvements in registration, resolution, and pigments have allowed screen-printed decals to reduce the quality gap between them and lithographic decals. Moreover, satisfactory trichromatic designs have been achieved by printing cyan, yellow, magenta, and black in a dot pattern. Trichromatic printing has been of interest for ceramic and glass decorating for some time because of the obvious advantages of printing a full color spectrum using only three or four colors. Unfortunately for glass and ceramics, satisfactory colors have been unavailable because of pigment limitations, and, although improvements have been made, magenta is still a problem, particularly with under-glaze ceramics (Ref 41). However, with on-glaze ceramics and glass, limited trichromatic printing has been demonstrated with decals, double offset, and even pad printing, although the registration capabilities of the latter are only marginally sufficient for trichromatic designs.

Double Offset. Decals were widely used for many years, despite being expensive, because there was no economically acceptable alternative for achieving high-quality multicolor decorations. This high decal cost eventually induced one glass products manufacturer, the world's largest user of ceramic heat-release decals, to seek an alternative method of multicolor decorating. Because of the large cost-saving potential, this manufacturer was able to undertake an extensive direct printing development project, which was successfully placed into production at the end of 1979. At first, this technology involved printing each color in succession from a series of silicone transfer rolls. The inks were thermoplastic (Ref 42), and the silicone rolls received the color images from etched plates in a similar manner to that employed for pad printing. Later, however, this technology evolved into a double offset or collector process, that in effect produced a multicolored decal on a silicone membrane or pad, which subsequently transferred the multicolored image to glass in a single application step (Ref 43). This was accomplished using high-tack thermoplastic inks (Ref 44) and two silicone compositions with balanced surface energies, so that the ink could be transferred from one to the other without leaving a shadow. This permitted a multicolored design to be printed onto a second, or collector, silicone composition from a series of transfer rolls composed of a lower surface energy silicone composition. This process is used both for wrap-around cup decorations and total-area-coverage tableware decorations (Ref 45).

Offset screen printing has been of interest for many years as a means to overcome the shape limitations of direct screening in which the ware surface must maintain near-parallel linear contact with the squeegee during printing. Consequently, direct screening was limited to either relatively flat surfaces or the exterior surfaces of cylindrical or conical holloware shapes.

Although offset screen printing machines were pioneered by companies using membranes and rolls for the offset surface, an improved offset screen printing process was developed in the early 1980s (Ref 46). The improved concept combines the advantage of silicone pads for printing over curved surfaces with the advantage of screening in producing print designs with large, solid-color areas. The process prints thermoplastic colors from a heated, stainless steel screen to a heated, flat silicone-coated platen. The print is picked up from the platen by a silicone pad and transferred to the ware. The pattern is built, color by color, on the ware by indexing through the machine. This process requires a unique thermoplastic medium, which allows the printing enamels to be first screened to the silicone-coated platen, transferred from the platen to the silicone pad, and subsequently transferred to the ware.

Process Comparison for Multicolored Designs

Several factors must be considered when selecting a process for printing a multicolored pattern onto ware. Although screening does not have the resolution and fine toning capabilities of the offset gravure processes, it is better able to apply solid color over larger areas, and can deposit heavier enamel thicknesses. Screening does have the capability of printing halftones ranging from 2.4 to 3.2 lines/mm (60 to 80 lines/in.). Flexography also can apply a solid enamel color over large areas, but enamel thickness is not as heavy as that which is screened; moreover, the resolution/toning capabilities are even more limited. However, flexography is often the cheapest to operate, because the flexographic typefaces (stamps) are easy to reproduce once the master pattern is made. Screens are somewhat more expensive to reproduce, whereas the etched plates needed for gravure are the most expensive of the three options. The gravure plates not only need to be etched, but must be made to tight tolerances in order to be satisfactorily doctored. Flexible doctor blades tolerate more variations in the plate surfaces, but compound the problem of color variation in the larger color areas because they are more prone to scooping color out of the open etched areas.

Process speeds for all three printing options can be quite high, and depend primarily on the decoration dimensions and contour of the surface to be printed. Consequently, printing process selection most often depends on the requirements of a particular decoration in terms of resolution, registration, toning, color intensity, shape of the article to be printed, and other factors. The indirect multicolor options of decal and offset collector processes have the greatest registration capabilities, because the registration becomes independent of ware shape and application speeds.

Photosensitive Glass

As has been reported in the literature (Ref 47), it is possible to induce a broad spectrum of colors in certain photosensitive glass compositions through ultraviolet (UV) exposures and heat treatments. Colored patterns can be created in transparent or opaque glass, or even in selectively opacified sections of glass. The depth of the color can also be controlled. The photocolors are complex functions of glass composition and the processing variables. The color is due to small, precipitated, subcolloidal silver particles, and color differences are due to variations in anisotropy.

Decorating Materials

Glass Enamels. Ceramic colors can be applied under-glaze, in-glaze, or on-glaze; on-glaze materials are similar to those used for glass. The principal materials used for decorating glass and on-glaze ceramics are the glass enamels, which comprise a low-melting glass frit, with about 5 to 30% of an oxide pigment. The glass frit is generally ground to an average particle size of less than 20 μm (0.8 mil); about 5 μm (0.2 mil) or less is common for gravure and screen-printing enamels. The coefficient of thermal expansion is critical and it is usually selected to be slightly below that of the substrate (Ref 48, 49). Under-glaze decorations use higher pigment levels, even up to 100%. The pigments must survive the glaze firing cycles, which limits those that can be employed and, consequently, the color palette. Details on frit and pigment compositions are provided in the article "Glazes and Enamels" in this Volume and in Ref 50–53, which discuss glass enamels, under-glaze pigments, and the decorative effects that can be produced through alterations of the glaze itself.

Metallics. Decorating materials are not limited to glass enamels and ceramic-oxide pigments. Precious metals and organometallic compounds have been used for years. Precious metals include particulate metals suspended in a medium, which must be burnished after firing to develop the metallic luster, as well as bright golds from a metallic resinate, which exhibit luster without the need to burnish (Ref 54, 55). Electroplating has even been used to increase the thickness of precious metal deposits, particularly for sil-ver, which is overplated with rhodium to prevent tarnishing (Ref 3). Beautiful iridescent, metallic sheen, and mother-of-pearl effects can also be produced on glass by using lusters, which are metallic resinates of such metals as tin, antimony, copper, and others.

Stains. Ceramic oxides or salts that stain the underlying glaze or glass during firing can also be employed as decorating materials but the color palette is limited. A cobalt oxide/titania stain is used for the cornflower blue decoration well known on Corningware. Silver chloride is an example of a well-known salt stain used to impart an amber color to glass (Ref 56).

Media. Glass enamels must be suspended in an organic vehicle, or medium, to be applied via the various decorating processes. A three-roll mill is commonly used to prepare the suspended colors, but many of the methods used for paints and inks can be employed (Ref 57). In many cases, the same glass enamel can be used for numerous application processes, but with different organic media which control the process compatibility.

The organic medium must provide satisfactory color dispersion, as well as adequate ware surface wetting and adhesion. It also must impart the rheological characteristics necessary for the application process. Through its effect on rheology, the medium can determine dot resolution, print definition, smoothness, and other parameters. The medium must also fire satisfactorily, which means it must either volatilize or burn uniformly and cleanly at a temperature below that necessary for the glass enamel to mature. Carbonaceous residue can either reduce the enamel or pigment oxides, thereby altering their characteristics, or result in enamel porosity, which can in turn result in poor durability and gloss.

In many cases, advances in decorating techniques have been made possible through the development of more sophisticated organic media. Readily available natural gums and resins were employed originally for such simple processes as hand painting, screening, and spraying. Later, improved media were developed using synthetic resins. Development tended toward aqueous-based and thermoplastic media. Aqueous media became preferred for spraying, in particular, because of the toxicological and flammability concerns associated with organic solvents.

Although aqueous binders still include natural materials, such as sugar, starches, gum arabic, and alginate, the synthetic water-soluble resins, such as the cellulosic ethers, often give superior results. Methyl and hydroxypropylmethyl cellulose, in particular, offer some unique benefits that are due to their thermal gelling properties. Polyvinyl alcohol, polyethylene glycol, acrylic acid polymers, and even acrylic latexes have been utilized. Besides the binders themselves, many other organic additives are often needed, such as dispersants, wetting agents, antisettling agents, and thixotropic agents. Organic solvent or oil-based media are still generally used for screening and other processes requiring lower-volatility solvents than water. Ethyl cellulose and polyvinyl butyral in pine oil are common media used for screening and many stamping processes. Alkyd resins are also widely used.

Thermoplastic media offer many advantages over liquid media, especially for screening multicolored patterns at high rates, because there is no need to dry between colors. These media generally contain high percentages of waxes, which often are fatty alcohol or polyethylene glycol based (Ref 58). Other waxes are used as modifiers, and the wax blend generally contains polymeric materials such as cellulosic ethers or acrylic resins. The acrylics are among the best-firing polymers, because they depolymerize to monomer at relatively low temperatures, and therefore do not require oxygen for clean removal.

The advent of silicone pad printing in the late 1960s required the development of new media to fully realize the shadowless release potential of the silicone pads. Key considerations were solvent selection to minimize pad absorption and saturation, as well as the affinity of the ink constituents for the silicone material. Thermoplastic media were particularly adaptable for use with silicone pads, and rapidly became the media of choice, especially for on-glaze ceramics and glass. The cohesive strength increase, which resulted from cooling on the pad surface, was very useful in attaining complete transfer. The thermoplastic media were often modified with rosin, plasticizers, and other materials to increase tack retention and thereby further aid in transfer. Thermoplastic media can also be modified for other application techniques, such as electrostatic spray, fluidized bed coating, and even electrophotographic processes such as xerography, for which constituents with the desirable triboelectric properties are used to develop the needed charge (Ref 59). However, these processes are used infrequently.

An alternative to thermoplastic media is the use of UV-curing binders (Ref 60). Recent formulation advances have improved the firing characteristics of these binders such that they can be employed satisfactorily for many multicolor applications. These binders are particularly useful for multicolor screening.

Organic Colors and Processes. When decorating on-glaze ceramics and glass, it is not always necessary to fire the ware. Organic polymers can be utilized as permanent binders, end-use conditions permitting. Thermoplastic lacquers or thermosetting organic enamels, including UV-curing enamels, can be formulated to provide reasonably durable decorations (Ref 61). These organic inks not only can be formulated to be compatible with the same processes that are used for glass enamels, but also are compatible with several processes that otherwise could not be used for glass decorating.

Two such processes are heat-shrinkable labeling and transfer printing with sublimable dyes. In the former process, decorations are printed onto heat-shrinkable plastic tubing or film, and the plastic is then shrunk around the glass, either in a heat tunnel or by using a hot air gun. This process has been used for decorating glass Christmas ornaments, wine bottles, and other items (Ref 62). Of even greater interest is transfer printing with sublimable dyes (Ref 63–65). This process requires the application of an organic coating onto the ware to act as a receptor for the dyes. The dyes are then transferred to the coating from a dye carrier film through the application of heat and pressure. In another variation, fine dot patterns can be printed onto the coated glass through the localized application of heat, high-intensity light, or laser. Extremely high-quality patterns are achievable by this means, and have excellent color reproduction.

The use of organic (cold) enamels also greatly expands the decal possibilities, and includes many processes that formerly were considered only appropriate for plastics, such as pressure-sensitive labels, hot transfer foils, and others.

The relative ease in formulating low-viscosity, nonparticulate inks also facilitates compatibility with such marking processes as ink jet printing (Ref 66), one of the few noncontact decorating processes. However, ink jet printing is not limited to organic colors. It has been employed with soluble inorganic stains and organometallics, but it has primarily been used to produce trademarks. The BCRA is currently researching the applicability of this technology for printing decorations on glass and ceramics. The key to success is the development of glass enamel inks that are compatible with this printing technique. The long-term goal is to have a noncontact multicolor method of decorating, with the potential of computer-generating the design directly on the ware. This process, if successful, will join the list of direct decorating methods.

REFERENCES

1. P. Freeman, *The Development of Ceramic Colours and Decorating Techniques in the Staffordshire Potteries*, Gladstone Pottery Museum, 1977
2. V.H. Remington, History and Evaluation of Screen Stencils, *Decorating in the Glass Industry*, A.G. Pincus and S.H. Chang, Ed., Books for Industry, 1977
3. V.H. Remington, The Art of Glass Decorating, *Glass Ind.*, May-Aug 1937
4. J. Zuckerman, Color: The 90's Controversy, *Ceram. Ind.*, Sept 1991, p 29–36
5. Color Turns to New Hues, Challenges as Decade Closes, *Ceram. Ind.*, Sept 1988, p 31–48
6. Beauty and the Beast: Colors Shine Without Lead, *Ceram. Ind.*, Feb 1989, p 21–22
7. F. Singer and S.S. Singer, Glazing and Decorating, Chapt. 9, *Industrial Ceramics*, Chapman and Hall, 1963, p 805–809
8. J.A. Stewart, Glass Decorating and Finishing, *Handbook of Glass Manufacture*, Vol II, Ashlee Publishing, 1984, p 843
9. G.L. Schneberger, *Understanding Paint and Painting Processes*, 3rd ed., Hitchcock Publishing, 1985
10. R.G. Sorice, Flat Glass Decorating: An Update, *Glass Int.*, Dec 1986, p 15–19
11. J. Zuckerman, Decorators Focus on Flat Glass, *Glass Ind.*, Aug 1990, p 10, 13
12. D.E. Hoover, Gaining Confidence with Machine Lining of Precious Metals, *Ceram. Bull.*, Vol 68 (No. 1), 1989
13. R.B. McKay, Glass Frosting with HCl: The HF Alternative, *Ceram. Ind.*, Dec 1987, p 38–39
14. New Decorating Styles Enrich Glass/Ceramics, *Glass Ind.*, Aug 1986, p 14–15
15. J. Wolf, Sand Blasted Decoration, *Decorating in the Glass Industry*, A.G. Pincus, *et al.*, Ed., Books for Industry, 1977
16. The Stylistic Trend Poses a Special Challenge for Decorators, *Glass Ind.*, Aug 1985, p 24–26
17. Technology: Meeting the Design Challenges, *Ceram. Ind.*, Sept 1987, p 39
18. Discovery in Decoration, New Riches for Glass, Ceramics, *Ceram. Ind.*, Sept 1984, p 31
19. J.J. Svec, Printing Four Colors…31 Pieces per Minute, *Ceram. Ind.*, June 1976, p 40–41
20. F. Singer and S.S. Singer, Glazing and Decorating, *Industrial Ceramics*, Chapman and Hall, 1963, p 809–811
21. F. Singer and S.S. Singer, Glazing and Decorating, *Industrial Ceramics*, Chapman and Hall, 1963, p 811–813
22. J. Tunicliffe, *et al.*, Screen Printing, *Vit. Enam.*, Vol 40, 1989, p 1–5
23. The Basics of Silk Screen Printing, *Sprechsaal*, Vol 115 (No. 8), 1982, p 682–691 (translated from German)
24. E. Messerschmitt, *The Ultimate Screen for Close Tolerance Screen Printing*, Academy of Screen Printing Technology, 1982
25. M.A. Coudray, Screen Printing Halftones, Part 1, *Screen Print.*, Oct 1989, p 124–129
26. M.A. Coudray, Screen Printing Halftones, Part 2, *Screen Print.*, Nov 1989, p 204–209
27. M.A. Coudray, Screen Printing Halftones, Part 3, *Screen Print.*, Dec 1989, p 110–128
28. M.A. Coudray, Screen Printing Halftones, Part 4, *Screen Print.*, Feb 1990, p 106–111
29. Buyers Guide, *Screen Print.*, 1990
30. J.E. Cooper, Underglaze Colors Improve Decorating Art, *Ceram. Ind.*, Nov 1982, p 42
31. A.F. Chapman, Silicone Pad Prints Multicolor Designs, *Ceram. Ind.*, July 1980,

p 36–37

32. W.D. Wells, Who in the World Wants Silicone Printing, *Brit. Ceram. Rev.*, No. 47, 1981, p 34–36

33. J.E. Cooper and W.G. Coulter, Silicone Pad Offset Printing, *Am. Ceram. Soc. Bull.*, 1981, p 516

34. W. Roberts, *et al.*, "Offset Printing: A Review of BCRA Developments 1977–1984," Research paper 729, British Ceramic Research Association, 1984

35. R.W. Pilcher, Decals for Ceramics, *Screen Print.*, Sept 1982, p 56–58, 102

36. J. Zuckerman, Yesterday—The Making of a Decal, *Ceram. Ind.*, Nov 1980, p 42–43

37. F.S. Child, Glass Decorating, *Glass Ind.*, Sept 1978, p 12

38. K.P. Heinbach, *et al.*, "Decalcomania," U.S. patent 4,322,467, Mar 1982

39. J. Zuckerman, Decals: Putting New Appeal on Ceramics, Glassware, *Ceram. Ind.*, Sept 1989, p 21–33

40. R.E. Johnson, *et al.*, "Decalcomania," U.S. patent 4,477,510, Oct 1984

41. W. Roberts and A.C. Airey, "Colour Reproduction," Special Publication 103, British Ceramic Research Association, 1981

42. R.E. Johnson, "Thermoplastic Ink Composition for Decorating Glass, Glass-Ceramic, and Ceramic Ware," U.S. patent 4,280,939, July 1981

43. C.E. Ford, *et al.*, "Article Decorating," U.S. patent 4,445,432, May 1984

44. R.E. Johnson, *et al.*, "Thermoplastic Inks for Decorating Purposes," U.S. patent 4,472,537, Sept 1984

45. Developments in Decorating Machines for Tableware, *Glass*, Feb 1985, p 58–59

46. Ceramic Research, *Ceram. Ind. J.*, Aug 1987, p 62, 65

47. S.D. Stookey, *et al.*, Full-Color Photosensitive Glass, *Ceram. Ind.*, June 1978, p 37–39

48. J.S. Nordyke, Glass Enamels and Colors, *Lead in the World of Ceramics*, American Ceramic Society, 1984, p 91–98

49. J.A. Stewart, Glass Decorating and Finishing, *Handbook of Glass Manufacture*, Vol II, Ashlee Publishing, 1984, p 850

50. D. Rhodes, *Clay and Glazes for the Potter*, Chilton Book Company, 1973

51. C.W. Parmlee, *Ceramic Glazes*, Cahners Books, 1973

52. R.A. Eppler, Glazes and Enamels, Chapt. 4, *Glass: Science and Technology*, Vol 1, Academic Press, 1983, p 301–338

53. R. Newcomb, Glazes and Decoration, Chapt. 8, *Ceramic Whitewares*, Pitman Publishing, 1947

54. A.N. Papzian, Gold Provides the Quality Image, *Glass Ind.*, Aug 1987, p 20–21

55. J.A. Stewart, Glass Decorating and Finishing, *Handbook of Glass Manufacture*, Vol II, Ashlee Publishing, 1984, p 838

56. V.H. Remington, Some Current Methods of Glass Decorating, *Decorating in the Glass Industry*, A.G. Pincus, *et al.*, Ed., Books for Industry, 1977, p 30

57. T.C. Patton, *Paint Flow and Pigment Dispersion*, John Wiley & Sons, 1979

58. D.R. Bush, *et al.*, "Composition for and Method of Applying Ceramic Color," U.S. patent 3,089,782, May 1963

59. C.D. Oughton, Decoration of Glass and Ceramic Articles by Xerography, Chapt. 25, *Decorating in the Glass Industry*, A.G. Pincus, *et al.*, Ed., Books for Industry, 1977, p 123–126

60. R. Sayer, UV Curing Update, *Glass Ind.*, Sept 1981, p 20–28

61. M.S. Kaye, A Look at the Advantages of Organic Coating for the Glass Decorator, *Am. Glass Rev.*, Nov 1977, p 12

62. C. Lodge, Creative Solutions to Tough Decorating, *Plast. World*, June 1990, p 48–51

63. *Glass Ind.*, Jan 1988, p 23

64. Reprint Process: Enhancing at 200 °C, *Interceram*, No. 6, 1986, p 4–5

65. K.H. Kramer, "Process for Imprinting Ceramic Articles and Apparatus," U.S. patent 4,943,684, July 1990

66. "The Basics of Better Product Marking and Coding," Video Jet Systems International Inc.

SECTION 7

Joining

Chairman: Ronald E. Loehman, Sandia National Laboratories

Introduction

MANY APPLICATIONS of ceramics and glasses require them to be joined to each other or to other materials such as metals. Frequently these joints escape the notice of the casual observer because, when they are made well, there is a smooth transition from one material to the other with no obvious demarcation of the joint. Yet the technologies that are used to join ceramics and glass are truly critical, because without them many uses of ceramics would be impossible or uneconomical. Electric light bulbs are one widespread example where glass/metal seals are essential to the function of the device. Other examples include cathode ray tubes, silicon nitride automotive turbocharger rotors, other automotive components such as valve guides and cam followers, vacuum devices such as laser and microwave tubes, feedthroughs, and connectors, hermetic packaging for hybrid microcircuits, lithium battery headers, and connectors for optical fibers.

The objective of this Section is to present a general introduction to ceramic joining that will be useful to a broad spectrum of readers. The presentation starts with an overview of the different applications of ceramic joining. This introductory article will give readers who are unfamiliar with ceramic joining a better idea of how the technology is used and its importance. The next article, "Wetting, Surface Energies, Adhesion, and Interface Reaction Thermodynamics," emphasizes the fundamental physical and chemical processes that form the basis for ceramic joining. The intent is to give the reader the necessary conceptual framework to understand the different joining processes. Thus, the choice of a particular firing atmosphere, surface pretreatment, or joining composition can be understood mechanistically, and the features common to different joining methods can be better appreciated.

The following articles are more detailed presentations of most of the joining processes used in industry. Joining of glasses, glass-ceramics, oxide ceramics, and non-oxide ceramics all are discussed by noted experts in their fields. Metallizing, which is important to many methods for joining ceramics to metals, is also discussed. The article on "Techniques of Seal Design" should be particularly helpful to potential users of ceramic joining because it discusses the design criteria for joints in terms of fundamental materials properties.

Together these articles form as comprehensive a description of modern joining technology as one is likely to find. They should be useful to a variety of readers from those who need only the most general of introductions to those who desire details on how to join a particular ceramic. Those of us who contributed to this Section hope that it will result in a wider appreciation of ceramic joining and in expanded areas of application.

Types of Ceramic Joining and Their Uses

Randall D. Watkins, Sandia National Laboratories

CERAMIC JOINING technologies can be broadly classified by material types. These technologies include traditional glass/metal sealing, glass-ceramic/metal joining, and ceramic/metal joining, ceramic/ceramic joining, and the more advanced joining of non-oxide ceramics. As glass and ceramic materials have evolved throughout the years, so also have the potential applications for these materials grown and evolved in many areas of electronics and high-temperature, high-strength structural design. The need for joining technologies has arisen as a result of the more frequent use of ceramics with other materials; in addition, complex ceramic shapes are often required that can be made only by joining together less complex shapes. Because the requirements of a ceramic joint can vary from application to application (that is, hermeticity, ductile strain relief, corrosion resistance, high-temperature strength), the ability to tailor the physical and mechanical properties of the seal interface is essential. The successful development of a wide variety of ceramic joining processes has had a fundamental impact on the field of engineered materials, often bridging the gap between conceptual design and practical use.

In addition to classification by material type, ceramic joining can also be classified by the bonding mechanism or process. Within each material classification introduced above, there exist sub-classes based on the type of joint. These types of joints include mechanical fastening, adhesive joining, and chemical bonding. Mechanical and adhesive joining of ceramics have useful application in low-temperature, low-stress environments. However, due to the frequent reliability and performance limitations encountered with these joining techniques, only chemical bonding processes, developed for high-reliability applications, will be addressed in this article.

Glasses and ceramics can be chemically bonded to metals, to themselves, or to other ceramics by a number of techniques. Below is a historical overview of the development of various existing chemical joining processes. Applications are described to illustrate the technological and practical significance of each joining technology. Due to the unique processing requirements as well as the technical significance of non-oxide ceramics relative to traditional oxide ceramics, these joining processes are discussed in greater detail in separate articles in this Section.

Glass/Metal Seals

Traditional glass/metal sealing is a fusion bonding technique. The glass is heated to a molten state and bonding occurs through wetting and chemical reactions between the molten glass and suitable metals or alloys. Hermetic (vacuum-tight) glass/metal seals have found use in the electronics industry for nearly a century. Edison used glass/metal seals in the incandescent lamp in the late 1800s, and patents for various types of glass seals were awarded around the turn of the century. As the applications became widespread around the 1950s and 1960s, attention was devoted toward developing a more fundamental understanding of the science of glass/metal sealing and sealing technology. Numerous classical papers were published on the subject between then and the late 1970s. Despite the declining scientific attention given to glass seals in recent years, due primarily to the limitations of glass in high-temperature, high-strength applications, these seals remain of interest to the glass technologist and component designer. Glass/metal seals are easy to manufacture in batch or continuous processes, are relatively inexpensive, and can provide long-term reliability under moderate thermal and mechanical loads. Billions of glass/metal seals are produced worldwide each year. Glass seals are fabricated for home use such as the common light bulb, for biomedical applications such as long-life lithium power supplies for pacemakers and insulin pumps, and seals are used in numerous military and weapon applications requiring long shelf life.

Much of the work reported in the literature prior to the mid-1960s is reviewed in some detail by Kohl (Ref 1). Kohl describes the technology of glass/metal sealing as developed and performed by glass blowing technologists, classifying glass seals by:

- *Fabrication technique*, including flame seals, induction seals, frit seals, and tape seals
- *Seal geometry*, such as bead seals, tube seals, disk seals, and Housekeeper seals
- *Relative coefficients of thermal expansion* between the glass and the metal, including "matched" and "unmatched" (that is, coaxial compression) seals
- *Materials*, describing terms such as hardglass, soft-glass, or quartz seals in reference to the type of glass used, and Kovar (iron-nickel-cobalt alloy), copper, or ironnickel alloy seals in reference to the type of metal used in the seal

Much of the technology reviewed by Kohl, although developed by glass blowers for one-piece-at-a-time, hand-made articles, is still used today by mass production sealing companies. Glass-to-metal sealing is a somewhat "forgiving" technology, as evidenced by the wide variety of processes utilized in industry today. Although the level of fundamental understanding, or the science, of glass sealing has matured steadily over the past 30 years, it is recognized that the techniques required to achieve near optimum seals can vary considerably and yet remain effective; companies often employ proprietary processing to preserve unique capabilities. A wide variety of commercially available glasses have been developed for sealing to numerous metals and alloys and semiconductors including silicon, tungsten, molybdenum, platinum, nickel-iron low-expansion alloys containing 42%, 46%, and 52% Ni, 400 series stainless steels, and cast iron. See Ref 2 and the article "Glass/Metal and Glass-Ceramic/Metal Seals" in this Volume for additional information.

Unique bonding processes have been developed for applications where traditional glass fusion-bonding has not been appropriate. An example is field-assisted bonding (Ref 3, 4), described as a "solid-state ceramic braze." Under the influence of an electric field, strong hermetic bonds between glass wafers and metals or semiconductors can be produced at temperatures well below the softening point of the glass. This technique is limited to the bonding of flat, wafer-like geometries and requires carefully prepared polished mating surfaces. It is for this reason that field-assisted bonding has found wide use in the silicon semiconductor packaging industry. Recently, field-assisted bonding has been used to join single-crystal quartz at 400 °C (750 °F) well below the alpha-beta quartz transformation temperature (Ref 5).

Glass-Ceramic/Metal Sealing

Glass-ceramics offer a number of engineering advantages over glasses and traditional polycrystalline ceramics (Ref 6). Primarily, glass-ceramics can be easily shaped and formed while still in the glassy, molten state, and following controlled nucleation and crystallization, they exhibit properties closely approaching those of ceramics. Thermal properties of glass-ceramics can be tailored over a wide range, permitting close expansion match to numerous alloys. As a result, for hermetic seal applications, the advantages of using glass-ceramics are clear. Ideally, interfacial reactions and bonding between the glass precursor and the metal occur during the initial high-temperature stage of the sealing process just as in glass/metal sealing (see the article "Wetting, Surface Energies, Adhesion, and Interface Reaction Thermodynamics" in this Section). Further heat treatment allows subsequent crystallization to proceed, yielding a glass/polycrystalline composite microstructure with greater strength and toughness, and often more suitable thermal expansion compatibility with the metal.

Significant attention has been directed to the development of lithium-silicate based glass-ceramics for hermetic sealing to high-strength superalloys such as Inconel 718 (Ref 7-9). These seals are used in electrical feed-throughs for pyrotechnic valve actuators requiring structural integrity. Because of their high strength and toughness, these glass-ceramic sealed components are able to withstand very high internal gas pressures.

Care needs to be taken in the preparation and use of glass-ceramic/metal seals. The bulk properties of the glass-ceramic can often change depending on the processing environment. For instance, glass-metal reactions are observed to disrupt the crystallization behavior in the interfacial region, and can lead to high residual stresses in the nominally "matched" expansion seal (Ref 10, 11). Despite this limitation, it is anticipated that glass-ceramic/metal seals will continue to gain significance in applications where the ease of glass sealing can be utilized and the mechanical, electrical, and/or thermal properties of ceramics are required. Other uses of glass-ceramic/metal seals include high-voltage insulators, which require high dielectric strength, and nuclear reactor feed-through seals

which require mechanical integrity at high temperatures and pressures. See the article "Glass/Metal and Glass-Ceramic/Metal Seals" in this Section for additional information.

Ceramic/Metal Joining

Ceramic seal components offer a number of advantages over the use of glass seals. Polycrystalline ceramics typically are mechanically stronger and less flaw-sensitive than glasses and are more resistant to damage from exposure to high temperatures and thermal shock. Ceramics also have superior dielectric properties and often are more corrosion resistant. The primary disadvantage of ceramic joining is the complexity of processing relative to glass sealing, as reviewed below. Both liquid-phase and solid-state oxide ceramic joining technologies have been available for many years (Ref 1, 3, and 12), but have received increased attention in recent years due to the need for seals that can survive in severe mechanical and thermal environments. Presently, Al_2O_3-to-Kovar or copper seals, commonplace in the electronics industry, are used for insulators, capacitors, discharge tubes, x-ray tubes, and heat sinks. Alumina joints to stainless steel, titanium and titanium alloys, and precious metals such as platinum or palladium are used in biomedical implants. See the article "Ceramic/Metal Seals" in this Section for additional information.

Liquid-Phase Joining. Joining ceramics to metals typically involves the use of some type of metal brazing alloy, often in conjunction with a metallization film. Metallization is used as an intermediate process to improve the wetting and adherence of the braze alloy to the base ceramic, and techniques described below for brazing can often be applied to metallizing (see also the article "Metallizing' in this Section). The primary classifications for brazes are:

- Refractory metal brazes, often synonymous with sintered metal-powder brazes
- Precious metal brazes
- Active metal brazes

Refractory metal brazing is the basis for the well-known "moly-manganese" (Mo-Mn) process reviewed in Ref 1 and the article "Joining Oxide Ceramics" in this Section. Pure metal or oxides of molybdenum and manganese are mixed in a slurry and deposited on a ceramic substrate, typically by screen printing, spraying, or transfer tape. The coating is fired in a H_2 atmosphere to promote controlled oxidation-reduction of the mixture and reactions between the braze and the ceramic (glass phase/metal migration in the case of Al_2O_3). Although the Mo-Mn process is the most widely used commercial brazing process, tungsten is also a fairly common refractory metal braze. Molybdenum-manganese brazes are used primarily as a metallization layer to promote improved wetting

and bonding of other suitable braze filler metals.

Although molten metals typically do not wet ceramics well in the absence of a metallization layer, precious metal braze joints have been produced between alloys of platinum, palladium, gold, and silver, and a number of oxide ceramics including Al_2O_3 and MgO (Ref 13). These metals are often alloyed with base metals such as copper or nickel. The improved wetting of these brazes has been attributed to the formation of precious-metal oxides when the metals are heated in air above their melting points.

Active metal brazing is an area of ceramic joining receiving recent scientific attention, in efforts to simplify processing while preserving or enhancing bond integrity. Active metal brazing is a technique where wetting and bonding of the base braze alloy is improved due to the presence of a small amount of active metal such as titanium or zirconium. Although early active metal processes involved the deposition of the active species as a metallization film (typically in salt or hydride form), recent advances in the processing and properties of ceramic/metal joints have been made by incorporating optimum amounts of the active metal directly into the braze alloy (see the article "Joining Oxide Ceramics" in this Section). This single-step process has obvious advantages over the multiple-step Mo-Mn process.

Metal brazes are often useful for joining ceramics to metals because the braze may provide ductile strain relief when the thermal expansion mismatch between the ceramic and metal is large. Glass bonding of ceramics to metals can be performed in cases where the coefficient of thermal expansion of the ceramic is closely matched to the metal, or where the geometry is such that the formation of large residual tensile forces in the glass and ceramic is prevented. Alumina has been joined to Kovar, stainless steel, and molybdenum using a MnO_2-Al_2O_3-SiO_2 glass bond (Ref 1), and a glass-bonded Al_2O_3/niobium envelope is being developed for sodium vapor lamps (Ref 14). A series of low-temperature sealing glasses have been developed for packaging ceramic and metal microelectronic devices (Ref 15).

Solid-State Joining. In the absence of a liquid phase, ceramic/metal joining can only be accomplished with temperature, pressure, and time sufficient to promote mobility and interdiffusion of reacting species, producing adequate interfacial contact between the ceramic and the metal. The nomenclature used to describe solid-state joining varies and includes reaction bonding, solid-state welding, diffusion bonding, and pressure bonding. These techniques are distinguished only by direct bonding (that is, where reaction and bonding between the ceramic and metal members are thermodynamically favorable) and indirect bonding (that is, where interlayer materials are required to improve reactivity between the ceramic and metal, or sometimes to introduce strain relief).

Direct bonding of Al_2O_3 to various metals including aluminum, copper, tantalum, iron, and niobium is reviewed by Nicholas and Mortimer in Ref 16. Pressures from 1 to 100 MPa (0.15 to 15 ksi), times from 1 to 4 h, and temperatures from 50 to 90% of the melting point of the metal were used. Results suggest that bond integrity is controlled by sufficient reaction, and that prolonged heating can cause degradation of the interface strength.

Indirect solid-state bonding of metal to ceramics (again primarily Al_2O_3) has been successful using an aluminum foil interlayer (Ref 12). Low melting aluminum is useful only when the temperature requirements of the component are not too severe. Besides foils, interlayer materials deposited using techniques similar to metallization films can be used for solid-state bonding. In fact, a Mo-Mn/Ti metallization layer has been used to bond molybdenum to Al_2O_3 (Ref 12). A third technique for indirect bonding involves application of mixtures of ceramic and metal powders in layer form; an expansion gradient results, minimizing residual stresses in the joint.

Ceramic/Ceramic Joining

Ceramic/ceramic joining is necessary in the fabrication of complex shapes from simpler, more producible subassemblies, and can be useful in the repair of components that have fractured during service. A common technique for ceramic/ceramic joining is solid state pressure/diffusion bonding. Processes have been developed for diffusion bonding, or welding, of numerous polycrystalline and single-crystal oxide ceramics including Al_2O_3, MgO, NiO, CaO, and ZrO_2 (Ref 3). Common in all techniques is the application of a layer of oxide powder between the two parts; the composition of the bonding medium depends on the materials being bonded. The assembly is then placed under pressure and brought to an appropriate temperature for bonding. Ceramics can also be joined by solid-state brazing. Alumina has been bonded to Al_2O_3 using aluminum, silver, copper, or nickel metal interlayers at temperatures slightly below the melting point of each metal (Ref 17).

Ceramics can be joined using liquid-phase active-metal techniques as described above for ceramic-to-metal joining. The high-temperature mechanical properties of the ceramic are typically compromised; as in any composite system, the weak link (in this case the metal layer) determines the bulk behavior. Finally, ceramics can be joined using a glass or glass-ceramic interlayer. Glasses wet and bond well to most ceramics. However, due to the brittle nature of glass, an adequate thermal expansion and elastic moduli match to the ceramic is critical. See the articles "Joining Oxide Ceramics" and "Joining Non-Oxide Ceramics" in this Section for additional information.

Non-Oxide Ceramic Joining

The use of non-oxide ceramics is being considered more and more due to their superior high-temperature mechanical properties, including fracture toughness and corrosion and wear resistance. Silicon carbide (SiC) and silicon nitride (Si_3N_4) ceramics are being developed for engine components such as turbocharger rotors (Ref 18), exhaust valves for marine diesel engines (Ref 19), and gas turbine stators and rotor blade rings and hubs (Ref 20). Such applications are discussed in the section in this Handbook entitled "Structural Applications for Technical, Engineering, and Advanced Ceramics." Aluminum nitride (AlN) is being considered as a replacement for Al_2O_3 as a substrate material in microelectronic devices due to its higher thermal conductivity and electrical resistivity, and better expansion match to silicon (Ref 21). In each of these cases, effective joining remains a primary requirement.

Many of the techniques used to join oxide ceramics can be used to join non-oxide ceramics. However, it is important to distinguish between the two classes of materials; due to differences in the chemical nature of non-oxide ceramics, the interfacial ceramic/metal reactions, bonding mechanisms, and resulting joint properties are often significantly different (see the article "Wetting, Surface Energies, Adhesion, and Interface Reaction Thermodynamics" in this Volume).

Liquid-phase joining techniques typically include some type of active metal brazing or glass sealing. Brazing of Si_3N_4 to molybdenum, titanium, and steel with Ag-Cu, Au-Ni, and Au-Ni-Pd filler metals has been accomplished by first vapor depositing a layer of titanium on the Si_3N_4 (Ref 22). Direct brazing SiC and Si_3N_4 was performed with titanium-activated copper and silver metal fillers (Ref 23, 24). A novel technique for joining Si_3N_4 involves the synthesis of a glass bonding composition that closely matches that of the grain-boundary phase in the ceramic (Ref 25). Wetting behavior, chemical reactions, and resulting high-strength joints have been characterized.

Numerous solid-state joining techniques are being developed for non-oxide ceramics. Indirect bonding, where a layer or layers of material are applied to the interface, has found widespread application. The so-called "composite interlayer bonding" (Ref 26), where a mixture of the metal and ceramic powder is applied to reduce the thermal expansion mismatch and thus the resulting stress at the interface, has been demonstrated in joints between molybdenum and AlN, TiN, BN, and Si_3N_4. Silicon nitride joints have been produced using nickel/niobium interdiffusing metal layers at temperatures below the melting temperatures of the metals (Ref 27), and in addition, ZrO_2 was shown to provide a suitable ceramic bonding layer for joining Si_3N_4 (Ref 28). Solid-state reaction bonding has been investigated for the joining of reaction-sintered Si_3N_4 (Ref 20), and involves the controlled application of a silicon-metal slip.

Direct solid-state bonding of non-oxide ceramics to themselves or metals has been demonstrated at elevated temperatures and pressures. Silicon nitride has been diffusion bonded to aluminum alloys A1050 and AC-8A (Ref 29) and to Nimonic 80A (a high creep-resistant nickel-base superalloy typically used in turbine engines) (Ref 19). In addition, direct bonding of hot pressed Si_3N_4 to reaction-sintered Si_3N_4 has been demonstrated in an effort to develop the "duodensity" turbine rotor concept (Ref 20). See the article "Joining Non-Oxide Ceramics" in this Section for additional information.

REFERENCES

1. W.H. Kohl, *Handbook of Materials and Techniques for Vacuum Devices*, Reinhold Publishing, 1967
2. J. Francel, Sealing Glasses, *Advances in Ceramics*, Vol 18, 1986
3. R.W. Rice, Joining of Ceramics, *Advances in Joining Technology*, J.J. Burke, Ed., Brook Hill Publishing, 1975, p 69–111
4. G. Wallis and D.I. Pomeranz, Field-Assisted Glass-Metal Sealing, *J. Appl. Phys.*, Vol 40 (No. 10), 1969, p 3946–3949
5. R.D. Watkins, *et al.*, Field-Assisted Bonding of Single Crystal Quartz, Proceedings of the 12th Quartz Devices Conference, Electronic Industries Association, 1989, p 6–17
6. P.W. McMillan, *Glass Ceramics*, 2nd ed., Academic Press, 1979
7. M.P. Borom, A.M. Turkalo, and R.H. Doremus, Strength and Microstructure in Lithium Disilicate Glass-Ceramics, *J. Am. Ceram. Soc.*, Vol 58 (No. 9–10), 1975, p 385–391
8. H.L. McCollister and S.T. Reed, "Glass-Ceramic Seals to Inconel," U.S. patent 4,414,282, 1983
9. T.J. Headley and R.E. Loehman, Crystallization of a Glass-Ceramic by Epitaxial Growth, *J. Am. Ceram. Soc.*, Vol 67 (No. 9), 1984, p 620–625
10. R.D. Watkins and R.E. Loehman, Interfacial Reactions Between a Complex Lithium Silicate Glass-Ceramic and Inconel 718, *Adv. Ceram. Mater.*, Vol 1 (No. 1), 1986, p 77–80
11. S.C. Kunz and R.E. Loehman, Thermal Expansion Mismatch Produced by Interfacial Reactions in Glass-Ceramic to Metal Seals, *Adv. Ceram. Mater.*, Vol 2 (No. 1), 1987, p 69–73
12. M.M. Schwartz, *Ceramic Joining*, ASM International, 1990

13. H.J. DeBruin, A.F. Moodie, and C.E. Warble, Ceramic-Metal Reaction Welding, *J. Mater. Sci.*, Vol 7, 1972, p 908–918

14. R.E. Loehman and A.P. Tomsia, Joining of Ceramics, *Am. Ceram. Soc. Bull.*, Vol 67 (No. 2), 1988, p 375–380

15. R.E. Hogan, Solder Glasses, *Chem. Tech.*, Jan 1971, p 41–43

16. M.G. Nicholas and D.A. Mortimer, Ceramic/Metal Joining for Structural Applications, *Mater. Sci. Technol.*, Vol 1 (No. 9), 1985, p 657–665

17. M.G. Nicholas, Diffusion Bonding Ceramics with Ductile Metal Interlayers, *Brazing Soldering*, Vol 10, 1986, p 11–13

18. K. Suganuma and Y. Miyamoto, Joining of Ceramics and Metals, *Ann. Rev. Mater. Sci.*, Vol 18, 1988, p 47–73

19. T. Yamada, *et al.*, Diffusion Bonding SiC or Si_3N_4 to Nimonic 80A, *High Temp. Technol.*, Vol 5 (No. 4), 1987, p 193–200

20. M.U. Goodyear and A. Ezis, Joining of Turbine Engine Ceramics, *Advances in Joining Technology*, J.J. Burke, Brook Hill Publishing, 1975, p 113–153

21. R.K. Brow, *et al.*, Interface Reactions During Brazing of AlN, *Adv. Ceram.*, Vol 26, 1989, p 189–196

22. M.L. Santella, Brazing of Ti-Vapor-Coated Silicon Nitride, *Adv. Ceram. Mater.*, Vol 3 (No. 5), 1988, p 457–462

23. M. Naka, *et al.*, Joining of Silicon Nitride to Metals or Alloys Using Amorphous Cu-Ti Filler Metal, *Trans. JWRI*, Vol 14 (No. 2), 1985, p 85–91

24. T. Yano, *et al.*, High-Resolution Electron Microscopy of a SiC/SiC Joint Brazed by a Ag-Cu-Ti Alloy, *J. Mater. Sci.*, Vol 23, 1988, p 3362–3366

25. M.L. Mecartney, R. Sinclair, and R.E. Loehman, Silicon Nitride Joining, *J. Am. Ceram. Soc.*, Vol 68 (No. 9), p 472–478

26. K. Suganuma, *et al.*, New Method for Solid-State Bonding Between Ceramics and Metals, *Comm. Am. Ceram. Soc.*, July 1983, p C117-C118

27. Y. Iino and N. Taguchi, Interdiffusing Metals Layer Technique of Ceramic-Metal Bonding, *J. Mater. Sci.*, Vol 23, 1988, p 981–982

28. P.F. Becher and S.A. Halen, Solid-State Brazing of Si_3N_4, *Am. Ceram. Soc. Bull.*, Vol 58, 1979, p 582–586

29. Y. Arata, *et al.*, Solid State Reaction Bonding of Non-Oxide Ceramics to Metals, *Trans. JWRI*, Vol 14 (No. 2), 1985, p 79–84

Wetting, Surface Energies, Adhesion, and Interface Reaction Thermodynamics

Joseph A. Pask and Antoni P. Tomsia, Lawrence Berkeley Laboratory and Department of Materials Science and Mineral Engineering, University of California

THE ACHIEVEMENT of tight, strong, and reliable seals of any two dissimilar materials depends on both chemical and physical (or application) factors. The chemical factors address those design concerns that are related primarily to the chemical bonding at all interfaces and the development of a favorable microstructure in the interfacial zone. Physical factors deal with the determination of seal strength, which is influenced by the strain and stress patterns that are controlled by the coefficients of thermal expansion (CTE) of the various phases in the interfacial zone.

Both chemical and physical factors play significant roles in developing acceptable assemblies with adequate strength and overall properties. This article, however, deals primarily with the chemical factors and chemical bonding in which one of the two materials is generally a ceramic (where a ceramic is considered any non-metallic and inorganic composition, whether crystalline or glassy). Considerable research on chemical factors is in progress. Although most of the research is qualitative or semi-quantitative in nature, it is providing a background of chemical data that is contributing to a basic understanding of the principles of chemical bonding.

The physical factors of chemical bonds are also a subject of considerable basic research. The physical approach is more complex and time-consuming but is necessary for the understanding and development of a basis for the interpretation of the chemical results (especially when overall strength of the assembly is a concern). Physical factors also are important in the analysis of assembly interfaces without a chemical bond (for example, some composites). Furthermore, in some designs adequate strength of an assembly is achieved by a design in which the joint is satisfactorily strengthened by the development of a compressive stress on the ceramic or roughening the interface in lieu of a chem-ical bond or developing a favorable interfacial microstructure by control of solid-state chemical reactions.

Basic Factors in Bonding

The objective of this article is to define and understand the principles and fundamentals that play a role in bonding dissimilar materials. In most cases one of the bonding phases is a liquid at the bonding temperature. Wetting and spreading (discussed in a later section) are thus important phenomena in that they play a critical role in distribution of the liquid and the formation of an intimate atomic interference without the application of pressure. These phenomena are best understood on the basis of surface and interface free energies (discussed below). The driving force for wetting is the reduction of the surface free energy of the solid by the liquid ($\gamma_{sv} - \gamma_{sl}$). Spreading requires the additional contribution to the driving force of the free energy of the interfacial reaction.

A chemical bond is formed at an interface when a balance of bond energies and a continuous electronic structure are present across the interface for any two dissimilar phases. This structure occurs when a thermodynamically stable chemical equilibrium exists at the interface and is achieved by chemical reactions at the interface. Generally, equilibrium compositions (which can be determined if an equilibrium phase diagram of the two phases being bonded is available) at the interface are attained at the reaction temperature very rapidly. Continuation of heating maintains equilibrium at the interface, but the reaction products increase in amount and increase the thickness of the interfacial zone, whose microstructure may also change. Two phases with a chemical bond are referred to as compatible phases.

Usually, there is an optimum thickness that maximizes the strength of the assembly. However, it is difficult to give detailed physical models and mathematical analyses of adhesive strength because of the limited data and the complexities in the developing microstructures that cause non-uniform strain and stress patterns. A chemical bond is represented by an electronic structure and a balance of bond energies across the interface whether the bonding is ionic, covalent, or metallic. These factors influence the bond microstructure. For example, when the two reacting phases (ceramic/ceramic or metal/metal) have no changes in valence, then the reactions are formation of a solid-solution alloy or compound by interdiffusion. In this case the solution rate at the interface would have to be faster than diffusion from the interface to maintain saturation and stable chemical equilibrium at the interface. When the two reacting phases (ceramic/metal) have changes in valence, the reactions are referred to as reduction-oxidation (redox). For example, active metals such as titanium and zirconium have high oxidation potentials and can reduce cations with lower oxidation potentials. The amount of preoxidation or the extent of the redox reactions also determines the microstructure.

At the present stage it is important to develop a fundamental understanding of the problems and difficulties arising in making joints and seals, and their sources. It should be recognized that a great deal of experimental and developmental work is being done not only to improve the strength of the interfacial bonding (which practically always occurs with the onset of reactions) but also, knowingly or unknowingly, to improve the strength of the assemblies through modifications of the microstructures in the interfacial zone. Furthermore, a fundamental understanding of the mechanisms of the solid-state chemical reactions provides directions and clues for analysis and further work.

Surface and Interfacial Free Energies. Surfaces and interfaces are descriptive terms. A surface is defined as an interface between a condensed phase and a vacuum or gas, and an interface generally refers to a two-dimensional contact between two condensed phases. A surface free energy denoted by γ is best described thermodynamically as the excess free energy per unit area at the surface over an equivalent layer in the bulk and is thus always positive (Ref 1). The cgs units used for surface free energy are ergs per square centimeter (ergs/cm^2) and the corresponding SI units are joules per square meter (J/m^2). The term surface tension was first applied to liquids as a measure of the work required to increase a unit area of the surface. The corresponding units are dynes per centimeter (dyn/cm) in cgs and newtons per meter (N/m) in SI units; dynes/cm and mN/m have the same numerical value.

When a tensile stress is applied to pull a specimen apart without deformation, the energy expended is the work of cohesion and is retained in the two newly formed surfaces. This energy is stored in the two surfaces because the atoms are not fully coordinated (as they are in an equivalent layer of atoms in the bulk). The tensile stress that provided the energy to break the bonds and transformed the surfaces to a higher energy state is the cohesive strength, which is expressed as (Ref 2)

$$W_c = 2\gamma_{s*} \qquad \text{(Eq 1)}$$

where γ_{s*} is the maximum or ideal surface free energy for pristine surfaces. It is important to remember that this relationship is based on the assumption that no bulk distortion occurred and that two pristine surfaces were created with no changes other than the breakage of bonds. These surfaces would have the maximum value of γ_{s*} surface energy equivalent to half the tensile breaking strength, but this condition is essentially impossible to achieve experimentally. The value of γ_{s*}, if it could be measured after formation, would actually always be less because a system is thermodynamically always progressing toward the lowest free energy state. This lowering would be due to physical relaxation of the atomic structure, adsorption of any gases that may be present in the ambient atmosphere, and to any distortion during measurement. Nevertheless, the concept is important because it contributes to our overall understanding which is adequate from a practical or technological viewpoint.

Some generalities can be made as to the relative magnitudes of surface free energies for different types of materials (Ref 3). Metals have the highest values because there is no screening of surface atoms. Ceramics or nonmetallic inorganic materials have lower surface free energies, because there is some screening of the cations by the anions at the

surface. Liquid metals generally have lower surface energies than the corresponding crystalline metals because there is some atomic structural relaxation at the surface due to the absence of long-range order. Glasses likewise have lower values than their crystalline counterparts for the same reason. Organic materials generally have the lowest surface free energy values because they are covalently bonded and their bonds are weaker than those in ceramics.

A similar analysis could also be used for an interface between any two dissimilar materials assuming that failure occurs exactly at the interface. For a crystalline solid or a glass, for example, the strength of the interface would now be equivalent to work of adhesion (Ref 2) and would be expressed as:

$$W_a = \gamma_{s*} + \gamma_{l*} - \gamma_{s*l*} \qquad \text{(Eq 2)}$$

where γ_{s*}, γ_{l*}, and γ_{s*l*} refer to the surface energies of the solid, liquid (or glass), and interfacial energy between the solid and liquid, respectively. The starred values again refer to pristine surfaces under the same assumptions of ideality as above. This equation for W_a is identified as *Dupre's equation*. This concept appears to be simple, but in reality it is complex. One additional factor in comparing work of adhesion with work of cohesion is the stress distribution in the interfacial zone due to differences in the coefficients of thermal expansion (CTE) of the two bonded phases. The formula as written assumes no stresses from differences in expansion. If fracture occurs close to the interface in one of the phases, then that phase is weaker than the interface or it is more highly stressed (see the article "Techniques of Seal Design" in this Volume). In this case, the interface is generally interpreted as being chemically bonded.

Adhesion and adherence are used interchangeably to define the physical strength of an interface between two regions of a material system (Ref 4). The term adhesion, however, is used more extensively with applications of organic polymers. Both terms generally refer to the molecular attraction exerted between the surfaces of bodies in intimate contact. When joining inorganic materials, however, the attraction can best be understood in terms of chemical bonding, as when a balance of bond energies exists across the interface between the two phases (Ref 5).

When the bond energies are balanced, a continuous electronic structure occurs across the interface and in the interfacial zone. If an interface forms in which the attraction is due to induced electronic dipoles, then van der Waals bonding is present. As a first approximation, ionic/covalent bonding can be distinguished from van der Waals by initiating fracture in the interfacial zone. If the fracture occurs in the interfacial zone near the interface, chemical bonding is present across the interface. Experimentally, with chemical bonding at the interface, fracture has not been

observed at the interface with both newly formed surfaces cleanly exposed. On the other hand, when fracture exposes both surfaces at the interface, then van der Waals bonding exists at the interface. This analysis assumes that the flaw populations in the bulk and the interface are similar and that the interface is unstressed.

Chemical bonding at the interface is realized when chemical equilibrium exists at the interface. This interfacial equilibrium is defined by equality in the chemical potentials of various species at the interface. Any system is driven to its lowest free energy state, which is defined as stable equilibrium. Systems away from equilibrium tend to undergo reactions at the interface. Because reactions, in principle, can occur between adjoining surfaces merely by electron transfer without chemical diffusion, an initial reaction may occur rapidly. The interface is driven toward equilibrium but may not be in its lowest energy state. With subsequent diffusion into the adjoining bulk phases the system as a whole tends toward equilibrium and reactions will continue at the interface to maintain equilibrium compositions until the system is in its lowest energy state or it is quenched by lowering its temperature.

Wetting, Non-Wetting, and Spreading

The first step in sealing requires an intimate interface so that the two phases are separated by only atomic distances with either chemical or van der Waals bonding between them. It is possible to obtain such interfaces by contacting two clean polished surfaces, applying a pressure, and heating the interface to a high enough temperature to provide sufficient activation energy for the chemical bonds to form. At the sealing temperatures, however, one of the phases frequently will form a liquid either by melting or a reaction. The advantage of a liquid is that an intimate interface can be achieved at the interface without the application of pressure. Wetting and spreading then become factors to be considered.

Wetting and spreading are used interchangeably to describe the liquid distribution on solids. They should, however, be distinguished on the basis of thermodynamics. Experimental measurements are most conveniently based on the configuration of a steady-state drop of liquid on a flat surface, which is referred to as a *sessile drop*. If the liquid drop is small enough so that the gravitational force is small relative to its surface tension, the drop will attempt to assume the shape of a sphere which represents its smallest area and thus the lowest total surface free energy. Cross sections formed by a plane through the vertical diameter of the drop and perpendicular to the substrate or base are shown in Fig 1. The top cross section shows a sessile drop formed by the truncating plane

Sessile Drop Configurations

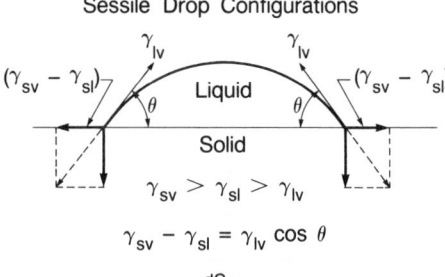

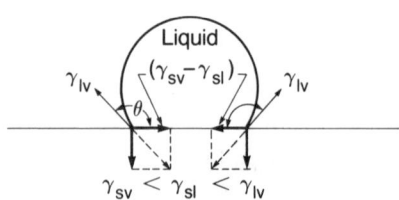

Fig 1 Sessile drop configurations: (top) wetting, and (bottom) non-wetting. See text for significance of equations.

passing through the upper hemisphere. The contact angle is measured inside the drop and is acute. The angle is formed within the vertical plane by the interface and a tangent to the drop's surface at its point of intersection with the substrate surface. The bottom cross-section shows a sessile drop formed by the truncating plane passing through the lower hemisphere. The contact angle in this case is obtuse.

The surface tensions and surface free energies of the liquid/vapor, solid/vapor, and the interfacial energy of the solid/liquid are represented by the symbols γ_{lv}, γ_{sv}, and γ_{sl}. The terms are used interchangeably. Use of surface free energies, however, provides a better understanding of the thermodynamic relationships. If the free energy of a solid surface is lowered when in contact with a liquid, it will tend to be covered by the liquid and is thus wetted. However, if the surface free energy of the solid is increased in contact with the liquid, then the solid/liquid area is minimized and the solid is thus thermodynamically not wetted by the liquid. Nevertheless, an interface still forms in this case if the surface free energy of the liquid is lowered by contact with the solid (that is, if the free energy of the system is decreased).

The rigid substrate determines the final configuration of the sessile drop. When γ_{sv} is reduced by the liquid, then a driving force equivalent to $\gamma_{sv} - \gamma_{sl}$ acts on the periphery of the drop and extends the drop/substrate interface. In the absence of a reaction this situation occurs when $\gamma_{sv} > \gamma_{lv}$, $\gamma_{sv} > \gamma_{sl} > \gamma_{lv}$, and $\theta < 90°$. A decrease of the contact angle causes an increase of the liquid drop surface area and thus the total liquid surface free energy. The total surface free energy of the solid decreases concurrently. A balance of

these two forces results in a steady-state condition represented by an acute contact angle. Mathematically, this balance is expressed as Young's equation acting at the periphery of the drop.

$$\gamma_{sv} - \gamma_{sl} = \gamma_{lv}\cos\theta \qquad (Eq\ 3)$$

The driving force for wetting thus is ($\gamma_{sv} - \gamma_{sl}$). The balancing resisting force is represented by the horizontal component of the surface tension of the liquid ($\gamma_{lv}\cos\theta$) as shown in Fig 1. A balance of vertical forces also exists, but they do not play a role in this problem.

When $\gamma_{sv} < \gamma_{lv}$ and $\gamma_{sv} < \gamma_{sl} < \gamma_{lv}$ (again in the absence of a reaction), a steady-state condition results with $\theta > 90°$. With a decrease of the obtuse contact angle, the liquid-drop surface area (and thus the total liquid surface free energy) also decreases. Concurrently, the total surface free energy of the solid decreases as the solid/liquid area increases. A balance of the respective driving forces results in a steady-state condition represented by an obtuse contact angle. Mathematically, this balance is also expressed by Eq 3.

Young's equation (Eq 3) represents a steady-state condition for a solid/liquid interface in stable or metastable thermodynamic equilibrium. However, there is no definite indication of whether chemical or van der Waals bonding exists other than that the contact angle is generally smaller with chemical bonding versus van der Waals bonding (because when γ_{sl} is smaller, the driving force for wetting is greater).

In either wetting ($\gamma_{sv} > \gamma_{lv}$) or non-wetting ($\gamma_{lv} > \gamma_{sv}$), if an intimate interface does not form between the phases to be bonded (as with contaminated solid surfaces such as adsorbed carbonaceous layers), γ_{sl} is larger than γ_{sv} and also γ_{lv}. Then, a large obtuse contact angle forms and approaches 180° with decreasing attractive forces, as indicated by absence of reduction of either the solid or liquid surface free energy and very weak adherence. An optimum 180° angle would indicate no attractive forces between the two phases.

Technologically, a non-wetting liquid is highly unfavorable with regard to the formation of an intimate interface due to its lack of capability of penetration of surface and grain boundary irregularities because of the lack of capillary behavior. Also, the liquid does not distribute itself uniformly.

In either case of a starting acute or obtuse contact angle, the characteristic of wetting can be achieved and enhanced by a reaction at the interface at an elevated temperature. Thermally activated reactions can occur because most systems are not at chemical equilibrium. However, the reactions that contribute to wetting are those that increase the driving force for wetting ($\gamma_{sv} - \gamma_{lv}$) which is acting at the periphery of the liquid drop. It has been found experimentally that the effective reactions are the ones in which the substrate's composition changes by dissolution of a com-

ponent of the liquid. On the other hand, a reaction which results in a change of the liquid's composition by dissolution of the solid substrate but with no change in the substrate composition was found not to contribute to the driving force for wetting. These two reactions are differentiated by defining the solid substrate in the former case as an active participant and in the latter case as a passive participant (in the sense that it does not initiate the reaction). The significance of this behavior is that an active participant has an energetic attraction for the second participant whether it is passive or active.

If the rigid substrate is an active participant in a reaction, the free energy of the reaction at the perimeter of the sessile drop will contribute to the driving force for wetting. As the perimeter of the drop expands, the advancing liquid maintains contact with unreacted solid and thus the free energy of the reaction continues to contribute to the driving force for wetting. Young's equation (Eq 3) for a nonreacting steady-state sessile drop can then be modified to include the contribution of the free energy of reaction:

$$\gamma_{sv} - \left(\gamma_{sl} + \frac{\Delta G_R}{dA_s dt}\right) \geq \gamma_{lv}\cos\theta \qquad (Eq\ 4)$$

The free energy required for the increase of the surface area of the drop as the perimeter expands provides the only resisting force to the expansion. It can be shown thermodynamically that in the absence of a reaction the driving force for wetting does not exceed γ_{lv}, resulting in a steady-state contact angle (Ref 6). The driving force with the contribution of the free energy of reaction in most cases exceeds the resisting force represented by γ_{lv} since θ is 0° during spreading. A dynamic contribution then exists until the liquid is consumed in the reaction, equilibrium compositions are attained, or the reaction is terminated by reducing the temperature. This condition of an expanding drop during a reaction is defined as spreading. It can be seen that the free energy of a reaction in which the substrate is a passive participant does not contribute to the driving force for wetting; thus spreading does not occur. The contact angle, however, adjusts to conform with the surface-energy changes of the liquid caused by composition changes due to the reaction. In the early stages of an experiment, spreading due to reaction should be distinguished from the spreading or slow rate of movement toward the equilibrium contact angle that a highly viscous nonreactive liquid drop undergoes in reaching its stable configuration.

The final example can be based on a reaction with a passive substrate when the starting surface energies would result in an obtuse contact angle (that is, $\gamma_{lv} > \gamma_{sv}$, in the absence of a reaction). The free energy of the reaction causes the interfacial energy of the liquid/solid to decrease with a decrease of the obtuse angle to 90°. This step is favorable because the liquid/vapor area decreases with a decrease in the total surface free energy of the liquid. The angle does not decrease below 90° to become acute because an increase of the liquid/vapor area would occur at this point which would make the process thermodynamically forbidden.

Young's equation (Eq 3) can be modified to represent a spreading coefficient, S (Ref 7), or a work of spreading, W_S, by taking the extreme case of wetting when the contact angle approaches zero and $\cos\theta$ approaches 1. The resisting force to the extension of the drop then is γ_{lv} as discussed earlier. Young's equation can then be expressed as

$$W_S = \gamma_{sv} - \gamma_{sl} - \gamma_{lv} \qquad (Eq\ 5)$$

W_S has to be positive in order to have spreading occur. Under these conditions the driving force for wetting $(\gamma_{sv} - \gamma_{sl})$ is greater than γ_{lv}. If the W_S is negative, then the driving force for wetting is smaller and spreading does not occur but an acute angle forms. If a reaction occurs in which the substrate is an active participant, then the free energy of the reaction, $\Delta G_R / dAdt$, contributes to the driving force for wetting, which practically always exceeds γ_{lv}. Spreading thus occurs.

Copper-Silver System. The equilibrium phase diagram for the copper-silver binary system (Fig 2) can be used to illustrate examples of wetting and spreading (Ref 8). The system has a eutectic at 780 °C (1435 °F) with 72 wt% of silver. At 900 °C (1650 °F), the solid-solution limit is 5 wt% Ag in copper and 8 wt% Cu in silver.

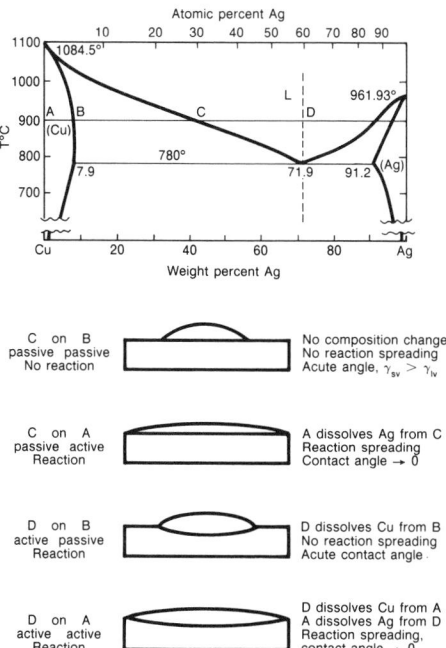

Fig 2 Stable phase equilibrium diagram for copper and silver with sessile drop examples. In all cases $\gamma_{sv} > \gamma_{lv}$. No reaction with C and B. Reaction in other shown cases with B as passive participant and A as active participant. See text.

Several compositions are identified in the phase diagram by letters A to D. When a drop of liquid C is placed on solid B at 900 °C (1650 °F), wetting occurs with a contact angle of 11° and no chemical reaction since the phases are in chemical equilibrium. This behavior corresponds to $\gamma_{sv} > \gamma_{lv}$ because in a given system the surface free energy of a liquid is less than that of a solid due to its lack of long-range order. The liquid thus has the opportunity to rearrange its surface structure to a lower free energy state. However, when liquid C is placed on solid A, spreading occurs because substrate A (as an active participant in the reaction) changes its surface composition toward B. Equation 4 applies in this case.

Another example is that of liquid D on solid B. Liquid D is not in equilibrium with B and dissolves some of the substrate to change its composition to C. Even though a reaction occurs, there is no spreading because B is a passive participant with no change in composition even though it is being dissolved. However, with liquid D on solid A, spreading occurs because both are active participants as they change to equilibrium compositions C and B, respectively. In both of the latter examples, liquid D is an active participant because it dissolves some of the substrate to reach equilibrium compositions. It does not, however, contribute to spreading, which is controlled by the active participation of the substrate.

The nickel-silver system has a eutectic at a fractional concentration of nickel (<1 wt% Ni) at 961 °C (1761 °F); the melting point is 962 °C (1763 °F) for silver and 1455 °C (2650 °F) for nickel. At a sessile drop temperature of 970 °C (1780 °F), liquid silver has a fractional concentration (in wt%) of nickel in solution, and the solid nickel has several percent of silver in solution (Ref 9). Energy dispersive spectroscopic (EDS) compositional traces across cross-sections of sessile drops of silver on nickel in gettered helium (top) and in air (bottom) are shown in Fig 3.

In helium, an intimate nickel/silver interface forms and the EDS traces indicate a small amount of intersolution but a greater amount of nickel in silver. This reversal of respective solutions can be attributed to the presence of some oxygen in the silver, which is common. (The surface energy of silver is sensitive to the amount of oxygen dissolved in the silver; the decrease in γ_{Ag} with increase of oxygen is shown in Fig 4.) The contact angle decreased to 9° in 30 min. It was acute because $\gamma_{sv} > \gamma_{lv}$. Because of the small amount of interdiffusion, the contribution of the free energy of the reaction to the driving force for wetting caused a reduction in the angle; however, it was not large enough to result in complete spreading of the liquid.

In air, a layer of NiO formed on the surface of nickel before the melting temperature of silver was reached (Ref 9). The EDS traces in Fig 3 indicate some solution of NiO in

nickel and in silver but no silver in NiO. On melting, the silver drop had an initial obtuse contact angle $(\gamma_{sv} < \gamma_{lv})$, and a steady-state angle of 90° was attained in less than 5 min. The substrate was a passive participant in the reaction because silver was not dissolved in the NiO, although the liquid was an active participant because NiO was dissolved in silver. As a result γ_{sl} was continuously reduced until it was equal to γ_{sv}, thus forming the 90° angle according to Young's equation. The contact angle does not decrease below 90° because γ_{sv} is not decreased by the liquid. Also, because the total area of the liquid surface decreases with a decrease of contact angle to 90° and would increase with further decrease of contact angle, it would be energetically unfavorable for the contact angle to decrease below 90°.

Chemical Reactions Leading to Equilibrium

Good adherence was observed at all of the interfaces discussed above because of the development of equilibrium compositions at the interfaces. The formation of an interface in intimate contact is a necessary first step in wetting and in making any kind of seal when one of the two members is a liquid at the sealing temperature. If the materials at the interface are at chemical equilibrium, a chemical bond will form when the temperature is sufficient to provide the necessary activation energy. If the two materials are not at chemical equilibrium but do not react because of kinetic or some other barriers, a thermodynamically metastable condition will exist and van der Waals bonding will be present at the interface. In both of these cases, a steady-state sessile drop contact angle, acute or obtuse, will be determined by the magnitudes of the interfacial energies $(\gamma_{sv}, \gamma_{lv}, \gamma_{sl})$ as discussed previously. On the other hand, if the materials are not at chemical equilibrium and the temperature is high enough, a reaction will occur at the interface to form equilibrium compositions. When the solid substrate is an active participant (that is, its interface composition changes either by dissolving a component of the liquid phase or by forming a compound with the liquid phase), the free energy of reaction contributes to the driving force for wetting, causing spreading associated with a dynamic decrease of contact angle. If the driving force $(\gamma_{sv} - \gamma_{sl})$ exceeds γ_{lv}, complete spreading occurs. This dynamic stage continues until the reactants are exhausted or until a steady-state equilibrium condition develops at the periphery of the drop. However, as discussed before, spreading will not occur if only the liquid phase is an active participant (that is, the solid only dissolves in the liquid).

In all cases, however, regardless of the sessile drop configuration, reactions lead to chemical equilibrium at the interfaces. The resulting phases at the interface become com-

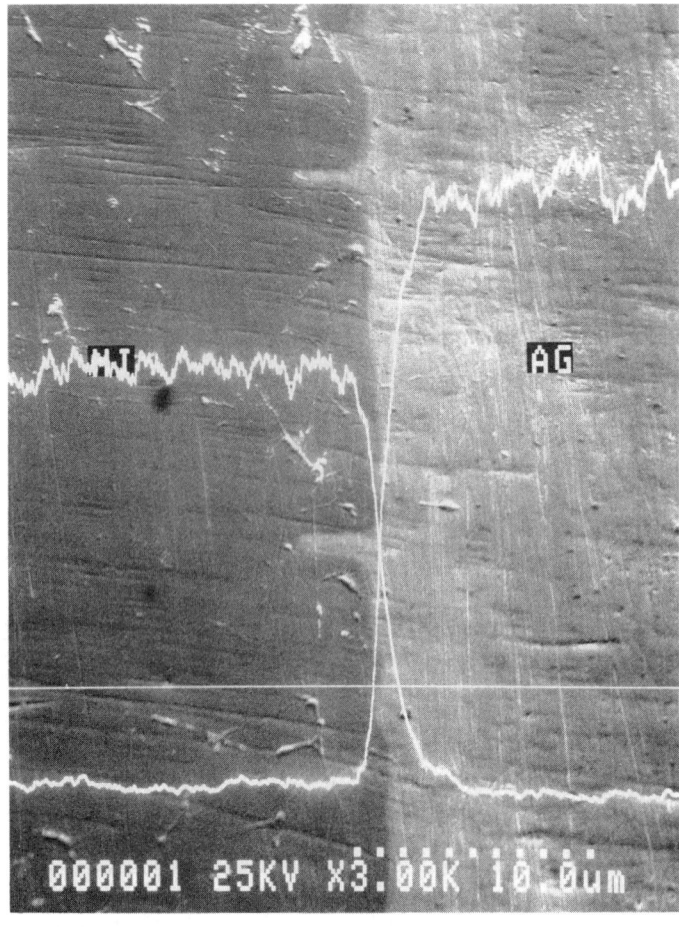

(a)

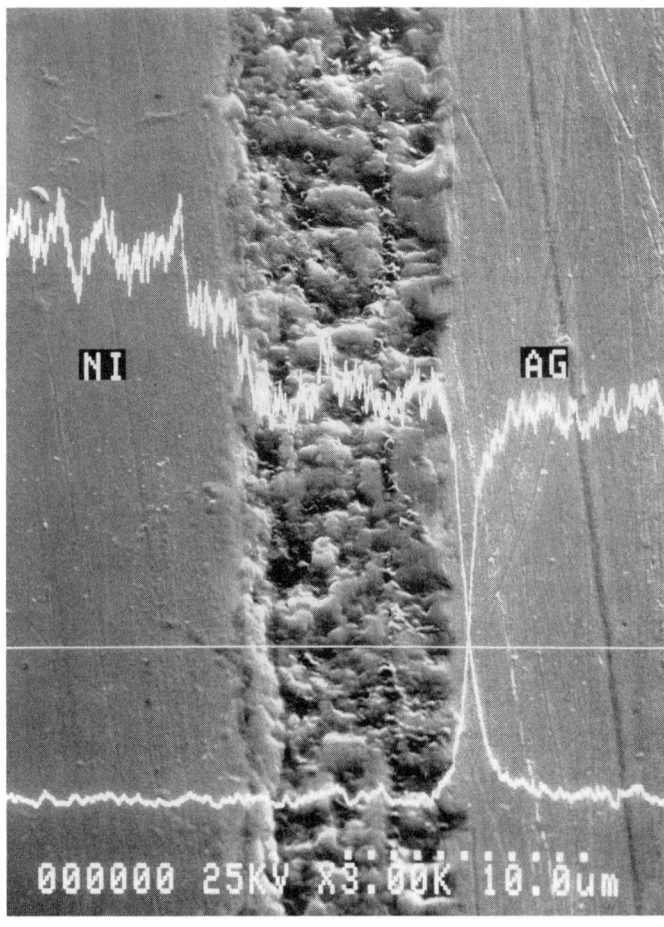

(b)

Fig 3 Cross-sections of sessile drops of silver on nickel in gettered helium (a) and in air (b) at 970 °C (1780 °F). The photos show EDS traces of nickel and silver along white horizontal line.

patible, permitting the formation of an electronic structure with a balance of bond energies across the interface which constitutes a chemical bond.

Interfacial chemical reactions can be distinguished according to whether or not electron transfer occurs. Without actual electron transfer reactions but with an electronic structure across the interface, the bonding can be

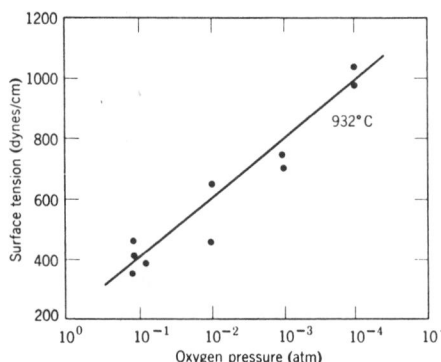

Fig 4 Effect of oxygen pressure on surface energy of solid silver at 932 °C (1710 °F). Source: Ref 10

classified simply as metallic or covalent; if electron transfer occurs the bonding is ionic. Ionic bonding frequently occurs in ceramic/metal joining through redox (reduction-oxidation) reactions. In simplest form there is electron transfer between two cationic species across the interface resulting in a valence change in each. In extreme cases a cation can be reduced all the way to a neutral metal precipitate.

Reactions with No Valence Change

When the two reacting phases have no changes in valence, then the reactions form solid-solution alloys or compounds by interdiffusion. In this case, the solution rate at the interface would have to be faster than diffusion from the interface to maintain saturation and a stable chemical equilibrium.

Diffusion Bonding. This type of reaction is applied to the joining of two clean and polished solids with similar electronic structures (for example, two metals). Polishing is performed to eliminate surface irregularities as much as possible. The usual sealing procedure is to hold the two surfaces together with sufficient pressure to form an intimate inter-

face at an elevated temperature ($\approx 0.9\ T_m$) and in a controlled atmosphere. Diffusion occurs across the interface until each phase is saturated with the other at the interface. As saturation at the interface is lost by diffusion into the bulk, further solution reaction takes place to reestablish saturation. Because the dissolution rate at the interface is faster than the diffusion rate into the bulk, saturation and chemical equilibrium are retained at the interface. The two phases at equilibrium thus remain compatible with a balance of bond energies, and chemical bonding is maintained across the interface.

An example is provided by the copper-silver system (Fig 2). Solid copper and silver in contact just below the eutectic temperature will react with interdiffusion. At the interface, copper will retain saturation with silver, and vice versa. The saturated phases are compatible and form a chemical bond.

Another example is provided by the NiO-CaO system (Ref 11). Polished specimens are held together under pressure at a temperature just below the eutectic temperature of 1695 °C (3085 °F). Equilibrium is attained at the interface by interdiffusion of nickel and calcium cations with maintenance of a charge balance until saturated NiO and CaO solid solutions form. The imbalance introduced by different solubility limits is avoided by a movement of the phase boundary (interface) in the direction of achieving a mass balance. At the reaction temperature, the phases at the interface also reach and maintain saturation equilibrium. Precipitates may form in one or both phases if the solubility limits decrease significantly on cooling from the bonding temperature. The equilibrium phases at the interface are compatible and form a chemical bond.

Solution Bonding. This category also applies to the bonding of two phases with similar electronic structures, one of which is in a molten state at the sealing temperature. The characteristics can be illustrated with a system of molten silicate glass on an oxide substrate, both of which have covalent-ionic electronic bonding. At equilibrium, the liquid glass will form a steady-state, acute-angle contact because $\gamma_{sv} > \gamma_{lv}$. If the glass is not in chemical equilibrium with the substrate oxide, it will dissolve sufficient oxide to reach equilibrium at the interface and the dissolved species will become part of the molten glass. The composition at the substrate side of the interface in most cases does not change extensively. The solubility limit for different glass constituents may be below the limit of detection. Spreading of the liquid will not occur unless the substrate is a sufficiently active participant in the reaction (that is, the substrate composition changes with reaction).

Another example can again be provided by the copper-silver system (Fig 2). If a small piece of silver is placed on a copper substrate saturated with silver at 900 °C (1650 °F) (equivalent to point B on the diagram) in a non-oxidizing atmosphere, it will dissolve copper at the interface by solution and diffusion. At saturation, it begins to melt and dissolve copper by solution until it becomes all liquid and saturated with copper (equivalent to point C in Fig 2). A steady-state, acute-angle contact is maintained at this point because B is a passive participant. If silver is placed on pure copper (equivalent to point A) under the same conditions, copper also begins to dissolve silver by diffusion. In this case, because the substrate is also an active participant in the reaction, spreading of the copper-saturated, silver liquid occurs.

Reactions with Valence Changes (Redox Reactions)

In general, metals will not bond chemically to ceramics because the metallic electronic structure is not compatible with the ionic-covalent electronic structure of ceramic materials. A phase, which theoretically could constitute just one molecular layer, is thus necessary at the interface as a compatible transition between the metal and the ceramic. With oxide ceramics, for example, the compatible phase is one of the oxides of the metal. In this case, the oxide maintains a continuous

electronic structure across the interface and a transition in bond type and bond energies between the metal and the ceramic when the interface is saturated with the metal oxide. The oxide is obtained by a redox reaction if it is not initially present.

Solid-State Joining (Reaction Bonding). Solid-state joining is usually performed by pressing the components together at a high temperature. Success depends on the formation of an adequate interfacial contact. The most extensively studied systems are alumina with niobium, aluminum, nickel, platinum, and titanium. In one study (Ref 12), transmission electron microscopy of single-crystal Al_2O_3 bonded to niobium at 1700 °C (3090 °F) for 1 h showed a layer of NbO_x (≈ 20 μm thick) at the interface and a solid solution of niobium and aluminum from the following reaction:

$$Al_2O_3 + 5Nb \rightarrow 3NbO + 2Al\text{-}Nb \qquad (Eq\ 6)$$

However, another somewhat similar set of experiments (Ref 13) gave contradictory results. In this study, single crystals of niobium and Al_2O_3 were bonded by heating at 1727 °C (3140 °F) for 2 h. The crystals were oriented so that the close-packed niobium (110) and Al_2O_3 (0001) planes were parallel. High resolution electron microscopy images showed the perfect lattices of each crystal extended to within 3 to 4 lattice planes of the interface. Within that region, the lattices were distorted, with most of the distortion in the niobium. No reaction layer was found at the interface, although complex redox reactions may be responsible. Precipitates of θ-Al_2O_3 were observed in the niobium, which indicates that Al_2O_3 reacts with niobium and dissolves in it. Aluminum diffusion profiles in niobium, measured by analytical electron microscopy, generally agreed with theoretical predictions. The reason for this discrepancy is not apparent, although it may be due to a lower oxygen partial pressure in the reaction atmosphere. A similar reaction to Eq 6 was reported to occur at an Al_2O_3-Ti interface (Ref 14). Other metals also react with Al_2O_3 to produce various interfacial compounds such as $NiAl_2O_4$ (wherein NiO is formed by a redox reaction and then reacts with Al_2O_3). A redox reaction is always a possibility (that is, Al_2O_3 is reduced and the metal is oxidized). For a spontaneous reaction the overall free energy change must be negative. The alloying of aluminum with the metal is favorable toward the reaction by contributing to a negative free energy. Also, the formation of a compound between the formed metal oxide and Al_2O_3 is favorable.

Preoxidation of Metal. An extensively used method to obtain the necessary substrate metal oxide at the interface is to preoxidize the metal at an elevated temperature in air or in an atmosphere with a controlled partial pressure of oxygen and/or water vapor. The metal at the interface oxidizes to form cations as it reduces the adsorbed oxygen from the

atmosphere to form anions. These combine to form the oxide whose thickness is kinetically controlled by continued oxidation:

$$(2x)Fe + O_2 = 2Fe_xO \qquad (Eq\ 7)$$

The glass is applied to the preoxidized metal as a powder coating or as a solid piece. The temperature is set so that the glass softens to become a liquid and dissolves the oxide. The solution rate usually is faster than the diffusion rate into the glass, so the glass at the interface remains saturated with the oxide as long as the oxide layer has not been completely dissolved (Fig 5). Thus, the necessary compatibility of phases for chemical bonding is achieved at both interfaces (metal-oxide and oxide-glass). The thickness and other properties of the oxide control the behavior of the interface such as vacuum tightness and inherent strength. Furthermore, as shown in the figure, the thicker oxide layers result in a more gradual decrease of the composition gradients from the interface into the glass.

Even precious metals, like platinum and silver, will chemically bond to a ceramic glass or oxide when heat-treated in air. Several molecular layers of metal oxide form on the metal surface, which interact with the ceramic to form a chemical bond at the sealing temperature. The small amount of metal oxide is sufficient to form the chemical bond because

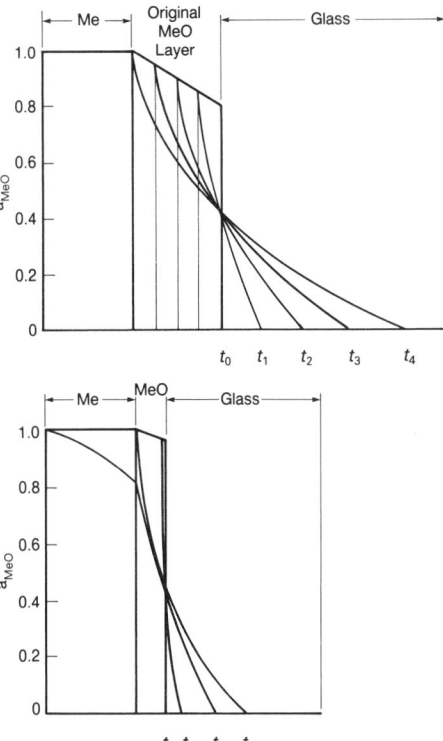

Fig 5 Schematic showing lengthening activity gradients of MeO in glass with continuing solution of oxide by glass. (top) Gradient extension proportional to thickness of oxide. (bottom) loss of saturation at interface with continued heating beyond complete solution of oxide layer. As time increases from t_0 to t_4, curves appear to rotate counterclockwise about a common origin located at $a_{MeO} \approx 0.4$

its solubility in the ceramic is extremely low and the diffusivity is extremely slow so that it is not easily depleted at the interface (Ref 15).

Spontaneous Redox Reactions in Glass/Metal Sealing. When the oxide layer on the metal is completely dissolved by the glass, then a redox reaction has to occur at the interface to form more metal oxide. The model system discussed below has been studied extensively. The deduced principles have served as the basis for developing an understanding and for establishing the conditions under which chemical bonding at interfaces is attained in all joining systems. The system to be discussed is Fe/CoO-containing, sodium-disilicate, glass. This is the model system for porcelain enameling that achieves chemical bonding at the interface through a spontaneous redox reaction (Ref 16, 17).

In this system an iron substrate is oxidized and the CoO in the glass is reduced. This reaction occurs spontaneously at 900 to 1000 °C (1650 to 1830 °F). The oxidation potential for iron is higher than that for cobalt ($Fe^{2+} = -0.44V$, $Co^{2+} = -0.28$ V). Thus, ΔG_R° under standard conditions is negative. The net reaction (Eq 8d) is represented by three step reactions (Eq 8a, b, c) as follows:

$$Fe + CoO_{(gl)} = FeO_{(int)} + Co \qquad (Eq\ 8a)$$

$$FeO_{(int)} = FeO_{(gl)} \qquad (Eq\ 8b)$$

$$Co + xFe = Fe_xCo \qquad (Eq\ 8c)$$

$$(1 + x)Fe + CoO_{(gl)} = FeO_{(gl)} + Fe_xCo \qquad (Eq\ 8d)$$

Equation 8(a) is the redox reaction forming FeO and occurs at the interface. Reaction 8(b) represents the solution of FeO into the glass; however, some of the FeO is also dissolved by the iron if the iron does not happen to be saturated with FeO at the interface. A layer of FeO thus continually remains at the interface and acts as the compatible oxide phase that provides the chemical bonding between the metal and the glass. Step reaction 8(c) represents the alloying of iron with the reduced cobalt to form dendrites in the interfacial zone by a microgalvanic cell mechanism shown schematically in Fig 6. This alloying step also

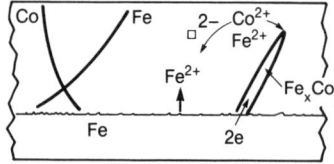

$$(x+y)\ Fe + Co_yNa_2Si_2O_{5+y} \rightarrow Fe_yNa_2Si_2O_{5+y} + Fe_xCo_y$$

Fig 6 Schematic cross-section showing glass containing the CoO reacting with iron substrate and resulting in indicated Co^{2+} and Fe^{2+} gradients and the formation of Fe_xCo_y dendrites by microgalvanic cell mechanism with an electronic circuit of electrons and negative vacancies. Electrons released as iron atom enters liquid glass as Fe^{2+}; negative vacancies form as Fe^{2+} and Co^{2+} precipitate on dendrite as atoms. Standard free energies are negative for redox reaction shown.

has a negative free energy, which contributes to the driving force for the net reaction (Eq 8d).

Non-Spontaneous Redox Reactions in Glass/Metal Sealing. If the glass does not have an oxide in its composition whose metal has a lower oxidation potential than that of the substrate metal, a redox reaction of a spontaneous nature would not take place. The formation of FeO would then depend upon a favorable equilibrium condition. A representative model system is the iron/sodium-disilicate system (Ref 18–20) held at 1000 °C (1830 °F). The net reaction (Eq 9c) consists of two step reactions: the formation of the metal oxide at the interface (Eq 9a) followed by its solution by the glass (Eq 9b):

$$Fe + Na_2O_{(gl)} = FeO_{(int)} + 2Na_g \uparrow \qquad (Eq\ 9a)$$

$$FeO_{(int)} = FeO_{(gl)} \qquad (Eq\ 9b)$$

$$Fe + Na_2O_{(gl)} = FeO_{(gl)} + 2Na_g \uparrow \qquad (Eq\ 9c)$$

$$\Delta G_R = \Delta G_R^\circ$$
$$+ RT \ln \frac{a(FeO)_{(int)} \cdot [a(Na \uparrow)]^2}{a(Fe) \cdot a(Na_2O)_{(gl)}} \qquad (Eq\ 9d)$$

The activity quotient (Eq 9d) for Eq 9(a) has to be sufficiently smaller than unity for the overall term to be more negative than the positive ΔG_R° (standard free energy). The reaction is thus favored by a low $a(Na)$, a low $a(FeO)_{int}$ and a high $a(Na_2O)_{gl}$. The term $a(Fe)$ is close to unity and can be neglected. The pressure of the sodium vapor produced must exceed ambient pressure to nucleate bubbles of sodium vapor at the interface and escape. Therefore, this type of reaction is potentially more favorable in a dynamic vacuum or a low-pressure atmosphere.

It is necessary to consider the step reactions in order to understand the nature of the net reaction (Eq 9c) fully and to interpret the kinetics correctly. The second step reaction (Eq 9b) has a negative ΔG_R° and thus occurs spontaneously. In any complex reaction, it is important to determine all of the step reactions in order to determine the nature of the complicating factors and their effect on processing control factors. In practice, however, it is more convenient to have compositions that lead to spontaneous redox reactions under regular processing conditions, generally in atmospheric air.

Complex Redox Reactions in Glass/Metal Sealing. In order to realize a favorable negative standard free energy ΔG_R° for a redox reaction, it is necessary to use a highly reactive metal as a substrate. Experiments have been made with chromium, titanium, and zirconium (Ref 21–23), which have sufficiently high oxidation potentials to reduce even silica in the glass with the formation of silicide or an alloy of the metal with silicon. A model system that illustrates the complexities that can arise is the chromium/sodium-disilicate system (Ref 21).

The reactions in this system are associated and dependent on the evolution of gaseous products, which are affected by the thickness of the glass layer and the atmospheric pressure and composition. Sessile drop experiments are excellent for determining the reaction mechanisms and their controlling factors, especially when several reactions can be going on concurrently. For example, the net reaction (Eq 10c) in the periphery of the drop corresponds to the formation of a thin layer of glass at 1000 °C (1830 °F) in a low ambient pressure. The most favorable reaction at the interface is the formation of CrO (Eq 10a, 10b, and the net reaction 10c) which is verified by observation of a blue coloration of the glass:

$$Cr^\circ + Na_2O_{(gl)} = CrO_{(int)} + 2Na_g^\circ \uparrow \qquad (Eq\ 10a)$$

$$CrO_{(int)} = CrO_{(gl)} \qquad (Eq\ 10b)$$

$$Cr^\circ + Na_2O_{(gl)} = CrO_{(gl)} + 2Na^\circ \uparrow \qquad (Eq\ 10c)$$

The first step (Eq 10a), whose free energy equation is indicated by Eq 10(d), is the controlling reaction. The next reaction (Eq 10b) takes place spontaneously because it has a negative ΔG_R°, as described in the previous example.

$$\Delta G = \Delta G^\circ$$
$$+ RT \ln \frac{a(CrO)_{(int)} \cdot p(Na^\circ)^2}{a(Cr^\circ)_{(s)} \cdot a(Na_2O)_{(gl)}} \qquad (Eq\ 10d)$$

When the ratio $a(CrO)/a(Na_2O)$ reaches some critical value, Eq 10(a) becomes unfavorable and Cr^{3+} becomes more stable, leading to the formation of Cr_2O_3 according to Eq 10(e), which colors the glass green:

$$2CrO_{(gl)} + Na_2O_{(gl)}$$
$$= Cr_2O_{3(gl)} + 2Na^\circ \uparrow \qquad (Eq\ 10e)$$

The sequential reactions (Eq 10(c) and Eq 10e) lead to the overall reaction Eq 10(f) as follows:

$$2Cr_{(s)}^\circ + 3Na_2O_{(gl)}$$
$$= Cr_2O_{3(gl)} + 6Na^\circ \uparrow \qquad (Eq\ 10f)$$

All of the indicated steps are dependent on the presence of a thin glass layer and a low ambient pressure to permit the evolution of the gaseous sodium. In this sequence CrO_{gl} is an intermediate product and the extent of its formation is dependent on the kinetics of the various steps. The practical significance is that the amount of CrO (blue glass) formed is not sufficient to saturate the glass at the interface because CrO is very soluble in molten silicates, and no adherence develops. On the other hand, the solubility of Cr_2O_3 in the glass (green color) is much less and thus it saturates the glass with development of adherence.

Another example involves the examination of the proposed sequence of reactions in the center of the sessile drop, equivalent to a thick glass layer, at the same temperature and ambient pressure. In this case, Eq 10(c) and

10(e) are severely retarded because the sodium gas cannot escape, as indicated by the initial absence of bluish glass and poor adherence. The most favorable reaction is between the chromium and SiO_2 in the glass to form $CrSi_x$ alloy as follows:

$$(1 + 2x)Cr° + xSiO_{2(gl)}$$
$$= CrSi_{x(s)} + 2xCrO_{(gl)} \qquad (Eq\ 11a)$$

The $CrSi_x$ forms at the interface as needles or whiskers as seen in Fig 7. It is proposed that the growth of these crystals occurred by a microgalvanic cell mechanism similar to that deduced for growth of Fe_yCo dendrites at the interface of the iron substrate and $NaSi_2$-glass containing CoO in solution. As CrO in solution increases, at some point Eq 11(b) becomes operable:

$$2CrO_{(gl)} + SiO_{2(gl)}$$
$$= Cr_2O_{3(gl)} + SiO_{(g)} \uparrow \qquad (Eq\ 11b)$$

This reaction is thermodynamically unfavorable under standard conditions but proceeds because of a favorable equilibrium constant as a result of low SiO gas pressure and a high heat formation of Cr_2O_3. As the SiO pressure builds up in the glass, it can alternatively react with chromium to form $CrSi_x$ plus Cr_2O_3 as follows:

$$(3 + 2x)Cr° + 3xSiO$$
$$= 3CrSi_{x(s)} + xCr_2O_{3(gl)} \qquad (Eq\ 11c)$$

The net reaction of Eq 11(a, b, and c) eliminates CrO and SiO and results in:

$$(3 + 4x)Cr° + 3xSiO_2$$
$$= 3CrSi_{x(s)} + 2xCr_2O_{3(gl)} \qquad (Eq\ 11d)$$

The interfacial zone becomes saturated with Cr_2O_3 as indicated by a green coloration of the glass and the formation of a chemical bond across the interface.

Redox Reactions in Metal Alloy-Glass Sealing. Far fewer scientific studies have been made on the bonding of glass to alloys because of the complexity of dissolution and oxidation reactions. This section describes initial studies on the bonding of a K-aluminosilicate glass to two 80Ni20Cr alloys (Ref 24). One was a 99.99% pure alloy prepared in the laboratory, and the other was a commercial alloy of 98% purity with a major impurity of 1.5 wt% Si.

The oxidation characteristics of the alloys were significantly different. For example, the weight gain at 1000 °C (1830 °F) in air for 2 min was 0.19 g/cm^2 for the pure alloy and 0.03 g/cm^2 for the commercial alloy (Fig 8). As expected, an auger analysis combined with argon ion sputtering showed different compositional profiles. The purer alloy shows a multilayer oxide scale with an outer layer of NiO. The commercial alloy shows a single-layer scale of a mixture of Cr_2O_3 and NiO, with the outer portion of the scale richer in Cr_2O_3. It can be deduced that the mobility of the nickel species relative to that of the chromium species is considerably reduced by the presence of silicon. Further work is necessary to develop an understanding of the oxidation mechanisms of 80Ni20Cr alloys with various additives.

Glass layers were fired onto platelets of both alloys at 1020 °C (1870 °F) for 10 min in air. The pure alloy was preoxidized for 1 min and the commercial alloy for 60 min at 1000 °C (1830 °F) in air. Energy dispersive spectro-

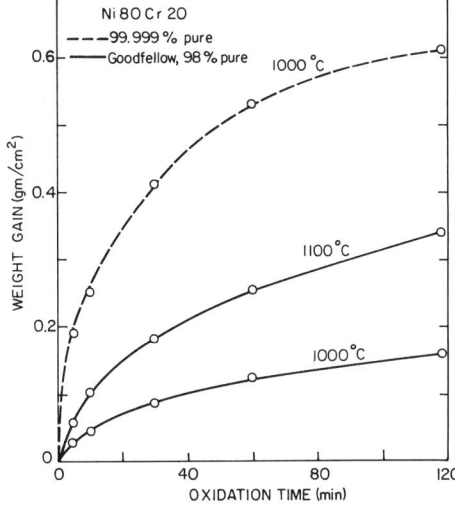

Fig 8 Weight gain versus time for two 80Ni20Cr alloys. The 98% pure alloy had 1.5 wt% Si.

scopic line scans of the cross-sections of the fired assemblies for chromium, nickel, and silicon are shown in Fig 9. The firing time was not long enough to dissolve the oxide scales. Thermal expansion mismatch with the pure alloy platelet assembly caused fracture to occur within the multilayer oxide scale, which was not penetrated by the glass. In the case of the commercial alloy assembly, fracture occurred in the glass close to the oxide/glass interface because the oxide was strengthened by glass penetration. The significant characteristic of the assemblies from a chemical viewpoint was that adherence was present at both metal/oxide and oxide/glass interfaces, indicating equilibrium compositions and compatibility at all of the interfaces. It should again be emphasized that adherence does not necessarily mean that the system is at equilibrium. From a physical or application viewpoint, the commercial alloy assemblies probably were satisfactory but the others were not because of fracture in the weaker oxide layer.

Metal Brazing

To avoid high bonding temperatures (especially in assemblies not exposed to high temperatures in use) and achieve greater reliability, metallic interlayers are used as a bonding agent. Intermediate metals that are used for making assemblies below 400 °C (750 °F) are identified as solders, and those that are used above 600 °C (1100 °F) are called brazing alloys. Most of the studies of brazing to date have focused on oxide and nitride ceramics. Brazing alloy compositions and their reactivity with oxide ceramics are well documented in the literature (Ref 25, 26).

A widely used brazing alloy is based approximately on the eutectic composition in the Cu-Ag system (Fig 2). It must react and reach chemical equilibrium at the interfaces with both the metal and ceramic components

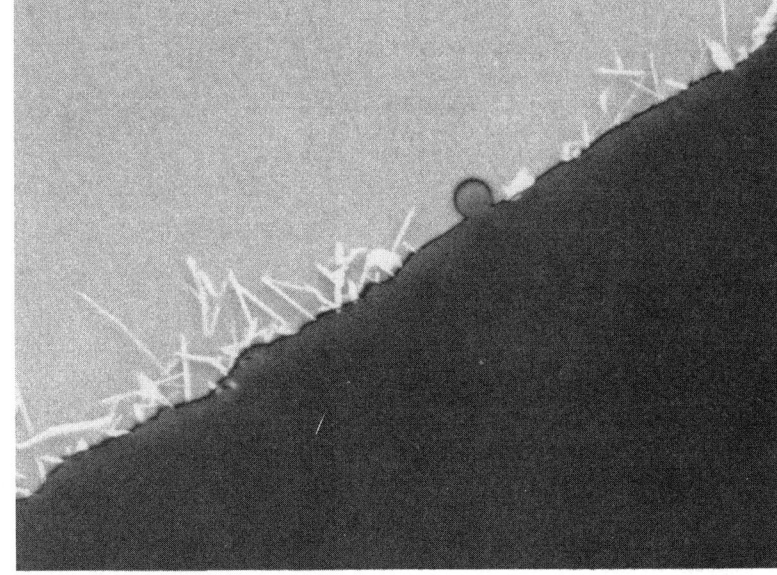

Glass

Cr silicide

resin

$\vdash\!\!-\!\!-\!\!\dashv$
50 µm

Fig 7 SEM photograph of cross-section of interfacial zone for chromium and glass showing dendritic growth of chromium silicide in the glass at the interface. Dendrites formed by microgalvanic cell mechanism similar to Fig 6.

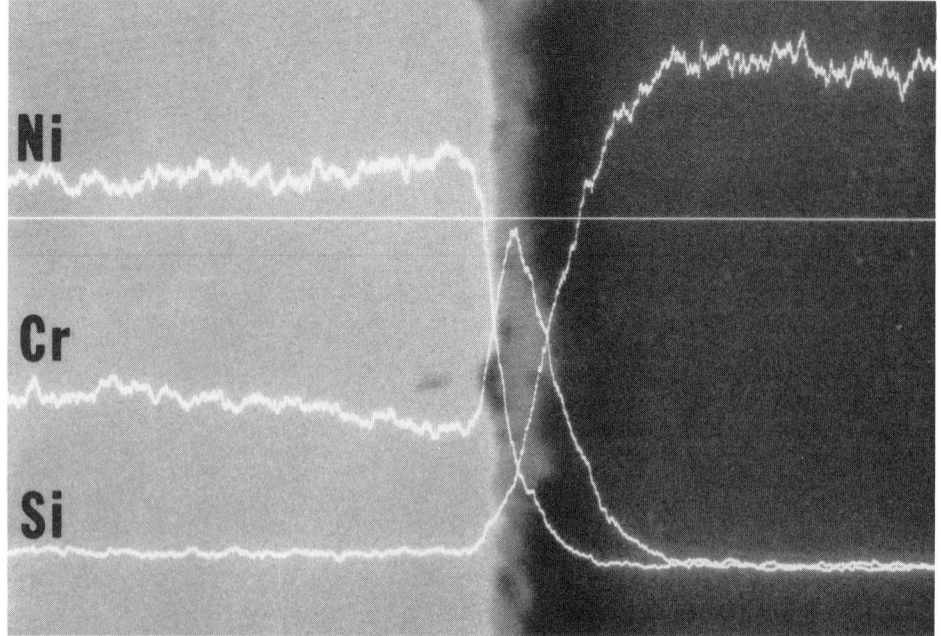

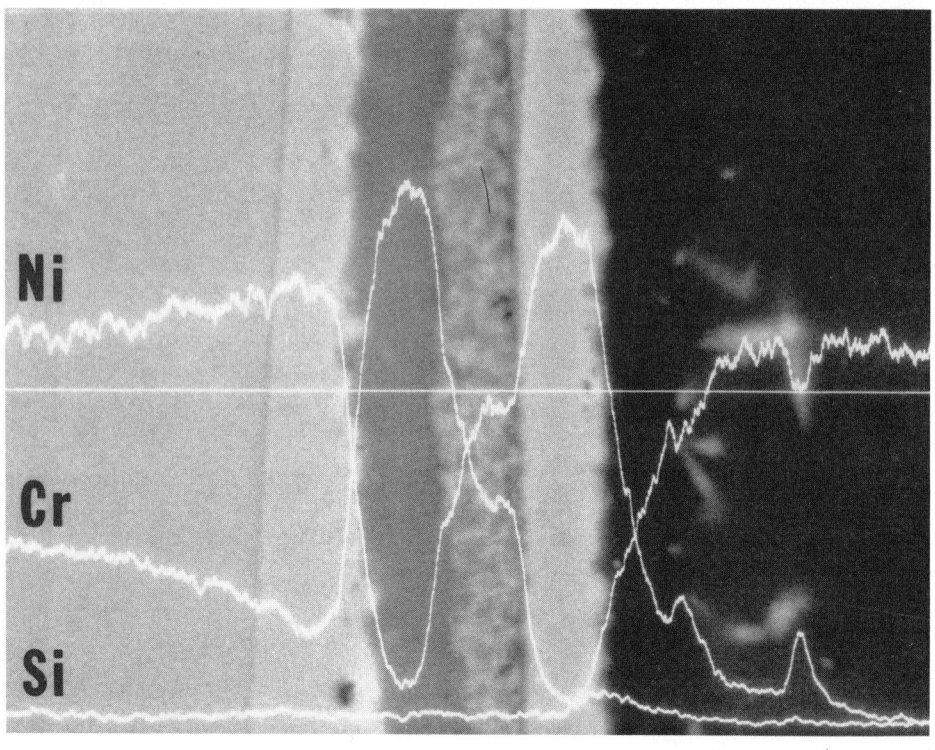

2 µm

Fig 9 X-ray scans for chromium, nickel, and silicon across the interfaces of (top) dental porcelain glassy phase on 80Ni20Cr type alloy with silicon additive preoxidized 60 min at 1000 °C (1830 °F) in air, and (bottom) an alloy with no additive preoxidized 1 min at 1000 °C (1830 °F) in air.

potential of the titanium (as discussed below) causes the element to undergo a redox reaction with the Al_2O_3, which causes the formation of an oxide compound that results in a driving force for spreading of the brazing alloy. The compound is compatible with both phases, producing a chemical bond at the interfaces. The wetting and solution/diffusion bonding of the brazing alloy with the metal component are unaffected.

Reaction at the interface with a reactive metal promotes chemical bonding. The net reaction is indicated as follows:

$$Al_2O_3 + 5Ti = 3TiO + 2AlTi \qquad \text{(Eq 12)}$$

The titanium reduces Al_2O_3 to form TiO_2 which reacts at the interface with Al_2O_3 to form a thin layer of Al_2TiO_5 at the interface. This thin layer is compatible with both Al_2O_3 and the brazing alloy. The aluminum and some of the oxides go into solution in the brazing alloy. A higher active metal concentration or the brazing temperature and time increases the extent of the reaction at the interface. The strength of the assembly at first increases and ultimately weakens (Ref 27). The strength changes are not due to variations in bonding at the interface, but due to changes in the microstructure of the interfacial zone, because of variations in strain and stress patterns. Consequently, there is an optimum reactive metal concentration in the brazing alloy. It has been pointed out in brazing alumina ceramics that the optimum reactive metal concentration for bonding by alloys is frequently less than that needed to achieve spreading (Ref 26).

Another example of the effectiveness of small amounts of additives was provided by sessile drop experiments of copper on sapphire substrates (Ref 28). With increasing additions of oxygen for Cu_2O formation, the contact angle decreased (for example, 0 wt% and 6.7 wt% showed contact angles of 163° and 27°, respectively). Bonding strength developed with the formation of $CuAlO_2$ at the interface.

Mo-Mn Metallizing Process

An extensively used procedure for joining is to first metallize the surface of the ceramic. A brazing alloy is then used to join the metallized ceramic with another, compatible metal component. The metallized ceramic is obtained through an intermediate glass phase that bonds to the metal constituent and the ceramic grains.

The molybdenum-manganese process is the most commonly used industrial technique for metallizing alumina. A paint containing powders of molybdenum and manganese (or their oxides) is applied to the ceramic, generally Al_2O_3 or BeO, and fired in H_2 with a controlled dew point so that manganese is present as MnO and molybdenum as a metal. The paint generally contains additional constituents to control the glass phase viscosity and

being bonded. Metal systems are generally compatible, as discussed earlier, resulting in wetting and solution/diffusion bonding of the brazing alloy with the metal component. The brazing alloy, however, usually does not wet the ceramic. It also does not react because of the oxidation potentials of copper and silver (in the case of Al_2O_3, for example, the oxi-

dation potentials of copper and silver are less than that of aluminum). Without a thermodynamically stable chemical equilibrium, bonding does not occur at the interfaces. Consequently, Cu·xAg brazing alloys to be used directly with ceramics generally have a small percentage of a reactive metal added (for example, titanium). The high oxidation

firing temperature. The MnO reacts with the ceramic grains and the liquid glassy phase in the ceramic to form a controlled amount of glassy phase at the ceramic/molybdenum interface. The glass migrates into the Al_2O_3 grain boundaries and into the metallizing layer and forms a dense glass/molybdenum composite structure. In addition, the glass at the interface dissolves MoO formed on molybdenum and also reacts with the molybdenum to become saturated with MoO, thus forming a chemical bond with molybdenum. Since the wetting of the molybdenum coating by the brazing alloy is critical, the coating is generally electroplated with nickel to provide a clean and continuous surface as well as one on which the braze spreads easily. A similar process is also followed for tungsten metallizing with an appropriate glass composition. There is extensive literature on various aspects of the process (Ref 29–33).

Although it is not normally recognized, the increased bond reliability obtained from the Mo-Mn process is suggested to be most likely due to a more favorable stress pattern across the interface because of the formation of a thicker interfacial zone with a more extended and graded microstructure.

Joining Non-Oxide Ceramics

Non-oxide ceramics are being developed for a wide variety of applications, ranging from use as structural components in heat engines to high-performance substrates for hybrid microcircuits. Many of these applications require development of suitable sealing or joining techniques. Most joining methods use brazing alloys (Ref 34) or nonmetallic materials such as reactive ceramics or glasses (Ref 35).

Much of the work has focused on use of metallic brazing alloys as the bonding medium, closely analogous to the brazing techniques used in conventional oxide ceramic joining. The same principles apply. Wetting and spreading are dependent on the same surface and interfacial free energy requirements. Lastly, chemical bonding at the interface is dependent on the existence of a thermodynamically stable phase equilibrium which results in the presence of compatible phases in the interfacial zone.

Figure 10 shows the wetting behavior of pure Sn and a CuAg alloy (Cusil) (Ref 36). The figure shows that tin did not wet Si_3N_4 as indicated by an invariant contact angle of 153° for the entire 60 min because $\gamma_{lv} > \gamma_{sv}$. On cooling, the tin formed a near spherical bead that rolled off the Si_3N_4 substrate when it was tilted. If the interfacial free energy is greater than either of the surface free energies ($\gamma_{sv} < \gamma_{sl} > \gamma_{lv}$), then van der Waals bonding exists at the interface, as indicated in this case. For the CuAg alloy the contact angle on as-polished Si_3N_4 was constant at 142° with no adherence to the substrate. The effectiveness of small amounts of a reactive metal (titanium) in a braze is indicated by actual experimental sessile drop data (Fig 10). Cusil ABA (1.5 wt% titanium) exhibited a low contact

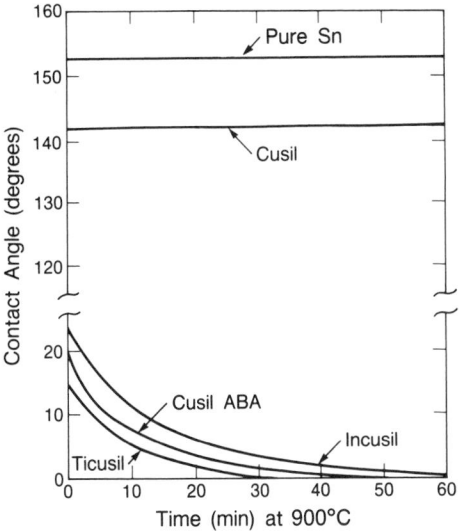

Fig 10 Variation of contact angle with time at 900 °C (1650 °F) of tin and copper-silver alloys on NC132 Si_3N_4 (Norton Co.). Partial pressure of O_2 was 10^{-16} atm during the experiments.

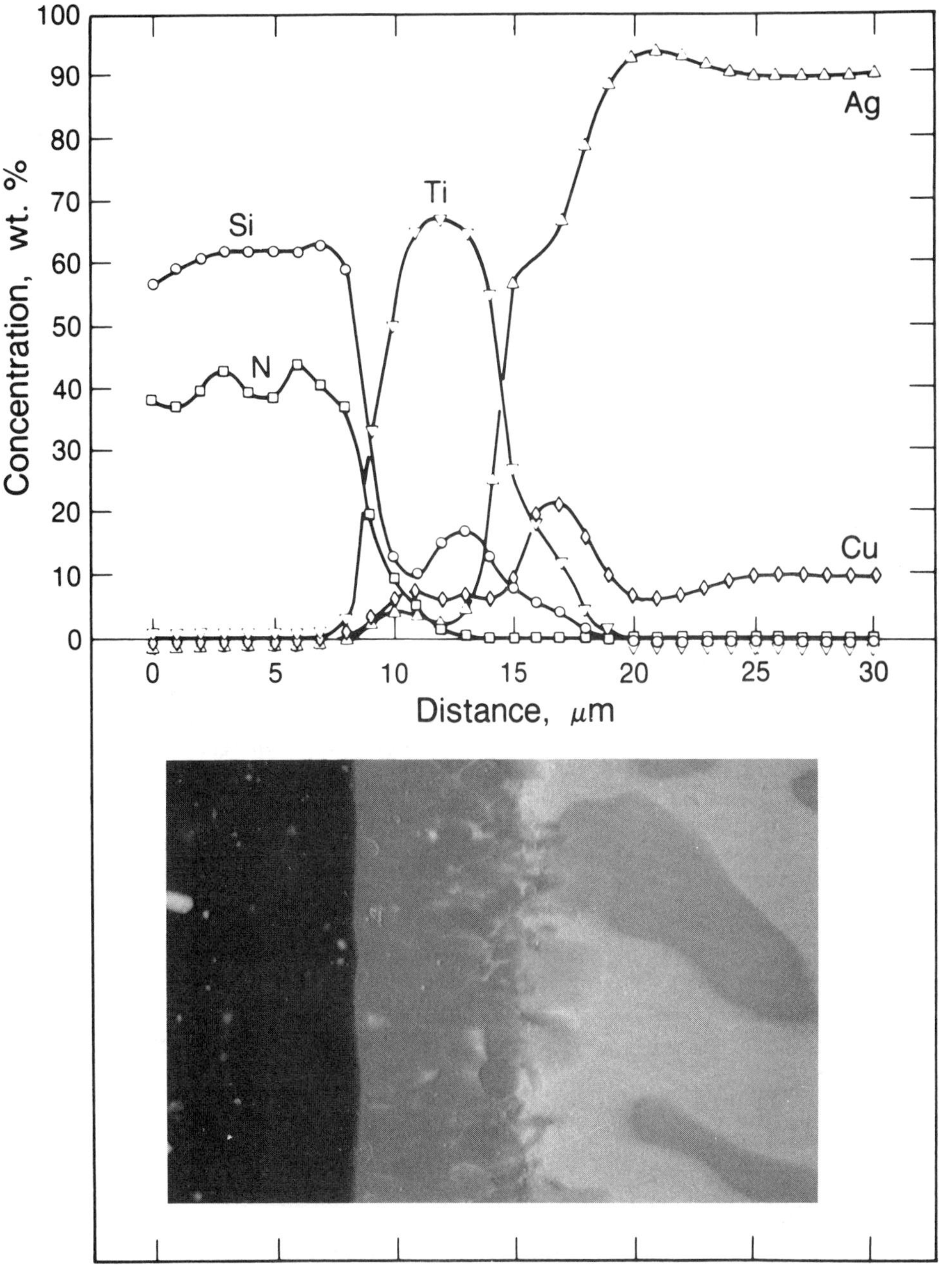

Fig 11 Interface from heating Cusil ABA (1.5 wt% Ti) (GTE Products Corp., Wesgo Division) on Si_3N_4 for 20 min at 900 °C (1650 °F) in argon. Four zones are visible: (1) unreacted Si_3N_4; (2) an ≈10 μm thick layer of TiN and titanium silicides; (3) a mixed zone containing titanium nitrides, silicides, and Cu-Ti eutectic structure; and (4) a Cu-Ag eutectic structure. Microscope scans show compositional variation through the interface.

angle and rapid spreading indicative of interfacial reactions. Complete spreading, as defined by a zero contact angle, was reached after 50 min. Ticusil (4.5 wt% titanium) reacted and spread more rapidly, reaching a zero contact angle in about 32 min.

On heating at 900 °C (1650 °F), the titanium segregates to the Si_3N_4 interface and forms a discernible reaction layer as seen in Fig 11 for Cusil ABA. The driving force for titanium migration is the energetically favorable reactions of titanium with Si_3N_4 such as

$$4Ti + Si_3N_4 = 4TiN + 3Si \qquad (Eq\ 13)$$

$$\Delta G^\circ_R(1200K) = -30.71\ kcal\ (-128.5\ kJ)$$

$$5Ti + Si_3N_4 = Ti_5Si_3 + 2N_2 \qquad (Eq\ 14)$$

$$\Delta G^\circ_R(1200K) = -9.80\ kcal\ (-41\ kJ)$$

The silicon and N_2 that are products of reactions (Eq 13 and 14) can react further with titanium to form nitrides and silicides. The titanium that has not reacted during the time the specimen is at temperature forms the phase-separated copper-titanium eutectic structure on cooling. The reaction compounds are compatible and in equilibrium with both the substrate and the braze as a solid. The details of this study appear in Ref 37.

Similar reaction mechanisms are being found for active metal brazing of AlN (Ref 38).

The next article "Glass/Metal and Glass-Ceramic/Metal Seals," discusses more specific examples of sealing, while the articles that follow cover physical principles which deal with the strength of the assemblies.

REFERENCES

1. J.W. Gibbs, *Collected Works*, Vol 1, Longmans-Green, 1928
2. W.D. Harkins, *Z. Phys. Chem.*, Vol 139, 1928, p 647; *The Physical Chemistry of Surface Films*, Reinhold, 1952
3. G.A. Somorjai, *Chemistry in Two Dimensions: Surfaces*, Cornell University Press, Ithaca and London, 1981, p 30–31
4. D.M. Mattox and D.A. Rigney, Adhesion Processes in Technological Applications, *Mater. Sci. Eng.*, Vol 83, 1986, p 189–195
5. J.A. Pask and R.M. Fulrath, Fundamentals of Glass-to-Metal Bonding: VIII, *J. Am. Ceram. Soc.*, Vol 45 (No. 12), 1962, p 592–596
6. I.A. Aksay, C.E. Hoge, and J.A. Pask, Wetting Under Chemical Equilibrium Conditions, *J. Phys. Chem.*, Vol 12 (No. 78), 1974, p 1178–1183
7. A.W. Adamson, *Physical Chemistry of Surfaces*, 4th ed., John Wiley & Sons, 1982, p 339
8. P.R. Sharps, A.P. Tomsia, and J.A. Pask, Wetting and Spreading in the Cu-Ag System, *Acta Metall.*, Vol 29 (No. 7), 1981, p 855–865
9. V.K. Nagesh and J.A. Pask, Wetting of Nickel by Silver, *J. Mater. Sci.*, Vol 18, 1983, p 2665–2670
10. F.H. Buttner, E.R. Funk, and H. Udin, *J. Phys. Chem.*, Vol 56, 1952, p 657
11. M. Appel and J.A. Pask, Interdiffusion and Moving Boundaries in NiO-CaO and NiO-MgO Single Crystal Couples, *J. Am. Ceram. Soc.*, Vol 54 (No. 13), 1971, p 152–158
12. S. Morozumi, M. Kikuchi, and T. Nashino, Bonding Mechanisms between Alumina and Niobium, *J. Mater. Sci.*, Vol 16, 1981, p 2137–2144
13. M. Ruhle, K. Burger, and W. Mader, Structure and Chemistry of Grain Boundaries in Ceramics and Metal/Ceramic Interfaces, *J. Microsc. Spectrosc. Electron.*, Vol 11, 1986, p 163–177
14. R.E. Tressler, T.L. Moore, and R.L. Crane, Reactivity and Interface Characteristics of Titania-Alumina Composites, *J. Mater. Sci.*, Vol 8, 1973, p 151–161
15. S.T. Tso and J.A. Pask, Wetting and Adherence of Na-Borate Glass on Gold, *J. Am. Ceram. Soc.*, Vol 62 (No. 9–10), 1979, p 543–544
16. M.P. Borom and J.A. Pask, Role of 'Adherence Oxides' in Development of Chemical Bonding at Glass-Metal Interfaces, *J. Am. Ceram. Soc.*, Vol 49 (No. 1), 1966, p 1–6
17. M.P. Borom, J.A. Longwell, and J.A. Pask, Reactions Between Metallic Iron and Cobalt Oxide-Bearing Sodium Disilicate Glass, *J. Am. Ceram. Soc.*, Vol 2 (No. 50), 1967, p 61–66
18. A.P. Tomsia and J.A. Pask, Kinetics of Iron-Sodium Disilicate Reactions and Wetting, *J. Am. Ceram. Soc.*, Vol 64 (No. 9), 1981, p 523–528
19. J.J. Brennan and J.A. Pask, Effect of Composition on Glass-Metal Interface Reactions and Adherence, *J. Am. Ceram. Soc.*, Vol 56 (No. 2), 1973, p 58–62
20. C.E. Hoge, J.J. Brennan, and J.A. Pask, Interfacial Reactions and Wetting Behavior of Glass-Iron Systems, *J. Am. Ceram. Soc.*, Vol 56 (No. 2), 1973, p 51–54
21. A.P. Tomsia, F. Zhang, and J.A. Pask, Reactions and Bonding of Sodium Disilicate Glass with Chromium, *J. Am. Ceram. Soc.*, Vol 68 (No. 1), 1985, p 20–24
22. A.P. Tomsia and J.A. Pask, Chemical Reactions and Adherence at Glass/Metal Interfaces: An Analysis, *Dent. Mater.*, (No. 2), 1986, p 10–16
23. J.A. Pask and A.P. Tomsia, "Wetting, Spreading and Reactions at Solid/Liquid Interfaces," *Surfaces and Interfaces in Ceramics and Ceramic-Metal Systems*, J.A. Pask and A.G. Evans, Ed., Plenum Publishing, 1981
24. A.P. Tomsia and J.A. Pask, Bonding of Dental Glass to Ni-Cr Alloys, *J. Am. Ceram. Soc.*, Vol 69 (No. 10), 1986, p C239–240
25. H. Mizuhara, "Step Brazing Using Active Alloys," presented at the 37th Pacific Coast Regional Meeting of American Ceramic Society, San Francisco, 28–31 Oct 1984
26. M.G. Nicholas, T.M. Valentine, and M.J. Waite, The Wetting of Alumina by Copper Alloyed with Titanium and Other Elements, *J. Mater. Sci.*, Vol 15 (No. 9), 1980, p 2197–2206
27. M.G. Nicholas, "Bonding Ceramic-Metal Interfaces and Joints," presented at *Ceramic Microstructures '86: Role of Interfaces Conference*, University of California, Berkeley, July 1986
28. A.C.D. Chaklader, A.M. Armstrong, and S.K. Misra, Interface Reactions Between Metals and Ceramics: IV, Wetting of Sapphire by Liquid Copper-Oxygen Alloys, *J. Am. Ceram. Soc.*, Vol 51 (No. 11), 1968, p 630–633
29. M.E. Twentyman, High Temperature Metallizing, Part 1, The Mechanism of Glass Migration in the Production of Metal-Ceramic Seals, *J. Mater. Sci.*, Vol 10, 1976, p 765–776
30. L.W. Bean, The Sintering of Molybdenum Metallizing, *Trans. J. Brit. Ceram. Soc.*, Vol 70 (No. 3), 1971, p 121–122
31. M. Hirota, Mechanism of Mo-Mn-Ti Metallizing and Cu Brazing in Metal-to-Ceramic Seals, *Trans. Jpn. Inst. Met.*, Vol 10 (No. 2), 1969, p 98–106
32. E.P. Denton and H. Rawson, The Metallization of High-Al_2O_3 Ceramics, *Trans. J. Brit. Ceram. Soc.*, Vol 59, 1960, p 25–37
33. D.M. Mattox and H.D. Smith, Role of Manganese in the Metallization of High Alumina Ceramics, *Am. Ceram. Soc. Bull.*, Vol 64 (No. 10), 1985, p 1363–1367
34. K. Suganuma, T. Okamoto, M. Koizumi, and M. Shimada, Effect of Surface Damage on Strength of Silicon Nitride Bonded with Aluminum, *Am. Ceram. Soc. Bull.*, Vol 1 (No. 4), 1986, p 356–360
35. P.F. Becher and S.A. Halen, Solid-State Bonding of Si_3N_4, *Am. Ceram. Soc. Bull.*, Vol 58 (No. 6), 1979, p 582–586
36. R.E. Loehman and A.P. Tomsia, Joining of Ceramics, *Am. Ceram. Soc. Bull.*, Vol 67 (No. 2), 1988, p 375–380
37. R.E. Loehman, A.P. Tomsia, J.A. Pask, and S.M. Johnson, Bonding Mechanisms in Silicon Nitride Brazing, *J. Am. Ceram. Soc.*, Vol 73 (No. 3), 1990, p 552–558
38. R.K. Brow, R.E. Loehman, A.P. Tomsia, and J.A. Pask, Interface Interactions During Brazing of AlN, *Advances in Ceramics*, Vol 27, 1989, p 189–196

Glass/Metal and Glass-Ceramic/Metal Seals

Antoni P. Tomsia and Joseph A. Pask, Lawrence Berkeley Laboratory and Department of Materials Science and Mineral Engineering, University of California
Ronald E. Loehman, Electronic Ceramics Division, Sandia National Laboratories

GLASS/METAL AND GLASS-CER-AMIC/METAL SEALS have found wide application in different engineering and electronic applications. Incandescent and vapor lamps, electron tubes, electrical feedthroughs, and housings for semiconductors are some of the many applications that use glass/metal or glass-ceramic/metal seals. In a broad sense the field also includes porcelain enameling, high-temperature protective coatings, joining ceramic to ceramic and/or metals through a glassy phase (high-pressure sodium lamps, magnetic recorder heads, and high-voltage feedthroughs), and composite materials.

In composite structures, such as lamps and electronic tubes, glass is used because it transmits light, can be easily formed into useful shapes, and provides electrical insulation and chemical inertness in corrosive or oxidizing environments. When used for protective coatings and in composites, the glass resists oxidation (especially at elevated temperatures). The primary focus of this article is glass/metal and glass-ceramic/metal seals. Many of the principles from a theoretical viewpoint, however, apply to glass/glass, glass/ceramic, and ceramic/metal joining. Glass and metal can also be joined by organic adhesives or by metallizing and soldering techniques, but they are less widely used and will not be covered in this article.

In many applications, one desires to make an electrical connection through a bulkhead that is hermetic, that is, one that is vacuum-tight. That requires strong adherence between glass and metal. The basic requirements for strong glass/metal seals are chemical bonding and favorable stress gradients in the interfacial zone. It is generally recognized that a number of factors must be controlled if a satisfactory bond or seal between two dissimilar constituents is to be produced. These factors can be grouped into three general categories: physical, chemical, and processing.

There have been vast numbers of reports and monographs on the general problem of bonding. Some are specific to the area of glass/metal seals, but others deal with related topics. The bibliography on this subject, covering the period from 1950 to 1990, contains nearly 2500 citations, and this does not include the extensive patent literature on the topic. The size of the available data base reflects the complexity of the subject and also its considerable industrial importance. The term bonding as used in this article refers to chemical bonding at the interface, whereas the term adherence also takes into account the effects of physical factors on the strength of the assembly.

Composition and Structure of Glass and Glass-Ceramic Materials

The most general definition of glass is that it is an amorphous solid with no long-range order. According to another widely used definition, glass is an inorganic product of fusion that has been cooled to a solid condition below its glass transition temperature, T_g, without crystallizing. Below T_g there are no further rearrangements of the atoms and the only contraction is a result of smaller thermal vibrations that are the origin of the thermal expansion or contraction of the corresponding crystal. These properties are illustrated schematically for a glass-forming oxide in Fig 1. The melt solidifies without the formation of crystals, has a higher specific volume, and thermodynamically is in a metastable state as compared with the corresponding crystal. This condition implies that if either a high enough temperature or a long enough time were available, crystallization to a thermodynamically more stable state would occur.

Not all oxides will, by themselves, form glasses. As shown in Table 1, glass-forming ability appears to be related to the viscosity of the melt at the melting point. This table shows that some materials can be either crystalline or a glass, depending upon the rate of cooling. The best example is SiO_2, which in nature generally occurs as quartz but which is used industrially as silica glass.

In addition to the glass- or network-forming oxides just described, there are two other classes of oxides—the network-modifying oxides and the intermediate oxides. A modifying oxide is one which is itself incapable of forming a continuous network, and its presence leads to a weakening of the polymeric glass structure. Typical examples of network modifiers are the alkali oxides, Na_2O and K_2O, and alkaline earth oxides, MgO and CaO. The intermediate oxides are capable of taking part in the glass network but are unable to form a glass by themselves at ordinary cooling rates. The most important of these is Al_2O_3.

In a glass there is a random polymeric network, which is formed by the oxygen and the glass-forming and intermediate ions, that is broken in places by the modifying ions. The network is irregular with no long-range order as originally described by Zachariasen in 1936. This absence of periodicity accounts for the very diffuse bands found in the x-ray diffraction patterns of glasses.

More recently a further refinement of the Zachariasen model has been suggested. The above model predicts a homogeneous structure throughout the glass but evidence now suggests that there is local inhomogeneity at about the 10 to 100 nm level. For example, in a sodium-silicate glass composition, electron microscopy shows that there is phase separation into soda-rich and silica-rich regions.

Glass-ceramics are another class of materials that are assuming increasing importance in glass/metal sealing. Glass-ceramics are materials that are processed and formed as glasses and then converted into a crystalline ceramic by a subsequent heat treatment. Whether a crystalline or glassy phase is formed depends upon many factors such as components present, the rate of cooling, and the presence or absence of nucleating agents. Crystallinity in glass-ceramic materials may exceed 90%. In a typical commercial glass-ceramic, grain sizes are generally smaller than 1 μm.

Classification of Glass/Metal Seals

Glass/metal seals can be classified in a number of ways (Ref 1). The following condensed listing is not complete, but it provides an indication of the terminology and also

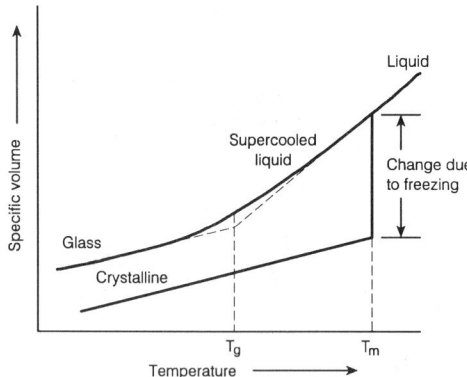

Fig 1 The difference in the relationship between a glass and a crystalline solid when the specific volume is plotted against the temperature

reflects the wide variety of approaches that can be used:

- Classification based on the geometry of the metal parts, for example, bead, tubular, disk, ribbon, and feather-edge (Housekeeper)
- Classification based on the type of metal or alloy used for the metallic parts, for example, molybdenum, platinum, copper, Kovar, and iron-nickel alloys
- Classification based on the type of glass used, for example, soft glass (soda-lime-silica or lead oxide-mixed alkali oxide-silica with a working temperature in the flame of 800 to 1000 °C, or 1470 to 1830 °F, with a coefficient of thermal expansion, CTE, of above 50×10^{-7}/°C); and hard glass (borosilicate and alumino-silicate glasses, or vitreous silica with working temperature in the flame of 1000 to 1300 °C, or 1830 to 2370 °F, and a CTE of below 50×10^{-7}/°C)
- Classification based on the type of joining, for example, mechanical, adhesive, fused glass, solder glass, and metallization
- Classification based on the technique used to make them, for example, flame, induction, frit, pressure-diffusion, tape, and high-frequency resistance welding
- Classification into two categories based on the level of stress in the glass, for example, matched or balanced, and unmatched or unbalanced (compression and ductile metal or Housekeeper)

Table 1 Glass forming ability of various compounds

| Oxide | Melting point | | Viscosity at melting point, poise |
	°C	°F	
As$_2$O$_3$	309	588	10^6
B$_2$O$_3$	450	842	10^5
GeO$_2$	1115	2039	10^7
SiO$_2$	1710	3110	10^7
H$_2$O	0	32	0.02
LiCl	613	1135	0.02
CdBr$_2$	567	1053	0.03

The final listing is a frequently used classification. In matched seals, the coefficient of thermal expansion of the glass and metal are similar over the entire temperature range of sealing. Sealing is the result of good chemical bonding between the glass and the metal at the interface. Matched glass seals can be made between glasses and metals, ceramics, or other glasses. Figure 2 shows a typical design of a matched glass/metal seal.

In unmatched seals, the coefficients of thermal expansion of the various parts are different over the temperature range of sealing. These seals are divided into two groups: compression seals and ductile metal, or Housekeeper, seals. In compression seals, joining is accomplished when the metal establishes a compressive hoop stress on the glass. With these seals, mechanical bonding may be sufficient, but the development of chemical bonding is always desirable. Compression seals are made with only the metals applying the compressive stress. Figure 3 shows a typical design of a compression seal.

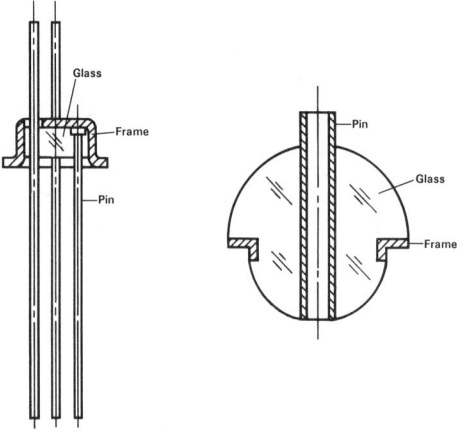

Fig 2 Typical design of matched glass/metal seal

Ductile metal, or Housekeeper, seals were introduced by G.W. Housekeeper to join many types of glasses to copper (Ref 2). Figure 4 shows typical steps in making a Housekeeper seal between copper and glass tubing. Copper has a relatively high thermal expansion (18×10^{-6}/°C) and does not match any of the usual commercial sealing glasses. Due to its ductility, however, copper can be joined to almost any glass, from Pyrex type to iron-sealing glasses, using the Housekeeper seal. Because of its high ductility, if copper is thin enough, it yields under the stresses caused by large CTE differences with sealed glasses, so that no cracking of the seal occurs. These types of seals are now used also with Kovar and several stainless steels. Housekeeper seals require good chemical bonding between glass and metal.

Matched glass/metal seals can be additionally divided into two groups: seals with hard glasses (low expansion glasses of the borosilicate type) and seals with soft glasses (high expansion glasses of special types).

Glass/metal compression seals can be divided into three subcategories (Ref 2):

- Standard matched compression seals where the glass and internal, coaxial pins have matched thermal expansion coefficients but with an external compression ring (frame) that has a considerably higher coefficient of thermal expansion. Typical materials include steel for the frame (CTE $= 13 \times 10^{-6}$/°C), soft, special glass, such as Schott 8422 (CTE $= 9 \times 10^{-6}$/°C), and Alloy 49 (50Ni-49Fe-1Cr and CTE $= 9 \times 10^{-6}$/°C) for the pin
- Low expansion, matched compression seals similar to the standard matched compression seals described above but made of hard glass whose coefficients of thermal expansion are lower by about 5×10^{-6}/°C. Typical materials include Alloy 49 for the frame, Schott 8250 glass

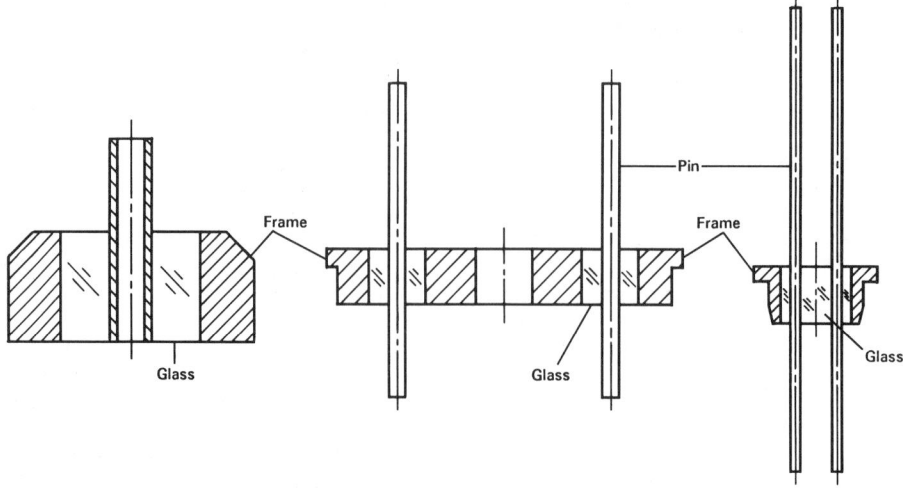

Fig 3 Typical design of compression glass/metal seal

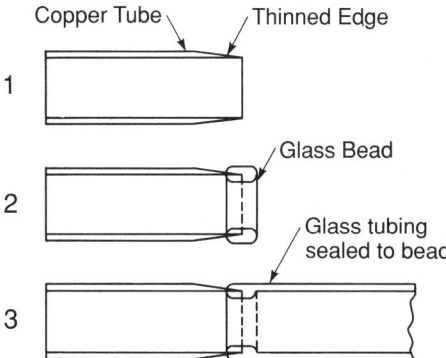

Fig 4 Steps in making a Housekeeper seal between copper and glass tubing. 1, The copper tube has to be thinned. 2, Glass bead is applied to the edge of the thinned copper tube. 3, The glass tubing is then sealed to the bead.

(CTE = 5×10^{-6}/°C), and Kovar (CTE = 5×10^{-6}/°C) for the pin
- Reinforced compression seals where the thermal expansion of the pin is lower than that of the sealing glass, resulting in reinforcement (amplification) of the radial compressive stress on the pin. Typical materials include steel for the frame, soft glass, and Kovar for the pin

Physical Factors in Making Glass/Metal Seals

The CTE of metals is usually considerably greater than that of glasses and ceramics. Thus, the most important requirement in developing a successful glass/metal seal is a proper matching of the thermal expansion or contraction characteristics of the bonded materials. Ideally, one strives to develop a slight compressive stress in the glass component of the seal during cooling from the firing or maturing temperature. The compressive strength of glasses usually exceeds the tensile strength by about 10 to 20 times and is approximately 20 to 80 MPa (2.9 to 11.6 ksi). Thermal expansion mismatch also results in shear stresses acting along glass/metal interfaces that tend to separate the glass from the metal. The bond developed between the glass and metal must be of sufficient strength to withstand these stresses, otherwise adherence failure will occur. If bonding is weak, a crack will propagate at the interface, causing separation of the bonded components. If bonding is strong and the CTE difference is large, a crack will nucleate in the glass very close and parallel to the interface.

In considering thermal expansion/contraction characteristics, the temperature range of interest is more extensive than sometimes appreciated. In order to obtain a correct indication of the interfacial strain, it is essential to consider thermal expansion behavior up to the set point temperature of the glass, which

is approximately the glass transition temperature, T_g, and not just to the 300 °C (570 °F) usually given. This point is illustrated in Fig 5 by the following example. If a seal is assumed to be stress-free at 300 °C (570 °F), on cooling a differential expansion of 0.00025 (unit/unit) would be predicted at room temperature. However, stress development initiates in the seal on cooling once the temperature drops below the set point of the glass (500 °C, or 930 °F in this example). If the entire thermal expansion curve is evaluated, a much lower (and more acceptable) differential expansion of only 0.00007 unit/unit results. By examining the offset between the curves representing the metal and the glass, it can be seen that the actual concern in this particular case would be the integrity of the system during cooling; the system undergoes a maximum (transient) expansion differential of 0.00032 unit/unit (at 400 °C, or 750 °F). If the system can be cooled through 400 °C (750 °F) without breaking, stresses at room temperature will be relatively modest.

Failure of glass/metal and ceramic/metal assemblies is often thought to arise at regions of high residual stress. The interfacial zone between metal and the glass is one region where the high residual stress and flaws can arise. These stresses are mainly due to reactions at the interface and diffusion of the metal ions into the glass. As a result of this process, the properties of the interfacial zone, such as the CTE, T_g, and elastic modulus will more than likely be different from those of the bulk glass.

Residual stresses in glass/metal systems arise from several sources, but the dominant one is the difference in thermal expansion coefficients of the metal and the glass, which results in different contractions during cooling from the softening or set point temperature to the use temperature. The magnitude of the residual stress is given by:

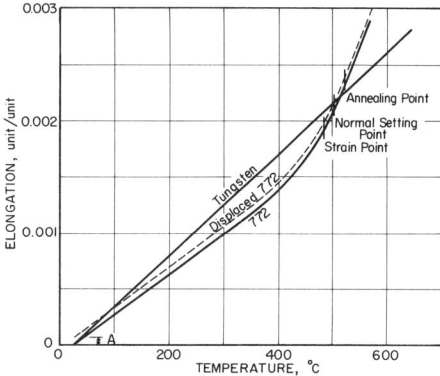

Fig 5 To determine differential expansion in a glass/metal seal, the glass expansion curve is displaced until the setting point for the conditions of cooling coincides with the metal expansion curve. The spread between these curves then indicates differential expansion, which governs the degree of stress on cooling. The indicated differential expansion, A, determines the internal stresses in the seal at room temperature.

$$\sigma_R = CE \int_{T_R}^{T_0} \Delta\alpha \Delta T \qquad \text{(Eq 1)}$$

where $\Delta\alpha$ is the difference in thermal expansion coefficients of the metal and the glass (in general, a function of temperature), T_R is the use temperature, T_0 is the softening temperature (or set point), E is the elastic modulus, and C is a constant which is dependent upon the geometry and relative thicknesses of the two materials. There is some sensitivity of σ_R to thermal history because the set point can be dependent upon cooling rate and thermal cycling (Ref 3), and because the thermal contraction of the glass can be influenced by crystallization, at a temperature below T_0, of phases with different densities than their parent glass. Such changes in thermal expansion may produce a serious thermal expansion mismatch. Thermal gradients in the glass, produced by rapid heating and cooling, can also induce either permanent residual stresses (if the gradient exists during cooling through the set point, that is, thermal tempering), or transient stresses (if the gradient is established at temperatures below the set point). Therefore, residual stresses can be minimized to some extent by control of thermal history. However, for a given glass/metal system, a minimum residual stress level at room temperature is dictated by the expansion coefficient difference, $\Delta\alpha$, and the geometry of the system (through the constant C). The dependence of σ_R on geometry provides a strong incentive for devising methods to measure residual stresses in actual components, rather than in specially constructed test specimens.

Another physical characteristic or factor that is important is the nature of surfaces being bonded. A highly irregular surface finish, produced either by abrasion or chemical erosion, can provide a mechanical contribution to adherence. There is evidence that surface roughness produced by sandblasting or chemical etching of the metal substrate can provide mechanical interlocking and also increase the surface area for glass attachment (Ref 4). Although such interfacial roughening contributes to adherence by producing a more diffuse (wider) transition zone and increasing the contact area, it is not necessary if a chemical bond is attained throughout the interface and the CTE difference is not too large.

The adherence between dissimilar phases that are brought into contact results from the interaction of surface (and possibly near-surface) atoms, ions, and molecules of the individual phases. These interactions stem from a difference between the environments of atoms on the surface and those in the interior, and the result is a net force of attraction between the contacting surfaces. Atoms in the interior of a solid or liquid are symmetrically coordinated (and bonded), while those on the surface are asymmetrically coordinated. As a result, surface atoms have an excess energy in comparison to bulk atoms. This difference is manifested as the surface energy. The high surface energy of metals (ranging from 480 ergs/cm^2 for lead to 1840 ergs/cm^2 for iron) produces a strong driving force for shape changes that reduce the interfacial area and chemical changes that lower the surface energy; both produce a lower energy state.

The surface free energy can be lowered through chemical (surface) interactions. For example, metal surfaces readily adsorb gases. On an atomic scale, dipole-dipole interactions, electron transfer, and covalent bond formation, all of which can be termed chemical or electrical interactions, may take place and result in bonding (Ref 5). Some gases, for example, oxygen, react chemically because of the strong electropositive nature of the metal atoms, and the thermodynamic driving force to form an oxide. Monolayers of oxygen or oxide persist, even at low pressures and high temperatures, on metal surfaces that otherwise appear clean, including the surfaces of noble (precious) metals. As a result, the high sensitivity of metal surfaces to oxygen, and possibly to other gases as well, must be taken into account during study of the adherence mechanism.

Chemical Factors in Making Glass/Metal Seals

The two most important chemical factors in the development of good adherence at glass/metal interfaces are (1) the formation of an intimate interface in contact on the atomic level and (2) the reactions to reach stable chemical equilibrium at the interface which results in chemical bonding. An intimate solid/liquid interface is formed when the liquid glass wets or spreads and thereby penetrates irregularities and distributes itself on the metal surface. Reactions invariably form more stable phases. At glass/glass, ceramic/ceramic, and metal/metal interfaces either solution or compound formation takes place readily in order to reach lower free energy compositions. The simplest reaction is the solution of one phase by the other, leading to and maintaining saturation in both phases at the interface. A continuation of the reaction is associated with diffusion into the bulk. The overall reaction is thus referred to as diffusion bonding. Such interfaces, as well as any compounds that form, are chemically compatible with both phases.

Glass/metal as well as ceramic/metal interfaces, however, generally undergo redox reactions at the interface, that is, oxidation of the metal and reduction of a cation in the glass or ceramic (Ref 6, 7). In the case of an oxide phase, the oxygen released by the reduction of the cation forms an oxide with the metal. Saturation of the interface with the metal oxide and subsequent formation of an oxide layer drive the interface toward equilibrium and result in chemical bonding at the interface. Necessarily, the oxide layer is compatible with both its parent metal and the glass or ceramic. Commercially, the necessary compatible oxide layer at the interface is normally obtained by preoxidizing the metal, which itself is a redox reaction, before the application of the glass (Ref 8).

Wetting and Spreading. In studying glass/metal interfaces by a sessile drop technique (Ref 9–11), the temperature must be high enough so that the glass will be in the liquid state. The surface energy of the solid is reduced when the liquid wets it. Wetting of the metal can be associated with either van der Waals bonding or chemical bonding at the interface. Liquids wet more strongly, however, when strong chemical bonds form at the interface, resulting in a smaller contact angle.

Very few systems are initially in chemical thermodynamic equilibrium. As a result, the possibility of a chemical reaction is almost always present. If the reaction does not occur because of unfavorable (sluggish) kinetics, then the system may be in a state of metastable chemical equilibrium. If a reaction does occur in response to the driving force towards stable chemical equilibrium, it initiates at the interface. During this transient stage, as equilibrium is approached, interfacial energies are modified and the measured contact angles are therefore affected. If the existence of such reactions is not recognized, as may be the case when the reactions are proceeding slowly, the measured and reported values of contact angles do not correspond to the stable equilibrium state and are in this respect incorrect. Unfortunately, many of the reported results in the literature are incorrect or suspect owing to this effect. If the reaction proceeds sufficiently rapidly, spreading will occur.

Extensive studies have been made using the sessile drop technique. Such experiments provide a means of studying wetting and spreading of a liquid on a rigid solid and determining relative surface and interfacial energies. Proper interpretation provides information on whether a reaction is taking place at the interface and on the occurrence of contamination of the surfaces and its effect. The technique also provides a convenient method of studying the reactions that occur at the interface by characterizing cross sections obtained by cutting the frozen sessile drop assembly perpendicularly to the interface.

In glass/metal systems, the contact angle, θ, formed by the liquid glass on the metal substrate is acute when the glass wets the metal. This arises because the glass surface energy is less than that of the metal. Surface energies of silicate glasses do not vary significantly and are generally in the range of 250 to 400 ergs/cm^2. The surface energies of clean metals always exceed those of glasses and are typically in the range 500 to 2000 ergs/cm^2. The acute contact angle is the result of wetting of the solid by the liquid. The surface energy of the metal in contact with the liquid glass (γ_{sl}) is less than that for metal in contact with the vapor (γ_{sv}). The reduction in surface energy stemming from coverage by the liquid ($\gamma_{sv} - \gamma_{sl}$) provides the driving force for wetting. The resistance to continued wetting of the solid by the liquid drop stems from the eventual increase of total surface energy of the liquid as further extension occurs or by completion of the reaction. Wetting and spreading are covered extensively in the article "Wetting, Surface Energies, Adhesion, and Interface Reaction Thermodynamics" in this Section.

Reactions at Glass/Metal Interfaces. It was stated above that chemical bonding at glass/metal interfaces is associated with the tendency toward stable chemical equilibrium. Equilibrium requires the presence of phase compatibility at the interfaces as given by the equilibrium phase diagram for the system. Equilibrium is achieved by saturating the interfacial zone with a low valent oxide of the substrate metal which results in a transition layer of the oxide at the interface that theoretically has to be only one molecular layer thick.

Preoxidation of Metal. The practical procedure for achieving chemical bonding in glass/metal assemblies is to preoxidize the metal and apply the glass as shown schematically in Fig 6. The $Na_2Si_2O_5$ glass and iron metal system is used as an example (Ref 9–11). At temperature, the molten glass wets and dissolves any oxide on the iron. The glass at the interface immediately becomes saturated with Fe^{2+} because the dissolution rate of the oxide is faster than the diffusion rate of the dissolved oxide into the bulk glass. Chemical bonding at the interface is thus realized. The oxide layer also bonds chemically to the metal, which in turn is saturated with oxygen, at least at its surface. The saturated interfaces and the oxide layer thus have a chemical activity of one for Fe_xO, which is a requirement for chemical bonding in this system.

As the dissolved oxide at the interface diffuses into the bulk glass, a concentration gradient is formed. As the oxide dissolves, the concentration gradients become more extended, but the equilibrium saturation at the oxide/glass interface is always maintained

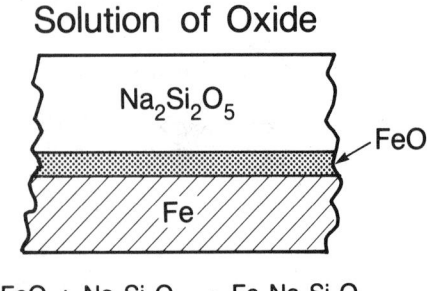

Solution of Oxide

$$xFeO + Na_2Si_2O_5 \rightarrow Fe_xNa_2Si_2O_{5+x}$$

Fig 6 Schematic cross sections of glass on preoxidized metal with equation representing solution reaction of oxide by the glass

because the reaction rate is faster. Dissolution of a metal oxide into the glass affects its coefficient of expansion because of an increase of its O/Si ratio and introduction of cations with different degrees of covalency. Thermal expansion coefficient gradients are proportional to compositional gradients, which generally results in more favorable stress gradients (Ref 12). It can be seen that the composition gradients become more extensive with increasing thicknesses of starting oxide. It is thus evident that many, if not most, experiments based on varying thicknesses of oxide that were presumably done to improve chemical bonding were actually strengthening the assembly by realizing more favorable stress gradients, which is similar to a graded seal.

Reactivity of Metals. In making glass/metal seals, glasses contact the metal substrate after dissolving any existing metal oxide. Redox reactions are then necessary to maintain saturation of the interface with the metal oxide. The reaction at the interface depends on the relative oxidation potentials of the metal and the cations in the glass. The metals can be arbitrarily grouped on the basis of reactivity with oxygen as low, intermediate, and high.

The low reactivity metals are those with relatively small negative Gibbs energies for oxidation and they include gold, platinum, and silver. Oxidation in air does not produce bulk oxides, although the metals can adsorb a surface oxide layer. Thicker layers generally do not form or, if formed, they dissociate on heating in normal vacuum and inert atmospheres, that is, at low $p(O_2)$. On firing gold or platinum/glass assemblies in air, the thin oxide layer that forms is sufficient to saturate the interfacial zone with the respective substrate oxide because of the very low solubility and diffusivity of these oxides in glasses. In case of silver, however, the Ag_2O forms a silver-silicate compound at the interface that is compatible with both silver and the Ag_2O-saturated glass. On firing gold or platinum/glass assemblies in low $p(O_2)$ atmospheres, however, no redox reactions occur at the interface and thus there is no chemical bonding. Redox reactions, however, occur in vacuum between silver and molten silicate glasses with the development of chemical bonding (Ref 13).

The group of metals that form intermediate strength oxides includes iron, nickel, and cobalt. The high reactivity group includes such metals as chromium, titanium, and zirconium. The general chemical behavior of these metals with molten silicate glasses have already been discussed.

The Making of Glass/Metal Seals

Processing factors are very important technically since they require reproducible control once laboratory scale procedures for making a glass/metal seal have been established. Processing variables also can control

the degree of oxidation of the metal surface, the chemical reaction behavior, and the nature of the strain development in the interfacial zone.

The metals commonly used in making glass/metal seals are platinum, tungsten, molybdenum, copper, iron, nickel, and alloys of various proportions of iron, nickel, cobalt, and chromium. Specific sealing glasses are needed for each metal or alloy primarily because of the need for a near-match of thermal expansions. Combinations of metal and glass recommended in commercial practice are listed in Table 2.

In making a particular seal design, the metal and glass are selected on the basis of specific properties and requirements. The metal is usually selected first based on electrical and thermal conductivity, thermal expansion, ability to be welded or soldered, mechanical strength, and chemical resistance requirements. The glass is then selected based primarily on electrical resistivity, dielectric strength, low gas permeability, environmental stability, and thermal expansion characteristics. In general, the CTE of the glass and metal should be within 10% of each other, with that of the metal being higher for conventional ring-geometry seals (see Fig 5). The contraction of interest is from the set point of the glass, often approximated as the glass transition temperature, to room temperature. If the difference in contraction is less than 100 ppm, the stress level will be quite safe, and even 500 ppm usually is satisfactory. A seal with a contraction difference between 500 and 1000 ppm can still be used but only for progressively smaller and thinner assemblies. A close match in expansion coefficients, however, is not as critical for thin glass coatings as for bulk specimens.

Glasses and glass-ceramics for sealing are usually prepared by melting oxides, carbonates, and hydroxides. One of the advantages of the glass and glass-ceramic materials is their ability to be made easily into a variety of shapes by casting, rolling, blowing, extrusion, and molding (see the section on "Glass Processing" in this Volume). The method used depends on the particular sealing technique.

Table 2 Recommended glass/metal seal combinations

Metal or alloy	Corning glass code No.	Schott glass No.
Tungsten	7720, 3320, 7750	8486, 8487
Kovar	7052, 7056, 7050, 7040	8454, 8436
Molybdenum	7052, 7056, 7050, 7040 1720, 1723	8250, 8245
Dumet, No. 4 alloy, or platinum	0120, 0010, 0080, 9010	8095, 8531, 8532
Fe-28%Cr	7570	
Fe-17%Cr	9019, 0129	8418
1010 steel	1990, 1991, 0110	8422

Specific examples of forms used for sealing are tubing, cast preforms of various shapes, and pressed powder preforms.

In preparation for sealing, the metal and glass parts are thoroughly cleaned, primarily to remove organics that normally interfere with wetting of the metal by the glass. Organics also form CO and CO_2 on reaction with oxides in the glass which produce bubbly seals that do not have a very good appearance, are weak, and may not be hermetic if the bubbling is severe. The usual method of removing organics is to oxidize them by heating in a wet hydrogen atmosphere to about 1100 °C (2010 °F). The cleaning may also achieve some etching or roughening of the metal surface. The clean metal is then preoxidized if the amount of oxidation that will occur during the sealing process is insufficient for good bonding. Figure 7 indicates the oxidation of Kovar, a cobalt-nickel-iron alloy that is used extensively in glass/metal seals (Ref 8).

The glass parts to be sealed frequently are provided as preforms. Those preforms are fixtured with the metal parts and then heated to a temperature above the softening point of the glass. During the thermal cycle it is essential that the glass should dissolve most of the surface oxide on the metal in order to form a bond with the metal. In most cases the amount of oxide, which is controlled by the time and temperature of its formation, is the principal factor in developing good adherence. If the oxide layer is too thin, upon its solution by the glass, chemical bonding is lost and the seal has poor strength (although it may be vacuum tight, especially under compressive stresses). If the oxide layer is too thick, a certain amount of the oxide may remain after sealing, and the strength of the seal is affected by the strength of the oxide and its adherence to the metal. Also, if the oxide layer is con-

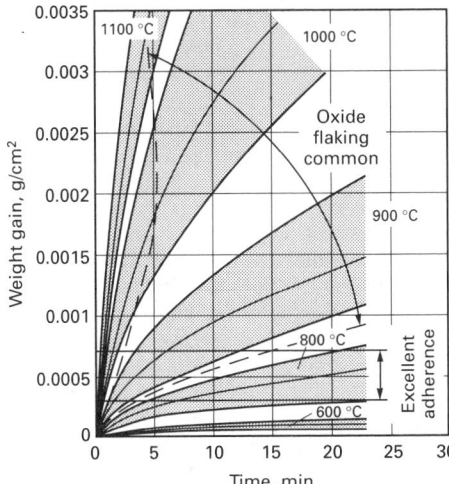

Fig 7 Oxidation as a function of time and temperature for small pieces of Kovar. The area inside the V-shaped dotted curves indicates conditions under which the greatest tendency for oxide flaking exists.

tinuous and porous, the seal may not be vacuum tight. Figure 7 also shows the temperature-time conditions under which flaking of the oxide on Kovar is most apt to occur. Such flaking results in poor adherence. It may be noted that the color of the glass at the interface at room temperature provides some indication of the quality of the seal. For example, in good Kovar-Corning 7052 glass seals with the proper amount of oxide, the color at the interface is dark bluish-gray. This color is not different if the oxide layer is too thick, but is lighter if there is insufficient oxide.

After the seal is made, the assembly is cooled to room temperature. The critical range for cooling is just below the transformation temperature of the glass. The assembly must be annealed or cooled slowly through this range in order to avoid destructive thermal strains in the glass. The importance of annealing increases with an increase in the thickness of the glass.

High-Pressure Sodium Vapor Lamp. The bonding of niobium to Al_2O_3 as used in sodium vapor lamps is a widespread example of glass/metal sealing. It represents probably one of the most technically sophisticated glass/metal seal products. Transparent alumina is used as the arc tube material in high-pressure sodium vapor lamps because it is unreactive with sodium vapor at operating temperatures ≥ 600 °C (≥ 1110 °F). The end cap and electrode structures are made of niobium to take advantage of similar CTEs to Al_2O_3 and to minimize mechanical stresses during thermal cycling (polycrystalline alumina CTE = 81 $\times 10^{-7}$/°C, niobium CTE = 78 $\times 10^{-7}$/°C from room temperature to 1000 °C, or 1830 °F). The Al_2O_3 tubes are sealed hermetically to the niobium end pieces using CaO-Al_2O_3 or CaO-MgO-Al_2O_3 glass frits that also contain small amounts of additives such as BaO, SiO_2, B_2O_3, or V_2O_3 (Ref 14). The assembly is fired in a vacuum furnace at 1450 to 1550 °C (2640 to 2820 °F) to melt the frit and to bond it to the niobium and Al_2O_3.

Two distinct interfaces exist in this system: Al_2O_3-glass and niobium-glass. In order to achieve equilibrium compositions that promote good bonding at both interfaces, the reacting phases must be compatible. At the Al_2O_3-glass interface, Al_2O_3 reacts to form a compound that is possibly a calcium-aluminate with an adjoining BaO-rich layer. Because the interface reactions are faster than diffusion of products away from the interface, near equilibrium conditions are maintained, which is favorable for formation of a strong bond. Depending on the cooling rate, some of the residual glass appears to crystallize and forms a glass-ceramic. The extent of crystallization on cooling is sensitive to the amount of Al_2O_3 dissolved, which in turn affects properties such as CTE (Ref 15).

The molten glass does not form strong chemical bonds to pure niobium. The interface thus is somewhat weak mechanically.

Proposed solutions to increase bond strength have included introducing an intermediate layer of WO_3 (Ref 16), coatings of titanium and zirconium hydrides (Ref 17), alloying niobium with 1 to 1.5% Zr (Ref 17), using an intermediate silicon layer (Ref 18), and using reactive metal brazes based on the Zr-V-Ti system (Ref 19). The objective of all these modifications is to increase the glass-metal reactivity in order to develop better bonding.

Figures 8 and 9 show two views of a cross section of a commercial Nb-Al_2O_3 high-pressure sodium lamp seal. Both figures show a scanning electron micrograph (SEM) of the glass-metal interface displayed below the electron microprobe elemental profiles of the same region. In Fig 8, niobium was alloyed with 1.5 wt% Zr to enhance bonding to the glass. The corresponding micrograph and compositional profile of the glass-Al_2O_3 interface is shown in Fig 9.

The microprobe data show that niobium and zirconium from the metal diffused 10 to 12 μm into the glass during sealing. The larger scale microprobe trace shows that zirconium is depleted for about 5 μm into the metal at the interface and that there is a symmetric diffusion profile of aluminum into the metal. The compositional data suggest a reaction sequence in which niobium and zirconium were oxidized by reaction with the glass at the metal interface to produce cations that diffused into the molten glass. These reactions reduced some Al^{3+} in the glass at the interface to $Al°$, which then alloyed with the niobium. The analytical techniques used are not sensitive to elemental valence so the exis-

tence of cationic species is only inferred. However, the proposed sequence is supported by the negative Gibbs free energy for the reaction:

$$3Zr° + 2Al_2O_3 = 4Al° + 3ZrO_2$$

$$\Delta G° \ (1527 \ °C) = -81.10 \ kJ \ (-19 \ kcal)$$

An additional driving force for this reaction could be provided by the free energy of mixing for alloying aluminum with niobium (Ref 7). Al_2O_3 is more stable than Nb_2O_5, so oxidation of niobium must be by reduction of some other species in the glass or by reaction with O_2 in the furnace atmosphere. The compositional variation in the calcium and aluminum in the glass is due to phase separation from partial crystallization.

Figure 9 illustrates the glass-Al_2O_3 side of the Nb-Al_2O_3 lamp seal. The main feature of the glass-Al_2O_3 interface is the 3 to 4 μm thick dark layer (indicating lower average atomic number in the backscattered electron image) between the bulk glass and the Al_2O_3. The compositional profiles show the dark layer is enriched in calcium and depleted in barium relative to the bulk glass. The reaction sequence inferred from these data is the reaction of Al_2O_3 with the molten glass with the formation of calcium aluminate and an enriched adjoining BaO layer. No oxidation-reduction reaction is necessary in this case because the calcium, barium, and aluminum remain as cationic species with unchanged valences.

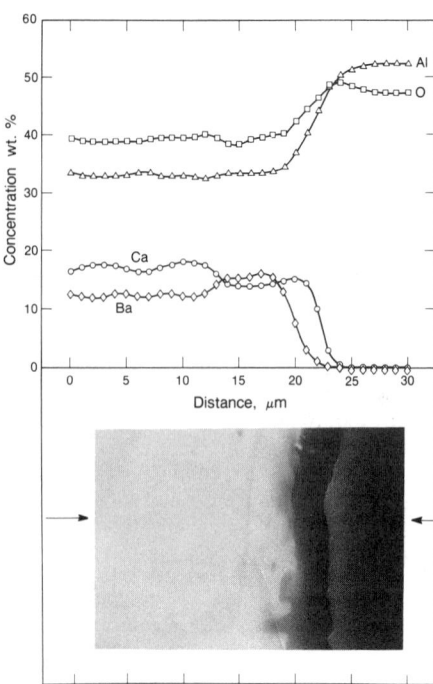

Fig 9 Cross section of Nb-Al_2O_3 lamp seal with microprobe compositional profiles of glass-Al_2O_3 interface

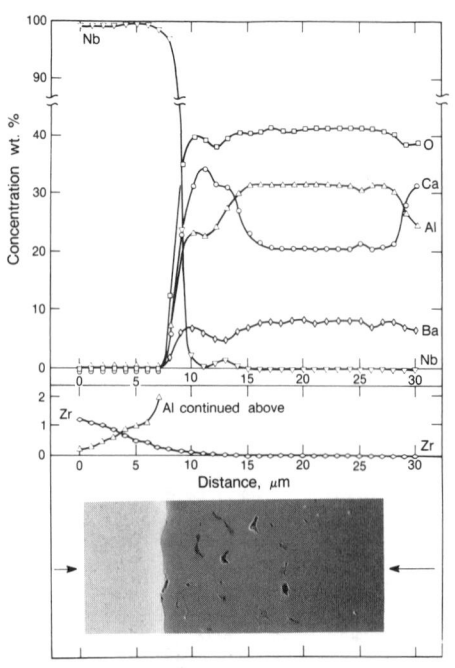

Fig 8 Cross section of Nb-Al_2O_3 lamp seal with microprobe compositional profiles of Nb-glass interface

Glass-Ceramic/Metal Seals

Glass-ceramics have several advantages over glasses and ceramics. Compared with ceramics, glass-ceramics can be formed into complex shapes more easily by methods common to glasses such as casting, rolling, and molding. Compared with glasses, glass-ceramics generally have superior mechanical properties and have better corrosion resistance because they are at least partly crystalline.

Glass-ceramics have another property that makes them particularly useful for metal-insulator seals. With proper control of crystallization, glass-ceramics can be made with a much wider range of CTEs than can be achieved with glasses or conventional ceramics, which allows them to be used as matched-expansion seals for many different alloys. For example, glass-ceramics can be made with expansion coefficients ranging from zero (β-eucryptite) to over $200 \times 10^{-7}/°C$ (mixtures of cristobalite and tridymite match the CTEs of copper and some stainless steels). The ability to fill re-entrant angles and complex internal shapes by viscous flow of the molten glass and then to crystallize it with subsequent heat treatment to match the CTE of the metal makes glass-ceramics particularly suited to applications where high strength of the system is important.

Glass-Ceramic to Metal Processing. The principles involved in designing and making seals with glass-ceramics are very similar to the making of the glass/metal seals. However, with glass-ceramics, seal properties can be varied over wide ranges by careful control of their crystallization.

Several commercial glass-ceramics are used for seals to metals. Pyroceram, Mycor (Corning), Mykroy, Re-X (GE), Zerodur (Schott) and S glass-ceramic (Sandia) are some examples. Although lithium silicate glass-ceramics are perhaps best known for very low expansion materials based on β-spodumene or β-eucryptite, other compositions in that family are capable of being crystallized to much higher CTE values. Many glass-ceramics in this family are useful for sealing to metals. McMillan reported a variety of compositions that use P_2O_5 as a nucleating agent to crystallize glass-ceramics with expansion coefficients ranging up to $160 \times 10^{-7}/°C$ (Ref 20). Borom *et al.* developed a P_2O_5 nucleated lithium silicate glass-ceramic for sealing applications that has a CTE of $110 \times 10^{-7}/°C$ (Ref 21). McCollister and Reed discovered an alternate route for the glass composition of Borom *et al.* that gives a glass-ceramic with a CTE of $145 \times 10^{-7}/°C$ that is useful for making high-strength hermetic seals to superalloys such as Inconel 718, Hastelloy C276, and several stainless steels (Ref 22). The particular mix of crystalline species obtained in this family depends on the glass composition and the heat treatment used, but typically they include one or more phases such as lithium metasilicate (Li_2SiO_3), lithium disilicate ($Li_2Si_2O_5$), cristobalite (SiO_2), or one of the lithium zinc silicates. Table 3 contains the CTEs of some of those individual compounds.

The thermal cycles used to process the vast majority of glass-ceramics employ an initial lower-temperature nucleation treatment, followed by a higher-temperature stage where the glass crystallizes. Somewhat more complex thermal cycles are used to make glass-ceramic/metal seals. For such seals, the glass initially must be heated to a relatively high temperature to make it flow into the metal part where it wets the surface and reacts to form an interface before the subsequent nucleation and crystallization portions of the cycle. In general, initially heating the glass to make it flow does not affect its later crystallization, so the same phases are produced regardless of whether the glass-ceramic is part of a seal or if it is made by the conventional nucleation and growth cycle.

The glass-ceramic composition developed by Borom, *et al.* (Ref 21) and listed in Table 4 is an exception to this generalization in that the sealing and conventional crystallization cycles produce two different phase mixtures. Specifically, a nucleation treatment at 550 to 650 °C (1020 to 1200 °F), followed by crystallization at 800 °C (1470 °F) gives mostly $Li_2Si_2O_5$ with a CTE of $110 \times 10^{-7}/°C$, whereas a sealing cycle with plateaus at 1000 °C (1830 °F), 550 to 650 °C (1020 to 1200 °F), and 800 °C (1470 °F) gives a mixture of Li_2SiO_3, $Li_2Si_2O_5$, and cristobalite with CTE of $145 \times 10^{-7}/°C$ (Ref 22). Heating this latter glass-ceramic longer at 800 °C (1470 °F) eventually leads to disappearance of the cristobalite, indicating that it is metastable at that temperature.

Figure 10(a) is a schematic representation of the thermal cycle used to fabricate a seal of S glass-ceramic to Inconel 718 and Hastelloy C276 and to crystallize the glass. The

Table 4 Examples of glass-ceramic compositions for sealing to Inconel 718, Hastelloy C276, and type 430 stainless steel (S glass-ceramics), to type 304 stainless steel (GC1014 and 1008), and to copper (SLM)

Compound/ property	Sealing glass/ composition, wt%			
	S glass-ceramics	GC1014	GC1008	SLM
SiO_2	67.1	67.0	71.0	69.3
Li_2O	23.7	24.5	25.0	20.2
Al_2O_3	2.8	...	3.0	2.2
K_2O	2.8	4.8	...	...
MgO	...	...	...	6.8
B_2O_3	2.6	2.2	...	...
P_2O_5	1.0	1.0	1.0	1.5
CTE $\times 10^{-7}/°C$	100	177	185	192

isothermal treatment at 1000 °C (1830 °F) serves several purposes. At that temperature the glass is fluid enough to flow into the Inconel 718 housing, to wet the housing and pins, and to bond to them chemically. On the ramp to 1000 °C (1830 °F) and during the 1000 °C (1830 °F) hold, a small amount of Li_3PO_4 crystallizes to form small (0.1 to 0.75 µm) crystals that serve as substrates for subsequent epitaxial growth of Li_2SiO_3, α-cristobalite (SiO_2), and $Li_2Si_2O_5$ that occurs during the later 650 and 800 °C (1200 and 1470 °F) thermal holds (Ref 23). The Li_2SiO_3 and α-cristobalite are desired high thermal

Table 3 Thermal expansion coefficients of crystalline species that may be present in glass-ceramics

Species	CTE $\times 10^{-7}/°C$
β-eucryptite ($Li_2O \cdot Al_2O_3 \cdot SiO_2$)	−86 at 20 to 700 °C (70 to 1290 °F)
β-spodumene ($Li_2O \cdot Al_2O_3 \cdot 4SiO_2$)	9 at 20 to 1000 °C (70 to 1830 °F)
Lithium disilicate ($Li_2O \cdot 2SiO_2$)	110 at 20 to 600 °C (70 to 1110 °F)
Quartz (SiO_2)	132 at 20 to 300 °C (70 to 570 °F)
	237 at 20 to 600 °C (70 to 1110 °F)
Cristobalite (SiO_2)	125 at 20 to 100 °C (70 to 212 °F)
	500 at 20 to 300 °C (70 to 570 °F)
Tridymite (SiO_2)	175 at 20 to 100 °C (70 to 212 °F)
	250 at 20 to 200 °C (70 to 390 °F)

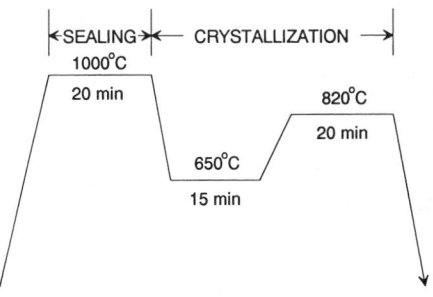

(a)

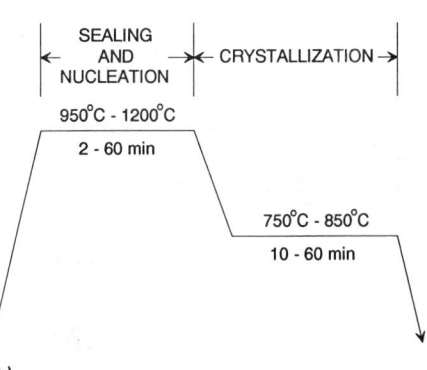

(b)

Fig 10 Typical thermal cycles for sealing glass-ceramics to Inconel 718 and Hastelloy C276 (a) and for sealing glass-ceramics to stainless steel and copper (b)

expansion phases that are the major crystalline species in the final glass-ceramic. The S glass-ceramic after this thermal cycle has a thermal expansion coefficient of 130 to 150 $\times 10^{-7}/°C$.

Figure 10(b) represents the thermal cycle used to produce the seal of S glass-ceramic to type 430 stainless steel. The crystallized glass-ceramic contains mostly lithium metasilicate ($LiSiO_3$) plus residual glass. It may also contain some lithium disilicate ($LiSi_2O_5$). The CTE is in the range of 90 to $100 \times 10^{-7}/°C$.

The thermal cycle used for joining other experimental lithium silicate glass-ceramic compositions to type 304 stainless steel and copper is similar to that shown in Fig 10(a). These glass-ceramics consist of various mixtures of cristobalite, tridymite, and quartz with lithium metasilicate, and sometimes lithium disilicate. Crystallization is by epitaxial growth on $LiPO_4$ crystallites formed during the 1000 °C (1830 °F) nucleation/flow step. The higher expansion results from having more of the various SiO_2 phases crystallize. The CTEs range from about $170 \times 10^{-7}/°C$ to over $210 \times 10^{-7}/°C$.

Glass-Ceramic/Metal Bonding Reactions. The general thermodynamic principles described in the previous article in this Section also apply to glass-ceramic interfacial reactions (see the article "Wetting, Surface Energies, Adhesion, and Interface Reaction Thermodynamics" in this Section). The main difference is that reaction products that form during the high-temperature sealing step can affect subsequent crystallization. Watkins and Loehman (Ref 24) studied the high-temperature sealing reactions between the molten S glass-ceramic and Inconel 718 and found that there is a distinct reaction zone adjacent to the metal, on the order of 25 to 100 μm thick, where the glass-ceramic microstructure is different from that of the bulk material (Fig 11). Crystallization in the reaction zone is altered because chromium from the Inconel 718 reacts at the 1000 °C (1830 °F) sealing temperature

with Li_3PO_4 crystallites in the melt to give $Cr_{12}P_7$ (the bright particles near the interface in Fig 11) (Ref 23). In the absence of Li_3PO_4, subsequent crystallization of the glass is surface nucleated, no cristobalite is formed, and the overall degree of crystallinity is lower than in the bulk glass-ceramic. The reaction-zone microstructure is similar to that of incompletely crystallized glass that has a thermal expansion coefficient 25% lower than the bulk glass-ceramic (Ref 25). That thermal expansion mismatch suggests that the interface may be significantly stressed.

The interfacial reactions that occur when the S glass-ceramic is sealed to Hastelloy C276 has also been investigated (Ref 26). Precipitates of molybdenum and phosphorus in the grain boundaries and along the glass-ceramic/metal interface were found. The microstructural observations can be rationalized by the following reaction:

$$6SiO_2 + Li_3PO_4 + 5CrO + xMo$$

$$= Mo_xP + 3LiCr(SiO_3)_2 + Cr_2O_3$$

Silica and Li_3PO_4 in the glass react with chromium and molybdenum from C276 to precipitate a molybdenum phosphide in the alloy grain boundaries at the glass-metal interface and $LiCr(SiO_3)_2$ particles in the reaction zone. $LiCr(SiO_3)_2$ has been identified in the interface by x-ray diffraction. The stoichiometry of the Mo_xP was not determined. The important conclusion to be drawn from these data on the reaction of the glass with both Inconel 718 and Hastelloy C276 is that crystallization at the glass/metal interface occurs differently than in the bulk.

In many glass-ceramic to metal seal applications the thin reaction zones described above are of no practical significance. However, for some demanding, high-strength uses those zones potentially can give rise to high interfacial stresses. The argument is as follows. Matching a given alloy CTE requires careful crystallization of the glass-ceramic. Interfacial reactions that change the crystallization alter the CTE. One then obtains a thin layer with a different CTE between metal and glass-ceramic layers with nominally matched CTEs. That difference in CTEs can lead to substantial interfacial stresses that should be considered in the mechanical design of high-performance devices (Ref 25).

Applications. Glass-ceramic/metal seals can be made by two general methods. In the first, the glass is crystallized independently and then joined to the metal in a second stage by ordinary ceramic/metal brazing techniques. The two advantages over conventional ceramic brazing are possible elimination of grinding steps because the glass can be cast to shape before it is crystallized, and the availability of a wider range of expansion coefficients with glass-ceramics than with conventional polycrystalline ceramics. In the second method for making glass-ceramic/metal seals, a glass preform is made to flow

into the metal part, then bonds to it, and finally is crystallized to the desired mixture of crystalline species all in one continuous process. As mentioned above, this second technique takes full advantage of glass-ceramic properties, especially when the seal requires the ceramic in a complex shape that is more easily obtained by glass flow into the part than by grinding or cutting. Also, the glass preform can be made from cast glass or from a sintered powder compact (pressed powder preform). Two less frequently used methods for joining glass-ceramics to metals are metallizing and brazing (see discussions of the moly-manganese process in the articles that follow in this Section), and bonding with an intermediate glass layer. In addition to their use in devices of various sorts, glass-ceramics can also be applied as coatings on metals or alloys.

REFERENCES

1. "Glass-to-Metal Seals," product information, No. 4830e, Schott, Jena Glasswerk Schott & Gen., Inc.
2. W.G. Housekeeper, The Art of Sealing Base Metals Through Glass, *J. Am. Inst. Elec. Eng.*, Vol 42, 1923, p 840–876
3. M.P. Borom and C.A. Johnson, Thermomechanical Mismatch in Ceramic-Fiber-Reinforced Glass-Ceramic Composites, *J. Am. Ceram. Soc.*, in press
4. R.L. Bertolotti, Porcelain-to-Metal Bonding and Compatibility, *Dental Ceramics*, Proceedings of the First International Symposium on Ceramics, J.W. McLean, Ed., Quintessence Publishing, 1983, p 415–431
5. M. Humenik, Jr. and T.J. Whalen, Physiochemical Aspects of Cermets, *Cermets*, J.R. Tinklepauch and B.W. Crandall, Ed., Reinhold Publishing, 1960, p 6–49
6. A.P. Tomsia and J.A. Pask, Chemical Reactions and Adherence at Glass/Metal Interfaces: An Analysis, *Dent. Mater.*, Vol 1 (No. 2), 1986, p 10–16
7. J.T. Klomp, Ceramic-Metal Reactions and Their Effect on the Interface Microstructure, *Ceramic Microstructures '86: Role of Interfaces*, J.A. Pask and A.G. Evans, Ed., Plenum Publishing, 1987, p 307–317
8. J.A. Pask, New Techniques in Glass-to-Metal Sealing, *Proc. of Inst. Radio Elec.*, Vol 26 (No. 2), 1948, p 286–289
9. A.P. Tomsia and J.A. Pask, Kinetics of Iron-Sodium Disilicate Reactions and Wetting, *J. Am. Ceram. Soc.*, Vol 64 (No. 9), 1981, p 523–528
10. J.J. Brennan and J.A. Pask, Effect of Composition on Glass-Metal Interface Reactions and Adherence, *J. Am. Ceram. Soc.*, Vol 56 (No. 2), 1973, p 58–62
11. C.E. Hoge, J.J. Brennan, and J.A. Pask, Interfacial Reactions and Wetting Behavior of Glass-Iron Systems, *J. Am.*

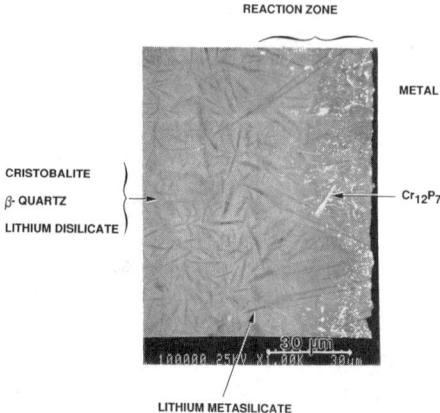

REACTION ZONE

METAL

CRISTOBALITE
β-QUARTZ
LITHIUM DISILICATE

$Cr_{12}P_7$

30 μm

LITHIUM METASILICATE

Fig 11 Inconel 718/S glass-ceramic reaction interface illustrating altered microstructure in reaction zone. Bright, elongated particles are $Cr_{12}P_7$.

Ceram. Soc., Vol 56 (No. 2), 1973, p 51–54

12. P. Mayer, J.A. Topping, and M.K. Murthy, Correlation Between Thermal Expansion of Glass and Glass-to-Metal Adherence, *J. Can. Ceram. Soc.*, Vol 43, 1974, p 43–46

13. V.K. Nagesh, A.P. Tomsia, and J.A. Pask, Wetting and Reactions in the Lead Borosilicate Glass-Precious Metal Systems, *J. Mater. Sci.*, Vol 18 (No. 11), 1983, p 2665–2670

14. C.I. McVey, High Pressure Sodium Lamps and Recent Improvements, *JIES*, Vol 8 (No. 2), 1979, p 72–77

15. R.E. Loehman and A.P. Tomsia, Joining of Ceramics, *Am. Ceram. Soc. Bull.*, Vol 67 (No. 2), 1988, p 375–380

16. W.C. Louden, "Niobium End Seal," U.S. patent 3,448,319, June 1969

17. P.J. Jorgensen, "Ceramic-Metal Bonding Composition and Composite Article of Manufacture," U.S. patent 3,598,435, Aug 1971

18. R.S. Bhalla, Improved End Seals for High-Pressure Sodium Lamp Arc Tubes, *JIES*, Vol 9 (No. 1), 1971, p 86–89

19. S.A.R. Rigden, B. Heath, and J.B. Whiscombe, "Closure of Tubes of Refractory Oxide Materials," U.S. patent 3,428,846, Feb 1969

20. P.W. McMillan, *Glass Ceramics*, 2nd ed., Academic Press, 1979, p 245–266

21. M.P. Borom, A.M. Turkalo, and R.H. Doremus, Strength and Microstructure in Lithium Disilicate Glass-Ceramics, *J. Am. Ceram. Soc.*, Vol 58 (No. 9–10), 1975, p 385–391

22. H.L. McCollister and S.T. Reed, "Glass-Ceramics Seals to Inconel," U.S. patent 4,414,282, Nov 1983

23. T.J. Headley and R.E. Loehman, Crystallization of a Glass-Ceramic by Epitaxial Growth, *J. Am. Ceram. Soc.*, Vol 67 (No. 9), 1984, p 620–625

24. R.D. Watkins and R.E. Loehman, Interfacial Reactions Between a Complex Lithium Silicate Glass-Ceramic and Inconel 718, *Adv. Ceram. Mater.*, Vol 1 (No. 1), 1986, p 77–80

25. S. Kunz and R.E. Loehman, Thermal Expansion Mismatch Produced by Interfacial Reactions in Glass-Ceramics to Metals Seals, *Adv. Ceram. Mater.*, Vol 2 (No. 1), 1987, p 69–73

26. R.E. Loehman, Processing and Interfacial Analysis of Glass-Ceramic to Metal Seals, *Technology of Glass-Ceramic, Glass, Ceramic or Ceramic-Glass to Metal Sealing*, M.D.-Vol 4, W.E. Maddeman, C.W. Mertin, and D.P. Kramer, Ed., American Society of Mechanical Engineers, 1987, p 39–45

Ceramic/Metal Seals

Howard Mizuhara and Toshi Oyama, GTE Products Corp.

CERAMIC-TO-METAL SEALS can be obtained using a variety of joining methods/processes. This article focuses on the most popular methods of obtaining seals between oxide ceramics and metals. Primary applications for oxide ceramic-to-metal seals are vacuum devices, such as laser and microwave tubes. Hermeticity and joint strength are emphasized in this article, because they represent basic requirements for vacuum applications, and the current vacuum industry standard for a hermetic seal, which limits the helium leak rate to no more than 10^{-9} cm^3/s, is utilized.

Process Evolution. A technique that predates most other joining processes used for vacuum devices involved oxidized Kovar and a glass envelope (Ref 1). The transition from the use of a glass envelope to an alumina ceramic envelope occurred during the 1950s. There are many advantages to using alumina rather than glass for the envelope (Ref 2). Compared to typical glass materials, alumina has better mechanical properties, greater thermal shock resistance, better dielectric properties, and higher temperature resistance. These advantages lead to longer life, lower noise, and higher vacuum in devices, because final pump-down can be carried out at higher temperatures.

The original development of alumina-to-metal joining technology (Ref 3) utilized a metallizing process, in which refractory metal powders were applied to the ceramic surface and sintered at high temperature ($\sim$1600 °C, or 2910 °F). The metallized ceramic was then brazed to a metal member using a conventional brazing filler metal. Nolte (Ref 4, 5) developed a metallizing mix consisting of molybdenum and manganese. The manganese addition decreased the sintering temperature to about 1400 °C (2550 °F), and provided greater latitude for processing. Again, the metallized ceramic was brazed to a metal member using a conventional brazing filler metal. Forty years later, the moly-manganese process is still the most common method of joining alumina ceramics to metals. Nearly all current moly-manganese processes are based on Nolte's development.

The active brazing process was also investigated extensively during the 1950s (Ref 6, 7). This process allowed direct brazing of ceramics and metals by using a brazing filler metal that was capable of wetting both ceramic and metal. Because this process involved only one step, it was more economical. However, the results obtained in the 1950s indicated that the moly-manganese process led to more consistent joint properties. Therefore, it became the predominant process.

Further development of the active brazing process occurred in the course of investigating appropriate bonding processes for non-oxide ceramics (Ref 8–21). It has now been established that reliable joints between ceramics and metals can be obtained by using the active brazing process (Ref 19). In retrospect, there are many reasons for the early failures that occurred when active brazing was applied to ceramic-to-metal seals, but the most important reasons are:

- Ductile active brazing filler metals were not available
- The requirement that vacuum brazing furnaces have low leak-up rates was not appreciated
- The importance of ceramic surface conditions was not well understood

In this article, both the moly-manganese and active brazing processes are reviewed in detail. Other joining methods that can be used to obtain ceramic-to-metal seals are mechanical attachment, adhesive bonding, solid-state (diffusion) bonding, fusion welding, soldering, electroless plating, and copper-based processes. Some of these techniques are not yet fully developed, and the others are not used as widely as the brazing techniques because of various limitations.

Materials

The materials that are commonly used in ceramic-to-metal joining are listed in Table 1. Of these, alumina is used most frequently and is therefore emphasized in this article. However, the same principles associated with its use equally apply to other oxide ceramics, such as zirconia and mullite, as well as to non-oxide ceramics. Also provided in Table 1 are important material properties, of which the coefficient of thermal expansion (CTE) represents the most significant material selection parameter. Because the CTE values for

metals are higher than those for ceramic materials, residual stress will develop upon cooling from the elevated temperatures that are typical of the joining process. High residual stress results in low joint strength and even fractured joints. High residual stress can also result in delayed fracture. Although the residual stress might not be high enough to result in fracture at the time of joining, it may be high enough to cause slow crack growth, sometimes called "static fatigue" (Ref 30). When the crack length reaches the critical length after some time, a catastrophic fracture will occur without warning. To minimize residual stress, the materials selected as the ceramic and metal members should have CTE values that are as close as possible. The use of the *average* CTE value for the temperature range involved in the joining process is recommended.

The elastic moduli of both ceramic and metal members can also affect residual stress. The lower the elastic modulus, the lower is the residual stress. Mechanical strength of the metal member is also important. Plastic deformation of the metal can accommodate a significant fraction of the thermal expansion difference between ceramic and metal and thereby decreases the residual stress (Ref 31). The lower the yield strength of the metal member, the lower is the residual stress.

Moly-Manganese Process Considerations

The moly-manganese process is the most common method of joining oxide ceramics to metals. The ceramic surface is initially metallized and is then brazed to metal using one of various brazing filler metals.

Process Mechanism

The mechanism involved in the moly-manganese process has been investigated extensively (Ref 32–39). Basically, a mixture of molybdenum and manganese powders (typically 10 at.% manganese) is applied on the ceramic surface, and then sintered to bond to the ceramic. The sintering temperature is usually over 1400 °C (2550 °F) in an atmosphere of hydrogen and nitrogen with a controlled dew point. During the sintering process, glass material that was originally in the alu-

Table 1 Typical materials used for ceramic-to-metal seals and important properties

| Material | Coefficient of thermal expansion, 10^{-6}/K | Temperature range | | Yield strength | | Tensile strength | | Elongation modulus, % | Elastic modulus | | Ref |
		°C	°F	MPa	ksi	MPa	ksi		GPa	10^6 psi	
Alumina	7.6	At 500	At 930	...	...	~5000		0	390	57	22, 23
Mullite	5.1–5.8	At 1000	At 1830	...	...	~4000		0	145	21	22, 23
Beryllia	7.6	At 500	At 930	...	...	~4000		0	380	55	22, 23
PSZ (2 wt% Y_2O_3)	7.5–13	20–1000	68–1830	...	...	~4000		0	207	30	24
Alloy 42 (UNS 94100)	7.9	20–500	68–930	295	45	550	80	43.7	145	21	25, 26
Kovar	5.2	20–200	68–390	410	60	534	77	25	138	20	27, 26
Nickel (Nickel 200)	13.2	25–100	77–212	148	20	462	67	27	204	30	25
Titanium (Grade 1)	10.1	20–815	68–1500	170–241	25–35	240–331	35–48	30	102.7	15	25
Copper (UNS C10100)	17.7	20–300	68–570	69	10	220	32	45	115	17	25
TZM (UNS R03630)	4.9	20–40	68–105	860	125	965	140	10	315	46	25
Steel (AISI-ASE 1010)	15	20–700	68–1290	115	17	310–380	45–55	33	205	30	28
410 stainless steel (UNS S41000)	11.6	0–538	32–1000	205	30	450	65	22	200	29	29

mina is drawn into the interstices of the molybdenum powder.

The effect of manganese was examined by Mattox and Smith (Ref 39). Upon heating, manganese is oxidized to form MnO, which enhances permeation of the glass material from the alumina to the molybdenum layer by lowering its viscosity. It also penetrates into the alumina grain boundary phase, and it changes the properties of the glass phase in the alumina, as well. These changes decrease both the thermal expansion mismatch between the molybdenum layer and alumina and also the glass transition temperature. As a result, there is less residual stress at the metallized surface.

Processing Steps

The chemical composition and properties of the glassy phase in the moly-manganese layer are very important. Glass or ceramic additions are made to this layer, based on the composition and properties of the alumina, in order to achieve better compatibility with the alumina. Although the basic mechanism in moly-manganese processing is understood, a detailed analysis is difficult because the processing of the alumina and the chemical composition of the moly-manganese paint are generally proprietary.

Typical steps in the moly-manganese process involve, for example, a ceramic vacuum envelope with a diameter over 25 mm (1 in.) that is dry-pressed and sintered. After the ceramic surface is ground to meet both flatness and parallelism requirements, moly-manganese paint is applied, dried, and sintered (Fig 1). To permit brazing with a conventional brazing filler metal, the moly surface is nickel-plated and then sintered. Finally, the metallized ceramic can be brazed to a metal member with a brazing filler metal. A variety of brazing filler metals are available. A typical microstructure at the ceramic-metal joint, obtained by the moly-manganese process, is shown in Fig 2.

The most important parameters in each processing step are described below.

Ceramic. The compatibilities of chemical compositions and properties between alumina substrate and moly-manganese layer are very

important. The success of the moly-manganese process depends on consistency in the ceramic manufacturing process. The chemical composition of the glass phase in the ceramic, which is a function of sintering temperature, affects the bonding process of the molybdenum layer. The ceramic grain size,

which also is a function of sintering temperature, has also been reported to influence the moly-manganese process (Ref 40). Therefore, if the ceramic sintering temperature is not controlled carefully, the chemical compositions of the glassy phase and grain size may vary in each sintering batch. Inconsis-

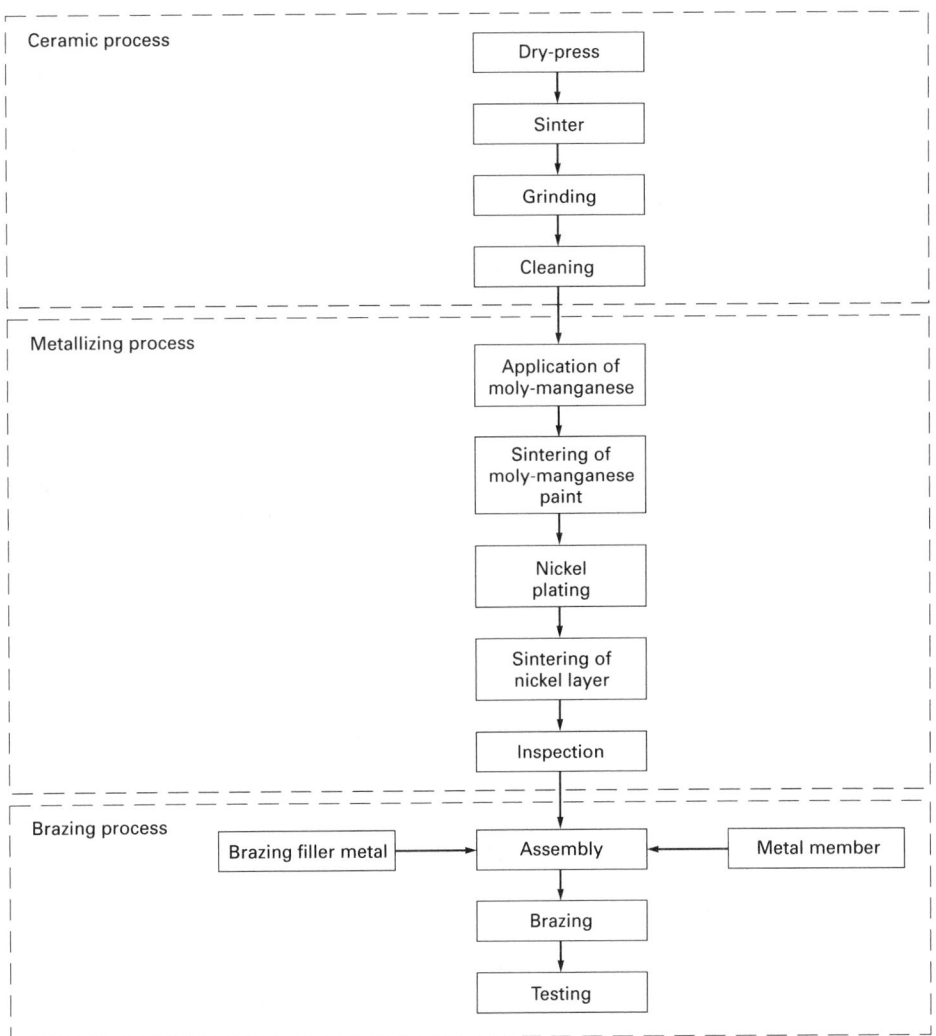

Fig 1 Typical steps in moly-manganese process

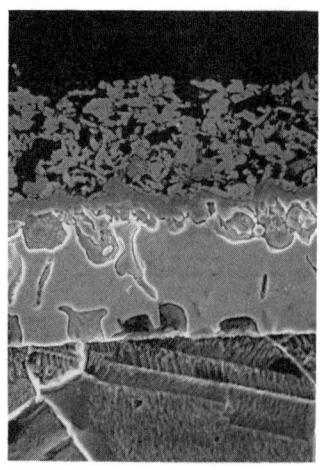

Alumina (94%)

Mo Kayer
Light area-Mo
Dark area-
glassy phase

Nickel layer

Brazing filler
metal (Ag-Cu
eutectic alloy)
Cu-rich phase
is etched

Alloy 42

10 μm

Fig 2 Scanning electron micrograph at the ceramic-metal joint (moly-manganese process)

tent bonding of the moly-manganese layer may result.

Painting. Initially, molybdenum powder, manganese powder, and, possibly, glass (or ceramic) powder are milled to form a pigment mix, which is then combined with an organic binder. A binder system can be added directly to the pigment mix and milled to create a brushing paint. However, to create screening paint, which requires high viscosity, the pigment mix should be prepared separately and combined with screening oil using a three-roll paint mill.

Flat surfaces can be coated by a screening operation. Key variables are screen size, screen tension, snap-off distance, squeegee design, and squeegee durometer, as well as moly-manganese paint properties and screening temperature. The coated alumina is dried at room temperature or in a heated oven, depending on the organic binder system.

Sintering. After the ceramic surface is coated with moly-manganese paint and dried, the coated ceramic is fired at about 1500 °C (2730 °F) under a hydrogen-nitrogen atmosphere with a controlled dew point of about 30 °C (85 °F). A typical sintering time is 30 min. The resulting thickness of the sintered molybdenum layer is typically about 25 μm (1 mil). The layer should be free of pin-holes and other defects.

Nickel Coating. To enhance its ability to be brazed, the molybdenum surface is next plated with either nickel or copper. The selection of the coating metal may depend on the brazing filler metal that will be used. Because nickel coating is more popular, three common nickel plating methods are described below. The same arguments can be applied to the copper coating.

Electrolytic nickel plating is used whenever a dense and thin layer of nickel is required. The primary requirement is that the

molybdenum coating must be electrically conductive when a barrel plating technique is used.

Electroless nickel plating is used in cases where there are many isolated moly-manganese tabs and electrical contact cannot be easily made for the barrel plating technique. Electroless plating is also excellent for coating the inside surface of a hole. However, strict control of the plating solution is required, which is a difficult process.

After either of these two plating processes, parts are heated to about 950 °C (1740 °F). During this heat treatment, the nickel layer sinters and bonds to the molybdenum layer through diffusion. In cases where copper is used as a brazing filler metal, a sintering step for the plated nickel layer may not be required. This is because the brazing temperature for copper is typically about 1120 °C (2050 °F), and sintering of the nickel can be carried out during the brazing cycle.

Nickel oxide paint is used in the third plating method, in a manner similar to the moly-manganese paint. The nickel oxide paint is applied over the metallized surface by screening or brushing. This layer is then sintered in a hydrogen-nitrogen atmosphere, typically at 950 °C (1740 °F). During sintering, nickel oxide is reduced by hydrogen to nickel metal, which bonds to the molybdenum layer through diffusion. Typical thickness of the nickel coating ranges from 2.5 to 7.5 μm (0.1 to 0.3 mil) for all three techniques.

Brazing. Many alloys can be used to braze the metallized alumina to metal, including pure copper. Step-brazing can be carried out by selecting brazing filler metals that have different melting temperatures. For example, a copper-gold brazing filler metal can be used in the first brazing step, and a lower melting point silver-copper eutectic brazing filler metal, in the second.

One of the major problems associated with brazing metallized ceramic is the penetration of the brazing filler metal through the molybdenum layer. For example, pure molten copper dissolves and consumes the nickel layer. Once the nickel layer is consumed, nickel in the molten filler metal can then attack the molybdenum layer. This can result in bond failure at the molybdenum-alumina interface. To minimize penetration, the brazing temperature and time should be controlled carefully. In the case of large components, where longer time is required for establishing uniform temperature, a copper coating is recommended because it will drastically minimize the penetration of a brazing filler metal.

Gold-copper brazing filler metals also show a strong tendency to attack and penetrate the molybdenum layer, but only if nickel plating is present. In contrast, silver-based brazing filler metals, such as the silver-copper eutectic alloy, do not readily attack because of the limited solubility between silver and the nickel that is plated over the molybdenum layer.

Active Brazing Process Considerations

In active brazing, the ceramic and metal are brazed directly, using an active filler metal, which contains element(s) that chemically reacts with the ceramic and enhances the wetting characteristics of the brazing filler metal on the ceramic. Table 2 lists some commercially available active brazing filler metals, all of which contain titanium as the active element. The availability of active brazing filler metals with different melting temperatures allows step brazing.

Process Mechanism

Wetting characteristics of metals on ceramic materials have been investigated extensively (Ref 41–43). It is well established that wetting of the ceramic material is accomplished by the chemical reaction of the active element with the ceramic. However, the exact chemical reactions are not always well understood. In the case of alumina ceramic and titanium-containing brazing filler metal, both TiO and Ti_2O_3 have been reported as reaction products. A $Cu_2(Ti,Al)_4O$ layer was also observed as the reaction layer on alumina (Ref 44, 45). Severian *et al.* investigated the reaction of pure titanium with sapphire substrates (Ref 46). In this case, the reaction product was mostly Ti_3Al. In addition, recent studies using advanced analytical techniques revealed more than one reaction product for CuAgTi alloys brazed on AlN and Si_3N_4 (Ref 47, 48). These results indicate that chemical reaction(s) involved in the active brazing process may be much more complex than previously believed. Further systematic studies are necessary to define the chemical reaction(s) responsible for wetting ceramic materials.

Techniques

Several techniques that can be used for the active brazing process are described below.

Coating. A simple approach is to place a layer of titanium on the ceramic surface which is then brazed to a metal member using a conventional brazing filler metal. There are many methods for coating the titanium layer on ceramic surfaces. Titanium powder with an organic binder can be screened on the ceramic surface or a very thin foil of titanium can be placed on the surface instead. Titanium can also be sputter-coated on the ceramic surface. After making a titanium layer using one of these methods, a conventional brazing filler metal can be placed on the coated surface, and brazing can be carried out in a vacuum furnace at the appropriate temperature. Because titanium readily reacts with oxygen and water vapor, brazing is typically carried out either in high vacuum or in gettered argon.

Mixed Powder. Titanium or titanium hydride powder is mixed with a powder of the conventional brazing filler metal. This mixed powder is combined with an organic binder and applied on the ceramic surface.

Table 2 Commercially available active brazing filler metals

Alloy system	Chemical composition, wt%	Solidus temperature °C	Solidus temperature °F	Liquidus temperature °C	Liquidus temperature °F	Brazing temperature range °C	Brazing temperature range °F	Coefficient of thermal expansion, 10⁻⁶/K	Density, g/cm³	Elastic modulus GPa	Elastic modulus 10⁶ psi	Yield strength MPa	Yield strength ksi	Tensile strength MPa	Tensile strength ksi	Elongation, %	Ref
AuNiTi	96.4Au-3Ni-0.6Ti	1003	1837	1030	1886	1025–1030	1880–1886	16.1	18.2	87	13	209	30	334	48	29	(a)
CuAlSiTi	92.75Cu-2Al-3Si-2.25Ti	958	1756	1024	1875	1025–1050	1880–1920	19.5	8.1	96	14	279	40	520	75	42	(a)
AgCuAlTi	92.75Ag-5Cu-1Al-1.25Ti	860	1580	912	1674	900–950	1650–1740	20.7	10	77	11	136	20	282	40	37	(a)
AgCuTi	63Ag-35.2Cu-1.75Ti	780	1436	815	1499	830–850	1525–1560	18.5	9.8	83	12	271	40	346	50	20	(a)
AgCuSnTi	63Ag-34.2Cu-1Sn-1.75Ti	775	1427	806	1483	810–860	1490–1580	8.7	9.7	83	12	260	38	402	58	22	(a)
AgCuInTi	59Ag-27.2Cu-12.5In-1.25Ti	605	1121	715	1320	700–750	1290–1380	18.2	9.7	76	11	338	50	455	66	21	(a)
TiCuNi	70Ti-15Cu-15Ni	910	1670	960	1760	960–1000	1760–1830	...	5.3	...	...	...	...	...	...	...	(a)
TiCuAg	68.8Ag-26.7Cu-4.5Ti	830	1526	850	1560	810–900	1490–1650	...	9.4	...	...	...	...	...	...	...	(a)
AgTi	96Ag-4Ti	970	1778	970	1778	1000–1050	1830–1920	...	...	...	...	...	...	...	...	...	(b)
AgCuTi	64Ag-34.5Cu-1.5Ti	770	1418	810	1490	850–950	1560–1740	...	...	...	...	...	...	...	...	...	(b)
AgCuTi	70.5Ag-26.5Cu-3Ti	780	1436	805	1481	850–950	1560–1740	...	...	...	...	...	...	...	...	...	(b)
AgCuInTi	72.5Ag-19.5Cu-5In-3Ti	730	1346	760	1400	850–950	1560–1740	...	...	...	...	...	...	...	...	...	(b)
PbInTi	92Pb-4In-4Ti	320	608	325	617	850–950	1560–1740	...	...	...	...	...	...	...	...	...	(b)
SnAgTi	86Sn-10Ag-4Ti	221	430	300	572	850–950	1560–1740	...	...	...	...	...	...	...	...	...	(b)
AgCuInTi	61.5Ag-24Cu-14.5In + (Ti)	620	1148	710	1310	~845	~1555	...	...	...	...	...	...	...	...	...	(c)
AgCuNiTi	71.5Ag-28Cu-0.5Ni + (Ti)	780	1436	795	1463	~927	~1700	...	...	...	...	...	...	...	...	...	(c)
AgCuNiTi	56Ag-42Cu-2In + (Ti)	770	1418	895	1643	~954	~1750	...	...	...	...	...	...	...	...	...	(c)

(a) GTE Wesgo. (b) Degussa AG. (c) Lucas-Milhaupt

Titanium hydride decomposes at about 500 °C (930 °F) into metallic titanium, well before the brazing temperature is reached. Because titanium is exposed to the atmosphere, either a high vacuum or inert gas atmosphere is required.

In a clad filler metal, the titanium layer is sandwiched between two layers of a regular brazing filler metal and diffusion bonded. For example, layers of the silver-copper eutectic alloy can be bonded with a titanium layer at the center. This filler metal has the advantage that the cladding alloy protects the titanium inner layer during the vacuum furnace heat-up cycle, when outgassing products such as water vapor may react with the titanium.

In a true alloy system, many active brazing filler metals that have different liquidus and solidus temperatures can be used (Table 2). Titanium is somewhat protected from the atmosphere by the diluting effect of the other alloy constituents.

Processing Steps

Figure 3 depicts the active brazing process, which has fewer steps than the moly-manganese process. Figure 4 shows the microstructure of a ceramic-metal joint that was made using the active brazing process. Important parameters in this process are described below.

Materials. Each of the three materials that are used in the active brazing process, that is, the ceramic, the brazing filler metal, and the metal member, is discussed separately in terms of parameters that must be considered.

The ceramic is the most critical part of the joint because of its brittle nature. Its surface preparation requires special attention in order to achieve a reliable joint between the ceramic and metal.

The importance of the quality of the ceramic surface on peel strength is shown in Table 3.

Peel test procedures are described in detail in Ref 8. In Table 3, the as-sintered, ground-and-lapped, and ground-and-resintered materials showed excellent peel strength (>100 N, or 22.5 lbf). However, the as-ground ceramic showed a low peel strength (58 N, or 13 lbf), because it contains microcracks introduced by the grinding operation.

Figure 5 shows microstructures of the brazed joints for the ground-and-resintered and as-ground material. No surface defects are observed for the ground-and-resintered material (Fig 5a), but microcracks are seen in the as-ground ceramic (Fig 5b). The ground-and-lapped and as-sintered ceramics also did not show any defects.

Table 3 also shows that surface roughness does not affect peel strength. Both the as-ground and as-sintered materials have approximately the same roughness, but the as-sintered ceramic gives a much higher strength value than the as-ground ceramic. Additionally, the ground-and-lapped and as-sintered ceramic materials have high peel strengths despite the fact that they have different surface roughness properties.

Alumina ceramic materials for vacuum devices are typically produced by dry-pressing, sintering, and then grinding to the finished dimension. The material in this condition is not suited for active brazing, as indicated above. The grinding defects should be healed

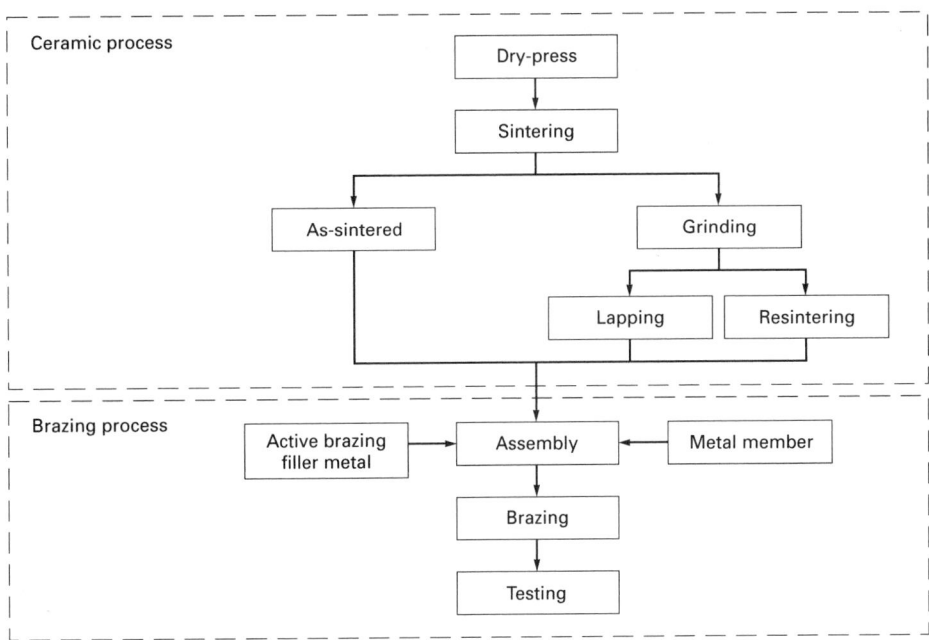

Fig 3 Typical steps in active brazing process

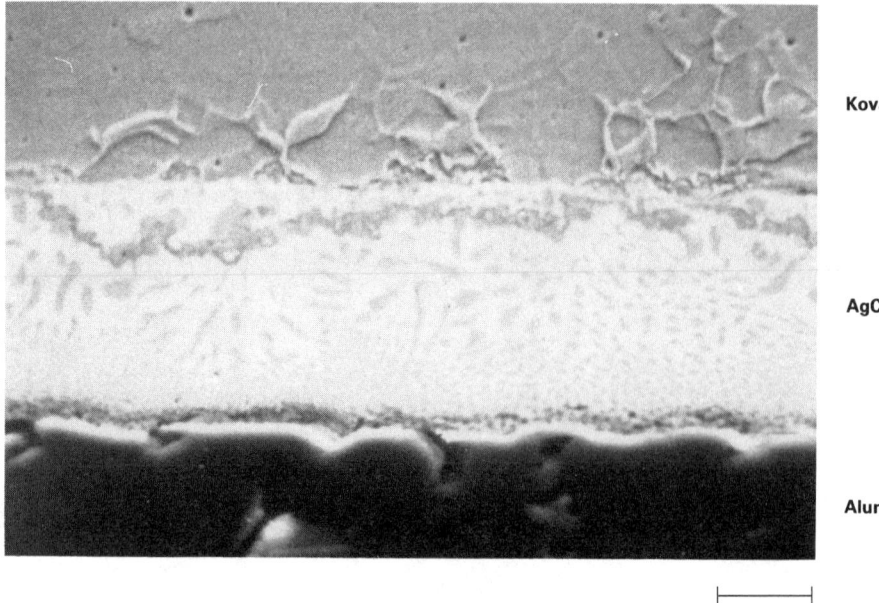

Kovar

AgCuTi

Alumina

|——————| 10 μm

Fig 4 Scanning electron micrograph of ground-and-lapped ceramic showing the ceramic-metal joint (active brazing process)

prior to the active brazing step. For example, the ground alumina can be resintered at about 1600 °C (2910 °F) for about 1 h in air to reduce the surface flaw population.

A simple dye test can be carried out to check whether the grinding defects have been completely healed. In this test, the ceramic material is immersed in either red or fluorescent dye and dried. It is then examined under either natural light, to see the red dye, or ultraviolet light, to see the fluorescent dye. Cracks that are not healed completely will be penetrated by dye and can thus be detected. The best and simplest way to determine surface suitability is to establish a standard material without any surface defects. This standard and the material to be tested should be dye-checked simultaneously and then compared to determine whether the grinding defects were completely healed.

It should be noted that in the moly-manganese process, ceramic material is heated to 1500 °C (2730 °F) to sinter the moly-manganese layer. This heat treatment will also heal any grinding defects.

Table 3   Effect of ceramic surface conditions on peel strength

Ceramic surface treatment	Surface roughness		Peel strength	
	μm	μin.	N	lbf
As-sintered	1.17	46	111	25
As-ground	1.04	41	58	13
Ground and lapped	0.10	4	111	25
Ground and resintered	0.94	37	107	24

Note: Ceramic, 99.5% Al₂O₃; Metal, Kovar (6.5 × 75 × 0.25 mm, or 0.25 × 3 × 0.01 in.); Brazing filler metal, Cusil ABA (50 μm, or 2 mils, thick); Brazing temperature, 825 °C (1515 °F); Brazing time, 10 min; Peel rate, 50 mm/min (2 in./min). Source: Ref 8

The brazing filler metal has several basic requirements. It should wet ceramic materials by the chemical reaction that occurs between the active element (typically, titanium) and the ceramic. The amount of titanium is important. For example, Mizuhara *et al.* investigated the effect of titanium concentration in the copper-silver filler metal on peel strength (Ref 11). Peel strength was maximum at a titanium concentration of about 1.5 wt%. In this alloy system, a titanium concentration below about 1.5 wt% is insufficient for complete wetting of alumina. When the titanium concentration is more than about 2 wt%, the titanium addition increases the strength of the brazing filler metal. Plastic deformation of the brazing filler metal during cooling from the brazing temperature accommodates thermal expansion mismatch (Ref 20). Therefore, the increase in strength of the brazing filler metal that is due to excess titanium leads to poorer accommodation of thermal expansion mismatch. A high residual stress can result and, hence, low joint strength.

Metal Member. Plastic deformation of the metal member during cooling accommodates the thermal mismatch between the ceramic and metal. For example, Mizuhara and Oyama (Ref 20) compared shear strengths of brazed joints between graphite-copper and graphite-molybdenum. Although the graphite material was chosen for its CTE match with molybdenum, the graphite-copper joint showed much higher shear strength because its greater plastic deformation better accommodated the thermal expansion mismatch. It has also been demonstrated that shear strength of the brazed joint between alumina and Alloy 42 with Cusil-ABA can be improved by annealing cold-worked Alloy 42 (Ref 19).

Brazing atmosphere, heating rate, and temperature parameters are described below.

Atmosphere. Titanium readily reacts with oxygen, nitrogen, or water vapor in the brazing atmosphere. Consumption of titanium by these reactions depletes the amount of titanium available to wet the ceramic. The reaction product also forms on surfaces of the brazing filler metal which may prevent physical contact between the titanium and the ceramic. As a result, bonding with the ceramic may be poor. One of the best atmospheres for active brazing is a vacuum. A level of 0.0013 Pa (10^{-5} torr) is usually sufficient. A vacuum furnace must have a low leak rate. Although high vacuum can be maintained by virtue of fast vacuum pumps, if the leak rate is high, then titanium can be eventually consumed by the reaction with air that continuously leaks into the chamber. The leak rate can be simply measured by pumping down to about 0.0013 Pa (10^{-5} torr) vacuum, isolating the chamber, and checking the change in vacuum level after 1 h. The required leak rate is less than 0.67 Pa/h (5 μm Hg/h).

In cases where vacuum is impractical, high-purity argon, helium, or hydrogen can be used. However, the dew point should be carefully controlled to be below about −70 °C (−95 °F), and the oxygen concentration should be less than about 10 ppm. Inert gases can be purified by passing them through gettering furnaces.

The heating rate is a critical factor, especially when the filler metal shows a strong alloying tendency with the substrate material. For example, when silver-copper eutectic (72Ag-28Cu) is used to braze a copper substrate, copper readily dissolves into the silver-copper alloy. This leads to an increase in the melting temperature of the alloy. A slow heating rate may result in a partial braze. Brazing of nickel-base alloys, such as Kovar, with silver-base brazing filler metal does not have this problem, because silver and nickel are not mutually soluble in the 800 to 900 °C (1470 to 1650 °F) temperature range used for brazing.

Overall, a rapid heating rate is desirable, although this may not be possible when the ceramic member has a much higher mass than the metal member. The difference in thermal conductivities can also be considerable. In this case, a slow heating rate is necessary to maintain uniform temperature. An example of such a case is the brazing of a light-gage metal cap to an alumina cylinder that has a mass about 50 times that of the metal member. When this brazing assembly is subjected to a rapid heating rate, the temperature of the metal member will be much higher than that of the ceramic when it reaches the brazing temperature. The active brazing filler metal then will be drawn on to the metal. When the ceramic member finally reaches the brazing temperature, the amount of filler metal available for wetting the ceramic will be insufficient. To eliminate this problem, a long soak

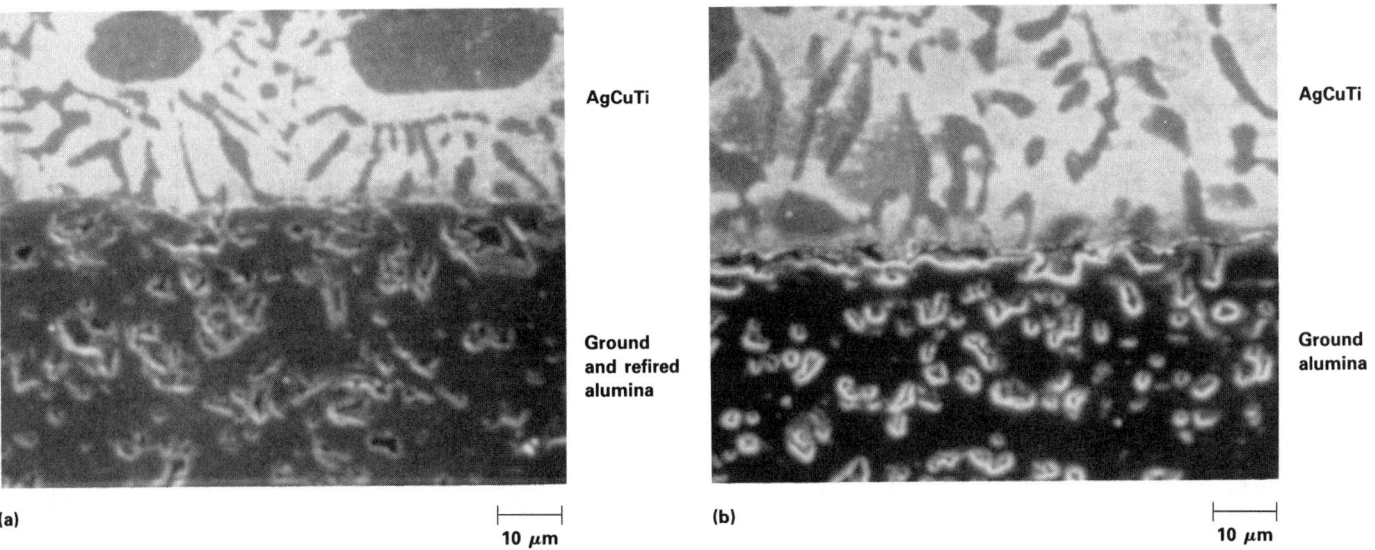

Fig 5 Scanning electron micrographs at the brazed joints by the active brazing filler metal. (a) Ground-and-resintered ceramic surface. (b) As-ground ceramic surface

should be carried out just below the solidus temperature of the filler metal to obtain uniform temperature both in the metal and ceramic members. This soak is then followed by a slow heating rate (for example, 5 °C/min, or 9 °F/min) to maintain the uniform temperature distribution between the metal and the ceramic until the brazing temperature is reached. In this case, the brazing filler metal and metal member combination must be carefully selected to minimize the reaction between them.

Brazing Temperature. Figure 6 shows the effects of brazing temperature on peel strength for Cusin-1-ABA. The brazing time was kept constant at 10 min. The peel test procedures have been described elsewhere (Ref 8). Peel strength is shown to increase as the brazing temperature increases from 810 to 840 °C (1490 to 1545 °F), and then decrease with the further increase in the brazing temperature. The increase of peel strength is explained below. The greater reaction of titanium that occurs at higher temperatures decreases the titanium concentration in the brazing filler metal. Correspondingly, the strength level of the filler metal decreases. As a result, the residual stress becomes smaller, and the joint strength becomes higher. However, when temperature is increased beyond 840 °C (1545 °F), the reaction between Kovar and filler metal becomes important. Primarily, nickel in Kovar dissolves into the filler metal. After this reaction from CTE mismatch, the filler metal is less ductile and the residual stress increases.

Hermeticity of the brazed joint has also been measured as a function of brazing temperature. Two alumina tensile test specimens (ASTM F19–16T) were brazed to a 250 μm (10 mil) thick Kovar washer with 50 μm (2 mil) thick preforms. Thus, one sample had two brazed joints. Four samples containing

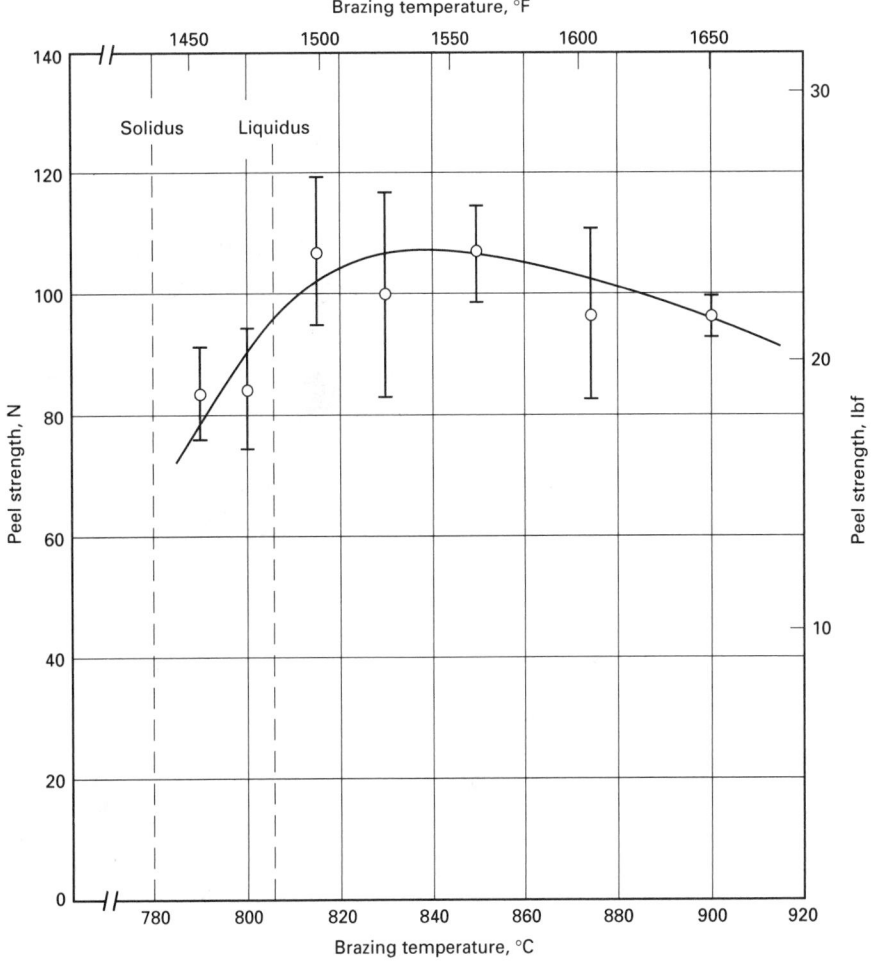

Fig 6 Peel strength as a function of brazing temperature, after 10-min braze applied to Cusin-1-ABA 50 μm (2 mil) foil and 6.5 × 75 × 0.25 mm (0.25 × 3 × 00.1 in.) Kovar and 94% Al$_2$O$_3$

eight joints were tested at each temperature. The helium leak rate was then measured by helium mass spectroscopy. The results showed that 100% hermeticity was obtained at all temperatures tested (810 to 900 °C, or 1490 to 1650 °F). These samples were then thermally cycled 10 times between 25 and 500 °C (77 and 930 °F) and retested. Hermeticity was maintained for all samples, indicating excellent joint property.

Brazing Fixture. Mechanical design, material selection, and the construction of brazing fixtures are important parts of the brazing process. The most desirable approach is to design components that are self-aligning, which would eliminate the need for brazing fixtures and simplify assembly. Because the brazing assembly does not have the extra mass of the fixture, furnace loading is more efficient. In many cases, however, self-aligning components either cannot be designed or may not be economical. When fixtures have to be utilized, the parameters described below should be considered.

Design. An example of a simple brazing fixture for an edge braze using an active brazing filler metal is shown in Fig 7. Edge brazing is described in the section "Active Brazed Joints." It should be noted that because an active brazing filler metal wets most materials, the design should not allow the filler metal to touch any part of the fixture. This brazing fixture set consists of a side fixture for centering the alumina cylinder and both top and bottom parts for centering the end caps. The fixture material can be mild steel or stainless steel. Machinable-grade steels should not be used because they typically contain selenium, in the case of stainless steel, or lead, in the case of mild steel, to enhance machinability. Both elements have high vapor pressure that is not acceptable for the vacuum devices. They may also contaminate the furnace. A brazed sample is also shown in Fig 7.

Fixture Material. Graphite is being used extensively as a brazing fixture material because of its cost and machinability. However, the use of graphite during the active brazing process requires attention. Typically, graphite materials are not totally dense and have a strong affinity for water and gas absorption. During brazing, these contaminants are outgassed at temperatures up to 1100 or 1200 °C (2010 or 2190 °F). When water vapor or oxygen is released from graphite, it reacts with titanium and leads to marginal properties in the brazed joint. Pyrolytic boron nitride also shows similar outgassing behavior. In this case, however, the boron nitride does not outgas until about 950 °C (1740 °F). Therefore, when the brazing temperature is below the outgassing temperature, no harm to brazing is incurred.

In cases where graphite must be used, a high-density vacuum grade is appropriate. Prior to brazing, the fixture should be heated in vacuum to about 1200 °C (2190 °F) for about 1 h to ensure complete outgassing. The brazing assembly should be made immediately upon removal of the graphite fixture from the vacuum furnace, and the assembly should be placed in the vacuum brazing furnace without delay.

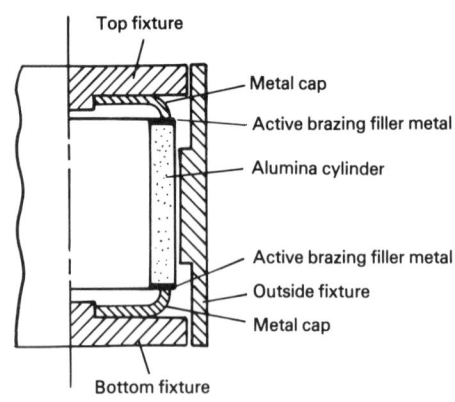

(a)

(b)

Fig 7 (a) Edge-braze fixture for active brazing process and (b) brazed sample

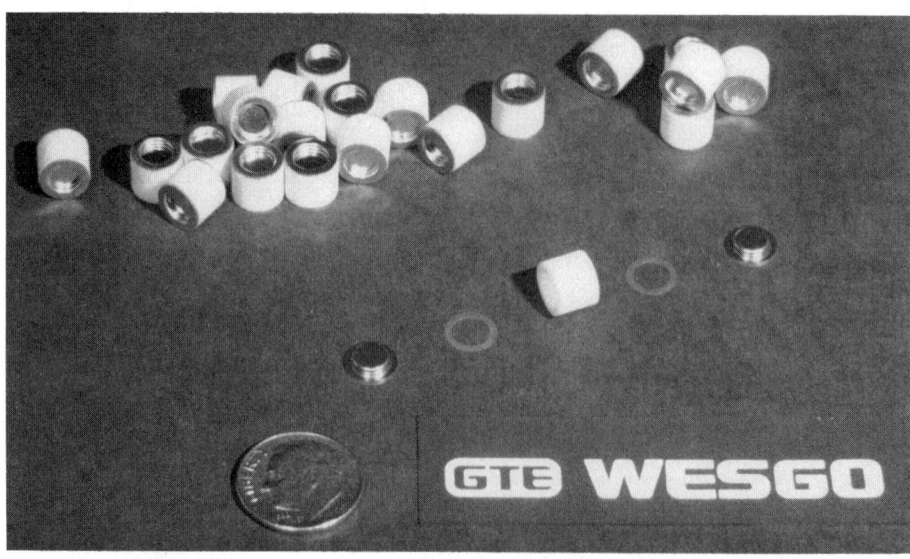

(a)

Cusil-ABA

Metal cap (Alloy 42)

(b)   Alumina cylinder (94%) |— 0.02 mm

Fig 8 (a) Butt seal design and (b) microstructure at the joint

Active Brazed Joints

The article "Techniques of Seal Design" describes in detail the importance of joint design when ceramics and metals must be joined. Briefly described below are two examples of successful joints made by the active brazing process.

The butt seal is one of the simplest designs. However, it cannot accommodate much residual stress. Therefore, this design is limited to components that have less than about a 15 mm (0.6 in.) diam. Examples of the butt braze and microstructure at the brazed joint are shown in Fig 8. Butt seals up to about a 100 mm (4 in.) diam can be made by using an alumina back-up ring. These sandwich seals distribute the shear stress between two alumina cylinders.

An edge braze is a compliant joint design in which the residual stress is accommodated by the elastic compliance of the joint. The edge-braze design can be applied to joint systems with a poor thermal expansion match between the ceramic and metal members, such as an alumina ring with a 430 stainless cap. This design can also be applied to joints of large size. The example in Fig 7 shows a successful joint between alumina cylinders that are over 100 mm (4 in.) in diameter and copper-clad 430 stainless steel end caps.

Fillet formation is an important element of an edge braze. A fillet can distribute the residual stress and eliminate stress concentrations at the joint. Figure 9 shows fillet formation at the brazed joint of an alumina cylinder and a mild steel part. A screenable paste of the active brazing filler metal (Cusin-1-ABA) was used as a brazing filler metal. The amount of the paste is important for fillet

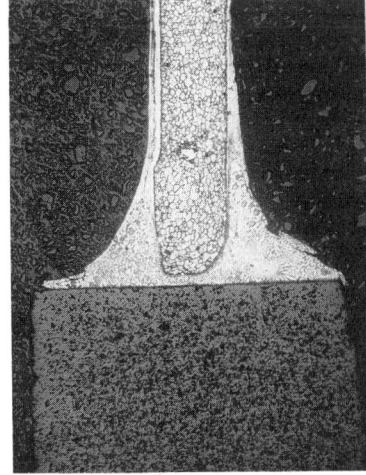

Mild steel

Brazing filler metal

0.5 mm

Alumina cylinder (94%)

Fig 9 Fillet formation at the edge-brazed joint by the active filler metal paint

Table 4 Comparison between foil preform and screenable paste for edge brazing using the active brazing filler metal

	Foil preform	Screenable paste
Organic vapor	None	Some
Rapid heating rate	Good performance	Best
Inverted edge braze	Poor	Good
Quantity of material placement	Very good	Good
Preform punching die	Expensive	Not required
Cost of material	High	Low

size control. It should be sufficient to form the fillet, without being excessive. Because the brazing filler metal typically has a higher CTE value than the ceramic, excessive brazing filler metal may cause high residual stress at the brazed joint.

When using an edge braze, paste can offer an advantage that a preform cannot. Table 4 compares the use of foil preform and screenable paste for edge brazing by the active metal process. In the case of paste, the organic binder should be designed to evaporate completely before the brazing takes place. Carbon residue or incomplete evaporation will lead to the reaction of titanium with carbon or unevaporated binder, to form titanium carbide. This consumes titanium and, hence, results in marginal brazing. The temperature gradient between a ceramic and the metal members is not disruptive in the case of the paste. This is because the metal powder attached to the ceramic surface will follow the ceramic temperature, and will melt and wet the ceramic when the ceramic temperature reaches the melting temperature of the brazing filler metal. The foil preform, under a similar situation, will melt via heat transfer through the metal cap, and the molten alloy will be drawn toward the metal member because the temperature of the metal member is higher than that of the ceramic. Therefore, there is too little brazing filler metal to wet the alumina ceramic. Also, a foil preform in the inverted position may sag during the heating cycle. When it sags, the preform loses physical contact with the ceramic and a fillet cannot form.

REFERENCES

1. W.G. Houskeeper, The Art of Sealing Base Metals Through Glass, *J. Am. Inst. Elec. Engr.*, Vol 42, 1923, p 870–876
2. W.H. Kohl, Materials and Techniques for Electron Tubes, *General Telephone and Electronics Technical Series*, Reinhold Publishing, 1960
3. H. Pulfrich, "Ceramic to Metal Seals," U.S. patent 2,163,407, June 1939
4. H.J. Nolte, "Method of Metallizing a Ceramic Member," U.S. patent 2,667,427, Jan 1954
5. H.J. Nolte, "Metallized Ceramic," U.S. patent 2,667,432, Jan 1954
6. C.S. Pearsall, New Brazing Methods for Joining Non-Metallic Material to Metals, *Mater. Meth.*, Vol 30, 1949, p 61–62
7. F.C. Kelly, "Metallizing and Bonding Non-Metallic Bodies," U.S. patent, 2,570,248, October 1951
8. H. Mizuhara and K. Mally, Ceramic-to-Metal Joining with Active Brazing Metal, *Weld. J.*, Vol 64 (No. 10), 1985, p 43–51
9. E. Lugscheider, H. Krappits, and H. Mizuhara, "Joining of Non-Metallized Ceramic with Metals by Insert of Ductile Active Brazing," paper presented at the Second International Colloquium, Joining of Ceramic, Glass, and Metal, Bad Nauheim, Federal Republic of Germany, 1985
10. A.J. Moorhead and H. Keating, Direct Brazing of Ceramics for Advanced Heavy-Duty Diesels, *Weld. J.*, Vol 65 (No. 10), 1986, p 17–31
11. H. Mizuhara and E. Huebel, Joining Ceramic to Metal with Ductile Filler Metal, *Weld. J.*, Vol 65 (No. 10), 1986, p 43–51
12. H. Mizuhara, Vacuum Brazing Ceramics to Metals, *Adv. Mater. Processes*, Vol 13 (No. 2), 1987, p 53–55
13. W. Weise, W. Malikowski, and W. Böhn, "Bonding Ceramics Together or to Metals Using Active Brazing Alloys in Argon or a Vacuum," Technische Keramik Jahrbuch, Vulcan-Verlag, Essen, Federal Republic of Germany, 1988, p 154–157
14. A.J. Moorhead, Direct Brazing of Alumina Ceramics, *Adv. Ceram. Mater.*, Vol 2 (No. 2), 1987, p 159–166
15. J.A. Pask, From Technology to the Science of Glass/Metal and Ceramic/Metal Sealing, *Ceram. Bull.*, Vol 66 (No. 11), 1987, p 1587–1592
16. R.T. Cassidy *et al.*, "Joining Alumina Ceramics with Titanium Containing Braze Alloys," MLM-3394, Monsanto Corp., 1986
17. M.G. Nicholas, T.M. Valentine, and M.J. Waite, The Wetting of Alumina by Copper Alloyed with Titanium and Other Elements, *J. Mater. Sci.*, Vol 15, 1980, p 2197–2206
18. R.E. Loehman and A.P. Tomsia, Joining of Ceramics, *Ceram. Bull.*, Vol 67 (No. 2), 1988, p 375–380
19. H. Mizuhara, E. Huebel, and T. Oyama, High-Reliability Joining of Ceramic to Metal, *Ceram. Bull.*, Vol 68 (No. 9), 1989, p 1591–1599
20. H. Mizuhara and T. Oyama, "Direct Brazing of Graphite using Copper-Based Active Brazing Filler Metal," paper presented at American Welding Society 71st Annual Convention, Anaheim, CA, April 1990

21. M. Ueki, M. Naka, and I. Okamoto, Wettability of Some Metals Against Zirconia Ceramics, *J. Mater. Sci. Lett.*, Vol 5 (No. 12), 1986, p 1261

22. E.A. Brandes, Ed., *Smithells Metals Reference Book*, 6th ed., Butterworths, 1983

23. M.F. Ashby and D.R.H. Jones, *Engineering Materials: An Introduction to Their Properties and Applications*, Pergamon Press, 1980

24. An Introduction to Zirconia, *Magnesium Electron*, Publication 113, 1986

25. *Properties and Selection, Non-Ferrous Alloys and Special-Purpose Materials*, Vol 2, *Metals Handbook*, 10th ed., ASM International, 1990

26. International Nickel Company data sheets, International Nickel Company

27. R.C. Gibbons, Ed., *Woldman's Engineering Alloys*, 6th ed., American Society for Metals, 1979

28. *Properties and Selection: Irons and Steels*, Vol 1, *Metals Handbook*, 9th ed., American Society for Metals, 1978

29. *Properties and Selection: Stainless Steels Tool Materials*, Vol 3, *Metals Handbook*, 9th ed., American Society for Metals, 1980

30. M.F. Ashby and D.R.H. Jones, Chapt. 18, *Engineering Materials 2: An Introduction to Microstructures, Processing and Design*, Pergamon Press, 1986

31. J.F. Burgess and S.A. Neugebauer, Direct Bonding of Metals to Ceramics for Electronic Applications, *Advances in Joining Technology*, Burke *et al.*, Ed., 4th Army Materials Technology Conference, Brook Hill, 1975, p 185–197

32. G.R. Van Houten, A Survey of Ceramic-to-Metal Bonding, *Am. Ceram. Soc. Bull.*, Vol 38 (No. 6), 1959, p 301–307

33. M.E. Twentyman, High-Temperature Metallizing, Part 1, The Mechanism of Glass Migration in the Production of Metal-Ceramic Seals, *J. Mater. Sci.*, (No. 10), 1975, p 765–776

34. M.E. Twentyman and P. Popper, High-Temperature Metallizing, Part 2, The Effect of Experimental Variables on the Structure of Seals to Debased Aluminas, *J. Mater. Sci.*, (No. 10), 1975, p 777–790

35. J.T. Klomp, Interfacial Reactions Between Metals and Oxides During Sealing, *Am. Ceram. Soc. Bull.*, Vol 59 (No. 8), 1980, p 794–802

36. S.S. Cole, Jr. and G. Sommer, Glass-Migration Mechanism of Ceramic-to-Metal Seal Adherence, *J. Am. Ceram. Soc.*, Vol 44 (No. 6), 1961, p 265–274

37. A.G. Pincus, Metallographic Examination of Ceramic-Metal Seal, *J. Am. Ceram. Soc.*, Vol 36 (No. 5), May 1953, p 152–158

38. A.G. Pincus, Mechanism of Ceramic to Metal Adherence of Molybdenum to Alumina Ceramics, *Ceram. Age*, Vol 63 (No. 3), March 1954, p 16–32

39. D.M. Mattox and H.D. Smith, Role of Manganese in the Metallization of High Alumina Ceramics, *Ceram. Bull.*, Vol 64 (No. 10), 1985, p 1363–1367

40. J.R. Floyd, Effect of Composition and Crystal Size of Alumina Ceramic on Metal-to-Ceramic Bond Strength, *Am. Ceram. Soc. Bull.*, Vol 42 (No. 2), 1963, p 65–70

41. J.V. Naidich, The Wettability of Solid by Liquid Metals, *Prog. in Surface and Membrane Science*, Vol 14, 1981, p 353–486

42. M. Nicholas, R.P.D. Forgan, and D.M. Poole, The Adhesion of Metal/Alumina Interfaces, *J. Mater. Sci.*, Vol 3, 1968, p 9–14

43. C.S. Kanetkar, A.S. Kacar, and D.M. Stefanescu, The Wetting Characteristics and Surface Tension of Some Ni-Based Alloys on Yttria, Hafnia, Alumina and Zirconia Substrate, *Metall. Trans. A*, Vol 19A, 1988, p 1833–1839

44. C. Peytour, F. Barbier, and R. Reve-colevschi, Characterization of Ceramic/ TA6V Titanium Alloy Brazed Joints, *J. Mater. Res.*, Vol 5 (No. 1), 1990, p 127–135

45. F. Barbier, C. Peytour, and R. Reve-colevschi, Microstructure Study of the Brazed Joint Between Alumina and Ti-6Al-4V Alloy, *J. Am. Ceram. Soc.*, Vol 73 (No. 6), 1990, p 1582–1586

46. J.H. Selverian, F.S. Ohuch, and M.R. Notis, Microstructures and Kinetics of the Interface Reaction Between Titanium Thin Films and (IT2) Sapphire Substrates, *Mater. Res. Soc. Symp. Proc.*, Vol 167, 1990, p 353–340

47 A.H. Carim, Identification and Characterization of $(Ti,Cu,Al)_6N$, a New η Nitride Phase, *J. Mater. Res.*, Vol 4 (No. 6), 1989, p 1456–1461

48. A.H. Carim and R.F. Loehman, Microstructure at the Interface Between AlN and a Ag-Cu-Ti Braze Alloy, *J. Mater. Res.*, Vol 5 (No. 7), 1990, p 1520–1529

SELECTED REFERENCES

- J.C. Ambrose, "Competing Joining Process," paper presented at BABS Autumn Conference, British Association of Brazing and Soldering, Bristol, 1990

- C.I. Helgesson, *Ceramic-to-Metal Bonding*, Boston Technical Publishers, 1968

- W.H. Kohl, Materials and Techniques for Electron Tubes, by *General Telephone and Electronics Technical Series*, Reinhold Publishing, 1960

- W. Kraft, *Joining Ceramics, Glass and Metal*, Compilation of papers presented at The International Conference, Bad Nauheim, Federal Republic of Germany, DGM Informationsgesellschaft Verlag, 1989

- R.E. Loehman, S.M. Johnson, and A.J. Moorhead, Ed., *The Proceedings of The International Forum on Structural Ceramics Joining*, Pittsburgh, 1987, The American Ceramic Society, 1989

- M.M. Schwartz, *Ceramic Joining*, ASM International, 1990

Joining Oxide Ceramics

A.J. Moorhead and Hyoun-Ee Kim, Oak Ridge National Laboratory

CERAMIC MATERIALS have traditionally been used at relatively low temperatures, or under low tensile stress at high temperatures, in applications that take advantage of such favorable properties as high hardness, wear and corrosion resistance, electrical resistivity, and low friction. These applications have included electrical insulators, crucibles, pump and valve components for the chemical industry, and in drawing and extrusion dies. Only in the past two decades have major efforts been made to use ceramics in structural applications that include a combination of conditions including significant tensile or flexure stresses, a range of temperatures, and corrosive environments. The ceramics that are being developed to meet these needs make up one segment of the broader class of materials that in Japan are referred to as "fine ceramics" or in the United States as advanced ceramics or high-performance ceramics. Materials of this type are generally produced by state-of-the-art technology and are therefore also known as high-technology ceramics. Advanced or fine ceramic materials also include the electronic ceramics such as those used as substrates and encapsulants, ionic conductors, piezoelectric devices, and high critical temperature superconductors; but these materials, and the technology to join them are outside the realm of this article. The primary reason for much of the interest in structural ceramics is the desire to improve the efficiency of various automotive engines. For example, in recent reviews of the current research on ceramic joining technology, emphasis has been placed on the research and development aimed at utilization of structural ceramics in advanced heat engines (Ref 1–3). In the United States, there are also major programs aimed at developing ceramic heat exchangers to recover the waste heat that is traditionally lost in the high-temperature, corrosive exhausts of industrial furnaces (Ref 4,5).

In recent years it has become more widely recognized that one of the key technologies that will enhance or restrict the use of ceramic materials in the above mentioned high-performance applications is the ability to:

- Reliably join simple-shape components to form complex assemblies
- Join unit lengths of material to form large systems
- Join ceramic components to metals

Although ceramic joining technology has been highly developed over the past fifty years or so, most of the effort has been expended in developing materials and techniques for applications at "low" temperatures and with low stress requirements. The development of technology for joining ceramics for use at elevated temperatures, at high stress levels, and in dirty environments has been much more limited.

For the purposes of this article, three major processes for joining oxide ceramics to themselves or to metals will be considered. These include diffusion welding (sometimes referred to as diffusion bonding or solid-state bonding), brazing with metallic filler materials, and brazing with nonmetallic filler materials, that is, glasses. As it is difficult for any one review of the technology for joining of ceramics to be all-inclusive, an attempt has been made to make this article complementary with two excellent reviews in the recent literature (Ref 6, 7).

Ceramic Materials

In general, as compared to metals, the structural ceramics have different atomic bonding (ionic and covalent versus metallic), have lower coefficients of thermal expansion, and are brittle rather than ductile materials (Ref 8). Typical physical and mechanical properties that may influence the joining behavior of some of the oxide ceramics considered for structural applications are given in Table 1. Monolithic ceramics have much lower fracture toughness than metal alloys. Even the advanced oxide-matrix composites have lower fracture toughness, ranging from 8 MPa$\sqrt{m}$ (7 ksi $\sqrt{in.}$) for SiC whisker-reinforced alumina (Al_2O_3-SiC_w) to 14 MPa$\sqrt{m}$ (13 ksi $\sqrt{in.}$) for the transformation-toughened zirconias (TTZs), which are less than those of cast aluminum alloys or cast iron. The structural ceramics tend to have greater thermodynamic stability than metals, so it is difficult to form in ceramic joints the strong chemical bonds that enhance adhesion.

Structural ceramics can be divided into monolithic materials and ceramic-matrix composites. The monolithic oxide materials include alumina, magnesia (MgO), zircon ($SiO_2 \cdot ZrO_2$), spinel ($Al_2O_3 \cdot MgO$), mullite ($3Al_2O_3 \cdot 2SiO_2$), cordierite ($2Al_2O_3 \cdot 2MgO \cdot 5SiO_2$), and lithium aluminosilicate. The composites can be divided into two broad classes: those containing randomly dispersed particles or whiskers to improve fracture toughness, and those containing semi-continuous fibers for sustaining load during matrix-microcracking under high stresses (Ref 9). Examples of oxide-matrix composites toughened by dispersion of a second phase or material include the transformation-toughened zirconias (Ref 10), zirconia-toughened aluminas (Ref 11), and SiC_w-reinforced alumina and mullite (Ref 12). Examples of continuous-fiber-reinforced composites include SiC fiber-reinforced glass ceramic composites (Ref 13), glass or glass-ceramic matrices reinforced with fibers ranging from SiC to alumina (Ref 14), and continuous- and discontinuous-fiber-reinforced oxide-matrix composites in which the matrix is formed from the hydrothermal decomposition of metal-organic precursors (Ref 15).

Oxide ceramic bodies, including some of the composite materials, are fabricated by consolidating a powder (or powder and reinforcing material) into a "green" body, then solidifying the body through coalescence of the powder particles during sintering. Sintering can be accomplished in some materials either with or without pressure. Sintering readily occurs in some materials via a totally solid-state process without the use of any additives (sintering aids). However, sintering aids are added to improve the sinterability of many materials even though they are not required. For example, high-purity alumina can be sintered without any additives, but MgO is typically added as a sintering aid. The role of the sintering aids varies from material to material, but generally they serve to promote densification by enhancing surface diffusion or restricting grain growth. In some systems the sintering aids remain as a solid throughout the process, but in others the sintering aids liquify at processing temperatures allowing for more rapid transport of material. In most cases at least some residue of the sintering aid remains at the grain boundaries after densification of the ceramic. These grain-boundary phases can substantially affect both the mechanical properties of the material (such as influencing creep resistance) and the behavior of the material during joining.

Most of the work to date on joining structural oxide ceramics has been done on three families of materials: the monolithic high-

Table 1 Data on the properties of various oxide ceramics

Ceramic	Flexural strength (σ_f)		Fracture toughness (K_{Ic})		Coefficient of thermal expansion (α), $\times 10^{-6}$/°C	Thermal conductivity (k), W/m · K	Ref
	MPa	ksi	MPa$\sqrt{m}$	ksi$\sqrt{in.}$			
Al_2O_3	200–600	29–87	3–4	2.7–3.6	7–9	15–40	87–91
PSZ(CaO, MgO)	600–1000	87–145	8–16	7.3–14.6	10	2.2	87–90, 92
TZP (Y_2O_3, CeO_2)	800–2000	116–290	7–8	6.4–7.3	10	2.2	88–90, 92
Al_2O_3-ZrO_2	1500–2000	218–290	8–13	7.3–11.8	...	...	88, 89, 92, 93
$3Al_2O_3 \cdot 2SiO_2$	200	29	2	1.8	5.3	3.5	87, 88, 90, 94
$2MgO_2 \cdot 2Al_2O_3 \cdot 5SiO_2$	150	22	1–2	0.9–1.8	2	3	87, 90

alumina ceramics, the transformation-toughened zirconias, and SiC$_w$-reinforced alumina composites. All three materials are commercially available and are being used, or considered for use, in a wide range of structural applications. Because they vary widely in physical and/or mechanical properties, and would probably behave quite differently during joining, more detailed information concerning these particular materials will be presented.

High-Alumina Ceramics. There are a wide variety of high-alumina ceramics ranging in Al_2O_3 content from 79 wt% (alumina-mullite) (Ref 16) to 99.9 wt% (Ref 17, 18). These materials all readily densify without pressure and thus are almost invariably fabricated by pressureless sintering. The aluminas not only vary widely in chemical composition (and thus in the amounts and composition of grain-boundary phases) but also in grain size and porosity. The various high-alumina ceramics can, therefore, be expected to respond very differently during joining or in service. Such differences during joining might include the formation of different interfacial reaction products depending on chemical composition of the ceramic, the wicking of filler metal away from a brazed interface in the case of a porous ceramic body, or the resistance to desirable microscopic deformation at the interface during diffusion welding if a large-grained material is selected.

Transformation-Toughened Zirconias. The considerable interest in the science and technology of the transformation-toughened zirconias over the past decade has been described as a "scientific revolution" (Ref 19). There are two types of material commercially available: the partially stabilized zirconias (PSZ), and tetragonal zirconia polycrystal (TZP) ceramics. A typical PSZ ceramic contains a stabilizer of MgO, CaO, or Y_2O_3 (for example, Y-PSZ containing 8 wt% Y_2O_3) (Ref 20) and has a microstructure consisting of 40 to 60 μm cubic grains containing finely dispersed precipitates of submicron tetragonal (t) and monoclinic (m) ZrO_2. The TZP materials contain a smaller amount of dopant (for example, 4 to 5 wt% Y_2O_3 or 2 to 3% CeO_2) as the stabilizer and have a microstructure consisting of 1 to 5 μm grains of predominantly tetragonal ZrO_2. If the amount and type of stabilizing agent, the precipitate size in the PSZs, and the grain size in the TZP ceramics are carefully controlled, the tetragonal phase

in both materials is metastable at low temperatures (<22 °C, or 72 °F). Applied tensile stresses can then initiate the t to m transformation in the zone around a propagating crack. This martensitic type of transformation is accompanied by a volume increase and shear strain which lower the stress intensity acting on the crack tip, thereby increasing the fracture toughness of the material (Ref 21). The TTZ ceramics are both strong and tough, and have relatively high coefficients of thermal expansion (CTE) that are favorable for joining to metals. The TTZ materials are generally fabricated by pressureless sintering, but also may be produced by hot pressing or hot isostatic pressing.

Although very promising structural ceramics, the transformation-toughened zirconias have several characteristics that can affect the joining process itself or the behavior of a joint in service. First, care must be taken during any thermal processing to avoid degradation of the controlled microstructure of the ceramic. For example, the PSZ materials can be degraded by a long thermal cycle in the 1000 °C (1830 °F) range (the TZP materials appear to be more thermally stable). During thermal treatments, the tetragonal precipitates or grains can grow and spontaneously transform on cooling (if a critical size is exceeded) with an accompanying loss in strength and toughness. A similar aging phenomenon occurs when the zirconia is exposed to humid environments at temperatures of 200 to 500 °C (390 to 930 °F) for extended periods of time. The process is similar to the overaging phenomenon observed in precipitation-hardened aluminum or steel. Also of concern, when brazing these materials, is a possible loss of strength and toughness from chemical reactions with an active metal (such as Ti) in a brazing filler metal. As will be seen in the section "Direct Brazing" in this article, there is a very noticeable darkening of zirconia ceramics in an area adjacent to the interface when these materials are brazed with filler metals containing active metals. Some, but probably not all, of this effect is caused by the loss of oxygen to the filler metal. Such loss is more likely to occur in zirconias than in other oxides due to the high transport rate of oxygen ions in the zirconias.

SiC$_w$-reinforced ceramic-matrix composites were invented less than ten years ago. Microscopically small (<1 μm in diameter by 25 to 50 μm in length) single-crystal whis-

kers of SiC were blended with ceramic powders using conventional ceramic processing techniques. Following densification, the resulting ceramic-matrix composites were twice as strong and tough as their monolithic counterparts (Ref 22, 23). Although the inclusion of these whiskers in many other matrices (including mullite and alumina-zirconia) has subsequently been explored (Ref 24), the most widely used material of this type is the SiC$_w$-reinforced alumina composite. In addition to the previously mentioned improvements in strength and toughness, the introduction of SiC$_w$ into alumina has resulted in significant improvements in resistance to slow crack growth at room temperature (Ref 25), toughness at elevated temperatures (Ref 9), and resistance to creep (Ref 26). The SiC$_w$-reinforced aluminas can be fabricated by pressureless sintering or by the application of pressure during densification. Either method causes the whiskers to have some degree of preferred orientation in the body. This orientation effect, and the fact that the composite can be fabricated with whisker loadings ranging from about 10 vol% (for pressureless sintered material) to 30 vol% or more in hot pressed materials, may affect the behavior of the composite during joining and the thermal stability of the joints.

Mechanical Properties of Joints and Their Measurement

Joints in ceramic materials generally fail in a brittle manner. The most likely mechanism is, as is the case for the ceramics themselves, the stress-induced formation and/or propagation of cracks from small defects in the joint area. In a ceramic, the originating defects may lie on the surface of the sample or component, or within the volume of the material. In the case of a joint, such defects may exist in interfacial reaction layers or in the ceramic adjacent to the joint. The defects in the ceramic itself may have been introduced during the original sintering process or from other fabrication steps, particularly machining. Defects in interfacial reaction layers may be the result of such factors as the generation of gaseous reaction products. Cracks may be created at such defects under the influence of external stresses or residual stresses resulting from mismatches in material properties such as coefficients of thermal expansion. The effects of such stresses may be exacerbated by the presence of stress concentrations in the joint region.

Because the failure of a ceramic joint is so dependent on the flaw population and distribution, the "strength" of a joint is statistical in nature, depending on the probability that a flaw severe enough to cause fracture at a given applied stress is present in the volume of material that is exposed to the peak stress. The relationship between the strengths of brittle materials and the effective volume of the sample, can be expressed as the Weibull failure probability function:

$$F(\sigma) = 1 - \exp\left[-V_e(\sigma_{max}/\sigma_0)^m\right] \qquad \text{(Eq 1)}$$

in which $F(\sigma)$, σ, σ_0, m, and V_e are the failure probability distribution function, tensile stress, material constant, Weibull modulus, and effective volume, respectively. The effective volume of a specimen is equivalent to the volume of that particular specimen which is under uniaxial tensile stress; therefore, a sample of given cross-section tested in three-point bending will yield a higher mean strength value than the same specimen tested in four-point bending, because the effective volume of the four-point bending test is considerably higher. For example, Katayama and Hattori (Ref 27) calculated (using the specimen dimensions, and a Weibull modulus of 10 for the material) the effective volumes for a 3 mm (0.12 in.) thick by 4 mm (0.16 in.) wide flexure specimen in a sintered silicon nitride (Si_3N_4) as 1.57 mm³ (9.57×10^{-5} in.³) when tested in three-point bending and 6.66 mm³ (4.06×10^{-4} in.³) in four-point bending. Because the tensile stress is applied to all parts of a tensile specimen, the effective volume of a tensile specimen is much greater. For example, Soma *et al.* (Ref 28), calculated the effective volume of a 10 mm (0.4 in.) diam tensile specimen (also in a Si_3N_4 with Weibull modulus of 10) as 3.14×10^3 mm³ (0.19 in.³). Statistically, there is a much greater likelihood that a critical flaw will be present in such a large volume. In other words, the average strength measured by a tensile test will be considerably lower, but more representative of the inherent strength of the joint. This relationship between test volume and the measured strength for a brittle material, including joints in brittle materials, is illustrated in Fig 1. This figure illustrates the point that samples with small effective volumes can significantly overestimate the strength of a ceramic material, and it is believed, that of a joint. This assumption is supported by the work of Suganuma *et al.* (Ref 29), who reported that the three-point bending strength of joints between Si_3N_4 and a low-expansion metal alloy (brazed with aluminum filler metal, at 800 °C, or 1470 °F, in flowing Ar) was 125 MPa (18 ksi), a value 2.5 times that of the tensile strength of similar brazements with the same cross-sectional area.

Because larger effective test volumes give a more accurate assessment of joint strength,

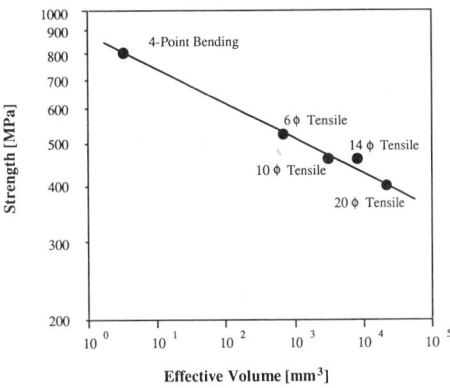

Fig 1 Relationship between the measured strength and effective volume of sintered silicon nitride specimens. The bend specimens were 3 mm (0.12 in.) thick by 4 mm (0.16 in.) wide. The gage diameters of the tensile specimens are given in millimeters. This relationship is similar for joints in structural ceramics. Source: Ref 28

one would think that a tensile test would be widely used in the development of joining procedures. However, in most work a flexure test is used rather than a tensile test for one of several reasons. First, a tensile specimen requires much more material than does a flexure specimen. Some ceramic materials are expensive and others may be under development themselves and, therefore, available in limited quantities. Second, because any surface flaws (such as scratches introduced during grinding) should be parallel to the tensile axis of the specimen, very specialized equipment is required to fabricate ceramic tensile specimens. Finally, as in all tensile tests conducted on ceramic materials, there is the ever present concern over alignment of the loading system to ensure that the specimen is loaded in uniaxial tension with a minimum of parasitic bending stresses. Some recent advancements, such as the development of self-aligning grips, have greatly alleviated this problem; but at this time they are available in only a limited number of facilities, and ceramic tensile testing is still not widely conducted.

There is a standardized tensile test for joints that has been used in a number of studies on both brazing and diffusion welding. Standard ASTM F 19 ("Method for Tension and Vacuum Testing Metallized Ceramic Seals") was first issued in 1961, revised in 1964 and reapproved without change in 1987. The specimen design is shown schematically in Fig 2. A metal washer is joined between the ceramic test pieces to evaluate the strength of ceramic-metal joints. Although tensile testing of joints is worthwhile, this particular standard appears to have some serious limitations. For example, the thickness of the joint is not controlled (such as might be done by the placement of metal wire or foil between the ceramic parts), and it has been shown in a number of brazing studies that the mechanical behavior of a joint is strongly dependent on joint thickness. Also,

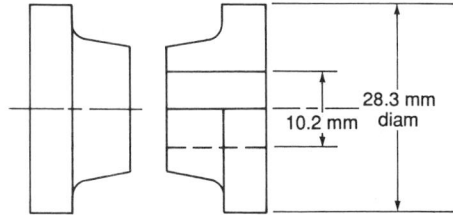

Fig 2 Design of specimen, specified in ASTM F 19, for determining tensile strength of metallized ceramic seals. A metal washer can be inserted between the test pieces to evaluate the strength of ceramic/metal joints.

the ASTM F 19 sample is tested in the as-joined condition. The specification calls for sufficient brazing material "to produce an even fillet at the joint," but a specimen that is not machined after joining is more susceptible to testing errors or uncertainties resulting from specimen-to-specimen variations in joint underfill or reinforcement. There are also concerns over the alignment of this specimen during testing. The standard recommends a compressible "sponge ring" and fluorocarbon washer to accommodate any misalignments in the specimen itself or in the load train, but these devices are presently considered to be unacceptable for generation of accurate tensile test data.

The strengths of joints in structural ceramics can also be characterized by flexure testing with either three- or four-point loading. One technique used for fabrication of brazed flexure specimens is illustrated in Fig 3. The tantalum spacers are used to control the joint thickness and prevent the joint from becoming too thin from the squeezing action of the molybdenum springs. The brazed assemblies are ground on both major surfaces to remove any build-up of filler metal, or remove material if the filler metal has not fully filled the joint. After grinding, the surface of the assembly that will eventually be the tensile surface of the test bars is polished as in preparation of a specimen for microscopic examination. The polishing operation is done to

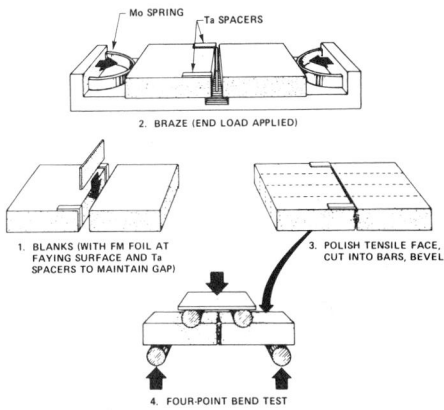

Fig 3 Fabrication of specimens for measuring the strength of ceramic/ceramic or ceramic/metal brazements in three- or four-point bending

further reduce the number and size of any surface defects (such as machining marks) that can influence the test results. Testing is done in four-point bending if the objective is to determine overall strength and weakest "link" of the brazement. Three-point bending flexural tests are conducted when the effects of processing parameters on interfacial reaction product formation are studied to ensure that failure occurs at the joint itself.

In recent years several investigators have suggested that the role of material and interfacial characteristics (the latter also being influenced by the joining parameters) on the mechanical behavior of ceramic joints can be more readily ascertained by a fracture mechanics test rather than by flexure or tensile testing (Ref 30). Although a valuable tool, the fracture mechanics test does not alleviate the eventual need for a flexure or tensile test to determine the strength of a joint. In the fracture mechanics approach, the quality of a joint is characterized by the amount of energy required to separate the joint as a single crack (from a sharp interface notch or precrack) propagates through the interface at the onset of catastrophic failure. The rationale is that such an approach eliminates the size or volume of the stressed region as a variable. According to the following relation (Ref 31):

$$\sigma = Y K_{Ic} c \qquad \text{(Eq 2)}$$

the strength, σ, of a brittle material, which includes all the structural ceramics as well as most joints that contain them (since most joints contain brittle reaction products as well as the intrinsically brittle ceramics), is a function of a parameter Y that depends on the specimen and crack geometry, the crack length c, and

the critical stress intensity factor K_{Ic}. The latter parameter, which is also referred to as the fracture toughness of the material, is the stress intensity at which a crack will propagate and lead to fracture. The critical stress intensity factor is considered to be a property of the ceramic or joint (independent of test volume), and is influenced by the microstructure of the interface region. Some of the test methods (Ref 30, 32, 33) that have been developed for measuring the fracture energy and toughness of bulk ceramics are also used for joints, and are shown schematically in Fig 4. Unfortunately, there is no standardized fracture mechanics test, and relatively little published test data. It is important to note that because of testing difficulties (such as the effect of crack length on the test results) it is generally not feasible to compare results obtained through the various fracture mechanics test methods. It is generally found in bulk ceramics, for example, that some tests, such as the single-edge-notched beam (SENB), result in inherently higher values of fracture toughness than others, such as those based on the double-cantilever beam (DCB) geometry.

Diffusion Welding

Diffusion welding or bonding is a solid-state process by which two prepared surfaces are made to coalesce through the application of heat and pressure. Both ceramic-ceramic and ceramic-metal joints have been fabricated using this process. Most of the work has been on ceramic-metal joints in which the metal portion of the joint is either a bonding material added to the interface between two ceramic components or is the second member of a

bimaterial joint. The more extensively studied ceramic-metal joint will be discussed first and in more detail. Ceramic-ceramic diffusion welding and a variation on this process in which ceramic powder compacts are simultaneously sintered and bonded (sinterbonding process) will be discussed at the end of this section.

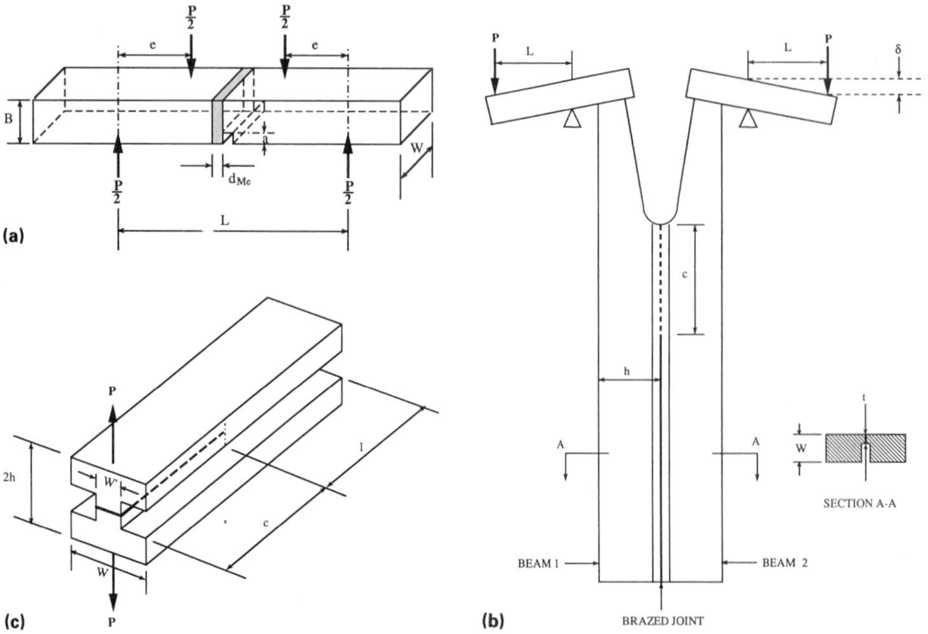

Ceramic/Metal Joints

Two types of interaction have been observed in diffusion welding of ceramic/metal joints. In the first type, bonding is driven by a decrease in surface energy as a new interface is formed to replace the two original surfaces. This physical interaction is indicated by an abrupt transition in microstructure between metal and ceramic at the interface. Examples of this type of bond include those between alumina welded under high vacuum to the metals Al, Cu, Ni, Pt, and Nb (Ref 34, 35). However, under certain conditions of temperature and excess oxygen activity, reaction products can form at some of these interfaces (Ref 36). In the case of the Al_2O_3-Nb couple, a model system which has been extensively studied, a reaction layer does not occur at the interface because Al_2O_3 dissolves in the Nb (Ref 35). The driving force for the formation of the physical type bond is the change in surface energy (ΔG_s) as given by the Dupre equation:

$$\Delta G_s = (\gamma_c + \gamma_m) - \gamma_{cm} \qquad \text{(Eq 3)}$$

where γ_c and γ_m are the surface energies of the ceramic and metal, respectively, and γ_{cm} is the surface energy of the interface formed during joining. The change in surface energy (ΔG_s) is the same as the thermodynamic work of adhesion (W_{ad}), the energy required to separate a unit area of the bonded interface.

In the second type of solid-state bond between ceramics and metals, chemical reaction further lowers the energy of the system and increases the strength of the bond. Chemical reaction product layers are observed in the joints after welding. These reaction products generally cause stronger bonds; however, weakening of the joint may occur if the reaction products grow to excessive thicknesses during welding or in service. Diffusion welds between oxide materials and metals that result in the formation of interfacial reaction products are alumina with Al (although it is suggested in Ref 37 that the reaction is between Al and the SiO_2 in the alumina and not with the Al_2O_3 itself); Al and Mg with SiO_2 and ZrO_2 (Ref 38), Ni with MgO, and Ni and Ti with Al_2O_3 (Ref 39).

The temperature required for diffusion welding to occur usually ranges from 0.5 to 0.8 of the absolute melting point of either the ceramic, the metal being bonded to the ceramic, or an interlayer that is added at the interface to promote bonding. A wide range of applied pressures has also been used, ranging from 15 kPa to 200 MPa (3 psi to 29 ksi) and higher. The surface finish of the mating

Fig 4 Test methods that have been developed for measuring the fracture energy and fracture toughness of ceramic-ceramic and ceramic-metal joints. (a) Single-edge-notched beam. Source: Ref 32. (b) Double-cantilever beam with constant moment loading. (c) Double-cantilever beam with tensile loading. Source: Ref 33

materials is also a critical variable, because the success of the process depends on the attainment of intimate interfacial contact. If all other variables are equal, the surface finish and pressure are related since rough surfaces require greater pressures to force the surfaces into contact for bonding. The observed effects of each of these variables (applied pressure, temperature, and surface finish) will be discussed below.

Applied Pressure. Diffusion welding is usually performed under conditions in which at least one component undergoes plastic deformation under the applied load. The material that deforms may be the ceramic, the metal portion of a ceramic-metal joint, or an intermediate material placed in the joint. Klomp suggests that applied pressure is critical in solid-state bonding of oxygen-active metals to oxide materials (Ref 37). The applied pressure creates an active metal surface by plastic deformation of the surface oxide layer that usually inhibits bonding, thereby allowing contact between clean metal and the ceramic during the bonding process. Such behavior was clearly observed in a study conducted by Arata and Ohmori on the solid-state bonding of CaO-stabilized ZrO_2 to itself in vacuum using 0.5 mm (0.02 in.) thick inserts of 5052 aluminum alloy (2.2 to 2.8 wt% Mg) (Ref 38). The surfaces of the ZrO_2 to be joined were polished with 1500 grade emery paper and degreased with acetone prior to bonding. The effect of bonding pressure on the room-temperature tensile strength of the joints is clearly shown in Fig 5. The bonding mechanism was described as a sequential process in which the oxide film on the aluminum alloy is broken, the alloy is forced into the pores of the zirconia, and finally the Al and Mg of the alloy react with the grain-boundary phases of the zirconia and with the zirconia.

Bonding Temperature. Temperature is the most significant variable affecting the strength of diffusion welded joints, yet the mechanisms involved in ceramic-ceramic or ceramic-metal diffusion welding have not been modelled. Formation of metal-metal joints (Ref

40, 41) are considered to involve the processes associated with pressure sintering, that is, plastic flow, creep deformation, and diffusion along surfaces and grain boundaries or through the bulk of a grain. If similar behavior occurs in joints containing ceramics, it can be concluded that increased temperature increases contact between the materials being joined by reducing their yield strengths, and increases the diffusion rates that control pore closure, grain growth, and chemical interactions.

The effect of temperature on mechanical deformation at the joint can be illustrated by the previously cited work of Arata and Ohmori (Ref 38). As shown in Fig 6, a 25-fold increase occurred in the room-temperature tensile strength of ZrO_2 to ZrO_2 welds (using 0.5 mm, or 0.02 in., thick inserts of 5052 aluminum alloy) when the bonding temperature was raised from 450 to 550 °C (840 to 1020 °F).

Similar behavior was observed by Nicholas and Crispin (Ref 42) in diffusion welding an alumina (97.5% Al_2O_3) to type 321 stainless steel using an Al interlayer as shown in Fig 7. The effects of variations in welding parameters were determined in a series of experiments utilizing a steel insert between standard ASTM F 19 test pieces. Fixed variables were: a 0.5 mm (0.02 in.) thick interlayer of commercial purity aluminum at both alumina-steel interfaces, an applied pressure of 50 MPa (7.25 ksi), a time at temperature of 30 min (2 to 3 h heat-up time), and a vacuum of a few mPa. The bonding surfaces of the alumina test pieces were in the as-received, fired condition. Increasing the welding temperature from 496 to 625 °C (925 to 1155 °F) resulted in progressively stronger joints. In room-temperature tensile testing, the welds made at 625 °C (1155 °F) failed at the stainless steel-aluminum interface, while all others failed at the alumina-aluminum interface. These authors concluded that the relationship between fabrication temperature and the strengths of the joints which failed at the alu-

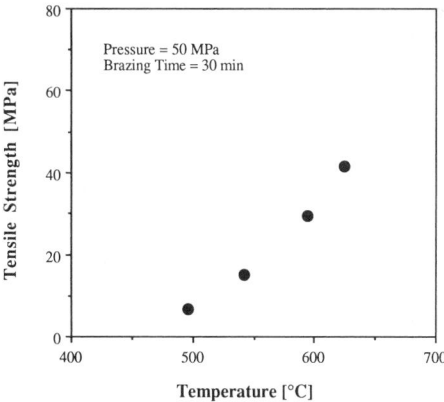

Fig 7 The effect of bonding temperature on the room-temperature tensile strength of diffusion welds between alumina and type 321 stainless steel (made using an aluminum interlayer). Source: Ref 42

mina-aluminum interface could be expressed as follows:

$$\sigma = \sigma_0 \exp\left(-Q/RT\right) \qquad (Eq\ 4)$$

where σ is the failure stress, in MPa, of samples bonded for a fixed time, at a given applied pressure, over a range of temperatures. In this equation, σ_0 is a constant (3.14×10^6 MPa), Q is an apparent activation energy (83.3 kJ mol^{-1}) for the bonding process, R is the gas constant, and T is the absolute temperature. Derby (Ref 43) observed that the activation energy derived from the data in this study was approximately the same as the 80.3 kJ mol^{-1} reached in a similar study by Dawihl and Klinger (Ref 44). Derby compared these values with the activation energies for self diffusion by lattice-, grain-boundary-, or surface-diffusion mechanisms and for vapor-phase transport in aluminum and alumina. Because of the close agreement with the published value for grain-boundary self diffusion in aluminum (84 kJ mol^{-1}), Derby concluded that grain-boundary diffusion in the aluminum was the most probable bonding route in these welds.

Crispin and Nicholas also conducted a similar study on diffusion welding the same alumina (97.5% Al_2O_3) to itself using a 0.5 mm (0.02 in.) thick copper foil interlayer (Ref 45). Welding was conducted in vacuum (<10 mPa) at temperatures ranging from 600 to 950 °C (1110 to 1740 °F), a time at temperature of 30 min, and a pressure of 50 MPa (7.25 ksi) which was applied when the sample reached the bonding temperature. They concluded that the most obvious correlation in the experimental data was between average joint strength and fabrication temperature. The room-temperature tensile strengths of the ASTM F 19 alumina specimens increased from essentially no strength when welded at 600 °C (1110 °F) to a maximum of 54 MPa (7.8 ksi) when welded at 875 °C (1605 °F). Further increases in welding temperature led to decreases in joint strength. The samples welded at temperatures

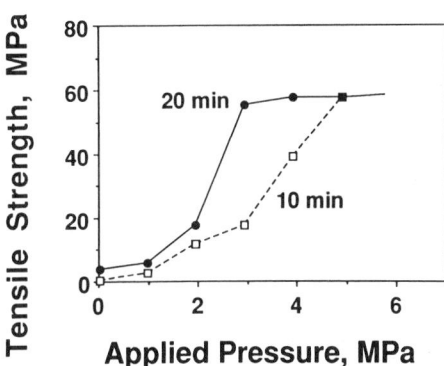

Fig 5 Effect of applied pressure (for two bonding times) on the tensile strength of solid-state welds of CaO-stabilized ZrO_2 joined to itself using an interlayer of 5052 aluminum alloy. Source: Ref 38

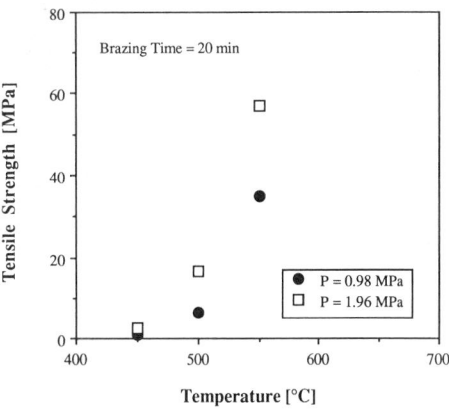

Fig 6 Effect of bonding temperature on the room-temperature tensile strength of solid-state welds of CaO-stabilized ZrO_2 joined to itself using an interlayer of 5052 aluminum alloy. Source: Ref 38

>700 °C (>1290 °F) failed within the alumina, but generally near to and influenced by the interface.

Surface Roughness. The effect of surface roughness on the strength of diffusion welds has been explored in several studies. In research by Derby (Ref 43), ASTM F 19 test pieces fabricated from alumina (99.5% Al_2O_3, 0.5% MgO + CaO) were joined using a 0.5 mm (0.02 in.) thick interlayer of 99.99% aluminum, an applied pressure of 50 MPa (7.25 ksi), and a temperature of 600 °C (1110 °F). The atmosphere during bonding was not specified, but the bonding conditions were reported to be similar to those of Nicholas and Crispin, who used a vacuum of <10 mPa. Two surface conditions were investigated: a ground surface with an average surface roughness of 1.4 μm and average peak-to-valley height of 5 μm, and a polished surface with values of 0.4 μm and 1 μm, respectively. A post-bonding heat treatment of 30 min at 400 °C (750 °F) in the bonding furnace was followed by a 4 h anneal at 200 °C (390 °F) in a furnace. The strength of the samples was negligible without these thermal treatments. For both surface conditions the bond strength increased with time. However a maximum of only 20 MPa (2.9 ksi) after 40 min bonding time was reached in the ground specimens, while the strength of the lapped specimens reached an average strength of 32 MPa (4.6 ksi) after a 20 min bonding time. An interesting footnote to this study is that much stronger joints, with tensile strengths averaging 50 MPa (7.25 ksi) were achieved at bonding temperatures above the melting point of aluminum (660 °C, or 1110 °F).

Suganuma and co-workers investigated the solid-state bonding of an alumina (99.7 wt% Al_2O_3) to niobium (99.5% pure) as a function of bond face grinding conditions (Ref 46). The bonding was done in a hot press under a vacuum of 2.7 mPa, at a temperature of 1500 °C (2730 °F) for 1 h, and an applied pressure of 20 MPa (2.9 ksi). The effects on room-temperature bond strength of three surface roughness conditions of both the alumina and the niobium were explored. The bonded specimens were ground to remove a 1.5 mm (0.06 in.) wide unbonded region that was sometimes present along the edges of the joint, and then cut into bars for four-point flexure testing. As in the study by Derby, these researchers concluded that significantly stronger joints resulted in welds made with the alumina having the smoothest surface (average surface roughness of 0.97 μm). However, as shown in Fig 8, the surface finish of the niobium had just the opposite effect, that is, the strongest bonds were produced with the roughest niobium (average surface roughness of 3.49 μm). In addition, the rough bond face on the niobium resulted in the elimination of the unjoined region at the edges of the original specimens. Both beneficial effects of niobium roughness were probably caused by plastic flow of the metal.

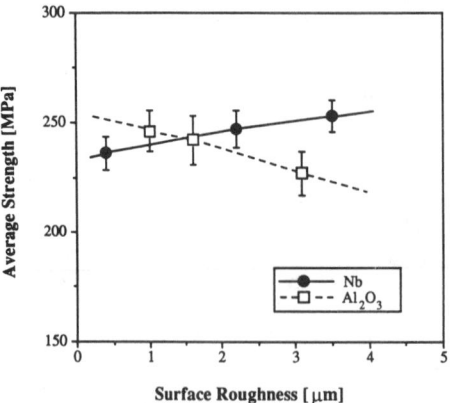

Fig 8 The effect of the roughness of the surfaces of the materials being joined when alumina was diffusion welded to niobium. Source: Ref 46

Ceramic/Ceramic Joints

As stated previously, diffusion welding is usually performed under conditions of temperature and pressure where at least one component plastically deforms under the applied load so that the surfaces to be joined are brought into intimate contact. Although the material that deforms in most cases is either the metal portion of a ceramic-metal joint, or an intermediate material placed in the joint, it may also be the ceramic. For example, Elssner et al. reported weak joints when a coarse-grained alumina (GS ~ 18 μm) was bonded to itself without an interlayer, because the ceramic was creep resistant and, therefore, resisted formation of an intimate interface (Ref 47). Conversely, bonds between the same coarse-grained alumina and an alumina with smaller grains (~1 μm) were much easier to produce as the fine-grained material deformed much more readily. The flexure strength (test method not given) for joints between the two aluminas reached a maximum value of 200 MPa (29 ksi) when welded at 1750 °C (3180 °F).

Scott and Tran studied the effect of the level of MgO sintering aid on diffusion welds in high-purity alumina (Ref 48). Dense, fine-grained alumina bodies (1 to 3 μm average grain size) were fabricated by pressureless sintering at 1500 °C (2730 °F), a series of compacts containing 0.0075, 0.075, or 0.75 wt% MgO. The sintered pellets were diffusion welded in a vacuum hot press at temperatures from 1200 to 1500 °C (2190 to 2730 °F) and applied pressures from 34 to 138 MPa (4.9 to 20 ksi). The bonded pellets were cooled to room temperature and then resintered in H_2 at 1875 °C (3405 °F) for 3 h for grain growth. As expected, the best results (as indicated by minimal interfacial porosity and grain growth across the original interface) were achieved at the higher temperatures and pressures. The study also showed that the level of MgO dopant in the alumina bodies had a significant effect on bond quality. Joints between samples with the highest

level of MgO (0.75 wt%) had fewer interfacial voids and a much finer grain size. The lower dopant levels allowed exaggerated grain growth to occur, leaving porosity trapped in the large grains and at the interface.

The effect of the superplastic behavior of Y_2O_3-stabilized tetragonal zirconia polycrystal ceramic (Y-TZP) on diffusion bonded joints has been investigated by Nagano and co-workers (Ref 49). The materials included fine-grained alumina (99.9 wt% Al_2O_3, 0.64 μm grain size), coarse-grained alumina (99.7 wt% Al_2O_3, 5.87 μm grain size), Y-TZP (3 mol% Y_2O_3, 0.59 μm grain size), and a series of Y-TZP/alumina composites with the alumina content ranging from 20 to 80 wt%, and grain sizes of about 1 μm. Bonding was conducted in air at temperatures ranging from 1450 to 1500 °C (2640 to 2730 °F) under an applied stress of 12.5 MPa (1.8 ksi). The welded samples were cut into bars that were tested in 4-point bending at room temperature. They concluded that the strengths of the bonds increased with the degree of deformation during welding (Fig 9), which resulted from higher bonding temperatures or increased amounts of the superplastic Y-TZP phase in at least one member of the joint. Flexural strengths greater than 1000 MPa (145 ksi) were achieved. Similar to Elssner (Ref 47), they found that it was difficult to bond the coarse-grained alumina to itself, but that it was possible to bond the coarse-grained alumina to materials that were more readily deformed during welding, such as the 80% Al_2O_3-20% Y-TZP composite or the fine-grained alumina.

Brazing with Filler Metals

Brazing with filler metals is another attractive process for joining structural ceramics for many applications. Wetting and adherence are the principal requirements for brazing; however, most ceramics are not wetted by con-

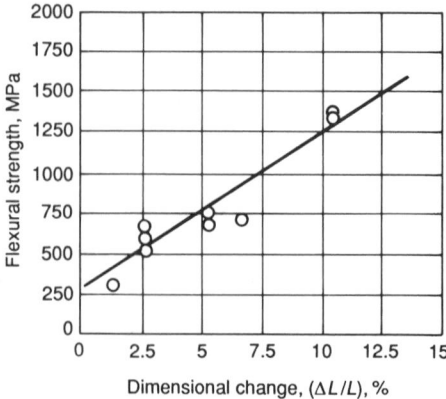

Fig 9 Relationship between the degree of deformation and the strength of solid-state welds. A series of Y-TZP/alumina composites containing 80 wt% ZrO_2, 3 mol% Y_2O_3, and 20 wt% Al_2O_3 were joined to each other at temperatures ranging from 1450 to 1500 °C (2640 to 2730 °F). Source: Ref 49

ventional brazing filler metals. This problem can be overcome either by coating the ceramic surface with a suitable metal layer prior to brazing (indirect brazing) or through the use of specially formulated filler metals that wet and adhere directly to an untreated ceramic surface (direct brazing).

Indirect Brazing. In this technique, the ceramic is first coated in the joint area with a material (usually a metal) that can be wetted by a filler metal that would not wet the untreated surface. Coating techniques include sputtering (Ref 50–53), vapor plating (Ref 54–56), and thermal decomposition of a metal-containing compound such as TiH_4 (Ref 57).

The most widely used brazing technique for joining alumina ceramics is the moly-manganese (Mo-Mn) process. In this process a slurry consisting of powders of Mo and/or MoO_3 (or W/WO_3), Mn or MnO_2, and various glass-forming compounds is applied to the ceramic in the form of a paint. The coated ceramic is fired in a wet H_2 atmosphere at a temperature of about 1500 °C (2730 °F). During firing, glassy material from the ceramic migrates into the Mo or W constituent of the powder layer bonding it to the ceramic surface. The process is dependent on the glass migration out of the ceramic and, therefore, is dependent upon such factors as amount and viscosity of the glassy phases, ratio of grain size in the ceramic to the pore size in the Mo powder, and processing temperature and atmosphere. The Mo-Mn process is thoroughly reviewed in a series of papers by Twentyman (Ref 58–60).

As mentioned previously, ceramics can also be coated by physical vapor deposition prior to brazing with conventional (non-active) brazing filler metals, and there are indications that this technique is superior to sputtering as the coating method. For example, Santella developed a technique for vapor plating a thin film of Ti onto PSZ materials prior to brazing, and in the process overcame a problem that arose in earlier work. In that work (see Ref 51), PSZ (MS grade, Nilcra Ceramics Pty. Ltd., Victoria, Australia) was coated with sputtered Ti and then brazed at 750 °C (1380 °F) to nodular cast iron (NCI) or to Ti with a commercial Ag-Cu-Sn filler metal (AWS BAg-18). Room-temperature shear strengths of both the PSZ/NCI and PSZ/Ti joints were about 140 MPa (20 ksi); however, at 400 °C (750 °F) the shear strength of the PSZ/NCI brazement was only 28 MPa (4 ksi). Santella found that Ti coatings deposited by sputtering contain microporosity (apparently from trapped plasma gases), that the gases trapped in such porosity were released during brazing, and that the high-temperature mechanical properties of such joints were adversely affected. Physical deposition by evaporation of the Ti in a vacuum solved this problem.

Direct Brazing. The wetting of an untreated ceramic by a molten metal and the associated adhesion in such a system increases with growing affinity of the metal constituents for the elements constituting the solid phase (Ref 61). Thus, chemically oxygen-active metals such as Ti, Zr, Al, Si, Mn, or Li, either pure or when alloyed with other metals, should enhance both wetting of and adherence to oxide ceramics, without the need for coating the ceramic surface. Similarly, the wetting of and adherence to SiC and Si_3N_4 are improved in filler metals that contain elements that strongly interact with Si, C, or N. Whether the ceramic is an oxide, carbide, or nitride, the active metal reacts with the ceramic surface, forming an interfacial layer that can be wetted by the bulk of the filler metal. Titanium is the most extensively studied and widely used active element addition to filler metals formulated to directly braze high-melting oxide ceramics (Ref 62–66). The critical interfacial reaction product in the case of oxide ceramics brazed with Ti-containing filler metals is either TiO or Ti_2O_3, with appreciably higher adhesion in systems that result in the formation of TiO (Ref 67).

There are three generally studied primary variables in direct brazing of ceramics: active element content of the filler metal, brazing time, and brazing temperature. Other variables such as applied pressure across the joint, thickness of the filler metal, purity of the atmosphere during brazing, and the roughness of the surfaces being joined, are potentially very important, but infrequently studied. The wetting or contact angle (θ) between the filler metal and ceramic generally decreases continuously as the primary variables are increased (Ref 62, 68, 69). The relationships between adherence and the primary variables, however, are not as clear. In general, as the primary variables are increased, the strength of a joint increases to a peak value and then decreases, as shown schematically in Fig 10 (Ref 57, 62, 70, 71). This behavior can be explained on the basis of the formation and growth of the reaction products at the interface between the ceramic and filler metal. These products are a necessary part of the wetting and adherence process, but if the reaction products grow excessively, the bond may be weakened. It is assumed that the stresses and strains associated with volume changes that occur when the reaction products are formed and grow, create flaws that result in the observed degradations of joint strength.

The type of oxide ceramic and factors such as compositional differences as the result of alloying additions or sintering aid compositions or amounts can also have a significant effect on wetting, adherence, and the sensitivity of the joint strength to changes in the primary variables, and to the behavior of the brazements in service. For example, in an as yet unpublished study on the direct brazing of TTZ ceramics, it was found that those stabilized with MgO or Y_2O_3 (whether partially-stabilized zirconias or tetragonal zirconia polycrystal materials) were readily wetted and

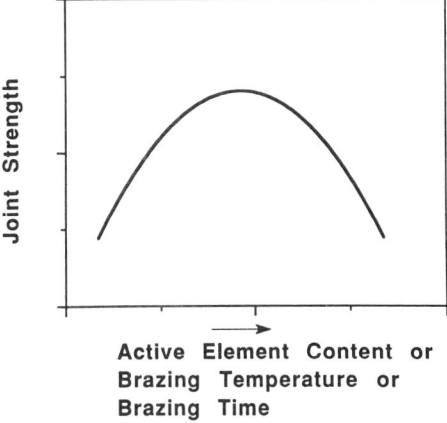

Fig 10 Relationship between the joint strength and the primary brazing variables for structural ceramics joined by active filler metals

strongly bonded by silver-copper base filler metals containing various amounts of titanium. However, when a TZP material stabilized with 12 mol% CeO_2 was brazed with active filler metal foils, the filler metals either did not wet the surface or were so poorly bonded that they were easily pulled intact from the ceramic as illustrated in Fig 11.

The darkened surface of the zirconia is typical of the effect of active filler metals on this type of ceramic and is a function of several variables, including titanium content of the filler metal and the brazing cycle variables such as time and temperature. The effect of brazing conditions on the degree of discoloration of a Mg-PSZ is apparent in Fig 12, which shows alumina and PSZ flexure bars machined from coupons brazed in vacuum of about 7 mPa with a filler metal of composi-

Fig 11 Tetragonal zirconia polycrystal ceramic containing 12 mol% CeO_2 brazed in vacuum of 7 mPa for 5 min at 950 °C (1740 °F) with two active filler metals. The filler metals have either withdrawn from the surface on melting (left) or were easily separated from the ceramic after brazing (right).

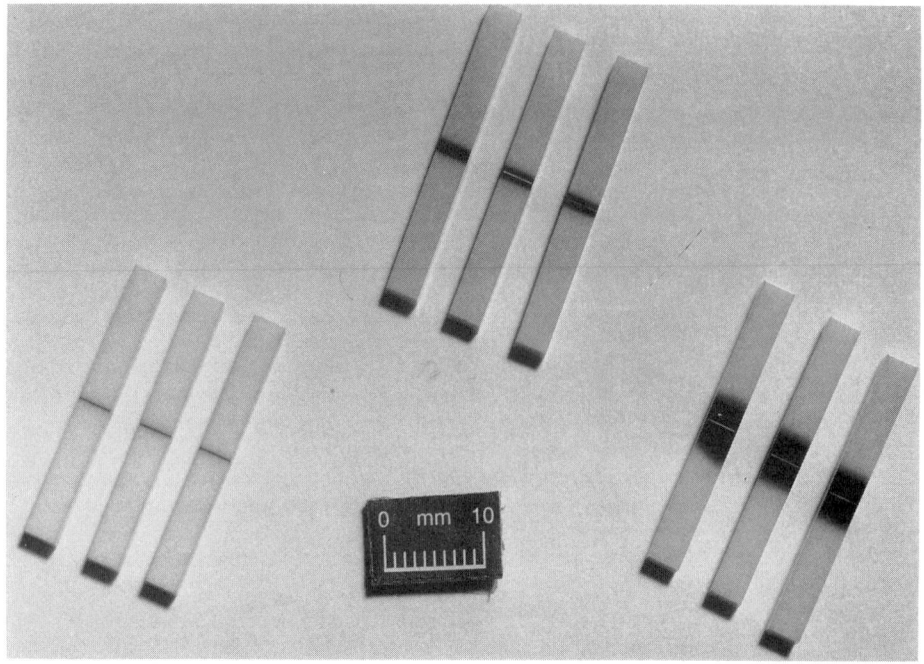

Fig 12 Flexure bars brazed in vacuum of 7 mPa with active filler metal of composition Cu-44Ag-4Sn-4Ti (at.%). Left: alumina brazed for 5 min at 800 °C (1470 °F). Center: Mg-PSZ brazed for 5 min at 800 °C (1470 °F). Right: Mg-PSZ brazed for 20 min at 900 °C (1650 °F)

tion Cu-44Ag-4Sn-4Ti (at.%). Note that there is little if any discoloration of the alumina bars shown on the left in Fig 12, but that the PSZ is darkened in the vicinity of the braze joint in the samples brazed for 5 min at 800 °C (1470 °F) (center) and much more extensively when brazed for 20 min at 900 °C (1650 °F) (right). Similar behavior has been observed when a Y-TZP material is brazed. In addition to the observed darkening of the PSZ, it has also been shown (Fig 13) that the

reaction product (TiO) at the interface between Mg-PSZ and an active filler metal (Cu-44Ag-4Sn-4Ti, at.%) grew significantly in thickness as the brazing conditions were increased from 5 min at 800 °C (1470 °F) to 20 min at 900 °C (1650 °F) (Ref 72). On the other hand, similarly brazed alumina had a very thin layer of TiO that could only be detected by transmission electron microscopy, and flexural strength was not affected by brazing conditions.

The darkening of zirconia ceramics during processing has been widely recognized, but attempts to determine the cause and the effect of such darkening on the performance of these ceramics has produced much controversy, with one suggestion being that the ZrO_2 is reduced to Zr_2O_3 (Ref 73). Whatever the cause of the darkening, active metal brazements in TZP materials behave very differently than those in the alumina ceramics. For example, the oxidation behavior of brazements of Cu-41.1Ag-3.6Sn-7.2Ti (at.%) on a Mg-PSZ are significantly different than those on a high-purity alumina (99.8% Al_2O_3). This behavior is evident in Fig 14 which shows that the specific weight gains of brazed coupons of Mg-PSZ exposed to an Ar-20% O_2 atmosphere at 400 and 500 °C (750 and 930 °F) are higher than similar brazements on alumina (Ref 74). Examination of polished cross-sections of the brazements before and after oxidation revealed that the increased weight could be attributed to oxidation of the alloy at the ceramic/filler metal interface by oxygen from the atmosphere being transported through the PSZ. The effect of such behavior on the strength of active metal brazements in a Mg-PSZ material is much more significant as can be seen in Fig 15. Note that the strengths of the brazements in alumina are relatively unaffected by 100 h exposures at elevated temperatures, but that the strengths of the PSZ brazements have dropped precipitously after exposure, particularly at 500 °C (930 °F).

Rathner and Green recently conducted a study on brazing of a Y-TZP material using filler metals of pure aluminum (99.9% Al) and Al-5.8 wt% Zr (Ref 75). Brazing was done in a high-purity argon atmosphere, with a 1 h hold at temperature. They found that a temperature of 1200 °C (2190 °F) was required to produce joints with the pure aluminum, but that strong bonds were formed with the Al-5.8 wt% Zr at ~900 °C (~1650 °F). The 4-point bending strength of the samples brazed with Al were 145 ± 29 MPa (21 ± 4 ksi).

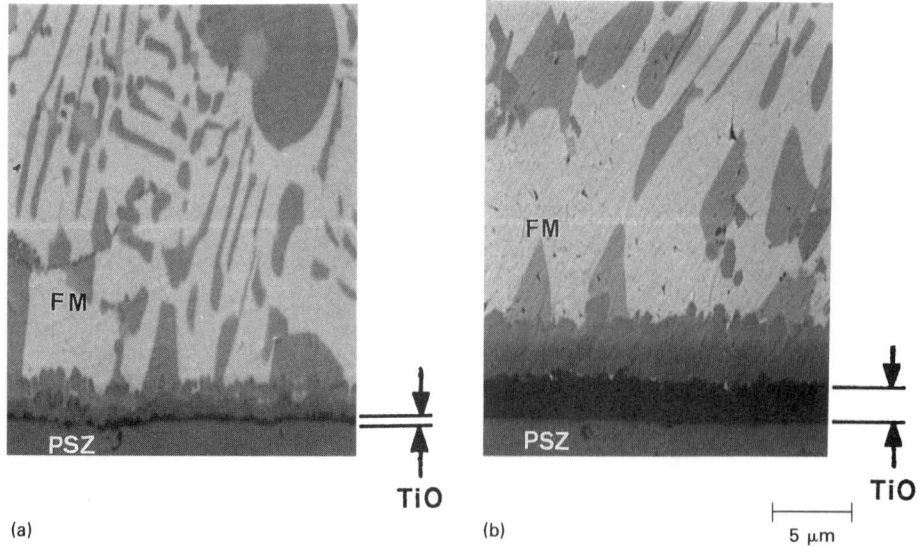

(a)   (b)   5 μm

Fig 13 Growth of interfacial reaction product (TiO) between Cu-44Ag-4Sn-4Ti (at.%) and Mg-PSZ as brazing conditions were increased from 5 min at 800 °C (1470 °F) to 20 min at 900 °C (1650 °F)

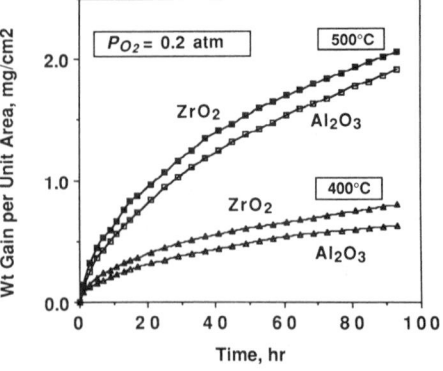

Fig 14 Comparison of weight gains during oxidation of Mg-PSZ and alumina brazements joined with Cu-41.1Ag-3.6Sn-7.2Ti (at.%). The increased weight change is attributed to oxidation at the filler metal/ceramic interface due to the high oxygen ion conduction through the zirconia.

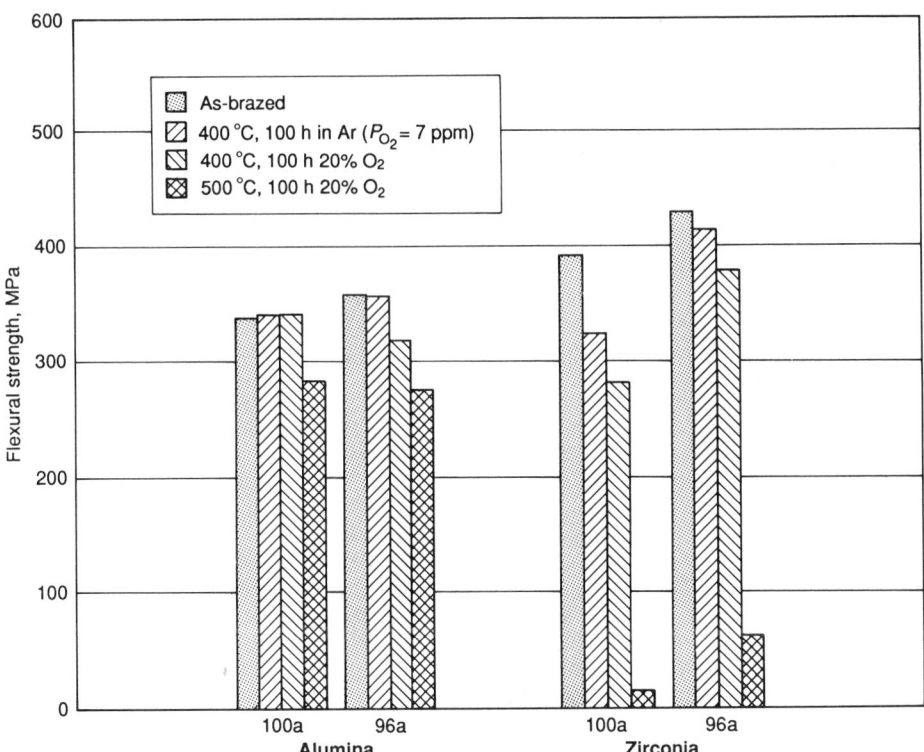

Fig 15 Effect of oxidizing environments on the room-temperature flexural strengths of active filler metal brazements in alumina and zirconia (Mg-PSZ). Filler metal compositions are: (96a) Ag-41.6Cu-9.7Sn-5.0Ti (at.%) and (100a) Cu-41.1Ag-3.6Sn-7.2Ti (at.%).

The strengths of the samples brazed with Al-5.8 wt% Zr were stronger than those with pure aluminum because of the formation of Al_3Zr precipitates in the former brazements. The strengths of the Al-5.8 wt% Zr brazements increased as the thickness of the joint decreased either as the result of gravity or applied pressure. The thinnest joint (25 μm) contained about 20 vol% of the Al_3Zr precipitates, and the joint was brittle but strong (402 ± 27 MPa, or 58 ± 4 ksi). The thicker joints were ductile, but the strength was significantly less (246 ± 31 MPa, or 36 ± 4.5 ksi). The photographs and micrographs shown in this article indicate that the TZP was not darkened by the brazing process, and the authors indicate that there was no evidence of interfacial reaction products.

The silicon carbide whiskers in an Al_2O_3-SiC_w composite affect the thermal stability of the joints when this material is brazed with an active filler metal. For example, brazements in Al_2O_3-SiC_w were significantly stronger than those in alumina, but decreased in strength to a value equal to that of the alumina brazements as the test temperature was increased (Fig 16a), or when aged at elevated temperatures (Fig 16b) (Ref 76). It has recently been determined that the orientation of the whiskers with respect to the interface also has an effect on the strength of the brazements (Fig 17).

One example of a very successful high-temperature application of brazements with a Ti-containing filler metal were sensors for measuring fluid flow phenomena in steam-water mixtures (T = 850 °C, or 1560 °F, and P = 0.6 MPa, or 87 psi) during loss-of-coolant accidents in simulated nuclear reactor experiments (Ref 77). Hundreds of the sensors shown in Fig 18, each consisting of a 12 mm (0.5 in.) diam alumina insulator separated from a type 446 stainless steel body by a platinum transition piece (to accommodate the differences in CTE), were brazed in vacuum at 970 °C (1780 °F) with a filler metal of composition 49Ti-49Cu-2Be (wt%). Test pieces had He leak rates <2 × 10^{-5} cm^3/s after being quenched 25 times from air at 500 °C (930 °F) into water at 80 °C (175 °F) (thermal transient >300 °C/s, or 540 °F/s). The sensors survived years of service.

Brazing with Glasses

A number of studies have been conducted to develop materials and techniques for joining structural ceramics—including alumina (Ref 78), PSZ (Ref 79), Si_3N_4 (Ref 80–84), and sialon (Ref 85, 86)—by brazing with nonmetallic glasses as filler materials. The rationale for this approach is that many ceramics have amorphous phases remaining at the grain boundaries after the sintering process, so they would be expected to be wet by, and compatible with, similar glassy materials introduced at the joint. Glass viscosity and flow properties can be controlled over a wide range of temperatures; however, glasses have much lower toughness (typically <1 MPa$\sqrt{m}$, or 0.9 ksi $\sqrt{in}$.) than even monolithic ceramics, have low Young's modulus, and are more susceptible than ceramics to stress-corrosion effects (slow crack growth). As will be discussed below, some of the glasses, even those formulated to match the grain-boundary composition of a specific ceramic, have CTEs significantly different from those of the ceramics, which causes high residual stresses and cracking in the joint.

A joint is usually prepared by applying a slurry of powdered glass with a binder to one or both of the mating components, which are then heated until the glass melts and reacts with the ceramics. In a variation of the process the surface of one of the components was first coated with molten glass, and this glazed surface was ground to precisely control the thickness of the joining material (Ref 82). In contrast to diffusion welding with high pressures, brazing with glasses and metals uses relatively low pressures ranging from that provided by the weight of one of the parts being joined up to 14 MPa (2 ksi). Brazing is typically conducted in air when using glasses to join the structural oxides.

The most extensive studies on brazing with glasses have been those on Si_3N_4. The glasses

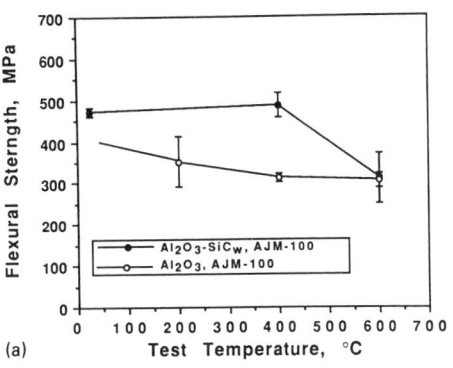

(a)

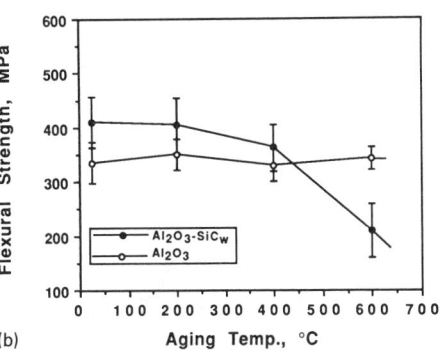

(b)

Fig 16 Effect of (a) testing temperature and (b) aging temperature on the strength of active filler metal brazements in Al_2O_3-SiC_w and alumina

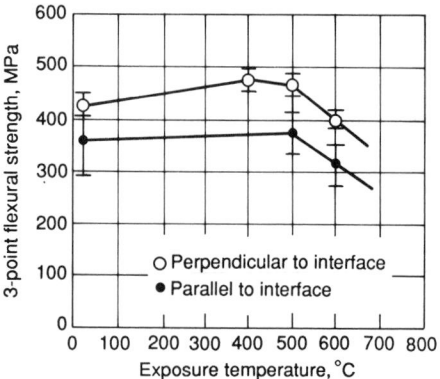

Fig 17 Effect of whisker alignment on the strength of active filler metal brazements in Al$_2$O$_3$-SiC$_w$ composites

include those similar to that found at the grain boundaries of a hot pressed Si$_3$N$_4$ (55SiO$_2$, 35MgO, 10Al$_2$O$_3$, wt%) (Ref 82, 84–86) and a very different glass such as CTS (35CaO, 25TiO$_2$, 40SiO$_2$) (Ref 83). A CTS glass was also investigated for joining Mg-PSZ (Ref 79). This type of glass was chosen because there are several eutectic compositions in the ternary system with melting temperatures below the 1400 to 1500 °C (2550 to 2730 °F) annealing range of the ceramic, so that the brazing step could be combined with the annealing treatment of the partially stabilized zirconia. An applied pressure of 14 MPa (2 ksi) was used to distribute the molten glass through the joint. Unfortunately, the differences in CTE of the materials (10 × 10^{-6}/ °C for the PSZ versus 3 to 4 × 10^{-6}/°C for the glass) resulted in extensive cracking through the glass and interface when the joint area was relatively large (about 20 cm^2, or 3 in.^{3}). Based on the work of Iwamoto, the CTE of the interlayer was made closer to that of the PSZ by adding 30 wt% ZrO$_2$ to the CTS interlayer powder. Using a thinner layer of

the modified powder and a longer pressing time at 1420 °C (2590 °F), joints were produced that had an average 4-point bending strength of 66 MPa (9.6 ksi).

Several glass compositions were investigated for joining an alumina (94% Al$_2$O$_3$) ceramic (Ref 78). This work differed from that of the preceding studies in that one of the goals of these researchers was to initiate and control crystallization of the glass in the joint to increase fracture toughness and resistance to stress corrosion. Zirconia powder was added to a MgO-Al$_2$O$_3$-SiO$_2$ glass as a nucleating agent. Unfortunately, spontaneous crystallization occurred at the glass-alumina interface limiting the desired degree of control over crystallization, and resulting in large voids at the interface. Accordingly, the fracture toughness of the joint as measured by the single-edge-notched beam technique (Fig 4) was only 1.8 MPa$\sqrt{m}$ (1.6 ksi$\sqrt{in.}$) versus a value of about 4 MPa$\sqrt{m}$ (3.6 ksi$\sqrt{in.}$) for the non-brazed alumina. The most successful joints (K_{Ic} = 2.3 MPa$\sqrt{m}$, or 2 ksi$\sqrt{in.}$) resulted when the alumina was joined at 1500 °C (2730 °F) with a mixture of 50 wt% talc (hydrated magnesium silicate) and 50% alumina. The higher toughness was attributed to extensive crystallization of the glassy phase in the joint. However, these joints also contained a substantial number of pores that degraded the mechanical integrity of the joints. The authors suggested that the pores were caused by the evolution of adsorbed gases from the alumina body; therefore, the porosity would be difficult to eliminate.

ACKNOWLEDGMENT

The authors appreciate the important technical contributions made to this manuscript during review by their colleagues J.I. Federer and T.N. Tiegs. Some of the research discussed in this chapter was sponsored by the U.S. Department of Energy under contract DE-AC05–84OR21400 with Martin Marietta Energy Systems, Inc.

REFERENCES

1. D.A. Parker, Technological Requirements for Designing Interfaces for Mechanical and Thermal Applications, *Designing Interfaces for Technological Applications: Ceramic-Ceramic, Ceramic-Metal Joining*. The European Colloquium, Petten, The Netherlands, 20–21 Apr 1988, S.D. Peteves, Ed., Elsevier, London, 1989, p 3–32

2. R.E. Loehman, Ceramic Joining in the U.S., *Designing Interfaces for Technological Applications: Ceramic-Ceramic, Ceramic-Metal Joining*, The European Colloquium, Petten, The Netherlands, 20–21 Apr 1988, S.D. Peteves, Ed., Elsevier, London, 1989, p 235–245

3. T. Suga, Current Research and Future Outlook in Japan, *Designing Interfaces for Technological Applications: Ceramic-Ceramic, Ceramic-Metal Joining*, The European Colloquium, Petten, The Netherlands, 20–21 Apr 1988, S.D. Peteves, Ed., Elsevier, London, 1989, p 247–263

4. B.D. Foster and J.B. Patton, Ed., *Ceramics in Heat Exchangers*, Vol 14, Proc. of three symposia on heat exchangers: First, 86th Annual Meeting of the American Ceramic Society, Pittsburgh, 29 Apr-3 May 1984; Second, Pacific Coast Regional Meeting, San Francisco, 28–31 Oct 1984; Third, 87th Annual Meeting, Cincinnati, OH, 5–9 May 1985, The American Ceramic Society, 1985

5. A.J. Hayes, W.W. Liang, S.L. Richlen, and E.S. Tabb, Ed., *Industrial Heat Exchangers*, Proc. 1985 Exposition and Symposium on Industrial Heat Exchanger Technology, Pittsburgh, 6–8 Nov 1985, American Society for Metals, 1985

6. M.G. Nicholas and D.A. Mortimer, Ceramic/Metal Joining for Structural Applications, *Mater. Sci. Technol.*, Vol 1 (No. 9), 1985, p 657

7. R.E. Loehman and A.P. Tomsia, Joining of Ceramics, *Ceram. Bull.*, Vol 67 (No. 2), 1988, p 375

8. D.W. Richerson, *Modern Ceramic Engineering: Properties, Processing, and Use in Design*, Marcel Dekker, 1982, p 6–16

9. M.H. Lewis, Ceramics To Be Joined 10 Years From Now, *Designing Interfaces for Technological Applications: Ceramic-Ceramic, Ceramic-Metal Joining*, The European Colloquium, Petten, The Netherlands, 20–21 Apr 1988, S.D. Peteves, Ed., Elsevier, London, 1989, p 271–290

10. P.F. Becher and M.K. Ferber, Mechanical Behaviour of MgO-Partially Stabilized ZrO$_2$ Ceramics at Elevated

Fig 18 The stages (from left to right) in the fabrication of brazed sensor assemblies consisting of a dispersed-platinum alumina insulator, platinum electrodes (arrow), and type 446 stainless steel body

Temperatures, *J. Mater. Sci.*, Vol 22 (No. 3), 1987, p 973–980

11. M. Rühle and N. Claussen, Transformation and Microcrack Toughening as Complementary Processes in ZrO₂-Toughened Al₂O₃, *J. Am. Ceram. Soc.*, Vol 69 (No. 3), 1986, p 195–197

12. P.F. Becher and T.N. Tiegs, Temperature Dependence of Strengthening by Whisker Reinforcement: SiC Whisker-Reinforced Alumina in Air, *Adv. Ceram. Mater.*, Vol 3 (No. 2), 1988, p 148–153

13. K.M. Prewo, Tension and Flexural Strength of Silicon Carbide Fibre-Reinforced Glass Ceramics, *J. Mater. Sci.*, Vol 21 (No. ID), 1986, p 3590–3600

14. J.J. Brennan, Glass and Glass-Ceramic Matrix Composites, *Fiber Reinforced Ceramic Composites*, K.S. Mazdiyasni, Ed., Noyes Publications, 1990, p 222–259

15. T. Mah *et al.*, Ceramic Fiber Reinforced Metal-Organic Precursor Matrix Composites, *Fiber Reinforced Ceramic Composites*, K.S. Mazdiyasni, Ed., Noyes Publications, 1990, p 278–310

16. J.B. Wachtman, Jr. and R.A. Haber, Advanced Ceramics Involving Alumina, *Alumina Chemicals: Science and Technology Handbook*, L.D. Hart, Ed., The American Ceramic Society, 1990, p 329–335

17. "Mechanical and Industrial Ceramics," CAT/2T8505TKU/3000E, Kyocera Corporation

18. "Coors Ceramics—Materials for Tough Jobs," Bulletin No. 953, Coors Porcelain Company

19. G. Fisher, Zirconia: Ceramic Engineering's Toughness Challenge, *Ceram. Bull.*, Vol 65 (No. 10), 1986, p 1355–1360

20. A.H. Heuer, R. Chaim, and V. Lanteri, Review: Phase Transformations and Microstructural Characterization of Alloys in the System Y₂O₃-ZrO₂, *Advances in Ceramics*, Vol 24, *Science and Technology of Zirconia III*, 1988, p 3–20

21. P.F. Becher and M.K. Ferber, Mechanical Behaviour of MgO-Partially Stabilized ZrO₂ Ceramics at Elevated Temperatures, *J. Mater. Sci.*, Vol 22 (No. 3), 1987, p 973–980

22. P.F. Becher and G.C. Wei, Toughening Behavior in SiC-Whisker-Reinforced Alumina, *Comm. Am. Ceram. Soc.*, Vol 67 (No. 12), 1984, p C267-C269

23. G.C. Wei and P.F. Becher, Development of SiC-Whisker-Reinforced Ceramics, *Am. Ceram. Soc. Bull.*, Vol 64 (No. 2), 1985, p 298–304

24. T.N. Tiegs and P.F. Becher, Whisker Reinforced Ceramic Composites, *Tailoring Multiphase and Composite Ceramics*, Vol 20, Proc. The Twenty-First Univ. Conf. on Ceramic Science, 17–18 July 1985, R.E. Tressler *et al.*, Ed., Plenum Press, 1986, p 639–647

25. P.F. Becher *et al.*, Toughening of Ceramics by Whisker Reinforcement, *Fracture Mechanics of Ceramics*, Vol 7, R.C. Bradt, A.G. Evans, D.P.H. Hasselman, and F.F. Lange, Ed., Plenum, 1986, p 61–73

26. H.-T. Lin and P.F. Becher, Creep Behavior of a SiC-Whisker-Reinforced Alumina, *J. Am. Ceram. Soc.*, Vol 73 (No. 5), 1990, p 1378–1381

27. Y. Katayama and Y. Hattori, Effects of Specimen Size on Strength of Sintered Silicon Nitride, *J. Am. Ceram. Soc.*, Vol 65 (No. 10), 1982, p C164-C165

28. T. Soma, M. Matsui, and I. Oda, Tensile Strength of a Sintered Silicon Nitride, *Non-Oxide Technical and Engineering Ceramics*, S. Hampshire, Ed., Elsevier, 1986, p 361–374

29. K. Suganuma, Relationship between the Strength of Ceramic/Metal Joints in Tensile and Three-Point Bending, *J. Am. Ceram. Soc.*, Vol 67 (No. 10), 1986, p C235-C236

30. A.J. Moorhead and P.F. Becher, Adaptation of the DCB Test for Determining Fracture Toughness of Brazed Joints in Ceramic Materials, *J. Mater. Sci.*, Vol 22 (No. 9), 1987, p 3297–3303

31. D.W. Richerson, *Modern Ceramic Engineering: Properties, Processing, and Use in Design*, Marcel Dekker, 1982, p 95

32. G. Elssner, Bond Strength of Ceramic/Metal Joints, *Designing Interfaces for Technological Applications: Ceramic-Ceramic, Ceramic-Metal Joining*, The European Colloquium, Petten, The Netherlands, 20–21 Apr 1988, S.D. Peteves, Ed., Elsevier, London, 1989, p 77–104

33. G. de With, Aspects of the Fracture Behaviour of Joints, *Designing Interfaces for Technological Applications: Ceramic-Ceramic, Ceramic-Metal Joining*, The European Colloquium, Petten, The Netherlands, 20–21 Apr 1988, S.D. Peteves, Ed., Elsevier, 1989, p 105–123

34. J.T. Klomp, Ceramic and Metal Surfaces in Ceramic-to-Metal Bonding, *Ceramic Surfaces and Surface Treatments, British Ceramic Proceedings*, No. 34, R. Morrell and M.G. Nicholas, Ed., 1984, Shelton, Stoke-on-Trent, Staffs, U.K., p 249–259

35. G. Elssner *et al.*, Fracture of Ceramic-to-Metal Interfaces, *J. De Physique*, Vol 46, 1985, p C4–597 to C4–612

36. M. Rühle and W. Mader, Structure and Chemistry of Metal/Ceramic Interfaces, *Designing Interfaces for Technological Applications: Ceramic-Ceramic, Ceramic-Metal Joining*, The European Colloquium, Petten, The Netherlands, 20–21 Apr 1988, S.D. Peteves, Ed., Elsevier, London, 1989, p 145–195

37. J.T. Klomp and A.J.C. VandeVen, Parameters in Solid-State Bonding of Metals to Oxide Materials and the Adherence of Bonds, *J. Mater. Sci.*, Vol 15, 1980, p 2483–2488

38. Y. Arata and A. Ohmori, Studies on Solid State Reaction Bonding of Metal and Ceramic (Report I), *Trans. JWRI*, Vol 13, 1984, p 41–46

39. H.J. DeBruin, A.F. Moodie, and C.E. Warble, Ceramic-Metal Reaction Welding, *J. Mater. Sci.*, Vol 7, 1972, p 909–918

40. B. Derby and E.R. Wallach, *Met. Sci.*, Vol 16, 1982, p 49

41. K. Nishiguchi and Y. Takahashi, Determination of Optimum Process Conditions in Solid Phase Bonding, by a Numerical Model, *Quart. J. Jap. Weld. Soc.*, Vol 3, 1985, p 65

42. M.G. Nicholas and R.M. Crispin, Diffusion Bonding Stainless Steel to Alumina Using Aluminum Interlayers, *J. Mater. Sci.*, Vol 17 (No. 11), 1982, p 3347–3360

43. B. Derby, The Influence of Surface Roughness on Interface Formation in Metal/Ceramic Diffusion Bonds, *Ceramic Microstructures '86*, J. Pask and A. Evans, Ed., Plenum, 1988, p 319–328

44. W. Dawihl and E. Klinger, Mechanical and Thermal Properties of Welded Joints Between Alumina and Metals (in German), *Ber. Deutsch. Keram. Ges.*, Vol 46 (No. 12), 1969

45. R.M. Crispin and M.G. Nicholas, Alumina-Copper Diffusion Bonding, *Ceram. Eng. Sci. Proc.*, Vol 10 (No. 11–12), 1989, p 1575–1581

46. K. Suganuma, T. Okamoto, and M. Koizumi, Effect of Surface Grinding Conditions on Strength of Alumina/Niobium Joint, *Ceram. Eng. Sci. Proc.*, Vol 10 (No. 11–12), 1989, p 1919–1933

47. G. Elssner *et al.*, Microstructure and Mechanical Properties of Metal-to-Ceramic and Ceramic-to-Ceramic Joints, *Surfaces and Interfaces in Ceramic and Ceramic-Metal Systems*, J. Pask and A. Evans, Plenum Publishing, 1981, p 629–639

48. C. Scott and V.B. Tran, Diffusion Bonding of Ceramics, *Am. Ceram. Soc. Bull.*, Vol 64 (No. 8), 1985, p 1129–1131

49. T. Nagano, H. Kato, and F. Wakai, Diffusion Bonding of Zirconia/Alumina Composites, *J. Am. Ceram. Soc.*, Vol 73 (No. 11), 1990, p 3476–3480

50. R.L. Bronnes, R.C. Hughes, and R.C. Sweet, Ceramic-to-Metal Bonding with Sputtering as a Metallization Technique, *Philips Tech. Rev.*, Vol 35, 1975, p 209–211

51. J.P. Hammond, S.A. David, and M.L. Santella, Brazing Ceramic Oxides to Metals at Low Temperatures, *Weld. J.*, Vol 67 (No. 11), 1988, p 227s-232s

52. P.O. Jarvinen, Novel Ceramic Receiver for Solar Brayton Systems, CONF-

790305–7, MIT Lincoln Laboratory, 1979

53. W.F. Heim and H.L. McDowell, Method of Metallizing a Ceramic Substrate, U.S. patent 4,342,632, Aug 1982
54. S. Weiss and C.M. Adams, Jr., The Promotion of Wetting and Brazing, *Weld, J. Suppl.*, Vol 46 (No. 2), 1967, p 49s–57s
55. E.F. Brush and C.M. Adams, Vapor Coated Surfaces for Brazing Ceramics, *Weld. J.*, Vol 47 (No. 3), 1968, p 106s–114s
56. M.L. Santella, Brazing of Titanium-Vapor-Coated Silicon Nitride, *Adv. Ceram. Mater.*, Vol 3 (No. 5), 1988, p 457–462
57. M.J. Ramsey and M.H. Lewis, Interfacial Reaction Mechanisms in Syalon Ceramic Bonding, *Mater. Sci. Eng.*, Vol 71, 1985, p 113–122
58. M.E. Twentyman, Part 1. The Mechanism of Glass Migration in the Production of Metal-Ceramic Seals, *J. Mater. Sci.*, Vol 10 (No. 5), 1975, p 765–776
59. M.E. Twentyman and P. Popper, Part 2. The Effect of Experimental Variable on the Structure of Seals to Debased Aluminas, *J. Mater. Sci.*, Vol 10 (No. 5), 1975, p 777–790
60. M.E. Twentyman and P. Popper, Part 3. The Use of Metallizing Paints Containing Glass or Other Inorganic Bonding Agents, *J. Mater. Sci.*, Vol 10 (No. 5), 1975, p 791–798
61. Yu.V. Naidich, The Wettability of Solids by Liquid Metals, *Progress in Surface and Membrane Science*, Vol 14, Institute of Material Problems, Academy of Sciences, Ukrainian S.S.R., 1981, p 333–484
62. M.G. Nicholas, Active Metal Brazing, *Br. Ceram. Trans. J.*, Vol 85, 1986, p 144–146
63. M.G. Nicholas, T.M. Valentine, and M.J. Waite, The Wetting of Alumina by Copper Alloyed with Titanium and Other Elements, *J. Mater. Sci.*, Vol 15 (No. 9), 1980, p 2197–2220
64. H. Mizuhara and K. Mally, Ceramic-to-Metal Joining with Active Brazing Filler Metal, *Weld. J.*, Vol 64 (No. 10), 1985, p 27–32
65. A.J. Moorhead and H. Keating, Direct Brazing of Ceramics for Advanced Heavy-Duty Diesels, *Weld. J.*, Vol 65 (No. 10), 1986, p 17–31
66. A.J. Moorhead, Direct Brazing of Alumina Ceramics, *Adv. Ceram. Mater.*, Vol 2 (No. 2), 1987, p 159–166
67. Yu.V. Naidich and V.S. Zhuravlev, Adhesion, Wettability and Reaction of Titanium-Containing Melts with High-Melting Oxides, trans. *Ogneupory*, Vol 1, 1987, p 50–55
68. T. Iseki, H. Matsuzaki, and J.K. Boadi, Brazing of Silicon Carbide to Stainless Steel, *Am. Ceram. Soc. Bull.*, Vol 64 (No. 2), 1985, p 322–324
69. M. Naka *et al.*, Joining of Silicon Nitride with Al-Cu Alloys, *J. Mater. Sci.*, Vol 22 (No. 12), 1987, p 4417–4421
70. M. Naka *et al.*, Observation of Al/Si_3N_4 Interface, *J. Mater. Sci. Lett.*, Vol 5 (No. 7), 1986, p 696–698
71. M. Naka *et al.*, Influence of Brazing Condition on Shear Strength of Alumina-Kovar Joint Made with Amorphous $Cu_{50}Ti_{50}$ Filler Metal, *Trans. JWRI*, Vol 12 (No. 2), 1983, p 341–343
72. A.J. Moorhead, H.M. Henson, and T.J. Henson, The Role of Interfacial Reactions on the Mechanical Properties of Ceramic Brazements, *Ceramic Microstructures '86*, S. Pask and A. Evans, Ed., Plenum, 1988, p 949–958
73. G. Ingo, Origin of Darkening in 8 wt% Yttria-Zirconia Plasma-Sprayed Thermal Barrier Coatings, *J. Am. Ceram. Soc.*, Vol 74 (No. 2), 1991, p 381–386
74. A.J. Moorhead and H.-E. Kim, Oxidation Behaviour of Titanium-Containing Brazing Filler Metals, *J. Mater. Sci.*, Vol 26 (No. 15), 1991, p 4067–4075
75. R.C. Rathner and D.J. Green, Joining of Yttria-Tetragonal Zirconia Polycrystal with an Aluminum-Zirconium Alloy, *J. Am. Ceram. Soc.*, Vol 73 (No. 4), 1990, p 1103–1105
76. A.J. Moorhead and H.-E. Kim, Comparison of Strengths of Active Metal Brazements in Alumina and SiC-Whisker Reinforced Alumina, *Ceram. Eng. Sci. Proc.*, Vol 10 (No. 11–12), 1989, p 1854–1865
77. A.J. Moorhead, Welding and Brazing of Film Probe Sensor Assemblies, *Weld. J.*, Vol 62 (No. 10), 1983, p 17–27
78. W.A. Zdaniewski *et al.*, Crystallization Toughening of Ceramic Adhesives for Joining Alumina, *Adv. Ceram. Mater.*, Vol 2 (No. 3A), 1987, p 204–208
79. S.L. Swartz *et al.*, Joining of Zirconia Ceramics with a $CaO-TiO_2-SiO_2$ Interlayer, *Proc. of the Twenty-Sixth Auto. Tech. Dev. Contractors' Coord. Mtg.*, P-219, Dearborn, MI, 24–27 Oct 1988, The Society of Automotive Engineers, p 149–154
80. M.E. Milberg *et al.*, The Nature of Sialon Joints Between Silicon Nitride Based Bodies, *J. Mater. Sci.*, Vol 22 (No. 7), 1987, p 2560–2568
81. S. Baik and R. Raj, Liquid-Phase Bonding of Silicon Nitride Ceramics, *J. Am. Ceram. Soc.*, Vol 70 (No. 5), 1987, p C105-C107
82. S.M. Johnson and D.J. Rowcliffe, Mechanical Properties of Joined Silicon Nitride, *J. Am. Ceram. Soc.*, Vol 68 (No. 9), 1985, p 468–472
83. N. Iwamoto *et al.*, *Ceramic Materials and Components for Engines*, Proc. of the Second International Symposium, Lübeck-Travemünde, FRG, 14–17 Apr 1986, W. Burk and H. Hausner, Ed., Deutsche Keramische Gesellschaft, FRG, 1986, p 467–474
84. M.L. Mecartney *et al.*, Silicon Nitride Joining, *J. Am. Ceram. Soc.*, Vol 68 (No. 9), 1985, p 472–478
85. R.M. Neilson, Jr. and D.N. Coon, Strength of Silicon Nitride-Silicon Nitride Joints Bonded with Oxynitride Glass, *Ceram. Eng. Sci. Proc.*, Vol 10 (No. 11–12), 1989, p 1893–1907
86. D.N. Coon, R.L. Tallman, and R.M. Neilson, Jr., Hot Isostatically Pressed $Si_3N_4-Si_3N_4$ Joints Bonded with Oxynitride Glass, *Adv. Ceram. Mater.*, Vol 3 (No. 2), 1988, p 154–158
87. W.D. Kingery, H.K. Bowen, and D.R. Uhlmann, *Introduction to Ceramics*, Wiley, 1976
88. R.W. Rice, Ceramic Composites; Future Needs and Opportunities, *Fiber Reinforced Ceramic Composites*, K.S. Mazdiyasni, Ed., Noyes Publications, 1990, p 451–495
89. P.F. Becher, Microstructural Design of Toughened Ceramics, *J. Am. Ceram. Soc.*, Vol 74 (No. 2), 1990, p 255–269
90. "Material Properties Data Sheet," Coors Ceramics Company
91. P. Chantikul, S.J. Bennison, and B.R. Lawn, Role of Grain Size in the Strength and *R*-Curve Properties of Alumina, *J. Am. Ceram. Soc.*, Vol 73 (No. 8), 1991, p 2419–2427
92. S. Somiya, N. Yamamoto, and Yanagida, Ed., *Science and Technology of Zirconia III*; A.H. Heuer and M. Ruhle, Ed., *Science and Technology of Zirconia II*; L.W. Hobbs and A.H. Heuer, Ed., *Science and Technology of Zirconia*, American Ceramic Society
93. D. Shin, K.K. Orr, and H. Schubert, Microstructure-Mechanical Property Relationships in Hot Isostatically Pressed Alumina and Zirconia-Toughened Alumina, *J. Am. Ceram. Soc.*, Vol 73 (No. 5), 1950, p 1181–1188
94. H. Schneider and E. Eberhard, Thermal Expansion of Mullite, *J. Am. Ceram. Soc.*, Vol 73 (No. 7), 1990, p 2073–2076

Joining Non-Oxide Ceramics

Katsuaki Suganuma, Department of Materials Science and Engineering, National Defense Academy, Yokosuka, Japan

NEW TYPES of ceramic materials that have excellent strength at elevated temperatures, light weight, and good thermal properties, among others, are in demand, because conventional oxide ceramics are inadequate for many engineering applications. Among new engineering ceramics, silicon nitride and silicon carbide represent two candidates that have outstanding potential for high-temperature structural applications. Silicon nitride, in particular, has already found application in some automobile engine components, because of its high strength and fracture toughness. Electronic components also need superior thermal properties than are available with conventional alumina ceramics. Aluminum nitride is under development as one of the best potential substrates for integrated circuits, because of its very large thermal conductivity and high electrical resistivity.

Conventional joining research and development for ceramics has concentrated on oxide ceramics. The use of the molybdenum-manganese metallizing method is well established in industry, and is described in detail in the article "Ceramic/Metal Seals" in this Section of the Volume. However, this method cannot be used for non-oxide ceramics because it is based on the reaction between glass-forming compounds in the paste and the glass sintering additive of the alumina ceramic. A glass phase suitable for this reaction does not exist in non-oxide ceramics. Thus, a considerable number of alternative joining techniques have been developed and reported for nitride and carbide ceramics since 1984.

Before a reliable process for joining ceramics to themselves or to metallic components can be established, two basic factors must be considered for every application. One factor is the strength of the interfacial zone that is obtained through reaction processes. In addition to an interfacial bond with acceptable lattice matching of the phases at the interface, the formation of macroscopic defects (such as the reaction layer itself) and the unjoined area have great influence as joint properties. The reaction conditions must be selected to optimize the interfacial strength for each combination of constituents. The second factor to consider is the means of accommodating the usual thermal expansion mismatch between ceramics and metals. Silicon-base ceramics are among the most difficult ceramics to join to metals because they have very small coefficients of thermal expansion, and the resulting expansion mismatch can lead to large interfacial stresses.

Interfacial Reactions

The interfacial microstructures of silicon nitride and metals are generally divided into three types, based on their reaction processes. The first type of microstructure is the interface made by active metal brazing, which is one of the most popular joining processes for ceramic to ceramic or ceramic to metal systems. It is also suitable for oxide ceramics. The second type of microstructure is formed by a eutectic melting reaction between silicon-base ceramics and metals. Silicon-base ceramics react with some transition metals to form eutectic liquids. The method utilizes this eutectic reaction by optimizing the reaction conditions. The third type of interface is formed by complete solid-state reactions under pressure.

All of these reactions, as well as the reaction process and products, can be partly predicted from thermodynamic data. For instance, suppose that the reaction between silicon nitride and a metallic element, M, produces a metal silicide and free nitrogen. The reaction is

$$Si_3N_4 + M \rightarrow M\text{-silicide} + \text{free } N_2 \uparrow \qquad (Eq\ 1)$$

The free energy of this reaction is obtained as a function of temperature, as shown in Fig 1 (Ref 1). The resultant free nitrogen may further react with the metal to form a metal nitride, or it may be released from the system as nitrogen gas, or it may remain in the joint in solution. In the temperature range in which the free energy change is negative, the reaction proceeds naturally in the direction shown. If the formation of a metal nitride becomes the rate-determining step, the reaction

$$Si_3N_4 + M \rightarrow M\text{-nitride} + \text{free } Si \qquad (Eq\ 2)$$

will take place. The free energy change of this reaction is also obtained from Fig 1. The resultant free silicon may further react with the metal. For niobium, titanium, tantalum, aluminum, zirconium, and other elements, the reactions can occur in the entire temperature range of interest because their nitrides are more stable than silicon nitride.

It should be noted that the minimum observable temperatures for both reactions shown in Eq 1 and 2 cannot be deduced solely from free energy changes because of kinetic constraints. If the temperature is too low, diffusion is limited. In addition, the formation of nitrides also depends on the partial pressure of nitrogen in the joining atmosphere. These parameters are discussed fully in the article "Wetting, Surface Energies, Adhesion, and Interface Reaction Thermodynamics" in this Section of the Volume. However, Fig 1 is useful in selecting the reactive elements to form silicon nitride. The joining of most silicon nitrides will also be influenced by the presence of grain boundary phases, that is, oxides and free silicon. Free silicon is reactive to many metals, forming silicides and oxides, and is expected to promote wetting by molten metals that are strong oxide formers.

These considerations also apply to silicon carbide and metals. The main difference is the different reactivity of metals with carbon, compared to nitrogen. This difference can be taken into consideration by replacing Eq 2 with:

$$SiC + M \rightarrow M\text{-carbide} + \text{free } Si \qquad (Eq\ 3)$$

and using the appropriate free energy data.

Joining Methods

Active metal brazing materials, which contain a small weight percent of active elements such as titanium, are available for silicon-base ceramics. Several kinds of active filler metals have been examined. Silver-copper eutectic alloy with a small weight percent of titanium or zirconium, as well as copper-titanium eutectic alloy (Ref 2–5), and pure aluminum and aluminum-titanium alloy (Ref 4, 6–8) have been examined for use with silicon nitride. Silver-copper-titanium alloy (Ref 9–11), nickel-titanium alloy (Ref 11), and aluminum (Ref 12) have been explored for use with silicon carbide.

Brazing with the silver-copper-titanium alloy is a typical process that is accomplished without any metallization. The constituents of the joint are fixed in a furnace and a slight

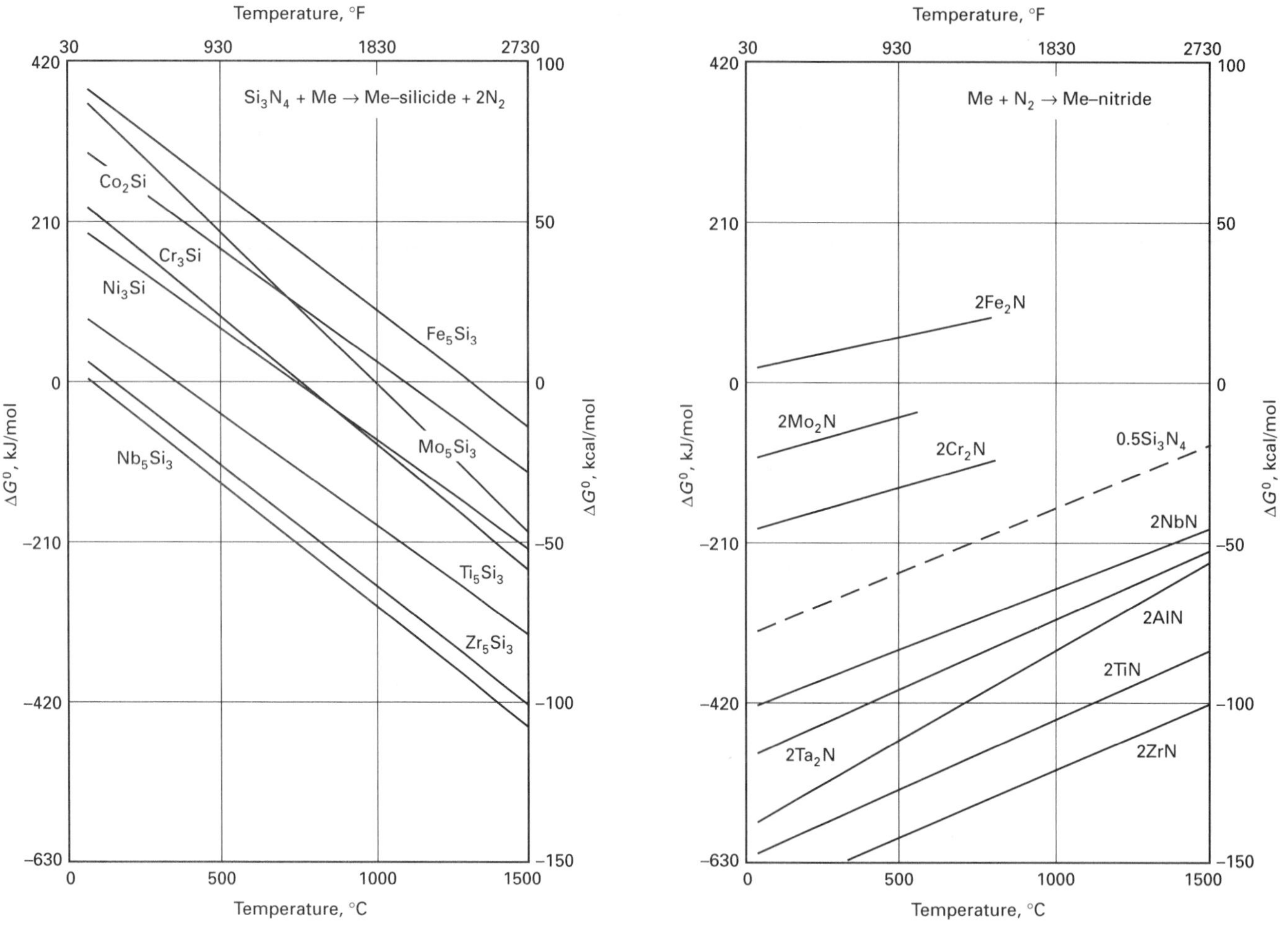

Temperature, °F

$Si_3N_4 + Me \rightarrow Me{-}silicide + 2N_2$

Co₂Si

Cr₃Si

Ni₃Si

Fe₅Si₃

Nb₅Si₃

Mo₅Si₃

Ti₅Si₃

Zr₅Si₃

Temperature, °F

$Me + N_2 \rightarrow Me{-}nitride$

2Fe₂N

2Mo₂N

2Cr₂N

0.5Si₃N₄

2NbN

2AlN

2TiN

2Ta₂N

2ZrN

Fig 1 Free energy changes of formations of metal silicide by reaction between silicon nitride and metals and of metal nitrides as a function of temperature. Source: Ref 1

load is applied. They are then held at the brazing temperatures for Ag-Cu-Ti filler metals, which range from 800 to 950 °C (1470 to 1740 °F). The brazing atmosphere is either a vacuum below about 0.0133 Pa (10^{-4} torr) or an inert gas, such as high-purity argon. The maximum interfacial strength obtained using the Ag-Cu-Ti filler metals exceeds 500 MPa (72.5 ksi), as measured by the four-point bend test. A representative interfacial microstructure is shown in Fig 2.

The main reaction products are titanium nitride adjacent to the silicon nitride, several types of Ti-Si-Cu-N compounds, and the eutectic layer (Ref 13). The formation of titanium nitride is predicted by the free energy diagrams shown in Fig 1. The very thin titanium nitride layer (~100 nm, or 4 μin.) consists of fine grains. The compound layer consists primarily of the orthorhombic Ti-Si-Cu-N phase with additional phases. The microstructure beyond it consists of the silver and copper eutectic. Most of the titanium in the braze alloy segregates to the interface to form the nitride and the silicide. In the case of silicon carbide, a titanium carbide layer is formed instead of the nitride layer (Ref 9, 10).

The layer is also very thin, and two compounds, Ti₅Si₃ and Ti₃SiC₂, are formed.

Several modifications of the active metal brazing process are possible. For example, Ag-Cu brazing can be combined with metallization of the silicon nitride by either vapor deposition of titanium film or brazing with a laminate of titanium in a Ag-Cu foil. Gold-nickel alloy used with titanium- or zircon-

Fig 2 Scanning electron micrograph of interfacial microstructure of silicon nitride/Ag-Cu-Ti braze reacted at 900 °C (1650 °F) for 10 min in vacuum. Although thin silicide layer is observed, titanium nitride is not because it is too thin.

ium-coated ceramics is also an effective way to make a heat-resistant interface (Ref 14). The microstructures obtained by these processes are basically the same as that shown in Fig 2.

Aluminum is another element that is active to silicon-base ceramics, as predicted by Fig 1. It forms a stable nitride and carbide, but its silicide is not stable. Figure 3 illustrates the interfacial microstructure of silicon nitride and aluminum that have been reacted at 800 °C (1470 °F) (Ref 15). Two reaction layers were recognized at the interface. The one adjacent to the silicon nitride consisted of fine grains of β′-sialon with a thickness of about 50 nm. The other was an amorphous alumina layer containing silicon that was about 450 nm (18 μin.) thick. About 1000 °C (1830 °F), 15R-AlN type sialon was formed instead of β′-sialon (Ref 16). The fact that sialon or amorphous alumina, both of which contain oxygen as an element, were formed by the reaction indicates the importance of the oxygen fugacity during brazing. Silicon nitride is well known as being nonwettable by aluminum liquid below about 1000 °C (1830 °F). The presence of oxygen in the reaction is expected

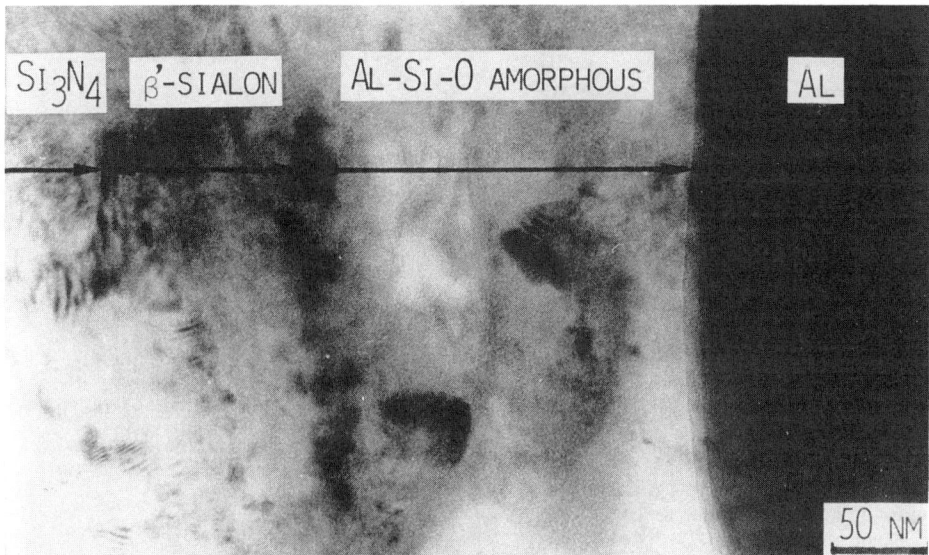

Fig 3 Transmission electron micrograph of the interface of silicon nitride/aluminum reacted at 800 °C (1470 °F) for 15 min in vacuum. Fine particle layer consists of β'-sialon grains smaller than 10 nm (0.4 μin.). Source: Ref 15

silicide layer is seen to grow into the metal. For silicon carbide, carbide layers of mainly the M_2C type are formed at the interface, as shown in Fig 5(b) (Ref 20). The microstructure contains multiple reaction layers with an irregular interface, which sometimes becomes the dispersion structure of a third phase in the reaction layer. Such complex phenomena can be explained by reference to the diffusion path in the ternary phase diagram (Ref 21). The M_2C carbide layer sometimes contains refractory metal elements in solid solution.

Because the thick silicide layer sometimes contains cracks, the strength of such a joint is lowered as the thickness of the silicide layer increases. The maximum interfacial strength is achieved by the optimization of temperature, time, and pressure. It is possible to achieve an interfacial strength above 300 MPa (45 ksi) for silicon nitride/refractory metal systems. In solid-state bonding, the pressure has a great influence on the completeness of the contact at the interface. Isostatically applied high pressure is the most effective means for obtaining full contact (Ref 19).

Joining by room-temperature bonding is carried out in an ultrahigh vacuum (Ref 22). A vacuum below about 130×10^{-10} Pa (10^{-10} torr) is required. The surface to be joined should be cleaned by ion bombardment in the same chamber before bonding. Because the metal should be soft to allow deformation during bonding under pressure, aluminum is an attractive candidate. Furthermore, no third phase is formed at the interface with aluminum when appropriate processing conditions are used and the interface strength with silicon nitride exceeds the strength of aluminum. However, only small parts can be joined using this method.

Friction welding has also been examined for joining silicon nitride to metals (Ref 23). This method is a simple, but effective, process. Joints can have a tensile strength beyond 100 MPa (15 ksi). However, because the joining torque must be low to prevent damage to the ceramic, only soft metals, such as

to promote wetting of the ceramic by molten metal. In fact, an argon atmosphere, which is slightly oxidizing, was found to be better for bonding than a vacuum (Ref 8). Preoxidation of silicon nitride to form a silica layer on its surface also promoted wetting (Ref 17). The maximum strength of Si_3N_4 joined by aluminum brazing was greater than 500 MPa (72.5 ksi), as measured in four-point bending.

A unique process using casting was developed in 1989. In this process, which is called SQ brazing, brazing materials are infiltrated under pressure, so that the lack of wettability is not a serious limitation (see Fig 4). This process can make strong joints in air without the oxidation problem of conventional brazing methods. The interface strength of a sil-

icon nitride joint made by this process is comparable to those made by conventional brazing.

Solid-State Bonding. Refractory metals, such as molybdenum, niobium, tantalum, zirconium, and tungsten, react with silicon ceramics by solid-state processes to form strong interfaces in the temperature range between 1300 and 1500 °C (2370 and 2730 °F). The main reaction products are silicides of stoichiometry M_5Si_3 (M is the refractory metal atom), as well as other nitrides and carbides (Ref 18, 19). Other types of silicides were also observed in some cases. The nitrides and carbides of those refractory metals tend to be more stable at elevated temperatures than are silicon-base ceramics, as seen in Fig 1.

A typical interfacial microstructure of the silicon nitride/metal system is shown in Fig 5(a). The nitride layer is not apparent, but the

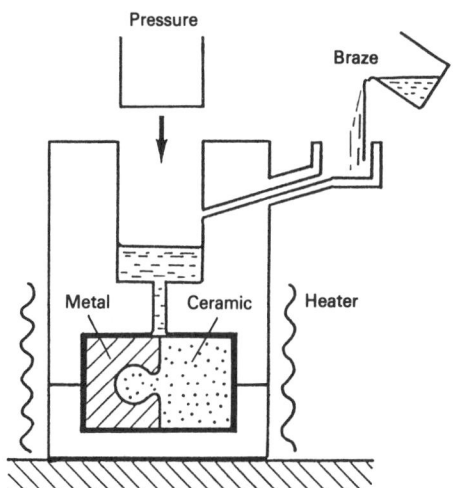

Fig 4 Schematic of SQ brazing process, where constituents are set in a die and the braze infiltrated through the interchannel, under pressure

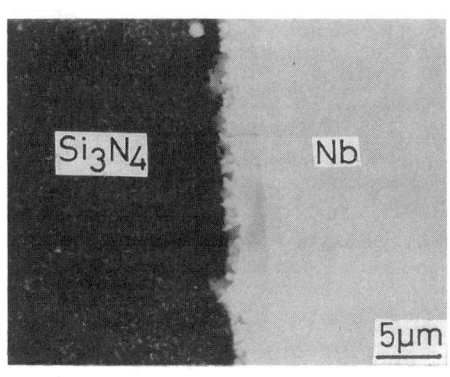

(a)

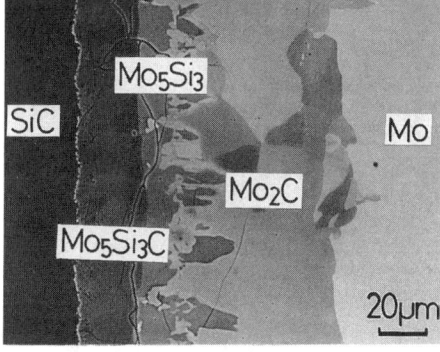

(b)

Fig 5 Interfacial microstructures. (a) Silicon nitride-niobium reacted in solid-state bonding at 1400 °C (2550 °F) for 30 min under a pressure of 100 MPa (15 ksi). (b) Silicon carbide-molybdenum reacted in solid-state bonding at 1700 °C (3090 °F) for 1 h under a pressure of 100 MPa (15 ksi)

aluminum, can be used as the metallic constituent.

Eutectic Joining. Most metallic materials, such as iron, nickel, and cobalt, and their alloys, such as low-alloy steels, austenitic and ferritic stainless steels, and nickel-base superalloys, react with silicon ceramics to form eutectic liquids between the silicon and the metal. Those reactions can be used for joining (Ref 24). The eutectic reaction temperatures, which are primarily determined from the two-component phase diagrams of metal/silicon systems, are about 1100 °C (2010 °F) for nickel and its alloys and about 1200 °C (2190 °F) for iron and its alloys with silicon nitride. For silicon carbide, the temperature ranges are lowered by 50 to 100 °C (90 to 180 °F) by the presence of carbon, and the reactions are extensive to silicon nitride. The presence of oxygen in the reaction atmosphere may lower the temperature of liquid formation (Ref 25). Reaction layers are formed at the interface in these systems. The reaction products are mainly metal silicides when silicon nitride is joined (Ref 24, 26). The reaction layer is thin and silicon diffuses deeply into the metal (Fig 6).

In addition, the joining of pressureless-sintered silicon nitride, which has a large amount of oxide boundary phases, forms complex phases of nickel and silicon with the elements of the oxide additives. Iron nitrides are not stable at the joining temperature in a vacuum. On the other hand, silicon carbide reacts with iron and its alloys to form carbon precipitates in either the silicide reaction layer or the metal-silicon alloy layer. Those structures are also observed in solid-state reactions for which the phases at the interface can be predicted from the appropriate phase diagrams. The interfacial strengths are beyond 300 MPa (45 ksi) for the joining of iron, nickel, and their alloys to silicon nitride.

Joint Properties

The interfacial strengths obtained by optimum joining conditions of silicon nitride to various metals mentioned above are summarized in Table 1. From these data, it could be deduced that these joining methods can produce satisfactory interfacial strengths if the joining conditions are appropriately selected. Unfortunately, because neither the testing methods for strength nor the metal layer thickness are consistent for every case, absolute comparisons of strength cannot be made. These data, however, still provide a suggestion of relative bond strengths.

Heat and oxidation resistances of silicon ceramic joints are two of the critical properties that are required for high-temperature applications. Figures 7 and 8 summarize those properties for the joining methods discussed here. Although brazing with aluminum cannot survive beyond about 300 °C (570 °F), it is a simple process that can produce strong joints at low temperatures. Silver-copper-titanium brazing can produce joints with heat resistance up to about 500 °C (930 °F), but it does not have much oxidation resistance. Eutectic joining can produce structures that can resist slightly higher temperatures than active metal brazing, and it can also provide oxidation resistance if high-nickel or high-chromium alloys are selected. Solid-state bonding using refractory metals can provide heat resistance beyond 1000 °C (1830 °F), but it does not have oxidation resistance beyond 500 °C (930 °F). Mechanical joining is also effective for high-temperature applications. Thus, each joining method has advantages and disadvantages. An appropriate method must be selected on the basis of particular requirements.

Thermal stress relaxation methods used for the butt-type joining of silicon-base ceramics and metals are summarized here. Thermal stresses are distributed nonuniformly in a joint along the interface. The key requirement is to reduce the tensile stress on the interface and in the ceramic, rather than in the metal parts for which local deformation can relax stress concentrations. Furthermore, the magnitude of residual stress depends on the shape and size of the interface (Ref 28). Thermal stresses increase as the size of the

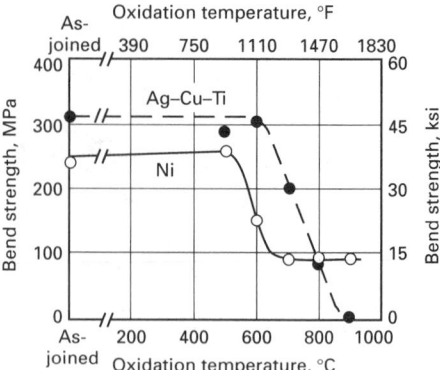

Fig 7 Oxidation resistance of silicon nitride joints with nickel interlayer and Ag-Cu-Ti braze. Joints exposed in air at various temperatures for 100 h and tested at room temperature. Source: Ref 27

part being joined increases. In addition, large thermal stresses sometimes induce flaws in a joint, which not only weakens the joint, but may increase the scatter in strength values (Ref 29). Therefore, it is very important to examine the degree of scatter in strength, especially for silicon-base ceramic and metal joints, because the thermal stress must be large in this system.

A wide variety of methods using particular interlayers have been developed for many ceramic-metal systems. They are also suitable for silicon-base ceramics. Soft metal interlayers (Ref 30), composite interlayers (particulate and fibrous (Ref 31), laminate interlayers (soft metal/low expansion and hard metal) (Ref 32), and fine-crack interlayers (Ref 8) are typical.

A soft metal interlayer, such as aluminum or copper, which sometimes involves harder pure metals, such as nickel, iron, or niobium, will reduce the thermal stress by the elastic, plastic, or creep deformation of the interlayers. However, such a single layer does not remove the stress effectively for silicon-base ceramic-metal systems. In addition, it cannot

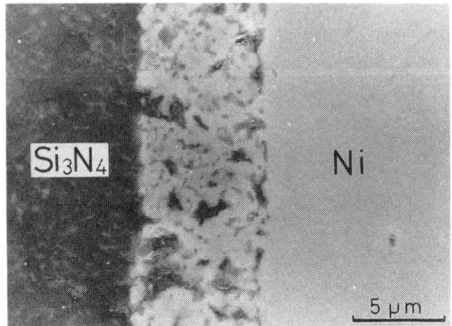

Fig 6 Scanning electron micrograph of interfacial microstructure of silicon nitride-nickel reacted at 1200 °C (2190 °F) for 30 min under a pressure of 100 MPa (15 ksi). Thick silicon diffusion layer is observed into nickel.

Table 1 Room-temperature bend strength of silicon nitride-metal joints

Joining techniques and conditions	Metal	Layer thickness μm	Layer thickness mil	Strength MPa	Strength ksi
Active metal brazing at 850 °C (1560 °F) and 3.3 MPa (0.5 ksi) for 15 min, except as noted in (a) and (c)	Ag-33.5wt%Cu-1.5wt%Ti(a)	10–15	0.4–0.6	650(b)	95(b)
	Ag-26.7wt%Cu-4.5wt%Ti(c)	15	0.6	650(b)	95(b)
	Al	<10	<0.4	380(d)	55(d)
	AA6061	<10	<0.4	470(d)	68(d)
	AA2017	<10	<0.4	400(d)	58(d)
Solid-state bonding at 1400 °C (2550 °F) and 100 MPa (15 ksi) for 30 min, except as noted in (e)	Mo(e)	100	4	420(f)	61(f)
	Nb	100	4	290(f)	42(f)
	Ta	100	4	380(f)	55(f)
Eutectic joining at 1200 °C (2190 °F) and 10 MPa (1.5 ksi) for 30 min	Fe	200	8	330(d)	48(d)
	Ni	200	8	250(d)	36(d)
	Type 410	200	8	320(d)	46(d)
	Type 304	200	8	220(d)	32(d)

(a) Joined at 950 °C (1740 °F) for 1 h in argon. (b) Data from Ref 5. (c) Joined at 875 °C (1605 °F) for 0.5 h in argon. (d) Specimen bar was 3 × 3 × 35 mm (0.12 × 0.12 × 1.4 in.) for four-point bend test; upper span, 10 mm (0.4 in.), lower span, 30 mm (1.2 in.); cross head speed, 0.5 mm/min (0.02 in./min). (e) Same conditions, but at 1500 °C (2730 °F). (f) Specimen bar was 2 × 2 × 15 mm (0.08 × 0.08 × 0.6 in.) for three-point bend test; span, 10 mm (0.4 in.), cross head speed, 0.5 mm/min (0.02 in./min). Source: Ref 27

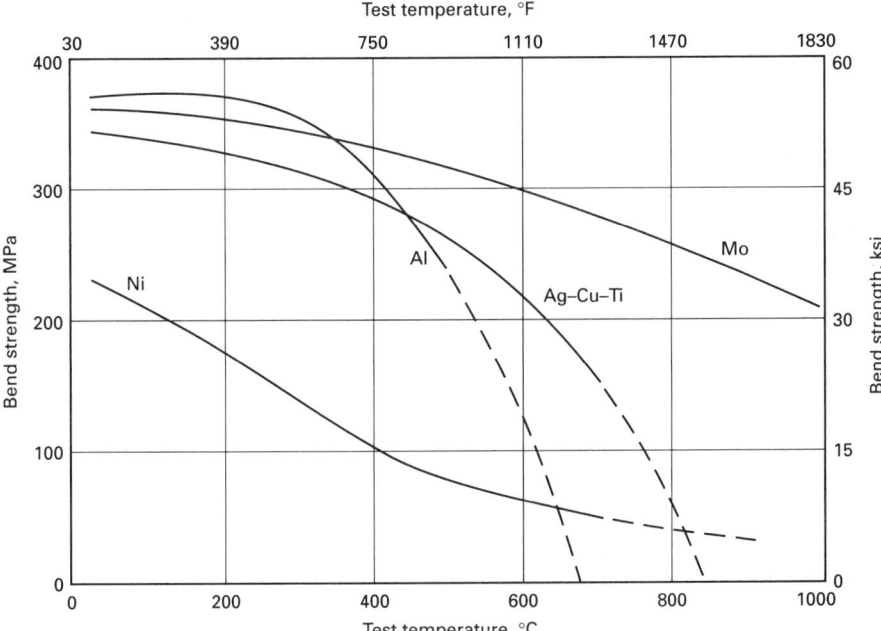

Fig 8 High-temperature strength of silicon nitride joint fabricated using various interlayers. Source: Ref 27

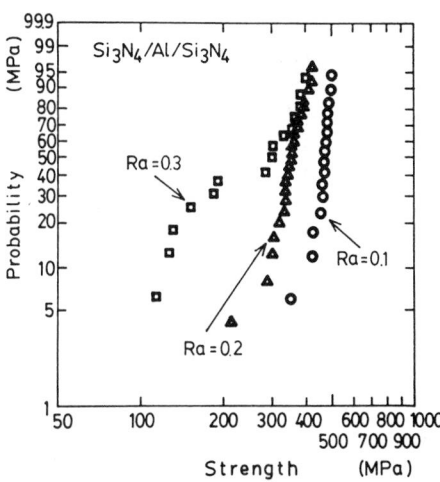

Fig 9 Weibull plots of three-point bend strength of silicon nitride joint with different bond face grinding conditions brazed with aluminum at 800 °C (1470 °F) for 7 min in argon flow. R_a is average surface roughness in μm. Source: Ref 36

survive a sudden temperature change or a severe thermal cycle because, even for soft metals, plastic or creep deformation cannot follow sudden temperature changes. However, such interlayers are useful for joints in low-temperature applications. The laminate interlayer structure improves resistance to thermal stress that is caused by sudden temperature changes (Ref 33). Because the soft metal is restricted on either side by a ceramic and a low-expansion hard metal, it deforms easily to accommodate the shrinkage of the ceramic. In addition, the harmful effect of the large expansion or shrinkage of the base metal is blocked by the hard metal layer. This type of interlayer has been used widely as a heat-resistant joining interlayer. A part of the commercial turbocharger rotors discussed later employ this interlayer structure. However, this interlayer technique applies to a limited range of joint sizes. The upper limit of size may be about 15 to 20 mm (0.6 to 0.8 in.) in diameter for silicon nitride-metal systems.

The composite interlayer, which is also called the powder-graded seal (Ref 34) or the functionally gradient material (Ref 35), is a potentially effective method that is limited by the strength and reliability of the interlayer itself. Finely cracked interlayers could produce a strong silicon ceramic-metal joint (Ref 8). The cracking of an interlayer occurs during joining to relieve thermal stress. It is possible to make a stable joint by carefully controlling crack structure. For this type of interlayer, the cracking layer must be sandwiched by ductile layers to prevent the growth of cracks, especially into silicon-base ceramics. In addition, the cracks must be aligned vertical to the interface.

Roughness, or the prepared condition of the surfaces to be joined, has some influence on joint properties, especially strength. A rough surface will prevent full contact at an interface under a limited pressure, and there will be a damaged layer near any ceramic interface that has deep scratches and severe residual stress induced by grinding. On the other hand, a rough bond face might have an anchoring effect that promotes joining by mechanical interlocking at an interface. In an actual case, these three effects will have a combined influence on the mechanical properties of a joint. Which effect is dominant cannot be defined generally, because every joining system differs.

Weibull plots of the bend strength of silicon nitride joints, which show the influence of grinding and polishing conditions of bond faces, are depicted in Fig 9. It is clear that the rougher bond face made the joint weaker, which can be explained by the fact that the rough grinding makes a damaged layer on the bond face that remains even after joining. In such a case, fracture of the joint proceeds in the damaged layer in the ceramic near the interface. The types of grinding or polishing also have an important influence on mechanical properties. Polishing, to achieve a particular level of surface roughness, is a promising method of controlling metal-ceramic interface properties (Ref 37). When reactions at the interface are extensive, the damaged layer may disappear and will not influence mechanical properties (Ref 38).

The weakening effect on the joint described above by the damaged layer is closely related to stress corrosion induced by the presence of water in the testing atmosphere. It is well known that silicon nitride is sometimes weakened by moisture due to stress corrosion. Scratches and flaws along the interface also reduce the strength of the joint by stress corrosion (Ref 39).

Joining Specific Materials

Joining a silicon-base ceramic to itself can be performed using metallic interlayers, as described above. The main advantage of using metal interlayers is that it becomes possible to reduce the temperature, pressure, and time required for joining. However, with interlayers, it is not possible to have both heat resistance and oxidation resistance in a joint at the same time. On the contrary, direct joining without any metallic interlayer gives joints that are stable above 1000 °C (1830 °F).

Hot isostatic pressing without any interlayer under isostatic pressures greater than 100 MPa (15 ksi) can give a nearly ideal interface. Complete contact at the interface is achieved and the interfacial region has the same structure as the matrix (Ref 40). When silicon nitride contains a large amount of sintering additive, the additive may concentrate in the interface region during joining, which may reduce high-temperature strength.

Hot pressing, which is uniaxial, is another way to achieve solid-state joining without an interlayer. In the hot pressing process, the pressures are limited to much lower values than in hot isostatic pressing because the uniaxial pressing can easily cause large deformation of components. The nature of the boundary phases is important for controlling interface quality. The effects of boundary phases were examined using two types of pressureless-sintered silicon nitride, one with MgO-Al_2O_3 and the other with Y_2O_3-Al_2O_3 (see Fig 10). The former is softer at elevated

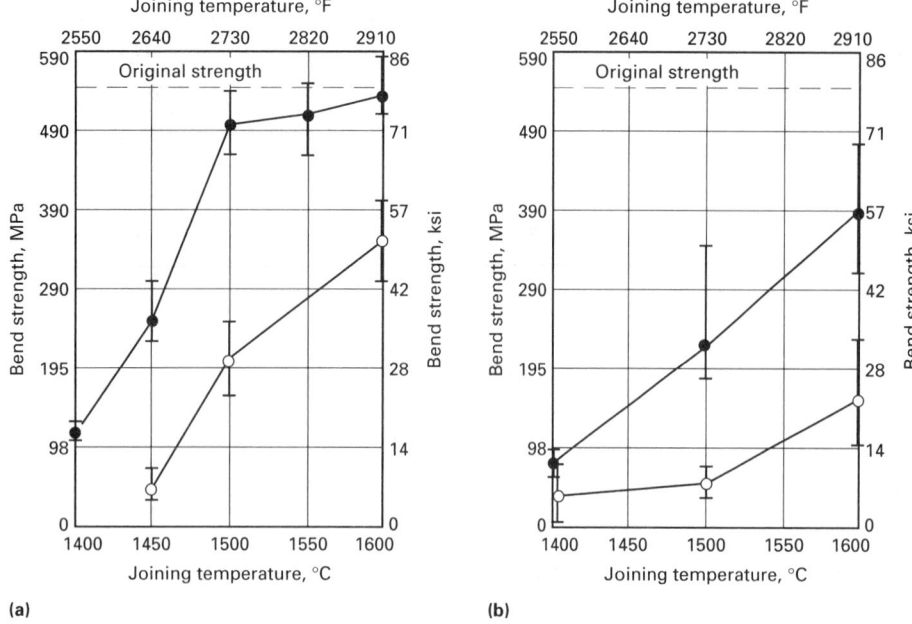

(a)

(b)

Fig 10 Bend strengths of silicon nitride joints without any interlayer as a function of joining temperature. Joining pressure was 20 MPa (3 ksi) (solid) and 1 MPa (0.15 ksi) (open). (a) With sintering additive MgO-Al₂O₃. (b) With sintering additive Y₂O₃-Al₂O₃

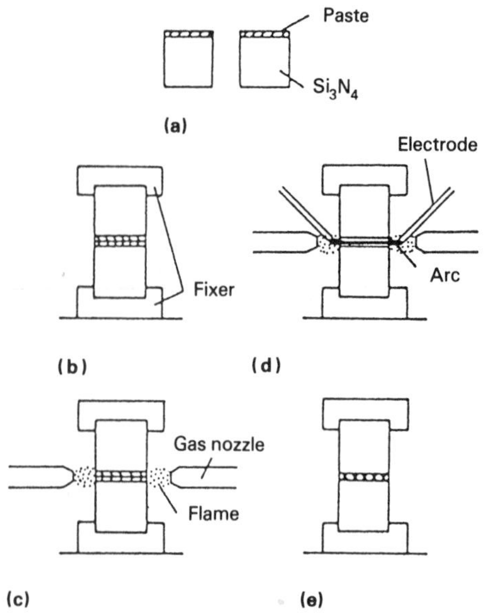

Fig 11 Process of electrical joining. (a) Coating samples with joining paste by screen printing. (b) Setting up in a joining fixer. (c) Preheating by gas flame to 700 to 800 °C (1290 to 1470 °F). (d) Joule heating by electric current through joining paste. (e) Cooling. Source: Ref 41

temperatures. When bonded above 1500 °C (2730 °F), the strength of the joint with the MgO-Al₂O₃ additive almost reached the strength of the original body. The other type of Si₃N₄ had much lower strengths. Thus, the viscosity of the grain boundary phase is very important to the ability to obtain a strong joint by maximizing the contact at the interface.

Because the high processing costs of solid-state bonding may prevent its wide application, several cost-saving improvements have been reported. Microwave heating of the interfacial region promotes joining and has been applied to the joining of silicon nitride (Ref 42). This method does not require conventional furnaces, and joining can be completed in a short time. A strength value of 390 MPa (57 ksi), measured by the bend test, was reported (Ref 42). The strength of the parent silicon nitride was 525 MPa (76 ksi).

Using ceramic agents as an interlayer is another way to achieve bonding. Bonding of hot pressed and reaction-bonded silicon nitride using ZrO₂ or ZrSiO₄ was examined at low pressures below 1.5 MPa (0.22 ksi) at 1400 to 1500 °C (2550 to 2730 °F) (Ref 43). The hot pressed high-purity silicon nitride exhibited a moderate strength of 170 MPa (25 ksi) with a ZrO₂ bond layer. Lowering the silicon nitride purity degraded strength because of the formation of a ZrN interface layer that depended on the ease of oxidation of silicon nitride.

An oxide glass has been used to join silicon nitride (Ref 44). The glass composition was chosen to approximate the oxide grain boundary phase in silicon nitride, for example, MgO-Al₂O₃-SiO₂. Joining was carried out in a nitrogen atmosphere in the temperature range between 1550 and 1650 °C (2820 and 3000 °F) without pressure. The oxide liquid reacted with silicon nitride and penetrated down the grain boundaries. The resultant interfacial phase was composed of Si₂N₂O, β-Si₃N₄, and a residual oxynitride glass. The maximum bending strength of the joint was beyond 450 MPa (65 ksi) when the joining interface was as thin as 30 μm (1.2 mils).

Fluorine compounds mixed with kaolin have been examined as the joining agent of sialon (Ref 45). The powder mixture of CaF₂-40 wt% kaolin was applied as a paste on a sialon surface. Joining was carried out at 1450 °C (2640 °F) without pressure, in air. A bend strength of 300 MPa (45 ksi) was obtained by this process (strength of the parent body was 570 MPa, or 83 ksi). Recently, this method has been modified to incorporate joule heating of the interface with electric current (Ref 46). Figure 11 is a schematic of the new process. Advantages of this method are that it does not require complex equipment, such as high-pressure and high-temperature furnaces; it can be carried out in air; and it can be completed within a short period. The strength of joined Si₃N₄ was 400 MPa (58 ksi) from room temperature to beyond 1000 °C (1830 °F).

Silicon carbide is a difficult material to join by solid-state methods. It has been reported that solid-state bonding, even by hot isostatic pressing, was not successful at 1950 °C (3540 °F) under a pressure of 138 MPa (20 ksi) (Ref 47). To promote joining, it is necessary to insert some appropriate interlayer, such as AlB₁₂, B₄C, and ZrB₂. Brazing with SiB₆, Al₄C₃, or MoSi₂ using hot isostatic pressing is also a possibility. Uniaxial hot pressing at 1950 °C (3540 °F) under a pressure of 13.8 MPa (57 ksi) was used to join silicon carbide when the plastic deformation of the silicon carbide became large. However, it is not a practical technique because the deformation of the constituents reach 25%. Furthermore, the periphery of the joined region was not bonded. If sintering powders such as aluminum, boron, and carbon were added, then solid-state bonding by hot pressing became possible, even at temperatures as low as 1650 °C (3000 °F) (Ref 48). The joint bonded with a 1 wt% Al-1 wt% B-1 wt% C interlayer under a pressure of 50 MPa (7.3 ksi) had the strength of 400 MPa (58 ksi), and it maintained a strength of 300 MPa (44 ksi) at 1500 °C (2730 °F).

Joining that utilizes reaction sintering of silicon carbide further reduces joining temperature (Ref 49). In this technique a powder mixture of silicon carbide and carbon was used as the joining interlayer and was siliconized at 1450 °C (2640 °F). The average strength of the joint reached 340 MPa (50 ksi), which was maintained to a temperature greater than the joining temperature.

Polycarbosilane is well known to convert to silicon carbide upon firing in an inert atmosphere, and a mixture of this polymer and silicon carbide powder can be used as a joining agent (Ref 50). Joining was carried out at 1500 °C (2730 °F) in nitrogen without any pressure, and this treatment gave strength of 40 MPa (6 ksi).

Joining Other Non-oxide Ceramics. Aluminum nitride has been proposed for the

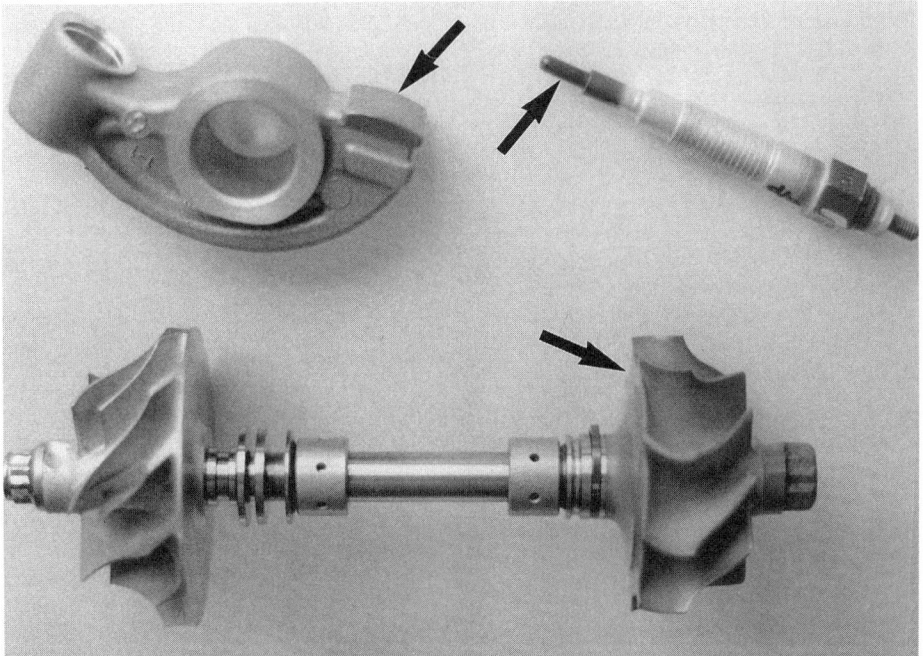

Fig 12 Silicon nitride automotive engine components (arrows indicate silicon nitride parts). Clockwise, turbocharger rotor with the silicon nitride blade, rocker arm with silicon nitride tip, and glow plug with silicon nitride head. (Courtesy of Nissan Motors Co., Ltd.)

Applications

Because automobile manufacturers have earnestly required the use of ceramics as high-temperature structural components, silicon nitride has been used for several components (Fig 12). Two typical products that utilize silicon nitride-metal joining are described below.

A rocker arm with a ceramic wear pad was introduced in 1983 (Ref 55). It was first made by aluminum die casting, which had been established for sintered alloy tips. The geometry of the silicon nitride insert is shown in Fig 13(a). The main advantage in replacing the sintered alloy tip with silicon nitride is to reduce stress concentration on the tip by rounding the tip edge. Since 1988, the rocker arm has also been fabricated by active metal brazing. The interlayer structure is shown in Fig 13(b) (Ref 56). One soft metal layer, that is, copper, is inserted, which is enough to maintain quality for low-temperature uses.

A silicon nitride turbocharger rotor was introduced in 1985 (Ref 57). The joint region is exposed to more severe heat conditions than the rocker arm. The highest operating temperature reached 550 °C (1020 °F). Two joining techniques, direct active metal brazing and shrunk-in insert, have been adopted (Fig 14). In brazing the silicon nitride rotor, a three-

next generation of integrated circuit substrates, because of its high heat conductivity and its coefficient of thermal expansion similar to that of silicon. Solid-state bonding of aluminum nitride to molybdenum was performed under an ultrahigh pressure, and thermal shock-resistant joining was shown to be possible using a composite interlayer with two constituents (Ref 31). However, solid-state bonding is not practical for use with commercial electronic devices, which require reliable and cost-saving metallization techniques for ceramics.

Active metal brazing of aluminum nitride is a practical alternative. Sufficient wetting can be obtained with commercial Ag-Cu brazes that contain a few wt% of titanium by brazing above 800 °C (1470 °F) (Ref 51, 52). A thin titanium nitride layer is formed at the interface. The shear strength of an aluminum nitride/molybdenum joint reached 190 MPa (28 ksi) (Ref 53). Successful copper-titanium braze joints require more than 7 wt% titanium to achieve enough wetting with aluminum nitride at 1150 °C (2100 °F) (Ref 52). Wetting can be achieved with a pure aluminum braze above 900 °C (1650 °F), but the interface strength, at about 35 MPa (5 ksi), is not as high.

There have been a few reports on joining methods used for other kinds of non-oxide ceramics. Nitrides, carbides, and borides of titanium have been established as cutting tool materials. Titanium nitrides and carbides are known as the interfacial reaction products for titanium-containing active metal brazes. It is reasonable to expect them to braze easily with their alloys, such as the Ag-Cu braze. How-

ever, the wettability of such carbides and nitrides with metal liquids may be influenced by their stoichiometry (Ref 54).

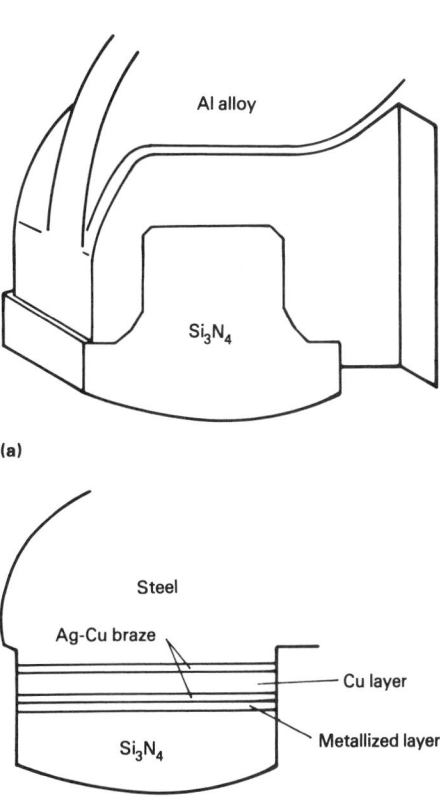

Fig 13 Schematic of silicon nitride rocker arm. (a) Aluminum die cast insert (Ref 55). (b) Active metal brazing with soft metal interlayer (Ref 56)

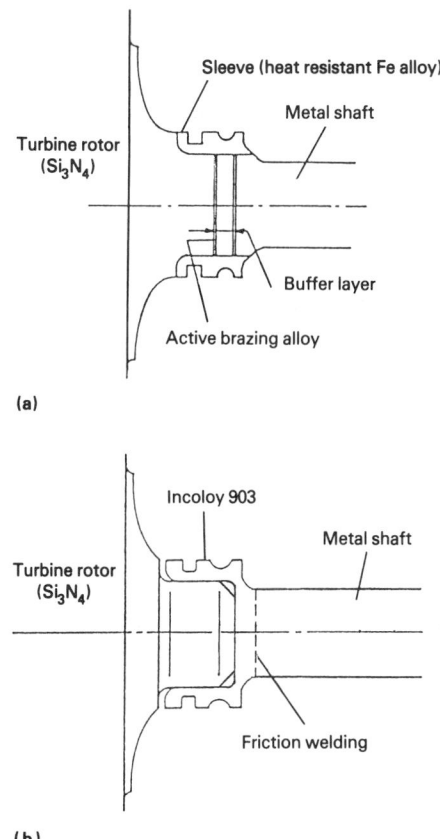

Fig 14 Schematic of silicon nitride turbocharger rotor. (a) Active metal brazed rotor with triple-laminated interlayers (Ref 57). (b) Shrunk-in inserted rotor, which has been renewed since 1989 (Ref 58)

layered interlayer structure, that is, nickel/tungsten/nickel, has been used to reduce thermal stress. The nickel layers fit the role of a soft metal and the tungsten represents the hard, low-expansion layer. The brazing material is the Ag-Cu alloy containing a few wt% titanium. The shrunk-in insert technique utilizes mechanical joining. In both cases, the bend strength of the joint region exceeds 200 MPa (30 ksi), and torsional strength of 50 N · m (37 lbf · ft) was maintained up to 500 °C (930 °F).

ACKNOWLEDGMENT

The author wishes to thank the following colleagues for providing information on the practical production of silicon nitride joining parts: M. Ito, NGK Spark Plug Co., Ltd.; T. Mizuno, NGK Insulators, Ltd.; and T. Nishi, Daihen Co.

REFERENCES

1. L.F. Voitovich, *Refractory Compounds*, Kiev, USSR, 1971, p 96–103

2. J.E. Siebels, Brazing of Silicon Nitride, *Progress in Nitrogen Ceramics*, F.L. Riley, Ed., Martinus Nijhoff Publishers, 1983, p 455–463

3. H. Mizuhara and K. Mally, Ceramic-to-Metal Joining with Active Brazing Filler Metal, *Weld. J.*, Vol 64, 1985, p 27–32

4. M.J. Ramsey and M.H. Lewis, Interfacial Reaction Mechanisms in Syalon Ceramic Bonding, *Mater. Sci. Eng.*, Vol 71, 1985, p 113–122

5. S.M. Johnson, Mechanical Behavior of Brazed Silicon Nitride, *Ceram. Eng. Sci. Proc.*, Vol 10 (No. 11–12), 1989, p 1846–1853

6. K. Suganuma, T. Okamoto, M. Koizumi, and M. Shimada, Method for Presenting Thermal Expansion Mismatch Effect in Ceramic-Metal Joining, *J. Mater. Sci. Lett.*, Vol 4, 1985, p 648–650

7. A. Kohno, T. Tamada, and K. Yokoi, Bonding of Ceramics to Metals with Interlayers of Al-Si Clad Aluminum, *J. Jpn. Inst. Met.*, Vol 49, 1985, p 876–883

8. K. Suganuma, T. Okamoto, M. Koizumi, and M. Shimada, Joining of Silicon Nitride to Silicon Nitride and to Invar Alloy Using Aluminum Interlayer, *J. Mater. Sci.*, Vol 22, 1987, p 1359–1364

9. T. Iseki, H. Matsuzaki, and J.K. Boadi, Brazing of Silicon Carbide to Stainless Steel, *Am. Ceram. Soc. Bull.*, Vol 64 (No. 2), 1985, p 322–324

10. S. Morozumi, M. Endo, M. Kikuchi, and K. Hamajima, Bonding Mechanism between Silicon Carbide and Thin Foils of Reactive Metals, *J. Mater. Sci.*, Vol 20, 1985, p 3976–3982

11. M. Naka, T. Tanaka, and I. Okamoto, Joining of SiC Using Amorphous Ti Base Filler Metal, *Kouon-Gakkai-Shi*, Vol 12 (No. 2), 1986, p 81–89

12. T. Iseki, T. Kameda, and T. Murayama, Interfacial Reactions between SiC and Aluminum during Joining, *J. Mater. Sci.*, Vol 19, 1984, p 1692–1698

13. A.H. Carim, Transitional Phases at Ceramic-Metal Interfaces: Orthorhombic, Cubic, and Hexagonal Ti-Si-Cu-N Compounds, *J. Am. Ceram. Soc.*, Vol 73, 1990, p 2764–2766

14. S. Kang, E.M. Dunn, J. Selverial, and H.J. Kim, Issues in Ceramic-to-Metal Joining: An Investigation of Brazing a Silicon Nitride-Based Ceramic to a Low-Expansion Superalloy, *Am. Ceram. Soc. Bull.*, Vol 68 (No. 9), 1990, p 1608–1617

15. X.S. Ning, K. Suganuma, M. Morita, and T. Okamoto, Interfacial Reaction between Silicon Nitride and Aluminum, *Philos. Mag. Lett.*, Vol 55, 1987, p 93–97

16. M. Naka, H. Mori, M. Kobo, I. Okamoto, and H. Fujita, Observation of Al/Si₃N₄ Interface, *J. Mater. Sci. Lett.*, Vol 5, 1986, p 696–698

17. M. Morita, K. Suganuma, and T. Okamoto, Effect of Preheat Treatment on Silicon Nitride Joining with Aluminum Braze, *J. Mater. Sci. Lett.*, Vol 6, 1987, p 474–476

18. S. Morozumi, M. Kikuchi, and S. Sugai, Reaction between Molybdenum and Several Nitrides, *J. Jpn. Inst. Met.*, Vol 45 (No. 2), 1981, p 184–189

19. M. Koizumi, T. Takagi, K. Suganuma, Y. Miyamoto, and T. Okamoto, Solid-State Bonding of Silicon Nitride to Metals using HIP, *High Tech. Ceramics*, P. Vinsenzini, Ed., Elsevier Science, Amsterdam, 1987, p 1033–1042

20. A. Horiguchi, K. Suganuma, Y. Miyamoto, M. Koizumi, and M. Shimada, Solid-State Interfacial Reaction in Molybdenum-Carbide Systems at High Temperature-Pressure, and its Application to Bonding Technique, *J. Soc. Mater. Sci. Jpn.*, Vol 35 (No. 388), 1986, p 35–40

21. F.J.J. van Loo, F.M. Smet, G.D. Rieck, and G. Verspui, Phase Relations and Diffusion Paths in the Mo-Si-C System at 1200 °C, *High Temp.-High Press.*, Vol 14, 1982, p 25–31

22. T. Suga, K. Miyazawa, and H. Takagi, TEM Observation of the Al and Cu Interfaces Bonded at Room Temperature by Means of the Surface Activation Method, *J. Jpn. Inst. Met.*, Vol 54 (No. 6), 1990, p 713–719

23. A. Suzumura, T. Onzawa, Y. Arata, A. Omori, and S. Sano, Friction Welding of Ceramics to Aluminum, *Kouon-Gakkai-Shi*, Vol 13, 1987, p 43–51

24. K. Suganuma, T. Okamoto, and M. Shimada, High Pressure Bonding of Nitrides and Iron, *High Temp.-High Press.*, Vol 16, 1984, p 627–635

25. G.J. Tennenhouse, A. Ezis, and F.D. Runkle, Interaction of Silicon Nitride and Metal Surfaces, *Comm. Am. Ceram. Soc.*, Vol 68 (No. 1), 1985, p 30–31

26. M.J. Bennet and M.R. Houlton, The Interaction between Silicon Nitride and Several Iron, Nickel, and Molybdenum-Based Alloys, *J. Mater. Sci.*, Vol 14, 1979, p 184–196

27. K. Suganuma, Recent Advances in Joining Technologies of Ceramics/Metals, *ISIJ Int.*, Vol 30 (No. 12), 1990, p 1046–1058

28. K. Suganuma, T. Okamoto, M. Koizumi, and J. Kamachi, Influence of Shape and Size on Residual Stress in Ceramic/Metal Joining, *J. Mater. Sci.*, Vol 22, 1987, p 3561–3565

29. K. Suganuma, T. Okamoto, M. Koizumi, and M. Shimada, Acoustic Emission from Ceramic/Metal Joints on Cooling, *Am. Ceram. Soc. Bull.*, Vol 65 (No. 7), 1986, p 1060–1064

30. M.G. Nicholas and M.R. Crispin, Diffusion Bonding Stainless Steel to Alumina Using Aluminum Interlayers, *J. Mater. Sci.*, Vol 17, 1982, p 3347–3360

31. K. Suganuma, M. Shimada, T. Okamoto, and M. Koizumi, New Method for Solid-State Bonding between Ceramics and Metals, *Comm. Am. Ceram. Soc.*, Vol 66 (No. 7), 1983, p 117–118

32. K. Suganuma, M. Shimada, T. Okamoto, and M. Koizumi, Effect of Interlayers in Ceramic-Metal Joints with Thermal Expansion Mismatches, *Comm. Am. Ceram. Soc.*, Vol 67 (No. 12), 1984, p 256–257

33. K. Suganuma, T. Okamoto, M. Shimada, and M. Koizumi, Solid-State Bonding of Alumina to Steel, *J. Nucl. Mater.*, Vol 133 & 134, 1985, p 773–777

34. H. E. Pattee, Joining Ceramics to Metals and Other Materials, *Bull. Ser. Weld. Res. Council*, Vol 178, 1972, p 1–43

35. A. Kawasaki and R. Watanabe, Finite Element Analysis of Thermal Stress of the Metal/Ceramic Multi-layer Composites with Controlled Compositional Gradients, *J. Jpn. Inst. Met.*, Vol 51 (No. 6), 1987, p 525–529

36. K. Suganuma, T. Okamoto, M. Koizumi, and M. Shimada, Effects of Surface Damage on Strength of Silicon Nitride Bonded with Aluminum, *Adv. Ceram. Mater.*, Vol 1 (No. 4), 1986, p 356–360

37. M. Coubire, "Study on Joining of Ceramic-Metal, Application for Copper-Alumina," Thesis, L'Ecole Centrale de Lyon, 1986

38. M. Coubire, M. Kinoshita, and I. Kondoh, Silicon Nitride/Carbon Steel Joining by HIP Technique, *Joining Ceramics, Glass and Metal*, W. Kraft, Ed., DGM Informationsgesellschaft, Oberursel, 1989, p 95–102

39. K. Suganuma, M. Morita, T. Okamoto, and M. Koizumi, Effect of Testing Atmosphere on Mechanical Properties of

Ceramic/Metal Joints, *Ceram. Eng. Sci. Proc.*, Vol 10 (No. 11–12), 1989, p 1908–1918

40. T. Kaba, M. Shimada, and M. Koizumi, Diffusional Reaction-Bonding of Silicon Nitride Ceramics under High Pressure, *Comm. Am. Ceram. Soc.*, Vol 66 (No. 8), 1983, p 135–136

41. H. Tabata, S. Kanzaki, and M. Nakamura, Solid-State Joining of Silicon Nitride Ceramics, *Proceedings of International Symposium on Ceramic Components for Engine*, 1984, p 387–393

42. H. Fukushima, T. Yamanaka, and M. Matsui, Microwave Heating of Ceramics and its Application to Joining, *Microwave Processing of Materials, Symposium Proceedings*, Vol 124, W.H. Sutton, M.H. Brooks, and I.J. Chabinsky, Ed., Materials Research Society, 1988, p 267–272

43. P.F. Becher and S.A. Halen, Solid-State Bonding of Si_3N_4, *Am. Ceram. Soc. Bull.*, Vol 58 (No. 6), 1979, p 582–586

44. R.E. Loehman, Transient Liquid Phase Bonding of Silicon Nitride, *Surface and Interfaces in Ceramic and Ceramic/Metal System*, J.A. Pask and A.G. Evance, Ed., Plenum Press, 1981, p 701–711

45. Y. Ebata and M. Kinoshita, Joining of Sialon with the Aid of Fluorine Compounds, *J. Ceram. Soc. Jpn.*, Vol 90 (No. 12), 1982, p 714–716

46. Y. Ebata, M. Kohyama, M. Iwasa, M. Kinoshita, K. Okuda, H. Takai, and T. Nishi, Electrical Joining of Silicon Nitride Ceramics, *J. Ceram. Soc. Jpn.*, Vol 97 (No. 1), 1989, p 88–90

47. T.J. Moore, Feasibility Study of Welding of SiC, *Comm. Am. Ceram. Soc.*, Vol 68 (No. 6), 1985, p 151–153

48. T. Iseki, K. Arakawa, H. Matsuzaki, and H. Suzuki, Joining of Dense Silicon Carbide by Hot-Pressing, *J. Ceram. Soc. Jpn.*, Vol 91 (No. 8), 1983, p 349–354

49. T. Iseki, M. Imai, and H. Suzuki, Joining of Dense Silicon Carbide Containing Free Silicon by Reaction-Sintering, *J. Ceram. Soc. Jpn.*, Vol 9 (No. 6), 1983, p 259–264

50. S. Yajima, K. Okamura, T. Shishido, Y. Hasegawa, and T. Matsuzawa, Joining of SiC to SiC using Polyborosiloxane, *Comm. Am. Ceram. Soc.*, Vol 60 (No. 2), 1981, p 253

51. R.E. Loehman, Joining and Bonding Mechanisms in Nitrogen Ceramics, *International Meeting on Advanced Materials*, Vol 8, Materials Research Society, 1989

52. M.G. Nicholas, D.A. Mortimer, L.M. Jones, and R.M. Crispin, Some Observations on the Wetting and Bonding of Nitride Ceramics, *J. Mater. Sci.*, Vol 25, 1990, p 2679–2689

53. M. Nakahashi, M. Shirokane, and H. Takeda, Characterization of Nitride Ceramic-Metal Joints Brazed with Ti-Containing Alloys, *J. Jpn. Inst. Met.*, Vol 53 (No. 11), 1989, p 1153–1160

54. M.G. Nicholas, Reactive Metal Brazing, *Joining Ceramics, Glass and Metal*, W. Kraft, Ed., DGM Informationsgesellschaft, Oberursel, 1989, p 3–16

55. Y. Ogawa, M. Machida, N. Miyakawa, K. Tashiro, and M. Sugano, "Ceramic Rocker Arm Insert for Internal Combustion Engines," Technical Paper Series, No. 860397, Society of Automotive Engineers, 1986

56. M. Itoh, private communication, 1991

57. M. Itoh, N. Ishida, and N. Katoh, "Development of Brazing Technology for Ceramic Turbocharger Rotors," Technical Paper Series, No. 880704, Society of Automotive Engineers, 1986

58. T. Mizuno, private communication, 1991

Techniques of Seal Design

E.K. Beauchamp and S.N. Burchett, Sandia National Laboratories

A PHILOSOPHY OF SEAL DESIGN and general criteria for use in the design process are described in this article. Issues of mechanical and electrical performance and of survival in the chemical environment are addressed. The use of finite-element analysis in determining the response to mechanical loading is treated briefly. Details are provided more to describe potential pitfalls than to prescribe specific design approaches.

The goal in design of any seal is to select a configuration and a fabrication process that will satisfy the functional requirements, survive the anticipated environmental conditions, and permit production of the seal at an acceptable cost. Usually, the functional requirements and the environmental conditions are set by someone other than the designer. Moreover, requirements on the size and shape of the seal structure may already have been dictated by constraints on the device that will contain the seal. It is an essential first step in any design to determine that the functional requirements and the constraints on size and shape are realistic and that there is a clear understanding of how the environment might affect the performance of the seal. It can be expensive to fabricate a seal whose performance substantially exceeds the real needs but catastrophe can result from underdesigning a seal. To ensure optimization of the final design, it is also essential to maintain close cooperation between the designer and the fabricator during design and prototype fabrication.

There is a very broad range of structures, from mechanical joints to vacuum-tight couplings, that can be classed as seals and a wide spectrum of processes that can be used to produce those seals including such diverse processes as fusion welding, diffusion bonding, glazing, adhesive bonding, metallizing and brazing, direct brazing, and mechanical attachment. However, many of the design problems and approaches are common to all of the structures and processes. Consequently, in describing the design process, the more commonly used procedures will be emphasized. These include metallize/braze, metallize/solder, and glass/metal sealing. Also, in many instances, the discussion will apply to either glasses or ceramic materials as the insulators. In those instances, the term "ceramic" is used as a generic descriptor

denoting either material. It should be clear from the context when the word "ceramic" is used to refer specifically to polycrystalline bodies.

The basic function of a seal or a joint is to provide a mechanical junction between a ceramic and a metal part or between two ceramics. In general, the combination of materials in the seal is intended to take advantage of the attributes of the ceramic while protecting against its limitations. In many applications such as in substrates or in pin seals, the role of the ceramic is to electrically isolate the metal components from one another. In other applications, the ceramic may be an active element such as a magnetic ferrite, a ferroelectric or an ion conductor and the function of the joint is simply to couple that active element, electrically and mechanically, to the rest of the structure. Obviously, the performance of the seal in the anticipated environment depends not only on the stability of the individual components of the seal in that environment, but also on the response of the combination of materials.

Mechanical Integrity

Ceramic Fracture. A successful seal or joint design must ensure against failure from either the residual stresses generated in fabrication or stresses generated in the use environment. These requirements imply prevention of failure in any element of the seal—the ceramic, the joint, or an adjacent metal member; however, it is usually the insulator that is the most vulnerable element. Therefore, the design philosophy for most seals is to maintain minimal tensile stress in the ceramic. For the purposes of clarifying the strategy of mechanical design of seals, some aspects of failure in ceramic materials (including glass) will be briefly summarized. A more complete description of the mechanical behavior of ceramics is given in the article "Crack Propagation in Ceramics" in this Volume. General design practices for monolithic ceramics and composite structures that will also be useful to the designer of seals are reviewed in the Section of this Handbook entitled "Design Considerations."

The key feature of ceramics used at operating temperatures for most seals is that they fail in a brittle manner, that is, they deform

elastically up to the failure stress and then fail catastrophically. Virtually all of the brittle failures occur by propagation of a crack in a tensile stress field. To provide a useful design criterion, it would be desirable to be able to specify a tensile stress limit for the ceramic. Unfortunately, failure stress is not a material constant for glasses and ceramics. Rather, all ceramics contain a population of flaws of varying severity and the worst of those flaws determines the strength. Because the size of the worst flaw varies from part to part, the failure stress varies. The flaw population depends on how the ceramic is initially formed, on subsequent machining processes, and finally on reactions with other components of the seal during seal formation.

For many ceramics, the parameter that can be regarded as a material constant describing the resistance to crack growth of a ceramic is the parameter K_{Ic}. K_{Ic} is the critical value of the stress intensity factor, K_I, which is a measure of the force acting on individual bonds at the boundaries of the flaw (at the tip of a crack) and is usually given by:

$$K_I = Y \cdot \sigma \cdot c^{1/2} \qquad \text{(Eq 1)}$$

where Y is a constant related to the geometry of the crack, σ is the applied stress and c is a measure of the size of the flaw (the radius of a circular or semicircular crack or a characteristic linear dimension for other types of flaws).

K_{Ic} for glasses ranges from 0.5 to 1.5 MPa$\sqrt{m}$ (0.45 to 1.4 ksi$\sqrt{in}$.) and for polycrystalline ceramics from 1 to 20 MPa$\sqrt{m}$ (0.9 to 18 ksi$\sqrt{in}$.). For even the highest K_{Ic} values, the flaws in materials of importance to seal applications are so small that nondestructive detection of them has not been very successful.

There are a number of ceramics whose resistance to crack growth is not a constant but increases with the size of the crack so that, in some circumstances, the failure stress can be insensitive to the crack size. These include materials such as partially stabilized zirconia, which displays transformation toughening, as well as aluminas in intermediate grain size ranges where crack interface bridging enhances the toughness. Usually, tabular K_{Ic} data for these materials are obtained at large crack size where the effective toughness is greatest. However, most failures in ceramics, including those in seals, involve very small originating cracks where the effective toughness is relatively lower.

The general concept of resistance to crack growth was formulated by Griffith (Ref 1) who introduced the relationship between failure stress and crack size that is embodied in Eq 1. That relationship was derived by assuming that the onset of crack growth occurred when the strain energy released by the extension of the crack was equal to the energy required to create the new surfaces. K_{Ic} is the value of the stress intensity factor corresponding to that energy equivalence. From Eq 1, it should be apparent that the failure stress for the ceramic will vary inversely as the square root of the crack size.

In many ceramics, particularly oxide ceramics and glasses, delayed failure can occur even though the initial K_I at the flaw is less than K_{Ic}, that is, the stress is less than the instantaneous failure stress. This phenomenon, described as static fatigue, results from the slow growth of the crack until it reaches critical size for the applied stress, at which point the part fails catastrophically. This subcritical crack growth is a consequence of the simultaneous action of the applied stress and a chemical reaction to break the bond. The most common reagent responsible for delayed failure in ceramics is water vapor from the air; however, ammonia, hydrazine, and other reagents with electron donor sites on one end of the molecule and proton donor sites at the other have similar effects. Michalske and Freiman described subcritical crack growth in terms of a model in which the strain energy introduced into the bond at the crack tip increases the reactivity of that bond with the reagent (Ref 2). The reactivity increases logarithmically with K_I. The rate of bond rupture and, hence, the velocity of crack growth increases with the concentration of the reagent, that is, the relative humidity in the case of water vapor from the air. Because the rate of growth can be extremely slow, failure can be delayed for days, weeks, or even years after application of a load. Consequently, to ensure very high reliability in a seal containing oxide ceramics or glasses, the stress intensity must be maintained at a low enough value that subcritical crack growth is effectively eliminated. For many materials, that means that K_I must be maintained less than $K_{Ic}/3$.

In general, ceramics contain many flaws and for any group of ceramic parts made by identical processes, the flaw population will vary from part to part so that the failure stresses of the ceramic parts will vary over wide ranges. As a consequence, it becomes very difficult to design a seal so that survival of the ceramic can be predicted. If very high reliability is required in an application, it will be necessary to guarantee a very low probability of failure. One way of achieving low failure probability is to design the seal so that stresses are limited to only a fraction of the average failure stress obtained on represen-

tative samples. That condition can be a severe constraint on a design.

An alternative to severely limiting stresses in the seal is to impose a proof test that assures essentially zero failure probability for any part that survives the test. To be effective, the proof test must accurately replicate the stresses anticipated for the structure in the use environment. In general, the stresses acting on a ceramic in a seal application are so complex that that condition is usually difficult, if not impossible, to achieve. Service and Ritter have described procedures that account for the possibility of subcritical crack growth (Ref 3).

It is convenient to describe the variation in failure stress in ceramics in terms of extreme value statistics and to analyze failure data using the formalism developed by Weibull (Ref 4) to describe the failure probability as a function of the stress. In general, the predictions for very low failure probability are much more accurate with Weibull plots than when failure stress data are assumed to follow normal statistics. In selecting a material for reliability, Weibull plots are a better indicator than normal statistics.

In spite of the advantage over normal statistics, the use of Weibull statistics for predicting failure in a seal assembly has the same limitation that any other treatment of strength test data has, namely, the strength tests are usually performed on samples with a different flaw population than that of the parts in the seal. Even if the methods of preparation of the ceramics in the test bars and the seals are identical so that the bulk flaw populations are very similar, the surface flaw populations may be very different. For alumina metallized with Mo-Mn and brazed, for example, the surface condition at the edge of the metallized zone where stresses are likely to be highest may have been modified by the glass from the metallizing paint. Consequently, the flaw population may be very different from that of as-sintered strength test samples.

The poor fracture resistance of ceramics and glasses is usually coupled with low thermal conductivity and often with a high coefficient of thermal expansion (CTE). The net result is a low resistance to thermal shock. That feature not only affects the mechanical integrity of a seal in the use environment but can severely restrict heating and cooling rates during processing. A useful equation for selecting limits on heating and cooling rates is

$$\Delta T_f = k\sigma_f(1 - \nu)[1/0.31r_mh]/E\alpha \qquad \text{(Eq 2)}$$

where ΔT_f is the temperature difference between the interior of the ceramic and the heat transfer medium, σ_f is the strength of the ceramic, ν is Poisson's ratio, E is the Young's modulus for the ceramic, α is the coefficient of expansion, k is the thermal diffusivity of the ceramic, r_m is the characteristic "radius" of the ceramic body, and h is the heat transfer coefficient at the ceramic surface (Ref 5).

Mechanical Behavior of Interfaces. The most critical element in a seal design is the metal/ceramic interface. In part, its critical character is related to the nature of the bond and the sensitivity of the bond strength to details of the fabrication process. In general, the interface involves a transition from the predominantly ionic and covalent bonding in the ceramic to metal bonding. The basic processes that govern the formation of bonding between dissimilar materials are described in the article "Wetting, Surface Energies, Adhesion, and Interface Reaction Thermodynamics" in this Volume. It will be clear from that discussion that, although there is a good general understanding of the chemistry and thermodynamics of the interface, it is not yet known how to optimize bond strength.

The other feature of the ceramic/metal junction in a seal which leads to its critical importance in mechanical performance of the seal is the very complex mechanical behavior of the structure surrounding the interface. That complexity exists, not only because the interface couples materials with very different properties, but also because the processing required to create the bond can substantially change the behavior of the constituents.

The failure mode for a seal will depend on the general character of the seal, but also on the details of the changes induced by the processing. For a braze-bonded or solder-bonded joint, the filler is intended to deform in a ductile manner. However, many of the braze and solder alloys solidify as intermetallics or eutectic structures that show only modest ductility prior to failure. Moreover, during fabrication, the filler materials can dissolve constituents of the adjacent materials (the metallizing material or the metal component of a metallized ceramic/metal seal, for example) that can change the mechanical response of the filler.

In general, the ceramic bond interface will behave in a brittle manner so that the strength will be determined by the fracture toughness of the interface and by the size of flaws in the interface. The energy required to propagate a crack through the interface will be related to the energy of adhesion. However, it will also include the energy expended in a number of other processes at the crack tip, including deformation of adjacent material, crack branching, bridging, blunting, and shielding. Thus, for a failure in the glass layer at the interface between a moly/manganese metallized layer and an oxide ceramic, the molybdenum, the nickel plating, and the adjacent filler material may deform plastically as the crack propagates. The energy expended in that deformation can raise the effective fracture toughness well above that for a monolithic glass structure in which no plastic deformation occurs. Cannon *et al.* have described some of the processes that affect the fracture toughness of the interface and suggested some techniques for enhancing those effects (Ref 6).

The presence of residual stresses and differences of elastic moduli for components of most seals makes it difficult to characterize the strength and fracture toughness of bond interfaces. In test specimens, Elssner and coworkers (Ref 7, 8) and Evans and coworkers (Ref 9, 10) have examined the problem of specifying interface bond strength and toughness and have developed a number of experimental procedures to obtain meaningful data. Much of the work by both groups was conducted on Nb/Al_2O_3 seals in which a good match in thermal expansion coefficients limited residual stresses in the seals. They discovered that, even in this relatively simple system, the resistance to crack growth in the interface (K_{Ic}) can depend markedly on details of the crack origin and the dynamics of crack growth. Turwitt *et al.* found that the initial grain size of the niobium was an important factor in determining the degree of bonding and, hence, the effective fracture energy (Ref 8). Fine-grained niobium yielded the highest fraction of bonded surface, apparently because the large amount of grain growth that occurred during the bonding operation permitted better accommodation of the mismatch at the interface during diffusion bonding.

Stress Minimization. Ensuring mechanical integrity of a seal implies not only selecting materials with high strength but also selecting a combination of materials that will result in minimum residual stress in the structure or in a stress state that protects the vulnerable elements (the ceramic and the bond material) of the seal. One method for limiting residual stress is to select combinations of materials whose thermal contractions after bonding are matched. Alternatively, materials with different CTEs can be coupled through filler materials whose high creep rate and low yield strength reduce the stresses generated by the differential contraction.

In making the selection for a CTE match, it is important to note that very few pairs of materials have essentially identical thermal expansion curves. Producing a stress-free structure often involves bonding a pair of materials at a very specific temperature so that their net contractions on cooling to room temperature are the same even though the contraction curves may not coincide. A particular example of this procedure is the use of FeNiCo alloys, including Kovar, in combination with alumina ceramic. As Fig 1 shows, the expansion curves for Kovar and alumina are very different. Yet, by forming a bond between these two materials that becomes rigid at the intersection of the two curves (620 °C, or 1150 °F), it is possible to produce a structure that is stress free at room temperature. Note, however, that the large separation of the two curves between the intersect point and room temperature implies a very substantial transient differential strain during cooling. For certain geometries, a tensile stress of 140 MPa (20 ksi) could be generated in the alumina in such a seal at about 400 °C (750 °F).

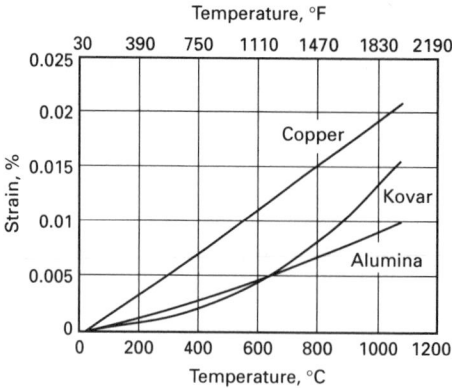

Fig 1 Thermal expansion curves for seal materials

Most Kovar/alumina seals are not truly matched seals but are produced by metallizing the alumina and using a braze filler to form the bond at some other temperature than the intersection of their CTE curves. The deformation of the filler by time-independent plastic flow or creep relaxation limits the stresses generated in the ceramic. In principle, it should be possible to measure the deformation behavior of a braze material and take account of that deformation in thermal stress analysis of a seal. As will be described later in this article, considerable progress has been made in translating creep and plastic deformation data obtained on test specimens to stress relaxation in analysis of a seal.

In any attempt to model the true behavior of the materials in the seal for stress analysis, it is essential to remember that the composition of the braze alloys or solder alloys may change as the seal is formed. Solution of constituents of the adjacent metal or the metallized layer into the alloy can drastically change the mechanical response of the bond layer. Even without a compositional change, the mechanical behavior of the braze alloy in the seal may be different from that of the starting material. For example, certain copper/gold alloys phase separate or even decompose spinodally during cooling. The rate of cooling can determine the size of the two-phase regions and, hence, the deformation behavior of the alloy.

Selection of matching materials for glass/metal and glass-ceramic/metal seals also requires careful scrutiny of the true behavior of the materials. The critical factor for thermal stress is the change in dimensions of the seal components from the temperature where the seal is effectively rigid down to room temperature. The relevant CTE would be the mean value for that contraction. Seals are often designed using tabular data obtained during temperature increase of the sample. However, the cooling curve, which is more relevant to the seal application, may be very different. Also, in some tables, the quoted

CTE data may be the average from room temperature to, for example, 300 °C (570 °F). Although most glasses have nearly linear expansion curves up to the transformation range, occasionally there is enough curvature so that the 25 to 300 °C (75 to 570 °F) value may not be truly representative of the contraction of the glass after the seal is formed.

Even when the contraction curve for a glass is available, it is not always clear what the net contraction of the glass will be after the glass becomes rigid. As noted in the article "Glass/Metal and Glass-Ceramic/Metal Seals" in this Volume, there is an abrupt change in the expansion curves for most glasses in the transformation range. Also, in the transformation range, the viscosity changes by many orders of magnitude in a small temperature range. The set point at which the glass is effectively rigid is nominally taken to be 20 °C (35 °F) below the annealing temperature (where viscosity = 10^{12} Pa · s). An annealed glass can be cooled rapidly below the set temperature without introducing residual stresses from temperature gradients during cooling. For glasses that behave normally, the set point is approximately the temperature at which straight lines drawn through the expansion curves above and below the transformation range intersect (see Fig 2). The net strain is then the change from that point to room temperature. However, a glass which is cooled rapidly through the transformation range can become effectively rigid at substantially higher temperatures so that the effective strain can be substantially higher, that is, the glass behaves in the seal as if it had a substantially higher CTE.

As in the case of braze alloys, the composition of the glass may be changed by dissolution of constituents from the metal or ceramic it contacts during bond formation. In most instances, the mass transport of these dissolved atoms will be slow so that the affected glass may be only a thin layer near

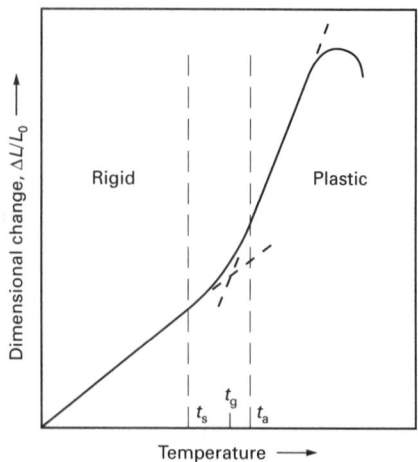

Fig 2 Dilatometer curve for a glass showing strain point, t_s, the glass transition (set point), t_g, and the annealing point

the bond interface. However, as noted by Pask *et al.*, the resultant change in viscosity and CTE for that layer may induce substantial stress into the bond (Ref 11).

Residual stresses can be produced not only by CTE mismatch, but by retaining thermal gradients in a structure to temperatures at which the resultant stresses can no longer relax. For glass/metal seals, these thermal gradient-generated stresses are minimized by slow cooling of glass/metal seals and by using annealing cycles. But ceramic/metal seals can also develop undesirable residual stresses.

The ultimate test of whether a set of materials has an adequate expansion match is, of course, whether the stresses in the structure are limited to acceptable values. For transparent insulators, stress optical measurements can be used to monitor the stress state. An indication that the match is satisfactory can also be obtained from the deformation of a bimorph consisting of thin lathes of pairs of the materials bonded together. For controlled geometry (both materials as thin, well-defined plates) a quantitative measure of the stress in the bimorph can be obtained from its final curvature. There is an important caveat to the use of this test to infer the stress state in a seal, namely, stress relaxation is likely to be very different in the seal configuration than in a flat plate so that the resultant stresses will be different.

Compression Seals. In some seals, residual stresses are generated deliberately to provide protective compressive stresses in the ceramic part and in the bond interface. Usually this is accomplished by selecting components with different CTEs. Many cylindrically symmetric glass/metal and glass-ceramic/metal pin seals are made as compressive seals in which the body of the seal (the metal surrounding the glass insulator) has a substantially higher CTE than that of the glass. The pin material is usually chosen to have the same CTE as, or slightly smaller than, the glass. The result of the differential contraction of the body is a radial compressive stress in the glass that is maximum at the body/glass interface and decreases in magnitude at smaller radii. Depending on the relative outer and inner diameter of the glass annulus, a sizeable fraction of that stress may be transferred to the glass/pin interface (one of the advantages of a compression seal is that it can help to ensure hermeticity at a metal/glass interface where good chemical bonding does not occur). The differential contraction in the axial direction also produces axial compressive and shear stresses in the glass which are maximum at the body/glass interface and a function of axial position.

Even though the compression seal leaves most of the glass with a protective compressive residual stress, the shear stresses can lead to very substantial tensile stresses in some zones of the glass. In particular, if the glass forms a meniscus on the metal body, local tensile stresses can cause cracking of that meniscus. In most instances, that cracking is of only cosmetic concern.

For a flat seal configuration as in a butt seal or a coating on an alumina substrate, higher expansion of the metal relative to the ceramic will put the ceramic within the bond zone in compression. However, at the periphery of the bond zone, very large tensile stresses can be generated in the ceramic. Those stresses are frequently singular (Ref 12) but decrease rapidly away from the singularity depending on the thickness of the metal. The maximum tensile stress is oriented at an angle to the bond interface so that a crack generated in the ceramic will usually enter at an angle of 30 to 45° to the interface. The very high values of tension which are generated in this seal configuration make it very easy to initiate a crack in the alumina even when the fracture toughness of the bond is lower that that of the ceramic. However, the very rapid decrease of the stress with distance from the edge often leads to an arrest of the crack before the seal is severely compromised. The crack stops growing because the stress intensity factor, K_I, drops below K_{Ic}, or in the event that subcritical crack growth can occur, it drops low enough that crack growth is effectively zero. In some designs, therefore, the practice has been to accept some cracking at the periphery of such seals provided it can be demonstrated that these will not extend in use.

Often, for butt seals, the diameter of the metallized zone on the ceramic is slightly larger than that of the metal washer. The filler metal wets the entire metallized zone so that a fillet of the filler extends beyond the washer. The differential contraction of the filler metal relative to the ceramic is then a more important factor than the CTE of the washer in the development of high local stresses at the periphery of the seal. Modification of the mechanical response of the filler by dissolution of adjacent components becomes especially critical in this event.

The high, but rapidly decaying, tensile stresses at the periphery of butt seals are also present at the edge of coaxial compression seals in which a high expansion metal surrounds a lower expansion insulator. In this geometry, however, there is rarely any concern about propagation of cracks through the insulator. Cracks may initiate at the edge of the braze fillet, but the radial compressive stress produced in the insulator usually prevents much growth.

One of the strategies for reducing the large stress concentrations at the periphery of a butt seal is to add a retainer (or backup) ring of the low expansion material (the alumina in a Kovar/alumina seal) to the outside of the seal as shown in Fig 3. This "sandwich" seal reduces the bending moment in the metal part and distributes the shear load between the two metal/ceramic interfaces.

Finite-Element Stress-Analytic Techniques

General Approach. Numerical stress analysis techniques have been and are increasingly being used in the design of both glass/metal and ceramic/metal seals. The most commonly used numerical analysis procedure is the finite-element technique which has gained wide acceptance in many fields of engineering analysis including heat transfer and structural mechanics.

In general, the problem for every joining technique is basically the same. Dissimilar materials are joined together at elevated temperature and subsequently cooled to room temperature. Unequal thermal contractions of the constituent materials generate residual stresses as the assembly is cooled. Unfortunately, the mechanical responses of many of the materials used in seals are nonlinear; therefore, nonlinear numerical-mechanics techniques are required to predict stresses developed in a joined assembly.

Finite-element analysis requires a base of information containing three critical items. First, the geometry of the assembly must be accurately described. Often, the analyst simplifies that geometry by imposing symmetry conditions that do not exist in the assembly. The analyst may also include in the model only the major components, omitting, for example, thin layers of materials that are deemed unimportant in the behavior of the structure. It is essential to justify those assumptions and to qualify the results of the analysis in terms of the simplified geometry. Secondly, the constituent materials must be identified, and the mechanical response of those materials must be known. In the analysis program, that mechanical response must be translated into an appropriate mathematical representation (material constitutive model). Finally, the loading of the assembly must be defined. For thermal stress analysis, the loading may be simply a statement of the

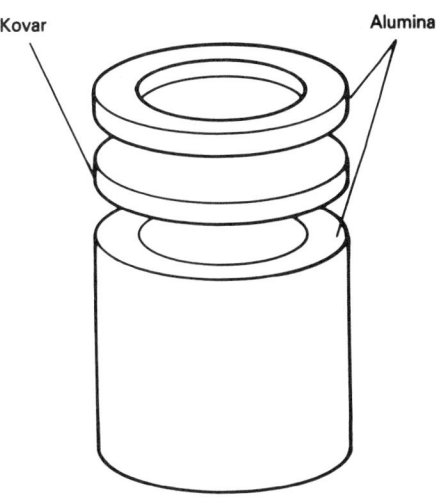

Kovar   Alumina

Fig 3 Sandwich seal with backup ring

temperature change in the assembly as it is cooled after the joint is formed.

In the analysis of the fabrication of glass or ceramic/metal seals, it is assumed that the assembly is stress free at some elevated temperature. The selection of the stress-free temperature is critical to the analysis and will be discussed in detail later. It is also generally assumed that the cooling rate is slow enough so that the assembly is uniformly cooled, that is, there are no temperature gradients. That assumption is not always appropriate and, because temperature gradients produce stresses even in a monolithic structure, it is sometimes necessary to include those gradients in the problem statement.

Glass/Metal Seals. In order to numerically predict the stresses in a glass/metal seal, current common practice is to assume that the assembly is stress free at the set point of the glass. The commonly assumed constitutive relationship for the glass is a linear elastic model. For the metal constituents of the seal assembly, however, a nonlinear, temperature-dependent, time-independent plasticity constitutive model is commonly assumed. The analysis proceeds from the stress-free temperature, and stresses in the assembly are predicted as a function of temperature using appropriate finite-element software (Ref 13, 14).

Once the stresses in the assembly have been calculated, the analyst must evaluate the predicted stress state in terms of the response of the structure, that is, is the glass likely to fracture? While some finite-element software packages incorporate fracture mechanics concepts into the analysis, this type of calculation is generally reserved for specific problems in which there is a pre-existing crack in a brittle material. These types of problems require a more detailed definition of the geometry and are often computationally expensive. Currently it is more common, particularly in design calculations, to use the maximum tensile stress in the glass as a measure of the severity of the stress state. For design purposes, various rules of thumb have been developed from past experience with seals. For instance, calculated tensile stresses in the glass less than 6.9 MPa (1000 psi) are judged to be acceptable. Calculated tensile stresses in the glass between 6.9 MPa and 34.5 MPa (1000 and 5000 psi) are judged to be marginal and calculated tensile stress states in the glass in excess of 34.5 MPa (5000 psi) are judged to be unacceptable.

Judgment of the acceptability of a calculated stress state in the glass is also dependent upon the location of the tensile stress. If tensile stresses are developed on the surface of the glass, extra caution must be exercised because there is a greater potential of severe flaws.

One of the useful questions that can be addressed with finite-element analysis is how to optimize pin seal geometry so that minimum tensile stress is generated in the glass.

Reference 15 describes a set of guidelines that were established, through analysis, for selecting relative dimensions of the central pin, the glass insulator, and the header body, including acceptable eccentricity in location of the central pin. The guidelines were developed for a compressive pin seal with a standard material combination consisting of a 52Ni-Fe alloy (Alloy 52) pin, Corning 9013 or Kimble TM-9 glass, and a type 304 stainless steel housing; however, the general principles can be applied to any set of seal materials. In this material combination, the alloy 52 and the glass have essentially the same thermal expansion/contraction characteristics while the expansion of type 304 stainless steel is much higher. While these design guidelines have not been confirmed in detail through experiments, their usefulness has been demonstrated in the manufacture of seals.

Figure 4 shows the model for the pin seal with the pin centrally located. Analyses of that model with various values for R_1, R_2, R_3, and t_p yielded maximum values for the tensile stress in the glass. That maximum stress is plotted in Fig 5 as a function of the ratio of the width and the thickness of the glass annulus. The tensile stresses are a consequence of the shear strain developed in the glass by the housing. The important feature of this curve is that there is a very definite minimum. The first design guide, obtained from that plot, is that for minimization of the tensile stress in the glass, the ratio $(R_2 - R_1)/t_p$ must be held between 0.25 and 0.33.

To account for possible eccentric location of the pin in the housing hole, analysis was performed on the geometry shown in Fig 6. That analysis showed that to minimize the tensile stress in the glass it was necessary to

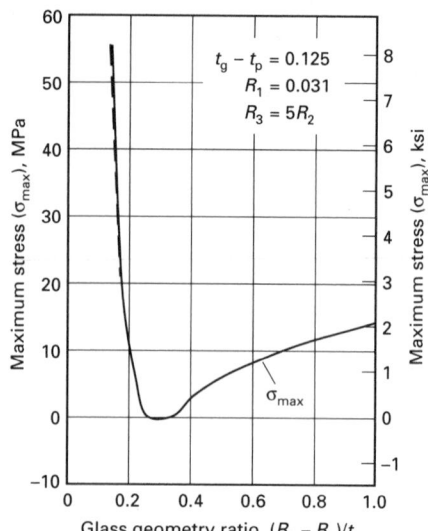

Fig 5 Maximum stress of a single-pin seal as a function of the ratio of the width and thickness of the glass annulus

limit the eccentricity so that $[R_2 - (R_1 + t_0)] \geq 0.25$ and that $[R_2 - (R_1 - t_0)] \leq 0.33$. Note that $[R_2 - (R_1 + t_0)]$ is the thinnest width of the glass annulus while $[R_2 - (R_1 - t_0)]$ is the thickest width.

Figure 7 shows the model used to calculate the optimum thickness of the glass annulus relative to the depth of the hole in the body. The analysis showed that stress was minimized when $t_g \cong 0.95\ t_p$.

In general, for this combination of materials the type 304 stainless steel will yield on cooling to room temperature from the bond-

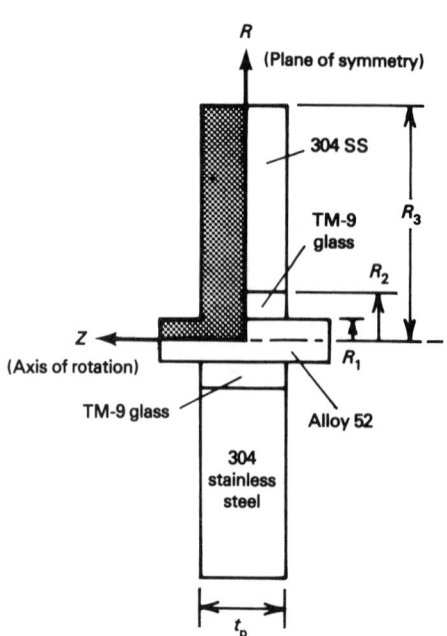

Fig 4 Single-pin seal geometry

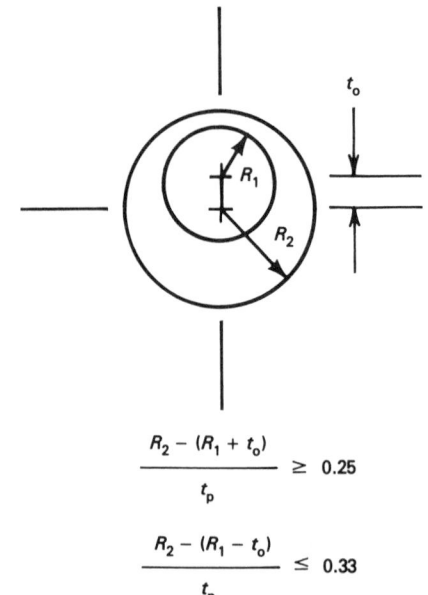

Fig 6 Design guidelines for a single-pin seal. The joint consists of Alloy 52/Corning glass 9013/304 stainless steel.

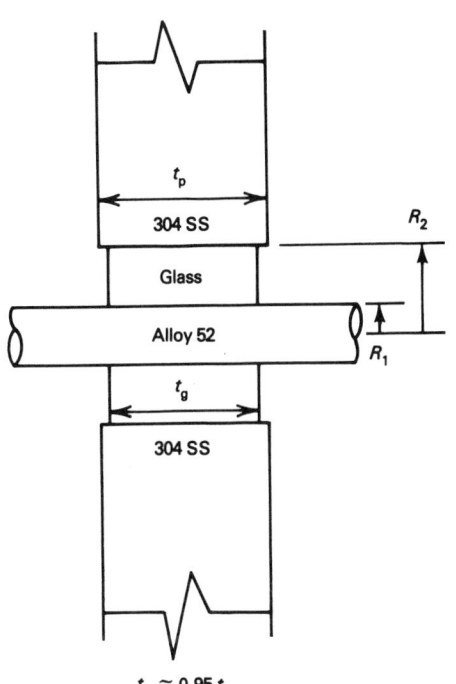

$t_g \cong 0.95 \, t_p$

Fig 7 Design guidelines for calculating the optimum thickness of the glass annulus in a single-pin seal. The joint consists of Alloy 52/Corning glass 9013/304 stainless steel.

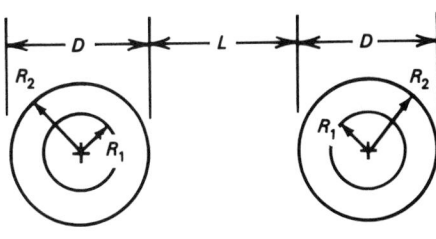

L/D < 0.5 = Poor
0.5 < L/D ≤ 0.75 = Marginal
L/D > 0.75 = Adequate
L/D > 1.0 = Desired

Fig 8 Design guidelines for a multiple-pin seal. The joint consists of Alloy 52/Corning glass 9013/304 stainless steel.

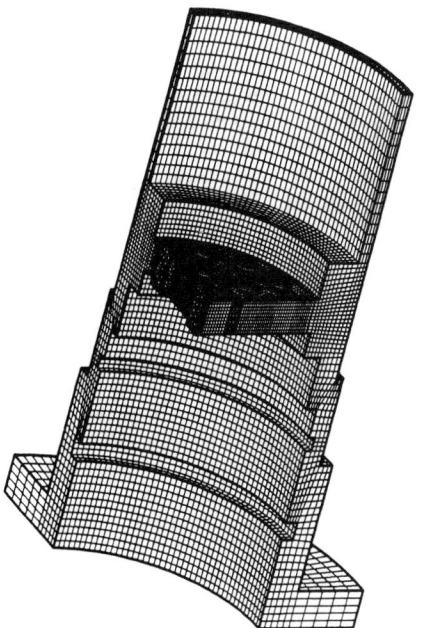

Fig 9 Complex glass/metal seal connector

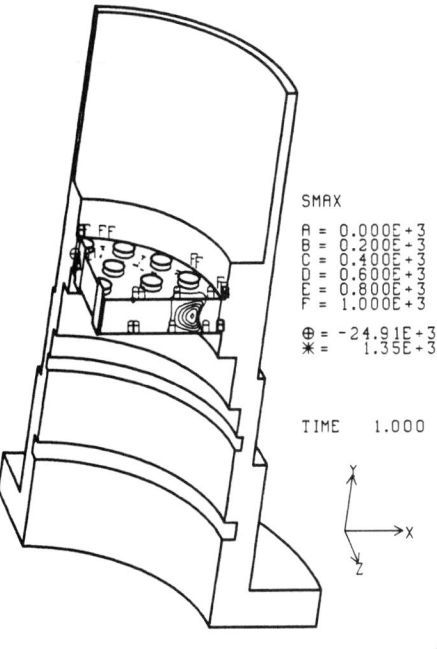

Fig 10 Calculated maximum tensile principal residual stresses developed in a glass insulator

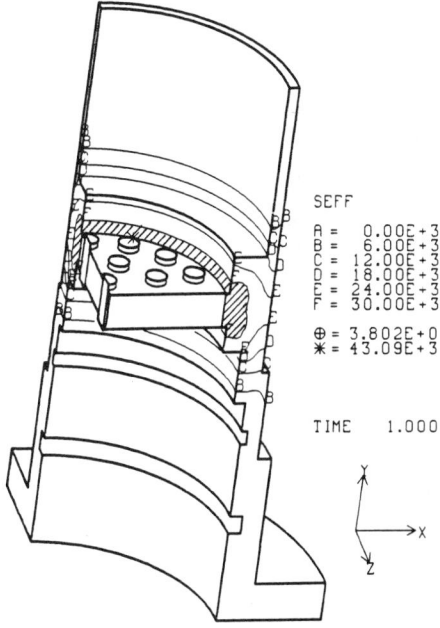

Fig 11 Calculated effective residual stress developed in a housing. The crosshatched areas denote where the stresses exceed the yield strength.

ing temperature. If the assembly is subsequently heated, damaging radial tensile stresses are developed in the glass at temperatures in excess of ~230 °C (~450 °F).

In multiple-pin seal geometries, there is an additional design guide related to the spacing between seals. For the multiple-pin geometry shown in Fig 8, the calculated flow stress in the type 304 stainless steel housing is enhanced when the geometry ratio L/D becomes too small. The enhanced yielding can produce damaging tensile stress in the glass. The guide included in Fig 8 shows the L/D ratios where those stresses are acceptable.

Figure 9 shows an example of a complex glass/metal seal whose design was aided through the use of finite-element analysis. This connector is a compression seal; however, the glass insulator has multiple metallic conductor pins extending through the glass insulator. In the finite-element idealization of this connector, two symmetry planes were assumed. The residual stress state in the connector was evaluated using the finite-element computer program JAC3D (Ref 13). In Fig 10, contours (lines of equal stress) of the calculated maximum principal stress developed in the

glass insulator at room temperature are plotted. The peak tensile stress is approximately 8.3 MPa (1200 psi) and occurs in the interior of the glass. Although this is in excess of the acceptable value quoted earlier, it occurs away from surfaces where flaws are expected to be severe. Therefore, this design was concluded to be acceptable. In Fig 11, contours of the calculated effective stress developed in the type 304 stainless steel housing are plotted. The stresses in the housing exceed the yield strength of type 304 stainless steel in the regions denoted by crosshatching. The yielding in the housing is not detrimental to the design of the glass/metal seal unless the seal is required to perform at temperatures in excess of 250 °C (480 °F).

As noted earlier, for most glass/metal seals, it is assumed that the glass responds linear elastically and that stress development starts at a stress-free temperature and is independent of time. While these assumptions appear to be appropriate and qualitatively accurate for most glasses, the response of phosphate-glass and some other low-melting glasses cannot be accurately predicted using this approach. A rigorous stress analysis of glass/metal seals must incorporate the complex and coupled changes in volumetric strain and stress relaxation which occur as glass passes through the liquid/solid temperature regime. Chambers has developed a viscoelastic material model for computing stresses in glass (Ref 16). This new model eliminates the need for all three of the previous assumptions and promises to open a new era of predictive capability in glass/metal seals. The model provides the capability to analyze and design actual manufacturing processes by predicting the effects of temperature history on the residual stress state. Unfortunately, at the present time, the data required by the model are available for only a few glasses.

Ceramic/Metal Seals. In the calculation of residual stresses in a ceramic/metal seal, it is generally assumed that the ceramic is a linear elastic material through the entire temperature range whereas the metal members, particularly the filler metal, can deform in response to the stresses developed during cooling. The complexity of that deformation makes the accurate numerical prediction of residual stresses developed in ceramic/metal seals more difficult than analysis of glass/metal seals. Part of the difficulty is that the mode of deformation changes as the temperature changes. At high temperature, when the metals are above about 0.6 of the homologous melting temperature, time-dependent nonlinear creep dominates the mechanical response of the filler material, whereas at lower temperatures, nonlinear time-independent plasticity is the dominant mechanism. Because the response is nonlinear, the loading or cooling rate must be accurately represented.

For most filler metals, the available mechanical data are too sparse to accurately describe their response through the entire temperature range of interest. Furthermore, most of the data are obtained on test specimens with relatively simple stress states and the translation of those data to an accurate mathematical description of the response of an element in the complex stress state of a seal is no simple task. Even for those materials whose behavior is well understood, the dissolution of adjacent metal constituents into the filler can change the response and render

an analysis based on the pure filler behavior useless.

Currently, the mechanical response of the filler material is treated in one of three ways. First, it is often assumed that the filler material can be characterized with a temperature-dependent, time-independent plasticity constitutive model. Secondly, it is assumed, especially for solder alloys, that the filler material can be characterized solely with a creep constitutive model. Finally, some finite-element software packages allow for both plasticity and creep in the filler material. The procedure involves taking a plasticity step followed by a creep step, and so on until the assembly is at room temperature. For the first option, the cooling rate does not enter into the analysis; however, for the second and third options, the cooling rate must be accurately known. Analysis using any of the three options is extremely valuable in comparing optional designs and for obtaining accurate values of the actual stresses developed in the assembly.

In spite of the caveats about the treatment of stress relaxation in the filler metals, numerical predictive analysis techniques are very useful in ceramic/metal seal design. As an example of the use of finite-element analysis in the design of ceramic/metal seals, a commonly used shear geometry pin seal, shown in Fig 12, was analyzed to estimate the room-temperature residual-stress state. The purpose of this design was to hermetically join a Kovar conductor pin to an alumina ceramic housing. In this design, a Kovar washer was used to make a shear seal to the ceramic body and the braze material was pure oxygen-free, high-conductivity (OFHC) copper. The actual geometry of this design is cylindrical; there-

fore, a two-dimensional axisymmetric geometry was used for the analysis. The melt temperature of the OFHC braze material is 1083 °C (1981 °F). As the assembly is cooled from this temperature, stresses are developed due to the mismatch in thermal contractions of the various materials. In this analysis, the OFHC copper braze material and the Kovar was characterized using a temperature-dependent, time-independent plasticity constitutive model. The residual stress state in the assembly was evaluated using the finite-element computer program JAC2D (Ref 13). In Fig 13, contours of the calculated maximum principal stress developed in the ceramic insulator at room temperature are plotted. A very severe stress concentration (approaching a singularity) is developed at the outer periphery of the braze material in the ceramic with a peak tensile stress of approximately 100 MPa (14,500 psi). The calculated magnitude of the stress at the concentration is, however, dependent on the size of the elements in the mesh. Once a baseline design has been established, the effect of various design parameters such as braze thickness, washer thickness, and diameter can be evaluated.

Electrical Integrity

Generally, except for applications involving high voltage, maintaining electrical integrity simply means ensuring that some minimum resistance exists between the metal components (for example, between the pin and the housing of a pin seal). For most of the insulators used in electrical feedthroughs, the bulk resistivity is high enough that ensuring a minimum resistance means limiting the

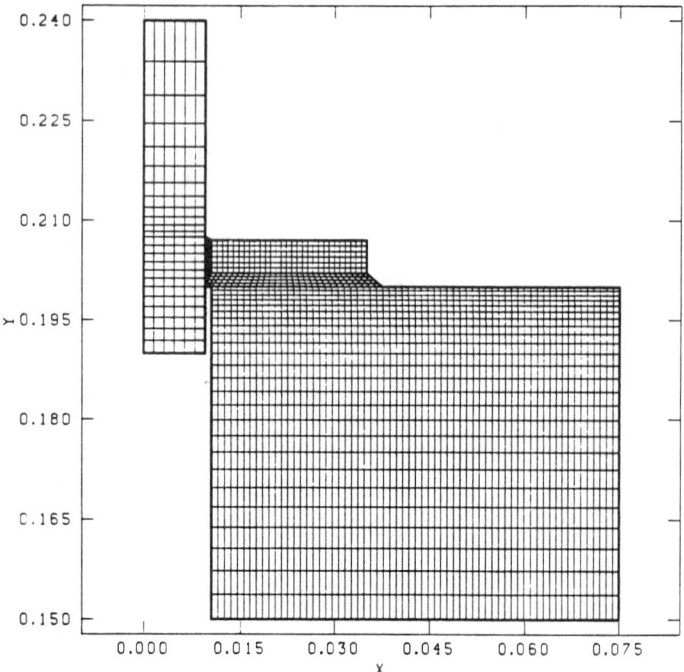

Fig 12 Geometry and finite-element idealization of a shear seal

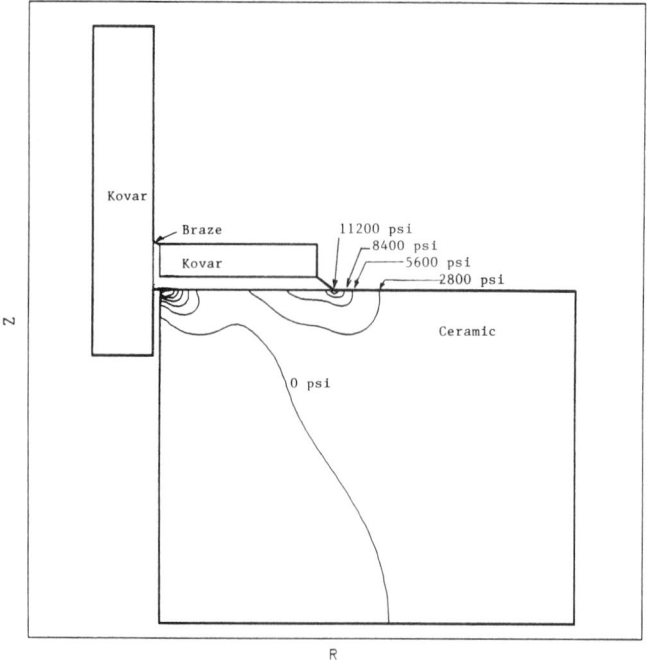

Fig 13 Calculated effective residual stress developed in a ceramic insulator

conductance across the surface of the insulator. Usually, that can be accomplished by preventing contamination of the surface with conducting materials. However, there are some commonly used insulators, such as the low-melting glasses containing lead oxide, whose conductivity can be greatly enhanced by firing them in a reducing atmosphere. For some of these materials, reduction does not even require the introduction of a reducing gas into the firing chamber. Graphite fixturing in contact or in close proximity with the insulator may be sufficient. Obviously, care needs to be exercised in processing such materials.

In some applications where the separation between electrodes is very limited, field-induced generation of metal dendrites (electromigration) over the insulator surface can cause delayed electrical failure. Similarly, a mechanical crack penetrating into a ceramic capacitor (or a delamination between plates) can lead to electrical failure through migration of metal over the fracture surface. Many of these failures involve the use of silver or silver alloys with moisture present. However, tin, lead, copper, aluminum, and chromium can also migrate in the presence of water (Ref 17). Precious metals can migrate in the presence of a halogen (Ref 18).

In high-voltage applications, the design goal is to maintain the local electrical field intensity below the dielectric strength of the insulator. For oxide ceramics, the tabular data typically indicate dielectric strengths of 100 kV/cm (255 kV/in.) for insulator thicknesses of 0.63 cm (0.25 in.), increasing to about 300 kV/cm (760 kV/in.) when thickness is decreased to 0.02 cm (0.008 in.). Theoretically, however, the dielectric strength of these materials ought to be 2 to 6 MV/cm (5 to 15 MV/in.) and experimental values approximating these have been measured on single crystals and very thin films (Ref 19, 20).

The relatively low tabular values and the dependence of dielectric strength on insulator thickness reflect, in part, the fact that in the standard dielectric strength test, local fields are much higher than the average value calculated by dividing the applied voltage by the dielectric thickness. Those high local fields are generated at the edge of the electrode contact on the flat dielectric specimen and at small electrical stress concentrators, for example, asperities on electrodes or cracks in insulators. The effect of those high fields is compounded by the presence of a low dielectric strength material such as air adjacent to the contact edge. In that event, a precursor to breakdown through the ceramic is the development of corona from the metal edge culminating in an arc that produces either surface breakdown or puncture of the insulator. In the standard test, that corona-induced breakdown is suppressed by immersing the specimen in oil and/or by using a guard ring. Nevertheless, the values obtained from the test often depend more on the electrical behavior

of the oil and the relative dielectric constants of the oil and the ceramic than they do on the intrinsic strength of the ceramic. For a true measure of the dielectric strength, most experimental studies attempt to eliminate local field concentrations by using special electrode configurations such as the Rogowski contour.

Even when the electrode geometry is controlled to eliminate high local fields, the dielectric strengths of engineering materials are not as high as those predicted and measured in single-crystal samples and films. Dielectric strengths in polycrystalline materials are lower for much the same reason that measured mechanical strengths of ceramics are substantially less than the theoretical values. In both cases, the microstructure of the ceramic results in local values of the stress (mechanical or electrical) that can greatly exceed the applied stress. Experimentally, it has been shown that, for rapidly increasing fields, dielectric strength of MgO ceramics decreased as grain-size and/or grain-boundary porosity increased (Ref 21). The grain-boundary pores produced local field-intensity enhancement and induced failure. Moreover, the presence of subsurface cracks produced by machining drastically reduced the dielectric strength of MgO single crystals, probably because of intrusion of electrically conducting material into the cracks to yield high field concentrations. With the removal of those cracks by etching, dielectric strengths of 2 MV/cm (5 MV/in.) could be obtained. Other studies suggest that the effect of microstructure on dielectric strength of ceramics is a general one (Ref 22, 23).

The high local-field concentration present in the standard test without a guard ring is also present in applications such as butt seals and other seal configurations where a sharp-edged electrode intrudes into the dielectric or is bonded to its surface. Where it is impossible, because of size constraints, for example, to make the electrical path through the insulator long enough to lower the average field (the ratio of applied voltage to thickness) below the dielectric strength, it may still be possible to survive higher fields than the tabular dielectric strength. The strategy is to eliminate the features which lead to high local fields and/or to prevent the development of corona which is the precursor to breakdown. Corona-induced breakdown can be suppressed by immersing the seal in oil or by encapsulating in a high dielectric strength material, for example, SF_6. Corona can also be suppressed by embedding the electrode in the electrolyte. This is commonly done with glass/metal seals but is difficult to achieve in ceramic/metal seals. The field condition around an embedded electrode can be further improved by eliminating sharp projections. Although special electrode contours such as the Rogowski contour can produce very uniform field conditions around the electrode, these are very rarely used in practice.

For vacuum applications, breakdown over the surface of the ceramic can reduce the voltage holdoff capability of the ceramic below that of a vacuum gap of the same dimensions. There are, however, a variety of techniques used to enhance that capability primarily by suppressing the electron avalanche process that is a precursor to the surface breakdown (Ref 24). Shielding the triple junction (the zone where the electrode edge contacts the dielectric in vacuum) to remove the high-field gradients is effective. Changing the angle of the insulator surface so that the secondary electrons do not return to the surface or roughening the surface so that the production of secondaries is reduced can also raise the breakdown voltage. One very effective method of increasing the surface flashover strength is to apply a slightly conductive or semiconducting coating to the insulator surface (Ref 25). The assumption is that the coatings reduce the effective secondary emission yield and/or reduce the fields that drive the electrons from the surface.

Chemical Integrity

The designer of a seal must select materials and materials combinations that will survive the anticipated environment. It is also essential to select materials, material combinations, and fabrication processes that will not result in adverse effects on performance because of changes during fabrication. In this discussion, high temperature is included as a parameter in the chemical environment.

Survival in the Environment. For most seal applications, the environment of concern is simply the ambient with, perhaps, some small temperature excursions. Survival is, therefore, predominantly a question of long-term stability in the presence of relatively benign reagents such as water (including salt water), oxygen, carbon dioxide, and other gases of lower concentration in the air. In some applications, such as chemical batteries, seals can experience much more aggressive environments.

Ceramics are generally considered to be much more stable than metals so that, in a seal, the primary concern is for the survival of the metal parts including the filler metals in the bond. Metal corrosion and other reactions of metals with the ambient are very broad topics and will not be addressed here. However, there are some genuine concerns for the survival of the insulator that need to be considered. In particular, for glass insulators, exposure to aqueous solutions can be a threat. It is a particular concern because many of the glasses that have desirable thermal expansion behavior may have poor durability in water. As an example, phosphate glasses can be made with expansion matches to stainless steels and other high-expansion metals, including aluminum. However, except for special compositions (Ref 26), they have poor durability in water. For all glasses, the dissolution rate is a strong function of the pH of the water.

Except for glasses with very low durability, the concern is not that the glass will dissolve and the seal will be destroyed. It is more likely that a surface layer of the glass will be modified by extraction of some of the constituents and that that layer may become porous. The result may only be a cosmetic concern. However, a porous layer can retain water plus alkalis extracted in the dissolution and develop enough electrical conductivity to affect the electrical performance of the seal.

Durability of glasses is often reported in terms of the weight loss per unit area exposed in a 5% HCl or NaOH solution at 100 °C (212 °F). However, in an application such as an insulator in a seal, where the concern is for the amount of alkali made available for surface conduction, a better measure of the durability is provided by ASTM test method P-W (see ASTM C 225). This test measures the quantity of a dilute solution of sulfuric acid required to titrate water exposed to a crushed 10 g (0.35 oz) sample of the glass for 30 min. A large value indicates poor durability. For most seal applications, a glass with a durability of 5 mL N/50 H_2SO_4 or less should be satisfactory.

In more reactive environments such as that in a Li/SO_2 cell, survival can only be assured by selecting special glass formulations designed for that specific application. Calcium boroaluminate glasses (CABAL glasses) have corrosion rates in Li/electrolyte environments low enough to project 25-year lifetimes for these batteries (Ref 27).

Applications at higher temperatures increase the reactivity of ceramics and glasses and make their survival more problematic. As an example, sodium/sulfur cells that are being developed for potential electric vehicle applications operate at about 450 °C (840 °F). In some Na/S cell designs the β″-alumina ceramic that provides the ion-conducting path is bonded to an α-alumina ceramic insulator with a glass of matching CTE. Similarly, seals in sodium vapor lamps are made with glass. Many of the early candidate glasses were severely degraded by the exposure to sodium and a number of new glasses are being developed for these applications (Ref 28, 29).

Processing Effects. For most seals, the only exposure to high temperature is during fabrication. It is important to recognize that conditions in the fabrication environment can be very aggressive and that both metals and ceramics can experience serious degradation in performance as a result of this exposure. There are two classes of reactions that need to be considered. One of these is the direct result of exposure of the individual components to the high temperature and the atmosphere in the processing environment. The other has to do with accelerated interactions between components of the seal.

Probably the most important factor in high-temperature exposure of the seal materials is

the change in kinetics of reactions. Diffusion is enhanced so that grains grow and chemical distribution is changed. In Kovar, for example, sustained exposure to temperatures necessary for copper brazing can convert a fine-grained structure in a pin lead to a grain size so coarse that a single grain boundary may extend across the part. The strain to failure can be dramatically decreased. It is, therefore, essential in circumstances where strength of metal parts must be maximized, that the high-temperature treatment be limited to that necessary to obtain wetting and flow of the braze alloy.

Changes in thermodynamic equilibrium are also important to the performance of seal materials. Often, the equilibrium phase is different at the process temperature than at room temperature. More importantly, the original phase relationship that was produced by precisely controlling the heat treatment of the metal to give optimum performance may be affected. To reproduce the desirable metal behavior in those cases, it is necessary to modify the heat-treatment schedule during cooling from the process temperature so that the appropriate phases are again present at room temperature.

For ceramics and glasses, the exposure to high temperature can also lead to microstructural changes. In the article "Glass/Metal and Glass-Ceramic/Metal Seals" in this Volume, the process of transforming an amorphous glass to a polycrystalline glass-ceramic is described in terms of a very carefully controlled thermal treatment. That heat treatment first nucleates the crystallization and then promotes the growth of the predominant crystal phase. In general, the thermal expansion of the glass-ceramic will not be the same as that of the glass. The CTEs may be different enough that cooling the seal to room temperature before crystallizing the glass could generate stresses that would crack the glass. As a consequence, it may be necessary to form the bond and crystallize the glass without cooling to room temperature. In the thermal cycle, the temperature would first be raised high enough to bond the glass to the metal parts, then lowered to a temperature where nucleation could proceed and finally raised to the temperature where the major phase would crystallize.

For the more refractory ceramics such as alumina, it is not likely that significant grain growth will occur at brazing temperatures or even at higher temperatures where metallizing paints are consolidated. However, high-temperature treatment may produce subtle changes that can affect the mechanical behavior to some extent. As an example, in alumina with a substantial percentage of glass phase, the glass phase can crystallize and the degree of crystallization and the crystalline phases produced can vary with the heat treatment. As noted in the discussion of mechanical integrity, fracture in alumina and, in particular, subcritical crack growth, is dom-

inated largely by the glass phase. A change in the microstructure of the glass phase might be expected to change its fracture behavior. Raynes has shown that crystallization of the glass phase in certain aluminas is sensitive to thermal treatment at temperatures in the range where Mo-Mn metallizing paints are processed and where high-temperature brazes melt (Ref 30). The rates of subcritical crack growth depend on the details of the crystallization process.

There is a particular concern for the effects of the seal fabrication environment on the stability of partially stabilized zirconia (PSZ) and PSZ-toughened aluminas. The very high strength and fracture toughness and the excellent thermal shock resistance of these materials make them very interesting candidates for seal application. However, the mechanical properties of these materials are critically dependent on their microstructures. Subtle changes in chemistry, which could be induced by diffusion of constituents of metallizing paint into the ceramic, could make the tetragonal zirconia phase so stable that no transformation could occur or so unstable that conversion to the monoclinic phase would occur spontaneously during cooling. Alternatively, for yttria-PSZ, exposure to water vapor at temperatures of 100 to 400 °C (210 to 750 °F) degrades the ceramic by promoting transformation to the monoclinic phase (Ref 31). Any projected use of these materials with conventional high-temperature metallizing paints and wet hydrogen firing requires analysis of the effects on the zirconia transformation.

Many of the changes in material behavior that are produced during fabrication result from reactions with other components in the seal. Some of those reactions are a necessary part of the fabrication process. In metallization of a 96% alumina ceramic with Mo-Mn, for example, the intent is to interpenetrate both the ceramic and the sintered molybdenum with a glass phase so that the metallization layer is mechanically coupled to the ceramic. The MnO (and, sometimes, TiO_2) addition acts as a network modifier to lower the viscosity of the glass phase and promote that interpenetration. Too little network modifier or too little glass can result in a weak interface. It is essential to match the metallizing paint to the ceramic and to avoid processing of the ceramic that removes glass that is necessary for seal formation.

Brazing and soldering processes also require a reaction between the filler metal and the metallizing layer to produce a metallurgical bond. However, as noted in the discussion of mechanical integrity, those reactions can change the mechanical response of the metals. The designer and the fabricator must cooperate in selecting materials combinations and process conditions that optimize the strength of that bond without seriously degrading the performance of the seal components. Where titanium coatings are used to metallize the ceramic, the filler metal must not completely dissolve the titanium. Where gold is used to provide a solderable surface on a Kovar or stainless steel part, care must be taken, by selection of the solder or by limiting the amount of gold dissolved in the solder, to prevent the formation of brittle gold intermetallics in the bond interface.

For most seals between metals and either glasses or glass-ceramics, there is some reaction at the interface during processing. Often, the metal is pre-oxidized to promote wetting. To maintain a strong interface it may be essential to prevent complete dissolution of that oxide layer into the glass. As noted in the discussion of mechanical integrity, the dissolution of constituents of the metal parts into the glass may change the thermal expansion of the glass in the vicinity of the interface and alter the residual stress condition. Certain types of redox reactions can also generate bubbles at the interface that can affect the bond strength. For glass-ceramic/metal bonds, the local chemical change may alter the crystallization process so that the phases or crystal morphologies produced at the interface are different from those in the bulk (Ref 32).

REFERENCES

1. A.A. Griffith, The Theory of Rupture, *Proceedings of the First International Congress on Applied Mechanics*, J. Waltman, Jr., Ed., Delft, 1924, p 55
2. T.A. Michalske and S.W. Freiman, A Molecular Mechanism for Stress Corrosion in Vitreous Silica, *J. Am. Ceram. Soc.*, Vol 66, 1983, p 284–288
3. T.H. Service and J.E. Ritter, Proof Testing to Assure Reliability of Structure Ceramics, *Fracture Mechanics of Ceramics*, Vol 7, 1986, p 255–264
4. W. Weibull, A Statistical Distribution Function of Wide Applicability, *J. Appl. Mech.*, Vol 18, 1951, p 293
5. W.D. Kingery, H.K. Bowen, and D.R. Uhlmann, *Introduction to Ceramics*, 2nd ed., John Wiley & Sons, 1976, p 822–823
6. R.M. Cannon, V. Jayaram, B.J. Dalgleish, and R.M. Fisher, Fracture Energies of Ceramic-Metal Interfaces, *Ceramic Microstructure '86: Role of Interfaces*, Plenum, 1987
7. G. Elssner, Bond Strength of Ceramic/Metal Joints, *Designing Interfaces for Technological Applications: Ceramic-Ceramic, Ceramic-Metal Joining*, Elsevier, 1989
8. M. Turwitt, G. Elssner, and G. Petzow, On the Fracture Behavior and Microstructure of Metal-to-Ceramic Joints, *Ceramic Microstructure '86: Role of Interfaces*, Plenum, 1987
9. A.G. Evans and M. Ruhle, On the Mechanics of Failure in Ceramic/Metal Bonded Systems, *Mater. Res. Soc. Symp. Proc.*, Vol 40, 1985, p 153–166
10. D.B. Marshall and A.G. Evans, Measurement of Residually Stressed Thin Films by Indentation. I. Mechanics of Interface Delamination, *J. Appl. Phys.*, Vol 56, 1984, p 2632–2638
11. J.A. Pask, From Technology to the Science of Glass/Metal and Ceramic/Metal Sealing, *Bull. Am. Ceram. Soc.*, Vol 66, 1987, p 1587–1592
12. D.B. Bogy, The Plane Solution for Joined Dissimilar Elastic Semistrips Under Tension, *J. Appl. Mech.*, Vol 42, 1975, p 93–97
13. J.H. Biffle and S.N. Burchett, Nonlinear Structural Analysis of Complex Electronic and Electromechanical Assemblies, *ATT Technical Journal*, Vol 66, 1987, p 51–62
14. *ABAQUS User's Manual*, Version 4.8, Hibbitt, Karlsson Sorenson, Inc., Providence, RI, 1989
15. J.D. Miller and S.N. Burchett, "Some Guidelines for the Mechanical Design of Coaxial Compression Pin Seals," SAND82–0057, Sandia National Laboratories Report, 1987
16. R.S. Chambers, F.P. Gerstle, Jr., and S.L. Monroe, Viscoelastic Effects in a Phosphate Glass-Metal Seal, *J. Am. Ceram. Soc.*, Vol 72, 1989, p 929–932
17. A. DerMarderosian, The Electrochemical Migration of Metals, *Proceedings of the International Society of Hybrid Microelectronics*, 1978, p 134
18. F.J. Grunthaner, T.W. Griswold, and P.J. Clendening, Migratory Gold Resistive Shorts: Chemical Aspects of a Failure Mechanism, *Proceedings of the 13th International Reliability Physics Symposium*, 1975, p 99
19. R. Cooper and W.A. Smith, Electric Breakdown of Sodium Chloride, *Proc. Phys. Soc.*, Vol 78, 1961, p 1734–1742
20. S.N. Koikov and A.N. Tsikin, Dielectric Strength of Thin Layers of Aluminum Oxide, *Sov. Phys. Tech. Phys.*, Vol 1, 1957, p 2180–2185
21. E.K. Beauchamp, Effect of Microstructure on Pulse Electrical Strength of MgO, *J. Am. Ceram. Soc.*, Vol 54, 1971, p 484–487
22. I.S. Kainarskii, E.V. Degtyareva, and I.G. Orlova, Interrelation Between the Electrical and Mechanical Strengths of Corundum Ceramics, *Dokl. Akad. Nauk SSSR*, Vol 157, 1964, p 168–170
23. M. Yoshimura and H.K. Bowen, Electrical Breakdown Strength of Alumina at High Temperature, *J. Am. Ceram. Soc.*, Vol 61, 1978
24. J.D. Cross, The Mechanism of Flashover Across Solid Insulators in Vacuum, *Proceedings of the VII International Symposium Disch. Elect. Insul. Vac.*, 1976, p 24–37
25. H.C. Miller, Improving the Voltage Holdoff Performance of Alumina Insulators in Vacuum by Quasimetallizing or Doping, *Physica*, Vol 104C, 1981, p 183–188
26. B.C. Bunker, G.W. Arnold, and J.A. Wilder, Phosphate Glass Dissolution in Aqueous Solutions, *J. Non-Cryst. Solids*, Vol 64, 1984, p 291–316
27. R.D. Watkins, B.C. Bunker, S.C. Douglas, R.K. Brow, and E.E. Hellstrom, Corrosion Resistant Glasses for Use in Lithium Ambient-Temperature Batteries, *Proceedings of the 22nd Intersociety Energy Conversion Engineering Conference*, 1987, p 1113–1118
28. T.I. Barry, G.S. Shajer, and F.M. Stackpool, "Glass Seals for Sealing Beta-Alumina in Electro-Chemical Cells or other Energy Conversion Devices, Glasses for Use in such Seals and Cells or Other Energy Conversion Devices with such Seals," U.S. patent 4,291,107, Sept 1981
29. D.S. Park, M.W. Breiter, B.S. Dunn, and L. Navias, "Sodium Resistant Sealing Glasses and Sodium-Sulfur Cells Sealed with Said Glasses," U.S. patent 4,341,849, July 1982
30. A.S. Raynes, "Mechanical Properties and Fatigue Behavior of Metallized Glass-Bonded Alumina Ceramics," Ph.D. thesis, Pennsylvania State University, 1989
31. H.-Y. Lu, H.-Y. Lin, and S.-Y. Chen, Autocatalytic Effect and Microstructural Development During Ageing of 3 mol% Y_2O_3-TZP, *Ceram. Internat.*, Vol 13, 1987, p 207–214
32. R.D. Watkins and R.E. Loehman, Interfacial Reactions Between a Complex Lithium Silicate Glass-Ceramic and Iconel 718, *Adv. Ceram. Mater.*, Vol 1, 1986, p 77–80

Metallizing

Jerry Steinberg, Cermalloy Division, Heraeus Inc.

CERAMIC BODIES have many properties that make them important components in complex systems (Ref 1). For example, refractory-type ceramics are useful as structural materials in the metallurgical industry, cements are essential to the architectural and building industries, and various special electrical and magnetic ceramics are essential to the development of computers and other electronic devices. Often, these ceramic-based components must be connected to other parts of a system. This connection can be accomplished by metallizing the ceramic body. The metallization usually connects the ceramic either mechanically or electrically. In a mechanical connection, the metallization acts as an intermediate layer. Another metal, applied usually by soldering or brazing, makes the connection to the other components of the system. An electrical connection relies on the transport of current or electrons. The ceramic can either be a functional device or simply a substrate. If it is a functional device, such as a sensor, then the metallization plays an important role in the final properties of the device (Ref 2).

For the metallization to be useful, it must have high adhesion to the ceramic, and retain its properties when subjected to the same environment as the ceramic. For instance, if the ceramic is to be used as a sensor of exhaust gases in an automobile, then the metallization must not corrode when exposed to these gases. The metallization must also retain its adhesion to the ceramic, even though it is under high stresses induced by thermal cycling and shock as the engine temperature varies.

Before choosing a metallization, each ceramic must be considered individually. Even within a given ceramic type, minor changes in the ceramic formulation can drastically alter its properties and, therefore, the choice of the metallization. Consider ferroelectric ceramics, either strontium or barium titanate (Ref 3), which are commonly used as capacitor dielectrics. Before these electrical components are connected to a printed circuit board, they are first metallized with either a precious or noble metal, such as silver, palladium, platinum, or gold. These electrodes are fired onto the ceramic at temperatures between 750 and 1100 °C (1380 and 2010 °F). The use of a copper electrode requires that the firing

atmosphere have an oxygen content of less than about 10 ppm oxygen. This reducing condition results in a change of the valence states of the titanium and a loss of the ferroelectric properties (Ref 4). If ruthenium is substituted for a small amount of the titanium, these particular ceramics become electrically conductive (Ref 5). They can then be utilized as resistive components. Firing the termination metallization in an oxygen-containing atmosphere results in a large contact resistance and poor electrical properties. In this case, copper is the best type of metallization. The electrical conductivity of the ceramic is not affected when fired in a reducing atmosphere.

Besides the availability of numerous ceramics that require metallization, there are a large number of techniques to apply the metallizations. Three methods can be utilized to apply a coating of tungsten metal to an alumina ceramic: screen printing an ink containing tungsten powder and then firing in an inert atmosphere (Ref 6), chemical vapor deposition of a tungsten-containing gas (Ref 7), and sputtering of a tungsten target (Ref 8).

The metallization technique is chosen on the basis of the final properties it provides, the shape of the ceramic, and economics. For instance, chemical vapor deposition can be used to metallize an oddly shaped ceramic body, whereas screen printing is the obvious choice to metallize a low-cost consumer electronic component.

Methods of Metallizing Ceramics

There are three steps in the formation of a metallic coating onto a ceramic body (Ref 7):

- Synthesis or creation of the depositing species
- Transport from source to substrate
- Film growth and interaction with the substrate

These three steps can be individually varied and controlled, giving a high degree of flexibility to the metallizing process.

The methods that are used to form coatings on various substrates are classified in Table 1, based on the dimensions and amounts of

the depositing species, that is, whether it is atoms/molecules, liquid droplets or solid particulates, or bulk quantities.

In the atomistic deposition process, atoms condense onto a ceramic substrate. A continuous film is formed by nucleation and growth. Atoms can be produced by various methods, such as thermal vaporization, sputtering, vaporized chemical species in a carrier gas, or ionic species in an electrolyte solution. Because the individual atoms do not always achieve their lowest energy states, metallizations formed by atomistic deposition can contain high concentrations of structural imperfections and residual stress. Annealing is sometimes used as a post-treatment to give the coating a chance to come to equilibrium.

All particulate depositions start with finely divided particles of the metal. These powders are either dispersed into some type of organic medium or transformed into the liquid state before they can be applied to the ceramic substrate. A common technique for applying dispersed particles is by screen printing (Ref 9). A continuous film of the metal is formed by using a controlled thermal process to first burn out the organics and then to sinter the particles. Flame spraying is an example of a liquid state being used to apply a metallization of the ceramic. In this process, a powder is melted while being carried by a gas stream to the ceramic substrate. The solidification of the molten metal forms the coating on the ceramic. In both cases, the thermal history of the coating is an important factor in its quality.

Bulk metallization involves the application of large amounts of material in a short period of time (Ref 10). Typically, the metal to be deposited is a powder suspended in an organic vehicle. The ceramic can be coated by techniques such as dipping, spraying, brushing, or spinning. These particular techniques of application can be used for ceramic objects that have complex shapes.

The choice of metallization technique to use in a particular application is not always clear. The overriding factor in the decision is the need to obtain a particular final property. For instance, if the ceramic is used in an electronic application where the metallized pattern requires high definition, then sputtering must be used rather than screen printing.

Table 1 Major ceramic metallization techniques

Atomistic	Particulate	Bulk
Plating	**Organic medium**	**Organic medium**
Electrolytic	Screen printing	Brushing
Electroless	Brazing	Roller
Evaporation	**Powder**	Dipping
Vacuum	Flame spraying	Spin
Flash		**Foil**
Electron		Cladding
beam		
Plasma		
Sputtering		
Vapor		
Chemical		
deposition		

Often, the final properties do not influence the choice of metallization technique. Consider a disk capacitor, which is a cylindrical ceramic dielectric body. To attach this capacitor onto a printed circuit board, leads are soldered to the metallized body. The metallization must provide good adhesion to the ceramic, solderability, and minimal degradation to the electrical characteristics of the capacitor. The metallization typically used for this application is silver, which can be deposited by either screen printing, spraying, sputtering, or even dipping. In this case, choice of technique is generally made on economics, available equipment, or technical expertise.

General selection criteria are:

- Form and purity of metal to be deposited
- Substrate limitations, such as ceramic shape, size, and maximum operating temperature
- Chemistry concerns, such as whether the ceramic forms compounds with certain metal oxides or is easily wet by glass
- Economics, in terms of how many parts must be produced (rate of deposition) and availability of apparatus and engineering expertise
- Final properties, based on type of and end-use environment, and the need for soldering, brazing, or other attachment to the metallization

Various Metallization Techniques. One application in which the choice of deposition techniques must be carefully considered, because of the complexity of the final product, is the production of electrical resistors. Resistive components can be manufactured by depositing a metal or metal alloy onto a ceramic substrate. One metal composition is an alloy of 80 wt% nickel and 20 wt% chromium (Ref 11). The deposition technique requires special attention because a multicomponent alloy is being deposited and the metallization must meet stringent electrical properties.

These resistors can be either atomistically or particulately deposited. Atomistic deposition is referred to as a thin-film process, whereas particulate deposition is a thick-film

process. For this alloy, thin-film resistors are the most common and give the best properties.

The most popular method of thin-film deposition is vacuum evaporation, in which the charge or material to be deposited in a vacuum is heated (Fig 1a). Vapors condense onto the cooler ceramic substrate. For this application, the charge is a multicomponent metal alloy, in which both constituents have different vapor pressures. At 1300 °C (2370 °F), chromium evaporates eight times faster than nickel. For this reason, a controlled metallization composition cannot be achieved by vacuum evaporation.

To overcome the problems that are due to the different vapor pressures of the nickel and chromium, two other methods of evaporation have been attempted: flash and electron beam. In the flash method, a stream of finely divided powder of the alloy composition is fed onto a preheated hot strip. The evaporation takes place in a short time. This method has been

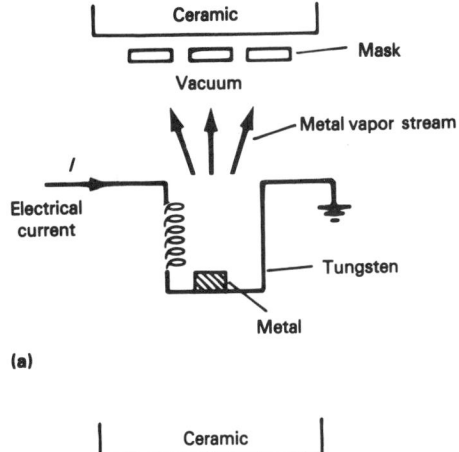

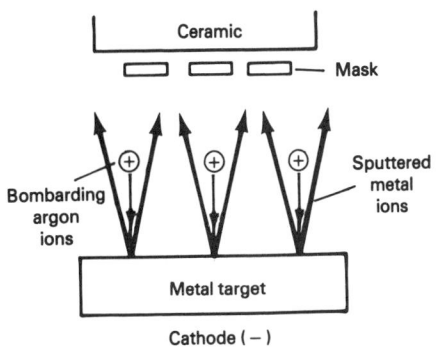

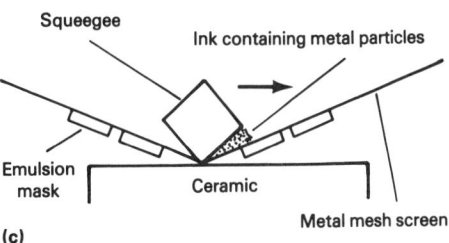

Fig 1 Common methods of metallizing ceramics. (a) Vacuum evaporation. (b) Sputtering. (c) Screen printing

reported to give metallizations of the appropriate metallurgical composition, but of diminished electrical quality. Flash evaporation is also difficult to implement in a production environment. Electron beam evaporation tries to overcome the large differences in evaporation rates of the two metals by supplying energy from an intense beam of electrons. This technique could not give reproducible resistance values.

To produce Ni-Cr resistive thin films consistently and efficiently, sputtering is the method of choice. Sputtering is the process by which particles from the top surface of a target are stripped off by bombardment of energetic positive ions. These ions are usually an inert gas, such as argon (Fig 1b). Sputtered films usually have higher adhesion than evaporated films because the particles have higher energy. Because the energy to produce the coating material is generated through momentum transfer (mechanical) rather than chemical or thermal (evaporation) processes, virtually any material, including alloys, can be a coating candidate. Reproducible Ni-Cr resistive films are made by this deposition technique.

The thick-film technique for making Ni-Cr resistors uses a screen printing process, in which an ink is transferred through a fine-mesh wire screen that is coated with a patterned emulsion (Fig 1c). The transfer of the ink is accomplished by the hydraulic pressure that is generated by a squeeze moving the ink across the screen (Ref 12).

The choice of using the thin- or thick-film technique to produce Ni-Cr metallizations depends on the demands of the final application. The thick-film technique is lower cost, but does not yield resistors with a quality equivalent to thin-film deposition. The range of resistance values in the thick-film process is from 1 to 3 Ω/square, and the thermal coefficient of resistance (TCR) is only about 200 ppm (Ref 13). For the thin-film case, these values are 10 to 400 Ω/square and less than 100 ppm. Since the thick-film resistors are made by the particulate deposition technique, the resistor is only formed after the applied ink is heat treated, which burns off the organics and sinters the nickel-chromium alloy powder. Other particulate components, such as glass powders, are added to the ink to give adhesion to the alumina substrate. These components affect the electrical properties of the fired resistor.

Metallization Adhesion

Regardless of the metallization end use, it must adhere to the ceramic body under all conditions. Adhesion is a macroscopic property that depends on the bonding across the interface of the two materials to be joined. A metallization with good adhesion can be defined as one that does not fail under assembly or service conditions nor at acceptably low stress levels under test conditions. The morphological and chemical properties of the interfacial reaction determine the adhesion, or mechanical integrity, of the metallization.

Interfacial regions can be classified as mechanical, compound, or chemical. Usually, the interface is not of a single type, but a combination. The type of interfacial region formed depends on the deposition process and the materials to be bonded. Figure 2 shows a metal-ceramic interface.

Mechanical Adhesion. A mechanical interface is characterized by interlocking of the metallic film with the ceramic body. The ceramic can be deliberately roughened or made porous to increase the mechanical interlocking. Chemical etching or abrasion are common techniques. The adhesion depends greatly on the surface preparation of the ceramic body. Electroless and electrolytical plating deposition techniques generally rely on this type of adhesion mechanism.

Although bonding by a mechanical formed interface can give high adhesion, it can be difficult to control. Changes in ceramic surface characteristics and grain size can be caused by small variations in the particle sizes of the ceramic powder or the thermal processing techniques. The etching time will then vary for maximum adhesion. The proper etchant must be experimentally determined for each type of ceramic.

An example of a mechanical bond is the metallization of a lead zirconate titanate (PZT) ceramic body by electroless deposition of nickel (Ref 14). To maximize the interfacial adhesion, the ceramic grain boundaries are etched in a HBF_4/HNO_3 solution. As the etching time increases, so does the adhesive strength and weight loss of the PZT. The plated nickel is anchored deeper into the grain boundaries. Eventually, a point is reached where the adhesion begins to decrease. This is where the ceramic grains are undercut by the etchant and nickel completely encapsulates the ceramic grains.

Compound Adhesion. A compound interface is formed by the chemical interaction of the metallization and the ceramic. The compound that is formed is a mixture of glass and crystalline phases.

For the best adhesion, the properties of the compound and the processing conditions to form the compound must be carefully considered. The compound should not be brittle, and it should contract less than the ceramic body during cooling and easily wet the particles of both the sintered metallization and the ceramic body. The amount and particle sizes of the metal powders and additives are critical. The processing parameters, such as the firing profile and atmosphere conditions, are also important. The metallization must be fully sintered and there must be sufficient compound formation at the interface. Figure 3 shows some of the conditions that must be avoided to maximize adhesion.

Examples. Each ceramic type requires special consideration if the metallization is to be bonded via compound formation. For every type of ceramic, there can be many different routes to form a compound bond. In the case of alumina, a common substrate material used in electronics, bonding can include brazing, direct-bond, and thick-film methods.

Brazing is a process in which a molten metal contacts the ceramic. To braze a silver metal-

lization onto alumina, an alloy consisting of Ag-Cu-Ti is commonly used (Ref 15). Compounds such as $(Ti, Al)_4Cu_2O$ have been identified as being responsible for the adhesive bond.

Direct-bond copper technology is a method of bonding copper to ceramic substrate materials using copper oxide as a chemical bonding mechanism (Ref 16). In the copper-oxygen phase diagram, a eutectic exists at 0.39% oxygen that is liquid at 1065 °C (1950 °F). This is 18 °C (32 °F) lower than the melting point of copper. By preoxidizing a copper foil, bonding can take place at 1065 °C (1950 °F) via wetting of the eutectic mixture on the alumina substrate. At the interface between the metallization and the alumina ceramic, a compound bond consisting of Cu_2O-Cu-$CuAlO_2$ forms.

In the thick-film process, a mixture of metal powders and additives dispersed in an organic vehicle can either be post-fired onto an alumina substrate or cofired with the green alumina body (Ref 9, 17). For the latter, various types of metallization, such as gold, platinum, silver, palladium, copper, and combinations thereof are commonly found. The firing temperature usually ranges from 850 to 1000 °C (1560 to 1830 °F). A compound bond is formed via oxide additives that react with alumina or any other components of the substrate. Typical oxides used are copper and cadmium (Ref 17). At a firing temperature of 850 °C (1560 °F), the kinetics of the reaction to form a chemical bond are slow. A combination of glass and oxides helps the compound formation. The glass helps by dissolving and transporting the oxides to the substrate, thus effectively increasing the kinetics of the reaction. Instead of a glass, a low-melting point oxide can be used as the flux. Typical glasses or fluxes are lead borosilicates or bismuth oxide (Ref 18). Compound bonds for postfired thick films can be classified as fritted, where only a glass additive is used; fritless, where only oxides are used; and mixed bonded, where both glasses and oxides are used (Ref 19).

When cofiring the metallization with the alumina, only refractory metals can be used because of the high temperatures (greater than 1400 °C, or 2550 °F) needed to sinter the alumina. Molybdenum and tungsten are good candidates because they have the lowest expansion and highest electrical conductivity of the refractory metals. To prevent oxidation, firing must be done in a wet H_2/N_2 atmosphere. To form an adhesive bond, it is advantageous to add a mixture of glass-forming and glass-modifying oxides to the refractory metal phase. For molybdenum, typical additions are SiO_2 and MnO (Ref 20). The MnO forms a spinel phase, $MnAl_2O_4$, with the alumina ceramic. The addition of SiO_2 results in a glassy phase of $MnO/Al_2O_3/SiO_2$. The glassy phase not only provides some mechanical bonding but also helps wetting of the refractory metal and increases the kinetics

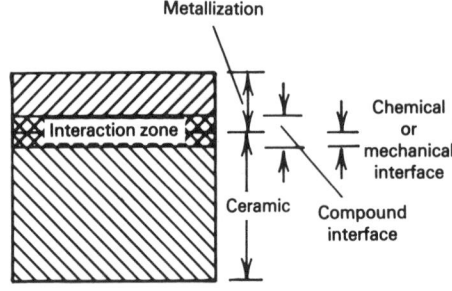

Fig 2 Ceramic metallization schematic showing interaction zone extending into ceramic and metal

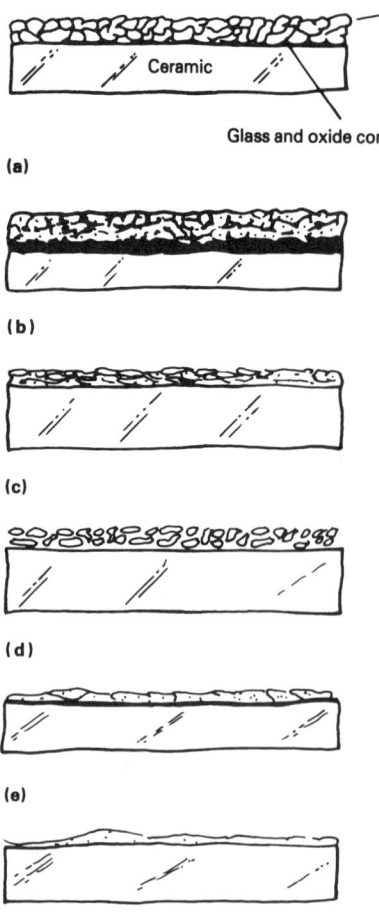

Fig 3 Conditions to avoid to maximize adhesion in a compound bond. (a) Too good wetting of the metal by the glassy phase, causing poor sintering. (b) Too much glass at the interface. (c) Inadequate amount of compound formation. (d) Inadequate sintering of the metallization. (e) Oversintering. (f) Lack of melt phase. Source: Ref 6

of the spinel formation. Both the spinel phase and the glassy phase are necessary for good adhesion. A typical metallization mixture is 80%Mo/15%MnO/5%SiO$_2$ by weight.

Chemical Adhesion. A chemical interface is formed by atom-to-atom bonding either by van der Waals or ionic attraction. This type of bonding usually is associated with atomistic deposition. During the deposition process, sufficient energy must be supplied to the depositing atoms in order to form a chemical bond.

Example. Electronic circuitry that is expected to operate in the microwave frequency range requires special properties to minimize the propagation losses and delay times. To meet these requirements, the metallization must have high electrical conductivity and be capable of being deposited with precise dimensions. Because the electrical signal propagates only in a thin top "skin" layer of the metallization, long-term inhibition of surface corrosion or oxidation is also important. The best choice for microwave circuitry is a gold metallization sputtered onto an alumina substrate. However, sputtering of gold onto alumina only provides a mechanical bond. Adhesion is not adequate for many applications. Therefore, an intermediate metallization is sputtered prior to gold metallizing. This adhesive layer can be chromium or titanium tungsten. Either of these metals can form chemical bonds with the alumina substrate (Ref 8).

REFERENCES

1. W.D. Kingery, H.K. Bowen, and D.R. Uhlmann, *Introduction to Ceramics*, 2nd ed., John Wiley & Sons, 1987
2. M. Taguchi, Applications of High Technology Ceramics on Japanese Automobiles, *Adv. Ceram. Mat.*, Vol 2 (No. 4), 1987, p 754
3. Ceramic Chip Capacitors, Technical brochure, Novacap Co.
4. I. Burn, MN doped Polycrystalline BaTiO$_3$, *J. Mater. Sci.*, Vol 14, 1979, p 2453
5. J. Steinberg, "Nitrogen Fireable Resistor Compositions," U.S. patent 4,814,107
6. L. Reed, "Ceramic to Metal Sealing," Special Publication No. 3, American Ceramic Society, 1969
7. R.F. Bunshah, Ed., *Deposition Technologies for Films and Coatings*, Noyes Publishers, 1982
8. T.S. Laverghetta, *Microwave Materials and Fabrication Techniques*, Artech House Inc., 1984, p 21
9. P.J. Holmes and R.G. Loasby, *Handbook of Thick Film Technology*, Electrochemical Publishers Ltd., Scotland, United Kingdom, 1976
10. T.C. Patton, *Paint Flow and Pigment Dispersion*, 2nd ed., John Wiley & Sons, 1979, p 551
11. K. Ramachandian and E. Giani, NiCr Thin Film Resistors, *Proceedings of the International Society for Hybrid Microelectronics*, 1981, p 269
12. D.E. Riemer, Ink Hydrodynamics of Screen Printing, *Proceedings of the International Society for Hybrid Microelectronics*, 1985, p 52
13. Technical brochure R8501.1, Heraeus Inc., Cermalloy Div.
14. C.E. Baumgarten, Adhesion of Electrolessly Deposited Nickel on Lead Zirconate Titanate Ceramic, *J. Am. Ceram. Soc.*, Vol 72 (No. 6), p 890
15. A.H. Carim, Transitional Phases at Ceramic Metal Interfaces: Orthorhombic, Cubic and Hexagonal Ti-Si-Cu-N Compounds, *J. Am. Ceram. Soc.*, Vol 73 (No. 9), p 2764
16. J.F. Dickinson, Direct Bond Copper Technology—Materials, Methods and Applications, *Proceedings of the International Society for Hybrid Microelectronics*, 1982, p 103
17. J.R. Blum and W.R. Cannon, Ed., *Advances in Ceramics*, Vol 19, American Ceramic Society, 1986
18. K.R. Bube and T.T. Hitch, "Basic Adhesion Mechanisms in Thick and Thin Films," Final Report, Contract No. N00019–77–0176, Naval Air Systems Command, 1978
19. S. Kadota and K. Shibata, Gold Conductor Pastes for High Density Circuits, *Proceedings of the International Microelectronics Conference*, 1980, p 127

SECTION 8

Testing, Characterization, and NDE

Chairman: H. David Leigh, Clemson University

Introduction

THIS SECTION is concerned with the testing, characterization, and nondestructive evaluation of advanced ceramic materials. The scope of this Section is confined to tests used to evaluate finished or formed ceramic materials. It should be recognized that the performance of ceramic materials is greatly influenced by the characteristics of the starting materials used and the processing of the ceramic. Tests used for the characterization of raw materials and the evaluation of process consistency are covered elsewhere in this Handbook. The tests and procedures discussed in this Section should provide the user with an understanding of the strengths and weaknesses of various tests. The details of special precautions or sample preparation techniques are discussed and the limitations of each method are covered. It is the aim of this Section to allow a more complete understanding and appreciation of the problems associated with interpretation of physical property data obtained on advanced ceramic materials.

For the purposes of this introduction, properties or characteristics may be divided into those which are intrinsic or inherent to the ceramic by nature and those which are extrinsic in nature, that is, related to a performance parameter. The purpose of testing is to provide a basis to evaluate materials properties, which in turn implies that predictions about performance may be precisely made. The understanding of the relationship between the inherent or intrinsic nature of the ceramic to the extrinsic or performance characteristics of the ceramic is still somewhat simplistic and inadequate. The state-of-the-art of testing ceramics of advanced engineering applications does not at present allow predictions of performance to be made with high precision. For example, the grain size distribution of ceramic microstructures can be determined with high precision. In turn it is known that strength of the ceramic will increase as the grain size decreases. It is not possible, however, to predict the change in strength that will occur in a ceramic with high precision by simple comparisons of average grain size. One of the purposes of this Section is to apprise users of ceramic materials of these limitations and provide necessary cautions against gross assumptions and extrapolations of extrinsic properties.

Many of the tests which are used to evaluate advanced ceramic products have their genesis from tests developed for traditional ceramic products such as refractories, whitewares, and structural clay products. Advanced ceramic materials are generally produced from chemically synthesized raw materials which are much finer in size and higher in purity than materials used for the production of traditional ceramic materials. The naturally occurring flaw size in advanced ceramics can therefore be much smaller than those found in traditional ceramics. As a consequence, handling and sample preparation are far more important to the observed property data obtained on advanced ceramics than traditional ceramics.

Due to the diverse and specialized nature of applications for advanced ceramic materials, a multitude of specific tests have been developed by workers in the field. This fact makes the development of meaningful comparisons among laboratories and existing experimental data, at the very least, difficult. The number of different tests used to evaluate ceramic materials can lead to different property data for identical materials. In addition to the influence that test methodology has on the property data that are obtained, minor changes in the micro- and macrochemical and microstructural nature of the ceramic can have an extensive influence on the resultant properties. These factors have frustrated workers trying to develop ceramics for sophisticated and critical applications.

Compromises must be made when selecting between tests which, on the one hand, offer speed and economy versus those tests which offer the attainment of a more rigorously "correct" property value. It is now becoming recognized that certain tests, for example, bend bar modulus of rupture tests which are used to determine strength of ceramic materials, are not adequate for design purposes. In the area of mechanical property testing of ceramic materials, it does not seem possible at this time to define the "best method" for particular property measurement. Due to these considerations, agreement upon standardized tests has been slow, but standardization of techniques is essential if ceramic applications are to be developed. These problems are recognized by the community of ceramic engineers and concerted efforts are now being mounted internationally and domestically to standardize test methodology to eliminate variations caused by variations in test procedures. This Section will discuss the strengths and weaknesses of methods which are, or appear to be, likely standard test procedures.

There are a large number of tests which have been developed over a number of years which are adequate for the evaluation of intrinsic characteristics of ceramic materials. Some examples are chemical analysis, phase analysis, microstructural analysis, and thermomechanical property evaluations. There are techniques and precautions which must be employed to obtain reproducible and reliable results of these inherent or intrinsic properties. In some cases, close attention must be paid to the homogeneity of specimens which requires special techniques to detect and evaluate characteristics on a microscopic scale. This is especially true for composite ceramic materials which contain dispersed reinforcing, and exsolved or separated phases. Examples of these are the silicon carbide whisker-reinforced alumina and partially-stabilized zirconia ceramics.

Although ceramics are by their nature resistant to change and can withstand relatively high temperatures and aggressive environments, they are subject to alteration during use. This is especially true of non-oxide based ceramics such as carbides and nitrides. These materials are prone to oxidation effects. Although the methodologies for the use of proof testing, high temperature and environmental testing are still being developed, detailed information on such test procedures is included in this Section.

Chemical Analysis

J.E. Enrique, E. Ochandio, and M.F. Gazulla, Instituto de Tecnologia Ceramica, University of Valencia and Asociación de Investigación de Industrias Cerámicas (Spain)

CHEMICAL ANALYSIS represents one of the most important tests for obtaining and maintaining a high level of quality in the production cycle. For many years, traditional wet chemical methods of analysis (Ref 1, 2) were the most common, precise, and often the only methods available for obtaining the chemical composition of ceramic materials.

The technological development of instrumental methods of analysis has been rapid since the early 1980s (Ref 3–5). These methods have virtually replaced traditional methods and have made it possible to routinely undertake some determinations that could not be attempted previously. To accommodate the wide range of analytes and different matrices, the analytical techniques used must be capable of providing both routine and nonroutine analysis.

The need to know more about the chemical composition of ceramic and glass materials is motivated by the need to:

- Consolidate industrial quality control at the heart of enterprises, especially in the areas of raw materials and finished products
- Improve the physical-chemical characteristics of the finished product and continue research on new bodies and glazes
- Develop new materials
- Search for the origin of defects, such as impurities, or problems, such as lack of homogeneity, in the intermediate or finished product
- Establish the possibilities of total process automation (as in the cement industry)
- Acquire knowledge of the competitor's products

Types of Materials

Considering the different raw materials involved in the ceramic process, such as clays, borates, zircon sand, feldspars, and silica sand (Ref 6), and the variety of manufactured products, such as ceramic bodies, refractories, glasses, frits, tiles, enamels, and bricks (Ref 7–9), there is a wide range of chemical compositions that can exist.

The oxide constituents of raw materials and bodies are listed in Tables 1 and 2, respectively, whereas frits and glazes are classified by their chemical compositions in Table 3.

Analytical Process Steps

The chemical composition of a material is defined by the quality and quantity of its constituent chemical elements present. These elements can be categorized according to their quantity, that is, major, minor, and trace amounts (Ref 4, 10). A simple scheme for delineating these categories is: major $> 1\%$ $>$ minor $> 0.1\%$ $>$ trace.

In terms of the chemical composition of ceramic and glaze materials, major and minor elements are usually expressed as a mass percentage of their oxides, whereas trace elements are usually expressed as parts per million.

Unlike many fields of analysis, in which the percentage of a given element is recorded, ceramic analysis reports the percentage of the oxide. Rather than being an empirical convenience, this convention represents a truer picture of the constitution of the material. Although the ceramic analyst may conduct the normal complete analysis, he should be aware that it ignores the possible presence of many other constituents in very small amounts, because their presence is of little consequence in practice (Ref 6). This is an important distinction between ceramic analysis and geological analysis, in which the presence of trace elements sometimes may be of major importance.

An analytical process generally consists of five steps:

- Problem identification and method selection
- Sampling
- Sample preparation according to selected method
- Measurement (including calibration graph)
- Evaluation of data and report preparation

Problem Identification and Method Selection. All analytical processes start with the presence of a problem that must be solved. In many cases, the history of the problem is so important that the method is chosen on the basis of:

- Time required for the analysis
- Precision and accuracy
- Number of analytes to be determined in one sample
- Diversity of the samples
- Sample size and concentration ranges to be covered
- Cost of analysis
- Analytical techniques available in the laboratory

When developing or choosing an analytical method, the selectivity or specificity achievable with that method should be emphasized. Selectivity or specificity refers to the ability of the method to determine the analyte without interference from the other components present in the analyte matrix.

Because it is relatively rare to find a truly specific method for the analysis of all the elements of interest, it is often necessary to combine several methods or instrumental techniques.

Sampling is one of the most important steps in the analytical process. In modern terminology, the general term "sample" has been replaced by the more precise term "representative sample." The sample must be of the same composition as the material from which it originated, which means that the portion of material selected possesses the essential characteristics of the bulk.

Sampling can be a very complex and difficult task (Ref 11, 12). It cannot be known whether the portion of material that is eventually analyzed truly represents the bulk of material from which it was taken. Even the method that is chosen to take the sample is the one that only appears to give the correct results. Sampling techniques are described in Ref 13 to 15, rather than herein, because the work of an analyst usually begins at a later stage in the process. It is sufficient to point out here that if the sample is not truly representative of the bulk, subsequent work will be a waste of time.

Sample preparation is required by most analytical techniques in order to achieve maximal quality with minimal work. With ceramic and glaze materials, solids should never be considered to be homogeneous until

Table 1 Typical oxide compositions of raw materials, wt%

Oxide	Silica sand	Clay	Feldspar	Borate	Talc	Zircon	Calcite	Dolomite
SiO_2	85–99	40–65	60–75	1–10	50–65	30–35	0–1	0–1
Al_2O_3	0–10	10–35	10–20	0–5	0–3	0–2	0–1	0–1
B_2O_3	<0.1	<0.1	<0.1	20–40	<0.1	<0.1	<0.1	<0.1
Fe_2O_3	0–1	1–10	0–2	0–1	0–3	<0.5	<0.5	<0.5
Na_2O	0–1	0–1	0–10	0–20	0–1	<0.2	0–1	0–1
K_2O	0–5	0–5	0–15	0–1	0–5	<0.5	0–1	0–1
CaO	0–5	0–15	0–3	0–25	0–5	<0.5	50–55	25–35
MgO	0–1	0–5	0–2	0–5	20–30	<0.5	0–5	15–25
TiO_2	0–1	0–4	0–2	0–1	0–1	<0.5	0–1	0–1
ZrO_2	<0.1	<0.1	<0.1	<0.1	<0.1	60–65	<0.1	<0.1
Other	0–1	0–1	0–1	0–1	0–1	0–1	0–1	0–1

Table 3 Typical oxide compositions of frits and glazes, wt%

Oxide	Transparent	Opaque	Lead	Borax	Mat(b)
SiO_2	50–60	50–60	30–40	10–40	50–60
Al_2O_3	4–10	4–10	0–5	0–5	5–10
B_2O_3	5–15	5–10	0–20	20–40	5–10
Na_2O	0–10	0–5	0–10	0–5	0–5
K_2O	0–10	0–5	0–5	0–5	0–5
CaO	0–10	0–15	0–5	0–5	5–20
PbO	0–10	0–5	30–60	20–50	0–5
ZnO	0–10	0–10	0–10	0–10	5–20
ZrO_2	0–3	5–15	0–5	0–5	0–5
Other(a)	0–10	0–10	0–10	0–10	0–10

(a) SrO, Fe_2O_3, SnO_2, Sb_2O_3, TiO_2, Li_2O, and others. (b) Met glazes or frits.

they are ground. Dry grinding of the sample is necessary for most chemical analyses, and a method that will not contaminate the sample with the elements of interest should be used. The degree of contamination will depend on both the medium used to reduce the particle size and the hardness of the material being ground.

It is important to realize that errors related to grinding cannot be eliminated completely, but can be reduced to a tolerable level if sufficient care is taken. The laboratory itself must be clean and free from sources of contamination.

After being ground, the materials should normally be dried at 110 °C (230 °F) for at least 2 h and then cooled in a desiccator, using a suitable drying agent. From this stage, sample preparation becomes specific to the analytical technique chosen, and is therefore discussed more fully in the section "Major Analytical Techniques" in this article.

Measurements. Nearly all modern instrumental analytical systems are indirect methods that are based on the measurement of a characteristic effect of a given element, the concentration of which is proportionate to the intensity of the effect itself, all other conditions being equal. Indirect methods are comparative, that is, they require calibration with standard substances, and are used to evaluate or quantify the concentration of unknown samples.

Correct operating conditions are, of course, a prerequisite in any analytical scheme. Errors that are due to the improper selection of operating conditions have been drastically reduced by automatic and semiautomatic equipment. Like sample preparation, operating conditions are specific for each analytical technique, as well as for the possible interferences.

For a single analyte, many interferences can be avoided by the correct choice of standards and conditions for a particular analysis.

Quantification Methods. Quantitative analysis almost always involves more than a straightforward comparison between samples and standards. Quantification procedures for analytical instrumental techniques have been considerably simplified since the advent of computers. Computing facilities enable the use of regression programs of various degrees of complexity in order to correlate the reported chemical analysis values of standards with their measured intensities.

Because the majority of instrumental methods are comparative, the obtained results can be no better than the quality of the standard substances used and the care taken in producing the calibration. The choice of standards should be such that, between them, they cover the expected concentration ranges in the sample batch. The use of certified reference materials must be considered in order to attain maximum quality.

A material or substance with one or more properties that have been sufficiently established can be called a reference material (Ref 16, 17). When the value of that property has been certified by a technically valid procedure and the sample of the material is accompanied by a certificate or other documentation, the material then can be defined as a certified reference material.

The certified value assigned to the material is often based on results obtained from a number of laboratories, at which different methods of analysis have been applied. The certified values represent the mean of the accepted results. Reference materials are widely used in analytical laboratories to calibrate analytical instruments, to validate ana-

lytical methods, and enable the transfer of analytical results between one laboratory and another (Ref 18–20).

Before analytical results can be transferred among laboratories with confidence, an analyst must be prepared to conduct a check on the analytical method via an independent standard. An adequate certified reference material can serve to justify reliability, both of the method used and the results supplied. These certified reference materials are available from the sources listed in Table 4.

If suitable standards are not available in cases where there is a need to cover a greater range of one or more elements, then the main approaches to take are:

- Use pure chemicals spiked into a suitable geological matrix
- Mix proportions of well-determined geochemical reference standards (this is fast and generally works well)
- Create a secondary standard that is either analyzed by different methods or has been very well analyzed many times

A synthetic standard is one that has been formulated from known quantities of constituents (pure chemical or reference materials) in order to represent as closely as possible the real samples to be analyzed. The disadvantages of using synthetic samples are the time necessary to prepare them and the loss of accuracy of the synthetic standard obtained.

Major Analytical Techniques

There is an extremely broad range of modern laboratory methods and instruments. The most common analytical techniques are described below.

X-ray fluorescence spectrometry (XRFS) is an instrumental technique of elemental analysis that assesses the presence and concentration of various elements by measuring the secondary radiation (fluorescence) emitted by a material that has been excited by an x-ray source (Ref 21–23). The selection of the radiation composing the emission spectrum can be obtained through either energy dispersion spectrometry (EDS) or wave-

Table 2 Typical oxide compositions of various ceramic bodies, wt%

Oxide	Earthenware	Porcelain	Stoneware	Brick	Refractory
SiO_2	30–60	40–60	45–65	45–65	20–60
Al_2O_3	5–20	15–30	10–20	10–20	25–90
Na_2O	0–1	0–1	0–1	0–1	0–2
K_2O	0–1	0–1	0–1	0–2	0–2
CaO	0–20	0–2	0–2	0–10	0–5
MgO	0–10	0–1	0–1	0–2	0–3
Fe_2O_3	0–1	0–1	0–1	0–5	0–2
TiO_2	0–1	0–2	0–2	0–3	0–3
Other	0–1	0–1	0–1	0–1	0–3

Table 4 Sources of certified reference materials

U.S. Department of Commerce, National Institute of Standards and Technology
   Office of Standard Reference Materials
   Washington, D.C. 20234

Bureau of Analysis Samples Ltd.
   Newham Hall, Newby
   Middlesbrough, Cleveland, UK TS8 9EA

Canadian Certified Reference Materials
   55 Booth St.
   Ottawa, Ontario
   Canada K1A 0G1

MBH Analytical Ltd
   Holland House, Queens Road
   Barnet, Herfordshire
   England EN5 4DJ

Community Bureau of Reference
   Commission of the European Community
   Rue de la Loi, 200
   Brussels, Belgium B-1049

Geostandards
   15 Rue, Notre-Dame des Pauvres, B.P.20
   54501 Vandoeuvre, lès-Nancy Cedex, France

CONOCO Specialty Products, Inc., CONOSTAN Division
   P.O. Box 1267
   Poca City, OK 74603

Helvent A.G.
   Sonnenbergstrasse 24
   8600 Dubendorf Switzerland

Brammer Standards Co.
   14603 Benfer Rd.
   Houston, TX 77069

Analytical Standards
   PL 2366
   S-43400 Kungsbaka Sweden

V. Kocman
   Domtar Research Centre
   P.O. Box 300
   Senneville, Quebec
   Canada H9X 3L7

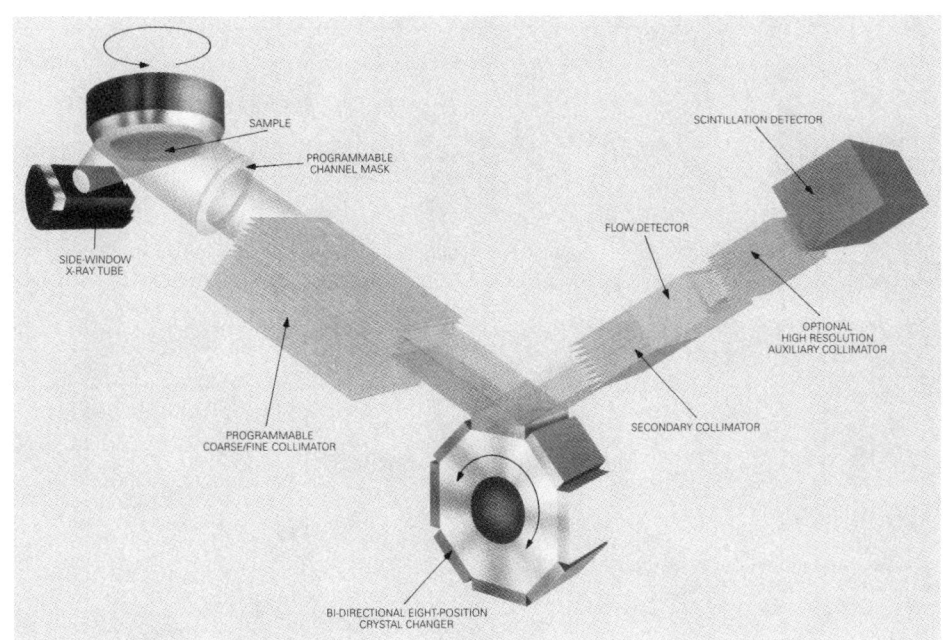

Fig 1 Typical wavelength-dispersion XRFS configuration

length-dispersion XRFS (WXRFS). Both EDS and WXRFS are two of the few techniques where analytical measurements are made directly on a solid matrix.

The EDS technique is normally used in tandem with the electronic microscope in the microanalysis of samples. The accuracy of this analytical technique is considerably less than the accuracy of WXRFS.

The characteristics of WXRFS are described in detail because it has become the method of choice for quality control (and for many other analytical purposes) in a wide variety of industries (Fig 1). The technique has also found considerable utility in research laboratories since the development of mathematical approaches for the conversion of x-ray intensity to a composition that does not require a large library of type standards.

One of the biggest advantages of XRFS is the automation of the spectrometer by a computer, which selects the instrumental conditions for the measurements, gathers the measurement data, and processes the data into calibration graphs or routine results. Another

important advantage is its ability to scan a sample and make a qualitative or semiquantitative estimation of any unusual element or even any usual element that is present in particularly large or small concentrations (Fig 2).

The ideal sample preparation for presentation to the x-ray beam is flat, homogeneous, infinitely thick with respect to the x-rays, and capable of withstanding a vacuum. Two basic types of sample preparation methods are used in XRFS analysis, depending on the precision, type of material, sample size, possible matrix effects of the elements that accompany the analyte in the sample, and other factors. One method utilizes fused beads, in which the sample powder is mixed with a suitable flux, fused into a glass, and is either cast or pressed into a disk. The other method utilizes pressed powder pellets, in which the sample powder (with or without a binder) is compressed to produce a solid disk of powder. Useful binding agents are polyvinyl alcohol, urea, cellulose, and boric acid, which are added in the minimum quantities possible to avoid sample dilution. Samples and standards must be pressed at the same pressure.

The advantages of using the fusion procedure to prepare samples for XRFS analysis, rather than pressed powder pellets, are:

● Both the mineralogical and particle size effects are totally eliminated
● Standards of almost any composition can be prepared from mixtures of pure oxides (Ref 9)
● The matrix effects can be reduced considerably
● Procedure is easy to use, rapid, and offers good reproducibility
● Cross-sample contamination is minimal

The fusion procedure that was first described in a 1957 paper (Ref 24) has not changed in its essential details (Ref 25, 26). The procedure essentially involves heating the fusion mixture to be fused at a temperature that melts the flux and dissolves the samples, agitating the melt to homogenize it, then pouring the molten glass into a mold, and cooling to obtain a homogeneous vitreous disk.

The fusion mixture consists of a powdered sample and a sufficient amount of flux. Other materials, such as nonwetting agents, are sometimes added to facilitate fusion and the subsequent casting process. To avoid introducing an extra variable, the nonwetting agent must be added in similar amounts to the samples, as well as to the standards. Several methods of fusion are discussed in detail by Hutchinson (Ref 27).

The flux is required to dissolve the sample to get a solid and homogeneous disk. A wide variety of fluxes can be used, but the most classical are lithium tetraborate, natrium (or sodium) tetraborate, and lithium metaborate. The flux also has to be free from spectral interferences.

In XRFS analysis, the flux-to-sample ratio in fused disks has been varied between 2:1 and 100:1 (Ref 28). The chosen ratio normally depends on the constituents, the nature of the sample, the interest in trace elements, and other factors.

Operating Conditions. The instrumental parameters for each analyzable element in XRFS are the crystal analyzer, 2θ angles (analyte peak and background), collimator, and detector.

The values of 2θ angles should be chosen so that peaks and backgrounds are free from overlap from any other spectral line.

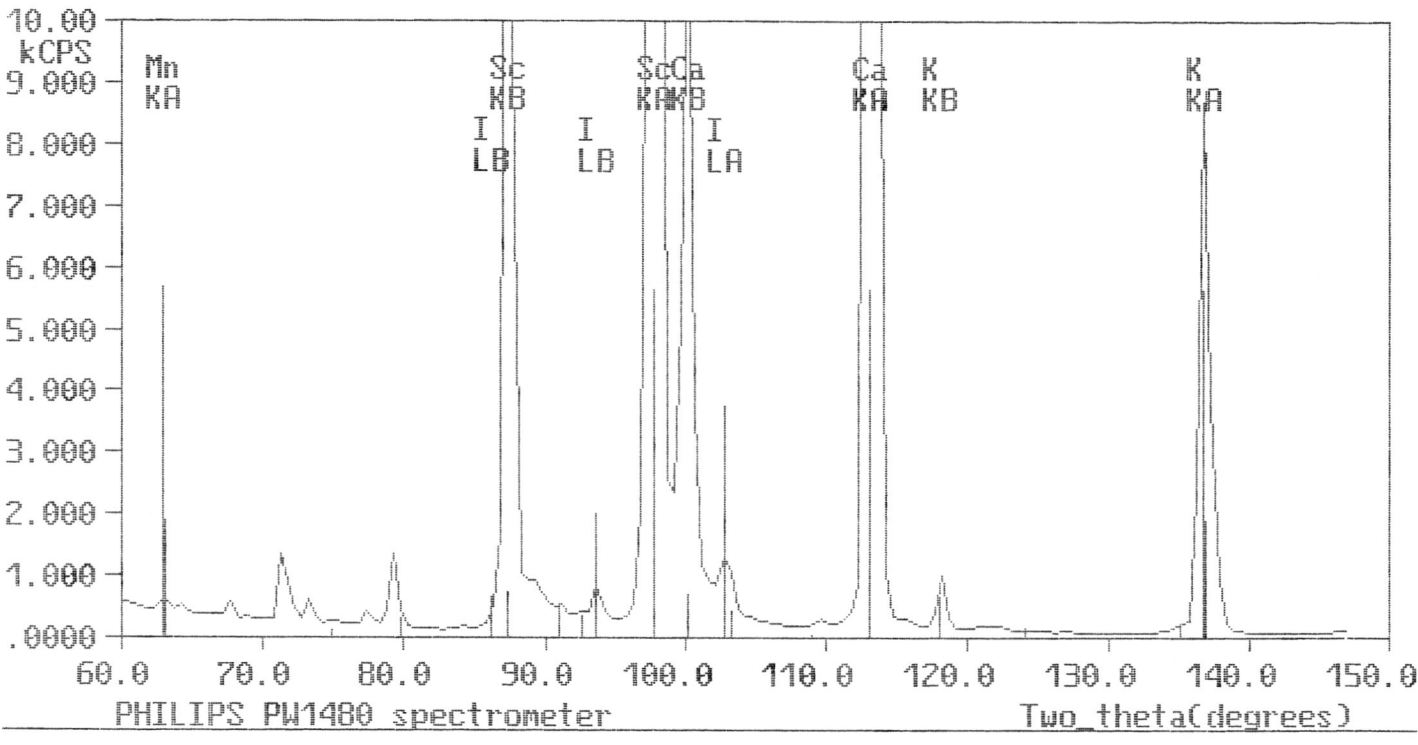

Fig 2 Qualitative scan of sample using XRFS. kCPS, kilocounts per second, or counts per second × 10³

Corrections for overlap can be made by means of factors introduced in the software of the computer that controls the spectrometer.

The effect of a spectral interfering element will be to increase the apparent concentration of the analyte by an amount that is linearly related to the concentration of the interfering element. Choice of crystals can reduce interferences, but some peaks cannot be resolved, for example, Pb (L_a) and As (K_a), and Sr (K_b) with Zr (K_a). First-order interferences can produce large errors, whereas high-order interferences can be reduced by the electronic device pulse height discriminator.

Operating conditions and limits of detection for elements being determined by XRFS vary with the matrix of the sample and with the equipment used.

Calibration Curves. Quantification of the most usual elements is carried out using the respective calibration curves obtained with standard reference materials that are as close as possible to the samples being analyzed.

Unfortunately, there is an absence of suitable standards for all types of materials analyzed (Tables 1–3). The recommended approach is to prepare synthetic standards by:

- Using pure chemicals spiked into a suitable geological matrix
- Mixing proportions of well-determined geochemical reference standards, covering the proper ranges of the different elements (Ref 6)

The use of a machine monitor sample to check the stability of the instrument can be very useful. The monitor needs to have a similar composition to the unknown sample.

In the XRFS technique, the count rate that is obtained from a particular element when a specimen is irradiated is proportional to the concentration of that element. Thus, if the count rate for the element in a sample is compared to the count rate of a standard of known composition, then the amount of the element can be estimated.

The calculation of element concentration from the measured count rate is based on the following relationship between count rate and concentration:

$$C_i = D_i + E_i R_i$$

where C_i represents the concentration of element i, R_i is the net count rate or a count rate ratio, and both D_i and E_i are constants that are dependent on the element, material type, and spectrometer.

However, factors that are caused by differences in composition in different specimens will influence the count rate and must be corrected. In many cases, a simple linear conversion is not possible and corrections for interelemental absorption and enhancement effects (matrix effects) must be made. Corrections for matrix effects can be made in a number of ways:

- By using standards of very similar composition to the samples
- By physical and/or chemical modification of the samples (and standards), for example, by fusion

- By using interelement influence coefficients or α-factors

A number of different mathematical models have been derived to compensate for matrix effects and to convert intensity into concentration at the same time. They have the general form:

$$C_i = D_i + E_i R_i (1 + A_i)$$

in which A_i represents the matrix correction. There are different mathematical models to apply for the correction of matrix effects (Ref 29–35). The typical calibration curve of an element in x-ray fluorescence analysis is illustrated in Fig 3.

Atomic absorption spectrophotometry (AAS) measures the variation of luminous energy that is produced by a monochromatic source because of absorption by nonionized atoms. Atomization of the sample to be analyzed is obtained by dispersion of the corresponding solution in a flame by means of field discharge (AAS with or without flame). The atomic concentration in the flame is proportional to the measured absorbance. From this, the element concentration in the sample solution can be found by standardization (Fig 4). Atomic absorption remains essentially a single-element technique. The relatively low cost of AAS and its high-performance characteristics have justified its widespread use (Ref 36–38).

Operating Conditions. For each elemental analysis, the best wavelength must be selected and the instrumental conditions must be opti-

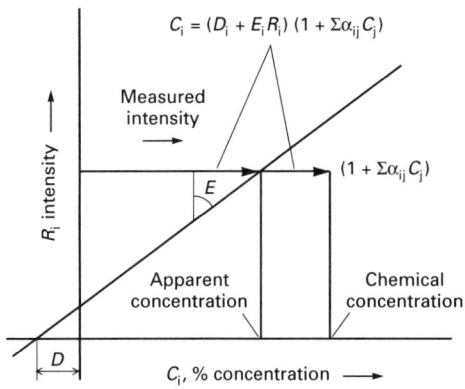

$$C_i = (D_i + E_i R_i)(1 + \Sigma \alpha_{ij} C_j)$$

Fig 3 Typical calibration curve of an element in XRFS

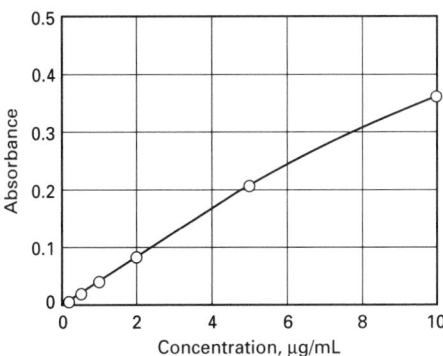

Fig 5 Typical calibration curve of element in AAS

mized. These conditions include the choice of flame type, fuel and oxidant flow rates, burner height, and slit width. Most manufacturers supply "cookbooks" that form a good starting point for establishing operating conditions.

Physical effects, such as viscosity of the sample solution, the presence of organic solvents, or high salt content, can also produce an interference effect. This usually can be overcome by dilution, or by matching the standard solution matrix composition to that of the sample, or by using the method of standard addition.

Sample Preparation. The three most common methods of bringing the sample into solution are (Ref 39):

- Direct dissolution in water (applicable in select cases)
- Dissolution in acids or mixes of acids, with digestion in microwaves or Teflon reac-

tors; acids most often used are HCl, HF, HNO_3, $HClO_4$, and H_2SO_4 mixed in different proportions
- Fusion techniques similar to those used in XRFS, but with a final dissolution of the melt in acids; fluxes most commonly used are Na_2CO_3, Ka_2CO_3, BO_3H_3, $Li_2B_4O_7$, and $Li_2B_4O_7$

Calibration Curves. The usual method of quantitative analysis entails the preparation of an analytical working curve by measuring the absorbance of a series of standard solutions of the element of interest and plotting absorbance versus concentration, as shown in Fig 5. Synthetic multielement calibration solutions are prepared to cover the expected compositional ranges. In preparing the solutions, chemical compatibility of the constituents must be taken into account in order to avoid precipitation.

The element concentration in the sample solutions can be obtained from this graph by interpolation. The curve is essentially linear over most of the range, with only a slight and quite typical curvature at high concentrations.

Calibration errors can be minimized by carefully choosing and preparing standards. Operationally, the process of standardizing the instrument to read concentration directly has been greatly facilitated by the incorporation of a computer in the instruments.

Plasma-emission spectrophotometry is an analytical technique that measures the luminous energy emitted by an excited atom in a plasma source. The plasma is usually a highly ionized gas, such as argon, extremely hot (more than 10,000 °C, or 18,030 °F), and is stable and chemically inert, in order to bring the atoms of any elements up to very high

excitation levels. There are two methods for obtaining the plasma: direct current and inductively coupling radio frequency (ICP) (Fig 6).

In addition to an extremely high temperature, the ICP source offers the possibility of determining elements such as phosphorus, sulfur, tungsten, tin, and boron, which are not normally determined by atomic absorption. Other analytical characteristics include a wide linear working range, low limits of detection, and freedom from chemical interferences.

Spectral interferences are a serious problem when analyzing a wide variety of materials made of different matrices. The nebulizing system is extremely delicate and is the source of many interruptions. Both the initial cost and operating costs are high, because of the argon gas, the changing of nebulizing systems, and other factors.

The sample preparation is very similar to that for atomic absorption, because it is necessary to bring the sample into solution. Only in exceptional instances can solid samples be analyzed, as in the case of slurries (Ref 40).

Other Instrumental Techniques. Ultraviolet (UV)/visible spectrophotometry measures the energy absorbed by a solution through which monochromatic radiation of known length has passed, within the UV or visible range. After choosing the best operating conditions, the quantitative analysis is made by comparison with a calibration curve.

Using electrochemistry, the difference in potential established between two specifically preset electrodes is measured. An example of this is fluoride analysis, in which an ion-selective electrode is used with a potentiometer. The determination of fluoride is carried out with a calibration curve that is obtained by using several standard samples (Fig 7).

Some traditional methods (Ref 41) are used for specific cases, such as the determination of boron by titration with NaOH 0.1 N (Fig 8). Sample preparation involves alkaline fusion at 950 °C (1740 °F) with an equimolecular mixture of Na_2CO_3 and K_2CO_3, followed by dissolution with HCl (1:1) and the elimination of chemical interferences.

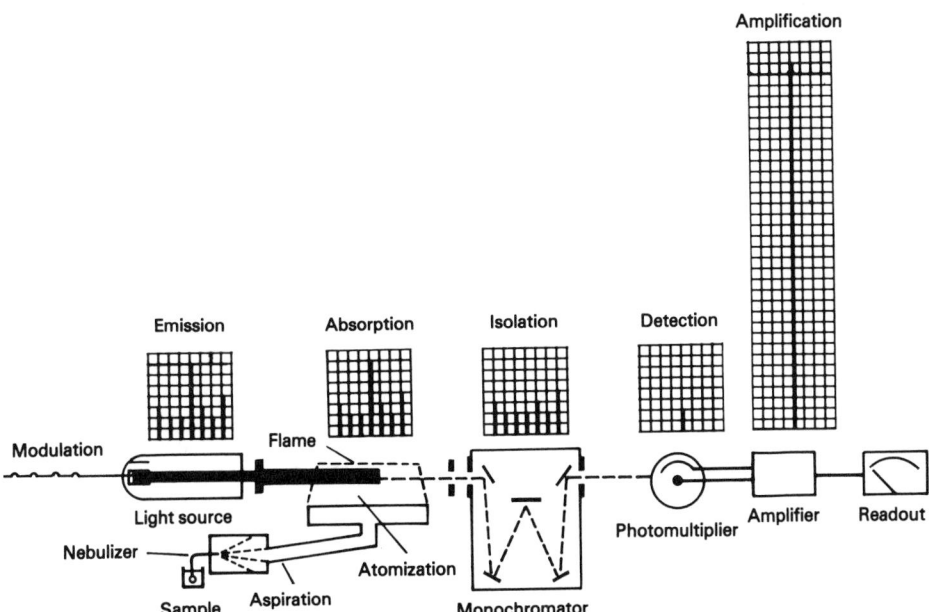

Fig 4 Basic configuration of atomic absorption spectrophotometer

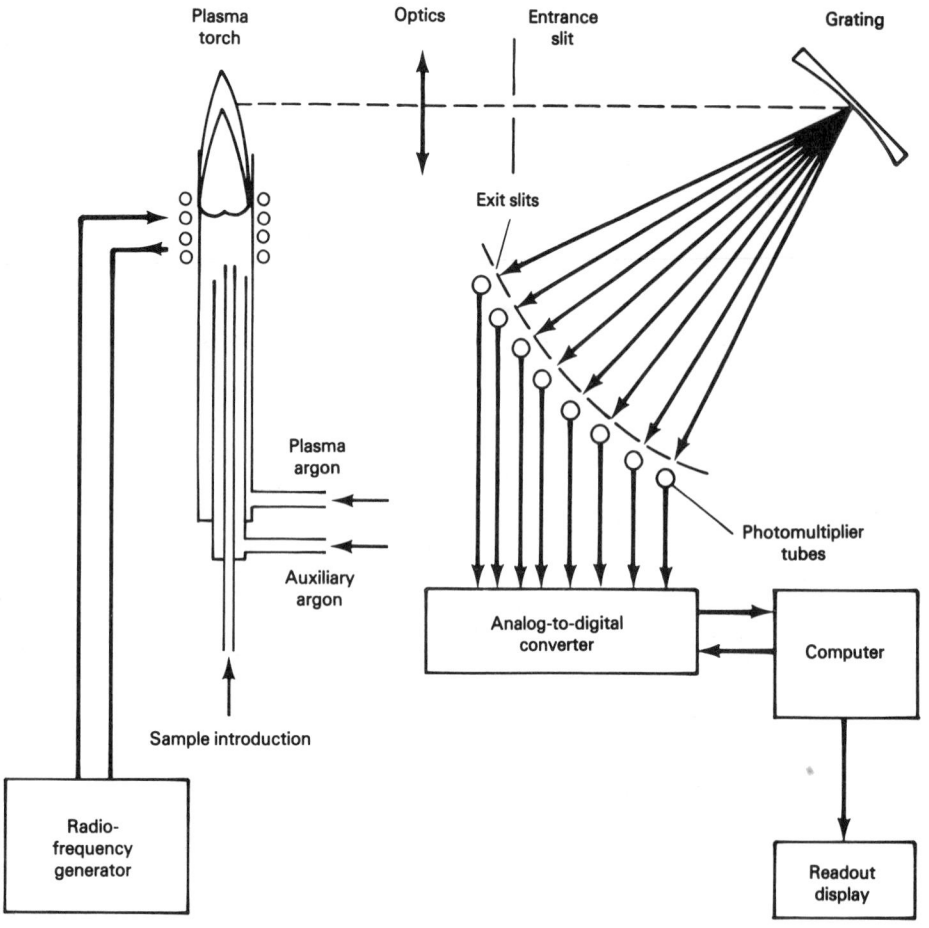

Fig 6 Typical ICP configuration

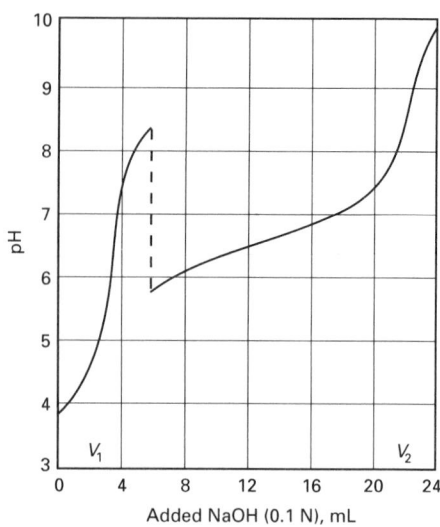

Fig 8 Titration curve of boric oxide. V_1 represents the end point of the titration of HCl, and V_2 represents the end point of boric acid.

Titration can be carried out automatically with a compatible PC hooked up to two automatic microburets (NaOH and mannitol) for the titration and excitation, respectively, of boric acid, a potentiometer for pH measurement, and a sampler.

Analytical Technique Selection

The traditional methods are now used infrequently because of their slowness, the possibility of human error during operation, difficulties in automation, and other concerns. Although they are used for specific problems, they essentially have been replaced by instrumental techniques.

Many factors, such as time savings, sample preparation, automation, precision and sensitivity, equipment cost, skills needed to operate the instrument, and other criteria, must be considered in order to select the most appropriate technique. However, not every laboratory needs a system of wide coverage and maximum performance, and many are satisfied with simpler and less-expensive equipment by either necessity or choice.

The possibility of conducting a qualitative analysis is important in those cases where samples of partially unknown composition frequently must be analyzed.

Bearing in mind that ceramics analysis usually involves solid samples, it should be noted that, in most cases, the material must

be brought into solution, because the instruments only take liquid samples. Only x-ray fluorescence works on solid samples, even though it is possible to work in the liquid phase, in some cases.

Plasma and x-ray fluorescence techniques enable the simultaneous determination of a large number of elements, which is not possible using either atomic absorption (whether it is used with a flame source or a graphite oven) or UV/visible spectrophotometry.

Considering the same analytical technique, the limit of detection is never equal for different elements. However, it is possible for all these analytical techniques to determine the major and minor elements in a sample, whereas the possibility of determining the trace elements is conditioned, to a considerable degree, by the need to dissolve the sample before it can be measured.

It should be noted that other factors, such as matrix or interference effects, have a major influence on the viability of a particular analysis, and that the quoted detection limits are for single-element solutions under optimal conditions. In many cases, the lowest concentration that can usefully be determined in practical sample solutions will be greater, by a factor of 5 or 10, than the ideal detection limits.

The linearity range of the calibration curve (intensity versus concentration) is defined by the number of orders of magnitude within which the intensity of the effect measured is directly proportional to the concentration of the element. The larger the linearity range, the lower the number of points necessary to define the curve; thus, the simpler and quicker the analysis. From this point of view, x-ray fluorescence is clearly superior to all other techniques.

Many instruments, including those with some degree of automation, require the inter-

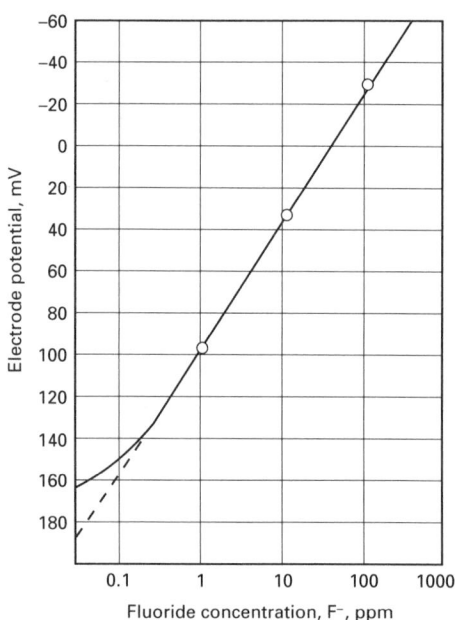

Fig 7 Typical calibration curve of fluoride

vention of the user. However, the major feature of the latest instruments is the ability to perform such functions as storing standard readings, calculating response curves and then sample values, subtracting backgrounds, and calculating the means and standard deviations of replicate samples, which avoids manual handling of the data and the associated risk of transcription error. Equipment costs vary considerably, because of the great variety of models on the market.

Precision and Accuracy

One of the principal objectives of an analytical laboratory should be the production of high-quality analytical results. However, there is the additional requirement to provide these results in a minimum period of time. This can be achieved by using accurate and reliable analytical methods that are appropriate for the analysis being conducted. However, deficiencies may occur when either the choice of an analytical method or its application has not received sufficient attention. The growing concern about poor laboratory practices has led to a proliferation of laboratory accreditation programs, government regulations governing good laboratory practices, and the development and application of quality control and quality assurance programs. The common purpose of all of these efforts is to ensure the accuracy, precision, and reliability of analytical results.

Various sources of error in an analysis can be distinguished either as those linked with the samples (sample preparation, contamination, nonhomogeneity, inaccurate chemical value) or those associated with the instrumentation. In many cases, the accuracy achievable is not determined by the precision of the measurements, but by factors associated with the sample itself.

Accuracy is defined as the closeness of the analytical result to the true figure, and should not be confused with precision, which is the mutual closeness of a replicate set of results for the same analysis obtained by the same analyst, on the same day, using the same equipment, and so on.

The obtainable precision of AAS is in the region of 0.5 to 2%. Greater precision and the capacity for multielement analysis can be achieved by ICP. Analysis by XRFS is capable of greater precision for most of the major elements, and is less tedious to use for a large group of elements. The quality control of analytical data in XRFS is potentially very rigorous, because of the nondestructive nature of the technique, its high precision, and the degree of automation that can be achieved.

Errors greater than 1% of the amount present are rarely permissible when dealing with analytical determinations. To obtain the requisite level of accuracy, more precautions are necessary, which prolongs the time of the determination.

REFERENCES

1. H. Bennet and R.A. Reed, *Chemical Methods of Silicate Analysis,* Academic Press, 1971
2. J. Debras Guedon, I.A. Voinovitch, and J. Louvrier, *L'Analyse des Silicates,* Hermann, 1962
3. R. Tertian and F. Claisse, *Principles of Quantitative X-Ray Analysis,* Heyden and Son Ltd., 1982
4. B. Fabbri, C. Fiori, and F. Donati, Modern Laboratory Instruments and Methods for Determining the Chemical, Mineralogical and Granulometric Composition of Ceramic Raw Materials, *Interbrick,* Vol 4 (No. 6), 1988
5. P.W.J. Boumans, *Inductively Coupled Plasma Emission Spectroscopy,* John Wiley & Sons, 1987
6. E. Ochandio, T. Gonzalez, and F. Povo, La Aplicación de la Fluorescencia de Rayos-X al Análisis de Materias Primas y Pastas Cerámicas, *Técnicas de Laboratorio,* No. 141, 1988
7. A.H. Jones, Analysis of Glass and Ceramic Frit, *Analyt. Chem.,* Vol 37, 1965, p 1761
8. J. Debras Guedon, L'Analyse de Émaux et Frittes. Application des Techniques Instrumentales d'Analyse, *L'Industrie Ceramique,* No. 765, 1982, p 759–762
9. E. Ochandio, J. Ballester, and F. Povo, La Aplicación de la Fluorescencia de Rayos-X al Análisis de Fritas y Esmaltes Cerámicas, *Técnicas de Laboratorio,* No. 149, 1989
10. J. Eastell and J.P. Willis, A Low Dilution Fusion Technique for the Analysis of Geological Samples. 1—Method and Trace Element Analysis, *X-Ray Spectrometry,* Vol 19, 1990, p 3–14
11. C.O. Ingamells and P. Switzer, A Proposed Sampling Constant for Use in Geochemical Analysis, *Talanta,* Vol 20, 1973, p 547–568
12. A.D. Wilson, The Sampling of Silicate Rock Powders for Chemical Analysis, *Analyst,* Vol 89, 1964, p 18–30
13. A.W. Kleeman, Sampling Error in the Chemical Analysis of Rocks, *J. Geol. Soc. Australia,* Vol 14, 1967, p 43–48
14. P. Gy, "Contribution a l'Étude de l'Heterogeneité d'un Lot de Matière Morcelée," Thèse de Docteur-Ingénieur, Université de Nancy, 1972
15. J. Visman, A General Sampling Theory, *Mater. Res. Stnd.,* Vol 9 (No. 11), 1969, p 8
16 S. Abbey, The Search for Best Values— A Study of Three Canadian Rocks, *Geostandard Newsletter,* Vol 5, 1981, p 13–26
17. S. Abbey, Studies in Standard Samples of Silicate Rocks and Minerals, 1969–1982, *Geol. Surv. Can. Paper,* Vol 83 (No. 15), 1983
18. F.J. Flanagan, Reference Samples in Geology and Geochemistry, *U.S. Geological Survey Bulletin,* No. 1582, 1986
19. G.Y. Tao, Z.Y. Zhang, and A. Ji, XRF Procedures for Analysis of Standard Reference Materials, *X-Ray Spectrometry,* Vol 19, 1990, p 85–88
20. V. Davison Vivit and Bi-Shia Wang King, The Determination of Major Oxide Trace Element Concentrations in Eighteen Chinese Standard Reference Samples by X-Ray Fluorescence Spectrometry, *Geostandards Newsletter,* Vol 12 (No. 2), Oct 1988, p 363–370
21. R. Jenkins, R.G. Gould, and D. Gedke, *Quantitative X-Ray Spectrometry,* Marcel Dekker, 1981
22. C. Whiston, *X-Ray Methods,* John Wiley & Sons, 1987
23. J. Bermudez Polonio, *Teoría y Práctica de la Fluorescencia de Rayos-X,* Editorial Alhambra, 1977
24. F. Claisse, "Accurate X-Ray Fluorescence Analysis without Internal Standards," Norelco Reporter III, No. 1, 1957, p 3
25. F. Claisse, "Automated Sample Preparation, X-Ray Fluorescence Analysis in the Geological Sciences, Advances in Methodology," Vol 7, short course, Geological Association of Canada, 1989
26. R. Le Houillier, S. Turmel, and F. Claisse, Loss on Ignition in Fused Glass Buttons, *Adv. X-Ray Anal.,* Vol 20, 1977, p 459–469
27. C.S. Hutchinson, *Laboratory Handbook of Petrographic Techniques,* Wiley, 1974
28. A. Moises, Glass Disk Fusion Method for the X-Ray Fluorescence Analysis of Rocks and Silicates, *X-Ray Spectrometry,* Vol 19, 1990, p 203–206
29. F. Claisse and M. Quintin, Generalization of the Lachance-Traill Method for the Correction of Matrix Effects in X-Ray Fluorescence Analysis, *Can. Spectrosc.,* Vol 12, 1967, p 129–133
30. S.D. Rasberry and K.F.J. Heinrich, Calibration for Interelement Effects in X-Ray Fluorescence Analysis, *Analyt. Chem.,* Vol 46, 1974, p 81–89
31. J. Lucas-Tooth and B.J. Price, A Mathematical Method for the Investigation of Inter-element Effects in X-Ray Fluorescence Analysis, *Metallurgia,* Vol 64, 1961, p 149–152
32. J.W. Criss, Fundamental Parameters Calculations on a Laboratory Microcomputer, *Adv. X-Ray Anal.,* Vol 23, 1980, p 93–97
33. J.W. Criss, L.S. Birks, and J.V. Gilfrich, Versatile X-Ray Analysis Program Combining Fundamental Parameters and Empirical Coefficients, *Analyt. Chem.,* Vol 50, 1978, p 33–37
34. G.Y. Tao, P.A. Pella, and R.M. Rousseau, NBSGSC—A Fortran Program for Quantitative X-Ray Fluorescence Analysis, U.S. Government Printing Office,

1985

35. G.R. Lachance, Defining and Deriving Theoretical Influence Coefficients in XRF Spectrometry, *Adv. X-Ray Anal.*, Vol 31, 1988, p 471–478

36. C.W. Fuller and J. Whitehead, The Determination of Trace Metals in High Purity Sodium Calcium Silicate Glass, Sodium Borosilicate Glass, Sodium Carbonate and Calcium Carbonate by Flameless Atomic Absorption Spectrometry, *Anal. Chimica Acta,* Vol 68, 1974,

p 407

37. W. Kilroy and C. Moyniham, Atomic Absorption Borosilicate Glass, *Anal. Chimica Acta,* Vol 83, 1976, p 389

38. R. Crow and J. Connolly, Atomic Absorption Analysis of Portland Cement and Raw Mix Using a Lithium Metaborate Fusion, *J. Test. Eval.*, Vol 1, 1973, p 382

39. D.K. Fung, J.K. Dubois, and B. Kubler, Comparison de Deux Méthodes de Mise en Solution des Silicates en Vue de Leur

Dosage par Spectrophotometrie d'Absorption Atomique, *Analusis,* Vol 11 (No. 6), 1983, p 291–294

40. L. Halicz and J.B. Brenner, Nebulization of Slurries and Suspensions of Geological Materials for Inductively Coupled Plasma-Atomic Emission Spectrometry, *Spectrochimica Acta,* Vol 42B (No. 1/2), 1987, p 207–217

41. E.P. Serjeant, *Potentiometry and Potentiometric Titrations,* John Wiley & Sons, 1984

Phase Analysis

Charles A. Sorrell, U.S. Bureau of Mines

THE PROPERTIES of a ceramic material are functions of the intrinsic properties of the constituent phases and of the boundaries between the phases. It is necessary, therefore, to know the complete chemistry, the identities and compositions of the phases, and the detailed microstructure of the material to understand the structure-property relationships and, in turn, the effects of processing on properties. Consequently, phase analysis is not an isolated activity but one essential part (see the articles "Chemical Analysis" and "Microstructural Analysis" in this Section) of the characterization necessary to enable production of a ceramic material with properties suitable for a given application.

Phase analysis requires detailed accurate determination of the composition, structure, and amount of each and every constituent phase in the ceramic. In addition, the sizes and size distributions of the grains or intergranular phases (including voids), phase morphologies, phase boundary structures, and surface features are also important. Collectively, these parameters are the microstructure; ideally, microstructural characterization, chemical analysis, and phase analysis are conducted simultaneously using a variety of techniques, some of which provide information about several important characteristics of the material. Careful planning of these integrated characterization procedures is very important to maximize efficiency and equipment utilization.

Because of the way understanding of a ceramic material is developed, phase analysis is important at several different stages. Ceramics are made by a combination of raw materials in predetermined amounts, followed by reaction and/or sintering at elevated temperatures. It is important, therefore, that the chemical compositions of the *raw materials* be accurately known and, in many cases, that the structures, particle sizes, and states of aggregation also be determined. Any ceramic material belongs to a chemical system in which the phase assemblage developed during processing is a function of chemical composition, temperature, pressure, and other variables. In most cases, it is essential that the *equilibrium phase assemblages* be determined for a wide range of chemical and physical variables and that the *sequence of phase changes* occurring during the approach to equilibrium be investigated as a function of time. This involves use of a wide variety of techniques, including phase identification by x-ray or electron diffraction, spectroscopic methods, and thermal analysis methods.

If equilibrium phase assemblages in a ceramic system have been determined experimentally, the information may be used (generally in the form of a phase diagram) to predict the physical structures, compositions, and percentages of the phases that will be present upon attainment of chemical equilibrium for any given mixture of raw materials. Phase diagrams also define the stability limits of the phases in terms of the physical variables and can be useful for prediction of the effects of contamination, contact with other materials, extended use at high temperatures, and so on. They do not, however, provide information about kinetics of the approach to equilibrium or other chemical change. Therefore, phase analysis remains essential for the determination of the identities and compositions of phases in the processed ceramic material to ensure that equilibrium has been attained or, in some cases, that a desired nonequilibrium assemblage has been produced. Phase analysis in the processed ceramic is most effectively accomplished in conjunction with microstructural analysis.

The first two requirements for phase analysis are the chemical compositions and the structures of the phases in the ceramic. Each of these can be considered at a number of different levels of detail. The identity of a crystalline phase, for example, can be easily determined in most cases by x-ray powder diffraction. Details of crystal perfection, lattice ordering, residual lattice strains, and so on require more detailed study by a wide variety of methods such as Mössbauer spectroscopy, nuclear magnetic resonance, or electron spin resonance. Similarly, even though identification of a crystalline phase by x-ray diffraction, coupled with detailed phase equilibria information, provides a reasonably accurate picture of the phase composition, chemical inhomogeneities may be present for a variety of reasons, and phases present in very small quantities may not be readily identifiable except by microchemical methods such as electron microprobe analysis, secondary ion mass spectroscopy (SIMS), electron spectroscopy for chemical analysis (ESCA), or any one of several other sophisticated techniques. These microchemical and microstructural techniques are not discussed in this article, but detailed descriptions may be found in other works (Ref 1). Rather, this article is restricted to those diffraction, spectroscopic, and thermal analysis methods that are particularly useful for phase analysis of ceramic materials and yield sufficient detail for the majority of needs.

Phase Identification by Diffraction

Diffraction is the coherent scattering of electromagnetic waves (for example, light or x-rays) or moving subatomic particles (for example, electrons or neutrons) by materials containing structural units with sizes similar to the wavelengths of the impinging radiation. Crystalline solids are regular periodic structures made up of millions of identical unit cells assembled along three crystallographic axes. All the atoms in the crystal are thus located on planes having many orientations in three-dimensional space relative to the edges of the unit cells (crystallographic axes). In the direction perpendicular to any plane there is a periodic repeat unit called the interplanar spacing or *d*-spacing. The identities and positions of all atoms within the periodic repeat unit are identical with those in all the other units stacked in that orientation. When radiation impinges upon a crystallographic plane, the electrons of the atoms scatter the radiation in all directions. At some angles (θ), where the path difference between the incident (impinging) radiation and the scattered radiation is an integral number ($n = 1, 2, 3, \ldots$) of wavelengths (λ), constructive interference among the scattered waves from all the atoms takes place, producing an intense diffracted x-ray beam. Destructive interference occurs at all other angles and scattered radiation is thus weak. The relationships among the angle of incidence (θ), the wavelength (λ), and the interplanar spacing (d) is given precisely by the Bragg equation:

$$n\lambda = 2d \sin \theta \qquad \text{(Eq 1)}$$

It is apparent, therefore, that measurement of the angles at which diffraction of radiation of a known wavelength occurs enables simple calculation of the interplanar spacings of a crystal by solution of the Bragg equation, rewritten as:

$$d = \frac{n\lambda}{2 \sin \theta} \qquad \text{(Eq 2)}$$

A variety of experimental geometries and detection methods can be used to obtain diffraction angles and intensities (diffraction patterns) for single crystals or polycrystalline materials containing one or more phases (Ref 2–5). Derivation of information from diffraction patterns is strictly a mathematical procedure. The diffraction angles are functions only of the interplanar spacings which are in turn functions of the size and shape of the unit cell (lattice parameters). On the other hand, the relative intensities of the diffracted beams are functions of the chemical identities and positions of the atoms within the unit cell and the mean amplitudes of vibration of the atoms about their positions (and several geometric factors dependent on diffraction angle but independent of crystal structure). Although many compounds may crystallize into the same structural arrangement and will have very similar diffraction patterns, no two compounds will have the same lattice parameters and relative intensities. The diffraction pattern of any crystalline material is therefore unique and may be used to identify that phase unequivocally.

X-Ray Diffraction. X-rays are electromagnetic radiation produced generally in an evacuated tube by electrons impinging on a metal target. The source of electrons is a fine wire (cathode), normally tungsten, that is heated electrically at about 20 A. Electrons are removed from the cathode and accelerated across an evacuated space by application of high voltage (about 50 kV) to the target (anode). Two types of radiation are produced as a result. Kinetic energy lost by electrons slowed or stopped by collision with electrons of the anode metal atoms is emitted as a continuous spectrum (white radiation or brehmsstrahlung) with wavelengths equal to or greater than that corresponding to the applied voltage (λ_{min}). Superposed on the continuous spectrum is a series of intense spectral lines corresponding to electronic transitions in the anode metal atoms. Because the wavelengths are functions of the energy level differences in the anode metal, the collection of lines is called the characteristic spectrum. The lines most commonly used for x-ray diffraction are those of the K-spectrum, which result from transitions from the L-shell (no. 2 quantum level) and the M-shell (no. 3 quantum level) to the K-shell (no. 1 quantum level). The transition from the M_{III} subshell to the K-shell produces the K_β line; transitions from the L_{II} and L_{III} subshells to the K-shell produce the intense closely spaced $K\alpha_1$

and $K\alpha_2$ lines. The wavelengths of the K spectral lines of target materials of commonly used x-ray tubes are shown in Table 1. Because the intense $K\alpha_1$ and $K\alpha_2$ lines of these target metals have wavelengths suitable for diffraction from crystallographic planes of most inorganic crystals, these lines are used almost exclusively for phase identification and crystal structure analysis.

A given crystal plane will diffract x-rays from the continuous spectrum over a continuous range of diffraction angles (θ) corresponding to solutions of the Bragg equation (Eq 2) and will also diffract $K\beta$ radiation as well as $K\alpha$ radiation. It is desirable, therefore, to eliminate as much of the continuous spectrum and the $K\beta$ radiation as possible from the x-ray beam emitted from the tube. This can be done in two ways. First, the beam may be directed to the surface of a single crystal at a suitable angle for diffraction of only the $K\alpha$ radiation, which is then directed to the sample to be examined. This produces a single wavelength (monochromatic) beam that eliminates background scatter from the continuous spectrum and redundant data from diffraction of the $K\alpha_2$ and $K\beta$ lines. Single crystals produce a monochromatic beam of much lower intensity than the $K\alpha_1$ radiation emitted from the x-ray tube, however, and data acquisition is consequently time-consuming. A satisfactory alternative for most diffraction work is the use of a filter to remove much of the continuous and $K\beta$ radiation. A thin sheet of a material with a K absorption wavelength between the $K\alpha_1$ and $K\beta$ wavelengths will absorb a high percentage of the $K\beta$ radiation and of the continuous spectrum with wavelengths less than that of the absorption wavelength of the filter. The absorption wavelength corresponds to the energy required to remove electrons from the K-shell of the filter. Therefore, a high percentage of the energy greater than that (shorter wavelengths) is used to produce the characteristic spectrum of the filter that radiation emitted in all directions so little impinges on the sample. The resulting spectrum from filtered radiation then consists of the $K\alpha_1$ and $K\alpha_2$ lines and relatively low-intensity continuous radiation. Because absorption of $K\alpha_1$ and $K\alpha_2$ radiation by the filter is small, the beam approximates monochromatic radiation suitable for most purposes. The $K\alpha_1$ and $K\alpha_2$ wavelengths are very close and thus are vis-

ible as separate diffraction lines only at high diffraction angles. The unresolved $K\alpha_1$-$K\alpha_2$ doublet is referred to as the $K\alpha$ line, whose wavelength is a weighted average of the wavelengths of $K\alpha_1$ and $K\alpha_2$. Data for filter materials used with commonly available x-ray tubes are shown in Table 1.

X-Ray Powder Diffractometry. There are many different geometric arrangements available for gathering x-ray diffraction data from single crystals or polycrystalline materials (Ref 2–5). The single most useful instrument for phase identification or quantitative analysis of mixtures of phases, however, is the x-ray powder diffractometer. It provides for rapid acquisition of data useful for a number of purposes:

- Identification of crystalline phases
- Quantitative analysis of mixtures of phases
- Precision measurement of lattice parameters
- Determination of the degree of preferred orientation in polycrystalline materials
- Estimation of mean size of very small crystalline particles, lattice strain, residual stresses, atomic vibration amplitudes
- Distinction between crystalline and amorphous phases
- With accessory instrumentation, determination of lattice parameters and structural changes at elevated temperatures
- Determination of distribution of atoms in solid solutions
- Crystal structure determination in some cases; provision of supplemental intensity data to be used in conjunction with single-crystal studies

A schematic diagram of the geometry of a powder diffractometer is shown in Fig 1. The material to be studied is ground to a fine powder, pressed into a recessed sample holder, and smoothed to produce a flat surface. Alternatively, powder may be sprinkled onto a glass slide coated with an adhesive or a solid polycrystalline sample may be cut with a flat surface. In any case, the sample should have a very large number of crystals randomly oriented with a flat surface. The sample is placed in a suitable clamping device in the center of the rotating, motor-driven goniometer. In the most common arrangement, the x-ray tube is mounted in a fixed position, the sample is at the center of rotation, and a detector (for example, a scintillation counter or a Geiger

Table 1 Spectral data for x-ray diffraction targets and filters

Target element	Spectral line wavelength, Å				Filter element	Absorption edge (K_{Abs}) wavelength, Å
	Kβ	Kα₁	Kα₂	Kα(a)		
₂₄Cr	2.0849	2.2897	2.2936	2.2910	₂₃V	2.2691
₂₆Fe	1.7566	1.9360	1.9400	1.9374	₂₅Mn	1.8964
₂₇Co	1.6208	1.7890	1.7929	1.7903	₂₆Fe	1.7435
₂₉Cu	1.3922	1.5406	1.5444	1.5418	₂₈Ni	1.4881
₄₂Mo	0.6323	0.7093	0.7136	0.7107	₄₀Zr	0.6888

(a) Wavelength is the weighted average of the $K\alpha_1$ and $K\alpha_2$ wavelengths.

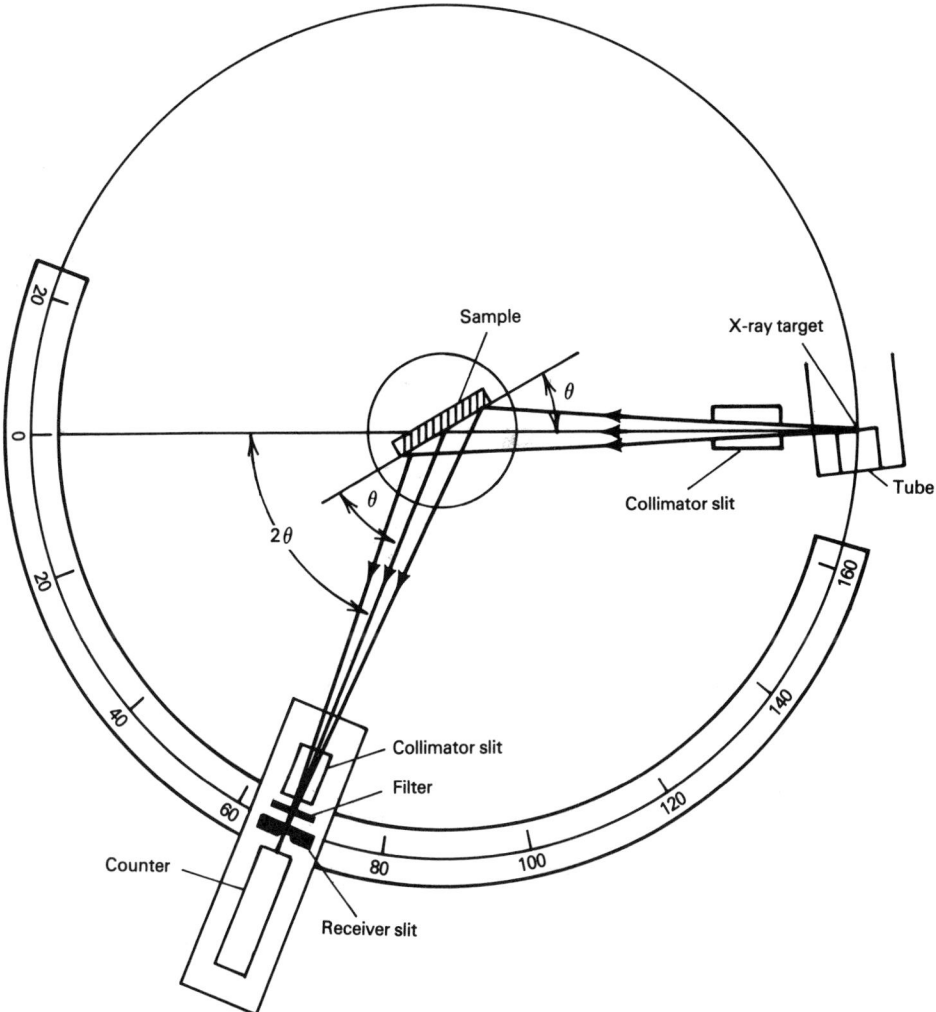

Fig 1 Schematic showing key components of an x-ray powder diffractometer

counter) is mounted onto the goniometer so it will rotate in a circular arc around the sample. The mechanism is calibrated so the sample rotates at half the speed (scanning rate) of the detector so that the angle at which the x-ray beam impinges on the sample is the same as the angle from which radiation is sensed by the detector. Angles are measured as 2θ, twice the angle of diffraction, in this arrangement. The filter or monochromator and collimating slits are placed conveniently in the beam paths, as shown. As the sample and detector are rotated from nearly 0° 2θ to near 180° 2θ, normally at rates between 0.125 and 4.0° 2θ/min, radiation sensed by the detector is amplified and recorded as a function of 2θ by means of a strip-chart recorder or is stored electronically. In the latter case, the data may be retrieved in a number of different graphical or numerical formats or fed directly to a computer for evaluation.

The powder diffraction method depends on the presence of many individual crystals, randomly oriented, so that all crystallographic planes will be, in some of the crystals, oriented at an angle with the incident beam in such a way as to satisfy the Bragg equation for the appropriate *d*-spacings and the wavelength of the x-rays. During the entire exposure of the sample to the x-ray beam, all the planes are diffracting radiation at the appropriate angles and the detector scans through the entire angular range to produce a picture of the intensity of radiation versus the angle of diffraction.

As indicated previously, the diffraction pattern, consisting of a number of diffraction peaks of different intensities, is diagnostic for a crystalline phase. In samples with a mixture of phases, the complete patterns of all the crystalline phases are present in the data set. Diffraction data for a crystalline phase are normally compiled as a list of *d*-spacings (in Angstrom units) in order of decreasing value, the intensities of the peaks relative to the intensity of the strongest peak (given the value of 100), and the indices for the crystallographic orientations of the diffracting planes. Powder diffraction patterns for thousands of crystalline phases have been so compiled (Ref 6) and are available in card or book form along with lists of the *d*-spacings of the strongest

lines arranged in such a way as to enable reasonably rapid identification of phases. Several programs for computer searching the powder diffraction files for matching patterns are also available. Figure 2 shows powder diffraction patterns for a single-phase material. Interpretation of patterns of mixtures of materials may be quite difficult for obvious reasons. Fortunately, patterns of the raw materials (reactants) for ceramics and metals may be obtained easily, and knowledge of the chemistry of the system and knowledge of the phase equilibria (even if only partial), in most cases, provide information about what phases should be present, thereby simplifying the task significantly.

Identification of phases may be complicated by errors in either of the kinds of acquired data—the diffraction angles or the relative intensities. Measured diffraction angles may be in error because of improper alignment of the sample and detector relative to the incident x-ray beam, failure to pack the powder in the sample holder properly, or even slippage or worn gears in the mechanism. Corrections for misalignment may be made by use of an internal standard. A single crystalline phase with precisely known interplanar spacings (for example, quartz or sodium chloride), with strong diffraction lines over a broad range of 2θ values and with minimum coincidence with the lines of the sample, may be ground and mixed with the powder sample. Differences between the correct and measured angles for the lines of the standard are used to plot a correction curve as a function of 2θ and suitable corrections are then made for each line of the sample.

Alternatively, the diffractometer may be aligned so that correct angles are obtained for the standard phase and then the entire range may be scanned. This procedure usually eliminates the need for corrections for the diffraction lines of the sample. Line intensities are most commonly in error because of preferred orientation of the individual crystals in the powder sample. If phases have a sheetlike or needlelike morphology or if there is well-developed cleavage, preparation of the sample surface commonly causes a disproportionate number of the crystals to be oriented with the sheets, needles, or cleavage planes parallel with the sample surface. Diffraction from those crystallographic planes will thus be more intense, and the diffraction from other planes will be less intense than if the crystals are randomly oriented. Even though the diffraction angles should be correct, this effect complicates identification by comparison with literature data which are tabulated according to the most intense lines. The effect may be minimized most effectively by mixing the sample with powdered material having isotropic morphology (for example, MgO or ground glass). The sample crystals tend to be oriented randomly in the interstices of the isotropic material rather than preferentially oriented. Preparing the mixture as a moist

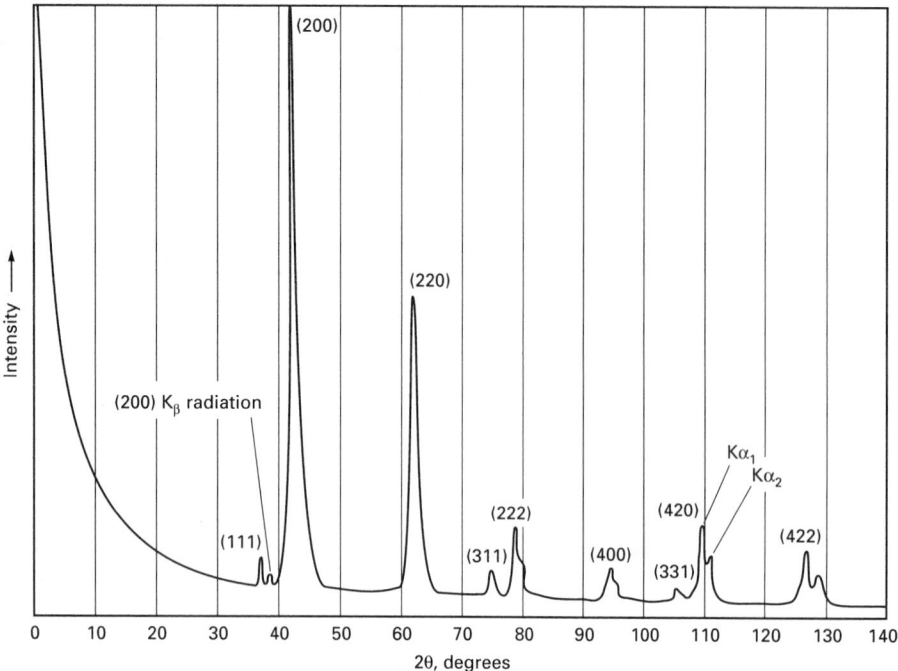

Fig 2 Plot of intensity versus twice the angle of incidence, 2θ, to show powder diffraction pattern obtained for face-centered-cubic (fcc) phase magnesium oxide using a copper target and a nickel filter. Acceleration voltage was 50 kV at a current of 20 mA. Numbers in parentheses are Miller indices (h k l), which denote the orientation of the diffracting planes relative to the crystal axes. Diffracted Kβ radiation shows only for strongest lines; the Kα doublet is resolved only at high angles.

paste, using water or organic liquid, enhances random orientation.

Quantitative analysis of mixtures of phases in a powder sample is accomplished by measurement of the intensities of one or more diffraction lines of each of the phases, taking care that there is no overlap of the lines with those of the other phases. Though simple measurement of the heights of diffraction lines is adequate in some cases, the true intensity of a line is the area under the curve (that is, above background) whether the area is measured from a strip chart or by other means from electronically stored data.

If a phase is composed of very small particles (<1 μm, or 40 μin.), has significant residual strain, compositional gradients, or high defect concentration, the height of the line is decreased relative to that of a phase with none of these structural features. These features produce line broadening so that the area will not be affected significantly. A variety of methods may be used to make quantitative measurements (Ref 4), but the most reliable is the comparison of the intensities of phases in the sample with those of standard mixtures. Pure samples of each of the phases found in the sample are mixed in varying percentages and scanned by diffractometer. Intensities of diagnostic lines are then measured and plotted versus the known percentages to provide calibration curves for comparison with intensities from the sample pattern. For two-phase mixtures, this procedure is straightforward, even if one of the phases is noncrystalline (amorphous or glassy).

For samples with three or more phases, however, many more standard mixtures must be prepared, more complex intensity plots are needed, and iterative measurements may be necessary (new standards, approximating the initial analysis are prepared and the procedure is repeated). In general, the phase percentages in complex mixtures may be determined within 1 to 5% of the correct values, depending on the intensities of the diagnostic lines and the relative x-ray absorption coefficients of the phases. In any case, results should be rationalized with calculations of phase contents from chemical composition of the sample and complemented with other experimental observations (for example, optical or electron microscopy).

Electron Diffraction. The quantum characteristics of moving particles (that is, electrons and neutrons) can be described not only in terms of particle behavior but also as waves. They can, therefore, be diffracted by periodic arrays of matter in a way completely analogous to x-rays. Electrons behave as waves with a wavelength determined by their energy, which is, in turn, determined by the accelerating voltage. At velocities well below that of light (nonrelativistic), the wavelength λ is very nearly:

$$\lambda = \frac{(1.5)^{1/2}}{V} \qquad \text{(Eq 3)}$$

where V is in electron volts and λ is in nanometers (1 nm = 10^{-9} m). Thus, electrons accelerated at 100 kV behave as waves with

a wavelength of 0.0037 nm (0.037 Å). The physics and geometry of the diffraction of electrons is identical with that of x-rays except that the atomic scattering factor depends on scattering by the nucleus as well as by the electrons of the atoms.

Electron diffraction is most commonly accomplished in a transmission electron microscope (TEM), in which resolutions in the range of 0.2 nm (2 Å) are attainable and the depth of focus is such that the image is, in effect, a two-dimensional projection of a three-dimensional object. The electrons can be focused to produce an approximately parallel, coherent beam of 2 to 6 μm (80 to 240 μin.) in diameter at the specimen so it is possible to obtain diffraction data from very small crystals without interference from the others in the sample. This is accomplished by use of a special diffraction lens and aperture that are not used for normal TEM examination to obtain a physical image.

The data are obtained as diffraction spots on the imaging screen of the microscope. The data constitute, in effect, a single-crystal diffraction pattern. The pattern produced is similar to both x-ray Laue and precession photographs in that it is a direct representation of a small portion of the reciprocal lattice sufficient to enable determination of the symmetry and orientation of the crystal as well as lattice parameters. Electron diffraction is a useful technique for identification of phases present in amounts too small to be detectable by x-ray diffraction or to determine the spatial relationships with other phases. Because all periodic structures can diffract, arrays of dislocations, phase boundaries, and precipitates on a small scale can be detected by electron diffraction analysis (Ref 5, 7–9).

Spectroscopic Methods

Atoms in molecules or crystals vibrate continuously at very high frequencies about a mean position. Because the vibration frequencies in atoms or molecules are in the range of 10^{13} to 10^{14} cycles per second (Hertz or Hz), which is of the same order of magnitude as infrared radiation (IR), interaction between infrared radiation and the chemically bonded atoms causes absorption of infrared radiation at wavelengths characteristic of the bond strengths and geometry. Infrared absorption spectroscopy is a valuable tool, therefore, to study structural detail and to identify phases in the solid, liquid, gaseous, or glassy state. Electromagnetic radiation in the visible range (light) also interacts with matter in another way. Energy may be transferred between atoms and photons, resulting in frequency shifts in scattered light. These Raman shifts are related to vibrational and rotational frequencies of the chemically bonded atoms and, like infrared radiation, can be used to determine structural detail and identify phases. Interpretation of IR and Raman spectra for purposes of phase identification is more complex than x-ray diffraction, but the methods are useful for determining structural features not easily detectable by x-ray diffraction (for example, the presence and orientation of hydroxyl units and bonding in glassy phases and for identifying very small quantities of phases undetectable by x-rays).

Infrared Absorption Spectroscopy. The most common type of IR spectrometer uses an optical grating to analyze the spectrum transmitted by the sample. Radiation from a source (metal coil or ceramic heater) is transmitted by the sample and directed by a system of mirrors to the grating which is rotated in order to scan the spectrum. Radiation from the grating is collimated, filtered, and directed to a thermocouple detector. In most spectrometers, the sample beam and a reference beam are alternately directed to the detector so that the net signal is a difference in the intensity of the two beams. The pattern is obtained as a plot of absorption, in percent transmission, as a function of wave number (in cm^{-1}) or of wavelength (for instrumental and procedural details, see Ref 10, 11). Identification of many phases can be accomplished by comparison of experimental spectra with those in several compendia (Ref 12–14).

Raman Spectroscopy. Raman shifts can be either positive or negative, depending on whether energy is lost or gained by interaction with the bonded atoms. Energy lost to the atoms results in shifts to greater wavelengths, producing Stokes lines. Energy gained causes shifts to smaller wavelengths, producing anti-Stokes lines. Stokes lines that result because of shifts from the ground state to an excited state are thus more common and produce much greater intensities.

Both IR and Raman spectra are shown in Fig 3. In this example, formation of La$_2$Sn$_2$O$_7$ from precursors that were amorphous to x-rays were studied at various stages of treatment. The patterns clearly show that both methods are sensitive to the crystallinity of the materials. Precipitated materials, which produced no x-ray diffraction pattern, produced broad IR and Raman lines. Increasing crystallinity produces more numerous and sharper lines. A comparison of IR and Raman suitability for various purposes is given in Table 2, and an indication of the versatilities of the methods is given in Table 3.

Thermal Analysis

In the broadest sense, thermal analysis refers to the measurement of any property change that occurs as a result of a change of temperature. Table 4 indicates some of the many types of analyses. In addition, other analytical techniques, such as x-ray diffraction, spectroscopy, or mass spectrometry may be adapted to study reactions at high or low temperatures. For ceramic materials, the two most widely used methods for phase identification and characterization are differential thermal analysis (DTA), which measures changes in

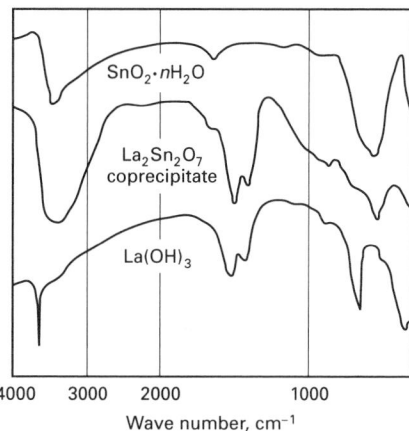

(a)

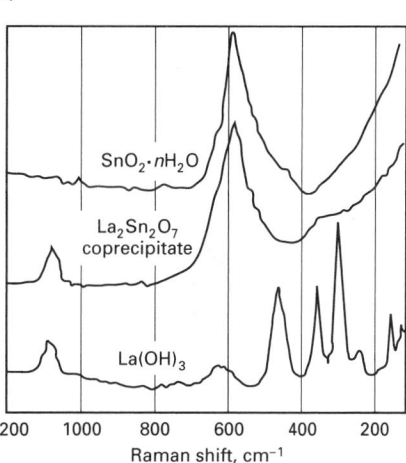

(b)

Fig 3 Formation of La$_2$Sn$_2$O$_7$ from precipitated ceramic precursor materials using spectroscopic methods. (a) Infrared absorption. (b) Raman spectra. Broad spectral lines are typical of poorly crystallized materials. Source: Ref 15

heat content, and thermogravimetric analysis (TGA) or differential thermogravimetric analysis (DTG), which measures changes in weight. Measurements are made during heating and cooling, usually at a constant rate.

Differential Thermal Analysis. Heat is absorbed (endothermic) or released (exothermic) whenever a chemical process results in

Table 2 Comparison of quality of information provided by Raman and infrared spectroscopy

	Spectroscopy	
	Raman	Infrared
Fingerprinting	Excellent	Excellent
Best vibrations	Symmetric	Asymmetric
Structural	Excellent	Very good
Group frequencies	Excellent	Excellent
Aqueous solutions	Excellent	Very difficult
Quantitative analysis	Fair	Good
Low-frequency modes	Excellent	Difficult
Libraries	Fair	Excellent

Source: Ref 16

Table 3 Specific material applications of IR spectroscopy

Product or service	Applications
Air pollution monitoring	Determination of ppm SO$_2$, SO$_3$, and H$_2$SO$_4$
	Studies of the chemistry of smog formation in a laboratory reactor
	Detection of hazardous vapors in the ppm and ppb range
Biomedical materials	Analysis of urinary stones or calculi
	Detection of allergenic substances on skin
	Analysis of anesthetic gases
	Characterization of intermediates in the CO/H$_2$ reaction on SiO$_2$ supported Ru
	Studies of cooking on zeolites
	Studies of chemisorption of H$_2$O and CO on Fe-Cr catalysts
Chemicals	Quality control of intermediates
	On-line process stream analysis
	Identification of impurities in products
Coatings, paints, varnishes, and resins	Identification of polymers, pigments, and solvents
	Quantitation of coating components
	Study of weathering effects on chemical structure
Drugs and pharmaceuticals	Quality control of products
	Analysis of competitive products
	Identification of antibiotics
Fibers and textiles	Analysis of cotton-polyester fiber blends
	Characterization of flame retardants, lubricants, and softeners
	Determination of crystallinity, orientation, and polymorphic form in fiber
Inorganic materials	Characterization of soils
	Identification of pigments (study of hydration reactions in portland cement)
Metals	Determination of boron nitride in transformer steel
	Determination of oxygen in sodium metal
	Characterization of coatings on metals
Petroleum	Tracing crude oil spills as to origin
	Determination of additives in lubricants
	Analysis of crude oil fractions
Polymers and elastomers	Identification of polymers, blends, and plasticizers
	Characterization of stereoregularity of isotactic and syndiotactic polypropylene
	Determination of hydroxyl groups in epoxy resins
Semiconductors	Analysis of silicon for ppb oxygen, boron, carbon, and phosphorus

(continued)

Table 3 (Continued)

Product or service	Applications
	Determination of epitaxial layer thickness
	Analysis of silicon halides used to produce silicon
Wood and paper	Analysis of paper for groundwood content
	Determination of lignin in birchwood pulp
	Determination of silicone release coating on paper

Source: Ref 17

a phase change. The amounts of heat exchanged (calories per mole), usually expressed as ΔH or enthalpy, and the temperatures at which the reaction occurs are fixed and characteristic for any given material. For purposes of phase identification, the most useful reactions are isothermal (that is, the temperature of the phase remains constant during the entire time the reaction is occurring). A change from a low-temperature structure to a high-temperature structure (polymorphic transformation), dissociation with loss of a volatile component (for example, loss of H_2O or CO_2), reduction of oxidation state, and fusion are all endothermic isothermal processes that are detectable by thermal analysis. The reverse of all of these processes is the exothermic isothermal process.

A schematic diagram of a DTA apparatus is shown in Fig 4. The sample and a reference material are placed in suitable holders inside a furnace that can be programmed to heat or cool on a predetermined schedule, usually at a constant rate. The reference material should undergo no phase changes, and its specific heat should be close to that

Table 4 Measurement capabilities of thermal analysis techniques

Technique	Parameter measured
Differential thermal analysis	Temperature difference
Calorimetry	Heat
Dilatometry	Length or volume
Thermomechanical analysis	Stress, strain
Thermogravimetry	Mass
Dynamic mechanical analysis	Stress, strain, time
Electrical conductivity analysis	Electrical resistance
Emanation thermal analysis	Release of radioactive material
Evolved-gas analysis	Pyrolysis with gas analysis
Thermal conductivity analysis	Thermal conductivity
Thermal diffusivity analysis	Thermal diffusivity
Thermoacoustimetry	Sound effects
Thermoluminescence analysis	Light emission
Thermomagnetic analysis	Magnetic susceptibility
Thermooptical analysis	Microscopy

Source: Ref 18

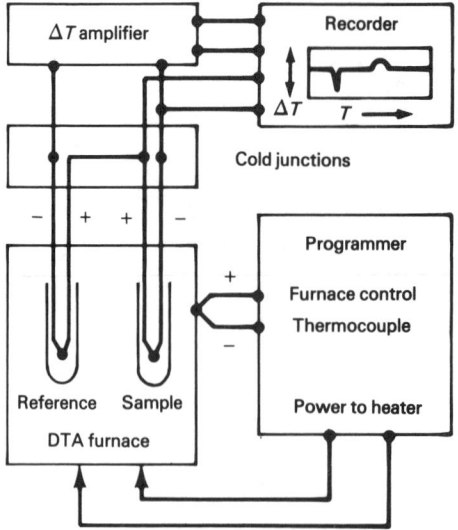

Fig 4 Schematic showing primary components of a DTA instrument

of the sample over the entire temperature range. The thermocouple circuitry is arranged so that the temperature of the reference material, T, and the temperature difference between the sample and the reference material, ΔT, are obtained, amplified, and recorded. The plot of ΔT versus T is called a thermogram. At the onset of a phase change during either heating or cooling, the temperature of the sample deviates from the reference temperature and remains constant until the reaction is complete. The temperature difference, ΔT, is negative for an endothermic reaction or positive for an exothermic reaction and is shown as a deviation from the baseline, or peak, on the thermogram. A typical thermogram is shown in Fig 5.

Thermogravimetric Analysis. When the weight of a sample changes as a result of a reaction (for example, loss of H_2O, CO_2, or other dissociation product, oxidation or reduction, or vaporization of the phase), thermogravimetric analysis or differential thermogravimetric analysis, may be useful in phase identification or study of the reaction process. The apparatus is similar to the DTA apparatus shown in Fig 4 except that an analytical balance is placed into the furnace and the weight change is measured electronically, amplified, and plotted versus temperature (TGA). To improve the precision with which the onset of the weight change can be measured, a plot of the slope of the TGA curve is prepared electronically. Both TGA and DTG curves are generally plotted simultaneously on the same chart. Figure 6 shows TGA and DTG data for two phases in the system ZnO-$ZnCl_2$-H_2O system.

Quantitative Analysis

As with most methods, quantitative analysis is possible if suitable standard samples

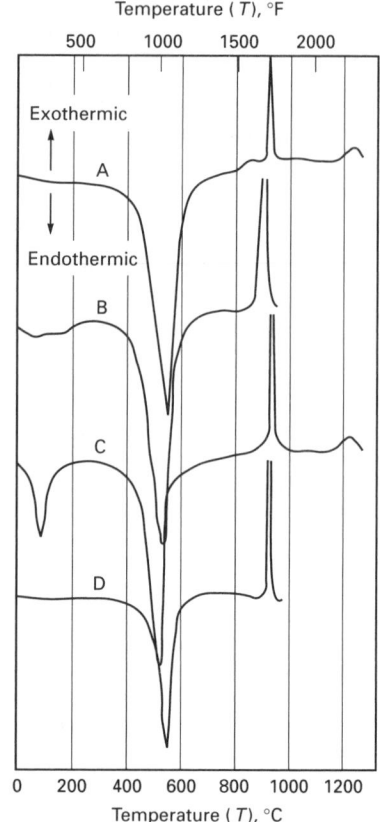

Fig 5 Typical thermogram showing temperature difference, ΔT, as a deviation from a baseline for four types of kaolinite $[Al_2Si_2O_5(OH)_4]$ clay; A, well crystallized; B, poorly crystallized, C, hydrated halloysite $[Al_2Si_2O_5(OH)_4 \cdot 2H_2O]$; D, anauxite $[Al_2(SiO_7)(OH)_4]$. Source: Ref 20

are prepared, and experimental variables are carefully controlled. Most DTA equipment provides, in most cases, only a fair approximation of the quantity of a phase in a mixture because the variables (notably sample packing) are difficult to control. Within the past few years, however, more sophisticated equipment has become available that makes possible reasonably accurate determination of enthalpy changes for specific reactions in single-phase samples. These newer methods, called differential scanning calorimetry (DSC), have proved to be quite accurate for absolute measurement of enthalpy changes, but quantitative analysis in mixtures of phases remains more easily and accurately done by other methods. Thermogravimetric analysis can be very reliable for quantitative analysis if the reactions involving weight changes in the various phases in the mixture are isolated (that is, no two reactions occur simultaneously or overlap over the temperature range to be investigated). For further information on DTA or TGA, see Ref 22 to 25. The most thorough compendium of thermophysical data for use in the interpretation of thermal analyses is Ref 26.

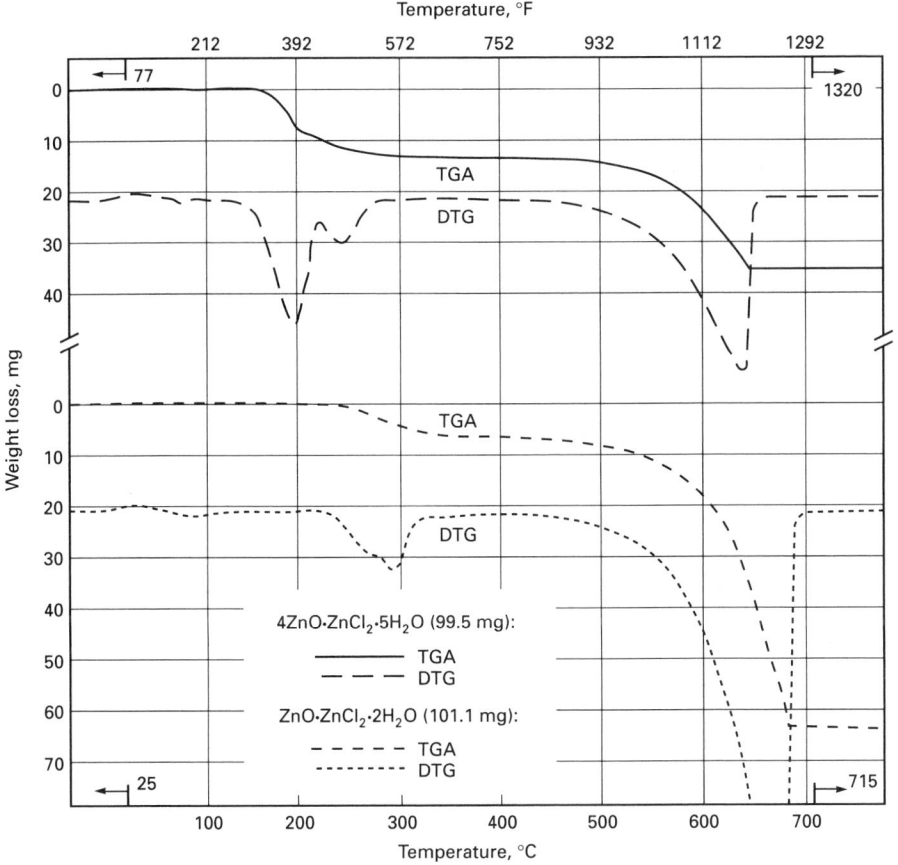

Fig 6 TGA and DTG curves for two compounds in the ZnO-ZnCl₂-H₂O system. Source: Ref 18

REFERENCES

1. M.B. Bever, *Encyclopedia of Materials Science and Engineering*, Pergamon Press, 1986
2. B.D. Cullity, *Elements of X-Ray Diffraction*, Addison-Wesley, 1978
3. L.V. Azaroff and M.J. Buerger, *The Powder Method in X-Ray Crystallography*, McGraw-Hill, 1958
4. H.P. Klug and L.E. Alexander, *X-Ray Diffraction Procedures for Polycrystalline and Amorphous Materials*, Wiley, 1974
5. L.H. Schwartz and J.B. Cohen, *Diffraction from Materials*, Academic Press, 1977
6. Joint Committee on Powder Diffraction Standards, Powder Diffraction File
7. S. Amelinckx, R. Gevers, and J. Van Landuyt, *Diffraction and Imaging Techniques in Material Science*, North-Holland, 1978
8. J.W. Edington, *Practical Electron Microscopy in Materials Science*, Van Nostrand-Reinhold, 1976
9. J.W. Steeds, Convergent Beam Electron Diffraction, J.J. Hren, J.J. Goldstein, and D.C. Joy, Ed., *Introduction to Analytical Electron Microscopy*, Plenum, 1979
10. *Manual on Practices in Molecular Spectroscopy*, Committee E-13, American Society for Testing and Materials, 1979
11. N.B. Colthup, L.H. Daly, and S.E. Wiberley, *Introduction to Infrared and Raman Spectroscopy*, Academic Press, 1975
12. C.D. Craver, Ed., *The Coblentz Society Deskbook of Infrared Spectra*, 1977
13. Infrared Spectroscopic Committee of the Chicago Society for Coatings Technology, *An Infrared Spectroscopic Atlas for the Coatings Industry*, Federation of Societies for Coating Technology, 1979
14. C.J. Pouchert, *The Aldrich Library of Infrared Spectra*, Aldrich Chemical Co., 1975
15. J. Takahashi and T. Ohtsuka, Vibrational Spectroscopic Study of Structural Evolution in the Coprecipitated Precursors to La₂Sn₂O₇, *J. Am. Ceram. Soc.*, Vol 72 (No. 3), 1989, p 426–431
16. H.J. Sloane, The Technique of Raman Spectroscopy: A State-of-the-Art Comparison to Infrared, *Appl. Spectrosc.*, Vol. 25, 1971, p 430–439
17. M.B. Bever, *Encyclopedia of Materials Science and Engineering*, Pergamon Press, 1986, p 2304
18. M.B. Bever, *Encyclopedia of Materials Science and Engineering*, Pergamon Press, 1986, p 4902
19. M.B. Bever, *Encyclopedia of Materials Science and Engineering*, Pergamon Press, 1986, p 4903
20. R.E. Grim, *Clay Mineralogy*, McGraw-Hill, 1968
21. C.A. Sorrell, Suggested Chemistry of Zinc Oxychloride Cements, *J. Am. Ceram. Soc.*, Vol 60 (No. 5–6), 1977, p 217–220
22. C. Duval, *Inorganic Thermogravimetric Analysis*, Elsevier, 1963
23. R.C. McKenzie, Ed., *Differential Thermal Analysis*, Interscience, 1970, 1972
24. F.D. Rossini and S. Summer, Ed., *Experimental Thermochemistry*, Vol 3, Pergamon Press, 1979
25. W.W. Wendlandt, *Thermal Methods of Analysis*, Wiley, 1974
26. Y.S. Touloukian and C.Y. Ho, Ed., *Thermophysical Properties of Matter: The TPRC Data Series. A Comprehensive Compilation of Data*, IFI/Plenum, 1970–1979

Structure and Properties of Glasses

Theodore D. Taylor, Department of Ceramic Engineering, Clemson University

UNLIKE CRYSTALLINE SOLIDS, glass lacks long-range order. It has only short-range order. This simplifies the measurement of macroscopic properties that have a directional character, but enormously complicates matters when scientists attempt to understand properties associated with microscopic structure. This article briefly reviews the properties of glass that are useful for understanding glass structure. Additional information can be found in the references and in the Section of this Handbook on "Properties" (see in particular the article "Oxide Glasses and Other Inorganic Glasses").

Glass Characterization

Statistical terms are often the most effective way to describe a glass. In order to help define the scope of this article a definition of glass will be helpful. The ASTM definition of glass is (Ref 1):

glass—an inorganic product of fusion that has cooled to a rigid condition without crystallizing.
(a) Glass is typically hard and brittle, and has a conchoidal fracture.
(b) A glass may be colorless or colored. It is usually transparent, but may be made translucent or opaque.
(c) When a specific kind of glass is indicated, such descriptive terms as flint glass, barium glass, and window glass should be used following the basic definition, but the qualifying term is to be used as understood by trade custom.
(d) Objects made of glass are loosely and popularly referred to as glass: such as glass for a tumbler, a barometer, a window, a magnifier, or a mirror.

This definition would exclude splat quenched glasses, glasses made under high pressure, sol-gel, and sputtered glasses. This definition is somewhat restrictive but does give a traditional definition to the term "glass." Many of the physical properties described in this article may be difficult to measure in these nontraditional glasses. Glass does not have a crystalline structure. Pauling describes the structure of ionic crystalline substances by setting down five rules (Ref 2):

1. A coordinated polyhedron of anions surrounds each cation. The cation-anion distance is determined by the radius sum. The coordination number is determined by the radius ratio
2. In a stable coordination structure, the electric charge of each anion tends to compensate for the strength of the electrostatic valence bonds reaching to it from the cations at the centers of the polyhedra of which it forms a corner
3. The presence of shared edges and, particularly, shared faces in a coordination structure decreases its stability
4. In a crystal containing different cations, those with high charge and small coordination number tend not to share polyhedron elements with each other
5. The number of essentially different kinds of constituents in a crystal tends to be small

At this point, it is useful to compare Pauling's rules with Zachariasen's rules (Ref 3):

1. An oxygen atom is linked to not more than two glass-forming cations
2. The number of oxygen atoms around a cation of a glass-forming cation must be small
3. The oxygen polyhedra share corners but not edges or faces
4. If the network is to be three-dimensional, at least three corners in each polyhedron must be shared

Both sets of rules are concerned with how cations and anions are linked together. Both also mention polyhedra with oxygen anions at the corners. In glass, the polyhedra must have relatively few corners (three or four). Both sets of rules describe how the polyhedra are to share common elements. It is energetically most favorable if the polyhedra share only corners but not edges or faces. Although Zachariasen's rules are not explicit, Zachariasen's polyhedra also would have to conform to Pauling's Rules 1 and 2, that is, there must be a charge balance and the central cations must be geometrically small enough to fit inside the oxygen polyhedra. Pauling's Rules state nothing about a three-dimensional requirement as does Zachariasen. Zachariasen makes no requirement regarding the number of "essentially different kinds of constituents" as does Pauling. This last point is important: glasses may contain literally hundreds of different constituents.

The structure of glass can be characterized as consisting of polyhedra of anions with cations at the centers. Most of the polyhedra have relatively few sides. These polyhedra are linked into a three-dimensional structure by having each polyhedron forming links to at least three other polyhedra. These links are essentially shared elements from each polyhedron. These elements are usually corners.

In the oxide glasses, it is customary to classify oxides as to whether they are glass formers, glass modifiers, or intermediate oxides. Glass formers will readily form a glass without the aid of other constituents. Examples are SiO_2, B_2O_3, and P_2O_5.

Glass modifiers, on the other hand, do not form glasses by themselves. These are usually oxides of large cations that have a small charge. These oxides furnish non-bridging oxygen ions to the glass structure. These ions are linked to only one glass-forming polyhedron. Examples are the alkali and alkaline earth oxides.

Intermediate oxides may form glasses, but only with assistance from other compounds. Examples of intermediate glass-forming oxides are Al_2O_3 and PbO.

It is difficult to describe the structure of glass in exact quantities. In contrast to crystalline solids, there are no repeating units (such as unit cells) or unique directions in glass. For example, in SiO_2 one may select a silicon cation. There is a very high probability of finding four oxygen anions surrounding this cation. The probability of finding each of these anions associated with another tetrahedron is still very high, yet it is still somewhat less than the first probability. Probabilities of finding a definite configuration become progressively smaller further from the central silicon cation. In other words, glass has good short-range order but poor long-range order.

Crystalline substances, on the other hand, have good short-range order and good long-range order.

Glass is a metastable solid, that is, there is a lower free-energy state that exists for a substance with this composition. As with most metastable substances, thermal history plays a very important role in establishing the physical and chemical characteristics of glass. In other words, besides the usual thermodynamic quantities, time is also a consideration. Figure 1 illustrates this for the density of an arbitrary glass as a function of cooling rate.

Properties of Glass

Glasses possess a wide range of properties; some are incredibly strong while others are readily soluble in water. Engineers and scientists usually quantify these qualities. Properties that are most important are dependent on the application. Usually, people who are working with glass are interested in:

- Optical properties
- Chemical durability
- Mechanical properties
- Electrical properties

Table 1 lists some of these properties.

Optical Properties

The most commonly measured optical properties are index of refraction, optical transmission, and optical dispersion.

The index of refraction of glass is the ratio of the speed of light in a vacuum to the speed of light in the glass. It manifests itself by causing a beam of light to bend as it crosses a glass surface at an angle. Snell's Law quantifies this effect as:

$$n_1 \sin \theta_1 = n_2 \sin \theta_2 \qquad \text{(Eq 1)}$$

where n_1 is the refractive index of air, n_2 is the refractive index of glass, θ_1 is the angle of incidence into glass surface, and θ_2 is the angle of incidence of light path in glass. It should be noted that the angles are measured with respect to the surface normal. There are a variety of methods for measuring the index of refraction. The choice depends on the degree of accuracy that is necessary.

The simplest and least accurate methods are the Becke line and other petrographic methods (Ref 4). A petrographic microscope is used to examine the powdered glass. One places a few grains of the glass powder (-100 mesh$+200$ mesh) on a microscope slide with a drop of index oil used for examining petrographic specimens. A cover slide is placed over the mixture of grains plus oil. The microscopist can tell by observing the behavior of the Becke line whether the oil has a higher or lower refractive index than the glass grains. The accuracy of these methods is usually ±0.004%.

Other methods of measuring index of refraction require the careful grinding of at least two surfaces of the glass specimen (Ref 5). Further, the most accurate results are obtained when at least one of the surfaces is ground flat to within one wavelength. Because the index of refraction of glass is a function of wavelength, it is usually necessary to use monochromatic light. Previous thermal history also profoundly affects the index of refraction.

Dispersion is a measure of the variation of refractive index with wavelength. The most common measure of dispersion is the Abbe number, v_d, which is given as:

$$v_d = \frac{N_d - 1}{N_f - N_c} \qquad \text{(Eq 2)}$$

where N_c is the index of refraction for the 656.27 nm line of hydrogen, N_d is the index of refraction for the 587.56 nm line of sodium, and N_f is the index of refraction for the 486.13 nm line of hydrogen.

The Transmission of Light. Various factors affect the fraction of light that gets through a piece of glass. The first factor is reflection. The fraction of light reflected, R, is:

$$R = \left(\frac{N_1 - N_2}{N_1 + N_2} \right) \qquad \text{(Eq 3)}$$

where N_1 is the refractive index of material No. 1 and N_2 is the refractive index of material No. 2. If material No. 2 happens to be air, $N_2 = 1$. If there is no light absorbed by the glass, the fraction of light, T, exiting from side No. 2 will be:

$$T = (1 - R_1)(1 - R_2) \qquad \text{(Eq 4)}$$

where R_1 is the reflection loss from surface No. 1 and R_2 is the reflection loss from surface No. 2. If the glass material contains compounds that absorb light, another cause of light loss is encountered. Ignoring the loss from reflections, the loss due to absorption by the glass is:

$$e^{-\epsilon_\lambda \cdot c \cdot t} \qquad \text{(Eq 5)}$$

where ϵ_λ is the molar absorptivity (in L/mole · cm), c is the concentration of colorant (in mole/L), and t is the light path length (in cm). Note that the molar absorptivity, ϵ_λ, is a strong function of wavelength.

Consider a piece of glass with light entering from the atmosphere ($N_2 = 1$). If the above equations are combined, an equation that accounts for the losses occurring when light traverses a thickness of glass is obtained:

$$T = \left[1 - \left(\frac{N_1 - 1}{N_1 + 1} \right)^2 \right]^2 \times e^{-\epsilon_\lambda \cdot c \cdot t} \qquad \text{(Eq 6)}$$

The method most commonly used to measure the transmission of light through glass is to use a double beam spectrophotometer. To obtain the most accurate measurements, the glass specimen should have two parallel well-polished surfaces. This technique is applicable to wavelengths from the ultraviolet to the infrared.

Photoelasticity. When placed under stress, glass becomes birefringent. The degree of birefringence is proportional to the difference between the stresses that are at right angles to the viewing direction. Polarized light passing through a piece of glass under stress splits into two components, one traveling faster than the other. Each component vibrates perpendicularly to the other. When these components pass through the analyzer, they recombine. Since they have traveled at different velocities, there is an interference effect that is directly proportional to the amount of strain (or stress) in the glass under examination. The details of this test are described in ASTM Standard C 770 (Ref 6).

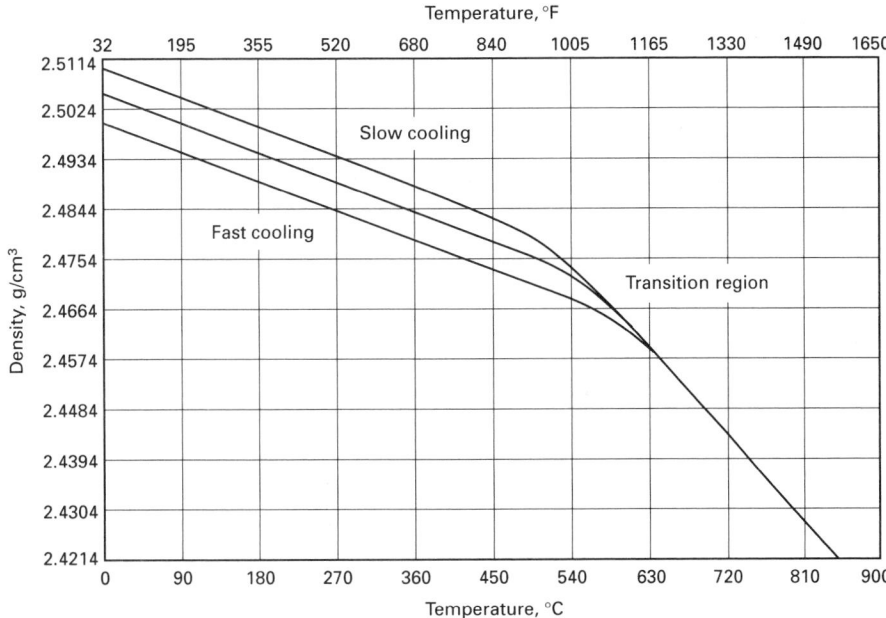

Fig 1 Density of an arbitrary glass as a function of cooling rates

Table 1 Properties of selected commercial glasses

| Glass | Compound, wt% | | | | | Thermal expansion, $\times 10^7/°C$ | Softening point | | Density, g/cm^3 | Refractive index | Young's modulus | |
	SiO$_2$	Na$_2$O + K$_2$O	CaO + MgO	B$_2$O$_3$	Al$_2$O$_3$		°C	°F			GPa	10^6 psi
Fused silica	100	. . .	. . .	. . .	. . .	5.5	1667	3033	2.2	1.458	69	10.0
Soda-lime (plate)	71–73	12–14	11–16	. . .	0.5–1.5	87	735	1355	2.5	1.51	69	10.0
Soda-lime (containers)	70–74	13–16	10–13	. . .	0.5–1.5	85	730	1345	2.5	1.51	69	10.0
Borosilicate	81	4	. . .	13	2	32	820	1510	2.23	1.47	68	9.8
Fiber glass (E)	54.5	0.5	22	8.5	14.5	60	830	1525	2.568	1.537	87	12.6

Source: *Glass Engineering Handbook*, 2nd ed., McGraw-Hill, 1958, p 17

Chemical Durability

There are many ways to measure glass durability, most of which are for specific glass applications. ASTM Standard Test Method C 225 specifies three tests to measure the durability of glass containers (Ref 7). The first test measures the durability of glass containers containing solutions with a pH less than 5. The second test measures the durability of glass containers containing solutions with a pH greater than 5. The third test measures the dissolution rate of powdered glass in distilled and deionized water. All three of these tests utilize an autoclave heated to 121 °C (250 °F). High temperature accelerates the dissolution process. Each of these tests carefully specifies the preparation of the water and reagents used in the test. This is because pH of the solution is extremely important. The presence of any buffering agents, such as CO$_2$, will affect the results.

In addition to pH and temperature, the glass surface area/volume ratio of solution is important. As glass constituents enter the solution, the pH changes. This in turn will cause a change in the dissolution rate. Further details can be found in Ref 8.

Density

Glass density is relatively easy to measure. Because there are no grain boundaries or pores, the usual procedures for measuring density (such as verifying that the pores are saturated) in ceramic materials are unnecessary. As with other glass measurements, previous thermal history is significant, especially when making precise measurements. Because density measurements are easy to make, they are often used in composition control.

The use of a density comparator is the usual method for measuring density in glass manufacturing plants (Ref 9). This method makes use of certain heavy fluids (such as a mixture of isopropyl salicylate and tetrabromoethane). The proportions of the fluid constituents are adjusted so that the fluid has a density approximate to that of the glass. Reference chips of glass are also needed and can be obtained from commercial sources. The measurement procedure starts by first cooling the heavy fluid to a point where the glass specimens float on top of the fluid. The tem-

perature of the fluid and specimens is then gradually raised. Eventually, the glass chips slowly start to fall. The temperature difference between the reference specimens falling and the unknown specimens falling is proportional to the density difference between the reference and unknown specimens.

Electrical Properties

Glass often serves as an electrical insulator, is used extensively in electrical lighting, and is an essential material in devices such as cathode ray tubes. The electrical characteristics of a glass play a prominent role for these and other applications.

Electrical conductivity for most oxide glasses is due to ionic diffusion under the influence of an electrical field. In silicate glasses, this conductivity is due to the motion of alkali cations that have high mobility in the glass structure. As these cations move toward the negative electrode, the layer next to the positive electrode becomes depleted of alkali ions. Therefore, direct current and low-frequency measurements require the use of nonpolarizing electrodes that supply (or take up) alkali cations adjacent to the electrodes. Surface conductivity may present a problem. Guard electrodes placed around the measuring electrodes can eliminate this effect. ASTM specifications provide further details (Ref 10). The conductivity of glass varies over a wide range, from 10^{-5} $\Omega \cdot$ m for a metallic glass to 10^{25} $\Omega \cdot$ m for calcium aluminoborate glass (Ref 11).

Dielectric constant of glass is measured in the same manner as other ceramic materials. An electrode material is applied to opposite faces of a glass wafer to form a capacitor. This capacitor is placed in a bridge circuit (such as a Schering bridge) at audio frequencies or a resonance circuit at radio frequencies. Balancing the bridge circuit or tuning the resonance circuit will facilitate the measurement of the capacitance and shunt resistance of this capacitor. The dielectric properties (permittivity and dissipation factor) of the glass may then be calculated (Ref 12).

The mechanisms responsible for the dissipation factors are dependent on frequency as well as temperature. At very high fre-

quencies, for example, cations cannot respond to a rapidly oscillating field and do not participate to any extent in the polarization that takes place in a dielectric. At lower frequencies, however, cations do take part. This accounts for the observation that the permittivity of a substance such as glass generally decreases with increasing frequency.

Mechanical Properties

Most regard glass as a weak, brittle substance. This is only partly true—it is brittle. Hillig (Ref 13) has reported glass fibers with tensile strengths of 1.35 GPa (2.0×10^6 psi). Like all brittle materials, glass is very susceptible to microflaws. Glass behaves as a truly isotropic substance, that is, the properties are independent of direction.

Elastic Properties. Because of the isotropic nature of glass, it is possible to define completely the elastic properties of glass by combining just two independent parameters. This makes the measurement of elastic properties comparatively simple. As with most materials, however, glass is subject to elastic relaxation processes. This means that the elastic constants will vary with frequency. As with other glass properties, thermal history is significant. Generally speaking, slowly cooling a glass causes the glass structure to become more compact and "stiffer." The elastic properties of glass do not vary over wide ranges as a function of composition (Ref 14). With increasing temperature, however, the elastic modulus of glass generally decreases (fused silica is an exception) (Ref 15).

Resonance (Ref 16) as well as flexural (Ref 17) methods are commonly used to determine elastic moduli. Identical results, however, are not obtained from both methods. The sonic method utilizes an oscillating stress whereas the flexural method does not.

Strength of Glass. Glass, like most other brittle solids, has an unpredictable strength. Glass will withstand far greater compressive stresses than tensile stresses. Strength depends primarily on the condition of any surface that is under tensile stress. Other factors that are important are previous thermal history, rate of load application, environmental conditions, and specimen size. When considering the strength of glass, it is best to associate the level of stress with a risk of failure. This is often done using a Weibull function:

$$R = 1 - \exp - \left(\frac{\left(\int_{\text{vol}} \sigma(v)dv - \sigma_0 \right)}{\sigma_u} \right) \quad \text{(Eq 7)}$$

where R is the risk of failure, $\sigma(v)$ is the stress as a function of its position in the specimen, σ_0 is the stress where no fracture occurs, σ_u is a normalizing factor, and N is the Weibull modulus. The σ_0 is often set to zero. This is equivalent to stating that there is no chance of fracture when the stress is zero. By rearranging this equation and taking logarithms twice, it is possible to linearize the equation. In the case of uniaxial tension, this equation reduces to:

$$R = 1 - \exp - \left(\frac{v \cdot \sigma - \sigma_u}{\sigma_0} \right)^N \quad \text{(Eq 8)}$$

Static Fatigue. Thus far, the rate of loading or environment has not been accounted for in this article. In order to use these equations properly, the rate of loading and environmental conditions must remain constant during both the testing and application phases of product development. The parameters by themselves are of limited value in understanding or characterizing glass properties.

The static fatigue process differs from the ordinary fatigue process associated with cyclical loading. Traditional fatigue is a mechanical process, not a chemical process. Fracture from ordinary fatigue is due to work hardening and subsequent embrittlement of the region around the crack tip. To the casual observer, fracture appears to occur at relatively low intensities of cyclical stress and after a period of time.

Static fatigue, on the other hand, is a chemical process associated with delayed fracture. Cyclical loading does not play a role. The necessary components for the static fatigue process are a low-level stress and a reagent in the environment. Typically, the reagent for silicate glasses is water vapor. With static fatigue, the glass appears to withstand the stress, but after a period of time, fracture occurs. Another manifestation of static fatigue is the higher strength values obtained when the stress is quickly applied rather than when the stress is slowly applied.

When a crack propagates through a solid, the velocity of the crack tip, v, is related to low levels of the stress-intensity factor K_I by:

$$v = a \cdot K_I^N \quad \text{(Eq 9)}$$

where a is a constant and $K_I^N = \sigma_{\text{applied}} \cdot Y \cdot c^{1/2}$, where σ_{applied} is the applied stress, Y is a geometric factor, c is the flaw depth, and N is an experimentally determined exponent.

The above equation describes subcritical crack growth or region I crack growth. The stress-intensity factor in Region I is not great enough to create crack-tip extension unless there are chemical agents present to help sustain the advance of the crack tip. These

chemical agents (such as water vapor) must be present at the crack tip to reduce the energy required to form the new surfaces as the crack propagates. The longest part of a crack's life is in region I. Beyond this region, there is region II that is relatively insensitive to the stress-intensity factor.

Freiman has shown that these relationships can be combined into a more general relationship that can account for rate of load application, environmental effects, stress magnitude, and sample size (Ref 18). Freiman also describes various techniques for making fracture toughness measurements (Ref 18).

Strength measurements are meaningful only if temperature, environmental conditions, loading rate, surface condition and preparation, specimen size, specimen configuration, and thermal history are specified. Most strength tests for glass specimens are either 3-point or 4-point flexural tests. The ASTM Standard Test Method C 674 describes a strength measurement technique for whiteware products. This test could be adapted for glass specimens (Ref 17).

Strength improvements to glass articles consist of either counteracting tensile stresses on the surface that result from applied bending moments or by removing flaws from the surface. Tensile stresses on the surface can be counteracted by placing the surface layers in compression. This is done by either quickly cooling the glass article or by exchanging smaller alkali cations in the glass surface for larger alkali cations in a salt bath (Ref 19).

Thermal Properties

Knowing the thermal properties of a particular glass is essential for efficient manufacturing and application. The thermal properties of glasses vary widely.

Viscosity. During the manufacturing process, the glass viscosity ranges from approximately 30 Pa · s (300 poise) during melting to greater than 10^{-19} Pa · s (10^{20} poise) at room temperature.

The viscosity of glass changes rapidly with temperature. Many equations have been proposed for predicting glass viscosity as a function of temperature. It is widely recognized that flow is a thermally activated process, yet when the logarithm of glass viscosity, η, is plotted against $1/T_{\text{abs}}$, where T_{abs} is the absolute temperature, there is a curve instead of a straight line. A closer approximation to experimental results is:

$$\log \eta = A + \frac{B}{T - T_0} \quad \text{(Eq 10)}$$

where A, B, and T_0 are experimentally derived constants and T is the temperature.

The viscosity of oxide glasses is profoundly affected by the number of non-bridging oxygen ions per glass-forming cation. There are several other factors such as bonding forces, polarizability of cations, and

coordination numbers of cations to consider (Ref 20). All of these factors are composition dependent. Lakatos *et al.* (Ref 21) have developed formulae that accurately compute the constants in Eq 10 for silicate glasses containing Al_2O_3, CaO, MgO, Na_2O, and K_2O. It is important to note that the viscosity of glass changes with time. As stated above, the structure of glass is dependent on thermal history.

Commercial glass-melting processes deal with glass viscosities ranging from around 10 to 10^2 Pa · s (10^2 to 10^3 poise). Glass is formed at viscosities ranging from 10^2 to 10^5 Pa · s (10^3 to 10^6 poise) (Ref 22). Rotating viscometers (such as the concentric cylinder viscometer) are suited for more fluid glasses in the range of from 0.1 to 10^5 Pa · s (1 to 10^6 poise) (see ASTM Test Method C 965 for measuring the viscosity of glass above the softening point as described in Ref 23). Fiber elongation methods are used for glass viscosities $10^{6.6}$ Pa · s ($10^{7.6}$ poise) or greater.

Viscosity of glass in the low-temperature range is measured by varying the temperature until a certain rate of fiber elongation is reached. Low-temperature viscosity is traditionally measured at three temperature points (Ref 1):

- *Softening point.* Temperature at which a glass fiber of uniform diameter elongates at a specific rate under its own weight when measured by ASTM Test Method C 338 (Ref 24). The viscosity at the softening point depends on the density and surface tension. For example, for a glass of density 2.5 g/cm^3 and surface tension 300 dynes/cm, the softening point temperature corresponds to a viscosity of $10^{6.6}$ Pa · s ($10^{7.6}$ poise)
- *Annealing point.* Temperature corresponding either to a specific rate of elongation of a glass when measured by ASTM Test Method C 336 (Ref 25), or a specific rate of midpoint deflection of a glass beam when measured by ASTM Test Method C 598 (Ref 26). At the annealing point of glass, internal stresses are substantially relieved in a matter of minutes
- *Strain point.* Temperature corresponding to a specific rate of elongation of a glass fiber when measured by ASTM Test Method C 336 (Ref 25), or a specific rate of midpoint deflection of a glass beam when measured by ASTM Test Method C 598 (Ref 26). At the strain point of glass, internal stresses are substantially relieved in a matter of hours

Annealing of Glass. After a glass object has been manufactured, it cools in such a way that residual stresses may develop. These stresses develop as a result of different regions of the part being at varying temperatures and therefore varying viscosities. The hotter regions can yield to stress caused by uneven cooling; the cooler more rigid regions cannot yield to such a stress. A consequence of this is a buildup of residual stress (Ref 27). In the case of safety plate glass, this stress buildup is desirable.

In many cases, no stress is desirable. Stress relief is closely approximated by the empirical relation (Ref 26):

$$\frac{d\sigma}{dt} = k \cdot \sigma^2 \qquad \text{(Eq 11)}$$

where σ is stress, t is time, and k is a constant. This equation accounts for the fact that at annealing temperatures, the glass structure relaxes (Ref 28). McMaster gives a practical account of how to set up an annealing process (Ref 29).

The primary objective when annealing optical glass is to stabilize and control the index of refraction. Chilled glass has a lower index of refraction than glass that is slowly cooled. Glass regions that have different thermal histories are likely to have different refractive indices.

At a constant annealing temperature, the index of refraction, N, varies according to the following relation (Ref 28):

$$\frac{1}{N_e - N} - \frac{1}{N_e - N_o} = A \cdot t \qquad \text{(Eq 12)}$$

where N_e is the equilibrium refractive index, N_o is the original refractive index, A is a rate constant and is a function of temperature, and t is time. This relation has a form that is very similar to the previous equation for the release of stress.

Thermal expansion varies from very low values (silica, for example, is $5 \times 10^{-7}/°C$) to values greater than $150 \times 10^{-7}/°C$. Low thermal expansion is necessary for glasses exhibiting high thermal shock resistance. Thermal expansion can be measured using ASTM Standard Test E 228 (Ref 30) or Standard Test C 372 (Ref 31). Both methods utilize dilatometers, which are devices which measure length very precisely. Test C 372 is a general description of dilatometric thermal expansion tests, while Test E 228 describes the use of a fused-silica dilatometer. The basic idea is to compare the known thermal expansion of a length of fused silica with that of the specimen. The specimens for measurement are usually rods that are typically 100 mm (4 in.) long.

Another method of measuring is the interferometric method (Ref 32). Three specimens are arranged between fused-silica optical flats. The length of the specimens (~5 mm, or 0.2 in.) is adjusted so that the number of fringes from optical interference can be conveniently seen. This assembly is placed in a furnace, and as the temperature of the assembly rises, the length of the specimens between the interferometer plates increases causing the distance between the interferometer plates to increase and the fringes to shift. By measuring the shift of the fringes, one can measure the linear length change, and therefore the

thermal expansion. The advantage of this method over the ASTM E 228 method is the small size of the specimens.

REFERENCES

1. Standard Definitions of Terms Relating to Glass and Glass Products, ASTM C 162–85, *Annual Book of ASTM Standards*, Vol 15.02, American Society for Testing and Materials

2. L. Pauling, *J. Am. Chem. Soc.*, Vol 51, 1929, p 1010; as cited by W.A. Weyl and E.C. Marboe, *The Constitution of Glasses, A Dynamic Interpretation: Vol I: Fundamentals of the Structure of Inorganic Liquids and Solids*, Interscience, 1962, p 96–98

3. W.H. Zachariasen, The Atomic Arrangement in Glass, *J. Am. Chem. Soc.*, Vol 54, 1932, p 3841–3851

4. F.D. Bloss, *An Introduction to the Methods of Optical Crystallography*, Holt, Reinhart, and Winston, 1961, p 48–63

5. H. Rawson, *Properties and Applications of Glasses*, Vol 3, *Glass Science and Technology*, Elsevier, Amsterdam, 1980, p 157–160

6. Standard Practices for Measurement of Glass Stress-Optical Coefficient, ASTM C 770–77, *Annual Book of ASTM Standards*, Vol 15.02, American Society for Testing and Materials

7. Standard Test Methods for Resistance of Glass Containers to Attack, ASTM C 225–85, *Annual Book of ASTM Standards*, Vol 15.02, American Society for Testing and Materials

8. A. Paul, Chapt. 4, *Chemistry of Glasses*, Chapman and Hall, 1982

9. Standard Test Method for Density of Glass by the Sink-Float Comparator, ASTM C 729–75, *Annual Book of ASTM Standards*, Vol 15.02, American Society for Testing and Materials

10. Standard Test Method for D-C Volume Resistivity of Glass, ASTM C 657, *Annual Book of ASTM Standards*, Vol 15.02, American Society for Testing and Materials

11. H. Rawson, *Properties and Applications of Glasses*, Vol 3, *Glass Science and Technology*, Elsevier, Amsterdam, 1980, p 238

12. D.G. Holloway, Chapt. 3, *The Physical Properties of Glass*, Wykeham Publications Ltd., London, 1973

13. W.B. Hillig, *J. Appl. Phys.*, Vol 32, 1961, p 41; as cited by R.H. Doremus, *Glass Science*, John Wiley & Sons, 1973, p 264

14. D.G. Holloway, *The Physical Properties of Glass*, Wykeham Publications Ltd., London, 1973, p 128

15. G.W. McLellan and E.B. Shand, *Glass Engineering Handbook*, 3rd ed., McGraw-Hill, 1984, p 219

16. Standard Test Method for Young's Mod-

ulus, Shear Modulus, and Poisson's Ratio for Glass and Glass-Ceramics by Resonance, ASTM C 623–71, *Annual Book of ASTM Standards*, Vol 15.02, American Society for Testing and Materials

17. Standard Test Methods for Flexural Properties of Ceramic Whiteware Materials, ASTM C 674–81, *Annual Book of ASTM Standards*, Vol 15.02, American Society for Testing and Materials

18. S.W. Freiman, Fracture Mechanics of Glass, *Glass Science and Technology*, Vol 5, D.R. Uhlmann and N.J. Kreidl, Ed., Academic Press, 1980, p 21–75

19. H. Rawson, *Properties and Applications of Glasses*, Vol 3, *Glass Science and Technology*, Elsevier, Amsterdam, 1980, p 134–137

20. W.A. Weyl and E.C. Marboe, *The Constitution of Glasses, A Dynamic Interpretation: Vol II: Constitution and Properties of Some Representative Glasses*, Interscience, 1964, p 669–678

21. T. Lakatos, L.-G. Johnasson, and B. Simminsköld, *Glass Technol.*, Vol 13, 1972, p 88–95; as cited by H. Rawson, *Properties and Applications of Glasses*, Vol 3, *Glass Science and Technology*, Elsevier, Amsterdam, 1980, p 157–160

22. H.H. Holscher, The Processing of Bottles and Other Hollow Ware Articles, *The Handbook of Glass Manufacture*, Vol 2, F.V. Tooley, Ed., Ashley Publishing, 1984, p 676

23. Standard Practices for Measuring Viscosity of Glass Above the Softening Point, ASTM C 965, *Annual Book of ASTM Standards*, Vol 15.02, American Society for Testing and Materials

24. Standard Test Method for Softening Point of Glass, ASTM C 338–73, *Annual Book of ASTM Standards*, Vol 15.02, American Society for Testing and Materials

25. Standard Test Method for Annealing Point and Strain Point of Glass by Fiber Elongation, ASTM C 336–71, *Annual Book of ASTM Standards*, Vol 15.02, American Society for Testing and Materials

26. Standard Test Method for Annealing Point and Strain Point of Glass by Beam Bending, ASTM C 598–88, *Annual Book of ASTM Standards*, Vol 15.02, American Society for Testing and Materials

27. S.R. Scholes and C.H. Greene, Chapt. 16, *Modern Glass Practice*, Cahners Book Publishers, 1974

28. O.S. Narayanaswamy, Annealing of Glass, *Glass Science and Technology*, Vol 3, D.R. Uhlmann and N.J. Kreidl, Ed., Academic Press, 1986, p 278–282

29. H.A. McMaster, Annealing and Tempering, *The Handbook of Glass Manufacture*, Vol 2, F.V. Tooley, Ed., Ashley Publishing, 1984, p 799–832–13

30. Test Method for Linear Thermal Expansion of Solid Materials with a Vitreous Silica Dilatometer, ASTM E 228, *Annual Book of ASTM Standards*, Vol 14.02,

American Society for Testing and Materials

31. Standard Test Method for Linear Thermal Expansion of Porcelain Enamel and Glaze Frits and Fired Ceramic Whiteware Products by the Dilatometer Method, ASTM C 372–88, *Annual Book of ASTM Standards*, Vol 14.02, American Society for Testing and Materials

32. Standard Test Method for Linear Thermal Expansion of Rigid Solids with Interferometry, ASTM E 289–70, *Annual Book of ASTM Standards*, Vol 14.02, American Society for Testing and Materials

Microstructural Analysis

Heinrich Mörtel, Institut für Werkstoffwissenschaften Glas und Keramik, Erlangen, Germany

CERAMIC MATERIALS are used today in nearly all areas of engineering. The properties of the materials are influenced significantly by their microstructures; therefore, the quality of these materials always must be carefully inspected. For this purpose the microstructure is examined with different microscopy techniques such as transmitted or reflected light microscopy using polarized light, scanning electron microscopy (SEM) or transmission electron microscopy (TEM) with additional analysis techniques such as energy dispersive analysis of x-rays (EDAX) or wavelength dispersive analysis of x-rays (WDS). The quality of preparation of the ceramics samples for these techniques is the determining factor in the quality of the materials analysis. The procedures used are similar to metallographic or petrographic methods but are not identical. Furthermore, for each type of ceramic material, a special procedure must be developed to obtain the best results for each sample.

Sample selection, cutting, infiltration, mounting, grinding, polishing, and methods of image contrasting are the procedures which are discussed in this article. The use of different microscopy techniques are also briefly reviewed. Additional information can be found in the Selected References that follow as well as in Volume 9, *Metallography and Microstructures*, and Volume 10, *Materials Characterization*, of the ASM Handbook.

Sample Preparation

Sampling should be carried out according to a predetermined plan. The investigator must be sure of what type of information can be expected from the microstructural investigation. Ceramic samples have a microstructure that is determined by their fabrication process. For example, the outer zone of an extruded rod is quite different from the inner zone of the same piece. In addition, the microstructure of the outer zone of a ceramic is different from its interior because the surface is exposed directly to the atmosphere during sintering. The sintering reaction is more severe on the exterior than in the interior of the part, as a result of exposure time. A map should always be drawn that shows where the sample under consideration was taken from and its exact orientation within the ceramic piece under consideration. The size of the sample should be as small as possible to minimize preparation time.

Sectioning of a ceramic sample must be done such that the microstructure or defect present is not altered. In the case of brittle materials such as ceramics, heat and mechanical damage are the most common. To perform a quick and economical sample removal from the bulk, sawing is used. Dry cutting should never be performed as it damages the microstructure of the ceramic and reduces the service life of the cutting wheel. Fluids such as water, mineral oil, or emulsions are recommended for wet sawing. Care must be taken for sufficient cooling. In the case that materials are soluble in the cooling agent, another lubricant must be used. For example, with water-soluble materials such as β-alumina, calcium oxide, magnesia, sintered magnesite or cement-clinker, alcohol may be used. Wet cutting results in a minimally damaged microstructure which saves time in subsequent steps of preparation.

The damage to the microstructure depends on the speed of the saw, the nature of the abrasive, and the grain size of the abrasive used in the consumable abrasive wheel. High-speed sawing saves time in sectioning of a sample, but more time may then be needed to grind and polish the sample properly down to the undisturbed microstructure. High-speed cutoff wheels are used for cutting a coarse piece. The subsequent sample should be sectioned from this coarse piece using a low-speed cutoff machine with a fine-grain abrasive in the wheel.

Abrasives such as diamond and corundum are commonly used for cutting ceramics such as silicon carbide and alumina. These abrasives are bonded with rubber, resin, or a metallic bond (normally brass). Diamond is the most effective abrasive for all kinds of ceramics because of its hardness. Other abrasives can be used only for materials whose hardness is less than their own. The type of material used to bond the abrasive is also an important consideration. Quartz, for instance, can only be sectioned with resin-bonded diamond, not with brass-bonded diamond. The reason is that a fresh, sharp edge of the abrasive particle must always be exposed. In the case of quartz, brass smears the cut, and no further progress in cutting may be achieved. With resin-bonded diamond, particles break out, and new surfaces are exposed. The consumption of such wheels is high, but the effect in terms of cutting is also reasonably high.

A smeared wheel can be sharpened again by cutting a piece of ceramic which is manufactured for this purpose, for example, silicon carbide for brass-bonded wheels. In general, hard wheels are used for soft materials and soft wheels (rubber-bonded) for hard ceramics.

The width of the cutting wheels is also important. Although thick wheels result in a higher loss of material and a quicker cutting, they also cause greater damage to the microstructure of the sample. Thin cutting wheels expend more time for cutting, but result in a lower loss of material and affect the microstructure less.

After sectioning, the samples must be cleaned carefully. In the case of porous samples, one must take into consideration that the cooling agent and/or fine debris can fill the pores of the sample. These must be flushed out, and the sample dried. Otherwise, coarse grains resulting from the cutting may escape from the pores during the subsequent grinding steps and cause scratches on the surface.

After each step of grinding and polishing a very careful cleaning is recommended for the above-mentioned reason. Ultrasonic cleaners are recommended for this purpose. Ultrasonic cleaning, however, may also cause some pullout in the case of soft materials or even porous materials. In these cases, the ceramics might be impregnated before fine cutting or the first grinding step.

Impregnation and Mounting. The handling of specimens during grinding and polishing is made easier when the samples are embedded in a polymer (Fig 1). Embedding also prevents pullout of grains, especially from the outer zone of the sample, and it prevents obtaining a curved surface, because the area of the sample is enlarged. Both grain pullout and surface curvature effects may also be reduced when the specimen is closely surrounded by a hard material, or when two pieces are placed face-to-face close together, such as when the glaze/body interface of tiles is to be studied.

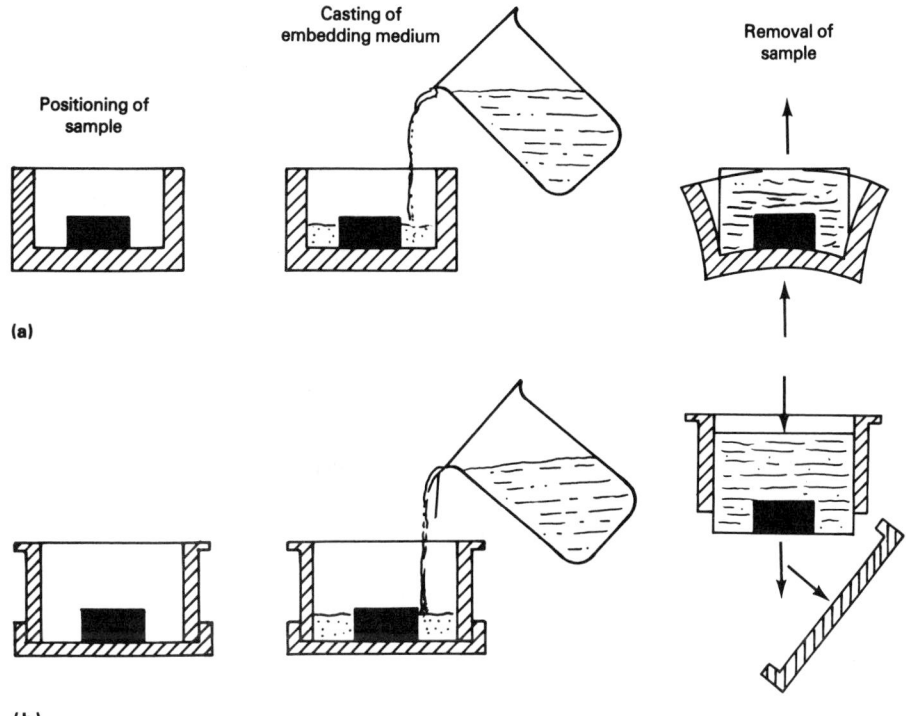

Positioning of
sample

Casting of
embedding medium

Removal of
sample

(a)

(b)

Fig 1 Embedding of ceramic samples in (a) silicone mold and (b) inflexible polymer mold

ber mold, or metallic mold) contains the samples (Fig 2). The device is evacuated to a pressure below which foaming of the polymer can start. This pressure should be determined experimentally for each polymer before starting any embedding procedure. Foaming pressures should be noted for all of the polymers used in a laboratory, because different polymers may be used for different applications. The vacuum should be held for at least 30 min. The valve to the reservoir containing the polymer is then opened and the estimated polymer volume for impregnation and/or molding is poured onto the sample. Care must be taken to ensure that some polymer is still in the container during the pour; otherwise air will interrupt the vacuum. The vacuum should be held for another 30 min. The container distributing the polymer should then be cleaned off before polymer hardening. After this, the exsiccator valve may be opened, but this should be done very slowly to smoothly lower the vacuum to normal pressure.

Impregnation or molding-agent resins are often used in combination with a hardener. Many of them harden at room temperature, although some need higher temperatures. In many cases, higher temperatures speed up the curing time. The temperature that should be used depends also on the nature of the material, that is, whether or not it is able to take heat without changing its microstructure. Suppliers of molding-agent resins, most of which are epoxies and araldites, offer lists with the exact data and applications.

Some samples require further treatment, that is, impregnation. This is recommended, for instance, for soft materials, green (unfired) ceramics, and porous materials. To determine the true porosity of a sample, one must ensure that no pullout occurs. Therefore, infiltration or impregnation with a fluid resin is recommended. This is also true after grinding or polishing steps when closed pores are opened up or pullout occurs. However, filling of the pores with a polymer prevents the porosity from filling up with abraded material during grinding or polishing, which may lead to flawed microstructural interpretation.

Before impregnation, the samples must be dried carefully. In most cases a temperature of 110 °C (230 °F) is applied for not less than 1 h. In some cases (for example, cement or calcium-silicate hydrates) temperatures as low as 70 °C (160 °F) are used; lower temperatures (40 °C, or 105 °F) are recommended for gypsum. Higher temperatures for such materials can alter their phase content and microstructure as well as cause them to lose water. The impregnation is done using low-viscosity polymers and is carried out under vacuum. This process removes most of the air from the open porosity and replaces it with the uncured polymer which, when hardened, forms a tight bond with the ceramic.

A simple device for this technique is a vacuum exsiccator with a small container for the curing agent on its top that is closed by a three-way valve.

In this exsiccator, a mold of some type (paper cup, polymer cup, polymer mold, rub-

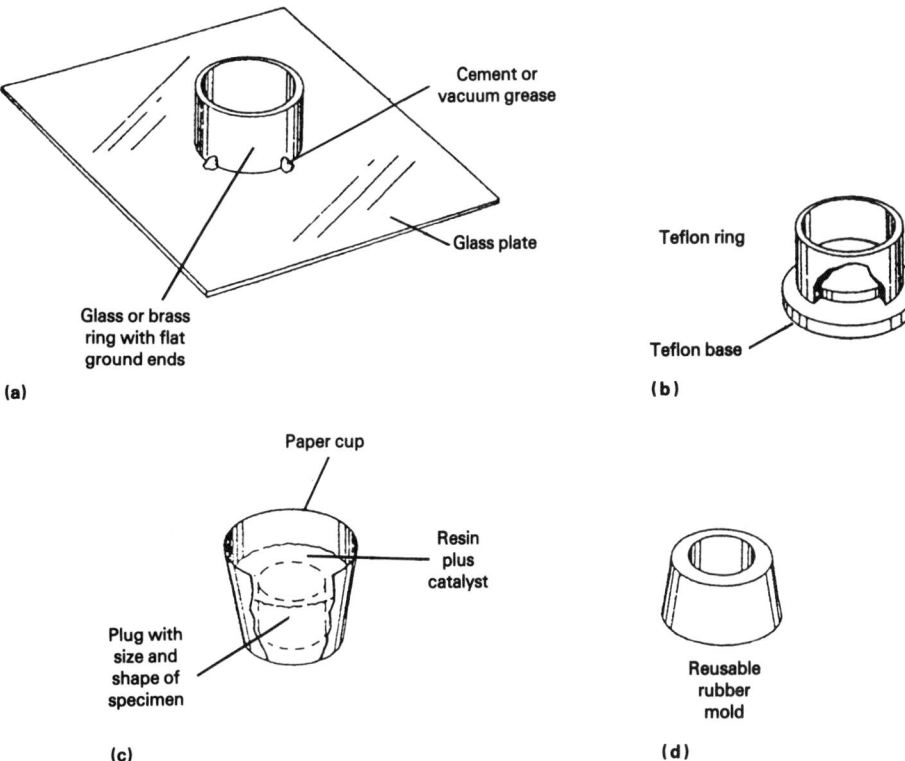

Cement or
vacuum grease

Glass plate

Glass or brass
ring with flat
ground ends

(a)

Teflon ring

Teflon base

(b)

Paper cup

Resin
plus
catalyst

Plug with
size and
shape of
specimen

(c)

Reusable
rubber
mold

(d)

Fig 2 Casting molds for the preparation of specimen mounts. (a) Glass plate and ring. (b) Teflon two-piece mold. (c) and (d) Rubber mold

Sometimes cracking or bursting occurs during hardening of the polymer, especially during thermosetting. Typical defects and advice for their prevention are shown in Fig 3.

Impregnation and embedding are very often carried out simultaneously. Independent steps are necessary only when different polymers for the infiltration and the mounting are required.

A filler may also be used in the embedding agent. Fillers enhance the hardness of the polymer close to the ceramic, making it easier to obtain a flat surface. Alumina is often used for this purpose. In other cases, an electrically conducting material may be used to produce the required conductivity in electron microscopes because most ceramics are insulators. Graphite, carbon, or silicon carbide are recommended.

Sometimes mechanical mounting devices can be utilized (Fig 4). Specimens are fixtured in a sample holder with screws or temporarily glued on a metallic plate (Fig 4). In most cases, however, samples are embedded and/or impregnated.

In special cases, a very thin layer or interlayer must be characterized. When a sample

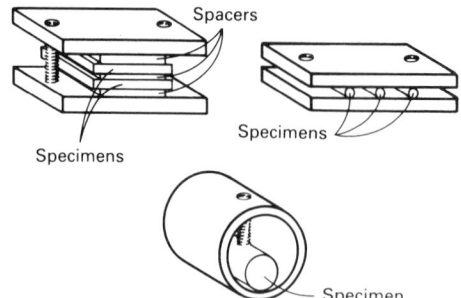

Fig 4 Clamps which may be used for the polishing of small specimens, thin sheets, or materials which react with the conventional media

like this is embedded under a very small angle with respect to this zone, the surface of the width can be enlarged significantly. This is referred to as taper mounting. For example, one can get an enlargement of 25:1 with an angle of about 2°, or 5:1 at an angle of about 11°. This is easy to perform as shown schematically in Fig 5.

Before the samples are embedded, they should be marked. This may be done with a piece of paper that contains all the important

remarks such as the date, source, and orientation. Careful numbering of the specimens is important because many of these samples may be stored after use as reference materials.

Mechanical grinding is performed by hand or more recently by semiautomated or fully automated machines. Because each ceramic material responds differently, only general features will be discussed. However, many parameters are similar, and standard methods can be set up for many ceramics.

The goal of grinding and polishing is to obtain a flat specimen surface for examination of its microstructure. The first step is the removal of irregularities (unevenness) perpendicular to the axis of the piece or mounting to enable further grinding and polishing with mechanical devices.

Grinding by hand is performed on solid, flat surfaces of metal or glass using loose abrasive grit dissolved in a liquid or emulsion. One other option is to use the abrasive bonded to cloth or paper. These are available for both rotary and fixed grinders.

All grinding of ceramics should be done wet, because dry grinding will cause local heating which may have a deleterious effect on the microstructure. Lubricants may be water, oil, or alcohol, as was discussed for cutting. A continuous flow of the coolant should be maintained in order to sweep away the abraded material.

Grinding abrasives include silicon carbide, alumina, diamond, and emery. The most common abrasive is silicon carbide, because of its effectiveness, long life, and low cost.

Grinding by hand is performed using a figure-eight motion and the sample turned slightly between each pass to ensure that another portion of the sample in the same plane is exposed to the abrasive. This holds for grinding on a solid plate with or without cloth and for a rotary grinder. Another possibility is to work on an inclined block with different grades of silicon carbide papers, moving the sample back and forth until the scratches from the former grade disappear and only scratches from the present grade are present. Then the piece is turned 90° and the same operation is performed until the former scratches disappear.

The progress of grinding should be monitored with a binocular microscope. Before examination under the microscope, the specimen must be carefully cleaned to remove abrasives and lubricants and dried. As stated earlier, the specimens must also be cleaned carefully when switching from one step to the next. If considerable pullout occurs, the grinding must be interrupted, and the specimen infiltrated again.

The procedure is similar for rotating grinders. The specimens should be turned slightly between steps until all scratches from the former step disappear. The grinding wheels have a diameter of 200 to 300 mm (8 to 12 in.). The rotation speed ranges from 50 to 1800 rev/min, and may be altered continuously or stepwise. The load on the specimen should

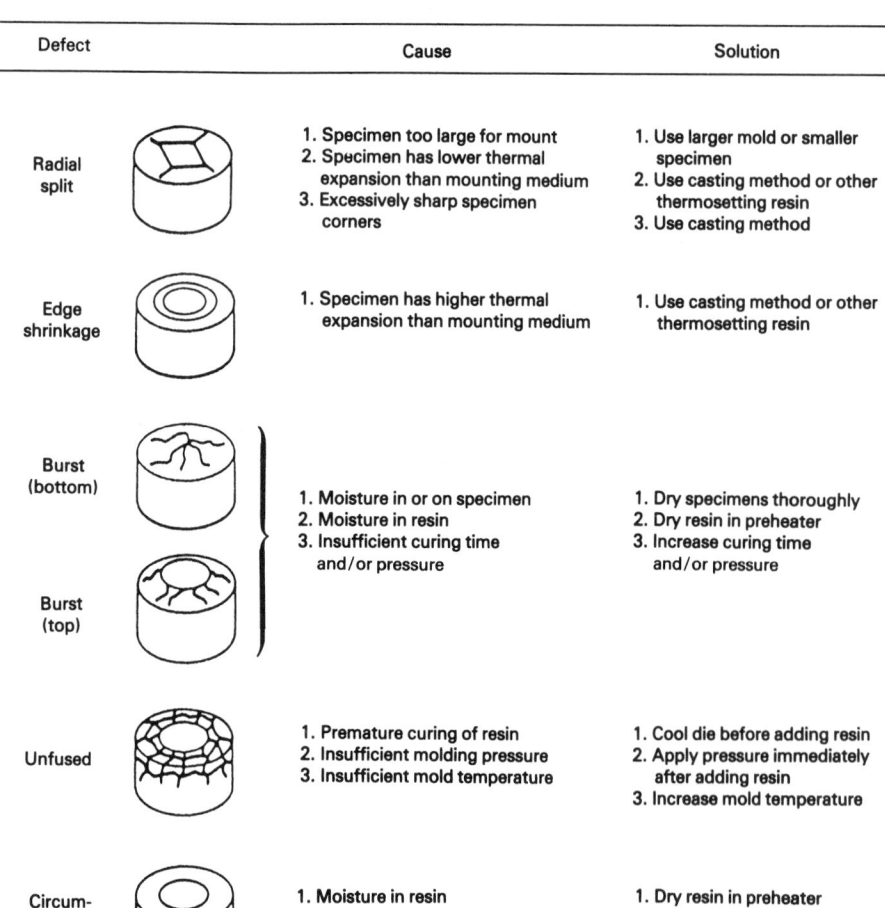

Defect		Cause	Solution
Radial split		1. Specimen too large for mount 2. Specimen has lower thermal expansion than mounting medium 3. Excessively sharp specimen corners	1. Use larger mold or smaller specimen 2. Use casting method or other thermosetting resin 3. Use casting method
Edge shrinkage		1. Specimen has higher thermal expansion than mounting medium	1. Use casting method or other thermosetting resin
Burst (bottom) Burst (top)		1. Moisture in or on specimen 2. Moisture in resin 3. Insufficient curing time and/or pressure	1. Dry specimens thoroughly 2. Dry resin in preheater 3. Increase curing time and/or pressure
Unfused		1. Premature curing of resin 2. Insufficient molding pressure 3. Insufficient mold temperature	1. Cool die before adding resin 2. Apply pressure immediately after adding resin 3. Increase mold temperature
Circumferential split		1. Moisture in resin 2. Mold initially too hot 3. Mold temperature too high	1. Dry resin in preheater 2. Cool die between mounts 3. Reduce mold temperature

Fig 3 Mount defects for thermosetting resins

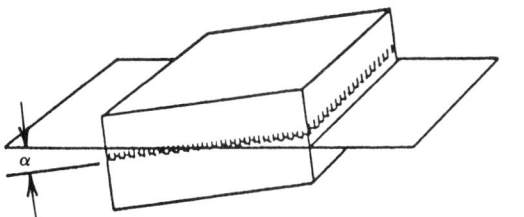 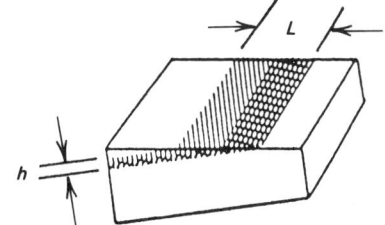

Fig 5 Geometry of the taper section. With a wedge angle, α, of 5.7° the apparent magnification of the layer structure $L/h = 10/1$

be selected according to the grade of the abrasive and the type of sample. The load commonly varies between nearly 0 and 300 N (0 and 7 lbf).

Semiautomated machines use a multiple specimen holder which can take 10 or more similar samples, that is, similar with respect to their hardness and grinding behavior. This sample holder sweeps over the rotating wheel. The cylindrical samples are free moving, so that a stepwise turn of the samples occurs when the holder sweeps back and forth. This ensures a continuous abrasion of all parts of the surface to provide a flat surface. The desired load on the samples can be obtained by putting a weight on each sample or by a spring system.

Another system uses samples glued to a metal plate which is then fixed on a magnet of the grinding machine. The grinding wheel is positioned on top of the magnet, which may be motor driven, and rotates or sweeps over the samples. This grinding wheel has the shape of a cut cone, and only the wide end of this cone that touches the samples has fixed abrasives (usually diamond or alumina).

Each grinding step causes destruction of the microstructure according to the grade of the abrasive. The subsequent grinding step must remove this damaged material. Therefore it makes no sense to start with a very coarse abrasive because too much time is needed to remove this deformed layer. The depth of this layer varies with different materials and is roughly proportional to hardness. However, it should be emphasized that in each of the grinding steps, the abrasive action and pressure are sufficient to exceed the elastic limit of the material, resulting in a newly generated layer of deformation just below the ground surface. Grinding must continue, using progressively finer abrasives, until one can be assured that all of the remaining deformation can be removed by subsequent polishing operations. This is shown schematically in Fig 6.

Semiautomated grinding, when compared to manual grinding, results in improved flatness of samples and less breakout, especially at the outer zone of the samples. Less time is also required for obtaining a high-quality finish on the samples. The reproducibility of the results is also enhanced, because parameters such as rotational speed, load, turning of the specimens, and time can be more accurately controlled.

With automatic machines, grinding parameters are experimentally determined and stored in a computer which manipulates the machine. These fully automatic machines are used when samples have to be prepared in great numbers, for example, quality control in a production situation. They are less advantageous for laboratories with frequently changing sample requirements. The primary benefit of these machines is their reproducibility, even with fewer trained or skilled personnel performing the work.

Coarse and fine polishing may be performed using the same or similar tools as mentioned in the section on grinding. The primary differences are only the grades and perhaps the type of abrasive, the kind of cloth or paper, and the lubricants. Other techniques, such as steps to remove small irregularities from the surface play no important role in ceramics.

One important point that should be mentioned is that grinding and polishing should be performed in separate rooms in order to keep samples free from coarse particles of abraded material. Cutting, embedding, and grinding may be performed in one room.

For polishing operations, alumina, cerium oxide (mainly for glass samples), and diamond are recommended, with diamond being the most common. Diamond is available in a paste or as a spray and is used on cloths of different hardness. Diamond is mainly used on rotating wheels. Grain sizes of 15, 9, 6, 3, 1, and 0.25 μm are available. These sizes are average values of grain size distributions, which are usually very narrow. The width of the grain size distribution determines the quality of the surface and the detectable microstructure of the sample, because the coarsest abrasive grain determines the depth of the scratches.

When using an abrasive-containing paste, small dabs are put on the cloth at several places. A lubricant, which is dipped onto the cloth during polishing, takes away the abraded material and retains the active diamond. This lubricant must be used carefully, otherwise aquaplaning can occur and the diamonds do not remove material. A simple test to check for the right amount is to rub a finger on the cloth. If a shiny film can be seen on the skin, the proper amount of lubricant is present.

When using a spray, one starts with uniform spraying of the cloth. Lubricant is dripped continuously onto the cloth during polishing and diamond spraying of the cloth is occasionally repeated. For refractories or zirconia, after polishing has been completed with diamond down to a 1 μm finish, fine alumina (0.3 μm) dispersed in distilled water is used as a last step.

The polishing cloths are fixed on polymer or metal wheels. Metal is preferred because of its thermal conductivity. The cloths used may be perforated to retain the abrasive. Hard cloths result in strong retention of the abrasive and in high abrasion of material, leading to a flat surface of the specimen. They are recommended for all types of ceramics. Other softer cloths with short naps abrade less aggressively and more slowly. They need a higher load on the sample to obtain a flat surface. The hard perforated cloths are preferred for the 6 and 3 μm finishing steps; the softer synthetic cloths, which are not perforated and contain short naps, are preferred for 3 μm finishes.

For final polishing, cloths from natural silk or synthetic silk are used. They are recommended for 1 μm and sometimes 0.25 μm

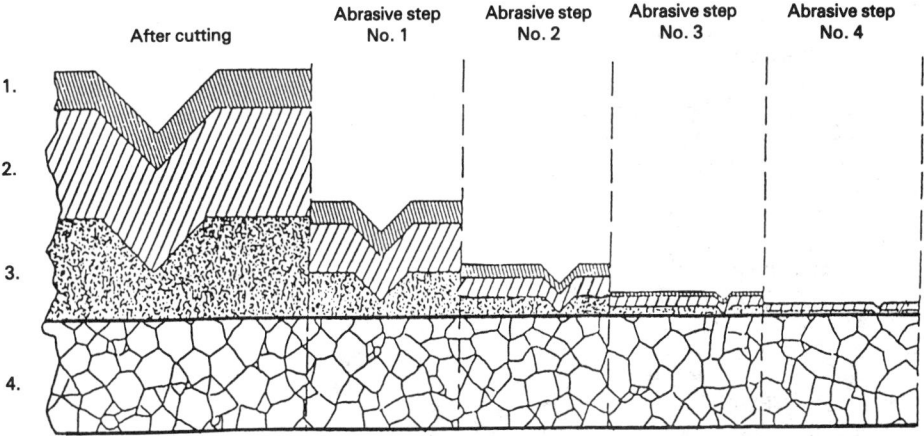

Fig 6 The effect of grinding with successively finer grit abrasive on the severely deformed material (1), the heavily deformed layer (2), the lightly deformed layer (3), and the undisturbed material (4)

finishes. This final polishing should be performed for only a short time or an undesirable relief starts to develop.

During polishing, a speed of 120 to 300 rev/min is recommended for 300 mm (12 in.) disks. The positioning of the sample holder with respect to the rotating wheel is the same as described in the discussion on grinding, and this always produces a difference in the rotating speed of the samples. Their turning and direction of movement ensures uniform material removal and results in a flat surface. The speed of polishing for ceramics is between 1 and 5 m/s (3 and 16 ft/s). The specimen load for diamond polishing is typically between 150 N (35 lbf) for coarse polishing and 30 N (7 lbf) for finishing, that is, about 0.08 to 0.02 MPa (1 to 3 psi). Hard ceramics need higher loads, which can lower the quality of the polished surface (see the discussion on grinding). For some ceramics standardized techniques are recommended, and some of these are described in Tables 1 to 3.

Vibratory polishing machines, where the amplitude of vibration is adjustable, are also available. Vibrations in both the vertical and horizontal directions induce the specimen to move about the cloth periphery. Because the polishing rate is low, several hours may be required, and thus the method is generally limited to final polishing.

Standard Methods of Sample Preparation for Incident Light Examination. Many ceramics have similar behavior with respect to grinding and polishing. It is possible, therefore, to standardize the preparation methods for SiC and/or diamond abrasive grinding and polishing. Silicon carbide is only recommended for materials with a hardness equal to or less than SiC; diamond, because of its superior hardness, can be used for all ceramics.

The recommended time for semiautomatic and manual grinding and polishing given in Tables 1 to 3 are valid for embedded samples with an area of about 1 cm (0.4 in.) square and a diameter of the embedding polymer between 25 and 30 mm (1 and 1.25 in.). Larger samples require longer preparation time (refer to Tables 1 to 3).

In general, three grinding steps are sufficient for preparation before polishing. In all cases wet preparation is recommended. The rotating speed of the grinding wheel should not exceed 120 rev/min. In manual preparation, the first step involves coarse grinding of the specimen using a grain size between 120 and 90 μm (diamond D 91 for example), followed by fine grinding using D 64 and D 15. In semiautomatic grinding, the coarse grinding is performed under a load of 120 to 150 N (25 to 35 lbf) at 120 to 300 rev/min. During fine grinding this load is reduced to 60 to 90 N (15 to 20 lbf) to avoid pullout of grains, especially at the sample border.

Polishing with diamond requires more time than grinding for both semiautomatic and manual methods; automated polishing requires

less time. The polishing is performed in three steps using diamonds with 6, 3, and 1 μm diameter. Hard cloths are recommended for coarse and fine polishing, soft cloths for the final polishing. Some of the cloths used for polishing ceramics include pellon (a rayon/nylon mixture), nylon, and silk. Handbooks describing the proper use of various grinding and polishing materials are offered by all metallography/ceramography equipment suppliers.

After each step, a thorough cleaning of the samples must be carried out with an ultrasonic cleaner. A detergent may be used to enhance the cleaning progress. In addition, the operator's hands must be washed between

steps, as they are an important source of cross contamination. *Do not wash the hands in the operating ultrasonic cleaner!*

Revealing of Microstructure

The previous treatment of specimens, correctly performed, yields samples which are flat, smooth, free from physical imperfections, and representative of the ceramic under consideration. Their surface should appear almost featureless in the microscope, the exception being polyphase materials which contain differences in hardness, anisotropy, color, surface defects, or reflectivity. Such

Table 1 Recommendations for grinding with SiC abrasives

Grinding method	Abrasive grain size	Lubricant	Speed, rev/min	Load, N (lbf)	Time, min(a)
Manual					
Parallel grinding	220 grade	Water	120	Strong	5
Coarse grinding	320 grade	Water	120–300	Strong	5
Fine grinding	800 to 1000 grade	Water	120–300	Strong	5
Semiautomatic					
Parallel grinding	220 grade	Water	120	30–150 (7–35)	5
Coarse grinding	320 grade	Water	120–300	30–90 (7–20)	5
Fine grinding	800 to 1000 grade	Water	120–300	30–90 (7–20)	5

(a) Time is dependent on material. It must be ground to a flat parallel surface.

Table 2 Recommendations for grinding with diamond wheels

Grinding method	Abrasive grain size, μm	Lubricant	Speed, rev/min	Load, N (lbf)	Time, min(a)
Manual					
Parallel grinding	120–90	Water	120	Strong	5–10
Coarse grinding	60–45	Water	120–300	Strong	5–10
Fine grinding	25–10	Water	120–300	Strong	5–10
Semiautomatic					
Parallel grinding	60–45	Water	200–300	120–150 (25–35)	5–10
Fine grinding	25–10	Water	200–300	60–90 (15–20)	5–10

(a) Time depends on material. A flat surface is required.

Table 3 Recommendations for diamond polishing

Polishing method	Cloth	Abrasive grain size, μm	Lubricant	Speed, rev/min	Load, N (lbf)	Time, min
Manual						
Coarse polishing	Pellon(a)	6	Alcohol-based	120–200	Strong	30–40
Fine polishing	Fine pellon	3	Glycerol	120–200	Medium	30–40
Final finishing	Soft cloth or silk	1	Oil	120–200	Medium	20–30
Semiautomatic						
Coarse polishing	Pellon(a)	6	Alcohol-based	120–200	120–150 (25–35)	20–30
Fine polishing	Fine pellon	3	Glycerol	120–200	90 (20)	10–20
Final finishing	Silk	1	Oil	120–200	30 (7)	5–10

(a) May be perforated or unperforated pellon

Table 4 Examples of chemical etching of selected ceramics

Material	Etchant	Temperature °C	°F	Time
Silicate ceramics	2 to 5% hydrofluoric acid (HF)	20	70	1–20 min
Calcia and magnesia ceramics	Conc HCl	20	70	35–6 min
Barium titanate	1/3 HF/conc HCl	20	70	7 min–2 h
Silicon nitride	NaOH melt	400–450	750–840	1 min–10 min
Silicon carbide	Modified Murakami's reagent (3 g KOH or NaOH, 30 g K-hexacyanoferrate (III), 60 mL water)	Boiling	Boiling	3 min–30 min
Zirconia ceramics	Concentrated phosphoric acid	Boiling	Boiling	5 min–10 min

Conc, concentrated

samples show the true pore-size distribution and should be used for quantitative analysis.

Chemical and Thermal Etching. In order to fully reveal the microstructure of a ceramic, further treatments, that is, etching methods, are required. Etching involves removal of material from the sample by acids (chemical etching), evaporation (sputtering), or thermal etching. During etching, atoms, ions, and molecules are removed. The bonding of atoms, ions, or molecules is weaker at exposed positions such as at grain boundaries and imperfections; therefore etching first initiates at these locations, yielding a delineation of the microstructure. Thermal etching takes place at least 100 K below the sintering temperature and is governed by surface tension between phases in contact with each other at equilibrium with the surrounding atmosphere.

The concentration of the chemical, the time of exposure, and temperature all play an important role during the etching process, and each procedure must be individually tailored for every ceramic. Because every ceramic must be considered on an individual basis, only a few examples can be given in this article. Table 4 contains some ceramic and chemical combinations used for chemical etching. Hydrofluoric acid etching may be used for silicate-containing ceramics, mainly if a glass phase is present. Proper safety precautions must be observed when performing chemical etching.

Table 5 contains some etchants and suggested times for proper treatment of ceramics containing 94 to 98.8 wt% Al_2O_3. Since chemical etching involves removal of material, the following complications must always be considered:

- Enlargement of all pores and grain boundaries
- Dissolution or loss of small particles
- Widening of scratches
- Etch pits and other artifacts on mineral surfaces
- Layers of chemical reactants on the surface
- Poor reducibility, particularly when using salt melts

Thermal etching is done on samples that are not embedded because the polymer will burn away. Porous materials, which must be infiltrated for grinding and polishing purposes, present a problem. The infiltration may burn off leaving a carbon layer on the surface of the thermally treated sample.

Another problem may be the sintering behavior of the material, which causes phase transitions, and subsequently influences the microstructure. Thermal etching should be carried out in a special furnace used for this purpose only in order to prevent contamination. Thermal treatment at excessively high temperatures will cause grain growth as sintering mechanisms may proceed. When carried out properly, however, thermal etching causes very small lines (grooves) to be produced at grain boundaries and phase boundaries, thereby making it a very effective technique for delineating grain structure in ceramics using optical and electron microscopes. Thermal etching parameters are outlined in Table 6.

Other Etching Methods. Cathodic vacuum etching (ion bombardment) may also be used for delineating a ceramic microstructure. Using this technique, selective removal of atoms takes place from the specimen sur-

face by positive bombardment in a glow-discharge environment. The glow discharge is initiated by applying a high direct current voltage between the specimen support and the anode in a partial vacuum. With such a device, etching of embedded specimens is also possible. The necessary voltage is in the range 1 to 7.5 kV; ion current is 60 to 400 µA. Argon or krypton gas is used for the partial vacuum control, the bombardment angle should be 5 to 60°, and the etching time is 20 to 90 min. Good results have been achieved with porcelain, silicon nitride, and zirconia ceramics.

Electrolytic etching may be carried out on ceramic semiconductors or conductors. Procedures are described in the literature for oxides, borides, and carbides. The grain size of the material is important. In some cases electrolytic etching is preferred instead of chemical etching, for example, for SiC with a coarse microstructure. For fine-grained materials chemical methods produce better results.

Another possibility for revealing the microstructure is the use of vapor deposited interference films for phase contrast. This method is based on an optical-contrast mechanism without chemical or morphological alteration of the specimen surface. The specimen is coated with a transparent layer whose thickness is small compared to the resolving power of the optical microscope. In interference layer microscopy, light that is incident on the deposited film is reflected at the air/layer and layer/specimen interfaces. Phases with different optical constants appear in various degrees of brightness and colors. The color of a phase is determined by its optical constants and by the thickness and optical constants of the interference layer.

In many cases, the reflectance of the ceramic phases is too similar to distinguish between them or to detect grain boundaries, even if they are properly developed. In these cases the contrast can be enhanced by sputtering a thin layer of a metal onto the surface. Gold or aluminum is normally used as the contrasting film.

One reason for coating with gold is the fact that most ceramics possess a high transparency for light. As a result, upon examination not only can the surface be viewed, but reflected light from the inner portions of the

Table 5 Chemical etching of alumina-containing ceramics (94 to 99.8 wt% Al_2O_3)

Etchant	Temperature °C	°F	Time
10% HF	20	70	15 min
Conc H_2SO_4	230	450	2–10 min
Conc H_3PO_4	250	480	1–10 min
$K_2S_2O_4$ melt	650	1200	1–10 min
V_2O_5 melt	900	1650	1 min
Borax melt	900	1650	15–45 s

Table 6 Thermal etching of selected ceramics

Material	Atmosphere	Time	Temperature °C	°F
α-Al_2O_3 (<96 wt% Al_2O_3)	Air	0.5–4.5 h	1250–1500	2280–2730
β-Al_2O_3	Air	1–5 min	1470(a)	2680(a)
Strontium and barium ferrites	Air	1 h	1050–1150	1920–2100
TiO_2	Air	1 h	1350	2460
Si_3N_4	Vacuum	15 min	1250	2280
	Nitrogen	5 h	1600	2910
SiC	Vacuum	1–3 h	1300–1500	2370–2730

(a) Thermal etching should be performed 200 °C (360 °F) below the sintering temperature.

sample, where the light is scattered and/or reflected from pore walls, grain boundaries or inhomogeneities of any kind, can also be seen. This scattered and reflected light yields a milky appearing surface with very low contrast.

The thickness of such gold layers should be between 5 and 30 μm. A layer of only 5 μm still allows the detection of the different reflectivities of the phases under investigation. Carbon is recommended as the conducting material when a quantitative chemical analysis with excited x-rays in a microprobe or an electron microscope shall be carried out. Interference light microscopy frequently uses samples with sputtered thin layers of oxides or non-oxides (differential interference contrast microscopy). This technique can be used to make visible small differences in the roughness of a surface depending on the preparation or the microstructure of a sample. The relief of such samples seems to be very smooth in ordinary incident-beam microscopy but becomes seemingly sharply contoured in the differential interference microscope. This technique is recommended for examining the quality of a polished surface, since tiny scratches become visible.

Preparation of Thin Sections for Transmitted Illumination

Thin sections of materials can be used successfully in investigations of translucent phases. With the polarizing microscope it is possible not only to determine the microstructure, that is, the size distribution and shape of the components, but also the types of phases present. It is then possible to determine many physical parameters that relate to the interaction with light such as refractive index, birefringence, optical angle, and color, which lead to a true identification of the material in question.

The thickness of the samples for these investigations must be on the order of the diameter of a grain, otherwise different "layers" of different phases are superimposed and disturb each other optically. In petrographic samples, this thickness is standardized to 30 μm. In ceramic samples, the thickness may have to be even thinner according to the diameter or fineness of the components.

The preparation of thin sections is generally very similar to the preparation of specimens for incident light microscopes. The sectioning, infiltration and embedding, grinding, and polishing are identical until the 7 μm polishing step.

In the standard procedure, the sample will be bonded to a standardized glass slide (Giessener format) with a resin of defined refractive index. The resin most often used is Canada balsam with a refractive index of 1.54. More recently, Lakeside 70 cement, transparent epoxy cements, or araldite have also been used. The refractive index of the latter materials runs between 1.55 and 1.61. The resin can be used as a reference for determinations of the refractive index of the phases in the specimens. The infiltration of the specimens should be carried out with the identical epoxy system for this reason.

The bonding of the specimen onto the glass slide using Canada balsam is performed on a hot plate which should be set at 150 ± 5 °C (300 ± 9 °F). Heating is complete when a small thread of the resin drawn from the slide is hard but flexible. To ensure that no bubbles are trapped, light pressure should be applied to the specimen slide segment as it is moved back and forth. Foaming of the resin should be prevented. The goal is a very thin film of resin without any air bubbles.

The slide is then placed on a cooling block under moderate load until the cement has hardened. In the event that air bubbles are present, the slide can be reheated and the procedure repeated.

The sample is then fixed to a thin-section saw, where the glass slide is attracted to the device by vacuum. The cutting wheels contain diamond. A slow cutting speed and fine diamonds disturb the microstructure less than a high speed and coarse grains. The distance between the sample holder and cutting wheel, which determines the thickness of the thin section, is controlled by a microscrew, which allows for an adjustment within a micrometer with high precision. After proper sectioning, further polishing down to about a 7 μm finish is needed. Using such cutoff devices, the desired thickness of the thin section can nearly be achieved. The final grinding should be performed with a slowly rotating (≤100 rev/min) or stationary 600- or 1000-mesh diamond lapping wheel.

The thickness of the thin section can be determined with a polarizing microscope under crossed Nicols making use of the interference colors and the Michel-Levy Cart (Fig 7). The last step is to cover the specimen with a cover slip, using a procedure similar to the attachment of the specimen to the glass slide. The function of this cover slip and its mounting medium is twofold: they substitute a smooth,

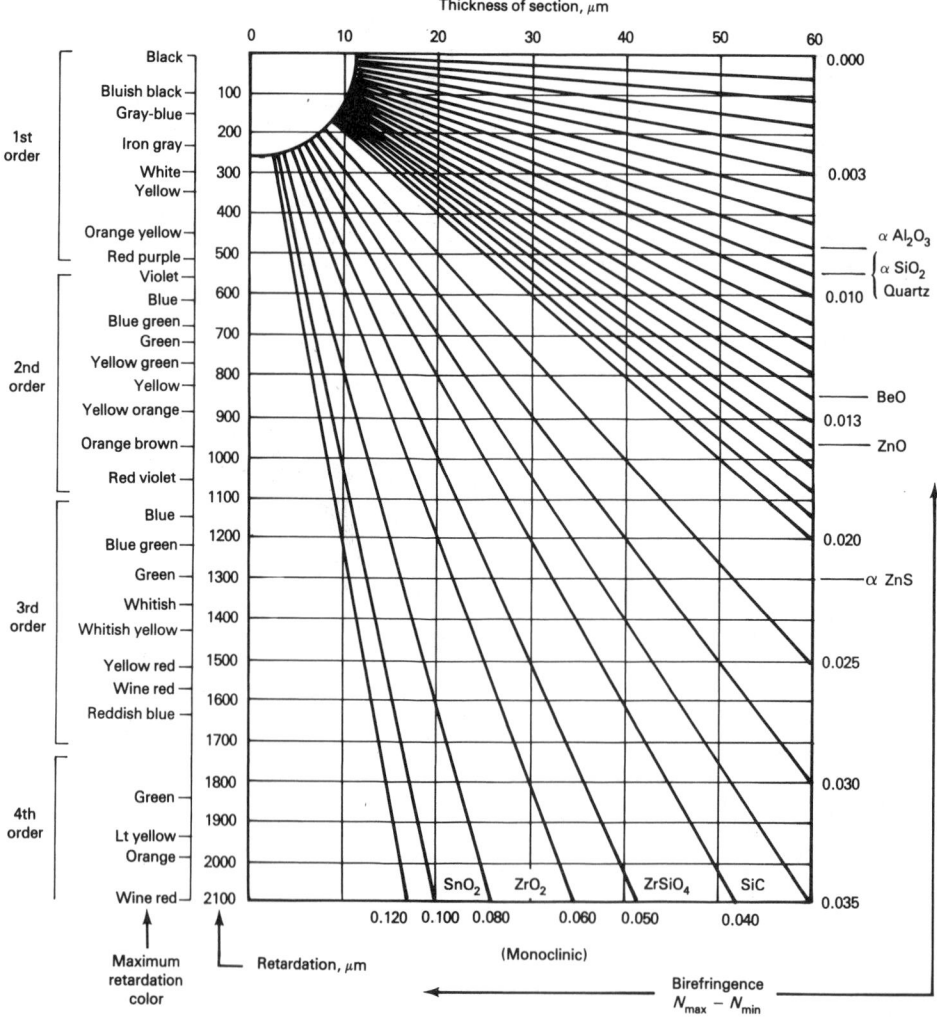

Fig 7 Diagram showing the relations of interference colors to thickness and birefringence. The birefringences of the more common non-cubic minerals are listed.

continuous, solid-air interface for the relatively rough segment surface which would otherwise scatter the illumination and ruin the image quality. Secondly, they provide the proper optical path for transmitted-light objectives at high magnifications.

Air bubbles must again be avoided when attaching the cover slips. One way is to perform the following procedure:

- Warm the thin section to a temperature appropriate for the resin being used—in most cases about 110 °C (230 °F)
- Place a small amount of cement on the ground section and allow it to cure slightly
- Press the clean and dry cover slip down into the mounting medium, thereby eliminating air bubbles from the mounting

After cooling, the thin section is ready for examination.

Cases where a final polishing step is performed instead of using a cover slip include samples examined with a polarizing microscope in transmitted light or incident-light illumination, examinations in a scanning electron microscope with x-ray analysis, and examinations with an electron microprobe. For electron microscope applications, thin sections must be covered with a gold or carbon layer.

Fracture Surface Specimens

The primary goal of examining fracture surfaces is to determine what kind of flaw (an aggregate, pore, microcrack, and so on) was the origin/cause of the failure. Even the path of fracture, whether through grain boundaries or through the grains themselves, provides useful information in connection with material testing, troubleshooting of fabrication problems, and service failures.

In many cases, a fracture surface of a ceramic is all that is needed to gain insight into the microstructure, particularly information about the pore size distribution, the grain size distribution, and the homogeneity of the material.

Small specimens can be examined without any cutting. For larger ones, the sample should be cut off carefully from the bulk. Suitable samples cannot be obtained by breaking off a piece with a hammer because this will produce additional cracks and microcracks which are usually impossible to distinguish from the original ones.

Sometimes the fracture surface should have a special location or orientation. One can ensure this by cutting a small slit to a certain depth into the ceramic with a very fine cutoff wheel at a very low speed, and then bending the specimen. The fracture initiates exactly along this slit.

After removing the specimen from the bulk, it can be mounted in a sample holder for further investigations with optical microscopes without any further treatment. Sometimes,

however, a chemical etching is carried out to develop the grain boundaries.

If observations with electron microscopes or an electron microprobe are planned, the specimen needs to be bonded to the sample holder with a conducting resin and thinly coated with carbon or a metal (gold or gold-palladium). This coating is done either by evaporation in a high-vacuum chamber or by sputtering in a low-vacuum chamber.

Replication. In order to preserve polished sections or fracture surfaces, the surface can be replicated as a thin film which preserves its topographical details. The original sample is available after replication for further treatments or other examinations.

The process may be as simple as pressing a plastic tape, softened with an appropriate solvent or heat, against the surface and stripping it off. Another technique is to use a silicone rubber, applied as a liquid, which polymerizes quickly and gives true results. The replica is a negative of the surface, that is, high points in the sample correspond to hollows in the replica. Such replicas may be examined with an optical microscope. Coating with gold may enhance the contrast when using incident illumination.

The replication process for electron microscopy usually involves several stages. The specimen is first coated with a film of Formvar as a 1 or 2% solution in chloroform. This is backed with a 7% solution of a soluble resin (Bedacryl) in benzene. The replica is then stripped from the surface, and carbon is evaporated onto it from a carbon arc in a vacuum chamber. The backing resin is then dissolved with acetone, and the Formvar removed with chloroform. The film is then ready for examination.

To enhance contrast in such replicas and to provide a relief effect giving three-dimensional information, the replicas are generally shadowed by evaporation of a metal such as gold at a glancing angle so that the metal accumulates on the near side of positive details, while the far sides are sheltered (Fig 8). The appearance of such shadowed replicas in the electron microscopes gives the impression of illumination at a glancing angle and throws details into strong relief.

Microstructural Analysis

The study of microstructures is concerned essentially with the ability to distinguish between materials having different chemical composition, structure, or orientation and to distinguish between details separated by only

small distances. Techniques for revealing microstructural features are almost all based either on the response of their details to incident radiation or on the quality and intensity of the radiation emitted by features when appropriately excited.

When a homogeneous object is viewed in transmission, that is, electromagnetic radiation is transmitted through it, details corresponding to greater thickness have greater absorption and will appear as dark regions in the image. At constant thickness, details having greater absorption coefficients also yield darker images and, if the absorptions differ sufficiently for different light wavelengths, they may show image color differences. The resolution, or resolving power, is a function of the wavelength, λ, of the radiation or incident light and the numerical aperture of the reflective lens.

Refraction effects on phase boundaries, for example, grain boundaries, may change the light path, brightening part of the structure and darkening the neighboring area. This effect is made more visible by being slightly out of focus and restricting the illumination aperture (for example, the iris diaphragm on a light microscope). Scattering effects on the grain boundaries can also be used for imaging. The intensity of this effect depends on the relative refractive indices. Using dark-field illumination, scattering on grain (phase) boundaries in the light path produces bright images on a dark background. On the other hand, in bright-field illumination, phase boundaries show up as dark lines owing to scattering and loss of light.

Polarization interference effects occur when an object is between crossed polarizing elements (polarizer and analyzer), with or without additional accessories. These effects may be useful in identifying minerals (phases). All minerals except those belonging to the cubic system and all transparent solids under stress split an incident linearly polarized light-ray into two rays vibrating perpendicular to each other that travel at different velocities through the material. When the two rays emerge, one lags behind the other to an extent determined by the wavelength of the light, the thickness of the material, and the birefringence (that is, the difference between the refractive indices of the two rays in the material). Under crossed polarizing elements, this path difference (retardation) is seen as color. The colors change if the mineral's orientation is changed. One can make use of this to determine this orientation quantitatively by viewing the rear surface of the objective lens when the light incident on the mineral is in the form of a strongly converging cone. Therefore, the polarizing microscope (petrographic microscope) is not only a magnification device, it is a tool for making physical measurements. This kind of observation in transmitted light (that is, using crossed polarizing elements) becomes more difficult with small particle sizes. Even thin sections that are only 10 μm

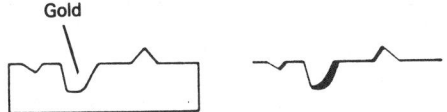

Fig 8 Replication technique using evaporated gold that produces a "shadowed" replica

thick may contain overlapping particles, which makes it impossible to perform analysis on individual grains in the material. The recourse is to use polished sections and work with reflected light. This is the common approach with most oxide and non-oxide ceramics today.

In both transmission and reflection conditions, a light ray passing through the grain or being reflected from the surface is retarded somewhat behind those which pass around it or which have encountered materials having other refractive indices. In phase-contrast microscopy, an annular diaphragm is inserted in the condenser, and a second annular ring is set in the conjugate plane of the condenser ring, thus introducing an additional small phase shift. Interference between these two in the image plane provides sharp contrast between details which originally varied only slightly.

Interference Microscopy. The phenomenon of interference is also applied in the so-called interference microscope. The most widely used technique is Nomarski differential-interference microscopy.

In the simplest arrangement of this type, a ray of light emitted from a light source is linearly polarized after it passes through the polarizer. It then enters the Nomarski biprism (Wollaston prism), which consists of two optically uniaxially doubly refracting crystals and is divided into two rays of linearly polarized light. The planes of vibration of these rays are perpendicular to each other. Upon passing through the objective, the rays become parallel and impinge on the specimen. After reflection from the specimen surface, they are recombined by the biprism. Interference is produced when these recombined rays pass through the analyzer.

As with normal polarized light microscopy, the analyzer is in a crossed relationship with respect to the polarizer. Phase differences resulting from the two spatially separated beams reflecting from the specimen are due to differences in height of the surface relief, which are modified by the optical properties of the specimen. These phase differences cause the light-dark or color interference contrast. Lateral displacement of the biprism allows an additional phase difference to be superimposed that varies color contrast. The achievable contrast depends on the local gradient of the phase difference. Therefore, this type of contrast is termed differential interference contrast. Images produced using this optical etching technique are characterized by their three-dimensional appearance.

Polarizing Microscopy. To the ceramist, the polarizing microscope is an essential tool for revealing ceramic textures, grain size, and phases. After some practice, most of the common phases can be recognized by their appearance using such obvious features as color, relative relief, typical inclusions, order of birefringence, and type of extinction. The polarizing microscope is restricted to grains having a diameter larger than 1 μm. Therefore, it can be used with raw materials, refractories, and other ceramics having a coarse microstructure.

Reflected-Light Microscopy. Microscopic examinations of ceramics by transmitted light often suffer from the fact that the thickness of single grains of constituents is much less than the thickness of the section. Examination with the reflected-light microscope eliminates the difficulty associated with overlap of constituents within the thickness of the section, since the image is formed essentially by reflection from the polished surface of the object.

Since reflectivity varies directly with the index of refraction, the low indices of most ceramic materials, which make the polarizing microscope so effective with transmitted light, constitute a decided handicap in the use of the metallograph. It is partly for this reason that the metallograph has found its greatest field of usefulness in the examination of materials that have relatively high refractive indices, such as the constituents of basic refractories, some of the special oxides, porcelains, or carbides. The reflectivity can be enhanced by sputtering thin layers of a metal onto the surface (interference staining) to increase the contrast. Quantitative measure of reflectivity can be combined with microhardness measurements in order to identify the phases present.

Image Analysis. The metallograph is often used for quantitative image analysis of polished thin sections that are observed using transmitted and reflected light (both utilizing polarization effects). Image analysis is also combined with results obtained using electron microscopes. Image analysis is used for quantitative evaluation of microstructure in terms of grain size distribution, pore size distribution, shape of constituents, paragenesis, and abundance of constituents. Constituents in the microstructure can be characterized by their contrast, color, and chemical composition. The use of etchants on the polished specimens is a universal practice to heighten the contrast between the phases and help in their identification. Stains can also be used, either following the etch, or in place of it.

Image analysis is carried out using any type of microscope combined with a video camera which converts the information into digitized signals which can be analyzed by a computer. The operator compares the calculated image on the screen with direct observation in the microscope. The best fit of the calculation procedure is then used as the basis for further mathematical treatment. The procedure depends on the ability of the system to distinguish contrasts. Therefore, many systems offer the possibility for an interactive procedure where the operator draws separating lines or unifies separated particles on-line. Image analysis systems may operate off-line using micrographs, or on-line, being directly connected to a microscope.

Electron Microscopy. The resolution of a microscope depends on the wavelength of the radiation used. A beam of electrons has a very short wavelength of about 0.015 nm. The electron microscope uses an incandescent filament as a source of electrons, accelerates them to high velocity by means of an electric field, and focuses them with magnetic or electrostatic lenses in much the same way an optical microscope focuses light beams. One significant difference compared to light microscopes is the very poor penetrating power of the electron beam, which limits objects to extremely thin sections unless a reflecting technique is employed. The effectiveness of the electron microscope is directly linked to the quality of specimen preparation, particularly the replica techniques. Useful magnifications up to about 2×10^5 are routinely employed, and under special conditions, dislocations themselves may be directly observed by using diffraction effects. Many instruments are equipped with the means for obtaining electron diffraction patterns of individual grains located in the field of view, and thus phase identification and orientation can be studied.

Scanning Electron Microscopy. In this instrument the electron beam is brought to a fine focus on a very small region of the sample, and the scattered electrons (which may be of several types) are collected on a fixed detector, the signal of which is fed to a cathode-ray tube (CRT). The beam is swept across the surface of the specimen and the CRT, which is synchronized with the beam, traces an image of the specimen. Such characteristics as atomic number and density of atoms in the specimen affect the contrast of the image. Useful magnifications of about 50,000$\times$ are possible, but magnifications in the same range as the optical microscope are more commonly used. The depth of focus, however, is much greater than that of the optical microscope at any particular magnification, and excellent photomicrographs of numerous ceramics have been obtained. This instrument is one of the more popular used by materials scientists today, especially when it is equipped with detectors for studying the excited x-rays for qualitative and quantitative chemical analysis. In this respect, it is very similar to the electron microprobe.

Electron probe microanalysis provides the means of determining the chemical composition of very small volumes at the surface of polished sections or polished thin sections of ceramic materials. The name derives from the essential feature of a fine electron beam which is directed at the point to be analyzed. The x-rays generated by the incident beam are characteristic of the elements, and the intensity of these x-rays is an approximately linear function of concentration.

In its simplest form, the electron microprobe consists of an electrooptical system which features an electron beam that is focused onto an area about 1 μm in diameter on the surface of the specimen, a stage on which the specimen and standards are mounted, a microscope which allows the area of interest to be selected, and a detection system which measures the intensity of the characteristic radiation of the elements to be determined. Scanning the beam over the specimen allows the distribution of chosen elements on the surface of the specimen to be displayed as an enlarged image on the screen of a CRT. If the electron beam is swept across the surface while the x-ray detector is set to record a wavelength characteristic of a particular element, a plot may be obtained on an oscilloscope; brightness in the oscilloscope image corresponds to points in the specimen where the element is found in abundance.

The characteristic x-ray spectra generated by the electron beam are identical to those produced in fluorescence excitation by x-ray radiation. In addition, the slowing of the electrons produces a continuous spectrum or "bremsstrahlung," which constitutes a background upon which the characteristic lines are superimposed. These x-rays can be analyzed with respect to wavelength (or energy) and intensity to yield spectra from which the constituent elements can be identified both qualitatively and quantitatively. The basic measurement is a comparison of the net intensity of a particular x-ray line generated in the specimen with that generated in a standard by the same incident current. However, many corrections are needed to obtain true results, for example, corrections related to atomic number, absorption, and fluorescence. Geometric factors that influence the results include orientation of the specimen and width of the beam. Under normal conditions, the resolution of this method is equal to that of the light microscope. The resolution is related to the size of the excited area which, in turn, depends on the thickness of the specimen. Only in the transmission mode using very thin specimens does the resolution become improved, approaching the diameter of the electron beam.

SELECTED REFERENCES

- K.M. Bowkett and D.A. Smith, Field Ion Microscopy, North Holland, 1970
- D.J. Clinton, *A Guide to Polishing and Etching of Technical and Engineering Ceramics*, The Institute of Ceramics, Stoke-on-Trent, England, 1987
- G. Elszner, *et al.*, Methoden zur Ansohliffpraperation Keramischer Werkstoffe, Fachausschußbericht No. 24, "Keramische Werkstoffe in der Technik" der Deutschen Keramischen Gesellschaft, Bad-Honnef, 1985
- V.D. Frechette, Experimental Techniques for Microstructure Investigations, *Microstructure of Ceramic Materials*, Proceedings of American Ceramic Society Symposium, Pittsburgh, 27–28 Apr 1963, NBS Publication 257, National Bureau of Standards, 1963, p 15–28
- N.H. Hartshorne and A. Stuart, *Crystals and the Polarising Microscope*, Edward Arnold Publishers Ltd., 1969
- H. Insley and V.D. Frechette, *Microscopy in Ceramics and Cements*, Academic Press, 1955
- E. Kordes, *Optische Daten zur Bestimmung Anorganischer Substanzen*, Verlag Chemie, 1960
- T.P. McKinley, Ed., *The Electron Microprobe*, John Wiley & Sons, 1966
- L. Reimer, *Scanning Electron Microscopy*, Springer Verlag, 1972
- J.R. Richardson, *Optical Microscopy for Material Sciences*, Marcel Dekker, 1971
- G. Thomas, Ed., *Electron Microscopy and Structure of Materials*, University of California Press, Berkely, 1972
- P.R. Thornton, *Scanning Electron Microscopy*, Chapman and Hall Ltd., 1968
- H.-R. Wenk, Ed., *Electron Microscopy in Mineralogy*, Springer Verlag, 1976
- J. Zussmann, Physical Methods in Determinative Mineralogy, Academic Press, 1967

Porosity, Density, and Surface Area Measurements

Joan E. Shields, Department of Chemistry, C.W. Post Campus, Long Island University

TRADITIONAL CERAMICS are composed of natural mineral substances, but an increasing number of specialized ceramics are based on chemically processed materials that may not originate from mined products. Therefore, these new composites require a higher degree of characterization than traditional ceramics, and a knowledge of such macroscopic properties as particle size, surface area, porosity, and density is critical.

The processing of ceramics usually involves compaction and firing at elevated temperatures to develop useful properties. Thus, it is significant to be able to characterize the changes that occur in these properties. A powder compact, upon formation and before firing, comprises individual particles that usually have between 25 and 60 vol% porosity, depending on the material and the process used. Generally, the most significant changes that occur during the firing process include an increase in particle size and changes in pore shape and size, which usually lead to a decrease in surface area. The elimination of porosity maximizes such properties as strength, translucency, and thermal conductivity.

Two general categories of ceramics characterization are atomic characterization, which defines the relative location of atoms, and macroscopic property characterization, especially using porosity, density, and surface area measurements. This article focuses on these particular macroscopic properties.

Porosity

Porosity is almost always present in ceramics that are prepared by powder compaction and heat treatment. Upon firing, pores initially present in a ceramic can change in shape and form either channels or isolated spheres, without necessarily changing in size. More commonly, however, both the size and shape of the pores are altered during the firing process. The pores become smaller and more spherical in shape as sintering (agglomeration by heating) is continued. As particles merge together, intraparticle pores, or voids, are transformed into interparticle pores.

Porosity is defined as the volume fraction of pores present. The amount of porosity in a ceramic can vary from zero to more than 90% of the total volume. In a powder compact, the void volume, V_v, which is the volume not occupied by solids, can be used to define the porosity, p, as:

$$p = \frac{V_v}{V_v + V_s} \qquad \text{(Eq 1)}$$

in which V_s is the volume occupied only by the solid, and $V_s + V_v$ represents the total volume of the compact.

The volume of a solid is open to several interpretations because of the possible existence of various types of pores (voids). The individual particles may contain pores (intraparticle) that are either open to the surface (open porosity) or totally enclosed within the particle (closed porosity). In addition, interparticle voids may be present in a solid compact, depending on the degree of packing.

Two types of porosity, open and closed, can be used to describe a solid. Because closed, or blind, pores usually cannot be detected, porosity is generally characterized as open, or apparent, porosity, which describes those pores that have access to the surface. It should be emphasized that the total, or actual, porosity includes both open and closed pores. Apparent porosity is not only easy to measure experimentally (compared to closed porosity), but is more important because it directly affects such properties as permeability, vacuum tightness, and surface availability for catalytic reactions. Closed pores have little or no effect on these properties because of their lack of accessibility.

Before firing, almost the entire porosity of a ceramic material is represented by open pores. During the firing process, the porosity of the sample decreases. Some open pores are totally eliminated, and some are transformed into closed pores. As a result, the volume fraction of closed pores increases initially and then decreases toward the end of the firing process. Open porosity is generally eliminated when the porosity has decreased to about

5% of its initial value. By the time 95% of the theoretical sample volume is reached, the ceramic is gastight, that is, open porosity has been essentially eliminated.

Open Porosity

Two commonly used experimental methods for determining porosity are mercury porosimetry and vapor sorption.

Mercury Porosimetry. Since the development of the first mercury porosimeter in 1945 (Ref 1), this technique has become an established method for the determination of open porosity in porous materials. The basis of this method is the Washburn equation (Ref 2), which relates the pressure, P, needed to force mercury into pores of radius, r, assuming cylindrical pore geometry:

$$Pr = -2\gamma \cos \theta \qquad \text{(Eq 2)}$$

In Eq 2, the surface tension and contact angle of mercury with the pore walls are expressed as γ and θ, respectively. The Washburn equation is derived by equating the force that drives mercury out of a pore to the force associated with the pressure that directs mercury into a pore.

Commercially available porosimeters operate at pressures that range from below ambient to 414 MPa (60 ksi). Assuming a value of 0.480 N/m (480 ergs/cm^2) for γ and 140° for θ in Eq 2, intrusion of mercury will occur in pores ranging in size from 7.25 μm (290 μin.) to 1.7 nm (0.07 μin.). A typical high-pressure mercury intrusion-extrusion curve is illustrated in Fig 1.

Because ceramic materials usually possess some pores that are larger than those that will intrude mercury at these pressures (that is, pores with more than a 7.25 μm, or 290 μin. radius), low-pressure mercury porosimetry may be more valuable in characterizing these materials. This technique involves the evacuation of the sample and subsequent measurement of the intrusion of mercury, from about 130 Pa (1 torr) up to ambient pressure, corresponding to pores that range from about 107 to 7.25 μm (4280 to 290 μin.). A typical

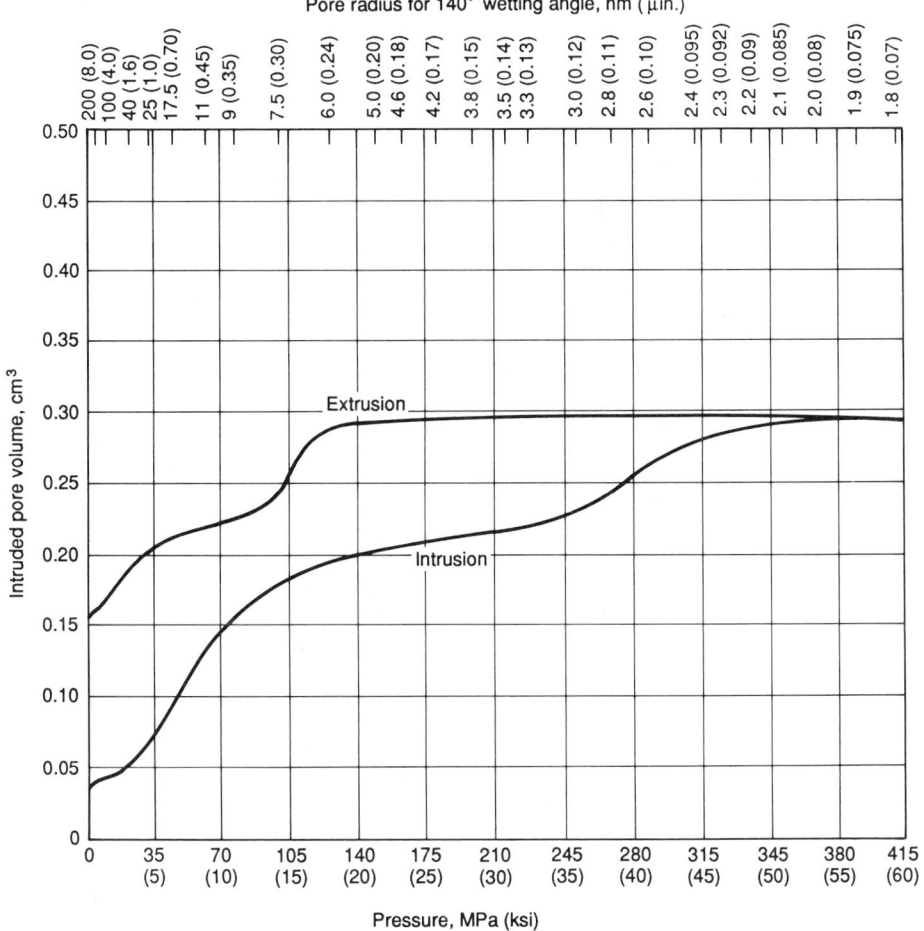

Fig 1 High-pressure mercury porosimetry curve obtained on an Autoscan 60. Courtesy of Quantachrome Corp.

Although the shape of intrusion curves depends on the size and volume of the pores, a characteristic feature of almost all porosimetry curves is that they exhibit uptake of mercury at low pressures as mercury intrudes into the larger interparticle voids, followed by intrusion into smaller intraparticle pores at higher pressures.

Limitations of mercury porosimetry are described below (Ref 3). First, the Washburn equation (Eq 2) governing the calculation of pore size assumes that the pores are cylindrical. This may not reflect their true geometry. Second, the assumption of a value for the contact angle θ in Eq 2 simplifies the experimental procedure, but could introduce an error. However, the availability of commercial anglometers for measuring the contact angle of mercury on a solid surface eliminates the need for making the assumption. On the other hand, a potential limitation is imposed by the possible variations in the contact angle and the surface tension over a very wide range of pressures, from below ambient pressure to 414 MPa (60 ksi).

A third limitation of mercury porosimetry is the possible influence of high pressure on the sample. Fortunately, most solid materials possess such low compressibility characteristics that the effect is negligible. However, some softer solids could show an appreciable compression at high pressures, which would appear as an intrusion of mercury. Henrion and Leurs (Ref 4) have reported the effects of sample compaction on compressibility, and the extent of pore deformation at high pressure has been studied by Brown and Lard (Ref 5).

Finally, because commercial porosimeters have an upper limit of 414 MPa (60 ksi), the lower limit of pores into which mercury can intrude is approximately 1.7 nm (0.07 μin.) in radius. Therefore, smaller pores (micropores) must be studied using a different technique.

Gas Sorption. There are two types of adsorbing interactions between gases and solids. Chemical adsorption, or chemisorption, involves strong chemical interactions between a gas (adsorbate) and a solid (adsorbent). The measurement of chemisorption is particularly important for the characterization of catalysts. Physical adsorption involves weaker van der Waals forces, similar to the liquefaction of vapors.

A widely used technique for determining porosity is based on the physical adsorption and subsequent desorption of an adsorbate on the surface of an adsorbent. The vapor sorption method is not only used for measuring pore size distributions but also is the basis for determining surface area.

When an adsorbent (solid) is allowed to interact with an adsorbate vapor, typically nitrogen at low temperatures, an amount of the vapor will physically adsorb on the surface and in pores as a function of the partial pressure of the vapor. Sorption studies lead-

low-pressure intrusion curve is illustrated in Fig 2. A significant advantage of mercury porosimetry is that relatively large pores can be measured.

The data obtained from the intrusion of mercury into a porous material can be reduced to provide the following functions:

- Percent pore volume plotted versus pressure and/or pore radius
- Derivative of the cumulative pressure-volume data, expressed as dV/dP versus pressure and/or pore radius
- Pore size or pore volume distribution function, $D_v(r)$, which is the volume change (dV) of mercury intruded per unit change in pore radius (dr), and is calculated from:

$$D_v(r) = \frac{P^2}{2\gamma \cos\theta}\left(\frac{dV}{dP}\right) \qquad \text{(Eq 3)}$$

- Surface area of interparticle and intraparticle pores, which is calculated from:

$$S = \frac{1}{\gamma \cos\theta}\int_0^v P\,dV \qquad \text{(Eq 4)}$$

and the pore surface area distribution [$D_s(r)$], or the surface area per unit pore radius, can

be obtained from the pore size distribution, as shown in Eq 5:

$$D_s(r) = \frac{2D_v(r)}{r} \qquad \text{(Eq 5)}$$

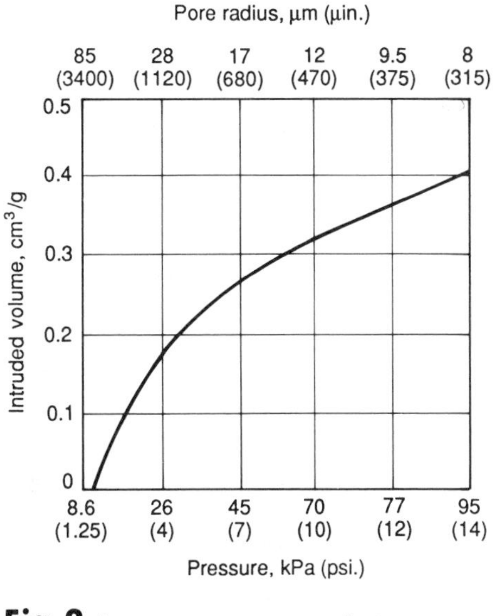

Fig 2 Low-pressure mercury porosimetry curve

ing to the calculation of pore size and pore size distributions generally make use of the Kelvin equation (Ref 6):

$$\ln \frac{P}{P_0} = \frac{-2\gamma \bar{V} \cos \theta}{rRT} \quad \text{(Eq 6)}$$

which relates the equilibrium vapor pressure, P, of a curved surface, such as that of a liquid in a capillary or pore of radius r, to the equilibrium pressure, P_0, of the same liquid on a plane surface. The remaining terms, γ, $\bar{V}$, θ, R, and T, represent the surface tension, molar volume, and contact angle of the adsorbate; the gas constant; and absolute temperature, respectively. According to the Kelvin equation, a vapor will condense into pores of radius r when the equality expressed in Eq 6 is realized. Because the contact angle of nitrogen is essentially zero, the vapor condenses at relative pressures P/P_0 substantially less than unity.

The amount of a vapor that will adsorb on a solid surface is a function of temperature, pressure, and the interaction potential between the adsorbate and the adsorbent. A plot of the quantity of vapor adsorbed or desorbed versus pressure at constant temperature is called an isotherm, portions of which are used for surface area and pore size distribution determinations.

A typical isotherm of a porous material is shown in Fig 3. For thermodynamic reasons (Ref 7), pore size distributions are usually calculated from the desorption isotherm, D, that is, the volume of gas desorbed from the porous material as the relative pressure is lowered. A common numerical integration method for the calculation of pore sizes is that of Barrett, Joyner, and Halenda (Ref 8), which takes into account the fact that a multilayer of adsorbed film remains on a pore wall when evaporation of an adsorbate occurs out of the pore.

Pore size calculations from a desorption isotherm are generally terminated at a relative pressure of 0.3, or even higher, if the hysteresis loop between the adsorption and

desorption curves is closed, an indication of the absence of pores below that relative pressure.

Below relative pressures of 0.3, the validity of the Kelvin equation is questionable because of the uncertainty in the value of the surface tension, contact angle, and molar volume of the adsorbate when only one or two molecular diameters are present in the pores. Because a relative pressure of 0.3 corresponds approximately to pores of 1.5 nm (0.06 μin.) radius, this size represents the lower limit of pore sizes that should be calculated from the Kelvin equation. However, the characterization of microporosity (pores smaller than 1.5 nm) by gas adsorption is possible by utilizing any of several theories, such as the t-method (Ref 9) and the α_s-method (Ref 10).

Pore size distribution measurements by nitrogen adsorption based on the Kelvin equation (Eq 6) are limited to pore radii smaller than 100 nm (4 μin.). Mercury porosimetry, in contrast, has the capability of measuring the volume of pores with radii as large as several hundred micrometers.

Closed Porosity

Because blind pores, that is, those not open to the surface, are inaccessible to fluids, mercury porosimetry and gas sorption techniques cannot detect them. If the presence of closed pores is suspected, the sample should be pulverized to attempt to remove them. In the event that the pores became closed during the firing process, a comparison of the true (theoretical) density before and after sintering would give an indication of the quantity of closed pores. After sintering, a ceramic usually reaches a value greater than 95% of the theoretical density. Thus, the volume of closed pores is very small.

Density

The total porosity of a sample can readily be determined by comparing its bulk density, ρ_B, that is,

$$\rho_B = \frac{\text{Sample weight}}{\text{Total volume (including pores)}} \quad \text{(Eq 7)}$$

with its true or theoretical density, ρ_T:

$$\rho_T = \frac{\text{Sample weight}}{\text{Volume of solids (excluding pores)}} \quad \text{(Eq 8)}$$

It is often convenient to express porosity as the fraction of porosity, that is,

$$f_p = \frac{\rho_T - \rho_B}{\rho_T} = \frac{1 - \rho_B}{\rho_T} \quad \text{(Eq 9)}$$

Bulk density is the density of a sample obtained from its bulk volume, or the volume of the sample including pores. For large regular solids, such as bricks, the bulk volume is readily obtained from the sample dimensions. For smaller and/or irregularly shaped

materials, bulk volume can be determined by measuring the amount of a nonwetting liquid (one that does not penetrate the pores) displaced by the sample (Archimedean principle).

Mercury porosimetry provides a convenient technique for this purpose. Because the size of pores into which mercury intrudes as a function of pressure is known from Eq 2, it is possible to determine the bulk density that corresponds to any given pressure or pore size. This technique gives the true density of samples that do not possess pores or voids smaller than those into which intrusion occurs at the highest pressure attainable in the porosimetry, that is, 1.7 nm (0.07 μin.) at 414 MPa (60 ksi). If pores smaller than this size are present, the density obtained from mercury porosimetry is an apparent density.

Determination of the bulk volume of a sample by mercury porosimetry at ambient pressure (0.101 MPa, or 14.7 psi) is accomplished by first weighing a sample cell filled with mercury and then weighing the same sample cell with both the sample and mercury present. The sample volume is simply the difference between the two mercury volumes. It should be noted that any pores larger than 7.2 μm (290 μin.) that fill with mercury at ambient pressure are excluded from the sample volume.

The experimental intruded volume of mercury from a porosimetry curve, such as that shown for a silica sample in Fig 4, can be used to calculate the apparent density as a function of pores remaining unfilled with mercury at the maximum pressure used to develop the curve. Table 1 illustrates the use of data from Fig 4 to calculate the sample volume, including pores not filled at different specified pressures. The calculated apparent densities are obtained by subtracting the intruded volume of mercury at each pressure from the bulk sample volume (at ambient pressure) and dividing the resulting value into the sample weight.

A plot of density versus pore radius from the data in Table 1 is shown in Fig 5. The horizontal line at a density of approximately 2 g/cm^3 represents the true density of the sample.

True, or theoretical, density is defined as the ratio of the mass of the sample to the volume occupied by that mass. Contributions to the volume by pores or void spaces must be excluded when measuring true density.

If a material is nonporous, its true density can be determined by displacement of a fluid in which the solid is chemically inert. However, if the solid contains pores, cracks, or crevices that are not completely penetrated by the liquid, then true density can only be measured using a gas as the displaced fluid. Helium is the most commonly used gas for measuring true density because it is inert and the small size of the helium atom permits it to penetrate even extremely small pores. A helium pycnometer is commonly used for measuring true density.

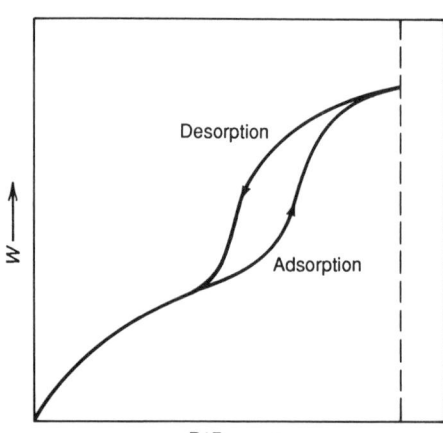

Fig 3 Gas sorption isotherm

Table 1 Density of silica at various pressures

Pressure		Radius		Intruded volume, cm³	Sample volume, cm³	Apparent density, g/cm³
MPa	ksi	nm	μin.			
0.10	0.015	7260	290	0	1.115	0.967
34	5	21.3	0.838	0.024	1.091	0.988
69	10	10.7	0.421	0.050	1.065	1.01
103	15	7.11	0.280	0.080	1.035	1.04
138	20	5.34	0.210	0.117	0.998	1.08
172	25	4.27	0.168	0.160	0.955	1.13
207	30	3.56	0.140	0.203	0.912	1.18
241	35	3.05	0.120	0.245	0.870	1.24
276	40	2.67	0.105	0.285	0.830	1.30
310	45	2.37	0.093	0.323	0.792	1.36
345	50	2.13	0.084	0.357	0.758	1.42
379	55	1.94	0.076	0.385	0.730	1.48
414	60	1.78	0.070	0.406	0.709	1.52

Note: Radius was calculated from Eq 2, assuming θ = 140°.

Effective Density. A sample containing blind pores that are totally encapsulated by the solid will give a volume measurement that includes these pores. The resulting density, which is less than the true density, is called the effective density.

Surface Area

A perfect sphere of radius r would exhibit a surface area of $4\pi r^2$. However, very few perfect spheres exist because of imperfections and flaws on the surface and pores within the particles that compose the solid material.

The most common method for measuring the surface area of a solid sample involves the adsorption of a vapor on the solid surface. The total surface area, S_T, of a solid material can be expressed as the product of the number of molecules of adsorbate in a monolayer, N_m, and the effective cross-sectional area of the adsorbate molecule, A_{cs}, that is,

$$S_T = N_m A_{cs} \qquad \text{(Eq 10)}$$

Thus, the measurement of surface area depends on the ability to predict the number of adsorbate molecules required to exactly cover the solid with a single molecular layer, or monolayer, and the cross-sectional area of the adsorbate molecule. Substituting a weight relationship for N_m, a useful expression for the calculation of surface area emerges:

$$S_T = \frac{W_m N A_{cs}}{M} \qquad \text{(Eq 11)}$$

where W_m is the weight of a monolayer of adsorbate on the surface of the solid, N is Avogadro's number (6.023×10^{23} molecules/mol), and M is the molecular weight of the adsorbate (grams/mol). Thus, surface area measurements depend on the success of the theory used to predict N_m and a knowledge of A_{cs}.

BET Theory. As pressure is increased during the process of adsorption of a vapor on a solid surface, the surface becomes progressively coated with adsorbate. However, vapor molecules will be adsorbed on previously bound molecules, as well as on the solid surface. As a result, second and higher adsorbed layers will be formed prior to complete surface coverage, that is, before the completion of a monolayer. The BET theory (Ref 11), proposed in 1938, is still the most generally used theory for estimating W_m.

The adsorbate cross-sectional areas of a number of different gases are known. However, nitrogen is the most universally used adsorbate because it adsorbs easily on most solid surfaces and its cross-sectional area is well established. The commonly accepted value for the cross-sectional area of a nitrogen molecule, calculated from its liquid density, is 16.2×10^{-20} m² (175×10^{-20} ft²).

The determination of surface area using the BET theory requires the application of the BET equation (Eq 12):

$$\frac{1}{W[(P_0/P) - 1]} = \frac{1}{W_m C} + \frac{C - 1}{W_m C}\left(\frac{P}{P_0}\right) \qquad \text{(Eq 12)}$$

in which W is the weight of vapor adsorbed at a relative pressure, P/P_0, C is the BET constant, and W_m is the weight of adsorbate in a monolayer.

A plot of $1/W[(P_0/P) - 1]$ versus P/P_0 should produce a straight line (as shown in Fig 6), usually in the range of $0.05 \leq P/P_0 \leq 0.35$. For most adsorbents, a linear plot is obtained in this relative pressure range because monolayer coverage will have been achieved. However, the linear region of the BET plot for microporous materials occurs at much lower relative pressures because of capillary condensation in very small pores at low relative pressures. Thus, for these materials, lower relative pressures should be used to acquire the data. The slope, s, and intercept, i, of a BET plot from the linear equation (Eq 12) can be expressed as:

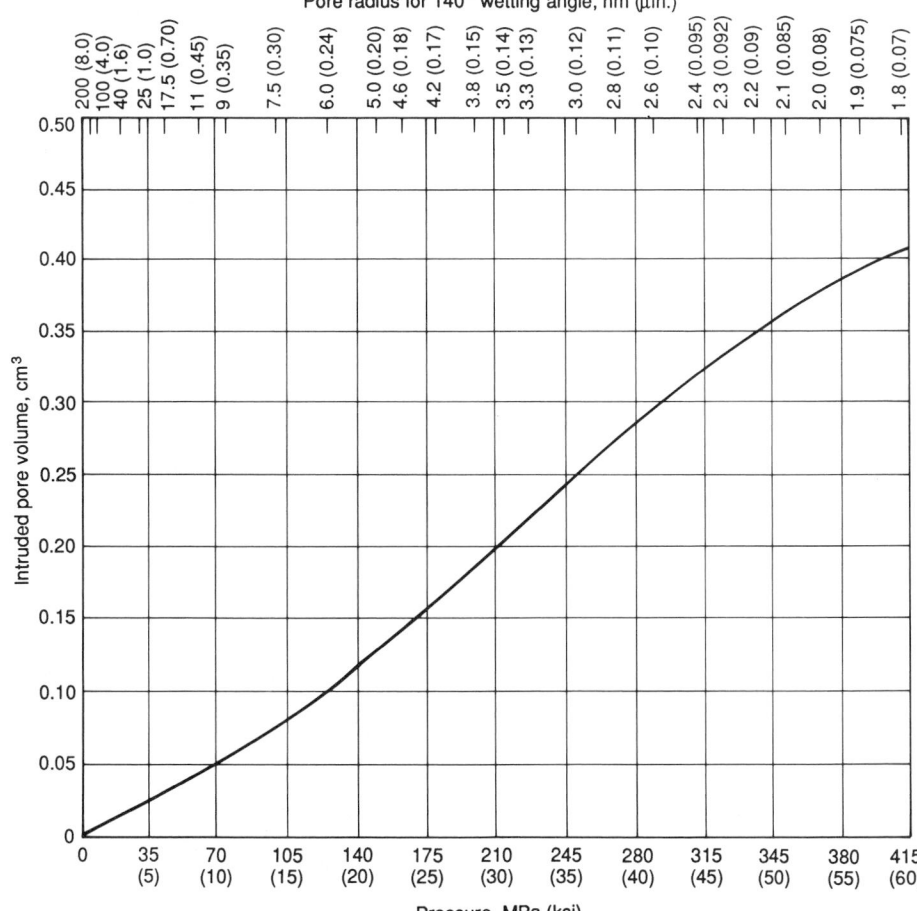

Pore radius for 140° wetting angle, nm (μin.)

200 (8.0) 100 (4.0) 40 (1.6) 25 (1.0) 17.5 (0.70) 11 (0.45) 9 (0.35) 7.5 (0.30) 6.0 (0.24) 5.0 (0.20) 4.6 (0.18) 4.2 (0.17) 3.8 (0.15) 3.5 (0.14) 3.3 (0.13) 3.0 (0.12) 2.8 (0.11) 2.6 (0.10) 2.4 (0.095) 2.3 (0.092) 2.2 (0.09) 2.1 (0.085) 2.0 (0.08) 1.9 (0.075) 1.8 (0.07)

Intruded pore volume, cm³ (vertical axis: 0 to 0.50)

Pressure, MPa (ksi): 0, 35 (5), 70 (10), 105 (15), 140 (20), 175 (25), 210 (30), 245 (35), 280 (40), 315 (45), 345 (50), 380 (55), 415 (60)

Fig 4 Mercury intrusion curve for a silica sample

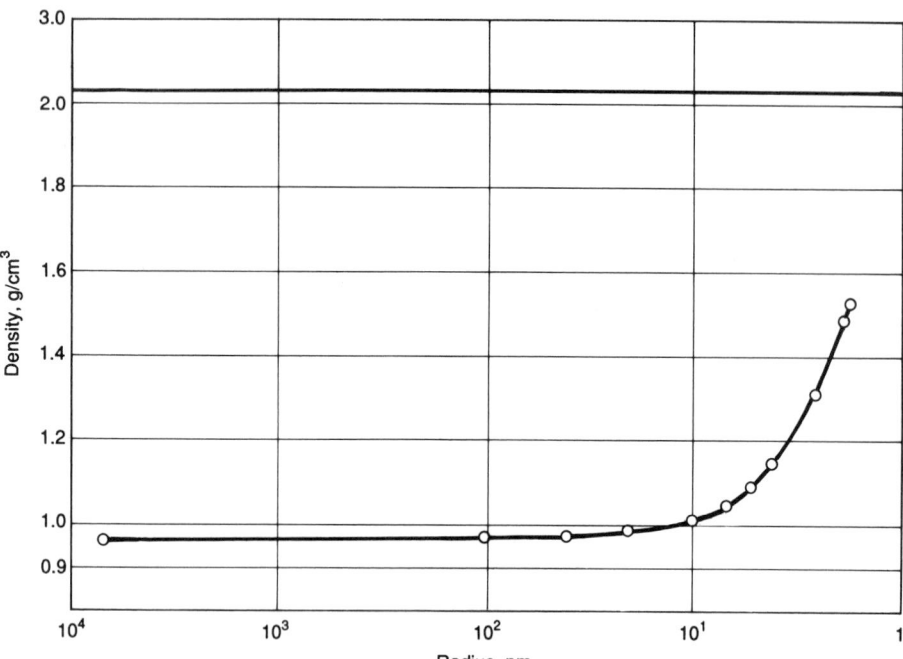

Fig 5 Density versus pore radius plot for a silica sample

$$s = \frac{C - 1}{W_m C} \qquad \text{(Eq 13)}$$

and

$$i = \frac{1}{W_m C} \qquad \text{(Eq 14)}$$

Solving Eq 13 and 14 for W_m can be calculated by:

$$W_m = \frac{1}{s + i} \qquad \text{(Eq 15)}$$

Substituting W_m from Eq 15 into Eq 11 gives the total surface area. The specific surface area, S, can then be obtained by dividing S_T by the sample weight.

The BET equation (Eq 12) requires that the amount of adsorbed vapor be determined at no less than three different relative pressures (multipoint method) so that a plot such as that shown in Fig 6 can be produced. An adaptation of the BET equation, called the single-point method (Ref 12), permits the surface area to be calculated from only one data point. The intercept, i, for reasonably high C values is small, compared to the slope, s, and may assumed to be zero. As a result, Eq 12 reduces to

$$W_m = W(1 - P/P_0) \qquad \text{(Eq 16)}$$

and the total surface area by the single-point method is expressed as

$$S_T = \frac{W(1 - P/P_0)NA_{cs}}{M} \qquad \text{(Eq 17)}$$

An error analysis of the single-point method, relative to the multipoint method, indicates that the error for a C constant of 100 and a relative pressure of 0.3 is 2%. On the majority of surfaces, the C constant is sufficiently high that the single-point error reduces to a negligible value.

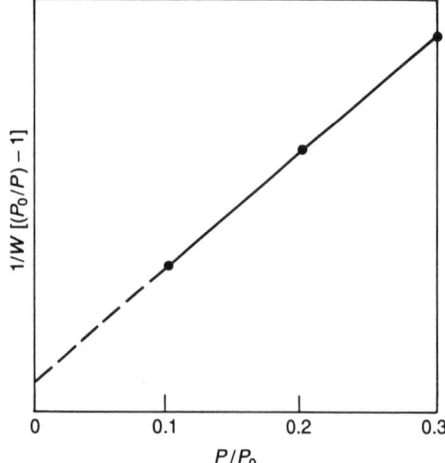

Fig 6 BET plot

Experimental Methods. There are two commonly used techniques for measuring the quantity of vapor adsorbed by a solid at various relative pressures.

The dynamic, or continuous flow method developed by Nelson and Eggertsen (Ref 13) involves the flow of mixtures of known concentration of adsorbate in a carrier gas through the sample. Usually, a mixture of nitrogen in helium is allowed to flow through the sample. At the same time, a thermal conductivity detector monitors the change in the effluent vapor concentration during adsorption and desorption when the sample is immersed in, and subsequently removed from, a liquid nitrogen bath. Calibration of the thermal conductivity signal with a known volume of nitrogen allows the calculation of the quantity of nitrogen adsorbed or desorbed by the sample. The procedure is repeated using different concentrations of nitrogen in helium to provide the various relative pressures needed for the BET equation.

In the vacuum volumetric technique, the volume of vapor adsorbed is determined by dosing an evacuated sample with adsorbate. After equilibrium has been reached, the resulting pressure value is used to calculate the amount of adsorption at that specific relative pressure. The dosing procedure is repeated until enough data points ranging from a P/P_0 of 0.05 to 0.35 have been obtained to make a BET plot.

REFERENCES

1. L.C. Ritter and R.L. Drake, *Ind. Eng. Chem. Anal. Ed.*, Vol 782, 1945, p 17
2. E.W. Washburn, *Proc. Natl. Acad. Sci.*, Vol 7, 1921, p 115
3. S. Lowell, *Powder Technol.*, Vol 25, 1980, p 37
4. P.N. Henrion and A. Leurs, *Powder Technol.*, Vol 145, 1977, p 17
5. S.M. Brown and E.W. Lard, *Powder Technol.*, Vol 9, 1974, p 187
6. W.T. Thompson, *Philos. Mag.*, Vol 42, 1871, p 448
7. S. Lowell and J.E. Shields, *Powder Surface Area and Porosity*, 3rd ed., Chapman and Hall, 1991, p 55
8. E.P. Barrett, L.G. Joyner, and P.P. Halenda, *J. Am. Chem. Soc.*, Vol 73, 1951, p 373
9. B.C. Lippens and J.H. DeBoer, *J. Catal.*, Vol 4, 1965, p 319
10. S.J. Gregg and K.S.W. Sing, *Adsorption, Surface Area and Porosity*, 2nd ed., Academic Press, 1982
11. S. Brunauer, P.H. Emmett, and E. Teller, *J. Am. Chem. Soc.*, Vol 60, 1938, p 309
12. G. Sandstede and E. Robins, *Chem. Ing. Tech.*, Vol 32, 1960, p 413
13. F.M. Nelson and F.T. Eggertsen, *Anal. Chem.*, Vol 30, 1958, p 1387

Strength and Proof Testing

George D. Quinn, National Institute of Standards and Technology

CERAMIC MATERIALS have been used in a variety of engineering applications that utilize their wear resistance, refractoriness, hardness, and high compression strength. Traditionally, they have not been used in tensile-loaded structures because they are brittle, that is, they experience catastrophic failure before permanent deformation (Fig 1). Nevertheless, their extreme refractoriness, chemical inertness, and favorable optical, electrical, and thermal properties are inducements to use ceramics in certain tensile load-bearing applications.

Strength and Fracture Phenomena

Glasses and ceramics are inherently brittle at ambient temperature because there is no mechanism to permit either general or localized yielding at cracks. The brittleness of crystalline ceramics at low temperatures is a consequence of the strong covalent bonding of many ceramics, or the lack of a sufficient number of independent slip systems for dislocation motion to provide general plasticity. At moderately high temperatures, some oxides can exhibit ductility, but covalently bonded ceramics, such as silicon carbide or silicon nitride, have negligible dislocation activity until very high temperatures.

Brittle failure in monolithic, linearly elastic materials can be well modelled by Griffith's criterion for fracture, where it can be shown that a critical balance of strain-energy release rate is offset by the energy to create new fractured surfaces (Ref 1, 2):

$$\sigma_f = \left[\frac{4\gamma_f E}{\pi(1 - v^2)c}\right]^{1/2} \qquad \text{(Eq 1)}$$

where σ_f is the failure stress, γ_f is the energy to create unit surface area, E is the elastic modulus, v is Poisson's ratio, c is the flaw dimension (radius, in this case), and the mathematical constants are those for a penny-shaped crack in a thick plate.

Alternatively, a fracture mechanics approach can be used, wherein the stress intensity K_I at the tip of sharp cracks in a linearly elastic material is expressed as:

$$K_I = Y\sigma c^{1/2} \qquad \text{(Eq 2)}$$

where σ is the applied far-field stress, and Y is a dimensionless shape factor that describes the severity of a defect and geometrical considerations, such as proximity to a surface or position within a stress field ($Y = 1.13$ for a penny-shaped defect in the bulk of a uniformly stressed specimen). Fracture occurs when the level of K_I reaches a critical level, K_{Ic}, which is a material property known as fracture toughness. The K_{Ic} value is a good index of resistance to fracture for many brittle, monolithic ceramics. Fracture toughness measurement methods are discussed in the article "Toughness, Hardness, and Wear" in this Section of the Volume.

The Griffith and fracture mechanics approaches are equivalent in linearly elastic materials and it can be shown that for plane strain,

$$k_{IC} = \left[\frac{2E\gamma_f}{(1 - v^2)}\right]^{1/2} \qquad \text{(Eq 3)}$$

Material scientists sometimes prefer to use γ_f, because it can be viewed as a fracture surface energy, whereas engineers and designers prefer K_{Ic} for its mathematical versatility in analyzing loadings on cracks.

Unfortunately, the low values of fracture toughness (0.5 to 3.0 MPa$\sqrt{m}$, or 0.46 to 2.7 ksi$\sqrt{in}$.) that are typical of traditional ceramics means that strength-limiting defects are extremely small. (Most metals have toughnesses of 20 MPa$\sqrt{m}$, or 18 ksi$\sqrt{in}$., or higher.) A strength of 700 MPa (100 ksi) is associated with a defect diameter of only 1 to 30 μm (0.04 to 1.2 mils). Traditional ceramics rarely exceed 350 MPa (50 ksi) in strength.

Small defects are randomly distributed in size and location throughout ceramic bodies, causing a range of strengths in otherwise identical ceramic parts or components. This scatter in defects leads to a size effect upon strength, that is, the larger the component, the greater the chance of a larger defect and, thus, a lower strength. Therefore, there is no deterministic strength for brittle monolithic ceramics, unless the defects are extremely uniform and consistent or unless particular R-curve phenomena apply. Factors of safety cannot be used. Strength varies significantly with specimen type or size and with process-

ing conditions. There are even batch-to-batch differences, a consequence of material inconsistency. These factors, coupled with the inherent brittleness, mean that extremely conservative design approaches must be used for tensile-loaded ceramics.

On the other hand, engineering ceramics are typically 8 to 20 times stronger in compression than in tension. This is because dislocation activity that could cause yielding is nil and the defects in the material are resistant to propagation in compression.

Recent developments in high-strength ceramics have allowed good opportunities to utilize engineering ceramics in tensile-loaded structural applications. New ceramic materials, notably silicon carbide and silicon nitride, as well as toughened zirconia are now available. Monolithic silicon carbides and silicon nitrides have higher fracture toughnesses (typically 3 to 4 MPa$\sqrt{m}$, or 2.7 to 3.6 ksi$\sqrt{in}$., for SiC and 4.5 to 5.5 MPa$\sqrt{m}$, or 3.8 to 5.0 ksi$\sqrt{in}$., for Si$_3$N$_4$) and appreciably lower thermal expansion coefficients, which means that they are less susceptible to thermal stress and thermal shock, the bane of many traditional ceramics. Flexural strengths that exceed 700 MPa (100 ksi) and 400 MPa (60 ksi) are common for silicon nitride and silicon carbide, respectively. They are able to sustain high tensile loads up to temperatures that exceed 1500 °C (2730 °F) for SiC and 1300 °C (2370 °F) for Si$_3$N$_4$. At these temperatures, most metals are either molten or have negligible creep resistance. In some instances, these ceramics have been able to sustain loadings for extended times at temperatures within 100 to 200 °C (180 to 360 °F) of their processing temperature, typically 1800 °C (3270 °F) for Si$_3$N$_4$, and ≥2000 °C (3630 °F) for SiC. Recent technical advancements by careful microstructural tailoring have pushed the fracture toughnesses to levels from 6 to 10 MPa$\sqrt{m}$ (5.5 to 9.1 ksi$\sqrt{in}$.) for Si$_3$N$_4$ (Ref 3–5) and as high as 10 MPa$\sqrt{m}$ (9 ksi$\sqrt{in}$.) for SiC (Ref 6, 7).

The otherwise mundane refractory material ZrO$_2$ has recently ascended to new levels of interest. Careful microstructural control and utilization of a peculiar phase transformation phenomenon have led to the attainment of fracture toughnesses between 10 and 15 MPa$\sqrt{m}$ (9 and 14 ksi$\sqrt{in}$.) and flexural

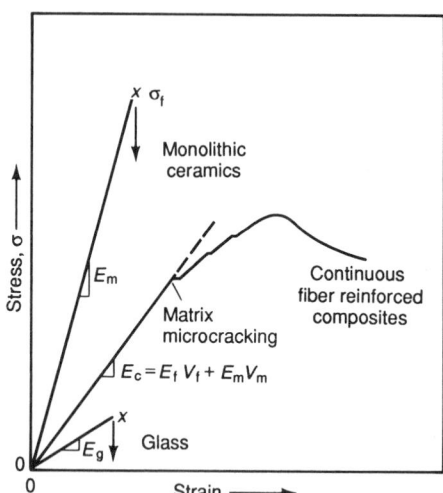

Fig 1 Glasses, monolithic ceramics, and fine scaled, isotropic ceramic composites behave brittlely and are elastic to failure whereas continuous fiber reinforced ceramics are quasiductile

strengths as high as 1500 to 2000 MPa (210 and 300 ksi). Zirconia is remarkably tough for a ceramic and has found some applications that exploit its resistance to fracture. Demonstrations with zirconia ball-peen hammers, knives, and scissors make quite an impression on skeptics who believe that ceramics are hopelessly brittle.

There also has been encouraging progress with glasses, which are notoriously brittle materials (K_{Ic} typically ranges from 0.7 to 0.9 MPa$\sqrt{m}$, or 0.6 to 0.8 ksi$\sqrt{in}$.). Chemical or thermal treatments are frequently used to put surfaces into compression, thereby reducing sensitivity to surface damage, which is the most common source of glass failures. Further advances in glass-ceramic technology permit bodies to be fabricated by glass-making methods and then fully crystallized by controlled thermal treatments. The fracture toughnesses of such crystallized glass-ceramics have been increased to 2.5 MPa$\sqrt{m}$ (2.3 ksi$\sqrt{in}$.). This technology has been utilized for applications that range from missile radomes to common kitchenware.

Analytical tools, such as the scanning electron microscope (SEM), have led to major advances in understanding the strength of ceramics. In the past, it was common simply to relate strength to the inverse square root of the average grain size. The more appropriate interpretation from the Griffith's or fracture mechanics criteria (Eq 1 or 2) have a square root dependence on the size of the largest, most highly stressed defect. Electron microscopy permits the characterization of these defects. Engineers can now relate strength directly to specific causes and factors. Processing is less haphazard or inferential and processors now focus much of their work on flaw management. Ultimately, as processors eliminate large, extrinsic defects, the strength of ceramics may revert to being controlled by

the mainstream microstructural features, for example, grains from the large end of the grain-size distribution.

Improved statistical analyses are available to better model the seeming randomness of strength. Strength scatter is now reasonably well understood and is manageable. The Weibull statistical model (Ref 8, 9) is applicable to uniaxial strength analysis, and is summarized below. More advanced analyses are available for multiaxial stressed components (Ref 10, 11).

The application of fracture mechanics analysis to ceramics has increased the fundamental understanding of the factors that control fracture phenomena. Fracture in monolithic ceramics is now understood to be a consequence of the stress intensification that occurs in the vicinity of defects. Unfortunately, it is difficult to apply fracture mechanics to practical ceramic engineering problems (other than laboratory specimens), because it requires detailed specific knowledge (the precise size, shape, and orientation around the defects). Stress intensity is the driving force for other phenomena, such as stress-corrosion crack growth, as discussed below.

The advent of fine-scale, isotropic ceramic composites with either second phases or particulate reinforcement has brought about the exciting prospect of tough and very strong ceramics. These composites employ a variety of means to impede or resist crack propagation, including crack deflection or branching, microcracking (to blunt the crack or dissipate the stress concentration), crack-particle interactions, and behind-the-crack-ligament reinforcement, as shown in Fig 2 (Ref 12, 13).

Particulate, second phase, or whisker-reinforced ceramics can have their toughnesses enhanced by a factor of two or more above the monolithic levels. Examples are partially stabilized zirconia, silicon carbide whisker-reinforced alumina, and titanium carbide particulate-reinforced silicon nitride. Toughnesses for these materials can range from 7 to 15 MPa$\sqrt{m}$ (6.4 to 13.7 ksi$\sqrt{in}$.). Even more exciting is that these materials exhibit so-called "R-curve" behavior, meaning that as a crack advances, it encounters greater resistance to propagation. A crack that extends from a microscopic defect may have several centimeters of impeded growth before the peak toughness has been achieved. There is no single value of fracture toughness, K_{Ic}, for such materials. Indeed, such R-curve behavior has been demonstrated for some common, coarse-grained ceramics, such as aluminum oxide. The R-curve phenomenon is a consequence of behind-the-crack-tip bridges that reinforce a crack (Ref 14). Such ceramics have a greater flaw tolerance than conventional, constant K_{Ic} materials (Ref 15).

Continuous fiber-reinforced ceramics have remarkable resistance to classical fracture (Fig 1) and, for practical purposes, can be considered ductile (Ref 16, 17). By the appro-

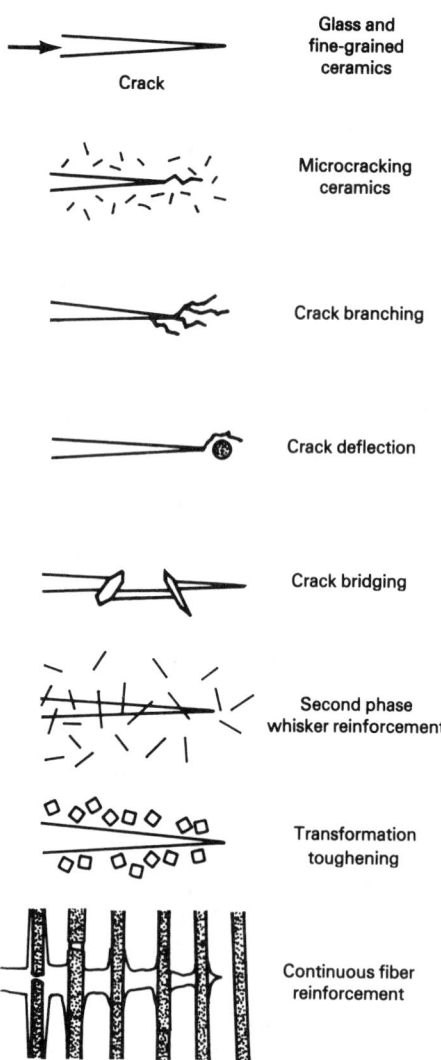

Fig 2 Cracks in glasses and fine-grained monolithic ceramics have simple, sharp form. Coarser-grained and composite ceramic cracks experience much more interaction with microstructure.

priate mismatch between fiber and matrix properties and careful microstructural control, such composites can completely disrupt crack propagation (Fig 2). Unlike organic-matrix composites, it is important that fiber-matrix bonding *is not too strong* or else the ceramic composite will fail brittlely. Fracture toughness typically cannot be measured for such materials, because they are so tough that cracks refuse to propagate in a brittle fashion. Conventional K_{Ic} measurement techniques are not valid.

Crack Growth and Creep. Many engineering ceramics, such as silicon carbide and silicon nitride, are extremely resistant or immune to crack growth at low temperatures (Ref 18), whereas glasses and many oxide ceramics are susceptible to environmentally assisted crack growth, which can lead to strength degradation or fracture (Ref 19, 20). Glasses and oxides are susceptible to stress corrosion, because of the chemical interaction between water (either in liquid or vapor form) and stressed cracks. Although water is the severest active species likely to be encountered, recent evidence shows that other compounds can cause similar stress corrosion (Ref 21). Substantial empirical evidence has indicated that there is a strong relationship between stress intensity, K_I and crack velocity, as shown in Fig 3(a).

At values of K_I below the critical fracture toughness K_{Ic}, there are three regions of crack growth. Region I, which is of the greatest concern to design engineers, because components will usually be stressed well below critical loading conditions, is a reaction-rate controlled region, where K_I and V can be related through a power function:

$$V = AK_I^n \qquad (Eq\ 4)$$

where A is a constant and n is called the slow-crack-growth exponent. A low n value indicates that the material is susceptible to slow crack growth over a broad K_I range, whereas a high value indicates susceptibility only over a narrow range. A plateau, region II, occurs where insufficient reactant is available to a crack, precluding higher velocities. Region III is a less well-characterized region that is intrinsic to a material, rather than dependent on a reactant species. Most of the lifetime of a crack that is growing under either stress-corrosion or slow-crack-growth conditions will be controlled by region I. Because the exponent n typically ranges from 5 to 100, the lifetime will depend very strongly on the initial stress or stress intensity, K_I. There is some evidence that a threshold stress intensity can exist, that is, a critical level, K_{Iscc}, below which crack growth will not occur (Ref 22–24). Such a value is of great engineering interest, but is difficult to measure, because the crack velocities are $\leq 10^{-10}$ m/s (3.3×10^{-10} ft/s).

Equations 2 and 4 can be combined and integrated to give a time-to-failure for a specimen or component under constant stress, σ_a:

$$t_f = constant\ \sigma_a^{-n} \qquad (Eq\ 5)$$

A stress-rupture graph with logarithmic axes of stress and time-to-failure will have a slope of $-1/n$, as shown in Fig 3(b). Such constant load, stress-rupture tests are also called static fatigue tests by ceramists. The rate of strength degradation is thus directly related to the slow-crack-growth exponent, n, for which high values are desired. Ceramists also often report strengths as a function of loading rate, as shown in Fig 3(c). The slower the rate of loading, the greater the opportunity for crack growth to occur, which leads to a lower measured strength. This test procedure is known as dynamic fatigue in the ceramics community, and the difference in strengths can once again be readily related to the slow-crack-growth exponent, as shown in Fig 3(c).

At higher temperatures, oxidation, diffusion, or cavitation processes, or, possibly, dislocation activity can lead to crack extension in ceramics. Engineering ceramics are often made by sintering powders with sintering aids that provide the necessary atom mobility for consolidation. This same atom mobility can recur during high-temperature mechanical loading and lead to crack nucleation and/or growth. Creep and sintering phenomena are closely coupled in ceramics.

Slow crack growth may be environmentally assisted at elevated temperature. Certain defects, such as pores or porous zones, that are located at or near the surface are more vulnerable to time-dependent environmentally assisted crack growth than are similar defects buried in the bulk of a ceramic specimen or component (Ref 25). Such crack extension is typically called slow crack growth in the ceramics community, but could alternatively be described as stress corrosion.

At high temperatures, intrinsic deformation processes such as grain boundary sliding, diffusion, cavitation, or dislocation activity can lead to crack nucleation, growth, and coalescence. At moderate temperatures,

deformation may be confined to the immediate vicinity of a preexistent crack (because of high concentrated stresses) and can lead to negligible overall deformation, but sufficient local deformation to cause crack growth. This mode of crack extension is also called slow crack growth. Bulk deformation in a ceramic body at high temperature, which involves nucleation of cavities, growth, and coalescence, causes creep fracture. Strength degradation can also be caused by the nucleation of new defects at the surface, such as surface pitting during oxidation exposure of silicon nitride at 1200 °C (2190 °F) (Ref 26).

Continuous fiber reinforced ceramic composites are extremely vulnerable to environmental degradation of strength, a consequence of attack of the fiber-matrix interface. As mentioned above, the interface bond must not be too strong, lest the composite be brittle. This controlled weakness may be an Achilles' heel that causes the environmental vulnerability, with or without applied stresses.

Defect Management in Ceramic Fabrication. Perhaps the greatest practical engineering impediment to the reliable use of monolithic or composite engineering ceramics in high tensile load-bearing applications, besides the inherent brittleness, is material inconsistency. There are often problems maintaining consistency on a day-to-day, batch-to-batch, or shape-to-shape basis. Monolithic ceramic strengths are extremely sensitive to this type of problem. Consistency here refers not only to bulk parameters, such as average grain size, grain size distribution, and density, but to the extremes of the microstructure, as well, that is, the strength-limiting defects. Consistency is difficult to assess, short of actually fracturing some specimens (Ref 27, 28).

Many ceramics that are sintered from powders are susceptible to subtle density and microstructural fluctuations, laminations, or gradients. Defects themselves are very sensitive to minor processing variations, and,

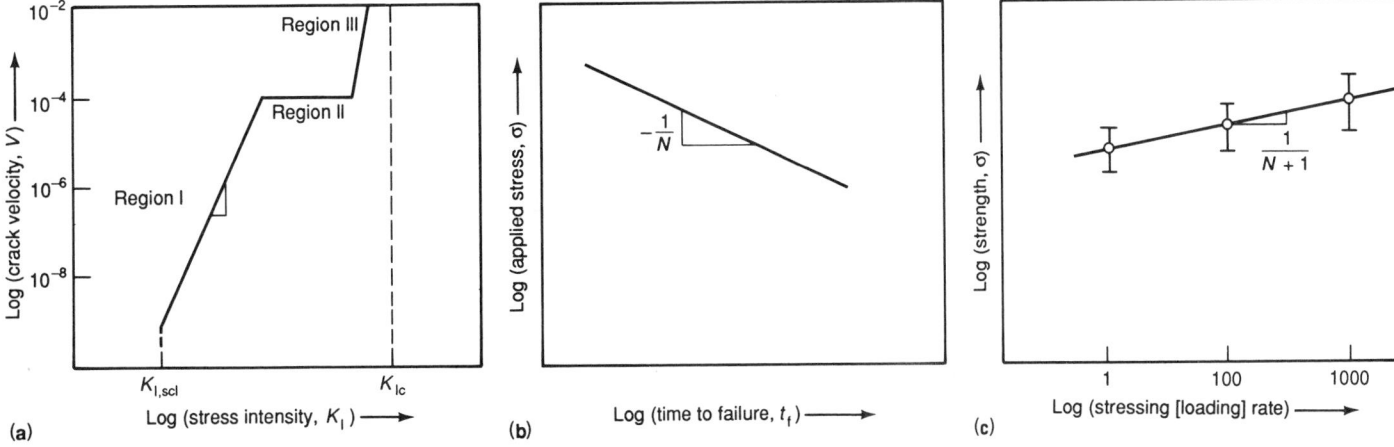

Fig 3 Cracks in glasses and ceramics can grow subcritically. Among the ceramics technical community, all such stable crack growth is regarded as occurring below K_{Ic}. If region I in (a) controls behavior, then stress rupture and dynamic fatigue experiments will give (b) and (c), respectively.

although the bulk properties may be consistent among components, the defects themselves are not. Thus, the skill of a processor is measured not only by his ability to maintain consistent bulk properties, but by his ability to manage defects.

Morrell (Ref 29) has discussed this topic in some detail, particularly in the context of whether any laboratory test bar data can ever be truly representative of components. A detailed review of this topic has recently been prepared and concludes that a number of assumptions must be met before the test data are useful (Ref 30). The accurate, routine correlation of test bars to components has been elusive, but the record shows a number of successes (Ref 30, 31).

The fundamental uncertainty about the reliability of monolithic ceramics has prompted many to turn to ceramic composites in the hope that their quasiductility will make them more forgiving. However, there will be a price to pay, both in material and fabrication costs and the need to develop a new design philosophy. Consistency in ceramic composites is still an uncertain issue. Failure analyses may need to be more complex (especially for other than unidirectionally reinforced composites). Composites will probably be much more susceptible to mechanical fatigue than monolithic ceramics, and many fatigue studies are now underway.

The rest of this article describes monolithic and particulate-reinforced ceramics separately from continuous fiber reinforced ceramics because the testing methodologies are different.

Room-Temperature Strength Test Methods for Isotropic Ceramics

Uniaxial Tensile Strength. The nonductile nature of monolithic ceramics and their high sensitivity to stress concentrators has meant that conventional direct tensile testing is difficult and expensive. Gripping with jaws, screw threads, or other conventional devices causes invalid test results, because of specimen breakage at the grips. The high stiffness (elastic modulus) of many ceramics means that a misalignment of only a few thousandths of a centimeter can lead to bending stresses with errors of 10% or more. Specimen preparation to exacting tolerances with minimal machining damage and careful tapers to avoid stress concentrators has been an expensive proposition. Considerable new work since the mid 1980s has focused on improving tensile test methods for ceramics, with the result that tensile testing is becoming more routine. Commercial equipment is readily available, and specimen costs are falling, but it will always be more difficult to conduct direct tensile tests for ceramics than for metals.

These experimental difficulties, coupled with the problems of fabricating sufficiently large specimens, have prompted ceramists to use alternative test methods (Fig 4). The most common is flexure testing, either in the so-called three-point or four-point configuration. The latter is usually further specified by a description of the distance from the outer support points and the inner points, such as $^1/_4$ or $^1/_3$ four-point loading. The small size, low cost, and easy preparation of a flexure specimen accounts for its popularity, but there are distinct drawbacks. The bending creates a stress gradient in the specimen and only a small volume is exposed to high tensile stress. The specimens are very sensitive to edge or surface machining damage. The test is deceptive in that it appears easy to set up and conduct, but misalignments and experimental errors can easily ruin it.

Through the years a variety of flexural test procedures proliferated and there was high scatter in strength results. This was due to material inconsistency, statistical effects from sampling, specimen size effects, and experimental error. Furthermore, it was difficult to discriminate between these sources. Some establishments even exploited the test by using small, highly polished specimens in three-point loading to produce very high strengths. Improved statistical analyses, fractography, and comprehensive error analyses (Ref 32, 33) have permitted refinements to flexure testing. Standard test methods are now available that permit accurate strength measurements for standard sizes and shapes, as shown in Fig 5.

Nevertheless, it is still preferable to perform direct tension testing. Current testing systems are designed with self-aligning features that limit the imposed bending stresses to approximately 1%. There is usually less extrapolation of the strength data from test specimen to component size. Tensile specimens are still expensive, however, as a con-

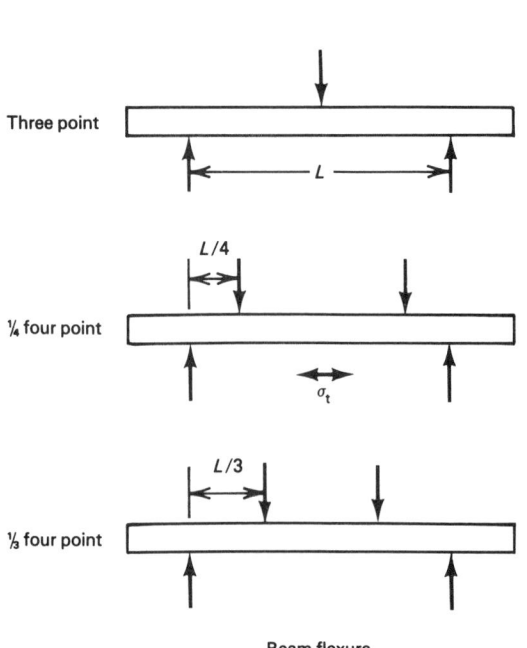

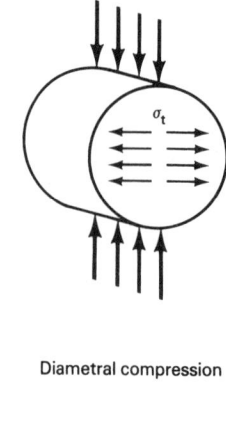

Diametral compression

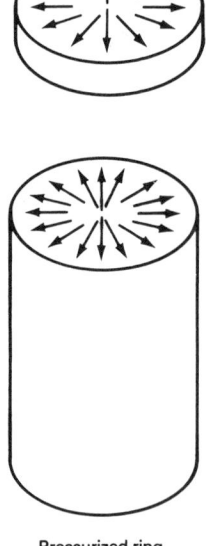

Pressurized ring or tube

Beam flexure

Fig 4 Alternative specimens for uniaxial strength measurements

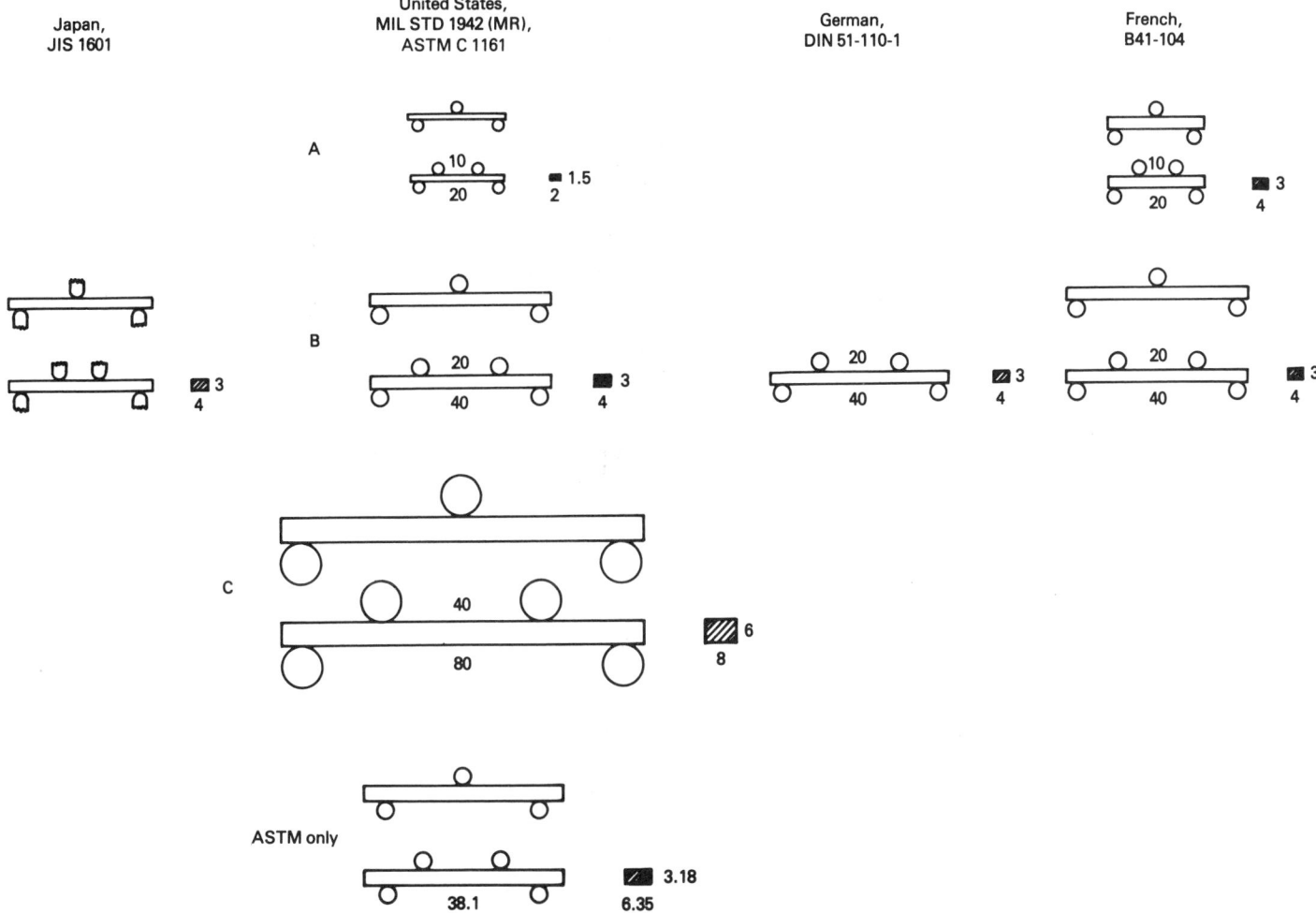

Fig 5 Flexure strength standard test methods; all dimensions in mm

sequence of costly fabrication and machining. They are inconveniently large, as well, because most systems are designed for high-temperature test rigs that use cold grips. Until recently, only a few laboratories had the ability to test or to even afford direct tensile experiments. A new emphasis on attaining accurate, quality data in support of ceramics in heat engine programs has led to rapid improvements in the field, and commercial test systems are now readily available. Different tensile specimen geometries that are being used are shown in Fig 6 (Ref 34–39).

Another occasionally cited test for engineering ceramics is the so-called diametral compression, or Brazilian disk test (Fig 4) wherein a circular cylinder is loaded at its ends (Ref 40, 41). The test is actually biaxial, because in addition to the tensile stresses that tend to laterally split the specimen, there are compressive stresses that are three times as great, which act axially through the specimen. However, compressive stresses of this magnitude are not likely to affect uniaxial strength, an effect peculiar to monolithic ceramics, as discussed below. The specimen loading is between two platens with pads of

compliant material (such as a metallic shim or paper) to avoid high shearing stresses. Careful machining of the end faces of the specimen is essential, once again to avoid damage that compromises the test. This point is often overlooked. This test is occasionally employed by ceramic processors for ceramics fabricated in cylindrical shapes.

Many ceramic materials have strengths that are specific to the shaping process being used, such as injection-molded turbocharger rotors or extruded heat-exchanger tubes. In such cases, it is not practical to cut tensile specimens from the part, but separately cast tensile specimens may not have the same microstructure or defects as the component, and therefore are irrelevant (Ref 29–31). It is optimal to test components in as close a configuration to the final component shape as is possible. Thus, in the case of a tube, a ring can be cut from the tube and pressurized to obtain a uniaxial hoop stress testing configuration (Ref 42, 43). Indeed, contrary to expectations, such a test can be conducted at high temperatures and one of the highest recorded strength test temperatures for a ceramic (2180 °C, or 3955 °F) was on a pres-

surized tube (Ref 44). Extreme care must be taken to ensure that the edges are not chipped or have excessive machining damage, lest the test merely become a measure of machining damage.

There is no simple answer to the question of what specimen is best for measuring strength data. The best practice is to test a configuration that most resembles the actual component in its service conditions, and to ensure that the test material accurately represents the component material. It is likely that the first available data will be flexure strength data, which is typically higher (10 to 50%) than tensile specimen data, because of the dependency of strength on test specimen size. Nevertheless, considering the trade-offs in cost, quantity of results, and difficulty in testing, it is likely that future engineering data bases will feature complementary flexure and tensile data. Indeed, it will be beneficial to have strength data from different sizes and shapes to permit an assessment of material consistency, flaw uniformity, and the veracity of strength-size scaling models.

Elastic Modulus. Several methods are used to evaluate the elastic moduli of monolithic

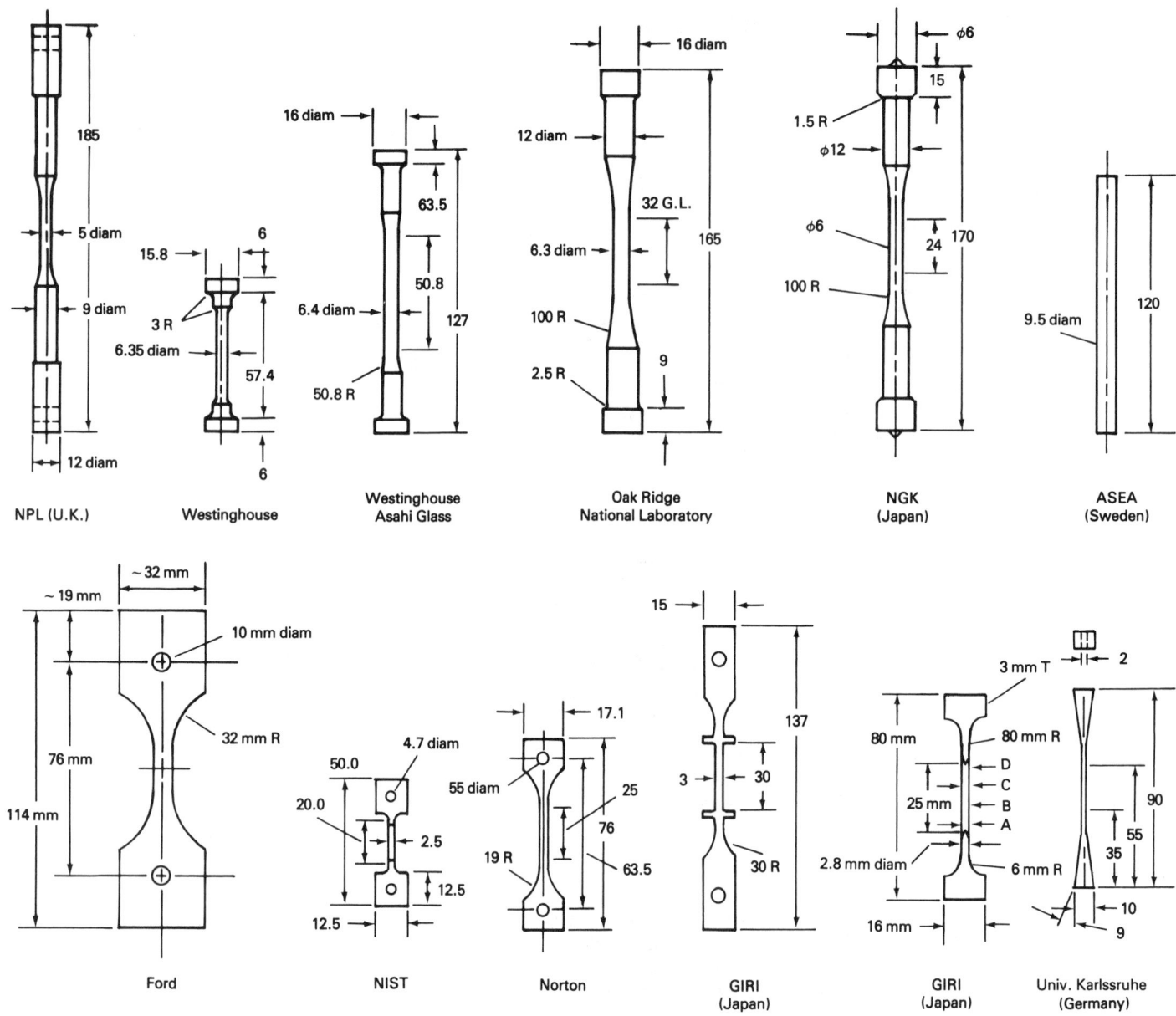

Fig 6 Tension specimens used for monolithic ceramics (each is in correct proportion to the others); all dimensions in mm. Upper row for round specimens; lower, for flat specimens. Adapted from Ref 37

or fine-scaled, isotropic composites. The most common are deflection measurements in flexural strength tests (with proper consideration of the test machine compliance) or strain gage experiments in flexure or direct tension. Dynamic measurements are also quite common, with either sonic excitation of prismatic specimens at their resonant frequency or time-of-flight measurements of ultrasonic waves.

Interpretation of Uniaxial Strength. The scatter in uniaxial strengths is well modelled by Weibull statistics (Ref 8, 9, 30). Weibull observed that the strength of brittle materials is controlled by the presence of randomly distributed defects, and that failure is controlled by the largest, most severely stressed defect.

Fracture occurs when a defect in one particular element of the body reaches a critical loading. This analysis is colloquially known as the weakest-link model, in direct analogy to the strength of a chain.

The Weibull strength distribution is:

$$F = 1 - \exp - \int_v \left(\frac{\sigma - \sigma_u}{\sigma_0} \right)^m dV \qquad \text{(Eq 6)}$$

where F is the probability of failure of a specimen or component, V is the specimen volume, σ is the maximum tensile stress at any given point, σ_u is the threshold stress (below which no failure will occur), σ_0 is the characteristic strength, and m is the Weibull

modulus. Equation 6 is known as the Weibull three-parameter distribution, because three parameters define the shape of the strength distribution. The Weibull modulus characterizes the scatter in strengths, and the characteristic strength is the location parameter (that is, how high or low the strengths are). Higher values of σ_0, σ_u, and, especially, m are preferred. This distribution is similar to a Gaussian distribution, except that there is a skew to lower strengths.

It is instructive to derive the above formula to show how simple it is, and to detect the underlying assumptions. For a chain of N elements, if any element fails, the component will fail. If F_i is the probability of failure of the ith element, and the probability of failure of the whole component is F, then the probability of survival of the component is:

$$1 - F = (1 - F_1)(1 - F_2)(1 - F_3) \ldots (1 - F_N)$$

This multiplication can be altered to an addition by taking the natural log of each side:

$$\ln(1 - F) = \ln(1 - F_1) + \ln(1 - F_2) + \ln(1 - F_3)$$

$$+ \ldots \ln(1 - F_N) = \sum_{i=1}^{N} \ln(1 - F_i)$$

The expression $\ln(1 - F_i)$ can be expanded with a MacLaurin series:

$$\ln(1 - F_i) = -F_i - F_i^2/2 - F_i^3/3 \ldots$$

and if F_i is small, then the higher-order terms can be deleted:

$$\ln(1 - F_i) \cong -F_i$$

Note that this is true for an individual element F_i, but not for F. For the whole component:

$$\ln(1 - F) = \sum_{i=1}^{N} \ln(1 - F_i) = \sum_{i=1}^{N} - F_i$$

At this point, Weibull chose an arbitrary, empirical expression for F_i:

$$F_i = \left(\frac{\sigma - \sigma_u}{\sigma_0}\right)^m$$

This is a critical point that should not be overlooked. This expression is quite simple and reasonable, and it satisfies boundary conditions at the limits of low or high stress. However, it is not founded upon any concept of flaw density or severity, or on fracture mechanics criteria. Continuing, for the whole structure:

$$\ln(1 - F) = \sum_{i=1}^{N} - F_i = \sum_{i=1}^{N} \left(\frac{\sigma - \sigma_u}{\sigma_0}\right)^m$$

and in the limit as each element approaches infinitesimal size:

$$\ln(1 - F) = \int_v \left(\frac{\sigma - \sigma_u}{\sigma_0}\right)^m dV$$

$$F = 1 - \exp - \int_v \left(\frac{\sigma - \sigma_u}{\sigma_0}\right)^m dV$$

This expression can be simplified further for the case of a test specimen experiencing a uniform tensile stress:

$$F = 1 - \exp\left[-V\left(\frac{\sigma - \sigma_u}{\sigma_0}\right)^m\right]$$

The threshold stress, σ_u, is a stress below which failure will not occur, and corresponds in concept to a maximum flaw size. In practice, it is usually set equal to zero. Although there are some recorded instances where it was not zero, these are rare and require extensive testing to verify (Ref 45). The Weibull modulus, m, has no units and is the factor that

determines the scatter in strength. High values are optimum. Traditional ceramics, such as whitewares and brick, may have values from 3 to 5. A good material has a value that exceeds 10. A ceramic with an m value ≥ 30 has very consistent strengths and could be practically considered to have a deterministic value of strength over a range of several orders of magnitude volume. The normalizing parameter, σ_0, is called the characteristic strength and has awkward units of stress times volume$^{1/m}$, but can be thought of as a location parameter that defines how low or how high strength is (analogous to the mean for a Gaussian distribution, but at the 63.2% probability of failure: $F = 1 - \exp[1.0]$) for a specimen of unit volume if σ_u is zero.

For the case where the threshold stress, σ_u, is set to zero:

$$F = 1 - \exp - \int_v \left(\frac{\sigma}{\sigma_0}\right)^m dV$$

This is the so-called Weibull two-parameter strength distribution, the most commonly used method for routine analyses.

Most ceramic bodies are nonuniformly stressed, and it is necessary to perform the stress-volume integration, Eq 6. A very useful concept of Weibull statistics is the effective volume, which permits an easy, rapid comparison of the strengths of different-sized specimens. For many bodies where the state of stress is known, it is a simple matter of basic calculus to perform the stress-volume integration and derive the effective volume. More sophisticated bodies can be analyzed with finite-element stress analysis codes.

A uniform tension specimen has, of course, an effective volume equal to the actual volume in the gage length, V. A three-point loaded, rectangular flexure specimen of total volume V within the support spans has an effective volume of (Ref 9):

$$V_E = \frac{V}{2(m + 1)^2}$$

This means that if a bend specimen has a maximum (outer fiber) stress of σ_R on it, it is equivalent (same probability of failure) to a tensile specimen with the same stress, but with a volume of only V_E. The effective volume varies with the Weibull modulus. The flexure specimen effective volume is always less than the volume of the specimen. If $m = 10$, then $V_E = V/242$. A three-point test specimen is thus very inefficient for stressing much of the material. A $1/4$ four-point rectangular specimen has an effective volume:

$$V_E = \frac{V(m + 2)}{4(m + 1)^2}$$

and for an m of 10, $V_E = 3/121\ V$. The four-point specimen loads a higher volume than the three-point specimen, but is still low compared to uniform tension.

Often, flaws are located at the surface, such as those resulting from machining, impact, or

chemical damage, in which case the Weibull integration should be over the stressed surface, and not the volume. This is especially true for glasses.

$$F = 1 - \exp\left(\frac{\sigma}{\sigma_0}\right)^m dS$$

and it can be shown that the effective surface is:

$$S_E = \frac{S(m + 2)}{4(m + 1)^2}$$

for square three-point specimens, and

$$S_E = \frac{S(m + 2)^2}{8(m + 1)^2}$$

for square $1/4$ four-point specimens.

Effective volumes or surfaces are used to compare the strengths of different sized or differently stressed specimens or components. For the two-parameter strength distribution, the strength of two different sized bodies are related as follows (Ref 9):

$$\frac{\sigma_1}{\sigma_2} = \left(\frac{V_{E2}}{V_{E1}}\right)^{1/m} \quad \text{or} \quad \left(\frac{S_{E2}}{S_{E1}}\right)^{1/m} \quad \text{(Eq 7)}$$

The strengths of two glass optical fibers of equal diameter, but different lengths, is thus:

$$\frac{\sigma_1}{\sigma_2} = \left(\frac{V_2}{V_1}\right)^{1/m} = \left(\frac{L_2}{L_1}\right)^{1/m}$$

As an example, if m is 15, and $V_{E2} = 4\ V_{E1}$, then:

$$\sigma_1 = 1.10\ \sigma_2$$

or, in other words, the shorter fiber would be 10% stronger on the average than the longer fiber. Another example would be the case of two batches of flexure specimens, size 1 being $4 \times 4 \times 40$ mm ($0.16 \times 0.16 \times 1.6$ in.), and size 2 being $1 \times 1 \times 10$ mm ($0.04 \times 0.04 \times 0.4$ in.), which is the size within the fixture support span, where

$$\frac{\sigma_2}{\sigma_1} = 64^{1/15} = 1.32$$

for volume flaws, and

$$\frac{\sigma_2}{\sigma_1} = 16^{1/15} = 1.20$$

for surface flaws.

These examples show that strength not only scales with specimen size, but that the magnitude of the change strongly depends on whether the defects are surface or volume. Obviously, it is essential to know whether flaws are of one or the other category if the laboratory strength data are going to be size scaled to predict component performance.

Weibull graphs are a convenient means to report strength data. The graph usually has special axes chosen to linearize the data. This is done in the same fashion that probability paper can be used to linearize data for a Gaussian distribution. If the data fit a two-parameter distribution, then:

$$F = 1 - \exp\left[-V_E\left(\frac{\sigma}{\sigma_0}\right)^m\right]$$

$$(1 - F) = -V_E\left(\frac{\sigma}{\sigma_0}\right)^m$$

$$-\ln(1 - F) = \ln(1 - F)^{-1} = \ln\frac{1}{1 - F} = V_E\left(\frac{\sigma}{\sigma_0}\right)^m$$

and taking one more natural-log:

$$\ln\left[\ln\left(\frac{1}{1 - F}\right)\right]$$

$$= \ln V_E - m \ln \sigma_0 + m \ln \sigma \quad \text{(Eq 8)}$$

The first two terms on the right side are constant for a batch of specimens of identical size and type. Equation 8 corresponds to a straight line if the axes on a graph are chosen to be $\ln \ln[1/(1 - F)]$ and $\ln \sigma$. It is typical to plot the former as the ordinate and the latter as the abscissa, as shown in Fig 7. This procedure is best illustrated by a simple example. Five specimens are tested with the following outcomes: 290, 315, 295, 300, and 340 MPa (42, 46, 43, 44, and 49 ksi). These strengths are arranged in ascending order in Table 1 and each strength is assigned a probability estimate of $F = (i - 0.5)/N$, where i is the ith specimen, and N is the total number of specimens. This estimator has been shown to have the least bias and lowest scatter for the present analysis (Ref 46). The values of $\ln \ln[1/(1 - F)]$ and $\ln \sigma$ are tabulated, and the graphed results are shown in Fig 7. A least squares curve can be fitted to the data, usually minimizing the vertical deviations. The all-important Weibull modulus is easily interpreted as the slope of the line on the graph. The location parameter, the "characteristic strength of the test specimen," is the stress

Table 1 Simple Weibull analysis of strength data

i	$F = (i - 0.5/N)$	σ	$\ln \ln[1/(1 - F)]$	$\ln \sigma$
1	0.10	285	−2.25	5.653
2	0.30	310	−1.03	5.737
3	0.50	335	−0.37	5.814
4	0.70	350	0.19	5.858
5	0.90	360	0.83	5.886

Note: Least squares fitting of $\ln \ln[1(1 - F)]$ versus $\ln \sigma$ gives a slope of 12.3, and an intercept of −72.02; for $\ln \ln[1/(1 - F)] = 0 = m \ln \sigma - 72.02$, then $\ln \sigma = 5.832$, or $\sigma = 341$ is the characteristic strength of the specimen.

for which the vertical axis is 63.2% (or $\ln \{\ln[1/(1 - F)]\} = 0$) and can be corrected to the "characteristic strength" for *unit volume* by using the appropriate size-scaling Eq 7.

This scheme is by far the most common means of depicting strength data of monolithic engineering ceramics. At a glance, it is possible to qualitatively assess whether the data fits the Weibull distribution. The process will only give estimates of the Weibull parameters, and for small sample sizes, these estimates can be quite deviant from the actual population parameters. It is usually recommended that no less than 30 specimens be tested. Curvature on the graph can result from several sources. If there is a threshold value of stress below which failure will not occur or if specimens have been proof tested, then the data will be truncated and will curve downward away from the straight line at the low-strength end. If several flaw types are present, then each will have a separate Weibull distribution, which can overlap and create a curved overall distribution.

Although this is the most common means of depicting strength data, it is by no means universal, and it does have some shortcomings. The least squares analysis places strong emphasis on the low-strength data point, but in many instances, this may be appropriate, because the low-strength end of the distribution is of most concern to a user. The maximum likelihood estimation scheme (Ref 46, 47) is a popular alternative means of estimating Weibull parameters and has virtues that make it appealing to the design community (Weibull analysis is discussed in several articles in the Section "Design Considerations" in this Volume). Because the different analyses will lead to different estimates of the Weibull parameters, particularly for small sample sizes (30 or less), it is important that it be explicitly stated how the data were analyzed, and it is preferable to give the actual strength values in tables, as well.

The Weibull analysis is adequate for multiaxially, tensilely loaded ceramics, provided that the second or third principal stresses are significantly less than the principal tensile stress. If this is not the case, then more sophisticated analyses that take into account the effect of multiaxial tensile stresses on defects are appropriate (Ref 10, 11). The Weibull analysis also has limitations if the defects are liable to grow subcritically during

a test or if the material exhibits an R-curve phenomena. A newly recognized phenomenon that could occasionally pose problems in strength analysis is *latent* defects caused by localized surface impact or contact stresses. Concentrated microdamage can occur that can lead to a larger microcrack popping-in after an incubation period (Ref 48, 49).

Fractography. Strength values by themselves are only half the picture. The defects that account for these strengths are equally important. Figure 8 shows some typical defects. A strength data base for a material that reflects defects of one class, such as pores, is misleading and possibly irrelevant if another specimen or component fails from other defects, such as inclusions or machining damage. Each flaw type will have its own Weibull distribution. Multiple flaw populations are common in ceramics, and it is essential that the defects be as clearly associated with the strength values as possible. A labelled Weibull graph (Ref 50), such as that shown in Fig 9, can be a helpful means to combine the data in an easy-to-appreciate format. Fractographic techniques are discussed in more detail in the Section "Failure Analysis" in this Volume, and a proposed standard guide to fractography for monolithic ceramics is being developed (Ref 51).

Although the Weibull analysis is empirical and relatively simple, it is quite versatile and has been repeatedly shown to model uniaxile strength data exceptionally well. Indeed, it can be shown that the Weibull distribution can be related to a plausible, power-law flaw size distribution (Ref 52). Flaw size distributions have only rarely, and usually incompletely, been characterized in ceramics. This is a topic that has received scant attention and warrants further study.

Uniaxial Compression Strength. The high compressive strength of ceramics is a consequence of the resistance of the material to plastic flow and the insensitivity of defects to compressive stress. Ancient structural applications of ceramics were columns and walls that capitalized on high compressive strength. The fact that ceramics fail at all in compression is a result of the distortion of the stress field in the immediate vicinity of the tip of a defect. This distortion causes a localized tensile stress concentration that, for defects at the worst orientation ($-30°$ to the axial stress) is about one-eighth of the concentration if the specimen is loaded in tension. Thus, a Griffith-type criterion for failure would predict that the compression specimen will fail at about 30° to the specimen axial direction when the compressive stress is eight times the tension strength, but this is an oversimplification.

The tensile stresses in the immediate vicinity of a defect will cause a crack to propagate stably for a slight distance (Ref 53, 54). The crack then aligns itself with the compression stress and is arrested (Fig 10). Progressively more defects grow until the damage that has

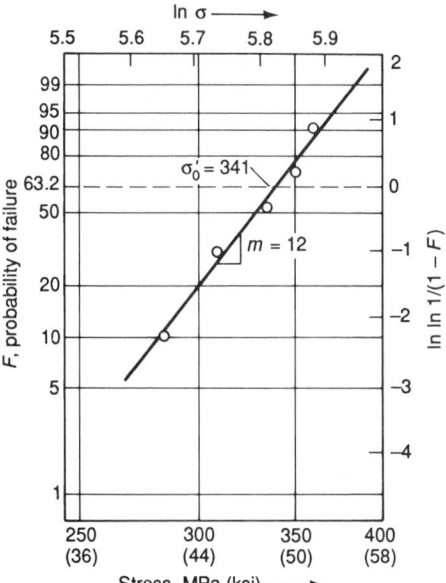

Fig 7 Simple Weibull graph linearizes strength-probability data to permit assessment of whether the data fit a Weibull distribution; strengths are from Table 1

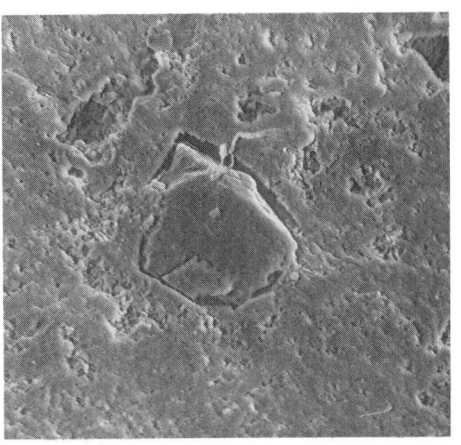

(a)

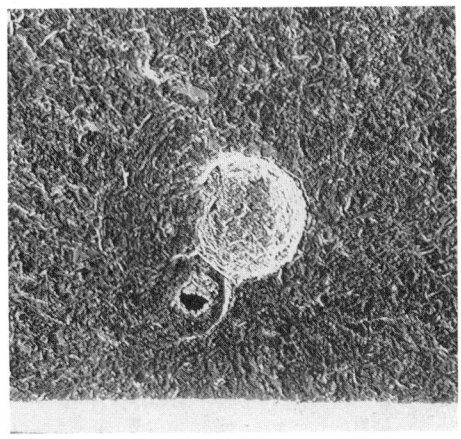

(b)

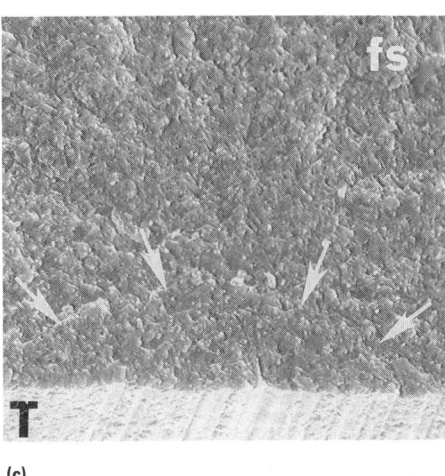

(c)

Fig 8 Scanning electron photomicrographs of strength-limiting defects in ceramics. (a) Silicon inclusion in reaction-bonded silicon nitride, 370×. (b) Powder agglomerate in sintered silicon carbide, 150×. (c) Machining damage in hot pressed silicon nitride. 370×

accumulated in the specimen reaches some limit, and the specimen virulently disintegrates into powder (often with a triboluminescent emission) (Ref 56, 57). Compression strength thus depends not on the largest, worst-oriented, highest-stressed defect, but on the entire defect population. The high compressive stresses can nucleate cracks due to twinning or dislocation activity, as well (Ref 58, 59). Compression strength may have a dependence on the square root of grain size, because the size of defects may scale with the average grain size or because of microplasticity in the grains. Weibull statistics are irrelevant, and compression strengths often have very low scatter (Ref 56, 57).

The compression test appears deceptively easy to conduct. However, it is extremely difficult to accurately measure compression strength, because slight misalignments can create bending stresses, and end-loading effects can cause parasitic tensile stresses that cause fracture. Mismatches of the elastic properties of the platens and test specimen can cause tensile stresses or frictional constraints. Buckling can occur if specimens are too long. Because the stresses being applied to a compression specimen are extremely high, the alignment errors may be exacerbated above what they would be for an equivalent tensile test specimen. True compression tests may be as difficult to conduct as direct tensile tests. Refined compression strength tests have been developed, as shown in Fig 11 (Ref 55, 56).

Multiaxial Strength. The small amount of available data on multiaxially stressed ceramics is primarily derived from research studies. A generalized failure criteria envelope is shown in Fig 12. Superimposing a modest stress, σ_2, onto a tensile stress, σ_1, usually changes the failure stress only a minor amount. In the tension-tension quadrant, the equibiaxial strength usually decreases by 0 to 15% (for example, Ref 60, 61), although there are even reported instances of strengthening, as discussed by Rosenfield *et al.* (Ref 61). Most equibiaxial data have been generated by disk flexure tests with ring-on-ring, ball-on-ring, or uniform pressure-on-ring configurations. The simplest model to account for the weakening is a Weibull model which assumes independent action of the principal stresses. Only those defects normal to the tensile stress in uniaxial tension are likely to cause fracture. In equibiaxial stress, many more flaws are active, causing reduced strength. However, this model overestimates the strength degradation. A more sophisticated analysis devised by Batdorf and Heinisch (Ref 10), which considers the precise nature of the defects and their shear sensitivity through a failure criterion, is currently in vogue (for example, Ref 60).

There is little or no weakening in the tension-compression stress quadrant for stress ratios σ_2/σ_1 of 0 to −4 (Ref 62–64). Uniaxial compression has been previously discussed. Because the tensile and compressive strengths are controlled by different phenomena, there is no simple ratio for σ_c/σ_t (such as the value of 8 predicted by the Griffith model).

Strengths in the compression-compression quadrant are not dramatically changed from the uniaxial compression strength, according to findings by Adams and Sines (Ref 53–55). However, they did develop a model that suggests strengthening could occur, depending on the character and number of defects that are activated (Ref 52).

Flaw Key		
A	Agglomerate	
P	Pore	
PS	Porous Seam	
PR	Porous Region	
LG	Large Grain	
?	Uncertain	

$\sigma_0 = 466$ MPa

Characteristic Strength of the Bend Bar

Stress	Flaw
562	A
543	?
500	?
499	?
499	PS
486	PS
485	?
480	PR*
466	P
465	P*
462	A
455	PS
449	PR
445	PR
445	A
441	PS
437	PS
435	PS*
434	A or LG
430	?
428	P*
411	PS
409	PS
407	PS
402	PS
393	PS
393	PR
380	PR
371	?
307	PS
Avg. 444 MPa	

m
Slope
10.2

466

Fig 9 Labelled Weibull graph that displays both strength and fractography data for a 99.9% alumina tested in three-point flexure at room temperature

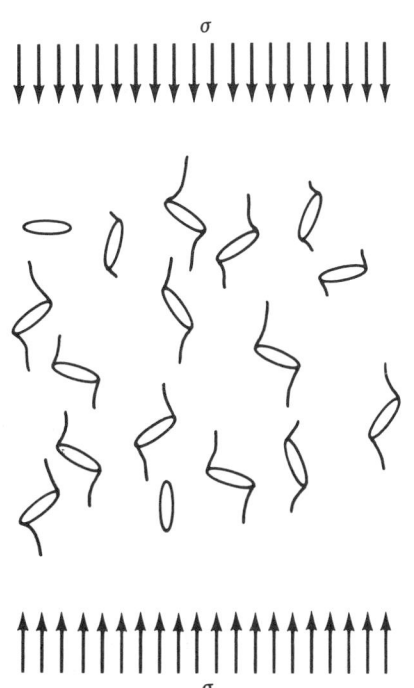

σ

σ

Fig 10 Crack extension generated by flaws in compression

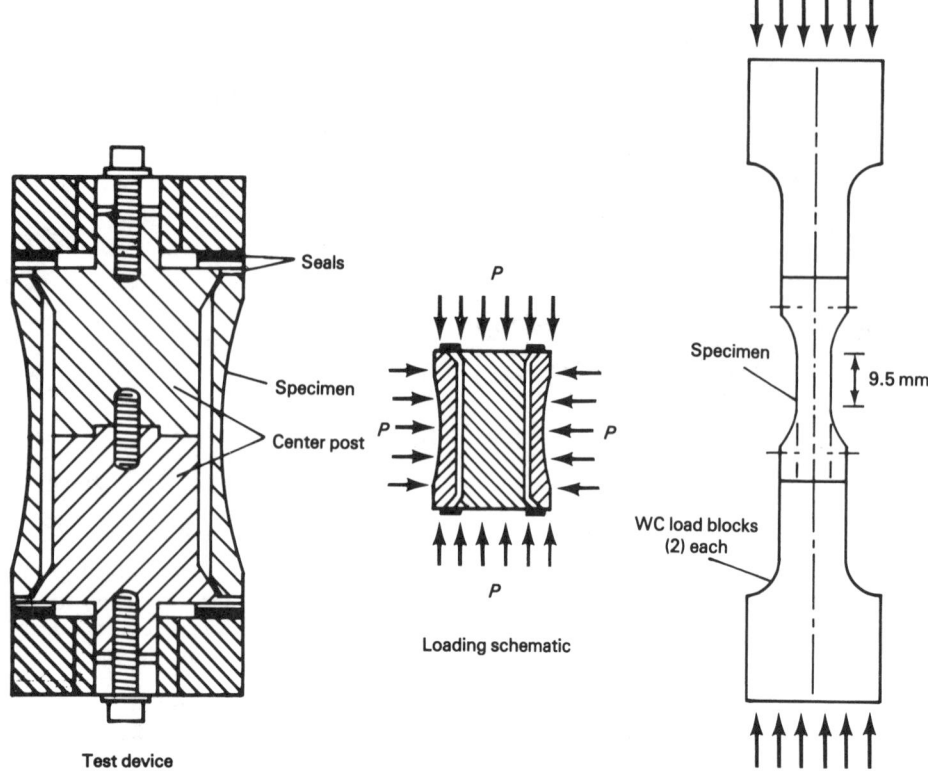

Fig 11 Compression test specimens. After Ref 55 and 56

High-Temperature Strength Test Methods for Isotropic Ceramics

Fast Fracture. The overwhelming majority of high-temperature strength tests have been done in four-point loading. Standards are being developed that are extensions of the low-temperature procedures. A variety of furnaces and environments can be used with temperatures typically up to 1600 °C (2910 °F) in air, but with some vacuum and inert gas systems up to 2000 °C (3630 °F). The test fixtures themselves must be dense ceramics, and are usually fairly pure forms of silicon carbide, although occasionally alumina fixtures are used at lower temperatures, and graphite fixtures are used in inert atmospheres.

The upsurge in tensile testing previously discussed had been driven in large part by new programs to use ceramics at high temperature in heat engines. As a result, most tensile test systems have been designed with high temperatures in mind (Ref 35–38). The gripping schemes must not only be elaborate enough to avoid stress concentrators and to align very precisely, but they must also be capable of being used in conjunction with furnaces. Most tension systems use cold grips with relatively long (for ceramics) specimens of 150 mm (6 in.). Such systems are commercially available and extensive testing is underway in the United States, Japan, and Germany.

Multiaxial strength tests are extremely rare at high temperature and are usually based on ring-on-ring loaded disks. The limited results for these experiments suggest the high-temperature equibiaxial strength is 15 to 30% less than uniaxial strengths (Ref 65, 66).

Creep and Stress Rupture. Direct tension tests of long duration are becoming more common, but once again most test systems are complicated and expensive. They typically are derivatives of the fast-fracture systems, using cold grips and long specimens (Ref 35–38). Most experiments are limited to a 1000 h duration. An economical alternative test system with hot grips and a flat dog bone specimen configuration has been developed, and is optimized for long duration, low-stress creep experiments (Ref 34). A short tapered specimen for similar experiments has been successfully used by Grathwohl (Ref 37). Strains must be measured with specialized extensometers because ceramic strains are extremely small, and resolutions of 1 μm (0.04 mil) over the course of hours must be recorded. The extensometers in use today are either delicate mechanical units (Ref 37) or lasers that monitor distance between flags (Ref 34) or diffract when passed through a narrow slit between two flags (Ref 38). Reference 67 is an excellent compilation of several tension testing papers that discuss these methods in detail.

Most investigators have at some time resorted to using flexure testing, which is much less expensive and allows strains to be readily measured from the curvature in the specimen. The bend specimen, in a sense, acts as a deflection magnifier, because the deflection associated with the integrated curvature is larger and easier to measure than the extension of a tension specimen. The drawback of the method is that there is a stress gradient in the specimen, which changes dramatically as the material creeps. Flexural creep testing is not a constant stress test. The strain is measured from the curvature, but this too must be adjusted for the proper constitutive equation.

In recent years, it has become painfully evident that flexural creep data can be misleading or even erroneous, a consequence of the stress gradient, the relaxation of such gradient, and the complicated constitutive equations that apply to ceramics. Analytical attempts to deconvolute the tensile and compressive creep behavior are usually tainted or compromised by the assumptions that have to be made about the constitutive equations. It is far more rational to conduct direct tension or compression experiments for careful creep work. Flexure tests can be used for qualitative assessments of conditions for the onset of creep.

Stress-rupture data require extremely long duration experiments. Some static-fatigue phenomena occur in the absence of bulk creep deformation and flexure testing may be eminently suitable in these cases. Failures out to 18,000 h have been reported (Ref 68) and data are readily accumulated, as shown in Fig 13. Perhaps the best approach is to conduct complementary direct tension and flexure experiments, as discussed in Ref 68 and 69. It then becomes feasible to construct fracture mechanism maps, as shown in Fig 14, patterned after those developed by Ashby and coworkers (Ref 70). In each case, fractography is essential for proper identification of the phenomena that cause failure.

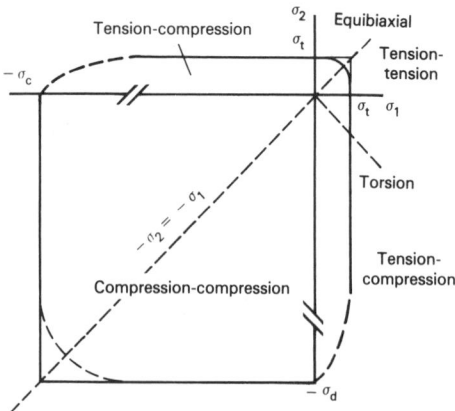

Fig 12 Approximate multiaxial failure envelope for monolithic ceramics. Uniaxial loadings lie on the axes. Tensile stresses are positive and compression stresses are negative. Failure occurs for any combination of σ_1, σ_2 that lies outside the envelope.

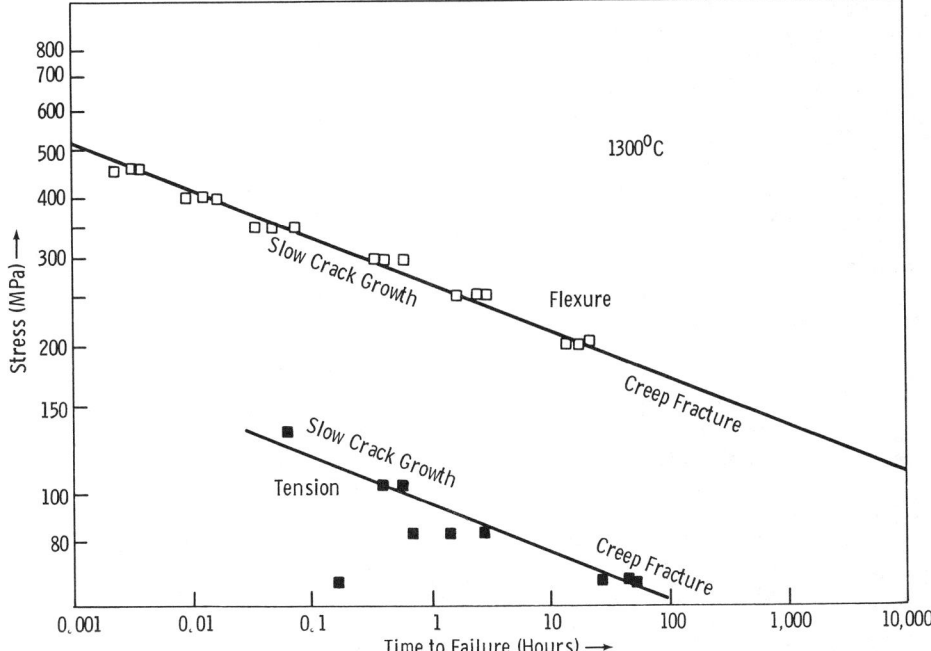

Fig 13 Complementary flexure and tensile stress rupture data for silicon nitride at 1300 °C (2370 °F). Stress levels are different for the two methods. Source: Ref 68, 69

Multiaxial stress-rupture experiments are exceptionally rare at high temperature. One recent study with ring-on-ring equibiaxially loaded specimens showed the strength degradation rate in silicon nitride that was due to crack growth was identical to the uniaxial rate (Ref 66), similar to other findings at room temperature (Ref 71).

Strength of Continuous Fiber Reinforced Composites

Continuous fiber reinforced ceramics open new vistas with respect to mechanical properties (Ref 16, 17, 72–76). The fibers and the matrix are separately quite brittle materials (for example, the glass K_{Ic} is ~0.7 MPa$\sqrt{m}$, or 0.6 ksi$\sqrt{in.}$, whereas SiC is ~3.0 MPa$\sqrt{m}$, or 0.27 ksi$\sqrt{in.}$), but together the toughness is so high as to be unmeasurable by conventional fracture mechanics tests. The fiber and matrix properties are often chosen such that the fiber elastic modulus is higher than that of the matrix in order to promote load transfer from the matrix to the fiber. The thermal expansion coefficient of the fiber is usually greater than that of the matrix, so that as they cool from the processing temperature (stress-free state) the fiber will tend to contract more, thereby placing the matrix into compression. A critical aspect is the fiber compatibility with the matrix, because severe chemical interactions in processing and in use should be avoided. Bonding should have low-to-medium strengths, so that cracks are redirected, branched, and dissipated by propagating along the matrix-fiber interface (Fig 2). Fibers remain intact and are load bearing

until they, too, are fractured. Extra energy is consumed in fracturing the fibers and pulling them out of the matrix. This gives rise to a fibrous fracture surface.

The stress-strain diagram shows linearly elastic behavior with a composite modulus equal to the volume fraction weighted average of the fiber and matrix moduli (Fig 1). The fracture strength of the matrix is usually much less than that of the fiber. Nonlinearity occurs after matrix microcracking occurs, causing permanent damage analogous to

yielding in metallic materials. Beyond the initial microcracking point, progressive damage/crack propagation occurs until, finally, the fibers themselves fracture. Depending on the specimen configuration and the ceramic composite itself, the stress-strain diagram may show either a sudden fracture or the stress level may gradually taper off (for example, in flexure loading).

Most ceramic-matrix composites being made are unidirectionally reinforced, and strengths normal to the fiber orientation are extremely low (~20 MPa, or 3 ksi, Ref 16). Two-dimensionally reinforced composites have also been developed, but their strengths are less than unidirectionally reinforced composites tested in their strongest orientation, because fewer fibers are reinforcing in each direction.

These ceramics must be regarded as prototype materials available in only small sizes and in research quantities. Issues of consistency, cost, mechanical fatigue, and environmental-exposure sensitivity remain open, but the allure of the "ductile" ceramic is strong, and it is likely that these advanced ceramics will find specialty applications where no other material can function, such as in very high temperature engine components.

Test methods for ceramic-matrix composites are typically quite different than for monolithics, and borrow heavily from the organic-matrix and carbon-carbon test procedures (Ref 16, 73, 74, 76). Uniaxial tensile strength testing is most commonly done in direct tension or flexure loadings. Flexure testing is not preferred, because the failure mechanism can be tensile, compressive, or interlaminar shear, depending on the composite components, the reinforcement architecture, and the loading geometry (Ref 72).

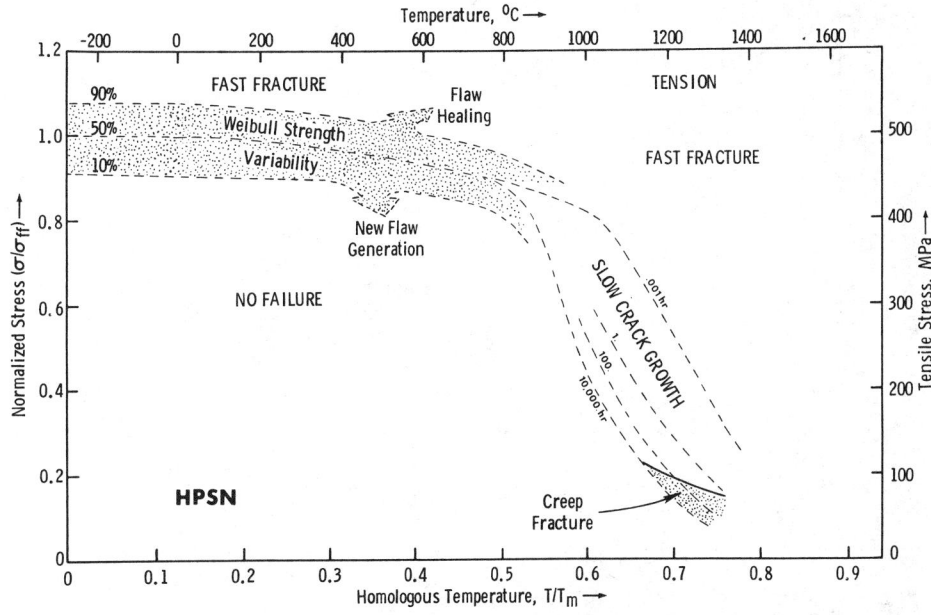

Fig 14 Stress rupture data used to create fracture mechanism maps. Loci of constant failure time are labelled in hours. Source: Ref 68

Flexure testing is acceptable for measuring the matrix microcracking stress, the shear strength of one-dimensionally reinforced composites (with the fibers perpendicular to the maximum stress), the effects of exposure or heat treatments, or certain high-temperature properties (Ref 72–74). Direct tensile loading is much easier to conduct for composites than for monolithics, because the former are more tolerant of slight misalignments. Flat specimens with glued tabs (to avoid gripping damage) are quite adequate with commercial test machine grips. Ordinary clip extensometers or strain gages are suitable for measuring strain. At high temperatures, glued tabs are not adequate, and specimens with holes or tapered-wedge shoulders are necessary (Ref 72–76). However, the low shear strength of unidirectionally reinforced composites can make testing of pin-loaded specimens difficult or impossible, because the pins shear through the specimen. It is actually easier to test two-dimensionally fiber reinforced composites for this reason.

Proof Testing

Ceramic users ultimately must have confidence that components will have a reliable minimum strength or performance level. The object of using nondestructive evaluation methods is to be able to discern defects to permit the culling of unacceptable components, but the state of the art is inadequate at the moment. Proof testing is a viable means of weeding out unacceptable parts in monolithic, brittle ceramics (Ref 77–79).

Proof testing entails stressing all components to a proof stress, σ_p, in order to cause fracture in the parts that are weaker than σ_p. This is illustrated in Fig 15, which shows the

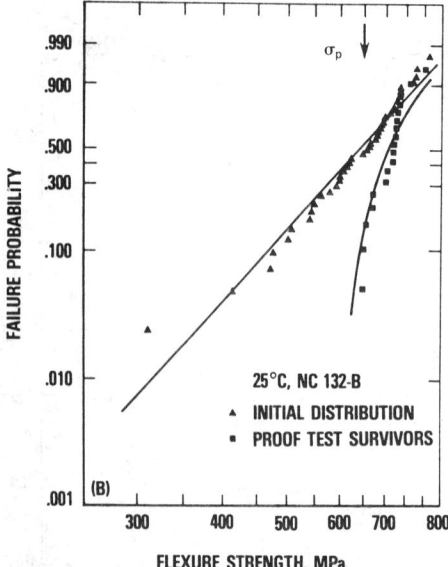

Fig 15 Proof testing at 650 MPa (94 ksi) truncates the Weibull strength distribution for hot-pressed silicon nitride and eliminates low-strength specimens. After Ref 26

strength distribution for hot-pressed silicon nitride flexure specimens in the as-received state, and after the application of a 650 MPa (95 ksi) proof stress (Ref 26). The proof strength distribution is truncated and curves downward in a fashion identical to a Weibull three-parameter distribution with a threshold stress, σ_u, equal to σ_p.

There are, however, severe restrictions on the utility of this method. Proof testing is only effective if the test precisely simulates the actual service conditions. Any deviation incurs risk. For example, the author once received a batch of flexure specimens for evaluation, which, unbeknownst to him, had been proof tested in three-point flexure in order to weed out "atypically weak specimens." This was ineffective, because the specimens were actually tested by the author in *four-point loading* and did have a wide strength scatter. The three-point flexure "proof tested" only a tiny portion of the specimen, and gave a false sense of assurance to the proof tester. Another example was in the final report of the Ceramic Applications in Turbine Engines Program (Ref 31):

"The CATE proof testing eliminated numerous flawed components . . . however, not all the flawed ceramic parts were eliminated by proof tests and engine failure resulted from one such part. This emphasizes the fact that proof tests are valuable; however they cannot exactly duplicate engine environments, and there is no substitute for engine testing as the ultimate demonstration of acceptability."

A further problem occurs if the material is susceptible to slow crack growth during the proof test. Defects can grow and weaken the specimen (Ref 2). For a proof test to be effective in narrowing a strength distribution, it can be shown that n must meet the criterion: $(n - 2)/m > 1.0$ (Ref 79). Ideally, the slow-crack-growth exponent n should be high under the conditions of the proof test. In the worst case, if unloading rates are low, it is even possible for some specimens to be weaker than σ_p after the test. Proof tests are typically done to stress levels commensurate with or somewhat higher than the stresses expected in service. If slow crack growth is anticipated in the service conditions, then it may be necessary to apply proof stresses much higher than the service levels. Analytical procedures exist that permit integration of the slow crack growth and statistical analyses, which allows the estimation of minimum lifetimes (Ref 2, 77), usually in the form of strength-probability-time diagrams (Ref 2).

The proof test only culls out specimens with defects at the time of the proof test. If new flaws subsequently develop, for example, during high-temperature exposure or service loadings, then the proof test may be negated (Ref 26).

To reiterate, effective proof testing must be conducted in as close a fashion as possible

to service conditions. It may be regarded as an expensive procedure, because every part must be tested, but it is routinely applied to the silicon nitride ceramic turbocharger rotors in Japanese automobiles and metallic rotors for gas turbine engines (Ref 78). Even if proof test conditions do not exactly match the application, the proof test will give some modicum of reassurance, and hopefully weed out the components with extraordinary, gross defects that can sometimes bedevil ceramic users.

REFERENCES

1. W.E. Creyke, I.E.J. Sainsbury, and R. Morrell, *Design With Non-Ductile Materials,* Applied Science Publishers, 1982
2. R.W. Davidge, *Mechanical Behavior of Ceramics,* Cambridge University Press, 1979
3. D.F. Carroll and A.J. Pyzik, "Microstructural/Mechanical Property Relationships in Self-Reinforced Silicon Nitride," presented at the 14th Annual Conference on Composites and Advanced Ceramic Materials, Cocoa Beach, FL, Jan 1990
4. C.W. Li and J. Yamanis, Super Tough Silicon Nitride With R-Curve Behavior, *Ceram. Eng. Sci. Proc.,* Vol 10 (No. 7–8), 1989, p 632–645
5. E. Tani, S. Umebayashi, K. Kishi, K. Kobayashi, and M. Nishijima, Effect of Size of Grains With Fibre-Like Structure of Si_3N_4 on Fracture Toughness, *J. Mater. Sci. Lett.,* Vol 3, 1985, p 1454–1456
6. R.S. Storm, An Advanced Silicon Carbide With Enhanced Strength and Fracture Toughness, *Structural Ceramics,* Proceedings of the Army's 37th Sagamore Conference, 3 Oct 1990
7. K.Y. Chia and S.K. Lau, "High Toughness Silicon Carbide," presented at the 15th Annual Conference on Composites and Advanced Ceramics, Cocoa Beach, FL, Jan 1991, to be published in *Ceram. Eng. Sci. Proc.*
8. W. Weibull, Statistical Theory of Strength of Materials, *R. Swedish Inst. Eng. Res. Proc.,* No. 151, 1939, p 1–45
9. D.G.S. Davies, The Statistical Approach to Engineering Design in Ceramics, *Proceedings of the British Ceramics Society,* Vol 22, 1973, p 429–452
10. S.B. Batdorf and H.L. Heinisch, Jr., Weakest Link Theory Reformulated for Arbitrary Fracture Criterion, *J. Am. Ceram. Soc.,* Vol 61 (No. 7–8), 1978, p 355–358
11. A.G. Evans and T.G. Langdon, Structural Ceramics, *Progress in Materials Science,* No. 21, 1976, Pergamon Press, p 171–425
12. A.G. Evans, Perspective on the Development of High-Toughness Ceramics, *J. Am. Ceram. Soc.,* Vol 73 (No. 2), 1980, p 187–206
13. R.W. Rice, Ceramic Matrix Composite

Toughening Mechanisms: An Update, *Ceram. Eng. Sci. Proc.*, Vol 6 (No. 7–8), 1985, p 589–607

14. R. Knehans and R. Steinbrech, Memory Effect of Crack Resistance During Slow Crack Growth in Notched Al_2O_3 Bend Specimens, *J. Mater. Sci. Lett.*, Vol 1, 1982, p 327–329

15. S.J. Bennison and B.R. Lawn, Flaw Tolerance in Ceramics With Rising Crack Resistance Characteristics, *J. Mater. Sci.*, Vol 24, 1989, p 3169–3175

16. R.W. Davidge and J.J.R. Davies, Ceramic Matrix Fibre Composites: Mechanical Testing and Performance, *Int. J. High Tech. Ceram.*, Vol 4, 1988, p 341–358

17. K.M. Prewo and J.J. Brennan, Silicon Carbide Yarn Reinforced Glass Matrix Composite, *J. Mater. Sci.*, Vol 17, 1982, p 1201–1206

18. G.D. Quinn, Review of Static Fatigue in Silicon Nitride and Carbide, *Ceram. Eng. Sci. Proc.*, Vol 3 (No. 1–2), 1982, p 77–98

19. R.J. Charles, Static Fatigue of Glass, II, *J. Appl. Phys.*, Vol 29 (No. 11), 1951, p 1554–1560

20. S.M. Wiederhorn and L.H. Bolz, Stress Corrosion and Static Fatigue of Glass, *J. Am. Ceram. Soc.*, Vol 53 (No. 10), 1970, p 543–548

21. T.A. Michalske and S.W. Freiman, A Molecular Mechanism for Stress Corrosion in Vitreous Silica, *J. Am. Ceram. Soc.*, Vol 66 (No. 4), 1983, p 284–288

22. T.A. Michalske, The Stress Corrosion Limit: Its Measurement and Implications, *Fracture Mechanics of Ceramics*, Vol 5, R. Bradt, A. Evans, D. Hasselman, and F. Lange, Ed., Plenum Press, 1983, p 277–289

23. B.J.S. Wilkins and R. Dutton, Static Fatigue Limit With Particular Reference to Glass, *J. Am. Ceram. Soc.*, Vol 59 (No. 3–4), 1976, p 108–112

24. E.J. Minford, D.M. Kupp, and R.E. Tressler, Static Fatigue Limit for Sintered Silicon Carbide at Elevated Temperatures, *J. Am. Ceram. Soc.*, Vol 66 (No. 11), 1983, p 769–773

25. G.D. Quinn and R.N. Katz, Time-Dependent High-Temperature Strength of Sintered α-SiC, *J. Am. Ceram. Soc.*, Vol 63 (No. 1–2), 1980, p 117–119

26. S.M. Wiederhorn and N.J. Tighe, Proof Testing of Hot Pressed Silicon Nitride, *J. Mater. Sci.*, Vol 13, 1978, p 1781–1793

27. D. Lewis III, Observations on the Strength of a Commercial Glass-Ceramic, *Ceram. Bull.*, Vol 61 (No. 11), 1982, p 1208–1214

28. M.L. Hanney and R. Morrell, Factors Influencing the Strength of a 95% Alumina, *Engineering With Ceramics*, Vol 12, *Proc. Br. Ceram. Soc.*, 1982, p 277–290

29. R. Morrell, Mechanical Properties of Engineering Ceramics: Test Bars Versus Components, *Mater. Sci. Eng.*, Vol A109, 1989, p 131–137

30. G.D. Quinn and R. Morrell, Flexure Testing for Design of Engineering Ceramics: A Review, *J. Am. Ceram. Soc.*, Vol 74 (No. 9), 1991, p 2037–2066

31. S. Thrasher, Ceramic Applications in Turbine Engines (CATE) Program Summary, *Proceedings of the 21st Contractors Coordination Meeting*, Report P138, Society of Automotive Engineers, March 1984, p 255–267

32. F.I. Baratta, G.D. Quinn, and W.T. Matthews, "Errors Associated With Flexure Testing of Brittle Materials," U.S. Army MTL Technical Report TR 87–35, July 1987

33. R.G. Hoagland, C.W. Marschall, and W. Duckworth, Reduction of Errors in Ceramic Bend Tests, *J. Am. Ceram. Soc.*, Vol 59 (No. 5–6), 1976, p 189–192

34. D.F. Carroll, S.M. Wiederhorn, and D.E. Roberts, Technique for Tensile Creep Testing of Ceramics, *J. Am. Ceram. Soc.*, Vol 72 (No. 9), 1989, p 1610–1614

35. T. Soma, M. Matsui, and I. Oda, Tensile Strength of a Sintered Silicon Nitride, *Nonoxide Technical and Engineering Ceramics*, S. Hampshire, Ed., *Proceedings of the International Conference*, Limerick, Ireland, 1985, p 361–374

36. T. Ohji, Towards Routine Tensile Testing, *Int. J. High Tech. Ceram.*, Vol 4, 1988, p 211–225

37. G. Grathwohl, Current Testing Methods—A Critical Assessment, *Int. J. High Tech. Ceram.*, Vol 4, 1988, p 123–142

38. K.C. Liu, H. Pih, and D.W. Voorhes, Uniaxial Tensile Strain Measurement for Ceramic Testing at Elevated Temperatures: Requirements, Problems, and Solutions, *Int. J. High Tech. Ceram.*, Vol 4, 1988, p 69–87

39. J. Nilsson and B. Mattsson, A New Tensile Test Method for Ceramic Materials, *Ceramic Materials and Components for Engines*, W. Bunk and H. Hausner, Ed., German Ceramic Society, Berlin, 1986, p 651–656

40. A. Rudnick, C.W. Marschall, W.H. Duckworth, and B.R. Emrich, "The Evaluation and Interpretation of Mechanical Properties of Brittle Materials," U.S. Air Force Technical Report TR 67–316, Air Force Materials Laboratory, April 1968

41. J.E.O. Ovri and T.J. Davies, Diametral Compression of Silicon Nitride, *Mater. Sci. Eng.*, Vol 96, 1987, p 109–116

42. R. Sedlacek and F.A. Halden, Method for Tensile Testing Brittle Materials, *Rev. Sci. Instr.*, Vol 33 (No. 3), 1962, p 298–300

43. R. Jones and D.J. Rowcliffe, Tensile-Strength Distributions for Silicon Nitride and Silicon Carbide Ceramics, *J. Am.*

Ceram. Soc., Vol 58 (No. 9), 1979, p 836–844

44. R. Charles and S. Prochazka, "Stress Rupture Testing of Silicon Carbide at Very High Temperatures," Technical Report 77CRD035, General Electric Co., Mar 1977

45. D. Lewis, Effects of Surface Treatment on the Strength and Thermal Shock Behavior of a Commercial Glass Ceramic, *Ceram. Bull.*, Vol 58 (No. 6), 1979, p 599–605

46. K. Trustrum and A. De S. Jayatilaka, On Estimating The Weibull Modulus for a Brittle Material, *J. Mater. Sci.*, Vol 14, 1979, p 1080–1084

47. R.A. Jeryan, Use of Statistic in Ceramic Design and Evaluation, *Ceramics for High Performance Applications II*, J. Burke, E. Lenoe, and R. Katz, Ed., Brook Hill Publishing, 1978, p 35–52

48. T.P. Dabbs, C.J. Fairbanks, and B.R. Lawn, Subthreshold Indentation Flaws in the Study of Fatigue Properties of Ultrahigh-Strength Glass, *Methods for Assessing the Structural Reliability of Brittle Materials*, STP 844, S.W. Freiman and C.M. Hudson, Ed., American Society for Testing and Materials, 1984, p 142–153

49. S.R. Choi, J.E. Ritter, and K. Jakus, Failure of Glass With Subthreshold Flaws, *J. Am. Ceram. Soc.*, Vol 72 (No. 2), 1990, p 268–274

50. G.D. Quinn, Fractographic Analysis and the Army's Flexure Test Method, *Advances in Ceramics*, Vol 22, *Fractography of Glasses and Ceramics*, V. Frechette and J. Varner, Ed., American Ceramic Society, 1988, p 319–333

51. G.D. Quinn, J.J. Swab, and M.J. Slavin, A Proposed Standard Practice for Fractographic Analysis of Monolithic Ceramics, *Proceedings of the Conference on Fractography of Ceramics and Glasses, II*, July 1990, V. Frechette and J. Varner, Ed., Alfred University, to be published

52. K. Trustrum and A. De S. Jayatilaka, Applicability of Weibull Analysis for Brittle Materials, *J. Mater. Sci.*, Vol 18, 1983, p 2765–2770

53. M. Adams and G. Sines, A Statistical, Micromechanical Theory of the Compressive Strength of Brittle Materials, *J. Am. Ceram. Soc.*, Vol 61 (No. 3–4), 1978, p 126–131

54. M. Adams and G. Sines, Crack Extension from Flaws in a Brittle Material Subjected to Compression, *Technophysics*, Vol 49, 1978, p 97–118

55. M. Adams and G. Sines, Methods for Determining the Strength of Brittle Materials in Compressive Stress States, *J. Test. Eval.*, Vol 4 (No. 6), 1976, p 383–396

56. C.A. Tracy, A Compression Test for High Strength Ceramics, *J. Test. Eval.*, Vol

15 (No. 1), 1987, p 14–19

57. C.A. Tracy, M. Slavin, and D. Viechnicki, Ceramic Fracture During Ballistic Impact, *Advances in Ceramics,* Vol 22, *Fractography of Glasses and Ceramics,* 1988, p 319–333

58. J. Lankford, Compressive Strength and Microplasticity in Polycrystalline Alumina, *J. Mater. Sci.,* Vol 12, 1977, p 791–796

59. J. Lankford, Uniaxial Compressive Damage in α-SiC at Low Homologous Temperatures, *J. Am. Ceram. Soc.,* Vol 62 (No. 5–6), 1979, p 310–312

60. M.N. Giovan and G. Sines, Biaxial and Uniaxial Data for Statistical Comparisons of a Ceramic's Strength, *J. Am. Ceram. Soc.,* Vol 62 (No. 9–10), 1979, p 510–515

61. A.R. Rosenfield, D.K. Shetty, and W.H. Duckworth, private communication, 1987

62. M.G. Stout and J.J. Petrovic, Multiaxial Loading Fracture of Al_2O_3 Tubes, I. Experiments, *J. Am. Ceram. Soc.,* Vol 67 (No. 1), 1984, p 14–18

63. I. Oda, M. Matsui, T. Soma, M. Masuda, and N. Yamada, "Fracture Behavior of Sintered Silicon Nitride Under Multiaxial Stress States, *J. Jpn. Ceram. Soc.,* Vol 95 (No. 5), 1988, p 539–545

64. G. Tappin, R.W. Davidge, and J.R. McLaren, The Strength of Ceramics Under Biaxial Stress, *Fracture Mechanics of Ceramics,* Vol 3, R.C. Bradt, D.P.H. Hasselman, and F.F. Lange, Ed., Plenum Press, 1978, p 435–449

65. M.N. Giovan and G. Sines, Strength of a Ceramic at High Temperatures Under Biaxial and Uniaxial Tension, *J. Am. Ceram. Soc.,* Vol 64 (No. 2), 1981, p 68–73

66. G.D. Quinn and G. Wirth, Multiaxial Strength and Stress Rupture of Hot Pressed Silicon Nitride, *J. Eur. Ceram. Soc.,* Vol 6 (No. 3), 1990, p 169–178

67. B.F. Dyson, R.D. Lohr, and R. Morrell, Ed., *Mechanical Testing of Engineering Ceramics at High Temperatures,* Elsevier, 1989

68. G.D. Quinn, Fracture Mechanism Maps for Advanced Structural Ceramics, *J. Mater. Sci.,* Vol 25, 1990, p 4361–4376

69. R. Govila, Uniaxial Tensile and Flexural Stress Rupture Strength of Hot-Pressed Si_3N_4, *J. Am. Ceram. Soc.,* Vol 65 (No. 1), 1982, p 15–21

70. C. Gandhi and M. Ashby, Fracture Mechanism Maps for Materials Which Cleave: F.C.C., B.C.C. and H.C.P. Metals and Ceramics, *Acta Metall.,* Vol 27, 1979, p 1565–1602

71. B.J. Pletka and S.M. Wiederhorn, A Comparison of Failure Predictions by Strength and Fracture Mechanics Techniques, *J. Mater. Sci.,* Vol 17, 1982, p 1247–1268

72. D.B. Marshall and A.G. Evans, Failure Mechanisms in Ceramic-Fiber/Ceramic-Matrix Composites, *J. Am. Ceram. Soc.,* Vol 68 (No. 5), 1985, p 225–231

73. D. Lewis, C. Bulck, and D. Shadwell, Standardized Tests of Refractory Matrix/Ceramic Fiber Composites, *Ceram. Eng. Sci. Proc.,* Vol 6 (No. 7–8), 1985, p 507–523

74. D.C. Phillips and R.W. Davidge, Test Techniques for the Mechanical Properties of Ceramic Matrix Fibre Composites, *Br. Ceram. Trans. J.,* Vol 85, 1986, p 123–130

75. A.G. Evans, The Mechanical Performance of Fiber-Reinforced Ceramic Matrix Composites, *Mater. Sci. Eng.,* A107, 1989, p 229–239

76. J. Mandell, D.H. Grande, and B. Edwards, Test Method Development for Structural Characterization of Fiber Composites at High Temperature, *Ceram. Eng. Sci. Proc.,* Vol 6 (No. 7–8), 1985, p 524–535

77. S.M. Wiederhorn, Reliability, Life Prediction and Proof Testing of Ceramics, *Ceramics for High Performance Applications,* J. Burke, A. Gorum, and R. Katz, Ed., Brook Hill Publishing, 1974, p 635–664

78. D.W. Richerson, *Modern Ceramic Engineering,* Marcel Dekker, 1982

79. J.E. Ritter, Jr., P.B. Oates, E.R. Fuller, and S.M. Wiederhorn, Proof Testing of Ceramics, Part I Experiment, *J. Mater. Sci.,* Vol 15, 1980, p 2275–2281

Toughness, Hardness, and Wear

Kamal E. Amin*, 3M Industrial and Electronic Sector, 3M Company

THE APPEAL OF CERAMICS as structural materials is based on their light weight, combined with their high-temperature resistance, high hardness values, chemical inertness, and anticipated unique tribological characteristics. A major goal of current ceramic research and development is to produce tough, strong ceramics that can provide reliable performance.

Ceramics fracture toughness is still poor, compared to that of metals and composites. Precise probabilistic design methodologies and modeling are therefore necessary to help predict the performance of ceramics. Such modeling requires statistically valid data on fracture toughness, wear, strength, hardness, and other properties, particularly crack-propagation phenomena. A number of these properties are discussed in this article.

Fracture Toughness

Fracture toughness values are used extensively to characterize the fracture resistance of ceramics/brittle materials (Ref 1–9). The ASTM Committee E24 has defined plane-strain fracture toughness as the crack-extension resistance under conditions of crack-tip strain. For brittle materials (ceramics), K_{Ic} is used to designate plane-strain fracture toughness at the onset of rapid crack extension under the conditions specified by the method. This means that K_{Ic} will be a function of the operating/service conditions, such as environment and loading rate. Addressed below are the major issues relating to definitions, theory, and the advantages/limitations of measurement techniques used to determine the fracture toughness of polycrystalline monolithic ceramics. Fracture toughness measurements on single crystals are not discussed in this article, but are described in Ref 10 and 11.

The fracture of brittle ceramics is usually controlled by the fracture toughness, mode I (Fig 1). A simple dimensional analysis (Ref 5) of a body containing a crack of length $2a$ subjected to an applied stress, σ_a, shows that the stress intensification at the crack, K_I, is:

$$K_I = \sigma_a Y \sqrt{a} \qquad \text{(Eq 1)}$$

where K_I is referred to as the stress-intensity factor, and Y is a dimensionless constant that depends on the geometry of the loading and the crack configuration. Fracture toughness of brittle materials is usually considered a material parameter (property), because plane-strain conditions occur over the entire length of the crack. At a critical level of stress intensity, K_{Ic}, fracture will occur.

For semicircular or semi-elliptical flaws, either at or near the surface of ceramics, K_{Ic} becomes (Ref 7):

$$K_I = 1.12\sigma_a \sqrt{\pi a}/E_K \qquad \text{(Eq 2)}$$

where E_K depends on the crack ellipsometry (ratio of major to minor axis). Crack orientation to the stress axis usually dictates the mode of failure or mode of loading (I, II, and III).

Under certain service and/or environmental conditions, stable crack extension or slow crack growth can occur at stress intensities that are less than the critical value, K_{Ic}. Under such conditions, K_I becomes dependent on the crack growth rate (crack velocity) and, hence, the characteristics of the system. In other words, the value of K_{Ic} for a given material is a function of environment, loading rate, and possibly test geometry. This behavior is often depicted graphically by V-K curves, that is, by plotting the log of crack velocity, V, versus the log of the applied stress intensity, K (Fig 2). Knowledge of the slow crack growth parameters is important in assessing the long-term strength prospects for a ceramic specimen. At high temperatures, chemical and viscous effects influence the crack-growth behavior.

A Griffith's criterion based on the energy balance necessary to create new fracture can be used to describe fracture of ceramics, and leads to a relationship between K_{Ic} and the thermodynamic surface energy, Y (Ref 6):

$$2E\gamma_i = \sigma_a^2 Y^2 a \qquad \text{(Eq 3)}$$

and

$$\gamma_i = K_{Ic}^2/2E \qquad \text{(Eq 4)}$$

where E is the Young's modulus. With the exception of ionic ceramics, the correlations in Eq 3 and 4 are not satisfied, and the surface energy underestimates the energy needed to cause fracture. Therefore, it is necessary to measure stress-intensity factors for direct failure prediction.

Fracture Toughness Measurements

There is no standard specimen type for determining fracture toughness of engineering ceramics, although numerous test techniques are available (Ref 1, 4, 12–23). The choice of technique is determined by the type of information needed, and is usually affected by the localized fracture energy around flaws (different for small cracks, compared to large cracks). Specimen geometry, preparation, and fabrication history is critical for correlating test specimen behavior to actual component fracture toughness.

At least five features of ceramic fracture mechanics usually can be observed. First, most fracture mechanics parameters must be obtained using precracked specimens. Second, the production of a sharp precrack is very critical and occasionally difficult to introduce. With ceramics, there is significant crack branching and crack-microstructure interactions. In some instances, an array of cracks, rather than a single crack, develops (Ref 9). Accurate methods of measuring crack length must be utilized. Third, techniques used at ambient temperatures, such as indentation fracture, may not be suitable at high temperatures. Fourth, limited or no material plasticity leads to linear elastic behavior. It is adequate to use the crack-propagation load to measure K_I. Fifth, crack growth in ceramics is usually enhanced by environment at ambient temperature (for silicate-base glasses, oxides, and compounds such as MgF_2, ZnSe, and GaAs, per Ref 12). Environment could range from water to chemical/corrosive species. This phenomena must be carefully examined when conducting fracture toughness measurements.

Extensive literature exists on the various techniques used for measuring the fracture toughness of ceramics. These techniques, which are described below, include:

*Formerly with Norton Company, Advanced Ceramics

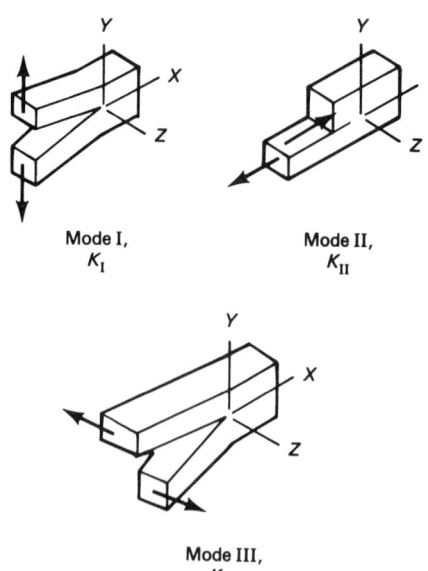

Fig 1 Modes I, I, and III of crack propagation during fracture toughness testing

- Double torsion (DT)
- Indentation crack length/fracture (IF)
- Indentation strength (IS)
- Chevron notch bend specimen (CNB)
- Double cantilever beam (DCB)
- Single-edge notched beam (SENB)
- Single-edge precracked beam (SEPB)
- Fractography approach
- Compression precracking

Double Torsion Technique (Ref 12–20). This type of specimen test is popular, and allows the use of a variety of specimen geometries. Basically, the specimen is a thin plate of about 75 × 25 × 2 mm (3 × 1 × 0.08 in.) (Fig 3). A variety of specimen width-to-length ratios can be used. The specimen sometimes has a side groove, which is usu-

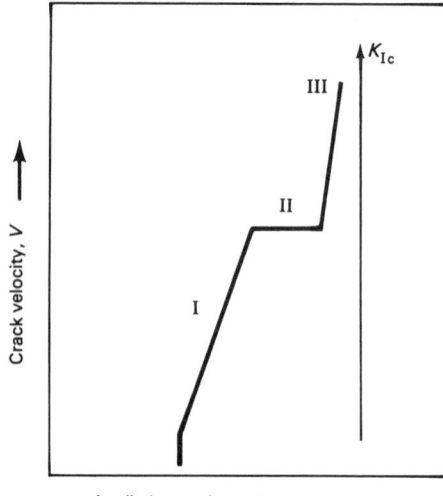

Fig 2 Three stages of crack growth based on crack velocity versus applied stress-intensity factor

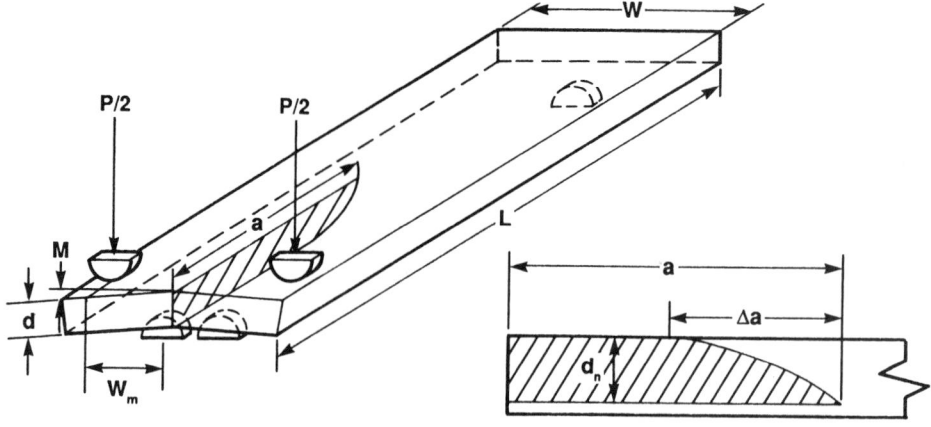

Fig 3 Double torsion test configuration

ally cut along its length to guide the crack. Recent work indicates that best results are obtained without a groove, provided that there is good alignment (Ref 10). The specimen is best loaded and supported by ball bearings. It can be studied in terms of crack-growth behavior, as well as fast fracture toughness measurements.

A compliance analysis of a precracked double torsion specimen indicates that the applied stress-intensity factor, K_I, is independent of crack length over a certain range of specimen length for a certain class of materials. Trantina (Ref 13), Shetty and Virkar (Ref 14), Fuller (Ref 15), and Pletka et al. (Ref 20) all suggest that the extent of the crack length independence of K_{Ic} is limited to only a portion of the specimen length (away from the ends). The effects of side grooves and changes in crack profile with crack length, have been indicated to be important (Ref 15).

A three-dimensional, elastic, finite-element stress analysis was conducted on double torsion specimens (Ref 17, 18). It was found that the stress intensity was nearly constant for crack lengths greater than 0.55 times the specimen width (W) and ligament lengths greater than 0.65 W.

Tsang and Berry (Ref 18) used finite-element analysis to derive Eq 5 for computing fracture toughness:

$$K_{Ic} = AP_cW_m\left[\frac{1}{8t_cI_p(1-v)}\right]^{1/2} \quad \text{(Eq 5)}$$

where P_c is the critical load, $W_m/2$ is the distance between the closed bottom support and the load, I_p is the polar moment of inertia of one half of the specimen, t_c is the thickness of the specimen at the groove, m is Poisson's ratio, and A is a modification factor of about $1.65 + 0.14 [\ln(W/2t)]$.

The double torsion technique requires a large amount (volume) of material, which may not always be available. The technique also suffers from the fact that the crack front is curved, which means that it is not under a uniform stress intensity. However, the tech-

nique is useful at high temperatures and severe environments and requires no particular fixtures (simple loading conditions).

A rigid machine is essential to conduct either precracking work or experiments where crack velocity is related to specimen compliance (during constant deflection or deflection rate trials). Some strain energy is stored in the test machine, because of its finite compliance. As the crack propagates, elastic energy is released from the machine, the specimen deflection increases, and crack velocity becomes excessive, compared to a true deflection experiment (Ref 16).

Double torsion specimens are generally used for slow crack growth studies using simple load-relaxation tests. An entire V-K_I curve may be generated using load relaxation of this technique geometry from the equation (Ref 18):

$$V = -a_iP_i\frac{dP}{dt} \quad \text{(Eq 6)}$$

where a_i is the initial crack length, P_i is the initial load, P is the instantaneous load, and dP/dt is the slope of the load-time trace. Usually, a precracked specimen is loaded at a relatively rapid crosshead speed (0.25 to 0.5 mm/min, or 0.01 to 0.02 in./min) to a load, P, that is somewhat less than the critical load, P_k, necessary to initiate fast fracture of the specimen. If the load is sufficient and the crack starts moving rapidly, that is, a large relaxation is observed, the crosshead is arrested and the load is monitored as a function of time. Minimum crack velocity, when accurately measured, is typically greater than 10^{-7} m/s (3.3×10^{-7} ft/s). After the test is completed, the specimen is removed and the final crack length is measured. If P and dP/dt are known, then K_I can be calculated using Eq 6. Evans has measured crack velocities as low as 10^{-11} m/s (3.3×10^{-11} ft/s) in glass using this technique (Ref 14). Microstructural details and materials with R-curve behavior appear to have measurable effect on the K_I data and the crack shape change during testing.

Indentation Fracture (Ref 21–25). Interest in this technique stems from its simplicity and the small volume of material required to conduct K_{Ic} measurements. A Vickers indentation is implanted onto a flat ceramic surface and cracks develop around the indentation in inverse proportion to the toughness of the material. By measuring crack lengths, it is possible to estimate K_{Ic}. The crack morphology formed during the elastic-plastic contact between a sharp indenter and a brittle medium consists of both median and lateral vent cracks (Fig 4). It must be noted that under small indentation loads, only small Palmqvist cracks form. The median vent cracks are used for fracture toughness computations.

Evans (Ref 21) derived an approximate analysis of the indentation stress fields and obtained the correlation between K_{Ic} and the size of the indentation cracks by the additive principle. Marshall *et al.* (Ref 22–24) considered residual stresses and the influence of geometric shape in elastic-plastic indentation. However, so far no exact solution of the indentation stress field has been achieved. Anstis *et al.* (Ref 25) employed a simplified two-dimensional fracture mechanics analysis and obtained:

$$K_{Ic} = 0.016 \frac{P}{C_0^{3/2}} \left(\frac{E}{H}\right)^{1/2} \qquad \text{(Eq 7)}$$

where P is the load in newtons, C_0 is the crack length from the center of indent to the crack tip, in meters; E is the Young's modulus, in GPa; and H is the Vickers hardness, in GPa.

The Palmqvist crack model equation can also be used to compute fracture toughness if only shallow cracks form:

$$K_{Ic} = 0.035 \left(\frac{H}{E}\right)^{-2/5} \left(\frac{I_0}{a}\right)^{-1/2} (H_a^{1/2}\phi^{-3/5}) \qquad \text{(Eq 8)}$$

where I_0 is the Palmqvist crack length from the corner of the indent to the crack tip at time $t = 0$, in meters; a is the indenter radius; in meters, H is the hardness, in GPa; P is the applied load, in newtons; E is Young's modulus, in GPa; and ϕ is a geometric constant equal to 3. The concept of using a crack resistance $w = P/l$ parameter has been suggested as a replacement for crack length because of its independence of both load and material hardness.

Crack dependence on sample preparation is well known for the Palmqvist techniques. The preparation of the sample surface, using effective polishing to achieve a stress state representative of the bulk, is recommended in order to achieve maximum crack length. Annealing also can be used. Crack length-to-indent radius ratios of about 23 or more were recommended in order to achieve consistent results. In addition, the crack length must be measured immediately after the indentation to minimize possible postindentation slow crack growth, especially in glasses, glass-ceramics, and ceramics that have a glassy grain boundary.

Kwon and Trogolo (Ref 26) have investigated the indentation K_{Ic} measurements on zirconia ceramics. Their major findings are:

- Shape of the crack by Vickers indent is a typical Palmqvist crack, instead of a half-penny median crack
- Crack length-to-indent radius ratio, C_0/a, is <2.5 at 490 N (110 lbf), a clear disagreement with the recommended ratio $C_0/a \geq 3$
- Cracks continue to grow rapidly after indentation; apparently, what has been measured is the time-dependent crack size, rather than the initial crack size. Therefore, K_{Ic} is expected to decrease as a function of the crack-length measurement. In practical terms, it takes around 30 min to measure both indents and crack size
- The postindentation slow crack growth is significantly retarded by indentation in oil. It is therefore assumed that stress-corrosion cracking contributes to crack growth in a moisture-containing environment
- All tested yttria-tetragonal zirconia polycrystals show relatively similar slow crack growth phenomena; zirconia-toughened aluminas show the same phenomena at a lower rate

Indentation Strength Method (Ref 24). In this method, Vickers indentations are induced on flexure strength bars that have a smooth surface finish. The bars, which are usually 3 × 4 × 40 mm (0.12 × 0.16 × 1.6 in.), are fractured in three-point bending. The bending strength is computed using:

$$\sigma_c = \frac{3P_c L}{2Wh^2} \qquad \text{(Eq 9)}$$

where P_c is the maximum load, L is the fulcrum distance, W is the specimen width, and h is the specimen height. The fracture toughness is then computed using (Ref 24):

$$K_{Ic} = 0.59 \left(\frac{E}{H}\right)^{1/8} (\sigma_c P_{1/3})^{3/4} \qquad \text{(Eq 10)}$$

where E is Young's modulus, H is Vickers hardness, σ_c is bending strength, and P is indentation load. The advantage of this method is that it is not necessary to measure crack

length. The measurement, however, shows indent load dependency.

Chevron Notch Method (Ref 27–34). This method is gaining popularity because it uses a relatively small amount of material (short rod or bar, Fig 5). The fracture toughness calculations are dependent on the maximum load and on both specimen and loading geometries. No material constants are needed for the calculations. The technique is also suitable for high-temperature testing, because flaw healing is not a concern. However, the technique requires a complex specimen shape that has an extra machining cost. A sawed notch is induced in a test bar that is usually 3 × 4 × 50 mm (0.12 × 0.16 × 2 in.). The notch angle varies from 30 to 50°. On subsequent testing of the specimen, a crack will develop at the chevron tip and extend stably as the load is increased, and is later followed by catastrophic fracture.

For a given specimen and chevron-notched geometry, maximum load always occurs at the same relative crack length, providing that the material has a flat crack growth resistance curve. It was shown (Ref 31, 32) that using one loading rate for the entire test sometimes does not produce valid results. A two-step technique has been successfully utilized. The specimen is precracked very slowly at a crosshead speed of 0.005 mm/min (0.2 mil/min) and loaded to failure at a rate of 0.05 mm/min (2 mils/min). A plot of the load versus crosshead motion is recorded, as is the maximum load of both the precrack and failure (Fig 6). Knoop indentations can be used to introduce a sharp starter crack at the chevron tip (Ref 32).

The fracture toughness is calculated from the maximum load and the specimen and the loading geometry using one of the following equations. First, for the short bar specimen:

$$K_{Ic} = \frac{P}{B\sqrt{w}} Y_m^* \qquad \text{(Eq 11)}$$

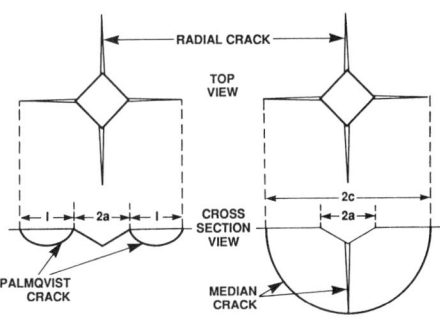

Fig 4 Indentation fracture configuration

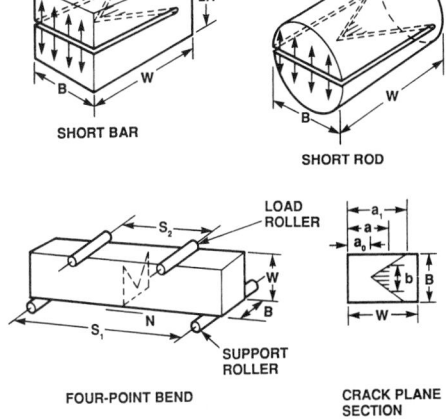

Fig 5 Chevron notch test configuration for both bar and rod

CRACK EXTENSION

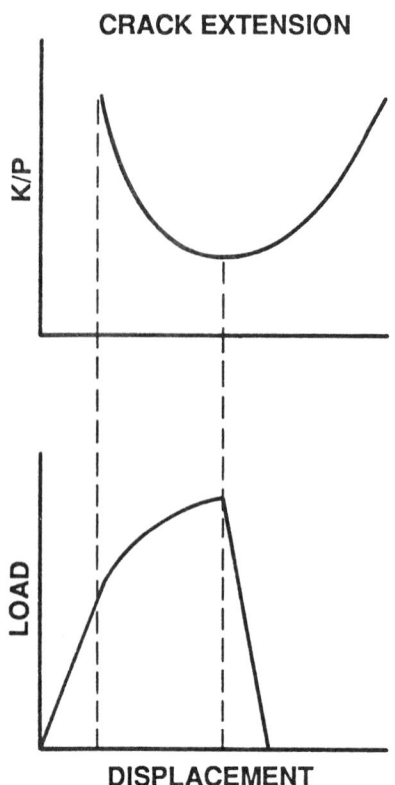

Fig 6 Load-displacement correlation showing maximum load at minimum Y^* for fracture toughness computations

where Y_m^* is a compliance function at the onset of unstable fracture where the maximum load, P_m, occurs (Ref 27, 33). The Y_m^* value may be obtained using either Munz's straight-through crack assumption (Ref 29, 33) or Blum's slice model (Ref 34). Using Blum's slice model:

$$Y_m^* = (3.08 + 5.00\,\alpha_0 + 8.33\,\alpha_0^2)\frac{(S_1 - S_2)}{W}$$

$$\left[1 + 0.007\left(\frac{S_1 - S_2}{W}\right)^{1/2}\right]\left[\frac{\alpha_1 - \alpha_2}{1 - \alpha_0}\right]$$

$$\text{(Eq 12)}$$

Using a straight-through crack approximation (Ref 25):

$$Y_m^* = (2.92 + 4.52\,\alpha_0 + 10.14\,\alpha_0^2)$$

$$\left(\frac{S_1 - S_2}{W}\right)\left(\frac{\alpha_1 - \alpha_2}{1 - \alpha_0}\right)^{1/2} \qquad \text{(Eq 13)}$$

Both the crack length-to-width ratio and Y_m^* are usually a function of both the crack and specimen geometry.

Recent studies (Ref 32) have demonstrated that an overload is necessary for crack initiation because of a nonideally machined notch tip in yttria-partially stabilized zirconia (Y-PSZ) ceramics. However, transformation-induced compressive residual stresses that result from the machining of Y-PSZ material usually deactivate surface microcracks as crack starters. Consequently, overloading may

exceed the maximum load and result in immediate catastrophic fracture without any stable crack growth. This is an area that requires more research. A load-displacement curve exhibiting both stable and unstable crack extension is shown in Fig 7.

Some ceramics are susceptible to slow crack extension in a corrosive environment at a stress intensity well below the critical value. Similar behavior exists at high temperatures. In both cases, the crack velocity is related to the stress-intensity factor, giving a three-region K_I-V curve (Fig 2). A lower limit threshold K_I exists, below which a crack will not grow. At high stress levels, the curve asymptotes to K_{Ic} beyond which fast fracture occurs. Between the threshold K_I and K_{Ic}, subcritical crack growth follows a power-law relation and the crack grows at a relatively constant speed over a considerable range of K_I. Based on these characteristics, in-test precracking of a chevron-notched bend bar can be simply carried out using the appropriate stress-intensity level to induce subcritical crack growth. The K_{Ic} test is conducted after the appropriate precracking. This subcritical precracking is facilitated by an appropriate displacement rate chosen experimentally without prior knowledge of the K_I-V curve for the test material.

Double-Cantilever Beam Method (Ref 28, 35–38). This method is discussed in detail by Freiman et al. (Ref 28), who introduced the applied moment version of the technique to eliminate the need for a tapered specimen section. This was subsequently modified by Wu et al. (Ref 35, 36). Three different loading configurations are usually applied (Fig 8). A tapered double-cantilever double beam has also been suggested (Fig 9).

Fracture toughness is usually computed using the following formula:

$$K_{Ic} = 3.45\,\frac{Pa}{bh^{3/2}}\left[1 + 0.7\left(\frac{h}{a}\right)\right] \qquad \text{(Eq 14)}$$

where h is the half-specimen height, a is the notch length, b is the specimen thickness, W is the specimen width, and P is the normal tensile load.

The technique possesses a number of advantages over other testing techniques.

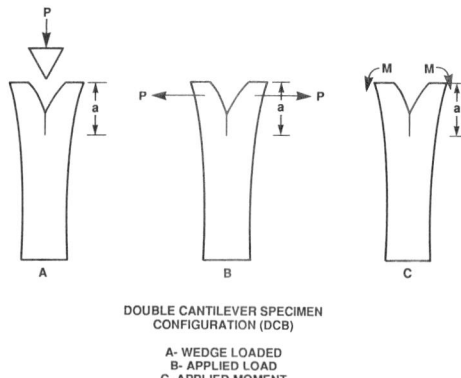

DOUBLE CANTILEVER SPECIMEN CONFIGURATION (DCB)

A- WEDGE LOADED
B- APPLIED LOAD
C- APPLIED MOMENT

Fig 8 Double-cantilever specimen configuration

Stress intensity is independent of the crack length, in the case of the constant applied moment loading, and sample preparation and the testing procedure are both relatively simple. The specimen must be precracked (sharp cracks emanating from a blunt notch) to ensure that failure initiates from a sharp flaw of the correct geometry. Most of the time, a number of very small cracks emanate from a blunt notch with a tip radius of about 15 nm (0.6 μin.) or more. This usually results in crack growth away from the notch tip (uncontrolled geometry) and produces anomalously higher fracture toughness values than those obtained from specimens that have sharp cracks with the appropriate geometry.

Several techniques for introducing a sharp starter crack tip into the specimen (for example, pop-in sharp crack at the end of a machined notch using a stiff machine) are described in the literature. Extreme care must be exercised to avoid either a premature failure of the specimen or uncontrolled crack propagation directions. Jessen and Lewis described a neat technique that avoids these problems (Ref 37). Ready et al. (Ref 38) pointed out that the use of a short specimen will ensure that the crack follows the centerline of the specimen without a side groove and will have a straight front.

Single-Edge Notched Beam (Ref 1, 39). This method has been commonly used because

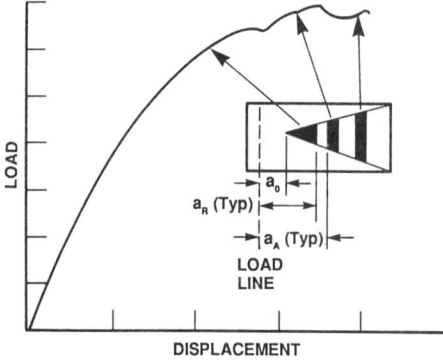

Fig 7 Load-displacement experiments showing both stable and unstable crack extension during loading of Chevron-type specimens

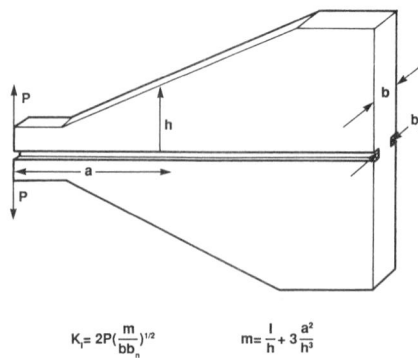

$$K_I = 2P\left(\frac{m}{bb_n}\right)^{1/2} \qquad m = \frac{l}{h} + 3\frac{a^2}{h^3}$$

Fig 9 Tapered double cantilever beam configuration

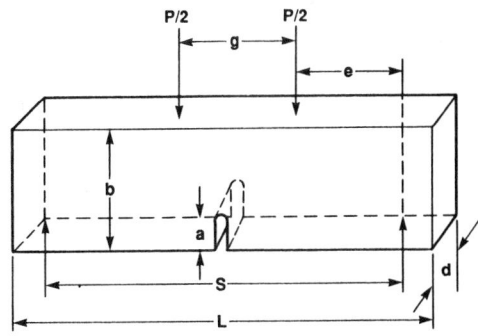

Fig 10 Single-edge notched beam specimen

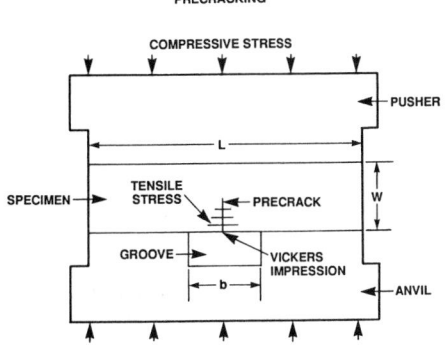

Fig 11 Loading fixture for precracking of the single-edge precracked beam specimen

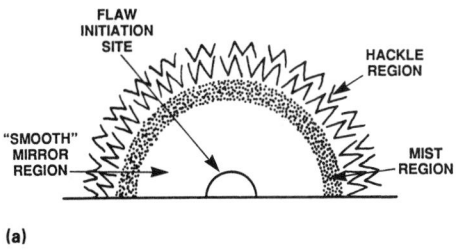

(a)

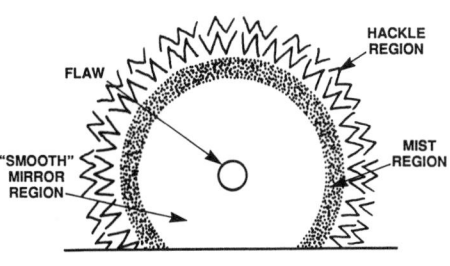

(b)

Fig 13 Fracture surface of a ceramic sample that failed from surface flaw (a) and volume flaw (b)

of its simplicity (Fig 10). The sharp crack requirement is replaced by a narrow notch, which is easy to introduce and can be measured more accurately. Fracture-toughness measurements are usually conducted using four-point bending apparatus and Eq 15:

$$K_{Ic} = \frac{P(S_1 - S_2)a^{1/2}}{2db^2} Y \qquad \text{(Eq 15)}$$

where P is the maximum load (fracture load), S_1 and S_2 are the outer and inner span, respectively, d and b are defined in Fig 10, a is the notch depth, and Y is defined as:

$$Y = 1.99 - 2.47\left(\frac{a}{b}\right) + 12.97\left(\frac{a}{b}\right)^2$$
$$- 23.17\left(\frac{a}{b}\right)^3 + 24.8\left(\frac{a}{b}\right)^4 \qquad \text{(Eq 16)}$$

Unfortunately, it has been reported that the results of this test are very sensitive to the notch width and depth, and either a precracked single-edge notched beam or a single-edge precracked beam is preferred.

Single-Edge Precracked Beam (Ref 40–42). The concept of this technique can be traced to Nunomura and Jitsukawa (Ref 40), who applied it to create and arrest a brittle crack in a bearing steel. A modification of the method was later termed the bridge indentation (BI) method by Warren and Johnson (Ref 41). The main feature of the method is the loading fixture for precracking (Fig 11), in which a beam-shaped specimen (flexure bend bar was also suggested) is compressively loaded against a centrally located groove in an anvil (Fig 12). This generates a local tensile field of a Vickers indentation (or a straight notch), which is placed in the center of the tensile surface of the specimen. Then, upon gradual loading, pop-in sound is detected and a median crack is induced from the indent, extending both inwards and sideways. Eventually, the crack front is arrested as a straight line through the thickness of the specimen.

This technique is a refinement of the single-edge notched beam technique for introducing a precrack. Nose and Fujii (Ref 41) emphasized the need to carefully prepare (grind/polish) the beam surfaces and edges 0.08 nm (3 mils) or better to help eliminate

undesirable crack starters. Careful parallelism and squareness of sample surfaces is essential for the success of precracking. The specimen is then loaded to failure in bend fixtures in the same fashion as a single-edge notched beam, and fracture toughness is computed by (Ref 42, 43):

$$K_I = \frac{3SP}{2BW^2} a^{1/2} F(\alpha) \qquad \text{(Eq 17)}$$

where $\alpha = a/w$,

$$F(\alpha) = \frac{1.99 - \alpha(1 - \alpha)(2.15 - 3.93\alpha + 2.7\alpha^2)}{(1 + 2\alpha)(1 - \alpha)^{2/3}}$$
$$\text{(Eq 18)}$$

when S is the fulcrum distance equal to 16, a is the precrack length, and W is the specimen height.

Fractography Approach (Ref 44–47). This method has gained acceptance recently. Most fracture surface observations yield substantial information that allows the calculation of the fracture toughness of the material. Moreover, the roughness of the fracture surface gives qualitative information on the extent of crack deflection, which is a toughening mechanism. Figure 13 shows the different features that can be observed on a fracture surface, originating from either a volume or surface flaw. It is difficult to directly measure the flaw-initiating site, especially in very

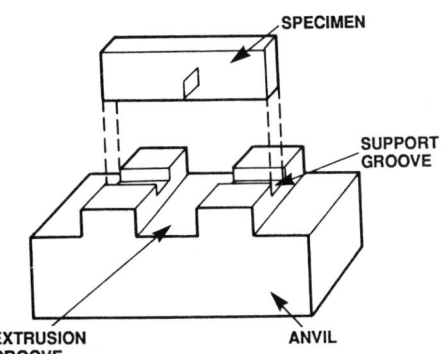

Fig 12 Loading anvil technique for generating a precrack in the SEPB method

high-strength, fine-grained ceramics and in cases where failure is caused by poor machining practices. However, where the flaw itself cannot be measured, the region from which the failure occurred can be determined by observing the patterns on the fracture surface. The original flaw usually grows subcritically, unless fracture occurs extremely rapidly. This means that, in most materials, the critical flaw boundary is not an observable feature.

Another feature of fracture surface arises during the propagation of the flaw at high velocities (after passing through the K_{Ic} point). Four distinct regions surrounding the original flaw may be observed: a smooth region, known as fracture mirror; a region of small radial ridges surrounding the mirror, known as mist; an even rougher region of larger ridges, known as hackle; and the crack-branching region. There is a quantitative relationship between the stress at fracture and the distance from the flaw to each of the boundaries of these regions:

$$\sigma r_j^{1/2} = A_j \qquad \text{(Eq 19)}$$

This equation can be combined with either Eq 1 or 2 to give:

$$K_{Ic} = \frac{Y A_j \sqrt{a}}{r_j^{1/2}} \qquad \text{(Eq 20)}$$

The equation most commonly observed for many ceramics is:

$$K_{Ic} = \frac{1.12 A_j \sqrt{\pi a}}{r_j^{1/2} E_K} \qquad \text{(Eq 21)}$$

Compression Precracking (Ref 48). Suresh *et al.* developed this procedure for measuring fracture toughness in either bending or tension after precracking notched specimens in uniaxial cyclic compression to produce a controlled and through-thickness fatigue flaw (stable mode I fatigue growth). A schematic of the technique is shown in Fig 14. After precracking, the specimen can be loaded in flexure in the single-edge notched beam configuration.

Transformation Toughening (Ref 49). In some brittle materials, such as zirconia-base ceramics, stress-induced phase transformation changes the fracture toughness. Lange (Ref 49) and others have analyzed this using both the Griffith and Irwin approaches. Both approaches result in an expression for the critical stress-intensity factor, K_{Ic}, of:

$$K_{Ic} = \left[K_0^2 + \frac{2RE_c V_i (|\delta G^c| - \delta Usef)}{(1 - \nu_c^2)} \right]^{1/2}$$

(Eq 22)

where K_0 is the critical stress intensity for the material without transformation phenomenon, δG^c is the chemical free energy change associated with the transformation, $(|\delta G^c| - \delta Usef)$ is the work done per unit volume by the stress field to induce the transformation, E_c and Y_c are the elastic properties, V_i is the volume fraction of the retained high-temperature phase, and R is the size of the transformation zone associated with the crack. The two expressions indicate that the contribution of the stress-induced transformation can be maximized by maximizing both the volume fraction, V_i, of grains (inclusions) in the untransformed state, the elastic modulus, E_c, of the composite, the quantity $(|\delta G^c| - \delta Usef)$, and the size of the transformation zone, R, associated with the propagating crack-front.

The R-curve phenomena influences the fracture-toughness measurements K_{Ic} or R (defined below) and their dependence on the measurement techniques (Ref 50). It is a consequence of most crack-tip shielding toughening mechanisms for ceramics, such as transformation toughening (as in zirconia ceramics, discussed above), microcrack toughening (alumina), ligament development resulting from some fiber (Ref 51), or other bridging and load-carrying processes behind the crack tip. All of these mechanisms tend to produce an increase in fracture toughness with continued crack growth, that is, resistance to crack growth is an increasing function of flaw size up to some steady-state value (R curve). A schematic of such behavior is shown in Fig 15. The R-curve behavior can also be caused by transition from single-crystal to polycrystalline fracture.

For materials exhibiting an R curve, the toughness K (MPa$\sqrt{m}$) or crack extension, and the crack resistance, R (N/m), is a function of a. Crack initiation occurs when K or R exceeds a threshold value, K_0 or R_0:

$$G \geq R_0 = \frac{\pi \sigma_i^2 a}{E}$$

(Eq 23)

The stability of the fracture process will depend on the point at which the curve in Fig 15 is intersected by the loading curve for the component (at R_c). Catastrophic fracture occurs when the rate of change of strain energy with crack length exceeds the rate change of crack resistance with crack length, or

$$\frac{\delta G}{\delta a} \geq \frac{\delta R}{\delta a}$$

(Eq 24)

Microstructural Effects (Ref 8). When the grain size of the material becomes large with respect to flaw size, the crack does not encompass enough grains for the full polycrystalline toughness to apply. This results in a reduction in both fracture toughness and fracture strength. A more detailed discussion of the effects of microstructural features is available in Ref 8, among others.

Guidelines

At the present time, fracture toughness values obtained through different techniques must be considered in a relative context, when comparing materials. Although a remarkable level of effort has been focused on this subject, it appears that a complementary number of techniques may have to be used to generate K_{Ic} values under different testing conditions.

To successfully standardize test measurements, very careful attention and consideration must be given to testing details (both operators and machine), sample preparation, and prehistory in terms of microstructure, thermomechanical processing, and composition.

Both indentation crack length (fracture) and indentation strength methods can be successfully used to measure K_{Ic} only at ambient temperature in ceramic materials where neither significant slow crack growth nor R-curve behavior is observed. Simple sample preparation and small sample size are needed for such techniques.

Double torsion is applicable at high temperatures when enough material is available and under conditions where notch/crack geometry is established to allow for nearly uniform K_I value at the crack front.

The R-curve phenomena, which significantly affects the contribution of crack size and character to fracture toughness measurements, has been experimentally observed. Some mechanisms/models have been suggested to explain it.

Various double-cantilever techniques are advantageous because they use a small amount of material. Analytical solutions are available for accurately computing K_I values from double-cantilever configurations. However, loading fixtures and details are difficult and cumbersome, especially for high-temperature use.

Another technique that has proven to be useful for fracture toughness measurements in ceramic multilayer electronic capacitors is the miniaturization of the double-cantilever beam. Space constraints preclude a more detailed discussion.

The Japanese Industrial Standards Association committee on the standardization of fine ceramics adopted the use of both indentation crack length/fracture and single-edge precracked beam techniques as standards in JIS R1607 in 1989. However, there are shortcomings to these methods, and it is not likely that they will be standardized in the United States in the very near future. Those and other techniques still need to resolve the meaning of average resistance to crack growth and the undesirable high cost of machining, sample preparation, and fixturing.

Fractography appears to reflect real crack-growth conditions and to simplify calculations. It was also found to be the only means by which to identify probable effects of

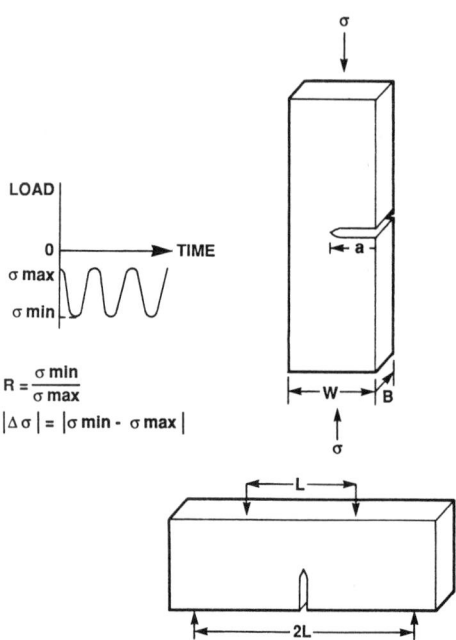

Fig 14 Cyclic compression technique for creating a sharp precrack

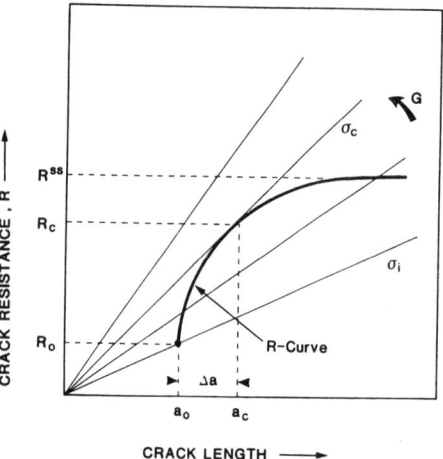

Fig 15 Plot of crack stability that is due to R curve of a material. Crack resistance, R, versus extension is plotted on the same graph, along with applied strain-energy release rate, G. Catastrophic failure occurs at the stress when $G = R_c$. (After Ref 49)

microstructural stresses and should be adopted on a wider scale. Several problems associated with the application of the fracture toughness measurement techniques discussed above to single crystals, a subject not addressed here, are discussed in Ref 47 to 49.

Hardness (Ref 22, 52, 53)

Hardness is defined in the conventional sense as a means of specifying the resistance of a material to deformation, scratching, and erosion. It is an important property for engineering applications that require good tribological resistance, such as seals, slurry pumps, rollers, and guides. Hardness tests are based on indenting the sample with a hard indenter, which may be spherical, conical, or pyramidal. There is a lack of experimental evidence to support the use of hardness for ceramics evaluation, because a combination of plastic flow, fragmentation, and cross cracking leads to considerable scatter between indentations and to differences between observers.

Common techniques for measuring hardness in ceramics are known as Vickers (HV), Knoop (HK), and Rockwell superficial (HR). The following guidelines should be observed when conducting hardness measurements:

- Based on a study sponsored by the Versailles project on Advanced Materials and Standards (VAMAS) (Ref 51), hardness tests can be used for engineering ceramics if it is recognized that errors (as high as 15%) and biases lead to high levels of uncertainty, and increase with increasing hardness level
- The indentation must be larger than the microstructural features. An adequate number of indentations must be used, preferably 10 or more of good geometry
- Badly damaged indentations must be ignored. Cracking from corners has to be accepted, but the impression of the corners must be undamaged
- The machine/observer combination must have a means of calibration, preferably a high hardness test block
- The geometry of the diamond indenter must be checked at intervals, especially in the high-load tests
- Hardness can vary with indentation load for small loads. Indentation loads greater than or equal to 9.8 N (2.2 lbf) are recommended for Knoop and Vickers indentations

Formulas used to compute hardness are designed below. For Vickers indentations:

$$H_v = 1.8544 \frac{P}{d_v^2} \qquad \text{(Eq 25)}$$

where d_v is the mean diagonal length in mm, and P is the applied force in kgf.

For Knoop indentations, the long axis of the indentation should be measured, and the hardness calculated as:

$$H_K = 14.229 \frac{P}{d_K^2} \qquad \text{(Eq 26)}$$

where d_k is the long diagonal length in meters and P is the applied force in kgf.

For Rockwell hardness, the superficial N scale, a 440 N (1 tonf) load is preferred for use (diamond N brale indicator). If this is not available, the A scale can be used. The hardness reading can then be directly obtained from the scale to the nearest 0.1.

From an overview perspective, observer biases and random error are the major sources of potential error in hardness measurements. One important source of variation occurs when the combination of indenter geometry and load, as well as specimen microstructure, no longer make the indent large compared to the microstructure or its variations. The statistical variations in microstructure are a result of variations in grain size, particle size, porosity, pore sizes, and their distributions. Such intrinsic variation in hardness is a common occurrence in ceramics because of their high hardness. Calibration blocks of high hardness are needed to reduce machine bias and help correct measurement error. It is important to understand the implications of using a hardness test. A review of Ref 51 and other conclusions and recommendations is advised before analyzing hardness data.

Wear of Ceramics (Ref 54–60)

Ceramics are attractive candidates for wear applications because of their high-temperature strength and hardness, chemical inertness (particularly in severe corrosive and oxidizing atmospheres), high contact fatigue resistance, and low inertia. Wear applications include low-heat rejection engines, high-efficiency gas turbines, high-temperature bearings, and high-speed, high-precision cutting tools.

Although a great deal of information exists on the wear testing of ceramics a variety of standard methods and a means of interpreting results are not currently well established. This stems from the fact that wear resistance is not a material property, but a system response or system-dependent. The tribosystem that governs this response is composed of the material in contact, the environment surrounding the contact, the type of relative motion at the interface, the loading pattern/severity, the stiffness of the testing machine, and the geometry of the interface. It is not unusual for one class of materials to behave differently in different types of wear tests. Because of system-related factors, wear testing is a complex subject and can be approached from several different aspects.

Before the service performance of ceramic wear components can be predicted using laboratory data, it is necessary to ensure that reproducible and reliable results are obtained in laboratory wear tests. ASTM Committee G2 for the standardization of tests such as abrasion, cavitation erosion, sliding wear, or slurry erosion has addressed the field in terms of materials, sponsoring several symposia and special technical publications (STP) on the subject. These include the selection and use of wear tests for metals (STP 615), plastics (STP 701), coatings (STP 769), and ceramics (STP 1010). Numerous other ASTM wear tests cover slurry abrasion, jet erosion, and many more topics.

Conventional techniques can be segregated into two categories, one of which is laboratory (simple geometry) wear tests, including:

- Single pin on disk
- Three pins on disk
- Ball on three flats
- Ball on three balls (ASTM D 2266–86)
- Ring on a block
- Thrust washer
- Liquid erosion

The other category is simulated field tests, which include:

- Lip seal
- Face seal
- Stick-slip
- Walking cam lubrication
- Three-ball microfilm
- Cyclic contact stress
- CV-joint test
- Rolling contact fatigue
- Vane blade assembly

Examples of some of these test configurations are shown in Fig 16 to 18. Other configurations can be found in the literature or by contacting commercial vendors.

Most of these tests have been adopted for metals and metal-ceramic combinations. Simple-geometry modifications for ceramic testing have been attempted in order to reduce the need for excessive fabrication and machining costs. The only ASTM standard test method that explicitly involves the sliding wear of ceramics is ASTM G 99 (1990), which employs a pin-on-disk apparatus. For this test, a pin with a radiused tip is positioned perpendicular to a flat circular disk. A fixed ball is often used as the pin specimen (see Fig 16). Different pin-on-disk system configurations, where either the pin revolves on the disk or vice versa are shown in ASTM G 99. This test configuration has been used for evaluating ceramic coatings. It also has been useful in providing data to understand tool wear in machining operations, such as turning and boring (Ref 53, p 43–57).

Another testing geometry that provides useful information for applications such as ball bearings, cams, and gears is known as the rolling contact fatigue test. One form of this is shown in Fig 18 and simply consists of a cylindrical rod specimen rotated against stationary surfaces (usually spheres) under a variety of loads, speeds, and, possibly, high temperatures and different atmospheres. The

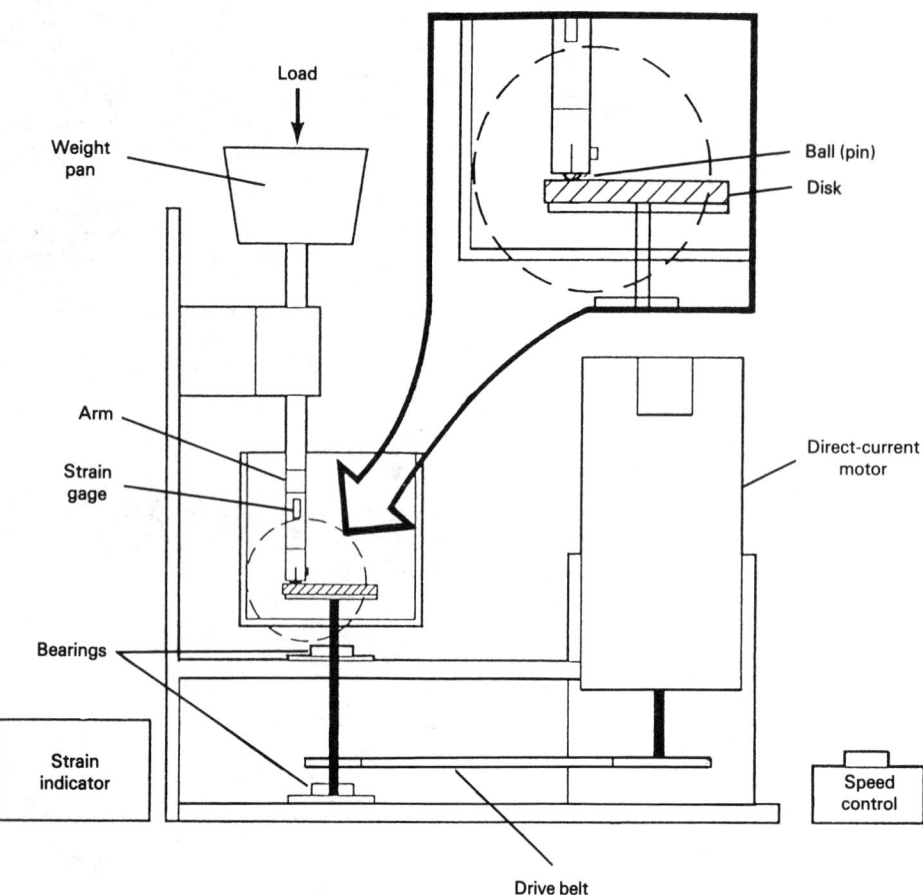

Fig 16 Pin-on-disk wear test configuration for ceramics

of ceramics because of their high elastic moduli and poor thermal conductivities. A combination of both high contact stresses and dry contact (temperature spikes) is usually more likely with ceramics than it is in metals. For example, in a pin-on-disk test, Hertzian stress at the interface can be very large, resulting in initiation of surface fracture. Similar situations arise in rolling contact fatigue situations. If a wear test is done in a manner to avoid crack initiation at localized areas (which can be due to inclusions, agglomerates, thermal mismatch, residual stresses, and voids and which results in accelerated wear and premature failure), then the test results may be meaningful. However, such structural and microstructural inhomogeneities lead to variation in wear and friction behavior for the same material.

Sliding Wear Test Geometry. A decision must first be made as to whether the investigator wants the test to be conformal or nonconformal. Conformal tests offer a constant apparent contact area (that is, constant nominal bearing pressure), but obtaining the proper initial mating of the test specimens can be difficult, leading to variations in the running-in characteristics from one test to another (Ref 56, 57). A nonconformal test, typified by the pin-on-disk (ASTM G 99) or block-on-ring test (ASTM G 77), is easier to set up, but suffers from very high contact stresses (similarly rolling contact fatigue tests). The choice of either test geometry depends on the specific applications for which the test is intended to provide data. For example, conformal tests make sense for cam followers, whereas non-

applied load can be either a steady-state force or a repetitive function (sinusoidal) with frequencies up to 50 MHz.

Specific issues in ceramic wear testing that focus mainly on material details and test methodology are described below.

Laboratory Test/Field Test Correlation. It is usually difficult to establish a direct correlation between a laboratory test and a field test. This is one of the wear technology areas that needs tremendous attention. There is a tendency to accelerate laboratory tests, which produces wear modes that are much more severe than would be expected in the application, except perhaps at the end of the life-

time of a component. It is recommended that laboratory tests be used on a comparative basis for material composition and microstructure development. Then, having limited the choices, actual field tests or simulated field tests can be conducted and the results carefully analyzed to try and establish possible correlations between the two types of tests.

Critical Parameters for Wear Testing. Surface hardness, fracture toughness, and tensile strength may be important properties to consider in certain types of wear applications, but they may be insignificant in others. This is a situation where the laboratory test sometimes fails and the field test proves to be valuable. For example, fracture toughness is a major issue in erosive wear, whereas in lubricated wear it is of minor significance. On the other hand, the grain-boundary purity and chemistry play a major role in increasing the wear resistance of ceramics under reactive lubricating conditions (oil reacts with certain chemical species at the grain boundaries, forming a beneficial adherent film) (Ref 60). The same grain-boundary features may adversely affect wear resistance in nonlubricated, humid, and erosive environments.

Hertzian Contact Stresses. The issue of contact stress and stress distribution must be considered seriously during the wear testing

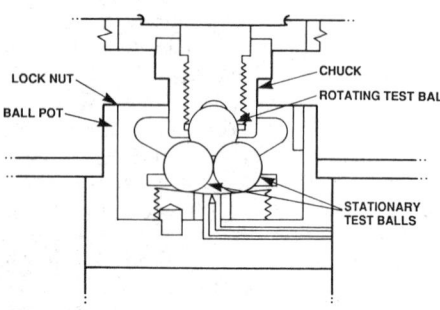

Fig 17 Four-ball wear test configuration

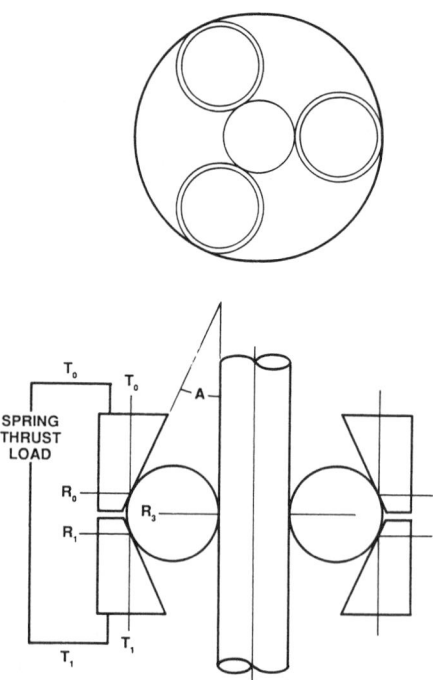

Fig 18 Rolling contact fatigue (RCF) configuration for testing ball bearings

conformal configurations are desirable for coatings and adhesion applications.

Specimen Fabrication Costs. Both the ball on three balls and ball on three flats configurations offer load alignment and stress distribution that are reproducible in all tests. However, machining spherical ceramic shapes with a high surface finish is very expensive and few facilities have this capability.

Material/Specimen History. The prehistory of a specimen dictates its wear performance. For example, machining induces different conditions of residual stress distribution (surface compressive stresses), which affects the mechanics and timing of crack initiation and surface damage. It also leads to differences in wear rates for samples made of the same material.

Dynamic Behavior of the Test Apparatus. Some researchers ignore the dynamic vibration changes/acceleration component in the direction perpendicular to the interface, which modifies both the shear force and stress distribution during the test and reduces the contact time. Such a condition ultimately modifies the process of surface crack initiation from sample to sample. Yust and Bayer address some of these issues in Ref 52.

Surface Cleanliness. The presence of residual material from the cleaning solution affects friction and wear to some degree through chemical reactions. The effect may vanish at high temperatures because of evaporation of such residue unless some reaction occurs, resulting in advancing slow crack growth. Therefore, a standard procedure must be adopted for ceramics cleaning before testing. Gates *et al.* have addressed this subject thoroughly (Ref 58).

Step loading versus constant applied load testing is an issue that was raised by a number of investigators. Some tests, such as the ASTM G 77 block-on-ring, allow for the step-loading sequence, and the ASTM D 2670 pin-on-V-block test uses a step loading. Others use a constant applied load. The idea here is to partially offset the increase in the area of contact with progressive wear damage and keep the normal pressure nearly constant, while allowing for the measurement of seizure load. The fact is that in a real-life situation, this pressure/contact area change may not be controllable because of temperature effects, slow crack growth, progressive wear, and other factors. In the tribology community, there is no obvious need for the step-loading approach.

Sliding Motion. The choice of such a motion becomes important when both reciprocating and nonreciprocating (unidirectional) sliding, fatigue, and fretting effects are introduced during the test. It is safe to say that the desired sliding motion must be chosen based on the purpose of the test or type of component to be simulated.

Sliding Speed. If the laboratory test is to produce a certain type of wear mechanism (effects) for the purposes of material selection and screening, then speeds that trigger thermoelastic instability and are sometimes critical should be determined. Such instability arises from the fact that most ceramics have poor thermal shock resistance, low thermal conductivity, and produce very high frictional heat build-up during testing under non-lubricating conditions.

The wear maps methodology examines a material under a full spectrum of tribological conditions to establish critical operating limits. The emphasis of this approach is on determining what material characteristics either limit or control the boundaries between critical and noncritical operating regions and finding conditions that enhance the useful wear operating regions.

The wear maps concept was first discussed by Tabor (Ref 57), based on the Ashby deformation maps concept. Lim and Ashby (Ref 59) followed this effort by publishing their own wear maps for the dry wear of steels. They identified four major wear mechanisms: seizures, melt-dominant wear, oxidation-dominant wear, and plasticity-dominant wear, including delamination. Based on these classifications, they identified the localized "flash" temperature as a unifying concept and were able to calculate a "temperature map" and derive the corresponding wear equations for the four dominant wear modes, which were proven experimentally. Hsu *et al.* (Ref 59) have extensively applied the concept to ceramics. Blau and Yust (Ref 56) indicated some reservations about such an approach. Their point is that long-term fatigue effects must be considered when safe and unsafe wear regions are determined from short-term experiments (Ref 56). Other types of wear maps have been successfully applied since the mid-1970s.

Standard Reference Samples. The VAMAS project was initiated as an international effort

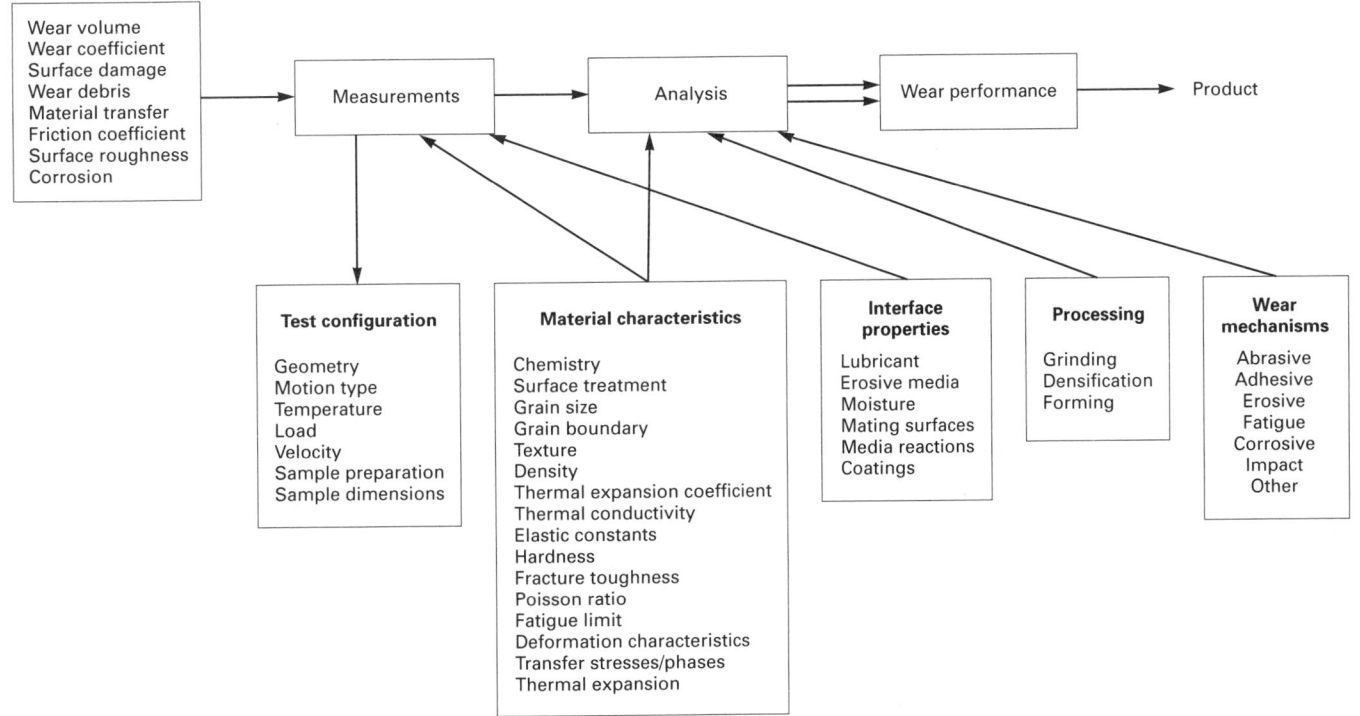

Fig 19 Measurements, test parameters, material processes, and interface details that should be considered in the analysis of wear performance of ceramics

to help standardize wear tests by producing a standard methodology for data analysis. The reduction of test errors (both systematic and random) and the minimization of data variability also were goals, among others. The task group on wear recommended the use of reference samples to help point out differences between the testing procedure of one laboratory and another, ensure that the test is accurately repeated within the same laboratory, and/or understand dominant wear degradation mechanisms under specific conditions of testing. However, many tribologists and engineers use wear tests as a tool to screen materials and for composition development. If design data is needed, then field tests should be conducted. As a consequence, the usefulness of the idea of testing reference wear samples becomes questionable.

Postmeasurement Handling. It is advisable to preserve the wear surface with its debris in place, that is, avoid the use of solvents in washing or cleaning. This helps extract additional information about wear degradation mechanisms, including thermochemical and mechanical effects.

Presentation of Wear Data. A very satisfactory methodology for presenting wear and friction data is described in Chapter 1 of ASTM STP 1010. Among the information that should be documented is the history of the specimen, including the full characterization of the specimen surface (topographic features, anisotropy, and residual stresses).

Guidelines. Wear testing is an essential part of the wear technology and science field. Component designers or users of engineering wear applications need to carefully define the field testing conditions and use these data to design laboratory wear tests that ensure the optimum material/product design for specific applications. Having said this, it becomes obvious that standard wear tests are very much needed to provide reliable data for such a goal. Therefore, a global approach aimed at understanding material behavior in relation to test conditions is a priority. All relevant factors are incorporated in Fig 19, which defines laboratory measurements and shows that they depend on test configuration, material characteristics, and interface properties. While analyzing such data, the processing history and dominant (desirable) wear mechanisms should be considered. Then, reliable conclusions can be drawn and correlated to product performance. Failure to consider wear technology in a global sense may lead to wrong conclusions. The VAMAS project aims to achieve international agreement on wear standards through multilateral results.

REFERENCES

1. S.W. Freiman, Brittle Fracture Behavior of Ceramics, *Ceram. Bull.,* Vol 67 (No. 2), 1988, p 392
2. A.G. Evans, Fracture Mechanics Determinations, *Fracture Mechanics of Ceramics,* Vol 1, *Concepts, Flaws, and Fractography,* R.C. Bradt, D.P.H. Hasselman, and F.F. Lange, Ed., Plenum Press, 1973, p 17
3. R.M. Anderson, Testing Advanced Ceramics, *Adv. Mater. Process.,* Vol 3, 1989, p 31
4. R.F. Pabst, K. Kromp, and G. Popp, Fracture Toughness—Measurement and Interpretation, *Proc. Brit. Ceram. Soc.,* Vol 32, Mar 1982, p 89
5. P.C. Paris and G.C. Sih, "Fracture Toughness, Testing and Its Applications," STP 381, American Society for Testing and Materials, 1965
6. A.G. Evans and R.W. Davidge, The Strength and Oxidation of Reaction Sintered Silicon Nitride, *J. Mater. Sci.,* Vol 5, 1970, p 314
7. G.R. Irwin and P.C. Paris, Fundamental Aspects of Crack Growth and Fracture, *Fracture III,* H. Liebowitz, Ed., Academic Press, 1971, p 2
8. E.R. Fuller, Jr. and R.M. Thomson, Lattice Theories of Fracture, *Fracture Mechanics of Ceramics,* Vol 4, R.C. Bradt, D.P.H. Hasselman, and F.F. Lange, Ed., Plenum Press, 1978
9. C.C.M. Wu, R.W. Rice, and P.F. Becher, *The Character of Cracks in Fracture Toughness Measurements of Ceramics,* STP 745, "Fracture Mechanics for Ceramics, Rocks, and Concrete," American Society for Testing and Materials, 1982, p 127
10. W. Rice, Fractographic Determination of K_{Ic} and Effects of Microstructural Stresses in Ceramics, *Proceedings of Alfred Fractography Conference,* 1990
11. J.L. Henshall, D.J. Rowcliffe, and J.W. Edington, Fracture Toughness of Single-Crystal Silicon Carbide, *J. Am. Ceram. Soc.,* July 1977, p 373
12. G.D. Quinn, Delayed Failure of a Commercial Vitreous Bonded Alumina, *J. Mater. Sci.,* Vol 22, 1987, p 2309
13. G.G. Trantina, Stress Analysis of the Double Torsion Specimen, *J. Am. Ceram. Soc.,* Vol 60, 1977, p 338
14. D.L. Shetty and A.V. Virkar, Determination of the Useful Range of Crack Lengths in Double Torsion Specimens, *J. Am. Ceram. Soc.,* Vol 61, 1978, p 93
15. E.R. Fuller, Jr., An Evaluation of Double Torsion Testing: Analysis, *Fracture Mechanics Applied to Brittle Materials,* STP 678, S.W. Freiman, Ed., American Society for Testing and Materials, 1979
16. A.V. Virkar and D.L. Johnson, Some Kinetic Consideration Regarding Double-Torsion Specimen, *J. Am. Ceram. Soc.,* Vol 59 (No. 5–6), 1970, p 197
17. G.G. Trantina, Stress Analysis of the Double-Torsion Specimen, *J. Am. Ceram. Soc.,* 1977, p 338
18. A. Tsang and J.T. Berry, A Three Dimensional Finite Element Analysis of the Torsion Test, Publication No. 78-PVP-96, American Society of Mechanical Engineers, 1978, p 1
19. J. Nakayama, Direct Measurements of Fracture and Adhesion in Heterogeneous Materials, *J. Am. Ceram. Soc.,* Vol 48 (No. 11), 1965, p 583–587
20. B.J. Pletka, E.R. Fuller, Jr., and B.G. Koepke, An Evaluation of Double Torsion Testing: Experimental, *Fracture Mechanics Applied to Brittle Materials,* STP 678, S.W. Freiman, Ed., American Society for Testing and Materials, 1979
21. A.G. Evans, *Fracture Toughness, the Role of Indentation Techniques,* STP 678, American Society for Testing and Materials, 1979, p 113
22. B.R. Lawn and D.B. Marshall, Hardness, Toughness and Brittleness: An Indentation Analysis, *J. Am. Ceram. Soc.,* Vol 62 (No. 7–8), July-Aug 1979, p 347
23. D.B. Marshall, B.R. Lawn, and A.G. Evans, Elastic/Plastic Indentation Damage in Ceramics—The Lateral Crack System, *J. Am. Ceram. Soc.,* Vol 65 (No. 11), 1982, p 561
24. P. Chantikul, G.R. Anstis, B.R. Lawn, and D.B. Marshall, A Critical Evaluation of Indentation Techniques for Measuring Fracture Toughness: I, Strength Method, *J. Am. Ceram. Soc.,* Vol 64 (No. 9), 1981, p 533
25. G.R. Anstis, P. Chantikul, B.R. Lawn, and D.P. Marshall, A Critical Evaluation of Indentation Techniques for Measuring Fracture Toughness: I, Direct Crack Measurements, *J. Am. Ceram. Soc.,* Vol 64, 1981, p 533
26. O.H. Kwon and J.A. Trogolo, "Postindentation Slow Crack Growth in Zirconia Ceramics," paper presented at the 91st Annual ACS Meeting, Indianapolis, 23 Apr 1989
27. J.A. Salem and J.L. Shannon, Jr., Fracture Toughness of Si_3N_4 Measured With Short Bar Chevron-Notched Specimens, *J. Mater. Sci.,* Vol 22, 1987, p 312–324
28. S.W. Freiman, D.R. Mulville, and P.W. Mast, Crack Propagation Studies in Brittle Materials, *J. Mater. Sci.,* Vol 8 (No. 11), 1973, p 1527–1533
29. D. Munz, R.T. Bubsey, and J.L. Shannon, Jr., Fracture Toughness Determination of Al_2O_3 Using Four Point Bend Specimens with Straight-Through and Chevron Notches, *J. Am. Ceram. Soc.,* Vol 63 (No. 5–6), 1980, p 300–305
30. T.B. Troczynski and P.S. Nicholson, Effect of Subcritical Crack Growth on Fracture Toughness and Work of Fracture Tests Using Chevron-Notched Specimens, *J. Am. Ceram. Soc.,* Vol 70 (No. 2), 1987, p 78–85
31. S.X. Wu, Stability and Optimum Geometry of Chevron Notched Three Point Bend Specimens, *Int. J. Fract.,* Vol 26 (No. 2), 1984, p R43–47
32. J. Sung, "Valid K_{Ic} Determination via In-

Test Subcritical Precracking of Chevron Notched Bend Bars," Norton Company Advanced Ceramics internal document, Jan 1989

33. D.M. Munz, J.L. Shannon, and R.T. Bubsey, *Int. J. Fract.*, Vol 16, 1980, p R137–142

34. J.J. Blum, Slice Synthesis of Three Dimensional Work-of-Fracture Specimens, *Eng. Fract. Mech.*, Vol 7 (No. 3), 1975, p 593

35. C.C. Wu, K.R. McKinney, and D. Lewis, Grooving and Off-Center Crack Effects on Applied Moment Double-Cantilever-Beam Tests, *J. Am. Ceram. Soc.*, Vol 67 (No. 8), 1984, p C166-C168

36. C.C. Wu, D. Lewis, and K.R. McKinney, Strength and Toughness Measurements of Ceramic Fiber Composites, *Fracture Mechanics of Ceramics*, Vol 7, R.C. Bradt, A.G. Evans, D.P.H. Hasselman, and F.F. Lange, Ed., Plenum Press, 1986, p 53–60

37. T.L. Jessen and D. Lewis III, Modified Specimen Preparation for Applied-Moment Double-Cantilever-Beam Testing, *J. Am. Ceram. Soc.*, Vol 72 (No. 11), 1989, p 2196.36.K

38. M.J. Ready, A.H. Heuer, and R.W. Steinbrech, Crack Propagation in Mg-PSZ, Vol 78, *Mater. Res. Soc. Sympos. Proc.*, 1987, p 107

39. R.F. Pabsi, Determination of K_{Ic}-Factors with Diamond Saw-Cuts in Ceramic Materials, *Fracture Mechanics of Ceramics*, Vol 2, Microstructure, Materials, and Applications, R.C. Bradt, D.P.H. Hasselman, and F.F. Lange, Ed., Plenum Press, 1974, p 555–565

40. S. Nunomura and S. Jitsukawa, Fracture Toughness S853 Evaluation for Bearings Steels by Indentation Cracking Under Multi-axial Stress, *Tetsu-to-Haganen*, Vol 64 (No. 5853), 1978 (in Japanese)

41. R. Warren and B. Johnson, Creation of Stable Crack in Hard Materials Using Bridge Indentation, *Powder Metal.*, Vol 27 (No. 1), 1984, p 25

42. T. Nose and T. Fujii, Evaluation of Fracture Toughness for Ceramic Materials by a Single-Edge Precracked-Beam Method, *J. Am. Ceram. Soc.*, Vol 71 (No. 5), 1988, p 3278

43. J.E. Strawley, Wide Range Stress Intensity Factor Expressions for ASTM E399 Standard Fracture Toughness Specimens, *Int. J. Fract.*, Vol 12, 1976, p 475

44. R.W. Rice, Perspective on Fractography, *Advances in Ceramics* Series, Vol 22, *Fractography of Glasses and Ceramics*, American Ceramic Society, 1988, p 3

45. J.J. Mecholsky and S.W. Freiman, Determination of Fracture Mechanics Parameters Through Fractographic Analysis of Ceramics, STP 678, American Society for Testing and Materials, 1980, p 136

46. J.J. Mecholosky, S.W. Freiman, and R.W. Rice, Fracture Surface Analysis of Ceramics, *J. Mater. Sci.*, Vol 11, 1976, p 1310

47. R.W. Rice and D. Lewis III, Limitation and Challenges in Applying Fracture Mechanics to Ceramics, *Fracture Mechanics of Ceramics*, Vol 5, R.C. Bradt, A.G. Evans, D.P.H. Hasselman, and F.F. Lange, Ed., Plenum Publishing, 1988, p 659

48. S. Suresh, L. Ewart, M. Maden, W.S. Slaughter, and M. Nguyen, Fracture Toughness Measurements in Ceramics: Precracking in Cyclic Compression, *J. Mater. Sci.*, Vol 22, 1987, p 1271

49. F.F. Lange, Transformation Toughening—Part 2 Contribution to Fracture Toughness, *J. Mater. Sci.*, Vol 17, 1982, p 235

50. M.V. Swain, R-Curve Behavior and Thermal Shock Resistance of Ceramics, *J. Am. Ceram. Soc.*, Vol 73, 1990, p 621

51. C.C. Wu, D. Lewis, and K.R. McKinney, Strength and Toughness Measurements of Ceramic Fiber Composites, *Fracture Mechanics of Ceramics*, Vol 7, R.C. Bradt, A.G. Evans, D.P.H. Hasselman, and F.F. Lange, Ed., Plenum Press, 1986

52. D.M. Butterfield, D.J. Clinton, and R. Morrell, The VAMAS Hardness Test Round-Robin on Ceramic Materials, Report No. 3, National Physical Laboratory, Teddington, England, Apr 1989, p 1

53. I.J. McColm, *Ceramic Hardness*, Plenum Press, 1990

54. C.S. Yust and R.C. Bayer, Ed., *Selection and Use of Wear Tests for Ceramics*, STP 1010, American Society for Testing and Materials, 1988

55. P.J. Blau, *Friction and Wear Transitions of Materials-Break-in, Run-in, Wear-in*, Noyes Publications, 1989, p 189

56. P.J. Blau and C.S. Yust, "Sliding Wear-Testing and Data Analysis Strategies for Advanced Engineering Ceramics," paper presented at ASTM Symposium on Wear Testing of Advanced Materials, San Antonio, 14 Nov 1990

57. D. Tabor, Status and Direction of Tribology as a Science in the 80's: Understanding and Prediction, in *Proceedings of the International Conference on Tribology in the 80's*, CP 2300, National Aeronautics and Space Administration, 1984, p 1

58. R.S. Gates, J.P. Yellets, D.E. Deckman, and S.M. Hsu, *Considerations in Ceramic Friction and Wear Measurements*, C.S. Yust and R.G. Bayer, Ed., STP 1010, American Society for Testing and Materials, 1988

59. S.M. Hsu, Y.S. Wang, and R.G. Munro, Quantitative Wear Maps as a Visualization of Wear Mechanism Transitions in Ceramics, *Proceedings of the ASME Wear of Materials Conference*, American Society of Mechanical Engineers, 1989, p 723

60. S.M. Hsu, NIST, private communication, 1989, 1990

Thermophysical Properties

William F. Hammetter, Sandia National Laboratories

METHODS FOR DETERMINING thermal expansion, thermal conductivity, heat capacity, and emissivity of ceramics and glass are described in this article, as is the measurement of these properties as a function of temperature. Quality measurements of any of these thermophysical properties require, in part, careful sample preparation; good experimental technique, instrumentation, and controls; precise calibration; attention to detail; and careful observation. Rather than discuss these important points, this article is intended to provide insight into the property itself, explain what is actually measured, show how the specific property is deduced from the measurement, and describe how it is expressed (units, representation) in scientific literature. Through necessity, only brief descriptions of techniques are presented. However, references to more complete reviews will be given. For some measurements, standard test methods that pertain to ceramics and glass have been established by ASTM, and are listed in Table 1.

Thermal Expansion

Almost without exception, the volume of a substance changes reversibly during heating and cooling. It is crucial that the volume changes of ceramics and glasses be quantified to ensure successful application when they are combined with different materials, such as metals, polymers, other ceramics, and other glasses. Examples of typical joining applications include composites, structural components, coatings, films, and refractory furnace linings. The importance of a quantitative ranking of expansion lies in the necessity to match the volume changes of a substance with that of other materials in contact with it. For example, when making hermetic glass-to-metal seals, the difference between glass and metal, in terms of thermal expansion, should be no greater than approximately 2% over the temperature range required to fabricate the seal. The two material properties that are used to quantify this phenomena are the coefficient of volume thermal expansion, α_v, and the coefficient of linear thermal expansion (CTE), designated simply as α.

The CTE is a measure of the volume expansion along only one dimension of a sample. It is defined (Ref 1) as: "The change in length, relative to the length of the specimen, accompanying a unit change of temperature, at a specified temperature." In practice, it is presented in two slightly different forms. First, following the strict definition, it can be presented as the slope of a plot of sample length versus temperature:

$$\alpha(T_i) = \left(\frac{\partial L(T)}{\partial T} \right)_{T=T_i} \frac{1}{L(T_i)} \qquad \text{(Eq 1)}$$

where $L(T_i)$ is the experimentally measured length of the sample at temperature T_i. Second, it can be presented as an average value between temperature limits T_1 and T_2 (where $T_2 > T_1$) as:

$$\alpha_{ave} = \frac{L(T_2) - L(T_1)}{L(T_1)} \frac{1}{(T_2 - T_1)} \qquad \text{(Eq 2)}$$

where $L(T_n)$ is the measured length at T_n ($n = 1,2$). The difference in expansion coefficients, as defined above, is shown in Fig 1. The average coefficient of thermal expansion, α_{ave}, can be used with sufficient accuracy when the temperature extremes are not too far apart, when the expansion is constant with respect to temperature, and when no polymorphic phase transitions (which are usually accompanied by a discontinuous, relatively large dimensional change) are contained within the temperature interval. Note that the crystallization of glasses is similar to a phase transition in which a discontinuous change in length is observed.

Thermal expansion is reported in the general units of (length)(length)$^{-1}$(temperature)$^{-1}$, or simply T^{-1} where the "change in length per unit length" is understood. The correct SI unit for thermal expansion is K^{-1}. Because the fractional length change for a wide variety of materials has been found to be on the order of 10^{-6}, thermal expansions are commonly expressed as parts per million per degree Celsius or Kelvin, in the form of $10^{-6}/$K, where the units can be μm/m or μin./in.

To a good approximation, the relationship between the average coefficient of volume thermal expansion, $(\alpha_v)_{ave}$, and the average coefficient of linear thermal expansion, α_{ave}, is:

$$(\alpha_v)_{ave} = 3\alpha_{ave} \qquad \text{(Eq 3)}$$

Thermal expansion is a tensor property, that is, the magnitude of the expansion is different along the different crystallographic axes. In fact, with increasing temperature, some substances like calcite, $CaCO_3$, will expand along one crystallographic axis and contract along another. However, most applications are concerned with the volume change of polycrystalline aggregates or amorphous glasses. For polycrystalline materials, the measured expansion will be some weighted average of the directionally dependent magnitudes. For aggregates of noncubic crystalline substances, variations in the expansion coefficient along the unique axes can cause restricted movement between randomly oriented crystallites during heating and cooling, resulting in residual stress. Actual cracks can form between the different crystal faces because of expansion mismatch in the extreme case. Such cracking that is due to thermal cycling will manifest itself as a hysteresis in apparent expansion measurements during repeated temperature cycling. Glasses, which are noncrystalline, exhibit isotropic expansion.

Dilatometry is by far the most widely used method to measure the thermal expansion of a wide variety of materials. Several ASTM standards are available to guide the interested experimentalist. Commercially available instrumentation and associated software have made the direct determination of both the temperature-dependent expansion coefficient and the average expansion coefficient an almost trivial experiment. A rod-shaped sample is placed between a fixed base and a moveable "push rod," the sample is heated slowly at a fixed rate (generally not exceeding 3 °C/min, or 5 °F/min), and the resulting expansion is transferred by the push rod to some external sensing device and recorded for later analysis.

Commercially available dilatometers are of either the single or double push rod type. When using the single type, a reference sample with known expansion is used to calibrate the length versus temperature signal under the environmental conditions and temperature range of interest. Subsequently, a separate experiment is conducted to measure the expansion of the sample and calculate the coefficient of expansion by comparing it with the previously measured standard. The cali-

Table 1 Relevant ASTM standards

Thermal expansion

Standard Practice for Making and Testing Reference Glass-Metal Bead-Seals to Determine Degree of Mismatch, ASTM F-14 (10.04), 1990

Standard Practice for Making and Testing Reference Glass-Metal Butt-Seals to Determine Degree of Mismatch, ASTM F-140 (10.04), 1990

Standard Practice for Making and Testing Reference Glass-Metal Sandwich-Seals to Determine Degree of Mismatch, ASTM F-144 (10.04), 1990

Standard Test for Linear Thermal Expansion of Porcelain Enamel Frit by the Interferometric Method, ASTM C-539 (02.05, 15.02), 1990

Standard Test for Linear Thermal Expansion of Porcelain Enamels, Glaze Frits and Fired Ceramic Whiteware Products by the Dilatometry Method, ASTM C-372 (15.02), 1990

Standard Test for Linear Thermal Expansion of Refractories Under Load, ASTM C-832 (15.01), 1990

Standard Test for Linear Thermal Expansion of Rigid Solids with Interferometry, ASTM E-289 (03.01, 14.02), 1990

Linear Thermal Expansion of Solid Materials by Thermomechanical Analysis, ASTM E-831 (08.03, 14.02), 1990

Standard Test for Linear Thermal Expansion of Solid Materials with Vitreous Silica Dilatometer, ASTM E-228 (03.01, 14.02), 1990

Standard Test for Linear Thermal Expansion of Vitreous Glass Enamels and Glass Color Frits by Dilatometry, ASTM C-824 (15.02), 1990

Thermal conductivity/diffusivity

Standard Test for Thermal Diffusivity of Carbon/Graphite by the Thermal Pulse Method, ASTM C-714 (15.01), 1990

Standard Test for the Thermal Conductivity of Ceramic Whitewares, ASTM C-408 (15.02), 1990

Standard Test for Thermal Conductivity of Electrical Grade Magnesium Oxide, ASTM D-2858 (10.02), 1990

Standard Test for Thermal Conductivity of Solids by Guarded-Comparative-Longitudinal Heat Flow Technique, ASTM E-1225 (14.02), 1990

Standard Practice, Requirements/Guidelines for Thermal Transmission Properties Calculated from Steady-State Heat Flux Measurements, ASTM C-1045 (04.06), 1990

Standard Test for Thermal Conductivity of Brick other than Insulating Firebrick, ASTM C-202 (15.01), 1990

Standard Test for Thermal Conductivity of Carbon and Carbon-Bearing Refractories, ASTM C-767 (15.01), 1990

Standard Test for Thermal Conductivity of Insulating Firebrick, ASTM C-182 (15.01), 1990

Standard Test for Thermal Conductivity of Unfired Monolithic Refractories, ASTM C-417 (15.01), 1990

Standard Test for Thermal Transmission Properties of Insulating Specimens by Heat Flow Meter, ASTM C-518 (04.06, 14.01), 1990

Standard Practice for Thermal Transmission Properties of Insulating Specimens Using a Guarded Hot Plate in One-Sided Mode, ASTM C-1044 (04.06), 1990

Standard Test for Thermal Transmission Properties of Insulating Specimens by Guarded Hot Plate Apparatus, ASTM C-177 (04.06, 08.01, 14.01), 1990

Specific heat capacity

Standard Test for Specific Heat Capacity of Materials by Differential Scanning Calorimetry, ASTM E-1269 (14.02), 1990

Standard Test for Specific Heat of Liquids/Solids, ASTM D-2766 (05.02), 1990

Emittance

Standard Test for Total Normal Emittance by Inspection-Meter Technique, ASTM E-408 (15.03), 1990

(continued)

Table 1 Continued

Standard Test for Spectral Normal Emittance at Elevated Temperatures, ASTM E-307 (15.03), 1990

Standard Test for Spectral Normal Emittance of Nonconducting Specimens at Elevated Temperatures, ASTM E-423 (15.03), 1990

Standard Test for Total Hemispherical Emittance by Calorimetric Determination, ASTM E-434 (15.03), 1990

Standard Test for Total Hemispherical Emittance of Metal/Coated-Metal Surfaces from 20° to 1400 °C, ASTM C-835 (04.06), 1990

General

Standard Definitions of Terms Relating to Thermophysical Properties, ASTM E-1142, 1990

bration procedure corrects for the expansion of the dilatometer materials, the existence of thermal gradients, and possible nonlinearity in heating rate.

In a dilatometer with dual push rods, the sample and standard are mounted side-by-side and heated together. The expansion of the sample is then determined by direct comparison with the standard.

Dilatometers can operate in different temperature ranges, depending on materials of construction and the type of external heating/cooling hardware. Common materials of construction include silica (with a very low inherent expansion), alumina, and sapphire. Under reducing conditions, common materials are graphite, molybdenum, tantalum, and tungsten. It is usually easy to construct or modify commercial dilatometers to operate in specific atmospheres or vacuum.

Interferometry is an optical technique based on the principle that two parallel reflecting surfaces a short distance apart and illuminated by a monochromatic light source will display interference fringes. These fringes will move as the displacement between the reflecting surfaces changes. As applied to thermal expansion, a sample is placed between the reflecting surfaces, and, as its length changes upon heating, the reflecting surfaces (usually optically flat, polished, fused quartz disks) move apart and the interference fringes move past some reference point on one reflecting plane. The change in length of the sample is related to the number of fringes, N, which pass the reference point by:

$$\frac{\Delta L}{L} = \frac{\lambda N}{2L} + \frac{A}{L} \qquad (Eq~4)$$

where ΔL is the change in sample length, L is the sample length, λ is the wavelength of the light source, and A is a correction factor for the measuring atmosphere ($A = 0$ for measurements in vacuum).

Although there are several variations of the above basic technique, the sample requirements for each are generally the same. For best results, the sample should be made such that it makes three-point contact with both top and bottom reflecting surfaces. In addition, the planes defined by the three top points and three bottom points of the sample must be

parallel, and the sample must be heated uniformly to maintain the parallel separation of the reflecting surfaces. These restrictions on sample geometry render this technique to be far less popular than dilatometry for routine measurements. However, for ceramics or glasses with some volatile component, such as lead in $Pb(Zr_xTi_{1-x})O_3$ ferroelectrics, it is possible to construct a closed, isothermal sample chamber in which the vapor pressure of the volatile species can be adjusted, controlled, or contained. Thus, the fact that the expansion is transferred to an external device optically rather than mechanically is the main advantage of this technique for ceramics and glasses.

X-ray diffraction (XRD) probes the crystalline nature of substances and, among other things, measures the spacing between different sets of planes in a crystalline substance. By measuring the changes in interplanar spacing as a function of temperature, the tensor components of thermal expansion can be calculated. Methods that use small single crystals and powder samples are possible, and one great advantage of this method is that the amount of sample necessary for a measurement is small.

The resolution of thermal expansion into its tensor components is sometimes important, for example, to estimate residual internal stress during thermal cycling. However, to combine the tensor components into a single value for a polycrystalline aggregate (especially for anisotropic expansions) requires that the volume expansion of the unit cell be calculated from the temperature-dependent tensor components, followed by the estimation of the average linear thermal expansion using Eq 3. It should be noted that the processing history of the polycrystalline aggregate may influence the expansion toward different extremes, depending on such factors as preferred orientation, or preferred crystallite growth morphologies. A separate expansion measurement of the polycrystalline aggregate is usually necessary.

Expansion measurements by XRD are not automated as such nor can they be made in a temperature-scanning mode. Interplanar spacings are usually measured at different temperatures and expansion coefficients are determined in a separate calculation.

Measuring Microscopy. One of the most direct methods for determining thermal expansion is to obtain a data set of sample length versus temperature by simply measuring the sample length (usually with a measuring microscope or telescope) at temperature. Sometimes, this may be the only viable method, either because measurements at extremely high temperatures are required or because there are serious compatibility problems with materials of commercially available instruments. Custom sample chambers can be made to withstand severe temperatures or environments, or they can be constructed from some exotic, but compatible, material.

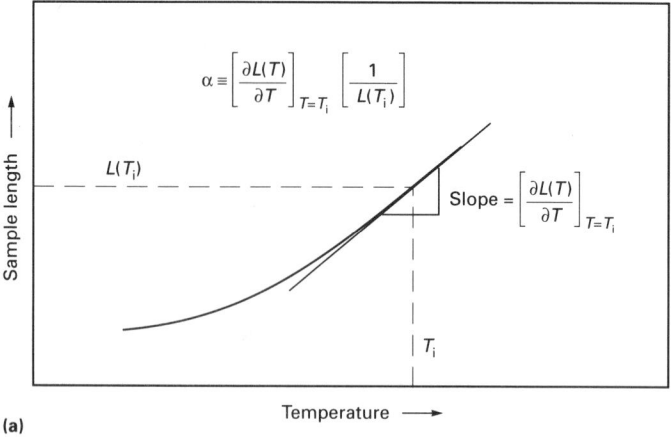

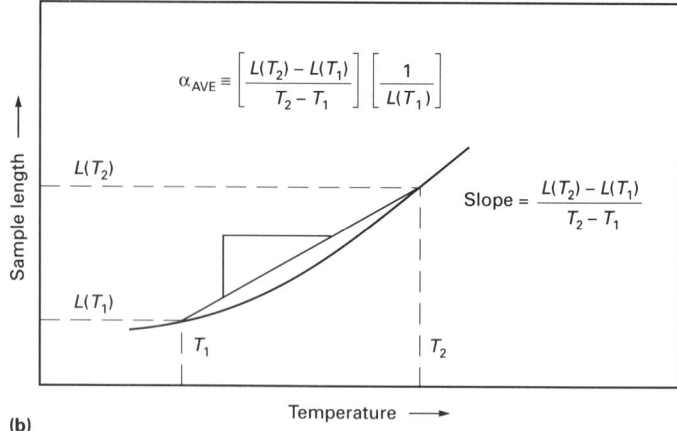

Fig 1 Different thermal expansion coefficients. (a) Linear. (b) Average

The sample must be visible from outside the isothermal environment. Usually, back illumination of the sample is necessary and there must be some arrangement to determine the sample length. For example, a reference line in a microscope or telescope can be focused on one end of a sample and then translated parallel to the sample expansion to the other end of the fixed sample. The distance traveled by the microscope (or by the sample relative to a fixed microscope) is directly related to the length of the sample.

For this method, the sample should be as long as possible (consistent with the ability of the furnace to maintain an isothermal zone) so that small changes in length can be detected. Usually, both ends of the sample must be visible and sample movement must be unrestricted in the expanding direction. If the sample is fixed on one end (for example, a sample placed vertically on some support), both ends must be visible to measure length changes directly. If one end is fixed and only the free end is visible, then the expansion must be compared to the expansion of a standard reference material measured under the same conditions (to correct for the expansion of the sample holder).

Films and Coatings. The *in situ* measurement of thermal expansion of ceramic or glass films or coatings on substrates can pose significant challenges that are not easily addressed by the previously discussed techniques. In many instances, the expansion of a bulk sample of the film or coating composition can be measured and successfully used to predict stresses, adhesion, and other important characteristics of the film-substrate interaction. In a few cases, however, the unique orientation within a film or even the uniqueness of the composition of the film itself cannot be reproduced in bulk samples, and an indirect measure of film thermal expansion is necessary.

If the film or coating is applied at or processed at elevated temperature, and if the substrate is not too stiff, the difference between film and substrate expansion will cause the coated substrate to exhibit some curvature upon cooling to ambient temperature. The curvature can be measured by several commercially available instruments and related to the stress in the film if the geometry and elastic properties of the substrate are known. That stress can be used, in turn, to estimate the thermal expansion of the film. Usually, if the film stress is known, the thermal expansion of the film is moot. But if one wants to identify a more compatible substrate (with respect to thermal stresses), then the estimate of film expansion will result in a more efficient search for an alternate.

Thermal Conductivity

Thermal conductivity provides a measure of how well heat is transferred through a substance by conduction. Empirically, it was found that the rate of heat flow through a material, dQ/dt, was proportional to the area through which the heat was conducted and to the temperature gradient across the sample, dT/dx. The coefficient of thermal conductivity, or simply, thermal conductivity, κ, is then defined by the equation resulting from these observations:

$$\frac{dQ}{dt} = -\kappa A \frac{dT}{d\chi} \qquad \text{(Eq 5)}$$

where the negative sign indicates that heat flows "down" the temperature gradient (from high to low temperature). In some publications, the symbol λ is used for thermal conductivity, but κ or K is more common. The SI unit of thermal conductivity is the W/m · K. Other common units are the W/cm · K, Btu/ft · h · °F (British thermal unit per foot per hour per degree Fahrenheit), and various mixed units.

Thermal conductivity is emerging as an important material property for microelectronic substrates and packaging materials. With the current trend toward developing higher-density circuits, energy dissipation becomes a critical and often limiting consideration. Even with clever electrical and mechanical designs and good thermal management, the heat generated by current carriers must be transported through substrate and packaging materials by conduction. Searching for electrical substrates and packaging materials with high resistivity, low dielectric constant, and high thermal conductivity is currently of active research interest.

Heat conduction takes place by interactions of lattice vibrations and by electron motion and interaction with the atoms. Thus, thermal conductivity is a second-ranked tensor property, $\kappa_{ij}(T)$ and:

$$\frac{dQ_i}{dt} = -\kappa_{ij}(T)A\frac{dT}{d\chi_j} \qquad \text{(Eq 6)}$$

which means that for nonisotropic crystalline solids, a temperature gradient in one direction, χ_j, will result in heat flow along that direction, and in orthogonal directions, dQ_i/dt, as well. This anisotropy of thermal conductivity may be important for heat flow in oriented films or coatings. Unlike thermal expansion, the anisotropy of thermal conductivity in itself is not usually observed in measurements on randomly oriented polycrystalline aggregates. But, if the thermal stresses that are due to anisotropic expansion are severe enough to cause separation between grains, the thermal conductivity can be drastically decreased by the cracks. The thermal conductivity of glasses is isotropic.

Another way of expressing the ability of a material to conduct heat is by its thermal diffusivity, usually denoted by α (not to be confused with thermal expansion). Thermal diffusivity arises from consideration of heat as an entity, which can diffuse through a material subject to different boundary conditions, causing both spatial and temporal variations of temperature. Thermal diffusivity, $\alpha_{ij}(T)$, is a temperature-dependent material tensor property defined from:

$$\frac{dT}{dt} = \alpha_{ij}(T) \, \nabla^2 T \qquad \text{(Eq 7a)}$$

or in Cartesian coordinates, as:

$$\frac{dT}{dt} = \alpha_{ij}(T)\left(\frac{\partial^2 T}{\partial \chi_1^2} + \frac{\partial^2 T}{\partial \chi_2^2} + \frac{\partial^2 T}{\partial \chi_3^2}\right) \qquad \text{(Eq 7b)}$$

where χ_1, χ_2, and χ_3 are generalized orthogonal axes. Thermal diffusivity is related to thermal conductivity by:

$$\alpha_{ij}(T) = \frac{k_{ij}(T)}{\rho(T)c_p(T)} \qquad \text{(Eq 8)}$$

where $\rho(T)$ and $c_p(T)$ are the temperature-dependent density and specific heat capacity, respectively. Thermal diffusivity is usually expressed in units typical of diffusion, cm^2/s; but the correct SI units are m^2/s.

The guarded hot plate technique is one of several comparative methods used for the measurement of thermal conductivity under steady-state conditions. In principle, comparative methods can be used to measure values of thermal conductivity over a wide range of values. For best results, a reference material with a conductivity similar to that of the sample is necessary. These methods can measure thermal conductivity from subambient temperatures to about 700 to 800 °C (1290 to 1470 °F), where excessive heat losses render the technique of little use.

Sample and reference materials are right circular cylinders with identical diameters in the range of 20 to 100 mm (0.8 to 4 in.) and usually have a height equivalent to the diameter. The experimental arrangement is shown in Fig 2. The thermocouples positioned near the faces of sample and reference cylinders measure the respective temperature gradients. The sample of unknown conductivity is sandwiched between two cylinders of the reference material with known conductivity. The mating surfaces are polished, and a thermally conductive paste is used to provide good thermal contact when possible. The entire apparatus can be placed in a furnace (not shown in Fig 2) to allow the measurement to be made at elevated temperatures. A temperature gradient and, subsequently, a steady-state heat flux are established between the upper and lower stack heaters. To eliminate radial heat losses from the stack, a similar temperature gradient is established in the "guard," as well. In addition, the measurement is frequently made in vacuum or reduced atmosphere to eliminate convective heat losses, especially at high temperatures.

When a steady state is reached, the heat flux through the stack is calculated from the known conductivity of the reference material and the measured temperature gradients in both the top and bottom reference samples using Eq 5. These two values of heat flux are averaged and used as the heat flux through the sample. The two independent calculations of heat flux also serve as a measure of heat losses in the stack. The thermal conductivity of the sample is then deduced from the known geometry, the measured temperature gradient along the sample length, and the average heat flux through the stack using Eq 5.

Many commercially available instruments use the comparative method to measure thermal conductivity. Each is designed to operate in a specific temperature range for materials within different thermal conductivity limits, and for different sample types and sizes.

The radial heat flow method is an example of a noncomparative technique used to measure thermal conductivity. An electrical heater, calibrated with respect to heat generation per unit length, is placed inside a long cylinder of the sample to be measured. Good thermal contact between the heating element and the inner sample diameter must be maintained throughout the measurement. At steady state, the known heat generated by the electrical element is used to determine the heat flux in the radial direction through some small volume element near the center of the cylindrical sample. End losses are neglected, and axial heat flow near the center of the cylinder is presumed to be small. The radial temperature gradient is measured by judiciously placed thermocouples, and the thermal conductivity is calculated from the solution to Eq 7(a) in a cylindrical coordinate system.

This method generally requires very long (to assure only radial heat flow at the center of the cylinder), large-diameter (to ensure a measurable gradient) samples and is not generally used for routine measurements unless the samples are normally in this configuration. Automation of the complete method is not generally available. This technique can be made into a "comparative" method by placing another cylinder of known conductivity around the sample, around the heater, or both, and then making the necessary additional temperature measurements.

The laser flash method for measuring thermal diffusivity is the most common of several "thermal pulse" techniques. Thermal diffusivity, the material property that describes heat flow under transient conditions $[\partial T(x,y,z,t)/\partial \neq 0]$, is related to thermal conductivity by Eq 8. This is a fairly simple experiment that can be used over a wide range of thermal diffusivities (conductivities) and sample temperatures. In fact, it is the preferred measurement technique at high temperatures, where steady-state heat flow conditions are difficult to maintain. The laser flash method derives its name from the fact that a single laser pulse, or flash, imparts a thermal pulse to the sample under measurement. However, other sources can also be used, such as xenon or quartz flash lamps, electron beams, and others. The following discussion applies to all of the thermal pulse techniques.

With the laser flash technique, a sample in the form of a thin disk held at constant, uniform temperature, T, is subjected to a thermal pulse, that is, a laser flash, on its front face, for example. Assuming that the energy is deposited uniformly over the surface, and that the heat flows along the thickness direction only, and that heat deposited by the pulse results in sample heating only (the rate of heat loss to the surroundings before the pulse is unchanged by the pulse), then the temperature history of the back face can be used to deduce the thermal diffusivity of the sample. A schematic of the experimental set-up and a typical data set are shown in Fig 3.

Thermal diffusivity is calculated from different values of time corresponding to different fixed points on the back-face temperature-time history. Some algorithms use the time required to reach the maximum temperature at the rear face, t_{max}. Some use the time to get to one-half the maximum temperature, $t_{1/2}$. Others use the extrapolated incubation time, t_i, to calculate thermal diffusivity. Each of these times is shown in Fig 3. Still other algorithms use a difference method based on two values of temperature and time (T_1,t_1) and (T_2,t_2) to calculate diffusivity. The algorithms referred to above are quite complex and would require too much space to describe adequately. For a detailed description of these algorithms, the review article by Righini and Cezairliyan (Ref 2) should be consulted. Note that the temperature rise due to a thermal pulse is usually small, on the order of less than 1 K to perhaps as much as 10 K, maximum, and, as such, the value of $\alpha(T)$ is a good representation of the diffusivity value at the nominal temperature of the sample. Note also that it is not necessary to know the temperature rise of the front face nor the quantity of heat deposited by the thermal pulse. The laser flash technique has been fully automated and,

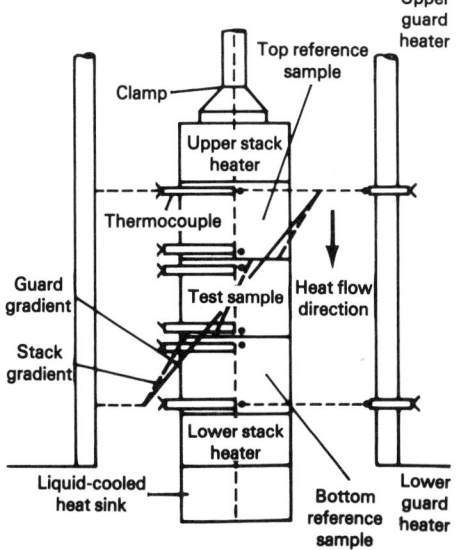

Fig 2 Typical test stack and guard system illustrating matching of temperature gradients. Source: Ref 1

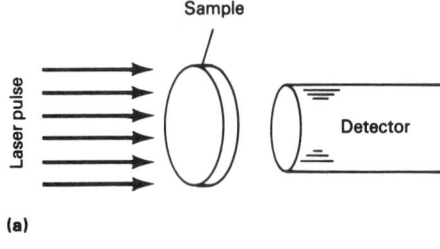

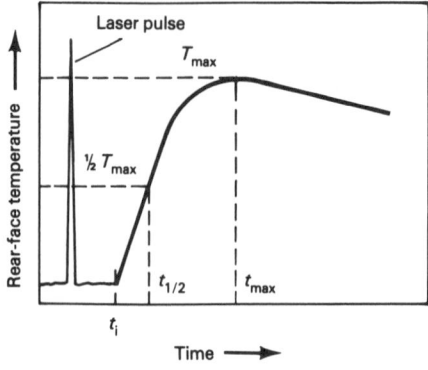

Fig 3 Laser flash experimental set up (a) and typical data set (b)

with some instruments, auxiliary information such as specific heat can be measured with little or no modifications.

In principle, the diffusivity (conductivity) of films or coatings can be measured using this technique by solving Eq 7(a) for a two-layer structure with the diffusivity of the film as the only unknown (the diffusivity of the substrate being either known or measured in a separate experiment). However, from a practical perspective, the assumption that the pulse source deposits energy on the front surface only may not be fulfilled if the film or coating is too thin or transparent to the pulse source.

Heat Capacity

Materials differ from one another in the amount of energy (actually, enthalpy) required to produce an identical temperature increase in identical masses. Thus, the total enthalpy or heat content of different materials at the same temperature is different. Heat capacity is a general term for an assortment of closely related, more definitive terms, each related to the quantity of heat that must be added to or rejected from a substance to raise or lower its temperature. The most commonly used, measured, and tabulated term in this assortment for condensed phases is the specific heat capacity at constant pressure, or, simply, the specific heat, designated by c_p. It is strictly defined (Ref 1) as the quantity of heat required to raise the temperature of 1 g (0.035 oz) of a substance by 1 K at constant (usually atmospheric) pressure. The specific heat of a substance is a scalar quantity derived from the

thermodynamic state property of enthalpy, H, as:

$$\left(\frac{\partial H}{\partial T}\right)_p = c_p \qquad \text{(Eq 9)}$$

where T is temperature and the subscript p denotes a constant pressure measurement. Although specific heat is normally presented in units of cal/g · C or Btu/lb · °F, the correct SI units are J/mole · K.

The variation of specific heat with temperature is discontinuous at second-order phase changes, such as order-disorder structural transformations, ferromagnetic transitions, ferroelectric transitions (Curie temperature), and the superconducting transition temperature (critical temperature). In fact, the experimental observation of a discontinuity in specific heat often heralds the discovery of a second-order transition. The most notable example is the discontinuity of specific heat at the glass transition temperature, T_g, of a glass or an amorphous substance. Above T_g the glass is viscous or rubbery, whereas below T_g the glass is hard and brittle.

Although specific heat and its variation with temperature can be measured by several techniques, in principle it is the variation of enthalpy with temperature that is actually measured, and specific heat is calculated from the slope following Eq 9. Like thermal expansion, the specific heat can be considered constant within a narrow temperature range if that range does not include discontinuities. Experimentally measured specific heats are usually tabulated as polynomials of the following form to represent as much of the nonlinearity as possible:

$$c_p(T) = a + bT + cT^2 - dT^{-2} + eT^{-1/2} \qquad \text{(Eq 10)}$$

The constants a through e are derived by fitting the measured values to the above polynomial. For many materials, one such relationship is insufficient to describe the data over a large temperature interval (especially one that contains first- or second-order phase changes) and several polynomials of the above form are used in different temperature intervals.

The enthalpy versus temperature relationship for a substance can be measured by various forms of calorimetry. Subsequently, the temperature-dependent specific heat is determined. Also, differential thermal analysis (DTA) can be used to estimate specific heat.

Differential scanning calorimetry (DSC) is the most commonly used technique for measuring specific heat capacity of a ceramic or glass. A DSC measures and records the rate of heat flow (dH/dt) to a sample as it is heated at a predetermined constant rate (dT/dt). In principle, the specific heat capacity at a series of temperatures, $c_p(T)$, can be calculated from:

$$\frac{dH(T)}{dt} = mc_p(T)\frac{dT}{dt} \qquad \text{(Eq 11)}$$

where m is the mass of the sample and $dH(T)/dt$ has been corrected for the presence of a sample holder. However, to eliminate systematic errors such as the effect of the thermal resistance between the sample and heat flow sensor, the measurement of an unknown is usually compared to that of a standard reference material (sapphire or silver) measured under identical conditions (same or similar sample holder, same heating rate, and so on). For a comparative measurement:

$$c_p(T)_u = c_p(T)_R \frac{m_R}{m_u}\frac{(dH(T)/dt)_u}{(dH(t)/dt)_R} \qquad \text{(Eq 12)}$$

where the subscripts R and u refer to the reference material and unknown, respectively.

Many DSCs are commercially available and the literature for each instrument usually describes the measurement of specific heat in detail. This method is suitable for both ceramics and glasses, as well as many other types of materials (metals, polymers, and others) including liquids. Sample sizes are on the order of 1 to 10 mg (35 to 350 × 10^{-6} oz). The major disadvantage to this method is that commercially available DSCs can only be used to about 800 °C (1470 °F).

Calorimetry. The heat content of a substance can be determined at a single temperature by "drop" calorimetry. A known weight of a substance is heated to some temperature, T_i, and "dropped" from the external furnace into a calorimeter at T_0. The function of the calorimeter is to quantify the heat rejected by the sample in cooling from T_i to the final temperature of the calorimeter and sample, T_f, which is simply the change in enthalpy or heat content between T_i and T_f. Measuring $H(T)$ at a series of temperatures (samples dropped into the calorimeter from different temperatures) allows a plot of $H(T)$ versus T to be made and $c_p(T)$ to be calculated from Eq 9.

Many different calorimeters have been described in the scientific literature and each measures the heat content of the sample in a slightly different way. A detailed description of each calorimeter is beyond the scope of this discussion, but some of the more general methods should be mentioned. They include the measurement of temperature rise in a well-stirred liquid of known heat capacity in an adiabatic container, the measurement of the temperature rise of a solid block of material with known heat capacity (usually copper or silver), or the measurement of the amount of vapor produced when the hot sample contacts some volatile liquid (liquid nitrogen, Freon, water) for which the heat of vaporization is known accurately.

Calorimetry allows specific heats to be calculated over a wide range of temperatures, and if any first-order transitions (that is, melting) are included, the enthalpy associated with the transition is measured, as well as the specific heats above and below the transition. However, this technique suffers from insensitivity to transitions with small enthalpy changes, such as magnetic transitions in which the specific heat is discontinuous.

Differential thermal analysis (DTA) is a fairly easy way to estimate the specific heat of a substance. This method measures the temperature difference, ΔT, between a sample and reference material as each is heated at the same rate in small, separate, side-by-side sample holders. If the instrument baseline is well established (by recording a scan of the empty sample holders), the ΔT signal recorded for a sample of an unknown specific heat, ΔT_u, can be compared with the ΔT signal recorded for a standard material of known specific heat, ΔT_k, and then:

$$c_p(T)_u = c_p(T)_k \frac{m_k (\Delta T_u - \Delta T_b)}{m_u (\Delta T_k - \Delta T_b)} \qquad \text{(Eq 13)}$$

where $c_p(T)|_u$ and $c_p(T)|_k$ are the specific heats of unknown and standard, respectively, m_k and m_u are the mass of the standard and unknown, and ΔT_b is the measured baseline signal. For this method to give even approximate results, the standard and unknown must be in the same physical state and must have similar emittance or be enclosed to prevent different radiation losses.

Emissivity

Any object held at temperature, T, above the absolute zero of temperature will radiate heat energy. The rate at which heat is radiated, dQ_R/dt, is given by:

$$\frac{dQ_R}{dt} = \epsilon A \sigma T^4 \qquad \text{(Eq 14)}$$

where A is the surface area of the object, ϵ is the emissivity, and σ is the Stefan-Boltzmann constant, numerically equal to 5.6686 $\times 10^{-8}$ W/m$^2 \cdot$ K^4 (10^{-5} erg/s $\cdot$ cm$^2 \cdot$ K^4). The emissivity of a "perfect emitter" or "blackbody" is unity, and values of emittance, which are dimensionless, for all real surfaces lie between zero and unity.

Emissivity is one of four thermal radiative properties. The others are reflectivity, absorptivity, and transmittivity. Quantities ending in "-ivity," by convention, refer to intrinsic material properties of samples that have clean, optically smooth surfaces. These properties are rarely measured in ordinary experiments. Quantities ending in "-ance," such as emittance, reflectance, absorptance, and transmittance, refer to properties of "real" samples, regardless of thickness or surface conditions. The "-ance" quantities are measured in routine experiments, but, very often, the distinction between the "-ivity" and "-ance" property is forgotten, and the terms are (incorrectly) used interchangeably.

Emittance, also designated by ϵ, is defined (Ref 3) as the ratio of the radiant flux per unit area leaving the surface of a body at a given temperature to that of a blackbody at the same temperature. The heat energy radiated by an object appears as a spectrum of electromag-

netic waves over a wide range of wavelengths. The spectral radiance of a blackbody, $L(\lambda)$, is given by:

$$L(\lambda) = \frac{C_1 \lambda^{-5}}{\exp\left\{\dfrac{C_2}{\lambda T}\right\} - 1} \qquad \text{(Eq 15)}$$

where T is absolute temperature, and the constants $C_1 = 1.1911 \times 10^{-5}$ erg/s $\cdot$ cm$^2 \cdot$ sr, and $C_2 = 1.4387$ K $\cdot$ cm are Planck's first and second radiation constants and give $L(\lambda)$ in terms of erg/s $\cdot$ cm$^3 \cdot$ steradian. This relation is plotted in Fig 4 for several temperatures. The greatest radiance is in the infrared region of the electromagnetic spectrum. It is interesting to note that significant amounts of radiant flux are emitted in the visible portion of the spectra only at temperatures above approximately 450 °C (840 °F).

The underlying principle of most emittance measurements is to compare the radiant energy spectrum from a real surface at known temperature to that derived from either Eq 15 or Fig 4, the spectrum for a blackbody (that is, perfect) emitter at the same temperature. Emittance, as well as the other thermal radiative properties, can be measured at a single

wavelength, termed the spectral emittance, $\epsilon(\lambda)$; over some specified portion of the electromagnetic spectrum, given by the integrated emittance, $\epsilon(\lambda)$; or over the entire wavelength spectrum, represented by the total emittance, $\epsilon(t)$. For each measurement, the necessary ratio must be formed with both numerator (radiant flux from the sample surface) and denominator (blackbody radiant flux) measured or calculated at either the same wavelength or over the same wavelength limits.

In some instances, it is easier to measure one of the related thermal radiative properties and infer the emittance. Reflectance, ρ, absorptance, α, and transmittance, τ, are defined as the ratios of the reflected, absorbed, and transmitted fluxes, respectively, to the incident flux. From a flux balance:

$$1 = \rho + \alpha + \tau \qquad \text{(Eq 16)}$$

In addition, it can be shown that for restricted measurement geometries (the illuminating geometry for α identical to the collecting geometry for ϵ), the emittance, ϵ, is numerically equal to the absorptance, α. Thus, emittance can be inferred directly from absorptance measurements or from reflec-

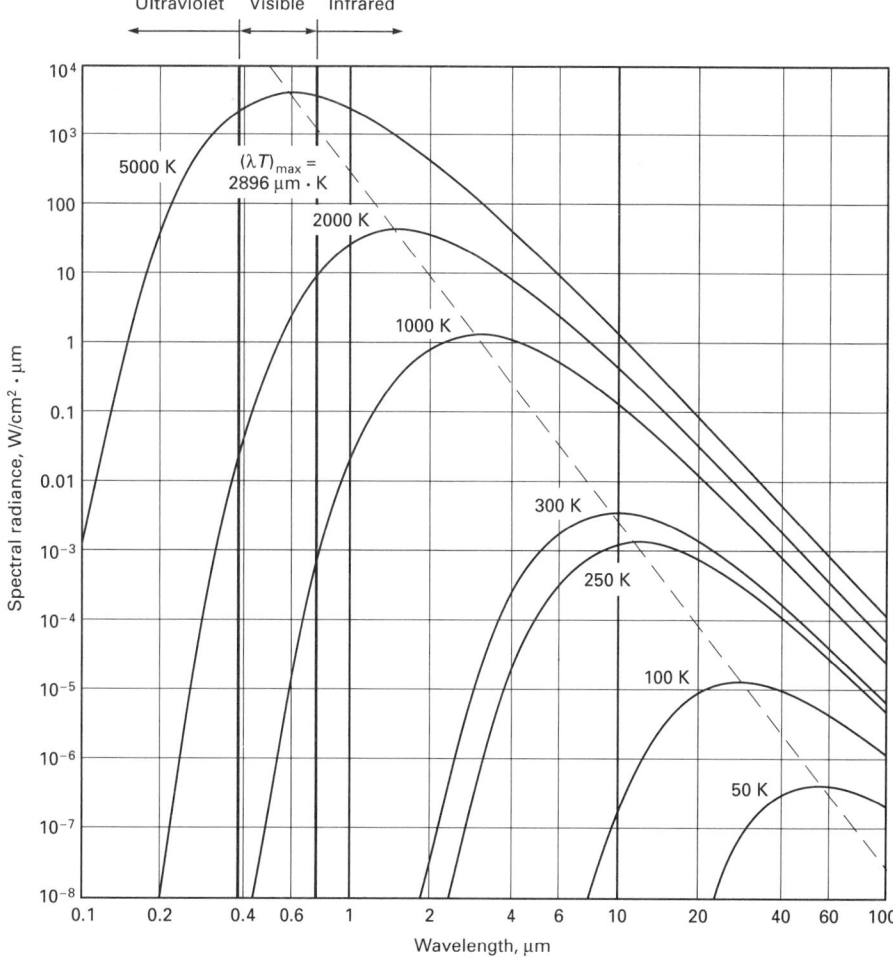

Fig 4 Planck distribution law, spectral radiance of blackbody radiation as a function of temperature and wavelength

tance measurements on opaque materials for which $\tau = 0$ and:

$$1 = \rho + \epsilon \qquad \text{(Eq 17)}$$

Direct Measure of Emitted Radiant Flux. Radiant thermal flux can be measured directly using different spectrometers, each usually sensitive to a portion of the thermal radiation wavelength range, such as ultraviolet, visible, or infrared spectrometers. At one extreme are fixed-wavelength or one-color instruments, such as some optical pyrometers, which measure radiant flux at a single, predetermined wavelength.

To estimate the emittance of an object at some temperature, T, the emitted radiant flux is measured directly and compared to that expected from a blackbody at the same temperature. For spectrometer measurements, some wavelength range must be selected for the comparison. Comparisons can be made on a wavelength basis, where the measured radiant flux at several different wavelengths is compared to the blackbody flux at the same set of wavelengths and temperature, as calculated from Eq 15. A comparison can also be on an integrated intensity basis, by which the total radiant energy between specified wavelength limits is measured and compared to the integrated form of Eq 15 with the same wavelength limits. Obviously, a fixed-wavelength instrument can make the comparison only at the predetermined wavelength.

The above techniques are used if the true temperature of the object can be measured independently (usually by using thermocouples). However, emittance values may often be required at temperatures above which an accurate, independent temperature measurement can be made. In this high-temperature region, the measured emitted radiant flux from the sample can be compared directly to that measured from an "approximate" blackbody. The radiation emitted through a hole in a hollow cavity maintained at some uniform temperature is a good approximation to the blackbody radiation spectrum at that temperature, described by Eq 15 and shown in Fig 4.

Such an approximation to a blackbody radiator is shown in Fig 5(a). Several blackbody reference sources are commercially available. A more practical (and thus a more suspect) approximation to a blackbody cavity is simply a thin, deep hole drilled into the sample itself, such as that shown in Fig 5(b). There are no generally accepted criteria for the aspect ratio of the hole, and different authors suggest different dimensions depending on the often conflicting requirements for the hole to appear "black" (usually satisfied by deep, thin holes) and yet remain at constant temperature over the entire cavity (satisfied by shallow depressions).

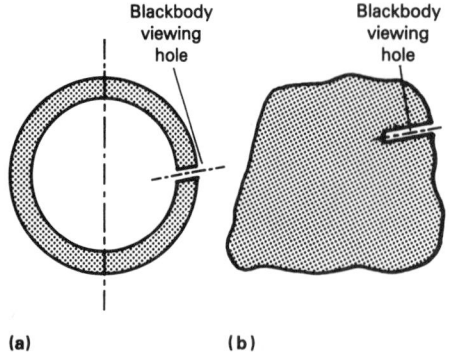

Fig 5 Approximations to blackbody radiation. (a) Radiant energy emitted from a hollow cavity heated uniformly to some temperature. (b) Thin, deep hole formed in actual sample

To make the comparative measurement, the emitted radiant flux is measured first from the sample surface and then from the hole approximating the blackbody. Care must be taken to maintain the sample and blackbody at the same temperature and to maintain that temperature uniformly over the cavity. Another precaution is that the measuring optics must collect radiant flux from an area entirely within the hole or depression approximating the blackbody, and that an area of the same size must be used to collect radiant flux from the sample surface. Emittance is then calculated from the ratio of flux emitted from the surface to flux emitted from the blackbody cavity.

For the special case of fixed-wavelength optical pyrometers, the result of a measurement on the sample surface is a temperature, T_S, characteristic of the true sample temperature and the sample emittance. To estimate the sample emittance, another pyrometer sighting is taken on the hole approximating the blackbody, giving the blackbody temperature, T_B, and the emittance is obtained from $\epsilon(\lambda) = T_S/T_B$.

The above fundamental principles take many forms when applied in commercially available instrumentation. Among the common variations in instrumentation are different detectors (as previously mentioned), vastly different optical paths and collection geometries for measuring the emitted radiant flux, and different methods of blackbody comparison. In principle, these measurements can be made at any sample temperature, but one must ensure that energy is lost from the sample by radiation only (conductive and convective losses must be minimized or eliminated).

Calorimetric Determinations. Two forms of calorimetric measurements are commonly used to measure emittance. In the first, a sample and embedded electrical heater are held at some temperature, T. The energy supplied to the heater to maintain the sample at constant temperature is equivalent to the radiant heat lost by the sample (provided that conduction and convection losses are eliminated). By comparing the energy emitted by the sample to the calculated total energy radiated from a blackbody at the same temperature, the emittance can be determined.

Emittance can also be inferred from a calorimetric determination of absorptance. A sample is placed in close contact with a calorimeter and irradiated with a known thermal flux. The calorimeter measures the change in heat content of the sample, which can be converted to a measure of the absorbed energy. Comparing the energy absorbed by the sample to the energy of incident radiation gives a measure of the sample absorptance that is numerically equal to the emittance for opaque materials.

Reflectivity Measurements. Emittance can be inferred from reflectance measurements on opaque materials. If a sample is irradiated with a known flux and the reflected radiation is measured (as discussed above for the direct measurement of emitted radiant flux), then emittance can be calculated from Eq 17. Instrumentation to measure reflectance, like that to measure emittance directly, can take many different forms. Principle variations are in the type and spatial distribution of incident flux, the optical path and solid angle by which reflected flux is collected, and the method of detecting and quantifying reflected flux.

REFERENCES

1. Standard Definitions of Terms Relating to Thermophysical Properties, ASTM E-1142, *General Methods and Instrumentation*, Vol 14.02, *Annual Book of ASTM Standards*, American Society for Testing and Materials, 1990
2. F. Righini and A. Cezairliyan, Pulse Method of Thermal Diffusivity Measurements (A Review), *High Temperature—High Pressure*, Vol 5, 1973, p 481–501
3. *Thermophysical Properties of Matter*, 13 Vol., Y.S. Touloukian *et al.*, Ed., IFI/Plenum, 1970–1977

SELECTED REFERENCES

- R. Berman, *Thermal Conduction in Solids*, Clarendon Press, Oxford, 1979
- G. Grimvall, *Thermophysical Properties of Materials*, North Holland, 1986
- O. Kubaschewski and C.B. Alcock, *Metallurgical Thermochemistry*, Pergamon Press, 1979
- G.S. Sheffield and J.R. Schorr, Comparison of Thermal Diffusivity/Conductivity Methods, *Am. Ceram. Soc. Bull.*, Vol 70 (No. 1), Jan 1991, p 102–106
- R. Siegel and J.R. Howell, *Thermal Radiation Heat Transfer*, 2nd ed., Hemisphere Publishing, 1981

NDE Testing and Inspection

L.A. Lott and D.C. Kunerth, Idaho National Engineering Laboratory

THE INCREASING USE of advanced monolithic and composite ceramic materials in critical applications has necessitated the development of nondestructive evaluation (NDE) techniques that can be used to ensure final component integrity. Although many NDE techniques are routinely and successfully applied to metals and other materials, the techniques cannot always be directly applied to ceramics. This is due to the unique physical properties of ceramics, which not only make them quite sensitive to very small defects, but also impede the application of conventional electromagnetic, ultrasonic, and radiographic techniques.

There is a large body of literature on NDE techniques, including several handbooks that can help solve NDE problems (Ref 1–3). Although most techniques described in these sources were developed primarily for application to metallic materials and structures, and therefore are generally not complete with respect to ceramics, a number are applicable to some degree and should be consulted.

The two characteristics that make advanced ceramic materials unique when compared to other structural materials are that components are often fabricated from fine powders and that they are typically brittle in nature. Defects that limit the utility of the materials are often generated by the processes used to consolidate the powders into useful monolithic components and finished structures. The brittle nature of the consolidated material means that under the appropriate conditions, extremely small defects (10 to 100 μm, or 0.4 to 4 mils) are critical to the material performance and must be prevented, either by careful process control or by inspecting finished components and rejecting those containing critical defects. Nondestructive evaluation has a critical role in both process control and final inspection.

Except for some ferro-oxide, carbide, and boride materials, ceramics are electrically nonconductive, with low magnetic permeabilities. Typical eddy current inspection techniques are generally not applicable. However, this does not exclude the use of microwave techniques, which are also sensitive to electrical permittivity. Microwave techniques have been used to characterize materials and identify anomalies, but are limited by wavelength (Ref 1, 4).

Using ultrasonic techniques to inspect ceramics is complicated by higher ultrasonic velocity and small critical flaw size. At the same test frequency, the larger ultrasonic velocities in ceramics, compared to those in metals, result in much longer wavelengths, which are directly related to the size of defect that can be detected. For this reason, the test frequencies used in ceramics must be significantly increased (20 to 100 Mhz), compared to the conventional test frequencies (1 to 15 MHz) commonly used for metals. This increases both the potential for attenuation problems and the sophistication of the hardware and techniques that are utilized. The efforts of the ultrasonic industry to address these complications has provided instrumentation and transducers that have much higher bandwidths, as well as various techniques directly applicable to ceramics.

Similar problems are encountered with radiography techniques, because of the small defect size of interest and low radiographic contrast. To overcome these problems, low-energy projection microfocus radiography techniques have been developed. These techniques improve contrast and permit direct magnification for high-resolution enlargements.

Radiography and ultrasonics are the most commonly used evaluation techniques in industry and have therefore been emphasized with respect to ceramic applications. Other techniques, such as acoustic emission, thermography, acoustic resonance, microwaves, nuclear magnetic resonance, and dye penetrant, are applicable to ceramics to various degrees. Some manufacturers of structural ceramics use dye penetrant on all critical parts because it detects flaws on curved surfaces that are not easily detectable by other means. As with most inspection needs, one technique cannot provide all the answers. It may be necessary to use multiple techniques that provide complementary information.

Radiography

Radiographic techniques use penetrating radiation to interrogate internal structures using the interaction between material and electromagnetic waves/particles, which attenuates and scatters the radiation. These techniques are primarily used to detect and characterize inclusions, voids, porosity, cracks, and density variations that are due to material inhomogeneities.

Two basic approaches are used to extract and process the information carried by the transmitted energy. The first approach is to use film or other full-field detectors to produce an image of the test object. This provides a two-dimensional image of the object, in which the film density or image intensity is representative of beam attenuation that is due to interaction with material. Cracks, delaminations, and similar features that are improperly oriented with respect to the propagation direction of the beam have significantly reduced detectability. For example, a planar interface between two bonded materials with different attenuation characteristics will be undetectable when oriented transverse to the beam, but quite obvious when oriented parallel to the beam.

The second approach is computed tomography, which uses multiple radial attenuation measurements and an appropriate algorithm to generate a spatial distribution of linear attenuation coefficient in the plane of the radiation. This process provides an image of a two-dimensional transverse slice of the test object, in which the localized intensity of the image is representative of the linear attenuation coefficient of the material in that area. Multiple transverse slices are generated and stacked to acquire information in the third dimension of the test object. This attribute makes tomography well suited for complex geometries and overcomes many of the orientation problems associated with the first approach. Although tomography is a more versatile technique, its disadvantages are the complexity and cost of the instrumentation and computational time needed to generate an image.

Both imaging approaches can be applied using gamma, x-ray (conventional and microfocus), and neutron radiation sources. Although useful for some ceramic applications (Ref 5, 6), such as penetrating thick, high-density, high atomic number materials, gamma radiography is limited by the high energy of the radiation and/or the physical size of the source needed for usable intensities of commonly used isotopes (Ref 2, 7).

X-ray radiography has been found to have greater versatility, especially with the advent of microfocus technologies. Under the appropriate conditions, increased spatial resolution and sensitivity can be achieved.

Neutron radiography is not as highly developed as gamma or x-ray technologies, but does hold some promise for ceramic applications, such as drying and binder distribution and/or removal (Ref 8, 9). The advantage of neutron radiography for these applications stems from the interaction of the neutrons with atomic nuclei instead of orbital electrons. This results in distinct differences in the transmission of neutrons through low atomic weight materials such as hydrogen, compared to techniques based on x-ray or gamma radiation. The disadvantages of neutron radiography are its high cost, safety issues, and lack of source assemblies of a practical size and intensity to permit easy portability.

X-Ray Microradiography. The physics that govern defect detectability and/or sensitivity is the same for all techniques that use a divergent x-ray beam to image an object. However, the practical aspects of an implementation often determine which techniques can be used. Ceramics, which inherently have small critical defects of low contrast, require radiographic techniques that provide high resolution and contrast. Microradiography techniques that use moderate to low energies (typically <150 keV) best meet these requirements.

Microradiography can be performed using two basic techniques. The first is contact radiography, in which the object to be imaged contacts a high-resolution imaging medium, such as fine-grain film. This results in the highest resolution, particularly when the 1:1 radiograph is enlarged to obtain a magnified image. This method works well with thin objects in contact with the imaging medium because geometric unsharpness (U_g, loss of image resolution due to source size) is small. It has been reported that magnifications up to 200× can be obtained using very slow films, whereas magnifications of several hundred times, yielding spatial resolutions of 5 to 10 nm (0.2 to 0.4 μin.), can be obtained using x-ray-sensitive resist material. The disadvantages are that image sharpness and contrast can suffer as a result of scatter; exposures can be long, depending on the speed of the imaging medium; and the additional step of image enlargement is required.

The other technique, microfocus radiography, uses small x-ray sources that are usually less than 100 μm (4 mils) in size (Ref 2, 4, 10, 11). The small source permits direct magnification of the image through geometric projection, while still maintaining small geometric unsharpness. Other advantages of this technique are:

- The size of the radiation source is variable within certain limits
- Magnification can be varied by simply moving the test object between the source and imaging plane
- Sharp images can be produced with very short source-to-film distances
- Scatter that reaches the detector from the test object is reduced, thereby improving both image sharpness and contrast

The disadvantages of the technique are:

- Output intensity is low because of source size
- Coverage is limited by source-to-film/detector distance
- Geometric distortion can result with thick specimens

Defect detectability is governed by the combination of contrast and image definition or resolution. Resolution defines the degree to which the two-dimensional structure of the imaged object can be reproduced, whereas contrast is related to the ability of the sample to modify an electromagnetic wave and the ability of the imaging medium or detector to measure or record the variation in the transmitted radiation. The physical limits of resolution are controlled by the total amount of unsharpness, U_t, in the image, of which the more important contributions are inherent unsharpness, U_i, determined by the characteristics of the imaging medium or detector, and geometric unsharpness, U_g, resulting from the physical size of the radiation source. Beam scatter and movement of the test object also contribute to total unsharpness, but are assumed to be negligible in this discussion.

Inherent unsharpness, U_i, results from the resolution limits of the imaging medium, which are determined by the physical characteristics of the detector material and its interaction with the x-ray energy. At low magnification, this contribution can limit the lowest U_t level that can be achieved. This results from the fact that U_t is related to U_i and U_g by

$$U_t = (U_i^2 + U_g^2)^{1/2} \qquad \text{(Eq 1)}$$

This indicates that the largest contributor will dominate. As a result, a very small U_g has no effect if U_i controls U_t. This is compensated for with low-resolution detectors, such as image intensifiers, by magnifying the image sufficiently to make the image of the smallest desired detectable defect greater than the inherent unsharpness at the input to the detector.

Geometric unsharpness, U_g, is defined as

$$U_g = F\left(\frac{T}{D}\right) \qquad \text{(Eq 2)}$$

where F is the size of the radiation source, T is either the thickness of the object to be imaged when using contact radiography or the distance from the front of the object to the imaging plane when using a projection technique, and D is the distance from the source to the object. Figure 1 illustrates both U_g and

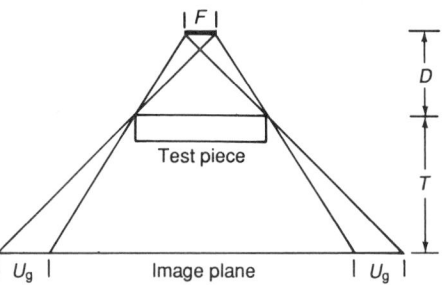

Fig 1 Projection radiography, where geometric unsharpness resulting from use of a finite size source yields ill-defined boundaries; separation of the test object from the imaging plane results in geometric enlargement

the principle of geometric enlargement. The magnification, M, that is achieved is defined as

$$M = \frac{D + T}{D} \qquad \text{(Eq 3)}$$

Combining Eq 2 and 3 yields

$$U_g = F(M - 1) \qquad \text{(Eq 4)}$$

which illustrates that with a constant F, U_g increases linearly with M. However, magnification does have a diminishing influence (Ref 12). The effective resolution at the detector plane must be compared to the size of a defect that is also magnified, as shown by the ratio:

$$\frac{U_g}{XM} = \frac{F}{X}\left(1 - \frac{1}{M}\right) \qquad \text{(Eq 5)}$$

As the magnification increases, the blur that is due to the finite source size never exceeds the focal spot size when compared to X, the actual defect size. Thus, increased U_g that results from increased geometric magnification does not adversely affect the detectability of a defect.

Contrast sensitivity is controlled by the ability of the test object to modify the propagating electromagnetic wave, as well as the ability of the detector to measure and record variations in the transmitted radiation. Test object contrast depends on the physical characteristics of the material and the spectral composition of the radiation. Ceramics, which are usually composed of low atomic number elements, do not attenuate x-ray radiation as strongly as metals. Figure 2 is a plot of calculated linear attenuation coefficients (Ref 2) for several ceramics, metals, and other constituents that are associated with ceramic flaws. The linear attenuation coefficient, μ_l, is a measure of the probability of x-ray absorption per centimeter of material traversed.

In Fig 2, it can be observed that as radiation energies decrease, μ_l increases, resulting in improved image contrast but reduced penetration capabilities, which are often needed for thick sections. This is a well-recognized trend in x-ray radiography, and can be under-

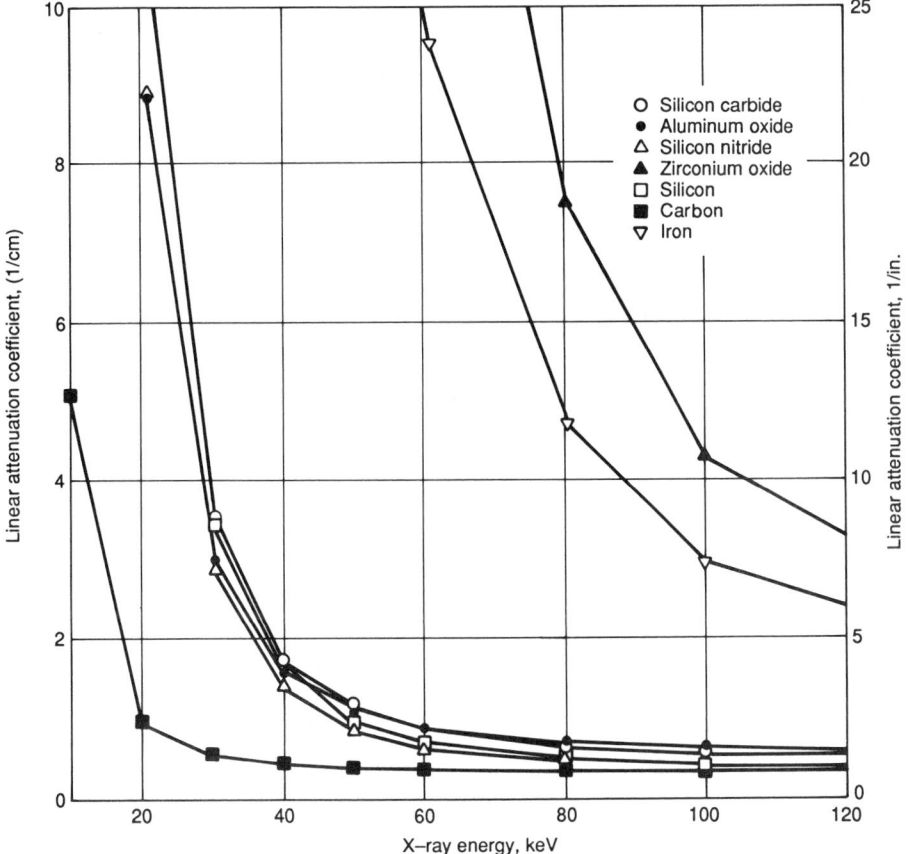

Fig 2 Calculated linear x-ray attenuation coefficients for various materials plotted as a function of beam energy

stood quantitatively by the definition of radiographic contrast:

$$\text{Contrast} = \frac{KX(\mu_{l1} - \mu_{l2})}{1 + I_s/I_d} \qquad (\text{Eq 6})$$

where K is the imaging medium constant, X is the thickness of the flaw traversed by the x-ray beam, μ_{l1} and μ_{l2} are the respective attenuation coefficients at a specific x-ray energy of the sample and flaw, and $(1 + I_s/I_d)$ is the build-up factor that takes into account the effect of scattered radiation produced in the sample (Ref 4, 11, 13). For projection radiographic techniques that separate the sample from the imaging plane, scatter is significantly reduced, making I_s/I_d relatively small, improving contrast as well as permitting Eq 6 to be simplified to

$$\text{Contrast} = KX(\mu_{l1} - \mu_{l2}) \qquad (\text{Eq 7})$$

The utility of Eq 7 is to allow the simple comparison of contrast differences between different defects, as well as different thicknesses of the test object. For example, it can be used to determine that, at 50 keV in SiC, an Fe inclusion will have approximately 12 times as much contrast as would a void of equivalent size. It can also be used to demonstrate that surface roughness or thickness variations, on the order of the small voids and

cracks usually of interest in ceramics, will give equivalent contrast variations as the defect, thereby potentially masking defects. However, typical x-ray sources are not monochromatic in nature. More precise computations of image contrast must treat μ_l and K as a function of beam energy over the functional x-ray spectrum of the source and detector.

Optimizing both contrast and resolution is necessary to obtain the best defect detectability. Figure 3 is a microradiograph of modulus of rupture (MOR) test bars that have metallic inclusions. Researchers (Ref 12, 14–16) have demonstrated that with microfocus radiography, detectability limits for voids in sintered SiC and Si_3N_4 are 1.5% of specimen thickness, with a 0.90 probability of detection at a 0.95 confidence level. This compares to 2.5% of specimen thickness for conventional contact radiography. Radiographic detectability of defects is typically reported as percent of thickness sensitivity (%TS) and defined as:

$$\% \text{ TS} = 100 \left(\frac{X}{T}\right) \qquad (\text{Eq 8})$$

where X and T are the thicknesses, respectively, of the defect and the specimen traversed by the interrogating x-ray beam.

Typically, penetrameters (Ref 2, 12) are used to indicate radiographic quality. This is determined by the ability of an inspector to detect penetrameter characteristics of known size and thickness in a through-sample radiograph. However, at this time there is no recognized standard for ceramic penetrameters, resulting in the need to develop a unique penetrameter for each application (Ref 3). It is sometimes convenient to use aluminum in place of a ceramic because of its similar absorption spectrum (Fig 2).

X-ray computed tomography (CT) is an established imaging technique that has been used in the medical field for many years and is now being developed for various industrial applications, including ceramics (Ref 5, 7, 17–19). The technique is based on a mathematical reconstruction of multiple one-dimensional transverse projections to a spatial map of x-ray attenuation in a slice of a test object with both high spatial and contrast resolution. Although CT is well suited for the detection and characterization of cracks, porosity, and long- or short-range density gradients in components with complex geometries, its applicability to ceramic inspection is at the stage of initial development. Presently, it is useful primarily as a research tool. In fact, issues such as defect sensitivity, required system hardware and software, inspection rates, and cost are only now being addressed. Nevertheless, the results that can be obtained with CT are impressive (Fig 4).

Like either conventional or microfocus radiography, the implementation of CT to ceramics requires that the technical issues controlling resolution and contrast be addressed (Ref 5, 7, 20). These include the typical parameters unique to the source (size, intensity, and energy content), detector (size/resolution, range, stability, and immunity to scatter), and artifacts produced by interaction of the x-ray beam with the sample (scatter and beam hardening). Because the physics are the same for all x-ray imaging techniques, many of these issues are addressed in a manner similar to those already discussed. For example, source size is maintained small to obtain better image resolution, whereas intensity and energy content are optimized and scatter minimized to obtain the best contrast at the detector. However, some issues are unique to CT and must be addressed differently.

Although film can be used, the detectors typically used for CT are either solid-state arrays or ionization chambers. The attributes required by these detectors are high stability, a large linear dynamic range, negligible crosstalk resulting from scatter (usually controlled by collimation of the incident beam), and high packing density for improved resolution. With ceramics, it is also important to control the thickness of the slice, z dimension, as well as the x and y dimensions, of the volume element (voxel) that is being examined in order to minimize its total vol-

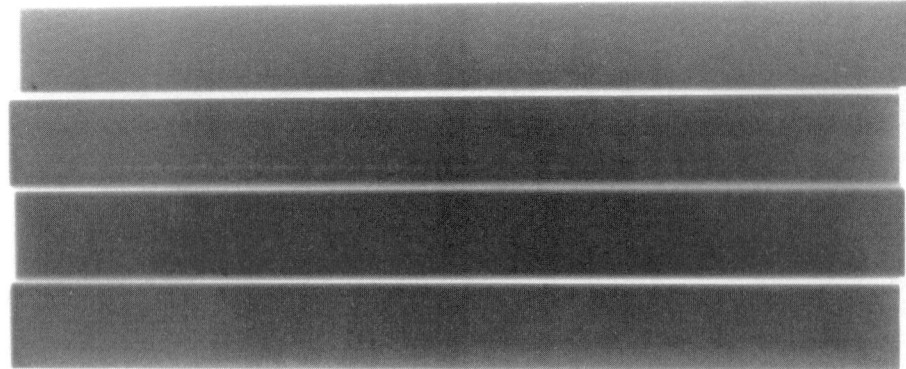

Fig 3 Microfocus radiograph of MOR test bars containing metallic inclusions. Courtesy of GTE Labs, Inc.

ume. The attenuation value for each imaged picture element (pixel) is an average of the attenuation measured for the voxel. Attenuation variations caused by large density gradients or defects whose volumes are only a small percentage of the voxel will be integrated over the entire voxel, thereby reducing detectability. This is called the partial volume effect and is controlled by collimation of the source and/or detector to reduce slice thickness.

Another effect that is significant for the application of CT to ceramics is beam hardening, which can result in errors for the x-ray attenuation values obtained, making an absolute density measurement impossible.

Beam hardening results when polychromatic sources (Bremsstrahlung) are used because x-ray attenuation varies along the path of the beam through the object. As shown in Fig 2, low-energy x-rays are attenuated faster than the higher-energy x-rays at the edge of the sample, resulting in an increase in average x-ray energy with path length through the sample. In the CT image of a uniform sample, this effect appears as an apparent higher attenuation at the periphery of the sample, which can be 10% or more in error of the overall attenuation value.

Various means of overcoming this problem are being used or developed (Ref 7, 21, 22). Methods to either control the amount of

beam hardening or correct for it include prefiltering of the source to reduce the amount of low-energy x-rays, the use of phantoms (test objects of known composition) to calibrate the detector response to a polychromatic source for a range of sample thicknesses, and analytical/hardware approaches, such as the dual energy and linearization methods. One approach to avoid beam hardening is to use monochromatic isotope sources, although its disadvantages are low intensities, larger source sizes, and stringent safety requirements.

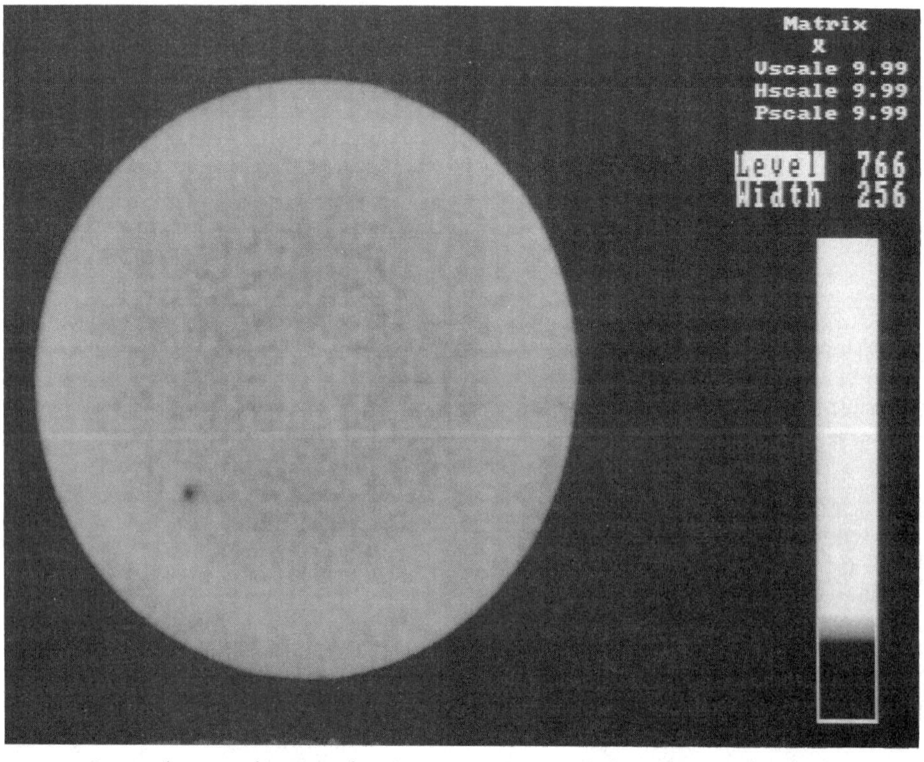

Neutron Radiography. With the introduction of smaller sources and improved imaging techniques (Ref 2, 23–26), neutron radiography is promising to become a more practical inspection technique with significant potential for ceramics applications. The difference between neutron radiography and x-ray or gamma ray radiography is that neutrons interact with the nuclei of the atoms, instead of orbital electrons, thereby resulting in significantly different transmission properties for materials. Some materials, such as hydrogen, lithium, and boron, strongly interact with neutrons. Others, such as aluminum and silicon, only weakly interact, thereby making neutron radiography potentially quite selective with respect to the atomic species within the sample being interrogated. This attribute makes neutron radiography ideal in evaluating materials or processes that inherently contain an active species whose presence, absence, or distribution is important.

The reported scattering and absorption coefficients for various elements found in ceramics are provided in Table 1. As can be observed, there is no correlation of these coefficients with the atomic number as there is with the x-ray attenuation coefficients.

An example of the selectivity is provided in Fig 5, a neutron radiograph (positive image) in which six different materials containing 750 μm (30 mil) diameter polystyrene beads are compared. The compacted powder disks are 25 mm (1 in.) in diameter, approximately 15 mm (0.6 in.) thick, and approximately 50% of full density. It is evident in Fig 5 that the SiC, Al_2O_3, and $Al_6Si_2O_{13}$ are relatively transparent, while the Si_3N_4, $Al_2O_3 \cdot 3(H_2O)$, and $LiAl_5O_8$ give greater contrast because of scattering by hydrogen and nitrogen and the

Fig 4 Computed tomography image of a 10 mm (0.4 in.) ceramic ball containing a 250 μm (10 mil) void. Image was produced using a 100 KeV source with a 50 μm (2 mil) focal spot diameter. Courtesy of Bio-Imaging Research, Inc.

Table 1   Neutron absorption and scattering coefficients

Material	Atomic no.	Scattering coefficient, cm²/g	Absorption coefficient, cm²/g	Total, cm²/g
H	1	23.0	0.196	23.2
Li	3	0.10	6.1	6.2
C	6	0.24	0.00018	0.24
N	7	0.43	0.081	0.51
O	8	0.16	$<7 \times 10^{-6}$	0.16
Al	13	0.031	0.0053	0.036
Si	14	0.036	0.0034	0.039
Fe	26	0.12	0.0282	0.15

Source: Ref 8, 26

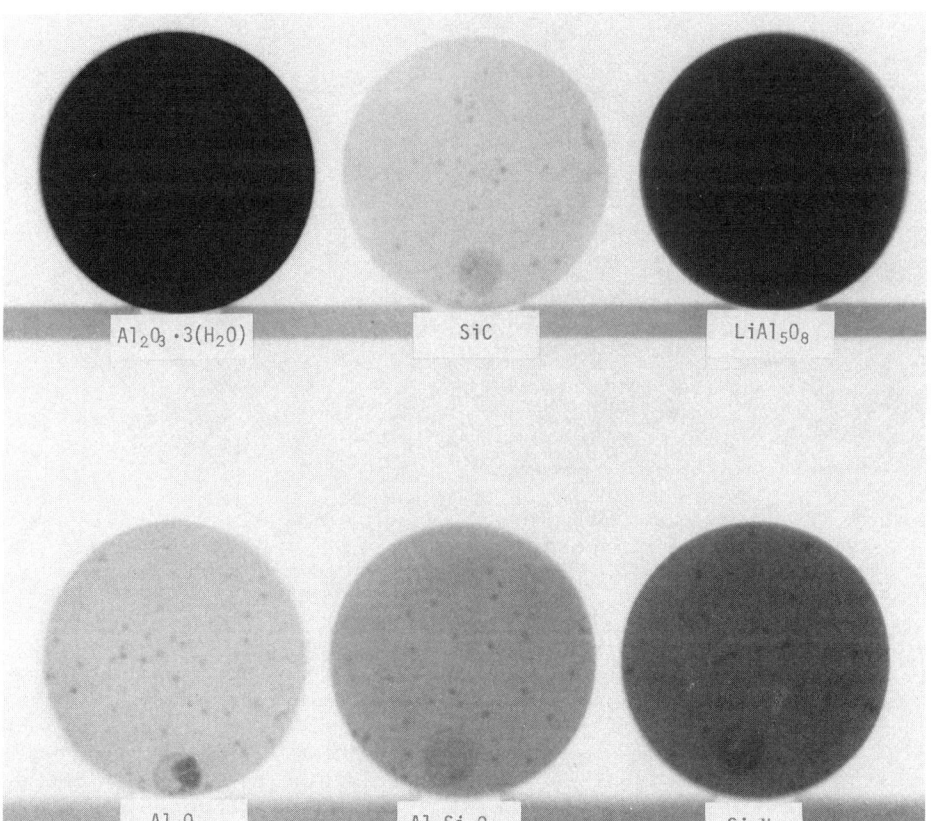

Fig 5 Positive print of a film neutron radiograph showing six different ceramic disks; each disk was seeded with 750 μm (30 mil) diameter polystyrene spheres. Courtesy of Materials Technology Laboratory

absorption by lithium. The polystyrene beads ($[-CH_2CH(C_6H_5)-]_n$), which are meant to simulate polymer inclusions, are readily detectable because of scattering by hydrogen.

Although this technique is capable of detecting porosity and cracks, the applications with the greatest potential are evaluations of injection molding, binder removal, and drying processes, which inherently contain hydrogen-rich constituents (that is, water or organic polymers). These processes, which are difficult to control, are important to the efforts of the ceramic industry to develop and utilize near-net-shape fabrication techniques. The unique ability of neutron radiography to image hydrogen in a dense solid matrix makes it a unique tool for studying, as well as controlling, these processes. Although these applications are not yet fully developed, feasibility has been demonstrated (Ref 8).

Ultrasonics

Ultrasonics is a versatile and relatively low-cost NDE technique for evaluating surface and bulk conditions in materials and structures. It can detect and characterize flaws and material conditions that are not possible using other techniques. Ultrasonic waves can be generated in, and made to propagate through, solid structures in a variety of wave modes and directions.

The fundamental measurable wave propagation parameters, such as attenuation and velocity, are directly determined by the elastic properties of the material. Information about material conditions, such as porosity and density, can be inferred by measuring these parameters. Additionally, ultrasonic waves are partially reflected or scattered at internal discontinuities, such as voids, cracks, and inclusions. These reflections, or echoes, can be detected, and will reveal the presence of these conditions, their location, and something about their size, shape, and/or orientation. Even in cases where ultrasonics cannot determine all the desired information, it can be valuable as a complement to other techniques.

The two main factors that make ultrasonic NDE of ceramics more difficult are the very small size of the critical flaws that must be detected and characterized and the typically high ultrasonic velocities, compared to those of metals. These factors can combine to force the use of focused transducers to concentrate the ultrasonic beam in a small volume and utilize higher ultrasonic wave frequencies than are normally required in metals applications, in order to resolve small defects. For example, critical defects in ceramics are often on the order of 100 μm (4 mils) or smaller, which means that ultrasonic waves with comparably sized wavelengths are necessary. This, in turn,

dictates the wave frequency that must be used. For a typical ceramic material such as sintered silicon carbide with a longitudinal wave velocity, v, of 12 mm/μs (0.5 in./μs), achieving a wavelength, λ, of 0.1 mm (4 mils) requires a frequency, f, of

$$f = \frac{v}{\lambda} = \frac{12 \text{ mm/μs}}{0.1 \text{ mm}} = 120 \text{ MHz} \qquad \text{(Eq 9)}$$

Frequencies lower than this can be useful for evaluating the bulk properties of ceramics. However, for applications in which high spatial resolution is required, frequencies of 100 MHz or higher are necessary, although the increasing attenuation of ultrasonic waves at higher frequencies limits the usable depth of penetration.

Measurement of Bulk Porosity. The physical properties of ceramic components fabricated from powders are generally greatly affected by the porosity of the material. Ultrasonic propagation measurements are capable of characterizing ceramic material porosity even though the individual pores are considerably smaller than the ultrasonic wavelength. Although individual pores are too small to be imaged (resolved) or even detected by ultrasonics, collectively, a large number of pores can affect the elastic properties of the material to the extent that the condition can be characterized by measurements of ultrasonic velocity and attenuation (Ref 27).

Ultrasonic Velocity. Procedures for measuring ultrasonic velocity in materials are well documented in ASTM E 494, "Standard Practice for Measuring Ultrasonic Velocity in Materials." Most of the techniques involve coupling a transducer to one face of a material specimen that has plane parallel faces and known thickness. The transducer is operated in a pulse-echo mode to measure the round-trip travel time of an acoustic pulse within the specimen. The practice is improved by observing several multiple round-trip echoes. For a specimen of thickness, d, for which a round-trip travel time, t, has been measured, the ultrasonic velocity, v, is given by

$$v = \frac{2d}{t} \qquad \text{(Eq 10)}$$

A linear relationship between porosity and ultrasonic velocity has been demonstrated for a wide variety of ceramic materials.

The relationship between velocity and porosity allows variations in porosity in a material specimen to be mapped with ultrasonic C-scans. Figure 6 shows a time-of-flight C-scan for a hot-pressed SiC plate that was intentionally processed to have incomplete densification (Ref 27). The pores were distributed in the shape of a disk, centrally located in the plate. The plate was sectioned transversely through the center, leaving a semicircular region of porosity in the sample. Figure 7 schematically shows the test setup and specimen, including the region of increased porosity.

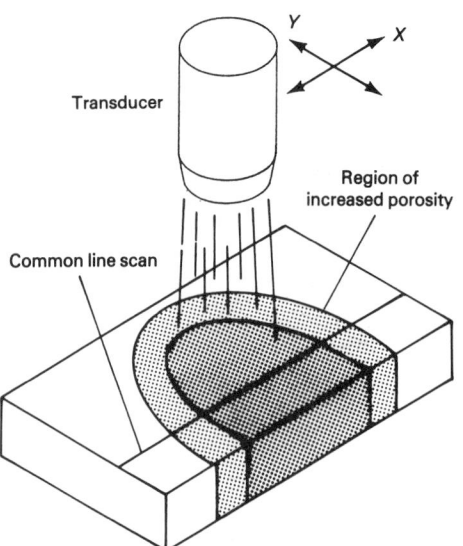

Fig 7 Schematic of SiC sample imaged in Fig 6 showing test setup and regions of porosity

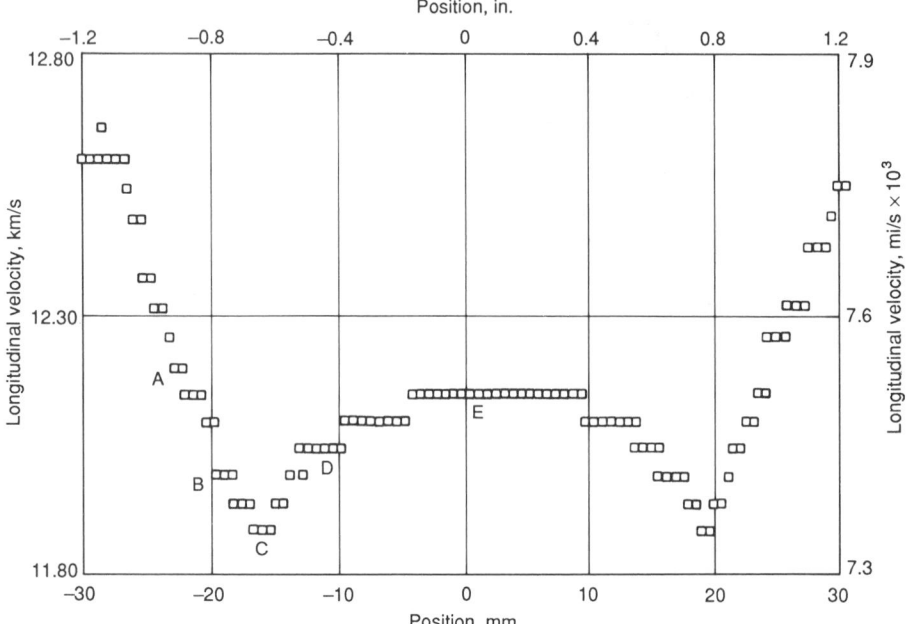

Fig 6 Time-of-flight C-scan of hot-pressed SiC plate specimen (66 × 33 × 14 mm, or 2.6 × 1.3 × 0.6 in.) depicting utility of porosity/ultrasonic velocity relationship to map porosity variations; graph shows longitudinal velocity for sample along the line shown in Fig 7. Source: Ref 27

The porosity distribution is clearly shown in the C-scan image. Each gray-level change represents an 80 ns time increment, corresponding to an approximate 3% step in velocity. For this scan, measurements were taken every 0.1 mm (4 mils) using a 50 MHz, 11.2 mm (0.45 in.) diameter, 55.9 mm (2.2 in.) water focal length, wideband transducer with the point of focus located about 6 mm (0.24 in.) from the sample surface.

The graph in Fig 6 shows the measured velocity along a line across the sample. Both the C-scan and the line scan clearly delineate the outer rim of high porosity, as well as indicate an overall reduction in velocity in the central region. These results are consistent with optically measured porosity values along the line scan. However, it should be understood that ultrasonic velocity measurements taken in this way cannot provide information about

the distribution of porosity through the thickness of the sample.

Ultrasonic Attenuation. As an ultrasonic beam propagates through a material, energy is lost by interactions with the material microstructure. Pores, voids, inclusions, and grain boundaries act as scattering sites, removing energy from the primary beam. The attenuated beam can be described by the relation

$$I = I_0 \exp(-\alpha x) \qquad \text{(Eq 11)}$$

where I is the intensity of the beam at the distance, x, in the material; I_0 is the intensity at $x = 0$, and α is the ultrasonic attenuation coefficient of the material. It is the magnitude of the attenuation coefficient and its frequency dependence that contains information about the material microstructure.

The apparent signal received by a transducer coupled to a material specimen for either a through-transmission or pulse-echo mode depends not only on the bulk attenuation in the material, but on reflection losses at the specimen surfaces and on geometrical beam spreading (diffraction) effects. The reflection losses and beam spreading effect must be accounted for in order to determine the bulk material attenuation. Methods of accomplishing this are described in detail by Papadakis (Ref 28).

A well-established pulse-echo method of measuring the ultrasonic attenuation as a function of frequency is to use a buffer rod coupled to the specimen, as shown in Fig 8. A broadband ultrasonic pulse from the transducer travels through the buffer rod. Reflection echoes from the front surface of the specimen, $(FS)_1$, and at least two multiple echoes from the back surface, B_1 and B_2, are recorded. The Fourier transform of each received waveform represents the amplitude and phase as a function of frequency and is denoted below by (f). The attenuation as a function of frequency is given by

$$\alpha(f) = \frac{1}{2X} \ln \frac{|B_1(f)| \, |R(f)|}{|B_2(f)|} \qquad \text{(Eq 12)}$$

$$R(f) = \frac{|(FS)_2 \, (f)|}{|(FS)_1 \, (f)|} \qquad \text{(Eq 13)}$$

where X is the sample thickness, $(FS)_1$ is the buffer rod echo without the sample in contact, and $(FS)_2$ is the buffer rod echo with the sample in contact.

An example of the effect of material microstructure on ultrasonic attenuation is shown in Fig 9, in which attenuation as a function of frequency is plotted for two groups of sintered α-silicon nitride samples (Ref 30). The samples are all of approximately the same density and show little difference in mea-

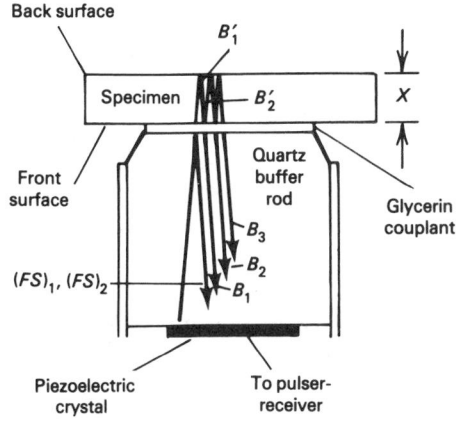

(a)

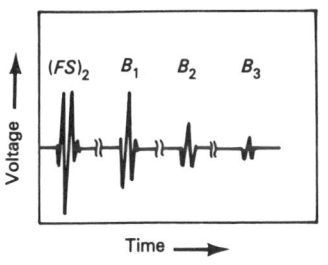

(b)

Fig 8 Pulse-echo attenuation measurement. (a) Path of front- and back-surface echoes. (b) Transducer output showing echoes. Source: Ref 29

sured ultrasonic velocity. However, the two groups of samples are significantly different in grain size and morphology. The most obvious difference is that the samples with the highest attenuation are large-grained, whereas those with the lower attenuation are

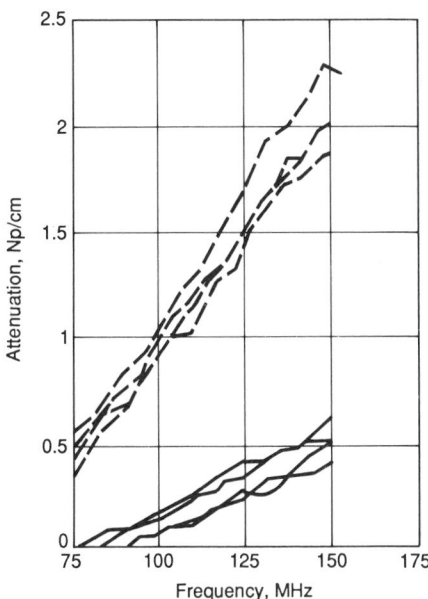

Fig 9 Attenuation versus frequency for two groups of sintered SiC samples differing in grain size. Source: Ref 30

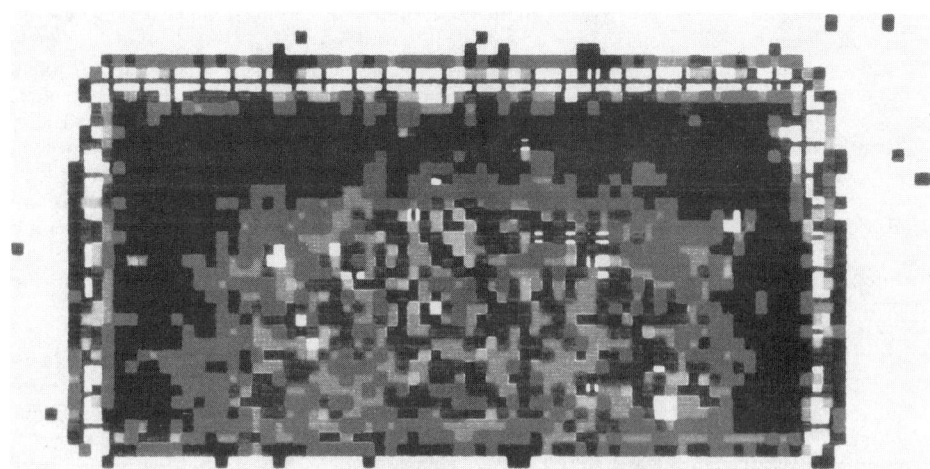

Fig 10 Back-surface echo amplitude image of SiC sample shown in Fig 7 illustrating the mapping of porosity by ultrasonic attenuation. Source: Ref 31

small-grained. Also, the less-attenuating sample has equiaxed grains, whereas the many grains in the more-attenuating sample are elongated. The void size and shape, as well as morphology, are also quite different. The example points out the difference in information obtained by ultrasonic attenuation and velocity measurements in ceramics. In general, ultrasonic velocity depends primarily on the porosity fraction, whereas in addition to porosity fraction, attenuation is highly sensitive to pore size and distribution, as well as grain size and morphology.

In many cases, it is not necessary to perform the complex calculations for ultrasonic attenuation described above to get useful information about ceramic porosity and microstructure. Ultrasonic measurements without corrections for reflection losses or beam spread, or even for frequency dependence, can give valuable information about the relative attenuation in samples. A method for determining the apparent attenuation by measuring the decay rate of multiple back reflections of longitudinal ultrasonic waves introduced into samples with flat, parallel surfaces, using the immersion technique, is described in ASTM E 664. Even simpler are C-scan measurements, in which the amplitude of the transmitted beam or the first back-surface reflection are plotted in a pseudo-color or gray scale to produce images of variations in attenuation in a single sample.

Figure 10 is a back-surface echo energy C-scan of the silicon carbide sample shown in Fig 6. For this scan, measurements were taken every 1.0 mm (0.04 in.) using a 20 MHz, 3.2 mm (0.128) diameter wideband transducer. The attenuation image is different from the velocity image because of a focusing effect that occurs in the transition region and is caused by the rapidly changing wave velocity. This effect of high apparent attenuation in the area of velocity gradients has also been observed by others (Ref 29).

Ultrasonic Coupling Techniques. Ultrasonic inspection requires some form of fluid coupling between the transducer and part being inspected in order to efficiently transmit ultrasound into and out of the part. This is normally accomplished by either completely immersing both part and transducer in water or placing the transducer and part in direct contact with a thin layer of viscous couplant material between them. Some ceramic materials, particularly highly porous components and green compacts, can absorb couplant materials and become contaminated and/or degraded. To prevent this, alternate ultrasonic coupling methods can be used.

A completely dry coupling can be obtained by simply forcing the transducer directly against the part. Kupperman and Karplus (Ref 32) successfully obtained ultrasonic coupling to green compacts with the application of direct pressure. However, the large contact pressures needed for efficient coupling can damage fragile components. Efficient coupling can be achieved with minimal pressure by placing an elastomer membrane between transducer and part. Jones et al. (Ref 33) achieved efficient shear wave coupling to ceramics with an elastomer thickness of 0.3 mm (12 mils) and a pressure less than 10 kPa (1.45 psi). Adhesive tape can also be used to protect the ceramic sample from contamination. Yamanaka et al. (Ref 34) achieved efficient coupling with both polypropylene and polyester tapes 28 μm (1.10 mils) and an adhesive thickness of 22 μm (0.87 mil). A frequency-dependent insertion loss as low as 0.7 dB at 0.5 MHz was obtained.

Roberts (Ref 35) extended this idea with a device in which a thin plastic film is affixed to the surface of a ceramic specimen by evacuating a container (in which the specimen has been placed) that is constructed of acoustically transparent film. Because the outer surface of the coupling film is open to the atmosphere, both conventional contact and immersion techniques can be used. For more critical applications, completely noncontacting measurements can be achieved using laser generation and detection of ultrasound (Ref 36, 37).

Standard Defect Specimens. Effective NDE techniques require standard reference specimens containing known defects of the same type, size, and location that the technique is required to detect. This is necessary for proper equipment calibration and performance checks and to assure that the technique itself is capable of the performance for which it is designed. Standard defects in metals often consist of machined holes or slots, which are usually not practical in ceramic materials.

A number of specialized techniques have been developed to produce ceramic reference specimens that contain known seeded defects. Sintered silicon carbide and silicon nitride specimens with seeded surface voids of diameters ranging from 40 to 165 μm (1.6 to 6.5 mils) have been successfully fabricated (Ref 38). The specimens are produced by pressing plastic microspheres into the surface of powder ceramic specimens during the initial stages of specimen fabrication. The microspheres are later burned out during heat treatment of the specimens, leaving craters, or voids, on the specimen surface.

Silicon nitride specimens that contain seeded internal inclusions of foreign phases that are representative of typical processing defects have been fabricated (Ref 16). Iron, tungsten carbide, and diamond particles ranging from 50 to 300 μm (2 to 12 mils) are placed in the interior of the specimens prior to cold and hot isostatic pressing.

Internal voids in silicon nitride and silicon carbide can be fabricated by a similar technique (Ref 39, 40). Plastic microspheres are embedded at known locations within green specimens and are later burned out to create voids ranging in size from 20 to 430 μm (0.8 to 17 mils) in sintered specimens. The voids tend to be ellipsoidal in shape, where the dimension of the void perpendicular to the specimen pressing direction is always larger than the dimension parallel to the pressing direction.

Defect Detection by Acoustic Microscopy. The ultrasonic detection and characterization of individual interior defects, such as voids, cracks, and inclusions in the size range of 20 to 100 μm (0.8 to 4 mils), in ceramics usually require the use of high-frequency, high-resolution techniques. These techniques are generally referred to as acoustic microscopy. In contrast to conventional ultrasonic inspection of metallic components, in which unfocused transducers in the frequency range from 1 to 10 MHz are used, ceramic inspection often uses focused transducers in the frequency range from 10 to 100 MHz. Although some acoustic microscopes operate at frequencies up to 1 GHz, the attenuation of structural ceramic materials generally limits the usable frequency range to 100 MHz and below.

The ultrasonic beam is generally scanned over the part being inspected to generate an image. For flaws larger than the focal spot size, flaw size and shape can be determined to some degree. Although flaws smaller than the focal spot can be detected, if two or more are spaced less than an ultrasonic wavelength apart, then they cannot be resolved. An excellent detailed description of acoustic microscopy is given in Ref 41.

Ultrasonic Focusing. Small ultrasonic focal spot sizes are usually obtained using plano-concave lenses attached to the face of a transducer. The resolution of an acoustic microscope is determined by the physical characteristics of the focal spot produced by the transducer in the test specimen. The focal spot size is determined by diffraction effects, so that point-focusing of an ultrasonic beam is never achieved. The −6 dB focal spot diameter, E_d, and depth of field, E_x, can be approximated by (Ref 42):

$$E_d = 1.22\,\lambda\,\frac{F}{D} \qquad \text{(Eq 14)}$$

$$E_x = 8.0\lambda\left(\frac{F}{D}\right)^2 \qquad \text{(Eq 15)}$$

where λ is the ultrasonic wavelength in the medium, F is the focal length of the lens in the medium, and D is the lens diameter.

To detect subsurface defects, the diameter and focal length of the lens are chosen to keep the angle of incidence of the entire beam at the part surface below the critical angle for longitudinal wave refraction. For a typical ceramic such as silicon carbide, this limits the aperture of the lens to approximately $F/4.0$. This corresponds to a focal spot diameter of 150 μm (6 mils) at 50 MHz in water. Because of the large difference between the speed of sound in water and the longitudinal wave velocity in silicon carbide, refraction at the interface greatly shortens the effective focal length in the solid. The maximum depth into silicon carbide at which the ultrasonic beam can come to a focus is about 1/8 (the ratio of the sound velocities) of the focal length in water. This and other practical considerations limit the effective depth at which flaws can be reliably detected to several millimeters.

C-Scan Acoustic Microscopy (C-SAM). Two variations of pulse-echo acoustic microscopy using focused transducers are capable of high-resolution flaw detection and materials characterization. The method most commonly used to detect internal defects in ceramics is sometimes referred to as C-SAM (Ref 41). In this technique, focused transducers in the frequency range from 10 to 100 MHz are coupled to the test part in a water-immersion tank in such a way that the focal spot is located at a desired depth below the surface of the part. Electronic gating is used to block out the reflected signal from the front surface of the part. The transducer is raster scanned over the area of the part of interest to generate an image of internal flaws based on the amplitude of the ultrasonic signal reflected from the defect. The image is displayed in either a gray-scale or pseudo-color scheme.

In a variation of the technique, the depth of field of the focused transducer is used advantageously, and the time of arrival of reflected signals is used to generate a C-scan of an object with internal defects, imaged in a gray level or false color scheme to denote flaw depth.

An example of such an image is shown in Fig 11 (Ref 31). For this scan, measurements were taken every 0.1 mm (4 mils) using a 50 MHz, 11.2 mm (0.45 in.) diameter, 55.9 mm (2.2 in.) water focal length, wideband transducer. The focal spot size in the material is approximately 200 μm (8 mils). The image shows individual voids and void clusters that are too few in number and small in concentration to be seen in the specimen velocity

Fig 11 High-resolution C-scan ultrasonic microscopic image of individual voids in SiC sample of Fig 7. Source: Ref 31

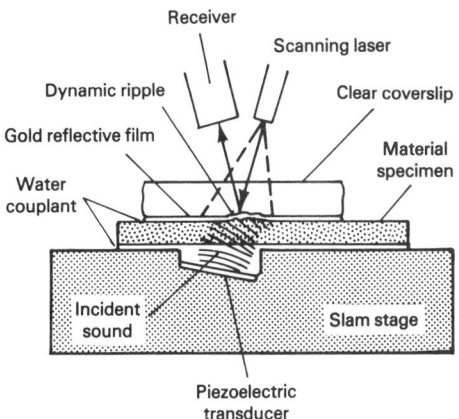

Fig 12 Scanning laser acoustic microscopy

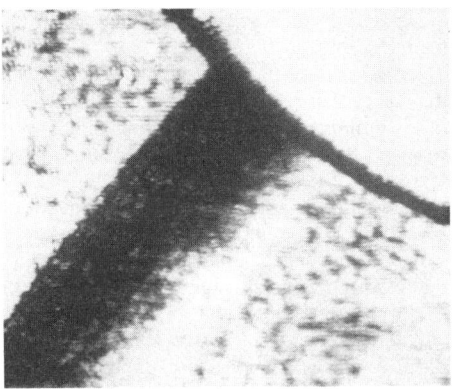

Fig 14 SLAM image showing crack in vicinity of laser-drilled hole in alumina substrate. Courtesy of Sonoscan, Inc.

and attenuation scans. The bulk of the porosity that produces the attenuation and velocity variations is too fine to be seen in the image of Fig 11.

Scanning acoustic microscopy (SAM), the other high-resolution technique, typically operates at higher frequencies (100 to 200 MHz) and is primarily used to detect surface and near-surface defects (Ref 41, 43, 44). A wide-angle acoustic lens produces ultrasonic rays beyond the critical angle to generate Rayleigh, or surface, waves in a circular pattern on the sample surface at the edge of the focused beam. Only that part of the beam that does not propagate into the bulk of the sample other than an evanescent wave that penetrates only approximately one-quarter of a wavelength, is used. The scanning acoustic microscope, when operating at very high frequencies, can achieve a resolution approaching that of an optical microscope. It has the advantage over optical microscopy of penetrating slightly into the material. In addition to generating images of surface and near-surface defects, it is also capable of generating images of material microstructure.

Scanning laser acoustic microscopy (SLAM), another useful technique, produces high-resolution images of internal discontinuities in a solid sample without the use of focused transducers (Ref 41, 45). The technique uses a high-frequency transducer (typically 100 MHz) operating in a continuous wave (CW), through-transmission mode, as shown in Fig 12.

The transducer, located beneath the sample, produces a collimated ultrasonic beam that is altered, as it propagates through the sample, by internal discontinuities, such as voids, cracks, and inclusions. A displacement pattern, or ripple, that contains information about the internal discontinuities is produced on the top surface of the sample. The surface pattern is converted to a visual magnified image by a laser beam that continuously scans an area of the surface. The reflected laser beam, which is angularly modulated by the surface ripple, is intercepted by a photodetector, which converts the modulated laser light to an electronic signal.

This signal is processed to create a real-time black and white or pseudo-color acoustic image of the test part on a video monitor at a magnification of typically 100. A clear plastic coverslip, coated on one side with a thin film of gold, is usually placed on top of the specimen and acoustically coupled to it with a thin layer of water. The coverslip provides a mirrorlike reflective surface for the laser beam, because typical ceramic surfaces are too rough to reflect sufficient light to form an image.

The capabilities of SLAM are shown in Fig 13 and 14. Depicted in Fig 13 is an alumina substrate containing laser-drilled holes at which microcracks are likely to occur. Because of the low contrast produced by tight cracks, optical inspection is very difficult and unreliable. In Fig 14, a 100 MHz SLAM acoustic image of a similar alumina substrate in the area of a laser-drilled hole clearly shows a crack. The crack appears as a dark band that extends diagonally across the image. The bright area in the top right corner of the image is the laser-drilled hole.

As another example, Roth *et al.* (Ref 38) reported 0.9 probability of detection of 100 μm (4 mil) voids at a 0.95 confidence level in polished samples of sintered silicon nitride. Similar detection reliabilities were not achieved for voids in as-fired samples with rough surfaces.

Other Techniques

Briefly described below are other inspection techniques that are applicable to ceramics to various degrees (Ref 46).

Liquid penetrants are used to detect surface-breaking flaws. The principle of this technique is to visually enhance surface defects by introducing dyes into the defect. The main drawback is that the typically rough or porous surfaces of ceramics also absorb the penetrant and tend to mask actual defects.

Acoustic emission (AE) detects the elastic stress waves emitted during the initiation and propagation of internal or surface flaws. This technique is usually associated with some mechanism or technique that initially stresses

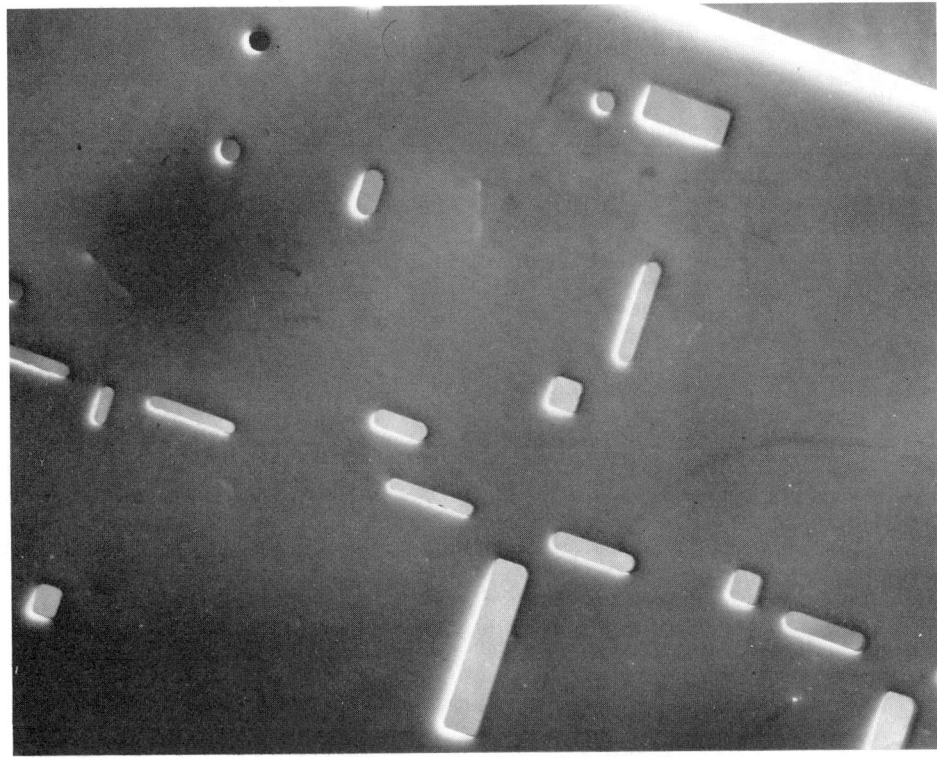

Fig 13 Alumina substrate containing laser-drilled holes. Courtesy of Sonoscan, Inc.

the material, such as thermal shock or proof testing.

Infrared inspection detects flaws or interfaces that alter normal thermal diffusion in a component. These inhomogeneities are detected through the measurement of surface temperature profiles with time.

Acoustic resonances are used to measure overall integrity of a component. Defects cause a change in the resonance spectrum of the defective component, compared to spectrum of a "good" component.

Microwaves at or above 100 GHz can be used to measure electromagnetic material properties and, at times, defects in low loss or nonconductive ceramics. However, the detection of small defects is limited by long wavelengths, which provide poor resolution. Ceramics such as carbides, borides, and ferrooxides, which are either electrically conductive or have high electromagnetic permeabilities, require much lower frequencies for inspection because of the lack of sample penetration.

Nuclear magnetic resonance (NMR), which is quite sensitive to hydrogen, is being used to spatially map hydrogen-rich binder and water distributions in green-body ceramics. This provides information quite similar to neutron computed tomography, but without radiation hazards and safety restrictions.

ACKNOWLEDGMENT

The preparation of this article was supported in part by the U.S. Department of Energy under DOE Contract No. DE-AC07–76ID01570.

REFERENCES

1. *Nondestructive Evaluation and Quality Control*, Vol 17, *Metals Handbook*, 9th ed., ASM International, 1989
2. L.E. Bryant and P. McIntire, Ed., *Radiography and Radiation Testing*, Vol 3, *Nondestructive Testing Handbook*, 2nd ed., American Society for Nondestructive Testing, 1985
3. *Nondestructive Testing*, Vol 03.03, *Annual Book of Standards*, American Society for Testing and Materials, 1990
4. K. Geobbels and H. Reiter, Non-Destructive Evaluation of Ceramic Gas Turbine Components by X-Rays and Other Methods, *Progress in Nitrogen Ceramics*, F.L. Riley, Ed., Martinus Nijhoff Publishers, 1983, p 627–634
5. T. Taylor, W.A. Ellingson, and W.D. Koenigsberg, Evaluation of Engineering Ceramics by Gamma-Ray Computed Tomography, *Proceedings of the 10th Annual Conference on Composites and Advanced Ceramic Materials*, American Ceramic Society, Jan 1986
6. H. Berger and D. Kupperman, Microradiography to Characterize Structural Ceramics, *Mater. Eval.*, Vol 43 (No. 2), Feb 1985, p 201–205
7. W.A. Ellingson and M.W. Vannier, "X-Ray Computed Tomography for Nondestructive Evaluation of Advanced Structural Ceramics," ANL-87–52, DE89 002075, Argonne National Laboratory, Sept 1988
8. N.D. Corbin, V.K. Pujari, J.J. Antal, and A.S. Marotta, A Preliminary Assessment of Neutron Radiography for Detecting Inhomogeneities in Ceramics, *Proceedings of Conference on Nondestructive Testing of High-Performance Ceramics*, American Ceramic Society, Aug 1987, p 114–127
9. D.S. Kupperman, H.B. Karplus, R.B. Poeppel, W.A. Ellington, and H. Berger, "Application of NDE Methods to Green Ceramics: Initial Results," CONF-8310205—8, DE84 006411, Argonne National Laboratory, Dec 1983
10. R.S. Peugot, Theoretical and Practical Considerations of Microfocus Radiography, *Mater. Eval.*, Vol 40 (No. 2), Feb 1982, p 151–152
11. R.W. Parish, Microfocus X-Ray Technology—A Review of Developments and Applications, *Review of Progress in Quantitative Nondestructive Evaluation*, Vol 5A, D.O. Thompson and D.E. Chimenti, Ed., Plenum Press, 1986, p 1–20
12. D.J. Cotter and W.D. Koenigsberg, Microfocus Radiography of High Performance Silicon Nitride Ceramics, *Proceedings of Conference on Nondestructive Testing of High-Performance Ceramics*, American Ceramic Society, Aug 1987, p 233–253
13. P.K. Khandelwal, Projection Microradiography of Ceramic Turbine Components, *Proceedings of Conference on Nondestructive Testing of High-Performance Ceramics*, American Ceramic Society, Aug 1987, p 254–267
14. G.Y. Baaklini and D.J. Roth, "Probability of Detection of Internal Voids in Structural Ceramics Using Microfocus Radiography," Technical Memorandum 87164, N86–13749, National Aeronautics and Space Administration, Nov 1985
15. G.Y. Baaklini, J.D. Kiser, and D.J. Roth, Radiographic Detectability Limits for Seeded Voids in Sintered Silicon Carbide and Silicon Nitride, *Adv. Ceram. Mater.*, Vol 1 (No. 1), 1986, p 43–49
16. K.E. Amin and T.P. Leo, Radiographic Detectability Limits for Seeded Defects in Both Green and Densified Silicon Nitride, *Proceedings of Conference on Nondestructive Testing of High-Performance Ceramics*, American Ceramic Society, Aug 1987, p 211–232
17. R.N. Yancey, S.J. Klima, and J.A. Smith, High Resolution Computed Tomography of Modern Ceramics, *Proceedings of Conference on Nondestructive Evaluation of Modern Ceramics*, American Society for Nondestructive Testing, July 1990, p 126–130
18. Y. Aiba, K. Oki, S. Matsuura, and M. Fujii, Development of Industrial X-Ray Computed Tomography and Its Application to Refractories, *Trans. ISIJ*, Vol 26, 1986, p 236–243
19. B.D. Sawicka, Density Gradients in Ceramics Investigated by Computed Tomography, *Proceedings of Conference on Nondestructive Evaluation of Modern Ceramics*, American Society for Nondestructive Testing, July 1990, p 63–68
20. B.R. Krohn and M.D. Silver, Optimizing Contrast Resolution in Industrial Computed Tomographic Scanners, *Mater. Eval.*, Vol 48 (No.10), Oct 1990, p 1296–1300
21. E. Segal, W.A. Ellingson, Y. Segal, and I. Zmora, A Linearization Beam-Hardening Correction Method for X-Ray Computed Tomographic Imaging of Structural Ceramics, *Review of Progress in Quantitative Nondestructive Evaluation*, Vol 6a, D.O. Thompson and D.E. Chimenti, Ed., Plenum Press, 1987, p 411–417
22. P. Engler and W.D. Friedman, Review of Dual Energy Computed Tomography Techniques, *Mater. Eval.*, Vol 48 (No. 5), May 1990, p 623–629
23. J.J. Antal, A Renaissance in Neutron Radiography via Accelerator Neutron Sources, *Materials Characterization for Systems Performance and Reliability*, J.W. McCauley and V. Weiss, Ed., Plenum Press, 1986, p 385
24. W.E. Dance and S.F. Carollo, High Sensitivity Real Time Imaging System for Reactor or Non-Reactor Neutron Radiography, *Proceedings of the 2nd World Conference on Neutron Radiography*, L.L. Person and H. Rottger, Ed., D. Reidel Publishing, 1987, p 415
25. Y. Ikeda, H. Sakai, K. Ohkudo, and G. Matsumoto, Neutron Computed Tomography with A High Speed Image Processor, *Mater. Eval.*, Vol 46 (No. 11), Oct 1988, p 1471–1476
26. P.V. Hardt and H. Rottger, Ed., *Neutron Radiography Handbook*, D. Reidel Publishing, 1981, p 1–170
27. D.C. Kunerth, K.L. Telschow, and J.B. Walter, Characterization of Porosity Distribution in Advanced Ceramics: A Comparison of Ultrasonic Methods, *Mater. Eval.*, Vol 47 (No. 5), 1989, p 571
28. E.P. Papadakis, Ultrasonic Diffraction from Single Apertures with Application to Pulse Measurements and Crystal Physics, *Physical Acoustics—Principles and Methods*, Vol 11, W.P. Mason and R.N. Thurston, Ed., Academic Press, 1975, p 151–211
29. E.R. Generazio, D.J. Roth, and G.Y.

Baaklini, Acoustic Imaging of Subtle Porosity Variations in Ceramics, *Mater. Eval.*, Vol 46 (No. 10), 1988, p 1338

30. S.J. Klima, NDE of Advanced Ceramics, *Mater. Eval.*, Vol 44 (No. 5), 1986, p 571

31. D.C. Kunerth and K.L. Telschow, "Advanced Ultrasonic NDE Methods for Characterizing Porosity in SiC," Presented at The Gas Turbine Conference and Exhibition, American Society of Mechanical Engineers, 31 May to 4 June 1987

32. D.S. Kupperman and H.B. Karplus, Ultrasonic Wave Propagation Characteristics of Green Ceramics, *Am. Ceram. Soc. Bull.*, Vol 63 (No. 12), 1984, p 1505–1509

33. M.P. Jones, G.V. Blessing, and C.R. Robbins, Dry-Coupled Ultrasonic Elasticity Measurements of Sintered Ceramics and Their Green States, *Mater. Eval.*, Vol 44 (No. 7), 1986, p 859

34. K. Yamanaka, C.K. Jen, C. Neron, and J.F. Bussiere, Improved Ultrasonic Evaluation of Green-State Ceramics with the Use of a Surface-Bonded Adhesive Tape, *Mater. Eval.*, Vol 47 (No. 7), 1989, p 828

35. R.A. Roberts, A Dry-Contact Coupling Technique for Ultrasonic Nondestructive Evaluation of Green-State Ceramics, *Mater. Eval.*, Vol 46 (No. 6), 1988, p 758

36. P. Cielo, X. Maldague, S. Johar, and B. Lauzon, Some Laser-Based Techniques for the Characterization of Sintered Ceramics, *Mater. Eval.*, Vol 44 (No. 6), 1986, p 770

37. J.F. Bussiere, Nondestructive Materials Characterization: An Important Tool in Manufacturing, *Proceedings of the 11th World Conference on Nondestructive Testing*, Vol 3, Nov 1985, p 1550–1564

38. D.J. Roth, S.J. Klima, J.D. Kiser, and G.Y. Baaklini, Reliability of Void Detection in Structural Ceramics by Use of Scanning Laser Acoustic Microscopy, *Mater. Eval.*, Vol 44 (No. 6), 1986, p 762

39. D.J. Roth, E.R. Generazio, and G.Y. Baaklini, Quantitative Void Characterization in Structural Ceramics by Use of Scanning Laser Acoustic Microscopy, *Mater. Eval.*, Vol 45 (No. 8), 1987, p 958

40. G.Y. Baaklini and D.J. Roth, Probability of Detection of Internal Voids in Structural Ceramics Using Microfocus Radiography, *J. Mater. Sci.*, 1986, p 456–467

41. L.W. Kessler, Acoustic Microscopy, *Nondestructive Evaluation and Quality Control*, Vol 17, *Metals Handbook*, 9th ed., ASM International, p 465

42. R.S. Gilmore, J.R.M. Vierth, and L.I. Halberg, Materials Characterization by Acoustic Microscopy, *Materials Technology, Journal of General Electric's Engineered Materials*, General Electric Company, Spring 1981

43. R.S. Gilmore, K.C. Tam, J.D. Young, and D.R. Howard, Acoustic Microscopy from 10 to 100 MHZ for Industrial Applications, *Phil. Trans. R. Soc.*, Vol A320, 1986, p 215–235

44. R.S. Gilmore, R.E. Joynson, C.R. Trzaskos, and J.D. Young, Acoustic Microscopy: Materials Art and Materials Science, *Review of Progress in Quantitative Nondestructive Evaluation*, Vol 6A, D.O. Thompson and D.E. Chimenti, Ed., Plenum Press, 1987, p 553

45. D.S. Kupperman, L. Pahais, D. Yuhas, and T.E. McGraw, Acoustic Microscopy Techniques for Structural Ceramics, *Am. Ceram. Soc. Bull.*, Vol 59, Aug 1980, p 814–816

46. D.R. Johnson, R.W. McClung, M.A. Janney, and W.M. Hanusiak, "Needs Assessment for Nondestructive Testing and Materials Characterization for Improved Reliability in Structural Ceramics for Heat Engines," ORNL/TM-10354, Oak Ridge National Laboratory, 1987

Failure Analysis

Chairman: John J. Mecholsky, University of Florida

Introduction

THIS SECTION is intended as a guide for the field engineer to determine the causes of failure in ceramic components. Ceramics here is a general descriptive term which includes glasses, polycrystalline, and single-crystal materials. The principles described within this Section are equally applicable to all these materials except where noted for specific characteristics (for example, the elastic anisotropy of single crystals). Since this field requires experience as well as specific principles, it is recommended that the reader study the references suggested and contact individuals for specific information.

The Section consists of four major subject areas: (1) overview, (2) terminology, (3) principles of failure analysis, and (4) case histories. The "Overview of Failure Analysis" article is intended to present all the principles involved in failure analysis without the supporting detail, to serve as a map of the rest of the Section. The article "Special Terminology Used in Fractography" is included because not every practitioner agrees on the same terminology for specific observed phenomena. While almost everyone familiar with failure analysis is aware of the different terminology, it is important for the novice to be aware of these differences. More important, the physical manifestations of these features and their origins should be understood. These matters will be described in detail in the article "Descriptive Fractography," which is part of the third subsection.

The Principles of Failure Analysis subsection covers three subjects: (1) descriptive fractography, (2) applied fracture mechanics, and (3) quantitative fracture surface analysis. The "Descriptive Fractography" article serves as a catalog of typical fracture origins and patterns for the reader, to help in deciding the appropriate analysis to apply. The article "Applied Fracture Mechanics" presents the available analyses (or references to them) for analysis of failure origins. The "Quantitative Fracture Surface Analysis" article is also intended as an analysis guide, for using the appearance of fracture surface characteristics surrounding the failure origin to quantitatively determine more information about the stress state and character of the failure. From the size, shape, and location of the fracture-initiating defect as well as from the size, shape, and location of the surrounding topography, the details of the failure history can be determined. In addition to describing the general principles, the article discusses the effects of:

- Microstructure (large grains, porosity, and so on)
- Stress-corrosion processes (slow crack growth due to stress-aided chemical reactions at the crack tip)
- Residual stresses (both local and global stresses which are present before the application of load)
- High-temperature effects such as oxidation and corrosion, to provide examples and to demonstrate methodology

The final subsection, Case Histories, is intended to apply all the principles described in the preceding subsection to the failure analysis of engineering components, so that the reader can appreciate not only the complexity of analysis but also the benefits of applying these principles. The examples have been selected to present general areas of interest rather than very specific examples that are not applicable elsewhere. Obviously, this subsection cannot be an exhaustive study of case histories. However, it should serve as a guide to applying the principles described in the other subsections.

Overview of Failure Analysis

V.D. Frechette, New York State College of Ceramics at Alfred University

FAILURE ANALYSIS, like any analysis, is directed toward acquiring specified information. For the field engineer this may include the suitability of a particular material for an intended application, the appropriateness of the design of a part, the competence with which the various steps of its manufacture have been performed, any abuse suffered by it in packing and transportation, or the severity of service under which failure has occurred. The test engineer will employ failure analysis to ensure that his test was properly designed and conducted to stress the part in such a way as to permit calculation of the relevant parameters.

In the extreme, failure analysis may be occasioned by the necessity to assign legal liability for a service failure. But whatever the aegis for failure analysis, it may be wise to satisfy nearly identical requirements with respect to positive identification of the evidence, specification of the methods employed in analysis, reasoned interpretation of observations, and careful records at each stage of the analysis. Certainly the questions submitted by the requester must be addressed, but questions that may arise later can often be anticipated as well.

This article briefly reviews the objectives and stages associated with failure analysis. More detailed information including specific case histories, can be found in the articles that follow in this Section.

Stages of an Analysis

Collection of Data. Description of the failed part should be specific enough to enable unambiguous identification. It typically includes the materials of construction, the intended function of the part, any serial numbers or other markings that may be present and specification of the person submitting it. The condition of the evidence, including the presence of any soil or other signs of past use, should be noted, together with a broad description of the failure itself.

Replicas. Since destructive treatments are often forbidden, and because the failed object may be unmanageably large, replication of critical fracture details may be useful. The techniques were first borrowed from electron microscopy, but the requirements are gener-

ally very much less severe. For example, replication materials may not be required to be transparent to an electron beam, if examination is to be limited to optical microscopy and scanning electron microscopy. For the same reason, replicas may not be limited with respect to thickness or restricted to those materials that have no details of their own on a scale that would confuse the microscope image. Replicas are especially useful in examining white ceramics, where the markings on fracture surfaces are often obscured by light scattered from subsurface details. (Occasionally this may be a disadvantage, however, as when subsurface stress modifiers are involved.) Finally, examination of plastic replicas in the optical microscope at high magnification poses a lesser danger to the objective lenses than does the ceramic specimen itself.

Photographs are useful in documenting the findings. They should, of course, include a reference scale as well as specification of the optical components of the photographic system, such as crossed polars and sensitive tint plates, where applicable, and mode of operation (in the case of scanning electron microscopy). Stereographic pairs of photos are helpful, particularly when the part under investigation cannot accompany the report. "Over-the-edge" photos of the origin showing the fracture surface at the same time as the neighboring free surface of the part, are often useful when it is desirable to relate the origin to abuse damage on the adjacent surface. Because of its great depth of focus, the scanning electron microscope is generally used to obtain these, but the optical microscope can also be effective at lower magnifications.

The history of the part and of the failure is useful for identifying the questions that may be raised, but it is of questionable value in the analysis itself. Witnesses—including factory personnel—are often unaware of all the details of the incident, and indeed failure analysis finds one of its greatest uses in identifying the unnoticed step in manufacture or in service. In addition, witnesses at law may have been led through their story so often that events are confused in their minds and in their testimony.

Part Inspection. Failure analysis begins in every case with a general inspection of the

failed part, an inspection that can be aided in routine cases by a checklist of features to be watched for. The relevant standards, both governmental and industrial, will suggest items to include. Irregularities of material and fabrication are noted, together with those markings that may serve to identify the part.

The fragments are then examined individually and each is marked with the direction and speed of cracking, the surface at which the crack front leads, that is, the surface under greatest tension as the crack front passes, the presence of cantilever stress, shear stress, arrests (also called dwell marks), flaking, Hertzian cone cracking, defects of material or fabrication, and signs of abuse.

The origin, that is, the site from which cracking originated, must be precisely located and it deserves particular attention. (Obviously if the crack had not started, there would have been no failure.) Its location and its topography can serve to distinguish failure due to a manufacturing defect from one introduced by abuse in service, or its appearance may indicate a simple case of failure under excessive loading. The dimensions of the origin flaw should be specified, although attempts to relate these quantitatively to the stress at failure have been less than entirely successful. More satisfactory means of estimating the stress at failure will be dealt with in a later section.

The fragments may then be reassembled, and the sequence of crack spread observed (see the case history in this Section of the Handbook on "Glass-Ceramic Cookware Failure Analysis"). The entire history of cracking, including the cracking pattern and the notations describing the information seen on the crack-generated surfaces, can serve to identify the kind and magnitude of stress responsible for failure and the role of such defects of material, design, manufacture, or abuse as may have contributed.

Unless the results constitute a pattern that is familiar to the analyst, it may be wise for him to undertake controlled experiments to reproduce it. Bursting pressure can be generated in duplicate containers by filling them completely with cold water or other fluid having high thermal expansion, sealing them, and warming them. Center heating (this is sometimes responsible for the failure of large windows from the heat of the sun and substrates or other plates subjected to localized heating) can be arranged by smoking the surface of such a sheet with soot, wiping the edge areas clean, and exposing it to a high-wattage lamp. Thermal shock, impact, and mechanical stresses can be applied by the standard materials-testing methods. It is profitable to be familiar with typical origins, and so specimens should be prepared by impacting, chill-checking, scratching, gouging, and otherwise abusing them to generate the common "seeds of failure." Typical manufacturing flaws should also be collected for study and reference.

The formal report should include all of the elements described, including the identity of any persons participating "under the analyst's direction and control." Because years may pass before there is need to review the matter, it pays to be thorough and explicit. Such a report commonly includes an abstract or summary that identifies the final conclusions of the analyst without requiring the reader to go through all of the details that were involved.

SELECTED REFERENCES

- T.J. Davies and I. Brough, General Practice in Failure Analysis, *Metals Handbook*, Vol 11, 9th ed., ASM International, 1986, p 15–46
- V.D. Fréchette, *Failure Analysis of Brittle Materials*, American Ceramic Society, 1990
- D.W. Richerson, Failure Analysis of Ceramics, *Metals Handbook*, Vol 11, 9th ed., ASM International, 1986, p 744–757

Special Terminology Used in Fractography

Compiled by John S. Wasylyk, Consultant

TERMS used to describe the features observed on the fracture surfaces of brittle materials have come to be used in different ways as a result of the varied interests and backgrounds of researchers in the ceramics and glasses field. Although a single system of nomenclature would be invaluable in aiding communication among fractographers, to date there is no universal acceptance of standard nomenclature. For this reason, this article on terminology associated with fractography is based on four primary sources, each of which is identifiable by the bracketed descriptor that accompanies each term. These include:

- [ASTM] ASTM Designation C 162, "Standard Terminology of Glass and Glass Products." Published under the jurisdiction of ASTM Committee C-14 on Glass and Glass Products, which is the direct responsibility of Subcommittee C-14.01 on Nomenclature and Definitions
- [FRECH.] V.D. Frechette, Markings on Crack Surfaces of Brittle Materials: A Suggested Unified Nomenclature, *Fractography of Ceramic and Metal Failures*, Special Technical Publication 827, ASTM, 1982, p 104–109
- [MECH.] V.J. Mecholsky, Jr., Department of Materials Science and Engineering, University of Florida, private communication, 1991
- [QUINN] G.D. Quinn, J.L. Swab, and M.J. Slavin, "A Proposed Standard Practice for Fractographic Analysis of Monolithic Advanced Ceramics," U.S. Army Materials Technology Laboratory, Watertown, MA, Nov 1990

Of the aforementioned sources, the ASTM-based material is specific to glasses. The terms taken from Frechette, Mecholsky, and Quinn *et al.* describe polycrystalline materials, with the emphasis of the latter source being fracture origins (both intrinsic volume and surface flaws).

These few pages should identify most of the common terms used in fractography. Additional fracture terms are discussed in the article "Descriptive Fractography," which follows in this Section.

anneal [ASTM]. To prevent or remove objectionable stresses in glassware by controlled cooling from a suitable temperature.

annealing [ASTM]. A controlled cooling process for glass designed to reduce thermal residual stress to a commercially acceptable level, and, in some cases, modify structure.

agglomerate [QUINN]. The clustering together of a few to many particles, whiskers, or fibers, or a combination thereof, into a larger solid mass.

arrest line [FRECH.]. Rib mark defining the crack front shape of an arrested crack prior to resumption of crack spread under an altered stress configuration. The duration of rest may be long to infinitesimal. See also *rib mark*.

arrest mark. See *dwell mark* [ASTM].

bending stress [ASTM]. A stress system that simultaneously imposes a compressive component at one surface, graduating to an imposed tensile component at the opposite surface of a glass section.

bifurcation [MECH.]. The separation of a material into two sections.

bump check. See *percussion cone* [ASTM].

butterfly bruise. See *percussion cone* [ASTM].

chatter sleek. See *frictive track* [ASTM].

cleavage crack [ASTM]. Damage produced by the translation of a hard, sharp object across a glass surface. This fracture system typically includes a plastically deformed groove on the damaged surface, together with median and lateral cracks emanating from this groove.

cleavage crack (crystalline) [MECH.]. A crack which proceeds across the grain, that is, a transgranular crack in a single crystal or in a single grain of a polycrystalline material.

contact stress [ASTM]. The tensile stress component imposed at a glass surface immediately surrounding the contact area between the glass surface and an object generating a locally applied force.

crack [QUINN]. A line of fracture without complete separation.

crack branching [MECH.]. The separation of a material into two or more segments.

crescent crack [ASTM]. Damage having the appearance of a crescent, produced in a glass surface by the frictive translation of a hard, blunt object across the glass surface. The crescent shape is concave toward the direction of translation on the damaged surface.

dwell mark [ASTM]. A fracture surface marking which resembles a pronounced ripple mark, the presence of which indicates that the fracture paused at the location of the dwell mark for some indeterminable length of time; also known as arrest mark.

fabrication traces [FRECH.]. Anomalous markings which may appear on crack surfaces where the developing crack encounters regions of unusual weakness, density, or elastic modulus, introduced, usually inadvertently, during fabrication.

feather [ASTM]. A striation having the appearance of a feather. See also *striation* [ASTM].

fine hackle. See *hackle* [ASTM].

flexure stress [ASTM]. The tensile component of the bending stress produced on the surface of a glass section opposite to that experiencing a locally impinging force.

forking [ASTM]. A phenomenon in which a propagating fracture in a fracture system branches into two or more new fractures, each separated from its immediate neighbor by an acute angle.

fracture surface markings [ASTM]. Fracture surface features that may be used to determine the fracture origin location and the nature of the stress that produced the fracture.

fracture system [ASTM]. That family of related fracture surfaces lying within an object, having a common cause and origin.

frictive track [ASTM]. A series of crescent cracks lying along a common axis, paralleling the direction of frictive contact; also known as a chatter sleek.

frosted area. See *hackle* [ASTM].

gray area. See *hackle* [ASTM].

hackle [ASTM]. A finely structured fracture surface marking giving a matte or roughened appearance to the surface, having varying degrees of coarseness. Finely structured hackle is variously known as fine hackle, frosted area, gray area, matte, mist and stippled area. Coarsely structured hackle is also known as *striation* [ASTM].

hackle [FRECH.]. A line on the crack surface, running parallel to the local direction of cracking, separating parallel but noncoplanar portions of the crack surface. See also *mist hackle*, *shear hackle*, *twist hackle*, and *wake hackle*.

hackle marks [ASTM]. Fine ridges on the fracture surface of the glass, parallel to the direction of propagation of the fracture.

Hertzian cone crack. See *percussion cone* [ASTM].

Hertzian stress. See *contact stress* [ASTM].

hinge stress [ASTM]. The tensile component of the bending stress generated on the same surface of a glass section, but not displaced from the site of a locally impinging force.

impact bruise. See *percussion cone* [ASTM].

inclusion [QUINN]. A foreign body from other than normal composition enclosed in the matrix.

intergranular fracture [MECH.]. The remnant of a crack which proceeded between a grain. Compare with *transgranular fracture* [MECH.].

intersection scarp [FRECH.]. A line, of any shape, which is the locus of intersection of two portions of a crack with one another. This is exemplified by intersection of a portion of a slow crack running wet with a portion not wetted. See also *transition scarp*.

lateral crack [ASTM]. A crack produced beneath and generally paralleling a glass surface during the unloading phase of mechanical contact with a hard, sharp object. See *cleavage crack*.

machining damage [QUINN]. Atypical or excessively large surface microcracks or damage resulting from the machining process; for example, striations, scratches, impact cracks. Note: Small surface and subsurface damage is intrinsic to the machining damage.

mirror region [MECH.]. The comparatively smooth region which symmetrically surrounds a fracture origin. The mirror region ends in a microscopically irregular manner at the beginning of the mist region.

median crack [ASTM]. Damage produced in glass by the static or translational contact of a hard, sharp object on the glass surface. The crack propagates into the glass perpendicular to the original surface. See *cleavage crack*.

mist hackle [FRECH.]. Markings on the surface of a crack accelerating close to the effective terminal velocity, observable first as a mist on the surface and with increasing velocity revealing a fibrous texture elongated in the direction of cracking and coarsening up to the stage at which the crack bifurcates. Velocity bifurcation or velocity

forking is the splitting of a single crack into two mature diverging cracks at or near the effective terminal velocity of about half the transverse speed of sound in the material. See also *bifurcation*.

percussion cone [ASTM]. Damage produced by contact stresses generated by mechanical contact of a hard, blunt object with a glass surface. Typically, it has the appearance of a semi-circular or circular crack on the damaged surface, propagating into the glass, flaring out with increasing depth into a cone-shaped crack; also called impact bruise, butterfly bruise, bump check, and Hertzian crack.

pit [QUINN]. A defect created by exposure to the environment, for example, corrosion, wear, or thermal cycling.

pore [QUINN]. (1) A small opening, void, interstice, or channel within a consolidated solid mass or agglomerate, usually larger than atomic or molecular dimensions. (2) An internal cavity which may be exposed by cutting, grinding, polishing, or fracture to become a pit, pock, or hole. (3) A discrete cavity or void in a solid material or a cavity or void larger than the typical porosity that might be present.

porous region [QUINN]. A three-dimensional zone of porosity or microporosity of higher concentration than is normally found in the matrix.

porous seam [QUINN]. A two-dimensional area of porosity or microporosity of higher concentration than is normally found in the matrix.

radial crack [MECH.]. Damage produced in brittle materials by a hard, sharp object pressed onto the surface. The resulting crack shape is semi-elliptical and generally perpendicular to the surface.

rib mark [FRECH.]. A curved line on the crack surface, usually convex in the general direction toward which the crack is running. The term is useful in referring to a mark of this shape until its specific nature is learned.

ripple mark. See *Wallner line* [ASTM].

river marks [MECH.]. Cleavage steps on individual grains of a polycrystalline material or on a single crystal. These markings spread out away from the point of origin. These are a special case of *twist hackle*.

scratch [ASTM]. Any marking or tearing of the surface produced in manufacturing or handling, appearing as though it were done by a sharp instrument.

scratch-resistant coatings [ASTM]. Coating applied to glass surfaces to reduce the effects of frictive damage. See also *frictive track*.

second-phase inhomogeneity [QUINN]. A microstructural irregularity related to the nonuniform distribution of a second phase, for example, an atypically large pocket of a second phase or a second-phase zone of composition or crystalline phase structure different than the matrix material.

sharks teeth [ASTM]. A striation consisting

of dagger-like step fractures starting at the scored edge and extending to or nearly to the compression edge.

shear hackle [FRECH.]. A *hackle* generated by interaction of a shear component with the principal tension under which the crack is running.

step fracture. See *striation* [ASTM].

stippled area. See *hackle* [ASTM].

strain [ASTM]. Elastic deformation due to stress.

stress [ASTM]. Any condition of tension or compression existing within the glass, particularly due to incomplete annealing, temperature gradient, or inhomogeneity.

striation [ASTM]. A fracture surface marking consisting of a separation of the advancing crack front into separate fracture planes. Also known as coarse hackle, step fractures, or lances. Striation may also be known as sharks teeth and whiskers.

surface void [QUINN]. A void which is located at the surface of a material and is a consequence of processing, that is, a surface reaction layer, as distinguished from a volume distributed flaw such as a *pore* or *inclusion*.

temper [ASTM]. (1) The degree of residual stress in annealed glass measured polarimetrically or by polariscopic comparison with a reference standard. (2) Term sometimes employed in referring to *tempered glass* [ASTM].

tempered glass [ASTM]. Glass that has been subjected to a thermal treatment characterized by rapid cooling to produce a compressively stressed surface layer.

thermal shock [ASTM]. A rapid change in temperature imposed on a glass body.

thermal stress [ASTM]. The stress produced by a temperature differential within a glass body.

transgranular fracture [MECH.]. The remnant of a crack which proceeded across a grain. Compare with *intergranular fracture*.

transition scarp [FRECH.]. A *rib mark* generated when a crack changes from one mode of growth to another, as when a wet crack accelerates abruptly from Region II (plateau) to Region III (dry) of a crack acceleration curve.

twist hackle [FRECH.]. A *hackle* that separates portions of the crack surface, each of which has rotated from the original crack plane in response to a twist in the axis of principal tension. In a single crystal, a twist hackle separates portions of the crack surface, each of which follows the same cleavage plane, the normal to the cleavage plane being inclined to the principal tension. In a bicrystal or polycrystalline material, a hackle is initiated at a twist grain boundary.

wake hackle [FRECH.]. A hackle line extending from a singularity at the crack front in the direction of cracking, as upon encounter with an *inclusion*.

Wallner line [ASTM]. A fracture surface

marking, having a wavelike profile in the fracture surface. Such marks frequently appear as a series of curved lines, indicating the direction of propagation of the fracture from the concave to the convex side of a given Wallner line. (synonymous with ripple mark).

Wallner lines [FRECH.]. Rib marks with wavelike contour caused by temporary excursion of the crack front out of plane in response to a tilt in the axis of principal tension induced by an elastic pulse. The Wallner line is the locus of interception of the spreading pulse with successive points along the running crack front. There are three classifications of Wallner lines. For a primary Wallner line, the elastic pulse is generated by the encounter of some portion of the crack front with a singularity in the specimen, such as a discontinuity at the free surface or within the specimen, or any localized stress field or elastic discontinuity. For a secondary Wallner line, the elastic pulse is generated by a discontinuity in the progress of the crack front, typically mist-hackle details. For a tertiary Wallner line, the elastic pulse, or train of pulses, is generated from outside the crack front by mechanical shock or by vibration of the specimen resulting from stress release by cracking.

whiskers. See *striation* [ASTM].

Descriptive Fractography

James R. Varner, New York State College of Ceramics, Alfred University

FRACTOGRAPHY is the examination of surfaces created by fracture and the interpretation of the fracture markings seen on these surfaces. It can be qualitative (for example, the direction of fracture propagation), or it can be quantitative (for example, the stress at the time of failure). This article will only deal with qualitative (descriptive) fractography; quantitative fractography is treated in the article "Quantitative Fracture Surface Analysis" in this Section of the Handbook. The formation of fracture markings is explained and examples given. Representative fracture surfaces of glasses, single crystals, and polycrystalline ceramics are shown. Fracture origins from a variety of causes are presented.

The science of fractography of glasses and ceramics has developed to the point where there is a considerable body of literature on the subject. It is beyond the scope of this article to provide an exhaustive bibliography, but several references are given which cover various aspects of descriptive fractography in detail (Ref 1–8).

Uses and Significance of Fractography

Fractography can be used to do many things relating to the fracture process, including:

- Locating fracture origins
- Determining direction of crack propagation
- Learning the sequence of crack propagation
- Deducing the stress state at the time of fracture
- Observing interactions between crack fronts and inclusions, grains, and so forth

The power of the technique in providing vital information about fracture makes it useful in understanding failure in production or in service. Fractography should be an integral part of strength testing, which means it is important in materials research and quality control. In-service failures often raise the issue of product liability, and fractography can provide factual information for comparison with eyewitness accounts. In short, fractography, both descriptive and quantitative, is essential in developing and producing glasses and ceramics, and in using them in any situation where stress-induced failure occurs.

Techniques of Fractography

The examination of fracture-exposed surfaces normally should be done using a sequence of observations starting at low magnification: unaided eye, hand lens, stereographic optical microscope, normal optical microscope, scanning electron microscope (SEM), transmission electron microscope (TEM). Not all of these steps will be necessary for every examination. Sometimes specialized techniques, such as replicating fracture surfaces or applying a reflecting coating, need to be employed.

Reconstructing a broken part by putting the pieces back together shows the overall crack pattern. This, in turn, usually reveals the location of the origin(s), and provides information on such things as the stress state at the time of failure and the sequence of crack development. When examining the origin, both fracture surfaces should be looked at, since the key visual clue, such as an inclusion, may be on only one of the surfaces. Also, post-fracture damage often happens right at the origin, causing part or all of the origin to be missing on one or both pieces. Looking at both will make this kind of post-fracture damage more obvious. Adjustment of the illumination when using an optical microscope is critical to seeing fracture markings. In the SEM, imaging techniques need to be matched to the specimen for best results, and stereographic pairs are extremely helpful in revealing the surface topography. Fractography should be well documented with notes, sketches, and photographs.

Fracture Markings

A fracture tends to propagate as a spreading front, much like a wave front, that is, as a line. Markings are formed when something happens to disturb the propagation of this front, for example, if the fracture stops, or if the local stress field changes direction. Typical markings on the fracture surfaces of glasses and ceramics are illustrated and explained in this Section.

Brittle materials, such as glasses and ceramics, fail in tension. Even in those situations where the applied stresses are not ten-sile, resolved tensile stresses act on discontinuities in the material (cracks, pores, and so forth). The absence of significant plastic deformation means that the only way for the material to relieve the stress concentration is through fracture. As a fracture moves through a glass or a ceramic, its propagation direction and velocity are affected by the stresses it encounters.

Central to understanding fracture markings and their interpretation in brittle materials is the so-called law of normal tension, which states that the fracture propagates normal to the direction of the local principal tension. In isotropic materials like glasses, the law of normal tension governs crack propagation. In single crystals and in polycrystalline materials at the microscopic level, strong cleavage tendencies will also play a role, particularly in the direction of propagation.

Mirror, Mist, and Velocity Hackle. Figure 1 shows a portion of the fracture surface of a glass rod broken in bending. The fracture origin is not seen in detail, but it is on the edge of the fracture surface, at the center of the smooth area. The fracture surface close to the origin is very smooth, and, when the illumination is just right, it will reflect light much like a mirror. Hence, this region is called the fracture mirror. A smooth surface like this indicates that the fracture was moving relatively slowly as a single spreading front. In fact, when glass is broken at low stress, for example, in many thermal failures, the entire fracture-exposed surfaces may be mirrorlike with no, or very few, markings.

In this example, typical of a strength test, the fracture was accelerating as it moved away from the origin. There is a considerable amount of stored elastic energy in the specimen at the time of failure, and much of this energy is used to create the fracture surface. However, when the velocity approaches the maximum for the material, the single crack front begins to break up on a microscopic scale. This is seen on the fracture surface as a roughening of the surface, giving it a misty appearance. This microscopic breaking up of the crack front allows an increased dissipation of the stored energy. This is often not sufficient, and the single crack branches (splits) into two cracks. The fracture at this point has reached its maximum velocity (about one-half

Fig 1 Fracture surface of a glass rod broken in bending. Fracture origin is at left; nearly semicircular region is the fracture mirror, bordered by mist and velocity hackle. Note the second region of mist and velocity hackle. SEM. 47×

to two-thirds the velocity of the transverse stress wave in the material). Branching is shown on the fracture surface by the presence of large, daggerlike marks called velocity hackle. In Fig 1, the mist and velocity hackle form the boundary of the fracture mirror. Velocity hackle is parallel to the direction of crack propagation.

Depending on the amount of stored elastic energy at the time of failure, the fracture may branch many times. This is the reason for the formation of the many fragments that occur in high-stress failures of glasses and ceramics. Multiple branching is shown in Fig 1, where the sequence of mist/velocity hackle occurs twice. The smooth region in between indicates the crack briefly slowed down before accelerating to the second branching.

The complexity of the fracture surface in the mist-hackle region is illustrated in Fig 2. From this photograph, it is seen that localized

crack branching and secondary cracking on a microscopic scale causes the formation of many finger-shaped markings, creating small-scale surface roughness.

The shape of the fracture mirror is influenced strongly by the stress state at the time of failure, as illustrated by the example shown in Fig 3. This glass rod was strengthened by an ion-exchange treatment, which produces a thin layer of high compression at the surface. The fracture origin is a large surface flaw, part of which is seen as the crescent-shaped feature in the mirror. The fracture evidently started at both ends of this crescent, resulting in a double mirror with a narrow mist-hackle region between the two halves. The extended parts of the mirror at the surface are due to the acceleration of the crack being hindered by the surface compressive stresses. The size of the mirror is inversely proportional to the stress at failure. For more information on this relationship, see the article "Quantitative Fracture Surface Analysis" in this Section.

Wallner lines (named for the scientist who first explained their formation) are lines on the fracture surface which are caused by disturbances of the fracture front by sonic waves. As shown in Fig 4, Wallner lines resemble waves on the surface of a still pond when a stone is dropped in. The blow which broke the glass in Fig 4 also created strong sonic waves which intersected the advancing crack front, causing momentary deviations in the direction of the local principal tension. The crack front follows these deviations, tilting out of its original plane momentarily, creating the wavy lines seen on this surface. It is important to note that Wallner lines do *not* give the shape of the fracture front, unless the sonic wave encounters the entire front at the same time, for example, when the fracture was moving slowly. However, the fracture origin is always on the *concave* side of Wall-

Fig 4 Fracture surface of a piece of glass broken by striking it with a hammer. Origin is at the lower left; the wave-like lines are Wallner lines. Optical microscope. 74×. Source: Ref 1

ner lines, and this fact can be used to determine the direction of propagation and the location of the origin.

Twist Hackle. Another form of hackle occurs when the stress field ahead of a propagating crack twists, that is, undergoes rotation about an axis normal to the crack front (parallel to the propagation direction). Obeying the law of normal tension, the fracture tries to follow this change by a corresponding twist; however, it is energetically very unfavorable for the entire crack front to twist as a single plane. Instead, the single crack front breaks up into a series of parallel, but noncoplanar, fronts which do line up with the new axis of local principal tension. A good analogy is a venetian blind, in which the slats are parallel, but not coplanar. Initially, these individual fracture segments are disconnected; but this cannot remain so for very long, and the material between the segments breaks through, forming the hackle.

On the fracture surface, twist hackle appears as very narrow, sharp daggers, such as seen in Fig 5. Twist hackle, like velocity hackle, is aligned parallel to the propagation direction. The example shown in Fig 5 is a portion of the fracture surface of a plate of glass broken in bending. The top edge of the photograph is at the surface that was in compression during the bending. The overall direction of propagation was from left to right, as can be seen from the curvature of the Wallner lines, but near the top of the picture the fracture was moving toward the top, nearly at a right angle to the compression surface. This appearance of twist hackle near the compression surface of a plate broken in bending is a characteristic feature for this kind of failure.

An examination of the other photographs in this article will reveal that twist hackle is a common occurrence on fracture surfaces of brittle materials. For example, twist hackle can be seen on the surface of the large hackle that runs diagonally through the region shown

Fig 2 High-magnification view of mist hackle on a glass fracture surface. Fracture direction is from lower left to upper right. Twist hackle (compare with Fig 5) is present at several places on the large hackle that runs diagonally through the picture. SEM. 670×

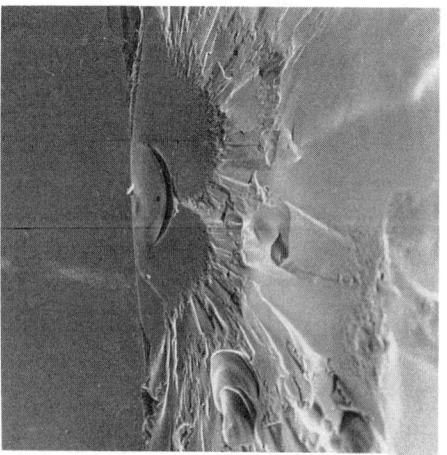

Fig 3 Double fracture mirror on the fracture surface of an ion-exchange-strengthened glass rod broken in bending. Specimen was tilted to show both the original surface (gray area at the left) and the fracture surface. Fracture started at the tips of the crescent-shaped flaw. SEM. 67×

Fig 5 Fracture surface of a glass plate broken in bending. Top of the picture is the compression side. Overall direction of propagation was from left to right, but nearly bottom to top in the region shown here. Twist hackle is seen at the top, and Wallner lines are also present. Optical microscope. 50×. Source: Ref 9

in Fig 2. Twist hackle is *not* associated with a particular velocity or range of velocities. In fact, twist hackle can form at very low velocities (for example, less than 1 m/s, or 3.3 ft/s).

Gull Wings and Wake Hackle. This set of fracture markings is actually a combination of two other markings—Wallner lines and twist hackle. The appearance of gull wings and wake hackle is shown in Fig 6. Part of the fracture surface of a glass capillary is shown; the dark circle is the hole of the capillary, and the fracture was moving from top to bottom, as can be seen from the intersecting Wallner lines. The mark at the bottom of the hole is twist hackle (called here wake hackle), and the two strong Wallner lines running away from the wake hackle are the gull wings. These markings are associated with inclusions in materials—pores, bubbles, solid particles—and are useful in determining the

direction of propagation. As seen here, the gull wings and wake hackle are on the side of the inclusion that is farthest away from the fracture origin. The wake hackle (sometimes called a "tail"), therefore, points back toward the origin.

Gull wings and wake hackle form when the fracture front encounters an inclusion. Even in the case of a pore or bubble, the front must move around the inclusion; that is, it does not simply pass through as though nothing was there. Instead, the front moves around the inclusion, effectively splitting into two fronts. When these two fronts meet on the opposite side of the inclusion, they nearly always are no longer travelling in *exactly* the same plane. Therefore, they overlap slightly, and the material separating them breaks through. This is the same mechanism that was described above for the formation of twist hackle. The sudden breaking of the material between the two fronts generates a sonic pulse which interacts with the moving fracture, creating the gull wings.

Intragranular porosity is often seen in polycrystalline ceramics. When the fracture is also intragranular, gull wings and wake hackle are generated at many of these pores. This can be the primary way in which the direction of fracture propagation is determined, thus aiding in locating and identifying the origin of failure.

An arrest line (sometimes called a rib mark) is a sharp line produced on the fracture surface when the crack stopped and subsequently restarted. Such a line can also be produced when the crack abruptly changes its plane of travel (as opposed to the wavelike, momentary change in plane associated with Wallner lines). For example, the edge of the crescent-shaped flaw in Fig 3 is an arrest line, because it marks the extent of the flaw before the load was applied which led to failure. Subsequent growth (that is, the main fracture) took place in another plane, leaving a sharp demarcation between the initial flaw and the fracture-exposed surface.

Arrest lines *always* give the shape of the fracture front, since they mark the position of the front at the time of an arrest or abrupt change in plane of travel. Recall that Wallner lines give the shape of the front only in certain circumstances. The origin will be on the concave side of arrest lines; thus, they are useful in determining fracture direction and in locating origins. They also define the boundary between a fracture-initiating flaw and the fracture surface, allowing the flaw to be identified and examined.

Scarps. In many glasses and ceramics, crack propagation at low velocities is affected by the environment, in particular by water or water vapor. If a slowly moving crack is running wet (that is, the entire crack front is exposed to water or water vapor), no unusual markings are generated. However, if part of the moving crack front is exposed to water and part is not, lines called scarps are gen-

erated at the intersections between the wet and dry parts of the front. The scarps that have been identified to date are discussed in detail (Ref 1). Two of these will be presented here.

A *deceleration scarp* (Fig 7) is formed when a crack front which has been running dry slows down enough to allow liquid water on the free surface(s) to be drawn in by capillary action. This exerts viscous drag on the edge of the crack front in contact with the water. Therefore, this part of the front is retarded with respect to the central part, which is still dry. As deceleration of the crack continues, the water reaches more and more of the front until the entire front is moving in contact with water. During this process, a line is generated on the fracture surface which represents the location on the crack front between the wet and dry parts. In Fig 7, the free surface which was wet is at the top of the picture. The nearly vertical lines are Wallner lines generated by pulses from a sonic generator. In this case, the Wallner lines *do* give the shape of the crack front. The deceleration scarp is the line running approximately horizontally across the picture. From the shape of the Wallner lines, it is clear that the outside (wet) part of the crack was lagging behind the inside (dry) part.

Sierra Scarp. A most interesting scarp (shown in Fig 8) is formed when, in the absence of water on the free surfaces, a wet-running crack accelerates, cavitates (that is, jumps abruptly in velocity, thereby leaving the water behind), decelerates, and then is overtaken by the water coming behind it. The water does not catch up to all parts of the dry crack front at the same time; instead, fingers of water touch parts of the front. At each of these positions, the water assists crack propagation, causing these parts of the front to bow out ahead of the dry portions. These bowed-out parts expand as more water catches up, until the entire crack front is again run-

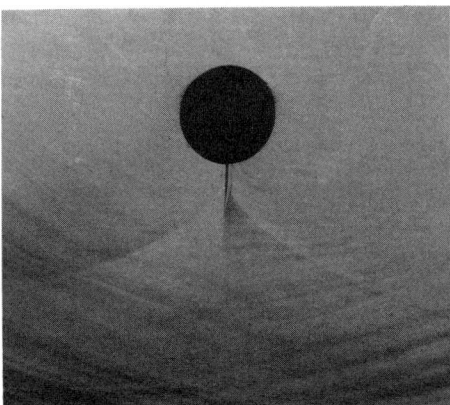

Fig 6 Fracture surface of a glass capillary broken in bending. Fracture propagated from top to bottom; gull wings and wake hackle formed when the front moved around the hole of the capillary (the black circle in the picture). Optical microscope. 50×. Source: Ref 9

Fig 7 Deceleration scarp and timing marks on the fracture surface of a glass plate. Fracture propagated from left to right. Edge of the fracture surface corresponds to the top of the picture. Optical microscope. 45×. Source: Ref 10

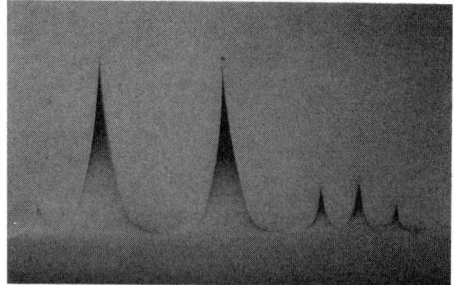

Fig 8 Sierra scarp on the fracture surface of a glass plate. Direction of fracture was from bottom to top. Optical microscope. 59×. Source: Ref 11

Fig 9 Wallner lines, mist hackle, and twist hackle on the fracture surface of a glass plate broken in bending. Fracture propagated from left to right; tensile surface at the bottom. Optical microscope. 35×. Source: Ref 1

ning wet. Again, the scarp, designated the Sierra scarp, is the locus of the boundary between the wet and dry parts of the front. Because this is happening at multiple locations, the scarps move toward each other, creating the line with multiple peaks seen in Fig 8. The name Sierra scarp was chosen because of this resemblance to a range of mountain peaks. In order for the Sierra scarp to form, a wet-running crack has to be subjected to a stress rise and decay. This may seem to be an unusual situation, but it occurs when there is fracture in a wet thermal downshock, as happens when a hot, wet drinking glass is touched by a cold metal utensil (knife, fork, and so forth).

Fracture Surfaces in Various Glass and Ceramic Materials

Glass. Fracture surfaces of glass have been shown in Fig 1 through 8. Glass is an ideal material for showing fracture markings, since it is homogeneous, isotropic, and has no microstructure to interfere with a spreading crack front. Crack propagation is governed solely by the stress state at the time of failure, whereby both applied and residual stresses (such as those induced by thermal or ion-exchange strengthening) play a role. Therefore, fracture markings can be seen easily and can be interpreted readily with respect to their implications regarding the stress state. Also, fracture surfaces will be smooth (mirrorlike) when generated by slowly moving cracks (for example, Fig 8), but will be rough and more complex when created by rapidly moving cracks (for example, Fig 1).

Two more examples of glass fracture surfaces will be presented here. The first, shown in Fig 9, is a fracture surface of a glass plate. An examination of the fracture markings reveals the stress state at the time of failure. The Wallner lines clearly indicate the crack was moving from left to right. The dark region at the bottom of the photograph is mist hackle; the short, straight lines near the top are twist hackle. Keep in mind that Wallner lines usually do not give the shape of the crack front

when it is moving at high velocity. Putting this information together, it can be deduced that the crack was moving at or near terminal velocity along the bottom edge and that the crack front was leading along this edge. On the other hand, the velocity of fracture was much lower toward the top, and the direction of propagation was nearly normal to the top edge. Therefore, there must have been a stress gradient in the plate during failure, with high tensile stresses along the bottom edge. In other words, the appearance of the fracture surface (the fracture markings) leads to the conclusion that the specimen failed in bending, which was, in fact, the case.

The second example, shown in Fig 10, presents a very different picture. The sharp lines on the fracture surface are arrest lines produced by slightly changing the direction of the applied tension at regular time intervals. The crack was moving at much less than

Fig 10 Arrest lines ("timing marks") on the surface of a glass plate broken in nearly uniform tension. Crack propagated from left to right. Only about one-half of the entire width of the fracture surface is shown in this photograph. Optical microscope. 75×. Courtesy of G. Caso, New York State College of Ceramics, Alfred University

1 m/s (3.3 ft/s), so these changes in the direction of applied tension produced slight, but sharp, changes in the plane of fracture. These deliberately induced arrest lines are called timing marks, since the velocity of propagation can be determined by measuring the distance between the arrest lines and dividing this by the time between the changes to the axis of applied stress. However, the purpose in showing this fracture surface here is to draw conclusions regarding the stress state at the time of failure from the information provided by the fracture markings. Since arrest lines show the shape of the fracture front, it is clear in this example that the crack was moving at a nearly uniform velocity along the length of the front shown. Only about one-half of the width of the fracture surface is seen in Fig 10. The slight lagging at the top is expected owing to the effect of the free surface. Therefore, the stress state at the time of failure must have been uniaxial tension. In fact, this was a double-cantilever specimen, and the stress at the crack tip is essentially uniform tension.

Single Crystals. When single crystals are broken, crack propagation is influenced by cleavage tendencies as well as by the stress state. Therefore, the appearance of fracture surfaces can be very different when cracks are moving in different crystallographic planes. Each material has to be investigated for its particular fracture behavior.

Rice points out some general observations about fracture in single crystals (Ref 4, 5). When the fracture occurs on primary cleavage planes, mist and velocity hackle occur usually without the crack branching mentioned above for glasses, or the branching occurs much farther away from the origin than would be expected in glasses. On the other hand, fractures which start on secondary cleavage planes usually branch without first exhibiting mist and velocity hackle. For cracks

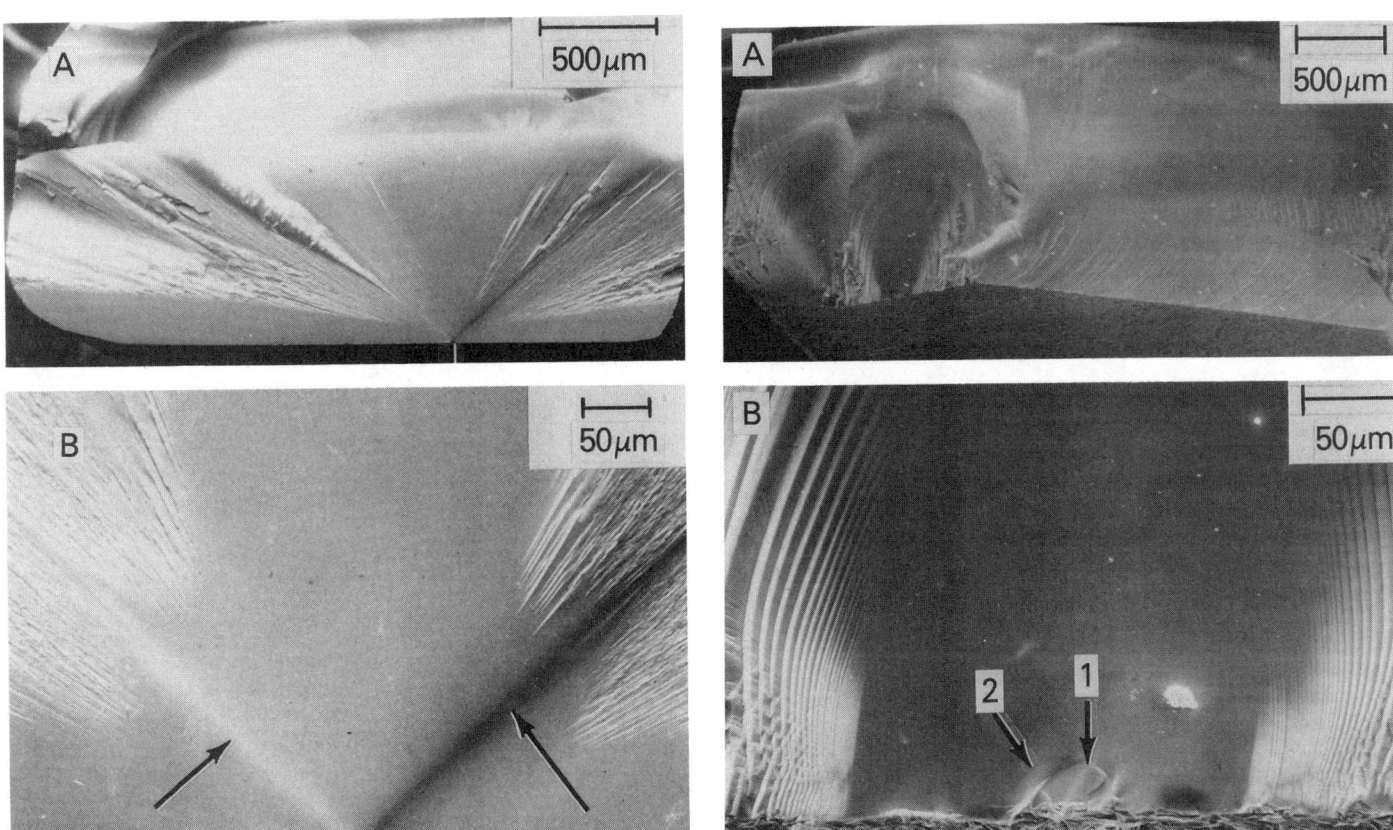

Fig 11 Views at lower and higher magnification of the fracture surface on a MgAl$_2$O$_4$ single crystal showing the mirror region with whisker-lance mist hackle (onset indicated by arrows); {100} tensile surface and ⟨100⟩ tensile axis. Source: Ref 5

Fig 12 Arc-rib mist hackle on the fracture surface of a single crystal of MgAl$_2$O$_4$; {100} tensile surface and ⟨100⟩ tensile axis. Numbers indicate the approximate extent of the fracture origin. Source: Ref 5

propagating on primary cleavage planes, mist occurs as radial ridges radiating away from the origin. These grow in density and size, with no clear distinction between mist and velocity hackle. Rice calls these markings whisker-lance mist hackle. Figure 11 shows fracture surfaces of single-crystal MgAl$_2$O$_4$ which exhibit this kind of hackle. Note that these markings occur only in certain directions, and, according to Rice, apparently only form when crack branching is energetically unfavorable, that is, when the fracture is on a primary cleavage plane.

When the fracture starts on a secondary cleavage plane, on the other hand, another type of hackle is formed—cathedral or arc-rib mist hackle, as shown in Fig 12. These marks increase in amplitude with increasing distance from the origin, until branching occurs. Arc-rib mist hackle also occurs only over certain portions of the fracture surface, that is, there are crystallographic limitations to the formation of these marks in certain directions.

The fracture surface of another crystal of MgAl$_2$O$_4$, shown in Fig 13, illustrates the influence of crystallographic orientation on fracture propagation. In this case, the tensile surface is a plane in the {110} system, and the tensile axis is ⟨100⟩. The arc-rib system

is very complex in the mirror region, and there are intersecting segments of whisker-lance mist hackle.

Figure 14 shows a broken single crystal of CaF$_2$ which has an internal origin (the crack curving to the left). This specimen was broken in bending, and this is one reason for the asymmetry of the mirror surrounding the origin. In particular, notice the absence of mist and hackle above the origin, that is, on the neutral-axis side. Also, the onset of the whisker-lance mist hackle is variable.

Space limitations prevent a more complete survey of fracture surfaces of single crystals. The reader is referred to the literature, particularly Ref 4 and 5 for more examples and more detailed information, including the effect of temperature.

Polycrystalline ceramics at the macroscopic level are isotropic; therefore, the stress state at the time of fracture governs the overall fracture patterns that develop in these materials. However, polycrystalline ceramics often present special challenges in fractography, because the microstructure strongly affects crack propagation at the microscopic level, causing perturbations all along the moving crack front. Fracture markings easily seen on glass fracture surfaces may be difficult or impossible to see on fracture sur-

faces of polycrystalline ceramics. In general, the problem is worse in coarse-grained ceramics. Markings in individual grains may be easier to see in larger grains, but the overall fracture surface is usually rougher, obscuring longer-range markings (that is, those that extend over many grains, such as Wallner lines, arrest lines, and velocity hackle).

On the other hand, fine-grained ceramics may have fracture surfaces that closely resemble those of glasses, as seen in the example shown in Fig 15. Mist, velocity hackle, and Wallner lines are clearly seen on this fracture surface of a piece of electrical porcelain. The shape of the Wallner lines and the fact that the mist and velocity hackle appear only along one edge indicate a stress gradient when failure occurred. It might be concluded that the piece failed in bending, as in Fig 9. However, this piece was a hollow cylinder which broke spontaneously while being cut with a diamond saw. Therefore, the stress gradient was present as a result of processing. In fact, the insulator was cooled too rapidly, resulting in compressive stress on the outside surface (at the top in Fig 15) and tensile stress on the inside surface (at the bottom in Fig 15).

The polycrystalline Al$_2$O$_3$ fracture surface in Fig 16 illustrates the point made above about

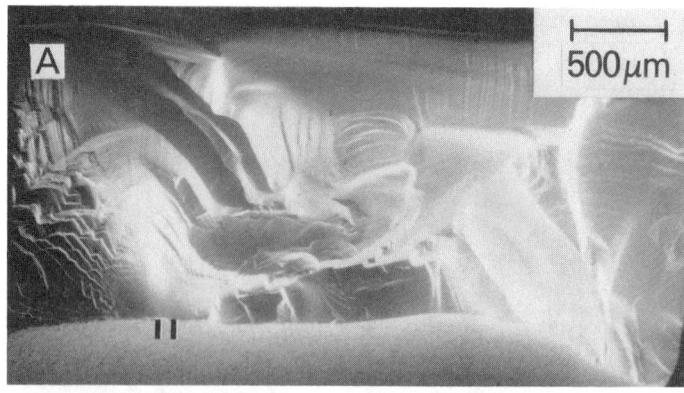

Fig 13 Complex arc-rib mirror and fracture surface of a single crystal of MgAl₂O₄. Note the intersecting segments of whisker-lance mist hackle; {110} tensile surface and ⟨100⟩ tensile axis. Source: Ref 5

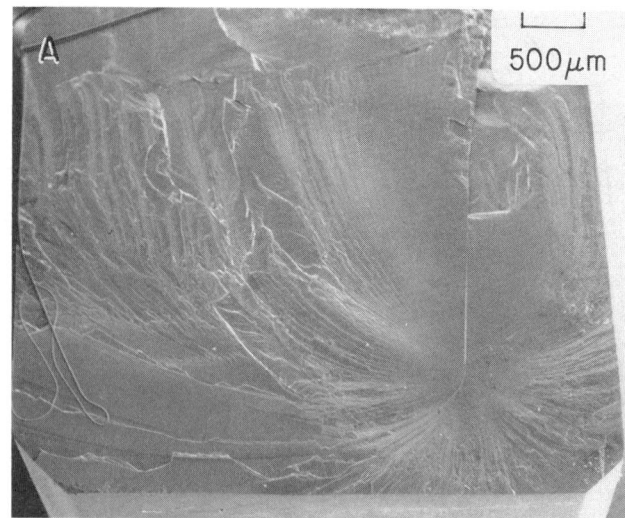

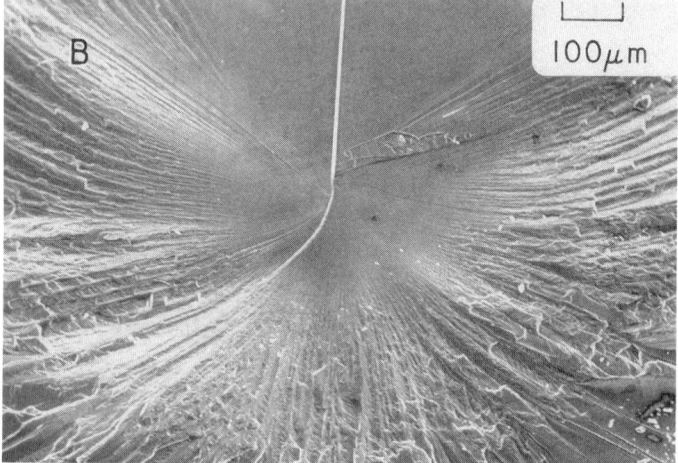

Fig 14 Internal origin in a bend specimen of single-crystal CaF₂. Origin is a crack, seen curving to the left; tensile surface was at the bottom. Source: Ref 5

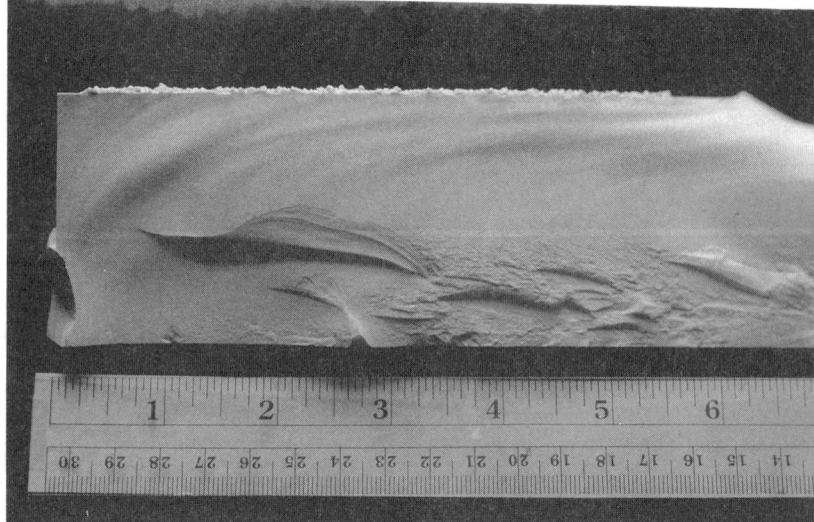

Fig 15 Fracture surface of a porcelain insulator which broke during cutting with a diamond saw. Fracture moved from right to left. Mist and velocity hackle and Wallner lines are readily seen in this fine-grained material. Optical microscope. Source: Ref 1

ceramics, even when the material has a small average grain size. In this photograph, and in several others to follow, the specimen was tilted in the SEM, allowing both the original surface and the fracture surface to be seen. The gray region at the left in Fig 16(a) and (b) is the original surface. In Fig 16(a) the origin is on the left, about halfway up from the bottom of the picture. The fracture surface close to the origin is relatively smooth compared to the region farther away. This is the fracture mirror, but it is difficult to define its extent with any accuracy. Outside of the mirror, there are ridges which run radially away from the origin, and these are presumably a form of velocity hackle. However, these ridges are also difficult to see.

A point can be made here regarding the level of magnification one should use in examining fracture surfaces of polycrystalline ceramics. Velocity-hackle ridges and the mirror region are often seen most readily using the unaided eye or a hand lens. This may be the quickest way to locate the fracture origin. Sometimes, however, it is virtually impossible to "read" the fracture surface well

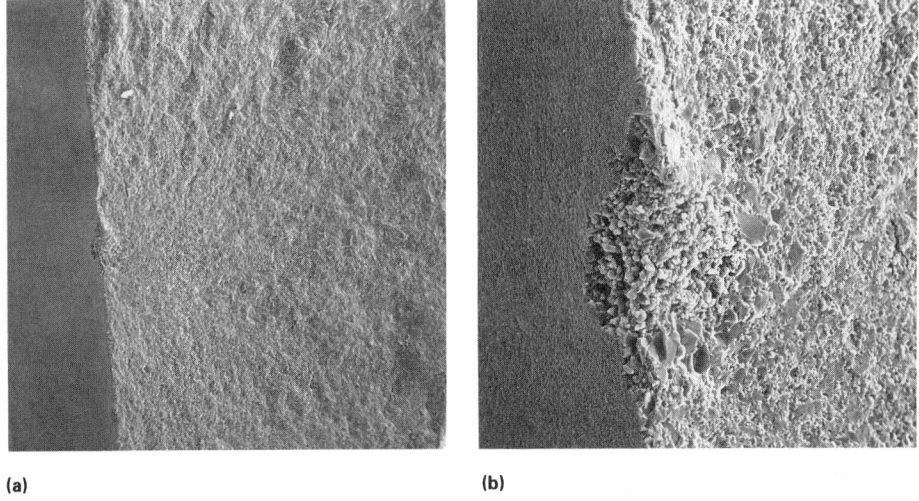

(a) **(b)**

Fig 16 Fracture surface of fine-grained Al_2O_3 shown at low and high magnifications. Specimen was tilted in the SEM allowing original surface (gray area at left) and fracture surface to be seen simultaneously. SEM. (a) 100×. (b) 500×

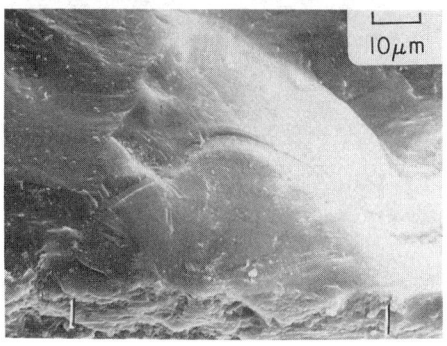

Fig 18 Two views at different magnifications of the fracture surface of large-grained B_4C. Source: Ref 5

the difficulty of seeing fracture markings in enough. In such cases, it may then be necessary to resort to very high magnifications using the SEM to try to see fracture markings

in individual grains (gull wings and wake hackle, in particular, at internal pores) in order to trace the way back to the origin. Of course, this is painstaking and time-consuming work.

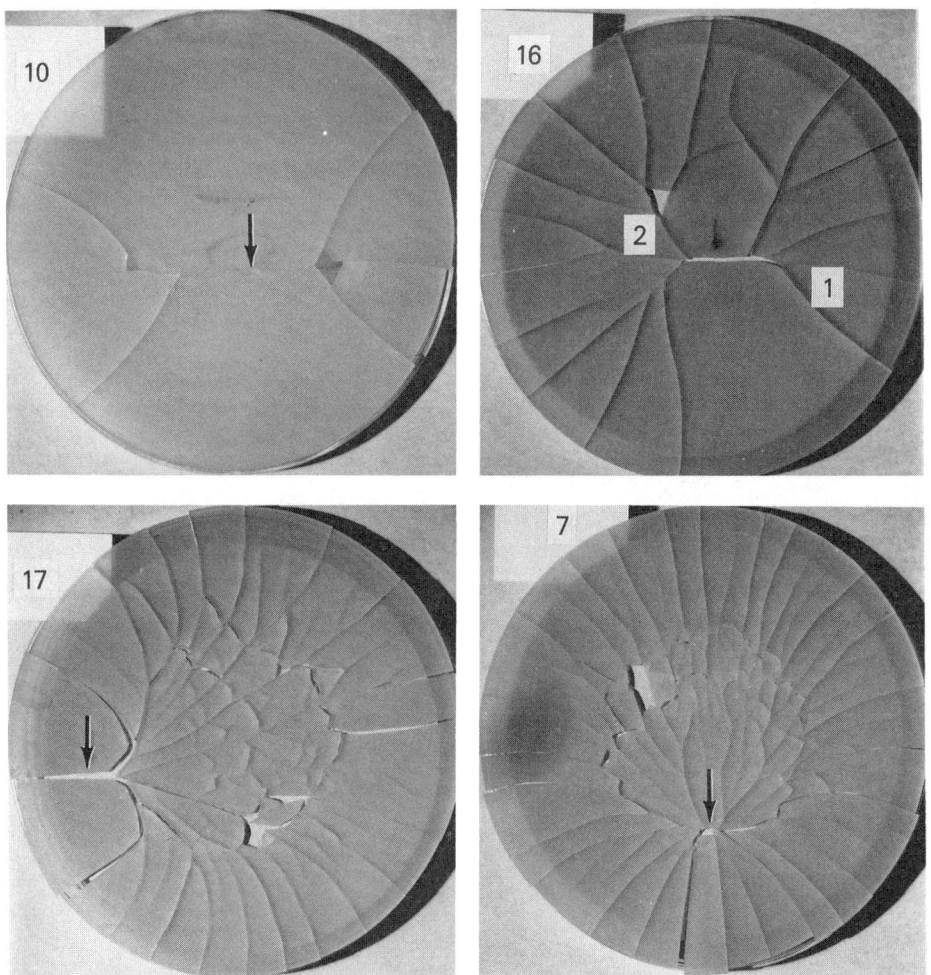

Fig 17 Crack patterns of four specimens of MgF_2 broken in biaxial flexure. The more extensive branching is associated with higher stress at failure. Arrows indicate locations of fracture origins. Disks are ≈50 mm (2 in.) in diameter. Source: Ref 5

Another way to locate the fracture origin, especially in strength tests, is to make use of the overall crack pattern. This can be facilitated in bend tests by applying tape to the compression side of the specimen before breaking it. The tape holds the pieces together without influencing the strength results. For example, Fig 17 shows four disks of MgF_2, broken in biaxial flexure, which failed at different stresses. The branching pattern is used to determine where the origin is located. Also, the branching is more extensive when the failure stress is greater.

Figure 18 shows the fracture surface of a specimen of large-grained B_4C. As in Fig 16, the mirror region is difficult to see, and velocity hackle shows up as ridges radiating away from the origin. Again, low magnification is more useful in observing these markings.

Fracture Origins

The key to understanding strength and failure of test specimens and manufactured parts is identifying the cause of fracture, that is, the fracture origin. Cracks, pores, bubbles, inclusions, grain boundaries—all of these discontinuities serve as stress concentrators in glasses and/or ceramics and can, therefore, be fracture origins. This section will present examples of fracture origins grouped together according to their mechanism of formation—indentation or impact, machining, processing. More examples can be found in the references listed earlier and in Ref 12 to 13.

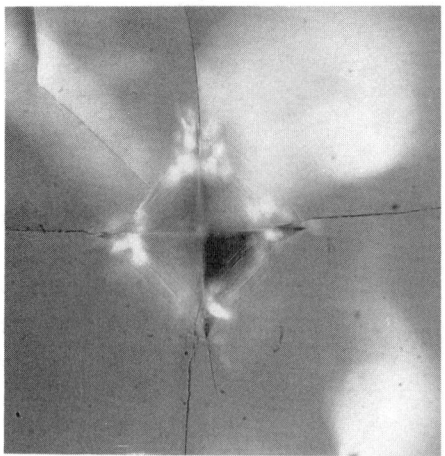

Fig 19 Indentation and associated cracks on a glass surface made by a Vickers diamond with a load of 2000 g. Bright areas are reflections from the surfaces of sub-surface cracks. Optical microscope. 500×. Courtesy of G. Klassen, New York State College of Ceramics, Alfred University

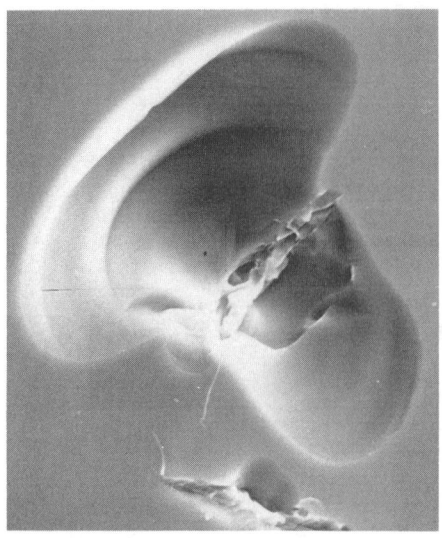

Fig 21 Impact site on a glass surface made by a 100-μm particle of SiC. SEM. 450×. Source: Ref 14

Fig 23 Impact fracture origin in glass caused by impact damage from a 100 μm SiC particle. Specimen was tilted in the SEM. Original surface is at left, fracture surface at right. SEM. 190×

Indentation and Impact Sites. Surfaces of brittle materials are particularly susceptible to contact damage caused by hard objects, either during indentation or impact (or machining, covered below). Figure 19 shows an indentation site caused by pressing a Vickers diamond onto the surface of a piece of glass. The cracks running from the corners of the indentation and the one coming off of one side are normal to the surface and extend some distance into the glass. These are the cracks that will become fracture origins when stress is applied to the piece. There are other cracks on the surfaces of the indentation, and the bright areas outside of the indentation are reflections from sub-surface lateral cracks that have not made it up to the surface. When they do, they cause material removal. The semicircular region shown in Fig 20 is a so-called half-penny crack that was generated from a similar indentation site by heating the glass and then thermally shocking it at the site with a copper block. The original indentation crack is seen on the edge of the fracture surface at the center of the half-penny crack.

Similar damage is produced when sharp particles impact glass or ceramic surfaces, as illustrated by the example in Fig 21. The lateral cracks reached the surface, causing material to spall off. Complex cracking and loss of material occurred at the center of the impact. Sub-surface cracking is not visible in this SEM photograph. However, Fig 22 shows a similar impact site after the specimen was broken in a strength test. As described earlier, the specimen was tilted in the SEM, allowing the original surface (at the left) and the fracture surface (at the right) to be seen simultaneously. The relationship between the impact site and the fracture origin is clear. The size of the origin flaw, as measured along the surface, is about the same as the size of the spalled-off region. This is typical, but there are many cases where the spalled-off region is much larger than the origin flaw, and vice versa. Note the twist hackle along part of the origin flaw.

Another impact origin is shown in Fig 23 to make the point that impact origins often have a more complicated morphology than indentation origins. Here, the origin looks like a truncated half-penny. In fact, the origin crack is intersected by a second crack (the narrow, bright feature at the impact site), and the origin flaw itself continues behind the fracture surface seen in the photograph. Nonetheless, this other crack influenced how the fracture started, that is, it affected the strength.

Blunt objects, such as spheres, also can produce contact damage, as shown in Fig 24. The elliptical lines on the left side of the photograph are contact cracks on the original surface of the specimen. (They are actually circular, but appear elliptical in this picture because of the tilting of the specimen in the SEM.) These contact cracks run normal to the

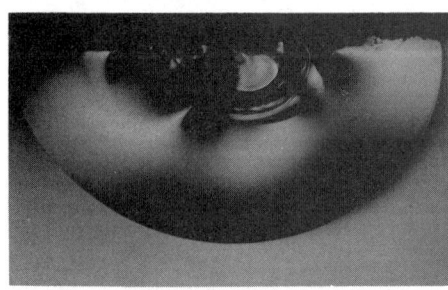

Fig 20 Half-penny fracture origin in glass. Crack at Vickers indentation site was extended using a thermal-shock technique to produce the large half-penny flaw. Optical microscope. 110×. Courtesy of V.D. Fréchette, New York State College of Ceramics, Alfred University

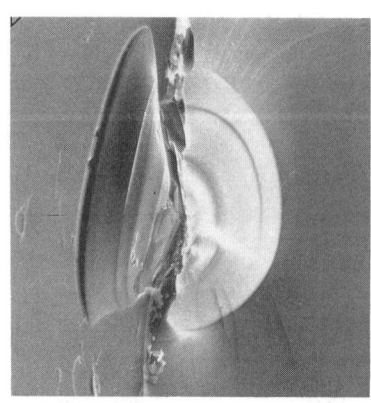

Fig 22 Impact fracture origin in glass caused by impact damage from a 100 μm SiC particle. Specimen was tilted in the SEM to reveal original surface (left) and fracture surface (right). SEM

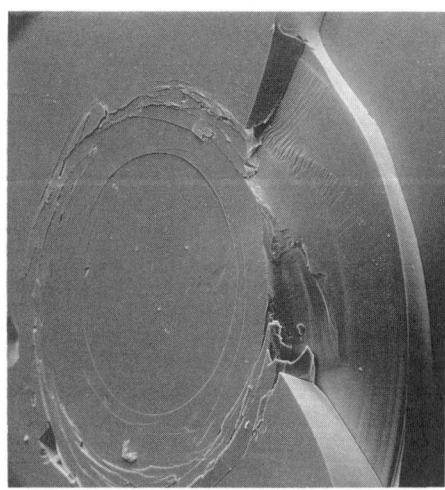

Fig 24 Hertzian impact site in glass. Specimen was tilted in the SEM to reveal original surface (left) and fracture surface (right). SEM. 145×

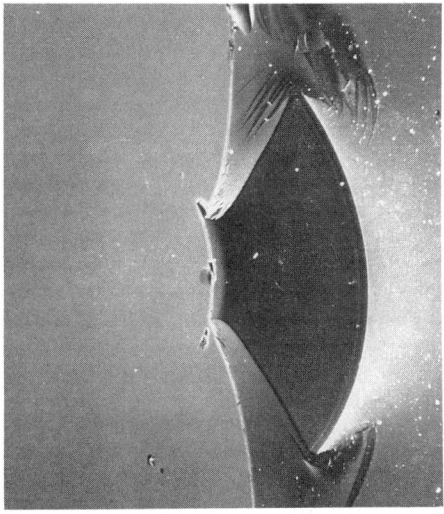

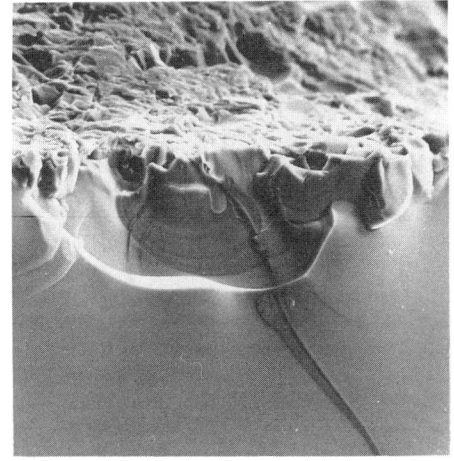

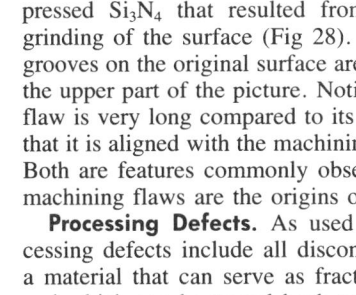

Fig 25 Hertzian fracture origin in glass. Original surface is at left, fracture surface at right. Main fracture started from base of Hertzian cone. SEM

Fig 27 Machining flaw as fracture origin in glass. Rough surface is the bottom of a groove cut by a diamond saw. SEM. 200×

surface for a short distance before flaring out into cone-shaped cracks (so-called Hertzian cones, named in honor of Hertz, the scientist who described the stress field associated with blunt-object contact). This is what happens when a "BB" hits a window, for example. Portions of two of these cones are seen in Fig 24. Notice the twist hackle and Wallner lines on the surface of the cone flaring out from the larger surface crack. Figure 25 presents a clearer picture of a Hertzian cone as a fracture origin. A single contact crack can be seen on the original surface, flaring out into the sub-surface cone. The twist hackle seen at two locations make it evident that the main fracture started from the base of the cone. This is not always the case, however, as shown in Fig 26. The cone in Fig 26 has been cut by the main fracture, which evidently (and sur-

prisingly) started at the surface and not from the base of the cone. This is shown by the mist and velocity hackle at the lower left of the picture, which indicate the fracture was moving from upper right to lower left and had reached terminal velocity.

Machining Flaws. When glasses and ceramics are machined or when they are cut using a diamond saw, extensive sub-surface damage is produced. The crack systems are similar to those seen at indentation or impact sites, but are still more complex, since many particles are involved instead of just one. Two examples of machining flaws are presented here. More examples can be found in Ref 4, 5, 8, and 12 to 13.

The first is a machining flaw in glass produced when a diamond saw was used to cut a groove in the glass (Fig 27). It appears that a series of sub-surface cracks linked up to produce the large flaw that caused failure of this piece when external loading was applied. The second shows a machining flaw in hot-

pressed Si_3N_4 that resulted from diamond grinding of the surface (Fig 28). Machining grooves on the original surface are evident in the upper part of the picture. Notice that this flaw is very long compared to its depth, and that it is aligned with the machining grooves. Both are features commonly observed when machining flaws are the origins of fracture.

Processing Defects. As used here, processing defects include all discontinuities in a material that can serve as fracture origins and which can be traced back to some step in manufacturing, except for those defects produced by contact damage. Examples include bubbles and crystalline inclusions in glasses, and pores, large-grained regions, and microcracks in polycrystalline ceramics.

Although pores are not as effective in concentrating stresses as cracks, they are a frequent cause of failure in polycrystalline ceramics. The example in Fig 29 shows a pore with an unusual shape that happened to be right on the tensile surface of this Al_2O_3 specimen when it was loaded in bending. Appar-

(a)

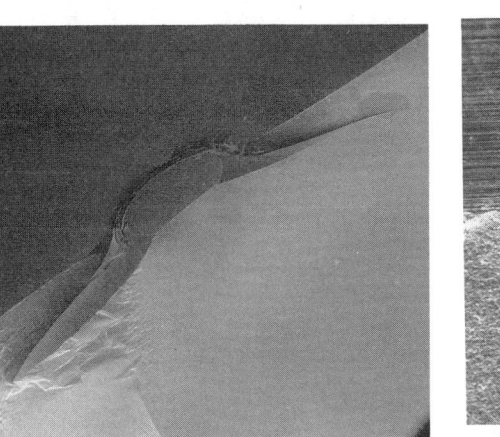

Fig 26 Hertzian fracture origin in glass. Line running diagonally through the picture is the edge between the original and the fracture surface. Main fracture started at the edge of the contact crack on the right-hand side of the cone. SEM. 40×

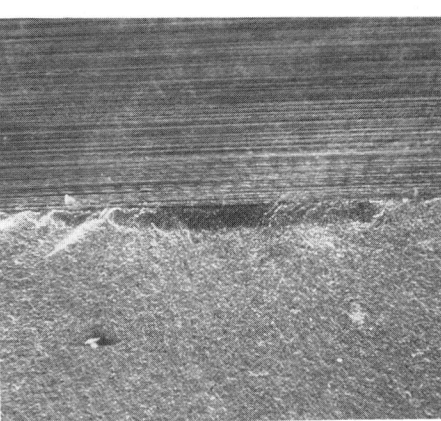

Fig 28 Fracture surface of Si_3N_4 with machining flaw as origin. Specimen was tilted in the SEM showing the machined surface at the top and the fracture surface at the bottom. Machining flaw is aligned with grooves on the original surface. SEM. 60×. Source: Ref 15

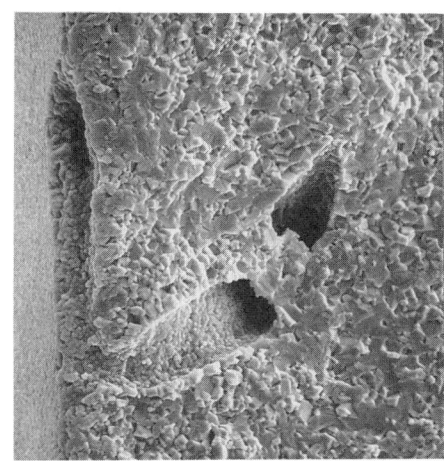

(b)

Fig 29 Lower and higher magnification views of fracture surface of fine-grained Al_2O_3 with a surface pore as fracture origin. SEM. (a) 69×. (b) 690×

ently something organic was in the green material (possibly a hair or a segment of poorly mixed binder) which burned out during sintering, leaving this large pore. Although the material overall is fine grained with only very small pores, this large pore is a processing defect which significantly reduced the strength of this particular piece.

Figure 16, discussed earlier, is an example of a large-grained region as fracture origin. In the high-magnification view of Fig 16(b), the origin is seen to be a region where the grains are larger than the average, and where densification is very incomplete. Notice the transgranular fracture surrounding the large-grained area. It is common to see transgranular fracture close to the origin, where the fracture was travelling slowly.

Glazed ceramics are interesting because origins can be in the glaze, at the glaze/body interface, or in the body. Locating and identifying fracture origins is, therefore, especially critical in understanding causes of failure. Figure 30 shows an example of an origin in the glaze. In this case, there was a

large grain of unreacted quartz located between two large bubbles. There will be cracking around such quartz particles owing to stresses that develop because of thermal-expansion differences between the quartz and the glaze. The large bubbles are also defects, but it was the quartz particle that caused fracture.

Fracture origins at the glaze/body interface are also very common. The glaze should be in compression and the body in tension. Reactions at the glaze/body interface may lead to local values of tension much higher than expected for the bulk of the body. Therefore, defects at the glaze/body interface are subjected to the combined effect of applied and residual stresses, making them frequent failure origins.

REFERENCES

1. V.D. Fréchette, *Failure Analysis of Brittle Materials*, Vol 28, *Advances in Ceramics*, The American Ceramic Society, 1990
2. V.D. Fréchette and J.R. Varner, Ed., *Fractography of Glasses and Ceramics*, Vol 22, *Advances in Ceramics*, The American Ceramic Society, 1986
3. V.D. Fréchette and J.R. Varner, Ed., *Fractography of Glasses and Ceramics, II*, Vol 17, *Ceramic Transactions*, The American Ceramic Society, 1991
4. R.W. Rice, Perspective on Fractography, *Fractography of Glasses and Ceramics*, V.D. Fréchette and J.R. Varner, Ed., The American Ceramic Society, 1986, p 3–56
5. R.W. Rice, Ceramic Fracture Features, Observations, Mechanisms, and Uses, *Fractography of Ceramic and Metal Failures*, STP 827, J.J. Mecholsky, Jr. and S.R. Powell, Jr., Ed., American Society for Testing and Materials, 1984, p 5–103
6. R.W. Rice, Fracture Topography of Ceramics, *Surfaces and Interfaces of Glass and Ceramics*, V.D. Fréchette, W.C. LaCourse, and V.L. Burdick, Ed., Plenum Press, 1974, p 439–472
7. J.J. Mecholsky, S.W. Freiman, and R.W.
Rice, Fractographic Analysis of Ceramics, *Fractography in Failure Analysis*, STP 645, B.M. Strauss and W.H. Cullen, Jr., Ed., American Society for Testing and Materials, 1978, p 363–379
8. R.W. Rice and J.J. Mecholsky, Jr., The Nature of Strength Controlling Machining Flaws in Ceramics, *The Science of Ceramic Machining and Surface Finishing II*, B.J. Hockey and R.W. Rice, Ed., Spec. Publ. No. 562, National Bureau of Standards, 1979, p 351–378
9. V.D. Fréchette, Fractology of Glass, *Introduction to Glass Science*, L.D. Pye, H. Stevens, and W. LaCourse, Ed., Plenum Press, 1972
10. T.A. Michalske, "Dynamic Effects of Liquids on Crack Growth Leading to Catastrophic Failure in Glass," Ph.D. thesis, Alfred University, 1979
11. V.D. Fréchette, Markings Associated with the Presence of H_2O During Cracking, *Fractography of Glasses and Ceramics*, Vol 22, *Advances in Ceramics*, V.D. Fréchette and J.R. Varner, Ed., The American Ceramic Society, 1986, p 71–76
12. R.W. Rice, J.J. Mecholsky, S.W. Freiman, and S.M. Morey, "Failure Causing Defects in Ceramics: What NDE Should Find," Memorandum Report 4075, National Research Laboratory, 1979
13. R.W. Rice, J.J. Mecholsky, Jr., and P.F. Becher, The Effect of Grinding Direction on Flaw Character and Strength of Single Crystal and Polycrystalline Ceramics, *J. Mater. Sci.*, Vol 16, 1981, p 853–862
14. J.R. Varner and H.J. Oel, Surface Defects: Their Origin, Characterization and Effects on Strength, *J. Non-Cryst. Solids*, Vol 19, 1975, p 321–333
15. E.S. Alfaro, J.V. Guiheen, and J.R. Varner, Insights Provided by Fractography in Strength Testing of Machined Si_3N_4 and Indented Al_2O_3, *Fractography of Glasses and Ceramics II*, Vol 17, *Ceramic Transactions*, V.D. Fréchette and J.R. Varner, Ed., The American Ceramic Society, 1991, p 485–508

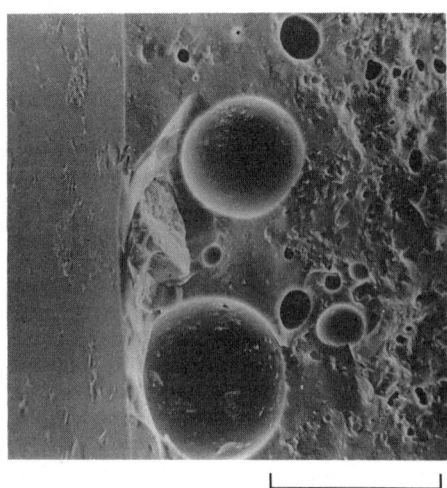

├─── 100 μm ───┤

Fig 30 Fracture surface of glazed porcelain. Fracture origin is a quartz grain between the two large bubbles in the glaze. SEM

Applied Fracture Mechanics

Isa Bar-On, Worcester Polytechnic Institute

FRACTURE MECHANICS can be used to obtain quantitative results from a failure analysis. Knowledge of crack characteristics and material properties is used in a typical fracture mechanics analysis to establish the fracture strength of a component containing cracks or flaws. In addition fracture mechanics analyses can provide insight into the role of loading conditions and crack geometry.

In ceramics, crack growth occurs by truly brittle fracture. The ceramic behaves linear elastically on a global scale as well as locally at the crack tip up to the point where atomic bond rupture occurs. As a result fracture mechanics analyses can be limited to linear elastic treatments.

This article discusses the basic concepts of linear elastic fracture mechanics and concentrates on models developed specifically for ceramic materials. The role of isolated flaws of different origins is discussed in some detail. The effect of specific flaw geometries on the fracture strength is presented. Lastly, the fracture behavior due to multiple flaws is mentioned briefly.

Linear Elastic Fracture Mechanics

Three modes of loading can be distinguished for a component containing a crack (Fig 1). Mode I is the opening or tensile mode, where the crack surface displacements are normal to the crack plane. For mode II, the sliding or in-plane shear mode, the displacement of the crack surfaces is in the plane of the crack and perpendicular to the leading edge of the crack. In mode III, the tearing or antiplane shear mode, the crack surface displacements are in the plane of the crack and parallel to the leading edge of the crack.

Mode I loading is technically the most important and will be treated in this article. Mode II and mode III solutions have been developed (Ref 1, 2). For many mixed mode loading conditions the solutions can be derived on the basis of these existing solutions.

Stress Field of Mode I Loading. The stress field in any body containing a crack under mode I loading can be found from the theory of elasticity. The stresses, σ_{ij}, at any location (r, θ) as shown in Fig 2 are given as:

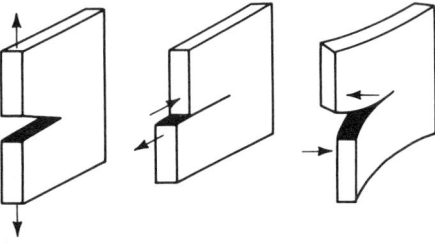

Mode I
Opening mode, tensile mode

Mode II
Sliding mode, shear mode

Mode III
Tearing mode

Fig 1 Modes of loading in fracture mechanics

$$\sigma_{ij} = \frac{K_I}{\sqrt{2\pi r}} f_{1ij}(\theta) + C_1 r^\circ f_{2ij}(\theta)$$
$$+ C_2 r^{1/2} f_{3ij}(\theta) + \dots \quad \text{(Eq 1)}$$

For small r (that is, for stresses close to the crack tip), the first expression in Eq 1 is dominant and the other terms can be neglected.

This approximation is justified because the fracture processes occur very close to the crack tip. The relevant stress expressions are then:

$$\sigma_{xx} = \frac{K_I}{\sqrt{2\pi r}} \cos\frac{\theta}{2} \cdot \left(1 - \sin\frac{\theta}{2}\sin\frac{3\theta}{2}\right) \quad \text{(Eq 2a)}$$

$$\sigma_{yy} = \frac{K_I}{\sqrt{2\pi r}} \cos\frac{\theta}{2} \cdot \left(1 + \sin\frac{\theta}{2}\sin\frac{3\theta}{2}\right) \quad \text{(Eq 2b)}$$

$$\tau_{xy} = \frac{K_I}{\sqrt{2\pi r}} \left(\sin\frac{\theta}{2}\cos\frac{\theta}{2}\sin\frac{3\theta}{2}\right) \quad \text{(Eq 2c)}$$

where the subscript I stands for mode I loading, and K is an expression of the stress intensification due to the crack.

Equations 2(a, b, and c) have the same general form for the three stress components. They consist of three parts: K_I, $(2\pi r)^{-1/2}$, and $f_{ij}(\theta)$. The combined expression $f_{ij}(\theta) \cdot (2\pi r)^{-1/2}$ depends only on the location (r, θ) with respect to the crack and will be the same for different K values.

The stress in the y-direction (σ_{yy}) along the plane $\theta = 0$ is given as:

$$\sigma_{yy} = \frac{K_I}{\sqrt{2\pi r}} \quad \text{for } \theta = 0 \quad \text{(Eq 3)}$$

because $f_{yy}(\theta = 0) = 1$. Therefore, the stress varies as a function of r (Fig 3). As r approaches zero the predicted elastic stress, σ_{yy}, approaches infinity. In reality, bond rupture will occur at some finite value of σ_{yy}. A similar argument can be made for the entire stress field.

The parameter K_I expresses the effect of a given crack and loading geometry on the stress field. K_I is defined as the stress-intensity factor for mode I loading. The magnitude of K_I differs for different cracked parts and its expression includes typically the effects of crack length, far field stress, and loading geometry. As such, the stress-intensity factor needs to be evaluated separately for any given specific loading geometry.

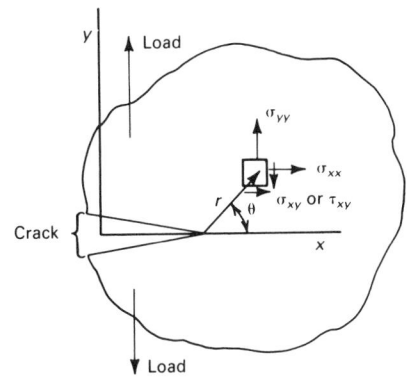

Fig 2 Coordinate system for load directions in a body

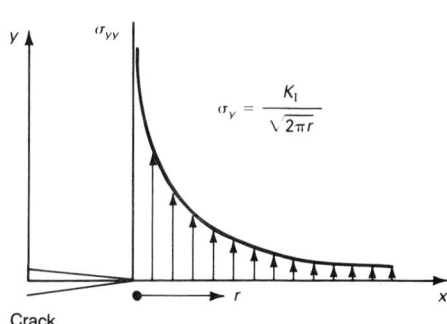

Fig 3 Crack-tip stress distribution

For a few special geometries K_I has been determined analytically. The majority of K_I expressions, however, have been derived by numerical procedures. Expressions for K_I can be found in handbooks and literature (Ref 1, 3, 4).

The stress solution for an infinitely wide plate containing a small center crack of length $2a$ (Fig 4), subjected at infinity to uniform tension, σ, is given as:

$$\sigma_{yy} = \frac{\sigma\sqrt{\pi a}}{\sqrt{2\pi r}} \quad \text{for } \theta = 0 \qquad \text{(Eq 4)}$$

where $K_I = \sigma\sqrt{\pi a}$.

This expression is frequently an excellent approximation for geometries involving small cracks in large parts loaded at some distance. The more general form of Eq 4 for plates of finite width and for alternate crack geometries includes a dimensionless expression Y. Y is a function of the crack length to width ratio, a/W. The more general form of Eq 4 accounting for finite dimensions and loading geometries is given as:

$$K_I = Y\left(\frac{a}{W}\right) \sigma \sqrt{\pi a} \qquad \text{(Eq 5)}$$

Analytical and empirical solutions have been determined for this geometry, which differ only slightly from each other (Ref 5, 6). An empirical solution for the center cracked plate of finite width due to Feddersen is:

$$K_I = \sigma\sqrt{\pi a}\,\sqrt{\sec\left(\frac{\pi a}{W}\right)} \qquad \text{(Eq 6)}$$

For the edge cracked plate loaded in uniform tension (Fig 5) the stress-intensity factor has been estimated as:

$$K_I = 1.12\,\sigma\,\sqrt{\pi a} \qquad \text{(Eq 7)}$$

More exact solutions are available including solutions for other geometries (Ref 3, 4, 7). The effect of the different geometries is expressed in the function $Y(a/W)$.

Fracture Toughness. Fracture in a component will occur when the applied stress intensity, K_I, reaches some critical value K_{Ic}, which is referred to as the plane strain fracture toughness. A standard method for the determination of the fracture toughness in ceramic materials has not been adopted yet. Several methods have been put forward, but none has gained general acceptance. This is due to the experimental difficulties associated with introducing a precrack in a controllable manner. The experimental fracture toughness values can thus vary depending on the testing method.

Two types of problems can be solved once the fracture toughness has been determined.

Either the crack length is known and the fracture strength is predicted or the stress is known and the critical crack length is estimated.

If $K_I = K_{Ic}$ (corresponding to the occurrence of fracture), then the following relationship is obtained from Eq 5:

$$K_I = Y\left(\frac{a}{W}\right) \sigma\sqrt{\pi a} = K_{Ic} \qquad \text{(Eq 8)}$$

This relates the fracture toughness, K_{Ic}, to the far-field stress, σ, and the crack length. The effect of the geometry is accounted for in the function $Y(a/W)$.

In the first problem, the crack length remains constant and is known as well as the fracture toughness. Equation 8 can then be reformulated to give the fracture strength, σ_f:

$$\sigma_f = \frac{K_{Ic}}{\sqrt{\pi a}\,Y\left(\dfrac{a}{W}\right)} \qquad \text{(Eq 9)}$$

This gives the fracture strength for a given flaw size and geometry.

In the second problem the applied stress remains constant, while the crack might grow to critical dimensions, a_c. If the fracture toughness is known then the critical crack length can be obtained by manipulating Eq 8 and using the condition $a = a_c$:

$$a_c Y^2\left(\frac{a_c}{W}\right) = \frac{K_{Ic}^2}{\sigma^2 \cdot \pi} \qquad \text{(Eq 10)}$$

Fracture Mechanics Analysis of Ceramics

For most ceramics the critical crack length will be quite small. In large-grained ceramics the crack could be smaller than a typical grain. In this case, the polycrystalline fracture toughness for a large crack is not applicable. Rather the single-crystal fracture toughness, which is typically a lower value, should be considered (Ref 8).

For several ceramic materials the fracture resistance, as measured by the fracture toughness, increases with stable crack extension. These materials are best characterized by a K_R or R-curve (Ref 9–11). A typical R-curve for alumina is shown in Fig 6. In those cases, care has to be exercised when inserting values for K_{Ic} in Eq 8 to 10.

Mechanical failure of ceramics does not occur from large straight cracks of the kind described in the previous section. Instead failure occurs from processing or handling defects. At low and intermediate temperatures, ceramics will fracture from the extension of these single flaws or the linking of several smaller flaws. At high temperatures the fracture process involves the formation and coalescence of cavities, which frequently is preceded by some nonlinear deformation (Ref 12).

Typically, flaws include internal or surface cracks, large voids, inclusions, and, at times, large grains. These can occur isolated or in

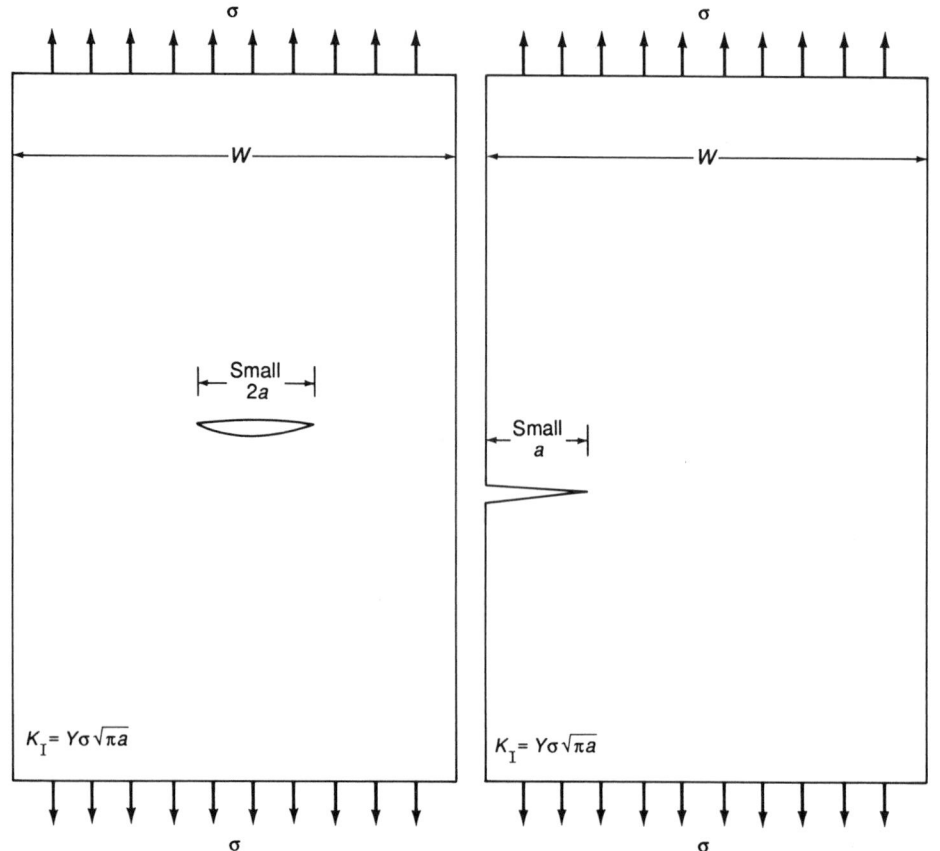

Fig 4 A center crack in a plate subjected to uniform tension

Fig 5 An edge crack in a plate subjected to uniform tension

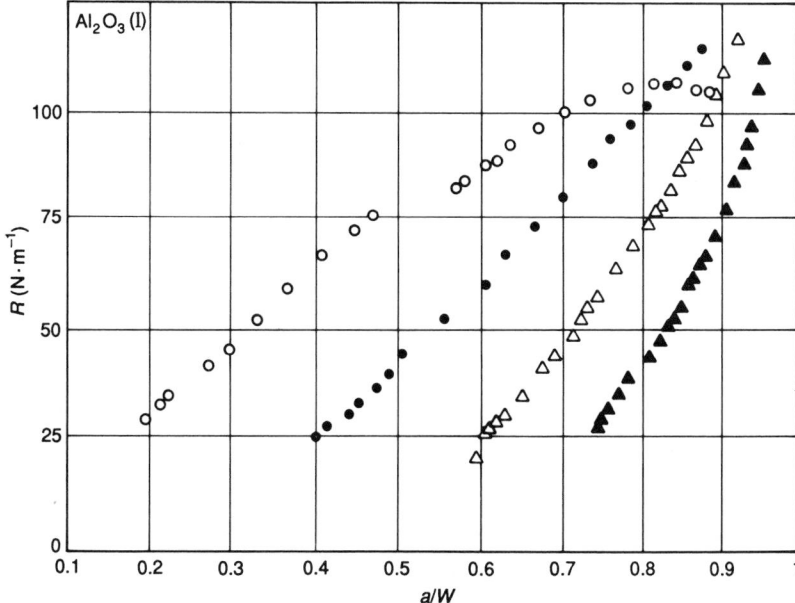

Fig 6 R curves for different notch depths in alumina. Source: Ref 9

$$K_I = \frac{\sigma\sqrt{\pi a}}{\Phi} \qquad \text{(Eq 12)}$$

For an elliptical flaw of a given size the fracture strength is then expressed as:

$$\sigma_f = \frac{K_{Ic} \cdot \Phi}{\sqrt{\pi a}} \qquad \text{(Eq 13)}$$

Equation 13 is the same as Eq 9, where $Y(a/w)$ has been replaced by the elliptical integral Φ.

Semielliptical Crack. A surface crack is a typical flat semielliptical crack. This geometry (Fig 8) requires a surface correction factor, M, in Eq 12 for the free surfaces, so that:

$$K_I = \frac{\sigma M \sqrt{\pi a}}{\Phi} \qquad \text{(Eq 14)}$$

where M has been estimated to be about 1.12. This equation is similar to Eq 7, but it considers the semielliptical flaw shape through use of the elliptical integral Φ.

For semielliptical cracks the K_I value changes along the circumference from a minimum at the end of the major axis to a maximum value at the minor axis. Raju and Newman (Ref 13) and Kobayashi (Ref 14) produced numerical stress-intensity factor solutions for shallow, nearly semicircular flaws loaded in tension. These analyses yielded M values varying from about 1.03 at the flaw depth to about 1.2 at the flaw surface. They also obtained values for similar flaws loaded in bending.

Irregularly Shaped Two-Dimensional Cracks. The stress-intensity expressions for elliptical, semielliptical, semicircular, and other crack shapes can be summarized in one expression through the introduction of a shape

Isolated Flaws in Two Dimensions

Elliptical Crack. A typical internal flaw can be modeled as a two-dimensional flat elliptical crack with minor half axis a, and major half axis c (Fig 7). This crack is located in an infinite body and is subjected to tensile loading. The derived stress-intensity factor for the elliptical crack is of the form

$$K_I = \frac{\sigma\sqrt{\pi a}}{\Phi}\left(\frac{a^2}{c^2}\cos^2\phi + \sin^2\phi\right)^{1/4} \qquad \text{(Eq 11)}$$

where σ is the applied stress, and ϕ is the angle defined in Fig 7(a). Φ is a complete elliptical integral of the second kind, its value increases as the a/c ratio of the elliptical crack increases, reaching a value of $\pi/2$ when $a/c = 1$. The value of Φ can be found either from tables or in a graph (Fig 7b). For this loading configuration the maximum stress intensity occurs at the end of the minor axis. The stress-intensity factor at this point is:

close proximity with each other. Fracture mechanics analyses exist for many of these different geometries, as discussed below.

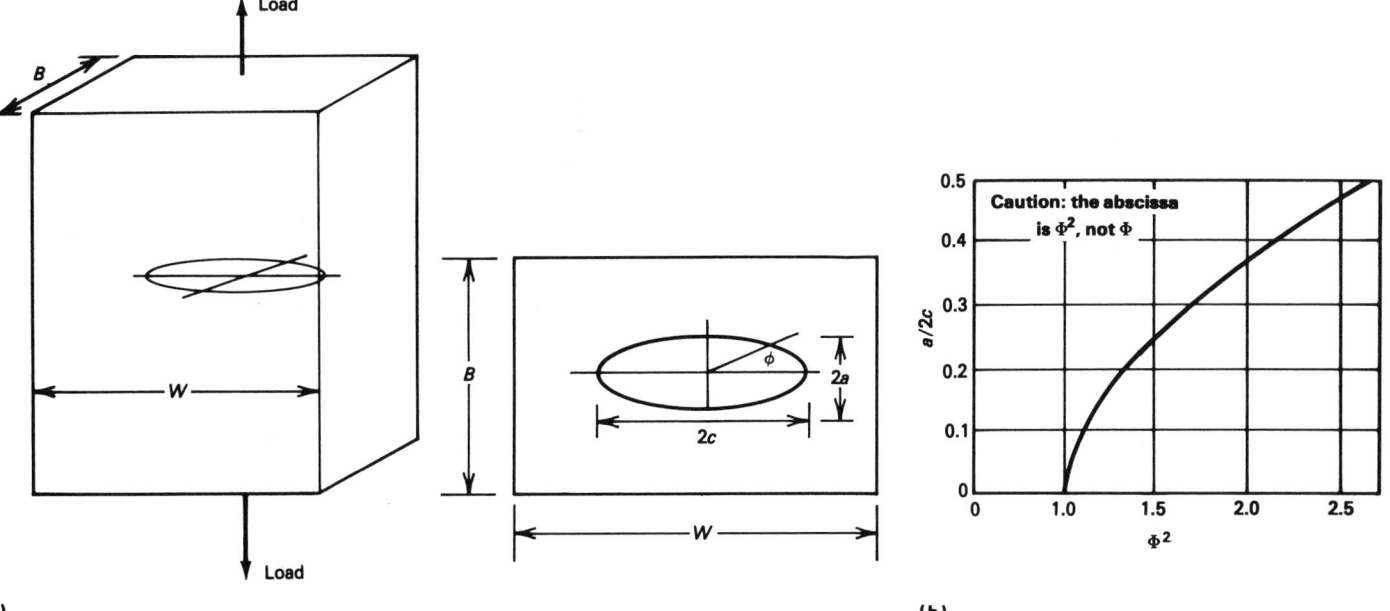

(a)   **(b)**

Fig 7 Flat elliptical model of an internal flaw. (a) Definition of terms. (b) Graph of the square of the elliptical integral (Φ^2) versus crack length

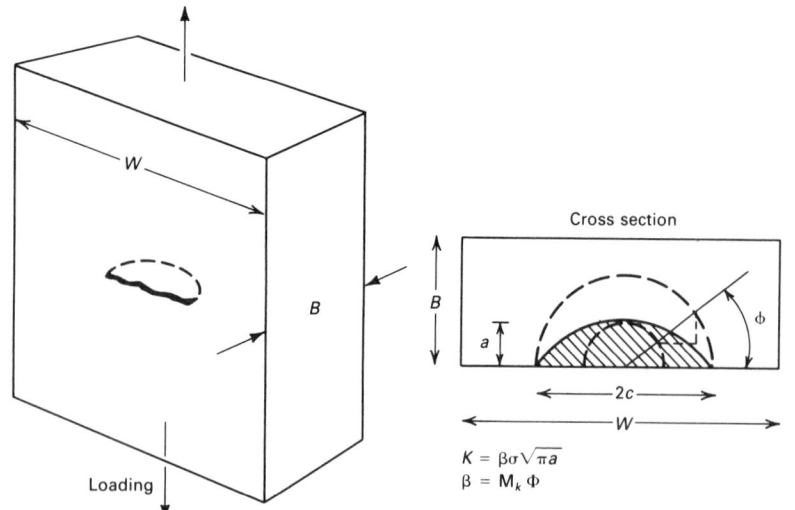

Fig 8 Semielliptical model of a surface flaw

$$K = \beta\sigma\sqrt{\pi a}$$
$$\beta = M_k \Phi$$

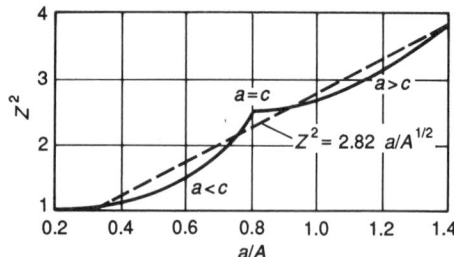

Fig 10 Relation of a flaw area (A) to Z. Source: Ref 15

factor Z. Equations 11, 12, and 14 are thus implied in the following form:

$$K_I = \frac{Y}{Z}\sigma\sqrt{a} \quad \text{(Eq 15)}$$

where Y is a numerical constant (that includes $\sqrt{\pi}$) dependent on the loading geometry and crack length and Z is the flaw shape parameter. Z is equal to 1 for a through-the-thickness crack. In a ceramic a typical crack has a flaw shape parameter, $Z > 0$. Z is a function of the ratio of minor to major axis, a/c (Fig 9).

For irregular as well as regular cracks, the ratio a/c can be substituted for by the crack area A (Ref 15). The relationship between the flaw shape parameter and the crack area is shown in Fig 10. This curve can be approximated by the expression

$$Z^2 = 2.82\frac{a}{\sqrt{A}} \quad \text{(Eq 16)}$$

where Z is the flaw shape parameter and A is

the crack area. The stress-intensity expression thus takes the form:

$$K_I = \frac{1}{1.68}YA^{1/4}\sigma \quad \text{(Eq 17)}$$

This relation, which gives the stress-intensity factor for different loading conditions as a function of the crack area instead of the crack length, permits the treatment of irregular crack shapes.

For a given flaw the fracture strength, σ_f, can then be determined as:

$$\sigma_f = \frac{1.68}{Y}\frac{K_{Ic}}{A^{1/4}} \quad \text{(Eq 18)}$$

Thus the fracture strength can be predicted for an irregular two-dimensional flaw, if the flaw area and fracture toughness are known.

Residual Stress Effects. Surface cracks created by the penetration of the surface by a hard particle (as in grinding or polishing, for example) develop in response to a residual stress created by the penetrating particle. The effect of this residual stress needs to be accounted for when determining K_I or the fracture stress (Ref 16–18).

Isolated Cracks in Three Dimensions

Voids do not cause fracture directly, because the stress intensification due to a void is usually not sufficient to cause fracture from the void surface. Fracture occurs when the stress field of the void interacts with defects in close proximity to the voids. For voids located close to the surface, these might be surface cracks. Alternatively, microcracking could be enhanced due to the stress concentration of the void.

Voids accompanied by circumferential cracks (Fig 11) have been observed by Bansal and Duckworth (Ref 19) and have been modeled by Baratta (Ref 20–22) and Evans

et al. (Ref 23) for uniaxial tension. The normalized stress-intensity factor for the case of uniaxial tension (Fig 11) is shown as a function of the crack length to void radius ratio. For short cracks, $L/R < 10^{-1}$, the stress-intensity factor due to the void is markedly higher. This is due to the presence of the void surface which allows a larger crack opening that will increase K_I. This effect decreases as the crack length approaches the size of the void.

The effect of multiaxial loading on the stress-intensity factor of voids with circumferential cracks has been approximated by Baratta and Parker (Ref 27) and Shetty et al. (Ref 28). The normalized stress intensity for these cases is summarized in Fig 12. The case of axial tension-biaxial compression ($\phi = 1$, $\lambda = \psi = -1$) results in the highest stress concentration followed by uniaxial tension ($\phi = 1$, $\lambda = \psi = 0$), triaxial tension ($\phi = \lambda = \psi = 1$) and finally biaxial hydrostatic compression ($\phi = 0$, $\lambda = \psi = -1$).

These stress-intensity calculations explain why materials containing cracked voids subjected to triaxial tension might appear stronger than those under uniaxial tension. In Fig 13 the fracture strength for two multiaxial stress states is compared with the strength under uniaxial tension for varying crack length to void ratios. Triaxial tension for short cracks can improve the strength by 34% when compared to uniaxial tension, whereas axial tension combined with biaxial compression decreases the strength by 20%.

Inclusions. The effect of inclusions on the fracture strength depends on the thermal-expansion coefficient mismatch between inclusions and matrix, and on the toughness and modulus of the inclusions with respect to the matrix. At present, these situations have been modeled either as voids (low effect on fracture) or as voids with circumferential cracks. Evans (Ref 12) has summarized the severity of these different conditions (Fig 14). It can be seen that the behavior of an inclusion can range from that of a void (severely contracting rigid inclusion) to that of a void with a radial crack.

Multiple Flaws

When flaws are located in close proximity, they may link up. Quantitative models for this situation are limited to three cases:

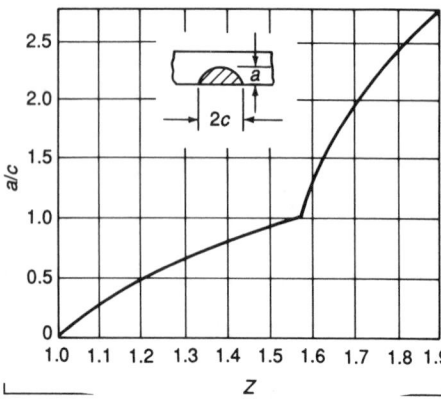

Fig 9 Effect of flaw shape on flaw shape parameter Z. Source: Ref 15

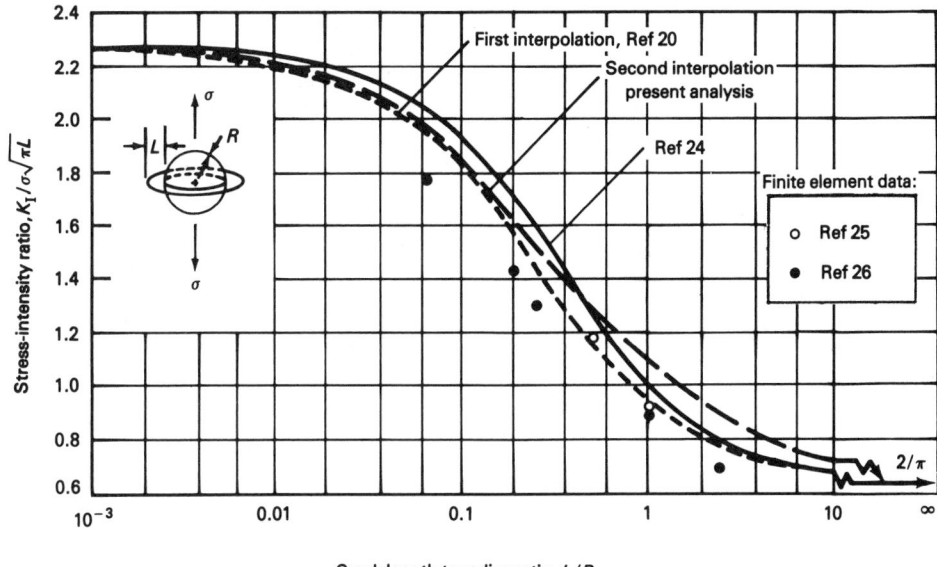

Fig 11 Stress-intensity ratio for a spherical void with a circumferential crack at its equator. Source: Ref 22

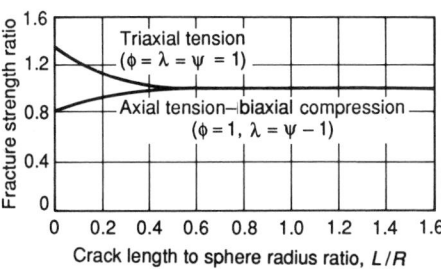

Fig 13 Ratio of multiaxial and uniaxial fracture strength of a spherical void with a circumferential crack in an infinite body. Source: Ref 27

the crack spacing is less than 50% of the crack size and decreases for larger spacing.

Similarly an approximate solution has been obtained for the stress necessary to link subsurface flaws with surface flaws (Ref 29). All three cases show that the linking stress intensity increases significantly only for configurations of closely spaced flaws. However, this linking stress might be considerably less than that required to extend in isolated flaw, especially when "easy fracture paths" are available.

REFERENCES

1. P.C. Paris and G.C. Sih, *Stress Analysis of Cracks,* STP 381, American Society for Metals, 1965, p 30–81
2. J. Eftis and H. Liebowitz, On the Modified Westergaard Equations for Certain Plane Crack Problems, *Int. J. Fract. Mech.,* Vol 8, 1972, p 383–392
3. H. Tada, P.C. Paris, and G.R. Irwin, *The Stress Analysis of Cracks Handbook,* Del Research, 1973
4. G.C. Sih, *Handbook of Stress Intensity Factors,* Lehigh University, 1973
5. M. Isida, On the Tension of a Strip with a Central Elliptical Hole, *Trans. Jpn. Soc. Mech. Eng.,* Vol 21, 1955
6. C.E. Feddersen, Discussion, *Plane Strain Crack Toughness of High Strength Metallic Materials,* STP 410, W.F. Brown, Jr. and J.E. Srawley, Ed., American Society for Testing and Materials, 1967, p 77–79
7. D.P. Rooke and D.J. Cartwright, *Compendium of Stress Intensity Factors,* Hillington Press, Uxbridge Middlesex, England, 1976
8. R.W. Rice, S.W. Freiman, and J.J. Mecholsky, Jr., The Dependence of Strength-Controlling Fracture Energy on the Flaw-Size to Grain-Size Ratio, *J. Am. Ceram. Soc.,* Vol 63, 1980, p 129–136
9. R. Steinbrech, R. Knehans, and W. Schaarwächter, Increase of Crack Resistance during Slow Crack Growth in Al₂O₃ Bend Specimens, *J. Mater. Sci.,* Vol 18, 1983, p 265–270
10. A.H. Heuer, M.J. Ready, and R. Steinbrech, Resistance Curve Behavior of Supertough MgO-Partially-Stabilized

- Multiple spherical voids with circumferential cracks
- The coalescence of a coplanar array of through-the-thickness cracks
- Linking of a subsurface pore with a surface flaw

In all three cases, the stress-intensity factor of a given flaw is increased due to the interaction between closely spaced flaws. Baratta and Parker (Ref 27) approximated the stress-intensity factors for an infinite body containing multiple spherical voids with circumferential cracks under uniaxial and multiaxial loading conditions. The results depend on the crack length to void radius ratio L/R and the spacing between voids (2 h in Fig 15).

Paris and Sih calculated the stress-intensity factor for an array of coplanar through-the-thickness cracks, as a function of the crack size, the distance between the cracks, and the number of cracks in the array (Ref 1, 7). The interaction effect becomes significant when

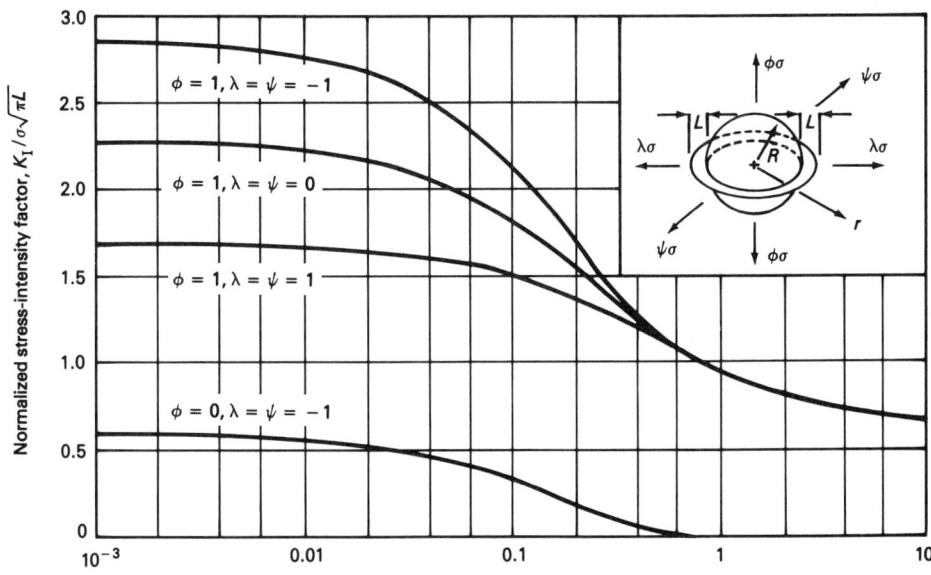

Fig 12 Normalized stress-intensity factor for a multiaxially stressed spherical void with a circumferential crack. Source: Ref 27

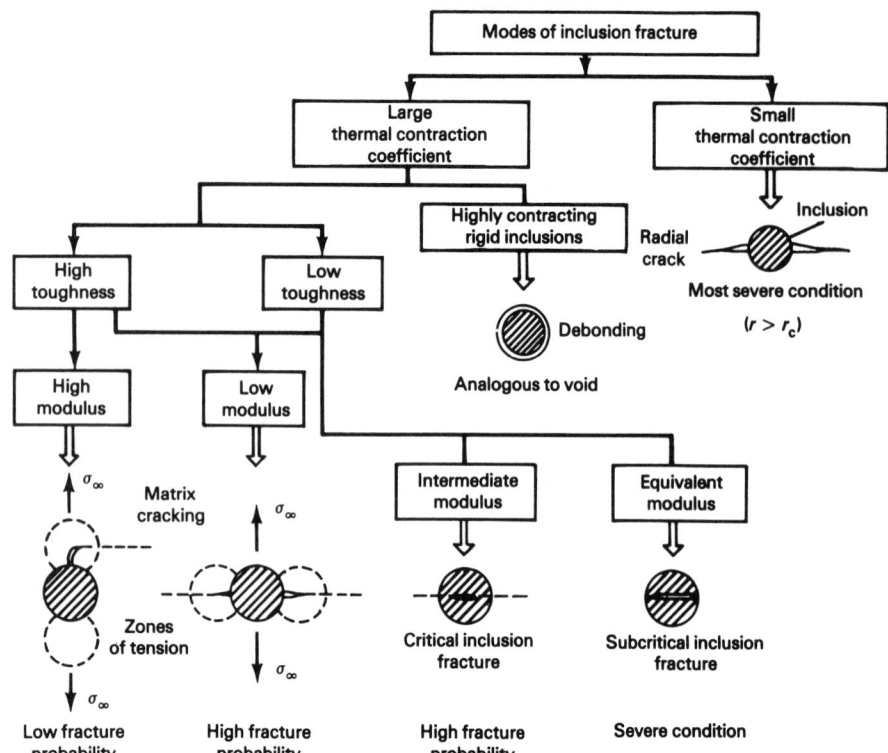

Fig 14 Various types of inclusion-initiated fracture processes. Source: Ref 12

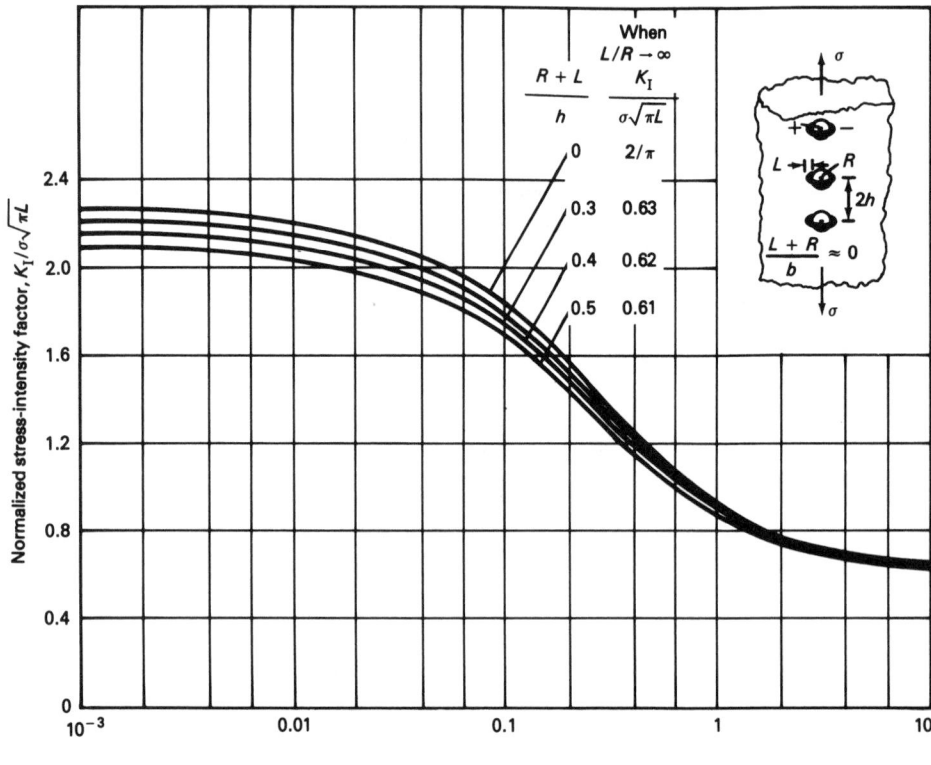

Crack length to shere radius ratio, L/R

Fig 15 Normalized stress-intensity factors for an array of spherical voids with circumferential cracks in an infinite body under uniaxial tension. Source: Ref 27

ZrO_2, *Mater. Sci. Eng. A.*, Vol 105, 1988, p 83–89

11. S.J. Bennison and B.R. Lawn, Flaw Tolerance in Ceramics with Rising Crack Resistance Characteristics, *J. Mater. Sci. Eng.*, Vol 24, 1989, p 3169–3175

12. A.G. Evans, Structural Reliability: A Processing-Dependent Phenomenon, *J. Am. Ceram. Soc.*, Vol 65, 1982, p 127–137

13. I.S. Raju and J.C. Newman, Stress Intensity Factors for a Wide Range of Semi-Elliptical Surface Cracks in Finite Thickness Plates, *Eng. Fract. Mech.*, Vol 11, 1979, p 817–829

14. A.S. Kobayashi, M. Zii, and L.R. Hall, Approximate Stress Intensity Factor for an Embedded Elliptical Crack Near to Parallel Free Surfaces, *Int. J. Fract. Mech.*, Vol 1, 1965, p 81–95

15. G.K. Bansal, Effect of Flaw Shape on Strength of Ceramics, *J. Am. Ceram. Soc.*, Vol 59, 1976, p 87–88

16. D.B. Marshall, B.R. Lawn, and P. Chantikul, Residual Stress Effects in Sharp Contact Cracking, II Strength Degradation, *J. Mater. Sci.*, Vol 14, 1979, p 2225–2235

17. B.R. Lawn, A.G. Evans, and D.B. Marshall, Elastic/Plastic Indentation Damage in Ceramics: The Medium/Radial Crack System, *J. Am. Ceram. Soc.*, Vol 63, 1980, p 574–581

18. D.B. Marshall, Surface Damage in Ceramics: Implications for Strength Degradation, Erosion and Wear, *Nitrogen Ceramics*, F.L. Riley, Ed., Nijhoff, The Netherlands, 1983, p 635–656

19. G.K. Bansal and W.H. Duckworth, Fracture Stress as Related to Flaw and Fracture Mirror Sizes, *J. Am. Ceram. Soc.*, Vol 60, 1977, p 304–310

20. F.I. Baratta, Stress Intensity Factor Estimates for a Peripherally Cracked Spherical Void and a Hemispherical Surface Pit, *J. Am. Ceram. Soc.*, Vol 61, 1978, p 490–493

21. F.I. Baratta, Correction—Stress Intensity Factor Estimates for a Peripherally Cracked Spherical Void and a Hemispherical Surface Pit, *J. Am. Ceram. Soc.*, Vol 62, 1979, p 527

22. F.I. Baratta, Refinement of Stress Intensity Factor Estimates for a Peripherally Cracked Spherical Void and a Hemispherical Surface Pit, *J. Am. Ceram. Soc.*, Vol 64, 1981, p C3-C4

23. A.G. Evans, D.R. Biswas, and R.M. Fulrath, Some Effects of Cavities on the Fracture of Ceramics: II Spherical Cavities, *J. Am. Ceram. Soc.*, Vol 62, 1979, p 101–106

24. D.J. Green, Stress Intensity Factor Estimates for Annular Cracks at Spherical Voids, *J. Am. Ceram. Soc.*, Vol 63, 1980, p 342–344

25. D.M. Tracey, private communication, 1980

26. R. Chang, Static Finite Element Stress Intensity Factors for Annular Cracks, Technical Report SC-PP 80–50, Rockwell International Science Center, 1980

27. F.I. Baratta and A.P. Parker, Mode I Stress Intensity Factor Estimates for Various Configurations Involving Single and Multiple Cracked Spherical Voids, *Fracture Mechanics of Ceramics,* Vol 5, R.C. Bradt, A.G. Evans, D.P.H. Hasselman, and F.F. Lange, Ed., Plenum Press, 1983, p 543–567

28. D.K. Shetty, A.R. Rosenfield, and W.H. Duckworth, Biaxial Stress State Effects on Strengths of Ceramics Failing from Pores, *Fracture Mechanics of Ceramics,* Vol 5, R.C. Bradt, A.G. Evans, D.P.H. Hasselman, and F.F. Lange, Ed., Plenum Press, 1983, p 531–542

29. A.G. Evans and G. Tappin, *Proc. Br. Ceram. Soc.,* Vol 20, 1972, p 275

Quantitative Fracture Surface Analysis

Terry A. Michalske, Sandia National Laboratories

FRACTURE SURFACE MARKINGS (FSM) can be used to draw quantitative conclusions about a fracture event because a passing fracture front leaves a permanent record of its history in the form of topographic markings (Ref 1–3). The FSM can be used to locate the failure origin and determine how the catastrophic crack developed under the influence of the applied and transient stress fields. The FSM also show how the microstructure and processing defects in the material influenced the catastrophic failure event.

The clues left behind by FSM are used to investigate the circumstances surrounding a failure event (during service life or testing) in hopes of answering questions such as:

- What was the nature of the defect that originated failure?
- What was the magnitude of the applied stress at failure?
- What was the stress state at failure (uniaxial tension, biaxial tension, and so forth)?
- Did factors such as residual stresses or the presence of stress-corrosion environments influence the failure event?

Quantitative answers to these questions are obtained by employing a combination of fracture mechanics-based and empirically based relations to convert measurements obtained from FSM to quantitative estimates for the pertinent fracture parameters. Fracture surface markings also are used to develop realistic models for failure mechanisms in a new material.

Fracture Features

This article discusses how important fracture features are developed, measured, and converted into a quantitative measure of the fracture conditions. For simplicity, the principle of quantitative fracture surface analysis will concentrate on isotropic, homogeneous brittle materials (for example, glasses) and then show how anisotropic crystalline or polycrystalline properties of a ceramic material may influence the characteristics of the FSM.

The basic feature of all failures begins at a specific point called the failure origin. Once

the failure origin has been identified, the origin flaw size can be used to estimate the magnitude of the stress at failure. As the fracture spreads away from the failure origin, the fracture velocity may increase. The formation of FSM such as mist, hackle, and branching, which indicate high-velocity fracture, can also be used to estimate the magnitude of stress at failure. Wallner lines offer a "snapshot" of the crack front as it passes through the material. The spacing between and the intersection of Wallner lines that are created during the fracture process can provide a quantitative, local measure of the stresses in the material. The fragmentation pattern created by the branching of the primary crack provides information about the magnitude and state of the applied stress (uniaxial or biaxial tension). Finally, effects such as residual stress and stress-corrosion cracking can be quantitatively assessed through measurements on the fracture surface. Each of these features are described in subsequent sections of this article.

Fracture Mechanics Relations. As indicated in the introduction, some of the quantitative aspects of fracture surface analysis are based upon continuum elastic fracture mechanics relations. The following discussions provide a brief description of the fracture mechanics relations that are used in quantitative fractographic analysis. Other articles in the Section "Failure Analysis" of this Volume provide more detailed discussions on the application of fracture mechanics to the failure of ceramic materials.

The most fundamental of fracture mechanics parameters is the crack tip stress intensity (K_I). When a continuous elastic material containing a crack is exposed to an externally applied stress, a concentrated field of stress is developed near the tip of a crack or flaw in the material. This stress field, which is concentrated by the crack in the material, produces the locally high stresses that lead to material failure and crack propagation. The magnitude of the stress field local to the crack tip is described by the K_I parameter. Fracture mechanics formulations show that the magnitude of K_I is given by:

$$K_I = \sigma Y \sqrt{a} \qquad \text{(Eq 1)}$$

where σ is the applied stress, a is the length of the crack, and Y is a parameter that accounts for the crack geometry. Equation 1 has important applications in failure analysis of glass and ceramic materials. If we can determine the size and shape of a crack and the magnitude of the external stress, we can identify a unique value of the crack tip stress intensity that will cause crack propagation in the material of interest. In practice, we use one of a number of controlled geometry fracture mechanics tests (Ref 4) to measure the value of K_I for crack propagation. The stress intensity value for catastrophic propagation is referred to as the critical stress intensity (K_{Ic}). The critical stress intensity is an extremely useful parameter in brittle materials failure since it provides a single parameter to describe the resistance of the material to crack propagation. In fact, Eq 1 can be rewritten as:

$$\sigma_f = \frac{K_{Ic}}{Y \sqrt{a}} \qquad \text{(Eq 2)}$$

where σ_f is the failure strength. Using this relation and the measured value for K_{Ic}, one can predict a practical strength for the ceramic material as a function of flaw size.

The fracture mechanics approach to materials strength allows us to understand and predict the effect of surface flaws on the strength of glasses and ceramic materials. We can easily see from Eq 2 that increasing the size of the surface flaws will greatly decrease the materials strength.

This limited discussion treats fracture as an on-off process. At applied stress intensities below K_{Ic}, no crack propagation occurs while above K_{Ic}, the material catastrophically fractures. In many engineering materials of interest, once a crack begins to propagate, the speed of the advancing crack will be dependent upon the magnitude of K_I. Because the crack speed increases very rapidly with increasing K_I, the magnitude of the error introduced by considering fracture as an on-off process is usually quite small. Only in special cases, particularly in the presence of chemically reactive environments, does this on-off approximation of the fracture event produce significant errors. In the section "Stress-Corrosion Fracture" of this article, the topic of stress corrosion in glass and ceramics and the fracture surface markings associated with that regime of failure is explored.

Origin Flaw

As its name implies, the origin flaw is the defect from which the entire fracture event develops. For this reason it is one of the most important features imprinted on the fracture surface. Location of the origin flaw is determined by the fractographic methods described in the article "Descriptive Fractography" in this Volume. Once the origin flaw is located, then its size can be used to quantitatively estimate the magnitude of stress at failure and to quantitatively interpret the event that produced the failure origin.

Failure origins are any topographic or compositional features that concentrate the applied stress to a magnitude that locally exceeds the fracture strength of the material. These features can be processing related as in the case of pores or inclusions, or they may be small surface cracks. In brittle ceramic or glassy materials, surface cracks are often generated by mechanical contact or surface abrasion events. In these hard materials, mechanical contact usually occurs at surface asperities. Localized contact generates large elastic stresses that are confined to the area of contact. When the local stress exceeds the materials strength, a surface flaw is created which develops only in the region of locally high stress. The resultant surface crack concentrates the applied stress (as previously discussed in the section "Fracture Mechanics Relations") and can serve to nucleate the catastrophic fracture event.

Identification and measurement of origin flaw size is possible because its boundary will be outlined by an arrest line. The arrest line that indicates the origin flaw is often easy to detect since the localized stress field that produced the origin flaw will not be oriented in exactly the same way as the applied stress that causes failure. Because cracks always propagate normal to the applied stress field, the plane of the origin flaw will be inclined with respect to the plane of catastrophic crack propagation and can easily be observed in reflected light conditions (Fig 1).

Estimating Failure Stress from Origin Flaw Size. After the size of the cracking origin is determined, linear elastic fracture mechanics (Eq 2) can be used to evaluate the applied stress that would have caused failure. Values for K_{Ic} are available for most glass and ceramic materials and can be obtained from materials properties handbooks (Ref 5, 6). The geometric factor (Y) for various idealized flaw shapes is available in linear elastic fracture mechanics complications (Ref 7). In the case of a well-defined surface crack, one should be able to use the technique described above to convert a measured crack size on the frac-

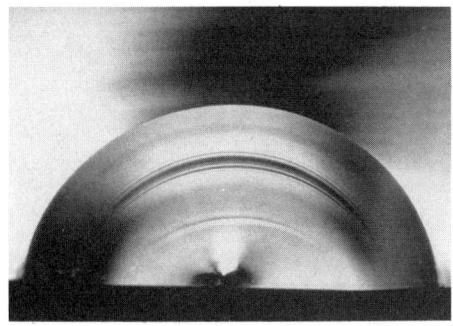

Fig 1 Semicircular fracture origin in glass surface. Slight differences in orientation of the origin flaw allow it to be distinguished from subsequent fracture plane using reflected light microscopy. 25×

ture surface into an estimate for the failure stress.

However, extreme caution should be used when applying the above method for estimating the stress at failure. Many of the surface flaws encountered in practice are far from idealized geometries (see Fig 2), and it may be difficult to determine an appropriate Y value to account for their stress-concentrating effect. For the case of well-defined flaws, the errors associated with the geometric factor are not large (less than a factor of 30%, Ref 8) and should not prevent a reasonable estimate of the failure stress. A more serious problem is that the origin flaw will most likely not be aligned with the direction of the applied stress that precipitates catastrophic failure. The relation given in Eq 2 only holds true for origin cracks that are aligned with the applied stress at failure. As the alignment between the origin flaw and the applied stress becomes worse, the stress-concentrating power of a given size crack will decrease and cause one to severely underestimate the failure stress.

When the flaw size is used to estimate the failure stress in polycrystalline materials, the above-mentioned precautions must be observed as well as a consideration for the effect of the microstructure. Mecholsky *et al.* (Ref 8) has reviewed the effect of microstructure on flaw

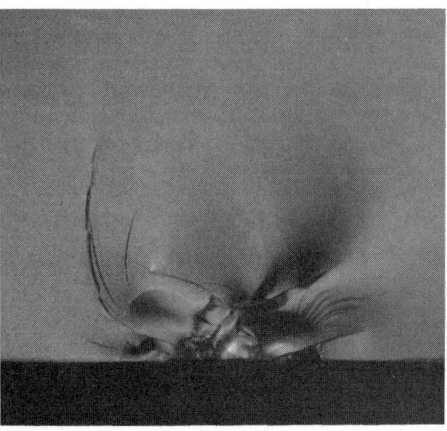

Fig 2 Typical failure origin in glass produced by asperity contact at glass surface. In practice, the complex geometry of failure origins can make flaw size determinations difficult. Reflected light, 50×

size estimates of failure stresses. They show that errors in failure stress determinations can be attributed to two separate effects: a microstructure dependent value for K_{Ic} and the generation of local residual stresses. Several studies have shown that the inherent fracture resistance of a material is affected by the ratio of the flaw size to the grain size (Ref 9, 10). In most cases, K_{Ic} increases as the flaw size/grain size increases. This effect would cause one to overestimate the failure strength that was determined from measurements of a small flaw, because values for K_{Ic} are normally determined from large-crack fracture mechanics test techniques (Ref 4).

Local residual stresses can arise from thermal expansion anisotropies in the grains of a polycrystalline material. In the case of flaws that are large with respect to the grain size, the flaw will span several grains and the effect of the local stresses are expected to average out. However, a flaw that encompasses only one or two grains may be subject to a large residual stress contribution. This stress contribution may either add or subtract from the applied stress and therefore can cause an under- or overestimate in the failure stress.

In the case of noncracklike origin flaws such as pores or inclusions, it is not generally possible to evaluate the magnitude of the applied stress from measurements of the defect size. In the case of a spherical pore the stress concentration is independent of the size of the pore (Ref 11). Therefore, the size of the pore alone cannot be used to estimate the failure stress during an investigation after failure. The stress concentration around an inclusion of a foreign material is extremely difficult to evaluate because it depends not only on the shape of the inclusion but also on the thermal expansion and elastic mismatch between the inclusion and the matrix, and on the strength of the interfacial bond between the inclusion and the matrix. All of these factors contribute to the stress concentration in a defect and make it extremely difficult to use the inclusion size to estimate failure stress.

Geometric Characteristics of Surface Cracks Caused by Contact Stresses. In the preceding section, the size of an origin flaw is used to estimate the failure stress. In the case of surface cracks caused by localized contact stresses, some quantitative aspects of the contact event can be determined from the size of a crack origin. Hertz (Ref 12), for example, studied the elastic problem of a sphere contacting a flat surface and showed that the largest tensile stresses occur at the flat surface just outside the radius of mutual contact. This stress distribution produces a cone crack that initiates just outside the zone of mutual contact (Fig 3). Lawn *et al.* (Ref 13) have shown that the size of the cone crack (R) can be related to the normal load (P) as follows:

$$P^2/R^3 = Eb \qquad \text{(Eq 3)}$$

where E is the elastic modulus of the substrate, and b is an empirically derived constant that must be evaluated for the material of interest. By measuring the diameter and depth of the Hertzian cone flaw, one can estimate the size and normal force of the projectile that contacted the surface.

For the case of highly angular contact asperities, fracture mechanics formulations for pyramidal indenters can be used to guide the quantitative fractographic analysis. Figure 4 shows the surface flaw geometry developed by contact with an angular asperity. Marshall *et al*. (Ref 14) examined the cracking around indenter tips and have shown that crack size can be related to the indenter force as follows:

$$P = \frac{(\sqrt{a})^3 K_{Ic}}{\chi} \qquad \text{(Eq 4)}$$

where P is the indenter load, a is the length of the radial cracks, and χ is an empirical parameter that must be determined for the specific material. Wiederhorn (Ref 15) has used this relation to examine erosion of glass and ceramic surfaces by angular particle impact. He has shown that the particle impact parameters can be estimated from inspection of the resulting surface cracks.

Mirror Region of Crack Propagation

Once the catastrophic crack has been nucleated, it will begin to spread from the origin under the influences of the applied stress field. The early stage of crack propagation produces a region which is very smooth and flat and is often referred to as the mirror. In glass, which has no microstructural detail, this region appears mirror smooth. In materials with microstructural details, however, the roughness of the mirror region will be dictated by the scale of the microstructure.

Wallner Lines. Within the mirror region of crack propagation, transient variations in the stress field will result in undulations in the crack path that will appear as Wallner lines (Ref 16) on the fracture-generated surface. As previously indicated, a crack will always propagate normal to the applied tensile stress field. When a transient stress wave acts to tilt the local applied stress, the crack will respond

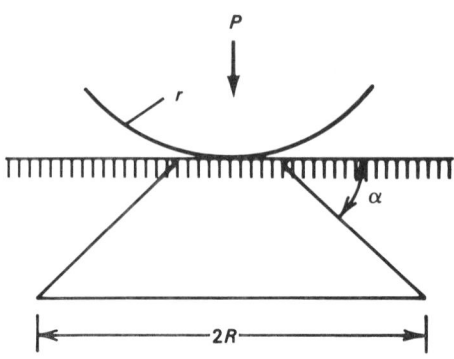

Fig 3 Hertzian cone fracture pattern produced by spherical particle contact with flat surface

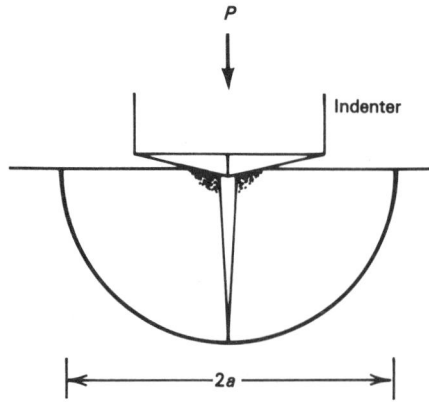

Fig 4 Radial fracture pattern developed by pyramidal indenter contact with flat surface

to the changing stress field by altering its plane of propagation during the transient event. As a transient stress wave sweeps past the propagating crack, the Wallner line that is developed will represent a snapshot of the crack front shape and position during the transient event. The stress transients that produce Wallner lines can be the result of:

- A crack front passing an inclusion or pore in the material that alters the local stress field
- The mechanical ringing of the sample as stress waves reflect across the interior of the sample during failure

Determination of Crack Velocity. Poncelet (Ref 17) has shown that the intersection of Wallner lines formed by two separate transient stress events can be used to determine the velocity of the advancing crack front. Figure 5 shows the geometric identities that are involved in determining the crack velocity from Wallner line intersection. This analysis provides an important means of quantitatively determining the local crack velocity from measurements of the FSM.

Ultrasonic fractography can also be used to assess the local crack velocity. In this technique, Wallner lines are produced by purposefully inducing a pattern of transient stress pulses during crack propagation. A piezoelectric or other electrochemical device is used to generate a series of controlled period stress transients. These stress transients produce a series of Wallner lines with a known temporal distribution (Fig 6). One needs only to measure the spacing between these lines on the fracture surface and divide by the period of the stress pulse signal to accurately determine the local crack velocity. This approach to crack velocity measurement is often referred to as ultrasonic fractography. The original studies (Ref 18–20) utilizing this approach were designed to investigate high-speed fracture events (crack velocity > 100 m/s), and so the frequency of stress pulses was necessarily in the MHz range. However, more recent applications of this technique (Ref 21–23) have shown that it can be used to examine crack

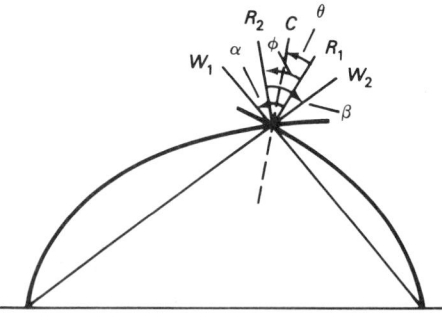

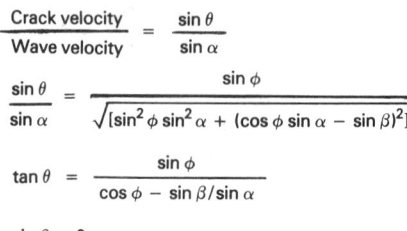

$$\frac{\text{Crack velocity}}{\text{Wave velocity}} = \frac{\sin \theta}{\sin \alpha}$$

$$\frac{\sin \theta}{\sin \alpha} = \frac{\sin \phi}{\sqrt{[\sin^2 \phi \sin^2 \alpha + (\cos \phi \sin \alpha - \sin \beta)^2]}}$$

$$\tan \theta = \frac{\sin \phi}{\cos \phi - \sin \beta / \sin \alpha}$$

$$\sin \beta < 0$$

Fig 5 Identities used to determine the crack velocity from the intersection of two Wallner lines on the fracture surface. R_2 and R_1 are lines drawn normal to each Wallner line at the point of intersection. C is a line drawn from the point of intersection to the failure origin.

velocities as low as 10^{-3} m/s (0.04 in./s) using low-frequency stress pulse signals.

Detailed information about the crack velocity can provide very much important information concerning the stress field responsible for the fracture event. Fracture mechanics studies have shown that a unique relation exists between the crack velocity (V) and the stress intensity (K_I). (The details of this relationship are discussed in the section "Stress-Corrosion Fracture" of this article.) By using the K_I versus V relationship and measuring the local crack velocity by fractographic techniques one can accurately measure the local stress intensity of a propagating crack. Michalske *et al*. (Ref 22–23) have used this approach to experimentally assess the

Fig 6 Wallner lines produced by imposing a periodic stress onto the static stress applied during crack propagation. Differential interference contrast, 50×

validity of fracture mechanics test methods and fracture mechanics analysis of various surface flaw geometries.

Mirror Radius

As a catastrophic crack continues to spread out from the origin under the influence of a constant applied stress, it may accelerate until it reaches its "terminal" velocity. Under static loading conditions, crack velocities in brittle materials have been observed by use of ultrasonic fractography and high-speed fractography to reach about six-tenths the Rayleigh wave speed (Ref 20). As the terminal velocity is approached, the fracture surfaces lose their mirror smooth condition and begin to show a gradual increase in surface roughness. The regions of increased surface roughness are referred to as mist and hackle. These regions on surface roughness are sometimes followed by crack bifurcation or branching. A microscopic examination of the fracture surface in the region of mist and hackle show that the observed roughness is a result of microscopic deviations of local sections of the crack away from the principal fracture plane.

As the surface goes from mist to hackle, the excursions become larger and proceed further from the principal fracture plane. At the point of crack branching, one of the deviations becomes large enough to nucleate a second crack which begins to propagate in a stable fashion. Several hypothesis have been put forth to explain the formation of the instability that generates the mist, hackle, and branching (Ref 24–26); however, no clear or definitive theories have yet emerged. In a qualitative sense, the formation of the instability is clearly associated with high crack velocity which accompanies a higher stress intensity. It has been suggested that either the increased velocity or the increased strain energy at the crack tip will produce the instability responsible for these FSM.

Figure 7(a) shows a schematic indicating the fracture mirror, mist, hackle, and branching pattern on a fracture surface. Figure 7(b) shows a photomicrograph of a representative fracture mirror in glass. The distance from the center of the fracture origin to the onset of mist is referred to as the mirror radius. It has been shown in several studies (Ref 27, 28) that an empirical relation of the form:

$$\sigma_f \sqrt{R_i} = A_i \qquad \text{(Eq 5)}$$

can be used to relate the mirror radius to mist, hackle, and branching to the magnitude of stress at failure (σ_f). In each case, R indicates the mirror radius and A is the empirically derived constant. A_M, A_H, and A_B correspond to the mirror constants for mist, hackle, and branching, respectively. Mecholsky (Ref 29) has examined the fracture mirror relationship for glass failures spanning nearly two orders of magnitude in failure stress. His results (Fig 8) indicate that a single mirror constant can be applied to the entire range of failure stresses.

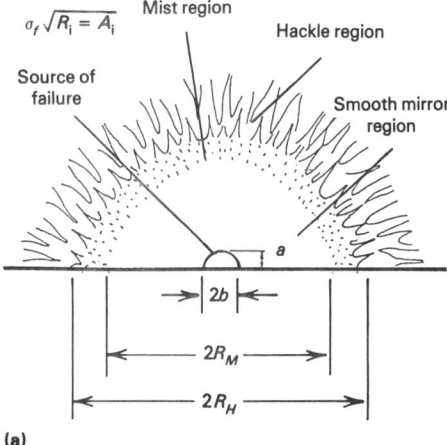

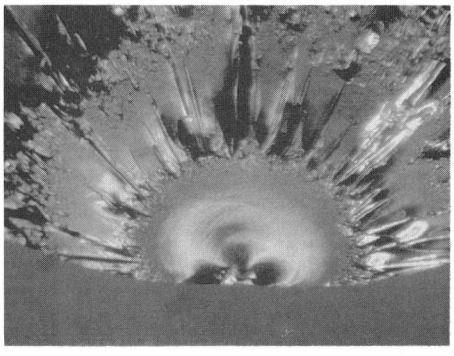

(a)

(b)

Fig 7 (a) Schematic representation of a fracture surface showing the formation of mist and hackle that define the mirror radius. (b) Photomicrograph of fracture surface in glass showing mirror, mist, hackle, and crack branching radii. Reflected light, 10×

Equation 5 is a very important relationship in quantitative fractography, because it can be used to estimate the failure stress with good precision over a wide range of applied stresses simply by measuring the mirror radius on a failed component. An important advantage of

the mirror radius method for estimating the failure stress over the use of the origin flaw size is that the mirror radius examines a portion of the fracture event where the crack has developed its equilibrium shape and is well aligned with the applied stress field. This removes some of the sources of error previously discussed in using the origin crack which may be highly complex in shape and poorly aligned with the applied stress at failure.

In many cases, the origin flaw will not be centered within the mirror boundary. This is often associated with complex flaws where the catastrophic crack preferentially nucleates from one portion of the flaw boundary. Since the mirror boundary is associated with a well-developed crack, the actual location of the origin flaw is unimportant and one should measure the complete mirror diameter and divide by two to obtain the radius.

In determining the failure stress from a mirror radius measurement, there are several important points which must be considered:

- The measurement technique
- The nature and uniformity of the stress field
- The presence of residual stresses

Measurement technique is important because the onset of gradual roughening associated with mist or hackle will be a subjective determination. That is, when viewed with a low-magnification stereoscopic microscope, the onset of mist or hackle may not be perceived at the same point as viewed with a high-powered optical microscope or a scanning electron microscope. For this reason, it is important to use the same observation technique during failure analysis as was used to produce the empirical mirror radius relation. It is also advisable to avoid measuring the mirror boundary along a free surface of the glass because surface damage can influence the nucleation of the mist and hackle feature

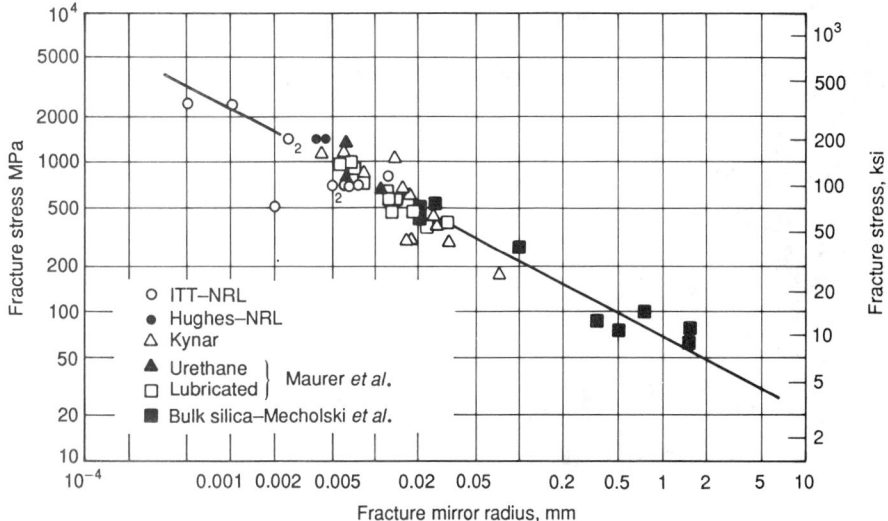

Fig 8 Mirror radius relation obtained for silica glass fracture. Mirror radius measurements made to mist boundary. Source: Ref 29

and may result in greater uncertainties in the estimate of failure stress (Ref 27).

Another important consideration in applying the mirror radius method for failure stress is the nature of the stress field both in the empirical calibration and in the actual failure event. Figure 9, for example, shows the mirror shape obtained in bending stress condition. In bending, the applied tensile stress is maximum at one surface and decreases linearly through the thickness of the sample, reaching zero stress at the centerline and producing an equal compressive stress on the opposite surface. Obviously, a flaw spreading under the influence of this nonuniform stress field will accelerate more rapidly in the regions of high tensile stress. In Fig 9, it is clear that the fracture mirror is not uniform throughout the thickness of the sample but is generally elongated in the direction of decreasing bending stress. It is obvious that a measurement of the mirror radius in such a case will be highly dependent upon the portion of the mirror boundary that is examined. For this reason, the results of a calibration obtained in a uniform stress field may be misleading if used to estimate the failure stress for a sample that fractured in bending. However, if the mirror region in the bending failure is small with respect to the specimen dimensions (<10% of the thickness), one may consider the stress field as a uniform tensile condition and therefore obtain results that are comparable to a uniform applied tensile stress.

Little work has been done to examine the mirror radius relation for more complex, multidimensional stress fields. Mecholsky and Rice (Ref 30) have examined the fracture surfaces of a glass ceramic and a glass specimen tested in biaxial flexure. Their results indicate that the mirror radius relation describing biaxial flexure and uniaxial flexure are indistinguishable. This is very important since it may not always be possible to determine the exact nature of the stress field involved in a field failure.

Residual stresses can also have an important effect on the formation of the mirror radius. Following from previous discussion, a residual tensile stress will add to the applied stress field and increase the rate of crack propagation. The effect of this will be to accelerate the development of mist, hackle, and branching. Alternatively, a residual compressive stress can be expected to suppress the formation of the mist, hackle, and branching features. Work by Kerper and Scuderi (Ref 31) has shown that residual stresses introduced by tempering can effectively shift the mirror radius relation if the residual stresses are uniform over the region of the mirror. If the residual stresses vary over the region of the mirror, the mirror shape can be distorted, making quantitative estimates of the residual stress and applied stress difficult or impossible by post-failure radius measurements.

The application of the mirror radius/failure stress relation to polycrystalline materials is subject to one major limitation.

The microstructure of the material can induce an instability in the plane of crack propagation that also creates fracture surface roughness. This inherent surface roughness can mask the features associated with mist and hackle. Wu *et al.* (Ref 32) has used x-ray microradiography to examine the crack path in various polycrystalline materials and has shown that crack path instabilities that result from the polycrystalline microstructure appear as microbranching of the primary crack and are similar to the character of mist and hackle features observed on an amorphous material. The increased "background" surface roughness can make it difficult to determine the precise onset of the crack plane instabilities associated with mist and hackle in polycrystalline materials. In a practical sense, this limitation means that the mist and hackle features must be coarser on a polycrystalline material before they can be distinguished. In extremely coarse-grained materials, where the scale of the microstructure is on the order of the mirror size it may be impossible to identify the mirror radius.

In cases where the microstructure elements are much smaller than the mirror radius, the mirror radius relation may be used to evaluate the stress at failure. Figure 10 shows the

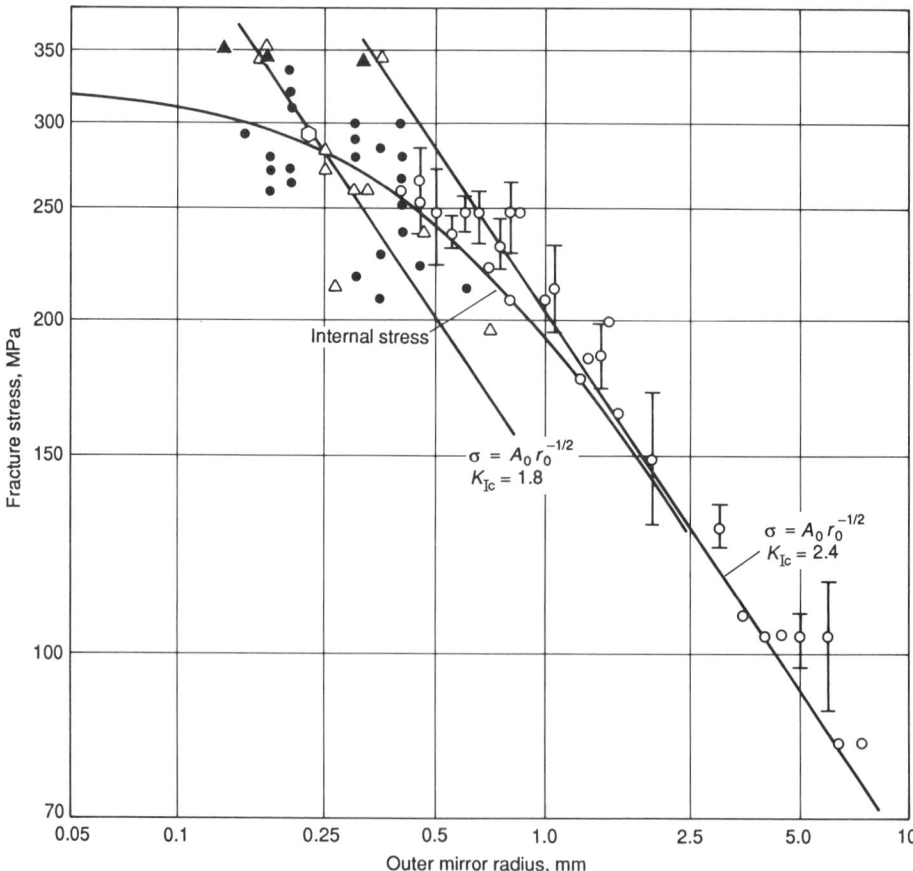

Fig 9 Fracture mirror in glass specimen exposed to bending stress. The asymmetric mirror shape is a result of a nonuniform stress field developed in bending. Reflected light, 14×

Fig 10 Mirror radius relation determined for polycrystalline glass ceramic. Internal stresses resulting from crystal anisotropy affect mirror radius relation at small mirror dimensions. Source: Ref 34

mirror radius relation determined for a polycrystalline glass ceramic material by Rice *et al.* (Ref 33). This work shows that the expected mirror relation (Eq 5) is observed for larger mirror radius conditions. However, at smaller mirror radius (larger failure stresses) the data indicate significant deviation from the strength/mirror radius relationship. Lewis (Ref 34) has attributed this behavior to the effects of microstructural stresses that result from anisotropic thermal expansion in the individual grains of polycrystalline material. These data indicate that caution must be applied when the mirror size is on the order of the scale of the microstructure.

Single-Crystal Fracture. The strength/mirror radius relation has been studied to a lesser extent for the case of single-crystal fracture. The limited results indicate that the extreme anisotropy in single crystals leads to fracture mirrors that are very complex and difficult to interpret. Rice (Ref 1) has recently reviewed the work done on the fracture mirrors in single-crystal ceramic materials and gives a very comprehensive account of the state of understanding in this area. His survey suggests that the complexity in the observed fracture mirror boundaries gives more clues toward the fundamental mechanisms of mist and hackle formation rather than providing a reliable approach to quantitative estimates for failure stress.

Crack Branching

As the developing crack continues to accelerate under the influence of the applied stress, it may branch and form multiple, stable primary fractures. The phenomenon of crack branching is a result of the same insta-

Table 1 Fragmentation patterns in thin walled structures

Specimen	Stress condition	Ratio of principal stresses, σ_y/σ_x	Maximum angle of branching, degrees
Tube	Torsion	−1	15
Lath	Crossbending	0	45
Container	Pressure	1/2	90
Sheet	Central pressure	1	180

bility that generates mist and hackle on the fracture surface, and a relation similar to the mirror radius relation established in the previous section can be used to estimate the failure stress from measurements of branching radius. In this section, however, two other quantitative fractographic relations that utilize measurements of crack branching are discussed. In one case, the total number of branches that form the fragmentation pattern

(a)

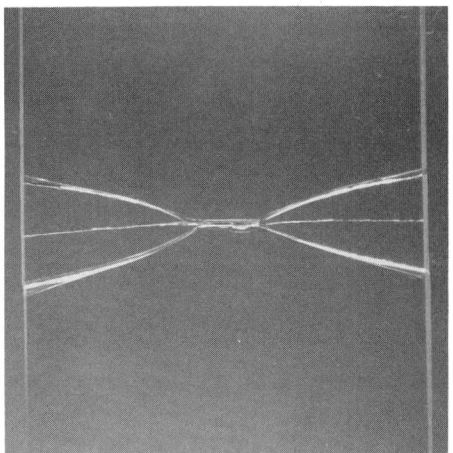

(b)

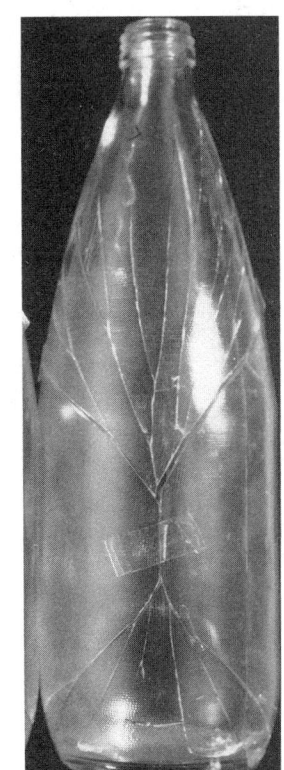

(c)

(d)

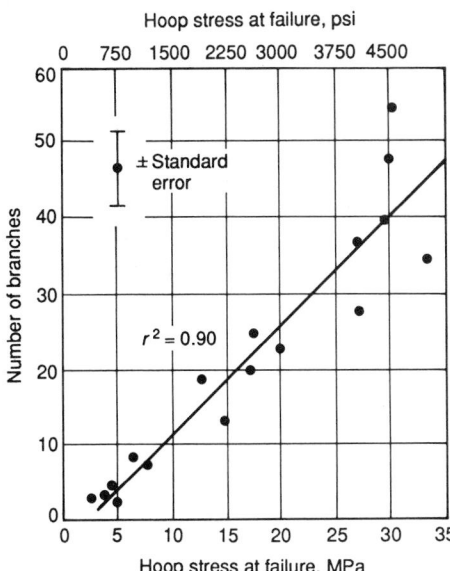

Fig 11 Failure stress versus the number of crack branching events observed for glass containers fractured under internal pressure. Source: After Ref 35

Fig 12 Fragmentation patterns observed under various stressing conditions. (a) Torsion. (b) Crossbending. (c) Internal pressure. (d) Drumhead tension

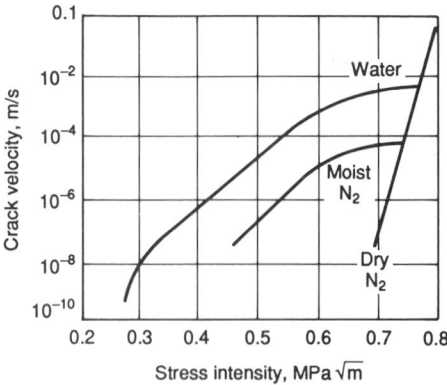

Fig 13 Schematic representation of crack velocity versus stress intensity relation for silicate glass in moist and dry conditions

can be used to estimate the stress at failure. In another quantitative fractographic technique, crack branching geometries are used to assess the nature (uniaxial, flexure, torsion) of the stress field that was acting at the time of failure.

Determination of Stress at Failure from Branching Patterns. At high enough crack tip stress intensity and resultant crack velocity, a single primary crack can branch into two stable propagating cracks. If there is enough stored energy in the system, each of the new cracks can also accelerate and experience the same instability that resulted in branching of the primary crack. In this way branched cracks can themselves branch, and this process can repeat as long as there is sufficient energy to continue driving the crack acceleration process. The process of multiple

crack branching results in a fragmentation pattern that can be used to determine the magnitude of the stress at failure.

Frechette and Michalske (Ref 35) studied the fragmentation patterns for glass containers that had failed by internal pressurization. In these experiments, the internal pressure at failure was measured and used to determine the hoop stress at failure in the glass container walls. Fractographic measurements of the mirror radius dimensions (to the mist and hackle boundaries) were correlated with the failure stress and were found to obey the empirical mirror radius/failure stress relation (Eq 5). In these experiments the fragmentation pattern was also examined, and the total number of branching events was correlated with the hoop stress at failure. The results given in Fig 11 show that a linear relationship between the number of branching events and the failure stress was discovered in these experiments. Although no models were presented to explain the existence of this empirical relation, a comparison of the scatter in the mirror radius data with the observed scatter in the fragmentation results indicated that the crack branching relation provided a better method for estimating the failure stresses in pressurized glass containers. It was reasoned that the smaller amount of scatter in the observed fragmentation relation could be attributed to the averaging process inherent to multiple crack branching events. That is, multiple crack branching samples a large portion of the glass container and may provide a better average of the overall thickness variations and other irregularities associated with glass beverage containers. Alternatively, the mirror radius measurement samples a small segment of the container and is therefore sensitive to the local distribution of geometric irregularities.

Estimating the Ratio of Principal Stresses at Failure. The fragmentation pattern generated by crack branching not only provides a means of estimating the magnitude of the applied stress at failure but can also be used to evaluate the ratio of the principal stresses acting at failure. Preston (Ref 36) shows that the included angle of branching observed in the fragmentation pattern fits an empirical relationship (Table 1). Here the included angle represents the entire branching pattern in the case of multiple branching failures. Figure 12 shows the fragmentation patterns observed for each of the four principal stress ratios considered by Preston. Although this is only a rough relation, it can be used to interpret fractographic observations to distinguish between the most commonly observed stressing conditions responsible for failure in glass and ceramic materials.

(a)

(b)

Fig 14 (a) Stages (1 to 5 described in text) in fracture front development from a dry starter crack in a partly wetted specimen. Stippled areas indicate liquid water trailing the crack front. Arrow shows initial area of water contact. (b) Fracture surface showing intersection scarp developed at water boundary. Reflected light, 25×. Source: Ref 38

Stress-Corrosion Fracture

It has long been recognized that glass and other ceramic materials crack more easily in the presence of water. In fact, glaziers often apply water to the shallow crack produced by their scribing tool. The water decreases the stress required to propagate the initial crack and causes failure at a lower applied stress. The effect of water is especially pronounced for cracks that are growing very slowly (<1 cm/s). Figure 13 shows a K_1 versus V plot for silicate glass indicating that the presence of water as a liquid or gas enhances the rate of crack propagation in the low stress intensity regime.

The extremely slow growth of a surface crack is very important because it can give rise to a phenomenon know as delayed failure. Delayed failure causes structures that have supported a load for long periods of time to suddenly and spontaneously fail. Longer flaws concentrate more stress and can result in catastrophic failure at smaller applied stresses. In the case of delayed failure, a small flaw can grow under the influence of the applied stress and water in the surrounding environment until it is large enough to cause catastrophic failure. Previous studies have shown that long-term failure stress may be as small as one-fourth the measured short-term (rapid) failure strength.

Because water can have such a profound effect on the strength and mechanical durability of glass and ceramic materials, it is often important to determine whether stress-corrosion fracture was responsible for a specific field failure. This section describes indirect and direct fractographic techniques that can be used to identify the presence of water during the fracture event. An indirect method for determining the stress-corrosion conditions involves the use of the mirror-to-flaw size ratio. It has also been shown that the presence of liquid water in the propagating crack will also produce unique FSM that can be used to directly assess the nature of the stress-corrosion conditions.

Indirect fractographic evidence of water in field failures utilizes fracture surface markings to assess the extent of slow crack growth that preceded the catastrophic failure event. In stress-corrosion assisted delayed failure, the initial flaw will slowly extend until the stress intensity (see Eq 2) reaches a critical value where catastrophic failure can occur. This means that a failure stress estimated from the initial flaw size (see the section "Origin Flaw") will overestimate the actual failure stress. In field failure, the actual magnitude of the failure stress is more than likely not known. However, the failure stress can be accurately estimated by measuring the mirror radius to mist or hackle. (During catastrophic failure event, the development of mist and hackle will be independent of the water in the surrounding environment because the crack will be moving too fast for the water to keep up with the moving crack. Therefore, the mirror radius only represents the stress level at failure according to the mirror relationship shown in Eq 5.) Using the mirror size to scale

the level of applied stress, one finds that the ratio of the mirror radius to the initial flaw size will be larger for the case of stress-corrosion fracture than for the case where no environmental effects influence the failure process.

Mecholsky (Ref 37) has examined the mirror-to-flaw-size ratio for glass and ceramic specimens that failed under long-term static loading. In an example of a silicate glass specimen that failed after 200 hours under load, a mirror-to-flaw-size ratio of 80 was observed. The same glass tested in rapid loading conditions (no stress-corrosion effects) gave a mirror-to-flaw-size ratio of 13. This and other results reported by Mecholsky (Ref 37) for the failure of polycrystalline ceramic materials such as MgF_2 indicate that the presence of stress-corrosion failure can be evaluated by a post-failure examination of the fracture surfaces. In addition, the authors present a fracture mechanics analysis that can

be used to relate the mirror-to-flaw-size ratio to the total time under load for the delayed failure event and found good correlation between the times predicted by this fractographic and the experimentally measured failure times.

Although this indirect fractographic method represents an important technique for identifying stress-corrosion failure conditions, its usefulness depends strongly upon ones ability to accurately measure the size of the initial flaw. As is discussed in the section "Origin Flaw," measurements of the initial flaw size are subject to the complexity of the surface flaw and its alignment with the direction of the applied stress field. The same cautions outlined in the discussion of origin flaws must also apply to the assessment of the mirror-to-flaw-size ratio measurements.

Intersection Scarp. In addition to indirect fractographic evidence for stress-corrosion failure, there are FSM that result directly from

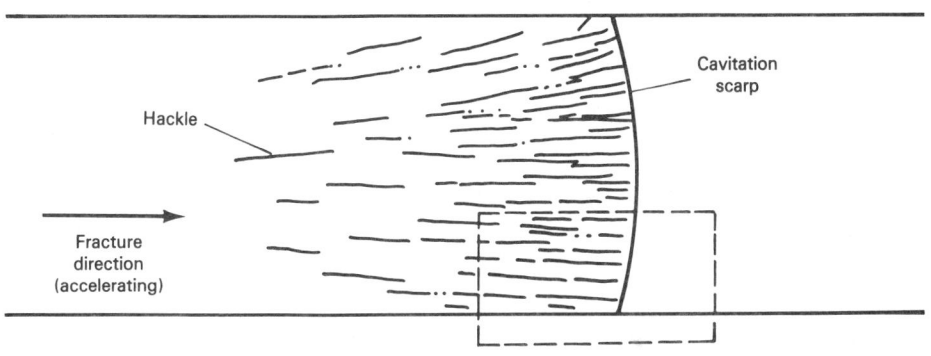

(a)

(b)

Fig 15 (a) Schematic of exposed surface of glass fracture accelerated in water. With increasing velocity, initially mirror-smooth surface (far left) shows generation of hackle steps, followed by a cavitation scarp and the return to mirror-smooth condition at greater velocity. (b) Micrograph of area within dashed lines of schematic. Differential interference contrast, 100×. Source: Ref 39

the presence of liquid water in the propagating crack. The first of these features is called an intersection scarp (Ref 38).

Intersection scarp is formed at the boundary between wet and dry portions of a crack front which is propagating in an environment with limited access to water. Specifically, crack fronts that are partly wetted by a source of liquid water are found to develop fracture surface markings at the boundary between wet and dry areas on the crack front. The fracture surface markings developed in this way are also called intersection scarps because they are ridges (scarps) that form at the intersection of the wet and dry portions along the crack front.

Figure 14(a) shows the sequence of events that is observed as a dry crack (free of liquid), propagating a glass plate, encounters water on the specimen surface. At position 1, the normal through thickness crack front shape is observed. At position 2, the propagating crack front encounters liquid water on the free surface of the glass plate. At this time, two events take place: capillary action draws some water into contact with the moving crack front, and the wetted portion of the crack front jogs slightly ahead of the dry section. Positions 4 and 5 represent the continued spread of water across the moving crack front. At position 5, the crack is completely wet and the normal crack front shape is reformed. Figure 14(b) is a photomicrograph showing a fracture surface in a soda-lime silica glass prepared in a manner previously described.

Formation of the intersection scarp can be interpreted on the basis of the K_I versus V diagram (Fig 13). At velocities below 0.2 mm/s (8 mil/s), the wetted portion of the crack front propagates more easily than the dry. (In this case, dry refers to the portion of the crack front exposed to room air.) Accordingly, the wet section requires less stress to propagate and is slightly ahead of the dry section. (The magnitude of this displacement is limited because the advance of one section will increase the mechanical loading on the other.) Because the conditions of fracture are different on the wet and dry portions of the crack front, the individual planes of fracture also differ. This difference in crack plane results in a sharp ridge or scarp at the line of intersection.

The intersection shows that water was present at the time of failure. This information would be of interest in a forensic case where it may be important to know if it was raining at the time of an incident or in questions of thermal shock fracture when the exact nature of the thermal shock event is unknown.

Cavitation Scarp. Another form of scarp marking is developed when liquid in a crack cavitates. Michalske *et al.* (Ref 39) examined the fracture surface markings that are generated as a crack is accelerated to catastrophic velocities (>1 mm/s) in a liquid environment. Figure 15(a) is a schematic of the fracture surface markings generated when an edge

cracked plate specimen accelerates in liquid water. At low velocity (left side of sketch), the crack surface appears mirror-smooth. As the velocity increases to approximately 10 mm/s (0.4 in./s), fine hackle ridges running parallel to the direction of fracture are observed on the fracture exposed surface, terminated by a distinct change in plane of the fracture surface. The fracture surface marking representing this change in plane is termed a cavitation scarp. After cavitation scarp formation, the fracture surface returns to a mirror-smooth condition. Figure 15(b) is a photomicrograph of the fracture surface of a soda-lime silica glass plate showing these features.

The process of liquid cavitation in the moving crack was modeled by Michalske and Frechette (Ref 39) by determining the negative pressure on the liquid due to viscous

effects. The calculated velocity for cavitation to occur was found to be in excellent agreement with the measured velocity of cavitation scarp formation for a number of different liquids tested. Because cavitation scarp results from the liquid cavitation phenomena, it marks the end of environment-affected, stress-corrosion crack growth.

Example 1: Interpretation of FSM after a Transition from Slow to Rapid Fracture in a Liquid Environment. Because the dynamic effects of liquids involve viscous drag effects, it is important to examine how those effects may operate in different crack geometries. Surface flaw geometry is of greatest practical importance since surface flaws are the most common strength-controlling defects in glass. Figure 16(a) is a schematic showing the fracture surface markings

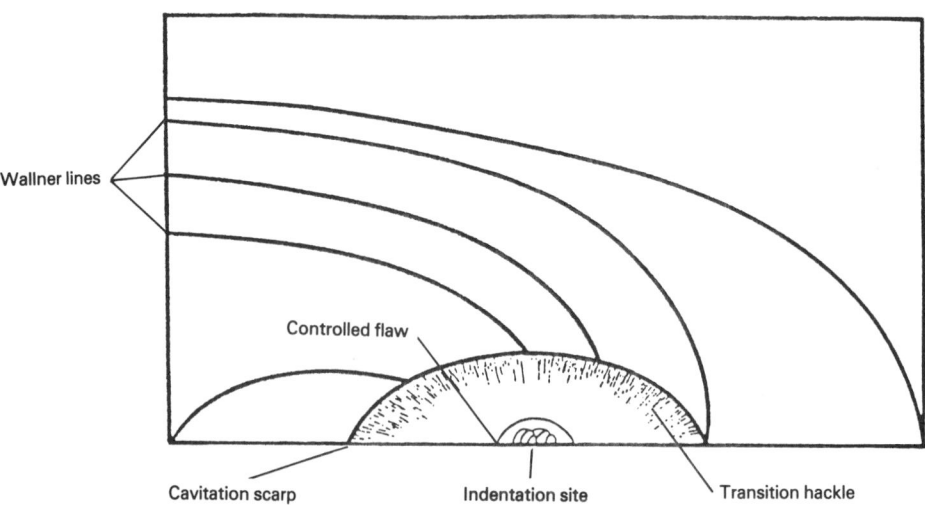

(a)

(b)

Fig 16 (a) Schematic of exposed surface of glass fracture resulting from bending in water. (b) Micrograph showing fracture surface represented in schematic. Reflected light, 15×. Source: Ref 40

Compressive surface

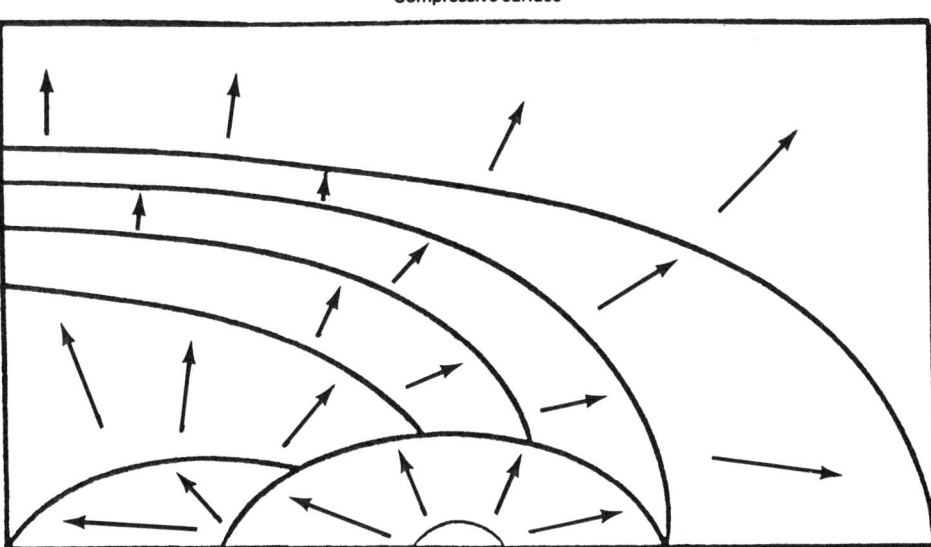

Tensile surface

Fig 17 Schematic showing the development of a surface flaw by application of bending stresses in water. Arrows indicate direction of local fracture.

observed when a surface flaw in soda-lime-silica glass is accelerated to critical velocity in a liquid water environment (Ref 40). The indentation flaw was used to serve as the origin of failure in this experiment and was present in the surface before the external stresses were applied. As external stress was applied to the sample, the fracture spread out into the thickness of the sample and a region of mirror-smooth surface was generated surrounding the origin flaw. Next, fine hackle ridges termed "transition hackle" developed as the origin crack accelerated under the applied stress. The transition hackle that developed in the surface flaw case was found to be identical to that observed on the edge cracked specimen. The surface flaw specimen also shows formation of a cavitation scarp which sharply terminates the transition hackle. Subsequent to the cavitation scarp are a series of Wallner lines indicating the growth pattern during the rapid fracture regime. Figure 16(b) is a photomicrograph of a soda-lime silica glass bar broken in such a manner.

In Fig 17, the arrows show the crack growth pattern that was deduced from the fracture surface markings. Initially, the origin flaw grows in all directions, spreading more rapidly along the tensile surface to form a semi-elliptical front. Following the cavitation scarp, the crack spreads preferentially from the left side of the ellipse formed by the cavitation scarp, sweeping in all directions to cause complete failure of the specimen. The fact that the crack spread preferentially (instead of evenly) from one side of the cavitation scarp is attributed to the effects of viscous drag. Along the tensile surface of the bend specimen, where the stress is highest, the crack

velocity is also greatest. The gradient in stress associated with the bending geometry is responsible for the elliptical shape of the growing surface crack.

Because the crack accelerates faster along the tensile surface, the velocity required for cavitation occurs first at the surface so that one end of the ellipse will serve as the site for cavitation. After cavitation has occurred on one portion of the crack front, the increased rate of crack propagation (~10-fold) allows that portion of the crack to sweep around, intersecting those portions still wet and subject to the effects of viscous drag. (The remaining wet section of the crack is unable to cavitate because its velocity is too low.) Therefore, the majority of the crack spreading from the origin flaw is retarded by the viscous drag of the liquid phase and remains "stationary," while the water-free section rapidly advances to catastrophic velocity.

This interpretation has several important implications. First, it provides an understanding of the mechanism responsible for the transition from slow to rapid fracture in liquid environments. Wiederhorn (Ref 41) has shown that this information is essential to the models used to predict the effect of proof testing on the strength. The cavitation scarp is also important in its role of defining the size of the flaw present at the onset of catastrophic failure. The critical flaw size is necessary to the calculation of the K_{Ic} value measured in a stress-corrosion environment. It is important to note here that critical flaw size as taken from the cavitation scarp must be measured from the center of the origin flaw to the point on the scarp where cavitation initiated. Finally, after the K_{Ic} has been mea-

sured by this technique, the stress at failure may be calculated (using Eq 2) from a post-failure measurement on the failed component.

REFERENCES

1. R.W. Rice, Perspective on Fractography, *Advances in Ceramics*, Vol 22, *Fractography of Glasses and Ceramics*, J.R. Varner and V.D. Frechette, Ed., The American Ceramic Society, 1988, p 3–56

2. V.D. Frechette, The Fractography of Glass, *Introduction to Glass Science*, L.D. Pye, H.J. Stevens, and W.C. LaCourse, Ed., Plenum Press, 1972, p 433–450

3. R.W. Rice, Ceramic Fracture Features, Observations, Mechanisms, and Uses, *Fractography of Ceramic and Metal Failures*, J.J. Mecholsky and S.R. Powell, Ed., American Society for Testing and Materials, 1984, p 5–102

4. S.W. Freiman, A Critical Evaluation of Fracture Mechanics Techniques for Brittle Materials, *Fracture Mechanics of Ceramics*, Vol 6, R.C. Bradt, A.G. Evans, D.P.A. Hassleman, and F.F. Lange, Ed., Plenum Press, 1983, p 27–45

5. *The Ceramic Source*, Vol 6, The American Ceramic Society, 1991

6. S.W. Freiman, T.L. Baker, and J.B. Wachtman, "A Computerized Fracture Mechanics Database for Oxide Glasses," Technical Note 1212, National Bureau of Standards, 1985

7. H. Tada, P. Paris, and G. Irwin, *The Stress Analysis of Cracks Handbook*, Del Research Corporation, 1973

8. J.J. Mecholsky and S.W. Freiman, Determination of Fracture Mechanics Parameters Through Fractographic Analysis of Ceramics, *STP 678*, American Society for Testing and Materials, 1980, p 136–150

9. R.W. Rice, S.W. Freiman, and P.F. Becher, Grain-Size Dependence of Fracture Energy in Ceramics: I, Experiment, *J. Am. Ceram. Soc.*, Vol 64 (No. 6), 1981, p 345–350

10. P.L. Swanson, Crack-Interface Traction: A Fracture-Resistance Mechanism in Brittle Polycrystals, *Advances in Ceramics*, Vol 22, *Fractography of Glasses and Ceramics*, J.R. Varner and V.D. Frechette, Ed., The American Ceramic Society, 1988, p 135–155

11. J.W. Dally and W.F. Riley, *Experimental Stress Analysis*, McGraw-Hill, 1965

12. H. Hertz, *Hertz's Miscellaneous Papers*, MacMillan, London, 1986

13. B.R. Lawn, S.M. Wiederhorn, and H.H. Johnson, Strength Degradation of Brittle Surfaces: Blunt Indenters, *J. Am. Ceram. Soc.*, Vol 58 (No. 9–10), 1975, p 428–432

14. D.B. Marshall and B.R. Lawn, *J. Mater. Sci.*, Vol 14, 1979, p 2001–2012

15. A.W. Ruff and S.M. Wiederhorn, Erosion by Solid Particle Impact, *Treatise on Plat. Sci. and Tech.*, Vol 16, C.M. Preece, Ed., Academic Press, 1979, p 69–126

16. V.H. Wallner, Linienstrukturen an Bruchflachen 2, *Phys.*, Vol 114, 1939, p 368–378

17. E.F. Poncelet, The Markings on Fracture Surfaces, *J. Soc. Glass Technol.*, Vol 42, 1958, p 279T

18. F. Kerhof, *Naturwissenschaften*, Vol 40, 1953, p 478

19. F. Kerhof, *Kurzzeitphysik*, K. Vollrath and G. Thomer, Ed., Springer-Verlag, 1967, p 498

20. F. Kerhof and H. Dreizler, *Glastechnische Berichte*, Vol 29 (1956), p 459

21. T.A. Michalske and V.D. Frechette, Modified Sonic Technique for Crack Velocity Measurement, *Int. J. Fracture*, Vol 17 (No. 3), 1981, p 251–256

22. T.A. Michalske, M. Singh, and V.D. Frechette, Experimental Observation of Crack Velocity and Crack Front Shape Effects in Double-Torsion Fracture Mechanics Tests, *Fracture Mechanics of Ceramics Rocks and Concrete*, E.R. Fuller and S.W. Freiman, Ed., American Society for Testing and Materials, 1981, p 3–12

23. T.A. Michalske and J.M. Collins, Fractographic Determination of Crack Tip Stress Intensity, *Advances in Ceramics*, Vol 22, *Fractography of Glasses and Ceramics*, J.R. Varner and V.D. Frechette, Ed., The American Ceramic Society, 1988, p 229–239

24. E.H. Joffe, The Moving Griffith Crack, *Philos. Mag.*, Vol 42 (No. 112), 1951, p 739–750

25. W. Doell, Investigations of Crack Branching Energy, *Int. J. Fracture*, Vol 11, 1975, p 184–186

26. H.P. Kirchner and J.W. Kirchner, Fracture Mechanics of Fracture Mirrors, *J. Am. Ceram. Soc.*, Vol 62 (No. 3–4), 1979, p 198–202

27. E.B. Shand, Breaking Stress of Glass as Determined from Dimensions of Fracture Mirrors, *J. Am. Ceram. Soc.*, Vol 42 (No. 10), 1959, p 474–477

28. N. Terao, Relation Between Resistance to Rupture and Mirror Surface in Glass, *J. Phys. Soc. Jpn.*, Vol 8, 1953, p 545–549

29. J.J. Mecholsky, S.W. Freiman, and S.M. Morey, Fracture Surface Analysis of Optical Fibers, *Fiber Optics: Advances in Research and Development*, B. Bendow and S.S. Mitra, Ed., Plenum, 1979, p 187–208

30. J.J. Mecholsky, Jr. and R.W. Rice, Fractographic Analysis of Biaxial Failure in Ceramics, *Fractography of Ceramic and Metal Failures*, ASTM STP 827, J.J. Mecholsky, Jr. and S.R. Powell, Jr., Ed., American Society for Testing and Materials, 1984, p 185–193

31. M.J. Kerper and T.G. Scuderi, Modulus of Rupture of Glass in Relation to Fracture Pattern, *Am. Ceram. Soc. Bull.*, Vol 44 (No. 12), 1965, p 953–955

32. C.Cm. Wu, S.W. Freiman, R.W. Rice, and J.J. Mecholsky, Microstructural Aspects of Crack Propagation in Ceramics, *J. Mater. Sci.*, Vol 13, 1978, p 2659–2670

33. R.W. Rice, R.C. Pohanka, and W.J. McDonough, Effect of Stresses from Thermal Expansion Anisotropy, Phase Transformations, and Second Phases on the Strength of Ceramics, *J. Am. Ceram. Soc.*, Vol 63 (No. 11–12), 1980, p 703–710

34. D. Lewis, III, Fracture Strength and Mirror Size in a Commercial Glass Ceramic, *J. Am. Ceram. Soc.*, Vol 64 (No. 2), 1981, p 82–86

35. T.A. Michalske and V.D. Frechette, Fragmentation in Bursting Glass Containers, *Bull. Am. Ceram. Soc.*, Vol 57 (No. 4), 1978, p 427–429

36. F.W. Preston, Angle of Forking of Glass Cracks as an Indicator of the Stress System, *J. Am. Ceram. Soc.*, Vol 18, 1935, p 175

37. J.J. Mecholsky and S.W. Freiman, Fractographic Analysis of Delayed Failure in Ceramics, *Fractography and Materials Science*, ASTM STP 733, American Society for Testing and Materials, 1981, p 246–258

38. T.A. Michalske, J.R. Varner, and V.D. Frechette, Growth of Cracks Partly Filled with Water, *Fracture Mechanics of Ceramics*, Vol 4, R.C. Bradt, D.P.H. Hassleman, and F.F. Lange, Ed., Plenum, 1978, p 639–649

39. T.A. Michalske and V.D. Frechette, Dynamic Effects of Liquids on Crack Growth in Glass, *J. Am. Ceram. Soc.*, Vol 63 (No. 11–12), 1980, p 603–609

40. T.A. Michalske, V.D. Frechette, and R. Hudson, Dynamic Effects of Liquids on Surface Crack Extension in Glass, *Advances in Fracture Research*, Vol 2, D. Francis, Ed., Pergamon, 1981, p 1091–1097

41. E.R. Fuller, S.M. Wiederhorn, J.E. Ritter, and P.B. Oates, Proof Testing of Ceramics, *J. Mater. Sci.*, Vol 15, 1980, p 2282–2295

Fracture Surface Analysis of Optical Fibers

John J. Mecholsky, Department of Materials Science & Engineering, University of Florida

OPTICAL FIBERS are used in many advanced technological applications such as telecommunications, photocopy machines, local military communication systems, sensors, and many more. In these uses, many of the fibers undergo severe loading conditions which require them to be very high in strength. It is therefore very important for them to be as flaw-free as possible. Fractographic analysis is an extremely useful technique for identifying the source(s) of failure and thus help eliminate it (them) during production or fabrication procedures. In addition, fractography can be used to determine the strength and stress state at failure and the time under load before failure.

It is the purpose of this article to demonstrate through examples, how the principles of fracture surface analysis can be applied to research, fabrication, and production problems for strong optical fibers. Although the examples will principally be directed towards optical fibers, the techniques and analyses are valid for most other fibers in different applications, for example, brittle fibers in composites, infrared fibers in various applications and brittle polymer fibers in lighting applications. The fibers that will be presented in this article, in most cases, were intentionally selected for their failure at low loads because the low-strength tail in the strength distribution is the controlling factor in production of long length, strong fibers. It will be shown how observations of the fracture surface can:

- Determine the failure stress
- Identify the size, shape, and type of fracture origin
- Aid in identifying modifications to production procedures to mitigate the flaw severity
- Estimate the time under load

Theoretical Background

Four definitive regions surrounding fracture-initiating flaws in silicate and non-silicate glasses have been observed (Fig 1) (Ref 1–3). The mirror (a flat, smooth region) is bounded by the onset of mist (a region of small radial ridges) which is bounded in turn by

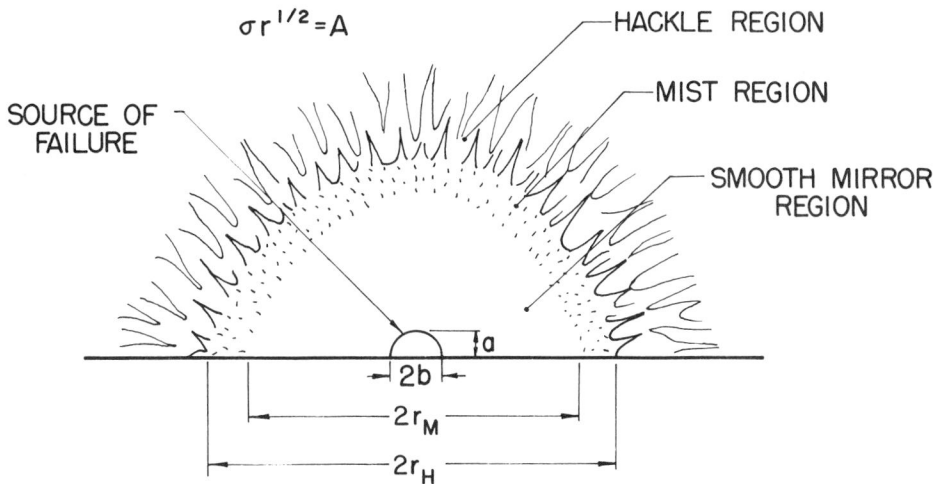

$$\sigma r^{1/2} = A$$

SOURCE OF FAILURE

HACKLE REGION

MIST REGION

SMOOTH MIRROR REGION

SHAPE AND GENERAL APPEARANCE OF FRACTURE MIRROR SURFACES WITH FLAW.

Fig 1 Schematic of fracture origin showing idealized semielliptical surface flaw and surrounding fracture features known as mirror, mist, and hackle. Crack branching is beyond the hackle.

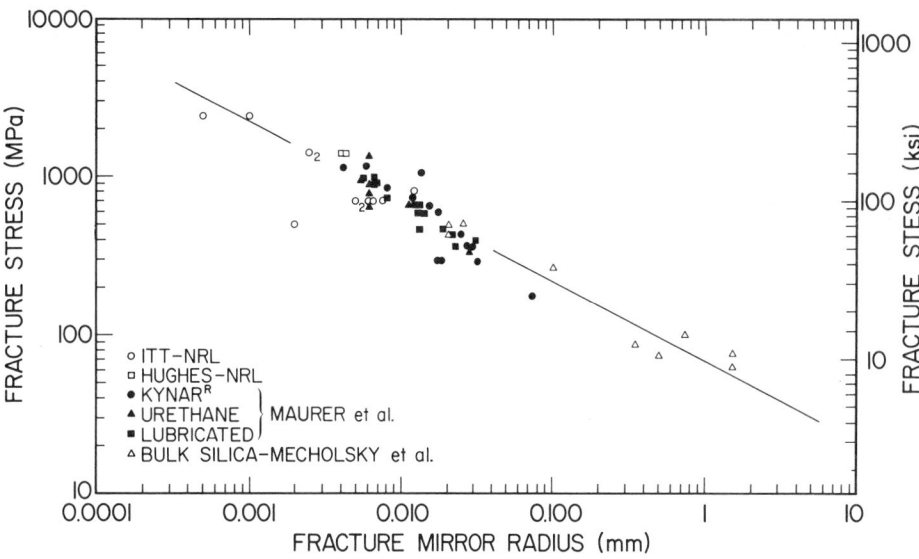

Fig 2 Fracture stress versus inner (mirror-mist) fracture mirror radius for optical fibers and bulk silica. The solid data were obtained from Ref 6 assuming Eq 1 is valid. The bulk silica data are from Ref 3. The other data are from Ref 10. The solid line is a linear least square fit with a slope of −0.5. A_M = 2.1 MPa$\sqrt{m}$.

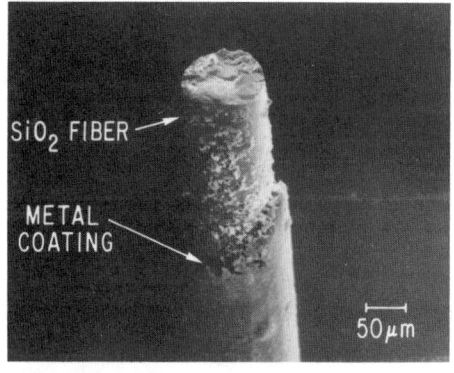

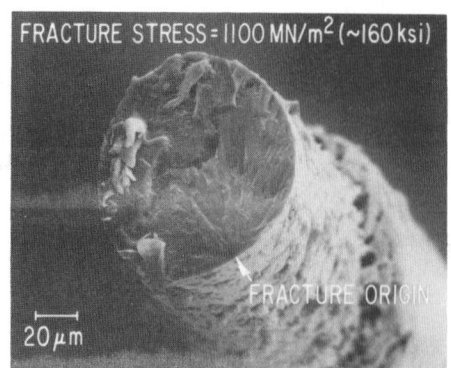

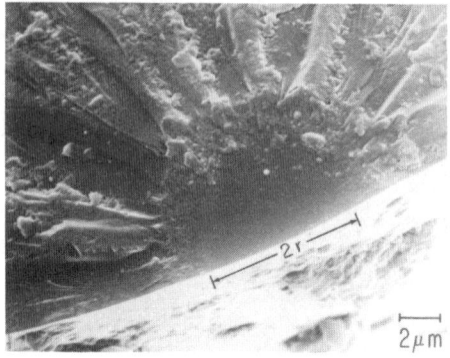

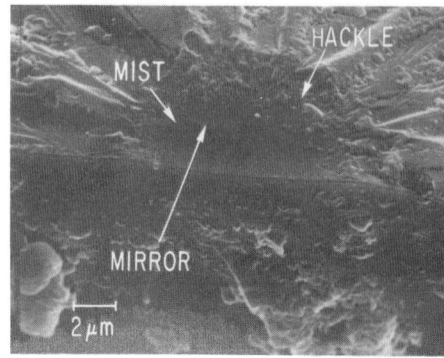

Fig 3 Scanning electron microscope (SEM) fractographs of a metal-coated fiber showing fracture demarcations surrounding the fracture origin (most likely a sharp crack—not visible on the surface)

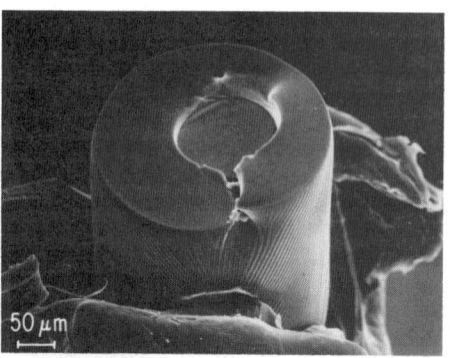

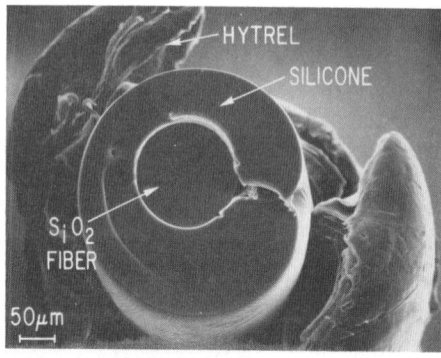

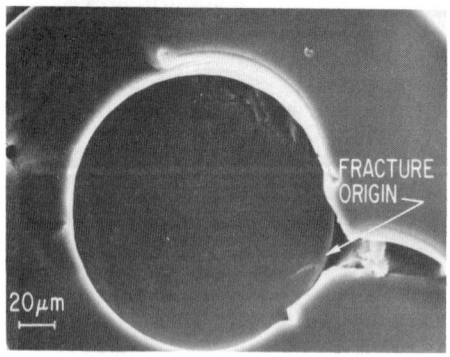

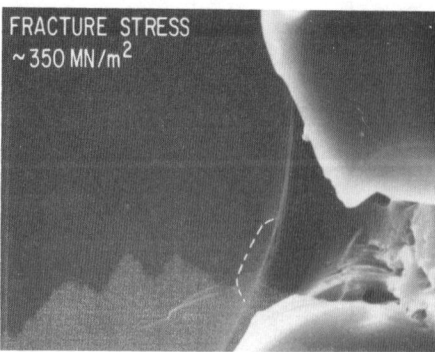

Fig 4 SEM fractographs of a fiber that failed from a sharp crack at the surface (dotted line in lower right fractograph). This fiber failed well below the proof stress of 1400 MPa (200 ksi) indicating failure on the edge of the drum before full stress was achieved.

hackle (a region of larger radial ridges) which is bounded by macroscopic crack branching. In a uniform stress field, these boundaries are circular arcs about the fracture origin. It has been extensively demonstrated that the products of the strength, σ, and the square roots

of the distances from the origin to the onset of mist (r_M), the onset of hackle (r_H), and the onset of crack branching (r_B) give three constant values for silicate glasses:

$$\sigma r_j^{1/2} = A_j \qquad (Eq\,1)$$

where j refers to the mirror-mist, mist-hackle, or crack-branching boundaries. It has been shown that these radii are related to the initial flaw depth, a, and half-width, b, through the combination of fracture mechanics and fracture surface analysis (Ref 3, 4):

$$\frac{c}{r_j} = K_{Ic}^2 Y^2/2A_j^2 \qquad (Eq\,2)$$

where $c = \sqrt{ab}$, Y is a factor which depends on the location and geometry of the crack ($Y = 1.12$ for a semicircular surface crack), and K_{Ic} is the fracture toughness (0.73 MPa$\sqrt{m}$, or 0.66 ksi$\sqrt{in.}$, for fused silica glass) (Ref 3). In circumstances where the flaw cannot be measured directly, its size can be inferred by that relationship (Ref 5). Even when origins are not macroscopically obvious, fracture markings point back to the area of the origin, indicating whether a surface or internal origin was the source of failure. Figure 2 is a graph of fracture stress versus inner mirror/mist boundary radius for fused silica obtained in various studies. The graph shows the validity of Eq 1 over a wide range of stress and mirror radius values. It also indicates that the mirror constant ($A_M = 2.1$ MPa$\sqrt{m}$, or 1.9 ksi$\sqrt{in.}$) is the same for bulk fused silica and fused silica fiber. The outer (mist-hackle) mirror constant, A_H, is 2.4 MPa$\sqrt{m}$ (2.2 ksi$\sqrt{in.}$).

An analytical approach to the measurement of the fracture mirror boundaries was introduced by Kirchner and Kirchner (Ref 7) and further developed by Kirchner and Conway (Ref 8). This approach assumes that the boundaries of the mirror-mist, mist-hackle, and crack branching each occur at a constant stress intensity value, that is, a different branching stress intensity for each boundary. For glasses and polycrystalline ceramics, then,

$$K_j = 2Q/(\pi)^{1/2}[\sigma r_j^{1/2}] \qquad (Eq\,3)$$

where σ and r_j are defined as before and K_j is a critical value of the stress intensity, that is, $j = 0,1,2,3$; for $j = 0$, $K_0 = K_{Ic}$ and r_0 is the critical crack size (in mode I loading). The $j = 0$ case is equivalent to the generally accepted fracture mechanics equation for a semicircular surface crack. The value Q is the value necessary to correct K_I for an internal penny shaped crack to obtain the stress intensity factor, K_j, for a semicircular surface crack. Tables are available for these values (Ref 7). The cases of $j = 1, 2$, and 3 correspond to the formation of the mirror-mist, mist-hackle, and crack-branching boundaries. The set of equations can relate all of the fracture demarcations that we have discussed. Notice that these sets of equations are, in principle, not any different than the previous set presented (Eq 1); however, the above equations

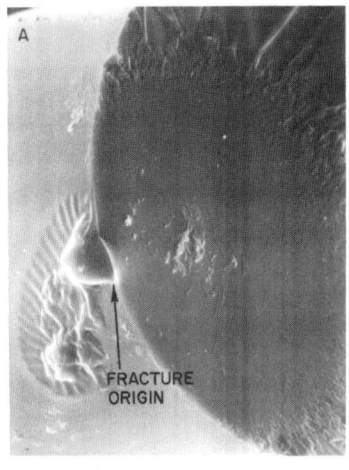

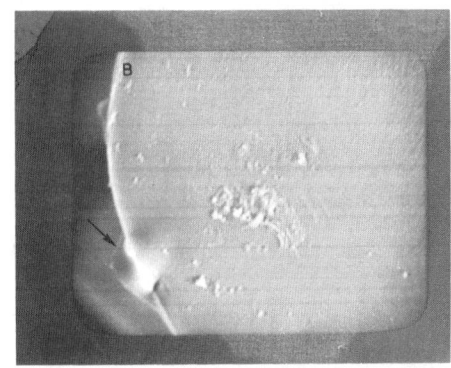

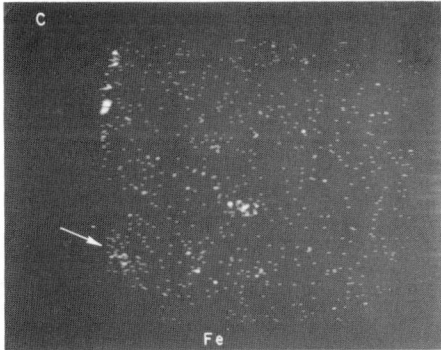

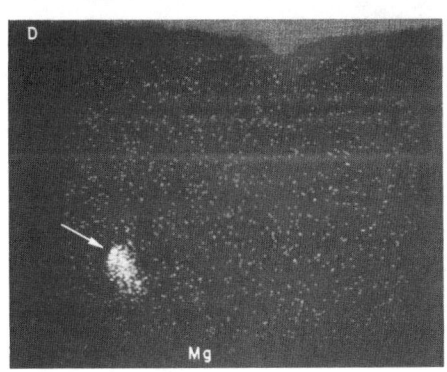

Fig 5 Fiber failure due to a foreign particle. (A) SEM fractograph of a polymer-coated fiber that failed in proof test at 1400 MPa (200 ksi). (B) Real image of A as given by microprobe unit. (C and D) Microprobe electron images showing relative concentrations of Fe and Mg, respectively. Arrows indicate fracture origin (crack between foreign "dust" particle and bulk SiO_2) for reference. The mirror-size measurements indicate a failure stress of ~315 MPa (~45 ksi). This means that this fiber most likely failed before the full proof-stress was achieved, that is, around the edge of the drum.

are based on well-formulated elasticity equations whereas the previous equations are empirical and only valid along the tensile surface. Also notice that the mirror-to-flaw size ratio naturally evolves from these equations, that is

$$r_1/r_0 = r_1/c = (K_1/K_{Ic})^2 = \text{constant} \qquad \text{(Eq 4)}$$

For the sake of completeness, it should be mentioned that Kirchner later modified his approach by requiring a constant strain intensity criterion (Ref 9):

$$K_j/E = \text{constant} \qquad \text{(Eq 5)}$$

where E is the elastic modulus. This will make a difference if single crystals are being analyzed. However, for our purposes there is no difference between Eq 5 and 3.

Experimental Procedure

The polymer-coated fibers used in this study consisted of a silica core with silicone coating and an exterior plastic coating (Hytrel). In order to examine the fractured surface of these fibers, it was sometimes necessary to strip the fiber of the outer Hytrel plastic coating. This was done manually by inserting a razor blade carefully around the fiber and then manually pulling off the severed plastic. The metal-coated fibers contained a metallic (aluminum) coating. When it was necessary to remove this coating, the fibers were placed in an aqua regia solution for 1 to 2 min and then rinsed in water. Before examination in a scanning electron microscope, both types of fibers were coated with gold or platinum.

Three conditions of previously tested fibers were examined. These include delayed failure specimens in which fibers were wrapped around a mandrel and times to failure in air or salt water were measured. This involved merely wrapping fiber around a mandrel of a certain radius. The stress induced in the fiber is related to the mandrel radius. In this case, a fully uniform tensile stress was not achieved, but rather the outer portion of the fiber was in tension and the inner portion of the fiber was in compression. The other conditions of tested fibers were those that were subjected to uniaxial tensile stresses, that is, either proof tested at a particular load or broken in tension on a test machine. Normally, proof testing is done by passing the fiber from one drum to another at a particular rate of speed with a drag on one drum and the load from the drag recorded. Fracture surface observations from

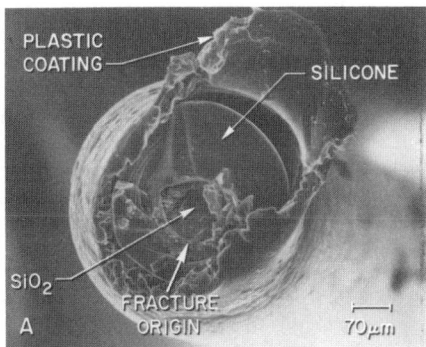

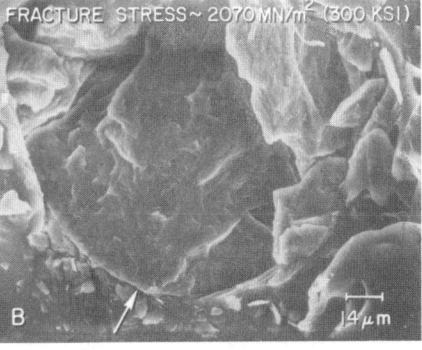

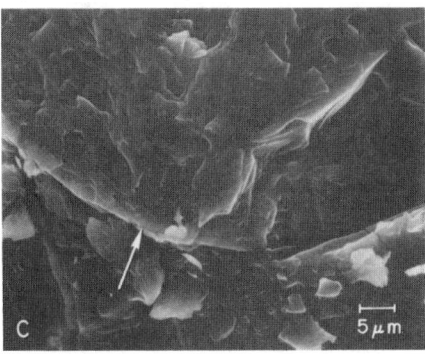

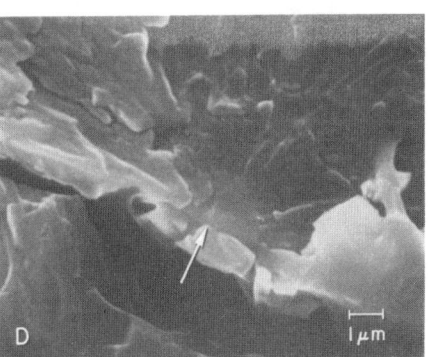

Fig 6 SEM fractographs of a fiber failed from the surface. The nature of the rough area surrounding the origin (arrow in B, C, and D) indicates a high stress (~2700 MPa, or 390 ksi). The exact nature of the fracture origin is unknown, but could be due to a thermal expansion mismatch between two phases (that is, glassy SiO_2 and crystalline SiO_2, or crystalline metal, or crystalline alumina).

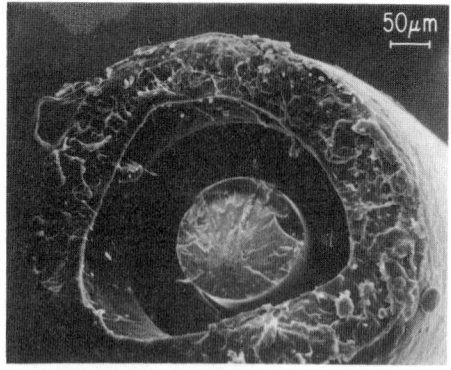
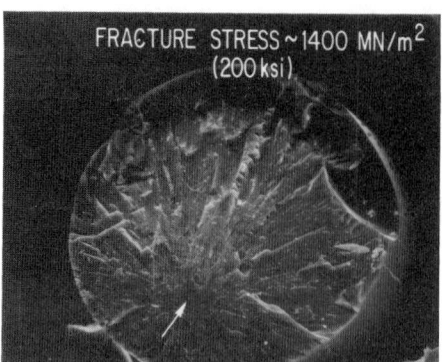

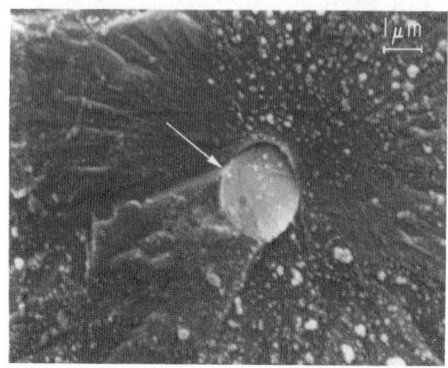

Fig 7 SEM fractographs of a polymer-coated fiber showing failure from a crack at the interface between a rare earth (Nd and La) inclusion and the bulk SiO_2 fiber. The size of the fracture "mirror" region surrounding the origin agrees with the 1400 MPa (200 ksi) proof-stress.

bending or tension tests will be similar and obey Eq 1 to 4 as long as stress-gradient effects are taken into account.

Analysis

More than 100 fibers were analyzed. There are basically two types of fibers that were examined. One type was an aluminum-coated silica-based fiber and the second was a silicone-coated, silica-based glass fiber with an outer organic coating (Hytrel). Since the purpose of this article is to illustrate the usefulness of fractographic analysis and its applications to fiber research, development, and production, only those fibers that illustrate a point will be presented. Further details can be obtained elsewhere (Ref 10).

Stress Analysis. Figure 3 shows a metal-coated fiber that demonstrated the classic mirror, mist, and hackle region schematically represented in Fig 1. The fracture origin in this fiber was most likely a sharp crack that could not be observed; however, by measurement of the mirror-mist boundary, Eq 1 was used to determine the stress (1100 MPa, or 160 ksi) which agreed with that recorded by a load cell. The critical defect size was estimated from Eq 2 to be 0.3 μm. The fiber shown in Fig 4 was fractured in a drum-to-drum proof stress at 1400 MPa (200 ksi). The failure stress inferred from the drag load

between the drums and the fiber dimension was 1400 MPa (200 ksi). However, the fracture surface measurements indicate a failure stress of 350 MPa (50 ksi). The difference is apparently due to the fact that failure occurred while that point of the fiber was still in contact with the drum, that is, before it had experienced the full drag load. Without fracture surface examination, it could only be concluded that a stress of 1400 MPa (200 ksi) or less was achieved on fibers failing during proof testing. The main point is that, regardless of the source of loading (mechanical or thermal), in most cases Eq 1 or 3 can be used to determine the stress at failure from the fracture surface.

Flaw Identification. Most of the fracture origins that could be identified (approximately 50) were surface failures. One of the surface failures was a result of a foreign particle (Fig 5A); the remainder resulted from cracks or mechanically induced chips (Fig 4) and "unidentified" sources of failure (Fig 6). It was suspected that at least four of the five fracture origins listed are from small crystallite formations, but this was not determined for certain. The handling or mechanically induced cracks were generally easy to identify and measure (Fig 4). There was generally good correlation between observation and that expected from Eq 1 and 2 (Ref 10).

However, other sources of failure were not as easily analyzed and necessitated other

techniques and deductive reasoning. For example, a foreign particle is shown attached to the silica fiber in Fig 5. The flaw severity was not so easy to determine for this attached foreign particle. The particle, which probably attached to the fiber during drawing, caused a small (~2 μm) crack, most likely upon cooling, which subsequently led to failure. The size of this crack is in good agreement with the calculated 3 μm from Eq 2. Chemical analysis by electron microprobe was taken to determine the origin of the particle. That analysis shows the presence of magnesium and iron as well as silica (Fig 5C and D) which implies a dust particle. This type of defect can be avoided by drawing the fibers in a clean room.

Although most fractures were from surface origins, there were seven cases with internal origins. One internal origin was an inclusion containing rare earth elements (neodymium and lanthanum) (Fig 7). This failure occurred due to a (0.2 μm) crack formed between the inclusion and bulk SiO_2. Failure from inclusions related to natural-forming elements can most likely be eliminated by using synthetic quartz material rather than natural quartz. This also applies to preform manufacture used in the chemical vapor deposition (CVD) process. If the outer tube or preform material is natural quartz, then defects such as bubbles and inclusions will be transferred to the final silica fiber. Higher magnification was required to determine the source of these failures because the size of the defect, in most cases, was less than 0.2 μm, and, consequently, detailed analysis was difficult.

Research and Production Aids. In addition to the last two examples given above (Fig 5 and 7), there are two other examples which point out the usefulness of fractography in developing strong optical fibers. One of these cases involves identification of fracture origins in several low-strength fibers. A semielliptical trough on the surface of most of the fibers examined was consistently the location of the failure (Fig 8). Because the ridge along the fiber surface was relatively smooth, it would appear that this defect was produced in the drawing process when the fiber was relatively hot. If the fiber were cool, one would expect a rough gouge if some object were pulled across the surface. Further communications between the failure analysis laboratory and the supplier determined that most probably a ZrO_2 particle adhered to the fiber drawing die orifice and scribed a trough on the hot silica fiber during drawing. The supplier eliminated this type of problem by removing the ZrO_2 tube from the drawing process.

Another example of fracture due to production procedures is shown in Fig 9. Failure is seen to originate at a "bubble" which is one of many along the surface of the fiber. The bubbles along the surface indicate that a large bubble in the preform was drawn out during fiber pulling. The large defect size at the origin (38 μm half width), resulted in

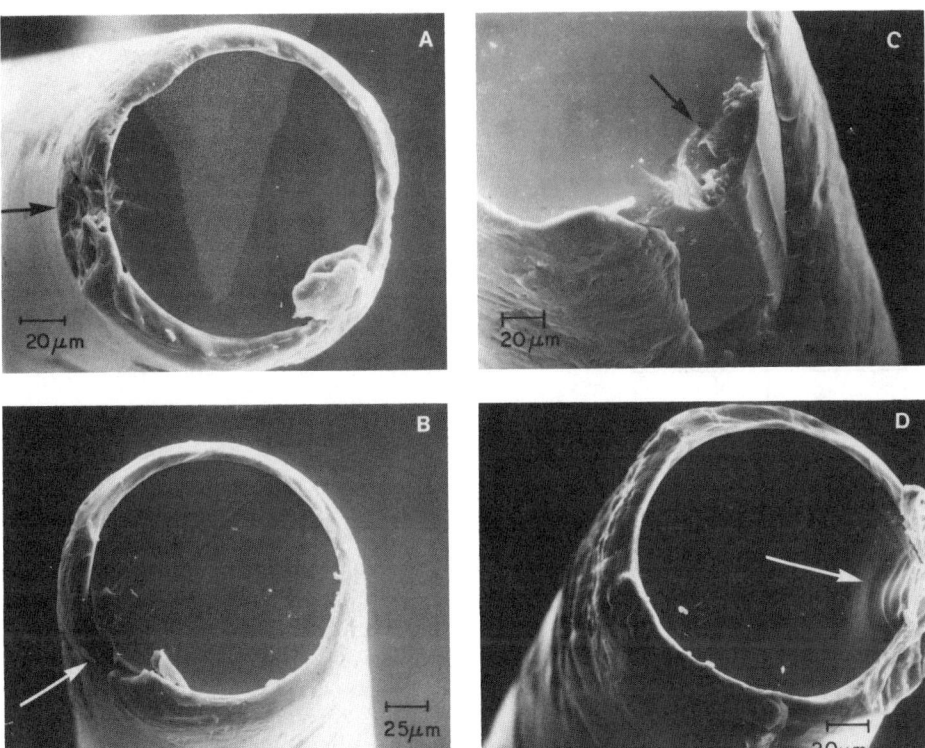

Fig 8 SEM fractographs of four metal-coated fibers failed in a proof test at relatively low strengths. Notice that all fibers indicate surface failure from an elliptical groove with ridge. The smoothness of the depression in B and C would indicate this occurred while the fiber was soft. Also, most likely, the metal is debonded from the fiber in the area around the defect in B and C.

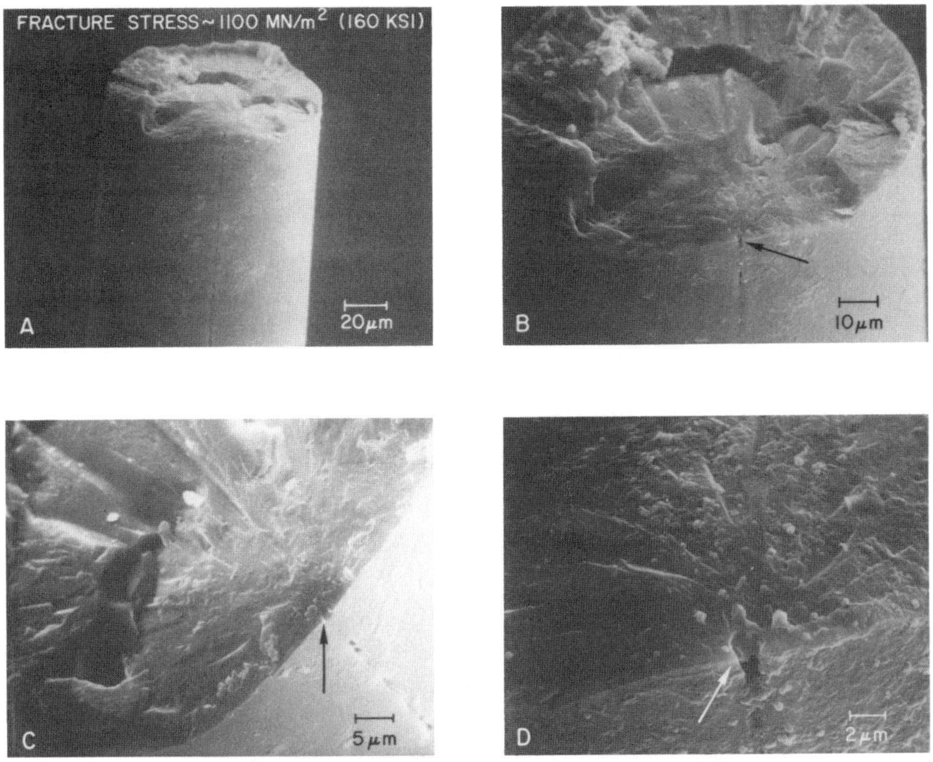

Fig 9 SEM fractographs of a fiber after removal of the metal coating. Source of failure (arrow) is from an elliptical defect, most likely from a bubble in the preform that was drawn out with the fiber.

approximately 1100 MPa (160 ksi) (from mirror-measurements) breaking stress in reasonable agreement with the approximately 1400 MPa (200 ksi) measured during the test. This emphasizes the need for good-quality preforms, if good-quality fibers are to be obtained.

Fractographic Analysis of Subcritical Crack Growth. Many fibers are subjected to subcritical stresses, that is, stresses which do not cause instantaneous failure for a period of time depending on the stress level, initial flaw size, c_i, and time under load, t. These may fail after some extended time under load. This delayed failure results from subcritical crack growth which will subsequently alter the appearance of the fracture surface shown in Fig 1 when failure occurs. The fracture surface produced in delayed failure will show evidence of that subcritical crack growth (Fig 10) where the initial flaw (solid curve) increases until it reaches the critical size (dashed curve) determined by the stress state and fracture toughness, at which time catastrophic fracture commences. The boundaries of the mirror, mist, hackle, and crack-branching regions occur just as they do in catastrophic failure. Thus, the ratio of the initial flaw size to the radius at which the mirror-mist or mist-hackle boundaries form is smaller for a longer time under load. It has been shown for bulk glasses that this was indeed the case both experimentally and theoretically (Ref 11). Only the results will be given here. The time under load and the initial flaw size are related by:

$$\frac{t}{c_i} = \left[\frac{(1.2\pi)^{1/2}}{\phi}\right]^{-n} \frac{(r_j/c_i)^{n/2}}{A_j^n A(1 - n/2)} \qquad \text{(Eq 6)}$$

where n and A are experimentally determined constants which describe subcritical crack propagation velocity (Ref 12) and t is the time under constant stress; r_j and A_j are defined in Eq 1; ϕ is an elliptical integral of the second kind and is a function of the geometry of the crack ($\phi = 1.0$ for a semicircular crack).

Equation 6 demonstrates that, to a reasonable approximation, the mirror radius to initial flaw size ratio should be a function of time under load. In addition, Eq 6 shows that the time to failure can be estimated through fracture surface analysis. A, n, and A_j are constants for a given material, loading, and environment so that measurement of the fracture mirror and initial flaw size can result in a calculation of the time under load. Obviously, this analysis can only be performed after the sample is broken, and thus cannot be used for time-to-failure predictions. However, for in-service failures, given no other information, this is a way to estimate the total time the fiber was subjected to load. An implied assumption is that the stress on the flaw is constant for this period of time. As yet, this analysis has not been tested on fibers, but has been shown valid for bulk glass (Ref 11).

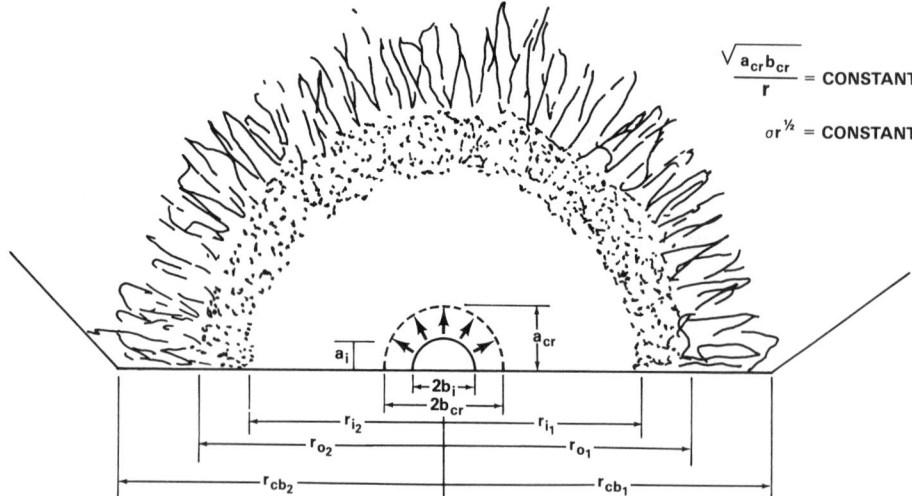

$$\frac{\sqrt{a_{cr}b_{cr}}}{r} = \text{CONSTANT}$$

$$\text{or } r^{1/2} = \text{CONSTANT}$$

Fig 10 Schematic of fracture surface of a brittle material subjected to a constant subcritical load, σ. The solid semielliptical curve at the center represents the initial flaw size and the dashed curve represents the outline of the critical flaw size before catastrophic failure. The inner, r_i, outer, r_o, and crack branching, r_{cb}, mirror radii are shown in the figure with subscripts 1 and 2 to indicate that there may be unsymmetric mirrors.

Although much research has been devoted to the study of stress corrosion in optical fibers, there are still many questions that remain unanswered, the greatest of which is: Can we predict the service life of fibers in a given environment? Fracture analysis can greatly aid in analyzing long-term results for comparison with theory (Ref 11). Use of fractography where possible in analyzing delayed failure and strain rate experimental results will be invaluable.

Conclusions

Based on the fractographic analysis described in this case history, the following conclusions were drawn:

- Fracture surface analysis can be used to identify sources for the low-strength (or unusual) failures. This knowledge can be used to correct processing or handling procedures to improve the fiber and get a higher yield of "good" fiber
- High-quality glass should be used in all phases of the fiber process

- Most of the low-strength failures of optical fibers are caused by mechanical damage during drawing or defects in the preform (such as bubbles) being drawn through in the fiber process, foreign particle or "stones" from naturally occurring elements or contamination during drawing. One or several of these causes could occur in any run
- The time-to-failure under constant stress may be estimated by observation of the fracture surface

ACKNOWLEDGMENT

The author thanks Dr. S.W. Freiman for much important discussion and encouragement, and S.M. Morey for excellent SEM fractographs. The author acknowledges the support of the Defense Advanced Research Projects Agency (DARPA Order No. 3285), and thanks Hughes Research Laboratory, ITT Research Laboratory, and Naval Ocean System Center, for cooperation in supplying fibers and discussing problems.

REFERENCES

1. E.B. Shand, *J. Am. Ceram. Soc.*, Vol 42 (No. 10), 1959, p 474–477
2. J.W. Johnson and D.G. Holloway, *Philos. Mag.*, Vol 14 (No. 130), 1966, p 731–743
3. J.J. Mecholsky, R.W. Rice, and S.W. Freiman, *J. Am. Ceram. Soc.*, Vol 57 (No. 10), 1974, p 440–443
4. J.J. Mecholsky, S.W. Freiman, and S.M. Morey, *Bull. Am. Ceram. Soc.*, Vol 56 (No. 11), 1977, p 1016–1017
5. D.A. Krohn and D.P.H. Hasselman, *J. Am. Ceram. Soc.*, Vol 54 (No. 8), 1971, p 411
6. R.D. Maurer, R.A. Miller, D.D. Smith, and J.C. Trondsen, "Optimization of Optical Wave Guides—Strength Studies," ONR contract No. N00014–73-C-0293, Corning Glass Works TR, Mar 1974
7. H.P. Kirchner and J.W. Kirchner, *J. Am. Ceram. Soc.*, Vol 62 (No. 3–4), 1979
8. H.P. Kirchner and J.C. Conway, Jr., *J. Am. Ceram. Soc.*, Vol 70 (No. 6), 1987, p 413–425
9. H.P. Kirchner, *Eng. Fract. Mech.*, Vol 10, 1978, p 283–288
10. J.J. Mecholsky, S.W. Freiman, and S.M. Morey, "Fractographic Analysis of Optical Fibers," DARPA Order No. 3285, National Research Laboratory Interim Technical Report from Defense Advanced Research Projects Agency, Nov 1977; also, J.J. Mecholsky, S.W. Freiman, and S.M. Morey, Fracture Surface Analysis of Optical Fibers, *Fiber Optics*, B. Bendow and S.S. Mitra, Ed., Plenum Press, 1979, p 187–208
11. J.J. Mecholsky, A.C. Gonzales, and S.W. Freiman, Fractographic Analysis of Delayed Failure in Soda Lime Glass, *J. Am. Ceram. Soc.*, Vol 62 (No. 11–12), 1979
12. S.M. Wiederhorn, *Fracture Mechanics of Ceramics*, Vol 2, R.C. Bradt, D.P.H. Hasselman, and F.F. Lange, Ed., Plenum, 1974

Glass-Ceramic Cookware Failure Analysis

P. Bruce Adams, Precision Analytical
S.E. DeMartino, Corning, Incorporated

FRACTOGRAPHY is the one indispensable tool for determining the cause of failure for a broken item. In a technical setting, that is, product design or engineering analysis, it is often the only tool necessary. However, in the context of a product liability case, where the jury is composed of lay people, the technical certainty of fractography may not suffice.

Assume that a fractographic analysis has unequivocally assigned the cause of breakage. In spite of this, there usually will be an opposing opinion, judged equally "expert" in the context of the legal proceeding, that presents a contrary view. The jury must choose. Yet they are rarely equipped to judge the validity of competing technical arguments.

Absent any additional information, the jury will make their choice based on their impressions of the credibility of witnesses. This may depend on academic or work-experience credentials, or, on who can best "teach" the jury.

However, there is another type of opportunity that is sometimes available and yet often overlooked. That is the breakage scenario itself.

The following is a case history in which two opinions resulted from the same fractographic evidence. Both experts were well qualified, credible, and fully capable of "teaching." However, one expert took advantage of corollary facts surrounding the incident; the other did not.

Accident Report

This case involves a glass-ceramic cooking vessel which was alleged to have broken spontaneously, spilling hot oil and causing severe burns to a housewife. She explained that she had ladled two tablespoons of shortening into a three quart glass-ceramic cooking vessel. She placed it on an electric stove to melt and left the kitchen. When she returned 15 min later the contents of the vessel were on fire. After opening the screen door leading to the outside and putting on oven mitts, she picked up the dish and started toward the door. Just before she reached the door, she stated that the dish spontaneously exploded.

As a result she was severely burned, necessitating an emergency transfer to a burn unit. She recovered but bore disfiguring scars.

Fractography

About 75 fragments, representing approximately 90% of the original item, were available for analysis. Two separate fractographic analyses were carried out.

Analysis No. 1. The initial analysis was focused on locating an origin and inspecting it under high magnification in order to elucidate the initiating flaw. The process involved:

- Visual and microscopic examination of the fragments using scanning electron microscopy
- Location and inspection of an origin
- Assessment of the features at, and in the vicinity of, the origin
- Interpretation of the relevance of these features as possible causes of breakage

During the course of the inspection of the fragments, an origin was identified on a piece from the rim near a handle of the dish (Fig 1a). Inspection of the detail of the origin under further magnification showed that the "flaw" location at the center of the origin was cratered (Fig 1b). Inspection of the as-formed free surface adjacent to this crater showed parallel tracks which appeared to lead toward the origin (Fig 1c). Finally, examination of the free surface showed certain irregularities (Fig 1d); these irregularities were attributed to the manufacturing process.

From the evidence as shown in Fig 1(a) through 1(d), it was concluded that the breakage was caused by an inhomogeneity, similar to that shown in Fig 1(d), which was present at the time of failure. It was assumed that it had been dislodged and lost during the fracture event. It was then concluded that this inhomogeneity was a manufacturing defect that so weakened the vessel as to allow it to fracture from relatively minor forces.

Analysis No. 2. The approach of fractographic analysis No. 2 was somewhat different. Each piece was inspected on all fracture surfaces and note was taken of the direction of crack propagation, the relative velocity of various cracks and whether they lead toward the internal or external as-formed surface. Notations were recorded with arrows marked on each piece. The product was then reconstructed using tape to hold the pieces together. Then the pattern of arrows was observed to determine the location of origins and the major forces which caused the product to break. All origins were inspected at various magnifications; mirror radius and initiating flaws were noted. The forces that initiated breakage were then considered as they related to the forces that generated further breakage. This analysis, then, involved:

- Visual and microscopic examination of fragments for fracture-generated features
- Reassembly of the vessel with available fragments
- Location and inspection of origins
- Interpretation of the stress system at the time of breakage
- Assessment and measurement of origin features
- Determination of stress at breakage

The reassembled vessel is shown in Fig 2(a). The pattern of fracture propagation shows that there were *two* primary origins. One is located at the rim near a handle (Origin A) as shown in Fig 1(a) through 1(d). The other is located near the center of the adjacent sidewall just below the rim (Origin B), as shown in Fig 3(a) and 3(b).

Origin A, which is the same as that identified by fractographic analysis No. 1, is on the rim of the dish. Within about 3 mm ($1/8$ in.), the crack forked to form six major cracks which fanned out across the bottom and sidewall of the dish. All of these cracks lead on the interior surface of the vessel.

Origin B is about 13 mm ($1/2$ in.) below the rim; it forked to form three cracks, all of which lead on the exterior surface of the vessel.

Cracks originating from the two different origins intersect and cross, as shown dia-

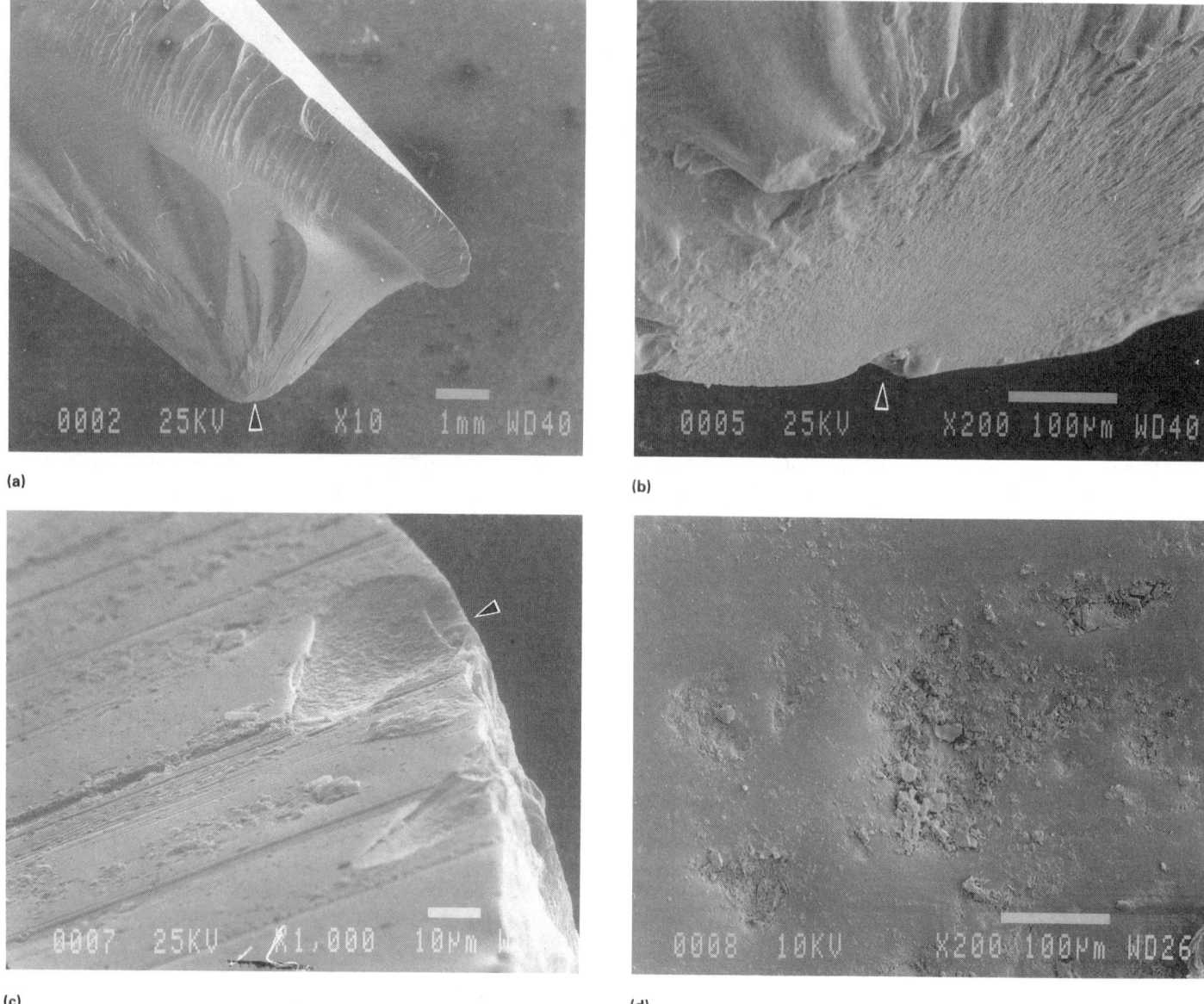

Fig 1 Fractographs of glass-ceramic cookware fragments. (a) Fracture-generated surface with arrow indicating Origin A location. (b) Oblique view of Origin A detail. (c) As-formed surface with arrow indicating Origin A location. (d) Irregular structures found on as-formed surface in vicinity of Origin A

grammatically in Fig 2(b). This fact dictates that at the instant of breakage there existed a rather complicated stress system that can be evaluated in terms of two major components. One component, manifested by a crack system leading on the interior, was the result of a severe force concentrated near Origin A and caused inward bending. The other component, manifested by a crack system leading on the exterior, was the result of a force causing the adjacent wall, about 90° from the first, to bulge outward. Cracks from Origin A, leading strongly on the inside surface (and trailing strongly on the outside surface) arrived at the intersection simultaneously with cracks from Origin B, leading strongly on the outside (and trailing strongly on the inside surface). Thus the two crack systems were able

to cross because the most advanced portion of one was on the opposite surface from the most advanced portion of the other when they intersected.

For this to occur, the origins which initiated the two crack systems were initiated almost simultaneously. Such a condition could only be created by a sudden impact causing the side of the dish to bend inward away from the impact and causing the vessel to deform such that the side which is about 90° away bent outward away from the center of the vessel.

Further, note that Origin A is smaller than Origin B and that there is significantly more crack branching associated with it. It follows therefore that Origin A was created by a larger breaking stress than Origin B. This in turn

means that the vessel was subjected to a force at least as large, if not greater than, the force required to create Origin A.

Discussion of Analyses. At this point there is agreement on the location of the principal origin and that the breakage resulted from mechanical forces producing tensile forces at the rim. But there is not agreement on the intensity of the initiating force. Furthermore, it was contended in analysis No. 1 that the manufacturer introduced a flaw at the location of Origin A which rendered the product defectively weak.

Although it is correct that such a "flaw" can often be identified and described, flaw size and the contribution of the flaw to crack initiation may be impossible to define and/or interpret. Fractographer No. 1 did not try

(a)

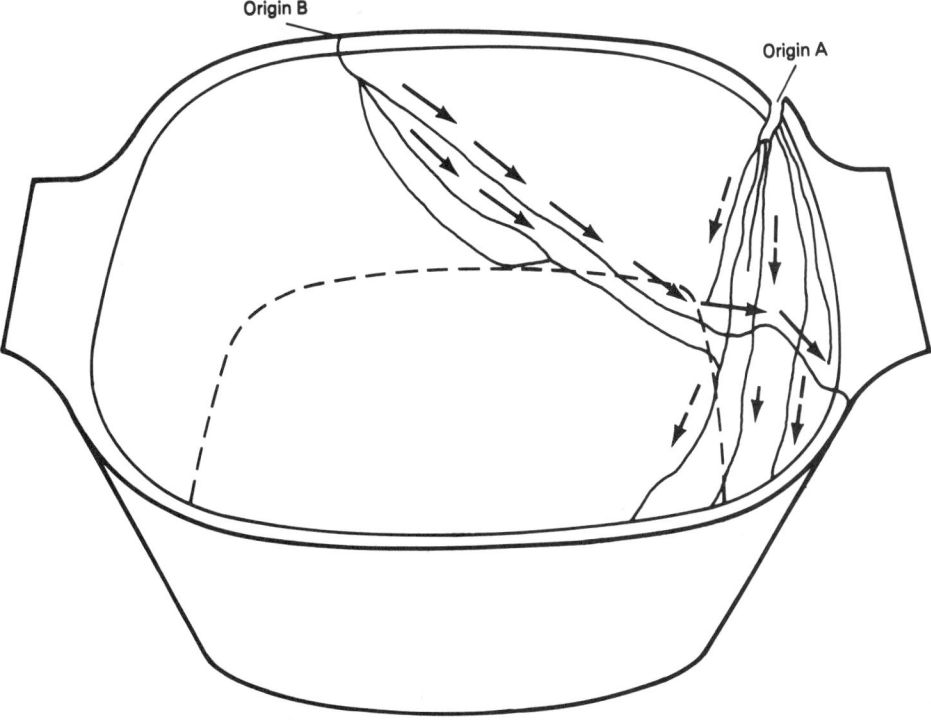

Origin B

Origin A

(b)

Fig 2 Fracture propagation pattern of broken vessel. (a) Reassembled product, showing two primary origins, the direction of cracks leading on the inside surface (dashed arrows) and the outside surface (solid arrows). (b) Sketch of the vessel showing the two major crack systems emanating from Origins A and B

Whether the "flaw" is identifiable or not, the analysis should include evaluation of the fracture-generated surfaces immediately surrounding the origin flaw, namely the mirror region. This relatively smooth surface area is bounded by an increased roughening, called mist-hackle. The mist-hackle is elongated in the direction of propagation and is produced when the crack accelerates from the origin flaw until it approaches terminal velocity. The development of a mirror region alone is an indicator of a high-stress breakage. It is also usually the most appropriate fracture-generated feature for a quantitative assessment of breaking stress. It is also very often available to the fractographer for measurement.

Measurement of Origin Mirrors. It has been repeatedly demonstrated that the product of the breaking stress, σ_F, and the square root of the distance from the origin to the onset of mist, which is the mirror radius (r_m), is equal to a constant for that material:

$$\sigma_F(r_m)^{1/2} = \text{constant} \qquad (\text{Eq 1})$$

The mirror radius, r_m, is half the distance as measured along the free surface edge between the onset of the mist-hackle on either side of the origin.

Figure 4 shows a plot of breaking stress versus the mirror size relationship for two commercial glass-ceramics (code 9608 being the one involved in this case). Having measured the mirror radius as 0.16 mm (0.0062 in.) at the impact origin (Origin A), and 0.46 mm (0.018 in.) for the other origin (Origin B), it is apparent that a force was imposed at the time of breakage which produced a breaking stress of at least 224 MPa (32.5 ksi). This force also caused the sidewall to bend outward producing a stress which allowed Origin B to be created at a breaking stress of 122 MPa (17.7 ksi).

Thus, analysis of the mirror dimensions provides a reliable indicator of the magnitude of forces. And, not incidentally, it provides a means of graphically conveying to a jury that the product was stressed far beyond the inherent strength of the material.

Investigation of Accident

The stage is set for a confrontation between two experts. Although the trained fractographer may be able to assess the relative technical merit, a lay jury is often in the position of having to make a decision on the basis of its assessment of the fractographer, rather than of his work. Therefore, it can be extremely useful to analyze corollary information that may be available.

In this case, fractographer No. 1 was unaware that such information existed. Only after intensive inquiry and effort was fractographer No. 2 able to ascertain that there was, in fact, a great deal of physical evidence that would help to determine what happened.

It turned out that many of the physical effects of various events which were part of

to do so in this case, and rightly so, because as is so often the case, any flaw that existed was destroyed by the fracture event.

However, without some indication of the breaking strength, the contribution of the

"flaw" cannot be quantified. Too often, then, as with analysis No. 1, there is simply a search for some physical difference to which to attribute a "weakness," such as the "inhomogeneities" identified in this case.

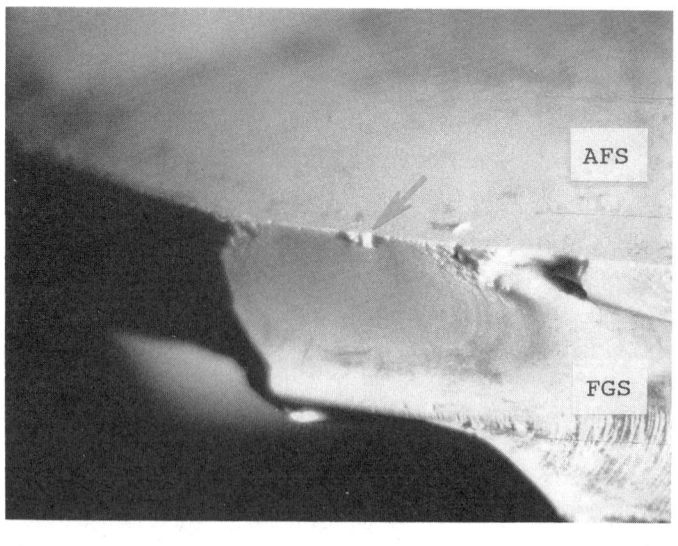

(a)

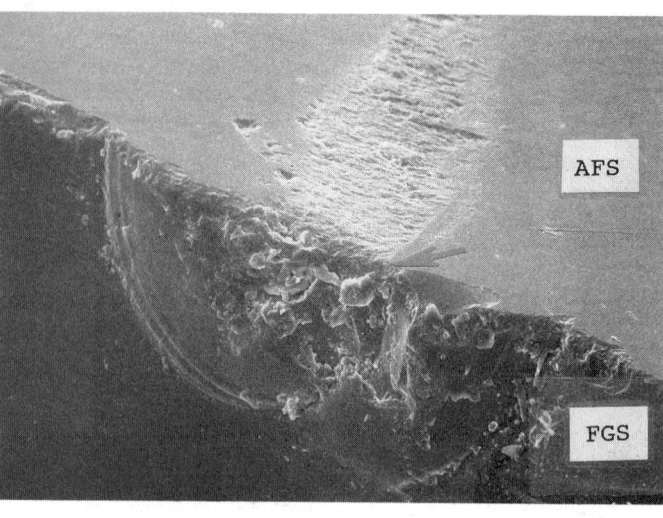

(b)

⊢———⊣
100 μm

Fig 3 Fractograph of fracture Origin B. (a) Fracture-generated surface (FGS) and as-formed surface (AFS). 8×. (b) Closeup showing pre-breakage damage on AFS and origin detail on FGS

the accident scenario were still evident. This evidence included charring in the stove area, burns on the rug, damage to a screen door as well as burn scars on the injured person.

A photograph of the hood immediately above the burner on which the vessel had been heated suggested that there had been a substantial grease fire in progress before the vessel was removed from the burner. Diagrams of the kitchen showed a pattern of burns on the rug in the area between the stove and the back door. The burns were a collection of points as would occur if burning material had been spattered on the rug. The character of the burned rug fiber was shown to be consistent with that resulting from hot grease.

Photographs of the back door taken after the accident showed there is a large hole in the screen and there is grease spattered about. In addition, remnants of shattered glass were found in the edge of the window frame.

Evaluation of the location and nature of the injured party's burns by a physician specializing in burns lead him to conclude that the burns about the face and neck were caused by flames, that those on the arms were caused by contact with hot grease, and those on the thighs were caused by spattered grease.

Accident Reconstruction

Based on the sum of the fractographic evidence and the corollary physical evidence surrounding the accident, it was concluded that the injured party had picked up the dish from the stove containing hot and flaming oil and had moved toward the back door. As she neared the back door, she hurled the vessel. It struck the door which was either closed or only partially open. The force shattered the glass in the door and punched a hole in the screen. The dish fell to the concrete steps below and broke.

The accident scenario was recreated in simulation and it was demonstrated that this reconstruction fit the details of the physical evidence. The simulation also demonstrated that burning grease produced a thick smoke that starved the fire of oxygen while the vessel remained on the burner; when moved rapidly toward the door, flames flared up. In an accident situation, this could easily be perceived as an explosive event.

Conclusion

When conducting an analysis for the purpose of understanding a breakage event, it is important to study all the evidence that is available. One cannot presume that the identification of a single fracture feature, such as an origin, will explain the fracture process.

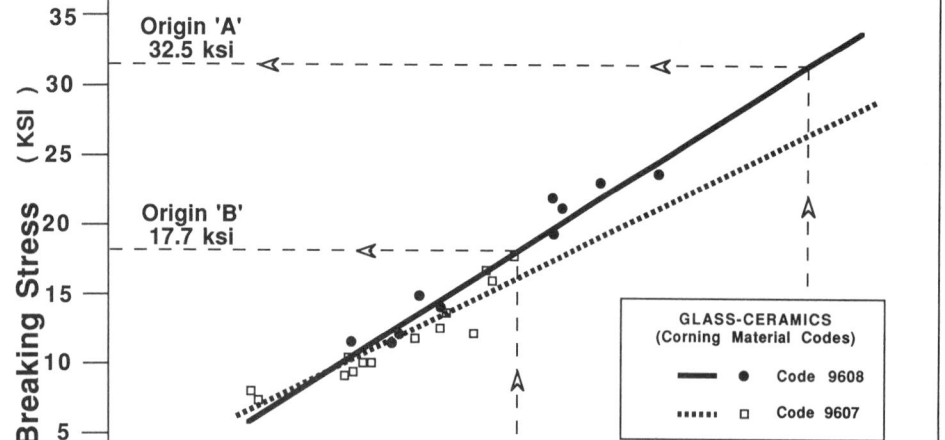

Fig 4 Breaking stress versus mirror radius for two glass-ceramic materials with the relationship shown for Origins A and B

Nor can one presume that the fractographic analysis can stand alone irrespective of other physical evidence. Certainly such other physical evidence must not bias the fractographic analysis; however, the fractographic analysis must be consistent with the entire breakage scenario.

This is especially true in the context of a product liability case where jurors may be ill-equipped to judge the relevance of conflicting technical testimony. The legal system does not require that either plaintiff or defendant "solve the crime." However, every jury wants to understand what happened and seeks a commonsense explanation.

SECTION 10

Design Considerations

Chairman: Raymond J. Bratton, Westinghouse Science & Technology Center

Introduction

THE DESIGN PROCESS for ceramic materials is generally more complex than metals owing to the ceramic characteristics of low-strain tolerance, low fracture toughness, and considerable scatter in strength properties, that is, all manifestations of a brittle material. These characteristics dictate the availability of consistent materials, flaw detection methods, and the appropriate design methodology in order to assure component reliability. This Section reviews brittle fracture behavior, the design approaches for ceramics, and in particular, the practical design methodology that is being developed to assure reliability of structural ceramics. Design practices for several applications of advanced ceramics, glass and glass fibers, and whisker-toughened ceramics are also discussed.

The first article, "An Overview of the Ceramic Design Process," summarizes the two major phases of design: conceptual design and reliability design. This article also includes a discussion of the following design approaches that are appropriate for ceramics:

- Empirical
- Deterministic
- Probabilistic
- Combined

The overview is followed by the article, "Ceramic Properties Data Base Systems," which describes the various sources of property information for engineers using ceramics and glasses. Emphasis is placed on existing computerized materials data systems. The article also discusses the need to use caution in the use of supplier property data, as well as other sources which do not provide data statistics or a detailed description of the material and test method used.

The article, "Crack Propagation in Ceramics" is presented next. Here the brittleness of ceramics and the use of linear elastic fracture mechanics to quantitatively explain fracture behavior, including environmentally enhanced crack growth, are described.

Having been introduced to the fracture behavior of ceramics, the design tools used for ceramics are discussed next. The comprehensive design methodologies being developed for ceramics focus on reliability. This, of course, is necessary because of the brittle behavior of ceramics. In the article, "Probabilistic Design of Ceramic Components with the NASA/CARES Computer Program," it will be shown that the design methodology uses three major elements: linear elastic fracture mechanics theory, which relates the strength of ceramics to the size, shape, and orientation of critical flaws; extreme value statistics to obtain the characteristic flaw size distribution function, which is a material property; and material microstructure.

Another article that contributes to design methodology, "Advanced Statistical Concepts of Fracture in Brittle Materials," reviews an ongoing project aimed at advancing the current understanding of fracture statistics. It describes in particular the development of confidence and tolerance bounds on predictions that use the Weibull distribution function. Four classes of problems involving Weibull estimation procedures for strength are analyzed.

A final series of articles in this Section describe design practices used for ceramic components for several different applications. These include:

- "Design Practices for Structural Ceramics in Gas Turbine Engines"
- "Design Practices for Structural Ceramics in Diesel Turbocharger Wheels"
- "Design Practices for Structural Ceramics in Gasoline Engines"
- "Design Practices for Whisker-Toughened Ceramic Components"
- "Design Practices for Glass and Glass Fibers"

An Overview of the Ceramic Design Process*

Arthur F. McLean, Consultant
Dale L. Hartsock, Ford Motor Company

IN DESIGNING engines or other engineering mechanisms, there are two major phases of design: conceptual design and detailed design. For both of these phases, the designer requires a knowledge of the key structural properties of the candidate materials being considered for the design. In general, his choice of materials will depend upon the material's ability to do the job, the material's general availability, and the material's competitiveness with other materials. Quite often, especially at the conceptual design stage, there are iterations between choice of materials and the resulting design concept, sometimes to the extent that the material choice has a significant bearing on the final design concept. Consider, for example, the contrast in design between two four-cylinder diesel engines, one with a four-cylinder, water-cooled, cast iron block and the other with four separate, uncooled, silicon nitride cylinders.

In considering the use of structural ceramics for engineering applications at the conceptual design stage, the designer needs to know the key physical and mechanical properties, principally by comparison with equivalent metal properties, with which he is generally familiar. At this stage, the designer is looking for the functional advantage to be gained by using ceramics.

Next comes the detailed design stage, where the designer will need more detailed information on the properties of the selected structural ceramics, along with an understanding of the methodology by which to design with ceramics. At this stage, the designer is seeking to design for reliability.

In summary, (1) conceptual design that is strongly related to design for function, and (2) detailed design that is strongly related to design for reliability are the key aspects of designing with structural ceramics and are the subjects of this article. Each of these phases of design, in differing degrees, requires a knowledge of the properties of the ceramic material under consideration.

Key Structural Ceramic Characteristics

Table 1 lists typical properties of the major structural ceramics that a designer would need as a basis for initial conceptual design of an engineering system.

Three grades of silicon nitride are shown: hot-pressed, sintered, and reaction-bonded. These are followed by three similar grades of silicon carbide, that is, hot pressed, sintered, and reaction-bonded. Next listed are three oxide ceramics: partially stabilized zirconia, lithium-aluminum-silicate, and aluminum titanate. For each material, some of the thermal, physical, and mechanical properties are listed and compared to the more common metals of construction: iron, steel, and aluminum. The following summarizes the key features of each of the ceramics listed.

- Hot pressed silicon nitride (HPSN) has the strongest specific strength (strength/density) at 600 °C (1110 °F) of any material. It has excellent thermal shock resistance
- Sintered silicon nitride (SSN) has high strength and can be formed into complex shapes
- Reaction-bonded silicon nitride (RBSN) can be formed into complex shapes with no firing shrinkage
- Hot pressed silicon carbide (HPSC) is the strongest of the silicon carbide family and maintains strength to very high temperatures (1500 °C, or 2730 °F)
- Sintered silicon carbide (SSC) has high-temperature capability and can be formed into complex shapes
- Reaction-bonded silicon carbide (RBSC) can be formed into complex shapes and has high thermal capability
- Partially stabilized zirconia (PSZ) is a good insulator and has high strength and toughness. It has a thermal expansion close to iron, facilitating shrink fit attachments
- Lithium-aluminum-silicate (LAS) is a good insulator and has very low thermal expansion
- Aluminum titanate is a good insulator with relatively low thermal expansion

These typical properties are useful in helping the designer generate a conceptual design to achieve a desired overall function.

Once the overall conceptual design is complete, detailed design, or design for reliability, is necessary. For this, the designer requires detailed property data for the selected structural ceramic as well as information on methodology for designing with ceramics.

At this detailed stage, the key difficulty with which the designer is faced in designing for reliability is that ceramics are brittle. This brittleness causes fracture to occur when the tensile stress exceeds a threshold causing cracks to propagate from inherent, minute flaws in the material, in a relatively catastrophic manner. The random size, shape, and distribution of these inherent defects causes a relatively large scatter in the strength of any group of material test specimens.

To facilitate a methodology to design for reliability, the strength of a ceramic is measured on many samples and the results quantified statistically. This strength testing is conducted in flexure, where the strength is the fracture stress calculated to pertain in the specimen outer fibers (tensile side) and is known as the modulus of rupture (MOR). Both three-point and four-point flexural testing are practiced, the latter being the favored approach. As mentioned, strength testing is conducted on many material specimens and the results are quantified statistically in terms of (1) characteristic MOR_0 (MOR where 63% of material specimens fracture) and (2) Weibull modulus, m (typically in the range 10 to 20). Note the higher the m value is, the less the scatter in strength.

Thus, for detailed design, the designer needs to know the material's statistical strength properties as well as detailed physical prop-

*Adapted, with permission, from *Treatise on Materials Science and Technology*, Vol 29, Academic Press, 1989, p 27–97

Table 1   Properties of structural ceramic materials

These properties are typical of these classes of materials, but in many cases, large variations may exist between various formulations. Strengths are for four-point bending spans of 19.05/9.525 mm (0.745/0.375 in.) and bar cross sections of 6.35 × 3.175 mm (0.25 × 0.125 in.).

Material	Young's modulus GPa	Young's modulus 10^6 psi	Poisson's ratio	Thermal conductivity, W/m·K RT	Thermal conductivity, W/m·K 600 °C (1110 °F)	Thermal expansion, $\times 10^{-6}$/K	Specific heat, J/g·°C	Density, g/cm³	Strength, MPa (ksi) RT	Strength, MPa (ksi) 600 °C (1110 °F)	Weibull modulus (m), at RT	Maximum use temperature °C	Maximum use temperature °F
Silicon nitride													
Hot-pressed (HPSN)	290	42.1	0.3	29	22	2.7	0.75	3.3	830 (120)	805 (117)	7	1400	2550
Sintered (SSN)	290	42.1	0.28	33	18	3.1	1.1	3.3	800 (116)	725 (105)	13	1400	2550
Reaction-bonded (RBSN)	200	29.0	0.22	10	10	3.1	0.87	2.7	295 (43)	295 (43)	10	1400	2550
Silicon carbide													
Hot-pressed (HPSC)	430	62.4	0.17	80	51	4.6	0.67	3.3	550 (80)	520 (75)	10	1500	2730
Sintered (SSC)	390	56.6	0.16	71	48	4.2	0.59	3.2	490 (71)	490 (71)	9	1500	2730
Reaction-bonded (RBSC)	413	59.9	0.24	225	70	4.3	1.0	3.1	390 (57)	390 (57)	10	1300	2370
Partially stabilized zirconia (PSZ)	205	29.7	0.30	2.9	2.9	10.5	0.5	5.9	1020 (148)	580 (84)	14	950	1740
Lithium-aluminum-silicate	68	9.9	0.27	1.4	1.9	0.5(a)	0.78	2.3	96 (14)	96 (14)	10	1200	2190
Aluminum-titanate	11	1.6	0.22–0.26	2	5	1.0	0.88	3.0	41 (6)	. . .	15	1200	2190
Common metals (Reference)													
Cast iron	170	24.7	0.28	49	40	12	0.45	7.1	620 (90)	100 (14.5)	. . .	500	930
Steel	200	29.0	0.28	38		14	0.45	7.8	1500 (218)	140 (20)	. . .	600	1110
Aluminum	70	10.2	0.33	160		22.4	0.96	2.7	370 (54)	0	. . .	350	660

(a) Maximum excursion from 350 to 800 °C (660 to 1470 °F), initial expansion is negative.

erties (methods to determine the mechanical and physical characteristics of ceramics are described in the Section "Testing, Characterization, and NDE" in this Volume). Assume, for example, the conceptual design indicates that sintered silicon nitride is preferred for a critical structural component. To conduct the required heat-transfer, stress, and reliability analyses, the designer will require such parameters as modulus of rupture (MOR_0), Weibull modulus (m), Young's modulus (E), thermal expansion coefficient (e), and thermal conductivity (k_t), all as functions of temperature (T). Figure 1(a) through 1(e) show curves of these properties for a typical grade of sintered silicon nitride. It should be noted that there is a large variation in strength and Weibull modulus for different grades of silicon nitride; these curves represent one particular case.

Conceptual Design

The application of structural ceramics to engineering systems hinges on the functional benefits to be derived and is first manifested in the conceptual design. The need to conceive new designs employing ceramics is often prompted by limitations of the metals being used in conventional designs. For example, the cast iron cylinder block of a conventional reciprocating diesel engine is water-cooled to limit material temperatures. In another example, the conventional gas turbine engine has to limit internal metal temperatures by consuming extra air, despite the use of costly, high-temperature, nickel-base superalloys. To appreciate the benefits from increasing material temperatures of these engines, we need to understand a little more about the functional aspects of the engine system under consideration. This can best be done by considering sample applications of structural

ceramics. In this article, this is accomplished through examples of diesel engine and turbine engine advanced concepts. More detailed information can be found in the following articles in this Section:

● "Design Practices for Structural Ceramics in Gas Turbine Engines"
● "Design Practices for Structural Ceramics in Diesel Turbocharger Wheels"
● "Design Practices for Structural Ceramics in Gasoline Engines"

Diesel Engines

Indirect-Injection Diesel Engines. Figure 2 shows the conceptual design of a ceramic precombustion chamber where combustion takes place. This concept of indirect injection is typical of high-speed diesels used for passenger cars and light trucks. The ceramic precombustion chamber shown is basically a cup containing initial combustion; the combustion gases exit through an angled hole in the bottom of the cup. Metal precombustion chambers are made of a high-temperature nickel alloy and represent one of the few uses of a relatively expensive alloy in the diesel engine. As such, this has been a target for the application of ceramics. Furthermore, the concept of a ceramic precombustion chamber offers the opportunity for improved insulation, which will tend to promote hotter and faster combustion, which, in turn, can improve noise and exhaust emissions. Clearly, to withstand initial ignition and continual thermal upshocks and downshocks, the ceramic selected must have excellent thermal shock resistance. Silicon nitride is an obvious choice for this application and Isuzu (Ref 2) and Toyota (Ref 1) have introduced silicon nitride precombustion chambers for their small automotive indirect injection (IDI) diesels.

Some of the attributes claimed for the better-insulating silicon nitride prechambers are:

● Increased engine power (10%) by adjusting fuel injection rate, compression ratio, and boost pressure
● Reduced idle noise resulting from reduced ignition delay (time from fuel introduction to ignition)
● Faster warm-up, resulting in reduced hydrocarbon emissions over the federal emissions driving cycle
● Reduced particulate emissions

Further design modifications for diesel engines include the direct-injection engine (Ref 3–6) and the turbo-compounded, turbocharged, direct-injection diesel engine (Ref 5).

Turbomachinery

Turbochargers are used in passenger cars, trucks, off-highway vehicles, and industrial applications to boost the intake pressure into gasoline or diesel engines and thereby achieve increased power output. To generate a reasonable degree of boost, the turbocharger has to reach a high speed. The turbocharger compressor or impeller, which boosts the intake pressure, is driven by a turbine powered by the exhaust gas of the engine; to tolerate hot exhaust-gas conditions, the turbine has been conventionally made of a high-temperature nickel alloy.

The conceptual design of a turbocharger incorporating a silicon nitride turbine is shown in Fig 3. The silicon nitride (or silicon carbide) turbine is only 40% of the density of the high-temperature nickel-base alloy and this translates into approximately 50% weight and inertia saving for the rotating assembly (that is, turbine, compressor, and shaft). Figure 4 shows test data comparing initial vehicle acceleration using a ceramic (silicon nitride) turbocharger versus a nickel-base alloy tur-

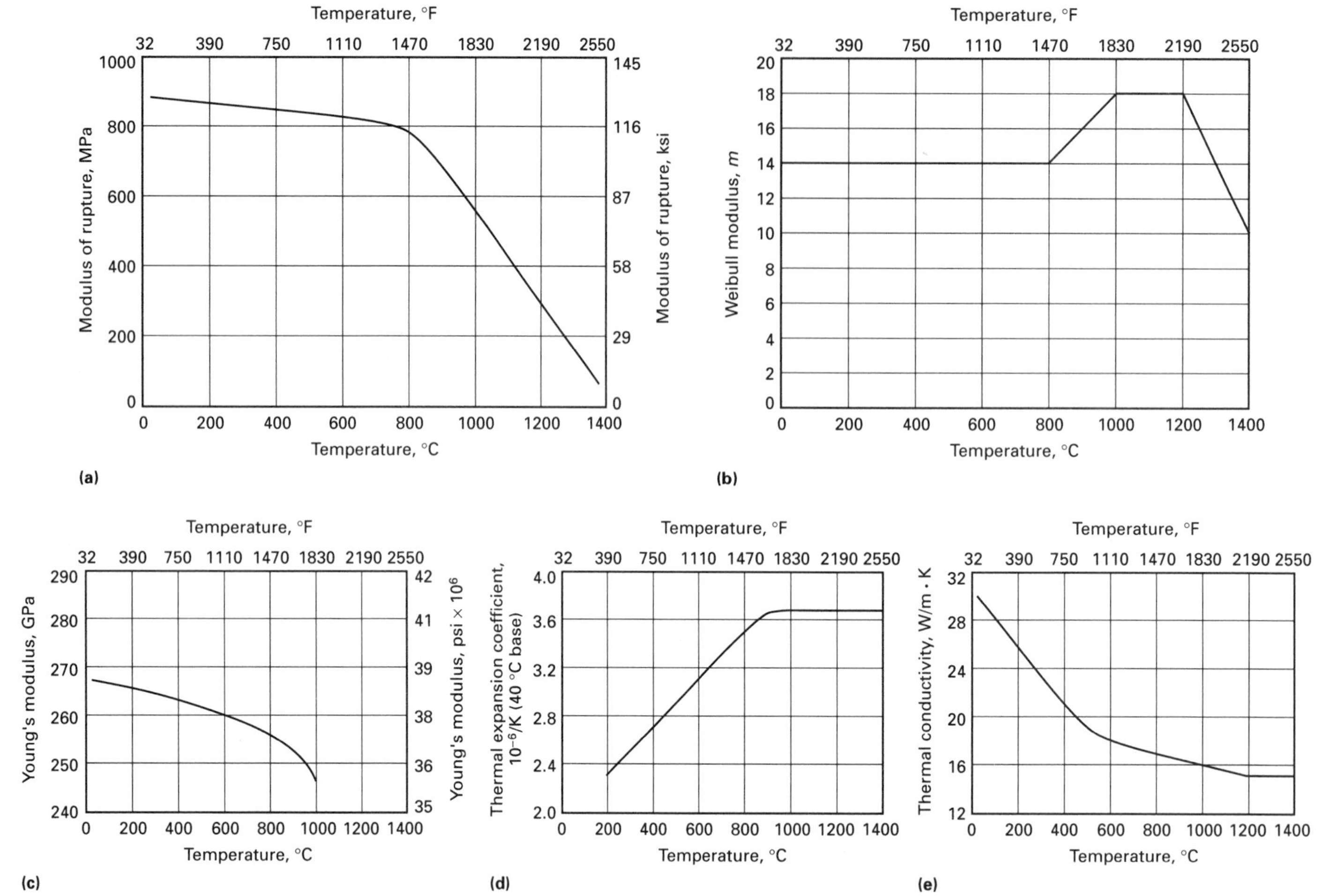

Fig 1 Properties of silicon nitride as a function of temperature. (a) Modulus of rupture. (b) Weibull modulus. (c) Young's modulus. (d) Coefficient of thermal expansion. (e) Thermal conductivity

bocharger; the turbocharger response time from idle to maximum is halved and this significantly improves the acceleration feel of the 1360 kg (3000 lb) test car in which it was installed. It is interesting to note from Fig 3 that the conceptual design utilized silicon nitride, not only for the turbine rotor itself, but also for half of the shaft. This was to facilitate a cooler ceramic-to-metal attachment location.

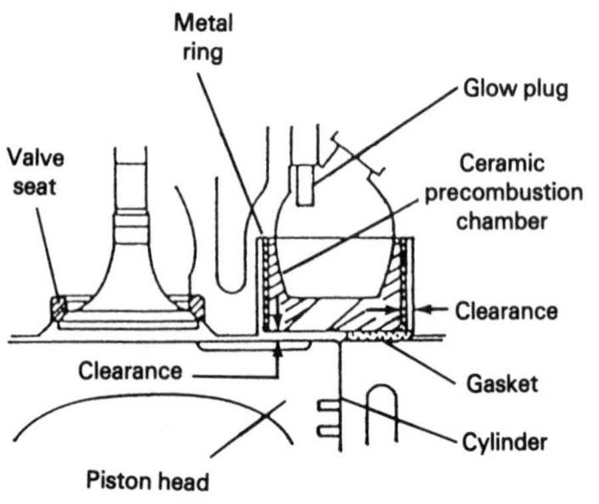

Additional advanced design concepts for turbomachinery include simple-cycle gas turbine engines (Ref 8, 9) and regenerative gas turbine engines (Ref 10–14).

Detailed Design

Following the conceptual design, the detailed design of the component is completed. The major aspect of this procedure is to design the component for an acceptable reliability. The means to do this are discussed in the following sections.

Fast Fracture Reliability

Ceramic materials have two basic properties that tend to make their successful application to structural components more difficult.

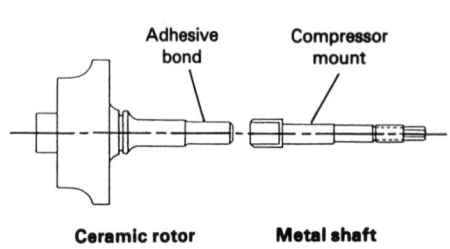

Fig 2 Ceramic precombustion chamber for the indirect-injection diesel engine

Fig 3 Ceramic turbocharger with metal shaft attached near the compressor and bearing by adhesive bonding

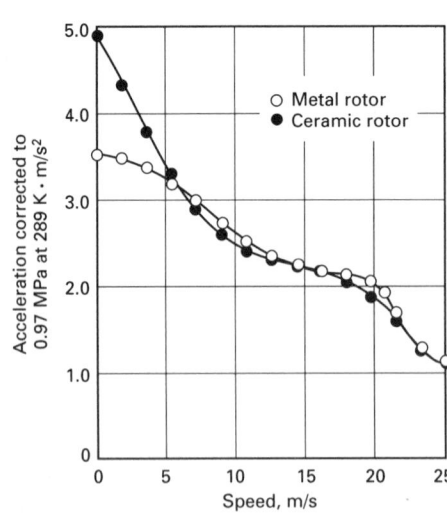

Fig 4 Comparison of wide-open throttle acceleration for ceramic and metal turbocharger rotors in a 2.3 L engine

First, the material has very low strain tolerance. Second, there is frequently a large scatter in the strength data. The first property manifests itself in that the material has a linear stress-strain relationship all the way from zero stress to failure. This linear relationship by itself is attractive to analysts because it makes the modeling process easier. However, the absence of plastic flow means that the mechanism by which the local stress concentrations are relieved around the ever-present flaws is eliminated. This means that the analysis will have to be refined enough to locate and eliminate local high-stress areas. This will frequently mean an analysis that is more complex than that which would be required for a similar part made of metal. The fact that the material strength shows considerable variability means that the design will have to be based on probabilistic methods.

For years, ceramics have been used in simple structures by designing to keep them in compression. This is because ceramics have been shown to be much stronger in compression than in tension. However, today there are many uses for which ceramics could offer performance benefits that are not conducive to this "design for compression" philosophy. Therefore, in order to confidently design components subjected to tensile loading, the design methodology had to be reevaluated. Weibull developed a probabilistic system of design based only on the tensile stresses in the component (Ref 15). Compressive failure is not considered in this technique because brittle materials usually fail from tensile stresses. His methodology has been used extensively in the design of ceramic components for engines (Ref 16–18) and has been shown to work quite well for a number of applications.

A number of articles within this Volume describe the use of Weibull analysis for determining the strength of ceramic components. These include the article "Strength and Proof Testing" (see the Section on "Testing, Characterization, and NDE") and the following articles in this Section of the Handbook:

- "Probabilistic Design of Ceramic Components with the NASA/CARES Computer Program"
- "Advanced Statistical Concepts of Fracture in Brittle Materials"

Alternative fast fracture reliability methods are also described in the aforementioned articles as well as in Ref 19 to 22.

Weibull Theory. Determining ceramic component reliability based on Weibull's weakest-link theory, very simply, is based on an assumption that a part is like a chain of many links. If any link (small element of the part) fails, then the whole chain (part) has failed. Similarly, if any small volume in a ceramic part is stressed sufficiently to cause a crack, the part will generally fail. Thus, the key is to determine if any of the elements in a part are likely to fail. Because ceramic

material properties are variable due to the random distribution of flaws in ceramics, there is a variation of strengths for various elements of a part. Thus, the strength of the various elements can be considered to have a statistical distribution with values above and below some characteristic strength. If an element is subjected to some value of stress, there is a certain probability that the local strength of the material will be exceeded. As the number of elements in a chain is increased, the probability that a weak link will occur and cause the chain to fail also increases. Similarly the probability of failure of a ceramic component is a function of the volume of material subjected to the various stress levels. By combining the probability of failure of all the elements, the probability of failure of the total part is determined.

Weibull established that the function that would describe the cumulative probability of failure of all the elements of a part of N elements is the following:

$$P_f = 1 - \exp\left[\sum_{i=1}^{N} \left(\frac{\sigma - \sigma_u}{\sigma_0}\right)^m \frac{V_i}{V_0}\right] \quad \text{(Eq 1)}$$

In Eq 1, σ_0 is a normalizing material strength and is defined as the characteristic stress at which a volume of the material (V_0) would fail in uniaxial tension. For convenience and to simplify the mathematics, Weibull chose to use a unit volume of material. Thus, since $V_0 = 1$, it is left out of the equation, but its existence must be remembered in order to obtain the correct units in the equation. Thus, the strength σ_0 should always be remembered to be a discrete value that is only valid for the unit volume of material for which it was calculated. One way to do this is to assign σ_0 the units of (force/length^2) × (length)$^{3/m}$. If these units are used, then it is possible to convert σ_0 to any unit system.

Using this simplification ($V_0 = 1$), and extending this to the infinitesimal elements of a part, Eq 1 can be written:

$$P_f = 1 - \exp\left[-\int_V \left(\frac{\sigma - \sigma_u}{\sigma_0}\right)^m dV\right] \quad \text{(Eq 2)}$$

Weibull recognized the fact that brittle materials are particularly susceptible to surface failures and thus there may be different σ_u, m, and σ_0 values for the surface and bulk material. This means that in general the exponent would include both an integral for the volume and a separate integral for the surface. Paluszny has demonstrated that by being careful in the preparation of the surfaces of test bars and components, the surface material parameters approximate the volume parameters, and using only the volume analysis gives valid reliability predictions (Ref 16). This is fortunate since, in reality, material property data broken down into separate values for surface and volume are usually not available. Thus the methodology given here utilizes only the volume analysis. However, it should be noted that as material surface

property data becomes available, the methodology could easily be extended to include the surface parameters.

The reliability of a component is defined as one minus the probability of failure.

$$R = 1 - P_f \quad \text{(Eq 3)}$$

The σ_u in Eq 2 is the stress below which the material will never fail. For a conservative design approach, this value is taken to be zero. Paluszny has shown that this assumption results in valid predictions of component reliability (Ref 16). Thus, the final equation used for prediction of reliability is:

$$R = 1 - P_f = \exp\left[-\int_V \left(\frac{\sigma}{\sigma_0}\right)^m dV\right] \quad \text{(Eq 4)}$$

Weibull chose to call the negative of the argument of the exponential function in this equation the risk of rupture (B). Thus, Eq 4 could be written:

$$R = \exp[-B] \text{ where } B = \int_V \left(\frac{\sigma}{\sigma_0}\right)^m dV \quad \text{(Eq 5)}$$

The risk of rupture is approximately equal to the probability of failure for small values ($B < 0.1$). Equation 5 is directly applicable for the case of uniaxial tension. In order to expand this equation for three-dimensional stress states, the concept of integrating the normal stress around the portion of the unit radius sphere where the normal stress is positive was introduced (Ref 15, 23). Only the positive portion is used since ceramics are much stronger in compression and usually fail in tension. The concept is shown in Fig 5 with the variables defined as follows:

$$\sigma_n = \cos^2 \phi (\sigma_1 \cos^2 \psi + \sigma_2 \sin^2 \psi) + \sigma_3 \sin^2 \phi \quad \text{(Eq 6)}$$

$$P_f = 1 - \exp\left\{-\int_V \left[k \int_A (\sigma_n)^m dA\right] dV\right\} \quad \text{(Eq 7)}$$

The k factor in Eq 7 is a compatibility term to force the equation to reduce to the original

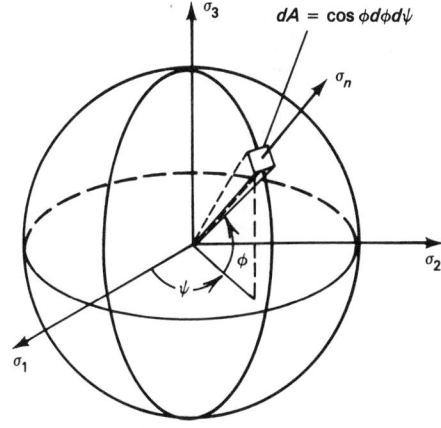

Fig 5 Geometric variables describing location on the unit sphere

equation for uniaxial tension (Eq 4). This gives a k value as follows:

$$k = \frac{2m + 1}{2\pi\sigma_0^m} \qquad (Eq\ 8)$$

Thus, the general equation for the probability of failure is:

$$P_f = 1 - \exp\left(\left(-\int_V \left\{\frac{2m + 1}{2\pi\sigma_0^m} \int_0^{2\pi} \int_0^{\pi/2}\right.\right.\right.$$

$$\cdot \left[\cos^2 \phi(\sigma_1 \cos^2 \psi + \sigma_2 \sin^2 \psi)\right.$$

$$\left.\left.\left.+ \sigma_3 \sin^2 \phi\right]^m \cos \phi\ d\phi\ d\psi\right\} dV\right)\right) \qquad (Eq\ 9)$$

Due to symmetry of the normal stress on the top and bottom of the sphere, it is only necessary to integrate the normal vector around the top half of the sphere. This is true since the area term from the integral is normalized by the 2π term in the denominator of the k term (Eq 8).

Weibull Material Property Generation. Before proceeding further, the method used to determine the material properties of ceramics must be discussed. This section will assist the designer in obtaining a meaningful and consistent data base for reliability analysis.

The first step necessary in designing with ceramics is to determine the material's unit volume characteristic strength and the Weibull modulus. First, the Weibull modulus is determined. By taking the logarithm of Eq 4 twice, it can be written:

$$\ln\left[\ln\left(\frac{1}{1 - P_f}\right)\right] = \ln(V) - m \ln(\sigma_0) + m \ln(\sigma)$$

$$(Eq\ 10)$$

which is of the form of a linear equation $y = a + mx$. From this equation, it is obvious that if test data is generated for σ, and the appropriate probability of failure is assigned for the ranked data, then the Weibull modulus m is just the slope of the best-fitting line through these data. The probabilities assigned for ranked data are simply the median ranks as defined in Table 2. The method of fitting the line is a matter of choice (for example, least squares, method of moments). One commonly used method of fitting the data is the maximum-likelihood estimator as described in Ref 25 and 26.

The most popular method of generating material data is to use test bars subjected to four-point bending. The bars are loaded to failure and the maximum outer-fiber stress σ_{max} is calculated. This stress is the one used in Eq 12 to determine the Weibull modulus. The characteristic modulus of rupture MOR_0 is defined as the maximum outer-fiber stress in the test bar that would cause failure 63.2% of the time. This corresponds to the stress at which the risk of rupture equals one.

The stresses at any point in the bar can be calculated in terms of the bar geometry, load points, and the σ_{max}. Using this, the risk of

rupture of the bar can be calculated by the following equation:

$$B = B_1 + B_2 \qquad (Eq\ 11)$$

where

$$B_1 = 2 \int_0^{L_2/2} \int_0^{h/2} \int_{-b/2}^{b/2} \left(\frac{2y\sigma_{max}}{h\sigma_0}\right)^m dx\ dy\ dz$$

$$B_2 = 2 \int_{L_2/2}^{L_1/2} \int_0^{h/2}$$

$$\int_{-b/2}^{b/2} \left[\frac{2y\sigma_{max}(L_1 - 2z)}{h\sigma_0(L_1 - L_2)}\right]^m dx\ dy\ dz$$

By solving this equation, the risk of rupture for test bars is:

$$B = \left(\frac{\sigma_{max}}{\sigma_0}\right)^m \frac{bh}{2} \left[\frac{mL_2 + L_1}{(m + 1)^2}\right] \qquad (Eq\ 12)$$

The last terms of this equation are the effective volume of the test bar. The effective volume is defined as the volume of material subjected to a uniaxial tension of σ_{max} that would have the same probability of failure as the sample.

$$V_{eff} = \frac{bh}{2} \left[\frac{mL_2 + L_1}{(m + 1)^2}\right] \text{ for test bars} \qquad (Eq\ 13)$$

As just mentioned, the characteristic strength of a material is defined as that strength that will have a probability of failure of 63.2% and a risk of rupture of 1. Thus the unit volume characteristic strength σ_0 can be obtained by substituting into Eq 12 the risk of rupture of 1 and using the characteristic strength MOR_0 for σ_{max}:

$$\sigma_0 = MOR_0 (V_{eff})^{1/m} \qquad (Eq\ 14)$$

This equation is shown graphically in Fig 6 for test bars of $3.175 \times 6.35 \times 31.75$ mm ($0.125 \times 0.25 \times 1.25$ in.) in height, width, and length, respectively, and load spans of 19.05 and 9.525 mm (0.75 and 0.375 in.).

As previously mentioned, the σ_0 and m values are constant for a given material and will not vary as a function of how they are generated (neglecting experimental error and accuracies affected by sample size). This being the case, the preceding equations show that the MOR_0 value must vary as a function of the test bar size and load spans. Thus MOR_0 data from different sources may not be comparable, and, if the actual test bar size and load spans are ignored in using the data, erroneous results can be obtained (Ref 27). The conversion of MOR data from one test bar size and load span (A) to another (B) can be done by the following equation (Ref 28):

$$\frac{MOR_B}{MOR_A} = \left(\frac{b_A h_A}{b_B h_B}\right)^{1/m} \cdot \left(\frac{mL_{A2} + L_{A1}}{mL_{B2} + L_{B1}}\right)^{1/m} \qquad (Eq\ 15)$$

During the development of the Weibull design methodology, companies utilized different statistical points for the strength parameter MOR. Some companies used the

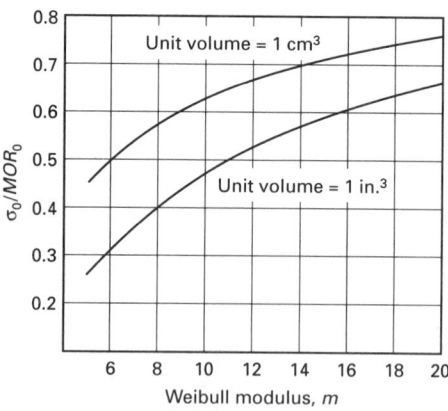

Fig 6 The ratio of the unit volume characteristic strength to the modulus of rupture for a given test bar size and load conditions versus the Weibull modulus

characteristic strength and some the mean strength. These different strengths continue to be used today and are related by the following equation (Ref 24):

$$MOR_0 = \sigma_{mean} \frac{1}{\Gamma\left(\frac{1}{m} + 1\right)} \qquad (Eq\ 16)$$

When generating material property data, it is necessary to use enough data points to get a good statistical representation of the material properties. A way of estimating the accuracy of a parameter is to look at the confidence intervals. Figure 7(a) shows the 90% confidence bounds for the Weibull slope, and Fig 7(b) shows the characteristic value. The bounds for the Weibull slope are a function of sample size only while, for the modulus of rupture, they are a function of both the sample size and the Weibull slope. As seen from this curve, using sample sizes of less than approximately 30 can lead to significant errors in the material parameters. The choice of sample size depends on many factors including the cost and timing of testing and the degree of conservation that is acceptable, but erroneous judgments may be made and unacceptable designs pursued if the sample sizes are too small.

Sensitivity of Reliability to Various Parameters. When designing with ceramics, it is often helpful for the designer to understand how sensitive the reliability calculations are to various parameters. This section will discuss how the stress state, volume, material strength, and Weibull modulus affect the reliability.

From Eq 9, the acceptable maximum stress in a component is a function of the multiaxial stress state in the component. In order to investigate the sensitivity of the risk of rupture to the relative magnitudes of the principal stresses, Eq 9 can be rewritten with C and D as the ratios of σ_2 and σ_3 to σ_1, as shown by Dukes (Ref 29):

Table 2   Median ranks

Rank order	Sample size																			
	1	2	3	4	5	6	7	8	9	10	11	12	13	14	15	16	17	18	19	20
1	.5000	.2929	.2063	.1591	.1294	.1091	.0943	.0830	.0741	.0670	.0611	.0561	.0519	.0483	.0452	.0424	.0400	.0378	.0358	.0341
2		.7071	.5000	.3857	.3138	.2645	.2285	.2011	.1796	.1623	.1480	.1360	.1258	.1170	.1094	.1027	.0968	.0915	.0868	.0825
3			.7937	.6143	.5000	.4214	.3641	.3205	.2862	.2586	.2358	.2167	.2005	.1865	.1743	.1637	.1542	.1458	.1383	.1315
4				.8409	.6862	.5786	.5000	.4402	.3931	.3551	.3238	.2976	.2753	.2561	.2394	.2247	.2118	.2002	.1899	.1806
5					.8706	.7356	.6359	.5598	.5000	.4517	.4119	.3785	.3502	.3258	.3045	.2859	.2694	.2547	.2415	.2297
6						.8909	.7715	.6795	.6069	.5483	.5000	.4595	.4251	.3954	.3697	.3471	.3270	.3092	.2932	.2788
7							.9057	.7989	.7138	.6449	.5881	.5405	.5000	.4652	.4348	.4082	.3847	.3637	.3449	.3280
8								.9170	.8204	.7414	.6762	.6215	.5749	.5349	.5000	.4694	.4423	.4182	.3966	.3771
9									.9259	.8377	.7642	.7024	.6498	.6046	.5652	.5306	.5000	.4727	.4483	.4263
10										.9330	.8520	.7833	.7247	.6743	.6303	.5918	.5577	.5273	.5000	.4754
11											.9389	.8640	.7996	.7439	.6955	.6530	.6153	.5818	.5517	.5246
12												.9439	.8742	.8135	.7606	.7141	.6730	.6363	.6034	.5737
13													.9481	.8830	.8257	.7753	.7306	.6908	.6551	.6229
14														.9517	.8906	.8364	.7882	.7453	.7068	.6721
15															.9548	.8973	.8458	.7998	.7585	.7212
16																.9576	.9032	.8542	.8101	.7703
17																	.9601	.9085	.8617	.8195
18																		.9622	.9132	.8685
19																			.9642	.9175
20																				.9659

$$P_f = 1 - \exp[-B]$$

$$B = \int_V \left\{ \frac{2m + 1}{2\pi} \int_0^{2\pi} \int_0^{\pi/2} \left(\frac{\sigma_1}{\sigma_0} \right)^m \right.$$

$$[\cos^2 \phi (\cos^2 \psi + C \sin^2 \psi) + D \sin^2 \phi]^m$$

$$\left. \cos \phi \; d\phi \; d\psi \right\} dV \qquad \text{(Eq 17)}$$

where:

$$C = \frac{\sigma_2}{\sigma_1} \quad \text{and} \quad D = \frac{\sigma_3}{\sigma_1}$$

From Eq 17, the risk of rupture can be written:

$$B = \int_V k_0 \left(\frac{\sigma_1}{\sigma_0} \right)^m dV \qquad \text{(Eq 18)}$$

where

$$k_0 = \frac{2m + 1}{2\pi} \int_0^{2\pi} \int_0^{\pi/2} [\cos^2 \phi (\cos^2 \psi + C \sin^2 \psi)$$

$$+ D \sin^2 \phi]^m \cos \phi \; d\phi \; d\psi$$

For cases of equal biaxial tension, the k_0 in the preceding equation reduces to:

$$k_0 = \frac{(2^m m!)^2}{(2m)!} \qquad \text{(Eq 19)}$$

For cases of equal triaxial tension, the k_0 equation is:

$$k_0 = 2m + 1 \qquad \text{(Eq 20)}$$

Utilizing the preceding equations, the effects of multiaxial stresses on the risk of rupture can be investigated. Figures 8 and 9 show how the value of k_0 and, hence, the risk of rupture vary with various ratios of σ_2 and σ_3 to σ_1. These figures show how k_0 rapidly increases when the second and third principal stresses are greater than approximately 80% of the maximum principal stress. Thus, when assessing the reliability of a component, it will be necessary to know the stress state, not just the maximum principal stress. If the second and third principal stresses are small percentages of the maximum principal stress, they will have a small effect on the risk of rupture. For approximate analysis, neglecting σ_2 for biaxial stress cases when σ_2 is less than 0.8 of σ_1 may be acceptable, and neglecting σ_2 and σ_3 for triaxial stress cases when σ_2 and σ_3 are less than 0.5 of σ_1 may be acceptable. For example, for the biaxial stress case, neglecting the second principal stress when it is equal to 0.8 of σ_1 will cause the reliability estimate to change from 0.99 to 0.977. Similarly, neglect of σ_2 and σ_3 when they equal 0.5 of σ_1 would cause the reliability to change from 0.99 to 0.97. The acceptability of these errors in final design would depend on the application, but they should be acceptable for early design calculations.

Another area of interest in ceramic design is the relative sensitivity of the risk of rupture to the volume and stress in the component. If two components a and b have the same risk of rupture, then from Eq 17, the following equation can be obtained:

$$k_{0a} \left(\frac{\sigma_a}{\sigma_{0a}} \right)^{m_a} V_a = k_{0b} \left(\frac{\sigma_b}{\sigma_{0b}} \right)^{m_b} V_b \qquad \text{(Eq 21)}$$

If the materials are the same, then the σ_0 and m values are the same for both components and

$$k_{0a}(\sigma_a)^m = k_{0b}(\sigma_b)^m V_b \qquad \text{(Eq 22)}$$

If the stress states within the components are the same, then the k_0 values are the same and the volume of material at stress σ_b that will have the same risk of rupture as a volume of material at σ_a is shown in Fig 10. This shows that for $m = 10$ and σ_b equal to 80% of σ_a, it takes 10 times as much material at stress σ_b to produce the same risk of rupture as one unit of material at σ_a. Thus, generally when the stress level drops below 80% of the peak stress, the contribution to the risk of rupture drops off rapidly. As the Weibull modulus increases, the contribution to the risk of rupture of stresses below the peak stress drops off even faster. The preceding example was for the condition where the states of stress at location a and b were the same, resulting in equal values of k_0. The case may exist where the peak stress is essentially a uniaxial stress, but somewhere else in the component there

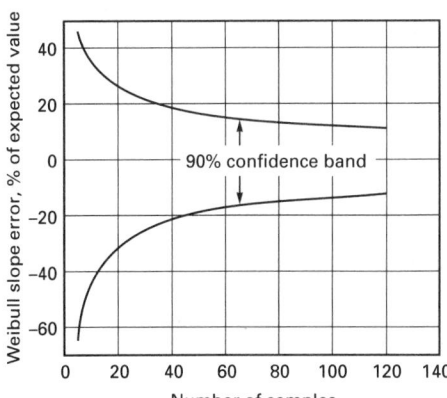

(a)

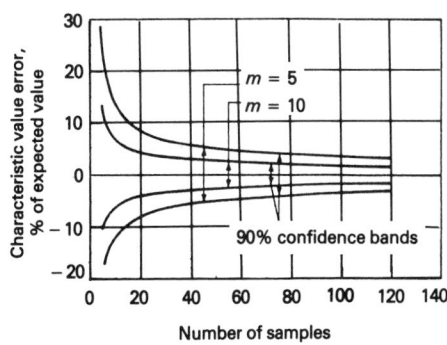

(b)

Fig 7   The possible error in (a) the predicted Weibull slope versus the sample size and (b) the Weibull characteristic strength value versus the sample size

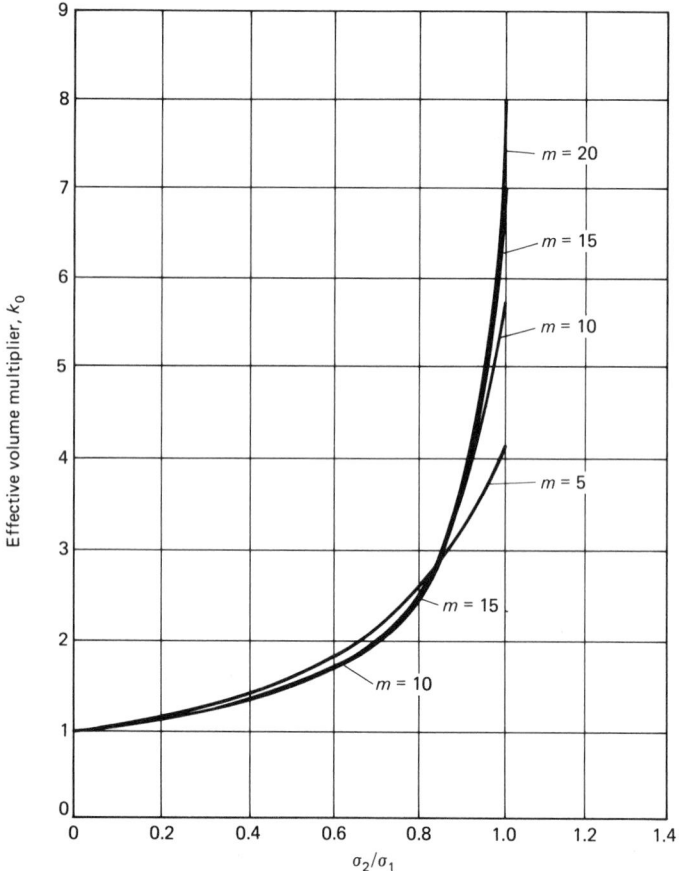

Fig 8 For biaxial stress states, the effective volume multiplier (k_0) is plotted versus the ratio of the second to the first principal stress for different Weibull moduli

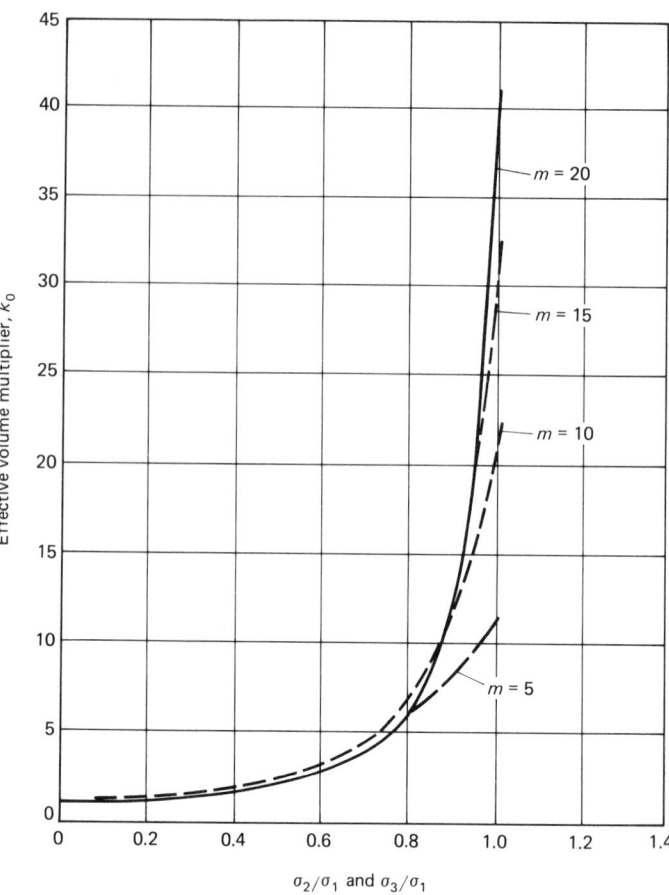

Fig 9 For triaxial stress states, the effective volume multiplier (k_0) is plotted versus the ratios of the second and third principal stresses to the first principal stress for different Weibull moduli

is a lower stress that is significant because it is biaxial. Such a case is shown in Fig 11 for a constant risk of rupture. Note here that for $m = 10$, that one unit volume of material in an equal biaxial stress state with a stress equal to 84% of the peak stress σ_a will have the same risk of rupture as a unit volume of material in uniaxial tension at a stress of σ_a.

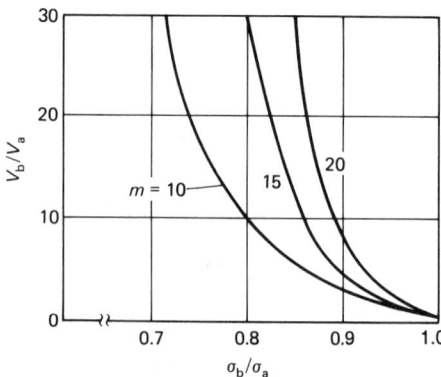

Fig 10 The ratio of volumes V_b/V_a versus the ratio of the relative stresses σ_b/σ_a for a constant probability of failure for various Weibull moduli. The volume V_b at stress σ_b will have the same probability of failure as volume V_a at stress σ_a

Thus, looking only at peak stresses can be misleading if a higher-order stress condition exists elsewhere with a maximum stress near that of the peak stress. With a triaxial stress state versus a uniaxial stress state, the situation is even worse and stresses as low as 60% of the peak stress are likely to be important.

When the ceramic material exhibits significant changes in strength and/or Weibull modulus with temperature, such as some of the partially stabilized zirconias (Ref 30), the analysis is complicated further. A volume of

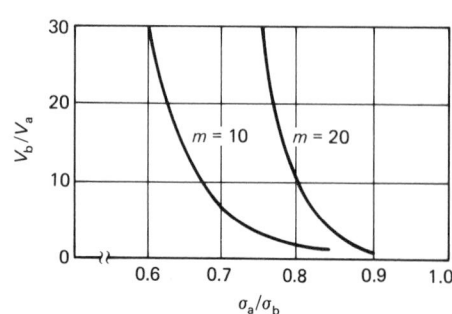

Fig 11 The ratio of volumes V_b/V_a versus the relative stresses σ_b/σ_a for a constant probability of failure when the stress state at σ_a is uniaxial tension and the stress state at σ_b is equal biaxial tension

material subjected to 50% of the peak stress but at a temperature where the ceramic has half the strength will have the same probability of failure as an equal volume at the peak stress. Similarly, if at some location in the part, there is a tensile stress nearly equal to the peak stress but, due to temperature, the Weibull modulus is half that exhibited at the peak stress, the probability of failure of this region may be several orders of magnitude higher than that of the peak stress region. Again, simply examining the peak stress can lead to problems.

The relative importance of several parameters can be illustrated by combining Eq 9 and 13. Figure 12 is a plot of various combinations of parameters for the special case of uniaxial tension. As seen, the allowable design stress is very sensitive to Weibull modulus and the allowable probability of failure. Since the allowable probability of failure is usually fixed, the "best" material from a given class will quite often be one with the highest Weibull modulus, assuming the modulus of rupture values are not greatly different.

From the previous discussion, it should be obvious why deterministic methods for ceramic design are usually not successful for high-performance ceramic components. Because

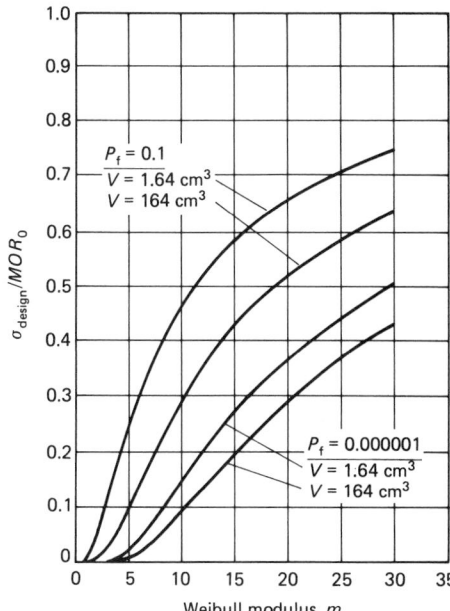

Fig 12 Ratio of design stresses to characteristic strength for a specific test bar size versus the Weibull modulus for different probabilities of failure and volumes stressed in uniaxial tension

of this, final design of components should be done using finite-element analysis coupled with a full-reliability calculation.

Simplified Structural Ceramic Design Technique. For preliminary design calculations or where accuracy is not as critical, a simplified structural design technique has been shown to be successful for approximate reliability calculations or alternatively as a means of estimating an allowable design stress (Ref 27). Note that this method is based on the Weibull theory and as such is sensitive to the same sources of errors as are present with the full Weibull analysis plus errors from volume and material property estimates. Consequently, this method should be used with caution and it is suggested that the user vary his estimates in the analysis to examine how sensitive his answer is to changes in the parameters. The two ways of using this technique will now be presented.

Estimating the reliability of a component involves the following steps:

1. Estimate the maximum stresses in the part
2. Estimate the volume of the component that is subjected to a tensile stress equal to or greater than 80% of the maximum tensile stress
3. Estimate the range of temperatures the tensile region of the part will pass through and the temperature of the part at the point of maximum stress
4. From the material property data, determine if the material strength and/or Weibull modulus drops off significantly in the temperature range of step 3. If so, use the minimum values and go to Step 6

5. From the material property data, get the MOR_0 and m value corresponding to the temperature at the maximum stress
6. From Fig 6, determine σ_0/MOR_0 for the given m value and calculate σ_0
7. Determine if the component is likely to be in a state of equal or nearly equal biaxial or triaxial tension. If so, determine the appropriate k_0 factor from Fig 8 or 9. If the stress is nearly uniaxial tension, then use k_0 equal to 1
8. Using the preceding information, substitute into the following equations to estimate the reliability:

$$B = \left[k_0 \left(\frac{\sigma_1}{\sigma_0} \right)^m V \right]_{\text{volume 1}} + \left[k_0 \left(\frac{\sigma_1}{\sigma_0} \right)^m V \right]_{\text{volume 2}} + \ldots$$

$$R = \exp(-B) \qquad \text{(Eq 23)}$$

The accuracy of the reliability estimate is directly dependent on the ability to determine the stresses within a component. The maximum stresses and also the stress state (for example, uniaxial, biaxial, or triaxial) are important. For a first estimate, it is frequently sufficient to use only the peak stress in the component. However, improved accuracy can be obtained by looking at all highly stressed areas in the component and including them in the analysis. In this case, steps 1 to 7 would be completed for each highly stressed region and the results used to calculate the reliability in step 8. A good rule of thumb is that any stressed region that has a higher-order stress state (for example, biaxial versus uniaxial), than that existing at the peak stress

region should be included in the analysis if its maximum stress is greater than or equal to 60% of the peak stress.

Estimating an allowable design stress involves the following steps:

1. Estimate the volume of the part that will be subjected to a stress equal to or greater than 80% of the peak stress in the part
2. From the data on the desired material, get the minimum Weibull modulus m for the temperature range where the tensile stresses occur in the part
3. Pick an acceptable probability of failure
4. Determine if equal or nearly equal triaxial or biaxial stresses are likely. If a biaxial stress state exists with $\sigma_2 > 0.8\sigma_1$, then use the biaxial curves. If a triaxial stress state exists with σ_2 and $\sigma_3 > 0.8\sigma_1$, use the equal triaxial stress curves
5. From the appropriate figure (13) or those generated at other P_fs, determine σ_1/MOR_0 and calculate σ_1. This is the maximum allowable design stress for the component-material combination and the chosen probability of failure

Figure 13 is obtained by utilizing Eq 9 and 14 and is valid only for the given size test bar and load span used in their generation. Other curves can be generated for any desired probability of failure or test bar configurations.

Finite-Element Models for Reliability Calculations. The use of finite-element analysis in the design of ceramic components has greatly improved the likelihood of suc-

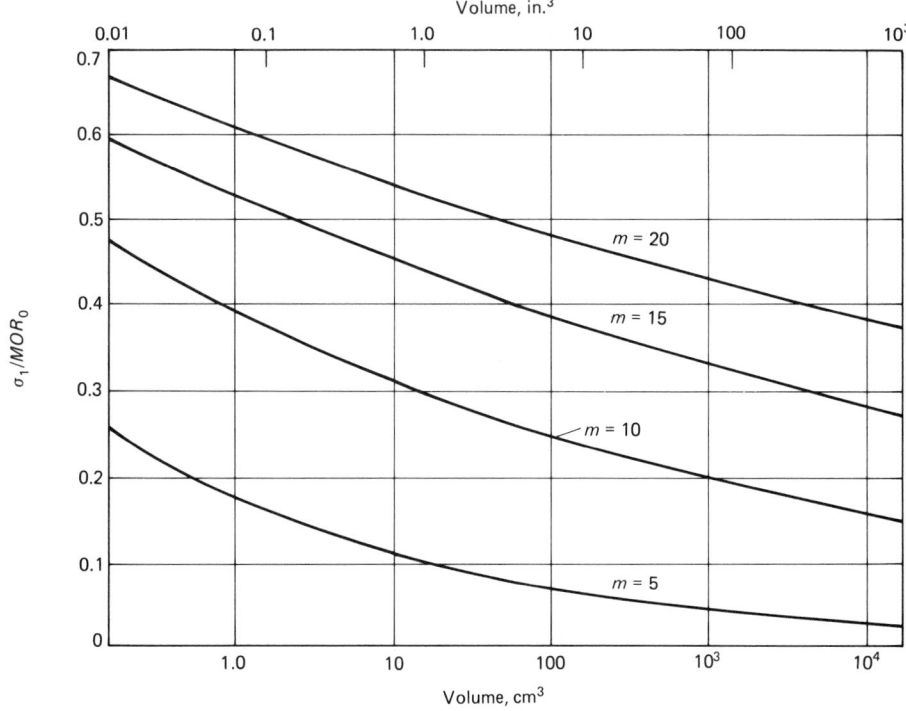

Fig 13 Ratio of design stress to standardized modulus of rupture versus the volume stressed for various Weibull moduli for uniaxial tension and a probability of failure of 0.01

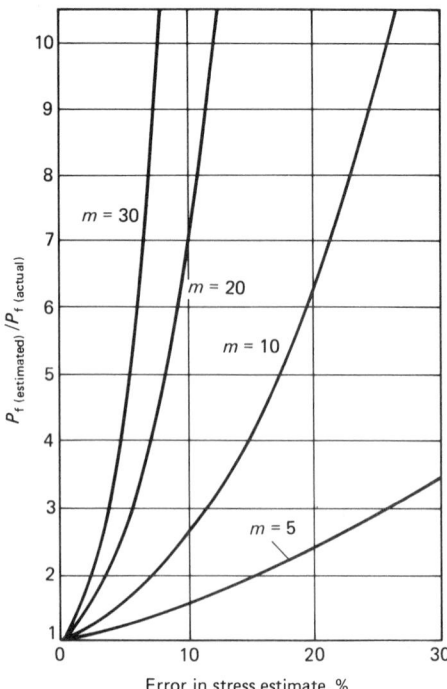

Fig 14 Ratio of the estimated to actual probability of failure versus the percent error in the stress estimate for various values of Weibull moduli

cess. Using this method, accurate temperatures and stresses can be determined for complex components. These values can then be fed into a Weibull reliability prediction program to assess the components' probability of success. However, even using finite-element analysis, one can get misleading results. The accuracy of the calculated probability of failure is very sensitive to errors in the estimated stress. As seen from Eq 17, the risk of rupture is a power function of stress, and, thus, the error resulting from the inaccuracies in the stress will also be a power function. The ratio of the estimated probability of failure to the actual is shown in Fig 14 for different values of Weibull modulus. Note that the error increases very rapidly as the Weibull modulus increases. Many ceramics today have a Weibull modulus of approximately 10, so an error of 20% in the stress will result in a factor of 6 error in the predicted probability of failure. Thus, the model geometry and boundary conditions must be such that correct values of stress and temperature are calculated.

There is another aspect of using reliability codes in conjunction with finite-element analysis that may not be so obvious as accuracy in the stress predictions. Most reliability codes in use today employ either the principal stresses at the centroid or the integration points of an element for the reliability calculation. This can lead to errors in the calculated reliability if there is a large stress gradient across an element. Take for example

an element in uniaxial tension with a stress gradient such that the maximum principal stress on one face of the element is higher than on the opposite face, such as exists in a test bar in four-point bending. The ratio of the side-to-side stresses is defined as R_s. For this case, the ratio of the actual risk of rupture (B_{act}) for the element to that predicted using the centroidal stress (B_{cent}) is given by the following formula:

$$\frac{B_{act}}{B_{cent}} = -\frac{2^m(1 - R_s^{m+1})}{(m + 1)(R_s - 1)(1 + R_s)^m} \qquad (Eq\ 24)$$

This is shown graphically in Fig 15. As can be seen from the figure for a Weibull modulus of 10 and a stress ratio of 2, the actual risk of rupture is more than three times higher than that predicted by using the centroidal stress. Today there are many higher-order elements (for example, 20 noded brick elements) that can give good thermal and stress predictions for a fairly coarse grid. However, if these stress results are used in conjunction with a reliability code that utilizes the element centroidal stresses, serious errors can result. By generating stresses at the integration points and using a reliability code based on the integration-point stresses, this error can be significantly reduced. The difficulty in using the integration-point stresses is that it can lead to very large output files being created by the finite-element analysis. This can be avoided by using a coarse grid to analyze the structure as a whole, and then breaking out a small section of the structure for reli-

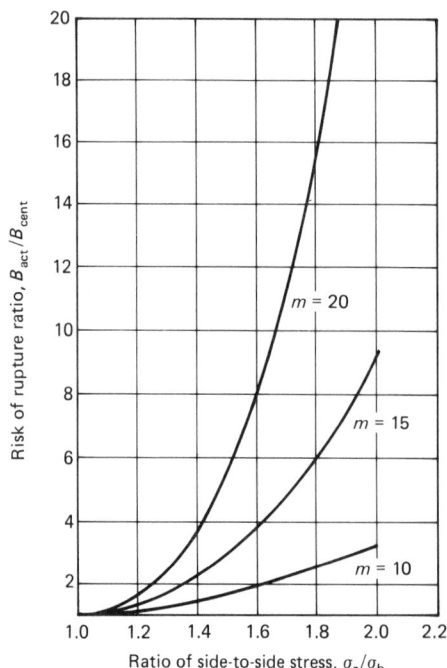

Fig 15 Ratio of actual to predicted risk of rupture versus the ratio of the side-to-side stress (σ_a/σ_b)

ability prediction. Obviously, this must be done with great care as indicated by the discussion in the previous section on sensitivity analysis.

Reliability Selection. The reliability of a ceramic component is a function of the material properties and the stress conditions within the part. The material properties are a function of the process controls established during the fabrication of the component starting with powder manufacturing. In general, the more detailed and more rigorous the process controls, the greater the consistency of the material in the component and, thus, the higher the Weibull modulus. A basic understanding of the material coupled with close process controls can lead to components having good microstructure and high strength. Obviously, high-strength materials with a high Weibull modulus coupled with good component design have the potential to produce the highest reliabilities. Producing the highest-quality components can also mean producing the lowest-cost components by elimination of waste (Ref 31). Thus, paradoxically, producing the highest-quality parts can also mean producing the most cost-effective parts. However, because of design constraints and material limitations, even the highest-quality parts may not demonstrate extremely high reliabilities in all applications. This can occur when the designer is trying to push a material design combination to the very limits of its capabilities. In these cases, the question arises as to what might be an acceptable reliability.

There is no clear-cut overall answer to the question of what is an acceptable reliability in these cases, and each will have to be determined on an individual basis. However, in order to give the reader some feel for acceptable reliabilities, some general cases are discussed next.

Choosing an acceptable reliability in these cases is dependent on the consequence of failure. For example, if failure could result in a life-threatening situation, then the reliability must be very high. Initially, design iterations must be incorporated to ensure the highest possible component reliability. Then, if possible, each component should be proof-tested to the full design stress condition to force the reliability to be 1 (naturally this assumes that tests have been performed to verify that the proof test does not damage the component). For cases where failure will result in factory downtime, warranty costs, loss of customers, or damage to the company image, high reliabilities will also be required. For these parts, a probability of failure of one in a million might be reasonable. For low-volume parts where replacement of the component would be easy and of little inconvenience to the user, a probability of failure of one in a thousand might be acceptable. For prototype parts for experimental evaluation to determine the benefits of using the component, a failure rate of one in ten could be acceptable.

Lifetime Reliability

In testing and evaluation of ceramics in actual operating environments, it has been discovered that in some cases, even though the component may survive initial tests, it may fail in subsequent tests. This occurs even though the severity of the test has not increased. This means that the fast fracture reliability determined for the new part is not an accurate prediction for the part during service. There are several reasons why the reliability may decrease during operation. These may include one or more of the following: physical change of the material properties, surface degradation from corrosion or erosion, or slow crack growth.

Material-Surface Degradation. A physical change in a material's properties can occur when the material goes through a compositional change or a phase change (Ref 32–34). An example of this was shown by Havstad (Ref 33) in the degradation of porous silicon nitride subjected to oxidation. In this case, the silicon nitride oxidized forming a new material with a higher thermal expansion and a lower strength. This material developed higher stresses because of the higher thermal expansion and this, combined with its lower strength, resulted in failure. When the material properties are changing with time, it is extremely difficult to get a good estimate of the component's reliability versus time. One approach to solving this problem would be to submit material to the same environment as it would see in operation for the desired life and then measure the material properties. These properties could then be used in the fast fracture reliability calculation. This approach is probably an improvement over using the "new" material properties, but it can still lead to difficulties. Since the material in the actual component may not change uniformly, there may be internal stresses that cannot be accounted for in the normal thermal and stress analyses. Also it has been shown that oxidation of ceramics can be enhanced by the addition of stress (Ref 35). Thus, simply soaking a material in a furnace will not lead to the same degradation of properties that may occur in the part that will be in a stressed state.

When the surface of the component is subjected to corrosion or erosion, the size and nature of surface flaws may be changed, which can significantly reduce the strength of the surface material. Once again, strength of test samples subjected to the operating environment could be determined and used for reliability calculations. This approach will again be better than using virgin material properties but may suffer from the same problem as mentioned in the previous paragraph, that is, damage to a part under stress may be more severe than that occurring in the unstressed condition. Generation of strength from test bars subjected simultaneously to stress and erosion or corrosion may give reasonable val-

ues for reliability calculations, but this is only conjecture and has not been proven by actual tests.

The study of slow crack growth in ceramics has received much attention and continues to be investigated (see the Section "Failure Analysis" in this Volume). This work is often discussed under the terminology of life prediction.

High-Cycle Fatigue and Slow Crack Growth. Life prediction in ceramics has concentrated in two areas, high-cycle fatigue and stress rupture, with by far most of the work being done on the latter.

For ceramics, significant high-cycle fatigue strength data are not available. Thus, methods for designing for high-cycle fatigue in ceramics have not been verified. Some researchers have been working in this area (Ref 36), and therefore design techniques are being developed.

Most work in life prediction of ceramics is for stress rupture based on the power relation of the crack velocity as a function of stress intensity:

$$V_c = A_c K_I^n \tag{Eq 25}$$

This equation was shown to accurately predict crack growth in steels at elevated temperature (Ref 37, 38), and later it was suggested that this equation would be valid for silicon nitride ceramics having slow crack growth as a result of grain boundary sliding (Ref 39).

This equation was combined with the Weibull probability equation (Ref 28, 40) to obtain a reliability at a certain number of hours.

$$R_t = R_{ff}^{(1+\sigma^2 t/B_p)^{m/n-2}} \tag{Eq 26}$$

In this equation, B_p and n are derived from experiments and calculation by two different procedures. These procedures incorporate either the double-torsion method (Ref 40) or the stress-rate technique (Ref 41). Unfortunately, these two techniques frequently give significantly different values of B_p and n that result in widely varying predictions of life. For example, in work on hot spin disks by Baker *et al.* (Ref 28), the reliability at 10 h varied by three orders of magnitude between 0.001 and 0.96 depending on which method was used. As a consequence of the variation, Swank has proposed an alternate equation based on empirical constants (Ref 42):

$$R_t = R_{ff} \exp[-\phi(t)] \tag{Eq 27}$$

The $\phi(t)$ in this equation is some function of the variables that describe the time-dependent behavior of the material. As of 1987 the variables that Swank has included in this function are the stress, volume, and time. This method has worked in some cases but not in others and needs further development.

Joints, Attachments, and Interfaces

Frequently, the most difficult problem facing the user of structural ceramics is the design of the ceramic/metal interface. The reason

for the difficulty is that usually the ceramic and the metal have very different thermal expansions. Thus, as the system changes temperature, the ceramic and metal components try to move relative to each other. This causes stresses in the ceramic and can lead to failure. This can occur even if the components are allowed to slide relative to each other. Richerson *et al.* (Ref 43) have shown that, for sliding components, if the contact stresses are high, a high tensile stress can exist just behind the contact point. These differential thermal-expansion-induced stresses can occur whenever components go through changes in temperature but are obviously a much more serious problem when large temperature fluctuations are involved.

For discussion purposes, attachment methods have been broken down into three categories: (1) systems using an interface material or compliant layer to allow for relative motion yet minimize contact stresses, (2) direct ceramic-to-metal contact with or without a friction-modifying interface, and (3) bonding or brazing of the ceramic to the metal. Each of these will be described below. More detailed information can be found in the Section "Joining" in this Volume.

Compliant Layers. One of the most common ways of handling the problem of the ceramic/metal attachment is to utilize a third material as a compliant layer (Ref 4, 44, 45). The interface layer has two functions. Depending on the application, one or both of the functions may be utilized. First, it distributes the contact loads to minimize stresses; second, it allows relative sliding motion with minimum resistance. This type of design obviously requires some external means of holding the pieces together. In some cases, the metal will be shrink-fitted over the interface material and the ceramic. In other cases, external bolts or clamps will be used to trap the ceramic between two metal components. In both of these cases, the design must be such that even with the differential thermal expansion, a sufficient load is maintained on the ceramic to hold it in the correct position over the operating temperature range of the component.

Direct Ceramic/Metal Interfaces. The second way of attaching ceramics to metals is with direct ceramic-to-base metal contact with and without friction modifiers. In this case, the ceramic may be held in place by a direct shrink-fit to the metal or there may be external retaining devices. In order to maintain an interference-fit, using the shrink-fit concept requires that the part does not operate over a large temperature range or that the ceramic and metal components have similar thermal expansions. Because of partially stabilized zirconia's and alumina's relatively high thermal expansion for ceramics (10.2×10^{-6}/ K and 7.4×10^{-6}/K, respectively), they can be used with low-expansion metal in shrink-fit designs. For the lower-expansion ceramics such as silicon nitride, silicon carbide, and lithium aluminum silicate, generally an alternative clamping mechanism will have to be used.

A prime example of the shrink-fit design is in the use of partially stabilized zirconia in metal extrusion dies. In this case, the PSZ is shrink-fitted into the metal holder that is then mounted to the extrusion apparatus. A key to successful shrink-fit construction is to avoid a design with the ceramic only partially inside the metal retainer. When this happens, high tensile stresses can occur in the ceramic where it leaves the metal component. These stresses can cause outright failure at assembly or later when combined with normal operating stresses. If a shrink-fit with the ceramic protruding out of the metal must be used, then the ends of the metal component should be tapered, the metal should be as thin as possible, the interference fit should be as low as possible, and the length of the joint should be kept to a minimum.

One of the more attractive methods of attaching ceramics to metals is to cast the metal around the ceramic. This has been done with aluminum and aluminum titanate port liners for reciprocating engine cylinder heads (Ref 46, 47). This type of assembly procedure may or may not utilize a crushable layer on the outside of the ceramic to absorb some of the stresses that would be caused by the high shrinkage of the metal as it cools from the casting temperature. Because of these high differential shrinkages, these casting methods usually require much development work and the resulting techniques are usually proprietary.

For cases where the load on the threads is not too high, the ceramic and metal can be threaded together. Here, the thread acts to hold the pieces in position, while the interface material carries most of the load. In general, the high contact loads and the stress concentrations in the thread roots can lead to problems. Thus, extreme care should be used in attempting to utilize threads in the design.

Other direct ceramic-metal interfaces can be successful by utilizing a friction-modifying interface. These interfaces may consist of a very thin (0.025 mm, or 0.001 in.) metal coating or a conventional lubricant (Ref 13, 48).

Bonding or Brazing. The third means of attachment is to bond or braze the ceramic to the metal. For relatively low-temperature attachments ($T < 200\ ^\circ C$, or 390 $^\circ F$), adhesives have been shown to work for attaching silicon nitride to metal shafts (Ref 7). In the design shown in Fig 3, the ceramic has been undercut in the center of the joint to allow sufficient room for the adhesive to flow while the ends of the joint are close-fitting for positioning.

When higher temperatures are encountered in the joint region, the components may be brazed together (Ref 49). In order to braze the ceramic, many different systems have been developed. Some systems metallize the ceramic and then braze to the metallized material. Other systems incorporate a reactive metal in the braze. This reactive metal reacts with the surface layer of the ceramic to wet it and the remaining braze constituents bond to this reacted layer.

When utilizing brazed joints, high residual stresses may be induced in the components during cooldown from the brazing temperature due to the differential thermal expansions. In order to minimize these stresses, usually a very slow cooling rate is specified to allow the braze material to creep. Even with a slow cooldown, there will always be residual stresses in the brazed system and these will have to be accounted for in the design calculations. Design methodologies for brazed ceramic/metal joints are described in the Section "Joining" in this Volume.

Thermal Shock Considerations

In designing with ceramics, thermally induced stresses are frequently critical. Of these, rapidly induced transient thermal stresses, known as thermal shock, are usually the worst, and thermal downshock, rapid cooling from a hot condition, is the worst of these. The reason that thermal downshock is so critical is that it induces a high tensile stress in a very thin surface layer. This can initiate a crack that can propagate through the part, causing catastrophic failure.

The best way to successfully design for thermal shock is by a complete transient thermal and stress analysis coupled with a reliability analysis (Ref 17, 50). This type of analysis is very expensive because of the many solution iterations that must be examined. Also, since it is frequently difficult to obtain accurate boundary conditions, this type of analysis needs to be calibrated with actual component data. However, this is still the best approach, especially if several components must be developed that will have different but similar designs.

In order to help with initial design, there are some guidelines on how to minimize thermal stresses. First, try to minimize any cross-sectional thickness changes in the component. This will tend to let the component respond to any thermal inputs more uniformly. Second, try to separate into two or more components any regions of the structure that see vastly different temperatures. This will tend to keep each part at a more uniform temperature. Third, avoid any sharp edges or very thin sections in regions subjected to high heat flux. This will prevent a rapid temperature change in a small region of the part.

When developing new components, it will sometimes happen that making actual hardware changes and evaluating on simple rigs can occur more rapidly than completing a revised thermal, stress, and reliability analysis. An example of this was the development of a turbine stator vane design. This component was subjected to the hot gases exiting from a combustor and was thus subjected to severe thermal transients. During startup-shutdown tests with the original stator design, cracks would frequently develop in the stator vanes. When segments of the stator were mounted on a thermal shock rig, similar cracks developed in the vanes (Ref 32). A design change was made to eliminate the thin trailing edge from the vane. When these stator segments were tested on the thermal shock rig, no vane cracks developed. Subsequent light-off qualification and engine testing verified this design change. Thus, rig tests can be useful in sorting out design changes.

Proof Testing

As mentioned earlier, the design of ceramics is based on probabilistic methods. A statistical distribution of the material strengths is generated from test bar data. This data is then used to predict the probability of failure of a component. In most cases it is desired that the probability of failure of the component be very much less than the failure probabilities used with the test bar strength data to generate the material characteristic strength and Weibull modulus. For example, if 30 data points were used, the lowest probability of failure used in the material property generation is 0.022. Thus, the failure distribution curve from the test data is extrapolated well beyond the limits of the data to the extreme tails of the distribution. This obviously limits the confidence in the projected reliabilities. One way of improving the confidence in the predicted reliabilities is to proof test the component. This is a form of testing that will weed out the weakest components and thus truncate the tail of the distribution.

Another reason why one might want to proof test the component is to raise the allowable design stress. As was shown in Fig 12, for very high desired reliabilities, the allowable design stress can become a small percentage of the test bar strength. By proof testing, the allowable design strength can be increased.

Effect of Proof Testing on Reliability. Weibull provided the general equation for the probability of failure of a component after it has been subjected to a proof-test stress of σ_p (Ref 15). The equation for this is:

$$P_{fp}(\sigma) = \frac{P_f(\sigma) - P_p(\sigma_p)}{1 - P_p(\sigma_p)} \qquad \text{(Eq 28)}$$

where $P_{fp}(\sigma)$ is the probability of failure at stress σ after being proof tested to a stress of σ_p. This assumes the directions of the stresses σ and σ_p are the same.

The corresponding reliability after proof testing is given by the following formula:

$$R_{ap}(\sigma) = \frac{R(\sigma)}{R_p(\sigma_p)} \qquad \text{(Eq 29)}$$

For a uniformly stressed element in uniaxial tension, this leads to the following equation if the element is subjected to a proof test at the same temperature and stress direction.

$$R_{ap}(\sigma) = \exp\left[-\left(\frac{\sigma^m - \sigma_p^m}{\sigma_0^m}\right)V\right] \qquad \text{(Eq 30)}$$

$$R_{ap}(\sigma_1) = \exp\left\{-\int_V \frac{2m+1}{2\pi}\right.$$

$$\left.\left[\int_0^{2\pi}\int_0^{\pi/2}(W_1 - W_2)\cos^2\phi\,d\phi\,d\psi\right]dV\right\} \qquad \text{(Eq 31)}$$

where

$$W_1 = \frac{[\cos^2\phi(\sigma_1\cos^2\psi + \sigma_2\sin^2\psi) + \sigma_3\sin^2\phi]^{m1}}{\sigma_0^{m1}}$$

$$W_2$$
$$= \frac{[\cos^2\phi(\sigma_{1p}\cos^2\psi + \sigma_{2p}\sin^2\psi) + \sigma_{3p}\sin^2\phi]^{mp}}{\sigma_{0p}^{mp}}$$

and the integral is evaluated only for cases where $W_1 - W_2$ is greater than zero.

For multiaxial stress states, the multiaxial Weibull equation must be used and the appropriate material parameters m and σ_0 used for the design conditions and the proof-test condition. This gives the following equation for the case of the proof test and design stress having their principal stresses in the same direction.

When the proof-test principal stresses are not in the same direction as the design principal stresses (the general case), then the equation must be modified to get the proof-test normal stress in the same direction as the design normal stress during evaluation of the integral.

As mentioned earlier, it is the stress state and the maximum stress that have the greatest effect on reliability. It follows that proof testing at a low stress level compared to the design conditions will have very little effect on the reliability. As an example, the effect of the proof-test load for the case of uniaxial tension where the proof test and the design stresses are in the same direction (that is, Eq 30 holds) is shown in Fig 16. Note that for a Weibull modulus of 10, reducing the risk of rupture from 0.1 to 0.05 requires that the component be proof tested to approximately 93% of the design stress. Also, to reduce the probability of failure by an order of magnitude, down to 0.01, the part must be proof tested to 99% of the design stress. As the Weibull modulus increases, the required proof test load is even closer to the design stress for a given reduction in the risk of rupture.

Another way of employing the proof test is to use it to increase the allowable design stress. To examine this, a uniaxially stressed component was held at a constant risk of rupture and the proof-test stress was varied to see the effect on the allowable design stress (σ_1). The results of this analysis are shown in Fig 17. As expected, the proof-test load must be very close to the desired design stress if there is to be any appreciable increase in the allowable design stress over the baseline

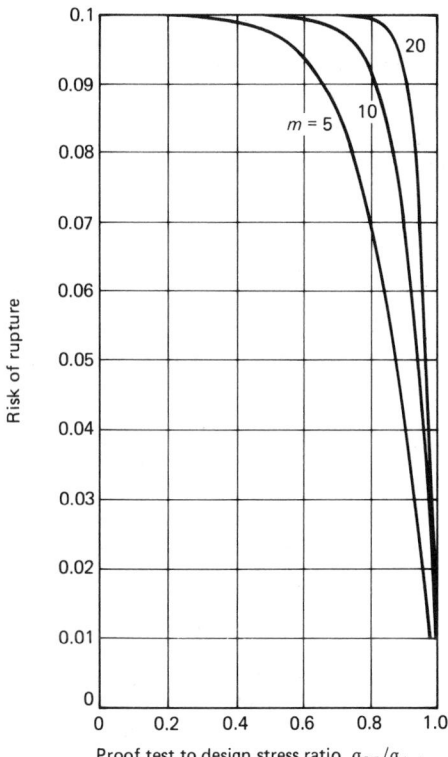

Fig 16 Risk of rupture after proof testing to the ratio of proof-test stress to design stress

condition. For example, from Fig 17, for $m = 10$, if reliability calculations indicate that for the desired risk of rupture, the allowable design stress for a given component is 690 MPa (100 ksi) (σ_1) but operating conditions

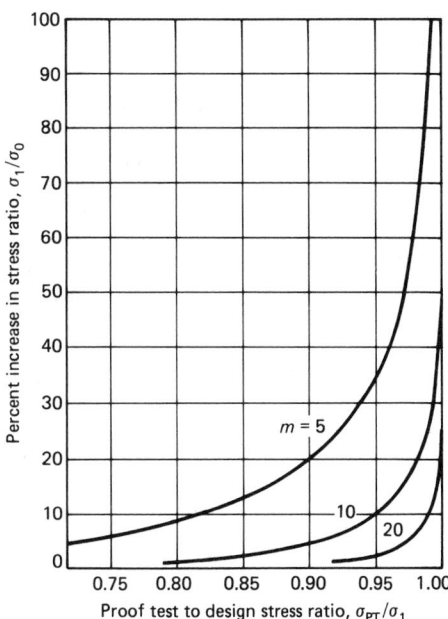

Fig 17 Percent increase in the ratio σ_1/σ_0 versus the ratio at proof-test stress to design stress for uniaxial tension at a constant risk of rupture

indicate stresses will be 760 MPa (110 ksi) (σ_1 must increase 10%) then the proof-test stress will have to be 722.35 MPa (104.765 ksi) ($\sigma_{pt}/\sigma_{design} = 0.952$ in order to obtain the given reliability). As shown, proof testing to above a given design stress essentially only raises the allowable design stress to a value approximately equal to the proof-test stress. Thus the proof-test stress is very close to the design stress.

In this discussion it was assumed that the proof test does not damage the part. Some researchers have indicated that in some cases, a proof stress does damage the part and therefore can be detrimental to the component (Ref 51, 52). In these cases the proof testing causes microcracking that over a long period of time causes failure by slow crack growth. An example of this is the failure of components at relatively low stress (less than 30% of fast fracture strength at the same temperature) when held at fairly high temperatures (1200 °C, or 2190 °F) for a period of time (Ref 53, 54). Since the damage to the part is a function of the stress application time, the proof-test time should be held to a minimum. As a further check to determine if the proof test is causing damage in the part, a series of components could be proof tested and then later tested to failure to see if the components passing the proof test fell on the predicted truncated distribution curve.

As just shown, the proof test can be used to increase the confidence for required reliability parts. In fact, proof testing the component to the actual stresses it will see in operation increases to 100% the confidence of 100% reliability for fast fracture and eliminates the statistics from the matter. In reality it is frequently impossible or impractical to develop a proof test to the identical stresses it will see in operation. This was shown to be true for an axial turbine rotor. In this case, the stresses in the hub region are the result of thermal and centrifugal loads. The stresses in the blade region are primarily centrifugal in nature. The most practical proof test was to cold spin the rotor to a sufficiently high load to induce the design stresses in the hub region. Since the design stresses in this region were a combination of both the rotational speed and the temperature gradient, the required proof test rotational speed was considerably above the 100% design speed of the rotor. Since the design stresses in the blades were primarily due to the rotational speed and were proportional to the square of the speed, the proof-test speed required for the hub region would significantly overstress the blades, causing failure or potential damage. This shows why it is necessary to be very careful in designing a proof-test to avoid damaging the component.

Another difficult area in designing a proof test is that of inducing the stresses in the component in the same direction in the proof test as those induced during actual operation. Obviously, if the proof-test stress of magnitude σ_p is in the x direction and the operational stresses are also of magnitude of σ_p but in the y direction, the proof test is not effective. Additional information can be found in the article "Strength and Proof Testing" in this Volume.

REFERENCES

1. S. Kamiya, M. Murachi, H. Kawamoto, S. Kato, S. Kawakami, and Y. Suzuki, "Silicon Nitride Swirl Lower-Chamber for Higher Power Turbocharged Diesel Engines," Publication 850523, Society of Automotive Engineers, 1985
2. H. Matsuoka, H. Kawamura, and S. Toeda, Development of Ceramic Pre-Combustion Chamber for the Automotive Diesel Engine, *Adiabatic Engines: Worldwide Review*, Publication 840426 in SP-571, Society of Automotive Engineers, 1984, p 1–7
3. W.R. Wade, P.H. Havstad, E.J. Ounsted, F.H. Trinker, and I.J. Garwin, "Fuel Economy Opportunities with an Uncooled DI Diesel Engine," Publication 841286, Society of Automotive Engineers, 1984
4. W.R. Wade and C.M. Jones, "Current and Future Light Duty Diesel Engines and Their Fuels," Publication 840105, Society of Automotive Engineers, 1984
5. W. Bryzik and R. Kamo, "TACOM/Cummins Adiabatic Engine Program," Publication 830314 in SP-543, Society of Automotive Engineers, 1983
6. A.F. McLean, Ceramics in Automotive Energy Systems, *Proceedings of the First International Symposium on Ceramic Components for Engine*, KTK Scientific Publishers, Tokyo, 1983
7. G.C. DeBell and J.R. Secord, "Development and Testing of a Ceramic Turbocharger Rotor," Paper 81-GT-195, American Society of Mechanical Engineers, 1981
8. J.P. Arnold, W. McGovern, J.C. Napier, and A.P. Batakis, "Demonstration of Ceramic Hot-Section Static Components in a Radial Flow Turbine," Publication 82-GT-184, American Society of Mechanical Engineers, 1982
9. A.F. McLean and D.A. Davis, "The Ceramic Gas Turbine—A Candidate Powerplant for the Middle- and Long-Term Future," Publication 760239, Society of Automotive Engineers, 1976
10. H.E. Helms and P.W. Heitman, "Ceramic Components for the AGT 100 Engine," Publication 84-GT-81, American Society of Mechanical Engineers, 1984
11. J.R. Kidwell and D.M. Kreiner, AGT101 Advanced Gas Turbine Technology Development Project, *Proceedings of the Twenty-Second Automotive Technology Development Contractors' Coordination Meeting*, Publication P-155, Society of Automotive Engineers, 1984, p 345–357
12. C.A. Fucinari, The Utilization of Data Relating to Fin Geometry and Manufacturing Processes of Ceramic Matrix Systems to the Design of Ceramic Heat Exchangers, *Ceramics for High Performance Applications—II*, J.J. Burke, E.N. Lenoe, and R.N. Katz, Ed., Brook Hill, 1977
13. A.F. McLean, "Ceramic Turbine Housings," Publication 82-GT-293, American Society of Mechanical Engineers, 1982
14. J.G. Lanning and D.A. Noll, "Casting Development of Ceramic Turbine Housings," International Gas Turbine Conference, London, Paper 82-GT-202, American Society of Mechanical Engineers, April 1982
15. W. Weibull, A Statistical Theory of the Strength of Materials, *Ing. Ventenstape Akad*, No. 151, 1939, p 1–45
16. A. Paluszny and W. Wu, Probabilistic Aspects of Designing with Ceramics, *J. Eng. Power*, Vol 99 (No. 4), 1977, p 617–730
17. J.C. Caverly and D.L. Hartsock, "Evaluation of Ceramics for Stator Applications—Gas Turbine Engines—Final Report—Reliability Analysis and Material Property Characterization," Report Cr 165270, National Aeronautics and Space Administration, 1980
18. R.A. Rackley, "Advanced Gas Turbine Power System Development Project—AiResearch/Ford," Highway Vehicle Systems Contractors' Coordination Meeting—Seventeenth Summary Report, U.S. Department of Energy, Conf.-791082, NTIS, 1979, p 143–164
19. J. Lamon and A.G. Evans, Statistical Analysis of Bending Strengths for Brittle Solids: A Multiaxial Fracture Problem, *J. Am. Ceram. Soc.*, Vol 66 (No. 3), 1982
20. S.B. Batdorf and H.L. Heinisch, Jr., Weakest Link Theory Reformulated for Arbitrary Fracture Criterion, *J. Am. Ceram. Soc.*, Vol 61 (No. 7–8), 1978
21. A.G. Evans, A General Approach for the Statistical Analysis of Multiaxial Fracture, *J. Am. Ceram. Soc.*, Vol 61, 1978, p 302–308
22. S.B. Batdorf and D.J. Chang, Relation Between the Fracture Statistics of Volume Distributed and Surface Distributed Cracks, *Int. J. Fract.*, Vol 15, 1979, p 191–199
23. O. Vardar and I. Finnie, An Analysis of the Brazilian Disk Fracture Test Using the Weibull Probabilistic Treatment of Brittle Strength, *Int. J. Fract*, Vol 11 (No. 3), 1975
24. L.G. Johnson, *The Statistical Treatment of Fatigue Experiments*, American Elsevier, 1964, p 46–47
25. A.F. McLean and E.A. Fisher, "Brittle Material Design, High Temperature Gas Turbine," Army Materials and Mechanics Research Center-CTR-77–20, Interim Report, Aug 1977, p 111–114
26. D.R. Thoman, L.J. Bain, and C.E. Antle, Inferences on the Parameters of the Weibull Distribution, *Technometrics*, Vol 11 (No. 3), 1969, p 445–460
27. D.L. Hartsock, "A Simplified Structural Ceramic Design," Paper 85-GT-100, American Society of Mechanical Engineers, 1985
28. R.R. Baker, L.R. Swank, and J.C. Caverly, "Ceramic Life Prediction Methodology—Hot Spin Disc Life Program," Army Materials and Mechanics Research Center TR 83–44, Interim Report, Contract No. DAAG 46–77-C-0028, 1983
29. W.H. Dukes, "Handbook of Brittle Material Design Technology," ADS-719–712, NTIS, 1971, p 16–24
30. D.L. Hartsock and A.F. McLean, What the Designer with Ceramics Needs, *Ceram. Bull.*, Vol 63 (No. 2), 1984, p 266–270
31. W.E. Deming, *Quality, Productivity, and Competitive Position*, Massachusetts Institute of Technology, Center for Advanced Engineering Study, 1982
32. D.L. Hartsock, R.R. Baker, P.H. Havstad, and J.J. Buechel, Test and Development of Ceramic Components, *Ceramics for High Performance Applications—II*, J.J. Burke, E.N. Lenoe, and R.N. Katz, Ed., Brook Hill, 1978, p 291–316
33. P.H. Havstad, "Silicon Nitride Life Prediction Method," U.S. patent 4,151,740, 1979
34. J.E. Ritter, S.M. Wiederhorn, N.J. Tighe, and E.R. Fuller, Jr., Fatigue Failure of Ceramic Components, *Ceramics for High Performance Applications—III*, E.N. Lenoe, R.N. Katz, and J.J. Burke, Ed., Plenum, 1983, p 521–522
35. R.K. Govila, J.A. Mangels, and J.R. Baer, Fracture of Yttria-Doped, Sintered, Reaction-Bonded Silicon Nitride, *J. Am. Ceram. Soc.*, Vol 68 (No. 7), 1985
36. A.G. Evans, Fatigue in Ceramics, *Int. J. Fract.*, Vol 16 (No. 6), 1980
37. M.J. Siverns and A.J. Price, Crack Growth under Creep Conditions, *Nature*, 21 Nov 1970
38. M.J. Siverns and A.J. Price, Crack Propagation Under Creep Conditions in a Quenched 2–1/4 Chromium 1 Molybdenum Steel, *Int. J. Fract.*, June 1973
39. A.G. Evans and S.M. Wiederhorn, Proof Testing of Ceramic Materials: An Analytical Basis for Failure Prediction, *Int. J. Fract.*, Vol 10 (No. 3), Sept 1974, p 6379–6392
40. D.D. Williams and A.G. Evans, Simple Method for Studying Slow Crack Growth, *J. Test. Eval.*, July 1973
40. A. Paluszny and P.F. Nicholls, Predicting Time-Dependent Reliability of Ceramic Rotors, *Ceramics for High Performance Applications—II*, J.J. Burke,

E.N. Lenoe, and R.N. Katz, Ed., Brook Hill, 1978

41. R.J. Charles, Dynamic Fatigue of Glass, *J. Appl. Phys.*, Dec 1978

42. L.R. Swank, "Correlation of Structural Ceramic Time Dependent Failure Data," Paper 86-GT-14, American Society of Mechanical Engineers, 1986

43. D.W. Richerson, L.J. Lindberg, and W.D. Carruthers, Contact Stress at Ceramic Interfaces, *Eighteenth Summary Report Automotive Technology Development Contractors' Coordination Meeting*, U.S. Department of Energy, Conference No. 801182, NTIS, 1981

44. G.S. Calvert and W.D. Carruthers, Ceramic Blade Attachments, *Ceramics for High Performance Applications—II*, J.J. Burke, E.N. Lenoe, and R.N. Katz, Ed., Brook Hill, 1978, p 839–860

45. R.L. Allor and J.C. Caverly, "Method of Attaching a Metal Shaft to a Ceramic Shaft and Product Produced Thereby," U.S. patent 4,499,646, 1985

46. R. Rocchio, Konstruktion und Eprobung Keramischer Bauteile für Pkw-Motoren, *Entwicklungslinien in Kraftfahrzeugtechnik und Strabenverkehr*, Verlag TUV Rheinland GmbH, Koln, West Germany, 1981, p 112–128

47. N. Zernig, Konstruction und Erprobung Keramischer Bauteile für Luftgekuhlte Lkw-Motoren, *Entwicklungslinien in Kraftfahrzeugtechnik und Strabenverkehr*, Verlag TUV Rheinland GmbH, Koln, West Germany, 1981, p 104–111

48. P.H. Havstad, J.C. Caverly, and R.R. Baker, Ceramic Turbine Rotors—Engine Test and Development, *Ceramics for High Performance Applications—II*, J.J. Burke, E.N. Lenoe, and R.N. Katz, Ed., Brook Hill, 1978, p 275–289

49. M.E. Twentyman, *The Joining of Engineering Ceramics—Final Report*, British Ceramic Research Association, Ltd., Stoke-on-Trent, England, 1978

50. C.R. Booher, Thermal Transient Stress Analysis of an Industrial Combustion Turbine Ceramic Stator Vane and Correlation with Test Results, *Ceramics for High Performance Applications—II*, J.J. Burke, E.N. Lenoe, and R.N. Katz, Ed., Brook Hill, 1978, p 731–783

51. A.G. Evans and S.M. Wiederhorn, Proof Testing of Ceramic Materials—An Analytical Basis for Failure Prediction, *Int. J. Fract.*, Vol 26, 1984, p 355–368

52. J.E. Ritter, Jr., D.C. Coyne, and K. Jakus, Failure Probability at the Predicted Minimum Lifetime after Proof Testing, *J. Am. Ceram. Soc.*, Vol 61 (No. 5–6), 1978, p 213–216

53. G.D. Quinn, "Static Fatigue of a Sintered Silicon Nitride," Report AMMRC TR 84–40, U.S. Department of Energy, 1984

54. R.K. Govila, "Ceramic Life Prediction Parameters," Report AMMRC TR 80–18, U.S. Department of Energy, 1980

Ceramic Properties Data Base Systems

R.G. Munro, Ceramics Division, National Institute of Standards and Technology

COMPUTERIZED DATA BASES have a significant role to perform in materials research and development. The processing capabilities of computerized systems allow data to be assessed more effectively to uncover the strengths and weaknesses of the data and to search for relations among property and performance variables. The immediate results can include guidance for new research and the optimization of materials processing or production methods. Such capabilities may be critical to the rapid and timely utilization of the technological advances in structural ceramics.

Ceramics Property Information

Important Properties. Advanced structural ceramics are of particular interest for applications involving high temperatures, typically above 1000 °C (1830 °F), or aggressive environments (Ref 1). Heat exchangers, valve components for combustion engines, turbine engine components, cutting tools, and roller bearings are some of the more prominent applications in which ceramics may enable, or significantly improve, performance at high temperatures (Ref 2). A generic list of the properties needed in these design areas is given in Table 1. Additional information on advanced structural materials can be found in the Section of this Handbook entitled "Structural Applications for Technical, Engineering, and Advanced Ceramics."

This set of properties is relatively easy to understand. Interest in thermal transport properties, heat capacity, and dimensional stability may be anticipated on the basis of the high temperatures involved in the applications. The presence of an aggressive environment dictates concern for corrosion and oxidation. The need for mechanical properties is mandated by the need to perform stress analyses and the concern about failure of the brittle material (Ref 4).

The full range of information important to the design engineer's application, however, extends significantly beyond the numeric values of these special properties and depends on how the material is to be used (Ref 5). The numeric values need to be supplemented by descriptive data, sometimes called "metadata," to identify details about the material's composition and structure and to indicate how the measured values were obtained (Ref 6). Such metadata are essential for understanding the appropriateness and reliability of the reported property data in the context of the design engineer's application.

Considerations in Data Reliability. Data reliability for advanced ceramics is closely associated with the complex relations among structures and properties (Ref 7). For example, a rather common property like density depends on what crystalline and glassy phases are present in the material and on the value of the total porosity. Such structure-property dependencies are reflected in the details of the chemical and phase compositions, the processing history, and the microstructure. As a result, comparisons of data that might be made to ensure reliable values require careful identification of the materials in question.

Data reliability is further compounded by a dependency on the details of the measurement procedure (Ref 8). Measures of the strength of a material, for example, can vary significantly as a result of differences in specimen preparation. It is especially important that specimen surfaces be prepared consistently because of the manner in which surface defects act as stress intensifiers. Such surface flaws are often identified as the fracture origins in fracture tests. Hence, variations in specimen preparation by different researchers may cause the data to be scattered in a manner that is more indicative of the inconsistent test details than of the variance of the strength of the material.

The dependence of certain materials properties on composition, structure, and measurement details cannot be overemphasized. Consider again fracture strength. The failure of a brittle ceramic is nearly always associated with a flaw, either internal to the specimen or on its surface (see the Section "Failure Analysis" in this Volume). The internal defects may be a result of porosity, cracks, dislocations, or inclusions. Inclusions can have a twofold influence on failure. First, the inclusion may function as a stress intensifier. Second, the coefficient of thermal expansion of the inclusion may be sufficiently different from the host material that a significant local stress field is generated from the thermal mismatch. The probability of failure is dependent on how many of these flaws are sampled during the test, that is, a larger test volume has a larger number of flaws and a greater chance for failure. Thus, measurement details relating to the volume of the specimen that is sampled during a test are important to consider when comparing values determined by different experiments. The latter is especially true when different techniques are used to measure what is nominally the same property, such as when flexural strength is measured by three-point bend, four-point bend, C-ring, and/or O-ring tests.

The effects of measurement details can be quite subtle. While it may be readily recognized that the presence of cracks affects the strength of a material, it may be less apparent that the propagation of a crack is dependent on the chemical environment at the tip of the crack (Ref 9). Chemical reactions at the tip can lead to changes in the local bonding and subsequent changes in the propagation rate. As a result, the measurement environment, such as air, water, or corrosive solution, can influence the measured values of a property such as strength or creep. Likewise, changes in the local environment as a function of temperature, such as the melting or softening of an intergranular phase, can result in unexpected temperature dependencies.

The general conclusion drawn from these observations is that the details of materials specifications and measurement methods are essential to the interpretation of materials property values for advanced ceramics. Materials specification and measurement method details must be considered integral parts of materials property data bases and of the use of the data bases for engineering designs.

Table 1 A summary of important materials properties for structural ceramics. Significant properties are denoted by an "X."

Property	HTEX	VALV	TURB	BRNG	TOOL
Thermal					
Conductivity	X	X	X	X	X
Diffusivity	X	X	X	X	X
Expansion	X	X	X	X	
Specific heat	X	X	X	X	X
Shock resistance	X	X	X	X	X
Emissivity	X				
Melting point	X	X	X		X
Maximum service temperature	X	X	X	X	X
Mechanical					
Elastic modulus	X	X	X	X	X
Poisson's ratio	X	X	X	X	X
Shear (torsion) modulus			X		X
Tensile strength	X	X	X	X	X
Flexural strength	X	X	X	X	X
Modulus of rupture	X	X	X	X	X
Weibull modulus	X	X	X	X	X
Hardness		X		X	X
Toughness	X	X	X	X	X
Creep	X	X	X	X	X
Impact resistance		X		X	X
Coefficient of friction		X		X	X
Wear resistance		X		X	X
Chemical					
Corrosion products	X	X	X	X	X
Corrosion rate	X	X	X	X	X
Oxidation rate	X	X	X	X	X

(a) HTEX, heat exchanger; VALV, valve components; TURB, turbine engines; BRNG, bearings; TOOL, cutting tools

Evaluated Data. Efforts to ensure that data are reliable can take numerous forms. Alternatives include reporting standard deviations and other statistics based on repeated measurements, making comparisons with theoretical models, and soliciting assessments based on an expert's extensive experience with nominally similar materials and measurement procedures. The end result of each of these efforts is a set of property values that are evaluated or validated.

A general set of assessment criteria may be summarized as follows:

- The material must be adequately identified
- The measurement method must be sufficiently described
- The quality of the data must be indicated in an acceptable form
- The property values must be reasonable
- The results must be documented

These criteria may serve as guidelines for the research community as well as for the data base development and the end user's application of the data. While the criteria, viewed individually, are somewhat vague, each provides an important safeguard toward the reliability of the data.

The last item, documentation, should not be overlooked. There is a fairly common practice in the literature of mentioning selected basic properties of a material without referencing their source or indicating how those values were determined. Often it is not clear whether these values are merely "typical" values that are assumed to be applicable to the material under study, or whether the authors actually measured the values. Documentation of the source of data provides the end user with an accountability for the values.

Alternative approaches for data evaluation include the development of consensus values and best judgment values. A consensus value can be obtained when each of a number of research groups agrees to measure a property value independently. The characterization of the property is then provided by the collective results after due assessments and deliberations have been completed by the group as a whole. International activities directed toward the establishment of standards may achieve consensus values. Best judgment values, in contrast, utilize the depth of experience of one or more noted experts who project or assign values to properties according to their best assessment, regardless of what has been measured. Best judgment values are recommended only when the alternatives are not practical. Such a situation has been encountered in the field of tribology (Ref 10).

In general, some form of data evaluation is essential for advanced ceramics. The properties of these materials are sufficiently susceptible to variations in processing, microstructure, and measurement procedures that to accept values without assessment is risky.

Existing Compilations of Property Data for Ceramics*

Studies of advanced ceramics have reached a level that will support sustained efforts to collect, organize, and evaluate property data. Several compilations of typical property data are available in printed collections, while evaluated sets in computerized formats are only slowly emerging.

Printed Compilations. Among the most commonly available collections of data are the more traditional compilations provided by research societies, trade journals, and handbooks. Each is a useful compendium of information on a broad range of ceramic materials. *Ceramic Source* (Ref 11), *Ceramic Industry* (Ref 12), and *Advanced Materials and Processes* (Ref 13) contain annual materials reviews that are especially noteworthy. These periodic publications provide a registry of companies, their products, services, and tradenames, and a survey of typical property

*Certain commercial names are identified in this article for the purpose of clarity in the presentation. Such identification does not imply recommendation or endorsement by the National Institute of Standards and Technology

values gleaned from manufacturer's literature and from selected reports in the technical literature. The materials are identified generically in most cases and occasionally by manufacturer. The intent of these compilations is to provide guideline or typical property values. Generally, specific information on the measurement methods or procedures are not discussed, nor is the quality of the data assessed explicitly. Several book sources provide similar information supplemented by annotations and discussions of typical trends. *Handbook of Properties of Technical & Engineering Ceramics* by R. Morrell (Ref 14), *Structural Ceramics* by J.B. Wachtman, Jr. (Ref 15), and *Ceramic Materials for Advanced Heat Engines* by D. Larsen *et al.* (Ref 16), as well as this Volume, are perhaps the most comprehensive sources of this type.

One of the first efforts to provide evaluated ceramics data was undertaken by the Metals and Ceramics Information Center (MCIC) at Battelle's Columbus Laboratories. The MCIC produced a three-volume series, *Engineering Property Data on Selected Ceramics*, which included nitrides (Vol 1, 1976), carbides (Vol 2, 1979), and single oxides (Vol 3, 1981) (Ref 17). Property values obtained from the technical literature and from manufacturers were reported along with appropriate references to the published sources. Measurement methods were usually identified, at least generically, although details were usually not given. An important assessment of the data was given in the form of plots showing bands of reported values, for generically identified materials. This series remains a good reference for older materials and for examples of the problems in data measurements.

A very useful complement to materials properties compilations is the series *Phase Diagrams for Ceramists*, which is a joint venture of the American Ceramic Society and the National Institute of Standards and Technology. This unique multi-volume series contains more than 8000 phase diagrams for binary, ternary, and higher order oxide, metal-oxide, metal-oxygen, halide, and other ceramic systems (Ref 18).

Computerized Data Bases. A significant step toward a standardized manner of collecting and evaluating data was provided by the report, "A Computerized Fracture Mechanics Database for Oxide Glasses" by S.W. Freiman, T.L. Baker, and J.B. Wachtman, Jr. (Ref 19). In this report, values of fracture toughness, fracture energy, subcritical crack-growth exponents, and Young's modulus were listed in tabular form for a wide variety of glasses. Values were presented in a standardized table which listed a generic material specification, the name of the manufacturer and the manufacturer's designation for the material, the composition of the glass, Young's modulus, the fracture mechanics data, a description of the test environment, and the reference for the data. A comment field was also included for supplementary information about the composition, test method, or data. It is interesting to note that the standardized presentation was the result of the preparation of the data in a computerized format for personal computers.

One of the first efforts to computerize data directly from the research laboratories is being pursued by the Ceramic Technology for Advanced Heat Engines program at Oak Ridge National Laboratory (ORNL) (Ref 20). Data are submitted from the contributing laboratories to a coordinating office at ORNL where the data are entered into a relational data base developed in dBASE IV (Ref 21). The results for nitrides, carbides, and oxides are principally research values obtained prior to technical peer review or other evaluation. The entries contain detailed materials identification including compositions, microstructures, physical properties, and names of manufacturers. Measurement methods are also identified.

The first program designed specifically to develop a user-friendly computerized data base for ceramics properties is the ongoing Structural Ceramics Database (SCD) project at the National Institute of Standards and Technology (NIST) (Ref 22). Developed with funding support from the Gas Research Institute, the Center for Advanced Materials at Pennsylvania State University, and the NIST, the SCD is designed to be used on personal computers. The SCD emphasizes ease of use as much as it stresses the evaluation of data. In Version 1.0, 127 sets of evaluated data on the high-temperature thermal and mechanical properties of silicon carbides and silicon nitrides are recorded. Materials specification, measurement method identification, specimen preparations, measurement procedures, and bibliographic information are included for all entries. Environmental effects and cautionary observations noted by the authors of the data or by the evaluators are also reported. Extension of the SCD to other properties, materials, and system features is ongoing.

As the recognition of the importance of computerized data bases of materials properties has increased, so also has the number of efforts increased. Several new programs are currently in the development stage and are expected to be significant in the future. A new effort, the Department of Defense Ceramics Information Analysis Center, will be pursued as part of the CINDAS program at Purdue University with an emphasis on composite materials (Ref 23). The Phase Diagrams for Ceramists program at NIST is developing a user-friendly system for personal computers to search and retrieve phase diagrams for on-screen plotting and manipulation (Ref 24). The New Glass Forum in Japan is developing a major resource for the materials property data of glasses (Ref 25). This system will operate on personal computers and will use CD-ROM (compact disk—read only memory) technology to hold approximately 90,000 data entries.

Issues for Future Developments

The evolution of computerized materials property data bases has a rather well defined future path. The current distributed data bases inevitably must be linked together into a broad knowledge base serving as the primary information resource for computerized expert systems.

This goal should be held firmly in mind whenever the requirements for new data base systems are considered. The more technology becomes dependent on diversely constructed sources of information, the more critical it will be that these sources meet the basic requirements for appropriate and reliable data (Ref 26). Likewise, the more interdependent systems become, the more critical it will be for the controlling system to have an easy to use, that is, user-friendly, design.

Efforts to link materials property data bases have already shown progress. MPD Network, Inc., in particular, is attempting to acquire property data bases for all types of materials (Ref 27). These data bases are accessible to remote users by means of the STN International computer network system. NASA is developing an integrated advanced materials data base, Materials and Processes Technical Information System (MAPTIS), as part of their advanced aerospace program (Ref 28). ACTIS, Inc. has indicated its plan to develop a gateway system in the future for tribological data which will include materials property data (Ref 29). CERABULL may provide an on-line system for ceramics data from the American Ceramic Society (Ref 30).

The potential for using the currently evolving structured query language (SQL) (Ref 31) to link materials property data bases distributed on personal computers in diverse formats has not yet been utilized. The SQL provides a basis for structurally different data bases, for example, different commercial packages, to communicate and exchange information. The SQL in this capacity serves as an interpreter or translator. Use of the SQL capabilities, therefore, could preserve the specialization needed for specific applications, while enabling data to be shared across imaginary disciplinary boundaries.

Progress toward expert systems for ceramics processing or manufacturing is slight at the present time (Ref 32). However, significant results that may be forthcoming soon in such areas as materials processing (Ref 33) and automated manufacturing (Ref 34) may provide a basis for the development of a collection of expert systems for the intelligent processing of ceramic materials also. It is well recognized that the microstructure and composition of a finished ceramic depends closely on its processing history. It follows that controlling the processing variables, such as temperature and temperature gradients, pressures and stress gradients, feed rates, and times of processing, is critical to limiting the vari-

ability of the finished material. As appropriate sensor devices are developed and reliable processing models are established, expert systems could provide the on-line decision-making capability needed to respond to processing variations in real time and, hence, to correct or to control errant deviations (Ref 35).

REFERENCES

1. J.A. Pask, Structural Ceramics, *J. Mater. Eng.*, Vol 11 (No. 4), 1989, p 267–274
2. D.R. Johnson and J.O. Stiegler, Structural Ceramics R&D, *Adv. Mater. Proc.*, Vol 138 (No. 3), 1990, p 55–61
3. R.G. Munro and C.R. Hubbard, Property Database for Gas-Fired Applications of Ceramics, *Ceram. Bull.*, Vol 68 (No. 12), 1989, p 2084–2090
4. R.G. Munro and E.F. Begley, *Materials Property Database Requirements for Gas-Fueled Ceramic Heat Exchangers*, STP 1106, American Society for Testing and Materials, 1990
5. R.G. Munro, F.Y. Hwang, and C.R. Hubbard, The Structural Ceramics Database: Technical Foundations, *J. Res. Natl. Inst. Stand. Technol.*, Vol 94 (No. 1), 1989, p 37–47
6. J.G. Kaufman, Standards for Computerized Material Property Data, ASTM Committee E-49, *Computerization and Networking of Materials Data Bases*, STP 1017, J.S. Glazman and J.R. Rumble, Jr., Ed., American Society for Testing and Materials, 1989, p 7–22
7. F.F. Lange, Relation Between Strength, Fracture Energy, and Microstructure of Hot-Pressed Si_3N_4, *J. Am. Ceram. Soc.*, Vol 56 (No. 10), 1973, p 518–522
8. B. Lawn and R. Wilshaw, Indentation Fracture: Principles and Applications, *J. Mater. Sci.*, Vol 10, 1975, p 1049–1081
9. T.A. Michalske and S.W. Freiman, A Molecular Mechanism for Stress Corrosion in Vitreous Silica, *J. Am. Ceram. Soc.*, Vol 66 (No. 4), 1983, p 284–288
10. S. Danyluk and S.M. Hsu, A Computerized Tribology Information System, ACTIS, *Managing Engineering Data: The Competitive Edge*, R.E. Fulton, Ed., American Society of Mechanical Engineers, 1987, p 133–144
11. *Ceramic Source*, W.J. Smothers, Ed., Vol 1–3, J.B. Wachtman, Jr., Ed., Vol 4, American Ceramic Society, 1986–1988
12. Data Book and Buyers Guide, *Ceram. Ind.*, Vol 135 (No. 4), 1990
13. Guide to Selecting Engineered Materials, *Adv. Mater. Proc.*, Vol 137 (No. 6), 1990
14. R. Morrell, *Handbook of Properties of Technical & Engineering Ceramics*, Her Majesty's Stationery Office, London,1985
15. J.B. Wachtman, Jr., *Structural Ceramics*, Academic Press, 1989
16. D.C. Larsen, J.W. Adams, L.R. Johnson, A.P.S. Teotia, and L.G. Hill,

Ceramic Materials for Advanced Heat Engines, Noyes, 1985

17. *Engineering Property Data on Selected Ceramics*, MCIC-HB-07, Vol I, 1976, Nitrides; Vol II, 1979, Carbides; Vol III, 1981, Single Oxides; Battelle Columbus Laboratories

18. *Phase Diagrams for Ceramists*, Vol I-VIII, American Ceramic Society, 1964–1990

19. S.W. Freiman, T.L. Baker, and J.B. Wachtman, Jr., "A Computerized Fracture Mechanics Database for Oxide Glasses," Technical Note 1212, National Bureau of Standards, 1985

20. B. Booker, "Ceramic Technology for Advanced Heat Engines Project Data Base," Technical Report DE89012888/HDM, National Technical Information Service, 1989

21. B. Keyes, CTAHE Test Data Archive, Oak Ridge National Laboratory, 1990

22. E.F. Begley and R.G. Munro, "Structural Ceramics Database 1.0," SRD No. 30, Standard Reference Data, National Institute of Standards and Technology, 1990

23. C.Y. Ho, Center for Information and Numerical Data Analysis and Synthesis, Purdue University, private communication

24. P.K. Schenck and J.R. Dennis, PC-Access to Ceramic Phase Diagrams, *Computerization and Networking of Materials Data Bases*, STP 1017, J.S. Glazman and J.R. Rumble, Jr., Ed., American Society for Testing and Materials, 1989, p 292–303

25. "INTERGLAD," International Glass Fact Data Base System, New Glass Forum, Tokyo, 1991

26. R.G. Munro, E.F. Begley, and T.L. Baker, Strengths and Deficiencies in Published Advanced Ceramics Data," *Ceram. Bull.*, Vol 69 (No. 9), 1990, p 1498–1502

27. G. Kaufman, "MPD Update," Materials Property Data Network Newsletter, Sept 1990, p 1–8

28. "Unified System of Data on Materials and Processes," Technical Report MFS-27212/TN, NASA Technology Transfer Division, 1989

29. S. Jahanmir, S.M. Hsu, and R.G. Munro, ACTIS: Towards a Comprehensive Tribology Data Base, *Computerization and Networking of Materials Data Bases*, STP 1017, J.S. Glazman and J.R. Rumble, Jr., Ed., American Society for Testing and Materials, 1989, p 340–348

30. G. Lewis and E.C. Skaar, A Look into the Future with CERABULL, *Ceram. Bull.*, Vol 67 (No. 12), 1988, p 1876–1878

31. R.F. van der Lans, *The SQL Standard: A Complete Reference*, Prentice Hall, 1989

32. S. Sengupta, Expert/Knowledge-Based Systems for Ceramic Glaze Technology, *Ceram. Bull.*, Vol 69 (No. 1), 1990, p 83–90

33. P. Palanisamy and T. Asokan, Intelligent Processing of ZnO-Based Ceramics, *Ceram. Bull.*, Vol 67 (No. 10), 1988, p 1695–1698

34. T. Sata, AI Tools for Manufacturing Automation, *Computers in Industry*, Vol 14 (No. 1), 1990, p 139–143

35. J. Schreurs and N. Pessall, Expert Systems Tackle Engineering Problems, *Adv. Mater. Proc.*, Vol 138 (No. 3), 1990, p 63–67

Crack Propagation in Ceramics

J.E. Ritter, University of Massachusetts, Amherst

INNOVATIVE PROGRAMS to develop new and improved ceramics for structural applications have been ongoing worldwide since 1970. The prime focus of these efforts has been to improve the fracture toughness, strength, and mechanical reliability of these brittle materials. Results from these programs, conducted in a broad range of industries, have considerably advanced the fundamental knowledge of brittle fracture.

Our understanding of crack propagation in ceramics has been developed using principles of fracture mechanics (Ref 1–6), since fracture mechanics concepts can be used to characterize the initial stress intensity of the crack, the conditions for slow crack growth, and the conditions for crack instability. This article describes crack propagation in ceramics, emphasizing the ways that stress and environment can influence crack propagation and the ways that microstructure can affect fracture resistance. The discussion concentrates on the concepts involved, rather than on the mathematical development of the models.

Fast Fracture

A measure of the magnification of the applied stress by a crack is given by the stress-intensity factor K_I (Ref 7). Failure (fast fracture) occurs when K_I equals a critical value K_{Ic}, which is generally termed fracture toughness and is considered to be a material constant. The relationship between failure stress or strength, σ_f, flaw size, a, and fracture toughness is given by (Ref 7):

$$\sigma_f = Y\left(\frac{K_{Ic}}{\sqrt{a}}\right) \qquad (Eq\ 1)$$

where Y is a well-documented crack/specimen geometry constant that generally ranges from 1.2 to 1.8.

Fast-fracture crack instability can also be derived by considering the energetics of crack growth, where the forces acting on the crack are expressed by the strain-energy release rate, G. This approach was originally proposed by Griffith (Ref 8) and was used recently by Cook (Ref 9) in explaining crack instability and thresholds in ceramics. This approach assumes that failure occurs if the incremental change in the net energy (work done—elastic strain energy) exceeds the energy required for crack

growth. The conditions for the onset of crack instability are equivalent to the stress-intensity approach, where the critical strain-energy release rate, often referred to as toughness, G_c, is given by:

$$G_c = \frac{K_{Ic}^2}{E} \qquad (Eq\ 2)$$

where E is the elastic modulus.

Equation 1 emphasizes the importance that K_{Ic} (or G_c) plays in the fracture of ceramics because K_{Ic} represents the material constant at which fast fracture commences for a critical combination of stress and crack length. Generally, the K_{Ic} level for ceramics is the point at which the crack is travelling at a high velocity, about 1 m/s (33 ft/s) (Ref 2). Actual crack growth can occur at stress-intensity levels less than K_{Ic}, because of environmental effects, which are discussed below. Therefore, a precise experimental definition of criticality is difficult. The value of K_{Ic} for a given material can be a function of environment, loading rate, and test geometry. The advantages and disadvantages of the various test techniques for measuring K_{Ic} are discussed in Ref 2 and 10.

Strength-Controlling Flaws. Failure in ceramics initiates at a single "worst" flaw (see Eq 1). These critical flaws may be an intrinsic feature of the microstructure: a pore, a weak grain boundary, an inclusion, or a large grain (Ref 11, 12). They may also be an extrinsic scratch, pit, or crack that is introduced either by surface finishing or abusive handling (Ref 13) or as a consequence of service environment (Ref 14). Because of the generally low fracture toughness of ceramics, these critical flaws are often 10 to 20 μm (0.4 to 0.8 mil) or less in size. Much of the processing research in the development of ceramic components is centered on identifying the sources of the strength-degrading flaws and then developing new processing techniques to eliminate or minimize the size and variability of these defects (Ref 15–18). However, several factors complicate the straightforward use of Eq 1.

First, actual flaws are generally complex, three-dimensional defects and the size of an equivalent, two-dimensional crack must be assumed (Ref 11). Second, a localized residual stress is often associated with the flaw

because of stresses generated by thermal expansion anisotropy, phase transformation, or contact stress damage. The residual stress around the flaw supplements the external applied stress and provides an additional crack driving force that must be considered (Ref 10–13). Third, the tensile stresses on a component are generally not perpendicular to the most severe flaw. Although there are several theories regarding mixed-mode loading, no one model has been shown to be universally acceptable (Ref 19–22).

Fractography. The examination of the fracture surface of ceramics can yield a great deal of information about the fracture process (Ref 2, 23). Fractography can show whether the flaw propagated transgranularly or intergranularly and, in the case of toughened ceramics, the possible mechanisms for the toughening. In addition, there are reproducible features that can be quantitatively related to the stress at fracture (Ref 2, 23).

The primary feature on the fracture surface is the strength-controlling flaw. Immediately surrounding the initial defect is a smooth region, commonly known as a fracture mirror, followed by a region of small radial ridges, known as mist, and then a rougher region of large ridges, known as hackle. There is a quantitative relationship between the failure stress and the distance from the initial flaw to each of the boundaries between the regions, r_j:

$$\sigma_f r_j^{1/2} = A_j \qquad (Eq\ 3)$$

where A_j is a constant. By combining Eq 3 and Eq 1, the boundary measurements can be related to the fracture toughness of the material:

$$K_{Ic} = YA_j\left(\frac{r_j}{a}\right)^{-1/2} \qquad (Eq\ 4)$$

Experimental measurements over a wide range of ceramics give (Ref 2, 23):

$$K_{Ic} = \frac{A_0}{2.9} \qquad (Eq\ 5)$$

where A_0 is the constant based on the mist/hackle boundary. By substituting Eq 4 into Eq 5, it can be shown that for a semicircular flaw, the ratio r_0/a is 13. Thus, using fractography, the failure stress and flaw size can be estimated based on the r_0 and K_{Ic} values or, using the failure stress and boundary determination, the K_{Ic} and flaw size can be estimated.

Environmentally Enhanced Crack Growth

Static Loading. Environmentally enhanced crack growth occurs at moderate temperatures in most ceramics at applied stress intensities less than the K_{Ic} value. The primary environmental constituent that causes subcritical crack growth is water. This phenomenon has been labelled as subcritical crack growth, stress-corrosion cracking, and static fatigue. The pioneering research of Wiederhorn (Ref 24) enabled the dependence of crack velocity on the applied stress intensity to be divided into three regimes: region I, where, above an initial threshold, the crack velocity is a strong function of K_I; region II, a central region, where the crack velocity is independent of K_I; and region III, where the crack velocity curve becomes quite steep and is independent of environment. Most of the research on subcritical crack growth has focused on region I, because the majority of the life of a ceramic is spent with the crack in this region. Characterizing the dependence of crack velocity on K_I in region I by a power-law relationship has led to the development of design-life prediction models (Ref 25, 26).

Michalske and Freiman (Ref 27) have proposed the mechanism shown in Fig 1(a) to explain subcritical crack growth in silica. At the crack tip, the strained Si—O—Si bonds alter the tetrahedral symmetry around the Si atoms, so that the Si atoms become strongly acidic and the bridging oxygens strongly basic. The water molecule orients itself so that a hydrogen bond forms at the bridging oxygen.

Electron transfer now occurs simultaneously from the O (water) to the Si with proton transfer to O (silica), forming two new bonds Si—O (water) and H—O (silica). Rupture of the weak hydrogen bond between O (water) and the transferred hydrogen then occurs. In place of the Si—O—Si crack tip, two Si—OH groups now form the surface behind the "new" crack tip. Crack propagation by this mechanism continues until the supply of water to the crack tip becomes limited. The plateau of region II of the crack velocity curve occurs because the diffusion of water to the crack tip becomes rate controlling. The success of the Michalske and Freiman model was demonstrated by showing that other molecules whose structure and bonding are similar to water, that is, can donate both electrons and protons, and are small enough to reach the crack tip, can promote subcritical crack growth. Molecules such as ammonia and hydrazine result in crack velocities similar to water, whereas N_2, CO, and acetonitrile are not effective (Ref 27–28).

Subcritical crack growth in many polycrystalline ceramics can be caused by the interaction of water with the silicate-base glassy phase present in the grain boundaries (Ref 2). However, subcritical crack growth can also occur in single-crystal and polycrystalline ceramics with no intergranular phase. Research by Michalske *et al.* (Ref 29) illustrates some of the basic features of stress corrosion in these ceramics. Sapphire (Al_2O_3) that has a similar ionic/covalent bonding to SiO_2 exhibited stress corrosion similar to SiO_2. However, in highly ionic MgF_2, the chemically active sites do not have a sufficient acid-base character for dissociative chemisorption to occur, and stress corrosion occurs by ion solvation.

An alternate description of the origin of subcritical crack growth is now emerging from recent research on the direct measurement of surface forces at crack tips (Ref 30–33). It is now well established, through direct observation, that cracks in ceramics are atomically sharp (Ref 1, 34, 35). Because of this sharpness, it is believed that an environmental molecule cannot penetrate to the crack tip and directly react with the crack-tip bonds. Instead, a "chemical wedge" forms when chemisorbing molecules enter the mouth of a brittle crack (Ref 31). Capillary action draws the chemisorbing molecules that form the wedge toward the crack tip, distorting the crack faces and causing a compensating elastic force on the molecules, which tends to eject them. Crack advance takes place when the Griffith energy condition is satisfied.

Thus, the energy approach emphasizes how the active environment affects the surface energy of the crack, rather than the corrosive bond-breaking mechanism. Figure 1(b) depicts a crack propagating through a solid under the action of a "chemical wedge" that generates atomic bonds in various states of rupture (Ref 36). Region 0 includes bonds that are in elastic equilibrium with the applied stresses. In region A, the bonds are elastically distorted by the crack-tip stress field. In region B, the cohesive forces are beyond their elastic limit, but nevertheless remain significant enough to provide primary bonding across the crack faces. Region C contains ruptured bonds that are due to the "wedging" that still interact through long-range, secondary bonds. In region D, the ruptured bonds are out of the range of the secondary forces, but are not yet in equilibrium with the environment in the crack, because of kinetic considerations. The final region E contains bonds (and surfaces) that are in equilibrium with the environment. Although crack kinetics have yet to be developed in the context of this energetic approach, this approach holds much promise, because

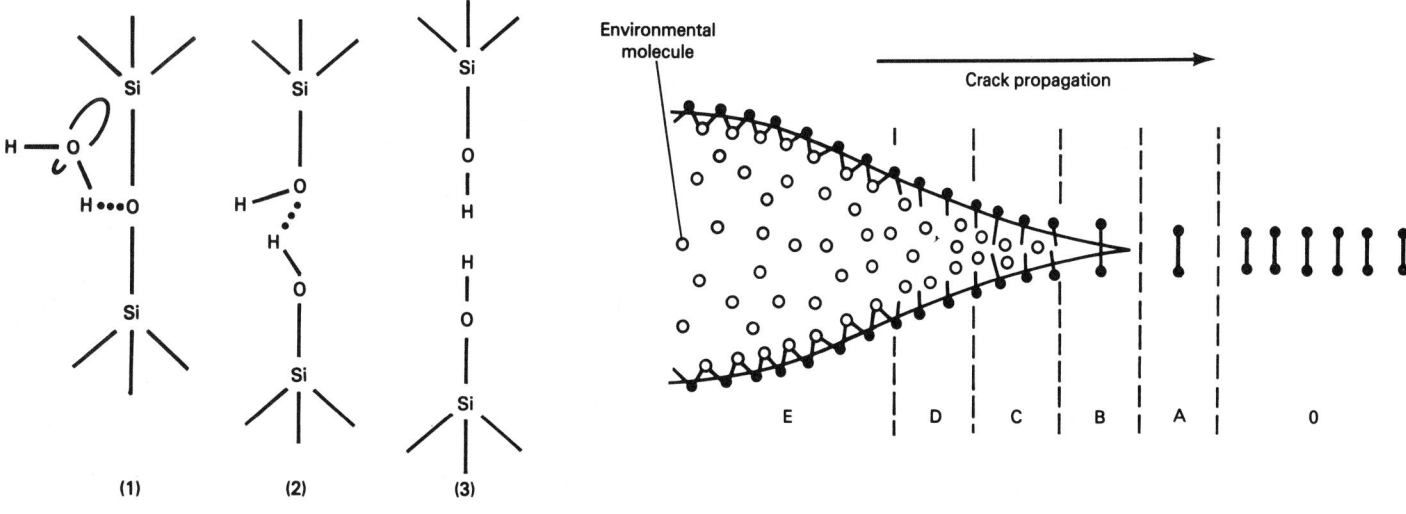

Fig 1 (a) Sequence of water molecule reacting with strained Si—O bonds at a crack tip in vitreous SiO_2 (after Ref 27). (b) Crack propagation in the presence of a chemical wedge (after Ref 36)

the relevant surface forces can be directly measured.

Cyclic Loading. Because ceramics have very limited crack-tip plasticity, the general belief has been that they are immune to cyclic stress and that apparent cyclic fatigue effects are simply manifestations of environmentally assisted crack growth under tensile loading (Ref 37). However, recent studies have shown conclusive evidence of true cyclic fatigue-crack growth (Ref 38–40). Figure 2(a) shows data for a Mg-PSZ, where growth rates are proportional to the cyclic range of stress intensity, rather than a constant, sustained stress intensity (Ref 38). The fact that growth rates are faster in water and moist air than in inert nitrogen indicates a marked corrosion-fatigue effect that involves the weakening of atomic bonds at the crack tip, which is due to water molecule adsorption. These results emphasize that cyclic fatigue in ceramics is a mechanically induced cyclic process that can be accelerated by the environment. In Fig 2(b), a comparison between crack growth measured under cyclic loads and crack growth measured under sustained loads at comparable stress intensities illustrates that cyclic crack growth rates in ceramics can be many orders of magnitude faster and can occur at stress intensities far lower than those for sustained-load cracking (Ref 38). This existence of cyclic fatigue effects is most troubling, because design-life predictions for ceramics are generally based on existing subcritical crack growth data generated under sustained-loading conditions that, based on the results in Fig 2(b), would result in extreme overestimates of lifetimes.

Although the detailed mechanism for cyclic crack growth has not been identified, data suggest that an intrinsic mechanism is operative in transformation-toughened ceramics because cyclic fatigue is not dependent on the degree of transformation (Ref 38). Potential intrinsic mechanisms include:

- Accumulation of damage ahead of the crack tip in the form of localized microcracking, particularly in grain-boundary and particle/matrix interface regions
- Mode II and III cracking that is due to the wedging action of crack surface asperities upon unloading
- Relaxation of residual stresses in specifically oriented grains, with subsequent development of additional tensile and shear stresses upon unloading because of the inability of the relaxed grains to be accommodated in their original position

Threshold stress intensity values for both static and cyclic crack growth have been demonstrated experimentally (Ref 10, 32, 36, 38, 41–43). This threshold, below which crack growth ceases, occurs at K_I/K_{Ic} values of 0.15 to 0.30 for static crack growth (Ref 41, 42) and approximately 0.50 for cyclic crack growth (Ref 38).

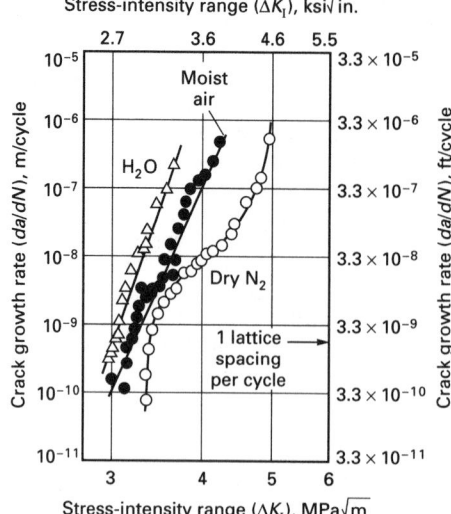

(a)

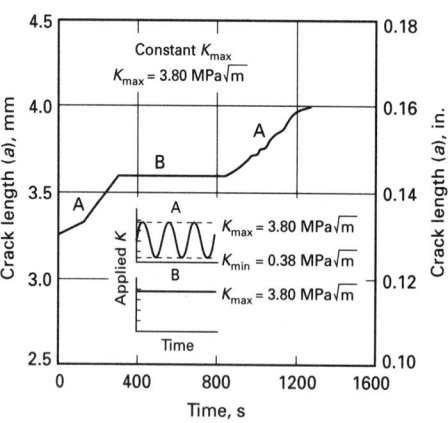

(b)

Fig 2 (a) Cyclic fatigue crack-growth rates in a low-toughness Mg-PSZ, illustrating the acceleration that is due to water (after Ref 38). (b) Effect of static and cyclic loading conditions on crack growth in a low-toughness Mg-PSZ (after Ref 38)

According to stress-corrosion theory (Ref 27, 28), a threshold in static crack growth arises naturally from the fact that, at a low K_I value, the crack opening is insufficient for the reactive molecule to reach the crack tip. The static fatigue threshold also arises quite naturally from surface force arguments (Ref 10, 32, 36, 43). Here, the threshold value is where the crack driving force falls to that given by the surface energy, that is, to the Griffith condition for crack equilibrium in the chemical environment of the fracture. For cyclic crack growth, the threshold apparently results from crack-tip shielding caused primarily by the wedging action of fracture surface asperities (Ref 38).

High-Temperature Crack Growth

Elevated Temperature. In contrast to fracture at low-to-moderate temperatures, high-

temperature failure of ceramics is much more complex. For example, oxidation-associated flaws can be generated in non-oxides that are exposed to oxidizing environments (Ref 14), and silicate flaws can develop in oxides that are contaminated by a silica impurity (Ref 44). In both cases, enhanced crack growth can also occur because of a stress-corrosion reaction where the amorphous phase wets the crack and enhances the local reaction at the crack tip (Ref 44). In addition, transport processes, such as viscous flow and diffusion, are greatly accelerated at high temperatures, leading to creep-induced crack initiation and growth. References 1 and 14 describe the range of fracture behavior exhibited by ceramics at elevated temperatures.

Figure 3 illustrates the general dependence of creep-induced crack-growth rate on the stress-intensity factor (Ref 45). Below a threshold stress intensity, K_{th}, pre-existing cracks will blunt and become innocuous, and failure will occur by the (Ref 1):

- Coalescence of creep damage that involves creep-nucleated cavities forming throughout the material
- Growth of some of these cavities across grain facets to become grain-sized cavities
- Coalescence of cavities that form on adjacent boundaries until such contiguous coalescence cavities constitute a distinguishable crack that is capable of propagating to failure by slow crack growth

At stress intensities above K_{th}, creep-induced crack-growth rates appear to be controlled by two distinct mechanisms (Ref 45). One mechanism involves the nucleation and linking up of voids ahead of the crack tip, and the other mechanism involves diffusion of atoms away from the crack tip. In general, the crack driving force is expected to be dependent only on K_I if the creep of the material is nearly linear ($\dot{\varepsilon} \propto \sigma^n$, where $n \approx 1$). In cases where creep is nonlinear, creep-induced crack-growth rates appear to be dependent on strain (Ref 46). At very high stress intensities, crack-growth rates are expected to become independent of transport processes and revert back to brittle behavior.

An important finding in a recent study (Ref 47) of crack growth in glassy alumina was that although a creep stress threshold existed for the growth of indentation cracks, cracks of an equivalent size under the same temperature and stress conditions, but nucleated by the creep process, exhibited no threshold stress. Furthermore, these creep-induced cracks grew at a faster rate than the indentation cracks. These results have important implications to design-life prediction models.

Toughening Ceramics

Toughening mechanisms in ceramics have been discussed in detail in Ref 1, 3, and 48. According to Evans (Ref 4), the known mechanisms can be conveniently considered

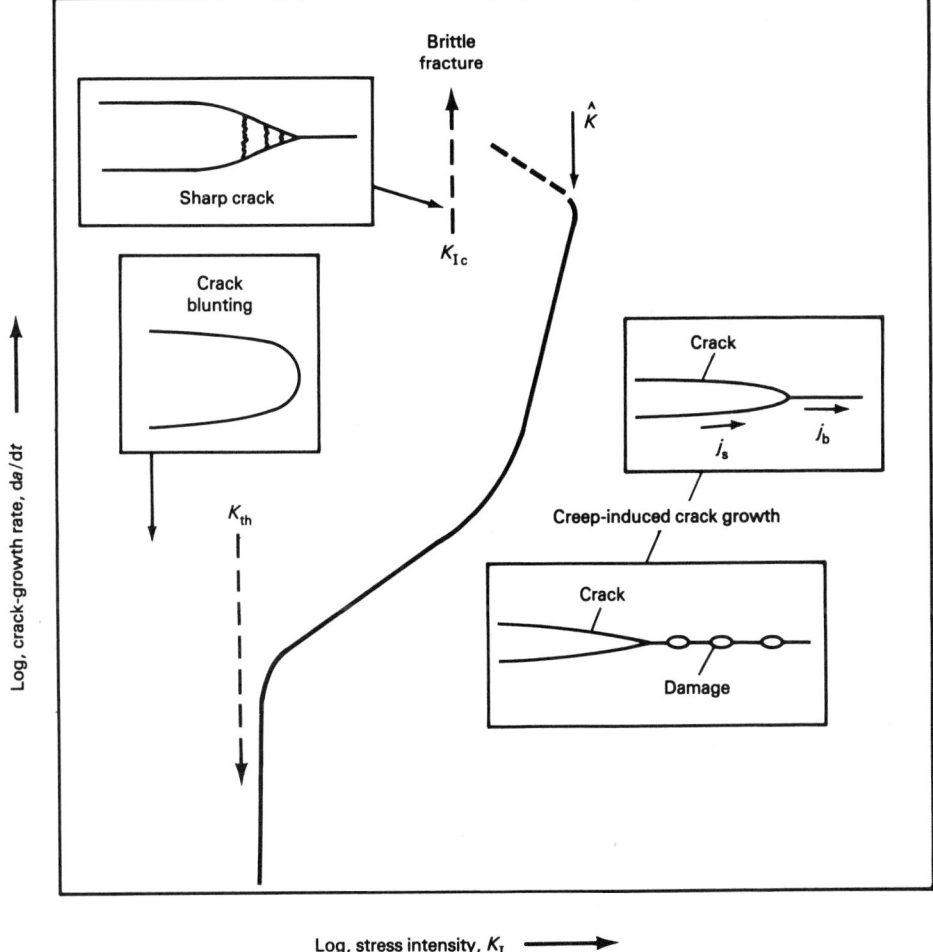

Fig 3 Generalized dependence of high-temperature crack-growth rates on stress-intensity factor (after Ref 45)

to involve either a process zone or a bridging zone. The former category exhibits a toughening that is fundamentally governed by a critical stress for the onset of nonlinearity, σ_0, in elements near the crack and by a dilation stress-free strain, ϵ_0. The resulting stress-strain hysteresis of those elements within the process zone then yields a maximum increase in toughness (ΔG_c) of (Ref 4):

$$\Delta G_c = 2 f \sigma_0 \epsilon_0 w \qquad \text{(Eq 6)}$$

where f is the volume fraction of the toughening agent and w is the width of the process zone in the steady state. Phase transformation and microcracking are mechanisms of this type. Partially-stabilized ZrO_2 is the most common ceramic that is toughened by phase transformation; toughnesses of about 20 MPa$\sqrt{m}$ (18 ksi$\sqrt{in}$.) have been achieved. Microcracking toughening up to about 10 MPa$\sqrt{m}$ (9 ksi$\sqrt{in}$.) has been obtained in materials such as Al_2O_3/ZrO_2, Si_3N_4/SiC, and SiC/TiB_2.

The bridging category exhibits toughening that is due to intact material ligaments that cause closure forces on the crack surfaces behind the crack tip, which results in a toughening (Ref 4):

$$\Delta G_c = 2 f \int_0^{u^*} t(u) \, du \qquad \text{(Eq 7)}$$

where $2u$ is the crack opening, $2u^*$ is the opening at the edge of the bridging zone, and t is the traction on the crack surfaces exerted by the intact ligaments. Ductile reinforcements, as well as whiskers, fibers, and large grains, toughen by means of bridging tractions. Ductile-reinforced ceramics include Al_2O_3/Al, Al_2O_3/Ni, and WC/Co, with toughening up to 25 MPa$\sqrt{m}$ (23 ksi$\sqrt{in}$.). Fiber reinforcing results in the highest toughnesses recorded for ceramics (>30 MPa$\sqrt{m}$, or >27 ksi$\sqrt{in}$.); examples are glass-ceramic/SiC, Al_2O_3/SiC, SiC/SiC, and SiC/C.

It should be emphasized that, in practice, it is likely that several toughening mechanisms occur concurrently. The interaction between mechanisms can be highly beneficial and can produce synergism (Ref 4, 48). Conversely, some interaction effects can be deleterious and reduce the efficiency of the individual mechanisms (Ref 4). Synergism is most likely to occur when bridging and process-zone mechanisms interact. The interactions between either two process-zone or

bridging-zone mechanisms are relatively unexplored.

Toughened ceramics are inevitably characterized by a crack-growth resistance curve (R-curve), rather than by a single-valued fracture toughness. The resistance to crack growth in such materials increases as the length of the crack increases, and strength is no longer related uniquely to an initial flaw size. Instead, over a certain range of flaw sizes (less than a critical value), stable crack growth precedes failure and strength is determined by the R-curve and loading configuration. This has important design implications, because the strength becomes damage-tolerant, that is, independent of initial flaw size.

Crack-resistance toughening can be modelled by assuming a power-law dependence between the fracture resistance, K_R, and crack length (Ref 49):

$$K_R = K_0 \left(\frac{a}{d} \right)^\tau \qquad \text{(Eq 8)}$$

where $a \geq d$. The term K_0 can be regarded as the base toughness in the absence of any toughening mechanism, and the term d is the crack size at which toughening begins. The toughening exponent τ characterizes the rate at which the toughness increases ($0 \leq \tau \leq 0.5$). A value of $\tau = 0$ corresponds to constant toughness, $K_R = K_0 = K_{Ic}$. A value of $\tau > 0.5$ is not physically possible, because it implies that there can be no catastrophic crack propagation.

By substituting K_R in Eq 8 for K_{Ic} in Eq 1, strength is given by:

$$\sigma_f = \frac{K_0 a^{(2\tau-1)/2}}{Y d^\tau} \qquad \text{(Eq 9)}$$

Increasing τ increases σ_f for a given value of a and decreases the variability in σ_f for a given variation in a (Ref 49).

For cracks that are generated by sharp particle contacts on the surface, strength may be written generally, in terms of the slope of the R-curve τ, as (Ref 49):

$$\sigma_f = \beta P^{(2\tau-1)/(2\tau+3)} \qquad \text{(Eq 10)}$$

where β is a material/indenter constant and P is the peak contact load. Increasing τ leads to a weaker dependence of strength on contact load with the slope of the $\sigma_f - P$ curve reducing from $-1/3$ for $\tau = 0$ towards 0 as τ approaches $1/2$. Thus, increasing the slope of the R-curve will decrease the range of measured strengths for a given range of contact loads, that is, the material becomes damage tolerant. It is important to note that Eq 10 is generally applicable to cracks generated around imbedded inclusions and phase-transformed particles, as well as at grain-boundary junctions on cooling ceramics with large thermal expansion anisotropies (Ref 9, 49).

As shown above, a range of flaw sizes or contact loads gives rise to a range of observed strengths. In terms of Weibull statistics (Ref 26), strength variability is inversely related to the Weibull modulus, m. It can be shown from Eq 9 that for a given range of flaw sizes, the Weibull modulus is (Ref 49):

$$m = \frac{m_0}{(1 - 2\tau)} \qquad \text{(Eq 11)}$$

and for a given range of contact loads, Eq 10 gives (Ref 49):

$$m = \frac{m_0}{3}\left(\frac{3 + 2\tau}{1 - 2\tau}\right) \qquad \text{(Eq 12)}$$

where m_0 is the Weibull modulus for $\tau = 0$. As can be seen, m depends only on the slope of the R-curve, not on the absolute value of toughness, because K_0 does not appear in either Eq 11 or 12. These equations show that significant increases in the Weibull modulus can be observed as the slope of the R-curve increases. For example, increasing τ from 0 to 0.3 will more than double the Weibull modulus. This type of increase in the Weibull modulus has been observed for transformation-toughened ZrO_2 (Ref 50).

Interfacial Crack Propagation

The mechanism of delamination along bimaterial interfaces has recently received considerable research interest because of the importance of the mechanical integrity of interfaces in ceramic matrix composites, electronic packaging, and thin films on ceramic substrates. For example, in the design of microelectronic components, combinations of semiconductors, ceramics, polymers, and metals are bonded together to form high-speed integrated circuits where interfacial debonding is invariably detrimental. Similarly, interfacial failure must be averted in the case of protective coatings on substrates. On the other hand, the design of ceramic-matrix composites requires controlled debonding along the matrix-fiber interface in order to provide high values of toughness.

The mechanics of cracks at bimaterial interface have been elucidated recently (Ref 51, 52) and test techniques to measure interfacial strength and toughness have been developed (Ref 53–55). An important technique for studying interfacial fracture is the four-point flexure specimen, because the crack driving force, as measured by the strain-energy release rate, G, is independent of crack size within the inner loading span (Ref 53). This test was recently used to study subcritical crack growth at polymer/glass interfaces (Ref 56). Figure 4 summarizes the subcritical crack velocity (da/dt) data as a function of applied G and relative humidity. These data were best fitted to a power-law relationship to give

$$da/dt = A\left(\frac{G}{G_c}\right)^{2.5} \qquad \text{(Eq 13)}$$

where $A = 1.82 \times 10^{-6}$ m/s (5.97×10^{-6} ft/s) for 20% relative humidity (RH), 37.7×10^{-6} m/s (124×10^{-6} ft/s) for 50% RH, and 136×10^{-6} m/s (446×10^{-6} ft/s) for 60%

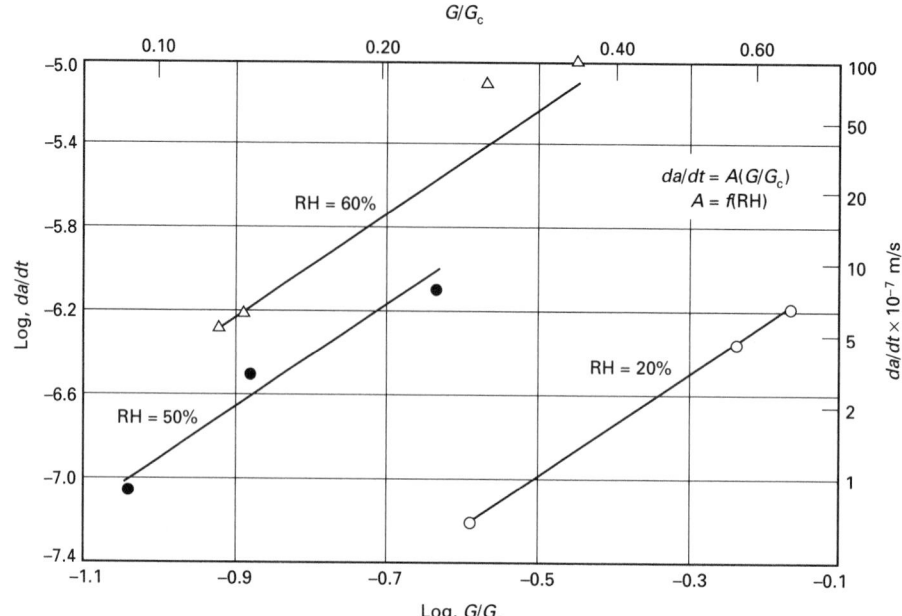

Fig 4 Dependence of crack-growth rates on strain-energy release rate at an epoxy acrylate/glass interface as a function of relative humidity (RH) (after Ref 56)

RH. This sensitivity of subcritical crack velocity on humidity indicates that the crack growth involves a stress-dependent interaction between moisture and the interfacial bonds at the crack tip, similar to that discussed previously for monolithic ceramics.

Because of subcritical crack growth, coating delamination can occur at an applied G less than G_c. The importance of this subcritical crack growth can be conveniently illustrated by considering a coating of thickness h under a residual tensile stress, σ_r, that forms during the curing of the polymer and provides the driving force for the interfacial crack to grow until the coating becomes detached. The strain-energy release rate for a crack propagating along the interface is (Ref 57):

$$G = \frac{(0.5\,\sigma_r^2\,h)}{E} \qquad \text{(Eq 14)}$$

where E is the elastic modulus of the coating. Substituting Eq 14 into 13 and integrating from an initial crack size, a_0, to a final crack size for complete delamination, a_f, gives the failure time for delamination, t_f:

$$t_f = \frac{1}{A}\left(\frac{2EG_c}{\sigma_r^2 h}\right)^{2.5}(a_f - a_0) \qquad \text{(Eq 15)}$$

Based on the data in Fig 4, it can be shown from Eq 15 that the time for delamination failure is increased by a factor of 75 by going from 60% RH to 20% RH. Decreasing the residual stress in the coating by one-half increases the failure time by a factor of 32.

Future Trends

Progress in understanding the fracture processes in ceramics is continuing and researchers are beginning to be able to ask fundamental questions about crack propagation in these materials. The knowledge base of crack propagation at low-to-moderate temperatures is far more complete than at high temperatures. Although a qualitative description of high-temperature fracture is now emerging, there is a need for mechanics analyses to describe creep-induced crack formation and growth and to predict the dependence of this formation and growth on microstructural features. Finally, it is now possible to envisage designing ceramic microstructures that can exhibit simultaneously two or more toughening mechanisms or that have different toughening mechanisms operating in different temperature regimes. This undoubtedly will open up new opportunities for ceramics as structural materials.

REFERENCES

1. D.R. Clarke and K.T. Faber, *J. Phys. Chem. Solids,* Vol 48, 1987, p 1115
2. S.W. Freiman, *Am. Ceram. Soc. Bull.,* Vol 67, 1988, p 392
3. D.B. Marshall and J.E. Ritter, *Am. Ceram. Soc. Bull.,* Vol 66, 1987, p 309
4. A.G. Evans, *J. Am. Ceram. Soc.,* Vol 73, 1990, p 187
5. A.G. Evans, *Mater. Sci. Eng.,* Vol 71, 1985, p 3
6. A.G. Evans and T.G. Langdon, *Progress in Mater. Sci.,* Vol 21, 1976, p 171
7. B.R. Lawn and T.R. Wilshaw, *Fracture of Brittle Solids,* Cambridge University Press, 1975
8. A.A. Griffith, *Phil. Trans. R. Soc. London,* Vol A221, 1920, p 163
9. R.F. Cook, *J. Appl. Phys.,* Vol 65, 1989, p 1902
10. S.W. Freiman, *Fracture Mechanics of*

Ceramics, Vol 6, R.C. Bradt, A.G. Evans, D.P. Hasselman, and F.F. Lange, Ed., Plenum Press, 1983, p 27–45

11. J.P. Singh, *Adv. Ceram. Soc.,* Vol 3, 1988, p 18

12. A.G. Evans, *J. Am. Ceram. Soc.,* Vol 65, 1982, p 127

13. D.B. Marshall, *Nitrogen Ceramics,* F.L. Riley, Ed., Martinus Nijhoff Publishers, 1983, p 635–656

14. S.M. Wiederhorn, *Mater. Sci. Eng.,* Vol 71, 1985, p 169

15. F.F. Lange, *J. Am. Ceram. Soc.,* Vol 72, 1989, p 3

16. J. Sung and P.S. Nicholson, *J. Am. Ceram. Soc.,* Vol 71, 1988, p 788

17. J. Neil, A. Pasto, and L. Bowen, *Adv. Ceram. Mater.,* Vol 3, 1988, p 225

18. S. Dutta, *Adv. Ceram. Mater.,* Vol 3, 1988, p 257

19. D.B. Marshall, *J. Am. Ceram. Soc.,* Vol 67, 1984, p 110

20. G.S. Glaesemann, J.E. Ritter, and K. Jakus, *J. Am. Ceram. Soc.,* Vol 70, 1987, p 630

21. L.-Y. Chao and D.K. Shetty, *J. Am. Ceram. Soc.,* Vol 73, 1990, p 1917

22. T. Thiemeier and A. Bruckner-Foit, *J. Am. Ceram. Soc.,* Vol 74, 1991, p 48

23. J.J. Mecholsky, S.W. Freiman, and R.W. Rice, *J. Mater. Sci.,* Vol 11, 1976, p 1310

24. S.M. Wiederhorn, *Fracture Mechanics of Ceramics,* Vol 4, R.C. Bradt, D.P.H. Hasselman, and F.F. Lange, Ed., Plenum Press, 1978, p 549–580

25. J.E. Ritter, *Yogyo-Kyoko Shi,* Vol 93, 1985, p 341

26. J.E. Ritter, *Fracture Mechanics of Ceramics,* Vol 5, R.C. Bradt, A.G. Evans, D.P.H. Hasselman, and F.F.

Lange, Ed., Plenum Press, 1984

27. T.A. Michalske and S.W. Freiman, *J. Am. Ceram. Soc.,* Vol 66, 1983, p 284

28. T.A. Michalske and B.C. Bunker, *J. Appl. Phys.,* Vol 56, 1984, p 2686

29. T.A. Michalske, B.C. Bunker, and S.W. Freiman, *J. Am. Ceram. Soc.,* Vol 69, 1986, p 721

30. K.-T. Wan, N. Aimard, S. Lathabai, R.G. Horn, and B.R. Lawn, *J. Mater. Res.,* Vol 5, 1990, p 172

31. R. Thomson, *J. Mater. Res.,* Vol 5, 1990, p 524

32. B. Lawn, D.H. Roach, and R.M. Thomson, *J. Mater. Sci.,* Vol 22, 1987, p 4036

33. R.G. Horn, *J. Am. Ceram. Soc.,* Vol 73, 1990, p 1117

34. H. Tanaka and Y. Bando, *J. Am. Ceram. Soc.,* Vol 73, 1990, p 761

35. B.R. Lawn, B.J. Hockey, and S.M. Wiederhorn, *J. Mater. Sci.,* 1980, p 1207

36. R.H. Cook, *J. Mater. Res.,* Vol 1, 1986, p 852

37. A.G. Evans, *Int. J. Fract.,* Vol 16, 1980, p 485

38. R. Dauskardt, D.B. Marshall, and R.O. Ritchie, *J. Am. Ceram. Soc.,* Vol 73, 1990, p 893

39. L. Ewart and S. Suresh, *J. Mater. Sci.,* Vol 22, 1987, p 1173

40. M.J. Reece, F. Guiu, and M.F.R. Sammur, *J. Am. Ceram. Soc.,* Vol 72, 1989, p 348

41. B.J.S. Wilkens and R. Dutton, *J. Am. Ceram. Soc.,* Vol 59, 1976, p 108

42. T.A. Michalske, *Fracture Mechanics of Ceramics,* Vol 5, R.C. Bradt, A.G. Evans, D.P.H. Hasselman, and F.F. Lange, Ed., Plenum Press, 1983, p 277–284

43. D.G. Roach, D.M. Heuckeroth, and B.R.

Lawn, *J. Colloid Interface Sci.,* Vol 114, 1986, p 292

44. H. Cao, B.J. Dalgleish, and A.G. Evans, *J. Am. Ceram. Soc.,* Vol 70, 1987, p 257

45. A.G. Evans and B.J. Dalgleish, *Advanced Materials for Severe Service Applications,* K. Iida and A.J. McEvily, Ed., Elsevier Applied Science, 1986, p 91–118

46. K. Jakus, C.E. Weigand, M.H. Godin, and S.V. Nair, *Ceram. Eng. Sci. Proc.,* Vol 10, 1989, p 1352

47. K. Jakus, S.M. Wiederhorn, and B.J. Hockey, *J. Am. Ceram. Soc.,* Vol 69, 1986, p 725

48. P. Becher, *J. Am. Ceram. Soc.,* Vol 74, 1991, p 255

49. R.F. Cook and D.R. Clarke, *Acta Metall.,* Vol 37, 1988, p 555

50. D.B. Marshall, *J. Am. Ceram. Soc.,* Vol 69, 1986, p 173

51. A.G. Evans, B.J. Dalgleish, M. He, and J.W. Hutchinson, *Acta Metall.,* Vol 37, 1989, p 3249

52. M.D. Thouless, *Thin Solid Films,* Vol 181, 1989, p 397

53. H.C. Cao, B.J. Dalgleish, and A.G. Evans, *Closed Loop,* Vol 17, 1989, p 18

54. J.E. Ritter, T.J. Lardner, L. Rosenfeld, and M.R. Lin, *J. Appl. Phys.,* Vol 66, 1989, p 3626

55. L.G. Rosenfeld, J.E. Ritter, T.J. Lardner, and M.R. Lin, *J. Appl. Phys.,* Vol 67, 1990, p 3291

56. J.E. Ritter, K. Conley, D. Steul, and T.J. Lardner, *Mater. Res. Soc. Symp. Proc.,* Vol 203, 1991, p 47

57. M.D. Thouless, H.C. Cao, and P.A. Mataga, *J. Mater. Sci.,* Vol 24, 1989, p 1406

Probabilistic Design of Ceramic Components with the NASA/ CARES Computer Program

Noel N. Nemeth and John P. Gyekenyesi, National Aeronautics and Space Administration, Lewis Research Center

THE UNIQUE PROPERTIES of advanced ceramics, such as high-temperature strength, environmental resistance, and low density, can contribute to greatly increased fuel efficiency and reduced emissions in aerospace and automotive engine applications. The potential use of these structural ceramics in high-temperature applications depends on their strength, toughness, and reliability. Components that use ceramics can be designed for high reliability in service, provided that the contributing factors causing material failure are defined. Consequently, research has focused on improving ceramic material processing and properties, as well as on establishing a sound design methodology.

The nature of ceramic failure is probabilistic, because of the variable severity of inherent flaws, and optimization of design requires the ability to accurately determine the reliability of a loaded component. Methods of quantifying this reliability and the corresponding failure probability have been investigated and refined by the authors. The result of this effort is a public-domain computer program known as CARES, or the Ceramics Analysis and Reliability Evaluation of Structures. This program was formerly known as SCARE, or Structural Ceramics Analysis and Reliability Evaluation (Ref 1). The CARES program (Ref 1–4) calculates the fast-fracture reliability of macroscopically isotropic ceramic components. These components may be subjected to complex thermomechanical loadings such as those found in heat-engine applications. The CARES framework will subsequently be built upon to include ceramic fatigue that is due to subcritical crack growth.

The design methodology used by CARES combines three major elements: (1) linear elastic fracture mechanics (LEFM) theory, which relates the strength of ceramics to the size, shape, and orientation of critical flaws; (2) extreme value statistics to obtain the characteristic flaw size distribution function, which is a material property; and (3) material microstructure. An inherent element of this design

procedure is that component integrity is a function of the entire field solution of the stresses and is not based only on the most highly stressed point. In addition, both the amount of the stressed material surface area and volume affect component strength (the so-called size effect).

Because ceramics fail at the weakest flaw, examination of their fracture surfaces can reveal the nature of the failure. Fractography of broken samples has shown that these flaws can be characterized as being either internal, or intrinsic, to the material volume (volume flaws) or extrinsic to the material volume (surface flaws). Intrinsic defects are a result of materials processing. Extrinsic flaws can result from grinding or other finishing operations, from chemical reaction with the environment, or from the internal defects intersecting the external surface. The different physical nature of these flaws results in dissimilar failure response to identical loading situations. Consequently, separate criteria must be employed to describe the effects of the applied loads on the component surface and volume.

Probabilistic component design requires the determination of the fracture strength distribution from simple-geometry flexural or tensile test specimens. The statistical material parameters are estimated as functions of temperature, specimen loading, specimen geometry, and type of defect. From these data the reliability for a complex component geometry and loading is then predicted. Appropriate design changes are made until an acceptable probability of failure has been reached.

Program Capability and Description

CARES, an integrated computer program written in FORTRAN 77, uses Weibull (Ref 5, 6) and Batdorf (Ref 7, 8) fracture statistics to predict the fast-fracture reliability of isotropic ceramic components. CARES has three primary functions:

- To analyze statistically the data obtained from the fracture of simple uniaxial tensile or flexural specimens
- To estimate the Weibull and Batdorf material parameters using these data
- To perform a fast-fracture reliability evaluation of a ceramic component experiencing thermomechanical loading

Component reliability is predicted using elastostatic finite-element analysis output from the MSC/NASTRAN or ANSYS computer programs. The CARES code and accompanying documentation can be obtained from the authors of this article. A PC-based version of CARES that performs statistical analysis and parameter estimation of specimen data only is also available (Ref 9).

The CARES code includes a number of fracture theories to predict material response due to multiaxial stresses. These methods are summarized in Table 1. The Batdorf method (Ref 8) is recommended because it couples LEFM with the Weibull weakest link theory (WLT) (Ref 5). The Weibull normal stress averaging method (Ref 6) and the principle of independent action (PIA) theories (Ref 10, 11) are included for comparison purposes and because of their previous popularity. All the fracture models available in CARES use the Weibull two-parameter probability of strength distribution.

Figure 1 shows the fracture criteria and crack geometries available to the user for both surface- and volume-distributed flaws. The PIA and Weibull normal stress averaging fracture theories do not require a crack geometry. Batdorf's fracture theory can be used with several different mixed-mode fracture criteria and crack geometries. For coplanar crack extension, CARES uses the total strain-energy release rate theory. Out-of-plane crack extension criteria are approximated by a simple semiempirical equation (Ref 12, 13). This equation involves a parameter that can be used to approximate various mixed-mode theories or experimental results. For comparison,

Table 1   Statistical fast-fracture models available with CARES

Weakest link fracture model	Size effect	Stress state effects	Computational simplicity	Theoretical basis
Weibull (1939)	Yes	Uniaxial	Simple	Phenomenological
Normal stress averaging (1939)	Yes	Multiaxial	Complex	Phenomenological
Principle of independent action (1967)	Yes	Multiaxial	Simple	Maximum principal stress theory
Batdorf: shear-insensitive (1974), shear-sensitive (1978)	Yes	Multiaxial	Complex	Linear elastic fracture mechanics

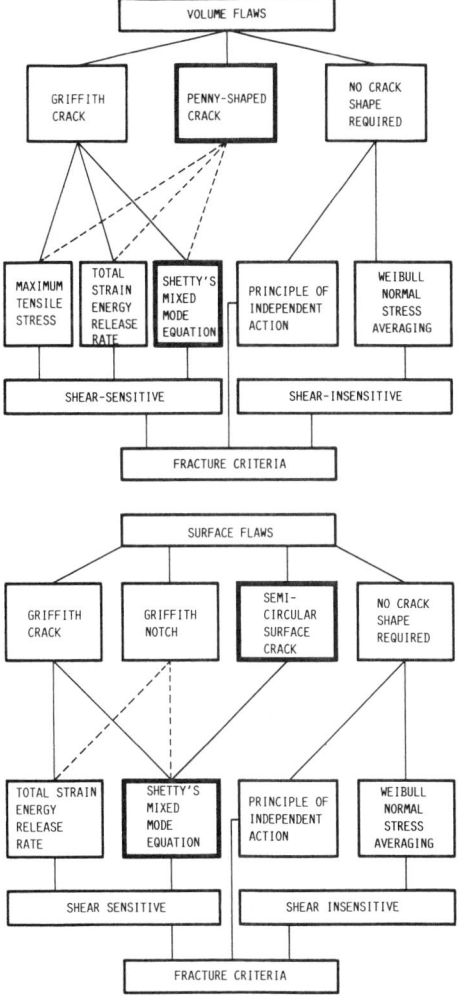

Fig 1 Available failure criteria and crack shapes (recommended criteria and shapes are highlighted)

Griffith's maximum tensile stress analysis for volume flaws is also included. The highlighted boxes in Fig 1 show the recommended fracture criteria and flaw shapes.

The statistical material parameters are obtained from the fracture stresses of many test specimens (ideally, 30 or more) of fixed geometry and loading. Solutions for the three- and four-point modulus of rupture (MOR) bending bar (Ref 14) and the pure tensile specimen (Ref 15) tested at a user-specified temperature have been incorporated in the CARES program. Because the statistical material parameters are a function of temperature, up to 20 data sets may be input at discrete temperature levels. Lagrangian polynomials are utilized to interpolate the parameter values at other temperatures. Each data set may consist of up to 200 specimen fracture stresses. Because each specimen can be identified by its mode of failure (volume flaw, surface flaw, or unknown), statistical parameters from competing failure modes can be estimated.

CARES can identify potential bad fracture stress data (outliers). The outlier test developed by Stefansky (Ref 16) and subsequently used by Neal, *et al.* (Ref 17) is employed. Although the technique is based on the normal distribution, and, therefore, its application to the Weibull distribution is not rigorous, it serves as a guideline to the user. The test can detect multiple outliers from a sample of up to 100 specimens at the 1%, 5%, or 10% significance levels. Data detected as outliers are flagged with a warning message, and any further action is left to the discretion of the user.

Weibull parameters are obtained as a function of the specimen surface and/or volume, as requested by the user. Biased estimates of the Weibull parameters are obtained from either least-squares analysis or the maximum-likelihood method for complete or censored samples (competing failure modes). CARES uses the Weibull log likelihood equations given by Nelson (Ref 18) and the rank increment adjustment method described by Johnson (Ref 19).

Because the estimates of statistical parameters are obtained from a finite amount of data, they contain an inherent uncertainty that can be characterized by bounds in which the true parameters are likely to lie. The *range* of these bounds is inversely related to the number of specimens. For the maximum-likelihood method, the 5- and 95-percentile confidence limits for the Weibull parameters are provided (Ref 20).

The ability of the probability distribution calculated from the Weibull parameters to reasonably fit the experimental data is measured with the Kolmogorov-Smirnov (K-S) and the Anderson-Darling (A-D) goodness-of-fit tests. These tests are discussed by D'Agostino and Stephens (Ref 21). The A-D test is provided because it is more sensitive to discrepancies at low and high probabilities of failure than the more commonly used K-S test. The Kanofsky-Srinivasan 90% confidence band values (Ref 22) about the Weibull line are given as an additional test of the fit of the data to the Weibull distribution.

Figure 2 illustrates the operational flow of the program. For a finite-element model, reliability calculations are performed at the element level, and the overall component reliability is then the product of all the element survival probabilities. Reliability due to the presence of volume flaws is calculated from the volume statistical material strength parameters and the output of the stresses, volumes, and temperatures from isoparametric brick, wedge, or axisymmetric finite elements. Reliability evaluation based on the presence of surface flaws is calculated from the surface statistical material strength parameters and the output of the two-dimensional surface stresses, areas, and temperatures from isoparametric quadrilateral and triangular shell elements. Solid elements are used for the structural modeling. Shell elements are only used to identify external surfaces of solid elements consistent with the component external boundaries that are required for the reliability analysis. CARES can compute reliability of finite-element

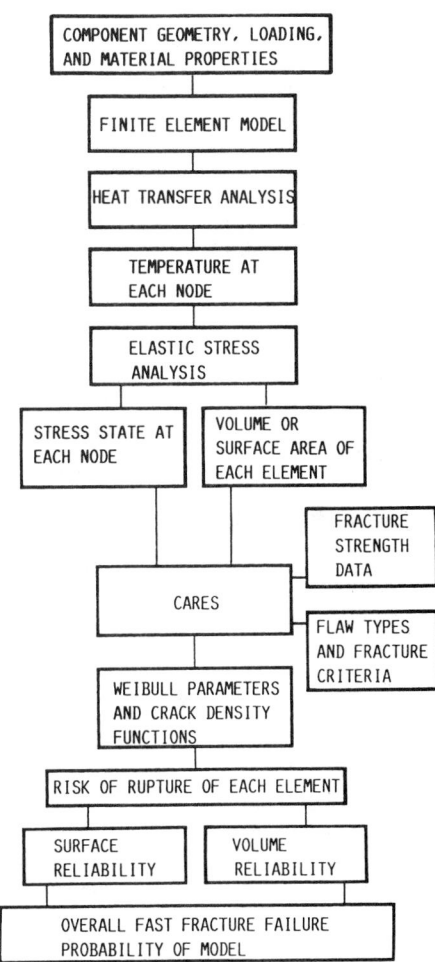

Fig 2 Block diagram for the analysis and reliability evaluation of ceramic components

models consisting of as many as 2000 solid and 2000 shell elements.

In CARES, if a principal compressive stress in an element is at least three times greater than the maximum principal tensile stress in absolute value, or if the maximum principal stress is compressive, then the corresponding element reliability is set equal to unity. Typically, brittle materials are much stronger in compression than in tension. It is assumed that the lower tensile strength limit will predominate over the higher compressive limit for a typical component design. If the compressive stresses are significant, they should be checked against limiting values by other methods. Note that when the PIA model is used, only tensile principal stresses can contribute to failure.

Provision is made in CARES to permit the use of cyclic symmetry modeling. CARES also has a multiple material capability, that is, a model can consist of up to 100 different materials (or up to 100 different statistical material characterizations). Elements not designated as brittle materials are ignored in the reliability computations. Temperature-dependent statistical material properties are interpolated at each individual element temperature. The risk-of-rupture intensity is also calculated for each element, and these values are sorted to determine the maximum values.

Two versions of the code, designated as CARES1 and CARES2, are available. The CARES1 version assumes that stress and temperature gradients within each element are negligible, and therefore, only element centroidal principal stresses are used in the reliability calculations. The CARES2 version takes into account element stress gradients by dividing each brick element into 27 subelements, and each quadrilateral shell element into 9 subelements. Subelement centroidal principal stresses are then computed and used in the subsequent reliability calculations. CARES2 enables the finite-element model to consist of fewer elements for the same level of convergence to the true solution as CARES1.

Input Requirements

To control the execution of the CARES program, an input file must be prepared. On the tape or disks provided with the program is a file called TEMPLET INP that can be used to construct an input file for a particular problem. Input to CARES is keyword driven. Data are input by the user under each keyword. Either an explanation of the input required or a list of input choices is provided next to the keyword.

The CARES program requires three categories of input: master control input; material control input, which includes temperature-dependent material data; and, optionally, MSC/NASTRAN or ANSYS output data files from finite-element analysis. The master control input is a set of control indices that directs overall program execution. The material con-

trol input consists of control indices and either the data required to estimate the statistical material parameters or direct input of the statistical parameter values themselves for various temperatures. This input category includes the choices of fracture criteria and flaw shapes shown in Fig 1. Both the master and the material control input are contained in the TEMPLET INP file. The third input category, MSC/NASTRAN or ANSYS output data files, includes finite-element analysis data files containing the element stresses, volumes/areas, and temperatures.

Output Information

The first part of the CARES output is an echo of the choices selected (or default values) from the master control input. If a finite-element model reliability analysis is not performed, then CARES proceeds to echo the material control input. If postprocessing of a finite-element model is done, then the centroidal, or subelement, principal stresses are listed for each element, with appropriate element area or volume and temperature. The printing of element stress tables in CARES is optional. In addition, two element cross-reference tables are printed. The first table lists the shell element number and gives the corresponding solid element to which it is attached. The second table lists the solid element identification number and up to six associated shell elements (for example, a brick element could have all of its six faces as external surfaces).

CARES echoes the user inputs for each section of the material control input. If statistical material parameters are not directly input, but are determined from experimental fracture data, then the output will identify the method of solution, the number of specimens in each batch, and the temperature of each test. In addition, the output echoes the sorted input values of all specimen fracture stresses with proper failure mode identification. Results from the statistical analysis of the fracture data are then printed.

The fracture strength and corresponding significance level are listed for detected outliers. Note that the lower the significance level, the more extreme is the deviation of the data point from the rest of the distribution. A 1% significance level indicates that there is a 1-in-100 chance that the data point is actually a member of the same population as the other data, assuming a normal distribution. The estimated statistical material parameters from least-squares or maximum likelihood analyses are listed following the results from the outlier test. The biased and unbiased values of the Weibull shape parameter, the specimen Weibull characteristic strength, the upper and lower bound values at 90% confidence level for both parameters, the specimen Weibull mean value, and corresponding standard deviation are printed for each temperature. For censored statistics, these values are generated

first for volume flaw analysis and subsequently for the surface-flaw analysis. It should be noted that not all the previously mentioned information is available for all methods of material parameter estimation.

The Kolmogorov-Smirnov (K-S) goodness-of-fit test is performed for each specimen fracture stress, and the corresponding K-S statistics D+, D−, and significance level are listed. Similarly, the K-S statistic for the overall sample set is printed, along with the significance level. This overall statistic is the absolute maximum of individual data D+ and D− factors. For the Anderson-Darling (A-D) goodness-of-fit test, the A-D statistic A^2 is determined for the overall population, and its associated significance level is printed. For the goodness-of-fit tests, the lower the significance level is, the worse the fit of the experimental data to the proposed distribution. For these tests, a 1% level of significance indicates that there is a 1-in-100 chance that the specimen fracture data were generated from the estimated distribution.

A table of data is generated to allow the user to construct Kanofsky-Srinivasan 90% confidence bands about the Weibull distribution. The table includes fracture stress data, the corresponding Weibull probability of failure values, the 90% upper and lower confidence band values about the Weibull line, and the median rank value for each data point.

The last table from the statistical analysis section of CARES summarizes the material parameters used in component reliability calculations. These parameters, which are listed as a function of temperature, include the biased Weibull modulus, the normalized Batdorf crack density coefficient, and the Weibull scale parameter or unit volume or unit area characteristic strength, depending on whichever is appropriate. The values printed correspond to the experimental temperatures input and five additional interpolated sets of values between each input temperature. The interpolated parameters are output for checking purposes. Information on the selected fracture criterion and the crack shape is printed as required.

If a component reliability analysis with finite-element data is being performed, then tables will be generated to summarize the reliability evaluation of each finite element. One table is provided for volume flaw analysis (solid elements), and one table is given for surface-flaw analysis (shell elements), as requested by the user. The tables list the element identification (ID) numbers and the corresponding element material ID, survival probability, failure probability, risk-of-rupture intensity (risk-of-rupture divided by element volume or area), and temperature-interpolated statistical material parameters. Following each table is a sorted list of the 15 most critical risk-of-rupture intensity values and corresponding element numbers. Also included is the probability of failure and survival for the component surface or volume, whichever is appropriate. Finally, the overall component probability of failure and the component probability of survival are printed.

Theory

The first probabilistic approach used to account for the scatter in fracture strength of brittle materials was introduced by Weibull (Ref 5, 6). This approach is based on the previously developed weakest link theory (WLT) (Ref 23–25), which is primarily attributed to Pierce, who proposed it while modeling yarn failure. The WLT is analogous to pulling a chain where catastrophic failure occurs when the weakest link in the chain breaks. The reliability of the chain is the product of the survival probabilities of the individual links.

Phenomenological observations indicate that monolithic ceramic failures behave in accordance with WLT. For a ceramic component that contains volume flaws and is loaded in uniaxial tension, the probability of survival is expressed as:

$$P_{sV} = \exp - \left[\int_V N_V(\sigma) dV \right] \qquad \text{(Eq 1)}$$

where V is the component volume and the subscript V denotes volume-dependent terms. The function $N_V(\sigma)$, referred to as the crack density function, represents the number of flaws per unit volume that have a strength equal to or less than the value of σ. The value of the integral is called the risk-of-rupture value.

Weibull introduced a three-parameter power function for the crack density function $N_V(\sigma)$:

$$N_V(\sigma) = \left(\frac{\sigma - \sigma_{uV}}{\sigma_{oV}} \right)^{m_V} \qquad \text{(Eq 2)}$$

where σ_{uV} is the threshold stress (location parameter), which is usually taken as zero for ceramics. The location parameter is the value of applied stress below which the failure probability is zero. When the location parameter is zero, the two-parameter Weibull model is obtained. The scale parameter σ_{oV} then corresponds to the stress level where 63.2% of specimens with unit volumes would fracture. The scale parameter has dimensions of stress(volume)$^{1/m}V$. The reciprocal of $\sigma_{oV}^{m_V}$ is called the uniaxial Weibull crack density coefficient, k_{wV}, and m_V is the shape parameter (Weibull modulus or Weibull slope), which is a measure of the degree of strength dispersion. These statistical parameters are material properties that are temperature and processing dependent. They are evaluated from uniaxial flexure or tensile specimen fracture data.

For surface-flaw-induced failure in ceramic structures, these expressions become a function of the component surface area, A. Herein the subscript S denotes analogous parameters that are a function of surface area.

To predict material response under multiaxial stress states, Weibull (Ref 6) proposed the normal stress averaging method. Although this approach is intuitively plausible, it is somewhat arbitrary. Subsequently, Barnett and Freudenthal (Ref 10, 11) proposed the PIA model. The PIA fracture theory is the weakest link statistical equivalent of the maximum stress failure theory and is only applicable for tensile states of stress. In the PIA model, the principal stresses $\sigma_1 \geq \sigma_2 \geq \sigma_3$ are assumed to act independently. If all principal stresses are tensile, then according to this approach, the probability of failure due to volume flaws is:

$$P_{fV} = 1 - \exp\left\{ -k_{wV} \int_V [\sigma_1^{m_V} + \sigma_2^{m_V} + \sigma_3^{m_V}] dV \right\}$$

$$\text{(Eq 3)}$$

Compressive principal stresses are assumed not to contribute to the failure probability. This equation yields nonconservative estimates of P_{fV} in comparison with the Weibull normal stress method. For surface-flaw analysis, σ_3 is zero and the integration in Eq 3 is evaluated about the component surface area to obtain P_{fS}.

The Weibull method of averaging the tensile normal stress and the PIA model have been the methods most widely used for brittle material design. However, they do not specify the nature of the defect causing failure. The Weibull normal stress averaging method and the Batdorf method with shear insensitive cracks yield identical reliability predictions.

Recognizing that brittle fracture is governed by LEFM, Batdorf (Ref 7, 8) proposed that reliability predictions should be based on a combination of the weakest link theory and fracture mechanics. Conventional fracture mechanics dictates that both the size of the critical crack and its orientation relative to the applied loads determine the fracture stress. In brittle ceramics, however, the small critical flaw size and the large number of flaws prevent determination of the critical flaw, let alone the determination of its size and orientation. What is calculated instead is the combined probability of the critical flaw being within a certain size range and being oriented so that it may cause fracture. Because flaw sizes correspond to strength levels, and because strength is easier to measure than size for these microscopic flaws, the probability of a crack existing within a critical strength range is determined.

The Batdorf theory assumes random flaw orientation and a consistent crack geometry. The component failure probability for volume flaws is expressed as:

$$P_{fV} = 1 - \exp\left[-\int_V \int_0^{\sigma_{e_{max}}} \frac{\Omega(\Sigma, \sigma_{cr})}{4\pi} \right.$$

$$\left. \frac{dN_V(\sigma_{cr})}{d\sigma_{cr}} d\sigma_{cr} dV \right] \qquad \text{(Eq 4)}$$

where σ_{cr}, the critical stress, is defined as the remote, uniaxial fracture strength of a given crack in mode I loading. The solid angle $\Omega(\Sigma, \sigma_{cr})$ is the area of a unit radius sphere that contains all the crack orientations for which $\sigma_e \geq \sigma_{cr}$ because of the existing stress state, Σ. The constant 4π is the surface area of a unit radius sphere and corresponds to a solid angle that contains all possible flaw orientations. The effective stress, σ_e, is defined as the equivalent mode I stress on the flaw. The limit of integration, $\sigma_{e_{max}}$, is the maximum effective stress. The Batdorf volume crack density function $N_V(\sigma_{cr})$, is approximated by the power function:

$$N_V(\sigma_{cr}) = k_{BV} \sigma_{cr}^{m_V} \qquad \text{(Eq 5)}$$

where the material Batdorf crack density coefficient k_{BV} and Weibull modulus m_V are evaluated from experimental uniaxial fracture data. In contrast to the Weibull coefficient, k_{wV}, which depends only on the specimen fracture data, the Batdorf coefficient requires a fracture criterion and crack shape.

The relationship between k_{BV} and k_{wV} is obtained by carrying out the integration in Eq 1 for a uniaxial stress state and equating the resultant failure probability to that of Eq 4. For shear insensitive cracks the result is:

$$k_{BV} = (2m_V + 1)k_{wV} = \bar{k}_{BV} k_{wV} \qquad \text{(Eq 6)}$$

Note that $\bar{k}_{BV}$ is the normalized Batdorf crack density coefficient for volume flaws. For shear-sensitive cracks, $\bar{k}_{BV}$ and $\bar{k}_{BS}$ are evaluated numerically (Ref 3) in the CARES program.

If a shear-insensitive condition is assumed, fracture occurs when $\sigma_n = \sigma_e \geq \sigma_{cr}$, where σ_n is the normal tensile stress on the flaw plane. However, for a flat crack, it is known from fracture mechanics analysis that a shear stress, τ, applied parallel to the crack plane (mode II or III) also contributes to fracture. Therefore, for polyaxial stress states, expressing the effective stress σ_e as a function of both σ_n and τ is more accurate than assuming shear insensitivity.

The equations derived by Batdorf and Heinisch (Ref 8) are based on self-similar (coplanar) crack extension. However, a flaw that experiences a multiaxial stress state usually undergoes crack propagation initiated at some angle to the flaw plane (noncoplanar crack growth). Shetty (Ref 13) performed experiments on polycrystalline ceramics and glass that considered crack propagation as a function of an applied far-field multiaxial stress state. He modified an equation proposed by Palaniswamy and Knauss (Ref 12) to empirically fit experimental data. This multimodal equation takes the form:

$$\frac{K_I}{K_{Ic}} + \left(\frac{K_\delta}{\bar{C} K_{Ic}} \right)^2 = 1 \qquad \text{(Eq 7)}$$

where K_I is the mode I stress intensity factor and K_δ is either K_{II} or K_{III}, whichever is dominant. The term $\bar{C}$ is a constant adjusted to best fit the data. Shetty (Ref 13) found a range of values of $0.80 \leq \bar{C} \leq 2.0$ for the materials he tested, which contained large induced flaws. As $\bar{C}$ increases, the response becomes progressively more insensitive to shear. Using this criterion and a penny-shaped crack, we obtain:

$$\sigma_e = \frac{1}{2}\left[\sigma_n + \sqrt{\sigma_n^2 + \left(\frac{4\tau}{\bar{C}(2-\nu)}\right)^2}\right] \quad (Eq\ 8)$$

where ν is Poisson's ratio.

For mixed-mode fracture that is due to surface flaws, the Batdorf failure probability equation is:

$$P_{fS} = 1 - \exp\left[-\int_A\int_0^{\sigma_{emax}}\frac{\omega(\Sigma,\sigma_{cr})}{2\pi}\right.$$

$$\left.\frac{dN_S(\sigma_{cr})}{d\sigma_{cr}}\,d\sigma_{cr}dA\right] \quad (Eq\ 9)$$

where $\omega(\Sigma,\sigma_{cr})$ is the total arc length on a unit radius circle on which the projection of the equivalent stress satisfies $\sigma_e \geq \sigma_{cr}$, and 2π is the perimeter of the circle. The cracks are assumed to be randomly oriented in the plane of the external boundary, with their planes normal to the surface.

The Batdorf surface crack density function $N_S(\sigma_{cr})$ has the same form as Eq 5. Following a similar procedure used to obtain Eq 6, the surface-flaw Batdorf crack density coefficient for shear-insensitive cracks is expressed as (Ref 26, 27):

$$k_{wpS} = \frac{m_S\Gamma(m_S)\sqrt{\pi}}{\Gamma\left(m_S + \frac{1}{2}\right)}k_{wS} = \bar{k}_{BS}k_{wS} \quad (Eq\ 10)$$

where Γ is the gamma function.

For noncoplanar crack growth and a semi-circular crack, we obtain for the effective stress:

$$\sigma_e = \frac{1}{2}\left[\sigma_n + \sqrt{\sigma_n^2 + 3.301\left(\frac{\tau}{\bar{C}}\right)^2}\right] \quad (Eq\ 11)$$

where, again, $\bar{C}$ is adjusted to best fit the data.

Selected statistical theories and equations for parameter evaluation are explained in detail in Ref 27. Typically, for brittle materials, the Weibull parameters are determined from simple specimen geometry and loading conditions, such as beams under flexure and either cylindrical or flat specimens under uniform uniaxial tension. The flexural test failure probability can be expressed in terms of the extreme fiber stress, σ_f, or modulus of rupture (MOR) using the two-parameter Weibull form:

$$P_f = 1 - \exp\left[-\left(\frac{\sigma_f}{\sigma_\theta}\right)^m\right] \quad (Eq\ 12)$$

where m is the Weibull modulus and σ_θ is the volume or area specimen characteristic strength (characteristic modulus of rupture, MOR_o). The characteristic strength σ_θ is defined as the uniform stress or extreme fiber stress at which the probability of failure is 0.6321. From the experimental fracture stresses, σ_f, and respective fracture origins, the Weibull parameters m and σ_θ are estimated using

maximum likelihood or least-squares analysis. Goodness-of-fit testing, calculation of confidence limits, and determination of confidence bands are also obtained from these data. Note that m and σ_θ are either volume-dependent or surface-flaw-dependent terms. The outlier test only requires σ_f for input data.

The Weibull scale parameter, σ_o, for volume and surface cracks is determined from σ_θ, m, the specimen geometry, and specimen loading. For volume flaws and the four-point MOR bar geometry, the scale parameter is determined by performing the integration in Eq 1 and equating the result with Eq 12:

$$\sigma_{oV} = \sigma_{\theta V}\left[\frac{wh}{2}\frac{(L_1 + m_vL_2)}{(m_v+1)^2}\right]^{1/m_V}$$

$$= \sigma_{\theta V}[V_e]^{1/m_V} \quad (Eq\ 13)$$

where L_1 is the outer load span, L_2 is the inner load span, w is the specimen width, and h is the specimen height. The term V_e is the effective volume. For uniaxial tensile loading, the effective volume is equal to the gage volume, V_g, which is the uniformly stressed region where fracture is expected to occur.

For surface flaws and the four-point MOR bar geometry, integrating Eq 1 about the specimen surface and equating it to Eq 12 yield the scale parameter:

$$\sigma_{oS} = \sigma_{\theta S}\left\{\left[\frac{(L_2/L_1)m_S + 1}{(m_S + 1)^2}\right]\right.$$

$$\left.\left(\frac{m_Sw}{w+h}+1\right)(w+h)L_1\right\}^{1/m_S}$$

$$= \sigma_{\theta S}(A_e)^{1/m_S} \quad (Eq\ 14)$$

where A_e is the effective area. For uniaxial tensile loading, the effective area is equal to the specimen gage area, A_g, which is the total specimen surface area of interest. Note in Eq 13 and 14 that when $L_2 = 0$, the solution for the three-point MOR bar is obtained.

Example 1: Statistical Material Parameter Estimation. To illustrate the methods used to estimate statistical material parameters, results from the fracture of four-point bend bars broken at the Lewis Research Center of the National Aeronautic and Space Administration (NASA) and analyzed by CARES are compared to results independently obtained by Bruckner-Foit and Munz (Ref 28) for the International Energy Agency (IEA) Annex II agreement (Ref 29). Two different materials were analyzed, namely, a hot isostatic pressed (HIP) silicon carbide (SiC) from Elektroschmelzwerke Kempten (ESK), Germany, and a HIP silicon nitride (Si_3N_4) from ASEA CERAMA, Sweden.

The flexure bars were distributed by Oak Ridge National Laboratory (ORNL) to the five participating U.S. laboratories, including NASA Lewis. The bars were fractured at these laboratories, and the fracture stress data sets were returned to ORNL as complete data without censoring for different failure modes. The number of specimens of a particular

material given to each U.S. participant was 80.

The results of the 80 SiC flexure bars tested at NASA Lewis were analyzed using the CARES code to calculate the least squares and maximum likelihood estimates (MLEs) of the Weibull parameters. The MLE values from CARES for a complete sample are compared in Table 2 to the values obtained from Ref 28. The SiC fracture data are plotted in Weibull coordinates in Fig 3, along with the proposed Weibull line and the Kanofsky-Srinivasan 90% confidence bands. Because no outliers were detected, all the data are within the 90% bands, and the goodness-of-fit significance levels are high, it is concluded that the fracture data show good Weibull behavior.

For the ASEA CERAMA hot isostatically pressed Si_3N_4 bars fractured at NASA Lewis and subsequently analyzed as a complete sample, the statistical material parameters that were estimated using CARES and Ref 28 are shown in Table 3. The comparison of the MLEs with Ref 28 is very good. When analyzed by CARES as a complete sample, the significance levels of 54% and 35% from the K-S and A-D goodness-of-fit tests, respectively, were relatively low, indicating a questionable fit to the proposed Weibull distribution. From the outlier test, the highest strength fracture stress was detected to be an outlier at the 1% significance level. Several of the lower strengths were flagged as outliers at various significance levels (1, 5, or 10%). Figure 4 shows a Weibull plot of the data. In this figure, it appears that the data are bimodal, with an outlier point at the highest strength.

Because of the observed trends, the data were reanalyzed, assuming a censored distribution and removing the highest strength value ($\sigma_f = 817.2$ MPa, or 118.5 ksi) as bad data. Although it is possible that both failure modes were surface induced, for the sake of this example it is assumed that the low-strength failures were predominantly due to volume flaws, and that the high-strength specimens predominantly fractured because of surface flaws. Because results from fractography of the individual specimens to identify the various failure modes were not available, the fracture origins had to be arbitrarily assigned prior to parameter estimation. Note that identification of individual specimen flaw origins is especially important for small sample sizes where a plot of the data does not yield clear trends. However, for the NASA Lewis Si_3N_4 data, the sample size is large, and clear trends can be observed, although extra care is required to determine if the trends are surface- or volume-flaw based.

Two censored distributions were analyzed. The first was based on an inspection of Fig 4, where the lowest nine strength values were assumed to be due to volume flaws, and the remainder to be due to surface flaws. The second censored distribution assumed 13 vol-

Table 2   Weibull parameters, confidence limits, Kolmogorov-Smirnov (K-S), and Anderson-Darling (A-D) test results for silicon carbide four-point bend bar fracture data

All estimates are biased estimates: 80 samples per material; complete sample analysis.

Method of analysis	Source of data	Shape parameter, m	90% confidence limits on m		Characteristic strength, σ_θ		90% confidence limits on σ_θ				K-S test statistic, D	K-S test significance level (α), %	A-D test significance level (α), %
							Upper		Lower				
			Upper	Lower	MPa	ksi	MPa	ksi	MPa	ksi			
Maximum likelihood	CARES	6.48	7.38	5.52	556	80.6	573	83.1	539	78.2	0.070	83	86
Maximum likelihood	Ref 28	6.59	7.65	5.61	556	80.6	574	83.3	539	78.2	0.063	...	...
Least squares	CARES	6.59	...	...	555	80.5	...	...	...	...	0.071	81	83

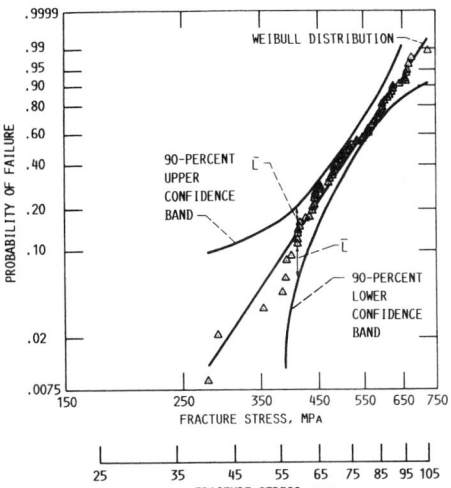

Fig 3 Weibull distribution and 90% confidence bands determined from fracture stress data for ESK hot isostatically pressed silicon carbide four-point bend bars broken at NASA Lewis (not all data points shown)

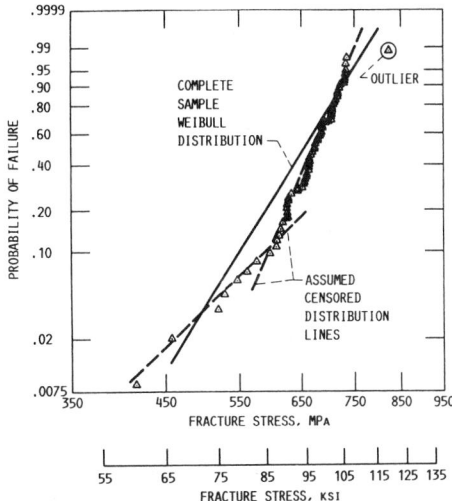

Fig 4 Complete sample, assumed censored sample Weibull distributions, and outlier determined from fracture stress data for ASEA CERAMA hot isostatically pressed silicon nitride four-point bend bars broken at NASA Lewis (not all data points shown)

ume flaws, where the particular volume flaw fracture strengths were assigned such that the MLEs more closely fit the experimental data.

Results of this analysis are shown in Table 3. The parameter estimates and goodness-of-fit statistics for the 80-specimen complete sample are provided, followed by a complete and censored sample analysis of the data with the outlier removed. The table shows the MLEs, along with the least square estimates.

When the goodness-of-fit scores are used as the basis of judgment, the censored sample that was estimated using the maximum-likelihood analysis and 13 volume flaws gives the best fit to the data. The high scores of 78% and 88% for the K-S and A-D tests, respec-

tively, indicate good Weibull behavior. Improvements in the goodness-of-fit scores may be gained by correctly identifying the location of fracture origins.

It should be noted from Fig 4 that the assumed volume-flaw distribution dominates the failure response at low probabilities of failure. In component design, therefore, it is essential to properly account for competing failure modes. Otherwise, nonconservative design predictions can result.

Example 2: Predicting the Reliability of Ceramic Components. Failure predictions that used the available fracture models from CARES for both volume and surface analysis were compared to failure predictions obtained by Swank and Williams (Ref 30) for a silicon nitride annular disk rotating at various speeds (Fig 5). The Weibull material parameters were independently evaluated from four-point MOR bar tests using a total of 85 specimens. The Weibull modulus, m, was 7.65, and the characteristic modulus of rupture, MOR_o, was 808 MPa (117 ksi). If fracture was assumed to be due to volume flaws, then from Eq 13, $\sigma_{oV} = 75$ MPa $\cdot$ m$^{0.3922}$ (45.8 ksi $\cdot$ in.$^{0.3922}$); if fracture was assumed to be due to surface flaws, then from Eq 14, $\sigma_{oS} = 232$ MPa $\cdot$ m$^{0.2614}$ (87.9 ksi $\cdot$ in.$^{0.2614}$). Swank and Williams assumed that both the bars and disks broke because of volume flaws.

Seven disks were fracture tested, and the Weibull modulus of 4.95 for the experimental disk was considerably different from the

Table 3   Weibull parameters, Kolmogorov-Smirnov (K-S), and Anderson-Darling (A-D) test results for silicon nitride four-point bend bar fracture data

Method of analysis	Source of data	Assumed distribution	Sample size	Shape parameter, m	Characteristic strength, σ_θ		K-S test statistic, D	K-S test significance level (α), %	A-D test significance level (α), %
					MPa	ksi			
Maximum likelihood	CARES	Surface	80	13.39	686	99.5	0.0901	54	35
Maximum likelihood	Ref 28	Surface	80	13.40	686	99.5	0.088	...	...
Least squares	CARES	Surface	80	11.74	691	100.2	0.128	15	11
Maximum likelihood	CARES	Surface	79	16.22	683	99.1	0.078	73	56
Least squares	CARES	Surface	79	11.98	688	99.8	0.124	18	10
Maximum likelihood	CARES	Volume	9	4.13	1128	163.6	0.081	69	58
		Surface	70	22.81	692	100.4	0.081	69	58
Least squares	CARES	Volume	9	6.74	830	120.4	0.112	28	35
		Surface	70	22.93	691	100.2	0.112	28	35
Maximum likelihood	CARES	Volume	13	6.79	876	127.1	0.074	78	88
		Surface	66	21.00	693	100.5	0.074	78	88
Least squares	CARES	Volume	13	6.84	864	125.3	0.085	62	38
		Surface	66	15.87	697	101.1	0.085	62	38

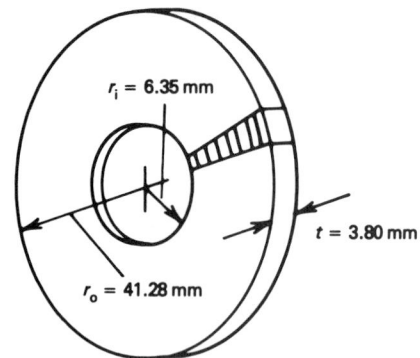

Fig 5 Rotating annular disk with 15° sector finite-element mesh containing eight brick elements (not to scale). Material was NC-132 hot pressed Si_3N_4; rpm range, 70,000 to 114,000.

7.65 value based on MOR specimen data. A better agreement between disk and MOR Weibull slopes would lead to improved predictions in failure probabilities for the entire data range. Estimating parameters from small sample sets greatly increases potential deviation from the true population parameters. Confidence limits are used to measure the intrinsic uncertainties in parameter estimates from finite sample sizes. If potential experimental errors are excluded, it is possible that the rotating disks and the MOR bars broke because of the same flaw population because their 90% confidence limits overlap between 6.56 and 6.98.

It should be noted that Swank and Williams (Ref 30) also spin tested and performed a volume-flaw reliability analysis on a contoured hub and a turbine blade ring geometry. Swank and Williams found that the Weibull moduli obtained from MOR bars cut from the hub and blade ring were in good

agreement with the Weibull moduli obtained from the hub and blade ring spin tests, respectively. They also noted that the material parameters used for the annular disk reliability analysis were obtained under less tightly controlled conditions than those used with the other geometries.

All reliability calculations were done with brick and quadrilateral shell elements. Because of symmetry, only 8 solid and 18 shell elements were used in one 15° sector for the model of the disk (Fig 5). Only one element spans both the thickness and circumferential directions. The shell elements were attached to solid elements consistent with the model external surfaces.

Experimental results are plotted in Fig 6, along with shear-insensitive (normal stress) and various shear-sensitive predictions from CARES2 for the volume-flaw analysis. Results were compared to data calculated by Swank and Williams, who used the Weibull normal stress averaging method and linear axisymmetric finite elements. The agreement between failure predictions is good; the small discrepancy is probably due to the different stress-volume data used in solving the reliability problem.

When using the Batdorf model with CARES, two methods are available to the user to describe the material shear sensitivity. With the input parameter, IKBAT, set equal to zero, k_B is calculated using only the σ_n stress component on the crack plane. This option assumes that mode I fracture is intrinsic to uniaxial loading. When IKBAT is set equal to one, then k_B is calculated using the user-selected fracture criterion and crack geometry. These two methods yield opposite trends relative to the shear-insensitive criterion, as shown in Fig 6. When IKBAT is set equal to zero, the subsequent reliability predictions are more con-

servative. The value of IKBAT is chosen so as to best fit the reliability predictions to the experimental data.

For this example, the laboratory measurements agree best with the shear-sensitive fracture models, using the IKBAT = 0 option. Note that results are given for an approximation of the maximum strain-energy release rate criterion, G_{max}, using a Griffith crack $\bar{C}$ = 0.82 and IKBAT = 1. Failure probabilities calculated from decreasing shear-sensitive effective stress equations move the probability of failure curves toward the shear-insensitive case. It is observed that Shetty's criterion for $\bar{C}$ = 0.80 and IKBAT = 0 with the penny-shaped crack gives the best agreement with experimental data.

The disks were reanalyzed assuming that fractures in the MOR bars, as well as the disks, were caused by surface flaws. Selected results are shown in Fig 7. The same trends are observed for both the shear-sensitive results and volume analysis results. However, for a given speed, failure probabilities are significantly less than those obtained by the volume-flaw analysis for all fracture models, indicating that material failure was most likely due to volume flaws. The main reason for the decreased failure estimates is the much higher equivalent surface Weibull scale parameter, σ_{oS}. The importance of postmortem fractography to identify the nature of the fracture-causing flaws is evident from the two different sets of answers in Fig 6 and 7.

In a more recent experiment involving alumina, Chao and Shetty (Ref 31) determined the failure strengths of three- and four-point MOR bars and disks in biaxial flexure. The disk specimens were supported by 40 freely rotating ball bearings located at a constant radius, and the disks were transversely loaded by uniform pressure. The bend bar and disk

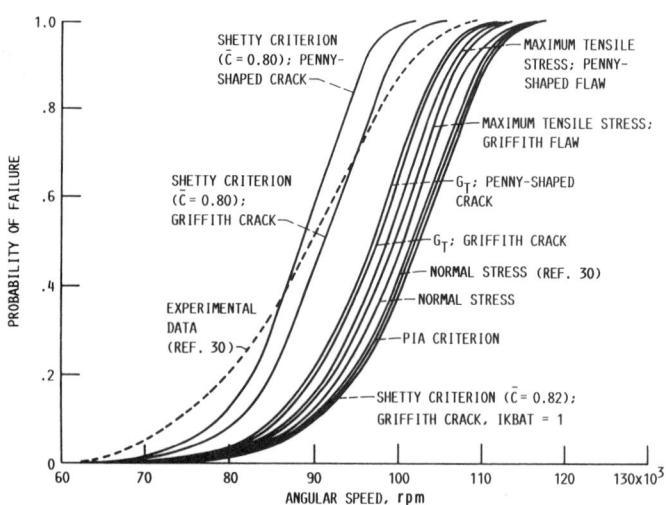

Fig 6 Comparison of experimental failure probabilities with those for various fracture models for a rotating annular disk, based on volume-flaw analysis, where m_V = 7.65; σ_{oV} = 75 MPa · $m^{0.3922}$; for IKBAT = 0, $\bar{k}_{BV}$ = 16.30; for IKBAT = 1, $\bar{k}_{BV}$ = 2.99 (only for Griffith crack, Shetty criterion, $\bar{C}$ = 0.82). G_T is the total strain energy release rate criterion.

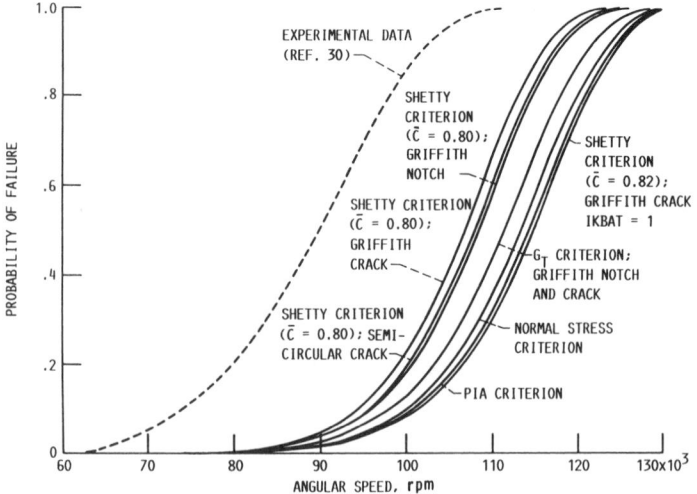

Fig 7 Comparison of experimental failure probabilities with those for various fracture models for a rotating annular disk, based on surface-flaw analysis, where m_S = 7.65; σ_{oS} = 232 MPa · $m^{0.2614}$; for IKBAT = 0, $\bar{k}_{BS}$ = 4.49; for IKBAT = 1, $\bar{k}_{BS}$ = 1.76 (only for Griffith crack, Shetty criterion, $\bar{C}$ = 0.82). G_T is the total strain energy release rate criterion.

specimens were made from the same powder lot, using the same pressing and sintering conditions.

The Weibull size effect was assessed in the three- and four-point bend tests, whereas the stress-state effect was evaluated from the biaxial flexure induced on the disks. Fractography was performed on these specimens and it was determined that all failures occurred on the material tensile surface. Careful surface preparation ensured that material strength was isotropic.

Table 4 shows the estimated Weibull parameters from least squares and maximum-likelihood analyses. The results show a consistently uniform Weibull modulus and reasonably high goodness-of-fit significance levels. Detected outliers were ignored.

The subsequent reliability analysis of the disks and bars was based on the MLEs obtained from four-point bend bar specimens. From Eq 14, the scale parameter for surface flaws was $\sigma_{oS} = 238.2$ MPa $\cdot$ m$^{2/m_S}$ (47.1 ksi $\cdot$ in.$^{2/m_S}$). A refined mesh of a 7.5° disk segment was prepared with shell and solid elements, and the results from the subsequent finite-element analysis compared well with strain gage measurements on the disks. Figure 8 is a Weibull plot of the fracture data and reliability predictions from CARES for the disks and MOR bars. As can be observed, the size effect is properly accounted for. Contrary to the rotating disk example, the normal stress criterion prediction of the disks is conservative in this case. It was found that the PIA criterion and the Batdorf method with the G_{max} fracture criterion and a semicircular crack ($\bar{C} = 0.82$; IKBAT = 1) match the disk data very well.

The significance of the pressure-loaded disk example and the rotating annular disk example is that different materials may not behave in a similar manner in multiaxial stress states. The Batdorf method has the flexibility to model the variety of responses observed,

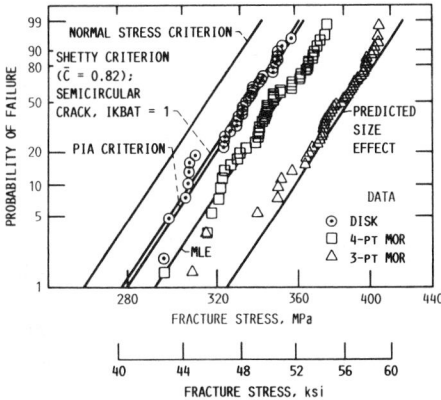

Fig 8 Weibull plot of fracture stresses compared to various fracture models (surface-flaw analysis) for an alumina ceramic tested in three-point, four-point, and biaxial flexure

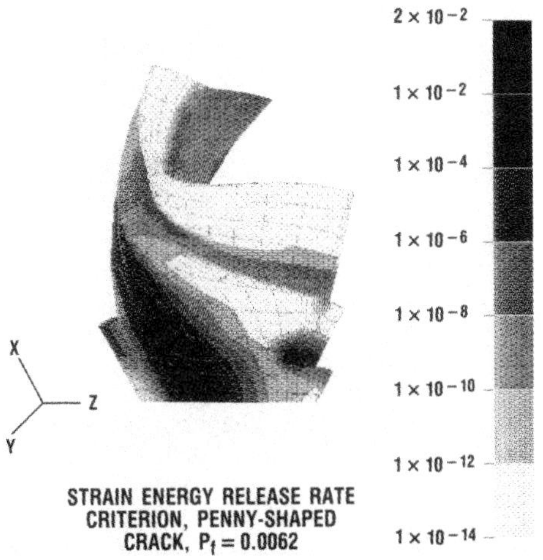

Fig 9 Mixed-flow rotor volume-flaw-based element risk-of-rupture intensities for the total strain-energy release rate criterion (G_T) with a penny-shaped crack geometry and IKBAT = 0

STRAIN ENERGY RELEASE RATE CRITERION, PENNY-SHAPED CRACK, $P_f = 0.0062$

unlike the PIA and Weibull normal stress methods.

The CARES code is used by numerous companies worldwide in the automotive, aerospace, nuclear, and computer software fields. At NASA Lewis, CARES has been used for the preliminary design of a silicon nitride mixed-flow rotor for application in small, high-temperature engines. A single blade and a section of the rotor hub were analyzed using the cyclic symmetry option of MSC/NASTRAN.

In Fig 9, the element risk-of-rupture intensities are plotted from the CARES volume-flaw analysis for the total strain-energy release rate criterion using IKBAT = 0 and the penny-shaped crack geometry. Note that the risk-of-rupture intensity is independent of the individual element geometry, unlike the probability of failure, and provides the designer with a means to visualize the critical stressed regions. This design was optimized to yield

a low probability of failure. It is observed in Fig 9 that the most critical region is at the rotor hub.

REFERENCES

1. J.P. Gyekenyesi, SCARE: A Postprocessor Program to MSC/NASTRAN for the Reliability Analysis of Structural Ceramic Components, *J. Eng. Gas Turbines Power*, Vol 108 (No. 3), July 1986, p 540–546
2. J.P. Gyekenyesi and N.N. Nemeth, Surface Flaw Reliability Analysis of Ceramic Components with the SCARE Finite Element Postprocessor Program, *J. Eng. Gas Turbines Power*, Vol 109 (No. 3), July 1987, p 274–281
3. N.N. Nemeth, J.M. Manderscheid, and J.P. Gyekenyesi, "Ceramics Analysis and

Table 4   Weibull parameters, confidence limits, Kolmogorov-Smirnov (K-S), and Anderson-Darling (A-D) test results for alumina bend bar and disk fracture data

Specimen	Method of analysis	Shape parameter, m	90% confidence limits on m Upper	90% confidence limits on m Lower	Characteristic strength, σ_θ MPa	Characteristic strength, σ_θ ksi	90% confidence limits on σ_θ Upper MPa	90% confidence limits on σ_θ Upper ksi	90% confidence limits on σ_θ Lower MPa	90% confidence limits on σ_θ Lower ksi	K-S test statistic, D	K-S test significance level (α), %	A-D test statistic, A^2	A-D test significance level (α), %	Sample size, N	Outliers MPa	Outliers ksi
Three-point bend bar	Maximum likelihood	25.42	29.91	20.47	385.9	56.0	389.9	56.6	382.0	55.4	0.0916	81.5	0.4316	81.7	48	307.7(a) 314.1(a)	44.6(a) 45.6(a)
	Least squares	20.41	...	...	387.0	56.1	...	...	...	...	0.1024	69.5	0.6718	58.4	48	307.7(a) 314.1(a)	44.6(a) 45.6(a)
Four-point bend bar	Maximum likelihood	23.79	27.99	19.15	353.4	51.3	357.3	51.8	349.6	50.7	0.1205	48.8	0.5379	70.9	48	294.8(b)	42.8(b)
	Least squares	23.34	...	...	353.4	51.3	...	...	...	...	0.1171	52.5	0.5013	74.6	48	294.8(b)	42.8(b)
Disk in flexure	Maximum likelihood	22.25	26.81	17.12	338.7	49.1	343.5	49.8	334.1	48.5	0.0858	99.0	0.3591	88.9	35	None	None
	Least squares	22.56	...	...	338.8	49.1	...	...	...	...	0.0877	99.0	0.3567	89.1	35	None	None

(a) At significance level of 1%. (b) At significance level of 5%.

Reliability Evaluation of Structures (CARES) User's and Programmer's Manual," NASA TP-2916, National Aeronautics and Space Administration, 1990

4. A. Pintz, G.H. Abumeri, and J.M. Manderscheid, "NASA/CARES Programs using ANSYS as a Pre-Processor Program," NASA Lewis Research Center, July 1989

5. W. Weibull, A Statistical Theory for the Strength of Materials, *Ing. Vetenskaps Akad. Handlinger*, No. 151, 1939

6. W. Weibull, The Phenomenon of Rupture in Solids, *Ing. Vetenskaps Akad. Handlinger*, No. 153, 1939

7. S.B. Batdorf and J.G. Crose, A Statistical Theory for the Fracture of Brittle Structures Subjected to Nonuniform Polyaxial Stresses, *J. Appl. Mech.*, Vol 41, June 1974, p 459–464

8. S.B. Batdorf and H.L. Heinisch, Jr., Weakest Link Theory Reformulated for Arbitrary Fracture Criterion, *J. Am. Ceram. Soc.*, Vol 61, 1978, p 355–358

9. S.A. Szatmary, J.P. Gyekenyesi, and N.N. Nemeth, "Calculation of Weibull Strength Parameters, Batdorf Flaw Density Constants and Related Statistical Quantities using PC-CARES," NASA TM-103247, National Aeronautics and Space Administration, 1990

10. R.L. Barnett, *et al.*, "Fracture of Brittle Materials under Transient Mechanical and Thermal Loading," AFFDL-TR-66–220, U.S. Air Force Flight Dynamics Laboratory, 1967

11. A.M. Freudenthal, Statistical Approach to Brittle Fracture, *Fracture, An Advanced Treatise*, Vol 2, *Mathematical Fundamentals*, H. Leibowitz, Ed., Academic Press, 1968, p 591–619

12. K. Palaniswamy and W.G. Knauss, On the Problem of Crack Extension in Brittle Solids under General Loading, *Mech. Today*, Vol 4, 1978, p 87–148

13. D.K. Shetty, Mixed-Mode Fracture Criteria for Reliability Analysis and Design with Structural Ceramics, *J. Eng. Gas Turbines Power*, Vol 109, July 1987, p 282–289

14. F.I. Baratta, W.T. Matthews, and G.D. Quinn, "Errors Associated with Flexure Testing of Brittle Materials," AD-A187470, U.S. Army Materials Technology Laboratory, 1987

15. K.C. Liu and C.R. Brinkman, Tensile Cyclic Fatigue of Structural Ceramics, *Proceedings of the Twenty-Third Automotive Technology Development Contractors' Coordination Meeting*, Society of Automotive Engineers, 1986, p 279–284

16. W. Stefansky, Rejecting Outliers in Factorial Designs, *Technometrics*, Vol 14 (No. 2), 1972, p 469–479

17. D. Neal, M. Vangel, and F. Todt, Statistical Analysis of Mechanical Properties, *Engineered Materials Handbook*, Vol 1, *Composites*, ASM International, 1987, p 302–307

18. W. Nelson, *Applied Life Data Analysis*, Wiley, 1982, p 333–395

19. L.G. Johnson, *The Statistical Treatment of Fatigue Experiments*, Elsevier, 1964, p 37–41

20. D.R. Thoman, L.J. Bain, and C.E. Antle, Inferences on the Parameters of the Weibull Distribution, *Technometrics*, Vol 11, 1969, p 445–460

21. R.B. D'Agostino and M.A. Stephens, *Goodness-of-Fit Techniques*, Marcel Dekker, 1986, p 97–193

22. P. Kanofsky and R. Srinivasan, An Approach to the Construction of Parametric Confidence Bands on Cumulative Distribution Functions, *Biometrika*, Vol 59 (No. 3), 1972, p 623–631

23. F.T. Pierce, Tensile Tests for Cotton Yarns, V. The 'Weakest Link'-Theorems on the Strength of Long and of Composite Specimens, *J. Textile Inst.*, Vol 17, 1926, p T355-T368

24. T.A. Kontorova, A Statistical Theory of Mechanical Strength, *J. Tech. Phys. (USSR)*, Vol 10, 1940, p 886–890

25. J.I. Frenkel and T.A. Kontorova, A Statistical Theory of the Brittle Strength of Real Crystals, *J. Phys. (USSR)*, Vol 7, 1943, p 108–114

26. B. Gross and J.P. Gyekenyesi, Weibull Crack Density Coefficient for Polydimensional Stress States, *J. Am. Ceram. Soc.*, Vol 72 (No. 3), Mar 1989, p 506–507

27. S.S. Pai and J.P. Gyekenyesi, "Calculation of the Weibull Strength Parameters and the Batdorf Flaw-Density Constants for Volume- and Surface-Flaw-Induced Fracture in Ceramics," NASA TM-100890, National Aeronautics and Space Administration, 1988

28. A. Bruckner-Foit and D. Munz, *Statistical Analysis of Flexure Strength Data*," Institut für Zuverlässigkeit und Schadenskunde in Maschinenbau, Universität Karlsruhe, Germany, 1988

29. V.J. Tennery, "IEA Annex II Management, Subtask 4 Results. Ceramic Technology for Advanced Heat Engines Project," ORNL/TM-10469, Semiannual Progress Report Oct 1986-Mar 1987, Oak Ridge National Laboratory, 1987

30. L.R. Swank and R.M. Williams, Correlation of Static Strengths and Speeds of Rotational Failure of Structural Ceramics, *Am. Ceram. Soc. Bull.*, Vol 60, No. 8, 1981, p 830–834

31. L.Y. Chao and D.K. Shetty, Reliability Analysis of Structural Ceramics Subjected to Biaxial Flexure, *J. Am. Ceram. Soc.*, Vol 74, 1991, p 333–344

Advanced Statistical Concepts of Fracture in Brittle Materials

C.A. Johnson and W.T. Tucker, General Electric Corporate Research and Development

CERAMICS are increasingly being considered as load-bearing structural materials that are suitable for use in both ambient and elevated temperature applications. These applications are driven by the need for specific combinations of critical properties (mechanical, thermal, chemical, optical, electrical, magnetic, and others) that are often only available in ceramic materials. The function of a load-bearing component, of course, is threatened by fracture from stresses generated during service. Therefore, an understanding of the unique characteristics of fracture in ceramics is important for successful implementation.

Many of the unique mechanical properties of ceramics can be traced to the difficulty of dislocation motion in their typical ionic and covalent bonded crystal structures (Ref 1). The absence of dislocation motion and the associated absence of inelastic deformation cause most monolithic ceramics to fail in a brittle manner, with relatively low energy absorption. With only elastic deformation available to accommodate crack tip stresses, conditions of crack stability are defined by the Griffith/Orowan criteria, where the fracture stress is inversely proportional to the square root of the flaw size. This inverse square-root relationship, combined with the nonuniform flaw sizes that are inevitable in ceramics, results in two common observations. First, multiple observations of fracture strength in otherwise similar specimens generally include a great deal of scatter or variability about the average strength. Second, the average strength is a function of specimen size such that larger specimens are lower in strength than smaller specimens.

Quantitative descriptions of the variability and size dependence of strength are necessary to design structural ceramic components, to predict component reliability, and to compare the merits of different ceramics for a given application. The purpose of this article is to review some recent advances in this area of fracture statistics.

Distributional Models

Quantitative descriptions of the statistical nature of fracture in brittle materials rely on a distribution function to define the cumulative probability of failure, P, as a function of all dependent variables that influence failure, such as stress level and component size and geometry. The majority of distribution functions considered for this purpose are based on the weakest link concept, which simply states that the entire body will fail when the stress at any defect is sufficient for unstable crack propagation of that defect. Examples of applicable weakest link distribution functions include the Weibull distribution (Ref 2, 3) and extreme value distributions (Ref 4).

The Weibull distribution has become very popular in descriptions of ceramic fracture for a number of reasons. The function is mathematically simple and easy to manipulate, and it is a weakest link distribution with the proper "limiting conditions." But most importantly, the distribution has been successful in describing a great deal of fracture data representing many different materials from numerous investigations. In many respects, the Weibull distribution is to weakest link problems what the Gaussian distribution is to situations where the central limit theorem applies. For these reasons, the remainder of this article concentrates on the Weibull distribution.

In his 1939 paper (Ref 2), Weibull introduced both a uniaxial model of probabilistic failure and a related multiaxial model that treats some aspects of failure produced by multiaxial stresses. His uniaxial model is given by

$$P = 1 - \exp\left[-\int^V \left(\frac{\sigma}{\sigma_0}\right)^m dV \right] \qquad \text{(Eq 1)}$$

where P is the cumulative probability of failure, σ is the stress of a unit volume dV at the time of failure, m is the Weibull modulus, and σ_0 is a normalizing parameter. (Note that the symbol F is more commonly employed for cumulatives in the statistical literature.) If failure is caused by defects that are distributed randomly throughout the volume of a specimen or component, then the integration is carried out over all elements of volume, dV. If the defects are restricted to surfaces (such as machining defects), then the integration is carried out over surface elements, dA.

For the general case where stress is a function of position in the body, the integration can be carried out to yield

$$P = 1 - \exp\left[-kV\left(\frac{\sigma_{max}}{\sigma_0}\right)^m \right] \qquad \text{(Eq 2)}$$

where k is a dimensionless "load, or structure factor" and σ_{max} is the maximum stress in the structure at the time of failure. For the case of uniform uniaxial tension, k is unity and Eq 2 reduces to the form of the two-parameter Weibull distribution commonly encountered in the statistical literature. For all other loading geometries, k is a function of m and, when evaluated, is always less than unity. The product of k times V is often termed the "effective volume" and, as the term implies, is the volume of material that is effectively under uniform uniaxial tension.

If the uniaxial Weibull model described above is valid for a given material, then the two adjustable parameters, m and σ_0, are material constants. The value of the Weibull parameters are estimated from experimental fracture strengths using an "estimator," such as linear regression or maximum likelihood (Ref 5). In the simplest practical situation, Eq 2 can then be used to estimate the probability of failure of a component of interest where the maximum stress, the component size, and the load factor are known for that structure. Conversely, Eq 2 can be solved for σ_{max} to estimate the maximum stress that can be survived in the component at a specified probability of failure (specified failure rate).

Unfortunately, the procedure outlined above is oversimplified. In most real-life situations, one or more complications are present and a more complex analysis is required. A partial list of such complications, with key references where available, includes:

- Multiple active flaw populations (Ref 6)
- Multiaxial stress states (Ref 7, 8)
- Censored and/or proof-tested data (Ref 9)
- In-service modification of flaws by oxidation, slow crack growth, and other factors (Ref 10)
- Directional anisotropy of strength

If not accounted for, the presence of any of these complications will reduce the accuracy of strength/probability estimates in unpredictable ways. Although many recent efforts in fracture statistics have addressed complications such as these, and progress has been made, many difficult problems remain. In particular, progress is needed in approaches that allow more than one complication to be present simultaneously.

In addition, there are three further requirements for a comprehensive probabilistic fracture methodology:

- The analysis should allow data to be combined or pooled from multiple specimen sizes and geometries, so that all available fracture data can be integrated into strength/probability estimates and size-scaling aspects of the model can be tested
- The analysis must be capable of providing a measure of the statistical uncertainty in estimates of strength/probability (confidence and tolerance bounds)
- The analysis must be compatible with goodness-of-fit tests to determine if the experimental data are consistent with all aspects of the assumed model of probabilistic behavior

A research effort funded by the U.S. Department of Energy, Office of Transportation Systems, which was subcontracted through Oak Ridge National Laboratory and carried out by the authors (Ref 11) has studied many aspects of the complications and requirements listed above. The following section reviews some results of that effort in the areas of combining data and confidence/tolerance bounds on estimates. The approaches developed for combined data and confidence/tolerance are described and demonstrated for the uniaxial model of Eq 2. Importantly, it has been shown that similar techniques are applicable to more comprehensive models of multiaxial failure, such as those of Ref 7 and 8. Thus, the form of Eq 2 is quite general.

Weibull Estimators for Combined Data

A Weibull estimator is a method, or algorithm, to analyze fracture data and estimate useful quantities. These quantities may include both point estimates, such as distribution parameters and predicted strengths, as well as

estimates of intervals, such as confidence and tolerance bounds. As mentioned above, there are advantages in efficiency and model validation that result from combining or pooling fracture data from multiple specimen sizes and geometries.

In the studies of combined data, it has been useful to categorize problems according to the type of loading and the presence or absence of size scaling. The following four "classes" of problem have been considered:

- Uniform tensile stress and single specimen size (Class I)
- Uniform tensile stress and multiple specimen sizes (Class II)
- Common load factor (k) and multiple specimen sizes (Class III)
- Differing load factors and multiple specimen sizes (Class IV)

In each, it is assumed that the Weibull distribution with a size-scaling term (volume, area, or edge length) adequately describes the strength to failure. Class I problems of strength/probability estimation are the easiest to analyze, but the least useful for practical applications. On the other hand, Class IV problems are the most useful of the four but, accordingly, are the most difficult to properly analyze. The progression of complexity from Class I to IV has helped in the development of flexible and rigorous Class IV estimators.

Example of Class IV Strength Data. To demonstrate the derivation and application of Weibull estimators for Class IV problems, strength tests were performed on specimens of boron-doped sintered silicon carbide (SiC) (Ref 12). Testing was done in six bending configurations. The A, B, and C specimen geometries of the MIL-STD-1942MR (virtually identical to ASTM Standard C1161, adopted in September 1990) were used in both 3-point and 4-point bending. Specimen A is defined to have a cross section of 1.5 by 2.0 mm (0.059 by 0.079 in.) and is tested on a 20 mm (0.787 in.) outer span (10 mm, or 0.394 in., inner span when 4-point testing is done). Specimen B doubles every dimension of specimen A, and specimen C doubles those dimensions yet again. The specimen volumes therefore vary by a factor of 64, and the areas vary by a factor of 16. The difference between 3-point versus 4-point bending contributes

another factor of approximately 10 to the range of effective volumes and areas. Therefore, the data span a range of 500 to 1000 in effective volume and a range of 100 to 200 in effective areas.

Specimens were prepared by isopressing and sintering billets from which the specimens were cut and ground to shape. The specimen matrix was planned to allow detection of complications, such as billet-to-billet differences and strength dependence on location within a billet. No such complications could be resolved and, therefore, all specimens were considered to have a common family of defects. A total of 137 bend specimens were tested. The results of analyzing the specimens, one group at a time, are provided in Table 1.

The Weibull parameters listed in Table 1 are the output of conventional maximum likelihood analysis. Group-to-group variations in the two Weibull parameters are typical of statistical sampling error on data sets of this size. Therefore, the variation in Weibull modulus, for instance, is not believed to indicate that different groups contained different fundamental flaw distributions.

Low-power (50×) stereo microscopy was used to identify the location of each fracture-initiating defect in the SiC and to determine whether it was a surface, subsurface, or edge-related defect. Because of the predominance of surface-related fracture origins observed in this data set, all Class IV analyses and all references to effective size discussed below assume that strength was controlled by surface defects.

In addition to the six estimates of m in Table 1, the strength as a function of specimen size can be used to derive a seventh, independent estimate of m. The seventh estimate adds to the confusion of determining the "best" estimate of the true Weibull modulus. Should the seven estimates be averaged? Should they be weighted, somehow, by the number of specimens contributing to each estimate? The Class IV maximum likelihood estimator described below is a more rigorous and statistically efficient approach to extracting all useful information in data sets such as the above example.

Maximum Likelihood Estimator. The two adjustable parameters of the Weibull distribution are derived from fracture data using a

Table 1 Analysis of strength tests performed on boron-doped sintered silicon carbide

Geometry	Number	Average strength		Standard deviation		Weibull modulus, m	σ_0	
		MPa	ksi	MPa	ksi		*MPa·mm$^{2/m}$	*ksi·in.$^{2/m}$
3-pt A	18	388.11	56.29	33.36	4.838	14.57	430.99	40.093
4-pt A	17	312.85	45.375	34.83	5.052	9.43	459.24	33.546
3-pt B	18	350.96	50.902	31.12	4.514	12.20	449.93	38.405
4-pt B	48	303.31	43.991	24.17	3.506	14.28	430.28	39.666
3-pt C	18	325.65	47.232	21.69	3.146	16.39	418.83	40.929
4-pt C	18	283.78	41.159	22.35	3.242	14.48	441.08	40.921

Weibull estimator. For the first three classes of problems described above, methods of estimating the parameters from fracture data are available in the literature. However, only a few methods have been developed for estimating parameters in Class IV problems (Ref 13, 14).

Two new Weibull estimators have recently been developed to analyze Class IV problems of combined data (Ref 12). One is based on linear regression and the other on maximum likelihood. Both require successive approximation methods and are therefore best suited for computer analysis. The estimator based on maximum likelihood has been found to be more efficient than linear regression in maximizing the use of information from a data set. Therefore, only the maximum likelihood method is described herein.

Derivation of the Class IV maximum likelihood estimator parallels the derivation of the Class I estimator as published by Trustrum and Jayatilaka (Ref 5), which in turn is based on maximum likelihood derivations, such as those of Lawless (Ref 15), Mann *et al*. (Ref 16), and Nelson (Ref 9). Unlike earlier derivations, the Class IV analysis must account for the loading factor, k, which is generally a function of the Weibull modulus.

Maximum likelihood finds the maximum, as a function of the unknown parameters, in a quantity known as the "log likelihood." The log likelihood, l, is defined as the sum of the logs of the probability density, f, for each observed specimen. The probability density of Eq 2 is:

$$f = \frac{dP}{d\sigma} = \frac{mkV}{\sigma_0^m}\sigma^{m-1}\exp\left[-kV\left(\frac{\sigma}{\sigma_0}\right)^m\right] \quad \text{(Eq 3)}$$

For simplicity in upcoming equations, the subscript "max" from Eq 2 has been dropped from the stress in the numerator of the exponential, but the term still represents the maximum stress in the structure. The log likelihood, l, for a group of n strength measurements is then:

$$l \equiv \sum_{i=1}^{n} \ln(f_i)$$

$$= n \ln(m) - nm \ln(\sigma_0)$$

$$+ \sum_{i=1}^{n} \ln(k_i V_i) + (m-1)\sum_{i=1}^{n} \ln(\sigma_i)$$

$$- \sum_{i=1}^{n} k_i V_i\left(\frac{\sigma_i}{\sigma_0}\right)^m \quad \text{(Eq 4)}$$

Partial derivatives of Eq 4 are then taken with respect to the two Weibull parameters, m and σ_0. The derivative with respect to m must account for the functional dependence of k_i on m. Each of these partial derivatives is then set equal to zero to find the combination of the two parameters that leads to the maximum in the log likelihood. The two partial derivatives can be combined such that σ_0 drops out, leaving the relationship:

$$0 = \frac{n}{m} + \sum_{i=1}^{n}\frac{dk_i/dm}{k_i} + \sum_{i=1}^{n}\ln(\sigma_i)$$

$$- \frac{n\sum_{i=1}^{n}[k_i V_i \sigma_i^m \ln(\sigma_i) + V_i\sigma_i^m\,dk_i/dm]}{\sum_{i=1}^{n}V_i k_i \sigma_i^m} \quad \text{(Eq 5)}$$

The maximum likelihood estimate of m is then evaluated by iteratively determining the value of m that satisfies Eq 5. Under usual conditions, only a single value of m satisfies Eq 5 for any given data set. After m is known, the second Weibull parameter can be determined without iteration by rearranging the partial of Eq 4 with respect to σ_0 that was earlier equated to zero:

$$0 = n\,\sigma_0^m - \sum_{i=1}^{n}k_i V_i \sigma_i^m \quad \text{(Eq 6)}$$

In order to evaluate the maximum likelihood estimates of the Weibull parameters, the dependence of both k and dk/dm must be known as a function of m for every specimen geometry tested.

The true parameters of the distribution are referred to as m and σ_0. By standard convention, estimates of these parameters from analysis of data are known as $\hat{m}$ and $\hat{\sigma}_0$. An efficient algorithm for the evaluation of maximum likelihood estimates for Class IV problems has been developed that requires only five to ten

iterations to determine $\hat{m}$ and $\hat{\sigma}_0$ to approximately five significant digits for most data sets that have been analyzed. The 137 strength measurements of SiC described above result in Weibull parameters of $\hat{m} = 14.22$ and $\hat{\sigma}_0 = 433.1$. These estimates assume that failure is controlled by a distribution of surface defects. The Weibull modulus is dimensionless. In order for the exponent of Eq 2 to be dimensionless, the dimensions of σ_0 must account for the surface area term in the exponential. Therefore, the dimensions of σ_0 are a function of the Weibull modulus and are $(MPa \cdot mm^{2/m})$.

Knowledge of the Weibull parameters allows the fracture strength to be estimated for any component size and shape that may be of interest at any probability of failure of interest using Eq 2. These estimated strengths can be displayed graphically, as in Fig 1, where log of fracture stress is plotted versus log of effective stressed area (again, fracture is controlled by surface-related defects). Figure 1 includes the complete set of 137 fracture strengths, with different symbols used to identify the various specimen geometries. The effective area used to plot each data set is the product of k times the surface area of the specimen. Because k is a function of $\hat{m}$, the plotting position is a function of the Weibull modulus.

Also included in Fig 1 are two parallel straight lines representing the strength versus

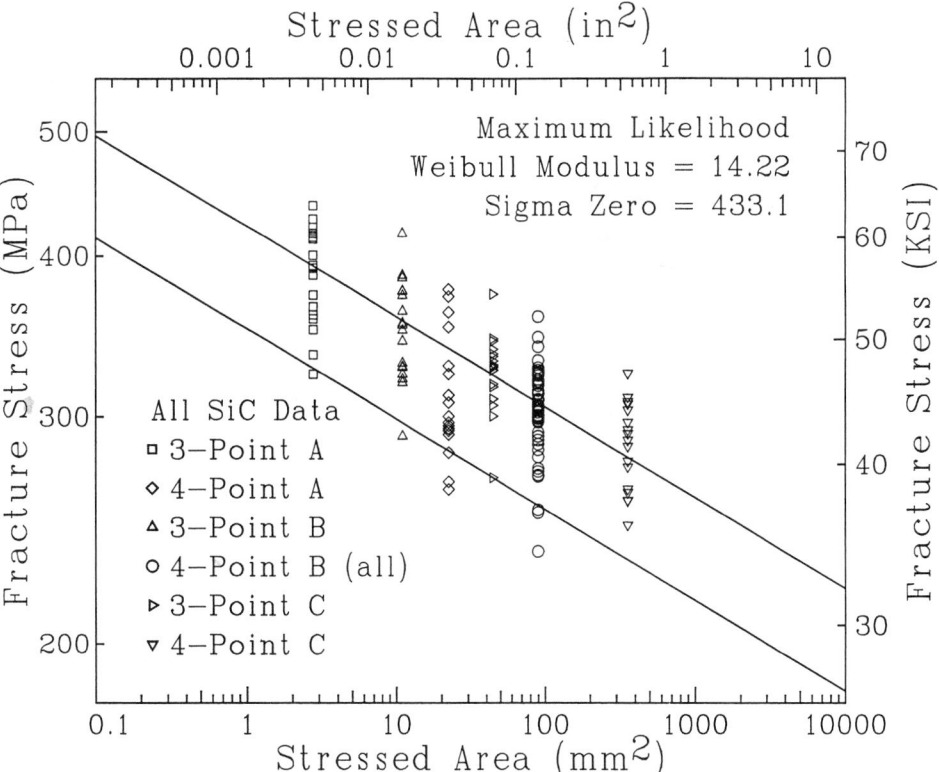

Fig 1 Dependence of strength on specimen size for boron-doped, sintered β-SiC tested in six configurations of specimen size and testing geometry. Upper line is 0.5 quantile predicted behavior; lower line is 0.05 quantile.

effective area at two different probabilities of failure. The upper straight line describes the median (or 0.5 quantile) behavior, where 50% of the specimens or components are expected to fail. The lower straight line describes the 0.05 quantile behavior where only 5% are expected to fail. Of course, an entire family of parallel straight lines exists that represents all quantile levels.

Confidence and Tolerance Bounds. Confidence bounds give a measure of statistical uncertainty associated with a quantity estimated from data sampled from a distribution. Using experimental data, one cannot exactly determine the complete distribution; this inexactness must be quantified in order to make sensible decisions. For instance, the exact location of the true 0.05 quantile line for the SiC data is unknown. Figure 1 only shows an estimate of the position of the line. Confidence bounds offer a way to quantify the inexactness and give an indirect measure of the risk associated with the use of an estimated value as the true but unknown value. In the above definition but confidence bounds, the term "statistical uncertainty" is the variation that is due to random sampling error and random measurement error; "estimated" means calculated from sampled data that were obtained from a distribution of all possible data values; and "quantity" is a parameter, quantile, or future value of a distribution.

There are three primary types of confidence bounds, which, by convention, are defined as:

- Confidence limits: bounds on a parameter of a distribution
- Tolerance limits: bounds on a quantile of a distribution
- Prediction limits: bounds on a future value sampled from a distribution

A distribution characterizes the population of strengths, for example, of all components made from a certain material using a given manufacturing process. Generally, there is only a sample of material available for use in specifying the distribution, and observations are taken in the presence of measurement error. The measurement error can either be of a random nature or occur as a bias (constant offset). In the developments herein, neither of these forms of measurement error are considered. Other sources of uncertainty, such as model "incorrectness," are not accounted for in the estimation procedures, either. The resulting bounds therefore only account for statistical sampling error.

Statistical sampling error arises from the fact that a sample quantity, for example, a sample mean, does not equal the true value associated with the population, for example, the population mean. Therefore, the underlying distribution cannot be *exactly* determined or specified with a finite number of observations. This inexactness must be quantified in order to make sensible decisions about design considerations and the ability of a

component to withstand a required service stress. One way to quantify the uncertainty is to employ confidence bounds, which give an indirect measure of the risk associated with using an estimated value as the true value. When a "high" confidence coefficient is used, then it is "safe" to act as if the true value is contained within the *observed* interval. But the terms high and safe are open to subjective interpretation; what is high and/or safe for one person may not be for another. Thus, the level of acceptable confidence must be determined for each individual problem.

The qualifier "observed" is emphasized above because the location and, often, the length of a confidence interval changes from one sample to the next under identical sampling conditions. The confidence coefficient reflects this kind of uncertainty by giving the long-run probability that an interval covers the true, but unknown, value of interest. This is illustrated schematically in Fig 2, where one of the 20 intervals shown does not cover the true value, whereas the others do. Each interval has been deduced by an independent sampling of the distribution. The behavior shown is analogous to that which would occur in the long run for a 95% confidence interval on the unknown true value. In practice, of course, the validity of a specific interval in covering the true value would not be known in advance. The probability of an interval missing the true value then gives a measure of the risk one is willing to take in making a decision about the true value. The future is indeed uncertain.

In the calculation of confidence and tolerance bounds, two overall approaches have evolved: exact bounds based on sampling theory and approximate bounds based on asymptotic considerations. Exact bounds arise by inverting statistical tests and/or conditional integration methods first described by Fisher (Ref 17). These methods can be shown, analytically, to have exact coverage probabilities, as opposed to an approximate coverage. Approximate bounds include those obtained by maximum likelihood asymptotics, some likelihood ratio methods, and some bootstrap (simulation) methods. Reference 18 gives an annotated bibliography that contains further references to the large literature on confidence bounds for lifetime distributions. Because a lifetime is analogous to a strength, lifetime methodologies can apply, in some cases, to the present situation.

To some extent, existing methods for normal theory and likelihood ratio approaches offer solutions up through Class III problems. However, current bootstrap procedures are generally not applicable, without some development, to any of the four classes. These two approaches are emphasized because they appear to be the most promising for use in obtaining confidence and tolerance bounds in Class IV problems. More details on this issue and references to other approaches are given in Ref 19.

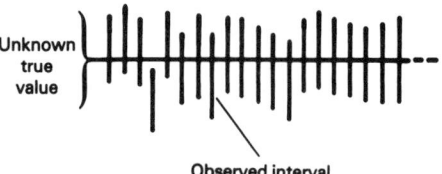

Fig 2 Projected confidence intervals from a sequence of independent estimates; for 95% bounds, 1 in 20 intervals on average will not cover unknown true value

Bootstrap Techniques for Confidence and Tolerance. As the term "bootstrap" implies, these methods pull themselves up by their own bootstraps; that is, they gather information about variability in estimates by evaluating only the data available in the data set of interest. A good review of bootstrap techniques and their basis can be found in Ref 20. These techniques are readily applied to a variety of distributions, but have not yet been applied to the Weibull distribution with size scaling.

Bootstrap techniques characterize a combination of the intrinsic uncertainty of estimation that is due to sampling error and any additional uncertainty that is due to inefficiencies of the estimator. These techniques are ultimately dependent on an estimator to digest the experimental measurements of fracture strength and estimate parameters of the distribution and/or estimates of the strengths of components with other sizes and loading geometries. The estimator used in this discussion is the Class IV maximum likelihood estimator described above.

Two types of bootstrap techniques have been studied: parametric and nonparametric. Both involve simulation and analysis of many "artificial" data sets that are derived either directly or indirectly from the experimental data. In the nonparametric technique, no assumption of the form of the strength distribution is made prior to generation of the simulated data sets. The simulated data are created by randomly choosing strengths from the original data "with replacements"; that is, after a strength is chosen from the list of experimental strengths, it will still be available to be chosen again within that same simulated data set. Therefore, each simulated data set is virtually guaranteed to have some repeat observations. In the parametric technique, the estimator is used to estimate the Weibull parameters for the experimental data. Simulated strengths are then generated by randomly choosing simulated specimens from the infinite population of possible specimens that are consistent with those distribution parameters.

In both the parametric and nonparametric techniques, the number of specimens generated for each size and geometry of test specimen is identical to that of the original data set. The estimator is then used to analyze the simulated data set. This simulation and analysis procedure is repeated a large number of times (typically, 1000 times). The variability in estimates from the many simulated data sets reflects the intrinsic variability of the estimator in analyzing data sets similar to the original experimental data set. For instance, if the $\hat{m}$'s resulting from the simulations are ordered from smallest to largest, then the range of $\hat{m}$ values from the 25th to the 975th position in this list of 1000 $\hat{m}$'s is an estimate of the 95% confidence interval on estimates of $\hat{m}$ using that estimator. For example, in 950 of 1000 trials, the estimated $\hat{m}$ value was found to fall in this interval when sampling error was the only source of error. Similar bounds can be placed on predicted strength values at a specific component size and probability of failure by estimating the component strength for each of the bootstrap simulations (using Eq 2) and carrying out a similar ordering to find the interval in strength that contains 95% of the estimated strength values. Because there is nothing magic about 95% intervals, the confidence interval could be estimated in this fashion for any level of confidence that may be of interest.

Figure 3 is very similar to Fig 1, but includes tolerance bounds on the 0.05 quantile. The curved lines bounding the 0.05 quantile straight line are the tolerance limits as a function of effective area as determined by the parametric bootstrap technique. The corresponding 95% confidence bounds on the two Weibull parameters are 13.01 and 15.81

for $\hat{m}$ and 421.0 and 444.5 for $\hat{\sigma}_0$. The solid triangular data points are also tolerance limits on the 0.05 quantile, but were calculated by the likelihood ratio technique, discussed in the next section of this article.

The shape of the tolerance limit lines are approximately hyperbolic, as should be expected. The hyperbolic shape results in a "pinch point," where the uncertainty in strength estimates is the smallest. Regardless of the quantile being considered, this pinch point is located near the average strength of the data. Thus, as one considers component sizes with strengths that are progressively farther from this average strength (either higher or lower), the width of the tolerance interval gets progressively larger. As one considers smaller quantile behaviors, the pinch point for the new quantiles will move to the left toward progressively smaller effective specimen sizes. Typically, strength predictions are made for components that are larger than the test specimens. Therefore, the width of the tolerance interval for most components will increase as one considers progressively smaller quantile behaviors.

Figure 3 does not include the tolerance limits for the nonparametric bootstrap technique. The results are very similar for this data set. Comparisons of parametric versus nonparametric bootstrap analyses have been made for many real and simulated data sets. These comparisons have shown that the nonparametric bootstrap is more prone to bias and

has a greater sensitivity to data sets that contain "apparent outliers." For these reasons, the parametric bootstrap is favored for most situations studied to date.

The bootstrap technique is computationally intensive, but it offers the potential of estimating confidence and tolerance bounds on a variety of very complex problems, even those beyond Class IV (such as those involving multiple flaw populations, multiaxial stresses, and strength degradation that is due to slow crack growth). If an estimator can be designed to estimate all the adjustable parameters of a given model, then the bootstrap technique is capable of estimating confidence bounds on all the parameters and tolerance bounds on estimated strengths. Of course, the better the quality of the estimator in using all the information available within the data set, the smaller the width of the confidence and tolerance bounds resulting from the analysis.

As with most estimators, the Weibull maximum likelihood estimator has the property of yielding offset, or biased, estimates of parameters, strengths, and confidence intervals. The magnitude of bias error often decreases as the number of samples increases. Unfortunately, the degree of bias error is difficult to predict in advance for complex estimators, such as this Class IV estimator. Bootstrap simulations for estimation of confidence and tolerance bounds contain information about the magnitude of bias introduced by the estimator. Reference 21 describes how such information can be used to correct bias errors. Although maximum likelihood estimates do generate bias, of all the Class IV estimator studies to date, maximum likelihood has generally been found to generate the least bias.

Likelihood Techniques for Confidence and Tolerance. Likelihood ratio methods offer another way to obtain confidence bounds. The development given by Cox and Oakes (Ref 22) as applied to the two-parameter (m, σ_0), Weibull distribution is closely followed. This method is based on the direct use of the likelihood ratio statistic

$$W(\sigma_0) = 2[l(\hat{m},\hat{\sigma}_0) - l(\hat{m}_{\sigma_0},\sigma_0)] \qquad \text{(Eq 7)}$$

where $l(.,.)$ is the log likelihood given by Eq 4, $(\hat{m},\hat{\sigma}_0)$ is the joint maximum likelihood estimate of (m,σ_0) and $\hat{m}_{\sigma_0}$ is the maximum likelihood estimate of m conditional on σ_0 (that is, taking σ_0 fixed). The function $l(\hat{m}_{\sigma_0},\sigma_0)$ of σ_0 is sometimes called the profile log likelihood for σ_0. Under the null hypothesis that σ_0 is the true or actual value of the second Weibull parameter, $W(\sigma_0)$ has, approximately, a χ^2 distribution with $p_{\sigma_0} = dim(\sigma_0)$ (= 1) degrees of freedom. Inverting the test yields a corresponding $1 - \alpha$ confidence region as:

$$\{\sigma_0 : W(\sigma_0) \le \chi^2_{p_{\sigma_0},\alpha}\} \qquad \text{(Eq 8)}$$

where $\chi^2_{p_{\sigma_0},\alpha}$ is the upper α point of the chi-squared distribution with p_{σ_0} degrees of freedom. The procedure of inverting a statistical

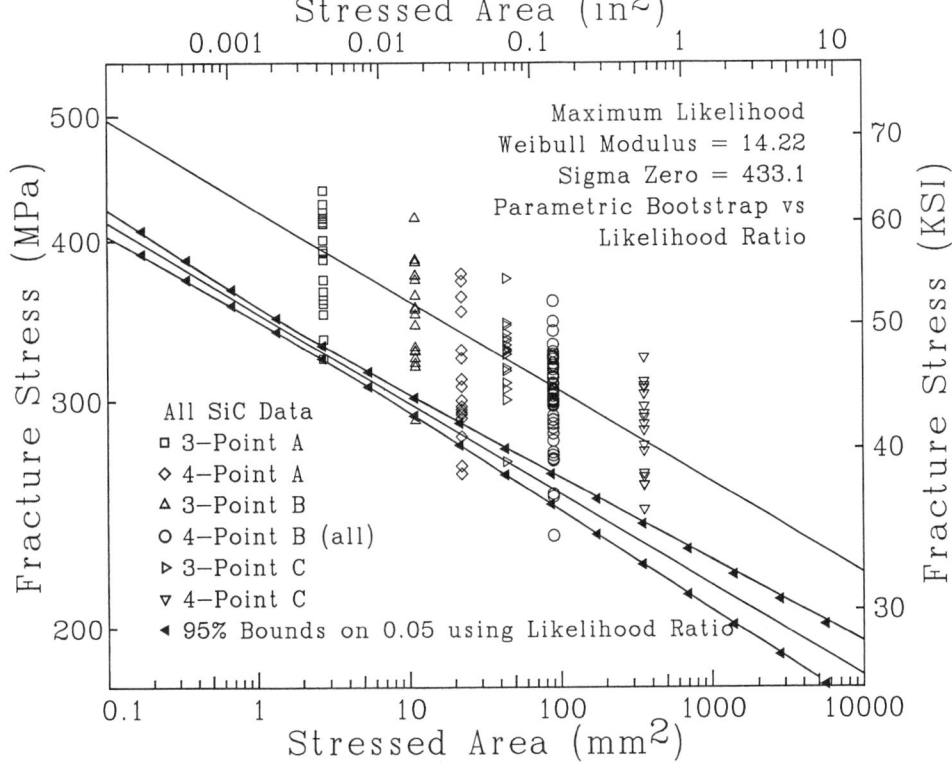

Fig 3 Plot similar to Fig 1, but includes 95% tolerance bounds on the 0.05 quantile as determined by parametric bootstrap (curved lines) and likelihood ratio (solid triangles) techniques

test to obtain a confidence region is common practice: The set of all σ_0 values that satisfy the acceptance criterion, that is, would not result in a decision of rejection if that particular value had been the null value, are associated with a test statistic value that has a probability of $1 - \alpha$ under the prescribed null hypothesis. But this just defines a $1 - \alpha$ confidence region. Thus, in order to obtain a $1 - \alpha$ confidence region on σ_0, one must find those σ_0 values that satisfy Eq 8.

The strength quantiles for a component, as shown in Fig 1, are denoted by σ_p. In practice, p is generally chosen to be "small," for example, 0.05 or 0.01. It is well known that with mild conditions, the maximum likelihood estimator (MLE) of a function of a parameter(s) is the function of the MLE(s). For example, if $\hat{\theta}$ is the MLE of θ, then the MLE of $f(\theta)$ is $f(\hat{\theta})$. This fact is commonly used to estimate σ_p; see, for example, Ref 9. However, because we also want to determine various kinds of confidence limits via the likelihood ratio method, the somewhat different course of redefining σ_0 into σ_p is followed. Thus, let σ_p be such that

$$p = 1 - \exp\left\{-kV\left(\frac{\sigma_p}{\sigma_0}\right)^m\right\} \qquad \text{(Eq 9)}$$

where k and V are associated with the component of interest. Then, solving for σ_0 yields

$$\sigma_0 = \sigma_p\left[\frac{1}{kV}\ln\left(\frac{1}{1-p}\right)\right]^{-1/m} \qquad \text{(Eq 10)}$$

and substituting Eq 10 into Eq 2 yields

$$P_i(\sigma_i) = 1 - \exp\left\{-r_iR_i\ln\left(\frac{1}{1-p}\right)\left(\frac{\sigma_i}{\sigma_p}\right)^m\right\} \qquad \text{(Eq 11)}$$

where $r_i = k_i/k$ and $R_i = V_i/V$. Equation 11 has the two unknown parameters m and σ_p and relates the component situation to the test situation. For example, the parameter σ_p is the pth quantile at size V associated with the component, appearing in the strength distribution of σ_i for the ith specimen. We note in Eq 11 that r_i is a (known) function of m and also R_i is known; m and σ_p must be estimated by taking data, σ_i. To this end, Eq 11 is employed in the remaining developments of the likelihood ratio procedure. In addition, this methodology can be employed to determine confidence bounds on other parameters. For example, it is straightforward to obtain confidence limits on either m or σ_0. Moreover, a reparameterization analogous to that given by Eq 9 to 11 can be employed to get confidence bounds on the reliability at a given strength. Thus, the likelihood ratio method is very flexible and can be employed to obtain various kinds of confidence limits. The trick is to make a suitable change of variables for the parameter(s) of interest.

The maximum likelihood solution for m is as given previously in Eq 5 with r_i replacing

k_i. In the present setup, the maximum likelihood solution for σ_p becomes

$$\hat{\sigma}_p = \left[\ln\left(\frac{1}{1-p}\right)\Sigma r_i(\hat{m})R_i\,\sigma_i^{\hat{m}}/n\right]^{1/\hat{m}} \qquad \text{(Eq 12)}$$

In order to obtain confidence limits on σ_p, Eq 8 must be solved with σ_0 replaced by σ_p. This requires evaluating $l(\hat{m}, \hat{\sigma}_p)$ and $l(\hat{m}_{\sigma_p}, \sigma_p)$. The method to obtain $\hat{m}$ and $\hat{\sigma}_p$ has already been described. Now, take σ_p fixed so as to determine $\hat{m}_{\sigma_p}$. Inspection of Eq 4 and employing Eq 11 reveals that the solution is

$$0 = \frac{\Sigma_{\hat{r}_i'}}{\hat{r}_i} + \frac{n}{\hat{m}_{\sigma_p}} + \Sigma\ln\sigma_i - n\ln\sigma_p$$
$$- \ln\left(\frac{1}{1-p}\right)\Sigma\left[\hat{r}_iR_i\left(\frac{\sigma_i}{\sigma_p}\right)^{\hat{m}_{\sigma_p}}\left\{\frac{\hat{r}_i'}{\hat{r}_i} + \ln\left(\frac{\sigma_i}{\sigma_p}\right)\right\}\right] \qquad \text{(Eq 13)}$$

where the circumflexes on the r's indicate that m_{σ_p} has been replaced by $\hat{m}_{\sigma_p}$. Equation 13 is solved in the same manner as Eq 5 in an iterative procedure to find the boundary values of σ_p such that

$$\{\sigma_p : W(\sigma_p) \leq \chi^2_{p\sigma_p,\alpha}\} \qquad \text{(Eq 14)}$$

is met, that is, replace σ_0 by σ_p in Eq 7 and 8 via the change in density from Eq 2 to Eq 11 to carry out the likelihood ratio procedure to obtain confidence limits on σ_p.

The results of applying the likelihood ratio method to the SiC data are shown in Fig 3. Using Eq 13 and 14 with a p of 0.05 and employing various component areas yields the 95% confidence bounds on the 0.05 quantile shown on Fig 3 as solid triangles. A comparison of the two techniques indicates (in a limited way) that the likelihood ratio and parametric bootstrap methods give confidence limits that practically agree. This is consistent with the general knowledge base for bootstrap methods. These Class IV approaches of analysis combine data from all SiC specimen sizes and geometries, and make strength predictions with confidence bounds on components with arbitrary sizes and geometries. The "component" loading of Fig 3 is that of uniform tension, because no allowance was made for the loading geometry in the calculation of the confidence bounds.

Unfortunately, the likelihood ratio method has one drawback. The limits obtained are generally biased in a similar manner to those described above for the bootstrap method. Note that if the asymptotic distribution used in deriving Eq 8 were exact, then the expected value of $W(\sigma_0)$ would be p_{σ_0}. Sometimes it is possible to find an expansion such that the expected value is $p_{\sigma0}[1 + c/n + o(1/n)]$. Then $(1 + c/n)$ is called a Bartlett correction factor and improved properties are obtained by replacing W with $W' = W/(1 + c/n)$ in Eq 7 and 8. Frequently, c must be estimated to carry out this correction procedure. As pointed out by Cox and Oakes (Ref 22), it is rarely feasible to carry out such calculations in the

presence of censoring. In addition, the increased complexity of a problem diminishes the ability to determine c.

As mentioned above, it is also possible to correct bootstrap limits for bias through information generated during the bootstrap simulation. However, a general study has not been carried out that compares all the methods of bias correction with the goal of determining a preferred way to obtain unbiased confidence limits. One can imagine correcting bootstrap confidence limits via Bartlett corrections, on the one hand, or, on the other, correcting likelihood ratio confidence limits via a bootstrap procedure.

Future Needs and Directions. A series of complications that are often present in real-life problems of probabilistic strength analysis was described in the section "Distributional Models" in this article. Since the early 1980s, a great deal of progress has been made in the development of methods to accommodate such complications. Unfortunately, these methodologies are not compatible with each other in many instances. Therefore, if several of the complications are present simultaneously, no overall approach is available. Additionally, in the presence of many of the complications listed earlier, methods of combining data, estimating confidence bounds, and testing for goodness-of-fit are either not available or are so computationally intensive that they may not be practical.

The five-year period from 1991–1996 is expected to be especially productive in addressing such problems of compatibility in probabilistic failure analysis. For example, as part of the DOE-sponsored program cited in Ref 11, two subcontracted efforts are independently addressing problems of probabilistic life prediction. One subcontract is with Garrett Auxiliary Power Division of Allied-Signal, Inc., and the other is with the Allison Gas Turbine Division of General Motors Corp. In both cases, the objective is to develop and test a ceramic life prediction methodology that simultaneously addresses many of the complications and requirements described in this article.

In the case of the Garrett effort (Ref 23), a comprehensive methodology is being developed that will address the entire list of complications. In addition, the methodology will have the desirable capabilities of analyzing pooled data and determining confidence bounds on estimates. This is a very ambitious undertaking, but there are no obvious roadblocks to the planned approach. In parallel with the development of probabilistic tools is a well-designed experimental testing plan. The resulting data will provide a challenging opportunity for the new tools to digest various types of test specimen data, predict the life of a simulated turbine component, and compare the predicted behavior with observed behavior. The ultimate success of comprehensive probabilistic life prediction tools will eventually require validation through numerous successful predictions of component behavior.

REFERENCES

1. W.D. Kingery, H.K. Bowen, and D.R. Uhlmann, Chapt. 14, *Introduction to Ceramics*, 2nd ed., John Wiley & Sons, 1976
2. W. Weibull, A Statistical Theory of the Strength of Materials, *R. Swedish Acad. Eng. Sci. Proc.*, Vol 151, 1939, p 1–45
3. W. Weibull, A Statistical Distribution of Wide Applicability, *J. Appl. Mech.*, Vol 18, 1951, p 293–297
4. E.J. Gumbel, *Statistics of Extremes*, Columbia University Press, 1958
5. K. Trustrum and A. De S. Jayatilaka, On Estimating the Weibull Modulus for a Brittle Material, *J. Mater. Sci.*, Vol 14, 1979, p 1080–1084
6. C.A. Johnson, Fracture Statistics of Multiple Flaw Distributions, *Fracture Mechanics of Ceramics*, Vol 5, Plenum Press, 1983, p 365–386
7. S.B. Batdorf and H.L. Heinisch, Weakest Link Theory Reformulated for Arbitrary Fracture Criterion, *J. Am. Ceram. Soc.*, Vol 61 (No. 7–8), 1978, p 355–358
8. J. Lamon and A.G. Evans, Statistical Analysis of Bending Strengths for Brittle Solids: A Multiaxial Fracture Problem, *J. Am. Ceram. Soc.*, Vol 66 (No. 3), 1983, p 177–182
9. W. Nelson, *Applied Life Data Analysis*, John Wiley & Sons, 1982
10. W.T. Tucker and C.A. Johnson, Modifications in Strength Distributions Due to Slow Crack Growth, *Proceeding of the 24th Automotive Technology Development Contractors' Coordination Meeting*, Publication P-197, Society of Automotive Engineers, Apr 1987, p 233–237
11. "Ceramic Technology for Advanced Heat Engines Project," Contract DE-AC05–840R21400 with Martin Marietta Energy Systems, Inc., Subcontract 86X00223X with General Electric Corporate Research and Development
12. C.A. Johnson and W.T. Tucker, Advanced Statistical Concepts of Fracture in Brittle Materials, *Ceramic Technology for Advanced Heat Engines Project Semiannual Progress Report for April 1987 Through September 1987*, ORNL/TM-10705, Oak Ridge National Laboratory, Mar 1988, p 200–208
13. C.A. Johnson and S. Prochazka, "Investigation of Ceramics for High Temperature Turbine Component," Final Report, Contract N62269–76–0243, June 1977
14. S.B. Batdorf and G. Sines, Combining Data for Improved Weibull Parameter Estimation, *J. Am. Ceram. Soc.*, Vol 63, 1980, p 214–218
15. J.F. Lawless, Construction of Tolerance Bounds for the Extreme-Value and Weibull Distributions, *Technometrics*, Vol 17, 1975, p 255–261
16. N.R. Mann, R.E. Shafer, and N.D. Singpurwalla, *Methods for Statistical Analysis of Reliability and Life Data*, John Wiley & Sons, 1974
17. R.A. Fisher, Two New Properties of Mathematical Likelihood, *Proc. R. Soc. A*, Vol 144, 1934, p 285–307
18. C.A. Johnson and W.T. Tucker, Advanced Statistical Concepts of Fracture in Brittle Materials, *Ceramic Technology for Advanced Heat Engines Project Semiannual Progress Report for October 1987 Through March 1988*, ORNL/TM-10079, Oak Ridge National Laboratory, Aug 1986, p 208–223
19. W.T. Tucker and C.A. Johnson, Confidence Bounds on Strength Estimates, *Proceeding of the 26th Automotive Technology Development Contractors' Coordination Meeting*, Publication P-219, Society of Automotive Engineers, Apr 1989, p 205–210
20. B. Efron and R. Tibshirani, Bootstrap Methods for Standard Errors, Confidence Intervals, and Other Measures of Statistical Accuracy, *Stat. Sci.*, Vol 1, 1986, p 54–77
21. C.A. Johnson and W.T. Tucker, Advanced Statistical Concepts of Fracture in Brittle Materials, *Ceramic Technology for Advanced Heat Engines Project Semiannual Progress Report for October 1989 Through March 1990*, ORNL/TM-11586, Oak Ridge National Laboratory, Sept 1990, p 298–316
22. D.R. Cox and D. Oakes, Chapt. 3.3, *Analysis of Survival Data*, Chapman and Hall, 1984
23. A.M. Comfort and J.S. Cuccio, Life Prediction Methodology for Ceramic Components of Advanced Heat Engines, *Proceeding of the 24th Automotive Technology Development Contractors' Coordination Meeting*, Publication P-230, Society of Automotive Engineers, Apr 1990, p 275–282

Design Practices for Structural Ceramics in Gas Turbine Engines

Leonard C. Lindgren, Robert L. Holtman, Kevin D. Kannmacher,
Jack D. Petty, Allison Gas Turbine Division, General Motors Corporation
George A. Costakis, Advanced Engineering Staff, General Motors Corporation

STRUCTURAL CERAMICS offer the potential for significant improvement in gas turbine engine performance. The U.S. government, through various agencies, has provided funding/support since the 1970s to pursue this potential. Successful short-term operation of engines with ceramic components such as combustors, vanes, blades, rotors, and others, has been demonstrated. Sophisticated analysis methods have been coupled with statistical characterization of material properties for a probabilistic approach in the design of these ceramic components.

This article first synopsizes the program history of ceramics research and development. It then describes design methodology, including reliability analysis, followed by guidelines as to materials selection, design tradeoffs, and component development. Finally, ceramic components are discussed in terms of their success or failure, and the lessons learned are summarized.

Program History

A project known as the Small Engine Components Technology Studies (Ref 1) has shown that the use of ceramics holds more promise than any other single technology improvement. This project, and most others related to gas turbine development, was directed by the National Aeronautics and Space Administration (NASA) Lewis Research Center.

One of the earliest applications of ceramics was in a portable 10 kW power pack designed for the U.S. Army, which used the Solar Gemini turbine engine. The use of ceramic components allows turbine inlet temperature to be raised approximately 200 °C (360 °F), with attendant reductions in fuel consumption (8%) and engine specific weight (35%) (Ref 2, 3).

The Defense Advanced Research Projects Agency (DARPA), which funds many defense-related programs, supported an extensive effort between 1972 and 1976 that focused on brittle materials (ceramic) design in high-temperature gas turbines. Ford Motor Company (Ref 4) and Westinghouse Electric Company (Ref 5) were the principal contractors. This program provided an important initial impetus for many later projects, even though it terminated prematurely. The Electric Power Research Institute (EPRI) subsequently funded the development of a rotor blade for large industrial gas turbines (Ref 6).

The U.S. Navy supported a demonstration of ceramic engine components on a Garrett axial flow turbine engine of 533 kW (715 hp) (Ref 7). Turbine inlet temperature was raised 200 °C (360 °F), producing 30% more power with a 7% improvement in SFC.

Serious work on the application of ceramics to civil transportation engines began in 1976 with a heavy-duty gas turbine engine (HDGTE) study, followed by the ceramic applications in turbine engines (CATE) program (Ref 8) in 1978. These programs were funded by the Department of Energy (DOE) Conservation and Renewable Energy, Office of Vehicle and Engine R&D, and administered by NASA Lewis Research Center. The principal contractor was the Allison Gas Turbine Division of General Motors Corporation. Other participants included Carborundum Company, Corning Inc., GTE Laboratories Inc., Pure Carbon Company, and Norton Company.

The CATE program accomplished 9398 hours of ceramic component testing at turbine inlet temperatures of 1038 °C (1900 °F) for 7814 hours and 1132 °C (2070 °F) for 1584 hours. The project demonstrated successful short-term application of a range of ceramic components, from rotor blades to supporting rings, but because it was curtailed by budgetary cuts, there was no testing at the planned target temperature of 1240 °C (2265 °F).

In 1977, the DOE Division of Energy Technology, Fossil Energy Supply R&D initiated a ceramics technology readiness development (CTRD) program with Westinghouse Electric Company (Ref 9) aimed at the possible application of ceramics in large utility combined-cycle power units. The planning phase was completed in 1979, but subsequent phases were not funded.

The U.S. government has also provided substantial direct support for the development of ceramic materials and component fabrication methods. Much of this work has been directed by Oak Ridge National Laboratory (ORNL).

The most important recent engine development programs for nonmilitary use have been sponsored by the DOE Office of Transportation Systems. The advanced gas turbine (AGT) program (Ref 10, 11) spanned from 1979 to 1987, and has been followed by the advanced turbine technology application project (ATTAP). The ATTAP program is planned to end November 1992. The aim of these programs has been the development of a 75 kW (100 hp) automotive gas turbine engine with 30% greater fuel economy than that of a gasoline engine and the capability to use a wider range of fuels with reduced emissions. Ceramic components that were designed, fabricated, and tested include combustors, radial inflow turbine rotors, vanes, scrolls, regenerators, and other associated parts. The ongoing ATTAP project is concerned with ceramic component technology, especially reliability and durability improvements (Ref 12, 13). The contractors in both phases of this work are Allison Gas Turbine Division of General Motors Corporation and Garrett Auxiliary Power Division of Allied-Signal Aerospace Company. These programs are also managed by the NASA Lewis Research Center.

ATTAP includes substantial subcontract development activities by the major domestic suppliers of ceramic components, managed by the engine manufacturers. The ongoing materials development and design programs coordinated by ORNL are also closely involved in the overall effort.

In Japan, direct government support for ceramic gas turbine development began only recently. Two major programs have emerged in the industrial and transportation engine fields. One is the "Moonlight" project (Ref 14) which focuses on a 300 kW (402 hp) turbine concept for cogeneration and industrial power generation. The other is a transportation engine project, which focuses on a 100 kW (134 hp) gas turbine. Both projects are supported by the Ministry of International Trade and Industry and cover a time span from 1988 to 1997.

In Europe, Daimler-Benz AG has developed ceramic components for an automotive research gas turbine. This two-shaft regenerative gas turbine engine is designed to use monolithic ceramic gasifier and power turbine wheels. The initial design for turbine inlet temperature was 1250 °C (2282 °F) for a 93 kW (125 hp) engine. The ultimate goal is 1350 °C (2462 °F) for a 110 kW (147.5 hp) engine. More than 20,000 km (12,400 miles) of road tests in the research car have been conducted since 1982. Foreign object damage has reportedly been a major problem for these axial turbine rotors. A new 7-yr project involving a 100 kW (134 hp) ceramic gas turbine is scheduled to begin in 1990. Its emphasis will be on fuel economy and exhaust emissions. The target turbine inlet temperature is greater than 1350 °C (2462 °F).

Design Methodology

The inherent characteristics of ceramic materials are sufficiently different from metallic structural materials to require special attention at all phases of the design effort. In some cases, specifically different practices are necessary.

The need for change in design practice is not a new phenomenon in gas turbine engine experience. Materials use has continually progressed to lower ductility levels through the incorporation of high-strength steel alloys, superalloy use, and high-strength titanium usage. The necessary changes in design methodology have addressed refined criteria definitions, improved fatigue characterization through finite-element modeling and combined loading conditions, use of fracture mechanics technology, and application of probability/reliability practices. Therefore, the advent of structural ceramic materials represents a logical extension of these changes in the design disciplines required for gas turbine components.

Probabilistic (as opposed to deterministic) design analysis is required for each ceramic turbine component. Finite-element models, either two- or three-dimensional, are generally required to adequately calculate the temperature and stress distribution necessary for assessing component reliability (probability of survival) at critical operating conditions. Normally, the critical operating condition occurs during rapid transient engine operation: either an acceleration from low to high power or a rapid shutdown from high to low power. Temperature gradients within the

components and, thus, thermal stresses, are most severe during this period. For this analysis, input data are derived from the engine duty cycle, material characteristics, reliability requirements, and proof-test definition.

This ceramic component design/analysis process is depicted in Fig 1. Thermal, aerodynamic, and mechanical loads for steady-state and transient operation in both rig and engine environments must be considered. Proof-test requirements are determined for each component. Proof-test failures can be predicted and measured against component operational reliability enhancement. The materials properties necessary for these analyses are continually updated from materials test data. In addition, component data obtained from rig and engine tests can be fed back through these analytic models for correlation of the analytic techniques. This feedback provides improved analytic predictive capability for evaluation of subsequent geometric configurations and environmental conditions.

As an example, the design analysis summary of the ceramic gasifier turbine scroll assembly for the AGT-100 engine in the AGT program (Ref 10) is presented. The general arrangement of the scroll assembly is depicted in Fig 2. Three-dimensional, cubic finite-element analysis of each component of the assembly was the basis for establishing design acceptability. This analysis was used to calculate temperatures, stresses, and failure probabilities, for steady-state and transient operating conditions. Figure 3 shows the transient cycle, along with the predicted reliability for each of the components for steady-state and the most severe transient conditions. Overall reliability is lowest 50 seconds into the transient conditions, where a failure rate of 9.1% is predicted.

Results of this type allow the designer to address those components with a low probability of survival. Options available are revised geometry to reduce stresses, stronger material to improve reliability, proof testing to screen flawed components, and finally, but least desirable, reduced operational severity.

Presently, ceramic gas turbine components are designed primarily to preclude "fast-fracture" failures. In the future, as the various structural ceramics mature, a data base for time-dependent failure modes will become available and influence the design of reliable components.

In the component fabrication and test phase, both ceramic specimens and components are initially subjected to materials and nondestructive inspection (NDI) testing for verification of desired properties, characteristics, and quality. However, an initial rig or engine test for some components may be performed on a metal part fabricated to simulate the characteristics of the ceramic component. This practice permits the evaluation of selected critical environmental situations without major risk to either the simulated ceramic component or other rig or engine parts. Proof tests should be defined to simulate the part-limiting operational conditions and parts proof tested for quality segregation prior to functional rig or engine testing.

Materials Selection, Design Tradeoffs, and Component Development

The material selection process for structural ceramic components involves consideration of a variety of material characteristics. Most obvious are strength and maximum operating temperature capability. Other char-

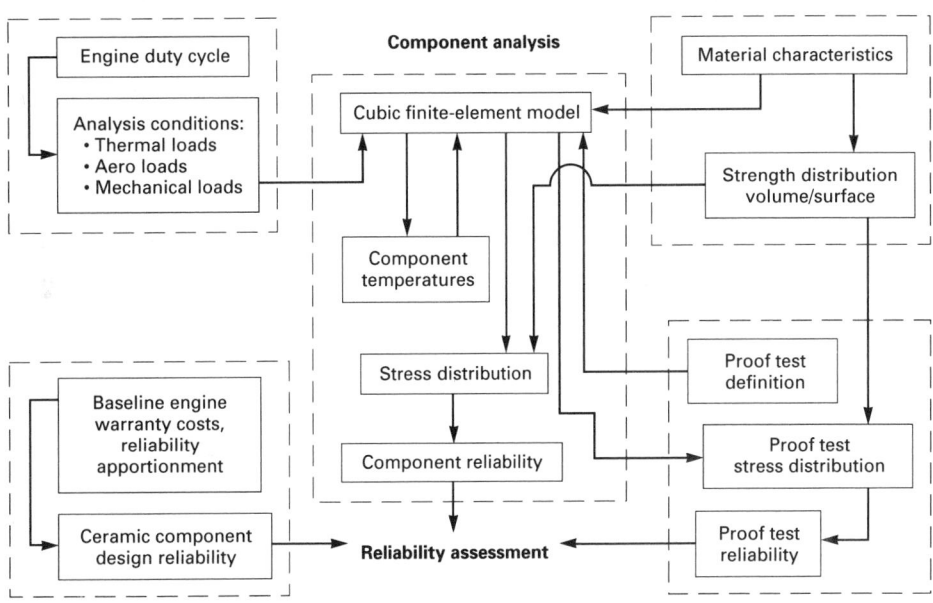

Fig 1 Ceramic component probabilistic design methodology

3-D FEM MODEL

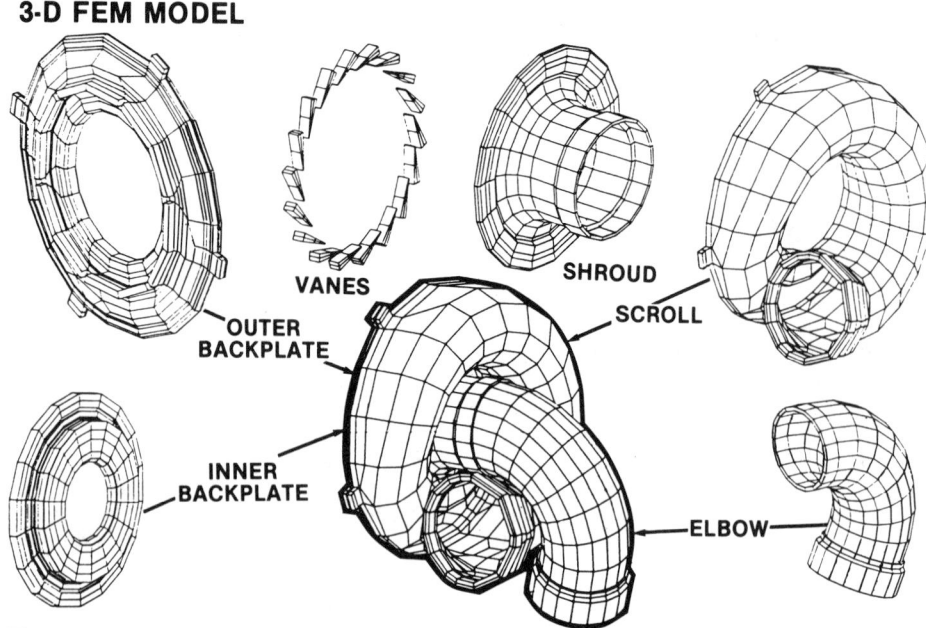

Fig 2 Three-dimensional finite-element model, AGT 100 gasifier turbine scroll assembly

acteristics that are also important to varying degrees include thermal conductivity, specific heat, Young's modulus, coefficient of thermal expansion, thermal shock resistance, method of fabrication, and cost.

Most ceramic component development activities to date have centered around high-strength structural ceramics, that is, vendor-specific silicon nitrides and silicon carbides. One exception is the low-strength zero-expansion material candidates that are suitable for the regenerator disks utilized in automotive-type engines. Development materials with moderate strength capabilities may be suitable for certain ceramic components, such as combustor bodies, turbine scrolls, rotor blade tip shrouds, and transition ducts. New ceramic materials are being introduced periodically. Some have been developed for specific applications, and others have increased strength, increased toughness, and/or increased operating temperature capability.

When designing structural ceramic components, it is extremely important to understand the operating environment. Maximum steady-state operating temperatures and/or temperature spikes/excursions, combined with the operating stresses, dictate the class of ceramic material required for each component. For components that operate at high temperature, emphasis should be placed on minimizing thermal gradients or, where this is not possible, on minimizing the stresses induced by the thermal gradients. This can be addressed by minimizing component stiffness where high thermal gradients exist and by designing components with features that have nearly uniform thermal inertia. Thermal stresses in the ceramic material are proportional to the product of elastic modulus, E,

thermal expansion coefficient, α, and temperature gradient, the latter of which is a function of the thermal conductivity, density, specific heat, and component geometry.

The manufacturability of a ceramic component should be examined early in the design process. Material formability and component-specific features, including size, shape, and thickness, determine the fabrication method. Techniques utilized to date for gas turbine component fabrication are slip casting, injection molding, die/isostatic pressing, and hot pressing. Slip casting offers relatively inexpensive tooling and the ability to form large and complex thin-walled structures, as well as thick-section turbine rotors. Slip-casting drawbacks include difficulty in controlling close tolerances, time required to cast thick sections, poor-quality surface finish, and low green strength. Injection molding offers increased dimensional control, near-net-shape components, and good surface finish. Drawbacks to injection molding include expensive tooling, part size restriction, difficulty in removing binder and limited ability to accommodate design modifications. Material contamination has also been a problem associated with injection-molded components. Die and isostatic pressing processes have similar benefits and drawbacks to injection molding; however, pressed geometries are further restricted to simple shapes such as cylinders, disks, blocks, and other shapes where simple die pull is possible. Fabrication of complex shapes from hot pressed "blanks" is possible, but not practical from both a cost and production standpoint.

Component evaluation of gas turbine ceramic parts is accomplished through nondestructive evaluation (NDE) techniques, proof tests, and by material verification of co-processed test bars. Available NDE methods for in-process and fully fabricated hardware include visual, dimensional, and fluorescent penetrant inspection (FPI), microfocus radiography, ultrasonic imaging, and x-ray computed tomography (CT). Unfortunately, the last three NDE techniques have thus far proven to be ineffective and/or cost prohibitive for most complex-shaped turbine components.

Subsequent to the successful completion of NDE, proof tests of turbine rotors include ambient spin tests that simulate operational stresses in rotors and/or blades. Static components undergo ambient mechanical loading (pressurization or flange loading) when practical. Following cold proof tests, development components receive hot rig evaluation at design operating conditions before becoming candidates for engine testing.

To further quantify the material characteristics in turbine rotors, spin tests to burst are performed. Figure 4 shows a radial inflow turbine rotor at the instant of burst at 130% of design speed. At the current state of structural ceramic development, it is imperative that NDE and proof testing be employed early and often in the fabrication process in order

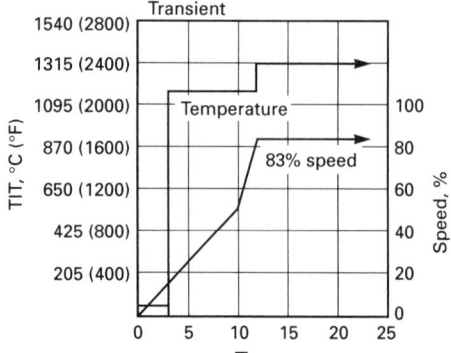

Component	Material	Steady state	Transient (50 s)
Scroll	SiC	0.999	0.952
Shroud	SiC	0.997	0.999
Elbow	SiC	1.000	0.999
Outer backplate	SiC	0.999	0.956
Inner backplate	Si_3N_4	0.999	0.998
Assembly	...	0.997	0.909

Fig 3 Probability of survival summary, AGT 100 gasifier turbine scroll assembly. TIT, turbine inlet temperature

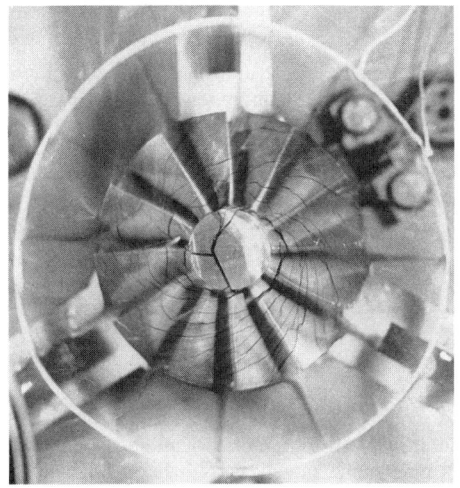

Fig 4 AGT 100 radial inflow turbine rotor burst at 130% design speed

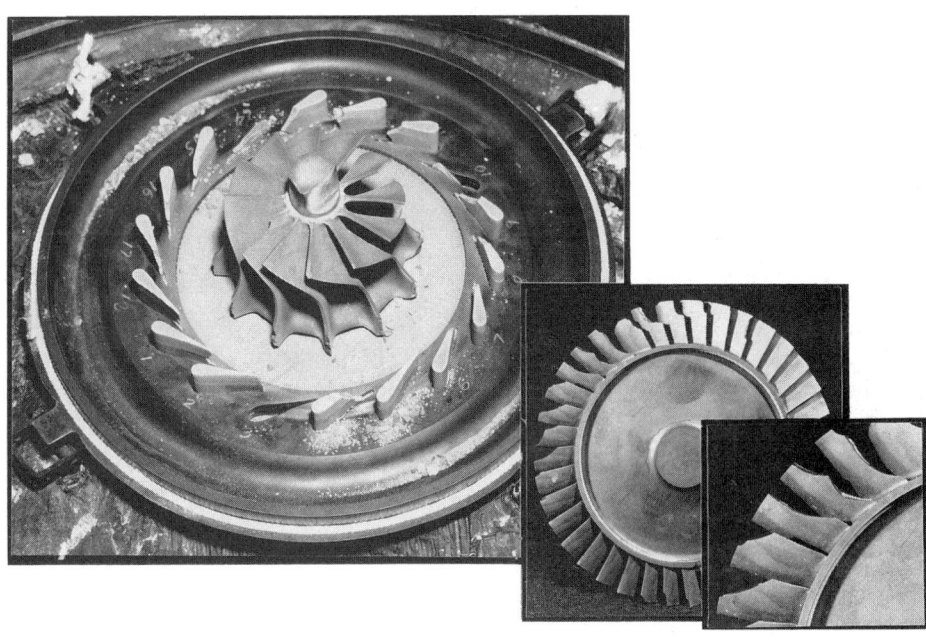

Fig 5 Foreign object damage to turbine rotors, radial inflow turbine (left), axial flow turbine (right)

to minimize the investment in substandard ceramic components.

Successful Ceramic Components

The successful performance of ceramic components has been demonstrated in gas turbine engines. The major efforts in programs subsequent to the CATE program have been directed toward turbine rotor development, specifically one-piece rotor/blade components of both radial inflow and axial flow configurations. Test experience in high-speed, high-temperature rigs and in full engine testing as well as design/manufacturing experiences have shown that there are major areas of concern in a gas turbine that utilizes monolithic ceramic components. These concerns include:

- Foreign object damage
- High-speed blade tip rubs
- High temperature and high rotational speeds
- Cyclic durability and time dependence
- Design/manufacturing tradeoffs

Foreign object damage (FOD) has proven to be a critical failure mode for ceramic blade materials. The inherent brittleness and low fracture toughness of ceramic materials can cause FOD to become a catastrophic failure mode for monolithic ceramic components, as shown in Fig 5. Sources of FOD are hard carbon from the combustor, loose pieces of insulation, and distressed metal fragments. This shortcoming has been addressed by utilizing materials of higher strength and toughness, such as silicon nitride (Si_3N_4) rather than silicon carbide (SiC), and by designing more robust blades with thicker leading and trailing edges and wider chords.

Performance of the radial inflow turbine rotors utilized by both contractors in the AGT programs (Ref 10, 11) did not meet design

requirements for a variety of reasons. The failures of many radial inflow turbine rotors during testing stem from: immature ceramic materials, inconsistency in material processing, low fracture toughness, and the propensity of the radial inflow turbine to self destruct when foreign objects are ingested, repelled, and reingested. Steps taken to increase the ruggedness of radial inflow turbine rotors were only partially successful. In ATTAP, the Garrett radial inflow turbine has been redesigned to combat failures that are due to FOD and to reduce tip speed and associated stress, whereas Allison now uses an axial flow configuration.

Initial testing of ceramic (SiC) axial flow turbine rotors was also disappointing because of blade damage that was due to FOD. However, improvements in Si_3N_4 materials and redesign of the rotor blades to greatly increase their ruggedness have resulted in rotors that have accumulated many hours of successful testing (Ref 15). Figure 6 shows the contrast between the initial turbine rotor design and current, more robust, version. Design iteration will continue as advanced ceramic materials are developed and cleaner engine operation (less FOD) is achieved, and will identify the best compromise in design parameters to meet design goals.

High-speed blade rubs have also proven to be troublesome for ceramic blades because of the low toughness of the high-temperature ceramic materials. Blade rubs result because tight clearances between rotating blades and stationary shrouds are desirable for performance reasons. Minimum cold tip clearances at assembly are established by performance and durability targets, test experience, and analyses that consider shaft dynamics excursions, centrifugal and transient thermal

growths, and tolerances. Higher-toughness materials, such as silicon nitride, and more robust blade design have significantly improved the ability of blades to sustain high-speed tip rubs without damage.

High Temperatures and Speeds. Ceramic rotors capable of running at high temperatures and high speeds have been recognized as the key to the success of the ceramic gas turbine. The ceramic materials for the blades must have sufficient as-fired surface and volumetric strength at temperature to survive at stress levels produced by centrifugal forces and thermal gradients. Thermal shock stresses at cold start conditions require the use of materials such as silicon nitride and silicon carbide for blades and vanes.

Cold starts, accelerations, decelerations, and shutdowns result in transient thermal gradients that can produce severe stresses in the disk of an axial turbine rotor and the back surface of a radial inflow turbine rotor. Present state-of-the-art ceramics have not demonstrated sufficient as-fired surface strength characteristics. Thus, the highly stressed regions of the rotor surfaces must be machined to remove surface imperfections and then heat treated to obtain the required material strength. Mass production cost considerations dictate that disk surfaces be as-fired or that a low-cost rapid machining technology be developed. Thermal stresses for the same gradient in the axial turbine wheel disk are higher for silicon carbide, relative to silicon nitride, because of its higher modulus of elasticity and higher coefficient of thermal expansion, resulting in a lower probability of survival for silicon carbide rotors. The higher strength of silicon nitride at moderate temperatures is an advantage because the transient engine oper-

Fig 6 Original design ceramic axial flow turbine rotor (left) and advanced, more rugged turbine rotor (right)

ation produces critical stresses in the disk at a time slice and in a region that is at a lower temperature than the gas temperature.

A time-dependent and cyclic-durability data base has not been sufficiently developed as yet, through hot rig and engine tests demonstrations, to influence the design of reliable components. Silicon nitride turbine rotors made by Kyocera Corporation have been run in rigs and engines to temperatures of 1400 °C (2550 °F). The accumulated time on individual rotors approaches 1000 h, and, on many rotors, is up to 3000 h. Cyclic durability testing is in progress.

Design/Manufacturing Tradeoffs. Integrally bladed axial turbine rotors have been successfully fabricated by injection molding and slip casting. Tooling requirements dictate that the airfoil passage be pullable, that is, airfoil sections must be stacked to permit the tooling inserts that form the ceramic blades to be pulled freely from between the blades without damaging the molded blades. This results in design compromises in aerodynamic performance. Disk rim configurations, blade-to-disk blending, and other parameters have been adjusted to facilitate the manufacture of a reliable ceramic rotor. Cold spin testing of rotors to stress levels to simulate operating conditions have been required as a proof test to establish engine-quality components.

Unbalance corrections on high-speed, highly stressed rotors have proven difficult for slip-cast components because of geometry variations caused by the soft tooling utilized to mold the part. To address this problem, it has been necessary to employ mass center balancing, in which the geometric center of the mount system is shifted to coincide with the mass center. Hard tooling, such as that used for injection molding, will minimize, but not necessarily eliminate, geometry variations that affect high-speed unbalance.

The attachment of a ceramic rotor to a metal shaft has been successfully accomplished by various means. Shrink-fit shafting has proven to be quite reliable, but requires precision machining, plating, and expensive low-expansion metals. Brazing is being explored, but at the present time does not have the load capacity at higher operating temperatures.

The encouraging test results of ceramic turbine rotors led to increased emphasis on the design of ceramic turbine static structure components required to permit operation of the engines to peak cycle temperatures above that possible with uncooled metallic static components, such as vanes. Mixed success was achieved with ceramic static structure components in the CATE and AGT programs. In the CATE program, the performance of vane and turbine rotor tip shrouds was improved as design revisions that reduced thermally induced stresses were implemented. This was accomplished by utilizing individually mounted vanes and by redesigning the flanges on circular components to reduce hoop stresses caused by radial thermal gradients.

In the AGT programs, the static components associated with the radial inflow turbines tended to be large in size, and, in the case of the AGT-100 engine (Ref 10), very complex in shape. The complex shape of the turbine scrolls required that several detail components be joined together to form the assembly (see Fig 2). Successful testing was accomplished with scrolls made of siliconized silicon carbide. No primary failures of ceramic-to-ceramic joints were experienced. However, failures in these joints were experienced in all scrolls tested that were fabricated of sintered silicon carbide and sintered silicon nitride. Unsatisfactory performance of axial flow turbine static structures in the ATTAP program has been attributed to deficient performance of ceramic-to-ceramic

joints. Consequently, elimination of ceramic-to-ceramic joints in component designs has been a major goal and has resulted in complex, one-piece component shapes. Figure 7 shows a turbine scroll/vane assembly for an axial flow turbine. The scroll is a one-piece casting. Loosely mounted vanes are held in place by a retaining ring. Development of successful ceramic-to-ceramic joining methods is essential to the manufacture of complex static structure components, particularly vane assemblies. Research at ORNL is presently directed toward this end.

Another aspect to the design of ceramic static components is the method by which they are attached to the metal engine structure. In contrast to rotor/shaft attachments, which are single-point attachments where no relative ceramic-to-metal movement is experienced, attachments of ceramic static structure components to metal engine structure are usually multipoint, and almost always experience relative movement. This movement is due to either differential temperatures or the rate of change in temperature, and the large differential in coefficient of thermal expansion that exists between ceramic and metallic materials. Successful designs that address these concerns through thermal isolation and freedom of relative movement have been demonstrated. These designs require closely controlled clearances, which, in ceramic components, translates into expensive post-sinter machining. A major effort is required to devise a rugged, easily assembled, and economical attachment system for mounting ceramic static structure components in an automotive engine.

All of the engine designs in the HDGTE, CATE, AGT, and ATTAP programs have utilized regenerators to improve engine cycle performance by capturing exhaust gas heat and transferring it to compressor discharge gases. Many hours of successful operation have been demonstrated with regenerator disks made of aluminum silicate (AS) material, primarily in the HDGTE and CATE programs. The achievement of design goals in the ATTAP programs has yet to be demonstrated. ATTAP program objectives require operation at

Fig 7 Ceramic turbine scroll/vane assembly for axial flow turbine

increased temperatures (to peaks of 1150 °C, or 2100 °F), which may necessitate additional material development.

Economical production of an AS regenerator disk is also a major issue. Current studies suggest that a regenerator disk fabricated of AS cannot meet the price requirements of high-production automotive applications. More economical materials, namely magnesium aluminum silicate (MAS), have not yet demonstrated the strength required for automotive application. A major effort is required to develop both a material and a manufacturing process for a ceramic regenerator disk.

Lessons Learned

A thorough analysis of a ceramic component design, with realistic operating boundary conditions, is essential to the development of a successful component. Analysis provides both insight as to why successful results are achieved during tests and a foundation on which to develop scenarios describing the factors that caused parts to fail. Analysis also can be used as an iterative tool to create an optimum design with the highest probability of success.

Difficulty of Large Components. Ceramic vendor response to questions regarding the effect of physical size of components on their probability of success has generally been that larger components present more difficulties than smaller components of the same configuration. Reasons cited for favoring components of reduced size are: less part distortion during demolding and sintering; thinner cross sections, resulting in less difficulty in removing binders; less volume of material in which failure sites may exist; and greater rigidity in the green state. However, there are practical limits to the miniaturization of components, such as fillet radii and blade trailing edge thickness.

Close coordination (simultaneous engineering) by ceramic component design engineers and ceramic processing engineers is essential to the development of a component that meets design goals. Lessons learned during development of past and current design components include the need to develop part shapes that are more conducive to simplified mold design, avoiding mold lock, a phenomenon in which the part becomes locked in the mold (introducing stress in the green part), because of part shrinkage during drying. In the case of turbine rotors, the design must allow blade passage mold segments to be pulled without distorting the air foils.

Thermally induced stresses are generally life limiting for ceramic components in an engine that operates under transient conditions, such as the automotive gas turbine. This requires that integral components be configured to have minimum thermal inertia and be free of significant constraint by associated components as they heat or cool (and expand or contract) at different rates. Materials that have high strength at moderately high temperatures, low coefficients of thermal expansion, and relatively low Young's moduli are preferable in such components.

REFERENCES

1. M.R. Vanco, *et al.*, "An Overview of the Small Engine Component Technology Studies," AIAA-86-1542, American Institute of Aeronautics and Astronautics, 1986
2. R.A. Mercure, "Small Gas Turbines for U.S. Army Auxiliary Power Systems," ASME 86-GT-282, American Society of Mechanical Engineers, 1986
3. J.C. Napier and J.P. Arnold, "Advancements in Application of Ceramics to the Gemini Radial Flow Gas Turbine," ASME 85-GT-183, American Society of Mechanical Engineers, 1985
4. A.F. McLean, Ceramics in Small Vehicular Gas Turbines, *Ceramics for High-Performance Applications*, J. Burke, A.E. Gowan, and R.N. Katz, Ed., Brookhill Publishing, 1974, p 9–36
5. R.J. Bratton and A.N. Holden, Ceramics in Gas Turbines for Electric Power Generation, *Ceramics for High-Performance Applications*, J. Burke, A.E. Gowan, and R.N. Katz, Ed., Brookhill Publishing, 1974, p 37–60
6. C.A. Anderson and R.J. Bratton, *et al.*, EPRI Ceramic Rotor Blade Program, *Ceramics for High-Performance Applications—III, Reliability*, E.M. Lenoe, R.N. Katz, and J. Burke, Ed., Plenum Press, 1983, p 783–803
7. D.W. Richerson and K.M. Johansen, "Ceramic Gas Turbine Engine Demonstration," Final Report, Navy Contract N00024-76-C-5232, May 1982
8. H.E. Helms, P.W. Heitman, L.C. Lindgren, and S.R. Thrasher, "Ceramic Applications in Turbine Engines," Final Report DOE/NASA/0017-6, NASA-CR-174715, Department of Energy, National Aeronautics and Space Administration, Oct 1984
9. R.J. Bratton and K.L. Rieke, Key Ceramic Development Issues for High-Temperature Utility Turbines, *Ceramics for High-Performance Applications—III, Reliability*, E.M. Lenoe, R.N. Katz, and J. Burke, Ed., Plenum Press, 1983, p 235–247
10. "Advanced Gas Turbine Technology (AGT) Project," Allison Gas Turbine Division, Final Report DOE/NASA 0168-11, NASA-CR-182127, Department of Energy, National Aeronautics and Space Administration, Aug 1988
11. "Advanced Gas Turbine (AGT) Development Project," Garret Auxiliary Power Division, Final Report DOE/NASA/0167-12, NASA-CR-180891, Department of Energy, National Aeronautics and Space Administration, Dec 1987
12. "Advanced Turbine Technology Applications Project," Allison Gas Turbine Division, 1989 Annual Report DOE/NASA/0336-2, NASA-CR-187039, Department of Energy, National Aeronautics and Space Administration, July 1990
13. "Advanced Turbine Technology Applications Project," Garrett Auxiliary Power Division, 1989 Annual Report DOE/NASA/0335-2, NASA-CR-185240, Department of Energy, National Aeronautics and Space Administration, Feb 1990
14. Y. Kiichiro, Y. Yamada, Y. Echizenyz, and S. Ishiwata, "Current Status of Ceramic Gas Turbine R & D in Japan," ASME 89-GT-114, American Society of Mechanical Engineers, 1989
15. *Proceedings of the Twenty-Eighth Automotive Technology Development Contractors' Coordination Meeting*, Dearborn, MI, 22–25 Oct 1990, Society of Automotive Engineers, April 1991, p 243

SELECTED REFERENCES

- "Ceramic Gas Turbine Engine Demonstration," Final Report, Contract N00024-76-C-5352, DARPA/U.S. Navy, 1976
- "Ceramic Rotor Blade Development," Vol 1, "Evaluation of MCrA/Y/CrO$_2$ Y$_2$O$_3$ Thermal Barrier Coatings Exposed to Simulated Gas Turbine Environments," Vol II, "Ceramic Combustor Design and Analysis," Vol III, Final Report, EPRI Report AP-1539-SY, Aug 1980
- S.M. DeCorso and D.E. Harrison, "Ceramic Industrial Gas Turbines," ASME 72-GT-94, American Society of Mechanical Engineers, 1972
- G.J. DeSalvo, "Theory and Structural Design Application of Weibull Statistics," WANL-TME-2688, Westinghouse Aeronautical Laboratory, May 1970
- K. Katayama, T. Watana, K. Matoba, and N. Katoh, "Development of Nissan High Response Ceramic Turbocharger Rotor," SAE Technical Paper 861128, Society of Automotive Engineers, 1986
- H. Masayo, N. Ishida, and N. Kato, "Development of Brazing Technique for Ceramic Turbocharger Rotors," SAE Technical Paper 880704, Society of Automotive Engineers, 1988
- G.W. Meetham, High Temperature Materials in Gas Turbine Engines, *Mater. Des.*, Vol 9 (No. 4), July/Aug 1988
- W.P. Parks, Jr., R.R. Ramey, D.C. Rawlins, J.R. Price, and M. vanRoode, "Potential Applications of Structural Ceramic Composites in Gas Turbines," Final Report ORNL/Sub/88-SA798/01 DOE Contract DE-AC05-840R21400, Oak Ridge National Laboratory, Sept 1989
- K.L. Reike, "Ceramics Technology Readiness Development Program Phase 1," Final Report, DOE Contract EF77-C-01-2786, Department of Energy, Nov 1989

Design Practices for Structural Ceramics in Automotive Turbocharger Wheels

Craig C. Baker and Donald E. Baker, Garrett Automotive Group, Allied-Signal, Inc.

TECHNICAL CERAMICS have generated a great deal of interest for various applications in internal combustion engines because of their excellent high-temperature strength and resistance to wear and oxidation.

One application for which ceramics are particularly attractive is turbocharger turbine wheels. These components must operate for long periods of time under conditions of high temperature and stress, which can result in creep-rupture failures in metallic materials.

However, the primary advantage of using ceramics for turbine wheels is their low density. A historic limitation of the automotive turbocharger is the time required to accelerate from the worst case, low-idle, no-load condition to some desired speed and load condition. The low density of ceramics offers a technological alternative that can significantly improve on this limitation by reducing the polar moment of inertia of the turbocharger rotating parts.

Turbocharger Operation. The modern automotive turbocharger is a device that uses the waste heat from an internal combustion piston engine to compress the intake air, thus allowing a larger quantity of fuel to be burned in the engine per unit time. This increases the power output relative to that of a nonturbocharged engine, and, when the system as a whole is properly designed, can improve either the specific fuel consumption or available torque, or reduce pollutants.

Hot exhaust gas from the engine is used to drive a turbine wheel mounted on one end of the shaft of the turbocharger (see Fig 1). On the opposite end of the shaft is a compressor wheel, which is driven by the turbine wheel and compresses the intake air. Depending on the specific application, turbine wheel temperatures can exceed 1000 °C (1830 °F). Rotational speeds range from 90,000 rev/min in large units used on large-displacement die-

sel engines to over 200,000 rev/min in smaller turbochargers used for passenger cars. At such speeds, centrifugal force can result in peak stresses of 260 MPa (38 ksi). These speeds are necessary so that the compressor wheel can provide air at rates ranging from 10 to 40 kg/min (22 to 88 lb/min) at pressures of 0.1 to 0.3 MPa (15 to 45 psi).

Advantages of Ceramics. Most internal combustion engines are required to adjust rapidly from one set of operating conditions (engine speed and load) to another. The turbocharger must respond quickly to changes in operating conditions in order to provide efficient operation of the engine system as a whole. The polar inertia of the rotating parts of the turbocharger presents the greatest limitation to turbocharger response to changes in engine operating conditions. A metal turbine wheel contributes about 70% of the total inertia. Reduced inertia, attained through lower density, is the most important advantage that ceramics offer to a turbine wheel application.

Ceramic turbine wheels were initially developed for the passenger car market, where the need for better responsiveness (to reduce so-called "turbo lag") was first recognized. In this kind of application, they contribute significantly to the driver's perception of vehicle responsiveness. They are also expected to be of benefit in commercial diesel applications, where their faster response to changes in engine operating conditions will assist in reducing the level of particulate emissions and unburned hydrocarbons during acceleration.

History

The early development of technical ceramics was primarily motivated by the need to improve the efficiency of gas turbine engines by raising their operating temperatures. By the 1960s, the potential for further

improvement in the temperature resistance of metals had begun to level off, and research efforts began to be focused on ceramic materials. During the 1970s and 1980s, a number of major programs sponsored by both United States and foreign governments (Ref 1–6) advanced the fabrication and use of non-oxide ceramics, such as silicon nitride and silicon carbide.

During this same time period, the need to reduce the rotating inertia of turbocharger turbine wheels began to be recognized. Faster response to changes in engine operating conditions, particularly throttle opening, would improve both the driveability of turbocharged passenger cars and the driver's sensation of acceleration. It was recognized that the most effective way to improve turbocharger response was to reduce the mass of the rotating parts of the turbocharger, particularly the turbine wheel. Because of their low density and excellent temperature resistance, ceramics were good candidates for such applications.

One development effort to use ceramics for turbocharger turbine wheels was initiated in 1977 by the Garrett Automotive Group of Allied-Signal Inc. Wheels were formed by slip casting, using reaction-bonded silicon nitride as the material. Initial samples were produced by the AiResearch Casting Company. Because material strength was expected to require considerable development, the first tests carried out were measurements of burst speed. Wheels were attached to the turbocharger shaft by means of an adhesive bond. The burst speeds measured during the early stages of the program (1978 to 1979) were considerably lower than those of metal wheels.

By 1982, several suppliers were involved in the program, and all were showing improved shape control and freedom from defects. As progress was made in material formulation and processing, the burst speeds

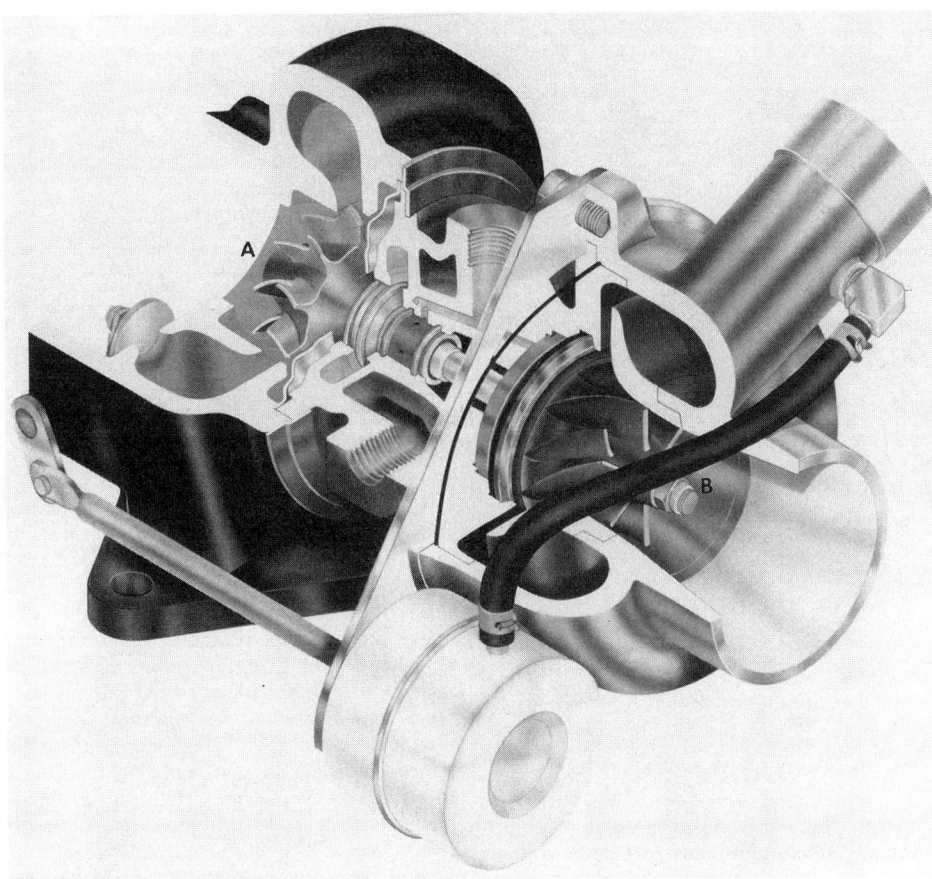

Fig 1 Diagram of turbocharger, showing principles of operation. Hot exhaust gas from engine turns turbine wheel A, which drives compressor wheel B, providing engine with increased quantity of intake air

gradually increased. Other turbocharger manufacturers began to announce plans to commercially release units with ceramic turbine wheels.

In 1983 it became clear that sintered silicon nitride was far superior to both silicon carbide and reaction-bonded silicon nitride, and efforts were concentrated on that material. The first wheel samples were received from suppliers such as NGK Insulators, Ltd. and Kyocera Corporation. Wheel quality became good enough to justify testing in vehicles.

In 1989, Garrett began to supply Nissan Motor Company with turbochargers with a ceramic turbine wheel. Designed by Garrett, the ceramic turbine wheel is manufactured by NGK Spark Plug Company, Ltd.

Technical Concerns. One of the first discoveries during vehicle testing was the sensitivity of ceramic turbine wheels to impact damage from very small particles in the engine exhaust gas. Oxide flakes, or weld spatter, from the exhaust manifold either eroded or, in some cases, shattered the ceramic wheels. This resulted in efforts to improve both the cleanliness of these engine parts and the design of the wheel to resist such damage.

Design efforts were concentrated not only on improving resistance to foreign object damage, but on minimizing the probability of failure under operating stresses. It was recognized that the brittleness and notch sensitivity of ceramics made it necessary, during design, to account for the presence of inevitable flaws by applying fracture mechanics and probability concepts. The problem of attaching the low-expansion-rate, brittle ceramic to the metallic turbocharger shaft with a joint that would support the applied loads and remain reliable under the temperature fluctuations of service in the field was solved in at least two different ways, with a graded metallic brazed joint (Ref 7) and a simple press-fit attachment (Ref 8).

Product Introduction. By 1985, ceramic suppliers had learned to produce material of high strength and Weibull modulus and high-quality, dimensionally accurate wheels. Manufacturing yields, which had been as low as 25% in the early years, were around 60% and continued to improve.

In October 1985, the first production turbocharger with a ceramic turbine wheel entered the automotive market in the Japanese domestic model Nissan 200Z.

Design Methodology*

The design of ceramic turbine wheels is heavily influenced by the properties of ceramic materials. Unlike metals, in which the effects of flaws and stress concentrations are reduced by plastic deformation, both of these factors are of critical importance to the performance of ceramic components.

The basic shape of the turbine wheel (Fig 2) is determined by aerodynamic, rather than mechanical, considerations. However, the thickness of the blades at any location is adjusted to satisfy three requirements: reducing stress concentrations, minimizing the probability of fracture because of flaws in the material, and resisting foreign object damage.

Minimizing Stress Concentrations. The thickness distribution of the blades is primarily determined to avoid stress concentrations and to make the stresses in both the blades and the hub as low and as uniform as possible, consistent with the need to provide acceptable aerodynamic performance. Preliminary hub stresses are determined using two-dimensional finite-element stress analysis on an axisymmetric model of the wheel, as shown in Fig 3. The axis of symmetry is parallel to the axis of rotation of the turbocharger. Although this model gives accurate stresses for the hub, it is not adequate for the blades because of their curvature. However, because all cross sections of the blades in a plane normal to the wheel axis are radial, they can be accurately modeled using a radial section, as shown in Fig 4.

This model is used to adjust the blade thickness at various axial locations to optimize the stress distribution in the blade and at the blade root fillet. An approximate thickness distribution for the whole blade is interpolated from these models. The resulting preliminary blade shape is then modeled in a more complicated (and computation-intensive) three-dimensional analysis.

The final three-dimensional finite-element model consists of about 650 eight-node brick elements. It is used only for the last design iterations because of the large amount of computer time required. The final refinements of wheel shape are made, and the analysis is repeated to obtain the final stress distribution.

The effect of thermal stresses is not included in the preliminary two-dimensional analysis. However, it is included in the three-dimensional model. The temperatures at various locations in the turbine wheel are calculated using the aerodynamic aerothermal solution as a boundary condition. An example of the resulting stress distribution, including the effect of temperature, is shown in Fig 5.

Minimizing Failure Probability. Four failure modes are considered when designing

---

*Most of this section was adapted from Ref 8 and therefore represents the practice and experience of the Garrett Automotive Group of Allied-Signal Inc. References to the published work of others are given as appropriate.

Fig 2 Several ceramic shaft-wheel assemblies, showing blade shapes

turbocharger turbine wheels: fast fracture, slow crack growth (static fatigue), creep, and oxidation fatigue. Of these, fast fracture has received the most theoretical attention. The discussion below focuses on only that mode to illustrate the design procedure. Details of the analysis for the other failure modes are available in Ref 9.

Ceramics are very sensitive to flaws, which are assumed to be randomly distributed either in the material or on its surface (with a different distribution in each case, of course). Therefore, under any condition of applied stress, there is a finite probability that fracture will initiate at any location in the wheel. The probability of fracture for the wheel as a whole is found by appropriately combining the probabilities of failure at a number of discrete locations in the wheel. This procedure has been described theoretically in several forms (Ref 10, 11), and has been incorporated into the computer program SCARE (later renamed CARES) (Ref 12).

Using this computer program, the probability of failure within any surface or volume element is computed and used to predict a total probability of fracture. CARES uses the principal stresses, as calculated by ANSYS at the centroid of the mesh element volume or surface area, as appropriate, to evaluate the probability of failure within any element. For example, the predicted probability of failure for the wheel depicted in Fig 5 is 0.03% at a rated speed of 177,000 rev/min and turbine inlet temperature of 955 °C (1750 °F).

Using this methodology, adjustments in the blade thickness distribution are then made, if necessary, to reduce the probability of wheel failure at rated speed.

Resisting Foreign Object Damage. Methods for analytically predicting the ability of a wheel design to resist damage from foreign objects in the engine exhaust are very limited, given the current state of the art. Therefore, the initial blade tip thicknesses are chosen by exercising engineering judgment, as a compromise between aerodynamic efficiency and durability. The resulting design is tested for resistance to foreign object damage, as described in the section "Vehicle Testing" in this article. Although some information has been gathered from ballistic impact measurements on ceramic materials (Ref 7), designing wheels to resist impact is largely empirical.

Joining Silicon Nitride to Metals. Two methods of joining silicon nitride and steel have been used in practice, namely active metal brazing and a purely mechanical attachment. The joint is located under the turbine end oil seal because the temperatures are

relatively low there, minimizing the stresses caused by thermal expansion mismatch between the ceramic and metal.

Finite-element analysis is used to analyze the possible configurations of both joint types in order to design the joint to minimize the stresses in the ceramic.

Brazed Joint Analyses. The turbine end oil seal must be metal. With a ceramic oil seal, fracture of the wheel would involve the risk of losing engine lubricant. This design requirement is satisfied by variations of the configuration shown in Fig 6. A low-expansion alloy sleeve is used between the ceramic wheel and the steel shaft, with the oil seal ring groove in the low-expansion sleeve. The finite-element model used to predict the stresses in each of the configurations is shown in Fig 7. This feature ensures that if the wheel fails, the engine oil will not be pumped out through the turbocharger (Ref 13).

In the analysis of this joint, the sleeve is made of the low-expansion alloy Incoloy 903 and is brazed to the ceramic along all mating surfaces. The brazing alloy was assumed to solidify at 771 °C (1420 °F). At this temperature the stresses in the assembly may be assumed to be zero. They increase as the temperature of the assembly is decreased toward room temperature. The result of finite-element analysis of this configuration after cooling from 771 °C (1420 °F) to room temperature is shown in Fig 8. The isopleths represent maximum principal stress, which is a more reliable predictor of failure in ceramic

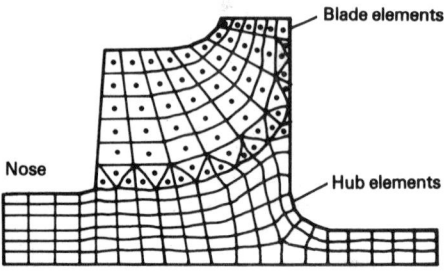

Fig 3 Two-dimensional finite-element mesh for axisymmetric modeling of a ceramic turbine wheel hub

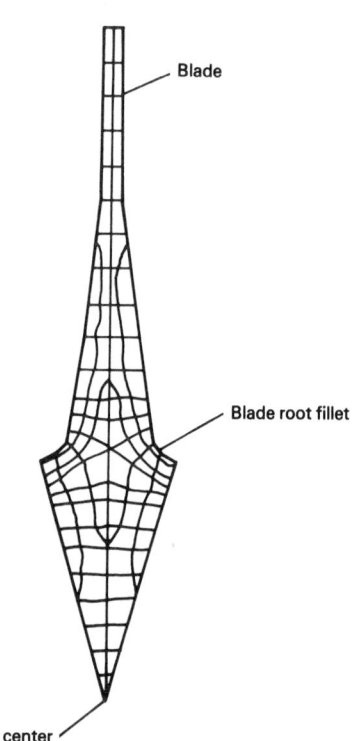

Fig 4 Two-dimensional finite-element mesh for radial modeling of a ceramic turbine wheel blade and blade/root fillet

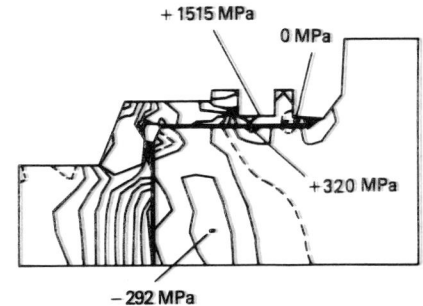

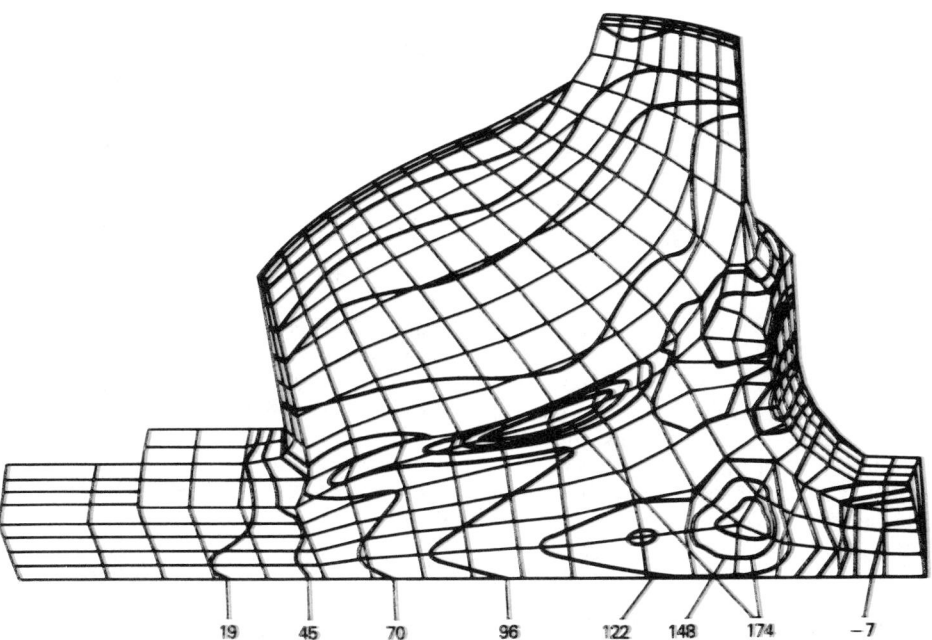

Fig 5 Distribution of maximum principal stress on the concave (high-pressure) side of a typical passenger car turbine wheel blade. Contours correspond to 177,000 rev/min and a turbine inlet temperature of 955 °C (1750 °F). Stress contours are shown in MPa.

Fig 8 Finite-element stress analysis results for joint configuration in Fig 6 at room temperature with all metal-ceramic contact surfaces brazed. Maximum and minimum values of stress in the ceramic and metal sleeve are indicated.

materials than the more frequently used equivalent stress.

The stresses predicted by the elastic analysis are about four times higher than the yield strengths of the metal parts. Accurate stress predictions for cooling to room temperature would require a plastic analysis. This was not done in the present example because plastic yielding in the metal sleeve would cause the actual stresses in the ceramic to be considerably below those values indicated by this analysis, which are already acceptable. The actual stresses in the ceramic at room temperature are estimated to be 50 to 75% as high as the analysis indicates.

Another way to accommodate the thermal expansion mismatch between metal and ceramic is with a "buffer layer," a disk of material of intermediate expansion coefficient placed between the ceramic and the steel shaft. This approach has been successful using a composite of nickel and tungsten (Ref 7) to produce the desired intermediate expansion coefficient.

Mechanical Attachment. An example of a mechanical attachment is a simple press fit using a sleeve configuration that is very similar to that shown in Fig 6, but with an interference fit between the low-expansion sleeve and the ceramic. The stress distribution for such a press fit is very similar to that for the brazed joint, if the sleeve can be assumed to not slide on the ceramic during thermal cycling. If sliding does occur, it will act to

reduce the stress in the ceramic, and the predicted stresses will be conservative.

The actual interference must be chosen to allow a sufficient attachment load at the operating temperature. The stress in the ceramic at the end of the sleeve, caused by the transition from compression under the sleeve to the free surface beyond it, is reduced by tapering the inside diameter of the sleeve.

Material Selection and Design Tradeoffs

The requirement for low mass density and the conditions of temperature and stress encountered in turbocharger turbine wheels restrict material choices to those that have high-temperature strength, oxidation resistance, and toughness. During the wheel development effort, the proper material could not be selected in a logical and systematic way because the materials were evolving, to a large extent, in support of the effort. The classes of ceramics that best satisfied the requirements of the application were the non-oxide ceramics, such as silicon nitride (in two forms), and silicon carbide. Table 1 summarizes the range of properties of each material, as of about 1983, and, for comparison, gives corresponding property values for sintered silicon nitride as of 1988.

Final Material Selection. At the beginning of the development program, all three of the materials listed in Table 1 were evaluated. Silicon carbide was eliminated early because its relatively low fracture toughness led to excessive chipping during machining. Sintered silicon nitride, rather than the reaction-bonded form, was chosen for further development because of its higher strength, especially after the development of improved sintering aids.

Because of intensive development efforts on the part of material suppliers, the characteristic strength of sintered silicon nitride improved from about 530 MPa (77 ksi) in 1983 to 900 MPa (130 ksi) in 1988. As stronger materials became available, they were used in turbine wheels. The quality of the molded

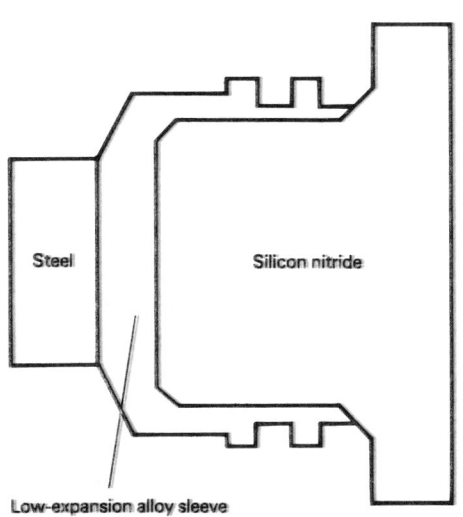

Fig 6 Configuration of braze joint that allows use of a metal oil seal

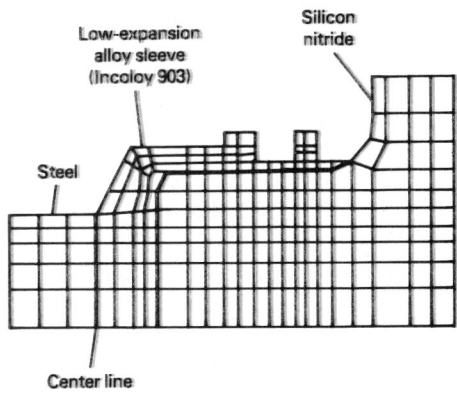

Fig 7 Finite-element model of joint configuration in Fig 6

Table 1 Properties of candidate ceramic materials

	Silicon nitride			
	Sintered		Reaction bonded, as of 1983	Silicon carbide, as of 1983
	1983	1988		
Modulus of rupture, MPa (ksi)				
at room temperature	600 (85)	900 (130)	350 (50)	500 (72.5)
at 1000 °C (1830 °F)	550 (80)	600 (85)	350 (50)	500 (72.5)
Fracture toughness, MPa$\sqrt{m}$ (ksi$\sqrt{in.}$)	7 (6.4)	7 (6.4)	6 (5.5)	4.5 (4.1)
Density, g/cm^3	3.15	3.20	3.12	3.18

parts available from suppliers also improved. Early in the program, the strength of bars cut from wheels was significantly lower than that measured on directly molded bars, but later data showed that they were virtually indistinguishable (Ref 8).

Most turbocharger turbine wheel design tradeoffs concern compromises between aerodynamic considerations and wheel durability. Aerodynamic performance is favored by making the blades as thin as possible. However, this increases susceptibility to particle impact damage and makes the wheel more difficult to produce.

Component Development

Burst Speed. The first functional test usually performed on a part such as a ceramic turbine wheel is a measurement of overall strength, such as burst speed. This measurement indicates the quality level of the manufactured part, as well as the material from which it is made. Burst speeds measured on early ceramic wheel samples were only about half the level required for satisfactory operation in a turbocharger, but they increased rapidly because the strength of the available materials improved and manufacturers quickly learned to eliminate small imperfections in the wheels. Wheels available now have burst speeds comparable to those of metal wheels (tip speeds of roughly 750 m/s, or 2460 ft/s).

Wheels that meet this fundamental requirement for burst margin can undergo a program of cyclic testing using either laboratory engines or gas stands (devices that simulate engines by supplying air at the same flow rates and temperatures). Early failures in such testing help to indicate the weak spots in the design of the turbocharger system as a whole.

Vehicle Testing. After the integrity and durability of the shaft/wheel joint and the wheel castings are established, over-the-road vehicle testing can begin. This type of testing provides a real-world experience base and can also indicate the need for design modifications that were not obvious in earlier testing.

For example, the earliest wheels tested in vehicles exhibited blade tip impact damage, which was traced to small particles of iron oxide from the engine exhaust manifold. The size of the particles that could cause this kind of damage was subsequently investigated by injecting steel spheres of various sizes into an operating turbocharger. Although the damage caused by a particle of a given size and weight depends greatly on the wheel design and the properties of the wheel material, the results of such testing indicated that particles weighing 0.5 mg (17.5×10^{-6} oz) could cause perceptible damage, and 1 mg (35×10^{-6} oz) particles sometimes completely destroyed the wheel. Although the incidence of particle impact damage has been reduced by thickening the blades of the wheel, vehicle testing also indicated the need to make changes in other parts of the engine, such as carefully cleaning exhaust systems prior to assembly.

However, before a ceramic turbine wheel could be used in large quantities in vehicles, much more extensive testing was necessary. For example, over 12.1×10^6 km (7.5×10^6 mi) were accumulated on four different turbocharger models in a wide variety of cars and trucks, both diesel and gasoline powered. This testing program was necessary to demonstrate that the newly developed ceramic wheels were a satisfactory replacement for standard metal wheels.

Nondestructive Evaluation

Because of the complicated shape, wide range of section thicknesses, and scarcity of flat surfaces in a turbocharger turbine wheel, standard nondestructive tests, such as radiographic and ultrasonic inspection, are difficult to carry out. For some special purposes, parts are inspected by real-time microfocus x-ray techniques.

Visual inspection for surface flaws is important, although after some manufacturing experience is acquired many of the defects found visually will have been eliminated by process modifications. The kinds of flaws that can be found visually include cracks, chips, pits, pocks, inclusions, fins, and flow lines.

With consideration to the difficulty and expense involved in applying most nondestructive inspection techniques, the most reliable and cost-effective test is a proof test, in which the completed shaft/wheel assembly is briefly run at its intended service rev/min and temperature.

Cost Considerations

Although silicon nitride turbine wheels were developed primarily to exploit the advantage of their low inertia and good temperature resistance, a potential for cost reduction was perceived, because the raw materials, sand and air, are cheap. However, the extreme purity required of the raw materials, long processing times, and low process yields experienced to date have kept the cost of finished parts considerably higher than that of conventional metal wheels.

The dilemma facing both users and manufacturers of ceramics and ceramic components is that the key to reducing cost is higher production volumes, and the key to higher volumes is reduced cost. The potential market for ceramic turbine wheels is probably not large enough to affect production volumes very much by itself. However, the increasing interest in ceramics for engine components other than turbocharger turbine wheels will help in this regard, and within the next few years all ceramic components may become much more economically attractive.

REFERENCES

1. C.C. Van Reuth, The Advance Research Project Agency's Ceramic Turbine Program, *Ceramics for High-Performance Applications*, J.J. Burke, A.E. Gorum, and R.N. Katz, Ed., Brook Hill Publishing, 1974, p 1
2. D.W. Richerson and K.M. Johanson, "Ceramic Gas Turbine Engine Demonstration Program," final report, Contract N00024-76-C-5352, sponsored by DARPA and prepared for the Naval Air Systems Command, May 1982
3. W. Bunk, "Overview of the German Program on Ceramic Components for Vehicular Gas Turbines," Paper I-2, International Symposium on Ceramic Components for Engines, Hakone, Japan, Oct 1983
4. H. Okuda, "National Projects on Research and Development of High-Performance Ceramics in Japan," Paper I-1, International Symposium on Ceramic Components for Engines, Hakone, Japan, Oct 1983
5. J.R. Kidwell and D.M. Kreiner, "AGT101-Advanced Gas Turbine Technology Update," Paper No. 85-GT-177, International Gas Turbine Conference, Houston, American Society of Mechanical Engineers, Mar 1985
6. L.E. Groseclose and R.R. Johnson, "Status of the AGT100 Advanced Gas Turbine Program," Paper No. 85-GT-178, International Gas Turbine Conference, Houston, American Society of Mechanical Engineers, Mar 1985
7. K. Matoba, K. Katayama, M. Kawa-

mura, and T. Mizuno, "The Development of Second Generation Ceramic Turbocharger Rotor—Further Improvements in Reliability," Paper No. 880702, International Congress and Exposition, Detroit, Society of Automotive Engineers, Feb-Mar 1988

8. C. Baker, R. Kobayashi, and D. Baker, "Garrett Experience in Ceramic Turbocharger Turbine Wheels," Paper No. 890426, International Congress and Exposition, Detroit, Society of Automotive Engineers, 1989

9. M. Matsui, T. Soma, Y. Ishida, and I. Oda, "Life Time Prediction of Ceramic Turbocharger Rotor," Paper No. 860443, International Congress and Exposition, Detroit, Society of Automotive Engineers, Feb-Mar 1986

10. H. Egli, "Development of Ceramic Turbine Wheels for Turbochargers with Emphasis on Burst Speed Prediction Methodology," International Conference on Supercharging Technology, Aachen, West Germany, Oct 1984

11. J.M. Manderscheid and J.P. Gyekenyesi, "Fracture Mechanics Concepts in Reliability Analysis of Monolithic Ceramics," TM-100174, National Aeronautics and Space Administration, Aug 1987

12. J.P. Gyekenyesi, SCARE: A Postprocessor Program to MSC/NASTRAN for the Reliability Analysis of Structural Ceramic Components, *J. Eng. Gas Turbines Power*, Vol 108 (No. 3), July 1986, p 540–546

13. J. Byrne, U.S. patent 4,749,334, June 1988

Design Practices for Structural Ceramics in Gasoline Engines

Ken Okajima, Kyocera Industrial Ceramics Corporation, and Ryuichi Matsuda, Kyocera Corporation, Japan

INTERNAL COMBUSTION ENGINES are facing global energy and environmental pressures. The energy situation presents a big hurdle for the gasoline engine as average fuel economy becomes more stringently regulated in the near future. Diesel engines also have to be improved significantly to pass 1991 and 1994 emission regulations, the latter of which is expected to be very tough to meet.

In an effort to resolve these issues, requirements for engine materials are becoming more demanding. Because ceramics have a lower density and excellent wear resistance compared to metals, as well as high hot strength properties, they are now being applied to internal combustion engines.

In terms of commercialization of ceramic parts, Japanese automotive manufacturers have been very aggressive and successful. Table 1 lists commercialized ceramic engine parts, all of which are made from silicon nitride. Although rocker arm tips and turbocharger rotors represent the only gasoline engine applications at present, valves, piston pins, and port liners are expected to be commercialized in the near future. This article describes material properties, the design procedure for structural ceramic engine parts, and the market expansion opportunities (Ref 1, 2).

Selection of Material

Table 2 identifies material properties of typical engineering (structural) ceramics. Figure 1 provides the useable operating temperatures for these materials and for aluminum oxide to allow comparison. Of course, price and weight considerations, in addition to stress and operating temperature characteristics, are important factors in the material selection process.

Silicon nitride is the most popular ceramic material being applied to engine parts. In conjunction with the light weight that is characteristic of many ceramics, its excellent thermal shock resistance, high hot strength, and relatively high fracture toughness offer the greatest potential for reducing the weight of engine parts used at high temperatures. Significant amounts of developmental work

have been conducted to apply it not only to gasoline engines, but to gas turbines as well. However, it is generally believed that the fracture toughness and strength of current silicon-nitride materials are not high enough to replace metal parts. In particular, its unreliability due to its brittleness has discouraged broad commercialization. Most efforts have been concentrated on strength and toughness improvement. As a result, material properties have been improved dramatically since the mid-1980s.

Basically, there are three major directions of development for silicon nitride. The first is to increase its strength at low temperatures based on application in highly stressed parts. The second is to increase its strength at high temperatures to allow the operating temperature of engines to be raised in order to obtain high thermal efficiency. The third development area is to improve fracture toughness to overcome its brittleness. There are some reports that fracture toughness can be improved by adding reinforcement material in the silicon nitride matrix. Figure 2 shows the development history and flexural strength improvements of silicon nitride.

Silicon carbide has a very high covalent bonding strength, like diamond. It exhibits high thermal conductivity and chemical durability, as well as excellent strength (creep resistance) at elevated temperature. However, its fracture toughness is so low that it

cannot be used in dynamically loaded parts. Consequently, its applications for engines has been to some static parts requiring high chemical durability, such as mechanical seal rings. Much developmental work to improve the fracture toughness has been performed using sophisticated doping and particle dispersion techniques. Although these techniques significantly increased fracture toughness values, high-temperature strength properties usually had to be sacrificed. As of 1991, silicon carbide has not been developed to fully overcome this obstacle.

Partially stabilized zirconia, in contrast to silicon carbide, has high strength at room temperature as well as excellent fracture toughness properties. However, its density is high and it loses a considerable fraction of its strength at moderate temperatures. Thus far, it has not been successful in applications requiring high temperatures and light weight. As a structural ceramic, it can be used for parts requiring high stress at low temperatures. The most feasible application is as a thermal barrier coating on piston heads, which utilize its low thermal conductivity. This coating technology has been used in heavy-duty diesel engines.

Design Procedure

One example of the iterative design process for ceramic components is depicted in

Table 1 Ceramic engine components under commercial production in Japan

Components	Production start	Engine	Manufacturer
Glow plug	1981	Isuzu	Kyocera
Swirl chamber	1983	Isuzu	Kyocera
Glow plug	1983	Mitsubishi	Kyocera
Intake heater	1983	Isuzu	Kyocera
Swirl chamber	1984	Toyota	Toyota
Rocker arm tip	1984	Mitsubishi	NGK
Glow plug	1985	Mazda	Kyocera
Glow plug	1985	Nissan	NTK
Turbocharger rotor	1985	Nissan	NTK
Rocker arm tip	1987	Nissan	NTK
Turbocharger rotor	1988	Isuzu	Kyocera
Link injector	1989	Cummins	Toshiba
Turbocharger rotor	1989	Toyota	Toyota
Turbocharger rotor	1990	Toyota	Kyocera
Rocker arm tip	1990	Mazda	Kyocera

Table 2 Comparison of ceramic material properties

Property	Silicon nitride	Silicon carbide	Zirconia
Bulk density, g/cm³	3.2	3.0	6.0
Flexural strength, MPa (ksi)			
at room temperature	676 (98)	490 (71)	980 (142)
at 1000 °C (1830 °F)	598 (87)	490 (71)	274 (40)
at 1200 °C (2190 °F)	392 (57)	490 (71)	167 (24)
Fracture toughness, MPa√m (ksi√in.)	5.9 (5.4)	3.4 (3.1)	6.7 (6.1)
Thermal shock resistance, water quench, ΔT, °C (°F)	670 (1240)	350 (660)	300 (570)
Thermal conductivity at room temperature, W/m·K (Btu·in./h·ft²·°F)	25 (175)	71 (490)	4 (30)

Note: Kyocera number for Si₃N₄, SN220M; SiC, SC221; ZrO₂, Z201N

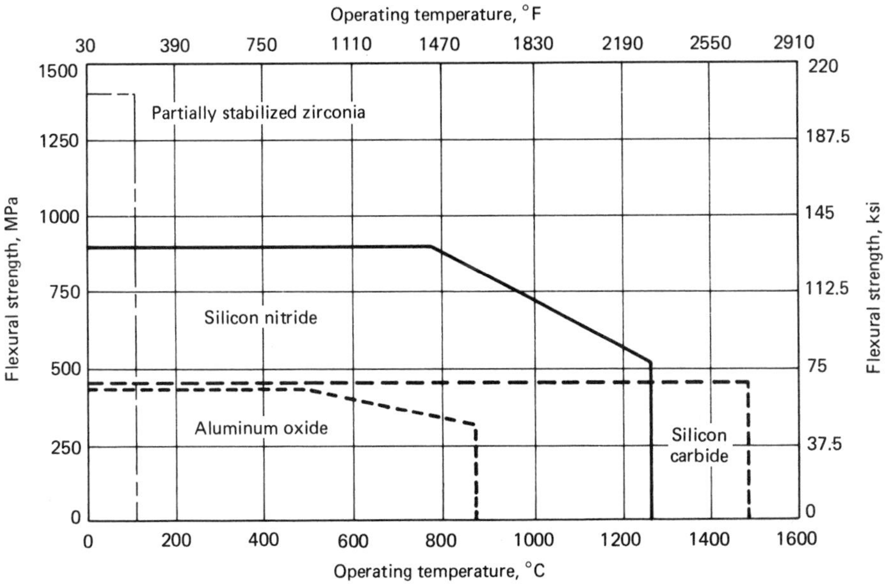

Fig 3. First, a certain material is selected based on the required operating conditions. Then, finite-element method (FEM) analysis is conducted to obtain the stress distribution within the part. The probability of failure of the part is calculated from the strength of the material and the stress distribution, utilizing linear fracture mechanics. In general, Weibull statistics can be applied to ceramics to describe fracture behavior. Probability of failure can be obtained using Eq 1:

$$F = 1 - \exp\left\{-\left[\left(\frac{S}{M}\right)\cdot\Gamma\left(\frac{m+1}{m}\right)^m\right]\right\} \qquad \text{(Eq 1)}$$

where S is the stress within the parts, M is the strength of the material, and the Weibull modulus m is a statistical measure of varia-

tion in the data. For example, Fig 4 shows correlations between the strength of ceramics and failure probability for various Weibull moduli in the case of a maximum principal stress of 390 MPa (57 ksi). It is seen that reducing the variation (increasing Weibull modulus) is as effective as increasing the strength of materials in order to make reliable ceramic parts. After this process is finished, a life prediction is conducted using characteristics of both the dynamic and static fatigue of the material. If the calculated results cannot meet the required operating conditions, the design and/or material is changed. The same process described above is then repeated.

Assurance of reliability is sometimes more important than the design when dealing with ceramic engine parts. Because ceramic is a brittle material, K_{Ic} (fracture toughness) is usually used to evaluate brittleness. It can be described as:

$$K_{Ic} = SYa^{1/2} \qquad \text{(Eq 2)}$$

where S is strength, Y is a constant determined by the shape of the defect, and a is the defect size.

Typical values of K_{Ic} for silicon nitride are between 5 and 10 MPa√m (4.6 and 9.1 ksi√in.), which is very small compared to values of 100 to 200 MPa√m (90 to 180 ksi√in.) for carbon steels. This means that ceramic parts have greater flaw sensitivity. In fact, x-ray and ultrasonic flaw detection methods are used in daily operation for metals. However, because the flaw size that must be detected for ceramic engine parts during nondestructive inspection is usually 10 μm (400 μin.), and the smallest detectable flaw size in daily operation is about 50 μm (2000 μin.), nondestructive techniques cannot provide enough assurance of reliability. Therefore, almost all ceramic parts are proof-tested in order to assure their reliability. Figure 5 indicates the importance of proof testing, in general, all sintered silicon nitride materials. It should be noted that the parts with undesirable strength values can be screened out by setting up appropriate proof conditions.

Fig 1 Useable operating temperatures for structural ceramic material

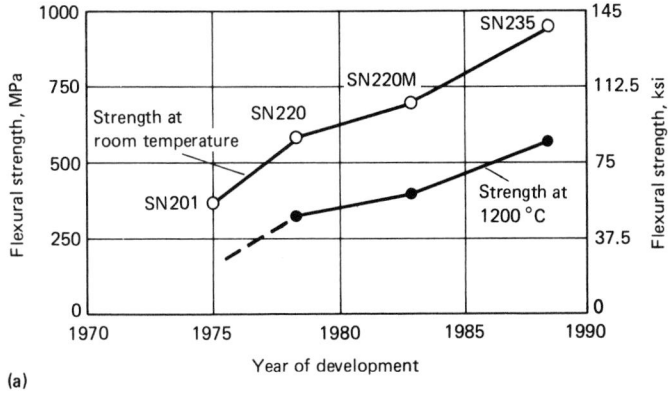

(a)

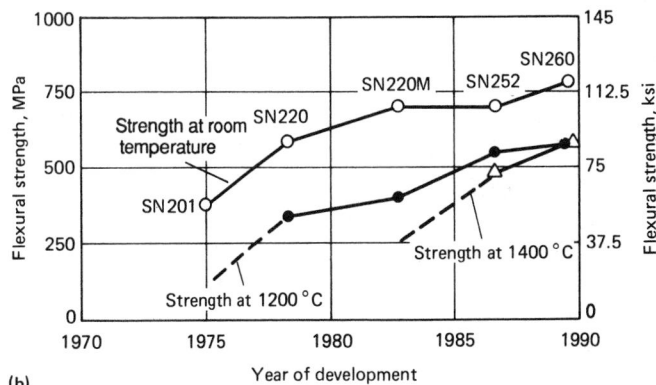

(b)

Fig 2 Flexural strength improvements of silicon nitride. (a) In high-stress applications. (b) In high-temperature applications. Courtesy of Kyocera Corporation

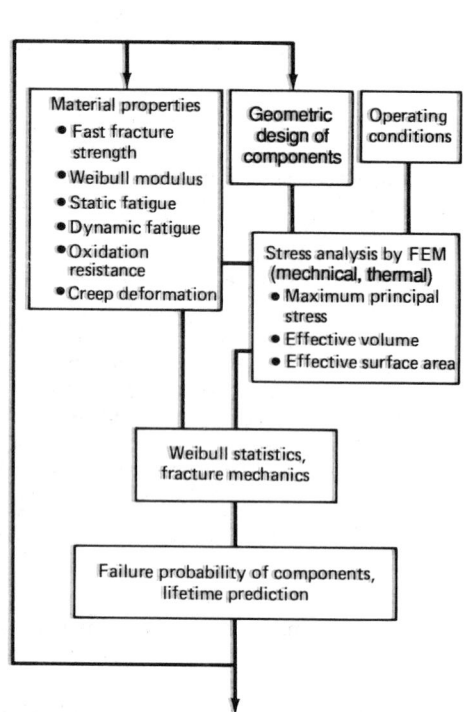

Fig 3 Design process for sintered Si₃N₄ ceramic components

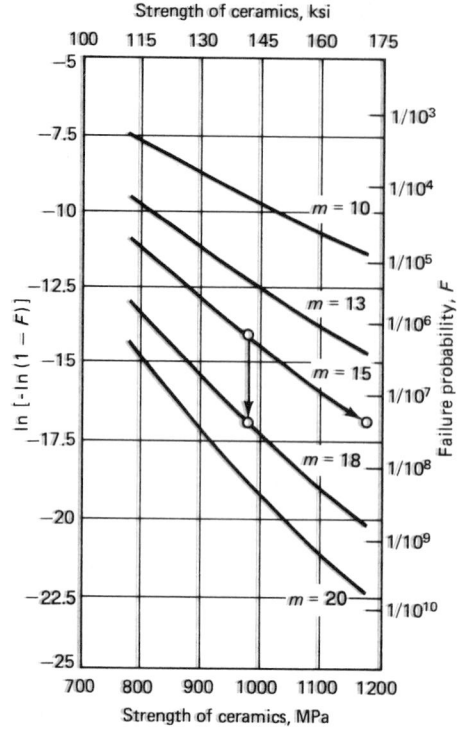

Fig 4 Correlation between ceramic strength and failure probability for various Weibull moduli. Example given at 390 MPa (57 ksi) for maximum principal stress

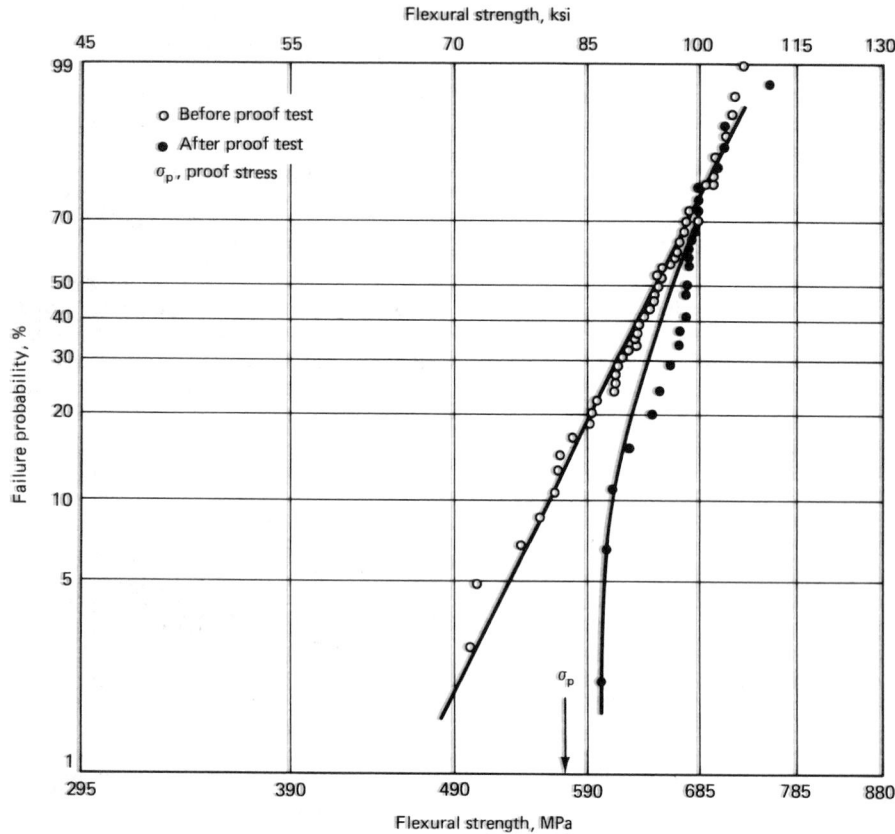

Fig 5 Positive effect of proof test in case of sintered silicon nitride

Design of Ceramic Valves

The resulting advantages of ceramic materials used in engine applications are shown in Fig 6. In valve applications, ceramics offer the valuable advantage of light weight. A lighter valve increases the maximum revolution limit. On a weight ratio basis, where a steel valve is represented as 100%, a hollow valve would have 88% of the weight of the steel valve, while a titanium valve would have 60%, and a silicon nitride valve, about 40%. When the maximum revolution limit of an engine with a metal valve is N_1, the limit associated with a ceramic valve, N_2, can be obtained from Eq 3:

$$N_2 = \left(\frac{W_1}{W_2}\right)^{1/2} N_1 \qquad \text{(Eq 3)}$$

where W_1 is the equivalent inertia weight of a ceramic valve train.

For example, in the case of valves for a passenger car engine with a stem diameter of 7 mm (0.3 in.), W_1/W_2 will be calculated to be 1.3. This means the revolution limit can be improved by 14%, theoretically. In fact, an 11% improvement was confirmed in the engine test. Load for the valve spring is reduced with a lightweight valve. This reduction in the friction loss corresponds to about 1% improvement in fuel efficiency.

Ceramics generally have better wear resistance than metal. In addition to this generic characteristic, a high Young's modulus prevents deformation of the valve head during combustion. This means that the friction between the valve head and valve seat is reduced significantly, thus improving the rate of wear.

Because ceramics are very sensitive to stress concentration, careful designing is required. In the case of valves, design of the lock groove and fillet becomes very important. Machining conditions also have to be considered because there is a strong relationship between the surface finish and the strength of the ceramic parts. Appropriate machining con-

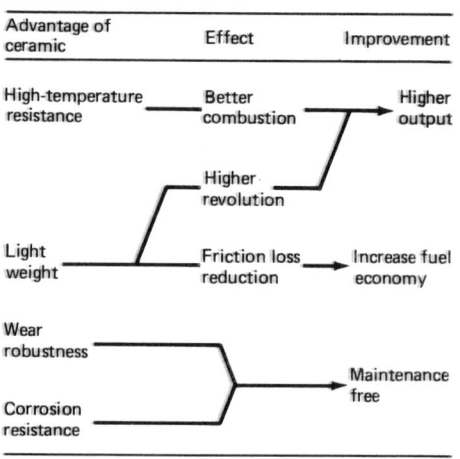

Fig 6 Advantages of ceramic engine valves

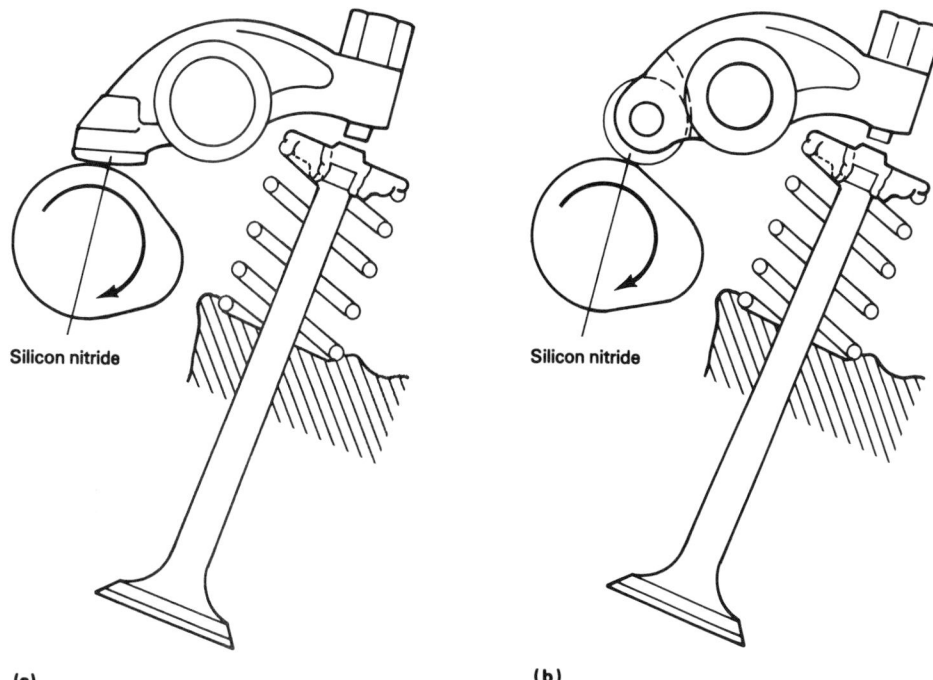

Fig 7 Rocker arms engine components. (a) Slipper type. (b) Roller type

ditions have been determined based on stress distribution. For example, careful attention has been paid to the selection of a diamond wheel and the machining direction for the highly stressed portion, such as the stem and groove areas.

Sliding Parts

Two typical rocker arms are shown in Fig 7. The commercial goal of using ceramics in the tips is to reduce weight and provide excellent wear resistance. For the slipper-type rocker arm, development efforts eliminated stress concentration by optimizing the shape. In addition, the grinding process was eliminated, except for the sliding face, in order to reduce cost. These kinds of design and process changes enabled the commercialization

of ceramic rocker arm tips that have low friction coefficients and long life, compared to metal tips. Figure 8 compares the wear rates of various materials used for rocker arm tips.

It should be noted that the wear rate of metal cams was reduced considerably.

Turbine Wheel

The great merit of a ceramic turbine wheel used for turbochargers is weight reduction. The density of the nickel alloy used for the turborotor is about 8.2 g/cm³, while the density of silicon nitride is about 3.2 g/cm³. This means that if a turborotor is made from silicon nitride, while keeping the same dimensions, about a 60% weight reduction becomes possible. Due to this low-inertia wheel system, the ceramic turborotor can reduce the turborotor lag significantly, even from a low rev/min.

As far as selection of ceramic material is concerned, low density and high strength at high temperatures are the key properties. Although both silicon nitride and silicon carbide meet these requirements, only silicon nitride was commercialized due to its superior thermal shock resistance and higher fracture toughness. Even for silicon nitride, the metal design cannot be applied directly because of its brittleness. When the metal design is utilized for the ceramic rotor, it usually fractures at low speed due to stress concentration. Silicon nitride, with high average strength and high Weibull modulus, should be selected in order to overcome this problem.

In addition to the proper material selection, redesign of the metal part using FEM is vital to eliminate stress concentration for ceramic applications and provide the necessary safety factor against brittle fracture strength by reducing the maximum principal stress. However, there is usually little freedom to redesign the original because the design has to reflect the aerodynamic requirements to obtain high

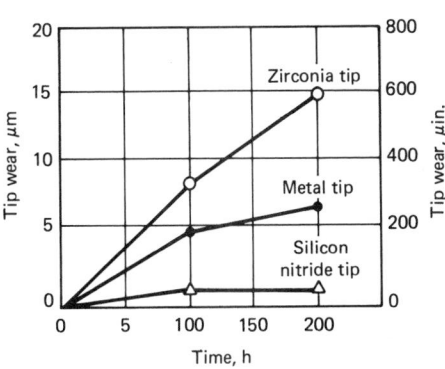

Fig 8 Wear rates for various materials used as rocker arm tips. Oil used was SAE 30 and 2% carbon; oil temperature was 100 °C (212 °F).

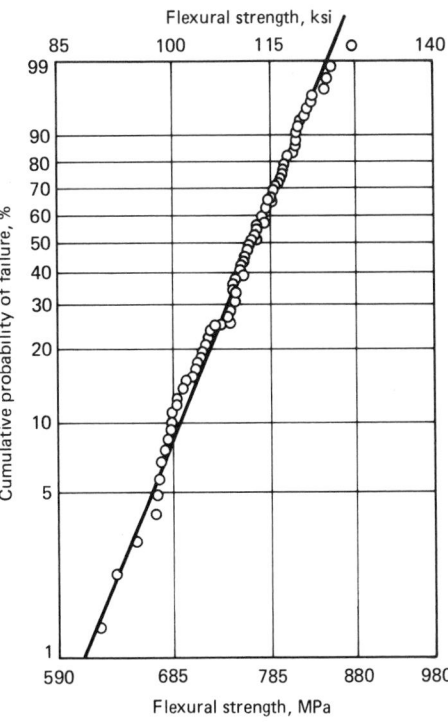

Fig 9 Weibull distribution of SN235 at 950 °C (1740 °F). Weibull modulus, 18.4; maximum strength, 870 MPa (125 ksi); minimum strength, 610 MPa (88 ksi); average strength, 760 MPa (110 ksi); number of specimens, 112

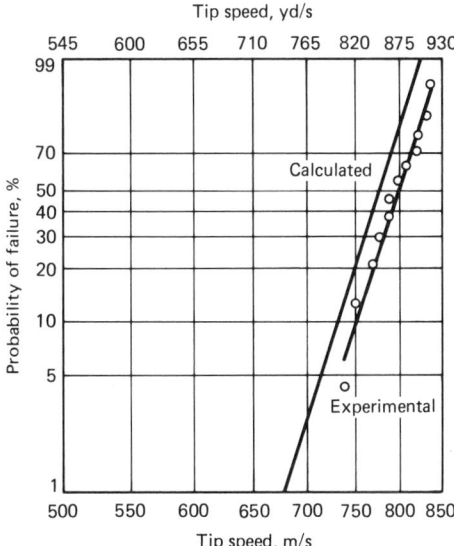

Fig 10 Analytical versus experimental test results comparison. Probability of failure at tip speed of 545 m/s (596 yd/s) was 8×10^{-4} for the experimental result and 1×10^{-3} for the predicted result.

efficiency in the turbocharger. Silicon nitride with a room-temperature strength of 900 MPa (130 ksi) was developed to meet this requirement.

Figure 9 shows the strength data at 950 °C (1740 °F) obtained from the specimens machined out of the rotor. It should be noted that material properties should always be evaluated on real parts. A turborotor with a maximum principal stress of 400 MPa (750 ksi) was designed with the high-strength silicon nitride. In this case, failure probability became 0.1% at a tip speed of 545 m/s (596 yd/s) and the burst speed was found to be 750 m/s (820 yd/s). Figure 10 shows the comparison between analytical results and experimental results for the burst test. Without this kind of positive agreement between analytical evaluation and rig test results, it is impossible to redesign ceramic engine parts.

REFERENCES

1. M. Yamamoto, Application of Ceramics to Automotive Components, Now and Future, *Innovation in Materials for the Transportation Industry Seminar*, Vol 2, Torino, ATA, 1987, p 95
2. M. Yoshida *et al.*, "Silicon Nitride for Automotive Applications," Paper 890424, Society of Automotive Engineers

Design Practices for Whisker-Toughened Ceramic Components

Stephen F. Duffy*, Cleveland State University
Atef F. Saleeb, The University of Akron

CERAMIC-MATRIX COMPOSITES, when compared with monolithic ceramics, have the potential advantages of increased fracture toughness and creep and corrosion resistance at very high service temperatures. Primary applications that are under consideration are advanced turbine engine components, cutting tool bits, heat exchangers, and aerospace components. Considering that these composites will be produced from nonstrategic materials, it is not surprising that concerted research efforts are underway, both in the materials science field (to advance processing techniques) and the engineering mechanics field (to develop design methodologies for these material systems).

The material systems of interest in this article are whisker-toughened ceramic-matrix composites. These systems exhibit a range of behaviors from essentially isotropic (random fiber orientation) to nearly orthotropic (aligned fiber orientation). The extent of anisotropy is often described by categories of material symmetry. Design practice must account for anisotropy in determining the stress states throughout a whisker-toughened ceramic component. Directional dependence can be characterized in terms of macroscopic stiffness parameters, which are determined experimentally.

Alternatively, constitutive relationships have been developed (Ref 1) that relate the fabrication-induced microstructure to thermoelastic properties. These relationships account for the constituent thermoelastic properties and the orientation, size, and shape of the whisker reinforcements, as well as the volume fractions of the constituents. Thus, when designing the material the orientation of the whiskers can be tailored for directions of high strength and stiffness that dramatically improve component performance. A method to predict thermoelastic properties based on the interaction of a number of these parameters that

characterize the composite microstructure is reviewed in this article.

Failure analysis of components fabricated from ceramic-matrix composites requires a departure from the design philosophy (that is, the factor of safety approach) that prevails when designing metallic structural components, which are more tolerant of flaws. Because failure of components fabricated from whisker-toughened ceramics is governed by the scatter in strength, statistical design approaches must be used. This article reviews several phenomenological failure models and reports on the development of a public domain computer algorithm that, when coupled with a general-purpose finite-element program, predicts the time-independent reliability of a structural component under multiaxial loading conditions. The present version of this algorithm, known as TCARES (toughened ceramics analysis and reliability evaluation of structures), is a direct offspring of the CARES, or SCARE, program (Ref 2), which is widely used in the design of monolithic ceramic components.

In addition to capturing the inherent scatter in strength, the reliability analysis of components fabricated from whisker-toughened ceramics must account for material symmetry imposed by whisker orientation. A noninteractive macroscopic model has been presented that accounts for the transversely isotropic material symmetry (Ref 3) often encountered in hot pressed and injection-molded whisker-toughened ceramics. A similar model (Ref 4) has been proposed for whisker-toughened ceramics with orthotropic material symmetry. This continuum approach excludes any consideration of the microstructural events that involve interactions between individual whiskers and the matrix.

Other authors have addressed fracture of ceramic-matrix composites on a more local scale. A model based on probabilistic principles has been developed to compute an increased energy absorption during fracture due to whisker pullout (Ref 5). The processes

of crack deflection (Ref 6) and crack pinning (Ref 7) have also been addressed. The latter two approaches are founded in deterministic fracture mechanics. Because these crack mitigation processes strongly interact, it is difficult to experimentally detect or analytically predict the sequence of mechanisms leading to failure.

A more feasible approach is to compute reliability in terms of macrovariables by using a continuum-based criterion. This underscores a fundamental difference that exists between the materials scientist and the engineer. The materials scientist focuses on mechanisms of failure at the microstructural level, whereas the engineer focuses on this issue at the component level. The failure models currently incorporated into the computer algorithm TCARES adopt the viewpoint of the engineer. This point of view implies that the material element under consideration is small enough to be homogeneous in stress and temperature, yet large enough to contain a sufficient number of whiskers, such that the element is a statistically homogeneous continuum. This does not imply that the microscopic and macroscopic levels of focus are mutually exclusive. Indeed, a close relationship must exist between the materials scientist and the design engineer to develop better failure models to facilitate the use of ceramic materials in structural components.

Estimating Thermoelastic Properties

Analysis of the stress state developed in a whisker-toughened ceramic component requires good knowledge of the thermoelastic properties. Indeed, accurate representation of the stress fields are critical to the reliability models presented here. Combining rules are necessary to relate the composition and the process-dependent microstructure to thermomechanical performance. Extensive research efforts have resulted in relationships that pre-

*NASA Resident Research Associate

dict thermoelastic properties for general heterogeneous systems (for example, continuous fiber, whisker/short fiber, or particulate-reinforced composites). Detailed accounts, including state-of-the-art reviews as well as critical assessments of major contributions at different stages of developments, are available in the literature (Ref 1, 8–10). The cited methods make use of mechanics of materials concepts, embedding or self-consistent fields, bounding theorems, and semi-empirical rules.

For the class of discontinuous-fiber-reinforced composites of interest here, a rather elaborate analytical approach that explicitly accounts for important microstructural features has been established. Referred to as the aggregate model (Ref 1, 8, 10), it has recently emerged as a promising tool in developing comprehensive relationships to predict thermoelastic properties for short-fiber polymeric composites. To date, specific applications of this model to whisker-reinforced ceramics are not available. However, the same degree of success is anticipated in view of the many qualitative similarities in the fundamental response characteristics of these two composite systems, as well as similarities in the dominant geometrical features of their microstructures. Consequently, the description of the computational procedures necessary for the application of the aggregate model are described below.

A matrix format is used throughout much of this section of the article. This is convenient in the subsequent development of computer algorithms linked with the overall reliability analysis outlined in the sections that follow. Mechanical properties are collectively described by the (6 × 6) arrays of elastic stiffness, C, and compliance moduli, S, with respect to a fixed material coordinate system (see Fig 1). Hence,

$$\sigma = C \cdot \epsilon \qquad (Eq\ 1)$$

and

$$\epsilon = S \cdot \sigma \qquad (Eq\ 2)$$

where

$$C \cdot S = I \qquad (Eq\ 3)$$

Here, I is the (6 × 6) identity matrix, and the familiar contracted notations are implied for the (6 × 1) vector representations of the stress tensor components, σ, and strain tensor components, ϵ. The engineering definition for shear strain components in ϵ is also assumed. In addition, the (6 × 1) vector α is used to denote coefficients of thermal expansion and is similarly associated with a fixed material coordinate system.

Expressions, in terms of homogenized engineering material descriptors such as macroscopic Young's moduli, shear moduli, and Poisson's ratios, for the independent entries in C or S are well known for transversely isotropic (Ref 11) and orthotropic material symmetries (Ref 12). However, the application of the aggregate model yields relationships in

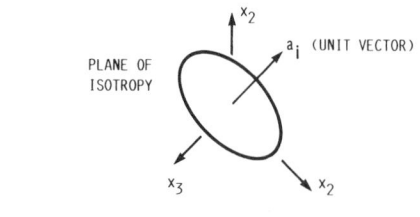

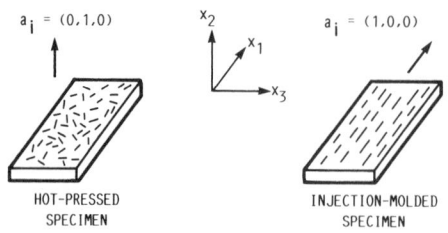

Fig 1 Examples of transversely isotropic whisker-toughened ceramics

terms of fiber geometry and orientation that are used to predict the effective composite stiffnesses (or compliance moduli) and coefficients of thermal expansions.

The aggregate model relationships are derived from an assumed interaction of a number of parameters that include the properties of the composite constituents and their volume fractions. A measure of the load transfer efficiency associated with different reinforcing geometries is represented by an effective aspect ratio, a_e. The parameter a_e is the effective length-to-diameter ratio of the reinforcing phase. In addition, intermediate states of fiber orientation require suitable measures that account for fabrication-induced directional dependency. This is accomplished through the use of four orientation parameters f_p, g_p, f_a, and g_a, which are referred to as integrated Hermans orientation parameters (Ref 13). The subscripts signify planar (p, two-dimensional) and axial (a, out-of-plane tilting, that is, three-dimensional) orientations, respectively. The specific roles that the aspect ratio and orientation parameters play in establishing the thermomechanical properties of the composite material will be presented.

Figure 2 depicts how these microstructural parameters and the features they represent are distinguished in various composite systems. From this perspective, continuous-fiber composites have the simplest microstructure to quantify. The continuity of perfectly collimated fibers ($a_e = \infty, f = 1$) reflects the physically appealing and simplifying condition that the strain field parallel to the aligned fibers is essentially uniform. With the ability to quantify the reinforcing geometry and orientation, composites with much more complex microstructures can be modeled. For example (refer to Fig 1 and 2), the orientation parameters f_p and g_p assume the value of zero for a totally random planar distribution found in hot pressed, whisker-reinforced ceramics.

On the other hand, $f_p = g_p = 1$ for the case of perfect collimation of whiskers, which is typical of injection-molded, whisker-toughened ceramics. Microstructures with partial fiber alignment are indicated by intermediate values of these parameters.

The general approach of the aggregate model necessitates the stipulation of combining rules. These rules (based on concepts found in Ref 14 through 16) center on the idea of partitioning the composite microstructure into subregions of characteristic material. The composite is then viewed as some combination (or aggregate) of these subregions. The subregions are partitioned in such a way that the properties (with respect to a local material coordinate system) of a characteristic subregion are equivalent for all subregions. These equivalent subregions are treated as an apparent homogeneous (but anisotropic) material with individual arrays of thermoelastic properties (C or S, and α). This abstraction reduces the analysis to that of a one-component material (similar to polycrystalline or directionally solidified metals) in which the subregions assume definite orientations with respect to the global axes of the component.

The application of the aggregate model includes two major steps. The first involves the estimation of the properties of a typical subregion. The second specifies the appropriate procedures for calculating averaged macroscopic properties that account for orientation dependence.

In estimating the properties of the typical subregion, one assumes that the reinforcing phase is composed of transversely isotropic, ellipsoidal inclusions embedded in an isotropic matrix. The starting point is a variational treatment (Ref 8), which leads to a set of generalized relations for the moduli and coefficients of thermal expansion of the characteristic subregion, that is,

$$C^* = C^0 + (M^{-1} + E^0)^{-1} \qquad (Eq\ 4)$$

where

$$\alpha^* = (C^*)^{-1}\Gamma^* \qquad (Eq\ 5)$$

$$M = \Sigma V_j m_j \qquad (Eq\ 6)$$

$$m_j = [(C_j - C^0)^{-1} - E^0]^{-1} \qquad (Eq\ 7)$$

$$\Gamma^* = \Sigma V_j \Gamma_j + (\tfrac{1}{2})[I + ME^0]^{-1}Q \qquad (Eq\ 8)$$

$$Q = [ME^0(\Sigma V_j \Gamma_j)] - \Sigma V_j m_j E^0 \Gamma_j \qquad (Eq\ 9)$$

$$\Gamma_j = C_j \alpha_j \qquad (Eq\ 10)$$

Here, V_j represents individual volume fractions of the constituents. The symbol Σ implies a summation on j from 1 to the number of constituents in the composite system. The possible existence of an interface between the whisker and the matrix is not recognized here, so that whisker-toughened composites are composed of only two constituents. Hence, the phase indicator $j = $ "w" refers to the whisker, and $j = $ "m" refers to the matrix phase.

Type	Symmetry	Microstructure
Continuous fibers 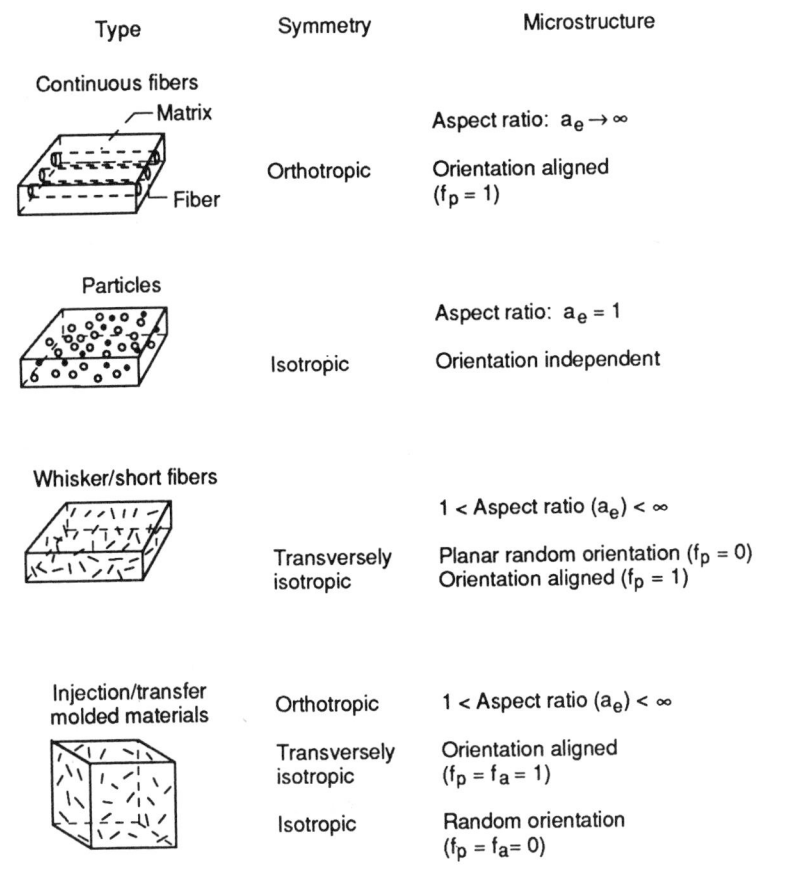	Orthotropic	Aspect ratio: $a_e \to \infty$ Orientation aligned ($f_p = 1$)
Particles	Isotropic	Aspect ratio: $a_e = 1$ Orientation independent
Whisker/short fibers	Transversely isotropic	$1 <$ Aspect ratio (a_e) $< \infty$ Planar random orientation ($f_p = 0$) Orientation aligned ($f_p = 1$)
Injection/transfer molded materials	Orthotropic Transversely isotropic Isotropic	$1 <$ Aspect ratio (a_e) $< \infty$ Orientation aligned ($f_p = f_a = 1$) Random orientation ($f_p = f_a = 0$)

Fig 2 Definition of the microstructural features distinguishing various classes of composite materials (after Ref 1)

The various C's, m's, and M are (6×6) arrays of elastic constants, and the various Γ's, α, and Q are (6×1) vectors of thermal coefficients. Note that an asterisk indicates the effective aggregate properties. The superscript zero signifies properties of an arbitrary reference material. The (6×6) array E^0 is often referred to as the Eshelby tensor, and it accounts for two-point correlation effects from adjacent constituents. Details are given in Ref 9 for the elements of the Eshelby tensor representing a transversely isotropic material in terms of C^0 and the apparent aspect ratio parameter a_e ($0 < a_e \leq \infty$). A range of 10 to 100 for a_e is typical of whisker-reinforced ceramics (Ref 17, 18).

If a size distribution is exhibited (Ref 19), one may estimate the apparent aspect ratio from the root mean square of the effective aspect ratio (Ref 1), that is,

$$a_e = \sqrt{\langle \bar{a}^2 \rangle} \tag{Eq 11}$$

where

$$\langle \bar{a}^2 \rangle = \int \bar{a}^2 p(a) \, da \tag{Eq 12}$$

and $p(a)$ is the relative number of inclusions having aspect ratio $\bar{a}$. Instead, one may use the following simplified relationship

$$a_e = [\langle \bar{a} \rangle^2 + s_a^2]^{1/2} \tag{Eq 13}$$

where $\langle \bar{a} \rangle$ is the average aspect ratio and s_a is

the standard deviation, both of which are determined experimentally. Typical values representing a Si_3N_4 matrix reinforced with SiC whiskers are provided in Ref 17. Note that a positive sign is selected in Eq 13 to ensure realistic values for any average/standard deviation combination.

The accuracy of the effective property estimates given by Eq 4 and 5 depends on the selection of appropriate reference elasticity properties in the C^0 array. The extreme cases of assigning zero and infinite rigidities to the elements in C^0 lead to the lower and upper bounds of Reuss and Voigt, respectively. Immediate improvement in these bounds may be obtained if C^0 is assigned the properties of the matrix phase ($C^0 = C_m$) or the whisker phase ($C^0 = C_w$). Unfortunately, these latter bounds are often too far apart.

A more reasonable choice for the reference elasticity properties is obtained on the basis of a "standard" composite of spherical inclusions ($a_e = 1$) with the same material properties and volume fraction concentration as those of the inclusions. A semi-empirical model, called the "S-combining" rule (Ref 10), is used to calculate the elastic constants found in the C^0 array, which are then adopted as the elastic properties for the reference material. The elements of this array are derived from the upper and lower bounds of the compli-

ance array, S_U and S_L, respectively. These compliance arrays are estimated by alternatively assigning $C^0 = C_w$ and $C^0 = C_m$ in Eq 4 and equating the inverse of the upper and lower bounds on the resulting stiffness arrays to S_U and S_L, that is,

$$S_L = C_L^{-1} \tag{Eq 14}$$

$$S_U = C_U^{-1} \tag{Eq 15}$$

The following S-combining rule is then used to obtain S^0:

$$S^0 = V_m S_L + V_f S_U + \gamma V_f V_m (S_L - S_U) \tag{Eq 16}$$

where

$$C^0 = [S^0]^{-1} \tag{Eq 17}$$

Here, γ is a parameter whose value depends on limiting aspects of phase continuity, maximum packing volume, and packing geometry. A value of $\gamma = 1/2$ was suggested in Ref 1 and 10 for polymeric particulate composites, which have vastly different values of constituent compliance moduli. If, in comparison, the compliance moduli of the constituents are reasonably close, as they are in whisker-toughened ceramics, the value of γ becomes relatively insignificant because the value of ($S_L - S_U$) approaches zero. For this case, S^0 is given by the simple rule of mixtures, that is, the sum of the first two terms of Eq 16.

Once the elastic properties of a typical subregion have been determined, the final step requires the computation of volume-averaged properties. The assumption is that the whiskers in a typical subregion are aligned such that a preferred direction of strength can be identified. The distribution of preferred directions throughout the material is then specified by an orientation density function, $n(\phi, \theta, \zeta)$ where ϕ, θ, and ζ are any set of three Euler angles. These angles identify the local preferred direction of a subregion with respect to the fixed global axes x_i ($i = 1, 2, 3$; again, resorting to index notation). The orientation density function is generally taken in the separable form

$$n(\phi, \theta, \zeta) = n_1(\phi) \, n_2(\theta) \, n_3(\zeta) \tag{Eq 18}$$

and is used as a weighting function to determine volume-averaged properties.

For the special case of a planar distribution of whiskers (see the hot pressed specimen in Fig 1), only one angle (for example, ϕ) is necessary for a complete description of the orientation density function. This function then must satisfy a set of normalization and symmetry conditions (Ref 10). In terms of ϕ, the symmetry conditions are

$$n_1(\phi) = n_1(-\phi) \tag{Eq 19}$$

and

$$n_1(\phi) = n_1(\pi + \phi) \tag{Eq 20}$$

The normalization condition requires

$$\int_0^\pi n_1(\phi) \, d\phi = 1 \tag{Eq 21}$$

The volume-averaged components of the effective moduli C^*, and the effective thermal expansion coefficients of α^* are obtained from

$$\langle C^* \rangle = \langle T_c \rangle C^* \qquad (Eq\,22)$$

and

$$\langle \alpha^* \rangle = \langle T_\alpha \rangle \alpha^* \qquad (Eq\,23)$$

where $\langle T_c \rangle$ and $\langle T_\alpha \rangle$ are matrix transformation operators weighted with respect to the orientation distribution function. In index notation, these matrix operators correspond to the following weighted transformation law for a fourth-order tensor

$$\langle C_{ijkl} \rangle = \langle \omega_{si}\omega_{tj}\omega_{uk}\omega_{vl} \rangle C_{stuv} \qquad (Eq\,24)$$

and a second-order tensor

$$\langle \alpha_{ij} \rangle = \langle \omega_{ki}\omega_{lj} \rangle \alpha_{kl} \qquad (Eq\,25)$$

A summation is implied over repeated indices in these expressions, and ω_{ij} is the second-order tensor of direction cosines. In the same sense that the stiffness array C is a contracted form of the fourth-order tensor C_{ijkl}, the weighted matrix transformation operators $\langle T_c \rangle$ and $\langle T_\alpha \rangle$ are the respective contractions of the weighted fourth-order tensor transformation operator

$$\langle \omega_{si}\omega_{tj}\omega_{uk}\omega_{vl} \rangle = \int_0^\pi n_1(\phi)\omega_{si}\omega_{tj}\omega_{uk}\omega_{vl}\,d\phi \qquad (Eq\,26)$$

and the weighted second-order tensor transformation operator

$$\langle \omega_{ki}\omega_{lj} \rangle = \int_0^\pi n_1(\phi)\omega_{ki}\omega_{lj}\,d\phi \qquad (Eq\,27)$$

Note that the second-order tensor of direction cosines is functionally dependent on ϕ for the special case of a planar distribution of whiskers.

Because of the symmetry conditions imposed on the orientation distribution function (Eq 19 and 20), the elements of the matrix transformation operators are linear combinations of the orientation parameters

$$f_p = 2\langle \cos^2(\phi) \rangle - 1 \qquad (Eq\,28)$$

and

$$g_p = (\tfrac{1}{5})[8\langle \cos^4(\phi) \rangle - 3] \qquad (Eq\,29)$$

where

$$\langle \cos^{2,4}(\phi) \rangle = \int_0^{\pi/2} n_1(\phi)\cos^{2,4}(\phi)\,d\phi \qquad (Eq\,30)$$

The orientation functions (which define the preferred directions in each subregion) have been constructed in such a way that $f_p = g_p = 1$ corresponds to the alignment of the whiskers. Alternatively, $f_p = g_p = 0$ characterizes a random distribution of whiskers. In general, the parameters f_p and g_p are independent.

However, if one takes the general form of the orientation distribution function in the following manner

$$n_1(\phi) = k^{-1}\cos^q(\phi)\,d\phi \qquad (Eq\,31)$$

where

$$k = \int_0^{\pi/2}\cos^q(\phi)\,d\phi \qquad (Eq\,32)$$

for any integer, q, then

$$g_p = 2f_p(7 - 2f_p)/5(4 - 2f_p) \qquad (Eq\,33)$$

This allows one to characterize whisker orientation in terms of the single orientation parameter, f_p. Thus for a planar distribution of whiskers, a specific form of $n_1(\phi)$ is not necessary. The assumptions given in Eq 31 and 32 then fix the value of g_p. Once f_p is stipulated, the elements of the tensor transformation operators can be constructed. Table 1 outlines the algorithm for establishing averaged elastic properties for a planar distribution of whiskers. The complete details of constructing the averaged elastic properties are given in Ref 1.

For the more general case where out-of-plane tilting occurs, whisker orientation is characterized in terms of two parameters. In reference to the orientation distribution function, this implies stipulating forms for two of the Euler angle functions [such as $n_1(\phi)$ and $n_2(\theta)$], where the behavior of $n_3(\zeta)$ is assumed a priori (Ref 10). For example,

$$n_3(\zeta) = \frac{1}{\pi} \qquad (Eq\,34)$$

Table 1 Computational algorithm for predicting the thermoelastic properties of two-component whisker-reinforced ceramic composites

Given the thermoelastic properties of fiber and matrix, such as E_f and E_m; the volume fraction concentration of fiber (or matrix); the effective aspect ratio, a_e; and the fiber orientation state, such as f_p and f_a, follow this procedure:

First, compute the properties of the reference material with spherical reinforcement
- Set the effective aspect ratio (a_e) to 1, which represents the reference material with spherical inclusions
- Assign C^0 the values of the matrix phase and compute the lower bound, C_L, from Eq 4; invert the results to obtain $S_L = (C_L)^{-1}$
- Assign C^0 the values of the fiber phase and compute the upper bound, C_U, from Eq 4; invert the results to obtain $S_U = (C_U)^{-1}$
- Use the results from preceding two steps in the S-combining rule of Eq 16 with an appropriate value of γ to obtain the properties of the material with spherical inclusions, S^0, invert to obtain stiffness properties $C^0 = (S^0)^{-1}$

Second, compute the properties of the characteristic subregion
- Set the parameter a_e to the current estimate for the effective aspect ratio
- Assign C^0 the values obtained in first set of steps and use these values to obtain C^* and Γ^* for the characteristic subregion

Third, determine the components of the volume-averaged stiffness array and thermal expansion coefficient vector. Using the special case of a planar distribution,
- Assume a value of f_p to $(0 \leq f_p \leq 1)$, and compute g_p from Eq 42
- Obtain $\langle C^* \rangle$ from Eq 31 and $\langle \alpha^* \rangle$ from Eq 32

for a random distribution in ζ. Again, because of the symmetry conditions imposed on $n_1(\phi)$ and $n_2(\theta)$, the elements of the transformation operators $\langle T_c \rangle$ and $\langle T_\alpha \rangle$ are functions of f_p and g_p, as well as

$$f_a = (\tfrac{1}{2})(3\langle \cos^2\theta \rangle - 1) \qquad (Eq\,35)$$

and

$$g_a = (\tfrac{1}{4})(5\langle \cos^4\theta \rangle - 1) \qquad (Eq\,36)$$

Note that

$$-\tfrac{1}{2} \leq f_a \leq 1 \qquad (Eq\,37)$$

and

$$-\tfrac{1}{4} \leq g_a \leq 1 \qquad (Eq\,38)$$

Taking $n_1(\phi)$ and $n_2(\theta)$ in forms given in Eq 31 and 32 leads to the previous relationship between f_p and g_p (Eq 33), as well as

$$g_a = \frac{3f_a}{5 - 2f_a} \qquad (Eq\,39)$$

Hence, the characterization of whisker distribution in the most general case requires the specification of f_p and f_a. The complete details of constructing the elastic properties for a three-dimensional distribution of whiskers are given in Ref 10.

Noninteractive Reliability Models

Here, a continuum is considered to be a chain composed of links connected in series. Therefore, the overall strength of the continuum is governed by the strength of its weakest link. It is further assumed that the events leading to failure of an individual link are not influenced by any other link in the chain. Defining f as the probability of failure of an individual link gives

$$f = \psi\Delta V \qquad (Eq\,40)$$

where ΔV denotes the volume of a link and ψ is a failure function per unit volume of material. By taking r as the reliability of an individual link, then

$$r = 1 - \psi\Delta V \qquad (Eq\,41)$$

If the failure of an individual link is considered a statistical event, and if these events are assumed to be independent, then the reliability of the continuum, denoted as R, is given as

$$R = \lim_{N \to \infty}\left\{\prod_{\lambda=1}^{N}[1 - \psi(x_i)\Delta V]_\lambda\right\} \qquad (Eq\,42)$$

where N denotes the number of links and $\psi(x_i)$ is the failure function per unit volume at position x_i within the continuum. Lowercase italic letter subscripts here and in the following expressions denote tensor indices with an implied range from 1 to 3. Greek letter subscripts and uppercase italic letters are associated with products or summations with ranges that are explicit in each expression. Alternatively, the reliability of the continuum is given by the following expression

$$R = \exp\left[-\int_V \psi \, dV\right] \qquad (\text{Eq } 43)$$

where the integral within the bracket is referred to as the risk of rupture.

Depending on fabrication, a whisker-toughened composite may have an isotropic, transversely isotropic, or orthotropic material symmetry. The principle of independent action (PIA) would be an appropriate first approximation macroscopic theory for isotropic whisker composites. In this instance, the failure function ψ would depend only on stress or the principal invariants of stress, that is,

$$\psi = \psi(\sigma_{ij}) = \psi(\sigma_1, \sigma_2, \sigma_3) \qquad (\text{Eq } 44)$$

where σ_{ij} is the Cauchy stress tensor and σ_1, σ_2, and σ_3 are associated principal stresses.

However, for transversely isotropic whisker composites, the failure function must also reflect material symmetry. This requires that

$$\psi = \psi(\sigma_{ij}, a_i) \qquad (\text{Eq } 45)$$

where a_i is a unit vector that identifies a local material orientation. This orientation, depicted in Fig 1, is defined as the normal to the plane of isotropy. The sense of a_i is immaterial, and thus its influence is taken through the product $a_i a_j$, that is,

$$\psi = \psi(\sigma_{ij}, a_i a_j) \qquad (\text{Eq } 46)$$

Note that $a_i a_j$ is a symmetric second-order tensor, the trace of which satisfies the identity $a_i a_i = 1$. Furthermore, the stress and local preferred direction may vary from point to point in the continuum. Thus, Eq 46 implies that the stress field and the unit vector field [that is, $\sigma_{ij}(x_k)$ and $a_i(x_k)$] must be specified to define ψ.

For orthotropic composites, the failure function must also reflect the appropriate material symmetry. This requires that

$$\psi = \psi(\sigma_{ij}, a_i a_j, b_i b_j) \qquad (\text{Eq } 47)$$

where a_i (a different vector than the one used for transverse isotropy) and b_i are unit vectors that identify local material orientations. These vectors are assumed to be orthogonal, such that $a_i b_i = 0$.

Because ψ is a scalar-valued function dependent on second-order tensors, the form of ψ must remain invariant under proper orthogonal transformations. This requires the function to be insensitive to the global coordinate system used to define the stress tensor and material directions. Through the use of invariant theory, a finite set of invariants known as an integrity basis can be developed for the isotropic, transversely isotropic, and orthotropic material symmetries (Table 2). The invariants of each integrity basis can be likened to a basis vector that helps to span a particular vector space (for example, the set of unit vectors that span the Cartesian space).

A slightly different set of invariants that correspond to physical mechanisms related to failure is constructed from each integrity basis. Table 3 provides a brief description of each

Table 2 Nonzero members of integrity basis corresponding to material symmetry

Material symmetry	Invariant
Isotropy	$I_1 = \sigma_{ii}$
	$I_2 = \sigma_{ij}\sigma_{ji}$
	$I_3 = \sigma_{ij}\sigma_{jk}\sigma_{ki}$
Transverse isotropy	$I_1 = \sigma_{ii}$
	$I_2 = \sigma_{ij}\sigma_{ji}$
	$I_3 = \sigma_{ij}\sigma_{jk}\sigma_{ki}$
	$I_4 = a_i a_j \sigma_{ji}$
	$I_5 = a_i a_j \sigma_{jk}\sigma_{ki}$
Orthotropy	$I_1 = \sigma_{ii}$
	$I_2 = \sigma_{ij}\sigma_{ji}$
	$I_3 = \sigma_{ij}\sigma_{jk}\sigma_{ki}$
	$I_4 = a_i a_j \sigma_{ji}$
	$I_5 = a_i a_j \sigma_{jk}\sigma_{ki}$
	$I_6 = b_i b_j \sigma_{ji}$
	$I_7 = b_i b_j \sigma_{jk}\sigma_{ki}$

invariant, and Fig 3, a graphical interpretation. These invariants can be identified with a principal stress or a component of the stress tractions coincident with a material direction.

The invariants used to form ψ are assumed to act independently in producing failure, such that ψ has the following general form:

$$\psi = \left(\frac{\hat{I}_1}{\beta_1}\right)^{\alpha_1} + \cdots + \left(\frac{\hat{I}_N}{\beta_M}\right)^{\alpha_M} \qquad (\text{Eq } 48)$$

where $N = 3$ and $M = 1$ for isotropy, $N = 4$ and $M = 3$ for transverse isotropy, and $N = M = 5$ for orthotropy (Table 4). In association with each invariant, the α's correspond to Weibull shape parameters and the β's correspond to Weibull scale parameters.

A variety of test methods could be used to determine these parameters. One approach is to obtain the data associated with the normal stress tractions from fast fracture of simple bend test specimens, often referred to as modulus of rupture (MOR) bars. The Weibull parameters associated with shear tractions would be obtained from appropriate shear strength tests. It is further assumed that compressive principal stresses and compressive stresses associated with a material orientation do not contribute to failure.

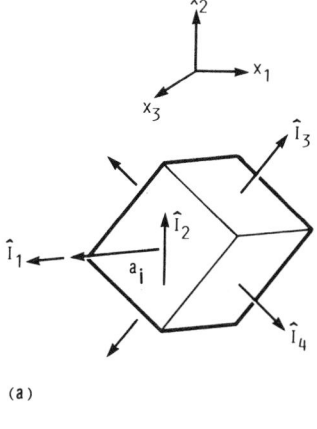

(a)

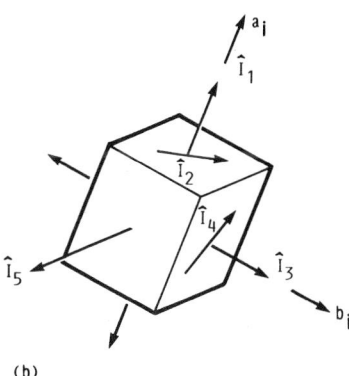

(b)

Fig 3 Invariants associated with transverse isotropy (a) and orthotropy (b)

TCARES Algorithm

The basic data requirements of TCARES (Fig 4) closely follow the structure of its parent code CARES. The algorithm requires the stress analysis from a general-purpose finite-element code. Currently, the preliminary version of TCARES is compatible with the MSC/NASTRAN (MacNeal-Schwendler Corporation/NASA structural analysis), although it

Table 3 Invariants associated with physical mechanisms directly related to failure

Material symmetry	Invariants used in ψ function	Comments
Isotropy	$\hat{I}_1 = \sigma_1$	Principal stresses; functionally dependent on the first three invariants of stress
	$\hat{I}_1 = \sigma_2$	
	$\hat{I}_1 = \sigma_3$	
Transverse isotropy	$\hat{I} = I_4$	Normal stress component of stress traction associated with a_i
	$\hat{I}_2 = [I_5 - (I_4)^2]^{1/2}$	Shear stress component of stress traction associated with a_i
	$\hat{I}_3 = \frac{1}{2}(I_1 - I_4) + R(a)$	Maximum normal stress in plane of isotropy
	$\hat{I}_4 = \frac{1}{2}(I_1 - I_4) - R(a)$	Minimum normal stress in plane of isotropy
Orthotropy	$\hat{I}_1 = I_4$	Normal stress component of stress traction associated with a_i
	$\hat{I}_2 = [I_5 - (I_4)^2]^{1/2}$	Shear stress component of stress traction associated with a_i
	$\hat{I}_3 = I_6$	Normal stress component of stress traction associated with b_i
	$\hat{I}_4 = [I_7 - (I_6)^2]^{1/2}$	Shear stress component of stress traction associated with b_i
	$\hat{I}_5 = I_1 - I_4 - I_6$	Normal stress in direction defined by cross product of a_i and b_i

(a) $R = [(\frac{1}{2})I_2 - I_5 + (\frac{1}{4})(I_4)^2 - (\frac{1}{4})(I_1)^2 + (\frac{1}{2}) I_1 I_4]^{1/2}$

Table 4 Functional forms of ψ corresponding to material symmetry

Material symmetry	Functional form of ψ
Isotropy $N = 3, M = 1$	$\psi = \left(\dfrac{\sigma_1}{\beta_1}\right)^{\alpha_1} + \left(\dfrac{\sigma_2}{\beta_1}\right)^{\alpha_1} + \left(\dfrac{\sigma_3}{\beta_1}\right)^{\alpha_1}$
Transverse isotropy $N = 4, M = 3$	$\psi = \left(\dfrac{\hat{I}_1}{\beta_1}\right)^{\alpha_1} + \left(\dfrac{\hat{I}_2}{\beta_2}\right)^{\alpha_2} + \left(\dfrac{\hat{I}_3}{\beta_3}\right)^{\alpha_3} + \left(\dfrac{\hat{I}_4}{\beta_3}\right)^{\alpha_3}$
Orthotropy $N = M = 5$	$\psi = \left(\dfrac{\hat{I}_1}{\beta_1}\right)^{\alpha_1} + \left(\dfrac{\hat{I}_2}{\beta_2}\right)^{\alpha_2} + \left(\dfrac{\hat{I}_3}{\beta_3}\right)^{\alpha_3} + \left(\dfrac{\hat{I}_4}{\beta_4}\right)^{\alpha_4} + \left(\dfrac{\hat{I}_5}{\beta_5}\right)^{\alpha_5}$

is anticipated that future versions compatible with the MARC, ADINA, and ANSYS finite-element codes will be available. The algorithm allows the user to specify temperature-dependent statistical material parameters for each material symmetry. Alternatively, the program has the capability to estimate statistical parameters from fracture data obtained from uniaxial tensile or flexural specimens. Details of this capability can be found in Ref 20.

The preceding section describing noninteractive theoretical models implies a volume flaw analytical approach. It is quite possible that the surface and volume of a structural component will fail because of distinctly different flaw populations. Accordingly, the TCARES program has the capability to separately conduct surface and volume reliability analyses. The program produces, as bulk output, a summary of input from the finite-element code, element statistical properties, element survival probabilities, and an overall component survival probability.

TCARES requires certain information from the finite-element structural analysis, including element volumes, nodal temperatures, centroidal or nodal stresses, element principal stresses (for PIA analysis), and element identification numbers. The current version of TCARES assumes that the nodal stresses from the finite-element code are provided relative to the local material orientation for transversely isotropic and orthotropic materials. This precludes having to input material orientation vectors for each finite element.

The TCARES user input requirements are grouped into three categories (Table 5). The first category, master control input, defines control indices for stress and graphics output,

the number of component materials, and information regarding the finite-element code and mesh. The second category, temperature-independent material control input, allows the user to designate the statistical parameters, and defines control indices for analysis (volume or surface) and material symmetry (isotropic, transversely isotropic, or orthotropic). The final category contains the temperature-dependent material symmetry statistical parameters. This includes the Weibull shape and scale parameters at each designated temperature.

The output from a reliability analysis conducted with TCARES can be grouped into eight broad categories. They include an echo of the master control input, a summary of element types, stress output, volume, and averaged temperature of each element. An echo of the material control indices and a summary of the analysis of each element are also provided. The latter includes material identification, failure probability, risk of rupture intensity, and the Weibull statistical parameters for each element. The program sorts and identifies 15 elements with the highest risk of rupture intensity to aid the design engineer in locating "hot" areas within a component. Finally, the overall component survival probability is given.

Examples

Multiaxial experiments are a necessity to assess the accuracy of the noninteractive modeling approach. One experimental avenue highlighted here is the application of mechanical loads to tubular specimens. Initially, a thick-walled tube subjected to an applied torque is considered. A second problem is presented, in which the same thick-walled tube is subjected to a simultaneous application of internal pressure and axial torque. In all applications considered, isothermal conditions are assumed. However, the algorithm is capable of nonisothermal analyses if the user specifies the values of the Weibull parameters at a sufficient (and appropriate) number of temperature values. Unfortunately, at the present time no data base exists to properly characterize the multiaxial statistical parameters for any anisotropic whisker-toughened ceramic. Thus, an assessment of the program output relative to actual structural component data is reserved for a later date. In the examples that follow, Weibull statistical parameters are assumed for the purpose of illustration. However, the values adopted are well within the range of the sparse data that can be found in the open literature (Ref 21–23).

To test the validity of the reliability calculations performed by the program, a comparison of output with a hand calculation is presented for the aforementioned simple structural problem, that is, a thick-walled tube subjected to an axial moment or torque. It is assumed that the cylinder is fabricated from

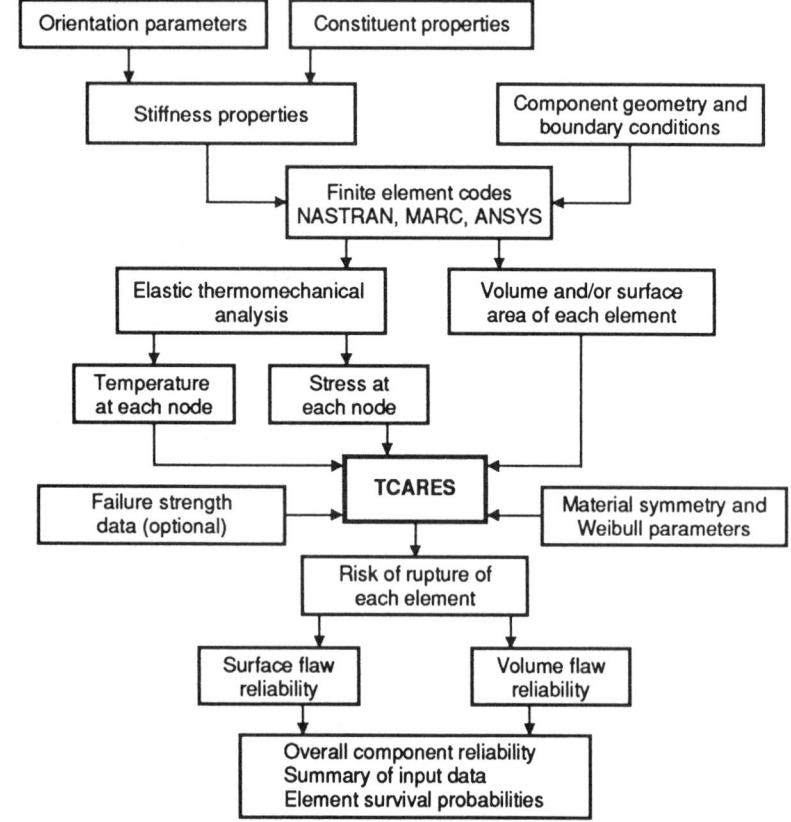

Fig 4 Data requirements of TCARES

Table 5 Description of input parameters

Category	Input	Description
Master control input	NE	Analysis option
		0—Experimental data analysis only
		1—MSC/NASTRAN analysis
		2—ANSYS analysis (option currently unavailable)
		3—MARC analysis (option currently unavailable)
	NMATS	Number of materials for surface flaw analysis
	NMATV	Number of materials for volume flaw analysis
	IPRNT	Control index for stress and/or fracture data output
		0—Suppress element stress and/or fracture data output
		1—Print element stress and/or fracture data output
	IGRPH	Control index for graphics output
		0—Suppress element risk-of-rupture intensity output
		1—Print element risk-of-rupture intensity output
	NS	Number of cyclic symmetry segments
Temperature-independent material control input	MATID	User-designated material identification number
	ID1	Control index for experimental data analysis
		1—Uniaxial tensile specimen data
		2—Four point bend test data
		3—All Weibull parameters specified by user
		4—Censored data for suspended item analysis (uniaxial tensile specimen)
		5—Censored data for suspended item analysis (four point bend test)
	ID2	Control index for material symmetry failure criteria
		1—Isotropic (PIA)
		2—Transversely isotropic
		3—Orthotropic
	ID4	Control index for volume or surface flaw analysis
		1—Volume
		2—Surface
Temperature-dependent material control input	TDEG	Temperature for specified data set
	PARAM	Weibull parameters corresponding to TDEG

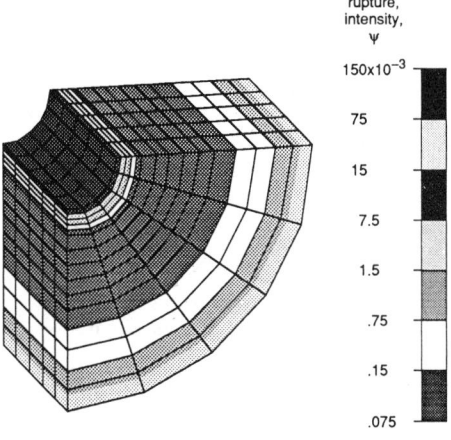

Fig 6 Risk-of-rupture intensity for a thick-walled tube subjected to a torque of 7500 N · m (66 × 10^3 lbf · in.) and internal pressure of 70 MPa (10 ksi)

a whisker-toughened ceramic material having an orthotropic material symmetry such that a_i = (0,0,1) and b_i = (0,1,0) at every point in the structure. A cylindrical coordinate system readily lends itself to this application. Hence, a_i is directed along the z-axis of the cylinder and b_i is oriented in the θ (circumferential) direction. With this geometry, material symmetry, and load condition, only the two terms (see Table 4) in the failure function associated with shear tractions are nonzero, and ψ takes the form

$$\psi = \left(\frac{Tr}{J\beta_2}\right)^{\alpha_2} + \left(\frac{Tr}{J\beta_4}\right)^{\alpha_4} \qquad (Eq\ 49)$$

where T is the applied torque, r is the radius, and J is the polar moment of inertia. Assuming an inner radius of 10 mm (0.40 in.), an outer radius of 50 mm (2 in.), and a length of 50 mm (2 in.) gives

$$R = \exp(-\iiint \psi r\ dr\ d\theta\ dz)$$

$$= \exp(-10\pi \int \psi r\ dr) \qquad (Eq\ 50)$$

For dimensionless reliability, the Weibull scale parameters ($\beta_1, \ldots, \beta_5$) have units of stress · (volume)$^{1/\alpha}$, and the Weibull shape parameters are unitless. With $\alpha_2 = 10$, $\beta_2 = 15{,}000$, $\alpha_4 = 6.5$, $\beta_4 = 10{,}000$ (the other Weibull parameters can be stipulated arbitrarily), and $T = 7500$ N · m (66 × 10^3 lbf · in.), this integration yields an overall reliability of 83.6%. The problem was also modeled by using MSC/NASTRAN to generate a numerical solution for the stress distribution. The model was composed of 1500 eight-node HEXA elements (the stresses were within

2% of the closed-form solution), which generated an overall component reliability of 81.8%.

Figure 5 is a plot of the variation of ψ in a quarter section of the component. Note that ψ (risk-of-rupture intensity) is a measure of reliability independent of the element geometry, and that it attains a maximum value along the outer edge of the tube.

Next consider the same tube, subject to the conditions stated in the preceding paragraph, with an additional applied internal pressure of 70 MPa (10 ksi). Here α_3 and β_3 can no longer be stipulated arbitrarily, and take values of α_3

= 7.5 and β_3 = 12,000. The overall component reliability decreases to 77.4%. For this load case, the circumferential stress is a maximum at the inner radius, and it decreases nonlinearly through the thickness. The shear stress from the applied torque is a minimum at the inner radius, and it increases linearly with the radius. Given this multiaxial stress distribution, one expects the maximum risk-of-rupture intensity to occur at some point midway though the thickness. However, this is not the case, as is evident in Fig 6. The maximum risk-of-rupture intensity occurs at the inner radius, and much of the inner volume of the tube remains relatively "cold." This underscores the need to not only consider overall component reliability, but also to consider where local "hot" spots occur within a component. If this particular component were to fail, one would expect the failure to originate in the vicinity of the inner radius.

Overall component reliability can be adversely affected by either a slight increase

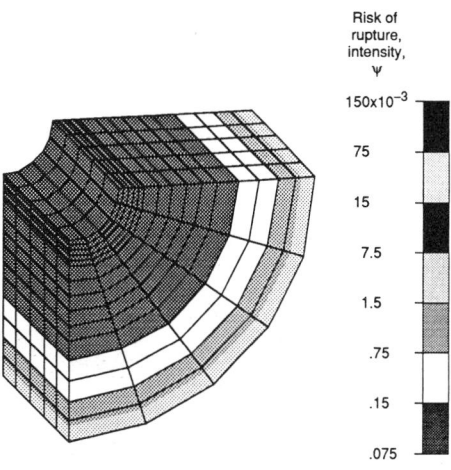

Fig 5 Risk-of-rupture intensity for a thick-walled tube subjected to a torque of 7500 N · m (66 × 10^3 lbf · in.)

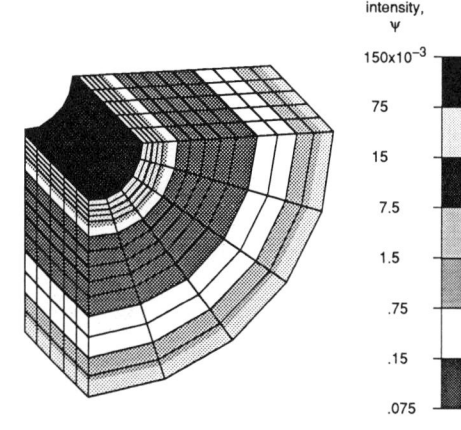

Fig 7 Risk-of-rupture intensity for a thick-walled tube subjected to a torque of 7500 N · m (66 × 10^3 lbf · in.) and internal pressure of 100 MPa (15 ksi)

in stress over large regions of the component (the so-called size effect), or by dramatically increasing the stress locally (thereby increasing the chance of failing a single link in the chain). Increasing the internal pressure of the previous example to 100 MPa (15 ksi) sharply decreases the component reliability to 36%. Figure 7 depicts the variation of ψ throughout the component. It appears that the additional affected region of the component is minimal. However, ψ changes two orders of magnitude along the inner radius. Although not depicted here, the overall component reliability of the tube subjected to only an internal pressure of 100 MPa (15 ksi) is 44%, indicating that the major source in the degradation of reliability is the internal pressure.

As a final note, the applications presented here represent very straightforward structural analyses. Similar calculations and graphical interpretations can be carried over into designs with much more complex geometry and boundary conditions.

REFERENCES

1. R.L. McCullough, Micro-Models for Composite Materials—Particulate and Discontinuous Fiber Composites, Chapt. 2.4, *Micromechanical Materials Modeling*, Vol 2, J.M. Whitney and R.L. McCullough, Ed., Technomic Publishing, 1990
2. J.P. Gyekenyesi, SCARE: A Postprocessor Program to MSC/NASTRAN for Reliability Analysis of Structural Ceramic Components, *J. Eng. Gas Turbines Power*, Vol 108 (No. 3), July 1986, p 540–546
3. S.F. Duffy and S.M. Arnold, Noninteractive Macroscopic Reliability Model for Whisker Reinforced Ceramic Composites, *J. Compos. Mater.*, Vol 24, 1990, p 293–308
4. S.F. Duffy and J.M. Manderscheid, Noninteractive Macroscopic Reliability Model for Ceramic Matrix Composites with Orthotropic Material Symmetry, *J. Eng. Gas Turbines Power*, Vol 112 (No. 4), Oct 1990, p 507–511
5. R.C. Wetherhold, Fracture Energy for Short Brittle Fiber/Brittle Matrix Composites with Three-Dimensional Fiber Orientation, *J. Eng. Gas Turbines Power*, Vol 112 (No. 4), Oct 1990, p 502–506
6. K.T. Faber and A.G. Evans, Crack Deflection Processes—I. Theory, *Acta Metall.*, Vol 31, Apr 1983, p 565–576
7. F.F. Lange, The Interaction of a Crack Front with a Second-Phase Dispersion, *Philos. Mag.*, Vol 22, 1970, p 983–992
8. C.T.D. Wu and R.L. McCullough, Constitutive Relationships for Heterogeneous Materials, *Developments in Composite-Materials*, G.S. Holister, Ed., Applied Science Publishers, London, 1977, p 119–187
9. B.W. Rosen and Z. Hashin, Analysis of Material Properties, *Composites*, Vol 1, *Engineered Materials Handbook*, ASM International, 1987, p 185–205
10. S.H. McGee, "Influence of Microstructure on the Elastic Properties of Composite Materials," Ph.D. Dissertation, University of Delaware, 1982
11. S.M. Arnold, "A Transversely Isotropic Thermoelastic Theory," TM-101302, National Aeronautics and Space Administration, 1988
12. B.D. Agarwal and L.J. Broutman, *Analysis and Performance of Fiber Composites*, John Wiley & Sons, 1980
13. R.C. Wetherhold and P.D. Scott, Prediction of Thermoelastic Properties in Short-Fiber Composites Using Image Analysis Techniques, *Compos. Sci. Technol.*, Vol 37 (No. 4), 1990, p 393–410
14. J.C. Halpin and N.J. Pagano, The Laminate Approximation for Randomly Oriented Fibrous Composites, *J. Compos. Mater.*, Vol 3, 1969, p 720–724
15. R.M. Christensen and F.M. Waals, Effective Stiffness of Randomly Oriented Fiber Composites, *J. Compos. Mater.*, Vol 6, 1972, p 518–532
16. A. Ghesquiere and J.C. Bauwens, Theoretical Model for the Elastic Behavior of Composites Reinforced with Short Fibers, *J. Appl. Polym. Sci.*, Vol 20 (No. 4), Apr 1976, p 891–902
17. P.D. Shalek, J.J. Petrovic, G.F. Hurley, and F.D. Gac, Hot-Pressed SiC Whisker/Si_3N_4 Matrix Composites, *Am. Ceram. Soc. Bull.*, Vol 65 (No. 2), Feb 1986, p 351–356
18. J.V. Milewski, Efficient Use of Whiskers in the Reinforcement of Ceramics, *Adv. Ceram. Mater.*, Vol 1, 1986, p 36–41
19. Y. Takao and M. Taya, The Effect of Variable Fiber Aspect Ratio on the Stiffness and Thermal Expansion Coefficients of a Short Fiber Composite, *J. Compos. Mater.*, Vol 21, 1987, p 140–156
20. N.N. Nemeth, J.M. Manderscheid, and J.P. Gyekenyesi, "Ceramic Analysis and Reliability Evaluation of Structures (CARES)—User's and Programmer's Manual," TP-2916, National Aeronautics and Space Administration, 1989
21. N. Claussen and G. Petzow, Whisker-Reinforced Zirconia-Toughened Ceramics, *Tailoring Multiphase and Composite Ceramics*, R.E. Tressler, *et al.*, Ed., Plenum Press, 1985, p 649–662
22. T.N. Tiegs and P.F. Becher, "Alumina-SiC Whisker Composites: Application to Heat Engines," presented at the 23rd Automotive Technology Development Contractors' Coordination Meeting, Dearborn, MI, 1985
23. J.F. Rhodes, C.J. Dziedzic, R.L. Beatty, and J.L. Cook, Ceramic Reinforced Ceramics, *PM Aerospace Materials*, Vol I, MPR Publishing, Shrewsbury, England, 1984, p 43–1 to 43–22

Design Practices for Glass and Glass Fibers

M. John Matthewson, Department of Ceramic Science and Engineering,
Rutgers, The State University of New Jersey

GLASS has been made and used by man throughout recorded history. Glass used for pottery glazes, small pieces of jewelry, and small containers was first found in Egypt and Mesopotamia and has been dated at around 7000 BC. Around 1500 BC, the Egyptians discovered the technique for making the first holloware, which involved dipping a ceramic core into molten glass.

A revolution in glass manufacture occurred around 200 BC, when Syrian craftsmen learned to use a pipe to blow glass objects so that thin-walled vessels with complex shapes could be made. Blowing the glass into a wooden mold allowed the duplication of large numbers of objects with standardized shapes and sizes. The glass-blowing technique also allowed production of flat glass by cutting and flattening cylinders of blown glass while still soft.

Improvements in processing conditions and raw materials enabled the first colorless glass to be made in the Roman Empire around AD 100. This gave impetus to a thriving industry producing, among other things, jewelry and art objects. In the Middle Ages, Venice led the world in the production of fine glass, aided in part by its use of high-quality crystal glasses.

Since the 17th century, there have been rapid improvements in the technology of glass manufacture and processing. In the 18th and 19th centuries, quality optical glasses were developed, primarily in Germany, and represented one of the earliest uses of glass for technical purposes. New manufacturing techniques, such as the tank furnace, allowed continuous mass production of glass. With these developments came an explosion in glass applications.

The most important current use of glass is in the manufacturing of containers, whereas the second largest use is as flat glass, primarily for window applications. Other uses include light bulb envelopes, optical lenses, vacuum tubes, radiation shields, jewelry, art objects, chemical apparatus, and electrical insulation. When made in fibrous form, glass is used as an acoustic and thermal insulator, as a reinforcement for high-strength, low-weight composites, and as optical waveguides for communications and sensing applications. In fact, with so many diverse applications, it is hard to envisage how current technology could be sustained without the use of glass.

The reason that glasses have wide application is that they have a unique and desirable combination of properties. The raw materials for silicate glasses are among the most abundant in the earth's crust. Glasses are easily made by firing the constituents in a furnace, a process that is readily adapted for bulk, continuous manufacturing processes. Because glasses soften before melting, glass articles can be made by a variety of techniques including pressing, blowing, combined pressing and blowing, casting, drawing, and spinning. Glasses can also be made via sol-gel processes, which allow components of complex shape to be made either directly or by fusing together smaller pieces.

Because glasses are usually mixtures of compounds, their compositions can be continuously varied to change their physical and chemical properties. The resultant glass articles are hard, transparent solids with good abrasion resistance, chemical durability, and thermal stability, and they are impervious to liquids and most gases, as well. However, the one major limitation that pervades glass science is that glasses are generally brittle, that is, the energy to fracture the material is low. Therefore, although glass generally has good compressive strength, it has low tensile strength.

With so many uses of glass, it is not possible to describe in this article the design methodology for all applications. However, because brittleness is a major limiting factor in most applications, this article concentrates primarily on the strength properties of glass. In fact, strength is one of the key factors in the two most important applications, namely container and flat glasses. This is especially true in the glass container industry, which is experiencing severe competition from plastics. The drive to produce lighter weight glass containers without compromising strength is an issue of great importance. Before discussing the implications of strength on design, the choice of glass composition and forming method is considered below.

Material Composition

The majority of glass manufactured today is based on silica, SiO_2, as the glass former, with various materials added to modify its properties. Boria, B_2O_3, is also a glass former but is most often mixed with silica to make borosilicate glass. Although the discussion in this article is limited to the silicate glasses, other materials that are used for specific applications can also form glasses, such as other oxides (particularly phosphate and germanate glasses), sulfides, tellurides, and halides (see Ref 1, p 12 to 13, for a comprehensive list). Although an enormous range of glasses can be prepared by using a number of different constituents in different proportions, several broad types of glass can be defined (Table 1).

Glass compositions that do not fall into one of these broad categories are generally known as specialty glasses. Selection of the glass composition for a particular application should be made after careful consideration of such requirements as durability, forming method, thermal expansion, conductivity, and any other relevant physical properties. Development of composition/properties data bases on computers will play an increasingly important role in making rapid and effective composition selections.

Forming Methods

The forming method chosen for making a glass component is primarily determined by the shape of the final article. Holloware is most conveniently formed by either blowing or combined pressing and blowing. Simple or quite complex shapes can be made by pressing. Complex shapes or those with tight dimensional tolerances can be made by sol-

Table 1 Main categories of silicate glasses

	Typical composition	Uses	Properties
Soda-lime	70–75% SiO_2 12–16% Na_2O 10–15% CaO	Bottles, glasses, windows	Optically clear, durable
Lead ("crystal") glasses	55–65% SiO_2 18–38% PbO 13–15% Na_2O or K_2O	Decorative items	High refractive index
Borosilicate	70–80% SiO_2 7–13% Ba_2O_3 4–8% Na_2O or K_2O 2–7% Al_2O_3	Chemical apparatus, lamp and tube envelopes	Chemical durability, low thermal expansion
Quartz glass	100% SiO_2	High-temperature uses	High softening temperature, low thermal expansion
Aluminosilicate glasses	52–58% SiO_2 15–25% Al_2O_3 4–18% CaO	High-temperature uses, thermometers, combustion tubes, cookware	High softening temperature, low thermal expansion

gel techniques. If cost is an important consideration, then expensive finishing techniques, such as machining, grinding, and polishing, should be avoided.

Intrinsic Strength of Glass

It has been found that the strength of pristine fused silica fibers in liquid nitrogen (which freezes out reactions with water) is ~14 GPa (2×10^6 psi) (Ref 2, 3). This represents a strain to failure of ~20%, which is close to the theoretical silicon/oxygen bond strength. It is also possible to produce multicomponent glass fibers with theoretical strength, although this is somewhat lower than the theoretical strength of pure silica because of the differences in the network structure (Ref 4). In general very high strength surfaces can be made from the melt (either directly or indirectly by flame polishing) or by etching a weaker surface. Thus, extremely high strengths can be obtained by careful preparation of the pristine glass surface, followed by protection of that surface from subsequent damage. In practice, however, glass components usually have a considerably lower strength, typically in the range of 1 to 100 MPa (0.15 to 14.5 ksi). This reduction in strength is due to the presence of defects, surface shapes, and elastic discontinuities that concentrate the stress locally.

Inglis (Ref 5) showed that when a plate containing an elliptical hole is subjected to a uniform tensile stress, σ, the tensile stress at the "tip" of the hole is given by

$$\sigma_t = \sigma\left(1 + \frac{2c}{b}\right) \qquad \text{(Eq 1)}$$

where c and b are the major and minor axis lengths of the ellipse (Fig 1). Because the radius of curvature at the tip of the ellipse is $\rho = b^2/c$,

$$\sigma_t = \sigma(1 + 2\sqrt{(c/\rho)}) \qquad \text{(Eq 2)}$$

If $c \gg \rho$, that is, the ellipse is sharp,

$$\sigma_t \approx 2\sigma\sqrt{c/\rho} \quad \text{or} \quad \sigma \approx \sigma_t/2\sqrt{(\rho/c)} \qquad \text{(Eq 3)}$$

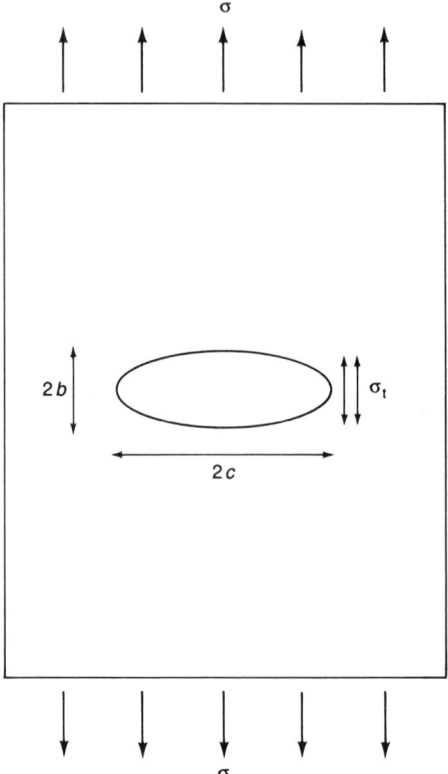

Fig 1 Stresses around an elliptical hole in a flat plate

If σ_t exceeds the local intrinsic strength of the glass, failure occurs, so that if $c \gg \rho$, the apparent strength, σ, is considerably lower than the intrinsic strength. The form of Eq 3 is valid for a wide variety of stress-concentration geometries, giving the general result that the apparent strength decreases with increasing size (c) and sharpness ($1/\rho$) of the stress concentrator. Stress concentrators take several forms, such as surface shapes, embedded impurities, or microscopic cracks. The successful design of a strong glass component lies in recognizing the nature of stress concentrators and reducing their effect by reducing either their size or the stress acting upon them.

Mechanisms of Strength Reduction

Component Shape. When a glass component is subjected to loading in service, a distribution of stress is set up within that body. The design process should produce a shape for the component that does not result in unduly high stresses anywhere. As a typical holloware example, a bottle is considered below.

The internal surface of the bottle has a sharp change of shape where the base meets the wall. The local stresses can be reduced by increasing the radius of curvature (Eq 3). Therefore, all reentrant angles should be as shallow as possible and have small curvature. The walls can be thickened in regions of high stress to reduce that stress. Equally, material can be removed in regions of low stress to reduce weight and cost. Certain areas of a component will be weaker than others. In the case of the bottle, its base and shoulder are expected to be weaker because they impact other objects; therefore, the walls in these regions can be made thicker to reduce the local stresses. Clearly, design of the component shape is an iterative process that can be greatly speeded by use of finite-element analysis (FEA) techniques.

Internal flaws in the glass are usually associated with processing. They may be small crystals that have formed from the glass (stones) and weaken it by having sharp edges, as well as local stresses that are due to elastic and thermal mismatches. Inhomogeneities in the glass (striae and cords) appear as strands and concentrate stress because of elastic discontinuities. Bubbles are also stress concentrators, but the effect is minor for nearly spherical bubbles, unless they are very close to the surface. However, bubbles produce tensile stress concentrations when subjected to compressive stress, thus compromising the compressive strength of the component, which would otherwise be high. Although internal defects are associated with processing, they can usually be controlled by improved processing techniques and thus are not usually a major design factor.

Surface flaws are by far the most important class of defects leading to strength reduction, and they can come from several sources:

- *Transfer of foreign material* onto the surface during the molding process; adhering particles deform the surface and are often associated with residual stresses and cracking due to thermal expansion mismatches
- *Local Sticking.* Small areas of the glass surface can stick to the mold and when adhesion is broken upon removal of the product from the mold, large local stresses can produce cracks in the glass surface
- *Thermal Damage.* The strength of glass is known to degrade with time when heated in air below the annealing temperature. This is thought to be possibly due to slight

surface devitrification, bonding of dust particles, or chemical attack by moisture

- *Mechanical Damage.* Throughout production and during subsequent use, the surface of the glass component can be damaged by contact with other objects. Because of frictional forces, high stress concentrations are produced by two surfaces that are in sliding contact with modest normal forces, which can lead to cracking. Additionally, dust particles between the surfaces can cause indentations with associated cracking

The first three mechanisms, and, to some extent, the fourth, can be controlled by careful design of processing conditions. However, because handling damage is usually unavoidable, various strategies can be used to strengthen the glass component.

Residual Stresses. When glass components are formed from the melt, residual stresses (that is, stresses that remain in the component when no external forces are applied) result and are due to differential cooling rates. Faster cooling areas (usually surfaces) "set" first and are pulled into compression by slower cooling areas (usually the interior) as they contract. The compressive stresses are balanced by tensile stresses in the slower cooling areas. If these tensile stresses appear on the surface of the component, then weakening occurs because less external loading is required to raise the stress to its critical value for failure. This effect is particularly noticeable for holloware that is blown into a mold. The outer surface cools rapidly in contact with the mold and goes into compression, leaving the internal surface in tension. It is usually necessary to remove these residual stresses by annealing the component. This is achieved by raising the temperature to the annealing point (defined when the viscosity is 10^{12} Pa · s) for some time, followed by slow cooling to the strain point (viscosity $10^{13.5}$ Pa · s), followed by fast cooling to room temperature.

Strength Testing Techniques

The strength of a glass component depends on many factors, particularly the history of the glass surface, that is, how it has been formed and handled. For this reason, it is not possible to assess the strength of a component by subjecting specimens with a simplified geometry to simplified stress distributions. For example, it is not possible to estimate the failure pressure of a bottle by measuring the bending strength of a lathe specimen of the same composition. It is therefore necessary to determine strength by tests on actual components that model the in-service stress conditions as closely as possible.

As an example, glass containers frequently fail because of either overpressurization during the filling process or thermal shock. These conditions can be modelled by subjecting the container to an increasing internal pressure until it bursts and by dropping a heated container into a cold bath. Both these test methods have been standardized and are described in the specifications ASTM C 149–50 and C 47–62, respectively.

Methods for Improving Glass Strength

Annealing and Etching. Although annealing removes any residual tensile surface stresses, it can also allow healing of surface defects (Ref 6). Surface damage can also be removed by acid etching (Ref 7). However, neither of these techniques result in a permanent strength increase unless that surface is protected from subsequent handling damage.

Tempering. Residual stresses can be used advantageously in strengthening the component, provided that the stresses are always compressive at the surface. Any applied loading must first overcome the compressive stress before surface defects can be placed in tension. This effect can increase the strength of glass components by a factor of approximately 5 and can be achieved by either thermal or chemical tempering.

Thermal tempering, or toughening, is achieved by rapidly quenching the surfaces of a component with jets of air or oil. An approximately parabolic distribution of stress is produced through the thickness of the specimen with a tensile region at the center of the section (Fig 2). Such stress profiles can be predicted from the thermal conductivity and thermal expansion of the glass, as well as from the starting and quenching temperatures, allowing such stresses to be tailored for a particular application (Ref 8).

Chemical tempering is achieved by immersing the glass in a molten salt that contains potassium ions. Ion exchange occurs between the potassium and sodium ions in the glass. The exchange can be accelerated by application of an electric field. Because potassium ions are somewhat larger than sodium ions, a compressive stress is set up in the surface. Unlike thermal tempering, the compressive layer is relatively thin (Fig 2). Therefore, if there are large surface defects that are somewhat deeper than this layer, the component is weakened rather than strengthened, because the defect experiences net tension.

Coatings. As discussed above, the pristine surface of glass formed from the melt has extremely high strength. One strategy for improving reliability is to attempt to preserve that strength by applying one or more protective coatings to the article as soon as possible after the forming process. The coating will then limit surface damage that can be caused by handling during subsequent manufacturing and use.

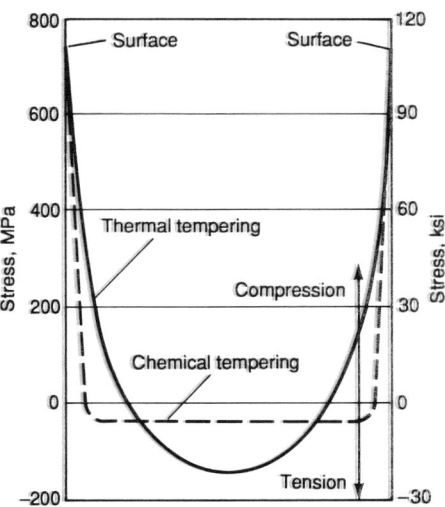

Fig 2 Stress distributions through the thickness of a tempered specimen

Perhaps the most successful application of protective coating technology is in the optical fiber industry. Optical fiber is drawn down to a 125 μm (5 mil) diameter from a preform in a furnace. At a distance from the furnace that allows the fiber to cool (usually ~2 m, or 6.6 ft) the fiber is pulled through a cup of liquid polymer, which coats it with a typically 50 to 100 μm (2 to 4 mil) thick layer. The polymer layer is then cured in-line by intense ultraviolet (UV) radiation. This in-line process ensures that the fiber surface is exposed to the environment for the shortest possible time. The resulting coated fiber is robust and can withstand substantial mistreatment without loss of the pristine strength.

The glass container industry uses coating technology as standard practice in order to produce stronger, lighter weight containers in response to competition from polymer containers. Glass containers usually receive two coatings. The "hot-end" coating is applied to the just-formed component as it enters the annealing furnace at high temperature, and the "cold-end" coating is applied as the component leaves the annealing furnace at lower temperature.

Hot-end coatings are usually either SnO_2 or TiO_2, which can be applied by passing the component through an atmosphere of volatile tin or titanium compounds. This type of coating is variously thought to improve scratch resistance, decrease friction, and promote adhesion of the cold-end coating.

Cold-end coatings are normally organic lubricants that are applied by spraying a solution or emulsion. The coating reduces the coefficient of friction of the glass surface, which reduces the level of contact stresses when the surface is in sliding contact with other objects, thus improving scratch resistance. The coating system should be carefully designed to avoid either overapplication or underapplication of the lubricant, which would lead to subsequent problems (Ref 9). The coating formulation needs careful design to ensure compatibility with adhesives that are used later to apply labels to the container (Ref 10).

Application of quite thick (mm) polymer coatings satisfies the dual role of both cushioning impact stresses and providing a labelling medium. This is illustrated by the increasing use of shrink-wrapped coatings on containers, which substantially improves their impact resistance. Design of the coating is simple for most applications—the thicker, the better. However, for more specialized applications, there are optimum values for the coating thickness and elastic properties, and, ideally, the coating should be rigidly bonded to the glass (Ref 11). Polymer coatings are compliant, compared to the glass itself, but much stiffer coating materials, such as diamond, can provide improved tribological properties and are a subject of intense current research (see, for example, the series of articles on diamond and diamond-like films in Ref 12).

Proof Testing. The strength of glass is a stochastic variable because the strength-degrading flaws are widely distributed in position, orientation, size, and form. For this reason, there is considerable variability in the measured strength. Although the median strength may be acceptable, for some applications the lower strengths encountered may be unacceptable. For example, failure of the occasional bottle (1 in 1000) may be tolerable, whereas the occasional failure of an aircraft window is not.

One method to combat this problem is to proof test all components by loading them up to a stress level that is somewhat higher than expected under service conditions. Any component that is weak enough to risk failing in service fails during the proof test and is necessarily discarded. Proof testing does not have a significant effect on the median strength of surviving components. Its importance is in truncating the strength distribution and in giving a minimum useful strength. After the important concepts of flaw statistics and time-dependent failure are considered below, proof testing will be discussed further.

Reliability Analysis

As stated above, strength is a stochastic quantity that is usually broadly distributed. This means that although most components may have sufficient strength, the survivability of all components cannot be guaranteed. Reliability analysis involves estimating the risk of failure, which may be small but is never zero. Clearly, risk analysis should be a key part of the design procedures in order to obtain components with an acceptably low probability of failure in service.

Flaw Statistics. The distribution of strength that results from the distribution of flaw position and severity can be characterized by

measuring the strength of several specimens, followed by fitting to a given probability distribution. A weakest-link model is assumed for brittle failure, namely, that failure of the weakest flaw (defined in terms of both the flaw severity and the local stress) leads to complete failure of the component. It is then found that the cumulative probability of failure by a stress, σ, is

$$F(\sigma) = 1 - \exp[-n(\sigma)A] \qquad \text{(Eq 4)}$$

where the risk function, $n(\sigma)$, is some monotonically increasing function of stress and A is the surface area of the specimen. (This assumes, as is usually the case for glass, that failure is dominated by surface flaws; A would be replaced by V, the volume, if internal flaws dominate.) The dependence on specimen size results from the greater probability of encountering a severe flaw in a larger specimen. The most commonly used form for $n(\sigma)$ is due to Weibull (Ref 13):

$$n(\sigma) = \frac{(\sigma/\sigma_0)^m}{A_0} \qquad \text{(Eq 5)}$$

where the Weibull modulus, m, is a measure of the distribution width and σ_0 is a stress-scale parameter. A_0 is an area of size unity. This parameter is usually not included but is included here to give σ_0 dimensions of stress (Ref 14).

For a general component, the stress, σ, is not uniform but is distributed over the component surface. It is then necessary to calculate the total failure probability by a suitable integration of Eq 4 over the entire specimen surface. This is most conveniently achieved by taking the stress distribution supplied by a finite-element analysis of the loading, and then calculating from that the total failure probability for a given flaw distribution using a program such as CARES. The article "Designing Ceramic Components with the NASA/CARES Computer Program" in this Section of the Volume provides details.

Time-Dependent Effects. All oxide glasses are subject to delayed failure or fatigue, that is, failure can occur after prolonged exposure to a stress level that is significantly less than the stress required to produce immediate failure. The mechanism for fatigue is chemical attack, usually by water, and the subsequent rupture of strained bonds at the tip of surface cracks (Ref 15). The crack advances slowly until it is sufficiently long to produce catastrophic failure. Assuming a power-law relationship between the crack growth velocity, v, and the stress-intensity factor, K_I,

$$v = AK_I^n \qquad \text{(Eq 6)}$$

Wiederhorn (Ref 16) showed that the time to failure, t_f, of a component of initial strength, σ_i, subjected to a constant applied stress, σ_a, is given by

$$t_f = 2\sigma_a^{-n} \frac{1}{AY^2(n-2)} \left(\frac{\sigma_i}{K_{Ic}}\right)^{n-2} \qquad \text{(Eq 7)}$$

where the exponent n is the stress-corrosion susceptibility parameter, K_{Ic} is the critical stress-intensity factor, or fracture toughness, and Y is a constant of order unity describing the crack geometry. This equation reduces to the form

$$t_f = B\left(\frac{\sigma_i}{\sigma_a}\right)^{n-2} \frac{1}{\sigma_a^2} \qquad \text{(Eq 8)}$$

The parameter, B, can be determined by performing accelerated static fatigue experiments with a high applied stress, σ_a, or by dynamic fatigue experiments in which the strength is measured as a function of loading rate. Equation 8 may then be used to predict the lifetime of a component of a given initial strength, subjected to a given service stress, σ_a. However, as already noted, the initial strength is statistically distributed, leading to a distribution of lifetime. Fatigue is difficult to avoid because it is almost always due to environmental water, which is ubiquitous. However, various strategies may be applied to reduce fatigue:

- Reducing the water concentration by using, for example, dessicants in a sealed environment
- Changing the pH in liquid water environments. Silica glass fatigues more slowly in a low pH environment, whereas soda-lime silicate glass fatigues more slowly at a high pH level
- Applying a fatigue-resistant coating, such as by incorporating nitrogen in the surface of silica fiber to produce a silicon oxynitride film, which increases the fatigue resistance (the value of n increases from ~10 to ~100) (Ref 17)
- Applying a hermetic coating that does not fatigue itself but acts as a barrier to exclude water from the glass surface. For example, silica fiber can be coated with a metal such as aluminum (Ref 18)

Proof Testing. Although proof testing guarantees a minimum strength, Eq 2 shows that this procedure also guarantees a minimum lifetime (Ref 19). If the initial strength, σ_i, is interpreted as the proof stress, σ_p, then Eq 8 may be used to calculate the minimum lifetime from a given maximum service stress, σ_a, or, inversely, the maximum allowable service stress for a given design life. This equation can be used to generate a proof test diagram, such as Fig 3, where the time to failure is shown as a function of applied stress for various ratios of the proof stress to applied stress.

Because fatigue of the component occurs during proof testing, the total loading time should be minimized to maximize the yield. It is especially important to maximize the unloading rate, because it is possible for fatigue effects to reduce the strength below the proof stress during unloading (Ref 20). Even at the fastest unloading rates, it is not possible to guarantee a strength, although the probability that a weak specimen will result

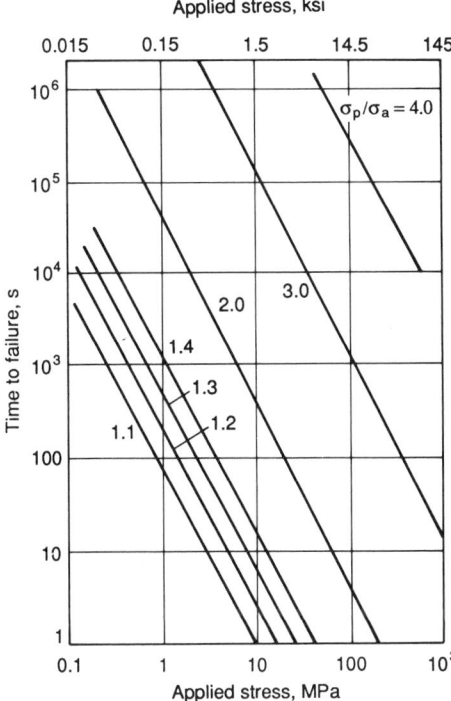

Fig 3 Proof test diagram for soda-lime glass in water. Number adjacent to each line represents the ratio of proof stress to applied stress. Source: Ref 19

can be made acceptably small by using a sufficiently fast unloading rate and by using an environment that is as inert as possible.

Lessons Learned

Because of its brittle nature, it is easy to introduce severe flaws into the surface of glass during handling. For this reason, glass has been traditionally thought of as a weak material. However, a pristine surface with a strength in excess of 1 GPa (0.15×10^6 psi)

can be prepared and, if that surface can be protected from damage, very high practical strengths can be realized. By careful design procedures aimed at lowering surface stresses and applying appropriate protective coating schemes, the old view of glass as a necessarily weak material must be abandoned. Perhaps the most impressive example of this new view of glass strength is provided by the optical fiber industry, in which multi-kilometer lengths of silica fiber are made cheaply and routinely with strength values of several GPa.

REFERENCES

1. R.H. Doremus, *Glass Science*, Wiley Interscience, 1973
2. B.A. Proctor, I. Whitney, and J.W. Johnson, *Proc. R. Soc. London A*, Vol 297, 1967, p 534–537
3. P.W. France, *et al.*, Liquid Nitrogen Strengths of Coated Optical Glass Fibres, *J. Mater. Sci.*, Vol 15, 1980, p 825–830
4. P.K. Gupta, Glass Fibers for Composite Materials, *Trans. Am. Ceram. Soc.*, Orlando, FL, Nov 1990, in press
5. C.E. Inglis, Stresses in a Plate due to the Presence of Cracks and Sharp Corners, *Trans. Inst. Nav. Archit. (London)*, Vol 55, 1913, p 219–230
6. P. Manns and R. Bruckner, Bending Strength of Flat Glass from Room Temperature to Littleton Temperature, *Strength of Inorganic Glass*, C.R. Kurkjian, Ed., Plenum, 1985, p 531–548
7. B.A. Proctor, The Effects of Hydrofluoric Acid Etching on the Strength of Glass, *Phys. Chem. Glasses*, Vol 3, 1962, p 7–27
8. H.A. Schaeffer, Thermal and Chemical Strengthening of Glass—Review and Outlook, *Strength of Inorganic Glass*, C.R. Kurkjian, Ed., Plenum, 1985, p 469–484
9. H.M. Snyder, Cold-End Coatings in Glass Container Manufacture, *Am. Ceram. Soc. Bull.*, Vol 69 (No. 11), 1990, p 1831–1833
10. L. Levene, How Container Coatings Effect Label Adhesion, *Glass Ind.*, Jan 1989, p 17–20
11. M.J. Matthewson, The Effect of a Thin Compliant Protective Coating on Hertzian Contact Stresses, *J. Phys. D: Appl. Phys.*, Vol 15, 1982, p 237–249
12. *J. Mater. Res.*, Vol 5 (No. 11), 1990, p 2273–2609
13. W. Weibull, The Phenomenon of Rupture in Solids, *Proc. R. Swedish Inst. Eng. Res.*, Vol 153, 1939, p 3–55
14. M.J. Matthewson, An Investigation of the Statistics of Fracture, *Strength of Inorganic Glass*, C.R. Kurkjian, Ed., Plenum, 1985, p 429–442
15. S.M. Wiederhorn, A Chemical Interpretation of Static Fatigue, *J. Am. Ceram. Soc.*, Vol 55 (No. 2), 1972, p 81–85
16. S.M. Wiederhorn, Subcritical Crack Growth in Ceramics, *Fracture Mechanics of Ceramics*, Vol 2, R.C. Bradt, D.P.H. Hasselman, and F.F. Lang, Ed., Plenum, 1974, p 613–646
17. R. Hiskes, Improved Fatigue Resistance of High Strength Optical Fibers, *Tech. Dig. Top. Meet. Opt. Fiber Comm.*, 1979, p 74–75
18. D.A. Pinnow, G.D. Robertson, Jr., and J.A. Wysocki, Reductions in Static Fatigue of Silica Fibers by Hermetic Jacketing, *Appl. Phys. Lett.*, Vol 34, 1979, p 17
19. A.G. Evans and S.M. Wiederhorn, Proof Testing of Ceramic Materials—An Analytical Basis for Failure Prediction, *Int. J. Fract.*, Vol 10 (No. 3), 1974, p 379–392
20. A.G. Evans and E.R. Fuller, Proof Testing—The Effects of Slow Crack Growth, *Mater. Sci. Eng.*, Vol 19, 1975, p 69–77

Properties

Chairman: Robert E. Moore, Department of Ceramic Engineering, University of Missouri—Rolla

Introduction

KNOWLEDGE OF ACCURATE VALUES for the properties of ceramics and glasses has become indispensable for their utilization in engineering design. Evidence of the reliability of the property values is also needed in the reporting of values which the design engineer will pass on to the pilot scale designer and ultimately to the production engineer. The need is no longer for characteristic numbers or typical values; it is for sets of quantified values for properties of interest which can be used now.

The current high-level need for useful property data for ceramics and glasses challenges many scientists and engineers working to produce such values because they are short on capabilities to do so and also short on understanding exactly what to produce. This uncertainty relates to the fact that only in the past decade have the materials engineer and the design engineer engaged in serious need-to-know discussions, especially in the areas of mechanical, thermal, and electrical properties.

Materials information needs for the design and production of automotive and aerospace ceramics, for fuel cells and ceramic batteries, as well as for advanced metallurgical processing and modern architecture require values of properties which are specific and which are accompanied by a listing of qualifications which will ensure the long-term integrity of the designs. The numerous examples of failed designs which were limited by materials serve to underscore the *sine qua non* role of materials in technology today. When ceramic materials with critical properties are not available, the design must often be scrapped. When the properties cannot be reliably realized, the pilot efforts must often be put aside. The cost effectiveness of production methodologies is increasingly tied to the levels and reliabilities of the properties of ceramic materials as was well illustrated by the failure to develop reliable ceramic materials for turbine rotors during the 1980s.

Properties of ceramics and glasses must increasingly be accompanied by qualifying information which the designer can comprehend. The designer often needs to know or be warned that the quoted values are valid over specified ranges of temperature, pressure, mechanical stress (including cycling), chemical environment, and presence of electromagnetic fields. Many times the designer also needs to be apprised of the reasons for the listed qualifications of engineering property values.

The use of property information by the experimentalist to elucidate mechanisms or the test development engineer to provide relatable data has accelerated with the explosion in the knowledge of materials and with the ensuing variety of advanced materials. The development of ceramic-matrix composites is, for example, requiring the development of new test methods to produce property information to aid in the testing of theories and models and for the certification of materials systems.

The coverage of the materials in the first eleven articles in this Section serves to exemplify the properties of greatest importance for each material, to connect the property levels to underlying mechanisms, and to give some insight into the utilization of property information in order to point to the directions for test development and to provide comment on the quality of data available. The emphasis is on the design engineering use of the data. The articles should serve to bring the users of property information up to date and to direct them to the sources of the most useful property data.

The final two articles have a similar purpose, that is, to provide the user-reader with examples of the utility of structural, thermodynamic and threshold values, and to point to sources for the most reliable and comprehensive information.

Engineering Properties of Single Oxides

Masaru Miyayama, Kunihito Koumoto, and Hiroaki Yanagida, University of Tokyo, Japan

CERAMIC MATERIALS owe much of their importance to the fact that they are generally more stable than other materials at elevated temperatures and under corrosive conditions. Among ceramics, the oxides are perhaps the most useful and common, because they are more easily fabricated and are more stable in air at high temperatures than non-oxide ceramics such as the nitrides, carbides, and sulfides.

When the configuration of the oxygen ions and the compositions of the included elements (and, hence, the ratio of the ion radii) has been determined, the crystal structure can be readily anticipated. The representative crystal structure of single oxides and examples of single oxides are shown in Table 1. Engineering properties of representative oxides and their groups with the same crystal structures are described below.

Properties of A_2O

Among A_2O-type oxides are Cu_2O, Ag_2O, Pb_2O, and others with a Cu_2O-type (cuprite) structure; Li_2O, Na_2O, K_2O, and others with an antifluorite structure; and Cs_2O, with an anti-$CdCl_2$-type structure.

Cu_2O is a well-known p-type semiconductor with a metal-deficient composition. The energy band gap is ~1.5 eV, and acceptor levels are usually formed 0.3 to 0.6 eV above the top of the valence band. The Seebeck coefficient of Cu_2O is rather large, ~700 μV/K at room temperature.

Alkali-metal oxides other than Li_2O are not stable themselves. There are few cases in which they alone are used as materials.

Properties of AO

AO-type oxides are generally classified into the two major groups that possess different crystal structures, that is, rock-salt and wurtzite. Among the oxides with rock-salt structure are those that consist of alkaline earth elements, such as MgO, CaO, SrO, and BaO, and transition metal oxides, such as TiO, VO, MnO, FeO, CoO, NiO, NbO, and ZrO. Oxides with a wurtzite structure include BeO, ZnO, and CdO. Other oxides, such as CuO, PbO, and PdO, have different structures because of the peculiarity of their electronic structures.

Alkaline-earth oxides (CaO, SrO, BaO) exhibit a tendency to become peroxides absorbing excess oxygen, because their cationic radii for octahedral coordination are too large. Magnesium oxide shows different chemical properties. Because all of these oxides are composed of typical element ions with closed-shell electronic structures, they are normally electric insulators and have high melting points (Table 2).

The first-transition metal oxides exhibit interesting electronic properties, because of the characteristic behavior of 3d electrons. These oxides (A = Ti, V, Mn, Fe, Co, Ni) would have partially filled 3d bands and would be expected to show metallic properties. Although TiO and VO show high metallic conductivity, MnO, FeO, CoO, and NiO are typical p-type semiconductors, based on comparatively large deviations from the stoichiometric compositions. This fact is closely related to the electronic structure and the localization of charge carriers in a crystal. The nearest-neighbor distance of cations varies with the number of 3d electrons, as shown in Fig 1, indicating that the degree of overlap of 3d orbitals is rather large in TiO and VO. However, it is small in MnO-NiO; hence, the carriers formed in the narrow 3d band would localize at cations, giving rise to low carrier mobility and low conductivity.

Among the oxides with a wurtzite structure, BeO is an electrical insulator, whereas ZnO and CdO are n-type semiconductors, based on small deviations from stoichiometry. BeO must be treated carefully because of its toxicity, but it is favored for integrated circuit (IC) substrate materials because it shows higher thermal conductivity than alumina substrates.

MgO is described below in terms of thermal, mechanical, electrical, optical, and chemical properties.

Thermal Properties. Molar heat capacity, C_p, of MgO can be expressed by the following equation for the temperature range from 298 to 2500 K:

$$C_p = 45.47 + 5.01 \times 10^{-3} T - 8.74$$
$$\times 10^5 T^{-2} \quad \text{(In J/mol} \cdot \text{K)} \qquad \text{(Eq 1)}$$

where T is the absolute temperature. Debye temperature is calculated to be 630 K. The thermal expansion coefficient of MgO is comparatively large, ranging from 12.0×10^{-6}/K to 17.3×10^{-6}/K between 200 and 2000 °C (390 and 3630 °F). Thermal conductivity as a function of temperature is shown in Fig 2. At room temperature, the thermal conductivity of MgO is rather low, ~60 W/m · K for a single crystal, compared to other typical ceramic materials.

Mechanical Properties. Young's modulus of a single crystal depends on crystallographic orientation and temperature. Its value at room temperature is reported to be ~300 GPa (45×10^6 psi) and decreases with increasing temperature. Shear modulus of single-crystal MgO is ~130 GPa (20×10^6 psi) at 300 K and decreases with increasing temperature. Porosity dependence of the shear modulus of polycrystalline MgO can be expressed as:

$$G = 20.24 \times 10^6 \exp(-4.90P) \qquad \text{(Eq 2)}$$

where P is the volume fraction of pores.

Compressive strength of a sintered body with the density of 3.48 g/cm^3 is ~1300 MPa (190 ksi), whereas flexure strength exhibits low values, such as 441 MPa (65 ksi) for a single crystal and ~90 MPa (15 ksi) for a polycrystal with 3.4 g/cm^3 density.

Vicker's microhardness slightly depends on crystal orientation: 8.9 GPa (1.29×10^6 psi) on (100), 9.7 GPa (1.40×10^6 psi) on (110), and 9.6 GPa (1.39×10^6 psi) on (111). Microhardness of polycrystalline MgO depends largely on grain size, porosity, purity, and temperature. The Mohs hardness that is usually observed ranges from 5.0 to 6.5.

MgO exhibits a brittle nature below ~1000 °C (1830 °F), but it becomes ductile above that temperature. A high-temperature deformation map is shown in Fig 3. Poisson's ratio below 1000 °C (1830 °F) ranges from 0.36 to 0.33, a value as high as ~0.47 in the ductile region (1100 °C, or 2010 °F).

Table 1 Representative crystal structures and single oxides

A/O ratio	Coordination number	Type of crystal structure	Examples of oxides
A_2O	2	Cuprite	Cu_2O, Ag_2O, Pb_2O
	4	Antifluorite	Li_2O, K_2O, Na_2O
AO	4	Wurtzite	ZnO, BeO, CdO
	6	Rock-salt	MgO, CaO, SrO, BaO
			FeO, CoO, NiO, VO
A_3O_4	4, 6	Spinel	Fe_3O_4, Mn_3O_4, Co_3O_4
A_2O_3	6	Corundum	α-Al_2O_3, α-Fe_2O_3, Cr_2O_3
	6	C-type rare-earth oxide	Y_2O_3, In_2O_3, Sc_2O_3
	7	A-type rare-earth oxide	La_2O_3, Nd_2O_3, Pr_2O_3
AO_2	4	Cristobalite	SiO_2
		Tridymite	SiO_2
		Quartz	SiO_2
	6	Rutile	TiO_2, SnO_2, GeO_2, VO_2
	8	Fluorite	ZrO_2, CeO_2, ThO_2, HfO_2
A_2O_5	...	...	V_2O_5, Nb_2O_5, Ta_2O_5
AO_3	...	...	WO_3, MoO_3, ReO_3

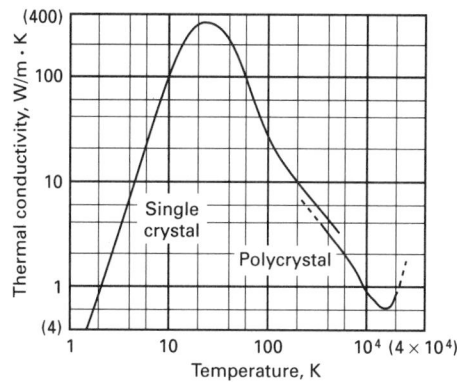

Fig 2 Thermal conductivity of MgO. Source: Ref 1

Electrical Properties. MgO is a typical insulator, usually showing low electrical conductivity ($<10^{-17}$ S/cm) at room temperature. However, it increases exponentially up to the order of 10^{-6} S/cm at 1200 °C (2190 °F), as shown in Fig 4. The contribution of ionic conduction is significant below ~1100 °C (2010 °F), and electronic conduction predominates at higher temperatures. The relative dielectric constant measured at 25 °C (77 °F) is about 9.65 for the wide frequency range of 102 to 108 Hz.

Optical Properties. A single crystal is highly transparent, showing over 80% transmittance for the wavelength between 7 and 300 μm. Refractive index in this wavelength range is ~1.75 and decreases with further increasing wavelength. Various values for the optical band gap have been reported in the literature, but the most reliable value may be 7.77 eV at 295 K.

Chemical Properties. MgO easily reacts with H_2O, CO_2, and acids, just as alkaline-earth oxides do. It shows high vapor pressure at elevated temperatures:

$$\log\left(\frac{P_{MgO}}{Pa}\right) = \frac{-30{,}100 \pm 2300}{T} + (13.82 \pm 2.56) \quad \text{(Eq 3)}$$

Table 2 Melting points of AO-type oxides

Oxide	T_m, K
BeO	2843
MgO	3098
CaO	2887
SrO	2727
BaO	2196
TiO	2023
MnO	2115
FeO	~1640
CoO	2208
NiO	2257
CuO	1300
ZnO	Evaporate
CdO	Evaporate
PbO	1159

Vapor pressure reaches the order of 10^{-3} Pa at 1700 °C (3090 °F). MgO decomposes into Mg vapor and O_2. Hence, the vapor pressure (or vaporization rate) depends on the ambient oxygen partial pressure.

NiO is described in terms of thermal, electrical, and chemical properties.

Thermal Properties. Nickel-oxide crystallizes into the rock-salt structure above the magnetic Neel temperature, ~525 K, and changes by squeezing ~0.15% in the [111] direction into the rhombohedral crystal below 525 K. Molar heat capacity, C_p in J/mol · K, shows three different temperature dependences:

$$C_p = 20.89 + 157.34 \times 10^{-3}\,T - 16.29 \times 10^5\,T^{-2}$$

(From 298 to 525 K)

$$C_p = 58.11$$

(From 525 to 565 K)

$$C_p = 46.81 + 8.46 \times 10^{-3}\,T$$

(From 565 to 2000 K) (Eq 4)

where T is the absolute temperature. The melting point is 2257 K.

The thermal expansion coefficient of NiO is comparable to those of typical ceramic oxides, such as MgO, Al_2O_3, and stabilized zirconia: 13.2 to 16.3×10^{-6}/K at temperatures between 200 and 1800 °C (390 and 3270 °F) for a sintered polycrystal. The thermal conductivity of a fully dense polycrystal

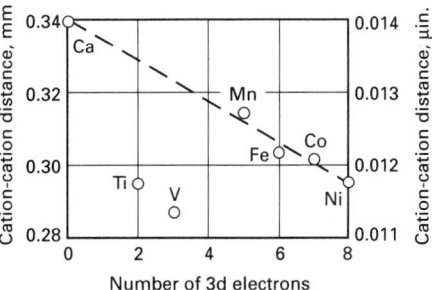

Fig 1 Nearest-neighbor distance of cations in rock-salt-type transition metal oxides as a function of number of 3d electrons

is ~2.4 W/m · K at 373 K and decreases down to ~4.5 W/m · K at 1273 K, which is comparable to that of stabilized zirconia.

Electrical Properties. NiO exhibits *p*-type semiconducting properties, based on a small amount of deficiency in cations. Electrical conductivity is very low at room temperature but reaches the order of 0.1 S/cm at 1000 °C (1830 °F), depending on both temperature and oxygen partial pressure, as well as on the purity of a crystal. The conductivity of a single crystal between 900 and 1400 °C (1650 and 2550 °F) and for $P_{O_2} = 10$ to 10^4 Pa (0.0001 to 1 atm) may be expressed as:

$$\sigma = 9.8 \times 10^2 \times P_{O_2}^{1/2} \exp(-0.81\,eV/kT)$$

(In S/cm) (Eq 5)

Conductivity can be increased by doping Li_2O and decreased by doping Al_2O_3. Thermally activated hopping of small polarons was believed to be the dominating process for conduction in NiO. Both theoretical and experimental investigations have shown that the mobility of holes introduced into the narrow 3d band is comparatively low, ranging from 0.2 to 0.5 cm²/V · s, as shown in Fig 5. However, recent research has shown that band-like conduction is rather predominant, and the conflict concerning the conduction mechanism is still an unsolved problem.

Chemical Properties. NiO can be easily reduced to a metal state at high temperatures, and the equilibrium decomposition pressure from 1000 to 1500 K is expressed as:

$$\log\left(\frac{P_{O_2}}{Pa}\right) = 13.91 - \frac{24{,}100}{T} \quad \text{(Eq 6)}$$

In the oxide-stable region, NiO shows nonstoichiometry, and the deviation from the stoichiometric composition, δ, in $Ni_{1-\delta}O$ has been measured by several methods and is shown in Fig 6.

The diffusion rate of cations in NiO is the lowest among the transition metal oxides with rock-salt structure, possibly because of its small deviation from stoichiometry. Tracer diffusion coefficients, D_{Ni^*} and D_{O^*}; a chemical diffusion coefficient, D_{NiO}, obtained by

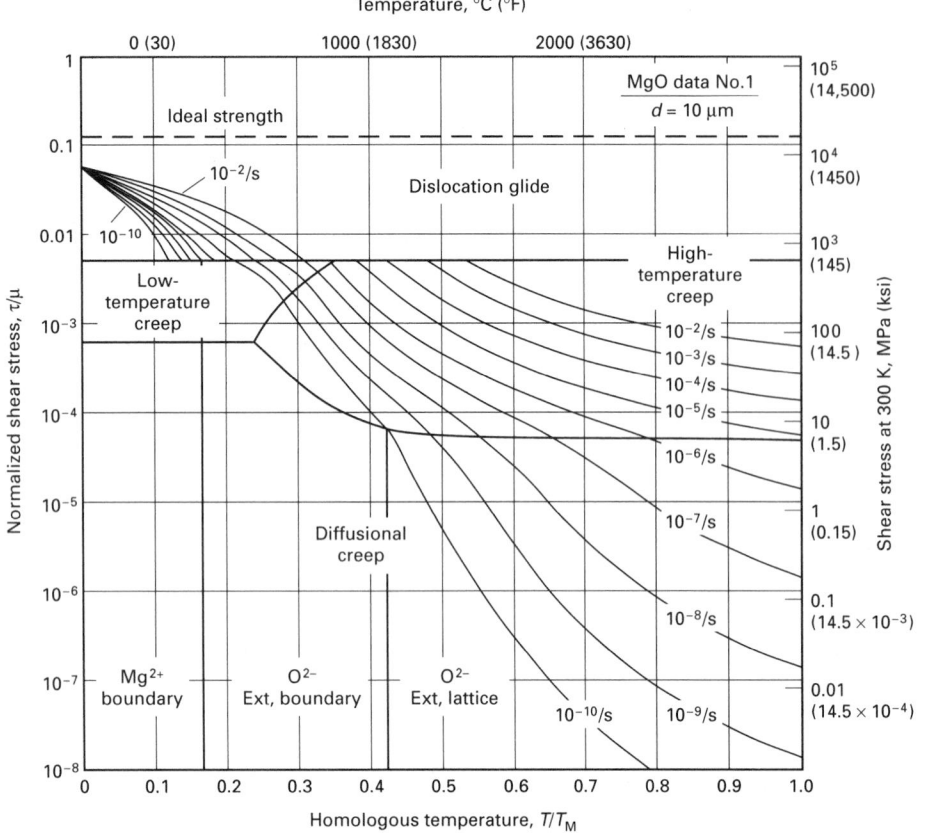

Fig 3 Tentative deformation map for MgO. Source: Ref 2

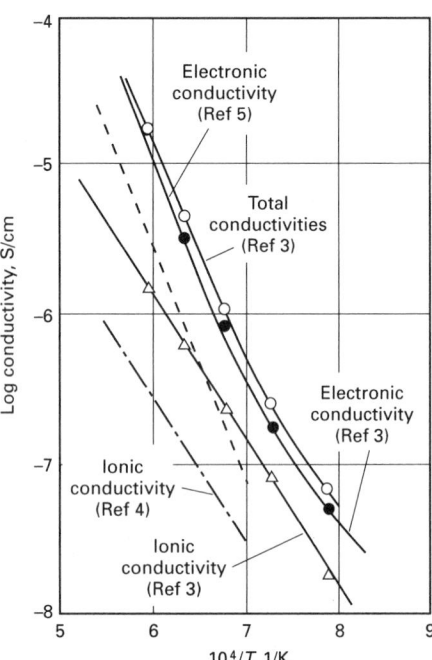

Fig 4 Total, ionic, and electronic conductivities of MgO single crystal as function of reciprocal temperature

a relaxation method; and impurity diffusion coefficients, D_{Co} and D_{Mg}, are expressed respectively, all in cm^2/s, as:

$$D_{Ni^*} = 1.5* 10^{-2} \exp\left(\frac{-2.51 \text{ eV}}{kT}\right)$$

(From 1028 to 1973 K, $P_{O_2} = 105$ Pa)

$$D_{O^*} = 6.24* 10^{-4} \exp\left(\frac{-2.49 \text{ eV}}{kT}\right)$$

(From 1273 to 1773 K)

$$D_{NiO} = 7.52* 10^{-4} \exp\left(\frac{-0.95 \text{ eV}}{kT}\right) \quad \text{(Eq 7)}$$

(From 1153 to 1623 K)

$$D_{Co} = 3.46* 10^{-2} \exp\left(\frac{-2.52 \text{ eV}}{kT}\right)$$

(From 1073 to 1973 K)

$$D_{Mg} = 9.2* 10^{-3} \exp\left(\frac{-2.56 \text{ eV}}{kT}\right)$$

(From 1073 to 1973 K)

ZnO is described below in terms of thermal, electrical, and chemical properties.

Thermal Properties. Zinc oxide has no clear melting point because of its high vapor pressure, which is remarkable at elevated temperatures. Molar heat capacity, C_p, is expressed as:

$$C_p = 49.03 + 5.11 \times 10^{-3} T - 9.13 \times 10^5 T^{-2}$$

(In J/mol · K)   (Eq 8)

Because ZnO has a wurtzite structure (hexagonal system), thermal expansion shows anisotropy, depending on crystallographic orientation. The thermal expansion coefficient parallel to the c-axis is ~5.0 × 10^{-6}/K and that which is perpendicular to the c-axis is ~5.5 × 10^{-6}/K between 0 and 400 °C (30 and 750 °F). Thermal conductivity of a fully dense ZnO polycrystal is a little lower than that of MgO, ranging from 17.05 W/m · K at 200 °C (390 °F) to 5.0 W/m · K at 800 °C (1470 °F).

Electrical Properties. ZnO shows n-type semiconductivity, because of its nonstoichiometric nature to generate excess zinc atoms, which behave as electron donors. The reported band gap is from 3.2 to 3.35 eV. Because of its anisotropic crystal structure, Hall mobility of electrons depends on crystal orientation, especially at low temperatures below 300 K. Hall mobilities parallel to the c-axis, $\mu_{H\parallel}$, and perpendicular to the c-axis, $\mu_{H\perp}$, both show the maxima in their temperature dependences. Maximum values are observed as 1350 cm^2/V · s at 60 K for the former and 2400 cm^2/V · s at 40 K for the latter. The effective mass of a conduction electron is reported to be 0.25 to 0.3 m_e where m_e is the mass of a free electron.

Chemical Properties. As already mentioned, ZnO shows nonstoichiometric composition with excess zinc atoms, and the solubility of zinc in zinc oxide is expressed by:

$$\log\left(\frac{X_{Zn}}{\text{ppma}}\right) = \frac{-4700}{T} + 5.6 \quad \text{(Eq 9)}$$

where T is the absolute temperature and 1 ppma = 4.21 × 10^{16} cm^{-3}.

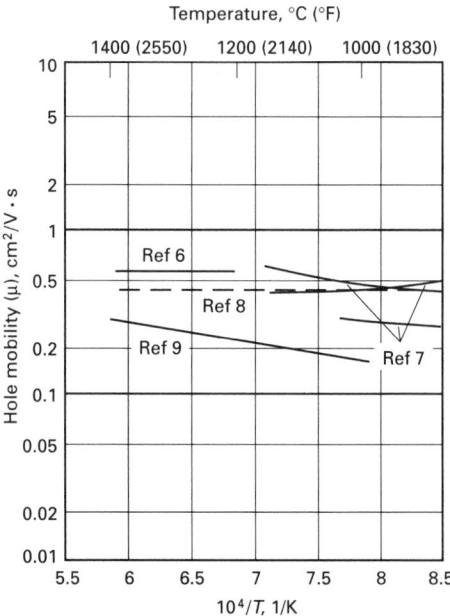

Fig 5 Mobility of holes in NiO

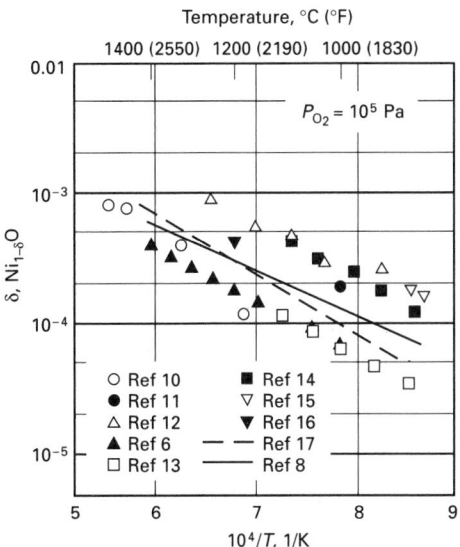

Fig 6 Temperature dependence of deviation from stoichiometry in NiO

Diffusion coefficients of zinc and oxygen ions have been reported by many workers, but they scatter in more than 2 orders of magnitude. Therefore, only example data are listed here (in cm^2/s):

$$D_{Zn} = 1.3 \exp\left(\frac{-3.19 \text{ eV}}{kT}\right)$$

(From 950 to 1370 K)

$$D_O = 6.5 \times 10^{11} \exp\left(\frac{-7.15 \text{ eV}}{kT}\right)$$

(From 1100 to 1300 K)   (Eq 10)

ZnO is one of the oxides with high vapor pressures. Vaporization becomes observable above ~1200 °C (2190 °F) in air. The total equilibrium vapor pressure (ZnO decomposes into zinc vapor and O_2) is shown in Fig 7.

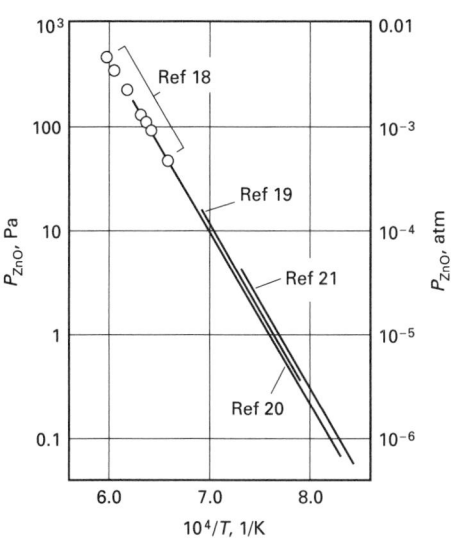

Fig 7 Temperature dependence of the total vapor pressure of ZnO

Properties of A_3O_4

In A_3O_4-type oxides, there are two kinds of valence states for A ions. Transition metal oxides, such as Mn_3O_4, Fe_3O_4, and Co_3O_4, consist of divalent and trivalent cations with the ratio of 1:2, as $A^{2+}B_2^{3+}O_4$. These oxides generally have spinel structures. Structural formulas are expressed as $A[B_2]O_4$, $B[AB]O_4$, or $A_{1-x}B_x[A_xB_{2-x}]O_4$, where [ ] denotes the octahedral sites. They are called normal spinel, inverse spinel, and intermediate (or random) spinel, respectively. Fe_3O_4 is an inverse spinel denoted as $Fe^{3+}[Fe^{2+}Fe^{3+}]O_4$, whereas Mn_3O_4 and Co_3O_4 are both normal spinels denoted as $Mn^{2+}[Mn_2^{3+}]O_4$ and $Co^{3+}[Co_2^{3+}]O_4$. Other oxides with the A_3O_4 composition, such as Pb_3O_4, have structures that differ from a spinel structure.

Iron oxide of engineering importance exists in four types: FeO, Fe_3O_4, α-Fe_2O_3, and β-Fe_2O_3. FeO has a wide nonstoichiometric composition. The x in $Fe_{1-x}O$ ranges from 0.05 to 0.15, depending on temperature and ambient oxygen partial pressure.

Stoichiometric composition has never been observed for FeO in which octahedral cation vacancies and interstitial Fe^{3+} ions at tetrahedral sites combine to form defect clusters dispersed in the rock-salt structure. $Fe_{1-x}O$ with a small x value is usually antiferromagnetic, having the antiparallel arrangement of magnetic moments along the (111) direction. However, in $Fe_{1-x}O$ with a large x value, defect clusters become large in size, with their structure similar to a spinel structure, and give rise to superparamagnetism.

Magnetite (Fe_3O_4) is a ferrimagnetic material in which the difference in magnetic moments between octahedral and tetrahedral Fe ions generates the spontaneous magnetization. Curie temperature for ferrimagnetic-paramagnetic transition ranges from 847 to 858 K. Electrically, Fe_3O_4 is a semiconductor at low temperatures but changes into a metallic conductor above 110 to 120 K (Fig 8). This phenomenon is called Verwey transition and is believed to be related to the order-disorder transition of Fe^{2+} and Fe^{3+} ions at octahedral sites, accompanied by the polymorphic transition from an orthorhombic to a cubic system.

Maghemite (γ-Fe_2O_3) has a defective spinel structure, and its cation distribution is expressed as $Fe^{3+}[Fe_{5/3}^{3+}\square_{1/3}]O_4$, where $\square$ denotes cation vacancies. Residual magnetic moment gives rise to ferrimagnetism in this material below the Curie temperature, 860 K.

Hematite (α-Fe_2O_3) has a corundum structure with octahedral Fe^{3+} ions. Magnetic moments of Fe ions, arranged parallel within (001) planes, are further arranged antiparallel along the [001] direction, giving rise to antiferromagnetism below the Neel temperature, 953 K, but with slight ferromagnetism along [001]. Below 260 K, hematite becomes completely antiferromagnetic, and this temperature is called Morin point.

Properties of A_2O_3

A_2O_3-type oxides are classified into those with a corundum structure and those with either a C-type or A-type rare-earth sesquioxide structure. As the ionic radius of the A ion increases, crystal structure changes from the corundum to the A-type rare-earth structure.

A representative oxide with the corundum structure is α-A_2O_3. Because it has excellent

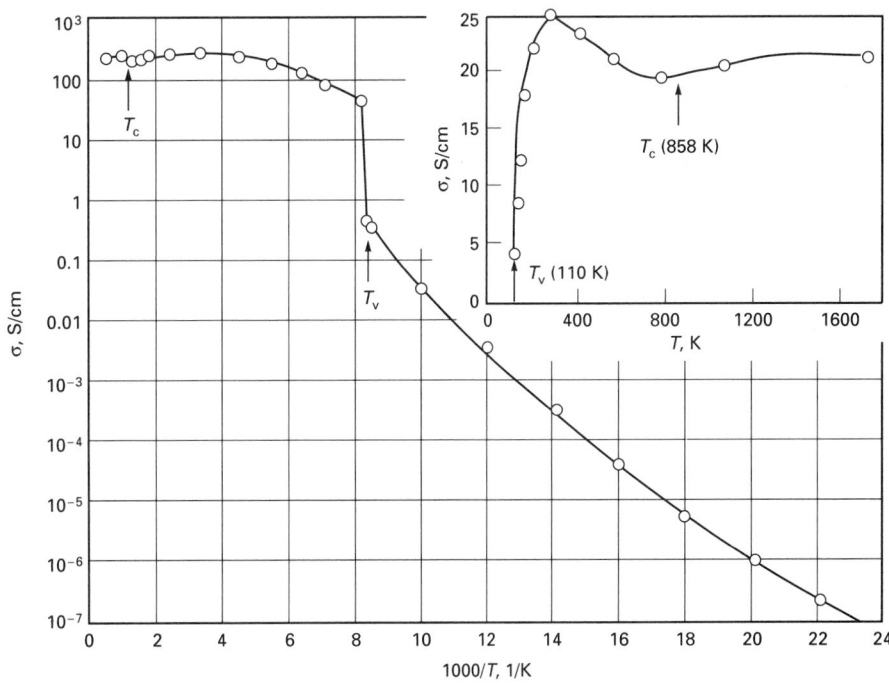

Fig 8 Electrical conductivity of Fe_3O_4. Source: Ref 22

properties in hardness, corrosion resistance, high electrical resistance, high thermal conductivity, and heat resistance, it has been used in a variety of structural and electrical applications as a typical oxide ceramic.

Rare-earth sesquioxides are stable materials at high temperatures. In oxides of scandium, yttrium, lanthanum, gadolinium, and lutetium, which are well-used materials, trivalence is extremely stable and no other valence can exist in those oxides. Because ionic radius is relatively large and continuously changing in a series of rare-earth sesquioxides, these are often applied as dopant materials to control various properties in doped materials.

Al₂O₃, also termed alumina, is produced by heating hydrates of alumina. A number of transitional Al_2O_3 structures can form initially with increasing temperatures, but all structures are transformed irreversibly to α-Al_2O_3 with a corundum structure of a hexagonal system. α-Al_2O_3, the only stable form above 1200 °C (2190 °F), is generally used for structural and electrical applications, except for γ-Al_2O_3, which is used for catalytic application. Hereafter, α-Al_2O_3 is represented simply as Al_2O_3.

Thermal and Mechanical Properties. Because of a strong chemical bond strength between the Al and O ion, as expected from the value of heat of formation (−400 Kcal/mol), Al_2O_3 has outstanding physical stability, such as a high melting point (2050 °C, or 3720 °F), the highest hardness among oxides, and high mechanical strength. Table 3 shows the physical, mechanical, thermal, and electrical properties of Al_2O_3. Specific heat is expressed as:

$$C_p = 27.43 + 3.06 \times 10^{-3} T - 8.47 \times 10^5 T^{-2}$$

(In cal/mol · K, from 298 to 1800 K)   (Eq 11)

Mechanical strength is high at room temperature but is lowered largely above 1100 °C (2010 °F) as shown in Fig 9 and 10. Thermal conductivity is relatively large among oxides. However, because the coefficient of thermal expansion is large, thermal shock resistance (Δ*T* of 110 to 270 °C, or 230 to 520 °F) is smaller than the values for Si_3N_4 and SiC, which are the representative high-strength materials. Fracture toughness at room temperature is from 3.85 to 3.95 MPa√m (3.50 to 3.59 ksi√in.) in sintered Al_2O_3 (95% relative density and 2 μm, or 80 μin., average grain size) and from 4.18 to 5.9 MPa√m (3.80 to 5.4 ksi√in.) in hot pressed Al_2O_3 (99.5% relative density and 2 μm average grain size). Some creep data are given in Table 4.

Electrical Properties. Al_2O_3 is a typical electrically insulating material. The volume resistivity of high-purity ceramics with a low

Table 3 Engineering properties of α-Al₂O₃

Property	Single crystal	99.9%	99.9%	99.5%	96%	90%
Physical						
Density, g/cm³	. . .	3.99	3.96	3.87	3.72	3.60
Grain size, μm (mil)	. . .	15–45 (0.6–1.8)	1–6 (0.04–0.24)	5–50 (0.2–2)	2–20 (0.08–8)	2–10 (0.08–0.4)
Surface finish, (authentic average), μm (mil)	. . .	62 (2.5)	50 (2)	87 (3.5)	162 (6.5)	162 (6.5)
Color	. . .	Translucent white	Ivory	Ivory	White	White
Mechanical						
Modulus of elasticity, GPa (10⁶ psi)	434 (63)	393 (57)	366 (53)	372 (54)	303 (44)	275 (40)
Modulus of rigidity, GPa (10⁶ psi)	. . .	162 (24)	158 (23)	151 (22)	124 (18)	117 (17)
Bulk modulus, GPa (10⁶ psi)	. . .	234 (34)	227 (33)	227 (33)	172 (25)	158 (23)
Poisson's ratio	. . .	0.22	0.22	0.22	0.21	0.22
Flexural strength, MPa (ksi)						
At 25 °C (77 °F)	634 (92)	282 (41)	551 (80)	379 (55)	358 (52)	337 (49)
At 1000 °C (1830 °F)	413 (60)	172 (25)	413 (60)	. . .	172 (25)	. . .
Compressive strength, MPa (ksi)						
At 25 °C (77 °F)	. . .	2549 (370)	3790 (550)	2618 (380)	2067 (300)	2480 (360)
At 1000 °C (1830 °F)	. . .	482 (70)	1929 (280)	. . .	. . .	. . .
Tensile strength, MPa (ksi)						
At 25 °C (77 °F)	. . .	206 (30)	310 (45)	262 (38)	193 (28)	220 (32)
At 1000 °C (1830 °F)	. . .	103 (15)	220 (32)	. . .	96 (14)	108 (16)
Transverse sonic velocity, 10³ m/s	. . .	9.9	9.9	9.8	9.1	8.8
Hardness (R45N)	. . .	85	90	83	78	79
Coefficient of thermal expansion, 10⁻⁶/K						
From 25–400 °C (77–750 °F)	. . .	7.4	7.4	7.6	7.4	7.0
From 25–1000 °C (77–1830 °F)	. . .	8.3	8.3	8.3	8.2	8.1
Thermal conductivity, W/cm · K						
At 20 °C (68 °F)	0.43	0.39	0.39	0.35	0.24	0.16
At 100 °C (212 °F)	. . .	0.28	0.27	0.26	0.19	0.13
At 400 °C (750 °F)	. . .	0.13	0.13	0.12	0.10	0.08
Dielectric constant at 25 °C (77 °F)						
At 1 kHz	. . .	10.1	9.9	9.8	9.0	8.8
At 1 MHz	. . .	10.1	9.8	9.7	9.0	8.8
At 10 GHz	. . .	10.1	9.8	9.7	8.9	8.7
Dissipation factor at 25 °C (77 °F)						
At 1 kHz	. . .	0.00050	0.0020	0.0002	0.0011	0.0006
At 1 MHz	. . .	0.00004	0.0002	0.0003	0.0001	0.0004
At 10 GHz	. . .	0.00009	0.0050	0.0002	0.0006	0.0009
Loss factor at 25 °C (77 °F)						
At 1 kHz	. . .	0.0050	0.020	0.002	0.010	0.005
At 1 MHz	. . .	0.0004	0.002	0.003	0.001	0.004
At 10 GHz	. . .	0.0010	0.005	0.002	0.005	0.008
Dielectric strength, AC, kV/cm (average RMS values at 60 Hz AC)						
0.63 cm thick	. . .	90.5	94.5	86.6	82.6	92.5
0.13 cm thick	. . .	200.7	181.1	169.3	145.6	177.1
0.02 cm thick	. . .	. . .	314.9	330.7	228.3	299.2
Volume resistivity, Ω · cm²/cm						
At 25 °C (77 °F)	. . .	. . .	>10¹⁵	>10¹⁴	>10¹⁴	>10¹⁴
At 500 °C (930 °F)	. . .	. . .	3.3 × 10¹²	. . .	4.0 × 10⁹	2.8 × 10⁸
At 1000 °C (1830 °F)	. . .	. . .	1.1 × 10⁷	. . .	1.0 × 10⁶	8.6 × 10⁵

Source: Ref 23

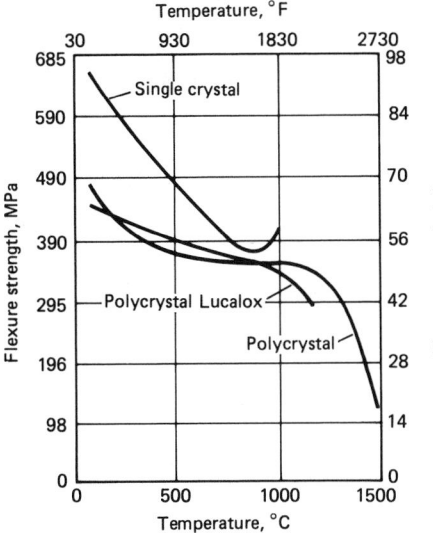

Fig 9 Flexure strength of single-crystal and polycrystal Al₂O₃. Source: Ref 24

Table 4 Creep data of sintered Al₂O₃

Material	Temperature		Stress		Creep rate, 10^{-6} cm/cm · h
	°C	°F	MPa	ksi	
Al₂O₃ (purity > 99%)	1300	2370	13.1	1.9	2.7
	1300	2370	24.9	3.6	11.5
	1300	2370	48.9	7.1	27.6
Al₂O₃ (purity > 99.9%) (density = 98%)	1000	1830	...	...	9
Al₂O₃ (density > 99.5%)	1300	2370	21.0	3.0	18.5
			56.0	8.1	115.0
Al₂O₃ + 1.38% Cr₂O₃ (density > 99.5%)	1350	2460	21.0	3.0	71.0
			56.0	8.1	375

Source: Ref 25

is a laser active material for a 694 nm wavelength.

Chemical Properties. Al₂O₃ is also chemically very stable and has high corrosion resistance. It is insoluble in water and very slightly soluble in strong acid and alkaline solution. Vapor pressure is small, even at high temperatures: 0.13, 1.33, and 13.3 kPa (1, 10 and 100 torr) at 2148, 2385, and 2665 °C (3900, 4325, and 4830 °F), respectively.

Properties of AO₂

AO₂-Type Oxides. Oxides with cristobalite, rutile, and fluorite structures are included in *AO₂*-type oxides. Because of a

alkali concentration is more than 10^{15} Ω · cm at room temperature. The electrical conductivity of single-crystal and polycrystal Al₂O₃ at high temperature is shown in Fig 11.

As shown in Table 3, values of elasticity, thermal conductivity, and electrical resistivity largely increase as purity decreases. Mechanical strength and dielectric strength also increase with increasing purity but are more dependent on density and ceramic microstructure.

Optical Properties. Al₂O₃ has a refractive index of 1.760. Dense Al₂O₃ ceramic prepared with high-purity, fine, and uniform powders shows good translucency (Fig 12). Both MgO and Y₂O₃ are often used as additives and effectively eliminate pores and protect against abnormal grain growth. A single crystal with a small amount of Cr₂O₃ (ruby)

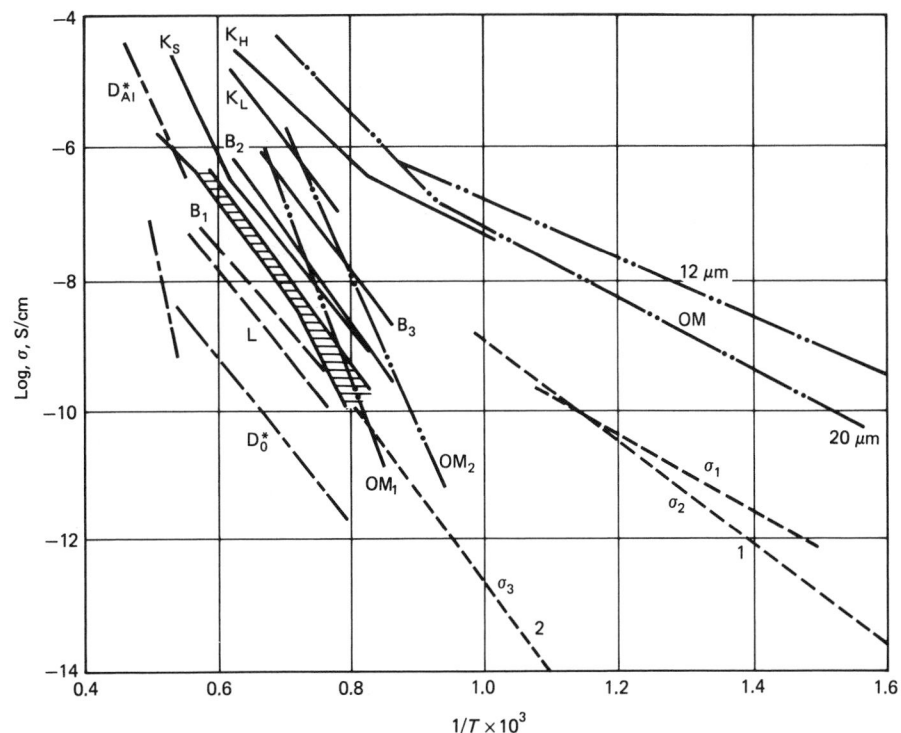

K_H: Hot-pressed PC, grain size 4 μm (160 μin.)
K_L: PC, Lucalox, grain size 30 μm (1.2 mils)
K_s: EFG-method sapphire (perpendicular c-axis)
OM: PC
B₁: SC (CZ), nondoped and Ti,Si doped
B₂: SC (CZ), Mg doped
B₃: SC (CZ), Co doped
L: SC (CZ), nondoped
OM₁: SC (CZ), σ, perpendicular to c-axis
OM₂: SC (CZ), σ, parallel to c-axis
1: SC (CZ), oxidized
2: SC (CZ), reduced
D*_Al: Conductivity calculated from Al diffusion constant
D*_O: Conductivity calculated from O diffusion constant

Fig 11 Electrical conductivity of single-crystal (SC) and polycrystal (PC) Al₂O₃. Source: Ref 26

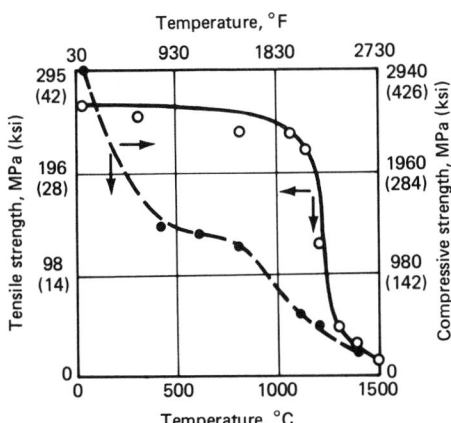

Fig 10 Tensile and compressive strengths of sintered Al₂O₃. Source: Ref 24

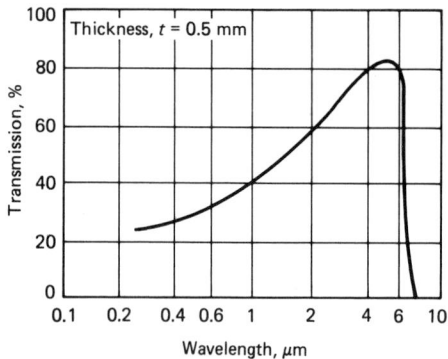

Fig 12 Optical transformation of translucent Al_2O_3 ceramics. Source: Ref 27

large flexibility in the stacking of AO_4 polyhedra, the cristobalite structure has many polymorphs. Tridymite and quartz structures are modified types of cristobalite structure.

SiO_2, a representative oxide with cristobalite structure, has peculiar engineering properties, such as low thermal expansion, low thermal conductivity, and glass-forming ability derived from the covalent nature of the Si–O bond.

Several oxides with the rutile structure, that is, TiO_2, VO_2, and MnO_2, easily form non-stoichiometric oxides and shear-structure oxides (magneli phase) derived from a large amount of oxygen vacancy. These materials have much potential for electrical applications.

Fluorite structure is observed in AO_2-type oxides with large cation radii, such as ThO_2, HfO_2, and CeO_2. Characteristic property of the fluorite-structure oxides is a high oxygen-ion conductivity because of a large space at the center of the cubic unit cell. A fluorite structure of pure ZrO_2 is stable only at high temperatures, because of the relatively small Zr^{4+} ion for the fluorite structure. It shows martensite transformation at low temperatures. This transformation is effectively applied for the toughening of ZrO_2-based ceramics.

SiO_2. Several crystalline phases of SiO_2 that are formed at atmospheric pressure are classified in Table 5. The predominant crystalline

phases of SiO_2, that is, quartz, tridymite, and cristobalite, are not well suited for use as the principal phase in refractory ceramics, because at relatively low temperatures they undergo phase modifications that are accompanied by a disruptive change in volume.

The liquid that is formed when crystalline SiO_2 is melted has an unusually high viscosity, which means that the liquid has a strong tendency to form a glass, rather than to crystallize upon cooling. This glass-forming ability, along with the widespread occurrence of quartz sands, has resulted in the use of SiO_2 as the basic raw material for the glass industry.

Two types of ceramics are made from fused silica. One type is referred to as bulk fused silica, and the other as polygranular fused silica. Bulk fused silica is made directly from the melt by more or less normal glass-processing methods. It can be either clear or translucent. Entrapped gas bubbles are present when it is translucent. Polygranular fused silica is prepared from fused-silica powder, either by sintering compacts or by hot pressing. The compacts are commonly formed by slip casting.

Thermal Properties. Specific heat (cal/mol) of SiO_2 can be expressed by:

Quartz (low) $C_p = 11.22 + 8.2 \times 10^{-3} T - 2.70$

$\times 10^5 T^{-2}$ ($T = 298$ to 848 K)

Quartz (high) $C_p = 14.41 + 1.94 \times 10^{-3} T$

($T = 848$ to 2000 K)

Fused silica $C_p = 13.38 + 3.68 \times 10^{-1} T - 3.45$

$\times 10^5 T^{-2}$ ($T = 298$ to 2000 K)   (Eq 12)

Thermal conductivity at 400 K is 7.9 W/m · K for a perpendicular c-axis and 4.8 W/m · K for a parallel c-axis in single-crystal quartz, 1.5 W/m · K in bulk fused silica, and 0.6 W/m · K in polygranular fused silica.

Linear thermal expansion of quartz increases up to 40×10^{-6}/K (perpendicular to c-axis) and 23×10^{-6}/K (parallel to c-axis) at 840 K with increasing temperature. Above that temperature, it rapidly decreases to a small negative value. Fused silica has a constant

coefficient of thermal expansion up to 1000 °C (1830 °F) but has a value as small as ~10% of that of single-crystal quartz.

Mechanical Properties. Young's modulus, shear modulus, and Poisson's ratio of bulk fused silica at room temperature are 74, 31, and 0.71 GPa (10, 4.5, and 0.1×10^6 psi), respectively. Table 6 shows the mechanical strengths of several kinds of SiO_2. Generally, strengths of SiO_2 are not large, except for single-crystal quartz.

Vickers hardness at room temperature ranges from 5 to 7 GPa (0.7 to 1.0×10^6 psi) in fused silica and from 9 to 11 GPa (1.3 to 1.6×10^6 psi) in single-crystal quartz.

Electrical and Chemical Properties. SiO_2 is a typical electrical insulator. Quartz has piezoelectricity, and its piezoelectric constants are 2.31×10^{-2} C/N(d_{11}) and 0.727×10^{-2} C/N(d_{14}). Artificial quartz is widely used for oscillators.

Quartz and fused silica have excellent resistance to ordinary weathering. No attack occurs at ordinary temperatures and pressures. However, both have relatively high solubility in a basic solution of pH > 9.

TiO_2 has three polymorphs: low-temperature stable anatase and brookite and high-temperature stable rutile.

Anatase and brookite transforms irreversibly to rutile at temperatures from 400 to 1000 °C (750 to 1830 °F), but the starting temperature and velocity of the transformation are largely affected by impurities, particle size, and treated condition due to reconstructive slow transformation.

Thermal and physical properties of TiO_2 are given in Table 7. Specific heat ($T = 298$ to 1800 K) is expressed as:

Rutile $C_p = 17.4 + 9.8 \times 10^{-4} T - 3.5$

$\times 10^5 T^{-2}$ (In cal/mol · K)

Anatase $C_p = 17.21 + 10.8 \times 10^{-4} T - 3.59$

$\times 10^5 T^{-2}$ (In cal/mol · K)   (Eq 13)

Mechanical properties of rutile are shown in Table 8. Reported creep rate (steady state) of rutile single crystal with 48.57 ppm impurities are 5.0×10^{-3}/in. · h (in O_2) and 7.2×10^{-3}/in. · h (in N_2) under a stress of 0.66 MPa (0.096 ksi) and 10.8×10^{-3}/in. · h (in O_2) and 21.6×10^{-3}/in. · h (in N_2) under 0.76 MPa (0.11 ksi).

Electrical Properties. High-purity rutile has low electrical conductivity ($\sim 10^{-7}$/Ω · cm at 500 °C, or 930 °F) and high dielectric constant (from 110 to 117 at room temperature). Anisotropy is observed in single crystals, where the perpendicular dielectric constant is 86, and the parallel, 170. Because rutile has a low tangent and a small negative value in the temperature coefficient of the dielectric constant (from 300 to 700 ppm/K), it has been applied to capacitor materials.

When reduced in a low oxygen pressure condition or when doped with a small amount of pentavalent cation (Nb^{5+}, and so on), rutile

Table 5   Crystalline phases of SiO_2 at atmospheric pressure

Classification of phases	Stability range, K	Crystal system	Theoretical density, g/cm³	Melting point, K	Remarks
Quartz, low	<845	Trigonal	2.65	...	Most common crystalline phase
Quartz, high	845–1140	Hexagonal	2.52	1743	High-temperature phase
Tridymite S	1140–1743	Orthorhombic	2.32	1933	Characteristic phase in refractories
S-I	<337				Polymorphic modification
S-II	337–391				Polymorphic modification
S-III	391–436				Polymorphic modification
S-IV	436–483				Polymorphic modification
S-V	483–748				Polymorphic modification
S-VI	748–1743				Polymorphic modification
Tridymite M	...	Hexagonal	2.30	...	Unstable, converts to Tridymite S
M-I	<391				Polymorphic modification
M-II	391–436				Polymorphic modification
M-III	>436				Polymorphic modification
Cristobalite, low	<545	Tetragonal	2.32	...	Low-temperature phase
Cristobalite, high	545–1998	Cubic	2.27	1998	Appears first when above phases are heated to 1672 K

Table 6 Strength value of fused silica and quartz at room temperature

| | | Strength | | | | | |
| | Relative density, % | Flexure | | Compressive | | Tensile | |
Type of material		MPa	ksi	MPa	ksi	MPa	ksi
Bulk fused silica							
Clear	100	104	15	690–1380	100–200	69	10
	100	48	7	1035	150	...	...
Translucent	100	45	6.5	276	40	21	3
Polygranular fused silica							
Hot pressed	~95	...	...	690–1380	100–200	...	...
Slip cast, sintered	34–98	3–120	(0.4–17)	17–490	2.5–71	5–32	0.7–4.6
Single-crystal quartz							
Parallel c-axis	100	...	...	~2070	~300	110–276	16–40
Perpendicular c-axis	100	...	...	<2070	<300	...	...

Source: Ref 28

shows an n-type semiconducting nature. Predominant lattice defects are reported to be interstitial Ti^{3+} or oxygen vacancies. The intrinsic band gap is 3.5 eV.

Chemical Properties. TiO_2 is a chemically stable material. However, it is soluble in hot H_2SO_4 and reacts rapidly with melt hydroxides and carbonates of alkalies. Titanium chloride, carbide, and nitride are formed upon contact with chloride at temperatures from 800 to 1000 °C (1470 to 1830 °F), carbon above 1100 °C (2010 °F), and nitrogen (+carbon) at 1200 °C (2190 °F). When annealed in a reducing atmosphere or when it contacts metals at high temperatures, TiO_2 is easily reduced to lower-valence oxides, such as Ti_2O_3, TiO, and magneli phases Ti_nO_{2n-1} ($n = 4$ to 10).

ZrO_2. The high-temperature stable form of ZrO_2 has a cubic fluorite structure. Because the ionic radius of Zr^{4+} is too small (0.084 nm, or 0.003 μin.) to satisfy the required ratio of cation and anion radius for an 8-coordinated polyhedron of the fluorite structure, pure ZrO_2 shows a polymorphic transformation:

$$\text{Monoclinic} \underset{900-1000\,°C}{\overset{1170\,°C}{\rightleftharpoons}} \text{Tetragonal} \overset{2370\,°C}{\leftrightarrow} \text{Cubic}$$
$$\overset{2680\,°C}{\leftrightarrow} \text{Liquid} \qquad \text{(Eq 14)}$$

As observed in the thermal expansion curve in Fig 13, the transformation between monoclinic and tetragonal forms is accompanied by a large volume change (~4.6%). This transformation is a martensite type and depends on dopant amount, grain size, and stress.

Electrical Properties. The cubic form of ZrO_2, stabilized down to room temperature by the doping of CaO, MgO, or Y_2O_3 (7 to 15 mol%), is called stabilized ZrO_2. Cations of dopants generally used for the stabilization have a lower valence than the tetravalent Zr ion. To maintain electrical neutrality when forming the solid solution, oxygen vacancies are generated according to the following reactions:

$$CaO \xrightarrow{ZrO_2} Ca''_{Zr} + O_0^* + V_\ddot{o}$$

$$Y_2O_3 \xrightarrow{2ZrO_2} 2Y'_{Zr} + 3O_0^* + V_\ddot{o} \qquad \text{(Eq 15)}$$

Stabilized zirconia shows high oxygen ion conductivity, because of the high concentra-

Table 7 Thermal and physical properties of polycrystal TiO_2

Property	Rutile	Anatase	Brookite
Density, g/cm^3	4.25	3.89	4.14
Melting point, °C (°F)	1855 (3370)	Transformed to rutile	
Coefficient of thermal expansion, 10^{-6}/K	7.14	10.2	11.0
Thermal conductivity, W/m·K	10.4(a)	...	...
	7.4(b)	...	...
At 300 °C	8.4	...	...
Magnetic susceptibility, cm^2/g	7.4 × 10^{-8}	...	...
Refractive index At 25 °C (77 °F) and 598 nm	n_ω 2.612 n_ϵ 2.899	n_ω 2.561 n_ϵ 2.488	n_α 2.583 n_β 2.584 n_γ 2.700
Reflectance, % At 400 °C (750 °F)	47–50	88–90	...
At 500 °C (930 °F)	95–96	94–95	...
Absorption of UV, at 360 μm %	90	67	...

(a) Parallel to c-axis, single crystal. (b) Perpendicular to c-axis, single crystal. Source: Ref 25

Table 8 Mechanical properties of rutile polycrystals

Young's modulus, GPa (10^6 psi)	282 (41)(a)
Shear modulus, GPa (10^6 psi)	111 (16)(a)
Bulk modulus, GPa (10^6 psi)	206 (30)
Poisson's ratio	0.278
Flexure strength, MPa (ksi)	340 (49)
	89 (13)(b)
Tensile strength, MPa (ksi)	69–103 (10–15)
Compressive strength, MPa (ksi)	800–940 (117–136)(b)
Hardness	10.4 (1.5)(a)(b)(c)
Knoop, 200 K g load	9.6 (1.4)(a)(b)(d)
	8.8 (1.3)(a)
	<2.0 (0.3)(b)(e)

(a) Room temperature. (b) Single crystal. (c) Parallel (110). (d) Perpendicular (110). (e) At 1000 °C (1830 °F)

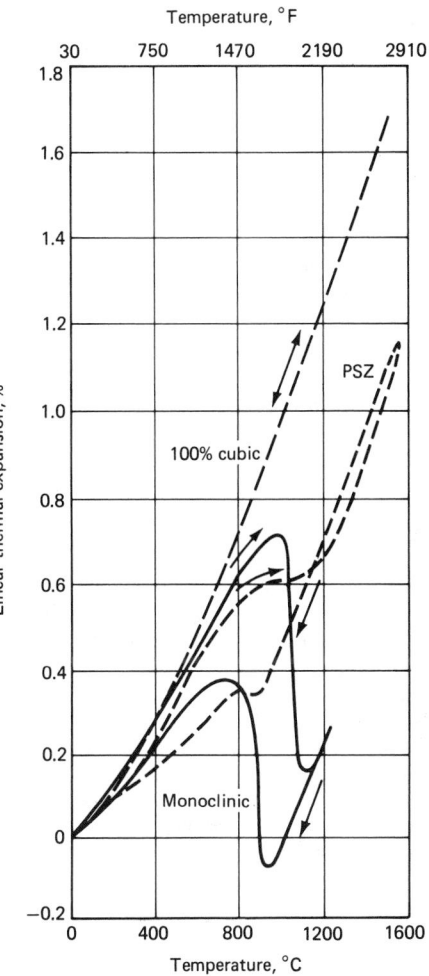

Fig 13 Thermal expansion curves of typical polymorphic forms of ZrO_2. Source: Ref 29

tion of oxygen vacancies, $V_\ddot{o}$. It has been widely applied as the solid electrolyte in oxygen concentration detectors or fuel cells.

Figure 14 shows oxygen ion conductivity at 800 °C (1470 °F) of cubic ZrO_2 stabilized

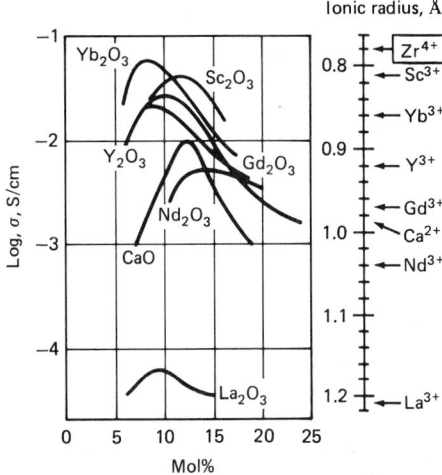

Fig 14 Electrical conductivity at 800 °C of cubic ZrO_2 stabilized with various oxides, and cationic radii of dopant oxides. Source: Ref 30

by various dopants. A maximum conductivity is given for the composition of the minimum dopant concentration required for stabilization. Generally, dopants with cations larger than Zr^{4+} give lower conductivities. Those phenomena have been explained to be caused by a formation of defect associates between oxygen vacancies and substituted cations.

Mechanical and Thermal Properties. The tetragonal form of ZrO_2 can be present at room temperature when the dopant amount is lower than that necessary for stabilization of the cubic form and when tetragonal particles are in very small (submicron) size.

Generally, multiphase ceramics with tetragonal and cubic forms are called partially stabilized zirconia (PSZ), and monophase ceramics of tetragonal form are called tetragonal zirconia polycrystal (TZP). Those ceramics show very high mechanical strength and fracture toughness. Some TZPs have flexure strength values that exceed 2000 MPa (290 ksi) or fracture toughness values greater than 20 MPa$\sqrt{m}$ (18 ksi$\sqrt{in}$.), which are the highest values among the available ceramics. Several properties of toughened TZP are showed in Table 9. Dopant concentrations that give maxima of strength and toughness are somewhat different, as shown in Fig 15. Fully stabilized ZrO_2 has flexure strength values from 360 to 660 MPa (52 to 96 ksi) and fracture toughness values of 2.4 MPa$\sqrt{m}$ (2.2

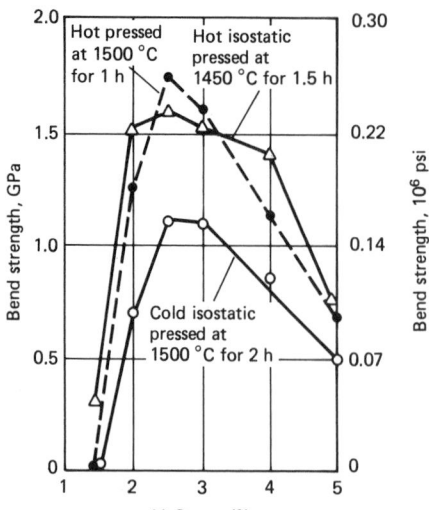

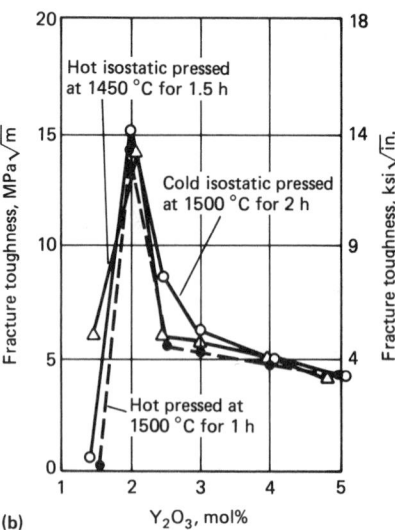

Fig 15 Composition dependence of bend strength (a) and fracture toughness (b) in Y_2O_3-added tetragonal ZrO_2 polycrystal. Source: Ref 31

Table 9 Several properties of tetragonal zirconia polycrystal (Y_2O_3, 3 mol%)

Color	Ivory
Density, g/cm³	6–6.05
Particle size, (tetragonal) μm (μin.)	0.2–1 (8–40)
Mechanical properties	
Flexure strength, MPa (ksi)	800–1500 (116–218)(a)
Compressive strength, MPa (ksi)	>2900 (420)(a)
Fracture toughness, MPa$\sqrt{m}$ (ksi$\sqrt{in}$.)	7–12 (6.4–10.9)(a)
Vickers hardness, GPa (10^6 psi)	12–13 (1.7–1.9)
Young's modulus, GPa (10^6 psi)	200–210 (29–30.5)
Poisson's ratio	0.31
Thermal shock (DT), K	360
Creep rupture, MPa (ksi)	300 (45)(b)
Thermal properties	
Specific heat, cal/g·K	0.11–0.12
Thermal expansion, 10^{-6}/K	9.6(c)
	10.4(d)
Thermal conductivity, W/m·K	2.0–3.3
Electrical properties	
Ionic conductivity, S/cm	~10^{-12}(a)
	3×10^{-4}(e)
Dielectric constant	15–32
Loss tangent	5–7 $\times 10^{-3}$

(a) Room temperature. (b) At 800 °C (1470 °F), 1000 h. (c) From 20–400 °C (68–750 °F). (d) From 20–1000 °C (68–1830 °F). (e) At 500 °C (930 °F)

ksi$\sqrt{in}$.), which are much lower than those of TZP. Specific heat (cal/mol · K) of MgO-PSZ is:

$$C_p = 18.78 + 3.67 \times 10^{-3} T - \frac{1310}{T + 273}$$

$$(T = 0 \text{ to } 900 \text{ °C}) \qquad \text{(Eq 16)}$$

The appearance of high strength and toughness is explained to be due to stress-induced phase transformation, absorption of fracture energy by martensite transformation from tetragonal to monoclinic form, or by microcracks produced by volume change at the transformation.

Chemical Properties. Because stabilized ZrO_2 has a high melting point (~2680 °C, or 4855 °F), low thermal conductivity (1 to 2 W/ m · K for fibers and porous ceramics), low vapor pressure (~0.1 Pa, or 10^{-6} atm at 1800 °C, or 3270 °F), and good corrosion resistance against alkaline and fused metal, it gives good performance as a refractory and a heat insulator for high-temperature use.

However, ZrO_2 reacts with HCl and HNO_3 more rapidly than with alkalies. Upon contact with carbon, nitrogen, or hydrogen at 2200 °C (3990 °F), ZrO_2 decomposes or forms zirconium carbide, nitride, or hydride.

Properties of AO_3

Typical AO_3-type oxides are WO_3, ReO_3, and MoO_3. Both WO_3 and MoO_3 are semiconductors, but ReO_3 is a metallic conductor whose electrical conductivity at room temperature is of the order of 10^5 S/cm, the highest value among oxides.

These oxides easily become the so-called bronzes, incorporating alkali metals, for instance, into their crystal lattices. Specifically, M_xWO_3 is called tungsten bronze, where M can be alkali metals, alkaline-earth metals, Cu, Ag, lanthanoid elements, and others, and $x = 0$ to 1. The bronze-type oxides have been extensively investigated because of their interesting properties from the viewpoint of solid-state chemistry.

WO_3 exists in five polymorphic forms, as shown in Table 10. Crystal structures of these polymorphs basically consist of [WO_6] octahedrons linked by sharing-corner oxygen ions. WO_3 usually shows a wide range of nonstoichiometry through forming oxygen vacancies. If the oxygen vacancy concentration becomes large, part of [WO_6] octahedrons would become arranged on certain shear planes being linked by sharing the edges. Hence, the oxygen vacancies would disappear. Increase

Table 10 Five polymorphs of WO_3

Crystal system	Space group	a nm	b nm	c nm	α, β, γ	Temperature region, K
Tetragonal	P4/nmn	0.525	...	0.391	...	>1013
Orthorhombic	...	0.7384	0.7512	0.3846	...	623–1013
Monoclinic	P2$_{1/n}$	0.7297	0.7539	0.7688	β = 90°91'	290–623
Triclinic	P2$_{1/n}$	0.730	0.752	0.769	α = 88°50'	233–290
		...	...	...	β = 90°55'	...
		...	...	...	γ = 90°56'	...
Monoclinic	...	0.527	0.516	0.767	β = 91°43'	<233

in the number of planar defects (shear planes) would then lead to their ordered arrangement, stabilizing a certain type of crystal phase just as the case of magneli phases in titanium oxide. When WO_3 is reduced to form non-stoichiometric phases, the corresponding amount of W^{6+} ions are reduced to W^{5+} ions to maintain the charge balance.

Tungsten oxide is anticipated for use in the electrochromic device (ECD) because of its capability to change colors by an electrochemical reaction. It is based on the following fundamental reaction:

$$WO_3 + xM^+ + xe^- \underset{\text{Oxidation}}{\overset{\text{Reduction}}{\longleftrightarrow}} M_xWO_3 \qquad \text{(Eq 17)}$$
(colorless) (blue)

The M^+ denotes monovalent ions, such as H^+, Li^+, Na^+, and Ag^+. The H^+ was used for an initial ECD, but Li^+ was most widely used recently as a solution containing H^+ in high concentration, which dissolves the WO_3 film.

REFERENCES

1. Touloukian *et al.*, Ed., *Thermophysical Properties of Matter*, Vol 1–13, 1970–1977
2. G.R. Terwilliger and K.C. Radford, *Am. Ceram. Soc. Bull.*, Vol 53 (No. 2), 1974, p 172
3. H. Murata, S.C. Choi, K. Koumoto, and H. Yanagida, *J. Mater. Sci.*, Vol 20, 1985, p 4507
4. D.R. Sempolinski and W.D. Kingery, *J. Am. Ceram. Soc.*, Vol 63, 1980, p 664
5. D.R. Sempolinski, W.D. Kingery, and H.L. Tuller, *J. Am. Ceram. Soc.*, Vol 63, 1980, p 669
6. C.M. Osburn and R.W. Vest, *J. Phys. Chem. Solids*, Vol 32, 1971, p 1331
7. A.J. Bosman and C. Crevecoeur, *Phys. Rev.*, Vol 144, 1966, p 763
8. K. Koumoto, Z.T. Zhang, and H. Yanagida, *J. Ceram. Soc. Jpn.*, Vol 92, 1984, p 83
9. I. Bransky and N.M. Tallan, *J. Chem. Phys.*, Vol 49, 1968, p 1243
10. S.P. Mitoff, *J. Chem. Phys.*, Vol 35, 1961, p 882
11. G.Z. Zintl, *Phys. Chem. N.F.*, Vol 48, 1966, p 340
12. M. Heward, Ph.D. thesis, University of Exeter, 1976
13. W.C. Tripp and N.M. Tallan, *J. Am. Ceram. Soc.*, Vol 53, 1970, p 531
14. Y.D. Tretyakov and R.A. Rapp, *Trans. AIME*, Vol 245, 1969, p 1235
15. R. Farhi and G. Petot-Ervas, *J. Phys. Chem. Solids*, Vol 39, 1978, p 1169
16. H.G. Sockel and H. Schmalzried, *Ber. Bunsenges. Phys. Chem.*, Vol 72, 1968, p 754
17. A. Atkinson, A.E. Hughes, and A. Hommou, *Philos. Mag. A*, Vol 43, 1981, p 1079
18. K. Koumoto, S. Mizuta, and H. Yanagida, *J. Ceram. Soc. Jpn.*, Vol 88, 1980, p 217
19. D.F. Anthrop and A.W. Searcy, *J. Phys. Chem.*, Vol 68, 1964, p 2335
20. M.S. Chandrasekharaiah, *The Characterization of High-Temperature Vapors*, J.L. Margrave, Ed., John Wiley & Sons, 1967, p 495
21. J.A. Kitchener and S. Ignatowicz, *Trans. Faraday Soc.*, Vol 47, 1951, p 1278
22. P.A. Miles, W.B. Westphal, and A. von Hippel, *Rev. Md. Phys.*, Vol 29, 1957, p 279
23. C.T. Lynch, Ed., *Handbook of Material Science*, Vol 2, CRC Press, 1974, p 360–362
24. Y. Shiraki, *Fine Ceramics: Sintered Ceramics 4*, Gihodo, Tokyo, 1976, p 337–353
25. *Fine Ceramics Dictionary*, Gihodo Tokyo, 1987
26. K. Kitazawa and R.L. Coble, *J. Am. Ceram. Soc.*, Vol. 57, 1974, p 245
27. R.L. Coble, U.S. patent 3,026,210, 1962
28. "Engineering Property Data on Selected Ceramics," MCIC Report, Battelle Columbus Laboratories, 1981
29. E.C. Subbarao, *Science and Technology of Zirconia*, Vol 3, *Advances in Ceramics*, A.H. Heuer and L.W. Hobbs, Ed., American Ceramic Society, 1981, p 1–24
30. Y. Saito, *Zirconia Ceramics 1*, S. Somiya, Ed., Uchida-Rokakuho, Tokyo, 1983, p 109–125
31. T. Masaki, *J. Am. Ceram. Soc.*, Vol 69, 1986, p 638

Engineering Properties of Multicomponent and Multiphase Oxides

Angus I. Kingon and Robert F. Davis, Department of Materials Science and Engineering, North Carolina State University
Michael M. Thackeray, Division of Materials Science and Technology, CSIR, South Africa

THE MULTICOMPONENT OXIDES refer to those oxides which contain more than one cation type in the structure. These oxides can be single phase, such as mullite ($3Al_2O_3 \cdot 2SiO_2$), or multiphase (mixture of oxides). This class therefore covers a huge number of compounds, for which it would clearly be impossible to present comprehensive property data. In general, with a few exceptions, mechanical properties of the multicomponent oxides are not as good as the single oxides (for example, Al_2O_3 and modified ZrO_2). However, the additional cations allow a wide range of structural types to be synthesized, allowing the selective optimization of a single property or a set of properties, including thermomechanical, thermal, chemical, dielectric, electrical, electronic, optical, and magnetic.

In this article the (arguably) most important structural classes (silicates, spinels, and perovskites) are used to illustrate the optimization of particular properties. Organization of the article is as follows:

- Mechanical and thermomechanical properties of the silicates are emphasized in the first section, while the most important aluminosilicate, mullite, is sufficiently important to be dealt with separately in a second section
- Optimized mechanical, electrical, and/or magnetic properties of the spinel-structured oxides are covered in the third section. Particular emphasis is placed upon the lithium-containing spinels which are being developed for new generation batteries
- The fourth section covers the perovskites, and briefly surveys the wealth of dielectric, optical, and electrical properties which can be controllably produced
- The fifth section covers two-phase Al_2O_3/ZrO_2 ceramics and reaction-sintered mullite/ZrO_2 which are two of the most important multiphase oxides utilized

because of their superior mechanical properties

Even more so than with the simple oxides, property data must be used with caution because of the variations in processing, compositions, microstructures, and perhaps the reported test methods. Property data for those oxides considered to be the most important are compared with alumina in Table 1.

Silicates

The majority of minerals in the earth's crust are silicates. Literally thousands of identified silicate structures and silicate structural variants exist. Most of the industrial silicates are not synthesized commercially, but are utilized as beneficial minerals and consumed in large volume by the refractories, brick, sanitaryware, china, and pottery industries. In general, mechanical properties of the silicates are inferior to the important single oxides such as Al_2O_3 and stabilized ZrO_2. This section briefly reviews the structural principles of the silicates and mentions important silicates in each structural category. Only two silicate materials are described in detail. These representative silicate ceramics, which are used for engineering applications (rather than as "traditional" ceramics), are cordierite and pyrophyllite. Cordierite is generally utilized in synthetic form while pyrophyllite is used in an as-mined or beneficiated form. For more information on the silicates, readers are referred to the text by Grimshaw (Ref 5).

Silicate Structures (Ref 6). The building block for the silicates is the unit tetrahedron consisting of a silicon ion placed at the center of four symmetrically arranged oxygen ions. These SiO_4 [or $(SiO_4)^{4-}$] units may be linked to other cations or to oxygen ions. Pauling has postulated a scheme giving a ranking of energy for the various linkage schemes. It is useful to categorize the silicates by the way

the tetrahedra link to each other as discussed below (Ref 6).

Island structures consist of $(SiO_4)^{4-}$ units which are linked to each other through additional (non-silicon) cations. This class is also called the "orthosilicates." Important materials in the class include zircon ($ZrSiO_4$), phenacite (Be_2SiO_4), willemite (Zn_2SiO_4), the olivines ($Mg_{2-x}Fe_xSiO_4$), and the aluminosilicate group. The last group includes (ideally) kyanite, andalusite, and sillimanite (all with ideal composition Al_2SiO_5), and mullite ($Al_6Si_2O_{13}$). For engineering applications, mullite is by far the most important material and is reviewed in a separate section of this article. With regard to the properties of the remaining materials, little reliable engineering property data are available. This is for the following reasons:

- The materials are predominantly used in impure, mineral form, and properties are significantly affected by both lattice and second-phase impurities
- The materials are usually selected for a combination of cost and high-temperature thermal properties rather than mechanical properties

Isolated group structures are two or more silica tetrahedra linked through a common corner. This can result in the $(Si_2O_7)^{6-}$ grouping, as well as the ring groupings $(Si_3O_9)^{6-}$, $(Si_4O_{12})^{8-}$, and $(Si_6O_{18})^{12-}$, shown in Fig 1. In all of these cases, the above groupings are isolated from each other. Charge balance and linkage between these groups must be maintained by additional cations. Structures with $(Si_4O_{12})^{8-}$ rings include the zeolites. These are important because the four-membered rings form channels of controllable size, which can be used to accommodate specific chemical species. They are therefore used in large volume as catalyst materials in the petrochemical industry. A mineral with the $(Si_6O_{18})^{12-}$ rings is beryl. Closely related to this is the

Table 1   Properties of selected multicomponent or multiphase oxides

	Materials					
Properties	99.9% α-Al₂O₃(a)	Mullite(b)	Spinel	Cordierite	Forsterite	Alumina/ zirconia(c)
Density, g/cm³	3.96	3.16	3.58	2.0–2.53	3.22	5.5
Melting point, °C (°F)	2015 (3660)	1830 (3325)	2135 (3875)	1470 (2680)	1885 (3425)	...
Thermal expansion coefficient from 25 to 1000 °C (77 to 1830 °F), 10⁻⁶/°C	8.3	4.5–5.3	7.6–8.8	1.4–2.6	1.5	9
Thermal conductivity, W/cm·K, at 100 °C (212 °F)	0.27	0.059	0.15	...	0.046	0.035
Young's modulus, Gpa (10⁶ psi)	366 (53)	50–220 (7–32) 150–270 (22–39)	240–260 (35–38)	139–150 (20–22)	...	260 (38)
Flexural strength at room temperature(d), MPa (ksi)	550 (80)	500 (72)(f)	110–245 (16–35)	120–245 (17–35)	...	2400 (348)
Volume resistivity, Ω·cm, at:						
100 °C (212 °F)	2 × 10¹⁵	>10¹⁴	>10¹⁴	>10⁴(?)	...	...
1000 °C (1830 °F)	10⁷	>10³	2 × 10⁵	...	10⁵	...
Relative dielectric constant, *K* and tanδ (1 MHz)	9.8(e) 0.0002(e)	6.6(e) 0.006–0.07(e)	... ...	5.0(e) ...	... ...	... ...
Composition	Al₂O₃	3Al₂O₃·2SiO₂	MgAl₂O₄	2MgO·2Al₂O₃·5SiO₂	2MgO·SiO₂	20.0 wt% Al₂O₃, 75.7 wt% ZrO₂, 4.2 wt% Y₂O₃

(a) Fine-grained α-Al₂O₃, for comparison purposes. (b) Typical of commercial solid state reaction synthesized mullite, with coarse grains. See text for comments on properties of chemically synthesized mullite, presently under development. (c) Most of the data on Al₂O₃/ZrO₂ is from ToyoSoda Mfg. Co., Ltd., referring to TSK Super-Z grade containing 80 wt% of Y₂O₃-stabilized zirconia and 20 wt% alumina. (d) conditions not generally specified. (e) At room temperature. (f) At 1300 °C (2370 °F). Source: Ref 1–4

ceramic material cordierite, which is widely used and discussed in this section.

Chain structures are formed by sharing of two oxygen atoms of every tetrahedron, thus forming extended parallel chains (Fig 2). This yields the general formula $n(SiO_3)^{2-}$, with charge balance again maintained by adjacent

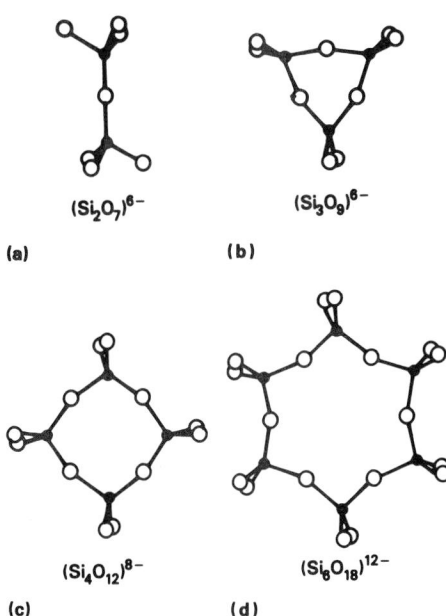

● Silicon ○ Oxygen

Fig 1 Silica tetrahedral groupings. (a) Two linked units. (b) Three unit ring. (c) Four unit ring. (d) Six unit ring. Adapted from Ref 6

cations. These cations also link the chains in three dimensions. These single-chain compounds are termed "pyroxenes." Alternatively, chains can be linked via shared oxygen ions (these silicates are known as "amphiboles"), and an example is shown in Fig 2(b). The general formula is now $n(Si_4O_{11})^{6-}$. Important materials of the class are the pyroxenes enstatite ($MgSiO_3$), wollastonite ($CaSiO_3$), and spodumene $LiAl(SiO_3)_2$. The last material has a low coefficient of thermal expansion, and is an important constituent of commercial glass-ceramics.

The sheet structures are formed by sharing three oxygens of each tetrahedron (Fig 3). The layer composition can be represented as $n^2(Si_2O_5)^{2-}$. Most clay materials, such as kaolinite, have these sheet structures, which affect both their behavior in aqueous suspension and their mechanical properties. The mineral pyrophyllite is discussed in more detail below.

Framework structures represent the extreme case where all four oxygen atoms of the silica tetrahedron are shared. The ideal examples are SiO_2 in its various structural modifications quartz, cristobalite, and tridyimite. Included in this group are the framework silicates in which some silicon atoms are replaced by aluminum. Further cations, usually alkali or alkali earth ions, then enter the lattice to maintain charge balance. The general formula is $(Si_{n-p}Al_pO_{2n})^{-p/2}$. An example is albite, a soda feldspar ($NaAlSi_3O_8$) in which $Al^{3+} + Na^+$ replace one Si^{4+} ion. The most important materials in this group are the feldspar minerals (a major constituent of igneous rock).

Cordierite (ideally $Mg_2Al_4Si_5O_{18}$) has an unusually low temperature coefficient of thermal expansion and excellent thermal shock resistance. Coupled with its reasonable mechanical properties, this has made cordierite useful in ceramic form as substrates for particular filters in diesel engines and spark plug insulators. It is also undergoing development for rotor vanes and heat exchangers in gas turbines.

Cordierite is isostructural with the mineral beryl. It is found in two structural forms: The common form in nature has orthorhombic symmetry ("low-temperature" form), while the modified form has hexagonal symmetry and is also being called "indialite." For engineering applications, it is usual to control purity and reproducibility by utilizing synthetic cordierite. Depending on process route, this generally forms the hexagonal structure.

A projection of the cordierite structure is shown in Fig 4. It is characterized by the six-membered rings. The structure is built up from (SiO_4) and (AlO_4) tetrahedra. These tetrahedra form six-(cation)-membered or four-(cation)-membered rings. The six cations forming the six-membered rings are designated T1 in Fig 4, and the tetrahedrally coordinated cations (which cross-link the large rings) are designated T2. The bridging oxygens contained in the large rings are designated O2; all the other oxygens are designated O1. There is one non-tetrahedrally coordinated cation site that is occupied by Mg ions (designated M in Fig 4) and cross-links the rings. The site occupancy in the low-temperature form is 4Si + 2Al in the six-membered rings (ordered) and 1Si + 2Al (ordered)

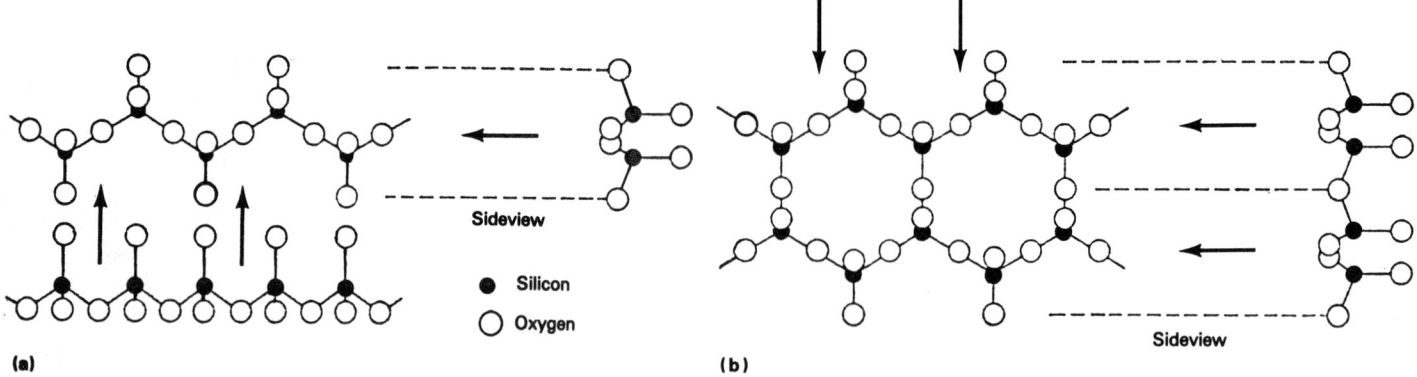

Fig 2 The silicate chain structures. (a) The pyroxene single chain. (b) The amphibole double chain. Adapted from Ref 6

in the cross-linking T2 sites. In the high-temperature form, there is some variation of aluminum and silicon site occupancies, as well as partial disorder. In the mineral forms, there are various impurities, which can substitute for each of the cations, as well as alkali, alkali earth, and hydroxy ion accommodation in the channels.

Typical thermal expansion curves are shown in Fig 5. The expansion is small and anisotropic. The c-axis expansion is negative, and the resultant volume expansion is therefore very small. The volume expansion is compared to other framework silicate structures (mineral form) in Table 2.

The relationship between thermal expansion and crystal structure of cordierite has been discussed by Hochella and Brown (Ref 8). With elevated temperatures, the increased thermal motion of the atoms is accommodated primarily by twisting and rotation of the rings. Selected properties of cordierite are compared to alumina and mullite in Table 3.

Pyrophyllite. This mineral, sometimes referred to as "wonderstone," is used for a surprisingly large number of applications. It is utilized because it is inexpensive, can be machined easily, and can then be "fired" to yield a strong, multiphase ceramic.

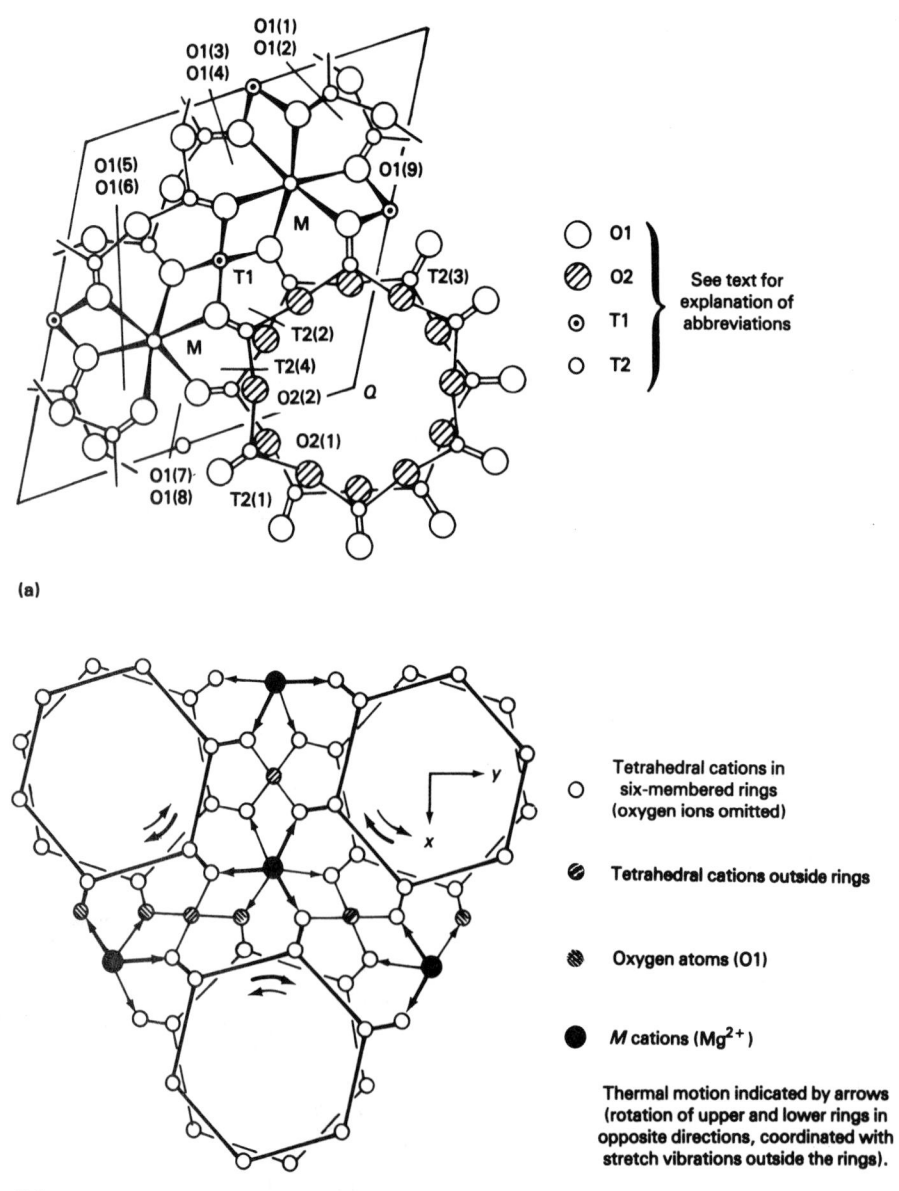

Fig 4 The structure of cordierite in its high-temperature form. (a) Unit all projected onto the 001 plane. (b) Schematic emphasizing the six-tetrahedral rings. Source: Ref 7, 8

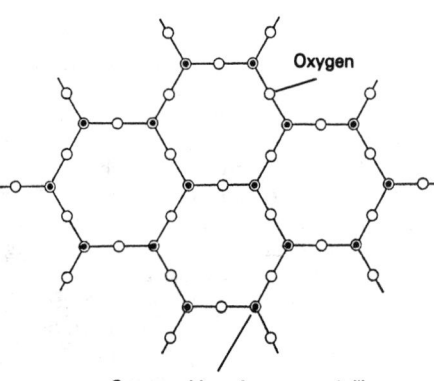

Fig 3 Sheet structure of silicates. Source: Ref 6

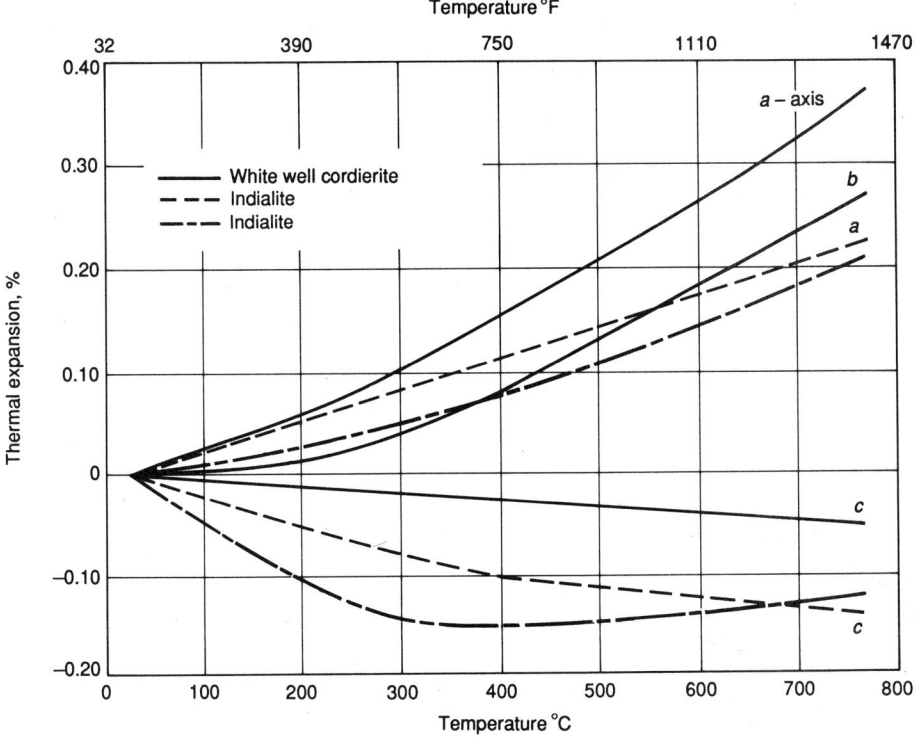

Fig 5 Selected thermal expansion curves for low- and high-temperature forms of cordierite. The low-temperature cordierite specimen is a natural single crystal. Source: Ref 8

Pyrophyllite is one of the layer structure silicates. It has the composition $Al_2(Si_2O_5)_2(OH)_2$ and contains a characteristic dioctahedral double-layer structure (Fig 6). The forces holding adjacent layers together (in c-direction) appear to be primarily Van der Waals type. Thus slippage occurs readily in the a-b plane, and the material is soft and can be readily machined. It is closely related to talc [$Mg_3(Si_2O_5)_2(OH)_2$].

Upon heat treatment, dehydroxylation occurs at about 800 °C (1475 °F) to yield a metastable phase. Upon further heating, a phase transformation occurs at ~1100 °C

(~2000 °F) to yield to a two-phase mixture of cristobalite (SiO_2) and mullite ($3Al_2O_3\cdot2SiO_2$). Importantly, the dimensional changes upon dehydroxylation and phase transformation are only ~2%. Furthermore, the mullite needles help to retain the dimensional stability at elevated temperatures. This implies that (small numbers of) components can be machined to near-net shape, to yield a product with reasonable properties after firing. Typical properties after firing are presented in Table 4.

Mullite

Mullite is the only chemically stable intermediate phase at atmospheric pressure in the SiO_2-Al_2O_3 system. It is rare in nature, the most important place of occurrence being the island of Mull, UK. Polycrystalline mullite has a lower thermal expansion coefficient than alumina. As a result it is more resistant to thermal shock, particularly from temperatures above 1000 °C (1850 °F). Reviews of this material have been presented by Davis and Pask (Ref 10) and Schneider (Ref 11). The proceedings of an international conference concerned with the science and technology of mullite have also been published (Ref 12).

Composition and Structure. The chemical composition of mullite is usually denoted as $3Al_2O_3\cdot2SiO_2$ (60 mole% Al_2O_3). In fact it is a nonstoichiometric solid solution having the structural formula $Al_2(Al_{2+2x}Si_{2-2x})O_{10-x}$, where x varies as a function of temperature over the total range from 0.17 to 0.50 (58 to 66.5 mole% Al_2O_3). This variation in the solid-solution range (especially near the melting point), the uncertainty associated with melting-point temperature, and the difficulty of nucleating a-Al_2O_3 from liquids within the mullite solid-solution range have caused considerable controversy in this portion of the phase diagram. Reference 10 and the first two chapters in Ref 12 discuss these arguments in terms of the experimental results and difficulties associated with the phase diagram studies. In recent studies, Aksay and Pask (Ref 13) have indicated that the compositional limits for mullite range from 60 to 63 mole% Al_2O_3 and are maintained, essentially to an incongruent melting temperature of 2101 ± 10 K (3322 ± 18 °F) above which Al_2O_3 and a liquid of another composition are formed. Prochazka and Klug (Ref 14) have confirmed that a peritectic occurs, at 2263 ± 10 K (3614 ± 18 °F) at 66.5 mole% Al_2O_3. All other compositions within the mullite phase field transform to a two-phase field of mullite plus liquid beginning at approximately 1860 °C (3380 °F), as shown in Fig 7. Their results also indicate that both solid solution limits shift through a continuum of increasing amounts of Al_2O_3 above ≈1600 °C (≈2900 °F).

The understanding of the solid solution range of mullite and the development of ever more sophisticated methods of structural

Table 2 Volume expansion coefficients for some framework silicates

Material	Stuffing cation	Volume expansivity, 10^{-6}/°C	Temperature range	
			°C	°F
Na feldspar (albite)	Na	29	24–750	75–1380
Mineral cordierite	Mg	7	24–775	75–1425
Commercial cordierite	Mg	3	24–600	75–1110
β-spodumene	Li	2	24–1200	75–2190
β-quartz	None	−6	600–900	1110–1650

Source: Adapted from Table in Ref 8

Table 3 Selected properties of alumina, mullite, and cordierite

Property	Material		
	Alumina (90%)(a)	Mullite(b)	Cordierite(c)
Thermal expansion coefficient (25–800 °C), 10^{-6}/°C	8.1	5.0	1.5
Dielectric constant, K (25 °C, 1 MHz)	8.8	6.6	5.0
Bend strength, MPa (ksi)	~330 (48)	270 (39)	245 (35.5)
Young's modulus, GPa (10^6 psi)	275 (40)	220 (32)	139 (20)
Shear modulus, GPa (10^6 psi)	117 (17)	87 (13)	45 (6.5)
Poisson's ratio	0.22	0.27	0.31
Vicker's hardness	~15	11.0	8.2
Fracture toughness, (K_{1c}), Mpa$\sqrt{m}$ (ksi$\sqrt{in.}$)	1.5–4.0 (1.4–3.6)	2.6 (2.4)	2.3 (2.1)

(a) Alumina data used for comparison is a low cost commercial alumina (90% Al_2O_3). (b) Data for mullite and cordierite primarily from Ref 1

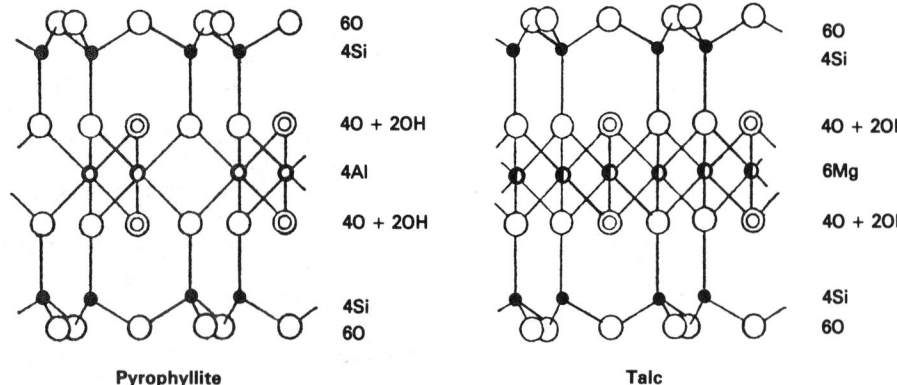

Fig 6 Schematic of the structure of pyrophyllite, emphasizing the layer structure. Talc is shown for comparison.

characterization have allowed an understanding of the crystal structure of this complex oxide to evolve. The close structural relationship between mullite and sillimanite (Al_2SiO_5) has been noted in virtually all the structural studies because of the similarity in their x-ray powder diffraction patterns. Taylor (Ref 15) and Warren (Ref 16) first proposed a structure for mullite based on that of sillimanite but with the introduction of oxygen vacancies. These vacancies allowed electrical neutrality to be preserved as a result of the substitution of Si^{+4} by Al^{+3}. Their existence also explained the smaller average density of mullite (≈ 3.16 g/cm^3) compared to that of sillimanite (≈ 3.22 g/cm^3). The statistical nature of the mullite structure was determined independently by Durovic (Ref 17) and Sadanaga et al. (Ref 18) using x-ray single-crystal diffraction data. Several related studies of the crystal structure of this material were also subsequently reported (Ref 19–22); however, the results are in good agreement with their earlier studies.

The average structure of the $2Al_2O_3 \cdot SiO_2$ mullite, commonly obtained by quenching a high Al_2O_3 melt, is shown in Fig 8. The structure is orthorhombic (space group Pbam) with the cell dimensions of $a \approx 7.6$ Å, $b \approx 7.7$ Å, and $c \approx 2.9$ Å. Mullite in all of its compositions consists of chains of slightly distorted Al—O octahedra which run parallel to the c-axis of the orthorhombic unit cell. These chains are linked by discontinuous

Table 4 Selected property data for fired pyrophyllite

Water absorption, %	2.5
Density, g/cm^3	2.3
Tensile strength, MPa (ksi)	~20 (~2.9)(a)
Compressive strength, MPa (ksi)	~180 (~26)(a)
Flexural strength, MPa (ksi)	~65 (~9)(a)
Softening temperature, °C (°F)	1600 (2900)
Dielectric breakdown field, V/m	4×10^6
Relative dielectric constant (at 1 MHz)	5.3
Loss factor (at 1 MHz)	0.053

(a) Test conditions unspecified. Source: Data for Lava Grade A supplied by the Maryland Lava Co. (Ref 9). After firing, consists of a multiphase mixture of cristobalite (SiO_2) and mullite

double chains of Al—O and Si—O tetrahedra with randomly distributed aluminum and silicon atoms. The discontinuities in the latter chains are caused by the movement of some of the aluminum ions (and possibly silicon ions) into normally unfilled tetrahedral positions because of insufficient oxygen atoms present to bond to them in the normal positions. The occupation of these new sites also increases the coordination of the remaining oxygen atoms and forces them into new positions that are slightly different from their original locations. The connections of octahedra and tetrahedra chains produce relatively wide structural channels that also run parallel to the c-axis. Moreover, long-range ordering of the oxygen vacancies results in a complex modulated or incommensurate structure of antiphase domains, as reported by Morimoto et al. (Ref 23). The modulated structure is constructed using the clusters of double Si—tetrahedra and triple Al—tetrahedra, which cross-link the edge-sharing AlO_6 octahedral chains running parallel to the c-axis.

The problems associated with the determination of the solid solution range of mullite formation, the occasional failure to recognize the existence of a solid solution, and the differences in mass transport of the Al- and Si-containing species during sintering has caused consternation in the fabrication of fully dense mullite ceramics free of an amorphous grain-boundary phase.

Properties and Processing. The recent phase equilibria research noted above, coupled with recent innovations in fine-particle synthesis, has resulted in advances in mullite technology. Nominal 3:2 mullite, produced by reaction sintering of a mechanical mixture of Al_2O_3 and SiO_2 powders at temperatures at or about 1600 °C (2900 °F) is usually characterized by low strength (≤ 200 MPa, or 29 ksi) and low fracture toughness (1 to 2 MPa$\sqrt{m}$, or 0.9 to 1.8 ksi$\sqrt{in}$.) caused by the presence of an amorphous boundary phase. It is known that, at these temperatures, the equilibrium phase field is mullite plus liquid, the latter of which may separate into two

phases and incompletely crystallize on cooling (Ref 24). Additional complications are caused by the much more rapid diffusion of Al_2O_3 into SiO_2 than vice versa. This results in the creation of porosity and the expansion of the sample caused by the formation of the mullite.

A new generation of single-phase (or virtually single-phase) higher strength (to 500 MPa at 1300 °C, or 72 ksi at 2370 °F) and higher toughness (2 to 4 MPa$\sqrt{m}$ or 1.8 to 3.6 ksi$\sqrt{in}$.) mullite ceramics is being produced by sintering or hot pressing submicron particles of intimately mixed and usually prereacted (at $T \leq 1600$ °C, or 2900 °F) materials produced via several unique routes. These routes include:

- Gelation from mixtures of colloidal suspensions such as of γ-Al_2O_3 and SiO_2, or AlOOH and $Si(OC_3H_7)$
- Hydrolytic or thermal decomposition of various metal alkoxides
- Coprecipitation from mixtures of $AlCl_3$ and $SiCl_4$ in diethyl ether and ammonia
- Spray pyrolysis of $AlNO_3$ and $Si(OC_3H_7)_4$
- Chemical vapor deposition (CVD) of selected gases
- Hydrothermal synthesis

Sacks et al. (Ref 25) have published an in-depth review of this powder preparation research as well as the densification methods used by the various investigators and the resulting microstructures and densities achieved.

Evaluation of the resulting materials has been obtained from mechanical property studies, usually bending strength. However, creep behavior also is used to evaluate the effects of composition and test temperature. These effects on mechanical properties are described for mullite produced by heating a conventional mixture of $Al(OH)_3$ and fused SiO_2, a sol-gel technique, and by sintering material derived via CVD in a combustion flame.

Monolithic and SiC-Reinforced Mullite. Hot pressed mullite produced by a conventional mixture of $Al(OH)_3$ and fused SiO_2 and having the approximate composition of $3Al_2O_3 \cdot 2SiO_2$ has been used to fabricate monolithic mullite as well as mullite composites containing 20 vol% SiC whiskers (Ref 26). Both materials contained an amorphous phase on the grain boundaries which were derived from the as-received powder. The composite also contained additional amorphous material derived from the oxide on the whiskers. The room-temperature strength and toughness values of the monolithic mullite were 200 MPa (29 ksi) and 2.2 MPa$\sqrt{m}$ (2 ksi $\sqrt{in}$.). With the addition of the whiskers, the corresponding values increased to 425 MPa (61.5 ksi) and 4.7 MPa$\sqrt{m}$ (4.3 ksi$\sqrt{in}$.). The strength was retained to approximately 1500 K (2240 °F); however, above this temperature, it decreased rapidly and was dependent on the amorphous grain-boundary phase.

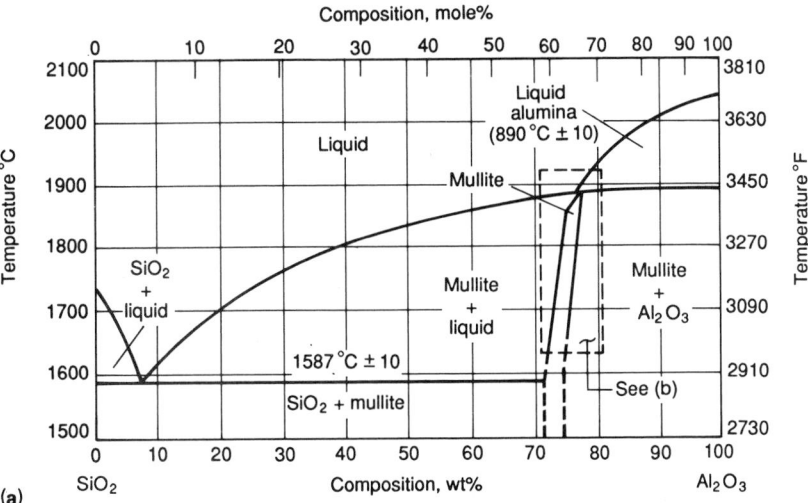

(a)

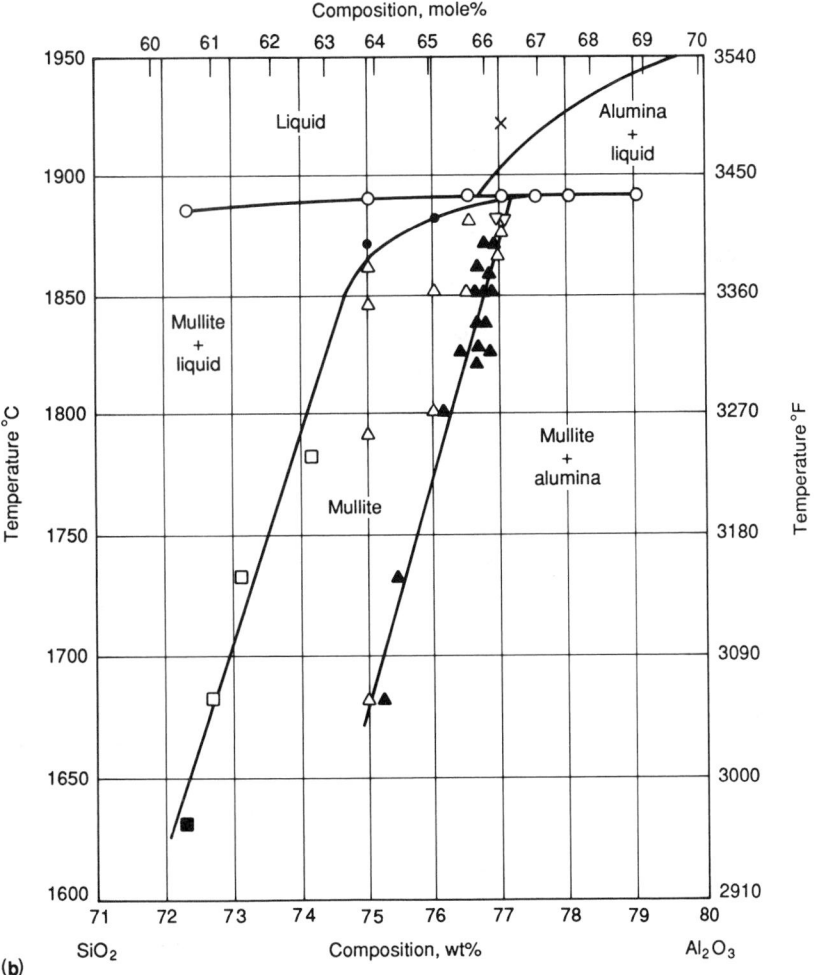

(b)

Fig 7 (a) Composite alumina-silica phase diagram and (b) expanded area inscribed by the dashed rectangle in (a) showing the melting behavior of the mullite solid solution and the various surrounding phase fields. Source: Ref 14

and 5 to 220 MPa (0.7 to 32 ksi). Both materials showed three regimes of temperature in which the activation energy of creep was different (Fig 9). Figure 10 reveals that the stress exponent *n* for the unreinforced mullite was approximately 1 for all temperature regimes; for the composite this parameter increased from approximately 2 in the lowest temperature regime to 3 at 1700 K (2600 °F). Transmission electron microscopy studies as well as the low-activation energy values obtained below 1560 K (2350 °F) indicated that creep in the lowest temperature regime was controlled by the viscous flow of the SiO_2-rich intergranular glass phase. With increasing temperature to 1630 K (2475 °F), extensive cavitation also occurred. Cavities were created in the glass phase at triple points and between mullite-mullite and whisker-mullite boundaries, leading to higher activation energy values as a result of the increase in free surface area. Above 1630 K (2475 °F), no cavities were observed, indicating that the viscosity of the glass decreased sufficiently such that the cavities created at lower temperatures were filled by the intergranular phase. The decrease in activation energy in this temperature regime supported this result.

Glass-Free Mullite Prepared by Sol-Gel Method. Nixon *et al.* (Ref 27) have also conducted creep research on a completely glass-free mullite prepared by Hirata *et al.* (Ref 28) using a sol-gel technique. In this process, a mixture of AlOOH and tetraethoxysilane was gelled by hydrolysis and subsequently crushed to form submicron grain size powder. This powder was pressed at room temperature and pressureless sintered in air at 1250 °C (2280 °F) to 2 h. The disk was subsequently annealed at 1600 °C (2900 °F) for 48 h in order to establish a grain size that would not change during creep experiments. For the creep research, the same experimental approach described above was employed with the temperature and stress ranges of 1600 to 1700 K (2420 to 2600 °F) and 20 to 300 MPa (3 to 43.5 ksi). An activation energy of 748 kJ/mole and stress exponents in the range of 1 to 1.2 were determined from these studies (Fig 11 and 12). This data as well as the results of scanning and transmission electron microscopies were consistent with a controlling mechanism of steady-state creep of grain boundary sliding accommodated by lattice diffusion and cavitation. The creep rate of this material was approximately an order of magnitude slower than the unreinforced mullite with the amorphous grain-boundary phase investigated by Nixon *et al.* (Ref 27).

Ultrafine Powder from CVD. Spherical ultrafine (grain size range = 40 to 70 nm) Al_2O_3-SiO_2 composite powders (67 to 80 wt% Al_2O_3) have been prepared by Hori and Kurita (Ref 30) via CVD using a combustion flame of approximately 1900 °C (3450 °F). These powders consisted of an amorphous phase and g-Al_2O_3. However, they transformed to mullite between 950 and 990 °C (1740 and

Nixon *et al.* (Ref 27) have investigated the steady-state constant stress compressive creep of these same materials. Because the crystal structure of mullite is orthorhombic, the five independent slip systems necessary for non-

diffusional dislocation mechanisms to deform a polycrystalline material are not present. Thus other mechanisms must be considered. The ranges of temperature and stress used by Nixon *et al.* were 1450 to 1700 K (2150 to 2600 °F)

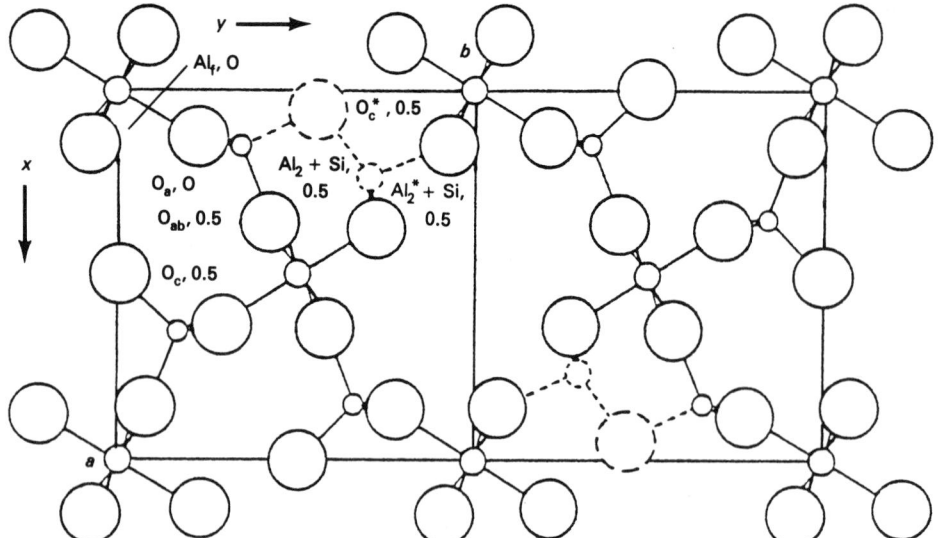

Fig 8 Projection on (001) for the 2:1 mullite structure. Source: Ref 20

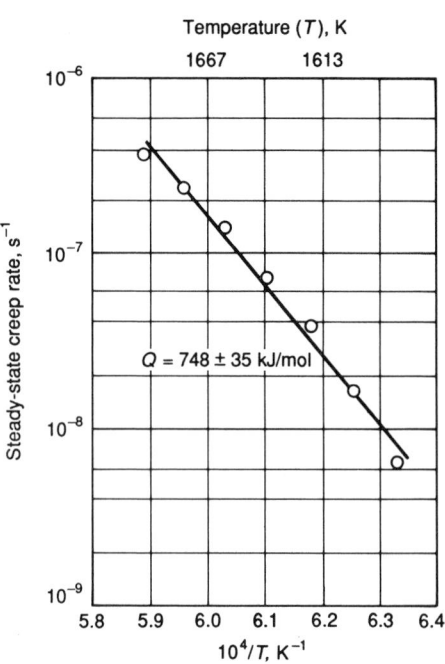

Fig 11 Steady-state creep rate as a function of reciprocal temperature obtained for a glass-free mullite at a constant stress of 100 MPa (14.5 ksi). Source: Ref 27

1815 °F) and densified to ≥98.8% of theoretical density by sintering at 1600 °C (2900 °F) for 2 h. Some glass was present on the grain boundaries, especially in the samples with the lower amounts of Al_2O_3, which allowed the formation of lath-shaped grains of mullite dispersed in the matrix of uniform grains of this material. Those samples with the higher concentrations of Al_2O_3 had a completely uniform grain size and contained both mullite and corundum. These differences in microstructure were reflected in the breaking strength (Fig 13). The increase in breaking strength was shown to be a direct linear function of the wt% of Al_2O_3.

Applications. Classical uses of mullite include refractories in the metallurgical industries, electric furnace roofs, and low-frequency induction furnaces. In the glass industries, these refractories are employed in the upper structure of the tank in which the glass is melted and for constructing the drawing chambers. Mullite is frequently used as kiln setting slabs and posts for firing ceramic ware. However, the spectrum of actual or potential employment also includes electronic substrates, protective coatings, parts for

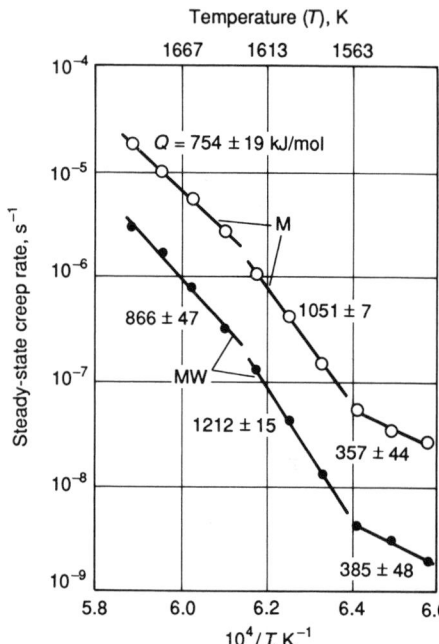

Fig 9 Steady-state creep rate as a function of (a) reciprocal temperature obtained for mullite (M) and mullite reinforced with SiC whiskers (MW) under a constant stress of 100 MPa (14.5 ksi). Q, activation energy. Source: Ref 27

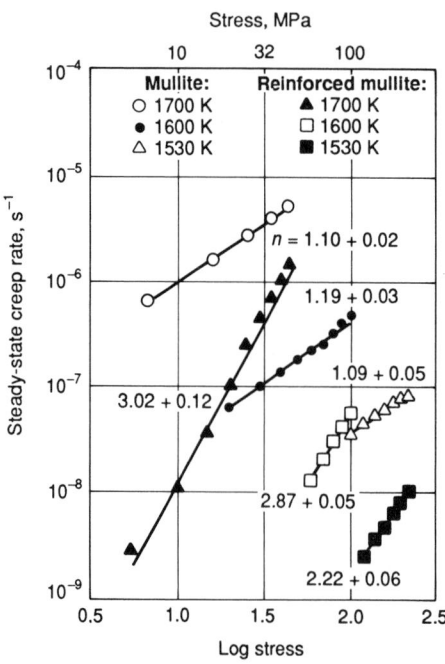

Fig 10 Steady-state creep rate as a function of log stress obtained for mullite and mullite reinforced with SiC whiskers at the temperatures of 1530 K, 1600 K, and 1700 K (2295 °F, 2420 °F, and 2600 °F). Source: Ref 27

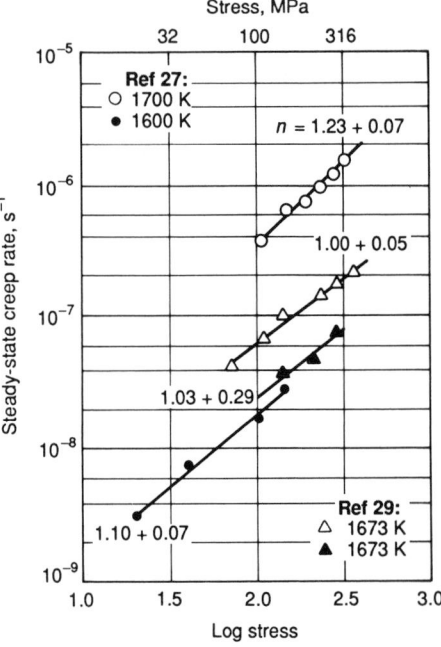

Fig 12 Steady-state creep rate as a function of log stress for a glass-free mullite (from Ref 27). Similar data for an analogous material studies by Dokko et al. (Ref 29) is also shown for comparison.

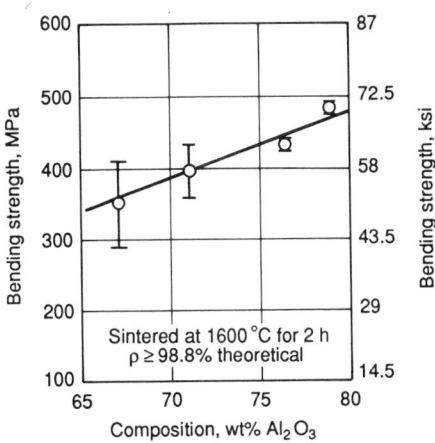

Fig 13 Room-temperature bending strength as a function of composition between 67 and 79 wt% Al_2O_3 of mullite-based ceramics produced via CVD processes in a combustion flame and sintered at 1600 °C (2900 °F) for 2 h. Source: Ref 28

turbine engines, and infrared transmitting windows.

The foregoing mechanical property results provide a basis regarding the fabrication and application of the newer mullite materials. Namely, if mullite is to be used in high-temperature ($\geq$1400 °C, or 2550 °F) load-bearing applications, the recent results of the high-temperature phase equilibria coupled with the newer processing procedures must be considered and applied to achieve a uniform and very fine-grained microstructure, free of an amorphous grain-boundary phase. Careful consideration of compositions and process routes will also have a dramatic effect on other properties such as thermal conductivity and thermal shock stability. Table 3 makes a direct comparison of selected properties of typical commercial mullite with those of cordierite and a high-purity Al_2O_3.

Spinels

Compounds having the spinel structure have been investigated for many years because of their interesting structural and physical properties. Many spinel compounds exist in nature; they have been formed at high temperature and/or pressure. Many synthetic spinels have been prepared in the laboratory by solid-state reactions of suitable precursor materials at temperatures above 800 °C (1470 °F). Others have been prepared at moderate temperatures, typically 300 to 500 °C (570 to 930 °F). Some spinel compounds have even been prepared at room temperature by using alternative chemical or electrochemical techniques.

General Characteristics and Properties. Stoichiometric spinels have the general formula $A[B_2]X_4$ where A refers to cations in tetrahedral sites, B to cations in octahedral sites, and X to the anions that are essentially cubic-close-packed (Ref 31). The prototype spinel is $Mg[Al_2]O_4$; it has cubic symmetry $Fd3m$. In this structure, with the origin of the

unit cell at the center ($3m$), the magnesium ions are located in the tetrahedral A-sites at the $8a$ positions (for example, $1/8,1/8,1/8$), the aluminum ions in the octahedral B-sites at the $16d$ positions (for example, $1/2,1/2,1/2$), and the close-packed oxygens at the $32e$ positions (u,u,u where the variable u is often referred to as the 'u'-parameter). In an ideal cubic-close-packed structure, $u = 1/4$. The coordination of the Mg, Al, and O ions is, respectively, MgO_4 (tetrahedral), AlO_6 (octahedral), and $OMgAl_3$ (tetrahedral). The interstitial space of the spinel structure is comprised of empty octahedra with sites at the $16c$ positions (for example, 0,0,0), and tetrahedra with sites at the $8b$ positions (for example, $3/8,3/8,3/8$), and $48f$ positions (for example, $u,1/8,1/8$). As in all close-packed structures, there are twice as many tetrahedral sites as octahedral sites.

In spinel structures the distribution of the cations over the A and B sites can vary. When the cations are arranged as in the formula $A[B_2]X_4$, the structure is referred to as normal. When 50% of the B cations are interchanged with the A cations to give the arrangement $B[AB]X_4$, the structure is referred to as inverse. The degree of inversion, g, is 0 for the normal structure and 1 for the inverse structure. Intermediate values can occur. For example, when $g = 0.67$, the cation distribution is random and the arrangement is $(A_{0.33}B_{0.67})_{tet}[A_{0.67}B_{1.33}]_{oct}X_4$.

The family of spinel compounds is a large one because A, B, and X can be represented by many different cationic or anionic types with various valence states, as illustrated by the examples given in Table 5. A comprehensive list of the oxide- and sulfide-spinels, in particular, have been catalogued in detail (Ref 32).

Properties. The mechanical properties of the spinels (Table 1) are not exceptional, with only the prototype $MgAl_2O_4$ utilized in high-purity, single-phase form, primarily for its large thermal expansion coefficient, and its high service temperature (usable to over 1900 °C, or 3450 °F, which is higher than

mullite). However, spinels are well known for their interesting magnetic and electrical properties. For example, the ferrites MFe_2O_4 where M represents a solvent metal cation such as Mn^{2+}, Fe^{2+}, Co^{2+}, Ni^{2+}, and Cu^{2+} are a family of commercially important magnetic oxides which all have magnetic spins which all have partial, or completely inverse structures. The ferric ions on the tetrahedral sites have magnetic spins which are antiparallel to those on the octahedral sites. This arrangement gives the ferrites either antiferrimagnetic or ferrimagnetic properties. Much has already been written about this subject in comprehensive treatises (Ref 37, 38).

The $[B_2]X_4$ framework of an $A[B]X_4$ spinel (Fig 14) provides good electrical conductivity and a stable three-dimensional interstitial space for lithium insertion and extraction reactions (Ref 39, 40). These properties have led to a concerted research effort to develop commercially attractive insertion electrodes for rechargeable lithium batteries. An insertion electrode acts as a host structure for the intercalation of guest ions during an electrochemical reaction; it must be a good ionic and electronic conductor.

The electrical conductivity of the $[B_2]X_4$ framework is largely dependent on the oxidation state of the B cations that reside in edge-shared octahedra of the framework and the degree of d-orbital overlap between them. For example, hausmannite, $Mn^{2+}[Mn^{3+}Mn^{3+}]O_4$, which is a normal spinel, is virtually an insulator, whereas magnetite, $Fe^{3+}[Fe^{3+}Fe^{2+}]O_4$, which is an inverse spinel with mixed-valence cations on the B-sites, has good electrical conductivity. The lithium spinels, $Li[M^{4+}M^{3+}]O_4$ (M = Mn, V, Ti), are also good electrical conductors (Ref 41). The degree of electrical conductivity can vary widely; $Li[Mn_2]O_4$ is a hopping semiconductor (Ref 42), $Li[V_2]O_4$ has metallic conductivity (Ref 43), and $Li[Ti_2]O_4$ is a superconductor below 12 to 13 K (Ref 44).

In general, lithium insertion reactions with oxide-spinels at room temperature result in the formation of a rock salt phase after the

Table 5 Examples of selected $A[B_2]X_4$ compounds illustrating the wide variety of cations and anions that are found in the spinel family

Spinel compound	A	B	X	Ref
Fe_3O_4	Fe	FeFe	O	32
$MgMn_5O_8$	$Mg_{0.5}Mn_{0.5}$	MnMn	O	32
$ZnAl_2O_4$	Zn	AlAl	O	32
$LiAl_5O_8$	Al	$Li_{0.5}Al_{1.5}$	O	32
$LiMn_2O_4$	Li	MnMn	O	32
$Li_4Mn_5O_{12}$	Li	$Li_{0.33}Mn_{1.67}$	O	32
$SiNi_2O_4$	Si	NiNi	O	32
Li_2MoO_4(a)	Mo	LiLi	O	33
$MnIn_2S_4$	$Mn_{0.33}In_{0.67}$	$Mn_{0.67}In_{1.33}$	S	32
$CuCo_2S_4$	Cu	CoCo	S	32
Ni_3S_4	Ni	NiNi	S	32
$CdFe_3Sn_2S_8$	$Cd_{0.5}Fe_{0.5}$	FeSn	S	32
$MgTm_2Se_4$	Mg	TmTm	Se	34
$CdYb_2Sc_4$	Cd	YbYb	Se	34
Li_2MgCl_4	Li	LiMg	Cl	35, 36
Li_2CdCl_4	Li	LiCd	Cl	35, 36

(a) High-pressure, high-temperature compound

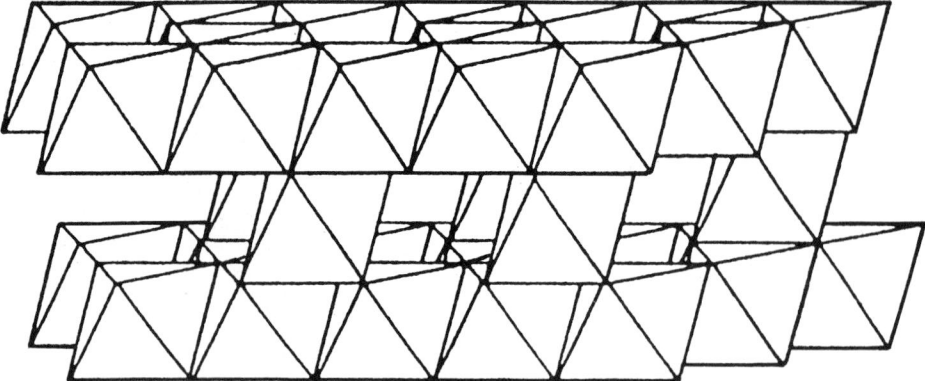

Fig 14 The $[B_2]X_4$ framework of an $A[B_2]X_4$ spinel

addition of one lithium ion per spinel formula unit and the concomitant reduction of an A- or B-type cation:

$$Li + A_{8a}[B_2]_{16d}O_4$$

At an early stage of the reaction, the A cations are cooperatively displaced from their tetrahedral positions into neighboring octahedral sites ($16c$) to yield a defect rock salt phase; this process allows further incoming lithium ions to percolate through a network of face-sharing $8a$ tetrahedra and $16c$ octahedra until all the $16c$ sites have been filled. The integrity of the $[B_2]O_4$ framework is maintained throughout this reaction.

When A is a relatively heavy, multivalent cation, Li^+ ion diffusion through the interstitial space of the $[B_2]O_4$ framework is hindered. Cyclic voltammetry studies of such spinel electrodes (for example, $Fe[Fe_2]O_4$, $Co[Co_2]O_4$, and $Mn[Mn_2]O_4$) have indicated that the lithium insertion reaction is electrochemically irreversible (Ref 45). These spinels therefore have limited application in room-temperature lithium batteries.

Spinels with lithium on the A-sites are of greater interest as insertion electrodes because the lithium ions can diffuse, unimpeded, through the network of face-shared $8a$ tetrahedra and $16c$ octahedra (Ref 46). Moreover, in some of these compounds it is possible to extract lithium from the tetrahedral sites and to simultaneously oxidize a B-cation, which significantly increases the working capacity of the spinel electrode.

Insertion Electrodes for Lithium Batteries. Lithium spinels having a $[B_2]O_4$ structure, as previously mentioned, have attracted interest as use for insertion electrodes in rechargeable lithium batteries. These batteries, with voltages typically between 3 and 4 V, are being developed as possible alternatives to nickel-cadmium batteries which have the disadvantages of a relatively low cell voltage (1.2 V), a poor shelf-life, and toxic electrode materials.

The spinel phases of interest for 3 to 4 V lithium battery applications are defined by $Li_xMn_{2-z}O_4$ ($0 \leq x \leq {}^4/_3$, $0 \leq z \leq {}^1/_3$), which

fall within the $Mn_3O_4 \cdot Li_4Mn_3O_{12} \cdot 1\text{-}MnO_2$ tie triangle of the Li-Mn-O phase diagram (Ref 47). A large number of stoichiometric and defect spinel structures exist in the Li-Mn-O system, but the various lithium spinels that provide stable $[B_2]O_4$ framework structures (see Table 6 for examples) in the Li-Mn-O system (Ref 47, 48) are represented by the general formula $Li_xMn_{2-z}O_4$ ($0 \leq x \leq {}^4/_3$, $0 \leq z \leq {}^1/_3$). In other lithium spinel systems, such as $Li_x[V_2]O_4$, the $[V_2]O_4$ framework is maintained over a limited range of x, $0.67 \leq x \leq 2.0$ (Ref 49, 50). For $x < 0.67$, the vanadium ions migrate irreversibly from the B-sites into neighboring octahedral sites ($16c$), which limits the utility of $Li_x[V_2]O_4$ electrodes in practical lithium cells. The x-ray data of $Li_{0.2}Ti_2O_4$ suggest that a similar process occurs during the delithiation of $Li[Ti_2]O_4$ (Ref 51).

Lithium insertion and extraction reactions with $Li[Mn_2]O_4$ have been studied extensively (Ref 39, 52–55). It is convenient to use $Li[Mn_2]O_4$ as the initial electrode in $Li/Li_x[Mn_2]O_4$ cells because this spinel composition is easily prepared by the reaction of lithium and manganese oxides, hydroxides, or salts at elevated temperature, typically $>600\ °C$ ($>1100\ °F$) in air. Lithium insertion into $Li[Mn_2]O_4$ induces a Jahn-Teller distortion, which reduces the crystal symmetry from cubic to tetragonal symmetry (Ref 39). In $Li/Li_x[Mn_2]O_4$ cells, this distortion produces a two-phase electrode and a constant open-circuit voltage response of 2.3 V over the range $1.1 \leq x \leq 2.0$. Although it is possible to remove the lithium from $Li[Mn_2]O_4$ by acid treatment at room temperature to yield the $[Mn_2]O_4$ spinel framework ($\lambda\text{-}MnO_2$) (Ref 52), electrochemical extraction of lithium under constant current conditions for $x \leq 0.54$ has been shown to be difficult, even at moderate rates (Ref 56). Under constant voltage conditions, however, a composition $Li_{0.27}[Mn_2]O_4$ has been achieved (Ref 55). It was also demonstrated that lithium extraction from $Li[Mn_2]O_4$ occurs at approximately 4 V versus Li/Li^+ in two stages, separated by only 170 mV. The two-step process has been attributed to the ordering of the lithium ions

on one-half of the tetrahedral A-sites at a composition $Li_{0.5}[Mn_2]O_4$ (Ref 41).

In practice, $Li[Mn_2]O_4$ electrodes synthesized at temperatures above $600\ °C$ ($1100\ °F$) (non-spiral phases) do not show good rechargeability in 3 V lithium cells. $Li/Li_x[Mn_2]O_4$ button cells, operating typically at $0.5\ mA/cm^2$ between voltage limits of 3.3 and 2.0 V lose 50% of their capacity over 10 cycles (Ref 47). However, spinel electrodes with superior performance to $Li[Mn_2]O_4$ have been synthesized by heating lithium and manganese salt at moderately high temperatures, typically at 350 to $450\ °C$ (660 to $840\ °F$); they are ideally represented by the system $Li_2O \cdot MnO_2$. The end members of this system are the stoichiometric spinel $Li_4Mn_5O_{12}$ ($y = 2.5$) with spinel notation $Li[Li_{1/3}Mn_{5/3}O_4$ and the defect spinel $\lambda\text{-}MnO_2$ ($y = \infty$). As y increases so does the theoretical capacity of the electrode which reaches $308\ mA \cdot h/g$ at $\lambda\text{-}MnO_2$ (see Table 6). For example, reaction of Li_2CO_3 with $MnCO_3$ in a 1:4 molar ratio at $400\ °C$ yields an intermediate member of the system, $Li_2Mn_4O_9$ ($y = 4$), which is a defect spinel $(Li_{0.89}\square_{0.11})[Mn_{1.78}\square_{0.22}]O_4$. However, strict control of the processing conditions is required to ensure full oxidation of the manganese cations, otherwise spinel phases, slightly deficient in oxygen, are produced. In particular, if $Li_2Mn_4O_9$ is heated above $400\ °C$ ($750\ °F$), it is reduced by loss of oxygen to the stoichiometric spinel $Li[Mn_2]O_4$:

$$Li_2Mn_4O_9 \rightarrow 2Li[Mn_2]O_4 + {}^1/_2\, O_2$$

$Li_2O \cdot yMnO_2$ spinel electrodes have demonstrated excellent rechargeability. Unlike $Li[Mn_2]O_4$, these electrodes have shown stable electrode capacities in excess of $130\ mA \cdot h/g$ in battery cells when cycled at $0.5\ mA/cm^2$ between voltage limits of 3.3 and 2.0 V (Fig 15) and hold promise for their practical application in rechargeable lithium batteries.

Perovskites

The class of ceramic materials with the perovskite crystal structure is characterized by a wealth of properties, which include ferroelectricity, piezoelectricity, pyroelectricity, electrostriction, high permittivity, and optical and electrooptic properties. They are not commonly utilized on the basis of unique mechanical properties. Instead, the above properties ensure their use in a wide range of commercial and consumer devices. This section attempts to demonstrate that variety, and in particular, attempts to elucidate principles of structure-property relations for the class.

The structure of the ideal ABO_3 perovskite is shown in Fig 16. It may be viewed, in its most symmetrical form, as a cube with the A-cations in the corners of the cube, the B-cations in the cube center position, and O-anions in the center of each face. The structure is densely packed, and it can be seen that

Table 6 Stoichiometric (S) and defect (D) spinels in the system Li$_x$Mn$_{2-x}$O$_4$ (O $\leq$ x $\leq$ $^4/_3$, 0 $\leq$ z $\leq$ $^1/_3$)

Spinel compound	x	z	Spinel notation	y in Li$_2$O·yMnO$_2$	Theoretical capacity(a), mA·h/g
LiMn$_2$O$_4$ (S)	1	0	Li[Mn$_2$]O$_4$	n/a	148
Li$_4$Mn$_5$O$_{12}$ (S)	$^4/_3$	$^1/_3$	Li[Li$_{1/3}$Mn$_{5/3}$]O$_4$	2.5	163
Li$_2$Mn$_3$O$_7$ (D)	$^8/_7$	$^2/_7$	(Li$_{0.86}$□$_{0.14}$)[Mn$_{1.71}$Li$_{0.29}$]O$_4$	3.0	184
Li$_2$Mn$_4$O$_9$ (D)	$^8/_9$	$^2/_9$	(Li$_{0.89}$□$_{0.11}$)[Mn$_{1.78}$□$_{0.22}$]O$_4$	4.0	213
λ-MnO$_2$ (D)	0	0	(□$_{1.0}$)[Mn$_2$]O$_4$	∞	308

(a) Based on a discharge reaction to a rock salt composition

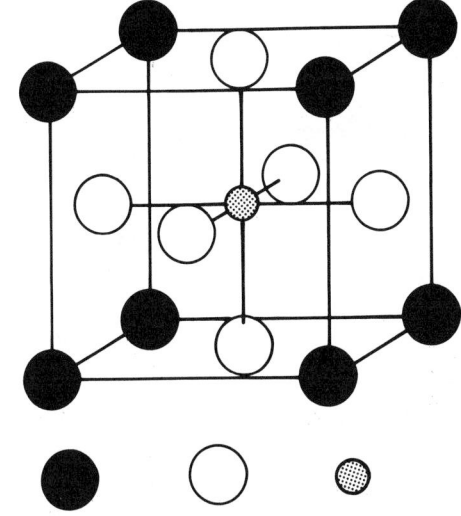

the A cation is coordinated to 12 oxygen ions, and the B-cation to 6. Predictably, therefore, the A-site cation is larger than the B-site cation. Goldschmidt has shown that the stability of the structure can be predicted on the basis of the ionic radii. He defined a tolerance factor, t:

$$t = \frac{R_A + R_O}{\sqrt{2}\,(R_B + R_O)}$$

Goldschmidt suggested that for values $0.8 < t < 0.9$, the cubic symmetry perovskite would form. The range is somewhat larger for distorted perovskite structures, where the symmetry falls to tetragonal, orthorhombic, or lower. As is shown below, the generally exploited properties of the perovskites are critically related to this distortion from the ideal cubic structure.

The commonly quoted prototype perovskite structure is that of BaTiO$_3$. This has cubic symmetry at subsolidus temperatures above ~130 °C (~265 °F), but undergoes a series of phase transitions as the temperature is lowered (Fig 17). The trend of phase transitions to lower symmetry structures with a decrease in temperature is typical of the perovskites.

Some small number of examples from the above perovskite families are listed in Table 7. These were selected from the Galasso text (Ref 58), in order to illustrate the categories. The number of identified oxides with the perovskite structure is astonishingly large. Galasso, as early as 1969, published a text which references and discusses hundreds of these compounds (Ref 58). It lists only limited property data, primarily unit cell parameters, and density. The text, however, is a valuable resource.

Galasso also further classifies the perovskites, in terms of the nominal oxidation states of the A- and B-site cations, as follows:

- $A^+B^{5+}O_3$
- $A^{2+}B^{4+}O_3$ (most commonly reported)
- $A^{3+}B^{3+}O_3$
- $A^{2+}(B_{0.67}^{3+}B_{0.33}^{6+})O_3$

Fig 16 The ABO_3 perovskite structure typified by BaTiO$_3$ above its Curie point (cubic symmetry). Source: Ref 57

- $A^{2+}(B_{0.33}^{2+}B_{0.67}^{5+})O_3$
- $A^{2+}(B_{0.5}^{3+}B_{0.5}^{5+})O_3$
- $A^{2+}(B_{0.5}^{2+}B_{0.5}^{6+})O_3$
- $A^{2+}(B_{0.5}^{1+}B_{0.5}^{7+})O_3$
- $A^{3+}(B_{0.5}^{2+}B_{0.5}^{4+})O_3$

In addition, there are variants which show a significant deviation from 1:1:3 stoichiometry, the most important of which are the tungsten bronzes. These are typified by

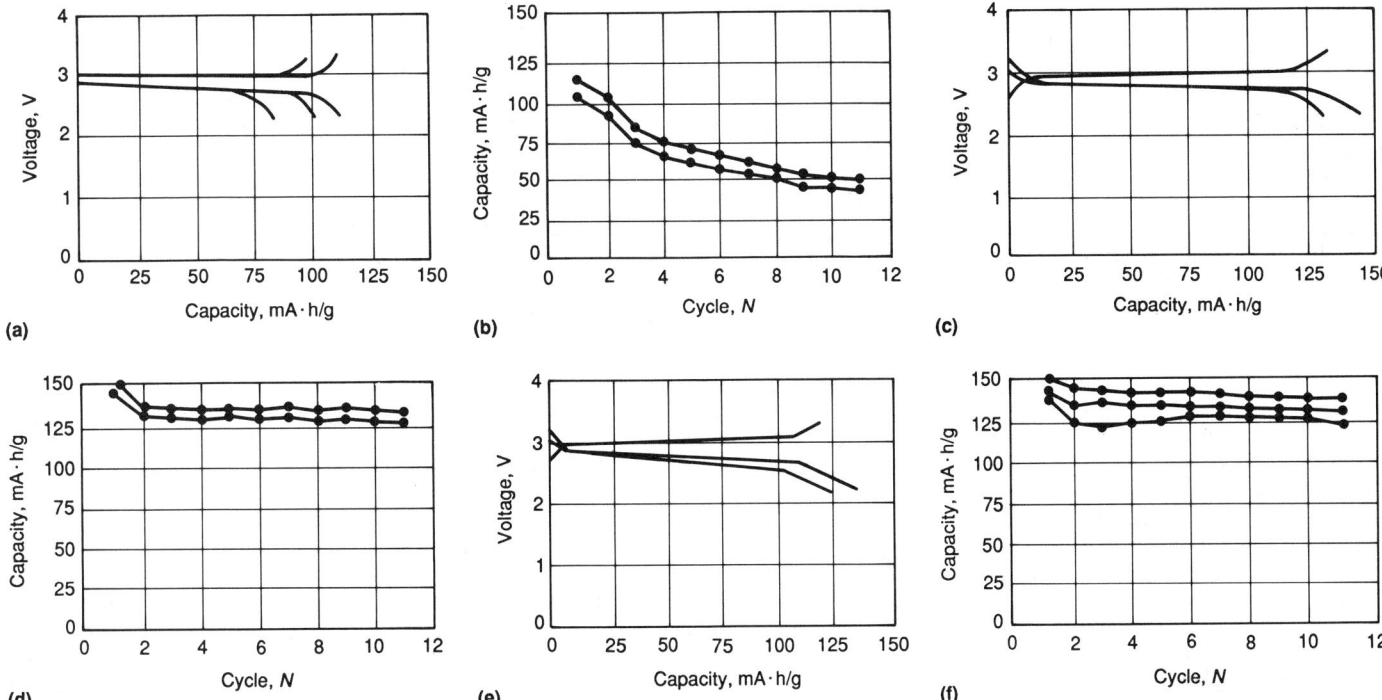

Fig 15 Electrochemical characteristics of rechargeable lithium cells. (a) and (b) Li/LiMn$_2$O$_4$. (c) and (d) Li/Li$_2$Mn$_4$O$_9$. (e) and (f) Li/Li$_4$Mn$_5$O$_{12}$. In (a), (c), and (e) typical discharge and charge profiles are shown only for the first few cycles. Source: Ref 47

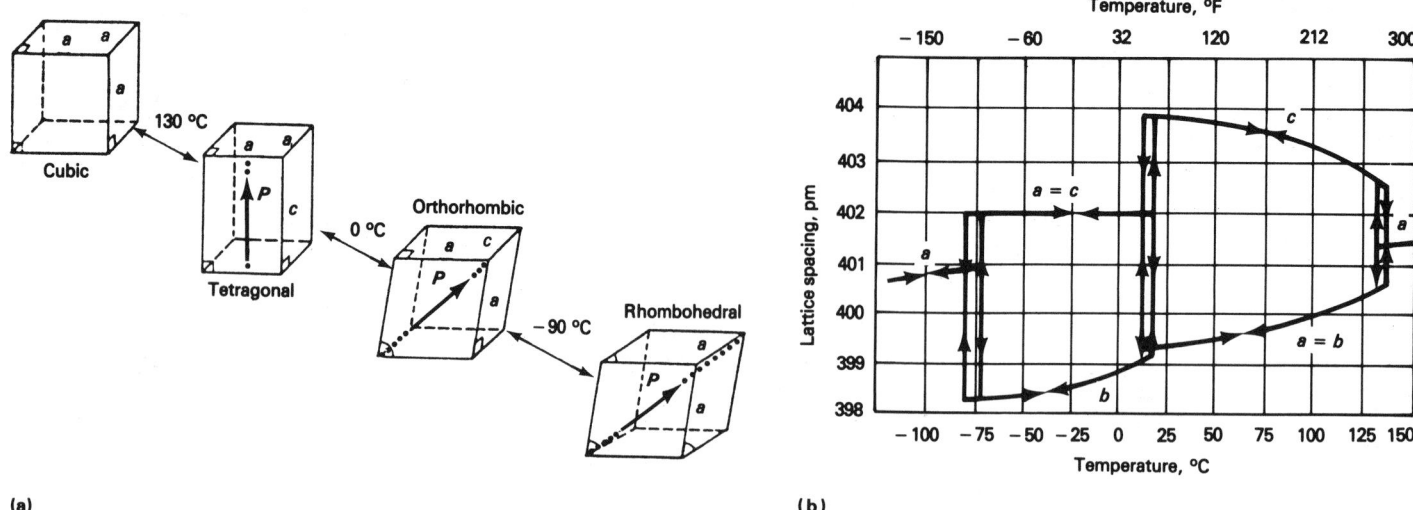

Fig 17 Phase transitions in BaTiO₃. (a) Schematic of structural changes. (b) Lattice parameters as a function of temperature. Source: Ref 57

Na_xWO_3 (0.3 < x < 0.95), Li_xWO_3 (0.35 < x < 0.57), and La_xVO_3 (0.66 < x < 1). These compounds are discussed in the article "Single Oxides" in this Volume.

Structure-Property Relationships. It should also be noted that properties can be modified by the addition of low concentrations of dopants (particularly to control point defects) and by forming disordered solid solutions of two or more perovskites. The relationships between structure and the dielectric, ferroelectric, piezoelectric, pyroelectric, and electrostrictive properties of the perovskites (discussed below) are often dependent upon deviation from cubic symmetry. Considering the classical 32 crystal classes into which all crystalline materials can be categorized, 21 lack a center of symmetry, and 20 of these 21 display the property of "piezoelectricity" (change in polarization on applied stress). Ten of the 20 contain a unique polar axis and are thus "pyroelectric" (change polarization upon uniform heating). The property of "ferroelectricity" can be displayed by all pyroelectric materials. Ferroelectricity is defined as the reversibility of the unique polarization (electric dipole) of a pyroelectric material by application of an electric field. It is determined only by experimentation and is an empirical classification.

These structure-property relationships are summarized in Fig 18 and are discussed below in terms of properties that are dominant in determining the application of perovskites. Only the most important perovskite(s) in the class are mentioned.

Ferroelectric Properties. Ferroelectricity is defined as the reversibility of the spontaneous polarization under application of an applied electric field. Ferroelectric perovskites therefore display hysteresis loops. Until recently, the property of ferroelectricity was only indirectly utilized in piezoelectric ceramics. In the past few years, however, thin-

film ferroelectric ceramics have been extensively investigated for their application in digital memory devices. The polarization direction in a film element can be sensed using a simple semiconductor circuit in close proximity, allowing the polarization direction to act as a digital "1" or "0." The advantages are: nonvolatility (information is not lost if power is lost), speed (intrinsic switching times have been demonstrated to be <1 ns), and radiation hardness. This application of perovskites as ferroelectric thin-film memories has been reviewed at recent symposia (Ref 59).

In polycrystalline ferroelectric ceramics, it is important to note that (for usual grain sizes) grains form multiple ferroelectric domains (in which there is a single polarization direction), in order to minimize stresses. The allowable polarization directions of each domain are determined by crystal symmetry. Domains are bounded by domain walls (Fig 19). Because the material is ferroelectric, domain walls can move on application of a small field or stress.

Dielectric properties are utilized in capacitors, with the most common perovskite materials being modified barium titanates. The large values of dielectric displacement or electric polarization which typically occur in these perovskite materials ensure that the materials have large relative permittivities, or relative dielectric constants. A newer class of perovskites has relative dielectric constants as high as 30 × 10³.

The anisotropic dielectric constants of single crystal, undoped BaTiO₃ as a function of temperature are shown in Fig 20. The maxima correspond to the phase transitions shown in Fig 17. For polycrystalline ceramics, the permittivities are higher (approximately double) than those one would predict from the single-crystal data. This effect is ascribed to a contribution to polarization from the move-

ment of domain walls upon application of the measurement fields. The domain wall movement also represents the major contribution to dielectric loss.

Most commercial capacitors are categorized in terms of the change in dielectric constant as a function of temperature. One can modify the dielectric constant and dielectric constant-temperature behavior by dopant addition and process procedure. The dopants control the phase transition temperatures, and dopants and processing together control the microstructure (grain size, domain structure). Examples of the effect of isovalent substitutions on the phase transitions including the ferroelectric-paraelectric transition temperature are shown in Fig 21. Aliovalent dopants in the A- and B-sites can affect properties by changing the domain wall mobility, grain-boundary phases, sizes, and the concentration of point defects. Common solution additives used to control the phase transition temperatures in commercial BaTiO₃ are shown in Table 8.

The relaxor ferroelectrics are a relatively new class of complex perovskites which should be mentioned. These are usually lead-based, with compositions such as $Pb(B_x'B_{1-x}'')O_3$, where B' is a low-valence cation such as Mg^{2+}, Sc^{2+}, Zn^{2+}, Fe^{3+}, and the B'' cation is of higher valence, such as Nb^{5+}, Ta^{5+}. The attraction of these materials is that they have relative dielectric constants (K) which can exceed 30,000. The characteristics of these "relaxors" are a diffuse ferroelectric-paraelectric phase transition (broad K versus T maxima), as well as a pronounced frequency dependence of permittivity. Dielectric properties of a typical relaxor ferroelectric, $Pb(Mg_{0.33}Nb_{0.67})O_3$, are shown in Fig 22. This relaxor behavior is believed to be related to compositional heterogeneity on a microscopic scale. There exist only local regions, or microdomains, a few nanometers in

Table 7 Examples of perovskites including unit cell parameters

Compound	a, Å	b, Å	c, Å	Remarks
$KNbO_3$	3.9714	5.6946	5.7203	Orthorhombic
$KTaO_3$	3.9885	...	...	Cubic
$NaNbO_3$	5.512	5.577	3.885	Orthorhombic
$NaTaO_3$	3.8851	5.4778	5.5239	Orthorhombic
$TlIO_3$	4.510	...	...	α = 89.34° rhombohedral
$A^{2+}B^{4+}O_3$				
$BaFeO_3$	3.98	...	4.01	Tetragonal
$BaTiO_3$	3.989	...	4.029	Tetragonal
$BaZrO_3$	4.192	...	...	Cubic
$CaTiO_3$	5.381	7.645	5.443	Orthorhombic
$PbSnO_3$	7.86	...	8.13	Tetragonal
$PbTiO_3$	3.896	...	4.136	Tetragonal
$PbZrO_3$	9.28	...	...	Pseudocubic, orthorhombic
$SrCeO_3$	5.986	8.531	6.125	Orthorhombic
$A^{3+}B^{3+}O_3$				
$BiAlO_3$	7.61	...	7.94	Tetragonal
$BiCrO_3$	3.90	3.87	3.90	α = γ = 90°35' triclinic β = 89°10'
$DyFeO_3$	5.30	5.60	7.62	Orthorhombic
$DyMnO_3$	3.70	...	...	Cubic
$GdAlO_3$	5.247	5.304	7.447	$GdFeO_3$ structure
$LaTiO_3$	3.92	...	...	...
$NdAlO_3$	3.752	...	...	Rhombohedral
$NdGaO_3$	5.426	5.502	7.706	$GeFeO_3$ structure
$PrCoO_3$	3.787	...	...	α = 90°13' rhombohedral
$SmVO_3$	3.89	...	...	
A_xBO_3 and ABO_{3-x}				
$Ce_{0.33}NbO_3$	3.89	3.91	7.86	Orthorhombic
$Ca_{0.5}TaO_3$	11.068	7.505	5.378	Orthorhombic
Li_xWO_3	(x = 1) 3.72	...	...	Cubic x = 0.35–0.57
$A(B'_{0.67}B''_{0.33})O_3$				
$Ba(Dy_{0.67}W_{0.33})O_3$	8.386	...	...	$(NH_4)_3FeF_6$ structure
$Ba(In_{0.67}U_{0.33})O_3$	8.512	...	...	$(NH_4)_3FeF_6$ structure
$Ba(Nd_{0.67}W_{0.33})O_3$	8.513	...	...	$(NH_4)_3FeF_6$ structure
$A^{2+}(B^{2+}_{0.33}B^{5+}_{0.67})O_3$				
$Ba(Cd_{0.33}Nb_{0.67})O_3$	4.168	...	...	...
$Ba(Fe_{0.33}Ta_{0.67})O_3$	4.10	...	...	...
$Ba(Sr_{0.33}Ta_{0.67})O_3$	5.95	...	7.47	Hexagonal ordered
$A^{2+}(B^{3+}_{0.5}B^{5+}_{0.5})O_3$				
$Ba(Bi_{0.5}Nb_{0.5})O_3$	8.630	...	...	$(NH_4)_3FeF_6$ structure
$Ba(Cu_{0.5}W_{0.5})O_3$	7.88	...	8.61	Tetragonal
$Ba(Ni_{0.5}Nb_{0.5})O_3$	4.1	...	...	...
$Ba(Sc_{0.5}Re_{0.5})O_3$	8.163	...	...	$(NH_4)_3FeF_6$ structure
$Ca(Gd_{0.5}Nb_{0.5})O_3$	4.03	4.04	4.03	β = 92°42' monoclinic
$A^{2+}(B^{2+}_{0.5}B^{6+}_{0.5})O_3$				
$Ba(Ca_{0.5}W_{0.5})O_3$	8.39	...	...	$(NH_4)_3FeF_6$ structure
$Ba(Cu_{0.5}U_{0.5})O_3$	8.18	...	8.84	Tetragonal
$Ba(Mg_{0.5}W_{0.5})O_3$	8.099	...	...	$(NH_4)_3FeF_6$ structure
$Ba(Sr_{0.5}W_{0.5})O_3$	8.5	...	...	$(NH_4)_3FeF_6$ structure

Source: Data selected from Ref 54, Table 2.2

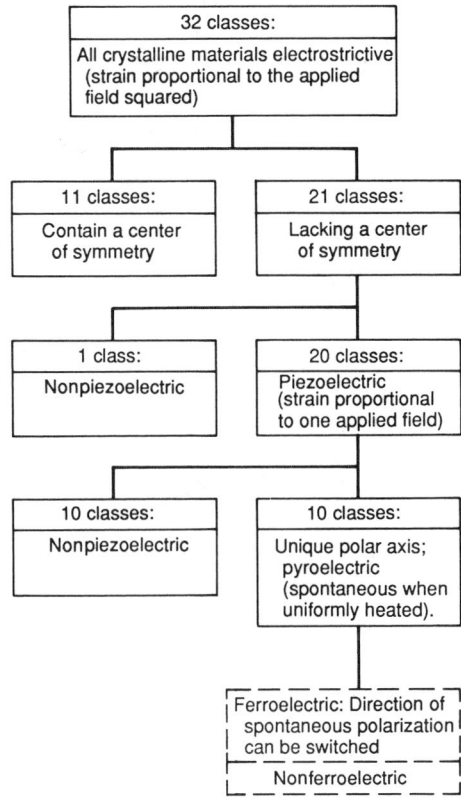

Fig 18 Classification scheme for properties based upon crystal symmetry

tion, temperature control, and heat dissipation measurement. For example, it has been adapted to form cost-effective icing sensors, flow rate monitors, and liquid level sensors (Ref 61).

Piezoelectric Properties. Piezoelectric materials change their electrostatic polarization on application of a mechanical stress, such that:

$$D_i = d_{ijk} T_{jk}$$

where i, j, and k are indices representing the cartesian axes, T_{jk} is the applied stress, D_i is the dielectric displacement or induced polarization, and d_{ijk} is the piezoelectric voltage coefficient. The converse piezoelectric effect is the change in the sample dimensions (strain) with applied field; that is,

$$S_j = d_{ijk} E_i$$

where S_j is the strain, and E_i the applied field. The strain and change in polarization can be related directly to ionic displacements at the unit cell level. It is clear from the above that the piezoelectric coefficient is a third rank tensor (as polarization is a vector, and stress a second ranked tensor). The applicable subscripts are dependent upon the symmetry of the system.

In order to synthesize a polycrystalline ceramic, it is useful to utilize the ferroelectric properties of the material. A polycrystalline ceramic, cooled down to room temperature

dimension, in which the polarization is ordered. These microregions change in size upon application of a field. The relaxor ferroelectrics are currently being developed for capacitor applications.

Positive temperature coefficient of resistivity serves as an illustration of how appropriate doping of the perovskite structure can yield useful properties. Lanthanum-doped $BaTiO_3$ displays the nonlinear resistivity depicted in Fig 23(a). La^{3+} substitutes into the Ba^{2+} site in the ABO_3 perovskite structure and therefore requires charge compensation. This is by promotion of electrons into the

conduction band and/or by formation of vacancies on the titanium site. With appropriate choice of processing conditions, these titanium vacancies (acceptor sites) can be concentrated at grain boundaries in the ceramic, as shown in Fig 23(b), resulting in potential barriers at the grain boundaries. The barrier is such that it increases exponentially with temperature above the ferroelectric-paraelectric transition temperature, consistent with the resistivity behavior shown in Fig 23(a).

This strongly nonlinear resistance or "thermistor" behavior has been utilized for a wide variety of devices, for temperature indica-

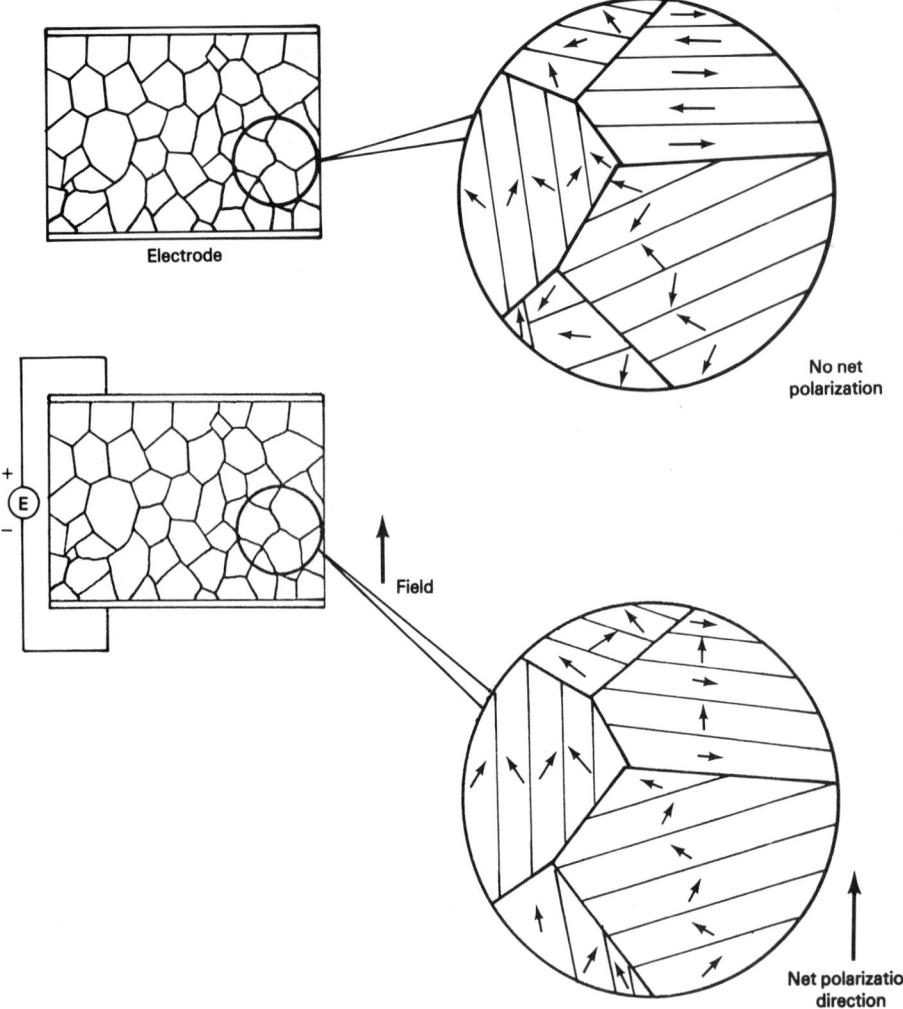

Fig 19 Schematic of domains within polycrystalline grains before and during application of an applied field

of Al_2O_3, while retaining the low cost and ease of processing of alumina ceramics. A large body of research has been undertaken, and it is clear that the properties of the Al_2O_3/ZrO_2 composites are dependent upon a number of parameters, including the $Al_2O_3{:}ZrO_2$ ratio, the dopants (stabilizers) in the ZrO_2, the grain size and distribution of each phase, and the processing conditions.

Typical properties of a commercially available grade of Al_2O_3/ZrO_2, containing 75.2 wt% Al_2O_3, 4.2 wt% Y_2O_3 (in solid solution with the ZrO_2), and 20 wt% Al_2O_3, are presented in Table 1. The flexural strengths are extremely high, and strength remains high even at elevated temperatures (see Fig 26).

The dependence of flexural strength upon Al_2O_3 content is shown in Fig 27. In addition, the Al_2O_3/ZrO_2 does not show the chemically driven degradation displayed by single-phase Y_2O_3-stabilized ZrO_2 (Fig 28). The material is also attractive from the point of view of ceramic component design, as the high toughness also confers upon the material a high Weibull modulus (Fig 29).

Mullite/zirconia appears to be an attractive candidate to replace single-phase or commercial Al_2O_3 in a number of applications. Its attraction lies in the mechanical properties which can be achieved at low cost. It can be produced by reaction-sintering, starting with zircon ($ZrSiO_4$) and alumina (Al_2O_3), and using the following reaction:

$$2ZrSiO_4 + 3Al_2O_3 \xrightarrow{1450\ °C} 2ZrO_2$$
$$+ 3Al_2O_3{\cdot}2SiO_2$$

This reaction is followed by densification, in a single process. The ZrO_2 is dispersed in the mullite matrix and provides a mechanism for "toughening" the mullite. Both alumina and zircon are available in large volume, satisfactory purity, and reasonable cost. The proc-

after sintering, has an essentially random orientation of crystallographic directions, with therefore no set or overall electric dipole for the sample. If the material is additionally ferroelectric, however, it is possible to achieve polarization reversal by application of a large field (the so-called "poling" field). Upon removal of the field, stress barriers stop a significant number of domains from switching back to their original direction. This results in a net polarization in the poling direction, and the ceramic is now piezoelectric. The relevant piezoelectric and elastic terms are described in Table 9.

The most commonly utilized piezoelectric ceramics are those from the lead zirconate titanate (PZT) solid-solution family. The subsolidus pseudobinary phase diagram is shown in Fig 24. The optimum properties are found for compositions with Zr:Ti ratio close to 1:1 (Fig 25); that is, close to the "morphotropic" phase boundary region of Fig 24. This behavior is due to the fact that ferroelectric switching (and thus "poling") is easier for compositions close to the boundary where the spontaneous polarization can be switched to one of 14 possible orientations (the eight $\langle 111 \rangle$ directions of rhombohedral unit cell, and the six $\langle 100 \rangle$ of the tetragonal cell).

Commercial PZT compositions are commonly doped in order to optimize particular piezoelectric properties. Isovalent substitution in the A-site, or aliovalent substitution in the B-site, can have a significant effect on the dielectric and piezoelectric properties (Table 10).

Multiphase Oxides

There are many combinations of the single- and multicomponent oxides which can yield multiphase oxides. This approach is usually attempted to optimize a particular set of properties (as discussed below for Al_2O_3/ZrO_2) or to utilize alternative raw materials (as discussed for mullite/zirconia). These two examples are important engineering ceramics.

Alumina/Zirconia. Multiphase $Al_2O_3/$ stabilized ZrO_2 ceramics have been developed in order to improve the toughness (K_{Ic})

Table 8 Effect of solid solution substitutions on the ferroelectric phase transitions of BaTiO₃ ceramics

Additive	Solid solution limit (mol%)	Change in transition temperature, Δ °C/mol%, for:		
		ΔT_c(a)	ΔT_1(b)	ΔT_2(c)
$PbTiO_3$	100	+3.7	−9.5	−6.0
$SrTiO_3$	100	−3.7	−2.0	0
$CaTiO_3$	21	0	−6.7	−6.0
$BaZrO_3$	100	−5.3	+7.0	+18
$BaSnO_3$	100	−8.0	+5.0	+16
$KNbO_3$	100	−9.0	+12	+35
$Ba(Fe_{1/2}Ta_{1/2})O_3$	100	−15	−2	−6
$Ba(Co_{1/2}W_{1/2})O_3$	50	−30		
$(K_{1/2}Nd_{1/2})TiO_3$	15	−10	−8	−6
$(K_{1/2}La_{1/2})TiO_3$	15	−15		
$La_{2/3}TiO_3$	15	−18		
$Ba_{1/2}NbO_3$	14	−26	+12	+6
$Ba_{0.5}TaO_3$	14	−29		

(a) T_c = tetragonal/cubic transition temperature. (b) T_1 = orthorhombic/tetragonal transition temperature. (c) T_2 = rhombohedral/orthorhombic transition temperature. Source: Ref 58 or 60

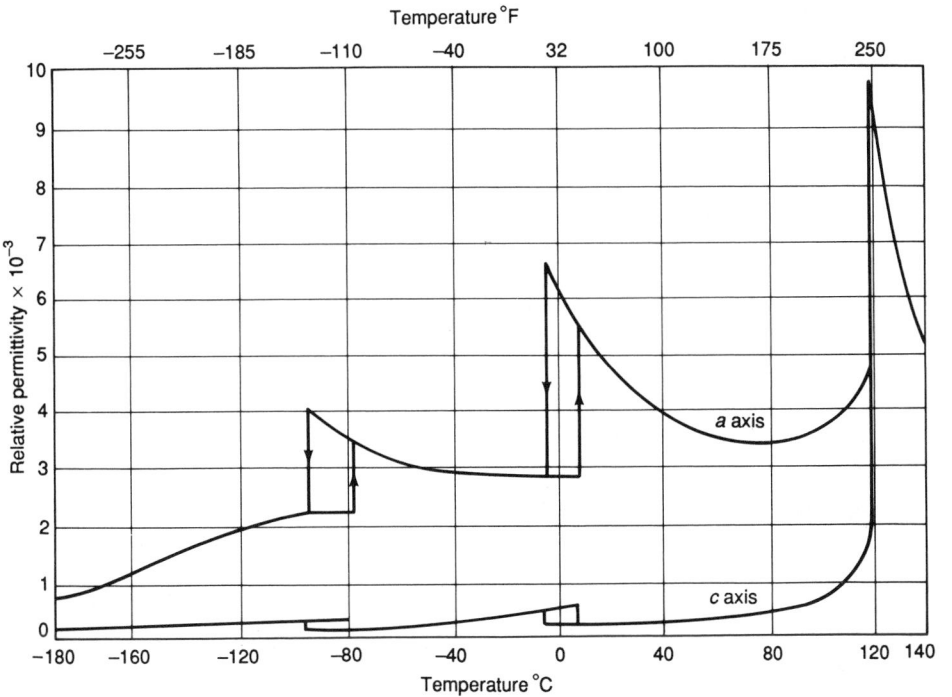

Fig 20 Anisotropic dielectric constants of single-crystal BaTiO₃ as a function of temperature. Compare to Fig 17. Source: Ref 57

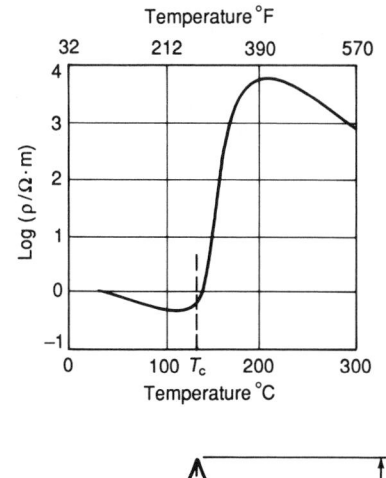

(a)

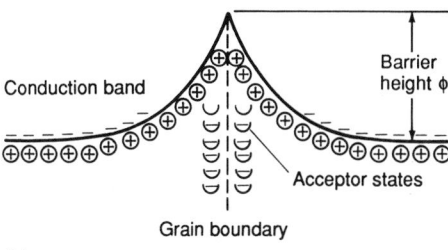

(b)

Fig 23 (a) Typical characteristics of a material having a positive temperature coefficient of resistivity based upon BaTiO₃. (b) Schematic of acceptors causing a potential barrier at the grain boundaries. Source: Ref 61

ess needs to be carefully controlled in order to optimize microstructures and thus properties. Comparative properties are tabulated in Table 11. It can be seen that flexural strengths are comparable to commercial Al_2O_3. The wear resistance under conditions of particle impact erosion appears to be significantly better than 90% Al_2O_3, making it of potential for large-volume, wear-resistant applications.

REFERENCES

1. B.H. Mussler and M.W. Shafer, *Bull. Am. Ceram. Soc.*, Vol 63 (No. 5), 1984, p 705–710

2. *Advanced Technical Ceramics*, S. Somiya, Ed., Academic Press, Tokyo, 1984, p 31–36

3. H. Salmang, *Ceramics: Physical and Chemical Fundamentals*, (trans. by M. Francis), Butterworths, London, 1961, p 294–298

4. J. Hiavac, *The Technology of Glass and Ceramics; An Introduction*, Elsevier Scientific, Amsterdam, 1983, p 232–237

5. R.W. Grimshaw, *The Chemistry and Physics of Clays and Allied Ceramic*

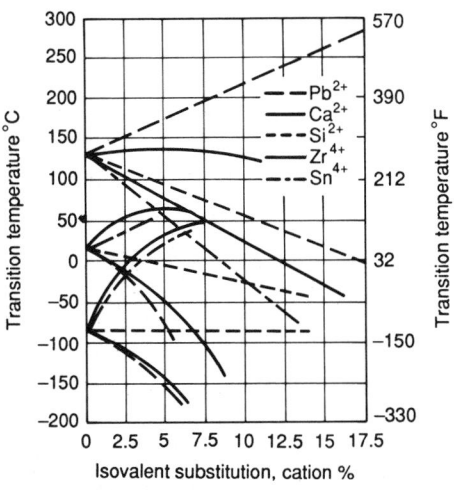

Fig 21 The effect of isovalent substitutions on the transition temperatures in BaTiO₃

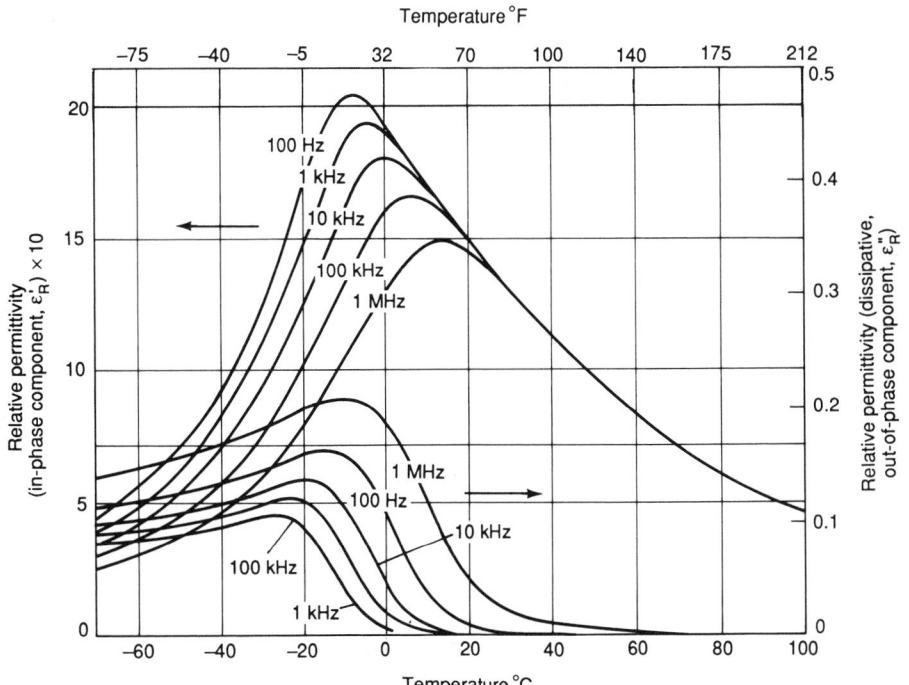

Fig 22 Dielectric properties of Pb(Mg₀.₃₃Nb₀.₆₇)O₃

Table 9 Equations of state for poled piezoelectric ceramics

General form of equations of state:

$$P = d \cdot T + \epsilon^T E$$
$$S = s^E T + d \cdot E$$

D: dielectric displacement or polarization (P)
T: stress
E: applied field
S: strain
s: elastic compliance
ϵ: constant stress permittivity (related to dielectric constant $K = \epsilon/\epsilon_0$)
d: piezoelectric charge coefficient

The complete form for a poled ceramic:

$$D_1 = \epsilon_1 E_1 + d_{15} T_5$$
$$D_2 = \epsilon_1 E_2 + d_{15} T_4$$
$$D_3 = \epsilon_3 E_3 + d_{13} T_1 + d_{31} T_2 + d_{33} T_3$$
$$S_1 = s^E_{11} T_1 + s^E_{12} T_2 + s^E_{13} T_3 + d_{33} E_3$$
$$S_2 = s^E_{11} T_2 + s^E_{12} T_1 + s^E_{13} T_3 + d_{31} E_3$$
$$S_3 = s^E_{13} T_1 + s^E_{13} T_2 + s^E_{33} T_3 + d_{33} E_3$$
$$S_4 = s^E_{44} T_4 + d_{15} T_2$$
$$S_5 = s^E_{44} T_4 + d_{15} T_1$$
$$S_6 = s^E_{66} T_6$$

Tensor conventions:

3: The "poling axis" (between poling electrodes)
1 and 2: orthogonal axes in plane normal to 3
4: abbreviation for 23 shear
5: abbreviation for 13 shear
6: abbreviation for 12 shear

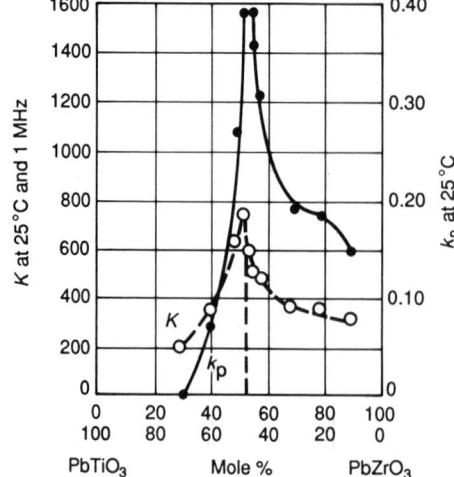

Fig 25 Dielectric constant (K) and piezoelectric (planar) coupling factor (k_p) as a function of Zr-Ti ratio. Source: Ref 57

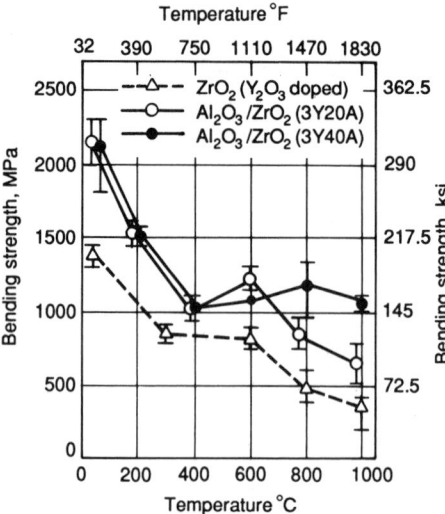

Fig 26 Flexural strengths of Al_2O_3/ZrO_2 3 mol% Y_2O_3-doped ZrO_2 as a function of temperature. Data on Al_2O_3/ZrO_2 is from ToyoSoda Manufacturing Co., Ltd., referring to TSK Super-Z grade containing 80 wt% of Y_2O_3-stabilized ZrO_2 and 20 wt% of Al_2O_3.

Materials, 4th ed., John Wiley & Sons, 1971

6. R.W. Grimshaw, Chapt. III, *The Chemistry and Physics of Clays and Allied Ceramic Materials*, 4th ed., John Wiley & Sons, 1971, p 97–169

7. H. Ikawa, *et al.*, *J. Am. Ceram. Soc.*, Vol 69 (No. 6), 1986, p 492–498

8. M.F. Hochella, Jr. and G.E. Brown, Jr., *J. Am. Ceram. Soc.*, Vol 69 (No. 1), 1986, p 13–18

9. J. Dinning, Maryland Lava Co., Inc., private communication

10. R.F. Davis and J.A. Pask, Mullite, *High Temperature Oxides*, Part 4, A. Alper, Ed., Academic Press, 1971, p 37–76

11. H. Schneider, "Formation, Properties and High-Temperature Behavior of Mullite,"

Inaugural Dissertation, University of Münster, Germany, p 1–148, 1987

12. S. Somiya, R.F. Davis, and J.A. Pask, Ed., *Mullite and Mullite Matrix Composites*, Vol 6, *Ceramic Transactions*, American Ceramic Society, 1990

13. I.A. Aksay and J.A. Pask, Stable and Metastable Equilibria in the System SiO_2-Al_2O_3, *J. Am. Ceram. Soc.*, Vol 58, 1975, p 507–512

14. S. Prochazka and F.J. Klug, Infrared-Transparent Mullite Ceramic, *J. Am. Ceram. Soc.*, Vol 61, 1983, p 874–880

15. W.H. Taylor, The Structure of Sillimanite and Mullite, *Z. Krist.*, Vol 68, 1928, p 503–521

16. B.E. Warren, The Role of Silicon and Aluminum in Complex Silicates, *J. Am. Ceram. Soc.*, Vol 16, 1933, p 412

17. S. Durovic, A Statistical Model for the Crystal Structure of Mullite, *Kristallografiya*, Vol 7, 1962, p 339

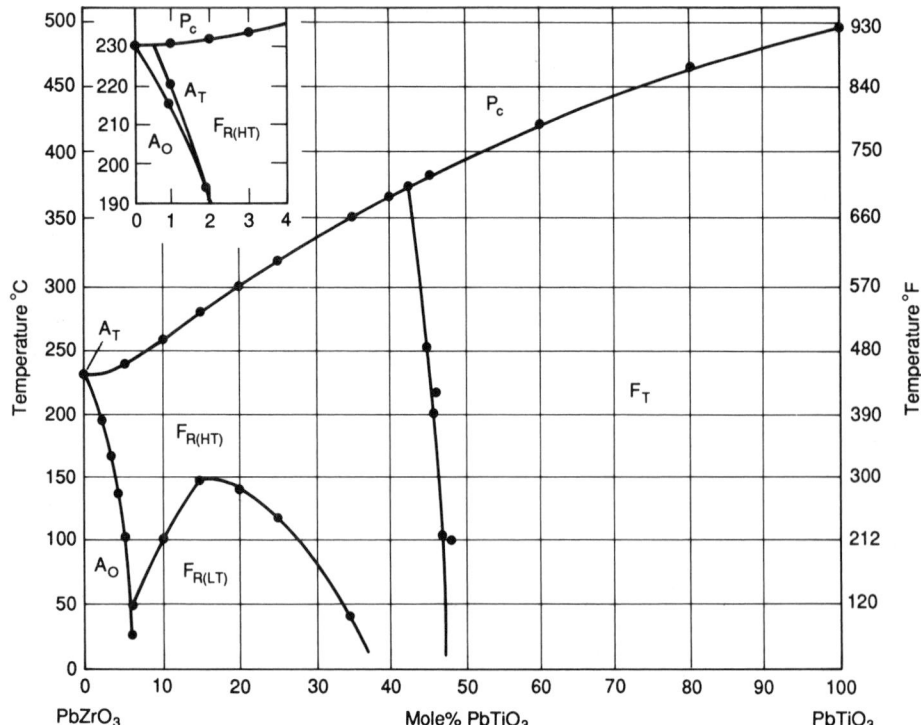

Fig 24 Pseudobinary, subsolidus $PbZrO_3$-$PbTiO_3$ phase diagram. P_c = paraelectric (cubic); F_T = ferroelectric (tetragonal); F_R = ferroelectric (rhombohedral); A_O = antiferroelectric (orthorhombic); A_T = antiferroelectric (tetragonal). Source: Ref 57

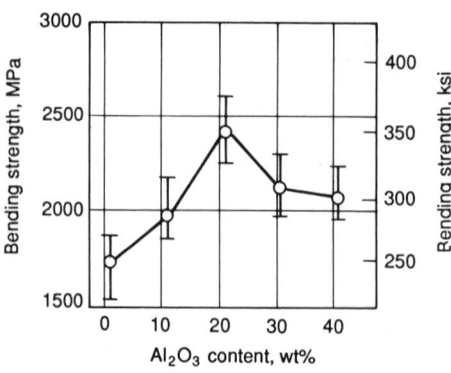

Fig 27 Dependence of the flexural strength of Al_2O_3/ZrO_2 upon Al_2O_3 content. Source: ToyoSoda Manufacturing Co., Ltd.

Table 10 Piezoelectric, elastic, and dielectric constants of several lead titanate zirconate compositions

	$Pb(Ti_{0.48}Zr_{0.52})O_3$	$(Pb_{0.94}Sr_{0.06})(Ti_{0.47}Zr_{0.53})O_3$	$Pb_{0.988}(Ti_{0.48}Zr_{0.52})_{0.976}Nb_{0.024}O_3$
Curie point, °C (°F)	386 (727)	328 (622)	365 (690)
Dielectric constants (unitless)			
K_1^T	1180	1475	1730
K_1^S	612	730	916
K_3^T	730	1300	1700
K_3^S	399	635	830
Dissipation factor	0.004	0.004	0.02
Piezoelectric coupling factors (unitless)			
k_p	0.52_9	0.58	0.60
k_{31}	0.31_3	0.33_4	0.34_4
k_{33}	0.67_0	0.70	0.70_5
k_{15}	0.69_4	0.71	0.68_5
Piezoelectric constants (units indicated)			
d_{31}, 10^{-12} C/N	−93.5	−123	−171
d_{33}	223	289	374
d_{15}	494	496	584
g_{31}, 10^{-3} Vm/N	−14.5	−11.1	−11.4
g_{33}	34.5	26.1	24.8
g_{15}	47.2	39.4	38.2
h_{31}, 10^8 V/m		−9.2	−7.3
h_{33}		26.8	21.5
h_{15}		19.7	15.2
e_{31}, C/m^2		−5.2	−5.4
e_{33}		15.1	15.8
e_{15}		12.7	12.3
Elastic compliances (*s*-constants), 10^{-12} m^2/N			
s_{11}^E	13.8	12.3	16.4
s_{33}^E	17.1	15.5	18.8
s_{12}^E	−4.07	−4.05	−5.74
s_{13}^E	−5.80	−5.31	−7.22
s_{44}^E	48.2	39.0	47.5
s_{66}^E	38.4	32.7	44.3
s_{11}^D	12.4	10.9	14.4
s_{33}^D	9.35	7.90	9.46
s_{12}^D	−5.38	−5.42	−7.71
s_{13}^D	−2.56	−2.10	−2.98
s_{44}^D	25.0	19.3	25.2
Elastic stiffness (constants), GPa			
c_{11}^E		139	121
c_{12}^E		77.8	7
c_{13}^E		74.3	75.2
c_{33}^E		115	111
c_{44}^E		25.6	21.1
c_{66}^E		30.6	22.6
c_{11}^D		145	126
c_{12}^D		83.9	80.9
c_{13}^D		60.9	65.2
c_{33}^D	134	159	147
c_{44}^D		518	39.7
Other elastic constants			
Q_M	500	500	75

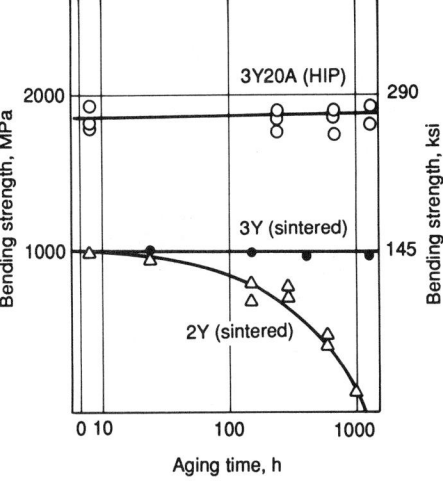

Fig 28 Flexural strength aging time at 300 °C (570 °F). Al_2O_3/ZrO_2 (3Y20A) and two grades of Y_2O_3-stabilized zirconia (2 and 3 mol% Y_2O_3). Source: ToyoSoda Manufacturing Co., Ltd.

18. R. Sadanaga, M. Tokonami, and Y. Takeuchi, The Structure of Mullite, $2Al_2O_3 \cdot SiO_2$ and Relationship with the Structure of Sillimanite and Andalusite, *Acta Crystall.*, Vol 15, 1962, p 65–68

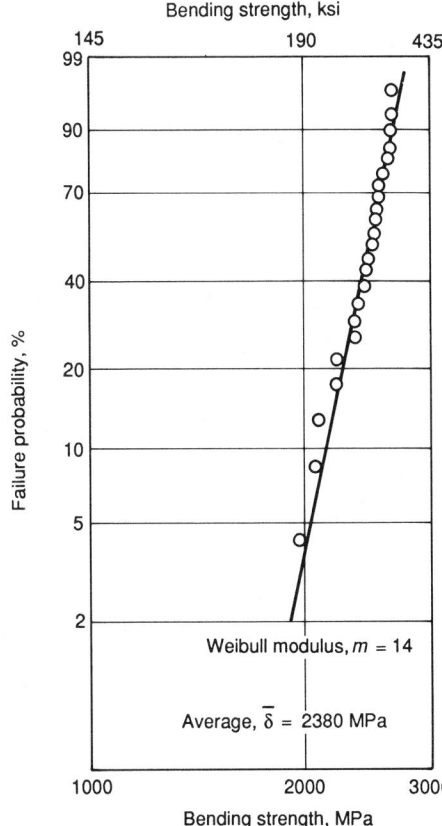

Fig 29 Weibull plot of the flexural strengths of Al_2O_3/ZrO_2 samples (20 wt% Al_2O_3/80 wt% yttria-stabilized ZrO_2, TSK Super-Z grade of ToyoSoda Manufacturing Co., Ltd.

Table 11 Selected property data for reaction-sintered mullite/zirconia, in comparison with commercial 90% Al$_2$O$_3$

	Mullite/zirconia(a)	90% alumina(b)
Four-point flexural strength, MPa (ksi)	190 (27.5)	200–330 (29–48)
Hardness, HV	10.1	10.0
Toughness, MPa$\sqrt{m}$ (ksi$\sqrt{in.}$)	4.0 (3.6)(c)	2.6 (2.4)(c)
Young's modulus, GPa (10^6 psi)	195 (28)	275 (40)
Density, g/cm^3	3.86	3.56
Particle impact erosion, mg/1000 g abrasive, for:		
SiC abrasive (100–150 μm)	450	600
SiO$_2$ abrasive (100–150 μm)	6	74

(a) Developmental grade of mullite/zirconia produced by Multotec S.A. Ltd, Kempton Park, South Africa. Data supplied by Nigel A. Stone, Mattek, CSIR, Pretoria, South Africa. (b) Data primarily from the article "Single Oxides" in this Volume. (c) Indentation method

19. S. Durovic, Refinement of the Crystal Structure of Mullite, *Slovak Acad. Sci., Cheicke Zvesti*, Vol 23, 1969, p 113–128

20. C.W. Burnham, *Crystal Structure of Mullite*, Carnegie Institution of Washington Year Book, Vol 62, 1963, p 223–227

21. C.W. Burnham, Composition Limits of Mullite and the Sillimanite-Mullite Solid Solution Problem, *Crystal Structure of Mullite*, Carnegie Institution of Washington Year Book, Vol 62, 1963, p 228–235

22. R.J. Angel and C.T. Prewitt, Crystal Structure of Mullite: A Re-examination of the Average Structure, *Am. Mineral.*, Vol 71, p 1476–1482

23. N. Morimoto, Y. Nakajima, and M. Kitamuro, *J. Am. Ceram. Soc.*, Vol 69 (No. 6), 1986, p 115–124

24. S.H. Risbud and J.A. Pask, Mullite Crystallization from SiO$_2$-Al$_2$O$_3$, *J. Am. Ceram. Soc.*, Vol 61, 1978, p 63–67

25. M.D. Sacks, H. Lee, and J.A. Pask, A Review of Powder Preparation Methods and Densification Procedures for Fabricating High Density Mullite, *J. Am. Ceram. Soc.*, Vol 69 (No. 6), 1986, p 167–207

26. T. Tiegs, P. Becher, and P. Angelini, Microstructures and Properties of SiC Whisker-Reinforced Mullite Composites, *J. Am. Ceram. Soc.*, Vol 69 (No. 6), 1986, p 463–472

27. R.D. Nixon, S. Chevacharoenkul, R.F. Davis, and T.N. Tiegs, Creep of Hot-Pressed SiC Whisker Reinforced Mullite, *J. Am. Ceram. Soc.*, Vol 69 (No. 6), 1986, p 579–603

28. Y. Hirata, I.A. Aksay, R. Kurita, S. Hori, and H. Kaji, Processing of Mullite with Powders Processed by Chemical Vapor Deposition, *J. Am. Ceram. Soc.*, Vol 69 (No. 6), 1986, p 323–338

29. P.C. Dokko, J.A. Pask, and K.S. Mazdiyasni, High Temperature Mechanical Properties of Mullite Under Compression, *J. Am. Ceram. Soc.*, Vol 60, 1977, p 150–155

30. S. Hori and R. Kurita, Characterization and Sintering of Al$_2$O$_3$-SiO$_2$ Powders Formed by Chemical Vapor Deposition, *J. Am. Ceram. Soc.*, Vol 69 (No. 6), 1986, p 311–322

31. A.F. Wells, *Structural Inorganic Chemistry*, 4th ed., Clarendon Press, Oxford, 1975, p 489

32. R.J. Hill, J.R. Craig, and G.V. Gibbs, *Phys. Chem. Minerals*, Vol 4, 1979, p 317

33. A.R. West, *Solid State Chemistry and its Applications*, John Wiley & Sons, 1984, p 44

34. P. de la Mora and J.B. Goodenough, *J. Solid State Chem.*, Vol 70, 1987, p 121

35. R. Kanno, Y. Takeda, K. Takada, and O. Yamamoto, *J. Electrochem. Soc.*, Vol 131, 1984, p 469

36. C.J.J. van Loon and J. de Jong, *Acta Crystall.*, Vol B31, 1975, p 2549

37. J.B. Goodenough, *Magnetism and the Chemical Bond*, Wiley Interscience, 1963

38. R.E. Newnham, Chapt. 6, *Structure-Property Relations*, Berlin, 1975

39. M.M. Thackeray, W.I.F. David, P.G. Bruce, and J.B. Goodenough, *Mater. Res. Bull.*, Vol 21, 1984, p 435

40. M.M. Thackeray and J.B. Goodenough, U.S. patent 4,507,371

41. J.B. Goodenough, M.M. Thackeray, W.I.F. David, and P.G. Bruce, *Rev. de Chim. Miner.*, Vol 21, 1984, p 435

42. I.T. Sheflel and Ya V. Pavlotskii, *Inorg. Mater.*, Vol 2, 1966, p 782

43. D.B. Rogers, J.B. Goodenough, and A. World, *Appl. Phys.*, Vol 35, 1964, p 1069

44. D.C. Johnston, H. Prakash, W.H. Zachariasen, and R. Viswanathan, *Mater. Res. Bull.*, Vol 8, 1973, p 777

45. L.A. de Picciotto and M.M. Thackeray, unpublished data

46. M.M. Thackeray, L.A. de Picciotto, A. de Kock, P.J. Johnson, V.A. Nicholas, and K.T. Adendorff, *J. Power Sources*, Vol 21, 1987, p 1

47. M.M. Thackery, A. de Kock, M.H. Rossouw, D. Liles, R. Bittihn, and D. Hoge, *Proc. Electrochem. Soc.*, Vol 91–3, 1991, p 326

48. M.H. Rossouw, A. de Kock, L.A. de Picciotto, and M.M. Thackeray, *Mater. Res. Bull.*, Vol 25, 1990, p 173

49. L.A. de Picciotto and M.M. Thackeray, *Mater. Res. Bull.*, Vol 20, 1985, p 1409

50. L.A. de Picciotto and M.M. Thackeray, *Solid State Ionics*, Vol 18 & 19, 1986, p 773

51. D.W. Murphy, R.J. Cava, S.M. Zahurak, and A. Santoro, *Solid State Ionics*, Vol 9 & 10, 1983, p 413

52. J.C. Hunter, *J. Solid State Chem.*, Vol 39, 1982, p 142

53. J.C. Hunter and F.B. Tudron, *Proc. Electrochem. Soc.*, Vol 85–4, 1985, p 444

54. A. Mosbah, A. Verbaere, and M. Tournoux, *Mater. Res. Bull.*, Vol 18, 1983, p 1375

55. T. Ohzuku, M. Klagawa, and T. Hilal, *J. Electrochem. Soc.*, Vol 137, 1990, p 769

56. M.M. Thackeray, P.J. Johnson, L.A. de Picciotto, P.G. Bruce, and J.B. Goodenough, *Mater. Res. Bull.*, Vol 19, 1984, p 179

57. B. Jaffe, W.R. Cook, and H. Jaffe, *Piezoelectric Ceramics*, Academic Press, 1971

58. F.S. Galasso, *Structure, Properties and Preparation of Perovskite-Type Compounds*, Pergamon Press, Oxford, 1969

59. See for example, Proceedings of the Second and Third International Symposia on Integrated Ferroelectrics, Published by the journal, *Ferroelectrics*

60. S.L. Swartz, *IEEE Trans. Elec. Insul.*, Vol 25, 1990, p 935–987

61. A.J. Moulson and J.M. Herbert, *Electroceramics: Materials, Properties, Applications*, Chapman and Hall, London, 1990

Engineering Properties of Zirconia*

R. Stevens, Department of Ceramics, University of Leeds, England

ZIRCONIA (ZrO₂) and zirconia-bearing oxides are utilized widely in the metallurgical and high-temperature chemical engineering industries because of their refractory nature and their relative inertness to many hostile environments. Zirconia-based solid solutions also exhibit oxygen ion conduction, and this property has been exploited in solid-state electrochemical oxygen monitors.

In this article, emphasis has been placed on the processing, properties, and applications of transformation-toughened zirconias (TTZ). During the past decade there has been considerable research into these materials, primarily for industrial applications requiring high-temperature capability, high strength, and an improvement in fracture toughness over existing ceramic materials.

Zirconia

Structure. Zirconia exhibits three well-defined polymorphs (Ref 1), the monoclinic, tetragonal, and cubic phases; it has also been shown that a high-pressure orthorhombic form exists (Ref 2). The monoclinic phase is stable up to about 1170 °C (2140 °F) where it transforms to the tetragonal phase, which is stable up to 2370 °C (4300 °F) when the cubic phase exists up to the melting point of 2680 °C (4855 °F). Crystallographic data as well as physical properties of pure ZrO₂ are given in Table 1.

Of greatest significance is the tetragonal to monoclinic transformation, unusual in that on cooling through the transformation temperature (the monoclinic), it is associated with a large volume change (3 to 5%). This is sufficient to exceed elastic and fracture limits even in relatively small grains of ZrO₂ and can only be accommodated by cracking. Consequently the fabrication of large components of pure zirconia is not possible due to spontaneous failure upon cooling from the sintering temperature. Additives such as calcia (CaO), magnesia (MgO), yttria (Y₂O₃), or ceria

(CeO₂) must be mixed with ZrO₂ to stabilize the material in either the tetragonal or cubic phase.

Part of the phase diagram for the ZrO₂-CaO system is shown in Fig 1. The essential features of this phase diagram are common to the other pseudobinary diagrams involving the stabilizing oxides and zirconia (Ref 3). The crosshatched area (C_ss) is the cubic solid-solution phase field, and materials prepared within this region are commonly referred to as fully stabilized zirconia. These materials are generally used as refractories and can be fabricated as ceramic bricks, foam, or wool. Stabilized zirconia ceramics are resistant to attack by most molten metals (with titanium and the alkali metals being the principal exceptions) and are therefore used as crucible materials.

Two-phase materials (T_ss + C_ss) which consist of precipitates of tetragonal and/or monoclinic phase (depending on the heat treatment) dispersed in a cubic matrix are known as partially stabilized zirconia (PSZ) as indicated by the shaded area in Fig 1. These materials have enhanced mechanical properties and are used in such applications as extrusion dies and high-temperature nozzles.

Transformation Toughening

Garvie et al. were the first to realize the potential of zirconia for increasing both the strength and toughness of ceramics by utilizing the tetragonal monoclinic phase transformation of metastable tetragonal particles induced by the presence of the stress field ahead of a crack (Ref 4). The volume change and the shear strain developed in the martensitic reaction were recognized as opposing the opening of the crack and therefore acting to increase the resistance of the ceramic to crack propagation.

A great deal of research has been expended since in attempts to devise theories and develop

mathematical frameworks to explain the phenomenon. Several thorough reviews of this area are available (Ref 5–9). It is generally recognized that apart from crack deflection which can occur in two-phase ceramics, the t-m transformation can develop significantly improved properties via two different mechanisms.

Microcracking. This can be induced by the incorporation of ZrO₂ particles in a ceramic matrix (cubic ZrO₂ or another ceramic such as Al₂O₃). On cooling through the transformation temperature (T_t-m), the volume expansion of 3 to 5% occurring in the ZrO₂ particles causes a crack to form (Fig 2).

Tangential stresses are generated around the transformed particle which induce microcracks in the matrix; these by their ability to extend in the stress field of a propagating crack or to deflect the propagating crack can absorb or dissipate the energy of the crack, thereby increasing the toughness of the ceramic. The optimum conditions are met when the particles are large enough to transform, but small enough to cause limited microcrack development. The ZrO₂ particle size can usually be controlled by milling time prior to sintering or aging conditions after sintering to develop the desired size range.

In order to obtain maximum toughness, the volume fraction of ZrO₂ inclusions must be at an optimum level (Ref 10). The influence of the ZrO₂ volume fraction is illustrated in Fig 3. The toughness will increase to a maximum above which microcracks generated by the ZrO₂ particles will interact with one another resulting in a decrease in strength.

Such microstructures are useful in resisting thermal shock conditions. There are two conditions, the first the resistance to the initiation of fracture by thermal shock, R':

$$R' = \frac{\sigma_f}{\alpha E} \qquad (Eq\ 1)$$

where σ_f is the fracture stress, α is the thermal expansion coefficient, and E is the elas-

*Adapted from *An Introduction to Zirconia*, Magnesium Elektron Ltd., 1986. With permission

Table 1 Crystallographic and physical properties of zirconia

Property	Value
Polymorphism, K	
Monoclinic to tetragonal	1273–1473
Tetragonal to cubic	2643
Cubic to liquid	2953
Crystallography, monoclinic	
a	5.1454 Å
b	5.2075 Å
c	5.3107 Å
β	99° 14'
Space group	$P2_1/c$
Crystallography, tetragonal	
a	3.64 Å
c	5.27 Å
Space group	$P4_2/nmc$
Crystallography, cubic	
a	5.065 Å
Space group	$Fm3m$
Density, g/cm³	
Monoclinic	5.68
Tetragonal	6.10
Thermal expansion coefficient, 10^{-6}/K	
Monoclinic	7
Tetragonal	12
Heat of formation, kJ/mol	−1096.73
Boiling point, K	4548
Thermal conductivity, W/m·K	
At 100 °C	1.675
At 1300 °C	2.094
Mohs hardness	6.5
Refractive index	2.15

Source: *Encyclopedia of Materials Science and Engineering*, Vol 7, The MIT Press, 1986, p 5130

tic modulus. R' may be increased by increasing σ_f or decreasing α and E. In practice it has been found possible to alter all three parameters beneficially.

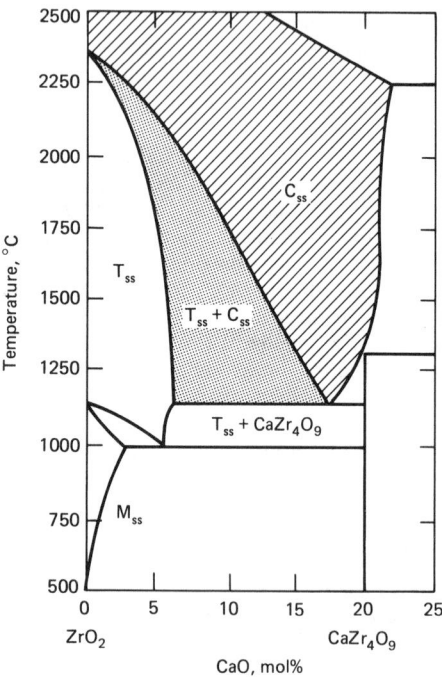

Fig 1 Part of the equilibrium phase diagram for the system ZrO₂-CaO. C_{ss} refers to the cubic solid-solution phase, T_{ss} to the tetragonal solid-solution phase, and M_{ss} to the monoclinic solid-solution phase. Source: Ref 3

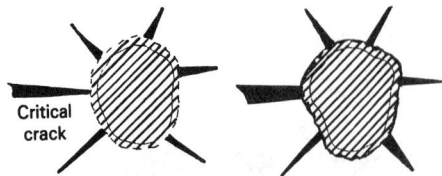

Fig 2 The martensitic transformation that occurs in ZrO₂ (tetragonal to monoclinic at 900 to 1100 °C, or 1650 to 2010 °F) with a 3 to 5% volume expansion, develops microcracks around the ZrO₂ particles (a). A crack propagating into the particle is deviated and becomes bifurcated (b), thus increasing the measured fracture resistance.

The second parameter R'' is considered to be the resistance to growth of a pre-existing flaw by a thermal stress and is given by:

$$R'' = \sqrt{\frac{E\gamma}{2\sigma_f^2}} \qquad \text{(Eq 2)}$$

In this case a tough but weaker ceramic is desirable to obtain higher values of R''.

Stress-Induced Transformation Toughening. On cooling ZrO₂ from above 1200 °C (2190 °F) to room temperature, the tetragonal to monoclinic transformation should occur. If, however, the ZrO₂ is finely divided, or a constraining pressure is exerted on it by the matrix, then the zirconia particle can be retained in the metastable tetragonal form. The particles can be introduced as a second phase during the initial fabrication, for example, zirconia in alumina, or may be developed as a second phase by heat treatment, during or after sintering.

The mechanism of toughening is considered to be a stress-induced transformation of the metastable tetragonal particles to the monoclinic form. If a crack is made to extend under stress (Fig 4), large tensile stresses are generated around the crack, especially ahead of the crack tip (Ref 11, 12). These stresses release the matrix constraint on the tetragonal

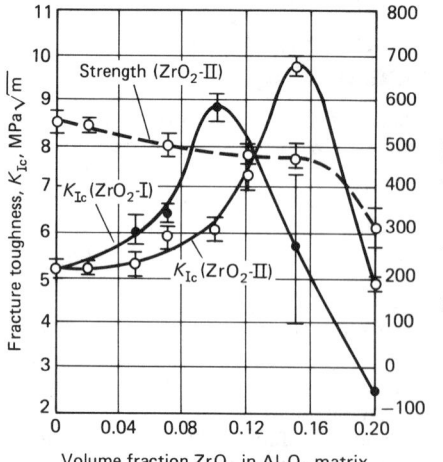

Fig 3 Fracture toughness and flexural strength as functions of the volume fraction of zirconia. Source: Ref 10

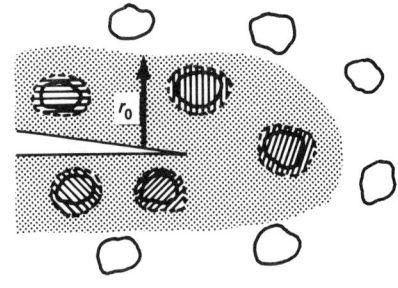

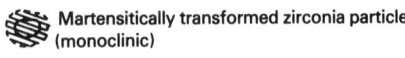
○ Original metastable zirconia particle (tetragonal)

Martensitically transformed zirconia particle (monoclinic)

Stress field around crack tip

Fig 4 Stress-induced transformation of metastable ZrO₂ particles in the elastic stress field of a crack

zirconia particles and if sufficiently large, could result in a net tensile stress on the particle, which under the new conditions will transform to monoclinic symmetry. The volume expansion (>3%) and shear strain (~1 to 7%) developed in the particle causes the martensitic reaction, with a resultant compressive strain being generated in the matrix. Since this occurs in the vicinity of the crack, extra work would be required to move the crack through the ceramic, accounting for the increase in toughness and hence strength.

A critical particle size range for zirconia exists, within which tetragonal particles can be transformed by stress. If the particles are less than a critical size they will not transform; if they are larger than a second critical size they will transform spontaneously. This critical size limit depends on the matrix constraint and the composition of the zirconia; as the cubic stabilizing oxide content is increased, the chemical free energy associated with the phase transformation decreases and hence larger particles can be induced to remain in the metastable tetragonal form.

Compressive Surface Layers. That compressive surface layers are developed in zirconia-based transformation-toughened ceramics is now a well known phenomenon (Ref 13–15). Such stresses develop as a result of the spontaneous t-m transformation of zirconia particles in or near the surface, due to the absence of the hydrostatic constraint near the free surface. The process is shown in Fig 5.

A considerable increase in the fracture strength of the ceramic can be obtained; up to double the value for the fracture strength of the matrix is not uncommon. Surface grinding has been found to be the most effective method of inducing the metastable tetragonal particles to transform, since the compressive stresses can be generated at some

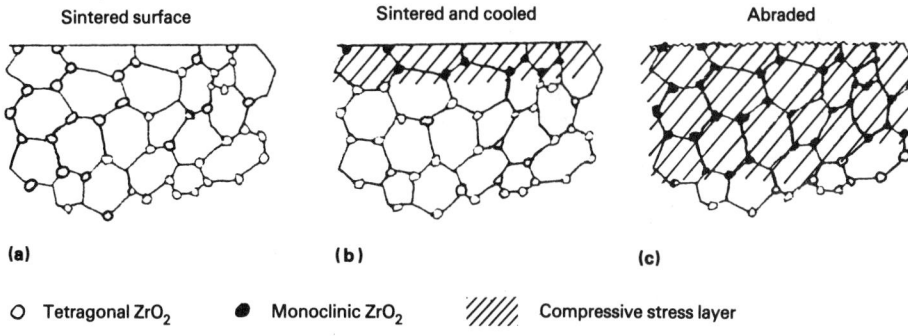

O Tetragonal ZrO₂   ● Monoclinic ZrO₂   ///// Compressive stress layer

Fig 5 Diagram of a section through a free surface at (a) the sintering temperature. On cooling, particles of ZrO₂ near the surface (b) transform due to reduced constraint, developing a compressive stress in the matrix. The thickness of this compressively stressed layer can be increased (c) by abrasion or machining.

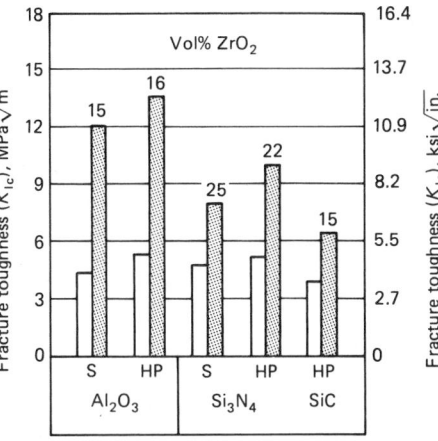

Fig 7 Increase in fracture toughness (K_{Ic}) observed upon inclusion of zirconia particles in the ceramic matrix. The volume of zirconia added is shown in the shaded histogram. The toughness of the matrix material is shown in the adjacent white histogram bars. S, sintered. HP, hot pressed

depth (10 to 100 μm) below the abraded surface. The strengthening effect is known to be dependent on the severity of the grinding. However it has also been shown that the amount of monoclinic phase present in the ground surface decreases rapidly when the surface is carefully polished away in steps (Fig 6). Maximum benefits result if the thickness of the transformed zone, that is, the thickness of the compressively stressed layer, is larger than the critical flaw size but small in comparison with the cross-section of the ceramic.

The development of a compressive surface layer in the ceramic when it is abraded or scratched is perhaps the most significant phenomenon associated with transformation-toughened ceramics. For the first time the engineer has a material which is less sensitive to small surface defects which can be introduced in handling. Any small abrasion, less than the critical flaw size, introduced into the surface is immediately placed into compression, the abrasion flaw effectively being removed from its possible role as a critical flaw. Since the maximum load is usually placed on the outer surface of a ceramic component, usually as a result of a bending stress due to alignment problems, then it is the population of surface flaws which is most likely to provide the critical flaw, hence the importance of the compressive surface stress.

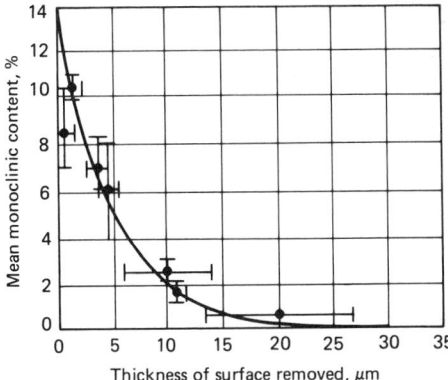

Fig 6 Apparent monoclinic content present in the surface layer as a function of depth

Fabrication of Toughened Ceramics

Five separate approaches have been made toward obtaining a dispersion of ZrO₂ particles in a ceramic matrix. They can be categorized as follows:

- Mixing of the precursor powders or chemicals, followed by sintering or hot pressing
- Heat treatment of partially stabilized zirconia (PSZ)
- Chemical reaction to produce matrix with dispersed zirconia particles
- Duplex structures
- Whisker-reinforced ceramics

Mixed Oxide Ceramics

One of the more common systems under investigation has been the Al₂O₃-ZrO₂ ceramic, when particles of pure zirconia or partially stabilized zirconia are added to Al₂O₃ powder, mixed, and sintered or hot pressed. A considerable increase in strength and toughness results, there being an optimum particle size for both the pure zirconia (~1 to 2 μm) and PSZ (2 to 5 μm) particles and also an optimum volume fraction of the second phase (~15% for ZrO₂ in Al₂O₃).

Zirconia particles have also been added to other matrix materials, such as spinel, mullite, and silicon carbide. Figure 7 illustrates the benefits of zirconia additions on the fracture toughness of several ceramics.

Silicon nitride has been shown to have enhanced properties, but the improvements are limited to relatively low temperatures, by high rates of oxidation, particularly grain-boundary oxidation which occurs in ceramics containing more than 10% ZrO₂.

The commercial application of this type of material is well under way; cutting tool tips of Al₂O₃-ZrO₂ are now in service and wear parts of many components are being manufactured and tested. For economic reasons this class of material is being investigated for internal combustion engine components. Seals, particularly where a combination of wear resistance and corrosion resistance is required, offer possibilities for this type of material.

Mixed oxide ceramics containing zirconia can also be prepared by routes other than mechanical mixing of the precursor oxide, notably sol-gel processing and solid-liquid mixing, followed by drying, calcining, forming, and firing. Care must be taken to optimize the firing cycle since the matrix is required to reach a high density in order to give the necessary mechanical constraint to the zirconia particles. On the other hand the zirconia must be grown to a sufficient diameter that is metastable and therefore transformable, but preferably does not exceed the critical diameter when spontaneous transformation occurs. In practice there is a range of particle sizes produced for a given heat treatment and for the Al₂O₃-ZrO₂ system the average size is considered to be ~0.75 μm. Inevitably some larger particles are produced and this leads to microcracking.

Partially Stabilized Zirconia

Several oxides will stabilize the cubic form of zirconia, of which three, MgO, CaO and, more recently, Y₂O₃, are commonly used to produce partially stabilized zirconia.

The phase diagrams of the binary oxide systems show that a PSZ, which is a mixture of cubic and tetragonal and/or monoclinic phases, can be produced providing the following conditions are met:

- The dopant oxide is present in a concentration less than that needed for complete stabilization of the cubic phase
- The cubic phase is heat treated to develop a two-phase ceramic

Mg-PSZ. A composition in the region of ZrO₂-3 wt% MgO allows the fabrication of a single-phase cubic ceramic at temperatures >1750 °C (>3180 °F). Quenching, to avoid large grain-boundary phase precipitation, followed by isothermal aging at ~1400 °C (~2550 °F) for an appropriate time, results in the development of small particles, lens shaped in section, of tetragonal symmetry, in a cubic matrix. The precipitation process starts with the development of tetragonal nuclei, initially rectangular. On aging for ~2 h the optimum size tetragonal precipitates are developed which are just below the critical size for spontaneous transformation. Overaging of the material leads to the further growth of the tetragonal precipitates and their spontaneous transformation to the monoclinic phase.

The martensitic shear transformation then causes the development of twins in the second phase, two distinct habit planes being apparent, one along and the other at an oblique angle to the long axis of the precipitate. The particles are not now able to contribute to the transformation toughening mechanism and are too small to develop lateral microcracks.

The commercially available Mg-PSZ has a rather more complex microstructure resulting from the sintering route employed. Fortunately, improved properties also result. The compacted green body is usually fired for several hours at temperatures of 1750 to 1800 °C (3180 to 3270 °F), either close to the phase boundary or just inside the single-phase cubic region of the phase diagram. The densified body, which now has a grain size of 40 to 70 μm, is then furnace cooled to the isothermal aging temperature. The furnace cooling results in phase separation, the tetragonal phase forming as a layer at the boundaries of the original cubic grain. On cooling to room temperature the majority of this grain-boundary phase would be expected to transform to the monoclinic phase.

However, two aging treatments are employed to develop the microstructure within the grain. Isothermal aging at ~1400 °C (~2550 °F) allows the development of the

oblate spheroids of metastable tetragonal, the optimum properties being reached when they have a maximum dimension of ~250 nm. The properties at room temperature can be further improved by a subeutectoid aging treatment at 1100 °C (2010 °F).

Ca-PSZ. Similar microstructural changes to those in the Mg-PSZ have been observed although the critical particle size at which transformation is spontaneous is smaller; for Mg-PSZ the critical size range is 25 to 30 nm, whereas for Ca-PSZ the range is 6 to 10 nm.

Garvie and co-workers have made an extensive study of the material (Ref 17). The as-fired PSZ was found to be stronger (200 MPa, or 29 ksi) than the single-phase fully stabilized cubic form (170 MPa, or 25 ksi), due to the presence of precipitates which acted as crack diverters. Aging at 1300 °C (2370 °F) developed a considerable increase in strength, ~650 MPa (~94 ksi) being attained (Fig 8).

Y-PSZ. The addition of Y_2O_3 to ZrO_2 as well as stabilizing the tetragonal and cubic forms, lowers the temperature of the tetragonal to monoclinic transformation. There is thus the practical advantage that in using yttrium oxide, larger particles of zirconia stabilized with this oxide may be retained in the metastable tetragonal form. The use of larger particles as the toughening agent in systems such as Al_2O_3-ZrO_2 considerably eases the fabrication problem.

Several groups are working on this system. Lange (Ref 18) has made important contributions to the science and technology of toughening ceramics. In particular he has shown that the toughening increases linearly with the amount of retained tetragonal phase. The logical consequence of this is the development of a ceramic that is wholly tetragonal, a feat which had been accomplished perhaps unknowingly several years earlier.

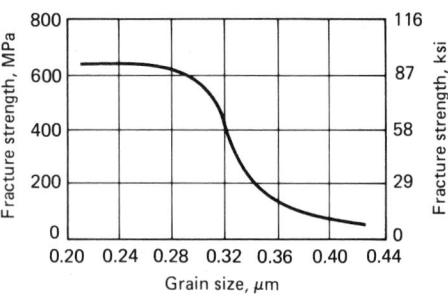

Fig 9 Influence of grain size on fracture strength

This class of material is now known as tetragonal zirconia polycrystal (TZP).

Tetragonal zirconia polycrystals (TZP) ceramics consisting of 100% tetragonal phase using the ZrO_2-Y_2O_3 system were first reported by Rieth (Ref 19) and subsequently by Gupta and co-workers (Ref 20–22). The ceramics produced were dense, fine grained, and stabilized by unspecified amounts of yttria and other rare-earth additives, and strengths of 600 to 700 MPa (87 to 102 ksi) were developed when sintering took place in the temperature range 1400 to 1500 °C (2550 to 2730 °F). When sintered above this temperature the fracture strength decreased due to the development of a transformed layer in the surface which was thought to contain microcracks. A critical grain size was reported above which spontaneous transformation occurred resulting in the lower strength values recorded (Fig 9).

Subsequently it has been found that the critical grain size is also dependent on the amount of stabilizer added and the degree of mechanical constraint. Such factors are interdependent, as illustrated by the relationship of the critical grain size with stabilizer content on an as-sintered surface in Fig 10.

Some doubt may be cast on the precision of this relationship due to the incomplete densification of the ceramic; however the important relationship, the increase in critical grain size from 2 mol% Y_2O_3 to 3 mol% Y_2O_3,

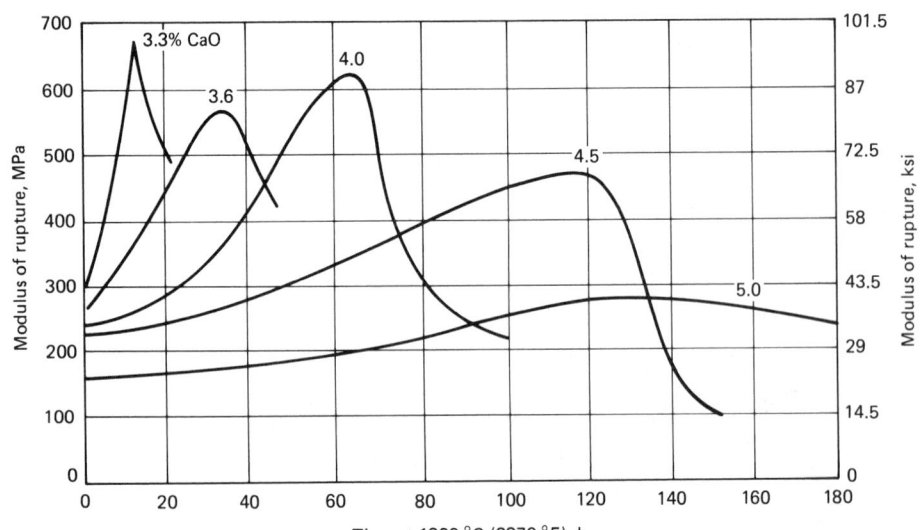

Fig 8 Strength/aging curves obtained by heat treatment at 1300 °C (2370 °F) for various compositions of CaO-PSZ materials

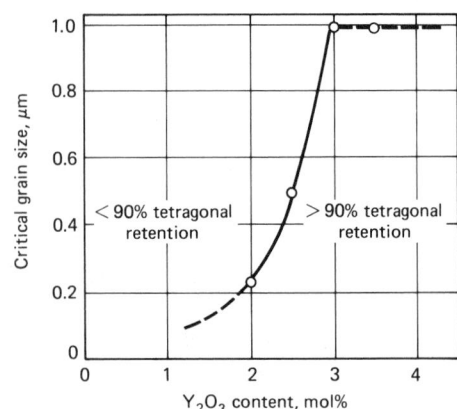

Fig 10 Dependence of the critical grain size on the yttria content

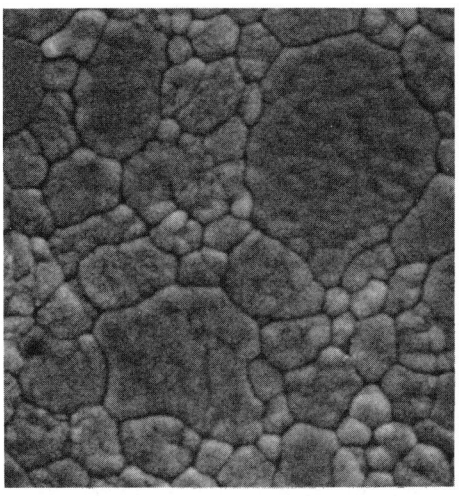

Fig 11 Scanning electron micrograph of a Y-TZP sample. The larger 3 to 5 μm grains are cubic (~5%); the smaller 1 to 2 μm grains are tetragonal (~95%). Source: Ref 40

remains valid. The effect of composition and sintering temperature on the mechanical properties of commercially available materials is now well established (Ref 23). It is possible to sinter the commercial powders to full density at temperatures below 1400 °C (2550 °F) to give a fine-grained tetragonal ceramic. There is usually a small percentage of the cubic phase present, readily identifiable as larger grains (Fig 11), typically 4 to 8 μm compared to the tetragonal phase grains which are usually less than 1.0 μm. The other feature of interest is the presence of the glassy phase at the grain boundaries, which would be expected to play a significant role in both the densification and in the properties, particularly those at high temperatures. The grain-boundary phase is thought to be a low-viscosity liquid in the Al_2O_3-Y_2O_3-SiO_2 system at sintering temperature, however its exact nature and effect on the TZP are not yet fully understood. A consequence of the fine grain size and the presence of the grain-boundary phase is the pseudo-superplasticity of this material at 1200 °C (2190 °F) with extension of >100% being measured in tension.

Of great importance is the provision of chemically homogeneous starting powders, since the relatively low sintering temperatures employed to densify these ceramics do not allow equilibrium to be attained by diffusion in the relatively short sintering schedules. It is considered that the concentration of yttria acts as the source of the cubic grains, yet concentration differences of yttria have been found in tetragonal grains (Ref 24).

The material has been attracting ever increasing interest and has found commercial application in many fields, notably those involving cutting or wear resistance. A detailed review of the TZP ceramics is available (Ref 25).

Alumina Toughened Zirconia. A ceramic having strength values of 2500 MPa (360 ksi) has become commercially available. Consisting of TZP (2 mole % Y_2O_3) with 20% Al_2O_3 addition, it is fabricated in the conventional manner but has the additional step of hot isostatic pressing (HIP) in the last stages of manufacture. This removes the processing flaws reducing the critical flaw size. The addition of alumina is considered to serve two purposes, the first to increase the elastic modulus, this having a minor effect on strength. More important, the fine particles of alumina act as grain-boundary pinning points preventing the development of large grains of the cubic phase which is inevitably present in TZP ceramics. The microstructure is thus made up of uniform fine-grained ceramic.

Since the strength of this ceramic is limited by the intrinsic flaw size, the lowest value of $K_{Ic} = 6$ MPa$\sqrt{m}$ (5.5 ksi$\sqrt{in}$.) being measured by the indentation technique, the high strength measured at room temperature is not degraded disproportionately at high temperatures, a value of 1000 MPa (145 ksi) at 1000 °C (1830 °F) being reported (Ref 26). Degradation on aging at 300 °C (570 °F) in water vapor did not occur, and the thermal shock resistance was superior to that of pure TZP ($\Delta T = 470$ °C, or 845 °F).

Chemical Reaction Zirconia

A distribution of zirconia particles can be produced in a mullite matrix by the reaction of zircon and alumina (Ref 27):

$$2ZrSiO_4 + 3Al_2O_3 \rightarrow 2ZrO_2 + 3Al_2O_3 \cdot 2SiO_2$$

$$(Eq\ 3)$$

Hot pressing mixtures of the above materials resulted in a material having a superior strength (400 MPa, or 58 ksi) to the hot pressed matrix material alone (269 MPa, or 39 ksi).

More recently enhanced toughening has been obtained in a ceramic electrolyte, β-Al_2O_3, used in the Na-S battery, by the incorporation of ZrO_2 particles (Ref 28). β-Al_2O_3, β″-Al_2O_3, and dispersed metastable tetragonal ZrO_2 have been produced by reacting sodium metazirconate with alumina (Ref 29):

$$Na_2ZrO_3 + xAl_2O_3 \rightarrow ZrO_2 + Na_2O \cdot xAl_2O_3$$

$$(Eq\ 4)$$

Under optimum conditions a doubling of the toughness (K_{Ic}) has been obtained.

Duplex Structures

By judicious design and use of a multiphase system, the properties of zirconia can be used to advantage, to an even greater extent than in the systems described earlier. A microstructure which used a polycrystalline agglomerate of metastable zirconia in a matrix of fine-grained alumina was first described by Stevens and Evans (Ref 30). Significant processing advantages were foreseen and the use of alumina as a matrix material resulted

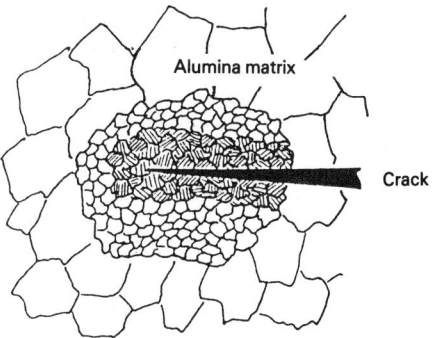

Y-PSZ agglomerate (t-ZrO_2)

Transformed Y-PSZ (monoclinic)

Fig 12 Schematic of the toughening mechanism in a duplex-toughened ceramic using agglomerated zirconia particles

in considerably lower raw material costs, as well as giving an improvement in mechanical properties. A schematic of the toughening mechanism associated with duplex-toughened ceramics is shown in Fig 12.

The microstructure has been developed further such that a toughened matrix is further toughened by the addition of polycrystalline zirconia. As with all toughened ceramics, the toughening is a result of microcracking and the amount of metastable tetragonal phase that can be transformed during the movement of a crack. Since the zirconia has a lower elastic modulus than the matrix, the cracks will tend to move toward the zirconia particles and the energy of the crack can be absorbed by the effective concentration of zirconia particles in the agglomerates. The microstructure is thus more efficient at stopping cracks than if the zirconia had been uniformly dispersed throughout the matrix.

The process may be further enhanced by causing the matrix to contain transformed zirconia particles, thereby placing the matrix in compression causing a further driving force for the propagating critical crack to move to the zirconia agglomerates.

Design and fabrication of duplex structures, which can be produced by conventional sintering, are described in Ref 31.

Whisker-Reinforced Zirconia

A limiting factor in the application of zirconia ceramics is the decrease in properties with increase in temperature—specifically the transformation toughening effect decreases as the transformation temperature is approached. Claussen (Ref 9, 32) had outlined the strategies available for microstructural control of properties at both low and high temperature. The reinforcement of TZP with SiC whiskers was thought to be the most attractive option for high-temperature properties (Ref 33). Whiskers were considered to be more suitable than fibers because of the higher strength and Young's modulus. They were considered to be more amenable to the processing route when fibers too would be mechanically and thermally degraded.

Compositions containing up to 30 vol% of SiC whiskers in TZP were fabricated by hot pressing at 1450 °C (2640 °F) in graphite dies. The mechanical properties showed interesting changes. Whereas the toughness at room temperature doubled for the 30 vol% SiC addition (6 to 12 MPa$\sqrt{m}$, or 5.5 to 11 ksi$\sqrt{in}$.), the fracture strength at room temperature decreased by a factor of 2 (1200 MPa to 600 MPa, or 174 to 87 ksi$\sqrt{in}$.), suggesting the critical flaw size increased by a factor of ~16 on the introduction of the whiskers, further suggesting possible improvements in the processing. However the significant result in the work was the doubling of the strength at 1000 °C (1830 °F) by the inclusion of the SiC whiskers. Claussen noted that the composite degraded in an oxidizing atmosphere at high temperature and suggested that reinforcement with Al_2O_3 whiskers might well be a more appropriate method of obtaining the desirable high-temperature long-term strength.

Properties of Zirconia-Toughened Ceramics

Three basically different classes of ZrO_2-toughened ceramics will be described in this section. These include:

- Tetragonal ZrO_2 polycrystals (TZP), which consist predominantly of fine (<1 μm) t-ZrO_2 matrix grains. The TZP ceramics contain a smaller amount of dopant (for example, 2 to 3 mol% Y_2O_3 or CeO_2) as the stabilizer and have a microstructure consisting of ~1 to 5 μm grains of predominantly t-ZrO_2
- Partially stabilized ZrO_2 (PSZ), in which t-ZrO_2 particles are coherently precipitated within a cubic (c) stabilized ZrO_2 matrix (precipitation-toughened ceramics). A typical PSZ ceramic contains a stabilizer of MgO, CaO, or Y_2O_3 (for example, Mg-PSZ containing 8 wt% MgO and has a microstructure consisting of 40 to 70 μm cubic grains containing finely dispersed precipitates of submicron t-ZrO_2 and m-ZrO_2
- ZrO_2-toughened ceramics (ZTC), where t- or m-ZrO_2 particles are dispersed in ceramic materials such as Al_2O_3, mullite ($3Al_2O_3 \cdot 2SiO_2$) and spinel ($MgAl_2O_4$) (dispersion-toughened ceramics). The critical ZrO_2 particle size varies with different matrices but is usually between 0.5 and 1.2 μm. Small amounts of Y_2O_3 are usually added to further strengthen the t-phase

Properties of TZP

Of particular importance for the TZP ceramics is their long-term stability and this very much depends on factors such as yttria content and grain size. In the presence of water at temperatures in the range 200 to 300 °C (390 to 570 °F), degradation of properties can be severe, to the extent that the topic warrants a separate section. In contrast the Mg-PSZ commercial ceramic appears not to be as sensitive to atmospheric moisture over this temperature range. This behavior might well be attributed to the low-volume fraction of the transformable tetragonal phase present in or adjacent to the grain boundaries.

The mechanism for the degradation of properties with time is still the subject of discussion. While it is agreed that the removal of elastic constraint at the grain boundaries is the cause of the phenomenon, the reason why this occurs is still open to question. There is one school of thought that attributes the action of water or polar liquids on the grain-boundary glassy phase as being significant (Ref 34, 35), while another suggests the removal of Y_2O_3 from solid solution in the ZrO_2 surface as being an important mechanism (Ref 36). The observation of the development of yttrium hydroxide crystals in the pores of thin films which had degraded was quoted as evidence in favor of this mechanism.

Although the matter has yet to be resolved, the importance of the nucleation event is appreciated, since once a monoclinic nucleus is formed, the transformation process can become self-generating in that adjacent grains can be triggered to transformation. The microcracks so generated can then provide ready access for the water vapor to continue the chemical attack on the TZP.

The yttria content of the zirconia is the significant controlling variable for property determination. In order to generate significant mechanical properties it is essential to have a microstructure free of any monoclinic phase which would act as a flaw, and this dictates the minimum level of stabilizer added. Thus approximately 1.8 mol% Y_2O_3 in solid solution results in a composition close to the phase boundary and in a ceramic where the metastable tetragonal is readily transformable. As a consequence high values of fracture toughness can be obtained with this composition. There is a propensity to spontaneous transformation on removal of constraint as would occur most readily at a free surface. Exposure to moisture at 200 to 300 °C (390 to 570 °F) enhances this process with the result that long-term exposure under these conditions produces a drastic decrease in fracture strength.

An indication of the microstructural metastability can be seen in the curves of aging time against surface monoclinic content where it is apparent that the surface is only stable when the composition of the solid solution increases to 2.8 mol% Y_2O_3. It is also apparent that an increase in grain size, equivalent to a decrease in elastic constraint, increases the degree of transformation, a grain size of 0.2 μm being required to retain metastable tetragonal in the crystallographic form required for the transformation-toughening process.

A maximum in the fracture strength of TZP is obtained at the 3 mol% Y_2O_3 composition and this must occur when the value of $\sqrt{\gamma/c}$, where γ is the fracture energy and c is the critical flaw length (Ref 37, 38), is optimized. If the intrinsic flaw size due to fabrication techniques can be considered constant over the range 2 to 4 mol% Y_2O_3 and γ is at a maximum at 2 mol% Y_2O_3, this would appear to be a paradox. This can be resolved by examination of the microstructure when at the 3 mol% composition relatively large grains (~5 to 8 μm) of the cubic phase are to be found often in fairly close association with one another. A relatively thick layer of glassy phase is often found at the cubic grain boundary which together with the lower value of the fracture toughness for the cubic phase ($K_{Ic} \simeq 2.2$ MPa$\sqrt{m}$, or 2.0 ksi$\sqrt{in}$.) does indicate a change in the flaw distribution.

The bend strength decreases with temperature, the rate of decrease fairly constant up to 800 °C (1470 °F); thereafter the decrease is at a lesser rate. The elastic modulus decreases with temperature and according to the Griffith equation (Ref 37) this would account for a small change in the fracture strength. Since the flaw size distribution should remain constant, the rapid initial decrease in strength can be explained by a decrease in the fracture toughness from the room-temperature value to that at the transformation temperature. The properties of TZP are summarized in Fig 13 to 16 and Table 2.

Properties of Mg-PSZ

The fracture toughness of the commercial Mg-PSZ can exceed 15 MPa$\sqrt{m}$ (14 ksi$\sqrt{in}$.) at room temperature, although the exact value depends on the method of testing. In addition, the fracture toughness of the ceramic decreases with temperature (Fig 17). The R-curve behavior of the material, which is of advantage when short microcracks are present in a stressed region, also decreases with temperature.

The fracture strength of the Mg-PSZ is not quite as high as that of TZP at room temperature of 650 to 800 MPa (94 to 116 ksi) being reported. However of great significance is the high value of the Weibull modulus m, which exceeds the value obtained for m for the yield stress of certain metals (Table 3). A gradual decrease is noted with temperature until at 800 °C (1470 °F) the fracture strength is 400 MPa (58 ksi). The decrease in strength with temperature can be attributed to the increased stability of the metastable tetragonal precipitates and also to the decrease in elastic modulus with temperature as shown below.

Temperature, °C (°F)	E, GPa (10^6 psi)
20 (70)	205 (29.7)
200 (390)	195 (28.3)
300 (570)	185 (26.8)
400 (750)	173 (25)
500 (930)	169 (24.5)
600–1100 (1110–2010)	169–165 (24.5–23.9)

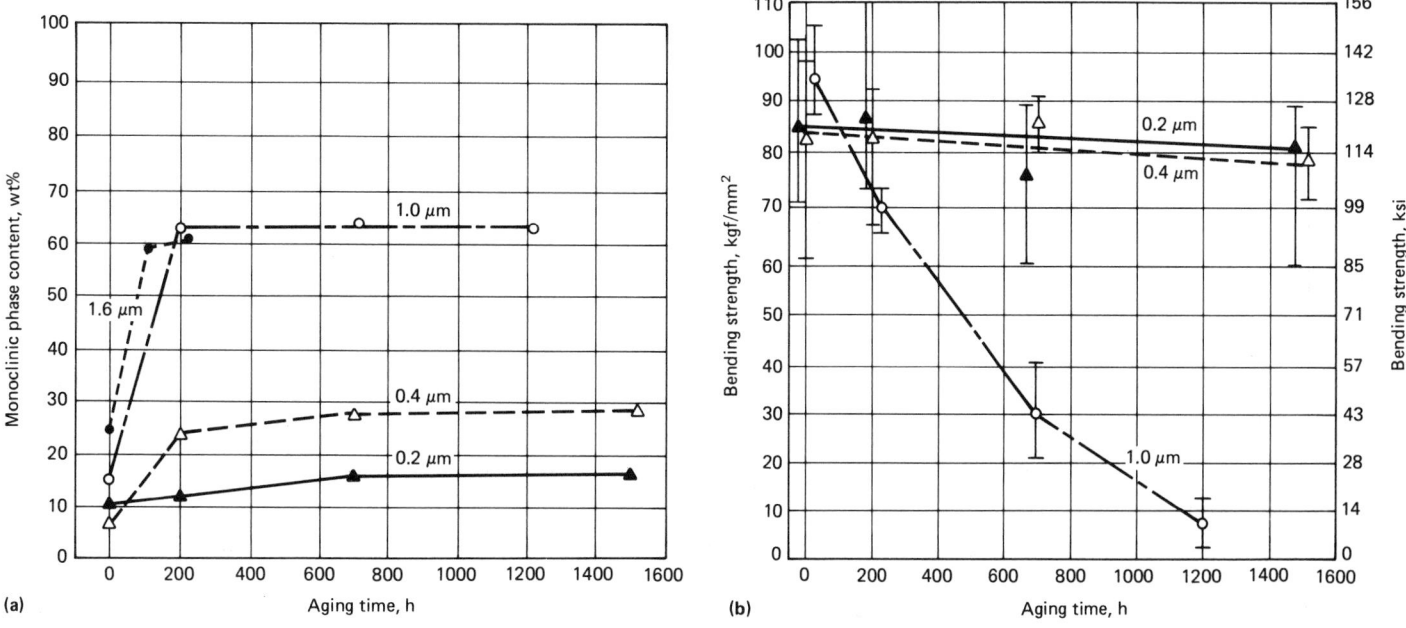

(a)

(b)

Fig 13 Relation between grain size and aging time at 230 °C (445 °F). (a) Monoclinic phase content versus aging time. (b) Bending strength versus aging time. 1 kg/mm² = 9.806 MPa

The hardness measured by indentation using a Knoop indenter also shows a decrease with temperature (Fig 18). Surprisingly the hardness when measured in argon using a Vickers indenter showed no decrease.

The testing of Mg-PSZ in compression has produced interesting results. A value of 2000 MPa (290 ksi) has been measured and of significance is the strain rate dependence, the failure strength increasing with strain rate. At the same time a permanent pseudoplastic strain of 1.3 to 1.4% at failure was measured. The

strain became apparent at loads of 65% of the failure load, giving further credence to the likelihood of transformation-modified critical flaws.

The thermal stability of the Mg-PSZ is reported to be good to temperatures of 800 °C (1470 °F) for periods of 500 h, only marginal reductions in strength being noted (~3%), well within the experimental variations one would expect between batches of specimens. Above 800 °C (1470 °F) small changes occur which are time dependent and a preliminary result

suggests permanent creep deformation can occur at 900 °C (1650 °F) under 200 MPa (29 ksi) loads.

Of significance is the thermal shock resistance of the ceramic when ΔT values in excess of 350 °C (630 °F) or higher are reported (Fig 19).

The thermal stability of several grades of Mg-PSZ and a TZP have been compared over a range of temperatures up to 1000 °C (1830 °F) and for aging times of 1000 h (Ref 39). Some change in the tetragonal to mon-

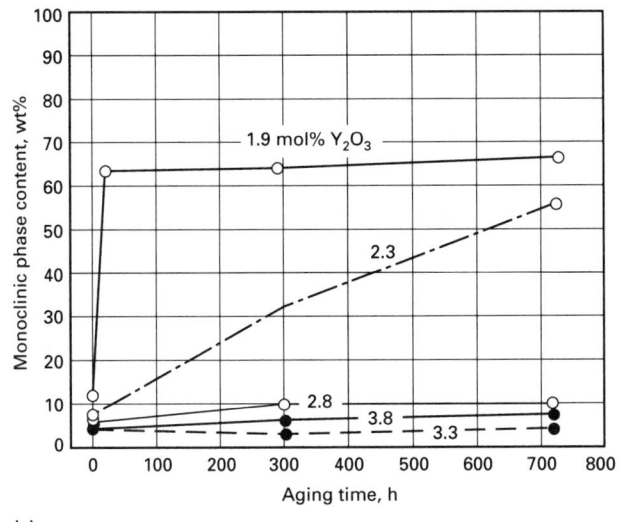

(a)

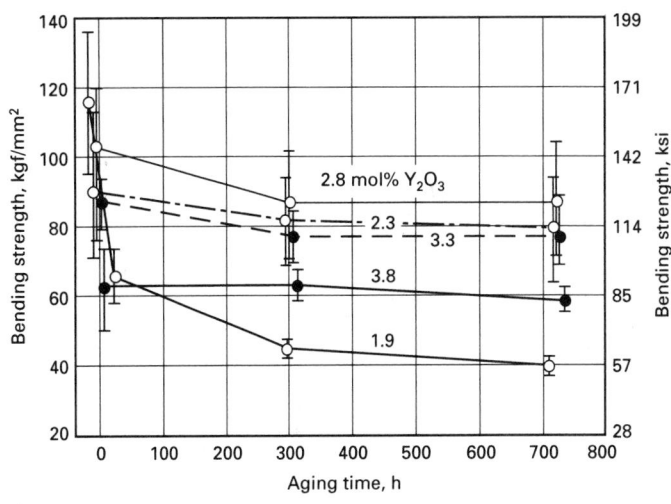

(b)

Fig 14 Relation between yttria content and aging time at 300 °C (570 °F). (a) Monoclinic phase content versus aging time. (b) Bending strength versus aging time. 1 kg/mm² = 9.806 MPa

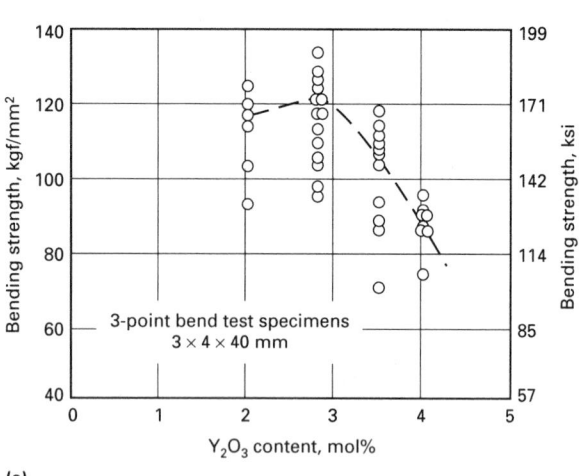

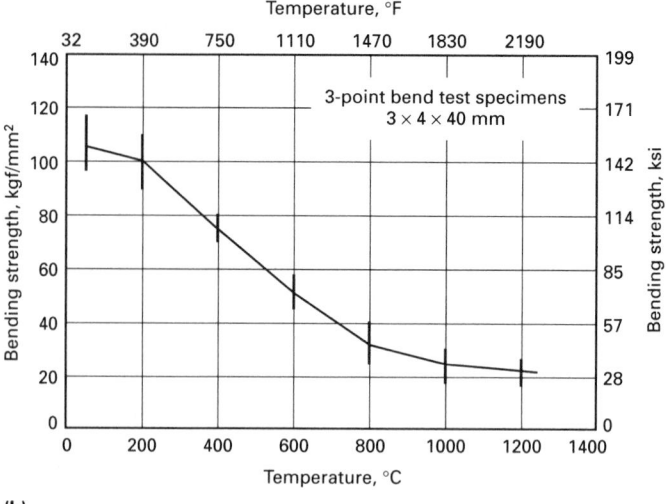

Fig 15 Bending strength of TZP. (a) Bending strength versus yttria content. (b) Bending strength versus temperature. 1 kg/mm² = 9.806 MPa

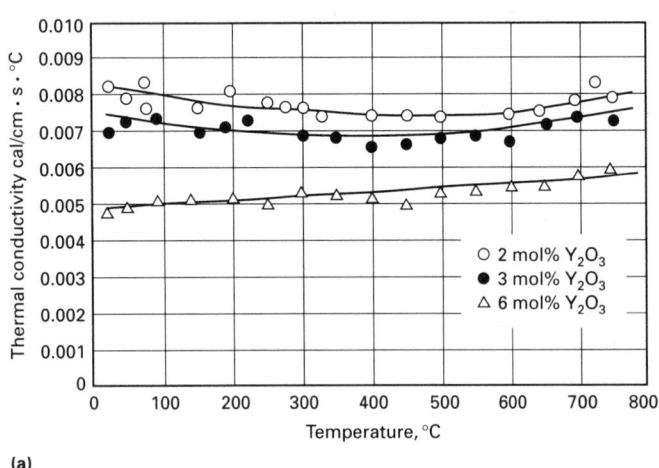

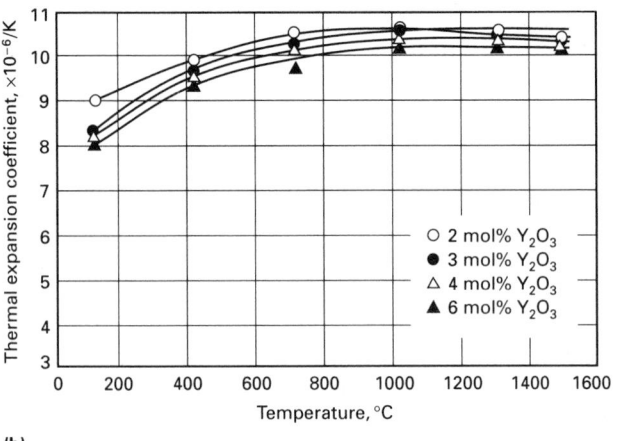

Fig 16 Thermal properties of TZP. (a) Thermal conductivity. (b) Coefficient of thermal expansion

oclinic phase ratio was noted and whereas there were only minor changes in the primary properties such as elastic modulus and thermal expansion, major changes in strength were reported, together with a considerable decrease in the thermal shock resistance ΔT. It would appear therefore that aging at 1000 °C (1830 °F) produces significant property changes whereas this is not the case at 800 °C (1470 °F), and that such behavior is consis-

tent with the instability of the microstructure at the higher temperature.

Thermal Properties. The thermal expansion behavior of the ceramic can be affected by its heat treatment, particularly the degree of metastability of the transformable tetragonal phase. Thus the MS grade shown in Fig 20(a) exhibits a nearly linear expansion and contraction, whereas the TS grade has a slightly lower expansion coefficient and can also exhibit a hysteresis indicative of spontaneous transformation in the microstructure. This is shown in Fig 20(b), the loop of the curve being due to the 4 vol% difference between the monoclinic and tetragonal phases. The thermal conductivity of the grades has also been found to differ, 2.2 W/m · K being reported for the TS grade and a lower value, 1.8 W/m · K, for the MS grade.

Zirconia-Toughened Alumina (Ref 40)

Zirconia-toughened alumina (ZTA) is the generic term applied to alumina-zirconia sys-

tems where alumina is considered the primary or continuous (70 to 95%) phase. Zirconia particulate additions (either as pure ZrO_2 or as stabilized ZrO_2) from 5 to 30%

Table 2   Typical properties of TZP

Property	Value
Density, g/cm³	6.05
Hardness, HR45N	83
Modulus of rupture(a), MPa (ksi)	900 (130)
Fracture toughness (K_{Ic}), MPa√m (ksi√in.)(b)	14 (12.7)
Elastic modulus, GPa (10⁶ psi)	200 (30)
Weibull modulus, m	14
Thermal conductivity, W/m · K	2

(a) Four-point bend test. (b) Single-edge notched beam test. Source: Coors Ceramics

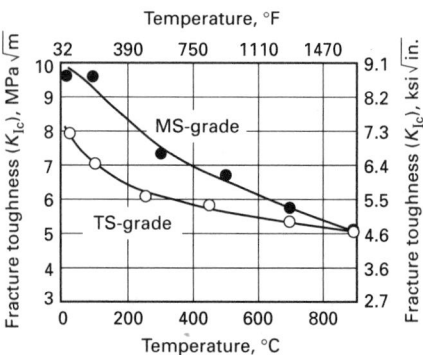

Fig 17 Temperature dependence of the fracture toughness of commercial Mg-PSZ grades. The MS-grade (low-temperature, mechanical shock resistant) and the TS-grade (thermal shock resistant) are manufactured by NILCRA Ceramics, Inc.

Table 3 Typical properties of Mg-PSZ (3 wt% MgO)

Property	Value
Density, g/cm³	5.75
Hardness, HR45N	74–79
Flexural strength, MPa (ksi)	
at 25 °C	634 (92)
at 500 °C	414 (60)
at 1000 °C	290 (42)
Tensile strength, at 25 °C, MPa (ksi)	352 (51)
Compressive strength, at 25 °C, MPa (ksi)	1758 (255)
Young's modulus, GPa (10⁶ psi)	200 (29)
Shear modulus, GPa (10⁶ psi)	69 (10)
Bulk modulus, GPa (10⁶ psi)	373 (54)
Poisson's ratio	0.22
Thermal expansion, at 25–1000 °C × 10^{-6}/K	10.1
Fracture toughness (K_{Ic}), MPa$\sqrt{m}$ (ksi$\sqrt{in.}$)	8–12 (7–11)
Weibull modulus, m	20

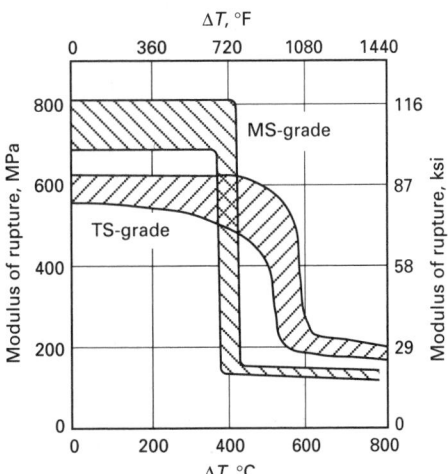

Fig 19 Thermal shock resistance of Mg-PSZ measured as retained strength after quenching through a temperature interval

represent the second phase (Fig 21). The solubility of ZrO_2 in Al_2O_3 and Al_2O_3 in ZrO_2 is negligible. The ZrO_2 is present either in the tetragonal or monoclinic symmetry. ZTA is a material of interest primarily because it has a significantly higher strength and fracture toughness than alumina.

The microstructure and subsequent mechanical properties can be tailored to specific applications. Higher ZrO_2 contents lead to increased fracture toughness and strength values, with little reduction in hardness and elastic modulus, provided most of the ZrO_2 can be retained in the tetragonal phase. Strengths up to 1050 MPa (152 ksi) and fracture toughness values as high as 7.5 MPa $\sqrt{m}$ (6.8 ksi $\sqrt{in.}$) have been measured. Wear properties in some applications may also improve due to mechanical property enhancement compared to alumina. These types of ZTA compositions have been used in cutting-tool applications.

Zirconia-toughened alumina has also seen some use in thermal shock applications. Extensive use of monoclinic ZrO_2 can result in a severely microcracked ceramic body. This microstructure allows thermal stresses to be distributed throughout a network of microcracks where energy is expended opening and/or extending microcracks, leaving the bulk ceramic body intact.

Friction and Wear

In comparison with other ceramics and metals, zirconia engineering ceramics offer a combination of properties which make them attractive for wear-resistant components. The desirable properties, high values of hardness, strength, and toughness together with chemical stability are to be found in commercially available material and can be used to advantage in both adhesive and abrasive wear conditions. However it still has to be remembered that the materials are brittle and catastrophic failure in tension is possible as a result of wear-induced mechanical or thermal shock. In addition the reduction in toughness of the tetragonal zirconia with temperature has to be taken into account and a further factor arises out of the sensitivity of the TZP ceramics to aqueous environments at 200 °C (390 °F) which are known to enhance the transformation process.

The literature to date on the friction and wear behavior of zirconia-based ceramics has been summarized in a comprehensive review by Wallbridge and Dowson (Ref 41). A great deal of the work has been carried out for specific applications and the results are often pertinent to the specific system. Furthermore the wide diversity of equipment and testing conditions makes the comparison of data from various sources difficult. In addition to the normal variables encountered in wear (load, interface speed, lubrication) with zirconia engineering ceramics, the effect of transformation of the tetragonal to the monoclinic must be taken into account, since it can directly affect the nature of the wear surface. A number of papers dealing with wear of ZrO_2-based ceramics have been presented (Ref 42–46) and the interest is likely to increase with further development of ceramic components for automotive engines.

The coefficient of friction of PSZ on hardened steel is 0.17 for unlubricated sliding which compares favorably with the value of 0.44 obtained with alumina. Low friction behavior of PSZ with aluminum alloys has also been reported. The wear behavior of Mg-PSZ is also very good in conjunction with itself and other ceramics, particularly at room temperature, although the wear rate does increase with temperature (Table 4).

In sliding wear the zirconia ceramics perform well due to a smearing of the surface to give a very smooth finish. In conjunction with this a compressive surface layer is developed due to the tetragonal-monoclinic transformation (see the Section "Compressive Surface Layers"), and this could well be a factor

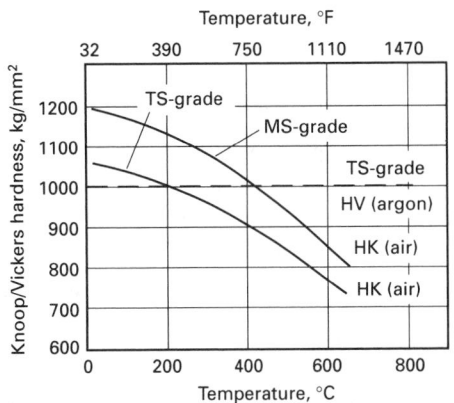

Fig 18 Temperature dependence of the hardness of commercial Mg-PSZ grades

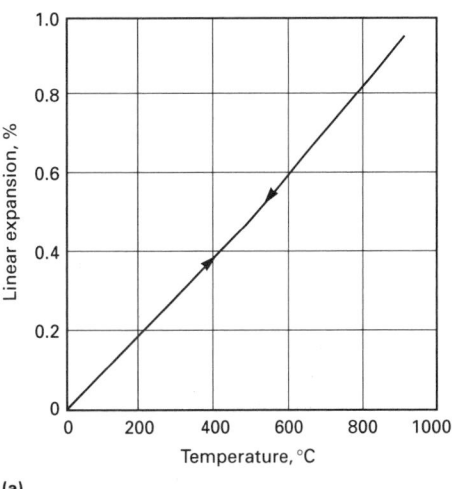

(a)

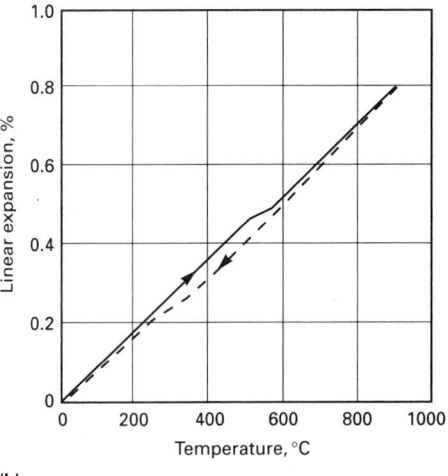

(b)

Fig 20 Thermal expansion behavior of (a) MS-grade Mg-PSZ and (b) TS-grade Mg-PSZ. See text for details.

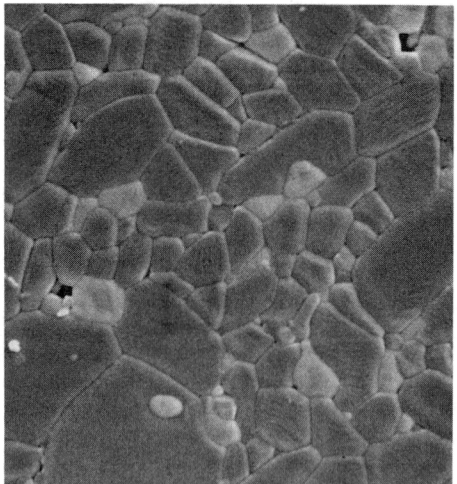

Fig 21 Scanning electron micrograph of high-purity, zirconia-toughened alumina showing dispersed zirconia phase (white) within an alumina matrix. Source: Ref 40

Table 5 Wear behavior of ceramics under abrasive conditions

Material	ASTM Test G 65 procedure A, volume loss, mm³
Standard tungsten carbide	5–20
Nilcra-PSZ-MS grade	10–12
Alumina	8–85
Stellite 1016	18–23
D2 tool steel	32–38
316 stainless steel	100–250
Carbon steel	200–250

in the eventual breakdown of the bearing surface.

Abrasive wear resistance is also high for the zirconia ceramics. The standard test consists of volume loss on abrasion with sand by means of a rubber wheel for 6000 revolutions. The results for several commercial engineering materials are shown in Table 5, where it can be seen that PSZ shows considerable promise.

Performance in the laboratory gives a good indication of wear behavior and of the wear mechanisms under a given set of controlled conditions. Industrial practice is often far more demanding due to irregular loadings and environmental changes with the result that wear behavior is less predictable. Field trials in the mining industry have shown that the Mg-PSZ outperforms previously used materials by a factor of 2 to 3 in applications such as belt scrapers and screw conveyor bearings.

Table 4 Wear behavior of ceramic materials

MS-grade PSZ against		Value of wear coefficient, k, at various temperatures(a)		
		20 °C (70 °F)	205 °C (400 °F)	425 °C (795 °F)
MS-grade PSZ	k pin	0.003	11.5	...
	k disc	...	25.5	222.0
Si₃N₄	k pin	9.3	17.8	95.1
	k disc	71.3	25.2	96.9
SiC	k pin	4.9	2.3	...
	k disc	19.1	0.77–3.70	29.0
Al₂O₃	k pin	67.6	36.3	107.2
	k disc	83.8	193.0	26.4

(a) k = wear coefficient = volume material removed/force × distance

Applications of Zirconia Ceramics

The properties of zirconia ceramics outlined earlier suggest that the combination of strength, toughness, and chemical resistance should allow application of the materials in harsh environments under severe loading conditions. In conjunction with the development of a surface compressively stressed layer, many novel applications in wear-resistant cutting devices are envisaged.

The alumina-zirconia ceramics have superior strength, toughness, and wear resistance when compared to conventional alumina, and consequently the composite ceramic has found use as a cutting tool tip (Ref 47). Alumina zirconia abrasion wheels outperform the pure alumina variety by a factor of 8 (Ref 48).

Many cutting applications have been found. Items such as scissors and shears have proved particularly successful for the cutting of difficult materials such as Kevlar, outlasting conventional tool steels. Cutting applications have also been found in the kitchen, the paper industry, and in hair salons, where clipper blades have been found to be advantageous due to their corrosion resistance.

Cutting and slitting of industrial materials have been exploited using zirconia blades, particularly for magnetic tape, plastic film, and paper items such as cigarette filters. The extra life obtained by the zirconia component allows longer runs of the machines and the resulting loss of down-time more than compensates for the higher initial cost of the component. Similarly zirconia wire-drawing dies and hot extrusion dies are proving themselves superior to conventional dies, particularly in the finishing runs where good dimensional tolerance and a high surface quality are required (Ref 49).

Seals in valves, chemical, and slurry pumps are also being made of zirconia ceramics and recently the impellers have been fabricated and tested. A pump fabricated in TZP is now commercially available.

Components requiring long life under low load conditions can also be made to advantage: thread guides and bearings and guides for dot matrix printers are two further applications.

The most attractive market to the manufacturers who are examining carefully the potential for zirconia is in automotive engine parts, in particular the diesel engine. Two types

of applications are envisaged. In the first the low thermal conductivity of zirconia would be used to advantage in components such as piston crowns, head face plates, and piston liners. As a result the heat loss from the combustion chamber would be reduced and the flame temperature increased resulting in increased engine efficiency. Wear would be reduced giving longer times between routine services and a considerably extended engine life.

There are also components in the engine which are limited by wear, particularly in the valve train, such as cams, cam followers, tappets, and exhaust valves. All these components have been fabricated and are under test (Ref 50). A rotary engine has been constructed using a silicon nitride casing and a zirconia rotor. Should significant advantages ensue in terms of fuel efficiency and/or long engine life, then the future of engineering ceramics would seem assured.

The possibility of zirconia as a biological implant material to replace worn joints is being actively investigated (Ref 51). *In vitro* experiments at 100 °C (212 °F) indicated a small loss of strength considered to be due to a stress-corrosion effect. It was suggested that the effect would be greatly diminished at body temperature.

In vivo experiments involving implants into rabbits to test biocompatibility showed the material to have considerable promise and that no decrease in properties of Mg-PSZ occurred on contact with living tissue for extended periods. Furthermore there was no adverse soft tissue response to the implants.

Other applications for zirconia-bearing ceramics include:

- Refractory components and bricks for high-temperature furnaces for metallurgical processing
- Refractory fibers that provide thermal insulation to separators in aerospace batteries, hot gas filters, and electrolysis diaphragms (Ref 52)
- Thermal barrier coatings for equiaxed, directionally solidified, and single-crystal superalloy turbine blades (Ref 53–56)
- Oxygen sensors for combustion control and atmosphere control in heat-treating furnaces
- Furnace heating elements
- Piezoelectric ceramics—solid solutions of Pb(ZrTi)O₃ (Ref 59, 60)
- Ferroelectric optically active transparent ceramic materials based on lead zirconate titanate system with lanthanum additions (PLZT ceramics) (Ref 61–63)
- ZrO₂-containing glasses and enamels (Ref 64, 65)
- Nucleating agents in glass-ceramics (Ref 66)
- Coloring agents used in the tile, tableware, and sanitaryware industries (Ref 67, 68)

- Single-crystal gemstones of zirconia (Ref 69, 70)

Many of the aforementioned applications for zirconia are described in greater detail in articles that appear in the Sections of this Handbook on "Applications for Traditional Ceramics," "Structural Applications for Technical, Engineering, and Advanced Ceramics," "Applications for Glasses," and "Electrical/Electronic Applications for Advanced Ceramics."

REFERENCES

1. R.C. Garvie, Zirconium Dioxide and Some of Its Binary Systems, *High Temperature Oxides Part II*, A.M. Alper, Academic Press, 1970, p 117
2. A.H. Heuer and L.K. Lenz, *J. Am. Ceram. Soc.*, 1982, p 192
3. V.S. Stubican and J.R. Hellman, Reported in Phase Equilibria in Some Zirconia Systems, *Science and Technology of Zirconia*, Vol 3, *Advances in Ceramics*, 1981, p 25
4. R.C. Garvie, R.H. Hannink, and R.T. Pascoe, *Nature*, Vol 258, 1975, p 703
5. A.G. Evans and R.M. Cannon, Toughening of Brittle Solids by Martensitic Transformations, *Acta Metall.*, Vol 34 (No. 5), 1986, p 761–800
6. A.G. Evans, Toughening Mechanisms in Zirconia Alloys, *Science and Technology of Zirconia—II*, Vol 12, *Advances in Ceramics*, N. Claussen, M. Rühle, and A.H. Heuer, Ed., American Ceramic Society, 1983, p 193–212
7. A.G. Evans, *Fracture in Ceramic Materials*, Noyes Publications, 1984
8. N. Claussen and M. Rühle, Design of Transformation Toughened Ceramics, *Science and Technology of Zirconia*, Vol 3, *Advances in Ceramics*, A.H. Heuer and L.W. Hobbs, Ed., American Ceramic Society, 1981, p 137
9. N. Claussen, Microstructural Design of Zirconia Toughened Ceramics (ZTC), *Science and Technology of Zirconia—II*, Vol 12, *Advances in Ceramics*, N. Claussen, M. Rühle, and A.H. Heuer, Ed., American Ceramic Society, 1983, p 325–351
10. N. Claussen, *J. Am. Ceram. Soc.*, Vol 59, 1976, p 49
11. A.G. Evans and A.H. Heuer, Transformation Toughening in Ceramics: Martensitic Transformations in Crack Tip Stress Fields, *J. Am. Ceram. Soc.*, Vol 63 (No. 5–6), 1981, p 241–248
12. R.M. McMeeking and A.G. Evans, Mechanics of Transformation-Toughening in Brittle Materials, *J. Am. Ceram. Soc.*, Vol 65 (No. 5), 1982, p 242
13. J.S. Reed and A.M. Lejus, Effect of Grinding and Polishing on Near Surface Phase Transformations in Zirconia, *Mater. Res. Bull.*, Vol 12, 1977, p 949–954

14. F.F. Lange and A.G. Evans, Erosive Damage Depth in Ceramics: A Study on Metastable Tetragonal Zirconia, *J. Am. Ceram. Soc.*, Vol 62 (No. 1–2), 1979, p 62–65
15. D.J. Green, F.F. Lange, and M.R. James, Residual Surface Stresses in Al_2O_3-ZrO_2 Composites, *Science and Technology of Zirconia—II*, Vol 12, *Advances in Ceramics*, N. Claussen, M. Rühle, and A.H. Heuer, Ed., American Ceramic Society, 1983, p 240–250
16. R.T. Pascoe and R.C. Garvie, *Ceramic Microstructures 1976*, R.M. Fulrath and J.A. Pask, Ed., Westview Press, 1977, p 774–785
17. R.C. Garvie, R.R. Hughes, and R.T. Pascoe, *Materials Science Research*, Vol II, H. Palmour III, R.F. Davis, and T.M. Hare, Ed., Plenum Press
18. F.F. Lange, *J. Mater. Sci.*, Vol 17 (No. 1–4), 1982, p 225–255
19. P.H. Rieth, J.S. Reed, and A.W. Naumann, Fabrication and Flexural Strength of Ultra-Fine Grained Yttria-Stabilised Zirconia, *Bull. Am. Ceram. Soc.*, Vol 55, 1976, p 717
20. T.K. Gupta, Sintering of Tetragonal Zirconia and Its Characteristics, *Sci. Sintering*, Vol 10, 1978, p 205
21. T.K. Gupta, J.H. Bechtold, R.C. Kuznickie, L.H. Cadoff, and B.R. Rossing, Stabilisation of Tetragonal Phase in Polycrystalline Zirconia, *J. Mater. Sci.*, Vol 12, 1977, p 2421
22. T.K. Gupta, F.F. Lange, and J.H. Bechtold, Effect of Stress Induced Phase Transformation on the Properties of Polycrystalline Zirconia Containing Metastable Tetragonal Phase, *J. Mater. Sci.*, Vol 13, 1978, p 1464
23. Technical bulletin TZ3Y Ceramics, TSK Ceramics, Tokyo
24. M. Rühle, N. Claussen, and A.H. Heuer, Microstructural Studies of Y_2O_3 Containing Tetragonal Zirconia Polycrystals (Y-TZP), *Science and Technology of Zirconia—II*, Vol 12, *Advances in Ceramics*, American Ceramic Society, 1983, p 352
25. I. Nettleship and R. Stevens, Tetragonal Zirconia Polycrystals (TZP)—A Review, *Int. J. High Tech. Ceram.*, Vol 3, 1987, p 1–32
26. K. Tsukuma, K. Ueda, and M. Shimada, Strength and Fracture Toughness of Isostatically Hot-Pressed Composites of Al_2O_3 and Y_2O_3 Partially Stabilised ZrO_2, *J. Am. Ceram. Soc.*, Vol 68 (No. 1), 1985, p C4-C5
27. E. Di Rupo, E. Gilbart, T.G. Carruthers, and R.J. Brook, *J. Mater. Sci.*, Vol 14, 1979, p 705–711
28. L. Vishwanathan, Y. Ikuma, and A.V. Virkar, Transformation Toughening of β Alumina by Incorporation of Zirconia, *J. Mater. Sci.*, Vol 18, 1983, p 109–113
29. J.G.P. Binner, R. Stevens, and S. Tan,

Toughening of Hot-Pressed Beta-Alumina Using Zirconia Additions Prepared from Zirconate, Tartrate and Carbonate Precursor Materials, *J. Microsc.*, Vol 140, Pt 2, 1985, p 183–194
30. R. Stevens and P.A. Evans, Transformation Toughening by Dispersed Polycrystalline Zirconia, *Br. Ceram. Trans. J.*, Vol 83, 1984, p 28–31
31. J. Wang and R. Stevens, "Design, Fabrication and Microstructure-Property Relationships of Duplex $Al_2O_3ZrO_2$ Ceramic," Second International Symposium on Ceramic Materials and Components for Engines, Lübeck, Apr 1986, Deutsche Keramische Gesellschaft
32. N. Claussen, Strengthening Strategies for ZrO_2-Toughened Ceramics (ZTC) at High Temperatures, *Mater. Sci. Eng.*, Vol 71, 1985, p 23
33. N. Claussen, K.L. Weisskopf, and M. Rühle, Tetragonal Zirconia Polycrystals Reinforced with SiC Whiskers, *J. Am. Ceram. Soc.*, Vol 69 (No. 3), 1986, p 288–292
34. T. Sato and M. Shimada, Transformation of Yttria Doped Tetragonal Zirconia Polycrystals by Annealing in Water, *J. Am. Ceram. Soc.*, Vol 68, 1985, p 356
35. T. Masaki, Mechanical Properties of Y-PSZ after Ageing at Low Temperatures, *Int. J. High Tech. Ceram.*, 1986, p 85–98
36. F.F. Lange, G.L. Dunlop, and B.I. Davis, Degradation during Ageing of Transformation Toughened ZrO_2-Y_2O_3 Materials at 250 °C, *J. Am. Ceram. Soc.*, Vol 69 (No. 3), 1986, p 237–240
37. A.A. Griffith, The Phenomena of Rupture and Flow in Solids, *Philos. Trans. R. Soc. London*, Vol A221, 1920, p 163
38. E. Orowan, Fracture and Strength of Solids, *Rep. Prog. Phys.*, Vol 12, 1949, p 185
39. D.C. Larsen and J.W. Adams, "Long-Term Stability and Properties of Partially Stabilised Zirconia," presented at 22nd DOE ATD Contractors Coordination Meeting, Dearborn, MI, Nov 1984
40. G.L. DePoorter, T.K. Brog, and M.J. Ready, Structural Ceramics, *Metals Handbook*, Vol 2, 10th ed., ASM International, 1990, p 1019–1024
41. N.C. Wallbridge and D. Dowson, The Wear Behaviour of Ceramics
42. R.H.J. Hannink, M.J. Murray, and M. Marmach, Magnesia Partially Stabilised Zirconias (Mg-PSZ) as Wear Resistant Materials, *International Conference on the Wear of Materials*, K.C. Ludema, Ed., Apr 1983, Reston, VA, American Society of Mechanical Engineers, p 181–186
43. R.H.J. Hannink, M.J. Murray, and H.G. Scott, Friction and Wear of Partially Stabilised Zirconia: Basic Science and Practical Applications, *Wear*, Vol 100, 1984, p 355–366

44. H.G. Scott, Friction and Wear of Zirconia at Very Low Sliding Speeds, *International Conference on the Wear of Materials*, K.C. Ludema, Ed., Apr 1985, Vancouver, Canada, American Society of Mechanical Engineers

45. R.W. Rice and C.Cm. Wu, Wear and Related Evaluations of Partially Stabilised ZrO₂, Ceramics for Automotive Engines, *Ceram. Eng. Sci. Proc.*

46. P.C. Becker, T.A. Libsch, and S.K. Rhee, Wear Mechanisms of Toughened Zirconias, *Ceram. Eng. Sci. Proc.*

47. R.C. Garvie, Microstructure and Performance of an Alumina-Zirconia Tool Bit, *J. Mater. Sci.*, Vol 3, 1984, p 315–318

48. R.C. Garvie, Structural Applications of ZrO₂ Bearing Materials, *Science and Technology of Zirconia—II*, Vol 12, *Advances in Ceramics*, N. Claussen, M. Rühle, and A.H. Heuer, Ed., American Ceramic Society, 1983, p 465–479

49. S.T. Gulati, J.D. Helfinstine, and A.D. Davis, Determination of Some Useful Properties of Partially Stabilised Zirconia and the Application to Extrusion Dies, *J. Am. Ceram. Soc.*, Vol 59 (No. 2), 1980, p 211–219

50. M. Marmach, D. Servent, R.H.J. Hannink, M.J. Murray, and M.V. Swain, Toughened PSZ Ceramics—Their Role as Advanced Engine Components, *MPR*, Jan 1984, p 7–12

51. R.C. Garvie, C. Urbani, D.R. Kennedy, and J.C. McNeuer, Biocompatibility of Magnesia Partially Stabilised Zirconia (Mg-PSZ) Ceramics, *J. Mater. Sci.*, Vol 19, 1984, p 3224–3228

52. B.H. Hamling and R.E. Lattimer, "Zirconia Fibres and Composites in Severe Environments," private communication, Zircar Products Inc.

53. R.J. Bratton and S.K. Lau, Zirconia Thermal Barrier Coatings, *Science and Technology of Zirconia*, Vol 3, *Advances in Ceramics*, A.H. Heuer and L.W. Hobbs, Ed., American Ceramic Society, 1981, p 226

54. P. Boch, P. Fauchais, B. Lombard, B. Rogeaux, and M. Vardelle, Plasma Sprayed Zirconia Coatings, *Science and Technology of Zirconia—II*, Vol 12, *Advances in Ceramics*, N. Claussen, M. Rühle, and A.H. Heuer, Ed., American Ceramic Society, 1983, p 488–502

55. D.S. Suhr, T.E. Mitchell, and R.J. Keller, Microstructure and Durability of Zirconia Thermal Barrier Coatings, *Science and Technology of Zirconia—II*, Vol 12, *Advances in Ceramics*, N. Claussen, M. Rühle, and A.H. Heuer, Ed., American Ceramic Society, 1983, p 503–517

56. A.S. Grot and J.K. Martyn, Behaviour of Plasma Sprayed Ceramic Thermal Barrier Coatings for Gas Turbine Applications, *Bull. Am. Ceram. Soc.*, Vol 60 (No. 8), 1981, p 807

57. B.C.H. Steele, J. Drennan, R.K. Slotwinski, N. Bonanos, and E.P. Butler, Factors Influencing the Performance of Zirconia Based Oxygen Monitors, *Science and Technology of Ceramics*, Vol 3, *Advances in Ceramics*, A.H. Heuer and L.W. Hobbs, Ed., American Ceramic Society, 1981, p 286

58. E.M. Logothetis, Zirconia Oxygen Sensors in Automotive Applications, *Science and Technology of Ceramics*, Vol 3, *Advances in Ceramics*, A.H. Heuer and L.W. Hobbs, Ed., American Ceramic Society, 1981, p 388

59. B. Jaffe, W.R. Cooke, and H. Jaffe, *Piezoelectric Ceramics*, Academic Press, 1971

60. J. van Randeraat and R.E. Setterington, Ed., *Piezoelectric Ceramics*, Mullard Ltd., London, 1974

61. G.H. Haertling and C.E. Land, *J. Am. Ceram. Soc.*, Vol 54 (No. 1), 1971, p 1

62. G.S. Snow, *J. Am. Ceram. Soc.*, Vol 56 (No. 2), 1973, p 91

63. A.D. James and P.F. Messer, *Trans. Br. Ceram. Soc.*, Vol 77, 1978, p 152

64. H. Rawson, Properties and Applications of Glass, *Glass Science and Technology—3*, Elsevier, 1980

65. K. Kamiya, S. Sakka, and Y. Tatemichi, *J. Mater. Sci.*, Vol 15, 1980, p 1765–1771

66. P.W. McMillan, *Glass Ceramics*, Academic Press, 1979

67. F.T. Booth and G.N. Peel, Preparation and Properties of Some Zirconium Stains, *Trans. Br. Ceram. Soc.*, Vol 61 (No. 7), 1962, p 359

68. R.A. Eppler, Zirconia Based Colours for Ceramic Glazes, *Bull. Am. Ceram. Soc.*, Vol 56 (No. 2), 1977, p 213

69. K. Nassau, Cubic Zirconia, the Latest Diamond Imitation and Skull Melting, *Lapidary J.*, Vol 31, 1977, p 900–904, 922–926

70. K. Nassau, Cubic Zirconia, an Update, *Gems and Gemnology*, Spring 1981, p 9

Engineering Properties of Borides

Raymond A. Cutler, Ceramatec, Inc.

BORON reacts with many elements in the periodic table to form a wide variety of compounds. The strong covalent bonding of most borides is responsible for their high melting points, moduli, and hardness values. Borides generally have high negative free energies of formation, which gives them excellent stability under many conditions.

The coefficients of thermal expansion for borides are in the moderate to high range for ceramics. The high thermal conductivity of borides, in general, gives many of them high thermal shock resistance. Borides have only moderate strength and toughness, compared to other ceramics, although recent research has suggested several approaches for enhancing strength and toughness. Most borides are excellent electrical conductors with resistivities in the 5 to 80 $\mu\Omega \cdot$ cm range. Borides have low work functions, and they generally have negative Hall coefficients. The magnetic properties of borides vary from diamagnetic to strongly ferromagnetic, with most borides being weakly paramagnetic at room temperature. The chemical resistance of borides is superior to most ceramics, and the oxidation resistance of HfB_2-SiC composites at elevated temperatures is excellent. Because borides have a unique combination of thermal, mechanical, electrical, and chemical properties, it is expected that the further development of ceramics for industrial applications will expand the use of these particular compounds.

The primary purpose of this article is to provide engineering data, where available, on binary boride compounds. Data on the refractory boron compounds B_4C and BN are covered in the articles "Engineering Properties of Carbides" and "Engineering Properties of Nitrides" in this Section of the Volume.

Figure 1 shows the known borides in the periodic system. Diagonal lines through certain elements in Fig 1 indicate that no boride formation has been reported. The compounds SiB_3 and SiB_6 are generally regarded as silicides and are not included in this discussion. Numerous reviews of borides are available in the literature (Ref 1–16). References 7, 8, and 12 are the basis for much of the property data

discussed below, but because these particular sources report data that is somewhat dated, this article uses recent literature, where available, to update properties data and to discuss property-structure relationships.

Classification and Structure

Binary metal borides range from compounds with low boron content, such as Be_5B, to those with primarily B—B bonds, such as LnB_{100}, where Ln represents the lanthanide elements Sm, Gd, Tb, Dy, Ho, Er, Tm, Yb, and Lu. Because of the larger atomic size of B ($\approx$0.91 Å, or 36 $\times$ 10^{-10} in.), compared to C ($\approx$0.77 Å, or 31 $\times$ 10^{-10} in.) or N ($\approx$0.71 Å, or 28 $\times$ 10^{-10} in.), interstitial substitution of boron in the undistorted octahedral site is rare, resulting primarily in boron—boron bonding for borides (Ref 14). Important factors that control the composition of borides are the ratio of atomic sizes of the boron to metal atoms; the electrochemical factor linked with electronic transitions; and electron concentration, determined by the degree of localization and delocalization of electrons (Ref 10).

During the formation of a boride, the outer electrons become redistributed and sp^2 and sp^3 electron configurations, which are characteristic of strong covalent bonds, are formed. The roles of sp^2 and sp^3 states change, depending on the donor ability of atoms reacting with boron. It is the various combinations of s^{2}p, sp, sp^2, and sp^3 electron configurations that account for the various boride structures (Ref 10).

Samsonov and Serebryakova (Ref 10) have categorized all borides into three classes: (1) borides formed by s elements having outer s electrons with completely filled or unfilled deeper electron shells, that is, alkaline (group IA) and alkaline earth (group IIA) metals; (2) borides formed by elements with incompletely filled d or f,d subshells, that is, transition metals (Groups IIIB-VIIIB), lanthanides (rare-earth metals), and actinides; and (3) borides formed by elements with valence s,p electrons (BN and BP). The greatest number

of different boride phases occurs in transition metal and lanthanide borides, because various valence states linked with overlapping d, f, p, and s orbitals are possible (Ref 10).

Borides may alternatively be classified into two groups based on crystallochemical considerations (Ref 10): boride structures of lower boron content (such as M_4B, M_3B, M_2B, M_3B_2, MB, and M_3B_4, where M is a metal) or boron-rich compositions (MB_2, MB_4, MB_6, MB_{12}, MB_{49}, MB_{66-100}). The structures of lower boron content borides are determined by their metallic lattices, whereas the structures of the boron-rich borides are determined by their boron atoms, which form covalent B—B bonds. Figure 2 shows how the B atoms change from isolated atoms to chains to three-dimensional networks.

Metal-rich borides with compositions between M_3B and MB_2 are composed of distorted trigonal prisms, whereas boron-rich borides are composed of rigid covalent boron lattices (Ref 14). As the degree of covalent bonding increases, so does the melting point, modulus of elasticity, and hardness. Many of the borides have exceptionally high melting points, as shown by the data in Table 1. Only borides with well-characterized x-ray diffraction patterns (Ref 17) are included in Table 1. Throughout this article, borides are listed in alphabetical order to facilitate locating a specific boride.

Most boride compounds exist over a compositional range (that is, they are nonstoichiometric). In monoborides and diborides, nonstoichiometry arises from either boron chain or lattice vacancies, not from metal vacancies (Ref 14). In boron-rich rigid-lattice borides, nonstoichiometry arises because of metal sublattice vacancies that are due to the strong covalent bonding in these compounds (Ref 14). Data on the homogeneity range of borides are available in Ref 12 and 18. Binary phase diagrams for the Ti-B, Zr-B, and Hf-B systems are shown in Fig 3(a) to (c).

Borides of similar crystal structure and lattice parameters (see Table 1) form solid solutions. The lanthanide hexaborides are an example of this (Ref 19). Rudy (Ref 18) gives phase diagrams for a number of ternary tran-

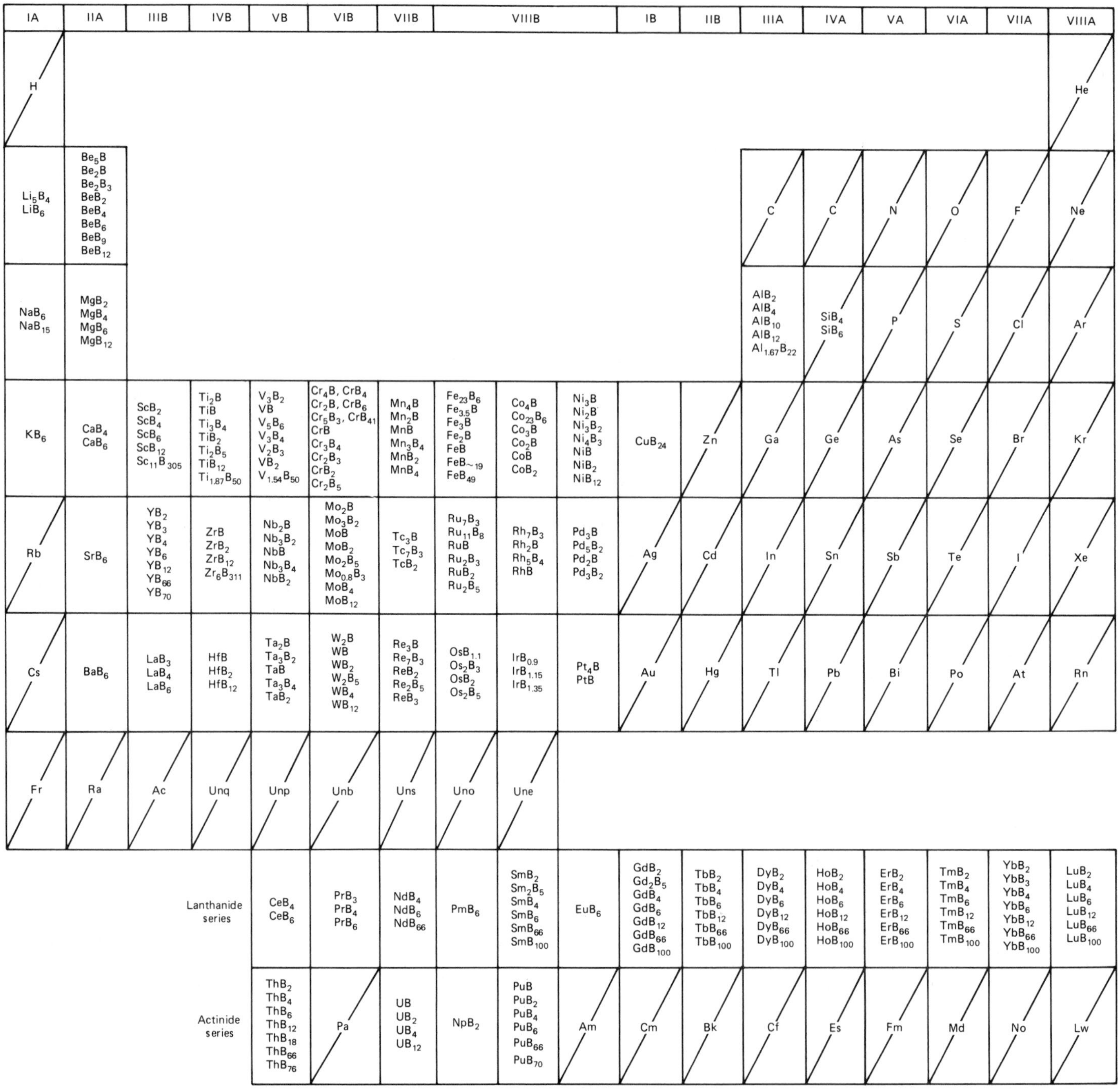

Fig 1 Known borides in periodic system. Elements with diagonals have no reported formation of borides. Source: Ref 12, 13

Main periodic table (borides listed per element):

IA	IIA	IIIB	IVB	VB	VIB	VIIB	VIIIB (Fe)	VIIIB (Co)	VIIIB (Ni)	IB	IIB	IIIA	IVA	VA	VIA	VIIA	VIIIA
H																	He
Li: Li_5B_4, LiB_6	Be: Be_5B, Be_2B, Be_2B_3, BeB_2, BeB_4, BeB_6, BeB_9, BeB_{12}											C	C	N	O	F	Ne
Na: NaB_6, NaB_{15}	Mg: MgB_2, MgB_4, MgB_6, MgB_{12}											Al: AlB_2, AlB_4, AlB_{10}, AlB_{12}, $Al_{1.67}B_{22}$	Si: SiB_4, SiB_6	P	S	Cl	Ar
K: KB_6	Ca: CaB_4, CaB_6	Sc: ScB_2, ScB_4, ScB_6, ScB_{12}, $Sc_{11}B_{305}$	Ti: Ti_2B, TiB, Ti_3B_4, TiB_2, Ti_2B_5, TiB_{12}, $Ti_{1.87}B_{50}$	V: V_3B_2, VB, V_5B_6, V_3B_4, V_2B_3, VB_2, $V_{1.54}B_{50}$	Cr: Cr_4B, CrB_4, Cr_2B, CrB_6, Cr_5B_3, CrB_{41}, CrB, Cr_3B_4, Cr_2B_3, CrB_2, Cr_2B_5	Mn: Mn_4B, Mn_2B, MnB, Mn_3B_4, MnB_2, MnB_4	Fe: $Fe_{23}B_6$, $Fe_{3.5}B$, Fe_2B, FeB, $FeB_{\sim19}$, FeB_{49}	Co: Co_4B, $Co_{23}B_6$, Co_3B, Co_2B, CoB, CoB_2	Ni: Ni_3B, Ni_2B, Ni_3B_2, Ni_4B_3, NiB, NiB_2, NiB_{12}	Cu: CuB_{24}	Zn	Ga	Ge	As	Se	Br	Kr
Rb	Sr: SrB_6	Y: YB_2, YB_3, YB_4, YB_6, YB_{12}, YB_{66}, YB_{70}	Zr: ZrB, ZrB_2, ZrB_{12}, Zr_6B_{311}	Nb: Nb_2B, Nb_3B_2, NbB, Nb_3B_4, NbB_2	Mo: Mo_2B, Mo_3B_2, MoB, MoB_2, Mo_2B_5, $Mo_{0.8}B_3$, MoB_4, MoB_{12}	Tc: Tc_3B, Tc_7B_3, TcB_2	Ru: Ru_7B_3, $Ru_{11}B_8$, RuB, Ru_2B_3, RuB_2, Ru_2B_5	Rh: Rh_7B_3, Rh_2B, Rh_5B_4, RhB	Pd: Pd_3B, Pd_5B_2, Pd_2B, Pd_3B_2	Ag	Cd	In	Sn	Sb	Te	I	Xe
Cs	Ba: BaB_6	La: LaB_3, LaB_4, LaB_6	Hf: HfB, HfB_2, HfB_{12}	Ta: Ta_2B, Ta_3B_2, TaB, Ta_3B_4, TaB_2	W: W_2B, WB, WB_2, W_2B_5, WB_4, WB_{12}	Re: Re_3B, Re_7B_3, ReB_2, Re_2B_5, ReB_3	Os: $OsB_{1.1}$, Os_2B_3, OsB_2, Os_2B_5	Ir: $IrB_{0.9}$, $IrB_{1.15}$, $IrB_{1.35}$	Pt: Pt_4B, PtB	Au	Hg	Tl	Pb	Bi	Po	At	Rn
Fr	Ra	Ac	Unq	Unp	Unb	Uns	Uno	Une									

Lanthanide series:

Ce	Pr	Nd	Pm	Sm	Eu	Gd	Tb	Dy	Ho	Er	Tm	Yb	Lu
CeB_4, CeB_6	PrB_3, PrB_4, PrB_6	NdB_4, NdB_6, NdB_{66}	PmB_6	SmB_2, Sm_2B_5, SmB_4, SmB_6, SmB_{66}, SmB_{100}	EuB_6	GdB_2, Gd_2B_5, GdB_4, GdB_6, GdB_{12}, GdB_{66}, GdB_{100}	TbB_2, TbB_4, TbB_6, TbB_{12}, TbB_{66}, TbB_{100}	DyB_2, DyB_4, DyB_6, DyB_{12}, DyB_{66}, DyB_{100}	HoB_2, HoB_4, HoB_6, HoB_{12}, HoB_{66}, HoB_{100}	ErB_2, ErB_4, ErB_6, ErB_{12}, ErB_{66}, ErB_{100}	TmB_2, TmB_4, TmB_6, TmB_{12}, TmB_{66}, TmB_{100}	YbB_2, YbB_3, YbB_4, YbB_6, YbB_{12}, YbB_{66}, YbB_{100}	LuB_2, LuB_4, LuB_6, LuB_{12}, LuB_{66}, LuB_{100}

Actinide series:

Th	Pa	U	Np	Pu	Am	Cm	Bk	Cf	Es	Fm	Md	No	Lw
ThB_2, ThB_4, ThB_6, ThB_{12}, ThB_{18}, ThB_{66}, ThB_{76}	Pa	UB, UB_2, UB_4, UB_{12}	NpB_2	PuB, PuB_2, PuB_4, PuB_6, PuB_{66}, PuB_{70}	Am	Cm	Bk	Cf	Es	Fm	Md	No	Lw

sition metal boron systems, including Ti-Zr-B, Ti-Hf-B, and Zr-Hf-B.

Preparation and Fabrication

Borides are generally prepared by high-temperature reactions, as described in detail by Schwarzkopf and Kieffer (Ref 1). These methods, which are used to make powders, *in situ* bodies, or coatings, include:

- Direct reaction of the elements by combustion synthesis (Ref 1, 20–23) or fusion (Ref 1)
- Reduction of the metal oxide and boron oxide with Al, Si, Mg, and/or B (Ref 1, 24)
- Reaction of the metal, oxide, or carbide with B_4C (Ref 1) or BN (Ref 25)
- Deposition by fused-salt electrolysis (Ref 1)
- Deposition from the vapor phase (Ref 1, 26–28)

Lower-temperature synthesis is also possible using methods such as single-crystal growth via aluminum fluxes (Ref 29, 30).

Borides have been fabricated by dry pressing, cold isostatic pressing, slip casting, tape casting, extrusion, injection molding, pressure casting, hot pressing, contained hot isostatic pressing, physical and chemical vapor deposition techniques (including ion implantation), explosive compaction, and reaction-forming processes. Although some of the data discussed below are for samples densified without external pressure (that is, pressureless sintering), the majority of the data have been obtained on samples fabricated by hot pressing. This is typical of the early stage of development of borides at the present time.

When comparing engineering property data, it should be realized that some properties are intrinsic to the material, that is, they are controlled by atomic structure. Theoretically, they can be calculated from a knowledge of atomic structure and deformation mechanisms, they are temperature and strain-rate dependent, and their magnitude is independent of specimen size and geometry, within a range of defined limits (Ref 31). Thermodynamic data, thermal expansion, melting point, theoretical density, Young's modulus, Poisson's ratio, fracture toughness, Curie temperature, and refractive index are examples of material properties. Other properties are very sensitive to stoichiometry and processing, including thermal conductivity, strength, hardness, creep, wear rate, electrical conductivity, optical transmission, magnetic susceptibility, and chemical resistivity.

Strength, for example, is dependent on specimen size and geometry, and is sensitive to processing-induced flaws. Improved methods of ceramic processing can be used to limit critical flaws that control strength (Ref 32). One can therefore expect improved mechanical properties for borides, similar to what has occurred for oxides, nitrides, and carbides, as more effort is focused on the processing of these materials. Composite materials that contain borides have already resulted in improved properties for many ceramic systems, as is discussed below, by relying on current methods for producing tough ceramics (Ref 33).

A review of factors that influence property values is given in the appendix of Ref 7. It cannot be overemphasized that many of the properties given below are dependent on the stoichiometry of the boride phase(s) present, grain size, porosity, impurities, test method, technique, and environment.

Engineering Property Data

Thermal Properties. Table 2 gives specific heat capacity, free enthalpy, and Gibbs free energy of formation (from the elements) data for selected borides. The data in Table 2 are from Pankratz et al. (Ref 34), who reviewed available data in the early 1980s and included only those data deemed reliable. The majority of the data are taken from Barin et al. (Ref 35). The enthalpies of formation are strongly correlated with Gibbs free energies of formation for borides, because the entropy term is small. This also means that the free energy is relatively insensitive to temperature, as shown in Table 2.

Enthalpies of formation at 298 K, reported by Samsonov and Vinitskii (Ref 12), are given in Table 3. It is clear that the stability of the

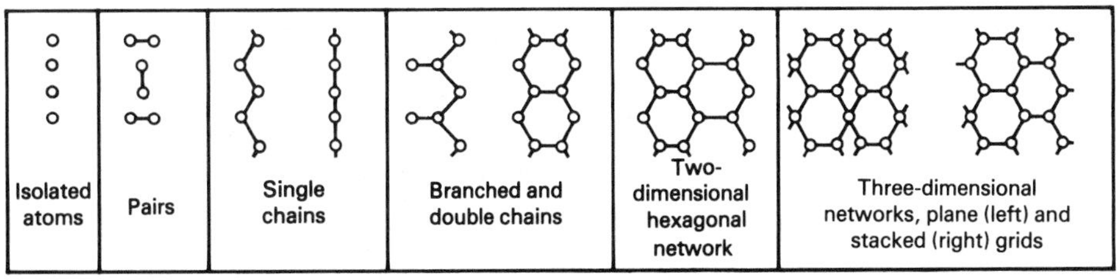

Isolated atoms	Pairs	Single chains	Branched and double chains	Two-dimensional hexagonal network	Three-dimensional networks, plane (left) and stacked (right) grids

Formula	Metal	Crystal system and structural type	Arrangement of boron atoms
M_4B	Pd, Pt	Cubic, Pt_4B-type	Isolated atoms
M_2B	Ta, Cr, Mo, W, Fe, Ni, Co	Tetragonal, $CuAl_2$-type	Isolated atoms
M_5B_3	Cr	Tetragonal, Cr_5B_3-type	
M_3B_2	V, Nb, Ta, Mo	Tetragonal, U_3Si_2-type	Isolated pairs
MB	V, Nb, Ta, Mo, Cr, W, Ni	Rhombic, CrB-type	Single chains
MB	Mo, W	Tetragonal, MoB-type	
MB	Ti, Hf, Fe, Co	Rhombic, FeB-type	
M_3B_4	V, Nb, Ta, Cr	Rhombic, Ta_3B_4-type	Double chains
MB_2	Sc, Y, Ti, Zr, Hf, V, Nb, Ta, Cr, Mo, W, U, Pu, Al	Hexagonal, AlB_2-type	Two-dimensional hexagonal network
M_2B_5	Mo, W	Hexagonal, Mo_2B_5-type	
MB_4	Y, lanthanides, actinides	Tetragonal, UB_4-type	Three-dimensional network
MB_6	Y, lanthanides, actinides	Cubic, CaB_6-type	Connected B_6 octahedrons
MB_{12}	Sc, Y, Zr, Mo, W, lanthanides, actinides	Cubic, UB_{12}-type	24 cross-linked boron atoms surround a central metal atom
MB_{66}	Y, Pu, lanthanides	YB_{66}-type	Three-dimensional network of B_{12} clusters

Fig 2 Structure of borides. Source: Ref 10, 13, 14

Table 1 Crystal structure, density, and melting point

Boride	Crystal structure	Lattice parameters a Å	b Å	c Å	β(°)	Theoretical density, g/cm³	Melting point °C	°F
AlB_2	Hexagonal	3.005	...	3.253	...	3.17	1654	3009
AlB_{10}	Orthorhombic	8.88	9.10	5.69	...	2.54	2421	4390
α-AlB_{12}	Tetragonal	10.16	...	14.28	...	2.58	2162	3924
β-AlB_{12}	Orthogonal	12.34	12.63	10.16	...	2.60	2212	4014
BaB_6	Cubic	4.262	...	...	...	4.35	2270	4118
δ-$Be_{5-x}B$	Tetragonal	3.368	...	7.050	...	1.94	1160	2120
Be_2B	Cubic	4.670	...	...	...	1.89	≈1500	≈2730
n-Be_2B_3	Tetragonal	7.25	...	8.46	...	...	...	...
BeB_2	Hexagonal	9.79	...	9.55	...	2.42	>1970	>3506
BeB_6	Hexagonal	10.16	...	14.28	...	2.33	2020–2120	3668–3848
δ-BeB_{12}	Hexagonal	5.46	...	12.42	...	2.42	2300	4170
CaB_6	Cubic	4.154	...	...	...	2.43	2235	4055
CeB_4	Tetragonal	7.202	...	4.093	...	5.74	2380	4316
CeB_6	Cubic	4.141	...	...	...	4.79	2550	4620
Co_4B	Orthogonal	4.408	5.220	6.630	...	...	...	...
$Co_{23}B_6$	Cubic	11.05	...	...	...	6.99	...	...
Co_3B	Orthogonal	4.408	5.223	6.629	...	8.13	1110	2030
Co_3B	Tetragonal	5.015	...	4.220	...	8.05	1260	2246
CoB	Orthogonal	3.948	5.243	3.037	...	7.32	...	...
Cr_4B	Orthogonal	4.26	7.38	14.71	...	6.28	1650	3002
Cr_2B	Orthogonal	7.409	14.712	4.250	...	6.58	1870	3398
Cr_5B_3	Tetragonal	5.370	...	10.188	...	6.61	1900	3452
CrB	Tetragonal	2.94	...	15.72	...	6.14	...	...
δ-CrB	Orthogonal	2.967	7.867	2.932	...	6.10	2100	3812
Cr_2B_3	Orthogonal	3.026	18.115	2.954	...	5.60	...	...
Cr_3B_4	Orthogonal	2.986	13.02	2.952	...	5.76	2070	3758
CrB_2	Hexagonal	2.973	...	3.071	...	5.20	2200	3992
CrB_4	Orthogonal	4.744	5.477	2.866	...	4.25	1400–1600	2552–2912
CrB_{41}	Orthogonal	10.964	...	23.848	...	2.65	...	...
CuB_{24}	Orthogonal	10.980	...	23.925	...	2.82	...	...
DyB_2	Hexagonal	3.287	...	3.847	...	8.49	2100	3812
DyB_4	Tetragonal	7.097	...	4.017	...	6.76	2500	4532
DyB_6	Cubic	4.097	...	...	...	5.49	2200	3992
DyB_{12}	Cubic	7.500	...	...	...	4.60	2100	3812
DyB_{66}	Cubic	23.45	...	...	...	2.71	2150	3902
ErB_2	Hexagonal	3.28	...	3.79	...	8.88	2185	3965
ErB_4	Tetragonal	7.068	...	3.993	...	7.01	2500	4532
ErB_{12}	Cubic	7.484	...	...	...	4.708	2080	3776
ErB_{66}	Cubic	23.33	...	...	...	2.73	2150	3902
EuB_6	Cubic	4.17	...	...	...	4.99	≥2580	≥4676
$Fe_{23}B_6$	Cubic	10.67	...	...	...	7.38	...	...

(continued)

diborides decreases in the order $HfB_2 > TiB_2$, $ZrB_2 \gg TaB_2$, $VB_2 > UB_2 \gg CrB_2$, MnB_2, $MgB_2 > AlB_2$. Bolgar *et al.* (Ref 36) report that the heat capacity of NbB_2 is 47.78 J/mol · K (11.4 Btu/lb · mol · °F), in accord with the data in Table 2 for VB_2 and TaB_2. The stability of lanthanum hexaborides, LaB_6, decreases in order of decreasing atomic number (that is, Lu > Yb > Tm > Er > Ho > Dy > Tb > Gd > Eu > Sm > Pm > Nd > Pr > Ce) or increases with increasing numbers of 4f electrons. Lanthanum hexaboride has similar stability to EuB_6. The alkaline hexaborides decrease in stability with decreasing atomic size ($BaB_6 > SrB_6 > CaB_6 > MgB_6$). Free energy of formation data for a number of borides, based on specific boro-thermic or carbothermic reductions of oxides, are tabulated by Samsonov and Vinitskii (Ref 12). The paucity of reliable thermodynamic data for borides makes thermochemical stability predictions impossible for many systems.

Thermal expansion and thermal conductivity data are given in Table 4. "Linear" coefficients of thermal expansion of ceramics are not strictly linear, and the coefficients generally increase with increasing temperature. It is therefore extremely important to use thermal expansion coefficients for the temperature range of interest and to compare similar temperature ranges when comparing thermal expansion coefficients for different materials. In addition, thermal expansion coefficients for anisotropic materials are dependent on orientation for single crystals.

The important implication for polycrystalline materials is that there is a critical size for microcracking (Ref 37). For example, hexagonal TiB_2 will microcrack when the grain size exceeds ≈ 15 μm (0.6 mil) (Ref 26, 38). Microcracking, which is generally deleterious to mechanical properties, decreases modulus, strength, and hardness. In some ceramic systems, microcracking increases fracture toughness, although multiple mechanisms are usually operative in such instances (Ref 33).

Thermal expansion coefficients for borides cover a wide spectrum of values, as seen in Table 4. A number of borides have thermal expansion coefficients in the 6 to 8×10^{-6}/K range at temperatures between 20 and 1000 °C (68 and 1830 °F). For reference, these thermal expansion values are greater than SiC and less than or equal to Al_2O_3. The borides with low thermal expansion values ($< 4 \times 10^{-6}$/K), shown in Table 4, do not have temperature ranges associated with the measurements, and such values should not be used for engineering design. Linear coefficients of expansion for the transition borides TiB_2, ZrB_2, and HfB_2 are shown in Fig 4 (Ref 7).

Thermal conductivities of borides are generally high, relative to many other ceramics. The high thermal conductivities result from both a lattice and an electronic contribution to phonon transport. The high-temperature (2027 K) data of Samsonov (Ref 12) suggest that the conductivities of transition metal and

Table 1 continued

| Boride | Crystal structure | Lattice parameters | | | | Theoretical density, g/cm³ | Melting point | |
		aÅ	bÅ	cÅ	β(°)		°C	°F
Fe₃.₅B	Tetragonal	8.62	...	4.28	...	...	...	...
Fe₃B	Tetragonal	8.674	...	4.313	...	7.30	...	...
Fe₂B	Tetragonal	5.132	...	8.532	...	7.34	1410	2570
FeB	Orthorhombic	4.059	5.503	2.947	...	6.73	1650	3002
FeB₄₉	Orthorhombic	10.951	...	23.861	...	2.35	...	...
GdB₂	Hexagonal	3.318	...	3.933	...	7.92	2050	3722
Gd₂B₅	Monoclinic	7.181	7.193	7.196	102.16	6.74	2100	3812
GdB₄	Tetragonal	7.133	...	4.047	...	6.47	2650	4802
GdB₆	Cubic	4.107	...	...	...	5.32	2510	4550
GdB₁₂	Cubic	7.524	...	...	...	4.475	...	...
GdB₆₆	Cubic	23.47	...	...	...	2.68	2150	3902
HfB	Orthogonal	6.517	3.218	4.919	...	12.19	2100 ± 20	3812 ± 36
HfB₂	Hexagonal	3.142	...	3.476	...	11.19	3380 ± 20	6116 ± 36
HfB₁₂	Cubic	7.377	...	...	...	...	...	...
HoB₂	Hexagonal	3.281	...	3.811	...	8.72	2200	3992
HoB₄	Tetragonal	7.087	...	4.008	...	6.87	2500	4532
HoB₆	Cubic	4.095	...	...	...	5.56	2180	3956
HoB₁₂	Cubic	7.492	...	...	...	4.67	2100	3812
HoB₆₆	Cubic	23.38	...	...	...	2.74	2025	3677
IrB₀.₉	Hexagonal	2.81	...	2.81	...	17.39	...	...
IrB₁.₁₅	Tetragonal	2.810	...	10.26	...	16.73	...	...
IrB₁.₃₅	Monoclinic	10.525	2.910	6.099	91.07	17.25	...	...
KB₆	Cubic	4.233	...	...	...	2.28	...	...
LaB₃	Tetragonal	3.82	...	3.96	...	4.92	...	...
LaB₄	Tetragonal	7.323	...	4.181	...	5.40	...	...
LaB₆	Cubic	4.157	...	...	...	4.71	2715	4919
LuB₂	Hexagonal	3.246	...	3.704	...	9.76	2250	4082
LuB₄	Tetragonal	7.036	...	3.974	...	7.37	2550	4622
LuB₁₂	Cubic	7.464	...	...	...	4.87	2170	3938
LuB₆₆	Cubic	23.412	...	...	...	2.76	2100	3812
MgB₂	Hexagonal	3.086	...	3.522	...	2.63	...	...
MgB₄	Orthogonal	5.464	7.472	4.428	...	...	...	...
MgB₆	Tetragonal	7.07	...	6.45	...	...	...	...
Mn₄B	Orthogonal	14.53	7.293	4.209	...	6.87	1285	2345
Mn₂B	Tetragonal	5.149	...	4.209	...	7.18	1580	2876
MnB	Orthogonal	4.147	5.561	2.977	...	6.36	1890	3434
Mn₃B₄	Orthogonal	3.302	12.86	2.960	...	5.98	1750	3180
MnB₂	Hexagonal	3.009	...	3.037	...	5.34	1988	3610
MnB₄	Monoclinic	5.503	5.367	2.949	122.71	4.45	2160	3920
β-MnB₄	Tetragonal	6.28	...	8.38	...	...	...	...
Mo₂B	Tetragonal	5.547	...	4.739	...	9.23	2280 ± 12	4136 ± 22
α-MoB	Tetragonal	3.105	...	16.97	...	8.77	...	...
β-MoB	Orthogonal	3.16	8.61	3.08	...	...	2600 ± 8	4712 ± 14
MoB₂	Hexagonal	3.04	...	3.07	...	7.99	2375 ± 15	4307 ± 27
Mo₂B₅	Orthogonal	3.012	...	20.937	...	7.45	2140 ± 15	3884 ± 27
Mo₀.₈B₃	Hexagonal	5.203	...	6.349	...	4.87	...	...
Mo₃B₂	Tetragonal	6.00	...	3.15	...	9.07	2070	3758
MoB₄	Hexagonal	5.214	...	6.358	...	6.18	1800	3272
MoB₁₂	Hexagonal	3.004	...	3.174	...	...	2020	3668
Nb₃B₂	Tetragonal	6.185	...	3.280	...	7.95	2080 ± 40	3776 ± 72
Nb₃B₄	Orthogonal	3.312	14.11	3.143	...	7.28	2935 ± 12	5315 ± 22
δ-NbB	Orthogonal	3.297	8.723	3.166	...	7.57	2917 ± 10	5283 ± 18
ε-NbB₂	Hexagonal	3.089	...	3.303	...	7.00	3036 ± 15	5497 ± 27
NdB₄	Tetragonal	7.217	...	4.102	...	5.83	2350	4262
NdB₆	Cubic	4.126	...	...	...	4.95	2610	4730
NdB₆₆	Cubic	23.42	...	...	...	2.63	2150	3902
Ni₃B	Orthogonal	5.211	6.619	4.389	...	8.20	1175	2147
Ni₂B	Tetragonal	4.991	...	4.247	...	8.05	1225	2237
Ni₄B₃	Monoclinic	6.430	4.882	7.818	103.3	7.43	...	...
Ni₄B₃	Orthogonal	11.96	2.98	6.57	...	...	...	...
NiB	Orthogonal	2.936	7.38	2.968	...	7.17	1590	2894
NiB₁₂	Cubic	7.385	...	...	...	...	2320	4208
NpB₂	Hexagonal	3.165	...	3.975	...	12.47	...	...
OsB₁.₁	Hexagonal	2.876	...	2.871	...	16.32	...	...
Os₂B₃	Hexagonal	2.910	...	12.910	...	14.48	...	...
OsB₂	Orthogonal	4.684	2.872	4.076	...	12.83	...	...
OsB₂	Hexagonal	2.876	...	2.871	...	17.10	...	...
Os₂B₅	Hexagonal	2.91	...	12.91	...	15.23	...	...
Pd₂B	Orthogonal	4.692	5.127	3.110	...	9.93	...	...
Pd₃B₂	Hexagonal	6.49	...	3.43	...	18.22	≈1020	≈1868
PmB₆	Cubic	4.128	...	...	...	...	...	...
PrB₄	Tetragonal	6.099	...	12.063	...	5.01	2350	4262
PrB₆	Cubic	4.132	...	...	...	4.84	2610	4730
PtB	Hexagonal	3.358	...	4.058	...	17.25	≈920	≈1688
PuB	Cubic	4.918	...	...	...	13.94	2050	3722
PuB₂	Hexagonal	3.19	...	3.90	...	12.67	1825	3317
PuB₄	Tetragonal	7.10	...	4.014	...	9.27	2050	3722
PuB₆	Cubic	4.120	...	...	...	7.57	2100	3812

(continued)

Table 1 continued

Boride	Crystal structure	aÅ	bÅ	cÅ	β(°)	Theoretical density, g/cm³	°C	°F
PuB_{66}	Cubic	23.43	...	...	...	...	...	...
Re_3B	Orthogonal	2.890	9.313	7.258	...	19.36	≈2150	≈3902
Re_7B_3	Hexagonal	7.50	...	4.88	...	18.63	≈2000	≈3632
ReB_2	Hexagonal	2.900	...	7.478	...	12.68	≈2400	≈4352
Rh_7B_3	Hexagonal	7.471	...	4.777	...	...	...	...
Rh_5B_4	Hexagonal	3.306	...	20.394	...	9.60	...	...
RhB	Hexagonal	3.309	...	4.224	...	...	...	...
Ru_7B_3	Hexagonal	7.467	...	4.714	...	...	≈1660	≈3020
$RuB_{=1.1}$	Hexagonal	2.852	...	2.855	...	...	1500	2730
$Ru_{11}B_8$	Orthogonal	11.60	11.34	2.83	...	10.69	...	...
$RuB_{1.1}$	Hexagonal	2.852	...	2.855	...	9.38	...	...
Ru_2B_3	Hexagonal	2.905	...	12.810	...	8.32	1550	2822
RuB_2	Hexagonal	2.852	...	2.855	...	10.14	1600	2912
Ru_2B_5	Hexagonal	2.89	...	12.81	...	9.19	...	...
ScB_2	Hexagonal	3.146	...	3.518	...	3.67	2250	4082
ScB_{12}	Tetragonal	5.22	...	7.35	...	2.89	2040	3704
$Sc_{11}B_{305}$	Orthogonal	10.965	...	24.087	...	2.51	...	...
SmB_2	Hexagonal	3.310	...	4.019	...	7.49	...	...
Sm_2B_5	Monoclinic	7.183	7.191	7.216	102.03	6.47	≈1980	≈3596
SmB_4	Tetragonal	7.167	...	4.070	...	6.15	2400	4352
SmB_6	Cubic	4.132	...	...	...	5.06	2580	4676
SmB_{66}	Cubic	23.48	...	...	...	2.66	2150	3902
SrB_6	Cubic	4.193	...	...	...	3.44	2230	4046
Ta_2B	Tetragonal	5.783	...	4.866	...	15.21	2417 ± 15	4383 ± 27
Ta_3B_2	Tetragonal	6.184	...	3.286	...	14.92	2180 ± 20	3956 ± 36
TaB	Orthogonal	3.280	8.671	3.156	...	14.19	3090 ± 15	5594 ± 27
$δ-Ta_3B_4$	Orthogonal	3.29	14.0	3.13	...	13.60	2990 ± 20	5414 ± 36
TaB_2	Hexagonal	3.098	...	3.227	...	12.54	3037 ± 20	5499 ± 36
TbB_2	Hexagonal	3.280	...	3.860	...	8.34	2100	3812
TbB_4	Tetragonal	7.119	...	4.029	...	6.58	2600	4712
TbB_6	Cubic	4.105	...	...	...	5.37	2340	4244
TbB_{12}	Cubic	7.504	...	...	...	4.54	2200	3992
TbB_{66}	Cubic	23.43	...	...	...	2.70	2100	3812
TcB_2	Hexagonal	2.900	...	7.475	...	7.30	...	...
ThB_4	Tetragonal	7.261	...	4.114	...	8.43	>2200	>3992
ThB_6	Cubic	4.112	...	...	...	7.09	2150	3902
ThB_{12}	Cubic	7.612	...	...	...	...	...	...
ThB_{66}	Cubic	23.46	...	...	...	2.92	...	...
TiB	Orthogonal	6.12	3.06	4.56	...	4.56	2190 ± 25	3974 ± 45
Ti_3B_4	Orthogonal	3.259	13.73	3.042	...	4.56	...	...
TiB_2	Hexagonal	3.030	...	3.230	...	4.52	3225 ± 20	5837 ± 36
Ti_2B_5	Hexagonal	2.98	...	13.98	...	4.63	...	...
$Ti_{1.87}B_{50}$	Tetragonal	8.830	...	5.072	...	2.65	...	...
TmB_2	Hexagonal	3.250	...	3.739	...	9.25	2250	4082
TmB_4	Tetragonal	7.0572	...	3.9882	...	7.10	2550	4622
TmB_{12}	Cubic	7.476	...	...	...	4.76	2180	3956
TmB_{66}	Cubic	23.433	...	...	...	2.73	2100	3812
UB	Cubic	4.88	...	...	...	14.22	...	...
UB_2	Hexagonal	3.131	...	3.987	...	12.69	2385	4325
UB_4	Tetragonal	7.075	...	3.979	...	9.38	2495	4523
UB_{12}	Cubic	7.473	...	...	...	5.85	2235	4055
V_3B_2	Tetragonal	5.746	...	3.032	...	5.75	1900 ± 12	3452 ± 22
VB	Orthogonal	3.060	8.048	2.972	...	5.60	2570 ± 15	4658 ± 27
V_5B_6	Orthogonal	3.058	21.25	2.974	...	...	...	...
V_3B_4	Orthogonal	3.030	13.18	2.986	...	5.43	2610 ± 12	4730 ± 22
V_2B_3	Orthogonal	3.061	18.40	2.984	...	...	...	...
VB_2	Hexagonal	2.998	...	3.056	...	5.07	2747 ± 15	4977 ± 27
$V_{1.54}B_{50}$	Tetragonal	8.824	...	5.072	...	2.60	...	...
W_2B	Tetragonal	5.568	...	4.744	...	17.09	2670 ± 16	4838 ± 29
$β-WB$	Orthogonal	3.19	8.40	3.07	...	...	2665 ± 16	4829 ± 29
$δ-WB$	Tetragonal	3.117	...	16.910	...	15.74	...	...
W_2B_5	Orthogonal	2.982	...	20.715	...	13.17	2365 ± 15	4289 ± 27
WB_4	Hexagonal	5.202	...	6.333	...	8.40	2020 ± 30	3668 ± 54
WB_{12}	Hexagonal	3.994	...	3.174	...	...	...	...
YB_2	Hexagonal	3.78	...	4.40	...	3.37	2100	3812
YB_4	Tetragonal	7.086	...	4.012	...	4.36	2800	5072
YB_6	Cubic	4.100	...	...	...	3.72	2600	4712
YB_{12}	Cubic	7.502	...	...	...	3.44	2200	3992
YB_{66}	Cubic	23.34	...	...	...	2.48	2100	3812
YbB_2	Hexagonal	3.250	...	3.732	...	9.47	≈1500	≈2732
YbB_4	Tetragonal	7.04	...	4.00	...	7.24	≈1850	≈3362
YbB_6	Cubic	4.146	...	...	...	5.55	2370	4298
YbB_{12}	Cubic	7.469	...	...	...	...	2200	3992
YbB_{66}	Cubic	23.30	...	...	...	2.75	2150	3902
ZrB	Cubic	4.65	...	...	...	6.48	2800	5072
ZrB_2	Hexagonal	3.169	...	3.530	...	6.10	3245 ± 18	5873 ± 32
ZrB_{12}	Cubic	7.408	...	...	...	3.63	2250 ± 40	4082 ± 72
Zr_6B_{311}	Orthogonal	10.956	...	24.020	...	2.60	...	...

lanthanide borides increase sharply at high temperature. These data should be treated with caution, because reliable diffusivity measurements at high temperatures are difficult, at best. Thermal conductivity data for TiB_2 are given in Fig 5 as a function of temperature (Ref 39). These data suggest that TiB_2 is an excellent choice for heat dissipation at high temperatures.

Mechanical Properties. Borides are known to be very hard because of their covalent bonding. Hardness is the only mechanical property that is well characterized for the majority of borides (Table 5a). The use of TiB_2 as an effective armor material is partly due to its high hardness and relatively low specific gravity, although armor materials are complex and difficult to model based on mechanical properties (Ref 41). Indentation loads used for hardness testing are given in Table 5(a), because hardness generally decreases with increasing indentation load. The wide scatter in hardness values from different researchers at the same load is primarily due to differences in grain size and porosity within materials of a given composition. Hardness data for single crystals are given in Ref 8. Hardness as a function of test temperature is shown in Fig 6 for HfB_2, TiB_2, and ZrB_2 (Ref 7).

The modulus of borides is generally high, as expected, because of the high hardness and high degree of covalent bonding. Molybdenum and tungsten borides have very high moduli values (675 to 775 GPa, or 98 to 112 $\times 10^6$ psi), comparable to tungsten carbide. The Young's moduli of TiB_2, ZrB_2, and HfB_2 are all very similar, linearly decreasing from 500 GPa (73 $\times 10^6$ psi) at room temperature to 460 GPa (67 $\times 10^6$ psi) at 1000 °C (1830 °F) (Ref 44). The shear modulus decreased from 225 GPa (33 $\times 10^6$ psi) at room temperature to 205 GPa (30 $\times 10^6$ psi) at 1000 °C (1830 °F) for the same three materials (Ref 44). Poisson's ratio of the same three transition diborides, calculated from Young's and shear moduli, increased slightly from room temperature to 1000 °C (1830 °F), with values between 0.10 and 0.13 (Ref 44). Bulk modulus, calculated from Young's modulus and Poisson's ratio, decreased from 210 GPa (30 $\times 10^6$ psi) at room temperature to 200 GPa (29 $\times 10^6$ psi) at 1000 °C (1830 °F) (Ref 44). Young's modulus data for high-purity TiB_2 as a function of temperature are shown in Fig 7.

The compressive strengths of borides are generally uncharacterized, but, like other ceramics, values of 3 to 10 times the flexural strength are expected. The compressive strength values of 1.3 to 1.6 GPa (0.19 to 0.23 $\times 10^6$ psi) reported by Samsonov and Vinitskii (Ref 12) for CrB_2, TiB_2, and ZrB_2 are more realistic than the 5.7 GPa (0.82 $\times 10^6$ psi) reported for TiB_2 (Ref 41).

The flexural strengths of borides vary widely, as expected, because of the variety of factors that affect strength measurements

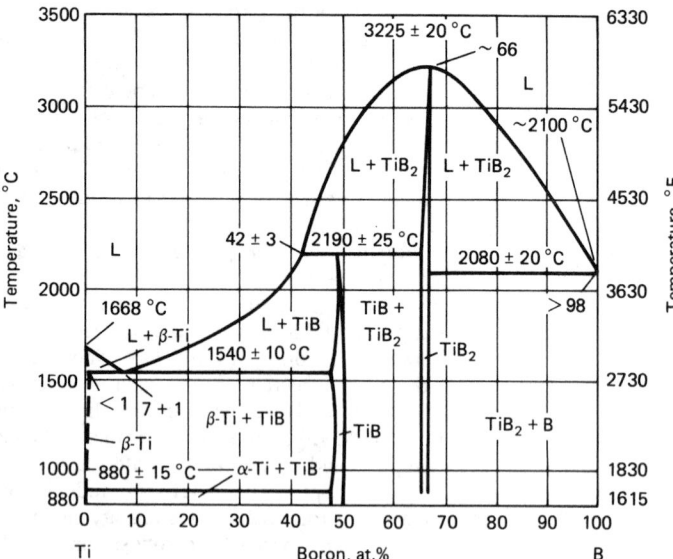

Fig 3(a) Binary phase diagram of Ti-B. Source: Ref 18

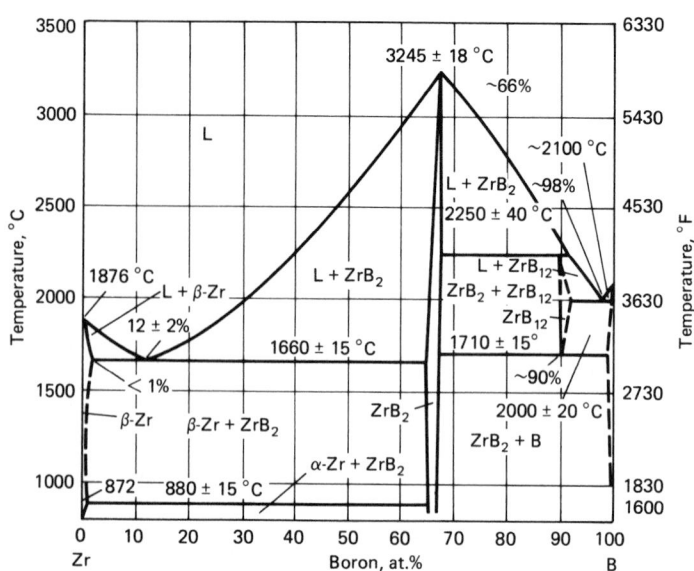

Fig 3(b) Binary phase diagram of Zr-B. Source: Ref 18

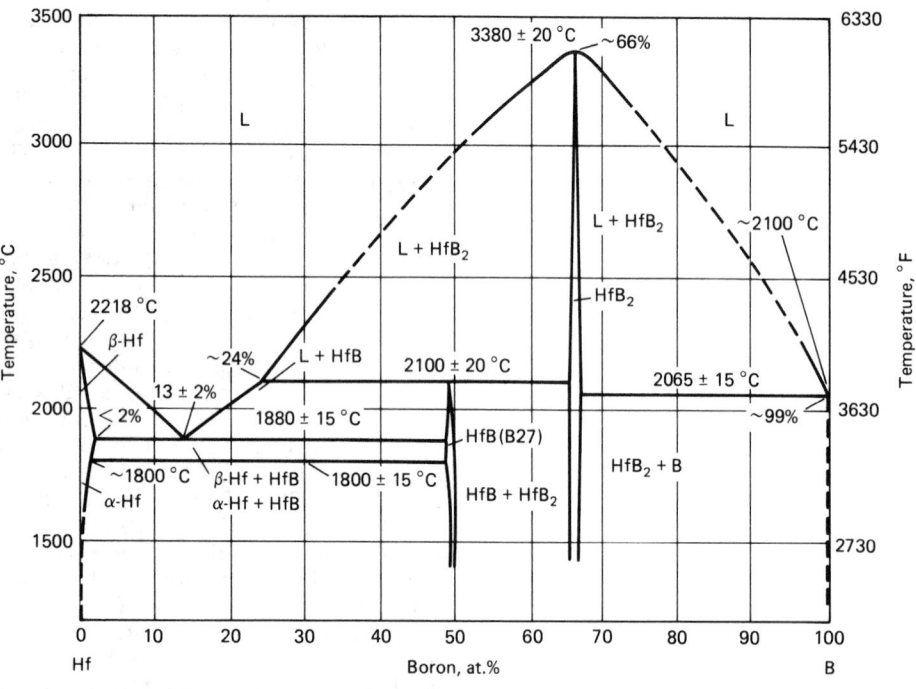

Fig 3(c) Binary phase diagram of Hf-B. Source: Ref 18

(Ref 7). Values for sintered or hot pressed TiB_2 are typically in the 300 to 400 MPa (44 to 58 ksi) range. The fact that significant improvements in flexural strength for TiB_2 can be made is obvious, based on the unpublished work of the author, which resulted in average four-point flexural strengths of 565 ± 75 MPa (82 ± 11 ksi) for pressureless sintered and unclad hot isostatically pressed material. The higher strength is due to smaller pores and fine grain size. It should be realized that significant improvements in the strengths listed in Table 5(b) will result from improved processing of these materials. The

strength of HfB_2, TiB_2, and ZrB_2, as a function of temperature, are shown in Fig 8.

Weibull moduli values of 11.3 and 28.7 were reported for sintered and hot pressed TiB_2, respectively (flexural strengths were 350 and 400 MPa, or 51 and 58 ksi, respectively) (Ref 45). Because flaw populations influence Weibull moduli, care should be taken in assuming high Weibull moduli ($m > 15$) for materials that have only been developed on a laboratory scale.

Zirconium diboride exposed to cyclic loading (2 to 8 million cycles) at 1800 Hz (at flexural stress levels of 85% of the average

static strength of the material) showed no degradation in strength (Ref 46). However, more recent data on a wide variety of ceramics has indicated that ceramics are susceptible to cyclic fatigue (Ref 47).

Dynamic fatigue data for TiB_2 are unclear, with large positive slow crack growth exponents reported for TiB_2 and both positive and negative exponents exhibited for TiB_2 in molten aluminum environments (Ref 43). Slow crack growth in the absence of molten aluminum is consistent with dynamic fatigue data for other ceramic materials.

Fracture toughness values have only been measured for TiB_2. Bulk toughness values range from 4.9 to 6.0 MPa$\sqrt{m}$ (4.5 to 5.5 ksi$\sqrt{in}$.) (Ref 26, 43), which are considerably lower than indentation or single-edge notched beam values of 6.7 to 8.0 MPa$\sqrt{m}$ (6.1 to 7.3 ksi$\sqrt{in}$.). Although borides are expected to be relatively brittle, recent work on nitride, carbide, and oxide ceramics have shown that considerable improvement in toughness can be achieved with focused research (Ref 33). Titanium boride is embrittled immediately upon exposure to liquid aluminum environments, with fracture toughness decreasing approximately 15% (Ref 43).

A coefficient of friction of 0.4 for fine-grained (0.8 μm, or 32 μin.) TiB_2 has been reported at room temperature, with a linear decrease to 0.2 at 1000 °C (1830 °F) (Ref 48). Borides generally have excellent wear and erosion resistance.

High-temperature deformation of TiB_2 under vacuum has been investigated by Ramberg *et al.* (Ref 49, 50). With the exception of high-purity SiC, the yield strength of TiB_2 is greater than any other ceramic compound for which such data exist. A high-temperature plastic yielding mechanism was observed, and a Hall-Petch dependency on grain size (Fig 9) resulted in an activation energy of 308 kJ/mol (292

Table 2 Thermodynamic data for borides

| | Heat capacity at constant pressure, C_p | | | | | | Enthalpy of formation, ΔH_f(a) | | | | | |
| | At 298 K | | At 1000 K | | At 2000 K | | At 298 K | | At 1000 K | | At 2000 K | |
Boride	J/mol·K	Btu/mol·°F	J/mol·K	Btu/mol·°F	J/mol·K	Btu/mol·°F	kJ/mol	Btu/mol	kJ/mol	Btu/mol	kJ/mol	Btu/mol
AlB$_2$	43.64	0.02298	78.21	0.04118	...	...	−66.94	−63.53	−81.04	−76.91	...	...
AlB$_{12}$	149.58	0.078769	317.82	0.16736	439.28	0.23132	−200.83	−190.58	−223.12	−211.74	−211.55	−200.76
CeB$_6$	103.31	0.054403	165.81	0.087316	198.65	0.10461	−351.46	−333.54	−360.62	−342.23	−389.18	−369.33
CoB	34.63	0.01824	56.45	0.02973	...	...	−94.14	−89.34	−96.14	−91.24	...	...
Co$_2$B	58.98	0.03106	89.28	0.04701	...	...	−125.52	−119.12	−128.68	−122.12	...	...
CrB	35.82	0.01886	57.35	0.03020	...	...	−75.31	−71.47	−74.86	−71.04	...	...
CrB$_2$	53.59	0.02822	85.07	0.04480	...	...	−94.14	−89.34	−94.02	−89.22	...	...
FeB	50.21	0.02644	54.72	0.02882	...	...	−71.13	−67.50	−72.96	−69.24	...	...
Fe$_2$B	75.33	0.03967	87.78	0.04622	...	...	−71.13	−67.50	−77.82	−73.85	...	...
HfB$_2$	49.45	0.02604	81.67	0.04301	...	...	−335.98	−318.85	−334.90	−321.62	...	...
MgB$_2$	47.82	0.02518	71.71	0.03776	...	...	−92.05	−87.36	−106.55	−101.12	...	...
MgB$_4$	70.17	0.03695	115.58	0.06086	...	...	−105.02	−99.66	−124.48	−118.13	...	...
MnB	35.82	0.01886	57.35	0.03020	...	...	−75.31	−71.47	−79.94	−75.86	...	...
MnB$_2$	53.62	0.02824	85.10	0.04481	...	...	−94.14	−89.34	−99.08	−94.03	...	...
NiB	34.63	0.01824	56.45	0.02973	...	...	−100.42	−95.30	−102.36	−97.14	...	...
Ni$_4$B$_3$	128.12	0.067468	201.32	0.10602	...	...	−311.71	−295.81	−318.79	−302.53	...	...
TaB$_2$	48.12	0.02534	76.77	0.04043	96.71	0.05093	−209.20	−198.53	−209.77	−199.07	−208.02	−197.41
TiB	29.67	0.01562	51.92	0.02734	53.40	0.02812	−160.25	−152.08	−161.85	−153.60	−187.00	−177.46
TiB$_2$	44.28	0.02332	76.89	0.04049	94.54	0.04978	−323.84	−307.32	−326.59	−309.93	−347.87	−330.13
UB$_2$	57.29	0.03017	87.75	0.04621	123.71	0.06516	−164.43	−156.04	−165.31	−156.88	−177.82	−168.75
UB$_4$	78.91	0.04155	137.49	0.07240	...	...	−245.60	−233.07	−248.34	−235.67	...	...
UB$_{12}$	169.58	0.089301	311.12	0.16384	...	...	−433.04	−410.95	−440.10	−417.65	...	...
V$_3$B$_2$	97.10	0.05113	142.09	0.074824	194.84	0.10260	−303.76	−288.27	−303.62	−288.14	−298.47	−283.25
VB	36.02	0.01897	55.74	0.02935	75.28	0.03964	−138.49	−131.43	−138.44	−131.38	−136.74	−129.77
V$_5$B$_6$	191.05	0.10061	303.80	0.15998	407.43	0.21455	−763.58	−724.64	−763.31	−724.38	−754.80	−716.31
V$_3$B$_4$	119.01	0.062671	192.32	0.10128	256.85	0.13526	−486.60	−461.76	−486.43	−461.62	−481.34	−456.79
V$_2$B$_3$	83.00	0.04371	136.59	0.071928	181.57	0.09561	−345.18	−327.58	−345.06	−327.46	−341.68	−324.25
VB$_2$	46.98	0.02474	80.85	0.04258	160.28	0.08440	−203.76	−193.37	−203.69	−193.30	−202.02	−191.72
ZrB$_2$	48.37	0.02547	71.99	0.03791	82.66	0.04353	−322.59	−306.14	−326.65	−309.99	−340.36	−323.00

| | Gibbs free energy of formation, ΔG_f(a) | | | | | |
| | At 298 K | | At 1000 K | | At 2000 K | |
Boride	kJ/mol	Btu/mol	kJ/mol	Btu/mol	kJ/mol	Btu/mol
AlB$_2$	−65.33	−62.00	−58.83	−55.83	...	...
AlB$_{12}$	−206.70	−196.16	−213.62	−202.73	−205.26	−194.79
CeB$_6$	−341.53	−324.11	−318.08	−301.86	−262.45	−249.06
CoB	−96.14	−91.24	−88.04	−83.55	...	...
Co$_2$B	−123.69	−117.38	−118.36	−112.32	...	...
CrB	−73.68	−69.92	−70.43	−66.84	...	...
CrB$_2$	−91.37	−86.71	−85.52	−81.16	...	...
FeB	−69.49	−65.95	−68.05	−64.58	...	...
Fe$_2$B	−71.75	−68.09	−68.19	−64.71	...	...
HfB$_2$	−332.20	−315.26	−324.49	−307.94	...	...
MgB$_2$	−89.52	−84.95	−80.40	−76.30	...	...
MgB$_4$	−103.72	−98.43	−94.66	−89.84	...	...
MnB	−73.68	−69.92	−68.68	−65.18	...	...
MnB$_2$	−91.37	−86.71	−83.79	−79.52	...	...
NiB	−98.73	−93.69	−93.25	−88.49	...	...
Ni$_4$B$_3$	−304.98	−289.43	−283.50	−269.04	...	...
TaB$_2$	−206.53	−195.00	−200.18	−189.97	−191.02	−181.28
TiB	−159.71	−151.56	−157.35	−149.33	−146.93	−139.44
TiB$_2$	−319.69	−303.39	−308.34	−292.61	−347.87	−330.13
UB$_2$	−162.34	−154.06	−161.46	−153.23	−149.39	−141.77
UB$_4$	−244.80	−232.32	−245.37	−232.86	...	...
UB$_{12}$	−438.63	−416.26	−455.88	−432.63	...	...
V$_3$B$_2$	−300.29	−284.98	−292.06	−277.16	−281.90	−267.52
VB	−136.84	−129.86	−132.94	−126.16	−127.93	−121.41
V$_5$B$_6$	−732.60	−695.24	−659.69	−626.05	−558.53	−530.04
V$_3$B$_4$	−480.12	−455.64	−464.89	−441.18	−444.84	−422.15
V$_2$B$_3$	−340.36	−323.00	−329.06	−312.28	−314.06	−298.04
VB$_2$	−200.60	−190.37	−193.19	−183.34	−183.20	−173.86
ZrB$_2$	−318.16	−301.93	−306.34	−290.72	−279.60	−265.34

(a) In standard state, pure phase, at 0.1 MPa (1 atm). Source: Ref 34

Btu/mol) (Ref 50). Up to five independent slip systems were observed using a constant stress of 900 MPa (130 ksi) at temperatures between 1600 and 2000 °C (2910 and 3630 °F) (Ref 50). The hardness of single-crystal diborides decreases nearly linearly with temperature, resulting in values that are one-half to one-third of their original values at 40% of their melting temperature (Ref 51, 52).

The thermal shock resistance of diborides is excellent, primarily because of their high thermal conductivity. Kalish et al. (Ref 42) used bend strength, elastic modulus, thermal expansion, and thermal conductivity measurements to determine thermal stress resistance parameters for ZrB$_2$ and HfB$_2$. Under steady-state heat flow conditions, the calculated thermal stress parameters were higher

Table 3 Enthalpies of formation for selected borides

Boride	Free enthalpy of formation(a) kJ/mol	Btu/mol	Boride	Free energy of formation(a) kJ/mol	Btu/mol
BaB$_6$	−333.0	−316	Mo$_3$B$_2$	−175.7	−166.7
Be$_4$B	−78.7	−74.7	MoB	−68.2	−64.7
Be$_2$B	−69.9	−66.3	MoB$_2$	−96.2	−91.3
BeB$_2$	−64.9	−61.6	Mo$_2$B$_5$	−209.2	−198.5
BeB$_4$	−87.0	−82.6	NbB$_2$	−246.9	−234.3
BeB$_6$	−108.8	−103.3	NdB$_6$	−430.1	−408.2
BeB$_9$	−161.5	−153.3	PmB$_6$	−442.2	−419.6
CaB$_6$	−119.7	−113.6	PrB$_6$	−416.7	−395.4
CeB$_6$	−338.9	−321.6	ScB$_2$	−264.8	−251.3
CrB$_2$	−125.5	−119.1	SmB$_6$	−454.4	−431.3
DyB$_6$	−502.1	−476.5	SrB$_6$	−210.9	−200.1
ErB$_6$	−527.2	−500.3	TaB$_2$	−192.5	−182.7
EuB$_6$	−469.4	−445.5	TbB$_6$	−492.9	−467.8
FeB	−38.5	−36.5	ThB$_4$	<−217	<−206
GdB$_6$	−479.9	−445.4	ThB$_6$	<−276	<−262
HfB	−196.6	−186.6	TiB$_2$	−292.9	−278
HfB$_2$	−358.1	−339.8	Ti$_2$B$_5$	<−439	<−417
HoB$_6$	−514.6	−488.4	TmB$_6$	−539.7	−512.2
LaB$_6$	−469.9	−445.9	VB	−129.7	−123.1
LuB$_6$	−560.7	−532.1	VB$_2$	−259.4	−246.2
Mg$_4$B	−70.3	−66.7	W$_2$B	−100.4	−95.3
Mg$_2$B	−59.0	−56.0	WB	−71.1	−67.5
MgB$_2$	−55.6	−52.8	W$_2$B$_5$	−146.4	−138.9
MgB$_4$	−73.6	−69.8	YB$_6$	−100.4	−95.3
MgB$_6$	−93.7	−88.9	YbB$_6$	−550.2	−522.0
MgB$_{12}$	−143.9	−136.6	ZrB	<−163	<−155
MnB$_2$	−79.5	−75.4	ZrB$_2$	−320.9	−304.5
Mo$_2$B	−106.7	−101.3	ZrB$_{22}$	<−502.1	<−476.5

(a) From the elements, in standard state, pure phase, at 0.1 MPa (1 atm) and 298 K. Source: Ref 12

than SiC, BeO, MoSi$_2$, and other refractory compounds at temperatures greater than 1000 °C (1830 °F) (Ref 42).

Electrical Properties. Electrical resistivity, work function, and Hall constants are shown for selected borides in Table 6 (Ref 12). The electrical conductivity of borides is high, with the exception of boron-rich beryllium borides. The electrical conductivity of alkaline-earth hexaborides increases with increasing atomic number. The low resistivity of most borides makes them machinable by electron discharge machining (EDM). The ability to apply this technique to TiB$_2$ has enabled the production of complex shapes of SiC-TiB$_2$ composites (Ref 53, 54), whereas the higher resistivity of SiC makes EDM of silicon carbide impossible. The electrical resistivity of borides increases with increasing temperature, as shown by the limited data for HfB$_2$, NbB$_2$, TaB$_2$, VB$_2$, and ZrB$_2$ in Table 6. As temperature increases, conduction electron scattering by lattice phonons and impurity atoms increases, lowering the electrical resistivity (Ref 55). It should be noted that the same boride from different suppliers will have a range of resistivities, rather than the specific resistivity shown in Table 6. Typical resistivities for TiB$_2$ range between 9 and 15

Table 4 Thermal expansion and thermal conductivity of selected borides

Boride	Thermal expansion, 10^{-6}/K α	α$_a$	α$_c$	Temperature range °C	°F	Ref	Thermal conductivity W/m·K	Btu/ft·h·°F	Temperature °C	°F	Ref
BaB$_6$	6.8 ± 0.5	...	...	20–800	68–1470	7,12	46.8 ± 1.7	27 ± 1	20	68	7
							36.4 ± 1.7	21 ± 1	20	68	12
CaB$_6$	6.5 ± 0.5	...	...	20–800	68–1470	7,12	39.3 ± 1.7	22.7 ± 1	20	68	7
							23.0 ± 1.7	13.3 ± 1	20	68	12
CeB$_6$	7.3 ± 0.5	...	...	20–800	68–1470	7,12	33.9 ± 0.9	19.6 ± 0.5	20	68	7,12
Co$_3$B	...	...	...	...	...	...	17.0	9.8	...	...	12
Co$_2$B	...	...	...	...	...	...	14.0	8.1	...	...	12
CoB	...	...	...	...	...	...	17.0	9.8	...	...	12
Cr$_4$B	...	...	...	...	...	...	11.0 ± 0.4	6.4 ± 0.2	20	68	12
Cr$_2$B	14.2	...	...	27–1027	80–1880	12	10.9	6.3	20	68	12
	15.0	...	...	1027–2027	1880–3680	12	...	...	...	...	...
Cr$_5$B$_3$	13.7	...	...	27–1027	80–1880	12	15.8	9.1	20	68	12
	14.2	...	...	1027–2027	1880–3680	12	...	...	...	...	...
CrB	12.3	...	...	27–1027	80–1880	12	20.1	11.6	20	68	12
	12.6	...	...	1027–2027	1880–3680	12	...	...	...	...	...
Cr$_3$B$_4$	11.8	...	...	27–1027	80–1880	12	20.5	11.8	20	68	12
	12.1	...	...	1027–2027	1880–3680	12	...	...	...	...	...
CrB$_2$	10.5	...	...	27–1027	80–1880	12	31.8	18.4	20	68	12
	11.8	...	...	1027–2027	1880–3680	12	34.0	19.7	1027	1880	12
	11.0	...	...	0–1200	30–2190	7	55.0	31.8	2027	3680	12
Cr$_2$B$_5$	...	...	...	...	...	...	18.0 ± 0.8	10.4 ± 0.5	20	68	12
DyB$_4$	5.9	...	...	25–1027	77–1880	12	118	68.2	27	80	12
DyB$_{12}$	4.2	...	...	...	...	12	...	...	...	...	...
ErB$_4$	7.6	...	...	20–1000	68–1830	12	...	...	...	...	...
ErB$_{12}$	3.7	...	...	...	...	12	...	...	...	...	...
EuB$_6$	6.9 ± 0.5	...	...	20–800	68–1470	7,12	23.0 ± 0.9	13.2 ± 0.5	20	68	12
							18.0	10.4	27	80	19
Fe$_2$B	...	11.8 ± 0.3	8.9 ± 0.6	20–800	68–1470	12	30.1	17.4	20	68	12
FeB	≈12	...	...	400–1000	750–1830	12	12.0	6.9	20	68	12
GdB$_4$	7.0	...	...	20–1000	68–1830	12	148.5	85.8	27	80	12
GdB$_6$	8.7 ± 0.5	...	...	20–800	68–1470	12	20.5 ± 1.3	11.8 ± 0.8	20	68	12
HfB$_2$	6.3	...	...	27–1027	80–1880	12	51.6	29.8	27	80	12
	6.8	...	...	1027–2027	1880–3680	12	60.0	34.7	1027	1880	12
	7.6	...	...	20–2205	68–4000	7	143.0	82.7	2027	3680	12
HoB$_4$	7.85	...	...	20–1000	68–1830	12	127.6	73.8	27	80	12
HoB$_6$	3.0	...	...	...	...	12	...	...	...	...	...
HoB$_{12}$	3.6	...	...	...	...	12	...	...	...	...	...
LaB$_4$	...	7.17 ± 1.16	8.36 ± 1.03	...	...	12	...	...	...	...	...
LaB$_6$	6.4 ± 0.5	...	...	20–800	68–1470	12	47.7 ± 4.2	27.6 ± 2.4	...	...	12
							45.0	26.0	...	...	19

(continued)

Table 4   continued

Boride	Thermal expansion, 10⁻⁶/K			Temperature range			Thermal conductivity		Temperature		
	α	α_a	α_c	°C	°F	Ref	W/m·K	Btu/ft·h·°F	°C	°F	Ref
LuB₁₂	3.4	...	...	...	...	12	...	...	...	...	...
Mn₄B	...	...	...	...	...	...	5.0	2.9	...	...	12
Mn₂B	...	...	...	...	...	...	6.6	3.8	...	...	...
MnB	...	...	...	...	...	...	7.7	4.5	...	...	...
Mn₃B₄	...	...	...	...	...	...	8.6	5.0	...	...	...
MnB₂	...	...	...	...	...	...	10.2	5.9	...	...	12
Mo₂B	5.0	...	...	25–500	77–930	7	...	...	...	...	...
MoB₂	7.7(Ref 12)	7(Ref 8)	10(Ref 8)	300–900	570–1650(a)	...	...	...	...	...	...
Mo₂B₅	8.6	...	...	27–1027	80–1880	12	≈50	≈28.9	20	68	12
	9.9	...	...	1027–2027	1880–3680	12	≈27	≈15.6	900	1650	12
MoB₄	6.5	...	...	...	...	12	...	...	...	...	...
Nb₃B₂	13.8	...	...	27–1027	80–1880	...	12.0	6.9	27	80	12
	13.9	...	...	1027–2027	1880–3680	...	...	...	...	...	...
NbB	12.9	...	...	27–1027	80–1880	...	15.6	9.0	27	80	12
	13.4	...	...	1027–2027	1880–3680	...	...	...	...	...	...
Nb₃B₄	9.9	...	...	27–1027	80–1880	...	20.5	11.8	27	80	12
	10.3	...	...	1027–2027	1880–3680	...	24.0	13.9	27	80	12
NbB₂	8.0	...	...	27–1027	80–1880	...	23.5	13.6	1027	1880	12
	8.5	...	...	1027–2027	1880–3680	...	40.3	23.3	2027	3680	12
	8.6	...	...	20–1650	68–3000(b)	...	...	...	...	...	...
NdB₄	5.84	...	...	27–1027	80–1880	12	...	...	...	...	...
NdB₆	7.30 ± 1.0	...	...	20–800	68–1470	12	47.3 ± 3.3	27.3 ± 2	20	68	12
Ni₃B	...	...	...	...	...	...	41.8	24.2	...	...	12
Ni₂B	...	...	...	...	...	...	54.8	31.6	...	...	12
NiB	...	...	...	...	...	...	21.9	12.7	...	...	12
PrB₄	5.0	...	...	27–1027	80–1880	...	...	...	...	...	...
PrB₆	7.50 ± 0.5	...	...	20–800	68–1470	...	41.0	23.7	20	68	12
ScB₂	...	6.8 ± 0.5	7.6 ± 0.5	20–600	68–1110	12	...	...	...	...	...
	...	4.4	4.9	30–1100	86–2010	8	...	...	...	...	...
ScB₄	4.1	...	...	...	...	12	...	...	...	...	...
SmB₆	6.8 ± 0.5	...	...	20–800	68–1470	12	13.8 ± 1.7	8.0 ± 1	20	68	12
SrB₆	6.7 ± 0.5	...	...	20–800	68–1470	12	26.4 ± 2.1	15.3 ± 1.2	20	68	12
TaB₂	8.2	...	...	27–1027	80–1880	12	16.0	9.2	27	80	12
	8.8	...	...	1027–2027	1880–3680	12	16.1	9.3	1027	1880	12
	8.4	...	...	20–1650	68–3000	7	36.2	20.9	2027	3680	12
	...	5.85	7.14	30–1000	86–1830	8	...	...	...	...	...
	...	7.82	8.40	1000–2300	1830–4170	8	...	...	...	...	...
TbB₄	6.55	...	...	27–1027	80–1880	12	126.4	73.1	27	80	12
TbB₆	7.8 ± 1.0	...	...	20–800	68–1470	12	20.1 ± 1.3	11.6 ± 0.8	20	68	12
TbB₁₂	3.2	...	...	...	...	12	...	...	...	...	...
ThB₄	7.9	...	...	20–1770	68–3220	12	≈25	≈14.5	20	68	12
	...	...	...	...	...	...	≈28	≈16.2	730	1346	12
	...	...	...	...	...	...	≈31	≈17.9	1230	2246	12
	...	...	...	...	...	...	41	23.7	1730	3146	12
ThB₆	7.8 ± 0.5	...	...	20–800	68–1470	12	44.8 ± 5.0	25.9 ± 3	...	...	12
TiB₂	...	6.63	8.65	30–1000	86–1830	8	64.4	37.2	27	80	12
	...	8.00	11.20	1000–2300	1830–4170	8	69.9	40.4	1027	1880	12
	4.6	...	...	27–1027	80–1880	12	122.2	70.6	2027	3680	12
	5.2	...	...	1027–2027	1880–3680	12	98	56.6	25	77	39
	8.6	...	...	20–2205	68–4000	7	115.2	66.6	27	80	26
	...	7.30	10.27	300–1300	570–2370	8	37	21.4	27–400	80–750	40
	...	...	...	27–1027	80–1880	12	158.2	91.4	27	80	12
TmB₄	6.5	...	...	...	...	12	...	...	...	...	...
TmB₁₂	3.85	...	...	...	...	12	...	...	...	...	...
UB₂	...	9	8	20–205	68–400	12	51.9	30.0	20	68	12
UB₄	7.0	...	...	20–1000	68–1830	12	4.0	2.3	20	68	12
UB₁₂	4.6	...	...	...	...	12	...	...	...	...	...
VB₂	7.6	...	...	27–1027	80–1880	12	42.3	24.4	27	80	12
	8.3	...	...	1027–2027	1880–3680	12	42.0	24.3	1027	1880	12
	...	...	...	...	...	...	67.4	39.0	2027	3680	12
W₂B	6.7	...	...	...	...	...	...	...	...	...	...
WB	≈6.9	...	...	20–2205	68–4000	7	...	...	...	...	...
W₂B₅	7.8	...	...	27–1027	80–1880	12	≈52	≈30.1	20	68	12
	8.8	...	...	1027–2027	1880–3680	12	≈26	≈15.0	800	1470	12
WB₄	5.8	...	...	...	...	12	...	...	...	...	...
YB₄	...	7.6 ± 0.5	6.4 ± 0.6	...	...	...	29.2	16.9	27	80	12
YB₆	6.02 ± 0.6	...	...	...	...	12	29.3 ± 4.2	16.9 ± 2.3	20	68	12
YB₁₂	6.6 ± 0.6	...	...	...	...	...	...	...	...	...	...
YbB₆	5.8 ± 0.5	...	...	20–800	68–1470	12	25.1 ± 1.7	14.5 ± 1	20	68	12
YbB₁₂	2.0	...	...	−196–20	−320–68	...	...	...	...	...	...
	3.9	...	...	20–300	68–570	...	...	...	...	...	...
	5.8	...	...	300–1000	570–1830	12	...	...	...	...	...
ZrB₂	5.9	...	...	27–1027	80–1880	12	57.9	33.5	27	80	12
	6.5	...	...	1027–2027	1880–3680	12	64.4	37.2	1027	1880	12
	8.3	...	...	20–2205	68–4000	7	133.9	77.4	2027	3680	12
							≈87	≈50.3	20–2000	68–3630	7

(a) Ref 8. (b) Ref 7

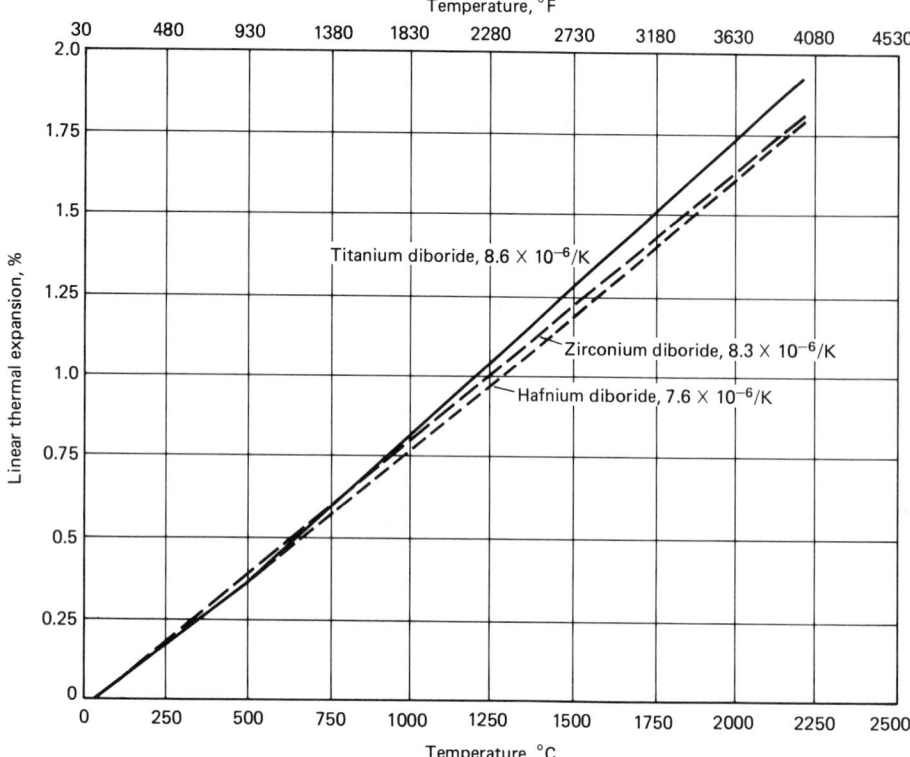

Fig 4 Linear thermal expansion for TiB₂, ZrB₂, and HfB₂. Source: Ref 7

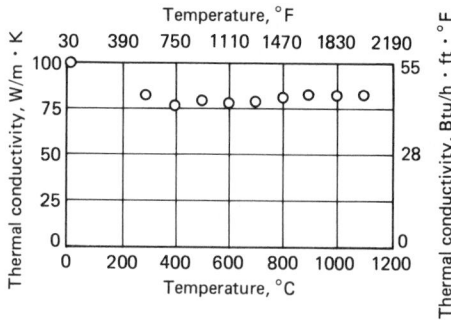

Fig 5 Thermal conductivity as a function of temperature for TiB₂, based on laser flash diffusivity measurements (Ref 39) and heat capacity data (Ref 34)

$\mu\Omega \cdot$ cm (Ref 13). The range of resistivities is very dependent on impurities in the borides.

Borides, in general, have low work functions (the lower the work function, the easier it is for electrons to escape from solid surfaces). The high electron emissivity of LaB_6 is the reason for its use in electron guns in scanning electron microscopes (Ref 56). The Hall constant is used for measuring the carrier concentration. It is negative for free electrons and positive for holes. The lower the carrier concentration, the greater the magnitude of the Hall constant (Ref 55). Hall coefficients are generally negative for borides, as expected. Transition metal borides are generally better electrical conductors than transition metal nitrides or carbides. A variety of borides are superconductors at temperatures less than 2 K (Ref 12). Niobium boride (NbB) has the highest superconducting transition temperature at 8.25 K (Ref 12).

Optical Properties. The optical properties of borides are relatively unexplored. Table 7 gives the color of a number of borides. Emission coefficients at a wavelength of 0.665 mm (in the infrared region) generally range between 0.7 and 0.8.

Magnetic Properties. Magnetic susceptibilities, effective magnetic moments, and Curie temperatures are given in Table 8. Magnetic susceptibility is a dimensionless unit obtained by dividing the magnetization by the magnetic field intensity. Susceptibilities in Table 8 are based on the molecular weight of each boride. Substances with a negative magnetic susceptibility are called diamagnetic, whereas substances with a positive susceptibility are called paramagnetic (Ref 55). A number of borides are diamagnetic, whereas the remainder range from weakly paramagnetic (TiB_2) to strongly paramagnetic (lanthanide hexaborides) to ferromagnetic (Fe_2B, FeB, and MnB) (Ref 14).

The magnetic moment of a free atom arises from the spin of electrons, the electron orbital angular momentum, and the change in orbital moment induced by an applied magnetic field (Ref 55). A ferromagnetic material has a magnetic moment even in zero applied magnetic field. The effective magnetic moments

Table 5(a) Ambient temperature mechanical properties of selected borides

Boride	Hardness		Load	
	GPa	10⁶ psi	N	lbf
AlB₂	9.6 ± 1.0	1.4 ± 0.15	1(a)	0.23(a)
AlB₁₀	26.0 ± 1.2	3.8 ± 0.17	1(a)	0.23(a)
α-AlB₁₂	23.5 ± 0.3	3.4 ± 0.04	1(a)	0.23(a)
β-AlB₁₂	22.5 ± 1.0	3.3 ± 0.15	1(a)	0.23(a)
BaB₆	29.4 ± 2.8	4.3 ± 0.41	1(a)	0.23(a)
Be₅B	6.1	0.9	...	...
Be₂B	8.7	1.3	...	...
BeB₂	31.2	4.5	...	...
BeB₆	25.3	3.7	...	...
CaB₆	26.9 ± 2.2	3.9 ± 0.32	1(a)	0.23(a)
	16.2	2.3	1(b)	0.23(b)
CeB₆	30.8 ± 1.9	4.5 ± 0.3	0.3	0.07
Co₃B	11.3	1.6	0.5	0.11
Co₂B	11.3	1.6	0.5	0.11
CoB	11.3	1.6	0.5	0.11
	23.0	3.3	1(b)	0.23(b)
CoB₂	25.3	3.7	0.5	0.11
Cr₄B	12.2 ± 0.6	1.8 ± 0.09	0.5	0.11
Cr₂B	13.2 ± 1.0	1.9 ± 0.15	0.5	0.11
Cr₅B₃	13.9–14.9	2.0–2.2	0.5	0.11
CrB	11.8–12.7	1.7–1.8	1	0.23
Cr₃B₄	13.7–14.7	2.0–2.1	1	0.23
CrB₂	20.6 ± 0.8	3.0 ± 0.1	0.5	0.11
	10.9	1.6	1(b)	0.23(b)
DyB₄	18.6	2.7	...	...
DyB₆	23.5 ± 1.0	3.4 ± 0.15	1	0.23
ErB₄	18.6	2.7	...	...
ErB₆	27.5 ± 1.0	4.0 ± 0.15	1	0.23
EuB₆	26.1	3.8	1	0.23
	18.0	2.6	1(b)	0.23(b)
Fe₂B	13.1 ± 0.5	1.9 ± 0.07	...	...
	17.7	2.6	1(b)	0.23(b)
FeB	16.2 ± 0.5	2.3 ± 0.07	...	...
	18.6	2.7	1(b)	0.23(b)
GdB₄	18.6 ± 0.5	2.7 ± 0.07	...	...
GdB₆	18.1 ± 0.5	2.6 ± 0.07	...	...

(continued)

Table 5(a) continued

Boride	Hardness		Load	
	GPa	10^6 psi	N	lbf
HfB_2	28.4 ± 5.0	4.1 ± 0.7	0.3	0.07
	23.5	3.4	1.6(a)	0.36(a)
	21.2	3.1	1(b)	0.23(b)
HoB_4	16.5	2.4	...	...
HoB_{12}	26.5 ± 1.0	3.8 ± 0.15	1	0.23
$IrB_{1.15}$	16.2	2.3	...	...
LaB_6	19.7	2.9(b)	...	...
LuB_{12}	28.4 ± 1.0	4.1 ± 0.15	1	0.23
Mn_4B	10.3 ± 0.5	1.5 ± 0.07	...	...
Mn_2B	17.7 ± 0.5	2.6 ± 0.07	...	...
MnB	20.1 ± 0.5	2.9 ± 0.07	...	...
Mn_3B_4	19.6 ± 0.5	2.8 ± 0.07	...	...
MnB_2	16.7 ± 0.5	2.4 ± 0.07	...	...
MnB_4	35.3 ± 1.0	5.1 ± 0.15	...	...
Mo_2B	24.5	3.6	0.5	0.11
α-MoB	23.0	3.3	0.5	0.11
β-MoB	24.5	3.6	0.5	0.11
MoB_2	11.8	1.7	0.5	0.11
Mo_2B_5	23.0	3.3	0.5	0.11
	29.5	4.3	1.0(a)	0.23(a)
	21.3	3.1	1.0(b)	0.23(b)
Nb_3B_2	22.5	3.3	0.5	0.11
NbB	21.5	3.1	0.5	0.11
Nb_3B_4	22.5	3.3	0.3	0.07
NbB_2	25.5	3.7	0.3	0.07
	20.9	3.0	1(b)	0.23(b)
NdB_4	19.1	2.8	...	...
NdB_6	24.9 ± 1.7	3.6 ± 0.25	0.7	0.16
Ni_3B	11.7	1.7	...	...
Ni_2B	14.0	2.0	...	...
Ni_4B_3	14.6	2.1	...	...
NiB	15.2	2.2	...	...
$OsB_{1.2}$	16.1	2.3	...	...
$OsB_{1.6}$	18.4	2.7	...	...
OsB_2	28.4	4.1	...	...
Pd_3B	4.6	0.7	...	...
Pd_5B_2	5.8	0.8	...	...
PrB_4	18.9	2.7	...	...
PrB_6	24.2 ± 1.0	3.5 ± 0.15	1	0.23
$PtB_{1.1}$	9.2	1.3	...	...
ReB_2	30.4	4.4	...	...
Rh_7B_3	7.6	1.1	0.5	0.11
$RhB_{\approx1.1}$	11.9	1.7	0.5	0.11
Ru_7B_3	11.1	1.6	...	...
$Ru_{11}B_8$	12.8	1.8	...	...
$RuB_{\approx1.1}$	13.8	2.0	...	...
Ru_2B_3	14.9	2.2	...	...
RuB_2	22.2	3.2	...	...
ScB_2	17.4 ± 0.3	2.5 ± 0.04	2	0.45
SmB_6	24.5 ± 3.0	3.5 ± 0.44	1	0.23
	13.6 ± 1.6(c)	1.9 ± 0.2(c)	...	...
SrB_6	28.6 ± 0.9	4.1 ± 0.1	0.3	0.07
Ta_3B_2	27.2	3.9	0.5	0.11
TaB	30.7	4.5	0.5	0.11
Ta_3B_4	32.9	4.8	0.5	0.11
TaB_2	24.5 ± 0.4	3.5 ± 0.06	0.3	0.07
	19.6	2.8	0.5(a)	0.11(a)
	24.5 ± 1(b)	3.6 ± 0.15(b)	...	...
TbB_4	18.6	2.7	...	...
TbB_6	22.6	3.3	1	0.23
TbB_{12}	25.5 ± 1.0	3.7 ± 0.15	1	0.23
ThB_4	...	...	...	...
ThB_6	17.1 ± 1.2	2.5 ± 0.17	0.2	0.05
ThB_{76}	22.7	3.3	...	...
TmB_4	25.5 ± 1.0	3.7 ± 0.15	1	0.23
TiB	22.7	3.3	...	...
TiB_2	33.0 ± 0.6	4.8 ± 0.09	0.3	0.07
	25.5	3.7	1(b)	0.23(b)
	...	...	...	...
	25.5	3.7	1(b)	0.23(b)
	20.8, 23.0	3.0, 3.3	5(c)	1.13(c)
UB_2	14.8	2.1	...	...
UB_4	11.1	1.6	1(b)	0.23(b)
UB_{12}	25.8	3.7	1(b)	0.23(b)
V_3B_2	22.4	3.2	0.5	0.11
V_3B_4	23.0	3.3	0.5	0.11
VB_2	27.5 ± 0.1	4.0 ± 0.01	0.3	0.07
	20.6	3.0	1(b)	0.23(b)

(continued)

in Table 8 have units of Bohr magnetons, μB. The Curie temperature is the temperature above which the spontaneous magnetization vanishes, separating the paramagnetic state from the ferromagnetic state. Iron boride (Fe_2B) has the highest Curie temperature of any boride, approximately 740 °C (1360 °F) (Ref 8).

Detailed information on the magnetic properties of all classes of borides is provided in the review article by Buschow ("Magnetic Properties of Borides"), which is contained in Ref 8. It should be noted that there is considerable disagreement between Ref 8 and 12 regarding the magnetic properties of borides.

Iron borides (Fe_2B and FeB) are strongly ferromagnetic. Lanthanide hexaborides show diverse magnetic properties, varying from diamagnetic to ferromagnetic (Ref 8). The magnetic properties of transition metal and actinide borides also vary widely. Detailed structure-property data are available for select borides (Ref 8).

Chemical Properties. Borides are known for their resistance to chemical attack, as well as their excellent oxidation resistance. Table 9 is a summary of an extensive compilation of data on the resistance of powdered borides to acids and alkalis (Ref 12). The resistance of a given compound to chemical attack is dependent on the surface area, the concentration of the reagent to which it is exposed, the temperature of exposure, and the time of exposure. A given compound can be subject to a widely varying level of attack in acids and bases, depending on the conditions to which it is exposed. Therefore, the values in Table 9 should only be used as a guide. Specific data on corrosion resistance are tabulated in Samsonov and Vinitskii (Ref 12). The data listed in Table 9 are for the most aggressive conditions reported, and are obtained at temperatures between 20 and 100 °C (68 and 212 °F). A quick perusal of Table 9 shows that no borides are resistant to attack by both acids and bases, although specific borides can be selected to resist attack under any given condition. Borides generally have better resistance to chemical attack than either their nitride or carbide counterparts. Both HfB_2 and ZrB_2 are more resistant to attack in elemental fluorine at 980 °C (1795 °F) than their carbide counterparts (that is, HfC and ZrC) (Ref 7). The diborides of Zr and Hf are attacked only at temperatures above the volatilization points of the metallic fluorides ($\approx$600 °C, or 1110 °F) (Ref 7).

The oxidation resistance of refractory diborides decreases in the order of HfB_2 > ZrB_2 > TiB_2 > TaB_2 > NbB_2 (Ref 57). The oxidation resistance of a given diboride decreases as the boron concentration increases (Ref 7). The addition of hexaborides decreases the oxidation resistance of a given diboride, whereas the addition of glass formers, such as SiC, results in composites with excellent resistance to oxidation (Ref 7). Composites of HfB_2 with SiC have excellent stability to

Table 5(a) continued

Boride	Hardness GPa	Hardness 10⁶ psi	Load N	Load lbf
	Hardness		**Load**	
Boride	GPa	10^6 psi	N	lbf
W$_2$B	23.7 ± 1.2	3.4 ± 0.17	1.5	0.34
WB	36.3	5.3	0.5	0.11
WB$_2$	26.1 ± 0.1	3.8 ± 0.01	0.3	0.07
W$_2$B$_5$	26.1 ± 0.1	3.8 ± 0.01	0.3	0.07
	24.5	3.6	1(b)	0.23(b)
WB$_4$	39.2 ± 2.0	5.7 ± 0.3	0.5	0.11
YB$_4$	27.9 ± 1.5	4.0 ± 0.22	...	...
YB$_6$	25.3 ± 1.0	3.7 ± 0.15	...	...
YB$_{12}$	24.5 ± 1.5	3.6 ± 0.22	...	...
YbB$_6$	26.1	3.8	1.0	0.23
	15.1 ± 0.3	2.2 ± 0.04	...	...
YbB$_{12}$	32.4 ± 1.0	4.7 ± 0.15	0.3	0.07
ZrB	34.3 − 35.3	5.0 − 5.1	0.3	0.07
ZrB$_2$	22.1 ± 0.2	3.2 ± 0.03	0.3	0.07
	17.9	2.6	1(b)	0.23(b)
ZrB$_{12}$	27.0 − 28.0	3.9 − 4.1		
	25.3	3.7	1(b)	0.23(b)

(a) Ref 7. (b) Ref 13. (c) Ref 41

2100 °C (3810 °F) (Ref 58). Appreciable oxidation of the borides of vanadium, niobium, and tantalum is reported to occur in the temperature range from 800 to 900 °C (1470 to 1650 °F) (Ref 7). Chromium borides are fairly resistant to oxidation; CrB$_2$ can withstand oxidation up to 1700 °C (3090 °F) (Ref 14).

Other reports suggest that noticeable oxidation in air begins at 1000 °C (1830 °F) for chromium borides, and that oxidation is appreciable by 1200 °C (2190 °F) (Ref 7). The oxidation resistance of molybdenum borides increases as the boron-to-metal ratio increases (Ref 7). Lanthanide hexaborides are less resistant to oxidation than transition metal borides (Ref 59–61).

Applications

Titanium boride has been investigated for some time as cathodes in Hall-Heroult cells (Ref 62). Despite the relatively high contact angle of molten Al on TiB$_2$ (θ = 60° at 1200 °C, or 2190 °F, per Ref 12), penetration by molten aluminum degrades mechanical properties (Ref 43, 63, 64), hampering the successful implementation of boride cathodes. Improved processing is still occurring for TiB$_2$, and the effects of impurities on properties are only beginning to be understood (Ref 26, 38, 39, 65–68). Further work on understanding processing-microstructure relationships, as well as systems design, is required if boride cathodes are to be used in Hall-Heroult reduction cells.

Titanium diboride is currently used as an armor material; SiC, AlN, B$_4$C, and Al$_2$O$_3$ represent competing materials. Diborides are used as crucibles for handling liquid metals, such as Al, Cu, Ag, Au, Zn, Cd, Mg, Ge, Pb, Bi, and Cr (Ref 16).

Titanium boride is mixed with BN and used as evaporation boats for coating capacitor foils with Al at 1400 to 1500 °C (2550 to 2730 °F) (Ref 16). The electrical conductivity of TiB$_2$ allows direct resistance heating of the boats, whereas BN gives the composites enhanced thermal shock resistance. Other composite materials based on TiB$_2$ additions include SiC-TiB$_2$ (Ref 53, 54), TiB$_2$-TaB$_2$-CoB (Ref 69), TiN-TiB$_2$ and Ti(C,N)-TiB$_2$ (Ref 70), TiB$_2$ with SiC whisker reinforcement (Ref 71), Al$_2$O$_3$-TiB$_2$ (Ref 72–75), TiB$_2$-B$_4$C (Ref 76, 77), and TiB$_2$-ZrO$_2$ (Ref 78, 79). Improvement in mechanical properties (that is, strength

Table 5(b) Ambient temperature mechanical properties of selected borides

Boride	Young's modulus GPa	Young's modulus 10^6 psi	Poisson's ratio	Flexural strength MPa	Flexural strength ksi	Compressive strength GPa	Compressive strength 10^6 psi	Fracture toughness MPa√m	Fracture toughness ksi√in.
BaB$_6$	385	56	...	...	...	...	...	...	...
CaB$_6$	451	65	...	138	20	...	...	...	...
CeB$_6$	379	55	...	...	...	...	...	...	...
CrB$_2$	211	31	...	607	88	1.3(a)	0.19(a)	...	...
EuB$_6$	...	...	...	183	27	...	...	...	...
Fe$_2$B	284	41	...	...	...	...	...	...	...
FeB	343	50	...	...	...	...	...	...	...
GdB$_6$	...	...	...	206(b)	30(b)	...	...	...	...
HfB$_2$	500(c)	72.5(c)	0.12(c)	350 ± 70(d)	51 ± 10(d)	...	...	...	...
LaB$_6$	479	69.5	...	126	18	...	...	...	...
α-MoB	...	...	...	345	50	...	...	...	...
Mo$_2$B$_5$	672	97.5	...	...	...	...	...	...	...
NbB$_2$	637	92.4	...	...	...	...	...	...	...
SmB$_6$	...	...	...	148	21	...	...	...	...
TaB$_2$	257	37	...	...	...	...	...	...	...
	248(c)	36(c)	...	...	...	...	...	...	...
TbB$_4$	...	...	...	137(c)	20(c)	...	...	...	...
ThB$_4$	148	21	...	137	20	...	...	...	...
TiB$_2$	551(e)	80(e)	0.11(e)	240	35	...	...	6.7 ± 0.2(e)	6.1 ± 0.2(e)
				300–370	49–54				
	545(f)	79(f)	0.093(f)	(f,g)	(f,g)	...	...	4.9 ± 0.4(f)	4.5 ± 0.4(f)
	500(c)	73(c)	0.10(c)	...	...	5.7 ± 0.17(h)	0.8 ± 0.02(h)	8.0 ± 0.3(e)	7.3 ± 0.3(e)
	530	77	...	...	...	1.3	0.19	6.0 ± 0.7(g)	5.5 ± 0.6(g)
	...	...	...	350–400	51–58	...	...	...	...
UB$_4$	440	64	...	413	60	...	...	...	...
VB$_2$	268	39	...	...	...	...	...	...	...
	262(c)	38(c)	...	...	...	...	...	...	...
W$_2$B$_5$	775	112	...	...	...	...	...	...	...
ZrB$_2$	343	50	...	305(d)	44(d)	...	...	...	...
	500(c)	73(c)	0.11	275(e)	40(e)	...	...	...	...
ZrB$_{12}$	...	...	...	...	...	1.6	0.23	...	...

(a) Ref 12. (b) 8% porosity. (c) Ref 7. (d) Ref 42. (e) Ref 41. (f) Ref 26. (g) Ref 43. (h) Ref 40

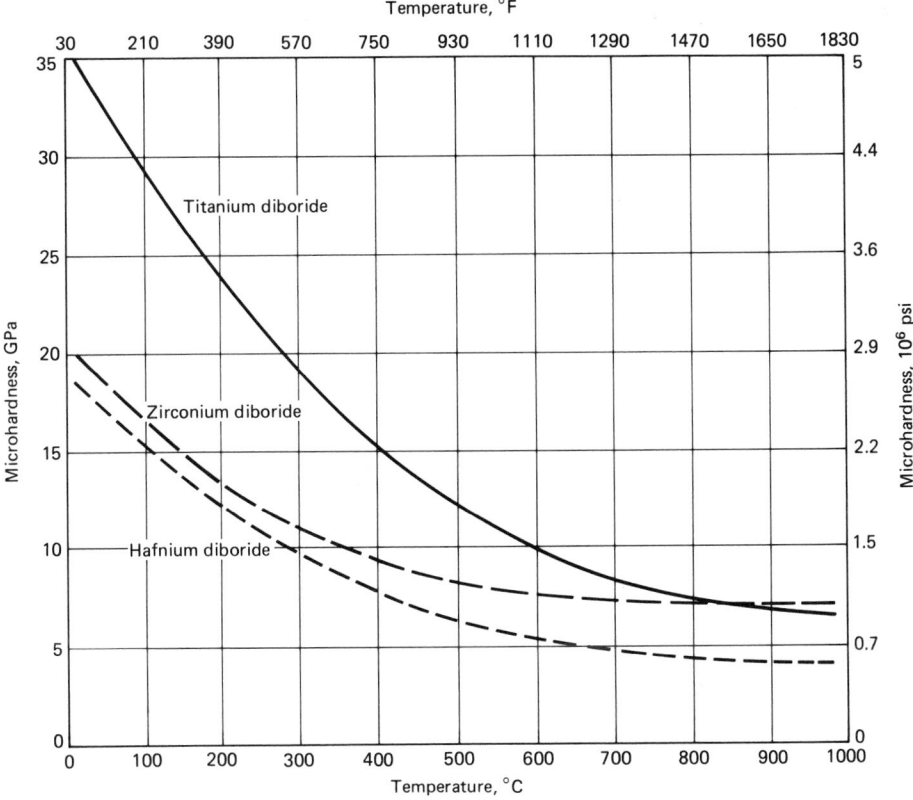

Fig 6 Hardness as a function of temperature for TiB₂, ZrB₂, and HfB₂. Source: Ref 7

and/or toughness) is characteristic of each of these systems. Strength values of 700 MPa (102 ksi) (Ref 76) and fracture toughness values of 7 to 8 MPa$\sqrt{m}$ (6.4 to 7.3 ksi$\sqrt{in.}$) (Ref 76) have been achieved in composite systems that can have Vicker's hardness values of 35 GPa (5 × 10⁶ psi) (applied load of 1 N, or 100 gf) (Ref 77). Relatively little research has been performed on boride systems, compared to oxide, nitride, and carbide ceramics. As other ceramics replace metals

and ceramic-metal (cermet) composites in wear and erosion applications, more attention will be focused on boride composites.

Boride-based composites synthesized by exothermic reactions appear promising (Ref 80–82), with excellent combinations of strength (800 to 900 MPa, or 116 to 130 ksi), toughness (10 to 15 MPa$\sqrt{m}$, or 9 to 13.6 ksi$\sqrt{in.}$), and thermal conductivity (70 W/m · K, or 40 Btu/ft · h · °F) reported for ZrB₂-ZrC-Zr composites (Ref 81, 82). Applica-

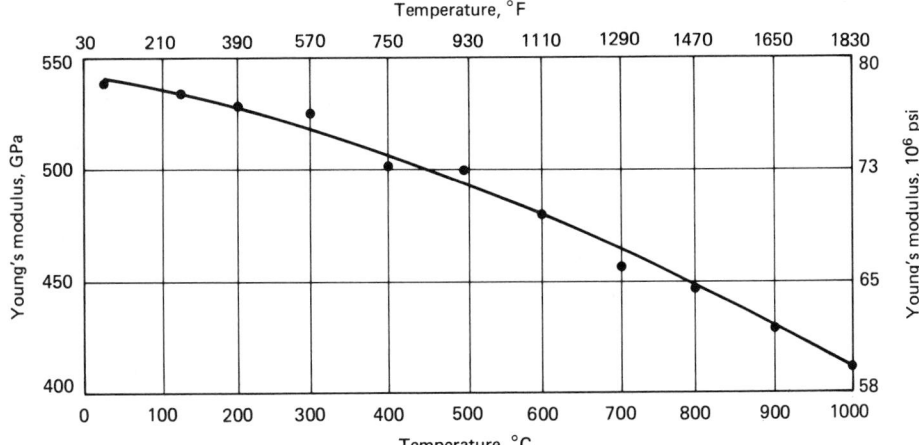

Fig 7 Young's modulus for high-purity TiB₂ as a function of temperature. Source: Ref 26

Table 6 Electrical properties at ambient temperature

Boride	Electrical resistivity, 10⁻⁶ Ω · cm	Work function 10¹⁹ J	Work function 10¹⁸ eV	Hall constant, 100 m³/K
BaB₆	77	5.52	34.4	−57.5
Be₅B	30	...	...	...
Be₂B	1 × 10³	...	...	...
BeB₂	2 × 10⁴	...	...	...
BeB₄	1 × 10¹²	...	...	...
BeB₆	1 × 10¹³	...	...	...
BeB₉	6 × 10¹³	...	...	...
CaB₆	222	4.58	28.6	−91.0
CeB₆	29	4.69	29.3	−4.18
Co₃B	28	...	...	...
Co₂B	33	...	...	...
CoB	76	...	...	−104
Cr₂B	107	4.13	25.8	−0.8
Cr₅B₃	49	4.39	27.4	−0.9
CrB	46	4.86	30.3	−1.0
Cr₃B₄	60	5.03	31.4	−1.0
CrB₂	30	5.12	31.9	−1.2
DyB₄	35	...	...	...
DyB₆	...	5.65	35.3	...
DyB₁₂	14	...	...	−4.6
ErB₄	50	...	...	...
ErB₆	...	5.39	33.6	...
ErB₁₂	16	...	...	−4.5
EuB₆	85	7.84	48.9	−50.2
Fe₂B	38	5.28	32.9	...
FeB	80	5.76	35.9	...
GdB₄	31	2.32	14.5	...
GdB₆	45	3.30	20.6	−4.39
HfB₂	11	5.76	35.9	−18.0
	46(a)	...	...	...
HoB₄	30	5.47	34.1	...
HoB₆	...	...	...	−5.5
HoB₁₂	15	...	...	...
LaB₄	24	...	...	...
LaB₆	15	4.29	26.8	−4.96
LuB₆	...	4.80	30.0	...
LuB₁₂	14	...	...	−4.8
Mn₄B	85	...	...	...
Mn₂B	40	...	...	...
MnB	57	...	...	...
Mn₃B₄	62	...	...	...
MnB₂	71	6.62	41.3	...
Mo₂B	40	...	...	...
α-MoB	45	...	...	...
β-MoB	25	...	...	...
MoB₂	45	...	...	...
Mo₂B₅	26	5.95	37.1	−6.6
Nb₃B₂	45	5.40	33.7	−0.46
NbB	40	5.78	36.1	−0.60
Nb₃B₄	34	6.24	38.9	−1.1
NbB₂	26	6.52	40.7	−1.5
	53(a)	...	...	...
NdB₄	39	...	...	...
NdB₆	20	6.35	39.6	−4.39
Ni₃B	21	...	...	0.0
Ni₂B	14	...	...	0.53
NiB	50	...	...	0.63
Pd₃B	10	...	...	...
Pd₅B₂	26	...	...	...
PrB₄	40	...	...	...
PrB₆	20	3.46	21.6	−4.33
Rh₇B₃	88	...	...	...
RhB₋₁.₁	880	...	...	...
ScB₂	7–15	4.64	28.9	...
SmB₆	207	7.04	43.9	1.54
SrB₆	111	4.27	26.6	−76.3
TaB	100	...	...	...
TaB₂	33	6.56	40.9	−2.1
	75(a)	...	...	...
TbB₄	32	...	...	...
TbB₆	37	5.22	32.6	−4.57
TbB₁₂	12	...	...	−4.2
ThB₆	15	4.67	29.1	−2.19
TiB	40	...	...	...
TiB₂	9	6.57	41.0	−19.6
TmB₄	35	...	...	...

(continued)

Table 6 continued

Boride	Electrical resistivity, $10^{-6}\ \Omega \cdot cm$	Work function		Hall constant, $100\ m^3/K$
		10^{19} J	10^{18} eV	
TmB_6	...	4.50	28.1	...
TmB_{12}	17	...	...	−4.7
UB_2	...	5.29	33.0	...
UB_4	98	5.42	33.8	...
UB_{12}	23	4.74	29.6	−0.24
VB	35	...	...	...
VB_2	23	6.45	40.2	−0.82
	71(a)	...	...	...
W_2B_5	22	6.12	38.2	−6.9
YB_2	39	...	...	−3.05
YB_4	29	3.33	20.8	−21.3
YB_6	40	5.52	34.4	−4.56
YB_{12}	95	7.36	45.9	−4.9
YbB_6	47	5.01	31.3	−83.6
YbB_{12}	185	...	...	−8.4
ZrB_2	10	5.99	37.4	−19.0
	32(a)	...	...	...

(a) At 1027 °C (1880 °F). Source: Ref 12

tions such as rocket engine components, wear parts, and biomaterials appear promising for these unique ZrB_2 platelet-reinforced ZrC-Zr cermets. Composites of HfB_2-SiC have also been used for rocket engine components, because of their excellent oxidation resistance and high thermal shock resistance (Ref 7).

Lanthanide and alkaline-earth hexaborides are used as cathodes in electron guns, because of their high electron emissivity. The preferred material is LaB_6, although solid solutions of lanthanide hexaborides have also been used for such applications.

Table 7 Optical properties

Boride	Color	Boride	Color
Be_2B	Gray with rose shade	MnB	Red-brown
BeB_2	Dark gray	MnB_2	Red-brown
BeB_6	Brick red	Mo_2B_5	Light gray
BaB_6	Black with violet shade	NbB_2	Gray
CaB_6	Black	NdB_6	Blue
CeB_4	Gray-brown	PrB_4	Gray-brown
CeB_6	Blue-violet	PrB_6	Blue
CrB_2	Gray	ScB_2	Gray
DyB_4	Gray-brown	SmB_4	Gray-brown
DyB_6	Blue	SmB_6	Blue
DyB_{100}	Black	SmB_{100}	Black
ErB_4	Gray-brown	SrB_6	Black with green shade
ErB_6	Blue	TaB_2	Gray
ErB_{100}	Black	TbB_4	Gray-brown
EuB_6	Dark gray	TbB_6	Blue
Fe_2B	Gray	TbB_{100}	Black
FeB	Gray	ThB_6	Red-violet
GdB_4	Gray-brown	TiB_2	Gray
GdB_6	Blue	TmB_4	Gray-brown
GdB_{100}	Black	TmB_6	Blue
HfB_2	Gray	TmB_{100}	Black
HoB_4	Gray-brown	UB_6	Gray-steel
HoB_6	Blue	UB_{12}	Black
HoB_{100}	Black	VB_2	Gray
LaB_6	Purple-violet	W_2B_5	Light gray
LuB_6	Blue	YB_4	Gray
LuB_{100}	Black	YB_6	Blue-violet
MgB_2	Dark brown	YbB_6	Black
MgB_6	Dark brown	YbB_{100}	Black
MgB_{12}	Dark brown	ZrB_2	Gray

Source: Ref 12

All borides can be used as neutron-absorbing materials. Lanthanide hexaborides are of interest for such applications, because of their high macroscopic capture cross-section for

Table 8 Magnetic properties

Boride	Molar magnetic susceptibility, × 10^6	Effective magnetic moment (Bohr magneton, μB)	Curie temperature, K
BaB_6	1526	1.9	...
CaB_6	813	1.4	...
CeB_6	...	2.5	344
CoB	Paramagnetic	...	477
CrB	Paramagnetic	...	...
CrB_2	390	...	...
DyB_4	...	10.4	...
DyB_6	...	10.6	...
ErB_4	...	9.5	...
EuB_6	...	8.1	...
Fe_2B	...	1.9	1013
FeB	...	1.8	598
GdB_4	...	8.1	...
GdB_6	...	8.0	60
HfB_2	−4.0	...	...
HoB_4	...	10.8	...
LaB_6	60	0.0	...
Mn_2B	Paramagnetic	...	...
MnB	...	2.7	578
Mn_3B_4	Antiferromagnetic	...	...
MnB_2	...	2.3	140
Mo_2B_5	61.5	...	...
NbB_2	8.0	...	...
NdB_6	...	3.5	455
Ni_3B	264	...	...
Ni_2B	94	...	...
NiB	−5.5	...	...
PrB_6	...	3.6	≈0
ScB_2	≈100	...	...
ScB_{12}	−65	...	...
SmB_6	...	≈1.5	...
SrB_6	0.0	0.0	...
TaB_2	−64.8	...	...
TbB_6	...	9.4	...
TiB_2	31.3	...	...
TmB_4	...	7.7	...
UB_2	550	...	...
UB_4	1390	...	...
UB_{12}	70	...	...
VB	Paramagnetic	...	...
VB_2	34.1	...	...
W_2B_5	506	...	...
YB_2	Diamagnetic	...	...
YbB_6	...	4.6	2
ZrB_2	−67.7	...	...

Source: Ref 12

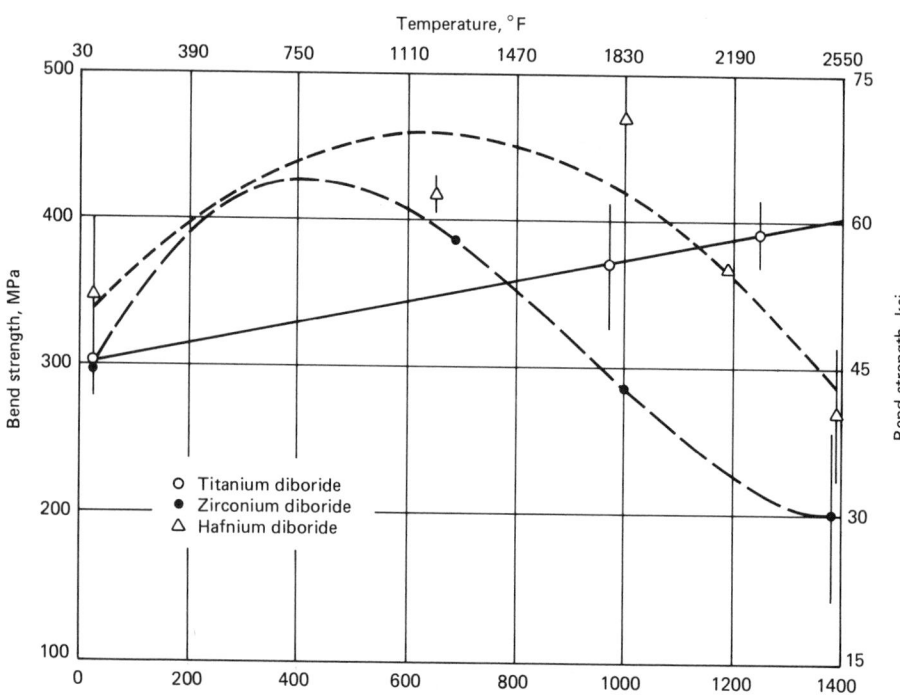

Fig 8 Strength for TiB_2 (Ref 26), ZrB_2 (Ref 42), and HfB_2 (Ref 42) as a function of temperature in inert environments

O Titanium diboride
● Zirconium diboride
△ Hafnium diboride

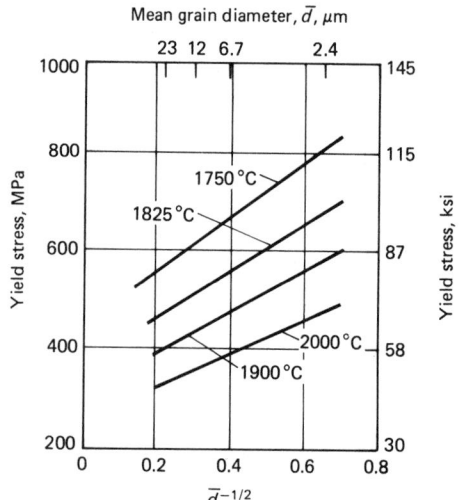

Fig 9 Hall-Petch relationships for yield stress data as a function of TiB_2 grain size in vacuum for temperatures between 1750 and 2000 °C (3180 and 3630 °F). Source: Ref 50

Table 9 Chemical properties of selected borides

Boride	Chemical resistance, % soluble					Oxidation resistance at 1000 °C (1830 °F), mg/cm²
	H_2O	HNO_3	HCl	H_2SO_4	NaOH	
Be_5B	...	...	...	...	...	258
Be_2B	...	...	...	...	...	132
BeB_2	...	...	...	...	...	22
BeB_4	...	...	...	...	...	30
BeB_6	...	...	...	...	...	64
BaB_6	...	100	3.0	0	1.9	45%(a)
CaB_6	...	100	0.5	0	2.6	37%(a)
CeB_6	...	100	5.0	2.5	1.1	...
Co_2B	...	100	100	...	...	...
CoB	...	...	0	...	...	...
Cr_4B	...	...	...	...	...	≈3.5
Cr_3B_2	...	...	...	...	...	≈15.5
CrB	...	...	...	...	...	≈1.0
Cr_3B_4	...	...	...	...	...	≈4.5(a)
CrB_2	...	59	97	97	12	≈7.8
FeB	...	100	0	0	...	...
GdB_6	...	100	9	100	0.6	6.0
HfB_2	...	97	94	98	...	(b)
LaB_6	...	100	3.8	100	1.5	...
MgB_2	100	100	100	100	...	...
MgB_{12}	0	100	0	0	...	...
MnB	...	100	100	100	...	...
Mn_3B	0	100	100	100	...	...
Mn_4B	0	100	100	100	...	...
Mo_2B_5	...	97	27	93	...	≈2.5
NbB_2	...	1	9	97	100	32.5
NdB_6	...	100	10.1	100	1.1	...
Ni_3B	100	...	...	...	...	...
PrB_6	...	100	9.2	100	0.7	...
ScB_2	...	84	75	67	...	...
SmB_6	...	100	22	100	0.6	...
SrB_6	0	100	1.5	0	1.9	43%(a)
TaB_2	...	0	2	97	100	2.52(a)
ThB_4	0	100	100	100	...	...
ThB_6	0	0	0	0	...	...
TiB_2	...	100	95	50	100	30(c)
UB_2	...	100	0	0	0	...
UB_4	...	100	100	100	...	...
UB_{12}	...	100	0	0	...	...
VB_2	...	99	97	93	39	...
W_2B_5	...	100	4	98	...	≈4
YB_4	...	100	0	100	0	...
YB_6	...	100	0	100	0	...
YB_{12}	...	>75	0	0	0	...
ZrB_2	...	100	98	100	100	30(d)

(a) At 900 °C (1650 °F). (b) 0.50 cm (0.02 in.) thick oxide layer after 100 h at 1300 °C (2370 °F) (Ref 7). (c) After 100 h. (d) After 150 h. Note: Chemical and oxidation resistance are very dependent on the nature of the attack. Above values represent worst possible scenarios. See Ref 12 for concentrations of reagents, temperature, and time of oxidation. Source: Ref 12

neutrons. One material that has been tested as neutron-absorbing pellets in control rods for fast-flux nuclear reactors is EuB_6 (Ref 16).

The metallurgical industry is the largest consumer of borides. Ferroboron is a grain refiner for steels, TiB_2 is used as a grain growth inhibitor for Al, and CaB_6 is a strong deoxidizer/defluxing agent for nonferrous metals such as copper. Welding alloys composed of Ni-Cr-B-Si are used to produce corrosion and wear-resistant surface coatings. The borides CrB, CrB_2, NiB, and FeB are among those used as hardfacing for wear parts (Ref 16).

ACKNOWLEDGMENT

The author gratefully acknowledges the help of Aaron Jones and Rachel Goeckeritz in preparing the manuscript. Discussions with Professor Anil V. Virkar of the University of Utah are appreciated.

REFERENCES

1. P. Schwarzkopf and R. Kieffer, *Refractory Hard Metals*, Macmillan, 1953, p 271–315
2. V. Mandorf, J. Hartwig, and E.J. Seldin, High Temperature Properties of Titanium Diboride, *High Temperature Materials*, Vol 2, G.M. Ault, Ed., Wiley-Interscience, 1963, p 455–467
3. P.T.B. Shaffer, *Materials Index*, Vol 1, *Plenum Press Handbooks of High Temperature Materials*, Plenum Press, 1964, p 6–77
4. G.V. Samsonov, *Properties Index*, Vol 2, *Plenum Press Handbooks of High Temperature Materials*, Plenum Press, 1964
5. J.L. Boone, Refractory Boron Compounds, in *Kirk-Othmer Encyclopedia of Chemical Technology*, Vol 3, 2nd ed., John Wiley, 1964, p 673–680
6. R. Thompson, *Borides: Their Chemistry and Applications, Lecture Series No. 5*, Royal Institute of Chemistry, London, 1965
7. J.F. Lynch, C.G. Ruderer, and W.H. Duckworth, Engineering Properties of Ceramics, *Borides*, American Ceramic Society, 1966
8. V.I. Matkovich, Ed., *Boron and Refractory Borides*, Springer-Verlag, 1977
9. R.H. Wentorf, Jr., Refractory Boron Compounds, *Kirk-Othmer Encyclopedia of Chemical Technology*, 3rd ed., Vol 4, John Wiley, 1978, p 123–129
10. G.V. Samsonov and T.I. Serebryakova, Classification of Borides, *Sov. Powder Metall. Met. Ceram.* (English transl.), Vol 17 (No. 2), 1978, p 116–120
11. G. Monkel, A. Lebugle, and H. Pastor, Manufacture and Electrotechnical Applications for Materials Containing Refractory Borides, Carbides and Nitrides, *Rev. Int. Hautes Temp. Refract.*, Vol 16 (No. 2), 1979, p 95–124
12. G.V. Samsonov and I.M. Vinitskii, *Handbook of Refractory Compounds*, Plenum Press, 1980
13. K.A. Schwetz, K. Reinmoth, and A. Lipp, *Production and Industrial Uses of Refractory Borides*, Vol 3, Radex Rundschau, 1981, p 568–585
14. I.C. McColm, *Ceramic Science for Materials Technologists*, Leonard Hill, London, 1983, p 330–343
15. F. Thevenot, Present Application of Borides, *Silic. Ind.*, Vol 51 (No. 1–2), 1986, p 17–22
16. H. Knoch, Borides, *Encyclopedia of Materials Science and Engineering*, Vol 1, M.B. Bever, Ed., Pergamon Press, 1986, p 398–402
17. *JCPDS Powder Diffraction File, Inorganic Phases*, International Centre for Diffraction Data, 1989, p 83–86
18. E. Rudy, "Compendium of Phase Diagram Data, Part V," Report No. AFML-TR-65–2, Wright-Patterson Air Force Materials Laboratory, 1969, p 198–215
19. M.I. Aivazov, T.I. Bryushkova, V.S. Mkrtchyan, and V.A. Rubanov, Thermal Conductivity of LaB_6-EuB_6 Solid Solutions, *High Temp.*, Vol 17 (No. 2), 1979, p 277–278
20. A.G. Merzhanov, V.M. Skhiro, and P. Borovinskaja, USSR patent 255,221, 1971
21. Z.A. Munir, Synthesis of High Temperature Materials by Self-Propagating Combustion Methods, *Am. Ceram. Soc. Bull.*, Vol 76 (No. 2), 1988, p 342–349
22. Z.A. Munir and V. Anselmi-Tamburnini, Self-Propagating Exothermic Reactions: The Synthesis of High-Temperature Materials by Combustion, *Mater. Sci. Rep.*, Vol 3, 1989, p 277
23. V. Hlavacek, Combustion Synthesis: A Historical Perspective, *Am. Ceram. Soc. Bull.*, Vol 70 (No. 2), 1991, p 240–243

24. J.D. Walton, Jr. and N.E. Poulos, Cermets from Thermite Reactions, *J. Am. Ceram. Soc.,* Vol 42 (No. 1), 1959, p 40–49

25. M. Shiota, M. Tsutsumi, and K. Uchida, Synthesis of LaB_6 from BN and Lanthanum-Citrate-Hydrate, *J. Mater. Sci.,* Vol 15, 1980, p 1987–1992

26. H.R. Baumgartner and R.A. Steiger, Sintering and Properties of Titanium Diboride Made from Powder Synthesized in a Plasma-Arc Heater, *J. Am. Ceram. Soc.,* Vol 67 (No. 3), 1984, p 207–212

27. J.G. Ryan and S. Roberts, The Formation and Characterization of Rare Earth Boride Films, *Thin Sol. Films,* Vol 135, 1986, p 9–19

28. C.J. McHargue, Ion Implantation in Metals and Ceramics, *Int. Met. Rev.,* Vol 31 (No. 2), 1986, p 49–76

29. S. Okaga, M. Sato, and T. Atoda, *J. Chem. Soc.,* Vol 4, Japan Chemical Industry Association, 1985, p 685–691

30. Y. Zhang, S. Okada, T. Atoda, T. Yamabe, and I. Yasumori, Synthesis of a New Compound WAlB by the Use of Aluminum Flux, *Yogy-Yyokai Shi,* Vol 95 (No. 4), 1987, p 374–380

31. E.A. Almond, Deformation Characteristics and Mechanical Properties of Hardmetals, *Science of Hard Materials,* R.K. Viswanadham, D.J. Rowcliffe, and J. Gurland, Ed., Plenum Press, 1983, p 517–557

32. F.F. Lange, Powder Processing Science and Technology for Increased Reliability, *J. Am. Ceram. Soc.,* Vol 72 (No. 1), 1989, p 3–15

33. A.G. Evans, Perspective on the Development of High-Toughness Ceramics, *J. Am. Ceram. Soc.,* Vol 73 (No. 2), 1990, p 187–206

34. L.B. Pankratz, J.M. Stuve, and N.A. Gokcen, Thermodynamic Data for Mineral Technology, Bulletin 677, U.S. Bureau of Mines, 1984, p 98–102

35. I. Barin, O. Knacke, and O. Kubaschewski, *Thermochemical Properties of Inorganic Substances,* Springer-Verlag, 1977, p 861

36. A.S. Bolgar, M.I. Serbova, T.I. Serebryakova, L.P. Isaeva, and V.V. Fesenko, High Temperature Enthalpy and Heat Capacity of Borides of the Niobium-Boron System, translated from *Poroshk. Metall.,* Vol 3 (No. 243), 1983, p 57–62

37. E.D. Case, J.R. Smyth, and O. Hunter, Grain-Size Dependence of Microcrack Initiation in Brittle Materials, *J. Mater. Sci.,* Vol 15, 1980, p 149–153

38. M.K. Ferber, P.F. Becher, and C.B. Finch, Effect of Microstructure on the Properties of TiB_2 Ceramics, *J. Am. Ceram. Soc.,* Vol 66 (No. 1), 1983, p C2-C4

39. V.J. Tennery, C.B. Finch, C.S. Yust, and G.W. Clark, Structure-Property Correlations for TiB_2-Based Ceramics Densified Using Active Liquid Metals, *Science of Hard Materials,* R.K. Viswanadham, D.J. Rowcliffe, and J. Gurland, Ed., Plenum Press, 1983, p 891–909

40. G.P. Peterson and L.S. Fletcher, On the Thermal Conductivity of Dispersed Ceramics, *Trans. AIME,* Vol 111, 1989, p 824–829

41. S.K. Chung, Fracture Characterization of Armor Ceramics, *Am. Ceram. Soc. Bull.,* Vol 69 (No. 3), 1990, p 358–366

42. D. Kalish, E.V. Clougherty, and K. Kreder, Strength, Fracture Mode, and Thermal Stress Resistance of HfB_2 and ZrB_2, *J. Am. Ceram. Soc.,* Vol 52 (No. 1), 1969, p 30–36

43. H.R. Baumgartner, Mechanical Properties of Densely Sintered High-Purity Titanium Diborides in Molten Aluminum Environments, *J. Am. Ceram. Soc.,* Vol 67 (No. 7), 1984, p 490–497

44. D.E. Wiley, W.R. Manning, and O. Hunter, Jr., Elastic Properties of Polycrystalline TiB_2, ZrB_2, and HfB_2 from Room Temperature to 1300K, *J. Less Comm. Met.,* Vol 18, 1969, p 149–157

45. C. Tracy, M. Slavin, and D. Viechnicki, Ceramic Fracture During Ballistic Impact, *Advances in Ceramics,* Vol 22, *Fractography of Glasses and Ceramics,* J. Varner and V.D. Frechette, Ed., American Ceramic Society, 1988, p 295–306

46. E.L. Strauss and T.F. Keiffer, Strength of Zirconium Diboride Ceramics After Cyclic Loading or Heating, *J. Am. Ceram. Soc.,* Vol 58 (No. 9–10), 1977, p 399–401

47. R.H. Dauskardt, D.B. Marshall, and R.O. Ritchie, Cyclic Fatigue-Crack Propagation in Magnesia-Partially Stabilized Zirconia Ceramics, *J. Am. Ceram. Soc.,* Vol 73 (No. 4), 1990, p 893–903

48. B.L. Mordike, The Frictional Properties of Carbides and Borides at High Temperatures, *Wear,* Vol 3, 1960, p 374–387

49. J.R. Ramberg, C.F. Wolfe, and W.S. Williams, Resistance of Titanium Diboride to High-Temperature Plastic Yielding, *J. Am. Ceram. Soc.,* Vol 68 (No. 3), 1985, p C78-C79

50. J.R. Ramberg and W.S. Williams, High Temperature Deformation of Titanium Diboride, *J. Mater. Sci.,* Vol 22, 1987, p 1815–1826

51. N. Nakano, H. Matsubara, K. Nakamura, and T. Imura, High Temperature Hardness of Titanium Diboride Single Crystals, *Nagoya Kogyo Gijutsu Shikensho Hokoku,* Vol 29 (No. 11), 1980, p 323–331

52. K. Nakano, T. Okubo, K. Nakamura, and T. Sugimura, High Temperature Hardness of NbB_2 and TaB_2 Single Crystals, *Yogyo Kyakai Shi,* Vol 90 (No. 6), 1982, p 285–289

53. C.H. McMurtry, W.D.G. Boecker, S.G. Seshadri, J.S. Zanghi, and J.E. Garnier, Microstructure and Mechanical Properties of SiC-TiB_2 Particulate Composites, *Am. Ceram. Soc. Bull.,* Vol 66 (No. 2), 1987, p 325–329

54. M.A. Janney, Mechanical Properties and Oxidation Behavior of a Hot-Pressed SiC-15 vol. % TiB_2 Composite, *Am. Ceram. Soc. Bull.,* Vol 66 (No. 2), 1987, p 322–324

55. C. Kittel, *Introduction to Solid State Physics,* John Wiley & Sons, 1975

56. R. Shimizu, Y. Kataoka, T. Tanaka, and S. Kawai, Application of LaB_6 in Electron Microscopy, *Jpn. J. Appl. Phys.,* Vol 14, 1975, p 1089

57. L. Kaufman and E.V. Clougherty, "Investigation of Boride Compounds for Very High Temperature Applications, Part I," ManLabs, Inc. Report No. RTD-TDR-63-4096, Dec 1963

58. L. Kaufman and E.V. Clougherty, "Investigation of Boride Compounds for Very High Temperature Applications, Part II," ManLabs, Inc. Report No. RTD-TDR-63-4096, Feb 1965

59. R.K. Islamgaliev, A.V. Zyrin, A.A. Semenov-Kobzar, O.I. Shulishova, and I.A. Shcherbak, Oxidation of Powders of Rare-Earth Element Hexaborides in Air, *Poroshk. Metall.,* Vol 12 (No. 300), 1986, p 980–983

60. V.A. Lavrenko, S.S. Chuprov, A.P. Umanskii, T.G. Protsenko, and E.S. Lugovskaya, High Temperature Oxidation of Composite Materials Based on Titanium Diboride, *Sov. Powder Metall. Met. Ceram.,* Vol 26 (No. 9), 1988, p 761–762

61. T. Sata, High Temperature Vaporization of Nonoxides, *Taikabutsu,* Vol 34 (No. 293), 1982, p 49–53

62. C.E. Ransley, Refractory Carbides and Borides for Aluminum Reduction Cells, *J. Met.,* Vol 14 (No. 2), 1962, p 129–135

63. R.C. Dorward, Aluminum Penetration and Fracture of Titanium Diboride, *J. Am. Ceram. Soc.,* Vol 65 (No. 1), 1982, p C6

64. C.B. Finch and V.J. Tennery, Crack Formation and Swelling of TiB_2-Ni Ceramics in Liquid Aluminum, *J. Am. Ceram. Soc.,* Vol 65 (No. 7), 1982, p C100-C101

65. P.F. Becher, C.B. Finch, and M.K. Ferber, Effect of Residual Nickel Content on the Grain Size Dependent Mechanical Properties of TiB_2, *J. Mater. Sci.,* Vol 5 (No. 2), 1986, p 195–197

66. S. Baik and P.F. Becher, Effect of Oxygen Contamination on Densification of TiB_2, *J. Am. Ceram. Soc.,* Vol 70 (No. 8), 1987, p 527–530

67. E.S. Kang, C.W. Jang, C.H. Lee, C.H. Kim, and D.K. Kim, Effect of Iron and Boron Carbide on the Densification and

Mechanical Properties of Titanium Diboride Ceramics, *J. Am. Ceram. Soc.*, Vol 72 (No. 10), 1989, p 1868–1872

68. J.D. Katz, R.D. Blake, and C.P. Scherer, Microwave Sintering of Titanium Diboride, *Ceram. Eng. Sci. Proc.*, Vol 10 (No. 7–8), 1989, p 857–867

69. T. Watanabe and K. Shobu, Strength of Hot Pressed TiB_2-5% TaB_2-1% CoB Bodies Using Fine Powders, *Yogyo Kyokaishi*, Vol 94 (No. 12), 1986, p 1213–1217

70. K. Shobu, T. Watanabe, Y. Enomoto, K. Umeda, and Y. Tsuya, Frictional Properties of Sintered TiN-TiB_2 and $Ti(CN)$-TiB_2 Ceramics at High Temperatures, *J. Am. Ceram. Soc.*, Vol 70 (No. 5), 1987, p C103-C104

71. A. Kamiya, K. Nakano, and A. Kondoh, Fabrication and Properties of Hot-Pressed SiC Whisker-Reinforced TiB_2 and TiC Composites, *J. Mater. Sci. Lett.*, Vol 8, 1989, p 566–568

72. E. Rudy, "Sintered Cermets for Tool and Wear Applications," U.S. patent 4,022,584, 10 May 1977

73. W. Stadlbauer, W. Kladnig, and G. Gritzner, Al_2O_3-TiB_2 Composite Ceramics, *J. Mater. Sci. Lett.*, Vol 8, 1989, p 1217–1220

74. J. Liu and P.D. Ownby, "Enhanced Mechanical Properties of Alumina by Dispersed Titanium Diboride Particulate Inclusions," paper 88-SIV-90 presented at the 92nd Annual Conference of the American Ceramic Society, Apr 1990

75. J. Liu and P.D. Ownby, "Boron Containing Ceramic Particulate and Whisker Enhancement of the Fracture Toughness of Ceramic Matrix Composites," paper presented at the 10th International Symposia on Boron, Borides and Related Compounds, Aug 1990

76. E.S. Kang and C.H. Kim, Improvements in Mechanical Properties of TiB_2 by the Dispersion of B_4C Particles, *J. Mater. Sci.*, Vol 25, 1990, p 580–584

77. A.K. Knudsen and W. Rafaniello, "TiB_2-B_4C with High Hardness and Toughness," U.S. patent 4,957,884, 18 Sept 1990

78. T. Watanabe and K. Shoubu, Mechanical Properties of Hot-Pressed TiB_2-ZrO_2 Composites, *J. Am. Ceram. Soc.*, Vol 68 (No. 2), 1985, p C34-C36

79. R. Telle, S. Meyer, G. Petzow, and E.D. Franz, Sintering Behavior and Phase Reactions of TiB_2 with ZrO_2 Additives, *Mater. Sci. Eng.*, Vol A105/106, p 125–129

80. R.A. Cutler, Synthesis, Sintering, Microstructure and Mechanical Properties of Ceramics Made by Exothermic Reactions, *Proceedings of the First US-Japanese Workshop on Combustion Synthesis*, Y. Kaieda and J.B. Holt, Ed., Japan National Research Institute for Metals, 1990, p 73–80

81. W.B. Johnson, T.D. Claar, and G.H. Schiroky, Preparation and Processing of Platelet-Reinforced Ceramics by the Directed Reaction of Zirconium with Boron Carbide, *Ceram. Eng. Sci. Proc.*, Vol 10 (No. 7–8), 1989, p 588–598

82. T.D. Claar, W.B. Johnson, C.A. Andersson, and G.H. Schiroky, Microstructure and Properties of Platelet-Reinforced Ceramics Formed by the Directed Reaction of Zirconium with Boron Carbide, *Ceram. Eng. Sci. Proc.*, Vol 10 (No. 7–8), 1989, p 599–609

Engineering Properties of Carbides

Peter T.B. Shaffer, Technical Ceramics Laboratories, Inc.

CARBIDES are typical brittle ceramic materials. As such they, unlike metals, show no significant ductility or "graceful" failure mechanisms. Where the properties of metals are given as yield strength, ultimate strength, and ductility, the strength of ceramics is presented as a Weibull plot, the probability of failure as a function of the applied load.

The mechanical properties of carbides and other ceramics are determined and limited by the presence of strength-limiting defects. At the strength levels where ceramics are effective, these strength-limiting defects are of the order of only a few microns in size. Thus the strength of a carbide will be determined by control of the fabrication processes, even those processes by which the powder itself was prepared.

In addition to these limitations, the situation with carbides is further complicated by the fact that their composition can vary without outward evidence of this in their structure. A sample of titanium carbide, for example, will appear the same as one with a 10% carbon deficiency, except for a slight shift in the lattice spacings (Ref 1).

Finally, the mechanical properties of all ceramic materials are strongly dependent on their surfaces and the presence of defects resulting from their finishing and handling.

Classes of Metal Carbides

The three general classes of carbides are:

- *Ionic carbides,* which form from the metals of Groups I (sodium, potassium, . . .), Group II (calcium, magnesium, . . .), Group III (aluminum, yttrium, . . .), and the lanthanides
- *Covalent carbides* consist of only two compounds of consequence, SiC and B_4C, both of which are well-established engineering ceramics due mainly to their excellent thermal and chemical stability and extreme hardness
- *Interstitial carbides,* which exist over a rather wide range of stoichiometry and form as stable compounds mostly with transition elements of the Groups

IVa (titanium, zirconium, . . .), Group Va (niobium, tantalum, . . .), and Group VIa (chromium, molybdenum, tungsten, . . .), and much less stably with the lightest metals in Group VIII (Fe, Co, Ni)

From an engineering standpoint, the ionic carbides can be ignored. They are unstable under atmospheric conditions, reacting with moisture, even atmospheric humidity, to produce various hydrocarbons (Al_4C_3 to methane, CaC_2 to acetylene, Mg_2C_3 to methylacetylene, for example).

Interstitial carbides encompass by far the greatest number of materials, but to date the primary system of engineering importance is based on tungsten carbide. Titanium carbide also has found commercial application.

This article focuses on the two main covalent carbides (B_4C and SiC) and on systems derived from tungsten carbides. Other typical interstitial carbides besides tungsten carbide are also briefly discussed. From a practical view other carbides may offer tremendous potential, but in general this potential has not been achieved much beyond the laboratory scale.

Boron Carbide

Most commercial boron carbide is not a compound but rather a composite of B_4C and carbon (in graphitic form). The presence of second-phase graphite is a major factor that limits the strength of commercial boron carbide.

The structure of the range of boron carbide compositions can be visualized as a rhombohedron, formed by the elongation of a cube parallel to one of the diagonals. Three carbon atoms are located along this axis. Regular dodecahedrons that each contain twelve boron atoms are located at the corners (Fig 1). This is essentially the structure of boron carbide ($B_{12}C_3$ or, for simplicity, B_4C).

The carbon sites along the diagonal are of two types, referred to as A-B-A. Each A carbon atom has one nearest carbon neighbor, while the B carbon atom (shaded in Fig 1) has two equally spaced nearest carbon neighbors.

The three carbon atoms can be substituted, all or in part, by several types of atoms, or vacancies. If all are replaced by boron atoms, the structure becomes that of elemental, rhombohedral boron. If one is replaced by a boron atom, the structure becomes that of $B_{12}C_2B$ or $B_{13}C_2$. If the two are replaced it becomes $B_{14}C$. Neglecting atoms other than boron, it is possible to form a range of compositions from pure boron to B_4C. With only minor changes in lattice dimensions, all have the same crystal structure.

Stoichiometry. Compound composition can range from B_4C at 78.25 wt% boron to at least $B_{6.5}C$ at 85.4 wt% boron. On the carbon-rich side, no pure single-phase boron carbide having less than four borons per carbon has been proved. Thus, any boron carbide product containing more carbon than corresponding to the B_4C phase (about 21.6 wt% carbon) exists as a mixture of boron carbide and graphitic carbon (Ref 2, 3). Even the best nuclear grade of boron carbide has a minimum boron specification of 76.5% boron. This corresponds to 2.25% carbon beyond the stoichiometric limit for B_4C.

Most commercial boron carbide has graphitic carbon as a second phase, which limits strength. To make the situation even worse, the graphite typically found in boron carbide exists in the form of large, extremely thin, easily cleaved lamellae. This is the worst possible shape for a strength-limiting discontinuity in a high-strength matrix, especially a brittle ceramic.

Manufacture. The majority of commercial boron carbide is produced by reacting and fusing a mixture of boric oxide and carbon (together with a quantity of partially reacted recycled material which has no bearing on the discussion here) in an electric arc furnace. The product in the arc furnace consists essentially of a partially fused "slush" having the composition approximating that of B_4C. Boron carbide does not melt congruently. That is to say, the liquid in contact with solid boron carbide does not have the same composition as the solid (Fig 2).

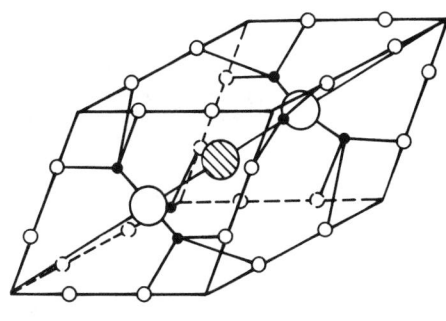

○ Boron

● Boron with links to carbon and other boron

◯ Carbon (type A, see text)

▨ Carbon (type B)

Fig 1 Structure of boron carbide. The shaded carbon atom represents the middle atom in the A-B-A sequence described in text.

The liquid in the arc furnace is solidified at a rate such that the solid and liquid do not reach an equilibrium. What happens is that the material left in the molten state becomes increasingly richer in carbon than in solid until, as the last liquid solidifies, its composition consists of that of a eutectic corresponding to a mixture of boron carbide and graphite. Thus, no matter how carefully boron carbide is prepared in practice, all arc furnace product will contain a quantity of free graphite.

A recent paper suggests that the incongruent melting of boron carbide may not be correct (Ref 4). This congruent melting is also shown in the boron carbon phase diagram in

Massalski (Ref 5), after Elliott (Ref 6). The fact remains, however, that commercial boron carbide typically contains lamellar crystals or graphite. Other processes such as the gas-phase reaction of boron suboxide and carbon yield a boron-rich boron carbide ($B_{4.1}C$) free of included graphite (Ref 7), but these syntheses do not constitute a significant fraction of the production.

Fabrication. Covalent ceramic materials such as boron carbide and silicon carbide are very difficult to fabricate in a fully dense, pure single-phase form except by hot pressing (including hot isostatic pressing). On a laboratory scale boron carbide has been sintered to quite high densities by additions of carbon to extremely finely milled boron carbide powder (Ref 8–11). As previously mentioned, however, the excess carbon yields a composite having less than ideal properties. Additional studies have been carried out to develop other metallic sintering aids that would eliminate the free carbon residual after sintering. To date these efforts have been only partially successful.

To date, all commercial boron carbide parts are produced by hot pressing or, in a few cases, by sintering followed by hot isostatic pressing. Typical hot pressing conditions of powder (<10 µm) are 2100 °C (3800 °F) at 35 MPa (5 ksi) for 30 min (Ref 8). Typical sintering and hot isostatic pressing of powder (<1 µm, 1 to 3 wt% C) consists of sintering at 2200 to 2250 °C (4000 to 4080 °F) for 30 min at low pressure (10 Pa, or 0.075 torr) followed by hot isostatic pressing at 2000 °C (3630 °F) and 200 MPa (29 ksi) in argon, for 120 min (Ref 8). Hot pressing limits the size and complexity of shapes that can be produced. Simple cylinders, blocks, flat plates,

axially symmetrical nozzles, and similar simple shapes can be formed directly by hot pressing.

Shaping these simple shapes requires very expensive diamond grinding. However, diamond grinding creates a multitude of microscopic flaws below the ground surface. These frequently are the source of strength-limiting defects within the machined shapes.

Properties. Typical mechanical and physical properties of boron carbide are listed in Table 1. Boron carbide has a low thermal conductivity and is very susceptible to thermal shock failure. Its outstanding hardness, however, is exceeded only by diamond and cubic boron nitride (Table 2). Mechanical strength can be influenced by processing (Table 3).

Hardness. In spite of the problems with stoichiometry and the presence of a free-graphite phase, dense boron carbide (even commercial hot pressed grades) is one of the hardest material mass produced on a commercial scale. Its hardness is significantly greater than that of silicon carbide, tungsten carbide and sapphire (Table 2). However, in spite of its extreme hardness, boron carbide

Table 1 Typical properties of boron carbide (B_4C)

Property	Value	Ref
General		
Density, g/cm³	2.51	
Boron content, %	78.28	
Carbon content, %	21.72	
Structure	Rhombohedral	
Lattice spacings of	$a = 5.60$	12
hexagonal cell, Å	$c = 12.10$	12
Color	Black	
Thermal		
Melting point, °C (°F)	2450 (4440)	
Boiling point, °C (°F)	3500 (6330)	
Specific heat, cal/g	0.226	12
Heat of formation, kcal/mol	−13.8	12
Entropy, kcal/mol · °C	6.5	12
Thermal expansion, 10^{-6}/°C, from room temperature to:		
800 °C (1470 °F)	5	12
800 °C (1470 °F)	4.5	13
500 °C (930 °F)	4.78	14
1000 °C (1830 °F)	5.54	14
1500 °C (2730 °F)	6.02	14
2000 °C (3630 °F)	6.53	14
Thermal conductivity, W/m · K, at:		
27 °C (80 °F)	28	15
975 °C (1790 °F)	16	15
20 °C (68 °F)	27	13
20 °C (68 °F)	29	16
425 °C (800 °F)	83	16
Mechanical(a)		
Young's modulus, GPa (10^6 psi)	445 (64.5)	Avg of Ref 8 and 17
Shear modulus, GPa (10^6 psi)	186.5 (27)	Avg of Ref 8 and 17
Bulk modulus, GPa (10^6 psi)	254 (37)	17

(continued)

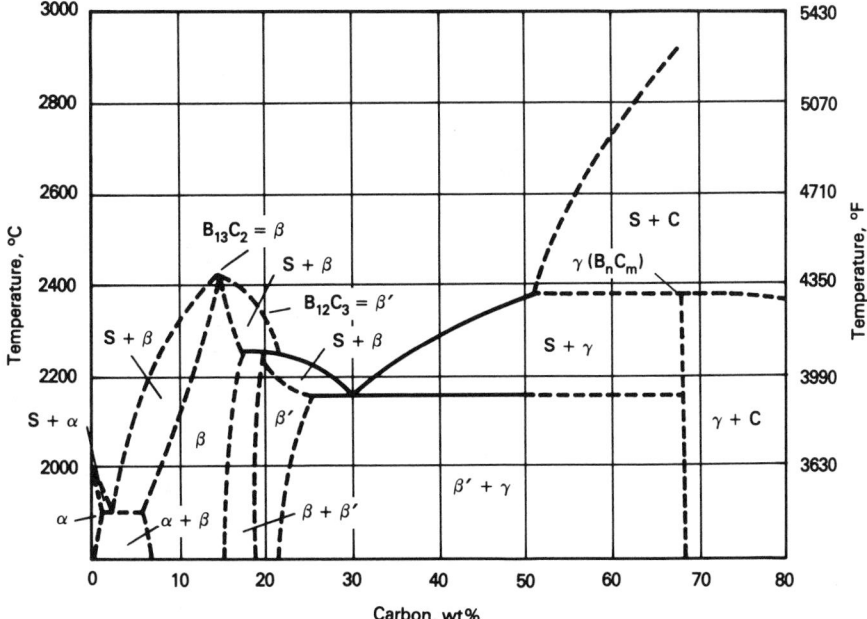

Fig 2 Phase diagram of boron-carbon system after Samsonov. Source: Ref 3

Table 1 continued

Property	Value	Ref
Speed of sound, km/s (miles/s)	14.5 (9)	17
Flexural strength, MPa (ksi), at:		
room temperature	480 (70)	8
room temperature	303 (44)	13
room temperature	345 (50)	18
650 °C (1200 °F)	290 (42)	18
1100 °C (2000 °F)	241 (35)	18
Tensile strength at 980 °C (1800 °F), MPa (ksi)	155 (22.5)	16
Compressive strength, MPa (ksi)	2855 (414)	13
Poisson's ratio	0.17	8
	0.207	17
Hardness numbers, kg/mm^2		
Knoop (100 g load)	3000	12
Knoop (100 g load)	3060	18
Knoop (100 g load)	2800	19
Knoop (1000 g load)	2230	19
Vickers	3700	19

Electrical and optical

Resistivity, $\Omega \cdot$ cm, at:		
room temperature	0.1–10	12
600 °C (1100 °F)	0.038	20
1000 °C (1830 °F)	0.030	20
2000 °C (3630 °F)	0.022	20
Coefficient of thermal emf, uV/°C	80	20
Band gap, eV	1.64	20
Emissivity of 0.65 μm energy at:		
880 °C (1615 °F)	0.76	21
1880 °C (3415 °F)	0.56	21
Total emissivity	0.85	20
Magnetic	Nonmagnetic	

Chemical stability(b)

In helium	Stable to 2250 °C (4080 °F)	16
In O$_2$	Stable to 540 °C (1000 °F)	16
In sulfur, 1200 °C (2200 °F)	Stable	22
In phosphorus, 1200 °C (2200 °F)	Stable	22
In iodine, 1200 °C (2200 °F)	Stable	22
In CO	Stable	22
In Cl$_2$, 600 °C (1100 °F)	Reacts	22
In metal oxides, >1500 °C (2730 °F)	Reacts(c)	22

Corrosion

Air, maximum-use temperature, °C (°F)	1000 (1830)	22
General corrosion weight loss (mg/cm^2) in:		
Concentrated HCl	0.095(d)	22
Concentrated H$_2$SO$_4$	0.130(d)	22
Concentrated HNO$_3$	0.269(d)	22
Concentrated H$_3$PO$_4$	0.086(d)	22
40% HF	0.095(d)	22
40% HF + concentrated HNO$_3$ + concentrated H$_2$SO$_4$	0.139(d)	22
25% NaOH	0.069(d)	22
25% KOH	0.068(d)	22

(a) Unless otherwise noted, room-temperature mechanical properties are given. (b) Stability of all carbide powders is a function of their surface area. (c) Boride reacts with many metal oxides at temperatures greater than 1500 °C (2730 °F). (d) General corrosion rate for a hot pressed product (2.49 g/cm^3 density) in the specified environment at 200 °C (390 °F) for 8 h

Table 2 Hardness of several common abrasives

Material	Knoop hardness (100 g), kg/mm^2
Sapphire (alumina)	2000–2050
Tungsten carbide (WC)	2050–2150
Silicon carbide (SiC)	2150–2950
Boron carbide (B$_4$C)	2900–3100
Cubic boron nitride (CBN)	4500–4600
Diamond (C)	8000–8500

is not useful as a bonded abrasive (grinding wheels, sharpening stones). The material is extremely brittle, and fractures into rounded fragments which are not effective in stock removal. All boron carbide abrasives are used instead in loose, slurry form.

Electrical. Boron carbide is a semiconductor, but it is not produced in a sufficiently pure form for its semiconductor properties to be studied in detail, nor is it available in a form in which it might be used as an electrical device. Table 1 gives a broad range of resistivity at room temperature, but its resistivity has been reported to be 0.3 to 0.8 $\Omega \cdot$ cm (Ref 13, 23).

Nuclear Properties. Because it has a good thermal-neutron capture cross-section of 3960 barns (Ref 24), one of the major applications for boron carbide is as a neutron absorber and shielding material. It is used in nuclear reactor control rods and as a shielding around spent fuel rods. Absorption of a thermal neutron liberates a low-energy particle. More importantly, however, decay of the $_5B^{10}$ atom after capture of the neutron yields no long-life daughter products or high-energy emissions.

$$_5B^{10} + {}_0n^1 \rightarrow {}_3Li^7 + {}_2He^4$$

Uses. Boron carbide is used in only a few applications, none of which utilize mechanical strength in the usual sense. It is used as a thermal-neutron absorber and as a loose abrasive. In hot pressed form, it is used in nozzles for abrasive slurries, and in wear applications in which the load is either compressive or sliding.

Silicon Carbide

Structures. Whereas boron carbide has a B$_4$C structure over a wide range of chemical compositions, several hundred structures (polytypes) of silicon carbide have been identified. The simplest of the polytypes of silicon carbide has the diamond structure in which alternate carbon atoms are replaced by silicon. This cubic structure is referred to as β. All other structures show either hexagonal or rhombic symmetry and are collectively called α.

Differences between these structures are simply a matter of stacking arrangements. All SiC structures are made up of hexagonal double layers; that is, one layer of carbon atoms in a hexagonal pattern placed above a layer of hexagonal silicon atoms. In this SiC struc-

ture, every carbon atom is surrounded tetrahedrally by four silicon atoms, and each silicon atom is surrounded tetrahedrally by four carbon atoms. Differences in the crystal symmetry arise from the different ways that the hexagonal layers are arranged one on the other.

In cubic SiC, each double layer is arranged above the next lower simply by translating it as shown in Fig 3(a). This arrangement gives rise to a cubic symmetry. The layer sequence arbitrarily is designated aaaa. Three layers develop a repetitive unit cell. This structure is referred to as 3C. (In this system of notation, the number refers to the number of layers in the unit cell, the letter to the crystallographic symmetry of the cell, C, cubic; H, hexagonal; R, rhombohedral.)

If the added layer is translated along a diagonal rather than parallel to the edge of the hexagon, the result is a structure in which the cubic symmetry disappears. If each layer is translated diagonally, alternate layers will be directly above one another (Fig 3b). Such a structure, which corresponds to the Wurtzite structure, is designated abab, and this is the simplest hexagonal structure. The structure is called 2H in the notation system described above.

In between these two most symmetrical structures are a large number of layer arrangements, some hexagonal in symmetry, some rhombohedral. Structures having repeat distances of many hundred layers have been identified. The various silicon carbide polytypes are described in a more detailed review (Ref 25). A summary of four basic structures is given in Table 4.

Manufacture. Neglecting films, surface coatings, and infiltrated structures of silicon carbide, all of which are prepared by chemical vapor deposition (CVD) or chemical vapor infiltration (CVI), silicon carbide in massive forms is derived from a powder or grain prepared by carbon reduction of silica. The original preparation of silicon carbide (Ref 27), and the production furnace devised by Acheson shortly thereafter (Ref 28) are largely unchanged during the century since their inception. Depending on the time and temperature of reaction, the resulting silicon carbide is either in the form of a fine powder or as a large bonded mass, which is crushed and sorted to provide typical abrasive powders and grains.

After crushing, milling, and separation into specific size fractions (submicron for sintering), the resulting silicon carbide powders are roasted to oxidize carbon and washed with hydrofluoric acid to remove silica to the required purity.

Fabrication. Commercial silicon carbide products are fabricated in one of three ways, depending on the requirements of the intended use. One method mixes the SiC powder with a second material such as resin, glass, silicon nitride, clay, or metal. This composite material is then treated to allow the second phase to develop a bond.

Table 3 Mechanical properties versus product condition of boron carbide

Atomic formula	Carbon content, wt%	Fabrication method	Density, g/cm³	Flexural strength(a)		Young's modulus		Shear modulus		Poisson's ratio
				MPa	ksi	GPa	10⁶ psi	GPa	10⁶ psi	
B₄C	21.7	Hot pressed	2.51	480 ± 40	70 ± 6	441	64	188	27	0.17
B₄C + 1C	22.5	Sintered + HIP	2.51	400 ± 20	58 ± 3	433	63	183	26.5	0.18
B₄C + 3C	24.8	Sintered + HIP	2.50	430 ± 63	62 ± 9	405	59	171	25	0.16

(a) Four-point bending. Source: Ref 8

Silicon carbide powder also can be mixed with carbon or silicon metal powder and formed into shapes that are reaction bonded. The carbon may react with silicon vapor to form a silicon carbide bond (self-bonded), or the silicon metal can react with nitrogen to form a silicon nitride bond. Either of these processes is referred to as reaction bonding.

In the early 1960s it was found that very fine silicon carbide powder (<1 μm average particle diameter) could be densified and sintered by additions of boron carbide (Ref 29, 30). In the mid-1970s Prochaska and his coworkers optimized the process and made high-density sintered silicon carbide practical (Ref 31, 32).

Each process has advantages as well as limitations. Bonding with second phases is standard for abrasives and most refractory applications. The products are limited in their applications, to a large extent by the properties of the bond phases. These are, however, the most economical products.

Self-bonded silicon carbide generally contains a quantity of free, unreacted silicon metal in the pores. This melts and exudes at temperatures above the melting point of silicon (1400 °C). It can be removed after the reaction bonding step by refiring in vacuum, or in the presence of excess carbon. Self-bonded silicon carbide exhibits better oxidation resistance than does sintered material.

Sintered silicon carbide, especially that material prepared with care to prevent the formation of exaggerated grain growth and other internal strength-limiting flaws, is the choice for mechanical applications where oxidation is not the limiting factor. Because sintered silicon carbide contains both boron and free carbon, it has less oxidation resis-

tance than reaction-bonded material. Free carbon oxidizes and develops pores in the body. Boron produces boric oxide during oxidation, which increases the fluidity of the protective silica layer and lowers its oxidation resistance.

Thermal Properties. Silicon carbide is well known in its refractory applications. It has excellent thermal conductivity and a low thermal expansion coefficient (Table 5), which together give rise to outstanding thermal shock resistance.

Thermal conductivity is adversely affected by the presence of impurities dissolved in the crystal structure. While the thermal conductivity of highly pure silicon carbide is 490 W/m · K (Ref 35), practical values of over 100 W/m · K are seldom achieved (Table 5). It is difficult to obtain high-purity silicon carbide commercially; impurities (boron) are added for sintering, or impurities are introduced from the silicon used in reaction bonding.

Mechanical Properties. Several values of the elastic moduli of silicon carbide in several forms of commercial material are included in Table 5. The data include values for reaction-bonded silicon carbide at ambient and elevated temperatures (Ref 36). The values were significantly higher, and much less temperature dependent than earlier reported (Ref 39). These differences are likely due to the increased purity of the materials used in the later study. McMurtry and coworkers later presented results of a study that covered pyrolytic silicon carbide, as well as addressing the effects of grain size and hot pressing (Ref 33). These are also included in Table 5.

Sintered silicon carbide is one of the strongest ceramic materials. Its strength is limited by the presence of a variety of flaws, crystallite agglomerates, oversize, elongated grains, and porosity. While the potential is there, the process and economics have not yet been optimized to the point where they can

be applied in mechanical applications except those where failure will not be catastrophic. In order to use ceramic materials in critical applications, it is necessary to introduce so many safety factors, that the ceramic materials are no longer economically viable.

Hardness. The property that allowed silicon carbide to be commercialized is its hardness. When Acheson (Ref 28) first discovered it, he noted this extreme hardness, and even spoke of its ability to cut diamonds. While this may be somewhat of an overstatement, the fact remains that silicon carbide is one of the most effective abrasives. It is not as hard as boron carbide, but it exhibits a conchoidal fracture that makes it extremely effective in material removal.

Hardness of silicon carbide varies depending on the crystallographic direction, the facet being exposed, the impurities present, and whether the surface has been polished. Even the test atmosphere has an effect. The hardness of silicon carbide and its relation to crystallographic direction are summarized in Table 6 from two papers.

Electrical. Silicon carbide is a semiconductor. Its energy gap is related to the structure (polytype). It ranges from 2.2 eV for the cubic material to 3.3 eV for the simple hexagonal polytype (2H, Wurtzite form) (Ref 42, 43).

Resistivity can vary by as much as seven orders of magnitude (Ref 44). If pure, silicon carbide is essentially an insulator. In contrast, the writer and coworkers have prepared dense polycrystalline parts, having resistivities approaching 0.001 Ω · cm (Ref 45). Of special interest, the nitrogen-saturated material exhibited a positive temperature coefficient from room temperature to 1000 °C (1830 °F). Such behavior is typical for metals, whereas silicon carbide semiconductors normally exhibit negative coefficients.

An extensive summary of the electrical and semiconducting properties of silicon carbide

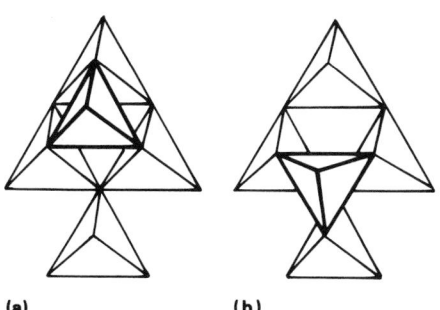

(a)   **(b)**

Fig 3 Hexagonal layers of silicon carbide arranged in (a) a type 3C (cubic) structure and (b) a type 2H structure. See text.

Table 4 Selected crystal structures of silicon carbide

Structure(a)	Structure type	Pearson symbol	Space group	Lattice spacings, nm
Cubic (3C)	ZnS/Zinc blende	cF8	F43m	a = 0.4358
Hexagonal (2H)	ZnS/Wurtzite	hP4	P6₃mc	a = 0.30763, c = 0.50480
Hexagonal (6H)	SiC	hP12	P6₃mc	a = 0.30806, c = 1.51173
Rhombohedral (15R)	. . .	hR48	R3m	a = 0.3082, c = 6.049

(a) See text for the description of the designators in parentheses. Source: Ref 26

Table 5 Summary of silicon carbide properties

Property	Value	Ref
Thermal properties		
Expansion coefficient, $10^{-6}/°C$, of:		
3.186 g/cm³ (sublimed)	4.51–4.73 (RT to 1250 °C)	33
3.190 g/cm³ (CVD at 25 °C)	4.78 (RT to 1250 °C)	33
3.10 g/cm³ (sintered α)	4.02 (RT to 700 °C)	34
Thermal conductivity, W/m·K, of:		
High-purity SiC	490	35
Sintered α	125.6	34
3.186 g/cm³ (sublimed)	110 (at 100 °C)	33
Sintered α	102.6 (at 200 °C)	34
3.186 g/cm³ (sublimed)	57 (at 1000 °C)	33
Hot pressed (2% alumina)	78 (at 100 °C)	33
Sintered α	77.5 (at 400 °C)	34
Hot pressed (2% alumina)	36 (at 1000 °C)	33
CVD	13.7 (at 100 °C)	33
CVD	9.95 (at 1000 °C)	33
Specific heat, J/g·K, of sintered α	0.67	34
Mechanical properties		
Young's modulus, GPa (10⁶ psi), of:		
Sublimed (0% porosity)	475 (69)	36
Sublimed (3.186 g/cm³)	459–476 (66.5–69)	33
CVD (3.190 g/cm³)	441–469 (64–68)	33
Hot pressed (2% alumina)	414–445 (60–64.5)	33
Sublimed (3.157 g/cm³)	420–438 (61–63.5)	36
Sintered α (3.10 g/cm³)	410 (59.5)	34
Reaction bonded (3.12 g/cm³)	382–394 (55.5–57)	36
Recrystallized (2.237 g/cm³)	143 (20.7)	36
Shear modulus, GPa (10⁶ psi) of:		
Sublimed (3.157 g/cm³)	181 (26)	36
Reaction bonded (3.12 g/cm³)	162 (23.5)	36
Recrystallized	59.5 (8.6)	36
Poisson's ratio (sintered α)	0.14	34
Weibull modulus (sintered α)	10	34
Fracture toughness, MPa√m (ksi√in.)	4.60 (42)	34
Flexural strength(a), MPa (ksi), of:		
Sublimed (3.186 g/cm³)	228–261 (33–37.8)	33
Hot pressed (2% alumina)	552–862 (80–125)	33
CVD (3.190 g/cm³)	255–465 (37–67.4)	33
Compressive strength (sintered α), MPa (ksi)	4600 (667)	34
Hardness	see Table 6	
Other physical properties		
Electrical resistivity	see text	
Refractive index	2.65–2.7	37
Birefringence ($N_e - N_o$)	0.042–0.097	37, 38
Dispersion (optical)	~0.085	37
General corrosion, mg/cm²·y, in:		
98% H_2SO_4 at 100 °C	1.8(b)	34
	55(c)	34

(continued)

Table 5 continued

Property	Value	Ref
50% NaOH at 100 °C	2.5(b)	34
	≥1000(c)	34
53% HF at 25 °C	<0.2(b)	34
	7.9(c)	34
85% H_3PO_4 at 100 °C	<0.2(b)	34
	8.8(c)	34
70% HNO_3 at 100 °C	<0.2(b)	34
	0.5(c)	34
45% KOH at 100 °C	<0.2(b)	34
	>1000(c)	34
25% HCl at 70 °C	<0.2(b)	34
	0.9(c)	34
10% HF + 57% HNO_3 at 25 °C	<0.2(b)	34
	>1000(c)	34

(a) Three-point bend test of flexural strength with a specimen 3 × 3 × 25 mm (0.12 × 0.12 × 1 in.). (b) Sintered α structure. (c) Corrosion in a wear environment from pin-on-disk friction

is appended in the last silicon carbide conference review (Ref 42).

Optical. It has been said that silicon carbide should be the perfect gem (Ref 46). It can be prepared in a wide range of colors, colorless (pure α/hexagonal), yellow (β/cubic), green (nitrogen or phosphorus doped), blue (aluminum doped), brown (boron doped), and black (heavily doped with aluminum). The refractive index (Table 5) is greater than that of diamond (2.65 versus 2.4 for diamond), the variation of refractive index with wavelength (dispersion) is more than twice that of diamond (0.10 versus 0.044 for diamond), and unlike diamond it can be made both isotropic (like diamond) and birefringent (unlike diamond).

Several papers have been published describing the optical properties of silicon carbide (Ref 37, 47, 48). Data from these papers are presented in Table 5.

Magnetic. Silicon carbide exhibits no magnetic properties of note. Its magnetic susceptibility has been reported (Ref 42, 49).

Tungsten Carbide

Tungsten and carbon form two simple compounds of interest, ditungsten carbide (W_2C) and tungsten monocarbide (WC). They also form another compound (pentatungsten tricarbide, W_5C_3) which is stable only above about 2535 °C, or 4595 °F (Ref 5). Villars (Ref 26) reports the formula $W_{31}C_{19}$. Of these two stable phases the monocarbide is the one of primary interest. It is the material on which the entire cemented carbide industry is based.

Cemented Carbides. Tungsten carbide parts, dies, and cutting tools are commonly referred to as tungsten carbide, although they are composites in which the tungsten carbide grains are bonded to one another by a metallic phase (typically cobalt). It is the combination of the rigid carbide network and the tougher metal substructure that provides these cemented carbides with their outstanding properties.

The amount of the metal phase, usually cobalt or an alloy of cobalt, inversely affects the hardness, corrosion resistance, and rigidity of the composite, and directly improves the toughness. More metal yields a softer, lower modulus, tougher composite that is more prone toward corrosive degradation. Grain size of the carbide phase also influences properties (Table 7).

There is a wide range of cemented carbide compositions (Tables 7 and 8). The metal binder can be varied in terms of the quantity and/or composition of the metal phase, while the carbide phase may contain carbides other than WC. For example, a number of cemented carbide grades contain tantalum carbide (TaC) and titanium carbide (TiC), either singly or in combination. However, the best combination of strength, wear resistance, and good toughness is provided by the straight (WC-Co) cemented carbides (Table 7). The wide range of cemented carbide products and producers is covered extensively in Brookes' *World Directory and Handbook of Hardmetals* (Ref 50).

Two processes for the production of these compositions (WC-TiC, WC-TaC, WC-TaC-TiC) predominate. The constituent carbides are intimately blended, typically by ball milling, then subjected to high-temperature annealing in hydrogen. They have also been reported to be produced by the reaction of the

Table 6 Effect of crystal orientation on hardness of silicon carbide

Crystal structure	Indentation test plane	Orientation of Knoop indenter	Knoop hardness(a), kg/mm²
6H polytype	0001 face	Parallel 1010	2917
6H polytype	0001 face	Parallel 1120	2954
6H polytype	1010 face	Parallel C axis	2129
6H polytype	1010 face	Perpendicular C axis	2700
6H polytype	1120 face	Parallel C axis	2391
6H polytype	1120 face	Perpendicular C axis	2717
Cubic (β)	100 face	Parallel edge	2525(b)
			2733(c)
Cubic (β)	100 face		2732(b)
		Perpendicular edge	2852(c)
Cubic (β)	111 face		2758(b)
		Parallel edge(d)	2878(c)
Cubic (β)	111 face		2772(b)
		Perpendicular edge(d)	2828(c)

(a) 100 g load. (b) Natural face. (c) Polished face. (d) Frequent cracking. Source: Ref 40, 41

Table 7 Properties of representative cobalt-bonded cemented carbides

Nominal composition	Grain size	Hardness, HRA	Density g/cm³	Density oz./in.³	Transverse strength MPa	Transverse strength ksi	Compressive strength MPa	Compressive strength ksi	Modulus of elasticity GPa	Modulus of elasticity 10⁶ psi	Relative abrasion resistance(a)	Coefficient of thermal expansion, μm/m·K at 200 °C (390 °F)	Coefficient of thermal expansion, μm/m·K at 1000 °C (1830 °F)	Thermal conductivity, W/m·K
97WC-3Co	Medium	92.5–93.2	15.3	8.85	1590	230	5860	850	641	93	100	4.0	...	121
94WC-6Co	Fine	92.5–93.1	15.0	8.67	1790	260	5930	860	614	89	100	4.3	5.9	...
	Medium	91.7–92.2	15.0	8.67	2000	290	5450	790	648	94	58	4.3	5.4	100
	Coarse	90.5–91.5	15.0	8.67	2210	320	5170	750	641	93	25	4.3	5.6	121
90WC-10Co	Fine	90.7–91.3	14.6	8.44	3100	450	5170	750	620	90	22	...	...	...
	Coarse	87.4–88.2	14.5	8.38	2760	400	4000	580	552	80	7	5.2	...	112
84WC-16Co	Fine	89	13.9	8.04	3380	490	4070	590	524	76	5	...	...	...
	Coarse	86.0–87.5	13.9	8.04	2900	420	3860	560	524	76	5	5.8	7.0	88
75WC-25Co	Medium	83–85	13.0	7.52	2550	370	3100	450	483	70	3	6.3	...	71
71WC-12.5TiC-12TaC-4.5Co	Medium	92.1–92.8	12.0	6.94	1380	200	5790	840	565	82	11	5.2	6.5	35
72WC-8TiC-11.5TaC-8.5Co	Medium	90.7–91.5	12.6	7.29	1720	250	5170	750	558	81	13	5.8	6.8	50

(a) Based on a value of 100 for the most abrasion-resistant material. Source: 10th Edition *Metals Handbook*, Volume 2, 1990, p 952

Table 8 Representative compositions and proprietary designations of various cemented carbide grades

WC	Co	TaC	TiC	Ni	Cr	Mo₂C	Adamas Carbide	Anderson Strathclyde	Carbidie	Carmet	Danit	General Carbide	General Electric Carboloy	GTE Valenite
96.5	3.0	0.5	...	...	...	...	GU2	CA	CD20	CA8	K04	GC003	999	VC3
94.0	5.5	0.5	...	...	...	...	PWX	CF	CD24	CA306	K10	GC005	895	VC2
94.0	6.0	...	...	...	...	...	A	CG	CD30	CA4	K20	GC106	883	...
91.0	9.0	...	...	...	...	...	B	...	CD35F	CA12	K30	GC009	...	VC152
87.0	13.0	...	...	...	...	...	BB	...	CD40	CA10	...	GC313	258	VC11
81.5	18.0	0.5	...	...	...	...	GU1	...	CD650	...	...	...	...	...
79.0	12.0	9.0	...	...	...	...	474	...	...	...	...	...	...	VC047
...	...	...	74.0	12.5	...	13.5	Titan 80	...	...	CA100	...	...	...	VC83
...	...	...	70.5	17.5	1.0	11.0	Titan 60	...	...	...	...	...	...	...
...	...	...	66.5	22.5	1.0	10.0	Titan 50	...	...	...	...	...	...	...
89.0	...	...	1.0	10.0	...	...	...	...	...	R10	...	...	...	VC099
85.0	15.0	...	...	...	...	...	HD15	CPM	CD50	CA11	DG30	GC315	268	VC12
80.0	20.0	...	...	...	...	...	HD20	...	...	CA20	DG40	GC320	...	VC13
75.0	25.0	...	...	...	...	...	HD25	...	...	CA225	DG50	GC325	190	VC14
75.0	20.0	5.0	...	...	...	...	HD20T	...	CD60	...	...	...	...	...
70.0	25.0	5.0	...	...	...	...	HD25T	...	CD70	...	...	...	...	...
94.0	6.0	...	...	...	...	...	575	CR	...	CA3	B030	GC206	44A	...
90.0	10.0	...	...	...	...	...	569	CM	...	2102	B050	GC410	90	...
89.0	11.0	...	...	...	...	...	783	...	CD337	CA411	B055	GC411	115	...
88.0	12.0	...	...	...	...	...	502	CT	...	CA412	B060	GC412	120	...

WC	Co	TaC	TiC	Ni	Cr	Mo₂C	Kennametal	Krupp Widia	Mefasa	Metallwerk Plansee	Mitsubishi	Sandvik	Sumitomo	Teledyne Firth Sterling
96.5	3.0	0.5	...	...	...	...	K11	THF	...	H03T	...	CS05	...	HF
94.0	5.5	0.5	...	...	...	...	K68	GT05	K1	H10T	GTi05	CS10	H1	HA
94.0	6.0	...	...	...	...	...	K6	GT10	K2	H16T	GTi10	HML	G10E	H6
91.0	9.0	...	...	...	...	...	K9	GT15	MK30	H30T	GTi15	H10F	G3	H8
87.0	13.0	...	...	...	...	...	...	GT3H	...	H40T	GTi20	R4	G5	H81
81.5	18.0	0.5	...	...	...	...	...	...	...	...	GTi40	...	...	...
...	...	...	74.0	12.5	...	13.5	K165	...	...	F05T	NX33	CN02	...	...
...	...	...	70.5	17.5	1.0	11.0	...	...	...	F10T	NX55	...	T12A	...
...	...	...	66.5	22.5	1.0	10.0	...	TTF	...	...	...	...	T12B	...
88.2	1.8	10.0	...	...	...	...	K602	...	...	...	...	...	...	...
85.8	10.1	...	...	...	4.1	...	K701	...	...	...	...	...	...	...
93.3	5.8	...	...	...	0.9	...	K703	...	...	...	...	...	...	...
88.4	6.1	4.5	1.0	...	...	...	K714	...	...	...	...	...	...	...
93.7	...	0.3	...	6.0	...	...	K801	...	...	WC6Ni	...	...	...	...
89.0	...	...	1.0	10.0	...	...	K803	...	...	TCR30	...	...	...	...
85.0	15.0	...	...	...	...	...	SP212	BT40	G3	B50T	...	CT60	G6	MPD160
80.0	20.0	...	...	...	...	...	...	...	G4	H60T	GTi40	CT75	G7	...
75.0	25.0	...	...	...	...	...	...	GT55	G5	H70T	...	CT85	G8	...
75.0	20.0	5.0	...	...	...	...	K91	...	...	...	...	...	...	ND20
70.0	25.0	5.0	...	...	...	...	K90	...	...	...	...	...	...	ND25
94.0	6.0	...	...	...	...	...	K3404	BT10	K3	B10T	...	CT30	...	HAN6
90.0	10.0	...	...	...	...	...	K3070	BT25	MK35	B30T	...	CT45	G3	MPD10
89.0	11.0	...	...	...	...	...	K3047	...	MK40	B36T	...	CT50	...	MPD11
88.0	12.0	...	...	...	...	...	K3030	BT30	...	B40T	...	...	G5	...

Note: This table lists approximate compositions and proprietary designations for a number of corrosion-resistant grades. These grades from 16 manufacturers worldwide are meant to be representative only in a general sense. There are well over 100 manufacturers throughout the world (over 25 in the United States alone); therefore, it is not feasible to include all. In addition, cross comparisons are not precisely possible. For example, a grade listed with an approximate composition of WC-25Co may be cross referenced with a comparable grade that contains only 24% Co. Reference 50 contains more complete data on any grade, and manufacturers can be consulted for more information. Source: 9th Edition *Metals Handbook*, Volume 13: *Corrosion*

Table 9 Crystal properties of tungsten carbide

Compound	Crystal structure	Structure type	Pearson symbol	Space group	Lattice parameters, nm
Monotungsten carbide (WC)	Face-centered cubic	WC	hP2	P6̄m2	$a = 0.29063$
Ditungsten monocarbide (W$_2$C)	Hexagonal(a)	CdI$_2$	hP3	P3̄m1	$a = 0.300$ $c = 0.4730$

(a) Low-temperature stable to 2753 K (4500 °F). Source: Ref 26

constituent carbides in a metal melt (Ref 51–53).

Recent research has shown that solid solutions of various carbides can be produced, free of the individual constituent phases, by direct reaction of the blended oxides with carbon in a continuous tube furnace (Ref 54).

Properties. Although pure carbides are never used as engineering materials, the properties of WC and W$_2$C pure phases are given in Tables 9 and 10. For properties of actual tungsten carbide products (that is, cemented carbides), the reader is referred to Ref 50 or to Volume 2 of the 10th Edition of *Metals Handbook*.

The use of W$_2$C is rather limited, primarily to applications in which the material will be fused during processing. Since the 1:1 ratio carbide does not melt congruently, its use under such conditions is prohibited. Ditungsten carbide is used in combination with metals such as nickel as a plasma spray hard facing material and as a powder filling for tubular hard facing welding electrodes.

Other interstitial carbides (Table 10) besides tungsten carbide show promise based on their refractory nature, high hardness, and comparatively low densities. However, none of these carbides have the widespread use of tungsten carbide.

Like tungsten carbide, the other interstitial carbides have been used as the hard phase in cermets, which is a general term for a heterogenous composite of a ceramic phase in combination with a metal alloy phase. In a broad sense, cemented carbides constitute the largest class of cermets.

Since the inception of ceramic technology, the dominant concept has been that of a material based on TiC as the primary hard and refractory constituent, with the bonding provided by any of a variety of lower-melting ductile metals or alloys (much the same as those used for cemented tungsten carbides). The TiC cermets have found use in tool and wear-resistance applications; in selected high-stress, high-temperature systems; and in corrosive environments. Cermets based on TiC are also used as cutting tool materials (see Volume 2, 10th Edition of *Metals Handbook*, p 995–1000). Cermets with a chromium carbide (Cr$_3$C$_2$) base have been used for a variety of corrosion-resistance applications and as gage blocks; however, they have apparently lost much of their industrial usage.

REFERENCES

1. P. Schwarzkopf and R. Kieffer, *Refractory Hard Metals—Borides, Carbides, Nitrides and Silicides*, MacMillan, 1953
2. Shuravlev *et al.*, *Nachr. Akad. Wiss. USSR*, Vol 1-S, 1961, p 113
3. G.V. Samsonov, *Metallphys. Metallkund. SSSR Ural Filial*, Vol 3-S, 1956, p 309
4. M. Bouchacourt and F. Thevenot, The Melting of Boron Carbide and the Homogeneity Range of the Boron Carbide Phase, *J. Less-Comm. Metals*, Vol 67 (No. 2), 1979, p 327–331
5. T.B. Massalski, *Binary Alloy Phase Diagrams*, American Society for Metals, 1986
6. R.P. Elliott, *Constitution of Binary Alloys*, McGraw-Hill, 1965
7. P.T.B. Shaffer and K.A. Blakely, Production and Properties of Sub Micron Type 1 Boron Carbide Powder, Conference Proceedings Series MMCIAC No. 696, *10th Annual Discontinuously Reinforced Metal Matrix Composites Working Group*, Park City, UT, 5 Jan 1988, p 141–159
8. K.A. Schwetz, W. Grellner, and A. Lipp, Mechanical Properties of Hot Isostatically Pressed Treated Sintered Boron Carbide, *Inst. Phys. Conf. Ser.*, No. 75, A. Hilger Publishing, 1986
9. S. Prochazka, S.L. Dole, and C.I. Hejna, Abnormal Grain Growth and Microcracking in Boron Carbide, *J. Am. Ceram. Soc.*, Vol 68 (No. 9), 1985, p C235-C236
10. S. Prochazka and S.L. Dole, Densification and Microstructure Development in Boron Carbide, *Ceram. Eng. Sci. Proc.*, Vol 6 (No. 7/8), 1985
11. K. Schwetz and W. Grellner, The Influence of Carbon on the Microstructure and Mechanical Properties of B$_4$C, *J. Less-Comm. Metals*, Vol 82, 1981, p 37–47
12. A. Graf von Matuschka, Technical Use of Boron Compounds, *Elektroschmelzwerk Kempten Spl. Reprint*, No. 98, 1974
13. G.R. Finlay, Refractories for 4000 F. and Higher, *Chem. Canada*, Vol 4 (No. 41), 1952
14. O.H. Krikorian, "Thermal Expansion of High Temperature Materials," UCRL Research Report 6132, Sept 1960
15. M. Bouchacourt and F. Thevenot, The Correlation between the Thermoelectric Properties and Stoichiometry in the Phase B$_4$C-B$_{10.5}$C, *J. Mater. Sci.*, Vol 20, 1985, p 1237–1247
16. W.G. Bradshaw and C.O. Mathews, "Properties of Refractory Materials: Collected Data and References," LNSD Report 2466, June 1958
17. S.M. Lang, "Properties of High Temperature Ceramics and Cermets—Elasticity and Density at Room Temperature," Monograph No. 6, National Bureau of Standards, Mar 1960
18. P.T.B. Shaffer, unpublished results
19. B.W. Mott, *Micro Indentation Hardness Testing*, Butterworths, London, 1956
20. G.V. Samsonov, *Handbook of High Temperature Materials*, Vol 2, *Properties Index*, Plenum Press, 1964

Table 10 Properties of selected interstitial carbides

Carbide	Hardness, HV (50 kg)	Crystal structure	Melting point °C	Melting point °F	Theoretical density, g/cm³	Modulus of elasticity GPa	Modulus of elasticity 10⁶ psi	Coefficient of thermal expansion, μm/m · K	Resistivity(a), μΩ · cm
TiC	3000	Cubic	3100	5600	4.94	451	65.4	7.7	68
VC	2900	Cubic	2700	4900	5.71	422	61.2	7.2	150
HfC	2600	Cubic	3900	7050	12.76	352	51.1	6.6	109
ZrC	2700	Cubic	3400	6150	6.56	348	50.5	6.7	63
NbC	2000	Cubic	3600	6500	7.80	338	49.0	6.7	74
Cr$_3$C$_2$	1400	Orthorhombic	1800(b)	3250	6.66	373	54.1	10.3	...
WC	(0001) 2200 (1010) 1300	Hexagonal	~2800(b)	5050	15.7	696	101	(0001) 5.2 (1010) 7.3	53
Mo$_2$C	1500	Hexagonal	2500	4550	9.18	533	77.3	7.8	97
TaC	1800	Cubic	3800	6850	14.50	285	41.3	6.3	30
W$_2$C(c)	2150(d)	Hexagonal	2785	5045 (Ref 5)	17.2 (Ref 1)	420	60.9 (Ref 55)	4 to 4.7(e)	...

(a) Resistivity values supplied by author. (b) Not congruently melting, dissociation temperature. (c) Data on W$_2$C obtained from Ref 1, 5, 13, 14, and 55. (d) Knoop hardness, 100 g load. (e) Expansion from room temperature to 1500 and 2000 °C (2730 and 3630 °F), respectively, per Ref 14. Source: Unless otherwise specified, data are from Ref 56.

21. W.D. Kingery, *Property Measurements at High Temperatures,* Wiley, 1959

22. A. Lipp, Boron Carbide—Production, Properties, Application, *Tech. Rundschau,* No. 14, 1965, p 28, 33; No. 7, 1966

23. P.T.B. Shaffer, *Handbook of High Temperature Materials,* Vol 1, *Materials Index,* Plenum Press, 1964

24. "Tetrabor"—Boron Carbide (B4C) Absorber Plates," Elektroschmelzwerk Kempten product bulletin

25. P.T.B. Shaffer, A Review of the Structure of Silicon Carbide, *Acta Crystall.,* Vol B25 (No. 3), 1969, p 477–488

26. P. Villars and L.D. Calvert, *Pearsons Handbook of Crystallographic Data for Intermetallic Phases,* American Society for Metals, 1985

27. E.G. Acheson, "Production of Artificial, Crystalline Carbonaceous Material," U.S. patent 492,767, Feb 1893

28. E.G. Acheson, "Electric Furnace," U.S. patent 560,291, May 1896

29. S.R. Billington, J. Chown, and A.E.S. White, The Sintering of Silicon Carbide, *Special Ceram.,* 1964, p 19–34

30. P.T.B. Shaffer, Growth and Densification of Silicon Carbide Crystalline Masses, *Conference of the International Committee on Silicon Carbide,* Vienna, Oct 1969

31. S. Prochazka, "Investigation of Ceramics for High Temperature Turbine Vanes," *Technical Report SRD-74–04,* General Electric Company, Dec 1973

32. S. Prochazka and R.M. Scanlan, Effect of Boron and Carbon on Sintering of Silicon Carbide, *J. Am. Ceram. Soc.,* Vol 58 (No. 1/2), 1975, p 72

33. C.H. McMurtry, M.R. Kasprzyk, and R.G. Naum, Microstructural Effects in Silicon Carbide, *Mater. Res. Bull.,* Vol 7 (No. 1), 1972, p 411–420

34. S.B. Lasday, Alpha Silicon Carbide Properties Advantageous for Automotive Water Pump Seal Faces Produced at New Facility in W. Germany, *Ind. Heat.,* Vol 35–39, Aug 1990

35. G.A. Slack, Nonmetallic Crystals with High Thermal Conductivity, *J. Phys. Chem. Solids,* Vol 34, 1973, p 321–335

36. C.K. Jun and P.T.B. Shaffer, Elastic Modulus of Dense Silicon Carbide, *Mater. Res. Bull.,* Vol 7 (No. 1), 1972, p 63–70

37. P.T.B. Shaffer, Refractive Index, Dispersion, and Birefringence of Silicon Carbide Polytypes, *Appl. Optics,* Vol 10 (No. 5), 1971, p 1034–1036

38. J.E. Field, *The Properties of Diamond,* Academic Press, 1979

39. J.B. Wachtman, Jr. and D.G. Lam, Jr., *J. Am. Ceram. Soc.,* Vol 42, 1959, p 254

40. P.T.B. Shaffer, The Effect of Crystal Orientation on the Hardness of Silicon Carbide, *J. Am. Ceram. Soc.,* Vol 47 (No. 9), 1964, p 366

41. P.T.B. Shaffer, The Effect of Crystal Orientation on the Hardness of Beta Silicon Carbide, *J. Am. Ceram. Soc.,* Vol 48 (No. 11), 1965, p 601–602

42. R.C. Marshall, J.W. Faust, and C.E. Ryan, "Silicon Carbide-1973," Proc. 3rd Intl. Conf. SiC, Miami Beach, FL

43. J.R. O'Conner and J. Smiltens, Silicon Carbide, A High Temperature Semiconductor, *Proc. 1st Intl. Conf. SiC,* Boston, 1960, Pergamon Press, 1960

44. J.H. Racette, Intrinsic Electrical Conductivity in Silicon Carbide, *Phys. Rev.,* Vol 107, 1957, p 1542–1544

45. J.E. Burlingame and P.T.B. Shaffer, "Dense High Conductivity Reaction Bonded Silicon Carbide," unpublished studies

46. J. DeMent, SiC May Be the Gem Men Once Sought, *Ind. Res.,* Vol 19, June 1970

47. P.T.B. Shaffer and R.G. Naum, Refractive Index and Dispersion of Beta Silicon Carbide, *J. Opt. Soc. Amer.,* Vol 59 (No. 11), 1969, p 1498

48. P.T.B. Shaffer, Use of the Microscope in the Observations of Silicon Carbide Structures, *Microscope,* Vol 18, 1970, p 179–191

49. D. Das, *Ind. J. Phys.,* Vol 40, 1966, p 684

50. K.L. Brooks, *World Directory and Handbook of Hardmetals,* 4th ed., Engineers Digest and International Carbide Data, Surrey, UK, 1990

51. P.M. McKenna, U.S. patents 2,113,333 and 2,113,335, 1935

52. P.M. McKenna, U.S. patent 2,124,509, 1938

53. K. Becker, *Z. Physik,* Vol 51, 1928, p 481

54. P.T.B. Shaffer, K.A. Blakely, W.E. Hauth III, S. Kumar, and C. Mroz, "Improved Materials for Ceramic Armor Fabrication," Advanced Refractory Technologies, Final Report to Naval Surface Weapons Center, Phase II, SBIR Contract No. N60921–88–C–A237, July 1990

55. V.S. Neshpor and G.V. Samsonov, *Fiz. Metal. Metalloved.,* Vol 4 (No. 1), 1957, p 181–182

56. H.E. Exner, *Int. Metall. Rev.,* Vol 24 (No. 4), 1979, p 149–173

Engineering Properties of Nitrides

Stuart Hampshire, Materials Research Centre, University of Limerick, Ireland

NITRIDE CERAMICS have a major advantage over metals in their high-temperature capability, reflected by the fact that these ceramics have higher strengths at temperatures above 1000 °C (1830 °F) and better oxidation and corrosion resistance. The greatest impetus to research and development on nitride ceramics has been the attempt to produce a ceramic gas turbine engine for which application silicon nitride has been a main contender.

Silicon nitride is the primary material in a family of nitride ceramics developed for engineering applications including a variety of types made by different processing routes or having different compositions or both. Silicon nitride has attracted much attention because of its unique combination of excellent high-temperature mechanical properties and resistance to oxidation and thermal shock. The thermal conductivity of Si_3N_4 is approximately double that of Al_2O_3-TiC, while the coefficient of thermal expansion is around one-half that of Al_2O_3. These thermal properties result in a much lower sensitivity to temperature changes and improved thermal-shock resistance.

The difficulty of utilizing these properties arises from the problems involved in sintering pure Si_3N_4. In Si_3N_4, the volume diffusivity is not large enough to offset the densification retardation effects of surface diffusion and volatilization phenomena; as a result, sintering is difficult. The covalent bond is believed to be the reason for such a low-volume diffusivity. Therefore, dense Si_3N_4-base material can be obtained only by alloying Si_3N_4 with additives that promote sintering. The properties of these Si_3N_4 alloys, especially at high temperatures, are to a large extent controlled by the additives. As a result, there is not just a single Si_3N_4 material, because the properties of Si_3N_4 alloys depend on the additives.

Various metal oxides and nitrides have been found to be effective sintering aids. The most commonly used solid-solution additives to Si_3N_4 are Al_2O_3, aluminum nitride (AlN), and silica (SiO_2). These additives result in the family of sialon ceramics that retain the structure of silicon nitride with incorporation of aluminum and oxygen in solid solution. These ceramics have already been successfully commercially exploited as cutting tool inserts. The other nitride ceramics within the Si-Al-O-N system include silicon oxynitride and aluminum nitride, but less research effort has been employed on these materials so far and their potential is still being assessed.

Silicon Nitride-Based Ceramics

Silicon nitride may be produced by various methods to give a range of products, each one named after the fabrication route employed. The major types are:

- Reaction-bonded silicon nitride (RBSN)
- Hot pressed silicon nitride (HPSN)
- Sintered (pressureless) silicon nitride (SSN)
- Sintered reaction-bonded silicon nitride (SRBSN)
- Hot isostatically pressed silicon nitride (HIPSN)

The latter four fabrication methods are used to make dense silicon nitride products.

Reaction-Bonded Silicon Nitride

Reaction-bonded silicon nitride (RBSN), which has been under development since the late 1950s (Ref 1), is made by first producing the required shape from silicon powder. The shaped powder form is then nitrided in molecular nitrogen, starting at 1150 °C (2100 °F) and slowly increasing the temperature to over 1400 °C (2550 °F). The resulting product consists of two different crystallographic forms of silicon nitride, alpha (α) and beta (β), and has 12 to 30% porosity. Pores are mostly in the sub-micron size range but with some occasional large pores (up to 50 μm) which are usually associated with the melting out of iron impurities in the original silicon powder. Flexural strength for RBSN falls in the range of 150 to 350 MPa (22 to 50 ksi), and this can be maintained to very high temperatures (>1400 °C, or 2550 °F) in an inert atmosphere.

One advantage of the reaction-bonding process is that the original dimensions of the silicon compact remain virtually unchanged during nitriding. A complex shaped component can be partially nitrided to give the piece strength and then easily machined to final size using conventional tooling before completion of the process.

Dense Silicon Nitride

Dense silicon nitride, in various forms, may be produced by sintering (with or without applied pressure) preformed silicon nitride with oxide additives which form liquid phases that allow more rapid diffusion of material and the attainment of near theoretical density. This may be achieved by hot pressing, by pressureless sintering of both silicon nitride powder compacts and reaction-bonded silicon nitride or by hot isostatic pressing of either powder compacts or all of the previously sintered types.

During firing, the following processes take place. Each particle of the original α-silicon nitride powder has around it a surface layer of silicon dioxide (SiO_2). An additive, such as magnesium oxide (MgO) or yttrium oxide (Y_2O_3), reacts with this silica and some of the nitride to form a liquid at the firing temperature which allows sintering to take place in three stages (Ref 2):

1. *Rearrangement*, induced by the liquid phase
2. *Solution* of material from α into the liquid, followed by *diffusion* of silicon and nitrogen and then, when conditions permit, the *precipitation* of the more thermodynamically stable β-silicon nitride
3. *Coalescence*, where the liquid acts to eliminate final closed porosity and to produce a more rounded grain morphology

After cooling, the liquid either forms a glass as with MgO additive or crystallizes to form various oxynitride phases as with Y_2O_3. The role of the additives is summarized by (Ref 2–4):

$$\alpha\text{-Si}_3\text{N}_4(+\text{SiO}_2) + M_xO_y$$

$$\rightarrow \beta\text{-Si}_3\text{N}_4 + M\text{-Si-O-N phase} \qquad \text{(Eq 1)}$$

The physical processes are shown schematically in Fig 1. With little or no β present in the original powder, a certain supersaturation must be reached before nucleation occurs and then growth of the fiber-like β grains takes place. The resulting morphology is rod-shaped, and the resulting microstructure after cooling consists of interlocking β-silicon nitride grains with a residual secondary phase at the grain boundaries. A scanning electron micrograph of the microstructure is shown in Fig 2.

Hot pressed silicon nitride (HPSN) is formed by the application of both heat and uniaxial pressure in graphite dies heated by induction to temperatures in the range 1650 to 1850 °C (3000 to 3360 °F) for 1 to 4 h under an applied stress of 15 to 30 MPa (150 to 300 bar). Boron nitride is applied as a coating to the graphite die and plungers to prevent reaction of these with the silicon nitride. Boron nitride powder is also used as a high-temperature solid lubricant to facilitate removal of the hot pressed material from the die. Surface contamination is usually a problem, but this can be minimized by pre-pressing the powder mix in a metal die to form a compact prior to its introduction into the graphite die. However, HPSN is limited to simple-shaped (cylindrical) billets and components must be machined from these using expensive grinding.

Sintered silicon nitride (SSN) allows a more cost-effective method of production of complex-shaped components in dense silicon nitride, without the requirement for machining to any extent. This involves firing the shaped component at 1700 to 1800 °C (3100 to 3275 °F) under a nitrogen atmosphere at 0.1 MPa (14.5 psi, or 1 atm). As with hot pressing, the additives provide conditions for liquid-phase densification (Ref 2–3) but, in the absence of applied pressure, the reduction in surface energy becomes the major driving force for sintering and so the use of high-surface-area powders is necessary. This increases the oxygen content of the powders which can also affect the quantity of liquid phase formed and, thus, the overall composition of the secondary phase (Ref 3).

Without pressure, dissociation of Si_3N_4 becomes a problem at high temperatures. During pressureless sintering at temperatures much above 1700 °C (3100 °F), density starts to decrease at longer times as a result of increasing weight losses (Ref 5). The use of so-called "powder beds" where the component to be sintered is surrounded by a mixture of powder of its own composition and inert

5 μm

Fig 2 Scanning electron micrograph showing β-silicon nitride grain morphology

boron nitride, has proved successful in reducing volatilization (Ref 6). This creates a local gas equilibrium immediately adjacent to the silicon nitride compact thus minimizing volatilization. An alternative is to increase the nitrogen pressure to higher levels (10 MPa, or 1500 psi), and development work by a number of manufacturers in Japan has demonstrated substantial improvements in properties by this method. Density values of 97 to 99% theoretical are routinely achieved with bend strengths of >1000 MPa (>145 ksi).

Sintered reaction-bonded silicon nitride (SRBSN) combines the technology of both RBSN and SSN. Silicon nitride powder compacts have low green densities (~45 to 55% theoretical) and, thus, sintering to high densities requires linear shrinkages of 15 to 18%. Consequently, control of the firing process for complex shapes becomes more difficult. Because reaction-bonded silicon nitride has a density within the range of 70 to 88% of theoretical, it is a suitable starting material for sintering (Ref 7, 8).

Additives such as MgO or Y_2O_3 are mixed with the silicon prior to shaping and then nitriding is carried out as for reaction-bonded silicon nitride. A further firing in the range of 1800 to 2000 °C (3275 to 3630 °F) under a nitrogen atmosphere (0.1 to 8 MPa, or 1 to 80 atm), using a protective powder bed to reduce volatilization, allows densification to 98% theoretical density with only 6% linear shrinkage.

Hot isostatically pressed silicon nitride (HIPSN) covers a number of different material types. The components are placed in an "autoclave" and subjected to high temperature and high pressure using argon or nitrogen as the pressure transmission medium to consolidate either a shaped silicon nitride

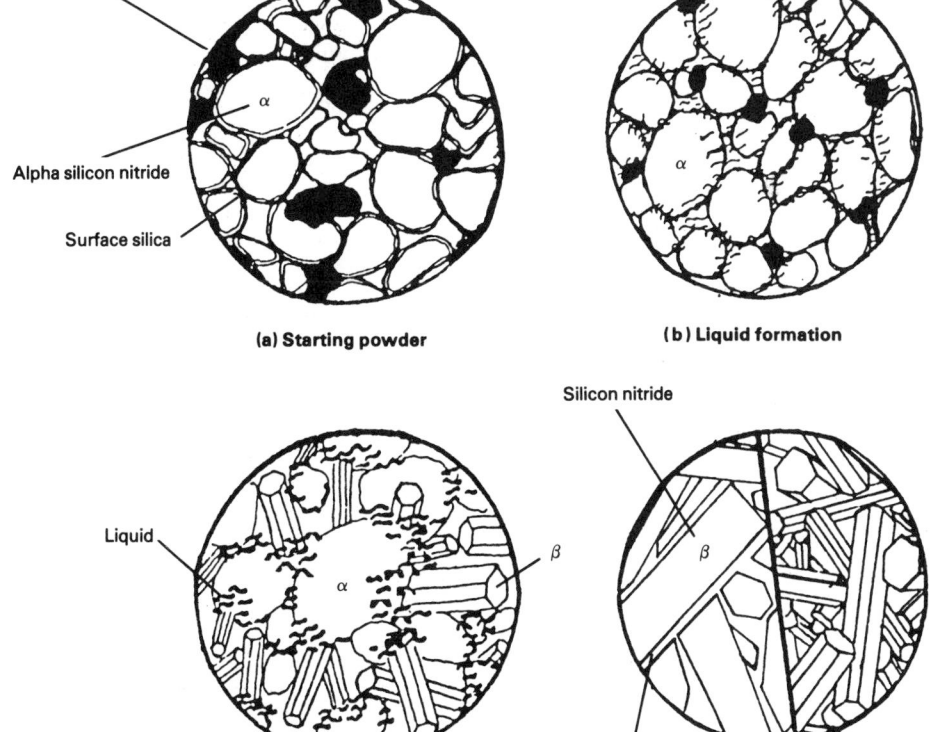

(a) Starting powder

(b) Liquid formation

(c) Precipitation of beta

(d) Morphology and microstructure after sintering

Fig 1 Physical processes occurring during sintering of silicon nitride ceramics

powder compact or to remove porosity from pre-fired RBSN, SSN, or SRBSN. In all cases, a small amount of sintering additive is required; however, because a lower quantity of additives may be used, properties should be superior to those of other forms of Si_3N_4.

Because powder compacts and reaction-bonded silicon nitride contain a large volume fraction of open porosity, an encapsulation technique is used to prevent penetration of the pressurized gas into the open pore network during hot isostatic pressing (Ref 9). During controlled cooling, the glass capsule cracks off and the hot isostatically pressed component is given a surface treatment such as sandblasting.

For pre-sintered materials such as SSN and SRBSN where there is no open porosity, encapsulation is unnecessary and the major reason for applying the HIP treatment is to remove residual porosity. Some studies suggest (Ref 10) that hot isostatic pressing of sintered silicon nitride actually results in a slight decrease of mean strength but with a substantial improvement in reliability (higher Weibull modulus).

Sialons

Sialons were a major breakthrough in silicon nitride technology in the 1970s with their concurrent discovery in England (Ref 11) and in Japan (Ref 12). Aluminum can replace silicon in the beta silicon nitride structure if, at the same time, nitrogen is replaced by oxygen giving a substituted solid solution referred to as β'-sialon with a composition range of $Si_{6-z}Al_zO_zN_{8-z}$ based on the β-Si_6N_8 unit cell.

Sialons are synthesized by reacting together silicon nitride with other constituents to achieve an appropriate β' composition as shown in Fig 3. References 13 to 15 give more details on interpretation of these behavior diagrams and their use in processing. Yttrium oxide is used as a liquid-phase sintering aid and, while it should be possible to produce single-phase β'-sialon, densification considerations require that the material chemistry is such that the final assemblage is multiphase. Two major material types have been developed commercially, one consisting of β' grains with a residual glass, the other consisting of β' grains and a semicontinuous intergranular crystalline phase of yttrium-aluminum-garnet (YAG).

A solid solution based on the alpha silicon nitride structure (Ref 14–16) has received less attention, but α'-sialon shows promise commercially because it has greater hardness than β'-sialon. In addition, it also offers an improved product with less grain-boundary glass as a result of the incorporation of the cation from the sintering additive (for example, Y_2O_3) into a solid solution of $M_x(Si,Al)_{12}(O,N)_{16}$ based on the α-$Si_{12}N_{16}$ unit cell. Single-phase α-sialons currently are only available in hot pressed or hot isostatically pressed forms, but composites of $\beta' + \alpha'$ are available produced by pressureless sintering.

Properties of Silicon Nitride

Key representative properties of silicon nitride in its various forms (including sialons) are given in Tables 1 and 2. Wide ranges of property values are observed which not only depend on the processing route but also on the microstructural characteristics of the particular material type. In addition to the normal dependence on porosity levels, the properties are also crucially influenced by the amount and morphology of the β-form of silicon nitride and the amount, characteristics, and distribution of the grain-boundary phase. The theoretical density of pure silicon nitride is 3.18 g/cm^3, but the values for the different types depend on the amount of secondary phase and its specific gravity.

The properties of silicon nitride-based ceramics must always be considered in relation to the processing route, the densification additives, and the resulting microstructure. The most desirable, in terms of strength and fracture toughness, consist of interlocking β grains with high aspect ratios, very little or no glass phase, and crystalline grain-boundary phases that are not oxidized at high temperatures. As shown schematically in Fig 4, as a crack moves through silicon nitride, it encounters a combination of fiber-like β grains and secondary grain-boundary material which tend to steer the crack along a tortuous path and, thus, more energy is required for propagation, resulting in an increase in toughness. A decreasing β-grain aspect ratio results in lower strength and lower fracture toughness.

Thus, there is an effective "microstructural toughening" in sintered silicon nitride and the concept of *microstructural engineering* (Ref 17, 18) has been applied to tailor the properties of silicon nitride ceramics to end applications. This involves manipulation of the chemistry of the silicon nitride/additive system and heat treatment, both during and after sintering, to give the most beneficial secondary phases. Because regions of uncrystallized glass residues after sintering allow easy creep cavitation and increased creep strain, it is important to minimize their presence. Materials which are hot pressed or hot isostatically pressed require much lower volumes of liquid for densification and, in some cases, have been produced almost glass-free (Ref 19), but these processes are expensive.

Thermal conductivity/diffusivity of silicon nitride (see Table 1) is controlled by the level of porosity, the β-phase content, the amount of secondary phases (particularly glass), the extent of any solid solution, and orientation effects (particularly for hot pressed materials). Reference 20 describes a systematic investigation of various microstructural effects on thermal diffusivity, a, of silicon nitride, defined as:

$$a = \lambda/\rho C_p \qquad \text{(Eq 2)}$$

where ρ is density in g/cm^3 and C_p is the spe-

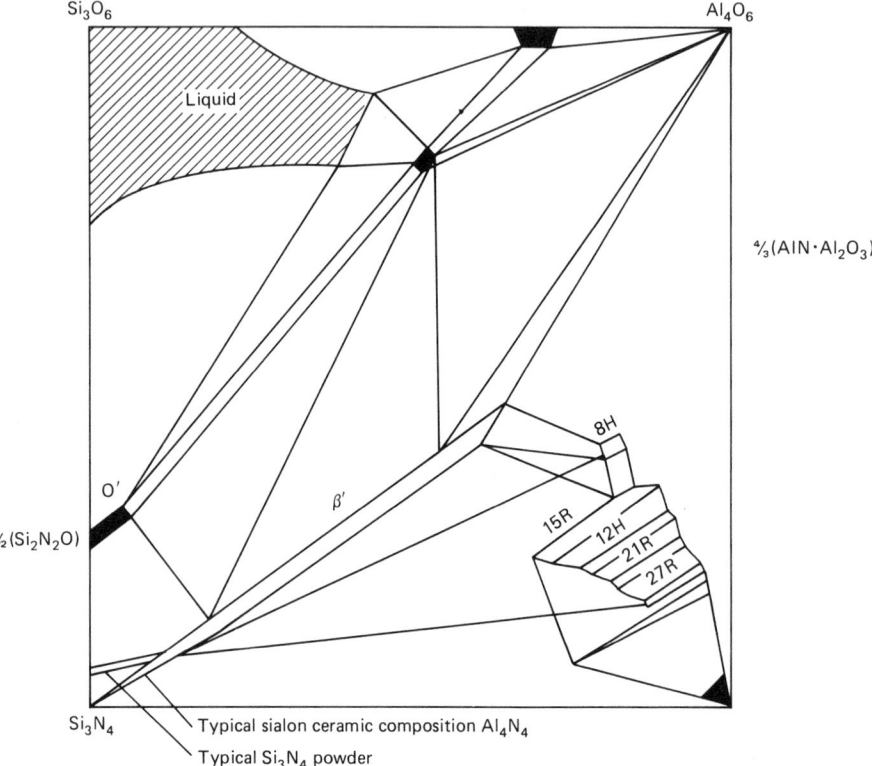

Fig 3 The Si-Al-O-N behavior diagram showing reaction of constituents to form β'-sialon ceramic. Source: Ref 15

Table 1   Thermal properties of silicon nitride ceramics

Property	Material type					
	RBSN	HPSN	SSN	SRBSN	HIP-SN	Sialon
Relative density, % theoretical	70–88	99–100	95–99	93–99	99–100	97–99
Coefficient of thermal expansion (α) from 25–1000 °C (77–1830 °F), 10^{-6}/°C	3.0	3.2–3.3	2.8–3.5	3.0–3.5	3.0–3.5	3.0–3.7
Thermal conductivity (λ) at:						
25 °C (77 °F), W/m · °C (Btu/ft · h · °F)	7–14 (4–8)	30–43 (17–25)	15–31 (8–18)	...	22 (12.27)	15–22 (8–13)
1000 °C (1830 °F), W/m · °C (Btu/ft · h · °F)	1.4–3 (0.8–1.7)	5–10 (3–6)	4–5 (2.3–3)	...	...	2.5 (1.4)
Specific heat (C_p) at:						
25 °C (77 °F), J/kg · °C (Btu/lb · °F)	700–1100 (0.17–0.26)	680–800 (0.16–0.19)	...	...	...	...
1000 °C (1830 °F), J/kg · °C (Btu/lb · °F)	1250 (0.30)	1200 (0.29)	...	...	...	...

Table 2   Mechanical properties of silicon nitride ceramics

Property	Material type							
	RBSN	HPSN	SSN	SRBSN	HIP-SN	HIP-RBSN	HIP-SSN	Sialon
Young's modulus (E), GPa (10^6 psi)	120–250 (17.4–36.2)	310–330 (44.9–47.8)	260–320 (37.7–46.4)	280–300 (40.6–43.5)	...	310–330 (44.9–47.8)	...	300 (43.5)
Poisson's ratio (v)	0.20	0.27	0.25	0.23	0.23–	0.27	...	0.23
Flexural strength (σ_f), MPa (ksi) at:								
25 °C (77 °F)	150–350 (21.7–50.7)	450–1000 (65.2–145)	600–1200 (86.9–173.8)	500–800 (72.5–115.9)	600–1200 (86.9–173.8)	500–800 (72.5–115.9)	600–1200 (86.9–173.8)	750–95 (108.7–137.7)
1350 °C (2460 °F)	140–340 (20.2–49.3)	250–450 (36.2–65.2)	340–550 (49.3–79.7)	350–450 (50.7–65.2)	350–550 (50.7–79.7)	250–450 (36.2–65.2)	300–520 (43.5–75.3)	300–550 (43.5–79.7)
Weibull modulus (m)	19–40	15–30	10–25	10–20	...	20–30	...	15
Fracture toughness (K_{Ic}), MPa$\sqrt{m}$ (ksi$\sqrt{in.}$)	1.5–2.8 (1.3–2.5)	4.2–7.0 (3.8–6.3)	5.0–8.5 (4.5–7.7)	5.0–5.5 (4.5–5.0)	4.2–7.0 (3.8–6.3)	2.0–5.8 (1.8–5.3)	4.0–8.0 (3.6–7.2)	6.0–8.0 (5.4–7.2)

cific heat in J/kg · °C. Thermal conductivity, λ, can be found from diffusivity data if density and specific heat are known. From calorimetric measurements, it was concluded that the specific heat of silicon nitride containing different sintering additives, phase assemblages, and grain morphology is independent of microstructural effects. Thermal conduc-tivity varies with microstructural effects as shown in Fig 5 for a range of RBSN grades and Fig 6 for a range of HPSN grades.

For reaction-bonded silicon nitride, thermal diffusivity changes with variations in porosity or pore size but these are also accompanied by changes in other microstructural features. Thermal diffusivity generally increases with increase in density, with increase in β content, and with increase in the amount of any unreacted silicon following the reaction-bonding process. Oxidation causes a decrease in thermal diffusivity due to the formation of amorphous silica film.

In commercial reaction-bonded silicon nitride, α may be the major phase with the β-phase content as low as 20%; in hot pressed silicon nitride, however, β is the major phase and usually full conversion of α to β has taken place. The thermal conductivity of the β-phase is higher than that for α, and the elongated β-silicon nitride grains also are orientated perpendicular to the hot pressing direction. With MgO as sintering additive, maximum values of diffusivity coincide with maximum glass content at 3 to 5 wt% MgO. With higher additive levels, a secondary crystalline phase, usually forsterite (Mg_2SiO_4) is formed. With Y_2O_3 as sintering additive, alumina is also used as a second additive. This allows formation of a larger volume of densification liquid, which is subsequently incorporated into a low z β'-sialon solid solution (where low z means low aluminum-substituted) causing a decrease in the thermal diffusivity.

Mechanical properties of silicon nitride ceramics are shown in Table 2. Even within each material type, the properties, particularly fracture strength and fracture toughness,

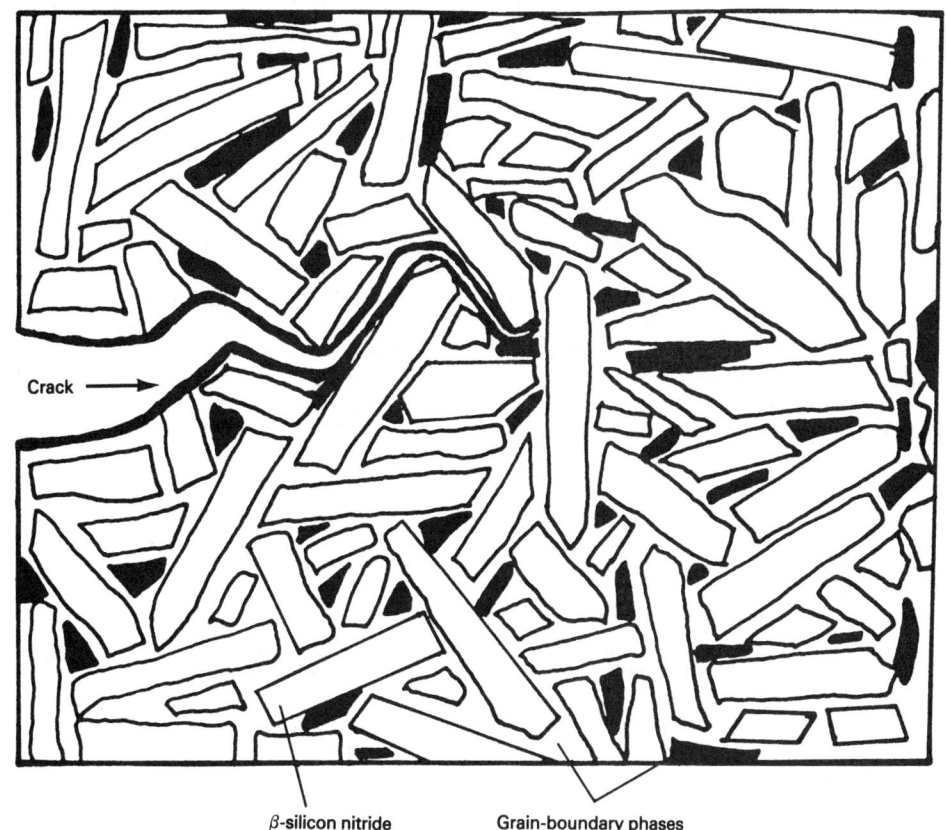

Crack →

β-silicon nitride   Grain-boundary phases

Fig 4   Impeded crack path by microstructural toughening in dense silicon nitride ceramics with secondary grain-boundary phase

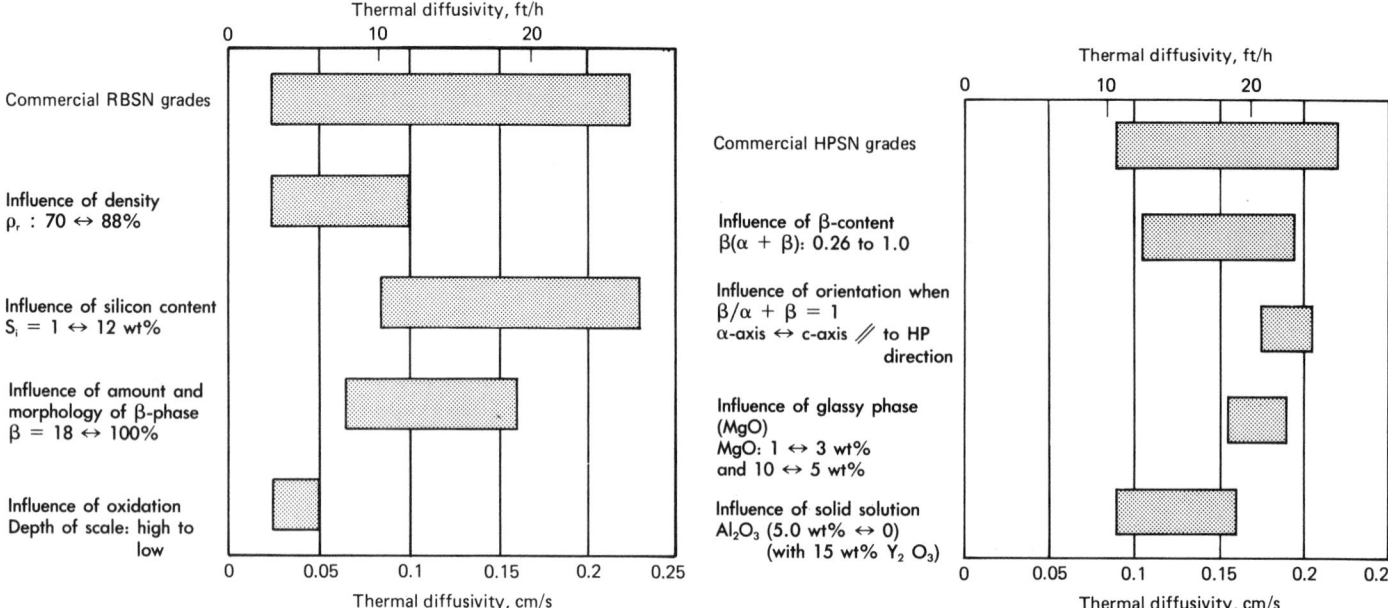

Fig 5 Influence of microstructural features on thermal diffusivity of reaction-bonded silicon nitride (RBSN). Source: Ref 20

Fig 6 Influence of microstructural features on thermal diffusivity of hot pressed silicon nitride (HPSN). Source: Ref 20

show large variations, again attributable to microstructural features.

Young's modulus of elasticity (E), for RBSN decreases with an increase in total porosity according to (Ref 21):

$$E = E_0 \exp(-3P) \qquad \text{(Eq 3)}$$

where P is the total volume fraction of porosity and E_0 is the Young's modulus for silicon nitride with zero porosity, taken as 300 GPa (43.5×10^6 psi). Materials with the same total porosity but different pore size distributions have very similar elastic moduli.

For dense silicon nitrides, values of Young's modulus vary between 260 and 330 GPa (38 and 48×10^6 psi), depending on the amount and orientation of other phases present in the material, including porosity.

Flexural strength usually is determined by four-point bending on specimens of dimensions 3.5 mm × 4.5 mm × 45 mm (0.14 × 0.18 × 1.8 in.). Some values quoted by various groups are inflated because they have been obtained on smaller specimens using three-point bending.

Strength of RBSN grades is dependent on the volume fraction of porosity but, more particularly, on the size of the largest pores which are generally created by the melting out of iron impurities present in the original silicon powders. Thus, for a given density, the fracture strength of reaction-bonded silicon nitride shows significant scatter. For the achievement of high strength, it is better to have a homogeneous microstructure with a narrow pore-size distribution at moderate densities than a high-density material with large voids present (Ref 4). The grain size, which is finer than the macro-pore size, has much less effect on strength.

Strength of Densified Silicon Nitride. Typical average fracture strength values at ambient temperature for HPSN grades are 600 MPa (87 ksi) with MgO additive and 800 MPa (116 ksi) with Y_2O_3 additive, the major difference being the morphology of the β grains in the microstructure. With MgO, the liquid phase during hot pressing allows easy densification but an equiaxed grain morphology. In contrast, the liquid formed with Y_2O_3 has a higher viscosity resulting in c-axis growth of the β grains and hence a higher aspect (length to diameter) ratio (Ref 2), giving higher strength and fracture toughness.

Improvements in the processing of sintered silicon nitride have resulted in values as high as or exceeding those of HPSN grades. However, Weibull moduli tend to be higher for the hot pressed silicon nitride and, in particular, the HIP routes. These processes result in flaw-healing and pore size reduction.

Microstructural and Processing Effects on Strength and Toughness. For any particular material type, the room-temperature mechanical strength and fracture toughness are dependent on, firstly, the aspect ratio of the β-silicon nitride grains and, secondly, the overall grain size. Figure 7 shows schematically how fracture strength and fracture toughness change during the sintering process for silicon nitride. As the α-form transforms to β-phase, the aspect ratio changes from equiaxed crystals to fine elongated crystals giving improved strength and toughness but, once the α-to-β transformation is complete, grain growth results in an increase in grain diameter with a consequent decrease in these properties. The reason for the improvement with higher aspect ratios is that the interlocking, elongated β grains have better resistance

to crack propagation because of crack branching and deviation as well as grain pullout, resulting in higher energy requirements for crack growth (Fig 3).

The high-strength values now achieved for sintered silicon nitrides are a result of process optimization including modifications to the type and composition of the grain-boundary phase by varying the amount and type of sintering additive. The use of mixed oxide additives (for example, $Y_2O_3 + Al_2O_3$, MgO + Nd_2O_3, and others) allows control of the properties of the sintering liquid (such as its volume and viscosity), which determine the growth of the β grains along the preferred c-axis direction and also the grain diameter (Ref 2).

The variation of strength with temperature for the different types of silicon nitride-based ceramics is shown in Fig 8. Because of the presence of porosity, reaction-bonded silicon nitride has the lowest flexural strength (150 to 350 MPa, or 22 to 50 ksi) at ambient temperature, but strength is retained up to very high temperatures (>1400 °C, or 2550 °F) because there is no glass phase.

With sintered silicon nitride, much higher strengths are achieved (600 to 1200 MPa, or 87 to 174 ksi) at ambient temperature. At temperatures exceeding 1000 °C (1830 °F), however, the strength may decrease rapidly due to the softening of the intergranular glass. Strength is retained to much higher temperatures when the secondary phase is crystalline (Ref 4, 17, 18).

Sintered sialons are of two types with different microstructural features: β'-sialon grains plus glass and β'-sialon grains plus crystalline YAG (Ref 14, 15, 17). The strength of Type 1 is similar to hot pressed silicon nitride

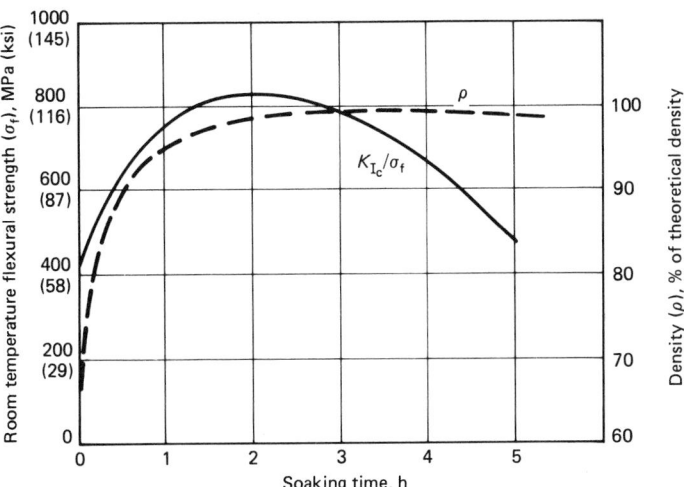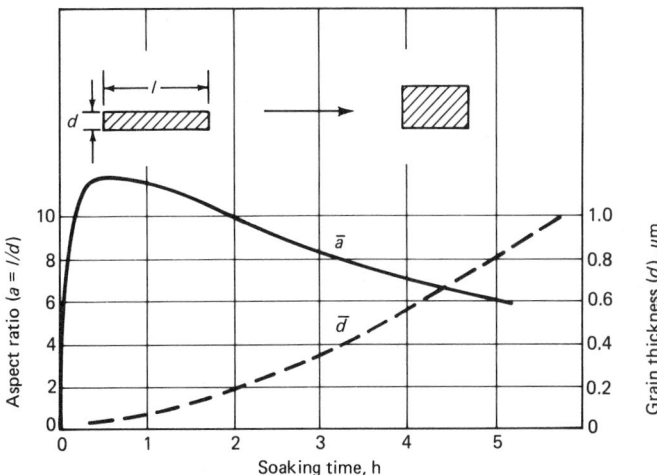

Fig 7 Changes in flexural strength and fracture toughness of silicon nitride as a function of aspect ratio and diameter of β grains. Values of K_{Ic} and σ_f vary directly with the microstructure.

at ambient temperatures, but its strength decreases as the intergranular glass softens above 1000 °C (1830 °F). Type 2 sialon has a lower ambient strength but retains strengths of 500 MPa (72.5 ksi) at temperatures of 1400 °C (2550 °F).

A characteristic property of silicon nitride ceramics containing intergranular glass phases is a transient rise in fracture toughness near the glass softening point. This is associated with the onset of subcritical crack growth (SCG) within a creep cavitation zone at the primary crack tip. The rise in K_{Ic} is due to energy absorption by plastic deformation as a result of viscous flow of residual glass bridging crack surfaces. In type 2 sialons containing the crystalline YAG phase, there

is no evidence for a zone of SCG preceding rapid fracture even at temperatures over 1450 °C (2640 °F) in an inert environment (Ref 22), while this phenomenon results in degradation at much lower temperatures of type 1 sialons and sintered silicon nitrides containing glass.

The HPSN and HIPSN grades, which have much lower levels of additives (hence less glass) and also zero porosity, generally have higher strengths than sintered silicon nitride or sialons at higher temperatures. In applications such as metalcutting, it is important to retain strength and hardness to high temperatures. Hardness values vary from 750 HV (100 g load) for RBSN with density of 2.7 g/cm^3 to 1600 to 1800 HV (100 g load) for

HPSN materials and 1350 to 1600 HV (100 g load) for sintered silicon nitrides. Hardness can be as high as 1800 HV in sintered sialons containing both β' and α' phases and the hardness differential due to the presence of α' in these sialon composites is retained up to high temperatures (Fig 9). At 1000 °C (1830 °F) these materials are much harder than alumina ceramics.

Creep behavior of silicon nitride ceramics is mainly controlled by grain-boundary sliding along the amorphous phase. In addition to the volume of glass, its viscosity is an important consideration. Many sintered silicon nitrides are densified with a combination of Y$_2$O$_3$ and Al$_2$O$_3$ additives. The β'-sialons are also sintered with Y$_2$O$_3$. This system (Y-Si-Al-O-N) offers a combination of moderately low liquidus temperature with a comparatively high residual glass viscosity and glass transition temperature, particularly compared with glasses in the Mg-Si-O-N system. Thus, the Y-Si-Al-O-N system exhibits far superior high-temperature properties. Ceramics have been developed which have survived very long times to failure at temperatures well in excess of 1300 °C (2370 °F) and stresses above 400 MPa, or 58 ksi (Ref 22). This has been achieved by care-

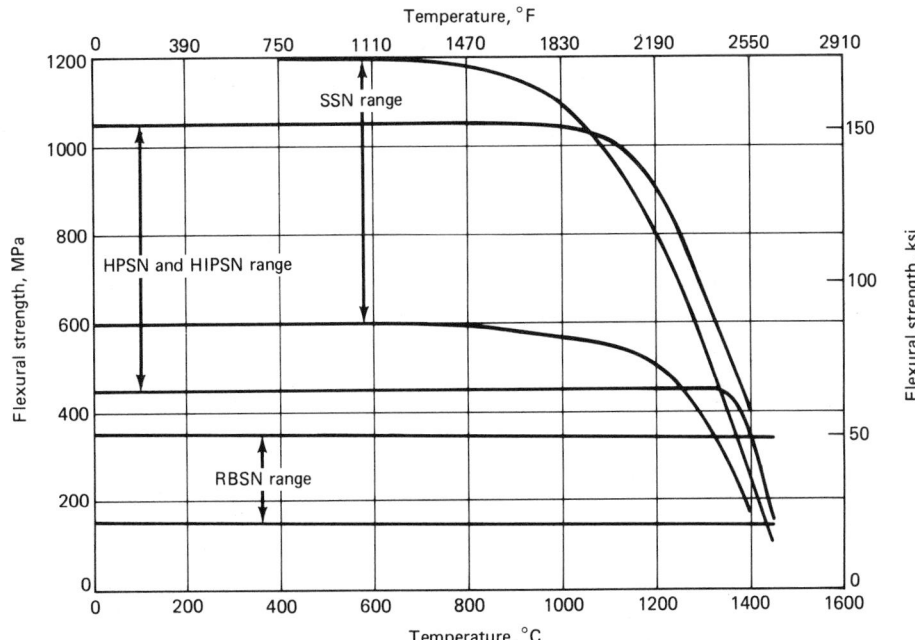

Fig 8 Variation of flexural strength with temperature for various types of silicon nitride ceramics

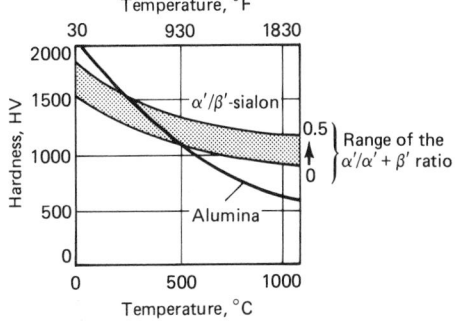

Fig 9 Hot hardness behavior of β'/α' sialon composites. Source: Ref 23

ful optimization of sintering and subsequent crystallization of the grain-boundary glass. This treatment suppresses creep cavitation and minimizes the creep rate at high applied stress levels.

Thermal shock resistance of silicon nitride ceramics varies widely and, as with the thermal and mechanical properties, depends on microstructural features in a complex way. The critical temperature difference, ΔT_c, after water and oil quenching for the various types of silicon nitride ceramics, is given in Table 3 (Ref 20). Unstable crack propagation was found to occur for all grades of silicon nitride. With ΔT_c equated to R, the thermal shock resistance parameter for instantaneous temperature changes:

$$R = \sigma_f(1 - \nu)/\alpha E \qquad \text{(Eq 4)}$$

For less severe quenching, a second parameter, $R' = R\lambda$ is defined. As the β content of silicon nitride ceramics increases, the thermal conductivity, λ, and the fracture strength, σ_f, also increase; hence there is an increase in R and R'. Other effects on R and R' are as follows:

- An increase in grain size decreases σ_f but has little effect on λ or Young's modulus, E; hence, R and R' also decrease with increasing grain size
- An increase in either glass content or in the extent of β'-sialon solid solution formation has little effect on R; however, because they result in a decrease in λ, then R' decreases

Thermal cycling behavior of silicon nitride ceramics has been reported (Ref 24–25). The retained fracture strength at ambient temperature, following thermal cycling from 1260 °C (2300 °F) is shown in Fig 10 for various grades of RBSN and HPSN as a function of the number of cycles. Quenching was carried out at a maximum cooling rate of 80 °C/s. Strength degradation occurs initially, and this is usually higher for RBSN than for HPSN grades. In spite of some loss of strength, silicon nitride retains a much higher strength after thermal cycling than other engineering ceramics (Ref 4, 25).

The interpretation of the thermal cycling behavior of the various types of silicon nitride is complex because a superposition of various

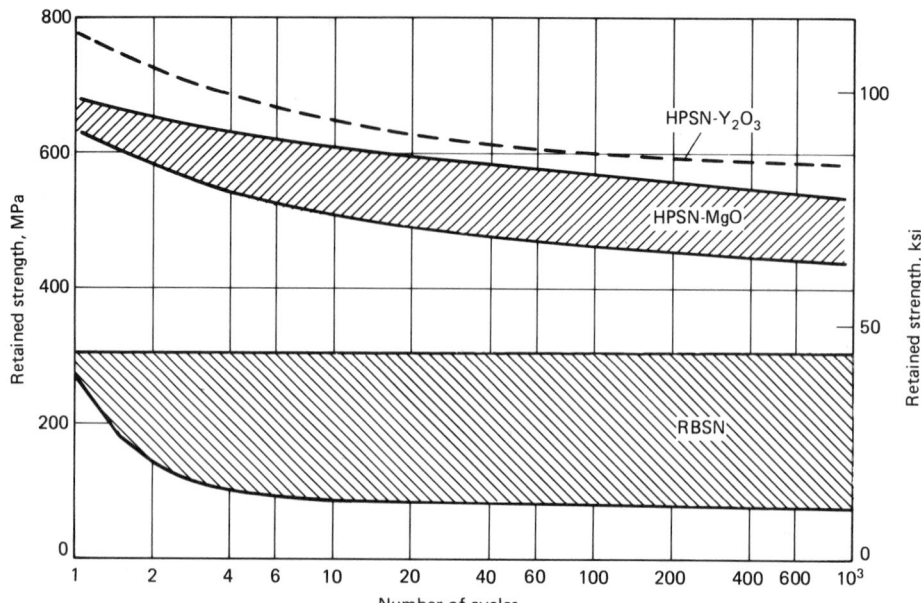

Fig 10 Thermal cycling behavior of various silicon nitride ceramics showing retained strength as a function of number of cycles from 1260 °C (2300 °F) to room temperature. Source: Ref 24

effects takes place. These effects include the induced thermal stress levels and plastic deformation (as well as oxidation) at high temperatures, which can result in crack healing and rounding of internal pores and flaws.

In all cases, the limiting temperature capability of silicon nitride-based ceramics appears to be around 1450 °C (2640 °F) in inert atmospheres. The added complication of using these ceramics in oxidizing atmospheres is a further limitation. Combined exposure with oxidation and high-temperature mechanical tests gives useful data on behavior (Ref 4, 25).

Oxidation of silicon nitride ceramics in air begins as low as 800 °C (1470 °F) and a thin protective layer of amorphous SiO_2 is formed on the surface of the silicon nitride according to:

$$Si_3N_4 + 3O_2 \rightarrow 3SiO_2 + 2N_2 \qquad \text{(Eq 5)}$$

This simple equation describes the situation for reaction-bonded silicon nitride and for dense silicon nitrides at lower temperatures. However, oxidation of these types at higher temperatures is much more complex. The rate of oxidation varies according to the amount

and type of densification additive used. In the case of materials containing a grain-boundary glass phase, various investigations (Ref 26–28) have shown that the oxidation reaction is diffusion controlled, the rate being limited by outward diffusion of metallic impurity or dopant ions into the SiO_2 scale.

In the case of materials containing crystalline grain-boundary phases which are oxynitrides, their oxidation products may have substantially different specific volumes giving rise to extensive cracking at the surface which then exposes fresh surfaces to further attack. This is particularly a problem in hot pressed silicon nitride doped with Y_2O_3 where the secondary phase is N-melilite (Ref 2–4, 14, 15); catastrophic oxidation in this case occurs in the range of 900 to 1200 °C (1650 to 2200 °F). Knowledge of phase relationships and potential alternative grain-boundary phases in the various M-Si-O-N systems is thus important (Ref 29).

Ceramic Composites Based on Silicon Nitride. The limitations outlined above have given the impetus to finding new ways of improving these materials and the present interest in applying composite principles to ceramic systems. For example, silicon nitride matrix composite ceramics employing SiC whisker or platelet reinforcement are under development. The main aim is to achieve improved toughness without the loss of high-temperature strength and other thermomechanical characteristics found in the monolithic ceramics.

During firing of the ceramic, the presence of the "inert" SiC whiskers inhibits shrinkage of the silicon nitride matrix and, thus, pressureless sintering is more difficult than with monolithic types. So far, only hot pressing

Table 3 Thermal shock resistance of silicon nitride ceramics

| | Critical temperature change (ΔT_c) for: | | | |
| | Water quench | | Oil quench | |
Material type	Δ °C	Δ °F	Δ °C	Δ °F
RBSN	200–600	360–1080	750–1250	1350–2250
HPSN	400–800	720–1440	>1400	>2520
SSN	600–750	1080–1350	>1400	>2520
HIP-RBSN	800	1440	...	...
Sialon	300–550	540–990	...	...

Source: Ref 20

has been successful in producing fully dense silicon nitride composites.

On cooling from the sintering temperature, stresses arise as a result of thermal mismatch between whisker and matrix. The thermal expansion coefficient for SiC is 4.4×10^{-6}/K while that for silicon nitride is 3.2×10^{-6}/K. Thus, the whiskers will be in tension and the matrix in compression, and a higher stress would be required for matrix cracking (although an overall decrease in strength for the composite may also result). Radially, the SiC whiskers should shrink away from the matrix resulting in a decrease in bond strength at the whisker/matrix interface allowing toughening via crack deflection and whisker pull-out.

Figure 11 shows the effects of different amounts of SiC whiskers on the fracture toughness of silicon nitride hot pressed by different workers (Ref 30–32) under different conditions. The variations in sintering schedules lead to differences in the composite matrix microstructure, which reflects the different results for fracture toughness. In case 1 (Ref 30) of Fig 11, samples are fully dense, fully transformed β matrices with highly elongated fibrous grains and toughness values of 10 MPa$\sqrt{m}$ (9 ksi$\sqrt{in.}$). Case 2 has a less well-developed β grain microstructure (Ref 31). In cases 3 and 4, there is evidence that fabrication-related defects were the origins of fracture initiation in these composites and composite toughness values are only fractionally better than those for monolithics.

Generally, results of mechanical testing reported for SiC whisker-reinforced Si₃N₄-based composites have shown a slight decrease in strength with whisker addition. However, simultaneous increases in fracture toughness

and the modulus of rupture with whisker addition also have been reported (Ref 30).

Initial results on the development of new SiC whisker-reinforced Si₃N₄-base composites are encouraging. Some improvements are achieved in fracture toughness, but only on fully dense hot pressed materials. A production route for economic fabrication of these ceramic composites still must be developed.

Other Nitride Ceramics

Key representative properties of silicon oxynitride, aluminum nitride, and boron nitride are given in Table 4.

Silicon oxynitride, Si₂N₂O, has received less attention as an engineering ceramic than silicon nitride and is still under development. It may be synthesized from mixtures of silicon nitride and silica in conjunction with densification additives. With Al₂O₃ present, some limited solid solubility occurs. Pressureless sintering (Ref 33) and pressure-assisted processes (Ref 34) may be employed. Mechanical properties of silicon oxynitride are inferior to those of silicon nitride, but the material may have potential in certain thermomechanical applications because of its lower Young's modulus and slightly higher thermal expansion coefficient (which make it more suitable for bonding to metals).

Aluminum nitride was developed as an insulating substrate for microelectronic applications where its high thermal conductivity allows good heat dissipation. In this regard, it is a cheaper, non-toxic alternative to beryllium oxide (see the article "Ceramic Substrates" in this Volume). The ceramic is manufactured from aluminum nitride powder mixed with densification additives such as CaO or Y₂O₃ by a process of tape casting into thin sheets followed by blanking out and sintering at 1650 to 1800 °C (3000 to 3270 °F) in a nitrogen atmosphere. With yttrium oxide as

sintering aid, the yttrium-aluminum-garnet (YAG) is formed at the grain boundaries and the thermal conductivity increases dramatically (Ref 35) with Y₂O₃ content. This is because at low yttrium levels (<0.8 wt%), YAG forms as an almost continuous grain-boundary phase that inhibits the conduction between AlN grains (thermal conductivity 50 to 90 W/m · °C). With increasing yttrium levels, YAG forms as larger grains (up to 15 μm) leaving "clean" contacts between individual AlN grains. At 4.2 wt% yttrium, thermal conductivity is 160 W/m · °C.

The mechanical properties at ambient temperatures would suggest that aluminum nitride has less potential in high-stress applications, though some limited studies have been carried out (Ref 36, 37) at temperatures up to 1200 °C (2200 °F). Oxidation at >800 °C (1470 °F) of AlN limits its usefulness.

Boron nitride is isoelectronic with carbon and exists as two structural modifications similar to graphite (hexagonal) and diamond (cubic). Hexagonal boron nitride (HBN) is soft and platelike and is manufactured as a ceramic by hot pressing. This material exhibits anisotropy because platelets are oriented in directions perpendicular to that of the applied pressure (Ref 35). Thin-walled shapes, for applications such as crucibles for special glasses, may be produced by a chemical vapor deposition (CVD) or a pyrolytic method. The boron nitride is deposited on a suitably shaped graphite substrate which is subsequently removed. Thermal (and electrical) conductivity parallel to and perpendicular to the plane of the substrate are substantially different (Ref 35).

The cubic form of boron nitride, with a much higher density, is very hard and is produced from the hexagonal form at high temperature and pressure, similar to the synthesis of diamond (Ref 38). Cubic boron nitride is used as grinding material in grit form or as a

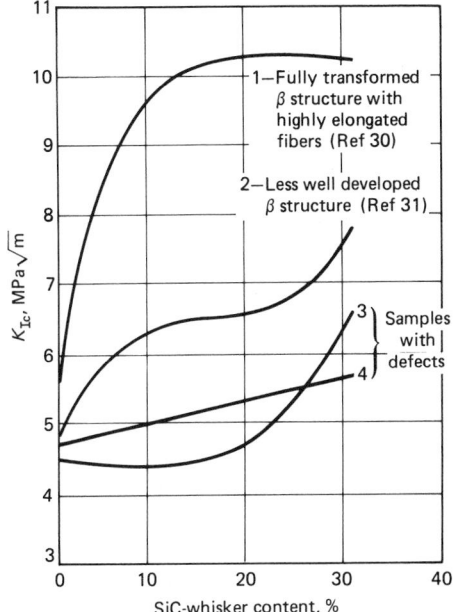

Fig 11 Effect of SiC whiskers on the fracture toughness of different silicon nitride matrix composites

Table 4 Properties of ceramic nitrides

Property	Silicon oxynitride (Si₂N₂O)	Aluminum nitride (AlN)	Hexagonal boron nitride Parallel to platelets	Perpendicular to platelets	Cubic boron nitride
Theoretical density, g/cm³	2.90	3.20	2.27	2.27	3.48
Coefficient of thermal expansion from 25–1000 °C (77–1830 °F), 10⁻⁶/°C	4.3	4.4–5.7	2–6	1–2	...
Specific heat (Cₚ), J/kg · °C (Btu/lb · °F), at:					
25 °C (77 °F)	...	800 (0.19)	...	780 (0.18)	...
1000 °C (1830 °F)	...	1570 (0.37)	...	1950 (0.47)	...
Thermal conductivity, W/m · °C (Btu/ft · h · °F), at:					
25 °C (77 °F)	8–10 (4.6–5.8)	50–170 (29–98)	20 (11.5)	33 (19)	...
1000 °C (1830 °F)	4 (2.3)	20–60 (11.5–35)	13 (7.5)	27 (15.6)	...
Young's modulus (E), GPa (10⁶ psi)	275–280 (39–40)	260–350 (37–51)	100 (14.5)	20 (2.9)	150 (22)
Flexural strength (σ_f), MPa (ksi)	450–480 (65–70)	235–370 (34–54)	Low	Low	High

solid ceramic for the manufacture of metal-cutting tools (see the article "Ceramic Cutting Tools" in this Volume).

ACKNOWLEDGMENT

The author wishes to thank Ms. D. O'Reilly, Mr. R. Flynn, and Mr. D. O'Sullivan for their help in the preparation of this article. Research on nitride ceramics at the University of Limerick has been sponsored by the European Community and EOLAS, the Irish Science and Technology Agency.

REFERENCES

1. N.L. Parr, G.F. Martin, and E.R.W. May, Preparation, Microstructure and Mechanical Properties of Silicon Nitride, *Special Ceramics*, Vol 1, P. Popper, Ed., Heywood, 1960, p 120–135
2. S. Hampshire, The Role of Additives in the Pressureless Sintering of Nitrogen Ceramics for Engine Applications, *Met. Forum*, Vol 7 (No. 3), 1984, p 162–170
3. F.F. Lange, Silicon Nitride Polyphase Systems: Fabrication, Microstructure and Properties, *Int. Met. Rev.*, (No. 1), 1980, p 1–20
4. G. Ziegler, J. Heinrich, and G. Wotting, Review: Relationships Between Processing, Microstructure and Properties of Dense and Reaction-Bonded Silicon Nitride, *J. Mater. Sci.*, Vol 22, 1987, p 3041–3086
5. G.R. Terwilliger and F.F. Lange, Pressureless Sintering of Si_3N_4, *J. Mater. Sci.*, Vol 10, 1975, p 1169–1174
6. G. Wotting and H. Hausner, Influence of Powder Properties and Processing Parameters on the Sintering of Silicon Nitride, *Progress in Nitrogen Ceramics*, F.L. Riley, Ed., Martinus Nijhoff, 1983, p 211–218
7. A. Giachello and P. Popper, Post-Sintering of Reaction Bonded Silicon Nitride, *Energy and Ceramics*, P. Vincenzini, Ed., Elsevier, 1980, p 620–631
8. J.A. Mangells and G.J. Tennenhouse, Densification of Reaction Bonded Silicon Nitride, *Am. Ceram. Soc. Bull.*, Vol 59, 1980, p 1216–1218
9. H.T. Larker, Hot-Isostatic Pressing of Shaped Silicon Nitride Parts, *High Pressure Science and Technology*, Vol 2, K.D. Tummerhans and M.S. Barber, Ed., Plenum Press, 1979
10. G. Ziegler and G. Wotting, Post-Treatment of Pre-Sintered Silicon Nitride by Hot Isostatic Pressing, *Int. J. High Technol. Ceram.*, Vol 1, 1985, p 31–58
11. K.H. Jack and W.I. Wilson, Ceramics Based on the Si-Al-O-N and Related Systems, *Nature Phys. Sci.*, Vol 238, 1972, p 28–29
12. Y. Oyama and O. Kamigaito, Solid Solution of Some Oxides in Si_3N_4, *Jpn. J. Appl. Phys.*, Vol 10, 1971, p 1637
13. N.E. Cother and P. Hodgson, The Development of Syalon Ceramics and Their Engineering Applications, *Trans. J. Brit. Ceram. Soc.*, Vol 81, 1982, p 141–144
14. K.H. Jack, Sialons—A Study in Materials Development, *Non-Oxide Technical and Engineering Ceramics*, S. Hampshire, Ed., Elsevier-Applied Science, 1986, p 1–30
15. K.H. Jack, Silicon Nitride, Sialons and Related Ceramics, *Ceramics and Civilization*, Vol III, *High Technology Ceramics—Past, Present and Future*, W.D. Kingery, Ed., American Ceramic Society, 1987, p 259–288
16. S. Hampshire, H.K. Park, D.P. Thompson, and K.H. Jack, α′-Sialon Ceramics, *Nature*, Vol 274, 1978, p 880–882
17. R.N. Katz and G.E. Gazza, Grain Boundary Engineering and Control in Nitrogen Ceramics, *Nitrogen Ceramics*, F.L. Riley, Ed., Noordhoff, 1977, p 417–431
18. M.H. Lewis, G. Leng-Ward, and S. Mason, Microstructural Design of High-Temperature Ceramics, *Engineering with Ceramics*, Vol 2, R. Freer, S. Newsham, and G. Syers, Ed., *Brit. Ceram. Proc.*, Vol 39, 1987, p 1–12
19. Y. Miyamoto, K. Tanaka, M. Shimada, and M. Koizumi, Survey for HIP Sintering Condition and Characterization of Dense Silicon Nitride without Additives, *Ceramic Materials and Components for Engines*, Proc. 2nd Int. Symp., W. Bunk and H. Hausner, Ed., Deutsche Keramische Gesellschaft, 1986, p 271–278
20. G. Ziegler, Thermal Properties and Thermal Shock Resistance of Nitrogen Ceramics, *Progress in Nitrogen Ceramics*, F.L. Riley, Ed., Nijhoff, 1983, p 565–588
21. A.J. Moulson, Review: Reaction Bonded Silicon Nitride: Its Formation and Properties, *J. Mater. Sci.*, Vol 14, 1979, p 1017–1051
22. M.H. Lewis, S. Mason, and A. Szweda, Syalon Ceramic for Application at High Temperature and Stress, *Non-Oxide Technical and Engineering Ceramics*, S. Hampshire, Ed., Elsevier-Applied Science, 1986, p 175–190
23. T. Ekstrom and N. Ingelstrom, Characterisation and Properties of Sialon Materials, *Non-Oxide Technical and Engineering Ceramics*, S. Hampshire, Ed., Elsevier-Applied Science, 1986, p 231–254
24. G. Ziegler, Thermal Cycling Behavior of Silicon Nitride, *Ceramic Components for Engines*, Proc. 1st. Int. Symp., S. Somiya, E. Kanai, and K. Ando, Ed., KTK Scientific, 1983, p 232–248
25. R.N. Katz and G.D. Quinn, Time-Temperature Effects in Nitride and Carbide Ceramics, *Progress in Nitrogen Ceramics*, F.L. Riley, Ed., Nijhoff, 1983, p 491–500
26. D. Cubicciotti, K.H. Lau, and R.L. Jones, Rate-Controlled Process in the Oxidation of Hot-Pressed Silicon Nitride, *J. Electrochem. Soc.*, Vol 124, 1977, p 1955–1956
27. M.H. Lewis and P. Barnard, Oxidation Mechanisms in Si-Al-O-N Ceramics, *J. Mater. Sci.*, Vol 15, 1980, p 443–448
28. M.J. Pomeroy and S. Hampshire, Oxidation Processes in Silicon Nitride Ceramics, *Mater. Sci. Eng.*, Vol A109, 1989, p 389–394
29. D.P. Thompson, Alternative Grain-Boundary Phases for Heat-Treated Si_3N_4 and β′-Sialon Ceramics, *Fabrication Technology*, R.W. Davidge and D.P. Thompson, Ed., *Brit. Ceram. Proc.*, Vol 45, 1980, p 1–13
30. S.T. Buljan and V.K. Sarin, *Silicon Nitride Based Composites*, Vol 18, 1987, p 99–106
31. P.D. Shalek, J.J. Petrovic, G.F. Hurley, and F.D. Gac, Hot Pressed SiC Whisker/Si_3N_4 Matrix Composites, *Am. Ceram. Soc. Bull.*, Vol 65, 1986, p 351–356
32. A. Bellosi and G. De Portu, Hot-Pressed Si_3N_4-SiC Whisker Composites, *Mater. Sci. Eng.*, Vol A109, 1989, p 357–362
33. M.H. Lewis, C.J. Reed, and N.D. Butler, Pressureless-Sintered Ceramics Based on the Compound Si_2N_2O, *Mater. Sci. Eng.*, Vol 71, 1985, p 87–94
34. H. Larker and L. Hermansson, Synthesis and Analysis of Silicon Oxynitride by Hot Isostatic Pressing, *Hot Isostatic Pressing—Theories and Applications*, CENTEK, 1988, p 375–381
35. C. Russel, T. Hofmann, and G. Limoner, Thermal Conductivity of Aluminium Nitride Ceramics, *Ceram. Forum Int.*, Vol 68, 1991, p 22–26
36. M. Billy and J. Mexmain, Processing and Properties of Aluminium Nitride, A New Candidate for High Temperature Applications, *World Ceram.*, Vol 2, 1985, p 91–95
37. F.P. Skeele, M.J. Slavin, and R.N. Katz, Time-Temperature Dependence of Strength in Aluminium Nitride, *Ceramic Materials and Components for Engines*, V.J. Tennery, Ed., American Ceramic Society, 1989, p 710–718
38. R. Morrell, *Handbook of Properties of Technical and Engineering Ceramics, Part 1, An Introduction for the Engineer and Designer*, Her Majesty's Stationery Office (UK), 1985

Engineering Properties of Diamond and Graphite

P. Darrell Ownby, University of Missouri—Rolla
Read W. Stewart, Consultant

CARBON is the sixth element in the periodic table with an average atomic weight of 12.011 in nature. Four of the six electrons in the neutral atom are in the outer L-shell and available for forming chemical bonds. In the solid state one s electron is excited into a p-state and covalent bonds are formed. Figure 1 shows the three general types of inorganic solid crystal structures which form from carbon.

In diamond, bonds of the type sp^3 are distributed tetrahedrally with lengths of 0.154 nm. This results in an extremely rigid, tight, and strong structure. In graphite, there are three coplanar bonds of the type sp^2 and of length 0.142 nm, with one electron available for a mixed role of coplanar and interplanar bonding (0.335 to 0.344 nm), as well as for electrical conduction. The carbon atoms lie at the corners of regular hexagons in monoplanes which are parallel to each other. Every successive plane is displaced such that every other plane is in alignment.

The third structure is spherical with its surface made up of from 28 to 540 carbon atoms in hexagonal and pentagonal arrays, with 60 atoms being the most stable as shown in Fig 1. This soccer ball shaped configuration is reminiscent of R. Buckminster Fuller's geodesic domes, with twelve pentagonal and twenty hexagonal sides, hence the name Buckminster Fullerene or "Buckyball" (Ref 3–6). The diameter of the sixty carbon structure is about 0.71 nm (Ref 1).

Diamond

Diamonds are the most highly valued gemstones and are characterized by their optical brilliance and resistance to chemical and abrasive attack. Yet ironically, diamond is metastable under standard temperature and pressure conditions, and only at very high pressures and temperatures is diamond thermodynamically stable.

In addition to possessing the highest hardness of any known substance, diamond exhibits the highest thermal conductivity of any material at room temperatures. The combination of these two properties renders it the ultimate cutting, grinding, and polishing material. Being the hardest, diamond not only cuts through all other hard substances but is able to rapidly dissipate the resultant heat from the cutting interface, thereby avoiding extreme temperature gradients and subsequent thermal shock that accompanies other hard materials.

Natural diamonds are produced deep within the earth's crust at extremely high temperatures and pressures. The carbon phase diagram is shown in Fig 2. The three stable phase fields are graphite at lower pressures, diamond at high pressures, and liquid at extremely high temperatures. The triple point where these equilibrium phase fields coincide occurs at about 4200 K and at 14 GPa (2×10^6 psi). Also depicted in Fig 2 are the regions where diamonds have been synthesized by man.

In the central portion of the diagram two regions are shown where diamonds are produced commercially by the simultaneous

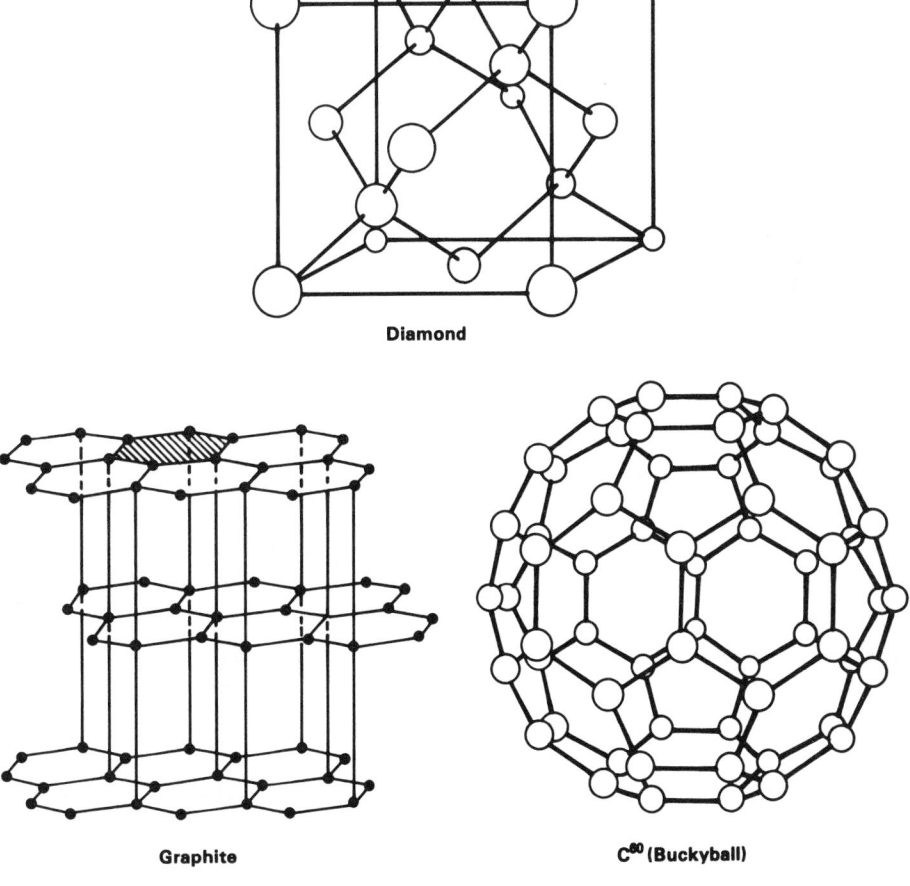

Fig 1 Crystal structures of diamond, graphite, and C^{60} (Buckyball). Source: Ref 1, 2

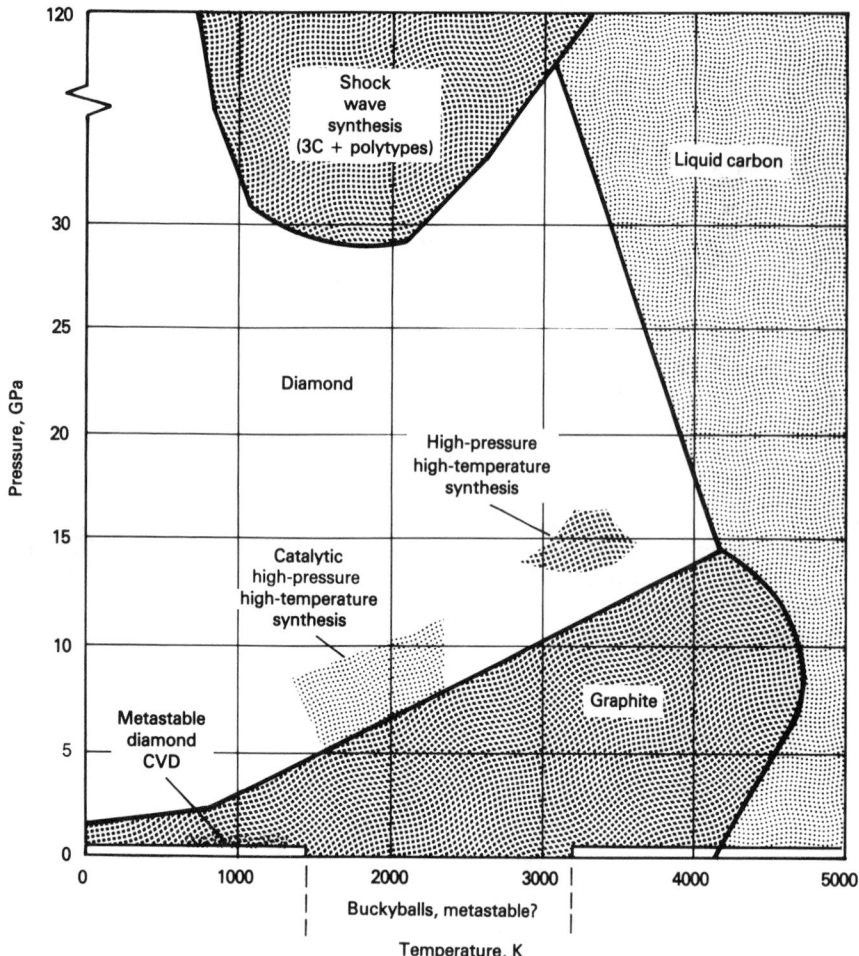

Fig 2 The carbon phase diagram. Source: Ref 7–9

application of static high pressure and temperature to graphite in specially designed tetrahedral, "belt," or cubic presses, with or without catalysts, such as nickel, iron, manganese, chromium, or cobalt. The catalysis method produces most of the synthesized commercial industrial diamond used throughout the world, in the micron to millimeter size range. A variation of this method utilizing the above catalysts as solvents is used to synthesize polycrystalline diamond (PCD), usually in the form of disks or hemispheres, 5 to 10 mm (0.2 to 0.4 in.) in diameter, for rock bit drilling media and bearing applications (Ref 10–15).

Near the top of the diagram a region is shown where dynamic explosive shock provides direct conversion of graphite to diamond, yielding diamonds in the nanometer to micron size range. The lattice planes of these diamonds do not all have sufficient time to stack in the normal cubic sequence and therefore consist of both cubic and non-cubic diamond polytypes (Ref 16).

At the bottom of the diagram, at pressure of one atmosphere or less, lies the region where diamond films can be synthesized metastably by chemical vapor deposition

(CVD). This thin film CVD process has been greatly advanced over the past decade *via* hot filament, ion beam, combustion moderation, sputter deposition, glow and microwave discharges, electron bombardment, direct current, radio frequency, laser, and reactive pulse plasmas, and microwave assistance to provide the required monatomic hydrogen partial pressure atmosphere at reduced total pressure of 10^3 to 10^5 Pa (Ref 2, 7, 9, 17–32). Various hydrocarbon CVD precursors have been used successfully including methane, ethanol, ethylene, benzene, and monosubstituted benzenes (Ref 26, 33). Once again, the kinetics of these CVD processes do not allow sufficient time for the ideal cubic lattice plane stacking sequence to always occur, thus creating, in general, a mixture of diamond polytypes in the deposited film (Ref 2).

Diamond Polytypes. The distinguishing crystallographic characteristics of the diamond polytypes are presented in Table 1. In the Ramsdell notation the number represents the number of unique lattice planes in a stacking repeat sequence and the letter denotes the cubic, hexagonal, or rhombohedral crystal structure of the polytype. The Jagodzinski notation emphasizes the type of structure

between lattice planes, k representing *kubisch* (cubic) and h, hexagonal. The percent hexagonal character of each polytype can be determined by inspection from this notation. It can be seen that while $3C$ diamond is 100% cubic and $2H$ is 100% hexagonal, all other polytypes contain mixed stacking characteristics. $2H$ diamond, also known as lonsdaleite, is found naturally in meteorites and bort as well as in synthetic forms, especially, thin films and explosively produced diamond as shown in Table 1. The new *ABC* notation distinguishes between two identical planes which are mirror images of each other (for example, AA').

Note that since the interatomic distance remains constant and only the stacking sequence varies between diamond polytypes, all of the theoretical polytype densities are the same (3.515 g/cm³).

Impurities. A great many properties of diamond are related in some manner to either the presence of impurities or the lack thereof. Impurities may exist substitutionally or as a second phase. Of all known impurities in diamond, only atomic nitrogen, boron, and the isotope C^{13} are known with certainty to be incorporated into the diamond lattice (Ref 37, 38). The types of impurities in diamond determine its "Type." Type Ia diamonds contain about 0.1% N atoms segregated into small aggregates and associated platelets. Most natural diamonds are of this type. Type Ib diamonds contain nitrogen dispersed in substitutional form. Most synthetic diamonds are of this type. Type IIa diamonds are very rare and have enhanced optical and thermal properties. They are virtually free of nitrogen. Type IIb diamonds are very pure, usually blue, with semiconducting properties due to the incorporation of boron. These are also extremely rare in nature.

Properties of Diamond

Chemical Inertness. An important property of diamond is its resistance to almost all forms of chemical attack. Diamond is not attacked by common acids at the temperatures and pressures commonly employed in chemical laboratories and can be safely cleaned in reagents such as boiling aqua regia or hydrofluoric acid (Ref 39). On the other hand, diamonds react comparatively easily in fluxes of caustic alkalis and various oxysalts (Ref 40). Such oxidation is often employed to produce an etched or roughened surface which is useful in revealing the microstructure of diamond. Prolonged oxidation, however, ultimately leads to the complete conversion of diamond to gaseous carbon dioxide (Ref 39). A one hour etching with potassium nitrate at 800 K is therefore common (Ref 39). Diamonds can also be oxidized to carbon monoxide in mixtures of sulfuric acid and potassium dichromate at temperatures in excess of 470 K (Ref 41). Diamonds have also been etched at high temperatures in monatomic gases such as oxygen, hydrogen, and chlo-

Table 1 Diamond polytype parameters

Crystal system	Crystallographic characteristics						
Ramsdell notation	3C	2H	4H	6H	8H	15R	21R
Jagodzinski notation	$(k)_3$	$(h)_2$	$(hk)_2$	$(hkk)_2$	$(hkkk)_2$	$(hkhkk)_3$	$(hkkhkkk)_3$
Zhdanov notation	(Infinity)	(11)	(22)	(33)	(44)	$(23)_3$	$(34)_3$
New ABC	ABC	AA'	AA'C'C	AA'C'B'BC	AA'C'B'A'ABC	AA'C'CABB' A'ABCC'B'BC	AA'C'B'BCABB'A' C'CABCC'B'A'ABC
Space group	FD3M	$P6_3/mmc$	$P6_3/mmc$	$P6_3/mmc$	$P6_3/mmc$	$R\bar{3}M$	$R\bar{3}M$
Factor group	O_h	$D6_h$	$D6_h$	$D6_h$	$D6_h$	$D3_d$	$D3_d$
Hexagonal, %	0	100	50	33	25	40	29
Lattice parameter, nm	$a = 0.35597$	$a = 0.25221$ $c = 0.41186$	$a = 0.25221$ $c = 0.82371$	$a = 0.25221$ $c = 1.23557$	$a = 0.25221$ $c = 1.64743$	...	...
Interatomic distance, nm	0.154	0.154	0.154	0.154	0.154	0.154	0.154
Atom positions	$(0,0,0)(\frac{1}{2},\frac{1}{2},0)$ $(0,\frac{1}{2},\frac{1}{2})(\frac{1}{2},0,\frac{1}{2})$ $(\frac{1}{4},\frac{1}{4},\frac{1}{4})(\frac{1}{4},\frac{3}{4},\frac{3}{4})$ $(\frac{3}{4},\frac{1}{4},\frac{3}{4})(\frac{3}{4},\frac{1}{4},\frac{1}{4})$	4 on $f(C_{3v})$	4 on $e(C_{3v})$ 4 on $f(C_{3v})$	4 on $e(C_{3v})$ 8 on $f(C_{3v})$	4 on $e(C_{3v})$ 12 on $f(C_{3v})$	30 on $c(C_{3v})$	42 on $c(C_{3v})$
Sources	Natural, HTHP, explosive, thin film	Natural, explosive, lonsdaleite, meteorites, bort	Thin film, explosive	Thin film, explosive	Thin film, explosive	Thin film, explosive	Thin film, explosive

Source: Ref 16, 34–36

rine as well as in carbon monoxide, carbon dioxide, and water vapor (Ref 42). A combination of energetic ions and physisorbed, oxygen-containing gas has been demonstrated to etch diamond surfaces as easily as those of any other semiconductor (Ref 88).

Chemical attack may occur in a manner which involves a group of avid carbide-forming metals. They include tungsten, tantalum, titanium, and zirconium. At high temperatures, they react chemically with diamond to form their respective carbides (Ref 43, 44).

Table 2 Mechanical properties of diamond

Properties	Natural diamond	Synthetic diamond	Polycrystalline diamond	Thin film Diamond	Ref
Density, g/cm³	3.51–3.52	3.20–3.52	3.00–4.00	1.80–3.50	2, 6, 9, 10, 13, 15, 17, 25, 30, 32, 34, 38, 40, 45–51
Elastic properties					
Elastic constants, GPa	$C_{11} = 950-1079$ $C_{12} = 120-330$ $C_{44} = 430-578$	...	...	...	2, 38, 44, 45, 52, 53
Young's modulus, GPa	700–1200	800–925	749–953	536–1035	2, 12, 15, 16, 20, 32, 38, 40, 45, 52–56
Bulk modulus, GPa	440–590	...	290–372(a)	...	2, 16–18, 38, 44, 45, 52–55, 57
Poisson's ratio	0.10–0.29	0.20	0.07–0.20(a)	...	2, 12, 13, 17, 38, 44, 45, 52, 53
Strength and brittleness					
Tensile strength, GPa	16.4–32.4	21.6–32.4	0.34–1.50(a)	...	13, 38, 44, 45, 52, 58
Flexure strength, MPa	1050	800–1400	389–1550(a)	...	12, 13, 15, 38, 52, 58, 59
Compressive strength, GPa	8.68–16.53	4.5–5.8	1.9–6.9(a)	...	10, 12, 13, 15, 38, 52, 58, 59
Fracture toughness, MPa√m	3.4	6.0–10.7	6.0–8.8(a)	...	12, 38, 52, 58, 60–62
Cleavage energy, J/m²	12(b) 10–20(d)	10–18	26–50(c)	...	37, 38, 54, 63, 64
Knoop hardness, GPa	...	54–84	49–78	65	2, 12, 15, 17, 20, 38, 44, 45, 52, 55, 65–67
(001) face	56–102	...	...	...	
(110),(111) faces	58–88	...	...	...	
Vickers hardness, GPa	...	95–131	25–98	29–118	
(001) face	88–147	...	...	...	
(111) face	98(e)	...	...	...	
Type 1b	...	88–108	...	...	
Type 11a	...	108–145	...	...	11, 15, 21, 30, 38, 45, 52, 68–71
Sliding coefficient of friction (f)					
In air	0.05–0.10	0.05–0.15	0.02–0.40	0.15–0.45	2, 9, 12, 37, 38, 52, 72
In vacuum	0.9	...	...	...	
Sound velocity, m/s	...	...	16,330	16,200	15, 40, 45
(100) face	17,500	...	...	...	
(110) face	18,200	...	...	...	

(a) Values dependent upon grain size and cobalt content. (b) Experimental values. (c) Fracture energy. (d) Theoretical values vary with planes. (e) Maximum value. (f) Varies with face, direction, surface roughness, and atmosphere

Another group of metals, though not distinguished for their carbide formation ability, are in the molten state true solvents for carbon and as such are capable of completely dissolving diamond crystals. This category includes the ferrous metals group such as iron, cobalt, manganese, nickel, and chromium, as well as the platinum group metals (Ref 7, 43, 44).

Mechanical Properties. Various mechanical properties, as well as their reported ranges of values are given in Table 2. The actual measured densities vary depending upon solvent content and degree of crystalline perfection. The mechanical properties of the various classes of diamond depend upon the manner in which they were synthesized. For some properties, crystallographic orientation or the presence of certain solvent(s) assumes greater importance.

The extremely high values for moduli, strength, and hardness are all related to the elastic constants of diamond which are the highest of any material. Like a ceramic, pure diamond is brittle and exhibits low fracture toughness. Conversely, the fracture toughness for polycrystalline diamond is considerably higher and can be attributable to the presence of a second phase.

In addition to its high strength and hardness, diamond also exhibits the lowest sliding coefficient of friction of any solid material, as well as the ability to propagate sound at high velocities.

Electrical Properties. For virtually every electronic property, diamond is unmatched. The electrical properties of diamond, which are listed in Table 3, are primarily dependent upon impurity type and concentration. It possesses the highest electrical resistivity of all

materials. Its variation with temperature and diamond type is shown in Fig 3. Theoretically, ideal diamond crystals should exhibit resistivities on the order of 10^{70} $\Omega \cdot$ cm (Ref 40, 87). However, impurities reduce this value considerably. In the case of polycrystalline diamonds, for instance, additions of transition metal solvents may render them as electrically conductive as some alkali, Group IIIa, and even some of the transition metals (Ref 37).

As semiconductors, diamonds exhibit the highest saturated electron velocity, the lowest dielectric constant, the highest dielectric strength, and the largest band gap of any semiconducting material (Ref 88).

The most prevalent acceptor impurity in both natural and synthetic Type IIb diamonds is boron. Both the boron concentration and the temperature range selected determine the activation energy. Generally, crystals with lower concentrations of uncompensated boron have higher activation energies, while those with higher concentrations tend to have lower activation energies due to the onset of impurity band conduction (Ref 89). Nearly all conductivity data for both natural and synthetic diamonds can be understood in terms of one acceptor—boron—at different concentrations and with varying degrees of compensation by nitrogen donors (Ref 45).

Thermal Properties. Selected thermal properties of diamond are presented in Table 4. Studies on the dependence of thermal conductivity on crystal symmetry revealed that conductivity in diamond is anisotropic and is characterized by isothermal surfaces which are ellipsoidal in nature with three different axes (Ref 40, 103). Figure 4(a) illustrates the temperature dependence of thermal conduc-

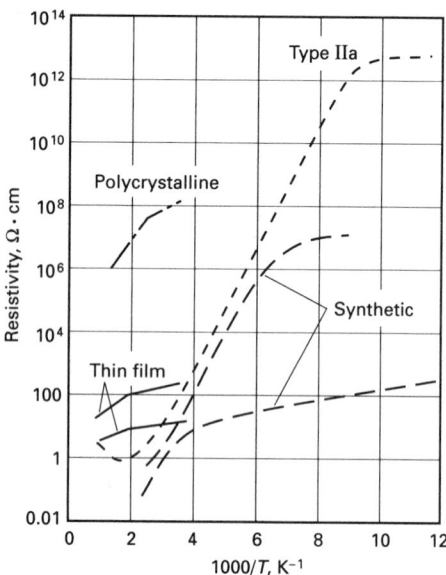

Fig 3 Temperature dependence of electrical resistivity for various types of diamond. Source: Ref 28, 68, 79, 85, 86

tivity for various types of diamond. With the exception of polycrystalline diamond, the maximum thermal conductivity for all diamonds occurs at approximately 80 K (Ref 2). At room temperature, the conductivity of polycrystalline diamond is five times that of copper. Although unsurpassed by all other materials at this temperature, substantially higher conductivity values as predicted for perfect crystals of diamond are not possible because of impurities.

The thermal conductivity of diamond is strongly sensitive to nitrogen impurities, an

Table 3 Electrical properties of diamond

Properties/Source	Natural diamond	Synthetic diamond	Polycrystalline diamond	Thin film diamond	Ref
Electrical resistivity, $\Omega \cdot$ cm, at 298 K(a)					
Type 1a	10^3–10^{16}	See Fig 3	See Fig 3	See Fig 3	
Type 1b	10^3–10^{16}				
Type 11a	10^3–10^{16}				
Type 11b	10–10^7				2, 9, 11, 14, 22, 24, 25, 32, 38, 40, 45, 68, 73–75
Dielectric constant (298 K, 1 MHz)	5.5–5.7	5.7	...	3.5–5.7	2, 9, 17, 18, 32, 38, 45, 57, 69, 74, 76–78
Breakdown strength (V/cm)	>10^7	...	...	...	2, 69
Activation energy, eV					
Type 11b (90–290 K)	0.29–0.37	0.015–0.37	0.42	...	14, 45, 48, 79–82
B-doped (270–710 K)	...	0.0029–0.087	...	...	
Al-doped (270–710 K)	...	0.32	...	...	
B-doped (270–710 K)	...	0.17–0.18	...	...	
Be-doped (270–710 K)	...	0.20–0.35	...	...	
Band gap energy, eV at 298 K	5.2–5.6	...	...	...	2, 9, 38, 69, 83
Effective mass					
Holes (m_h/m_o) Type 11b	<1.0	...	...	...	45, 80, 81, 84
Electrons (m_e/m_o) Type 11b	0.2	...	...	...	
Carrier mobility, cm^2/V $\cdot$ s					
Hole	1200–1600	...	...	...	38, 69, 77, 83
Electron	1800–2200	...	...	...	
Dissipation factor	0.0002	...	...	0.0140	33
Carrier lifetime, s	10^{-10}	...	...	...	69
Electron velocity, cm/s	2.7×10^7	...	...	...	69

(a) Variation with temperature is shown in Fig 3.

Table 4 Thermal properties of diamond

Properties/source	Natural diamond	Synthetic diamond	Polycrystalline diamond	Thin film diamond	Ref
Thermal conductivity, W/m · K at 298 K	See Fig 4(a)	See Fig 4(a)	543	See Fig 4(a)	2, 7, 9, 12, 15, 17, 22, 38, 40, 41, 45, 50, 54, 57, 75, 90–95, 102
Thermal diffusivity, cm²/s	...	...	...	0.3–8.1	93, 102
Type 11a at 77 K	4800	3300	...	...	
Type 11a at 298 K	10	10	...	...	
Coefficient of thermal expansion, 10^{-6}/K, at 298–373 K	See Fig 4(b)	...	1.5–3.8	...	2, 12, 17, 18, 29, 33, 38, 40, 41, 44–46, 50, 94, 96, 97
Specific heat, J/mol · K, at 300 K	See Fig 4(c)	...	...	33.2–41.8	17, 38, 45, 50, 94, 98
Heat of sublimation, kJ/mol	669	...	...	...	98
Debye temperature, K	See Fig 4(d)	...	...	...	2, 45, 91, 99
Thermal shock resistance parameter, W/m	3.8×10^8	...	...	1.9×10^8	20, 100
Oxidation rate, g/cm² · s	...	...	...	0.356–0.879	32
(111) plane	2.38	...	...	...	
(100) plane	0.167	...	...		
Activation energy of oxidation, kJ/mol · K					
823–1023 K	172–184	...	...	...	27, 101
923–1023 K	...	...	...	303–318	

effect which may be caused by the tendency of nitrogen atoms to cluster (Ref 110). Natural diamonds of the Type IIa variety contain very little nitrogen impurity whereas Type Ia diamonds are characterized by the presence of appreciable amounts of nitrogen. As a result, the thermal conductivity of Type IIa is about three times greater than that of Type Ia diamonds (Ref 2, 92).

One explanation for diamond's high thermal conductivity is that phonon heat transfer proceeds readily across its tetrahedral crystal lattice of low-atomic-mass atoms and stiff covalent bonds (Ref 111). Enhancement of the lattice quality can be attained if it is composed exclusively of a single isotope—C^{12} (Ref 92, 107). Natural diamond is composed of 98.9% C^{12} and 1.1% C^{13} (Ref 107). From earlier studies on the isotope effect in both germanium and lithium fluoride, it is presumed that these occasional C^{13} atoms (which are 8% heavier than C^{12} atoms) in the diamond lattice impede the transmission of heat by interfering with the passage of phonons much like the manner in which speed bumps slow traffic on an otherwise smooth highway (Ref 107, 111, 112). The elimination of so small a contamination has an effect on the thermal conductivity that is far greater than its abundance would seem to suggest. The removal of C^{13} atoms and its resulting isotope effect has been realized in diamonds composed almost entirely of the C^{12} isotope. They have been found to conduct heat 50% better than naturally mined diamonds (Ref 111). Predicted values of thermal conductivity for isotopically pure C^{12} diamonds with respect to temperature are shown in Fig 4(a) and are expected to approach 10^5 W/m · K (Ref 107).

Elements of the diamond-structure type group which include germanium, silicon, and diamond are characterized by very low ther-

mal expansions and isotropic linear expansion coefficients (Ref 45). The temperature dependence of the coefficients of thermal expansion is given in Fig 4(b). Diamond obeys Grüneisen's law between 420 and 1200 K (Ref 44, 91, 99, 107, 113). At very low temperatures, the variation of thermal expansion coefficients among the different classes of diamonds is relatively minor. However, with increasing temperature, the permissible range of values increases in an almost linear manner.

Diamond exhibits the lowest specific heat of any solid between 0 and 800 K. Figure 4(c) illustrates the variation of specific heat with temperature. Corresponding to its low specific heat, diamond also exhibits the highest Debye temperature of all solid substances. Since the heat capacity of diamond does not conform exactly to Debye theory, its Debye temperature varies somewhat with temperature as shown in Fig 4(d) (Ref 2, 45, 92).

Optical Properties. Exhibiting high transparency throughout the visible, near infrared, and thermal infrared, pure diamond has the widest spectral transmission range of all known solids (0.225 to 25 μm) (Ref 39, 43, 51, 56, 77, 114–116). Theoretically, ideal diamond crystals should be absolutely transparent to visible light, but impurities and other defects in the crystal structure affect absorption in the visible region. The dispersion of diamond is high (0.062) and as a result, its refractive index, n, is highly wavelength, λ, dependent (Ref 40). This can be expressed as an empirical equation of the form (Ref 117):

$$n^2 - 1 = \frac{\epsilon_1 \lambda^2}{\lambda^2 - \lambda_1^2} + \frac{\epsilon_2 \lambda^2}{\lambda^2 - \lambda_2^2}$$

where $\epsilon_1 = 0.3306$, $\epsilon_2 = 4.3356$, $\lambda_1 = 175.0$ nm, and $\lambda_2 = 106.0$ nm.

This dependence is illustrated in Fig 5. Furthermore, the refractive index decreases

when subjected to hydrostatic pressure and increases upon heating (Ref 40, 44).

Due to its high refractive index, the amount of incident light reflected off a diamond surface is quite high, usually between 17 to 18%, depending upon the wavelength of light used (Ref 44, 45, 51).

The coloration of diamond is caused by absorption bands in the visible region (400 to 700 nm) or in the case of Type IIb, to the tail of the infrared absorption spectrum associated with the acceptor center. The yellow coloration of Type Ib diamonds results from the tail of the ultraviolet absorption spectrum associated with a substitutional nitrogen impurity (Ref 99).

Many diamonds are birefringent due to inclusion-induced strain. The birefringence of many inclusion-free Type IIa natural diamonds is distinguished by a crosshatch of light and dark lines (Ref 45).

Furthermore, many diamonds also luminesce and, depending on the diamond, this luminescence can be excited electrically, by phonon irradiation, or by particle bombardment. The luminescence may take the form of fluorescence, phosphorescence, or thermoluminescence (Ref 45, 114, 121). Varying degrees of luminescence are observed when diamonds are subjected to ultraviolet radiation. Synthetic Type Ib and Type IIb diamonds are relatively inert to long-wave ultraviolet radiation while exhibiting various responses to short-wave radiation, namely, changes in color. In contrast, natural diamonds show a stronger response to long-wave than to short-wave ultraviolet radiation. This optical phenomenon has been utilized as a time-saving tool in the separation of natural from synthetic diamonds (Ref 122).

Applications for Diamond

The many remarkable mechanical properties of diamond make it uniquely qualified for many special applications. In addition to its aforementioned superiority in bearing, wire drawing, cutting, grinding, and polishing applications which utilize its extreme hardness, wear resistance, and low friction coefficient, many very new applications have been discovered. Several examples of enhanced properties of ceramic-matrix composites have been reported by the addition of diamond and diamond polytypes. These ceramic matrices include zinc sulfide (Ref 123), aluminum oxide (Ref 124–126), silicon nitride (Ref 127), and silicon carbide (Ref 16).

The great range in electrical resistivity possible in diamond makes it amenable to be tailored for an extremely wide range of electronic applications. The high saturated electron velocity, largest intrinsic band gap, lowest dielectric constant, and highest dielectric strength make diamond unmatched by other materials for specialized electrical applications. If the recently discovered mechanisms by which erbium impurities in III-V compound semiconductor junctions can be made

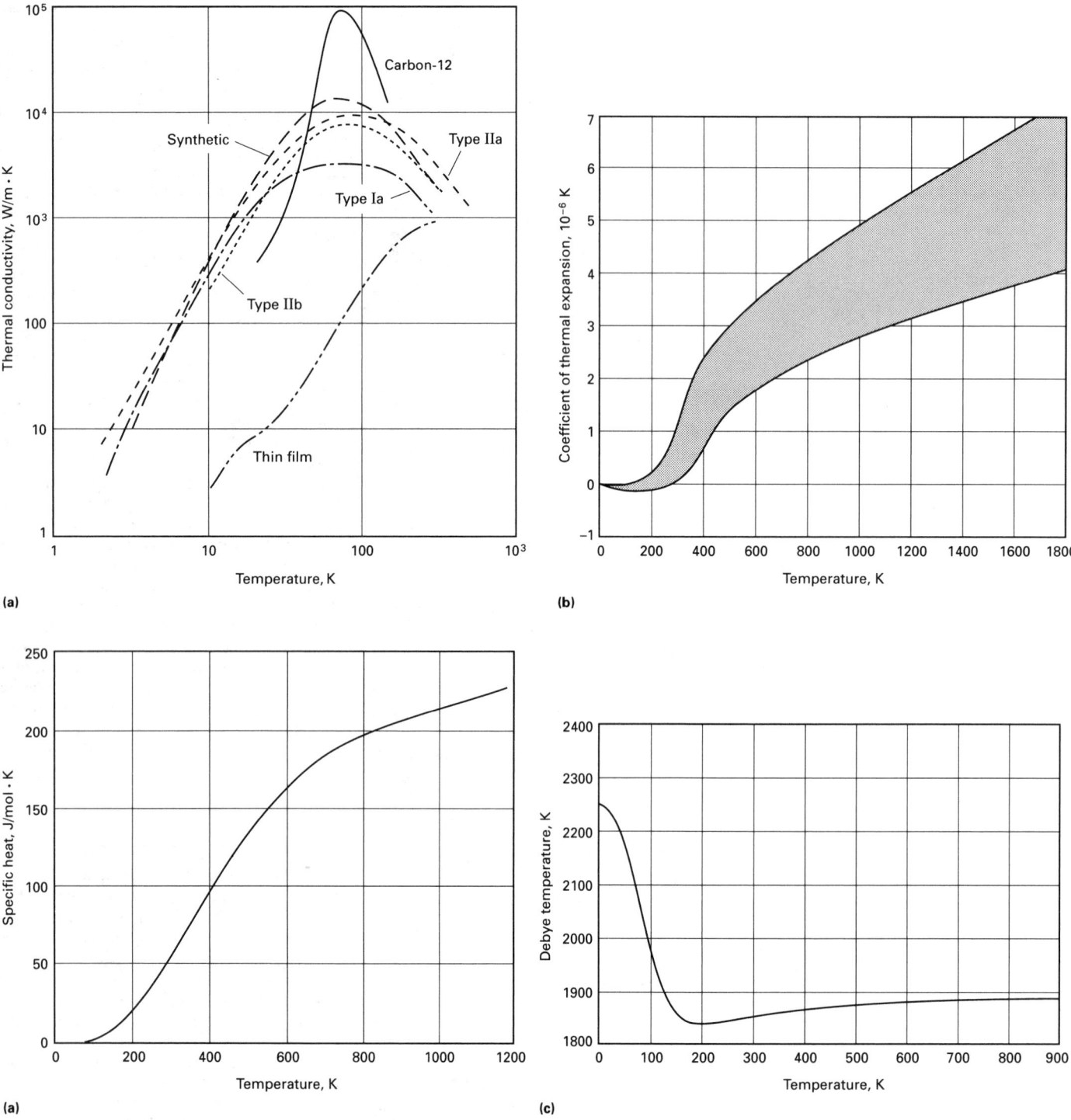

Fig 4 Thermal properties of diamond. (a) Thermal conductivity as a function of temperature for various types of diamond. Source: Ref 90, 91, 104–107. (b) Range of thermal expansion coefficients. Source: Ref 33, 40, 44–46, 50, 90, 94, 96, 97, 108, 109. (c) Temperature dependence of specific heat. Source: Ref 50. (d) Experimental variation of Debye temperature as a function of temperature. Source: Ref 40, 90, 92

to produce lasers from indirect gap semiconductors, then efficient diamond lasers which operate in attractive portions of the optical spectrum may become a distinct possibility (Ref 88).

In many applications, it is the unique combination of properties of a material which make them most useful. An excellent example is the use of diamond for scalpel blades in microsurgery (Ref 39). Edges can be polished on diamond which are at least fifty times sharper than those of the best razor blades or surgical steel scalpels. Diamond blades are thus used in delicate microsurgery because they cut cleanly with less force, less distortion and displacement of the surrounding tissue, and less cell trauma in the wound region. A transparent diamond blade may also be used as a light pipe to bring either illumination or laser energy to the cutting edge. The fact that diamond is the most corrosion resistant of cutting edges also contributes to the effectiveness of this application.

The advent of CVD diamond coatings has resulted in the first commercial consumer product based on diamond thin film coatings: a line of diamond-clad audio speaker tweeters

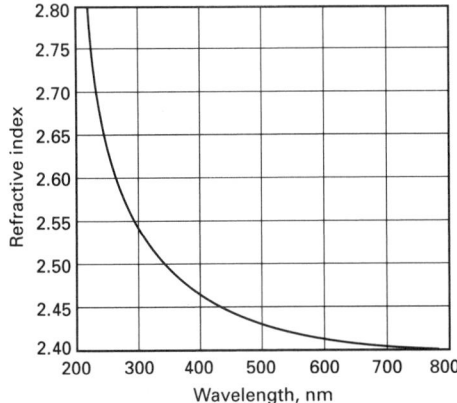

Fig 5 Refractive index of diamond as a function of wavelength. Source: Ref 2, 9, 17, 22, 38, 40, 43, 45, 57, 68, 82, 103, 114, 117–120

which take advantage of CVD diamond's high sound propagation velocity, high stiffness, high dissipation factor, and low weight (Ref 7, 33).

Graphite

Although "amorphous" and graphitized carbon have basically similar crystal structures, the external properties of the two materials are quite distinct. This can be explained by taking a closer look at the structures; namely the differences in the spatial arrangement of the carbon atoms in relation to each other and to those adjoining them.

In graphite, the carbon atoms lie at the corners of regular hexagons, and all atomic layers or planes are parallel to each other (Fig 1). In carbon black, the hexagon atomic arrangement is somewhat definite in one plane, but the arrangement of these to other planes is quite random. Single graphite crystals consist exclusively of hexagonal planes piled on each other, with crystalline order and closer spacing between the monoplanes and stacks.

This degree of order is found only in crystals where the planes have linear dimensions larger than 70 atom spacings. Such dimensions can only be achieved by heating the material to temperatures between 2773 and 3273 K, at which the monoplanes grow and become piled into stacks. Called graphitization, this process can be continued until crystallites are formed with linear dimensions of the order of 1000 atom spacings.

Displacement and rearrangement of layer planes and small groups of planes to achieve three-dimensional ordering also occur. However, carbons that have strongly bonded structures (cross-linked chains and stacks) cannot be graphitized. Materials are required that have atoms with high mobility; in other words, carbonaceous materials that go through a liquid stage during carbonization can usually be graphitized.

Besides the structure of the precursor material, the maximum temperature and soak time at temperature are important factors in achieving graphitization or crystallinity. The crystallite interlayer spacing has been shown to have a limited value at each temperature, although decreasing at every temperature, even after long soak times. When the maximum temperature is increased, the rate of approach to the limiting value is also increased.

Other research has shown that crystallite formation can be enhanced in the presence of oxidizing gases, which results in the elimination of cross-linking bonds, single-layer planes, or other amorphous regions (Ref 55, 128).

Because of graphite's unique structure, it is highly anisotropic or directional in properties. This is due to the differences in separation distances of atoms between the layer planes (0.335 nm). Hence, electrical conductivity is good along the direction of the lattice planes, and poorer perpendicular to the lattice planes. A similar result is seen with strength. The good lubricity of graphite products also results from the hexagonal arrangement of layer planes. Because each plane is weakly bonded to those above and below it, each is relatively free to slip or slide past one another.

In contrast, amorphous carbons have smaller monoplanes piled in turbostatic stacks. In such a structure, the *a* axes of different stacks are randomly oriented and hydrocarbon groups and chains can be attached to the periphery of the monoplanes. Heat treatment removes volatile groups, leaving cross-linked bonds between the different monoplanes of one stack and also connecting any stack with others.

Another major difference between carbon and graphite is the crystallographic size of the individual crystallites. In the case of carbon, typically very small crystallites are present on the order of 4 nm long and 2 nm thick. Graphites, on the other hand, can have crystallite sizes greater than a hundred nanometers and over 45 nm in thickness (Ref 129).

Therefore, a variety of different carbon and graphite grades can exist, depending on the crystal regularity and the presence of cross-linked bonds. Properties can also be varied

Table 5 Manufacturers, codes, and grades reported in Tables 7 to 10

Code	Manufacturer/address	Grades represented in this article
(1)	Superior Graphite Company 120 S. Riverside Plaza Chicago, IL 60606	MR(e) SFR(e) SR(e) SCR(e) SFH(e) SRH(e)
(2)	Tokai Carbon Co., Ltd. Aoyama Building 2–3, Kita-aoyama 1-Chome, Minato-ku, TOKYO 107, Japan	G250(1) G457(i) G540(i) G151A(e) G114(m)
(3)	Poco Unocal 76 1601 South State Street Decatur, TX 76234	PGCS-1 PGCS-2 PGCS-3 PXM ZXF-5Q AXM-5Q
(4)	Stackpole Carbon Company Stackpole Street St. Marys, PA 15857	Y20 331 2000 2019 2020 2191 2916 2917 6026 6077(e)
(5)	Carbon/Graphite (AIRCO) 800 Theresia Street St. Marys, PA 15857	H SX-4 873S 873RL 875G(e) 3499
(6)	Great Lakes Carbon 320 Old Briarcliff Road Briarcliff Manor, NY 10510	HPC(e) HLM(e) LHM-50(e) H489(e) B-404 H-050
(7)	Union Carbide 39 Old Ridgebury Road Danbury, CT 06817-0001	ATJ(i) AGSR(e) CS-312(e) ZTA CMB(e) EBP ECL ECV
(8)	Rinsdorf-Werke GmbH Drachenburgstrasse 1 D-5300, Bonn 2	EK40(m) EK462(i) EK47(e) EK78(i) EK599(e) EK98(i)
	Sigri Corporation Route 202-206 North P.O. Box 922 Somerville, NJ 08876	MKS-R(e)
(9)	Pechiney Group Carbon and Graphite Division 500 Plaza Drive Secaucus, NJ 07094	PH(e) PHA(e) PHAI(e) PHA2(e) PHAD(e)
	Ultra Carbon Division Div. of Carbone of America 900 Harrison Street Bay City, MI 48707	2122, 2123, and 2318
(10)	Toyo Tanso Co., Ltd. 7-12, 5-Chome, Takeshima, Nishiyodogawa-ku, OSAKA 555, Japan	IG-15(i) ISEM-1(i) ISO-88(i) IG-710(i)

Note: (e) = extruded, (i) = isostatically pressed, (m) = molded.

by using different binders and fillers, and as has already been mentioned, by changing process parameters. Such changes can result in different grain sizes, crystalline structure, and other physical properties. In turn, these physical characteristics have considerable effect on the final mechanical, thermal, electrical, and other properties, as will be presented below.

Properties of Shaped Graphite

The purpose of this compilation is to present a broad overview of the range and limits of the properties of several grades of products taken from the published commercial data of a selected number of producers of shaped graphite. Not all grades produced by these manufacturers are reported here, only representative grades showing typical ranges of properties. Additional detailed information can be obtained directly from the manufacturers, which are listed in Table 5, and in Ref 128 to 134.

There are principally three forming techniques in producing shaped graphite: extrusion, molding, and isopressing. Properties achieved will depend on the method of preparation, however, no special notation is provided here with the exception, of where possible, showing where the grades are either extruded (e), molded (m) or isopressed (i), and strength values, where with-grain (w), against-grain (a), and isotropic (i) designations are given.

The principal source of commercial grades of shaped graphite is petroleum cokes. Typical compositions of raw, calcined, and thermally purified petroleum coke used in synthetic or artificial graphite manufacture are shown in Table 6.

Grain Size. Depending on the manufacturer's methods, the shaped graphite will exhibit a characteristic grain size, as shown in Table 7, which can range from 2 μm up to 8 mm. The engineering properties will in general depend on this characteristic. The smaller the grain size, the higher cost of the graphite, but mechanical, thermal, and electrical properties are enhanced.

Bulk densities and ash content are shown in Tables 8 and 9, respectively. As-formed bulk densities can range from as low as 1.30 g/cm³ to more than 1.95 g/cm³; the theoretical density is recognized to be between 2.24 and 2.26 g/cm³. The range of ash content is from about 10 ppm up to 4%. Porosity in extruded bodies can reach as high as 30%.

Thermal Properties. Table 10 is arranged in ascending value of the coefficients of thermal expansion (CTE). For each grade shown, the corresponding value of thermal conductivity is listed. The range covers CTE values from 1.2 to 8.2 × 10⁻⁶/K, with variation if the measurement is taken with the grain or against the grain of extruded materials. Values of thermal conductivity range from 25 W/m · K to more than 470 W/m · K. The specific heat for these types of shaped graphite ranges between 0.17 and 0.20 cal/g · °C.

Mechanical Properties. The variability of flexure strength, which depends on the method of forming, is shown in Fig 6. Compressive strength, tensile strength, and the modulus of elasticity do not have as wide a range and fall respectively in the following values:

- Compressive strength: 16 to 186 MPa (2.3 to 27 ksi)
- Tensile strength: 6 to 70 MPa (0.85 to 10 ksi)
- Modulus of elasticity: 8 to 15 GPa (1.2 to 2.15 × 10⁶ psi)

Hardness, as measured by the Shore-scleroscope can be as low as 30 and as high as 100.

The electrical properties are usually reported as resistivity, and values will range between 5 and 30 μΩ · m.

Applications for Graphite

The layered crystal structure of graphite gives it some of its most important characteristics and one of the lowest coefficients of friction of any material. Over a wide temperature range, graphite is a sectile, flexible material exhibiting low coefficient of thermal expansion, high thermal and electrical conductivity, and is almost chemically inert. Consequently, graphite is commonly used for the following attributes: lubricity, electrical conductivity, refractoriness, thermal conductivity, chemical inertness, low thermal expansion, and as a carbon source.

Table 6 Nominal compositions and properties

Constituent/property	Raw coke	Coke calcined to 1300 °C	Coke purified to 2500 °C
Moisture, %	0.5–2	Negligible	Not detectable
Volatile matter, %	6–15	<0.2	<0.1
Fixed carbon, %	87–97	97–99	99.8+
Real density, g/cm³	1.2–1.8	2.08–2.13	1.8–1.9
Ash, %	0.1–1.0	0.2–1.5	<0.1
Sulfur, %	0.2–5.0	0.2–2.5	0.015–0.03
Hydrogen, %	3–4.5	<0.1	2–60 ppm
Nitrogen, %	0.1–0.5	<0.1	7–70 ppm
Oxygen, ppm	...	...	6–65
Iron, ppm	50–2000	50–2000	<25
Vanadium, ppm	5–500	5–500	1–50
Boron, ppm	0.1–0.5	0.2–0.7	0.1–0.5
Titanium, ppm	2–60	2–60	0.05–1
Silicon, ppm	50–300	50–300	20–60
Aluminum, ppm	15–100	15–100	<5
Manganese, ppm	2–100	2–100	<1
Nickel, ppm	10–100	10–100	<1
Calcium, ppm	25–500	25–500	<10
Magnesium, ppm	10–250	10–250	<0.15

Table 7 Reported average grain size versus manufacturer's product code and grades (MC&G). See also Table 5

Average grain size, mm	1	2	3	4	5	6	7	8	9	10
2 μm	...	G540 G457	...	...	...	...	...	...	...	...
4 μm	...	...	...	...	...	...	...	...	2318	...
0.0046	...	G250	...	...	...	...	...	...	...	...
0.0050	...	...	...	2000 2019	...	...	...	...	...	...
0.01	...	...	ZXF-50	2917	...	...	...	...	...	...
0.015	...	...	...	...	...	...	ATJ	...	...	...
16 μm	...	...	...	...	...	...	...	...	2123	...
0.043	...	...	...	2020 2191 2916, 331	...	...	...	...	...	...
20 μm	...	...	...	...	...	...	...	...	2122	...
0.05	...	...	AXM-50	...	...	...	...	...	...	...
0.076	...	...	...	Y20	3499	H489 H-050	...	...	...	...
0.10	...	...	PGCS-2	...	...	...	...	...	...	...
0.13	...	...	...	...	H	...	...	...	...	...
0.15	...	...	PGCS-1	...	...	...	ECV, ZTA	...	...	...
0.20	...	...	...	...	...	B-404	...	...	...	...
0.229	...	...	PGCS-3	...	...	...	...	...	...	...
0.71	...	...	...	...	...	...	...	MKS-R	...	...
0.76	SFH, SCR SFR	...	...	...	...	HLM-50 HLM	CS-312 ECL	...	All grades	...
0.813	...	...	...	...	873S 875G 873RL SX-4	...	...	...	...	...
1.02	...	...	...	6077	...	...	...	...	...	...
1.52	SR, MR SRH	G151A	...	6022	...	HPC	...	...	...	...
3.3	SRH	...	...	...	...	...	...	...	...	...
6.35	SCR	...	...	...	...	...	AGSR	...	...	...
8.0	...	...	PXM	...	...	...	...	...	...	...

Table 8   Apparent bulk density versus manufacturer's product code and grades (MC&G). See also Table 5

Bulk density range, g/cm³	MC&G									
	1	2	3	4	5	6	7	8	9	10
1.35–1.37	...	G114	...	...	...	...	...	...	...	...
1.38–1.40	...	G250	...	...	...	...	...	...	...	...
1.56–1.61	...	...	...	...	...	...	AGSR	...	PH	...
1.62–1.64	MR, SR SFR SCR	...	...	6077 Y20	875G H	HLM-50	...	EK48	...	...
1.65–1.67	SCR, SR SFR	...	...	...	3499	...	...	...	...	...
1.68–1.70	...	...	...	...	...	B-404	ECL ECV	EK5996	PHA PHAI	ISEM-1
1.71–1.73	SFH	...	...	6026 2020 2191	873RL SX-4	HPC H489	CS-312 ATJ EBP	EK40	PHA2	...
1.74–1.76	SFH SRH	...	PGCS-3 PXM AXM-5Q	331	873S	HLM	CMB	MKS-R	2122	...
1.77–1.79	...	...	PGCS-1	2916	...	...	...	...	PDAD	...
1.80–1.82	...	G457	ZXF-5Q	2000 2019	...	...	...	...	2123	...
1.83–1.85	...	G540	...	...	...	...	...	EK462 EK986	...	IG-710
1.86–1.88	...	...	PGCS-2	2917	...	H-050	...	EK78	...	...
1.89–1.92	...	...	...	...	...	...	...	...	2318	ISO-88
1.93–1.95	...	...	...	...	...	...	ZTA	...	...	IG-15

Although the graphite reported in this article focused principally on graphite shapes for electrodes used in the metallurgical industry, graphite is utilized in a broad range of commercial products. These include batteries, friction materials (brake shoes), carbon parts (brushes and bushings), lubricants (wire drawing dies), paints, packings, pencils, and seals. The highly purified graphite is used in analytical procedures such as in emission spectroscopy.

The following data published by the U.S. Department of the Interior shows the natural graphite use in the United States:

- Refractories 25%
- Foundries 20%
- Lubricants 10%
- Brake linings 9%
- Steelmaking 8%
- Batteries 6%
- Crucibles and stoppers 5%
- Pencils 5%
- Other 12%

The major use of artificial or synthetic graphite in extruded or molded shapes is for electrodes or machined parts, although some of the above applications will also use synthetic graphite instead of the natural mineral.

Buckyballs

These recently discovered structures (Fig 1) are apparently as common as soot and may have played a critical role in the formation of the planet while being the source of mysterious spectral lines emanating from distant stars (Ref 5). Having a truncated icosahedral (or soccer ball) structure with one dangling bond from each of the sixty carbon atoms, there has been much speculation as to the possible chemical and industrial uses of these fascinating structures. The perfection of this roundest of molecules stems from the fact that sixty is the largest number of proper rotations in the icosahedral point group making it the most symmetric possible molecule. If each of its sixty dangling bonds could be bonded to an atom like fluorine, it could serve as an ideal micro ball-bearing lubricant. This symmetry also allows electrons to move about freely through, and be stored in, the network structure, possibly leading to a new class of rechargeable batteries. Inside the spherical structure, the large space may provide a multitude of possibilities. Encapsulated atoms may be the basis for new classes of metallofulleroniums.

ACKNOWLEDGMENT

The authors would like to acknowledge the assistance of Timothy Lin, Jenq Liu, and Jyung-Lyang Shen of the University of Missouri—Rolla Ceramic Engineering Department, and Daniel McKean of Superior Graphite Company for their assistance in the compilation of the data and in the preparation of the text, figures, and tables.

REFERENCES

1. I. Amato, Buckeyballs Get Their First Major Physical, *Science News*, Vol 138 (No. 23), 1990, p 357
2. K.E. Spear, Diamond—Ceramic Coating of the Future, *J. Am. Ceram. Soc.*, Vol 72 (No. 2), 1989, p 171–191
3. W. Krätschmer, L.D. Lamb, Fostiropoulos, and D.R. Huffman, Solid C⁶⁰: A New Form of Carbon, *Nature*, Vol 347 (No. 6291), 1990, p 354–358
4. A.L. Mackay, Carbon Crystals Wrapped Up, *Nature*, Vol 347 (No. 6291), 1990, p 336–337
5. P.E. Ross, Buckeyballs, *Sci. Am.*, Vol 263 (No. 7), 1991, p 114–116
6. C. Vaughan, Tracking an Elusive Carbon, *Sci. News*, Vol 135 (No. 1), 1989, p 56–57
7. P.K. Bachmann and R. Messier, Emerging Technology of Diamond Thin Films, *Chem. Eng. News*, Vol 67 (No. 20), 1989, p 24–39
8. F.P. Bundy, The P,T Phase and Reac-

Table 9   Ash content versus manufacturer's product code and grades (MC&G). See also Table 5

Ash content, wt %	MC&G									
	1	2	3	4	5	6	7	8	9	10
≤20 ppm	...	...	...	...	...	...	...	...	2122PT 2123PT 2318PT	...
≤200 ppm	...	...	...	...	...	...	...	...	2122PN 2123PN 2318PN	...
500 ppm	...	...	...	...	...	...	...	...	2122 2123 2318	...
0.001–0.005	...	...	...	...	...	...	ECL ECV	EK5996 EK986	...	IG-710
0.01–0.059	...	...	PGCS-3	...	873RL	...	...	MKS-R	PHAI	ISEM-1 ISO-88 IG-15
0.06–0.099	...	G250 G457, G540	PXM	...	...	...	ATJ	...	...	...
0.1–0.299	...	G151A G114	...	...	SX-4 3499	H-050	ZTA AGSR	EK40 EK47 EK78	PH, PHA PHA2	...
0.3–0.699	MR	...	...	2191	...	...	...	...	...	...
1.0–1.3	...	...	...	...	873S, H 875G	...	...	...	...	...
3–4	...	...	...	...	...	...	...	...	PHAD	...

Table 10 Coefficient of thermal expansion, thermal conductivity, and specific heat versus manufacturer's product code and grades (MC&G). See also Table 5

CTE, 10^{-6}/K	Thermal conductivity, W/m · K	Specific heat, cal/g · °C	MC&G									
			1	2	3	4	5	6	7	8	9	10
1.20	93	...	...	...	...	...	...	...	...	...	...	ISEM-1
1.26	428	...	...	...	...	...	...	HLM-50	...	...	...	...
1.26	454	...	...	...	...	...	...	HPC	...	...	...	...
1.35	...	...	...	...	...	...	...	...	CS-312	...	...	...
1.58	190	...	...	...	...	...	...	HLM	...	...	...	...
1.6	70	...	...	...	...	...	...	...	...	EK50	...	...
1.88	...	...	...	...	...	...	...	...	AGSR	...	...	...
1.9–2.9	...	...	MR	...	...	...	...	...	...	...	...	...
2.0	473	...	...	...	...	...	...	...	...	MKS-R	...	...
2.2–3.8	100	...	...	...	...	...	...	...	...	...	PH	...
2.3	...	...	...	...	...	6077	...	...	...	...	...	...
	354	...	...	...	...	...	...	H489	...	...	...	...
2.5	437	0.20	...	...	PXM	...	...	...	...	EK5996	...	...
	...	...	...	...	...	6026	...	...	...	...	...	...
2.8	115.7	...	...	...	...	2191	...	...	...	...	...	...
2.8	...	...	...	...	...	2916	...	...	...	...	...	...
2.9–3.5	120–130	...	...	...	...	...	...	...	...	...	PHA, PHA2, PHAI, PHAD	...
3.00	58	...	...	G250	...	...	...	...	...	...	...	...
3.00	140	...	...	G457	...	...	...	...	...	...	...	...
3.08	125	...	...	...	...	...	...	...	ATJ	...	...	...
3.08	110	...	...	...	...	...	...	...	...	EK47	...	...
3.3	...	...	...	...	...	331	...	...	...	...	...	...
3.3	70	...	...	...	...	...	...	...	...	EK78	...	...
3.4	100	...	...	...	...	...	...	H-050	...	...	...	...
3.4	80	...	...	...	...	...	...	...	...	EK986	...	...
3.42	...	...	...	...	...	...	...	...	EBP	...	...	...
3.5	382	0.20	...	...	PGCS-3	...	...	...	...	...	...	...
3.7	25	...	...	...	...	...	...	...	...	EK40	...	...
3.77	25	...	...	...	...	...	...	...	ECL, ECV	...	...	...
3.8	95	...	...	...	...	...	...	...	...	EK462	...	...
4.0	93	...	...	G540	...	...	...	...	...	...	...	...
4.1	...	...	...	...	...	Y20	...	...	...	...	...	...
4.2	...	...	...	...	...	2000, 2020	...	...	...	...	...	...
4.4	361	0.20	...	...	PGCS-2	...	...	...	...	...	...	...
4.6	128	...	...	...	...	...	...	...	...	...	...	IG-710
4.7	...	...	...	...	...	2019	...	...	...	...	...	...
4.8	353	0.20	...	...	PGCS-1	...	...	...	...	...	...	...
4.8	139	...	...	...	...	...	...	...	...	...	...	IG-15
4.8	80	...	...	...	...	...	...	...	...	...	2122	...
5.0	...	...	...	...	...	...	...	...	CMB	...	...	...
5.2	90	...	...	...	...	...	...	...	...	...	2123	...
5.39	...	...	...	...	...	...	...	B-404	...	...	...	...
5.7	...	...	...	...	...	2917	...	...	...	...	...	...
6.4	70	...	...	...	...	...	...	...	...	...	2318	...
6.5	81	...	...	...	...	...	...	...	...	...	...	ISO-88
7.7	394	0.17	...	...	ZXF-5Q	...	...	...	...	...	...	...
8.2	377	0.17	...	...	AXM-5Q	...	...	...	...	...	...	...
8.2	...	...	...	...	...	...	...	...	ZTA	...	...	...

tion Diagram for Elemental Carbon, 1979, *J. Geophys. Res.*, Vol 85 (B12), 1980, p 6930–6936

9. R.C. DeVries, Synthesis of Diamond under Metastable Conditions, *Annual Rev. Mater. Sci. 1987*, Vol 17, 1987, p 161–187

10. M. Akaishi, K. Hisao, Y. Sato, N. Setaka, T. Ohsawa, and O. Fukunaga, Sintering Behavior of the Diamond-Cobalt System at High Temperature and Pressure, *J. Mater. Sci.*, Vol 17, 1982, p 193–198

11. M. Akaishi, S. Yamaoka, J. Tanaka, T. Oshawa, and O. Fukunaga, Synthesis of Sintered Diamond with a High Electrical Resistivity and Hardness, *J. Am. Ceram. Soc.*, Vol 70 (No. 10), 1987, p C237-C239

12. M.K. Keshavan, B. Liang, and M.

Russell, Tribological Properties of Polycrystalline Diamond and Its Application, *Finer Points*, (No. 9), 1990, p 21–27

13. A. Lammer, Mechanical Properties of Polycrystalline Diamonds, *Mater. Sci. Tech.*, Vol 4 (No. 11), 1988, p 949–955

14. R.H. Wentorf, Jr. and H.P. Bovenkerk, Preparation of Semiconducting Diamonds, *J. Chem. Phys.*, Vol 36 (No. 8), 1962, p 1987–1990

15. R.H. Wentorf, Jr., R.C. DeVries, and F.P. Bundy, Sintered Superhard Materials, *Science*, Vol 208, 1980, p 873–880

16. P.D. Ownby and J. Liu, Nano-Diamond Enhanced Silicon Carbide Matrix Composites, *Ceramic Engineering and Science Proceedings*, Vol 12 (No. 9–

10), 1991

17. J.C. Angus, F.A. Buck, M. Sunkara, T.F. Groth, C.C. Hayman, and R. Gat, Diamond Growth at Low Pressures, *MRS Bull.*, Vol 14 (No. 10), 1989, p 38–47

18. J.C. Angus and C.C. Hayman, Low-Pressure, Metastable Growth of Diamond and 'Diamondlike' Phases, *Science*, Vol 241, 1988, p 913–921

19. T.R. Anthony, Metastable Synthesis of Diamond, *The Physics and Chemistry of Carbides, Nitrides and Borides*, R. Freer, Ed., Kluwer Academic Publishers, Dordrecht, 1990

20. C.P. Beetz, Jr., C.V. Cooper, and T.A. Perry, Ultralow-Load Indentation Hardness and Modulus of Diamond Films Deposited by Hot-Filament-Assisted CVD, *J. Mater. Res.*, Vol 5

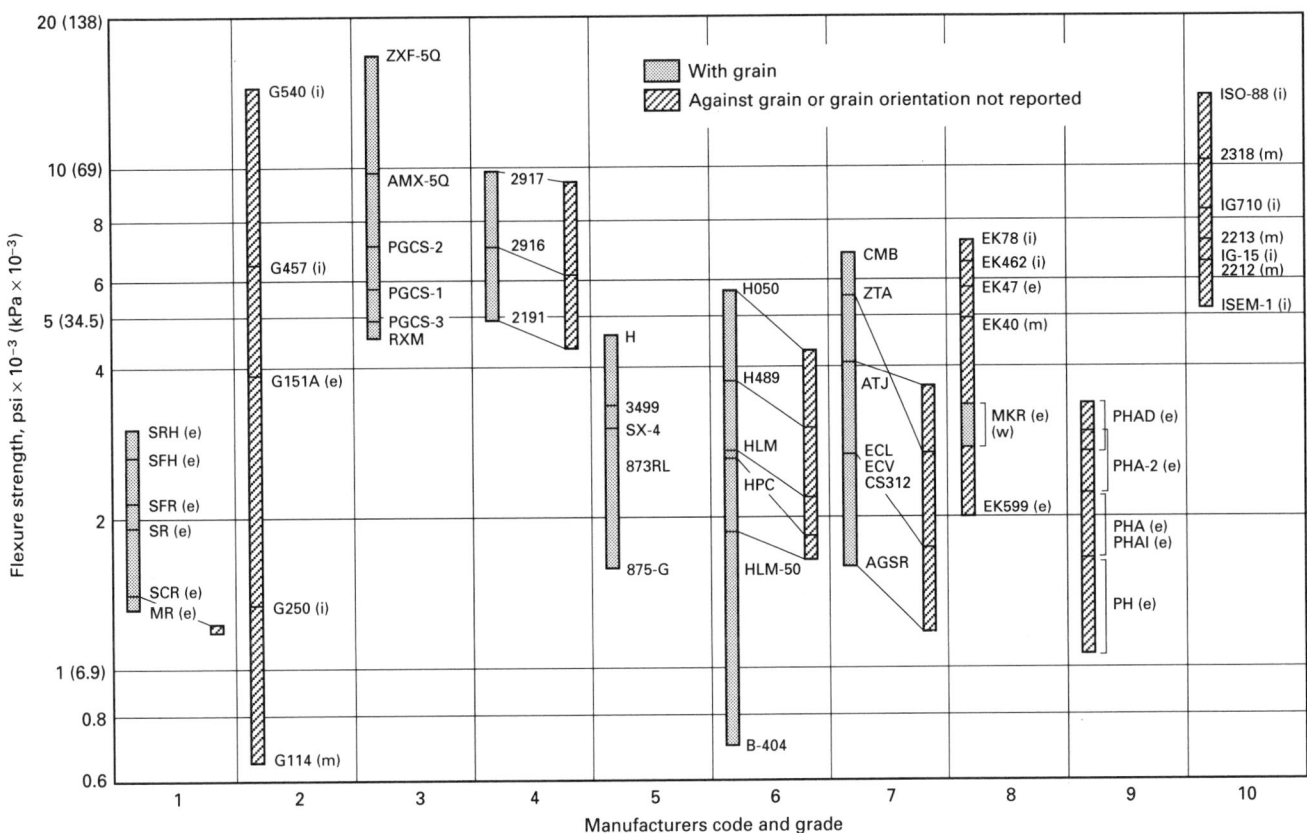

Fig 6 Flexure strength of various graphite grades. See Table 5 for an explanation of the manufacturer's code and grade. e = extruded, i = isostatically pressed (isotropic—no grain orientation), m = molded

(No. 11), 1990, p 2555–2561

21. C.B. Collins, F. Davanloo, E.M. Juengerman, D.R. Jander, and T.J. Lee, Preparation and Characterization of Thin Films of Amorphic Diamond, *Diamond Optics II—SPIE Proceedings—The International Society for Optical Engineering*, Vol 1146, A. Feldman and S. Holly, Ed., San Diego, CA, Aug 1989, p 37–47

22. A.H. Deutchman, R.J. Partyka, and J.C. Lewis, Dual Ion Beam Deposition of Diamond Films on Optical Elements, *Diamond Optics II—SPIE Proceedings—The International Society for Optical Engineering*, Vol 1146, A. Feldman and S. Holly, Ed., San Diego, CA, 7–8 Aug 1989, p 124–134

23. N. Fujimori, T. Imai, and A. Doi, Characterization of Conducting Diamond Films, *Vacuum*, Vol 36 (No. 1–3), 1986, p 99–102

24. A. Joshi, S.A. Gangal, and S.K. Kulkarni, Structure and Properties of Diamondlike Carbon Coatings Deposited in RF Plasma from Benzene and Monosubstituted Benzene, *J. Appl. Phys.*, Vol 64 (No. 12), 1988, p 6668–6672

25. M. Kawarada, K. Kurihara, K. Sasaki, A. Teshima, and K. Koshino, Diamond Synthesis by DC Plasma Jet CVD, *Diamond Optics II—SPIE Proceedings—The International Society for Optical Engineering*, Vol 1146, A. Feldman and S. Holly, San Diego, CA, 7–8 Aug 1989, p 28–36

26. S. Matsumoto, Y. Sato, M. Kamo, J. Tanaka, and N. Setaka, Chemical Vapor Deposition of Diamond from Methane-Hydrogen Gas, *Proceedings of the 7th International Conference on Vacuum Metallurgy*, Iron and Steel Institute of Japan, Tokyo, 1982, p 386–391

27. R.R. Nimmagadda, A. Joshi, and W.L. Hsu, The Role of Microstructure on the Oxidation Behavior of Microwave Plasma Synthesized Diamond and Diamond-Like Carbon Films, *J. Mater. Res.*, Vol 5 (No. 11), 1990, p 2445–2450

28. K. Nishimura, K. Das, J.T. Glass, K. Kobashi, and R.J. Nemanich, Electrical Properties of B-Doped CVD Grown Polycrystalline Diamond Films, *The Physics and Chemistry of Carbides, Nitrides and Borides*, R. Freer, Kluwer Academic Publishers, Dordrecht, 1990

29. R.W. Pryor, R.L. Thomas, P.K. Kuo, and L.D. Favro, Recent Developments in Growth and Characterization of Thin Diamond Films, *Diamond Optics II—SPIE Proceedings—The International Society for Optical Engineering*, Vol 1146, A. Feldman and S. Holly, San Diego, CA, 7–8 Aug 1989, p 68–77

30. A. Sawabe and T. Inuzuka, Growth of Diamond Thin Films by Electron Assisted Chemical Vapor Deposition and Their Characterization, *Thin Solid Films*, Vol 137 (No. 1), 1986, p 89–99

31. M. Sokolowski, A. Sokolowska, B. Gorkieli, A. Michalski, A. Rusek, and Z. Romanowski, Reactive Pulse Plasma Crystallization of Diamond and Diamond-Like Carbon, *J. Cryst. Growth*, Vol 47 (No. 3), 1979, p 421–426

32. K. Tankala, T. DebRoy, and M. Alam, Oxidation of Diamond Films Synthesized by Hot-Filament Assisted Chemical Vapor Deposition, *J. Mater. Res.*, Vol 5 (No. 11), 1990, p 2483–2489

33. T. Obata and S. Morimoto, Free-Standing Diamond Films—Plates, Tubes and Curved Diaphragms, *Diamond Optics II—SPIE Proceedings—The International Society for Optical Engineering*, Vol 1146, A. Feldman and S. Holly, San Diego, CA, 7–8 Aug 1989, p 208–216

34. C.E. Holcombe, Calculated X-Ray Diffraction Data for Polymorphic Forms of Carbon, *U.S. Atomic Energy Commission Publication Report Y-1887*, Oak

Ridge Y-12 Plant, 23 July 1973

35. D.S. Knight and W.B. White, Characterization of Diamond Films by Raman Spectroscopy, *J. Mater. Res.*, Vol 4 (No. 2), 1989, p 385–393

36. K.E. Spear, A.W. Phelps, and W.B. White, Diamond Polytypes and Their Vibrational Spectra, *J. Mater. Res.*, Vol 5 (No. 11), 1990, p 227–285

37. F.P. Bowden and D. Tabor, Deformation, Friction and Wear of Diamond, *Physical Properties of Diamond*, R. Berman, Ed., Clarendon Press, Oxford, 1965, p 184–218

38. *Properties of Diamond*, DeBeers Industrial Diamond Division Publication, 1982

39. M. Seal and W.J.P. van Enckevort, Applications of Diamond in Optics, *Diamond Optics—SPIE Proceedings—The International Society for Optical Engineering*, Vol 969, A. Feldman and S. Holly, Ed., San Diego, CA, 16–17 Aug 1988, p 144–152

40. Y.L. Orlov, *The Mineralogy of the Diamond*, John Wiley & Sons, 1977

41. P. Grodzinski, *Diamond Tools*, Anton Smit & Company, Inc., 1944

42. F.C. Frank and K.E. Puttick, Etch Pits and Trigons on Diamond: II, *Philos. Mag.*, Vol 3 (No. 35), 1958, p 1273–1279

43. H.B. Dyer, Physical and Mechanical Properties of Diamond, *Proceedings: The Industrial Diamond Revolution—A Technical Conference*, Industrial Diamond Association of America, Columbus, OH, 13–15 Nov 1967, p I-XI

44. *The Properties of Diamond*, J.E. Field, Ed., Academic Press, London, 1979, p 641–653

45. R.M. Chrenko and H.M. Strong, Physical Properties of Diamond, General Electric Report 75CRD089, 1975, p 1–46

46. K. Lonsdale, Divergent-Beam X-Ray Photography of Crystals, *Philosophical Transactions of The Royal Society of London*, A240, 1947, p 219–250

47. C.T. Lynch, *Practical Handbook of Materials Science*, CRC Press, Inc., Boca Raton, 1989, p 12–13

48. N.F. Mott and W.D. Twose, The Theory of Impurity Conduction, *Adv. Phys.*, Vol 10 (No. 37–40), 1961, p 107–163

49. W.L. Roberts, G.R. Rapp, Jr., and J. Weber, *Encyclopedia of Minerals*, Van Nostrand Reinhold Company, New York, 1974, p 172

50. *Thermophysical Properties of High Temperature Solid Materials Volume I: Elements*, Y.S. Touloukian, Ed., The MacMillan Company, New York, 1967, p 392–401

51. S. Tolansky, *The Microstructures of Diamond Surfaces*, N.A.G. Press, Ltd., London, 1955

52. J.E. Field, Mechanical and Physical Properties of Diamond, *Science of Hard Materials*, Adam Hilger, Ltd., Bristol, 1986, p 181–205

53. *Handbook of Tables for Applied Engineering Science*, 2nd ed., R.E. Bolz and G.L. Tuve, Ed., CRC Press, Boca Raton, 1987

54. P.J. Heath, Ultrahard Tool Materials, *Metals Handbook*, Vol 16, 9th ed., ASM International, 1989, p 105–117

55. C.L. Mantell, *Carbon and Graphite Handbook*, Robert E. Krieger Publishing Company, Melbourne, 1979

56. M.G. Peters, J.L. Knowles, M. Breen, and J. McCarthy, Ultra-Thin Diamond Films for X-Ray Window Applications, *Diamond Optics II—SPIE Proceedings—The International Society for Optical Engineering*, Vol 1146, A. Feldman and S. Holly, Ed., San Diego, CA, 7–8 Aug 1989, p 217–224

57. J.C. Angus, C.C. Hayman, and R.W. Hoffman, Diamond and 'Diamondlike' Phases Grown at Low Pressure: Growth, Properties and Optical Applications, *Diamond Optics—SPIE Proceedings—The International Society for Optical Engineering*, Vol 969, A. Feldman and S. Holly, Ed., San Diego, CA, 16–17 Aug 1988, p 2–13

58. J.E. Field, Strength and Fracture Properties of Diamond, *The Properties of Diamond*, J.E. Field, Ed., Academic Press, London, 1979, p 282–324

59. P. Gigl, *High Pressure Science and Technology*, Vol 1, K.D. Timmerhaus and M.S. Barber, Ed., 6th AIRAPT Conference, Boulder, CO, Plenum Press, New York, 1979, p 914–922

60. L.N. Devin, A.L. Maistrenko, E.S. Simkin, S.J. Sklyar, and N.V. Tsypin, Resistance to Cracking of Hard-Metal Composite Materials, *Sov. Powder Metall. Met. Ceram.*, Vol 21 (No. 5), 1982, p 419–423

61. S.N. Dub, Fabrication and Application of Superhard Materials, *Sverkhtverdye Materialy (Superhard Materials)*, (No. 5), 1983, p 75–78

62. A.A. Kukharenko, *Diamonds of the Urals*, Gosgeoltekhizdat, Moscow, U.S.S.R., 1955

63. *Handbook of Materials and Processes for Electronics*, C.A. Harper, McGraw-Hill Book Company, New York, 1970, p 7–45 to 7–48

64. V.R. Howes, Ring Cracks on Diamond Surface, *Physical Properties of Diamond*, R. Berman, Ed., Clarendon Press, Oxford, 1965, p 174–183

65. C.A. Brookes, Plastic Deformation and Anisotropy in the Hardness of Diamond, *Nature*, Vol 228 (No. 5272), 1970, p 660–661

66. F. Knoop, C.G. Peters, and W.B. Emerson, A Sensitive Pyramidal-Diamond Tool for Indentation Measurements, *J. Res. National Bureau of Standards*, Vol 23 (No. 1), 1939, p 39–61

67. C.G. Peters and F. Knoop, Metals in Thin Layers—Their Microhardness, *Metal Alloys*, Vol 12 (No. 3), 1940, p 292–297

68. M. Akaishi, S. Yamaoka, J. Tanaka, T. Oshawa, and O. Fukunaga, Synthesis of Sintered Diamond with a High Electrical Resistivity and High Hardness, *Mater. Sci. Eng.*, Vol A105–106 (No. 1–2), 1988, p 517–523

69. S. Albin, A. Cropper, L. Watkins, C.E. Byvik, P. Buoncristiani, K.V. Ravi, and S. Yokota, Laser Damage of Diamond Thin Film Windows, *Diamond Optics—SPIE Proceedings—The International Society for Optical Engineering*, Vol 1146, A. Feldman and S. Holly, Ed., San Diego, CA, 7–8 Aug 1988, p 186–193

70. E.O. Barnhardt, Über die Mikrohärte der Festoffe im Grenzbereich des Kick'schen Ähnlichkeitssatzes, *Zeitschrift für Metallkunde*, Vol 33 (No. 3), 1941, p 135–144

71. J.F. Spanitz, paper presented at Diamond, Partner in Productivity, Symposium of Industrial Diamond Association of America, Nov 1979

72. M. Casey and J. Wilks, The Friction of Diamond Sliding on Polished Cube Faces of Diamond, *J. Phys. D*, Vol 6 (No. 15), 1973, p 1772–1781

73. M.I. Landstrass and K.V. Ravi, Resistivity of Chemical Vapor Deposited Diamond Films, *Appl. Phys. Lett.*, Vol 55 (No. 10), 1989, p 975–977

74. S. Whitehead and W. Hacket, Measurement of the Specific Inductive Capacity of Diamonds by the Method of Mixtures, *Proceedings of the Physics Society*, Vol 51 (No. 1), 1939, p 173–190

75. S. Yazu, S. Sato, and N. Fujimori, Some Properties of Synthetic Single Crystal and Thin Film Diamonds, *Diamond Optics—SPIE Proceedings—The International Society for Optical Engineering*, Vol 969, A. Feldman and S. Holly, Ed., San Diego, CA, 16–17 Aug 1988, p 117–123

76. D.F. Gibbs and G.J. Hill, The Variation of the Dielectric Constant of Diamond with Pressure, *Philos. Mag.*, Vol 9 (No. 99), 1964, p 367–375

77. L.L. Hench and J.K. West, *Principles of Electronic Ceramics*, John Wiley & Sons, New York, 1990, p 111, 187, 357

78. E.J. Rymaszewski and R.R. Tummala, Microelectronics Packaging—An Overview, *Microelectronics Packaging Handbook*, R.R. Tummala and E.J. Rymaszewski, Van Nostrand Reinhold, New York, 1989, p 36

79. A.T. Collins and E.C. Lightowlers, Electrical Properties, *The Properties of Diamond*, J.E. Field, Ed., Academic

Press, London, 1979, p 79–105

80. P.J. Dean, E.C. Lightowlers, and D.R. Wight, Intrinsic and Extrinsic Recombination Radiation from Natural and Synthetic Aluminum-Doped Diamond, *Phys. Rev.*, Vol 140 (No. 1A), 1965, p A352-A368

81. P.T. Wedepohl, Electrical and Optical Properties of Type IIb Diamonds, *Proc. R. Soc.*, Vol B70 (No. 2), 1957, p 177–185

82. W.B. Wilson, Evidence for Hopping Transport in Boron-Doped Diamond, *Phys. Rev.*, Vol 127 (No. 5), 1962, p 1549–1550

83. *ASM Engineered Materials Reference Book*, ASM International, Metals Park, 1989

84. C.D. Clark, P.J. Dean, and P.V. Harris, Intrinsic Edge Absorption in Diamond, *Proc. R. Soc.*, Vol 277A (No. 1370), 1964, p 312–329

85. S. Ashley, Diamond-Hard Heat Sinks, *Mech. Eng.*, Vol 112 (No. 10), 1990, p 54–56

86. A.T. Collins and A.W.S. Williams, Nature of the Acceptor Centre in Semiconducting Diamond, *J. Phys. C: Solid State Phys.*, Vol 4 (No. 13), 1971, p 1789–1800

87. F.C. Champion, *Electronic Properties of Diamond*, Butterworth & Co., Ltd., London, 1963

88. M.N. Yoder, Artifact Diamond, *Diamond Optics—SPIE Proceedings—The International Society for Optical Engineering*, Vol 969, A. Feldman and S. Holly, Ed., San Diego, CA, 16–17 Aug 1988, p 106–113

89. A.W.S. Williams, E.C. Lightowlers, and A.T. Collins, Impurity Conduction in Synthetic Semiconducting Diamond, *J. Phys. C: Solid State Phys.*, Vol 2 (No. 8), 1970, p 1727–1735

90. R. Berman, E.L. Foster, and J.M. Ziman, The Thermal Conductivity of Dielectric Crystals: The Effect of Isotopes, *Proc. R. Soc. London*, Vol A237, 1956, p 344

91. R. Berman, Thermal Properties, *Physical Properties of Diamond*, R. Berman, Ed., Clarendon Press, Oxford, 1965, p 371–393

92. R. Berman, Thermal Properties, *The Properties of Diamond*, J.E. Field, Ed., Academic Press, London, 1979, p 3–22

93. A. Feldman, H.P.R. Frederiske, and X.T. Ying, Thermal Wave Measurements of the Thermal Properties of CVD Diamond, *Diamond Optics II—SPIE Proceedings—The International Society for Optical Engineering*, Vol 1146, A. Feldman and S. Holly, Ed., San Diego, CA, 7–8 Aug 1989, p 78–84

94. A. Goldsmith, T.E. Waterman, and H.J. Hirschhorn, *Handbook of Thermophysical Properties of Solid Materials, Volume I—Elements*, Pergamon Press, New York, 1961, p I-C-3-m

95. D.T. Morelli, C.P. Beetz, Jr., and T.A. Perry, Thermal Conductivity of Synthetic Diamond Films, *J. Appl. Phys.*, Vol 64 (No. 6), 1988, p 3063–3066

96. E.N. Bunting and A. Van Valkenburg, Some Properties of Diamond, *Am. Mineralogist*, Vol 43 (No. 1–2), 1958, p 102–106

97. B.J. Skinner, The Thermal Expansions of Thoria, Periclase and Diamond, *Am. Mineralogist*, Vol 42 (No. 1–2), 1957, p 39–55

98. C.A. Coulson, *Valence*, Oxford University Press, London, 1961, p 185

99. R.A. Swalin, *Thermodynamics of Solids*, 2nd ed., John Wiley & Sons, New York, 1972, p 62

100. C.E. Byvik, P. Hobsen, P. Buoncristiani, S. Albin, and V. Lakdawala, 2nd Annual Diamond Technology Initiative Seminar, SDIO.IST-ONR, 7–8 July 1987

101. T. Evans, Transmission Electron Microscopy of Diamond, *Physical Properties of Diamond*, R. Berman, Ed., Clarendon Press, Oxford, 1965, p 116–134

102. S. Albin, W.P. Winfree, and B.S. Crews, Thermal Diffusivity of Diamond Films, *Diamond Optics II—SPIE Proceedings—The International Society for Optical Engineering*, Vol 1146, A. Feldman and S. Holly, Ed., San Diego, CA, 7–8 Aug 1989, p 85–94

103. I.S. Rozhkov, K.K. Abrashev, and A.F. Konstantinova, Some Thermal Conductivity Features of Diamonds from the Mir and Aikhal Deposits, *Geoloigiya Geofizika Novosibirsk*, Vol 3, 1964

104. R. Berman, P.R.W. Hudson, and M. Martinez, Nitrogen in Diamond: Evidence from Thermal Conductivity, *J. Phys. C: Solid State Phys.*, Vol 8 (No. 21), 1975, p L430–434

105. P. Hawtin, J.B. Lewis, N. Moul, and R.H. Phillips, The Heats of Combustion of Graphite, Diamond and Some Non-Graphitic Carbons, *Philos. Trans. R. Soc. London*, Vol A261, 1966, p 67–95

106. R. Seitz, U.S. patent 3,895,313, 15 July 1975

107. S. Singer, The Potential of Diamond as a Very High Average Power Transmitting Optical Material, *Diamond Optics—SPIE Proceedings—The International Society for Optical Engineering*, Vol 969, A. Feldman and S. Holly, Ed., San Diego, CA, 16–17 Aug 1988, p 168–177

108. G.A. Slack and S.F. Bartram, Thermal Expansion of Some Diamondlike Crystals, *J. Appl. Phys.*, Vol 46 (No. 1), 1975, p 89–98

109. J. Thewlis and A.R. Davey, Thermal Expansion of Diamond, *Philos. Mag.*, Vol 1 (No. 5), 1956, p 409–414

110. M.N. Yoder, Synthetic Diamond, Its Properties and Synthesis, *Novel Refractory Semiconductors*, D. Emin, T.L. Aselage, and C. Wood, Ed., Materials Research Society, Pittsburgh, 1987, p 315–326

111. S. Ashley, Diamond-Hard Heat Sinks, *Mech. Eng.*, Vol 112 (No. 10), 1990, p 54–56

112. T.R. Anthony, W.F. Banholzer, J.F. Fleischer, L. Wei, P.K. Kuo, R.L. Thomas, and R.W. Pryor, Thermal Diffusivity of Isotopically Enriched C^{12} Diamond, *Phys. Rev. B*, Vol 42 (No. 2), 1990, p 1104–1111

113. D. Tabor, Adhesion and Friction, *The Properties of Diamonds*, J.E. Field, Ed., Academic Press, London, 1979, p 325–350

114. C.D. Clark, Optical Properties of Natural Diamonds, *Physical Properties of Diamond*, R. Berman, Ed., Clarendon Press, Oxford, 1965, p 295–324

115. P.J. Kemmey and P.T. Wedepohl, Semiconducting Diamond, *Physical Properties of Diamond*, R. Berman, Ed., Clarendon Press, Oxford, 1965, p 325–355

116. K.A. Snail, L.M. Hanssen, A.A. Morrish, and W.A. Carrington, Hemispherical Transmittance of Several Free Standing Diamond Films, *Diamond Optics II—SPIE Proceedings—The International Society for Optical Engineering*, Vol 1146, A. Feldman and S. Holly, Ed., San Diego, CA, 7–8 Aug 1989, p 144–151

117. F. Peter, Refractive Indices and Absorption Constants of the Diamond Between 644 and 226 μm, *Zeitschrift für Physik*, Vol 15, 1923, p 358–368

118. I.G. Austin and R. Wolfe, Electrical and Optical Properties of Semiconducting Diamond, *Proc. R. Soc. B*, Vol 69 (No. 3), 1956, p 329–338

119. A.A. Kukharenko, *Diamonds of the Urals*, Gosgeoltekhizdat, Moscow, U.S.S.R., 1955

120. J. Walton, *A Pocket Chart of Ornamental and Gem Stones*, Sir Isaac Pitman & Sons, Ltd., London, 1954, p 2–3

121. P. Denham, E.C. Lightowlers, and P.J. Dean, Ultraviolet Intrinsic and Extrinsic Photoconductivity of Natural Diamond, *Phys. Rev.*, Vol 161 (No. 3), 1967, p 762–768

122. E. Fritsch, J.E. Shigley, and J.I. Koivula, The Separation of Natural from Synthetic Diamonds on the Basis of Their Optical and Microscopic Properties, *Diamond Optics—SPIE Proceedings—The International Society for Optical Engineering*, Vol 969, A. Feldman and S. Holly, Ed., San Diego, CA, 16–17 Aug 1988, p 114–116

123. L.A. Xue and R. Raj, Effect of Dia-

mond Dispersion on the Superplastic Rheology of Zinc Sulfide, *J. Am. Ceram. Soc.*, Vol 73 (No. 8), 1990, p 2213–2216

124. J. Liu and P.D. Ownby, Normal-Pressure Hot-Pressing of α-Alumina-Diamond Composites, *J. Am. Ceram. Soc.*, Vol 74 (No. 10), 1991, p 2666–2668

125. T. Noma and A. Sawaoka, Toughening in Very High Pressure Sintered Diamond-Alumina Composite, *J. Mater. Sci.*, Vol 19, 1984, p 2319–2322

126. T. Noma and A. Sawaoka, Effect of Heat Treatment on Fracture Toughness of Alumina-Diamond Composites Sintered at High Pressures, *J. Am. Ceram.*

Soc., Vol 68 (No. 2), 1989, p C36-C37

127. T. Noma and A. Sawaoka, Fracture Toughness of High-Pressure Sintered Diamond/Silicon Nitride Composites, *J. Am. Ceram. Soc.*, Vol 68 (No. 10), 1985, p C271-C273

128. W.N. Reynolds, *Physical Properties of Graphite*, Elsevier Science Publishing Company, Inc., New York, 1968, p 1

129. "Carbon/Graphite Properties," The Stackpole Corporation, 1983, p 3–8

130. D.E. Carmichael, W.C. Chard, and P.D. Ownby, Dense, Isotropic Graphite Fabricated by Hot-Isostatic Compaction, *Carbon*, Vol 6 (No. 2), 1968, p 218

131. D.E. Carmichael, P.D. Ownby, and E.S. Hodge, Hot-Isostatic Compaction of Graphite, Battelle Memorial Institute Report, No. BMI-1746, 5 Oct 1965

132. G.S. Brady and H.R. Clauser, *Materials Handbook*, 12th ed., McGraw-Hill Book Company, New York, 1986, p 257–260

133. L.M. Ligget, Carbon-Baked and Graphitized Products & Manufacture, *Encyclopedia of Chem. Tech.*, 2nd ed., Vol 4, Kirk-Othmer, Ed., John Wiley & Sons, 1964

134. "Manufactured Carbon Products," The Stackpole Corporation, unpublished private communication

Engineering Properties of Carbon-Carbon and Ceramic-Matrix Composites

G. Ziegler, Institut für Materialforschung, Universität Bayreuth, Germany
W. Hüttner, Schunk Kohlenstofftechnik GmbH, Germany

CARBON-CARBON composites began replacing fine-grained graphite as nose tips in rockets in the mid-1960s, because they represented significant improvements in thermoshock behavior and erosion resistance. They are currently being introduced in fields that require their high specific strength and stiffness, in combination with their thermoshock resistance, chemical resistance, and fracture toughness, especially at high temperatures.

Carbon-carbon composites could be an ideal structural ceramic if they did not experience severe oxidation at temperatures above 500 °C (930 °F). Intensive research and development work is being conducted worldwide to solve this problem. The material needs for future-generation hypersonic aircraft and aeropropulsion systems are driving forces. Structural ceramics that have high damage tolerance and retain strength and stiffness up to 1800 °C (3270 °F), are urgently required. The two principal routes that are under investigation to achieve long-term resistivity are the surface protection of carbon-carbon composites on the basis of SiC (multilayer coatings of the inner/outer surface) and the development of fiber-reinforced ceramic-matrix composites with oxidation-resistant matrices, such as SiC.

For applications in the temperature range up to about 1800 °C (3270 °F), new compounds that are based on matrices with extremely high thermal stability must be developed. Combinations with other ceramics are necessary, in this case, for surface and bulk protection. Until now, no general solution has been developed. A specially tailored solution has to be devised for each requirement. Key problems are the control of fiber-matrix interfaces (damage tolerance) and chemical attack by atomic oxygen.

Recently, research and development efforts have focused on other high-performance matrix composites. The primary goal of this line of development is to improve the crack propagation and damage tolerance of ceramic

materials. Characteristic features of ceramic materials are their only elastic behavior without plastic deformation during loading and catastrophic fracture by overloading the materials. Therefore, monolithic ceramics exhibit fracture strain values of only a few tenths of a percent and fracture toughness values of a maximum 8 MPa$\sqrt{m}$ (7 ksi$\sqrt{in.}$) (for monolithic materials without initiating any toughening mechanism).

To improve crack propagation problems in ceramic materials, ceramic-matrix composites that incorporated continuous fibers, short fibers, whiskers, particles, and platelets, began to be developed. In the case of continuous fiber reinforced ceramics, investigations aim to improve high-temperature properties. Moreover, the incorporation of short reinforcements should result in improvements in wear resistance, impact strength, and thermal shock/cycling behavior. The promising aspects of this type of material fosters the continuing development and characterization of ceramic-matrix composites worldwide.

Carbon-Carbon Composites

Manufacturing routes generally follow the scheme depicted in Fig 1. Carbon fibers are impregnated by a suitable matrix precursor and fixed by winding or lamination in defined directions (not necessary in cases of freestanding three-directional billets). The next step is a pressing and/or curing cycle, followed by carbonization between 800 to 1200 °C (1470 to 2190 °F) in an inert atmosphere. After the carbonization treatment, further heat treatment to graphitization temperatures of greater than 2500 °C (4530 °F) can be performed. The result is a porous carbon-carbon composite that has approximately 30% open porosity, which will be decreased to less than 5% by densification cycles. These den-

sification cycles can consist of impregnation followed by recarbonization or impregnation, recarbonization, and then regraphitization. Instead of densification via liquid precursors, chemical vapor impregnations (CVI) of pyrolitic carbon into the porous skeleton can be performed to fill the interior spaces. The pyrolitic carbon matrix is formed by the cracking of hydrocarbons at temperatures ranging from 1000 to 2000 °C (1830 to 3630 °F). Suitable liquid matrix precursor mixes are polyaromatic thermoplastics, such as coal tar-based and petrol-based pitches, and thermosets, such as phenolics, polyimides, and others.

High-carbon yields are required from the precursors for two reasons. First, the weight loss and density change cause severe shrinkage of the precursors during carbonization and induce cracks. High-carbon yields reduce the shrinkage and, especially, the shrinkage stresses (Ref 1). Second, high-carbon yields reduce the numbers of redensifications, which translates to cost and time savings.

A different way to build up a dense, high-carbon-containing matrix was developed by hot pressing thermally pretreated pitch to 800 °C (1470 °F) (mesophase pitch) (Ref 2). Already, one densification cycle is sufficient to achieve porosities of less than 10% and a fiber yield strength of 80% in model unidirectionally (UD) reinforced composites. Progress in developing commercially available appropriate mesophase pitches will allow the continuation of this manufacturing method and the reduction of both processing time and costs, as well as the reduction of shrinkage stresses (Ref 3).

All types of carbon fibers can be used as reinforcements in many different architectures: random fibers; two-directional (2-D) fabrics in stacked, stitched, or pierced configurations; three-directional (3-D) geometries (cartesian or cylindrical coordinates) to increase the off-axis strength; or 3-D and multidirectional (4 to 11) weaves to minimize

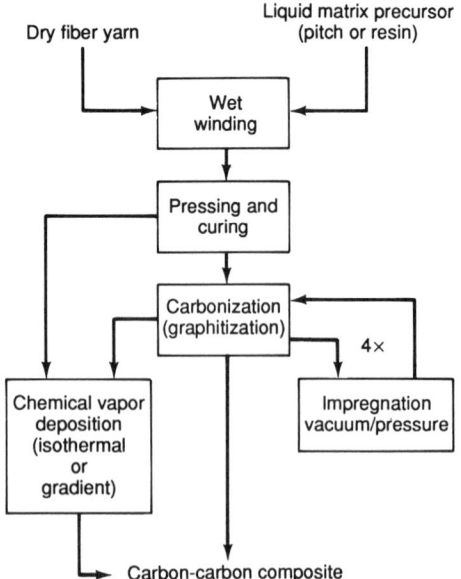

Fig 1 State-of-the-art processing of carbon-carbon composites

Table 1 Processing techniques for different geometries

Geometry	Axial pressing, 2-D	Vacuum-sack, 2-D	Autoclave, 2-D	Winding, 2-D	Weaving, 2.5/3-D
	Easy	Easy	Easy	Possible	Possible
	Easy	Easy	Easy	Not possible	Possible
	Not possible	Possible	Possible	Easy	Possible
	Not possible	Possible	Possible	Easy	Possible
	Not possible	Possible	Easy	Not possible	Possible
	Not possible	Possible	Possible	Easy	Possible
	Not possible	Possible	Possible	Easy	Possible
	Not possible	Possible	Easy	Not possible	Possible
	Not possible	Possible	Easy	Not possible	Possible
	Not possible	Not possible	Possible	Not possible	Possible

the empty spaces between the rod junctions (3-D unit cell can be filled up to a maximum 59 vol%). Unidirectional orientation has no importance for technical applications, but many basic investigations have been performed using UD-reinforced composites as models (Ref 1, 2, 4–7).

Both 3-D and multidirectional composites are exciting materials, but are generally too expensive and unnecessary for most industrial applications. The 2-D and multiaxial fiber architectures usually suffice. Different manufacturing techniques are listed in Table 1 with respect to the processibility of different geometries and shapes. In some cases, combinations of the main processing techniques could be reasonable for manufacturing complex structures.

The new developed 2.5-D technique allows the weaving of very complex integrated structures, such as double-T bars and wing leading edges with integrated stiffeners. Automated 3-D preform manufacturing facilities have been available since the 1980s, and can reduce the high costs to one-third and manufacturing time to one-half, compared to manual labor (Ref 8, 9).

Properties. Carbon-carbon composites are a family of materials with a wide range of different properties. Decisive influences on the mechanical, thermal, and electrical properties come from the fiber type, fiber volume fraction, fiber architecture, precursor, and processing cycle (Tables 2–4). With an increasing number of redensification cycles, porosity decreases and density, strength, and stiffness increase (Fig 2). The final heat-treatment temperature (HTT) influences fracture behavior and physical properties such as the coefficient of thermal expansion (CTE), resistivity, and conductivity (Table 4).

All composite properties are anisotropic. This originates from the carbon fibers, which are extremely anisotropic because of their graphitic crystal lattice. Along the fiber axis, stiffness, strength, and both electrical and thermal conductivity are excellent. Across the axis, these properties are poor. In the case of 2-D reinforcement, the ratio of anisotropy amounts to between 5/1 and 10/1.

The flexural strength of industrially manufactured carbon-carbon composites with different fiber architectures varies between 100

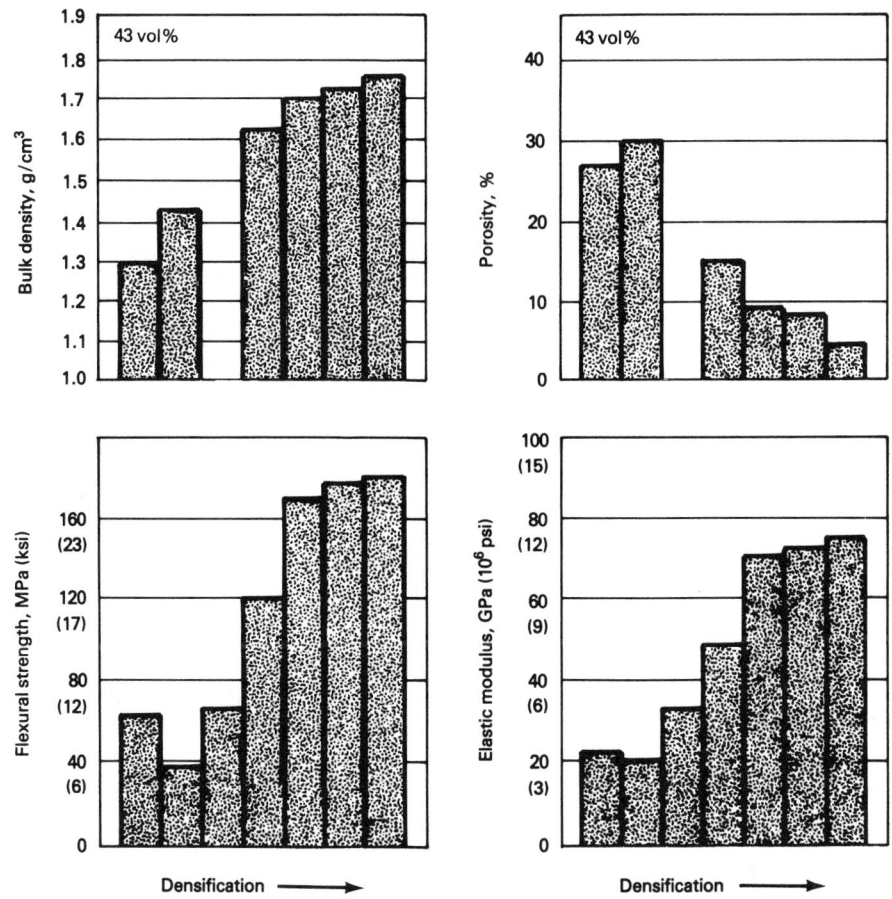

Fig 2 Change of properties with number of redensifications [2-D weave, 8 harness/satin (H/S)]

rials significantly lose strength at temperatures greater than 1300 °C (2370 °F), carbons and graphites keep their mechanical properties up to 2000 °C (3630 °F) (Ref 12). Ongoing tests clearly show increasing values up to the testing temperature of 1600 °C (2910 °F) in an order of magnitude of 40 to 50% (Ref 13), as shown in Fig 3. The interlaminar shear strength (ILSS) increases about 70% (from room temperature to 1800 °C, or 3270 °F) and the modulus, approximately 10%.

The physical and thermophysical properties are also of great interest for specific applications. In one direction, along the fibers or laminates, the material acts as a heat conductor, but across, as an insulator. The processing history of the composites is important to the final property values. Graphitization treatment increases conductivity and decreases resistivity (Table 4). This is still more pronounced if graphitizing matrices, such as pitch-based carbon or pyrolitic graphite, are used. In 3-D and multiaxial structures, the properties are balanced to a more isotropic behavior. However, the ratio of isotropy is dependent on the balance of fiber volume fraction in the x, y, and z directions. At high temperatures, electrical resistivity decreases and the composites show negative thermal conductivity (NTC) behavior. Thermal conductivity also drops with higher temperatures.

Thermal expansion parallel to the laminates of 2-D carbon-carbon composites is negative up to 800 °C (1470 °F). Across the laminates, a positive CTE is present. In general, however, the CTE data are low and can be tailored by the fiber architectures to a balanced expansion of nearly zero in the plane (Fig 4).

Fracture Behavior. High fracture toughness values and pseudoplastic fracture behavior are most attractive for numerous applications. Both brittle, catastrophic failure by overstressing and critical stress concentrations at notches can be excluded. This high damage tolerance can be demonstrated by nailing a composite. No catastrophic, brittle failure occurs as one might expect for ceramic materials. Only the area around the nail shaft is delaminated. Bend tests on 3-D composites have shown enormous strains up to 5% (Ref 23). The reasons for this extremely high damage tolerance of the brittle fiber-brittle matrix structures originates from the complex minimechanical and micromechanical behavior of the several interfaces that are present in a carbon-carbon composite. These are fiber intrabundle and interbundle fiber-fiber fiber-matrix and matrix-matrix interfaces (Ref 24). The crystalline structure of the matrix itself also influences toughness (Ref 25, 26). The micromechanics are not yet fully understood, but weak interfacial bondings seem to be a precondition for producing tough carbon-carbon composites (Ref 1, 27). Bonds between fiber and matrix that are too good induce excessive damage and promote fast crack propagation through the fibers. The compos-

and 600 MPa (15 and 90 ksi) (parallel to the laminates/fibers). A characteristic feature of carbon-carbon materials is that the ratio of flexural/tensile strength is 1/1.4. Wound tubes with fiber angles of ±15° and ±45° (normal to the axis) exhibit tensile strength values of 540 MPa (80 ksi) and 430 MPa (60 ksi), respectively. The strength values of 3-D reinforced weave structures range from 150 to 300 MPa (22 to 44 ksi). The strengths of felt and randomly oriented short fiber reinforcements are between 50 and 100 MPa (7.5 and 15 ksi).

Fatigue behavior is very attractive. Under alternating flexural loads, 70 to 80% of the ultimate flexural strength is still available (Ref 10, 11). Moduli values range from 40 to 150 GPa (6 to 22 × 10⁶ psi) for 2-D and 3-D weaves, respectively. Besides the fiber type and orientation, the type of matrix precursor and final HTT decisively influences the modulus value. If the graphite layers of the matrix are oriented parallel to the fiber axis, then they can contribute to the composite modulus nearly in the same order of magnitude as do the fibers.

A high-temperature strength capability is most attractive for many structural applications. Although other high-temperature mate-

Table 2 Mechanical properties of 2-D carbon-carbon at various final heat-treatment temperatures (HTT)

Fiber orientation	HTT		Density, g/cm³	Flexural strength		Modulus		ILSS		Strain to failure, %
	°C	°F		MPa	ksi	GPa	10⁶ psi	MPa	ksi	
±15° tow	1200	2190	1.55	520	75.4	90	13.0	18	2.6	0.4
±15° tow	2200	3990	1.60	470	68.2	130	18.9	12	1.7	0.3
Cloth, 8 H/S	1200	2190	1.50	250	36.3	75	10.8	16	2.3	0.4
Cloth, 8 H/S	2200	3990	1.60	230	33.4	80	11.6	12	1.7	0.3
Cloth, 8 H/S	2500	4530	1.65	210	30.4	85	12.3	10	1.5	0.3

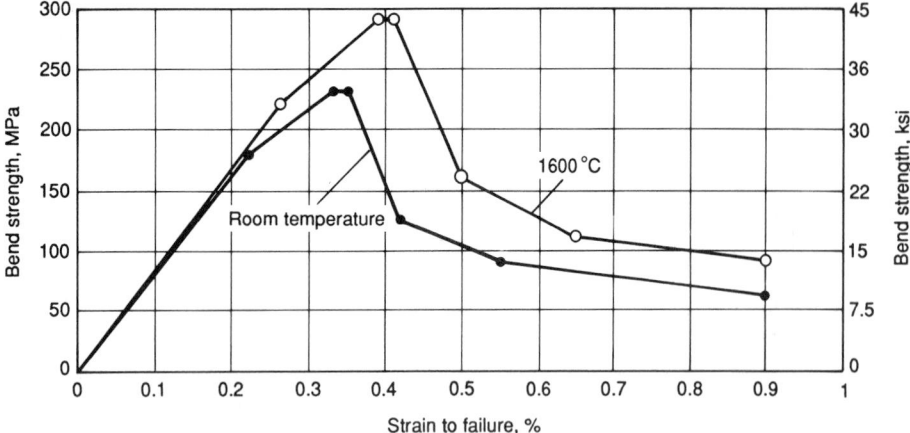

Fig 3 Comparison of bend strength at room temperature and 1600 °C (2910 °F) (2-D weave, 8 H/S in inert atmosphere)

ite fails in a brittle mode, and no strengthening effect is achieved. The mechanism of load transfer has been only poorly understood until now. If the interfacial forces are low, however, no brittle fracture occurs. Because the shear and transverse strengths are very low, one could assume that the loads are transferred across the interfaces primarily by friction. In addition, mechanical interlocking of cracked matrix parts and particles can occur.

One can also assume that microcracking is still in service when loads are applied on the composite. However, the debonded and weak interfaces avoid the crack propagation through the fibers. The energy dissipation is additionally promoted by poorly bonded matrix-matrix interfaces, which act as a type of "duplex mechanism" (Ref 28) and increase the fracture toughness.

The amount of influence of the graphitic matrix structure on toughness values was demonstrated in model experiments (Ref 29). Three-directional billets were graphitized to different extents to produce matrix structures from slightly to highly graphitic states. In Fig 5, the ability of the matrix to absorb energy

is expressed as relative toughness. The maximum is reached at 2400 °C (4350 °F). The explanation for this is given by different lamellar structures of the pitch-based matrix. Less-ordered microstructures allow crack propagation over long distances along the filaments, and energy dissipation is low with consequent low toughness. At 2400 °C (4350 °F), an intermediate stage exists, where the microstructure is sufficiently ordered to accommodate some slip from shear forces, but is disordered enough to prevent long-range slip. Energy absorption and, therefore, toughness is high. A good graphitic order of the matrix decreases against toughness, because there is more extensive microcracking in the matrix and the chance for multiple fracturing is the greatest.

Applications. As already noted and demonstrated by the mechanical and physical data parallel and perpendicular to the laminations, composite properties are controlled by the anisotropic properties of the carbon fibers. This has to be taken into account for every application. In most cases, the fabrication of structural parts must be fabricated near the

final shape to orient the fibers during the manufacturing process in the technically required design.

The main applications of carbon-carbon composites, in terms of money and mass, are in the military, space, and aircraft industries. In fields of general mechanical engineering, these materials are still new. The material is used where high temperature, high strength, high stiffness, and corrosion-resistant materials are needed. The material has been successfully used as fastening elements. Bolts, screws, nuts, and washers are used where high temperature and severe chemical conditions are present. Strength and stiffness at high temperatures guarantee high fastening stability. If graphite parts are screwed together, only low fastening moments are necessary. The system is self-fastening at the application temperature, because of the anisotropic CTE. Parts are applied in the semiconductor industry, in furnace constructions, and other high-temperature equipment.

Interesting applications are as heaters, heat shields, and furnace walls in vacuum furnaces. Liners, plates, tubes, crucibles, sleeves, and other auxiliary aids are applied in the field of apparatus construction. The properties of carbon-carbon composites are also useful as tool segments, pressure plates, and resistance elements in hot sintering applications. In some industries, these composites are applied to replace asbestos.

Ceramic-Matrix Composites

Toughening Mechanisms. Various toughening mechanisms are described in the literature (Ref 14, 31–35): microcracking, phase transformation, crack deflection, crack bowing, plastic deformation by incorporating metal dispersoids, and, particularly, whisker- and fiber-reinforcement initiating mechanisms, such as load transfer by crack bridging or matrix prestressing, interfacial friction, and fiber pullout. In many cases in ceramic materials and ceramic-matrix composites, a combination of various mechanisms occurs.

Microcracks occur within regions of local residual tension, and are caused by thermal expansion mismatch and/or by transformation. The microcracks locally relieve the residual tension and, thus, cause dilatation governed by the volume displaced by the microcrack. Furthermore, the microcracks reduce the elastic modulus within the microcrack process zone. In monolithic ceramics, microcracking is frequently caused by local stresses in the microstructure, because of differences in the CTE between adjacent grains (Ref 36). These local stresses are composed of both tensile and compressive components, which must average to zero over the volume of the body. However, the tensile components, on a scale of a grain or less, can lead to microcracking between grain facets.

Table 3 Fracture toughness values achieved to date utilizing various toughening mechanisms

Toughening mechanism	Highest toughness values achieved		Exemplary material-composite systems
	MPa$\sqrt{m}$	ksi$\sqrt{in.}$	
Microcracking	~10	~9.1	Al_2O_3-ZrO_2
Transformation	~20	~18.2	ZrO_2 (MgO)
Particle	~8	~7.3	Si_3N_4-SiC
Platelet	~8	~7.3	HIPRBSN-SiC
	~14	~12.7	Si_3N_4-SiC
Whisker	~8.5	~7.7	Al_2O_3-SiC
	~11	~10.0	Si_3N_4-SiC
Fibers	>30	>27.3	Glass-SiC
	>25	>22.8	Glass-ceramic/SiC
	>30	>27.3	SiC-SiC
	~16	~14.6	Si_3N_4-SiC
Metal dispersion	~25	~22.8	Al_2O_3-Al, Al_2O_3-Ni

Source: Ref 14, 15, 16–20, 21, 22

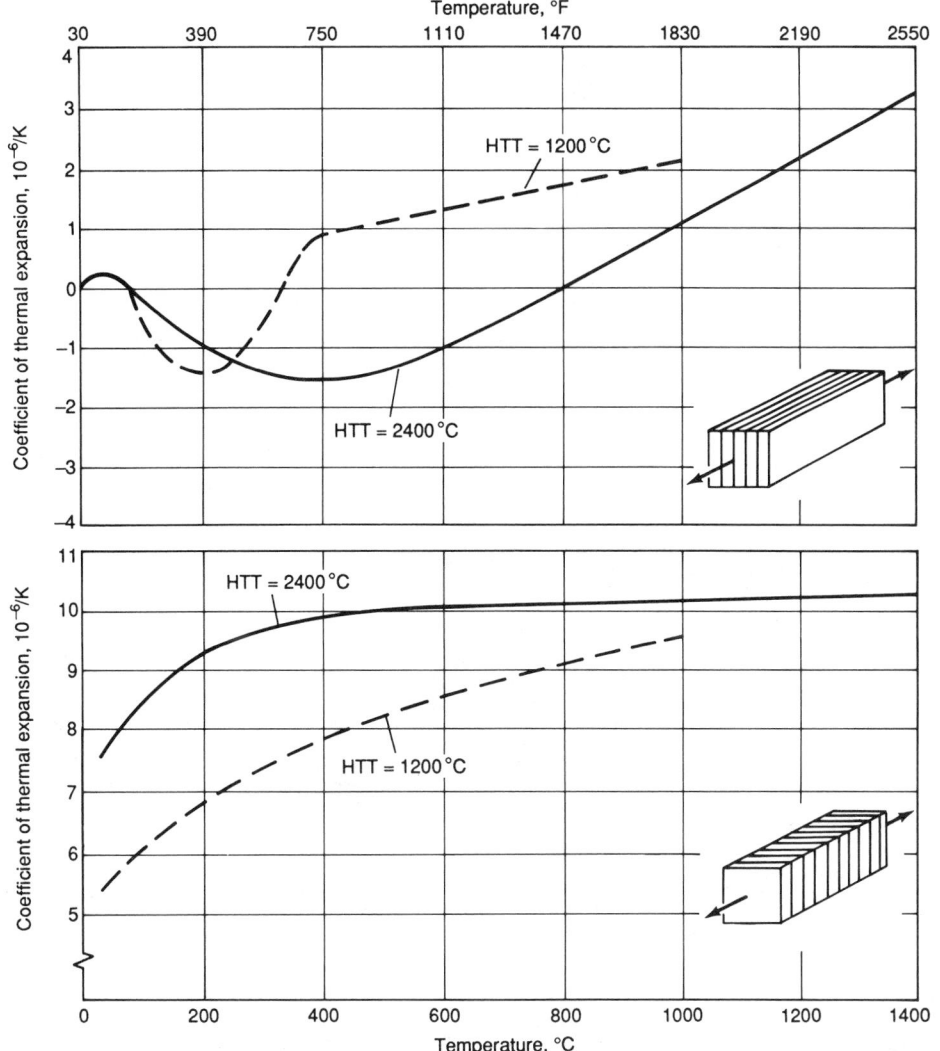

Fig 4 Coefficient of thermal expansion, parallel and perpendicular to laminates of 2-D weave carbon-carbon composite with final HTT of 1200 and 2400 °C (2190 and 4350 °F)

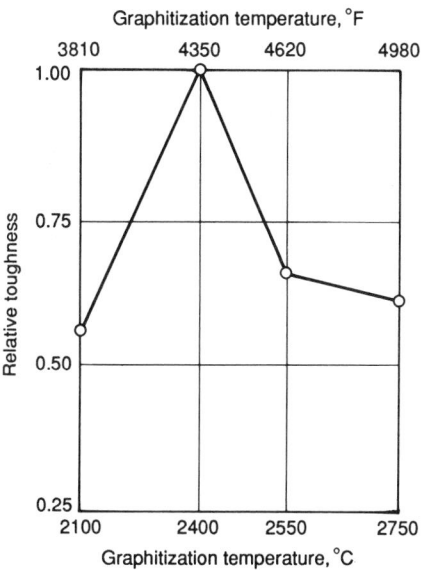

Fig 5 Relative toughness as function of graphitization temperature (pitch-based matrix, 3-D, carbon-carbon). Source: Ref 30

Microcracking appears to produce frequent toughening in conjunction with other mechanisms, such as phase transformation, for example, in ZrO_2 or in ZrO_2-containing matrix materials, like Al_2O_3-ZrO_2. Phase transformation, as a toughening mechanism, is caused by dispersoid volume change that is due to phase transformation, creating a process zone to shield the crack from applied stress. The

most common material in which phase transformation is the primary toughening mechanism is partially stabilized zirconia. In this material, tetragonal ZrO_2 transforms to the monoclinic form in the stress field of the crack tip. The transformation induces compressive stresses on the crack surface behind the crack tip.

The deflection mechanism originates in the presence of a stress field surrounding the dispersoid along the matrix-dispersoid interface, which is caused by either thermal expansion or elastic modulus mismatch (Fig 6a). In the case of crack bowing, dispersoid particles act as obstacles for crack propagation, particularly if the dispersoids have a higher fracture toughness than the matrix material (Fig 6b). The line tension effect is analogous to dislocation motion past obstacles. The stress needed to propagate the bowed segments of the crack front is greater than that needed to propagate the crack in a dispersoid-free material. Hence, an increase in toughness results.

Ductile reinforcements can profoundly increase toughness (Fig 6c). One contribution to toughness derives from crack trapping, while another involves crack bridging, and yet another involves crack shielding and plastic dissipation associated within a plastic zone. Experience and analysis have indicated that crack bridging is usually the most potent of these mechanisms. The material systems that exhibit plasticity-induced toughening can have three distinct microstructures: isolated ductile reinforcements, interpenetrating networks, and a continuous ductile phase.

In the case of whisker reinforcement, crack deflection, crack bridging, and, to a minor extent, fiber pullout are all effective (Fig 6d). Platelet toughening of ceramics (Fig 6e) includes contributions from debonding, crack deflection, crack bridging and pullout, and, in some cases, modulus load transfer in a manner similar to that described for whiskers (Ref 15). The most important parameters are orientation, aspect ratio, and interfacial properties. Crack deflection is the most effective contribution to toughening for randomly oriented platelets.

Based on these mechanisms, the resistance of brittle solids to the propagation of cracks can be strongly influenced by microstructure and by the use of various reinforcements. In most cases, toughening results in resistance-curve characteristics, wherein the fracture resistance systematically increases with crack extension. The resulting material strength values then depend on the details of the resistance curve and the initial crack lengths, such that toughness and strength optimization usually involves different choices of microstructure. The known mechanisms can be conveniently considered to involve either a process zone or a bridging zone.

Table 4 Electrical resistivity and thermal conductivity parallel and perpendicular to the laminates of 2-D weave carbon-carbon composites

HTT		Electrical resistivity, $\mu\Omega \cdot m$		Thermal conductivity, $W/m \cdot K$	
°C	°F	Parallel	Perpendicular	Parallel	Perpendicular
200	390	33–37	98–114	36–43	4–7
800	1470	8–12	68–81	127–134	39–46

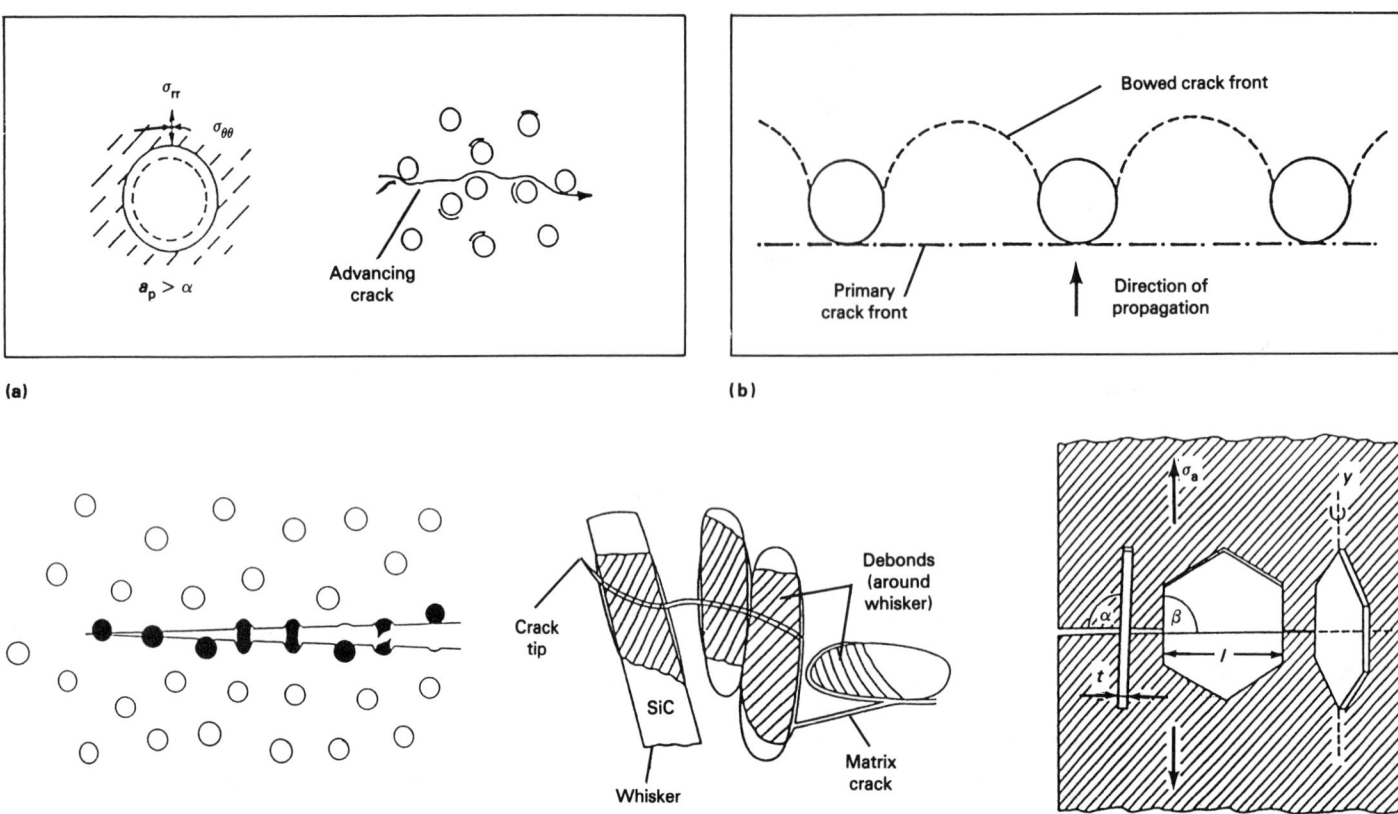

(a)

(b)

(c)   **(d)**   **(e)**

Fig 6 Various reinforcing mechanisms in ceramic-matrix composites. (a) Residual stress state around the dispersoid, having a higher thermal expansion than the matrix (left); resulting crack deflection (right). Source: Ref 32. (b) Crack bowing between dispersoid particles. Source: Ref 32. (c) Incorporation of metallic particles, where the possibility exists for crack to run around the inclusion; certain fraction of particles is therefore relatively undisturbed by fracture process. Source: Ref 37. (d) Whisker reinforcement implying debonding, deflection, and bridging. Source: Ref 30. (e) Platelets bridging a crack. Source: Ref 15

Until now, essential improvement in toughness has been achieved by fracture toughening/phase transformation and by the increase of crack surfaces that are due to crack deflection and crack branching. Regarding the mechanisms of crack deflection, toughness increase becomes higher the more the incorporated dispersoids differ from the spherical morphology (Fig 7). On a theoretical basis (Ref 38), rod-shaped particles are predicted to be more effective toughening agents than disk-shaped particles, which are more effective than spheres. However, the highest toughness values have been observed by incorporating continuous fibers.

Toughness of continuous fiber reinforced materials is mainly caused by the pullout of fibers in the crack wake, which absorbs energy (Fig 8). The pullout is strongly influenced by the sliding resistance of the interface and the properties of the fibers, particularly the statistical distribution of strength data. Thus, the characteristics of the interface and the stresses at the interface are of decisive importance. Moreover, fracture energy increases if the Weibull modulus of the fibers is small (Ref 33). A typical stress-strain curve of continuous fiber reinforced composites is shown in Fig 9.

Composite Systems and Properties. Many composite combinations based on different oxide, carbide, and nitride matrices have been investigated. Until now, the compounds investigated have primarily been selected according to these criteria: properties and availability of fibers or whiskers, chemical and physical compatibility of the reinforcements with the matrix material, and mechanical and physical characteristics of the matrix, as well as the technological realization. Primarily glass, glass-ceramics, alumina, mullite, zirconia, silicon carbide, and silicon nitride have been selected as matrix materials (Ref 30, 16–22, 40–45). The processing techniques used recently are powder method/slurry infiltration (Ref 15, 19, 20, 40, 42–44), gas phase reaction bonding (Ref 45), melt infiltration (Ref 41), chemical vapor infiltration (Ref 22, 46), the sol-gel route, the polymer pyrolysis technique (Ref 44, 47, 48), and the Lanxide process (directed metal oxidation) (Ref 49). Based on the selection criteria mentioned above, there are many cases where SiC whiskers, SiC fibers, as well as uncoated and coated carbon fibers, have been used (unidirectional and cross-woven bidirectional layers of continuous fibers or textile-like multidimensional prewoven structures). In addition to technical requirements, the availability of reinforcements plays a decisive role. Typical examples are the problems in obtaining fibers of high thermal stability, as well as platelets and whiskers of the desired morphology.

Improvements in toughness and the stress-strain behavior have been obtained in many composite systems (Table 3 and Fig 10), whereby the extent of property improvement is dependent on the type of composite, the morphology of the reinforcement, and on a variety of parameters, such as raw materials, processing techniques, and processing conditions. The toughening effect increases from particle to whisker to continuous fiber reinforcement.

Combinations of toughening mechanisms seem to be promising. For example, in the case of a combination of ZrO_2-transformation toughening and whisker reinforcement, very high toughness values of about 13.5 MPa$\sqrt{m}$ (12 ksi$\sqrt{in.}$) were found to be accompanied by strength values of ~700 MPa (100 ksi) (Ref 20). Positive results are reported by a combination of toughening mechanisms in mullite (SiC whiskers and tetragonal ZrO_2) (Ref 20) and Si_3N_4 (SiC whiskers and SiC particles, Ref 50). The most promising sys-

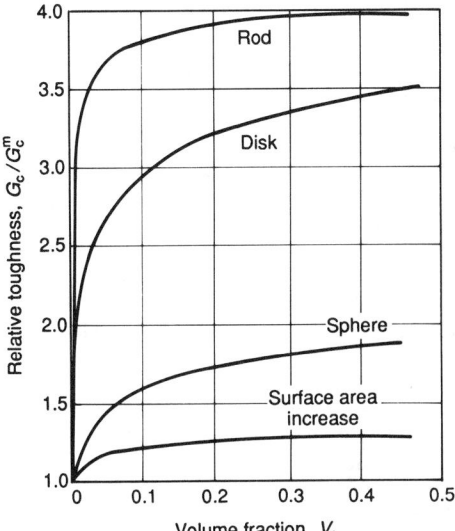

Fig 7 Theoretical predictions of composite toughening by dispersoids of various shapes caused by crack deflection. Source: Ref 38

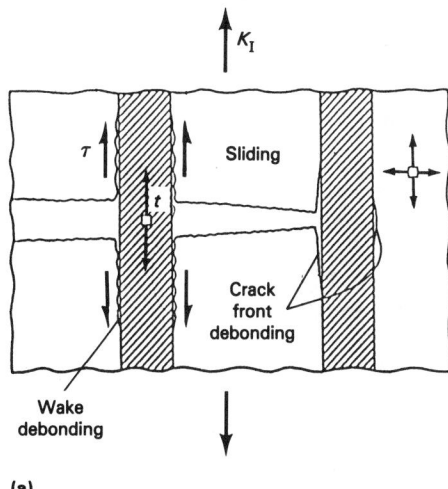

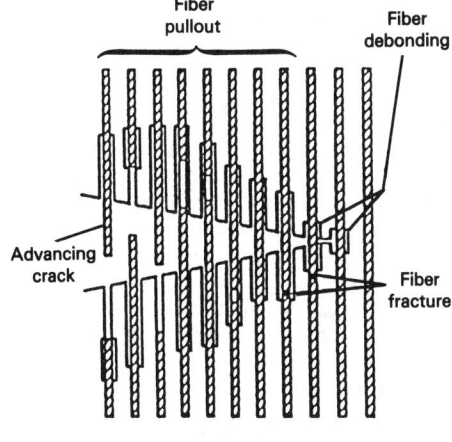

Fig 8 Toughening mechanisms in continuous fiber reinforced ceramic composites. (a) Initial debonding of fibers at the crack front, as well as fiber debonding and sliding in the crack wake (Ref 14). (b) Fiber pullout (Ref 39)

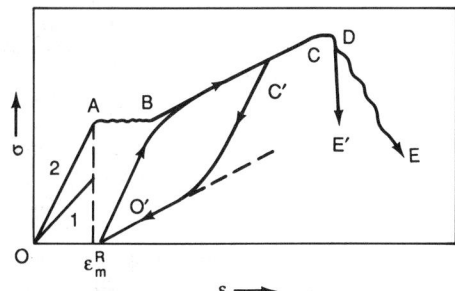

Fig 9 Typical stress-strain curve of a unidirectional continuous fiber reinforced composite under tensile load, compared to monolithic ceramic. OA, elastic region; AB, crack formation in matrix; BC/CD, fiber dominant region (load transfer from matrix to fibers, debonding and start of friction between fiber and matrix, fiber fracture); DE', fracture in case of strong bonding between fiber and matrix; DE, controlled fracture caused by fiber pullout dependent on properties of the fibers and interface. Source: Ref 31

tems at present are the fiber-reinforced glasses, glass-ceramics (Ref 16–18, 21) and SiC fiber reinforced SiC matrices produced by the CVI technique (Ref 22). In all these materials, high fracture toughness values greater than 30 MPa√m (27 ksi√in.) (these values are only a qualitative indicator for the toughness) and an essential improvement in stress-strain behavior have been obtained (see Fig 10). Continuous fiber reinforced reaction-bonded Si_3N_4 also has potential (Ref 45).

In various materials, fracture stress values also have been increased. This is particularly true in fiber-reinforced glasses and glass-ceramics. Flexural strength data up to 1000 MPa (145 ksi) and 750 MPa (109 ksi), respectively, have been measured (40 vol% fibers) (Ref 16–18). Depending on the type of reinforcements, improvements can be achieved in the scatter of strength data, impact strength, thermal shock and thermal cycling, as well as wear resistance. In some cases,

improvements in high-temperature strength and creep behavior have been reported. However, until now, the experimental data are partly contradictory, especially for short reinforcements.

Because ceramic-matrix composites are still in the development stage, commercial applications remain limited. One commercially available application is a cutting tool made from whisker-reinforced materials based on Al_2O_3 and Si_3N_4. In the case of SiC- or carbon-fiber reinforced SiC (CVI), one company in France succeeded in producing large, complex-shaped components, for example, high-temperature oxidation-resistant turbine wheels, liquid rocket engine nozzles, leading edges designed for multiple earth reentries, and relatively large spacecraft thermomechanical protection shingles.

State-of-the-Art and Problem Areas. Depending on the type of reinforcements, the various classes of ceramic composites result in different values of toughness, fracture strain, fracture stress, residual strength after deloading, and damage tolerance, which indicate large differences in future potential. On the other hand, the various classes of composites exhibit large differences in the efforts expended for processing, technological problems, and fabrication complexity and cost.

Continuous fiber reinforcement has the highest potential for improving stress-strain behavior and damage tolerance, but has the highest fabrication complexity and cost. One serious problem in processing continuous fiber reinforced ceramic composites is the limited temperature stability of the fibers. Those fibers that are of interest for most matrix materials because of their sufficient chemical stability, such as SiC fibers, lose their strength during processing at temperatures higher than about 1100 °C (2010 °F), primarily because of crystallization effects. Only the carbon fibers keep their strength up to very high temperatures, but only in an oxygen-free atmosphere.

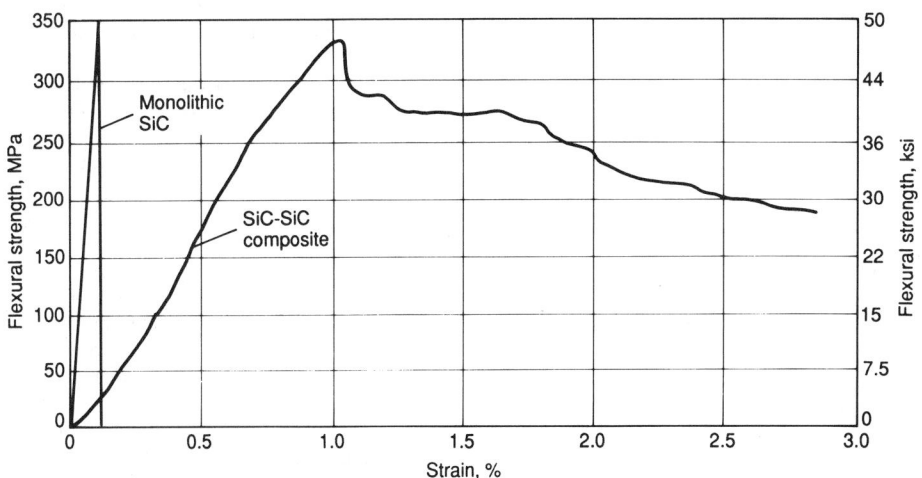

Fig 10 Stress-strain curve of SiC (Nicalon) fiber reinforced SiC (CVI-technique), compared to monolithic ceramic. Source: Ref 22

An additional consideration is the chemical attack of the fibers, which is due to reactions with the matrix material or/and with gaseous media of the sintering atmosphere. A typical example is the infiltration of carbon or SiC fiber preforms by liquid silicon. Mechanical damage of reinforcements, particularly of coated fibers, must be avoided during processing. Moreover, the interfacial characteristics have to be optimized in such a way that, during loading in the first step, load transfer from the matrix to the fibers is possible. However, in a second step, debonding and post-debond friction with subsequent fiber pullout can be initiated (Ref 33, 51, 52).

To meet these requirements, strong efforts are undertaken to avoid the high processing temperatures of the powder route. Consequently, some chemical techniques become more and more important, such as the pyrolysis of polymer precursors, the sol-gel technique, liquid phase infiltration, and directed metal oxidation. Additional promising routes are reaction sintering of silicon nitride at low temperatures, nearly without shrinkage, and the infiltration of porous ceramic preforms by liquid metals. Typical examples are the infiltration of reaction-bonded Si_3N_4 and Al_2O_3.

Using chemical techniques, the matrix materials often still exhibit a certain amount of residual porosity, which is not as critical as in the case of monolithic ceramics, because of the high toughness values of the composites. Nevertheless, serious problems may arise during high-temperature loading, because of oxidation.

In the case of whisker reinforcement, the main problem is the carcinogenity of the very fine whiskers ($\phi \sim 0.1$ to 2 μm, or 4 to 80 μin.). In Germany, for example, technological work with whiskers was stopped for this reason. Furthermore, to raise the effectiveness of the toughening mechanism, the whisker morphology has to be optimized. From the technological point of view, these issues need to be solved:

- Improve dispersion of whiskers in order to avoid hard agglomerates that can result in strength reduction (Ref 53)
- Avoid mechanical damage of the whiskers during processing
- Maintain stability of whiskers during sintering
- Sinter the compounds without applying pressure

For example, mechanical dispersion has to be replaced by chemical processing to avoid either surface or coating damage or the reduction of the aspect ratio of the reinforcements (Ref 53).

The alternative to whisker reinforcement is the incorporation of short fibers, particles, or platelets. The powder route is well suited for particle and platelet reinforcement. In general, property improvement for these types of reinforcements is limited, because of the less-energy-absorbing mechanisms acting in these composites. This is especially true if the reinforcing component is approaching spherical shape (see Fig 7).

In the case of chopped fiber reinforced composites, the main problems are the need to incorporate a high amount of chopped fibers (high enough to initiate reinforcing mechanisms) and to achieve proper dispersion of the fibers without causing artificial flaws by their introduction (Ref 20, 53). Positive effects can be expected if the fibers are oriented perpendicular to crack propagation. Depending on the composite system, problems with the thermal and chemical instability of the fibers during processing may arise, such as with nitrogen during sintering of Si_3N_4 (Ref 43).

Regarding hardness and wear resistance, and, to a minor extent, fracture toughness, promising results have been obtained for various particle-reinforced composites. The main advantage of this type of composite is the ease of processing. The microstructure-property relationship is important, particularly with respect to the size of the particles, the volume content, and the thermodynamic relation of particle-matrix. For example, toughness increase is only observed in various systems for coarse dispersoids. In this connection, the size and morphology of the dispersoids in relation to those characteristics of the matrix material have to be considered. Based on these facts, recently promising results regarding toughness have been observed by a combination of whisker and particle reinforcement (Ref 50). The results are interpreted by utilizing different reinforcing mechanisms caused by the particles and by the whiskers.

The introduction of platelets has not yet been very successful. The slight toughness increase is frequently connected with strength reduction (Ref 15, 40). The problem at the present time is that only large-sized platelets of relatively high impurity content are available.

Nanocomposites, in which very fine second-phase particulates ($\phi < 0.1$ μm, or 4 μin.), are dispersed within the matrix grains, have obvious potential to improve fracture strength even at high temperatures of about 1400 °C (2550 °F) (Ref 54).

The incorporation of ductile metal phases into ceramic materials has been shown to enhance toughness effectively (Ref 14, 37). An example of recent investigations is the toughness increase by incorporating nickel (13 vol%) in alumina by a factor of two, compared to monolithic alumina, suggesting a toughening mechanism involving plastic deformation of the metal particles. Perhaps further improvements in crack propagation of ceramics can be obtained by the new techniques for fabricating metal-reinforced ceramic-matrix composites, such as directed metal oxidation, as well as metal infiltration of reaction-bonded Si_3N_4 (RBSN) and Al_2O_3 (RBAO) by gas-pressure infiltration or squeeze casting to fill the pores of RBSN and RBAO with various metals (such as Al, Si, inter-metallics, or superalloys) (Ref 55). However, for all these novel techniques, more systematic basic and technological work has to be done.

It is quite clear from the reinforcing mechanisms that the interface of reinforcement-matrix plays a decisive role in improving and controlling the mechanical behavior of ceramic composites. This is particularly true for continuous fiber reinforced composites. Mechanical properties of these materials are strongly influenced by the debonding and sliding resistance of the interfacial region. Thus, the relationship between the characteristics of the interface and the mechanical behavior of the composites should be known in more detail for all important composite combinations. This problem area includes:

- The microanalytical and microstructural characterization of interfaces by means of high-resolution and analytical transmission electron microscopy, in order to understand the microchemistry and microstructure
- The analytical approach from the fracture mechanics point of view
- The measurement of mechanical and physical characteristics of the interface

Based on these results, experimental efforts have to be made in order to optimize the interface by coating. In some cases, not only one layer, but a double layer should be developed in order to protect the fibers and to control the interface.

Future Trends

Carbon-carbon composites have a clear potential for mechanical engineering applications in hot load-bearing structures and corrosive surroundings. They seem to be a useful addition to the fine-grained carbon and graphite grades. However, growth of applications will be slow, because of the cost situation and a lack of knowledge about several material items. The data base is currently small, and problems exist in comparing known and published data, because there is no test standard. The number of available publications is limited, because carbon-carbon composites are classified in the United States and France, where much of the work has been performed. In addition, open questions exist concerning the understanding of the influence of process conditions on properties, and many questions regarding the role and control of the complex interface situation still need to be addressed.

Carbon-carbon composites are also of interest from another point of view. The need is tremendous for damage-tolerant, tough, structural ceramic-ceramic composites in service at high temperatures above 1000 °C (1830 °F). Behind this demand are future considerations to construct hypersonic aircraft and spacecraft. One way to develop such urgently needed materials is to start from carbon/carbon composites and achieve oxidation stability, either by coating techniques or by transferring the carbon matrix partly or completely into an oxidation-resistant ceramic. As a spin-off from these developments, one can hope that oxidation-resistant carbon-carbon-silicon carbide composites will become available for use at temperatures ranging from 1000 to 1400 °C (1830 to 2550 °F). Such oxidation-inhibited carbon-carbon-based composites would have a chance for a much wider use in industry as one can imagine today (Ref 56–58).

Based on the state of the art, it can be concluded that, even under critical aspects, other high-performance ceramic-matrix composites also have high future potential. By selecting the type of reinforcement, it will be possible to tailor materials with regard to a property profile. Besides improvements in stress-strain behavior, damage tolerance, and the other properties mentioned above, two perspectives are of future interest. One is the reduction of the scatter of mechanical property data for ceramics, and the other is the significance of these materials for improving high-temperature mechanical properties, utilizing the mechanisms of crack bridging and fiber pullout in a glass-free matrix material, for example. However, a number of problems have to be solved before these materials can be produced either in large amounts or for complex-shaped components with the high reproducibility required. In this context, only some problems are mentioned here:

- Specific technological problems, depending on the processing route
- Further development of low-temperature processing techniques
- Lack of thermal stability of fibers
- Optimization and control of interface by coating
- Changes in microstructure and microchemistry at the interface after thermal exposure
- Fracture mechanics of interfaces
- Efforts to deepen the understanding of toughening mechanisms, including modeling
- Characterization of mechanical and thermomechanical properties
- Design of components
- Development of nondestructive techniques for improving quality control

Depending on the gradual solution of these problems, the various classes of ceramic-matrix composites will be used in the short, intermediate, or long term as damage-tolerant materials for structural applications, from room temperature up to extremely high temperatures.

REFERENCES

1. W. Huettner, doctorate thesis, University of Karlsruhe, 1980
2. H. Brueckmann, doctorate thesis, University of Karlsruhe, 1979
3. J.L. White and P.M. Shaeffer, Mesophase Stabilization by Oxidation Processing, *Carbon 88*, B. McEnaney and T.J. Mays, Ed., IOP Publishing Ltd., 1988, p 455–457
4. E. Fitzer and M. Heym, *Z. Werkstofftech.*, Vol 8, 1976, p 269–279
5. C.R. Thomas and E.J. Walker, *High Temp.-High Press.*, Vol 10, 1978, p 79
6. E. Fitzer and W. Huettner, *Sprechsaal*, Vol 6, 1980, p 451
7. E. Fitzer and W. Huettner, Structure and Strength of C/C-Composites, *J. Phys. G, Appl. Phys.*, Vol 14, 1981, p 47–71
8. M. Steinberg, *Spektrum der Wissenschaft*, 1986
9. P.G. Rolincik Jr., *SAMPE J.*, Sept/Oct 1987, p 40–47
10. Schunk Kohlenstofftechnik GmbH, Giessen, Germany, product information carbon-carbon
11. U. Soltesz, private communication, 1988
12. J.R. Strife and J.E. Sheehan, *Am. Ceram. Soc. Bull.*, Vol 67 (No. 2), 1988, p 369–374
13. R. Meistring, private communication, 1989
14. M. Rühle and A.G. Evans, High Toughness Ceramics and Ceramic Composites, *Progr. in Mater. Sci.*, Vol 33, 1989, p 85–167
15. N. Claussen, Ceramic Platelet Composites, in *Proceedings of the 11th Risø International Symposium on Metallurgy and Materials Science*, 1990, Risø National Laboratory, Denmark, p 1–12
16. K.M. Prewo, J.J. Brennan, and G.K. Layden, Fiber Reinforced Glasses and Glass-Ceramics for High Performance Applications, *Am. Ceram. Soc. Bull.*, Vol 65, 1986, p 305–322
17. K.M. Prewo, Fiber-Reinforced Ceramics: New Opportunities for Composite Materials, *Am. Ceram. Soc. Bull.*, Vol 66 (No. 2), 1989, p 395–400
18. H. Hegeler and R. Brückner, Fiber-Reinforced Glasses, *J. Mater. Sci.*, Vol 24, 1989, p 1191–1194
19. T.N. Tiegs and P.F. Becher, Sintered Al₂O₃-SiC-Whisker Composites, *Am. Ceram. Soc. Bull.*, Vol 66 (No. 2), p 330–333
20. J. Lehmann and G. Ziegler, Oxide-Based Ceramic Composites, *Proceedings of the 4th European Conference on Composite Materials ECCM-4*, Stuttgart, Sept 1990, p 425–434
21. G. Ziegler, Fiber-Reinforced Ceramic-Matrix Composites, *Proc. Materialforschung 1988 des BFMT*, Projektleitung Material-und Rohstofforschung (PLR) der KFA Jülich GmBH, 1988, p 765–786 (in German)
22. P.R. Naslain, Inorganic Matrix Composite Materials for Medium and High Temperature Applications: A. Challenge to Materials Science, *Development in the Science and Technology of Composite Materials*, A.R. Bunsell et al., Ed., 1st ECCM, Bordeaux, Elsevier Science Publishers Ltd., 1985, p 34–35
23. L.E. McAllister and W.C. Lachmann, *Handbook of Composites*, Vol 4, A. Kelly and S.T. Mileiko, Ed., North Holland, New York, p 139
24. J. Jortner, *Carbon*, Vol 24 (No. 5), 1986, p 603–613
25. J.E. Zimmer et al., *Molecular Crystals, Liquid Crystals*, Vol 38, 1977, p 188
26. J.E. Zimmer et al., *Advances in Liquid Crystals*, Vol 5, H. Brown, Ed., 1983, p 157
27. E. Fitzer, K.-H. Geigl, and W. Huettner, *Carbon*, Vol 18 (No. 6), 1980, p 383
28. J.G. Morley et al., *J. Mater. Sci.*, Vol 9, 1974, p 1171
29. R. Meyer et al., *Proceedings of the International Symposium on Science and Applications of Carbon Fibers*, Toyohashi University, Japan, 1984
30. G.H. Campell, M. Rühle, B.J. Dalgleish, and A.G. Evans, Whisker Toughening: A Comparison Between Aluminium Oxide and Silicon Nitride Toughened with Silicon Carbide, *J. Am. Ceram. Soc.*, Vol 73 (No. 3), 1990, p 521–530
31. D. Rouby, Ceramic-Matrix Composites, *Ceram. Forum Int.*, Vol 66, 1989, p 208–216
32. S.T. Buljan and V.K. Sarin, Silicon Nitride-Based Composites, *Composites*, Vol 18 (No. 2), 1987, p 99–106
33. M.D. Thouless, O. Sbaizero, L.S. Sigl, and A.G. Evans, Effect of Interface Mechanical Properties on Pullout in a SiC-Fiber-Reinforced Lithium Aluminium Silicate Glass-Ceramic, *J. Am. Ceram. Soc.*, Vol 72 (No. 4), 1989, p 522–532
34. A.G. Evans, Perspective on the Development of High-Toughness Ceramics, *J. Am. Ceram. Soc.*, Vol 73 (No. 2), 1990, p 187–206
35. R. Naslain and B. Harris, Ceramic Matrix Composites, *Compos. Sci. Technol.*, Vol 37, 1990
36. S.W. Freiman, Brittle Fracture Behavior of Ceramics, *Am. Ceram. Soc. Bull.*, Vol 67 (No. 2), 1988, p 392–402
37. W.H. Tuan and R.J. Brook, The Toughening of Alumina with Nickel Inclusions, *J. Eur. Ceram. Soc.*, Vol 6, 1990, p 31–37
38. K.T. Faber, Toughening of Ceramic Materials by Crack Deflection Processes, Ph.D. thesis, University of California, 1982
39. D.C. Philipps, *Survey of the Technological Requirements for High Temperature Engineering Applications*, Office for Official Publications of the European Communities, Luxembourg, 1985, p 48–73
40. K.-H. Heussner and N. Claussen, Yttria-

and Ceria-Stabilized Tetragonal Zirconia Polycrystals (Y-TZP, Ce-TZP) Reinforced with Al_2O_3 Platelets, *J. Eur. Ceram. Soc.*, Vol 5, 1989, p 193–200

41. E. Fitzer and R. Gadow, Fiber-Reinforced Silicon Carbide, *Am. Ceram. Soc. Bull.*, Vol 65 (No. 2), 1986, p 326–335

42. S.T. Buljan, J.G. Baldoni, and M.L. Huckabee, Si_3N_4-SiC-Composites, *Am. Ceram. Soc. Bull.*, Vol 66 (No. 2), 1987, p 347–352

43. W. Braue, A. Hölscher, B. Saruhan, and G. Ziegler, Effect of SiC-Whisker Characteristics of Interface and Mechanical Properties of Silicon Nitride Matrix Composites, *Euro-Ceramics*, Vol 3, *Engineering Ceramics*, G. de With, R.A. Terpstra, and R. Metselaar, Ed., Elsevier Science Publishers Ltd., 1989, p 3.253–3.267

44. R. Lundberg, R. Pompe, and R. Carlsson, Fiber Reinforced Silicon Nitride Composites, *Compos. Sci. Technol.*, Vol 37, 1990, p 165–176

45. R.T. Bhatt and R.E. Phillips, Thermal Effects on the Mechanical Properties of SiC Fiber-Reinforced Reaction-Bonded Silicon Nitride Matrix Composites, *J. Mater. Sci.*, Vol 25, 1990, p 3401–3407

46. D.P. Stinton, T.M. Besmann, and R.A. Lowden, Advanced Ceramics by Chemical Vapor Deposition Techniques, *Am.*

Ceram. Soc. Bull., Vol 67 (No. 2), 1988, p 350–355

47. D.C. Carlson, J.D. Cooney, S. Gauthier, and D.J. Worsfold, Pyrolysis of Silicon-Backbone Polymers to Silicon Carbide, *J. Am. Ceram. Soc.*, Vol 73 (No. 2), 1990, p 237–241

48. F. Sirieix, P. Goursat, A. Lecomte, and A. Dauger, Pyrolysis of Polysilazanes: Relationship between Precursor Architecture and Ceramic Microstructure, *Compos. Sci. Technol.*, Vol 37, 1990, p 7–19

49. M.K. Aghajanian, N.H. MacMillan, C.R. Kenney, S.J. Luszcz, and R. Roy, Properties and Microstructures of Lanxide Al_2O_3-Al Ceramic Composite Materials, *J. Mater. Sci.*, Vol 24, 1989, p 6587–6670

50. H. Kodama and T. Miyoshi, Fabrication and Properties of Si_3N_4 Composites Reinforced by SiC Whiskers and Particles, *J. Am. Ceram. Soc.*, Vol 73 (No. 3), 1990, p 678–683

51. R.J. Kerans, R.S. Hay, N.J. Pagano, and T.A. Parthasarathy, The Role of the Fiber-Matrix Interface in Ceramic Composites, *Am. Ceram. Soc. Bull.*, Vol 68 (No. 2), 1989, p 429–441

52. H.C. Cao, E. Bischoff, O. Sbaizero, M. Rühle, A.G. Evans, D.B. Marshall, and J.J. Brennan, Effect of Interfaces on the

Properties of Fiber-Reinforced Ceramics, *J. Am. Ceram. Soc.*, Vol 73 (No. 6), 1990, p 1691–1699

53. J. Lehmann, B. Müller, and G. Ziegler, Optimization of Slip-Casting Techniques for Short-Fiber and Whisker-Reinforced Ceramic Composites, *Euro-Ceramics*, Vol 1, *Processing of Ceramics*, G. de With, R.A. Terpstra, and R. Metselaar, Ed., Elsevier Science Publishers Ltd., 1989, p 1196–1200

54. K. Niihara, K. Izaki, and A. Nakahira, The Si_3N_4-SiC Nanocomposites with High Strength at Elevated Temperatures, *J. Jpn. Soc. Powder Powder Metall.*, Vol 37 (No. 2), 1990, p 352; *J. Mater. Sci.*, Vol 9, 1990, p 598–599

55. N. Claussen, "Novel Ceramic/Metal Composites," Seventh Cimtec World Ceramics Congress 1990, Montecatini Terme, Italy

56. W. Huettner, R. Weiss, G. Dietrich, and R. Meistring, Space Applications of Advanced Structural Materials, *Proceedings of the ESA Symposium*, ESA-SP-303, June 1990, p 91–95

57. W. Huettner *et al.*, *Ext. Abstr. 19 Bienn. Conf. on Carbon*, 1989, p 482–483

58. W. Hüttner, *Carbon Fibers, Filaments and Composites*, J.L. Figueiredo *et al.*, Ed., Kluwer Academic Publishers, Netherlands, 1990

Engineering Properties of Oxide Glasses and Other Inorganic Glasses

J.E. Shelby, W.C. Lacourse, and A.G. Clare, Institute for Glass Science and Engineering,
New York State College of Ceramics, Alfred University

THE EFFECT OF COMPOSITION on many properties of glasses is most easily understood by considering the structural roles of the components. Many elements have similar structural roles and thus affect properties in similar ways. Therefore, the discussion below is based on a direct comparison of components that have similar effects on the structure of glasses.

Traditionally, the structural role of a component in an oxide glass is designated as that of a glass former, a modifier, or an intermediate. Glass formers are those oxides that readily form a glass on their own, such as SiO_2, B_2O_3, and GeO_2. Modifiers include those oxides that are incapable of forming a glass by themselves but are frequently added in large amounts to alter, or modify, the properties of the glass, such as alkalies and alkaline earth oxides. Intermediate oxides are defined as those that are incapable of forming a glass by themselves (at least if traditional glass-forming techniques are used) but are capable of substituting for a glass former in the vitreous network. In other words, the behavior of these oxides is "intermediate" between those of the glass formers and the modifiers.

Although it is usually assumed that an intermediate has the same coordination number as the glass former it replaces, this is not a necessary condition for the classification of an ion into this category. It is only necessary for the ion to act to increase the connectivity of the vitreous network, as opposed to modifier ions, which may either increase or decrease the connectivity of the network.

The structural role of components in halide, chalcogenide, and oxyhalide glasses is not as clearly defined. However, the roles of components in these systems are also often discussed in terms of glass former, modifier, and intermediate. This nomenclature is used throughout this article, with the obvious qualification that the exact definition of these terms is more loosely defined for the non-oxide systems.

The specific structural role of an ion is a direct function of its size and charge, that is, its field strength. The oxides most commonly recognized as glass formers, for example, are based on cations such as boron and silicon, which have very high field strengths. At the other extreme, the most common modifier ions are the alkali ions, which have low charge and modest size, that is, low field strengths. Ions of intermediate field strength exhibit behavior intermediate to these ions. Changes in the field strength of the cation present will increase or decrease the value of properties as appropriate for the changes in bond strength and ionic size (coordination number) that result. An example of this effect is observed whenever gallium replaces aluminum in glasses. The decrease in field strength results in decreases in viscosity and glass transformation temperature.

Of course, the field strength of ions is not the only parameter that must be considered when discussing the properties of glasses. The density of a glass depends on both the structure and the mass of the ions present. Structural arrangements that increase the free volume of a glass have different effects on density from those that decrease the free volume. A discussion of refractive index involves not only consideration of changes in density of the material but also changes in polarizability of the ions, especially the anions, which are major contributors to the refractive index. Changes in the field strength of the cations often primarily alter the refractive index through associated changes in the polarizability of the adjacent anions, rather than through large changes in the polarizability of the cation itself.

Other properties are dependent on still other aspects of the structure. Viscosity and glass transformation temperature are determined by the strength of the bonds, by the connectivity of the structure (nonbridging versus bridging species), and by the coordination number of the ions present. Thermal expansion coefficients are complex functions of the degree of asymmetry of the structural units, the bond strength, and the connectivity of the network. Electrical conductivity is influenced by changes in the identity of the carrier ion, structural density of the network, and the strength of the bond binding the carrier ions to their sites. Gas mobility is determined by structural density, sizes of openings between interstices, elastic modulus of the glass, and other factors. It follows that one must include field strength, mass, ionic polarizability, connectivity of the network, coordination of ions, symmetry of building units, and possible intermediate range units, in any suggested explanation for trends in the properties of glasses.

Overview of Glass Structures

In addition to defining the general structural roles of components into glass formers, modifiers, and intermediates, it is necessary to discuss the effect of each of these in various glass-forming systems. Hence, the general discussion below of the structures of the primary glass-forming systems will simplify the later discussions of the properties of specific systems.

Each of the common oxide glass formers forms a random network structure. This structure may be composed of a two-dimensional network, such as that formed by vitreous boric oxide or a three-dimensional network, as is the case for vitreous silica and germania; or it may be composed of polymeric chains, as in vitreous phosphoric oxide. The introduction of modifiers to these structures can alter them by either forming nonbridging anions, as in the alkali-silicate glasses, or by producing two types of glass-former sites, for example, $(BO_3)^{3-}$ triangles and $(BO_4)^{4-}$ tetrahedra in alkali borate or $(GeO_4)^{4-}$ tetrahedra and $(GeO_6)^{6-}$ octahedra in alkali-germanate glasses.

Because the connectivity of the structure is severely altered by either the formation of nonbridging species or by changes in the coordination number of the glass former, the properties of the resultant glasses will also be severely altered by these changes. In general, structural changes that lead to the production of nonbridging species decrease the connectivity of the structure and result in changes in properties which are in the opposite direction from those structural changes which result in an increase in connectivity.

The structural changes that result from the addition of an intermediate ion to a glass are often opposed to those introduced by the addition of a modifier ion, that is, the reduction in the concentration of nonbridging anions or the reconversion of the glass-former ion to its coordination number in the unmodified glass. It follows that the addition of intermediate ions to a glass will usually (but not always) lead to trends in properties that are the opposite of those produced by the addition of modifier ions.

Although historically only cations were considered in the discussion of the structural roles of ions, the recent study of halide and oxyhalide glasses has forced consideration of the changes in properties that result from the replacement of one anion by another. It appears that the same arguments can be used to explain these trends. Thus, it is necessary to consider the effect of an added anion on the connectivity of the structure, the bond strengths between polyhedra, and the coordination number of the glass-former cation. Alterations of these and other parameters will lead to the same types of changes in properties as do those that are due to the replacement of one cation by another.

Finally, the structures of chalcogenide glasses are significantly different from those of the oxide, halide, and oxyhalide glasses. These glasses can display either one-, two-, or three-dimensional structures, depending on the overall composition of the glass. Chalcogenide glass structures are generally described by an "average coordination number," usually designated as R. The elements Se, S, and Te have coordinations of 2; As, Sb, and Bi have coordinations of 3; and Ge and Si, 4. The value R for a glass of the type $Ge_xAs_ySe_z$ is determined from:

$$R = \frac{4x + 3y + 2z}{x + y + z} \qquad \text{(Eq 1)}$$

Values of R between 2 and 2.4 correspond to chains ($R = 2$) or cross-linked chains. For As_2Se_3, R is 2.4, which corresponds to a layer structure with As surrounded by 3 Se, and each Se bonded to 2 As. For $GeSe_2$, $R = 2.66$, which corresponds to a three-dimensional structure similar to that of SiO_2.

Those properties that depend on the "tightness" of the structure will change in predictable ways. For example, for selenides of the type Ge-As-Se, the glass transformation temperature, hardness, Young's modulus, and strength increase, while the thermal expansion coefficient and molar volume decrease with increasing R. It must be noted that when using R values to predict properties, it is assumed that there is sufficient Se and/or S to provide the metal cations, such as As and Ge, with their desired coordination (that is, only chalcogen-chalcogen or metal-chalco-

gen bonds form, and no As-As, Ge-Ge, Sb-Sb bonds exist). Because Ge must share 4 Se and As must share 3 Se, the compositions GeSe, $Ge_{40}As_{30}Se_{30}$, and $As_{45}Se_{55}$ all are chalcogen deficient, and the use of R values to predict property changes would not be appropriate. The As_2Se_3, As_2S_3, Sb_2S_3, and $GeSe_2$ compounds define the minimum chalcogen contents at the limiting compositions. Discontinuities in property-composition relationships often occur when such compositions are reached.

The properties of these glasses also depend on the type of chalcogen or metal present. Therefore, although property trends would be similar for glasses of the type Ge-As-S, Ge-Sb-Se, and Ge-Sb-Se, property values would differ.

Physical and Thermal Properties

The density of glasses is controlled by the free volume of the vitreous network and by the masses of the ions present. In general, the addition of species that enter the interstices of the vitreous network will tend to increase the density by reducing the free volume. The effect of additions of other ions to a glass will alter the density in proportion to the mass of the added ions relative to that of the ions already present in the glass. These effects are further complicated by the possibility that the addition of another ion to a glass will result in an alteration in the coordination number of the glass-forming cation.

The variation of density of silicate glasses with composition is perhaps the most direct example of the effect of the placement of other ions into the interstices of the structure. The addition of alkali ions to vitreous silica will increase the density monotonically as the concentration of the alkali ion increases. However, the increase in density does not scale simply with the atomic mass of the added ion. Although sodium-silicate glasses are significantly more dense than the corresponding lithium-silicate glasses, potassium-silicate glasses have densities that are almost identical to those of the corresponding sodium-silicate glasses.

The densities of the alkali borate and germanate glasses exhibit trends with atomic weight of the alkali ion present that are similar to those of the alkali-silicate glasses. However, the anomaly in the trend for the sodium- and potassium-germanate glasses is even more pronounced, with density increasing in the order Li < K < Na < Rb < Cs. However, the simple trends in the density with increasing concentration of alkali ions observed for the silicate glasses are not observed in these systems. This is particularly true for the alkaligermanate glasses, where the density passes through a maximum and actually begins to decrease with increasing alkali content (Fig 1). The differences between the behavior of the alkali silicate glasses and that of other

systems is the direct result of the differences in structure discussed earlier.

The density of silicate, borate, and germanate glasses is decreased by the partial substitution of fluorine for oxygen, because the fluorine ions increase the free volume of the structure. The density of lead-silicate glasses is increased by the partial substitution of chlorine, bromine, and iodine for oxygen.

Densities of fluoride glasses range from BeF_2 (3.59 g/cm^3) and aluminum-fluoride-based glasses (3.9 g/cm^3) to thorium-fluoride-based glasses (6.5 g/cm^3). Heavy metal-fluoride-based glasses are quite dense, because of the presence of massive ions, as well as the high packing densities of these glasses. Modifiers such as barium are particularly useful in increasing the density through the elimination of large interstices.

Because of the greater mass of S, Se, and Te, as compared to oxygen, chalcogenide glasses generally have higher densities than those of oxide glasses (Table 1). On the other hand, molar volumes are also somewhat greater, because of structural and bonding considerations.

The thermal expansion behavior of glasses can be segregated into two primary regions that are separated by a third, smaller region known as the glass transition region. Vitreous materials behave as elastic solids for temperatures below the transition region and are properly termed glasses. At temperatures above this region, vitreous materials behave as viscous fluids and are more properly termed melts. These materials exhibit intermediate (viscoelastic) behavior within the transition region and are commonly called glasses, even though they actually are in transition between solid and liquid phases.

The thermal expansion of a solid glass is primarily controlled by the expansion of the bond length, which results from increased thermal energy. Structural changes that increase the number of nonbridging bonds will increase the thermal expansion coefficient of glasses, whereas changes in the coordination number of the glass-former cation may either increase or decrease the thermal expansion coefficient, α.

Because most of the thermal expansion coefficients reported in the literature are determined by conventional dilatometry, the accuracy of the values is usually of the order of $\pm 0.2 \times 10^{-6}$/K. Values reported to more than two significant figures should only be based on measurements obtained using greater care than is done in the average study. It is possible to obtain much more accurate thermal expansion coefficients by other methods, especially using interferometers, in which case values can be obtained that are accurate to several significant figures. Values quoted by the National Institute of Standards and Technology for SRM-731 are given to $\pm 0.01 \times 10^{-6}$/K.

The addition of alkali oxides to silica creates nonbridging oxygen ions in the structure,

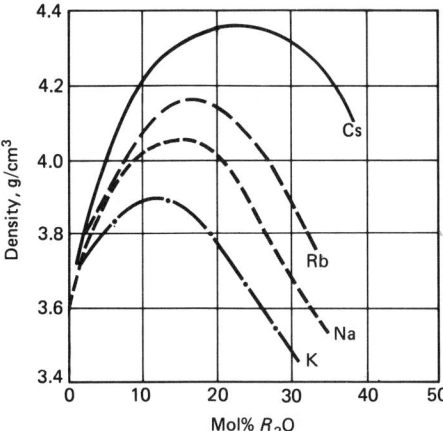

Fig 1 Effect of alkali-oxide concentration on the density of alkali-germanate glasses. Source: Ref 1

which results in an increase in α. The α for a given alkali oxide concentration (mol%) increases in the order Li < Na < K < Rb < Cs, that is, with increasing size or decreasing field strength of the alkali ion present. Reduction of the concentration of nonbridging oxygens by the addition of an intermediate, such as alumina or gallia, results in the restoration of the connectivity of the network and reduces the α of alkali-silicate glasses. Ions that cause the formation of nonbridging oxygens, but act to partially bridge the break in the network as a result of their own significant field strengths (for example, Mg^{2+}), increase the α of silicate glasses to a lesser extent than do monovalent ions.

The addition of alkali oxides to borate glasses results in the well-known "borate anomaly." The two-dimensional structure of vitreous boric oxide, which is based on a sheet-like structure consisting of $(BO_3)^{3-}$ triangles and exhibits a large α, increases in connectivity upon the addition of alkali oxides via the conversion of the $(BO_3)^{3-}$ triangles to $(BO_4)^{4-}$ tetrahedra. These tetrahedra are linked at all four corners and thus form a three-dimensional network. The α of the network decreases as connectivity increases. The structure eventually begins to break down as a portion of the alkali oxide begins to create nonbridging oxygen ions, the connectivity of

the network decreases, and the α increases with the further addition of alkali oxide. This model is an oversimplification of the actual structure, which involves a number of intermediate range order units. Although it does not explain the behavior of alkali-borate glasses in detail, it does serve as a first approximation for the purpose of this review. The α of the alkali-borate glasses varies in the order Li < Na < K < Rb < Cs.

A minimum in α also occurs for alkaline earth-borate glasses. However, the properties of these glasses are dominated by the fact that they exhibit stable immiscibility in the range of the borate anomaly. Thus, the trends in their properties are difficult to compare with those of homogeneous glasses. The addition of intermediates to alkali-borate glasses also decreases their thermal expansion coefficients, but to a lesser extent than is observed for alkali-silicate glasses. The borate anomaly is not eliminated by the addition of either alumina or gallia to alkali-borate glasses.

The thermal expansion behavior of alkali-germanate glasses is intermediate between that of alkali-silicate and alkali-borate glasses. A small "germanate anomaly" is observed in the α at 2 to 5 mol% alkali oxide. This effect is much less striking than that observed for alkali-borate glasses. On the other hand, the α of alkali-germanate glasses is almost independent of the identity of the alkali ion present. The addition of alumina to these glasses suppresses, but does not eliminate, the germanate anomaly and decreases the thermal expansion coefficient.

Partial replacement of oxygen by halide ions increases the thermal expansion coefficient. This effect has not been studied in detail but is tentatively attributed to the formation of nonbridging halide anions and a consequent reduction in the connectivity of the network.

The thermal expansion coefficients of heavy metal-fluoride-based glasses typically range from 15 to 20×10^{-6}/K, whereas the BeF_2-based glasses exhibit thermal expansion coefficients of 10 to 20×10^{-6}/K. The high thermal expansion coefficients of fluoride glasses relative to those of many oxide glasses are a direct result of the weak bonding of the singly charged fluoride ion. In many cases, the product of the α and the glass transition temperature, T_g, of these glasses has been found

to be a constant, that is, the α is inversely proportional to the T_g.

The thermal expansion coefficients for chalcogenide glasses range from about 12×10^{-6}/K for a commercial Ge-As-Se glass to as high as 30×10^{-6}/K. Most values are in the range of 12 to 20×10^{-6}/K. The α values decrease with the average coordination number.

The data available for the α of glass-forming melts are quite limited, usually to values determined in the viscosity range between the T_g and dilatometric softening temperature. The α always increases as the temperature passes through the transition region. The α of the melt increases by as little as a factor of 2 or less for some highly connected three-dimensional network glasses to as great as 50 to 100 for glasses such as vitreous boric oxide and the alkali borates. In general, thermal expansion coefficient trends in melts are similar to those of glasses.

Heat capacities of glasses are relatively constant over the range from room temperature to the glass transition region, varying according to the expression

$$C_p = a + bT + cT^2 \qquad (Eq\ 2)$$

where a, b, and c are constants and then increase rather abruptly as the glass transforms into a melt. Molar specific heats of glasses are typically of the order of 20 J/mol $\cdot$ K (4 Btu/lb $\cdot$ mol $\cdot$ °F), which is close to the value of $3R$, that is, the Dulong and Petit value for vibrational heat capacity. Measurement of the change in heat capacity in the glass transition region via a differential scanning calorimeter (DSC) or differential thermal analyzer (DTA) is often used to determine the T_g, which is simply a reference temperature within the transition region used to define the onset of the region.

There is little difference in the heat capacity of glasses as a function of composition. Values generally lie in the range from 628 to 1256 J/kg $\cdot$ K (0.15 to 0.30 cal/g $\cdot$ K). The heat capacity increase observed upon passing through the transition region is somewhat greater for borate glasses than for silicate and germanate glasses. This change in heat capacity is greater for glasses that either contain a large number of nonbridging anions or exhibit structures with low connectivity in three dimensions.

Vitreous SiO_2, GeO_2, and BeF_2, which essentially consist of three-dimensional, fully linked networks, exhibit very small changes in heat capacity in the transition region. As a consequence, it is very difficult to determine the T_g by using the DSC and DTA methods. Vitreous boric oxide, which has a two-dimensional structure, exhibits a very large heat capacity change, and the T_g is readily apparent in DSC and DTA curves. The addition of alkali oxides to silica decreases the connectivity of the network and increases the magnitude of the heat capacity change at the glass transition. The addition of inter-

Table 1   Properties of chalcogenide glasses

Glass(a)	R	Density, g/cm³	Molar volume, m³/mol × 10⁻⁶	T₁₂(c) °C	T₁₂(c) °F	E_η(d) kJ/mol	E_η(d) 10³ Btu/lb · mol
Se	2.0	4.25	18.6	30	85	...	...
As₂₀Se₈₀	2.2	4.46	17.5	82	180	235	101
As₄₀Se₆₀	2.4	4.60	16.8	165	330	280	120
As₄₀S₆₀	2.4	3.20	15.4	200	390	300	129
Ge₁₀As₁₀Se₈₀	2.3	4.36	17.8	...	...	...	...
Ge₁₀As₃₀Se₆₀	2.5	...	...	212	415	370	159
Ge₄₀As₂₀Se₄₀(b)	3.0	4.59	16.4	376	710	585	252

(a) Atomic basis. (b) Chalcogen deficient. (c) $T_{12} = 10^{12}$ P $= 10^{11}$ Pa $\cdot$ s isokom. (d) E_h, activation energy for viscous flow

mediates, such as alumina to alkali-silicate glasses, restores the connectivity of the network and leads to a decrease in the difference between the heat capacity of the glass and melt. Heavy metal-fluoride-based glasses exhibit large changes in heat capacity in the glass transition region.

The thermal conductivity of glasses is a weak function of composition, with values that range generally from 0.149 to 2.09 W/m · K (0.001 to 0.005 cal/cm · s · K) at room temperature. Values gradually increase with temperature but remain within a factor of two of the room-temperature value. Thermal diffusivity is also a relatively weak function of temperature, with typical values of 0.5 mm²/ s (0.02 ft²/h).

In most applications, the transmission of heat through glasses is dominated by radiation, rather than by conduction. Most common glasses are relatively transparent in the infrared (IR), with cutoff wavelengths in the range of 4 to 5 μm. The IR transmission of glasses is further discussed in the section "Optical Properties" in this article.

The viscosity of glass-forming melts can either be considered from the perspective of the entire viscosity curve or by considering specific fixed viscosity points. Historically, the viscosities of melts that can be cooled to form glasses have been measured between 10 and 10^{13} Pa · s. Measurements over such a broad range require the use of multiple instruments, each of which has its own associated errors and limitations. The measurement of the viscosity-temperature curve for a commercial material might utilize the fiber elongation technique for the range from 10^{13} to 10^5 Pa · s and a rotating spindle viscometer for the range from 10^5 to 10 Pa · s. Measurements for materials that crystallize readily might involve use of the beam-bending method from 10^{13} to 10^9 Pa · s, the parallel plate method from 10^9 to 10^5 Pa · s, and a rotating spindle for the remaining portion of the curve. These data are then fit to an empirical expression, and the fitting parameters are listed. Unfortunately, there is little agreement regarding the best empirical model. Therefore, a number of equations are quoted in the literature. Furthermore, the difficulty in producing the entire viscosity-temperature curve means that a limited number of compositions have been studied in detail.

The problems associated with the production of the entire viscosity-temperature curve usually are avoided either by only measuring the viscosity within a limited range or by using other parameters, such as the T_g, the dilatometric softening temperature, T_d, or the Littleton softening point, to compare the viscosity of different compositions. Even those studies that do measure the viscosity-temperature curve over a wide temperature range frequently avoid the problems associated with fitting the data to a single equation and list the results in a table, giving the temperature for a given viscosity. This isoviscosity tem-

perature is referred to as an isokom. Although it is also possible to specify the viscosity at a given temperature (the isothermal viscosity), this is rarely done for studies that cover a broad range of compositions.

The viscosity of a melt is controlled by a number of factors. However, the primary factors are the connectivity of the structure, the strength of the bonds connecting the short range order units, and the rate of change of the melt structure with temperature. Vitreous silica consists of a strongly bonded, three-dimensional network, which only slowly breaks down with increasing temperature. As a result, silica exhibits a viscosity-temperature curve with a small slope. This curve ranges from a T_g of around 1100 °C (2010 °F) to melting temperatures in excess of 2000 °C (3630 °F). The addition of a small amount of alkali oxide to vitreous silica results in a rapid decrease in viscosity and an increase in the slope of the viscosity-temperature curve. Melts that contain as little as 10 mol% alkali oxide can be fined at temperatures below 1600 °C (2910 °F). Further additions of alkali oxide rapidly decrease the temperatures required for forming a usable melt, until the processing temperature drops below 1300 °C (2370 °F). The T_g of the alkali-silicate glasses also decreases in a similar manner. The replacement of alkali oxides by alkaline earth oxides, which do not disrupt the network as greatly, results in more viscous melts and a consequent increase in the temperature required to form a satisfactory glass.

As might be expected from the earlier discussion, the effect on viscosity of adding alkali oxides to boric oxide is quite different from that found when adding these oxides to silica. The increase in the connectivity of the network leads to an increase in viscosity at the high-viscosity end of the viscosity-temperature curve, as evidenced by an increase in T_g

(Fig 2). The viscosity passes through a maximum at 25 to 30 mol% alkali oxide and then decreases as the formation of nonbridging oxygen ions disrupts the structure. The viscosity increases in the order Cs < Rb < K < Na < Li for these glasses. The addition of intermediates has a relatively small effect on viscosity, because the network does not contain a large concentration of nonbridging anions, even in the absence of the intermediate ion. The effect of the intermediate ion is larger at high alkali oxide contents, where a significant number of nonbridging oxygens do exist.

The alkali-borate systems offer a good illustration of the problems associated with only measuring viscosity in a limited temperature range. Whereas the T_g increases with increasing alkali-oxide contents up to about 25 to 30 mol% alkali oxide, the temperature needed to produce a homogeneous, bubble-free melt actually decreases as a result of a concurrent increase in the slope of the viscosity-temperature curve. In other words, these curves for the alkali borate melts cross with increasing temperature. The difference in slope is the result of the rate at which the structure of the melt depolymerizes with increasing temperature.

The viscosity of alkali-germanate melts is an even more complicated function of composition. The arguments presented above, combined with the simple structural model in which the initial addition of alkali oxides to germania result in the conversion of $(GeO_4)^{4-}$ units to $(GeO_6)^{6-}$ units, suggest that the viscosity of alkali-germanate melts should act in a manner similar to that of the alkali-borate melts, that is, increase with increasing alkali-oxide concentration.

In fact, the opposite occurs. The initial addition of alkali oxides to germania results in a very sharp decrease in viscosity, with a

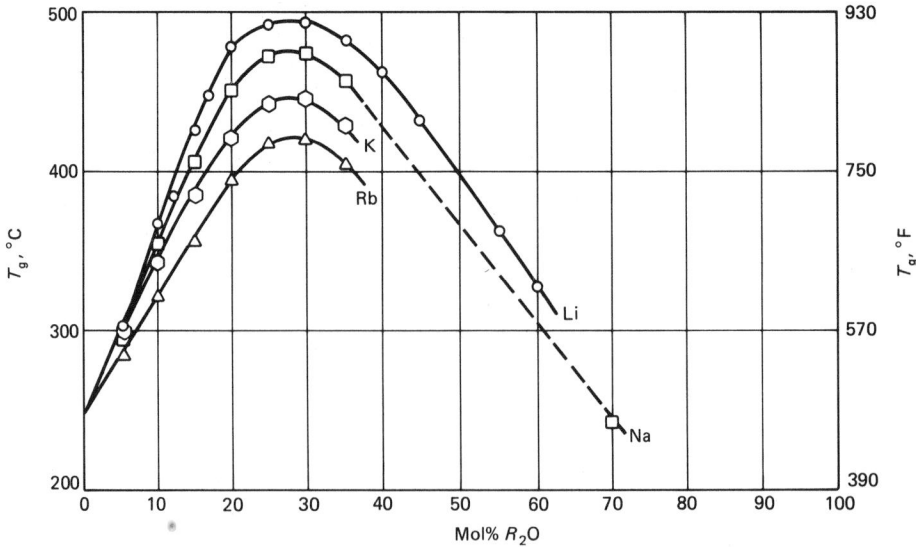

Fig 2 Effect of alkali-oxide concentration on the glass transition temperature of alkali-borate glasses. Source: Ref 2

decrease in T_g of over 100 K for a 1 mol% alkali oxide addition. The viscosity passes through a minimum at 2 to 5 mol% alkali oxide and then increases until it passes through a broad maximum at 15 to 20 mol% alkali oxide. The broad maximum is consistent with the general structural models for these glasses, which suggests that the maximum octahedral germanium concentration occurs in this composition range, with increasing formation of nonbridging oxygen ions for greater additions of alkali oxide. Nothing in the structural models for germanate glasses currently proposed explains the minimum in viscosity at around 2 mol% alkali oxide, that is, current structural models are inadequate to explain the details of the behavior of the alkali-germanate glasses.

Many of the glasses of interest for high-technology applications exhibit very steep viscosity-temperature curves. These materials form very fluid melts. As a result, melting times are measured in minutes for small batches. These melts are quite fluid at the liquidus and therefore are marginal glass formers, compared to the traditional silicate glasses. These materials include such diverse systems as the aluminates, galliates, and rare-earth aluminosilicates. The glass transformation temperatures of the glass formed from these melts can be very high, with values above 900 °C (1650 °F) for the rare-earth aluminosilicates.

The partial replacement of oxygen by a halogen results in a decrease in viscosity. Thus, the oxyfluoride glasses exhibit viscosities that are less than those of the corresponding oxide glasses.

Halide glasses are quite fluid and low melting, compared to silicate glasses. The shape of the viscosity-temperature curve is quite different from that observed for commercial silicate glasses. At low temperatures, the viscosity behavior is essentially Arrhenian, with activation energies that range from 600 to 800 kJ/mol (260 to 344 × 10³ Btu/lb · mol). The slope of these curves changes sharply in the region around 10^5 Pa · s, with a very small temperature dependence (as low as 70 kJ/mol, or 30 × 10³ Btu/lb · mol, for fluorozirconate melts) in the melt region. Although the Fulcher equation is often used to describe the temperature dependence of the viscosity of these materials, it has been shown that the Cohen-Grest model provides a better fit to the total viscosity-temperature curve.

The glass transformation temperatures of BeF_2 (250 °C, or 480 °F) and the fluorozirconate glasses (250 to 310 °C, or 480 to 590 °F) are quite low. The glass transformation temperatures of other glasses, such as the fluorozircohafnates, thorium fluoride based glasses, and aluminum fluoride-based glasses, can exceed 400 °C (750 °F), with the highest values approximating 450 °C (840 °F). In contrast, cadmium fluoride-based glasses can have T_g values as low as 110 °C (230 °F). The replacement of fluorine by other halogens

decreases the T_g of otherwise identical glasses in the order F > Cl > Br > I.

The addition of alkali fluorides to fluorozirconate and thorium fluoride-based compositions decreases the viscosity and the T_g. The activation energy for viscous flow also decreases with increasing alkali-fluoride concentration. The reduction in viscosity for corresponding compositions increases in the order Cs < K < Na < Li. Additions of trivalent and quadrivalent fluorides increase viscosity.

Glass transition temperatures of chalcogenide glasses can be even lower than those of halide glasses. Values range from 30 °C (85 °F) for pure Se, to 180 to 200 °C (355 to 390 °F) for As_2Se_3 and As_2S_3, to as high as 300 to 400 °C (570 to 750 °F) for glasses containing Ge, Si, As, and Se or S. Viscosities can be related to structure, with T_g increasing as the degree of cross-linking of the Se or S chains increases. The temperature dependence of viscosity varies in the same manner.

The T_g varies with the average coordination number R in a predictable way. Figure 3 shows the variation in T_g for As-S and As-Se glasses R between 2 (pure Se and S, that is, a chain structure) and $R = 2.4$ (40% As, that is, a layered structure). Note the break in the curves at the 40% As composition. Beyond this value, the glasses are chalcogen deficient, which leads to discontinuities in the property curves.

The discussion above is based on the assumption that the melt is homogeneous. In reality, many glass-forming melts exhibit immiscibility, which can lead to the complete separation of very fluid melts into two layers, each with the composition of one of the immiscible liquids. More commonly, the immiscibility is of the metastable type, which means that the melt is homogeneous at low viscosities and only separates into two liquids during cooling through the glass transition range. The microstructures resulting from this process consist of either droplets of one liquid dispersed in a matrix of the other or two continuous interconnected phases.

Measurement of the viscosity of these materials within the transition region can be dominated by the particular microstructure and

compositions of the two liquids in the sample. This effect is particularly evident for silicate systems, such as the lithium and sodium silicate and borosilicate systems, in which the viscosities of the two liquids differ radically. Phase separation actually dominates the measurement of T_g, so that it may appear to be independent of glass composition. It is also possible to alter the viscosity of a bulk sample by several orders of magnitude by varying the thermal history in such a way that the connectivity of the high-viscosity liquid changes from a droplet to a connected morphology, or vice versa.

Mechanical Properties

The elastic modulus, *E*, of a glass determines how the material responds to a particular stress. On a macroscopic scale, this response is due to the combined response of the individual bonds of the material. Therefore, to a first approximation, the moduli of homogeneous, single-phase glasses depend on the bond type, concentration of each type, and the network structure.

Glasses that have a chain structure generally have low moduli. Although primary intrachain bonds can be strong, interchain bonds are of the van der Waals type, or weak ionic links through network modifiers. Glasses with layered structures also have moduli that are lower than those of the more three-dimensional silicate glasses, because of weak bonding between layers. The concentration of nonbridging oxygens is important in three-dimensional, random network glasses, such as silicates, aluminosilicates, borosilicates, and soda-lime silicates. The type of modifier and intermediates present can also influence the value of the moduli.

The presence of Al_2O_3, which has a high modulus (380 GPa, or 55 × 10⁶ psi) in its crystalline state, also leads to high-moduli glasses. Glasses that contain Al_2O_3 also tend to have low nonbridging oxygen contents. In a similar manner, glasses containing Mg^{2+} or Be^{2+}, rather than Na^+ as a modifier have higher moduli. Special glasses developed to reinforce composites can have moduli as high as 140 GPa (20 × 10⁶ psi) (Table 2). Note that the high-moduli glasses are low in SiO_2, although Si may be present as Si_3N_4.

Young's modulus of most commercial silicate glasses varies from about 50 GPa (7 × 10⁶ psi) for low-T_g glasses to about 75 GPa (11 × 10⁶ psi) for fused SiO_2 and E-glass fibers. A special glass fiber (S glass, a magnesium aluminosilicate composition) has a modulus ranging from about 85 to 90 GPa (12 to 13 × 10⁶ psi). Other high-T_g aluminosilicate glasses and glass-ceramics also have Young's moduli of about 90 GPa (13 × 10⁶ psi).

Poisson's ration, ν, varies only slightly for multicomponent silicate glasses, ranging from about 0.2 to 0.3. Pure vitreous silica, which has a Poisson's ratio of 0.17, is an exception

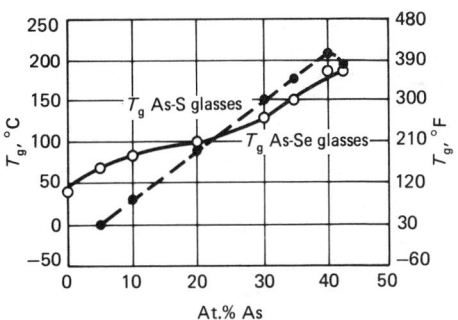

Fig 3 Effect of As concentration on the glass transition temperature of As-Se and As-S glasses

Table 2 Mechanical properties of glasses

Glass	Young's modulus GPa	Young's modulus 10^6 psi	Poisson's ratio	Surface energy N/m	Surface energy lbf/ft	Fracture toughness MPa$\sqrt{m}$	Fracture toughness ksi$\sqrt{in.}$
SiO_2	73	10.6	0.17	4.4	0.30	0.79	0.72
GeO_2	43	6.2	0.21	...	...	...	
B_2O_3	17	2.5	0.26	...	...	...	...
Se	10	1.5	0.33	...	...	...	...
As_2Se_3	17	2.5	0.29	...	...	...	...
$Ge_3As_4Se_3$	29	4.2	0.26	...	...	...	...
Ge-As-Se	18	2.6	...	...	...	0.25	0.23
$PbO-4B_2O_3$	60	8.7	0.26	...	...	...	...
Aluminosilicate	75	10.8	...	4.0	0.27	0.91	0.83
Borosilicate	60	8.7	...	4.6	0.31	0.77	0.70
Soda-lime-silica	66	9.6	...	3.9	0.26	0.75	0.68
$20La_2O_3-30Al_2O_3-50SiO_2$	100	14.5	...	...	...	...	...
$37.5Y_2O_3-8.6Al_2O_3-$ $37.5Si_3N_4-6.4SiO_2$	183	26.5	0.28	...	...	...	...

to this rule. The high-moduli glasses also have relatively high Poisson's ratios of 0.28. Because glasses are isotropic, the shear and bulk moduli can be calculated from the usual formulas.

Silica-free glass compositions, such as the phosphates, can have linear chain structures in which phosphate chains are bonded via alkali or alkaline earth cations. The weak bonding between chains leads to lower values of elastic modulus and higher values of v.

Pure B_2O_3 has a more two-dimensional structure, with boron forming planar triangular polyhedra with oxygen. This structure also leads to low elastic modulus and higher v values. As alkali and/or some divalent ions are added to boric oxide, the structure becomes more three dimensional, with boron in tetrahedral coordination. As a result, the moduli increase. In the lead-zinc borate system, which is important in electronic applications, the modulus can reach about 90 GPa (13×10^6 psi). Chalcogenide glasses can have chain, layered, or three-dimensional structures, and the moduli vary accordingly, as shown in Table 2. Halide glasses also have moduli that vary over large ranges.

Fracture toughness values for glasses are related to elastic modulus and fracture surface energy. Table 2 lists values for several commercial glasses. Note that the fracture toughness values do not vary much for commercial glasses.

The strength of a flawless glass in vacuum is determined by the strength of the chemical bonds. Calculated strength values are approximately 12 GPa (2×10^6 psi) for pure SiO_2. These high strengths are rarely attained in practice, because of the existence of surface flaws.

Two equations can be used to describe the influence of flaws on strength. Griffith's equation relates the observed fracture stress under tension to the existence of semi-elliptical cracks in the surface. The Inglis equation relates the stress at the tip of an elliptical flaw to the applied stress, the flaw length, and the radius of curvature at the crack tip.

Failure occurs when the stress at the crack tip reaches the strength of the chemical bonds in the crack-tip region. Long, sharp cracks are more detrimental to strength than short, rounded cracks.

As shown previously, the elastic modulus for most silicate glasses varies by only about a factor of 2. Similar variations are expected for the fracture surface energy. On the other hand, the crack length can vary from a few nanometers to 1000 μm (40 mils) or more. Therefore, strength variations that are due to differences in surface flaws often overwhelm compositional effects. Additional complications occur because of environmentally assisted crack growth under stress. For the same glass, the crack length (and, possibly, the radius of curvature at the crack tip) will change with time under stress at a rate that depends on the stress, the test environment, and the glass composition. Strengths in water are less than those tested in air. Humidity variations can also play a role in determining strength. As a result of these complications, few studies of compositional effects on strength have been carried out.

As might be expected from the Griffith and Inglis equations previously mentioned, and as demonstrated by the data of Table 3, samples tested under similar conditions fracture at quite different stresses. The type of test used is very important. Tests that tend to place the highest stress on the largest area produce the most reasonable values for strength. Thus, either three-point bend tests on bars or ball-on-ring tests on plates tend to yield higher average strengths with much lower Weibull moduli than do four-point bend tests (bars) or ring-on-ring tests (plates). Samples tested in pure tension (rarely done, except for fibers) show the lowest average strengths.

Of the silicate glasses, pure SiO_2 has the highest strength. Values near 10 GPa (1.5×10^6 psi) are often found for pristine optical fibers. On the other hand, pristine E-glass fibers, which are typically used in reinforcement applications, have strengths approximating 3.0 GPa (0.4×10^6 psi). Although it

might be expected that glasses with high Young's moduli would be stronger than those with low moduli, this is not necessarily the case.

Glass also exhibits a time dependence of strength. Samples with flaws that are not large enough to satisfy the Griffith equation may initially support an applied stress. However, in most glasses, the flaw geometry can change with time under stress. Flaws elongate until the Griffith equation is satisfied by the applied stress. The crack velocity, V, is given by

$$V = A\sigma_{tip}^n \qquad \text{(Eq 3)}$$

where σ_{tip} is the stress at the crack tip. The parameter n is material and atmosphere dependent and describes the sensitivity of the material to slow crack growth in a particular environment. The time to failure, t_f, at a given applied stress is given by

$$t_f = B\sigma_a^{-n} \qquad \text{(Eq 4)}$$

where B depends on the temperature and initial flaw geometry and σ_a is the applied stress.

Although it is not obvious from Eq 4, a high value of n is desirable, because it implies that once a crack begins to propagate, it will reach the critical value for failure. The time to failure is thus predictable. On the other hand, small values of n mean that cracks can grow for long periods of time prior to failure. Values of n range from 12 to 18 for soda-lime silicate glasses in water to 35 to 40 silica.

Tempered Glass. Glass strengths can be increased by placing surface flaws under compressive stresses using one of two methods. In thermal tempering, the glass is heated into the transition range and is then cooled rapidly. The differential cooling experienced by the surface and bulk glass causes the surface to be pulled into compression by the interior glass, which is, therefore, in tension. Compressive stresses ranging from 100 to 300 MPa (15 to 45 ksi) can be generated.

In ion-exchange strengthening, a glass that contains a small mobile ion, such as Li^+ or Na^+, is placed in a molten salt bath containing larger ions, such as K^+. The large ions from the bath replace the smaller ones from the glass, resulting in an increased volume in the surface. This places the surface under a compressive stress.

The strength of surface-compressed glasses is greater than that of normal glass by an amount approximately equal to the surface compression. Strengths of over 700 MPa (100 ksi) are attainable by ion exchange, and values near 400 MPa (60 ksi) are reached by thermal tempering. Thermal tempering is generally preferred to ion-exchange strengthening, because it is a rapid process that produces a thick (mm) surface compression layer in minutes. The ion-exchange method requires many hours to develop even a 25 to 50 μm (1 to 2 mil) thick layer.

Thermally tempered glass also has the ability to "dice" upon fracture, that is, breaks into numerous harmless pieces, rather than a

Table 3 Observed strengths of silicate glasses

Glass	Diameter		Surface	Temperature		Atmosphere	Load rate		Observed strength					
									Minimum		Mean		Maximum	
	mm	in.		°C	°F		mm/s	in./min	GPa	10^6 psi	GPa	10^6 psi	GPa	10^6 psi
SiO$_2$	1	0.04	Pristine	25	77	Air	1.7	4.0	0.55	0.08	4.48	0.65	5.52	0.80
	0.03	0.001	Pristine	25	77	Air	1.7	4.0	2.76	0.40	5.86	0.85	9.65	1.40
	0.03	0.001	Pristine	25	77	Vacuum	1.7	4.0	4.14	0.60	7.59	1.1	11.0	1.60
Soda-lime	0.5	0.02	Pristine	25	77	Air	...	...	0.89	0.13	3.93	0.57	5.38	0.78
	0.5	0.02	After 1 h at 125 °C (255 °F)	25	77	Air	...	...	0.42	0.06	1.17	0.17	1.72	0.25
7740	0.5	0.02	Pristine	25	77	Air	...	...	0.69	0.10	2.93	0.43	4.69	0.68
	4	0.16	Etched	25	77	Air	...	...	1.21	0.18	1.73	0.25	2.62	0.38
Soda-lime	0.25	0.01	Abraded	25	77	H$_2$O gas	0.02	0.05	0.1	0.014	0.15	0.02	0.21	0.03
	0.25	0.01	Abraded	−170	−275	H$_2$O gas	0.02	0.05	0.07	0.010	0.09	0.013	0.10	0.015
E-glass	0.0075	0.0003	Pristine	−170	−275	N$_2$	0.17	0.39	5.0	0.72	5.79	0.84	6.07	0.88
	0.0075	0.0003	Pristine	200	390	N$_2$	0.17	0.39	1.38	0.20	1.52	0.22	1.72	0.25
Soda-lime	1.3	0.05	Slight abrasion	25	77	Air	...	...	0.35	0.05	0.45	0.07	0.55	0.08
	0.7	0.03	Etched	25	77	Air	...	...	1.73	0.25	2.14	0.31	2.62	0.38
	Rect.	Rect.	Abraded	25	77	Air	(a)	(a)	0.04	0.007	0.05	0.008	0.06	0.008
	Rect.	Rect.	Abraded	25	77	Air	(b)	(b)	0.03	0.004	0.03	0.005	0.04	0.006
	Rect.	Rect.	Abraded	−170	−275	N$_2$	(a)	(a)	0.05	0.007	0.12	0.018	0.19	0.027

(a) 5.5 MPa/s (0.80 ksi/s) (b) 6.9 kPa/s (1 psi/s)

few sharp shards. The ion-exchange method does not generally yield materials that dice. However, the limits of thermal tempering are that neither thin nor complex shapes can be treated and optical quality is often reduced.

Hardness. Unlike metals, in which hardness can be varied over wide ranges, either by heat treatments or inclusion of second-phase particles, the hardness of silicate glasses does not vary dramatically. The Vickers microindentation hardness of pure SiO$_2$ is 6.0 GPa (0.9 × 10^6 psi), and the values for most silicate glasses are within about 30% of this value. Somewhat softer glasses are obtained by substituting B$_2$O$_3$, PbO, and/or alkaline earth and alkali ions for SiO$_2$. Harder glasses are generally obtained when they are alkali and alkaline earth free. Additions of Y$_2$O$_3$ or Al$_2$O$_3$ increase hardness to over 7.0 GPa (1.0 × 10^6 psi) (Table 4).

The highest hardness values are obtained on glasses that contain nitrogen. Replacing oxygen with nitrogen, either by the reaction of glass with N$_2$ at high temperatures or by co-melting oxides and nitrides, such as Si$_3$N$_4$, is extremely effective for increasing hardness.

Nonsilicate glasses are generally much softer. The open structure of B$_2$O$_3$ is characterized by a hardness of about 2.0, whereas the more compact lead-borate and alkali-borate glasses are much harder, because of the conversion of boron from three-fold to four-fold coordination. The hardness of chalcogenide glasses ranges from 0.3 GPa (0.04 × 10^6 psi) for Se (chain structure) to 1.4 GPa (0.20 × 10^6 psi) for As$_2$Se$_3$ (two-dimensional) to more than 2.0 GPa (0.3 × 10^6 psi) for Ge-As-S glasses.

Thermal Shock. The maximum thermal stress that can be generated in a glass due to a temperature difference can be estimated from

$$\sigma = \frac{\Delta T \alpha E}{1 - \nu} \qquad \text{(Eq 5)}$$

Table 4 Hardness of glasses

Glass	Hardness	
	GPa	10^6 psi
SiO$_2$	6.2(a)	0.90(a)
	4.7(b)	0.68(b)
GeO$_2$	2.4(b)	0.35(b)
B$_2$O$_3$	2.0(a)	0.29(a)
Soda-lime-silica	4.5(b)	0.65(b)
12.5Na$_2$O-17.5CaO-70SiO$_2$	5.5(b)	0.80(b)
37MgO-13Al$_2$O$_3$-50SiO$_2$	6.6(a)	0.95(a)
18Y$_2$O$_3$-24Al$_2$O$_3$-58SiO$_2$	8.1(a)	1.17(a)
37.5Y$_2$O$_3$-18.6Al$_2$O$_3$-37.5Si$_3$N$_4$-6.4SiO$_2$	11.4(b)	1.65(b)
30Na$_2$O-70B$_2$O$_3$	4.7(b)	0.68(b)
Sodium borosilicate	4.1(b)	0.59(b)
10BaF$_2$-30ZnF$_2$-30YF$_3$-30ThF$_4$	3.0(a)	0.44(a)
57ZrF$_4$-36BaF$_2$-3LaF$_3$-4AlF$_3$	2.5(a)	0.36(a)
Se	0.3(a)	0.04(a)
As$_2$Se$_3$	1.3(a)	0.19(a)
Ge$_25$Se$_{75}$	1.9(a)	0.28(a)
Ge$_{40}$As$_{15}$S$_{45}$	2.6(a)	0.38(a)
Ge$_{30}$Sb$_{10}$Se$_{60}$	1.9(a)	0.28(a)

(a) Vickers. (b) Knoop

where ΔT is the temperature difference and α is the linear thermal expansion coefficient. A 100 K temperature difference within vitreous silica ($\alpha = 0.5 \times 10^{-6}$/K, $E = 70$ GPa, or 10×10^6 psi, and $\nu = 0.16$) will create less than 10 MPa (1.5 ksi) of stress. The same temperature difference will produce about 22 MPa (3 ksi) of stress in a low-expansion borosilicate glass. However, the same temperature difference will produce from 75 to 80 MPa (11 to 12 ksi) of stress in a typical soda-lime silicate glass, which is often sufficient to cause catastrophic failure.

Electrical Properties

The electrical conductivity of most common oxide glasses results from ionic transport of monovalent cations, whereas the electrical conductivity of some halide and oxyhalide glasses is controlled by the ionic transport of monovalent anions. Electrical conductivities can vary by many orders of magnitude, from those of glasses that are effectively free of monovalent ions and act as excellent resistors to those of glasses that contain very large concentrations of silver, which exhibit conductivities similar to those of many metals. Electrical conductivity is also strongly affected by phase separation, especially when the connectivity of the more-conductive glass is altered by small changes in either glass composition or thermal history.

Glasses that are free of monovalent ions exhibit very low electrical conductivities. These glasses obviously include those consisting solely of a glass-former oxide, such as vitreous silica and vitreous boric oxide. They also include such diverse glasses as the alkaline earth silicates, aluminates, aluminosilicates, lead silicates, borates, borosilicates, calcium aluminoborates, rare-earth aluminosilicates, aluminoborates, and aluminogermanates, and essentially any other systems that are free of monovalent cations and anions.

The addition of alkali oxides to any of the glass formers results in a dramatic increase in the cationic conductivity of the glass (Fig 4). The conductivity of alkali silicate and borate glasses increases monotonically as the concentration of the alkali oxide increases. The effect of phase separation is superimposed on these trends, as indicated by the sharp change in the conductivity of the sodium-silicate glasses between 4 and 8 mol% Na$_2$O, as shown in Fig 4. The conductivity is often found to be greatest for glasses containing sodium, with conductivity decreasing in the order Na > Li > K > Rb > Cs. Although one might expect that the small Li$^+$ ion would be more mobile than the somewhat larger Na$^+$ ion, the increase in the binding energy between the Li$^+$ ion and its neighboring nonbridging oxygen ion par-

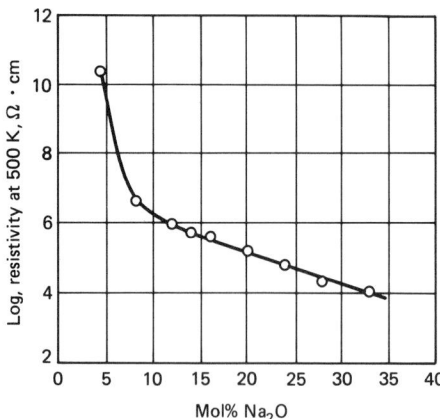

Fig 4 Effect of sodium-oxide concentration on the electrical conductivity of sodium-silicate glasses

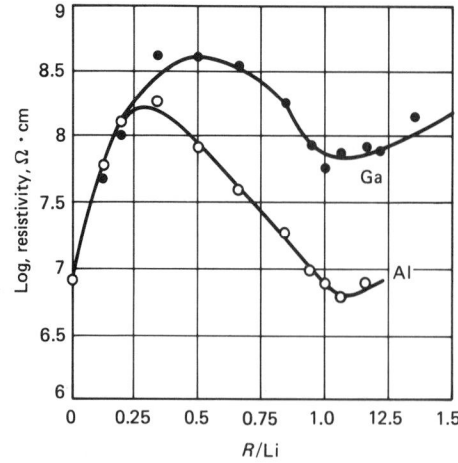

Fig 5 Effect of intermediate/alkali ratio on the electrical resistivity of lithium-aluminosilicate and lithium-galliosilicate glasses; $T = 100\ °C\ (212\ °F)$. Source: Ref 3

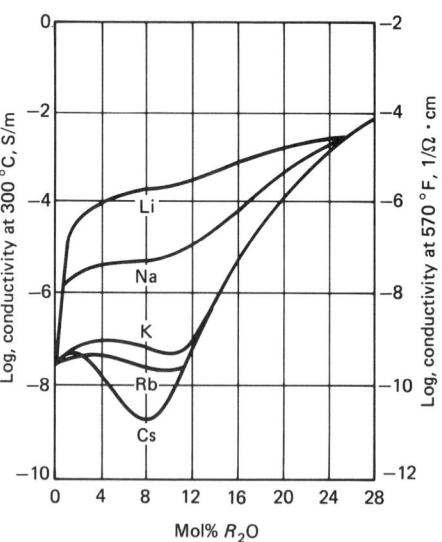

Fig 6 Effect of alkali-oxide concentration on the electrical conductivity of alkali-germanate glasses

tially offsets the lower energy needed to move the ion through the vitreous network.

These and other observations have led to various modifications of the Anderson and Stuart model for the activation energy for conduction. This model contains two terms, the first of which involves the binding energy of the ion to its equilibrium site in the vitreous structure. The second term involves the energy needed to dilate the doorway between adjacent interstices sufficiently to allow passage of the monovalent ion. Apparently, the balance between these two terms is such that a maximum in mobility is observed for sodium ions when the conductivity is plotted against the radius of the migrating species. In fact, this maximum does not actually occur for the sodium ion when all monovalent ions are considered. The silver ion, which is intermediate in size between sodium and potassium ions, actually appears to be the most mobile in many glass-forming systems.

The addition of an intermediate ion, such as aluminum or gallium, to an alkali-silicate glass results in a complicated behavior for the electrical resistivity as a function of the intermediate/alkali ratio in the glass (Fig 5). The addition of the intermediate eliminates the nonbridging oxygen ions from the structure. The alkali ions, which were all associated with the nonbridging oxygen ions, are now associated with the $(RO_4)^{5-}$ tetrahedra. Because the alkali ion associated with the intermediate ion/oxygen group is less localized than one associated with a nonbridging oxygen, the binding energy decreases. This effect is used to explain the high conductivity for glasses containing equimolar concentrations of alkali and aluminum ions. Unfortunately, arguments explaining the overall curve, including the initial decrease in conductivity upon the addition of a small amount of alumina to these glasses, are not universally accepted.

Any consideration of the conductivity of glasses containing alkali ions must include the infamous "mixed-alkali" effect. Many glass properties can be explained by simple addi-

tive behavior, that is, glasses can often be treated as almost ideal solid solutions. The electrical conductivity of glasses containing more than one type of alkali ion can definitely not be considered to be additive. In fact, simply varying the concentration ratio of the two types of alkali ions present in a glass, where one might expect a linear variation in conductivity from that of one single alkali glass to that of the other single alkali glass, can often result in a large minimum (several orders of magnitude) in electrical conductivity. This effect is observed in all glass-forming systems, regardless of the identity of the glass former or of the presence of other ions. To date, at least a dozen or more models have been offered to explain this behavior, although not one of these models has found general acceptance.

The electrical conductivity of the alkali-germanate glasses exhibits trends that are somewhat different from those of the corresponding silicate and borate glasses. The initial addition (1 mol%) of either lithium or sodium oxide to germania results in an increase in conductivity of two to four orders of magnitude. A similar addition of potassium, rubidium, or cesium oxide yields very little change in conductivity (Fig 6). At high alkali oxide contents (above 25 to 30 mol%), the conductivity of these glasses is almost independent of the identity of the alkali ion. This behavior is very different from that observed for many other properties of these glasses and has yet to be explained.

The electrical conductivity of oxyhalide glasses containing monovalent cations, such as sodium fluoroborate or fluorogermanate glasses, is almost unaffected by the presence of the monovalent fluorine ion. Although there is a small minimum in electrical conductivity with increasing fluorine content, the effect is not particularly striking. On the other hand, the replacement of PbO by PbF_2 in the lead

fluorosilicate, fluoroborate, and fluorogermanate systems results in increases of as much as eight orders of magnitude in electrical conductivity (Fig 7).

Similar effects have been reported for alkali-free phosphate glasses containing halogen ions. The behavior of fluorine appears to be somewhat different from that of the other halogens (Fig 8) in every system where other halogens have been studied. The existence of a true "mixed anion" effect has been reported for lead-halosilicate glasses containing mixtures of any two of the four halogen ions. Note that the term mixed anion effect is often used erroneously to describe the effect of the presence of two different anions in a glass on cationic conductivity.

The electrical conductivity of fluoride glasses appears to be the direct result of fluorine diffusion. An increase in the nonbridging fluorine concentration slightly increases

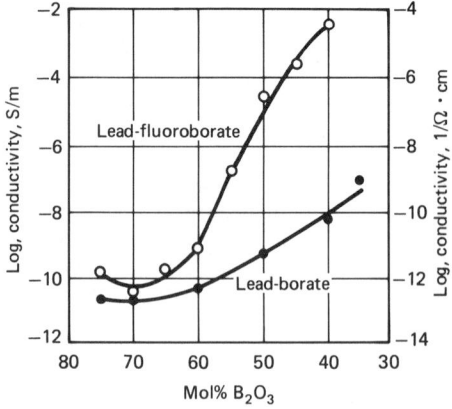

Fig 7 Effect of boric-oxide concentration on the electrical conductivity of lead-borate and lead-fluoroborate glasses. Source: Ref 4

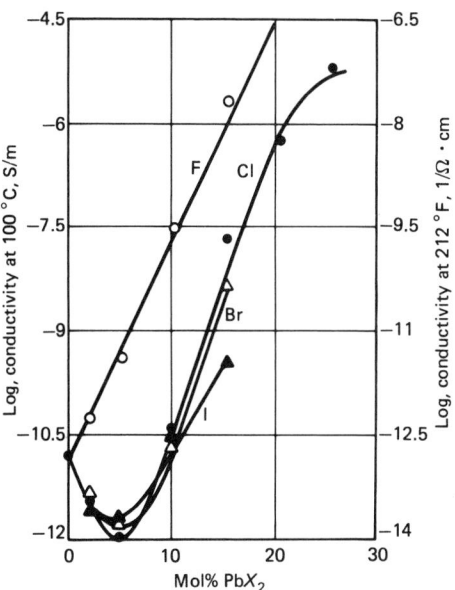

Fig 8 Effect of lead-halide concentration on the electrical conductivity of lead-halosilicate glasses. Source: Ref 5

Table 5 Color commonly produced by various transition and rare-earth elements

Element	Color
Titanium	Yellow
Vanadium	Green
Chromium	Green
Manganese	Purple
Iron	Yellow-green/blue-green
Cobalt	Blue
Nickel	Yellow/brown/purple
Copper	Green/blue
Cerium	Yellow/brown
Praseodymium	Green
Neodymium	Purple-red/blue
Samarium	Pale yellow
Europium	Dark yellow
Terbium	Very pale yellow
Dysprosium	Yellow
Holmium	Peach
Erbium	Pink
Thulium	Green

the electrical conductivity of these glasses. The addition of small amounts of alkali fluorides to fluorozirconate and thorium fluoride-based glasses actually decreases the electrical conductivity of these glasses. Interestingly, the electrical conductivity of the oxygen-free halogen glasses is significantly less than that observed for the lead fluorosilicate, borate, and germanate glasses discussed above.

Optical Properties

The presence of first-row transition and rare-earth ions is by far the most common cause of color in glasses. Table 5 lists the most common colors that result from the common coloring cations. The optical absorption bands that result from the presence of these ions are a direct result of electronic transitions within the 3d and 4f levels, respectively. These transitions result from the splitting of these levels as a result of the electric field generated by the surrounding anions. The degree of splitting is primarily determined by the coordination number of the cation, the valence of the cation, and the field strength of the surrounding anions. A Co^{2+} ion, for example, will yield a different set of absorption bands if it is in tetrahedral coordination from the set of bands that occur if it is in octahedral coordination. The spectrum that is due to Fe^{2+} ions will differ from that which is due to Fe^{3+} ions, even though the matrix material remains the same. Changes in coordination number most frequently result from changes in the glass structure, whereas changes in the valence state of the cation causing the coloration usually result from changes in the redox state of the glass.

Changes in the identity of the surrounding anions will change the degree of splitting of the levels, at least for the less-shielded 3d ions, which results in a shift in the position of the absorption bands. The greater shielding of the 4f electrons in the rare-earth ions results in an almost composition-independent position of the absorption bands. In general, decreases in the field strength of the anions will shift the absorption bands toward the infrared. The anions surrounding the coloring cation need not be the majority anion in the glass. The addition of very small concentrations of halide ions to an oxide glass will result in the complete association of the coloring cation with the halide in preference to oxygen ions. Thus, the addition of a very small amount of NaCl to a sodium-borate glass containing a small amount of Co^{2+} will change the characteristic cobalt blue color of this glass to a color with strong green contributions.

Other ions can also cause colors in glasses. Hexavalent uranium, for example, has been used to produce a characteristic yellow-green color. Other transition metals, such as platinum and rhodium, can cause colors in glasses under special conditions. These colors are usually a result of improper melting and are not normally desirable.

The presence of metallic colloids distributed in a glass can lead to coloration by either Mie or Rayleigh scattering processes. The Mie scattering, or plasma resonance, process is responsible for the red color of gold-ruby glasses and the yellow color of silver-stained glasses. These glasses are melted in such a manner that the gold or silver is present as ions, which do not impart any color to the glass. Lowering the temperature into the transition region results in the reduction of the gold or silver ions to the metallic state. The atoms then agglomerate into small (10 to 30 nm, or 0.4 to 1.2 μin.) spherical colloids, which exhibit characteristic absorption bands within the visible region of the spectrum. The positions of these bands are a function of the

refractive index of the glass and the shape of the colloids. It is therefore possible to vary the color produced by these colloids, either by processing variations that lead to elliptical colloids or by large changes in the refractive index of the glass.

Silver colloids can also be formed in the outer few micrometers of a glass surface by a combination of diffusion and reduction processes. The base glass is free of silver but contains a significant amount of sodium. A source of silver is applied to the surface and the glass is heated to allow the silver to exchange with the sodium in the glass. Reduction of the silver, either by reaction with ions in the glass, such as stannous ions in the surface of float glass, or by exposure to hydrogen gas, produces atomic silver, which agglomerates to form colloids.

If the colloids become too large, plasma resonance no longer occurs. In this case, the scattering is more aptly described by the Rayleigh formalism, and the scattering becomes dependent on the reciprocal of the fourth power of the wavelength of the incident light. This form of scattering essentially generates an optical loss curve, which exhibits a long tail extending from the ultraviolet into the visible wavelengths. This tail eventually overwhelms the absorption band, and the glasses become brown.

Only a few metals interact with light via the plasma resonance mechanism to yield specific absorption bands. Most colloids only exhibit Rayleigh scattering, even when quite small. The $1/\lambda^4$ dependence of the scattering curve produces yellow-brown to black glasses for other colloids, for example, lead, arsenic, antimony, bismuth, and other metals. This process is usually the result of the exposure of glasses containing the desired ion to hydrogen gas. Because the depth of penetration of the hydrogen, and hence the subsequent formation of the colloids, is a diffusion-controlled process, it is possible to form a strongly colored surface region while the interior of the glass remains colorless.

The IR transmission of glasses has recently become of major importance in the areas of optical fibers for communication systems and bulk optical devices operating in the IR region of the spectrum. For some applications, such as communication optical fibers, the presence of hydroxyl in the glass and the absorption bands that result from this hydroxyl control the use of the material. For other applications, such as IR windows, only the location of the IR cutoff (multiphonon edge) is important.

Hydroxyl is the most commonly occurring impurity in inorganic glasses. The primary absorption bands that are due to hydroxyl occur between 2.5 and 5.0 μm, with overtone bands throughout the near-infrared. The location and shape of the absorption bands are a function of the composition of the glass and the local environment of the hydroxyl within the structure.

The multiphonon edge is a function of the strength of the bonds in the glass, that is, the force constant, and the mass of the ions involved. Glasses that consist of relatively light ions, with strong interionic bonds, such as borates and silicates, exhibit IR edges below 5 μm. The replacement of silicon with the heavier germanium ion results in a shift of about 1 μm to an IR edge of about 6 μm. The most IR-transparent oxide glasses are obtained by using heavy ions, which also typically exhibit relatively weak bonding. Thus, the lead galliates exhibit IR edges as great as 7.8 μm. Other glasses containing heavy metals, such as lead germanates and bismuthates, and rare earth galliogermanates, also exhibit IR edges beyond 6 μm.

The replacement of oxygen with a halogen reduces the force constant and thus increases the wavelength for the IR edge, even though the mass of the halogen, that is, fluorine, may not be significantly different from that of oxygen. Thus, the replacement of a portion of the oxygen by fluorine in the various oxyfluoride glass-forming systems results in a slight shift in the IR edge to longer wavelengths. This effect is rather small, because the oxygen ions still outnumber the fluoride ions in these glasses.

A much more radical shift in the IR edge can be obtained by completely eliminating oxygen from the glass. The most common examples of such glasses are the zirconium-barium-lanthanum-aluminum fluoride (ZBLA) glasses, which routinely display infrared edges in the vicinity of 10 μm. Replacement of fluorine by the heavier, lower field strength chlorine, bromine, or iodine ions shifts the edge to even longer wavelengths, with reported values well beyond 15 μm. Unfortunately, the parameters that yield the large value for the IR edge wavelength lead to poor glass-forming behavior, low T_g values, and a strong tendency for the glass to dissolve in water.

The refractive index of oxide glasses varies from about 1.45 to more than 2.4. The addition of large, highly polarizable ions to a glass dramatically increases the refractive index, as does the conversion of bridging oxygen ions to nonbridging ions. Once again, glasses with strong bonds and highly connected networks (vitreous silica and boric oxide) exhibit low refractive indices, whereas heavy metal oxide glasses exhibit very high refractive indices.

The addition of any alkali oxide to either silica or boric oxide results in a monotonic increase in the refractive index. The refractive index of the alkali-silicate glasses increases in the order Na < K < Rb < Li < Cs. Because the refractive index depends not only on the polarizability of the ions, but also on the molar volume of the glass, the increase in the polarizability of these ions with increasing atomic number is partially offset by the changes in the molar volume. This is particularly striking for the lithium-silicate glasses, where the strong reduction in the molar volume of these glasses results in a much higher refractive

index than would be expected. These trends are shown in Fig 9 for the sodium-borate glasses.

The deviations from simple behavior as a function of composition observed for the alkali-borate glasses are much more pronounced for the alkali-germanate glasses. In fact, in these systems the refractive index actually passes through a maximum as a function of composition, first increasing and then decreasing with increasing alkali-oxide concentration (Fig 10). These trends are generally attributed to the competing effect of the change in coordination number of the germanium ion from 4 to 6, which produces a more dense structure and to the later formation of nonbridging oxygen ions. Additionally, the reconversion of germanium to fourfold coordination creates a less-dense structure (the densities of these glasses exhibit similar trends). Of course, because the formation of the more polarizable nonbridging oxygen ions should also increase the refractive index, there appears to be some question regarding the interpretation of these data. It should also be noted that the refractive index at higher alkali-oxide concentrations increases in the reverse order of the atomic number, that is, K < Na < Li. This effect is again attributed to the corresponding trend in the molar volume of the glasses.

The replacement of oxygen by fluorine to form oxyfluoride glasses results in a decrease in the refractive index. The partial replacement of oxygen by fluorine can even reverse the trend in the data, as shown in Fig 9. The refractive index of a high sodium content fluoroborate glass is actually less than that of vitreous boric oxide.

Although many of the halide glasses contain relatively large concentrations of heavy metals, they typically exhibit very modest refractive indices. The atomic compositions of most glasses consist of 60 to 80 at.% of anions. Typical ZBLA glasses, for example, contain over 70 at.% fluorine. Because fluorine has a relatively low polarizability for an

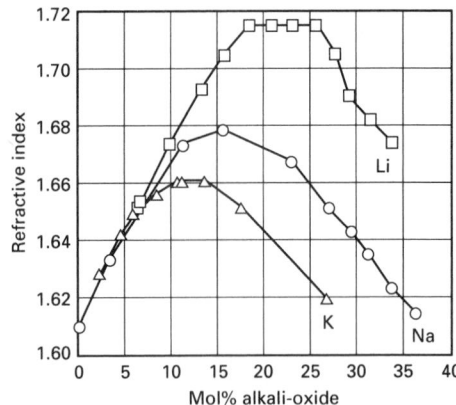

Fig 10 Effect of alkali-oxide concentration on the refractive index of alkali-germanate glasses

anion, these glasses exhibit moderate refractive indices (1.48 to 1.53), even though they contain large amounts of the heavy metals zirconium, barium, and lanthanum. The fluoroaluminate glasses, which contain lighter cations, exhibit significantly lower refractive indices in the range of 1.4 to 1.42, whereas BeF_2 has the lowest refractive index of any inorganic glass (1.28). Replacement of fluorine by chlorine, bromine, or iodine will significantly increase the refractive index, because of the increasing polarizability of these ions with increasing atomic number.

Magnetic Properties

In general, glasses are diamagnetic. The only exceptions are those glasses that contain transition metal, rare earth, and some transuranic ions. The behavior of the magnetic ion in those glasses controls their overall magnetic behavior. The magnetic susceptibility is controlled by the contribution of the specific paramagnetic ion present to the overall susceptibility of the glass (the bulk of the ions in typical glasses only contribute to the diamagnetic susceptibility). If the glass contains a high concentration of the paramagnetic ion, the paramagnetic contribution to the overall susceptibility is much greater than that of the diamagnetic contribution. The magnetic susceptibility of glasses that contain rare earth and/or transition metal ions is almost independent of the bulk composition of the glass. An example of the trend observed for the magnetic susceptibility for rare-earth aluminosilicate glasses is shown in Fig 11.

Although glasses are not frequently used specifically for their magnetic properties, they are often used in devices based on the Faraday effect. In this case, the interaction of the rare-earth ions with an external magnetic field results in a rotation of the plane of polarization of polarized light passing through the material. The angle of rotation is given by the product of the magnetic field strength, the optical path length within the glass, and a material constant known as the Verdet coefficient. Only the Verdet coefficient depends

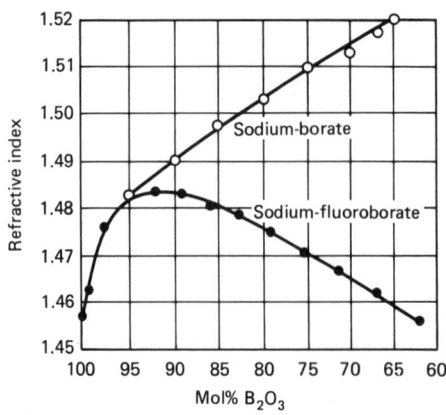

Fig 9 Effect of boric-oxide concentration on the refractive index of sodium-borate and sodium-fluoroborate glasses. Source: Ref 6

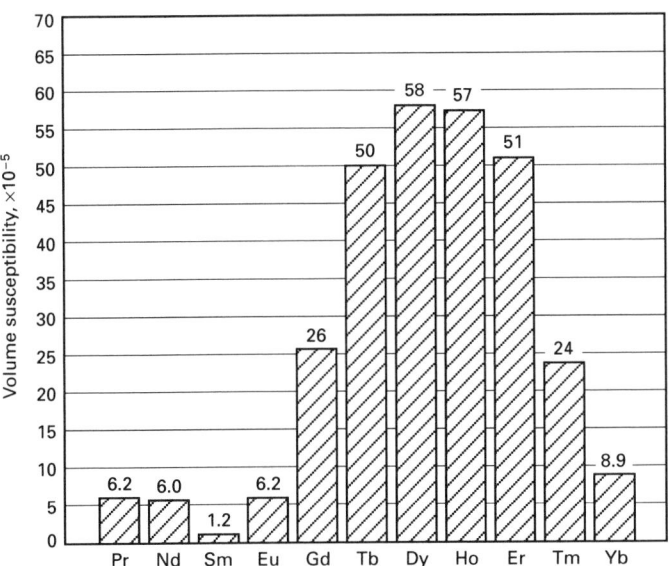

Fig 11 Effect of rare-earth identity on the magnetic susceptibility of rare-earth aluminosilicate glasses of the general formula $20R_2O_3$-$20Al_2O_3$-$60SiO_2$.
Source: Ref 7

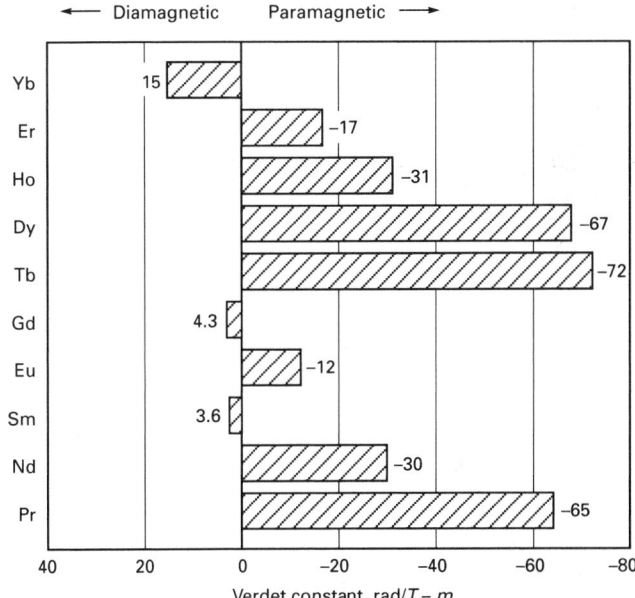

Fig 12 Effect of rare-earth identity on the Verdet constant of rare-earth aluminosilicate glasses of the general formula $20R_2O_3$-$20Al_2O_3$-$60SiO_2$. Source: Ref 7

on the glass composition. Once again, as in the case for magnetic susceptibility, the Verdet coefficient is more dependent on the specific properties of the rare-earth ion present than on the overall composition of the glass, so that no simple trends can be reported. It should also be noted that the Verdet coefficient is often highly dependent on the wavelength of the light. The Verdet coefficients of a series of rare-earth aluminosilicate glasses are shown in Fig 12.

Diamagnetic glasses can also exhibit the Faraday effect. In this case, the rotation of the light will be in the opposite sense from that of the so-called paramagnetic rotators. The Verdet coefficients of diamagnetic glasses are usually lower in absolute magnitude than those of highly paramagnetic glasses. Furthermore, while the Verdet coefficients of paramagnetic rotators are temperature dependent, those of diamagnetic rotators are usually temperature independent. It should also be noted that some glasses exist in which the magnetic susceptibility is paramagnetic, whereas the Verdet coefficient is diamagnetic.

Chemical Properties

Although volatilization of individual components during glass melting can be a major problem in producing many glasses, very few systematic studies have been published. The most volatile species include the alkali metals, the halogens, lead, and boron. The volatility of the alkali metals and the halogens increase with increasing atomic number. As a result, it can be very difficult to produce glasses that have, for example, well-controlled concentrations of cesium or iodine. Because volatilization losses are enhanced by

elevated temperatures, problems that are due to the high vapor pressure of these species are particularly troublesome in high-melting-temperature silicate glasses containing relatively small amounts of these ions. Volatilization losses are controlled by lowering the processing temperature, covering the melt to increase the vapor pressure of the volatile species directly over the melt, or by changing the composition of the melt by substituting other less-volatile components wherever possible.

The open network structure of many glasses results in relatively high diffusivities for many atomic, molecular, and ionic species. Because the diffusion of these different species is controlled by different aspects of the glass structure, it is common to discuss diffusion in terms of either extrinsic species, such as gases, or intrinsic species, such as alkali ions.

The most mobile gaseous species in any glass is always helium. Molecular species such as hydrogen do not dissociate at the surface of a glass and, hence, diffuse as molecules. The diffusion coefficient of gaseous species in glasses is a direct function of the size of the diffusing species. It is always observed that the diffusion coefficient decreases in the order $He > H_2 > Ne > Ar > Kr$ in any given glass. This observation has led to models for gas diffusion in glasses, which emphasize the size of the diffusing species relative to the size of the openings between the interstices in the vitreous network. Changes in glass composition that reduce the size of these openings, increase the energy required to dilate these openings, or fill the interstices with other ions reduce the diffusion coefficient for all gases.

The addition of alkali oxides to silica results

in a decrease in the diffusion coefficients of helium and hydrogen isotopes in those glasses. Although the trends are overwhelmed by the effects of phase separation for the lithium and sodium glasses, it can be stated that the diffusion coefficient for helium increases in the order $Li < Na < K = Rb = Cs$. Similar trends with alkali identity occur for the alkali-borate and alkali-germanate glasses, although the differences between the effects of the heavier alkali ions are more pronounced in the germanate glasses, where the helium diffusivity increases in the order $Li < Na < K < Rb < Cs$. However, in each of these systems, the more complex variation of glass structure with glass composition results in nonmonotonic trends in the helium diffusion coefficient with overall glass composition. Thus, for example, the helium diffusion coefficient passes through a minimum with increasing alkali oxide concentration for each of the alkali-germanate systems (Fig 13). These data clearly refute the often stated misconception that the diffusion coefficient for gases always decreases as the concentration of modifier ions increases.

The effect of intermediate ions on the diffusion coefficient of helium is more complex than is that of the modifier ions. The addition of alumina or gallia to silica initially leads to a decrease in the helium diffusion coefficient. This trend reverses as the intermediate-to-modifier concentration ratio increases, and eventually reaches a maximum when this ratio is near unity. The trend then, once again, reverses and the helium diffusion coefficient decreases with continuing increases in the concentration of the intermediate ion.

Very little is known regarding the diffusion of gases in halide and chalcogenide glasses. Efforts to make such measurements have not

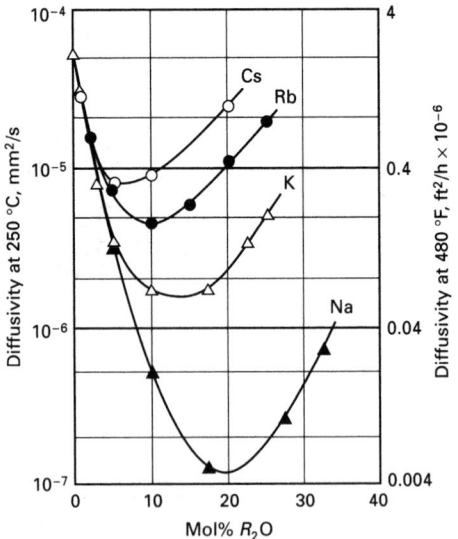

Fig 13 Effect of alkali-oxide concentration on the helium diffusivity at 250 °C (480 °F) of alkali-germanate glasses. Source: Ref 1

proved successful. Unfortunately, the techniques usually used for such measurements require that the glass exhibit a measurable permeability for the gas in question and cannot differentiate between materials where the low permeability is due to a very low diffusion coefficient and those where it is due to a very low solubility coefficient.

Study of the diffusion of cations in glasses has primarily focused on the monovalent ions for the very practical reason that the diffusion coefficients for ions having a valence number greater than 1 is usually very low. In general, the trends observed in the diffusion coefficients for these ions are identical to those discussed earlier for electrical conductivity. This is not surprising, because ionic diffusion is largely responsible for the electrical conductivity of many glasses. The reader should refer to the section "Electrical Properties" in this article for further information.

Chemical Durability. Chemical attack of glass surfaces occurs through either dealkalization or dissolution. Glasses that contain mobile ions, such as Na^+ or K^+, are subject to leaching processes in which the mobile ions in the surface region diffuse into the surrounding liquid. The alkali ions are replaced by protons, leading to the formation of Si—OH groups in the surface region via the reaction

$$(Si—O\ Na^+)_{glass} + H_2O =$$
$$(Si—OH)_{glass} + (NaOH)_{solution} \qquad (Eq\ 6)$$

Both the glass surface and the liquid composition are changed by this reaction. Dis-

solution of silicate glasses occurs via chemical attack by OH^- ions on the Si—O—Si bonds via the reaction

$$Si—O—Si + OH^- = Si—OH + Si—O^- \qquad (Eq\ 7)$$

Silica enters the solution as $Si(OH)_4$, which is known as silicic acid. Subsequent reaction of the $Si—O^-$ with water produces another OH^-, so that the reaction continues without depleting the OH^- concentration of the solution. Because the reaction depends directly on the concentration of hydroxyl, the rate of dissolution increases with increasing pH. Furthermore, the solubility of silica is only about 100 ppm at pH values below 7, but increases strongly above a pH equal to 8. It follows that most silicate glasses can be used at low pH but will dissolve in strong bases.

Several standard tests have been designed in an attempt to compare the durability of various glasses in solutions. These tests are summarized in Table 6. In general, these tests simply divide glasses into categories, with a low category number designating good durability.

Weathering of glasses occurs either when alkali is leached from the glass under high humidity or when the glass surface is exposed

Table 6 Durability tests

ISO 719 water resistance test

Sample: 2 g (0.07 oz.) powder; 300–500 μm (12–20 mil) diameter
Conditions: 50 mL water immersed in boiling water bath for 1 h
Measurement: alkali concentration by titration

Ratings:

Class	Alkali leached, μg/g	Interpretation
1	<31	Very high resistance
2	31–62	High resistance
3	62–264	Medium resistance
4	264–620	Low resistance

DIN acid resistance test

Sample: bulk glass of known geometry
Conditions: boiling 6 mol/L HCl solution for 6 h
Measurement: weight loss expressed as mg/100 cm²

Ratings:

Class	Half of weight loss	Interpretation
1	<0.7	Highly acid resistant
2	0.7–1.5	Acid resistant
3	1.5–15	Slight acid attack
4	>15	High acid attack

ISO 695 alkali resistance test

Sample: bulk sample of known geometry
Conditions: boiled in aqueous solution containing 1 mol/L of NaOH and 0.5 mol/L of Na_2CO_3 for 1 h
Measurement: weight loss expressed as mg/100 cm²

Ratings:

Class	Half of weight loss	Interpretation
1	<75	Low alkali attack
2	75–175	Slight alkali attack
3	>175	High alkali attack

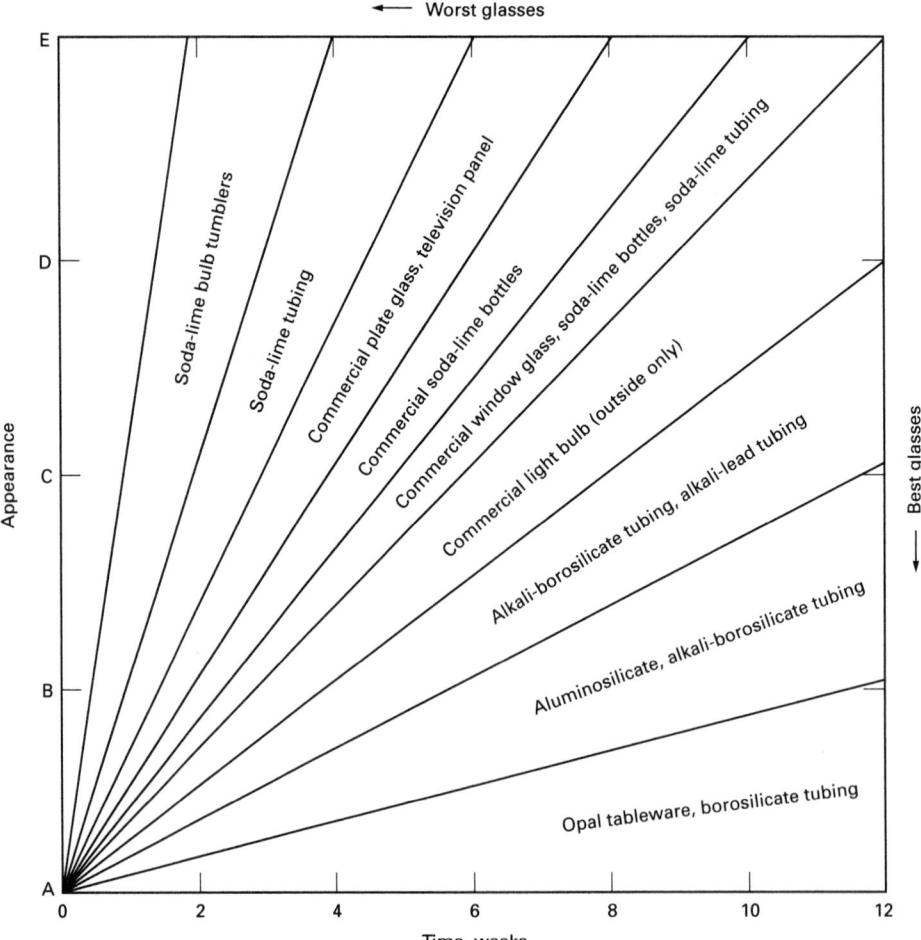

Fig 14 Average weatherability of various glasses under high humidity. Source: Ref 8

to alternating cycles of moisture condensation and evaporation. The alkali leached from the surface reacts with the water to form hydroxides and thus creates a very high pH environment on the surface of the glass. The high pH environment causes the silicate network to dissolve. Formation of visible deposits on the surface eventually reduces the transparency of the material.

The weathering resistance of several types of commercial glasses is summarized in Fig 14. Those that have good water resistance (ISO 169 test) also typically exhibit good weathering resistance. Because the leaching of alkali ions initiates the weathering process, glasses that are low in alkali generally offer good weathering resistance.

Vitreous silica is among the very best glass in terms of chemical durability. Even though significant chemical attack occurs at pH values greater than 8, silica is resistant to all acids, except hydrofluoric and hot phosphoric acids. Weathering of vitreous silica is negligible, because the glass is free of alkali ions.

Commercial borosilicate glasses exhibit excellent acid, water, and alkali resistance. However, borosilicate glasses that contain more than about 15 mol% boric oxide exhibit poor chemical durability in most environments. High-borate sealing glasses often rank in the lowest durability class in each type of test.

Few commercial glasses exhibit class 1 behavior for alkaline solutions. Although the addition of alkaline earths or alumina to glasses improves the alkaline-solution durability, glasses that contain alumina have poor acid resistance. Glasses that contain zirconia can be extremely resistant to both acid and alkaline solutions, with a commercial fiber composition used to reinforce cement, where the pH can exceed a value of 11 during curing.

REFERENCES

1. J.E. Shelby, Helium Migration in Alkali Germanate Glasses, *J. Appl. Phys.*, Vol 50 (No. 1), 1979, p 276–279
2. J.E. Shelby, Thermal Expansion of Alkali Borate Glasses, *J. Am. Ceram. Soc.*, Vol 66 (No. 3), 1983, p 225–227
3. J.L. Piguet and J.E. Shelby, Electrical Conductivity of Lithium Aluminosilicate, Galliosilicate, and Alumino/Galliosilicate Glasses, *Adv. Ceram. Mater.*, Vol 1 (No. 2), 1986, p 192–197
4. C.A. Gressler and J.E. Shelby, Lead Fluoroborate Glasses, *J. Appl. Phys.*, Vol 64 (No. 9), 1988, p 4450–4453
5. J. Coon and J.E. Shelby, Properties and Structure of Lead Halosilicate Glasses, *J. Am. Ceram. Soc.*, Vol 73 (No. 2), 1990, p 379–382
6. J.E. Shelby and L.K. Downie, Properties and Structure of Sodium Fluoroborate Glasses, *Phys. Chem. Glasses*, Vol 30 (No. 4), 1989, p 151–154
7. J.T. Kohli and J.E. Shelby, Magneto-Optical Properties of Rare Earth Aluminosilicate Glasses, *Phys. Chem. Glasses*, to be published
8. H.V. Walters and P.B. Adams, Effect of Humidity on Weathering of Glass, *J. Non-Cryst. Solids*, Vol 19, 1975, p 183–199

SELECTED REFERENCES

- O.V. Mazurin, M.V. Strel'tsina, and T.P. Shvaiko-Shvaikoskaya, *Properties of Glasses and Glass-Forming Melts*, Vol I-IVB, Izdatel'stvo Nauka, Leningrad, 1973–1981
- G.W. Morey, *The Properties of Glass*, Reinhold Publishing, 1938
- J.E. Shelby, Thermal Expansion of Alkali Borate Glasses, *J. Am. Ceram. Soc.*, Vol 66 (No. 3), 1983, p 225–227
- J.E. Shelby, Viscosity and Thermal Expansion of Alkali Germanate Glasses, *J. Am. Ceram. Soc.*, Vol 57 (No. 10), 1974, p 436–439
- J.E. Shelby, Property/Morphology Relations in Alkali Silicate Glasses, *J. Am. Ceram. Soc.*, Vol 66 (No. 11), 1983, p 754–757

Engineering Properties of Glass-Matrix Composites

R.A. Haber and R.M. Anderson, Rutgers, The State University of New Jersey

AS THE REQUIREMENTS for high-performance ceramics increase, the need to optimize material properties, such as strength and reliability, has placed greater emphasis on understanding the mechanisms that govern material behavior. Glass-matrix composites are high-performance materials that comprise a large portion of those ceramics that are technologically important (Ref 1).

Ceramic composites are composed of two or more phases that differ in physical and/or chemical properties. By this definition, the incorporation of any second phase, such as porosity or sintering aids, constitutes a composite. A more useful definition is to identify the additional phases as phases that are intentionally added to improve some property of the single-phase matrix. The resulting properties provided by the multiphase composite are often not attainable by single-phase materials (Ref 2–9).

Traditionally, glass-matrix composites have been classified as either dispersed-phase composites or fiber-reinforced composites. The basic requirement of the dispersed, or second, phase added to a matrix, whether it is a discrete particle or a void, is to improve the aggregate material properties, such as the coefficient of thermal expansion (CTE), dielectric constant, thermal shock resistance, strength, and toughness. Fiber-reinforced composites, in which either continuous fibers or discrete, chopped fibers are added to the matrix, were developed to provide low-density materials with both high strength and toughness. For the purpose of this discussion, attention is focused on dispersed-phase composites only.

The concept of engineering a microstructure to provide property enhancements is not new. One of the most common engineered ceramics, alumina, has often been modified by additions of glass. The resulting microstructures, containing up to 16% of a continuous glassy phase, have shown improved mechanical properties. Particulate-reinforced glass-matrix composites have since been used as model systems to study physical properties. Reviews of prior investigations are provided in Ref 10 and 11. The typical properties

of different matrix and particulate materials commonly used in dispersed-phase composites are listed in Table 1.

In a study conducted by Nicholson, a glass matrix was chosen because it represents an ideal brittle material for crack path studies (Ref 12). Others have used glass because it does not contain grain boundaries and does not have internal surfaces to contend with (Ref 13, 14). Internal stresses can be introduced by altering physical properties, such as thermal expansion and elastic modulus (Ref 12, 15, 16). Finally, depending on fabrication conditions, glass will develop a strong interfacial bond with a dispersed phase that, again, influences composite properties (Ref 10, 14, 17).

As an extension of the theories developed from low-solids composites, Davidge and Green hypothesized that as the volume fraction of glass present in a ceramic-glass matrix composite decreases, ceramic-sphere glass composite model systems are applicable in predicting mechanical behavior (Ref 18). More recent studies by Kumar et al. and Mussler and Shafer have shown that by properly engineering a dispersed-phase glass-matrix composite, that is, cordierite and glass, controlled dielectric properties and thermal expansion characteristics could be achieved (Ref 19, 20).

Composite Thermal Expansion

When a composite is connected or joined to other materials, it is important to match the expansion coefficient of the two materials. An important function of reinforcing particulate fillers in metals is the reduction and control of thermal expansion. With glass-bonded alumina and glass-bonded cordierite composites used in dielectric substrate applications, a mismatch in thermal expansion between the substrate and chip material leads to solder pad strain and eventual fatigue failure.

The thermal expansion coefficient of a composite material is defined as the average strains that result from a unit temperature rise in an unconstrained material. In considering the response of multiphase materials to changes

in temperature, the basic stress-strain relations used to define the elastic behavior of a material have been modified to include the thermal expansion coefficient, α (Ref 21, 22).

Effect of Composition. Several equations have been published in the literature to predict the thermal expansion behavior of composite bodies. Turner derived the following equation (Ref 23):

$$\alpha_{comp} = \frac{\Sigma\, \alpha_i K_i F_i / \rho_i}{\Sigma\, K_i F_i / \rho_i} \qquad \text{(Eq 1)}$$

where α_i, K_i, F_i, and ρ_i are the thermal expansion coefficient, bulk modulus, weight fraction, and the density of component i. The derivation of this equation assumes that the thermal stresses are not large enough to disrupt the body, that each component is constrained to change dimensions at the same rate as the aggregate, and that there is negligible shear deformation. Kingery has shown this relationship to give good agreement for Al_2O_3/SiO_2 glass composites (Ref 22). When the Poisson's ratio of the constituents is the same, the bulk modulus K in Eq 1 can be replaced with Young's modulus, E. Kerner developed a model that incorporates shear effects. An equation capable of predicting the thermal expansion coefficient of a two-phase composite is given in Kingery as (Ref 23):

$$\alpha_{comp} = \alpha_1 + V_1(\alpha_2 - \alpha_1)$$
$$\left\{ \frac{K_1(3K_2 + 4G_1)^2 + (K_2 - K_1)(16G_1^2 + 12G_1K_2)}{(4G_1 + 3K_2)[4V_2G_1(K_2 - K_1) + 3K_1K_2 + 4G_1K_1]} \right\}$$
$$\text{(Eq 2)}$$

A bound on composite expansion is obtained by reversing the role of matrix and inclusion in this equation. When all of the components have the same bulk modulus, a simple rule of mixtures would be expected to hold. Tummala et al. found reasonably good correlation between expansion values predicted by the Kerner relation for glass-particulate zirconia composites (Ref 24). Schapery, utilizing energy principles of thermoelasticity, derived the basic relationship for an isotropic composite containing two isotropic phases (Ref 25):

Table 1 Comparison of properties of selected reinforcement and matrix materials

Material	Young's modulus		Fracture strength		Coefficient of thermal expansion, 10^{-6}/K	Density, g/cm^3
	GPa	10^6 psi	MPa	ksi		
Matrix						
Borosilicate glass	60	8.7	100	14.5	3.5	2.3
Soda-lime glass	60	8.7	100	14.5	8.9	2.5
Lithium aluminosilicate glass-ceramic	100	14.5	100–150	14.5–21.8	1.5	2.0
Magnesium aluminosilicate glass-ceramic	100	14.5	110–170	16–24.7	2.5–5.5	2.6–2.8
Particulate						
Alumina	360–400	52–58	250–350	36.3–50.8	8.5	3.9–4.0
Zirconia	200	29	200–500	29–73	8	5.6
Mullite	220	32	200–300	29–44	5.3	3.1
Cordierite	132	19	245	36	1.4	2.5
Silicon carbide	400–410	58–59.5	310	45	4.8	3.2
Tungsten	340–410	49–59.5	3800(a)	550(a)	4.8	19.3
Stainless steel	150–210	22–30.5	2100(a)	305(a)	...	7.7–8.0
Kovar	135	19.6	...	...	5.6	8.2

(a) Values represent τ_T, where τ is the tensile yield point

$$\alpha^* = \bar{\alpha} + \frac{\alpha_1 - \alpha_2}{[(1/K_1) - (1/K_2)]}\left(\frac{1}{K^*} - \frac{\bar{1}}{K}\right) \quad \text{(Eq 3)}$$

where α_1, α_2 are the linear thermal expansion coefficients of the two phases, K_1, K_2 are their bulk moduli, K^* and α^* are the effective bulk modulus and thermal expansion of the composite, respectively, and the bars over the symbols indicate average values. Figure 1 compares the measured thermal expansion coefficients of a mullite-cordierite composite with prediction based on the Turner and Kerner models. In this system, good correlation is obtained because the thermoelastic properties of the two phases are not drastically different. In addition, the phases are well bonded, which is required to justify using these models.

Effect of Microstructure. Grain size and other microstructural features play a secondary role in determining composite thermal expansion. Both Tummala and Hunter *et al.* found reasonable agreement between experimental data and predictions resulting from Kerner's model (Ref 24, 26). Deviations were noted when the expansion between the dispersed phase and the matrix was so great that microcracking resulted. The extent of microcracking is particle-size dependent. Therefore, particle size can affect composite thermal expansion. Deviations can also be expected when the bond between the phases is poor.

Dielectric Properties

When a dielectric material is substituted for air between the plates of a capacitor, the effective capacitance is increased because of the ability of the dielectric to neutralize part of the applied field. This is a result of polarization within the dielectric. In general, four polarization mechanisms exist: electronic, atomic, orientation, and space charge. For a given frequency of applied voltage, the dielectric constant will be determined by the number of acting polarization mechanisms and the number of "dipoles" for each mechanism. Because relaxation times are associated with each mechanism, the dielectric behavior will depend on the applied frequency. The decrease in dielectric constant with increased frequency is a result of a decreased ability of the dipoles to keep up with the applied field. The speed at which the dipoles are able to orient to the applied field is related to the displacements involved. Space charge, which requires motion of ions over grain dimensions, can only contribute at low frequencies. Electronic polarization, which involves small displacements in the electronic cloud, can respond to frequencies in the optical range. For ceramics, the most important contribution to loss below a frequency of 10^{10} Hz is ion migration. Other mechanisms can occur, and are characterized by peaks in the loss versus frequency curve when the applied frequency is equal to the relaxation time of the mechanism (Ref 27).

Effect of Constituents. Many models exist for predicting the effective dielectric constant of a composite. The more complex models require the specification of phase distribution as well as particle shape.

Without this knowledge, it is impossible to predict an exact value of the composite dielectric constant, although it is possible to predict a bound for it, provided that the volume fraction of each phase is known (Ref 28). The widest bounds are based on the assumption that all of the individual grains act as capacitors, either in parallel or in series. The parallel mixing rule is given by:

$$\epsilon_{comp} = V_1\epsilon_1 + V_2\epsilon_2 \quad \text{(Eq 4)}$$

For series mixing, the expression is:

$$\frac{1}{\epsilon_{comp}} = \frac{V_1}{\epsilon_1} + \frac{V_2}{\epsilon_2} \quad \text{(Eq 5)}$$

where V and ϵ are the volume fraction and dielectric constant, respectively, of the individual phases.

An empirical mixing rule originally attributed to Lichtencker is the logarithm rule (Ref 29). This is expressed as:

$$\log \epsilon = \Sigma V_i \log \epsilon_i \quad \text{(Eq 6)}$$

A closer bound on ϵ is provided by Hashin and Shtrikman (Ref 30). The upper and lower bounds are given by the following expressions:

$$\epsilon(+) = \epsilon_2 + \frac{V_1}{[1/(\epsilon_1 - \epsilon_2)] + (V_2/3\epsilon_2)} \quad \text{(Eq 7)}$$

$$\epsilon(-) = \epsilon_1 + \frac{V_2}{[1/(\epsilon_2 - \epsilon_1)] + (V_1/3\epsilon_1)} \quad \text{(Eq 8)}$$

These equations assume isotropy within the components. A comparison of values predicted by Eq 4 and 5, along with those of Eq 7 and 8 for a two-phase composite with a 10:1 ϵ ratio is given in Fig 2. When the difference in ϵ between the two phases is large, the bounds predicted by the preceding equations are too wide to provide an effective description of ϵ. However, when the ϵ ratio is not too large (that is, less than 3), the mixing rules can provide a good estimate for the composite dielectric constant. This is illustrated in Fig 3 for a mullite-cordierite composite with a dielectric constant ratio of ~1.5. To provide closer bounds for ϵ, statistical information of the phase assemblage is required, which entails determining shape factor and spatial correlation functions. A detailed discussion of this is provided in Ref 28. Van Beek has critically revised many of the composite ϵ equations. For mixtures of glass spheres suspended in carbon tetrachloride ($\epsilon = 2.278$), Van Beek found that the following expression derived by Looyenga adequately described the variations in ϵ with an increase in the volume fraction of spheres (Ref 31, 32).

$$\epsilon = [\epsilon_1^{1/3} + V_2 [(\epsilon_2^{1/3})] - \epsilon_1^{1/3}]^3 \quad \text{(Eq 9)}$$

Microstructure. The two most important microstructural features are phase distribution and porosity. As noted above, the extremes in phase distribution, that is, series or parallel mixing, will yield different composite ϵ values. The effect of phase segregation or local scale agglomeration as deviations to idealized geometries has not been addressed by any of the authors discussed previously. Particle shape has been shown to have an important effect on predicted ϵ values. However, particle size should have a negligible effect on ϵ in cases where interfacial polarization is not a main contributor to ϵ. A secondary grain size effect will occur if the grains are larger than the critical size necessary for microcracking (Ref 31).

Porosity will provide the major microstructural effects in many systems. Porosity lowers the effective ϵ values for a dielectric medium. For a low volume fraction of porosity, the effect is linear. Weiner's expression is:

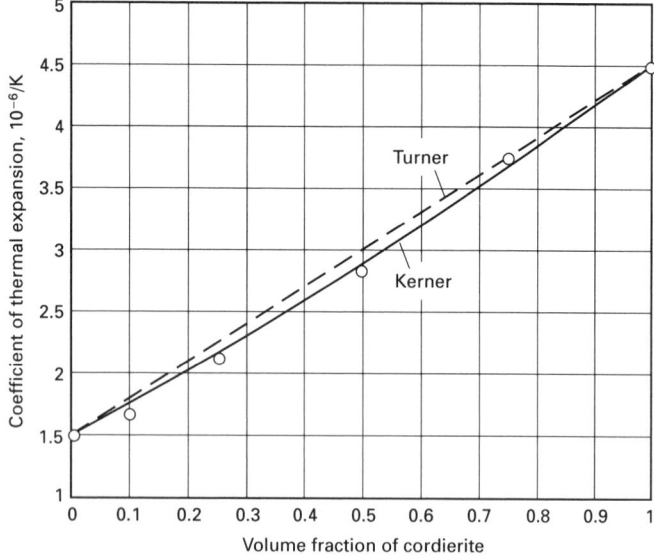

Fig 1 Thermal expansion of mullite-cordierite composites

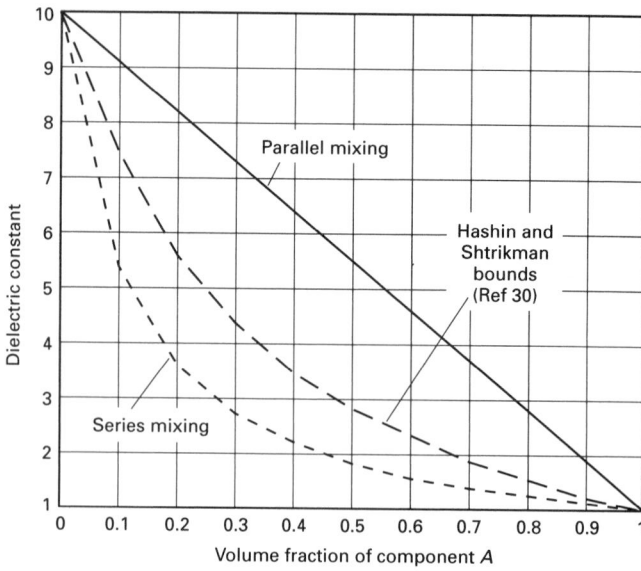

Fig 2 Dielectric constant of a two-phase composite. Dielectric constant of component A = 1 and dielectric constant of component B = 10

$$\epsilon = \frac{\epsilon_1}{1 + 1.5 \, V_2} \qquad \text{(Eq 10)}$$

where ϵ is the 0 porosity value and V_2 is the volume fraction of the porosity (Ref 33). This expression describes the effect of porosity on the ϵ of a BaTiO$_3$ material with up to 20 vol% porosity. Other authors (Ref 20) assume:

$$\frac{\epsilon}{\epsilon_1} = (1 - P)^{20} \qquad \text{(Eq 11)}$$

Composite Young's Modulus

The primary factors affecting the Young's modulus of a composite body are composition (phases), phase distribution, porosity, and grain or particle size. Depending on the match of these factors, stress distributions can be created, which can lead to reduction in composite properties.

Constituent Effect. For structural applications, it is important to know the Young's modulus value because it dictates the elastic strain of a body subject to a uniaxial stress. The measured Young's modulus of a single crystal is dependent on the orientation of the crystal. However, a body made up of many randomly oriented grains will behave isotropically on a macroscopic level. The measured modulus will be some average of all the possible values for different orientations of the grains. A similar situation exists when a body consists of two or more phases. Complicated phase shape, orientation, and interactions led to the conclusion that an exact solution for the composite modulus for arbitrary grain shape and distributions is not possible. A more tenable approach is to calculate upper and lower bounds for the moduli. Two early attempts at calculating composite moduli were

those of Ruess and Voigt (Ref 23). Voigt considered a phase geometry in which the strain in each constituent is the same. This leads to the expression:

$$E_c = \Sigma E_i V_i \qquad \text{(Eq 12)}$$

where E_c is the composite modulus and E_i and V_i are the modulus and the volume fraction, respectively, of component i. The Ruess model assumes that the stress in each component is the same. This consideration leads to the expression:

$$\frac{1}{E_c} = \frac{\Sigma V_i}{E_i} \qquad \text{(Eq 13)}$$

for the composite modulus.

Paul used energy theorems from elasticity theory to calculate upper and lower bounds for an arbitrary phase geometry (Ref 28). The lower bound was obtained utilizing the theorem of least work, whereas the upper bound was obtained through use of the minimum energy theorem. The low bound is equal to the Ruess equation discussed above. Paul's upper bound is equivalent to Voigt's expression when Poisson's ratio of the two phases are equal. An illustration of Voigt and Ruess model predictions, along with experimental values for a cordierite-mullite composite, is given in Fig 4. In this case, the elastic moduli of the two components are close enough to yield a narrow bound using this simple approach. Hashin and Shtrikman used a variational approach to provide the following upper and lower bounds for bulk and shear moduli (Ref 30):

$$K_L = K_1 + \frac{V_2}{1/(K_2 - K_1) + [3(1 - V_2)]/(3K_1 + 4G_1)} \qquad \text{(Eq 14a)}$$

$$K_U = K_2 + \frac{1 - V_2}{1/(K_1 - K_2) + 3V_2/(3K_2 + 4G_2)} \qquad \text{(Eq 14b)}$$

$$G_L = G_1 + \frac{V_2}{1/(G_2 - G_1) + [G(K_1 + 2G_1)(1 - V_2)]/[5G_1(3K_1 + 4G_1)]} \qquad \text{(Eq 15a)}$$

$$G_U = G_2 + \frac{1 - V_2}{1/(G_1 - G_2) + [6(K_2 + 2G_2)V_2/5G_2(3K_2 + 4G_2)]} \qquad \text{(Eq 15b)}$$

Bounds for Young's modulus can be obtained by combining the values for the shear and bulk moduli in the equation:

$$E = \frac{9KG}{(3K + G)} \qquad \text{(Eq 16)}$$

It has been shown that these expressions provide a good description of the behavior of a WC-Co alloy. The Hashin equations provide tighter bounds on the composite moduli. When the modular ratio $m = E_{part}/E_{matrix}$ is small, that is, $0.5 < m < 3$, the separation between these boundary solutions is small enough to obtain an estimate within 10% of the true value (Ref 34). Fulrath reported that his data for alumina dispersions in glass were well described by Hashin's equation (Ref 35). Several other equations for predicting the effective composite moduli are reviewed by Rice (Ref 36).

Effect of Porosity. It is well known that the effective elastic moduli of a given body decreases with increased pore content. Several authors have proposed theoretical as well as empirical relationships to describe this behavior (Ref 37, 38). Several of these equa-

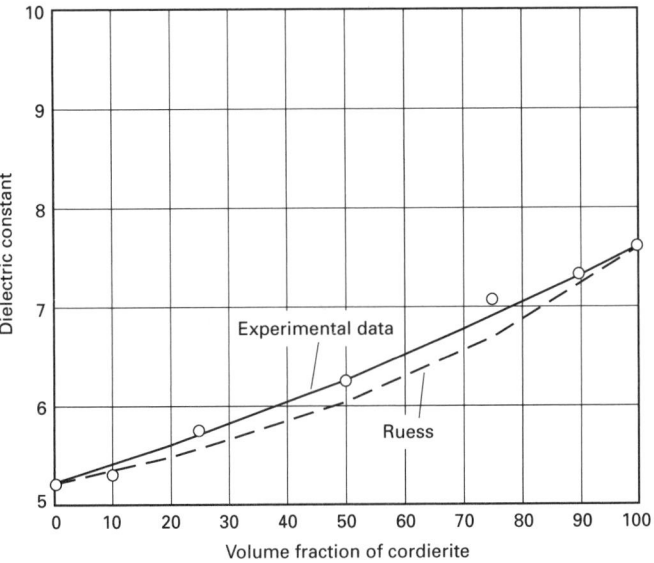

Fig 3 Dielectric constant of mullite-cordierite composites

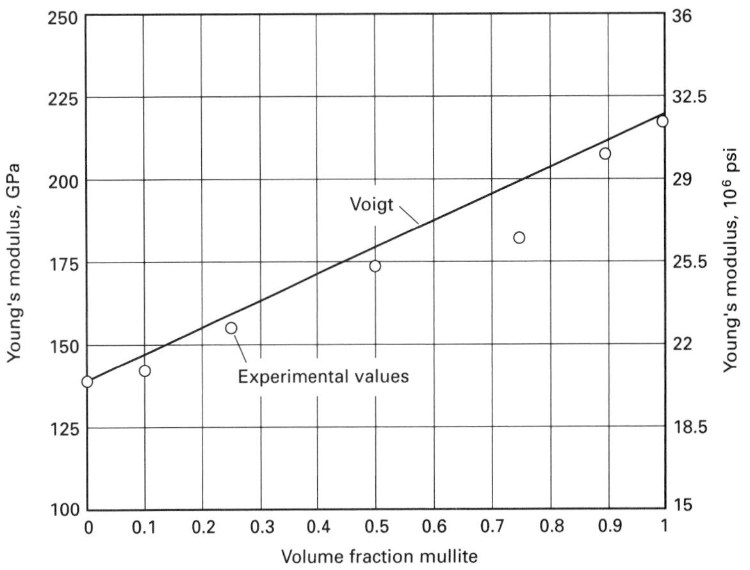

Fig 4 Voigt and Ruess model predictions and experimental values for cordierite-mullite composite

tions have been reviewed by Rice and Wachtman (Ref 36, 39). Wachtman indicates that a number of theoretical equations reduce to the form of $E/E_0 = 1 - bp$ for small pore fractions. For a Poisson's ratio of 0.2, b should equal 2. Dean *et al.* have compared this linear equation, as well as the equations of Spriggs and Hasselman, to actual data for a number of materials (Ref 40). The authors felt that the linear equation can be made to empirically fit the data better than the other expressions. Both E and b were considered to be adjustable parameters.

Many of the equations derived for two-phase materials are not appropriate to describe the effect of porosity (by setting $E_2 = 0$). For bounding equations, the size of the bounds increases significantly as the modular ratio increases. Paul's equations, for example, would yield bounds of $E = (1 - P)E_0$ and $E = 0$ for all fractions of porosity. Of all the analytical expressions investigated, Rice found Hashin's equations to best fit the experimental data (Ref 36). These equations, which are valid for low concentrations of uniformly dispersed spherical pores, are:

$$G = G_0 \left[\frac{1 - 15(1 - V_0)P}{7 - 5V_0} \right] \qquad \text{(Eq 17)}$$

$$K = K_0 \left[\frac{1 - 3(1 - V_0)}{2(1 - 2V_0)} \right] P \qquad \text{(Eq 18)}$$

In addition to theoretical equations, several empirical equations have been proposed. Hasselman utilized the expression (Ref 23):

$$\frac{E}{E_0} = \frac{1 - AP}{1 + (A - 1)P} \qquad \text{(Eq 19)}$$

This equation is based on Hashin's composite moduli formula. The value A is treated as an empirical constant that has been shown to vary

for different types of material and porosity. However, most values of A range from 2 to 4.

Spriggs used an expression of the form:

$$\frac{E}{E_0} = e^{-bP} \qquad \text{(Eq 20)}$$

where b is an empirical constant (Ref 41). This expression has some theoretical basis in that the exponential relationship describes the contact area between grains. Spriggs found that the exponential relationship described data well for alumina. Kingery lists an equation attributed to MacKenzie as (Ref 23, 42):

$$\frac{E}{E_0} = 1 - 1.9P + 0.9P^2 \qquad \text{(Eq 21)}$$

Most of the published expressions fit the data well for porosity values less than 10%. Discrepancies are mainly due to variations in pore shape and distribution. Rossi has shown that the porosity dependence of elastic moduli is affected drastically by the axial ratio of spheroidal pores (Ref 43). When more than one pore type is present, that is, there are different values of b, Rice has shown that a weighted average of b can be used to describe the total porosity content (Ref 44). He also states that the pore shape is the dominant factor in determining the porosity effect on elastic properties.

Rice lists an equation attributed to Ishai and Cohen, in which porosity and compositional effects are combined (Ref 36):

$$E = E_0(1 - P^{2/3})\{1 + \phi[E_1/E_1 - E_0$$

$$(1 - P^{2/3}) - \phi^{1/3}]^{-1}\} \qquad \text{(Eq 22)}$$

Other Microstructural Features. Although grain size has no direct effect on Young's modulus, it can produce secondary

effects, of which the most important is microcracking. Rice and Evans have shown that the microcrack density present after sintering a ceramic has a strong dependence on grain size and thermal expansion anisotropy. Evans has also shown that the presence of microcracks severely diminishes Young's modulus. Young's modulus can also show some dependence on grain size if there is some correlation between grain size and pore distribution (Ref 35).

Composite Mechanical Properties

The strength of a homogeneous body can be described in two essentially equivalent ways. Griffith utilized a thermodynamic least energy method to obtain the often quoted expression:

$$\sigma = Z \left(\frac{E\gamma}{c} \right)^{1/2} \qquad \text{(Eq 23)}$$

where Z is a flaw geometry parameter, E is Young's modulus, γ is the fracture energy, and c is the crack length (Ref 29). The second approach is to equate the stress at the crack tip to the theoretical bond strength of the material. Because the applied stress is greatly magnified at the crack tip, macroscopic applied stress levels that are several orders of magnitude lower than theoretical strength, $E/6$, can cause fracture. Both approaches yield the same dependence of E, γ, and c. A rigorous stress analysis shows that the stress-intensity factor, K_I, is related to the applied stress and crack length by:

$$K_I = \sigma(Yc)^{1/2} \qquad \text{(Eq 24)}$$

where Y is a dimensionless constant. Comparison with the Griffith equation yields the expression:

$$K_1 = \sqrt{2E\gamma} \qquad \text{(Eq 25)}$$

which relates the stress-intensity factor, or fracture toughness, to fracture energy.

This can further be extended by considering the crack extension force, G, which governs the advancement of the crack front. The value G can be expressed in terms of the elastic properties of the material such that:

$$G = \left(\frac{\pi c\sigma}{E}\right)^{1/2} \qquad \text{(Eq 26)}$$

Equation 25 can then be expressed as:

$$K_1 = (GE)^{1/2} \qquad \text{(Eq 27)}$$

The preceding equations are based on the assumption that the body acts as a perfectly elastic continuum. For a typical polycrystalline ceramic, many microstructural heterogeneities exist that can make this assumption invalid. Second phases that are either intentionally or unintentionally incorporated into the microstructure can either act as crack impediments or produce internal strain fields that can alter the crack path. These effects are intentionally produced to improve the mechanical properties of a single-phase matrix by incorporating secondary phases to form a composite. The manner in which these effects are manifested in strength and fracture toughness is the subject of the remainder of this article.

Effects of Adding a Second Phase

It is well established that internal stress results from thermal expansion mismatch within a ceramic body. Both multiphase and noncubic single-phase bodies are subject to this stress. The stress distribution for "real" systems is complicated, because of anisotropy within individual grains and particle-particle interactions. Expressions for simplified systems have been derived to calculate the magnitude of the internal stress (Ref 45). For a single spherical crystal within a glass matrix subject to a change in temperature, the stress resulting from the temperature change is given by (Ref 46):

$$P = \frac{\Delta\alpha\Delta T}{(1 + v_1)/2E_1 + (1 - 2v_2)/E_2} \qquad \text{(Eq 28)}$$

The difference in thermal expansion coefficients is given by $\Delta\alpha$, and v_1 and v_2 are Poisson's ratio of the glass and crystal, respectively.

The radial (P_r) and tangential (P_t) stresses within the matrix can be obtained from the expression:

$$P_r = -P_t = \frac{-PR^3}{r^3} \qquad \text{(Eq 29)}$$

where P is the same P value calculated from Eq 28, R is the radius of the particle, and r is the distance from the center of the particle. When calculating the stress developed upon cooling for an anisotropic ceramic, the ΔT of Eq 28 is not simply equal to the temperature that represents fabrication temperature minus room temperature. Coble has shown that stress relaxation will occur at high temperatures, so the upper temperature is the temperature at which the rate of stress relaxation is negligible (Ref 47).

The nature of the stresses depends on the relative thermal expansions of the phases. When the expansion of the dispersed phase is greater than the matrix, a radial tensile component is developed, along with a hoop compressive stress. If the expansion of the dispersed phase is lower than the matrix, the signs of the stresses are reversed. These effects are illustrated in Fig 5. Davidge and Green (Ref 18) demonstrated that for ThO_2 spheres dispersed in a glass matrix, the nature of the observed cracking depended on the relative expansion of the glass, α_g, to that of ThO_2, α_p. When $\alpha_g > \alpha_p$, circumferential cracking developed around the particle.

Table 2 provides property data for a number of glasses with ThO_2 and the calculated stress, P, from Eq 28. Fulrath (Ref 16) calculated that an increase in internal strain was observed for particles of Al_2O_3 in glass as the particle size of the Al_2O_3 decreased from 300 μm (12 mils) to less than 8 μm (0.3 mil). Borom (Ref 48) found that the stresses surrounding particles diminish as the distance from the glass-crystal interface increases. This produces a state of variable compressive stress throughout the matrix, which opposes crack propagation. Therefore, in the presence of higher-modulus particles coherently bonded in a glass matrix, the load to fracture increases in proportion to the increase in system modulus.

For a microstructure that consists of grain-grain contacts, the stress distribution along the grain boundaries is more complicated. For a single-phase anisotropic material, the stress along the grain boundary is given by (Ref 49):

$$\sigma_{yy} = [(\alpha_i^n + \alpha_2^n - \alpha)E\Delta T/1 + v] \qquad \text{(Eq 30)}$$

and

$$\sigma_{xy} = \left(\frac{\alpha_1^2 + \alpha_2^2}{2}\right)\left(\frac{E\Delta T}{1 + v}\right) \qquad \text{(Eq 31)}$$

where α_1^n equals the expansion coefficient normal to the boundary and α_i^2 equals the expansion coefficient tangential to the boundary.

A more rigorous analysis yields a logarithmic variation of stress along the grain boundary, with a maximum at a junction of facets. Along a single grain, it is also expected that there will be alternating compressive and tensile stresses along the grain facets. However, this does not preclude the presence of local regions of net tensile or compressive stresses around several grains.

A second source of internal stress is anisotropic elastic moduli. Numerous authors have considered the evolution of internal stresses that are due to elastic moduli differences in multiphase materials (Ref 12, 15, 35, 50, 51). Hasselman and Fulrath considered the evolution of internal stresses that are due to an elastic moduli mismatch for an alumina-glass system (Ref 52). These stresses arise during mechanical loading of the body and lead to inhomogeneities in stress distribution, commonly referred to as stress concentrations. During loading, the alumina is placed under stress, which leads to an increase in the load-bearing capacity of the composite.

Frey and Mackenzie studied the fracture strength of alumina-glass and zirconia-glass composites (Ref 15). The use of these crystalline materials allowed the evaluation of the effect of elastic moduli mismatches. Table 3 provides property data for the composite components. Composites containing alumina as the secondary phase were consistently stronger than the zirconia composites (Fig 6, 7). This strength difference was attributed to the higher elastic moduli of the alumina composites. The analysis stated that crystal-glass systems behave as constant strain systems, because of the observed relation between composite moduli and fracture strength (Ref 15). Considering that both types of composites were highly brittle, the strains measured prior to failure for both systems should be approximately equal. Because the alumina composites exhibited a higher composite

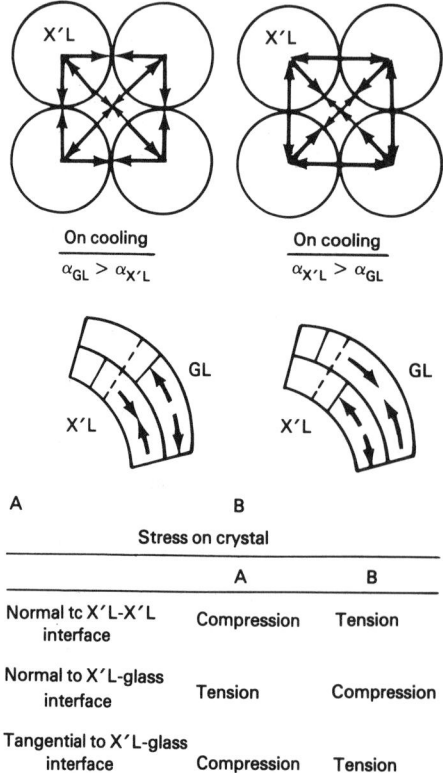

On cooling $\alpha_{GL} > \alpha_{X'L}$	On cooling $\alpha_{X'L} > \alpha_{GL}$

	Stress on crystal	
	A	B
Normal to X'L-X'L interface	Compression	Tension
Normal to X'L-glass interface	Tension	Compression
Tangential to X'L-glass interface	Compression	Tension

Fig 5 Stress distribution arising from crystal-glass thermal expansion mismatch

Table 2 Property data for glasses and thoria

Material	Density, g/cm³	Softening point °C	Softening point °F	Coefficient of thermal expansion, 10⁻⁶/K	Stress MPa	Stress ksi
Corning 7740 pyrex	2.23	820	1510	3.6	−267	−38.7
Borosilicate glass	2.46	765	1410	7.9	−45	−6.5
Soda-lime glass	2.49	715	1320	10.5	88	12.8
ThO₂ spheres	9.90	…	…	8.5(a)	…	…

(a) At 0–500 °C (30–930 °F)

Table 3 Property data of composite components

Material	Density, g/cm³	Softening point °C	Softening point °F	Elastic modulus GPa	Elastic modulus 10⁶ psi	Coefficient of thermal expansion, 10⁻⁶/K
Corning 6810 soda-zinc glass	2.65	770	1420	6.6	0.96	6.9
Al₂O₃	3.91	…	…	385	55.8	8.85
ZrO₂	5.65	…	…	142	20.6	5.47

elastic modulus than the zirconia composites, the former should fracture at a higher stress, all other things being equal.

Nicholson used ultrasonic fractography to observe crack propagation in brittle composite materials. This technique allowed for postfracture analysis of the crack-particle interactions (Ref 12). The systems studied were nickel in glass and aluminum in glass. The results indicated that a strong interfacial bond between the high-modulus particle, nickel, and the matrix leads to crack deflection. In the aluminum-glass system, the elastic moduli of both phases were equal, and, with a good interfacial bond, the crack intercepted the particle. For ductile inclusions, this type of behavior was desired, because it allows ductile failure of the inclusions to occur, which requires more energy for crack propagation than for the deflection process.

Crack-Second-Phase Interactions

It is well known that the fracture toughness of an amorphous material can be increased by the introduction of a hard, discrete second phase. This second phase may take the form of fibers, whiskers, or particulate (Ref 10, 12, 17, 18, 40, 53–56). For particulate systems, the increase in fracture toughness is associated with the interaction of the moving crack front with each discrete particle (Ref 57). Crack interaction mechanisms are those by which a crack deviates from planarity due to direct interaction between the crack and a microstructural feature (Ref 58). These interactions are affected by the properties of the matrix phase and the particulate phase, both of which determine/influence the resultant properties of the composite. Four processes which describe the interaction of a crack front with a second-phase particle are crack-interface traction (mechanical interlocking and ligamentation) (Ref 29, 59, 60), frontal zone microcracking (Ref 61), crack deflection (Ref 55), and crack bowing (Ref 10).

Crack-interface traction has been identified as a mechanism by which resistance to crack propagation is increased in brittle polycrystalline ceramics (Ref 62). Traction is associated with two mechanisms that promote an increase in measured fracture toughness. The first mechanism is due to a frictional or mechanical interlocking of opposing fracture surfaces as a crack propagates and the crack opening displacement increases. Figure

8(a) illustrates the mechanical interlocking which decreases the applied stress intensity factor at the crack tip. This mechanism has been observed to operate as far back as 100 microns behind the crack tip. The criteria for a successful interlocking is a microstructurally rough fracture surface which occurs in large-grained polycrystalline materials (Ref 63). The second mechanism involves the formation of ligamentary bridges, as shown in Fig 8(b). These ligamentary bridges, sometimes referred to as grain bridges, consist of intact islands of material left behind the moving crack front. These islands must then be fractured for the two opposing fracture surfaces to separate. These two mechanisms act as restraining forces that hold the fracture surfaces together, much in the same way that continuous fibers traverse the crack front in fiber-reinforced composites.

Frontal Zone Microcracking. For brittle materials, the most important effect of internal stress is microcracking (Ref 50). The occurrence of microcracking can increase the fracture energy, but generally decreases the strength, depending on the extent of microcracking (Ref 64). The extent of microcracking depends on the magnitude of the residual thermal stress and the grain size. Because the magnitude of the residual stress is independent of grain size, the grain size dependence must be rationalized in another manner. One approach is to equate the strain energy necessary to form a spherical crack with the surface area produced by the formation of the crack (Ref 12, 29, 46). The strain energy has an R^3 (R-particle size), whereas the surface energy has an R^2 dependence. This leads to a critical particle size radius for spontaneous microcracking, given by:

$$Rc \geq \frac{8\gamma_s}{\{P^2[(1 + \nu_1)/E_1 + 2(1 - 2\nu_2)/E_2]\}} \quad \text{(Eq 32)}$$

This equation was derived for a spherical particle in a glass matrix. Rice gives an expression for the critical grain size to fracture a grain boundary as (Ref 53):

$$Gs = \frac{12E\gamma_b(1 - \nu)}{\sigma_i^2} \quad \text{(Eq 33)}$$

A similar equation was derived by Davidge (Ref 18). Evans utilized a fracture mechanics

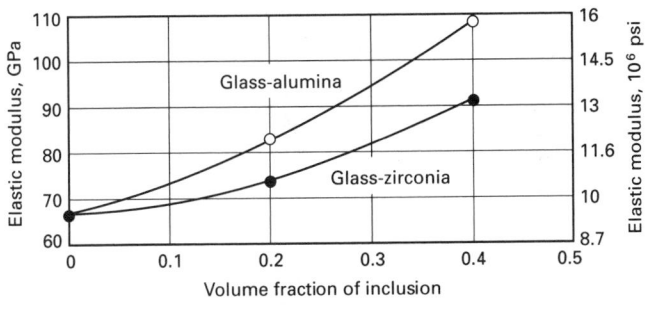

Fig 6 Experimental elastic moduli of glass-alumina and glass-zirconia composites

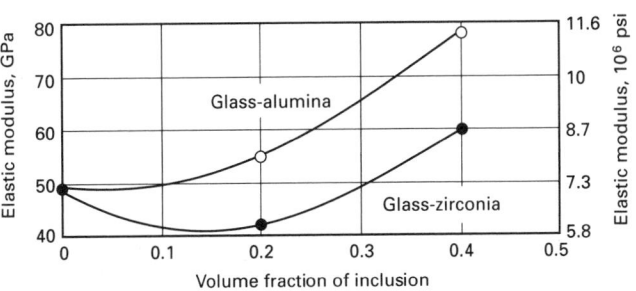

Fig 7 Experimental fracture strength of glass-alumina and glass-zirconia composites

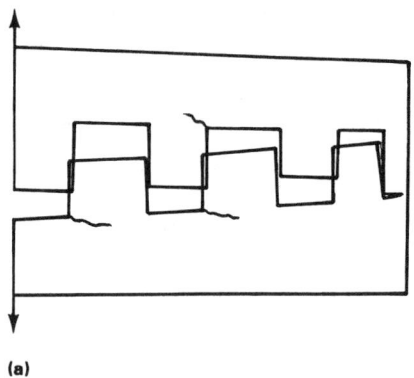

(a)

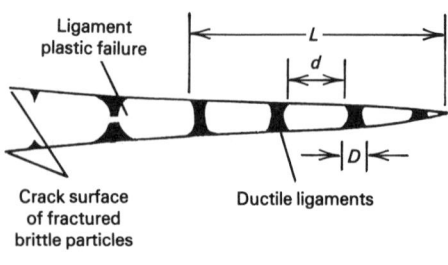

(b)

Fig 8 (a) Origin of mechanical interlocking, which leads to crack-interface traction. (b) Crack bridged by ductile ligaments

approach to develop an expression for the critical facet length (Ref 49):

$$lc^s = \beta\left(\frac{K_c^{GB}}{\sigma}\right)^2 \qquad \text{(Eq 34)}$$

where β = 3 to 4.

The preceding equations are used to predict the critical grain size for spontaneous fracture of grain facets. Microcracking can also occur in the presence of an applied stress. An externally applied stress can superimpose on the stress produced by thermal expansion mismatch to increase the number of microcracks and to decrease the residual stress (and grain size) necessary for microcrack formation (Ref 65). This is particularly true near a crack tip, where the applied stress is greatly magnified. Associated with this microcracking that occurs within a process zone around a crack tip is an increase in fracture energy. The increased fracture energy is associated with expenditure of the strain energy that is necessary to produce the microcracks. Evans attributes microcrack toughening to both elastic modulus diminution and dilation within the process zone (Ref 49).

Rice and Freiman have developed a model to explain the variation in fracture energy with grain size in noncubic materials (Ref 53). An increased grain size initially increases the fracture energy, but after passing through a maximum level, fracture energy drops precipitously with further increases in grain size. The increase in fracture energy is a result of increased process zone microcracking. However, once the grain size reaches a critical

value, the amount of spontaneous microcracking increases, thereby making the article more susceptible to failure that is due to microcrack linkup. Rice *et al.* give the following expression for the variation in fracture energy, γ, as a function of grain size (Ref 53):

$$\gamma = \gamma_{pc}\left(\frac{1-G}{Gs}\right) + M\Delta\epsilon[(9E\gamma_B)^{1/2} - \Delta\xi GE]$$

$$\text{(Eq 35)}$$

The same effect could also be expected for increased thermal expansion anisotropy within the microstructure as a result of adding more of a second phase. As the amount of second-phase particles increases, the number of unlike continuous particle boundaries increases. This should provide an increased propensity toward microcracking, especially in the presence of a crack tip, which could result in an increase in fracture energy.

Below the spontaneous microcracking threshold, the presence of residual stress seems to have a minor effect on fracture stress when the flaws are large, compared to the grain size. Frey *et al.* found that the strength of glass-alumina composites was not degraded by the thermal stress present, provided premature cracking did not occur (Ref 15). Swearingen *et al.* have shown that the presence of residual stress has little influence on the fracture toughness of alumina-glass composites (Ref 9). Varying the expansion of the glass (greater and lower than alumina) caused changes in the nature of crack propagation and particle interaction, but did not influence the composite K_{Ic}.

In terms of composite behavior, the preceding discussion illustrates the need to thoughtfully choose the constituent phases. In general, it is necessary to choose phases with similar thermal expansions in order to minimize the residual stress. However, when this is not possible, grains that are as fine as possible should be used to suppress spontaneous microcracking (Ref 45).

Crack Deflection. Deflection toughening arises whenever interactions between the advancing crack front and an inclusion cause the crack to deviate from planarity (Ref 66). Figure 9(a) illustrates the deflection of a crack from planarity during impingement of a single particle. Multiple deflection processes are also possible, as illustrated in Fig 9(b). The consequence of such an interaction is the reduction in the stress-intensity factor at the crack tip. Deflection processes can occur in polycrystalline bodies, where an amorphous grain-boundary phase gives rise to a weakened interface between particles (Ref 67). These non-planar cracks are also caused by residual strains in the body due to thermal and elastic mismatches (Ref 68). The sign of the residual strain determines the direction in which the rack is deflected. Experimental results reported by Binns (Ref 68) showed that although residual stresses determine the

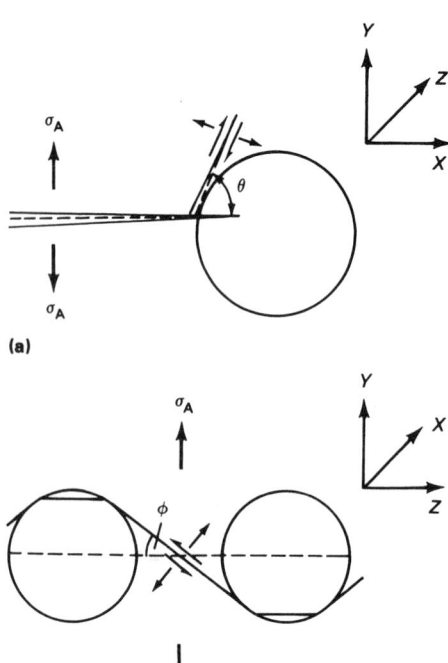

(a)

(b)

Fig 9 Typical crack deflections. (a) Tilted at angle θ about Z axis. (b) Twisted at angle ϕ about X axis. Source: After Ref 66

direction of crack propagation, there is no effect on the overall toughness measured.

The approach taken by Faber and Evans (Ref 58) to evaluate the deflection process was to consider the local stress-intensity factor associated with the advancing crack. The increase in fracture toughness that is due to deflection is obtained by averaging the local stress-intensity contributions at twist and tilted portions of the crack front. Assuming that the crack front is subjected to mode I, mode II, and mode III type fracture processes, the following expression can be written:

$$EG = k_1^2(1-\nu^2) + k_2^2(1+\nu^2) - k_3^2(1+\nu)$$

$$\text{(Eq 36)}$$

where ν is Poisson's ratio. Using this equation, they defined the term $\langle G \rangle$, as the average crack extension force across the crack front. This term represents the net driving force for fracture. Using $\langle G \rangle$, predictions were made for the incremental toughening produced by the inclusions, $\langle G_c \rangle$. The term $\langle G_c \rangle$ was formally defined as the relative toughness of the composite, which is the ratio of composite toughness to the parent matrix toughness. The expression derived was:

$$G_c - \left(\frac{G^m}{\langle G \rangle}\right)G_c^m \qquad \text{(Eq 37)}$$

After determining the local stress-intensity factor associated with a tilted crack and formulating a weighting function for different particle morphologies, particle morphology

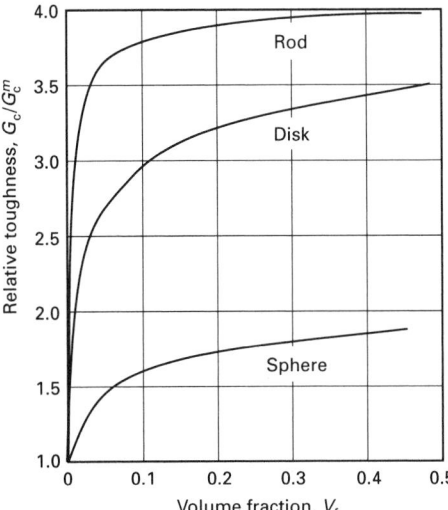

Fig 10 Relative toughening versus volume fraction of second phase present for spheres, disks, and rods. Source: After Ref 66

was shown to have a strong influence on the crack driving force. Figure 10 shows the toughening predictions for three particle morphologies: disk, rod, and sphere.

It was concluded that increases in fracture toughness were a function of the volume fraction of particulate and particle morphology. Favorable morphologies are ones of high aspect ratio, which allow for maximization of the twist type of fracture. The twist type of fracture will consume more energy than the initial crack tilt. Furthermore, after exceeding 10 to 20 vol% of inclusions, the relative toughness becomes asymptotic for all particle morphologies. It was also noted that the toughness of materials containing spherical particles is significantly influenced by the interparticle spacing distribution. The greatest toughening is afforded when the spheres are close, such that twist angles approach 90°.

Experimental results were provided to support the crack-deflection theory. Particle morphology effects were evaluated in a series of hot pressed silicon nitrides, and the effect of particle size was determined using a lithium-aluminosilicate glass-ceramic. The toughness of the glass-ceramic system was evaluated after various heat treatments, which promoted the growth of crystals of different sizes. Of specific interest was that fracture toughness was found to be invariant with particle size, when considering crack deflection as the toughening process. This has been a major source of conjecture.

Crack Front Bowing. Bowing can be described as the interaction of a particle that has a higher resistance to fracture with a moving crack front. The intersection causes the crack front to be temporarily pinned, and with the addition of more energy, the planar crack bows between particles. The resultant stress required to propagate the bowed segment of the crack front is greater than that

needed to extend the unpinned crack (Ref 69). Using the geometrical configuration in Fig 11, a relationship was derived using the concept of a line tension to describe the pinned crack front.

Lange (Ref 70) proposed that the stress, σ, necessary for a crack to propagate through a series of particles could be expressed as:

$$\sigma = \left[\frac{2E}{\pi c'(\gamma_0 + T/2c)} \right]^{1/2} \qquad \text{(Eq 38)}$$

where E is the elastic modulus of the material, γ_0 is the fracture surface energy of the material, c' is the depth of the primary crack through the thickness crack, T is the line tension, and $2c$ is the particle spacing.

Using Eq 38, it can be seen that the resistance of a material to fracture should increase as the spacing between homogeneities decreases. This equation also indicates, assuming that the fracture strength is proportional to the square root of the crack extension force, that the fracture strength is inversely proportional to the square root of the inclusion spacing. Lange concluded that this model is very simple in nature, in that no other particle-matrix interactions are considered, such as thermal and elastic mismatches. These mismatches would create residual stresses in the composite, which would affect correlation between experimental results and theoretical predictions. Qualitatively, Lange compared the experimental strength data of Hasselman and Fulrath (Ref 52) to this analysis and found good agreement over the range of volume fractions of particulate evaluated (Fig 12).

Evans considered the effect of second-phase dispersions on the strength of composite materials (Ref 54). The approach taken was to consider the energetics of the crack front upon impinging a singular inclusion or a series of inclusions. From this, the magnitude of the line tension, as discussed by Lange, can be determined for various obstacle configurations (Ref 70). Green (Ref 71) described the notion of particle penetrability. He defines the ratio $r/r_0 = 1$ as the particle at its maximum penetrability. When $r/r_0 = 0$, penetrability was at its minimum. By using the concept of an effective particle penetrability, Green was able to fit a polynomial expression to Lange's strength data.

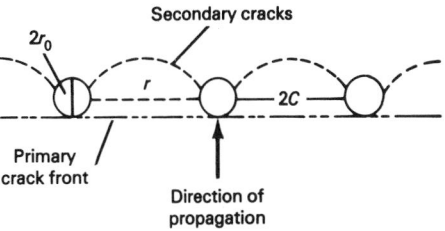

Fig 11 Interaction of advancing crack front with array of particles gives rise to secondary cracks. Crack front temporarily pinned at particle intersection

$$\frac{\sigma_A}{\sigma_c} = \sum_1^7 A_n \left(\frac{r'}{C} \right)^n \qquad \text{(Eq 39)}$$

Miyata (Ref 72) investigated the fracture properties of alumina-glass composites containing up to 30 vol% alumina. The two particle morphologies evaluated were spherical and angular. The experimental results were contrasted with theoretical predictions based on crack bowing, crack deflection, and frontal zone microcracking. All three mechanisms were found to occur in the composites tested. The experimental results could be explained in terms of a combination of mechanisms. In thermally matched systems, crack bowing and crack deflection are the predominant mechanisms. It was assumed that the dispersed particles acted as "weak" obstacles for the crack bowing. In systems where thermal stresses were generated, the increases in toughness were a combination of all three mechanisms. Miyata stated that, at the present level of development of fracture theory, the fractional contribution of each of these processes cannot be determined. However, it was hypothesized that at higher concentrations of dispersed phase, the contribution that was due to crack bowing was predominant. This was supported by Hannon (Ref 73) who showed that deflection was dominant in Al_2O_3-glass composites containing less than 30 vol% particulate, whereas bowing contributions dominated at higher particulate loadings.

Effect of Porosity

In general, the presence of porosity will degrade the mechanical properties of a ceramic article. The magnitude of its effect will depend on the nature (size, shape, and distribution) of the porosity. Knudsen used an exponential relationship of the form (Ref 74):

$$\sigma = \sigma_0 e^{-bP} \qquad \text{(Eq 40)}$$

to describe the strength of porous alumina. The exponential relationship has some physical significance in that it is related to a minimum cross-sectional area. All other things being equal, a crack will generally seek out a path or a least-solid area. Rice has shown that the fracture energy can be adequately described by Eq 40 for low levels of P (Ref 36). Mussler et al. used the exponential relationship to relate the fracture toughness of porous cordierite-mullite composites to their dense counterparts with some success (Ref 20).

The actual effect of porosity on fracture stress is more complicated than Eq 40 indicates. Pores affect strength in three ways. They are stress concentrators, they decrease E and γ, and they act as flaws. These points were reviewed by Rice (Ref 36). Cracks emanating from the equatorial plane of the pore can act as the critical flaw. The severity of this effect is dependent on the grain and pore size, along with pore locations. Pores located at a triple grain junction are more severe than grain-boundary pores and pores within grains.

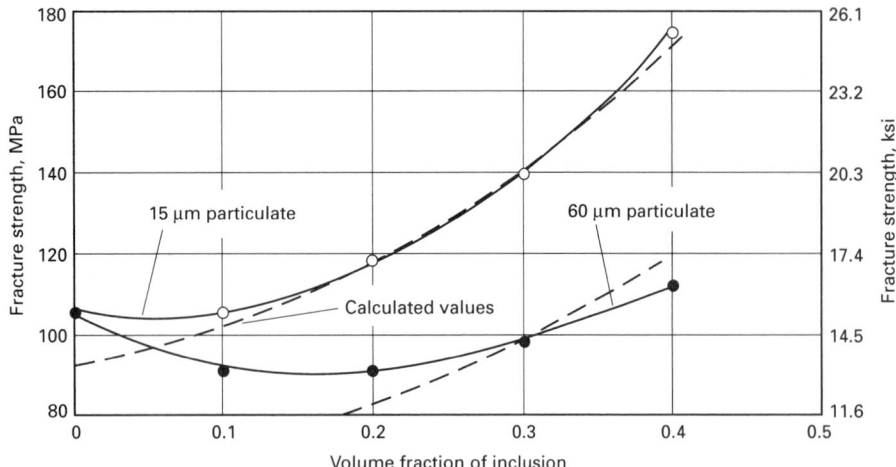

Fig 12 Comparison of fracture strength data of Hasselman and Fulrath (Ref 52) and calculated values for alumina-glass composites

Cracks associated with the pores can propagate one grain dimension into the matrix. The combined pore radius and crack length can be substituted for c in the Griffith equation to obtain the amount of strength degradation caused by the pore. The decrease in E and γ must also be taken into account. In some cases, the decrease in E and γ will be the controlling factors in the functional dependence of P on σ. In others, the size of the flaw associated with the pore will be strength controlling.

Effect of Volume Fraction and Size of Dispersed Phase

Borom (Ref 48) demonstrated that, for Al_2O_3-glass composites, the flaw size was associated with particle size additions of the dispersed phase. Hasselman and Fulrath (Ref 75) proposed that a dispersion of a hard crystalline second phase within a brittle glass matrix strengthens the composite by limiting the Griffith flaws. The distribution of a dispersed phase can have one of two effects on flaw size, depending on whether the intrinsic flaw size of the matrix is greater or smaller than the interparticle spacing (Ref 75).

In the first case for which the average distance between particles is greater than the flaw size, the size of the effective flaw will be reduced, at least on a statistical basis, by becoming entrapped or pinned by the particles. The reduction will be equal to the volume fraction of the dispersed phase:

$$C_c = C(1 - \phi) \qquad (Eq\ 41)$$

where C_c is the flaw size within the composite, C is the flaw size within the matrix prior to dispersed phase addition, and ϕ is the volume fraction of the dispersed phase.

Substituting for C_c in Eq 41 and rearranging, the effect of dispersion on the strength, σ_c, of the composite can be written as:

$$\sigma_c = \sigma(1 - \phi)^{-1/2} \qquad (Eq\ 42)$$

where σ is the intrinsic strength of the matrix for flaw size, C.

In the second case, the flaws will be larger than the distance between particles, either at higher volume fractions or a smaller particle size of the dispersed phase. The flaw size is now limited by the average mean free path between particles. Using the expression of mean free path, λ, between spherical particles of uniform size, R, distributed uniformly throughout a matrix, Fullman (Ref 76) determined their relation to be:

$$\lambda = \frac{4R(1 - \phi)}{3\phi} \qquad (Eq\ 43)$$

Substituting Eq 43 into Eq 23 shows that the strength of a composite strengthened by a flaw-limitation mechanism is:

$$\sigma_c = \frac{(3\gamma E\phi)^{1/2}}{\pi R(1 - \phi)} \qquad (Eq\ 44)$$

In case 1, when interparticle spacings are greater than flaw size, Eq 42 and 44 show that strength is a function of volume fraction only. Relatively large volume fractions of dispersed phase are required before strength increases appreciably. In case 2, when flaw size is large compared to particle spacing, strength is a function of volume fraction as well as particle size.

Fulrath and coworkers found good experimental agreement for alumina particles, ranging in size from 15 to 500 μm (0.6 to 20 mils), dispersed in a sodium borosilicate glass (Ref 10, 75). Lange extended the importance of particle spacing to show that the functional relation proposed by Hasselman and Fulrath concurred with that developed to describe the interaction of a crack front and a dispersed phase (Ref 54). Lange later experimentally observed that the fracture energy of a glass could be significantly increased by incorporating a second-phase dispersion of Al_2O_3 (Ref 77). The fracture energy of composites containing larger particle sizes of Al_2O_3 appeared

to result in larger values. However, these larger particles tended to increase crack size. As a result, a compromise had to be made in the choice of the particle size of the dispersed phase.

Nivas and Fulrath showed that for a tungsten sphere-glass composite with two particle sizes or a wide distribution about one average size, the average mean free path decreased more than could be achieved with a single-particle size dispersion (Ref 78). The resulting composite strengths indicated that the decrease in mean free path for a mixed, fine particle size, dispersed phase resulted in the maximum composite strengths. Hannon and Kowzan showed that for alumina-glass composites that vary in volume fraction and size of dispersed phase, both fracture strength and toughness increased with loading and decreased with increasing particle size (Ref 73, 79). Figures 13 and 14 show that up to a volume fraction of 0.5, composite fracture strength and toughness were grain size independent. Above this level, the finer size particulate provided greater fracture strength and toughness.

Effect of Interfacial Reaction

Aside from the apparent physical effects of mixing dispersed-phase particles in a continuous matrix, increased chemical bonding between particle and matrix can lead to composite strengthening (Ref 10, 17, 77).

Nason examined interfacial bonding in an attempt to disperse tungsten and nickel spheres in glass matrices for which particle-matrix thermal expansions were mismatched (Ref 80). It was observed that when the thermal expansion of the glass was much less than the expansion of the spheres, the spheres shrank away from the matrix glass upon cooling, and formed pseudoporosity. Upon examination, he observed that no bonding between phases occurred (Ref 80).

Stett and Fulrath (Ref 74), in a series of investigations, showed that if nickel spheres were preoxidized prior to forming, a bond developed between the nickel and glass, resulting in no pseudoporosity that was due to mechanical stresses. Figure 15 shows the composite strength of 30 vol% nickel spheres within a sodium borosilicate glass (Ref 10, 14, 74). They showed that a bond develops between a glass and metal when the glass is saturated with the oxide of the metal at the glass-metal interface. If the available oxide is entirely dissolved by the glass before the glass attains the saturation concentration, the resultant contact with the purely metallic surface results in a weak bond (Ref 78). However, if the available oxide layer is too thick and the glass is saturated, the resultant bond strength is also low, because fracture is apt to initiate within the oxide layer.

Studt and Fulrath examined a mullite-glass composite and found that interfacial wetting

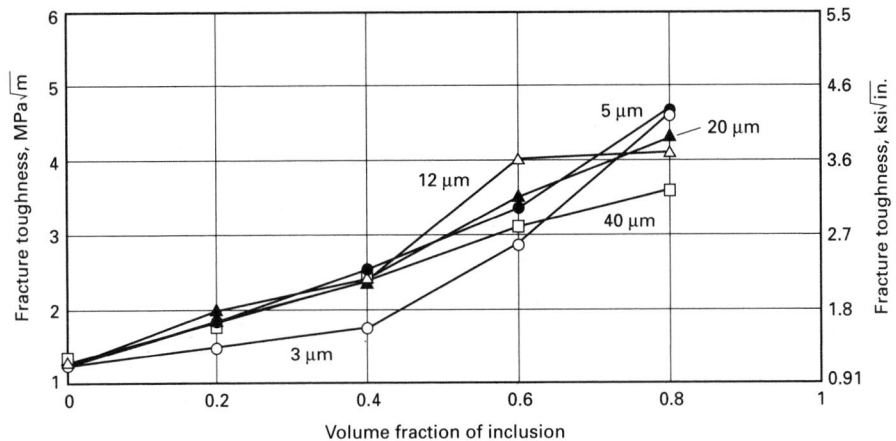

Fig 13 Experimental fracture toughness of hot pressed glass-alumina composites varying in volume fraction and particle size

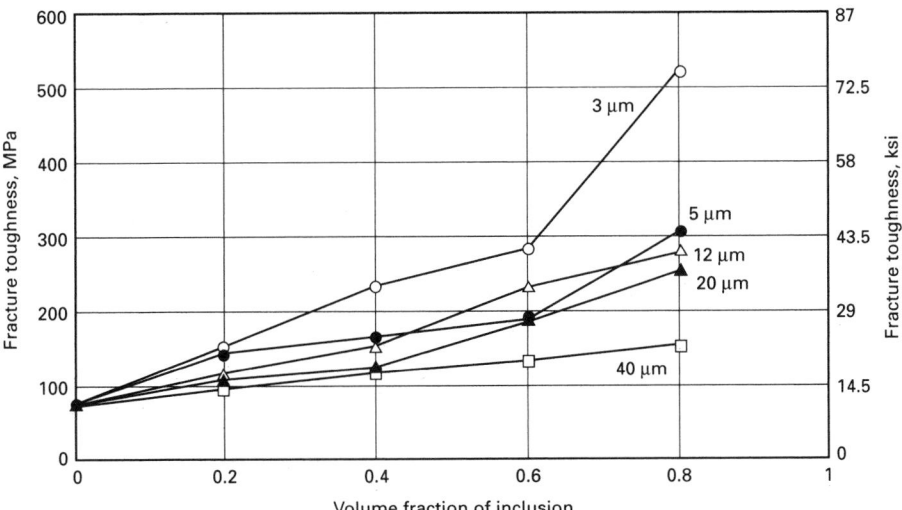

Fig 14 Experimental fracture strength of hot pressed glass-alumina composites varying in volume fracture and particle size

studies indicated particle-matrix chemical reaction (Ref 77). Upon the analysis of the composite as a function of particle size, and maintaining a constant solid volume fraction, it was observed that for similar degrees of interfacial reaction, the smaller particles provided the greatest strengths.

Hardness

Rice has reviewed the effect of microstructural features on microhardness (Ref 36). Limited data were presented for second-phase inclusions on microhardness. Hardness increased in softer materials, such as MgO,

when a second phase was added, presumably as a result of dislocation pinning (Ref 36). No models were found to predict the hardness of a two-phase particulate composite. However, porosity was found to lower hardness, as predicted by the exponential relationship used for fracture energy and Young's modulus. Mussler (Ref 20) found that the relationship

$$\frac{H}{H_0} = e^{-6P} \qquad \text{(Eq 45)}$$

reasonably described the effect of porosity on the hardness of mullite and cordierite bodies.

Kowzan (Ref 79) showed that for a series of hot pressed alumina-glass composites, Knoop hardness increased as a function of particulate loading. Figure 16 shows that there is a slight grain size dependence on hardness. Four different particle sizes of alumina were tested. In general, the finer the particle size, the greater the hardness value (Ref 79).

REFERENCES

1. J.R. McLaren and R.W. Davidge, The Combined Influence of Stress, Time and Temperature of the Strength of Polycrystalline Alumina, *Proc. Brit. Ceram. Soc.*, Vol 25, 1975, p 151–167
2. A.G. Evans and G. Tappin, Effects of Microstructure on the Stress to Propagate Inherent Flaws, *Proc. Brit. Ceram. Soc.*, Vol 20, 1972, p 275–297
3. H. Meredith, C.W.W. Newey, and P.C. Pratt, The Influence of Texture on Some Mechanical Properties of Debased Polycrystalline Aluminas, *Proc. Brit. Ceram. Soc.*, Vol 20, 1972, p 299–316
4. R.W. Davidge and G. Tappin, The Effects of Temperature and Environment on the Strength of Two Polycrystalline Aluminas, *Proc. Brit. Ceram. Soc.*, Vol 15, 1970, p 47–57
5. B.R. Steele, F. Rigby, and M.C. Hesketh, Investigations on the Modulus of Rupture of Sintered Alumina Bodies, *Proc. Brit. Ceram. Soc.*, Vol 6, 1966, p 83–95
6. S.H. Risbud, Influence of Surface Condition on the Strength of Alumina Glass

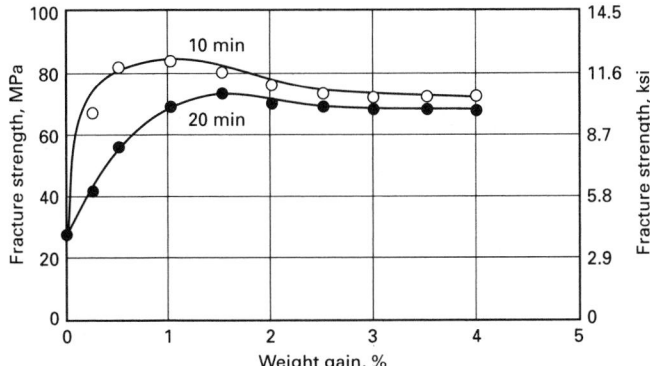

Fig 15 Strength as a function of weight gain for sodium borosilicate glass-preoxidized nickel sphere composites hot pressed for various times

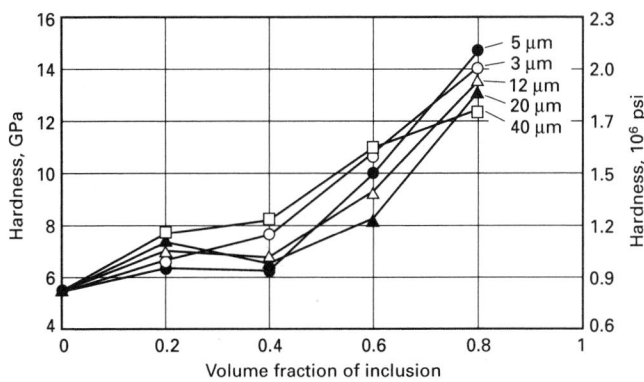

Fig 16 Experimental hardness of hot pressed glass-alumina composites varying in volume fraction and particle size

Ceramics, *Ceramic Proceedings*, American Ceramic Society, Jan-Apr 1980, p 702–711

7. M. Kunugi, N. Soga, and A. Konishi, Mechanical Properties and Microstructure of Glass-Ceramic Composite (Part 2), Asahi Garasu Kogyo Gijutsu Shoreikai Kenkyu, Vol 24, Japanese Ceramic Society, 1972, p 97–104

8. D. Binns, *Science of Ceramics*, Vol 1, G.H. Stewart, Ed., Academic Press, p 315–334

9. J.C. Swearingen, E.K. Beauchamp, and K.J. Egan, *Fracture Mechanics of Ceramics*, Vol IV, R.C. Bradt, D.P.H. Hasselman, and F.T. Lange, Ed., Plenum, 1978, p 973–987

10. M.A. Stett and R.M. Fulrath, Mechanical Properties and Fracture Behavior of Chemically Bonded Composites, *J. Am. Ceram. Soc.*, Vol 53 (No. 1), 1970, p 5–13

11. J.S. Nadeau and J.I. Dickson, Effects of Internal Stresses Due to a Dispersed Phase on the Fracture Toughness of Glass, *J. Am. Ceram. Soc.*, Vol 63 (No. 9–10), 1980, p 517–523

12. P.S. Nicholson, Crack Paths and the Toughening of Brittle Materials by Second-Phase Particles, *High Temp. Sci.*, Vol 13, 1980, p 279–297

13. G.R. Anstes, P. Chantikul, B.R. Lawn, and D.B. Marshall, A Critical Evaluation of Indentation Techniques for Measuring Fracture Toughness: I, Direct Crack Measurements, *J. Am. Ceram. Soc.*, Vol 64 (No. 9), 1981, p 533–538

14. R.L. Bertolotti and R.M. Fulrath, Effect of Micromechanical Stress Concentration on Strength of Porous Glass, *J. Am. Ceram. Soc.*, Vol 50 (No. 11), 1967, p 558–562

15. W.J. Frey and J.D. MacKenzie, Mechanical Properties of Selected Glass-Crystal Composites, *J. Mater. Sci.*, Vol 2, 1967, p 124–130

16. R.M. Fulrath, Internal Stresses in Model Ceramic Systems, *J. Am. Ceram. Soc.*, Vol 47 (No. 9), 1959, p 423–429

17. M.A. Stett and R.M. Fulrath, Chemical Reaction in a Hot-Pressed Al_2O_3-Glass Composite, *J. Am. Ceram. Soc.*, Vol 50 (No. 12), 1967, p 673–676

18. R.W. Davidge and T.J. Green, The Strength of Two-Phase Ceramic/Glass Materials, *J. Mater. Sci.*, Vol 3, 1968, p 629–634

19. A. Kumar, P. McMillan, and R.R. Tummala, "Glass-Ceramic Structures and Sintered Multilayer Substrates thereof with Circuit and Patterns of Gold, Silver or Copper," U.S. patent 4,301,324, 1980

20. B.H. Mussler and M.W. Shafer, Preparation and Properties of Mullite-Cordierite Composites, *Am. Ceram. Soc. Bull.*, Vol 63 (No. 5), 1984, p 705–714

21. R.K. Kirby, Thermal Expansion of Ceramics, *Mechanical and Thermal Properties of Ceramics*, Publication 303, 1968, National Bureau of Standards, p 41–62

22. W.D. Kingery, Note on Thermal Expansion and Microstresses in Two-Phase Compositions, *J. Am. Ceram. Soc.*, Vol 40 (No. 10), p 351–352

23. W.D. Kingery, H.K. Bowen, and D.R. Uhlman, *Introduction to Ceramics*, John Wiley & Sons, 1976, p 307–317

24. R.R. Tummala and A.L. Firedber, Thermal Expansion of Composites as Affected by the Matrix, *J. Am. Ceram. Soc.*, Vol 53 (No. 7), 1970, p 376–380

25. R.A. Schapery, Thermal Expansion Coefficients of Composite Materials Based on Energy Principles, *J. Comp. Mater.*, Vol 2 (No. 3), 1968, p 380

26. O. Hunter and W.E. Brownell, Thermal Expansion and Elastic Properties of Two-Phase Ceramic Bodies, *J. Am. Ceram. Soc.*, Vol 50 (No. 1), 1967, p 19–23

27. A. Von Hipple, *Dielectric Materials and Applications*, John Wiley & Sons, Inc., 1954, p 1–30

28. D.K. Hale, Review: The Physical Properties of Composite Materials, *J. Mater. Sci.*, Vol 11, 1976, p 2105–2141

29. R.W. Davidge, *Mechanical Behavior of Ceramics*, Cambridge University Press, 1979, p 1–50

30. A. Hashin and S. Shtrikman, A Variational Approach to the Theory of the Elastic Behavior of Multiphase Materials, *J. Mech. Phys. of Solids*, Vol 11, 1963, p 127–140

31. L.K.H. Van Beek, Dielectric Behavior of Heterogeneous Systems, *Prog. Dielectrics*, Vol 7, 1967, p 69–114

32. E.J. Smoke, "Inorganic Dielectrics Research," Engineering Research Bulletin, No. 50, Rutgers University, 1970, p 120–140

33. D.A. Payne and L.F. Cross, Microstructure-Property Relationships for Dielectric Ceramics: Mixing of Isotropic Homogeneous Linear Dielectrics, *Ceramic Microstructures*, R.M. Fulrath and J.A. Pask, Ed., Vestview Press, p 584–597

34. F.F. Lange, Fracture of Brittle Matrix Particulate Composites, *Fracture and Fatigue*, Vol 5, LaBroutman, Ed., Academic Press, 1974, p 2–43

35. A.G. Evans, The Role of Inclusions in the Fracture of Ceramic Materials, *J. Mater. Sci.*, Vol 9, 1974, p 1145–1152

36. R. Rice, Microstructure Dependence of Mechanical Behavior of Ceramics, Properties and Microstructure, *Treatise on Materials Science and Technology*, Vol 11, Academic Press, 1977, p 199–381

37. D.P.H. Hasselman and R.M. Fulrath, Effect of Alumina Dispersions on Young's Modulus of a Glass, *J. Am. Ceram. Soc.*, Vol 48 (No. 4), 1965, p 218–219

38. D.P.H. Hasselman, On the Porosity Dependence of the Elastic Moduli of Polycrystalline Refractory Materials, *J. Am. Ceram. Soc.*, Vol 45 (No. 9), 1962, p 452–453

39. J.B. Wachtman, Jr., "Elastic Deformation of Ceramics and Other Refractory Materials," Special Publication 303, 1968, National Bureau of Standards, p 139–162

40. E.A. Dean and J.A. Lopez, Empirical Dependence of Elastic Moduli on Porosity for Ceramic Materials, *J. Am. Ceram. Soc.*, Vol 66 (No. 5), 1983, p 366–370

41. N.M.H. Jinno, Theoretical Approach to the Fracture of Two-Phase Glass-Crystal Composites, *J. Mater. Sci.*, Vol 7, 1972, p 973–982

42. R.M. Spriggs, Expression for the Effect of Porosity on the Elastic Modulus of Polycrystalline Refractory Materials, Particularly Aluminum Oxide, *J. Am. Ceram. Soc.*, Dec 1961, p 628–629

43. R.C. Rossi, Prediction of the Elastic Moduli of Composites, *J. Am. Ceram. Soc.*, Vol 51 (No. 8), 1968, p 433–439

44. R. Rice and T.J. Donahue, Effect of Inhomogeneous Porosity Distribution on Elastic Moduli of Ceramics, *J. Am. Ceram. Soc.*, Vol 62 (No. 5–6), 1979, p 306–307

45. R. Rice, Effect of Stresses from Thermal Expansion Anisotropy and Phase Transformations on the Strength of Ceramics, *J. Am. Ceram. Soc.*, Vol 63 (No. 11), 1980, p 703–710

46. J. Selsing, Internal Stresses in Ceramics, *J. Am. Ceram. Soc.*, Vol 8, 1961, p 419

47. J.E. Blendell and R.L. Cobel, Measurement of Stress Due to Thermal Expansion Anisotropy in Al_2O_3, *J. Am. Ceram. Soc.*, Vol 65 (No. 3), 1982, p 174–178

48. M.P. Borom, Dispersion-Strengthened Glass Matrices-Glass-Ceramics, A Case in Point, *J. Am. Ceram. Soc.*, Vol 60 (No. 1–2), 1977, p 17–21

49. A.G. Evans, *Fracture in Ceramic Materials*, Noyes, 1984, p 154–170

50. R.L. Bertolotti and R.M. Fulrath, Effect of Micromechanical Stress Concentrations on Strength of Porous Glass, *J. Am. Ceram. Soc.*, Vol 50 (No. 11), 1967, p 558–562

51. V.D. Krstic, Fracture of Brittle Solids in the Presence of Thermoelastic Stresses, *J. Am. Ceram. Soc.*, Vol 67 (No. 9), 1984, p 589–593

52. D.P.H. Hasselman and R.M. Fulrath, Micromechanical Stress Concentrations in Two-Phase Brittle-Matrix Ceramic Composites, *J. Am. Ceram. Soc.*, Vol 50 (No. 8), 1967, p 399–404

53. R.W. Rice, Ceramic Matrix Composites, *Proceedings of the American Ceramic Society*, 1980, p 661–681

54. F.F. Lange, The Interaction of a Crack Front with a Second-Phase Dispersion, *Philos. Mag.*, Vol 22 (No. 179), 1970, p 983–992

55. A.G. Evans, The Strength of Brittle Materials Containing Second Phase Dispersions, *Philos. Mag.*, Vol 26, 1972, p 1327–1344
56. A.G. Evans and A.H. Heuer, Review Transformation Toughening in Ceramics: Martensitic Transformations in Crack-Tip Stress Fields, *J. Am. Ceram. Soc.*, Vol 63 (No. 5–6), 1980, p 241–248
57. A.G. Evans and E.A. Charles, Fracture Toughness Determinations by Indentation, *J. Am. Ceram. Soc.*, Vol 59 (No. 7–8), 1976, p 179–182
58. K.T. Faber and A.G. Evans, Intergranular Crack-Deflection Toughening in SiC, *Comm. Am. Ceram. Soc.*, Vol 66, 1983, p C94-C96
59. F.F. Lange, M.R. James, and D.J. Green, Determinations of Residual Surface Stresses Caused by Grinding in Polycrystalline Al_2O_3, *Comm. Am. Ceram. Soc.*, 1973, p C17
60. A.A. Griffith, Phenomena of Rupture and Flow in Solids, *Philos. Trans. R. Soc.*, Vol 221A (No. 4), 1920, p 163–198
61. E.M. Passmore, R.M. Spriggs, and T. Vasilos, Strength-Grain Size-Porosity Relations in Alumina, *J. Am. Ceram. Soc.*, Vol 48 (No. 1), 1965, p 1–5
62. P.L. Swanson, C.J. Fairbanks, B.R. Lawn, Y.W. Mai, and B.J. Hockey, Crack-Interface Grain Bridging as a Fracture Resistance Mechanism in Ceramics: I, Experimental Study on Alumina, *J. Am. Ceram. Soc.*, Vol 70 (No. 4), p 279–289

63. H. Wieninger, K. Kromp, and R.F. Pabst, Crack Resistance Curves of Alumina and Zirconia at Room Temperature, *J. Mater. Sci.*, Vol 21, 1986, p 411–418
64. S. Weiderhorn, Brittle Fracture and Toughening Mechanisms in Ceramics, *Ann. Rev. Mater. Sci.*, 1984, p 373–402
65. R. Rice, Grain Size Dependence of Fracture Energy in Ceramics, *J. Am. Ceram. Soc.*, Vol 64 (No. 6), 1981, p 351
66. K.T. Faber and A.G. Evans, Crack Deflection Processes: I, Theory, *Acta Metall.*, Vol 31 (No. 4), 1983, p 565
67. K.T. Faber, Toughening Mechanisms for Ceramics in Automotive Applications, *Proceedings of the 12th Automotive Materials Conference*, American Ceramic Society, 1984, p 409–439
68. D.B. Binns and P. Popper, Mechanical Properties of Some Commercial Alumina Ceramics, *Proc. Brit. Ceram. Soc.*, Vol 6, 1966, p 71–82
69. R.A. Sack, *Proc. Phys. Soc.*, 1946, Vol 58, p 729
70. F.F. Lange, Fracture Energy and Strength Behavior of a Sodium Borosilicate Glass-Composite, *J. Am. Ceram. Soc.*, Vol 54 (No. 12), 1971, p 614–620
71. D.J. Green, P.S. Nicholson, and J.D. Embury, Fracture of a Brittle Composite, *J. Mater. Sci.*, Vol 13 (No. 5), 1979, p 1657–1661
72. N. Miyata, S.I. Chikawa, H. Monji, and H. Jinno, Fracture Behavior of Brittle Matrix, Particulate Composites with Thermal Expansion Mismatch, *Fracture*

Mechanics of Ceramics, Vol 7, R.C. Brandt, A.G. Evans, D.P.H. Hasselman, and F.F. Lange, Ed., Plenum Press, 1985
73. G.E. Hannon, "Alumina-Glass Composites: The Effect of Microstructure on Mechanical Properties," Ph.D. thesis, Rutgers University, 1988
74. M.A. Stett and R.M. Fulrath, Strengthening by Chemical Bonding in Brittle Matrix Composites, *J. Am. Ceram. Soc.*, Vol 51 (No. 10), 1968, p 599–600
75. D.P.H. Hasselman and R.M. Fulrath, Proposed Fracture Theory of Dispersion-Strengthened Glass Matrix, *J. Am. Ceram. Soc.*, Vol 49 (No. 2), 1966, p 68–72
76. R.L. Fullman, Measurement of Particle Size in Opaque Bodies, *Trans. AIME*, Vol 197 (No. 3), 1953, p 447–452
77. P.L. Studt and R.M. Fulrath, Mechanical Properties and Chemical Reactivity in Mullite-Glass Systems, *J. Am. Ceram. Soc.*, Vol 45 (No. 4), 1962, p 182–188
78. Y. Nivas and R.M. Fulrath, Limitations of Griffith Flaws in Glass-Matrix Composites, *J. Am. Ceram. Soc.*, Vol 53 (No. 4), 1970, p 188–191
79. L. Kowzan, Alumina-Glass Composites: The Effect of Grain Size on the Mechanical and Tribological Properties, M.S. thesis, Rutgers University, 1989
80. D.O. Nason, "Effect of Interfacial Bonding on Strength of a Model Two-Phase System," M.S. thesis, University of California, Berkeley, 1962

Engineering Properties of Glass-Ceramics

Anna E. McHale, Consultant

GLASS-CERAMICS are polycrystalline solids produced by the controlled crystallization of glass. Their use as engineering materials is well established, and new applications are constantly under development. Glass-ceramics have a thermodynamic advantage in stability over their parent glasses and can usually be used at temperatures near the solidus temperatures of the crystalline phases, which are generally much higher than the softening points of the parent glasses.

To produce glass-ceramics, glasses are melted, fabricated to shape and then converted to ceramic (that is, "ceramed") by specific heat treatment. A typical heat-treatment cycle is shown in Fig 1. The microstructure is characteristically fine grained ($\sim$1 μm, or 40 μin.) and is generally randomly oriented in bulk pieces, with zero or near-zero porosity.

Numerous commercial compositions are available in formed ware, as sheet and cane stock, or as fritted glass. Within each general chemical system, minor variations in chemistry, nucleation agent, and heat-treatment cycle can be used to "tailor" properties such as transparency and coefficient of thermal expansion (CTE). This capability is particularly useful in sealing and substrate applications for the electronics industry.

Most commercial compositions are based on silicate crystalline phases. However, the glass-ceramic process is suitable for the manufacture of many nonsilicate phases where suitable glass-forming and nucleation agent systems can be formed. Examples include $SrTiO_3$ glass-ceramics (Ref 2), other titanates for piezoelectric and pyroelectric applications (Ref 3), and ceramic superconductors in the $BaO-Y_2O_3-CuO$ system (Ref 4). A specialty glass-ceramic cement that contains a mixture of α-alumina and zirconia, which is used in the hot repair of glass tanks (Ref 5), is available.

Numerous tables that define many types of glass-ceramics, in terms of both commercial and noncommercial compositions and their physical properties are provided throughout the article. The noncommercial, or research formulation, examples obtained from the recent literature indicate the range of properties available to the materials designer. Because glass-ceramics are considered to be composite materials, final properties often follow simple additivity rules. Therefore, new compositions and heat treatments to create desired phase assemblies and microstructures allow an ever-increasing scope for the application of these versatile materials.

Chemical Systems

The $Li_2O-Al_2O_3-SiO_2$ (MgO, ZnO) System. The glass-forming region of the $Li_2O-Al_2O_3-SiO_2$ system is shown in Fig 2. The majority of commercial compositions are found in this system (with MgO and ZnO as additional components, in many cases). The final phases in these glass-ceramics determine their overall properties. They are primarily dependent on Al_2O_3-content and the heat-treatment schedule. The glass-ceramics can be either transparent or opaque, with excellent thermal shock resistance and CTE values that vary from 0 to 20×10^{-6}/K. Tables 1 to 3 describe glass-ceramics that are based on lithium disilicate, β-spodumene, and β-quartz, respectively.

With a near-zero or low Al_2O_3 content, the dominant phase is lithium disilicate ($Li_2Si_2O_5$). These glass-ceramics can be "activated" by ultraviolet (UV) exposure. Crystallization occurs only in exposed areas (Ref 9). The remaining glass can be removed by chemical etching, resulting in complex shape formation, along with extremely high tolerances and excellent surface smoothness. The product is suitable for magnetic and optical data storage media. For decorative uses, color can be varied from white through yellow to brown by varying exposure and heat treatment (Ref 10).

With increasing Al_2O_3 content, the stable phase becomes β-spodumene ($Li_2O-Al_2O_3-4SiO_2$) or β-eucryptite ($Li_2O-Al_2O_3-2SiO_2$). These glass-ceramics are durable and extremely resistant to thermal shock, having a near-zero CTE. β-spodumene glass-ceramics have general applications in domestic cookware and appliances. Mixtures of β-spodumene and mullite that have improved strength for thermomechanical applications in turbine heat exchangers can also be formed.

Using controlled nucleation and heat treatment, similar compositions can be formed into glass-ceramics that have β-quartz, a metastable phase, as the dominant crystalline phase. These glass-ceramics can be highly transparent. Upon the transformation β-quartz $\rightarrow$ β-spodumene (at about 1000 °C, or 1830 °F), the nucleant TiO_2 also undergoes transformation to rutile, resulting in the opacification of the glass-ceramic. If ZrO_2 is selected as the nucleant, transparency can be maintained (Ref 8). β-quartz glass-ceramics have excellent thermal stability and are therefore used in optical applications such as telescope mirror blanks, optical benches, and fiber-optic communications.

The $Na_2O-Al_2O_3-SiO_2$ System. Glass-ceramics in this system generally have high CTE values ($\sim$10 $\times$ 10^{-6}/K) and are suitable for glazing to enhance surface strength. The dominant phase is nepheline ($NaAlSiO_4$) (Ref 9).

The $Cs_2O-Al_2O_3-SiO_2$ System. Glass-ceramics in this system are of interest for their high refractoriness and chemical durability. They have been suggested for applications related to nuclear waste disposal. The dominant phases are mullite ($3Al_2O_3-2SiO_2$) and/or pullucite ($CsAlSi_2O_6$) (Ref 9).

The $MgO-Al_2O_3-SiO_2$ System. The glass-forming region of this system is shown in Fig 3. The dominant phases are cordierite ($2MgO-2Al_2O_3-5SiO_2$) (Table 4) or (clino)enstatite ($MgO-SiO_2$) (Table 5). These materials are characterized by higher-temperature stability under load than lithium aluminosilicates, good thermal shock resistance, and good mechanical shock resistance. They are under investigation for numerous advanced ceramics applications. In addition, their low permittivity and high electrical resistivity properties lead to wide application as microwave component materials. They are also used in radomes.

The $ZnO-Al_2O_3-SiO_2$ (B_2O_3) Systems. The glass-forming region in the $ZnO-Al_2O_3-SiO_2$ system is shown in Fig 4. In compositions that do not contain B_2O_3, the primary phase is willemite (Zn_2SiO_4), with minor

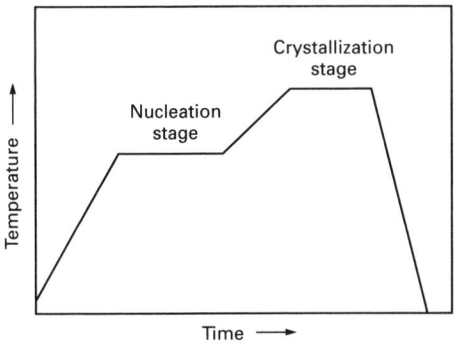

Fig 1 Typical two-stage heat-treatment cycle for bulk-glass ceramics. Source: Ref 1

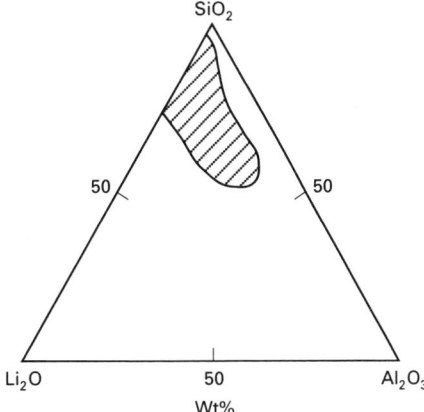

Fig 2 Glass-forming region in the $Li_2O-Al_2O_3-SiO_2$ system. Source: Ref 1

amounts of cristobalite (Table 6). In contrast to the systems previously discussed, P_2O_5 is often used as a nucleant. These zinc-containing systems have a significantly lower dielectric loss than lithium aluminosilicate-based glass-ceramics.

Glass-ceramics that do contain B_2O_3 possess crystalline zinc aluminates, zinc borates, and quartz as their crystalline phases. These compositions have been specifically developed for substrate applications (Ref 1).

Machinable glass-ceramics are based on crystalline phases such as phlogopite micas, which have a fine-grained "house-of-cards" microstructure (Table 7). Recently, fine-grained acicular $Ca_4Y_6O(SiO_4)_6$ has been formed in the system $CaO-Al_2O_3-Y_2O_3-SiO_2$. The phlogopite micas are considered to be easy to machine (Ref 11), whereas the latter materials require the use of carbide cutting tools.

Very fine surface finishes can be obtained on precisely machined parts.

Nepheline-, celsian-, and pollucite-based glass-ceramics are also said to be machinable using tungsten carbide tools (Ref 13). Their high chemical resistance makes them suitable for precision valves and nozzles in the chemical industry (Ref 13).

Machinable phlogopite micas and apatite-phlogopite glass-ceramics are biocompatible. Other biocompatible glass-ceramics are based on oxyfluorapatites, wollastonite ($CaSiO_3$), calcium phosphates, or leucite ($KAlSi_2O_6$) (Ref 5).

Other Systems. Specialty glass-ceramics with specific optical and dielectric properties have been produced for technical applications. These include transparent glass-ceramics for laser applications and new Sialon glass-ceramics for electronic applications (Table 8).

Other transparent glass-ceramics for laser applications are reported in the systems $RO-Al_2O_3-SiO_2$ (where $R = Mg$, Ca, Zn) having mullite, petalite, and/or spinel crystalline phases (Ref 5).

The glass-ceramic process is also used to produce architectural materials, such as synthetic granite or marble used in the decorative facing of buildings, especially in Eastern

Table 1 Primary phase lithium disilicate-based glass-ceramics

Product type	Manufacturer	Density, g/cm³	Coefficient of thermal expansion, 10⁻⁶/K	Modulus of rupture, MPa (ksi)	Ref	SiO₂	Al₂O₃	Li₂O	Na₂O/K₂O	Ag	Au	CeO₂	SnO₂	Sb₂O₃	Total alkali	Total alkaline earth	Nucleation
									Composition, wt%								
Commercial Fotoform	Corning	...	...	...	6	79.6	4.0	9.3	1.6/4.1	0.11	0.001	0.014	0.003	0.4	...	...	...
Noncommercial L4	...	2.40	9.3	380 (55)	1	(a)	5	...	...	...	...	...	...	...	15	5(BaO)	P₂O₅

(a) Balance is SiO₂.

Table 2 Primary phase β-spodumene-based glass-ceramics

Commercial product type	Manufacturer	Density, g/cm³	Coefficient of thermal expansion, 10⁻⁶/K	Ref	SiO₂	Al₂O₃	B₂O₃	Li₂O	Na₂O/K₂O	MgO/CaO	ZnO	As₂O₃/Sb₂O₃	Fe₂O₃	TiO₂/ZrO₂
								Composition, wt%						
C101 Cervit	Owens-Illinois	...	...	7	66.4	21.4	...	3.9	.../0.1	.../3.6	...	.../0.4	...	1.8/1.9
9608	Corning	2.5	0.4–2	7	69.7	17.1	...	2.5	0.4/0.1	2.8/...	...	.../0.2	...	4.8/0.1
9617	Corning	...	...	7	67.4	20.4	...	3.5	0.3/0.1	1.6/...	1.2	0.4/...	...	4.8/... (+0.2 F)
9455	Corning	...	...	7	71.8	22.9	...	5.1	...	...	...	...	...	0.1/...
0336	Corning	...	...	7	64.6	...	2.0	3.5	0.6/0.2	1.8/...	2.2	0.8/...	...	4.4/...
Corningware	Corning	...	...	6	69.7	17.8	...	2.8	0.4/0.2	2.6/...	1.0	0.6/...	0.1	4.7/0.1
Cercor	Corning	...	...	6	72.5	22.5	...	5.0	...	...	...	...	...	...

Noncommercial product type	Density, g/cm³	Coefficient of thermal expansion, 10⁻⁶/K	Modulus of rupture MPa	Modulus of rupture ksi	Ref	SiO₂	Al₂O₃	Total alkali	Total alkaline earth	Nucleant
								Composition, %		
L1	2.44	5.2	255	37	1	(a)	10	14	...	P₂O₅
L2	2.47	4.5	230	33	1	(a)	10	12	...	P₂O₅
L3	2.51	4.5	160	23	1	(a)	10	10	2(b)	P₂O₅
L5	2.59	5.0	175	25	1	(a)	10	10	9	P₂O₅

(a) Balance at SiO₂ assumed. (b) BaO only

Table 3 Primary phase β-quartz-based glass-ceramics

Product type	Manufacturer	Coefficient of thermal expansion, 10^{-6}/K	Modulus of rupture, MPa (ksi)	Ref	Composition, wt%									
					SiO_2	Al_2O_3	Li_2O	Na_2O/K_2O	MgO/CaO	BaO/ZnO	As_2O_5	P_2O_5	TiO_2/ZrO_2	Other
Commercial														
Vision	Corning	...	...	6	68.8	19.2	2.7	0.2/0.1	1.8/...	0.8/1.0	0.8	...	2.7/1.8	0.1 Fe_2O_3, 50 ppm CoO 50 ppm Cr_2O_3
Zerodur	Schott	0.03	...	6	56.5	25.3	3.7	0.5/...	1.0/...	.../1.4	0.5	7.0	2.3/1.9	0.03 Fe_2O_3
Narum 1	Nippon Electric and Glass	...	...	6	65.1	22.6	4.2	0.6/0.3	0.5/...	...	1.1	1.2	2.0/2.3	0.03 Fe_2O_3
9623 (T1)	Corning	0–0.5	...	8	65	23	3.8	...	1.8/...	.../1.5	0.9	...	2.0/2.0	...
Noncommercial														
T2	...	1.55	62 (9)	8	74	19.5	2	...	4.5/...	...	...	...	.../4	...
T3	...	3.1	83 (12)	8	65	25	...	...	10/...	.../6	1.0	...	.../10	...
T4	...	1.5	69 (10)	8	74	16.5	...	...	3.5/...	.../6	...	...	.../4	...

Note: All products are metastable phase and convert to β-spodumene if heated above 900–1000 °C (1650–1830 °F).

Table 4 Primary phase cordierite

Product type	Manufacturer	Density, g/cm³	Coefficient of thermal expansion, 10^{-6}/K	Modulus of rupture		Ref	Composition, wt%							Nucleant
				MPa	ksi		SiO_2	Al_2O_3	MgO	CaO	As_2O_3	Fe_2O_3	TiO_2	
Commercial														
9606	Corning	2.6–2.612	6.5	123–370(a)	18–54(a)	6	56.1	19.8	14.7	0.1	0.3	0.1	8.9	...
"Cordierite type"	...	2.61	1.1–5.7	240–350(b)	35–51(b)	5	...	...	...	...	...	...	...	...
Noncommercial														
M1	...	...	5.3	350	51	1	(c)	20	15	...	...	...	...	TiO_2
M2	...	...	3.2	162	24	1	(c)	26	12	...	...	...	...	TiO_2
M3	...	...	2.4	127	18	1	(c)	30	14	...	...	...	...	TiO_2
M4	...	...	3.1	270	39	1	(c)	26	12	...	...	...	...	TiO_2, ZrO_2

(a) Strength dependent on surface finish. (b) Higher if chemically strengthened. (c) Balance is SiO_2.

Table 5 Primary phase enstatite noncommercial glass-ceramics

Product type	Density, g/cm³	Coefficient of thermal expansion, 10^{-6}/K	Modulus of rupture		Ref	Composition, wt%					Comments
			MPa	ksi		SiO_2	Al_2O_3	MgO	Li_2O	ZrO_2	
E1	...	6.8	193 ± 15	28 ± 2	6	58	5.4	21	0.9	10.7	Enstatite, β-spodumene, and zirconia
E2	...	8.0	200 ± 15	29 ± 2	6	54	...	33	...	13.0	Enstatite, zircon, and zirconia
M5	3.18	8.2	750	109	1	(a)	22	22	...	(b)	(Clino)enstatite

(a) Balance is SiO_2. (b) Nucleation

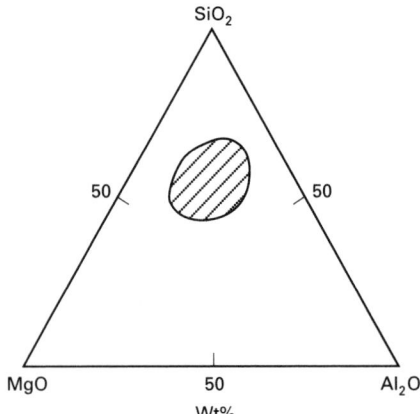

Fig 3 Glass-forming region in the MgO-Al_2O_3-SiO_2 system. Source: Ref 1

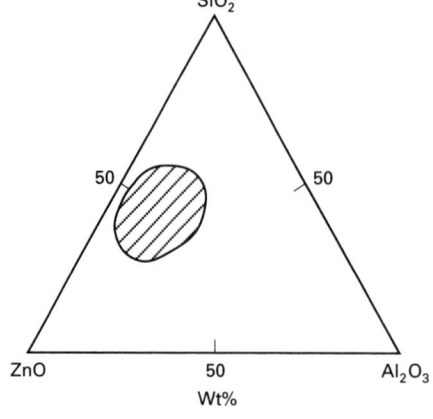

Fig 4 Glass-forming region in the ZnO-Al_2O_3-SiO_2 system. Source: Ref 1

Europe and Asia. The primary phases are wollastonite ($CaSiO_4$) or diopside ($CaMgSi_2O_6$) (Table 9). These materials may contain up to 99% residual glass and may have large crystals for aesthetic purposes. However, they are strong and durable and can be economical, because the raw materials are often the slag by-products of smelting operations. The β-spodumene glass ceramic Corning 0336 (see Table 2) is also used in architectural applications.

Piezoelectric and Pyroelectric Systems. Several systems have been investigated for potential application in advanced electromechanical devices (Table 10). Utilization of materials in this application requires that the crystallization be oriented, generally in a thin or surface film.

Table 6 Primary phase willemite (Zn_2SiO_4) and other zinc-containing phases of noncommercial glass-ceramics

Product type	Density, g/cm³	Coefficient of thermal expansion, 10^{-6}/K	Modulus of rupture MPa	Modulus of rupture ksi	Ref	SiO₂	Al₂O₃	B₂O₃	BaO	ZnO	Nucleant	Other
Z1	3.46	5.0	125	18	1	(a)	5–30	...	15	30	TiO₂	...
Z2	3.79	6.0	...	...	1	(a)	5–30	...	0	57	P₂O₅, TiO₂	...
Z3	3.70	3.0	95	14	1	(a)	5–30	...	15	42	P₂O₅, TiO₂	...
Z4	3.72	3.1	...	...	1	(a)	5–30	...	20	37	P₂O₅, TiO₂	...
Z5	3.68	3.8	...	...	1	(a)	5–30	...	5	47	P₂O₅, TiO₂	...
ZB1	...	4.1	180	26	1	20	20	30	...	20	...	Zn aluminate, Zn borate
ZB2	...	4.3	130	19	1	40	20	10	...	30	...	Quartz
ZB3	...	4.8	150	22	1	30	17	15	...	30	...	5 CaO, 3 P₂O₅
ZB4	...	4.9	115	17	1	26	16	15	...	35	...	5 BaO, 3 P₂O₅
ZB5	...	3.7	75	11	1	30	17	15	...	36	...	4 ZrO₂

(a) Balance is SiO₂.

Table 7 Machinable glass-ceramics

Product type	Manufacturer	Density, g/cm³	Coefficient of thermal expansion, 10^{-6}/K	Modulus of rupture MPa	Modulus of rupture ksi	Elastic modulus GPa	Elastic modulus 10^6 psi	Ref	SiO₂	Al₂O₃	Al₂O₃·Y₂O₃	CaO	B₂O₃	K₂O	MgO	CeO₂	F	ZrO₂
Commercial																		
Macor	Corning	2.52–2.63	6.3–9.7	60–102	8.7–14.8	~6.0	~8.7	5, 6	47.2	16.7	...	...	8.5	9.5	14.5	...	6.3	...
Dicor	Dentsply	...	...	...	...	...	...	6	56–64	0–2	...	...	...	12–18	15–20	0.06	4–9	0–5
Noncommercial																		
CAYS-20	...	3.23	5.3	...	...	105	15.2	12	50	...	30	20	...	...	...	...	...	...
CAYS-25	...	3.14	5.7	...	...	103	14.9	12	50	...	25	25	...	...	...	...	...	...
CAYS-30	...	3.10	5.8	...	...	100	14.5	12	50	...	20	30	...	...	...	...	...	...

(a) Commercial compositions given in wt%, noncommercial compositions given in mol% and are approximate

Table 8 Other glass-ceramics

Product type	Coefficient of thermal expansion, 10^{-6}/K	Ref	SiO₂	Al₂O₃	BaO	ZrO₂	MgO	ZnO	Cs₂O	Y/Mg	S	Al	O	N	Comments
Mullite															
T7	1.61	8	77	23	...	...	...	...	...	...	...	...	...	...	Transparent, heat treat 1000 °C (1830 °F)
T8	3.65	8	50	40	10	...	...	...	...	...	...	...	...	...	Transparent, heat treat 1000 °C (1830 °F)
Gahnite (ZrO₂ − t)															
T6	3.2	8	70	17	...	6	...	13	4	...	...	...	...	...	Transparent, heat treat 800–1000 °C (1470–1830 °F)
Spinel (ZrO₂ − t)															
T5	3.3	8	70	19	...	8	5	6	...	...	...	...	...	...	Transparent, heat treat 800–950 °C (1470–1740 °F)
Sialon															
YG2	...	14	...	...	...	...	...	...	...	15.2/...	14.6	8.7	54.6	6.9	(b)
Y3	...	14	...	...	...	...	...	...	...	15/...	15	10	45	15	(b)
E″	...	14	...	...	...	...	...	...	...	.../20	13.6	10	44.2	12.1	(c)
D	...	14	...	...	...	...	...	...	...	.../18.2	12	11.5	42.3	11.5	(c)

(a) Except for Sialon products, which are specified as at%. (b) 1050 °C (1920 °F) process temperature, Y₂SiAlO₅N phase. (c) 900 °C (1650 °F) process temperature, β″-Sialon.

Table 9 Primary phase wollastonite or diopside commercial slag-ceramics

Product type	Manufacturer	Density, g/cm³	Coefficient of thermal expansion, 10^{-6}/K	Modulus of rupture MPa	Modulus of rupture ksi	Ref	SiO₂	Al₂O₃	Na₂O/K₂O	MgO/CaO	ZnO	MnO	Fe₂O₃	S
Slagsital (wollastonite)	USSR	~2.6	7.2–9	80–120	12–17	5, 6(a)	55.2	8.3	5.4/0.6	2.2/24.8	1.4	0.9	0.3	0.4
Sigran	USSR	2.6–2.8	8–8.5	28–70	4–10	6	...	...	...	...	...	...	...	...
Minelbite (diopside)	Hungary	...	...	...	...	6	60.9	14.2	3.2/1.9	5.7/9.0	...	2.0	2.5	0.6
Neoparies(a)	Japan	2.7	6.2	50	7.3	5	...	...	...	...	...	...	...	...

(a) Composition given for white product. (b) Similar material, phase unknown

Table 10 Piezoelectric and pyroelectric glass-ceramics

Primary phase	Density, g/cm³	Dielectric constant	Tan δ	Specific heat, J/cm³·K	Pyroelectric coefficient, p, μCuries/m²·K	M₁(a), μCuries/m²/K	Ref	Composition, mol%							
								SiO₂/GeO₂	B₂O₃	BaO/SrO	Li₂O	CaO	TiO₂	PbO	ZnO
Ba₂TiSi₂O₈(b)	3.87–4.0	9–10.5	0.001	2.04	4.5–6.0	0.46–0.60	3	48–50/...	...	25–33/...	...	0–6.9	16–17	1.7	...
Ba₂TiSi₂O₈	4.05	10	0.001	2.04	8.0	0.80–0.88	3	50/...	...	32/...	...	...	17	...	...
Ba₂TiGe₂O₈(c)	4.78	15	...	...	–2.0	0.13	3	.../33	...	33/...	...	...	33	...	...
Sr₂TiSi₂O₈(b)	3.53–3.63	10.5–11.5	0.001	...	7.0–8.0	0.66–0.70	3	47–50/...	...	0–3.3/30–33	...	0–1.7	17	...	...
Li₂Si₂O₅(c)	2.47	6.5	0.008	2.36	–3.0	0.46	3	62/...	...	...	31	...	...	...	6
Li₂Si₂O₅	2.32–2.35	7.0–7.5	0.05	2.39	–10 to –11	1.33–1.57	3	53–60/...	6–13	...	33	...	...	...	...

(a) M₁ equals p/K and is used for rapid evaluation of material. (b) Tetragonal, 4 mm (0.16 in.). (c) Orthogonal, 2 mm (0.08 in)

Special Applications and Techniques

Fiber-Reinforced Glass-Ceramic Composites. Glass-ceramics in several chemical systems can be formed as silicon carbide fiber reinforced composite materials (Table 11). In contrast to conventional methods of melt-forming glass-ceramics, the glass is used as powder (often with additives such as Nb₂O₅ to promote interfacial reaction with fiber) and is mixed with a binder/solvent system to impregnate the fiber plies. The composites are laid up and densified using combined heat and pressure. A further heat-treatment step without applied pressure crystallizes the glass-ceramic.

Glass-Ceramic-to-Metal Seals. Glass-ceramics are suitable for many complex sealing applications. They can be formed using all conventional methods for obtaining glass-to-metal seals, including vacuum injection molding. A typical heat-treatment cycle is shown in Fig 5.

The high temperatures required for forming the seals and the thermomechanical properties of the resulting glass-ceramic seal are compatible with superalloys, although some changes in the alloy properties may be observed after processing. Glass-ceramic seals are stronger than conventional glass seals and extremely impervious to gases and environmental attack and may have varied properties, obtained by modifying the heat-treatment cycle (Table 12).

General Properties

Glass-ceramics are useful well above the softening point of the parent glass composition, but they are subject to creep deformation via diffusion through the residual glassy phase at higher temperatures (Ref 21), generally >600 °C (>1110 °F). Creep rate is inversely proportional to crystallite size. Thus, the use temperature limit, if limited permanent deformation is acceptable, is dependent on thermal process history (Table 13).

Like most ceramic materials, glass-ceramics are generally brittle and fail with lower stress in tension than in compressive loading (Table 14). Performance in service is affected by both a size effect (Ref 22) and an effect of stress loading rate, the latter not generally noted in brittle materials (Ref 23). The observed size effect correlates with fracture origin, either at surface flaws (most common) or at interior pores (rare). Lower strengths are observed in larger samples where the statistical probability of the existence of an interior pore in the sample volume becomes significant.

An approximate 30% reduction in strength may be expected in large pieces because of this effect. Statistical studies of failures in Corning 9606 (MAS-cordierite type) have indicated a narrow distribution of failure stress in most cases, which is highly dependent on surface quality (Ref 24).

Hardness and fracture toughness have been studied in several commercial glass-ceramic compositions, as well as in fiber-reinforced glass-ceramic composites. Generally, both nonreinforced and fiber-reinforced glass-ceramics that exhibit poor matrix-fiber interaction have fracture toughness (K_{Ic}) values of about 2, except for MAS-(clino)enstatite (E1 or E2) (Table 15).

The thermal and thermal-mechanical properties of selected glass-ceramics are listed in Tables 16 to 18, whereas dielectric properties at room temperature are provided in Table 19. Glass-ceramics are generally excellent elec-

Table 11 Fiber-reinforced (46% SiC fiber) noncommercial glass-ceramics

Product type	Primary phase	Major components	Minor components	Ref
LAS I	β-spodumene	SiO₂-Al₂O₃-Li₂O-MgO	ZnO, ZrO₂, BaO	15
LAS II	β-spodumene	LAS I + Nb₂O₅	ZnO, ZrO₂, BaO	15
LAS III	β-spodumene	LAS I + Nb₂O₅	ZrO₂	15
MAS	Cordierite	SiO₂-Al₂O₃-MgO	BaO	15
BMAS	Barium osmullite	SiO₂-Al₂O₃-BaO-MgO	...	15
Ternary mullite	Mullite	SiO₂-Al₂O₃-BaO	...	15
Hexacelsian	Hexacelsian	SiO₂-Al₂O₃-BaO	...	15

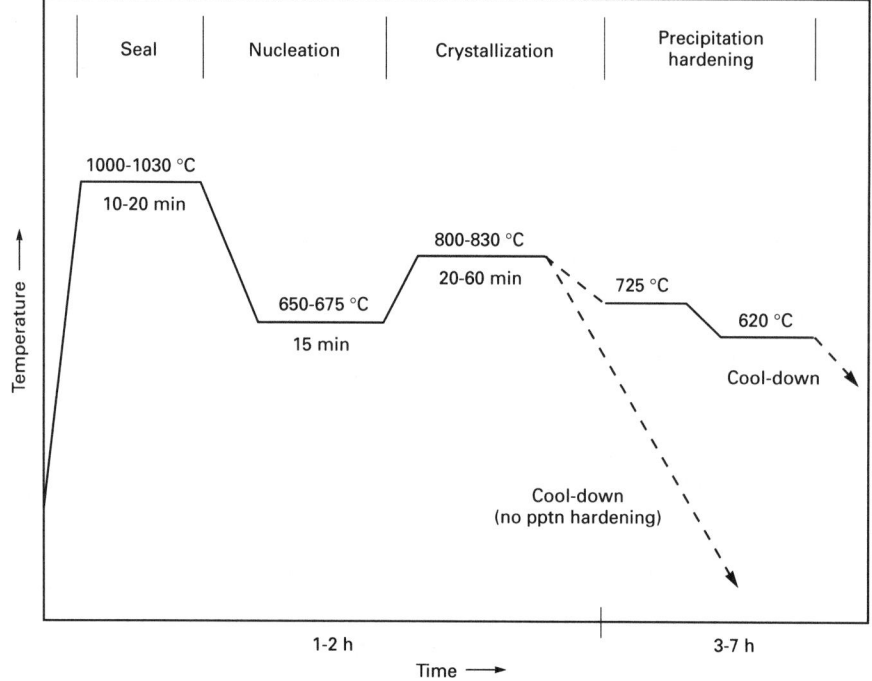

Fig 5 Heat-treatment cycle for forming glass-ceramic-to-metal seals (LAS type). Source: Ref 12

Table 12 Glass-metal seals

Product type	Manufacturer	Density, g/cm³	Coefficient of thermal expansion, 10⁻⁶/K	Modulus of rupture, MPa (ksi)	Ref	SiO₂	Al₂O₃	Li₂O/MgO	BaO/ZnO	Na₂O/K₂O	CaO	P₂O₅/B₂O₃	Fe₂O₃	Comments
Commercial														
7583	Corning	6.0	8.4	...	16	...	...	...	...	...	...	...	...	...
CVIII	Owens-Illinois	5.9	7	41–98 (6–14)(a)	16	...	...	...	...	...	...	...	...	...
Noncommercial(b)														
Austenitic SS (4–5 Al)	...	...	...	...	17	(c)	...	...	...	...	...	...	...	Preoxidize metal
304L SS	...	...	~20	...	17	(c)	...	...	...	...	...	...	...	...
304L SS	...	...	~17.5	...	12	83	<0.9	5.3/...	...	.../4.9	...	2.9/2.6	...	Proprietary heat treat
304L SS	...	...	13.5–17.5	...	12	80	5.1	5.3/...	...	.../4.5	...	2.6/1.4	...	CTE depends on heat treat
Molybdenum	...	...	...	...	17	47	9.75	...	4.75/32.5	3.75/...	...	2.2/...	...	...
Chromed steel	...	...	12.3–14.2	...	18	66	3.0	8.5/...	.../7.0	5.5/...	3.0	2.8/...	2.2	650–750 °C (1200–1380 °F) heat treat
Inconel 718	...	...	...	...	17(c)	...	...	...	...	...	...	...	...	...
Inconel 718	...	...	~14	...	19	67.1	2.8	23.7/...	...	.../2.8	...	1.0/2.6	...	Li₂Si₂O₅ type
Inconel 718	...	...	...	...	20	71.7	5.1	12.6/...	...	.../4.9	...	2.5/3.2	...	...

(a) Dependent on heat treatment, inversely proportional to crystallite size. (b) For sealing to metal. (c) Composition is LAS type. (d) Also recommended for Hastelloy C-276

Table 13 Maximum use temperature (no load)

Note: High-temperature use is generally limited to below 600 °C (1110 °F) under loaded conditions, but somewhat higher temperatures are possible for cordierite-type (9606, MAS) materials.

Product type	Manufacturer	Maximum temperature (no load) °C	°F	Ref	Comments
Zerodur	Schott	800	1470	13	β-quartz
LAS-type	...	800–900	1470–1650	5	β-eucryptite
Fluormicas					
Macor	Corning	950	1740	13	Creep over 750 °C (1380 °F)
Macor-type	...	800–1000	1470–1830	5	...
β-spodumene					
9608	Corning	1000	1830	13	...
LAS I	...	1000	1830	15	46% SiC fiber-reinforced composite
LAS II (Nb)	...	1100	2010	15	46% SiC fiber-reinforced composite
LAS III (Nb,Zr)	...	1200	2190	15	46% SiC fiber-reinforced composite
LAS-type	...	1200–1300	2190–2370	5	...
Cordierite					
9606	Corning	1100	2010	13	Creep over 900 °C (1650 °F)
MAS-type	...	1250–1300	2280–2370	5	...
MAS	...	1200	2190	15	46% SiC fiber-reinforced composite
Enstatite					
E1	...	1250	2280	6	Enstatite, β-spodumene, tetragonal zirconia
E2	...	1500	2730	6	Enstatite, zircon, tetragonal zirconia, cristobalite
Barium osmullite	...	1250	2280	15	Fiber-reinforced composite
Ternary mullite	...	~1500	~2730	15	Fiber-reinforced composite
Hexacelsian	...	~1700	~3090	15	Fiber-reinforced composite

Table 14 Elastic constants

Product type	Manufacturer	Modulus of elasticity, E GPa	10⁶ psi	∂E/∂T × 10⁵/K	Shear modulus, G GPa	10⁶ psi	Poisson's ratio, ν	Ref	Comments
9606	Corning	120	17.4	−13	...	...	0.22	13	Cordierite
9606	Corning	114–120	16.5–17.4	...	...	...	0.244	24	Cordierite
9608	Corning	88	12.8	...	...	...	0.25	13	β-spodumene
EE1087	GEC Ceramics Ltd.	92	13.3	...	37	5.4	0.245	13	Unknown
Zerodur	Schott	84	12.2	...	...	...	...	13	...
Macor	Corning	65	9.4	−6	30	4.4	0.27	13	Fluormica
Macor-type	...	59–64	8.6–9.3	...	...	...	...	5	Fluormica
LAS-type	...	52–98	7.5–14.2	...	...	...	...	5	β-eucryptite
LAS-type	...	49–87	7.1–12.6	...	...	...	...	5	β-spodumene
CIDI cervit	Owens-Illinois	92	13.3	...	...	...	...	1	β-spodumene
MAS-type	...	100–120	14.5–17.4	...	...	...	...	5	Cordierite

Table 15 Hardness and fracture toughness

Product type	Manufacturer	Ref	Hardness (load)(a)	K_{Ic}, MPa$\sqrt{m}$ (ksi$\sqrt{in.}$)
9606	Corning	13	660 (HK 0.1)	1.6–2.1 (1.5–1.9)
9608	Corning	13	590 (HK 0.1)	...
			650 (HK 0.05)	...
Zerodur	Schott	13	950 (HV 0.5)	...
Macor	Corning	13	420 (HK 0.02)	...
CAYS-20	...	25	~725 (HV 0.2)	...
CAYS-25	...	25	~700 (HV 0.2)	...
CAYS-30	...	25	~675 (HV 0.2)	...
MAS 46% SiC fiber reinforced [cordierite-type(b), zero additives]	...	26	...	4.7 ± 0.8 (4.3 ± 0.7)
MAS 46% SiC fiber reinforced [cordierite-type(b), 5% Nb_2O_5]	...	26	...	4.9 ± 1.3 (4.5 ± 1.2)
MAS 46% SiC fiber reinforced [cordierite-type(b), 5% Cr_2O_3]	...	26	...	2.0 ± 1.3 (1.8 ± 1.2)
E1	...	6	...	3.5 ± 0.4 (3.2 ± 0.4)
E2	...	6	...	4.6 ± 0.6 (4.2 ± 0.55)

(a) Load given as Vickers (HV) or Knoop (HK) indent values in kgf (9.8 N). (b) Ferro MAS8659 powder

trical insulators (bulk resistivities of about 10^8 to 10^{10} $\Omega \cdot$ cm or larger), although special compositions that are electrically conducting or superconducting can be formed. Dielectric constants are dependent on phase assembly and can be varied through the selection of a heat-treatment schedule.

Optical data are generally not reported on glass-ceramics, which are generally considered to be opaque materials. However, the MAS-type and LAS-type materials exhibit significant transparency between wavelengths of 1 and 4 μm (Ref 27). Transparent β-quartz glass-ceramics will transmit all shorter wavelengths, as well (Ref 8). Additional data on the spectral emittance, reflectance, and hemispherical transmission of MAS and LAS compositions for radome applications can be found in the literature (Ref 27).

Glass-ceramics possess good to excellent durability and weathering characteristics, consistent with comparable silicate-based ceramics (Table 20). The ASTM rankings shown in Table 20 are determined using evaluative criteria designed for conventional glasses and may not be meaningful, except as a guide (Ref 7). For instance, "weathering" is based on the appearance of surface clouding after exposure. Because most glass-ceramics are initially opaque, no discernible change is detected, and therefore the highest ranking is given. For specific applications, specialized testing methods are described in

Table 16 Coefficients of thermal expansion, 10^{-6}/K

Product type	Manufacturer	Ref	From 25 to −50 °C (77 to −60 °F)	From 25 to 100 °C (77 to 212 °F)	From 25 to 500 °C (77 to 930 °F)	From 25 to 1000 °C (77 to 1830 °F)
9606	Corning	13	1	6.5	4.8	4.5
9608	Corning	13	...	0.7–2	1–3	1.5–4
EE1087	GEC	13	...	...	10–11.5	...
Zerodur	Schott	13	0.05	0.03	...	...
Macor	Corning	13	...	8	10	13

Table 17 Thermal conductivity at T_c

Product type	Manufacturer	Ref	At −50 °C (−60 °F) W/m·K (Btu/ft·h·°F)	At 25 °C (77 °F) W/m·K (Btu/ft·h·°F)	At 100 °C (212 °F) W/m·K (Btu/ft·h·°F)	At 500 °C (930 °F) W/m·K (Btu/ft·h·°F)	At 1000 °C (1830 °F) W/m·K (Btu/ft·h·°F)
9606	Corning	13	4.5 (2.6)	3.6 (2.1)	3.6 (2.1)	3.3 (1.9)	3.1 (1.8)
9608	Corning	13	...	2.2 (1.3	2.3 (1.3)	2.4 (1.4)	...
EE1087	GEC Ceramics Ltd.	13	...	...	2 (1.2)	...	...
Zerodur	Schott	13	...	1.63 (0.9)	...	...	...
Macor	Corning	13	...	1.3 (0.8)	1.3 (0.8)	1.4 (0.8)	...
Slagsitall	USSR	5	...	0.9 (0.5)(a)	...	...	...
Sigran	USSR	5	...	0.9–1.2 (0.5–0.7)(a)	...	...	...
Neoparies	Japan	5	...	1.4 (0.8)	...	...	...

(a) Temperature not specified

Table 18 Thermal diffusivity at T_c

Product type	Manufacturer	Ref	At 25 °C (77 °F) 10^{-6} m²/s	At 25 °C (77 °F) 10^{-3} in.²/s	At 100 °C (212 °F) 10^{-6} m²/s	At 100 °C (212 °F) 10^{-3} in.²/s	At 500 °C (930 °F) 10^{-6} m²/s	At 500 °C (930 °F) 10^{-3} in.²/s	At 1000 °C (1830 °F) 10^{-6} m²/s	At 1000 °C (1830 °F) 10^{-3} in.²/s
9606	Corning	13	1.8	2.8	1.6	2.5	1.1	1.7	0.9	1.4
9608	Corning	13	1.1	1.7	1.0	1.6	0.8	1.2	...	...
EE1087	GEC Ceramics Ltd.	13	...	...	0.9	1.4	...	...	...	...
Zerodur	Schott	13	...	...	0.79	1.2	...	...	...	...
Macor	Corning	13	0.7–8	1.1–12.4	0.6–7	0.9–10.9	0.5–6	0.8–9.3	0.5–6	0.8–9.3

Table 19 General dielectric properties

Product type	Ref	Dielectric constant 1 MHz	Dielectric loss, tan δ, 1 MHz
9606(a)	(b)	5.6	0.017
9608(a)	(b)	6.9	0.023
LAS(c)	5	6.9–9.1	0.006–0.02
LAS(d)	5	5.5–7.8	0.0026–0.0090
	1	5.2–6.0	0.007–0.014
ZAS(e)	1	6.4–7.4	0.0005–0.0022
MAS(f)	5	5.6–7.5	0.0015–0.003
	1	5.6–8.5	0.0003–0.0005
MAS(g)	1	8.5	0.006
Macor type	5	5.9–6.1	0.0006–0.003
	14(h)	4.7(h)	0.0024(h)
SrTiO$_3$(i)	2	~35	0.005–0.01
MgSiAlON-type			
E″	14	9.1(h)	0.0029(h)
D	14	8.6(h)	0.003(h)
YSiAlON-type			
YG2	14	10.5–11.3(h)	0.0031–0.004(h)

(a) Manufactured by Corning. (b) Product literature. (c) β-eucryptite type. (d) β-spodumene type. (e) Broad composition range. (f) Cordierite type. (g) Enstatite type. (h) Data are measured at 1 to 20 kHz. (i) 35% glass former.

Table 20 Chemical durability, rated on ASTM C-225 basis where 1 is best, 4 is worst

Product type	Manufacturer	Ref	In 5 wt% HCl, 95 °C (205 °F), 24 h	In water, 90 °C (195 °F), 4 h	At 98% relative humidity, 50 °C (120 °F), 3 mo
0335	Corning	7	3	1	1
9455	Corning	7	4	1	1
9606	Corning	7	4	1–2	1
9608	Corning	7	2	1	1
9617	Corning	7	2	1	1

Product type	Manufacturer	Ref	In 5 wt% HCl, 95 °C (205 °F), 24 h	In 5% NaCO$_3$	In 5% NaOH	In H$_2$SO$_4$, varying concentrations and conditions	In HCl, varying concentrations and conditions
LAS(a)	...	5	3	0.04(b)	...	...	...
LAS(c)	...	5	3	0.1–0.2(b)	0–0.03(b)	...	...
MAS(d)	...	5	4	0.1–1.0(b)	0.01–0.1(b)	...	...
CVIII	Owens-Illinois	16	...	...	...	6.45(e)	...
		16	...	...	...	0.08(f)	...
		16	...	...	...	1.4(g)	...
		16	...	...	...	0.88(h)	...
		16	...	...	...	74.4(i)	...
		16	...	...	...	...	3.4(j)
		16	...	...	...	...	6.4(k)

(a) β-eucryptite type. (b) Given in units of mg/cm^2 material loss. (c) β-spodumene type. (d) Cordierite type. (e) In 50% H$_2$SO$_4$, 95 °C (205 °F), 1 h; units in mg/cm^2 material loss after washing. (f) In 10% H$_2$SO$_4$, 20 °C (68 °F), 10 min; units in mg/cm^2 material loss after washing. (g) In 18 N H$_2$SO$_4$, 50 °C (120 °F), 5 min; units in mg/cm^2 material loss after washing. (h) In 50% H$_2$SO$_4$ and 25% HNO$_3$, 25 °C (77 °F), 5 min; units in mg/cm^2 material after washing. (i) In +50% H$_2$SO$_4$ for 15 min and 5% HNO$_3$ for 5 min, at 95 °C (205 °F); units mg/cm^2 material loss after washing. (j) In 1 N HCl, 25 °C (77 °F) 5 min; units in mg/cm^2 material loss after washing. (k) In 10% HCl, 20 °C (68 °F), 10 min; units in mg/cm^2 material loss after washing

the industrial literature and in the publications of the International Glass Commission (ICG).

ACKNOWLEDGMENT

The assistance of the faculty and library staff of the New York State College of Ceramics at Alfred University is gratefully acknowledged.

REFERENCES

1. G. Partridge, C.A. Elyard, and M.I. Budd, Glass Ceramics in Substrate Application, *Glass: Science and Technology*, D.R. Uhlmann and N.J. Kreidel, Ed., Academic Press, 1984, p 226–271

2. S.L. Swartz and A.S. Bhalla, Dielectric Properties of SrTiO$_3$ Glass Ceramics, *Ferroelectrics*, Vol 87, 1988, p 141–154

3. A. Halliyal, A.S. Bhalla, R.E. Newnham, and L.E. Cross, Glass Ceramics for Piezoelectric and Pyroelectric Devices, *Glass: Science and Technology*, D.R. Uhlmann and N.J. Kreidel, Ed., Academic Press, 1984, p 297–315

4. A. Bhargava, A.K. Varshneya, and R.L. Snyder, On the Stability of Superconducting Y$_1$Ba$_2$Cu$_3$O$_{(6+\delta)}$ in a Borate Glass-Ceramic Matrix, *Mater. Lett.*, Vol 8 (No. 12), 1989, p 41–45

5. P.D. Sarkisov, The Modern State of Technology and Application in Glass Ceramics, *Glass '89, Proc. XV Int. Conf. Glass*, Leningrad, 1989

6. G.H. Beall, Design of Glass Ceramics, *Rev. Solid State Sci.*, Vol 3 (No. 3–4), 1989, p 333–354

7. D.E. Campbell and H.E. Hagy, Glasses and Glass Ceramics, *CRC Handbook of Materials Science*, C.T. Lynch, Ed., CRC Press, 1975

8. G.H. Beall and D.A. Duke, Transparent Glass Ceramics, *J. Mater. Sci.*, Vol 4 (No. 4), 1969, p 340–352

9. G.H. Beall, Property and Process Development in Glass-Ceramic Materials, *Glass…Current Issues*, A.F. Wright and J. Dupuy, Ed., Martinus Nijhoff, 1985, p 31–48

10. N.D. Stephenson, Photosensitive Glass Ceramics—A New Generation in Decoration, *Ceram. Eng. Sci. Proc.*, Vol 7 (No. 9–10), 1986, p 1359–1361

11. C. Saso, Machinable Glass Ceramics, *Ceram. Eng. Sci. Proc.*, Vol 3 (No. 7–8), 1982, p 405–409

12. W.E. Moddenan, R.E. Pence, R.T. Massey, R.T. Cassidy, and D.P. Kramer, Sealing 304L to Lithia-Alumina-Silica (LAS) Glass Ceramics, *Ceram. Eng. Sci. Proc.*, Vol 10 (No. 9–10), 1989, p 1394–1402

13. R. Morrel, *Handbook of Properties of Technical and Engineering Ceramics. Part I: An Introduction for the Engineer and the Designer*, National Physical Laboratory, H.M. Stationery Office, London, 1985

14. D.S. Perera, D.P. Thompson, and J.S. Thorpe, Nitrogen Glass Ceramics in the Mg-Si-Al-O-N and Y-Si-Al-O-N Systems, *Mater. Sci. Forum*, Vol 34–36, 1988, p 633–637

15. K.M. Prewo, Fibre Reinforced Glasses and Glass Ceramics, *Glass: Science and Technology*, D.R. Uhlmann and N.J. Kreidel, Ed., Academic Press, 1984, p 337–368

16. N.N. SinghDeo, Solder Glass Processing, *Glass: Science and Technology*, D.R. Uhlmann and N.J. Kreidel, Ed., Academic Press, 1984, p 170–206

17. L.F. Taswell and M.W. Jones, Glass Ceramic to Metal Seals, *Glass Technol.*, Vol 31 (No. 2), 1990, p 44

18. S.G. Dzhavuktsyan, High Temperature Glass Ceramic Coatings, *Glass Ceram.*, Vol 44 (No. 3–4), 1987, p 177–179

19. R.E. Loehman, S.C. Kunz, and R.D. Watkins, Reactions at Glass Ceramic to Metal Interfaces, *Ceram. Eng. Sci. Proc.*, Vol 7 (No. 7–8), 1986, p 721–726

20. R.E. Loehman et al., Effects of Atmosphere and Dew Point on the Wetting Characteristics of a Glass Ceramic on Two Nickel Based Superalloys, *Ceram. Eng. Sci. Proc.*, Vol 4 (No. 9–10), 1983, p 740–750

21. R. Raj and C.K. Chang, Solution Precipitation Creep in Glass Ceramics, *Acta Metall.*, Vol 29 (No. 1), 1981, p 159–166

22. G.K. Bansal, W. Duckworth, and D.E. Niesz, Strength-Size Relationships in Ceramic Materials: Investigation of a Commercial Glass Ceramic, *Am. Ceram. Soc. Bull.*, Vol 55 (No. 3), 1976, p 289–292

23. L. Lankford, Strength of Monolithic and Fiber-Reinforced Glass Ceramics at High Rates of Loading and Elevated Temperatures, *Ceram. Eng. Sci. Proc.*, Vol 9 (No. 7–8), 1988, p 843–852

24. D. Lewis III, Observations on the Strength of a Commercial Glass Ceramic, *Am. Ceram. Soc. Bull.*, Vol 61 (No. 11), 1982, p 1208–1214

25. A. Makishima, M. Asami, and Y. Ogura, A Machinable Calcia-Alumina-Yttria-Silica Glass Ceramic, *J. Am. Ceram. Soc.*, Vol 72 (No. 6), 1989, p 1024–1026

26. R. Chaim, D.G. Brandon, and L. Baum,

Mechanical Properties and Microstructural Characterization of SiC Fiber-Reinforced Cordierite Glass Ceramics, *Ceram. Eng. Sci. Proc.*, Vol 9 (No. 7–8), 1988, p 695–704

27. Y.S. Touloukian and C.Y. Ho, Thermophysical Properties of Selected Aerospace Materials, Part I: Thermal Radiative Properties, Candas, Purdue University, 1976

SELECTED REFERENCES

- J.F. Bednarik and P.W. Richter, A Machinable Glass Ceramic Based on a Crystalline Phase Other Than Fluorophlogopite, *Glass Technol.*, Vol 27 (No. 2), 1986, p 60–68
- T.R. Dinger, Microstructure Development during Controlled Crystallization of M-Si-Al-O-N Glass Ceramics, *Mater. Sci. Forum*, Vol 47, 1989, p 119–131
- G. Partridge, A Review of Surface Crystallization in Vitreous Systems, *Glass Technol.*, Vol 28 (No. 1), 1987, p 9
- G. Partridge, Glass Ceramic Material for Use in Substrate and Packaging Applications, *Glass Technol.*, Vol 30 (No. 6), 1989, p 215
- J.L. Plawsky, G.E. Williams, and P.A. Sachenik, Photochemically Machined Glass Ceramic Optical Fiber Interconnection Components, *Proc. SPIE Int. Opt. Eng.*, Vol 994, 1989, p 101
- K.M. Prewo, Tension and Flexural Strength of Silicon Carbide Fibre-Reinforced Glass Ceramics, *J. Mater. Sci.*, Vol 21 (No. 10), 1986, p 3590–3600
- N.E. Priestley, Glass Ceramic Substrates in Advanced Microwave Integrated Circuits, *Glass Technol.*, Vol 31 (No. 1), 1990, p 7–10
- A.J. Sturgeon, D. Holland, G. Partridge, and C.A. Elyard, Reactions at the Interface Between a Glass Ceramic and a Metal and Their Effect on Devitrification, *Glass Technol.*, Vol 27 (No. 3), 1986, p 102–107

Crystal Structures

Dawn A. Bonnell, The University of Pennsylvania

THE CRYSTAL STRUCTURES of ceramic compounds are best described in terms of structural classifications. A wide range of structures can be classified according to cation-anion (metal-nonmetal) ratios and are usually designated AX, AX_2, AB_2X_4, and so forth, where A and B represent cations and X the anion (which is often oxygen in the case of ceramics) (see Table 1). The particular structure adopted by a compound will depend on the ratio of the radii of the constituent elements and the type of bonding, that is, the degree of covalency. Most simple structures can be built upon the basis of a nearly close-packed anion lattice, with cations residing in available interstices. An illustration of the cubic and hexagonal close-packed lattices indicating the positions of tetrahedral and octahedral interstices is shown in Fig 1.

In this article, the name associated with the structure is given along with its space group designation, the structure is described in terms of close-packed lattices, and then the compounds exhibiting the structure are listed. Several compounds of special interest are discussed at the end of the article.

AX Structures

The Rock Salt Structure ($Fm3m$). Ionically bonded binary compounds with radius ratios between 0.73 and 0.41 favor octahedral coordination and crystallize in the cubic rock salt structure. The structure consists of a cubic close-packed anion lattice in which all octahedral interstitial positions are filled with cations (see Fig 1b). Both anions and cations are octahedrally coordinated.

Compounds exhibiting this structure include MgO, CaO, SrO, BaO, VO, MnO, FeO, CoO, NiO, MgS, CaS, SrS, BaS, MnS, CaS, SrS, BaS, PbS, LiF, LiCl, LiBr, LiI, NaF, NaCl, NaBr, NaI, KF, KCl, KBr, KI, RbF, RbCl, RbBr, RbI, CsF, AgF, AgCl, AgB, and transition metal nitrides.

The Nickel Arsenide Structure ($P6_3/mmc$). Binary compounds with radius ratios between 0.73 and 0.41 possessing some degree of covalent bonding will adopt the nickel arsenide structure. The structure consists of a hexagonal close-packed lattice of anions with cations filling octahedral interstices (Fig 1b).

Both anions and cations are octahedrally coordinated.

Compounds exhibiting this structure include NiAs, NiS, FeS, CoS, VS, and transition metals combined with Sn, Pb, As, Sb, Bi, Se, or Te.

Wurtzite Structure ($P6_3mc$). Binary compounds with radius ratios between 0.22 and 0.41 having a strong covalent contribution to bonding will favor tetrahedral coordination for both constituents. In wurtzite, two hexagonally close-packed interpenetrating lattices satisfy this requirement (Fig 2a).

Structures exhibiting this structure include ZnS, BeO, ZnO, SiC, AlN, GaN, and InN.

Zinc Blende Structure ($F43m$). Binary compounds with radius ratios between 0.22 and 0.41 and a strong covalent contribution to bonding can also be stable in the zinc blende structure. This structure is similar to diamond except that each atom is tetrahedrally coordinated by the other element. The structure can also be thought of as two interpenetrating cubic close-packed lattices (Fig 2b).

Compounds exhibiting this structure include ZnS, BeO, and SiC.

The Cesium Chloride Structure ($Pm3m$). For binary compounds with radius ratios greater than 0.73, stability requires that the atoms be eight-fold coordinated. This is accomplished in a simple cubic lattice of one element with the large interstice at the body-center position filled by the other element (Fig 3a).

Compounds exhibiting this structure include CsCl, CsBr, and CsI.

AX₂ Structures

The Fluorite Structure ($Fm3m$). For binary compounds in which the charge of the cation is twice that of the anion, charge neutrality requires twice as many anions as cations in the structure. For compounds with radius ratios greater than 0.73, eight-fold coordination is favored. This is realized in the fluorite structure which consists of a simple cubic anion lattice with half of the body-center positions occupied by cations. This structure is the same as cesium chloride with the exception that only half of the cation sites are filled (Fig 3b). A related structure, antifluorite, exists in which the anion and cation sites are reversed.

Compounds exhibiting this structure include HfO_2, ZrO_2, HfF_2, ZrF_2, SrCl_2, PrO_2, CeO_2, UO_2, CaF_2, CdF_2, ThO_2, BaCl_2, HgF_2, SrF_2, PbF_2, and BaF_2.

The Rutile Structure ($P4_2/mnm$). For AX_2 compounds with radius ratios between 0.41 and 0.73, the stable structure may be rutile in which cations are octahedrally coordinated by anions and each anion is bonded to three cations. Cations fill only half the available octahedral sites, and the closer packing of oxygen ions around the filled cation sites leads to the distortion of the nearly close-packed anion lattice.

Compounds exhibiting this structure include GeO_2, PbO_2, SnO_2, MnO_2, VO_2, RuO_2, TiO_2, IrO_2, OsO_2, NbO_2, WO_2, CrO_2, MoO_2, MgF_2, FeF_2, ZnF_2, PbO_2, MnF_2, TaO_2, NiF_2, NbO_2, CoO_2, and CaBr_2.

The Structures of SiO₂. Crystalline silica is stable in three phases: quartz, tridymite, and cristobalite, all three of which exist in high- and low-temperature modifications. The ideal cristobalite structure can be described in terms of the zinc blende lattice. The A atoms of cristobalite occupy both the A and X sites of the zinc blende structure with one X atom midway between each pair of A atoms (compare Fig 2b and 4). Each A atom is then surrounded by four X atoms which form a tetrahedron. Each X atom is coordinated by two A atoms. The ideal tridymite structure is related to wurtzite in the same way that cristobalite is to zinc blende. If all A and X positions in wurtzite are replaced by A and X atoms are introduced midway between the A atoms, the tridymite structure results (Fig 2a). The A atoms are again tetrahedrally coordinated while the X atoms are surrounded by two A atoms. In quartz the atoms are coordinated in the same manner and the silica tetrahedra are arranged in a hexagonal array.

The most stable structures are:

- Low quartz—below 573 °C (1063 °F)
- High quartz—573 to 867 °C (1063 to 1593 °F)
- High tridymite—867 to 1470 °C (1593 to 2678 °F)

Table 1 Summary of ceramic crystal structures

Class	Name	Anion lattice	Coordination	Cation occupation
AX	Rock salt	Cubic close-packed	6:6	All octahedral sites
AX	Nickel arsenide	Hexagonal close-packed	6:6	All octahedral sites
AX	Cesium chloride	Simple cubic	8:8	All cubic sites
AX	Wurtzite	Hexagonal close-packed	4:4	$\frac{1}{2}$ octahedral sites
AX	Zinc blende	Cubic close-packed	4:4	$\frac{1}{2}$ tetrahedral sites
AX_2	Fluorite	Simple cubic	8:4	$\frac{1}{2}$ cubic sites
AX_2	Rutile	Tetragonal close-packed	6:3	$\frac{1}{2}$ octahedral sites
AX_2	Silica	Connected tetrahedra	4:2	
A_2X	Antifluorite	Cubic close-packed	4:8	All tetrahedral sites
A_2X_3	Corundum	Hexagonal close-packed	6:4	$\frac{2}{3}$ octahedral sites
ABX_3	Perovskite	Cubic close-packed	12:6:6	$\frac{1}{4}$ octahedral sites
ABX_3	Ilmenite	Hexagonal close-packed	6:6:4	$\frac{2}{3}$ octahedral sites
AB_2O_4	Spinel	Cubic close-packed	4:6:4	$\frac{1}{8}$ tetrahedral sites (A) and $\frac{1}{2}$ octahedral sites (B)
$B(AB)O_4$	Inverse spinel	Cubic close-packed	4:6:4	$\frac{1}{8}$ tetrahedral sites (B) and $\frac{1}{2}$ octahedral sites (A, B)
AB_2O_4	Olivine	Hexagonal close-packed	6:4:4	$\frac{1}{2}$ octahedral sites (A) and $\frac{1}{8}$ tetrahedral sites (B)

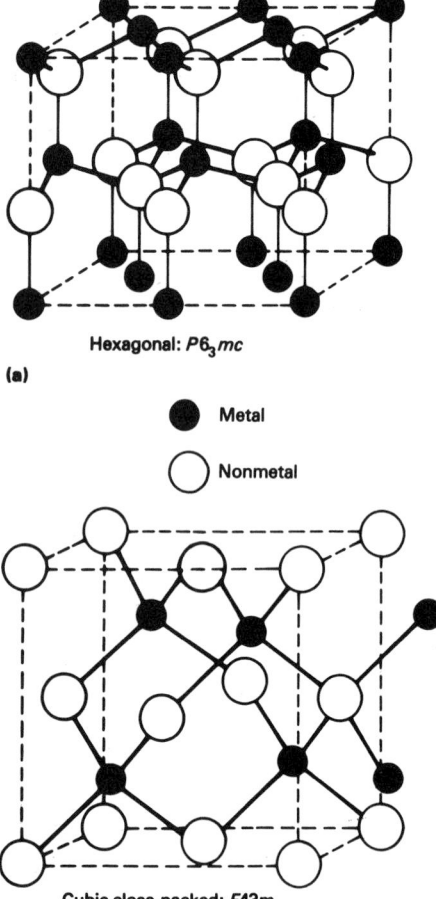

Hexagonal: $P6_3mc$

(a)

● Metal

○ Nonmetal

(b)

Fig 2 The wurtzite (a) and zinc blende (b) structures illustrating two arrangements of atoms which result in tetrahedral coordination of both constituents

- High cristobalite—1470 to 1710 °C (2678 to 3110 °F)

The low-temperature modifications are distorted versions of the high-temperature forms.

A_2X Structures

Antifluorite Structure. An oxide structure consisting of a cubic close-packed lattice of oxygen atoms with cations arranged in the tetrahedral sites has cation and anion posi-

Cubic close-packed: $F\bar{4}3m$

tions exactly reversed from the normal fluorite lattice.

Compounds exhibiting this structure include Li_2O, Na_2O, K_2O, Rb_2O, and sulfides.

A_2X_3 Structures

The corundum structure ($R\bar{3}c$) consists of a nearly hexagonal close-packed anion lattice with cations filling two-thirds of the octahedral sites (Fig 1d). The cations are in approximate octahedral coordination. When cations fill adjacent octahedra they repel each other so that neither is centered in its site. The anion positions are also distorted by 2 to 4%.

Compounds exhibiting this structure include Al_2O_3, Fe_2O_3, Cr_2O_3, V_2O_3, Ti_2O_3, Rh_2O_3, and Ga_2O_3.

ABX_3 Structures

Perovskite Structures ($Pcmn$). To a first approximation, the perovskite structure consists of a cubic close-packed anion lattice in which one-fourth of the ions are replaced by

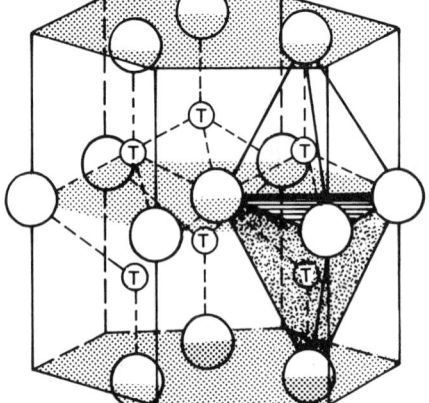

(a)

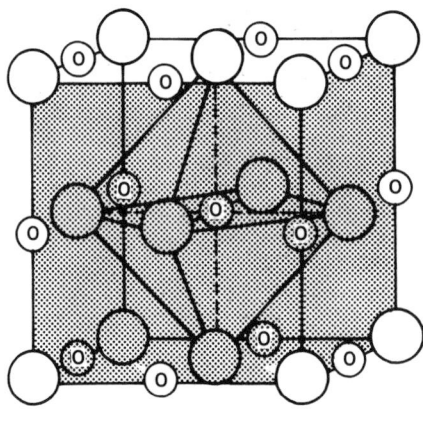

(b)

(c)

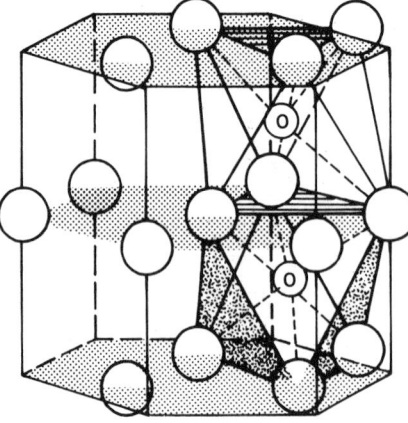

(d)

Fig 1 Cubic closed-packed (a and b) and hexagonal close-packed (c and d) lattices indicating the positions of tetrahedral (a and c) and octahedral (b and d) interstices

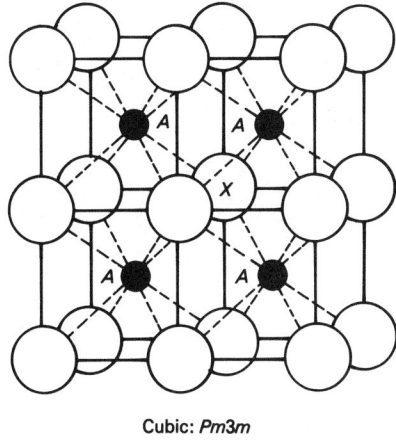

Cubic: *Pm3m*

(a)

● A (cation)

○ X (anion)

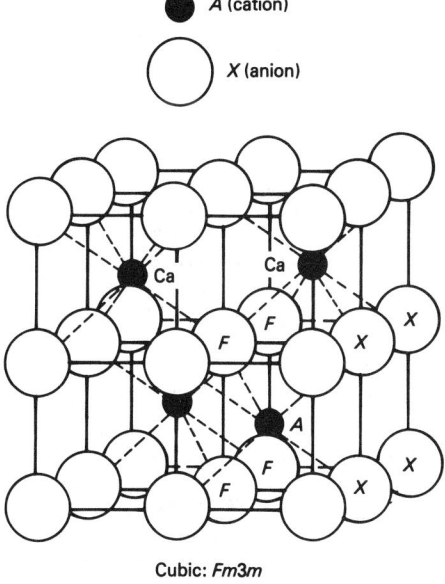

Cubic: *Fm3m*

(b)

Fig 3 The cesium chloride (a) and fluorite (b) structures, which differ only in the number of cations per unit cell. In cesium chloride all body-centered positions are occupied while in fluorite only half are occupied.

a large cation and a smaller highly charged cation occupies one-fourth of the octahedral interstices. The structure is illustrated for the $CaTiO_3$ structure in Fig 5. Each oxygen ion is surrounded by four calcium ions and eight oxygen ions, each calcium ion is surrounded by twelve oxygen ions, and the titanium ion is octahedrally coordinated. The stability of this ideal structure depends critically on the relative sizes of the cations. Most compounds form distorted perovskite structures in which the atoms are slightly displaced from their ideal positions. This reduces the symmetry of the structure so that most perovskites are orthorhombic or monoclinic.

The perovskite structure leads to the stability of a wide range of solid solutions. For example, combining titanates with cations of

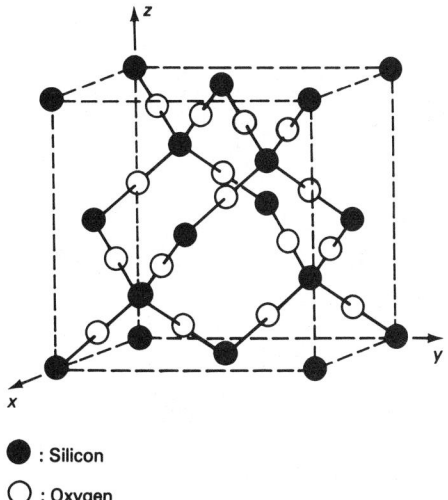

● : Silicon

○ : Oxygen

Fig 4 The unit cell of the idealized cristobalite structure

the same valence, such as $BaTiO_3$ and $SrTiO_3$, will lead to a stoichiometric solid solution. It is also possible to substitute an element of a different valence, in which case the resulting charge imbalance is accommodated through the formation of oxygen vacancies. Some of the compounds can accommodate a large number of oxygen defects and remain stable. The cuprate-based oxide superconductors are of this class of structures.

Compounds exhibiting this structure include $CaTiO_3$, $BaTiO_3$, $SrTiO_3$, $CdTiO_3$, $PbTiO_3$, $SrSnO_3$, $CaZrO_3$, $SrZrO_3$, $PbZrO_3$, $BaZrO_3$, $CaSnO_3$, $SrSnO_3$, $BaSnO_3$, $CdCeO_3$, $CaCeO_3$, $SrCeO_3$, $PbCeO_3$, $BaCeO_3$, $SrHfO_3$, $BaHfO_3$, $BaPrO_3$, $BaThO_3$, $KNbO_3$, $NaNbO_3$, $LaAlO_3$, $YAlO_3$, $LaMnO_3$, $LaFeO_3$, $LaCrO_3$, $KMgF_3$, $PbMgF_3$, $KZnF_3$, $KNiF_3$, $NaWO_3$, $LaBa_2Cu_3O_{7-x}$, and other oxide superconductors.

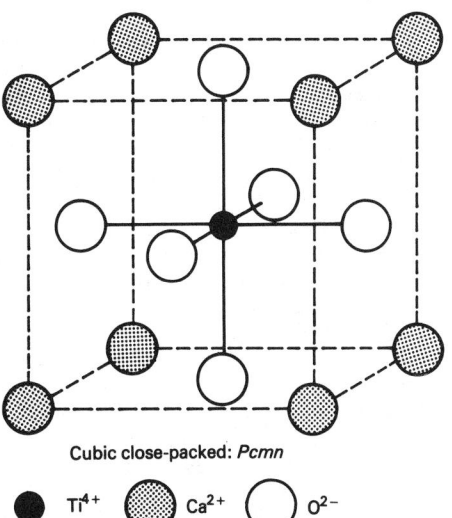

Cubic close-packed: *Pcmn*

● Ti^{4+} ◍ Ca^{2+} ○ O^{2-}

Fig 5 The unit cell of perovskite showing the relative positions of the tetravalent (Ti^{4+}) and divalent (Ca^{2+}) cations

The Ilmenite Structure ($R3$). Compounds in which the relative sizes of the cations do not allow the perovskite structure to be stable will adopt the ilmenite structure. This structure is similar to corundum except that the two cations reside in the octahedral sites in alternating layers.

Compounds exhibiting this structure include $MgTiO_3$, $NiTiO_3$, $CoTiO_3$, $FeTiO_3$, $MnTiO_3$, $NiTiO_3$, $CdTiO_3$, $LiNbO_3$, Al_2O_3, Ti_2O_3, V_2O_3, Fe_2O_3, Rh_2O_3, and Ga_2O_3.

AB_2X_4 Structures

The Spinel Structure ($Fd3m$). The normal spinel structure consists of a cubic close-packed oxygen lattice with divalent metal cations occupying one-eighth of the tetrahedral sites and trivalent metal cations occupying one-half of the octahedral sites (Fig 6). The unit cell contains 32 oxygen ions, 16 octahedral cations, and 8 tetrahedral cations.

Compounds exhibiting the normal spinel structure include $ZnFe_2O_4$, $CdFe_2O_4$, $MnFe_2O_4$, $MgAl_2O_4$, $FeAl_2O_4$, $FeCr_2O_4$, $CoAl_2O_4$, $NiAl_2O_4$, $MnAl_2O_4$, $ZnAl_2O_4$, MgV_2O_4, ZnV_2O_4, $MgCr_2O_4$, $MnCr_2O_4$, $FeCr_2O_4$, $CoCr_2O_4$, $NiCr_2O_4$, $CuCr_2O_4$, $ZnCr_2O_4$, $CdCr_2O_4$, $HgCr_2Se_4$, $CdCr_2Se_4$, Mn_3O_4, $ZnMn_2O_4$, Co_3O_4, $ZnCo_2O_4$, $MgRh_2O_4$, and $CaGa_2O_4$.

In a related structure, inverse spinel, the divalent ions switch places with half of the trivalent ions. Consequently, the A^{2+} ions and half the B^{3+} ions are on the octahedral sites; the other half of the B^{3+} are on tetrahedral sites, for example, $B(AB)O_4$. The presence of iron, indium, or gallium will tend to stabilize the inverse spinel.

Compounds exhibiting the inverse spinel structure include: $FeMgFeO_4$, $FeTiFeO_4$, Fe_3O_4, $ZnSnZnO_4$, $FeNiFeO_4$, $MgTiMgO_4$, $MgVMgO_4$, $FeCoFeO_4$, $FeCuFeO_4$, $CoSnCoO_4$, $ZnTiZnO_4$, $GaMgGaO_4$, $InMgInO_4$, $InFeInO_4$, $InCoInO_4$, and $InNiInO_4$.

Silicate Structures. Silicates, compounds containing silicon and oxygen, are a major classification of minerals, some of which are the basis of traditional ceramics. The radius ratio for Si—O is 0.29, corresponding to tetrahedral coordination, and four oxygen ions are almost invariably arrayed around a central silicon. Silicates can be classified according to the manner in which the silica tetrahedral are linked together.

Orthosilicates are the simplest structural arrangement in which the silica tetrahedra with the formula $(SiO_4)^{4-}$ are linked only through other cations. This structure results in an oxygen-silicon ratio of 4:1. Compounds of this type include the olivenes, $Mg_{2-x}Fe_xSiO_4$, forsterite, Mg_2SiO_4, phenacite, Be_2SiO_4, and garnets, $A_3B_2Si_3O_{12}$.

When two silica tetrahedra share a single oxygen ion to form a structural unit with the formula $(Si_2O_7)^{6-}$ the compound is referred to as a pyrosilicate. Compounds of this type include melilite, $Ca_2MgSi_2O_7$.

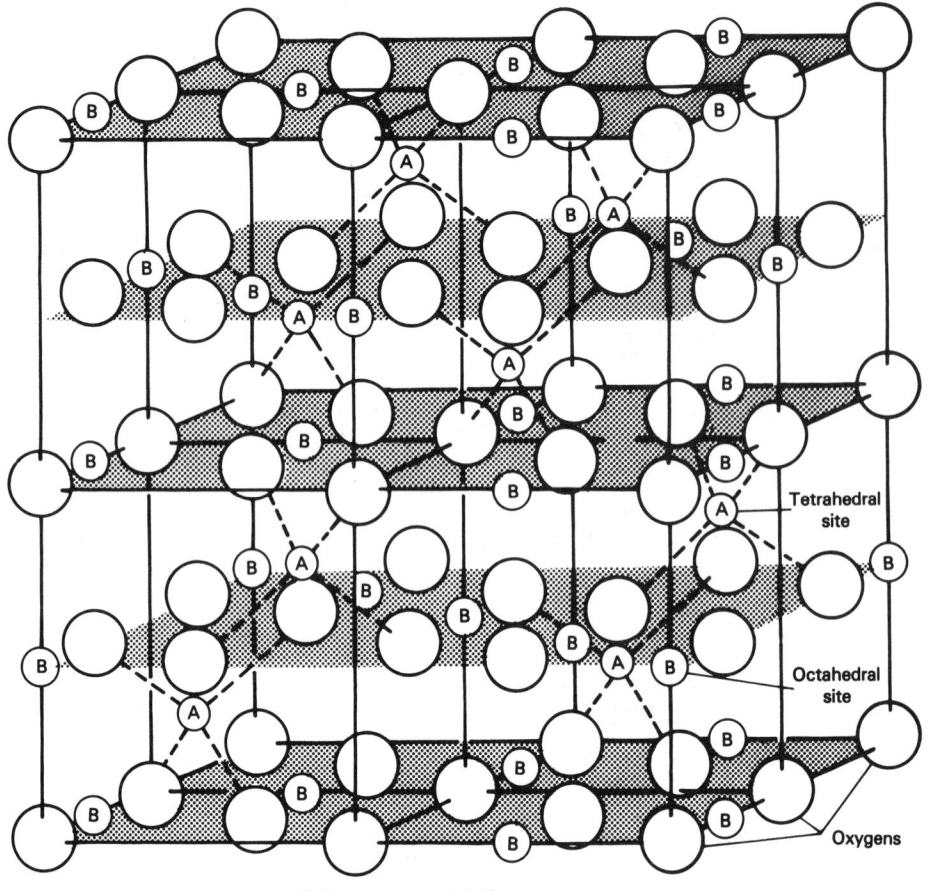

Cubic close-packed: *Fd3m*

Fig 6 The cubic close-packed lattice indicating the octahedral and tetrahedral positions normally occupied by divalent (A) and trivalent (B) ions in the ideal spinel structure

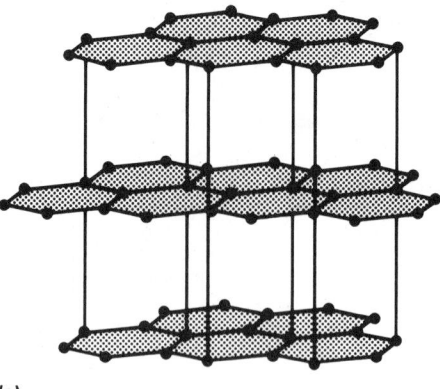

Fig 7 Structures of carbon. (a) Graphite. (b) Diamond

Metasilicates are constructed from closed rings with the composition $(SiO_3)_n^{2n-}$ which are formed when neighboring tetrahedra share two oxygen. The structures correspond to an oxygen-silicon ratio of 3:1. Compounds of this type include wollastonite, $CaSiO_3$, beryl, $Be_3Al_2Si_6O_{18}$, and cordierite, $Al_4Mg_2Si_5O_{18}$.

When the silica rings degenerate into chains, the structures are referred to as pyroxenes if they contain single-ring chains and amphiboles if they contain double-ring chains. Compounds of this type include diopside, $CaMg(SiO_3)_2$, enstatite, $MgSiO_3$, spodumene, $LiAl(SiO_3)_2$, and asbestos.

Structures in which the silica tetrahedra form a three-dimensional network are called framework silicates. Compounds of particular interest result from substitution of aluminum into the silica network. This replacement results in a net charge which is balanced by large ions on interstitial positions. Albite, $NaAlSi_3O_8$, anorthite, $CaAl_2Si_2O_8$, and orthoclase, $KASi_3O_8$, and zeolites are based on this type of structure.

The open structure of the silicate compounds allows a considerable degree of solid solution; therefore, the examples listed above

actually represent families of compounds. An important class of materials are based on the incorporation of water into a layered silicate structure. Hydrated aluminosilicates or clay minerals are in this class of structures and include kaolinite, $Al_2(Si_2O_5)(OH)_4$, and mica.

Silicon carbide is of special interest due to its rich polymorphism. Each carbon and silicon atom is tetrahedrally coordinated by four atoms of the other kind. Silicon carbide is stable in both zinc blende and wurtzite structures (Fig 2). Higher order modifications are constructed from layers of these two structures. Structures with unit cells containing 600 to 1000 layers have been reported; however, the most prevalent compounds contain fewer than 40 layers.

Carbon-Graphite. The *sp*2 hybridized bonding in graphite results in a hexagonal layered structure with strong σ bonds between carbon atoms in the plane and weaker interaction between π bonds between the planes. The structure shown in Fig 7(a) illustrates the anisotropy of the structure.

Elements/compounds exhibiting this structure include carbon and boron nitride.

Carbon-Diamond Structure (*Fd3m*). When carbon bonds consist of hybrid *sp*3 orbitals, the atoms must be tetrahedrally coordinated. This requirement defines the carbon crystalline structure to be a periodic array of carbon atoms with the structure shown in Fig 7(b). Note that the structure is essentially the same as zinc blende except that only one element is present.

Elements exhibiting this structure include carbon (diamond).

SELECTED REFERENCES

- F.D. Bloss, *Crystallography and Crystal Chemistry*, Holt, Rinehart, and Winston, 1989
- G. Burns and A.M. Glazer, *Space Groups for Solid State Scientists*, Academic Press, 1990
- R.C. Evans, *An Introduction to Crystal Chemistry*, Cambridge University Press, 1964
- W.D. Kingery, H.K. Bowen, and D.R. Uhlmann, *Introduction to Ceramics*, John Wiley & Sons, 1975

Melting Points, Crystallographic Transformation, and Thermodynamic Values

Mark E. Schlesinger, Department of Metallurgical Engineering, University of Missouri—Rolla

THE IMPORTANCE OF A GOOD DATA BASE for determining the ranges of temperature and chemical environment in which a material (ceramic or otherwise) can exist in a given state is usually so obvious as to go unstated. However, as the variety of materials used in ceramic processes continues to grow, the location of the necessary information for a particular compound grows increasingly difficult. In large part, this is due to the lack of experimentally obtained thermodynamic data for many compounds; in other cases, the availability of available information has not been disseminated widely enough. This article lists the best sources of thermodynamic information for compounds of interest to ceramists, and provides several tables designed to demonstrate the kind of information available for some of the more common compounds.

Types of Invariant Point

The Ti-TiO$_2$ phase diagram (Fig 1) provides an ideal illustration of the various types of transformation which a given compound may go through as its temperature increases. These transformations can be divided into solid-state and melting reactions.

Solid-state transformations include allotropic transformations, in which a single-phase material takes on a different crystal structure of the same composition (TiO, Ti$_2$O$_3$, and TiO$_2$ are examples), and peritectoid decomposition to two entirely different phases in the system (such as the formation of α-Ti and high-TiO solid solutions from δ phase). Compounds may also be formed by the solid-state eutectoid reaction of two adjacent phases in the phase diagram (for example, the formation of Ti$_3$O$_5$). For "line" compounds (those without significant solid solubility ranges), the temperatures at which these transformations occur are invariant at a given pressure; for nonstoichiometric compounds, these temperatures vary with the overall system composition.

Melting Reactions. The possible melting reactions in a system closely parallel the list of solid-state reactions. Congruent melting to an identical-composition liquid (Ti$_2$O$_3$, Ti$_3$O$_5$) is comparable to the allotropic transformation; peritectoid melting to a liquid and solid phase of separate compositions (TiO solid solution) is also a common occurrence. Again, the temperatures at which these reactions occur vary more as the stoichiometry range of the particular compound widens. Some compounds may sublime to an equilibrium gas phase instead of melting, especially if the solid-state phase is stable to very high temperatures; sublimation can also occur either congruently or incongruently.

Thermodynamic Properties

Heat Capacity. The heat capacity of a substance is defined as the amount of energy required to a raise the temperature of a given amount of that material by one degree. The generally tabulated unit is the molar heat capacity (J/mol · K), although listings of heat capacity per unit weight are also seen. The *specific heat* of a substance is the ratio of its heat capacity to that of liquid water (1.0 cal/g · K), and as a result is often tabulated as well.

Heat capacities are the most easily obtained thermodynamic property for most compounds, and as a result are also the most widely available. Their availability and reliability are generally better if the temperature of interest approaches ambient, and if the state of the substance in question is the most stable solid. For most solids, the constant-pressure heat capacity (C_p) as a function of temperature is expressed by a formula of the type (Ref 2),

$$C_p = a + b \times 10^{-3} T + c \times 10^5 T^{-2} \qquad \text{(Eq 1)}$$

where a, b, and c are constants, and T is the absolute temperature. For substances showing unusual discontinuities in heat capacity-versus-temperature curves (that is, disordering), higher-order expressions are used. The measured heat capacities of liquids (especially at high temperatures) are less reliable and usually considered a constant (Ref 2, 3).

Another general rule in the use of heat capacity data is that the most reliable data exist for the simplest compounds (for example, C_p for the oxide of a particular metal is more likely to be available than that of its aluminosilicate). In fact, heat capacity data are unavailable for many ceramic compounds, which has in turn led to the development of several estimating techniques. The most widely known technique is the *Neumann-Kopp rule* (Ref 2), which states that the heat capacity of a given compound is equal to the sum of its components in the same standard state at the same temperature. For example:

$$C_p \text{ (Cr}_3\text{Si, solid, 298 K)} = 3 \, C_p \text{ (Cr, solid, 298 K)}$$

$$+ \, C_p \text{ (Si, solid, 298 K)} \qquad \text{(Eq 2)}$$

This implies, of course, that ΔC_p for the formation of such a compound (and thus the entropy of formation as well) is zero. This assumption is less likely to be valid for compounds that are more ionic. A variation of the Neumann-Kopp rule useful for estimating the heat capacities of multiple oxides is to presume that the entropy of formation from the simple oxides is zero (Ref 3):

$$C_p \, (M_x\text{Si}_y\text{O}_{2y+x}) = x \, C_p \, (MO) + y \, C_p(\text{SiO}_2) \quad \text{(Eq 3)}$$

Because the heat capacity data for most simple oxides is reliable, this approach has some value; again, the more ionic the nature of the multiple oxide, the less likely predictions of this type are to be accurate. Obviously, actual experimental data are preferred if available for the particular compound.

The heat-capacity estimating method first derived by Kellogg and later revised by Kubaschewski and Ünal (see Ref 2) also deserves mention. This model expresses the heat capacity of a substance as the sum of "contributions" from the ionic groups which

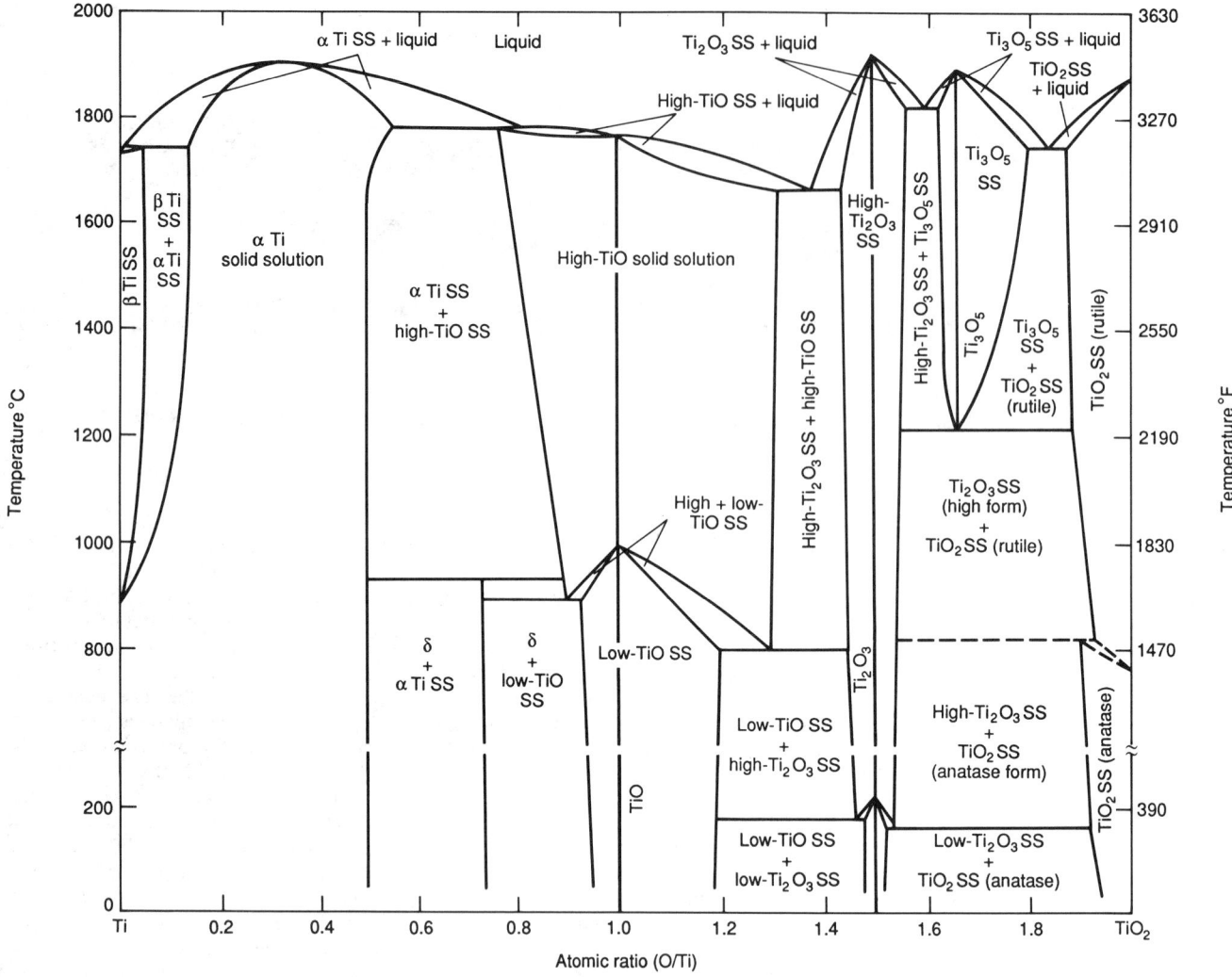

Fig 1 The Ti-TiO₂ phase diagram (SS = solid state). Source: Ref 1

make up the compound. Using calcium hydroxide as an example,

$$C_p \, [\text{Ca(OH)}_2, 298 \text{ K}] = \Xi_{\text{Ca}} + 2 \, \Xi_{\text{OH}} \qquad (\text{Eq 4})$$

where Ξ is the ionic group's particular "contribution." Tables 1 and 2, from the monograph by Kubaschewski and Alcock (Ref 2), list the values of Ξ for various cations and ionic "groups." Ünal also noticed that the heat capacity of most solids is about 30.23 J/g-atom · K at the melting point, *if* the substance undergoes a solid-solid transformation before melting; the heat capacity at the first transformation point (solid-state or melting) averages about 29.28 J/g-atom · K. Furthermore, Ünal pointed out that the derived value for c from most heat capacity measurements, although highly uncertain, was roughly equal to −4.18 J/g-atom · K. Using this information, Kubaschewski and Alcock have generated formulas for estimating the values of a, b and c in Eq 1:

$$a = \frac{\begin{array}{c}[T_m \times 10^{-3}(\Sigma\Xi + 1.12524n) \\ - 0.298n \times 10^5 \, T_m^{-2} - 2.16n]\end{array}}{T_m \times 10^{-3} - 0.298} \qquad (\text{Eq 5})$$

$$b = \frac{6.125n \times 10^5 \, nT_m^{-2} - \Sigma\Xi}{T_m \times 10^{-3} - 0.298} \qquad (\text{Eq 6})$$

$$c = -n \qquad (\text{Eq 7})$$

where T_m is the melting point, and n is the number of atoms per molecule. Using these equations, it is possible to estimate C_p as a function of temperature for any stable solid ceramic phase to its melting point, given its composition.

Less information is available for estimating the heat capacities of liquid ceramic phases, which are generally considered independent of temperature. Turkdogan (Ref 3) quotes a value of 29.2 (±10%) J/g-atom · K for liquid aluminosilicates, and Kubaschewski and Alcock (Ref 2) suggest that the heat capacity of a given solid phase is not substantially different from that of its liquid. Obviously, actual data should be used when available.

Heat of transformation is defined as the energy input necessary to change a given amount of a particular compound from one state to another at a single given temperature.

If that temperature is the equilibrium transformation point (the invariant points discussed above), this quantity is known as a *latent heat of transformation* and assigned the symbol ΔL. The hypothetical heat of transformation is assigned the symbol ΔH. More specifically, ΔH_{tr} is defined here as the symbol for the heat of solid-state transformations, ΔH_m or ΔH_f for the heat of melting or fusion, ΔH_s the heat of sublimation, and ΔH_v the heat of boiling or volatilization. As with heat capacity, the usual units use a molar basis (J/mole). Heat of transformation is also in units of J/g-atom, where g-atom is one mole of atoms in a compound.

Generally, the only heats of transformation that can be measured experimentally are the ΔL values. Determination of ΔH values requires knowing the heat capacity as a function of temperature of both the stable low- and high-temperature phases; the integral of the difference between the two as a function of temperature (ΔC_p) completes the equation

$$\Delta H_{tr} = \Delta L_{tr} + \int_{T_{tr}}^{T} \Delta C_p dT \qquad (\text{Eq 8})$$

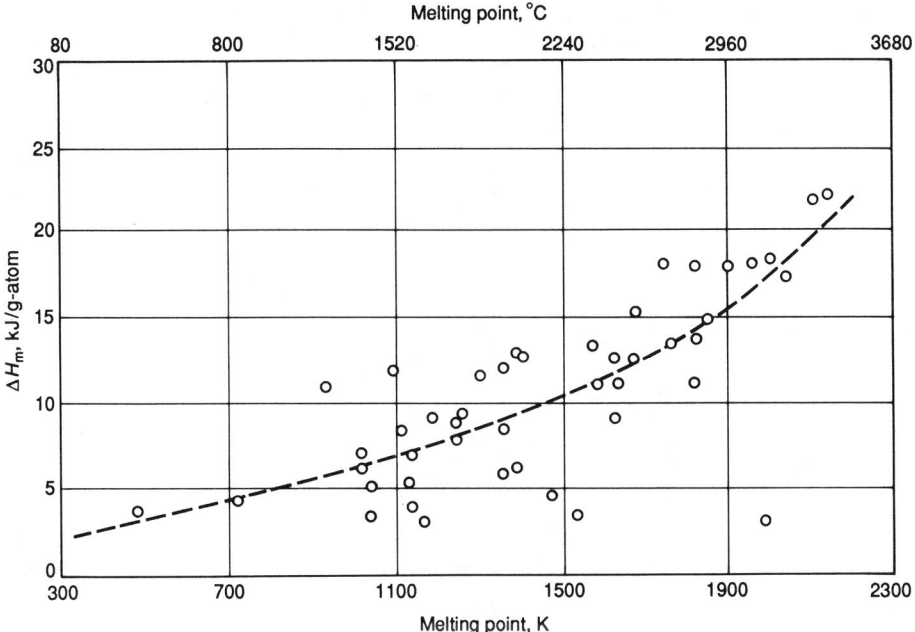

Fig 2 Enthalpy of fusion of network-forming oxides as a function of melting point. Source: Ref 3

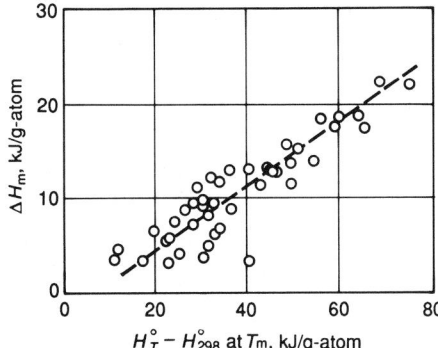

Fig 3 Enthalpy of fusion of network-forming oxides as a function of sensible heat at their melting points. Source: Ref 3

The availability and reliability of heat-of-transformation values are similar to those for heat capacities in that they improve with both lower temperature and compound simplicity. However, the high melting points and complexity of many ceramic materials make the compilation of a data base in this area especially difficult. The deriving of relationships which might make their estimation possible is also more difficult, as a result. Figures 2 and 3 show the measured latent heats of fusion for network-forming crystalline oxide compounds as a function of melting point; no such relationships are readily available for other types of ceramic compounds, and the reliability of ΔL_m estimates made by simply using the melting point (Fig 2) is questionable. Another approach for estimating heats of transition is to assume that the entropies of transition are equal for compounds of similar crystal structure, allowing the calculation of ΔL values for the unknown at its transformation temperature. The required heat inputs for solid-state transformations are generally small and can assumed to be zero if no better estimating approach is available.

Figure 4, also from Turkdogan's work (Ref 3), shows some "rules" for estimating the heat of crystallization (ΔH_v) of simple glasses (that is, those with equivalent compositions to the crystalline material). The figure is based on two rules: the heat capacity of a glass below T_g is equal at a given temperature to that of the corresponding crystalline solid and the heat capacity of the supercooled liquid above T_g is equal to that above the melting point T_m (generally considered a constant).

Heat of formation is defined as the heat input necessary (usually a negative number) to form a given compound or phase from the elements at a given temperature. If both the compound and the elements from which it is formed are in their standard state at the same temperature, the resulting enthalpy change is known as a *standard heat of formation* and denoted by the symbol $\Delta H°$. As with heat capacities, a standard mathematical expression is used to express $\Delta H°$ (in units of kJ/mole) as a function of temperature:

$$\Delta H° = a - b \times 10^{-3} T - c \times 10^{-6} T^2 + dT^{-1}$$

$$\text{(Eq 9)}$$

The temperature limits for equations of this type are set by the stability ranges of equilibrium phases of both the particular com-

Table 1 Cationic contribution to the heat capacity at 298 K (25 °C)

Metal	Ξ_{cat} J/°C	cal/°F	Metal	Ξ_{cat} J/°C	cal/°F
Ag	25.7	3.4	Mg	19.65	2.6
Al	19.7	2.6	Mn	23.4	3.1
As	25.1	3.33	Na	25.9	3.45
Ba	26.4	3.5	Nb	23.0	3.05
Be	(9.6)	(1.3)(a)	Nd	24.25	3.2
Bi	26.8	3.55	Ni	(27.6)	(3.67)
Ca	24.7	3.3	P	14.2	1.9
Cd	23.0	3.05	Pb	26.8	3.55
Ce	23.4	3.1	Pr	24.25	3.2
Co	28.0	3.7	Rb	26.35	3.5
Cr	23.0	3.05	Sb	22.2	3.15
Cs	26.4	3.5	Se	21.3	2.83
Cu	25.1	3.33	Si	...	...
Fe	25.9	3.45	Sm	25.1	3.33
Ga	(20.9)	(2.8)(a)	Sn	23.4	3.1
Gd	23.4	3.1	Sr	25.5	3.4
Ge	20.1	2.67	Ta	23.0	3.05
Hf	25.5	3.4	Th	25.5	3.4
Hg	25.1	3.33	Ti	21.75	2.9
Ho	23.0	3.05	Tl	27.6	3.67
In	24.25	3.2	U	26.8	3.55
Ir	(23.85)	(3.15)(a)	V	22.2	2.95
K	25.9	3.45	Y	(25.1)	(3.33)
La	(25.5)	(3.4)(a)	Zn	21.75	2.9
Li	19.65	2.6	Zr	23.8	3.15

(a) Values in parentheses are estimates. All values were converted from cal/K.

Table 2 Anionic contribution to the heat capacity at 298 K (25 °C)

Anion	Ξ_{an} J/°C	cal/°F	Anion	Ξ_{an} J/°C	cal/°F
H	8.8	1.15	SO_4	76.55	10.2
F	22.8	3.03	NO_3	64.4	8.55
Cl	24.7	3.3	P	(23.4)	(3.1)(a)
Br	25.9	3.45	CO_3	58.6	7.78
J	26.35	3.5	Si	(24.7)	(3.3)(a)
O	18.4	2.45	CrO_4	90.8	12.05
S	24.5	3.25	MoO_4	90.4	12
Se	26.8	3.55	WO_4	97.5	12.95
Te	27.2	3.6	UO_4	107.1	14.2
OH	30.95	4.1			

(a) Values in parentheses are estimates. All values were converted from cal/K.

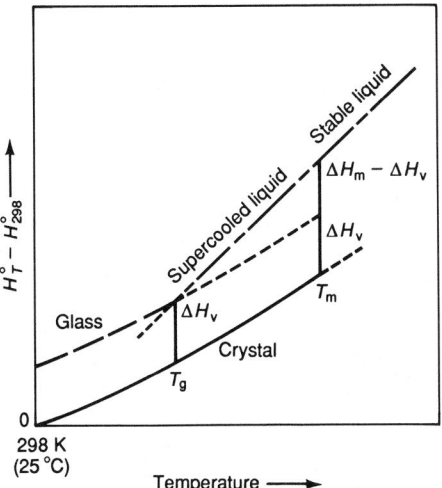

Fig 4 Depiction of "guidelines" for calculating the heat of crystallization of glasses as a function of temperature. Source: Ref 3

pound and the elements from which it is formed; for example, separate sets of coefficients are required for expressing $\Delta H°$ for solid $CuFeO_2$ in the temperature ranges 298 − 1043 K (the Curie temperature of iron), 1043 − 1091 K (a solid-state transformation point for $CuFeO_2$), 1091 − 1185 K (the ferrite-austenite transition temperature of iron), 1185 − 1357.6 K (the melting point of copper), and 1357.6 − 1470 K (the melting point of $CuFeO_2$).

A variety of methods exist for the experimental determination of heats of formation; the difficulty of obtaining accurate values generally increases with the complexity of the compound. Once a value of $\Delta H°$ is known at a particular temperature (ΔH_T), the difference in heat capacities between the particular compound and its constituent elements (all in their standard states and proper stoichiometric ratios) can again be used to calculate the heat of formation as a function of temperature, using *Kirchhoff's equation*:

$$\Delta H_T = \Delta H_0 + \int_0^T \Delta C_p \, dT$$

$$= \Delta H_{298} + \int_{298}^T \Delta C_p \, dT \qquad \text{(Eq 10)}$$

Equation 10 is first used backwards with a known value of ΔH_T to calculate ΔH_0, and then used forward to calculate heats of formation at other temperatures. If they occur in the appropriate temperature range, heats of transformation for the compound are subtracted from the right-hand side of Eq 10, and those for the elements (or compounds) from which the new compound is being formed are added to it.

Few, if any, universal rules exist for the estimation of unknown compound heats of formation. Kubaschewski and Alcock (Ref 2) discuss some of the approaches that have been tried, which point out that $\Delta H°$ seems to be related to the relative ionicity of the particular compound, along with perhaps the atomic number of the particular cation and/or the change in volume resulting from compound formation from the elements. However, no hard-and-fast rules exist for the prediction of $\Delta H°$ of simple compounds, much less for complex phases such as multiple oxides.

Free Energy of Formation. The free energy of a given reaction is most simply defined as the change in the chemical potential of a system resulting from its occurrence under set conditions. As with heats of formation, the free energy change resulting from the formation of a particular compound in its most stable state at a particular temperature from the elements in their most stable state at the same temperature is denoted as the standard free energy of formation of that compound, and denoted by the symbol $\Delta G°$. A standard-form equation of the type,

$$\Delta G° = a + b \times 10^{-3}\, T \ln T + c$$

$$\times 10^{-6}\, T^2 - (d/2)T^{-1} + e \times 10^{-3}\, T \qquad \text{(Eq 11)}$$

Table 3 Transformation temperatures and heats of transformation of various ceramic compounds

Compound	Molecular weight	Solid-state (SS) transformations:		Melting temperature (T_m), K	Latent heat of melting (ΔL_m), J/mole
		SS Transformation temperature (T_{tr}), K	SS Latent heat (ΔL_{tr}), J/mole		
Elements					
Al	26.98	...	...	933	10,711
C	12.01	...	...	[4070](a)	...
Mo	95.94	...	...	2897	39,099
Si	28.08	...	...	1685	50,208
W	183.85	...	...	3680	35,397
Single oxides					
Al_2O_3	101.96	...	...	2327	111,085
CaO	56.08	...	...	3200	79,496
Cr_2O_3	151.99	...	...	2603	129,704
Fe_2O_3	159.69	906	0	*1730*(b)	...
Fe_3O_4	231.54	850	0	1870	138,072
MgO	40.30	...	...	3105	77,822
MnO	70.94	...	...	2115	43,932
MoO_3	143.94	...	...	1075	48,911
NiO	74.69	525	0	2228	54,392
		565	0		
$^1/_2 P_4 O_{10}$	141.94	...	...	[632](a)	53,000
SiO_2	60.08	847 and 1079	728 and 1996	1996	9566
TiO_2	79.88	...	...	2130	66,944
ZnO	81.39	...	...	2248	54,392
ZrO_2	123.22	1478	5941	2950	87,027
Multiple oxides					
$3Al_2O_3 \cdot 2SiO_2$ (mullite)	426.05	...	...	2023	...
$BaTiO_3$	233.20	395 and 1733	201 and 0	1889	...
$CaSiO_3$ (pseudowollastonite)	116.16	1398	5400	1817	56,066
Ca_2SiO_4 (olivine)	172.24	1121 and 1712	13,987 and 14,188	2403	71,128
$Ca_3(PO_4)_2$	310.18	1423 and 1743	18,828 and 0	2083	167,360
$CaO \cdot Al_2O_3 \cdot 2SiO_2$ (anorthite)	278.21	...	...	1823	81,170
$CaO \cdot TiO_2 \cdot SiO_2$ (sphene)	196.04	...	...	1673	123,846
$CaWO_4$	287.93	...	...	1853	...
$CuFe_2O_4$	239.24	675 and 795	753 and 0	1358	13,054
Fe_2SiO_4 (fayalite)	203.78	...	...	1490	92,174
$LiAlO_2$	65.92	...	...	1883	87,864
$LiAlSiO_4$ (eucryptite)	126.01	1300	1255		
$MgAl_2O_4$	142.27	...	...	2408	196,648
$MgCr_2O_4$	192.30	...	...	2623	...
$MgSO_4$	120.37	...	...	1400	14,644
Mn_2SiO_4	201.96	...	...	1618	89,621
$MnTiO_3$	150.82	...	...	1677	...
$NaAlSiO_4$ (nepheline)	142.05	467 and 1180	0 and 0	...	...
Na_2CO_3	105.99	723	690	1123	29,665
$NiAl_2O_4$	176.65			2383	...
$PbTiO_3$	303.08	763	5648	1559	...
Zn_2SiO_4	222.86	...	...	1785	83,680
Zirconates					
$CaZrO_3$	179.30	...	...	2623	...
$SrZrO_3$	226.84	1003, 1133, and 1408	... ...	3023	...
$ZrSiO_4$	183.31	*1811*(b)	...		
Borides					
AlB_{12}	156.71			2423	...
CoB	69.74			1733	...
CrB	62.81			2343	...
CrB_2	73.62			2437	...

(continued)

is used to express free energies of formation (in kJ/mole) as a function of temperature, with a, b, c, and d being the same values as those used to calculate the heat of formation of the same compound in the same temperature range.

$\Delta G°$ is usually the most difficult thermodynamic property to accurately obtain for a particular compound and thus the property for which the smallest experimental data base is available. Nevertheless, enough data is available to allow for generalizations which can be used to predict free energies for some "unknowns." In particular, Turkdogan (Ref 2) has shown that for many solid alkaline-earth multiple oxides,

$$\Delta G°/N = -332.750 + 0.05933\,T \qquad \text{(Eq 12)}$$

and for alkali- and several transition-metal multiple oxides,

$$\Delta G°/N = -285.290 + 0.05933\,T \qquad \text{(Eq 13)}$$

where N is the number of atoms per molecule. For simple oxides and non-oxide compounds, correlations of this type are largely unavailable.

Data Sources

Invariant Points. Obviously, where thermodynamic data are available for a pure compound, the invariant points for the substance are given as part of the thermodynamic data listing. However, the number of ceramic compounds for which no thermodynamic data listing exists exceeds the number of those which have one. For these compounds, as well as for composite materials and nonstoichiometric phases, the best place to look for phase-transformation and melting points is the relevant phase diagram.

For simple compounds (one or two elements), the best phase-diagram compilation is the *Handbook of Binary Phase Diagrams* (Ref 4, 5). The *Journal of Phase Equilibria* (formerly *Bulletin of Alloy Phase Diagrams*) contains more thorough discussion on a growing number of these systems, in addition to a regularly updated bibliography on other review work. The series of phase-diagram assessments edited by Moffatt (Ref 6) is also a good source of information.

The data base for invariant points and thermodynamic properties shrinks dramatically as the complexity of the material in question increases. For multiple compounds (that is, three or more elements), the most thorough source of transformation and melting temperatures is the ongoing *Phase Diagrams for Ceramists* collection (Ref 1, 7–12) published by the American Ceramic Society. Unlike the binary-systems collections, the format of the diagrams reproduced in *Phase Diagrams for Ceramists* varies with the type of system under consideration, and the information available. Critical assessment is also less common, largely because so little data are sometimes available to assess.

Table 3 Transformation temperatures and heats of transformation of various ceramic compounds (continued)

| Compound | Molecular weight | Solid-state (SS) transformations: | | Melting temperature (T_m), K | Latent heat of melting (ΔL_m), J/mole |
		SS Transformation temperature (T_{tr}), K	SS Latent heat (ΔL_{tr}), J/mole		
FeB	66.66		...	1923	...
MnB₂	76.56		...	2261	...
Ni₄B₃	267.19		...	1853	...
TaB₂	202.57		...	3373	83,680
TiB₂	69.50		...	3193	100,416
VB₂	72.56		...	3020	...
ZrB₂	112.85		...	[4466](a)(b)	...
Carbides					
Al₄C₃	143.96		...	2500	...
B₄C	55.26		...	2743	104,600
Be₂C	30.04		...	2400	75,312
CaC₂	64.10	720	5523	...	...
Cr₂₃C₆	1267.98	...	...	1793	...
Fe₃C	179.55	485	7029	1500	51,463
HfC	190.50	...	...	4103	...
Nb₂C	197.82	1503 and 2723	...	3259	...
SiC	40.10	...	...	3245	...
TiC	59.89	...	...	3290	71,128
ZrC	103.24	...	...	3805	79,496
Nitrides					
AlN	40.99	...	...	[2790](a)	...
BN	24.82	...	...	[2600](a)(b)	...
Be₃N₂	55.05	...	...	2473	129,286
Cr₂N	118.00	...	...	[1810](a)(b)	...
Fe₄N	237.40	753	8368	...	...
Li₃N	34.83	...	...	1086	...
Mg₃N₂	100.93	823 and 1061	920 and 1088	...	...
NbN	106.91	1643	4184	2323	46,024
Si₃N₄	140.28	...	...	[2151](a)(b)	...
TiN	61.89	...	...	3223	62,760
VN	64.95	...	...	[2619](a)(b)	...
ZrN	105.23	...	...	3225	67,362
Halides					
AlCl₃	133.34	...	...	466	35,350
CaF₂	78.08	1424	4770	1691	29,706
CaCl₂	110.98	...	...	1045	28,543
CsI	259.81	...	...	900	25,606
KBr	119.00	...	...	1007	25,522
LiF	25.94	...	...	1121	27,087
MgF₂	62.30	...	...	1536	58,702

(a) Sublimation reactions are distinguished from melting reactions by the use of brackets. (b) Temperatures of incongruent melting or solid-state decomposition are italicized to distinguish them from congruent transformations. Source: Ref 18

Transition-point data are also available for some ceramic systems in older compilations. The series of binary phase-diagram compilations begun by Hansen and continued by Elliott and Shunk (Ref 13–15) contains a few reviews of metal-metalloid systems, all of which have since been reproduced or updated by the phase-diagram program of ASM International (Ref 4, 5). The index periodically published in the *Journal of Phase Equilibria* also lists other sources of elemental-system phase-diagram assessments. Two Soviet phase-diagram compilations for binary oxide and halide systems are also available in translation (Ref 16, 17).

Thermodynamic Data. By far the most complete compilation of thermodynamic data for inorganic compounds is the two-volume

Thermochemical Data for Pure Substances compiled by Barin (Ref 18). This set, a new version of the previous compilation of Barin and Knacke (Ref 19, 20), features an updated data base in a more convenient format than the previous work; being largely based on other compilations, it is also far more thorough.

The U.S. Bureau of Mines over the previous decade has also updated its thermodynamic collections, resulting in the publication as part of the *Bulletin* series of several excellent compilations. *Thermodynamic Properties of Elements and Oxides* (Bulletin 672, 1982; Ref 21), is a complete collection of data listing for those condensed-state single oxides and elements for which data are available; data are also given for several nonstable states and vapor species (but no multiple oxides). *Ther-*

modynamic *Properties of Halides* (Bulletin 674, 1984; Ref 22) also reviews vapor species as well as condensed-state compounds, as does *Thermodynamic Properties of Sulfides* (Bulletin 689, 1987; Ref 23). A particular advantage of these three compilations is their listing of coefficients for calculating heat capacities and heats and free energies of formation as a function of temperature for the various compounds, a task which requires interpolation between the tabulated values of Barin's work. For those without access to Barin's compilation, *Thermodynamic Data for Mineral Technology* (Bulletin 677, 1984; Ref 24) is a good "second-choice" overall collection; in addition to the compound classes covered by Bulletins 672, 674, and 689, reasonably thorough data listings are given for most classes of metalloid compounds, and for a few classes of binary oxides (carbonates, silicates, sulfates).

Both Barin's compilation and the U.S. Bureau of Mines publications draw heavily on the data listed in the various JANAF publications (Ref 25–30). The somewhat scattered nature of the JANAF publications makes them harder to use, and the list of tabulated compounds of interest to ceramists is shorter than in the previously mentioned general compilations; however, the JANAF tables have the advantage of a critical data assessment for each tabulation, describing the nature of the estimating techniques and assumptions that often accompany thermodynamic "data." Smaller general data compilations also exist, often accompanying thermodynamics textbooks; the most notable of these is the listing at the end of Ref 2. In addition, several compilations specific to a specific class of compounds have been published; these are listed among the Selected References.

Data Listings

Three types of tabulation of sample thermodynamic data are given in the following pages. The first (Table 3) is a listing of the solid-state transformation points and types and heats of transformation of various ceramic compounds, taken from the compilation of Barin (Ref 18). The temperatures of incongruent melting or solid-state decomposition reactions are italicized to distinguish them from congruent transformations; sublimation reactions are distinguished from melting by the use of brackets. The table is far from complete; data for other compounds can be obtained from the various references described above or from the Selected References listed at the end of this article.

The second data table (Table 4) lists coefficients for calculating the heat capacities of these compounds as a function of temperature, using the form of Eq 1. Since Barin (Ref 18) does not list these values, they are taken where possible from lists in other sources, notably Pankratz (Ref 21 and 23) for elements, halides, and oxides, and Kubas-

Table 4   Coefficients for calculation of heat capacity C_p (in J/mole) $= a + b \times 10^{-3} T + c \times 10^5 T^{-2}$

Compound	Temperature range(a), K	a	b	c
Elements				
Al (solid)	298–933	19.212	14.451	0.728
Al (liquid)	933–2791	31.748	...	...
C (graphite)	298–2000	14.719	6.410	−7.209
	2000–4100	23.601	1.121	−30.123
C (diamond)	298–1200	8.033	7.627	−146.015
Mo (solid)	298–2897	18.312	9.916	2.477
Mo (liquid)	2897–4978	40.350	...	...
N_2 (gas)	298–2000	27.266	4.929	0.331
	2000–5000(a)	36.169	0.469	−45.687
O_2 (gas)	299–2000	30.249	4.209	−1.891
	2000–5000(a)	34.893	1.749	−26.358
Si (solid)	298–1685	23.760	2.937	−4.121
Si (liquid)	1685–3505	27.196	...	...
W (solid)	298–3680	24.329	3.104	−0.879
W (liquid)	3680–5931	35.564	...	...
Oxides				
Al_2O_3 (α, solid)	298–2327	115.493	11.539	−35.487
Al_2O_3 (γ, solid)	298–2290	112.415	22.383	−32.324
Al_2O_3 (liquid)	2327–2500	46.000	...	...
CO (gas)	298–2000	28.065	4.627	−0.259
	2000–3000(a)	34.207	1.038	...
CO_2 (gas)	298–2000	45.365	8.686	−9.619
	2000–3000(a)	61.381	0.619	−90.563
CaO (liquid)	298–3200	50.741	3.682	−8.702
CaO (liquid)	3200–3500	(62.76)(b)	...	...
Cr_2O_3 (solid)	298–2603	109.645	15.455	...
Cr_2O_3 (liquid)	2603–3000	(156.90)(b)	...	...
Fe_2O_3 (solid)	298–960	98.194	80.613	−16.430
	960–1800	142.417	...	...
Fe_3O_4 (solid)	298–850	79.756	225.390	34.014
	850–1870	49.850	72.514	855.382
H_2O (liquid)	298–373	75.480	...	...
H_2O (gas)	373–3000	304.246	10.426	−0.393
MgO (solid)	298–3105	47.515	4.309	−10.347
MgO (liquid)	3105–4000	(66.94)(b)	...	...
MnO (solid)	298–2115	46.921	7.774	−4.569
MnO (liquid)	2115–2500	(60.67)(b)	...	...
MoO_3 (solid)	298–1075	75.505	32.081	−8.974
MoO_3 (liquid)	1075–1400	126.231	...	...
NiO (solid)	298–525	−6.322	131.229	10.221
	525–565	−34.249	168.440	...
	565–2228	−39.913	12.367	21.885
NiO (liquid)	2228–2500	(54.39)(b)	...	...
$^1/_2P_4O_{10}$ (solid)	298–632	70.275	169.754	13.374
$^1/_2P_4O_{10}$ (gas)	632–2500	158.131	4.157	90.238
PbO (solid)	298–762	43.578	15.204	−2.042
	762–1159	45.202	12.526	−1.916
PbO (liquid)	1159–1500	57.887	...	...
SO_2 (gas)	298–2000	47.377	6.661	−8.439
	2000–3000(a)	58.331	0.577	−50.147
SiO_2 (solid)	298–847	40.495	44.599	08.322
	847–1079	67.589	42.577	−1.364
				−23.731
	1079–1996	67.639	4.192	
SiO_2 (liquid)	1996–3000	(85.77)(b)	...	...
TiO_2 (solid)	298–2130	70.765	5.104	−15.317
TiO_2 (liquid)	2130–3000	(100.41)(b)	...	...
ZnO (solid)	298–2000	45.336	7.288	−5.771
ZrO_2 (solid)	298–1478	70.116	7.020	−14.242
	1478–2500	74.745	...	...
ZrO_2 (liquid)	2950–3300	(87.86)(b)	...	...
Multiple oxides				
$3Al_2O_3 \cdot 2SiO_2$	298–2023	498.60	31.664	−162.326
Al_2SiO_5 (kyanite) (solid)	298–2000	168.073	28.139	...
$BaTiO_3$ (solid)	298–395	61.938	136.566	−0.058
				−26.501
	395–1733	130.633	5.523	
	1773–1889	(134.93)(b)	...	...
$CaSiO_3$ (solid)	298–1817	103.748	18.529	...
$CaSiO_3$ (liquid)	1817–2200	146.44	...	...
Ca_2SiO_4 (solid)	298–1121	133.122	51.547	−19.413
	1121–1712	126.621	51.224	...
	1712–2403	205.016	...	...
Ca_2SiO_4 (liquid)	2403–2800	(209.20)(b)	...	...
$Ca_3(PO_4)_2$ (solid)	298–1423	183.657	148.055	...

(continued)

Table 4 Coefficients for calculation of heat capacity C_p (in J/mole) = a + $b \times 10^{-3} T$ + $c \times 10^5 T^{-2}$ (continued)

Compound	Temperature range(a), K	a	b	c
	1423–1743	(376.56)(b)	...	...
	1743–2083	(376.56)(b)	...	...
Ca₃(PO₄)₂ (liquid)	2083–2500	(380.74)(b)	...	...
CaAl₂Si₂O₈ (solid)	298–1823	(260.54)(b)	(63.43)(b)	...
CaAl₂Si₂O₈ (liquid)	1823–2100	380.74	...	...
CaO·TiO₂·SiO₂ (solid)	298–1673	177.35	23.18	−40.29
CaO·TiO₂·SiO₂ (liquid)	1673–2000	279.491	...	...
CaWO₄ (solid)	298–1000	110.787	45.771	...
CuFe₂O₄ (solid)	298–675	139.621	117.779	−23.431
	675–795	227.191	...	...
	795–1358	166.021	41.003	...
CuFe₂O₄ (liquid)	1358–1500	225.936	...	...
Fe₂SiO₄ (solid)	298–1490	152.751	39.160	−28.031
Fe₂SiO₄ (liquid)	1490–1700	240.580	...	...
LiAlO₂ (solid)	298–1883	92.337	12.163	−25.007
LiAlO₂ (liquid)	1883–2500	133.89	...	...
LiAlSiO₄ (solid)	298–1300	154.337	28.449	−43.934
	1300–1600	129.701	50.21	...
MgAl₂O₄ (solid)	298–2408	153.964	26.776	40.918
MgAl₂O₄ (liquid)	2408–2500	230.12	...	...
MgCr₂O₄ (solid)	298–2623	167.436	14.894	40.081
MgSO₄ (solid)	298–1400	112.449	41.076	−25.330
MgSO₄ (liquid)	1400–2000	158.992	...	...
2MgO·2Al₂O₃·5SiO₂ (solid)	298–1700	(612.19)(b)	(109.97)(b)	...
Mn₂SiO₄ (solid)	298–1618	159.074	19.500	−31.128
Mn₂SiO₄ (liquid)	1618–3000	243.09	...	...
MnTiO₃ (solid)	298–1677	(121.97)(b)	(9.29)(b)	(−21.88)(b)
Na₂CO₃ (solid)	298–723	29.234	226.225	...
	723–1123	50.087	129.072	...
Na₂CO₃ (liquid)	1123–2000	189.54	...	...
NaAlSiO₄ (solid)	298–467	27.739	0.295	...
	467–1180	112.087	67.113	...
	1180–1525	171.997	5.528	...
NiAl₂O₄ (solid)	298–2000	(159.20)(b)	(23.35)(b)	(−30.75)(b)
PbTiO₃ (solid)	298–763	119.534	17.911	−18.198
	763–1559	109.075	22.804	−13.337
ZnFe₂O₄ (solid)	298–1000	(136.48)(b)	(54.15)(b)	...
Zn₂SiO₄ (solid)	298–1785	(144.88)(b)	(36.94)(b)	(−30.29)(b)
Zn₂SiO₄ (liquid)	1785–2000	213.384	...	...
Zirconates				
BaZrO₃ (solid)	298–2000	(122.80)(b)	(8.79)(b)	(−21.96)(b)
				−22.945
CaZrO₃ (solid)	298–2623	119.194	10.702	...
Li₂Zr₂O₇ (solid)	298–1500	(139.20)(b)	(25.39)(b)	(−33.12)(b)
Nd₂Zr₂O₇ (solid)	298–1400	(255.22)(b)	(44.77)(b)	(−40.17)(b)
SrZrO₄ (solid)	298–2000	(124.06)(b)	(12.22)(b)	(−21.61)(b)
Y₂Zr₂O₇ (solid)	298–1500	(273.45)(b)	(9.54)(b)	(−54.45)(b)
ZrSiO₄ (solid)	298–2000	131.706	17.656	−33.805
Borides				
AlB₁₂ (solid)	298–2423	(211.29)(b)	(115.06)(b)	(−85.35)(b)
CoB (solid)	298–1300	(42.43)(b)	(14.64)(b)	(−11.26)(b)
CrB (solid)	298–1200	42.342	16.024	−10.041
CrB₂ (solid)	298–1200	9.63	10.7	...
FeB (solid)	298–1923	(48.29)(b)	(6.44)(b)	...
MnB₂ (solid)	298–1200	(40.25)(b)	(44.85)(b)	...
Ni₄B₃ (solid)	298–1400	(37.28)(b)	(11.74)(b)	(−9.03)(b)
TaB₂ (solid)	298–3373	59.452	18.785	−15.062
TaB₂ (liquid)	3373–3500	125.52	...	...
TiB₂ (solid)	298–3193	70.724	10.884	−26.511
TiB₂ (liquid)	3193–3500	108.784	...	...
VB₂ (solid)	298–2500	50.578	28.784	−10.828
ZrB₂ (solid)	298–3323	66.062	17.656	−14.685
ZrB₂ (liquid)	3323–3800	96.232	...	...
Carbides				
Al₄C₃ (solid)	298–2500	154.675	28.726	−41.922
B₄C (solid)	298–2743	96.186	22.593	−44.850
B₄C (liquid)	2743–3500	135.980	...	...
Be₂C (solid)	298–2400	36.894	21.364	...
Be₂C (liquid)	2400–3000	(92.05)(b)	...	...
CaC₂ (solid)	298–720	68.614	11.882	−8.660
	720–1200	66.571	6.520	...
Cr₂₃C₆ (solid)	298–1793	638.474	240.968	−73.072
Fe₃C (solid)	298–485	(82.18)(b)	(83.68)(b)	...
	485–1500	107.189	12.551	...
Fe₃C (liquid)	1500–2000	(121.34)(b)	...	...
HfC (solid)	298–3900	47.448	5.439	−13.021

(continued)

chewski and Alcock (Ref 2) for some intermetallic compounds and multiple oxides. Other values have been derived by multiple regression against the values listed by Barin; in some cases, parameters derived from heat capacities estimated by Barin have been parenthesized. Temperature limits for these equations are generally defined by the stability limits of each phase of the particular compound.

The third data table (Table 5) lists values for the standard heats and free energies of formation at 298 K (77 °F) of the compounds listed in Tables 3 and 4, using the values given by Barin. For elements, this value is of course equal to zero, excluding metastable phases (such as diamond for carbon); as a result, these are not listed. Liquid and gaseous standard states are noted separately; all other compounds listed in Table 5 have a solid standard state. $\Delta H°$ and $\Delta G°$ values can be found as a function of temperature either by interpolating between the tabulated values available in the data sources described above, or by using the equations listed by Pankratz *et al.* (Ref 21–23) and by Kubaschewski and Alcock (Ref 2).

Invariant Points and Thermodynamic Properties of Carbides, Nitrides, and Borides. This article applies to many of the compounds involved in glass making; the

Table 5 Heats and free energies of formation at 298 K (25 °C)

Compound(a)	$\Delta H°_{298}$, kJ/mole	$\Delta G°_{298}$, kJ/mole
Elements		
C (diamond)	1.90	2.90
Oxides		
Al₂O₃ (α)	−1675.69	−1582.27
Al₂O₃ (γ)	−1656.86	−1563.85
CO (gas)	−110.54	−137.18
CO₂ (gas)	−393.50	−393.36
CaO	−635.09	−603.51
Cr₂O₃	−1139.70	−1058.07
Fe₂O₃	−824.25	−742.29
Fe₃O₄	−1118.38	−1015.23
H₂O (liquid)	−285.83	−237.14
H₂O (gas)	−241.83	−228.62
MgO	−601.24	−568.94
MnO	−385.22	−362.90
MoO₃	−745.09	−667.99
NiO	−239.70	−211.54
½P₄O₁₀	−1504.97	−1361.67
PbO	−219.40	−189.28
SO₂ (gas)	−296.81	−300.10
SiO₂	−910.86	−856.44
TiO₂	−944.75	−889.41
ZnO	−350.46	−320.48
ZrO₂	−1097.46	−1039.72
Multiple oxides		
3Al₂O₃·2SiO₂	−6820.46	−6443.05
Al₂SiO₅ (kyanite)	−2594.08	−2443.88
BaTiO₃	−1659.80	−1572.44
CaSiO₃	−1628.40	−1544.74
Ca₂SiO₄	−2315.22	−2198.59
Ca₃(PO₄)₂	−4120.80	−3884.97
CaAl₂Si₂O₈	−4227.89	−4002.22
CaO·TiO₂·SiO₂	−2603.30	−2461.78
CaWO₄	−1645.15	−1538.42
CuFe₂O₄	−967.97	−863.24

(continued)

Table 4 Coefficients for calculation of heat capacity C_p (in J/mole) = $a + b \times 10^{-3} T + c \times 10^5 T^{-2}$ (continued)

Compound	Temperature range(a), K	a	b	c
Nb₂C (solid)	298–1800	(66.44)(b)	(12.55)(b)	(−8.58)(b)
SiC (solid)	298–3245	49.628	2.238	−20.728
TiC (solid)	298–3290	49.494	3.347	−14.978
TiC (liquid)	3290–3500	(62.76)(b)	...	...
ZrC (solid)	298–3805	50.905	3.386	−124.547
ZrC (liquid)	3805–4000	(62.76)(b)	...	...
Nitrides				
AlN (solid)	298–2000	(34.39)(b)	(16.94)(b)	(−8.37)(b)
BN (solid)	298–1900	33.889	14.727	−23.053
Be₃N₂ (solid)	298–2473	132.005	−10.470	−59.824
Be₃N₂ (liquid)	2473–3000	(133.89)(b)	...	...
Cr₂N (solid)	298–1810	63.761	28.450	...
Fe₄N (solid)	298–753	(112.29)(b)	(34.14)(b)	...
	753–953	(138.07)(b)		
Li₃N (solid)	298–1086	57.565	84.028	−6.527
Mg₃N₂ (solid)	298–823	95.432	30.542	...
	823–1061	123.840		
	1061–2000	123.595		
NbN (solid)	298–1643	46.674	7.443	−8.803
	1643–2323	(62.76)(b)		
NbN (liquid)	2323–3000	(62.76)(b)	...	...
Si₃N₄ (solid)	298–2200	70.539	98.738	...
TiN (solid)	298–3323	49.829	3.933	12.384
TiN (liquid)	3323–3500	(66.94)(b)	...	...
VN (solid)	298–2619	45.771	8.786	−9.246
ZrN (solid)	298–3225	46.940	7.029	−7.196
ZrN (liquid)	3225–3500	(58.58)(b)	...	...
Halides				
AlCl₃ (solid)	298–466	64.933	87.860	...
AlCl₃ (liquid)	466–1000	125.520	...	...
CaCl₂ (solid)	298–1045	69.836	15.388	−1.594
CaCl₂ (liquid)	1045–2206	122.263	−14.903	0.703
CaF₂ (solid)	298–1424	41.056	55.460	8.497
				−729.609
	1424–1691	154.098	0.000	...
CaF₂ (liquid)	1691–2804	99.914	...	...
CsI (solid)	298–743	52.820	7.439	−2.297
	743–900	66.230		
CsI (liquid)	900–1400	74.221	...	...
KBr (solid)	298–1007	46.808	16.267	0.364
KBr (liquid)	1007–1671	69.873	...	...
LiF (solid)	298–1121	43.281	16.727	−5.711
LiF (liquid)	1121–2000	64.347	...	...
MgF₂ (solid)	298–1536	73.137	8.878	−12.660
MgF₂ (liquid)	1536–2600	94.922	...	...
SiCl₄ (gas)	298–2000	105.260	1.699	−13.786
SiF₄ (gas)	298–2000	96.232	6.786	−21.902

(a) For some gases, separate coefficients are listed above 2000 K (3140 °F) to more accurately reflect actual values over a wide temperature range.
(b) Parentheses indicate estimated values.

Table 5 Heats and free energies of formation at 298 K (25 °C) (continued)

Compound(a)	ΔH°_{298}, kJ/mole	ΔG°_{298}, kJ/mole
Fe₂SiO₄	−1479.90	−1378.98
LiAlO₂	−1188.67	−1126.31
LiAlSiO₄	−2124.20	−2010.11
MgAl₂O₄	−2299.90	−2175.01
MgCr₂O₄	−1784.06	−1660.50
MgSO₄	−1284.90	−1170.58
MgO·2Al₂O₃·5SiO₂	−9161.70	−8651.34
Mn₂SiO₄	−1730.50	−1632.13
MnTiO₃	−1358.55	−1279.38
Na₂CO₃	−1130.77	−1048.00
NaAlSiO₄	−2094.66	−1980.01
NiAl₂O₄	−1915.90	−1797.12
PbTiO₃	−1198.72	−1111.85
ZnFe₂O₄	−1171.58	−1065.79
ZnSi₂O₄	−1262.57	−1179.50
Zirconates		
BaZrO₃	−1779.46	−1694.68
CaZrO₃	−1766.90	−1681.06
Li₂ZrO₃	−1760.20	−1666.84
Nd₂Zr₂O₇	−4046.92	−3844.62
SrZrO₃	−1767.30	−1681.68
Y₂Zr₂O₇	−4121.99	−3917.87
ZrSiO₄	−2023.80	−1909.32
Borides		
AlB₁₂	−266.10	−272.24
CoB	−94.14	−92.55
CrB	−75.31	−77.00
CrB₂	−94.14	−95.22
FeB	−71.13	−69.52
MnB₂	−94.14	−91.41
Ni₄B₃	−311.71	−305.05
TaB₂	−209.20	−206.57
TiB₂	−323.80	−319.65
VB₂	−203.76	−200.65
ZrB₂	−322.59	−318.24
Carbides		
Al₄C₃	−208.80	−196.46
B₄C	−71.13	−70.55
Be₂C	−116.98	−114.51
CaC₂	−59.80	−64.89
Cr₂₃C₆	−328.44	−338.57
Fe₃C	25.10	20.03
HfC	−251.04	−248.63
Nb₂C	−190.00	−185.66
SiC	−73.22	−70.85
TiC	−184.50	−180.84
ZrC	−196.65	−193.28
Nitrides		
AlN	−317.98	−287.00
BN	−254.39	−228.50
Be₃N₂	−588.27	−532.87
Cr₂N	−125.52	−102.20
Fe₄N	−11.09	3.72
Li₃N	−164.56	−128.64
Mg₃N₂	−460.66	−400.50
NbN	−235.10	−205.97
Si₃N₄	−744.75	−647.34
TiN	−337.86	−309.16
VN	−217.15	−191.08
ZrN	−365.26	−336.70
Halides		
AlCl₃	−705.63	−630.00
CaCl₂	−795.79	−748.11
CaF₂	−1225.91	−1173.54
CsI	−346.60	−340.59
KBr	−393.80	−380.43
LiF	−616.93	−588.66
MgF₂	−1124.24	−1071.11
SiCl₄ (gas)	−662.75	−622.76
SiF₄ (gas)	−1614.94	−1572.70

(a) Room-temperature values of the compound in the solid state, unless specified otherwise as either as gas or liquid

description and data lists described above are useful in this regard. However, when the carbides, nitrides, and, to a lesser extent, the borides are considered, the approach is too limited. These compounds exist over a wide composition range in most applications. Consequently, melting points and thermodynamic properties cannot be assigned to a characteristic, fixed stoichiometry. Furthermore, the melting point of the nitrides cannot be defined without also defining the nitrogen pressure. Most stoichiometric nitrides melt at very high and normally unattainable nitrogen pressure. References 31 to 36 can be consulted for melting point and thermodynamic information about the carbides and nitrides. Although some of these references are rather old, they have not yet been replaced by better data. Engineering properties of carbides,

nitrides, and borides are discussed in separate articles in this Volume.

REFERENCES

1. E.M. Levin, C.R. Robbins, and H.F. McMurdie, *Phase Diagrams for Ceramists*, Vol I, American Ceramic Society, 1964
2. O. Kubaschewski and C.B. Alcock, *Metallurgical Thermochemistry*, 5th ed., Pergamon Press, 1979, p 179–209
3. E.T. Turkdogan, *Physiochemical Properties of Molten Slags and Glasses*, The Metals Society, London, 1983, p 125–140
4. T.B. Massalski *et al.*, *Binary Alloy Phase Diagrams*, 1st ed., Vol 1 and 2, ASM International, 1986

5. T.B. Massalski *et al.*, *Binary Alloy Phase Diagrams*, 2nd ed., Vol 1–3, ASM International, 1991

6. G.W. Moffatt, *The Handbook of Binary Phase Diagrams*, General Electric Company, 1978

7. E.M. Levin, C.R. Robbins, and H.F. McMurdie, Ed., *Phase Diagrams for Ceramists*, Vol II, American Ceramic Society, 1969

8. E.M. Levin and H.F. McMurdie, Ed., *Phase Diagrams for Ceramists*, Vol III, American Ceramic Society, 1975

9. R.S. Roth, T. Negas, and L.P. Cook, Ed., *Phase Diagrams for Ceramists*, Vol IV, American Ceramic Society, 1981

10. R.S. Roth, T. Negas, and L.P. Cook, Ed., *Phase Diagrams for Ceramists*, Vol V, American Ceramic Society, 1983

11. R.S. Roth, J.R. Dennis, and H.F. McMurdie, Ed., *Phase Diagrams for Ceramists*, Vol VI, American Ceramic Society, 1987

12. L.P. Cook and H.F. McMurdie, Ed., *Phase Diagrams for Ceramists*, Vol VII, American Ceramic Society, 1989

13. M. Hansen, *Constitution of Binary Alloys*, 2nd ed., McGraw-Hill, 1958

14. R.P. Elliott, *Constitution of Binary Alloys (First Supplement)*, McGraw-Hill, 1965

15. F.A. Shunk, *Constitution of Binary Alloys (Second Supplement)*, McGraw-Hill, 1969

16. G.V. Samsonov, Ed., *The Oxide Handbook*, 2nd ed. (English trans.), IFI/Plenum, 1982

17. V.I. Posypaiko and E.A. Alekseeva, *Phase Equilibria in Binary Halides*, IFI/Plenum, 1987

18. I. Barin, *Thermochemical Data of Pure Substances*, Parts I and II, VCH Publishing, 1989

19. I. Barin and O. Knacke, *Thermochemical Properties of Inorganic Substances*, Springer-Verlag, Berlin, 1973

20. I. Barin, O. Knacke, and O. Kubaschewski, *Thermochemical Properties of Inorganic Substances (Supplement)*, Springer-Verlag, Berlin, 1977

21. L.B. Pankratz, "Thermodynamic Properties of Elements and Oxides," U.S. Bureau of Mines Bulletin 672, U.S. Department of the Interior, 1982

22. L.B. Pankratz, A.D. Mah, and S.W. Watson, "Thermodynamic Properties of Sulfides," U.S. Bureau of Mines Bulletin 689, U.S. Department of the Interior, 1987

23. L.B. Pankratz, "Thermodynamic Properties of Halides," U.S. Bureau of Mines Bulletin 674, U.S. Department of the Interior, 1984

24. L.B. Pankratz, J.M. Stuve, and N.A. Gokcen, "Thermodynamic Data for Mineral Technology," U.S. Bureau of Mines Bulletin 677, U.S. Department of the Interior, 1984

25. D.R. Stull and H. Prophet, *JANAF Thermochemical Tables*, 2nd ed., U.S. Department of Commerce, 1971

26. M.W. Chase, Jr., *et al.*, JANAF Thermochemical Tables, 1974 Supplement, *J. Phys. Chem. Ref. Data*, Vol 3, 1974, p 311–480

27. M.W. Chase, Jr., *et al.*, JANAF Thermochemical Tables, 1975 Supplement, *J. Phys. Chem. Ref. Data*, Vol 4, 1975, p 1–175

28. M.W. Chase, Jr., *et al.*, JANAF Thermochemical Tables, 1978 Supplement, *J. Phys. Chem. Ref. Data*, Vol 7, 1978, p 793–940

29. M.W. Chase, Jr., *et al.*, JANAF Thermochemical Tables, 1982 Supplement, *J. Phys. Chem. Ref. Data*, Vol 11, 1982, p 695–940

30. M.W. Chase, Jr., *et al.*, JANAF Thermochemical Tables, 3rd ed., Parts I and II, American Institute of Physics, 1985

31. E.K. Storms, Phase Relationships and Electrical Properties of Refractory Carbides and Nitrides, *Solid State Chem.*, Vol 10, *MTP International Review of Science*, University Press, 1972, p 37

32. E.K. Storms, *The Refractory Carbides*, Academic Press, 1967

33. E.K. Storms, Atom Vacancies and Their Effects on the Properties of NbN Containing Carbon, Oxygen or Boron. Part I: Phase Boundary, Thermodynamic and Lattice Parameters, *High Temp. Sci.*, Vol 7, 1975, p 103

34. E.K. Storms and J. Griffin, The Vaporization Behavior of the Defect Carbides. Part IV: The Zr-C System, *High Temp. Sci.*, Vol 5, 1973, p 291

35. F.B. Baker, E.K. Storms, and C.E. Holley, Enthalpy of Formation of Zirconium Carbide, *J. Chem. Eng. Data*, Vol 14, 1969, p 244

36. E.K. Storms, B. Calkin, and A. Yencha, The Vaporization Behavior of Defect Carbides. Part I: The Nb-C System, *High Temp. Sci.*, Vol 1, 1969, p 431

SELECTED REFERENCES

- Y.A. Chang and N. Ahmad, *Thermodynamic Data on Metal Carbonates and Related Oxides*, TMS-AIME, 1982

- C.W. DeKock, "Thermodynamic Properties of Selected Metal Sulfates and Their Hydrates," U.S. Bureau of Mines Information Circular 9081, U.S. Department of the Interior, 1986

- D. Garvin, V.B. Parker, and H.-J. White, Jr., *CODATA Thermodynamic Tables*, Hemisphere Publishing, 1987

- R. Hultgren *et al.*, *Selected Values of the Thermodynamic Properties of Binary Alloys*, American Society for Metals, 1973

- O. Kubaschewski, "The Thermodynamic Properties of Double Oxides," DCS Report 7, National Physical Laboratory, London, 1970

- V. Medvedev *et al.*, *Thermicheskie Konstanty Veshchestv* (Thermal Constants of Substances), Vol 1–10, Akademy Nauk SSSR, Viniti, Moscow, 1965–1981

- K.C. Mills, *Thermodynamic Data for Inorganic Sulphides, Selenides and Tellurides*, Butterworths, London, 1974

- P.J. Spencer, "The Thermodynamic Properties of Silicates," Report Chem. 21, National Physical Laboratory, Teddington, UK, 1973

- D.D. Wagman *et al.*, "Selected Values of Chemical Thermodynamic Properties: Compounds of Uranium, Protactinium, Thorium, Actinium, and the Alkali Metals," National Bureau of Standards Technical Note 270-8, U.S. Department of Commerce, 1981

- D.D. Wagman *et al.*, "The NBS Tables of Chemical Thermodynamic Properties," National Bureau of Standards, 1982

Applications for Traditional Ceramics

Chairman: Gordon Lewis, Department of Ceramic Engineering, Clemson University

Introduction

EARLY MANKIND used native stones as tools and implements and often recorded the development of their culture on cave walls. Improved tools resulted as mankind discovered how to shape the rough stone by chipping, grinding, and polishing. Even today some cultures still prepare tools and implements in this manner.

At some time before recorded history, mankind discovered that the soft clay exposed along river banks would mold into utensils and brick that became useful after drying. Next mankind discovered the added benefits and improved performance associated with heat treating the clay at higher temperatures compared to simple drying of ceramic articles in air or the sun. That point in time can be taken to represent the start of what is termed "traditional" ceramics today.

New ceramic products and applications form an evolutionary chain leading up to the present time. It seems reasonable to assume that following the discovery of the enhanced value of ceramic products over natural materials in some applications, other difficult materials problems were solved with ceramic products produced for completely different applications. It is equally possible that the discovery of new ceramic products followed from a defined need or application before the product existed. This need undoubtedly set off a search for new products.

Assuming the ceramic industry started with two defined product fields—clay vessels and bricks—it grew over the years from these to many other specialty subfields. This process of discovery of new subfields is still active today. For example, there was a significant expansion in the electrical ceramic materials field during the latter half of the 1980s with the discovery of ceramic superconductors. It should be noted, however, that electrical ceramic materials first appeared in the whitewares product line. The future range of ceramic products and applications seems as boundless today as it must have hundreds, or even thousands, of years ago.

The above discussion points out how difficult it is to draw boundaries between ceramic products or applications using terms such as ancient, traditional, modern, and advanced. This is because of the evolutionary nature of the field. Advances in preparing or beneficiating starting materials, processing operations, and quality control used in manufacturing result in new product lines with a new spectrum of applications. Demands from industry for special materials with designed properties to meet special applications lead to research into new products.

The raw materials used to form certain ceramic articles may represent a more satisfactory way to distinguish between todays and yesterdays products than a concept attached to an ill-defined time frame, a product application, or a processing method. Therefore in this Section, traditional ceramics will be classified as those ceramic products that use clay or have a significant clay component in the batch. Clay-based products were the only representatives of the ceramic field, other than glass, until about some one hundred years ago. At that point new classes of starting materials, such as silicon carbide for abrasives, started to enter the picture.

The five product areas represented by refractories, cement, whitewares, porcelain enamel, and structural clay products fit into this classification scheme. In their early forms, all contained clays as a batch component. Today, many of the product lines contain no clay at all, yet these materials continue to be classified into the family of traditional ceramics. Obviously, a single descriptive modifier is inadequate to give a complete picture of the product's capabilities.

To demonstrate the process of ceramic evolution in starting materials that leads to the concept of advanced materials, consider refractories. Refractory fire clay was the major starting material for refractory brick. Discoveries in the metals industry led to demands for new and special refractory types. Part of the answer in some cases required starting materials that represented a single chemical oxide, rather than the complex chemistry associated with clays. This led to silica brick, alumina brick, magnesia brick, zirconia products, carbon brick, and others. The refractory industry rightly classed these products as advanced refractories at the time of discovery. They were more refractory, stronger, and more corrosion resistant. They represented an advance over previous products.

Today the refractory user is moving toward monolithic linings for furnaces and vessels rather than brick. As manufacturers of these products gained insight into particle size distribution requirements, advances in refractory life and performance resulted. Again they produced an advanced ceramic, but not based on a chemical composition change.

Some advanced furnaces today have no brick or monolithic refractory in them; the thermal barrier lining is fiber refractory. These furnaces lose heat through conduction less rapidly and contain much less heat resulting from heat capacity. This makes the furnace more fuel efficient and allows faster heat up and cool down. The advanced concept is from a change in product internal structure.

The point at issue is that the traditional ceramic industry continues to grow and mature using recently discovered concepts or techniques to meet advanced needs. As such, traditional does not mean old-fashioned or obsolete.

In this Section, the authors will describe how each major ceramic product classification and the various products within that line, meet application requirements in other industries, such as manufacturing, energy production, transportation, construction, and even in our daily lives. Each article presents the considerations that should precede selecting a specific material or product in a given application.

The goal of this Section is not just to acquaint the reader with the products in the field and how to use them in traditional ways. The broader goal is to stimulate the reader to search for further applications for segments of the industry whose roots connect directly back to the beginning of time.

Refractories

REFRACTORIES are materials of construction that can withstand high temperatures and maintain their physical properties within a furnace environment. They are therefore widely used to build structures that can resist the thermal shock, physical wear, and corrosive chemicals associated with iron and steel production, copper and aluminum smelting, glass and ceramics manufacturing, and similar processes.

The most common refractories are made from natural materials. Most industrial refractories are composed of metal oxides or carbon, graphite, or silicon carbide. Newer materials, such as carbide, nitrides, borides, and silicides, are also used when the expense is warranted. Generally, the use of chemically processed synthetics or high-purity oxides has increased in response to the higher service temperatures required in many applications.

Numerous refractory compositions in a wide variety of shapes and forms can be used in a broad range of applications. Two-thirds of all refractories used by industry are preformed bricks and other fired shapes. The remainder take the form of monolithic materials, that is, castables, plastics, and gunning or ramming mixes. These materials are placed directly in a furnace to form a refractory lining upon being fired *in situ*. Mortars, which consist of grains with plasticizers, are also considered refractory monoliths, and are used for laying refractory brick. Refractory insulating materials are often used in high-temperature applications to reduce heat losses and conserve fuel. They can be made in the form of insulating brick, refractory fiber board or blanket, or special vacuum cast shapes.

This article first classifies refractory materials and describes the types that belong in each class. Various physical properties that should be considered in the refractory selection process are then discussed, followed by numerous examples of typical applications. Emphasis is placed on refractory bricks and shapes. A companion article on "Monolithic and Fibrous Refractories" follows in this Section.

Refractory Types (Ref 1)

Refractories are generally categorized as being either clay or nonclay products. The types of raw materials that constitute the various refractories in each of these two main categories are described below. The most commonly used refractory oxides are SiO_2, Al_2O_3, MgO, CaO, Cr_2O_3, and ZrO_2. Generally, the refractories that contain SiO_2 or ZrO_2 are considered to be acidic; those with MgO, CaO, Al_2O_3, or CrO_3, basic. This grouping scheme is convenient for describing refractory reactions with high-temperature slags, although it is not strictly true, chemically. A chemical analysis of typical refractory materials is provided in Table 1.

Clay Refractories (Ref 1, 2)

The primary types of clay refractories are fireclay and high alumina. Each type is used to produce bricks, as well as insulating refractories, which are manufactured to be less dense. Fireclay is also used to produce two types of ladle brick, both of which require high density and low porosity.

Fireclay generally consists of the mineral kaolinite ($Al_2O_3 \cdot 2SiO_2 \cdot 2H_2O$), with minor amounts of other clay minerals and impurities in quartzite, iron oxide, titania, and alkali forms. Clays are used either as mined or after calcination, a process that breaks down the clay into mullite and siliceous glass. The refractoriness of a product made of fireclay will increase with increasing amounts of alumina and decreasing amounts of impurities, such as alkalis and iron oxide. The fireclays from various deposits are often combined to produce products with different properties. Deposits are widely distributed throughout the world.

Fireclay refractories can be classified as being low-, medium-, high-, or super-duty, based on their resistance to high temperature or on their refractoriness. Bricks made of fireclay have an alumina content ranging from 25 to 45%, and offer moderately high resistance to thermal stress, low thermal expansion, and adequate thermal insulation. They also offer some resistance to attack by acidic materials, but fail rapidly at high temperatures when exposed to chemically basic materials.

High-alumina refractories are generally composed of bauxite or other raw materials that contain from 50 to 87.5% alumina. Those compositions with more than this level of alumina are classified as nonclay, extra-high alumina refractories, which are described in the section "Nonclay Refractories."

Bauxite is a naturally occurring raw material consisting mainly of the mineral gibbsite, $Al(OH)_3$, with varying amounts of kaolinite, and minor amounts of iron oxide and titania impurities. Because the loss on ignition of bauxite is high, it is first calcined at high temperatures to convert it to a dense grain consisting mainly of corundum (Al_2O_3) and mullite ($3Al_2O_3 \cdot SiO_2$). The most widely used refractory-grade calcined bauxite comes from Guyana and Surinam in South America and, more recently, from China. Deposits of kaolinitic bauxites containing 50–75% alumina on a calcined basis also occur in Alabama and Georgia in the United States.

Depending on its composition and impurities, a high-alumina refractory is generally multipurpose, and offers fair-to-excellent resistance to chipping and somewhat higher volume stability than most other clay refractories.

Nonclay Refractories (Ref 1, 2)

The types of refractories included in this category are those designated as basic (magnesia, dolomite, chrome, and combinations thereof), extra-high alumina, mullite, silica, silicon carbide, and zircon, each of which is described below. Each of these materials is used to produce brick forms. In addition, magnesite-chrome and alumina refractories are formed into special shapes using a fused casting process, whereby the materials are fused in electric furnaces and then poured, while molten, into molds.

Other nonclay refractories that are used in specific applications that benefit from their special properties are alumina-zirconia-silica, chrome-alumina, forsterite, graphite, magnesia-alumina, pyrophyllite-andalusite, and zirconia. These are not discussed in this article.

Basic refractories are produced from a composition of dead-burned magnesite, dolomite, and chrome ore, as well as mixtures of magnesite-dolomite, magnesite-carbon, dolomite-carbon, and magnesite-chrome.

Magnesia brick consists mainly of the mineral periclase (MgO) and is available in chemically bonded, pitch-bonded, burned or fired, and burned and pitch-impregnated forms. Historically, natural magnesite ($MgCO_3$) that was calcined provided the raw material for this brick, but with the increased demands for higher temperatures and the introduction of fewer process impurities, more high-purity magnesia from seawater or underground brines has been used. In the seawater and brine processes, the magnesia is obtained by calcining precipitated $Mg(OH)_2$, which provides MgO of purity up to 98%.

Seawater or brine plants are located in the United States, United Kingdom, Soviet Union, Mexico, Italy, and Japan. Natural magnesite deposits of refractory grade are found in Austria, Greece, China, and Brazil, as well as Washington and Nevada. Typically, the natural magnesites contain more impurities, but the Chinese deposits are reported to contain ≥95% MgO on a calcined basis.

Table 1 Chemical analysis of some typical refractory raw materials

Constituent	Refractory material composition, wt%							
	Magnesia(a)	Chromite(b)	Dolomite(c)	Bauxite(d)	Alumina(e)	Kyanite(c)	Flint clay(f)	Quartzite
Silica (SiO$_2$)	0.8	5.1	0.7	6.5	0.06	40.0	51.0	99.5
Alumina (Al$_2$O$_3$)	0.4	29.9	0.3	88.3	≥99.5	56.6	44.0	0.14
Titania (TiO$_2$)	trace	trace	trace	3.2	trace	1.0	2.5	0.03
Iron oxide (Fe$_2$O$_3$)	1.4	13.0	0.9	1.8	0.06	1.0	0.9	0.17
Lime (CaO)	0.9	0.5	57.7	...	...	0.5	0.2	...
Magnesia (MgO)	96.3	18.7	40.4	...	trace	0.5	0.3	0.01
Chromia (Cr$_2$O$_3$)	ND(g)	32.2	ND	ND	ND	ND	ND	ND
Alkalis (Na$_2$O + K$_2$O + Li$_2$O)	ND	ND	ND	0.02	0.05	0.2	0.6	0.03
Loss on ignition	...	0.3	47.0	0.25	...	14.2	13.2	...

(a) Seawater, U.K. (b) Philippine lump. (c) United States. (d) Refractory grade, Guyana. (e) Tabular. (f) Missouri. (g) No Data

Dolomite brick is made from a highly calcined natural mineral of the composition CaCO$_3$·MgCO$_3$. Because the lime component hydrates readily, unfired brick made with dolomite grain is usually tar- or pitch-bonded; burned dolomite brick is often impregnated with tar or pitch to extend its service life.

Chrome brick is made from naturally occurring chrome ore, which is a complex solid-solution series of spinel-type minerals, including spinel (MgO·Al$_2$O$_3$), picrochromite (MgO·Cr$_2$O$_3$), hercynite (FeO·Al$_2$O$_3$), ferrous chromite (FeO·Cr$_2$O$_3$), magnesioferrite (MgO·Fe$_2$O$_3$), and magnetite (FeO·Fe$_2$O$_3$). Silicate phases, such as serpentine, talc and enstatite, are associated with the spinel grains as gangue materials.

Refractory-grade chrome ores are found in the Philippines, Cuba, Turkey, Greece, and China. The African ores from Transvaal and Rhodesia, as well as some from the Soviet Union, are high in iron oxide content.

Combinations of chrome-magnesite, which have a larger proportion of chrome than magnesite, exhibit less thermal expansion than high-magnesia compositions. High-purity magnesite and chrome-free compositions of high-purity seawater or brine-well magnesia provide maximum refractoriness and resistance to iron oxides. These compositions, as well as dolomite and magnesite-dolomite, can be coal tar bonded and used in basic oxygen process (BOP) furnace applications. Magnesite-carbon bricks are used in the most severe areas of BOP vessels and electric arc furnaces. Carbon contents typically range from 5 to 35% and are provided by additions of natural flake graphite. Magnesite-carbon brick offers outstanding resistance to corrosion by steel-making slags.

Extra-high alumina refractories are made predominately from bauxite or alumina (Al$_2$O$_3$) that has been fused or densely sintered. Extra-high alumina refractories contain from 87.5% to slightly less than 100% alumina, and offer good volume stability at temperatures up to 1815 °C (3300 °F).

Mullite (3Al$_2$O$_3$·2SiO$_2$) refractories are made of kyanite, sillimanite, andalusite, bauxite, or mixtures of aluminum silicate materials, resulting in compositions of about 70% alumina. Any of these materials must be sintered at high temperatures or melted in electric furnaces to bring about the formation of mineral mullite. This is the most stable of any alumina-silica combination. These refractories are noted for their low level of impurities and their excellent resistance to loading at high temperatures.

Silica is a naturally occurring mineral that is abundant in the earth's crust. Fine-grained deposits with low deposits of alumina and alkali make excellent refractories with high-temperature load-bearing capabilities.

Brick made of quartzites, conglomerates of silica gravel, or novaculite is bonded by adding 3 to 3.5% of CaO, which forms a small amount of glass upon firing. The balance of the quartzite raw material is converted to tridymite and cristobalite. Silica brick is characterized by a very high coefficient of thermal expansion between room temperature at 500 °C (930 °F) and therefore must be heated and cooled extremely slowly through this temperature range.

Silica brick is available in three grades: super-duty, which has very low alumina and alkali contents; regular, or conventional, duty; and coke oven quality. Recently, silica compositions made from the fused or glassy form have been used for hot patching and other applications, such as shrouds, where thermal shock is significant. The vitreous form of silica has a much lower coefficient of expansion than the crystalline forms.

Semisilica brick is made from siliceous clays and contains mainly cristobalite bonded with a glassy phase. It typically contains 18 to 25% alumina and 72 to 80% silica. This brick has excellent load-bearing ability to 1300 °C (2370 °F), but, like regular silica brick, has a comparatively high thermal expansion from room temperature to 500 °C (930 °F).

Silicon carbide (SiC) is produced by the reaction of sand and coke in an electric furnace. It is used to make special shapes, such as kiln furniture, which are used to support and separate pieces of ceramicware as they are fired in kilns. Silicon carbide has exceptionally high thermal conductivity, good load-bearing characteristics at high temperatures, and good resistance to rapid changes in temperature.

Zircon is a naturally occurring zirconium silicate (ZrO$_2$·SiO$_2$) mineral. It has good volume stability for extended periods of exposure to high temperature. Bricks can be made by combining zircon with high-alumina materials and either firing them in a conventional manner or fusing and casting them from the melt. When refractories are made from synthetic high-purity zirconia (ZrO$_2$), they must be stabilized with small quantities of lime, magnesia, or yttria. Zircon refractories are widely used for glass tank construction, which requires resistance to certain molten glasses.

Properties (Ref 2)

The significant properties of any refractory, including its high-temperature strength, depend on its mineral makeup, the particle size distribution of the minerals, and the way these materials react to high temperatures and furnace environments. Particles that vary in size from 6 mm (0.24 in.) to less than 74 μm (3 mils) constitute the unfired refractories. Upon firing, the fines form a ceramic bond between the larger particles. The fired refractory consists of bonded crystalline mineral particles and glass or smaller crystalline particles, depending largely on the composition of the refractories.

A variety of physical properties must be considered when choosing a refractory for a particular application. The most significant properties are those that enable the refractory to withstand the conditions found in a furnace. A refractory not only must withstand the maximum service temperature of the furnace, but, in many cases, must also resist spalling, heavy loads, abrasion, and erosion or corrosion caused by liquids and gases. Tests that indicate performance under service conditions are the pyrometric cone equivalent, reheat, load, hot modulus of rupture, spalling, slag, and carbon monoxide resistance tests. Typical physical properties are listed in Table 2.

The bulk density of a material is its mass per unit volume, usually expressed in g/cm^3.

Table 2 Physical properties of some typical fired refractory brick

	Magnesia (95% MgO)	Chrome (30% Cr$_2$O$_3$)	90% alumina	70% alumina	Zircon	Fireclay (Missouri superduty)	Silicon carbide	Silica (superduty)
Bulk density, g/cm^3	2805–2950	3060–3140	2900–2965	2530–2600	3605–3720	2310–2370	2565–2660	1780–1875
Porosity, %	15–19	16–20	14–18	17.5–21.5	19–23	11–14	11–15	20–24
Cold crushing strength,		35–55	62–95	27–48		12–21	69–83	27–41
MPa (ksi)	48–70 (7–10)	(5–8)	(9–14)	(4–7)	48–76 (7–11)	(1.7–3)	(10–12)	(4–6)
Modulus of rupture, MPa	17–24	14–21	17–21	7.6–11	15–23	4.8–6.9	21–24	4.1–6.9
(ksi)	(2.5–3.5)	(2–3)	(2.5–3)	(1.1–1.6)	(2.2–3.3)	(0.7–1)	(3–3.5)	(0.6–1)
Reheat test, % permanent linear change after heating to:								
1600 °C (2910 °F)	...	...	...	+3.5 to 6.0	...	0.0 to 0.9	...	...
1650 °C (3000 °F)	...	...	...	...	0.0	...	...	...
1725 °C (3135 °F)	−0.2 to 1.0	...	+0.1 to 1.0	...	...	...	−0.1 to 0.1	...
Load test at 170 kPa (25 ksi); withstands load to temperature, °C (°F)	1620 (2950)	1400 (2550)	1760 (3200)	1450 (2640)	1600 (2910)	1450 (2640)	1650 (3000)	1680 (3055)

Apparent porosity is the ratio of the open void volume to the total volume. Open voids are those pores that are open to the surface of the refractory. Refractories with low apparent porosity have greater resistance to penetration by slags and fluxes. They also resist corrosion and erosion and usually have a lower gas permeability than those with high porosity. Thermal conductivity is also influenced by porosity. Typically, insulating refractories are very porous.

Cold crushing strength is the ability of a refractory to withstand handling/shipping and impact/abrasion at low temperatures. It does not indicate strength at service temperatures.

The pyrometric cone equivalent (PCE) test measures the refractoriness, or softening point, of a material by comparing the behavior of a test cone with reference cones of standard composition designated by values between 12 and 42. There is no distinct melting point, because a refractory is a multicomponent material. In this test, the cones are mounted on a plaque and heated until the test cone softens and bends. The PCE value assigned to the test specimen is that of the reference cone whose tip touches the plaque at the same time as the test cone.

The reheat test is used to evaluate a refractory for permanent shrinkage or expansion. Test bricks are measured, and then heated in a furnace at a specified rate to a temperature that is held for 5 h. After cooling, the brick is measured to determine the changes in linear dimensions and volume.

The load test measures the ability of a refractory to withstand a load at elevated temperatures. The most common method involves placing the brick on end in a furnace, applying a vertical load of 170 kPa (25 psi), and heating the refractory to a maximum temperature that is held for 90 min. From measurements made before and after testing, the percentage subsidence is calculated.

Refractoriness under Load. Because the changes that take place in load tests during the heating period are seldom recorded, these tests are being replaced by a refractoriness-under-load, or creep, test. In this test, the expansion and/or subsidence are recorded during the heating period and a 24 h hold period at a temperature of 1480 °C (2700 °F). This not only yields a plot of expansion during heating, but measures the amount of creep under load at the elevated temperature. This test gives information about how the refractory will behave under load for extended periods of time. Typically, a uniform creep rate for a refractory is established after 20 to 24 h.

Hot modulus of rupture is a measure of the strength of a refractory at high temperatures. The test specimen, supported at both ends across a span in the testing apparatus, is heated to a specified temperature and then broken in the middle. The force required to fracture the specimen is used to calculate the modulus of rupture.

Thermal Shock or Spalling Resistance. Three types of spalling are commonly found: thermal spalling; mechanical, or pinch, spalling; and alteration, or structural, spalling. In the case of thermal spalling, stresses are created in the brick by the temperature gradient and are due to thermal expansion. Upon failure, the brick tends to cube. Mechanical, or pinch, spalling is due to improper allowance for thermal expansion, resulting in a high stress concentration at the hot face. This type of spalling can be recognized by shearing of the corners of the brick at the hot face. Alteration, or structural, spalling results from the reaction of other materials or slag with the brick to produce crystalline phases that have different thermal expansion characteristics from those of the base brick. This type of spalling can be caused by vitrification of the hot face.

A panel spalling test developed many years ago has been found to be effective in evaluating fireclay refractories. However, it does not produce definitive results for high-alumina brick or basic brick, because these are much more refractory and do not readily densify or vitrify at the hot face. When bricks of these types undergo the panel spalling test, the nature of the spalling is purely thermal, in contrast to fireclay compositions, in which some alteration occurs.

Recently, a prism spalling test has been devised to evaluate these more refractory compositions. In this test, small prisms of refractory are heated to 1200 °C (2190 °F) and cooled by air and/or water spray for 30 s. The spalling resistance is either measured by the number of cycles that the brick is able to withstand or, in the case of extremely good spalling resistance, by the appearance or number of cracks visible. The ribbon spalling test is a variation of this test.

Because silica brick has a very high coefficient of expansion from room temperature to 500 °C (930 °F), a special hotplate method of test has been devised. The ends of six bricks are heated on a hotplate to 800 °C (1470 °F) at a rate between 75 and 550 K/h. The specimens are then allowed to cool to 650 °C (1200 °F) at a rate at least 50% slower than the heating rate, and then are cooled by natural radiation to 120 °C (250 °F), after which they are evaluated.

Slag Tests. There are numerous methods for evaluating the reaction of slag with a refractory material. In one very common method, the cup slag test, a cup that has been cut or drilled into the refractory is filled with slag. The specimen is then heated in a furnace to a specified temperature for a given length of time. The reaction between the slag and the refractory may be observed by cutting the refractory through the cup to determine the depth of reaction.

In a more dynamic test, the drip slag test, slag rods are fed through a water-cooled jacket into a hot furnace, where the slag melts and drips on a sample of refractory tilted 30° from the horizontal. Depending on the extent of wetting of the refractory by the slag, the slag either cuts a narrow gorge or spreads over a wide area on the refractory.

In another dynamic slag testing method, test refractories line a cylinder tilted at a slight angle from the horizontal. As this cylinder rotates, it is heated by a burner at the lower end and the slag is fed into the upper end, causing slagging and reaction over the length of the refractory lining. This test measures the depth of loss by corrosion.

Abrasion Resistance. In the abrasion test, the face of a refractory is abraded by 1000 g (2.2 lb) of silicon carbide grain, which is entrained in air and strikes the test sample at 90°. The grain sizing of the silicon carbide abrasive and the duration of the test must be carefully controlled for reproducible results. The test samples are weighed before and after, and the volume loss of the refractory is calculated to measure abrasion resistance.

Permeability. The rate at which a gas flows through a refractory is important in instances where this gas will react with other materials. To measure permeability, a 50 mm (2 in.) cube of the refractory is placed in a rubber form that, when compressed, permits the gas to flow through the refractory in one direction. By using various gases and flow rates controlled by flow meters, it is possible to calculate the permeability of the refractory.

Carbon Monoxide Resistance. If a refractory that contains unreacted iron oxide is exposed to carbon monoxide over extended periods of time, solid carbon may be catalytically deposited. The deposited carbon may develop internal stresses great enough to destroy the bonds of the refractory. Usually, the critical temperature range for carbon deposition ranges from 495 to 505 °C (925 to 940 °F). The test for carbon monoxide resistance is conducted over five 100 h periods, during which the test samples are exposed to a 95% carbon monoxide atmosphere in a heated test chamber. The specimens are inspected for degradation at the end of each test period.

Chemical composition is another factor that indicates how well a refractory material will stand up under certain conditions. For example, in fireclay and high-alumina refractories, as the alumina-silica ratio increases, the refractoriness usually increases. The presence of certain impurities or accessory oxides, such as soda, potash, lime, and iron oxide, which promote the formation of low-melting glasses, also tends to reduce refractoriness.

Chemical composition can also be used to predict the extent of corrosion, that is, the destruction of refractory surfaces by the chemical action of external agents. Acid refractories, for example, contain a substantial amount of silica that reacts chemically with basic refractories, basic slags, or basic fluxes.

Thermal Behavior. Refractory materials, like all solids, expand upon heating, the degree of expansion being related to the chemical composition of the refractory. Magnesium oxide has the highest coefficient of expansion of the refractory materials commonly used. Silicon carbide has a comparatively low coefficient of expansion, and fused or vitreous silica has the lowest expansion of all the refractory materials commonly used. Materials with high coefficients of expansion typically have poor resistance to thermal shock.

Thermal expansion curves are shown in Fig 1 and 2.

Another thermal property that must be considered in service applications is thermal conductivity (Fig 3 and 4). Of the refractory materials normally used in tonnage quantities, silicon carbide exhibits the highest thermal conductivity. Generally, it has been noted that thermal conductivity depends on chemical composition, porosity, temperature, and the type of crystalline composition, porosity, temperature, and the type of crystalline or glassy phase present. As a general rule, thermal conductivity decreases with increasing temperature for crystalline oxides but increases with increasing temperature for glassy materials. Castable and gunning materials that contain hydrated calcium aluminate

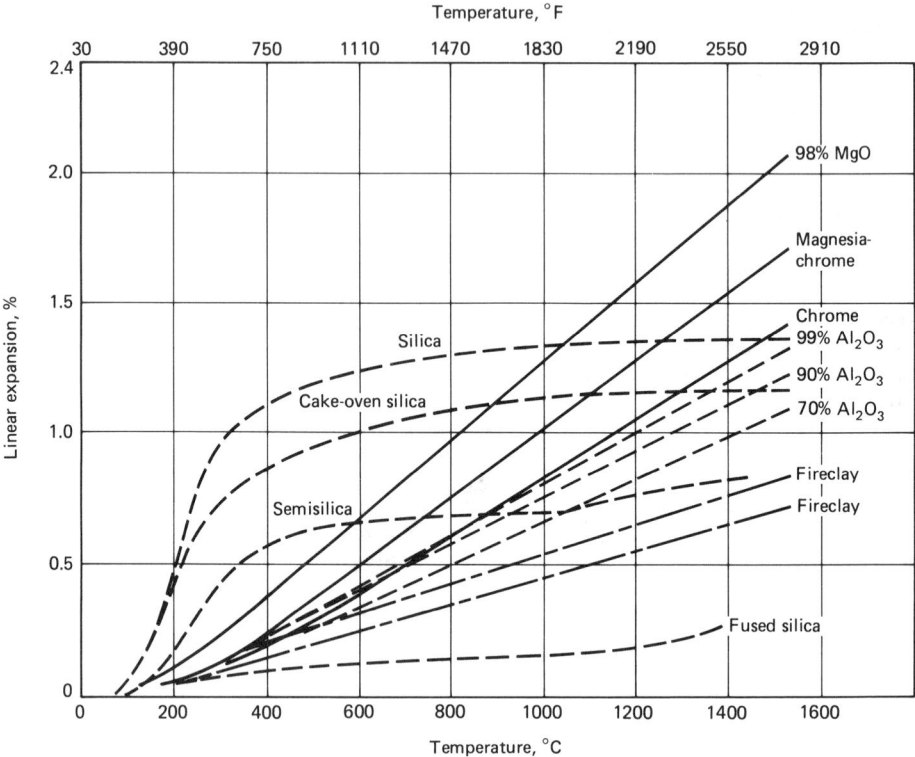

Fig 1 Linear thermal expansion of magnesia and chrome brick (solid lines) and silica-alumina brick ranging in composition from 96% silica to 99% alumina (dashed lines)

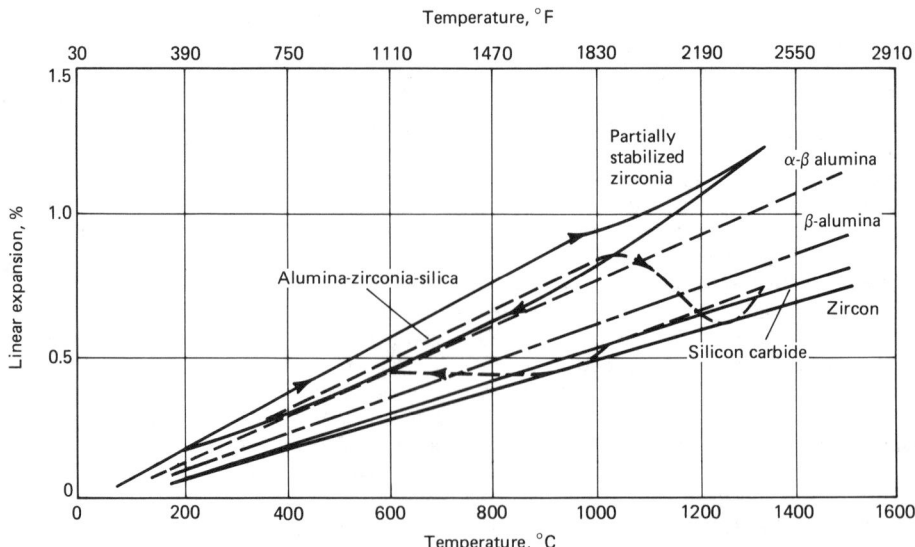

Fig 2 Linear thermal expansion of zircon, silicon carbide, and 4% CaO-stabilized zirconia brick (solid lines) and fused cast brick composed of α-β alumina, β-alumina, and alumina-zirconia-silica (dashed lines)

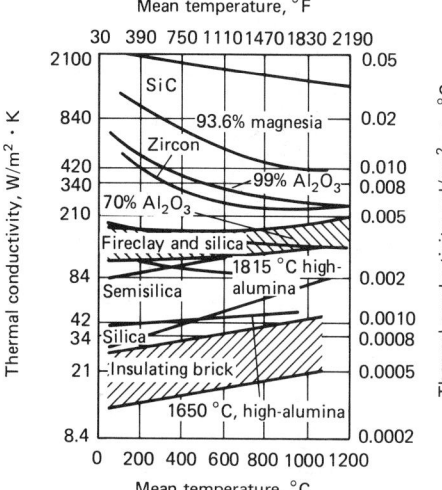
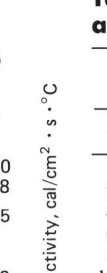

Fig 3 Thermal conductivity of various types of refractory brick

Table 3 Mean specific heats of refractory materials between 0 °C (32 °F) and indicated temperature

Temperature		Specific heat (a)				
°C	°F	Fireclay brick	Silica brick	Magnesia brick	Chrome brick	99% alumina brick
0	32	0.808	0.708	0.871	0.712	0.716
200	390	0.863	0.883	0.971	0.762	0.896
400	750	0.913	0.988	1.043	0.808	0.976
600	1110	0.976	1.051	1.097	0.850	1.034
800	1470	1.022	1.080	1.130	0.879	1.063
1000	1830	1.063	1.110	1.168	0.909	1.093
1200	2190	1.097	1.139	1.206	0.930	1.118
1400	2550	1.122	1.164	1.239	0.942	1.139
1500	2730	1.143	1.177	1.256	0.950	1.164

(a) Units of kJ/kg · K.

compounds may differ considerably in thermal conductivity. After the water of hydration is driven from the calcium aluminate cement component, thermal conductivity is appreciably reduced.

The specific heat capacity of refractories is very important when they are used in heat-exchange applications, such as blast furnace stoves or glass tank regenerators. However, the overall range of mean specific heat capacity of all refractories is not large, ranging from about 0.7 to 1.2 kJ/kg · K (Table 3).

Refractory Applications

Depending on the application, refractories must resist chemical attack, molten metal and slag erosion, thermal shock, physical impact, catalytic heat, and similar adverse conditions, generally within a high-temperature environment. Because the ingredients of refractories impart a variety of characteristics, many refractories have been developed for specific applications. This section describes the refractory products used by the iron and steel, nonferrous metals, and glass-melting industries. Also discussed are specialty refractories based on cordierite, zirconia, and $Li_2O \cdot Al_2O_3 \cdot SiO_2$ ceramics, which are used by several industries. Metallurgical industries in the United States consume about 72% of all refractories used, of which ferrous applications account for 63%. The glass industry

consumes 6.2%, and the ceramics industry, 11.8%. The balance is consumed by other industries.

Products Used in Iron and Steel Industries (Ref 3)

The structures that utilize refractory products include coke ovens, blast furnaces, blast furnace stoves, basic oxygen vessels, electric furnaces, open-hearth furnaces, argon-oxygen decarburization process vessels, pouring pit refractories, continuous casting vessels, soaking pits and reheating furnaces, and steel refining process vessels.

Coke Ovens. Operation of coke ovens is basic to the steel industry, because the coke is used as a raw material in the blast furnace, together with limestone and iron ore, to produce pig iron. Whereas the early production of coke was in beehive ovens constructed of silica brick, most production today occurs in by-product ovens that recover many of the chemicals given off as the coal is heated to produce coke. The by-product battery consists of a row of individual ovens, usually 24 or more, placed side by side. The ovens are separated by refractory walls that enclose the heating flues. High-duty and medium-duty fireclay brick and shapes are used for walls, flues, and regenerator checkers as conditions require. The chamber walls of the oven where coal is heated to produce coke are made of silica brick, which must be free of iron spots, chips, and other defects. Coke ovens are a highly specialized application requiring over 1000 shapes for construction.

Blast Furnaces. The operation of blast furnaces has undergone a considerable change over the last few years, with a trend toward larger-diameter furnaces with appreciably increased working volume. Blast furnaces with hearths 12 to 14 m (40 to 46 ft) in diameter are not uncommon, and are capable of producing up to 10,000 t (22×10^6 lb) of molten pig iron per day.

In the upper portion of blast furnaces (Fig 5), wear patterns are principally due to the abrasion of the charge. Between the bosh (the tapering section below the widest diameter) and the upper stack, carbon monoxide resis-

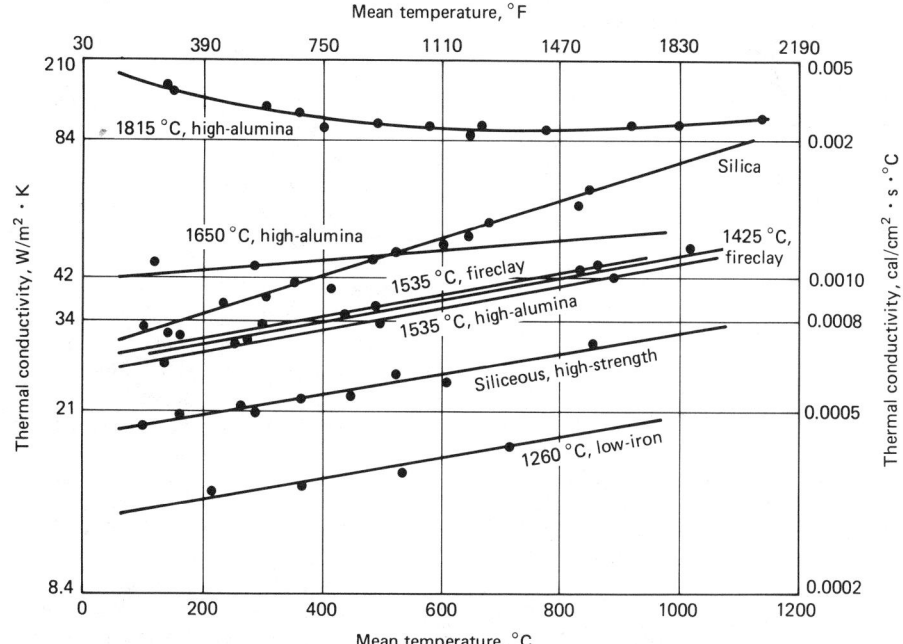

Fig 4 Thermal conductivity of various types of insulating brick

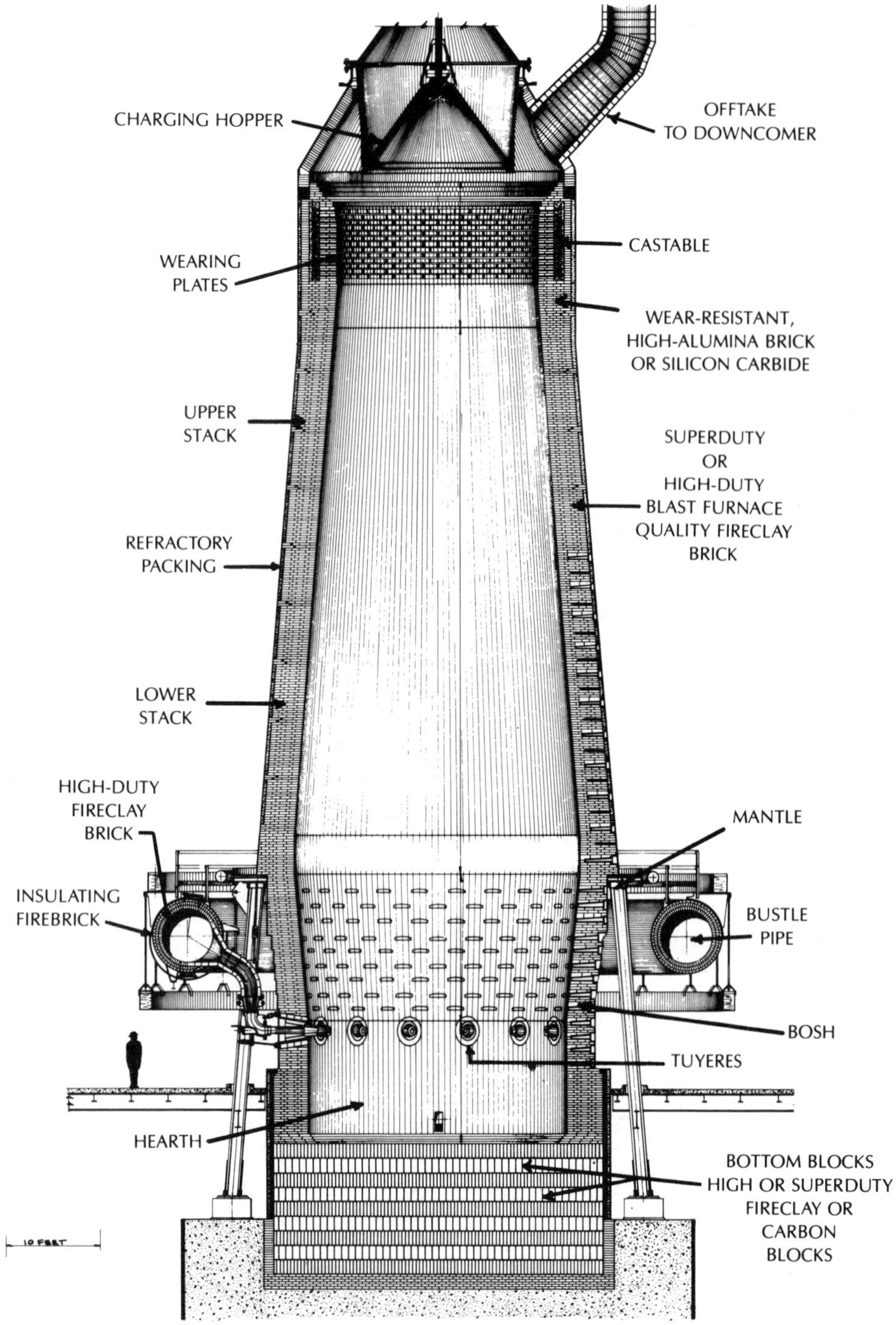

CHARGING HOPPER

OFFTAKE
TO DOWNCOMER

WEARING
PLATES

CASTABLE

WEAR-RESISTANT,
HIGH-ALUMINA BRICK
OR SILICON CARBIDE

UPPER
STACK

SUPERDUTY
OR
HIGH-DUTY
BLAST FURNACE
QUALITY FIRECLAY
BRICK

REFRACTORY
PACKING

LOWER
STACK

HIGH-DUTY
FIRECLAY
BRICK

MANTLE

INSULATING
FIREBRICK

BUSTLE
PIPE

BOSH

TUYERES

HEARTH

BOTTOM BLOCKS
HIGH OR SUPERDUTY
FIRECLAY OR
CARBON
BLOCKS

10 FEET

Fig 5 Typical blast furnace. Courtesy of The Refractories Institute

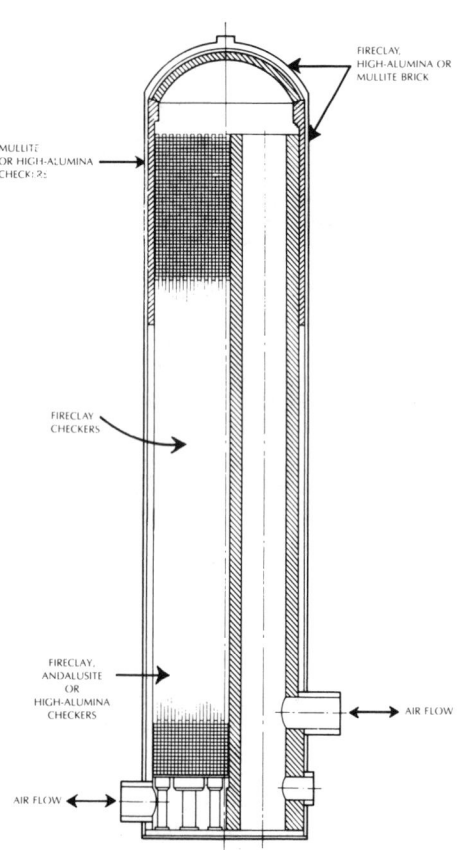

FIRECLAY,
HIGH-ALUMINA OR
MULLITE BRICK

MULLITE
OR HIGH-ALUMINA
CHECKERS

FIRECLAY
CHECKERS

FIRECLAY,
ANDALUSITE
OR
HIGH-ALUMINA
CHECKERS

AIR FLOW

AIR FLOW

Fig 6 Typical blast furnace stove. Courtesy of The Refractories Institute

tance is important. In the bosh and upper bosh area, the effects of alkalis, principally potassium species, and zinc oxide are very important. Therefore, blast furnace hearth walls are typically composed of carbon blocks, and the bottom of the hearth is composed of either carbon or super-duty blast furnace brick. Recently, a number of experiments have been conducted using nitride-bonded silicon carbide, which has very high thermal conductivity, as well as good resistance to alkalis and

zinc oxide. A number of experimental linings that utilize silicon carbide are being evaluated.

Blast Furnace Stoves. The need for greater production from blast furnaces has necessitated higher hot-blast temperatures. Increased temperatures, in turn, have required many changes in the refractories used in the stoves that preheat the blast (Fig 6). The lower hot-blast temperatures are readily accommodated by using semisilica brick in the walls, dome, and top checkers. However, with increasing

blast temperatures, more refractory high-alumina brick and silica brick have been used. When silica brick is used, heating and cooling must occur within the temperature range where the coefficient of thermal expansion is least critical (between 600 and 1300 °C, or 1110 and 2370 °F). This is usually accomplished by locating thermocouples strategically in the combustion chamber near the top of the chamber and in the upper walls and dome of the stove.

The lower checkers in the stove construction may be 38 to 55% alumina refractories. Frequently, mullite refractories are used in the burner area. A very important aspect in the operation of a blast furnace stove is the removal of flue dust from the blast furnace gas. Typically, this dust contains appreciable alkali, lime, and iron oxide, which have a deleterious effect on the stove refractories.

Basic Oxygen Vessels. Since the advent of the basic oxygen furnace process (Fig 7) in the early 1950s, its use has grown very rapidly. Furthermore, vessels used in this process have grown in size up to 300 t (660 $\times$ 10^3 lb) capacity of molten metal. Typically, the vessels have a permanent lining of burned magnesia brick against the shell and an inner working lining that may range in thickness from 450 to 900 mm (18 to 36 in.). Early vessels use tar-bonded dolomite brick

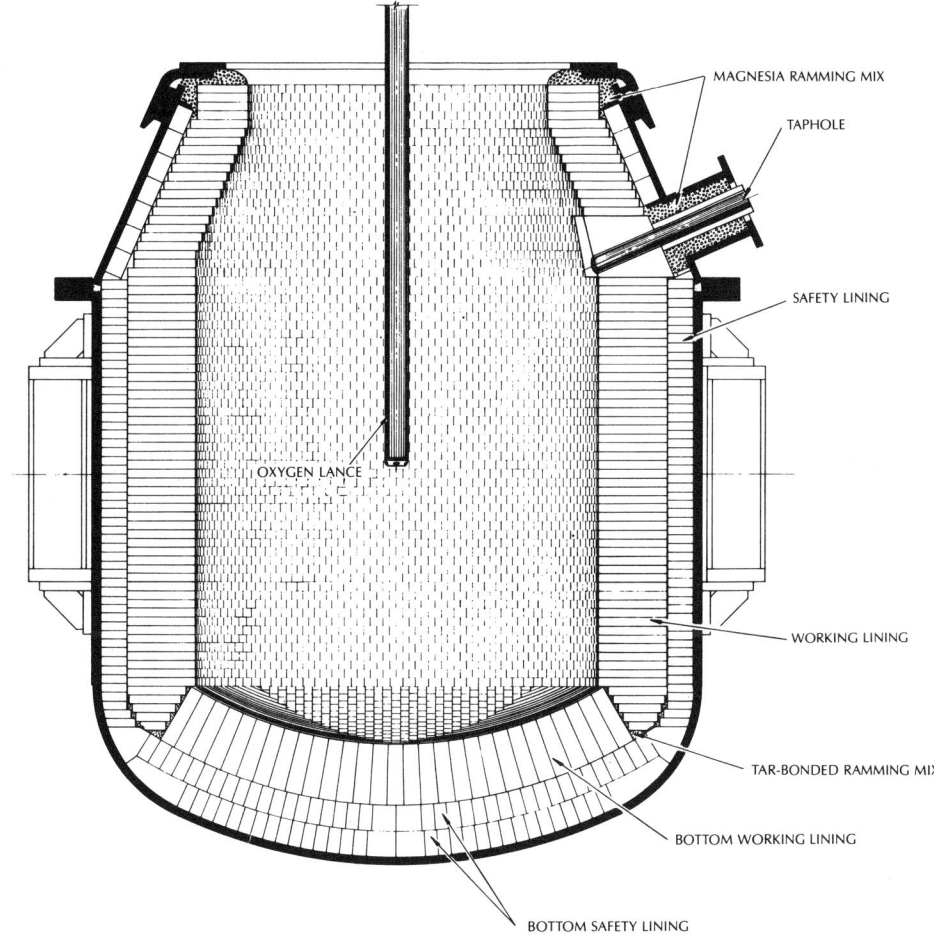

Fig 7 Typical basic oxygen furnace. Courtesy of The Refractories Institute

Labels in figure: MAGNESIA RAMMING MIX, TAPHOLE, SAFETY LINING, WORKING LINING, TAR-BONDED RAMMING MIX, BOTTOM WORKING LINING, BOTTOM SAFETY LINING, OXYGEN LANCE

in the working lining. The tar provides the bond in the brick for shipment, protects the dolomite against hydration, and is later converted to carbon when the lining is burned-in. The presence of carbon in this refractory helps prevent the metal and slag from penetrating the refractory during service. Typically, the altered zone of the refractory during service is seldom greater than 1.5 to 3 mm (0.06 to 0.12 in.) in depth.

To improve the storage life of these tar-bonded refractories, magnesia is often substituted in the matrix or fine portion of the refractory because it does not hydrate as readily as lime. Tar-bonded brick is also available in a tempered version that has been heat-treated to remove the lighter fractions of tar and thus provide greater stability through the softening range during burn-in. For the extreme high-wear areas in the basic oxygen furnace vessel, burned and impregnated refractories have been shown to have superior wear resistance. Special magnesia compositions based on both natural and synthetic magnesia have been developed to provide extremely high strength at elevated temperatures. Most vessels today are lined with a combination of refractories, described above, because the best lining life-

time, in terms of tonnes of steel produced, is achieved when the entire lining is consumed at an equal rate.

Electric furnaces are useful for making steel because they permit the product to be based entirely on scrap, thus eliminating the need for coke oven and blast furnace operations to produce pig iron. The capacity of electric furnaces has increased steadily. In furnaces that use basic slag, the furnace bottom is composed of magnesia brick or ramming material. The side walls are composed of either burned or chemically bonded magnesia-chrome brick, with fused cast compositions being used in the high-wear areas (Fig 8). Typically, the roof construction will utilize 70 to 90% alumina brick. Recent trends, particularly in Germany and Japan, have been to utilize water-cooled panels in the roof and sidewalls of the electric furnaces, thus eliminating many of the refractories.

The open-hearth furnace, which tends to be a more costly method of making steel, is still utilized to produce a considerable tonnage on a worldwide basis. These large vessels may contain up to 250 t (550×10^3 lb) of molten metal. The amount of scrap that can be utilized in an open hearth is not lim-

ited, as in the case of the basic oxygen furnace process. The conventional open-hearth process required 8 h for the refining of steel, but this time has been reduced to 4 h through the use of oxygen lances, making the process somewhat more competitive with other steel-making processes.

The refractories used for basic open-hearth furnaces are typically burned magnesia brick for the bottom, topped with a magnesia ramming mix, and either burned or chemically bonded metal-encased brick for the side walls and roof. The basic roof construction is typically suspended with a hold-down system that maintains the contour of the roof. The checkers in the regenerator can be basic or super-duty fireclay. The end walls in the ports are often metal-encased chrome-magnesia or magnesia-chrome brick.

The argon-oxygen decarburization (AOD) process for treating and producing low-carbon ferrochrome alloys conserves chromium. The vessels are typically lined with either burned magnesia refractories or fused cast magnesia-chrome refractories. They are often zoned using direct-bonded magnesia-chrome brick, rebonded fused grain brick, or fused cast brick. The process is used to make stainless steel from electric furnace melts.

Pouring Pit Refractories. The ladles used to transport molten steel to either the ingot casting line or the continuous caster are typically bloating brick made from a siliceous fireclay. Although this practice has provided low-cost refractories, longer ladle times and the need for cleaner steel have led the trend toward using 50 to 70% alumina made from special low-alkali bauxitic kaolins. Phosphate-bonded 75% alumina brick is often used at the slag lines. In Japan, ladle practice has moved to either zircon or zircon in combination with pyrophyllite or fireclay. Special ladles, such as those used for desulfurization processes, may require the use of basic linings to permit the occurrence of metallurgical reactions. Lime used as a refractory would be ideal, except for its tendency to hydrate.

Flow control of hot metal from the ladle has historically been achieved by using stopper rods composed of fireclay sleeves and clay-graphite stopper heads. However, with continuous casting, longer holding times in the ladle and higher temperatures are required, which has tended to force the practice toward ladle slide gates. Slide gate practice has increased steadily, and the use of 90% alumina and magnesia-graphite slides are now common.

Continuous Casting. The design and lifetime of tundishes used in the continuous casting of steel vary widely, depending on the type of operation (Fig 9). The refractory linings can be phosphate-bonded 85 to 90% alumina plastics of all-monolithic design or high-alumina brick. Pour pads are used in most tundishes in the area where the ladle stream impinges on the tundish bottom. Dense alumina and alumina-chromia block, as well as

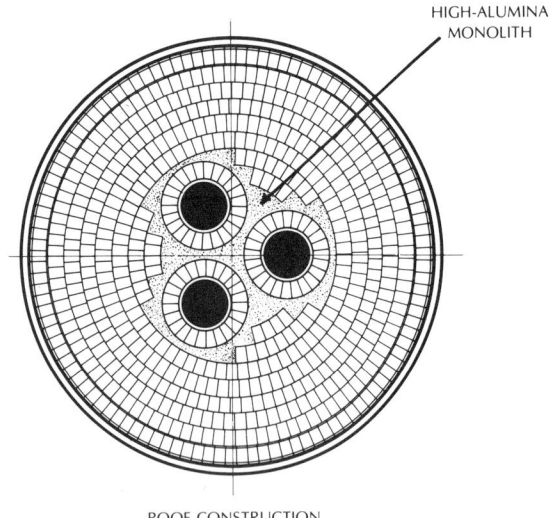

ROOF CONSTRUCTION

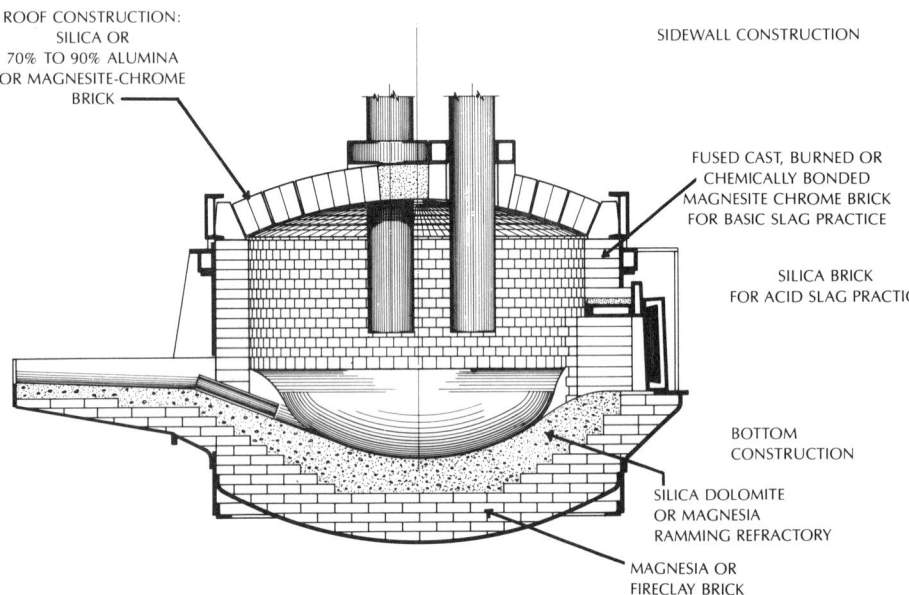

Fig 8 Typical electric arc furnace for steel production. Courtesy of The Refractories Institute

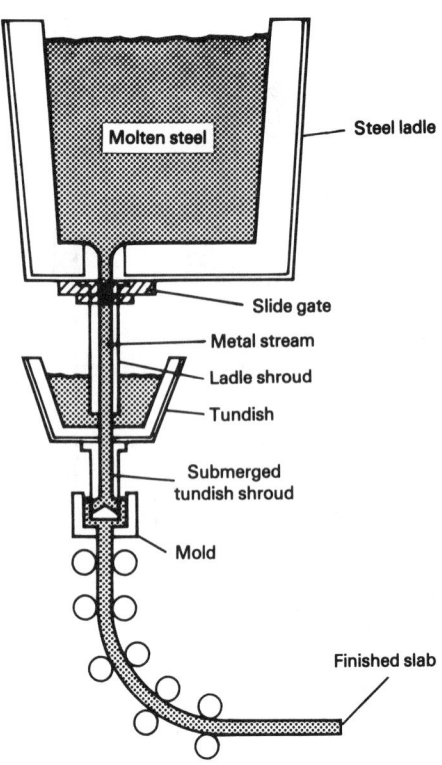

Fig 9 Continuous casting components

high-strength MgO block, have been used successfully.

Insulating boards are frequently used to line portions of the tundish so that hot metal charged to it remains hot longer. Tundish covers have been constructed utilizing insulating brick, plastics, and castables. However, new lightweight designs employing refractory fiber insulation appear to offer many advantages.

To control the flow of the metal from the tundish, single-piece isostatically formed stopper rods of alumina-graphite composition have been used for short strings of heats, but it appears that slide gate valves are more effective for long strings of heats.

Because continuous casting requires a nearly constant flow of steel from the tundish to the casting mold, nozzle materials of zircon and zirconia have been developed. Zirconia noz-

zles are used in conjunction with conventional tundish stopper rods and as nozzle inserts in slide gate valves.

Protection of the metal from oxidation as it flows from the tundish to the mold is accomplished by refractory shrouds. These are affixed to the bottom of the tundish nozzle and extend into the continuous casting mold several centimeters below the molten steel level. Rebonded fused silica shrouds have been used successfully for pouring many steels, because the low coefficient of expansion of fused silica contributes to the excellent thermal shock resistance necessary for this application. It has been found that alumina-graphite shrouds perform better when high-manganese steels are being poured. However, it is necessary to preheat these shrouds before pouring to prevent thermal shock damage by the initial stream of hot metal. The use of shrouds

from tundish to mold is accepted as standard practice. Shrouding from ladle to tundish is becoming more prevalent as the demand for cleaner steel is met.

Soaking Pits and Reheating Furnaces.
A decrease in the amount of steel cast into ingots generally reduces the number of new soaking pits constructed and causes many old pits to be removed from service. Refractory materials used in pits are many and varied but recently have tended toward forms that can be more easily installed, such as large preformed plastic blocks. Soaking pit recuperator tubes made of silicon carbide or high-alumina materials offer desirable crack resistance.

Other materials innovations include steel-wire-reinforced concretes. The steel fibers increase resistance to cracking and thereby increase furnace lifetime. They are useful for hearths, copings, cover edges, and other areas of physical abuse.

A large energy saver in pit construction is to use natural sandstone wherever possible, because thermal energy is expended during its preparation. Proper expansion calculations are essential when using sandstone, to allow for silica conversions.

In reheating furnaces, the current high cost of energy has brought more attention to construction that minimizes heat losses through the refractory walls and skid pipe systems. Because skid pipe insulation systems must resist vibration and even some impact, the refractory designs used must allow for some flexibility, such as that possible with composite constructions of fiber refractories and hard refractory shapes of fireclay and high-alumina compositions.

Heat losses through brick or monolithic walls (fireclay and high-alumina compositions) can be contained by simply applying exterior insulation, although the resulting increase in heat contained in the walls requires a corresponding increase in refractory quality.

Steel refining processes are used to produce specialty steels, tool steels, and superalloys. In both vacuum stream and vacuum ladle degassing, a vacuum chamber receives molten steel from a steel-making furnace and a vacuum pump removes unwanted gases before the metal solidifies. So-called DH and RH vessels are typical equipment for such treatment. Basic bricks of direct-bonded and rebonded fused grain qualities are most commonly used to line them.

Products Used in Nonferrous Metallurgical Industries (Ref 4)

Depending on the properties and reactivity of the metal being produced, applications can be very different from those used in the iron and steel industries, although as in the steel industry, the trend has been toward the use of continuous processes. This section describes the refractory products used in aluminum and copper production.

Aluminum Melting and Reverberatory Furnaces. Aluminum is being used increasingly throughout the world because of its light weight, oxidation resistance, and good conductivity. Unlike many other metals, aluminum is not reduced directly from the bauxite ore from which it is extracted. The bauxite is chemically treated with caustic soda under pressure to produce sodium aluminate, from which aluminum hydroxide is precipitated. The aluminum hydroxide is calcined in rotary kilns at the temperature required to drive off the combined water and to form commercially pure alumina (Al_2O_3). The temperatures involved and the highly abrasive character of the charge as it penetrates the joints necessitate the use of the proper refractories and the correct installation procedure for lining these kilns. Super-duty fireclay brick is most widely used for lining the high-temperature zone and for building the dams that are installed to provide longer retention time and prevent surging. In the cooler sections of the kiln, high-duty fireclay brick gives long service.

Reduction of alumina to the metal is accomplished in electrolytic reduction cells referred to as pots, which are usually lined with carbon blocks or carbon paste. The pot bottoms usually consist of carbon paste rammed into place, with an insulating layer of alumina or other material in powder form between the steel shell and the carbon. Insulating firebrick is used to good advantage in some installations.

The carbon anodes for the pots may be of the self-baking (Søderberg) or prebaked types. When prebaked anodes are used, they are fired in ring-type furnaces or continuous tunnel kilns. High-duty and super-duty fireclay brick have been used for many years with excellent results and continue to be widely regarded as the standard of quality for these firing installations.

The refractories that are best suited for lining aluminum melting and holding reverberatory furnaces depend on the design of the furnace, the melting rate, the kinds of metal or scrap charged and the desired alloys, and other operating conditions. Bottoms and sidewalls up to about 350 or 400 mm (14 or 16 in.) above sill level are the most critical areas. Careful selection of refractories is necessary for best economy. Resistance to corrosion reactions and penetration by molten aluminum and its alloys is vital, as is the ability to withstand mechanical abuse.

Certain types of phosphate-bonded 85% alumina brick have been developed primarily for aluminum melting furnaces. They are exceedingly resistant to reaction with aluminum alloys, have excellent resistance to dross build-up, and have good strength at operating temperatures. An 85% alumina mortar is the best choice for laying all the various types of brick in the bottoms and lower sidewalls of aluminum melting furnaces. Where monolithic linings are preferred for the bottom and lower sidewalls, phosphate-bonded plastic refractories and castables have been used in many installations.

Copper Reverberatory and Converter Furnaces. Reverberatory furnaces are used in copper production to:

- Smelt ore or its concentrates
- Refine blister copper, which is a relatively pure form, from converter furnaces
- Melt copper for casting or copper scrap refining

Each of these processes is performed in dedicated furnaces. Increased rates of production have been accompanied by higher operating temperatures, increased activity of chemical reactions, and larger volumes of metal, slag, and gases.

In reverberatory furnaces that are used to smelt copper ores, temperatures range from 1480 to 1650 °C (2700 to 3000 °F) in the smelting zone and at least 1090 to 1260 °C (1995 to 2300 °F) at the skimming end (Fig 10). The product of the furnace is copper matte, which consists mainly of cuprous sulfide, with appreciable proportions of ferrous sulfide and small proportions of other sulfides. At one time, silica brick was used almost exclusively for the complete construction of copper reverberatory furnaces. However, higher production rates have imposed increasingly severe conditions on refractories, which has led to the increased use of basic brick for the entire furnace structure.

Suspended arch construction is employed for roofs built entirely of basic brick. At furnace operating temperatures, the metal at the exposed ends oxidizes and combines with the magnesia of the brick to form a magnesioferrite bond. When wet concentrates are charged, spalling or hydration of the basic brick around fettling holes may dictate the use of high-alumina brick. The furnace walls above the slag line are advantageously built entirely of magnesia brick. The walls below the slag line are built of fired basic brick backed with fireclay brick. The reverberatory furnace bottom usually consists of magnesia ramming mixes backed by basic or fireclay brick.

In converter furnaces (Fig 11), the matte product from the reverberatory furnaces, which generally contains from 20 to 50% copper, is converted to blister copper of 96–99% purity. The slag that is formed in the converter, which primarily contains iron oxide and silica, is returned to the reverberatory furnace. Temperatures in the converter range from 1150 to 1315 °C (2100 to 2400 °F).

The lining of the furnace bottom and the tuyere zone, which is the zone of greatest wear, is usually fused cast magnesia-chrome brick. The lining of the holding furnaces in which molten blister copper from the converter is kept hot for casting into slabs and cakes is usually fired basic brick.

New continuous processes offer advantages in energy savings, materials handling, and capital costs when blister copper can be produced directly from ore concentrates. In these continuous processes, refractory selection is based on the same principles as for conventional smelting.

The Noranda process uses a furnace resembling an elongated conventional converter with oxygen-enriched air introduced into the matte layer through 50 or more tuyeres along one side. The tuyeres remain submerged during the entire process and are rotated out of the liquid batch only in emergencies.

The Worcra process is not unlike the Noranda process but utilizes a reactor of the hearth furnace type, more like the reverberatory furnace. Conventional processing introduces air into the molten matte through lances extended from the roof or upper sidewalls and provides a settling zone in the furnace so that slags can be directly tapped.

The Mitsubishi process is a continuous, but multiple-step process. Three furnaces are connected in cascade so that matte, slag, and, finally, blister copper flow by gravity through the system. The three units are a smelting furnace, an electric settling furnace, and a converting furnace.

Products Used in the Glass-Melting Industry (Ref 5)

Refractories used in glass-melting processes serve as containers for the molten glass, provide thermal insulation, and, conversely, function as a medium for heat exchange.

ROOF CONSTRUCTION
SUSPENDED OR SPRUNG ARCH — BASIC, HIGH-ALUMINA OR SILICA BRICK

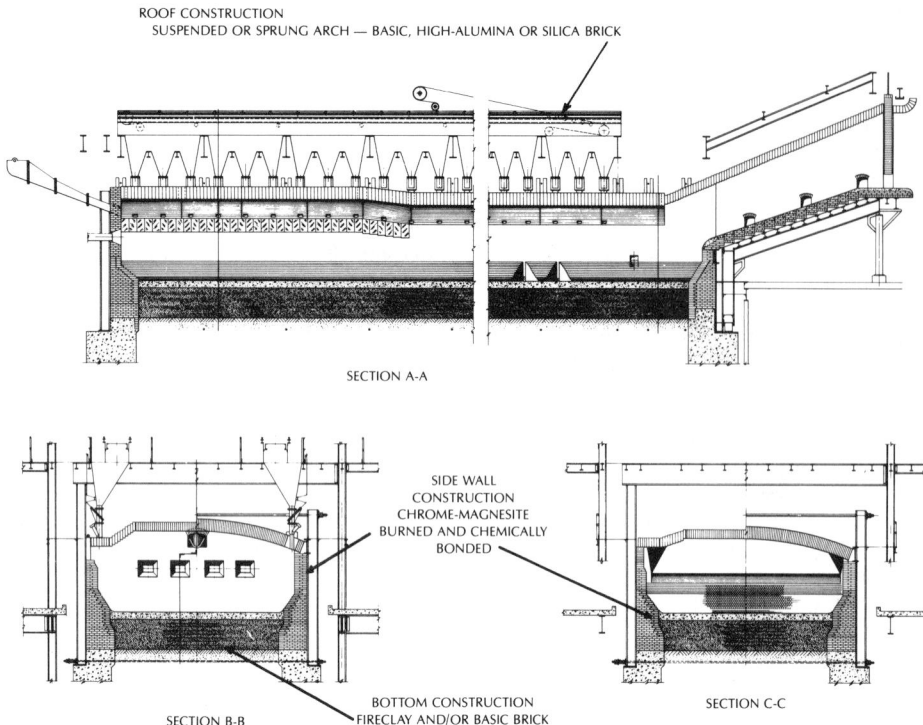

SECTION A-A

SIDE WALL
CONSTRUCTION
CHROME-MAGNESITE
BURNED AND CHEMICALLY
BONDED

SECTION C-C

SECTION B-B

BOTTOM CONSTRUCTION
FIRECLAY AND/OR BASIC BRICK
UNDER MAGNESITE HEARTH

Fig 10 Typical copper reverberatory furnace. Courtesy of The Refractories Institute

Glass-tank refractories also function as key structural elements, not just as liners. Figure 12 depicts the numerous refractory products used in a glass-tank melting furnace. The articles in the Section "Glass Processing" in this Volume fully describe the relevant processes.

The primary consideration for a glass-contact refractory is its resistance to corrosion by the glass. Unlike metallurgical applications, where refractory corrosion products are usually disposed with the slag, glass refractory corrosion products are retained in the glass and can adversely affect product quality. Crystalline inclusions, called stones, and glassy striae, called cord or ream, are defects that can result from excessive refractory corrosion.

The corrosion process in modern, dense glass refractories is usually a frontal chemical attack. In most cases, the alkali ions of the glass diffuse into the refractory and ultimately achieve surface concentrations sufficient to flux the refractory hot face. Porous refractories, or those exposed to glass on more than one face, will have a greater surface area of attack and poorer resistance to corrosive wear.

The most rapid corrosion of glass refractories occurs at the top surface of the melt, where the glass surface, refractory, and atmosphere all meet. Two forms of accelerated corrosion that are unique to the glass industry are related to melt-line corrosion. Upward drilling occurs on either horizontal faces of refractories, such as throat covers, or in horizontal joints when air or gas bubbles become entrapped there. Similarly, rapid downward drilling of bottom refractories results from metals at the glass-refractory interface. Metals can be introduced either as contaminants in recycled glass or by the reduction of glass constituents (for example, lead). Alkali enriched at the bubble or inclusion surface tends to reduce the surface energy, but accelerates the corrosion at the points of refractory contact.

The magnitude of the thermal gradient through a refractory determines the depth to which glass and alkali penetration can occur. Small gradients, such as those occurring in thicker or insulated walls, promote rapid corrosion. Measurements of refractory thickness in the melt line of tanks indicate that the rate of wear actually decreases as the wall thickness decreases and the thermal gradient becomes steeper. Forced air or water cooling of the cold face enhances this effect and is often used to extend furnace life.

Glass-Contact and Fused AZS Refractories. The temperatures within the specific operational zones of the tank determine the severity of corrosion and the resulting choice of refractory. A soda-lime glass melter operating at 1500 °C (2730 °F) will have a corrosion rate substantially greater than the refiner zone of the same tank operating at 1350 °C (2460 °F). A forehearth operating at 1200 °C (2190 °F) will have an even lower rate of corrosion. In the higher-temperature zones,

therefore, fewer materials offer the necessary combination of corrosion resistance and low defect potential.

The most common melter basin refractories are the fused-alumina-zirconia-silica (AZS) materials. Dense zircon is used in some low-expansion borosilicate melters. Clay, fused alumina, bonded AZS, and dense sintered alumina have limited use in lead and other specialty glasses, particularly those melted at lower temperatures.

Insulating fiberglass, a corrosive, high-alkali, borosilicate glass, is melted electrically using fused chrome-AZS or fused alumina-chrome ore refractories. Textile, or reinforcing, fiberglass, a low-alkali borosilicate, is melted using dense sintered chromic oxide throughout the melter basin and zircon and chrome oxide in the distribution channels. The coloring potential of chrome ions, which is of no concern for fiberglass, has limited the application of these materials to high-wear locations in other glasses. The higher cost of chrome-containing materials is usually justified by their markedly improved performance relative to fused AZS. Improved operating practices have largely overcome the thermal shock sensitivity common to chrome-containing materials.

In the conditioning zone of soda-lime glass tanks, fused AZS and fused α-β alumina are used for sidewalls and bottoms. The use of bonded AZS, zircon, and clay in refiner bottoms is also common. Fused α-β alumina has slightly lower corrosion resistance than fused AZS at conditioning temperatures, but many believe that it introduces potential defects to the product.

In the forehearth, fused α-β alumina or fused or sintered AZS are used most frequently for soda-lime glass channels, bowls, and skimmers. Zircon or fused AZS is required for borosilicate glasses. Expendable stirrers and other feeder parts are typically made of mullite or bonded AZS, because of the thermal shock encountered in their installation and use.

Tank sidewalls are normally constructed of large full-height, fusion-cast blocks in a soldier course construction. Multiple course construction using diamond-ground joints to prevent upward drilling is sometimes used to permit replacement of only the top course. Less durable, less expensive materials can be used lower on the wall, where less corrosion occurs. The latter construction is common in large, flat glass tanks. Paving materials are preferably placed inside the sidewalls to prevent upward drilling at the base of the sidewall.

The objective in any refractory application is to balance the refractory wear and thus extract maximum value from the refractories. In effect, all bricks that require replacement prior to a subsequent furnace campaign should fail simultaneously. There are two approaches to balancing refractory wear: zoning by thickness and zoning by composition. In the first case, thicker refractories are used where the

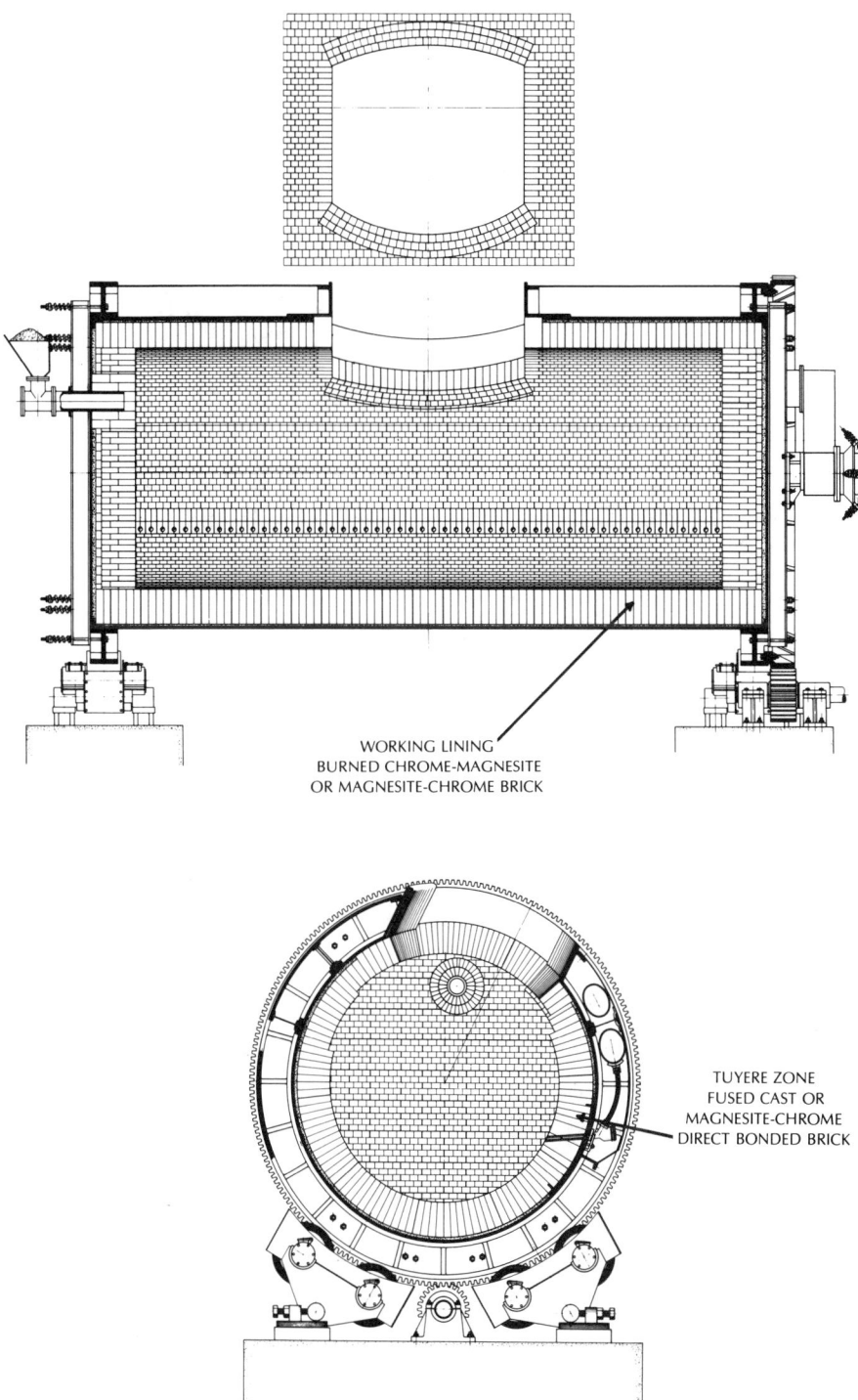

WORKING LINING
BURNED CHROME-MAGNESITE
OR MAGNESITE-CHROME BRICK

TUYERE ZONE
FUSED CAST OR
MAGNESITE-CHROME
DIRECT BONDED BRICK

Fig 11 Typical copper converter. Courtesy of The Refractories Institute

Throat covers are subject to upward drilling corrosion and are so critical to both tank life and operating efficiency that fused chrome-containing refractories are sometimes used. Sidewall or bottom blocks drilled for electrodes, and adjacent blocks are subject to excessive temperatures, high rates of convective flow, and a several thermal gradient to water-cooled electrode holders. Electrode blocks also benefit from the higher electrical resistivity of the premium refractory.

In general, the higher the corrosion-resistant zirconia content, the higher the corrosion resistance of the product. A 35% improvement is usually claimed between 34 and 41% ZrO_2 composition extremes. Prior to the introduction of oxidized or conditioned products in the late 1960s, the oxidation during service of carbon and other residual reduced contaminants resulted in sufficient internal pressure to cause the exudation of the glassy phase. This viscous exuded glass was an important source of product defects. Reduced AZS, still available from some suppliers, also has a markedly higher tendency to generate bubbles, seeds, or blisters in the glass and offers no compensating performance advantage.

Superstructure and crown refractories, despite being above the melt line, are subject to corrosive attack. Corrosive reactions occur with both the vapor species of batch components and with batch carryover (solid or molten dusts of batch components carried by the combustion gas stream). Direct attack occurs by the diffusion of the vapor species into the hot face of the refractory. The chemical changes in this reaction zone result in mechanical stresses and flaking, peeling, or spalling of the hot face. Vapor penetration into joints with subsequent condensation and corrosion can endanger the mechanical integrity of the superstructure or crown. Vapor attack, to different degrees, occurs in all of the operating zones of the tank. Superstructure insulation and the shallow thermal gradient that results are often beneficial, preventing spalling and causing condensation to occur in less critical backup refractories. Batch carryover problems are usually confined to the melter and regenerator sections of a glass tank. Fortunately, these problems can be controlled by the careful selection of raw materials, and by slightly wetting the batch.

Modern, high-pull furnaces often operate at superstructure temperatures that are 60 to 100 °C (110 to 180 °F) higher than the molten glass. Typically, melter breast walls, ports, and end walls are made of fused AZS up to the furnace hot spot, with fused β-alumina used downstream. Some sintered AZS is also used at the front end and near the back of the port necks. The horizontal port floors and sills, as well as the tuckstones that collect batch carryover, are almost always constructed of dense, void-free material. The crown is made of silica brick. Despite its low resistance to vapor attack and alkali condensate corrosion

wear is more severe. The latter concept involves the use of lower- or higher-performance refractories, depending on the relative wear at a specific location. Refractory prices generally parallel performance.

In the glass-contact zone of a melter, high-wear areas typically require premium, void-free, high-zirconia (41% ZrO_2) fused AZS, rather than the 34% ZrO_2 variety. One such

high-wear area involves doghouse refractories, which are contacted by the earliest melting and most corrosive batch constituents, and are also subject to thermal shock by the periodic introduction of cold batch. Structures having exposed corners, such as the doghouse, throat entrance or exit, and any steps in the bottom paving, corrode from two faces and often require the premium refractory.

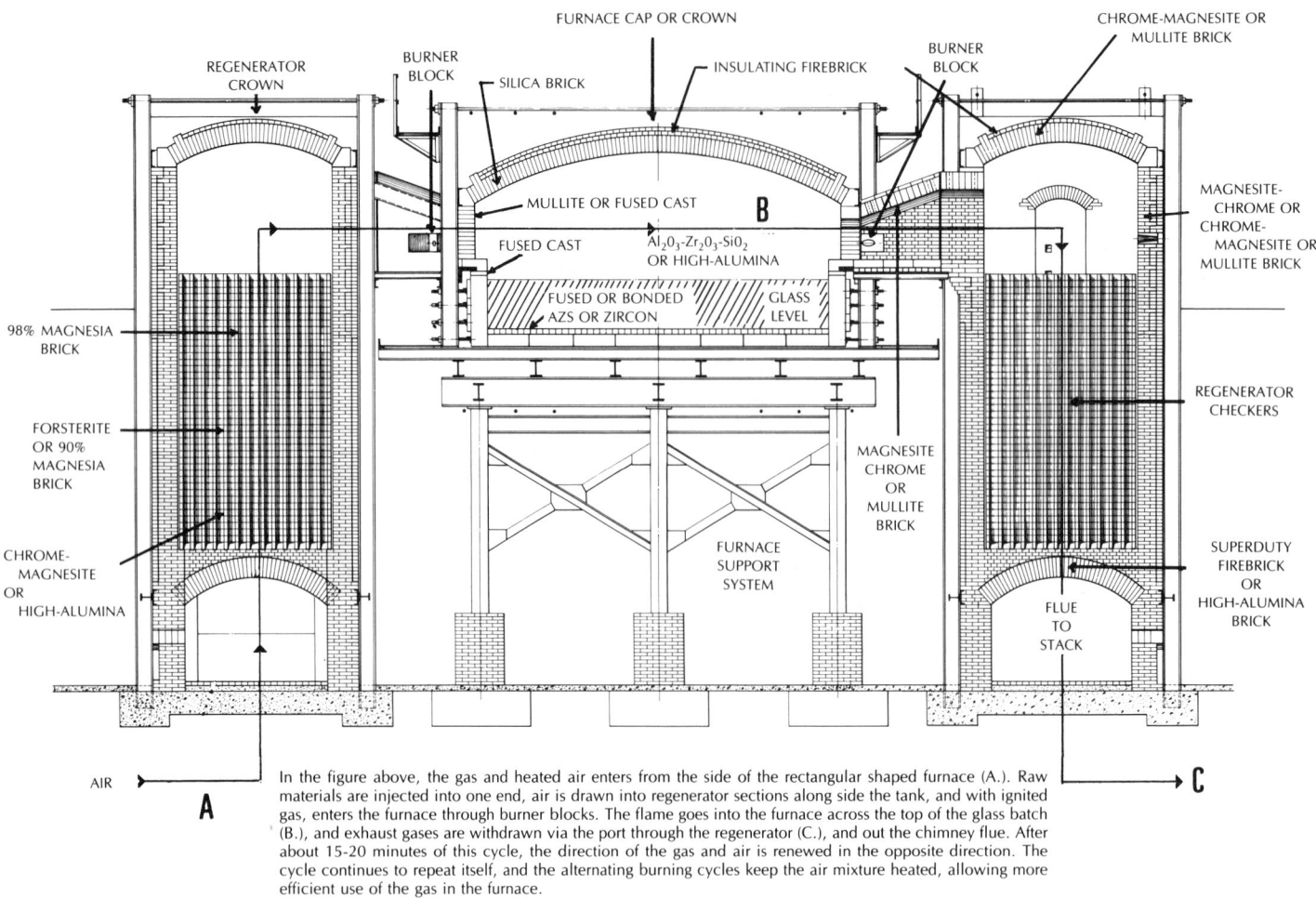

In the figure above, the gas and heated air enters from the side of the rectangular shaped furnace (A.). Raw materials are injected into one end, air is drawn into regenerator sections along side the tank, and with ignited gas, enters the furnace through burner blocks. The flame goes into the furnace across the top of the glass batch (B.), and exhaust gases are withdrawn via the port through the regenerator (C.), and out the chimney flue. After about 15-20 minutes of this cycle, the direction of the gas and air is renewed in the opposite direction. The cycle continues to repeat itself, and the alternating burning cycles keep the air mixture heated, allowing more efficient use of the gas in the furnace.

Fig 12 Typical regenerative side-port glass tank melting furnace

(ratholing), any drips that occur will be highly siliceous and easily homogenized in the melt. Some suspended back walls, shadow walls, or curtain walls also incorporate silica brick. Silica brick should never contact alumina-containing materials because of a potential melting reaction. Zircon is used as a buffer at the crown skewline, where the silica crown contacts the aluminous breast walls.

The refiner and forehearth superstructures, especially if isolated from the combustion atmosphere of the melter, can often be made from mullite, sillimanite, high alumina, or even fireclay materials. Silica refiner crowns predominate. Forehearth covers, however, are normally mullite or high alumina in soda-lime glasses, because vapor attack is very mild at the low temperature involved. In borosilicates, considerable vapor attack persists into the working end, and zircon or silica are used. In all-electric melters, which operate with a complete, constantly replenished batch cover, vapor species are absorbed in the solid batch and reintroduced into the melt. This, combined with the insulating efficiency of the floating batch, makes ordinary fireclay bricks suitable for the superstructure.

The regenerative heat exchanger is the predominant type of heat exchanger used in the glass industry. Two types of regenerative furnaces are used. End-port furnaces have two ports and regenerator chambers located at the back (upstream) end of the furnace. Side-port furnaces, like the example in Fig 12, have multiple ports and banks of regenerators on either side of the melter.

In regenerative furnaces, exhaust gases are passed through a massive open stack of refractory bricks, called checkers, which absorb heat. Regenerator refractories, particularly the checker bricks, must resist the corrosive action of carryover, vapors, and condensates. In addition, high thermal conductivity and heat capacity are necessary to maximize thermal-exchange efficiency. Fortunately, regenerator refractories rarely, if ever, generate quality defects during routine operation. This means that a broader selection of materials are suitable for use. The checker settings are supported by arches of super-duty fireclay, called rider arches. Rider tiles span these arches and provide the base for laying the checker bricks. Checker bricks are typically laid in a basketweave setting to

the level of the port floor, a height that ranges from 5 to 7 m (16 to 23 ft).

In soda-lime glass tanks, the top third of the checkerwork, the hottest section and that most directly affected by vapors and lime and silica carryover, is usually constructed of high-purity, direct-bonded magnesite (MgO) bricks. The middle third of the checker setting has traditionally been forsterite brick, but the trend is increasingly toward periclase or lower-purity magnesite brick. In the bottom third of the checkers, temperatures from 850 to 1000 °C (1560 to 1830 °F) promote the condensation of alkali and sodium sulfate. Sintered chrome, chrome-magnesite, or magnesite-chrome are used to resist these powerful fluxes.

In tanks fired with fuel oil, forsterite brick is sometimes used to resist vanadates and other fluxes introduced with the oil. The structure surrounding the checkers and the walls dividing checkers for individual ports are usually chrome-magnesite or magnesite-chrome. Essentially, the same materials are used to construct chambers for the recuperative heat exchangers. In borosilicate glasses, despite experimentation with chrome-containing materials, mullite, fireclay, or silica checkers are used. The corrosive attack of borates, particularly sodium tetraborate, is so severe that these lower-cost materials are replaced periodically.

A novel concept in checker design has been the introduction of fused AZS and fused high-alumina shapes in Europe. These shapes have a geometry promising higher efficiency, improved resistance to plugging and collapse, and materials with proven corrosion resistance. Rising energy and labor costs have rapidly offset the higher initial price of such installations.

Tin Oxide Electrodes. Much electrical melting of glass can be satisfactorily performed using water-cooled molybdenum electrodes. However, molybdenum is not chemically compatible with many lead-containing crystal or optical glasses. Dense iso-pressed or slip-cast tin oxide (SnO_2), with minor additions to increase its conductivity, is suitable as an electrode material in such high-quality applications. Tin oxide has a lower limiting current density than does molybdenum, and thus requires larger electrode cross-sections. Tin oxide electrodes, however, require no water cooling, which results in local heat losses. Tin oxide is second only to Cr_2O_3 in its corrosion resistance to most glasses and is extremely compatible with fused AZS refractories.

Specialty Refractories (Ref 6)

The unique properties of cordierite, zirconia, and $Li_2O \cdot Al_2O_3 \cdot SiO_2$ ceramics, which are considered to be specialty refractories, have been very useful in meeting current requirements and hold promise for future high-performance applications.

The choice of which refractory material performs best in a particular application depends on a number of properties and material characteristics. High-temperature capability, thermal shock resistance, high strength, and other properties, separately or in combination, can meet many energy-efficiency requirements. Other requirements that can be satisfied include resistance to molten-metal erosion and slag attack. Operating temperature, chemical compatibility, and cost also can be important characteristics in material selection.

Cordierite, zirconia, and $Li_2O \cdot Al_2O_3 \cdot SiO_2$ refractories not only offer a wide variety of desirable properties, but can be modified to accentuate one or more of these properties for a particular application. Zirconia, for example, has excellent resistance to molten metal erosion, is extremely refractory, and is chemically inert in most environments, although it has poor resistance to thermal shock stresses when fully stabilized. However, thermal shock resistance can be tailored over a wide range by varying the degree of stabilization. Glass-ceramic components fabricated from oxides in the $Li_2O \cdot Al_2O_3 \cdot SiO_2$ system can be made impervious with good strength and excellent thermal shock resistance, but are generally limited to use at temperatures of 1000 °C (1830 °F) or less because of residual glass. Range-top cookware is therefore an appropriate application. Increased temperature capability can be achieved by sintering naturally occurring minerals, such as petalite or spodumene, or by solid-state reaction of Li_2CO_3, Al_2O_3, and SiO_2, but at the expense of density and strength. Similar flexibility in forming and property development has been demonstrated with cordierite refractories.

Markets served by these specialty refractories include production of ferrous and nonferrous metals and glasses, and the processing of chemicals and ceramics. Products used in these markets include kiln furniture, crucibles, nozzles, pouring shrouds, tundishes, and wear-resistant parts.

Cordierite, a naturally occurring mineral, has the idealized composition $2MgO \cdot 2Al_2O_3 \cdot 5SiO_2$. Refractory and ceramic components containing cordierite as a major constituent are generally used where low thermal expansion and good thermal shock resistance are required.

Axial thermal expansion studies have determined that expansion is positive in the a crystallographic direction and negative in the c direction, with an aggregate thermal expansion of about 0.9×10^{-6}/K from room temperature to 800 °C (1470 °F). Glass-ceramic shapes with thermal expansion values approaching this theoretical limit have been made through controlled crystallization of articles shaped from cordierite glass powders. The combination of low thermal expansion and the high strength resulting from additives make this material attractive for heat exchangers used in gas-turbine engines.

Zirconia forms in nature as the mineral baddeleyite, which is used mainly as a raw material, either in fused cast refractories or in zirconia-containing abrasive grain. Most ceramic-grade zirconia is produced from zircon. It can be used in a wide variety of applications. Its refractoriness and resistance to both thermal shock and molten metal erosion are useful for tundish nozzles and other specialty parts in the casting of molten steel. It is also used for foundry crucibles. The chemical stability of zirconia is important in its utilization as setter plates for firing barium titanate capacitors. High density and excellent wear resistance are the properties that enable zirconia milling media to perform with high efficiency. Exceptional toughness, which results from controlled heat treatment of the appropriate starting material, provides excellent performance in abrasive wear applications, such as extrusion dies for metal rods and as thread guides. Its low thermal conductivity makes fibrous zirconia shapes extremely effective high-temperature insulation materials.

Unusually tough ceramics have been developed from partially-stabilized zirconia compositions through the transformation toughening mechanism. Zirconia composi-tions containing tetragonal zirconia within a matrix of the cubic phase have been developed using controlled heat treatment. Surface transformation of tetragonal zirconia results in compressive stresses at the surface and increased strength, while increased toughness results from transformation at crack tips.

The thermal shock resistance of fully stabilized zirconia refractories is very low because of their high thermal expansion and low thermal conductivity. There are many applications, however, where good to excellent thermal shock resistance is required. Requirements can be met by optimizing the level of stabilization for each application. As the level of stabilization is reduced, so are modulus of rupture, elastic modulus, and thermal expansion, until, at some point, the strength becomes too low for practical utilization. The very severe thermal shock stress requirements of crucibles used for induction melting of refractory metals, especially while pouring the molten metal over a relatively cold crucible spout, are met by zirconia compositions that have a cubic zirconia content of about 40%. A cubic zirconia content of 65% results in good performance of tundish nozzles in the casting of molten steel, whereas an 80% cubic zirconia content provides long life for zirconia setter plates for firing barium titanate capacitors. This flexibility, combined with the capability of microstructural control through heat treatment, provides the means by which the properties of zirconia can be engineered to fit a wide variety of special refractory applications.

Zirconia ceramics have found unique application in areas other than high-temperature environments. For example, zirconia milling media used for attrition milling and dispersion are often preferred because of increased milling efficiency and low wear rates. High density and superior toughness combine to yield these attractive properties. Low wear rates not only result in longer life of the media, but in less contamination of the milled product. High- and low-temperature oxygen ion conductors for oxygen sensors in automotive and industrial applications are other areas where the unique properties of zirconia ceramics have found acceptance. Solid electrolytes of both CaO- or Y_2O_3-stabilized zirconia have been used in the steel industry as oxygen probes, in galvanic cells, and in a variety of fossil-fuel systems.

The versatility of zirconia is evidenced by the wide variety of applications where they have found acceptance. Recent technical developments in transformation-toughening mechanisms may enable the use of zirconia components in applications such as high-performance ceramics used in adiabatic turbo-compound diesel engines.

The three ternary mineral compounds in the $Li_2O \cdot Al_2O_3 \cdot SiO_2$ system are of particular interest to ceramists. These three compounds, as well as their high-temperature modifications and solid solutions of these modifications, are found along the $Li_2O \cdot Al_2O_3 \cdot SiO_2$ join in a continuous compositional series of molar ratios from 1:1:2 to about 1:1:10.

The $Li_2O \cdot Al_2O_3 \cdot 2SiO_2$ compound is found in nature as the mineral α-eucryptite. This mineral can be synthesized hydrothermally in the temperature range from 450 to 970 °C (840 to 1780 °F) at pressures greater than 3.45 MPa (0.5 ksi), but the high-temperature modification, β-eucryptite, always forms when the oxide composition is fired to equilibrium in air or when α-eucryptite is heated to temperatures higher than 970 °C (1780 °F).

The axial thermal expansion of β-eucryptite is highly anisotropic, with positive expansion in the *a* crystallographic direction and negative expansion in the *c* direction. Useful ceramic products can be developed from β-eucryptite and β-eucryptite-SiO_2 solid-solution compositions by controlling stresses and stress-relief mechanisms caused by the anisotropic thermal expansion. Such controls can include SiO_2 content and crystal size. Because unit-cell volume contraction decreases when increased quantities of SiO_2 are taken into solid solution, thermal-expansion anisotropy stresses are reduced. Smaller crystal sizes reduce localized stresses at grain boundaries. Control of crystal size, then, results in control of grain-boundary stresses, which can enhance or inhibit microcracking and thereby influence strength and thermal shock properties.

The amount of SiO_2 that can be taken into solid solution in the β-eucryptite structure is limited by the formation of $Li_2O \cdot Al_2O_3 \cdot 4SiO_2$, the high-temperature form of which is β-spodumene. Like eucryptite, spodumene has a high-temperature and a low-temperature form. α-Spodumene, the low-temperature, naturally occurring variety, is commonly found in pegmatites. When heated in air to about 900 °C (1650 °F), the α form inverts irreversibly to β-spodumene, with large volume expansion, as indicated by a change in specific gravity from 3.2 to 2.4. β-Spodumene can also be synthesized by solid-state reaction of Li_2CO_3, Al_2O_3, and SiO_2, formulated to yield a 1:1:4 composition, at about 1300 °C (2370 °F) or by recrystallization of a glass of the same composition. The β-spodumene structure, like β-eucryptite, can accept SiO_2 in solid solution. The limit of solubility in compositions crystallized from glasses is about 81 wt% SiO_2, or slightly higher than a 1:1:9 composition. Within the limits of solubility, increasing levels of SiO_2 in solid solution with β-spodumene result in continually reduced thermal expansion of the crystal lattice. The volume expansion of $LiO_2 \cdot Al_2O_3 \cdot 8SiO_2$ is zero, from room temperature to 1000 °C (1830 °F).

The third mineral occurring in this system that is of interest to ceramist is petalite, which has the composition $LiO_2 \cdot Al_2O_3 \cdot 8SiO_2$, and dissociates to β-spodumene-SiO_2 solid solution when heated to temperatures greater than 1000 °C (1830 °F). Thermal expansion prop-erties of ceramic bodies made from naturally occurring petalite and fired at 1250 °C (2280 °F) for 1 h are comparable to synthetic 1:1:8 compositions. Commercially useful thermal-shock resistant ceramic products are prepared from natural, synthetic, or mixtures of both materials by firing at the temperature range from 1200 to 1300 °C (2190 to 2370 °F). Commercially available products have also been prepared by recrystallization of glass shapes. These ceramic products can be dense and impervious, such as cookware, or permeable, such as shapes formed from glass powders.

Failure of a β-spodumene-silica solid-solution composition for ceramic heat exchangers in experimental automotive and industrial gas turbine engines illustrates some limitations in the chemical stability of $LiO_2 \cdot Al_2O_3 \cdot SiO_2$ materials (Rahnke and Vallance, 1978). Failure resulting from chemical attack occurs in two ways, both relying upon substitution for the relatively mobile Li^+ ion. Sodium, which enters the engine through ingestion of road salts, from saltwater or from impurities in the fuel, reacts with the ceramic at the hot face. Substitution of Na^+ for Li^+ in the ceramic results in bloating and increased thermal expansion. At the cold face (~450 °C, or ~840 °F) of the heat exchanger, the second reaction is observed. Here, hydrogen first reacts with sulfur from fuel impurities to form sulfuric acid. The reaction of sulfuric acid and the heat exchanger at the cold face results in substitution of H^+ for Li^+ and volume contraction. The stresses resulting from contraction at one surface and expansion at the other result in the formation of shear cracks within the ceramic. Utilization of the H^+ for Li^+ substitution has led to a new family of low-expansion ceramics based on the keatite structure.

REFERENCES

1. *Refractories,* The Refractories Institute, Pittsburgh, PA, 1987
2. E. Ruh, Metallurgical Refractories: Manufacture and Properties, *Encyclopedia of Materials Science and Engineering,* Vol 4, MIT Press, 1986, p 4153–4157
3. E. Ruh, Refractories for the Iron and Steel Industries, *Encyclopedia of Materials Science and Engineering,* Vol 6, MIT Press, 1986, p 4140–4145
4. E. Ruh, Refractories for the Nonferrous Metallurgical Industries, *Encyclopedia of Materials Science and Engineering,* Vol 6, MIT Press, 1986, p 4145–4150
5. A.D. Davis, Jr., Refractories for Glass Melting, *Encyclopedia of Materials Science and Engineering,* Vol 6, MIT Press, 1986, p 4134–4140
6. D.F. Beal, Refractory Materials, Specialty, *Encyclopedia of Materials Science and Engineering,* Vol 6, MIT Press, 1986, p 4153–4157

SELECTED REFERENCES

Iron and Steel Refractories:

- J.H. Chesters, *Refractories—Production and Properties,* Iron and Steel Institute, London, 1973
- J.H. Chesters, *Refractories for Iron- and Steelmaking,* Metals Society, London, 1974
- C.R.V. DaCruz, Ceramic Refractories Produced in Brazil, *Am. Ceram. Soc. Bull.,* Vol 62, 1974, p 1019
- *Modern Refractory Practice,* 4th ed., Harbison-Walker Refractories Company, 1961
- K.K. Kappmeyer and D.H. Hubble, Steelplant Refractories in the Seventies, *Am. Ceram. Soc. Bull.,* Vol 51, 1972, p 568–573
- J. Mackenzie, A Review of the Developments in Iron and Steelmaking Technology and the Use of Refractories, *Proceedings of the Silver Jubilee Conference, Institute of Ceramics,* H.W.H. West, Ed., Institute of Ceramics, Stoke-on-Trent, England, 1981, p 52
- A. Muan and E.F. Osborn, *Phase Equilibrium Among Oxides in Steelmaking,* Addison-Wesley, 1965
- R.J. Rait, *Basic Refractories—Their Chemistry and Performance,* Interscience, London, 1950
- E. Ruh, The Refractories Industry Today—Japan and the U.S., *Am. Ceram. Soc. Bull.,* Vol 63, 1984, p 1140–1142
- E. Ruh, *et al.,* Thermal Conductivity and Thermal Expansion of Basic Oxygen Furnace Refractories, *Am. Ceram. Soc. Bull.,* Vol 45, 1966, p 643–645
- K. Shaw, *Refractories and Their Uses,* Halsted, 1972
- G.H. Workman and I.S. Kessell, Steelplant Refractories Face the Challenging Future, *Ind. Heat.,* Vol 48 (No. 4), 1981, p 8

Nonferrous Metallurgical Refractories:

- A.K. Biswas and W.G. Davenport, *Extractive Metallurgy of Copper,* 2nd ed., Pergamon, Oxford, 1980
- J.M. Cigan, T.S. Mackey, and T.T. O'Keefe, *Lead-Zinc-Tin,* Metallurgical Society, American Institute of Mining, Metallurgical, and Petroleum Engineers, 1980
- M. Goto and T. Echigoya, Effect of Injection Smelting Jet Characteristics on Refractory Wear in the Mitsubishi Process, *J. Metall.,* Vol 32 (No. 11), 1980, p 6–11
- M. Goto and K. Kanamori, Converting Furnace Operation of Mitsubishi Process, *Copper and Nickel Converters,* R.E. Johnson, Ed., Metallurgical Society, American Institute of Mining, Metallurgical, and Petroleum Engineers, 1979, p 210–224

- J.E. Hatch, *Aluminum—Properties and Physical Metallurgy*, American Society for Metals, 1984
- D.W. Hopkin, Ed., *Extractive Metallurgy of Copper*, Vol 20, Pergamon, Oxford, 1979
- J.A. Kozma, Jr., Refractory Practice for Reverberatory Furnaces—Aluminum, *Foundry*, Vol 101 (No. 7), 1973, p 36–39
- W. Lueking, Choosing Refractory for Aluminum Melting, *Foundry Man. Technol.*, Vol 54 (No. 6), 1968, p 100–103
- M. Shima and Y. Itoh, Refractories of Flash Furnaces in Japan, *J. Metals*, Vol 32 (No. 11), 1980, p 12–16
- H.Y. Sohn, D.B. George, and A.D. Zunkle, Ed., *Advances in Sulfide Smelting*, Vol 1; *Basic Principles*, Vol II; *Technology and Practice*, Metallurgical Society, American Institute of Mining, Metallurgical, and Petroleum Engineers, 1983
- R.W. Sundeen, Refractory Practice for Induction Melting—Nonferrous, *Foundry*, Vol 101 (No. 3), 1973, p 46–48
- T. Suzuki, T. Yanagida, M. Goto, T. Echigoya, and M. Kikumoto, Recent Operation of Mitsubishi Continuous Copper Smelting and Converting, *Copper and Nickel Converters*, R.E. Johnson, Ed., Metallurgical Society, American Institute of Mining, Metallurgical and Petroleum Engineers, 1979

Glass-Melting Refractories:

- R.S. Arrandale, Furnaces, Furnace Design and Related Topics, *The Handbook of Glass Manufacture: A Book of Reference for the Plant Executive, Technologist and Engineer*, F.V. Tooley, Ed., Books for Industry, 1974, p 249–386
- E.R. Begley, *Guide to Refractory and Glass Reactions*, Cahners, 1970
- E.R. Begley, Refractories, *The Handbook of Glass Manufacture: A Book of Reference for the Plant Executive, Technologist and Engineer*, F.V. Tooley, Ed., Books for Industry, 1974, p 401–454
- E.R. Begley, Critical Phenomena in the Use of Refractories in Contact with Glass, *Glass*, Vol 55, 1978, p 335–340
- R.W. Brown, Fused Cast Refractories, *Glass Ind.*, Vol 43, 1962, p 68–75, 94
- T.S. Busby, Flux-Line Corrosion, *Glass*, Vol 39, 1962, p 182–189
- T.S. Busby, *Tank Blocks for Glass Furnaces*, Society of Glass Technology, 1966
- T.S. Busby and J. Barker, Simulative Studies of Upward Drilling, *J. Am. Ceram. Soc.*, Vol 49, 1966, p 441–446
- F. Walther and J. Kivala, Prescription Basic Checkers for Glass Tank Regenerators, *Ceram. Ind.*, Vol 81 (No. 5), 1963, p 60–63

Specialty Refractories:

- S. DeAza and J.E. Monteros, The Mechanism of Cordierite Formation in Ceramic Bodies, *Bol. Soc. Esp. Ceram.*, Vol 11, 1972, p 315–321
- R.C. Garvie, R.H. Hannink, and R.T. Pascoe, Ceramic Steel?, *Nature*, Vol 258, 1975, p 703–704
- D.G. Grossman, "Process for Producing Aluminous Keatite," U.S. patent 4,033,775, 1976
- F.A. Hummel, Thermal Expansion of Some Lithia Minerals, *J. Am. Ceram. Soc.*, Vol 34, 1951, p 235–239
- H.L. Johns and A.G. King, Zirconia Tailored for Thermal Shock Resistance, *Ceramic Age*, Vol 86 (No. 5), 1970, p 29–31
- A.G. King and P.J. Yavorsky, Stress Relief Mechanisms in Magnesia- and Yttria-Stabilized Zirconia, *J. Am. Ceram. Soc.*, Vol 51, 1968, p 38–42
- M.E. Milberg and H.D. Blair, Thermal Expansion of Cordierite, *J. Am. Ceram. Soc.*, Vol 60, 1977, p 372–373
- A. Miyashiro, T. Iiyama, M. Yamasaki, and T. Miyashiro, The Polymorphism of Cordierite and Indialite, *Am. J. Sci.*, Vol 253, 1955, p 185–208
- W.C. Mohr, Development of Properties in Cordierite Kiln Furniture, *Am. Ceram. Soc. Bull.*, Vol 56, 1977, p 428–429
- C.J. Rahnke and J.K. Vallance, Reliability and Durability of Ceramic Regenerators for Gas Turbine Applications, *Ceramics for High Performance Applications—II*, J.J. Burke, E.N. Lenoe, and R.N. Katz, Ed., Brook Hill Publishing, 1978, p 335–347
- V.D.N. Rao and D.F. Beal, Ceramic Materials Development for Low Expansion, High Strength MAS Bodies to be Used in Ceramic Matrix Systems, *Ceramics for High Performance Applications—II*, J.J. Burke, E.N. Lenoe, and R.N. Katz, Ed., Brook Hill Publishing, 1978, p 369–384
- H. Schulz, Thermal Expansion of Beta Eucryptite, *J. Am. Ceram. Soc.*, Vol 57, 1974, p 313–318
- E.R. Segnite and A.E. Holland, Formation of Cordierite from Clinochlore and Kaolinite, *J. Aust. Ceram. Soc.*, Vol 7 (No. 2), 1971, p 43–46

Monolithic and Fibrous Refractories

Leonard P. Krietz, Plibrico Company

MONOLITHIC AND FIBROUS RE-FRACTORIES have been manufactured for decades, and the application range of these materials has seen considerable growth within the past twenty years. This has largely been due to refractory system design changes resulting from technical improvements in materials, physical properties, and installation techniques.

The term monolithic refractories includes a wide variety of material types and compositions using various bonding systems. While there are both acidic and basic types of monolithic refractories, the acidic aluminosilicate/alumina group is the most varied in both product type and application. Basic monolithic materials have their main applications as metal contact refractories for process vessels in the steel and ferrous foundry industry.

Applications for fibrous refractory products based on refractory ceramic fibers (RCF) have increased, spurred in many cases in recent years by the need for increased energy efficiency in many combustion processes. Ceramic fiber refractories are currently produced in a wide range of compositions designed for various service temperatures and conditions. They are also produced in a wide variety of physical forms to meet different installation and application requirements.

Monolithic Refractory Materials

Monolithic refractory materials had their "formal" origin in 1914 when the first commercial monolithic was produced. This first monolithic, a plastic, was a simple blend of crushed firebrick and fireclay; the intended application was to repair boiler settings (Ref 1).

Although slow to be accepted, the monolithic concept eventually grew and new materials were developed to meet the increasing application needs. One of the most notable of these, and of major importance today, was refractory castables based on calcium aluminate cements. These were first developed and applied in the 1920s, initially as field

mixes and later as premixed packaged products (Ref 2). Through technological advances in both composition development and installation methods, monolithic refractories have become increasingly important in modern refractory system designs and repairs.

While refractory product sales are cyclic and decreased dramatically in the early 1980s due to the restructuring of the steel industry, the ratio of the tonnage produced of monolithic refractories to refractory brick has been steadily increasing. In the early 1970s, approximately 3 tons of refractory brick was produced for every ton of monolithic refractory. In 1980, the ratio had decreased to 2:1 in favor of brick. However, toward the end of the 1980s this ratio was less than 1, indicating the importance of monolithic refractories in current refractory design and practice. The volume produced of monolithic refractory materials has increased from approximately 0.9×10^6 Mg/yr (1×10^6 tons/yr) in 1970 to approximately 1.2×10^6 Mg/yr (1.3×10^6 tons/yr) in the late 1980s, while refractory brick production has decreased from 3.2×10^6 Mg/yr (3.5×10^6 tons/yr) in 1970 to approximately 1×10^6 Mg/yr (1.1×10^6 tons/yr) in the late 1980s (Ref 3).

The growth in monolithics has been largely due to advances in technology and the inherent advantages in monolithic refractories. These include ready availability, faster installation time, reduced lining joints, and excellent physical properties. The main potential disadvantages of monolithics are increased care during installation and bake-out, although these factors have been addressed by the increase in skill level of refractory installers/contractors, and the use of auxiliary heating equipment for bake-out or contracted heating specialists. Table 1 is a listing of the various types of monolithic refractories available.

Acidic (aluminosilicate) monolithic refractory castables make up the most diverse group of these materials. Refractory castables are dry granular materials which require a water addition. Installation is by casting, vibratory placement or troweling. The majority of castables contain a calcium aluminate cement as the binder, although a few use portland cement.

Other specialized castables utilize cementless binder systems or a chemical bond. While conventional castables—those containing >10% cement—make up the greatest percentage of castables produced, the reduced cement variety, both low- and ultralow, have had the highest growth in castable applications over the past 10 years.

These reduced cement materials have their origins in the Prost Patent of 1970 (Ref 4), and were commercialized in the 1980s as materials that offer superior physical properties to conventional castables and, in many cases, to brick and shapes. To illustrate this, Fig 1 shows hot strength comparisons for super-duty class conventional castables, low- and ultralow cement castables, and brick. Insulating castables are also predominantly calcium aluminate cement bonded and come in a variety of types based on service limit and density. Their main purpose is to provide thermal insulation.

Gunning mixes, both dense and insulating, differ from castables in that they are designed to be applied by pneumatic guns in a process similar to that of conventional civil shotcreting. They are introduced to the gun either dry or predampened, and hydrating water is added at the nozzle by a skilled installer.

Gunning mixes are formulated to minimize rebound and slumping and maximize stickiness and compaction. Typically, they have lower physical properties than their castable counterparts, but installation is easier and faster because forming is eliminated.

Aluminosilicate gun mixes may be installed as primary construction materials or for repairs. For repairs they may be installed while the process vessel is on or off line. Basic gunning materials (which make up the greatest volume of basic monolithics produced) are applied as repair materials while the process vessel is at elevated temperatures. Basic gunning mixes differ from most acidic gunning mixes in that they are chemical bonded rather than cement bonded. The most common chemical binders used are sodium silicate, sodium phosphate, or an organic acid.

Plastic refractories are soft, moldable materials and are ready for installation as

Table 1 Types of monolithic refractories

Acidic materials (alumino-silicate-alumina)			Basic materials
Castables			**Gunning mixes**
Dense		Insulating	Acid bond
Conventional			Phosphate bond
Reduced cement			Silicate bond
(low and ultralow)			
No cement			**Dry bottom mixes**
Chemical bonded			
Gunning mixes			**Castables**
Dense		Insulating	Acid bond
Conventional			Phosphate bond
Clay bonded			Silicate bond
Chemical bonded			
Plastics			**Ramming mixes**
Ramming, gunning, extruded block, or vibrating			
Clay bond			
Air bond			
Chemical bond			
Ramming mixes			**Mortars**
Wet	Dry	Other	
Clay bond	Ramming	Mortars	
Air bond	Vibrating	Service coatings	
Chemical bond		Injection mixes	

received. They come in a variety of compositions, and while clay is used in almost all plastics for plasticity and bonding, auxiliary bonds may also be used.

The most notable of these is phosphate bonding by the use of monoaluminum phosphate. Phosphate-bonded plastics are typically stronger and more abrasion-resistant than their clay and air-bonded counterparts (Fig 2).

The majority of refractory plastics are installed by ramming with pneumatic ramming tools. In the early 1980s, gunning grade refractory plastics were developed (Ref 5). These materials are similar in composition to ramming plastics but are placed by the use of specialized pneumatic gunning equipment. The placement technique is similar to that of conventional gunning mixes, though there is no water addition at the nozzle.

Gunning grade plastics offer installation advantages of speed, uniformity, and simplified logistics over their rammed counterparts, as well as equivalent physical properties.

Two other varieties of plastic refractories are also available, although neither is used as widely today as it once was. Extruded plastic blocks are similar standard refractory plastic formulations but are extruded into block shapes and installed as large unfired "brick." Vibrating plastics are plastic formulations modified to flow and compact under vibration and are typically installed behind forms.

Refractory ramming mixes are granular materials and are available either wet or dry. Wet ramming mixes employ the same bond systems as conventional plastics. They differ from plastics in that they typically contain a finer grain size distribution and less moisture. They are therefore less workable and are installed by ramming behind forms.

Dry ramming mixes may require a water addition (as in the case of some basic varieties) or may be used totally dry. This second category includes basic granular mixes, which are installed by tamping and are carefully sized for maximum compaction (dry mixes), as well

as a class of materials commonly known as dry vibrating materials. These materials come in a wide variety of compositions and commonly employ a boron compound as the binder phase. They are precision sized to achieve flow and maximum density. Since they possess no mechanical strength after installation, initial heating occurs with the forms in place until the bond phase begins to set, which can occur at relatively low temperatures (200 to 1100 °C, or 390 to 2010 °F).

Other Monolithics. There are other varieties of monolithic refractories manufactured, the most common category of which is mortars. Mortars are used primarily to bond brick but may be used for repairs to help bond monolithics to old refractories.

Service coatings are also used for this purpose, as well as to seal cracks, act as parting compounds, repair worn surfaces, and to generally "dress up" a refractory lining.

Injection mixes are a relatively new idea in furnace maintenance to repair worn linings and insulation. They are applied from outside the furnace through injection nipples, and allow repair while the furnace is operating or still hot.

Fibrous Refractories

Fibrous refractory products are based primarily on refractory ceramic fibers. The initial refractory ceramic fibers were developed in the 1940s and have grown in use since then. Industry production and sales statistics are hard to obtain since the U.S. Census Bureau did not start tracking ceramic fibers as a separate product class until 1987. From the limited data available, ceramic fiber production in the U.S. has increased considerably. In 1987, six companies reported sales of 16×10^3 Mg (17.6 $\times 10^3$ tons) of 1090 to 1315 °C (1995 to 2400 °F) grade ceramic fiber. This had grown to nine companies reporting sales of 29×10^3 Mg (32×10^3 tons) in 1989 (Ref 6, 7).

Fiber Types. Refractory ceramic fibers are available in two basic types: amorphous and polycrystalline. Amorphous fibers account for the greatest amount produced and used in refractory applications. Ceramic fibers are low-alkali aluminosilicate compositions which may contain other oxides (such as zirconia or chromia) to increase service temperatures and modify physical properties. Table 2 lists some compositions and service limits of commercially available ceramic fibers.

Properties. Ceramic fiber products are used mainly for their low thermal conductivity and low heat storage (resulting from their low density) as compared to other types of available insulating materials. Figure 3 illustrates this in comparing the thermal properties of ceramic fiber to insulating brick and castables.

Low thermal conductivity and heat storage give these materials a major advantage over other insulation, especially in cyclic furnace operations. Ceramic fiber products' main dis-

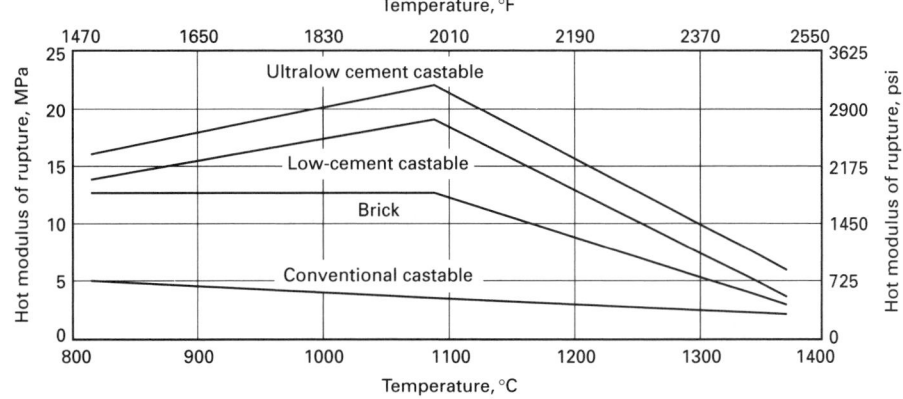

Fig 1 Hot strength comparison of several super-duty monolithic ceramics

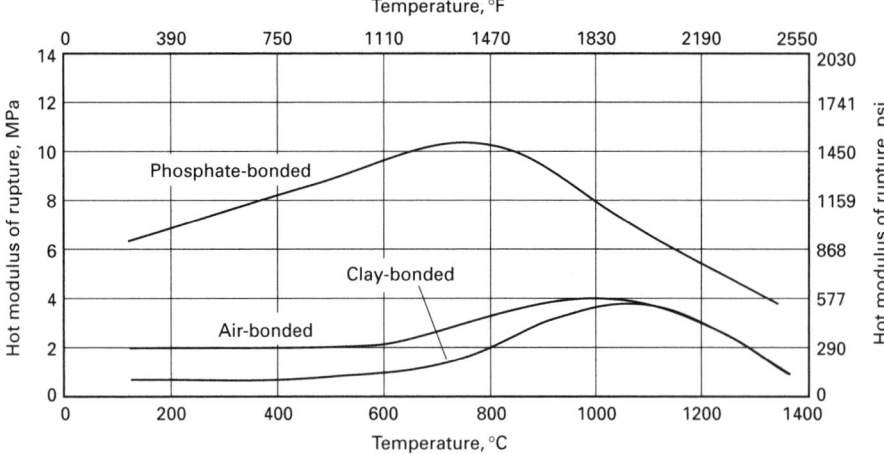

Fig 2 Effect of bonding method on hot strength of superduty monolithic ceramic materials

Table 2 General categories of ceramic fibers

Service temperature limit, °C (°F)	Composition, %					
	Al₂O₃	SiO₂	Fe₂O₃	TiO₂	Zr₂O₃	Cr₂O₃
670–700 (1260–1315)	44–51	47–53	0.1	1.0–1.8	...	...
1425–1480 (2600–2700) Type A	41–48	52–55	...	...	...	0.3–4
1425–1480 (2600–2700) Type B	32–36	44–48	...	...	14–21	...
1650 (3000)	96	4	...	...	...	...

advantages are their poor physical strengths, shrinkage, and their susceptibility to attack by corrosive environments and combustion atmospheres due to their high porosity.

Production Methods. Amorphous ceramic fibers are produced by melting the raw materials and forming the fibers by a blowing or spinning process. Polycrystalline fibers are produced from a solution chemistry or a sol-gel process (Ref 8). The fibers are then used

as is or further processed into a variety of forms such as blanket, board, strip, rope, paper, and vacuum-formed shapes.

Installation technique or application depends on the form of ceramic fiber used. RCF paper, rope, and strips are commonly used materials for expansion joint allowances or to service cracks. Modules may be attached by various methods directly to the furnace shell or veneered by use of a mortar over existing

refractory. Boards and blankets can be used as back-up insulation or as hot face lining with the use of proper attachment systems. Vacuum-formed shapes have a wide range of applications, including combustion chambers, burner cones, heating element holders, aluminum launders, and tap-out cones.

Applications

Monolithic and fibrous refractory products have been used in virtually every industrial sector as well as in common domestic applications (such as in fireplaces, water heaters, electric stove element holders, and home heating furnace combustion chambers).

When considering the application of refractories, it is common to find very similar applications which use totally different refractory materials, designs, and construction techniques. This is due to the engineering company or customer design preferences, operating conditions, installation requirements and/or economic considerations. Therefore, monolithic and fiber refractory products may compete with each other and also with brick and other products in many applications. However, most systems employ various types of refractories in conjunction with each other to make maximum use of the advantages of each type and for the economics of the total installation.

In repair situations or design modifications to an existing furnace lining, monolithics and fibrous products are used more often than competitive brick products due to their ease of installation and flexibility of application.

The application and construction design criteria for monolithics and fibrous refractories are much too numerous to list and discuss in this article. However, a review of some of the major applications of these materials by industrial sector can be useful in understanding their widespread use and flexibility.

Steel

The domestic steel industry is the largest single consumer of refractories, accounting for 50% of all refractory products produced. Monolithic refractories are used as primary construction materials in many applications and as repair materials in virtually all processing vessels and furnaces in the steel industry. Fibrous materials also have significant areas of application. Following are descriptions of the main application areas for fibrous and monolithic ceramics in iron and steel production in an integrated steel mill or mini mill.

Coke oven batteries are for the most part constructed of silica brick. Monolithic castables do find use as the primary materials for coke oven door plugs. Superduty castables are generally used for this application. In many cases, a low iron content cement is used as a binder because reducing conditions are experienced in the oven. Since thermal shock is a major problem in door plug applications,

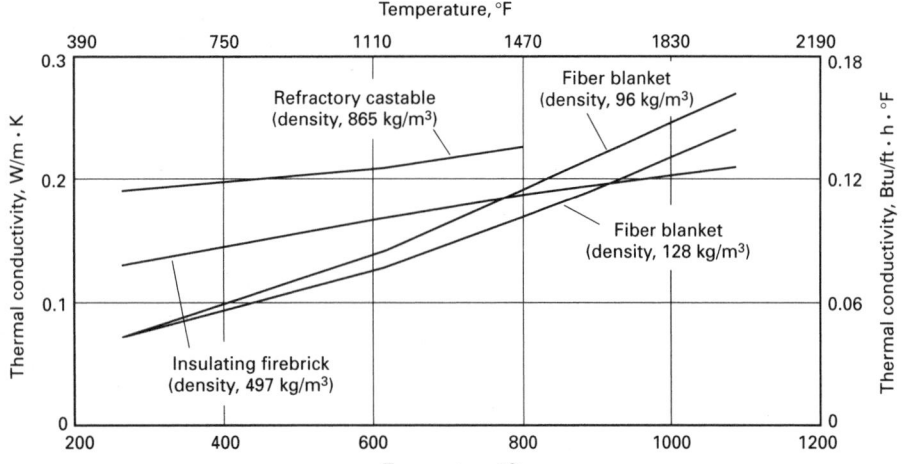

Fig 3 Comparison of thermal conductivities of ceramic fibers with those of other materials

stainless steel fiber is often used to reinforce the castable and reduce spalling when cracking occurs. In recent years, fused amorphous silica-based castables possessing excellent thermal shock resistance have been used in this application with success.

To repair the wear of coke oven door jambs, high-silica mixes are sprayed on periodically. These materials are applied on hot surfaces and are either silicate- or phosphate-bonded. Bulk, strip, or module ceramic fiber products are also used to pack and seal cracks around the door openings.

Blast Furnaces. While some castables and plastics are used in the initial or new lining of a blast furnace (mainly as back-up or "fill" materials), the primary working lining will consist of various grades of brick. However, as a blast furnace campaign progresses and the brick linings wear or deteriorate, monolithics become a major repair material. This is especially true in the stockline and upper and lower shafts.

During major repair the furnace is drained and cold gunning repairs of the stockline and shaft are carried out. Abrasion-resistant superduty gunning mixes are often used. These materials typically possess low iron content ($<$1.0%) to resist deterioration in the reducing atmosphere of the furnace and have good resistance to alkali attack.

For intermediate and on-line repairs, hot gunning or injection techniques are employed (Ref 9, 10). Both techniques have been used with success to significantly extend blast furnace campaigns.

For hot gunning repairs, the burden in the blast furnace is lowered below the area to be repaired, the furnace is taken off wind, and ports are opened above the area to be repaired. Special gunning formulations designed to be applied to warm linings are installed by the use of conventional guns with 10 to 20 m (33 to 66 ft) long nozzles. In addition to having resistance to abrasion, alkali vapors, and reducing atmospheres, gunning mixes used for this purpose must be able to bond and adhere to the old refractory at temperatures up to 600 °C (1110 °F).

Another technique used to repair the shaft is injection. In this method, injection nipples are installed in the area to be repaired. The furnace is taken off wind and a special injection mix or grout is pumped into the nipple. The injection mix is a heat set material, typically silicate-bonded, which begins to harden and form scabs when it mixes with the hot burden. The scabs formed during the injection then become the working lining.

Another area of the blast furnace where monolithics are used is in the trough and runner system. The monolithics used in this area have evolved from the original clay-sand-pitch mixes to the very sophisticated ones in use today.

Trough refractories must stand up to both corrosion and erosion by metals and slag and possess good resistance to both shock and oxidation. Another desired property, especially in single tap hole blast furnaces, is quick installation time. Materials for the trough and runner system are based on alumina, silicon carbide, and carbon. The types of materials currently in use are ramming mixes, ultralow or no-cement castables, and dry vibration materials (Ref 11). As wear occurs, the troughs and runners are maintained by the same types of materials. With proper maintenance, a trough system can last for 0.9 to 1.8 $\times$ 10^6 Mg (1 to 2 $\times$ 10^6 tons) of iron throughput.

Torpedo transfer ladles are originally brick lined, with high-alumina phosphate-bonded plastic used in the throat area. Life is sometimes increased by a gunning maintenance program. Typically a 50 to 70% alumina gun mix is applied at regular intervals. Ceramic fiber materials are often used to form a cap over the ladle opening to reduce heat loss during metal transfer.

Steelmaking vessels such as the basic oxygen furnace (BOF), Q-BOF, and electric furnace are lined with basic brick and shapes. However, basic gunning materials are extensively used to maintain the vessel lining and "protect" the more expensive bricks. The consumption of these basic maintenance materials during a vessel campaign far exceeds the amount of bricks originally used on a tonnage basis. Consumption of basic gunning mixes averages approximately 1 kg (2.2 lb) per 1000 kg (2200 lb) of steel in a BOF type vessel and approximately 5 kg (11 lb) per 1000 kg (2200 lb) of steel in an electric furnace. Basic gunning materials must possess good resistance to corrosion and erosion due to the molten metal and slag, as well as good gunning properties such as stickiness, minimum rebound, reactivity for fast sintering, and cost effectiveness.

Electric furnace bottoms are installed and repaired with so-called dry bottom basic mixes. These materials can be completely dry or slightly oiled (to retard hydration) and contain little if any binder. They are designed to be tamped or vibrated to maximum density. These products tend to have a high lime ($>$15%) and iron ($>$2%) content. The iron acts to aid in initial sintering and the lime retards slag penetration.

Steel transfer ladles have historically been lined with bloating fireclay or 70% alumina brick. In recent years, there has been success with ladles lined with vibratory cast low- and ultralow 70% alumina castables. This practice provides a lining with reduced cracks, excellent physical strength, and abrasion and corrosion resistance. For maximum life the slag line is in some cases cast with a basic castable (or lined with basic brick) to minimize corrosion in this area.

Soaking pits have historically been high-volume monolithic refractory consumers. While both castables and plastics have been used, the most typical construction utilizes superduty plastic or block refractory for the side and division walls backed by insulating brick, castables and/or mineral wool board insulation, and superduty castables for the pit floor. Depending on the slagging practice used, the slag line can be lined with high-alumina phosphate-bonded plastic to reduce corrosion in this area.

Covers are typically lined with superduty castables or plastic backed by insulating castable or mineral wool blocks. Though soaking pit use is declining with the increase in continuous casting, some monolithic innovations have been introduced such as large modular sections made of plastic, plastic block, or low-cement castables to reduce installation cost and time.

Continuous Casting. In the area of continuous casting, monolithic refractories have become very important in current tundish practice. Small tundishes (such as those used in mini-mill operations) use high-alumina phosphate-bonded plastics or either conventional or low-cement castables as the tundish lining.

It is only in the past few years that large tundish lining practice has begun shifting from brick to low-cement castables due to the improved durability of the castable tundish. Although 70% alumina brick tundish costs approximately 40% less to install than a tundish cast with 70% alumina low-cement castable, the average life of a brick tundish is between 200 and 500 heats. A low-cement castable tundish averages between 700 and 1200 heats. The integrity of the monolithic tundish also results in reduced maintenance costs over its campaign, typically requiring approximately one-third the maintenance expenditures of a brick tundish coupled with a campaign 2 to 3 times longer.

Another important monolithic application in the tundish is the increasing use of basic coating materials to coat the tundish lining to insulate and provide for easy desculling (Ref 12). Aluminosilicate coatings were originally used in this application but eventually gave way to board systems to form the tundish lining. In recent years, basic spray coatings and vibratables are replacing boards in many applications, providing excellent service, and reducing operating costs. These basic mixes are magnesite-based (containing between 70 and 92% MgO) and may be applied manually or by sophisticated robotic equipment.

Reheat Furnaces. Modern lining practice in the various types of reheat furnaces (pusher, in and out walking beam, and rotary hearth) utilizes monolithic refractories as the primary construction materials, although both brick and fiber products are used in conjunction with the monolithics. In reheat furnace applications, the refractory system must provide thermal insulation and maintain structural integrity for extended time at elevated temperatures. In addition, it must withstand, in certain zones, slag attack, flame impingement, and/or attack by combustion by-products. New designs, specifically the current generation of walking beam furnaces, stress thermal efficiency, and thus rely on increased insulation. Figure 4 illustrates typical hot face refractory use on a pusher-type reheat furnace. The other types of reheat furnaces will also use similar type refractories.

A typical sidewall construction in a reheat furnace consists of 240 mm (9.5 in.) of plastic refractory (grade varies depending on hot face temperature) backed by 114 mm (4.5 in.) of insulating fire brick or castable and 100 mm (4 in.) of mineral wool board insulation. Ceramic tile anchors and metallic wall supports are used to hold the lining back to the shell and support it. The suspended roof will typically consist of 225 mm (9 in.) of plastic backed by 50 mm (2 in.) of insulating castable and 50 mm (2 in.) of block insulation. The roof is supported from I-beams by ceramic tile anchors.

Soak zone hearths in pusher-type furnaces historically have been lined with 85 to 90% alumina phosphate-bonded plastic (for its strength and slag resistance) in the area between the skid blocks (typically fused cast refractory). In recent years, precast, prefired, ultralow cement castable block has been used to replace some varieties of fused cast skid blocks with success. Also low- and ultralow cement castables, because of their low porosities, excellent slag resistance, and high strengths, are being used in place of the phosphate-bonded plastic in the soak zone hearths.

The water-cooled pipe systems in pusher-type furnaces are covered with either preformed monolithic shapes made from castable or phosphate-bonded plastic or are cast *in situ* with castable refractory. Stainless steel

fibers are commonly used in pipe covering applications to minimize spalling due to thermal shock.

In walking beam type furnaces, the moving beams which take the place of the pipe and skid systems of the pusher furnace are lined with either a high-strength 70 to 85% alumina phosphate-bonded plastic or a high-strength low-cement castable to withstand the abuse and slag attack. Similarly, the support piers and hearths of in and out reheat furnaces and hearths in rotary reheat furnaces are installed with either phosphate-bonded plastic or high-strength conventional or low-cement castables.

While ceramic fiber refractories are used extensively in reheat furnace construction for expansion joint packing, door seals, or back-up insulation, a major application is the use of ceramic fiber modules as a veneer on the roof and sidewall of old or inefficient furnaces to increase efficiency and reduce energy consumption. The modules are applied directly to the existing refractory, provided the surface is not too irregular or spalled, and attached by high-temperature mastic or mortar. The type of fiber and thickness of the module are determined by the service conditions expected and may vary from zone to zone in the furnace. The use of veneering modules can be expensive, but they can reduce heat loss from the shell and roof areas by 30 to 50%.

Boilers and Incinerators

Boilers. Since monolithic refractory materials were originally designed for boiler applications, it is only natural that they play

a predominant role in modern boiler and incinerator practice. Boilers and incinerators vary widely in size, design, purpose, and fuel burned; therefore, many different refractory systems can be used.

In package, firebox, and utility boilers, refractory plastic and castables are used for burner cones, combustion chambers, walls, and as insulation behind tangent tubes. These materials are generally superduty plastics or castables and are backed up by an insulating castable, insulating firebrick, and/or block insulation. The hot face materials may be upgraded to higher-alumina types depending upon operating conditions (temperature, combustion by-products, slagging, abrasion, and so forth). This is especially true in waste heat units which utilize waste products such as wood, agricultural, and industrial waste as a primary fuel source.

Incinerators also vary from small package units which burn a constant waste product to large municipal and industrial units in which waste varies widely, resulting in temperature excesses, slagging, and corrosive gases. All of these factors can affect refractory choice and design.

Many large incinerators also use the waste heat to produce steam. Stringent environmental controls and the use of scrubbers and quench chambers cause further refractory problems and result in more frequent maintenance in these areas.

Monolithics used in incinerators include the standard superduty class plastics and castables, but more and more high-alumina phosphate-bonded plastics and low-cement

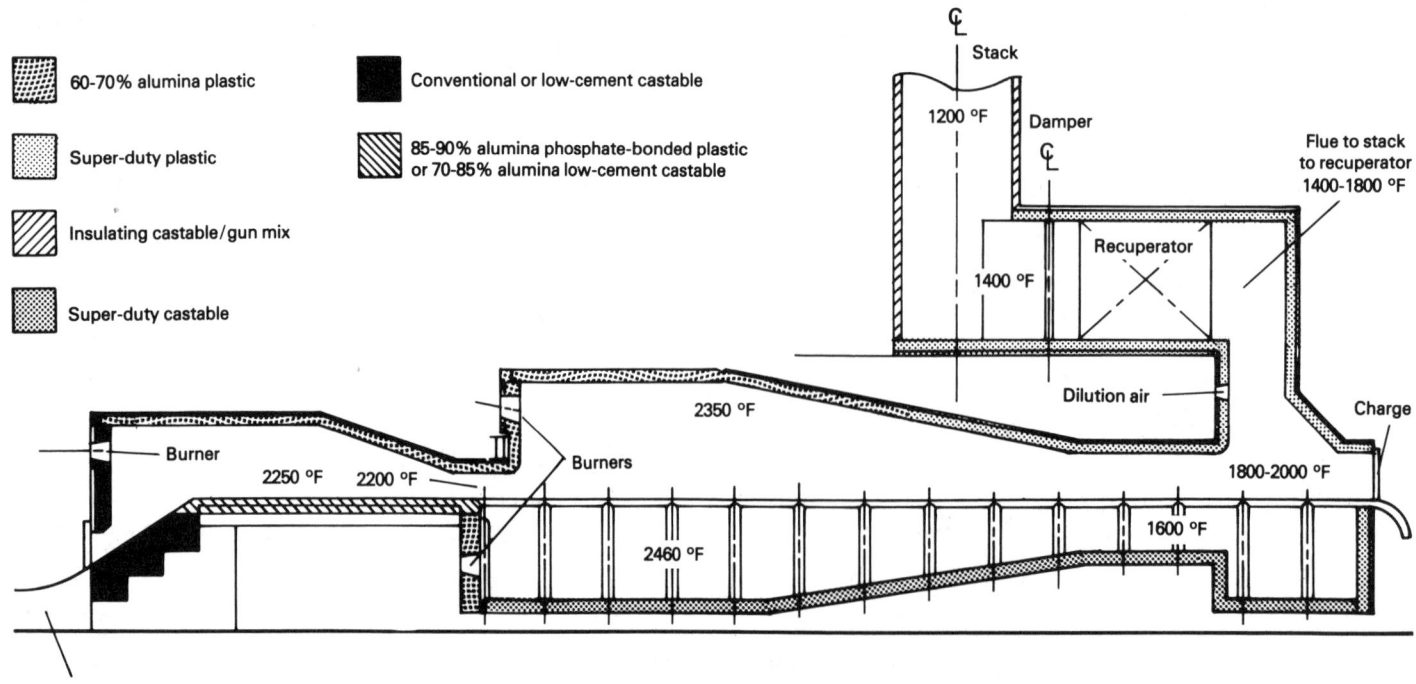

Fig 4 Schematic showing typical hot-face refractories used in a three-zone pusher furnace

castables are used to resist the high temperatures and slagging encountered. Silicon carbide based monolithics (castables, gunning mixes, and plastics) are also seeing increasing use in areas requiring slag resistance and high heat transfer, such as over water tubes (Ref 13). Acid-resistant gunning mixes and castables are being used more in scrubber sections, where acid solutions may form from treated combustion gases.

New Products and Application Techniques. Newly developed monolithic products and/or application techniques are continuously being applied in boiler and incinerator applications to increase refractory life and unit reliability. An example of this is in the area of wet ash hoppers. These large hoppers are located below utility boilers and are used to periodically quench the ash clinker produced in the boiler. The hoppers are filled with water and experience thermal shock and mechanical abuse during clinker removal. While some hoppers are brick lined, most are installed with a low-duty gunning mix. Annual repairs are the norm.

In recent years, thermal-shock-resistant, fused-silica castables, as well as high-strength abrasion-resistant castables reinforced with stainless steel fibers, have been used with great success, especially when coupled with a controlled low-temperature bake-out. Installations such as these have increased lining life and reduced maintenance repairs by more than 300%.

In cyclone burner systems, a coal-burning cylindrical combustion chamber is part of some utility boilers. The steam tubes have historically been protected from slagging and abrasion by a 25 to 35 mm (1 to 1.4 in.) layer of phosphate-bonded high-alumina plastic or a high-alumina gunning mix applied on studs over the internal steam tubes. Recent monolithic developments have led to new materials based on silicon carbide (both as plastics or gunning mixes) to resist slagging and abrasion and to increase heat transfer to the tubes. The use of plastic gunning techniques also has decreased installation time and cost in these units.

Circulating fluidized bed combustors are a relatively new technology gaining increased acceptance in many industrial applications. These units allow economical use of high-sulfur coal as well as the ability to burn other wastes (such as solvents, paints, and so forth) with reduced emissions. The fuels are burned with limestone in a circulating fluidized bed, which absorbs some combustion by-products.

Conditions in these units vary, with both oxidizing and reducing atmospheres experienced in different zones, as well as areas of high abrasion and temperature cycling. Depending on manufacturer, monolithic materials are currently competing with brick systems. A general monolithic lining would consist of an abrasion-resistant, high-strength material such as a low-cement castable or

phosphate-bonded plastic in the combustion chamber. The tube wall of the boiler may be studded in certain areas and covered with high-alumina or silicon carbide phosphate-bonded plastic or silicon carbide gun mix, while the cyclone section is installed with an abrasion-resistant gun mix backed with an insulating gun mix and block insulation (Fig 5).

Maintenance. In the area of maintenance of boilers and incinerators (and other furnaces), ceramic fiber-based injection mixes are becoming increasingly popular to repair degraded insulation and alleviate hot spots on the vessel shell. These injection mixes are paste like materials with good insulating properties which are pumped through nipples at hot spots while the unit is on line.

Petroleum and Petrochemical Applications

Monolithics and fiber-based refractories have been used extensively in refractory and petrochemical applications for many years. The stringent reliability requirements of refinery applications have led to the development of new monolithic materials, installation techniques, and production and field material controls which are now routinely used in many other industries. This latter category involves prequalification of the refractory materials at the production plant and during the installation process by independent inspection firms to assure the quality of the final installation.

Process Vessels. There are many types of refractory lined process vessels in a typical refinery and most operate at fairly low temperatures (<800 °C, or 1470 °F). Some, such as incinerators, boilers, and ammonia reformers, can operate at temperatures up to 1650 °C (3000 °F). The "heart" of the refinery is the fluidized catalytic cracking unit (FCCU) vessel, which consists of a regenerator and reactor vessel and their associated cyclones, catalyst transfer lines, and heat exchangers (Fig 6). In the regenerator vessel, low temperatures (<760 °C, or 1400 °F) and oxidizing conditions are encountered. In the reactor, reducing conditions occur in the same temperature range. The moving catalyst in these units produces mild erosion and sometimes thermal spalling. The use of a single-component, low-iron, semi-insulating gun mix on V anchors is common in these units, although two-component systems may be used.

The cyclones and transfer lines are subject to more erosion from the catalyst; therefore, erosion-resistant materials are used to line these sections. The cyclones are covered with

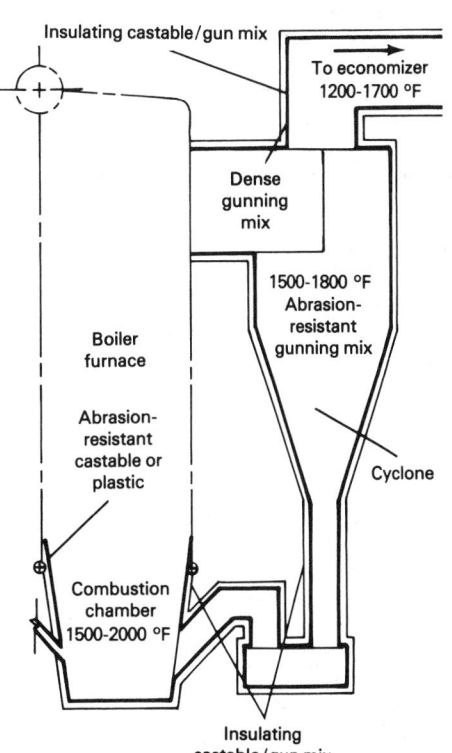

Fig 5 Schematic showing use of monolithic ceramic refractories in a circulating fluidized bed combustor

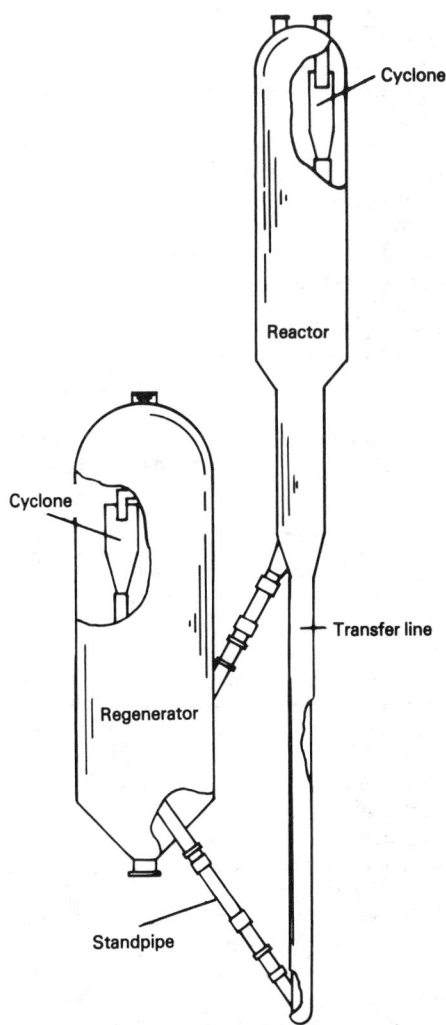

Fig 6 Schematic of a typical refinery fluidized catalytic cracking unit

approximately 25 mm (1 in.) of erosion-resistant castable (either cement- or phosphate-bonded) or phosphate-bonded plastic anchored by hex mesh or S bar anchors.

Transfer lines are critical sections in which the catalyst transfers back and forth between the reactor and regenerator. In the section of line from the regenerator to the reactor, the catalyst mixes with the feed and the cracking reaction begins. These sections have traditionally been lined with approximately 25 mm (1 in.) of abrasion-resistant phosphate-bonded castable or plastic anchored to the shell by hex mesh. In some cases, failures occur due to carbon (or coke) penetrating between the refractory and shell, causing the hex mesh to pull away from the shell. Other anchoring systems have been used to help alleviate this problem.

A new system of a vibration cast one-component lining has seen increasing use in recent years to line transfer lines (Ref 14). In this method, the lines are cast off the unit and later reassembled after a controlled bake out. An internal form is placed in the riser line and an erosion-resistant castable is carefully vibration cast in place. Continuous casts more than 12 m (39 ft) high have been successfully performed.

Ceramic fiber in blanket or module form also finds considerable use in various refinery heaters and process furnaces due to its thermal efficiency, ease of installation, and resistance to thermal shock. Insulating castables and insulating firebrick are also used in some cases.

Hydrogen synthesizer units and ammonia reformers operate at high temperatures (up to 1400 °C, or 2550 °F) with process atmospheres containing high amounts of hydrogen gas. These units are lined with both dense and insulating castables containing extremely low silica content, since silicon dioxide (SiO_2) can be reduced to silicon monoxide (SiO) gas by hydrogen at the high operating temperatures of these units (Ref 15).

Foundry Applications

In the foundry industry, many different types of melting, holding, and processing vessels and furnaces are employed depending on the size of the foundry, the scope of its production, and the type of metals processed. Some large foundries operate a fairly continuous operation, and most small ones have a very cyclic batch process in which various alloys may be processed in the same equipment. Monolithic materials are used quite extensively in melting, holding, and transfer operations, while ceramic fiber materials find use in heat-treating furnaces. Two specialized types of monolithics are used in foundry applications, namely graphitic plastic refractories (which resist slag and metal penetration) and dry vibratable ramming mixes.

The primary melting vessels in foundries are the cupola, electric arc furnace, and coreless and channel induction furnaces.

Electric arc furnaces in foundries may operate with either a basic or acidic practice. In basic arc furnaces, refractory practice is similar to that used in steel industry arc furnaces. Acidic arc furnace practice uses silica and/or high-alumina brick for initial lining, with fireclay gunning mixes and/or phosphate-bonded plastics used for maintenance.

The cupola operates under reducing conditions, and wide temperature variations occur. Brick or monolithics may be used either separately or in conjunction to line the cupola, with monolithics being used extensively in maintenance. The preheat and charge zones can be lined with low-iron, superduty castables or gunning mixes; the melt zone is lined with high-alumina graphitic plastic, castables or fireclay gunning mixes. Fireclay gunning mixes are used for maintenance in this area. The tap hole and trough are typically lined with graphitic high-alumina plastic refractory, although in recent years silicon carbide materials similar to those used in blast furnace troughs have also been applied.

Coreless induction furnaces may be lined with wet ramming mixes or low-cement castables, but the primary materials used are the dry vibratable materials, which are ideally suited for this application (Ref 16). The dry vibratable is applied behind a form, and bake-out may begin immediately with the form in place. After an initial set of the working surface of the lining occurs, the form is removed. The lining then has a hard surface but remains granular through most of its thickness. This allows for a more flexible lining with better stress relief. If a crack occurs, the granular nature of the lining minimizes its penetration, thus protecting the induction coils. Grades of the lining refractory vary with the type of alloy or metals melted. The caps and spouts of these furnaces are typically rammed with high-alumina phosphate-bonded plastic. Channel induction furnaces contain an upper case and a lower induction box. The upper case may be lined with various grades of brick, refractory plastic, conventional or low-cement castables. The lower induction box is lined with dry vibratable ramming mixes, wet ramming mixes, or low-cement castables.

Foundry ladles may be lined with a variety of refractory types, from brick to monolithics to board systems (similar to steel tundish practice), graphitic plastics, phosphate-bonded plastics, ramming mixes (both wet and dry vibratable), and castables (conventional and low-cement). All have merits and are used extensively. In some steel foundries, basic dolomitic ramming mixes are also used. These dolomitic ladles must be kept hot, since dolomite has a tendency to hydrate and lining degradation would result if cooled.

Heat-treating furnaces for hardening, normalizing, and tempering may have different refractory construction due to atmosphere and operating practices but are generally lined with insulating materials in the roof and sidewall and dense refractory in the hearth or car

top. Ceramic fiber blankets and modules, insulating castables, and insulating firebrick are all used. Ceramic fiber systems in this application have the advantage of thermal shock resistance, thermal efficiency, and low heat storage, especially in cyclic operations where rapid heating and cooling are required. Car tops and hearths, while they can be made of brick, are more typically lined with high-strength castables backed by insulating castables. Castable type is dictated by furnace operating conditions, temperature, and atmosphere.

Aluminum Processing Applications

The aluminum industry consists of three segments: primary, secondary, and remelt. The primary segment produces aluminum and its alloys from raw bauxite ore for use in the remelt segment. The secondary segment produces aluminum alloys from scrap also for use by the remelt segment. The remelt segment purchases aluminum in the form of ingots, billets, sows, and hot metal of known alloy, and produces finished products for sale to industrial and commercial markets (Ref 17).

In these segments, there are numerous applications for the use of monolithics and ceramic fiber based products. These range from the use of precast shapes of castable refractories and ceramic fiber pastes in carbon baking furnaces to castable and plastic refractory linings in dross reclaiming furnaces to the use of insulating castables and fiber products in various heat treating and normalizing furnaces.

Melting and Holding Furnaces. One of the important areas of monolithic application in recent years has been in the development and application of new monolithic materials for use in aluminum melting and holding operations. Common furnaces used are reverberatory, tilting, and barrel types, with the reverberatory being one of the most common. Reverberatory furnaces are used in every aluminum segment and range in size from the large 90,000 kg (200,000 lb), gas-fired melters and holders in the primary and secondary segment to small 200 kg (440 lb) melters (both gas-fired and electric) in the die casting industry. A typical gas-fired reverberatory furnace is shown in Fig 7. Before 1970, most of these furnaces were lined with 85% alumina brick (both burned and phosphate-bonded) in the metal contact area and superduty brick in the upper sidewalls. In the early 1970s, 85% alumina phosphate-bonded plastics began to be applied to the lower sidewalls with superduty plastics and castables in the upper sidewalls and roof. Brick was still in considerable use, especially in the metal holding zones (Ref 18).

In the early 1980s, radical changes occurred in monolithic compositions and applications for aluminum melting and holding furnaces. The first was the development and application of 70% alumina mullite-based phos-

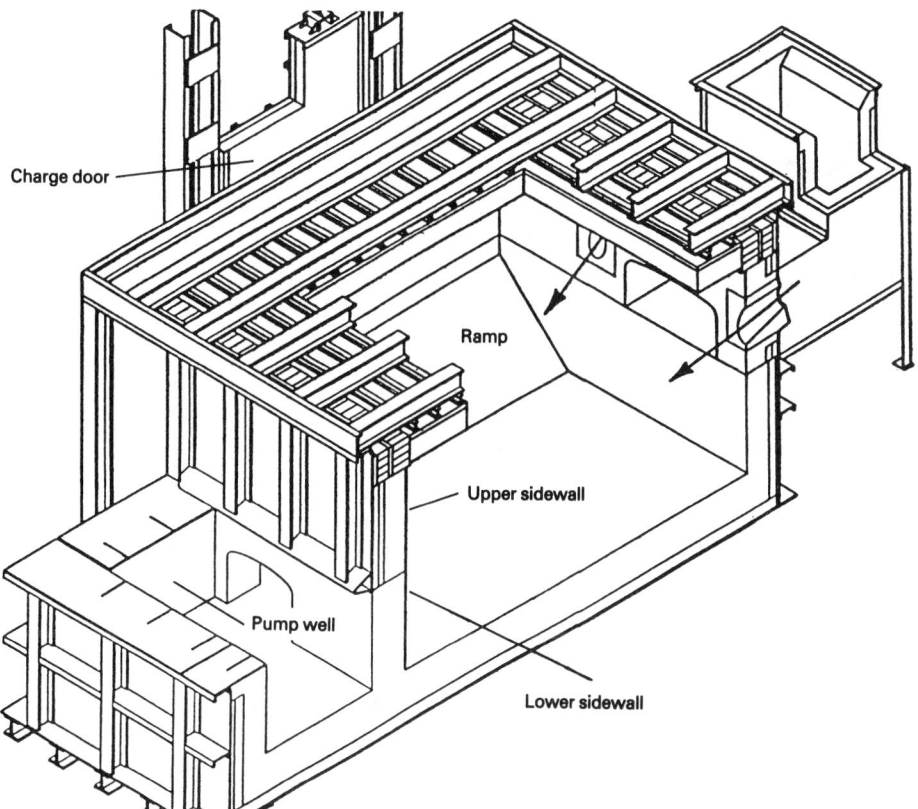

Fig 7 Schematic of a gas-fired reverberatory furnace used in the aluminum industry

phate-bonded plastics for metal contact areas. These materials proved just as resistant to aluminum penetration and oxide growth as the 85% materials they replaced and were more energy efficient due to their lower thermal conductivity. This lower conductivity had an added benefit in that the freeze plane location (the zone where molten aluminum will solidify) was moved closer to the hot face, thus providing extra protection from the possibility of metal runout if a crack developed in the lining. Following this was the development of nonwetting additives to monolithics for metal contact areas, further reducing aluminum wetting and oxide growth (Ref 18). This advance was followed immediately by the introduction of mullite-based, 70% alumina high-strength, low-cement castables for metal contact applications. This concept, along with the development of improved large-scale mixing and installation techniques, has revolutionized reverberatory furnace design and construction.

Today, nonwetting low-cement castables in large continuous pours (up to 60 tons, or 54 Mg, to minimize joints) are a standard installation in the aluminum industry. Both the nonwetting castables and plastics are backed up by nonwetting insulating materials such as modified insulating castables and ceramic fiber board. Nonwetting castables (both dense and insulating) are also being used with great success in aluminum transfer and over the road ladles.

Ceramic fiber products, which are very nonwetting to aluminum, are also used in various metal contact applications. Ceramic fiber paper is a standard expansion joint material, as are vacuum-formed ceramic fiber shapes used for tap-out cones. Doors are also lined with ceramic fiber modules or blankets to reduce heat loss and door weight.

Other Applications

Monolithics and ceramic fiber based products are also used extensively in other industry segments. Monolithics are applied in various areas of cement production and in kilns for processing lime, aggregates, ore, coal, and so forth. Ceramic fiber based products find considerable use in periodic, shuttle, and tunnel kilns in the ceramic industry and also in glass furnace applications.

REFERENCES

1. *Technology of Monolithic Refractories,* Plibrico Japan Co., Ltd., Tokyo, 1984
2. D.R. Lankard, Evolution of Monolithic Refractory Technology in the United States, *Advances in Ceramics,* Vol 13, *New Developments of Monolithic Refractories,* American Ceramic Society, 1985, p 46–65
3. E.S. Wright, Manufacturing and Marketing Trends in the U.S. Refractory Industry, *Am. Ceram. Soc. Bull.,* Vol 69 (No. 7), 1990, p 1155–1162
4. French patent No. 6,934,405, 1970
5. L.P. Krietz, G. Wilson, D. Hofman, and M. Tuskind, A Review of International Experience in Plastic Gunning, *Advances in Ceramics,* Vol 13, *New Developments in Monolithic Refractories,* American Ceramic Society, 1985, p 165–174
6. *Current Industrial Reports—Refractories,* U.S. Dept. of Commerce, Bureau of the Census, 1988
7. *Current Industrial Reports—Refractories,* U.S. Dept. of Commerce, Bureau of the Census, 1990
8. R.D. Smith, Refractory Ceramic Fiber, *Chemicals Science and Technology Handbook,* American Ceramic Society, 1990, p 385–391
9. L.P. Krietz, R. Woodhead, S. Chadhuri, and A. Egami, The Use of Monolithic Refractories in Blast Furnaces, *Advances in Ceramics,* Vol 13, *New Developments in Monolithic Refractories,* American Ceramic Society, 1985, p 323–330
10. L.P. Krietz, Refractory Injections for Blast Furnace and Maintenance, *Iron Steel Eng,.* Vol 64, (No. 12), 1987, p 31–34
11. D.H. Hubble, The Use of High Alumina Refractories in the U.S. Steel Industry, *Alumina Chemicals Science and Technology Handbook,* D.H. Hubble, Ed., American Ceramic Society, 1990, p 447–470
12. T. Morimoto, K Ogasahara, A. Matsuo, and S. Miyagawa, Introduction of Automatic Gunning Machines for Tundish Linings, *Advances in Ceramics,* Vol 13, American Ceramic Society, 1985, p 139–148
13. Refractory Construction for Waste-Burning Power Plants, *Am. Ceram. Soc. Bull.,* Vol 7, 1987, p 1107–1108
14. N. Heard, Vibration Cast Lining Procedures in the Petrochemical Industries, *Alumina Science and Technology Handbook,* American Ceramic Society, 1990, p 471–488
15. M.S. Crowley and R.E. Fisher, Petroleum and Petrochemical Applications of Refractories, *Aluminum Chemical Science and Technology Handbook,* American Ceramic Society, 1990
16. J.L. Turner and D.M. Myers, The Development of Dry Refractory Technology in the United States, *Advances in Ceramics,* Vol 13, *New Developments in Monolithic Refractories,* American Ceramic Society, 1985, p 161–164
17. J. Wunch, A Refractory Look at the Aluminum Industry, *Plibrico Company Market Place Innovators,* Issue 1, 1982
18. G.E. Graddy and D.A. Weirauch, Refractories Used for Aluminum Processing, *Alumina Science and Technology Handbook,* American Ceramic Society, 1990, p 489–493

Portland Cements and Concrete

Geoffrey Frohnsdorff and James R. Clifton, National Institute of Standards and Technology

PORTLAND CEMENTS are inorganic powders consisting predominantly of calcium silicates which, when mixed with a limited amount of water to form a paste, react slowly at ambient temperatures to produce a coherent, hardened mass with valuable engineering properties. The hardened product is porous and consists primarily of a poorly crystalline calcium silicate hydrate (Ref 1, 2). The main uses of these cements are in civil engineering for which, since the late 19th century, they have become indispensable (Fig 1). In these uses, cement is the essential ingredient in a castable or moldable mixture of cement with water and aggregates; the mixture may also contain either or both of chemical admixtures and mineral admixtures. If the maximum size of the aggregate is less than about 2 mm (0.08 in.), as in a sand, the mixture is called a mortar; if the maximum size is greater than about 5 mm (0.2 in.), as in a river gravel, it is called a concrete. A typical mortar consists of cement, fine aggregate (sand), and water in the ratios of about 1:3:0.5 by mass, while a typical concrete consists of cement, fine aggregate, coarse aggregate (for example, river gravel or crushed rock), and water in ratios of about 1:2:3:0.5 by mass. Common uses of portland cement mortars and concretes are listed in Table 1, together with some possible alternative materials.

Characteristics of Cements

To understand portland cements and the closely related blended cements, and to appreciate how the systems containing them can be designed for specific purposes, it is necessary to know their physical and chemical characteristics and how they interact with other ingredients of mortars and concretes.

Portland Cement. In the United States, but not in all countries, portland cements are used more extensively than blended cements. Portland cements are synthetic products made in great quantity (see Table 6). The first patent describing their manufacture was issued in 1824. The inventor, Joseph Aspdin, named them after a natural stone quarried at Portland on the south coast of England. In spite of a long history of use and evolutionary development, recent studies (Ref 3, 4) have shown that portland cement concrete has great potential for improvement in terms of durability, strength, and ease of placement (Ref 5).

The extensive use of portland and blended cements is possible, in part, because of the abundance of their raw materials—sources of the oxides CaO, SiO_2, Al_2O_3, Fe_2O_3, and SO_3, such as limestone, clay, iron ore, and gypsum (Ref 2, 6). These sources also contain lesser quantities of many unwanted elements and, for reasons of durability (Ref 6), a limit is always put on MgO, and frequently on ($Na_2O + K_2O$). In the industry, the common oxides are usually designated by single letters; thus, $C = CaO$, $S = SiO_2$, $A = Al_2O_3$, $F = Fe_2O_3$, $f = FeO$, $H = H_2O$, $\bar{S} = SO_3$, $M = MgO$, $N = Na_2O$, and $K = K_2O$. Almost all portland cements contain the same five main compounds or phases. These are:

- Alite (an impure tricalcium silicate, C_3S)
- Belite (an impure dicalcium silicate, C_2S)
- The aluminate phase (an impure tricalcium aluminate, C_3A)
- The ferrite phase (often referred to as tetracalcium aluminoferrite, C_4AF, or ferrite solid solution, fss, phase)
- Gypsum (calcium sulfate dihydrate, $C\bar{S}H_2$) or other hydrated forms of calcium sulfate

The ferrite phase is unusual among the cement compounds in forming an extensive solid solution series which extends from C_2F to C_6A_2F.

In the United States, portland cements are manufactured to comply with the ASTM Standard Specification for Portland Cement, ASTM C 150 (Ref 7). The specification defines five main types of portland cement. Type I, which is made in the greatest quantity, is intended for general use when the special properties of the other types are not required. The special properties of the other types when used in concrete are: Type II, moderate sulfate resistance or moderate heat of hydration; Type III, high early strength; Type IV, low heat of hydration; Type V, high sulfate resistance. The chemical and physical differences between the types that produce their special properties lie in the proportions of the cement compounds and in the fineness to which the cement is ground (Table 2).

Blended cements are used extensively in many countries, but not much in the United States. They are made by blending or intergrinding other materials (usually siliceous) with the normal ingredients of portland cements (Ref 6). In the United States, the cements are manufactured to comply with the ASTM Standard Specification for Blended Cements, ASTM C 595 (Ref 7). The blending materials that are currently used are listed in Table 3. They are finely divided materials which, while being almost inert to water when by themselves, contribute to the cementing reactions in the presence of portland cement; if added separately to a mortar or concrete mixture, they are termed mineral admixtures or, in the cases of granulated blast-furnace slag and high-lime fly ashes, supplementary cementing materials.

Although the blending materials are generally wastes or by-products from other industries, they should not be looked on as adulterants, since they usually have beneficial effects on the performance of the cements in which they are used (Ref 9). They have the characteristic that, in the presence of portland cement and water, the calcium hydroxide (CH) produced by the reactions of the calcium silicates in the portland cement promotes their reaction to form a calcium silicate hydrate.

Other Related Cements. Some specially formulated cements should be mentioned as being closely related to portland cement in that their main ingredient is portland-cement clinker (Ref 1, 8). Examples are masonry cements plasticized with hydrated lime or ground limestone for use in masonry mortars, expansive cements, and shrinkage-compensated cements formulated to cause controlled expansion during setting and hardening, and some oil-well cements (Ref 10).

Chemistry and Physics of Cements

The essential ingredient in the cements discussed in this article is ground portland-cement clinker. Clinker is the nodular product formed at about 1450 °C (2640 °F) in a rotary kiln fed with a carefully proportioned, finely ground, homogenized mixture of minerals which are sources of the required oxides; often, the ingredients are a low-magnesium limestone (source of CaO), a low-alkali clay

The chemistry of the hydration (that is, reactions with water) of portland cements is complex, and while the main products of the reactions are well-known, the exact mechanisms of the reactions are still debated (Ref 6, 12). The stoichiometries of the reactions of the C_3S and C_2S which make up about 80% of the typical portland cement (Table 2) can be represented roughly as follows:

$$C_3S + 3.3H \rightleftarrows C_{1.7}SH_2 + 1.3CH$$

$$C_2S + 2.3H \rightleftarrows C_{1.7}SH_2 + 0.3CH$$

Here, $C_{1.7}SH_2$ is the approximate composition of the nonstoichiometric calcium silicate hydrate (usually written C-S-H) formed, and CH represents calcium hydroxide, $Ca(OH)_2$.

Common reactions of the other main compounds can be represented as:

$$C_3A + 3C\bar{S}H_2 + 26H \rightleftarrows C_3A \cdot 3C\bar{S} \cdot 32H$$
$$\text{(ettringite)}$$

$$C_3A + C\bar{S}H_2 + 10H \rightleftarrows C_3A \cdot C\bar{S} \cdot 12H$$
$$\left(\begin{array}{c} \text{tetracalcium aluminate} \\ \text{sulfate 12-hydrate or ``monosulfate''} \end{array} \right)$$

$$C_3A + CH + 10H \rightleftarrows C_4AH_{10}$$
$$\left(\begin{array}{c} \text{tetracalcium} \\ \text{aluminate 10-hydrate} \end{array} \right)$$

$$C_4(A,F) + 6C\bar{S}H_2 + 2CH + 18H \rightleftarrows$$
$$2C_3(A,F) \cdot 3C\bar{S} \cdot 32H$$
$$\text{(iron-containing ettringite)}$$

Ettringite and its iron-containing analogs are often referred to as AFt phases, while the

Fig 1 Concrete is an essential element of the infrastructure of a modern industrial society. Courtesy of American Concrete Institute

Table 1 Uses of portland cement mortars and concretes and some alternative materials

Concrete products	Alternative materials
Factory-fabricated concrete products:	
Unreinforced products:	
Block	Clay brick, adobe brick, stone
Brick	Clay brick, adobe brick, stone
Tile	Clay tile, slate, wood shingles, fiber-reinforced cement
Steel-reinforced products:	
Pipe	Steel, fiber-reinforced cement, plastic, clay
Panels	Clay brick, fiber-reinforced concrete, plywood
Beams	Steel, lumber
Railroad ties	Wood
Boat hulls	Steel, wood, fiber-reinforced plastic
Other products:	
Fiber-reinforced products	Aluminum, steel, wood, plastic, glass, fired clay
Extruded products	Aluminum, steel, wood, plastic
Field-cast concrete structures:	
Predominantly unreinforced concrete:	
Highway pavements	Bituminous concrete
Canal linings, dams	Earth
Foundations	Steel, wood
Steel-reinforced concrete:	
Columns	Steel
Floor slabs	
Tunnel linings	Steel
Bridge decks	Bituminous concrete
Marine structures	Steel
Nuclear pressure vessels	Steel
Railroad structures	Steel
Other:	
Terazzo	Resin formulations
Stucco	Resin formulations
Masonry mortar	Resin formulations
Oil-well grouts	
Concrete patching	Resin formulations, bituminous materials
Roof decks	Steel, plywood

(source of SiO_2 and Al_2O_3), and iron ore (source of Fe_2O_3). The main compounds in the clinker are the previously-mentioned impure forms of C_3S, C_2S, C_3A, and C_4AF (Ref 2, 6). Areas of each of these can be seen in the scanning electron micrographs in Fig 2.

Portland cement is made by grinding portland-cement clinker with about 5% gypsum, or other form of calcium sulfate, the main function of which is to prevent too-rapid setting ("flash setting").

Table 2 Approximate composition and fineness ranges for ASTM standard types of portland cement

Compound/property	Composition, wt%/property value				
	Type I	Type II	Type III	Type IV	Type V
C_3S	42–65	35–60	45–70	20–30	40–60
C_2S	10–30	15–35	10–30	50–55	15–40
C_3A	0–17	0–8	0–15	3–6	0–5
C_4AF	6–18	6–18	6–18	8–15	10–18
$C\bar{S}H_2$	3–6	3–6	3–6	3–6	3–6
Specific surface area, m^2/kg (ft^2/lb)	300–400 (1465–1955)	280–380 (1365–1855)	450–600 (2195–2930)	280–320 (1365–1560)	290–350 (1415–1710)

Source: Ref 8

Table 3 Ingredients other than portland-cement clinker used in blended cements or added separately to concrete as mineral admixtures or supplementary cementing materials

Ingredient	Composition, wt% (major oxides only)				Typical % used(a)
	CaO	SiO_2	Al_2O_3	Fe_2O_3	
Pozzolans:					
Natural pozzolan	0.7–9.3	45–89	3–19	2.5–10	15–25
Fly ash	1–13	35–62	12–23	3–45	15–25
Silica fume	0.1–0.6	86–98	0.1–0.6	0–1	10–15
Ground, granulated blast-furnace slag(b)	37–45	32–37	10–16	0.3–9.5 (FeO)	15–50

(a) By mass of the portland cement. (b) Slag also contains 3.5–8.5% MgO. Source: Ref 6 with the exception of the fly ash data which is from Ref 14

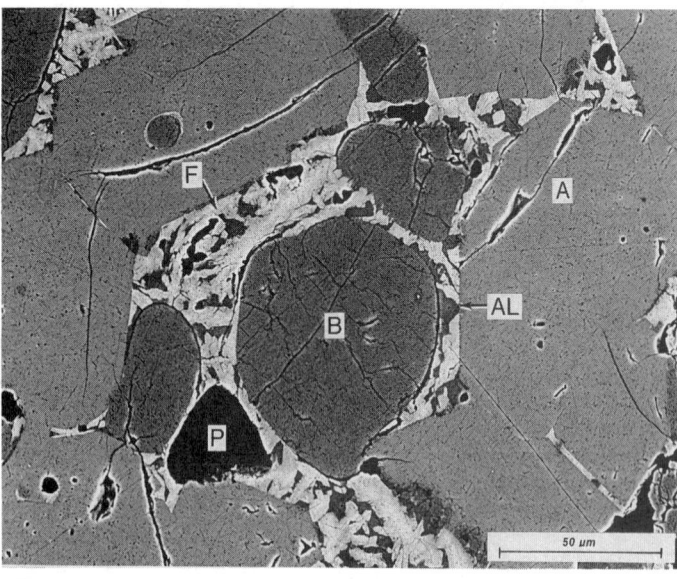

(a)

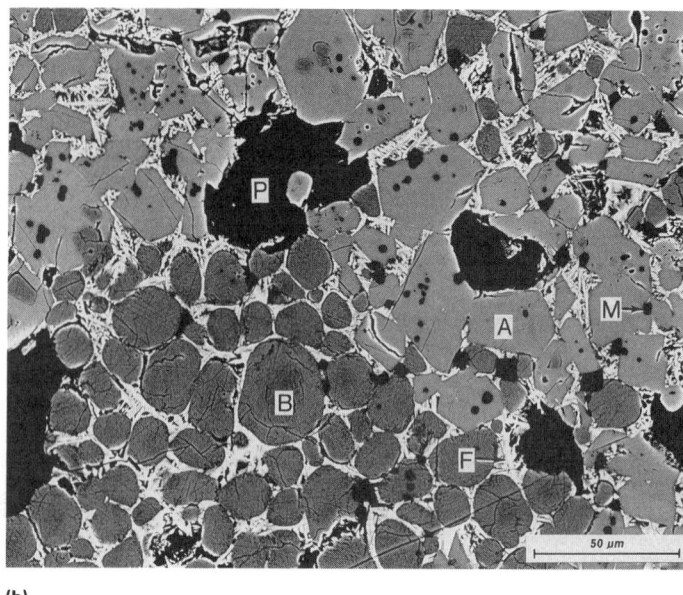

(b)

Fig 2 Scanning electron micrographs (backscattered electron image) of polished sections of ASTM Type I portland cement clinkers. (a) Clinker with large silicate crystals. A, alite (C_3S); B, belite (C_2S); AL, aluminate; P, pore. (b) Clinker with medium-sized silicate crystals. F, ferrite, M, periclase (MgO). Courtesy of P. Stutzman, NIST

monosulfate (tetracalcium aluminate sulfate 12-hydrate) and its analogs are referred to as AFm phases.

The formation of new phases as portland cement hydrates varies with the cement, but a typical course of the reactions is as shown in Fig 3.

Setting and hardening of the paste occurs as the original water-filled space becomes filled with solid products. The C-S-H appears to be the most effective binder among the hydrate phases, but the reasons for this have not been fully explained; it has been attributed to its layer structure, high specific surface area, and low degree of crystallinity. The microstructure of a hardened cement paste is shown in Fig 4.

In the case of a blended cement, the silicates in the portland cement component will react as outlined above to form C-S-H and

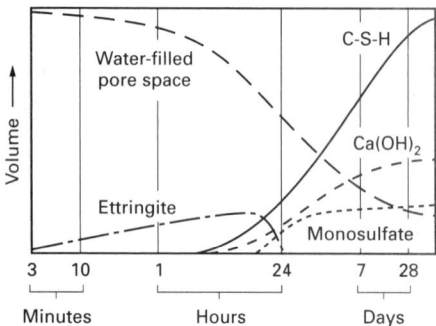

Fig 3 Formation of reaction products during the hydration of a Type I portland cement. With a Type II or Type V portland cement, little or no monosulfate would be formed at 24 h. Source: Ref 8

CH, then some or all of the CH formed and the water will react with SiO_2 in the siliceous phases in the blending material to form additional C-S-H. The reaction can be represented roughly as follows:

$$S + 1.4CH + 0.6H \rightleftarrows C_{1.4}SH_2$$

Because the main reaction product from the blending materials is similar to the C-S-H formed from portland cement and water, the blending materials do not cause great changes in the chemistry of the hardened product from the product obtained with portland cement. Rather, they change the proportions of the main reaction products, and they frequently reduce the rates of the reactions which cause hardening and strength development but without reducing the strength at late ages.

In addition to the four main compounds present in almost all portland-cement clinkers, minor compounds are always present due to one or more of:

● Impurities (particularly magnesium and alkali metal compounds) in the raw materials
● Ash or sulfur compounds from the fuel, particularly if the fuel is coal
● Incomplete reaction, that is, lack of attainment of high-temperature equilibrium, in the kiln
● Weathering by H_2O and CO_2 during storage of the clinker prior to grinding in the final stage of cement manufacture

The most common minor compounds in the clinker are CaO (free lime), MgO (periclase), sulfates of sodium and potassium, SiO_2 (insoluble residue), and products of hydration and carbonation on exposed surfaces of the

major compounds (loss on ignition). Specifications require each of these to be kept within limits in portland cement manufacture. In addition to the minor compounds which occur as discrete phases in the clinker, the major clinker phases always contain many impurities in solid solution in minor or trace quantities. In some cases, the impurities cause an identifiable new phase with a different stoichiometry to form, for example, $KC_{23}S_{12}$ in place of C_2S, or NC_8A_3 in place of C_3A. In other cases, the impurities may stabilize a different crystal form of the compound, for example, C_3S may occur in at least six crystallographically distinct forms, and C_2S in four (Ref 6).

Differences in the contents of minor elements between otherwise chemically similar clinkers from different cement plants can cause significant differences in the performances of the cements. Large interplant differences in clinker microstructure (Fig 2), which may be attributable to differences in raw materials, raw material preparation, kiln systems, and the operating conditions, are also believed to cause differences in performance.

Admixtures

Mineral Admixtures. It has already been mentioned that the finely divided materials blended with portland cements to make blended cements can also be added to the concrete mixture separately from the cement as mineral admixtures. The most commonly used ones are pozzolans—siliceous, or siliceous and aluminous, materials which react with calcium hydroxide and water at ordinary temperatures to form compounds with

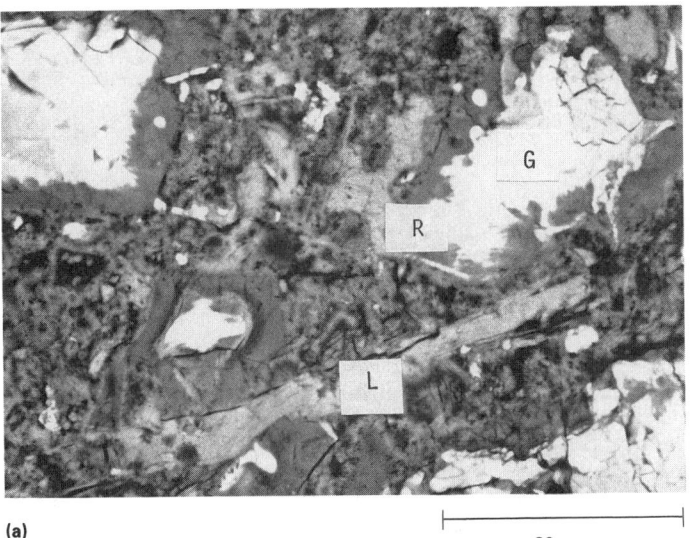

Fig 4 The microstructure of a hardened portland cement paste as seen in a scanning electron micrograph of (a) a polished section and (b) a fracture surface. The cement specimen (ASTM Type I) had a water/cement ratio of 0.45, was aged for 7 days, and was stored under water at 20 °C (70 °F). G, cement grain, with reaction rim; R, L, calcium hydroxide crystal. Most products cannot be identified definitively. Courtesy of P. Stutzman, NIST

cementitious properties. Typical levels of addition are between 5 and 30% by mass of the cement. In the United States, mineral admixtures are normally required to comply with ASTM Standard Specification for Fly Ash and Raw or Calcined Natural Pozzolan for Use as a Mineral Admixture in Concrete, ASTM C 618, or ASTM Standard Specification for Ground Blast-Furnace Slag for Use in Concrete and Mortars, ASTM C 989 (Ref 7). If they are properly selected and used, they can have substantial beneficial effects on concrete durability and strength (Ref 13–16), with the effects of silica fume being particularly noteworthy.

Chemical Admixtures. Another technologically important category of material which can be added to the concrete mixture is that of chemical admixtures (Ref 1, 17). These water-soluble materials may be organic compounds or inorganic salts; while they are only added in small quantities, for example, 0.1 to 2% by mass of the cement, they can have important effects on the performance of the mixture. In the United States, they are manufactured to comply with ASTM Standard Specification for Air-Entraining Admixtures for Concrete, ASTM C 260, or ASTM Standard Specification for Chemical Admixtures for Concrete, ASTM C 494 (Ref 7). The main use of air-entraining admixtures is to increase the frost resistance of concrete (Ref 1, 18). Their function is to cause small air bubbles to be entrained in the cement paste matrix of the concrete during mixing. If the volume of bubbles is sufficiently large and the bubble spacing is sufficiently small, they provide a mechanism for reducing the otherwise disruptive stresses associated with the crystallization of ice at sub-0 °C (32 °F) temperatures in water-filled pores of concrete which are not completely water-saturated. The uses of the

types of chemical admixtures covered by ASTM C 494 are indicated by their names:

- Type A, water-reducing admixtures
- Type B, retarding admixtures
- Type C, accelerating admixtures
- Type D, water-reducing and retarding admixtures
- Type E, water-reducing and accelerating admixtures
- Type F, high-range, water-reducing admixtures ("superplasticizers")
- Type G, water-reducing, high range, and retarding admixtures

Water-reducing admixtures are anionic surface-active agents that are adsorbed on the particles in the concrete. If enough admixture is present, the electrostatic charges of the adsorbed anions cause forces of attraction between particles to be overcome and interparticle repulsion to result. The immediate macroscopic effect is increased fluidity of the cement paste matrix and a reduction in the amount of mixing water required to achieve a given "workability" of the concrete. If the amount of mixing water is reduced, the ultimate strength of the concrete will normally be increased, and high-range water reducers are essential ingredients of very high strength concrete.

Retarding admixtures are added to the concrete mixture to increase the time of set, as in extending the working time of concrete in hot weather. The precise mechanism of their action is not known but, in most cases, a delayed nucleation of cement hydration products is thought to be involved. Accelerating admixtures cause the time of set to be reduced and the rate of early-age strength development to be increased. Calcium chloride, the best known accelerator, was formerly used extensively in concrete for cold weather con-

creting, but its use in reinforced concrete has diminished rapidly because of awareness that chlorides often accelerate corrosion of reinforcing steel.

Properties and Performance Characteristics of Mortars and Concretes

Concretes are composite materials consisting of some ingredients, for example, the aggregates (and reinforcement), which are intended to be inert and to retain their form and mechanical properties in the hardened product, and some, for example, the cement and mineral admixtures, which are intended to react to form new phases to act as the binder filling the spaces between the inert materials. Typical ranges of compressive and tensile strengths of concretes are shown in Table 4.

The tensile strengths of plain concrete are usually too low and variable to be relied on. When tensile properties are important, the concrete is usually reinforced with embedded steel in the form of rods ("rebars") or mesh (Ref 1, 19).

The combination of properties and performance characteristics desired of a mortar or concrete vary with the application. Table 5, which is based on a report from the National Materials Advisory Board (Ref 20), lists many characteristics of concrete materials and indicates, for each application mentioned in Table 1, which characteristics are limiting or, at least, could bring substantial benefits from improvement.

A concrete mixture represents a balance between many competing factors. For example, a large amount of mixing water increases the ease of placement but reduces the strength,

Table 4 Typical ranges of compressive and tensile strengths of moist-cured concretes

Property	Strength range MPa	psi
Compressive:		
1 day	2–40	300–6000
3 days	6–60	900–9000
7 days	10–65	1500–9500
28 days	20–70	3000–10,000
Tensile:		
1 day	0.2–4	30–600
3 days	0.6–6	60–900
7 days	1–6	150–900
28 days	2–7	300–1000

because the excess water increases the porosity of the hardened product. As another example, the $Ca(OH)_2$ formed in the hydration of the calcium silicates in the cement paste is often considered undesirable because it may be mechanically weak and the high pH it causes, particularly in the presence of alkali metal compounds, is a factor in promoting deleterious alkali-aggregate reactions in certain siliceous aggregates (Ref 1). On the other hand, the high pH provides important protection for reinforcing steel by passivating it against corrosion, and the $Ca(OH)_2$ can react with mineral admixtures to produce products and microstructures that are more desirable from the points of view of strength, permeability, and resistance to sulfate attack.

It is difficult to optimize a concrete for a given application because of the large number of performance characteristics which should be considered (Table 5), and the lack of knowledge of the precise relationships among them. As a result, mixture design is still heavily dependent on experience, rules of thumb, and trial and error (Ref 1, 8, 19). Concretes are usually formulated to achieve specified minimum compressive strengths at specified ages, while having suitable rheological properties (workability) to facilitate placing in the period prior to setting. Considerations of cost normally suggest that, consistent with attainment of the required strengths, the maximum amount of aggregate should be used and the minimum amount of cement. On the other hand, durability considerations suggest that the water/cement ratio should be a minimum, but that the amount of paste should be slightly more than is required to fill the spaces between the aggregate particles and provide adequate workability. A prototype knowledge-based expert system for aiding the design of durable concrete has recently been developed (Ref 21).

It is a goal of much current research to make concretes more predictable materials so that their performance can be optimized for specific purposes. This is through improved understanding of the materials science of concrete and other cementitious materials (Ref 12), development of mathematical models for simulating the development of concrete microstructure and of degradative phenomena (Ref 22), and development of expert systems for efficient delivery of current knowledge of concrete technology to the user (Ref 21). It is expected that one of the resulting products will be guidelines for optimization of the composition and microstructure of concretes and other cementitious materials for specific applications. Figure 5 shows the typical microstructure of a concrete specimen.

The chemistry and physics of concretes are largely governed by the cement hydration reactions that result in the formation of the

Table 5 Characteristics of concretes and mortars in which improvements could bring benefits to specific applications

Application	Compressive strength	Flexural or tensile strength	Bond strength	Volume stability	Controlled expansion	Uniform appearance	Color	Low density	Flow properties	Young's modulus	Impact resistance	Ductility	Energy absorption	Fracture toughness	Early strength	Quick setting	Low heat liberation	Low permeability	Freeze-thaw resistance	Sulfate and salt resistance	Low thermal expansion	Abrasion resistance	Stain resistance	Low thermal conductivity	High temperature resistance	Low cost	Estimate of fraction of total quantity of cement use
Block	X		X	X	X	X	X				X				X			X						X			4
Brick	X		X	X							X													X			0.2
Pipe	X	X	X								X		X	X	X			X		X		X					2
Panels		X	X		X	X	X				X			X	X			X				X		X			2
Beams	X	X	X							X		X		X	X	X						X					2
Tile		X		X			X	X			X				X	X		X	X				X				0.5
Extruded products		X	X	X					X		X		X	X	X	X		X									0.2
Fiber-reinforced products		X	X					X			X	X	X	X	X	X		X		X		X					2
Boats		X	X					X			X							X		X	X	X					0.1
Railroad ties		X									X	X	X	X	X	X		X		X		X				X	0.5
Foundations	X	X															X	X								X	40
Missile silos	X	X		X							X		X	X		X		X	X	X					X		0.2
Columns	X	X			X					X																	8
Slabs		X	X					X										X	X	X							15
Highways		X	X						X									X	X	X		X	X			X	15
Canal linings		X	X						X					X				X	X	X		X				X	2.5
Tunnel linings	X	X	X						X	X	X				X	X		X	X	X		X				X	2.0
Bridge decks		X	X															X	X	X		X	X				1.5
Desalination plants				X														X	X	X		X	X				0.1
Dams	X			X													X	X	X	X						X	0.7
Marine construction	X	X		X														X	X	X		X					1.1
Nuclear pressure vessels	X	X		X										X			X					X			X		0.1
Terazzo		X	X	X		X									X	X		X				X	X	X			0.1
Stucco		X	X	X		X			X		X							X						X			0.5
Masonry mortar	X		X	X					X										X	X							4.5
Oil well grouts			X	X					X													X			X		1.4
Concrete patching			X	X											X	X		X	X	X		X	X				0.1
Refractory linings			X	X				X	X				X	X	X	X						X	X		X	X	0.1
Roofing		X	X	X	X			X	X		X							X				X					0.1
Elevated railroad structures	X	X	X			X							X	X	X			X	X								0.5
Hardened MX missile sites	X	X	X								X	X	X	X	X	X										X	Large

Source: Ref 20

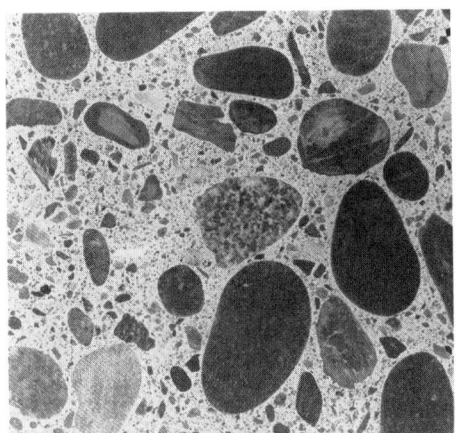

Fig 5 Reflected light micrograph of a polished section of concrete. The aggregate particles are embedded in a matrix of hardened cement paste. The width of the field is ≈120 mm (≈5 in.). Source: Ref 13

calcium silicate hydrate and other hydrate phases which cause the binding of the aggregate particles into a coherent mass (Ref 6).

In reality, none of the phases present in hardened concrete, not even the aggregates, is completely inert. Any phase might participate in degradative reactions to an extent which will depend on the composition and structure of the concrete and on the external environment. While it is relatively easy to recognize the more rapid degradative processes, the slower processes cannot be ignored if the concrete is intended to have a long life. The main degradative processes are (Ref 1, 23):

- Sulfate attack
- Frost attack (freezing and thawing)
- Alkali-aggregate reaction
- Corrosion of reinforcing steel
- Thermal stresses resulting from dissimilar coefficients of thermal expansion
- Acid attack (including carbonation)
- Temperature-induced phase changes

Among the external environmental factors which may cause, or accelerate, degradation of plain or reinforced concrete are:

- High temperatures, and temperature cycles
- Wetness, cycles of wetting and drying, and moisture gradients
- Wetness accompanied by freezing temperatures
- Wetting by solutions of salts, for example, sulfates, alkali metal salts, chlorides, and complexing agents
- Carbon dioxide
- Differences in electrical potential

The microstructural factors which affect the rate of degradation are:

- The porosity and pore distribution, particularly within the paste
- The distribution of aggregate and reinforcement

- The quantities and distributions of phases within the paste and the aggregate

Limitations of Concretes in Specific Applications. The generally recognized limitations of concretes are apparent from Table 5. They are:

- Low tensile and flexural strengths
- Brittleness
- Need to avoid cracking due to excessive temperature gradients caused by heat evolution during hardening of large masses of concrete
- Variability of cement, aggregates, and mixing and placing conditions
- Susceptibility to alkali-aggregate reaction
- Susceptibility of reinforcing steel to corrosion
- Susceptibility to sulfate attack
- Susceptibility to frost attack
- Susceptibility of strength development to temperature and to availability of moisture

These limitations are usually taken into account in the design process. As examples, steel reinforcement is widely used to provide needed tensile capability, and the selection of one of the special types of cement may be made so as to reduce the likelihood of alkali-aggregate reaction or attack of the concrete by high concentrations of sulfate in the environment. However, if the limitations could be reduced further, there would be significant economic benefits. As mentioned previously, current research programs are attempting to learn how to make concrete a more predictable material so that the limitations can be reduced.

Competitive Materials. The large amounts of concrete used in many different applications show that concrete competes effectively with other construction materials. Some materials that compete with concrete in its various applications are shown in Table 1. For the most part, they are steel, wood, asphalt, fired clay and, to a lesser extent, plastics. Characteristics other than cost which can give these materials a competitive advantage over concrete can be deduced from Table 5. However, it must be pointed out that concrete is not a single material, but a class of materials with a wide range of properties. Further, the limits of the properties of concrete have not yet been fully explored and defined. The changes in the markets for some of these materials and cement in construction since 1950 are indicated in Table 6.

Emerging New Materials and Applications

Cement and concrete technology have shown important advances in recent years. Two remarkable new products which are still in the developmental stage have been termed densified with small particles (DSP) cement (Ref 4) and macro-defect-free (MDF) cement (Ref 3). Both DSP and MDF cements are

Table 6 U.S. consumption of cement and other construction materials from 1950 to 1987 (millions of short tons)

Material	Consumption, tons × 10⁶					
	1950	1960	1970	1980	1987	Ref
Portland cement(a)	42	58	73	76	90	24
Ready-mixed concrete(b)	68	264	378	406	504	25
Steel(c)	12	14	19	18	14	26
Lumber	42	36	38	48	24	24
Brick(d)	16	16	16	15	18	24

(a) Tons of cement (calculated from figures given in barrels). (b) Tons of ready-mixed concrete calculated from number of cubic yards assuming 2 tons/yd³. The number of cubic yards for 1950–1970 are calculated from cement shipments assuming 5 sacks cement/yd³ and others from actual concrete shipments. (c) Total steel production is about 90 million tons/yr. The figures for construction steel given here include structural shapes, steel piling, steel plates, and reinforcing bars. (d) Tons of bricks calculated from numbers of brick assuming the average mass of a brick to be 5 lb

cement pastes rather than concretes, in that they do not contain aggregates. The DSP cements are low water/cement ratio pastes containing portland cement, fly ash, silica fume, water, and a superplasticizer. Low porosity is an important factor in helping them achieve compressive and flexural strengths that are unusually high for cementitious materials. The MDF cements are also low water/cement ratio pastes, but they include a small quantity of polymeric material. They require special processing to avoid the inclusion of macro-defects such as air bubbles, and the best performance appears to be obtained using a calcium aluminate cement rather than a portland cement.

With the stimulus of the properties attained with DSP and MDF cements, attention has turned to the more general problem of how to optimize concretes for specific purposes to produce high-performance concretes with desirable combinations of properties outside the range now obtainable. The outline of a national plan for high-performance concrete has been prepared and should stimulate research activity and development of standards and codes needed to make the United States a leader in high-performance concrete technology (Ref 27).

The mechanical properties of concretes reflect the inhomogeneities within them, for example, large pores, weak solid phases, and stress concentrators, due to interfaces between phases with different moduli. Removal of the inhomogeneities may increase the strength but at the expense of increased brittleness. Much interest in the possibilities for enhancing the properties and performance of concrete is now being expressed because of the recognition of its importance to the rebuilding of the civil works infrastructure.

New opportunities for designing concrete for specific applications are emerging because of the capabilities of computers to simulate structure formation and degradation in cementitious systems (Ref 22). Figure 6 shows

Fig 6 Simulated structure of cement paste in concrete near the interface with a cubic aggregate particle. In agreement with the experiment, the simulation shows enhanced porosity of the paste close to the aggregate surface. The width of the field is ≈1 mm (≈0.04 in.). Courtesy of E. Garboczi, NIST

an example of a simulated structure. By comparing the results of simulations of structure formation with different formulations of the concrete, it can be seen that fine mineral admixtures can cause improvement in the homogeneity of the cement paste near the aggregate interface, with its important implications for improvement in concrete performance.

Tests and Standards

In the United States, cement and concrete standards are developed and published by the ASTM and the American Concrete Institute (ACI). The ASTM standards deal with material properties, while the ACI addresses issues of structural safety and reliability. ASTM and ACI documents are referenced in national and local building codes concerned with public health and safety. Progress in cement and concrete technology in the 20th century is well indicated by the progress in standards development. However, new test methods and standards are needed to provide a strong materials science basis for cement and concrete technology. For example, although the capability to develop them exists, there are no standards for determining the compound (phase) compositions of cements nor for the hardened cement paste in concrete. Similarly, while particle size distribution is an important factor in determining the rheological properties of cementitious systems, there is no adequate standard method for determining the particle size distributions of cements and mineral admixtures. In general, the specifications for cements are not performance based. Before they can be, there must be standard tests for the various important aspects of performance, with particular emphasis on durability because of its importance in selecting materials. A particularly important need has arisen with recent developments in high-strength concretes. Few concrete testing laboratories have testing machines with the capacity to break the standard 150 mm (6 in.) diameter by 300 mm (12 in.) high compressive strength specimens of 100 MPa (14.5 ksi) concrete, and the reliability of tests on smaller specimens is not known.

ACKNOWLEDGMENT

The authors are grateful to Paul Stutzman of NIST for the scanning electron micrographs shown in Fig 2 and 4, and to Dr. Edward Garboczi, also of NIST, for the computer simulation of cement paste microstructure around an aggregate particle in Fig 6. They also wish to thank the American Concrete Institute for permission to reproduce Fig 1 from *Concrete International*, May 1989, Prentice-Hall for permission to reproduce Fig 5, and the National Materials Advisory Board for permission to use Table 5.

REFERENCES

1. S. Mindess and J.F. Young, *Concrete*, Prentice-Hall, 1981
2. F.M. Lea, *Chemistry of Cement and Concrete*, 3rd ed., Chemical Publishing, 1971
3. J.D. Birchall, Cement in the Context of New Materials for an Energy-Expensive Future, *Philos. Trans. R. Soc. (London)*, Vol A310, 1983, p 31–42
4. L. Hjorth, Development and Application of High-Density Cement-Based Materials, *Philos. Trans. R. Soc. (London)*, Vol A310, 1983, p 167–173
5. J.P. Skalny, Ed., "Concrete Durability: A Multi-Billion Dollar Opportunity," NMAB-437, National Materials Advisory Board, National Research Council, 1987
6. H.F.W. Taylor, *Cement Chemistry*, Academic Press, London, 1990
7. *Annual Book of ASTM Standards*, Vol 04.01, *Cement; Lime; Gypsum*; and Vol 04.02, *Concrete and Aggregates*, American Society for Testing and Materials, 1990
8. "Guide to the Selection and Use of Hydraulic Cements," ACI 225R-85, American Concrete Institute, 1985
9. G. Frohnsdorff, Ed., *Blended Cements*, STP 897, American Society for Testing and Materials, Philadelphia, 1986
10. E.B. Nelson, Ed., *Well Cementing*, Schlumberger Educational Services, 1990
11. P.E. Stutzman, S. Lenker, D. Campbell, L.J. Struble, H. Kanare, and F. Tang, Standard Cement Clinkers for Phase Analysis, *Proceedings of the 11th International Conference on Cement Microscopy*, 1989, p 154–168
12. J.P. Skalny, Ed., *The Material Science of Concrete*, Vol I, American Ceramic Society, 1989
13. P.K. Mehta, *Concrete: Structure, Properties, and Materials*, Prentice-Hall, 1986
14. E.E. Berry and V.M. Malhotra, "Fly Ash in Concrete," Publication SP-3, CAN-MET, Energy, Mines and Resources Canada, Ottawa, Feb 1986
15. V.M. Malhotra, Ed., *Supplementary Cementing Materials for Concrete*, CANMET, Ottawa, 1987
16. R. Helmuth, *Fly Ash in Cement and Concrete*, Portland Cement Association, 1987
17. V.S. Ramachandran, Ed., *Concrete Admixtures Handbook*, Noyes Publishing, 1984
18. S. Popovics, *Concrete-Making Materials*, McGraw-Hill, 1979
19. A.M. Neville, *Properties of Concrete*, 2nd ed., John Wiley & Sons, 1973
20. D.M. Roy, Ed., "The Status of Cement and Concrete R&D in the United States," NMAB-361, National Materials Advisory Board, National Research Council, 1980
21. J.R. Clifton and L.J. Kaetzel, Expert System for Concrete Construction, *Concrete Int.*, Vol 10 (No. 11), 1988, p 19–24
22. E.J. Garboczi and D.P. Bentz, Digital Simulation of the Aggregate-Cement Paste Interfacial Zone in Concrete, *J. Mater. Res.*, Vol 6 (No. 1), 1991, p 196–201
23. *Manual of Concrete Practice*, 5 Vol, American Concrete Institute, 1990
24. A.H. Ulrich, "U.S. Timber Production, Trade, Consumption, and Price Statistics, 1958–87," Miscellaneous Publication No. 1471, U.S. Dept. of Agriculture, Dec 1989
25. Statistics Provided by the National Ready Mixed Concrete Association, Silver Spring, MD
26. Annual Statistical Reports, American Iron and Steel Institute, Washington, D.C., 1960, 1970, 1980, 1989
27. N. Carino and J.R. Clifton, "Outline of a National Plan for High-Performance Concrete," NISTIR 4465, National Institute for Standards and Technology, Dec 1990

Tile Whiteware

Tiziano Manfredini and Gian Carlo Pellacani, Department of Chemistry, University of Modena, Italy

CHARACTERISTICS AND APPLICATIONS of ceramic tiles (tile whiteware) are reviewed in this article. The body formulations, the effects of raw materials, and the industrial preparation processes are briefly considered as these influence the characteristics (and consequently the applications) of the products obtained. Processing equipment is not considered in detail and specific plant designs are also not included because these depend on the raw materials being used and the scale of the operation. The considerations that should be made before selecting a specific material or product in a given application are presented, and emphasis is placed on the properties of the final product and to the type of applications they may have in the building industry.

Development of the Industry

The production of dried and fired clay tiles has very ancient origins dating back to the Middle Eastern populations (Ref 1). The tile whiteware industry developed significantly in Europe and by the beginning of the twentieth century, floor and wall tile production achieved industrial scale. Further development in this field occurred after World War II. Europe (Italy and Spain, in particular), Latin America, and the Far East are now the most important areas of industrial tile production.

The floor and wall tile sector of the whiteware industry has seen a great deal of development during the last ten years with the introduction of new technologies, automation, and integration production flow into the manufacturing process. Subsequently, productivity and efficiency have increased, while energy consumption and costs have been reduced. Tile manufacturing is now continuous in both wet and dry tile production, and many plants today have nearly 100% automation. The major innovations in the tile industry during the last decade include wet grinding, spray drying, high-pressure dry pressing, roller drying, and fast-firing technologies. The development of the tile whiteware industry is reviewed in Ref 2. Currently, Italy, Spain, Brazil, and the Far East are the primary producers of ceramic tile in the world. Figure 1 compares the use of various tile whiteware for architectural and building design applications.

Definitions

Ceramic floor and wall tiles include all ceramic products used as wall and floor coverings, tiles, and components for swimming pools, as well as relevant accessories (Ref 3). Definitions of specific types of tile whiteware are given below.

Ceramic Wall and Floor Tiles. European Standard EN 87, approved in November 1981 (Ref 4), specifies that "Ceramic wall and floor tiles are building materials that are generally designed for use as floor and wall coverings, both indoors and outdoors, regardless of shape and size. They are manufactured by standard ceramic processes. Ceramic wall and floor tiles are prepared from a mixture of ball clays, sand, fluxes, coloring agents, and other mineral raw materials, and they undergo processing such as milling, screening, blending, and wetting. They are shaped by pressing, extrusion, casting, or other process normally at room temperature and are subsequently dried and finally fired at high temperature. Ceramic wall and floor tiles are not combustible and are light fast. Wall and floor tiles may be glazed, unglazed, or engobed. Glazes are glasslike, impervious coatings, and engobes are matte, clay-based coatings that may also be porous. Glazed wall and floor tiles are produced either by single or two-stage firing... "

American National Standards Institute specification for ceramic tile (ANSI A 137.1) contains the definitions given below.

Ceramic mosaic tile is formed by either the dust-pressed or plastic method usually 6.4 to 9.5 mm ($1/4$ to $1/8$ in.) thick and has a facial area of less than 39 cm^2 (6 in.^{2}). Ceramic mosaic tiles may be either porcelain or natural clay composition, and they may be either plain or with an abrasive mixture throughout.

Decorative wall tile is glazed tile with a thin body that is usually nonvitreous and suitable for interior decorative residential wall use where breaking strength is not a requirement.

Paver tile is glazed or unglazed porcelain or natural clay tile formed by the dust-pressed method having 39 cm^2 (6 in.^{2}) or more facial area.

Porcelain tile is ceramic mosaic tile or paver tile that is generally made by the dust-pressed method with the resulting tile composition that is dense, impervious, fine-grained, and smooth with a sharply formed face.

Quarry tile is glazed or unglazed tile, made by the extrusion processes from natural clay or shale usually having 39 cm^2 (6 in.^{2}) or more facial area.

Wall tile is glazed tile with a body that is suitable for interior use and usually nonvitreous and is not required to withstand excessive impact or be subject to freezing and thawing conditions.

Individual tile whiteware grades include unglazed tiles (ceramic mosaic tile, quarry tile, paver tile) and glazed tiles (glazed wall tile, glazed ceramic mosaic tile, glazed quarry tile, glazed paver tile).

Classification of Ceramic Tiles

Redware and Whiteware. It is common knowledge that many types of ceramic tile are available on the market. They differ according to the condition of the surface, color of the body (white or red), manufacturing technology, raw materials, and end use (Ref 5). The difference between "red" and "white" tiles lies in the amount of iron minerals contained in the body. By reacting with the other body components, they can give more or less coloration and modify the behavior of the body during firing (Ref 6).

Many classifications have been proposed to define tiles. Some are classifications of use or based on the characteristics of the starting materials or firing cycle. A complete and exhaustive classification is very difficult owing to the extreme heterogeneity of tile products, their processing, and subsequent characteristics. In this article, European (EN) and ASTM standards are considered.

EN standards classify ceramic tiles as a function exclusively of water absorption (which directly correlates to the porosity) and shaping method (extrusion or pressing). The shaping methods are classified as:

- *Shaping Process A* (extruded floor tiles). This process includes split tiles and individually extruded tiles
- *Shaping Process B* (dry-pressed floor and wall tiles)

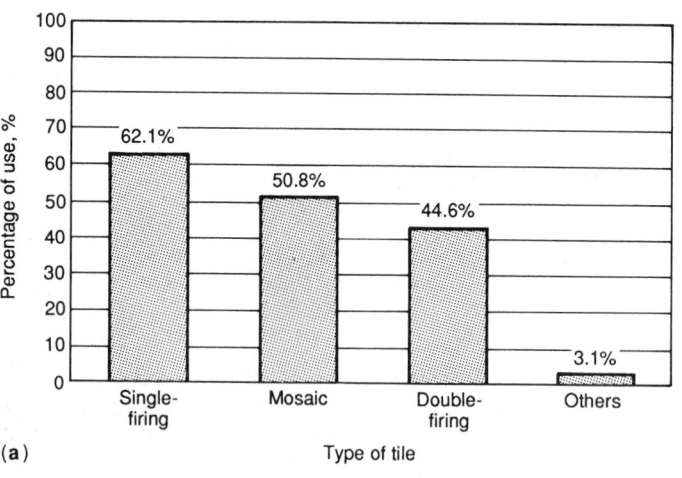
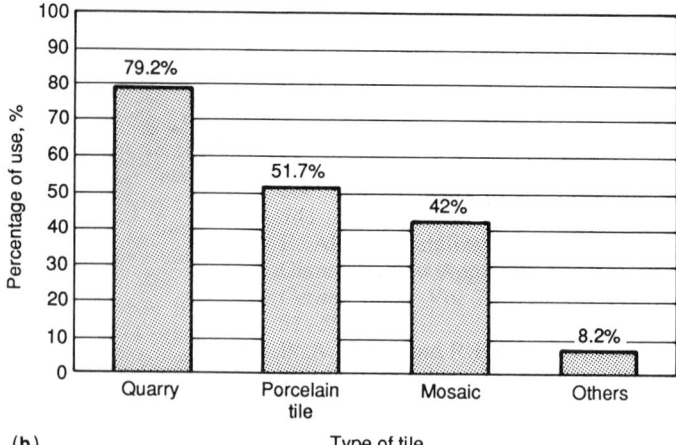

Fig 1 Types of ceramic tiles used for building in the United States. (a) Glazed products. (b) Unglazed products

- *Shaping Process C* (cast floor and wall tiles)

Water absorption, E, is expressed as a mass fraction given as a percentage and is classified as follows:

- *Group I* (low water absorption) where $E \leq 3\%$
- *Group II* (medium water absorption) where $3\% < E \leq 6\%$ (Group IIa) or where $6\% < E \leq 10\%$ (Group IIb)
- *Group III* (high water absorption) where $E > 10\%$

The EN classifications of tiles according to the shaping processes and water absorption are given in Table 1.

These parameters are directly related to the physical, chemical, mechanical, and microstructural properties such as:

- Modulus of rupture of fired tiles
- Abrasion resistance
- Chemical resistance
- Linear coefficient of thermal expansion
- Frost resistance

In addition to defining and classifying tile whiteware as well as outlining sampling pro-

cedures, EN standards provide thirteen specifications concerning the testing methods for the different products and eight concerning the requirements of the products and reference limits for the different properties. Table 2 lists these standards. Properties measured according to existing ASTM methods are provided in Table 3. Table 4 reports the technical features of vitrified, half-vitrified, and porous products which have been in compliance with the EN 176, EN 177, EN 178, and EN 159 standards for pressed tiles. A common classification scheme based on trade names for glazed and unglazed tiles is shown in Fig 2. The water absorption and the corresponding EN groups are also reported.

All these characteristics clearly point out that these materials are most suitable in specific floor and wall applications. Depending

on the type of glaze used, they can resist wear and high mechanical stresses.

Characteristics of Ceramic Tile

Ceramic tile is a stable, inexpensive building material of 5 to 15 mm (0.2 to 0.6 in.) thickness and a wide range of dimensions, as reported in Table 5.

The types of tiles presently on the market can be classified as single-fired or twice-fired. In both cases, the basic components of the raw materials used can be classified into three main types:

- Clay minerals
- Inert materials, such as feldspars, quartz, and pegmatite
- Others (calcite, dolomite, other oxides)

Some specialized bodies for specific applications contain other components, such as alumina or zircon in certain porcelains and bone ash in bone china. Tile bodies based on talc, pyrophillite, and wollastonite are also used in the United States. Raw materials (kaolin, clays, quartz, feldspars, calcium carbonate, and iron oxides), powder processing and compaction, and firing variables produce modifications of porosity (water absorption) and linear shrinkage and may alter the properties of the product (Ref 8).

The majority of tile and floor whiteware bodies are prepared today by a single firing technology (Ref 9) in the following steps (Ref 10):

- Batching
- Grinding
- Spray-drying
- Pressing
- Drying
- Glazing
- Firing

Each of these processing steps is described in detail elsewhere in this Volume.

Table 1 Classification of ceramic tiles in compliance with EN 87

| Tile type | Tile classification/water absorption, E, wt% | | | |
	$E \leq 3$ (Group I)	$3 < E \leq 6$ (Group IIa)	$6 < E \leq 10$ (Group IIb)	$E > 10$ (Group III)
Extruded tiles (a)	AI	AIIa	AIIb	AIII
Pressed tiles (b)	BI (d)	BIIa	BIIb	BIII
Cast tiles (c)	CI	CIIa	CIIb	CIII

(a) Includes split tiles and individually extruded tiles. (b) Includes dry-pressed floor and wall tiles. (c) Includes cast floor and wall tiles. (d) Water absorption for BI unglazed tiles is <1.5 wt%. For unglazed and very vitrified tiles, $E < 0.5$. Source: Ref 3

Table 2 European (EN) standards for ceramic tiles

Standard type	Standard No.	Description/classification
General	EN 87	Definition, classification
	EN 163	Sampling
Testing method	EN 98	Format, appearance
	EN 99	Water absorption
	EN 100	Bending strength
	EN 101	Mohs hardness
	EN 102	Wear resistance of unglazed tiles
	EN 103	Linear thermal expansion
	EN 104	Thermal shock resistance
	EN 105	Crazing resistance
	EN 106	Resistance of unglazed tiles to chemicals
	EN 122	Resistance of unglazed tiles to chemicals
	EN 154	Wear resistance of glazed tiles
	EN 155	Steam expansion
	EN 202	Frost resistance
Product requirements	EN 121	Group AI
	EN 159	Group BIII
	EN 176	Group BI
	EN 177	Group BIIa
	EN 178	Group BIIb
	EN 186	Group AIIa
	EN 187	Group AIIb
	EN 188	Group AIII

Table 3 Properties of ceramic tiles measured according to ASTM methods

Property	Tile type			
	Flat glazed wall tile	Flat ceramic mosaic tile	Flat quarry tile	Flat paver tile
Non-destructive tests				
Thickness	X	X	X	X
Facial dimension	X	X	X	X
Spacers	X			
Warpage	X	X	X	X
Wedging	X	X	X	X
Color uniformity	X			
Electrical properties		X		
Destructive tests				
Water absorption	X	X	X	X
Crazing	X	X	X	X
Thermal shock	X	X	X	X
Bond strength	X	X	X	X
Breaking strength	X	X	X	X
Abrasive hardness		X	X	X

Table 4 Selected technical features of vitrified, half-vitrified, and porous ceramic tiles

Test method standard	Property	Tile type/product requirement standard			
		BI (EN 176)	BIIa (EN 177)	BIIb (EN 178)	BIII (EN 159)
EN 100	Bending strength, MPa (ksi)	≥27 (≥3.9)	≥20 (≥2.9)	≥16 (≥2.3)	≥12–15 (≥1.7–2.2)
EN 101	Scratch hardness (Mohs scale)				
	Glazed tiles	≥5	≥5	≥5	...
	Unglazed tiles	≥6	≥6	≥6	...
	Wall tile	...	...	...	≥3
	Floor tile	...	...	...	≥5
EN 102	Abrasion resistance, mm³(a)	≤205	≤345	≤540	...
EN 103	Coefficient of thermal expansion at 20 to 100 °C	≤9 × 10⁻⁶ K⁻¹	≤9 × 10⁻⁶ K⁻¹	≤9 × 10⁻⁶ K⁻¹	≤9 × 10⁻⁶ K⁻¹
EN 155	Steam expansion, mm/m	...	...	≤0.6	...

(a) The abrasion resistance for unglazed tiles. For glazed tiles, the abrasion resistance is specified by the manufacturer.

Glazes

Glazes are thin, glassy coatings usually between 0.15 and 0.50 mm thick that are formed in place on a ceramic body. After blending the raw materials, the glaze mixture is spread on the surface and fired at high temperature. Glazes are usually applied to make the body nonporous, smooth, glossy, mechanically stronger, and chemically resistant. Glazes also improve the aesthetic appearance of the ceramic ware. A review of glaze materials and properties can be found in Ref 11. See also the article "Glazes and Enamels" in this Volume.

The composition of a glaze is adjusted to:

- Ensure adhesion to the surface of the ceramic body
- Improve some mechanical and physical properties
- Obtain different aesthetic effects (glossy glaze, matte glaze, satin glaze, opaque glaze)
- Provide color (Ref 12)

Tile Types

Technical and aesthetic characteristics of tile materials determine how they may be functionally classified (Ref 13).

Wall tiles (both medium and high porosity) are classified as decorated wall tiles, faience tiles, stove tiles, cottoforte, porous single-fired tiles, and exterior wall tiles. Floor tiles, which are low-porosity materials, include quarry tiles, stoneware tiles, klinker tiles, and mosaic tiles. Porcelain tiles are very low-porosity materials.

Wall Tiles

Wall tiles are mass-produced thin tiles with a porous body and a white or colored glossy or matte glaze. A wide variety of shapes and sizes are available and the great majority of the tiles are square or rectangular (Ref 14).

Wall tiles are used in domestic and institutional kitchens, bathrooms, washrooms, and swimming pools, as well as in decorative work such as tabletops and fireplace surrounds. Wall tiles are not used in places that are exposed to frost attack or severe abrasion.

The aesthetic appeal is the most important feature for a wall tile product used for interior decoration. As far as the technical characteristics are concerned, the tiles for wall decoration should match the following criteria:

- Excellent dimensional stability in the manufacturing process, usually obtained using body compositions with very low firing shrinkage (less than 0.5%). All wall tile product must meet these requirements
- Very narrow dimensional range to ensure easy setting and vertical applications
- The glaze must be easy to clean with no dust retention because one of the most important requirements in bathroom or kitchen applications is hygiene

Mechanical strength, abrasion resistance, impact resistance, and frost resistance are much less important for interior wall tiles than for floor tiles.

Decorated Wall Tiles. Earthenware and common pottery are some of the terms applied to fairly fine or even fine porous bodies manufactured in a wide range of decorated wall tile products.

Faience tiles are decorative tiles produced in a number of shapes with relief designs and in a wide range of sizes. The glaze is opaque and sometimes mottled. Often matte is used as a decorative treatment for walls. Other applications include window sills and fireplace surrounds.

Stove tiles (majolica, earthenware) are large and thick decorative tiles of considerable porosity that possess good thermal shock resistance and that are thermally efficient (both thermal insulation and heat storage). Although made of inexpensive raw materials, the surface is made attractive with engobes, clear or opaque glazes, and decoration.

Cottoforte include glazed semivitreous products for internal applications. This low-porosity product may be used for floor applications.

Porous single-fired ("Monoporosa") tiles are used for coverage of interiors and are suitable substitutes for classic tilings (faience, majolica). Unlike the vitrified or semivitrified single-fired tiles that occupy a different segment of the market and find a different use, this porous product finds application at a medium-high aesthetic level, with relatively contained costs (Ref 15). Table 6 gives data concerning the classification of the product according to EN specifications and the average properties of the products as compared with those of traditional majolica and cottoforte.

In view of its modulus-of-rupture values and the types of glaze applied (the product is always glazed), porous single-fired tiles can also be used as floor tiles.

Exterior wall tiles are glazed tiles with a vitreous or semivitreous tile body that resists

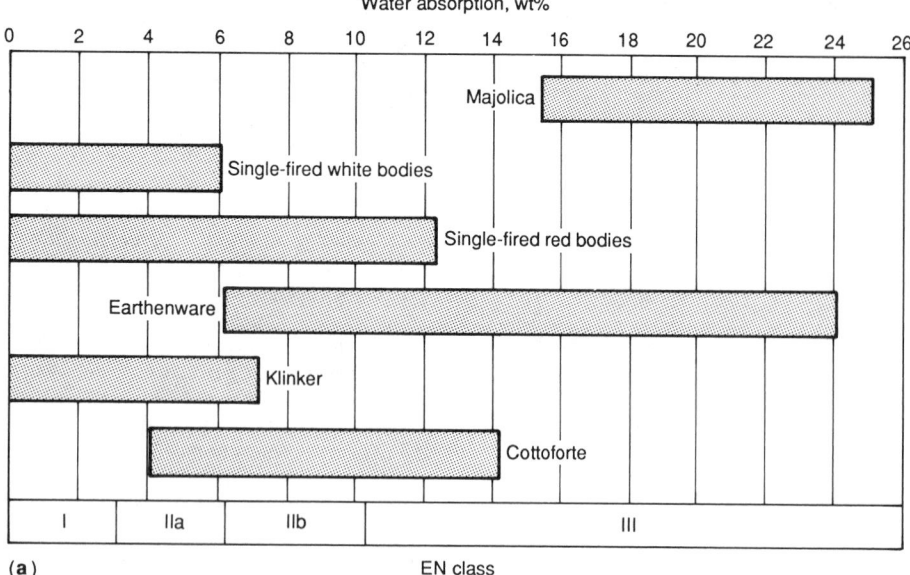

(a)

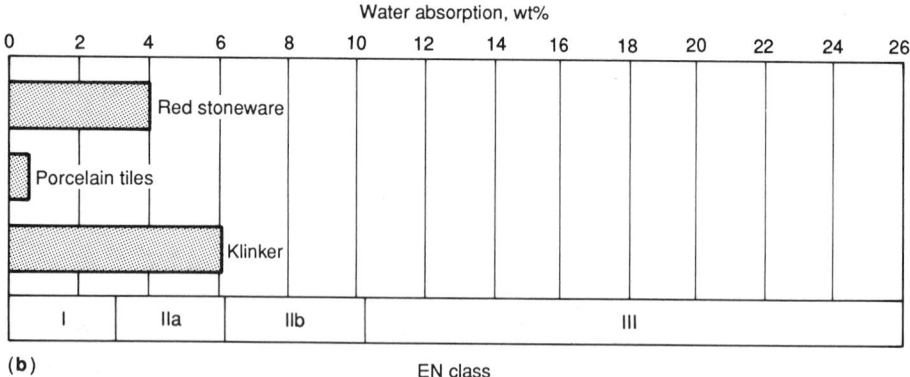

(b)

Fig 2 Water absorption characteristics of glazed (a) and unglazed (b) ceramic tiles. Source: Ref 7

Table 5 Common shapes and sizes of ceramic floor and wall tiles. Dimensions are given in centimeters. 1 cm = 0.4 in.

Square		Rectangular	
Floor tiles	Wall tiles	Floor tiles	Wall tiles
10 × 10	5 × 5	7.5 × 13.3	7.5 × 15
15 × 15	10 × 10	7.5 × 15	10 × 20
20 × 20	15 × 15	10 × 20	10 × 15
21.6 × 21.6	20 × 20	15 × 20	20 × 30
25 × 25		16.5 × 33	20 × 25
30 × 30		15 × 30	30 × 40
31.6 × 31.6		20 × 25	
33 × 33		20 × 30	
40 × 40		20 × 33.3	
50 × 50		20 × 40	
60 × 60		30 × 40	

Stoneware Tiles. This class comprises low-porosity products with a structure that consists of a glassy matrix with some crystalline phases (quartz is most prevalent with some mullite). Stoneware products can be glazed or unglazed and are used for exterior applications or applications in cold rooms (frost-resistant stoneware). Some varieties are:

- Classical stoneware (glazed or unglazed floor tiles)
- Chemical stoneware (glazed or unglazed chemical-resistant tiles)
- Fine stoneware (high-quality tiles that are normally glazed)

The single-fired stoneware products on the market have superior properties than specified by EN standard 176 (see Table 4). These include:

- Water absorption, 1 to 3 wt%
- Bend strength, ~40 MPa (5.8 ksi)
- Scratch hardness, 6 Mohs scale
- Abrasion resistance, ~180 mm^3
- Coefficient of thermal expansion at 20 to 100 °C (70 to 212 °F), 6 to 8 × 10^{-6} K^{-1}

Klinker tiles are sintered products with uniform microstructure with low porosity, resistance to chemical and atmospheric agents, high resistance to compression, abrasion, and frost. Klinker tiles are generally produced by extrusion and are widely employed for paving sidewalks, workshops, and floors subjected to high wear and chemical agents. They may also be produced in the glazed condition.

Mosaic tiles are small dense multicolored tiles that come in square, oblong, hexagonal, rhomboidal, diamond, and triangular shapes. These are either unglazed or glazed. There is a wider range of colors in the glazed type. Mosaic tiles are used as decorative wall finishers.

Porcelain Tiles

Porcelain tiles are prepared using kaolin or pure kaolinitic clays, quartz, and high amounts of feldspars as fluxing agents to increase sintering. The raw materials are more highly milled than those used for other types of products and the tile body is fired at about 1200 °C (2190 °F).

---

frost action. They are generally larger and thicker than interior tiles.

Floor Tiles

Ceramic floor tiles must be not only attractive and aesthetic but also strong. Floor tiles should have:

- *High mechanical strength* (breaking resistance, also referred to as modulus of rupture). Tile thickness must also be considered when determining mechanical strength
- *High abrasion resistance* is related to the hardness of the surface, the gloss of the surface, and the presence of cavities. A shiny product is intrinsically less resistant than a matte product with the same surface hardness. If the surface reveals pores after abrasion, they will retain the dirt and the tile becomes difficult to clean
- *High impact resistance* is related to the strength of the body and the thickness of the glaze. Unglazed porcelain tiles are the

best from this point of view. Tiles with a thick, dense glaze, such as those manufactured by hot glazing, display optimum properties

- *Frost resistance* is more important for floor tiles than for wall tile in exterior applications

Dimensional stability is less important for floor tiles than for wall tiles. It is easier to compensate for small changes in dimension in laying tiles on the floor. These materials are dense high-vitrified tiles. Many are smooth surfaced and some slightly self-glazed. Others are made with additions of abrasive grain or are produced with a corrugated surface. Floor tiles must also resist weathering and staining.

Quarry tiles are thicker and frequently larger than other floor tiles and have a slightly coarser texture. Their shapes are precise, and they are used as floors in kitchens, dairies, slaughterhouses, public buildings, balconies, terraces, porches, yards, and sills, and also as wall linings.

Table 6 Property comparisons of several types of ceramic wall tiles

Test method standard	Property	Group BIII (EN 159)	Porous single-fired	Cottoforte	Majolica
		Tile type			
EN 99	Water absorption, E, wt%	$E > 10$	$E = 13$ to 18	$E = 16$ to 19	$E = 18$ to 24
EN 100	Bending strength, MPa (ksi)	≥ 12 (≥ 1.7)	≥ 20 (≥ 2.9)	≥ 15 (≥ 2.2)	≥ 13 (≥ 1.9)
EN 101	Scratch hardness (Mohs scale)				
	Wall tiles	≥ 3	4	4	4
	Floor tiles	≥ 5	5	5	5
EN 154	Abrasion resistance(a)		Specified by the manufacturer		
EN 103	Coefficient of thermal expansion at 20 to 100 °C	$\leq 9 \times 10^{-6}$ K^{-1}	7 to 7.5×10^{-6} K^{-1}	6.5 to 7×10^{-6} K^{-1}	7 to 7.5×10^{-6} K^{-1}

(a) Glazed tile requirement

Table 7 Selected properties of porcelain tile

Standard	Property	Value prescribed by the standards	Real value of the products
EN 99	Water absorption, E, wt%	≤ 0.5	< 0.2
EN 100	Bending strength, MPa (ksi)	≥ 27 (≥ 3.9)	≥ 50 (≥ 7.25)
EN 101	Scratch hardness (Mohs scale)	≥ 6	7/8
EN 102	Abrasion resistance(a)	≤ 205 mm^3	< 130 mm^3
EN 103	Coefficient of thermal expansion at 20–100 °C	$\leq 9 \times 10^{-6}$ K^{-1}	$< 7 \times 10^{-6}$ K^{-1}

(a) Unglazed tile requirement

The tile is not glazed and may be colored. Its structure consists of a glassy matrix with mullite as the prevalent crystalline phase. Table 7 provides data on the classification and property values associated with porcelain tile products (Ref 16).

Porcelain tiles are used for antistatic environments where hygiene is paramount, as they combine sufficient cleanliness with suitable electrical properties. Since the product is unglazed, it may in certain cases be subject to dirt and staining if contaminants are not removed before they penetrate to any significant depth.

Emerging Materials

The change in firing technology from the traditional double firing to fast single-firing processes has enabled some ceramic tiles to be produced with a glass-ceramic glaze (Ref 17). Such dense coatings, which have a high percentage of crystal phase without voids in the glaze, allow the surface to be abraded without penetration of dirt in surface pores and discoloration with time.

Disadvantages of Tiles

Although tiles may have no obvious defects (either dimensional inconsistencies or visible cracks), they can be damaged when objects are dropped on them. Because of small variations in the composition of the raw materials

and industrial process, some tiles may be weaker than others. In addition, some types of tiles can suffer alteration by chemicals or may rupture due to exposure to cold or extreme fluctuations of temperature. It should be noted, however, that unacceptable service performance is due primarily to either selection of the wrong type of material or poor fitting characteristics, rather than defects in the product itself.

Alternative/Competitive Materials

The most important alternative materials to ceramic tiles in the building industry are natural stone, wood, moquette, and linoleum. With respect to the above-mentioned materials, ceramic tiles are easier to clean, more hygienic, easy to fit, and superior physically and mechanically. Alternative materials also have the following limitations:

- *Natural stone* has high absorption (difficult to polish), is very expensive, and must be smoothed
- *Wood* is expensive, is slippery, is subject to insect/worm infestation, fireproof, absorbs water, is easy to scratch
- *Moquette* is not very hygienic, holds dirt, burning generates toxic gases, is more expensive than ceramic floor, and has a short service life

- *Linoleum* has a low Mohs hardness (2 to 3) and burns very easily

REFERENCES

1. G. Biffi and E. Mas, Italia, Brasile e Spagna: Paesi Maggiori Produttori di Piastrelle Ceramiche a Confronto, *Ceramurgia*, Vol 17 (No. 5), 1987, p 171–182
2. A. Russel, Whiteware Developments. A European Perspective, *Int. Ceram. J.*, Vol 45, 1990, p 32–39
3. M. Drews, Wall and Floor Tiles, *Ceramic Monographs—Handbook of Ceramics*, Verlag Schmid GmbH, Freiburg, 1983
4. UNI "Piastrelle di Ceramica per Rivestimento di Pavimenti e Pareti," Norme Europee, UNI, Milano, 1985
5. G. Timellini, Caratteristiche Prestazionali delle Piastrelle Ceramiche, *Ceramurgia*, Vol 17 (No. 5), 1987, p 165–170
6. G. Nassetti, Thermal Energy Consumption in the Production of Single Fired Red Ware. Tendencies and Considerations, *Interceram*, Vol 34 (No. 4), 1985, p 40–41
7. C. Palmonari, "Le Piastrelle di Ceramica-Guida all'impiego," EDI.CER. Sassuolo, 1984
8. A.E. Benlloch, J.L. Amoros Albaro, and J.E.E. Navarro, Estudio de Gres para Pavimento, *Boll. Soc. Es. Ceram. Vid.*, Vol 20, 1981, p 17–24
9. R.C.P. Cubbon and J.R. Till, Preparation of Ceramic Bodies, *Ceramic Monographs—Handbook of Ceramics*, Verlag Schmid GmbH, Freiburg, 1983
10. A. Brusa, M. Dardi, and A. Bresciani, Single Fired Stoneware Floor Tiles, *Int. Ceram. J.*, Vol 45, 1990, p 22–31
11. S. Stefanov, Ceramic Glazes and Frits, *Ceramic Monographs—Handbook of Ceramics*, Verlag Schmid GmbH, Freiburg, 1983
12. R.A. Haber, K.R. Wilfinger, and R.A. McCauley, Stains and Coloring Agents, *Ceramic Monographs—Handbook of Ceramics*, Verlag Schimd GmbH, Freiburg, 1983
13. G. Ruini and J.V. Arnal, La qualità estetica della Piastrella, *Ceram. Inf.*, Vol 293, 1990, p 461–464
14. F. Singer and S.S. Singer, *Industrial Ceramics*, Chapman and Hall, London, 1963, p 1075–1084
15. A. Brusa, M. Dardi, and R. Raccagni, Rivestimenti e Pavimenti Porosi Monocotti, *Ceram. Inf.*, Vol 288, 1990, p 146–159
16. D. Gardini, Gres Fino Porcelanico, *Ceramica Informacion*, Vol 153, 1990, p 12–18
17. T. Manfredini and G.C. Pellacani, *et al.*, LiO$_2$-Al$_2$O$_3$-SiO$_2$-Me(II)O Glass Ceramic Systems for Tile Glaze Applications, *J. Am. Ceram. Soc.*, Vol 74 (No. 5), 1991, p 983–987

Ceramic Whiteware

Russell K. Wood, American Standard Inc.

WHITEWARES is the name given to a group of ceramic products characterized by a white or light colored body with a fine-grained structure. Most whiteware products are glazed—in whole or in part—and whiteware glazes may range from clear to completely opaque, white, or colored. Many whiteware products, such as tableware, are decorated with patterns or designs to enhance their beauty and appearance. Examples of whiteware products are sanitaryware, tableware, electrical porcelain, artware, stoneware, and tile. This article will be concerned only with the first three groups of products.

Whiteware Body Materials

Most ceramic whitewares are made from a combination of plastic, nonplastic, and fluxing materials, with some usage, depending on the application, of other materials such as bone ash, talc, limestone, dolomite, wollastonite, and pyrophyllite. Plastic materials include clays (ball and china) and kaolins. Nonplastic materials are silica in one form or another and occasionally alumina. Feldspar, feldspathic sands, and nepheline syenite are examples of fluxing materials.

Clays are needed in a whiteware body to develop plasticity and workability for the unfired body, permitting it to be shaped or cast into the desired article. Clays also serve as the starting point, during firing, of the desired crystalline structure of the fired body. Clays can be classified into three general groups: ball clays, china clays, and kaolins.

Ball clays are characterized by moderate kaolinite content, some organic material such as lignite, considerable silica in the form of quartz (sand) in varying degrees of fineness, illite, montmorillonite, and other minerals such as mica. As a group, ball clays tend to be fine grained. This characteristic, together with the organic and mica content, makes these clays plastic and workable with good strength when dry. Ball clays are always sedimentary clays, that is, they have been formed in one location and transported, usually by water, to another location where they settle, along with sand, vegetation, and other materials. As the deposit ages, the decaying vegetation forms complex organic compounds which, when dispersed in the clay, contribute significantly to the properties of the clay. Ball clays, depending on their mineral composition, burn from a cream to brown or gray color and therefore can be used only to a limited extent in bodies where whiteness and translucency are critical. While many ball clays can be used as mined, producers today are offering blends and combinations of clay (either in dry or slurried form) which have more stable and predictable properties in addition to convenience in handling when in the slurry state.

Table 1 shows the composition of several ball clays used in ceramic whitewares. The particle size distribution of ball clays is a critical characteristic affecting behavior in a ceramic body and representative distribution curves are shown in Fig 1.

China clays are made up primarily of the minerals kaolinite and mica, and occasionally significant amounts of halloysite. They have little if any organic content. China clays are mined extensively in southwestern England where large deposits occur as the result of weathering of pegmatites; china clays are also found to a limited extent in the United States (primarily in North Carolina). China clays contribute moderately to the plasticity of a whiteware body, but their usefulness is more that of a white-burning filler. Their unfired strength is very low, and so they contribute little if anything to the unfired strength of the ceramic body. China clays are classified as primary clays, that is, they are found in the location where they are formed. The as-found ore almost always requires processing (usually wet) to remove sand, partially weathered pegmatite and feldspar, coarse mica, and so on, before the clay is suitable for use.

Kaolins are composed mainly of the mineral kaolinite, with small amounts of montmorillonite silica, mica, and other minerals. They differ from china clays mainly in the absence of significant amounts of anything other than kaolinite. Kaolins occur extensively in many parts of the world; United States deposits are concentrated mainly in Georgia and South Carolina.

Kaolins are sedimentary in formation, that is, they have been formed in one location and transported to another, usually by water. Unlike ball clays, most commercial deposits of kaolins have not acquired significant amounts of other minerals (other than silica) or organic matter during their formation; occasionally small amounts of mica and montmorillonite will be present. As a result they will have a white or cream fired color. Kaolins, because of the absence of organic matter and minor contents of "binding" minerals such as montmorillonite, have very little strength when dry, and their contribution to a whiteware body, like china clays, is more that of a filler which improves the whiteness of the fired article and serves as the starting point for the desired crystalline makeup of the fired body.

Table 2 shows the chemical composition of several china clays and kaolins and Fig 2 the particle size distribution of representative materials.

Nonplastic materials are predominately silica in one form or another. Much of the chemical silica in a ceramic body is supplied by the clays and fluxing materials, but in most instances the formula for the body will call for direct quartz (often referred to as "flint") additions. While a chemical analysis of the unfired body will show a large amount of silica, much of this is combined silica from clays and fluxes. Occasionally alumina is added to the body, partially or totally replacing quartz, where properties such as thermal expansion and strength must be enhanced. Quartz functions in the body as a filler and as a modifier of fired-body properties such as thermal expansion. A paste of quartz powder has little strength when dry, and quartz contributes nothing to the plasticity of the ceramic body or to the strength of the body when dry (hence the classification as a "nonplastic"). It helps control the firing warpage of the body by forming a relatively nonfusible network of grains, counteracting to some extent the effect of feldspar.

Fluxing materials include such materials as feldspar, feldspathic sand, nepheline syenite, spodumene, and so on. The function of these materials is to form a glass in the body during firing which acts as a bond for other materials present, or which will dissolve other materials and allow formation of crystalline products in the body which enhance the fired properties of the body. The best example of the latter is the formation of mullite and a silica-rich glass in the body as the

Table 1 Typical ball clay compositions

	Location/content, %					
Material	Kentucky (US)	Kentucky (US)	Tennessee (US)	North Devon (UK)	South Devon (UK)	Dorset (UK)
Compound						
SiO$_2$	55.6	51.6	61.0	63.0	52.0	54.0
Al$_2$O$_3$	28.6	28.6	24.5	24.0	30.0	30.0
Fe$_2$O$_3$	1.0	0.9	1.0	0.9	1.2	1.4
TiO$_2$	1.8	1.7	1.3	1.5	1.0	1.2
CaO	0.1	0.1	0.1	0.2	0.2	0.3
MgO	0.4	0.5	0.1	0.4	0.4	0.4
K$_2$O	1.1	0.5	1.7	2.4	2.1	3.1
Na$_2$O	0.1	0.2	0.4	0.5	0.3	0.5
Ignition loss, %	11.4	15.9	9.7	6.8	12.2	8.8
Mineral						
Kaolinite	...	...	...	45.0	64.0	57.0
Mica	...	...	...	28.0	22.0	32.0
Quartz	...	...	...	26.0	11.0	11.0
Feldspar	...	...	...			
Other	...	...	...	1.0	3.0	...

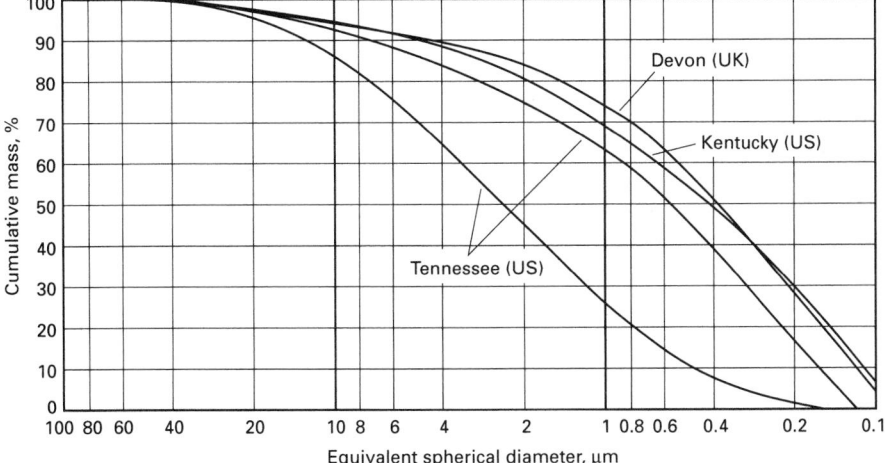

Fig 1 Particle size distribution for ball clays

Table 2 Typical china clay and kaolin compositions

	Location/content, %					
Material	Georgia (US)	North Carolina (US)	South Carolina (US)	Florida (US)	Cornwall (UK)	Cornwall (UK)
Compound						
SiO$_2$	44.4	47.3	44.6	46.1	48.0	50.2
Al$_2$O$_3$	39.6	36.4	39.5	38.8	36.9	34.1
Fe$_2$O$_3$	0.4	1.0	0.5	0.7	0.7	0.8
TiO$_2$	1.7	0.1	1.4	0.4		0.1
CaO	...	...	0.1	0.2	0.1	
MgO	...	...	0.1		0.3	0.3
K$_2$O	0.1	1.3	0.5	0.3	1.6	4.1
Na$_2$O	0.1	0.3			0.1	0.2
Ignition loss, %	13.7	13.6	13.3	13.5	12.3	10.2
Mineral						
Kaolinite	...	...	...	...	84.0	66.0
Mica	...	...	...	...	13.0	23.0
Quartz	...	...	...	...		
Feldspar	...	...	...	...	1.0	9.0
Other	...	...	...	...	2.0	2.0

result of kaolinite, having broken down during heating into primary mullite and quartz, dissolving in the feldspathic glass to form secondary mullite whose needle-like crystals contribute to the strength and other properties of the fired body. Table 3 lists the chemical composition of several fluxing materials for ceramic whitewares.

Fluxing materials are widely dispersed throughout the world. Both Canada and Norway have large deposits of nepheline syenite; in the United States large deposits of feldspar and feldspathic sand occur in North Carolina.

Other materials used in ceramic whiteware bodies are talc, wollastonite, bone ash, limestone, pyrophyllite, and dolomite. Talc and wollastonite are widely used in ceramic tile. Bone ash is the major ingredient in bone china. Limestone and dolomite are occasionally used as auxiliary fluxes in whiteware bodies. Pyrophyllite is sometimes used as a partial replacement for silica. The composition of several of these materials is shown in Table 4.

Whiteware Glaze Materials

Ceramic glazes are nothing more than glasses applied to a ceramic substrate. Glazes serve several functions:

- They provide a smooth, easily cleanable surface
- They enhance the appearance of the ceramic through the use of texture and color
- They conceal or mask undesired body qualities
- They provide protection for decorations that are applied underneath the glaze
- They improve the electrical properties of the ceramic over which they are applied

While there are many materials common to bodies and glazes (feldspar, silica and kaolin, for example) the family of glaze materials includes many that are unique to glazes only. This includes zinc oxide, barium carbonate, calcium carbonate, zircon and zircon-based compounds, ceramic pigments, lead compounds, and a variety of frits that are glasses which render soluble and toxic materials harmless and, in addition, help the appearance of the glaze by the "prefusion" of these materials. Glaze oxides such as soda, potash, calcia, lithia, strontia, boric oxide, and so on, are either glass formers of themselves, or act as modifiers for the silica lattice in the glaze. They are formulated to develop the desired maturing temperature and properties of the glaze such as thermal expansion and hardness.

Whiteware Manufacturing

While most whiteware products have very similar processing, one difference is in the way the product is formed. For example, almost all sanitaryware is cast. Tableware is

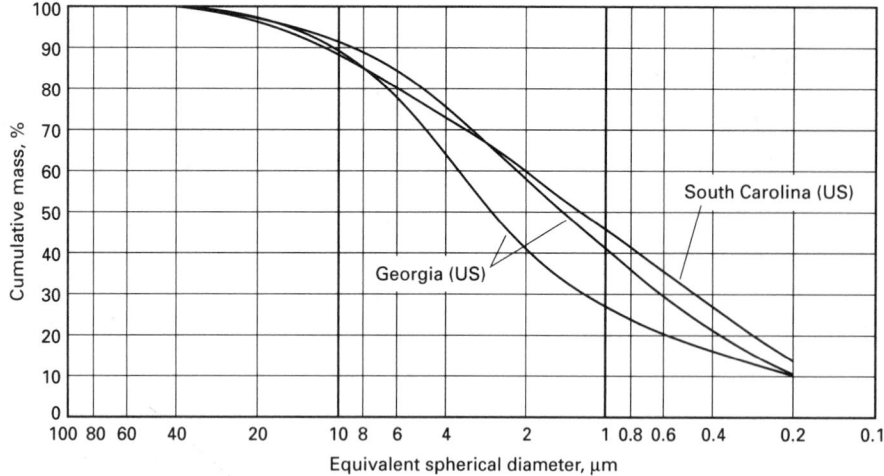

Fig 2 Particle size distribution for kaolins

usually "jiggered," that is, formed from a plastic mass, or pressed from powder or plastic mass, or cast. Electrical porcelain can be pressed, either dry or plastic, or extruded and then turned (machined) to the final shape. Tile is dry pressed or extruded. Artware and stoneware are usually formed from plastic mass or cast. Another difference in processing concerns the number of firings the product receives. Sanitaryware, for example, has only one firing (other than refiring to repair cos-

metic defects). Tableware (china) has two firings; the first where the body is matured and the desired porosity (if any) is developed, and the second where the glaze is fused onto the body. There may be a third firing to repair glaze defects. Hard porcelain also has two firings, but the first (bisque) firing is at a relatively low temperature that is sufficient to develop adequate strength in the product so that it can be handled. The second firing is at a much higher temperature which matures

the body and the glaze at the same time. In addition, especially with fine china or other highly decorated ware, there may be additional firings after that of the glaze to fuse the decoration onto or into the glaze. Electrical porcelain always has only one firing.

A simplified flow sheet for the production of whitewares is shown in Fig 3.

Fired Properties

Ceramic whitewares are sometimes called "triaxial" bodies, referring to the usual makeup of clays, nonplastics, and fluxes. Triaxial can also refer to the chemical composition of the fired product where the system is alkali-silica-alumina (ignoring minor constituents such as iron, titania, calcia, magnesia, and so forth). Whitewares in the traditional sense will fall within this system. In addition, most whiteware compositions lie within a relatively small compositional range in this system, as shown in Fig 4. Within this range, products can be distinguished by such physical properties as absorption, translucency, cleanability, compressive strength, thermal expansion, thermal conductivity, resistance to crazing, impact strength, hardness, acid resistance, thermal shock resistance, dielectric strength, power factor, and oxide composition. It is normal to change the oxide composition to develop the desired properties, but it should be remembered that the manufacturing process greatly affects these properties as well.

The physical properties of whitewares are listed in Table 5

If the list of properties were classified as to their importance to sanitaryware, cleanability, compressive strength, resistance to crazing, absorption, and thermal shock resistance would be included. For tableware, cleanability, translucency, resistance to crazing, impact strength, absorption, acid resistance, and thermal shock resistance are important. For electrical porcelain, compressive strength, impact strength, absorption, dielectric strength, and power factor are basic considerations.

Sanitaryware

The word sanitaryware, in its broadest sense, would include all products associated with residential and commercial bathrooms and facilities. Following this definition, such things as bathtubs (porcelain enameled and plastic), accessories such as towel bars and soap dishes, and shower surrounds would be included. In normal usage, however, sanitaryware refers to a group of products made from vitreous china which are used in residential or commercial bathrooms. This group includes lavatories and washbasins, toilets, bidets, urinals, and shower receptors.

Strictly speaking, china refers to the process wherein the body is fired by itself, followed by a second firing at lower temperature for the glaze. China is distinguished from

Table 3 Typical flux compositions

Compound	Location/content, %				
	Feldspar (US)	Feldspar (US)	Feldspar (Norway)	Nepheline syenite (Canada)	Cornish stone (UK)
SiO_2	68.2	66.3	69.2	60.7	79.5
Al_2O_3	18.9	18.6	18.7	23.3	12.0
Fe_2O_3	0.1	0.1	0.1	0.1	0.1
TiO_2	...	...	...	...	Trace
CaO	1.6	1.0	1.8	0.7	0.2
MgO	Trace	Trace	...	0.1	0.1
K_2O	4.5	10.5	2.8	4.6	3.8
Na_2O	6.7	3.2	7.2	9.8	3.9
Ignition loss, %	0.1	0.2	0.2	0.7	0.5

Table 4 Typical compositions of miscellaneous materials used in ceramic whitewares

Compound	Location/content, %				
	Talc (New York, US)	Talc (Montana, US)	Pyrophyllite (US)	Wollastonite (US)	Spodumene (US)
SiO_2	56.3	62.6	77.9	50.1	64.4
Al_2O_3	0.3	0.3	16.1	1.7	27.7
Fe_2O_3	0.1	1.5	0.2	0.3	0.5
TiO_2	...	Trace	0.1	...	...
CaO	7.0	Trace	0.1	44.0	0.1
MgO	30.7	30.3	0.1	1.5	...
K_2O	...	0.1	1.9	...	0.3
Na_2O	...	0.2	0.3	0.2	0.5
Li_2O	...	...	...	...	6.0
Ignition loss, %	5.6	5.0	3.2	2.2	0.5

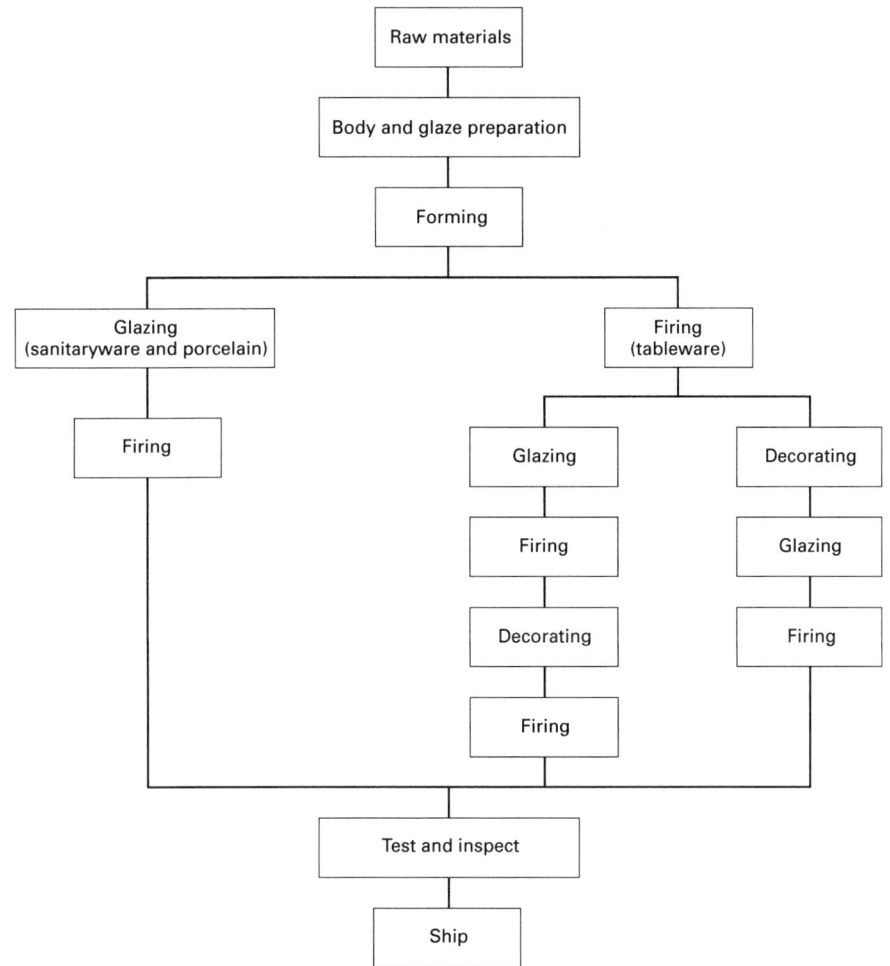

Fig 3 Flowchart of whiteware production

porcelain where the first firing does not vitrify the body, and the second firing at a higher temperature not only matures the glaze but vitrifies the body as well. In fact, sanitaryware originally was made by the china or two-fire process. Today, sanitaryware production more closely resembles the process for porcelain, where the body and glaze are matured together in only one firing, but with no pre-firing. The term vitreous china is still used, however, by most producers and users.

Essential Properties. The basic requirements for any material to be used for sanitaryware are:

- Cleanability
- Mechanical strength
- Thermal shock resistance
- Low, or zero, porosity

Cleanability refers to the ease of cleaning the fixture and to the resistance of the surface to cleansing agents—both mild and harsh—with or without abrasive content.

Mechanical strength refers to the compressive, flexural and impact strength of the product which must be sufficient to withstand stress during installation and usage (usage

meaning a service life which may extend for scores of years).

Thermal Shock Resistance. The fixture must have sufficient thermal shock resistance to withstand temperature cycling, for example, filling a lavatory with hot tap water or steam cleaning a commercial urinal. The resistance of the fixture to thermal shock is a function not only of the material but of its design and manufacturing process.

Porosity. The requirement for low, or zero, porosity results in the fixture having good mechanical strength, resistance to moisture expansion which can cause glaze crazing and loss of strength, and resistance to staining and discoloration if the glaze should be damaged in service. Porosity has a great influence on long term durability of the product.

Another requirement, not listed above, is the ability of the material to be formed into the desired shape on an economical basis.

Development of the Industry. While sanitaryware has been produced for many years, the vitreous china segment of the industry can be divided into two areas of commerce—that of developed nations, and that of developing countries and areas. In the

former, industry production is greatly influenced by remodeling and renovation, general building activity (principally housing starts for residential building) and commercial construction, exports, and the availability of products made of competitive materials. In the latter, industry production is more dependent on residential and commercial building, and the extent of imports from other areas, particularly where local manufacturing is limited. For developing countries, remodeling and renovation are less of a factor. Fixtures with strictly utilitarian application are a significant part of production.

The current capacity of the industry in the United States today is on the order of 20,000,000 pieces, where "piece" means individual parts (a residential toilet, with attached tank, would be considered as two pieces). This figure has not changed significantly for several years. What has changed is the company structure and arrangement of the industry, where plant closings and consolidations have been more or less balanced by plant expansions. The trend generally is to expand existing facilities, rather than build completely new plants.

The dollar value of United States vitreous china sanitaryware shipments for recent years is shown in Fig 5.

Competitive Materials. Of the factors affecting industry activity, the availability of competitive materials has had some influence on production volume, particularly in countertop lavatories. The demand for large countertops with integral basin has increased the use of polyester or acrylic resins. Vitreous china in this application is at some disadvantage because of the difficulty in making large, dimensionally uniform pieces with thin sections at acceptable cost. While the synthetic materials allow these countertops to be made easily, they do not have the cleanability and durability of china.

The volume of countertop and under-the-counter basins, installed on or under countertops of other materials, has been influenced by products made of porcelain enameled cast iron or steel, polyester, and Melamine. In spite of the availability of these competitive materials, the volume of vitreous china articles has been maintained, and this is attributed to some extent by the superior appearance and durability of china.

Vitreous china continues to dominate the market for toilets, bidets, and urinals, not only because of its cleanability and durability but because the production method of forming (slip casting) allows economical production of complex fixtures with extensive internal arrangements such as trapways. Occasionally a toilet bowl will be made of a combination of stainless steel and plastic (airplane toilets, for example) where weight and recirculating flushing media are important. There have been a limited number of all plastic toilets manufactured, but they have had modest acceptance in the market.

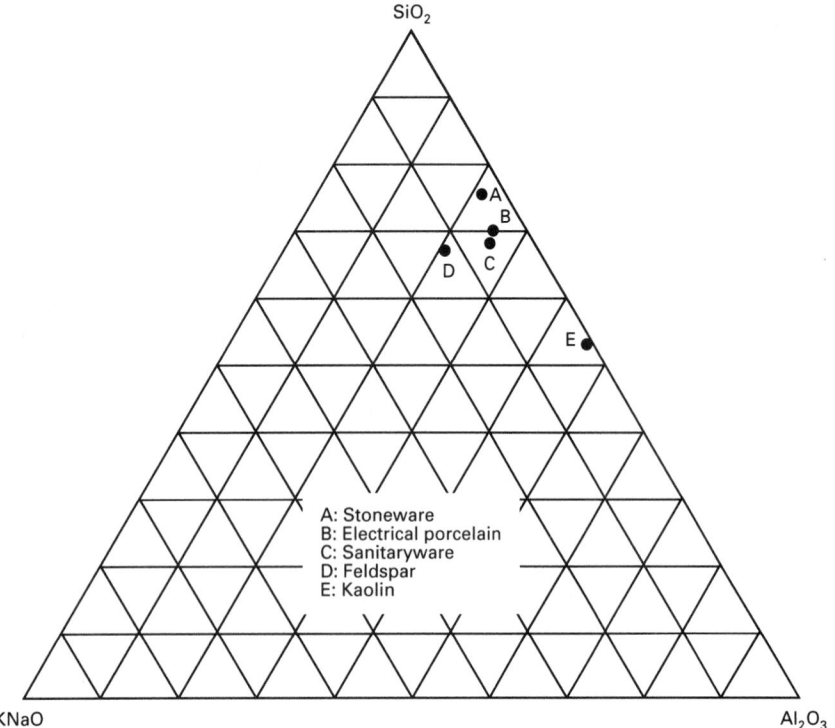

Fig 4 The alkali-alumina-silica system

A: Stoneware
B: Electrical porcelain
C: Sanitaryware
D: Feldspar
E: Kaolin

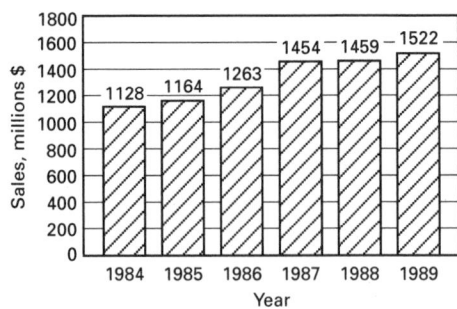

Fig 5 Annual sales of sanitaryware

Future Development. There is not much likelihood of a major change in materials used for sanitaryware or in the nature of the product itself. At the same time, the use of a traditional vitreous china sanitaryware body for other purposes is not foreseen. Major changes in manufacturing have and will continue to result in improvements in productivity of labor and utilization of facilities. Fixture development has been directed toward fixtures, both residential and commercial, with reduced water requirements. Many years ago, water consumption for a residential toilet was commonly 16 to 20 L (4 to 5 gal). This was reduced throughout the industry to 12 to 14 L (3 to 3.5 gal). More recently, many toilets operate on 6 L (1.6 gal) of water. The 6 L (1.6 gal) standard being adopted today by many localities has required major changes in trapway design and geometry.

Fixture design and appearance changes continue, with continued emphasis on more attractive fixtures in a wide variety of colors. A number of fixtures are available today with decorative effects and patterns.

Tableware

The word tableware includes a wide variety of products, including china (or chinaware), porcelain, earthenware, semi-vitreous ware, bone china, hotel or restaurant china, and some varieties of stoneware. All of these products are in the ceramic whitewares group, and it is convenient to classify them as being vitreous or porous (semi-vitreous). On this basis, china, porcelain, bone china, and restaurant china would be classified as vitreous, earthenware classified as semi-vitreous ware, and stoneware classified as porous.

Essential Properties. By the nature of its usage, tableware must have good resistance to impact, cutlery marking, and abrasion, be easily cleaned, resistant to acids and mild alkalis, and reasonable in thickness and weight. With semi-vitreous ware, there is some sacrifice of these properties (particularly in impact resistance) to permit a lower-cost product. Because design and appearance are important factors in tableware's acceptance by the customer, the product must be capable of accepting the desired decoration.

Impact resistance is a function of both the body composition and the design of the product. A vitreous body is required and occasionally some alumina is substituted for quartz in the body to improve its strength. The design of the ware is important—very thin, lightweight pieces obviously will be less durable, even though this may be required to develop the desired aesthetics. Sometimes the design can be modified to enhance durability. In restaurant china, for example, a "rolled edge" on plates is often used to increase resistance to chipping.

Cutlery marking is a function of the surface and composition of the glaze. A bright, glossy glaze is by nature more resistant to marking than a dull or matte glaze. Some glaze oxides are well known for their harmful effect in lowering the resistance of the glaze to cutlery marking.

Table 5 Physical properties of whitewares

Property	Earthenware	Hard porcelain	Bone china	Hotel china	Normal electrical porcelain	High-strength electrical porcelain
Water absorption, %	6–8	0.0–0.5	0.0–1.0	0.1–0.3	0.0	0.0
Specific gravity	2.6	...	2.75	2.6	2.4	2.8
Bulk density, Kg/m³	2200	2400	2700	2600	2400	2770
Compressive strength, MPa (ksi)	...	400 (58)	...	...	700 (102)	700 (102)
Tensile strength, MPa (ksi)	...	23–34 (3–5)	...	...	35 (5)	56 (8)
Modulus of rupture, MPa (ksi)	55–72 (8–10)	39–69 (5.5–10)	97–111 (14–16)	82–96 (12–14)	105 (15)	175 (25)
Modulus of elasticity, GPa (psi × 10³)	55 (8)	69–79 (5.5–11.5)	96 (13.9)	82 (12)	...	...
Linear thermal expansion (μm/m·K) 20–500 °C	7.3–8.3	...	8.4	7.3–8.3	5.7	6.7
20–1000 °C	...	3.5–4.5	...	...	...	...
Thermal conductivity, W/m·K 20–100 °C	1.26	...	1.26	...	...	...
Dielectric constant at 1 megacycle	...	...	...	...	5.6	6.9
Power factor at 1 megacycle	...	...	...	...	0.8	0.7

Abrasion resistance is required simply because tableware in usage rubs and scrapes upon itself and in addition may be subject to abrasive cleansers. This is one disadvantage of porcelain tableware—the unglazed foot of a plate will tend to scratch or abrade the surface of the plate underneath.

Cleanability is another requirement for tableware, particularly for ware used in commercial applications such as restaurants and hotels. The glaze must be free of pinholes and other surface imperfections. Nonglazed areas must be minimized.

Acid resistance refers mainly to resistance to fruit and vegetable acids. Here the composition of the glaze, its maturing temperature, and the texture of the glaze all have an effect. Modern ceramic technology and manufacturing processes have virtually eliminated the possibility of glaze ingredients being leached by glaze contact with fruit or vegetable acids.

Development of the Industry. The tableware industry, particularly in the United States, has undergone significant changes as the result of imports and competitive materials. First of all, semi-vitreous tableware, made at one time in large quantities by U.S. potteries, is now mainly imported. Vitreous china tableware for hotel and restaurant usage continues to be produced, but here again imports and competitive materials have forced consolidations and mergers in the industry with some reduction in capacity. Fine china continues to be produced but a large volume of ware is imported; a small volume of bone china is made in the United States but most of the ware marketed is imported from England. Tableware accessories such as mugs, serving dishes, and candlestick holders are often from overseas producers.

The emphasis in tableware production in developed countries has been to install more and more mechanization to reduce the labor content in the ware. At all stages of production—body preparation, forming, firing, decorating, and so on—the effort to mechanize has resulted in the development of highly sophisticated machines and equipment available from a variety of manufacturers.

Figure 6 shows the value of tableware shipments for the past several years.

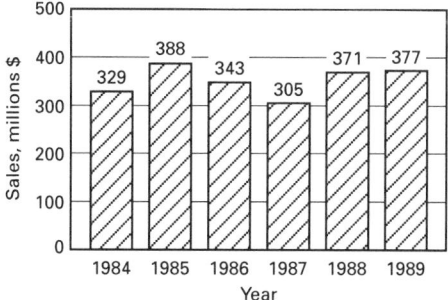

Fig 6 Annual sales of tableware

Competitive Materials. There are materials other than ceramic whitewares used for tableware today. Plastic ware, usually Melamine, is used for tableware for both households and restaurants. It has the advantage of being virtually unbreakable but is subject to staining and surface damage from knives and other sharp objects. Surface scratches and cuts are not desirable for ware in commercial service because of the difficulty in cleaning and sterilization. The material is not as easily wet (mildly hydrophobic) as a ceramic glaze, and this adds to the difficulty in cleaning.

Glass-ceramics are frequently used for tableware, both household and commercial. The process essentially is to shape the item as glass (pressed or blown), using a specially formulated composition. When the ware is reheated to a specified temperature, crystallization takes place with the result that the ware changes from a clear, transparent material to one that is fairly opaque. In addition, the mechanical strength is greatly enhanced, with the result that the strength of the final product is much greater than a product made of ordinary glass. A well known brand of glass-ceramics is Pyroceram. Since the product has no glaze as such, decorations are fused on the surface of the ware.

In addition to very good resistance to breakage, glass-ceramic surfaces resemble whiteware glazes in that they are hydrophilic. This greatly facilitates cleaning and maintaining a germ- and bacteria-free surface. Being a ceramic material, glass-ceramics are not as subject to scratching and cutting damage as are plastics.

Both plastic and glass-ceramic tableware have the advantage over whiteware products of greater mechanical strength and resistance to breakage. Whiteware manufacturers, particularly those supplying the commercial market, have developed body formulations with improved mechanical strength. There is a potential problem, however, in overdoing this in that the glaze may be abraded and scratched to the point of unserviceability before the ware is broken or cracked. Restaurants tend to discard only ware that is actually broken, not ware where the glaze has become unsightly. The fact that the body itself is vitreous, however, minimizes the danger of bacteria or germ contamination.

Whiteware products have a distinct advantage over competitive materials in the great variety of decorations that are possible, both over and under the glaze. Decorative processes such as silk screen and lithographic decals, air brush, intaglio, rubber stamping, banding and lining, and hand painting are all commonly used for tableware. In the case of bone china, fine china, and hard porcelain, the product has visual characteristics such as translucency and luster and texture of the glaze that are not available with competitive materials.

The possibility of still other materials significantly affecting the market for tableware

is remote. Conversely, it is difficult to imagine any of the present variety of tableware whiteware materials being used for other applications.

Future Development. Tableware has been, and continues to be, marketed on the basis of decoration and appearance, and to a lesser extent, durability and cost. Manufacturing efforts continue to improve productivity of labor and achieve better utilization of facilities. Radical change in the manufacturing process, however, is not anticipated.

Electrical Porcelain

Electrical porcelain is traditionally a dry- or wet-process material used for low-frequency electrical insulators for both low- and high-voltage service. This material is almost always a feldspathic porcelain, occasionally with added alumina in the body where the application calls for greater tensile strength. Porcelain insulators and bushings are always glazed to improve surface resistivity. Dry-press products, such as lamp sockets, fuse holders, tubes, and so forth, are usually unglazed.

The primary electrical usage today for wet-process porcelain is for suspension and pin-type insulators and high-voltage tubes and bushings. Figure 7 shows an example of high-tension porcelain insulators.

Essential Properties. Among the properties important for electrical porcelain, mechanical strength is critical because of the stresses involved in service. Suspension insulators, for example, must support considerable weight from transmission wire or cable. Good resistance to electrical stress is required

Fig 7 High-tension porcelain insulators. Courtesy of Swindell Dressler International

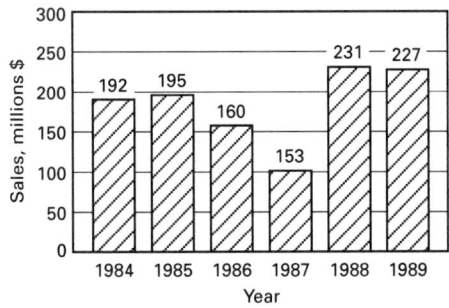

Fig 8 Annual sales of electrical porcelain

for insulators to avoid flashover and electrical leakage. Low porosity is essential for resistance to moisture penetration and for development of the desired mechanical strength. Resistance to thermal shock is required so that the insulator can withstand all types of weather without failure.

Development of the Industry. Electrical porcelain production is dependent upon the growth and development of the electrical generation and distribution industries. Construction of generating plants, installation of long-distance transmission lines and local distribution networks, and commercial building activity all influence the electrical porcelain industry. The use of competitive materials, such as glass and polymers, also affects porcelain production.

Sales of electrical porcelain (high-tension) in recent years is shown in Fig 8.

Competitive Materials. Both glass and polymeric materials are being used for electrical insulators. Polymeric materials have some advantage in strength and resistance to impact but currently their usage is restricted to smaller sizes of insulators.

Porcelain Enamel

Jeffrey F. Wright, Ferro Corporation
Clifton G. Bergeron, University of Illinois at Urbana-Champaign
John C. Oliver, Consultant, Porcelain Enamel Institute, Inc.

PORCELAIN ENAMEL can be defined as a very durable alkaliborosilicate glass coating bonded by fusion to a metal substrate at temperatures above 425 °C (800 °F). Porcelain enamel differs from organic paint coatings in its inorganic composition and the temperature at which the coating is fused to the metal substrate.

Porcelain enameling predates the birth of Christ, with the earliest articles being primarily decorative. Jewelry and art pieces produced in Egypt and other Middle Eastern countries back to the fourth century B.C. are found in museums throughout the world today.

Commercial porcelain enameling using sheet iron or steel and cast iron substrates began in the mid-1800s in Germany and Austria. By the end of the century several porcelain enameling plants had started up in the United States, and by 1910 some thirty to forty plants were in operation in the United States and Canada. Cooking utensils, tabletops, and signs made up much of the industry's early volume.

The period 1910 to 1940 saw the development of new, improved processes for making sheet iron and steel used for porcelain enameling.

Concurrently, commercial frit manufacturing companies were established to provide uniform, quality-controlled glass coating materials, known as frits, for use by the porcelain enameling industry. The steel producers and the frit manufacturers continued to use their research and development capabilities to provide improved base metals and coating materials for the industry. This 1910 to 1940 period marked the beginning of porcelain enamel's extensive application in products for the kitchen, bathroom, and laundry areas.

Traditional and newer porcelain enameling process involve grinding a mixture of frit, water, and various suspending agents to form a slurry of creamlike consistency that can be applied to the cleaned and prepared metal part by spraying, dipping, or flow coating. The applied coating is then dried to remove the water. In recent years, the new electrostatic dry powder process for application of porcelain enamel has emerged and is in wide use for high-volume production of similarly shaped components. For both systems, the metal with its coating is subjected to high temperatures (approximately 800 °C, or 1500 °F, for a steel substrate) to fuse the coating to a smooth, glossy surface and to bond the coating to the metal.

Utilizing the electrostatic dry porcelain enamel process, specially designed powder porcelain enamel systems have become commonplace for mass produced components such as range tops. Among its many advantages, the dry process uses up to 99% of all material, and little or no direct labor is required for the application operation.

In this dry process, the fluidized powder is propelled through the powder feed tubes, providing a steady, controlled flow of powder to the spray guns. The powder is directed toward the workpiece in the form of a diffused cloud. A high-voltage, low-amperage power unit supplies current to the charging electrode, causing powder to seek out and attach itself to the electrically grounded workpiece. The recovery booth collects and returns the powder that is not deposited on the workpiece. The powder is then recirculated through a closed loop system (with the use of multiple filters), and it is mixed with virgin powder before being pumped to the spray guns. Virtually none of the powder escapes into the environment.

Another major recent advance is the "clean only" metal preparation system now widely used throughout the industry. This simplified alkaline clean-rinse-neutralize method offers time and materials savings over the old clean-acid etch-nickel deposition process (Fig 1). It also greatly reduces the cost of materials used for metal preparation, environmentally related waste disposal costs, and maintenance costs.

Coating Materials

The frits for porcelain enamel are designed and manufactured to meet specific end-use applications and processing requirements. Porcelain enamel frits for sheet steel and cast iron are classified as either ground coat or cover coat types. Ground coat enamels contain metallic oxides that promote adherence to the metal substrate and may be used as a single functional coat or as a base coat for additional cover coats. Table 1 gives the composition of a typical ground coat enamel. Cover coat enamels (Table 2) are normally applied over ground coats for appearance and/or to improve the chemical and physical properties of the coating. Cover coats may also be applied directly to properly prepared decarburized steel substrates.

For aluminum, neither ground coats nor adherence-promoting oxides are required. Single-coat systems are used for most applications. Porcelain enamels for aluminum can be pigmented and opacified inorganically to produce the desired appearance.

Because all porcelain enamels are variations of borosilicate glass, they are characterized by end use and not by chemical composition. Some common designations for those with particular characteristics are acid-resistant, alkali-resistant, heat-resistant, glossy, low-gloss, and matte.

Metal Substrates

Porcelain enamels are applied primarily to components made of sheet iron or steel, cast iron, aluminum, or aluminum-coated steel.

Steel Substrates. Typical compositions of the various grades of low-carbon sheet iron or steel that are commercially available for porcelain enameling are listed in Table 3. Cold-rolled sheet steels used for porcelain enameling can be divided into three groups:

- Extra-low carbon steels (a maximum of 0.008% C), including those in which the carbon is stabilized by the addition of titanium or niobium
- Low-carbon steels containing about 0.02% C (these steels are suitable for ground or two-coat enameling
- Conventional cold-rolled sheets with higher carbon contents of about 0.06% (such sheets have a tendency toward primary boiling and sagging and are used in less critical ground coat and two-coat enameling applications)

Hot-rolled steels are generally used for porcelain enameled water heater tanks and for other applications where thickness and strength requirements dictate their use. When hot-rolled steels are used, components should be coated

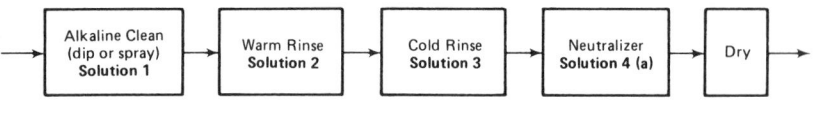

No.	Solution	Composition	Temperature °C	Temperature °F	Cycle time, min Dip	Cycle time, min Spray
1	Alkaline cleaner	Cleaner, 3.7-60 g/L (½-8 oz/gal)	Ambient to 100	Ambient to 212	6-12	1-3
2	Warm rinse	Water	49-60	120-140	½-4	½-1
3	Cold rinse	Water	Ambient	Ambient	½-4	½-1
4	Neutralize(a)	2/3 Na₂CO₃ and 1/3 borax, 0.60-2.10 g/L (0.08-0.28 oz/gal) as Na₂O	49-71	120-160	3-6	½-1

(a) Some systems may not require a neutralizer

Fig 1 Process for preparing sheet steel for porcelain enameling using the "clean only" system

Table 1 Melted-oxide compositions of frits for ground coat enamels for sheet steel

Constituent	Regular blue-black enamel	Alkali-resistant enamel	Acid-resistant enamel	Water-resistant enamel
SiO_2	33.74	36.34	56.44	48.00
B_2O_3	20.16	19.41	14.90	12.82
Na_2O	16.74	14.99	16.59	18.48
K_2O	0.90	1.47	0.51	...
Li_2O	...	0.89	0.72	1.14
CaO	8.48	4.08	3.06	2.90
BaO	9.24	8.59	...	...
ZnO	...	2.29	...	...
Al_2O_3	4.11	3.69	0.27	...
ZrO_2	...	2.29	...	8.52
TiO_2	...	...	3.10	3.46
CuO	...	...	0.39	...
MnO_2	1.43	1.49	1.12	0.52
NiO	1.25	1.14	0.03	1.21
Co_3O_4	0.59	1.00	1.24	0.81
P_2O_5	1.04	...	...	0.20
F_2	2.32	2.33	1.63	1.94

Table 2 Melted oxide compositions of frits used for cover coat enamels for sheet steel

Constituent	Titania white enamel	Semi-opaque enamel	Clear enamel
SiO_2	44.67	44.92	54.26
B_2O_3	14.28	16.40	12.38
Na_2O	8.27	8.67	6.55
K_2O	6.99	8.12	11.32
Li_2O	0.98	0.45	1.14
ZnO	...	0.74	...
ZrO_2	1.98	3.34	1.40
Al_2O_3	0.31	0.16	...
TiO_2	18.49	13.05	10.04
P_2O_5	1.32	0.88	...
MgO	0.5	...	...
F_2	2.21	3.27	2.91

Source: Porcelain Enamel Institute

on one side only to minimize processing defects such as fishscales.

Cast iron for enameling usually has a composition within the limits given in Table 4. Total carbon and silicon should vary in opposite directions within the ranges shown. If both are low, the iron tends to be brittle and to blister during porcelain enameling. If both are high, the iron is soft and warps easily when reheated for porcelain enameling. Manganese and sulfur should range in the same direction, so that all of the sulfur is converted to manganese sulfide. Within the normal range, phosphorus has a negligible effect on the strength of the cast iron at porcelain enameling firing temperatures.

Aluminum Alloys. The common aluminum alloys used for porcelain enameling are sheet alloys 1100, 3003, and 6061; extrusion alloy 6061; and casting alloys 43, 344, and 356. Of wrought alloys, only 6061 is heat treatable; it has better handling characteristics and is stronger after porcelain enameling. The nonheat-treatable alloys are used for small parts in which the amount of distortion and low strength encountered after firing are acceptable. Casting alloys to be porcelain enameled must be formed by permanent mold casting. Die-cast forming is normally unacceptable.

Service Properties

Porcelain enamels are designed to provide special properties for specific environments. These include chemical and weather resistance, temperature resistance, and special mechanical and electrical properties.

Chemical Resistance. Porcelain enamel is used extensively because of its resistance to household chemicals and foods. Mild alkaline or acid environments are generally involved in household applications. Table 5 gives examples of corrosive environments in which porcelain enamels are widely used. Special enamels or glass compositions are available to resist most acids—except for hydrofluoric or concentrated phosphoric—to temperatures of 230 °C (450 °F). Compositions may be formulated to resist alkali concentrations up to a pH of 12 at temperatures as high as 95 °C (200 °F).

Weather Resistance. The important characteristics that determine the weather resistance of porcelain enamels are chemical durability, color stability, cleanability, and continuity of coating. Gloss and enamel texture do not necessarily affect weather resistance. For indoor exposure, where attractive appearance is the principal requirement, porcelain enamel selection and processing are directed toward providing reproducible color along with optimum gloss and smoothness. Appearance standards are established according to the component, its use, and its location within the end product.

Service Temperatures. The temperature to which porcelain enamels can be exposed is limited by the softening of the glassy matrix. Surface temperature limits for porcelain enamels are listed in Table 6. Thermal shock intensifies the effect of elevated temperature, as does operation under severe temperature gradients. Enamels are formulated so that thermal expansion characteristics place the enamel in compression under service conditions.

Mechanical Properties. The hardness of porcelain enamels ranges from 3.4 to 6.0 on the Moh's scale. Porcelain enamels show a high degree of abrasion resistance. Abrasion resistance can be increased by adding crystalline particles to the enamel composition or by a devitrification heat treatment.

Table 3 Compositions of low-carbon steels used for porcelain enameling

Type of steel	C	Mn	P	S	Al	Ti	Nb	B
Replacement steel for enameling iron	0.02–0.05	0.15–0.3	0.015(b)	0.015(b)	0.03–0.07	...	...	0.006
Decarburized	0.005	0.2–0.3	0.01	0.02	(c)	...	...	...
Titanium-stabilized	0.05	0.30	0.01	0.02	0.05	0.30	...	...
Interstitial-free	0.005	0.20	0.01	0.02	...	0.04	0.09	...
Cold-rolled	0.06	0.35	0.01	0.02	(c)	...	...	...

(a) All compositions contain balance of iron. (b) Maximum. (c) Some steels may be supplied as aluminum-killed products. Data from Porcelain Enamel Institute

Table 4 Typical composition ranges of cast irons used in porcelain enameling

Constituent	Amount, %
Total carbon	3.20–3.60
Silicon	2.30–3.00
Manganese	0.30–0.60
Sulfur	0.05–0.12
Phosphorus	0.40–0.80

Electrical Properties. Porcelain enamels are electrical insulators. Electrical resistance per unit area is a function of enamel thickness and composition, as well as temperature. Other electrical properties (dielectric constant, dissipation factor, and dielectric strength) vary with temperature. For electronic applications such as porcelain enameled substrates for hybrid circuits, special electronic grade enamels have been designed. These compositions have considerably higher electrical resistivities and dielectric strengths and are less sensitive to temperature changes than conventional porcelain enamels.

Major Uses of Porcelain Enamels

Household Applications. The major applications for porcelain enamel remain today, as has been the case for more than half a century, in the household product area such as appliances and bathroom fixtures. These include the kitchen range, the dishwasher, the sink, the washer and dryer, the bathtub, and the lavatory. In addition, the so-called glass-lined water heater tank, which dominates the water heater market, is porcelain enameled.

These applications—for ranges, dishwashers, washers, dryers, water heaters, bathtubs, sinks, and lavatories—make up about 90% of the total volume (measured in square footage) for porcelain enamel usage in the United States and Canada. The rest of the world's produc-

tion of porcelain enamel products and components, estimated to be approximately 150% of the North American market in size, follows the same general end-product application pattern.

Other Applications. In addition to these primary uses, porcelain enamel is used in a wide variety of applications ranging from architectural panels, column covers, chalkboards, signage, specially executed murals, and electronic applications to chemical processing vats, piping and pump components, farm silos, storage tanks, and barbecue grills.

Porcelain enamel is normally specified where there is a need for one or more special performance requirements that only porcelain enamel can provide, such as chemical resistance, corrosion protection, abrasion resistance, thermal shock protection, weatherability, special appearance needs, permanent color, ease of cleaning, special mechanical requirements, or unique electrical properties.

Product Categories

For each major product category where porcelain enamel is utilized, a special family of porcelain enamel frit materials has been designed and developed. Each has its own special chemical and physical attributes. Figure 2 shows the individual portion of the total square footage of porcelain enamels applied in 1990 according to major consumer and commercial product lines.

Kitchen Ranges

Porcelain enamel is the dominant material of choice for four primary components of the range: the oven cavity, the inner door, the range cook top, and the burner box.

Oven Cavities. Today's oven interior may be one of three different types. The first is the conventional or standard, smooth surfaced porcelain enamel coating (gray/blue/

Table 6 Service temperature limits for porcelain enamels

Service temperature		Limiting conditions
°C	°F	
425	800	Usual limit for enamels maturing at about 815 °C (1500 °F)
540	1000	Maximum for enamels maturing at about 815 °C (1500 °F) without reboil
760	1400	Operating limit for special high-temperature enamels
1095	2000	Refractory enamels useful for short periods for protection of stainless steels and special alloys

Source: Porcelain Enamel Institute

black, or gray/blue/black speckled in color) that has been typical of oven interiors for more than 50 years. The porcelain enamel frit developed for the conventional household oven application is typically a combination of standard ground coat-type frits and mill addition materials engineered to provide a smooth, acid-resistant surface. It must be unaffected by normal baking and broiling soils, and must withstand repeated temperature recycling up to 260 to 290 °C (500 to 550 °F). In addition, typical porcelain enamel hardness and mild alkali-resistant properties are needed to resist abrasion and alkaline attack when cleaning the oven with commercial cleaners.

A second type of oven in common use, and growing in popularity, is the "self-cleaning" or pyrolytic oven. During its self-cleaning operation, when the oven door is locked for safety, temperatures in the 455 to 550 °C (850 to 1025 °F) range are reached. This cleaning cycle burns off (pyrolyzes) the accumulated organic soils, and after the oven cools any remaining ash particles can be easily wiped away with a damp sponge.

The special porcelain enamel frit systems for this application are designed, of course, to provide all the service requirements for normal baking and broiling functions as listed above for conventional ovens. In addition, the system must ensure an impervious porcelain enamel coating for the oven liner that will withstand the temperature recycling related to the cleaning operation. A combination of ground coat frits, with mill additions or additives, is utilized to design and formulate the porcelain enamel coating to meet these performance requirements.

A third type of oven is the "continuous cleaning" or catalytic oven. This brown/black non-smooth coating makes it possible for accumulated organic soils to "burn off" (oxidize) during normal baking and broiling operations. Both time and temperature are important factors in the oxidation of soils. The cleaning process in this system works most efficiently when the oven is operated at higher temperatures for longer periods of time.

The porcelain enamel frit system required for catalytic ovens contains various frits with

Table 5 Applications in which porcelain enamels are used for resistance to corrosive environments

Application	Temperature		pH	Corrosive media
	°C	°F		
Bathtubs	≤49	≤120	5–9	Water, cleansers
Chemical ware	≤100	≤212	12	Alkaline solutions
	≤100	≤212	1–2	All acids except hydrofluoric
	175–230	350–450	1–2	Concentrated sulfuric acid, nitric acid, and hydrochloric acid
Home laundry equipment	≤71	≤160	11	Water, detergents, and bleach
Dishwashers	≤82	≤180	8–12	Water, strong detergents
Range exteriors	21–66	70–150	2–10	Food acids, cleaners
Range oven liners:				
Conventional	20–315	70–600	2–10	Food acids, cleaners
Pyrolytic	20–540	70–1000	2–10	Food acids, cleaners
Range burner grates	20–595	70–1100	2–10	Food acids, cleaners
Refrigerators	−18–20	0–70	2–10	Food acids, cleaners
Kitchen sinks, lavatories	≤71	≤160	2–10	Food acids, water, and cleansers
Water heaters	≤71	≤160	5–8	Water

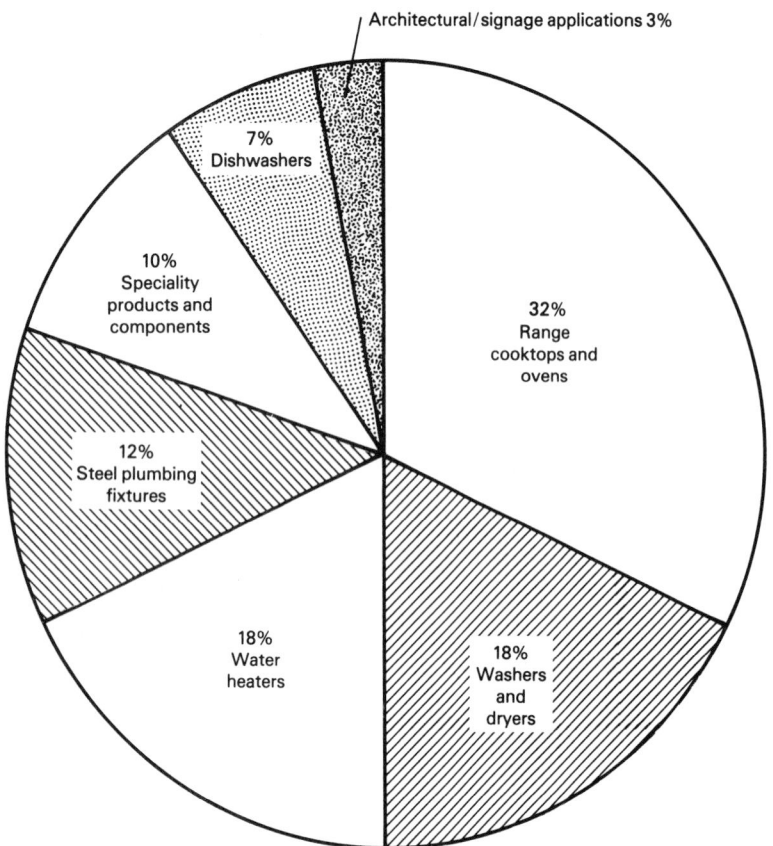

Architectural/signage applications 3%

7% Dishwashers

10% Speciality products and components

32% Range cooktops and ovens

12% Steel plumbing fixtures

18% Water heaters

18% Washers and dryers

Fig 2 Breakdown of porcelain enamels used in 1990 in major appliance and commercial product lines based on total square footage of enamels applied

a high loading of transition metal oxides to provide the oxidizing function. In addition, the coating is porous by nature, designed to ensure a greatly expanded surface area of microscopic "mountains and valleys" for increased oxidation action when applied to the oven liner.

Oven Doors. The inner door of the oven is coated with the same specially designed porcelain enamel frit used for the oven liner, whether it be a conventional oven, a continuous cleaning (catalytic) oven, or a self-cleaning (pyrolytic) oven. The service requirements for the inner door are, of course, similar to that of the oven liner itself.

Range Burner Box. The burner box catches grease and spills beneath the range top and is unseen except during cleaning of ranges with hinged tops. Most burner boxes are porcelain enameled. The porcelain frit requirements call for a ground coat-type utilitarian coating providing fire resistance and protection against rusting of the component.

Range Tops. Another component of the domestic range that utilizes porcelain enamel is the range top (or cook top). Porcelain enamel holds more than 95% of the total market for range tops, with stainless steel and glass/ceramic holding the rest. For many years, porcelain enamel has dominated the range top market, utilizing either a wet two-coat system

or a wet direct-on cover coat system. During the last decade, there has been rapid movement toward an innovative two-coat, one-fire electrostatic powder porcelain enamel system. Range top production has proved to be the leading application for porcelain enamel powder. It is estimated that approximately 80% of all range tops produced in the United States now utilize the porcelain enamel powder system; the remaining volume of porcelain enamel range tops still use wet system porcelain enamel.

The porcelain enamel frit systems for range tops—whether for use in the older conventional wet system or for the newer electrostatic powder method—are designed to provide a hard, easy-to-clean surface that has excellent acid resistance (food stains) and thermal shock resistance. Color reproducibility is essential for range tops because of cosmetic and decorative reasons. Porcelain enamel frit systems are formulated from one or more frits with additives to meet these service requirements. In wet system applications, one or more frits, with mill addition materials and pigments, are used to provide similar performance properties. During the last decade, the volume of porcelain enamel used for ranges has grown as total range production increased from 4,223,000 units in 1980 to 5,283,000 units in 1990. The range market uses an esti-

mated 32% of the total square footage for products that are porcelain enameled.

Home Laundry Equipment

Washers. Porcelain enamel has been used for key components of both the household washer and dryer for many years. An important application for porcelain enamel is the washer (spinner) basket. Most commonly, a wet system coating (blue/gray/black or blue/gray/black with speckles) is used, but white spinner baskets—usually porcelain enameled using the new electrostatic powder system—are being produced today. For this application, the wet porcelain enameling system uses one or more specially formulated ground coat-type frits with selected mill additions. They are designed to meet performance requirements including good resistance to alkali/caustic attack from laundry soaps and detergents, hot water resistance, and hardness to provide abrasion resistance from buttons, buckles, and zippers on clothing being washed. The specially formulated frit systems used for washer baskets coated by the new electrostatic powder system provide the same service characteristics. In addition, color permanence and a good continuity of coating are also important.

Another important application for porcelain enamel is the outer tub of the washer. This component, unseen by the homemaker while washing, contains the water that is used during the washing cycle. Polypropylene and other plastics are being introduced by some manufacturers for this component. The primary porcelain enamel frit requirement for this application is alkali resistance. The frit system must provide good continuity of coating for the component to prevent rusting and leaking, along with water- and alkali-resistant properties.

The remaining use of porcelain enamel for washer applications is for the washer tops and lids. Here, porcelain enamels are selected to provide maximum abrasion resistance; an alkali-resistant finish, and a permanent color and gloss that will not change over the years.

Dryers. For the accompanying clothes dryer, porcelain enamel's most important application is for the dryer drum. Its use here provides a smooth surface for excellent abrasion resistance and resistance to heat and moisture. To achieve the special characteristics of the porcelain enamel used for dryer drums, one or more ground coat-type frits with mill additions are used. In addition to ensuring a hard, abrasion-resistant surface, a good continuity of coating to protect against rusting is provided. Other materials and finishes competing for the dryer drum market include organic paint finishes and galvanized or aluminum coated steels.

Porcelain enamel is used for top-of-the-line dryer tops to provide maximum abrasion resistance and permanent color and gloss. It is often used for "twin" washer and dryer sets, and also on commercial units at coin-operated laundry centers. The special porcelain enamel frit systems for these applications are quite similar to those for washer tops and lids. They utilize two or more ground coat-type frits and mill additions to ensure a glossy, permanent color that is resistant to both alkali attack and abrasion. Dryer production increased from 3,176,000 units in 1980 to 4,595,000 units in 1990. Automatic washer volume rose from 4,426,000 units in 1980 to 6,192,000 units in 1990. Currently, washers and dryers together consume 18% of the porcelain enamel market based on total square footage used.

Dishwashers

Dishwashers evolved during the 1950s and 1960s from a luxury item to a necessary household appliance. Volume grew from 2,354,000 units in 1980 to 3,637,000 in 1990. From their earliest days, dishwashers utilized durable porcelain enamel extensively for the tub interior and inner door liner. In recent years, plastics (primarily polypropylene) have made gains for these applications, while stainless steel holds a small share of the market.

The special porcelain enamel frit systems formulated for the dishwasher interior draw upon two or more frits with strong alkali-resistant properties and selected mill addition materials to achieve needed resistance to chalking, hot water, and alkali attack from dishwasher detergents. Full continuity of coating is provided to protect against rusting. Porcelain enamels' hardness protects against cuts and gouges from knives and other sharp utensils. At present, 7% of all porcelain enamel total square footage goes into dishwasher tanks and inner doors.

Water Heaters

The domestic water heaters have utilized porcelain enamel almost exclusively as tank interior coatings for shells and flue pipes since World War II. The market for "glass-lined" water heaters has steadily increased from approximately 5,269,000 units in 1980 to nearly 7,132,000 units in 1990. Water heaters today account for approximately 18% of the total porcelain enamel market. The special porcelain enamel systems for water heater tanks are ordinarily a combination of two or more frits with a loading of high refractory mill additions to provide excellent resistance to corrosion by hot water and good coating continuity.

Plumbing Fixtures

Porcelain Enamel on Steel. In the period following World War II, an entire new category of plumbing fixture products emerged. Using automobile body metal stamping techniques, bathtubs and lavatories for the bathroom, along with kitchen sinks, were stamped from sheet steel; they were then porcelain enameled using a two-coat, wet application system.

For porcelain enamel on steel bathtubs, a two-coat system (ground coat under cover coat) is required to ensure a continuity of coating without pinholes. The cover coat is both acid-resistant and abrasion-resistant to withstand household cleaning compounds. The special porcelain enamels are formulated from two or more frits with mill additions. Good color stability and high gloss are important properties.

Where slip-resistant surfaces are required, special refractory materials are applied to the surface coating, and then the bathtub is fired in the enameling furnace. In 1980, 843,000 porcelain enamel units for steel bathtubs were produced. This figure increased to 913,000 units in 1990. Porcelain enamel on steel plumbing fixtures presently account for about 12% of the total square footage of porcelain enamel.

Porcelain Enamel on Cast Iron. For nearly a century, porcelain enamel on cast iron has been identified with high-quality bathtubs. These units, produced by the few remaining cast iron bathtub manufacturers in the United States (cast iron enameling is more extensively used in Europe), utilize a dry process cover coat. The frit for this application is made primarily by the individual bathtub manufacturers, rather than being purchased from commercial frit companies.

Special requirements for the dry process porcelain enamel frit application to cast iron bathtubs call for good coverage, good acid resistance, and close color control. Also, special attention to matching the coefficient of thermal expansion of the coating to that of the cast iron substrate is essential for a good fit and to eliminate crazing or chipping.

In 1990, 465,000 cast iron bathtubs were produced—down from 566,000 units in 1980. In addition to bathtubs, 700,000 kitchen sinks of porcelain enameled cast iron were produced in 1990, up from 433,000 in 1980. Also, 699,000 cast iron lavatories were shipped that year, an increase from 579,000 units in 1980. Porcelain enamel frit systems for cast iron sinks and lavatories are quite similar to those used for bathtubs.

Architectural Porcelain Enamel

For more than 50 years, porcelain enamel has been specified by architects for a wide variety of creative uses in modern building structures. Its time-tested performance under a wide range of environmental conditions has firmly identified architectural porcelain enamel as a durable, reliable material. Architectural applications represent about 3% of porcelain enamel's total square footage.

Porcelain enamel is used widely for exterior panel sections, fascias, and column covers for high-rise buildings. It is also used for interior paneling, decorative murals, signs, chalkboards, and other writing surfaces. Most commonly, panels and components are cus-tom designed, the sheet steel sections are fabricated, and then are porcelain enameled in plants specializing in this work. In addition to steel, some porcelain enameling of aluminum is done for architectural and signage applications.

Weatherability heads the list of requirements for these applications, with acid resistance being used as an accurate laboratory predictor of weathering characteristics of the porcelain enamel coating. Permanent colors and hues and a broad range of gloss from dull matte to high gloss are key properties.

Porcelain enamel systems for architectural applications use two or more frits with selected mill additions to achieve weatherability requirements. Modifications are made as needed to meet specific color and gloss specifications.

Specialty Products and Components

About 90% of porcelain enamel's total volume, measured in square feet, is concentrated in the application categories already discussed. The remaining share covers a wide range of specialty applications. These are usually porcelain enameled in contract or jobbing plants concentrating on these smaller volume markets.

In recent years, there has been a rapid growth of applications drawing upon porcelain enamel's properties of heat resistance, low coefficient of friction, and ease of cleaning. Barbecue grills, racks for these grills, grates for ranges, and range drip bowls are among the products and components being porcelain enameled.

The porcelain enamel systems used for these applications usually consist of one or more frits with mill additions or additives formulated to provide the utilitarian-type service demanded for the specific application. Color-fastness and smooth, easy-to-clean surfaces, plus heat resistance, are among the identified characteristics.

Another growth area in the specialty sector has been in the coating of pipe interiors and pump components for use in corrosive environments such as waste treatment facilities and chemical processing. The porcelain enamel frit systems for these uses, necessarily specially formulated for the particular service, are made up of two or more frits with selected mill additions that emphasize lubricity and easy-to-clean properties, along with chemical resistance, gloss retention, and permanence of the coating.

For harsh environments requiring resistance to heat, acid attack, and abrasion, porcelain enamel is specified for a wide range of industrial applications. These include heat exchangers, air preheater panels, and other power generation components. The porcelain enamel frit systems, each designed for the specific application, are usually two or more ground coat-type frits plus mill additions for increased heat resistance, abrasion resistance, and resistance to acidic atmospheres.

Another growth area is in electronic applications such as porcelain enameled substrates for printed circuit boards. Here, special porcelain enamel frit formulations optimize the desired electrical properties.

Additional specialized applications for porcelain enamel include:

- Porcelain enameled cookware
- Farm silo, grain storage facilities, municipal water storage tanks, beer and other beverage storage tanks, and agricultural slurry storage systems
- Conveyor equipment
- Automotive exhaust system components
- Frit systems as a thermal guard, providing oxidation protection during metal heat-treating operations
- Frit-based adhesives for joining metals

Future Outlook

During the coming decade and beyond, porcelain enamel can be expected to continue to serve sectors of the familiar home appliance, plumbing fixture, and water heater market. At the same time, competition will be intense in these markets as lower-cost materials claim to be "good-enough-for-the-application" for certain components.

A major role for porcelain enamel is likely to be for those appliance and related product applications viewed as requiring "critical performance," that is, where porcelain enamel's special properties cannot be readily matched by competitive materials. In addition, certain higher quality, top-of-the-line units can be expected to utilize and feature porcelain enamel.

The emergence of highly efficient porcelain enameling facilities—probably as separate contract or jobbing facilities that concentrate on niche markets using porcelain enamels—appears to be a good possibility.

The major growth areas are expected to be in specialized applications where one or more of porcelain enamel's unique service properties fill a need. Particularly important will be harsh environment applications where porcelain enamel provides the most cost-effective protection against high temperatures and attack by corrosive chemicals.

Furthermore, the growth opportunities in the electronics field appear to be promising as special porcelain enamels make it feasible for electronic components to operate under environmental conditions unfavorable to other coating materials.

SELECTED REFERENCES

- A.I. Andrews, *Porcelain Enamels*, 2nd ed., Garrard Press, Champaign, IL
- *Annual Book of ASTM Standards*, Vol 02.05, American Society for Testing and Materials, Philadelphia, PA
- "Appliance Finishes," A Brochure of Sheet Committees, American Iron and Steel Institute, Washington, D.C.
- *Appliance Mag.*, Annual Statistical Issues, Dana Chase Publications, Oak Brook, IL
- *Metals Handbook*, 9th ed., Vol 5, ASM International, Materials Park, OH, p 509–531
- "1939 Exposure Test of Porcelain Enamels on Steel, 30-Year Inspection," Building Science Series, National Bureau of Standards, Washington, D.C.
- Porcelain Enameling, *Appliance Mag.*, special product supplements, Dana Chase Publications, Oak Brook, IL, Mar 1988 and Apr 1990
- *Proceedings of the Technical Forum*, Annual Editions, Porcelain Enamel Institute, Washington, D.C.
- "Properties of Porcelain Enamel," Data Bulletin Series: PEI 501, 502, 503, 504, and 505; Porcelain Enamel Institute, Washington, D.C.
- *Tool and Manufacturing Engineers Handbook*, 4th ed., Vol 3, Society of Manufacturing Engineers, Dearborn, MI, p 23–1 to 23–10
- "Weather Resistance of Porcelain Enamels, 15-Year Inspection of the 1956 Exposure Test," Building Science Series 50, National Bureau of Standards, Washington, D.C.
- "Weatherability of Porcelain Enamel," Porcelain Enamel Institute, Washington, D.C.
- "Weathering of Porcelain Enamels on Aluminum, 12-Year Inspection, Exposure Period 1964–1976," Department of Ceramic Engineering, University of Illinois at Urbana-Champaign, IL

Structural Clay Products

Gilbert C. Robinson, Clemson University

STRUCTURAL CERAMIC PRODUCTS are used in construction. They are distinguished by their raw materials and manufacturing methods. The raw materials are naturally occurring clays or shales. The distinguishing manufacturing characteristic is the exposure to elevated temperatures (firing). This must be sufficient to develop a fired bond between the particulate constituents. The bond may result from the partial fusion of some of the constituents followed by cooling to develop a glassy bond. It can also come from interparticle reaction with or without the occurrence of fusion. The classification of a product as a structural ceramic product requires that the product be fired to at least the point of initial fired bonding (incipient fusion). Usually the heat treatment must be carried much further in order to develop sufficient bonding to meet the specifications for the product. Pore structure is a common property used to evaluate the extent of the heat treatment.

Products

There is a wide variety of structural ceramic products as shown by the listing of products in Table 1. Products are used for facing buildings, surfacing highways, making containers for corrosive acids, as aggregate for low-density concrete, as conduits for sewage, as structural arches supporting bridges, as roofs, and as chimney liners. Each product has its specific property requirements, and these will be considered in the discussion of individual products.

Common Properties

Although each product has its own distinguishing properties, there are two properties common to all the products—pore structure and fired bond. These properties are keys to defining the use of the various products.

Pore structure is basic to all the other properties of structural ceramic products. There are four qualities that define pore structure: quantity of pores, size distribution, interconnection of pores, and pore shape.

The quantity of pores is defined commonly by the weight percentage water absorption of the product. Sometimes the volume percentage of pores is expressed. Products may be exposed to rain water or may be asked to contain fluids. The quantity of pores will indicate the quantity of liquid that will enter within the volume of the object. This may influence the rate of corrosion of the object and the resistance to cyclic freezing when saturated with water. The quantity of pores relates inversely to the strength of the object and its thermal conductivity (Fig 1).

The size distribution of pores is particularly significant to the frost resistance and the capillarity of the product. Pores of small size are particularly damaging to frost resistance. The definition of small varies with the source. Generally harmful pores are considered 3 to 0.9 μm or smaller. Products of poor frost resistance will have a majority of their pores in the small category, while products with good frost resistance will have a preponderance of their pores larger than 1 μm. Small pores favor high height of capillary rise but a slow rate of rise.

The interconnection of pores influences the permeability of the object to fluids. The pores may be sealed against fluid penetration, or they may be open to penetration. The passageways between pores may be tortuous and inhibit flow or may provide easy paths for flow. High permeability in an object may prevent its use as a container for fluids or in preventing rain water from entering a structure, but high permeability may be desirable in an application such as drain tile.

Pore quantity, pore size distribution, and pore connection provide sufficient information to define key properties of the products. Pore shape may be useful in identifying the prior firing practice used in manufacturing the object.

Fired Bond. A distinguishing characteristic of all structural ceramic products is the requirement that they be processed at high temperatures to achieve interparticle bonding. Firing of the object causes interparticle reaction and partial fusion of the mass. The molten substance assists in pore elimination through drawing together of the particles and filling void spaces. The undesirable small pores are the first to disappear. Thus, the extent of fusion determines pore quantity and influences pore size distribution. The molten material solidifies on cooling and produces bonding of the constituents; thus, it determines the strength of the unit.

A product with a high amount of fused material is called vitreous, while products with lesser amounts are termed semivitreous. The boundary between vitreous and semivitreous is indefinite and changes with product type. It may be 6% absorption for vitreous sewer pipe or 0.5% for chemically resistant units.

Facing Materials

Variety of Products. Facing units provide the surface skin of structures and their primary functions are to provide a pleasing appearance and to isolate the interior from the exterior environment. The emphasis on appearance has resulted in a wide variety of products of various colors, textures, and shapes. Brick raw materials will produce fired colors of white, buff, red, brown, and black. Coated brick extend the variety of available colors. Textures can range from smooth to sandy to scratched and to imprinted designs. Shapes may be the familiar brick prismatic shape to almost any shape that the building designer conceives. The brick may be curved for construction of columns (Fig 2), be ornate blocks of intricate design (terra cotta) as shown in Fig 3(a) and (b), or be handcrafted brick for sculptured panels (Fig 3c).

The brick may be used as a veneer for wood and metal stud structures and facing for concrete and concrete masonry units. Cavity walls may be formed with an air gap between the facing brick and the interior wall of load-bearing brick or concrete masonry units. Furthermore, face brick are used in fireplace construction, chimneys, fences, garden walls, entry walls, outdoor signs, soundbarrier walls, and even supports for rural mail boxes. In addition, brick may be used for load-bearing structures.

Brick facing is prized for its beauty, variety, fire-proofness, low maintenance, and long life. Savings in maintenance and insurance costs make it lower in cost than its competition when calculated over the period of the mortgage.

There are certain distinctive properties that need to be considered in selecting a product for a particular application. These properties

Table 1 Structural ceramic products

Product	ASTM standard
Facing materials	
Face brick	C 216
Terra cotta	. . .
Thin brick veneer	C 1088
Sculptured brick	. . .
Special shapes	. . .
Glazed brick	C 126
Glazed structural tile	C 126
Load-bearing units	
Building brick	C 62
Hollow brick	C 652
Face brick	C 216
Structural tile facing	C 212
Structural tile floor	C 57
Structural tile wall	C 34
Paving units	
Light traffic pavers	C 902
Quarry tile	. . .
Paving brick	. . .
Chemically resistant units	
Chemically resistant masonry units	C 279
Industrial floor brick	C 410
Chemical stoneware	. . .
Sewer pipe	C 700
Chimney brick	C 980
Filter block	C 159
Roofing tile	. . .
Miscellaneous	
Flue lining	C 315
Lightweight aggregate	C 331
Drain tile	C 4

Fig 2 The Kellogg building in Battle Creek, Michigan, showing the use of circular brick columns

include color, texture, coatings, durability, efflorescence, moisture expansion, size tolerance, and chipping.

Color. Appearance is achieved through the provision of a variety of colors and textures. Color can be achieved through the selection of the raw material or blend of raw materials. Kaolins and fire clays tend to produce the buff and other light colors. Clays and shales with higher iron oxide contents will produce the brick red colors. Iron oxide and other iron compounds are the major natural colorants for structural ceramic products.

The colors produced by a given raw material can be changed by changing the maturing temperature and the time at temperature. Colors tend to deepen in shade as the temperature and/or time increase. Kiln atmosphere can change brick red colors into black, brown, and golden colors. The process called "flashing" purposely introduces an oxygen-deficient atmosphere (reducing atmosphere) to create the non-red colors. A color assortment can be achieved through controlling the exposure time and the time in the firing schedule where the reducing atmosphere is introduced. Variety can be extended through controlled blending of different colored brick.

Texture is produced during the shaping step of manufacture. Three shaping methods are employed: extrusion, molding, and pressing. Extrusion is the predominant process. The extrudate has a die-slickened surface and this smooth texture is one of the available finishes. Other textures can be achieved by cutting off a thin layer and leaving a wire cut or roughened surface. The extrudate may be scratched, brushed, or otherwise roughened. Special rolls may press various designs into the column surface. Sand may be applied to the surface to achieve a distinctive texture. The surface may be purposely roughened by a coating application or by cutting the extrusion with a superimposed paper layer.

The molding process shapes brick by forming wet, soft clay into a brick-shaped mold box. Originally, this was a hand-molding operation but today machinery is available for imitating the hand-molding process. Hand molding produced irregular shapes and the machine molding of today imitates this irregularity. This provides a brick wall of distinctive and prized texture. The molding process may coat the interior of the box with sand prior to introduction of the clay. This assists release of the clay from the mold and provides a desired sandy texture (sand-struck) on the exposed faces of the brick. In other instances the mold interior will be wet with water to assist release. This provides a water-struck brick texture.

Pressed brick are formed with low water content or a stiffer consistency than molded brick or extruded brick. As a consequence, these units are more regular in shape and sharper in outline than the others. Usually pressed brick have only the die-slickened exterior surface.

Coatings. The colors and textures of the base raw materials may be altered by the application of coatings of mineral mixtures to

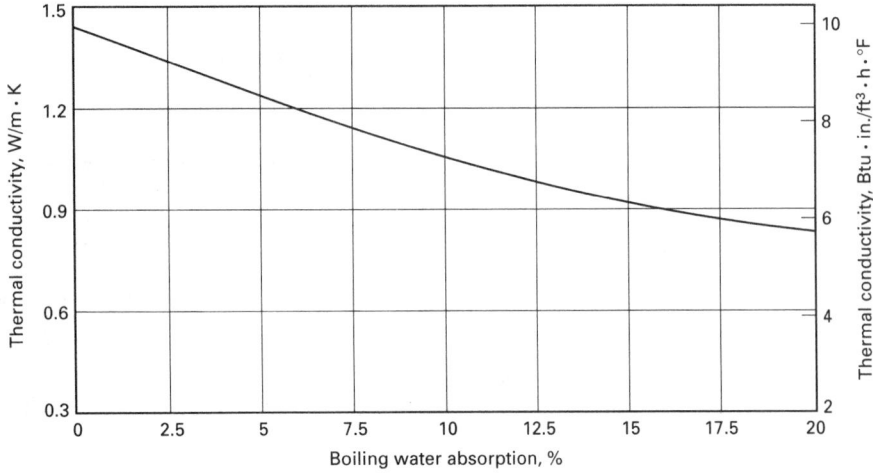

Fig 1 The relationship between thermal conductivity and boiling water absorption of brick. Source: Ref 5

(a)

(b)

(c)

Fig 3 Examples of decorative brick. Terra cotta exterior of the historic Barr building in St. Louis, Missouri. Note the assortment of shapes above and beside the doorway in (a). The repeating design of the cladding that decorates most of the multi-story building can be seen in (b). (c) Sculptured brick mural by Jay Tschetter measuring 4 × 12 m (14 × 40 ft) for Lincoln Station in Lincoln, Nebraska. Courtesy of Light Impressions

the extrusion column. The coating may be granular and uneven in application. Another type may be a continuous coating that completely masks the color of the base material.

Sands, colored sands, and clay/sand mixtures may be sprayed or rolled onto the column surface. Wet, runny slurries may be applied. The slurries may be clay based in an attempt to completely cover the brick with a different color coating. These coatings that mature to a semivitreous or vitreous but non-glossy state may be called "engobes." A coating that matures to a glass-like substance is termed a glaze. The glaze is impervious to water and easy to keep clean. The glaze has to be carefully formulated to match the thermal dimensional movements of the base body.

A mismatch can result in the separation of the glazed surface or the development of crazed cracks. In addition, the glaze must resist the dimensional movement from moisture expansion without developing crazed cracks. This resistance is evaluated by exposing units to steam at 1035 kPa (150 psi) pressure (Ref 1). Glazed units require the special attention to avoid crystallization of salts behind the glaze or crystallization of water to ice at this location. These processes can create an expansive force that will separate the glazed face from the brick.

Sometimes spots of color are produced on the brick as a result of naturally occurring auxiliary minerals in the clay. These fuse on firing and make dark colored spots. Other

times these spots are created by applying fusible granules to the surface of the brick column. Some manufacturers coat fired brick with sands or other substances using either portland cement or asphalt as the bonding constituent.

Efflorescence is a type of surface discoloration of brick which occurs during storage of the finished brick or after it has been placed in construction. The discoloration is a deposit of soluble substances on the surface of the unit. Water soluble constituents are present in mortars and may be present in brick and in water used in preparing the mortar. Furthermore exterior sources such as fertilizers and deicing salts can be the source of efflorescence. The soluble constituents are dissolved by water entering the masonry, and consequently surface evaporation of the water leaves behind a deposit of its contained salts. Salts may remain water soluble and wash away in following rains or by water washing. On the other hand, some salts are converted to insoluble constituents during passage through the masonry or after exposure to the atmosphere. These salts require careful and aggressive cleaning procedures. Improper cleaning can increase the severity of the problem. The cleaning process may be complex, and there are several references that provide cleaning instructions (Ref 2). The removal of mortar drippings appearing during wall fabrication uses chemical assistance to loosen the mortar. This process in itself can produce efflorescent salts if improperly conducted.

Durability. Bricks have outdistanced all other materials in their record of durability. Brick structures that have existed for more than 1000 years are documented, and there are numerous examples of brick structures still serviceable after hundreds of years of age. These may be observed in many locations in the United States, including Williamsburg, Virginia, a church at Jamestown, Virginia, and the home of Thomas Jefferson at Monticello. Despite this enviable record, there persists a small percentage of jobs where bricks fail after several years exposure. The vexing part of this situation is an inability to predict with high precision the possibility of this occurrence. The most reliable current method is submission of brick to 50 cycles of freezing and thawing when saturated with water (Ref 3). However, this procedure requires some 70 days for its performance and so is undesirable for routine control. Furthermore, there are cases of brick that passed the test and failed in service as well as the converse, brick satisfactory in service that failed the test. Considerable research is underway internationally to develop test procedures that are more precise predictors of durability and which can be performed with reasonable rapidity.

The appearance of durability failure takes many forms. There may be a splitting away of a thin layer of the face of the brick, which sometimes is termed "spalling." The brick may crack into several pieces, and the cracks may resemble cooling cracks. Craze cracks may appear on the surface of the brick, and these will gradually open followed by delamination of chips away from the surface. Another appearance is the crumbling and disintegration of the brick.

The problem of predicting durability is complicated by a variety of contributory causes. The source of failure that is usually considered is the expansive force of water converting to ice during freezing of a water-saturated brick. This force can break water pipes, heads of automobile engines, as well as brick, stone, and concrete. There are other causes equally damaging, but they receive less attention. Salt crystallization within the units can cause failure even in nonfreezing climates. Poor design practice that fails to allow adequate expansion joints for dimensional movements can cause failure. Dimensional movement may come from thermal expansion, or moisture expansion, or the expansion of mortar freezing. Mismatch in movement between building components can stress the outer skin to failure if unrelieved.

The place in the structure is also significant to durability failure. A unit in a vertical wall may exhibit no sign of failure after a hundred years, whereas the twin unit placed in the top of a chimney may fail in one year. The most demanding exposures are considered to be chimney tops, parapet walls, window sills, and garden walls. Horizontal brick paving, when poorly drained and capable of pooling water, is another severe exposure location. Adequate allowance for movement and good drainage of a supporting base is the difference between superior performance and poor performance.

There are indices that can predict probability of failure much as in weather forecasting. Two of these used in ASTM standards (Ref 4) are percentage water absorption and saturation coefficient. Water absorptions are calculated as the percentage weight increase after immersing a dry test sample in water. Two soaking procedures are specified (Ref 3). The first requires the sample to remain submerged in room-temperature water for 24 h (cold absorption), while the second calls for submersion in boiling water for 5 h (boiling absorption). Generally a cold water absorption below 8% indicates a high probability of success (Ref 5). However, brick with 14% absorption or higher may perform adequately provided that they have additional attributes. One such attribute is the saturation coefficient, often referred to as the C/B ratio. This is the ratio of the cold absorption to the boiling water absorption. It is presumed that the room water absorption represents the amount of water that the brick can pick up from extended exposure to rain water. Furthermore, it is presumed that any additional absorption attained in the boiling water exposure represents pores that can be intruded only upon the application of some superior force to rain water application. Such an example

would be water moving in advance of a freezing front that may be under elevated pressure. The surplus pores above those saturated by room water serve as pressure relief vessels to accept this higher pressure water. The brick with a saturation coefficient of one has no surplus pores and probably is of poor durability. The lower the saturation coefficient, the better the predicted durability. The ASTM standards use a value of 0.78 (Ref 4) as the maximum permissible saturation coefficient for brick intended for use where they may be frozen while saturated with water.

The saturation coefficient provides useful information but again is an imperfect index with numerous illustrations of products with a 0.9 coefficient that performed satisfactorily and, on the other hand, products with 0.70 coefficient that failed. It appears that this index becomes more reliable as the percentage of cold water absorption of the subject units increases above 10%, whereas it becomes less significant as the water absorption decreases below 8%. Furthermore, there are errors in the popular concept of saturation coefficient. One mistake is the assumption that the 24 h room absorption equals the maximum absorption by rain water. Actually, extended periods of exposure to water can cause the absorptions to climb above that of the 24 h absorption. The pressure relief concept needs qualification since this would imply the presence of small pores that could be intruded only by a fluid exposed to above atmospheric pressures. This might be true for a nonwetting fluid, but water will wet brick and capillary action will cause penetration of the small pores without application of superior pressures. Instead, the saturation coefficient, when it works, is probably indicating the presence of the larger pores which are difficult to keep saturated when opened to drainage channels.

Compressive strength is another index of durability employed in ASTM standards. A minimum compressive strength of 21 MPa (3 ksi) is required for brick that may be frozen while saturated with water (Ref 4). A compressive strength of below 14 MPa (2 ksi) as determined by the procedures of ASTM C 67 (Ref 3) is a fairly reliable indicator that the product will be nondurable. However, the strengths between 21 and 69 MPa (3 and 10 ksi) have very poor correlation with product durability.

Several authors have presented evidence that pore size is significant to durability. Maage has produced a formula of predicting frost resistance based on the total porosity of the unit and the percentage of porosity with pores greater than 3 µm (Ref 6). Robinson has shown that pores smaller than 1 µm are damaging to frost resistance (Ref 7). He further suggests that fired bond must be considered as well as pore size distribution.

Moisture Expansion. Brick exhibits a gradual and permanent expansion as a consequence of reaction with atmospheric water vapor. The expansion occurs rapidly in the

first month after emergence from the kiln. The pattern of expansion with time varies according to different researchers and probably with different raw materials. Nevertheless, most of the expansion is accomplished during the first year of exposure and with very little occurring thereafter. In the United States, brick exhibit moisture expansions between 0.02 and 0.09%. Higher expansions have been observed in other countries, particularly in Australia (Ref 8). The brick moisture expansion may show some reversal on heating through 110 °C (230 °F). The majority of expansion is nonreversible under this drying treatment (Ref 9).

Tests have been made to show that the moisture expansion does not damage the strength of brick or otherwise harm its performance as long as the expansion is unrestrained. Applications that do not provide adequate expansion joints will cause failure of units and wall segments in a similar fashion to restrained thermal expansion. Thus, the design of expansion joints should accommodate about twice the movement anticipated from thermal expansion.

Thermal Expansion. All structural materials including brick exhibit a reversible thermal expansion. This expansion can be accommodated with adequate expansion joints. Failure to provide expansion joints can result in breakage of brick or building components. The thermal expansion coefficient will be specific to a particular brand of brick. In the United States, thermal expansion values range from 4.5 to 9.0 µm/m · K.

Chipping. An aggravating problem is the tendency for brick to chip during packaging and transportation to the job site. The allowable number of chips and extent of chipping is specified in ASTM C 216 (Ref 4). Chips may become particularly noticeable on brick coated with engobes of different color from that of the brick.

Load-Bearing Products

Bearing Wall. Brick has traditionally supplied the needed structural strength of building components. It formed the walls for ancient monuments, coliseum, pyramids, castles, and homes. It was used in arch configurations to provide domes of cathedrals, bridges, and aqueducts. Times changed in about 1900, and brick began losing its share of the structural market. Steel, skeleton-frame structures started their rise to prominence in multi-storied buildings, and brick usage changed to being primarily a facing material.

Recent years have seen renewed interest in brick as the structural component of multi-story buildings. Work in Europe started this interest and demonstrated the feasibility of these structures. The renewed interest has been supported by demonstrated cost savings with brick. Furthermore, brick provides improved thermal performance, resistance to corrosion, and resistance to fire.

The contemporary bearing wall is based on the concept of engineered brick masonry (Ref 10). This provides a system of design based on material properties rather than the ancient concept of empirical design. This has made possible cost effective structures, with thinner walls providing superior structural performance.

The contrast in design methodology can be illustrated by comparing the Monadnock Building in Chicago with the Episcopal House of Reading, Pennsylvania. The Monadnock Building was designed by the old method which prescribed starting with a thickness of 300 mm (12 in.) on the top story and adding a specified amount of the thickness to each underlying floor. The building is 16 stories high and this formula produced walls 1.8 m (6 ft) in thickness on the bottom floor. The building was constructed in 1893 and was typical of multi-story structures of that era.

The Episcopal House of Reading, Pennsylvania, is 15 stories tall. Walls at the bottom floor are 300 mm (12 in.) thick with 200 mm (8 in.) walls on all the higher floors. This reduction in wall thickness was accomplished with engineered brick masonry. The design utilizes both walls and floors working together to effect the desired structural system. The design is based on the strength qualities of the particular materials to be used in the building. Brick and mortar are selected and controlled to provide the needed structural qualities.

Facing applications demand very little of brick strength. In contrast, brick strength is a major characteristic of brick to be used in load-bearing structures. The requirements vary with local building codes, and these should be consulted. Additional information is presented in "Building Code Requirements for Masonry Structures" (Ref 11). Comprehensive information on design is available from the Brick Institute of America (BIA) (Ref 10).

Design of load-bearing walls is based on the strength of individual brick and/or specimens of mortar that will be used in the structure. There is a major difference between the strength of an individual brick and/or mortar specimen and the unit strength of a composite wall of brick and mortar. Nevertheless, procedures are published to predict allowable wall stresses from strengths of individual brick and mortar specimens (Ref 10, 11). Another approach is to test a small wall section made from the brick and mortar. The section may be a prism constructed of five brick stacked together with mortar, but sometimes other shapes can be used.

An investigation by Watstein developed information on the relation of unit brick strength to prism strength (Ref 12). An investigation by West et al. compared the unit strengths of 225 mm (9 in.) test wall sections made from masonry cubes to those of large single-wythe walls measuring 2.6 m high by 1.4 m long (8$\frac{1}{2}$ ft high by 4$\frac{1}{2}$ ft long). Some of their results are shown in Table 2.

Table 2 The influence of brick and mortar strength on wall strengths

Sample no.	Brick compressive strength		Type M mortar						Type N mortar					
			Masonry cube strength		Wall strength		Mortar strength		Masonry cube strength		Wall strength		Mortar strength	
	MPa	psi	MPa	psi	MPa	psi	MPa	psi	MPa	psi	MPa	psi	MPa	psi
B3	96	14,000	35.5	5,147	20.0	2,900	14.8	2,140	30.6	4,440	14.2	2,053	4.8	709
C3	90	13,000	37.9	5,507	25.0	3,633	11.7	1,703	38.6	5,592	17.3	2,513	3.8	553
A3	83	12,000	32.6	4,730	21.3	3,090	12.4	1,795	30.6	4,440	15.6	2,260	5.3	775
F3	52	7,500	20.3	2,940	15.7	2,280	18.6	2,693	14.6	2,120	11.0	1,603	5.8	840
E3	48	7,000	19.5	2,833	16.1	2,340	13.1	1,900	17.8	2,583	10.9	1,583	5.4	784
E16	45	6,500	24.0	3,483	15.3	2,213	11.1	1,615	20.2	2,933	9.7	1,412	5.0	725
G14	31	4,500	12.7	1,848	13.2	1,915	16.7	2,427	10.0	1,452	8.9	1,293	4.9	720
G5	31	4,500	14.9	2,160	12.8	1,860	14.7	2,135	14.9	2,165	10.3	1,487	6.4	919

Source: Ref 13

The strength characteristics of brick used in the United States are shown in Table 3. It is possible to specify strengths within these limits, although all manufacturers may not be able to comply with the values.

Mortar Bonding. Bond between mortar and brick is essential to providing flexural strength of the wall and providing an effective barrier against water penetration. Most mortars are composed of portland cement, sand, and a constituent to provide workability. The workability is provided by hydrated lime in the portland cement-lime (PCL) mortars and by some proprietary constituent in the masonry cement mortars. The masonry cements are prepackaged blends of portland cement, a workability constituent, and are ready for mixing with sand. Portland cement-lime mortars require acquisition of separate bags of the cement and lime and proportioning these two constituents with sand. Thus,

Table 3 The general range of property values for commercial brick used in the United States

Property	Property value
Compressive strength:	
Extruded brick, MPa (psi)	41–110 (6,000–16,000)
Molded brick, MPa (psi)	21–45 (3,000–6,500)
Modulus of rupture, MPa (psi)	4.8–27.6 (700–4,000)
Modulus of elasticity, GPa (psi × 10^6)	9.7–34.5 (1.4–5.0)
Bulk density of solid volume, g/cm^3	1.65–2.08
Absorption for room-temperature water, %	0.5–10.0
Saturation coefficient	0.6–0.9
Initial rate of absorption, kg/m^2 · min (g/min · 30 in.^{2})	0.05–3.6 (1–70)
Thermal expansion coefficient, μm/m · K (μin./in. · °F)	4.5–9.0 (2.5–4.5)
Moisture expansion, %	0.02–0.09
Shrinkage in service, %	0.0
Corrosion resistance	Can contain all acids hot or cold except hydrofluoric; resistant to all alkalies
Thermal conductivity, W/ m · K (Btu · in./ft^2 · h · °F)	0.43–1.44 (3–10)

the masonry cement is more convenient for job-site mixing, but some users feel that the PCL mortars give superior performance (Ref 14) while others dispute this claim (Ref 15). Furthermore, there is a recent trend to supply prepackaged PCL mortars. Mortars are specified in ASTM C 270 (Ref 16). Mortars come in different types based on the type and percentage of cement in the composition. The prevalent types are Type M with the highest cement content, S and N with the lower cement content. There is a consequent change in the compressive strength of the mortar as indicated by the values shown in Table 2; however, corresponding changes in bond strength are not as great. Some research suggests the improvement in workability of the lower cement mortar may compensate for loss in compressive strength and provide equivalent bond strength.

The mortar bond strength is determined by applying a tensile or flexural stress to the mortar/brick interface. It relates directly to the flexural strength of masonry. Bond strengths with mortar usually measure between 207 and 965 kPa (30 and 140 psi). The performance of cement-based mortars can be enhanced by chemical additives and non-portland cement mortars are used for chemically resistant units and other applications.

The bond strength developed in a given system is very dependent on the workmanship used to build the assembly. Workmanship includes mortar preparation, selection of mortar consistency, pressure applied by the mason, and fullness of mortar joints. Beyond workmanship, there are certain brick properties that influence bond strength, and these are initial rate of absorption (IRA) and surface roughness. The IRA refers to the capillary suction power of brick. Capillary action can pull mortar into the larger open surface pores of the brick and then withdraw water from the mortar into the smaller capillaries. The IRA is determined by placing the mortar bed face of a brick to a depth of 3 mm ($\frac{1}{8}$ in.) into water. The weight gain in one minute for a solid standard size brick is the IRA. The bed surface of a standard size unit is 195 cm^2 (30 in.^{2}) in area. Measurements made with units with other bed areas are converted to the 195 cm^2 (30 in.^{2}) standard (Ref 3). Some suction is desirable since it provides prompt initial stiffening of the mortar. Excessive suction will stiffen the mortar before it can flow uniformly into the surface of the brick and before the upper brick can be placed on the mortar bed. This prevents adhesion to the upper unit.

It has been found that best bonding is obtained with IRAs of 6 to 30 g (0.2 to 1.0 oz) (however, pre-wetting brick prior to mortar placement produces satisfactory bonding with units as high as 60 g, or 2 oz). Mortar additives can also extend the IRA range. Brick of 0 g IRA can be bonded with lower water content mortars or specially designed mortar compositions. The mortar should be designed to be compatible with the IRA of the brick.

Mortar bonding is accomplished primarily by mechanical gripping. A brick bed surface with undercut indentations or sawtooth projections produces superior bonding to a slick surface. Adequate bond can be produced even with slick surfaces by designing the mortar for this purpose.

Water leakage through masonry walls can be another problem. Wind driven rain can be successfully repelled by use of a double-wythe wall with either a drainage gap between the sections or a moisture barrier. A wall section composed of a single masonry unit presents a greater challenge to make it leak-proof. The primary means of water penetration is through a crack at the interface between brick and mortar or through unfilled gaps in the mortar joints. The cracks may arise from movement of the brick after initial placement, incomplete bonding as a consequence of high IRA, or from excessive shrinkage of the mortar. Avoiding high IRAs or correcting for them with such techniques as pre-wetting can reduce water leakage. Shrinkage can be controlled by avoiding overuse of cement, by limiting the quantity of water in the mortar, and by proper aggregate particle sizing.

Bond strengths of conventional mortars are usually less than 758 kPa (110 psi). However, mortars modified with organic polymers can achieve bond strengths of 2070 kPa (300 psi) and give comparable wall flexural strengths. Caution should be exercised in the use of such additives in masonry containing steel. There have been instances of distress in such walls with the additive contributing to the corrosion of the metal. The consequent expansion of the metal has caused cracking and distortion of the masonry.

Reinforced brick masonry (RBM) has a long history of success. It increases the structural capability of masonry. Brick masonry is strong in compression but weak in tension. The inclusion of steel reinforcing bars within the masonry gives a great increase in its ability to resist tensile and shear forces. The reinforcing bars are surrounded by grout or mortar to form a composite element. The grout or mortar bonds the steel to the brick, and both

materials work together to resist applied loads. Hollow brick provide an economical means of constructing reinforced brick masonry (Fig 4). Lintels prepared from RBM can replace steel lintels at a saving in cost and without the tendency to corrosion exhibited by the steel member. Reinforced brick masonry has been used in a wide variety of structures, but its accomplishments are particularly noteworthy in retaining walls, preassembly panels, high-rise construction, and earthquake-resistant structures. The BIA gives instructions on the design and applications of RBM (Ref 17).

An interesting variation in RBM is the use of prestressed masonry. A tensile force is applied to the reinforcing bar or cable while the grout obtains its set. The setting grout or other system anchors the stretched bar to the brickwork, and the release of the tensile force imposes a compressive force on the brickwork. This compressive force greatly increases the resistance of the masonry to flexural or shear loads. The high compressive strength of brick makes it ideal for prestressed applications.

Prefabricated Panels. An important development of the 1950s and 1960s was the introduction of factory-assembled brick panels. It was anticipated that factory assembly would reduce the erection cost and improve the competitive position of brick. A number of assembly systems were developed in the United States, but the results were disappointing. The cost of panels persisted in being higher in many applications than hand-laid brick at the job site. There are a number of reasons for the increase in cost. One source of extra cost is the present system of shipping brick from manufacturer to another location

for panel assembly and then finally to the job site. This may double the transportation cost when compared with the hand-laid system wherein the manufacturer ships directly to the job site. This and other factors may be corrected, and it is anticipated that someday a system will be improved and factory preassembly will be the predominant method of construction.

Despite the disappointment in prefabricated panel cost, this type of construction persists in the United States (Ref 18). Panels offer a number of advantages over hand-laid masonry. They can provide speed of construction, can be produced in intricate designs (Fig 5a), eliminate the need for scaffolding at the job site, and offer improved quality control of the finished wall. Hollow brick offer ease of constructing reinforced brick panels. Thin brick are prefabricated with insulation to supply economical facing for structures (Fig 5b).

Other Load-Bearing Products. Structural tile, structural glazed tile, and hollow brick are other load-bearing products. All of these products contain core holes or open cells and the extent of open area helps to define the product. One puzzling term is "solid brick" as used in ASTM Standards C 62 and C 216. The term "solid" applies to any brick with 25% or less of its bearing surface area attributed to core holes. This means that users of brick that want a truly solid brick must be very specific in their instructions to the supplier.

Hollow brick have an increase in the permissible open area. The core-hole or cell areas may be up to 40% of the gross area for one class and 60% for a second class of hollow

Fig 4 Pouring of grout around a steel reinforcing rod in hollow brick wall

(a)

(b)

Fig 5 Examples of prefabricated brick. (a) Brick panel using high-bond mortar. (b) Thin brick veneer panels with plastic foam backing attached to masonry wall

brick. Structural tile overlaps the open area requirement of the hollow brick but can extend to still higher percentages.

Structural tile were once popular for backup or interior walls of structures. They could be scored to directly accept plaster coatings, or they could provide the finished surface of interior walls. They were an effective method of providing fireproofing in steel frame structures. They were used extensively until about 1946 when they began to be replaced by concrete block and other materials. Today they are no longer manufactured in the United States, although they continue to be favored in Europe. Many different size units were made with 300 mm (12 in.) being the frequent designated length. Wall thicknesses and unit heights both varied from 100 to 300 mm (4 to 12 in.). Structural tiles were used for floors as well as walls. The Europeans have extended the concept of hollow tile by producing ceramic planks 3.7 m (12 ft) in length (Ref 19). One size unit measures 300 × 300 mm × 3.7 m (12 × 12 in. × 12 ft) and weighs 22 kg (48 lb). It has 50% cored area. Thus one plank is capable of making a one-story-high wall section 0.3 m (1 ft) in width.

Hollow brick are enjoying a growing market in contrast to the disappearing structural tile. Hollow brick provides through-the-wall units (units that make up the entire wall thickness with a finished surface on both sides). They are larger in size than standard brick, and one unit may produce as much structure as 6 regular size brick. These features reduce wall cost. The design of the units and their higher strength versus hollow tile makes them excellent for load-bearing structures and reinforced structures. The larger cell areas permit grouting around in-place metal reinforcing bars (Fig 4). The cell areas permit placement of piping, wiring, and insulation inside the wall.

Glazed structural tile provides the added feature of a finished glazed surface. They are provided in a wide assortment of colors and give a permanent finish that is easy to keep clean.

Paving Units

Paving Brick (Pavers). Brick was the paving of choice between 1890 and 1910. Highways and streets were paved with brick. Many are still in use today, although most have been covered with asphalt or concrete. Marder reports on Florida highways still in use with sections of exposed brick surface, such as U.S. 19 near Palm Harbor and U.S. 92 between Deland and Daytona (Ref 20). Brick paving was popular until 1930 and then started a rapid decline in use as concrete and asphalt were used extensively. Paving brick production became so small that ASTM discontinued its standard C 7 for paving brick in 1980.

Recent years have seen a resurgence in the market for masonry paving units. The market increased from 4200 m² (45,000 ft²) in 1980 to 1,022,000 m² (110,000,000 ft²) in 1988. The initial growth was not for highways but instead for walkways, driveways, and patios. These applications did not require the strength and abrasion resistance of the former highway units, and so a new ASTM standard, C 902, was adopted for pedestrian and light traffic paving brick (Ref 21).

In addition, there has been a recent return to paving city streets and other heavy traffic areas with paving brick. Duluth, Minnesota has paved large areas of streets with brick (Ref 22). A number of other cities, including New Orleans, Louisiana, and Washington, D.C., have returned to the paving of streets with brick. This rapidly growing market has encouraged ASTM to develop a new standard to replace the old C 7, and this development is underway.

Highway pavers need abrasion and impact resistance. The old C 7 standard required a "Los Angeles rattler" test to evaluate these properties. This test impacted the brick with tumbling steel balls and measured the loss in weight of the pavers and the breakage. The equipment for the performance of this test disappeared and was no longer available at the time of discontinuance of ASTM C 7. These qualities are now evaluated by alternate procedures. A sand blast test is in use, and furthermore a ratio of the 24-h water absorption multiplied by 100 divided by the compressive strength is used as an abrasion index (Ref 21). Units previously used for highway pavers had less than 3% absorption, moduli of rupture of at least 17 MPa (2.5 ksi), and were slow fired to produce toughness. This procedure produced a paving material that remains unchallenged for its durability and wear resistance.

There are two systems for assembling paving bricks—flexible and rigid. They differ in the base for the pavers and the presence of mortar. Both systems start with a base of compacted earth and a base of several inches or so of washed gravel. The flexible paving covers the gravel with a 13 mm (½ in.) sand cushion and dry mounts the paving brick on this sand. The brick are placed as close together as possible, and fine sand is brushed in the gaps between the brick. No mortar is used (Fig 6a).

A concrete slab is placed over the gravel in the rigid system. Steel reinforcing is used in the concrete as needed. The setting bed above the concrete is mortar, and the brick are spaced to provide mortar joints in between the units (Fig 6b).

Both systems require good drainage of water in the base system. The rigid process may collect and pool water beneath the pavers and cause frost damage to the units. Proper drainage will prevent this problem.

The flexible system is lower in cost and offers advantages in that each brick is free to expand and contract somewhat independently of the others and independently of the base material. Narrow joints between units will result in some transmission of movement from one brick to the other, and so expansion joints are still needed, however, the demand for expansion joints is considerably less than for the rigid paving. The independence of the base in the flexible system frees the paving from stresses arising from the differential movement between concrete base and rigid paving. The flexible system can accommodate as heavy or heavier traffic than the rigid paving system. Machines have been developed to assist in the laying process.

Quarry tile overlaps the use area of paving brick. Quarry tile preceded the development of light traffic pavers. It is a flooring material that is thin and frequently in a square pattern such as $150 \times 150 \times 13$ mm ($6 \times 6 \times \frac{1}{2}$ in.). It is manufactured from the same raw materials and procedures as brick. Both brick manufacturers and tile manufacturers may produce thin rectangular units.

Quarry tile is usually fired to less than 4% cold absorption, thus providing a smooth impervious unit that is easy to keep clean.

Other Flooring. Floor brick are produced for industrial applications in a variety of types. The boiling absorption may vary between 1 to 10%, and there is also a variation in their resistance to chemical attack and thermal shock resistance. One application for these units is providing the flooring in chemical manufacturing factories. Resistance to chemical attack is important, and so the units must show less than 8% loss in weight after exposure to hot sulfuric acid for 48 h. Flooring for food preparation plants must provide cleanliness and in this application, boiling absorptions of less than 2% are required. Flooring for molten aluminum plants must emphasize thermal shock resistance and impact resistance. Four prescribed types of industrial flooring as described in ASTM C 410 (Ref 23) are used for these various applications.

Roofing Tile

Long life and beauty are the distinctions of ceramic roofing tile. These units are made in a variety of shapes from shingles to s-shaped spanish tile (Fig 7). They may include intricate patterns to assist in overlapping and drainage. They come in the natural colors of fired clays to a variety of applied colored coatings.

The tile must be made to provide good frost resistance, and they achieve this through the same methods described in the discussion on durability earlier in this article. In addition, they must shed water but avoid sweating on their underside. This is accomplished by limiting maximum and minimum water absorption. They must have sufficient flexural strength to support workers walking on the

Fig 7 S-shaped clay roofing tile with glazed terra cotta decorative pieces below the roof line

roof. An ASTM standard for clay roofing tile is under development and will probably appear in future issues of ASTM standards.

Aggregates

Certain clays and shales will bloat during firing and produce an expanded product of reduced density. This can be crushed and sized to produce a lightweight aggregate for concrete. This can produce as much as a 30% reduction in concrete density as compared to stone aggregate concrete. The high strength of the fired aggregate accomplishes the reduction in density with minimal sacrifice in strength. Units with a 15% reduction in density meet the same strength requirements as the heavier units. Thus the clay aggregate concretes have good strength but reduced dead load. The lower weight of masonry units made with lightweight aggregate increases mason productivity. The clay aggregate usually contains the larger size pores which are conducive to good frost resistance. The aggregate provides a method of adding desirable large pores to concrete. Expanded aggregate concrete provides savings in transportation costs.

Control of firing and raw material can produce aggregates ranging in lump bulk density from the 1.9 g/cm^3 (120 lb/ft^3) of unexpanded material down to 0.48 g/cm^3 (30 lb/ft^3); however, excess bloating reduces strength and produces floating aggregates in the concrete mixture. Specifications for aggregate call for loose-fill densities of mixed sizes of 1.0 g/cm^3 (65 lb/ft^3) (Ref 24). The loose-fill method includes pores in the space between aggregate particles and thus is considerably lower than the lump density of a particle.

Chemically Resistant Units

Many brick and other shapes are used in applications requiring resistance to chemical attack. This includes linings for industrial

(a)   (b)

Fig 6 Examples of (a) flexible dry brick paving and (b) rigid brick paving with mortar joints

chimneys, vats to contain chemicals, industrial flooring, and sewer pipe.

The distinctive specifications of chemically resistant units are low absorption and resistance to acid attack. Fired brick are resistant to all acids except hydrofluoric. Brick intended for these applications are tested for resistance to hot 78 wt% acid (Ref 25). The brick are crushed to provide fragments that are larger than 4.8 mm (0.2 in.) but smaller than 6.8 mm (0.27 in.). These fragments are immersed in the acid for 48 h, and then the loss in weight evaluated. This is limited to a maximum of 8% for corrosive resistance.

Vitrified Clay Pipe

Vitrified clay pipe was the predominant sewer pipe material until 20 years ago. Plastic pipe has replaced clay pipe for much of this market. Plastic pipe is lower in density and easier to lay and join. Nevertheless, clay pipe have proven records of service life in excess of 100 years and high resistance to corrosion by industrial sewage. Recent improvements have been made in joining systems to equal the convenience of plastic pipe. One system uses a stainless steel collar on one end and a ground surface with rubber gaskets on the opposite end (Fig 8). This permits slipping together of the pipe sections to form a water-tight joint. This pipe is applicable to tunneling under roadways or other such places. The pipe is jacked through the tunnel as it follows the drilling device.

Modest improvements in pipe strength would permit thinner wall sections and reduced weight. Perhaps application of knowledge on

Fig 8 Clay pipe being jacked into position. Note stainless steel collar (foreground) and rubber gasket on opposite end.

toughening mechanisms and composites would improve the competitive position of clay pipe and allow it to recapture part of its lost market.

Passive Solar Brick

Brick walls can capture solar energy without any moving parts. They can cut energy bills in winter and summer. In winter time, brick walls are heated by the sun and store the heat for release during the cooler temperatures of the night. In summer time, the walls can block the inflow of daytime heat and vent the heat to the atmosphere during the evening. The result is a saving in energy for both heating and air conditioning.

There are many designs of solar structures and information can be obtained from BIA (Ref 26). The lower the absorption of the brick, the better it performs in passive storage. Brown, red, or black colors perform slightly better than buff or white colors. Rough textures have a slight advantage over smooth textures. The solar properties of brick are reviewed in Ref 27.

Terra Cotta

The term terra cotta has several definitions and can include a variety of products. A popular usage in the United States describes a decorative facing material for buildings (see Fig 3). It can be made to imitate granite and limestones or it can look like itself, a fired clay material. It can be unglazed or finished in glazes of plain to vivid colors. The units may be sculptured to provide figures or other designs. The units are larger than brick and can be made in a near limitless variety of sizes and shapes. They usually provide for a 100 mm (4 in.) thick wall section and are fastened to the backup walls with metal anchors.

Terra cotta was a popular building product between 1880 and 1929 and was used extensively in the cladding of the Flatiron and Woolworth Buildings in New York and the Wrigley Building in Chicago. This industry started a rapid decline in 1930, and only a few plants remained in 1946. Recent years have seen a renewed market for terra cotta. Part of the demand is for the repair and maintenance of 1920 era buildings. Also architects are once again specifying terra cotta finishes for buildings (Ref 28, 29).

Most terra cotta buildings are now over 60 years old. Some of the terra cotta exhibits damage and needs repair. The damage comes from corrosion of the metal anchors and frost damage to the surface. The frost damage came from an ill-advised change in raw material composition of terra cotta. The newer composition of the 1920s showed excessive moisture expansion which caused a crazed pattern of glaze cracks. These cracks permitted entrance of water and subsequent splitting away of surface fragments. Modern technology teaches how to prevent this problem, and so replacement terra cotta should not exhibit

this trouble. Furthermore, corrosion-resistant metals are now available for anchoring terra cotta to the wall.

ACKNOWLEDGMENT

Appreciation is expressed to J. Gregg Borchelt of the Brick Institute of America and John Grogan of BIA-Region Nine for their review and contributions to this work.

REFERENCES

1. Standard Specification for Ceramic Glazed Structural Clay Facing Tile, Facing Brick and Solid Masonry Units, C 126–86, *Annual Book of ASTM Standards*, Vol 04.05, American Society for Testing and Materials, 1989, p 77–81

2. *Cleaning Stone and Masonry*, STP 935, James R. Chilton, Ed., American Society for Testing and Materials, 1983

3. Standard Methods of Sampling and Testing Brick and Structural Clay Tile, C 67–87, *Annual Book of ASTM Standards*, Vol 04.05, American Society for Testing and Materials, 1989, p 38–45

4. Standard Specification for Facing Brick, C 216–90a, *Annual Book of ASTM Standards*, Vol 04.05, American Society for Testing and Materials, 1989, p 104–107

5. G.C. Robinson, J.F. Edwards, and J.R. Holman, Durability of Structural Ceramic Products, *Am. Ceram. Soc. Bull.*, Vol 56 (No. 12), 1977, p 1017–1076

6. M. Maage, Frost Resistance and Pore Size Distribution of Bricks, Part 2, *Ziegelind. Int.*, Vol 43 (No. 10), 1990, p 582–588

7. G.C. Robinson, The Relationship between Pore Structure and Durability of Brick, *Am. Ceram. Soc. Bull.*, Vol 63 (No. 2), 1984, p 295–300

8. R.C. DeVekey, Moisture Expansion of Clay Masonry, *Trans. Brit. Ceram. Soc.*, Vol 82 (No. 2), 1983, p 55–59

9. G.C. Robinson, Reversibility of Moisture Expansion, *Am. Ceram. Soc. Bull.*, Vol 64 (No. 5), 1985, p 712–715

10. The Contemporary Bearing Wall, *Technical Notes on Brick Construction*, Vol 24, Brick Institute of America, 1970

11. "Building Code Requirements for Masonry Structures," ACI 530–88/ASCE-5–88, American Society of Civil Engineers, 1988

12. D. Watstein, Relation of Unrestrained Compressive Strength of Brick to Strength of Masonry, *J. Mater.*, Vol 6 (No. 2), 1971, p 304–320

13. H.W.H. West, H.R. Hodgkinson, D.G. Beech, and S.T.E. Davenport, The Performance of Walls Built with Wirecut Bricks: II Strength Tests, *Proc. Brit. Ceram. Soc.*, Vol 17 (No. 14–40), Feb 1970

14. J.H. Matthys, *Concrete Masonry Prism and Wall Flexural Bond Strength Using Conventional Masonry Mortars*, STP

1063, John H. Matthys, Ed., American Society for Testing and Materials, 1990, p 350–362

15. J.W. Ribar and V.S. Dubovoy, *Masonry Cements—A Laboratory Investigation*, STP 1063, John H. Matthys, Ed., American Society for Testing and Materials, 1990, p 85–96

16. Standard Specification for Mortar for Unit Masonry, C 270–88a, *Annual Book of ASTM Standards*, Vol 04.05, American Society for Testing and Materials, 1989, p 125–134

17. Reinforced Brick Masonry, *Technical Notes on Brick Construction*, Vol 17 I-IV, Brick Institute of America, 1988

18. J.J. Bailey, Prelaid Masonry Panels on Multistory Buildings, *J. TMS*, Vol 1 (No. 1), 1981, p G7-G11

19. G. Shellbach, Opening Up of New Dimensions in Building by Use of Plank Bricks, *Ziegelind. Int.*, Vol 39 (No. 9), 1986, p 448–460

20. Florida's Brick Roads of Historical Significance, *BIA News*, Vol 3 (No. 12), 1990, p 8–9

21. Standard Specification for Pedestrian and Light Traffic Paving Brick, C 902–87, *Annual Book of ASTM Standards*, Vol 04.05, American Society for Testing and Materials, 1989, p 539–541

22. "Duluth" 3 Million Brick Transform a City, *BIA News*, Vol 2 (No. 4), 1989, p 4–5

23. Standard Specification for Industrial Floor Brick, C 410–60, *Annual Book of ASTM Standards*, Vol 04.05, American Society for Testing and Materials, 1989, p 213–214

24. Standard Specification for Lightweight Aggregates for Concrete Masonry Units, C 331–87, *Annual Book of ASTM Standards*, Vol 04.05, American Society for Testing and Materials, 1989, p 159–161

25. Standard Specification for Chemical-Resistant Masonry Units, C 279–88, *Annual Book of ASTM Standards*, Vol 04.05, American Society for Testing and Materials, 1989, p 135–136

26. Passive Solar Heating with Brick Masonry, *Technical Notes on Brick Construction*, Vol 43, Brick Institute of America, 1981

27. G.C. Robinson, "Passive Solar Properties of Brick," Center for Engineering Ceramic Manufacturing, Clemson University, 1979

28. S. Tunick, Terra Cotta: A Legacy in Clay, *National Council on Education for the Ceramic Arts Newsletter*, Vol 9 (No. 1), 1985, p 4–7

29. E.E. Ryser, A Terra Cotta Primer, *Masonry*, July/Aug 1988, p 13–14

Ceramic Coatings

Richard A. Eppler, Eppler Associates

CERAMIC COATINGS are being used effectively in various elevated-temperature applications, providing energy savings, extending the service life of components, and increasing production. Addition of a ceramic coating to a substrate imparts physical and mechanical properties to the substrate not normally possessed by the substrate itself. Ceramic materials possess several desirable properties for use as coatings (Ref 1). Vitreous oxide coatings can provide a surface that is chemically inert and easy to clean, smooth, sufficiently hard to resist abrasion and scratching, and aesthetically pleasing. Crystalline ceramic coatings can provide thermal protection, high levels of abrasion and wear resistance, and oxidation and corrosion resistance (Ref 2).

A glass-matrix coating applied to a ceramic substrate is called a glaze (Ref 3). The coating may be completely vitreous or it may contain a dispersion of crystals that alter the coating properties. A vitreous coating applied to a metal is called a porcelain enamel. A coating applied to a glass substrate is called a glass enamel. In most cases, glass enamels are crystalline materials (primarily pigments) bonded by a low-melting glass. Plasma-spray coatings are usually refractory crystalline materials bonded to the substrate by means of temporary melting during the application process.

Ceramic coatings are applied to their substrates by one of several powder processing techniques. The raw materials are either crystalline oxides or frits. In wet processes, the raw materials are dispersed in a slip for application. In dry processes, the powders are applied directly although their surface characteristics may be adjusted for application.

The same raw material formulations can also be used to form intricate shapes unobtainable by conventional glass-forming techniques (Ref 4). The processing closely resembles that for powdered metals.

Raw Materials

Primarily for economic reasons, ceramic coatings are not usually prepared by mixing the simple oxides (Ref 5). Instead, as far as possible, naturally occurring minerals are used. Many of these minerals are the source of more than one of the constituent oxides. Table 1 is a compilation of the more common raw materials used in the formulation of ceramic coatings. Because water is used in applying many coatings, no materials that are soluble in water are included. Such materials are used only after fritting (that is, after being incorporated in a powdered glass).

The factors to be considered in selecting raw materials are:

- Chemical composition and uniformity of natural mineral sources
- Cost
- Mineral impurities
- Grain size
- Behavior in storage
- Behavior in processing
- Location and availability of the source
- Behavior in water suspension
- Environmental effects

Frits are glasses prepared by melting suitable raw materials in a gas-fired or oil-fired furnace or in an electric melting unit (Ref 6). The primary reason for the use of frits is that some of the ingredients required in coatings are soluble in water. If such materials were used directly, coating preparation by wet grinding and application as an aqueous suspension would not be possible. The solubility is a particular problem with respect to boron because there are no insoluble sources of that element (Ref 3). Other difficulties are encountered with the alkalis, where the alumina content of felspathic minerals limits the amount that can be used.

An advantage of frits is that when coating raw materials have been fritted, the reactions between them have already been largely carried to completion. Therefore, less heat work is required to fire the coating. As a result, the surface of a high frit coating is superior to that of a raw coating of the same composition fired at the same temperature.

A wide variety of frit compositions are available commercially. It is customary to formulate frits to meet the needs of specific applications (see the article "Glazes and Enamels" in this Volume).

The frits used for porcelain enamels are essentially alkali borosilicates to which ingredients appropriate to their application (Ref 1) are added. Ground coat frits (that is, base coats applied directly to the metal substrate) contain small quantities of transition metal ions to improve their adherence to the metal substrate. Alkali-resistant enamels contain zirconia to improve the corrosion resistance. Opaque cover coat enamels contain large quantities of titania, added as the opacifier.

A wide variation is found in the frits used for glazes. Some examples are given in Table 2.

The frits used for electronic applications and as the fluxes in glass enamels are very high lead content silicates and lead-zinc silicates.

Frit-Melting Furnaces

In a superficial way, frit processing is similar to the glass-melting processes described in the Section "Glass Processing" in this Volume (Ref 6). The principal difference is that frits do not require fining, which is that portion of the glass-making process whereby batch and molten glass are freed of bubbles. Another important difference is that frit melters must be usable for a whole family of related compositions. Compositional changes are made regularly without shutting down the melter.

Frit-melting furnaces are either of the box or rotary type. The rotary melter (Fig 1) is used where a number of frits are required in lots of about 450 kg (1000 lb). In the rotary process, a charge of material is placed in the preheated melter, brought up to temperature, and held for the time necessary to achieve reasonably complete solution of the batch (typically 15 min to 2 h). The furnace is slowly rotated during melting to enhance mixing and heat distribution. After melting, the burners are shut off and subsequently the molten frit is discharged and quenched.

In the continuous frit melter, raw material is continuously added to a pile at one end of the preheated box melter by a screw feeder (Fig 2). Burners directed at the raw material pile melt the frit, and the molten material flows by gravity to the other end of the melter where it is continuously discharged and quenched.

The quenching and breakup of the frit may be carried out by pouring the molten frit into the water. More commonly, the stream of molten frit is poured through a pair of water-cooled rolls and then broken up by an auto-

Table 1 Common raw materials used as sources of oxides for selected ceramic coatings

Oxide coating desired	Possible raw materials	Other oxides introduced
Al_2O_3	Clay	SiO_2
	Corundum	...
	Feldspars	SiO_2, Na_2O, K_2O
	Nepheline syenite	SiO_2, Na_2O, K_2O
BaO	Barium carbonate	...
CaO	Dolomite	MgO
	Whiting	...
	Wollastonite	SiO_2
Li_2O	Spodumene	SiO_2, Al_2O_3
MgO	Dolomite	CaO
	Heavy magnesium oxide	...
	Talc	SiO_2, CaO
Na_2O	Feldspars	SiO_2, Al_2O_3, K_2O
	Nepheline syenite	SiO_2, Al_2O_3, K_2O
PbO	Lead bisilicate(a)	SiO_2
K_2O	Feldspars	SiO_2, Al_2O_3, Na_2O
	Nepheline syenite	SiO_2, Al_2O_3, Na_2O
SiO_2	Clay	Al_2O_3
	Feldspars	Al_2O_3, Na_2O, K_2O
	Nepheline syenite	Al_2O_3, Na_2O, K_2O
	Quartz sand	...
	Talc	MgO, CaO
	Wollastonite	CaO
	Zircon	ZrO_2
SrO	Strontium carbonate	...
ZnO	Zinc oxide	...
ZrO_2	Zircon	SiO_2

(a) Lead bisilicate, as sold commercially, is a frit rather than a crystalline raw material.

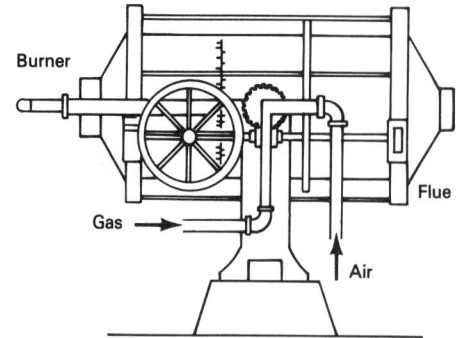

Fig 1 Key components of a rotary-type frit melting furnace

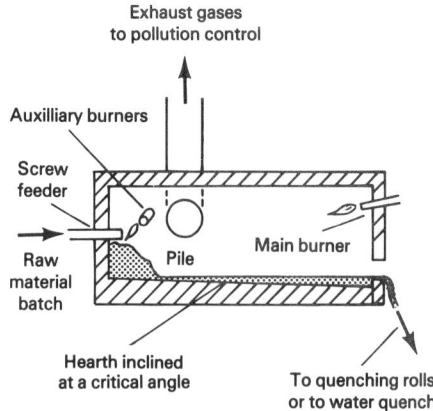

Fig 2 Schematic showing movement of raw material in a continuous frit melter

matic hammer. This latter process eliminates a water pollution problem inherent with water quenching.

Wet Application Processes

Glaze or enamel preparation involves the production of a finely ground and complete mixture of the selected raw materials. To attain this objective, most ceramic coatings are prepared as a dispersion of the raw materials in water or occasionally alcohol. This mixture, called a slip, consists of the medium (usually water), the coating ingredients, plus small quantities of minor materials such as flocculants, deflocculants, binders, and electrolytes that serve to keep the solids suspended in the liquid and to improve the adherence of the unfired coating to the substrate.

Slip Preparation

Processes for slip preparation typically involve mixing the ingredients, particle size reduction, dispersion in water, and the addition of minor amounts of additives to modify the rheological properties of the slip (Ref 7, 8). These processes are carried out together by wet grinding. Grinding involves the pulverization of the raw materials by impact and abrasion. The machinery most frequently used is a ball mill comprising a rotating cylinder partly filled with freely moving, impact-resistant shapes. An alternative method of energizing a mill is to vibrate it.

In a ball mill, particle size reduction arises from the cascading action of the media on the grain or lumps of the batch as the mill rotates. The factors that control mill efficiency are (Ref 9–11):

- Type of mill lining
- Type and size of grinding media
- Weight and ratio of grinding media, coating components, and water in the mill
- Mill speed
- Particle size of feed
- Consistency and hardness of the feed

The mill loading with grinding media, dry batch charge, and water is critical (Ref 7). The grinding media and intervening voids should occupy about 55% of the mill volume. The dry batch charge should occupy 11 to 18% of the total mill volume. The amount of water charged is usually about 50 to 55% of the total slip volume.

The mill speed must be controlled within specified limits (Ref 12). At very low rotational speeds, the media and coating components remain at the bottom of the mill. If the mill speed is too fast, centrifugal forces drive the contents outward onto the mill lining, where they remain stationary relative to the lining. The highest useful mill speed is called the critical mill speed. It can be estimated from:

$$21D^{-1/2} \leq \text{critical speed} \leq 30D^{-1/2} \qquad \text{(Eq 1)}$$

where the critical speed is in revolutions per minute (rev/min) and D is the mill diameter in meters. A mill speed of about 60% of the critical speed gives the best results.

Grinding times average 6 to 18 h in production-scale ball mills. In vibroenergy separator mills, $1^1/_2$ h would be a typical grinding time. The time of milling is controlled by periodically measuring the particle size.

Mill Additives

It is rare for a mixture of ground frit and raw materials suspended in water to be useable for a coating as is, particularly with any degree of reproducibility. Additions of rheology modifiers are required to (Ref 7):

- Control slip viscosity
- Alter thixotropy
- Overcome sedimentation
- Improve wetting properties

Table 2 Composition and application of selected frits used for glazes

Frit	Cone classifier	Composition, wt%														Glaze application
		Al_2O_3	B_2O_3	BaO	CaO	F	K_2O	Li_2O	MgO	Na_2O	PbO	SiO_2	SrO	ZnO	ZrO_2	
A	5	5.4	12.3	27.7	...	...	...	...	...	5.5	...	49.1	...	...	...	Partially fritted matte
B	4	5.1	9.2	...	12.1	...	3.6	...	...	1.7	...	59.9	8.4	...	...	All-fritted leadless
C	4	6.2	8.0	...	10.4	...	2.7	...	...	1.8	17.1	53.8	...	...	...	All-fritted lead
D	01	8.0	18.2	...	1.3	...	1.6	0.5	0.3	15.2	...	53.4	...	1.3	...	Partially fritted tile
E	01	12.1	12.7	...	0.4	2.3	5.0	...	...	11.9	...	43.6	...	...	12.1	High opacity tile
F	06	...	23.3	...	20.0	...	...	...	...	10.4	...	46.4	...	...	...	Partially fritted, alkali borosilicate flux
G	06	...	21.3	...	...	...	...	...	...	...	44.7	34.0	...	...	...	All-fritted lead for artware

- Control drying time
- Improve unfired or green strength

Binders. After drying but prior to firing, most ceramic coating layers are friable and can be easily damaged in the process of preparing the ware for firing. The addition of binders or hardeners is therefore necessary. The binder acts as a temporary cement holding the glaze particles on the surface until the firing operation. The amount of binder added can range up to 3% by weight, but 0.5% is considered typical.

Cellulose Ethers. The most commonly used binders are cellulose ethers. Cellulose is a partly amorphous and partly crystalline solid that is insoluble in water. By etherification, derivatives are prepared that are water soluble (Ref 13). A choice can be made between sodium carboxymethyl cellulose (CMC) and methyl cellulose (Methocel). These two products are available with a range of viscosities depending on the degree of polymerization. In water, polymerized cellulose swells to give clear solutions with viscosity dependent on the concentration and the molecular weight of the polymer. Lower viscosity grades are preferred for glaze hardening.

Gum tragacanth is a natural hydrophilic gum derived from a bush found in much of Asia. It is only partially soluble in water, in which it swells to form first a gel and then a sol. These sols have low surface tension and are useful as coating stabilizers (Ref 8).

Polyvinyl Alcohol (PVA). There are numerous PVA compounds that are efficient binders. Low molecular weight versions disperse more readily. Additions of up to 1% to the slip produce tough coherent layers. Wetting agents also improve the performance of polyvinyl alcohol binders.

Electrolytes. Another large group of additives are the electrolytes. In a slip, the solid particles can either be individually dispersed or agglomerated into flocs (Ref 14). Hence, control of the dispersion of the particles is critical.

Deflocculants. Materials that disperse solids as individual particles are called deflocculating electrolytes or deflocculants (Ref 7) and are of two types: the polyanion and the alkali cation. Polyanion deflocculants are complex salts of sodium and phosphoric acid and include:

- Sodium tripolyphosphate
- Tetrasodium pyrophosphate
- Sodium metaphosphate

In very small amounts, tetrasodium pyrophosphate will reduce the viscosity of pastes to fluid suspensions.

Alkali cation deflocculants include monovalent salts such as sodium nitride, borax, sodium aluminate, NH_4OH, Na_2CO_3 or K_2CO_3, sodium silicate, and KCl.

Flocculants can be used to control coating density and thinning of the slip during storage. Flocculants, which are divalent or high cation salts, include $CaCl_2$, $MgSO_4$, $CaSO_4$, $Ca(OH)_2$, alum, and NH_4Cl (Ref 15).

Suspending Agents. Most coating formulations contain a proportion of colloidal material, usually clay, that provides support in suspension for the inert pseudospherical components. Coatings without such colloidal material or with a high concentration of coarse constituents require additional suspending agents. Bentonite refers to a class of montmorillonite clays that are up to five times more effective as a suspending agent than normal clay. Bentonite is effective at concentrations of 0.5 to 2.0%. Most binders also have some suspending power and can be considered as suspending agents.

Wetting agents promote application by lowering the surface energy of liquids. They are of particular interest when underglaze decoration with organic pastes is specified. Adding a wetting agent to the water-based coating permits coating over the underglaze printing. The wetting agent additions are usually less than 1%.

Foam is deleterious to ceramic coatings. Phosphate ion is an effective foam-control agent, particularly if it can be added so that the cation acts as a deflocculant (Ref 7).

Organic Color-Code Dyes. All coatings, even those with pigments added, appear white in the green, or unfired, surface. Hence, organic dyes can be added to color-code the coating.

Bactericides. Many of the organic additives will undergo degradation in storage due to bacterial action. Hence, bactericides are needed to prevent their action. One example is a material called tris nitro, added at a concentration of ~0.01%.

Application Techniques

By the time a ceramic coating is ready for application, substantial value has been added to each component. Hence, the application process must be straightforward, foolproof, reproducible, cost-effective, and flexible (Ref 7).

For each coating application, there is an optimum application thickness. Usually, it is the thinnest coating that will give a smooth and uniform coverage. Substrate smoothness is the major parameter determining minimum coating thickness because it is difficult for a ceramic coating to heal across thin patches or to level over irregularities in the substrate. Normally, a coating thickness of 0.15 to 1 mm (6 to 40 mils) wet or 0.075 to 0.5 mm (3 to 20 mils) dry is applied.

Selection of the application technique is one of the most important decisions the coatings engineer must make. Criteria for this selection include:

- Type of ware
- Shape and size of ware
- Throughput required
- Energy and labor costs
- Space availability

Dipping is a simple, efficient, and rapid coating technique that requires no capital equipment. The dipping technique depends on the size and shape of the ware. The ware is immersed in the coating slip, moved around in a controlled way, removed from the slip, shaken to remove excess slip, and set down to drain and dry. Any bare spots are touched up with a finger wetted with coating material.

The surface texture of the fired glaze that has been dipped is affected by many interrelated factors:

- Slip density
- Slip viscosity
- Thixotropy of the slip
- Particle size of the coating
- Porosity of the ware
- Thickness of the ware
- Ware temperature
- Immersion time
- Slip additives
- Operator skill level

Of the parameters listed, operator skill is of paramount importance.

Spraying is a process where a coating slip is broken down to a cloud of fine particles that are transferred to the substrate by either pneumatic, mechanical, or electrical forces. Spraying requires a gun, a container or feed mechanism, an impelling agent, and a properly designed hood or booth maintained under negative pressure (Ref 16). Thick or thin coatings can be applied to substrates of varying porosity. Large or intricate shapes can be uniformly coated.

The quality of a fired ceramic coating that has been sprayed depends on numerous factors:

- Slip density
- Slip viscosity
- Thixotropy of the slip
- Fineness of the grind of the coating material
- Air pressure
- Air flow rate
- Coating feed rate
- Nozzle aperture
- Operator expertise
- Substrate porosity
- Substrate temperature
- Number and position of the spray guns

Spraying lends itself to high-volume automated systems (Ref 17). The workpieces are continuously fed via a wire belt through a preheat tunnel before the coating is applied from a battery of angled spray guns. Coating reclaim is an essential part of automated systems.

Slip can also be applied mechanically with a rotating atomizer. Slip is passed through a hollow spindle onto a set of closely spaced rotating discs. Centrifugal force throws the coating into a fan of droplets. The primary use of this technique is to produce textured coatings on tile.

Electrostatic Spray Coating. If the substrate is conductive (that is, a metal), the surface quality and uniformity of a ceramic coating can be improved by using the electrostatic spray coating technique (Ref 18, 19). In this system, the slip is broken into droplets either by air atomization or by centrifugal force from a sharp edged rotating surface. The drops acquire a high negative charge, are dispersed as a fine mist, and are then driven forward to the grounded substrate while following the lines of force. Hence, coating material can reach the underside of the workpiece and full edge coverage is achieved as a homogeneous cloud develops around the workpiece.

Additional Methods of Applying Coatings. There are other techniques for specific applications. Tile requires only one face to be glazed, but it must be covered with a very smooth coating. This suggests the waterfall or curtain technique may be utilized (Ref 7). A continuous feed of tiles is carried under a curtain of fluid slip. The slip continuously recirculates in the machine as it flows from an overlying reservoir in a controlled curtain to a lower sump from which it is pumped through a filter to the upper reservoir.

Painting and brushing are seldom used except by the craft potter for special effects and for applying glaze to areas inaccessible to the spray gun. One limitation of brushing is streaks, which cannot be entirely eliminated.

Silk-Screen Process. For substrates used in electronics, which require precisely positioned areas of coating, the silk-screen process can be used (Ref 20). Silk screening is also used for decorative effects on tiles. Finely powdered dry coating material is dispersed in a medium (usually a pine-oil based organic medium) to form a smooth paste. Using a squeegee, this paste is pressed through the open areas of a fine-mesh screen stretched on a frame. Once made of silk, the screen is currently made of metal or polyester. A substrate is positioned under the screen so that as they come in contact, the paste adheres to the substrate in the precise pattern of the design on the screen.

Doctor Blade Method. Another technique used primarily in electronics is the doctor blade. In this technique, a strip of unfired substrate is passed through a container of slip and positioned under a precisely set blade or several blades. The blades are set for a precise thickness of coating material across the surface of the substrate. The strip is allowed to dry before being cut to length and then fired in a kiln.

Dry Application Techniques

There are a few application techniques that do not require the preparation of a slip, including flame spraying, dry-powder cast iron enameling, and electrostatic dry-powder enameling.

Flame spraying can be used to apply ceramic coatings in the molten state to heat-sensitive or massive substrates. Most ceramic coating materials used currently can be applied by flame spraying (Ref 2). Silicates, silicides, carbides, oxides, and nitrides have all been deposited via this process.

In this process, the coating material is melted and then projected as heated particles onto the substrate where it instantaneously solidifies as a coating. Three methods of heating and propelling the particles in a plastic condition to the substrate surface are used:

- Combustion flame spraying (used for coating materials that melt readily)
- Plasma-arc flame spraying (used for very refractory materials, such as metal carbides)
- Detonation gun spraying (used for hard wear-resistant materials such as tungsten carbide)

Flame spray coatings generally lack smoothness and are usually porous. They are, therefore, limited to applications such as thermal barrier coatings (where porosity is desired) and wear-resistant coatings (where the materials cannot be applied readily by any other technique).

Dry-Powder Cast Iron Enameling. In dry-powder cast iron enameling, a very thin coat of ground coat enamel slip is applied to the cold casting, usually by wet process spraying (Ref 21). After drying, the casting is heated in a furnace to red heat, withdrawn from the furnace and, while still hot, dusted with dry powdered frit by means of a vibrating sieve placed over the surfaces to be coated. The powdered frit melts and adheres as it falls on the hot surface. Application of the frit continues until the casting has cooled sufficiently to the point that the powder will no longer adhere. The casting is then returned to the furnace to be reheated prior to an additional dusting of enamel. From one to five such dusting sequences are required to apply a coating of 1 to 1.5 mm (40 to 60 mil) thickness.

The principal limitation of dry-powder cast iron enameling is the high skill level required of the dusting equipment operators to obtain a uniform coating. The recent introduction of robotics into the dusting operation, in which the techniques and movements of skilled human operators served as a model for the programming of the robots, indicates that the machines have the potential to produce high-quality coatings.

Electrostatic Dry-Powder Coatings. The most important dry application method and the one most recently introduced, is dry-powder electrostatic application of all-fritted coatings to conductive substrates. This technique, first developed for application of organic paints, involves charging individual coating particles at a high voltage and then spraying them towards the substrate surface. Charging of particles requires very high resistance and low density, both of which are better for organic particles than for ceramics (Ref 22). Hence, while a charged organic particle will be attracted to a grounded substrate and will adhere to it without loss of charge, a ceramic particle will be deposited on the workpiece but will lose its charge and subsequently fall off. To solve this problem, frits that are to be electrostatic dry-powder coated are first encapsulated in an organic silane to provide the electrical properties of organic powders.

After grinding of the frit down to 30 μm (0.002 in.), the powder is dried and then injected with up to 0.5% of encapsulant. This coated material is suspended in clean compressed air into a fluid-like state in a fluidized-bed container (Ref 21). The fluidized powder is siphoned by high-velocity air flowing through a venturi and is propelled through powder feed tubes to special electrostatic powder guns for low-pressure application. The powder leaves the spray gun as a diffuse cloud and is propelled towards the substrate. The powder carries a potential of up to 100 kV, which causes the powder to seek out and to adhere to the grounded workpiece. The deposited coating is fragile compared to other application methods so that transfer to the firing kiln is difficult.

Spray plant manufacturers are now offering complete recirculating electrostatic powder spray machines, claiming a 95 to 99% efficiency in powder usage. Optimum conditions of temperature and humidity are recommended, to be controlled within narrow limits such as 40 to 50% humidity with temperatures ranging from 20 to 25 °C (70 to 75 °F).

Firing

After application, most ceramic coatings must be fired to convert the adhering particles to a continuous coating (Ref 5). Equally important, the coating materials must react with the substrate to form a firmly adherent ceramic-to-substrate chemical bond.

During the firing process, there will be considerable evolution of gases. This evolution must be completed before the coating materials melt sufficiently to heal over the surface of the substrate. The coating materials must then react to form a molten glass and flow sufficiently to produce a smooth surface before the workpiece is cooled. For coatings that devitrify or crystallize on cooling, that reaction must be completed during the cooling portion of the firing.

It is not possible to describe in this article all the different types of furnaces currently used for ceramic coating (see the article "Furnaces and Related Equipment" in this Volume). They can be classified into two basic types: batch or box furnaces and continuous furnaces or tunnel kilns.

Box furnaces come in all shapes and sizes. They are used mainly in laboratories and for shorter production runs. Although most box furnaces are loaded from a door on the front some are top loading and others bottom loading, requiring the use of a hydraulic lift. They can be instrumented to provide control as precise as desired that is consistent with the mass of the kiln. Obviously, a large kiln reacts more slowly to changes than a small unit. Modern low-mass insulation materials have greatly alleviated this limitation.

Continuous Furnaces. The work horses of the industry, however, are the continuous furnaces. These furnaces vary widely in size and are often circular or U-shaped in configuration for efficient loading and unloading and to minimize heat loss. They consist of three zones (preheat, fusing, and cooling), and the length of the fusing zone determines the maximum throughput. Energy savings of up to 50% are possible because of the use of low thermal mass insulation in the construction of these furnaces.

Fuel Costs. Energy consumption is a major concern in determining the cost of ceramic coatings. Two factors have to be considered: the cleanliness of the furnace atmosphere and the cost of the fuel. Electricity is the cleanest fuel, but it can be the most expensive. Most kilns in the United States today are fired with natural gas.

Single-Fire and Two-Fire Processing. There are two firing processes, both of which are extensively utilized (Ref 7). In the single-fire process, the coating is applied to a green (unfired) substrate and both the coating and the substrate are fired simultaneously in the same firing operation. In the two-fire process, the substrate is formed and fired prior to application of the coating. The coating is then fired at a temperature and time at which the substrate will not deform during the coating fire sequence.

A single-fire process is substantially lower in cost than a two-fire process. However, it is much more difficult to eliminate bubble defects, many of which arise in the maturing of the substrate from the coating. The maturing of the substrate and the coating must be carefully matched, a condition that often leads to difficult compromises. Some substrates cannot be single-fired because they must be supported during initial firing to prevent distortion. Recycling of defective workpieces is difficult to impossible with single-firing.

Coating Fit

Ceramic coatings are brittle materials. Thus, although they are very strong in compression, they are readily subject to tensile failures (Ref 5). When a ceramic coating is subjected to excessive tensile stresses, it develops fractures that are called crazing. The reverse case, in which a glaze with a low expansion coating causes tensile failure of the body or the glaze-body interface, is a condition called shivering or peeling.

Satisfactory ceramic coatings must not craze or shiver (also known as peel), either in manufacture or subsequently in service. A coating which meets these requirements is said to fit.

To evaluate fit, information is required on the thermal expansion of both the coating material and the substrate. Stress in a coating layer develops as a result of a differential thermal expansion between the coating and the substrate as they cool from the setting temperature to room temperature. The magnitude and sign of the stress will determine whether the coating will craze, be serviceable, or shiver.

Consider a ceramic body coated with a glaze at high temperature with the glaze-forming reactions complete. As the workpiece moves through the cooling cycle, the coating begins to solidify until a temperature is reached where the coating and substrate are rigidly bonded. At this point, the effect of any difference in the coefficients of thermal expansion of the glaze and of the substrate will become apparent (Ref 7).

Shivering. If the coefficients are equal, both coating and body will contract in unison and no strain will be generated. If the expansion coefficient of the substrate is greater than the coefficient of expansion of the coating, the substrate would contract a greater amount during cooling if it were not bonded to the coating. As a result, the coating will be compressed and the substrate will be permanently stressed in tension. At modest stress levels, this condition is acceptable and even desirable. However, when the difference in contraction between substrate and coating is much larger, the strength of adhesion between glaze and substrate may be exceeded, thus producing a stress-relieving fracture at the interface known as shivering or peeling.

Crazing. When the coefficient of expansion of the coating is greater than that of the substrate, the coating will contract more than the substrate. Because the final lengths of the two parts must be equal in the bonded state, the coating will compress the substrate and must therefore carry a balancing tensile stress. Even when the coating expansion is only slightly greater than the substrate expansion, the coating fails in tension because ceramic coatings cannot withstand tensile forces. A pattern of cracks in the glaze, called crazing, is the result.

The relative shape of the thermal expansion curves of the coating and the substrate can also affect the coating fit. The setting point (that is, the temperature below which coating-substrate interface can no longer relieve stress) is usually somewhat higher than the temperature at the top of the linear portion of the thermal expansion curve of the coating. A coating with a large expansion between the setting point and the top of the linear portion of the expansion-temperature curve will be more apt to craze than a coating with less expansion in this critical temperature zone.

Minimizing Coating Defects

To meet product specifications, a ceramic coating must be free of defects. A knowledge of the cause of these imperfections and the methods available to eliminate these defects from the process is essential to the operation of any coating process (Ref 23).

Crazing is the formation of a network of fine cracks within the glaze (Ref 7). Both primary and delayed crazing are caused by tensile forces on the glaze that are greater than it can withstand. Primary crazing refers to a craze pattern observed when the workpiece is removed from the firing kiln and is caused by poor glaze fit. Delayed crazing refers to crazing that appears after the workpiece is in service. This type of crazing is due to moisture expansion of the substrate (Ref 24).

Crazing resistance is improved by:

- Reducing the coefficient of thermal expansion of the coating
- Increasing the setting point of the coating
- Decreasing the thickness of the coating applied
- Increasing the coefficient of thermal expansion of the substrate
- Increasing the flux content of the body, which generally increases its thermal expansion coefficient

Shivering or Peeling. When peeling or flaking does result, the coating flakes or peels away from the surface of the body, usually along the interface (Ref 7). Peeling is controlled by:

- Increasing the thermal expansion of the coating
- Decreasing the setting point of the coating
- Decreasing the thermal expansion of the body

Bubbles. All vitreous coatings contain some bubbles (Ref 25). Most of the smaller bubbles have minimal effect upon the coating quality and can be ignored. However, larger bubbles disturb the coating surface and cannot be ignored. There are numerous sources of the gases that lead to bubble formation:

- Water trapped in frit
- Air trapped in the coating slip, arising either from excessive additions of wetting agents or from too vigorous agitation of the slip
- Air in the voids between the particles of the coating in the unfired layer
- Decomposition and interaction of coating components (notably carbonates, fluorides, clays, and organic compounds)
- Water vapor trapped by too rapid sealing of the coating
- In porous bodies, from water vapor and other gases released from the system and also entrapped in the glaze during melting
- Breakdown of organic binders

Blisters are large bubbles within the glaze that destroy the surface smoothness (Ref 7). Many factors beside the amount of gas evolved from the coating affect blistering. Blistering can be prevented by

- Reducing the quantity of those constituents that decompose with the evolution of gas
- Reducing the quantity of those mill additions that release gases
- Completely drying the coated workpiece before firing
- Reducing the maximum temperature reached in glost firing (the process of glazing and firing ceramic ware that has previously been fired at a higher temperature)
- Raising the maturing temperature of the coating
- Adjusting the firing cycle to give a slower rate of temperature rise
- Reducing the thickness of the glaze layer

Dimples and pinholes are bubbles that have burst and only partially healed (Ref 7). Where the surface is extensively marred by numerous dimples or pinholes, the defect is called eggshell or orange peel. There is no single solution for this fault. Four key factors determine the severity of dimples and pinholes in the workpiece:

- Thorough melting of any frits used
- Melting temperature of the coating, which must not heal over too early in the firing cycle
- Application technique
- Presence of any soluble vanadium salts

Specking. A speck is a discrete particle of unreacted or unwanted material and is most often noticed when it is a dark color. Although there are many causes of specking, the defect is usually caused by the contamination of the workpiece by extraneous material. Sources of contamination in the workpiece include:

- Impurities in the raw materials
- Wear in processing equipment during milling and storage
- Inadequate cleanliness during application
- Dirt from the firing kiln

Crawling is the exposure of areas of substrate. It is the result of a lack of wetting or of the absence of bonding between the glaze and the body (Ref 26). This defect can be overcome by:

- Keeping the clay content of the glaze as low as is consistent with proper slip suspension

- Avoiding overgrinding of the slip
- Keeping the quantity of opacifiers to a minimum
- Keeping the glaze thickness to the minimum consistent with good coverage

Metal Marking. Ceramics cannot be cut but they can be scratched with hard metals (Ref 27). Any metal cutlery will damage a coating surface to some degree, but on some occasions there is a trace of metal left behind on the coating. Another type of marking is the scuffing or scarring observed when the workpieces are stacked on top of each other without an intervening spacer. Placing cardboard between the workpieces solves this problem.

REFERENCES

1. R.A. Eppler, Glazes and Enamels, *Advances in Ceramics*, Vol 18, J.F. MacDowell and D. Boyd, Ed., American Ceramic Society, 1986
2. T.A. Taylor, C.G. Bergeron, and R.A. Eppler, Ceramic Coating, *Metals Handbook*, Vol 5, 9th ed., American Society for Metals, 1982, p 532–547
3. R.A. Eppler, Chapt. 4, Glazes and Enamels, *Glass Science and Technology*, Vol 1, Academic Press, 1983, p 301–337
4. W.M. Simpson, Intricate Glass Parts from "PM" Techniques, *Mach. Des.*, 8 Sept 1977
5. P. Rado, *An Introduction to the Technology of Pottery*, 2nd ed., Pergamon Press, Oxford, 1988
6. L.D. Gill and R.A. Eppler, Chapt. 7, Lead Frits, J.S. Nordyke, Ed., *Lead in the World of Ceramics*, American Ceramic Society, 1984, p 99–105
7. J.R. Taylor and A.C. Bull, *Ceramics Glaze Technology*, Pergamon Press, Oxford, 1986
8. J.S. Reed, *Introduction to the Principles of Ceramic Processing*, John Wiley & Sons, 1988
9. A. Laurs, Segregation of Glaze Suspensions during the Formation of Glaze Layers on Ceramic Bodies, *Keram. Z.*, Vol 24 (No. 9), 1972, p 490–492
10. C.W. Parmalee and C.G. Harmon, *Ceramic Glazes*, 3rd ed., Cahners Publishing, 1973
11. M. Melandri, Factors Affecting the Milling of Ceramic Glazes, *Ceram. Inf.*, Vol 4 (No. 2), 1969, p 70–73
12. K.A. Maskall and D. White, *Vitreous Enamelling*, Pergamon Press, 1986
13. T.A. Smith, Organic Binders and Other Additives for Glazes and Engobes, *Trans. Brit. Ceram. Soc.*, Vol 61 (No. 9), 1962, p 523–549
14. M. Ish-Shalom, I. Yaron, and D. Gans, Thinning of Slurries in Wet Process Manufacture of Cement, *Am. Ceram. Soc. Bull.*, Vol 50 (No. 9), 1971, p 733–741
15. E. Alston, Stabilization and Binding of Glazes—a New Approach, *Trans. J. Brit. Ceram. Soc.*, Vol 73 (No. 2), 1974, p 51–55
16. W.A. Bloor and R.E. Eardley, Environmental Conditions in Sanitary Whiteware Shops, II. Glaze Spraying Shops, *Trans. J. Brit. Ceram. Soc.*, Vol 77 (No. 2), 1978, p 65–69
17. M. Whitmore, Spraying of Earthenware Flatware, *Trans. J. Brit. Ceram. Soc.*, Vol 73 (No. 4), 1974, p 125–129
18. K. Hebberlein, Electrostatic Glazing of Tableware, *Ber. Deut. Keram. Ges.*, Vol 53 (No. 2), 1976, p 51–55
19. M. Lambert, Industrial Application of Electrostatic Enameling to Parts in Sheet Steel and Cooking Equipment, *Vit. Enameller*, Vol 24 (No. 4), 1973, p 107–109
20. W.S. Young, Chapt. 7, Multilayer Ceramic Technology, *Ceramic Materials for Electronics*, R.C. Buchanan, Ed., Marcel Dekker, 1986, p 403–442
21. ASM Committee on Porcelain Enameling, Porcelain Enameling, *Metals Handbook*, Vol 5, 9th ed., American Society for Metals, 1982, p 509–531
22. H. Emlemdi, The Electrostatic Application of Porcelain Enamel Powder, presented at the 91st annual meeting, American Ceramic Society, Indianapolis, Apr 1989, to be published
23. F. Singer and W.L. German, "Ceramic Glazes," Borax Consolidated, London, 1960
24. W. Lehnhauser, The Effect of Humidity on Moisture Expansion of Glazed Ceramics, *Sprechsaal*, Vol 102 (No. 23), 1969, p 1061–1068
25. T.J. Harper, Solubility of Gases in Glass-Review, *Glass Technol.*, Vol 3 (No. 5), 1962, p 171–175
26. D.W. Budworth, Theoretical Approach to the Crawling of Glazes, *Trans. J. Brit. Ceram. Soc.*, Vol 70 (No. 2), 1971, p 57–59
27. J.J. Bailey, The Scratch-Resisting Power of Glass and Its Measurement, *J. Am. Ceram. Soc.*, Vol 20 (No. 2), 1937, p 42–52

Structural Applications For Technical, Engineering, and Advanced Ceramics

Co-Chairmen: M.K. Ferber and V.J. Tennery, Oak Ridge National Laboratory

Introduction

THIS SECTION discusses applications for technical, engineering, and advanced ceramics. The application areas include mineral processing, machine tools, wear components, heat exchangers, automotive products, aerospace components, and medical products. While many of these areas have been identified as critical growth markets for ceramic materials, the current utilization of ceramics is still limited by technical issues pertaining to fabrication technology and long-term structural reliability. Once these issues are solved, the engineering ceramics market in the U.S., which is currently estimated to be slightly over $700 million, could approach $2 billion within 10 years.

The application areas cited above make use of critical properties of ceramics such as high-temperature stability, high-temperature strength, wear resistance, corrosion resistance, and chemical inertness. In many instances the application involves the substitution of ceramics for more traditional metallic components. Here the advantages to be gained include (1) lower cost of the raw materials, (2) less dependence upon strategic foreign raw materials often required for high-temperature alloys, and (3) cost savings arising from the use of lighter weight ceramic components. The paragraphs that follow provide a brief discussion of the motivation for using ceramics in many of these application areas as well as the associated technical issues remaining to be solved. The articles in this Section then provide a complete discussion of each application.

Ceramics have found use in minerals processing for nearly 30 years as a result of their substantially better wear performance. Chute linings, pipe linings, cyclones, and pumps and valves are the four primary areas of application where high volume exists. Sintered alumina, silicon carbide, and fused ceramics (alumina-zirconium-silicate and basalt) are the most commonly used ceramics for handling and processing equipment.

Ceramics for cutting tools represent a fairly well developed market. Ceramic cutting tools offer increased productivity in metal cutting through decreased wear and higher temperature capability as compared with their metallic counterparts. Consequently, ceramic cutting tools may be operated at higher feed rates or higher cutting speeds thereby reducing the overall cutting time while increasing tool life. Traditional ceramic material systems have been based upon aluminum oxide (alumina). However, because of the low toughness of this ceramic, alumina cutting tools have been limited to uninterrupted cuts at relatively low feed rates and depths. Currently, more advanced material systems are based upon cubic boron nitride, Sialon ceramics, and alumina which is reinforced with either silicon carbide whiskers or titanium diboride particulates.

There are a number of general applications which make use of the high wear resistance offered by many ceramic materials. These include dies for hot extrusion of metal rods and tubes, wear-resistant inserts for tablesetting and paper lapping dies, dies for cold extrusion of metals such as aluminum and copper, and seals. A prime candidate for many of these applications is transformation-toughened zirconium oxide (TTZ). TTZ ceramics generally exhibit high strength and toughness at temperatures less than 500 °C (930 °F). These properties result from the operation of a stress-induced martensitic transformation which increases the fracture resistance. Because the driving force for this transformation decreases with temperature, the toughness also drops as the temperature is raised.

The utilization of ceramic heat exchangers in waste heat recovery has received extensive interest over the last 5 to 10 years. Commercial products now include plate-fin and tubular recuperators. Two other potential application areas are process heat exchange and power generation heat exchange. The former involves both liquid-to-liquid and liquid-to-vapor heat transfer. In the case of power heat exchange, the ceramic heat exchanger transfers heat to a working fluid. The advantages of ceramics over their metallic counterparts are higher temperature capability, greater corrosion resistance, and longer-term mechanical and thermal stability. Problems remaining to be solved include (1) the development of high-temperature seals, and (2) heat exchanger size limitations arising from processing restrictions.

The utilization of ceramics in automotive applications is still in the early stages of development. Ceramic materials are currently being considered for application in (1) conventional internal combustion engines, (2) adiabatic diesel engines, and (3) advanced gas turbines. In the case of the internal combustion engine, commercialization of ceramic components such as intake and exhaust valves, valve seats, bearings, cam followers, exhaust port liners, and turbocharger rotors is becoming a reality. The major advantages to be gained include reduced inertia, friction and fuel consumption, lower mass, and improved wear resistance. Sialon, silicon carbide, and silicon nitrite are the primary material candidates in these engines.

Both ceramic coatings and monolithics have been considered for use in the adiabatic diesel engine. Critical components include ceramic cylinder liners and piston caps. Silicon carbide, aluminum oxide, and TTZ represent candidate material systems.

Applications for ceramic materials in the gas turbine engines include both stationary components (for example, turbine vanes, shroud rings, combustor body, and scrolls) and moving components (for example, turbine rotor and regenerator disks). Candidate material systems are typically based upon silicon nitride, silicon carbide, and aluminosilicates. When compared with conventional spark ignition engines, the ceramic gas turbine is expected to offer greater fuel efficiency, lower gaseous emissions, lower particulate levels, and multiple fuel capability. However, commercialization of the ceramic gas turbine still faces many challenges including improvement of component properties (particularly for the rotor) and development of design methodologies capable of describing the reliability of brittle ceramic materials.

Ceramics for aerospace applications have traditionally focused upon coatings used for their thermal insulating properties. However, in recent years considerable effort has been devoted to the development of ceramic gas turbines for aerospace vehicles. Here the materials issues are similar to those for the automotive gas turbine. Furthermore, the development of fiber-reinforced, ceramic-based composites has raised the possibility of using these materials as structural components in advanced aerospace vehicles.

The utilization of ceramics for medical applications (primarily hard tissue replacement) has received considerable attention over the last 10 to 20 years. A critical advantage of using ceramic materials over more traditional metallic components is that of greater biocompatibility which reduces toxicity problems often encountered with metals. Furthermore porous ceramics such as aluminum oxide have shown the ability to allow bone ingrowth thus creating a strong mechanical bond. Biocompatible glass-ceramics have also been developed which are capable of forming a chemical bond with bone. Specific application areas for ceramics include hip prostheses, dental implants, heart valves, and middle ear implants.

Mineral Processing

Jerry Weinstein, Alanx Products L.P.

IN THE MINERAL PROCESSING IN-DUSTRY, large volumes of ores, tailings, and other materials are handled. The abrasive nature of most of these materials results in extensive wear to both handling and processing equipment. Ceramics have found use in minerals processing in the past 30 years as a result of substantially better wear performance, both in extended life and ultimate economics realized, over previously used materials such as metals (Ref 1).

Ceramic materials differ considerably from metallic and polymeric materials with regard to wear mechanism. The causes for this are their atomic and molecular structure. For ceramic materials, the ionic or covalent bond always dominates. This type of bond is the reason for good corrosion resistance, stability at elevated temperatures, and high hardness. The high hardness increases resistance to penetration, scratching, and deformation, thus reducing wear. However, a major disadvantage of ceramics is their brittleness (Ref 2).

Materials

While wear-resistant ceramics find large-volume applications in the minerals processing industry, the extended use of ceramics is still in a stage of continuing development and will increase in the future. As ceramic processing technology and additional suppliers bring costs down with improvements in mechanical and wear properties, the application areas will increase.

Sintered alumina, silicon carbide based, and fused ceramics represent the high-volume ceramic materials used in the material processing industry for wear and erosion-corrosion resistance.

Table 1 shows a few basic physical and relative wear properties of the ceramic materials used in the minerals processing industry. The bulk of the ceramics used in minerals processing fall into the following categories (Ref 3):

- *Oxide ceramics,* which include sintered aluminas and transformation-toughened ceramics (such as partially stabilized zirconia and zirconia-toughened alumina)
- *Silicon carbide (SiC) ceramics* such as silicon nitride-bonded SiC, reaction-bonded SiC, and sintered SiC

- *Silicon carbide-based ceramic/metal composites (CMCs)* containing 75 vol% SiC, 15 vol% alumina, and 10 vol% metal
- *Fused cast materials* such as alumina zirconium silicate (AZS) and basalt
- *Ceramic-filled polymers, elastomers, or adhesives*

Oxide Ceramics. Currently the largest percentage of the wear applications get served with sintered alumina bodies containing 85 to 96% alumina, although bodies are available up to 99.9% for usually a higher cost along with increased processing limits. Wear resistance of alumina tiles usually increases with higher alumina content.

Sintered aluminas have high strength, good chemical resistance, heat resistance (although relatively poor thermal shock), and toughness (Ref 4). Alumina ceramics are produced in bulk quantities as tile segments or simple shapes and are applied as linings in chutes, grinding mills, and material handling operations. The material is generally inexpensive and is supplied by a wide variety of companies. In North America, major suppliers include Coors, Diamonite, PAKCO-Norton, Kyocera, and AlSiMag.

Transformation-toughened ceramics include partially stabilized zirconia (PSZ) and zirconia-toughened alumina. PSZ has received attention for various wear applications where greatly improved toughness and impact resistance is needed.

Properties of PSZ Ceramics. The mechanical and wear properties of PSZ are strongly controlled by the phase composition and stability in the microstructure, which is related to the stabilizers used as well as the processing conditions. While PSZ has high strength, toughness, and excellent corrosion resistance, it has poor thermal properties and only moderate hardness (Ref 5). The major North American suppliers include Coors, Nilcra, Kyocera, Zircoa Corning, Diamonite, and Ceramatec.

Zirconia-Toughened Alumina. By adding zirconia to alumina, increased toughness, strength, and wear resistance can be obtained, while retaining good chemical and heat resistance. The toughening mechanism is similar to PSZ (Ref 6). Unfortunately, the relative processing difficulties and increased costs have limited the extensive use of this ceramic for

abrasion resistance. North American suppliers include Diamonite.

Silicon carbide (SiC) ceramics have a higher thermal conductivity and lower thermal-expansion coefficient than alumina. These characteristics improve thermal shock resistance, which is why alumina-based cutting tool materials are reinforced with SiC whiskers. Silicon carbide is also used as the predominant base material in various ceramic products as described in the following three subsections.

Silicon nitride-bonded SiC (SNBSC) is the largest volume SiC-based ceramic used in minerals processing. Generally, SNBSC can be formed into more complex shapes and larger sizes than are capable in oxide ceramics. Silicon nitride-bonded SiC has low mechanical strength, toughness, good chemical resistance to caustic environments, excellent thermal shock properties, and heat resistance. The material is relatively inexpensive, especially for larger shapes. North American suppliers include Carborundum, Norton, Coors, and Ferro.

Reaction-bonded silicon carbide (RBSC) is a nonporous SiC-Si composite (Ref 7) having high strength, chemical resistance to caustic environments, heat resistance, and good thermal shock resistance and toughness. Processing difficulties give RBSC limits in size and geometry capabilities as well as a very high cost. Leading suppliers in North America include Carborundum, Norton, and Coors.

Sintered silicon carbide (SSC) is a fully dense, very strong, fine-grained SiC with good thermal properties and toughness, and excellent corrosion resistance to acids and caustic environments. However, due to special processing conditions and material costs, SSC has the highest cost of all materials discussed in this article. Size and geometry of fabrication are also limited. In North America, the leading suppliers include Carborundum, Coors, and Norton.

Ceramic/Metal Composites (CMCs). A fairly new CMC product is an SiC-alumina ceramic matrix containing a small percentage of metal for toughness. The fully dense, monolithic products consist of a coarse-grained ceramic phase of about 75 vol% SiC and 15 vol% alumina. The metal phase (~10 vol%) is not a binder but does improve properties

Table 1 Property comparison of various ceramic materials and abrasive-resistant steel

Product	Density		Color	Modulus of rupture		Fracture toughness, MPa	Knoop hardness, kg/mm²	Elastic modulus		Thermal expansion from 25 to 1000 °C (77 to 1830 °F)		Maximum thermal shock		Dry erosion resistance test(a); volume lost, cm³/h	Slurry erosion resistance(b)	Sliding abrasion(c), mm³/h
	g/cm³	lb/ft³		MPa	ksi			GPa	psi × 10⁶	ppm/°C	ppm/°F	Δ °C	Δ °F			
Sintered alumina, 85%(d)	3.41	213	White	317	46	3–4	960	221	32	7.2	4.0	300	540	0.65	0.85	16,000
Sintered alumina, 96%(d)	3.72	232	White	358	52	3–4	1088	303	44	8.2	4.6	250	450	0.50	1.0	14,000
Partially stabilized zirconia(e)	5.74	358	Ivory	820	118	8–12	1120	205	30	10.2	5.7	375	675	1.33	9.0	6200
Zirconia-toughened alumina(f)	4.15	259	Ivory	585	85	5.5	1380	335	49	8.0	4.4	445	800	0.70	1.8	10,200
Silicon nitride-bonded silicon carbide(g)	2.54	159	Gray	48	7	1.5	620	152	22	3.9	2.2	400	720	2.48	9.0	3700
Reaction-bonded silicon carbide(g)	3.09	193	Black	280	40	4.93	1880	380	55	5.0	2.8	350	630	0.10	40.0	1600
Sintered silicon carbide(g)	3.10	193	Black	460	67	4.6	2800	410	59	4.0	2.2	325	585	0.10	91.0	1800
Silicon carbide ceramic/metal composites(h)	3.26	203	Black	90	13	5.5	800	313	45	5.4	3.0	600	1080	0.20	45.0	1500
	3.28	205	Black	140	20	6.0	800	313	45	5.4	3.0	600	1080	0.30	25.0	2300
AZS fused cast(i)	3.60	225	Ivory	110	16	2.0	1000	...	...	7.5	4.2	300	540	...	...	...
Basalt fused cast(j)	2.90	181	Brick	64	9	1.5	880	104	15	10.1	5.6	100	180	...	...	...
NiHard IV abrasive-resistant steel(k)	7.4	462	Gray	650	94	2.7	650	104	15	7.3	4.1	>600	1130	9.62	4.5	12,200

(a) Grit blast testing with a stream of 150–300 μm silica particles at a pressure of 275 kPa (40 psi) directed at a stationary specimen at an angle of 30 degrees for five minutes. (b) Values listed are a ratio of volume loss for a reference pin (96% alumina) divided by volume loss of test specimen. Test pins rotated in an aqueous slurry of 40 wt% silica (300–600 μm) at 1750 rpm for 20 h. (c) Miller test, see text for test parameters. *Mechanical data values from product literature of:* (d) Coors Ceramic Company. (e) Nilcra Ceramics Inc. (f) Diamonite Products. (g) The Carborundum Company. (h) Alanx Products L.P. (i) Cohart Refractories Corporation. (j) Abresist Corporation. (k) Steel Casting Companies

in much the same way as cermets do. No shrinkage occurs during densification, enabling extremely large and complex monolithic shapes to be fabricated (Ref 8). This silicon carbide CMC has fair strength, chemical resistance, heat resistance, and excellent thermal shock resistance and toughness. The only current supplier of components for minerals processing is Alanx Products.

Fused cast materials include fused alumina zirconium silicate (AZS) and basalt.

Fused AZS is a relatively inexpensive fused cast material with a fine crystal size that imparts relatively high toughness to the material. Fused AZS is produced in standard shapes and large sizes including the direct casting of large monolithic pipes and elbows. Alumina zirconium silicate has good strength, toughness, chemical resistance, heat resistance, and reportedly better thermal shock resistance than sintered alumina. Hardness is lower than alumina, which results in lower wear resistance in some applications. In North America, leading suppliers include Corhart and Carborundum.

Basalt is a fused cast product with fair-to-poor mechanical and thermal properties but relatively good erosion resistance (Ref 9). However, because it is simply prepared from inexpensive raw materials (basalt rock) by a relatively low-temperature, low-cost process, fused basalt offers a significant cost advantage over other wear-resistant ceramics. It is generally fabricated in large standard tile and tubular shapes for lining chutes and pipes (Ref 9). In North America, major suppliers include Abresist.

Ceramic-Filled Polymers, Elastomers, or Abrasives. Cement, epoxy, and adhesive systems have incorporated everything from alumina beads, chips, and powders, to SiC powders. These are applied by trowelling and can be used up to 50 mm (2 in.) thick. The usual application of these materials is to repair worn patches of metal parts.

Ceramic-filled elastomer (urethane and rubber) systems are also commercially available for prefabricated wear-resistant parts. Parts made from silicon carbide-filled urethane have shown excellent slurry and dry grit blast results in chutes and hydrocyclones. Ceramics bonded to elastomer backing plates are used for chute and pipe liners to increase the impact resistance of the wear tiles. Ceramic cyclones and pump parts have been cast into urethane sleeves to increase toughness and reduce costs and ease installation in the field.

Wear of Ceramics

The most commonly encountered types of wear seen in ceramics which are used in the mining and minerals industry are abrasion, erosion, and corrosion. This wear behavior of ceramic materials is determined by three interdependent categories (Ref 10):

- The properties of the wear material
- The properties of the abrasive material
- The nature and severity of the interaction between abrasive and wear materials

Individual factors which influence wear behavior within these categories are classi-

fied in Table 2. For both abrasive and wear materials, the majority of factors relate to their mechanical properties. However, chemical processes (in particular corrosion and oxidation) also influence wear behavior, especially in the presence of elevated temperatures. The conditions which exist in mining and mineral processing equipment can be described in terms of the factors listed in Table 2, although a simpler, and more commonly used, method is to use the following general classifications (Ref 10):

- Gouging abrasion
- High stress or grinding abrasion
- Low stress or scratching abrasion
- Erosion-corrosion

These four classifications are illustrated schematically in Fig 1, along with examples of where they apply in minerals processing.

Laboratory Wear Testing. There are a large number of wear tests developed to simulate wear conditions in field applications. Three key tests (besides hardness) that have direct implications on the types of wear found in minerals processing are reviewed below. Comparative data from these three tests are listed in Table 1 (along with Knoop hardnesses) for the previously described ceramic materials.

Grit blast testing is run by directing a stream of abrasive particles onto a stationary target specimen. The target material is weighed both before and after grit blast testing and the resulting weight loss is used to calculate the erosion rate (volume loss per unit time). A

Table 2 Factors affecting wear of ceramics

Wear factor	Variables affecting wear factors	
Nature of the abrasive	Size	Wear increases with increasing particle size
	Shape	Angular particles cause about twice the wear of rounded ones
	Density	Dense materials create greater wear since kinetic energy of the particles is higher
	Hardness	Wear increases rapidly when the particle hardness exceeds that of the surface being abraded
	Concentration	In slurry systems, wear increases with concentration
Nature of contact	Velocity	The rate of wear increases rapidly with increasing velocity
	Impact angle	Wear increases when the impingement angle is increased towards 90°
	Load	Increased throughput causes an increase in wear rate
	Corrosivity	Complex interaction, but greater severity corrosion, lower wear resistance
Wear material	Fracture toughness	Greater wear resistance is usually obtained with higher fracture toughness
	Corrosion resistance	Very complex, but wear best for inert materials
	Porosity	Wear resistance best for materials with zero porosity
	Coefficient of friction	Lower coefficient of friction generally gives better wear resistance
	Grain size	For many sintered ceramics, fine grain size usually gives better wear resistance. However, with SNBSC, SCC/MC, and other multi-phase composites, the coarser grain size gives better wear, especially in abrasive slurries
	Hardness	Higher hardness ceramics usually have better wear resistance

Source: Ref 11

higher rate indicates increased susceptibility to erosive attack.

Rotating-Pin, Slurry Erosion Test. Slurry erosion characteristics can be obtained with

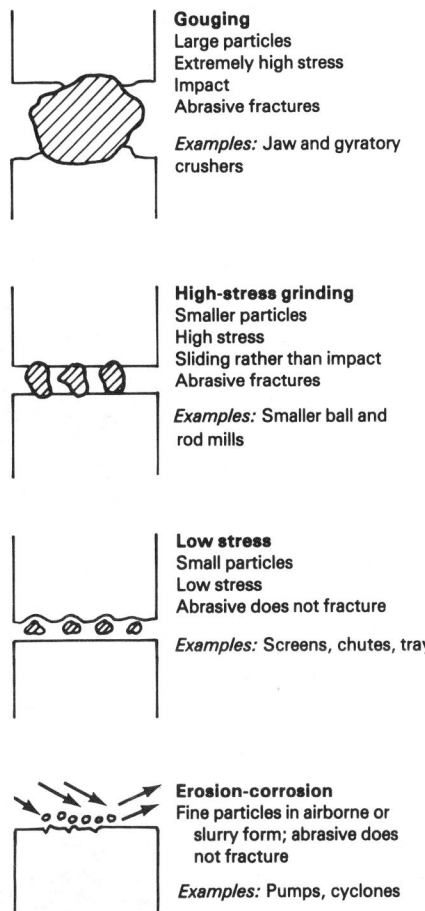

Gouging
Large particles
Extremely high stress
Impact
Abrasive fractures

Examples: Jaw and gyratory crushers

High-stress grinding
Smaller particles
High stress
Sliding rather than impact
Abrasive fractures

Examples: Smaller ball and rod mills

Low stress
Small particles
Low stress
Abrasive does not fracture

Examples: Screens, chutes, trays

Erosion-corrosion
Fine particles in airborne or slurry form; abrasive does not fracture

Examples: Pumps, cyclones

Fig 1 Classification of wear mechanisms commonly experienced in mineral processing operations

cylindrical test pins attached to a flywheel. The test pins and flywheel are then immersed into a pot containing an aqueous slurry containing abrasive particles. After testing, the pins are weighed and volume loss calculated for each material. A reference pin (for example, a 96% sintered alumina pin) is included in every test for comparative purposes. The slurry erosion resistance of a material is characterized by its wear resistance factor (that is, the volume loss of the alumina standard divided by the volume loss of the test material).

Sliding Abrasion Test. Sliding abrasion involves wear from reciprocating motion in the presence of a normal load (with respect to the direction of sliding). In the sliding abrasion test of Table 1, the material to be evaluated (wear block) is affixed to a mechanical arm and then immersed in a tray containing a 50% solids silica slurry (200 to 300 μm). With a 5 lbf load added, the wear block is then slid through the abrasive slurry at 0.16 m/s (0.5 ft/s) via the reciprocating motion of the mechanical arm. The weight loss on the wear block is used to calculate the wear rate in terms of volume loss per unit time. The magnitude of the wear rate indicates its susceptibility to abrasive attack.

Applications

High-volume usage of ceramic materials has four main application areas in mineral processing:

- Chutes
- Pipelines
- Hydrocyclones
- Pumps and valves

All of these applications involve predominantly low-to-moderate stress abrasion or erosion-corrosion conditions.

Chute Linings. The lining of chutes with alumina, AZS, basalt, nitride-bonded SiC, and SiC-based CMCs represents the greatest usage of ceramics as wear materials, in terms of both total volume and money spent on ceramics (Ref 11). Under dry conditions, sintered alumina, AZS, and basalt have been found to be cost effective in terms of a high wear life-to-cost ratio. If downtime and other associated costs are included, then the life/cost ratio of alumina is expected to be higher than basalt (Ref 11). A combination lining, using a number of different thicknesses and materials (Fig 2), makes it possible to achieve the most cost-effective selection of materials for wear resistance. Lower cost fused basalt may be used when the higher wear properties and impact resistance of alumina are not warranted, for example, under wet sliding conditions (Ref 12). In severe wet conditions, SiC-based CMCs have shown dramatic wear improvements over alumina and basalt (Ref 8). In the hard rock mining, mineral processing, and coal industries, sintered alumina tiles (mostly 85 and 90% alumina, Ref 12) and AZS tiles are the predominant ceramic used in chutes handling −300 mm ore (Ref 13). Problems in

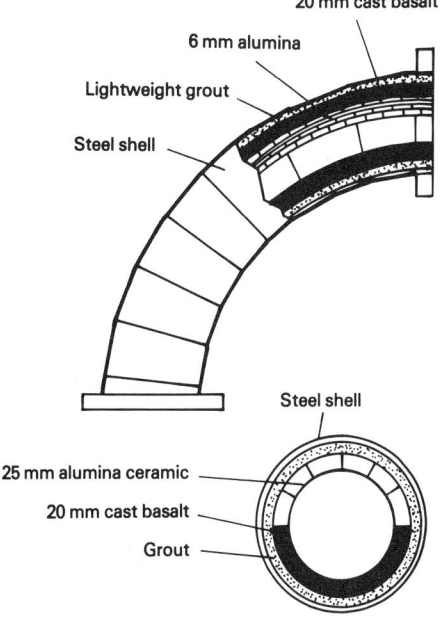

20 mm cast basalt
6 mm alumina
Lightweight grout
Steel shell

Steel shell
25 mm alumina ceramic
20 mm cast basalt
Grout

Fig 2 Example of pipe lined with different combinations of ceramic materials

having skilled personnel fit the tiles on site have largely been avoided by buying complete ready-to-install chute systems (Ref 10).

The new silicon carbide SiC-based CMCs offer improvements in impact resistance. These are available in large monolithic slabs, which have shown improved life as a liner in a primary feed line at a nickel ore mine. Larger tiles with fewer seams resulted in less wear (Ref 8). Another approach to improve the impact resistance of ceramic wear liners is the use of a ceramic filled polymer.

Pipe Linings. Ceramic pipe liners have been successfully used in both hydraulic and pneumatic piping systems. Pipes lined with alumina are available in a number of forms. Alumina bore segments (up to 300 mm, or 12 in., in diameter) are encased in either steel or fiberglass and are used in chemical processing industries. Larger bore pipes lined with alumina are produced by mortaring in tiles by a method similar to that used for lining chutes. Tapered alumina tiles specifically shaped for use in curved areas such as pipe bends are available from a number of suppliers (Ref 1, 4, 10, 12).

Basalt and AZS pipe liners can be cast directly into the steel outer pipe shell as well as cast into standard tile shapes to allow custom fits. Alumina zirconium silicate has been used for refuse handling in coal preparation plants as well as in limestone dredging operations. Basalt pipe liners are widely used in coal preparation plants and power generating stations. Silicon nitride-bonded silicon carbide finds extensive use as pipe and flue liners in coal-fired power generation stations for coal feed and fly ash handling due to its combination of wear and high-temperature resistance as well as good thermal shock properties. Standard tile shapes such as bends and straights are also produced to allow for custom installation in pipes (Ref 10).

Silicon carbide-based CMCs have been used successfully as elbow wear backs in sizes up to 300 mm (12 in.). These were manufactured as custom-made monolithic parts.

Cyclones. The operating conditions of a cyclone produce an environment which causes severe erosion of lining materials due to impingement by sharp, angular particles at very high velocities. Differences in particle sizes and the wear modes of different flow conditions in a cyclone produce markedly different erosion conditions, which can make material selection more difficult. At present, ceramics have been successfully used for the lower cone and apex liners, and for spigots, with more limited use in the upper cyclone housings (Ref 10).

In the coal industry, the use of nitride-bonded silicon carbide and SiC-based composites has resulted in good service life when raw coal sizes are finer than 2 mm (0.08 in.), while high-purity alumina (90%) ceramics have proved more cost effective for the coarser sizes. Fused AZS has also been used as complete cyclone liners and vortex finders in heavy

media cyclones in coal preparation plants (Ref 10, 14).

Flat alumina "hex" tiles (Fig 3) can be used to refurbish worn cyclone parts in coal preparation plants for approximately one-third of the cost of a new component (Ref 10). Silicon carbide-based CMCs and silicon nitride-bonded SiC have been used extensively in the hard rock mining and ceramic mineral process industries for lower cones, apex liners, and vortex finders (Ref 8). Upper cyclone and head liners up to 1.5 m (60 in.) diameter are made of silicon nitride-bonded SiC and of SiC-filled polymer systems.

Pumps and Valves. Ceramic-lined pumps are used for aggressive fluids where high abrasion, corrosion, and/or temperature degrade metal and polymeric-lined pumps (Ref 15). The ceramic surface is commonly a solid, self-supporting casting that is either cast in place or clamped inside a split housing. The liner provides wear and corrosion resistance but depends on the pump casing for mechanical strength to withstand internal hydraulic and external mechanical forces (Ref 15).

Clam shell split ceramic liners with matching ceramic impellers (Fig 4) of silicon nitride-

bonded silicon carbide and SiC-based CMCs have been marketed by several slurry pump manufacturers. In addition, ceramic shaft sleeves of alumina and SiC-based CMCs have shown lower wear and extended packing life over metal sleeves in abrasive slurry pump applications. Some pumps use ceramic inserts in only part of the inner wear surface where the erosion and abrasion causing wear are concentrated, such as inserts in the suction side wear plates and impellers or at the inlet throat and cutwater areas (Fig 5).

Breakage due to impact or installation stresses is a common complaint of abrasive slurry pump manufacturers and end users, whose main experience in ceramic pump parts has been with silicon nitride-bonded silicon carbide. The new SiC-based CMCs have shown less breakage and improved wear life in slurry pump suction side wear plates, impellers, and inlet throat liners.

Other materials such as sintered SiC, reaction-bonded SiC, and alumina are presently restricted by their processing methods to fabricating only small sizes and are in common use as shaft sleeves and pump face seal rings in place of hard alloys. Ceramic materials have been successfully used in flow control valves, such as ball valves, knife gate valves, spool

Fig 3 Vortex finder of a cyclone refurbished with alumina hex tiles. Source: Ref 10

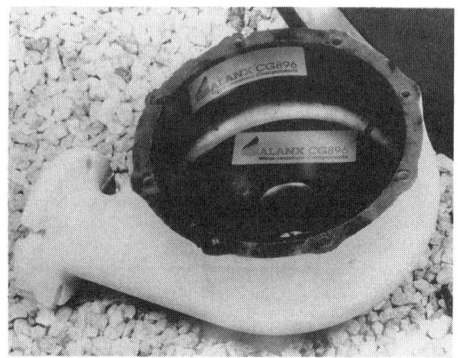

Fig 5 Wear plate in a hard-metal abrasive slurry pump replaced by a SiC-based CMC material

Fig 4 Ceramic pump liner cast into urethane backing in a clam shell design with matching impeller

Fig 6 Examples of ceramic valve components. Courtesy of Nilcra Ceramics Inc.

valves (Ref 16), and diaphragm valves (Ref 17).

Complete ceramic-lined valves and ceramic balls and valve trim (see Fig 6, for example) in metal or polymer housings have been available on the market for a number of years, but extremely high costs and long delivery times have limited their use in minerals processing. They are used successfully in gold ore autoclave processing.

REFERENCES

1. P. Verhey, Major Savings and Production Efficiences with Wear Resistant Ceramics, *Can. Min. Metall. Bull.*, Vol 74 (No. 834), 1981, p 88–91

2. D.W. Richardson, What are Ceramics?, *Chem. Eng.*, Vol 89 (No. 19), 1982, p 123–126

3. A.E. Horn, Abrasion-Resistant Ceramics—Application to the Mineral Processing Industry, *Mater. Aus.*, Vol 18 (No. 5), 1986, p 6–7

4. H. Hoppert, Aluminum Oxide Ceramics—for Wear Protection in Mineral Processing, *Aufbereit.-Tech.*, Vol 31 (No. 3), 1990, p 157–164

5. C.F. Lewis, Zirconia: The Tough Contender, *Mater. Eng.*, Vol 106 (No. 3), 1989, p 43

6. O.T. Sorensen, Wear Resistant Ceramic Materials Toughened with Zirconia Dioxide, *Structure*, Vol 3, 1989, p 5–7

7. D.W. Richardson, *Modern Ceramic Engineering*, Marcel Dekker, 1982, p 251–252

8. J. Weinstein, R. Webb, and M. Schreiner, "ALANX Ceramic/Metal Composites Applied to Mining Material Systems for Increased Life and Improved Maintenance Performance," Society for Mining and Metallurgical Exploration, Preprint No. 90–421, 1990

9. M.K. Aghajanian, E. Breval, J.S. Jennings, and N.H. MacMillan, The Erosion of Abresist, *Mater. Sci. Eng.*, Vol 91, 1987, p 257–264

10. A.M. MacDonald, P.J. Mutton, and W.J. Sinclair, Abrasion Resistant Materials for the Australian Minerals Industry, *AMIRA*, Vol 2, 1988, p 15–25, 29–41

11. P. Verhey, Mineral Processors Can Benefit from Long Life, High Efficiency with Ceramic Liners, *Can. Min. J.*, Vol 100 (No. 11), 1979, p 43

12. B. Drake, Resisting Abrasion, Downtime and Maintenance Costs, *Pit and Quarry*, Vol 81 (No. 2), 1989, p 24–25

13. R.E. Dial, How Super Materials Are Meeting Multi-Industry Needs, *Min. Proc.*, Vol 14 (No. 1), 1973, p 15–20

14. P. Darling, Cyclones—Silent Slaves of Coal Prep, *Coal*, Vol 9, 1990, p 46–52

15. R. Cunningham, Lined Pumps for Aggressive Materials, *Plant Eng.*, Vol 39 (No. 5), 1985, p 44

16. S.C. Lattin, Tough Valves with Ceramics, *Mach. Des.*, Vol 59 (No. 18), 1987, p 83–85

17. No Wear to Be Found with Ceramic-Lined Valves, *Mater. Eng.*, Vol 107 (No. 7), 1990, p 10

Ceramic Cutting Tools

James H. Adams, Bob Anschuetz, and Gene Whitfield, GTE Valenite Corporation

CERAMIC CUTTING TOOLS have four properties that distinguish them from traditional steel and tungsten-carbide cutting materials:

- They are more chemically inert
- Have a higher resistance to abrasive wear
- Have a higher hot hardness
- Are capable of superior heat dispersal during the chip-forming process

Collectively, these properties permit the user to increase his rate of metal removal while often obtaining longer tool life. Reduced operating time to produce a finished part, and reduced machine downtime through less frequent insert indexing and replacement result in an overall improvement in productivity and part cost reduction. Ceramic cutting tools can contribute substantially to meeting these objectives.

History of Ceramic Cutting Tools

Aluminum-Oxide-Based Ceramics. Traditional aluminum-oxide-based ceramic cutting tools have been available since the 1920s and, although very hard and chemically resistant, they had an inherent brittleness that limited their use. Thus, they remained an oddity until the 1960s. In the 1960s, material scientists improved these monolithic aluminum-oxide cutting tools by refining and alloying the heretofore brittle materials and gave them increased toughness and reliability. Although improved, these materials were still normally limited in application to uninterrupted cuts at moderate to light depth of cuts and feed rates.

In the 1970s and 1980s, however, process improvements, sintering aid refinements, and additions of zirconium oxide and other toughening/alloying agents were used to improve overall strength and fracture toughness. Simultaneously, cutting tool manufacturers learned how to protect and strengthen the cutting edge through the use of hones, edge chamfers (commonly referred to as T or K lands), or a combination of both hones and edge chamfers. These developments have led to a far broader usage of the aluminum-oxide and transformation-toughened oxide cutting tools. These products are no longer limited to finish turning of cast iron but have broadened their usage to machining of mild steels, and in some cases, nickel-base alloys. Normally, the monolithic aluminum-oxide-based cutting tools are manufactured by cold pressing and pressureless sintering, a process that is analogous to the manufacturing of cemented carbide cutting tools. A few producers today, however, subsequently hot isostatic press (HIP) their product to maximize density and to ensure product reliability.

Metal-Oxide Composites. In the early 1970s, it was discovered that aluminum oxide admixed with a refractory metal particulate (for example, titanium carbide) could produce a cutting tool with superior hardness and fracture resistance. These hot pressed metal-oxide composites consist of approximately 70% Al_2O_3 with 30% TiC particulate. This hot pressed grade typically worked well on materials with hardness ratings between 35 and 40 HRC and could be used for interrupted cuts, roughing and finishing of hardened steels, chilled cast iron, and cast iron with abrasive scale. Although the hot pressed metal-oxide composite was superior to the straight aluminum-oxide-based products, the costs associated with hot pressing placed a premium on this type of product.

With improvements in raw materials and processing techniques, the metal-oxide composite grade has undergone several improvements since its inception. Until recently, however, the metal-oxide composite grade was produced only as a hot pressed product. Technologies of the 1990s, however, may allow producers to manufacture this product outside of the hot press. Work is under way to utilize HIP equipment for consolidation in sintering (both encapsulated and unencapsulated processes) (see the article "Hot Isostatic Pressing" in this Volume).

Silicon-Nitride Ceramics. In the early 1980s, ceramic tool producers turned to a new base material, silicon nitride, for the development of advanced ceramic cutting tool products. Scientists have long been aware of the physical properties of silicon nitride. Only in the late 1970s and early 1980s, however, were the raw materials, technologies, and high-temperature/high-pressure facilities combined to enable the production of silicon-nitride-based cutting tools on a commercial basis. These innovations have proven themselves in actual production use over the past decade, providing reliable performance with predictability in the high-speed machining of cast irons. Silicon nitride cutting tools consistently show a greater strength, wear resistance, and fracture toughness than the traditional aluminum oxide cutting tools when machining the tenacious cast materials. However, silicon nitride is generally not recommended for the machining of steel or materials that produce a continuous chip.

Today, various compositions of silicon nitrides are available for a variety of applications. One, for example, is a silicon-aluminum oxynitride, called Sialon. This is essentially a silicon-nitride-based ceramic in which some of the silicon has been replaced with aluminum and some of the nitrogen with oxygen. Sialons have the ability to machine certain exotic materials at four to eight times the speeds typically used with carbide cutting tools.

Another composition uses silicon nitride as a base, with additives of yttrium oxide, aluminum oxide, and titanium carbide. Still other inserts are compositions of silicon nitride, yttrium oxide, magnesium oxide, and aluminum oxide. Gray cast irons, for example, can now be machined with silicon nitride tools at speeds of 460 to 1500 m/min (1500 to 5000 sfm) or more. Of these silicon-nitride-based ceramic materials, the sialons are generally cold pressed and sintered, while most of the other silicon nitrides are consolidated by hot pressing, over-pressure sintering, or HIPing.

Whisker-reinforced alumina is the latest development in advanced ceramic cutting tool materials. In this material, silicon carbide whiskers are added to a matrix of aluminum oxide. These whiskers reinforce the hard and somewhat brittle aluminum oxide, allowing it to better withstand high mechanical stresses. As a result, the composite is now widely used in the machining of high nickel-base aerospace materials such as Inconel, as well as severe machining applications in hardened alloy steels and chilled cast nodular iron. Whisker-reinforced alumina cutting tools are consolidated by hot pressing or HIPing techniques.

Cost-Effectiveness of Machining with Ceramics

Manufacturing efficiency is critical in today's world economy. Given the unprecedented scope and intensity of competition, every manufacturer is under pressure to maintain acceptable profit margins. Fortunately, efficiency improvements required are generally well within reach. This is especially apparent when the United States metalworking industry is considered as a whole. Because an estimated $115 billion are spent annually in labor and overhead costs and from 2 to 3 billion each on new machine tools and standard cutting tools, it is evident that even modest reductions in labor costs and other operating expenses can result in major savings and a significant improvement in global competitiveness.

Several approaches can be taken toward improved manufacturing efficiency. One approach is to acquire the most modern machinery and equipment and to institute new manufacturing methods. This approach can undoubtedly be effective; it has been estimated, for example, that, whereas conventional machine tools actually produce chips and make parts only about 30% of the time, the average level of uptime improves to 70% in the case of numerical control (NC) machines, which permit major reductions in nonproductive adjustment and setup time. On the other hand, investment in new capital equipment is very expensive and does not always provide the expected return. In many cases, a manufacturer will purchase an expensive new machine with which to improve efficiency but continues to waste time and money by failing to update other critical manufacturing.

Still another approach to increased manufacturing efficiency is to improve production methods along with an eye to maximizing the performance of existing equipment. This approach is often successful. Experience shows that manufacturers who are innovative, change methods, and drop operations that do not add value to the product and utilize ceramic tools, usually do increase their productivity, lower costs, and add profit at minimum expense.

A third approach is to utilize ceramic tooling on existing equipment to reduce machining cycle times by permitting faster feed rates and cutting speeds.

Individual Components of Total Machining Costs. Total costs for a machining operation comprise the combination of individual costs associated with the machine tool, labor, tools, and tool replacements. Specifically, these individual costs include:

- Overhead or burden rate for the machine tool. This includes both production (actual cutting) time and idle time, which covers rapid traverse, load and unload, and tool change. The production cost decreases with increasing cutting speed while the idle cost remains constant with changes in cutting speed
- Direct labor costs of tending to the machine
- Initial tool purchase costs prorated over the number of pieces produced by the tool
- Tool replacement costs, including costs associated with the repair or reconditioning of the original tool and the time required to install the replacement tool. Tool reconditioning costs generally increase with increasing cutting speed because this usually is accompanied by a higher tool wear rate

Generally, total machining costs are lowest at an intermediate cutting speed and rise sharply as cutting speeds are increased. The associated production rate in pieces per hour rises to a maximum with increasing cutting speed, then falls off as the higher tool wear rate necessitates more frequent downtime for tool changes. Machining with ceramic cutting tools also follows this pattern. However, because of their higher speed capability, the pattern is strongly shifted toward higher production rates.

Effect of Work Material on Machining Costs. Another factor affecting machining costs is the machinability of the work material, which is basically a product of its metallurgical composition and hardness. For example, if a machining cost of $1.00/piece is assumed for turning a typical aluminum alloy, a table of correspondingly higher costs can be projected for completing the same operation on more difficult-to-machine materials. Given the $1.00/piece for turning aluminum, these would work out to $2.50/piece for turning 1020 steel, $10.00/piece for turning 4340 steel quenched and tempered to 52 HRC, and $30.00/piece for turning Inconel 700.

Effects of Finish and Tolerance Requirements on Machining Costs. Two additional factors affecting machining costs are the workpiece finish and tolerance requirements. With respect to finish requirements, costs will generally increase in relation to the stringency of the requirements because additional machining passes will usually be needed to meet them. In a typical turning operation, for example, a sequence of roughing, semifinishing, and finishing cuts might be required. In some cases, moreover, where finish specifications are even finer than those readily obtained by a chip-removal process, a secondary operation such as grinding or even a tertiary operation such as honing may be required.

Machining costs will similarly increase with respect to more stringent tolerance requirements because these are closely associated with and usually dictate the achievement of finer surface finishes. Today's ceramic cutting materials enable the manufacturer in many applications to reduce some or all of the individual components of total costs for a machining operation.

Machining of Irons and Steels with Oxide-Based Ceramic Inserts

For a growing number of applications on irons and steels, today's oxide-based ceramic cutting materials can provide a key to an improved return on investment in expensive capital equipment.

Now both tougher and harder than when first introduced, the oxide-based ceramics can be run at speeds several times faster than those appropriate for coated carbides, thus resulting in dramatically reduced cycle times. The oxide-based ceramics provide better surface finishes, often eliminating the need for secondary operations such as grinding. When properly applied, the oxide-based ceramics also provide longer tool life because they reduce the frequency of insert indexing and replacement. Unlike carbide inserts, ceramic inserts are also well-suited to machining high-hardness and other difficult-to-machine materials.

Any job currently run with carbide or coated carbide on a machine capable of higher machining rates may be a candidate for the higher speeds and faster cycle times obtainable with today's oxide-based ceramics. Oxide-based ceramics can be applied most easily in single-point finishing or semifinishing operations and are especially productive in taking light finishing cuts at high cutting speeds because the excellent surface finish typically obtained often eliminates the need for secondary grinding.

Two basic types of oxide-based ceramic cutting materials are available to meet a broad range of application requirements:

- Cold pressed and sintered
- Hot pressed metal/oxide composite

Cold pressed oxide-based ceramic grades are composed mostly of aluminum oxide with small amounts of other oxides and compounds that promote toughness. Typically, the cold pressed grade is used for work materials with a hardness rating less than 35 HRC and performs well in light finishing applications on cast iron, carbon and alloy steels, and annealed tool-and-die steels. Normally, a cold pressed oxide-based ceramic is not used on workpieces that require interrupted cutting.

By comparison, HIP metal/oxide composite grade combines approximately 70% Al_2O_3 and 30% TiC. The hot HIPped grade typically works well on materials with hardness ratings between 35 and 70 HRC and can be used for interrupted cuts, and for roughing and finishing of hardened steels, chilled cast iron, and cast irons with an abrasive scale (see Table 1).

Table 1   Parameters for machining of selected materials with HIP metal-oxide composite grade ceramic insert cutting tools

Application	Insert(a)	Machining operation	Workpiece material	Hardness, HRC	Speed m/min	Speed sfm	Feed mm/rev	Feed in./rev	Depth of cut mm	Depth of cut in.
Steel mill	RBMA-106T	Turn outside diameter	Nodular iron	54	46–76	152–250	0.30	0.012	2.54–10.2	0.100–0.400
		Contour outside diameter	52100	56–62	85–98	280–320	0.13	0.005	10.2(b)	0.400(b)
							0.13	0.005	0.51(c)	0.020(c)
	RBMT-1012T	Plunge groove outside diameter	Nodular iron	56	30	100	1.5–2.0	0.060–0.080	(d)	(d)
	GR-500T	Contour outside diameter	Tool steel	52–55	98	320	0.51	0.020	2.54	0.100
Automotive	CNGA-433T	Hard turn and face	Alloy steel	58–60	85	280	0.2	0.008	0.15	0.006
Ordnance (shells)	RNG-43A	Turn outside diameter	1045	26–34	460	1500	0.2	0.008	0.25	0.010
Railroad (axles)	SNG-434T	Turn outside diameter	Alloy steel	38–42	206–351	676–1150	0.1	0.004	2.0	0.080

(a) GTE Valenite Q-32 grade with 70 Al_2O_3/30 TiC composition. (b) Rough cut. (c) Finish cut. (d) 31.75 mm (1.25 in.) wide by 102 mm (4 in.) deep

Alumina-based cutting tools have been used in industry for several decades and their performance standards are well documented. A typical application involving the alumina/titanium carbide composite will illustrate the cost savings and productivity improvements that are possible with oxide-based ceramics.

Example 1: Reduction of Machining Time for Reworking of Primary and Secondary Backup Rolls with the Use of Alumina/Titanium Carbide Ceramic Inserts. A major steel company had a problem reworking primary and secondary backup rolls for their rolling mill. For years they had reworked their used and damaged rolls using regrinding techniques. It would require 14 to 18 h to refurbish each roll and would consume a $700.00 vitrified grinding wheel in the process. The roll refurbishing capacity was therefore one roll per day (based on a two-shift operation). The company had only one grinder capable of supporting and doing the rework (the rolls were approximately 4.3 m, or 14 ft, long and 760 mm, or 30 in., in diameter). Because the mill's production generated more than one roll for rework a day, their roll shop and maintenance areas contained over 1000 rolls awaiting reworking. The company had to find a more effective way of refurbishing rolls or invest in another grinding station. The steel company had an older but rigid lathe in their maintenance shop. At the recommendation of a supplier of alumina/titanium carbide cutting tools, steel company personnel implemented a refurbishing procedure that cut the total man hours required to rework a typical roll to less than 6 h, reduced time on the grinder to less than 3 h, and extended the life of the vitrified grinding wheel over 4 rolls.

Using the maintenance lathe, the damaged rolls were rough-turned using three alumina/titanium carbide inserts in a single holder. The holder was designed so that the first insert would scarf the scale, rust, and trash (pieces of metal that welded to the roll during operation). The second insert was staggered so that an additional 1.5 mm (0.060 in.) of stock was removed and it could also do the job of rough-turning if the first insert was damaged or bro-

ken. The third insert was offset by 0.8 mm (0.030 in.) and was the finishing tool. Using this three-tool setup, the rolls could be roughed and trued in one pass, in about 3 h on the lathe (machining hours). The roll was then sent to the grinding area where the finishing could be done in about 2 h (actual machining time). Utilizing the alumina/titanium carbide insert, the steel mill has not only been able to keep up with new requirements for refurbished rolls but has also all but eliminated their backlog. The savings in vitrified wheel costs alone was many times more than the cost of the alumina/titanium carbide inserts.

In the early 1980s, alumina/titanium carbide composite cutting tools became the benchmark material for roughing and finishing of steel mill rolls. It is also the tool of choice in many other steel and cast alloy applications, especially for machining of hardened materials. Prior to the introduction of alumina/titanium carbide, most machining was done at a fraction of the speed of the ceramic inserts by carbide tools or grinders (if too hard or tenacious for carbide). Since the introduction of these composite cutting tools, the machining of difficult-to-machine materials has become commonplace.

Carbide Cutting Tools Versus Oxide-Based Cutting Tools. Generally, speeds may be increased 200 to 300% over carbide operations when using oxide-based cutting tools. The harder the work material, the slower the recommended cutting speed. Lower feed rates may also be required for harder materials to reduce insert damage. Table 2 gives general guidelines which can be used when using a HIPped alumina/titanium carbide ceramic insert.

Toolholder Systems. Good performance of the oxide-based ceramic cutting grades depends on a proper toolholder system. The holder must be sturdy with a minimum of overhang, pockets must be absolutely clean, and damaged shim seats must be replaced.

Insert geometry is another important consideration. The strongest shape compatible with the operations should always be used. These rank from strongest to weakest as fol-

lows: round, pentagon, square, 80° diamond, triangular, 55° diamond, and 35° diamond (Fig 1). In addition, the largest permissible corner radius) are often used to strengthen the cutting edge in light or finishing applications, inserts used to perform similar operations. Negative-rake tooling is preferred for most applications.

Edge Preparation. Still another important factor in obtaining successful performance from the oxide-based ceramics is edge preparation. In general, chamfer widths should be proportional to the feed rate. For turning operations, this typically means 50 to 75% of the feed rate; for milling operations, 1.2 to 1.5 times the feed rate per tooth. Hones (edge radius) are often used to strengthen the cutting edge in light or finishing applications, while edge chamfers with a hone are used together in severe applications. Figure 2 illustrates the various edge preparation options. Recommended edge preparation guidelines for selected machining operations are listed in Table 3. The edge radius is designated as:

Hone designation	Radius mm	Radius in.
A	0.013–0.076	0.0005–0.003
B	0.076–0.13	0.003–0.005
C	0.13–0.18	0.005–0.007

Edge preparations on inserts used for turning of steel-mill rolls are:

- Steel rolls: 0.060 in. × 30° with an A hone (T06025A)
- Cast rolls: 0.060 in. × 30° with a B hone (T06025B)

Coolant Applications. Water or oil coolants are *not* recommended for use with cold pressed oxide-based ceramics because they may cause the insert to crack. If carbide tooling is used to machine a part run with coolant and a subsequent operation is planned using a cold pressed oxide-based ceramic, the residual coolant should be blown away from the part. Coolants *can* be used with metal/oxide composite ceramics, but only if the insert can be flooded.

Table 2 Recommended machining specifications for rough and finish turning of selected steels and cast irons with HIP metal-oxide composite grade ceramic insert cutting tools

Workpiece material	Hardness	Machining operation(c)	Speed		Feed		Depth of cut	
			m/min	sfm	mm/rev	in./rev	mm	in.
High-temperature alloy(a)	...	Roughing	90–150	300–500	0.08–0.15	0.003–0.006	>1.3	>0.050
		Finishing	120–185	400–600	0.08–0.15	0.003–0.006	<1.3	<0.050
Carbon and alloy steels	20–30 HRC	Roughing	245–460	800–1500	0.13–0.25	0.005–0.010	>1.3	>0.050
		Finishing	305–490	1000–1600	0.08–0.30	0.003–0.012	<1.3	<0.050
Hardened steels(b)	30–50 HRC	Roughing	150–245	500–800	0.10–0.20	0.004–0.008	>1.3	>0.050
		Finishing	185–305	600–1000	0.08–0.25	0.003–0.010	<1.3	<0.050
Hardened steels and chilled iron	50–60 HRC	Roughing	60–120	200–400	0.05–0.15	0.002–0.006	>1.3	>0.050
		Finishing	90–185	300–600	0.05–0.15	0.002–0.006	<1.3	<0.050
Gray cast iron	140–260 HB	Roughing	150–365	500–1200	0.13–0.30	0.005–0.012	>1.3	>0.050
		Finishing	150–490	500–1600	0.13–0.30	0.005–0.012	<1.3	<0.050
Nodular cast iron	150–300 HB	Roughing	150–305	500–1000	0.10–0.30	0.004–0.012	>1.3	>0.050
		Finishing	150–425	500–1400	0.10–0.30	0.004–0.012	<1.3	<0.050

(a) Includes Inconels, Waspaloy, Hastelloys, René alloys, Haynes Stellites, Incoloys, and others; excludes titanium alloys, precipitation hardening (PH) stainless steels, and 300-series stainless steels. (b) Includes tool and die steels, carbon and alloy steels, and 400-series stainless steels. (c) Ceramic insert used is GTE Valenite grade Q-32 with 70 Al_2O_3/30 TiC composition

Recommended Machining Practices. Observing the following recommended machining practices can help achieve maximum productivity with oxide-based ceramic cutting tools:

- Use only rigid machine tools with adequate horsepower and speed capabilities to utilize ceramic cutting rates
- Ensure workpiece/clamping rigidity
- Be sure mechanical chipbreakers are adjusted to produce chips in an open C or figure 6 shape
- Prechamfer the workpiece, allowing the cutting tool to enter on a clean and uniform surface. A rough workpiece frequently requires a 50% reduction in feed for entry and exit
- Avoid dwelling in the cut in order to reduce the chance of thermal damage to the insert or the workpiece
- Carefully establish machining feeds and speeds with reference to the ceramic cutting grade, the hardness of the part material, and the depth of cut

Following these general guidelines can provide an important first step toward achieving full productivity benefits from the use of oxide-based ceramic cutting materials. Because of the elevated machining rates involved, however, as well as significant variables that may be present in specific applications, additional assistance from a tooling specialist can be helpful.

High-Productivity Machining of Cast Iron with Si₃N₄-Based Ceramics

The development and introduction of silicon-nitride-based ceramic cutting grades have made possible a genuine breakthrough in the machining of cast irons. Representing an advanced generation of ceramic cutting materials, the silicon nitrides are capable of machining cast irons at speeds ranging from 110 to 1500 m/min (350 to 5000 sfm) and beyond while providing exceptional tool life with remarkably reliable performance.

A silicon-nitride-based ceramic grade is formulated to combine the characteristic high hardness of ceramic cutting materials with superior high-temperature strength, oxidation resistance, fracture toughness, and thermal shock resistance. With such properties, the silicon nitrides provide for higher machining rates and improved reliability in both roughing and finishing operations on cast irons. Their increased fracture toughness makes them especially well suited for milling applications. In general, the wide range of cutting speeds of which these materials are capable makes possible levels of manufacturing productivity and economy previously unobtainable in the machining of cast irons.

The following examples document production applications of silicon-nitride-based ceramic grade to illustrate the effectiveness of the silicon nitrides in providing for more productive machining of cast iron.

Example 2: Application of Silicon-Nitride-Based Ceramic Inserts to Face Mill Gray Cast Iron Gear-Case Housing Components for Oil Pumps. A manufacturer of oil pumping units uses the silicon-nitride-based ceramic grade in a series of computer numerical control (CNC) face-milling operations on

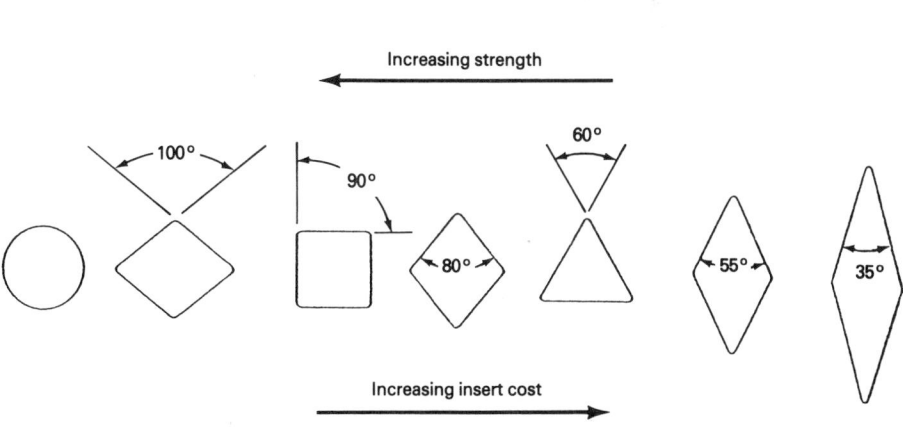

Fig 1 Effect of insert configuration and geometry on the strength and cost of oxide-based ceramic cutting tools

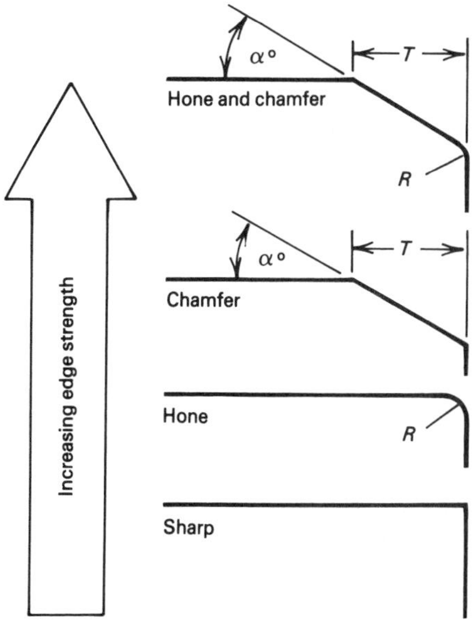

Fig 2 Relationship of typical edge preparation to edge strength for oxide-based ceramics

Table 3 Recommended edge preparation specifications for oxide-based ceramics used in selected machining operations

| Machining operation | Insert | Edge Preparation | | | |
| | | Chamfer (T) | | Angle (α), degrees | Hone |
		mm	in.		
Finishing	T00325	0.08	0.003	25	A
General purpose	T00820	0.20	0.008	20	...
Roughing	T01525	0.38	0.015	25	...
Turning of hard irons and steels	T00630B	0.15	0.006	30	B
Milling	T00630	0.15	0.006	30	...

gray cast iron gear-case housings. Productivity results represent a dramatic improvement from those previously obtained with a titanium-nitride-coated carbide tool material.

For example, in rough and finish passes on three bearing retainer surfaces, cutting speeds were increased from 275 to 610 m/min (900 to 2000 sfm), while the depths of cut (3.17 mm, or 0.125 in., roughing and 1.5 to 2.0 mm, or 0.060 to 0.080 in., finishing) and feed rate (0.004 ipt, or inch per tooth, roughing and finishing) were kept the same. This resulted in a reduction by half in overall cycle time, an increase in tool life from one to six parts per edge, and a 50% reduction in average insert costs per part. Although not generally recommended for maximum tool life, a water-based coolant is liberally used in these high-speed operations without fear of insert fracture, because, among other advantageous properties, the silicon-nitride-based ceramic is highly resistant to thermal shock.

Additional benefits of the silicon nitride in this application is the dramatic reduction in insert flank wear that made it possible to use the same face mill for both rough and finish passes. Previously, with the coated carbide, two different cutters had to be used to ensure the accuracy of the finishing pass. Furthermore, the sharp reduction in cycle time served to free up valuable machine time for additional production operations, including a growing number of outside jobs involving a wide range of cast-iron parts.

Yet for all these advantages, the productivity potential of the silicon-nitride-based ceramic grade actually *exceeds* the results obtained. The end user believes the material could be run much faster still without any problems, but it is already pushing the limits of the 19 kW (25 hp) Mazak horizontal machining center on which it is used.

Example 3: Silicon-Nitride-Based Ceramic Inserts Used to Machine Outside Diameter of Cast Iron Cylinder Liners for Truck Diesel Engines. Silicon-nitride-based ceramic cutting tool inserts are being used with special cartridge-style internal milling cutters to machine the straight outside diameter of cast iron cylinder liners for truck diesel engines. By replacing previous block-style cutters using an oxide-based ceramic cutting grade, the new tooling has improved produc-

tivity dramatically in machining the scaled, often out-of-round cylinder-liner castings. Run in this application without coolant, the silicon-nitride inserts have increased tool life from a previous 110 to 150 pieces per index to an average of 1200 pieces per index. They have also totally eliminated insert breakage, so common in the previous tooling that, on average, only three of ten available cutting edges of the pentagon-shaped inserts could be used before replacement was required. In contrast, all eight available edges of the square-shaped silicon-nitride inserts can be used, raising overall tool life from a maximum 450 pieces per cutter load of inserts to an average 9600 pieces per cutter load of inserts. These improvements represent not only a twenty-fold increase in tool life but also a major reduction in downtime for cutter replacement. With increased machine uptime, parts production per shift has increased 25%.

Even more impressive than these productivity improvements, however, are the cost savings now being realized. The dramatic increase in tool life, together with the elimination of cartridge destruction caused by broken inserts, has reduced per-piece perishable tool costs in the operation by an astounding 95%! And when now-eliminated labor costs involved in tool setup and delivery from crib to line are included, total annual savings can be calculated in the tens of thousands of dollars.

Example 4: Silicon-Nitride-Based Ceramic Used to Rough-Bore Internal Surfaces of 196 to 288 HB Cast Iron Truck Brake Drums. Silicon-nitride-based ceramic grade is used to rough-bore the heavily-scaled internal surfaces of truck brake drums made from centrifugally-produced cast iron hardened to 196 to 288 HB. The bore has an ID of 420 mm (16 1/2 in.) and a depth of 200 mm (8 in.).

The manufacturer had been machining the drums in a single pass with a special boring cutter holding two coated carbide inserts running in tandem at a speed of 150 m/min (500 sfm) and a feed of 0.8 mm/rev (0.030 in./rev). Because of the heavy scale, however, inserts broke frequently, causing tool damage and stopping production.

A special 5° lead toolholder was designed to clear the bottom of the drum. With a single

silicon-nitride insert taking a full 9.5 to 13 mm (3/8 to 1/2 in.) depth of cut, the operations were run, as before, in one pass but at a speed of 335 m/min (1100 sfm) and a feed of 0.41 mm/rev (0.016 in./rev). As a result, the manufacturer was able to:

- Reduce total cycle time per workpiece
- Increase tool life significantly
- Reduce downtime for tool changes
- Produce better-quality chips for disposal by the machine filtration system
- Lower tool cost per part
- Obtain predictable tool life with no catastrophic failures

Example 5: Comparison of Machining Capabilities of a Silicon-Nitride-Based Ceramic Versus an Oxide-Based Ceramic when Taper-Facing a Sheave-Half Movable Fan. A silicon-nitride grade ceramic was compared against a competitive oxide-based ceramic grade in taper-facing a sheave-half movable fan. Two criteria were established to evaluate tool performance:

- Notching at the depth of cut
- Part surface finish

As the comparative test operations proceeded, the oxide-based ceramic inserts exhibited severe notching after only five to eight pieces while the silicon-nitride inserts showed no notching at all. In addition, parts machined by the oxide-based ceramic material began to deviate from surface-finish specifications at an average of seven pieces per cutting edge. In contrast, parts machined by the silicon-nitride-based ceramic material maintained surface finish within specified limits even at twice the number of pieces per edge.

Additional Machining Guidelines. Silicon-nitride-based ceramics can be used for roughing, semifinishing, and finishing on cast iron. They have high resistance to oxidation as well as thermal shock. And unlike aluminum-oxide ceramics, silicon nitride has great fracture toughness and works best under moderate to heavy feed rates. In addition, silicon-nitride-based ceramics are well-suited to interrupted cuts and milling operations.

Tables 4, 5, and 6 provide general guidelines that can be used to machine selected cast irons with silicon-nitride-based ceramic inserts.

Difficult-to-Machine Materials

While the improved oxide-based ceramics and high-performance silicon nitrides offer cost-saving performance on a wide range of the most common work materials, a new advanced ceramic has long been sought for more productive machining of materials that can be described as especially difficult to machine. These include severe applications in traditional materials as well as the latest high-nickel and high-temperature/high-strength superalloys.

Table 4   Advantages of silicon-nitride-based ceramic inserts versus oxide-based ceramic inserts when machining selected cast irons

Workpiece material	Machining operation	Hardness, HB	Speed m/min	Speed sfm	Feed mm/rev	Feed in./rev	Depth of cut mm	Depth of cut in.	Al_2O_3 coated carbide	TiC coated carbide	Sialon	WC	Number of pieces	Si_3N_4-based(b), number of pieces	Improvement in production of Si_3N_4-base material over competitor materials, %
Gray cast iron	Turning	190	365	1200	0.330	0.0130	0.475	0.187	X	...	...	...	6	10	67
		230	700	2300	0.254	0.0100	3.18	0.125	X	...	...	...	75	475	533
		235	215	700	0.389	0.0153	6.35	0.250	...	X	...	...	18	25	39
	Boring	170	915	3000	0.635	0.0250	6.35	0.250	X	...	...	...	8	61	663
		185	565	1860	0.457	0.0180	3.18	0.125	X	...	...	...	11	24	118
		200	1100	3600	0.254	0.0100	2.54	0.100	...	...	...	X	13	150	1054
		214	440	1443	0.508	0.0200	3.81	0.150	X	...	...	...	18	30	67
		235	76	250	0.432	0.0170	4.78	0.188	...	...	...	X	55	102	85
	Milling	200	610	2000	0.10(a)	0.0040(a)	5.08	0.200	X	...	...	...	1	6	500
		245	503	1649	0.12(a)	0.0049(a)	3.18	0.125	X	...	...	...	600	1765	194
Nodular cast iron	Turning	207	336	1104	0.381	0.0150	4.75	0.187	...	...	X	...	100	140	40
		221	610	2000	0.406	0.0160	4.75	0.187	X	...	...	...	25	45	80
		250	287	943	0.305	0.0120	3.18	0.125	X	...	...	X	2	17	750
		300	122	401	0.330	0.0130	3.18	0.125	...	...	...	...	166	1689	917

(a) Feed per tooth. (b) GTE Valenite Q-6 ceramic insert

Today, the new whiskered ceramic cutting grades seem to be providing an effective solution. Not long removed from the confines of the cutting-tool industry's most advanced metallurgical laboratories, they are already providing practical performance benefits in production use on a wide range of difficult-to-machine materials.

Several successful whisker-reinforced ceramic grades now on the market use reinforcing silicon-carbide whiskers to add properties such as high edge strength, increased fracture toughness, good thermal shock resistance, and hot hardness to an abrasion-resistant aluminum-oxide base. With these properties, the whiskered material has already demonstrated its ability to improve productivity greatly in the roughing or finishing operations on high-nickel alloys and on severe applications of ferrous and cast alloys. The whiskered material takes even interrupted cuts at high machining rates with or without coolant.

The following examples demonstrate the machining capabilities of whisker-reinforced ceramic inserts.

Example 6: Machining of a 16-Pitch V-Style Thread in a 56 to 58 HRC 300M Alloy Steel Landing Gear Lever Arm with Whisker-Reinforced Ceramic Inserts. Threading inserts in the whisker-reinforced ceramic cutting grade are being used by a major aircraft manufacturer to machine a 75 mm (3 in.) long thread form on a landing-gear lever arm premachined from hardened (56 to 58 HRC) 300M alloy steel. The 16-pitch V-style thread form had previously been produced in a grinding operation which took approximately 1¼ h to complete. The form was especially difficult to machine, moreover, because of several severe interruptions, including three drilled holes (5.26 mm, or 0.207 in., in diameter and 12.7 mm, or 0.500 in., deep) and two keyways (6.4 mm, or ¼ in., wide and 6.4 mm, or ¼ in., deep). Despite these challenges, the whisker-reinforced threading inserts, running dry and taking 14 passes at 24 m/min (80 sfm) and a 0.13 mm (0.005 in.) depth of cut, are now producing finished thread forms in less than 10 min each, thus saving 65 min, or more than 85%, of costly machining time per workpiece. The surface finish is equal to or better than that previously obtained by grinding.

Example 7: Rough and Finish Turning of a D2 Tool Steel Flow-Turn Die Forging with a Whisker-Reinforced Ceramic Insert Cutting Tool. In an effort to reduce machining time, a die manufacturer substituted whisker-reinforced inserts for carbide in rough and finish turning operations on a heavily-scaled 280 mm (11 in.) diameter flow-turn die forging of hardened 60 HRC D2 tool steel.

Table 5   Recommended starting conditions for the rough and finish machining of selected cast irons with silicon-nitride-based ceramic insert cutting tools

Lead angles, thickness, and insert size are similar to those used for carbide inserts.

Workpiece material	Hardness, HB	Machining operation	Speed m/min	Speed sfm	Feed(b) mm/rev	Feed(b) in./rev	Depth of cut mm	Depth of cut in.
Gray cast ion	180–240	Finishing	490–1500	1600–5000	≥0.25	≥0.010	≤1.5	≤0.060
		Roughing(a)	365–1070	1200–3500	0.38–0.64	0.015–0.025	≤1.3	≤0.050
Nodular cast iron	220–260	Finishing	275–460	900–1500	≥0.25	≥0.010	≤1.5	≤0.060
		Roughing(a)	185–460	600–1500	0.25–0.50	0.010–0.020	≤9.5	≤0.375

(a) Lead angles improve tool life by reducing the tendency for breakout and burrs. (b) For improved tool life, avoid dwelling and light feed rates.

Table 6   Recommended edge preparation specifications for silicon-nitride-based ceramics used in selected machining operations

Machining operation	Application	Insert(b)	Chamfer(T) mm	Chamfer(T) in.	Angle (α), degrees
Turning	Light	T00515	0.13	0.005	15
	General(a)	T00820	0.2	0.008	20
	Heavy	T01230	0.3	0.012	30
Boring	Light	T00515	0.13	0.005	15
	General	T00820	0.2	0.008	20
	Heavy	T01230	0.3	0.012	30
Milling	Light	T00515	0.13(c)	0.005(c)	15(c)
	General	T00830	0.2	0.008	30

(a) Stocked edge preparation. (b) GTE Valenite Q-6 grade ceramic insert. (c) A hone

Table 7 Recommended machining parameters for roughing and finishing of selected difficult-to-machine materials with whisker-reinforced alumina ceramic insert cutting tools

Workpiece material	Machining operation	Speed		Feed		Depth of cut	
		m/min	sfm	mm/rev	in./rev	mm	in.
Superalloys	Roughing	120–305	400–1000	0.10–0.25	0.004–0.010	≤0.51	≤0.200
	Finishing	185–610	600–2000	0.10–0.25	0.004–0.010	≤1.5	≤0.060
Hardened steels, chilled cast iron (60–65 HRC)	Roughing	46–305	150–400	0.10–0.25	0.004–0.010	≤3.81	≤0.150
	Finishing	90–305	300–1000	0.10–0.30	0.004–0.012	≤1.5	≤0.060
Gray cast irons	Roughing	≤610	≤2000	0.25–0.75	0.010–0.030	≤6.35	≤0.250
	Finishing	≤610	≤2000	0.25–0.75	0.010–0.030	≤1.5	≤0.060
Nodular cast irons	Roughing	≤245	≤800	0.20–0.64	0.008–0.025	≤0.51	≤0.200
	Finishing	≤305	≤1000	0.13–0.50	0.005–0.020	≤1.5	≤0.060

Table 8 Recommended ceramic grade inserts for turning and milling of selected steels, cast irons, and high-temperature alloys

Workpiece material	Hardness	Machining operation				Ceramic insert grade(a)	
		Turning			Milling	Preferred	Alternate
		Rough	Semifinish	Finish			
Low carbon and mild steel	<30 HRC	X	...	...	...	Q-32	V-44
		...	X	...	...	V-44	Q-32
		...	...	X	...	V-44	Q-32
		...	...	...	X	Q-32	Q-32
Alloy and hardened steel	32–65 HRC	X	...	...	...	WRC	Q-32
		...	X	...	...	Q-32	V-44
		...	...	X	...	Q-32	V-44
Cast iron	140–260 HB	X	...	...	...	Q-6	Q-32
		...	X	...	...	Q-6, Q-32	V-44
		...	...	...	X	Q-6	Q-32
Nodular iron	150–300 HB	X	...	...	...	WRC	Q-32
		...	X	...	...	V-44	Q-32
Chilled cast iron	50–65 HRC	X	...	...	...	WRC	Q-32
		...	X	...	...	Q-32	WRC
High-temperature alloys (nickel-, cobalt-, and iron-base)	...	X	...	...	...	Sialon, WRC	Q-32
		...	X	...	...	Sialon, WRC	Q-32
		...	...	X	...	Sialon, WRC	Q-32
		...	...	...	X	WRC	WRC

(a) Q-6, GTE Valenite monolithic silicon nitride; Q-32, GTE Valenite alumina/titanium carbide composite; V-44, GTE Valenite transformation-toughened alumina; Sialon, silicon-aluminum oxynitride; WRC whisker-reinforced alumina ceramic

In the roughing operation, the rough insert was indexed four times in completing the required 200 mm (8 in.) long cut, averaging a 50 mm (2 in.) length of cut per index. In the follow-up finishing operation, the entire piece was machined with a single-insert edge. With the completed machining of eight of the dies, it was estimated that reduced downtime for insert indexing and replacement had saved approximately three weeks of production time.

Example 8: Whisker-Reinforced Ceramic Inserts Used in Semifinish Turning of 38 mm (1 1/2 in.) Diameter 54% Co Alloy Component Tripled Production Compared to Previously Used Carbide Inserts. In test semifinish turning operations on 38 mm (1 1/2 in.) diameter parts made from

a 54% Co alloy, a square insert machined six pieces with each of four corners of the insert, as compared to only two per each of four corners of the insert by a previously used carbide insert. Because of the dramatic reduction in downtime for insert indexing, it was decided to switch to the whisker-reinforced ceramic grade after exhausting the supply of carbide.

Example 9: Finish Turning of 79.25 mm (3.12 in.) Diameter 56 to 58 HRC Forged 300M Alloy Steel Aircraft Landing-Gear Lever Using Whisker-Reinforced Ceramic Insert Cutting Tools. In finish-turning operations on an aircraft landing-gear lever forged from 300M alloy steel and heat-treated to a hardness of 56 to 58 HRC, the whisker-reinforced ceramic grade cut machining time per

part from the 12.5 h required with a grinding operation to just 1.5 h. The 150 mm (6 in.) long cut over the 79.25 mm (3.12 in.) diameter workpiece, run at a feed of 0.13 mm/rev (0.005 in./rev), a speed of 27 m/min (89 sfm) and a 0.18 mm (0.007 in.) depth of cut, is severely interrupted. However, a single-insert edge machines an entire part and still has usable tool life when it is indexed routinely regardless of condition. The whisker-reinforced ceramic grade also improved both the accuracy of the cut and the surface finish. Runout was reduced to just 0.0003 total indicator reading (TIR) from the 0.003 TIR experienced with the diamond grinding wheel and surface finish was improved from 35 to 30 rms.

General Guidelines for Machining Exotic Materials. Table 7 provides machining parameters for whisker-reinforced ceramic insert cutting tools. The following are general guidelines that can be used when applying whisker-reinforced alumina:

- Prechamfer the workpiece or generate a progressive entry into the work
- A chamfer on the exit of the tool path from the work will help eliminate a burr on the workpiece
- Use the greatest lead angle that is practical
- Use round inserts whenever possible
- Generally, the depth of cut should not exceed 25% of the insert diameter when using round inserts. However, greater depths are possible
- Vary the depth of cut on repetitive cuts to decrease the development of a notch at the depth of cut line
- Ramping cuts can increase machine uptime while increasing tool life per index
- Flood coolant is generally recommended for maximum performance

Table 8 is a summary of recommended ceramic insert grades for selected machining operations.

Wear Applications

Jack D. Sibold, Coors Ceramics Company

CERAMICS, because of their ionic and covalent bonding, have the hardness and corrosion resistance necessary for wear applications. With the development of new ceramic formulations, improved ceramic fabrication technologies, ceramic joining strategies, and ceramic design criteria, the range of wear applications for ceramics has broadened considerably. Applications have grown from thread guides and wear-resistant seal rings to water faucet valves and soft drink dispenser valves to thoughts of perhaps spikes for golf shoes and buttons for men's dress shirts. These applications have required a broader base of knowledge of wear and friction (tribology) for ceramic materials. This is, in fact, being met. A recent literature search of wear in ceramics listed over 150 articles and papers from Jan 1980 to Aug 1990.

Wear can probably be organized into two categories, impingement wear and rubbing wear. Impingement wear involves particles entrained in fluids or gases striking a surface of a wear plate. Examples would be coal, gravel, ore, or grain striking a chute, an elbow, or a fan blade in a material handling system. Rubbing wear involves sliding abrasion of two materials in contact with each other, with or without abrading particles between the sliding pairs. The sliding pairs may or may not be the same materials. Examples of rubbing wear would be rotating shaft seals for automotive water pumps and wire drawing capstans.

Parameters Affecting Wear of Ceramics

Hardness. In wear applications, one usually thinks of hardness as the critical property, that is, the harder the material, the lower the wear rate. This is, of course, true. However, hardness is measured by the resistance of one material to the penetration of another material. This is usually measured by forcing a small point of a hard material into a flat surface of the subject material and measuring the depth of penetration. Typical hardness testing methods used for ceramics are Rockwell, Knoop, and Vickers with diamond indenters.

Unfortunately, if some of the applications already listed are considered, the wear is not just a function of indentation depth of a specifically prepared surface; rather, it is wear by the literal removal of material that has been subjected to corrosion, adhesion, material transfer, and fracture due to thermal shock, thermal stress, and Hertzian loads. Properties such as strength, fracture toughness, thermal expansion, thermal conductivity, modulus of elasticity, and corrosion resistance therefore must be considered in wear applications. Table 1 compares some properties of technical ceramics with those of other materials.

Ceramics, because of their high modulus of elasticity and lack of plastic flow, concentrate stresses where those stresses are applied and readily crack. In wear applications, if sharp particles are loaded against a ceramic surface with enough force, microcracking will take place at this location of concentrated stress. This is easily observed with harness indenters if a heavy load is used. The presence of these microcracks will increase the wear rate.

Fortunately, most naturally occurring abrading materials are softer than technical ceramics, so it takes quite large loads to cause microcracking and loss of wear resistance. Nevertheless, this must be taken into account in designing with ceramics. Sometimes instead of a uniformly hard and brittle ceramic such as fine-grained sintered silicon carbide, a coarse-grained silicon carbide (SiC) bonded by a more resilient (lower modulus of elasticity) matrix of silicon, silicon nitride (Si_3N_4), and/or glass results in a better wearing material. The large SiC grains resist penetration by impinging particles, while the matrix resists microcracking. This composite approach to mixing hard ceramic materials with more plastic ceramics yields a better wearing material than the fine-grained, single-phase silicon carbide (by most standards, an advanced engineering ceramic).

Thermal Conductivity. Since most ceramics have lower thermal conductivity than metals, localized heat build-up in rubbing or sliding applications must be considered. Too much heat build-up from friction can lead to localized thermal stress and thermal shock microcracking, resulting in poorer than expected wear. Thermal checking on a microscopic scale can be observed if localized loading and heating is not evaluated and controlled in the design. Surface finish, particularly asperities, must be considered, since too rough a surface can lead to localized heating from friction and microcracking.

Fracture Toughness. Because of the possibilities of creating microcracking and reduced wear resistance, localized as well as bulk fracture toughness should be considered. Most ceramics, of course, are brittle when compared to metals and engineered plastics. However, with their superior hardness, technical ceramics excel in wear. The range of fracture toughness of technical ceramics is from approximately 1 to 12 MPa$\sqrt{m}$ (0.9 to 11 ksi$\sqrt{in.}$) for microlithic ceramics and may be up to 20 MPa$\sqrt{m}$ (18 ksi$\sqrt{in.}$) for composites.

Some examples show that the ability of a ceramic with higher fracture toughness to withstand microcracking from localized loads and thermal shock and stress can offset lower hardness. Toughened zirconia (Zr_2O_3) compositions outwear harder alumina (Al_2O_3) in some applications, and zirconia-toughened Al_2O_3 can significantly outwear normal Al_2O_3 compositions. Silicon carbide whisker-toughened Al_2O_3 demonstrates excellent wear rates. On a microstructural level, glass-bonded Al_2O_3 and silicon-bonded SiC can limit microcracking to the Al_2O_3 or SiC grain, making the wear rate better than if a microcrack involved multigrains. Material selection therefore is not based solely on hardness but on other properties as well.

Wear Method. The manner in which a ceramic wears can also offset a lower hardness. For instance, most ceramics wear when individual grains pop out. Some other ceramics wear by losing microparticles from a grain or surface. These microparticles are much smaller than a grain and are submicronsize. Alumina, the most commonly used wear-resistant technical ceramic, wears most often by grain pluck-out and pop-out. This is mainly because of the anisotropy of properties such as thermal expansion in different crystal directions. This leads to cracking along grain boundaries, particularly in coarser grain formulations. Since the average grain size of Al_2O_3 is 3 to 30 μm with the largest grains up to 60 μm, fairly large particles are being lost to wear even if the microcracks go through the grain.

This leads to two potential wear problems. The first is the hole left on the surface, which allows a high angle of attack for the next

Table 1   Properties of technical ceramics and tool steel

Material	Vickers hardness		Fracture toughness		Thermal conductivity		Thermal expansion, $\times 10^{-6}/°C$	Flexural strength	
	GPa	ksi	MPa$\sqrt{m}$	ksi$\sqrt{in.}$	W/m·K	Btu/ft·h·°F		MPa	ksi
Al_2O_3	14	2030	4	3.6	16–39	9.3–22.5	8	50	7.2
Zr_2O_3	12	1740	11	10.0	2	1.2	10.5	90	13.0
SiC	25	3626	3	2.7	110	63.6	4.3	65	9.4
WC	18	2610	20	18.2	100	57.8	5.1	300	43.5
Tool steel	10	1450	80	72.8	25	14.5	6	550	79.7

abrading particle. Impingement wear tests show that impingement angles above 60° lead to much higher wear rates (Fig 1). Thus, once the initial grain pluck-outs begin, the wear rate can increase.

The second problem is that the particle being removed can become an abrading particle itself. This removed particle is usually sharp, as large as the surface grain size and as hard as the surface grains, making it a far more effective abrasive particle than the quartz dust or gravel particle that is part of the wearing medium.

In a material such as partially stabilized zirconia (PSZ) or a glass-ceramic, the particles being removed from the surface of the wear plate are submicronsize because of their engineered microstructures. PSZ is toughened by submicron metastable tetragonal precipitates in a (60 μm) cubic grain matrix. Theoretically, wear-induced damage causes the surface precipitates to transform to a lower-density monoclinic crystalline form, the surface goes into compression, the microcracks are blunted by the precipitates, and only submicron-subgrain size microparticles are worn away. This leads to surprisingly good wear rates for a material less hard than, for example, Al_2O_3. Simple evidence to support this theory is to examine the worn surfaces of Al_2O_3 and PSZ. Even polished aluminum oxide surfaces turn matte after exposure to wear, while matte as-fired PSZ often becomes

polished in a wear application indicating a submicron surface damage.

Corrosion resistance of ceramics is potentially as important as hardness in many wear applications. This corrosion effect is not duplicated by normal testing procedures. In the real world, there always seem to be corrosive chemicals present. These chemicals can range from water to aggressive acids and bases. In wear applications, this corrosion is particularly important since fresh surface is being created for chemical reactions to take place. In wear applications, these chemical reactions may alter the surface to a less hard compound, enhance slow crack growth, or make microcracks grow, any of which will increase the wear rate. For example, in a geothermal well application, a 99.9% Al_2O_3 bearing surface was dissolved by clean circulating hot water. This left a chalky bearing surface. Another quite different Al_2O_3 formulation works well as a shaft and sleeve bearing in a hot water recirculation pump.

In some cases, the corrosion of a surface can have beneficial effects. Silicon nitride used in high-temperature oxidizing applications forms a silicon dioxide glassy layer that acts as a lubricant and lowers friction and blunts microcracks. This leads to lower wear rates.

Joining Methods. Since ceramics are traditionally only placed in the area of high wear because of cost and fabrication considerations, they must be joined to other materials of construction. Significant thermal expansion differences exist between ceramics and metals and engineering plastics; therefore care should be taken to put the ceramic wear surface in compression if at all possible. Any induced tensile stresses can lead to slow crack growth, enhanced erosion, and premature failure. If a "resilient join" can be made that allows wear loads to be distributed, improved results can be expected because surface microcracks are minimized. If a design can allow the use of many small ceramic pieces attached to a support structure, a cost-effective approach can be achieved. For example, large fan blades for coal-fired power plants are "armored" with 25 mm (1 in.) squares of Al_2O_3 and SiC bonded to cast iron plate. This allows the most wear-resistant (but most expensive) ceramic, SiC, to be placed on the blade tips while less expensive Al_2O_3 is used on the balance of the blade. This application has extended fan blade life three times, thus

reducing costly maintenance and down times. A review of some successful wear applications of technical ceramics will further illustrate the effects of properties and demonstrate that technical ceramics are the solution to some wear problems.

Example Applications

Rotary Seals. One of the early uses of technical ceramics in wear applications was in seal faces for rotary pumps. Alumina ceramics have long been used for this application. Rubbing against a carbon-phenolic seal face, alumina provides the primary liquid seal in these pumps. In fact, an 85% Al_2O_3 seal ring has been standard in automotive water pumps for over 20 years. These rings, approximately 25 mm (1 in.) in diameter, are mounted on a 16 mm ($^5/_8$ in.) shaft and spring-loaded against a carbon-phenolic ring. The ceramic ring is press fitted into a metal cup which is attached to the shaft. The application requires corrosion resistance to engine cooling fluids (water/ethylene glycol mixtures) and erosion by any slag created in the cooling system. A major problem is to obtain a surface finish on the seal face that does not allow leakage yet is not too smooth as to raise the torque too high for efficient pump operation. High torque is also a problem for cold or dry starts. The solution is to control the grinding and lapping of the seal face such that the surface imperfections are too small for leakage but large enough to allow a liquid film to form and act as a lubricant. Typical seal rings can be seen in Fig 2.

Other large-volume seal face applications are in automotive air conditioners, clothes washers, and dishwasher pumps. Direct sintered SiC is being evaluated for use in automotive seals to meet 100,000 mile warranty specifications; however, the use of SiC would

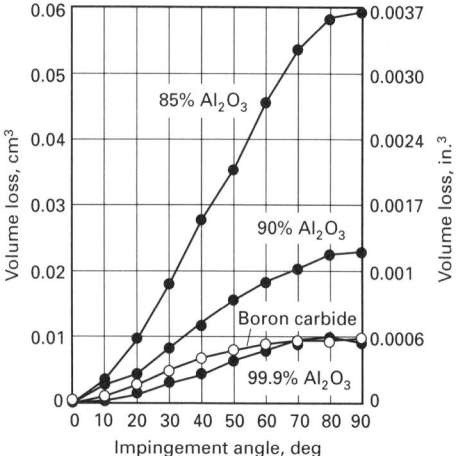

Fig 1   Impingement wear of several technical ceramics as a function of impingement angle

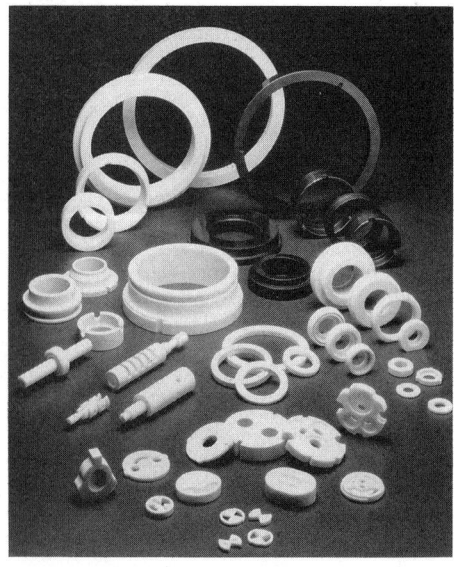

Fig 2   Alumina and SiC seal rings used in various types of valves and pumps

cost nearly three times as much as Al_2O_3. Improved Al_2O_3 compositions are also being evaluated to meet this longer life requirement.

There is a low-volume segment for pump face seals in chemical, petrochemical, petroleum refining, food pharmaceuticals, pulp and paper, mining, and power generation pump applications. These are often larger diameter rings for larger shafts. The seal materials are usually Al_2O_3, SiC, and tungsten carbide (WC). The Al_2O_3 compositions are normally 99.5% or purer. The SiC materials are either reaction bonded or direct sintered, and are replacing WC for cost and performance reasons.

Alumina is used primarily for its lower cost. An Al_2O_3 seal costing $10 would be about $30 in SiC and about $40 in WC. Alumina is not recommended with hydrofluoric acid and sodium hydroxide, and caution is recommended with caustic cyanogen, potassium hydroxide, and hydofluorosilic acid.

Silicon carbide is increasingly the premium seal face material of choice. It is much more acid-resistant than WC, and has a relatively high pressure-velocity (P-V) ratio, which is the performance measurement for seal faces. Figure 3 illustrates the relative P-V ratios for various seal couples. The higher the P-V ratio, the better the performance for most seal applications.

Figure 3 shows that SiC gives enhanced P-V ratios over WC in all seal pairs. However, the SiC user must overcome the lower cost of Al_2O_3 and the higher strength and toughness of WC. The cost issues were overcome by evaluating part cost versus down time, maintenance costs, inventory costs, and backup pump costs, since the lower corrosion and erosion resistance of Al_2O_3 shortens its life. The strength and toughness issue required design modifications around chamfers and notch radiuses. Reaction-bonded SiC, which contains some free silicon, is recommended for tougher applications. Direct sintered SiC is the most corrosion-resistant.

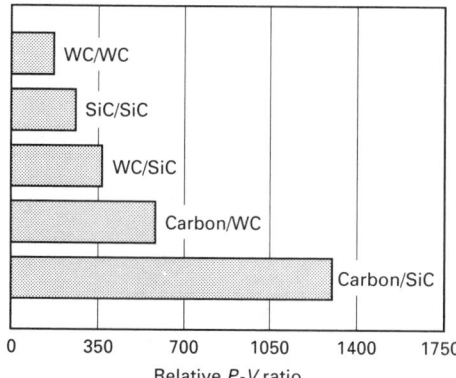

Fig 3 Relative pressure-velocity ratios for several pump seal couples. The higher the *P-V,* the better the seal performance. Data courtesy Carborundum Corporation

Water Faucet Valves. An obvious application similar to rotary seal faces is in water faucet valves. Again, this is a seal face which rotates to close and open water valves. Examples are seen in the foreground of Fig 2. Some are single valves for hot and cold water, while the larger valves with two holes are used in single handle faucets. For cost reasons, the ceramic materials used are usually various grades of Al_2O_3, although other material options might be SiC, Si_3N_4, or Zr_2O_3. The water faucet application demands corrosion resistance and resistance to various waters as well as sealing capabilities. Dissolved materials in water in various locations change the erosive and corrosive characteristics of water systems.

The competing materials are polymers and metals—primarily brass and bronze. Polymers usually degrade with time, especially the hot water side. Corrosion deposits scratch hard plastics, causing leaks. Resilient polymers "cold flow" and change the valve opening as water temperature changes. This can lead to cooling or heating of water in the morning shower. Brass and bronze are easily scratched by corrosion products and mineral deposits, which results in leaks. Alumina solves these problems but had to overcome some problems of its own.

There is a cost disadvantage due to manufacturing cost of the many intricate shapes. Additionally, diamond grinding and lapping are required to obtain flatness and surface finishes required for sealing. Other problems were high torque and noisy operation after some period of use. The use of dissimilar ceramic compositions for the mating faces was one solution, while control of surface condition proved very important. Initially the thought was to have the best surface finish that cost would allow would make the smoothest operating faucet. After much trial and error, it was found that a diamond grinding to ensure flatness was necessary. This creates many peaks and valleys. If a subsequent lap and polish is done, a very high surface contact area is obtained and the frictional forces become very high. If the surface contact area is reduced but the surface defects are small enough to allow the surface tension of the water to preclude leakage, then high torque and noisy operational problems are solved.

This still left a final problem of dimensional tolerance and alignment. Since ceramics have a very high modulus of elasticity, they are very rigid. If alignment of the ceramic in the faucet assembly is not very good, then scuffing wear will occur in one area of the seal face, leading again to high torque and noisy operation. This required redesign of the faucet assemblies to include tighter tolerances in the non-ceramic components as well as some self-aligning designs.

Rotary and Gate Valves. Other valve applications for ceramics are rotary and gate valves for handling corrosive and erosive liquids such as brine water, coal slurries, and

drilling muds. Examples of these types of valves can be seen in the right side of Fig 4. These Al_2O_3 valve trims are generally replacing cast iron and coated metal in the most difficult slurry handling applications. Because of their large size, they are generally made of Al_2O_3 for cost and fabricability reasons. They are most successful when used in an on/off application rather than as a throttling valve. Of course, metal and other materials do not survive very long with this valve design for throttling. Throttling with this type of valve leads to high-velocity fluid flow, which in turn leads to "wire line cutting" and leakage. However, in normal on/off applications, ceramics have given useful life many times that of metal components they are replacing, but the reduction in maintenance costs, down time, and backup valve costs usually outweighs the added part cost of ceramics.

Like water faucet valves, these applications were not without design problems. Again, the surface condition of the ceramic is important for low friction operation. This is especially important for remote control operation of valves. It was further found that inserting ceramic seats in the sealing surfaces of these valves improved life and performance. Tolerances of the valve assembly must be tight to avoid chattering of the valve faces, which can cause cracking and erosion on the ceramic faces. Another design problem encountered was in the actuating method for rotating or sliding the valve. Traditional approaches for metal are notches or key ways. In brittle materials such as ceramics, these become stress concentrators which lead to fracture and failure. The solutions were to greatly increase the radius of the notches or key ways or replace them completely with metal bands around the outer perimeters of the valve bodies.

Other valve applications include some ball valves, ball and seat chokes, and metering valves. The metering valves are generally small piston and sleeve assemblies (lower center in Fig 4) for precise metering of acidic

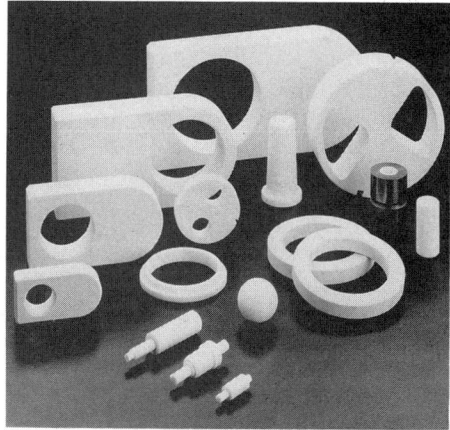

Fig 4 Alumina components for rotary valves and gate valves

or basic liquids in medical to chemical applications. This includes some oxygen valves.

Pump Components. In addition to valve components for handling corrosive and erosive liquids, ceramics are being successfully utilized as pump components—shaft sleeves, linings, and plungers, also seen in Fig 4. Ceramic pump plungers have been used for many years in the secondary recovery of oil. Water is pumped into oil formations to displace oil so it can be recovered. Since waters used for these operations are usually brine waters or contain sand and sediments, a tremendous erosion and corrosion problem exists. This is solved by Al_2O_3 plungers.

Another application in the oil industry is in the handling of drilling muds. In this application, a fairly high-viscosity dispersion of clay in water is pumped to lubricate and cool the drill bit. The linings of these pumps last only a short time even though they are fabricated from chromium steels, nickel-base alloys, and other hard coatings. Again, Al_2O_3 ceramics have been successfully used. As in previous application examples, redesign was required to overcome alignment problems, surface finish problems, and attachment problems. Self aligning designs and improved inside diameter honing solved the first two problems, while thermal shrink fitting is the best method for attachment. However, since these liners can be 150 mm (6 in.) in diameter and 750 mm (30 in.) long, handling these brittle tubes requires care and training.

Papermaking Machines. Another fluid handling application for ceramics is in the wet end of high-speed papermaking machines. Again, abrasion and corrosion resistance allow ceramics to stand up to acidic or basic suspensions of pulp as dewatering elements in the papermaking process. Figure 5 illustrates the shapes of ceramics used. These include segmented and continuous forming boards, foils, suction box covers, and foil cleaning devices. Ceramics have replaced polymers and metals on the higher speed machines for over fifteen years. Since the ceramic is supporting

fast-moving continuous plastic filament screen, a fine surface finish must be achieved and maintained on the support surface of the ceramic. This is important to the life of the screen as well as to minimize power requirements from friction effects. Rough surfaces erode the screen and raise the coefficient of friction. Many of these paper machines are up to 10 m (400 in.) wide, meaning the designers faced a difficult task as ceramic is not readily made in such lengths. The solution was to bond segments of fired ceramic shaped like piano keys to either metal or plastic/fiberglass composite strips. Then diamond grinding and finishing machines were designed and built to achieve the required surface finish. These machines are also capable of maintaining the extreme flatness and tolerance over the whole width demanded by the high-speed papermaking process. The leading edge of the dewatering element also has some very specific shape requirements that these grinding machines are capable of creating across the width of the element.

For many years the ceramic of choice for paper machine applications was either a 96% or a 99.5% Al_2O_3 formulation. In the last five years, transformation-toughened zirconia (TTZ) has become increasingly popular. Because of its high toughness, zirconia has excellent wear and withstands installation with few failures. The toughness also allows it to withstand thermal shocks associated with equipment failure, allowing the paper machine to run dry and hot followed by a rapid cooling by water. Also, because zirconia wears by a submicron microparticle method as opposed to the grain pluck-out wear method of Al_2O_3, a smoother, improved surface is maintained. The improved surface increases screen life, reducing down time. Thus, although zirconia is more expensive, it increases paper machine capacity, allowing the paper mill to produce more paper without a large capital investment.

Silicon carbide also is being tried on papermaking machines. Silicon carbide is even more corrosion-resistant and harder than TTZ, and thus may show improved life over both Al_2O_3 and zirconia. However, SiC is more brittle than Al_2O_3 and more expensive than

zirconia. Therefore, it must provide even more added value than zirconia in order to be viable.

Yet another approach being considered is combining the toughness attributes of zirconia with the hardness of Al_2O_3 to get zirconia-toughened alumina (ZTA). ZTA appears to be an excellent wear material. Although its toughness is about 50% greater than Al_2O_3, it seems to be very notch sensitive and is reported to be difficult to grind without chipping.

Another wear application in the paper industry is slitting and sizing knives. These are basically used on the finished end of the papermaking machine for slitting the paper in widths. Since papers may contain some clays as well as other hard particles from the pulp, erosion on the metal slitters dulls the edge quickly. This is compounded by corrosion, since paper may still have some moisture with the associated acids or bases left from the pulp dispersion. Again, transformation-toughened zirconia (TTZ) has greatly increased slitter longevity.

The design was a critical issue for success. Since most of these slitting operations are rotary shear type or slitting (see lower left, Fig 5), initially both wheels were made of TTZ. However, since the wheels do not run completely true, localized rubbing caused local heating which degraded the ceramic. This led to premature failure. It was found that with very precise slitting operations, extremely long runs could be achieved. However, a compromise of running TTZ against metal yields satisfactorily longer life without the need for precise setup of the slitting equipment. The ceramic seems to keep a sharp edge on the metal wheel.

Wire-Drawing Machines. Another industry where improved product quality, improved component service life, and lower-cost manufacture can be attributed to ceramics is in wire drawing. Here Al_2O_3 and partially stabilized TTZ capstans, pulleys, rolls, guides, and some dies provide extended wear times over steel and carbide-coated components. As wire is drawn over the guides, pulleys, and capstans, wear of these components by the wire degrades the component surface, which in turn degrades the wire. Ceramics with sur-

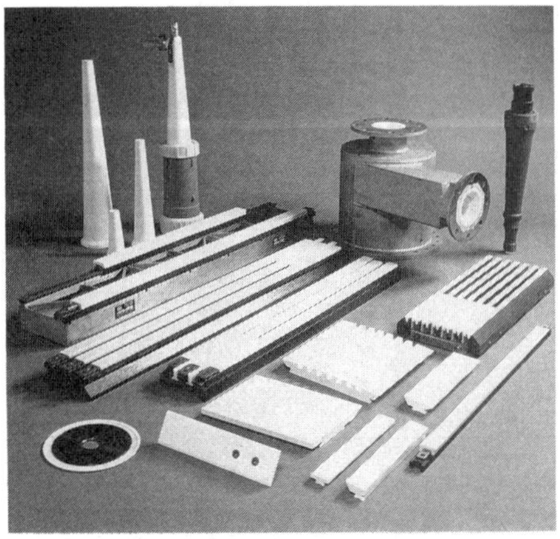

Fig 5 Ceramic components used in large, high-speed papermaking machines

Table 2 Characteristics and properties of ceramic and steel materials

	90% Al_2O_3	TTZ	SiC(a)	SiC(b)	Steel	Coated steel
Characteristics(c)						
Impact resistance	G	VG	P	P	E	P
Wear resistance	E	E	E	E	P	G
Corrosion resistance	E	E	E	E	P	G
Inertia (weights)	E	VG	E	E	G	G
Surface finish	VG	E	E	E	G	G
Properties						
Density, g/cm^3	3.6	5.75	3.1	3.14	...	...
Hardness, HR 45N	79	75	93(d)	...	...	...
Compressive strength, MPa (ksi)	2480 (360)	1760 (255)	2480 (360)	...	...	...
Tensile strength, MPa (ksi)	220 (32)	352 (51)	303 (44)	394 (57)	...	...

(a) Reaction sintered. (b) Sintered. (c) E, excellent; VG, very good; G, good; P, poor. (d) HR 30N

Fig 6 Ceramic components used to extend the life of wire-drawing equipment

face finishes as fine as 2 rms do not promote adhesive wear and are not affected corrosively by the wire-drawing coolants. The ceramic materials of choice are Al_2O_3, Zr_2O_3, and SiC, with TTZ playing an increasing role with its toughness and fine wear. Table 2 compares characteristics and properties of these materials.

While many designs exist for the various wire-drawing machines, most capstan and pulleys are a bolt-together design for flexibility and cost reasons. The flanges are usually steel, aluminum, or a polymer, depending on surface conditions. Some monolithic designs (Fig 6) are available, but these are more difficult to manufacture in ceramics and therefore more costly.

Ceramic wire-drawing capstans typically last 7 to 10 times longer than carbide-coated capstans. An example of such increased life was exhibited in a molten salt bath. A steel guide wore out within months, causing the wire being drawn to break. Ceramic guides endured two years without measurable wear.

In another example, the smooth surface of TTZ guide blocks reduced galling damage to aluminum strips passing through the blocks.

Metal Extrusion Dies. Another application in the wire industry is in extrusion dies. Life ten to fifteen times longer than steel and reduced stress cracking are claimed for aluminum and copper/brass extrusions using ceramic dies. The die material is partially stabilized or transformation-toughened zirconia (PSZ or TTZ). PSZ extrusion dies were used in a 1970s experiment for extruding steel rod. Although the project itself proved uneconomical, the PSZ dies proved capable of withstanding the stresses of steel extrusion. Once again, the design and attachment are critical. The unsupported strength of PSZ is about 445 MPa (65 ksi), which would be marginal at best. However, by shrink fitting PSZ blanks into steel holders and applying a hoop stress, the sum of the strength can approximate the 1890 MPa (274 ksi) of the H-13 die steel.

A PSZ die insert may cost 4 times as much as a steel die and twice as much as a WC die.

However, when PSZ inserts in an aluminum extrusion outlast five carbide tools, the economics look good. The application has spread beyond the wire industry to tubing and other metal extrusions. Up to 15% faster extrusion rates are claimed possible because of reduced galling and metal pickup. In addition, since the PSZ holds its dimension longer, less metal is used because the tools stay closer to nominal dimension for a longer time. PSZ dies maintain surface finish longer, increasing the visual quality of the extruded shapes.

SELECTED REFERENCES

- Application Guide Bulletin No. 980, Coors Ceramics Company
- Ceramic Material Solutions for Wire and Cable Manufacturing Bulletin No. 608, Coors Ceramics Company, 1987
- M.G. Gee and E.A. Almond, The Effect of Surface Finish on the Sliding Wear of Alumina, *J. Mater. Sci.*, Vol 25 (No. 1A), Jan 1990, p 296–310
- S.T. Gulati, *et al.*, Zirconium Oxide: A New Die Material for Hot Extrusion, *Metal Progr.*, Feb 1984
- J.N. Pennington, Ceramics Add Life to Extrusion Dies, *Modern Metals*, May 1990
- J.L. Spodnik and J.J. Wert, The Effect of Surface Finishing on the Tribological Behavior of a 94% Alumina Ceramic Material, *New Material Approaches to Tribology: Theory and Application*, Materials Research Society, 1989, p 309–318
- S.M. Wiederhorn and B.J. Hockey, Effects of Material Parameters on the Erosion Resistance of Brittle Materials, *J. Mater. Sci.*, Vol 18, 1983, p 766–780

Heat Exchangers

Scott L. Richlen and William P. Parks, Jr., Office of Industrial Technologies, U.S. Department of Energy

HEAT EXCHANGERS are found in nearly every industrial process and can be critical to the viability and effectiveness of a process. While its operation is relatively simple (exchanging heat between two streams), its effects on the thermodynamics of a process offer significant opportunities to improve the performance and efficiency of industrial processes. Many of these opportunities relate to improvements in heat-exchanger materials such as ceramics (Ref 1) that can increase both the operating temperature and corrosion resistance. Increases in thermal transfer can also be influenced by the use of enhancement devices or reduction in fouling impact (Ref 2); however, in many applications heat-exchanger performance depends on better materials. This may be accomplished by using ceramics, either as a monolithic material or as a composite.

The heat exchangers currently on the market are overwhelmingly of conservative design and metal manufacture. Ceramics must penetrate a heat-exchanger manufacturing community where metals have a long-accepted role; where the fabrication process of cutting, forming, and joining of metal sheet and components provides a clear "value-added" economic contribution; and where performance and economics of metal designs and applications are well understood.

The design of a ceramic heat exchanger must not rely upon conventional metal-based design approaches. While ceramics have unique properties that lend themselves to greatly improved heat-exchanger performance, such as high-temperature strength and corrosion resistance, other properties can significantly detract from heat-exchanger performance when not taken into consideration. At least one early attempt to apply ceramic materials in the fabrication of a heat exchanger ended in catastrophic failure of the ceramic component because the nature of the material (such as brittleness or thermal shock susceptibility) was not properly taken into account during the design process (Ref 3). Ceramics were simply viewed as direct substitutes for metal in the conventional design and little or no redesign effort was made. The best service performance of ceramic heat exchangers was obtained when the properties of the ceramic material were key considerations in the design effort.

Heat Transfer Considerations. Design of a heat exchanger is based upon its purpose (to transfer heat) and its application (use in industry). Because heat can be transferred in three distinct modes and because there are numerous ways the heat exchanger is used in industry, the applied design of the heat exchanger is an extensive subject. However, an explanation in generic terms of heat transfer and application can provide the basis for understanding ceramic heat-exchanger design for the specific, individual application. The three modes in which heat transfer occurs are conduction, convection, and radiant.

Conduction refers to heat transfer which occurs in a medium due to the motion of the atoms or molecules which comprise the medium. It is the only way that heat is transferred through an opaque solid. The rate of heat conduction through a plane wall of unit area is primarily dependent upon the thermal gradient through the wall and the thermal conductivity of the wall. The thermal conductivity is a property of the material through which the heat is being transferred and for some materials can change significantly with temperature.

Because ceramics, at their best, have thermal conductivities similar to stainless steel, the substitution of certain ceramics (for example, SiC) for metal is little cause for concern in terms of thermal conduction. What has caused concern is when materials with very low conductivities, such as alumina, are proposed. Generally, the use of alumina would only occur when no other material would survive exposure to the use environment. However, there are situations when other modes of heat transfer limit the overall heat exchange, thus making the low conductivity of alumina no longer a factor.

Convection refers to heat transfer between a surface and a fluid moving over the surface. The heat transfer rate for a unit area of surface is a function of the difference between the surface and fluid temperatures, and the average convection heat-transfer coefficient for the surface. The manner in which the convection coefficient is determined and the definition of the fluid temperature depends upon the specific nature of the convection process. With respect to the surface, the convective heat-transfer coefficient is dependent upon surface properties (such as roughness) and is not dependent upon the material properties of the surface.

Convection dominates the transfer of heat to air. Most current ceramic heat-exchanger designs are targeted for high-temperature use where another method of heat transfer, radiation, predominates. In this case, the air side represents the limiting factor in the flow of heat. The heat simply cannot be transferred into the air as fast as heat is being transferred out of the hot fluid. This frequently makes the low conductivity of many ceramics irrelevant in the weighing of materials selection criteria.

Radiation heat transfer differs significantly from that of conduction or convection. Radiation is the transfer of heat energy by electromagnetic emission and absorption. The rate of radiation heat transfer to a surface of unit area is a function of the difference in the fourth powers of the absolute temperatures of the surfaces involved and of the surface condition of the radiation surfaces and their areas and positions. Although radiation is usually associated with solid surfaces, certain gases can emit and absorb radiation. These include H_2O, CO_2, CO, SiO_2, NH_3, and the hydrocarbons. Some of these gases are present in every combustion process.

Heat Transfer in Exchangers. Although all three types of heat transfer are active to some extent in all heat exchangers, one mode of heat transfer often predominates. For example, the recovery of waste heat energy from a high-temperature furnace is often dominated by radiation on the hot, exhaust side of the exchanger and convection on the air side (if air is being heated). On the exhaust side of the exchanger, the high temperatures increase the extent of radiative transfer, which is proportional to the fourth power of absolute temperature (T). Convection is also present on the exhaust side, but it is not as significant as radiation at high temperatures. Instead, convection is dominant on the air side because air is relatively "transparent" to radiant heat. For these reasons, a ceramic heat exchanger for high-temperature use would be designed to optimize radiant heat transfer on the exhaust side of the exchanger and convection on the air side.

Application Areas

Three major industrial application areas of the heat exchanger are process heat exchange, power generation heat exchange, and industrial waste heat recovery.

Process Heat Exchangers. Two major industries that employ process heat exchange are the chemical industry and the primary metals industry. For process heat exchange in the chemical industry, primary uses include liquid-to-liquid and liquid-to-vapor heat exchange. Most applications are generally in medium-to-low temperature ranges where the ceramic heat exchanger would have no advantage over metallic heat exchangers with respect to temperature. However, many process streams, because of their corrosive nature, require that the metal heat exchanger have a protective coating. In addition to the use of glass coatings and glass-tubed heat exchangers, a ceramic tubed heat exchanger is available. Higher temperature-applications in this industry include steam reforming and other chemical reaction processes.

For process heat exchange in the primary metals industry, uses include the indirect-fired radiant tube furnace and the immersion heater. The radiant tube furnace is used by industry in heating processes, where the product must not be contaminated by the burner combustion gases. The combustion flame transfers its heat to the interior of the radiant tube, which in turn radiates heat to the product. Because ceramic materials have a higher melting temperature than metals, by replacing the metal radiant tube with a ceramic tube a higher gas firing rate into the tube and higher heat flux from the tube can be achieved along with an extended tube lifetime (Ref 4).

The immersed-tube melter is a new concept (Ref 5) which may be an efficient alternative to the reverberatory furnace for industrial metal melting operations. While the reverberatory furnace operates primarily by radiant heat transfer from a combustion flame firing over the surface of the furnace load, the immersed-tube melter transfers heat by conduction from an internally heated tube immersed in the furnace load. In this manner, transfer of heat to the load is not hindered by any dross insulating the surface of the load nor is the chamber refractory degraded from high chamber temperatures. By isolating the combustion flame from the load, the immersed-tube melter eliminates oxidation and entrainment from the melt surface, thus reducing metal loss and allowing corrosive/fouling-free recuperation of heat from the combustion exhaust gases. Finally, the lower temperatures required by an immersion melter should lead to lower NO_x emissions and longer life from less expensive furnace chamber refractory.

Power generation heat exchangers, such as indirect-fired gas turbines, transfer heat from a combustion gas or high-temperature exhaust to a working fluid. This often occurs in a high-temperature, potentially corrosive use environment. Because the working fluid (typically air) can be under high pressure, development of the ceramic heat exchanger for this market must include development of high-pressure seals and joint configurations.

Heat Recovery Exchangers. The industrial waste heat recovery environment is typically medium-to-high temperature and can often be corrosive. In the less corrosive streams, use of metallic heat exchangers is widespread. For industry with corrosive waste streams, a large market exists for the heat exchanger that proves viable (for example, in the past only about 5% of aluminum remelt furnaces were recuperated) (Ref 6).

Heat Exchanger Operating Modes and Effectiveness. By examining the operating modes (and therefore the change in temperature of the fluid streams as they exchange heat within the heat exchanger), the performance of the individual heat exchanger can be better understood. The two simplest operating modes of any heat exchanger are single-pass parallel flow (also called co-flow) and single-pass counter flow.

The parallel-flow exchanger (Fig 1), with its high initial temperature difference to drive heat transfer, requires less area and can potentially be the cheapest unit. On the other hand, this limits the highest temperature reachable by the "cold" fluid being heated to the lowest temperature of the hot fluid being cooled. The axial temperature gradient of the conducting wall separating the two fluid streams is also relatively small during this operation. Thus, parallel flow minimizes the stresses induced by the temperature-gradient along the length of the heat-exchanger surface.

The counter-flow exchanger (Fig 1) permits the temperature of the heated fluid to equal or exceed the exit temperature of the cooled fluid. This would normally require more transfer area compared to parallel flow because the counter-flow temperature differences are usually smaller than the differences across the parallel-flow exchanger. In this mode of operation, the conducting wall separating the two fluid streams can be exposed to:

- Very high temperatures at the hot-fluid inlet (thus accelerating any corrosion or oxidation of the wall material)
- Very low temperatures at the hot-fluid outlet (and if sufficiently low to reach the hot-fluid dewpoint, to cause acid corrosion)
- A steep axial temperature gradient, inducing high stresses along the length of the conducting surface

A third mode of operation is termed cross-flow. This system can be configured into relatively compact heat exchangers. Although it heats the "cold" fluid to a higher potential temperature than a parallel-flow exchanger, it has the potential liabilities of the counter-flow exchanger including hot and cold corners (with their attendant accelerated corrosion/oxidation and acid condensation, respectively), and

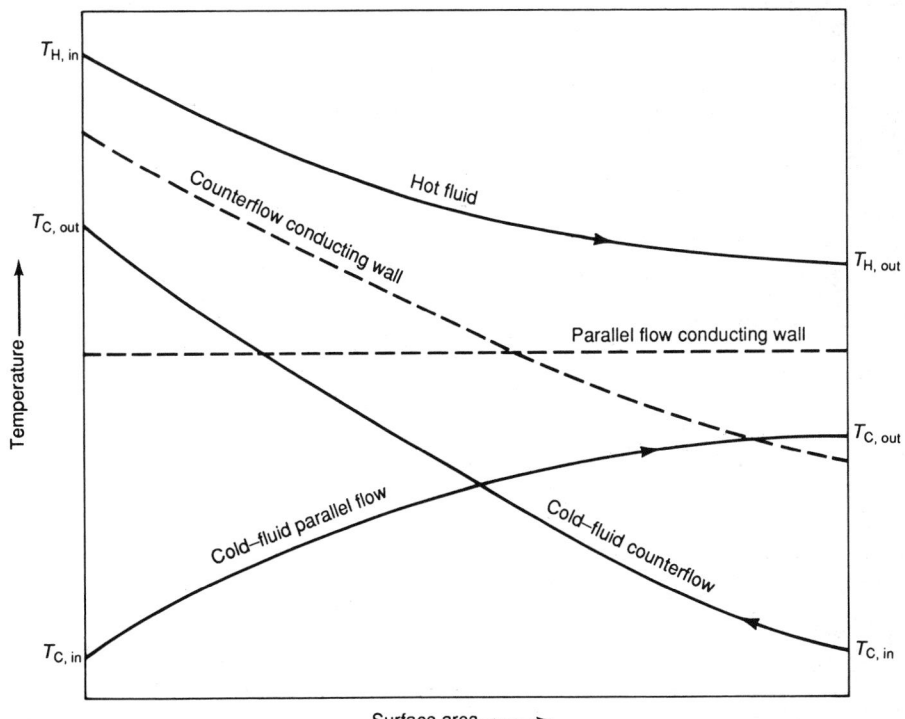

Fig 1 Temperature profiles of single-pass parallel-flow and counter-flow heat exchangers

temperature-gradient-induced stresses across the surface of the conducting wall.

Heat Transfer Effectiveness. A heat transfer term of particular importance to the user is heat transfer effectiveness. The effectiveness compares the actual heat transfer rate to the thermodynamically limited, maximum possible heat transfer rate as would be realized only in a counter-flow heat exchange of infinite heat transfer area. Heat exchanger effectiveness can give the potential user an approximate idea of recovery that would be realized from installation of a recuperator. The relationships between effectiveness versus the heat transfer area and the overall thermal conductance has been determined for various heat-exchanger operating modes. For exchangers optimized for convection, cost is almost directly proportional to the heat exchange area, while the cost of operation is a complex function of the conductance. Increases in effectiveness become nearly asymptotic with area and conductance increases, which is why metallic heat exchangers are so often manufactured in standard sizes and often as modular units (Ref 7). Only for the larger systems do special designs become economically worthwhile. Therefore, the one-time item manufactured of ceramic is not likely to be economically viable in the heat-exchanger market.

Heat-Exchanger Configurations. The basic geometries or configurations of heat exchangers (Ref 8) include the plate-type (also called the plate-fin), the tubular (which includes the tube-in-shell, the U-tube, and the tube-in-tube), the shell-in-shell (also referred to as the stack or radiant), and miscellaneous (including fluidized beds, spiral, heat pipes, and regenerative heat exchangers such as heat wheels and checkerworks).

The plate-type heat exchanger is typically a compact geometry (a high ratio of heat-exchange surface area to heat-exchanger volume) that is operated in a cross-flow mode. It is often designed with enhancement devices (usually fins or corrugations) to provide high effectiveness, frequently greater than 50% and sometimes as high as 75%. It is usually applied to relatively clean environments.

The tubular heat exchanger is typically operated in a combination of cross-flow and either counter- or parallel-flow modes. It is available in numerous variations and is applied throughout industry. Because of this wide variation, effectiveness ranges from around 30% (parallel-flow, low-temperature application) to as high as 90% (counter flow, high-temperature application). This type of metallic heat exchanger has by far the largest sales because it includes boilers and condensers used in mid-temperature applications throughout industry. This type of heat exchanger is also operated in high-temperature environments for waste heat recovery.

The shell-in-shell heat exchanger is used primarily for waste heat recovery. This exchanger consists of two concentric shells

that make up an inner shell through which the exhaust flows and an annulus between the shells. The combustion air flows through the annulus. This exchanger can be operated either in counter-flow or parallel-flow modes; however, parallel flow is often used to avoid overheating of the lower portions of the inner shell wall. Effectiveness is usually less than 50%.

Fluidized beds operate by ducting a high-temperature gas stream either through a series of individual nozzles or through the fine perforations in a distributor plate to fluidize a bed of small particles. This fluidization enhances the heat-transfer to in-bed products or to tubes (carrying the heated fluid) by a factor of up to 8 over non-fluidized convection. For exhaust gas heat recovery, the system would use a brush or air/steam jet to dislodge fouling particles from the under side of the distributor plate.

Ceramic Materials in Heat Exchangers

A variety of ceramic materials have been considered for use in heat exchange systems. They have been used primarily in the critical zones in an effort to minimize cost over metallic counterparts. Ceramics have been used for the main heat transfer surfaces (tubes, plates, and so forth), seals, tubesheets, and insulation. Heat-exchanger designs have ranged from all ceramic to ceramic-metal hybrids with varying numbers of ceramic components.

Silicon-Based Ceramics. A major effort has been directed toward development of non-oxide silicon-based materials in tubular form for the major heat transfer surfaces. These materials are produced commercially by slip-casting, extrusion, and isopressing. Table 1 and Fig 2 show mechanical properties of some of these materials produced by Norton, Carborundum, and Coors (Ref 9–12). They have been made in a variety of shapes such as plain surface, finned, or cruciform. Complex shapes

(Fig 3) including U-bends and headers with tube holes (Ref 13, 14) have also been produced and tested. Properties vary from shape to shape for the same composition due to shrinkage during processing (often up to 20%), batch variation, and manufacturer. In development efforts to date, performance of simple shapes has been superior to complex shapes. Two highly characterized materials are Norton's silicon-infiltrated silicon carbide (NC-440/CS-101K) and Carborundum's sintered α (SA) silicon carbide. The NC-440 type silicon carbide is based on slip-cast β-SiC material which is fired and infiltrated with 12 to 17% silicon metal. The SA material is 100% α-SiC produced by extrusion and/or isostatic pressing. Various other silicon nitride, Sialon, and silicon carbide materials have been evaluated as shown in Table 1.

Ceramic composites based on continuous ceramic fibers and CVD/CVI β silicon carbide have more recently been developed for heat exchangers by Thermo-Electron Company, Amercom, RCI, and 3M (Table 1). Thermo-Electron Company, using SiC (Nicalon) or graphite fibers, produced a plate-type heat exchanger with corrugated surfaces (Ref 15). Amercom and RCI developed tubular shapes based on Nicalon SiC fiber (Ref 16). The 3M Company licensed Amercom CVI technology for use with their Nextel family of ceramic fibers for radiant tube applications. The 3M materials have been used in several radiant tube field test installations (Ref 17).

Oxide ceramics have also been used for heat-exchanger components (Table 1). Plate-type heat exchangers have used a cordierite-based composition (Ref 18) and a proprietary aluminosilicate composition (Ref 19). Both of these materials can be produced through reasonably low-cost slip-casting or extrusion methods. They have low thermal expansion and moderate refractory properties, but are susceptible to corrosion due to glass formation in the presence of elements such as alkalis, calcium, and lead. Thin-walled complex

Table 1 Selected ceramics used for heat exchanger applications

Material	Processing method	Manufacturer	Brand	Density, g/cm³	Strength MPa	Strength ksi
SiC-Si	Extruded, slip cast	Norton	NC-430/CS101-K	3.07	160(a)	23(a)
SiC (100%α)	Extruded, slip cast	Carborundum	Hexaloy SA	3.10	320(b)	46(b)
					246(c)	35.7(c)
SiC	Extruded, slip cast	Norton	CS-101	2.7	91(a)	13(a)
					100–200(b)	14.5–29(b)
Si₃N₄-bonded SiC	Extruded	Norton	CX-589	2.51	84(a)	12.2(a)
SiC	Extruded, slip cast	Coors	SC-2	3.10	525(d)	76(d)
					175(c)	25(c)
SiC	Slip cast	Coors	RBSC-205	3.05	287(d)	41.5(d)
SiC	Chemical vapor infiltration	Amercom	...	2.7–3.2	300	43.5
Al₂O₃/ZrO₂	Sol-gel winding	B&W	...	3.7	350(c)	50(c)
SiC whisker-reinforced alumina	DIMOX(e)	Lanxide	...	...	460(c)	66.5(c)

(a) O-ring strength test. (b) Four-point bending flexural strength. (c) C-ring strength test. (d) Three-point bending flexural strength. (e) Proprietary process

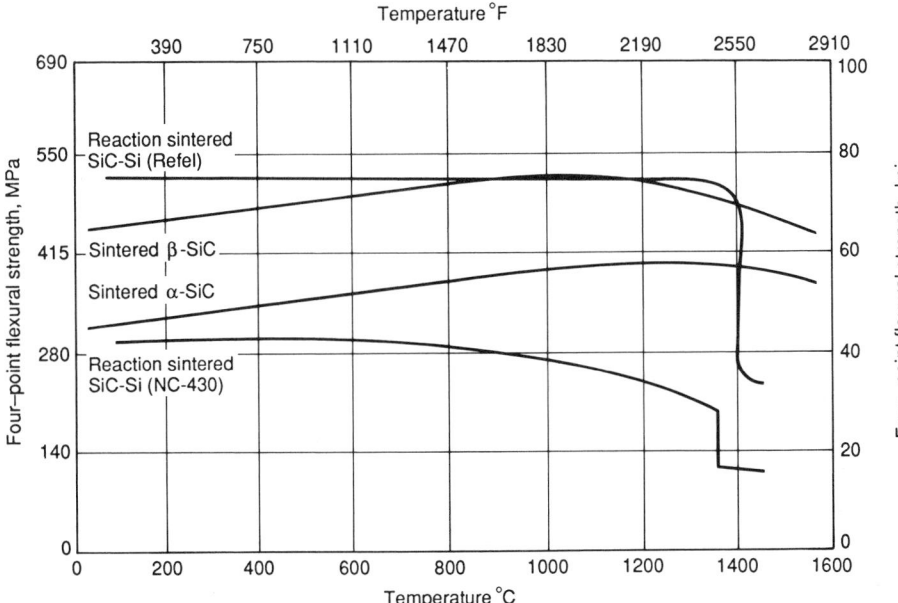

Fig 2 Strength of SiC ceramics as a function of temperature. Source: Ref 1

shapes have been made from these materials. Oak Ridge National Laboratory (ORNL), Lanxide, and Babcock & Wilcox (B&W) have developed ceramic composites for use in heat exchangers. Oak Ridge developed extruded SiC whisker-reinforced alumina tubes (Ref 20). These materials show enhanced thermal shock resistance compared to alumina tubes. Based on the ORNL work, MER and Vesu-vius/McDanel are developing a slip cast/pressureless sintering method of tube production. Lanxide has produced alumina-silicon carbide tubes for testing based on their DIMOX process (Ref 21). Babcock & Wilcox has developed continuous-fiber-reinforced composites based on alumina and zirconia. Tube sizes up to 10 cm (4 in.) in diameter and 1 m (39 in.) long have been produced. These materials have performed well in corrosion tests and a prototype heat exchanger will be tested in 1993 (Ref 22).

Ceramic Seals. Several types of ceramic seals or ceramic-containing seals have been developed or tested in ceramic heat exchangers. Very early designs used metallic parts to aid the sealing process such as the compression spring design used by the Hague Industrial heat exchanger (Ref 23). Low-pressure units have used glass fiber parts, such as Koawool, for seals. These seals typically show increasing leakage with greater differential pressures. Glass seals have also been studied by Solar Turbine, AiResearch, C&H, B&W, and others (Ref 22, 24, 25) using refractory glasses with controlled viscosity at temperature to seal between gas streams. Glasses have been characterized from 800 to 1650 °C (1470 to 3000 °F) and up to 1.38 MPa (200 psi).

Tubesheet or headers have also varied in design as well. All-ceramic tubesheets have had mixed success in field trials. AiResearch used commercial-grade silicon nitride bonded plates which failed in service due to insufficient densification (Ref 26). Carborundum uses a teflon-encapsulated steel tubesheet for their low-temperature process heat exchanger (Ref 27). Babcock & Wilcox used an insulated metallic tubesheet. Solar Turbine uses a metallic header for moderate temperatures with

a ceramic/metal transition piece to the tubes (Ref 24). They also produced a preliminary design for a composite header using Amercom's CVI silicon carbide materials.

Test Methods

Critical to the successful usage of ceramic in heat exchangers has been the continuing development of design methodology and test methods. This has been a trend for all structural ceramics, but the performance of heat-exchanger materials under potentially corrosive conditions has received substantial attention.

A variety of mechanical tests have been employed to characterize ceramic materials for heat exchangers. Strength tests have been run using C-ring or slit-ring tests (on tubes), flexural tests, and a lesser amount of burst or tensile testing. Compressive C-ring tests are simple tests that rely on tensile failure of outer surfaces where the most serious property degradation would occur. Several materials have been well characterized, including Weibull statistics such as those shown in Fig 4. Creep testing has also been performed on α-SiC and silicon-infiltrated silicon carbide. Creep theories have been postulated, including void formation and coalescence, plastic flow of the free silicon, and grain-boundary sliding.

Results of nondestructive evaluation (NDE) and mechanical property tests on heat-exchanger components have been reported by several laboratories including ANL, Penn State, and Virginia Tech (Ref 28–30). As part of the overall design of the heat exchanger, considerable effort has been placed on NDE and/or proof testing of the ceramics. Major test programs have been conducted on heat-exchanger tubular shapes (Ref 31, 32). Tech-

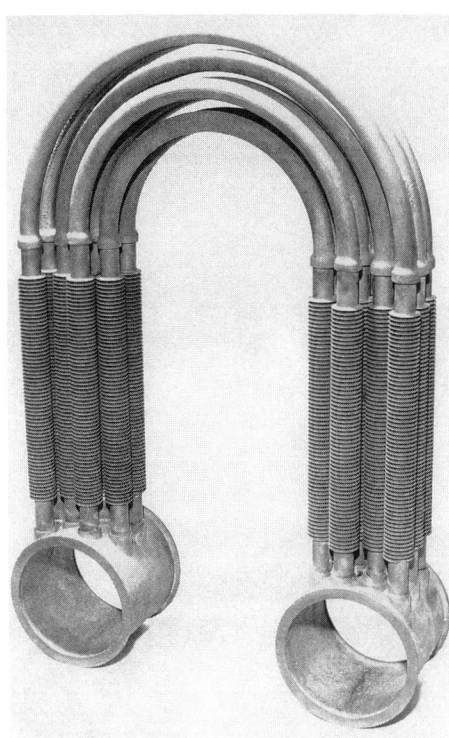

Fig 3 Ceramic heat exchanger test module. Courtesy of Allied-Signal Aerospace Co.

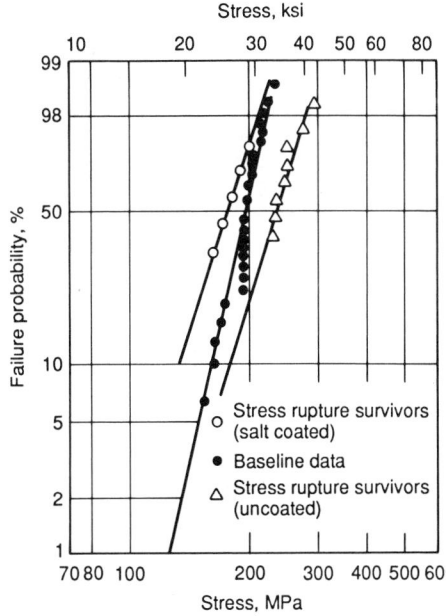

Fig 4 Weibull plots for NC-430 SiC before and after 500 h stress rupture

niques have been developed using computer tomography, scanning laser acoustic microscopy, acoustic emission, and radiography for flaw detection and prediction of performance. Several of these tests have been correlated to destructive methods such as C-ring and burst testing. Predictive models have been or are being developed. Several companies have used proof testing to screen materials before use.

Corrosion Tests. A number of corrosion tests have been run at universities, national laboratories, and private industry using both simulated conditions and actual field exposure tests. Major targeted processes include steel reheating furnaces, various coal-fired processes such as gasifiers and combined cycles, and aluminum remelt furnaces. Typical of the industrial sector, the extent and chemistry of the corrosive species vary in each application and control material performance. Corrosion testing in several applications are reviewed to present some of the variations in performance observed. Results of many additional studies are available in the literature.

In a simple combustion environment (for example, reheat furnaces or radiant furnaces) the primary species are oxygen, hydrogen, carbon, and nitrogen. This can produce free or reactive oxygen given sufficient temperature, and oxygen partial pressure. Silicon carbide materials have performed well in these simple environments in a number of tests due to the formation of a silicon oxide layer at the surfaces of these materials which protects and retards oxidation and diffusion of oxygen through the surface glasses. In contrast, oxide compositions are not externally attacked by the combustion environment in the presence of free oxygen. Therefore, performance of these materials is controlled by thermal shock, internal recrystallization, creep, or other mechanisms.

Some soaking pits use topping compounds on the steel billets which can become airborne and deposit on the ceramic parts. These can be iron, silicon, and alkali containing. At sufficient temperatures, these components can react with the surface layers of the ceramics. This is most detrimental in non-oxide, silicon-based systems where the glassy surface layer becomes less viscous and more reactive. This results in the formation of lower refractory glasses and produces much faster oxidation rates. Testing by B&W and ORNL (Ref 33) has shown that SiC-based materials will perform well at temperatures up to 1150 °C (2100 °F).

In coal processes the primary corrosive species are iron, alkalies, calcium, aluminum, and silicon (Ref 34–36). The chemistry can vary significantly whether the exhaust is acidic or basic. This requires that design be based on specific coal types in order to predict performance. On non-oxide materials, these slags can produce increased wear due to glass formation. The oxides vary in reac-

tivity to these species. In one set of tests done by ORNL (Ref 37), an experiment was run in a laboratory test facility at furnace temperatures of 1440 °C (2625 °F) and wall temperatures of 1240 °C (2265 °F) for 260 h. The major ash constituents included silica, iron oxide, lime, and alumina. All tubes tested showed significant erosion. Of the eight silicon carbide samples, the chemical vapor infiltrated materials showed the least erosion. Alumina, the only oxide tested showed 15% erosion. Sialon materials exhibited moderate to excessive erosion. Strength loss after exposure was significant in all but one silicon carbide material. For practical application, service life for these materials needs to approach a minimum of 10,000 h.

The results in aluminum remelt furnaces have varied from site to site. In general the most corrosive species are alkali fluoride and chlorides including cryolite (NaCaAlF$_6$). In one series of tests (see Fig 5) on oxides and carbides at Reynolds Metal at 1260 °C (2300 °F), the carbides performed much worse than the oxide compositions. Oxides with silica performed worse than oxides with little or no silica.

Battelle-Columbus (Ref 39) performed a study on materials for heat-condensing application where there were products of combustion of oil or gas. This can create H$_2$SO$_4$, HCl, and nitrous oxides. The materials were exposed to cyclic conditions up to 290 °C (550 °F) for 900 h (Table 2). Silicon carbide compositions exhibited little corrosion, while

Si$_3$N$_4$ and oxides exhibited moderate to significant erosion.

Ceramic Prototype Systems

Over the last fifteen years various companies and research organizations have attempted to develop ceramic heat exchangers for industrial use (Table 3). Some have seen more success than others; a number have been successfully tested in industrial environments and have been commercialized. This section reviews some of these prototype ceramic heat exchangers with emphasis on their unique features and their performance under industrial testing.

Table 2 Ceramic corrosion in the presence of combustion products

Material	Corrosion depth data(a)		Projected corrosion	
	µm	mils	µm/y	mils/y
Cordierite	9.4	0.37	58.0	2.3
Mullite	0.85	0.034	8.2	0.32
Reaction-bonded SiC	0.20	0.0080	1.9	0.08
	0.16	0.0063	1.5	0.06
	0.18	0.0069	1.7	0.07
Sintered α-SiC	0.37	0.014	3.5	0.14
Reaction-bonded Si$_3$N$_4$	0.90	0.035	8.6	0.34
Glass-enamel	1.4	0.054	13.3	0.52

(a) Data are based on a heat condensing application where products were exposed to combustion products (H$_2$SO$_4$, HCl, and nitrous oxides) under cyclic conditions up to 290 °C (550 °F) for 900 h. Source: Ref 39

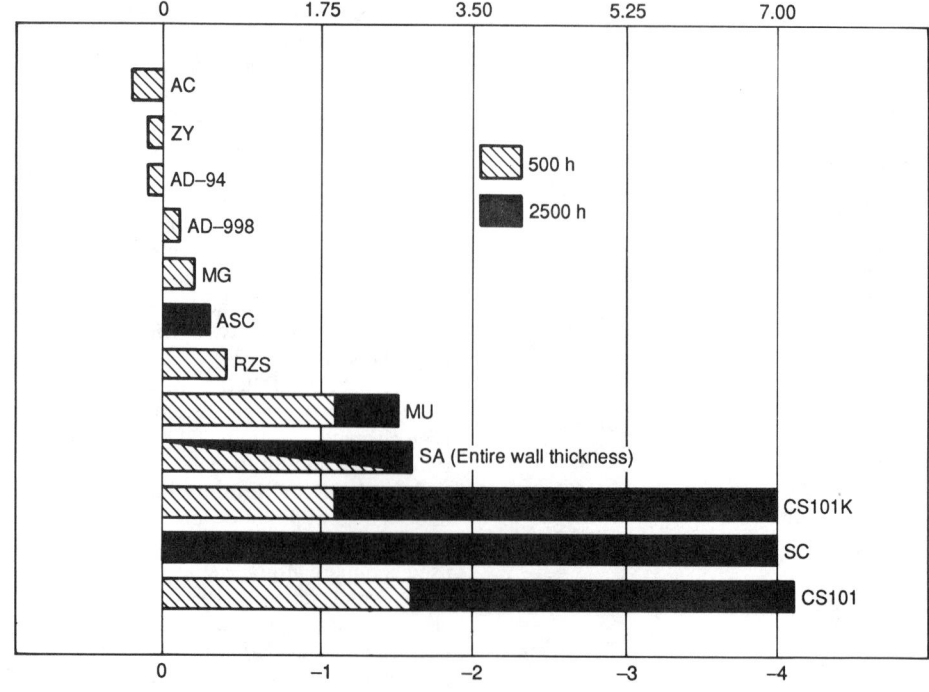

Fig 5 Projected annual recession rates of materials in an aluminum remelt furnace. Source: Ref 38

Table 3 Summary of ceramic heat exchanger systems

Type system	Designer	Capacity, GJ/h	Ceramic dimensions(a), cm	Test duration, h	Usage	Temperature °C	°F	Corrosive species
Tubular (shell and tube)	AiResearch	13	. . .	2000	Forging furnace	1150	2100	Combustion
	C&H	. . .	5 OD × 120 long	Various	Aluminum remelt, casting furnace	Various		
	Hague Industrial	4	120 long	Various	Various	1430	2600	. . .
	Carborundum	. . .	1.3–2.5 OD 270–430 long	Various	Condensing streams	200	390	Acids
Tubular (tube-in-tube)	B&W	22	9 OD × 195 long (50 tubes)	1400	Soaking pit	870–1230	1600–2250	Fe, Si, alkalis
	Solar Turbines	. . .	4–8 OD × 180 long (440 tubes)	6000	Aluminum melt furnace	1010	1850	Alkali salts
Plate fin	GTE	0.6, 1.6	25–46	10,000	Multiple	870–1370	1600–2250	Clean (good), alkalis (poor)
	Coors	0.25, 1.0	30 × 30 × 46	4000–10,000	Multiple			Clean (good), alkalis (poor)
Radiant tubes	INEX	. . .	5–15 OD × 180 long	17,000	Radiant furnace	1315	2400	Alkalis, HF, HCl
	3M	. . .	2.5–20 OD × 2	. . .	Radiant furnace	1450	2640	Alkalis, HF, HCl
Fluidized beds	Aerojet	7	. . .	8000	Aluminum melt furnace	600	1110	Oxygen
	TECO	17	. . .	8000	Aluminum melt furnace	1200	2190	. . .
High-pressure designs	Solar Turbines	41	5 OD × 180 long (864 tubes)	. . .	Hazardous waste	1260	2300	Na, Cl
	Stone & Webster	580	9 OD × 1200 long (650 tubes)	. . .	Gas reformer	1260	2300	Oxygen, hydrogen

(a) OD = outer diameter

Tube-in-shell type heat exchangers for waste heat recuperation (Ref 40) have been field tested by several companies such as AiResearch Manufacturing, C&H, and Hague Industrial. These designs have all used silicon carbide tubes of varying composition and manufacture.

The most advanced of these designs, by AiResearch Manufacturing (Ref 25), used silicon carbide tubes with an internal cruciform shape supported in silicon nitride-bonded silicon carbide header plates (Fig 6). The cruciform shape served as internal finning to increase the heat transfer area inside the tube (air side) and to minimize the number of tubes required. The tubes were cemented to the header sheets using a high-alumina mortar containing calcium-alumina cement in the low-temperature end of the heat exchanger and a phosphate-bonded alumina cement in the high-temperature end. The recuperator comprised four modules with 110 tubes per module.

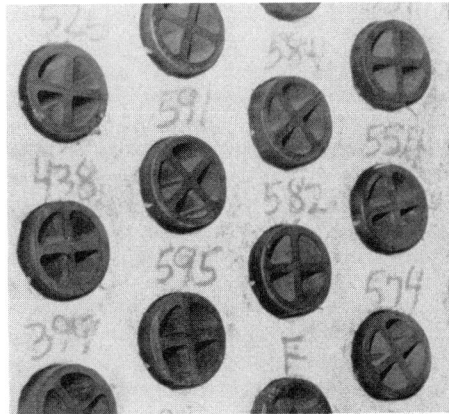

Fig 6 Cruciform tubes in ceramic header

During field testing on a steel forging furnace, the system provided over 2000 h of service. A post-test inspection of the cruciform tubes found no corrosion or other signs of degradation. The cements filled the joint well and sealed against leakage. However, the silicon nitride-bonded silicon carbide header plates experienced preferential oxidation which caused severe cracking. Air preheat temperatures as high as 1040 °C (1900 °F) from flue gases at 1150 °C (2100 °F) were achieved. Calculated fuel savings were in the range of 22 to 47%, depending on furnace operation. The furnace, designed for a total burner capacity of 37 GJ/h unrecuperated, matched original design performance with a total burner capacity of 13 GJ/h recuperated.

A tube-in-shell recuperator designed by C&H uses 5 m (2 in.) OD tubes with a 1.2 m (47 in.) length (Ref 41). The tubes are set in a refractory block wall and use a compressive fiber seal. Several field tests in different furnaces produced negligible wear on the silicon carbide parts. A design developed by Hague Industrial used ceramic tubes in a metallic shell structure (Ref 23).

Bayonet-style heat exchangers have been field tested by Solar Turbines and B&W. The B&W design, tested in a steel soaking pit, utilized silicon-infiltrated silicon carbide tubes (Ref 33). The outer tubes were 9.5 cm (3.7 in.) OD and 1.95 m (77 in.) long with one end flanged and one end closed. The primary corrosive species were iron silicides derived from the topping compounds at temperatures up to 1230 °C (2245 °F) for 1400 h. The tubes exhibited minimal wear in surface with minor oxidation and surface bubble formation.

A modular, bayonet-style recuperator has been designed by Solar Turbines (Ref 24). A typical module contained 12 to 16 silicon car-

bide tubes ranging from 4 to 8 cm (1.5 to 3 in.) in diameter and 1.5 to 1.8 m (60 to 70 in.) long. The ceramic tubes are braised to a metallic sleeve, which in turn is joined to a metallic header. Only the ceramic tube is exposed to the corrosive gas stream. Field tests on a 384 tube unit have been conducted for over 6000 h on a new aluminum remelt furnace, which used sodium and potassium chloride fluxing agents. Peak temperatures were 620 °C (1150 °F) with stack gases at 1010 °C (1850 °F) entering the recuperator. After 4300 h, salt deposits began to reduce the effectiveness. The tubes were cleaned and no detectable corrosion had occurred. Fuel savings of 32% were achieved. A service life of five additional years was projected.

Ceramic prototypes of tubular heat exchangers cover a variety of intended applications and use different ceramic materials. For refuse applications, for example, B&W has designed but not tested a tubular, oxide-ceramic composite based on DuPont PRD-166 alumina-zirconia fiber with alumina and/or zirconia matrices (Ref 42). This material has greater thermal shock resistance than the simple oxides with the benefit of their good corrosion resistance.

Other tubular designs have used either silicon carbide or ceramic composites. Solar Turbine Inc. used U-shaped silicon carbide by Carborundum and a complex header component. These shapes are more susceptible to thermal shock than the tubular counterparts and have been abandoned. Other companies have produced tubular or complex shapes based on various ceramic fibers (such as Nicalon, Nextel 312, or graphite) with chemical vapor deposition/infiltration (CVD/CVI) β-silicon carbide matrices. These materials exhibit greater thermal shock resistance with similar corrosion resistance to monolithic silicon carbide. Three other tubular designs are described below.

A low-temperature heat exchanger aimed at the corrosive chemical processing market has been developed by Carborundum (Ref 27). The unit is a tubular design using α-SiC tubes in a teflon-encapsulated steel tubesheet. Maximum operating conditions are 0.79 MPa (115 psi) and 200 °C (390 °F). The SiC tubes range from 1.3 to 1.9 cm (0.5 to 0.75 in.) in diameter and 0.9 to 4.3 m (35 to 170 in.) long. The units compete with exotic metals such as Hastalloy and tantalum-based systems. In this application the high thermal conductivity of the silicon carbide adds to the performance benefits.

Steam Reformers and High-Pressure Heat Exchangers. Preliminary studies by Stone & Webster project that a steam reformer using ceramic tubes could achieve significantly higher performance than conventional reformers of the radiant-box type (Ref 43). This high-temperature/pressure unit would convert methane (natural gas) and steam into a synthesis gas (H_2, CO, CO_2, H_2O, and CH_4) at 2.07 MPa (300 psi) and 1040 °C (1900 °F). Higher reaction temperatures increase feedstock conversion (from around 82% to as high as 98%), reduce energy requirements (by about 12%), and reduce plot area (the higher conversion efficiency allows a more compact reformer and a size reduction of downstream processes). High-pressure heat exchangers may also economically recover heat from high-temperature corrosive exhaust streams (for example, from municipal or hazardous waste combustion). Prototype development is to be completed in 1996.

Another bayonet design by Solar Turbines is a high-pressure heat exchanger. The heat exchanger (Ref 44) uses a continuous fiber ceramic composite header with monolithic ceramic tubes to heat 315 °C (600 °F), 1.03 MPa (150 psi) air to 900 °C (1650 °F) from an exhaust stream in excess of 1260 °C (2300 °F). The header uses a three-dimensional weave with Amercom CVI silicon carbide matrix. In the preliminary design of this system, 5 cm (2 in.) diameter ceramic tubes are brazed into the composite header to eliminate leakage. The composite header is further protected from the corrosive constituents of the waste stream by being cast into a refractory casing.

Compact plate-type cross-flow oxide recuperators have been developed by GTE and Coors. Over 1400 units of the GTE system (Fig 7) have been installed on new or retrofitted sites (Ref 18). The units consisted of cordierite plates stacked and mortared together. A resilient fibrous seal was used to prevent leakage between the unit and its metallic housing. The preheat air made three passes through the unit while the flue gas made one. Three sizes were available: 0.6, 1.0, and 1.6 GJ/h. The units operated at 870 to 1370 °C (1600 to 2500 °F) with 12 to 61% fuel savings (38% average). Eighty-seven percent of the units in exhausts below 1100 °C (2000 °F) operated four years or longer, while above 1100 °C (2000 °F), 57% survived. Plugging and corrosion occurred if sufficient alkalis were present and/or the temperature exceeded 1150 °C (2100 °F). Coors also developed a cross-flow heat exchanger based on a proprietary low-expansion aluminosilicate composition (Ref 45). The bonded ceramic plates were housed in a stainless steel shell with a ceramic fiber gasket to prevent leakage. Fuel savings of 35 to 50% were achieved. Over ten trials were run with up to 10,000 h-unit life obtained. These units were rated at 0.25 to 1.0 GJ/h.

Fluidized Beds. To allow a higher operation temperature of a fluidized bed, Aerojet Energy Conversion Company designed and tested a fluidized bed that replaced the metal distributor plate and cleaning brushes with a ceramic distributor plate and ceramic hot-gas lance assembly. By not requiring dilution of the furnace exhaust gases, the system was more compact (a 70% reduction in volume) and operated at higher efficiency (a 32% increase in heat recovery).

The ceramic distributor plate was made up of adjacent rows of alumina ceramic rods spaced, such that the exhaust gas passing between them became jets that fluidized the particle bed. A 0.8 by 0.5 m (31 by 20 in.) bench-scale unit of the ceramic distributor plate was tested at inlet exhaust temperatures up to 1370 °C (2500 °F) using No. 2 fuel. During these tests there was some breakage of the rods; however, no degradation in performance occurred. It was concluded that while the concept and design would likely perform well in full-scale, it would require the advent of reinforced alumina to avoid rod breakage.

Immersion Heaters. In a project conducted by Babcock & Wilcox (Ref 6), two small aluminum holding furnaces were modified to use immersion tube melting. These furnaces operated for over 7500 h in a production environment and over 185 Mg (410,000 lb) of scrap were processed.

Previous to the testing of the furnaces, coupon samples of various ceramic materials were exposed for a $5^{1}/_2$ month duration to molten aluminum. Sintered α-SiC performed well with only a very slight penetration of aluminum and flux. Reaction-bonded silicon nitride also performed well, while nitride-bonded silicon carbide performed third best in the tests. In later tests, sintered silicon nitride appeared to perform as well as sintered α-SiC.

During the testing of the modified holding furnaces, immersed-tube burners were operated continuously at 221 MJ/h to 264 MJ/h heat input. During one period of operation the melter obtained 0.2% greater metal recovery with 60% lower NO_x emissions than that produced by production melters.

Economics of Ceramic Heat Exchangers

Presently, ceramic heat exchangers cost more than their metallic counterparts. For example, a simple tube-in-shell heat exchanger with ceramic tubes and header of silicon carbide would cost approximately 68% more than the metal counterpart (INCO alloy 600) (Ref 46). The costs of the ceramic components in tubular recuperators are compared in Fig 8.

Various projections indicate that the cost of the ceramic components would decrease significantly with sufficient volume of sales. Silicon carbide tubes were estimated to cost $197 dollars each at a production rate of 2500/yr and to drop to $76 each at a production rate of 25,000/yr, a 62% drop in cost. Increased production rates could thus make the ceramics cost competitive with metals. However, first time installations of ceramic recuperation systems can require upgrades for burners, ducts, and controls to accommodate the higher temperature of the preheated air.

Fig 7 Compact, plate-type, cross-flow, oxide recuperator

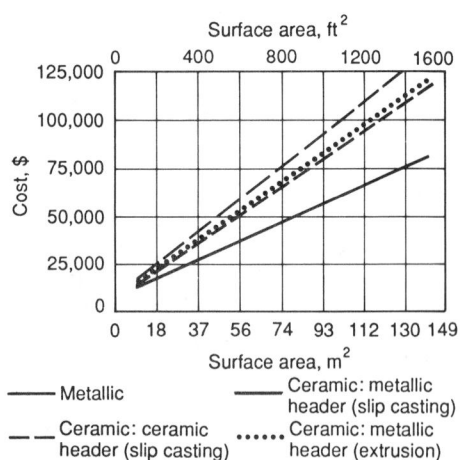

Fig 8 Sensitivity of tubular recuperator costs to changes in surface area

Clearly, unless the ceramic components provide benefit (such as higher performance, longer life, or entry into new market areas) that offsets their increased cost, the ceramic units will not replace metals. A performance benefit such as high-temperature capability is obviously required. For example, the replacement of a metallic unit operating in a 1260 °C (2300 °F) waste gas stream diluted to 800 °C (1490 °F) at the heat exchanger inlet to protect the metal is not as efficient as a ceramic unit which would require no dilution and yield much higher preheated combustion air temperatures. Many furnaces, especially unrecuperated furnaces, would benefit from ceramics.

Therefore, in many of the ceramic heat-exchanger projects the cost differential drove the designers to seek a performance benefit to justify the use of the ceramic unit. For example, the small size and high effectiveness of the compact plate-type recuperator (Fig 7) provides a niche market. Ceramic tubular units have typically doubled the fuel savings over their metallic counterparts.

Market Share

The slow acceptance and usage of the ceramic heat exchanger in the 1980s has been controlled by several factors. First, development and testing has been relatively recent with a number of successful trials not conducted until the mid-1980s. Advances in materials performance have been rapid, and two major silicon carbide producers now offer two year warranties on all ceramic tubes as an indication of their confidence in tube performance. There has been enough success to warrant use of ceramic heat exchangers, but the industrial sector is conservative and demands multiple successful demonstrations of a new technology before acceptance.

The second major market barrier has been the low price of fuel in the latter part of the 1980s. For example, natural gas prices peaked in 1984 and have decreased since then (Ref 47). This has also depressed the metallic recuperator market. Gas prices in the $4 to 6/GJ range would make the current systems cost effective with 1 to 2 year payback.

The third major market barrier has been the reluctance of a conservative industry to make major capital expenditures and accept long-term payback scenarios. Simple payback scenarios have shown potential payback in the range of 2 to 6 years depending on assumptions used and plant conditions. However, many industrial firms were spending little on capital improvements and expected these to have extremely short payback periods. A 3.5 year payback at one potential installation site was unacceptable because the company wanted a 2 to 2.5 year payback and lower capital expenditures. In other cases, firms had set maximum allowable payback periods as low as 6 months.

The market barriers listed above have created a lag in the further development of ceramic heat exchangers. A number of optimized design options have been reported or investigated, but lack of market penetration in the first generation systems curtail research and development spending. For example, various enhancement methods have been proposed to obtain greater heat transfer while minimizing the use of expensive structural ceramic tubes. A potential market area that has been little explored involves improving industrial processes through the incorporation of ceramic components. For example, higher temperatures can significantly increase chemical reaction rates, while higher pressures can increase product throughput (Ref 48, 49).

ACKNOWLEDGMENT

Much of the progress in ceramic heat-exchanger technology is the result of research projects funded by the Office of Industrial Technologies, U.S. Department of Energy, the Gas Research Institute, or EPRI. Significant contributions to our knowledge of corrosion effects on ceramics has been provided by researchers at the Oak Ridge National Laboratory.

REFERENCES

1. S.L. Richlen, A Survey of Ceramic Heat Exchanger Opportunities, *Ceramics in Heat Exchangers*, Vol 14, *Advances in Ceramics*, American Ceramic Society, 1985, p 3–14
2. W.J. Rebello, "Assessment of Heat Transfer Enhancement and Fouling in Industrial Heat Exchangers," DOE/CE/40716–3, U.S. Department of Energy, Nov 1987
3. D.M. Kotchick, private communication, Allied-Signal Aerospace Company, Feb 1991
4. M.A. Lukasiewicz, Radiant Heating, *Center for Advanced Materials Newsletter*, Vol 4 (No. 1), 1990, p 73–78
5. D.L. Hindman, "Gas-Fired Immersed Ceramic Tube Aluminum Melter—Phase 2, GRI-89/0098, Gas Research Institute, Feb 1989
6. M. Coombs, D.M. Kotchick, and H. Strumpf, "High Temperature Ceramic Recuperator and Combustion Air Burner Programs Semi-Annual Report April-October 1981," GRI-81/18550, Gas Research Institute, Nov 1981
7. W.M. Kays and A.L. London, *Compact Heat Exchangers*, McGraw-Hill, 1964
8. S.L. Richlen, Factors in Choosing Heat-Recovery Systems, *Eng. Dig.*, Vol 18 (No. 12), 1990, p 34–36
9. J.I. Fererer, private communication, Oak Ridge National Laboratory, Jan 1991
10. M.L. Torti, N.I. Paille, and J.J. Litwinowich, Materials and Process Development of Heat Exchanger Tubes, *Ceramics in Heat Exchangers*, Vol 14, *Advances in Ceramics*, American Ceramic Society, 1985, p 291–300
11. J.M. Halstead, Ceramic Recuperator Tube Manufacturing Method Comparison, *Ceramics in Heat Exchangers*, Vol 14, *Advances in Ceramics*, American Ceramic Society, 1985, p 279–289
12. D.W. Roy, K.E. Green, and M. Rivkin, Development of Cost Effective Ceramic Tube Process for High Temperature Heat Recovery Systems, *Ceramics in Heat Exchangers*, Vol 14, *Advances in Ceramics*, American Ceramic Society, 1985, p 269–278
13. K.O. Smith, T.E. Easler, and R.B. Poeppel, Experimental Assessment of an Axially Finned Ceramic Tube Heat Exchanger, *Ceramics in Heat Exchangers*, Vol 14, *Advances in Ceramics*, American Ceramic Society, 1985, p 79–96
14. D.M. Kotchik, M.G. Coombs, and W.T. Bakker, Development of a High Pressure Ceramic Heat Exchanger, *Ceramics in Heat Exchangers*, Vol 14, *Advances in Ceramics*, American Ceramic Society, 1985, p 227–241
15. W.E. Cole *et al.*, "Research and Development of a CVD Ceramic Composite Heat Exchanger for Industrial Waste Heat Recovery," Final Report DOE-ID-12544–3, U.S. Department of Energy, May 1988
16. J.E. Snyder and W.P. Parks, "Research and Development of a Ceramic Composite Heat Exchanger," Phase 1 Final Report DOE-ID-12536–1, U.S. Department of Energy, July 1985
17. M.A. Lukasiewicz, private communication, Gas Research Institute, Jan 1991
18. J.M. Gonzalez, private communication, GTE, Oct 1990
19. R.N. Kleiner, L.E. Coubrought, and L.R. Stasbaugh, Design of a Durable Compact Heat Exchanger, *Ceramics in Heat Exchangers*, Vol 14, *Advances in Ceramics*, American Ceramic Society, 1985, p 115–126
20. M.A. Janney, "Extrusion of SiC Whisker-Toughened Alumina Tubes," presented at DOE Advanced Heat Exchanger Program Review, Washington, D.C., U.S. Department of Energy, Oct 1987
21. A. Patel, private communication, Lanxide Corporation, Jan 1990
22. W.P. Parks, "Research and Development of a Ceramic Composite Heat Exchanger," Phase 2 Final Report, DOE-ID-12536–2, U.S. Department of Energy, Sept 1989
23. R.M. Woodward and R.G. LaHaye, Ceramic Tubular Heat Exchangers: A Summary of Seven Years of Operating Experience, *Proceedings of Industrial Heat Exchanger Symposium*, American Society for Metals, 1985
24. M.E. Ward, A.D. Russell, and W.W. Liang, Ceramic Recuperator Design for

an Aluminum Reclamation Furnace, *Ceramics in Heat Exchangers*, Vol 14, *Advances in Ceramics*, American Ceramic Society, 1985, p 71–79

25. M.G. Coombs, "High Temperature Joints/Seals for Ceramic Heat Exchanger Components," presented at DOE Advanced Heat Exchanger Program Review, Washington, D.C., U.S. Department of Energy, Oct 1987

26. D.M. Kotchick *et al.*, "Development of a High Pressure Ceramic Heat Exchanger," Final Report DOE ID 12170–1, U.S. Department of Energy, Dec 1987

27. M. Sohal, private communication, EC&G Idaho, Jan 1991

28. W. Ellingson, private communication, Argonne National Laboratory, Jan 1991

29. R. Tressler, "Cooperative Program in High Temperature Materials Research," Semiannual Report, Center of Advanced Materials, Mar 1990

30. T.J. Dunyak, K.L. Reifsneider, and W.W. Stinchcomb, "An Examination of Selected NDE Methods for Ceramic Composite Tubes," ORNL/Sub/87-SA946/01, Oak Ridge National Laboratory, Dec 1989

31. W. Rueter and D. Kotchik, "Assessment of Strength Limiting Flaws in Ceramic Heat Exchanger Components," presented at DOE Advanced Heat Exchanger Program Review, Washington, D.C., U.S. Department of Energy, Oct 1989

32. J. Bower *et al.*, "Assessment of Strength Limiting Flaws in Ceramic Heat Exchanger Components," Final Report DOE-ID-12566–3, U.S. Department of Energy, Jan 1989

33. J.I. Federer, "Corrosion of Materials by High Temperature Industrial Combustion Environments—A Summary," Report TM-9903, Oak Ridge National Laboratory, Feb 1986

34. R.A. Perkins and H.W. Lavendel, "Reaction of Silicon Carbide with Fused Coal Ash," Final Report EPRI AF-294, Electric Power Research Institute, 1976

35. T.E. Easler, C. Tan, and L.M. Putz, "Influence of Oxidizing and Reducing Environments on Coal-Slag-Induced Corrosion of Silicon Carbide Ceramics," Final Report ANL/FE-85–11, Argonne National Laboratory, 1985

36. T.E. Easler, "Comparison of Results of 200 and 500 Hour Exposures of Silicon Carbide to a Slagging Coal Gasification Environment," Final Report ANL/FE-85–11, Argonne National Laboratory, 1985

37. M.K. Ferber and V.J. Tennery, "Evaluation of Tubular Ceramic Heat Exchanger Materials in Basic Coal Ash from Coal-Oil-Mixture Combustion," Report TM-8385, Oak Ridge National Laboratory, Oct 1982

38. J.I. Federer, Corrosion Behavior of Heat Exchanger Materials, presented at Advanced Heat Exchanger Review, Washington, D.C., U.S. Department of Energy, 1987

39. I. Sekercioglu, R. Razgaitus, and J. Lux, Evaluation of Ceramics for Condensing Heat Exchanger Applications, *Ceramics in Heat Exchangers*, Vol 14, *Advances in Ceramics*, American Ceramic Society, 1985, p 359–369

40. W.P. Parks *et al.*, "High Temperature Burner Duct Recuperator System Evaluation," Final Report DOE-ID-12296–5, U.S. Department of Energy, July 1988

41. R.C. Graham, Development of an Air-to-Air Heat Exchanger with All Ceramic Internals, *Ceramics in Heat Exchangers*, Vol 14, *Advances in Ceramics*, American Ceramic Society, 1985, p 43–48

42. W.P. Parks *et al.*, "High Pressure Heat Exchange System for Combined Cycles for Power Generation," Final Report DE-FC07–88ID12798, U.S. Department of Energy, Oct 1989

43. J.J. Williams, "Development of a High Pressure Heat Exchange System for Reforming of Methane," Final Report DE-FC07–88ID12797, U.S. Department of Energy, May 1990

44. B. Harkins and W. Ward, "High Pressure Heat Exchange System," Final Report DE-FC07–88ID12799, U.S. Department of Energy, Mar 1990

45. L.C. Hoffman and A.H. Kreeger, "Research and Development of a Ceramic Tubular Distributor Plate for Advanced FBHX," Final Report DOE-ID-12535–2, U.S. Department of Energy, Aug 1987

46. S. Das, T.R. Curlee, and R.A. Whitaker, "Ceramic Heat Exchangers: Cost Estimates Using a Process-Cost Approach," ORNL/TM-10684, Oak Ridge National Laboratory, Aug 1988

47. "Annual Energy Review, 1988," Energy Information Administration Report DOE/EIA-0384 (88), U.S. Department of Energy

48. S.L. Richlen, "Overview of DOE's Office of Industrial Programs Development Program in Continuous Fiber Compositions," 34th International SAMPE Symposium and Exhibition, Reno, Nevada, 8–11 May 1989

49. Assessment of High Pressure Heat Exchange Systems Technology," Vol I and II, DOE/CE/40716–2, U.S. Department of Energy, Mueller Associates, Oct 1987

Adiabatic Diesel Engines

Roy Kamo, Adiabatics, Inc.

ADIABATIC DIESEL ENGINES, as the name implies, are designed to reduce heat loss during engine operation. In practice, heat loss from an engine can be reduced by:

- Insulating the combustion chamber (cylinder liner, piston crown, and cylinder head), exhaust and intake ports, and the exhaust manifolds
- Eliminating the cooling system and its associated parasitic losses
- Utilizing exhaust heat by turbocompounding (which involves a turbine system geared to the crankshaft)

The simplest type of "adiabatic" engine may involve just insulation of the combustion chamber. In this case, material selection for hot engine components is not severely affected because the insulated adiabatic engine would still include a cooling system. In a water-cooled diesel engine, for example, temperatures at the cylinder liner seldom exceed 200 °C, or 400 °F (Fig 1).

On the other hand, an uncooled, adiabatic engine, without a constant, low-temperature sink on the cylinder liner, experiences an extremely high temperature on the cylinder liner wall. The piston and the piston rings also experience high temperatures (Fig 2). The uncooled adiabatic engine also suffers higher compression work in the cycle, but, due to the higher temperature and pressure after combustion, the expansion work is greater.

With cylinder wall temperatures exceeding 1000 °C (1830 °F), uncooled adiabatic engines have been considered a potentially useful application for emerging high-performance ceramics such as pressureless-sintered silicon nitride, partially stabilized zirconia, and stabilized magnesia. These successful new ceramic products, however, are still not considered reliable enough for an uncooled, adiabatic diesel engine of 1500 kPa (220 psi) BMEP output. Most of the materials will not meet all the requirements of an adiabatic, turbocompounded diesel engine (Table 1). Compromises thus have been made in terms of materials, engine performance, and life.

Soon the monolithic ceramic designs were being replaced in the adiabatic turbocompound engine program. Focus was directed to thick ceramic coatings that are applied by plasma spray, detonation gun, or the slurry, chemical-bond process. Graded coatings, new bond coats, and many new coating developments resulted from these investigations.

Adiabatic Engine Designs

The designation of so-called "adiabatic" engines is a misnomer in thermodynamic terms, as the word adiabatic means that no heat is added or lost during a thermodynamic process. In practice, however, the basic intent of an adiabatic engine is to reduce heat loss from the combustion chamber and its intake and exhaust ports. Uncooled adiabatic engines also may have turbocompounding systems, which transform exhaust heat into chemical energy for the crankshaft system.

Uncooled (waterless) adiabatic engines offer the best thermodynamic efficiency, because the elimination of the cooling system (radiator, fan, coolant, and pump) raises engine temperatures and removes a built-in heat loss from cooling. A well-designed, uncooled adiabatic engine also insulates the exhaust and intake ports. The exhaust ports are insulated to minimize heat loss and maximize the exhaust-energy availability to the turbocompounding system. The intake ports are insulated to prevent heat addition to the intake charge for maintaining good volumetric efficiency.

Types of Adiabatic Engines. Table 2 indicates how today's adiabatic engines are categorized and the organizations associated with each one. The two general categories are cooled and uncooled designs, although the engine performance in either category is not necessarily comparable.

The water-cooled adiabatic engines include a monolithic ceramic engine designed by NGK Ltd, several engine designs containing ceramic coating, and engines using low-conductivity metals (such as Nimonic) with air-gap insulation. The ceramic-coated engines dominate the current development of insulated engines and are sometimes referred to as low-heat-rejection engines (LHREs). Water cooling maintains the temperature of engine components within required limits. In the case of the monolithic ceramic engine, however, the only reason for water cooling was to enable adequate lubrication on the cylinder liner and piston ring (see the section "Materials for Adiabatic Engines" for discussion of lubrication problems).

Uncooled adiabatic engines also include monolithic ceramic engines and ceramic-coated engines (Table 2). These engines also had turbocompounders. A simplified schematic of an adiabatic turbocompound engine is shown in Fig 3. Following the flow paths in the figure, air enters the turbocharger, is compressed, and then enters the insulated, high-temperature combustion chamber. The insulated, combustion-chamber components include the cylinder head, valves, cylinder liner, and piston. Combustion occurs, and useful energy is extracted by the piston. The high-temperature, high-pressure exhaust gas is then expanded through two turbine wheels to extract as much as possible of the remaining energy. One wheel is used to drive the compressor; the second is connected by gears (turbocompounding the system) to the engine crankshaft to further increase the useful power output of the engine.

The key to energy recovery is an efficient turbocompounding system. An overall turbocharger efficiency of 0.64 is desired. The free-power turbine geared to the engine crankshaft should have an efficiency of greater than 80% and a mechanical reduction gear efficiency of greater than 92% (Ref 3). Turbocompounding also reduces the in-cylinder peak pressures when the engine timing of an adiabatic engine is retarded. Retarded timing can be used without sacrificing fuel economy because the ignition delay in the adiabatic engine is reduced. With a shorter ignition delay and retarded injection timing, the emissions and fuel economy trade-off, as well as the peak cylinder pressure of an adiabatic turbocompound engine are lower when compared to a conventional, turbocharged engine.

Advantages of an adiabatic turbocompounded diesel engine include:

- Reduced fuel consumption
- Reduced emissions and white smoke
- Multifuel capability
- Reduced noise level
- Improved reliability and reduced maintenance
- Longer life
- Smaller installed volume and lighter weight

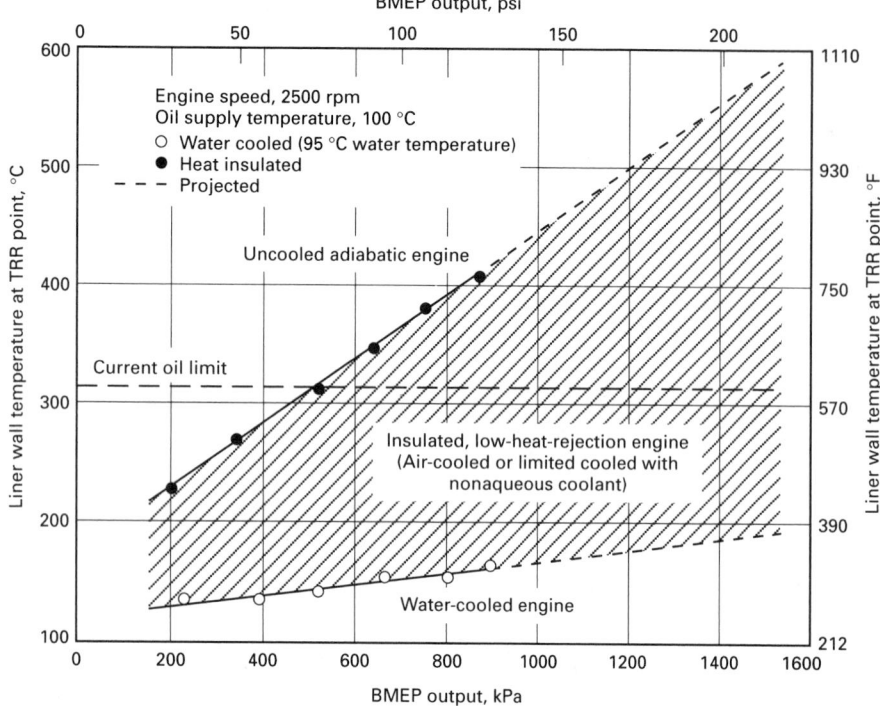

Fig 1 Top ring reversal (TRR) temperature versus diesel engine output (inclusive of pumping losses). BMEP = brake mean effective pressure. The TRR point is the top dead center ring position.

Table 1 Typical desired material properties for an adiabatic diesel engine

Temperature limit, °C (°F)	1350 (1800)
Fracture toughness (K_{Ic}), MPa$\sqrt{m}$ (ksi$\sqrt{in.}$)	>8.0 (7)
Flexural strength, MPa (ksi)	>800 (115 ksi)
Thermal conductivity (k), W/m·K	<0.01
Thermal shock resistance (ΔT), °C (°F)	>500 (900)
Coefficient of expansion, 10^{-6}/°C	>10
Creep rupture strength, MPa (ksi)	>97 (14)
Friction coefficient (dry)	<0.2

Most of the above advantages have been demonstrated in the laboratory and in the 5-ton U.S. Army vehicle.

Reduced fuel consumption, as previously mentioned, is attributed to heat recovery (turbocompounding), a higher temperature and pressure environment, a more favorable expansion ratio during the powerstroke, and a lack of parasitic power without a cooling system. The breakdown of fuel improvements is estimated to be: in-cylinder work improvement, 5%; turbocompound improvement, 7%; cooling fan, 7%; water pump, 1%; and friction reduction, 8%. The total is 20%. Other factors leading to lower fuel consumption are improved heat release rates and a smoother pressure diagram. Prototypes developed by Cummins, Hino, and Komatsu (Table 2) achieved fuel consumptions of less than 0.13 kg (0.29 lb) of fuel per brake horsepower-hour.

Reduced Emissions. In the area of emissions, the shortened ignition delay in a hot combusting adiabatic engine provides a more desirable brake-specific trade-off of NO_x emissions and fuel consumption trade-off. Hydrocarbons and CO emissions are also reduced due to the higher temperatures in the oxidizing environment. Smoke and particulate reductions have been noted during emission investigations. Particulate reductions of 60 to 80% reduction in particulates has been reported by other investigators in the engine field.

Multifuel Applications. The adiabatic engine is better suited for alternative fuels because its hot environment allows better combustion. If engine temperatures are not high enough, then low-cetane fuels do not burn or cause a long delay for ignition. This delay in ignition time can increase the pressure-rise rate in the combustion chamber and subsequently increase engine noise.

Reduced Noise Levels. The typical diesel knocking at low power is not present in adiabatic diesels because its shorter ignition lag provides smoother operation. Elimination of cooling fans also reduces engine noise.

Improved Reliability. The adiabatic engine can also reduce maintenance costs and repair time. Expensive aluminum or copper radiators are no long required; because the cooling system represents over 50% of the field war-

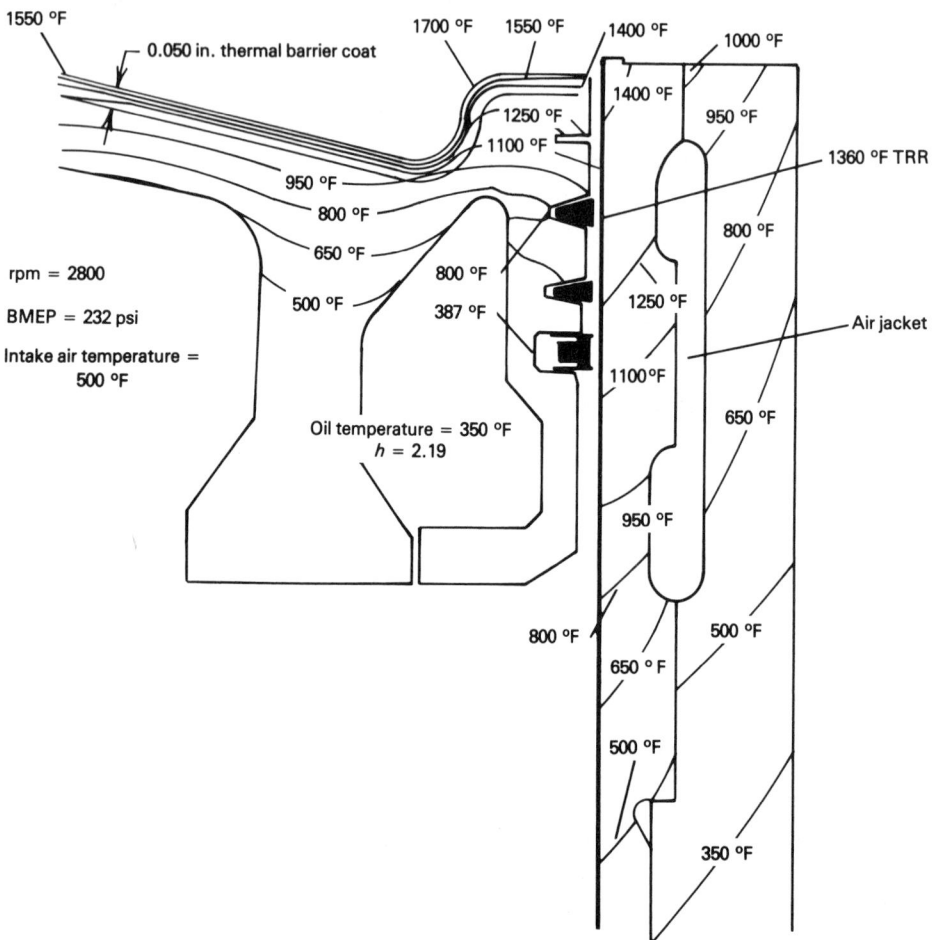

Fig 2 Predicted cylinder and piston temperatures of an uncooled, adiabatic diesel engine. Source: Ref 1

Table 2 Developers of cooled and uncooled adiabatic diesel engines

Engine material	Organizations with waterless designs	Organizations with water-cooled designs
Monolithic ceramics	Kyocera, Ford, Cummins Engine Co., and Isuzu	NGK
Monolithic/coating	. . .	GM
Ceramic coating	Adiabatics Inc., Cummins, and Komatsu	Hino Motors, Cummins, Caterpillar, Detroit Diesel, British Leyland, and Daimler Benz
Low-conductivity metal (with air gap)	. . .	Woschni (Ref 2) and GM

ranty problems, repair time also is greatly reduced. In addition, the smoother pressure rise offers improved noise characteristics and longer engine bearing life (Ref 4).

Engine Temperatures and Loads. Heat-transfer models in a diesel-cycle simulator usually assumes a mean, constant-cycle temperature during the entire cycle. In an actual engine, swing in wall temperatures occurs even in a water-cooled iron engine. When insulation material is applied in the combustion chamber and the cooling water removed, the swing in wall temperatures is significantly higher. The ideal case would be a thin insulation material possessing no specific heat and an infinite resistance to heat flow. The insulated cases with various thickness of zirconia (ZrO_2) begin to show this trend toward combustion gas temperatures with a consequent reduction in heat loss out of the combustion chamber.

The environment within the insulated combustion chamber of the adiabatic engine depends on the load, speed, air-fuel ratio, intake-air temperature, and injection timing. In an adiabatic engine, the temperature rises rapidly and lubrication of cylinder liners and piston rings are no long possible. Combustion chamber materials begin to deteriorate because of the elevated surface temperature and thermal fatigue failure.

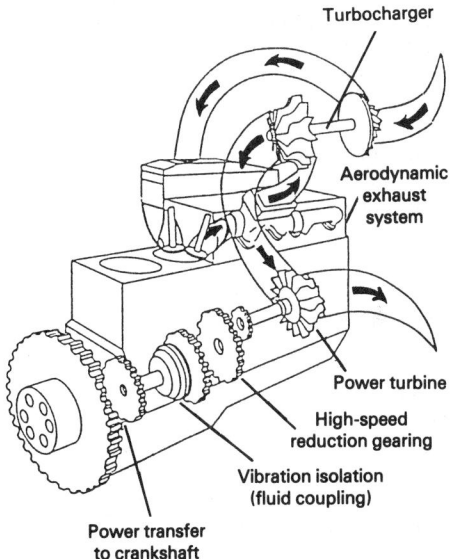

Fig 3 Schematic of adiabatic, turbocompounded diesel engine

Figure 1 shows the increase in the liner's wall temperature at the top-ring reversal (TRR) point as the diesel engine output continues to climb. The two lines show the temperature region for all engines between the uncooled, adiabatic and the watercooled diesel engines. The relative temperature between the cylinder liner and the piston along the axis of the piston travel shows that both the cylinder wall and piston temperatures drop rapidly along the axis from the top, dead-center position downward. Although the temperature profiles are similar, the liner temperatures are higher than the piston temperatures by approximately 50 °C (90 °F) (Ref 5). The piston and the piston rings also encounter high temperatures (Fig 2).

Although the stresses of the engine components are a combination of mechanical loading and thermal stress, the mechanical loadings are usually not the cause of the principal stresses (Ref 1). However, with future diesel engines going to higher outputs, material strength must be capable of handling at least 1035 kPa (150 psi) BMEP for turbocharged passenger cars, and at least 1515 kPa (220 psi) BMEP for turbocharged diesel truck engines. Under these load conditions, cylinder pressures will exceed 14 MPa (2 ksi). The combustion diagram of a waterless diesel engine having a 140 mm (5$^{1}/_{2}$ in.) bore and 150 mm (5 in.) stroke and operating at full load and speed shows that the maximum piston load is 220 kN (25 tonf).

Due to the fluctuating load and temperatures experienced in piston engines, significant thermal stresses are set up with temperature gradients caused by insulation and the expansion coefficient of the material. For example, in the headplate of an adiabatic engine, temperatures approaching 980 °C (1800 °F) have been measured on the exhaust valve seat area, while a minimum temperature of 173 °C (343 °F) has been measured behind the cylinder headplate by the cooler intake side (Ref 6).

Materials for Adiabatic Engines

Some of the preferred materials used by various investigators of adiabatic engines are shown in Table 3. Ceramic materials are dominant due to their high-temperature strength and relative availability as compared to superalloy materials that require strategic metals such as molybdenum, chromium, nickel, and titanium (Ref 7). Of the metallic materials, however, titanium offers excellent potential for use in adiabatic engines because of its high-temperature strength, light weight, low thermal conductivity, and (above all) ductility (Ref 1).

When selecting materials for adiabatic engines, the basic considerations are:

- Material strength and Weibull modulus for high reliability in ceramic components operating at high stress and temperature levels
- Stability at high temperatures
- Thermal shock resistance
- Low thermal conductivity to reduce heat losses
- Low thermal expansion so that the piston-to-cylinder bore clearances can be maintained at acceptably small levels over a wide range of piston-to-cylinder temperature differences

Material selection is also influenced by the particular component function and engine type. For example, aluminum titanate offers excellent application with aluminum components such as cylinder head port linings but is not reliable for use with iron cylinder heads.

A summary of ceramic applications for adiabatic engines is shown in Table 4. There is also current interest in ceramic applications for other engine components such as valve trains. For high-performance engines, lightweight silicon nitride is being seriously considered to reduce the valve train dynamics problems for greater output. The light weight, hardness, strength, and good tribological properties and strength of silicon nitride make reliable and durable valve train components having low friction and low dynamic stress. Light weight and hardness also give rise to other important applications such as prechamber bowls, rocker arm tips, cams, tappets, turbochargers, and bearings.

For adiabatic engines, the application of ceramics focuses on their insulation characteristics and high-temperature properties. High-temperature materials for insulating heat engines are described in the next section "Insulation for Engines." Insulation is used for pistons, cylinder heads, exhaust/intake ports, and cylinder liners.

Materials for the insulated adiabatic engine are plentiful. However, most of the new, high-performance ceramics (Table 3) will not meet all the requirements (Table 1) of an uncooled, turbocompounded diesel engine. Experience has shown at high-use temperatures, or high-unit loading, the toughened ceramics will change phase and subsequently fail. Consequently, in spite of the new, successful ceramic products, monolithic ceramics are still not reliable and durable enough for the uncooled adiabatic diesel engine having a 1515 kPa (220 psi) BMEP.

To compound matters, the lubricant for uncooled adiabatic engines is also inade-

Table 3 Properties of various materials used in adiabatic engines

Material	Silicon nitride	Silicon carbide	Zirconia	TiC	Alumina	Plasma-sprayed alumina	Plasma-sprayed Cr_2O_3	Cr_2O_3 densified coating	Hard chrome plate	Cast iron (flake graphite)
Bulk density, g/cm³	3.2	3.1–3.2	6.0	6.0	3.8	3.5	5.0	3.3	7.0	7
Hardness, HV	1500	2000–2800	1128–1250	1630	1170	850	1260	1800	900	23
Linear thermal expansion coefficient, 10^{-6}/°C (10^{-6}/°F)	3.2 (1.8)	4.0–4.4 (2.2–2.4)	10–11 (5.5–6.0)	7.4 (4.1)	8.0 (4.4)	7.4 (4.1)	7.4 (4.1)	13.3 (7.4)	9.4 (5.2)	11 (6.1)
Thermal conductivity, W/m·K	20.9	62.8–75.4	2.9–3.8	16.7	20.9	2.7	2.6	3.7	19.2	33
Specific heat, kJ/kg·K	0.67	0.63	0.46–0.50	0.71	0.92	...	...	0.59	0.46	0.5
Melting point, °C (°F)	1900 (3450)	2350–2700 (4260–4890)	1600–2677	...	2400 (4350)	2035 (3695)	2400 (4350)	1610 (2930)	1850 (3360)	...
Thermal shock resistance, Δ °C (Δ °F)	600 (1080)	380–400 (685–720)	300–350 (540–630)	350 (630)	150 (270)	...	...	800 (1440)	...	...

quate. Durability tests for adiabatic engines have encountered cylinder liner scuffing, ring sticking, and a host of other problems. Today's lubricating oil has a TRR temperature capability of about 288 °C (550 °F), which is woefully inadequate when the higher TRR temperatures of adiabatic engines are incurred (Fig 1). Problems associated with lubrication are discussed in the section "High-Temperature Tribology" in this article.

Compromises in materials thus have led to compromises in engine performance and life. No commonality of materials used by the various investigators has been noted, but monolithic ceramic designs have been replaced in some adiabatic-turbocompounded engine programs. Focus is now on the thick ceramic coatings. These ceramic coatings were applied by plasma spray, detonation gun, or the slurry, chemical-bond process. Graded coatings, new bond coats, and many new coating developments have resulted from these investigations (see the section "Ceramic Coatings" in this article).

Insulation for Engines

A ceramic material capable of providing the insulation characteristics and thermal expansion properties compatible with an iron engine was found to be partially stabilized zirconia (PSZ). The thermal conductivities of PSZ are low, and its thermal expansion is similar to iron. The material also possesses excellent fracture toughness (K_{Ic}, 8 to 15 MPa $\sqrt{m}$, or 7 to 14 ksi $\sqrt{in.}$), good thermal shock resistance, and a flexural strength of 1020 MPa (148 ksi) at room temperature (Ref 6). However, at higher engine combustion temperatures, a phase change occurs which thus limits the high-temperature use of PSZ.

Glass ceramic was included in the early phase of the adiabatic engine development solely for its insulation effectiveness. Three basic materials were considered during the initial developmental phase. They were the high-performance ceramic and glass monolithic ceramic and strategic metals.

The metallic engine required 10 layers of roughened, 0.25 mm (0.010 in.) thick stainless steel shims for insulation of the piston crown. Hot pressed silicon nitride was used for the high-performance ceramics. The glass ceramic materials were inadequate in strength to withstand the diesel operating environment. Also, the hot pressed silicon nitride with its low coefficient of expansion and lack of adequate insulation presented design problems when used in conjunction with the metal engine. The metallic design presented no problem except that it is difficult to maintain the degree of adiabaticity required in an adiabatic engine. In addition, the temperature limitations at 65% adiabaticity will soon result in failure by thermal fatigue. Better materials were needed.

High-Temperature Tribology

When ceramics are used in an adiabatic engine, care must be exercised in mating parts such as valves, valve seat inserts, and cylinder liner and piston ring. In general, high-performance ceramics are not the best tribological materials, especially in like pairs. For example, zirconia on zirconia or silicon nitride on silicon nitride are poor from the standpoint of friction and wear.

A case in point is the selection of valve seat and valve seat insert materials for gas engines. Stellite 6 on sintered-silicon-nitride valve seat inserts showed the best results in laboratory rig tests. Full-size engine tests of the mating pairs gave threefold increases in wear properties.

Probably the most important factor in the success of the adiabatic diesel engine can be attributed to high-temperature lubrication. Today, the best available commercial synthetic lubricant is the polyolester formulated oil. Polyolester oil shows a typical friction characteristic with onset of thermal oxidation (stability) at about 300 °C (570 °F). This is the temperature limit of this class of oil before scuffing occurs at the TRR point.

Table 4 Summary of ceramics applications for adiabatic engines

Adiabatic components	Desired characteristics							High-technology ceramics
	Low friction	Light weight	Insulation	Wear resistance	Heat resistance	Corrosion resistance	Expansion coefficient	
Piston		X	X		X	X	X	Si_3N_4, PSZ, TTA
Piston ring				X	X			SSN, PSZ, coating
Cylinder liner	X			X	X	X	X	Si_3N_4, PSZ, coating
Prechamber			X		X	X		PSZ, Si_3N_4
Valve		X	X	X	X	X		SSN, PSZ, composite
Valve seat insert			X	X	X	X		PSZ, SSN
Valve guides	X		X		X	X		PSZ, SSN, SiC
Exhaust/intake ports			X		X	X		ZrO_2, Si_3N_4, $TiO_2Al_2O_3$
Manifolds			X		X	X		ZrO_2, Si_3N_4, $TiO_2Al_2O_3$
Tappets		X		X		X		PSZ, SiC, Si_3N_4
Mechanical seals	X			X		X		SiC, Si_3N_4, PSZ
Turbocharger								
Turbine rotor		X	X		X	X	X	Si_3N_4, SiC
Turbine housing			X		X	X	X	LAS
Heat shield			X		X	X	X	ZrO_2, LAS
Ceramic bearings	X	X		X	X	X	X	SSN

Nomenclature: SSN, sintered silicon nitride; PSZ, partially stabilized zirconia; LAS, lithium alumina silicate; TTA, transformation toughened alumina

There are many ways of lubricating the surfaces between the cylinder liner and piston. The important ones are:

- Hydrodynamic liquid lubricant
- Solid lubricant
- Vapor-phase lubrication
- Catalytic generation of lubricants
- Gas lubricants

The important properties for high-temperature lubrication can be itemized in their order of importance as follows.

- Deposit formation
- Thermal stability/volatility
- Oxidative stability
- Corrosion control
- Wear and friction control
- Viscometrics

Hydrodynamic Lubrication. With available liquid lubricants there are many piston ring and cylinder liner materials. Table 5 shows the coefficient of friction, μ_f and scuffing temperature, T_s, of various metallic and ceramic material. Tests were conducted with conventional SAE 30-weight petroleum-base lubricating oil. Densified Cr_2O_3 shows significant superiority over cast iron versus cast iron for a ring and liner combination.

For one reason or another, hydrodynamic liquid lubricant has been preferred over the years. However, with the operating temperatures of the adiabatic diesel engine, the hydrodynamic liquid lubricant is no longer adequate to perform its tribological function. The "knee" of the coefficient of friction curve rises suddenly at around 300 °C (570 °F) with polyolester base oil. This temperature is also known as the "scuffing" temperature. Therefore, polyolester-base lubricants limit the BMEP output of uncooled adiabatic engines (Fig 1).

Polyolester-base synthetic lubricating oil is the current limit of the hydrodynamic liquid lubricant. Development work is going on with aromatic esters, polyophenolesters, and others, but none appears viable from the standpoint of performance and cost.

The probable approach to engine lubrication over the next generation can be predicted for the future uncooled adiabatic diesel engines. The petroleum-base oils will suffice to 230 °C (450 °F) TRR temperature. Synthetic polyolester can furnish the necessary lubrication to 315 °C (600 °F). A 315 to 430 °C (600 to 800 °F) temperature represents the next major challenge to any liquid lubricant in terms of performance and cost. Polyphenol ether-type oil may fulfill this temperature area, but the cost becomes prohibitive for a commercial lubricating oil.

Solid Lubricants. Beyond 430 °C (800 °F), solid lubricants are necessary. Much work is going on in the field of solid lubricants, especially for high-temperature application. The problem with solid lubricants to date has been its high coefficient of friction and high wear rates. Low friction has been reported once in a while but only for certain temperature regimes. However, the research in solid lubricant continues, and many good, high-temperature, solid lubricants are becoming available. The solid lubricants may bridge the gap between solid lubricant materials and polyolester-base oils (Ref 5).

There are many solid lubricant materials investigated for friction and wear. Most solid lubricants indicate a high coefficient of friction of about 0.2. Wear has also been excessive. Recently, solid lubricants have been developed capable of approaching coefficient of friction value of hydrodynamic lubrication. Among these are gallium/indium/tungsten/selenium ($Ga/In/WSe_2$) versus Cr_2O_3 and organometallic compound with Cr_2O_3 intermediates and elemental carbon achieving coefficient of friction values of 0.03 to 0.06 (Ref 5). Figure 4 shows the test results on the block and roller tester for organometallic compounds with intermediate Cr_2O_3 and elemental carbon, which demonstrates low friction and is low in cost.

A solid lubricant developed by NASA combines piston rings made of Stellite 6B and cylinder liners made of PS 200. The PS 200 material is a composition of silver, calcium

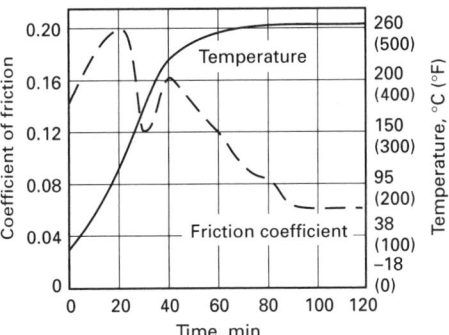

Fig 4 Dry friction test of organometallic compound with intermediate Cr_2O_3 and elemental carbon

fluoride, and barium fluoride in a chrome carbide matrix. The NASA-developed solid lubricant resulted from a need to raise the ring seal in a Stirling engine so that the dead air space between the piston and cylinder liner is reduced. The Stirling engine cannot tolerate oil. The Stirling engine test was conducted for 25 h and 7% improvement in thermal efficiency was reported.

Lubrication above 300 °C (570 °F). Because no lubricant is capable of exceeding a TRR temperature of 300 °C (570 °F), one must resort to other means of lubrication. Three important undergoing developments are:

- Vapor-phase lubrication as reported by Detroit Diesel Corporation and Pennsylvania State University
- Catalytic generation of lubricant for high-temperature lubricant by Professor Lauer, *et al.*, of Rensselaer Polytechnic Institute
- Hybrid piston by Adiabatics Inc. and the U.S. Army TACOM

The vapor-phase lubrication pursued by Detroit Diesel obtains its lubrication by introducing TCP onto the hot walls to obtain its lubrication. Tricresylphosphate (TCP) is being introduced in a chamber space between the top piston crown and the lower piston. It is suggested that TCP can also be introduced through the injector or through the intake system.

The catalytic approach uses ethylene gas impinging on a hot tribological surface with a nickel catalyst to generate a solid-lubricant film. Drastic reduction in friction is noted. The technique has reduced the coefficient of friction from 0.4 (dry) to 0.04. This approach developed by Professor Lauer is still in the laboratory stage with no known application for high-temperature diesel engines.

The hybrid piston being developed by Adiabatics Inc. uses solid lubrication at the top ring of the piston where the temperature is highest. On the same piston, a hydrodynamically lubricated lower ring is located where the temperature is low enough so that the oil is thermally stable. The piston crown is lubricated with solid-lubricant rings such as Stellite 6B (like the NASA design) or chrome

Table 5   Scuffing temperatures (T_s) and coefficients of friction (μ_f) between ring and cylinder liner materials

	Piston ring								
Cylinder liner	Plasma-sprayed alumina	Silicon carbide	Zirconia	TiC cermet	Cast iron	Silicon nitride	Plasma-sprayed Cr_2O_3	Hard chrome plate	Densified Cr_2O_3
Silicon nitride									
μ_f	0.014	0.022	0.005	0.008	0.019	0.010	0.007	0.004	0.009
T_s (°C)	166	160	177	200	240	242	260	268	286
Zirconia									
μ_f	...	0.026	0.012	0.012	...	0.012	...	0.004	0.006
T_s (°C)	...	127	177	255	...	223	...	250	250
Cast iron									
μ_f	...	0.034	0.013	0.039	0.038	0.023	0.065	0.013	0.013
T_s (°C)	...	195	160	165	137	138	179	155	222
Cr_2O_3 densified									
μ_f	...	0.037	...	...	...	0.007	...	0.007	0.007
T_s (°C)	...	199	...	...	...	250	...	238	288

oxide. A TRR temperature capability of 430 °C (800 °F) is obtained with this design. The proposed U.S. Army TACOM hybrid piston design also requires stainless steel cylinder liners and cylinder heads because of the elevated temperatures encountered in the uncooled "adiabatic" design.

Ceramic Coatings

There are many coating processes available for coating the combustion chamber and parts of an adiabatic diesel engine. Some of the more important thermal spray and thermomechanical ceramic coatings are listed below:

- Thermal spray processes such as plasma spray, flame spray, Jetkote, Gator gard, water plasma, plasma transfer arc, detonation gun, high-velocity oxygen flame, and spark discharge
- Chemical ceramic coating by sol-gel or slurry methods
- Chemical vapor deposition (CVD)
- Physical vapor deposition (PVD)
- Anodizing
- Ion implantation

Ceramic coatings act as thermal barriers in an adiabatic engine. They may also provide wear resistance, friction reduction, corrosion/erosion resistance, inertness, and non-magnetic properties.

Exhaust ports, intake ports, cylinder heads, and piston crowns require thermal-barrier insulation. However, components such as pistons, rings, and cylinder liners which experience relative motion must have good tribological (friction and wear) properties. Coating materials which exhibit good insulative properties at high engine combustion chamber temperatures are the zirconia groups such as:

Material	Thermal conductivity, W/m·K
Monolithic zirconia	2.3
Slurry-densified zirconia	1.4
Low-porosity zirconia	0.95
High-porosity zirconia	0.78
Chrome oxide	3.1
Titanium	12.0

Most ceramic coatings can be applied with adequate surface smoothness so that machining is not necessary. However, for tribolog-ical components, machining must be considered in application and cost analysis. Usually, sufficient substrate material is removed to allow for the coating thickness. One promising type of coatings, namely slurry coatings, is usually smooth and the grinding operation can be eliminated. Honing or surface polishing one time suffices. Because coatings of ceramic parts are very hard and approach monolithic hardness, the cost of the coating process should be carefully considered in total overall cost. Ion implantation is a coating process that eliminates the need for subsequent machining. Thus far, ion implantation has shown promise with hard surfaces such as M2 steel. However, for soft material such as aluminum, more work is needed to overcome its wear problems.

Two important tests are needed to qualify coatings for engine applications, namely, thermal-fatigue tests and bond-strength tests at elevated temperatures. For thermal-fatigue tests, a rig comprised of a high-temperature natural gas burner, a coupon specimen of the coating, and a reservoir of water for quenching is used. A test cycle consists of heating the thermal-sprayed coating sample with the gas flame, followed by quenching in the water. This is a severe test, because the coating material is heated to a temperature of 760 °C (1400 °F), and subsequently quenched in room-temperature water. When the data of specimen thickness is plotted against rig cycle, it is found that thick coating material crack and fail rapidly. When specimen thickness is less than 1 mm (0.04 in.), the number of cycles to failure rises rapidly. Coatings less than 1 mm (0.04 in.) in thickness are adequate for thermal insulation, as will be shown in the subsequent section "Thin Thermal-Barrier Coatings."

Bond strength of ceramic coating is necessary for its mechanical integrity. In the above list, most of the ceramic coating processes are mechanical bond. Slurry coating, chemical vapor deposition, physical vapor deposition, ion implantation, plasma transfer arc, and anodizing are molecular bond and generally stronger. Bond-strength measurement can be made by a pull test on a coated specimen with a known bond surface area. These tests normally are performed per ASTM C 633.

Thin Thermal-Barrier Coatings. The wall temperature profiles during a complete engine cycle is a function of zirconia coating thickness as follows:

- The mean temperature of an insulated component keeps rising with insulation thickness
- The cyclical transient temperature swings around the mean temperature are almost the same for thin coatings as for thick coatings

Assanis (Ref 9) has shown in earlier work that the temperature swing ΔT_s in the combustion chamber surface of an engine is inversely proportional to the square root of the product of thermal conductivity k, density ρ, and specific heat C.

$$\Delta T_s \approx 1/(k\rho C)^{1/2} \qquad \text{(Eq 1)}$$

Equation 1 is not influenced by the insulated wall thickness.

The low-conductivity plasma-sprayed zirconia coating with a thermal conductivity, k, of one-fifth to one-tenth of the conductivity of denser monolithic ceramics shows the largest temperature swings, approaching 240 °C (430 °F). Table 6 tabulates the important thermophysical properties and expected temperature swings for several conducting and insulating materials.

The next important factor is the penetration of the thermal wave into the material. The skin depth, δ, often defined as the distance from the surface at which the temperature swings decay to 1% of their value at the surface, is proportional to the square root of the ratio of the thermal diffusivity over the engine speed N:

$$\delta \sim (\alpha/N)^{1/2} \qquad \text{(Eq 2)}$$

where

$$\alpha = k/(\rho C) \qquad \text{(Eq 3)}$$

A factor of 10 difference can be observed in the penetration of transients between zirconia and iron material. Table 6 also summarizes skin depth for various materials at an engine speed of 1900 rpm. Clearly, temperature swings in the material are dampened very quickly in low-thermal diffusivity materials when compared to high-diffusivity materials.

Table 6 Thermal properties of engine wall insulator lining materials

Material	Thermal conductivity W/m·K	$\rho \cdot C$ (J/m³·K)	$k \cdot \rho \cdot C$	Temperature swing Δ °C	Temperature swing Δ °F	Diffusivity (α), m²/s	Skin depth mm	Skin depth in.
Cast iron	54	3.46×10^6	186.6×10^6	18.0	32.4	15.62×10^{-6}	2.75	0.108
Aluminum	155	2.52	389.9	12.5	22.5	61.61	5.46	0.215
Reaction-bonded silicon nitride	7.5	1.78	13.31	67.4	121	4.22	1.43	0.056
Partially stabilized zirconia	2.0	3.14	6.28	98.1	176.5	0.64	0.55	0.022
Plasma-sprayed zirconia	0.6–1.2	1.74–3.81	1.04–4.57	115–240	207–430	0.31–0.34	0.39–0.41	0.015–0.016

Source: Ref 10

The low thermal inertia of the thin coating ensures that the component surface responds to the gas temperature transients. This could result in improved volumetric efficiency and thermal efficiency. Because the mean temperatures of a component are conservative, high-temperature lubrication will not become a problem as with thick insulation. The merit of thin insulation becomes apparent.

Thin insulating materials can also be used in gasoline engines where the above performance benefits can be realized. In particular, engine knock will not develop. Unburned hydrocarbon and carbon monoxide emissions will drop due to reduced flame quenching as the walls try to follow the gas temperatures. Higher available exhaust gas energy could be recovered through the use of turbocharging. Other benefits would be lower friction, improved erosion/corrosion resistance, longer component life, and improved reliability.

Design Considerations

Methodology (Refill). Designing reliable engine components with ceramics is considerably more difficult and, unquestionably, different than designing with ductile materials. The probability of failure of a metallic design is quite predictable, while ceramic designs are considerably more unpredictable. To avoid failures, very low component stresses and ceramics of good quality and strength must be used. Many iterative processes in design and testing are necessary before an acceptable and viable adiabatic engine design is available.

A totally integrated adiabatic engine design methodology is necessary to ensure that each adiabatic engine component is synergistically analyzed in the context of the total engine system. Computer-aided modelling and simulation are used extensively today in the design phase of the adiabatic diesel engine. The steps taken when conducting a design analysis of an adiabatic engine are shown in Table 7.

Cylinder Liner Analysis. A design analysis assessment of the composite zirconia cylinder liner is presented to determine the various stress distributions. Three different conditions which cause stresses in the liner and combination thereof were considered:

- The first condition was assembly, which involves the interference-fit between the zirconia liner and cast iron sleeve
- The second condition was the internal pressure-induced stresses associated with combustion
- The third condition was the thermal stresses associated with the large temperature variations between the various parts and within the parts

The thermal distribution itself can be divided into four parts: (1) quasi-steady state axisymmetric, (2) transient axisymmetric, (3) quasi-steady state circumferential variation portion, and (4) transient circumferential variation

Table 7 Design methodology for the components of adiabatic engines

Design step	Design task
1.	Select design approach, materials, type of construction, and performance goals. Gather all the data required to prepare input files for the cycle simulation code and geometry file for the finite element analysis code
2.	Conduct thermodynamic analysis and cycle simulation. Engine-cycle simulation program is employed to estimate engine performance, peak pressure, peak bulk-gas temperature, and the heat release for each crank angle position
3.	Calculate equivalent steady-state heat transfer. Dynamic heat transfer data are used to calculate the equivalent steady-state (bulk combustion) gas temperature and combustion gas heat flux, which are needed for the steady-state finite element analysis (FEA)
4.	Build FEA models of engine and components. Several files usually are needed for heat transfer calculations of engine assembly and heat transfer and stress analysis of each component and design option to be studied. The engine assembly heat-transfer FEA model should include all of the engine components surrounding the combustion chamber; that is, ceramic piston caps, metal base piston, ceramic fire deck, all air gaps, piston rings, liner, ceramic coatings, engine block
5.	Calculate a steady-state temperature map of engine assembly. The temperature map for the entire engine assembly determines the interrelationships between engine components
6.	Calculate a steady-state temperature map of each engine component. The temperature and heat flux output of step 5 is employed as input in step 6 to ensure that the analysis of each component is consistent with the overall engine assembly temperature map. The final output of step 6 is the temperature at each node which is saved and employed to calculate the thermal stress in step 7
7.	Calculate a stress and deflection map of each component due to pressure, temperature, and inertial loading. Calculated stress can be used to assess probability of failure if test data is available
8.	Modify component design (revise material properties, revise geometry). After the completion of the first finite element analysis, several iterations are usually required to optimize the design. In the case of an adiabatic engine design, both the heat transferred through the component and stress are important. Revisions to the FEA geometry and material input files are made, and the sequence returns to step 5. The analysis (steps 5, 6, 7) is repeated until the stress and heat transfer goals are met
9.	Final component design (design goals met)
10.	Prototype fabrication. The adiabatic engine components are ready for prototype fabrication after successful completion of steps 1 through 9
11.	Ceramic components need careful screening for manufacturing flaws. Optical, ultrasonic, and vibration bench tests are employed to screen unsatisfactory samples. (Return to step 10 and build new prototypes, if necessary)
12.	Dynamic engine test. After all the components have successfully passed all bench tests, the engine is assembled and dynamically tested. Test failures will result in design revisions which will cause revisions to the FEA model
13.	Modifications. Return to step 5 for design revisions
14.	Final design. After successful completion of the dynamic engine test, the final step in the design methodology is the documentation and storage of the final validated design

portion. Only the axisymmetric quasi-steady state case was considered. In order to calculate the thermal stresses, it was first necessary to develop a detailed temperature profile for the piston, cylinder liner, sleeve, and the engine block. The boundary conditions were obtained from the single-cylinder test.

In the thermal plot of the important areas of the systems modeled, the piston was modeled as all cast iron, consistent with the liner test configuration. Conductance resistance between the zirconia liner and the cast iron sleeve is assumed to be zero. Also, at the interface between the sleeve and the engine block (close fit at top), the conductance resistance is assumed to be zero. Convection and radiation heat transfer in the gap between the sleeve and block is modeled along with convection and radiation from the engine block to the test cell.

In carrying out the various stress analyses involving the quasi-steady state temperature distribution, it is convenient, with respect to interpretation, to include the assembly conditions. The pressure-induced stresses, which come on and off during the engine cycle, need not be added because basically only the hoop stress changes. The major concern, based upon failure-mode data, is the axial stress near the top of the cylinder liner assembly.

Figure 5 presents the contact and liner stresses, respectively, for the case where the six mil radial interference is applied and then the thermal is added with a friction coefficient of infinity. The case of zero friction was also calculated. The maximum axial stress increases to about 90 MPa (13 ksi). This increase in axial stress is due to the coefficient of thermal expansion mismatch with the cast iron sleeve expanding faster than zirconia. With the above analysis and considerations, the composite zirconia cylinder liner was designed, fabricated, and tested. Weibull analyses also indicated acceptable durability.

Piston Design Considerations. The piston probably represents the most difficult design challenge because it is heavily loaded, exposed to high temperatures, and is constantly moving up and down.

Currently, the best compromise of a material for the piston is low-pressure-sintered silicon nitride. It possesses high-temperature strength, a low coefficient of expansion, and a moderate thermal conductivity. For application in a diesel engine, a material with high-temperature strength properties of silicon nitride and the thermal conductivity of zirconia is desired. In the absence of the desired thermal conductivity in silicon nitride, one must resort to design techniques to build insulation into it by air gaps or multiple layers.

Whatever course is taken, one must always make sure that the piston surface itself is insulated. Otherwise the conducting surface undergoes heat storage and heat release during engine operation because of the mass and specific heat of the material. The temperature swing or the so-called adiabatic effect is strictly

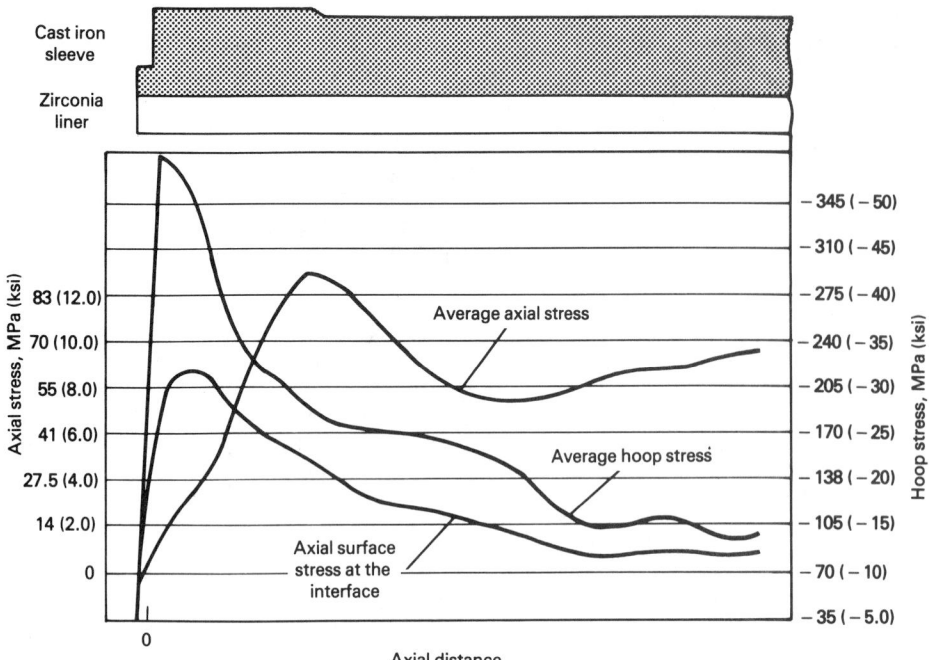

Fig 5 Stresses in a ceramic-lined, cast-iron sleeve

a function of material property. It was shown in the section "Ceramic Coatings" that the insulation need not be thick (Ref 10).

Economics of the Adiabatic Engine. The commercialization of structural ceramics in heat engines with their consequent low fuel consumption will have a wide range of benefits for both the manufacturers and consumers. The development of reliable structural ceramics in the transportation sector also should reduce the nation's dependence on several strategic materials. Structural ceramics need not be limited only to the transportation sector; applications could become widespread.

Considering the adiabatic engine for the transportation sector, one must look at the power plant on a total system basis. Also,

designing a completely new adiabatic engine should be considered rather than merely substituting metal components with carbon-copy structural ceramic components. With the elimination of the cooling system, for instance, new designs for the smaller-sized system can prove to be even more energy efficient on vehicles.

Reductions in engine noise have not been documented, but it is well known that fan noise is one of the greatest vehicle noise sources. However, noise level and emissions advantages are quite subjective and economic trade-offs cannot readily be made. Table 8 shows various advantages of the adiabatic turbocompound compared with an advanced metallic diesel and a turbocompounded con-

ventional diesel engine in the 340 kW (450 bhp) size range. Note the cost of the adiabatic turbocompound engine is no greater than the advanced metallic engine (Ref 4). These advantages of an adiabatic turbocompound engine are projected over interstate highway use by road transport.

REFERENCES

1. M.W. Woods, P.C. Glance, and E. Schwarz, "Advanced Insulated Titanium Piston for Adiabatic Engine," Paper No. 900623, Society of Automotive Engineers, 1990
2. G. Woschni, W. Spindler, and K. Kolsea, "Heat Insulation of Combustion Chamber Walls—A Measure to Decrease the Fuel Consumption of I.C. Engines?," Paper No. 870339, Society of Automotive Engineers, 1987
3. R. Kamo and W. Bryzik, "Adiabatic Turbocompound Diesel Engine," 15th International Congress on Combustion Engines, Paris, France, June 1983
4. R. Kamo, Adiabatic Diesel Engine Technology in Future Transportation, *Energy*, Vol 12 (No. 10/11), 1987, p 1073–1080
5. W. Bryzik, E. Valdmanis, and R. Kamo, "High Temperature Tribology for Adiabatic Diesel Engines," Energy Society of Detroit Conference, Dearborn, MI, June 1990
6. R. Kamo, "Status of Advanced Materials for Heat Engine Components," MRS International Meeting on Advanced Materials, Tokyo, Japan, Materials Research Society, Feb 1988
7. W. Wade, P.H. Havstad, V.D. Rao, M.G. Aimone, and C.M. Jones, "A Structural Ceramic Diesel Engine—The Critical Element," Paper No. 870651, Society of Automotive Engineers, 1987
8. Y. Wakuri, M. Soejima, T. Kitahara, and K. Kinoshita, Evaluation of Friction and Scuffing Resistance of Ceramics for Cylinder Liner and Piston Ring, 5th International Congress on Tribology, June 1989, Vol 4, Espoo, Finland
9. D.N. Assanis and E. Badillo, "Transient Heat Conduction in Low-Heat Rejection Engine Combustion Chambers," Paper No. 870156, Society of Automotive Engineers, 1987
10. R. Kamo and D.N. Assanis, "Thin Thermal Barrier Coatings for Engines," Paper No. 89-ICE-14, American Society of Mechanical Engineers, 1989
11. P. Glance, "Computer Aided Engineering Analysis of Adiabatic Engine Components," Paper No. 850361, Society of Automotive Engineers, 1985

Table 8 Estimated advantages of an adiabatic turbocompound engine

Engine and components	Weight		Cost, U.S. dollars	Size		Brake specific fuel consumption		Power	
	kg	lb		cm	in.	g/kJ	lb/bhp·h	kW	bhp
Advanced metallic diesel	163	360	Base	180	71.0	0.057	0.340	300	400
Conventional turbocompound diesel	68	150	0.15 base	2.8	1.1	0.003	0.017	315	425
Conventional turbocompound diesel	1545	3410	1.15 base	183	72.1	0.545	0.323	315	425
Advanced adiabatic turbocompound diesel	1165	2565	100.4 base	153	60.1	0.047	0.276	360	485

Advanced Gas Turbines

Jay R. Smyth, Garrett Auxiliary Power Division, Allied-Signal Aerospace Company

THE CONFIGURATION AND OPERATING CONDITIONS of gas turbine engines provide the capability for greater fuel efficiency and lower gaseous and particulate emissions compared to conventional internal combustion (IC) engines (both spark ignition and diesel). Also, the gas turbine has multifuel capability and fewer moving parts than IC engines. These attributes give the gas turbine engine the potential to be more flexible, reliable, and cost-effective than conventional IC engines. The application of ceramics in gas turbine engines enhances these benefits by increasing the allowable operating temperature.

As reported by Carruthers and Lindberg (Ref 1), the potential benefits of ceramic materials in gas turbine engines have been documented numerous times during the past decade. Early studies indicated that the use of ceramic components in an automotive gas turbine operating at 1370 °C (2500 °F) could double horsepower and improve fuel economy by 20% compared to a similar size all-metallic engine. Updates of these studies projected diesel fuel economy of 5.4 L/100 km (42.3 mpg) for a gas turbine engine in a 1361 kg (3000 lb) vehicle over the federal combined urban and highway driving duty cycle.

Benefits have also been projected for non-automotive gas turbine applications. During joint NASA/U.S. Army-funded studies, use of ceramics in auxiliary power units (APUs), and commuter, helicopter, and missile engines was evaluated. These studies concluded that the use of ceramics and/or ceramic composites could account for 23 to 65% of projected fuel and cost savings achievable in the year 2000 using advanced technologies. In addition, improved turbine nozzle life during sand ingestion conditions with the use of ceramic turbine stators has been projected. Under certain conditions, an advantage of up to 300% was observed for ceramics, compared to superalloys.

This article discusses the background of ceramic gas turbine development in the United States, the challenges of applying ceramics in gas turbines, current global development efforts in the area of ceramic gas turbine engines, and ceramic materials of interest.

Background

Richerson (Ref 2) presented a review of the history of ceramic gas turbine development in the United States. During the early 1970s, the emphasis was on determining whether gas turbine components could be successfully designed with brittle ceramic materials. This program demonstrated that ceramic turbine components could successfully be designed and could survive engine operating conditions.

The emphasis during the second half of the 1970s shifted toward incorporation of ceramic components into existing turbine engines to demonstrate improvement in performance and other characteristics. One of the first successful applications was ceramic stator vanes and shrouds in a turbine-powered 10 kW generator set. In this case, the objective was to utilize ceramics to decrease dust erosion of the stators, rather than achieving increased performance.

A separate program was directed toward short-time demonstrations of increased operating temperature and improved performance. A turboprop engine containing 104 separate ceramic parts in the hot section was successfully operated during two-hour cycles to 693 kW (930 shaft hp). This represented a 30% increase in power and a 7% decrease in fuel consumption compared to the metal engine, achieved through a 200 °C (360 °F) increase in the turbine inlet temperature.

A third program evaluated ceramic stator vanes and rotary regenerator heat exchangers in a truck engine operating at a turbine inlet temperature of 1040 °C (1900 °F). The engine with ceramic components was operated in a truck, accumulating 92 h operation and 2961 km (1840 mi), 1698 km (1055 mi) of which were over Belgian block and truck rough-road test courses at an automotive proving grounds. This program demonstrated that ceramic components could survive both thermal and mechanical (road-imposed shock and vibration) stresses in a realistic vehicle environment.

These programs, evaluating ceramic components in existing turbine engines, successfully demonstrated that substantial improvements in performance could be achieved through the use of ceramics. However, design and materials deficiencies requiring further development were also identified.

The next step to be achieved in the evolution of ceramic turbine engine technology in the United States was development of new engines designed specifically for ceramics. Two Advanced Gas Turbine (AGT) technology programs were conducted from 1979 through 1987 by the Allison Gas Turbine Division of General Motors, and Garrett Auxiliary Power Division of Allied-Signal Aerospace Company. Under these programs, both companies successfully designed engines with ceramic hot sections and demonstrated the feasibility of ceramic gas turbines.

These programs demonstrated that it is possible to achieve successful designs with brittle turbine engine ceramic materials and that ceramic components can operate in the severe environment of a gas turbine, but the programs have failed to demonstrate the degree of ceramic component reliability adequate for industry to commit to production. To achieve the reliability and durability goals requires continued extensive efforts in design, material property improvement, component fabrication, and quality control, and the accumulation of thousands of additional hours of engine testing.

The Challenges

Before addressing the engineering challenges for successful application of ceramics, it is beneficial to summarize the demands placed on ceramics by the gas turbine engine environment. These demands are quite severe and can vary considerably, depending upon the application. The ceramic gas turbine applications to be considered include missile, automotive, military and commercial propulsion, and auxiliary power engines. Each of these applications has a different duty cycle, life requirement, operating temperature, and stress envelope. The range of engine lifetimes and power requirements is illustrated in Fig 1. Design lives can vary from 15 to 10,000 h, with duty cycle time at maximum power ranging from 1 to 40%. The majority of the engines operate most of the time at power

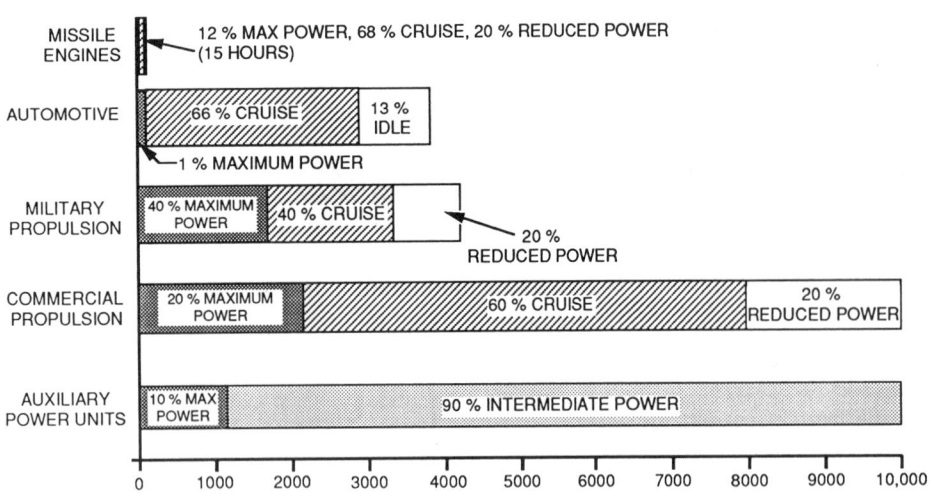

Fig 1 Duty cycles and lifetime requirements for different types of gas turbine engines. Lifetime given in hours

levels below maximum power, as illustrated in Fig 1. The specified duty cycle and engine life will dictate the required strength, crack growth and creep rates, and environmental stability of the selected materials.

Although thermal transients also vary with turbine engine design and application, peak lightoff temperatures are generally reached in 1 to 15 s. The severity of shutdown transients depends on the ram air or rolldown air temperature and differs significantly for regenerated and nonregenerated engines. Since thermal transients are often responsible for the peak stresses in ceramic components, expected transients typically dictate the degree of thermal shock resistance needed in the material and component designs.

Development Sequence. The introduction of advanced materials such as ceramics into gas turbines generally has followed a sequence of development activities aimed at establishing feasibility, performance, reliability, and competitive cost. These efforts typically overlap.

First, the feasibility of both the design and the material to survive the severe duty cycle of the gas turbine must be established. As discussed previously, this duty cycle is application-dependent, but in general, critical engine conditions include the thermal gradient induced stresses of lightoff, shutdown, and cyclic use, as well as sustained stresses and temperatures encountered during maximum and part-power engine operation. Primary material properties affecting feasibility include strength, thermal expansion, thermal conductivity, elastic modulus, fatigue, oxidation, and corrosion. Feasibility also depends on how well the component design minimizes the thermal and mechanical stresses.

Extensive analytical and experimental evaluation efforts are required to confirm all aspects of feasibility, including transient response, steady-state, peak power, and extended life. The evaluations typically address these issues in a progressive manner.

Once feasibility of a new material has been established, its performance, reliability, and cost competitiveness must be demonstrated. Performance includes meeting power, fuel consumption, and/or specific power goals. Reliability includes demonstrating predictable engine durability under normal and adverse conditions. Cost competitiveness includes the initial cost as well as the cost of operation, which will vary with the nature of the application.

Although the development efforts of the past two decades have focused heavily on establishing the feasibility of using ceramics in gas turbines, recent work is also addressing the issues of performance, reliability, and cost. As experience in each of these four aspects is gained, so is a clearer understanding of the critical technical challenges which remain (Ref 1).

Current Development Programs. The U.S. Government sponsored efforts in the 1970s to evaluate the potential of candidate heat engines to meet national goals for energy conservation and environmental impact. These efforts identified the ceramic gas turbine engine as meriting further development as an automotive power source, due to significant potential advantages including:

- Significantly reduced fuel consumption
- Ability to meet emission standards
- Ability to operate on alternate fuels

Currently, the U.S. Department of Energy (DOE) is sponsoring a major program continuing the efforts to develop gas turbine technology for automotive use. The Advanced Turbine Technology Applications Project (ATTAP) involves two U.S. industrial contractors and their ceramics suppliers. Along with the DOE-sponsored automotive gas turbine program, several U.S. company-funded programs are addressing the application of ceramics to gas turbines.

In addition, very aggressive government- and industry-funded programs supporting the development of ceramic gas turbine engines exist in both Japan and Europe. These worldwide development programs are discussed in detail in the following sections.

The DOE-ATTAP Program

The ATTAP is a continuation of activities sponsored by the U.S. Department of Energy (DOE) to develop the technology for an improved automobile propulsion system. Two contractors, Allison Gas Turbine Division of General Motors Corporation and Garrett Auxiliary Power Division of Allied-Signal Aerospace Company were selected for this project. The project goal (Ref 3) is to develop and demonstrate ceramic technology that has the potential of operating for 3500 h in an automotive gas turbine engine duty cycle at temperatures up to 1371 °C (2500 °F).

ATTAP is intended to advance the technological readiness of an automotive ceramic gas turbine engine, based on previous efforts begun in the DOE-sponsored AGT Project. Although ATTAP focuses strongly on the continued development of ceramic hot section component technology, its overall intent remains that of bringing the automotive turbine engine to a state at which industry can make commercialization decisions. The declared ATTAP objectives are the following:

- To enhance the development of analytical tools for ceramic component design utilizing the evolving ceramic properties data base
- To establish improved process for fabricating advanced ceramic components
- To develop improved procedures for testing ceramic components
- To evaluate ceramic component reliability and durability in an engine environment

Allison Gas Turbine Division

Haley (Ref 4) provided the following information on the Allison ATTAP program. The Allison engine, developed by the General Motors Advanced Engineering staff, is a two-shaft, regenerative configuration with axial-flow gasifier and power turbines. An engine computer model was created, based on the measured performance of the AGT-5 engine. The model was then modified to increase the turbine inlet temperature (TIT) to 1371 °C (2500 °F), taking advantage of the ceramic flow-path to be incorporated during the ATTAP program. Since this TIT increase in the AGT-5 resulted in a substantial horsepower gain, the engine model was then scaled back (in size and mass flow) to better match the ATTAP acceleration target. Engine component aerodynamic performances were scaled accordingly, based on data from small compressor and turbine characteristics.

The projected fuel economy gain of the ceramic turbine over a baseline 2.5 L (153 in.³) spark ignition IC engine was dramatic: 35% (based on volumetric flow), exceeding the DOE-AGT program fuel economy target of 30% improvement.

The ATTAP emissions and alternate fuels goals are considered achievable, based on demonstrated General Motors experience with turbines such as the AGT100 engine developed during the earlier AGT program. In the AGT100 engine, a premixed, prevaporized, variable-geometry design was used to achieve steady-state emission levels of nitrous oxides (NO_x), carbon monoxide (CO), and unburned hydrocarbons (HC) that were within U.S. Federal Emission Standards. This performance was achieved in tests using DF-2 diesel fuel, JP-5 jet fuel, and methanol.

During the first years of the Allison ATTAP program, a low-aspect-ratio ceramic turbine rotor design was successfully engine-demonstrated at 1205 °C (2200 °F) and 100% speed, including survival of particle impacts and other hostile flowpath conditions. Turbine flowpath components have been designed for the 1371 °C (2500 °F) ATTAP cycle using improved monolithic ceramics, and major development/fabrication efforts have been subcontracted with the Carborundum Company, GTE Laboratories, Corning Glass Company, Garrett Ceramic Components, and the Manville Corporation.

Garrett Auxiliary Power Division

During the AGT project, the focus was on proving the concept of an all-ceramic gas turbine hot section. That objective was successfully met while also demonstrating the potential for achieving the intended fuel economy, emissions, and multifuel operation goals. Specifically, the Garrett AGT project demonstrated the feasibility of an all-ceramic engine, identified technologies that needed further development, and provided a 1371 °C (2500 °F) ceramic gas turbine test bed.

During the AGT project, Garrett was the first U.S. company to successfully run an all-ceramic hot section gas turbine engine. The AGT101 engine ran for a total of 91 h, with 85 h of continuous operation at a shaft speed of 60,000 rpm and a temperature of 1205 °C (2200 °F).

Figure 2 shows the Garrett AGT101 engine, which is being used as the test bed engine for ATTAP. The AGT101 is a single-shaft, regenerated gas turbine engine utilizing an all-ceramic hot section. The engine is flat-rated at 74.5 kW (100 hp), with a specific fuel consumption (SFC) goal of 0.18 kg/kW · h (0.3 lb/hp · h). The single-shaft rotating group is composed of a radial turbine, centrifugal compressor, and output spline. This rotating group is supported by an air-lubricated foil/journal bearing on the turbine end, and a conventional ball bearing and an oil-film thrust bearing on the compressor end. The maximum (steady-state) engine speed is 100,000 rpm, and idle speed is approximately 50,000 rpm. The maximum TIT is 1371 °C (2500 °F).

Fig 2 Cutaway view of the Garrett AGT101 ceramic gas turbine test engine

Design Methodology. In the ATTAP program, Garrett is accomplishing an extensive "up-front" analysis of thermal and stress conditions, combined with increased understanding of ceramic material properties, behavior, and failure criteria, pioneering a design methodology that addresses the peculiar requirements of ceramic materials. This endeavor has involved a wide diversity of participants, including U.S. Government laboratories, universities, ceramics manufacturers, and gas turbine designers.

The brittle nature of ceramic materials, often leading to catastrophic failure, requires sophisticated and rigorous methods capable of addressing the design requirements of ceramic components. To provide robust component designs for gas turbine use, methods are needed to address the numerous failure modes of ceramics including fast fracture, slow crack growth, creep, impact, contact, oxidation and corrosion, and thermal fatigue.

Garrett has formulated a comprehensive plan to develop verified methods to address both near-term and future ceramic design needs. The most urgently needed design systems are being developed now at Garrett under both the ATTAP and the DOE-sponsored Oak Ridge National Laboratory (ORNL) Life Prediction Methodology Project, which is part of the Ceramic Technology for Advanced Heat Engines programs. The DOE/ORNL program efforts are addressing fast fracture, slow crack growth, and creep properties for ceramics.

The Garrett ATTAP program is focusing on design needs for impact and contact damage reduction. Turbine rotors and stators are vulnerable to impact from debris and/or combustion products, and nearly all ceramic components are vulnerable to "contact damage" at interfaces due to loading and relative motion. Design methods development at Garrett under both the ATTAP and the DOE/ORNL ceramic life prediction programs are utilizing analytical and experimental work to develop models of material behavior, stress states, and material failure criteria.

Design Improvements. Previous durability testing of the Garrett AGT101 engine showed the radial turbine design to be susceptible to impact damage from combustor-generated carbon. To address this problem, efforts were initiated to improve the impact resistance of the turbine rotor. To maintain compatibility with the existing AGT101 engine and minimize the number of components requiring modification, the new designs were constrained to operate within the 100,000 rpm speed and existing compressor pressure ratio and mass flow characteristics. The design for an improved, impact-resistant turbine included:

- Reduction in turbine blade tip speed
- Redesigned blades with impact-resistant geometry
- Ceramic material strength improvements

Reduction of the turbine blade tip speed was fundamental to improving the impact resistance of the design. Because impact occurs when a slow-moving carbon particle is impacted by a very-fast-moving blade, reducing tip speed from 701 to 564 m/s (from 2300 to 1850 ft/s) reduces the relative impact velocity by 20%. To redesign the turbine blade geometry for improved impact resistance, preliminary techniques under development for impact analysis were integrated into the design procedures.

Further improvements in particle impact resistance are expected, through the use of ceramic materials with increased strength. As ATTAP component development progresses, materials with surface strengths as high as 700 MPa (100 ksi) are projected. New design criteria developed with the higher-strength materials will be reflected in improved impact resistance.

The improved turbine was designed to meet the ATTAP criteria for aerodynamic performance, mechanical integrity, reduced tip speed, and fabricability. A partial engine cross-section (Fig 3) shows the redesigned turbine and its surrounding structures. The turbine stator, turbine shroud, transition duct, and combustor baffle were directly affected by the reduced turbine diameter, though these components retained the same general structural configuration of the previous AGT101 design.

In operation, the impact-resistant ceramic turbine rotor will experience stress and temperature levels similar to the AGT101 radial ceramic turbine rotor. The highest rotor stress is predicted to be 319 MPa (46 ksi). The predicted temperature at this peak stress location is 394 °C (741 °F). The peak rotor operating temperature is 1261 °C (2302 °F) while operating at the flat-rated, steady-state condition.

The ceramic engine structures surrounding the turbine stage also were redesigned to accommodate the necessary changes in the gas flowpath. A design analysis was performed to evaluate these components for temperature, stress, and deflection levels during a

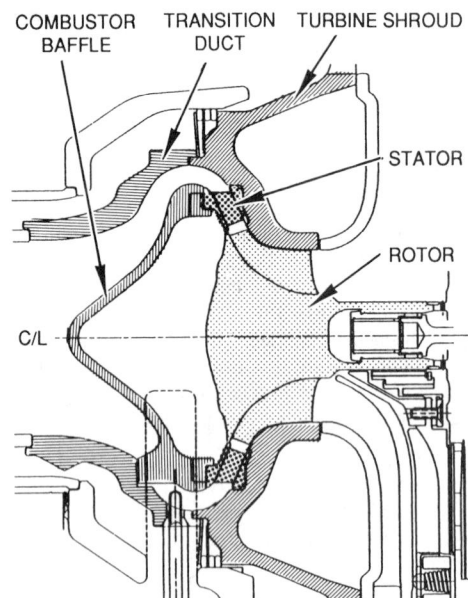

Fig 3 Cross-section of impact-resistant Garrett ceramic turbine and supporting structures showing five modified components (shaded)

Fig 4 Metallic seals located between the AGT101 regenerator core and flow separator housing

simulated start to flat-rated power transient, and at the steady-state operating condition. The highest stressed component was the turbine shroud. A two-dimensional analysis of the shroud shows peak stress of 276 MPa (40 ksi). This peak stress coincides with a predicted material temperature of 1021 °C (1870 °F).

Based on the earlier AGT101 testing, regenerator hot seal durability is one of the major focuses of the Garrett ATTAP program. The AGT101 regenerator system consists of a ceramic regenerator core, hot and cold metallic seals, and the drive system. The rotating ceramic regenerator core recirculates exhaust gas heat to the compressor discharge air to provide the energy recovery needed to meet fuel consumption goals. Metallic seals are located between the ceramic honeycomb regenerator core and the metallic exhaust housing (cold seal) and the ceramic lithium aluminum silicate (LAS) flow separator housing (hot seal), providing separation between the low-pressure and high-pressure regions in the engine (Fig 4). Both seals consist of a metallic diaphragm seal and metallic "shoe" which rides against the ceramic core surface. The wear surface of the metallic shoe is coated with a low-friction, high-temperature-resistant ceramic coating.

AGT101 test experience identified durability problems with the metallic hot seal due to mechanical distortion and creep, and spalling of the wearface coating. Garrett's current efforts in seal development are directed at improvements to the metallic seal. The seal is designed to meet the initial goal of operating in a ceramic engine for 100 h at a regenerator inlet temperature of 982 °C (1800 °F).

The improved configuration, now being evaluated in a hot regenerator test rig, consists of the baseline AGT101 coating applied on a new metallic substrate material and a new mechanical design allowing for thermal expansion without creating distortion. This seal has been exposed at 982 °C (1800 °F) for 50 h in a test rig, with insignificant wear and no signs of coating spalling or mechanical deformation.

Fabrication Technologies. The Garrett ATTAP program also places major emphasis on developing high-volume, near-net-shape ceramic fabrication technology, which must be demonstrated before the gas turbine engine can find acceptance in the automotive marketplace. A ceramic material properties data base is being established to provide feedback for fabrication process development and design methods development. The ceramic materials characterization activity to establish the data base depends upon destructive test data acquired from test specimens cut from components.

In recognition of the need to maintain a competitive position for domestic U.S. suppliers in critical ceramic technologies, the ATTAP program has placed heavy emphasis on the role played by the ceramics subcontractors. The development of fabrication techniques to produce high-quality, reliable ceramic components is critical to the continued growth of ceramic applications. The ATTAP U.S. suppliers are focusing on the technologies to fabricate the large and complex shapes needed for gas turbines. This fabrication technology must not sacrifice the temperature capability, and the strength, reliability, and durability properties which make ceramic materials so desirable.

Three subcontractors have been selected by Garrett to develop fabrication methods for high-quality ceramic ATTAP components: Norton/TRW Ceramics (NTC), Garrett Ceramic Components (GCC), and the Carborundum Company (CBO). These suppliers have demonstrated process and fabrication capabilities to produce components with material properties required for potential gas turbine use. The average flexure strength of NTC's NT-154 and GCC's GN-10 hot isostatically pressed (HIPped) silicon nitrides (Si_3N_4) are above 825 MPa (120 ksi) at ambient conditions. As the temperature approaches 1371 °C (2500 °F), this strength is reduced, but each material retains adequate strength to withstand component stresses at AGT101 operating conditions. A significant material data base was present for CBO's Hexoloy SA sintered alpha silicon carbide (αSiC) at the inception of ATTAP. The strength of this material is virtually constant from room temperature to the AGT101 engine maximum operating temperature. Shape-forming capability for these materials was also evaluated before selecting the firms to pursue ceramic fabrication development under subcontract.

Process Control. The ceramic component subcontractors are using in-process nondestructive evaluation (NDE) to identify flaws as they appear during fabrication. This takes the "guesswork" out of determining where flaws are being introduced in the process, so that applicable improvements can then be appropriately incorporated. High-resolution microfocus x-ray radiography and fluorescent penetrant inspection (FPI) are two NDE techniques in use by all three subcontractors. In addition, acoustic microscopy, ultrasonic testing, and other advanced techniques are being investigated by Garrett and the ATTAP subcontractors.

As the relationship between various process parameters and final product quality is determined, Statistical Process Control (SPC) is also being implemented. NTC is targeting their process development towards the fabrication of ATTAP rotors and stators from NT-154 HIPped Si_3N_4. GCC is developing a fabrication process for the ATTAP rotor using GN-10 HIPped Si_3N_4, and CBO is fabricating Hexoloy SA sintered αSiC into three structural components: the pilot combustor support, transition duct, and combustor baffle. Some of the components are shown in Fig 5.

Durability Testing. As an integral part of the ATTAP effort, durability testing with the Garrett AGT101 ceramic engine test bed provides a final measure of verification of the ceramic components as well as the design methods and processes that went into their making. This activity is planned in three, sequential phases: a 1370 °C (2500 °F) short-term test, a 100 h steady-state test, and a 300 h cyclic test.

The durability testing planned for the AGT101 test bed is a demanding demonstration of the technologies developed under ATTAP. The tests will be a rigorous evaluation of the design methodologies, improved designs, regenerator seal system, combustor, and ceramic component integrity (Ref 3).

U.S. Development for Auxiliary Power Units

Garrett Auxiliary Power Division is applying the ceramics technology developed

(a)

(b)

(c)

Fig 5 Ceramic turbine engine components being developed for the ATTAP program. (a) Turbine rotor. (b) Transition duct. (c) Pilot combustor support

under ATTAP, other U.S. Government-funded programs, and company-sponsored efforts to the development of advanced auxiliary power units (APUs).

One of the most significant new features of advanced APU designs will be a marked improvement in power density over today's designs. APUs for 21st-century commercial transport aircraft will have to meet a power section power density goal of more than 10.5 MW/m^3 (400 hp/ft^3). To achieve this power density, ceramics will replace conventional metal parts in the combustion and turbine sections, allowing operation up to 1371 °C (2500 °F) with an uncooled configuration.

The results of a NASA study (Ref 5) showed that the direct operating cost benefits of ceramics would far exceed the cost benefits expected from the anticipated improvement in engine component efficiency. The results of this cost benefit study are shown graphically in Fig 6. The operating cost reduction associated with ceramics is due to the increased cycle efficiency afforded by higher turbine inlet temperature capability without the parasitic loss of cooling flows. Also, significant weight-saving advantages over metals can be realized with ceramics. The higher strength-to-weight ratios of ceramics allow higher rotating component tip speeds for smaller, lighter rotors and reduced containment weight (Ref 6).

A Garrett program currently underway will demonstrate the use of a ceramic turbine stator in a commercial APU. Garrett has designed a ceramic stator and redesigned the surrounding metal hardware for the GTCP85–129 APU (used on the Boeing 737 and Douglas DC-9) to evaluate the durability of ceramics in gas turbine engine service (Fig 7). As a replacement for the current metal stator, the ceramic part should provide greater than three times the life. This is due to the improved erosion and corrosion resistance of ceramics and the elimination of low-cycle fatigue (LCF) as a failure mode, the primary causes of limited life of metallic components in this engine. A GTCP85 engine equipped with the ceramic stators and modified metal components is currently undergoing laboratory testing. To date, the ceramic stators have successfully accomplished over 300 h of engine testing, both steady-state and cyclic, up to a maximum inlet temperature of 980 °C (1800 °F). Upon completion of laboratory testing, the ceramic stators will have accumulated over 1200 h of testing, including accelerated cyclic endurance testing. Late in 1991, the program will transition to field testing, in which GTCP85 engines with ceramic stators will be evaluated in service by both military and commercial customers.

Sundstrand Power Systems. Bornemisza and Napier (Ref 7, 8) have reported on the evaluation of structural ceramic materials for APU components at Sundstrand Power Systems. Although the original intent was improved erosion resistance, the favorable high-temperature properties of new ceramic structural materials offered the potential for higher turbine operating temperatures, thus increasing power output and efficiency. Over the years, a number of Si$_3$N$_4$ and SiC turbine components have been designed and tested

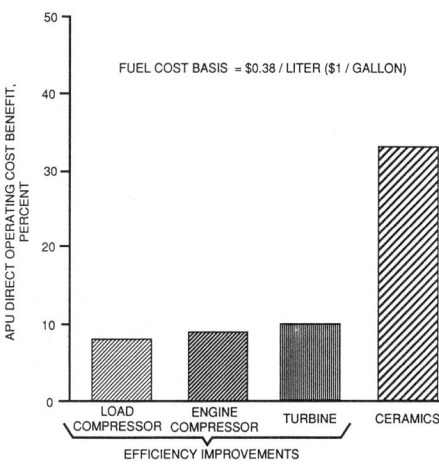

Fig 6 Reduction of APU direct operating costs

by Sundstrand to evaluate these continually improving structural ceramic materials.

The progression of ceramic hardware demonstrations in the Sundstrand Gemini 29.8 kW (40 hp) engine is summarized in Fig 8. The results of these tests demonstrated the future potential of ceramic components in small gas turbine engines and the need for improved materials, designs, and manufacturing methods.

Efforts have continued at Sundstrand in design and testing of ceramic static and rotating components operating at temperatures in excess of 1090 °C (2000 °F). The ceramic technology development tasks include the development of new design and analytical techniques, nondestructive and destructive material evaluation methods, and the fabrication and use of ceramic engine hardware, in close cooperation with the suppliers of the ceramic parts.

In 1985, a long-range Ceramic Technology Development Program was initiated by Sundstrand, with the objective of developing ceramic turbine components. During the first phase of this company-funded effort, a 112 mm (4.4 in.) diameter radial in-flow ceramic turbine wheel was designed and tested in the Gemini engine. Two Si$_3$N$_4$ rotors were tested at full speed (93,880 rpm) up to 1037 °C (1900 °F) TIT.

In 1988, Sundstrand successfully demonstrated ceramic turbine components in an engine at 1205 °C (2200 °F) TIT. The preliminary test results indicated that the ceramic hardware was capable of operating under the high thermal shock condition of fast starting to full speed, as well as full load/no-load cycling without failure.

Ceramic Turbine Engine Development in Japan

There are presently two ceramic gas turbine (CGT) programs underway in Japan (Ref 9). A 300 kW (402 hp) industrial CGT project is being conducted by the Moonlight Proj-

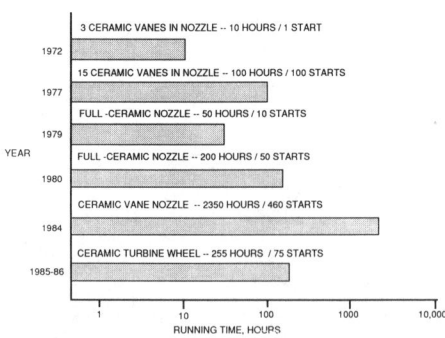

Fig 8 Ceramic components demonstrated by Sundstrand Power Systems. Source: Ref 8

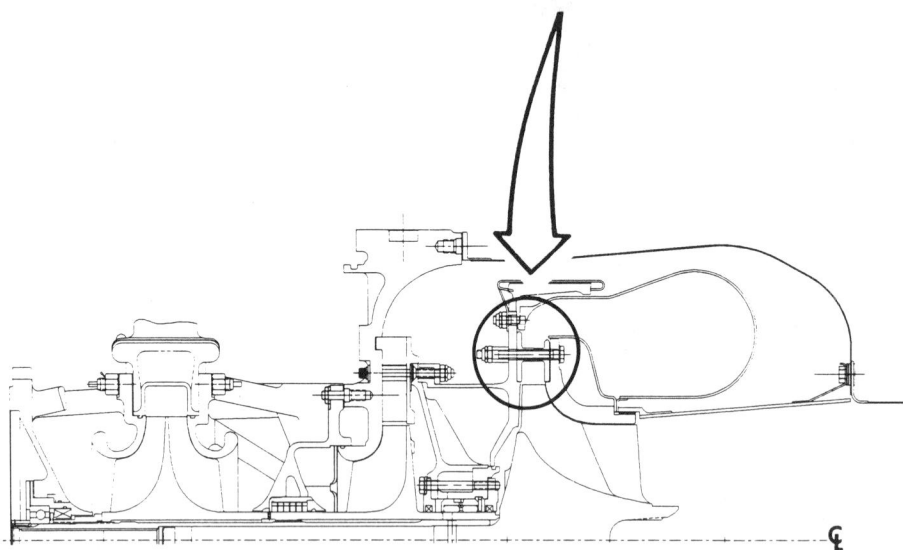

Fig 7 Ceramic stator for Garrett GTCP85 APU to evaluate component durability in field service

ect Office of the Japanese Ministry of International Trade and Industry (MITI). A nine-year program was initiated in 1988, similar in approach and goals to the U.S. DOE-sponsored AGT/ATTAP program. Three teams of engine manufacturers and ceramics companies are developing three different engines incorporating ceramics in the hot section. The Moonlight Project is in its third year, having concentrated on ceramic technology, component technology, and engine design. Utilizing knowledge gained during the first three years and from the U.S. ceramic gas turbine programs, the program is proceeding with fabrication and design of a metal version of each engine.

The second Japanese program is a 100 kW (134 hp) automotive CGT project being conducted by the Petroleum Refinery Division of MITI. The program is led by Nissan, Toyota, and Mitsubishi Motors, with several Japanese oil companies participating. Following a two-year preliminary design effort, they have selected a single-shaft radial turbine regenerative engine similar in design and performance goals to the Garrett AGT101. A full-scale engine development program was initiated in 1990.

Ceramic Gas Turbine Engine Development in Europe

Since the 1970s, Daimler-Benz in Germany has been conducting research in ceramics for gas turbine engines (Ref 10). A gas turbine engine has been operated in the Daimler-Benz Research Automobile since 1982. At present, the research gas turbine is being operated in a Mercedes-Benz mid-size series car. Through 1989, operating experience had been gathered with ceramic components in excess of 600 h in the engine and several hundred h on component test benches, and over 20,000 km (12,000 mi) of road testing had been accumulated.

Monolithic HIPped Si_3N_4 turbine wheels of three different designs were manufactured and have undergone testing in component and engine tests. During the initial phase in the early 1980s, a turbine wheel accumulated a total running time of 58 h at temperatures up to 1350 °C (2462 °F) and speeds up to the maximum design speed of 61,000 rpm. Several starts were performed with another turbine wheel that completed a total running time of 10 h, 9 h of which were at 1250 °C (2282 °F) and a speed of 61,500 rpm. These operating conditions correspond to the current full-load operation of the Daimler-Benz research gas turbine.

One injection-molded HIPped Si_3N_4 rotor was tested under engine operating conditions for 10 h without failure and with only minor damage to the blade trailing edges, operating for 9 h at TIT = 1250 °C (2282 °F) and 61,500 rpm. A slip-cast Si_3N_4 turbine rotor was run for 32 h at 1350 °C (2462 °F) in an engine installed in a current production car. An engine with a slip-cast turbine rotor installed in a Mercedes-Benz automobile was shown at the 1990 ASME International Gas Turbine Congress in Brussels, Belgium, after performing 6.5 h of demonstration driving between Stuttgart and Brussels, a distance of 660 km (410 mi).

Ceramic Materials

While a number of unknowns still need to be addressed, the ceramic materials available for gas turbine use have improved dramatically during the past twelve years. Not only have shape and size fabrication capabilities been improved, but fast fracture and stress-rupture properties have also been steadily improved.

The fast fracture flexure strength of Hexoloy SA sintered αSiC (Fig 9) was increased from 310 MPa (45 ksi) in 1976 to 415 MPa (60 ksi) in 1988. Sintered SiC materials exhibit excellent strength retention at temperatures up to 1540 °C (2800 °F) and continue to retain most of their initial strength after extended cyclic exposure to combustor conditions.

Dense Si_3N_4 materials have higher strength at low temperature than SiC materials, but the strength of Si_3N_4 decreases with increasing temperature. In 1976, fully dense Si_3N_4 was only available in hot pressed forms such as Norton NC-132. NC-132 has room-temper-

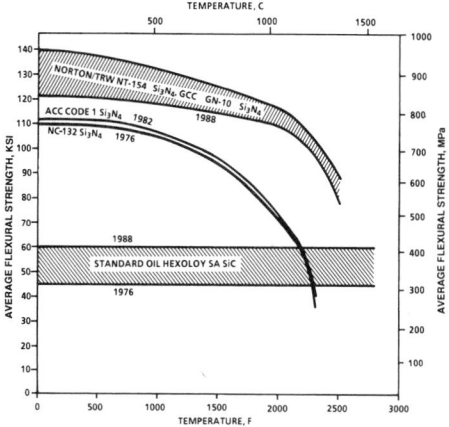

Fig 9 Strength versus temperature properties of advanced ceramics

ature flexure strength of over 690 MPa (100 ksi), but strength decreases rapidly above 1000 °C (1800 °F). Fabrication of turbine components from NC-132 required grinding from billet material. Net-shape sinterable Si_3N_4, such as AiResearch Casting Company (ACC) Code 1 sintered Si_3N_4, became available in the late 1970s. These initial sinterable materials had the advantage of near-net-shape capability but exhibited similar strength loss at high temperatures (Fig 9). Current state-of-the-art materials, such as Norton/TRW NT-154 and Garrett Ceramic Components (GCC) GN-10 Si_3N_4, have the potential for net-shape capability, room-temperature strength of 828 to 966 MPa (120 to 140 ksi), and retain this high strength to higher temperatures.

The stress-rupture life of Si_3N_4 materials has also been improved considerably (Fig 10).

ACC Code 1 achieved an improvement in stress-rupture life over NC-132; and current state-of-the-art materials, for example, GN-10 and NT-154, exhibit large increases in both stress-rupture and temperature capability. The early materials such as NC-132 and ACC Code 1 did not have adequate stress-rupture properties to satisfy the requirements of the AGT101 gas turbine rotor at 1370 °C (2500 °F) TIT. Current improved materials, such as NT-154 and GN-10, appear capable of withstanding the temperatures and stresses of a turbine rotor at that temperature.

Strength retention during long-term cyclic exposure to combustion environments has also improved significantly for Si_3N_4. In the mid-1970s, only the reaction-bonded form of Si_3N_4 was available in complex net shapes. This low-density Si_3N_4 maintained its strength after 1200 °C (2200 °F) cyclic exposure, but the strength was insufficient for most turbine rotor applications. Initial versions of sinterable, fully dense Si_3N_4 showed significant reductions in strength retention due to pitting and thick, coarse oxide formation on exposed surfaces. The more refractory (and stronger) sintered materials of the 1980s retain a greater portion of their initial strength following cyclic exposure. Sintered SiC material has consistently exhibited excellent strength retention after exposure to 1370 °C (2500 °F) and this durability has been maintained as higher-strength SiC has become available (Ref 1).

The current silicon-based ceramics appear to meet the needs of the ceramic gas turbine engines under evaluation. Table 1 shows the material properties required to provide 3500 h of life at the operating conditions of the Garrett AGT101 engine.

Table 1 AGT101 ceramic turbine engine material requirements
Materials also should have Weibull modulus >12.

Temperature		Flexural strength		100 h stress-rupture strength	
°C	°F	MPa	ksi	MPa	ksi
RT	RT	897	130	...	...
1204	2200	620	90	483	70
1371	2500	483	70	...	...

Silicate Materials. A second class of ceramic materials that are essential to the application of ceramic gas turbine engines in vehicles (automobiles/trucks) and some stationary power systems are based on silicates. These materials are used in regenerator cores that provide heat recovery for improved fuel consumption. Presently, variations of magnesium-aluminum-silicate (MAS) and aluminum-silicate (AS) are being evaluated for this application. The materials are formed into honeycomb structures similar to the ceramic substrate in automotive catalytic converters. The requirements for use in a regenerator core far exceed those needed for catalytic converters. The materials must withstand more severe mechanical and thermal loads, and the honeycomb pattern must provide for many more channels per unit area. This is an area where improvements in material capability and fabrication technology are needed.

ACKNOWLEDGMENT

The author would like to acknowledge the support of the management of Garrett Auxiliary Power Division, Allied-Signal Aerospace Co. in the preparation of this article. The author would also like to extend special recognition to GAPD staff members W.D. Carruthers, M.L. Easley, R.E. Morey, R.W. Schultze, R.L. Strong, and M.W. Rettler for their contributions to the development of ceramics technology at GAPD, and to G.A. Lucas for his skills in editing this manuscript.

REFERENCES

1. W.D. Carruthers and L.J. Lindberg, Critical Issues for Ceramics for Gas Turbine Engines, *Proceedings of Third International Symposium on Ceramic Materials and Components for Engines*, Las Vegas, 27–30 Nov 1988, V.J. Tennery, Ed., American Ceramic Society, 1989

2. D.W. Richerson, Evolution in the U.S. of Ceramics Technology for Turbine Engines, *Ceram. Bull.*, Vol 64 (No. 2), 1985

3. W.D. Carruthers and J.R. Smyth, "Advanced Ceramic Engine Technology for Gas Turbines," Paper 91-GT-368, American Society of Mechanical Engineers, 1991

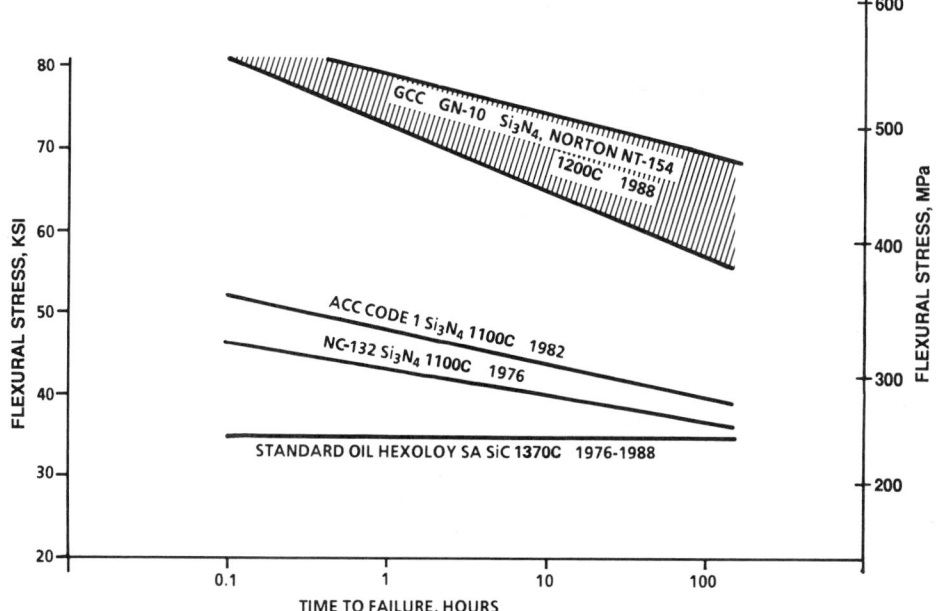

Fig 10 Stress-rupture characteristics of advanced ceramics

4. P.J. Haley, "Advanced Turbine Technology Applications Project (ATTAP)—Overview, Status, and Outlook," Paper 89-GT-118, American Society of Mechanical Engineers, 1988

5. M. Early, *et al.,* "Small Engine Component Technology Study Final Report (Contract No. NAS3–24544)," Report No. CR-175078, National Aeronautics and Space Administration, U.S. Army AVSCOM Report TR-86-C-11, Garrett Report No. 21–5776–2A, Mar 1986

6. J.L. Elliott and J.L. Munier, "Auxiliary Power Units and Their Technologies for the 21st Century," Garrett Report No. 31–8870, presented at Institute of Mechanical Engineers 21st Century Aero Engine Design 1990 Scenario, Dowty, England, 17–18 May 1990

7. T. Bornemisza, "Ceramic Small Gas Turbine Technology Demonstrator," Paper 90-GT-306, American Society of Mechanical Engineers, June 1990

8. T. Bornemisza and J. Napier, "Comparison of Ceramic Versus Advanced Superalloy Options for a Small Gas Turbine Technology Demonstrator," Paper No. 88-GT-228, American Society of Mechanical Engineers, June 1988

9. Y. Tsutsu, "The Status of Ceramic Gas Turbine Programs in Japan," Presented at 28th DOE Automotive Technology Development Contractors Coordination Meeting, Detroit, Oct 1990

10. K.D. Moergenthaler, "Gas Turbine Activities at Daimler-Benz," Presented at 28th DOE Automotive Technology Development Contractors Coordination Meeting, Detroit, Oct 1990

Aerospace Applications

Stanley R. Levine and Thomas P. Herbell, NASA Lewis Research Center

AEROSPACE NEEDS for ceramics and ceramic matrix composites are quite broad, encompassing electronics, wear, and thermostructural applications. This article primarily addresses the status and needs for thermostructural ceramics, which implies applications that have a high-temperature environment and attendant application-generated stresses. Thermostructural ceramics have many potential aerospace applications, ranging from relatively nondemanding uses (such as the thermal protection system on the Space Shuttle and high-temperature electrical insulators in space-based nuclear power systems like the SP-100) to the far more demanding and distant uses such as an integrally bladed turbine rotor in an aircraft engine.

Potential thermostructural applications for ceramics are outlined in the areas of aeropropulsion, space propulsion, space power, aerospace vehicles, and space structures. Several prototypes for specific component applications are described, and the payoffs and technical needs associated with ceramic implementation outlined. As an example, consider high-temperature thermal or electrical insulation. This type of application may seem minimally demanding. However, important application-related factors such as acoustic loads and loads associated with accelerations and decelerations on launches and landings must be considered in addition to loads generated by temperature gradients.

These factors are also important in structural, load-bearing applications (where the payoffs from ceramics and the barriers to their use are far greater). For highly loaded, very high temperature applications requiring composites, a number of key issues (Table 1) need to be resolved for ceramics to have a significant impact on aerospace systems. First, and foremost, the required base for fiber technology is not yet in place. Use temperatures are limited to about 1100 to 1300 °C (2000 to 2370 °F) depending on specific fibers and applications. Fiber coatings (interphases) also need further development. In addition, materials processing, basic understanding of materials behavior, and design and life prediction methodologies are far from mature.

Benefits and Limitations of Thermostructural Ceramics

Ceramics certainly have the potential to exceed the thermostructural, environmental durability, and weight limitations of superalloys given the desirable characteristics and potential benefits of ceramics (Table 2). Specific materials that often display the first five benefits (Table 2) include SiC, Si_3N_4, Al_2O_3, and mullite either in monolithic form or as a constituent in composites. In addition, ceramics generally offer low thermal conductivity, which makes them useful as thermal insulators but also detrimental in terms of thermal shock resistance. Silicon carbide (SiC), aluminum nitride (AlN), and beryllia (BeO) are examples of ceramics with relatively high thermal conductivities, which contribute to their good thermal shock resistance. Electrical properties for ceramics cover a broad spectrum—from insulators to superconductors.

Ceramics, in their present stage of development, still have major liabilities for many of the demanding aerospace thermostructural applications. The brittle nature of ceramics makes them unforgiving of small flaws that arise from impurities, processing, or in-service use. As a result, monolithic ceramics generally exhibit low fracture toughness and broad distributions of strength. Recent efforts have been able to narrow the strength distribution for SiC and Si_3N_4 to acceptable levels for applications where catastrophic failure is an acceptable consequence. Of course, catastrophic failure behavior limits their application in the aerospace sector. One example of the use of a monolithic ceramic in a low-risk application is the silicon-infiltrated SiC spacer in the Pratt and Whitney F-100 engine (Ref 1). The spacer replaced a metal part that was no longer available. The lightly loaded spacer separates a platinum part from a superalloy part in a catalytic ignitor support.

Considerable effort has been expended to lessen the sensitivity of ceramics to process- and service-generated flaws by increasing fracture toughness. Whisker reinforcement, particulate toughening, and microstructural optimization can raise fracture toughness by as much as 2 to 4 times (Ref 2). Catastrophic fracture, however, remains an issue. Therefore, for most aerospace applications, continuous fiber-reinforced ceramics (FRC) are expected to have the broadest use for demanding thermostructural applications. If the toughening mechanisms of crack deflection around fibers and crack bridging by fibers remain operable over the life of the system, brittle catastrophic failure can be avoided (Fig 1). However, fiber-reinforced ceramics are a long way from maturity and much needs to be done (Ref 3) to develop stable, high-temperature fibers, durable fiber-matrix interfaces, cost-effective processing, and protection schemes for oxidation-prone matrices. In addition, advances in fabrication technology and processing science are critical to the development of a viable industrial base. The base of understanding for design and life prediction, although receiving considerable attention, is presently inadequate to develop capable tools for confident design. Test methods also need further development and standardization, and a data base needs to be established. The technology for joints, attachments, and interfaces needs further development as well.

Applications

Ceramics and ceramic composites are considered an enabling technology for advanced aircraft and space propulsion engines and space power systems. Because of their potential to allow increased engine operating temperature, the benefits are quite significant.

Aircraft propulsion applications are outlined in Table 3. For small engines, below 750 kW (1000 hp), small component size limits or excludes the ability to cool effectively. As pressure ratios climb from present levels of about 25:1 in "large" engines to 50:1 to 100:1, engine core sizes will diminish and components will shrink in size. Smaller cores are beneficial from an overall efficiency standpoint, but again will make blade and vane cooling more difficult and less efficient.

Table 1 Key technical issues confronting the application of thermostructural ceramics in future aerospace systems

Issue	Primary solution	Other contributors
Reliability	Fiber reinforcement	Fiber/interphase stability, matrix reliability
Toughness	Fiber reinforcement	Small-diameter fibers, tough matrix
High-temperature capability	Stable fibers	Matrix strength, environmental durability
Environmental durability	Fiber coatings (interphases)	Matrix, surface coating
Design		
Joining		
Manufacturing base		

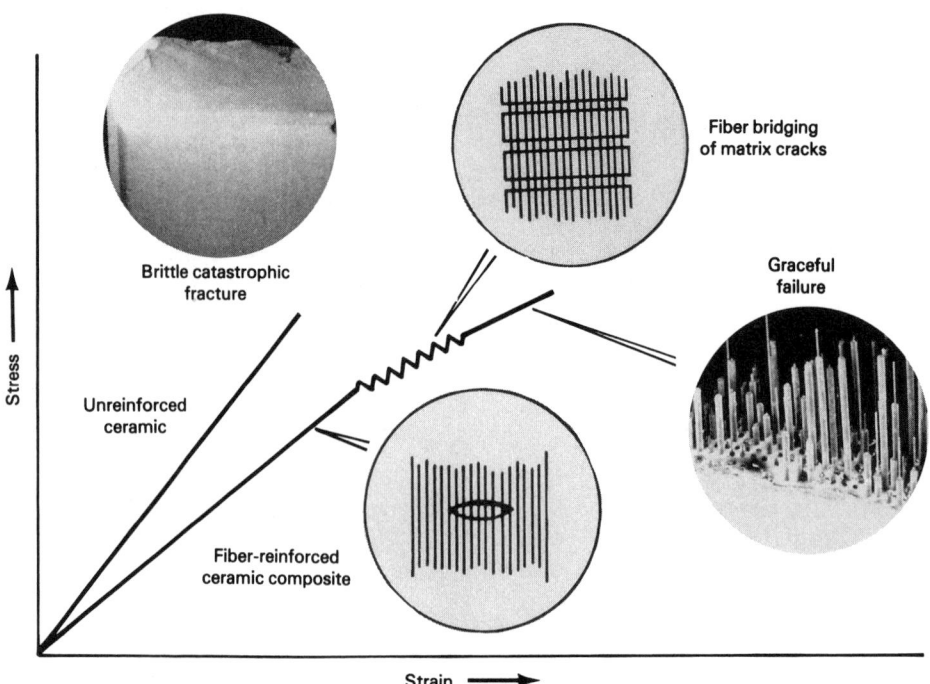

Fig 1 Stress-strain behavior of reinforced and unreinforced ceramics. Fiber reinforcement of ceramic composites avoids the brittle catastrophic fracture characteristic of monolithic materials. However, the matrix cracking stress of reinforced ceramics is generally lower than the fracture stress of monolithic ceramics.

Engine efficiency does not increase monotonically with increased cycle temperature if increased cooling air is required. Thus, the impetus exists for ceramics, which can be run uncooled or minimally cooled.

The potential advantages of ceramics in aircraft propulsion were examined in a 1986 Small Engine Component Technology Study (Ref 4). Five small U.S. engine manufacturers examined various engine/airframe/mission combinations and estimated the benefits possible from various technologies. Technology rankings based on direct operating cost (DOC) or life cycle cost (LCC) were made. For the rotorcraft or commuter aircraft applications, the introduction of ceramics was projected to yield the largest benefits ranging from 40 to 65%. These projected benefits exceeded the potential benefits from improved aerodynamics and other design optimizations, and are up to an order of magnitude greater than the benefits projected from utilization of advanced metallic materials. Further improvements are possible with recuperated cycles if improved lightweight ceramic heat exchangers can be developed.

Currently, under the NASA-sponsored High Temperature Engine Materials Program, studies are underway examining the potential

Table 2 Benefits of ceramics in aerospace systems

Ceramic properties	System benefits
Low density	Reduced system weight
High specific stiffness and strength	High thrust-to-weight ratio
Property retention at high temperatures	Thermal efficiency
Low coefficient of thermal expansion	Dimensional stability
Environmental durability	Long life
Thermal conductivity and electrical properties	Application and material specific

benefits of advanced composites for large subsonic and supersonic civil transport aircraft (Ref 5). Preliminary results indicate that fiber-reinforced ceramics provide significant benefits in applications like advanced, subsonic-engine components (Fig 2).

Technology for high-speed (~Mach 2.5) civil transport with entry into service by 2005 is also being pursued by NASA. Environmental acceptability from an emissions and noise standpoint are critical to the viability of such a vehicle. NO_x emissions at 20 km (12 miles) altitude must be at a low level to avoid ozone layer depletion. Fiber-reinforced ceramic matrix composites are considered a

Table 3 Potential applications of ceramics in aircraft propulsion systems

Engine system	Components application
Small engines	
Propulsion	Regenerator/recuperator, combustors
Auxiliary power units (APUs)	Turbine blades/bladed disks
Large engines	
Subsonic	Vanes, seals, other static components, tail cone, and nozzle
Supersonic	Bearings, vanes, seals, and other static components
Hypersonic	
Air-turbo-ramjet	Compressor, recuperator, ducting, seals
Scramjet	Uncooled ducting, seals

potential enabling technology for low NO_x combustor designs. Noise generation during taxi, take-off, climb, and cruise must also be minimized. Engine generated noise will be partially handled through nozzle design. Here too, fiber-reinforced ceramics are considered a potential enabling technology along with a lightweight, intermetallic matrix composite structure. Critical issues include high thermal conductivity to minimize thermal gradients and combustor inner-wall temperature, durability in combustor environments that may be reducing as well as oxidizing, and long life (20,000 h desired) at near maximum-use temperature. Potential materials applications in a second-generation, Mach 3.2 civil transport are also being evaluated.

Detailed studies of SiC fiber-reinforced lithium-aluminum-silicate (LAS) shingle combustor liners have already been carried out (Ref 6). The initial study showed technical feasibility and a projected 25% combustor life-cycle cost savings based on life equivalent to that of a metallic shingle concept. A follow-on effort involved combustor sector rig testing. While significant material degradation was observed as one might expect based on the known limitations of SiC-reinforced LAS, the results confirmed the potential of fiber-reinforced ceramics for this application (Ref 7).

Jeal (Ref 8) has predicted that ceramic matrix composites use in jet engines will grow to nearly 30% of engine weight by the year 2010. Ceramic composites will be introduced in less demanding applications first. These tend

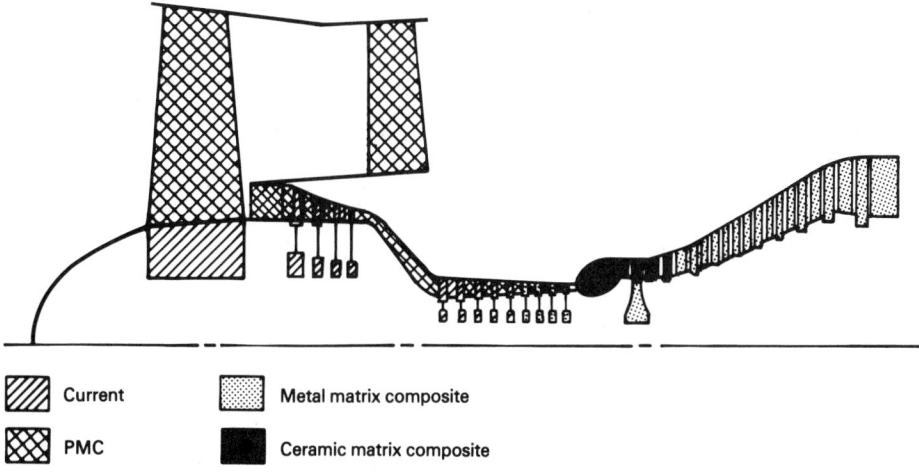

Fig 2 Potential applications for advanced composites in a conceptual 21st century ultra-high bypass turbofan engine

Legend:
- Current
- PMC
- Metal matrix composite
- Ceramic matrix composite

to be in the aft end of the engine, where temperatures and loads are relatively low. Examples are exhaust nozzle components and the tail cone. Weight reductions in these applications yield significant benefits. Recently, fiber-reinforced ceramic components fabricated by chemical vapor infiltration have been demonstrated. Inner exhaust nozzle flaps of SiC/SiC and outer flaps of C/SiC were installed in a SNECMA M53 turbo-fan engine on a French Air Force/Dassault-Breguet Mirage 2000 flown at the 1989 Paris Air Show (Ref 9).

Ceramics also appear to be enabling technology for air-turbo-ramjet engines, where compressor temperatures are extremely high and use of an uncooled approach via ceramics is desirable. For scramjet engines being developed for the National Aerospace Plane (NASP), the preferred approach is to use hydrogen-cooled, high-conductivity metals. The liquid or slush hydrogen fuel acts as a heat sink and requires reheating to as high a temperature as possible within materials constraints to optimize engine efficiency. Ceramics may find use in engine panel seals and uncooled ducting. The latter application may become a critical one if insufficient hydrogen is available for engine and airframe cooling.

Diesel engines have been given some attention as an alternative to small gas turbines. One concept involves a compound cycle with a diesel core (Ref 10). Ceramics could play an important role here from the standpoint of weight savings and efficiency. In thermal barriers for diesel engines, ceramics have applications for pistons, exhaust port liners, wear parts, and so forth.

Space Propulsion. There are many potential applications for ceramics in space propulsion systems. Thrust chambers for small chemical rockets are a relatively low-risk application, which may afford some durability and weight advantages. For large chemi-

cal rockets, the impetus for ceramics use is an expansion of the turbopump operating envelope from the current 820 °C (1500 °F) level set by uncooled superalloys to about 1200 °C (2200 °F) or higher. The large temperature gain translates into enormous savings in dollars per pound of payload-to-orbit because the impact cascades throughout the vehicle, reducing engine weight, fuel and liquid oxygen requirements, and tank sizes.

The benefits of using ceramics in rocket engine turbopumps have been described in several reports (Ref 11–13). Early studies (Ref 11, 12) focused on monolithic ceramic materials, while the potential of fiber-reinforced ceramics are evaluated in Ref 13. Fiber-reinforced ceramics are currently being considered in preference to monolithic ceramics for reasons of reliability and improved thermal shock resistance. The first study (Ref 11) projected gains in thrust and durability for a Space Shuttle Main Engine (SSME). For example, a 2.5% increase in thrust was projected for about a 200 °C (360 °F) increase in fuel turbopump temperature. This translates to a 6% increase (about 4500 kg) in payload. A later study (Ref 12) examined advanced blading candidates for SSME including ceramics. Large system benefits compared to development cost (<20:1) were projected for a 200 °C (360 °F) increase in engine temperature based on ceramics. Additional benefit studies addressing fiber-reinforced ceramics are currently in progress. Prototype subcomponents have been built using reinforced ceramic composites, and they have survived tests in an environment that simulates or exceeds the thermal shock transient of the SSME (Ref 13).

Although tests of full-scale turbopump components are planned, the full implementation of ceramics in turbopumps is at least a second generation away from today's engines. However, selective use of ceramics in large chemical rocket components may be closer at

hand for valves, bearings, thrust chambers, exit cones, and static turbopump hot section components.

Applications for structural ceramics also exist in solar thermal propulsion (for example, collectors and thermal insulators). In nuclear propulsion systems, uncooled turbine and reactor components are possible ceramic applications. High-temperature superconductors made from ceramics may enable the development of high-field magnets required for plasma confinement in magneto-plasma-dynamic thrusters, if upper critical field strengths can be raised.

Space Power. Two of the space power applications for ceramics are closed-cycle analogs of the space propulsion systems (that is, solar thermal and nuclear, high-temperature gas reactor). In space-based nuclear power generators like the SP-100, limited use is made of ceramics in the current thermoelectric power conversion system. The most demanding use is for Al_2O_3 as a high-voltage, high-temperature insulator with requisite thermodynamic stability in a reducing environment. Space radiator structures are an emerging important application for fiber-reinforced ceramics.

An alternative power conversion system is a space Stirling engine. A recent assessment for a space-based 1050 K Stirling engine (Ref 14) has identified areas where ceramic use could provide beneficial enhancements to a "superalloy" engine. In the displacer and power pistons and cylinders, SiC/SiC is a potential alternative to beryllium because of mass and environmental durability of SiC. For the alternator support structure, SiC/SiC is an alternative to a titanium alloy support structure and a graphite/polymer composite plunger.

Zirconia fuel cells and high-temperature superconductors for power transfer lines and magnetic energy storage are ceramic applications where unique electrical properties are the enabling factors, but where (thermo)structural considerations will certainly play an important role.

Aerospace Vehicles. The primary application for ceramics in aerospace vehicles like the Space Shuttle is in the passive thermal protection systems developed to withstand reentry heating. Here, primary loads are thermal and acoustic. More aggressive structural applications are being considered for hypersonic vehicles, such as the National Aerospace Plane. While carbon/carbon is the primary nonmetallic candidate for load-bearing use on the airframe and control surfaces in some designs, the issue of oxidation protection system reliability in a load-bearing structure must be considered. Significant expenditures by the Department of Defense have been made to solve the problem of oxidation for carbon/carbon with unsuccessful results. Fiber-reinforced ceramics are considered a less mature back-up material that may emerge as the eventual choice, as in the proposed French Hermes vehicle. Windows and radomes are examples of current important applications. As speed and life requirements increase, higher temperature materials are required.

Space Structures. Applications for ceramics in space structures are driven by factors such as stiffness, dimensional stability and durability in the space environment. Atomic oxygen stability is an important, across-the-board virtue for low-earth-orbit applications. Damping by passive means is a possibility via fiber/matrix sliding and high-damping fibers. Low coefficients of thermal expansion are an important consideration for precision structures, and fiber-reinforced ceramics with near zero coefficient of thermal expansion are available. Resistance to fatigue due to thermal cycling also makes low thermal expansion important. Space shields to protect vehicles against orbital debris and natural and man-made projectiles will be an increasingly important consideration for future manned space vehicles as the amount of "space junk" in orbit increases.

REFERENCES

1. R. Hilsdorf, Ceramic Usage in Turbines Advances, *Am. Met. Mark.*, 5 June 1987, p 1, 8

2. K.T. Faber, Toughening Mechanisms for Ceramics in Automotive Applications, *Ceram. Eng. Sci. Proc.*, Vol 5, 1984, p 408–439

3. "New Structural Materials Technologies: Opportunities for the Use of Advanced Ceramics and Composites—A Technical Memorandum," OTA-TM-E-32, U.S. Congress, Office of Technology Assessment, Sept 1986, p 11–42

4. M.R. Vanco, W.T. Wintucky, and R.W. Niedwiecki, "An Overview of the Small Engine Component Technology (SECT) Studies," NASA TM 88796, National Aeronautics and Space Administration, 1986 (also AIAA-86-1542)

5. J.R. Stephens, Intermetallic and Ceramic Matrix Composites for 815 to 1370 °C (1500 to 2500 °F) Gas Turbine Engine Applications, *Matrix Composites: Processing, Modeling & Mechanical Behavior*, R.B. Bhasat, *et al.*, Ed., The Mineral, Metals & Materials Society, 1990

6. M.A. Sattar and R.P. Lohmann, "Advanced Composite Combustor Structural Concepts Program," NASA CR 174733, United Technologies Corp., 1984

7. D.J. Dubiel, R.P. Lohmann, S. Tanrikut, and P.M. Morris, "Energy Efficient Engine Pin Fin and Ceramic Composite Segmented Liner Combustor Selector Rig Test Report," NASA CR 179534, United Technologies Corp., 1986

8. R. Jeal, Moving Towards the Non-Metallic Aero Engine, *Metallurgia*, Vol 55 (No. 8), 1988, p 371–374

9. P.J. Lamicq, G.A. Bernhart, M.M. Dauchier, and J.G. Mace, SiC/SiC Composite Ceramics, *Am. Ceram. Soc. Bull.*, Vol 65 (No. 2), Feb 1986, p 336–338

10. J.G. Castor, "Compound Cycle Engine for Helicopter Application—Executive Summary," AVSCOM TR-86-C-15, Garrett Turbine Engine Co., 1986

11. H.W. Carpenter, "Ceramic Turbine Elements," RI/RD 84–164, Rocketdyne Div., Rockwell International, 1984

12. W.T. Chandler, "Materials for Advanced Rocket Engine Turbopump Turbine Blades," NASA CR 174729, Rocketdyne Div., Rockwell International, 1983

13. T.P. Herbell and A.J. Eckel, "Ceramic Composites for Rocket Engine Turbines," SAE Technical Paper 911108, Society of Automotive Engineers, Apr 1991

14. C.M. Scheuermann *et al.*, "Materials Technology Assessment for a 1050 K Stirling Space Engine Design," NASA TM 101342, National Aeronautics and Space Administration, 1988

Medical and Scientific Products

Larry L. Hench, Department of Materials Science and Engineering and Bioglass Research Center, University of Florida

CERAMICS, GLASSES, AND GLASS-CERAMICS have been essential for a long time in the medical industry for eyeglasses, diagnostic instruments, chemical ware, thermometers, tissue culture flasks, fiber optics, and so on. Insoluble porous glasses have been used as carriers for enzymes, antibodies, and antigens, offering the advantages of resistance to microbial attack, pH changes, solvent conditions, and temperature (Ref 1). Ceramics are also widely used in dentistry as restorative materials such as gold porcelain crowns, glass-filled ionomer cements, dentures, and so forth. These applications are called dental ceramics and are discussed in Ref 2.

This article is limited to ceramics, glasses, and glass-ceramics used as biomedical implants. Although dozens of compositions have been explored in the past, relatively few have achieved human clinical application. This article concentrates on these compositions, examines their differences in processing and structure, describes the chemical and microstructural basis for their differences in physical properties, and relates properties and tissue response to particular clinical applications. Reference 3 provides a historical review of these biomaterials.

Tissue Attachment Mechanisms

No one material is suitable for all biomaterial applications. As a class of biomaterials, ceramics, glasses, and glass-ceramics are generally used for repair or replacement of musculoskeletal hard connective tissues. Their use depends on achieving a stable attachment to connective tissue. Carbon-based ceramics are also used for replacement heart valves, where resistance to blood clotting and mechanical fatigue are essential characteristics.

The mechanism of tissue attachment is directly related to the type of tissue response at the implant interface. No material implanted in living tissues is inert; all materials elicit a response from living tissues. Four types of response are possible (Table 1). These types of tissue responses allow four different means of achieving attachment of prostheses to the musculoskeletal system. Table 2 summarizes the attachment mechanisms with examples.

A comparison of the relative chemical activity of the different types of bioceramics, glasses, and glass-ceramics is given in Fig 1.

Relative reactivity correlates very closely with the rate of formation of an interfacial bond of ceramic, glass, or glass-ceramic implants with bone (Fig 2). Figure 2 will be discussed in more detail in the section "Bioactive Glasses and Glass-Ceramics" in this article.

The relative level of reactivity of an implant influences the thickness of the interfacial zone or layer between the material and tissue. Analysis of failure of implant materials during the last 20 years generally shows failure originating from the biomaterial-tissue interface. When biomaterials are nearly inert and the interface is not chemically or biologically bonded, there is relative movement and progressive development of a fibrous capsule in soft and hard tissues. The presence of movement at the biomaterial-tissue interface eventually leads to deterioration in function of the implant or the tissue at the interface or both. The thickness of the nonadherent capsule varies depending upon both material and extent of relative motion.

Inert and Nearly Inert Materials. The fibrous tissue at the interface of dense alumina (Al_2O_3) implants is very thin. Thus, if alumina devices are implanted with a very tight mechanical fit and are loaded primarily in compression, they are very successful. In contrast, if a nearly inert implant is loaded such that interfacial movement can occur, the fibrous capsule can become several hundred micrometers thick and the implant can loosen very quickly.

The concept behind nearly inert microporous materials is the ingrowth of tissue into pores on the surface or throughout the implant. The increased interfacial area between the implant and the tissues results in an increased inertial resistance to movement of the device in the tissue. The interface is established by the living tissue in the pores. Consequently, this method of attachment is often termed biological fixation. It is capable of withstanding more complex stress states than implants with morphological fixation. The limitation associated with porous implants, however, is that for the tissue to remain viable and healthy it is necessary for the pores to be greater than 50 to 150 μm in size. The large interfacial area required for the porosity is due to the need to provide a blood supply to the ingrown connective tissue; vascular tissue does not appear in pores less than 100 μm in size. If micromovement occurs at the interface of a porous implant, tissue is damaged, the blood supply

may be cut off, the tissues will die, inflammation ensues, and interfacial stability can be destroyed. When the material is a metal, the large increase in surface area can provide a focus for corrosion of the implant and loss of metal ions into the tissues. This can be mediated by using a bioactive ceramic material such as hydroxylapatite as a coating on the porous metal. The fraction of large porosity in any material also degrades the strength of the material proportional to the volume fraction of porosity. Consequently, this approach to solving interfacial stability is best when used as coatings or when used as unloaded space fillers in tissues.

Resorbable biomaterials are designed to degrade gradually over a period of time and be replaced by the natural host tissue. This leads to a very thin interfacial thickness. This is the optimal solution to the biomaterials problem if the requirements of strength and short-term performance can be met. Natural tissues can repair themselves and are gradually replaced throughout life. Thus, resorbable biomaterials are based on the same principles of repair which have evolved over millions of years. Complications in the development of resorbable bioceramics are:

- Maintenance of strength and the stability of the interface during the degradation period and replacement by the natural host tissue
- Matching resorption rates to the repair rates of body tissues

Some materials dissolve too rapidly, and some too slowly. Because large quantities of material may be replaced, it is also essential that a resorbable biomaterial consists only of metabolically acceptable substances. Porous or particulate calcium phosphate ceramic materials such as tricalcium phosphate (TCP) are successful materials for resorbable hard tissue replacements when low loads are applied to the material.

Bioactive Materials. Another approach to solving problems of interfacial attachment is the use of bioactive materials. The concept of bioactive materials is intermediate between resorbable and bioinert. A bioactive material is one that elicits a specific biological response at the interface, which results in the formation of a bond between the tissues and the material. This concept has now been expanded to include a large number of bioactive materials with a wide range of rates of bonding

Table 1 Possible tissue responses to biomedical implants

Implant material characteristics	Tissue response
Toxic	Surrounding tissue dies
Nontoxic, biologically inactive	Fibrous tissue of variable thickness forms
Nontoxic, bioactive	Interfacial bond forms
Nontoxic, dissolves	Surrounding tissue replaces material

Table 2 Tissue attachment mechanisms for bioceramic implants

Type of attachment	Example
Dense, nonporous, nearly inert ceramics attached by bone growth into surface irregularities by cementing the device into the tissues, by press-fitting into a defect, or attachment via a sewing ring (morphological fixation)	Al_2O_3 (single-crystal and polycrystalline) LTI (low-temperature isotropic carbon)
For porous inert implants, bone ingrowth occurs, which mechanically attaches the bone to the materials (biological fixation)	Al_2O_3 (polycrystalline) Hydroxylapatite-coated porous metals
Dense, nonporous, surface-reactive ceramics, glasses, and glass-ceramics attach directly by chemical bonding with the bone (bioactive fixation)	Bioactive glasses Bioactive glass-ceramics Hydroxylapatite
Dense, nonporous (or porous), resorbable ceramics are designed to be slowly replaced by bone	Calcium sulfate (plaster of paris) Tricalcium phosphate Calcium phosphate salts

and thickness of interfacial bonding layers. They include bioactive glasses such as Bioglass; bioactive glass-ceramics such as Ceravital, A/W glass ceramic, or machinable glass-ceramics; dense hydroxylapatite such as Durapatite or Calcitite, or bioactive composites such as Palavital or stainless steel fiber-reinforced Bioglass.* All of the above bioactive materials form an interfacial bond with adjacent tissue. However, the time dependence of bonding, the strength of bond, the mechanism of bonding, and the thickness of the bonding zone differ for the various materials.

Relatively small changes in the composition of a biomaterial can dramatically affect whether it is bioinert, resorbable, or bioactive. These compositional effects on surface reactions are discussed in the section "Bioactive Glasses and Glass-Ceramics" in this article.

*Bioglass is a trademark of University of Florida; Ceravital and Palavital are trademarks of E. Leitz Company; Durapatite is a trademark of Sterling Winthrop; and Calcitite is a trademark of Calcitek.

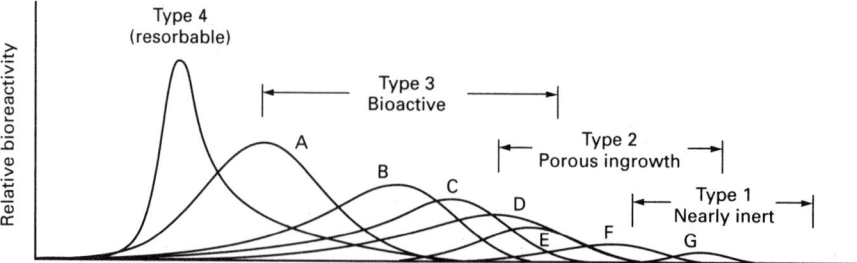

Fig 1 Relative rates of bioreactivity for ceramic implant materials. A, 45S5 Bioglass; B, KGS Ceravital; C, 55S4.3 Bioglass; D, A/W glass-ceramic; E, hydroxylapatite; F, KGX Ceravital; G, Al_2O_3, silicon nitride (see Table 4 for compositions)

Nearly Inert Crystalline Ceramics

High-density, high-purity ($>99.5\%$) Al_2O_3 is used in load-bearing hip prostheses and dental implants because of its combination of excellent corrosion resistance, good biocompatibility, high wear resistance, and high strength (Ref 3–5). Although some dental implants are single-crystal sapphire, most Al_2O_3 devices are very fine-grained polycrystalline α-Al_2O_3 produced by pressing and sintering at temperatures of 1600 to 1700 °C (2910 to 3090 °F). A very small amount ($<0.5\%$) of magnesia (MgO) is used as an aid to sintering and to limit grain growth during sintering.

Strength, fatigue resistance, and fracture toughness of polycrystalline α-Al_2O_3 are a function of grain size and purity. Al_2O_3 with an average grain size of <4 μm and $>99.7\%$ purity exhibits good flexural strength and excellent compressive strength. These and other physical properties are summarized in Table 3, along with the International Standards Organization (ISO) requirements for Al_2O_3 implants. Extensive testing has shown that Al_2O_3 implants which meet or exceed ISO standards have excellent resistance to dynamic and impact fatigue and also resist subcritical crack growth (Ref 6). An increase in average grain size to >7 μm can decrease mechanical properties by about 20%. High concentrations of sintering aids must be avoided because they remain in the grain boundaries and degrade fatigue resistance.

Methods exist for lifetime predictions and statistical design of proof tests for load-bearing ceramics. Applications of these techniques show that specific prosthesis load limits can be set for an Al_2O_3 device based upon the flexural strength of the material and its use environment. Load-bearing lifetimes of 30 years at 12,000 N (2700 lbf) loads have been predicted (Ref 5). Results from aging and fatigue studies show that it is essential that Al_2O_3 implants be produced at the highest possible standards of quality assurance, especially if they are to be used as orthopedic prostheses in younger patients.

Alumina has been used in orthopedic surgery for nearly 20 years, motivated largely by two factors:

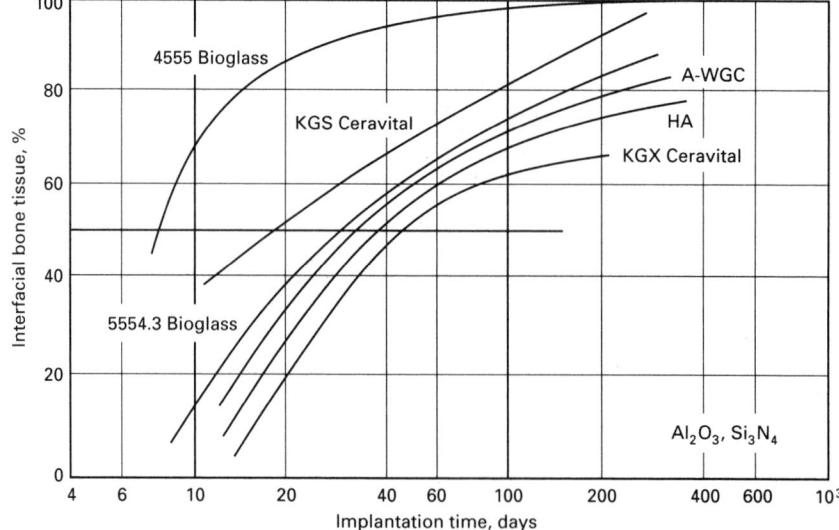

Fig 2 Time dependence of formation of bone bonding for the materials shown in Fig 1

- Excellent biocompatibility and very thin capsule formation, which permits cementless fixation of prostheses
- Exceptionally low coefficients of friction and wear rates

The superb tribologic properties (friction and wear) of Al_2O_3 occur only when the grains are very small (<4 μm) and have a very narrow size distribution. These conditions lead to very low surface roughness values ($R_a \leq$ 0.02 μm; see Table 3). If large grains are present, they can pull out and lead to very rapid wear of bearing surfaces due to local dry friction.

Alumina on alumina load-bearing wearing surfaces, such as in hip prostheses, must have a very high degree of sphericity produced by grinding and polishing the two mating surfaces together. An Al_2O_3 ball and socket in a hip prosthesis are polished together and used as a pair. The long-term coefficient of friction of an Al_2O_3-Al_2O_3 joint decreases with time and approaches the values of a normal joint. This leads to wear of Al_2O_3 on Al_2O_3 articulating surfaces being nearly 10 times lower than the metal-polyethylene surfaces (Fig 3).

Low wear rates have led to widespread use in Europe of Al_2O_3 noncemented cups press fitted into the acetabulum of the hip. The cups are stabilized by bone growth into grooves or around pegs. The mating femoral ball surface is also of Al_2O_3, which is bonded to a metallic stem. Long-term results in general are excellent, especially for younger patients. However, stress shielding due to the high elastic modulus of Al_2O_3 may be responsible for cancellous bone atrophy and loosening of the acetabular cup in older patients with senile osteoporosis or rheumatoid arthritis (Ref 5). Consequently, it is essential that the age of the patient, nature of the disease of the joint, and biomechanics of the repair be considered carefully before any prosthesis is used, including Al_2O_3 ceramics.

Other clinical applications of Al_2O_3 prostheses reviewed in Ref 3 include: knee

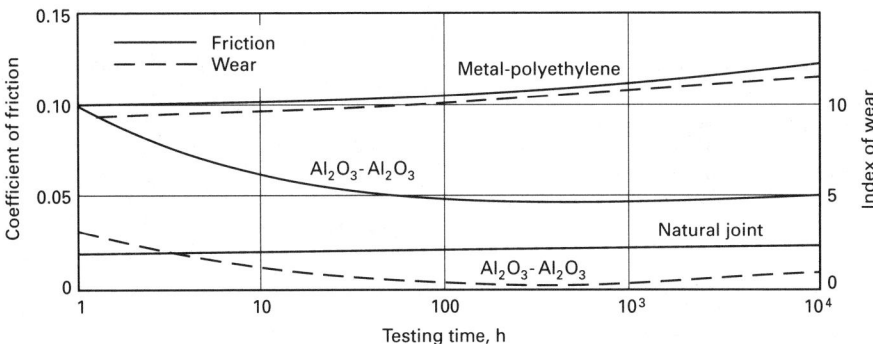

Fig 3 Friction and wear of Al_2O_3-Al_2O_3 hip joint compared to a metal-polyethylene prosthesis and a natural joint in *in-vitro* tests

prostheses, bone screws, alveolar ridge and maxillofacial reconstruction, ossicular bond substitutes, keratoprostheses (corneal replacements), segmental bone replacements, and blade and screw and postdental implants.

Porous Ceramics

The potential advantage offered by a porous ceramic implant is its inertness combined with the mechanical stability of the highly convoluted interface developed when bone grows into the pores of the ceramic. Mechanical requirements of prostheses, however, severely restrict the use of low-strength porous ceramics to nonload-bearing applications. Studies reviewed in Ref 1 and 3 have shown that when load bearing is not a primary requirement, nearly inert porous ceramics can provide a functional implant. When pore sizes exceed 100 μm, bone will grow within the interconnecting pore channels near the surface and maintain its vascularity and long-term viability. In this manner, the implant serves as a structural bridge and model or scaffold for bone formation. The microstructures of certain corals make an almost ideal investment material for the casting of structures with highly controlled pores sizes. White *et al.* (Ref 7) developed the replamineform process to duplicate the porous microstructure of corals that have a high degree of uniform pore size and interconnection. The first step is to machine the coral with proper microstructure into the desired shape. The most promising coral genus, *Porites*, has pores with a size range of 140 to 160 μm with all the pores interconnected. Another interesting coral genus, *Goniopora*, has a larger pore size, ranging from 200 to 1000 μm. The machined coral shape is fired to drive off carbon dioxide from the limestone ($CaCO_3$), forming calcia (CaO) while maintaining the microstructure of the original coral. The CaO structure serves as an investment material for forming the porous material. After the desired material is cast into the pores, the CaO is easily removed from the material by dissolving in dilute HCl.

The primary advantages of the replamineform process are that the pore size and microstructure are uniform and controlled and there

is complete interconnection of the pores. Replamineform porous materials of α-Al_2O_3, titanium dioxide (TiO_2), calcium phosphates, polyurethane, silicone rubber, polymethyl methacrylate (PMMA), and cobalt-base alloys have been used as bone implants, with the calcium phosphates being the most acceptable (Ref 7, 8).

Porous ceramic surfaces can also be prepared by mixing soluble metal or salt particles into the surface or using a foaming agent such as $CaCO_3$, which evolves gases during heating. Pore size and structure is determined by the size and shape of the soluble particles that are subsequently removed with a suitable etchant. The porous surface layer produced by this technique is an integral part of the underlying dense ceramic phase. Porous materials are weaker than the equivalent bulk form in proportion to the percentage of porosity. Much surface area is also exposed, so that the effects of the environment on decreasing the strength become much more important than for dense, nonporous materials. The environmental sensitivity of the high-strength ceramics and the loss of strength of porous ceramics with aging are reviewed in Ref 1.

Bioactive Glasses and Glass-Ceramics

Certain compositions of glasses, ceramics, glass-ceramics, and composites have been shown to bond to bone (Ref 1, 9–11). These materials have become known as bioactive ceramics. Some even more specialized compositions of bioactive glasses will bond to soft tissues as well as bone. A common characteristic of bioactive glasses and bioactive ceramics is a time-dependent, kinetic modification of the surface that occurs upon implantation. The surface forms a biologically active hydroxylapatite layer which provides the bonding interface with tissues.

Materials that are bioactive develop an adherent interface with tissues that resists substantial mechanical forces. In many cases the interfacial strength of adhesion is equivalent or greater than the cohesive strength of

Table 3 Physical characteristics of Al_2O_3 bioceramics

	High-alumina ceramics	ISO Standard 6474
Alumina content, %	<99.8	≥99.50
Density, g/cm³	>3.93	≥3.90
Average grain size, μm	3–6	<7
Surface roughness (R_a), μm	0.02	...
Vickers hardness	2300	>2000
Compressive strength, MPa (ksi)	4500 (653)	...
Bending strength, MPa (ksi) (after testing in Ringer's solution)	550 (80)	400 (58)
Young's modulus, GPa (psi × 10⁶)	380 (55.2)	...
Fracture toughness (K_{Ic}), MPa√m (ksi√in.)	5–6 (4.5–5.5)	...

the implant material or the tissue bonded to the bioactive implant.

Glasses. Bonding to bone was first demonstrated for a certain compositional range of bioactive glasses which contained silica (SiO_2), soda (Na_2O), calcia, and phosphorus oxide (P_2O_5). There were three key compositional features to these glasses that distinguished them from traditional soda-lime-silica glasses: 1) less than 60 mol% SiO_2, 2) high Na_2O and CaO content, and 3) high CaO/P_2O_5 ratio. These compositional features make the surface highly reactive when exposed to an aqueous medium.

Many bioactive silica glasses are based upon the formula called "45S5," signifying 45 wt% SiO_2 (S = the network former) and 5:1 molar ratio of CaO to P_2O_5. Glasses with lower molar ratios of CaO to P_2O_5 do not bond to bone. However, substitutions in the 45S5 formula of 5 to 15 wt% boron oxide (B_2O_3) for SiO_2 or 12.5 wt% calcium fluoride (CaF_2) for CaO, or ceraming the various bioactive glass compositions to form glass-ceramics, has no measurable effect on the ability of the material to form a bone bond. However, addition of as little as 3 wt% Al_2O_3 to the 45S5 formula prevents bonding.

Glass-Ceramics. Gross *et al.* (Ref 10) have shown that a range of low-alkali (0 to 5 wt%) bioactive silica glass-ceramics (Ceravital) also bonds to bone. They find that small additions of Al_2O_3, tantalum, titanium, or zirconium inhibit bone bonding. A two-phase silica-phosphate glass-ceramic composed of apatite [$Ca_{10}(PO_4)_6(OH_1F_2)$] and wollastonite ($CaO·SiO_2$) crystals and a residual silica glassy matrix termed A/W glass-ceramic (Ref 11) also bonds with bone. Addition of Al_2O_3 or TiO_2 to the A/W glass-ceramic inhibits bone bonding, whereas incorporation of a second phosphate phase, B-whitlockite ($3CaO·P_2O_5$) does not.

Another multiphase bioactive phosphosilicate containing phlogopite [$(Na,K)Mg_3$ $(AlSi_3O_{10})F_2$] and apatite crystals bonds to bone even though Al_2O_3 is present in the composition (Ref 12). However, the Al^{3+} ions are incorporated within the crystal phase and

do not alter the surface reaction kinetics of the material. Compositions of these various bioactive glasses and glass-ceramics are compared in Table 4.

Surfaces characteristic of bioactive glasses and glass-ceramics form a dual protective film rich in CaO and P_2O_5 on top of an alkali-depleted SiO_2-rich film. When multivalent cations such as Al^{3+}, Fe^{3+}, or Ti^{4+} are present in the glass or in the solution, multiple layers form on the glass as the saturation of each cationic complex is exceeded. This leads to a surface which does not bond to tissue.

A general equation describes the overall rate of change of glass surfaces and gives rise to the interfacial reaction rates and time dependence of bone bonding profiles shown in Fig 2. The reaction rate R depends on at least five terms (for a single-phase glass). For polycrystalline ceramics, or glass-ceramics, which have several phases in their microstructures, each phase will have a characteristic reaction rate, R_i, which must be multiplied times its areal fraction exposed to tissue in order to describe overall kinetics of bonding:

$$R = \underset{\text{Stage 1}}{-k_1t^{0.5}} \underset{\text{Stage 2}}{- k_2t^{1.0}} \underset{\text{Stage 3}}{+ k_3t^{1.0}} \underset{\text{Stage 4}}{+ k_4t^y} \underset{\text{Stage 5}}{+ t_5}$$

(Eq 1)

The first term describes the rate of alkali extraction from the glass and is called a Stage 1 reaction.

In Stage 1, the initial or primary stage of attack is a process which involves ion exchange between alkali ions from the glass and hydrogen ions from the solution, during which the remaining constituents of the glass are not altered. During Stage 1, the rate of alkali extraction from the glass is parabolic in character.

Stage 2 is interfacial network dissolution whereby siloxane bonds are broken, forming a large concentration of surface silanol groups. Stage 2 kinetics are linear. A resorbable glass experiences a combination of Stage 1 and Stage 2 attacks.

Stages 3 and 4 result in a glass surface with a dual protective film. The thickness of the secondary films can vary considerably, from

as little as 0.01 μm for Al_2O_3-SiO_2 rich layers on inactive glasses to as much as 30 μm for CaO-P_2O_5-rich layers on bioactive glasses. The formation of the dual films is due to a combination of the repolymerization of SiO_2 on the glass surface (Stage 3) by the condensation of the silanols (Si—OH) formed from the Stage 1 and 2 reactions, for example:

$$Si—OH + OH—Si \rightarrow Si—O—Si + H_2O \quad \text{(Eq 2)}$$

Stage 3 protects the glass surface. The SiO_2 polymerization reaction contributes to the enrichment of surface SiO_2 characteristic of bone-bonding glasses. It is described by the third term in Eq 1. This reaction is interface controlled with a time dependence of $+k_3t^{1.0}$. The interfacial thickness of the most reactive bioactive glasses is largely due to this reaction. The fourth term in Eq 1, $+k_4t^y$ (Stage 4) describes the precipitation of an amorphous calcium phosphate film which is characteristic of bioactive glasses.

In Stage 5, the amorphous calcium phosphate film crystallizes to form hydroxylapatite crystals. The calcium and phosphate ions in the glass or glass-ceramic provide the nucleation sites for crystallization. Carbonate anions (CO_3^{2-}) substitute for OH^- in the apatite crystal structure to form a hydroxyl-carbonate apatite (HCA) similar to that found in living bone. Incorporation of calcium fluoride (CaF_2) in the glass results in incorporation of fluoride ions in the apatite, resulting in a hydroxyl carbonate fluorapatite which matches dental enamel. Crystallization of HCA occurs around collagen fibrils present at the implant interface and results in interface bonding.

In order for the material to be bioactive and form an interfacial bond, the kinetics of reaction in Eq 1, and especially the rates of Stages 4 and 5, must match the rate of biomineralization that normally occurs *in vivo*. If the rates in Eq 1 are too rapid, the implant is resorbable; if the rates are too slow the implant is not bioactive.

By changing the compositionally controlled reaction kinetics (Eq 1), the rates of formation of hard tissue at a bioactive implant

Table 4 Compositions and structures of bioactive glasses and glass-ceramics

All compositions in wt%

Material / Constituent	45S5 Bioglass	45S5F Bioglass	45S5.4F Bioglass	40S5B5 Bioglass	52S4.6 Bioglass	55S4.3 Bioglass	KGC Ceravital	KGS Ceravital	KGy213 Ceravital	A-W-GC	MB-GC
SiO_2	45	45	45	40	52	55	46.2	46	38	34.2	19–52
P_2O_5	6	6	6	6	6	6	...	...	...	16.3	4–24
CaO	24.5	12.25	14.7	24.5	21	19.5	20.2	33	31	44.9	9–3
$Ca(PO_3)_2$	...	...	...	...	...	...	25.5	16	13.5	...	...
CaF_2	...	12.25	9.8	...	...	...	...	...	...	0.5	...
MgO	...	...	...	...	...	...	2.9	...	...	4.6	5–15
MgF_2	...	...	...	...	...	...	...	...	...	...	...
Na_2O	24.5	24.5	24.5	24.5	21	19.5	4.8	5	4	...	3–5
K_2O	...	...	...	...	...	...	0.4	...	...	...	3–5
Al_2O_3	...	...	...	...	...	...	...	...	7	...	12–33
B_2O_3	...	...	...	5	...	...	...	...	...	...	...
Ta_2O_5/TiO_2	...	...	...	...	...	...	...	...	6.5	...	...
Structure	Glass	Glass	Glass	Glass	Glass	...	Glass-ceramic	Glass-ceramic	...	Glass-ceramic	Glass-ceramic

interface can be altered, as shown in Fig 2. Thus, the level of bioactivity of a material $t_{0.5bb}$ can be related to the time for more than 50% of the interface to be bonded: Index of Bioactivity $I_B = (100/t_{0.5bb})$. It is necessary to impose a 50% bonding criterion for an index of bioactivity because the interface between an implant and bone is irregular (Ref 10). The initial concentration of cell at the interface varies as a function of the frit of the implant and the condition of the bony defect. Consequently, all bioactive implants require an incubation period before bone proliferates and bonds, with the length of the incubation period varying over a wide range depending on composition.

Bioactive implants with intermediate I_B values do not develop a stable soft tissue bond; instead, the fibrous interface progressively mineralizes to form bone. Consequently, there appears to be a critical boundary beyond which bioactivity is restricted to stable bone bonding. Inside the critical boundary, the bioactivity includes both stable bone *and* soft tissue bonding depending on the progenitor stem cells in contact with the implant.

The thickness of the bonding zone between a bioactive implant and bone is proportional to its index of bioactivity I_B. The failure strength of a bioactively fixed bond appears to be inversely dependent on the thickness of the bonding zone. For example, 45S5 Bioglass with a very high I_B develops a gel bonding layer 200 μm thick which has a relatively low shear strength. In contrast, A/W glass-ceramic, with an intermediate I_B value, has a bonding interface in the range of 10 to 20 μm and a very high resistance to shear. Thus, the interfacial bonding strength appears to be optimum for I_B values of ~4. However, it is important to recognize that the interfacial area for bonding is time-dependent. Therefore, interfacial strength is time-dependent and is a function of morphological factors such as the change in interfacial area with time, progressive mineralization of the interfacial tissues, and resulting increase of elastic modulus of the interfacial bond as well as shear strength per unit of bonded area. A comparison of the increase in interfacial bond strength of bioactive fixation of implants bonded to bone with other types of fixation is given in Fig 4.

Clinical applications of bioactive glasses and glass-ceramics are reviewed in Ref 9 to 11 and shown in Table 5. The eight-year history of successful use of Ceravital glass-ceramics in middle ear surgery (Ref 11) is especially encouraging as is the four-year use of A-W glass-ceramic in vertebral surgery (Ref 11) and five-year use of 45S5 Bioglass in endosseous ridge maintenance.

Calcium Phosphate Ceramics

Calcium phosphate based bioceramics have been in use in medicine and dentistry for nearly twenty years, as reviewed in Ref 3, 14 to 16.

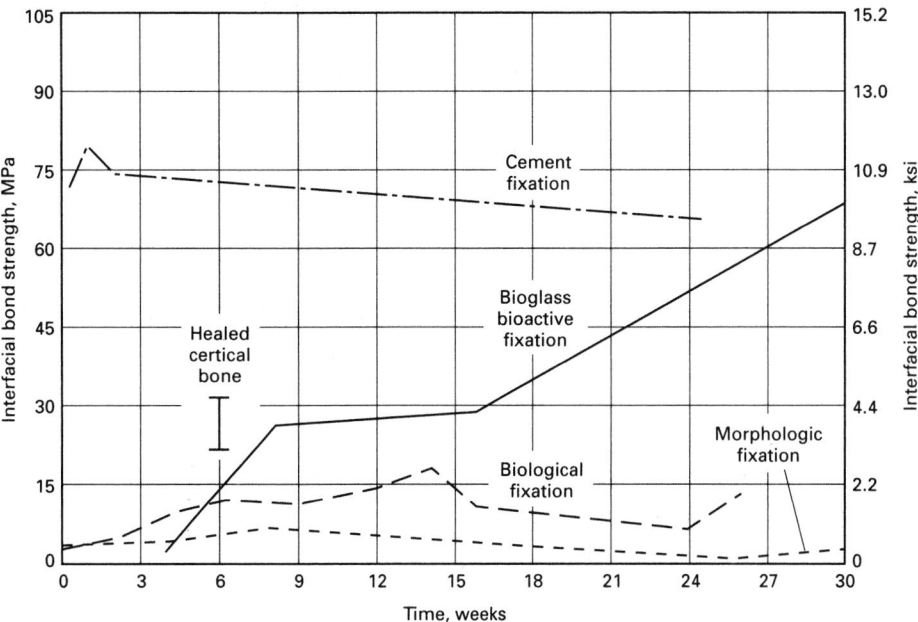

Fig 4 Time dependence of interfacial bond strength of various fixation systems in bone. Source: Ref 13

Applications include dental implants, periodontal treatment, alveolar ridge augmentation, orthopedics, maxillofacial surgery, and

Table 5 Present uses of bioceramics

Application	Material(s) used
Orthopedic load-bearing applications	Al$_2$O$_3$
Coatings for chemical bonding (orthopedic, dental and maxillary prosthetics)	HA, surface-active glasses and glass-ceramics
Dental implants	Al$_2$O$_3$, HA, surface-active glasses
Alveolar ridge augmentations	Al$_2$O$_3$, HA, HA-autogenous bone composite, HA-PLA composite, surface-active glasses
Otolaryngological applications	Al$_2$O$_3$, HA, surface-active glasses and glass-ceramics
Artificial tendons and ligaments	PLA-carbon fiber composites
Coatings for tissue ingrowth (cardiovascular, orthopedic, dental, and maxillofacial prosthetics)	Al$_2$O$_3$
Temporary bone space fillers	Trisodium phosphate, calcium and phosphate salts
Periodontal pocket obliteration	HA, HA-PLA composites, trisodium phosphate, calcium and phosphate salts, surface-active glasses
Maxillofacial reconstruction	Al$_2$O$_3$, HA, HA-PLA composites, surface-active glasses
Percutaneous access devices	Bioactive glass-ceramics
Orthopedic fixation devices	PLA-carbon fibers, PLA-calcium/phosphorus-base glass fibers

otolaryngology (Table 5). Different phases of calcium phosphate ceramics are used depending upon whether a resorbable or bioactive material is desired.

The stable phases of calcium phosphate ceramics depend considerably upon temperature and the presence of water, either during processing or in the use environment. At body temperature, only two calcium phosphates are stable in contact with aqueous media, such as body fluids; at pH < 4.2 the stable phase is $CaHPO_4 \cdot 2H_2O$ (dicalciumphosphate or brushite), while at pH ≥ 4.2 the stable phase is $Ca_{10}(PO_4)_6(OH)_2$ (hydroxylapatite, HA). At higher temperatures, other phases, such as $Ca_3(PO_4)_2$ (β-tricalciumphosphate, C_3P, or TCP) and $Ca_4P_2O_9$ (tetracalcium phosphate, C_4P) are present. The unhydrated high-temperature calcium phosphate phases interact with water, or body fluids, at 37 °C (98.6 °F) to form hydroxylapatite. The HA forms on exposed surfaces of TCP by the following reaction:

$$4Ca_3(PO_4)_2 + 2H_2O \rightarrow$$
$$\text{(Solid)}$$
$$Ca_{10}(PO_4)_6(OH) + 2Ca^{2+} + 2HPO_4^{2-} \quad \text{(Eq 3)}$$
$$\text{(Surface)}$$

Thus, the solubility of a TCP surface approaches the solubility of HA and decreases the pH of the solution, which further increases the solubility of TCP and enhances resorption. The presence of micropores in the sintered material can increase the solubility of these phases.

Sintering of calcium phosphate ceramics usually occurs in the range of 1000 to 1500 °C (1830 to 2730 °F) following compaction of the powder into the desired shape. The phases formed at high temperature depend not only on temperature but also the partial pressure of water in the sintering atmosphere. This is because with water present HA can be formed and is a stable phase up to 1360 °C (2480 °F). Without water, C_4P and C_3P are the stable phases. The temperature range of stability of HA increases with the partial pressure of water as does the rate of phase transitions of C_3P or C_4P to HA. Due to kinetics barriers which affect the rates of formation of the stable calcium phosphate phases, it is often difficult to predict the volume fraction of high-temperature phases that are formed during sintering and their relative stability when cooled to room temperature.

Starting powders can be made by mixing in an aqueous solution the appropriate molar ratios of calcium nitrate and ammonium phosphate, which yields a precipitate of stoichiometric HA. The Ca^{2+}, PO_4^{3-}, and OH^- ions can be replaced by other ions during processing or in physiological surroundings; for example, fluorapatite, $Ca_{10}(PO_4)_6(OH)_{2-x}$ with $0 < x < 2$; carbonate apatite, $Ca_{10}(PO_4)_6(OH)_{2-2x}(CO_3)_x$ or $Ca_{10-x+y}(PO_4)_{6-x}$ $(OH)_{2-x-2y}$, where $0 < x < 2$ and $0 < y < \frac{1}{2}x$ can be formed. Fluorapatite is found in dental enamel, and hydroxyl-carbonate apatite is present in bone. For discussion of the structure of these complex crystals, see Ref 16.

The mechanical behavior of calcium phosphate ceramics strongly influences their application as implants. Tensile and compressive strength and fatigue resistance depend on the total volume of porosity. Porosity can be in the form of micropores (<1 μm in diameter, due to incomplete sintering) or macropores (>100 μm in diameter, created to permit bone growth). The dependence of compressive strength σ_c and total pore volume V_p is described in megapascals (Ref 17):

$$\sigma_c = 700 \exp{-5V_p} \qquad \text{(Eq 4)}$$

where V_p is between 0 and 0.5. Tensile strength σ_t (in megapascals) depends greatly on the volume fraction of microporosity V_m:

$$\sigma_t = 220 \exp{-20V_m} \qquad \text{(Eq 5)}$$

The Weibull factor n of hydroxylapatite implants is low ($n = 12$) in physiological solutions, which indicates low reliability under tensile loads. Consequently, in clinical practice calcium phosphate bioceramics should be used as:

- Powders
- Small, unloaded implants such as in the middle ear
- With reinforcing metal posts, as in dental implants
- As coatings (for example, composites)
- As low-loaded porous implants where bone growth acts as a reinforcing phase

The bonding mechanisms of dense hydroxylapatite (HA) implants appear to be very different from that described above for bioactive glasses. Evidence for the bonding process for HA implants is described in Ref 15. A cellular bone matrix from differen-

tiated osteoblasts appears at the surface, producing a narrow, amorphous, electron-dense band only 3 to 5 μm wide. Between this area and the cells, collagen bundles are seen. Bone mineral crystals have been identified in this amorphous area. As the site matures, the bonding zone shrinks to a depth of only 0.05 to 0.2 μm. The result is normal bone attached through a thin epitaxial bonding layer to the bulk implant. Transmission electron microscope (TEM) image analysis of dense HA bone interfaces has shown an almost perfect epitaxial alignment of the growing bone crystallites with the apatite crystals in the implant.

A consequence of this ultrathin bonding zone is a very high gradient in elastic modulus at the bonding interface between HA and bone. This is one of the major differences between the bioactive apatites and the bioactive glasses and glass-ceramics. The implications of this difference on the implant interfacial response is discussed in Ref 1.

Resorbable Calcium Phosphates

Resorption or biodegradation of calcium phosphate ceramics is caused by: physiochemical dissolution, which depends on the solubility product of the material and local pH of its environment; physical disintegration into small particles due to preferential chemical attack of grain boundaries; and biological factors, such as phagocytosis which causes a decrease in local pH concentrations.

All calcium phosphate ceramics biodegrade at increasing rates in the following order: $TCP > \beta\text{-}TCP \gg HA$. The rate of biodegradation increases as:

- Surface area increases (powders > porous solid > dense solid)
- Crystallinity decreases
- Crystal perfection decreases
- Crystal and grain size decrease
- Ionic substitutions of Co_3^{2-}, Mg^{2+}, and Sr^{2+} in Ha increase

Factors which tend to decrease rate of biodegradation include:

- F^- substitution in HA
- Mg^{2+} substitution in β-TCP
- Lower β-TCP/HA ratios in biphasic calcium phosphates

Carbon-Base Implant Materials

Primarily three types of carbon are used in biomedical devices: the low-temperature isotropic (LTI) variety of pyrolytic carbon, glassy (vitreous) carbon, and the ultralow-temperature isotropic (ULTI) form of vapor-deposited carbon (Ref 18, 19).

These carbon materials in use are integral and monolithic materials (glassy carbon and LTI carbon) or impermeable thin coatings

(ULTI carbon). These three forms do not suffer from the integrity problems typical of the other available carbon materials. With the exception of the LTI carbons codeposited with silicon, all the clinical carbon materials are pure elemental carbon. Up to 20 wt% silicon has been added to LTI carbon without significantly affecting the biocompatibility of the material. The composition, structure, and fabrication of the three clinically relevant carbons are uniquely compared with the more common naturally occurring form of carbon (graphite) and other industrial forms produced from pure elemental carbon.

Subcrystalline Forms. The LTI, ULTI, and glassy carbons are subcrystalline forms and represent a lower degree of crystal perfection. There is no order between the layers as there is in graphite, so the crystal structure of these carbons is two-dimensional.

The densities of these carbons range between about 1.4 and 2.1 g/cm^3. High-density LTI carbons are the strongest bulk forms of carbon, and their strength can be further increased by adding silicon. ULTI carbon can also be produced with high densities and strengths, but it is available only as a thin coating (0.1 to 1 μm) of pure carbon. Glassy carbon is inherently a low-density material and as such is weak. Its strength cannot be increased through processing. Processing of all three types of medical carbons is discussed in Ref 19.

The mechanical properties of the various carbons are intimately related to their microstructures. In an isotropic carbon, it is possible to generate materials with low (20 GPa, or 2.9×10^6 psi) elastic moduli and high flexural strength (275 to 620 MPa, or 40 to 90 ksi). There are many benefits as a result of this combination of properties. Large strains ($\sim 2\%$) are possible without fracture. Carbon materials are extremely tough compared with ceramics such as Al_2O_3. The energy to fracture for LTI carbon is approximately 5.5 MJ/m^3 (115×10^3 ft · lbf/ft^3) compared with 0.18 MJ/m^3 (3.8×10^3 ft · lbf/ft^3) for Al_2O_3; that is, the carbon is more than 25 times as tough. The strain to fracture for the vapor-deposited carbons is greater than 5.0%, making it feasible to coat highly flexible polymeric materials such as polyethylene, polyesters, and nylon without fear of fracturing the coating when the substrate is flexed. For comparison, the strain to failure of Al_2O_3 is approximately 0.1%, about one fiftieth that of the ULTI carbons.

These carbon materials have extremely good wear resistance, some of which can be attributed to their toughness—that is, their capacity to sustain large local elastic strains under concentrated or point loading without galling or incurring surface damage.

The bond strength of the ULTI carbon to stainless steel and to Ti-6Al-4V exceeds 70 MPa (10.2 ksi) as measured with a thin-film adhesion tester. This excellent bond is, in part, achieved through the formation of interfacial

Table 6 Properties of biomedical carbons

Property	Material		
	Low to high-density LTI carbon	Silicon-alloyed LTI carbon	ULTI carbon
Density, g/cm^3	1.5–2.2	2–2.2	1.5–2.2
Crystallite size (L_c), nm	3–4	3–4	0.8–1.5
Flexural strength, MPa (ksi)	275–550 (40–80)	550–620 (80–90)	345 to >690 (50 to >100)
Young's modulus, GPa (psi × 10^6)	17–28 (2.5–4.1)	28–41 (4.1–5.9)	14–21 (2–3)
Fatigue limit/flexural strength ratio	1	1	1
Poisson's ratio	0.2	0.2	
Diamond pyramid hardness	150–250	230–370	150–250
Thermal expansion coefficient, 10^{-6}/K	4–6	5	
Silicon content, wt%	0	5–12	0
Strain energy to fracture, MJ/m^3 (ft·lbf × 10^3/ft^3)	2.7–5.5 (56.4–115)	5.5 (115)	9.9 (207)
Strain to fracture, %	1.6–2.1	2.0	>5.0
Impurity level, ppm	<100	<100	<100

Table 7 Successful applications of glassy, LTI, and vapor-deposited ULTI carbons

Application	Material
Mitral and aortic heart valves	LTI
Dacron and Teflon heart valve sewing rings	ULTI
Blood access device	LTI/titanium
Dacron and Teflon vascular grafts	ULTI
Dacron, Teflon, and polypropylene septum and aneurism patches	ULTI
Pacemaker electrodes	Porous glassy carbon-ULTI-coated porous titanium
Blood oxygenator microporous membranes	ULTI
Otologic vent tubes	LTI
Subperiosteal dental implant frames	ULTI
Dental endosseous root form and blade implants	LTI
Dacron-reinforced polyurethane aloplastic trays for alveolar ridge augmentation	ULTI
Percutaneous electrical connectors	LTI
Hand joints	LTI

Source: Ref 17

carbides. The ULTI carbon coating generally has a lower bond strength with materials that do not form carbides.

Another unique characteristic of the carbons is that they do not fail in fatigue; unlike metals, ultimate strength does not degrade with cyclical loading. The fatigue strength of these carbon structures is equal to the single-cycle fracture strength. It appears that unlike other crystalline solids, these forms of carbon do not contain mobile defects which at normal temperatures can move and provide a mechanism for the initiation of a fatigue crack.

Some of the more important known properties of biomedical carbons are listed in Table 6.

The most important biomedical application is the cardiovascular area, such as in heart valves, the first of which was implanted in 1969. Since then, more than 600,000 pyrolytic carbon valve components have been produced for implantation. The cardiovascular application is particularly demanding. Early attempts at producing successful heart valves failed because the materials used were either thrombogenic or suffered from high wear and mechanical failure. Thrombus, wear, distortion, and biodegradation have been virtually eliminated because of the biocompatibility and mechanical durability of pyrolytic carbon, clearly establishing it as the material of choice for heart valves.

Carbon surfaces are not only thromboresistant but also compatible with the cellular elements of blood. The materials do not influence plasma proteins or alter the activity of plasma enzymes. In fact, one of the proposed explanations for the blood compatibility of these materials is that they adsorb blood proteins on their surface without altering them.

Table 7 summarizes the successful uses of glassy, LTI, and ULTI carbons in various medical areas.

REFERENCES

1. L.L. Hench and E.C. Ethridge, *Biomaterials: An Interfacial Approach*, Academic Press, 1982
2. J.D. Preston, Properties in Dental Ceramics, *Proceedings of the IV International Symposium on Dental Materials*, Quintessa Publishers, 1988
3. S.F. Hulbert, J.C. Bokros, L.L. Hench, J. Wilson, and G. Heimke, *High Tech Ceramics*, P. Vincenzini, Ed., Elsevier Science, Amsterdam, 1987, p 189–213
4. P.M. Boutin, *Ceramics in Clinical Applications*, P. Vincenzini, Ed., Elsevier, 1987, p 297
5. P. Christel, A. Meunier, J.M. Dorlot, J.M. Crolet, J. Witvolet, L. Sedel, and P. Boritin, Biomechanical Compatibility and Design of Ceramic Implants for Orthopedic Surgery, *Bioceramics: Material Characteristics Versus In-Vivo Behavior*, P. Ducheyne and J. Lemons, Ed., Vol 523, *Annals of NY Academy of Science*, 1988, p 234
6. E. Dörre and W. Dawihl, Ceramic Hip Endoprostheses, *Mechanical Properties of Biomaterials*, G.W. Hastings and D.F. Williams, Ed., John Wiley & Sons, 1980, p 113–127
7. E.W. White, J.N. Webber, D.M. Roy, E.L. Owen, R.T. Chiroff, and R.A. White, *J. Biomed. Mater. Res. Symp.*, Vol 6, 1975, p 23–27
8. R.T. Chiroff, E.W. White, J.N. Webber, and D. Roy, *J. Biomed. Mater. Res. Symp.*, Vol 6 (No. 29), 1975, p 45
9. L.L. Hench and J.W. Wilson, Surface-Active Biomaterials, *Science*, Vol 226, 1984, p 630
10. V. Gross, R. Kinne, H.J. Schmitz, and V. Strunz, *CRC Crit. Rev. Biocompatibility*, Vol 4 (No. 2), 1988
11. T. Yamamuro, L.L. Hench, and J. Wilson, *Handbook on Bioactive Ceramics*, Vol I, *Bioactive Glasses and Glass-Ceramics*, Vol II, *Calcium-Phosphate Ceramics*, CRC Press, 1990
12. W. Hohland, W. Vogel, K. Naurnann, and J. Gummel, Interface Reactions between Machinable Bioactive Glass-Ceramics and Bond, *J. Biomed. Mater. Res.*, Vol 19, 1985, p 303
13. L.L. Hench, Bioactive Ceramics, *Bioceramics: Materials Characteristics Versus In-Vivo Behavior*, P. Ducheyne and J. Lemons, Ed., Vol 523, *Annals of NY Academy of Science*, 1988, p 54
14. K. de Groot, *Bioceramics of Calcium-Phosphate*, CRC Press, 1983
15. M. Jarcho, Calcium Phosphate Ceramics as Hard Tissue Prosthetics, *Clin. Orthop. Relat. Res.*, Vol 157, 1981, p 259
16. R.Z. Le Geros, Calcium Phosphate Materials in Restorative Dentistry: A Review, *Adv. Dent. Res.*, Vol 2, 1988, p 164–180
17. K. de Groot, C.P.A.T. Klein, J.G.C. Wolke, and J. de Blieck-Hogervorst, Chapt. 1, *Handbook on Bioactive Ceramics: Vol II*, CRC Press, 1990
18. J.C. Bokros, Carbon Biomedical Devices, *Carbon*, Vol 15, 1977, p 355–371
19. A.D. Haubold, R.A. Yapp, and J.C. Bokros, *Concise Encyclopedia of Medical and Dental Materials*, D. Williams, Ed., Pergamon Press, 1990

Applications for Glasses

Co-Chairmen: Thomas P. Seward III and Paul S. Danielson, Corning Incorporated

Introduction

GLASS in its variety of forms is almost ubiquitous, yet its use is oftentimes unrecognized as many of the example applications in this Section will show. The applications range from the comparatively unsophisticated, not unlike the drinking vessels and containers used thousands of years ago, to the high technology of opto-electronics, telecommunications, magnetic disc recording, and biological implants.

In terms of sales, the glass business in the United States may be divided into four roughly equal parts, each of which represents several billions of dollars annually. (Ref *Ceramic Industry*, Vol 133 (No. 2), Aug 1989, p 33–36). The first three categories are:

- Flat glass for architectural and transportation windows
- Container glass for food, beverages, and cosmetics
- Glass fibers for insulation, reinforcement, and textiles

The fourth part is comprised of all the remaining glass applications, such as television, lighting, consumer products, electronic glasses, and so on. The term "specialty glasses" is often applied to this last group.

In terms of composition type, the clear majority of glass products, particularly those that may be regarded as commodities, are represented by soda-lime-silica glasses. Typical products include most of the flat glass and containers, much of the fiberglass, and many of the lighting products. This workhorse glass type varies only slightly in composition from one application to the next, and is more likely to be influenced by regional raw material price and availability than other factors. As a result, the compositional and, therefore, physical and chemical property diversity of interest to the applications engineer will most likely be found among the more highly engineered glasses required for the specialized products.

In this Section, glass and glass-ceramic applications along with the key properties which influence the materials selection process will be described. Little attention will be paid to the processes that go into making the materials, since these have been dealt with in earlier Sections of this Volume (see the Section on "Glass Processing"). Exceptions will be in areas where the applications and the processes are intimately related.

Selection of a material, be it glass or glass-ceramic, generally depends on the properties needed for the end-use application. These properties often include mechanical (strength, toughness, and hardness), chem-ical (acid, base, and weather durability) and physical (thermal expansion, softening temperature, density, refractive index, light transmittance as a function of wavelength, and light scattering) properties. Often, the properties which influence materials selection depend on the manufacturing process, as well as on the end use. Properties affecting the manufacturability of the material include batch melting characteristics, liquidus temperature, viscosity as a function of temperature, visible light transmittance, and near infrared transmittance. Cost-driven compromises are often necessary.

Some applications require precise control of one or more mechanical, physical, or chemical property. Other applications require very little control, broad ranges of properties being acceptable. In many cases, close control of composition and properties is required in order to maintain process control during manufacture. Again, cost-driven compromises are often necessary.

Article authors will briefly describe their topic applications. They will then address the key processing or application requirements which influence the choice of glass or glass-ceramic composition. Typical example properties will be given. The authors will discuss possible tradeoffs in properties that may be required and any secondary processing that may be involved—for example, grinding, polishing, coating, and/or sizing. In some cases, potentially competitive non-glass materials will be mentioned with their relative advantages and disadvantages.

The Section is intended to provide the applications engineer with examples of the wide applicability of glass and glass-ceramics and with examples of the types of materials (glass or glass-ceramic) most suitable to those applications. By comparing properties with applications, it should be possible for the engineer to at least identify a family of glasses which might suit his or her needs, and perhaps even to find a specific composition which is currently available from one or more suppliers or manufacturers worldwide.

To list all commercially available glasses, along with only a few of their possible applications, is well beyond the scope of this Handbook. The list of useful compositions is continually growing. Oftentimes, new compositions must be developed for specific applications. This need has sometimes led non-glass manufacturers to develop their own glass chemistry expertise and even to start producing their own proprietary glasses.

Aerospace and Military Applications

Herbert A. Miska, Corning Incorporated

GLASS and glass-ceramics products are encountered in some of the most unusual and often most severe applications in the aerospace and defense industries. The applications range from the seemingly mundane such as aircraft windshields to the exotic, such as giant space telescope mirrors and the windows for manned spacecraft. Environmental and use conditions are typically extreme, and the need for sophisticated design to ensure minimum weight with safety requires that the designer have a very thorough understanding of the properties of these materials. Often the design approach and philosophy employed with conventional engineering materials do not apply.

This article describes a number of very diverse applications and attempts to show how the wide range of key mechanical, thermal, electrical, and optical properties of specialty glasses and glass-ceramics can be used to optimum advantage. Special processes are discussed to the extent that they are uniquely developed for unusual applications.

Mirrors for Space

Telescope mirror substrates, whether they are used in ground-based instruments or in space-based applications, have some common requirements, such as extreme dimensional stability and the ability to be finished to an ultrasmooth surface. The mirrors often also must be able to be fabricated in very large sizes. When they are to be used in space, weight is another extremely important attribute. To the extent that light weight also implies a low thermal mass and therefore the ability to reach thermal equilibrium with the environment faster, it is also important for ground-based instruments. Light weight also allows a less massive and therefore less expensive support structure, but nowhere is it more important than in space applications.

Two distinctly different materials can be made to meet the demanding requirements for space-born mirror substrates: titania-doped silica glass (Corning Code 7971 ULE) and certain lithium aluminosilicate glass-ceramics (Schott Zerodur). Important properties of these two types of mirror materials are given in Table 1.

Ultralow-Expansion Glass. Corning's ultralow-expansion (ULE) glass achieves a zero coefficient of thermal expansion over the 5 to 35 °C (40 to 95 °F) temperature range by means of the addition of 7.5% titania to fused silica. It is manufactured using a chemical vapor deposition (CVD) process as opposed to melting from a batch. Extreme purity and composition control can be achieved, which ensures that the coefficient of thermal expansion, which is a direct function of the amount of titania in the glass, can be very exactly controlled. Values of zero $\pm 3 \times 10^{-8}/°C$ ($\pm 1.7 \times 10^{-8}/°F$) over a temperature range of 5 to 35 °C (40 to 95 °F) are achieved.

A zero-expansion glass, as opposed to a glass-ceramic, also offers the ability to be fabricated into complex configurations by fusion bonding various components. This operation is analogous to welding in metals and can be carried out in a room-temperature environment. The residual stresses due to nonuniform cooling resulting from this operation are so low that they do not lead to fractures at the time of fabrication. Complex cores can be fabricated and then annealed in their entirety after completion.

Lightweight mirrors for space applications typically have a "honeycomb" configuration in which a hexagon, square, or triangular cell core, made by the fusion seal method, is fitted with top and bottom plates. These can be attached either by fusion or by the use of a devitrifying solder glass (frit) that matches the expansion of the glass (Fig 1). The greatest weight savings are achieved when the plates are assembled to the core using the frit technique. Assemblies weighing less than 10% of the equivalent solid volume have been successfully made. Figure 2 shows the lightweight blank for the primary mirror of the Hubble telescope.

Other designs range from thin meniscus plates to machined core light-weights. Thin meniscus mirrors are mounted on a series of actuators which control the optical figure in the finished instrument. Machined cores for honeycomb configurations offer great flexibility in design but are typically not as light as fused cores.

Thermal expansion characterization is critical for materials which are to be used in astronomical and other space-based instruments. Ultralow-expansion glass provides the opportunity to take advantage of a unique relationship of properties and composition. Both sonic velocity in the material and the coefficient of thermal expansion (CTE) have been found to be directly related to titanium content. Expansivity can therefore be indirectly measured by testing the ultrasound velocity in the material nondestructively. Large pieces can be fully characterized for their exact CTE, and assemblies can be optimized to eliminate any unfavorable residual stresses and distortions.

Glass-ceramic materials are made by conventional glass melting and forming techniques from compositions that contain nucleating agents such as titania. A secondary heat treatment converts them to a largely crystalline structure. Mirror blanks made from these materials can be lightened by machining techniques but in general do not offer the same flexibility in fabrication method, and therefore design, as glass. Nevertheless, careful control of composition and the subsequent crystallization heat treatment can ensure that the thermal properties required for extreme dimensional stability can be achieved. The ultrasound method of characterizing the expansion properties is not applicable in these rather complex, multiphase materials.

Spacecraft Windows

The earliest encounters with the design problems associated with windows for spacecraft came even before orbital flight was achievable. The North American (now Rockwell International) X-15 rocket-powered aircraft of the late 1950s and early 1960s achieved such high levels of performance that some of the same design issues were encountered that later had to be addressed for manned orbital spaceflight. In the case of the X-15, the problem was primarily one of high levels of mechanical stress due to pressure, shock,

Table 1 Properties of materials for space-based mirrors

Property	Ultralow-expansion glass (Corning 7971)	Lithium-aluminosilicate glass-ceramic (Schott Zerodur)
Average coefficient of thermal expansion in ppm/°C for:		
5–35 °C (40–95 °F)	0	−0.05
20–300 °C (70–570 °F)	0.03	0.05
Density, g/cm³ (lb/ft³)	2.20 (137)	2.53 (158)
Knoop hardness (200 g load), kg/mm²	460	580
Modulus of rupture, MPa (ksi)	50 (7.25)	76.5 (11)
Young's modulus, GPa (10⁶ psi)	67.6 (9.8)	91 (13.2)
Stress-optical coefficient, nm/cm/kg/cm²	4.15	2.94
Service temperature:		
Maximum operating temperature, °C (°F)	800 (1470)	150 (300)
Short-term operating temperature, °C (°F)	1050 (1920)	625 (1160)
Thermal conductivity, W/m·°C	1.31	1.64
Thermal diffusivity, cm²/s	7.9 × 10⁻³	7.9 × 10⁻³

vibration, and the thermal stress from aerothermal heating at high Mach numbers. The problem was solved by the use of very high strain point aluminosilicate glass (Corning Code 1723), which while quite low in thermal expansion (4.5×10^{-6}/°C), could nevertheless be fully tempered. The mechanical and thermal requirements were therefore able to be met with a single glass.

Early orbital manned space flight always involved a single launch and reentry. Again, pressure loading, shock, and vibration in launch and orbit followed by the extreme thermal conditions of reentry were major requirements. New combinations of materials had to be employed. Because the tempered aluminosilicate glass had been found to be well suited for the X-15, it continued to be employed as a pressure pane in orbital spacecraft. This pane is always located in the innermost position. A thermal pane, separate from the pressure pane, is fitted to the outside of the craft. It is not subjected to the cabin pressure, and its primary requirement is the ability to withstand both the extreme temperature (over 700 °C, or 1300 °F) and thermal shock of reentry. The most practical glasses, which are both highly refractory and thermal-shock resistant (by virtue of very low thermal expansion), are those consisting almost entirely of silica.

Both Vycor brand 96% silica glass and fused silica have been successfully employed. Usually a redundant pane is fitted between the pressure and thermal panes. This is designed to pick up the function of either the pressure or thermal panes, should either fail. Fortunately, this design philosophy was never put to a test in an actual mission, because no window failures have ever occurred.

With the advent of the space shuttle, additional design elements became critical. For example, the new requirement of reusability demanded an assessment of static fatigue (slow crack growth) because glass, for all practical purposes, must always be considered flawed. In the presence of moisture, even relatively low levels of tensile stress can cause these flaws to propagate, leading ultimately to failure. A fracture mechanics approach was used to analyze the design of the outer thermal panes, and it was determined that in fused silica (which was the material of choice for its superior thermal properties) any flaw deeper than 0.045 mm (0.0018 in.) in the glass surface might propagate under the anticipated loads. It was assumed that any visible flaw in the glass surface might have associated with it an invisible, subsurface flaw three times deeper than the one actually measured. All flaws deeper than 0.015 mm (0.0006 in.) are therefore rejectable in space shuttle panes.

The fatigue characteristics of fused silica appear to be the most favorable of all silica-base glasses. This allows stresses of nearly 20 MPa (3000 psi) to be sustained for very long periods of time without the risk of slow crack growth, when surface flaws are limited to those described above. All outer panes are inspected for new flaws after each flight and the panes are replaced if any are found which fail to meet the original design criteria.

Prolonged exposure to the extreme radiation of outer space leads to yet another design issue which was much less critical in previous, shorter-duration missions. Most glasses experience considerable darkening when subjected to high-energy radiation, particularly deep ultraviolet rays, x-rays, and gamma radiation. Fused silica is unique among glasses in its ability to resist such darkening. On the other hand, fused silica passes ultraviolet (UV) rays rather than absorbing them like most glasses, which means the crew needs to be protected from the harmful effects of UV radiation. Because the shuttle window system has many of the same structural requirements as earlier spacecraft, tempered aluminosilicate glass is still used as the pressure pane. Fortunately its UV absorption characteristics are such that it can serve a dual function as UV shield and pressure pane.

Additional protection from harmful radiation is provided by UV and infrared reflective coatings on the pressure and redundant panes. As in earlier designs, the redundant pane is also fused silica. It is considerably thicker than both the thermal and pressure panes.

Table 2 lists the properties of fused silica and aluminosilicate glasses, which have been commonly applied to spacecraft windows. Figure 3 shows the space shuttle window assembly.

Missile Nose Cones (Radomes)

High-speed, radar-guided, air-to-air, and surface-to-air missiles require an aerodynamic fairing over the radar antenna, which is mounted near the front of the missile. The material requirements for this fairing are extremely demanding in that they must provide an optimum combination of mechanical, thermal, and electrical properties to ensure the neccessary level of performance for the missile.

Radomes for relatively low-performance, subsonic systems are typically made from materials such as glass fiber-reinforced resin or other combinations of organic and inorganic materials. High-performance, supersonic systems generate temperatures that demand the use of ceramic or glass-ceramic materials. Table 3 lists the properties of a

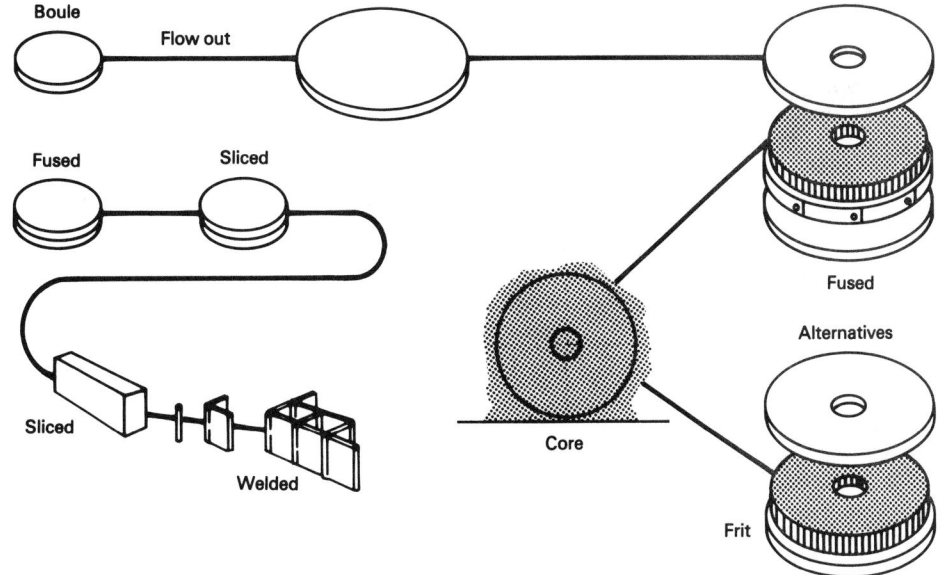

Fig 1 Assembly steps to fabricate a lightweight space mirror

Fig 2 Mirror blank prior to finishing for the Hubble telescope. The 2388 mm (94 in.) mirror blank is constructed from Corning Code 7971 ultralow-expansion glass.

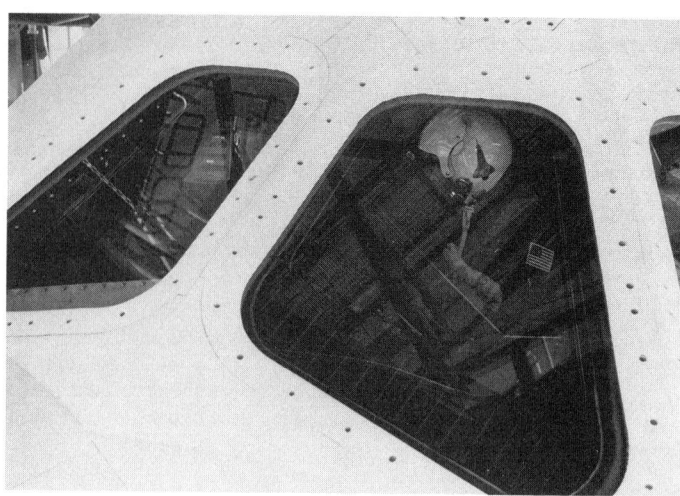

Fig 3 Space shuttle windows consisting of fused silica and aluminosilicate glass panes

ceramic-glass material (Pyroceram) and a fused silica ceramic that are used for radomes.

In order to transmit a strong, undistorted radar signal through a radome, the material needs a low dielectric constant (which is related to the refractive index in glass). A low dielectric constant minimizes reflective losses as well as "boresight error" (which is equivalent to aberration in optical systems). The loss tangent, which is a measure of signal absorption, also needs to be kept to a minimum. The best radome materials exhibit very little change in electrical properties with increasing temperature, so that optimum performance of the guidance system is not impaired when the radome material heats up in flight.

Mechanical and thermal properties of radome materials are closely linked because the primary source of stress in radomes in flight is due to the high temperature which develops during flight. As the outside of the dome heats up due to friction with the air, the inside temperature lags and in any material with a positive expansion coefficient, tensile stress results on the inside surface. Thermal shock failure can be prevented either by the use of very strong materials, by means of materials having very low thermal expansion, or by combinations and compromises of both properties.

Rain erosion resistance is another desirable characteristic, which can be difficult to achieve in brittle materials. The failure mechanism in a high-speed encounter with rain is very similar to that involved in contact stresses. Two measurable properties that provide some indication of the relative rain erosion resistance of glass or glass-ceramic materials are modulus of rupture and hardness. Fracture toughness provides a measure of the readiness of a material to propagate cracks and therefore also provides a clue as just how readily cracks initiated by rain impact might result in catastrophic failure of the radome.

When glass-ceramic materials were discovered by Dr. Donald Stookey in the Corning research laboratories in the 1950s, it almost immediately became obvious that the properties of certain members of this new family of materials might be well suited to meet the demanding requirements of newly developed high-speed, radar-guided missiles. A joint program with the U.S. Navy and the Applied Physics laboratory at Johns Hopkins University led to the use of a magnesium-aluminosilicate glass-ceramic (Pyroceram Code 9606) as the standard radome material for a long series of radar-guided missiles. More than 30 years later, Pyroceram is still the material of choice on most of these systems. The most recent application is the Advanced Medium-Range Air-to-Air Missile (AMRAAM).

Pyroceram Code 9606 features a moderately low dielectric constant, a very low loss tangent, high mechanical strength, and sufficiently low thermal expansion. Its rather complex crystal structure is developed at over 1200 °C (2200 °F) and consists mostly of cordierite with lesser amounts of cristobalite and various magnesium titanates. A very small amount of glassy phase also remains after the crystallization heat treatment.

It was discovered early in the radome development program that the apparent

Table 2 Properties of materials for spacecraft windows

Property	Fused silica (Corning Code 7940)	Aluminosilicate glass (Corning Code 1723)
Mechanical (at room temperature)		
Young's modulus, GPa (10⁶ psi)	73 (10.5)	85.5 (12.2)
Modulus of rupture, MPa (ksi)	60 (8.7)	170 (24.5) (tempered)
Poisson's ratio	0.17	0.26
Knoop hardness (100 g load), kg/mm²	500	595
Optical		
Refractive index	1.459	1.547
10% UV cut-off(a), nm	165	350
Birefringence constant, nm/cm/kg/cm²	3.45	2.40
Thermal		
Strain point, °C (°F)	990 (1815)	670 (1240)
Coefficient of thermal expansion from 0–200 °C, ppm/°C	0.57	4.6
Thermal conductivity, W/m·°C	1.38	1.29
Physical		
Density, g/cm³ (lb/ft³)	2.2 (137)	2.63 (164)

(a) Measured through 10 mm (0.4 in.) of glass

Table 3 Materials for radomes

Property	Pyroceram 9606	Slip-cast fused silica
Physical properties		
Density, g/cm³ (lb/ft³)	2.6 (162)	1.9–2.1 (119–131)
Water absorption, %	0.00	1–6 (ASTM-C373-56)
Mechanical properties (at room temperature)		
Young's modulus, GPa (10⁶ psi)	120 (17.4)	34–55 (4.9–8)
Modulus of rupture, MPa (ksi)	241 (35)	48 (7)
Poisson's ratio	0.245	0.15
Knoop hardness (100 g load), kg/mm²	698	...
Electrical properties		
Dielectric constant	5.5(a)	3.32(a)
Loss tangent	0.00033(a)	0.00002(b)
Thermal properties		
Coefficient of thermal expansion, ppm/°C	5.7(c)	0.55(d)
Thermal conductivity, W/m·°C	3.3	0.8
Thermal diffusivity, cm²/s	0.0127	0.006

(a) At 25 °C (77 °F) and 8.5 GHz. (b) At 25 °C (77 °F) and 10 kHz. (c) From 20 to 320 °C (68 to 610 °F). (d) From 25 to 950 °C (77 to 1740 °F)

strength of Pyroceram Code 9606 could be considerably enhanced by means of a chemical etch in a boiling 5% solution of sodium hydroxide. This is called fortification. A twofold benefit results from this treatment. Residual cracks and flaws resulting from the surface grinding operations, required to produce a dimensionally accurate dome, are rounded so that their stress-raising effect is minimized. A porous layer results which protects the actual load-bearing surface underneath from further damage. Modulus of rupture is increased from about 160 MPa (23 ksi) to over 240 MPa (35 ksi) by fortification.

Pyroceram Code 9606 radomes are manufactured by casting "blanks" from optical-quality molten glass to near-net shape using a centrifugal casting process. The typical ogive shapes are extremely well suited to this process, as is the fact that the particular glass composition employed has a rather steep viscosity curve (that is, it is rather fluid at the delivery temperature). Highly viscous glasses do not lend themselves to this forming process.

Domes are then partially finished in the glassy state, heat treated to convert them to the glass-ceramic form, and then finished to the final configuration by a series of grinding and lapping operations. Fortification is the final processing step.

Radome contours and wall thickness (typically half a wavelength at the particular microwave frequency employed) are held to within less than ±0.05 mm (±0.002 in.). All radomes made in this manner are proof tested by plunging them into a bath of molten nitrate salts at about 500 °C (930 °F) to simulate the aerodynamic heating encountered in flight. Figures 4 and 5 show a Pyroceram radome blank and finished dome assembly, respectively.

Fused Silica. The thermal and electrical properties of fused silica in its slip cast form are nearly ideal for radome applications. Its dielectric constant and loss tangent are both very low, and the change of these properties with increasing temperature is negligible. The inherently low coefficient of thermal expansion ensures that thermal stresses are at a minimum even when large temperature gradients are experienced. The less-than-optimum mechanical properties, and to some extent the high cost of fused silica, limit its usefulness as a radome material. It has found application on the Patriot surface-to-air system and is being considered for the latest generation of the Standard Missile series.

In the area of rain erosion, the poor mechanical properties of slip cast fused silica limit its applicability. Numerous attempts have been made to enhance rain erosion resistance by means of fiber reinforcement, and some limited success has been achieved.

Solar Cell Covers

Solar cells, which are used to power most satellites, need to be protected from the dam-

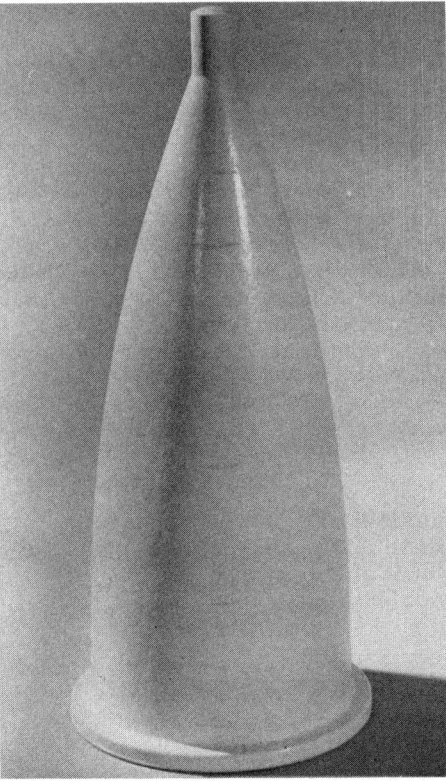

Fig 4 As-cast radome blank made of Pyroceram prior to finish grinding

Fig 5 Finished Pyroceram radome including guidance section shell

aging environment of space by means of glass covers. The primary function of these covers is to protect the solar cells from the damaging effects of ultraviolet radiation. There is also some concern about purely mechanical damage from the impact of micrometeorites and the glass covers also protect the solar cells from attack by atomic oxygen. Two distinct families of materials have been used successfully to produce solar cell covers: fused silica and various cerium-doped glasses.

Fused Silica. The material traditionally used has been fused silica because of its outstanding resistance to radiation darkening. By itself, however, it cannot do the job because of its inherently high UV transmittance. It requires highly specialized proprietary UV reflective coatings in order to serve its primary function. Fused-silica solar-cell covers are manufactured by slicing very thin plates from blocks and then finishing these slices in a series of lapping and polishing steps followed by coating. The leading specialists in this technology are Optical Coating Laboratories Inc.

Cerium-Doped Glasses. The addition of cerium to glass inhibits darkening. The mechanism of protection is not particularly well understood, but the presence of iron and other transition metal impurities is probably involved.

Cerium-doped solar-cell cover glasses are available from Corning in the United States, Asahi in Japan, and Pilkington in the United

Kingdom. The Corning product is designated Code 0213. It is a borosilicate composition with an expansion coefficient of $7.5 \times 10^{-6}/$ °C ($4.15 \times 10^{-6}/$°F) and is suitable for silicon solar cells. Pilkington has three varieties:

- CMX, which is similar to Corning 0213
- CMZ, which matches the thermal expansion of silicon more closely
- CMG, which is intended to match the expansion characteristics of gallium arsenide

The Asahi product is based in soda lime glass and is designated AS.

Infrared Glasses

Most silica-base glasses allow very good transmission of wavelengths less than 2.5 μm. If they are melted under conditions which allow the inclusion of water in the structure, as is the case in gas-fired melting units, there is a strong absorption band centered at about 2.8 μm. The silicon-oxygen bond itself begins to absorb at about 3.5 μm, so that beyond that wavelength, specialized infrared (IR) glasses need to be used. Table 4 lists the properties of some important IR glasses.

It is very difficult to formulate glasses with good IR transmittance while at the same time maintaining the other desirable properties of optical glasses. The wavelength at which glasses begin to absorb IR radiation is determined by the frequency of vibration of the

Table 4 Properties of infrared transmitting glasses

Property	Calcium aluminate (Barr & Stroud BS 37A)	Alkaline earth germanate (Corning, 9754)
Mechanical		
Young's modulus, GPa (10⁶ psi)	. . .	84 (12)
Poisson's ratio	. . .	0.29
Modulus of rupture, MPa (ksi)	83 (12)	45 (6.5) (abraded)
Hardness	6 on Moh scale	590 HK
Density, g/cm³ (lb/ft³)	. . .	3.581 (223.5)
Thermal properties		
Average thermal expansion (25–300 °C), ppm/°C	8.35	6.2
Service temperature, °C (°F)	700 (1300)(a)	650 (1200)
Optical properties		
10% IR cut-off(b), μm	5.45	5.7
Index of refraction at:		
3 μm	1.627	1.625
4 μm	1.607	1.606

(a) Moisture protective coating good to 500 °C (930 °F). (b) Wavelength for 10% cutoff in 2 mm (0.08 in.) thickness

ionic bonds. This frequency is generally lower for larger cations and anions because they have lower field strengths. Unfortunately, this also implies weaker chemical bonding and therefore less durable glasses.

The best of the silica-base glasses are the calcium aluminosilicates, which transmit out to about 4.7 μm (10% cut-off through a 2 mm section). Corning Code 9753 is an example of this type of glass, and it has the advantage (like most silica-base glasses) of being hard and durable, much like optical glasses used in the visible.

Barr and Stroud, a division of Pilkington, manufactures calcium aluminate glasses designated BS 37 A and BS 37 B, which transmit to almost 6 μm, but are somewhat less durable than the calcium aluminosilicate. Specially developed coatings improve the durability, so that transmittance is not degraded upon exposure to water.

Germanium-base glasses feature excellent transmittance from the ultraviolet at 0.33 μm to about 5.33 μm in the infrared. These glasses are very durable but are less stable than silica-base glasses, that is, they have a much greater tendency to devitrify unless cooled very rapidly from the molten state. Forming methods have been devised to allow the manufacture of germanate glass without crystallization; one composition, Corning Code 9754, is successfully employed on the Stinger Post heat-seeking antiaircraft missile.

There is also considerable interest in glasses for use in the 8 to 12 μm band. These are typically non-oxide glasses with very marginal mechanical properties and poor durability.

Frangible Glasses

Glasses and glass-ceramics may be chemically treated to induce high levels of residual stress, which, when released, will cause the materials to break almost explosively. Frangible materials that are made in this manner and breakable on command offer desirable characteristics for a number of defense-related applications. For example, a single frangible glass component can sometimes be substituted for complex mechanical devices and pyrotechnic systems.

Frangibility in glass results from an ion exchange that induces the prestress used for strengthening the material. The intent of ion

Table 5 Properties of Corning Code 0313 sodium-aluminosilicate for high-strength and frangible applications

Property	Property value
Mechanical properties (at room temperature)	
Young's modulus, GPa (10⁶ psi)	71 (10.3)
Shear modulus, GPa (10⁶ psi)	29.5 (4.3)
Poisson's ratio	0.21
Modulus of rupture, MPa (ksi)	275 (40)
Knoop hardness, (100 g load), kg/mm²	480
Physical properties	
Density, g/cm³ (lb/ft³)	2.46 (153)
Softening point, °C (°F)	870–880 (1600–1615)
Annealing point, °C (°F)	622–631 (1151–1168)
Strain point, °C (°F)	574–583 (1065–1080)
Water absorption, %	0.00
Porosity, %	0.00
Permeability	Impermeable under vacuum conditions
Thermal properties	
Coefficient of thermal expansion (25–300 °C), ppm/°C	8.8
Thermal conductivity (25 °C), W/m · °C	1.03
Specific heat, J/kg · °C	803
Optical properties	
Refractive index (20 °C)	N_f = 1.5129 N_d = 1.5068 N_c = 1.5043
N_u value	58.9
Transmittance (visible), %	91.6(a)
Electrical properties	
Loss tangent, %	0.012(b)
Dielectric constant	7.38(b)
Loss factor, %	0.088(b)
Volume resistivity, Ω · cm, at:	
25 °C	14.55
250 °C	6.8
350 °C	5.4

(a) Transmittance of visible wavelength through 2 mm (0.08 in.). (b) At 25 °C (77 °F) and 1 MHz

exchange or chemical strengthening is to induce a layer of compressive stress in the surface of the glass object, usually by the substitution of potassium ions for sodium ions. The compressive stress which is caused by the crowding of the larger ions in the smaller sites vacated by the sodium ions is balanced by an interior tensile stress. When a crack is deliberately caused to run into the tensile area of the glass, it accelerates and begins to branch with increasing frequency. The glass breaks up in a multitude of small fragments.

Frangible glass is most often applied as protective covers for missile components, such as sensitive infrared optical elements, that cannot withstand the flight environment. When the missile is ready to be launched, the glass is broken by impacting the edge or surface with a tungsten carbide stylus which penetrates the compression layer. The frangible glass disintegrates into fine particles the size of sugar grains and is harmlessly dispersed in the airstream. Other applications include deicing devices fitted to the Hellfire missile and tamper-proof containers.

A specialized sodium-aluminosilicate glass (Code 0313 developed by Corning) is the only glass routinely used for these applications, because it has the ability to reach very high levels of interior tension in the ion-exchange process. It is amenable to standard glass-forming processes and can therefore be furnished in a wide variety of shapes and sizes. However, all forming and finishing must be completed before the ion-exchange treatment can be carried out.

Table 5 lists the properties of Code 0313 sodium-aluminosilicate glass. Its strength and fracture characteristics can be varied over a wide range depending on the time and temperature of the ion-exchange treatment.

SELECTED REFERENCES

- P.S. Danielson, Vitreous Silica, *Kirk-Othmer: Encyclopedia of Chemical Technology,* Vol 20, 3rd ed., John Wiley & Sons, 1982
- W.H. Dumbaugh, Infrared Transmitting Glasses, *Opt. Eng.,* Vol 24 (No. 2), Mar/Apr 1985
- H.A. Miska, Understanding the Basics of Chemically Strengthened Glass, *Mater. Eng.,* June 1976
- D.A. Noll, Ceramic Radomes Materials and Design, *Electro-Technol.,* June 1962
- R.W. Smith, *The Space Telescope,* Cambridge University Press, 1989, p 69, 222–225, 239, 413
- J. Spangenberg-Jolley and T. Hobbs, Mirror Substrate Fabrication Techniques of Low Expansion Glasses, *Advances in Fabrication and Metrology for Optics and Large Optics,* Vol 966, Proceedings, Society of Photo-Optical Instrumentation Engineers, 17–19 Aug 1988
- J.D. Walton, *Radome Engineering Handbook,* Marcel Dekker, 1970

Architectural, Transportation, and Construction

Darryl J. Costin, Libbey-Owens-Ford Co.

GLASS is a unique material because it incorporates the properties of light transmission, impact strength, thermal and acoustic insulation, and corrosion resistance in a single material. It is this combination of properties that make it an ideal material for use in architectural, automotive, aircraft, and construction applications.

Architectural Glass

The history of the window in architectural applications has been one of multiple, and frequently conflicting, requirements. Windows have been used to permit the passage of light in and allow people to see out. They can allow free solar heating to enter yet must prevent heating ventilation and air-conditioning (HVAC) heating and cooling from escaping. During the summer, they should prevent excessive solar heat gain yet not impede the view to the exterior. Windows must withstand all loads imposed on them by wind and weather yet must on occasion be openable to allow room ventilation or emergency egress and entrance. Windows are required to be almost invisible and still provide a high level of acoustic insulation.

Today's windows, while meeting the above requirements, offer the architect a wide variety of colors to choose from when designing the exterior appearance of a building. Glass can be dark and unobtrusive with a low degree of reflectivity, or it can be highly reflective and mirror the changing sky and the surrounding landscape.

Glass Strength Properties

Wind loads and thermal stresses in heat absorbing and spandrel (part of a wall between the sill of a window and the head of the window below it) glasses often exceed the allowable limits for ordinary annealed glass. The strength of glass can be doubled or quadrupled by heat-strengthening or tempering, but this does not change the modulus of elasticity or stiffness. Where increased strength is needed in large glazing expanses, then the thickness of the glass plate is often increased to prevent the glass from visibly flexing under wind loads. Glass is now readily available in thicknesses from 2.4 to 25 mm (0.095 to 1 in.). Heat treating glass to provide additional strength induces some extra visible distortions beyond that induced from the original production of the flat annealed sheet of glass. Chemical tempering will eliminate this thermal distortion but has yet to be used in the architectural market because of the relative economy of production and because the layer of compression is considered too thin for most industrial applications.

Tempered Glass. It must be recognized that thermally tempered glass has a large amount of strain energy locked in by the quenching process. In the unlikely event that specific types of small stone inclusions are present, some glasses have an affinity to spontaneous shattering. This phenomenon has been known to occur many years after manufacture. A heat soaking process is available that can uncover most such inclusions. However, heat soaking can be time consuming and does entail some small loss of temper level imparted to the product. Proper glass design and selection ensures that tempered glass is only used where such unlikely breakage will not cause major problems.

Laminated glass is no more resistant to initial breakage than annealed product, but it does have the unique property of being highly resistant to penetration and fallout after breakage. For this reason, it is widely used in overhead glazing. Laminated glass also gives the maximum possible protection against UV (ultraviolet)-induced fading of furnishings and organic-based materials without any major reduction of visible light transmission. Further reduction of UV transmission requires tints and coatings with consequently reduced light transmission.

Wired glass has a unique strength property. It is actually weaker than annealed glass, but the enclosed wires hold broken glass pieces together in the event of a fire and prevent the passage of smoke and flames. New laminated glasses with intumescent interlayers offer the same or better fire ratings without the presence of the wire grid.

Regulation of Heat and Light

The control of heat loss through windows has now advanced to the stage where the thermal resistance of single glazing, R1, has been increased to R4 by the addition of a second layer of glass with a transparent low emissivity coating and an insulating gas such as argon in the cavity. Multiple layers and coatings can give R8 performance or even higher.

Half of the potential solar heat gain can be controlled by blocking the invisible infrared (IR) component of the sun's spectrum with light green tinted glasses. To further reduce the solar energy transmission, the visible light transmission must also be reduced. This is done by tinting the glass in bronze, green, gray, or blue colors and then further adding reflective coatings until the solar and visible transmissions are as low as 8%.

Figure 1 shows the theoretical and practical limits that can be achieved when controlling light transmission and solar heat gain through glass.

Table 1 gives the typical functions over which control is desired in residential and commercial glazing.

As new glass developments in passive and active switchable glazing become available, windows will take on a dynamic function and be capable of optimizing free passive solar heating when desired while also controlling excessive summer heat gain, glare, and privacy requirements. Estimated cooling and lighting energy requirements for different glass types are shown in Fig 2.

Transportation Glass

The primary uses of glass in the transportation industry are for automotive and aircraft window applications.

Automotive Glass

Glass is one of the basic elements for automotive design. Larger glass surfaces contribute to aerodynamics, aesthetics, comfort, and increase visibility and contact with the outside world.

Safety Requirements. United States federal regulations relative to the safety requirements of glass used in motor vehicles are specified in Federal Motor Vehicle Safety Standard Number 205 (FMVSS 205). Two basic types of safety characteristics are normally monitored: visible light transmission and injury potential upon breakage. The visible light transmission (illuminant A) in the United

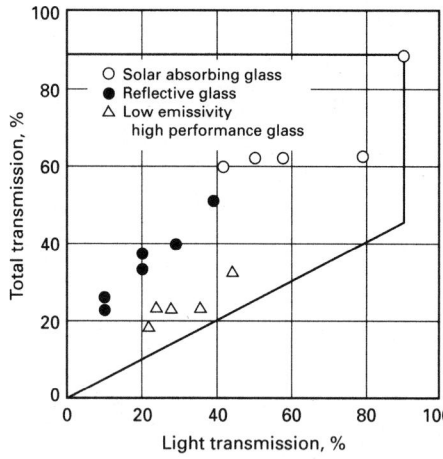

Fig 1 Plot of total transmission versus light transmission of selected 3.2 mm (¹/₈ in.) thick architectural glasses

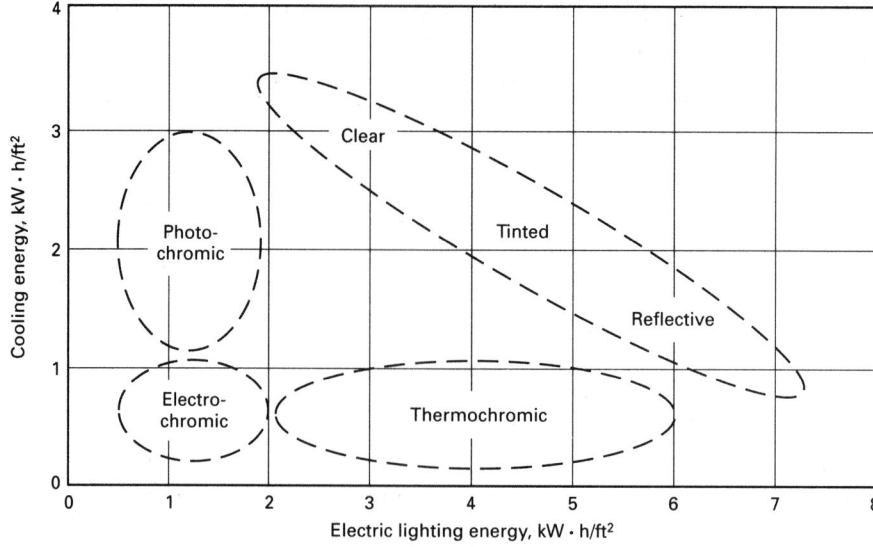

Fig 2 Plot of cooling energy versus electric lighting requirements to compare three types of optical switching materials with conventional glazings

States must be 70% in all passenger car windows. Nonpassenger cars (that is, vans, trucks, and so on) may have any glass visible transmission levels behind the B pillars.

Safety glass can be divided into two basic groups designated by the amount of internal stress present in the part:

- Annealed glass (not a safety glazing material) has very low internal stress, is brittle, and can be easily broken. When it breaks, long sharp dangerous shards can result. If used in a vehicle, these shards of glass could be propelled through the passenger compartment on impact to create a very severe hazard. For this reason,

unlaminated annealed glass cannot be used in motor vehicles

- Tempered glass has very high internal stresses that increase impact resistance, and upon breakage, small, blunt-edged particles result

Laminated glass, used primarily for windshields, consists of glass with a plastic-like material, typically polyvinyl butyral (PVB), adhered between the glass layers. Windshields absorb energy on impact and reduce penetration of objects from both external and internal sources. Special high-penetration

resistant (HPR) plastics are used in virtually all commercially available windshields to prevent occupant head penetration for speeds up to 40 km/h (25 mph).

The glass used for windshields is annealed. Glass used for other laminated parts could be heat strengthened (semitempered) or tempered.

Glass laminates may be colored by using dyed plastic or by applying a coating to the glass or plastic surface prior to lamination. Laminated glass is manufactured by subjecting the glass/plastic/glass sandwich to high pressure and temperatures (typically 150 °C, or 300 °F, and >700 kPa, or 100 psi).

Antilacerative Windshields. An antilacerative windshield, which consists of a conventional laminated windshield with a polyester of PVB plastic material laminated to the occupant surface of the part, significantly reduces lacerations. Upon breakage, the PVB plastic remains intact providing a protective layer over the broken glass.

Tempered Glass. Glass can only be broken in tension. The strength of glass can be significantly increased by having the exterior surfaces protected with a layer of compression. The process of creating this protective compressive stress on the surface of glass is called tempering.

There are several techniques for tempering glass. Chemical tempering involves creating surface compression by replacing surface sodium ions with larger potassium ions. This creates very high, thin compression layers. Chemical tempering is expensive and is seldom used for automotive glass.

A unique glass property permits glass to be thermally tempered, which is the most common strengthened glass. Glass is an amorphous material in a liquid state. The viscosity of soda-lime glass, which is used in most architectural and automotive applications, is

Table 1 Typical parameters that can be controlled by specific residential and commercial glazing

Capability or control function	Control requirement(a)								
	Residential			Commercial					
	Windows	Skylights	Sunspaces	View windows	Daylight windows	Envelope systems	Skylights	Atria	Interiors
Glare (visual performance)	...	...	...	S	P	P	P	P	S
Glare (visual comfort)	S	S	P	S	P	P	S	P	S
Privacy	P	...	P	S	...	...	...	...	P
Control of interior fading	P	...	P	S	S	S	...	...	S
Daylight control	...	...	...	S	P	P	P	P	S
Thermal comfort	P	S	P	S	S	P	S	S	...
Solar gain control (cooling)	P	P	P	P	P	P	P	P	...
Chiller and HVAC size	...	...	...	S	P	P	P	P	...
Peak demands control	...	...	...	...	P	P	P	S	...
Winter solar gain control	S	...	S	...	...	S	...	...	...
Control responsiveness	...	...	...	...	P	P	P	P	...
Control reliability	...	...	...	...	P	P	P	P	...

(a) P, primary; S, secondary

such that it behaves as an elastic solid at temperatures below the glass transition, which for soda-lime glass is approximately 515 °C (960 °F) and becomes less viscous as the temperature is raised. It is this unique ability of the glass to act as an elastic solid or a highly viscous liquid, coupled with its thermal expansion characteristics, that allow it to be thermally tempered.

Thermally tempered glass is manufactured by heating the glass part to a temperature in excess of 565 °C (1050 °F) and then rapidly cooling the part. This quenching creates temporary stresses that result in a permanent compression layer on the surface at room temperature.

When tempered glass parts are broken, small, blunt-edged particles result. FMVSS 205 requires that the largest particle from a piece of broken tempered glass must weigh no more than 4.3 g (0.15 oz). Tempered glass is used in sidelights, backlights, and sunroofs. It is not used in windshields because it does not offer sufficient penetration resistance required to protect occupants from ejection.

Solar Control. Glass provides the opportunity to reduce vehicle wind drag, increase design expression, and increase visibility to improve safety. In doing this, however, the solar load through the glass may be increased. This greenhouse effect, combined with the reduced efficiencies of the environmentally safer fluorocarbons used in modern automotive air-conditioning systems, have created a need for better solar reduction glass. Solar control automotive glass must reduce the solar energy transmitted without reducing the visible light transmission below the legal limit of 70%. Solar glass either reflects or absorbs the sun's energy.

Solar reflective parts are manufactured by depositing an extremely thin molecular layer of a metal or metal-oxide material on the glass or plastic interlayer. These coatings do not have the durability of glass and must be protected by lamination.

A second technique for reducing the solar energy transmission is through the use of a chemical change to the glass batch. These products absorb solar energy and have the durability of glass. Both the total solar energy and the ultraviolet light transmission can be significantly reduced with unique glasses.

The spectral properties of a typical commercial solar windshield are shown in Table 2. Note that the visible reflection is changed only slightly, which is an important safety concern.

Heated glass products reduce the amount of time required to defog and deice glass by passing an electric current through a conductor placed on the glass surface. Tempered back windows have a silver paste conductor fired onto the glass surface.

Laminated glass, used largely in Europe, may have very fine wires embedded into the PVB interlayer or have an extremely thin

Table 2 Comparison of spectral properties of commercial solar glass windshields with a standard tinted glass windshield

Property	Standard tinted	Solar Reflective	Solar Batch
Illuminant A transmission	77%	71%	71%
Visible reflection	7%	10%	6%
Solar transmission	54%	41%	43%
Solar reflectance	6%	31%	5%
UV transmission	20%	16%	33%(a)

(a) 17% for windshields due to PVB interlayer

conductive metallic-oxide coating over the entire glass surface.

Tinted Glass. The use of batch tinted glass is very popular with more than 90% usage in passenger vehicles and vans. Green tint is the most common. However, blue, gray, and bronze are also available.

Colored PVB can also be laminated between two plies of glass to provide tinting. Tinting that results in a visible light transmission of <70% can only be used behind the front door (B pillar) in vans and utility vehicles and only in sunroofs for passenger cars.

Decorative Glass. Painting glass, to make it more aesthetically pleasing or to hide interior body components, is a common practice. The paint used to accomplish this is a pigmented glass frit paste that is screen printed on the glass and then fired at approximately 565 °C (1050 °F) (during the bending and tempering operations).

Bent Glass Manufacture. Bent glass is either gravity sag bent or press bent at temperatures where the glass is sufficiently soft to flow or deform to a contoured surface. Press bending, which has rapidly grown in popularity over the last few years, is the only technique currently available to produce the more sophisticated parts required by future automotive designs.

Encapsulation. Encapsulated glass parts dramatically reduce the number of steps required to assemble products into a vehicle. Encapsulated parts are produced primarily by two basic techniques: polyvinyl chloride (PVC) injection molding and reaction injection molding (RIM), where a urethane type material is used. The fabrication technique is to clamp a glass part in a machined mold and inject the liquid encapsulation material.

Aircraft Window Glazings (Ref 1–12)

Aircraft window glazings (known as transparencies) vary in construction and use across the aviation industry, as defined in the following major topics:

- Cross-sectional design (compound curvature and weight reduction)
- Interlayers

- Testing (elevated-temperature testing, pressurization, bird impacts, and anti-ice/fog protection)

Cross-Sectional Design. Most of the aircraft transparency designs used in the commercial and military aircraft industry are based upon aerodynamics, weight reduction, and compound curvature. The majority of the designs are constructed from annealed or tempered glass (temper is obtained by either chemical or thermal processing), interlayer, and polymeric sheeting with abrasion-resistant coatings.

Optical Properties. On parts that have very rapid changes in radii, some problems with double vision (DV) (a condition when a glazing creates two images from a single object viewed by transmission) may exist. Therefore, to eliminate the DV effects, annealed aircraft glazings are cut to a size that is slightly larger than the final product. The glass blank is then shaped (generally to a compound curvature) by either gravity or press bending. The perimeter of the bent annealed part is cut to the desired shape, thereby reducing the DV and edge distortion caused by the shaping process.

Glass Temper. The same methods used to produce tempered automotive glass (chemical tempering and thermal tempering) are also used to produce tempered aircraft window glazings (see the section "Automotive Glass" in this article).

Polymers. Polymeric sheeting is made to simulate the appearance and overall optical properties of glass. Acrylic monomer is thermally polymerized between two glass blanks to obtain a large thick plate. This plate has holes drilled around the periphery and is subjected to heat and multiaxis tension to produce thin sheeting. The thin sheets are then placed onto grinding and polishing machines to obtain the desired thickness and optical properties. These optical grade blanks are then cut and subjected to heat and molds (which may or may not include vacuum) to obtain the desired compound curvature shape. Polymers (acrylic and polycarbonate) have better impact resistance than glass and are generally lighter in weight. They do, however, suffer from poor abrasion resistance and therefore require abrasion-resistant coatings.

Weight Reduction. One of the key variables considered when designing an aircraft transparency is weight. With the current emphasis on weight reduction and fuel expenditures, the lighter the design the better. Polymeric materials allow the engineers to design complex bends while maintaining properties and reducing weight. Table 3 shows a design comparison of various configurations.

Interlayers (layers of polymeric material that absorb impact energy while adhering to the substrate) are placed between the glass, polymer, or glass/polymeric constructions to provide the desired properties. Interlayers are

Table 3 Weight saving of glass-faced acrylic and polycarbonate composite aircraft window glazing configurations compared to an all-glass and polycarbonate aircraft window glazing configuration

Configurations were designed to withstand 1.8 kg (4 lb) bird weight impact at a 41.5° angle at a speed of 670 km/h (360 kt).

Configuration	Weight kg	Weight lb	Weight savings, %
All glass	46.3	102	0
Glass-faced acrylic composite	42	92	10
Glass-faced polycarbonate composite:			
A	24	52	49
B	29	65	36

Source: Ref 7

generally categorized as being either cast-in-place or sheeting.

Cast-in-Place. In cast-in-place systems, the interlayer is created by reacting the monomer and other additives *in situ* between the substrates (which are separated by a perimeter gasket). The most common forms of cast-in-place interlayers are from the silicone and polyurethane families.

Sheeting. Interlayers such as polyvinyl butyral (PVB) arrive at the manufacturer in sheet form. These sheets are laid between the substrates and subjected to pressure and temperature to achieve the desired adhesion and optical properties.

Testing. An aircraft glazing is subjected to a large number of tests to determine its effectiveness. Repeated temperature cycling between −70 to 45 °C (−90 to 110 °F) is not uncommon for laminate verification. Additionally, these tests are carried out in conditions of partial vacuum (7 kPa, or 1 psi) to simulate the conditions that supersonic aircraft experience at elevations of 15,000 m (50,000 ft) and at temperatures 117 °C (242 °F). Table 4 shows a portion of the remaining tests that focus on the polymeric material performance.

Pressure Loading. This testing is performed to indicate the ability of the laminate to withstand elevation changes and is considered insignificant in terms of the energy imparted by a bird impact. Creep testing also shows how a polymer (either interlayer or substrate) will deform at elevated temperatures and reduced pressures.

Bird Impact. One of the most important properties that an aircraft glazing must have is an ability to resist penetration during bird impact. Generally, a 1.8 kg (4 lb) bird is projected to the glazing at 822 km/h (444 kt). Various studies do show, however, that the speed at which the bird is projected may change as a function of the speed and impact of the aircraft or installation angle of the glazing. Results of this testing are generally expressed as supports and penetrations with the amount of the tearing being recorded.

Anti-ice and Antifog Protection. Commercial and military aircraft incorporate advanced filming technology to achieve anti-ice and antifog protection. This is achieved by application of an electroconductive layer (by chemical vapor deposition or CVD, sputter coating, and so on) onto the outermost substrate. These films must exhibit adequate adhesion to the interlayer, not degrade when subjected to testing with the interlayer, and provide high visible transmission without producing haze. The goal of these coatings is to dissipate 7.0 kW/m^2 (4.5 W/in.^{2}) of power density during operation. By applying a current to the bus bars, which are connected to the electroconductive film, resistive heating (and therefore anti-ice and antifog properties) are obtained.

Very advanced coatings protect the pilots against radioactive particles, radio frequency (rf) fields, and other unique military situations. Additional information is available in the article "Aerospace and Military Applications" in this Volume.

Future Outlook. By controlling the variables that are used in their design, manufacture, and testing, high-tech aircraft glazings can be brought into the 21st century. It must be kept in mind that many of the properties of aircraft glazings are at their maximum performance, and that new novel approaches are required if additional advancement is to be made.

Construction Glass

There are applications of glass in the construction industry that play an important role but are less obvious to the consumer than the

Table 4 Mechanical, thermal, and physical properties of transparent structural materials

Material	Military specification	Tensile strength MPa	Tensile strength ksi	Compressive strength(a) MPa	Compressive strength(a) ksi	Modulus of elasticity(b) GPa	Modulus of elasticity(b) 10⁶ psi	Design stress MPa	Design stress ksi
Polycarbonate	MIL-P-83310	68–72	9.8–10.5	86	12.5	2.1–2.4	0.3–0.35	28	4
Stretched acrylic	MIL-P-25690	79–82	11.5–11.9	120	17	3.1–3.4	0.45–0.49	21	3
Chemically tempered compound	MIL-G-25661-V	240–360	35–52	...	...	72	10.5	...	...

Material	Military specification	Coefficient of thermal expansion(c) 10⁻⁶/K	Coefficient of thermal expansion(c) 10⁻⁶/°F	Specific heat(d) kJ/kg·K	Specific heat(d) Btu/lb·°F	Coefficient of thermal conductivity(e) W/m·K	Coefficient of thermal conductivity(e) Btu·in./ft²·h·°F	Density(e) g/cm³	Density(e) lb/in.³	Poisson's ratio	Specific gravity
Polycarbonate	MIL-P-83310	6.25	3.47	1.17	0.28	0.22	1.5	1.19	0.043	0.36(f)	1.2
Stretched acrylic	MIL-P-25690	...	...	1.47	0.35	0.17	1.15	1.19	0.043	0.35	1.19
Chemically tempered compound	MIL-G-25661-V	0.88	0.49	0.84	0.20	...	...	2.52	0.091	0.22	2.44

(a) ASTM D 694 and FTMS 406–1021. (b) ASTM D 747 and FTMS 406–1011. (c) ASTM D 696 and FTMS 406–2031. (d) ASTM C 351–54. (e) ASTM C 177. (f) AFFDL-TR-25-2 (Air Force Dynamics Laboratory)

more familiar use as windows, doors, furniture, and partitions. Prominent among them are foam glass, glass blocks, and glass-ceramic building cladding. Use of these products allows the architect more freedom in design concepts with potential long-term savings in building maintenance costs.

Foam Glass. One of the major applications of foam glass, sometimes referred to as cellular glass, is insulating panels. The insulation is a light weight, rigid material composed of millions of closed glass cells, each acting as an insulating space.

The all glass closed-cell structure results in panels having moisture resistance, dimensional stability, and high compressive strength. Other advantages include being totally incombustible, the capacity to reduce transmission of airborne sound, and the total resistance to rodents, vermin, insects, and other organisms. With these properties, foam glass becomes an ideal material for a multitude of building insulation applications. For example, the material provides a firm, stable base for modified bitumen or elastomeric membrane roofing systems and can be used for insulating exterior walls (Ref 13).

The advantages of using foam glass insulation compared to other types commonly used are listed in Table 5. The comparisons are based upon criteria established by the National

Roofing Contractors Association (NRCA).

Glass Block. The wide range of patterns and sizes of glass block available allow the architect freedom of design, both functional and aesthetic. Examples of glass block usage include facades, windows, interior dividers, partitions, skylights, floors, walkways, and stairways. The block, made from two pieces of glass fused together with a partial vacuum in the interior, can be either transparent or translucent.

Glass block is used for environmental control of light (night/day and natural/artificial), noise, dust, drafts, and thermal transmission. Glass block has an insulating value equivalent to 3.2 mm ($^1/_8$ in.) double glazed glass units or 305 mm (12 in.) of concrete. Other options available include insertion of fibrous material in the block for further insulating value and the application of thermally bonded oxide surface coating for solar reflectance. This oxide coating reduces light transmission up to 95% and solar heat gain up to 80% compared to conventional 3.2 mm ($^1/_8$ in.) float glass (Ref 14, 15).

Ceramic Building Cladding. The use of glass/ceramic material for exterior cladding of buildings and interior applications is relatively new in the U.S. market. An alternative to natural stone, the material has a density similar to granite and marble but has three

times the bending strength. As a result, the material can be used in panels thinner than natural stone, thus reducing the dead load imposed on a building's structure.

The material is produced by melting, at elevated temperatures, silica, feldspar, calcium carbonate, zinc oxide, and barium carbonate into a glass and then submerging the molten glass into water to create granulated particles. The granulated glass particles are then placed in molds and subjected to a post heat treatment to fuse the particles and induce crystallization. The resulting product is then ground flat in panels ≥16 mm (≥$^5/_8$ in.) thick.

The crystallized material, approximately 40% crystalline, can then be heat softened and bent into curved panels if desired. The panels, flat or bent, are extremely hard and are difficult to abrade or scratch. The material is dimensionally stable due to an extremely low coefficient of thermal expansion.

The glass/ceramic material, being nonporous, does not stain as readily as natural stone and is not subject to damage by freeze-thaw cycles. It is less susceptible to damage by acids and alkali than granite, thus becoming a desirable substitute in areas where severe pollution or acid rains exist. A comparison of the properties of the glass/ceramic material to those of marble and granite is given in Table 6.

Table 5 Properties of foam glass insulation compared to other selected insulation materials

Insulation material	Noncombustible	Moisture resistant	Low K value(c)	Constant K value	Dimensionally stable	Compatible with bitumen	Strong, rigid, impact-resistant	Resists deterioration	Surfaces provide secure attachment	Compatible with membrane
Foam glass	Y	Y	I	Y	Y	Y	Y	Y	Y	Y
Perlite	N(c)	N	I	N	Y	Y	Y	N(e)	Y	Y
Glass fiber	N(c)	N	L	N	Y	Y	LR	N(e)	Y	Y
Polystyrene	N	Y	L	N	N	Y	LS	Y	N	N
Polyisocyanurate	N	Y	VL	N	N	C(d)	LS	Y	Y	C(d)
Phenolic	N	Y	VL	Y	N	Y	Y	Y	Y(f)	Y

(a) Y, yes; N, no; I, intermediate; VL, very low; C, caution; LR, low rigidity; LS, low strength. (b) Organic burns per NRCA criteria. (c) Very low, <0.20; low, 0.20–0.30; intermediate, 0.30–0.36; high, >0.37. (d) May cause blistering under BUR membranes. (e) Organic portions deteriorate with moisture. (f) If covered with acceptable insulation layer

Table 6 Properties of glass-ceramic compared to those of marble and granite used as building materials

Material	Bending, kg/cm²	Compressive, ton/cm²	Charpy shock, kg·cm/cm²(a)	Young's modulus, kg/cm²	Specific heat (at 50 °C, or 120 °F), cal/°C	10⁻⁶/°C	10⁻⁶/°F	W/m·K	Kcal/m·h·°C	Water absorption rate, %	Hardness, Mohs scale	g/cm³	lb/in.²	Acid resistance 1% H₂SO₄(b)	Alkali resistance 1% NaOH(b)	Freeze resistance(c)
Glass-ceramic	510	1.2–5.6	2.5	5.2	0.19	6.2	3.3	1.6	1.4	0.00	6.5	2.7	0.098	0.08	0.05	0.028
Marble	170	0.9–2.3	2.1	2.8–8.4	0.19	8.0–26.0	4.4–14.5	2.2–2.3	1.9–2.0	0.30	3–5	2.7	0.098	10.3	0.30	0.23
Granite	150	0.6–3.0	2.0	4.3–6.1	0.19	5.0–15.0	2.8–8.3	2.1–2.4	1.8–2.1	0.35	5.5	2.6	0.094	1.0	0.10	0.25

(a) Energy needed for rupture by instantaneous load. (b) Weight loss of test piece 15 × 15 × 10 mm (0.59 × 0.59 × 0.39 in.) after 650 h immersion of 25 C (75 °F). (c) Weight loss of testpiece 15 × 15 × 10 mm (0.59 × 0.59 × 0.39 in.) after 25 cycles: immersion of testpiece in water of 25 °C (75 °F) for 2 days exposure for 4 h in a temperature of −20 °C (−5 °F) (JISA5209, Japan Industrial Standard). Source: Ref 16

REFERENCES

1. J. Ondrejaus and F. Zitz, "Repair Laminate (Aircraft Glazings)," U.S. patent 4,855,182

2. G. Calhoon, "Window Liner for Use in Aircraft," U.S. patent 4,877,658

3. D. Voss, H. DeCamp, and G. Culp, "Electroconductive Film System for Aircraft Windows," U.S. patent 4,874,930

4. N.S. Corney, Report on the Conference on Transparent Aircraft Enclosures, Las Vegas, 5–8 Feb 1973, National Technical Information Service, U.S. Department of Commerce, p 3–28

5. W.P.C. Soper and R.A. Brown, "Concorde Glazings—5 Years of Mach 2 Service," S.B.A.C. Transparencies Symposium

6. P. Bain, Design Guidelines for Subsonic Transport Windshields, *Conference on Aerospace Transparent Materials and Enclosures*, 18–21 Nov 1975

7. B.G. Hinds, "The Certification of Polycarbonate Transparencies—An Appeal for Reasonable Requirements," AD POO3225, p 871

8. B.D. Gibbs, Operational Aspects and Performance of Transparencies Fitted to Current and Future Generation Civil Aircraft, *Conference on Aerospace Transparent Materials and Enclosures*, 1975, p 2–22

9. R.W. Wright, High Strength Glass in Service—A Status Report, *Conference on Aerospace Transparent Materials and Enclosures*, 1975, p 39–66

10. H.U. Reppermund and W.W. Hornsey, General Aviation Business Transparencies—An Update, *Conference on Aerospace Transparent Materials and Enclosures*, 1975, p 67–78

11. J.B.R. Heath and A.J. Bosik, Bird Impact Test Program for Windshields of Small, Light Aircraft, *Conference on Aerospace Transparent Materials and Enclosures*, 1975, p 851–886

12. Damping, Static, Dynamic, and Impact Characteristics of Laminated Beams Typical of Windshield Construction," Technical Report AFFDL-TR-76–156, Air Force Dynamics Laboratory, Air Force Wright Aeronautical Laboratories, Air Force Systems Command

13. 1990 Sweet's Catalog File, 07220/PIR, Vol 5, McGraw-Hill

14. 1990 Sweet's Catalog File, 08810/PIT, Vol 11, McGraw-Hill

15. B. Pennycook, *Building with Glass Blocks*, Doubleday & Co., 1987

16. N.B. Solomon, Glass: Cladding Innovations, *Architecture*, Aug 1990, p 101–102

Glass Fibers

P.F. Aubourg, C. Crall, J. Hadley, R.D. Kaverman, and D.M. Miller, Owens-Corning Fiberglas Corporation

GLASS FIBERS have evolved from a laboratory curiosity to a multibillion dollar market; applications have grown from a single filtration product to a very broad range of products including thermal and acoustical insulation, reinforcement, filtration, and other applications. The production aspects of glass fibers as well as optical fibers have been discussed in this Volume (see the Section on Glass Processing and specifically the article "Optical Fibers"). This article will focus on applications of glass fibers. Glass fibers can be divided in two broad groups depending on the geometry of the fibers: (1) the fibers used for yarn, fabric, and reinforcement, which typically are continuous or chopped fibers and (2) the discontinuous fibers used as batts, blankets or boards for insulation, and filtration. For more detailed information on such applications, see Ref 1 and 2.

Continuous Filaments

Compositions and Properties. Continuous glass fiber filaments are used for a wide variety of applications such as reinforcement for plastics, concrete, and gypsum board, fabrics for draperies, fabric structures, and filtration. The properties of the final product are controlled by the glass properties, fiber diameter, filament and fabric construction, and the surface finish of the glass as well as in the case of composites, the matrix. Depending on the properties desired, a variety of glass compositions have been developed; however, the large majority of the continuous fibers used now are of the E-glass composition. Traditionally, glass compositions have been given letter names; E for electrical. Other common textile glass compositions are S (strength), C (chemical), and E-CR (E-type, corrosion resistant).

E-glass is a family of glasses with a calcium aluminoborosilicate composition and a maximum alkali content of 2.0%. E-glasses are used as a general-purpose fiber when strength and high electrical resistivity are required.

S-Glass fibers have a magnesium aluminosilicate composition, which demonstrates high strength and is therefore used in applications where very high tensile strength is required. S-Glass and S-2 Glass fibers have the same glass composition (Ref 3) but dif-

ferent coatings. More stringent quality control procedures are necessary with S-Glass fibers to meet military specifications. S-Glass fibers are also used in high-temperature applications.

C-glass has a soda-lime-borosilicate composition that is used for its chemical stability in corrosive environments. Therefore, it is often used on composites that contact or contain acidic materials.

E-CR Glass fibers have a calcium aluminosilicate composition with a maximum alkali content of 1.0%. E-CR Glass fibers are used for applications similar to E-glass when improved acid corrosion is required.

Other compositions are or have also been used. These include:

- A-glass, a standard soda-lime silicate, was the first composition used but its poor corrosion resistance and high electrical conductivity and dielectric constant have limited its use
- A.R. glass (alkali-resistant), a zirconia-titania containing soda-lime silicate is used for cement reinforcement
- L-glass (lead) has been used for radiation shielding or as a tracer in composites
- D-glass (dielectric)
- M-glass (high-modulus)

Most of these glasses have, however, only limited market use.

When discussing the chemical composition of a specific glass type, the range of each of its oxide components must be presented. This is necessary because each manufacturer, and even different manufacturing plants for the same company, may use slightly different compositions for the same glass. These variations result from differences in the available glass batch (raw materials), or in the melting and forming processes, or from different environmental constraints at the manufacturing site. These compositional fluctuations do not significantly alter the physical and chemical properties of the glass type. It should be mentioned that while the compositions may vary from location to location, very tight control is maintained within a given production facility to achieve consistency in glass composition and maximize production efficiencies. Compositions and properties of E, S, C, and E-CR glasses are given in Tables

1 and 2. A more detailed discussion of properties can be found in Ref 4 and 7 to 10.

Textile Fibers. There are two basic forms of fiberglass textile fibers: continuous fibers and staple fibers. Continuous fiber strands consist of fibers having a length that extends throughout the entire length of the strand. Staple fiber strands are composed of individual fibers, 200 to 400 mm (8 to 16 in.) long, formed by pulling the molten glass filaments onto a revolving vacuum drum, where they are subsequently gathered into a strand. The majority of the textile fibers produced in the United States are of the continuous type. Staple fiber yarn provides bulkiness for filling, cushioning, and filtering and improves bond with impregnate. Some of these properties can also be obtained with continuous fibers by texturizing the strand, that is, separating the individual fibers.

Typical fiber diameters range from 3 to 20 μm. The individual filaments are combined into a strand, which is the basic building block for glass fiber products. Glass filaments are highly abrasive to each other. To minimize an abrasion-related degradation of filament strength, "size" coatings are applied before the strand is gathered. The size may be temporary, as in the form of a starch-oil emulsion that is subsequently removed by heating and replaced with a glass-to-resin coupling agent known as a finish. On the other hand, the size may be a compatible treatment that performs several necessary functions during the subsequent forming operation and which, during impregnation, acts as a coupling agent to the resin being reinforced.

Extensive research in this area has resulted in the development of coupling agents tailored to specific applications and resins. For example, silane coupling agents have an asymmetrical molecule that bonds to the organic polymer and the glass surface.

Reinforcement Fibers. The strands are processed in various forms to meet specific needs. For reinforcement, the products described below are used.

Continuous filament yarns are used in woven and nonwoven reinforcing fabrics. These yarns are made by twisting and/or plying a number of fiber glass strands together. A proprietary yarn is engineered for the braiding of three-dimensional reinforcement structures. This special yarn does not require

Table 1 Compositional ranges for glass fibers used in composite materials

Compound	Composition range, wt%			
	E-glass(a)	S-glass(a)	C-glass(a)	E-CR glass(b)
Silicon dioxide	52–56	65	64–68	54–56
Aluminum oxide	12–16	25	3–5	9–14.5
Boric oxide	5–10	...	4–6	...
Sodium oxide and potassium oxide	0–2	...	7–10	0–1
Magnesium oxide	0–5	10	2–4	0–4
Calcium oxide	16–25	...	11–15	17–25
Barium oxide	...	...	0–1	...
Zinc oxide	...	...	...	0–5
Titanium oxide	0–1.5	...	...	0–4
Zirconium oxide	...	...	...	0–0
Iron oxide	0–0.8	...	0–0.8	...
Fluorine	0–1	...	...	0–0.8

(a) Source: Ref 4. (b) Source: Ref 5, 6

the user to burn-off the sizing for handleability before adding a second sizing for resin compatibility. A single, dual-purpose sizing protects the fiber during braiding and still provides compatibility with polyester and epoxy resins.

Chopped strands are available in various lengths for compounding with resins and additives. Fiber length varies by compound but is usually in the 3 mm (¹/₈ in.) to 50 mm (2 in.) range. Chopped strands are available with several different surface treatments, each designed for a specific thermosetting or thermoplastic resin.

Continuous roving is bundled, untwisted strands. It provides excellent mechanical properties for reinforcement. For some molding processes, continuous roving is chopped into specified lengths to provide isotropic strength properties within the resin matrix. For other processes, the fiber is used continuously within the matrix for exceptional strength in the fiber's longitudinal direction. "Bulked" continuous roving has inherent loops in both axial and transverse directions. The result is multiaxial reinforcement with uniaxial input.

Woven roving is a heavy, drapeable fabriclike product, woven from continuous rovings. It gives high strength to large molded parts and is lower in cost than conventional woven fabrics.

Reinforcing mats are made with randomly dispersed chopped fibers or continuous fiber strands, laid down in a swirl pattern. In either case, fibers are held together with resin-compatible binder. Mats can also be produced by a technique similar to that used to make paper. Mats and woven rovings are sold in various densities or weights per unit area. They are used in hand lay-up processes in the manufacture of large structural objects such as boats.

Reinforcement fabrics are woven from untextured or textured glass fiber yarns in a variety of weaves and weights per unit square. Non-woven fabrics are made by knitting layers of fiber glass yarn together in unidirectional patterns. Combination products are made by stitching, needling or bonding two or more different types of roll goods together. A typical combination is a chopped strand mat/woven roving product. The benefit is the ability to lay-up two different layers of reinforcing materials in a single step.

Milled fibers, as the name suggests, are made by hammermilling glass fiber. The fibers are supplied in length categories by running them through different screen sizes. Milled fibers do not provide as good a stiffness and strength properties as chopped strands, which are longer. The function of milled fibers is more to control heat distortion of the resin and to improve surface finish of molded parts.

Glass flakes are used in resinous coatings to increase resistance to permeability from moisture, vapors, and solvents. Glass flakes have also been used in urethanes to enhance the surface finish of reaction injection molded parts.

Applications for Reinforcement Fibers. As mentioned at the beginning of this article,

Table 2 Typical properties for glass fiber types

Material	Density, bulk annealed, g/cm³	Tensile strength								Young's modulus of elasticity at 538 °C (1000 °F)		Elongation, %
		at −190 °C (−310 °F)		at 23 °C (72 °F)		at 371 °C (700 °F)		at 538 °C (1000 °F)				
		MPa	ksi	MPa	ksi	MPa	ksi	MPa	ksi	GPa	10⁶ psi	
E-glass	2.62	5310	770	3445	500	2620	380	1725	250	72.3	10.5	4.88
S-glass	2.50	8275	1200	4585	665	4445	645	2415	350	88.9	12.9	5.7
C-glass	2.56	5380	780	3310	480	...	...	...	...	...	...	4.8
E-CR glass	2.76	5310	770	3445	500	...	...	...	...	81.3	11.8	...

Material	Chemical resistance (percent weight loss)									Relative permittivity		Dissipation factor	
	in H₂O		10% HCl		10% H₂SO₄		1% Na₂CO₃		10% NaOH	at 1 MHz	at 60 Hz	at 1 MHz	at 60 Hz
	24 h	186 h	24 h	168 h	24 h	168 h	24 h	168 h	168 h				
E-glass	0.7	0.9	42	43	39	42	2.1	2.1	20	6.6	6.7	0.0025	0.0034
S-glass	0.5	0.7	3.8	5.1	4.1	5.7	2.0	2.1	66	5.3	5.4	0.0034	0.0129
C-glass	1.1	2.9	4.1	7.5	2.2	4.9	24	31	...	6.9	...	0.0085	...
E-CR glass	0.7	0.7	5.4	7.7	6.2	10.4	...	1.8	16	6.9	7.2	0.0028	0.0031

Material	Volume resistivity, Ω · m	Surface resistivity, Ω	Dielectric strength		Viscosity softening point		Viscosity annealing point		Viscosity strain point		Thermal expansion, 10⁻⁶/K(a)	Specific heat		Refractive index, bulk annealed
			KV/cm	V/mil	°C	°F	°C	°F	°C	°F		at 23 °C (72 °F) kJ/kg · K (Btu/lbf · °F)	at 200 °C (392 °F) kJ/kg · K (Btu/lbf · °F)	
E-glass	0.402 × 10¹⁵	0.42 × 10¹⁶	103	262	846	1555	657	1215	615	1140	5.4	0.810 (0.193)	1.03 (0.247)	1.562
S-glass	0.905 × 10¹³	0.886 × 10¹³	130	330	970	1778	810	1490	760	1400	1.6	0.737 (0.176)	...	1.525
C-glass	...	...	...	...	750	1382	588	1090	552	1025	6.3	0.787 (0.188)	0.90 (0.215)	1.537
E-CR glass	0.384 × 10¹⁵	0.116 × 10¹⁷	98	250	882	1619	...	...	...	...	...	...	0.97 (0.232)	1.583

(a) From −30 °C (−20 °F) to 250 °C (480 °F) Source: Ref 4

glass fibers are used in a wide variety of applications. Those applicable to reinforcement fibers are described below.

The automotive market has been, and is projected to continue to be, the largest consumer of reinforced plastics and composites. The most visible applications are exterior body panels. A growing market for reinforced plastics is emerging in structural automotive applications. Examples are composite drive shafts, wheels, and leaf springs.

Other automotive applications of reinforced plastics include headlamp housings, spare tire covers, and bumper beams. In addition to cars and light trucks, reinforced plastics are widely used for other ground transportation vehicles. Examples include hood/fender assemblies and engine components for heavy-duty trucks, body panels and engine components for recreational vehicles, and pultruded body panels and airducts for buses.

The marine market has been a large user of reinforced plastics for years (glass-reinforced polyester). Some high-performance boats are made with enhanced performance materials, such as S-2 Glass fibers and vinyl esters. In addition to pleasure boats and sailboards, plastics reinforced with glass are specified for the hulls and decks of commercial fishing boats and military minehunters.

Corrosion-resistant materials cover a variety of different markets. Major user-industries that benefit from special reinforced plastics are chemical processing, oil and gas, pulp and paper, waste water, air pollution control, and electroplating. Probably the most well-known corrosion-resistant application is the underground gasoline storage tank. Other corrosion-resistant applications include grating, handrails, work platforms, walkways, structural beams, even nuts and bolts.

Major construction market applications are paneling and skylights, bathtubs and shower stalls, doors, and windows.

The aircraft/aerospace/military market is low in volume but makes stringent demands on the material. Both commercial and military aircraft make extensive use of reinforced plastic materials.

Advanced composites are also used for helicopter rotor blades, aircraft wing and tail components, and engine ducts. Also, because of their radar transparency, advanced composites play a key role in radar-evading "stealth" technologies. In another military use, properly engineered composites work well as protective armor against ballistics attack. For example, the U.S. Navy uses shipboard armor made of an S-2 Glass fiber-reinforced thermoset polyester.

Electrical/electronic market applications vary widely. Printed circuit boards, composite ladders, poles, industrial circuit breaker housings, and conduit for power cables provide protection against electrical shock.

Consumer Goods and Business Equipment. New assembly techniques create opportunities for increased use of reinforced plastics in the appliance market. The business equipment market turns to reinforced composites for parts consolidation, close molding tolerances and heat resistance.

Many composite applications in the consumer goods market are related to sports and recreation. Composite bicycle frames are lightweight and stiff. Reinforced plastic skis made with braided preforms have exceptional torsional stiffness. Another growing area of applications is for power and hand tool housings and components.

Other Applications. In addition to uses for markets specifically mentioned above, reinforced plastics are made into stadium and institutional seating, power saw tables, road repair plates and decking for platform tennis. Medical uses include artificial joints and bones, casts, and equipment housings. Materials handling and industrial machinery are also potential markets for the growth of reinforced plastics because of light weight, durability, and vibration dampening.

Other fibers can be used for some applications; for example aramid fibers and carbon fibers offer higher specific, mechanical properties, but this is achieved at a high cost penalty.

Continuous glass fibers are used in many applications other than the reinforcement applications described above. Glass fabrics are used for filtration, as decorative fabrics and wall covering, and curtains and clothing when fire resistance is desired. Teflon-coated fabrics are used for domes and building covers (for example, stadiums). Coarse industrial fabrics are used for carpet and floor tile backing and roofing mats are used for reinforcing shingles. Glass fibers are also used to reinforce paper, tapes, and for insect screening. Glass yarn is also used to reinforce rubber for tires. Many other applications exist where glass imparts strength, dimensional stability, and fabrication efficiency.

To improve composite properties, shaped fibers and hollow fibers have also been developed.

Glass Wool

Compositions and Properties. In contrast to textile and reinforcement fibers, glass wool fibers are produced by processes which generate noncontinuous fibers of random lengths. Glass wool products are best known for their properties as thermal or acoustical insulation and as filtration media. Wool products, due to the inherent variation in the manufacturing processes, possess a distribution of fiber diameters and lengths which, along with the amount and type of binders added, determine the product properties.

Various glass compositions have been used for wool products. Factors affecting the choice of a glass composition are the availability and cost of the raw materials, the melting cost, the forming process used (glass liquidus and viscosity requirements), and the product properties (durability, resilience). For example, the low-cost rock and slag wool compositions have a high liquidus and hence cannot be formed by the rotary process. These compositions are fiberized instead by an air blast process or a rotating disk or multiple rotating drum process resulting in a high-shot (droplets of glass that have not been fiberized) content. Rock wool and slag wool were the first glass insulation products. The terms rock wool and slag wool denote similar products differentiated by the raw materials from which they are made. Due to the high surface area of the wool fibers, the glass must be resistant to water attack and, for some applications, have good chemical durability. Attempts to use a standard soda-lime silicate composition resulted in the product crumbling under humid conditions and vibrations. Typical compositions for glass wool and rock and slag wool are given in Table 3. As in the case of textile fibers, even though a broad range of composition exists, within a given plant the composition is tightly controlled.

Wool glass fibers have virgin tensile strength of about 2410 MPa (350 kpsi) (versus about 3450 MPa, or 500 kpsi, for E-glass). They have higher acid durability and a comparable or slightly lower water durability than E glass.

Applications. The most familiar application for wool products is for thermal insulation. The random intertwining of many small fibers effectively traps air within the insulation pack thus providing the insulating properties. In addition, the fibers tend to block radiative heat transfer due to their optical properties. The inherent fire resistance, chemical stability, and resistance to moisture attack make glass an ideal material for this application.

The same ability to trap air makes glass fiber an excellent acoustical insulation. This has led to the use of glass fiber as ceiling and wall panels in buildings and vehicles and to its use in air handling systems where noise control is important. The large amount of surface area per unit volume make the material useful as a filtration medium.

Processing versus Performance. The fiber diameter of the wool products is an important indicator of its performance. In general, finer fiber products are more costly to produce but provide better thermal and acoustical performance per pound of glass. Current technology can produce average fiber diameters that range from about 1 to 25 μm. In a wool pack, the fiber diameter will vary considerably from the average value. While the fiber diameters within a wool pack are not necessarily normally distributed, the standard deviation is still a useful measure of the spread of the distribution. Standard deviations tend to about 50 to 70% of the mean fiber diameter.

In most glass wool products, a phenolic binder is added during the manufacturing process to bond the mat together and to pro-

Table 3 Compositional ranges and properties for insulation-type glasses

Compound/property	Composition range, wt%			
	Rock wool made from basalt melted in a furnace	Rock wool made from basalt and other materials melted in a cupola	Slag wool made from slag melted in a cupola	Wool glass
Silicon dioxide	45–48	41–53	38–52	55–70
Aluminum oxide	12–13.5	6–14	5–15	0–7
Iron oxide(a)	5–6	3–8	0–2	0.1–0.5
Boric oxide	...	...	5–15	3–12
Sodium oxide	2.5–3.3	1.1–3.5	0–1	13–18
Potassium oxide	0.8–2	0.5–2	0.3–2	...
Magnesium oxide	8–10	6–16	4–14	0–5
Calcium oxide	10–12	10–25	20–43	5–13
Barium oxide	...	...	0–8	0–3
Titanium oxide	2.5–3	0.9–3.5	0.3–1	0–0.5
Sulfur(b)	0–0.2	0–0.2	0–2	0–0.5
Fluorine	...	...	...	0–1.5
Phosphorus pentoxide	...	...	0–0.5	
Viscosity = 1000 P at temperature, °C (°F)	...	...	...	915–1085 (1680–1985)
Liquidus temperature, °C (°F)	>Viscosity	>Viscosity	>Viscosity	880–955 (1615–1755)

(a) In rock and slag wool produced from materials melted in a cupola with coke as fuel, all iron oxide is reduced to FeO. During the spinning process, a surface layer may form in which the iron is oxidized to Fe_2O_3. Typically, 8 to 15% of the iron is oxidized to Fe_2O_3. In an electric furnace melting basalt, up to 50% of the iron is in the form of Fe_2O_3 and is more evenly distributed throughout the entire volume. (b) In wool glass, sulfur is oxidized to sulfate. Source: Ref 11

tect the glass fibers from abrasion. Depending on the application, the amount of binder can vary between 4 and 15% by weight. The light-density insulation products tend to use only enough binder to give the product the integrity required for installation of the product in its end application. For the heavier density products like glass fiber insulation for roof decks, higher levels of binder are required to provide the compressive strength and dimensional stability required. The amount of binder is usually measured by determining the loss on ignition of the product when brought to a temperature of around 540 °C (1000 °F).

Bulk density is a major factor in determining the performance of the product and will vary depending on the application. Insulation intended for residential buildings tends to be the lowest density with values in the neighborhood of 8 kg/m^3 (0.5 lb/ft^3). So-called board products tend to range from 16 to 80 kg/m^3 (1 to 5 lb/ft^3). The physical properties of the product (stiffness, compressive strength, and so forth) will tend to increase with increasing density.

In some applications, wool products are faced with materials which are intended to provide a moisture barrier or to provide a decorative surface. In other applications, the material may be molded (under heat and pressure) to produce various shapes. An example is the top liners used for automobile ceilings. Other fabrication processes are possible, including quilting (where additional integrity is required), die cutting (to produce complex shapes), or total encapsulation (to prevent fiber fly in clean room applications).

Specialty products such as the very fine microfibers are also produced for filtration or other special applications.

Health Aspects

Glass Filaments. The technology used to produce continuous glass filaments results in the production of fibers with near uniform diameters. Most reinforcement glass fibers have diameters considerably larger than 3 μm and as such are not respirable. Being too large to be inhaled into the lower lung effectively precludes chronic pulmonary effects. The size of the fibers does result in the fibers causing mechanical irritation to the skin and less frequently the upper respiratory tracts. On rare occasions, eye irritation may occur. Proper work procedures, as outlined by the manufacturers, will minimize possible irritation.

Though glass fibers will not split into finer fibers during machining or grinding of glass-reinforced composites, nonfibrous respirable dust from the resin system may be generated. For this reason, engineering controls to minimize respirable dust or, if these are impractical, respiratory protection should be used to prevent inhalation of these dusts.

Insulation Wool. Glass fibers used in most insulation products have nominal diameters from about 2 to 9 μm. Their method of manufacture is such that glass wool-type insulation products contain some fibers with diameters less than 3 μm which are respirable. Measurements of airborne glass fibers in both the manufacture and use of glass wool-type fibers have consistently shown that the airborne fiber levels are low. The World Health Organization concluded that, but for a few specific instances such as blowing wool applications, levels are less than one fiber per cubic centimeter (Ref 12). These fiber levels can be contrasted to the levels of airborne asbestos found historically which measured in the tens to hundreds of fibers per cubic centimeter (Ref 13).

Since their commercialization more than 50 years ago, extensive research on the health effects of glass fibers has been conducted (Ref 12). This work has included both experimental animal studies as well as studies of glass fiber manufacturing workers. The research has not demonstrated significant adverse effects of airborne glass fibers.

Human Studies. Large scale epidemiology studies of workers in both the United States and Europe have been conducted. These studies have involved thousands of past and current workers in the glass manufacturing industry. It has been consistently reported that these studies have not established a cause and effect relationship between airborne glass fibers and the development of disease in man. Glass wool insulations can, however, cause mechanical irritation of the skin, upper airways, and eyes. Appropriate work practices, as found in manufacturer's material safety data sheets will minimize this irritation. Additional information can be found in Ref 12.

Animal Research. A number of studies have been conducted to determine the effects of inhaled glass fibers on experimental animals. These studies have exposed animals to high levels (some experiments have used fiber levels in excess of 1000 fibers/cm^3) of glass fibers throughout much of their lives. These inhalation studies have consistently reported no significant adverse lung disease in the test animals. In addition to the inhalation studies, a number of investigators have directly injected or implanted large quantities of glass fibers directly into experimental animals. These routes of exposure bypass all of the body's natural defense systems. In some of these studies which have used fine diameter glass fibers designed for use in specialized applications (for example, filtration), animals have developed tumors following the injections. It is important that when these same fibers were tested by inhalation, no significant adverse pulmonary effects were found. Despite the lack of effect of airborne glass fibers, the results of studies using artificial injection or implantation of glass fibers directly into sterile body cavities of test animals have been considered by the International Agency for Research on Cancer (IARC) to provide "sufficient evidence of carcinogenicity" in animals (Ref 14). This finding coupled with "inadequate" evidence of carcinogenicity from human studies led IARC to classify glass wool fibers as Group 2B, or possible carcinogenic to man. It is important to note that only studies using artificial routes of administration have produced positive results in experimental animals.

REFERENCES

1. J.G. Mohr and W.P. Rowe, *Fiber Glass*, Van Nostrand Reinhold, 1978
2. "Textile Fibers for Industry," Publication 5–700–8285-B, Owens-Corning Fiberglas

3. R.W. Fulmer, "S-2 Glass Fiber Bridges a Gap in the Reinforcement Spectrum," Paper presented at the National SAMPE Symposium, Society for the Advancement of Material and Process Engineering, May 1980
4. D.M. Miller, Glass Fibers, *Composites*, Vol 1, *Engineered Materials Handbook*, ASM International, 1987, p 45–48
5. T.D. Erickson and W.W. Wolf, U.S. patent 3,847,627, 1974
6. T.D. Erickson and W.W. Wolf, U.S. patent 3,876,481, 1975
7. P.K. Gupta, Glass Fibres for Composite Materials, *Fibre Reinforcements for Composite Materials*, A.R. Bunsell, Ed., Elsevier Science, 1988, p 19–71
8. P.F. Aubourg and W.W. Wolf, Glass Fibers, *Commercial Glasses*, Vol 18, *Advances in Ceramics*, The American Ceramic Society, 1984, p 51–63
9. W.W. Wolf and S.L. Mikesell, Glass Fibers, *Encyclopedia of Materials Science and Engineering*, 1st ed., 1986
10. R.E. Lowrie, Glass Fibers for High-Strength Composites, *Modern Composites Materials*, Addison-Wesley, 1967
11. TIMA, Inc. Health and Safety Aspects of Man-Made Vitreous Fibers, submitted to NIOSH 10 July 1990
12. "Health Hazard Assessment of Non-Asbestos Fibers," Health and Environmental Review Division. Office of Toxic Substances, U.S. Environmental Protection Agency, Final Draft, 30 Dec 1988
13. "Asbestos and Other Natural Mineral Fibers," IPCS (1986) Environmental Health Criteria 53, World Health Organization, Geneva
14. *IARC Monograph on the Evaluation of Carcinogenic Risks to Humans: Man-Made Fibers and Radon*, Vol 43, World Health Organization/International Agency for Research on Cancer, Lyon, France, 1988

Lighting*

Peter R. Prud'homme van Reine, Willem J. van den Hoek, and Alexander G. Jack, Philips Lighting B.V., The Netherlands

GLASS is an important material for the lighting industry. The most important function of glass in lamps is its use as an envelope for the gas discharge or to keep the glowing filament in the appropriate atmosphere. The glass envelope has to transmit the light generated in the lamp with maximal, or controlled, transmission in the visible part of the spectrum. Silica glass was used for the first incandescent lamps and is still used on most lamp types. The introduction of new gas-discharge lamps, and the trends to miniaturization and optimization of light output, increased the demands on properties and dimensional accuracy of the glass. Glass remains an integral part of the lamp system, and its thermal, mechanical, and chemical properties influence the lamp behavior.

This article will review the various types of glasses used in lighting application, describe the manufacturing steps for glass lamps, and highlight the properties of importance for lighting. In the final section, emphasis is placed on the primary types of light sources.

Glasses Used for Lighting

The glasses that are used in lamps may be classified into five categories according to their basic compositions. Typical compositions and properties of glasses used for lamps are listed in Tables 1 and 2.

High-silica glasses (containing 96 to 100% SiO_2). Quartz glass has extremely high purity (>99.9% SiO_2) and high resistance to heat and thermal shock. The high thermal shock resistance is due to the low thermal expansion coefficient. Quartz glass is chemically stable and has high electrical resistance. The drawback of pure quartz glass is its high processing temperature. Therefore, two types of high-silica glass have been developed which have many of the favorable properties of quartz glass but require lower processing temperatures. These include Vycor and doped quartz glass. Vycor is made by melting a borosilicate glass (containing SiO_2, 4% B_2O_3, and 0.1% Na_2O), leaching the $Na_2O \cdot B_2O_3$ phase, and sintering to a high density. The final product consists of about 96% SiO_2, 4% B_2O_3, and 0.1% Na_2O. Doped

quartz glass is made by the addition of a few hundred parts per million of oxides such as BaO, Al_2O_3, and K_2O to a pure SiO_2 melt. See Tables 1 and 2 for compositions and properties of high-silica glasses.

Aluminosilicate glasses are alkali-free glasses that contain a high proportion of Al_2O_3, have high electrical resistivities, low expansion coefficients, and can withstand relatively high temperatures (Tables 1 and 2).

Borosilicate glasses have good chemical durability and high electrical resistivity (Tables 1 and 2). The addition of B_2O_3 reduces the viscosity (borosilicate glasses can be melted at 1500 °C, or 2730 °F) without increasing the expansion coefficient, thus maintaining a high thermal shock resistance.

Soda-Lime Glasses. The addition of alkali oxides (Na_2O and K_2O) causes a decrease in viscosity, electrical resistivity, and chemical durability and an increase in thermal expansion coefficient and in the tendency to devitrify (Tables 1 and 2). The addition of alkaline earth oxides (CaO, MgO, and sometimes BaO and SrO) is necessary to improve the durability. A small addition of Al_2O_3 further improves the chemical durability and reduces the tendency to devitrify.

Lead Glasses. The replacement of CaO and MgO by PbO increases the electrical resistivity and reduces the viscosity of lead glasses (Tables 1 and 2).

Glass Manufacture

Glass Melting. The first step in the manufacture of glass for lamps is the fusion of the raw materials. The basic glass melting process is melting of the mixed raw materials (sand, carbonates, oxides, and cullet, which is glass scrap of the same composition to facilitate the melting of the glass batch) in a furnace, and maintaining it at the melting temperature for a sufficient time to obtain a homogeneous glass that is free from bubbles. The furnace is heated by oil or gas flames above the molten glass bath and often additionally by the passage of electric current through the glass by immersed electrodes. Refining agents (such as antimony oxide, arsenic oxide, and sodium sulfate) are often

added to accelerate the removal of small bubbles in the molten glass. The temperature of the glass is reduced during the passage to the "working end" of the furnace where it is delivered to machines so that it can be formed in the shape of bulbs and tubes.

Glass Forming. Lamp envelopes can have the shape of bulbs or tubes. The electric feedthroughs for the glow wires or the electrodes are often sealed in a separate stem glass and made of tube glass with a flare on it. Narrow bore exhaust tubes, which are necessary for the degassing process, are sealed to the stem or directly to the bulb. The stem and either the bulb or tube are sealed together by applying gas flames. Close control of bulb dimensions and wall distribution is essential. For the bulb-blowing process, several types of machines are used. The most common is a fast ribbon machine that can produce 100,000 bulbs/h. Using this machine, a continuous stream of glass from the furnace passes between two water-cooled rollers which flatten the stream into a ribbon (Fig 1). These rolls are shaped to form heavier sections at intervals. The ribbon is supported on a moving chain of steel plates, each with an orifice hole to match one of the heavier areas in the ribbon. The glass sags through the orifices to form bubbles. A second chain carrying blowing tips is brought down on the ribbon of glass from above, each tip registering with the hole in an orifice plate. These tips supply puffs of air that expand the bubbles. A series of wet paste molds, traveling under the orifice plates, close around the bubbles of glass and rotate so that each bulb is blown by the final puffs from the blowing tips against a steam cushion. After the bulb is blown, the mold opens and falls away; a tap from a small hammer cracks the bulb from the ribbon into a conveying system that carries it to the annealing furnace.

For tube drawing, two different processes are used. In the Danner process, a continuous stream of hot glass flows over the surface of a hollow rotating mandrel, the axis of which is inclined slightly to the horizontal. The tubing is drawn from the free end of this mandrel, while air blown through the center of the mandrel maintains bore and diameter

*Adapted from *Ullman's Encyclopedia of Industrial Chemistry*, Vol A15, VCH Publishers, Inc., 1989, p 123–127, 133–136. With permission

Table 1 Typical compositions of glass types used in lamps

Glass	Composition, wt%								
	SiO_2	Na_2O	K_2O	B_2O_3	Al_2O_3	MgO	CaO	BaO	PbO
Soda-lime glass	72	16	1	...	2	4	3	...	...
Lead glass	63	7	7	...	2	...	...	...	21
Borosilicate glass	78	5	...	15	2	...	...	...	...
Aluminosilicate glass	61	...	...	1	16	...	10	12	...
Vycor	96	...	...	4	...	...	...	...	...
Quartz glass	>99.9	...	...	...	...	...	...	...	...

dimensions. A conveyer, consisting of heat-resistant rollers, carries the tubing to the cutting machine, where it is cut into lengths. The speed of the drawing machine, the flow rate of the glass, the air pressure, and the temperature of the glass determine the dimensions of the tube. The Danner process is the oldest continuous tube draw process still in use today.

The second type of tube drawing process is the Vello process. In this process, molten glass flows downward through the annular space between a vertical mandrel and a refractory ring set in the bottom of a special forehearth, called a "bowl." Air blown through the mandrel maintains dimensions of both bore and diameter. The soft tubing assumes the shape of a catenary as it is drawn onto a horizontal conveyer that carries the tubing to the cutting machine.

This process has gained favor over the Danner process because of its flexibility. It can accommodate a wide range of glass compositions as well as production rates varying from a few kilograms per hour to over 45 kg/min (100 lb/min). Dimensional control is good over diameters ranging from nearly 0 to 64 mm (2.5 in.). Under carefully controlled conditions, the standard deviation of diameter variation can be held to 0.3 to 0.5% of the nominal dimension. Thermometer tubing, in which the fine bore must be controlled precisely, is made by the Vello process.

Manufacture of Quartz Glass Tubing. Conventional glass melting techniques are not suitable for the manufacture of quartz glass because of the high temperatures required and the high viscosity of silica at such high temperatures. Raw materials can be natural quartz rock crystal or pure quartz sand. In one process, purified quartz sand is fed into a cylindrical refractory crucible mounted vertically in a furnace with a coaxial tungsten heater. The molten quartz glass flows through an orifice in the base of the crucible and over a central tubular mandrel, through which nitrogen is passed. A helium/hydrogen gas stream is fed through the crucible to prevent the formation of bubbles that would lead to defects in the drawn product. The glass tubing is then fired in a dry atmosphere or in vacuum to eliminate residual water and gases.

Properties of Glasses

Mechanical Properties. Glasses are brittle materials with a relatively high modulus of elasticity and reasonable tensile strength, which are almost independent of composition. Characteristic for glasses is that the practical tensile strength (30 MPa, or 4 ksi) is much lower than the theoretical value (10 GPa, or 1.5×10^6 psi). This is due to surface defects which act as stress raisers. Surface flaws may result from contact with glass-forming machine parts, other glass products, and/or from weathering. Weathering is caused by reactions of alkalis from the glass with water and CO_2 from the atmosphere. Flaws from contact with metal during forming operations can be diminished by the addition of lubricants. The strength of glass can be improved by protective coatings, by introducing a compressive stress at the outer surface by tempering, or by ion-exchange techniques (see the Section on Glass Processing in this Volume).

Another characteristic of glass is the occurrence of stress-corrosion cracking. Glass will fracture under a sustained stress of 40% of the mean instantaneous breaking stress. At a stress of 30% of this value (~10 MPa, or 1.5 ksi), fracture will not occur. This factor is taken into consideration in the design of glass/glass and glass/metal seals in lamps.

The resistance to thermal shock is directly dependent on product shape, the strength of the glass, and the reciprocal of the expansion coefficient. Quartz glass has a superior thermal shock resistance in comparison with other glasses, due to its low thermal expansion coefficient.

Thermal Properties. The maximum temperature a glass can withstand without deformation and the temperature at which glass can be formed are dependent on the viscosity/temperature characteristics. Viscosity-temperature curves of glasses are often identified by specifying the following four viscosity, η, points:

- *The strain point*: the temperature at which $\eta = 10^{13.5}$ Pa · s
- *The annealing point*: the temperature at which $\eta = 10^{12}$ Pa · s
- *The softening point*: the temperature at which $\eta = 10^{6.5}$ to 10^7 Pa · s
- *The working point*: the temperature at which $\eta = 10^3$ Pa · s

It is the working point that glass-forming processes like blowing, drawing, and sealing can be performed. At the softening point, glass deforms by its own weight and shaping in a gas flame can be performed. The annealing point and the strain point mark the "transformation range" of the glass, at which stress release in the glass by atomic rearrangement is possible. Due to temperature differences in a glass part during cooling after processing (bending of tubes, flanging, pinching), stresses arise. These stresses can be reduced by annealing in the transformation range. At the annealing point, stresses are reduced to 5% of their initial values in about 15 min. At the strain point, stresses are reduced to 5% of their initial values in about 4 h.

The thermal expansion coefficient is important for glass/glass and glass/metal seals. A thermal expansion mismatch leads to

Table 2 Properties of glasses used in lamps

	Density, g/cm^3	Strain point		Annealing point		Softening point		Working point		CTE at 0–300 °C (32–570 °F), 10^{-7} K^{-1}	Young's modulus at 20 °C (70 °F)		Electrical resistivity at 350 °C (660 °F), $\Omega \cdot cm$
		°C	°F	°C	°F	°C	°F	°C	°F		GPa	10^6 psi	
Soda-lime glass	2.5	490	915	520	970	700	1290	1015	1860	94	72	10.4	10^5
Lead glass	2.8	410	770	445	835	635	1175	1000	1830	93	61	8.8	10^7
Borosilicate	2.3	520	970	570	1060	800	1470	1200	2190	40	64	9.3	10^7
Aluminosilicate	2.6	770	1420	810	1490	1025	1875	1250	2280	45	88	12.8	10^{11}
Vycor	2.2	890	1635	1020	1870	1530	2785	...	...	7.5	72	10.4	10^8
Quartz glass	2.2	1070	1960	1140	2085	1670	3040	...	...	5.5	73	10.6	10^{10}

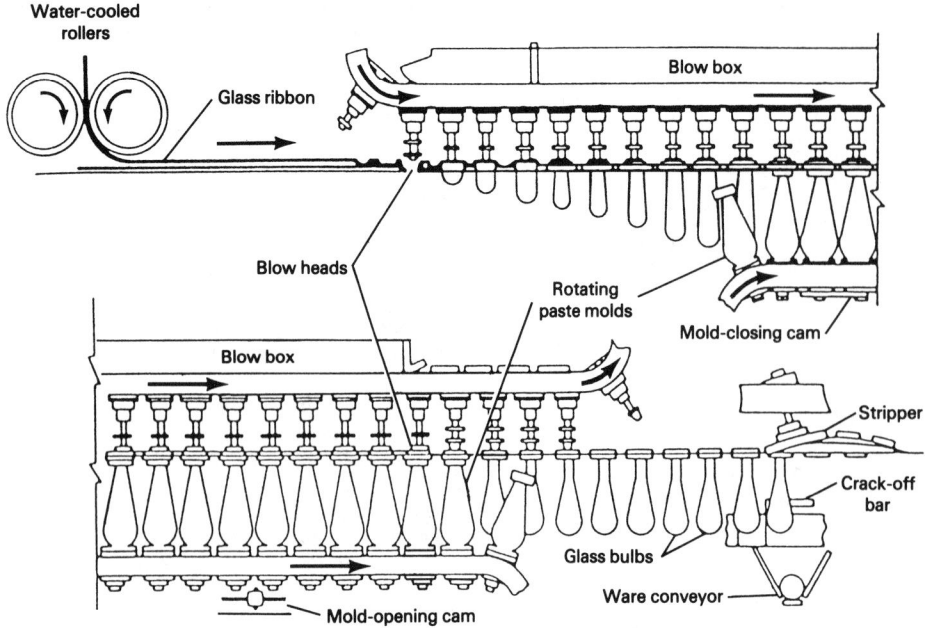

Fig 1 Ribbon machine for electric bulb manufacture

stresses in the glass parts after cooling to room temperature. With respect to thermal properties, the glasses used in lamps can be divided into three main groups:

- Soft glasses (soda-lime glasses and lead glass) with a thermal expansion coefficient of 90 to 100 × 10^{-7} K^{-1}
- Hard glasses (borosilicate and aluminosilicate glasses) with a thermal expansion coefficient of 35 to 50 × 10^{-7} K^{-1}
- High-silica glasses with a thermal expansion coefficient of 5 to 7 × 10^{-7} K^{-1}

Optical Properties. Pure quartz glass has a wide light transmission range from 0.18 μm (ultraviolet) to 4 μm (infrared). Glasses may be made colored, translucent, or opaque by the presence of dissolved, colloidal, or crystalline material. Oxides of iron, nickel, chromium, and manganese are some of the materials used to color glass. Iron contamination gives many soda-lime glasses their characteristic green color. A controlled amount of iron is often required in glasses to absorb unwanted ultraviolet radiation. Special ultraviolet transmission demands can be met by the addition of TiO_2.

Electrical Properties. Glasses are insulators at room temperature. At higher temperatures, soda-lime glass shows appreciable electrical conduction with Na^+ and K^+ as the charge carriers. Addition of PbO increases the electrical resistance by hindering the mobility of Na^+ and K^+. Alkali-free glasses (aluminosilicate, quartz glass) have high electrical resistance.

Chemical Properties. Resistance to the lamp atmosphere (sodium, mercury, halogen compounds) at high temperatures imposes special demands on glasses for lamps. Chem-

ical stability of glass at elevated temperature during lamp operation is necessary. Traces of water vapor and hydrogen must be eliminated. Manufacturing methods and subsequent thermal treatment are important factors. Quartz glass can be used up to 1100 °C (2010 °F) although devitrification (that is, crystallization) may occur at lower temperatures. The onset and the rate of devitrification can be accelerated by the presence of impurities (especially alkali metal ions) in the glass or on the surface. The product of devitrification is cristobalite (a crystalline phase of SiO_2) that has a much higher expansion coefficient than quartz glass. This acts as a stress raiser and can cause lamp failure.

Application of Glass in Lamps

The main application of glass in lamps is for lamp envelopes. General requirements for lamp envelopes are:

- Good transmission at all wavelengths in the visible spectrum, and in some cases the ultraviolet
- Resistance to the temperatures and pressures that occur during lamp manufacture and subsequent operation of the lamp
- Impermeability to gases during lamp operation
- Good degassing characteristics so that unwanted gases are not evolved during lamp operation
- Suitability for making gastight seals to metals for current leadthroughs
- Adequate electric resistivity at the maximum temperature of the seal during lamp operation

- Chemical inertness to the atmosphere in the lamp
- Economic forming to the required shapes and to close tolerances

Requirements for various types of lamps are summarized in Tables 3 and 4.

Incandescent Lamps. The heart of the incandescent lamp is a tungsten filament that is coiled either singly or in multiples and enclosed in a hermetically sealed glass envelope (Fig 2). A special copper-clad (Dumet) wire is used for the glass/metal seal. In order to prevent oxidation of the tungsten filament, the impurity level in the glass vessel must be very low. Substances known as getters are used to absorb traces of water. Usually the vessel is filled with an inert fill gas (Fig 2) to decrease the evaporation of tungsten from the filament. A more diffuse light source is obtained by etching the inside of the bulb or coating it with a powder layer of TiO_2. Colored lamps can be made by using pigments in the powder layer or an exterior heat-resistant paint. The lamp cap is attached to the vessel by means of a cement.

Glasses for the envelopes of incandescent lamps must be suitable for large-scale, high-speed production to keep down manufacturing costs. Impurities in the lamp atmosphere such as water vapor, oxygen, carbon dioxide, and hydrocarbons should not be present at levels exceeding a few parts per million. Soda-lime glass is adequate when an effective degassing process is provided during the lamp manufacture. This glass is very similar to the glass used for containers and window glass but is designed to have optimum viscosity/temperature characteristics so that bulbs can be made and processed on high-speed machines.

Soda-lime glass is not recommended for use in the feedthrough because its electrical resistivity is not sufficient. For this section and for the components carrying the filament-support wires, it is necessary to use lead glass.

For incandescent lamps subjected to high-temperature loading, borosilicate or aluminosilicate bulbs and stem tubes are used.

Halogen Incandescent Lamps. In halogen incandescent lamps, the inert gas pressure is higher than that of a normal incandescent lamp to further diminish the evaporation of tungsten. For a given design life, this means that the lamp can operate at a higher tungsten filament temperature and thus has a higher luminous efficacy. Due to the higher pressures, a smaller vessel is necessary. Halogens are used to keep the quartz vessel free from tungsten which evaporates from the filament.

Due to the higher wattage loading of the halogen lamp bulbs in comparison with incandescent lamps, soda-lime glass cannot be used. The choice between pure quartz glass, doped quartz glass, Vycor, aluminosilicate, and borosilicate glass has to be considered carefully for each halogen lamp type. Impor-

Table 3 Properties of typical high-pressure and low-pressure sodium and mercury discharge lamps

Property	High-pressure sodium	High-pressure mercury	Low-pressure sodium	Low-pressure mercury (fluorescent)
Lamp power, W	400	400	130	36
Discharge tube material	Polycrystalline alumina	Quartz glass	Soda-lime glass and borate layer	Soda-lime glass
Discharge tube inner diameter, mm	7.5	19	19	24
Electrode spacing, mm	82	72	1890	1120
Gas/vapor pressure in operating lamp				
Na, Pa	1×10^4		0.5	
Hg, Pa	8×10^4	4×10^5		1
Noble gas pressure (Ar, Kr, Ne, Xe), Pa	2×10^4	2×10^4	730	200
Lamp current, A	4.4	3.2	0.62	0.44
Lamp voltage, V	105	140	245	102
Discharge input power per unit length, W/m	4500	5500	60	30
Electron density at axis, m^{-3}	10^{21}–10^{22}	10^{21}–10^{22}	10^{18}–10^{19}	10^{18}–10^{19}
Gas temperature at axis, K	4000	6000	530	320
Electron temperature at axis, K	4000	6000	10,000	13,000

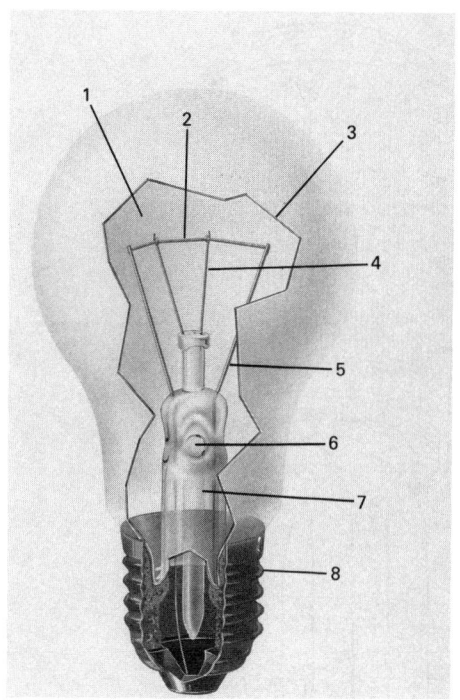

Fig 2 Diagram of an incandescent lamp. 1, fill gas; 2, filament; 3, bulb; 4, support wires; 5, lead-in wires; 6, stem; 7, fuse; 8, lamp cap

tant factors are the operating temperature and the demands on purity (hydrogen, water vapor, alkali impurity content) and the required service life. Typical values for the maximum operating temperature of these glasses are:

Glass	Temperature	
	°C	°F
Borosilicate	500	930
Aluminosilicate glass	650	1200
Doped quartz glass	800	1470
Vycor	900	1650
Quartz glass	1100	2010

Fluorescent (Low-Pressure Mercury) Lamps. The tubular fluorescent lamp consists of a glass tube with electrodes sealed in at each end (Fig 3). These coiled tungsten electrodes are coated with a metal oxide mixture which improves electron emission from the electrodes. The lamp is based on the gas discharge in a low-pressure mercury vapor (≈ 1 Pa) and argon gas (≈ 400 Pa) which primarily generates ultraviolet mercury radiation. The inner surface of the tube is coated with a fluorescent powder which converts the generated ultraviolet radiation into light. By varying the

composition of the fluorescent powder, lamp types with different color appearance and color rendering properties can be made.

In compact fluorescent lamps, the relatively long tube must be folded because simple shortening of the tube leads to an inefficient gas discharge. Examples of compact fluorescent lamps are shown in Fig 4.

As in almost all gas discharge lamps, the lamp voltage of fluorescent lamps decreases as the current increases. This means that a current limiter (a ballast) must be placed between the lamp and the electric main supply. Also the main voltage is usually insufficient to ignite the gas discharge so a starter is required. The starter and the ballast are often separate from the lamp but may also be integrated into a single unit (Fig 5), which can be directly connected to the main. In such an integral lamp, the mercury vapor pressure is very high due to the high wall temperature and an amalgam is used to obtain the optimum mercury vapor pressure.

For normal fluorescent tubes, a soda-lime glass with a composition that closely resembles that of incandescent bulbs is adequate. The composition (especially iron content) is

controlled in such a way that radiation below 300 nm is absorbed.

Blackening of fluorescent lamps is a complicated interaction process between the mercury discharge, the fluorescent powder, and the glass, in which the mobility of sodium ions in the glass is an important factor. For lamps with high wall loading (compact fluorescent lamps with narrow tubes) a special

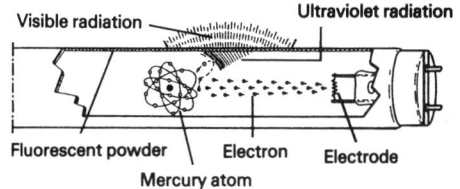

Fig 3 Schematic of a fluorescent lamp

Table 4 Important lamp parameters

Lamp type	Power, W	Luminous flux, lm	Luminous efficacy, lm/W	Color temperature, K	Color rendering index	Lamp life, h
Incandescent						
General	10–1500	50–35,000	5–25	2500–3000	100	100–2000
Halogen	10–2000	150–60,000	15–30	2800–3300	100	50–2000
Fluorescent	4–200	150–15,000	40–100	2500–6500	60–95	10,000–20,000
High-pressure mercury	50–1000	1500–60,000	30–60	3000–6000	25–60	10,000–20,000
Metal halide	30–2000	1500–250,000	50–125	3500–6500	50–85	2000–10,000
High-pressure sodium	30–1000	1500–150,000	50–150	2000–2500	20–80	10,000–25,000
Low-pressure sodium	15–200	1500–35,000	100–200	1700		10,000–15,000

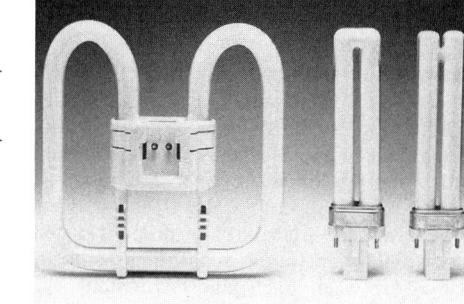

Fig 4 Examples of compact fluorescent lamps

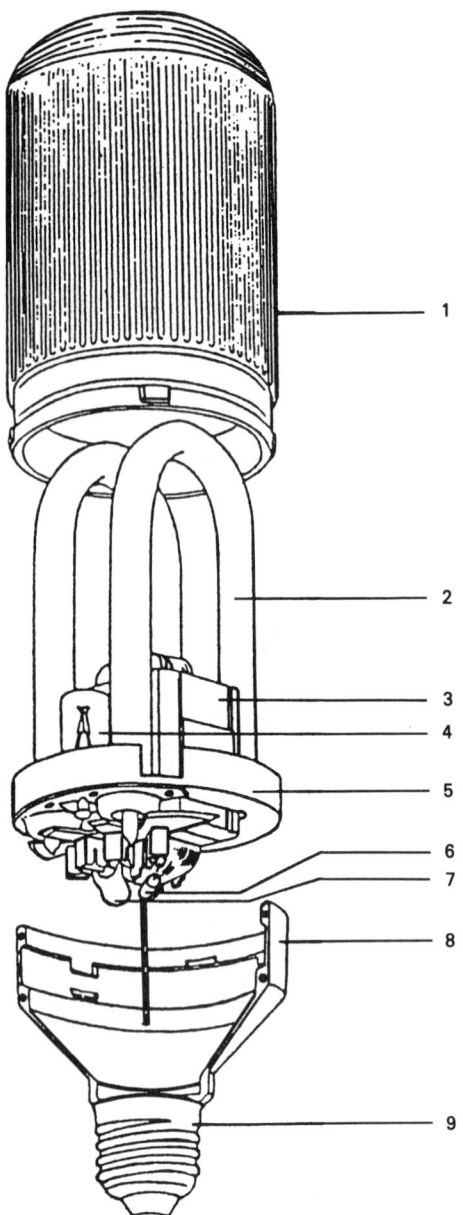

Fig 5 Exploded view of a compact fluorescent lamp. 1, outer bulb; 2, discharge tube; 3, ballast; 4, starter; 5, mounting plate; 6, thermal cutout; 7, capacitor; 8, housing; 9, lamp cap

soda-lime glass, or lead glass with a low-alkali ion mobility must be used. Germicidal lamps (high transmission at 254 nm) require a soda-barium-silicate glass. The composition is controlled in such a way that under the influence of ultraviolet radiation, the ultraviolet transmission is hardly reduced (low solarization effect). Special lamps for use in solariums (low transmission of <280 nm) and therapeutic equipment require glass with special ultraviolet transmission properties for which TiO_2-containing soda-barium-silicate glasses are used. Each tube has a stem and exhaust tube at either end for which lead glass is used.

Low-Pressure Sodium Lamps. Like the fluorescent lamp, the low-pressure sodium lamp (Fig 6) is based on a low-pressure discharge but in a sodium/neon gas mixture rather than mercury/argon gas. The fact that sodium is used in place of mercury has significant consequences. Since sodium radiates in the visible region, a fluorescent powder is not needed. In order to obtain a sodium pressure of about 0.5 Pa, however, the gas discharge tube must operate at a temperature of about 260 °C (500 °F). To minimize heat losses, the discharge tube is folded and placed in an evacuated outer bulb with a getter to ensure a good vacuum. The inside of the outer bulb is coated with a film (typically indium oxide with a tin dopant) which reflects in the near-infrared and is transparent in the visible. At 260 °C (500 °F), sodium attacks normal soda-lime glass so the inside of the gas discharge tube is coated with a thin film of sodium-resistant borate glass (20% B_2O_3, 50% BaO, 10% CaO, 10% Al_2O_3, 5% SiO_2, 5% MgO). The dimples, as shown in Fig 6, prevent the sodium from forming a film over the entire discharge tube surface.

High-pressure mercury lamps (Fig 7) are based on a discharge in mercury vapor with a typical pressure of a few hundred kilopascal. The presence of argon gas at low pressure facilitates ignition. In order to withstand the high wall temperatures (800 °C, or 1470 °F), the discharge tube is made from quartz glass. The tungsten electrodes are coated with an emitter material and are sealed into the quartz discharge vessel using molybdenum foil seals. The discharge generates visible radiation and some UV radiation. The quartz glass transmits the ultraviolet radiation emitted from the discharge, which can be converted into visible light by a fluorescent powder (phosphor) coating on the outer bulb. When special ultraviolet transmission demands on the quartz glass are imposed (lamps for use in solariums or photocopy lamps), TiO_2-doped quartz can be used to shift the absorption edge from 162 nm to higher wavelengths.

The outer bulbs of high-pressure mercury lamps for the higher wattages are made of borosilicate glass and sometimes aluminosilicate glass because of the required temperature and thermal shock resistance. Ultraviolet transmission demands are also imposed on the outer bulbs. Stem and exhaust tube of these outer bulbs are of borosilicate glass. For the

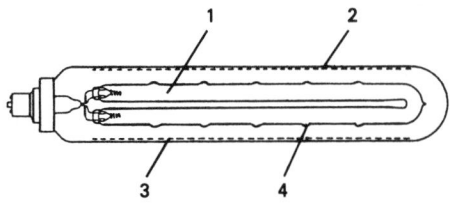

Fig 6 Schematic diagram of a low-pressure sodium lamp. 1, discharge tube; 2, outer bulb (evacuated); 3, indium oxide film; 4, dimples filled with sodium

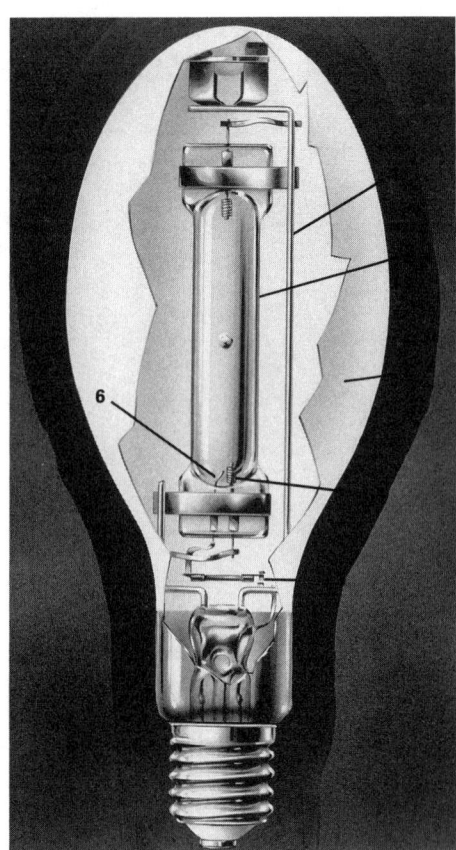

Fig 7 Diagram of a high-pressure mercury lamp. 1, support for discharge tube; 2, discharge tube; 3, outer bulb; 4, main electrode, 5, resistor; 6, auxiliary electrode

lower wattages, soda-lime glass bulbs are used, with lead glass stem and exhaust tubes.

The metal halide lamp is similar in construction to a high-pressure mercury lamp, the major difference being that the discharge tube contains one or more metal halides in addition to mercury. The use of metal halides has two advantages:

- The vapor pressure of the metal halide is usually higher than that of the metal
- The metal halides are chemically less aggressive toward quartz glass than the metals

The halides are partially or wholly vaporized when the lamp reaches its normal operating temperature. The halide vapor dissociates in the hot central region of the arc into the halogen and the metal, and the vaporized metal radiates its characteristic spectrum. The tungsten halogen cycle also operates in some of these lamps. Various metal halide systems have been developed, examples being Na-In-Tl iodides, Na-Sc iodides, Na-Tm iodides, and Sn iodides and bromides. Glass requirements are similar to those for high-pressure mercury lamps; discharge tubes for these lamps are made from quartz glass. The inclusion of metal halides results in requirements for quartz glass components of higher quality. It is essential to exclude the presence of water vapor or hydrogen during lamp manufacture because these increase the striking voltage of the lamps. The quartz glass must also have a low hydroxyl content (<5 ppm) so that evolution of hydrogen or water vapor does not occur during lamp operation. For lamps that contain sodium iodide, measures have to be taken to reduce photo electric currents which otherwise can cause sodium loss by electrolysis through the wall of the arc tube into the outer envelope. A typical metal halide discharge tube is shown in Fig 8.

High-pressure sodium lamps are based on a high-pressure sodium discharge where the sodium vapor pressure is typically 10 kPa. Mercury is also present with a pressure of 80 kPa. Due to the high pressure, the yellow sodium resonance lines are broadened, pro-

Fig 8 A quartz glass 400 W metal halide discharge tube during runup of the discharge. Salt deposits are visible on the wall.

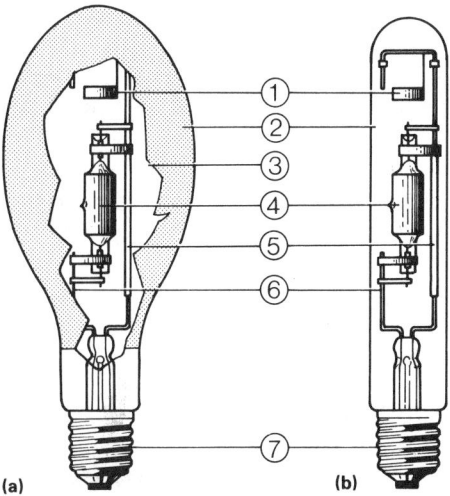

(a)　　　　　　　　(b)

Fig 9 Diagrams of a diffuse ovoid (a) and a clear tubular (b) high-pressure sodium lamp. 1, support springs to maintain discharge tube alignment; 2, lead-in wire; 3, hard glass outer bulb; 4, translucent polycrystalline alumina discharge tube; 5, diffusion coating; 6, end cap of discharge tube; 7, lead-in wire; 8, getter rings for maintaining high vacuum; 9, screw base

ducing a golden yellow light rather than the monochromatic yellow given off by the low-pressure sodium lamp. The production of the high-pressure sodium lamp (Fig 9) became possible after the development of a material that was resistant to high sodium pressures at high temperatures (≈1500 K). The arc tube consists of polycrystalline alumina and the electrodes are sealed using a niobium tube and a sealing frit. The arc tube assembly is placed in an evacuated outer bulb which contains a getter.

Glass/Metal Seals

Glass/metal seals are important for all lamp types. The electric lead wires can be sealed in a stem glass or sealed directly in the bulb by a pinch seal. General requirements for a gas-tight glass/metal seal are:

- The expansion of the stem glass should match the expansion of the metal from the sealing temperature to room temperature
- There should be good adhesion between the glass and the metal. Good adherence is often accomplished by oxidizing the metal prior to sealing
- The glass should retain its electrical resistance at high temperature during operation of the lamp

For soft glass seals, copper-clad wire (Dumet) and alloys of chromium, nickel, and iron are used, mostly sealed into the lead glass stem. The wires are preoxidized in order to obtain either a 3 to 5 nm thick Cu_2O layer or a 1 to 2 nm thick Cr_2O_3 layer that ensure good adhesion to the glass. For hard glass seals (borosilicate and aluminosilicate glasses) tungsten or molybdenum leadthrough wires are used. These wires can be sealed in a separate stem glass or pinched directly in the bulb. The tungsten or molybdenum wires are sometimes enameled to ensure adequate adhesion. When tungsten is sealed into an alkali-containing glass, the adhesion is obtained by the formation of reaction products of alkali oxides and tungsten oxide.

For quartz glass seals, there is no metal that can meet both the refractory and thermal expansion coefficient requirements. Tungsten and molybdenum have much higher thermal expansion coefficients (45×10^{-7} K^{-1} and 54×10^{-7} K^{-1}, respectively) than quartz glass (5.5×10^{-7} K^{-1}). Seals can be made using an intermediate sealing glass with a thermal expansion coefficient between quartz glass and molybdenum or tungsten, but seals are usually made by pinching the quartz glass tube over a thin molybdenum foil with a thickness/width ratio of 1:100 and a sharp foil edge. As the temperature is lowered after making the seal, the thermal expansion mismatch between the quartz glass and the molybdenum causes stresses to develop rapidly. When these stresses reach the yield point of the molybdenum, they are relaxed by the plastic flow of the metal and further stress development does not occur. More detailed information on joining glasses and metals can be found in the article "Glass/Metal and Glass-Ceramic/Metal Seals" in this Volume.

SELECTED REFERENCES

- M.A. Cayless and A.M. Marsden, *Lamps and Lighting*, 3rd ed., Edward Arnold, London, 1983
- C. Meyer and H. Nienhuis, *Discharge Lamps*, Kluwer Technical Books, Deventer, 1988
- *IES Lighting Handbook*, Illuminating Engineering Society, 1981
- G.W. McLellan and E.B. Shand, *Glass Engineering Handbook*, 3rd ed., McGraw-Hill, 1984

CRTs and TV Picture Tubes

John H. Connelly, Corning Incorporated

Donald J. Lopata, Corning Asahi Video Products Company

A CATHODE-RAY TUBE (CRT) is basically an electron tube in which a beam of electrons can be focused to a small area and varied in position and intensity on the surface. Cathode-ray tube applications include capital electronic measuring equipment such as oscilloscopes and the general field of print communications. To the typical consumer, the most familiar form of the cathode-ray tube is the television picture tube found in the home's television receiver.

This article will focus on the requirements of the various glasses that are found in today's color television (CTV) or black and white (B&W) set. Figure 1 shows a view of the CTV picture tube. The major glass components are a panel, a funnel, and a neck. Other glass parts are the frit (solder sealing glass), which seals the panel to the funnel, and two or three small pieces of glass known as multiform gun mounts or multiform pillars that are part of the gun assembly. The final piece of glass is the stem glass that seals the electrical wiring and provides the exhaust connection to the tube.

Direct View Cathode-Ray Tubes

Color Panel or Faceplate Glass. The glass requirements for the CTV faceplate have been and continue to be in a state of change depending on the ever-changing needs of the tube maker in response to the demands of the customer. From the mid-1960s until about 1982, there was an ever-increasing demand for a brighter and brighter picture. Subsequently, the more recent requirement has been for greater contrast. The brighter picture was achieved by increasing the voltage used to accelerate the electron beam. However, this required an increase in the x-ray absorption coefficient of the glass and a relatively high luminous transmittance of the glass. Since 1982, the emphasis has been on contrast and most of the glass produced has been of a darker tint. The primary requirements for the panel glass from the customer's (CTV set manufacturer) point of view are:

- Color
- Thermal expansion
- Sufficient x-ray absorption
- Freedom from defects
- Dimensions within specifications

Secondary requirements include:

- Browning protection
- Nonsolarizing
- Durability
- Viscosity such that the glass can be processed

Another customer is the glass plant's melting and forming department. Glass that cannot be melted and formed without defects is of no commercial value.

Dimensions of the color panel are carefully controlled. The inside face contour is critical to the manufacture of a high-quality picture. The dimensions of the contour of this surface are typically held to a tolerance of ± 0.25 mm (± 0.010 in.). This is a requirement because the distance of the metal aperture mask from the glass surface (the Q space) must be uniform to ensure color purity when the phosphor screen is excited.

The panel periphery is controlled to a tolerance of ± 1.0 mm (± 0.04 in.). This is to ensure that the panel seal edge and the funnel seal edge will mate properly for sealing. The flatness of the seal edge itself is controlled to 0.15 mm (0.006 in.) maximum to prevent gaps in the seal.

Metal pins are glass-to-metal sealed into the inside panel skirt. They are precisely controlled for height from the inside face surface and x and y location in the panel wall. The metal pins support the aperture mask when the tube is constructed.

The outside face surface is ground and polished. The surface is shaped to fit the TV set producer's bezel.

The thermal expansion coefficient (TEC) or CTE (coefficient of thermal expansion) of the panel glass is critical because it must match the funnel and the frit to seal the parts together, and the expansion match with the metal pins must be sufficient for a good glass-to-metal seal.

Funnel Glass. Requirements for the funnel glass are a little less stringent than those for the panel glass. Because the funnel is not the portion of the set viewed, there is no color specification and the inclusion level for seeds and blisters is less strict. However, because the funnel glass is much thinner, the x-ray absorption coefficient, μ, required is much higher (62 cm^{-1} versus 28 cm^{-1}) for the panel. This higher μ value is obtained through the use of lead oxide in amounts approaching 23

wt%. It is also required that the glass be capable of being spun or pressed for the forming process. The expansion coefficient of the funnel glass is closely controlled so that it can be sealed to the panel and the neck glass. The funnels are designed for structural integrity of the vacuum vessel, sufficient clearance for the electron beam to scan the entire screen on the panel, and enough thickness for x-ray absorption.

The yoke area of the funnel is precision pressed to allow the magnetic yoke to be placed as close to the electron beam as possible on the outside of the glass while providing adequate clearance to prevent the beam from striking the glass on the inside. An anode button is glass-to-metal sealed through the funnel wall to provide electrical contact to the inner surface of the tube.

For tube construction, the funnel is referenced in the x and y planes by three alignment points or alignment pads that are molded into the outside surface of the funnel wall near the seal edge. A defined diameter on the outside glass surface of the funnel yoke is the z reference plane.

Neck Glass. The relatively thin CTV neck glass requires an even higher x-ray absorption coefficient (92 cm^{-1}). Another important requirement is the liquidus-viscosity (viscosity at which devitrification occurs) of the glass. There are several forming processes to make tubing (for example, Vello process, Danner process, and updraw). It should be noted that the process which makes the best tubing also requires the most stable glass. A liquidus-viscosity approaching 10^5 Pa$\cdot$s (10^6 P) has been determined to be required for suitable Vello processing, while a liquidus viscosity of 5×10^3 Pa$\cdot$s (5×10^4 P) is satisfactory for Danner or updraw processing.

The neck is glass-to-glass sealed to the funnel body to form a graded seal between the funnel and a gun assembly. The stem of the electron gun is typically a G-12 glass.

Solder Sealing Glass. A typical solder sealing glass composition is shown in Table 1(a). Properties for this glass composition are shown in Table 1(b). The glass, which has a very low softening point, is initially made as sheet glass and then ball milled with various concentrations of zircon to produce a frit for suspension in an appropriate vehicle for extrusion application. This frit permits relatively low sealing temperatures to be used such

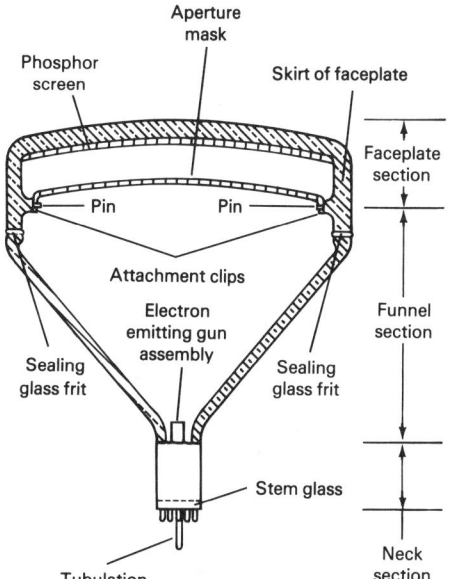

Fig 1 Schematic showing cross section of components of a conventional color television tube. The glass envelope consists of a funnel section, a faceplate section, and a neck section. The tubulation is used to evacuate the tube and is removed after vacuum processing.

that no distortion of the sealed parts occurs. During the sealing cycle, the glass melts, devitrifies (crystallizes), and provides a seal with higher temperature resistance and suitable electrical and mechanical properties (Ref 1, 2).

TV Gun-Mount Pillars. The gun-mount pillars support and insulate the electron gun electrodes. They are made from a high-stability multiform glass with very high electrical resistivity and very low thermal expansion coefficient.

Multiform is powdered glass that is coated with a binder and a coloring agent. It is dry pressed into the compensated shape and sintered in a furnace to produce the required gun mount pillars in various colors.

The TV gun manufacturer partially softens the pillars and embeds the metal gun parts into the softened multiform glass. The glass quickly hardens and mechanically locks the metal parts into place; it is not a glass-to-metal seal. The multiform gun-mount pillars maintain the critical spacing of the electron gun parts and electrically insulate them from each other.

The various color multiform gun-mount pillars are used by the manufacturer for gun identification.

B&W TV Picture Tubes

The glass requirements for a B&W TV picture tube are not as stringent as for a color TV tube. The panel and funnel are glass-to-glass sealed; therefore, the maximum tube processing temperature is not as high as for a CTV tube. A lower strain point glass is used.

The panel and funnel are made from the same glass. This is acceptable because the tube operating voltages are much lower than in a CTV tube; therefore, fewer x-rays are generated inside the tube. The linear absorption coefficient of this glass is also lower (that is, μ is 21.5 cm^{-1} at a wavelength of 0.06 nm, or 0.6 Å).

The neck glass on a B&W TV tube is G-12 (code 0120), which is the same glass used for the electron gun stem.

Projection CTV Tubes

Projection TV receivers have three monochrome type tubes, a red tube, a green tube, and a blue tube. Each tube has a lens system attached to the face and the three independent monochrome pictures are converged on the screen to produce the resultant color picture.

The tubes in a projection TV are monochrome, but the glass requirements are different from a standard B&W TV picture tube. The projection tubes are operated at much higher voltages; therefore, nonbrowning glass is required and x-ray absorption must be provided either by the glass or by means of an absorption system outside of the projection tube.

The liquidus-viscosity requirement is less stringent than that required for the large TV panel faceplates.

Specific Properties Imparted by Various Oxides Used in CTV Tubes

The following oxides are used in the production of CTV tubes because of the specific properties they impart to the CTV tubes.

SiO$_2$. Basic glass forming oxide (comes primarily from sand).

Al$_2$O$_3$. Network modifier generally used to adjust viscosity, to improve liquidus-viscosity to aid durability; sourced from feldspar, nepheline syenite, sandspar, Al$_2$O$_3$, Al(OH)$_3$, petalite, spodumene, and, at one time, lepidolite.

Alkalis. At one time or another all of the alkali metal oxides have been used in the manufacture of television faceplate glasses. In today's marketplace, the Rb$_2$O and Cs$_2$O are too expensive to consider as raw materials for the television industry. The alkalis Li$_2$O, Na$_2$O, and K$_2$O are the primary fluxes in the melting process and are considered network modifiers affecting viscosity, resistivity, dielectric properties, and expansion. Sources for these oxides are the carbonates, nitrates, and to some extent the alumina-bearing minerals listed above.

Alkaline Earths. The oxides of calcium and magnesium have been part of the glassmaker's repertoire for many years. They are also network modifiers and are important in that they contribute durability and increased electrical resistivity to the glass. These oxides

are commonly sourced from dolomitic limestone or high-calcium (calcite) limestone.

Barium oxide and strontium oxide are two other alkaline earth oxides that, in addition to the above effects on the glass composition, also have relatively high mass absorption coefficients at the wavelengths associated with x-ray radiation found in television tubes. Sources for these oxides are carbonates.

Lead Oxide. Also considered a network modifier, lead oxide has the highest mass absorption coefficient of any of the oxides used in television glasses. There is one drawback with the use of lead oxide in the faceplate and that is browning. Electrons and x-rays striking a glass containing lead oxide will cause a phenomenon known as electron browning or x-ray browning. While this is of no consequence in the funnel and neck compositions, its appearance in the panel is undesirable. For this reason, the upper limit of lead in the panel is 3 to 5 wt% Pb. At this level in a properly screened tube, the browning is not a problem. Lead is sourced as the oxide or silicate. Also, the handling of lead compounds in the batch mixing area requires proper hygiene procedures because of its toxicity.

Zirconium Oxide. A recent addition to the panel composition, zirconia is considered a network former. It has a mass absorption coefficient similar to strontium oxide and will improve the hardness of the glass. A fine particle size is required to ensure meltability. Zirconia is sourced as the oxide or the silicate.

Minor Ingredients. The following oxides are added in seemingly small amounts, but their use is important to the final product.

Cerium oxide enhances the fading of x-ray browning when the television set is turned off. Usual concentration is 0.2 to 0.3 wt% CeO$_2$. Source is usually the least expensive that can be found.

Titanium oxide enhances the effect of cerium and also makes the glass resistant to solarization. Typical concentration is 0.4 to 0.5 wt% TiO$_2$.

Antimony trioxide is a fining agent (aids in removing gaseous inclusions from the glass) usually batched at 0.35 to 0.50 wt% Sb$_2$O$_3$. Source is sodium antimonate or the oxide.

Arsenic trioxide is also a fining agent that is now being used in lower amounts due to environmental concerns. When used, it is usually batched at 0.1 to 0.2 wt% As$_2$O$_3$. The source is arsenic acid or the oxide.

Fluorine. This anion, which can replace oxygen in the glassy matrix, has a very powerful effect on the viscous properties of the glass. Its use, which is less common today than several years ago because of environmental manufacturing concerns, is as an aid to melting and fining of the glass. Today it is used at ≤0.25 wt% in some panel glasses. Prior usage was as high as 1.2 wt%. Presently, fluorine is sourced from fluorspar or sodium silica fluoride. Lepidolite was at one time used as a source of fluorine.

Table 1(a)   Composition of selected television component glasses

Component	Glass type (a)	SiO$_2$	Al$_2$O$_3$	B$_2$O$_3$	Na$_2$O	K$_2$O	Li$_2$O	CaO	MgO	PbO	SrO	Sb$_2$O$_3$	As$_2$O$_3$	ZnO	BaO	ZnO$_2$	TiO$_2$	CeO$_2$	F$^-$
Envelope																			
Stem	0120	56	2	...	4	9	...	...	...	29	...	0.5	...	...	...	...	...	...	...
Funnel	0138	54	2	...	6	8	...	3.5	2.5	23	...	0.1	...	...	...	...	...	...	...
Neck	0137	52.5	1.0	...	0.8	12.0	...	...	...	28.0	5.0	0.5	...	...	...	...	...	...	...
Multiform	7761	80.0	...	17.5	...	2.5	...	...	...	...	...	...	...	...	...	...	...	...	...
Frit	7580	2.0	0.1	8.4	...	...	...	...	...	75.4	...	...	...	12.2	1.9	...	...	...	...
Panel																			
Black and white	9008	55.1	3.6	...	6.8	6.4	0.5	...	.	1.7	...	0.4	0.2	...	12.0	...	...	...	1.0
Color	9061	62.0	2.0	...	7.0	9.0	...	1.7	0.7	2.2	10.3	0.4	0.2	...	2.4	...	0.5	0.2	0.3
Projection	9039	57.5	1.7	...	6.3	5.8	1.0	...	...	...	8.7	0.4	...	...	15.0	3.0	...	0.7	...

(a) Corning Glass Works products

Table 1(b)   Properties of selected television component glasses

Property	Stem (0120)	Funnel (0138)	Neck (0137)	Multiform (7761)	Frit (7580)	B&W (9008)	Color (9061)	Projection (9039)
Thermal expansion coefficient at 0–300 °C (30–570 °F), 10^{-7}/°C	89.5	97.0	97.0	28.0	98.0	89.0	99.0	96.4
Thermal stress resistance, °C	20	17	19	...	...	18	15	17
Viscosity data:								
Strain point, °C (°F)	395 (745)	435 (815)	436 (817)	468 (874)	293 (559)	406 (763)	460 (860)	458 (856)
Annealing point, °C (°F)	435 (815)	474 (885)	478 (892)	517 (963)	311 (592)	444 (831)	501 (934)	500 (930)
Softening point, °C (°F)	630 (1165)	654 (1210)	661 (1222)	820 (1510)	374 (705)	646 (1195)	689 (1270)	680 (1255)
Working point, °C (°F)	986 (1805)	965 (1770)	978 (1792)	1300 (2370)	890 (1635)	1004 (1839)	1000 (1830)	983 (1800)
Density, g/cm^3	3.05	2.980	3.18	2.16	6.47	2.64	2.70	2.90
Young's modulus (kg/cm^2) $\times$ 10^6	0.6082	0.7038	0.6609	...	...	0.7122	...	...
Poisson's ratio	0.22	0.24	0.23	0.20	0.25	0.24	0.23	0.24
Log$_{10}$ of dc volume resistivity:								
25 °C (75 °F)	>17	>17	>17	>17	13.6	>17	>17	>18
250 °C (480 °F)	10.1	9.7	10.1	12.5	6.7	9.4	9.2	9.6
350 °C (660 °F)	8.0	7.6	8.3	10.1	8.2	7.4	7.5	7.7
Dielectric properties at 1 MHz and 20 °C (70 °F):								
Dielectric constant	6.97	7.92	8.60	4.0	...	6.3	6.97	5.44
Loss tangent	0.0013	0.0011	0.0011	0.0009	...	0.0017	0.0013	0.0007
Refractive index at Na D line (0.5893 μm)	1.560	1.565	1.550	1.46	1.65	1.506	1.518	1.553
Birefringence constant	317	236	279	...	...	277	258	...
Linear absorption coefficient at 0.06 nm (0.6 Å) minimum, cm^{-1}	75.0	62.0	90.0	8.0	40.0	20.0	28.0	35.0

Coloring Oxides. Typical coloring oxides are cobalt, nickel, chromium, and iron. Their particular effect on color is discussed in the section "Color and Transmittance" in this article. Concentrations of these oxides in the glass are a function of the desired luminous transmission.

Iron oxide is rarely purposely added because it is found as a contaminant in many of the raw materials.

Cobalt, nickel, and chromium can be sourced as the oxide in which case they are made into a color mix with sand or feldspar so that precise concentrations can be exactly weighed. Dissolving cobalt nitrate, nickel nitrate, and potassium dichromate in water allows a color solution to be made that permits an even more precise control of colorant to be achieved.

Neodymium Oxide. In recent years, there has been some use of neodymium oxide as part of the color package (Ref 3). This oxide has the unique capability of absorbing light at the 588 nm (5880 Å) wavelength and pro-vides an added measure of contrast to the viewed picture.

Manganese dioxide is another coloring oxide that is used in the black-and-white compositions to mask the effect of x-ray browning. In this case, the manganese dioxide causes a discoloration to the exposed glass to such an extent that the net effect is to hide the browning caused by the x-rays.

Nitrates. Sodium nitrate or potassium nitrate is generally part of the batch make-up. This material is used to supply oxygen to the melt for the oxidation of any stray metal or carbonaceous material present in the cullet. The amount is generally 18 to 23 kg (40 to 50 lb) per batch that will yield 450 kg (1000 lb) of glass.

X-Ray Radiation Limits for TV Tubes

The U.S. Federal Standard for x-rays from television receivers under Public Law 90–602 (Radiation Control for Health and Safety Act of 1968) established the maximum permissible radiation level for the complete receiver. Appropriate controls and test methods were developed for the TV tube as a receiver component. When the picture tube in a TV receiver is operating, it is a source of x-rays. The glass envelopes used for TV picture tubes are designed to provide the required attenuation of x-rays. This is accomplished by designing and controlling both the glass composition and the minimum thickness of the glass parts.

The maximum allowable x-ray output from a TV receiver is 0.13 µC/kg · h (0.5 milliroentgens per hour or mR/h). The TV tubes are designed to attenuate x-rays so the receiver will remain within this limit under failure mode conditions. Such abnormal conditions often result in the picture tube being subjected to anode voltages that exceed its maximum ratings. Prior to 1970, the requirements were for TV receivers under normal operating conditions.

The total x-ray energy, E, at the target and inside the tube depends on the accelerating voltage, V, the current, i (a measure of the number of electrons), and the atomic number of the target, Z:

$$E = ciV^2Z \qquad \text{(Eq 1)}$$

where c is a constant.

A monochrome tube operates at approximately 16 kV and the target material would be the aluminum, the phosphor, and the face glass. A color TV tube operates at approximately 25 kV, and the primary target is the iron in the aperture mask. Because the iron aperture mask intercepts 80% of the electrons, the aluminum, phosphor, and face glass are only minor contributors to the x-rays. With much higher operating voltage and a better x-ray target, a color TV tube generates many more x-rays than a monochrome tube.

The absorption of x-rays by an elementary material, such as iron, is dependent upon the wavelength of the radiation and the density, thickness, and atomic number of the material. The mass absorption coefficient, ω, is the unit used to express the absorbing characteristics of the material. In the case of x-rays, the mass absorption coefficient is independent of the physical state of the material and can be applied to gases, liquids, and solids.

In a compound or a mixture such as glass, each component element absorbs independently of the others. The total absorption is the summation of these separate independent absorptions. Thus, the mass absorption coefficient of a mixture, $\omega_{mixture}$, is determined by summation of the contributions of the components, as follows:

$$\omega_{mixture} = \Sigma(f_c \cdot \omega_c) \qquad \text{(Eq 2)}$$

where f_c is the weight fraction of each component element, and ω_c is the mass absorption of each component element.

For glasses, the absorption is calculated using the coefficients for the oxides used in the glass. The coefficients are published in various tables over a range of wavelengths. The wavelengths of interest for television are in the range of 0.4 to 0.06 nm (0.4 to 0.6 Å). This is because many TV tubes are rated at 27.5 kV and with present color TV glasses, x-rays are trivial at <20 kV.

Planck's constant appropriate for energy in keV and wavelength in Angstrom units is 12.393. The relationship between minimum wavelength, λ_{min}, and accelerating voltage is defined below:

$$\lambda_{min} = \frac{12.393}{keV} \qquad \text{(Eq 3)}$$

where keV is numerically equal to the accelerating voltage, kV.

$$\omega_{glass} = \Sigma(f_{oxide} \cdot \omega_{oxide}) \qquad \text{(Eq 4)}$$

where f_{oxide} is the weight fraction of component oxide, and ω_{oxide} is the mass absorption of the oxide.

The linear absorption coefficient of the glass, μ_{glass}, is given by:

$$\mu_{glass} = \omega_{glass} \cdot \rho \qquad \text{(Eq 5)}$$

where ρ is the density.

A given glass can be characterized by specifying the absorption coefficient at a single wavelength. For purposes of control and specification of TV glass, the industry has arbitrarily chosen 0.06 nm (0.6 Å). For control of absorption, the linear absorption coefficient is calculated from total chemical analysis. A minimum value is established, which is based on the making of x-ray dose rate measurements on tubes with parts of known thickness and composition. A sample calculation is shown in Table 2.

Minimum values have been established for the linear absorption coefficient at 0.06 nm (0.6 Å) for TV glasses as follows:

TV component	Linear absorption coefficient (μ), cm^{-1}
Color set	
Panel glass	28.0
Funnel glass	62.0
Neck glass	90.0
Black and white set	
Glass	21.5

The design of TV tubes takes into consideration the glass μ value and the thickness, t, for absorption of x-rays generated inside the tube. The face glass of a TV tube is much thicker than the funnel or neck glass and for this reason has a lower required μ value.

Color and Transmittance

Maintaining the chromaticity of the television faceplate glass is one of the more interesting challenges facing the glass technologist. Chromaticity is a measure of the color of transmitted light through the glass panel face and can be graphically represented by x and y coordinates. This property is specified in terms of illuminant C (the common or standard light source under average daylight conditions) or I.C.I. (x and y) coordinates defining a polygon (usually a hexagon) with the center or bogie color point indicated at 11.43 mm (0.450 in.). The light source for the measurement of this property for CTV glass is identified as P22 standard phosphor illuminant (at 8600 K or, 15,000 °F, and 25 minimum perceptible color difference, or MPCD) or in terms of I.C.I. (x and y) coordinates with x value of 0.2869 and y value of 0.3159. For monochrome (B&W) panel glass, the chromaticity is measured at 10.16 mm

Table 2 Calculation of linear absorption coefficient at 0.06 nm (0.6 Å) for an actual CTV panel glass

Oxide	Wt%	Mass absorption of oxide at 0.06 nm (0.6 Å), ω_{oxide}	Weight fraction of component oxide, f_{oxide}	$f_{oxide} \cdot \omega_{oxide}$(a)
SiO$_2$	62.86	2.34	0.6286	1.4709
Al$_2$O$_3$	2.06	2.11	0.0206	0.0435
TiO$_2$	0.50	9.12	0.0050	0.0456
CaO	1.76	8.81	0.0176	0.1551
MgO	0.84	1.92	0.0084	0.0161
SrO	10.43	53.40	0.1043	5.5696
BaO	2.32	25.10	0.0232	0.5823
PbO	2.35	82.90	0.0235	1.9482
Na$_2$O	7.02	1.69	0.0702	0.1186
K$_2$O	8.81	8.45	0.0881	0.7444
As$_2$O$_3$	0.17	33.20	0.0017	0.0564
Sb$_2$O$_3$	0.38	18.20	0.0038	0.0692
CeO$_2$	0.20	25.30	0.0020	0.0506
F$_2$	0.30	1.20	0.0030	0.0036

(a) $\omega_{glass} = \Sigma(\omega_{oxide} \cdot f_{oxide}) = 10.8741$. Density = $\rho = 2.697$. μ_{glass} at 0.06 nm (0.6 Å) = $\rho \cdot \Sigma(\omega_{oxide} \cdot f_{oxide}) = 29.33$

(0.40 in.) with P4 standard phosphor illuminant (at 7000 K, or 12,140 °F and 25 MPCD) or in terms of I.C.I. coordinates with values at 0.3058 for x and 0.3171 for y at MPCD.

Figure 2 shows the color polygon that is used for the 52% transmittance glass (now the most commonly produced tint) at 11.43 mm (0.450 in.) of thickness. In terms of illuminant P22, the center point of this polygon is at x value of 0.2835 and y value of 0.3160. The effect of small variations in the oxide content of the glass shown in Fig 2 on x and y values is listed in Table 3.

For glasses with a transmittance of 68% and 72%, the center point of the polygon is an x value of 0.2840 and a y value of 0.3160, while the 85% transmittance glass has a center at an x value of 0.2845 and a y value of 0.3160.

Luminous transmittance ($\bar{Y}$ or T) is the measure of the intensity of the emergent beam versus that of the incident beam. Transmittance follows the Beer-Lambert law and can be mathematically expressed as follows:

$$T = I/I_0 = (1 - r)2 \cdot e^{-kt} \qquad \text{(Eq 6)}$$

where I is the emergent intensity; I_0 is the incident intensity; r is the reflection loss at 4% for a glass having a refractive index of 1.5; k is the sum of the absorption factors for the coloring oxides times their concentration; and t is the thickness of the glass.

Table 3 Effect of small variations in oxide concentration of an 11.43 mm (0.450 in.) thick 52% transmittance glass with a P22 illuminant light source shown in Fig. 2

Typical center, wt%	Oxide	Additional oxide concentration ppm(a)	Coordinates x	Coordinates y	Transmission ($\bar{Y}$), %
0.0120	NiO	10	+0.0010	+0.0010	−1.4
0.0020	Co$_3$O$_4$	1	−0.0010	−0.0010	−0.5
0.0006	Cr$_2$O$_3$	1	−0.0001	+0.0003	−0.1
0.04	Fe$_2$O$_3$	100	+0.0002	+0.0004	−0.1

(a) 1.0 ppm = 0.0001 wt%

The usual coloring oxides in a television faceplate glass are the oxides of Co_3O_4, NiO, Fe_2O_3, and Cr_2O_3. The cerium and titanium oxides have an effect, but because their concentration is usually held constant, no effort is made to measure their effect in terms of color adjustment. To precisely measure the effect of the various coloring oxides, it is necessary to make a number of experimental melts of a given composition wherein the colorant concentrations are varied to determine their individual contributions to the color of a given glass. It is these authors' experience that the general effects are similar but slightly different from one composition to another. Figure 2 and Table 3 show the typical effect of adding the various coloring oxides to a glass that is in the nominal center of the polygon. The dotted line inside the solid polygon designates the area where color is maintained so that, if there are any small measurement errors, the color will still be within specifications.

Properties of Glasses

The chemistry and typical physical properties for the various glasses found in a color or B&W television set are shown in Table 1. Normally, only a few of these properties and the chemistry are routinely measured in production. Because the properties are a function of composition, if the chemistry is on target, the properties will also be on target. Typically, the softening point, thermal expansion, and density are routinely measured in addition to the chemistry. These are relatively easy to do on a daily basis and provide the glass technologist an excellent indication of how the glass is performing.

Effects of Glass Composition. Table 4 shows the chemistry for a variety of color panel compositions (Ref 3–5). Many approaches have been used to achieve the desired physical properties. In general, most of the panel glasses produced are interchangeable so that the tube maker can purchase glass from a number of different manufacturers and use the parts in his process without changing his processing procedures. Table 4 lists the mass absorption coefficients at 0.06 nm (0.6 Å) for the various oxides and the typical range of properties of these oxides.

Thermal Expansion. One property that has created a bit of confusion over the years is thermal expansion. The coefficient of expansion listed in this article refers to the 0 to 300 °C (30 to 570 °F) expansion that is a reference target which the glass technologist maintains for the given glass. This property is measured in the laboratory with an expansion dilatometer making measurements on a standard piece of glass at 0 °C (32 °F) and 300 °C (572 °F). Other companies report expansion from 30 to 300 °C (86 to 572 °F). Figure 3 shows the expansion curves for the glasses listed in Table 1. As indicated in Fig 3, the expansion curve is not linear and increases as the temperature increases. The expansion curves for the pin metal (430 Ti) and button metal (Sylvania 4) are also shown.

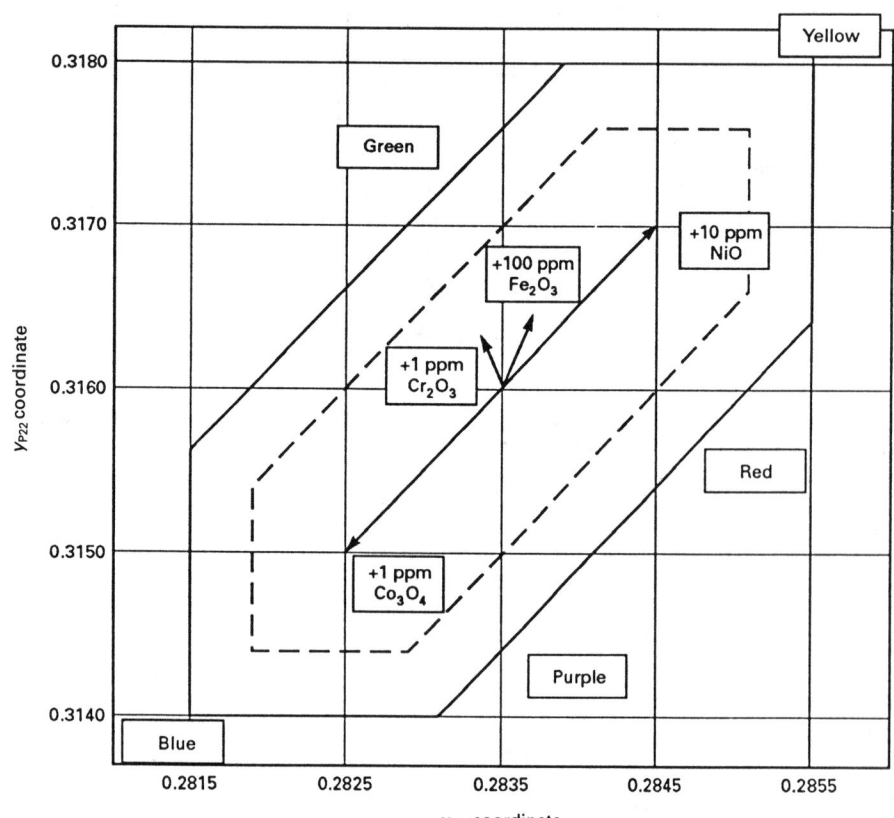

Fig 2 Color polygon for an 11.43 mm (0.450 in.) thick 52% transmittance glass with a P22 illuminant light source. Center point of polygon is at $x = 0.2835$ and $y = 0.3160$.

Table 4 Compositions and properties of selected glasses used for color television faceplate applications

Constituent	Composition, wt%(a)								Mass absorption coefficient of oxide constituent at 0.06 nm (0.6 Å), ω_c
	9068(b)	TL-28(c)	G999(d)	8051(e)	PT-28C(f)	5008(g)	9075(h)	Nd₂O₃(f)	
SiO_2	62.90	64.39	62.61	62.16	61.43	59.48	61.30	62.60	2.34
Al_2O_3	2.10	1.31	1.83	2.10	2.06	2.27	2.06	1.00	2.11
Fe_2O_3	0.04	0.04	0.04	0.03	0.04	0.04	0.04	...	...
TiO_2	0.52	0.47	0.42	0.48	0.51	0.39	0.51	0.30	9.12
CaO	1.79	2.80	2.21	0.84	0.12	2.44	1.88	0.50	8.81
MgO	0.85	0.05	1.53	0.56	0.03	0.52	0.04	0.50	1.92
SrO	10.31	9.42	10.13	9.99	9.61	7.93	9.51	9.00	53.40
BaO	2.37	2.09	2.03	7.38	8.71	8.80	5.51	7.20	25.10
PbO	2.31	2.84	2.48	0.05	0.04	0.05	0.05	0.50	82.90
Na_2O	6.99	7.59	7.81	7.70	7.48	7.96	8.43	7.20	1.69
K_2O	8.90	8.08	7.92	7.85	7.58	6.77	6.69	7.20	8.45
As_2O_3	0.16	0.15	0.31	0.01	0.01	0.01	0.22	...	33.20
Sb_2O_3	0.38	0.40	0.43	0.60	0.32	0.31	0.40	...	18.20
CeO_2	0.21	0.22	0.25	0.25	0.29	0.26	0.31	0.50	25.30
F^-	0.29	0.26	...	...	...	...	...	...	1.20
ZrO_2	...	...	...	...	1.28	2.58	2.59	1.00	54.5
ZnO	...	...	...	...	0.49	...	0.00	1.00	28.5
Li_2O	...	...	...	...	...	0.19	0.46	0.50	0.55
$^-O_2$	−0.12	−0.11	...	...	...	...	...	...	0.86
Nd_2O_3	...	...	...	...	...	...	...	1.00	29.4
Total	100.00	100.00	100.00	100.00	100.00	100.00	100.00	100.00	
Density, g/cm³	2.696	2.696	2.708	2.770	2.769	2.797	2.759	...	
Linear absorption coefficient, μ_{glass}, at 0.06 nm (0.6 Å), cm⁻¹		29.1	28.9	29.2	29.3	29.4	29.0	28.9	29.0

(a) Glass properties include: softening point, °C (°F): 685–705 (1265–1300); annealing point, °C (°F): 510–525 (950–975); strain point, °C (°F): 455–470 (850–880); thermal expansion, 10^{-7}/°C: 98.5–99.5; $\log_{10}$ of dc volume resistivity: >9.0 [at 250 °C (480 °F)], >7.0 [at 350 °C (660 °F)]. (b) Corning Glass Works. (c) Owens-Illinois. (d) Radio Corporation of America (now Thomson Consumer Electronics). (e) Schott. (f) Nippon Electric Glass. (g) Asahi. (h) Samsung Corning

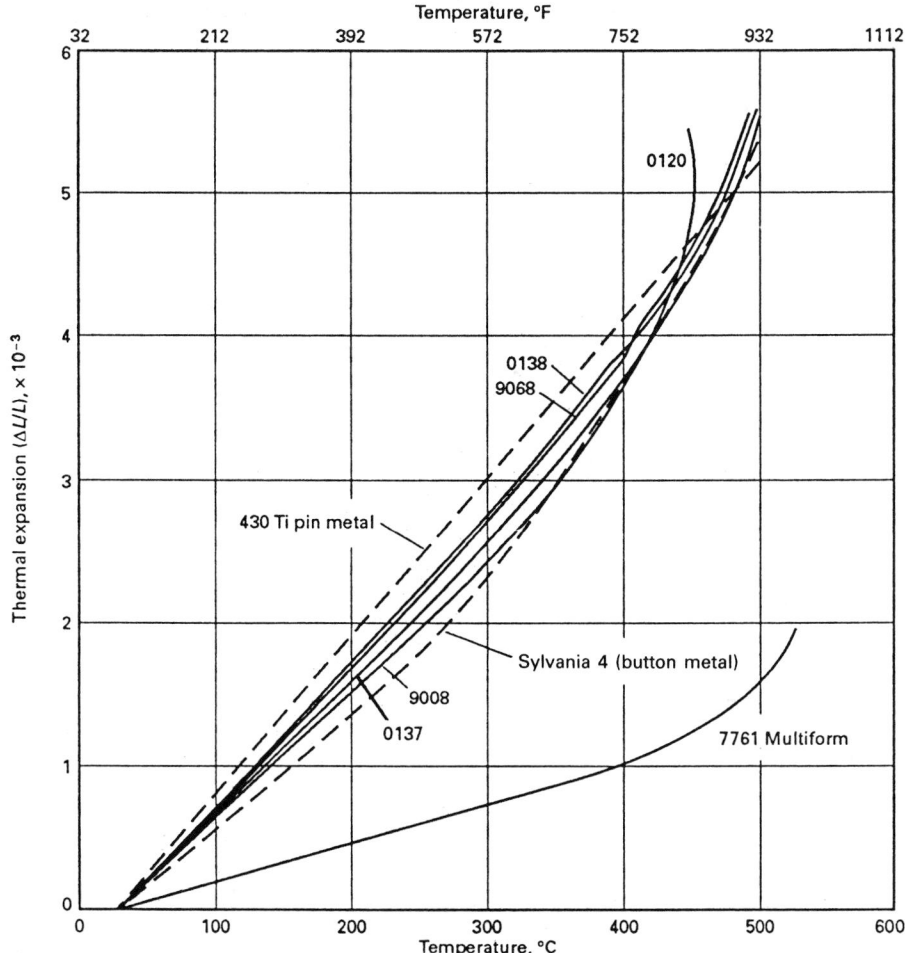

Fig 3 Comparison of thermal expansion coefficients of six selected CTV component glasses and two pin metals

If one compares the 0 to 300 °C (32 to 572 °F) expansion coefficient of the panel and funnel glass (that is, 99.0 versus 97.0), it would appear that these expansions are too far apart to achieve a satisfactory seal. However, when the expansions of the two glasses at the set point (defined as 5° above the strain point of the softer glass) are compared, it can be seen that the expansion curves are very close together. The ideal situation would be to have an exact match at this temperature. It is at this temperature that no additional viscous flow of the glasses is assumed to occur and that when the glasses, which have been sealed cool to room temperature, will have no permanent stress built into the seal. During cooling, there is a temperature stress generated by the different coefficients of the two glasses over the temperature range between the set point (440 °C, or 825 °F) of the softer glass and room temperature. An excellent discussion of this phenomenon and an approach to making the measurements is given by Hagy (Ref 6).

Glass Property Nomenclature. Physical property definitions are given below for those properties with which the reader may not be familiar.

Viscosity is the flow due to the force of shear and induced most commonly by gravity. The English unit of viscosity measurement is the poise, P, and the metric unit of measurement is Pa · s.

Strain point is the temperature at which the internal stresses are reduced to low values after 4 h. Viscosity of the glass is $10^{13.5}$ Pa · s ($10^{14.5}$

P). At this viscosity, glass is substantially rigid.

Annealing point is the temperature at which the internal strains are reduced to an acceptable commercial limit in 15 min. Viscosity of the glass at the annealing point is typically 10^{12} Pa·s (10^{13} P).

Softening point is the temperature at which glass will deform under its own weight. Viscosity of the glass at the softening point is typically $10^{6.6}$ Pa·s ($10^{7.6}$ P).

Working point is the temperature at which glass flows sufficiently for the glass to be formed or sealed. Viscosity of the glass at the working point is typically 10^3 Pa·s (10^4 P).

Melting temperature is the temperature at which the glass has a viscosity of 10 Pa·s (100 P) and is commonly used as a reference point when comparing differences in meltability between two glasses.

Future Trends

High-definition television (HDTV), which has scanning rates of 1125 lines per frame (more than twice the 525 lines per frame in United States-receivers) at 60 fields per second will most likely be the next major change in the television industry. The technology is currently available, and once the politics are worked out this system will be offered. Some interesting challenges will face the glass manufacturer. Because this system runs at even higher voltages than those currently used, the x-ray absorption coefficient requirement will increase and the dimensional specifications will be even more precise than today.

With the environmental concerns now being legislated, it appears that the old TV set soon will no longer be disposed of at the local landfill. It is expected that an industry devoted to reclaiming glass for use as cullet by the glass manufacturer will evolve. This new and variable raw material will provide interesting challenges to the field of glass technology in the future.

REFERENCES

1. F.W. Martin and F. Zimar, "Fusion Seals and Their Production," U.S. patent 3,258,350, June 1966
2. S.A. Claypoole, "Composite Article and Method," U.S. patent 2,889,952, June 1959
3. N. Daiku, "X-Ray Absorbing Glass for a Color Cathode Ray Tube Having a Controlled Chromaticity Value and a Selective Light Absorption," U.S. patent 4,390,637, June 1983
4. JT-32 Linear Absorption Coefficient Round Robin on International TV Glasses, Electronic Industries Association, 1985
5. JT-32 Linear Absorption Coefficient Round Robin, Electronic Industries Association, 1984
6. H.E. Hagy, Trident Seal—A Rapid and Accurate Expansion Differential List, *J. Am. Ceram. Soc.*, Vol 62 (No. 1–2), 1979, p 60–62

SELECTED REFERENCES

- W.H. Armistead, "Soft Glass Having Wide Working Range," U.S. patent 2,527,693, Oct 1950
- D.C. Boyd and D.A. Thompson, "Lead Free Glasses of High X-Ray Absorption for Cathode Ray Tubes," U.S. patent 4,277,286, July 1981
- J.H. Connelly and G.B. Hares, "X-Ray Absorbing Glass Compositions," U.S. patent 3,464,932, Sept 1969
- J.H. Connelly, "Glass for Projection Cathode Ray Tube Faceplate," U.S. patent 4,830,990, May 1989
- W.J. Cook, Spoils of a Good Air War, *U.S. News & World Report*, 10 Sept 1990, p 75–80
- J. De Gier and J.A.M. Smelt, "Cathode-Ray Tube," U.S. patent 2,477,329, July 1949
- J. De Gier, "Cathode-Ray Tube Having a Non-Discoloring X-ray Absorptive Display Window," U.S. patent 3,422,298, Jan 1969
- *Kirk-Othmer Encyclopedia of Chemical Technology*, Vol 11, 3rd ed., p 826–827
- R.L. Mathew, Without Glass, It's Not a Tube, *Information Display*, Vol 14 (No. 6), June 1988, p 18–21
- "Properties of TV Glasses," Television Glass Catalog Section B, Corning Glass Works, 1 July 1979
- J.A. Shell, "Zirconia Containing Glass Compositions for Cathode Ray Tubes," U.S. patent 3,987,330, Oct 1976
- B.L. Steierman, "Cathode-Ray Tube," U.S. patent 4,065,697, Dec 1977
- TEPAC Publication No. 194, Electronic Industries Association, July 1986
- J. Van der Geer, W. Van Pelt, M. Ploeger, W.J. Spoor, and G.H.A.M. Van der Steen, "Cathode-Ray Tube Face-Plate," U.S. patent 4,337,410, June 1982

Information Display

Peter L. Bocko and Roger K. Whitney, Corning Incorporated

INFORMATION DISPLAY technologies that employ glass as a critical component include cathode-ray tubes (CRTs), flat panel displays, and to a lesser extent, hard-copy devices (printers). Flat panel displays, especially, are becoming a major new application for high-performance glass because these devices offer advantages over CRTs in terms of less weight and smaller size for a given display size as well as providing better display performance in some aspects.

Glass as used in the construction of video and other cathode-ray tubes (CRT) is covered in the article "CRTs and TV Picture Tubes" in this Volume. The focus here will be on glass as used for information display in such non-CRT flat panel display applications as liquid crystal displays (LCD), thin-film electroluminescent (TFEL or EL) displays, light-emitting diode (LED) displays, plasma display panels (PDP), and vacuum fluorescent displays (VFD), all of which use flat glass as an integral part of the display. Hard copy generating devices (for example, dot matrix printers and ink-jet printers) use specialty glass components in the writing function. Each of these display technologies relies on some of the inherent or unique properties of glass for effective performance.

The application of glass to these various information display technologies will be reviewed from the point of view of basic device construction and the associated glass property requirements. These property requirements can be grouped under general headings such as environmentally protective (which, in some instances, may mean hermetically sealed or vacuum tight), mechanically strong, optically transparent, thermally and dimensionally stable, electrically insulating, and chemically inert.

Flat Panel Display Technologies

Substrate needs for flat panel displays are defined by the display characteristics. Therefore, in order to discuss substrate requirements and properties for these applications, a brief description of display types is necessary. Detailed information on these products is available in Ref 1.

Liquid Crystal Displays

The LCD cell is most commonly constructed of two glass substrates encapsulating a liquid crystal material. The two substrates are usually sealed together with epoxy, acrylic, or silicone adhesives. Liquid crystal displays are passive displays (that is, dependent on external sources of light for illumination). Two basic types of matrix configurations are intrinsic matrix addressed (relying on the threshold properties of the liquid crystal material) and extrinsic matrix or active matrix (AM) addressed in which an array of diodes, ferroelectric switches, metal-insulator-metal (MIM) devices, or thin-film transistors (TFTs) provide an electronic switch for each pixel or picture element. There are different substrate requirements for the two matrix addressing schemes. Intrinsically addressed LCDs are fabricated using thin-film deposition at temperatures ≤ 350 °C (≤ 660 °F) followed by photolithographic patterning. As a result, soda-lime silicate glass with a silica barrier layer is adequate for most needs. A high-performance version of intrinsically addressed LCDs, the super twisted nematic (STN), has an added substrate requirement of extremely precise flatness for the purpose of holding the gap dimensions uniform. Therefore, the soda-lime silicate glass used for these displays may require polishing to achieve flatness requirements.

Extrinsically addressed LCDs can be further subdivided into categories according to the material used for the addressing element. These include MIM, CdSe (cadmium selenide), amorphous silicon (a-Si), and polycrystalline silicon (poly-Si) devices. For MIM and a-Si devices, glasses with low alkali content (of the order of 0.1 wt%) and temperature capability above 550 °C (1020 °F) are preferred. Alkaline earth aluminoborosilicates made by commercial sheet-forming processes are available to meet these requirements.

Poly-Si based displays are processed at higher temperatures than those used with a-Si TFTs. Substrates capable of use temperatures (taken to be 25 °C, or 45 °F, below the strain point of the glass) ranging from 600 to 800 °C (1110 to 1470 °F) are required. The actual temperature is determined by the particular process used in fabricating the TFTs. Those with deposited gate dielectrics require 600 to 650 °C (1110 to 1200 °F) while those with thermal oxides require approximately 800 °C (1470 °F). In the former case, alkaline earth aluminoborosilicates have been developed to meet these requirements. For the 800 °C (1470 °F) application, alkaline earth aluminosilicates or vitreous silica (called fused silica or fused quartz, depending upon the source of raw materials) can be used. However, this application is somewhat hampered because commercial large sheet-forming processes (such as fusion draw, float process, slot draw, and so on) cannot be applied to these extremely high-temperature glasses; therefore, their cost is high.

Both a-Si and poly-Si processes require precise alignment of successive photolithographic patterns; therefore, the thermal shrinkage of the substrate should be low.

Sodium contamination of TFTs by alkali from the substrate is a major concern during processing. It is normal practice to coat the substrate glass with a barrier layer to block the migration of trace amounts of alkali. There is near zero alkali vitreous silica available; however, its high price, as noted above, precludes its commercial use.

Thin-Film Electroluminescent Displays

TFEL displays emit light and hence are called active. Their substrate needs are not as stringent as those of active matrix LCDs (AMLCDs). Films of conductors, dielectrics, and phosphors are deposited on the glass sheet in a layered structure. The display is constructed on a single piece of glass and a second piece of glass is sealed with adhesive to the first for protection of the various thin-film layers of the display. The limiting process temperature is the annealing temperature of the phosphor. If annealing is done at high temperature (up to 550 °C, or 1020 °F), the same alkaline earth aluminosilicate compositions used for AMLCD are preferred over soda lime. High annealing temperatures may be advantageous in raising phosphor efficiency.

Plasma Display Panels, Vacuum Fluorescent Displays, and Light-Emitting Displays

PDPs, VFDs, and LEDs are all active displays. Compared to the LCD and TFEL applications, they have a lower temperature requirement and require sealability with low-temperature frits. PDP and VFD requirements are generally met by soda-lime glass made by the float process. Compatibility with glass-sealing frits is also important for creating a hermetic (for PDPs) or a vacuum (for VFDs) enclosure. Frit/substrate compatibility is needed to provide a seal with minimal residual stress. Typically, seals between two glasses that exhibit a differential elongation <100 ppm at the setting point (taken as 5 °C, or 9 °F, above the strain point of the lower viscosity glass) is deemed to be a satisfactory expansion mismatch (Ref 2).

PDP devices typically utilize two glass-sealing frits. A dielectric frit, which is a higher sealing temperature glass frit than the final closure glass frit, covers the multiple electrical terminations. In addition, a soda-lime glass tubulation provides a means to evacuate and/or back fill the cavity of the PDP and VFD devices.

LED displays do not use precision flat glass as a substrate or sealing frits in their construction. They use powdered glass as a light-diffusing medium within the molded polymer encapsulation.

Safety Precautions for High-Lead Frits

Sealing-glass frits used in PDPs and VFDs typically contain high levels of lead oxide (≤80%) and should be handled with care. The following is a summary of some of the safety precautions that should be taken during use and storage of high-lead frits.

- Employees should be familiar with the hazards of inorganic lead as directed by the Hazard Communication Standard
- Lead can be an acute (short term) or chronic (long term) poison if too much is inhaled or ingested. To understand the hazards and symptoms of lead poisoning, the material safety data sheet (MSDS) should be reviewed carefully
- Other sources should also be reviewed. For example, the Occupational Safety and Health Administration (OSHA) has a specific standard concerning occupational exposures to inorganic lead. This standard, found at 29 CFR 1910.1025, sets forth specific controls, air and medical monitoring, work practices (such as the use of respirators, protective clothing, shower rooms, and so on) and training for employees exposed above certain airborne concentrations of lead. The appendices also contain useful information concerning lead hazards
- It is recommended that good personal hygiene practices be maintained when working around lead material, even when lead is not in the air. Employees working with a lead-containing material should wash their hands and face before eating, drinking, or using any tobacco product. Also, care should be used to ensure that the lead-containing material is not carried home incidentally (for example, on work clothing)

Flat Panel Display Substrate Requirements

The three critical substrate attributes for flat panel displays are dimensional precision, temperature capability, and thermal shrinkage. To this list should be added two additional requirements: freedom from significant faults (both internal and surface) and resistance to chemical reagents used during display fabrication. This section focuses on how these five requirements relate to flat panel display fabrication and function.

Because of the extremely stringent requirements of present AMLCD applications, the discussion below is largely oriented towards those requirements. As other flat panel display technologies evolve, many of these same substrate issues may become increasingly important for those applications.

Dimensional Precision

High-performance flat panel display fabrication includes multiple steps of precision photolithography. Therefore, the substrate has to be precise in size, shape, and edge finish to a tolerance on the order of 0.1 mm (0.004 in.). In the case of AMLCDs, deviation of the substrate from a plane due to either warp, surface roughness, or thickness variation must be minimal. Cell spacing (distance between the two display substrates) is a critical variable in display construction. If glass faults cause localized cell spacing variation, the electric field in that region of the cell will deviate from that in the surrounding pixels. This field variation is made manifest in the final display as nonuniformity in gray scale or color. The effect is noticeable for surface deviations with an amplitude as small as 0.01 μm (0.5 μin.) and an extension of a few millimeters. Flatness deviations that occur over larger scales, such as warp, can be compensated for by the now standard use of plastic beads or glass rods mixed with the liquid crystal material as spacers to maintain cell thickness. The cell thickness deviation occurs only when the wavelength of the substrate deviation is so small that the cell's other substrate is unable to conform to the contour and keep cell spacing constant across the cell.

Substrate deviations from the plane can cause an additional problem in fabrication because the photolithographic exposure may not be in focus across the substrate display, thereby creating defects in display circuitry.

Also, if proximity printing is used, warped substrates may damage the photomask.

Thermal Requirements

The two thermal requirements, temperature capability and thermal shrinkage, are interrelated. Rigidity of the glass substrate (that is, resistance to viscous flow) is an issue during thermal processing only when the peak temperature is above the annealing point (temperature at which the viscosity of the glass is approximately 10^{13} P or 10^{12} Pa · s). However, at processing temperatures approaching the strain point (temperature at which the viscosity is approximately $10^{14.5}$ P or $10^{13.5}$ Pa · s) of the glass, a more significant issue is mechanical stress buildup due to thermal gradients that can result if the glass is cooled too rapidly. Limiting upper use temperatures of display substrates to 25 °C (45 °F) below the strain point will generally prevent unacceptable substrate stress buildup in display processing. Higher temperature processing requires care in designing thermal ramps to prevent thermal breakage.

Even at temperatures well below the strain point, the substrate can display dimensional changes by volume relaxation. The substrate glass, as it is cooled to room temperature during manufacture, achieves a final density that is a function of the rate at which it was cooled through the glass transformation region. (The transformation range of a glass is the temperature range in which cooling rate can affect the structure-sensitive properties such as density, refractive index, and volume resistivity. This temperature range generally is in the vicinity of the strain and annealing points of the glass under consideration.) If a quickly cooled glass is reheated and cooled in a manner that allows relaxation of the glass structure during a subsequent thermal process such as display fabrication, the glass densifies. This process is sometimes referred to as thermal shrinkage or compaction. The resultant change in linear dimensions can cause mismatch between photolithographic patterns during display processing. Permissible substrate shrinkage during display processing depends upon the nature of the display circuitry design as well as the size of the display. In the case of AMLCDs, this means that the shrinkage can amount to no more than a fraction of the smallest feature (such as the width of an addressing line) across the maximum dimension of the display. This can be as small as a few microns over hundreds of millimeters (that is, a few ppm).

Glass viscosity and thermal history affect the thermal shrinkage behavior of glass. To illustrate this, Fig 1 represents laboratory shrinkage data for three substrate glasses after 1 hour exposure at various temperatures. Curve A represents a glass with a strain point of 593 °C (1100 °F) that has not been subjected to a preshrinkage treatment. Curve B is for the same glass after a preshrinkage treatment. Curve C is the shrinkage of a higher strain

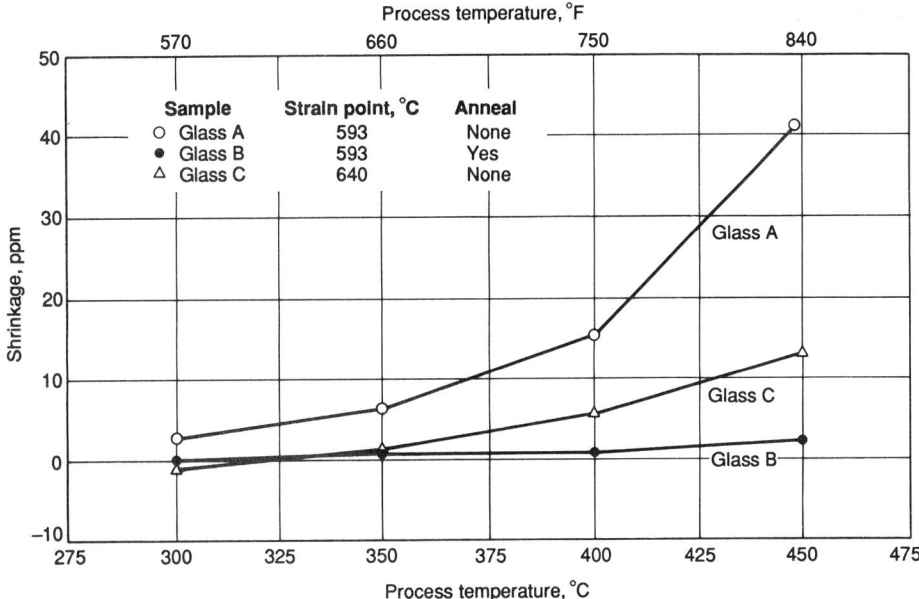

Fig 1 Plot of thermal shrinkage versus process temperature for three glasses with different thermal histories. Glasses have been subjected to 1 h treatment at process temperatures ranging from 300 to 450 °C (570 to 840 °F).

point substrate (640 °C, or 1185 °F) without a preshrinkage treatment. Thermal shrinkage is dependent upon viscosity. The shrinkage of the higher strain point glass (curve C) is less than that of the lower strain point glass (curve A) in the case where neither has undergone preshrinkage. However, a properly designed preshrinkage (compaction) schedule can greatly reduce the level of shrinkage, as evidenced by curve B.

Corrosion Resistance

The chemistry of the display fabrication process, especially that of AMLCDs, may well be the most diverse one that glass has had to endure for any application (Ref 3). An a-Si AMLCD substrate can be host for over seven thin-film patterns and as many etching steps, consisting of treatments by complex mixtures of acids, bases, and fluorides.

The chemical durability requirement for display substrates can be stated as follows: the various etch solutions should not corrode the glass enough to compromise the visual appearance of the display or to interfere with subsequent film deposition.

Table 1 lists weight loss data for several substrate materials after various chemical treatments. The reagent conditions are more severe than is typical in display processing; therefore, the data are intended for comparison of the relative chemical resistance of typical substrate materials.

Surface and Bulk Defects

Flat panel display substrates must be of very high quality with respect to surface and internal faults. The surface upon which the circuitry is fabricated must be free from scratches or other surface blemishes and faults as small as a few microns in size to avoid display circuitry failures. The question of when a microscopic substrate surface fault is merely cosmetic or when it is a potential display-killing defect is an open question.

Internal or bulk faults, such as gaseous inclusions (often called seeds) and particulate inclusions, do not interfere with display operation. Therefore, the issue is simply one of visibility. As long as the inclusion is less than a certain fraction of the pixel size, it may be acceptable. A reasonable fault size limit may be 25% of the surface area of a single pixel. Therefore, a display with a 100 μm (0.004 in.) pixel pitch would have a 50 μm (0.002 in.) inclusion size limit.

Dot Matrix Printers and Ink-Jet Printers

Glass, when used for these two applications, may not seem as dramatic nor as spec-tacular in that the glass component is not seen as in flat panel displays.

Dot matrix printers utilize a series of solenoid actuated tungsten wires striking an inked ribbon to impart ink spots selectively on a page and thus to create images. These tungsten wires are held in a suitably fixed array by a wire guide. The wire guide is often produced from a glass-ceramic. This alkali zinc silicate glass-ceramic or ceram has the capability of being produced with precisely located holes utilizing a photographic imaging process, coupled with a heat treatment, and followed by an acid etching process. In addition, the material exhibits a hardness characteristic that provides for high wear resistance for many millions of impressions.

Ink-jet printers can also utilize the same photochemically machined alkali zinc silicate glass or glass-ceramic to fashion very precise channels or orifices through which ink is directed to form images. (For more information on photochemically machined ceramics see Ref 4.) The precision of the hole locations and the durability of the material to potentially corrosive ink solutions dictate the use of this material. In some circumstances, channels and holes are laboriously machined in soda borosilicate glass parts to perform the same function.

Display Glass Properties

Glasses used in non-CRT display applications are grouped according to application in Table 2. Properties listed include thermal (viscosity and expansion), density, index of refraction, electrical and mechanical characteristics. Several notations and properties in Table 2 bear clarification. Strain and annealing points correspond approximately to the temperatures at which the glass viscosity is $10^{13.5}$ Pa · s ($10^{14.5}$ P) and 10^{12} Pa · s (10^{13} P), respectively. The softening point, defined in terms of the elongation of a glass fiber under its own weight, corresponds to the temperature at which the viscosity is $10^{6.6}$ Pa · s ($10^{7.6}$ P).

Table 1 Chemical corrosion of selected substrate glasses by hydrochloric acid, sodium hydroxide, or water at 95 °C (205 °F)

| | Corrosive | | | | | |
| | 5% HCl 24 h | | 5% NaOH 6 h | | H₂O 24 h | |
Glass composition type	Weight loss, mg/cm²	Visual appearance(a)	Weight loss, mg/cm²	Visual appearance(a)	Weight loss, mg/cm²	Visual appearance(a)
Soda-lime silicate	0.01	1	0.7	3	0.01	1
Soda borosilicate	<0.01	1	1.5	3	<0.01	1
Barium aluminoborosilicate	11	4	3	2	0.03	1
Alkaline-earth boroaluminosilicate	4	1	2	3	0.02	1
Alkaline-earth aluminosilicate	<0.01	1	1	3	<0.01	1

(a) Visual appearance key: 1, no change; 2, slight surface darkening or iridescence; 3, slight or moderate surface frost; 4, heavy surface frost or surface crazing

Table 2 Properties of glasses used in non-CRT applications

Application(a)	Glass code	Glass composition type(b)	Strain point °C	Strain point °F	Annealing point °C	Annealing point °F	Softening point °C	Softening point °F	Maximum use temperature °C	Maximum use temperature °F	Thermal expansion at 300 °C (570 °F) 10⁻⁷/°C	Thermal expansion at 300 °C (570 °F) 10⁻⁷/°F
L	0211	Alkali zinc borosilicate	508	946	550	1020	720	1330	483	901	74	41
L	1724	Alkaline-earth boroaluminosilicate	674	1245	726	1340	926	1699	649	1200	44	24
L	1733	Alkaline-earth boroaluminosilicate	640	1180	689	1270	928	1702	615	1140	37	21
L	1729	Alkaline-earth aluminosilicate	799	1470	855	1570	1107	2025	774	1425	35	19
E, L	7059	Baria aluminoborosilicate	593	1100	639	1180	844	1570	568	1055	46	26
F, P	7555	Lead borate V	326	619	370	700	415	779	301	574	88	49
F, P	7568	Lead zinc borate D	...	...	...	...	320	610	...	...	88	49
F, P	7570	Lead borosilicate V	358	676	376	708	447	837	333	631	83	46
F, P	7575	Lead zinc borosilicate D	...	...	...	...	380	720	425	800	89	49
F, P	7599	Lead zinc borate D	...	...	...	...	320	610	400	750	89	49
J, S	7740	Soda borosilicate	510	950	560	1040	821	1510	485	905	32.5	18
L	7913	96% silica	890	1635	1020	1870	1530	2785	865	1590	7.5	4.2
L	7940	100% silica	990	1815	1075	1965	1585	2885	965	1770	5.6	3.1
J	8603	Alkali zinc silicate (glass)	...	...	...	...	...	...	450	840	84	47
P	8603	Alkali zinc silicate (opal)	...	...	...	...	...	...	550	1020	89	49
J	8603	Alkali zinc silicate (ceram)	...	...	...	...	...	...	750	1380	103	57
F, P	AS	Soda-lime silicate	511	951	554	1030	740	1365	486	907	81(c)	45(c)
L	AX	Borosilicate	522	972	568	1055	789	1450	497	927	49(c)	27(c)
L	AN	Alkaline-earth aluminoborosilicate	616	1140	661	1220	862	1585	591	1095	45(c)	25(c)
L	AQ	Vitreous silica	1000	1830	1120	2050	1600	2910	975	1785	6(c)	3(c)
E, L	NA 35	Borosilicate	650	1200	700	1290	940	1725	625	1155	37(d)	21(d)
E, L	NA 40	Alkaline-earth aluminosilicate	656	1215					631	1170	43(d)	24(d)
E, L	NA 45	Baria aluminoborosilicate	610	1130	658	1215	859	1580	585	1085	46(d)	26(d)
L	BLC	Alkali borosilicate	535	995	575	1065	775	1425	510	950	51(e)	28(e)
E, L	OA 2	Alkaline-earth aluminoborosilicate	635	1175	685	1265	895	1645	610	1130	47(e)	26(e)
F, P	LS-7105	Lead borate devitrifiable sealing glass D	...	...	...	...	400	750			85(f)	47(f)
F, P	GA-8	Lead borosilicate	...	...	...	...	490	915	...	...	81(f)	45(f)
F, P	GA-9	Lead borosilicate	...	...	...	...	430	805	...	...	90(f)	50(f)
F, P	GA-12	Alkali borosilicate	...	...	...	...	560	1040	...	...	73(f)	41(f)
F, P	GA-21	Lead borosilicate	...	...	...	...	450	840	...	...	83(f)	46(f)
F, P	PLS-2401	(VFD glaze)	...	...	...	...	545	1015	...	...	77(f)	43(f)
F, P	PLS-3108A	(VFD glaze)	...	...	...	...	590	1095	...	...	65(f)	36(f)
P	PLS-3130	(PDP dielectric)	...	...	...	...	485	905	...	...	81(f)	45(f)
F, P	CV-455	Devitrifying solder glass (for soda lime) D	...	...	...	...	365	670	...	...	86	48
P	PP-200	Alkali free dielectric glass	...	...	...	...	340	645	...	...	125	70
F, P	SG-100	Vitreous solder glass V	...	...	...	...	362	684	...	...	77	43
E, L	AF45	Borosilicate	627	1160	663	1225	876	1610	602	1115	45	25
L	D263	Alkali-zinc borosilicate	...	...	557	1035	736	1355	...	...	73	41
P	G 017-340	Lead borate (for float) V	...	...	...	...	360	680	...	...	70(g)	39(g)
F, P	8596	Devitrifying solder glass D	...	...	...	...	370	700	...	...	87(g)	48(g)
F, L, P		Soda-lime silicate (Fourcault process)	510	950	536	995	687	1270	485	905	91	51
F, L, P		Soda-lime silicate	490	915	...	...	720	1330	465	870	84	47
F, L, P		Soda-lime silicate (Float process)	523	973	545	1015	737	1360	498	930	77.5	43

Application(a)	Glass code	Glass composition type(b)	Density g/cm³	Density lb/in.³	Volume resistivity Ω·cm 250 °C (480 °F)	Volume resistivity Ω·cm 350 °C (660 °F)	Dielectric constant at 1 MHz	Loss tangent at 20 °C (70 °F),%	Young's modulus GPa	Young's modulus 10⁶ psi	Index of refraction at 589.3 nm (5893 Å)	Poisson's ratio
L	0211	Alkali zinc borosilicate	2.57	0.0929	8.3	6.7	6.7	0.46	74.5	10.8	1.523	0.22
L	1724	Alkaline-earth boroaluminosilicate	2.64	0.0954	13.8	11.6	6.6	0.001	82.7	12.0	1.540	...
L	1733	Alkaline-earth boroaluminosilicate	2.49	0.0890	13.4	11.4	5.3	0.09	66.9	9.7	1.516	0.235
L	1729	Alkaline-earth aluminosilicate	2.56	0.0925	13.1	11.0	5.9	...	80.6	11.7	1.52	0.216
E, L	7059	Baria aluminoborosilicate	2.76	0.0997	13.1	11.0	5.84	0.10	67.6	9.8	1.533	0.280
F, P	7555	Lead borate V	5.70	0.206	10.5	13.74(j)	...	...	...	...	...	...
F, P	7568	Lead zinc borate D	...	...								
F, P	7570	Lead borosilicate V	5.46	0.197	10.6	8.7	15.0	0.22	53.5	7.76	1.860	...
F, P	7575	Lead zinc borosilicate D	3.80	0.137	8.6	7.05	20.4(j)	0.91	51.3	7.44	...	0.25
F, P	7599	Lead zinc borate D	5.78	0.209	9.3	7.7	17.7(j)	1.1	44.8	6.5	...	0.27
J, S	7740	Soda borosilicate	2.23	0.0806	8.1	6.6	4.6	0.4	62.7	9.1	1.473	0.20
L	7913	96% silica	2.18	0.0788	9.7	8.1	3.8	0.04	66.2	9.6	1.458	0.19
L	7940	100% silica	2.20	0.0795	12.3	10.7	3.8	0.00	72.4	10.5	1.458	0.16
J	8603	Alkali zinc silicate (glass)	2.365	0.0854	6.27	4.90	7.62(k)	0.80(k)	76.9	11.15	...	0.22
P	8603	Alkali zinc silicate (opal)	2.380	0.0860	8.81	7.23	5.73(k)	0.40(k)	82.7	12.0	...	0.21
J	8603	Alkali zinc silicate (ceram)	2.407	0.0870	8.76	7.07	5.63(k)	0.30(k)	86.9	12.62	...	0.19

(continued)

Table 2 continued

Application(a)	Glass code	Glass composition type(b)	Density		Volume resistivity Ω · cm		Dielectric constant at 1 MHz	Loss tangent at 20 °C (70 °F), %	Young's modulus		Index of refraction at 589.3 nm (5893 Å)	Poisson's ratio
			g/cm³	lb/in.³	250 °C (480 °F)	350 °C (660 °F)			GPa	10⁶ psi		
F, P	AS	Soda-lime silicate	2.49	0.0900	7.7(h)	5.9(i)	7.5	0.9	71.7	10.4	1.52	0.21
L	AX	Borosilicate	2.42	0.0874	8.0(h)	6.6(i)	5.9	0.9	68.9	10.0	1.50	0.18
L	AN	Alkaline-earth aluminoborosilicate	2.72	0.0983	15.5(h)	12.0(i)	6.3	0.07	73.8	10.7	1.54	0.22
L	AQ	Vitreous silica	2.20	0.0795								
E, L	NA 35	Borosilicate	2.50	0.0903	...	...	5.3	0.07	70.3	10.2	1.516	0.24
E, L	NA 40	Alkaline-earth aluminosilicate	2.87	0.104	...	...	6.7	...	92.4	13.4	1.574	0.26
E, L	NA 45	Baria aluminoborosilicate	2.78	0.100	...	...	5.6	...	68.9	10.0	1.533	0.24
L	BLC	Alkali borosilicate	2.36	0.0853	6.9	...	5.7	0.60	...	...	1.493	...
E, L	OA 2	Alkaline-earth aluminoborosilicate	2.76	0.0997	13.0	...	6.3	0.10	...	...	1.54	...
F, P	LS-7105	Lead borate devitrifiable sealing glass D	6.37	0.230	...	...	...	...	...	...	...	...
F, P	GA-8	Lead borosilicate	5.37	0.194	...	...	...	...	...	...	...	...
F, P	GA-9	Lead borosilicate	5.77	0.208	...	...	...	...	...	...	...	...
F, P	GA-12	Alkali borosilicate	2.95	0.107	...	...	...	...	...	...	...	...
F, P	GA-21	Lead borosilicate	5.74	0.207	...	...	...	...	...	...	...	...
F, P	PLS-2401	(VFD glaze)	...	...	12.2	...	12.2	0.266	...	...	...	...
F, P	PLS-3108A	(VFD glaze)	...	...	14.0	...	9.2	0.40	...	...	...	...
P	PLS-3130	(PDP dielectric)	...	...	12.2	...	11.7	0.264	...	...	...	...
F, P	CV-455	Devitrifying solder glass (for soda lime) D	5.915	0.214	...	8.3	18.3	0.68	...	...	...	...
P	PP-200	Alkali free dielectric glass	6.62	0.239	...	...	...	...	...	...	...	...
F, P	SG-100	Vitreous solder glass V	6.698	0.242	...	9.0	31.8	0.80	...	...	...	...
E, L	AF45	Borosilicate	2.72	0.0983	13.8	11.5	6.2	0.015	66.0	9.57	1.5276	0.235
L	D263	Alkali-zinc borosilicate	2.51	0.0907	...	...	...	...	...	...	1.5225	...
P	G 017-340	Lead borate (for float) V	4.80	0.173	...	...	...	...	...	...	...	...
F, P	8596	Devitrifying solder glass D	6.43	0.232	...	...	...	...	...	...	...	...
F, L, P		Soda-lime silicate (Fourcault process)	2.47	0.0892	...	...	8.3	...	71.0	10.3	1.513	0.22
F, L, P		Soda-lime silicate	2.483	0.0897	...	...	7.75	...	69.6	10.1	...	0.22
F, L, P		Soda-lime silicate (Float process)	2.498	0.0903	...	...	...	...	73.1	10.6	...	0.22

(a) E, TFEL or EL; F, VFD; J, ink-jet printer; L, LCD; M, dot matrix printer; P, PDP; S, LED. (b) V, vitreous; D, devitrifying. (c) 50-200 °C (120-390 °F). (d) 100-300 °C (210-570 °F). (e) 30-380 °C (85-715 °F). (f) 30-300 °C (85-570 °F). (g) 20-250 °C (70-480 °F). (h) At 200 °C (390 °F). (i) At 300 °C (570 °F). (j) KHz at 20 °C (68 °F). (k) 100 KHz at 21 °C (70 °C).

Table 2 designates several flat glasses by their method of manufacture:

- Float process, in which a ribbon of molten glass is delivered upon a bath of molten tin, upon which the glass becomes flat and parallel as it is slowly cooled
- Fourcault process, a version of a drawn sheet process in which molten glass is drawn from a specially designed orifice to the final desired thickness
- Fusion (overflow) process, used for display substrate manufacture

A detailed discussion of flat sheet manufacturing methods can be found in Ref 5.

Listed in Table 2 are three forms of photochemically machinable alkali zinc silicate glass (glass code 8603) used in wireguides for dot-matrix printers and cell sheets in plasma display panels. The three forms listed (glass, opal, and ceram) are the same composition but differ in the nature of their crystallinity. The glass form is noncrystalline. With appropriate ultraviolet (UV) exposure and heat treatment, the glass converts to the dark purple or black opal form, having approximately 28% crystalline phase (lithium metasilicate). A different heat treatment results in the conversion of the glass to the ceram form that has approximately 50% crystalline phase of lithium disilicate (Ref 4).

Glass frits listed in Table 2 are classified as vitreous and devitrifying. Vitreous sealing frits are thermoplastic materials that melt and flow at the same temperature each time they are processed. Devitrifying sealing frits are thermosetting materials that crystallize during heat treatment. Due to the crystalline nature of the material, a devitrified sealing glass has thermal and chemical stability greater than that of the original glass.

REFERENCES

1. L. Tannas, *Flat-Panel Displays and CRTs*, Van Nostrand Reinhold, 1985
2. H.E. Hagy, Thermal Expansion Mismatch and Stress in Seals, *Elec. Pack. Prod.*, 1978, p 182
3. P.L. Bocko, C.A. Rosenblatt, and L.R. Morse, Chemical Durability and Cleaning of Flat Panel Display Substrate Glasses, *1990 Society for Information Display International Symposium Digest of Technical Papers*, Vol 21, 1990, p 73
4. R.K. Whitney, Precision Photochemical Machining of a Glass Without Photoresist, *Proceeding of the Technical Program, National Electronic Packaging and Production Conference, NEPCON West '86*, 1986, p 495
5. D.C. Boyd and D.A. Thompson, Glass, *Kirk-Othmer: Encyclopedia of Chemical Technology*, 3rd ed., John Wiley & Sons, Vol 11, 1980, p 807

SELECTED REFERENCES

- W.H. Dumbaugh and P.L. Bocko, Glass Substrates for Flat Panel Displays, *1990 Society for Information Display International Symposium Digest of Technical Papers*, Vol 21, 1990, p 70
- W.H. Dumbaugh, P.L. Bocko, and F.P. Fehlner, Glasses for Flat Panel Displays, *High Performance Glasses*, M. Cable and J. Parkers, Ed., Blackies & Son, Ltd., London, 1990
- C.K. Edge, Flat Glass Manufacturing Processes Update, *The Handbook of Glass Manufacture*, F.V. Tooley, Ed., Ashlee Publishing Company, 1984
- W.C. Hynd, Flat Glass Manufacturing Processes, *Glass Science and Technology*, Vol 2, *Processing*, D.R. Uhlmann and N.J. Kriedl, Ed., Academic Press, 1984

Telecommunications and Related Uses

Suzanne R. Nagel, AT&T Bell Laboratories

TELECOMMUNICATION SYSTEMS, along with other communication and information networks, are undergoing revolutionary transformations in information handling capacity, due to the use of lightwave transmission rather than the more conventional electrical and microwave approaches. A critical element in this revolution has been the development and use of high-silica glass optical fibers as the transmission media of choice. This article will restrict its discussion to optical fibers of this type. Polarization-maintaining fibers are a special class of fiber that have emerged due to the technological advances in fiber fabrication. They will be briefly reviewed, although their utility has been more driven by sensor and component applications rather than their use in transmission systems.

Other specialized applications for silica glass involve couplers and splitters in fiber or planar waveguide form, as well as specific bulk components in communications systems. A more recent breakthrough has been the practical realization of active devices based on silica glass, in particular, optical fiber amplifiers. Research and development continues to examine other nonlinear glass devices for applications such as four-wave mixing, soliton transmission, and frequency upconversion. This article will focus on high-silica fibers as the transmission media for communication applications and highlight the status of other communications uses of glass.

Applications of High-Silica Fibers for Communications

The potential information-carrying capacity of a communication system increases directly with frequency: The use of light thus represents a natural extension of frequency relative to conventional approaches based on electrical and radio waves. The advent of the laser in 1959 provided the modern day impetus for lightwave communications systems, as it represented a practical source of coherent light which could be modulated to encode information. However, a practical, reliable, and flexible transmission medium was needed to economically exploit the advantages of such a system. In 1966, fiber lightguides based on silica glasses were proposed to have the theoretical optical transparency required in the near-infrared region of the spectrum (0.6 to 1.6 μm), if impurity effects on transmission loss could be overcome (Ref 1). Subsequent worldwide research and development resulted in silica-base glasses, modified with small amounts of glass-forming dopants, becoming the materials of choice for high-performance communications lightguides. Not only was theoretical transparency achieved, but such fibers had intrinsic high strength, excellent chemical durability, and most importantly, there was the evolution of processing technology to realize low loss, high bandwidth, and optimal designs at low cost. Concurrent with advances in fiber technology came advances in propagation theory, light sources, detectors, system design, and packaging technology, resulting in today's very high-bandwidth communications systems (Ref 2–6). Some of the important fiber performance characteristics that must be met for their use in communications are described below.

Light Guidance, Fiber Design, and Dispersion. An important element in realizing the high bandwidth (information-carrying capacity) of a system operating at optical frequencies is the design of fibers which guide light over long distances with minimal distortion of the light pulses they carry. The refractive index, n, of glasses is the key property controlled to achieve lightguiding structures; it is a measure of speed of light in vacuum, c, relative to its speed in a material. The higher the index, the more light is retarded, or the slower it travels. In addition, when light travels from one medium to another, it is refracted by an angle proportional to the relative refractive indices of the two media: this is known as Snell's law of refraction. Light guidance is based upon the principle of total internal reflection, which states that for light travelling in a higher refractive index medium, if it impinges on the interface of a lower refractive index medium at an angle less than a critical angle, θ, it will be reflected rather than refracted and thus guided in the high index medium.

In simplest form, a fiber lightguide consists of a circular core of material whose refractive index, n_2, is higher than the index of a surrounding cladding material, n_1, as illustrated in the top portion of Fig 1. When a light pulse is injected into the fiber core, certain modes of light at particular angles will meet the critical angle criterion for total internal reflection and will be guided; others will be refracted into the cladding. A basic parameter which measures the fiber light-collecting capacity is the numerical aperture, NA, which defines the maximum angle, θ_0, of incident light that can be totally internally reflected:

$$NA = \sin \theta_0 = (n_2^2 - n_1^2) \approx n_2(2\Delta) \qquad (\text{Eq 1})$$

where

$$\Delta = (n_2 - n_1)/n_2 \qquad (\text{Eq 2})$$

The number of modes of light that can propagate in a fiber are governed by Maxwell's electromagnetic field equations and are related to a dimensionless parameter V:

$$V = \frac{2\pi a}{\lambda} (NA) \sim \frac{2\pi a n_2}{\lambda} (2\Delta) \qquad (\text{Eq 3})$$

where a is the core radius. When V is less than 2.405, only a single mode of light can be propagated; all other modes are cut off.

Optical fibers for communications are made from silica ($n = 1.456$ at 589.3 nm). Small amounts of GeO_2 or P_2O_5 are used to increase the index, while additions of B_2O_3 or F can be used to decrease the index, thus allowing lightguide core and cladding compositions to be realized.

In addition to guiding light, fibers for communications applications must transmit encoded information. Typical lightwave communications systems are digital, with transmitted information encoded into a binary format of ones and zeros: a "pulse" of photons of light from a laser or light-emitting diode (LED) injected into the fiber represents a "one"; the absence of a pulse represents a zero. At the receiving end of such a system, a photodetector is used to detect the presence

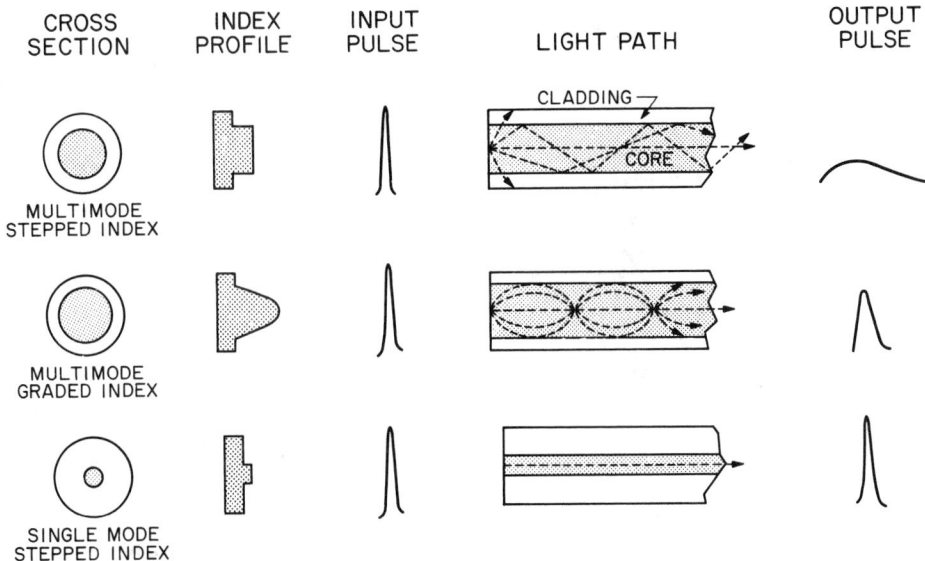

CROSS SECTION	INDEX PROFILE	INPUT PULSE	LIGHT PATH	OUTPUT PULSE

MULTIMODE STEPPED INDEX

MULTIMODE GRADED INDEX

SINGLE MODE STEPPED INDEX

Fig 1 Characteristics of the three basic types of optical fibers

or absence of such signals within a prescribed time slot, related to the bit rate of the system. Typically, telecommunications systems operate at Mbit/s to Gbit/s rates.

It is thus important to examine fiber designs in the context of pulse dispersion or broadening. The ray diagram provides a useful construct, although more rigorous propagation theory must be considered, as discussed in Ref 2 to 3 and 5 to 8. When a temporally narrow light pulse is injected into the fiber depicted in the top portion of Fig 1, each mode or guided angle of light has a different path length through the fiber, leading to modes initially injected at approximately the same time arriving at the far end of the fiber at different times. This is referred to as intermodal dispersion and limits the bandwidth, or rate at which individual pulses can be injected into the fiber and be individually detected, without overlapping, after travelling some length through the fiber.

One approach to minimizing intermodal dispersion is to use fibers with graded refractive index cores, such as depicted in the middle portion of Fig 1. In such graded index multimode fibers, the different angles of light injected into the fiber are refracted, resulting in gentle periodic paths of propagation. For any radial position, r, the refractive index $n(r)$ is described by:

$$n(r) = n_2 [1 - \Delta(r/a)\alpha] \qquad \text{(Eq 4)}$$

where α is the profile parameter. The optimal profile parameter is a function of both wavelength and glass composition (for more detail, see Ref 2, 3, and 5–8). While the detailed propagation of each mode is complex, in general the higher-order modes travel over longer path lengths than lower-order modes but travel on average faster as they proportionately reside more in the lower index portion of the core. Thus, near parabolic ($\alpha =$

2) profiles are optimal for minimizing intermodal dispersion. Use of such graded index profiles can reduce the pulse broadening by over three orders of magnitude relative to step index fibers of equivalent NA.

The most typical multimode communications fibers are called standard and high NA fibers. The standard fiber has an $NA = 0.20$ to 0.23, core diameter = 50 μm and outer diameter of 125 μm. High NA fibers typically have a 62.5 μm core, $NA = 0.3$, and outer diameter of 125 μm.

Another source of pulse spreading in glasses is material dispersion, resulting from the variation in the refractive index with wavelength. Figure 2 illustrates the relationship among refractive index, pulse delay, and material dispersion. Thus, there is an additional source of pulse spreading associated with

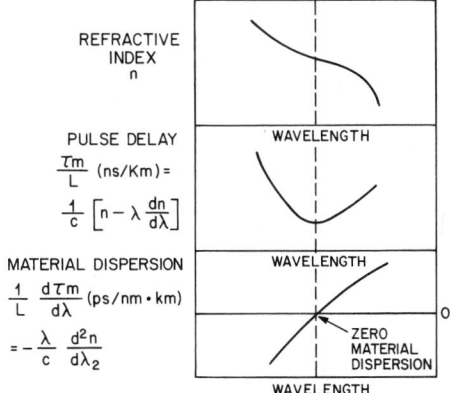

REFRACTIVE INDEX n

WAVELENGTH

PULSE DELAY

$\dfrac{Tm}{L}$ (ns/Km) =

$\dfrac{1}{c} \left[n - \lambda \dfrac{dn}{d\lambda} \right]$

WAVELENGTH

MATERIAL DISPERSION

$\dfrac{1}{L} \dfrac{dTm}{d\lambda}$ (ps/nm · km)

$= -\dfrac{\lambda}{c} \dfrac{d^2 n}{d\lambda_2}$

ZERO MATERIAL DISPERSION

WAVELENGTH

Fig 2 Relationship of refractive index versus wavelength behavior of glasses to the resultant material dispersion. At any wavelength, a pulse will be delayed by a time T_m per unit length L according to the equation shown, where c is the speed of light. The material dispersion is a measure of the delay with respect to wavelength and is minimized at the zero material dispersion wavelength.

the spectral width, $\Delta\lambda$, of the light source used. Typical LEDs have $\Delta\lambda \approx 20$ to 100 nm width, while lasers range from 2 nm to extremely narrow line sources, depending on the application. For high-silica glass compositions, material dispersion is minimum near 1.3 μm. Its effect on bandwidth in multimode fibers is small relative to intermodal effects but increases as the spectral width increases. It is a major factor in single-mode fibers, and is exploited to realize the very highest bandwidth designs.

The bottom portion of Fig 1 depicts a single-mode waveguide in which the core size, index difference, and profile shape are chosen such that only the fundamental mode of light is guided. This mode, as it propagates through the fiber, is actually characterized by a near Gaussian power distribution with some fraction of the power travelling in the cladding. The wavelength, λ_c, at which a fiber becomes single mode, is termed the cutoff wavelength, and occurs in ideal step-index fibers at $V < 2.405$, as described in Eq 3.

Pulse spreading in single-mode fibers results from chromatic dispersion which consists of two terms: material dispersion and waveguide dispersion. This is illustrated in Fig 3 for two single-mode fiber designs. The zero dispersion wavelength, λ_0, is the wavelength at which these two effects counterbalance each other, and represents the maximum bandwidth wavelength for such a fiber. The values of Δ, a, and profile can be chosen to tailor λ_0 to a desired operating wavelength. Typically, 1.3 μm fiber designs have step-index cores with $\Delta \approx 0.3$ to 0.4% and core diameters of 8 to 10 μm. In contrast, 1.55 μm designs have triangular or graded-index cores with 6 to 8 μm core diameters and $\Delta \approx 0.7$ to 1.0%. The standard outer diameter of such fibers is 125 μm.

As this discussion indicates, the dispersion of a fiber is determined by the specific fiber design characteristics, glass composition, and wavelength of operation. Single-mode fibers have orders of magnitude lower dispersion than multimode graded index fibers, but in each case the resultant bandwidth is determined by the spectral width of the source. The small core size of the single-mode fiber in general favors its use with a laser in high bandwidth applications, while multimode fibers are used with both LEDs and lasers for low-to-intermediate bandwidth systems.

Optical Transmission Loss. Another critical property of optical fibers for communications is the optical transmission loss at the wavelength(s) of the intended application, as it affects the system distance over which a signal can be transmitted. The optical loss, measured in units of dB/km, is defined as:

$$\text{Loss} = 10 \log [P_i/P_o]/L \qquad \text{(Eq 5)}$$

where P_i and P_o are the input and output intensity in dB for a fiber length L.

The achievement of very low losses in long fiber lengths represents one of the important

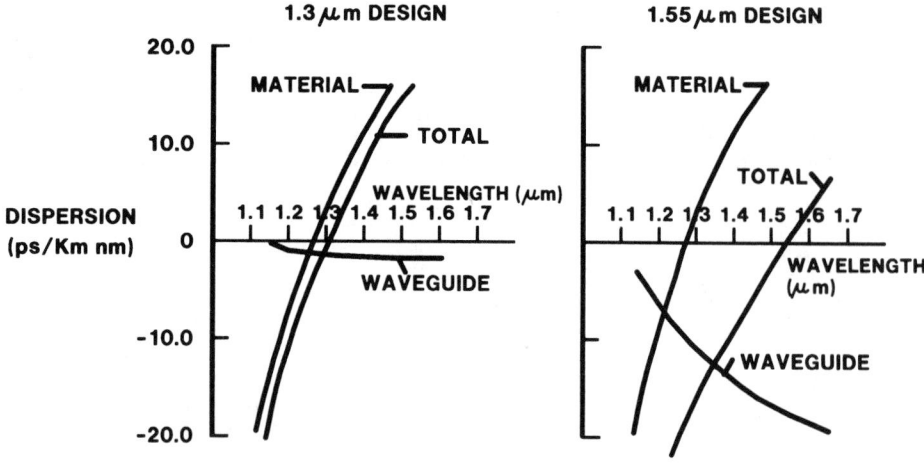

Fig 3 Dispersion in single-mode lightguides. Fiber designs optimized for 1.3 μm (left) and 1.55 μm (right) transmission are shown.

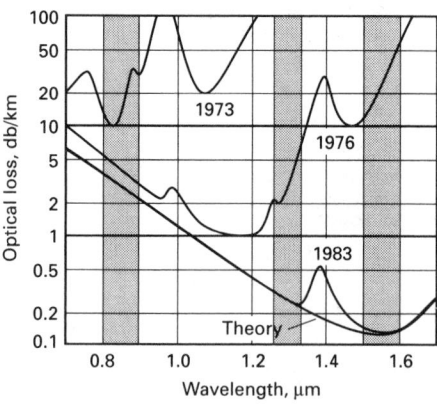

Fig 5 Optical fiber loss as a function of time. The remaining loss is due to trace amounts of OH, giving rise to Si-OH absorption at 1.39 μm.

triumphs in the evolution of lightguide technology. Figure 4 schematically illustrates the theoretical transmission window for silica-base lightguides, where the total theoretical loss is the sum of contribution from the ultraviolet edge associated with electronic transitions, the infrared edge associated with molecular vibrations, and the Rayleigh scattering of light caused by density and compositional fluctuations. Rayleigh scattering varies as $A\lambda^{-4}$, where A is a material constant which depends both on glass composition and the glass preparation method. It should also be noted that the absorption edge associated with B_2O_3 occurs at shorter wavelengths; this dopant has been eliminated in low-loss long-wavelength designs. In addition to these intrinsic loss mechanisms, however, a variety of extrinsic effects limit the achievement of low attenuation. Impurity absorptions are extremely important, and transition metal cations and water in the form of OH must be controlled to the part per billion level (ppb). Typically, the OH level in manufactured fibers is on the order of 5 to 10 ppb by weight. Figure 5 demonstrates the historical reduction in loss which has enabled communications fibers to operate in three "windows": short wavelengths (0.8 to 0.9 μm), 1.3 μm, and 1.55 μm, where solid-state laser sources and detectors are available. This has been achieved through improved processing, optimal compositions and fiber designs, and the ability to achieve very low water levels.

Fiber design can also impact the sensitivity of a given fiber to micro- and macrobending induced losses. Such effects are discussed in Ref 2 to 8. In addition, two long-term reliability issues relating to the interaction of radiation and hydrogen with glass fiber lightguides have been extensively studied. Reviews of such effects are summarized in Ref 5.

Mechanical Strength and Fatigue. The mechanical properties of optical fibers are critical to allow handling, cabling, installation, and high reliability over the lifetime and in the environment of their application. The theoretical strength of silica is on the order of 20 GPa (3000 ksi), but under ambient temperature and moisture conditions this strength is reduced to ~5.5 GPa (~800 ksi) due to dynamic fatigue during testing. When very long lengths are tested, lower breaking strengths are observed, where fracture occurs at the weakest link. This reduction in strength is attributable to chemically or mechanically induced flaws which act as stress concentrators and lead to local failure. The longer the fiber length, the greater the statistical probability of encountering increasingly larger flaws. The fracture stress is described by the Griffith criteria of failure, which relates the crack size that will lead to failure to the failure stress. The stress causing failure at a given flaw is inversely proportional to the square root of the flaw size. For example, a 0.8 μm flaw will cause a fiber to fail at a stress of 690 MPa (100 ksi); a 0.2 μm flaw results in failure at 1380 MPa (200 ksi). While plastic coatings are applied to fibers as they are formed to protect the pristine surface of the fiber, as the length of the fiber is increased, the probability of encountering increasingly

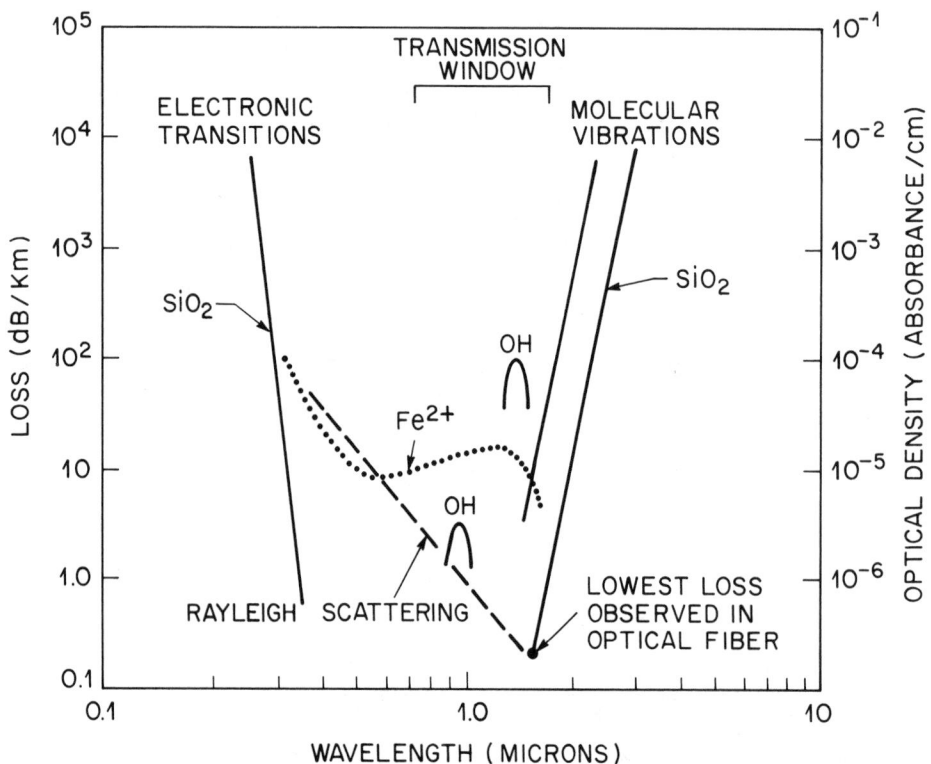

Fig 4 Schematic representation of loss mechanisms in silica-base lightguides as a function of wavelength. The lowest cost fiber reported has a loss of 0.16 dB/km at 1.55 μm.

larger flaws related to the overall process sequence increases.

In practice, all fiber is proof-tested or exposed to a short-term tensile strain test in order to eliminate weak links or larger flaws. This establishes a minimum strain capability for the fiber and thus prevents subsequent cabling and installation failures.

The long-term mechanical strength of a fiber is impacted by fatigue effects, which result in time-dependent failure attributed to the slow growth of flaws in an active environment. This is frequently termed stress corrosion where stress, particularly at cracks, enhances the interaction of water with the silica surface via the reaction:

$$Si\text{-}O\text{-}Si + H_2O \geq 2SiOH \qquad (Eq\ 6)$$

To mitigate against such effects in practical applications, simple design rules are used to set the proof-test level relative to the stress the fiber might experience in a given application. For example, the proof-test level should be 5 times higher than any long-term stress and 1.5 times higher than any short-term stress the fiber might experience during handling and use.

Basically, the long-term strength of fiber is assured by using in-line applied coatings to protect the fiber surface from abrasion, 100% proof-test screening, cable designs which minimize strain, and conservative static-fatigue design diagram approaches. More recently, hermetic coatings to prevent moisture from reaching the fiber surface are being explored to make fibers more fatigue resistant. Amorphous carbon coatings have achieved the best practical results to date. More detailed treatment of the mechanical properties of fiber are contained in Ref 2 and 9.

Fiber Processing and Cost. Fibers for communications applications are fabricated in a two-step process where first a glass precursor is made, followed by drawing a light-guiding structure into coated fiber. Vapor-phase processing is the basis for all current manufacturing techniques, as it has enabled very high purity, highly controlled core and cladding glass compositions to be fabricated with optimal dimensional, optical and mechanical properties with high yield. Fiber drawing and coating techniques have evolved to allow long lengths of high-strength, dimensionally controlled coated fibers to be realized. These techniques have been reviewed in Ref 2, 4, and 5 and in the Section of this Volume entitled "Glass Processing."

Typically, precursor glass preforms capable of yielding 15 to 100 km (9 to 60 miles) of fiber are made, then drawn at speeds of 5 to 10 m/s (16 to 33 ft/s). As a result, coated and fully characterized fibers are available in the commercial market place at prices ranging from \$0.06 to \$0.40/m (\$0.02 to \$0.12/ft), depending on the specific specifications. Specialized enhancements or fiber types are priced somewhat higher.

It is the availability of multikilometer lengths of fibers in this price range that have resulted in fiber deployment extensively throughout the long haul telecommunication network as well as from central offices to remote nodes. All new undersea systems are based on lightwave communications. As prices of other components come down, fiber is increasingly being used for Local Area Networks (LANs), Metropolitan Area Networks (MANs), and computer data links. In addition, there are many field trials underway to deploy fiber-to-the-home (FTTH), curb and business place systems.

Polarization-maintaining fibers refer to a class of single-mode fibers that can maintain the state of linear or circular polarization of light over a given length. Ideal circular single-mode fibers support two orthogonal polarizations of light that propagate independently and at the same speed. However, real fibers are often birefringent as a result of deviations from perfect circular symmetry, anisotropic stresses, and external factors such as pressure, temperature, twists, and bends. As a result, the polarization state of light varies in an unpredictable manner and the phase information is lost. In addition, the delay difference between the two orthogonally polarized modes can reduce the bandwidth of the fiber due to the polarization mode dispersion. Excellent reviews of the origin and control of such effects are contained in Ref 10 to 14.

Two basic fiber design approaches have been used to address such effects. Low birefringent fibers can be made by using highly controlled fabrication techniques, resulting in highly circular cores with minimal built-in anisotropic stresses. Conventional single-mode fibers have low modal birefringence, but the output polarization is typically unstable due to environmental and external perturbations. These changes typically occur with very slow time constants and can be compensated for by a polarization controller at the output end of the fiber for applications such as coherent detection, which require such control.

A second approach, typically used for sensors and other specialized applications, is to deliberately make fibers with sufficiently high birefringence so that there is minimal coupling between the two linear modes, even when subject to external perturbations. In the extreme, short lengths of fiber can be made which only support one of the two fundamental modes; these are called single polarization fibers or polarizers and represent useful fiber components.

Polarization-maintaining fibers based on silica glass compositions are currently commercially available in multikilometer lengths with excellent loss, strength, and polarization characteristics. They are typically fabricated using variations on conventional processing approaches. More details on the use of such fibers in sensor applications are contained in the article "Ophthalmic and Optical Applications" in this Section.

Glass Couplers, Splitters, and Components. Glass fibers have enabled cost-effective high-bandwidth communications systems based on point-to-point transmission to be realized, particularly for telecommunications systems. As applications for lightwave systems have evolved, a need for a wide range of passive components to handle the routing of optical signals has emerged. Such components include couplers, splitters, taps, wavelength multiplexers/demultiplexers, filters, isolators, polarizers, and lenses. It is far beyond the scope of this article to review all of these devices. Some will be covered in the article "Ophthalmic and Optical Applications" in this Section, while reviews are also contained in Ref 5, 6, and 15 to 19. A subset of such devices is based on glass and fiber and represent an important area of application.

The evolution of communications architectures, as well as the use of fibers in sensor applications, increasingly requires the ability to couple, split, or tap optical signals of the same or different wavelengths. In point-to-point communications systems, capacity can be upgraded by two approaches: Additional transmission wavelength signals can be added by multiplexing them on to the same fiber; fibers can also be used in a bidirectional manner, with signals of the same or different wavelengths travelling in opposite directions over the same fiber. As systems are designed to bring services directly to subscribers, such as to the home or business or to network many users together, such as in a Local Area Network, there is a need to split signals off the fiber "bus." Other architectures require multiple signals to be joined or separated in star-type configurations. Very high-capacity applications contemplate densely packed signals simultaneously transmitted over a single fiber. All of these interconnection approaches require specialized devices to meet specific transmission requirements.

Couplers generically refer to a general class of devices that allow photons to be split or joined. Fiber couplers include: star couplers, T couplers, bidirectional couplers, splitters, combiners, wavelength sensitive couplers, active couplers, passive couplers, and $n \times m$ couplers, where n and m are integers. Such couplers are characterized by such features as:

- The number of ports
- The sensitivity to the direction in which light is transmitted
- Wavelength sensitivity
- Single versus multimode
- Passive versus active
- Power requirements and/or losses
- Connection sensitivity to loss
- Cost considerations

Couplers based on fibers are commercially available and quite effective for handling modestly complex interconnect requirements. As more complex and demanding interconnect routing schemes evolve, the possibility of integrating a number of functions into a single device is being explored by the use of planar fabrication techniques, which offer the potential to be more cost-effective and compact for complex systems. Glass devices of this type have been achieved by using ion exchange on glass substrates and, more recently, by depositing and etching glass films on silicon or silica substrates.

Fiber Amplifiers and Lasers. One of the most exciting new areas of fiber research and development for communications is the field of rare earth doped glass-fiber amplifiers and lasers. By doping the single-mode fiber core with rare earth ions then injecting light of an appropriate wavelength, the fiber can act as a lasing or amplifying device. An excellent tutorial on this general subject is contained in Ref 20. Such devices are of particular interest for applications in telecommunications, medicine, sensing, and spectroscopy.

Of particular interest for telecommunications applications are erbium-doped fiber amplifiers, which allow amplification of signals in the very important 1.5 to 1.6 μm communications window. Specific characteristics of such amplifiers are reviewed in Ref 21, while Ref 20 to 22 discuss various approaches to fabricating silica-base fibers containing rare earth dopants.

There are a number of advantages to erbium-doped fiber amplifiers. They are compatible with existing transmission fibers and can thus be used as in-line devices. They have demonstrated low coupling losses, high gain with a large gain bandwidth (30 nm), polarization insensitivity, and low interchannel cross talk. While a number of pump wavelengths are possible, two laser diode pumps have been used to obtain excellent amplifier characteristics: 0.98 μm and 1.48 μm. The specific application and characteristics dictate the preferred source. The combination of simple laser-diode pumps combined with short, easily assembled fiber components also offers the potential for an inexpensive, easily manufacturable amplifier.

There are a variety of communications applications for such amplifiers. As a power amplifier, longer spans can be achieved before the need for reamplification. As a preamplifier, greatly improved signal-to-noise ratios can be achieved in high bit-rate long-span systems. In-line amplifiers are of particular interest and utility in telecommunications systems. Not only does this configuration allow very long-distance systems to be achieved before electronic regeneration of the signal is necessary, but a number of other features come into play. Once optical amplifiers are installed in-line in a system, additional channels can be added at the front end without the need to change the optical amplifiers, thus allowing very cost-effective upgradeability of systems. Most recently, soliton transmission experiments using in-line fiber amplifiers have demonstrated transmission distances of up to 10,000 km (6200 miles) without the need to electronically reshape the transmitted pulses.

While erbium-doped fiber amplifiers are currently commercially available, research and development continues to explore other rare earth dopants in silica and other host glasses which might be used to optically amplify signals at other wavelengths.

Future Directions/ Summary

High-silica glass fibers have gone from a research concept to a large commercial market in a relatively short time frame. Further work continues to develop a variety of special fibers for telecommunications, sensor, and other communications applications. Standard fiber designs are commercially available in the worldwide marketplace, and advances continue to further improve the price and performance of such fibers. Increasingly, a wide range of fiber devices in the form of couplers, splitters, sensors, and active devices are available. As interconnection complexity grows to build large optical networks, planar waveguide devices based on similar glass compositions are being developed and marketed. While the last twenty years have shown dramatic progress in the development of fiber technology, the future will continue to extend the performance limits and applications for such guided-wave devices based on high-silica glass.

REFERENCES

1. K.C. Kao and G.A. Hockham, Dielectric Fiber Surface Waveguides for Optical Frequencies, *Proc. IEEE*, Vol 133 (No. 7), 1966, p 1151–1158
2. S.E. Miller and A.G. Chynoweth, Ed., *Optical Fiber Telecommunications*, Academic Press, 1979
3. J.E. Midwinter, *Optical Fibers for Transmission*, John Wiley & Sons, 1979
4. T. Li, Ed., *Optical Fiber Communication*, Vol 1, *Fiber Fabrication*, Academic Press, 1985
5. S.E. Miller and I. Kaminow, Ed., *Optical Fiber Telecommunications II*, Academic Press, 1988
6. F.C. Allard, Ed., *Fiber Optics Handbook for Engineers and Scientists*, McGraw-Hill, 1989
7. A.W. Snyder and J.D. Love, *Optical Waveguide Theory*, Chapman and Hall, 1983
8. M.J. Adams, *An Introduction to Optical Waveguides*, John Wiley & Sons, 1981
9. C.R. Kurkjian, Ed., *The Strength of Inorganic Glass*, Plenum Press, 1985
10. T. Okoshi, Single Polarization Single Mode Optical Fibers, *IEEE J. Quant. Electron.*, Vol QE17 (No. 6), 1981, p 879–884
11. I. Kaminow, Polarization in Optical Fibers, *IEEE J. Quant. Electron.*, Vol QE17 (No. 1), 1981, p 15–22
12. D.N. Payne, A.J. Barlow, and J.T. Ramskov-Hansen, Development of Low and High Birefringent Optical Fibers, *IEEE J. Quant. Elec.*, Vol QE18 (No. 4), 1982, p 477–488
13. S.C. Rashleigh, Origins and Control of Polarization Effects in Single Mode Fibers, *J. Lightwave Technol.*, Vol LT1 (No. 2), 1983, p 312–321
14. J. Noda, K. Okamoto, and Y. Sasaki, Polarization Maintaining Fibers and Their Applications, *J. Lightwave Technol.*, Vol LT4 (No. 8), 1986, p 1071–1089
15. W.J. Tomlinson, Passive Components, paper FF1, Proceedings, *Optical Fiber Comm. Conference*, 1990, p 438–500
16. R.H. Stolen, Single-Mode Fiber Components, *Proc. IEEE*, Vol 75 (No. 11), 1987, p 1498–1511
17. D.B. Keck, A.J. Morrow, D.A. Nolan, and D.A. Thompson, Passive Components in the Subscriber Loop, *J. Lightwave Technol.*, Vol 7 (No. 11), 1989, p 1623–1633
18. T. Miyashita, S. Sumida, and S. Sakaguchi, Integrated Optical Devices Based on Silica Waveguide Technologies, *SPIE*, Vol 993, 1988, p 288–294
19. Y. Yamada, T. Miya, M. Kobayashi, S. Sumida, and T. Miyashita, Optical Interconnections Using Silica-Based Waveguide on Si Substrate, *SPIE*, Vol 991, 1988, p 4–11
20. P. Urquhart, Review of Rare Earth Doped Fiber Lasers and Amplifiers, *IEEE Proc.*, Vol 135 (No. 6), 1988, p 385–406
21. D.N. Payne and R.I. Laming, Optical Fiber Amplifiers, paper THFI, *Proceedings Optical Fiber Comm. Conference*, 1990, p 331–353
22. J. Simpson, Fabrication of Rare-Earth Doped Glass Fibers, *SPIE*, Vol 1171, 1989, p 2–7

Electronic Processing and Electronic Devices

Robert R. Shaw, Advanced Module Technology, IBM East Fishkill

GLASS is used in an extremely wide range of applications in electronics. Glass sometimes has a stand-alone function (such as carrier racks in semiconductor processing). More often, it is an important, and sometimes indispensable, component in electronic systems and processes. This article will focus on providing examples of these numerous applications; additional information is available in Ref 1 to 3.

Glass Applications in Electronic Processing

Silicon Crystal Growth. Extremely pure silicon is melted in high-purity silica glass crucibles in induction-heated crystal-pulling furnaces. The pure SiO_2 glass is commonly referred to as quartz or fused quartz in the industry. The hot zone of the pulling furnace is also usually surrounded by silica glass tubes and fittings, which permit easy monitoring and atmosphere control. Table 1 shows the purity levels available from commercial material.

Planar Processing. Integrated circuits are commonly formed by planar processing, in which silicon transistors are fabricated using glasses in the form of diffusion tubes, photomask substrates, and many types of carriers and fixtures. These are generally made of fused quartz or other low-expansion low-alkali compositions that avoid contamination of the silicon wafers. Wafer doping is done within quartz tubes followed by a diffusion anneal at a higher temperature. Currently, fused quartz tubing is drawn from the melt, and is available from numerous suppliers (Ref 4). It is available in sizes up to 350 mm (13.8 in.) OD (Ref 5). Impurities are in the 5 to 30 ppm range (Table 1). Recent work in centrifugal casting of colloidal gels and silica suspensions, followed by sintering, shows promise for producing relatively inexpensive and high-quality tubing (Ref 8), with impurities in the ppb range (Table 1). Tube OD is currently only 24 to 26 mm ($^{15}/_{16}$ to $1^1/_{32}$ in.).

Photomasks are used to pattern the metal wiring on the devices. These consist of chrome deposited on glass substrates that must have low thermal expansion, good transparency to ultraviolet (UV) light, and freedom from alkalis which can cause pinholing in the single-layer chrome used for the highest-resolution patterns (Ref 9). Soda-lime glass can be used in some cases, but aluminoborosilicates have better properties for most applications (Ref 10). Fused quartz provides the ultimate in transparency and stability (Fig 1). Mask substrate properties are shown in Table 2. Another glass substrate process use is in gallium arsenide wafer grinding and polishing. Fotoform glass with photomachined through-holes allows wafer separation with backside solvent application (Ref 22). Glass properties of Fotoform glass are listed in Table 2.

Packaging. Many ceramic packages that support electronic devices use glass components in their processing. Glasses are used as sintering aids to speed up densification and decrease the peak temperatures. Glass bonding occurs both between particles and at interfaces. Other uses of glasses as densification and bonding aids occur in conductive pastes, capacitor compositions, silver-glass die-attach materials, and other applications. Some of these applications are listed in Table 3.

Glass solder dams were useful in controlling the position of molten solder used in early high-density flip-chip interconnects (Ref 29). This function is now provided by metals and polymers in most current practice.

Glass Applications in Electronic Devices

Substrates. Glass substrates offer excellent combinations of smoothness, purity, planarity, and low cost for a large number of electronic applications. Polycrystalline silicon thin films, used in active-matrix liquid-crystal displays (LCDs), are deposited onto fused quartz substrates at temperatures above 600 °C (1110 °F). Borosilicates and soda-lime glass substrates are used at temperatures as low as 425 °C (795 °F) (Ref 30). Polycrystalline deposition at 250 °C (480 °F) was achieved on borosilicate glass (Ref 31). Techniques include evaporation, molecular beam deposition, reactive ion beam deposition, and several varieties of chemical vapor deposition (CVD). The silicon films are usually amorphous when deposited below about 500 °C (930 °F) and crystallize with strong preferred orientation at higher temperatures (Table 4).

Magnetic and Optical Storage Applications. Glass substrates fill many of the critical needs in memory storage applications where smoothness, flatness, and stability are important. Magnetic storage predominates in the field of memory storage. Current high-performance Winchester disk drives have a read/write head operating at high speeds only a few thousand Angstroms above the disk. This head clearance is being decreased further to increase the bit storage density, and new high-coercivity magnetic materials are being explored. Soda-lime glass substrates are often used in this demanding work (Ref 37). The dimensional requirements for optical storage technology are much more relaxed, but for the important case of write once read many (WORM) storage, there is still a need for the smoothness and stability of a glass backing substrate in high-performance applications (Ref 38). Magneto-optical storage technology typically depends on glass substrates in its current phase of development (Ref 39).

Solar Energy Applications. Glass also plays a useful role in solar energy technology as a substrate for solar cells, as protective glazing, and as a thermal absorber. Glass substrates are usually coated with a conducting layer of indium-doped tin oxide (ITO), evaporated aluminum, or screen-printed molybdenum, titanium, or nickel (Ref 40) before depositing a variety of active layers. The following table lists sources of information on glass substrates used in thin-film solar cells:

System	Ref
Si:H	41
$ZnSe-Zn_3P_2$	42
In_2Te_3-CdS	43
Cd(Zn,Mn)Te-CdS	44
CdTe-CdS	45
FeS_2	46
CdTe-CdZnS	47

Table 1   Impurity levels in fused quartz ware used for semiconductor processing

Material	Commercial quartz product	Impurity level, ppm						
		Alkali	Al	M^{2+}	OH	Zr	Cl	Ref
Beneficiated sand	Melt-drawn crucibles and tubes	2–5	15–20	2–10	30	2–3	...	4, 5
Plasma-torch SiO_2	Melt-drawn tubes	1	1	1	1000	...	100	4
Sol-gel/colloid	Sintered tubes	0.39	...	0.009	...	...	...	6, 7

Glass is the preferred protective glazing material because of its superior weathering resistance and screening of deep harmful ultraviolet radiation. Thin UV-protective solar covers have been developed (Ref 13 and Fig 1). The glass thickness strongly affects the cell efficiency in some cases (Ref 48). Spacecraft solar cells need radiation protection, particularly from proton bombardment, that can be provided by thin glass sheets containing up to 8 wt% CeO_2 (Ref 49). Borosilicate glass is also used in this application (Table 2). Solar collectors use absorber plates that must remain stable against ultraviolet radiation while maintaining a high absorbance-to-emittance ratio. A value of 0.93/0.05 was obtained from vacuum-coevaporated quartz glass and copper (Ref 50).

Other glass substrate applications are in electrochromic devices, in which the basic glass substrate is covered with ITO or other transparent, conductive films. Soda-lime glasses can contribute to alkali poisoning under some conditions (Ref 51).

Printed Circuit Boards. Another important substratelike function for glass is as a component of printed circuit boards (PCBs). Glass filaments as fine as 25 μm (1000 μin.) diameter are bundled into yarns and made into plain-weave glass cloth. This fabric is incorporated into epoxies, polyimides, and polyethylenes to make PCBs and cards. Low-alkali borosilicate (E-glass) and aluminosilicate (S-glass) compositions having the properties listed in Table 2 are most often used. The finished board properties are listed in Table 5, with the most commonly used material being the epoxy-glass FR-4.

Planarization and Passivation. Multilevel electronic devices and wiring must be kept reasonably planar during processing and must also be protected from chemical and mechanical damage after fabrication. Spin-on glasses (SOG) are popular in providing planarity and consist of organic siloxane- or phosphosilicate-based liquids that are spun on to flood surface irregularities. After thermal curing at 150 to 425 °C (300 to 795 °F) and dry-proc-

ess etchback (sometimes in the reverse order), the resulting surface is considerably more planar. Films typically have only parts per billion (ppb) concentration of alkalis and yield crack-free films about 0.15 to 0.5 μm (6 to 20 μin.) thick. Pure SiO_2 or phosphosilicate glass remains after removal of the organic components during bakeout. The films can incorporate about 10% H_2O on a molecular basis (Ref 52). Several modeling studies are available in the technical literature (Ref 53, 54).

Other types of glasses are used for interlevel dielectrics and passivating layers. These are applied with sedimentation and screening techniques and deposition techniques, that is, evaporation, sputtering, and several varieties of chemical vapor deposition (Ref 55). Silica, borosilicate, phosphosilicate (PSG), and borophosphosilicate (BPSG) compositions are used. Because of microcracking and other coverage problems, passivation glasses are often used in conjunction with an organic encapsulant to extend the passivation lifetime. PSG and BPSG types are the most popular, with the P_2O_5 content not more than a few percent in order to avoid corrosion problems. Application by plasma-enhanced chemical vapor deposition (PE-CVD) shows some advantages (Ref 56). A detailed discussion of passivation glass properties is available in Ref 2. A silica spin-on glass also provided a barrier layer against alkali migration in tungstic oxide (WO_3) electrochromic devices (Ref 57).

Electron Tubes. The vacuum tube industry, now nearly obsolete, commonly uses soda-lime glass as the vacuum envelope or bulb and lead silicates for the stem. The lower-resistance soda-lime composition allows charge to bleed off the inner bulb wall to minimize voltage buildup, and the high-resistance lead glass minimizes electrolysis. Large complex tubes with elaborate electrode structures and multiple feedthroughs might also have complex electrode spacers and other structures of glass. The latter are inexpensively prepared by molding glass powder and then sintering the powder to the final shape (Corning's multiform process) (Ref 58). High-power tubes, which operate at higher temperatures, require borosilicate glasses. All tubes need glass-to-metal sealing (see the article "Sealing Glasses" in this Volume). An excellent overview of vacuum tube technology and materials is provided by Ref 59. Although the need for standard vacuum tubes has languished, a recent variation is the fluorescent indicator panel (FIP) that was developed by NEC (Ref 60). This device combines vacuum tube technology with semiconductor technology (Fig 2). A glass vacuum envelope holds a filament, a patterned phosphor anode, and up to six integrated circuit (IC) chips. The chips control the illumination pattern of the dot character panel.

Microchannel Plates (MCPs) are disks made of parallel open glass microtubes that

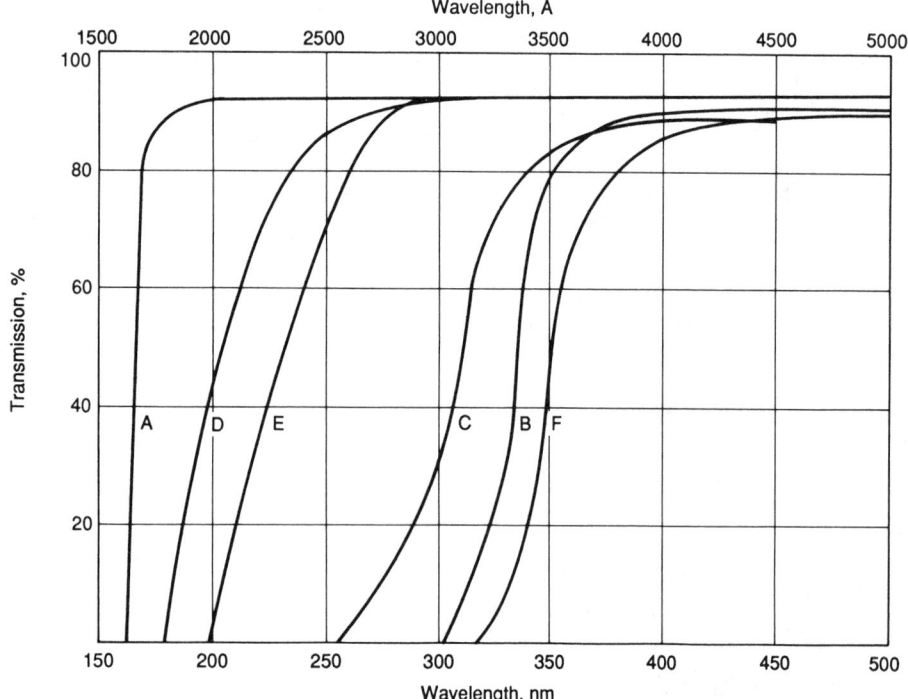

Fig 1   UV transmission curves for selected glasses used in photomask applications: A, 2.3 mm (0.091 in.) thick fused quartz; B, 2.3 mm (0.091 in.) thick soda-lime glass; C, 2.3 mm (0.091 in.) thick Hoya LE-30 aluminoborosilicate; D, 1 mm (0.039 in.) thick Corning 9741 EPROM window; E, 1 mm (0.039 in.) thick Fotoform glass; F, 0.125 mm (0.00492 in.) thick Corning code 0213 solar cell window. Source: Ref 9, 11–13

Table 2 Properties of selected glasses used in electronics

Application	Glass composition type	Thermal expansion coefficient, 0–300 °C (32–570 °F)		Softening point		Density		Log₁₀ volume resistivity at 350 °C (660 °F)	Refractive index, n_d	Dielectric constant 1 MHz, 20 °C (70 °F)	Ref
		10^{-7}/K	10^{-7}/°F	°C	°F	g/cm³	lb/in.³				
Substrates	Fused quartz	5.5	3.1	1580	2875	2.200	0.079	10.2	1.459	3.8	
	Aluminosilicates	31–43	17–24	924–1070	1695–1960	2.550–2.870	0.0921–0.104	10.1–11.6	...	...	10
	Fotoform glass (Li silicate)	84	47	450	845	2.365	0.0854	4.90	...	7.62(b)	12
	Fotoform opal	89	49	550	1020	2.380	0.0860	7.23	...	5.73(b)	12
	Fotoceram	103	57.3	750	1380	2.407	0.0870	7.07	...	5.63(b)	12
Solar cell cover	Alkali borosilicate	32	18	...	...	2.130	0.0770	11.2(a)	1.469	4.1	14
		74	41	718	1325	2.600	0.094	...	1.5281	6.69	13
Fiberglass	E-glass (Mg-Ca-Al borosilicate)	50	28	843	1550	2.520–2.620	0.0910–0.0947	...	1.546–1.560	5.8	15, 16
	S-glass (Mg aluminosilicate)	29–30	16–17	860	1580	2.470–2.490	0.0892–0.0900	...	1.524–1.528	4.53	15, 16
Microchannel plates	Tube (K-Rb lead silicate)	90.3	50.2	600	1110	3.970	0.143	...	...	...	17
	Carrier (K-Na lead silicate)	89.5	49.8	630	1165	3.850	0.139	...	...	...	18
EPROM windows	NaAl borosilicate	38	21	705	1300	2.170	0.0784	7.6	1.47	4.70	19
Diode enclosures	K lead silicate	91	51	578	1070	4.280	0.155	9.5	1.7	9.1	11
	Ba lead borosilicate	45	25	720	1330	2.880	0.104	11.3	1.55	6.0	11
Recording heads	Fotoceram	103	57.3	750	1380	2.407	0.0870	7.07	...	5.63(b)	12
	Gap glass (Ba-Ca borosilicate)	80	44	770	1420	...	...	...	...	...	3
	Bonding glass (lead borosilicate)	80	44	490	915	...	...	...	...	...	3
	Na lead silicate	95	53	495	925	5.100	0.184	...	...	...	20
	Na-Zn borosilicate	104	57.8	580	1075	2.980	0.108	...	...	...	20
Resistors	Aluminosilicate	44	24	926	1700	2.640	0.0954	13.5(a)	1.547	6.6	21

(a) At 250 °C (480 °F). (b) At 100 kHz

extend from one face of the disk to the other face, and have a high voltage impressed across the metallized disk faces (Fig 3). The inner surfaces of the lead silicate tubes are partially reduced to form an emissive layer along the walls. In operation, an incoming photon or charge particle (for example, an electron derived from a photocathode) strikes the wall and liberates secondary electrons. These cascade down the tube under the applied field and can generate a gain of up to 10^6 for a curved-channel plate. Paired plates can yield up to a 10^7 gain and even more can be obtained with triplet plates (Ref 62). These devices are used for image and signal amplification, photon and particle detection, and for measuring extremely fast events. The disks consist of thousands to millions of tubes, with bores as small as 4 μm (160 μin.) spaced on 5 μm (200 μin.), centers (Ref 63). Disk operation is in high vacuum with response times on the order of picoseconds (Ref 64). Glass properties of the disks are listed in Table 2. The sensitivity is improved by minimizing radioisotopes to reduce the background noise from spontaneous β-emission. A composition yielding only 0.08 counts/cm² · s (0.52 counts/in.² · s) background is, in weight percent, 30 to 50% SiO_2, 50 to 57% PbO, 2 to 10% Cs_2O, 0 to 5% Mo, and 0.1 to 1% (Al_2O_3/ZrO_2/TiO_2/Nb_2O_5) (Ref 19).

EPROM Windows. Erasable programmable read-only memory (EPROM) devices are erased by UV radiation. A borosilicate glass tailored for good UV transmission is shown in Table 2 and in Fig 1 (curve D). During processing, the optimum transmission is obtained by using a slightly reducing atmosphere (Ref 11).

Diode Tubing. Discrete diodes of axial-lead or radial-lead geometries are encapsulated in glass that makes a hermetic seal to the lead wires (Ref 58). Lead silicates with up to 6 wt% alkali are used, depending on the diode sensitivity to sealing temperatures and alkali poisoning (Ref 65). Selected properties of diode tubing are shown in Table 2.

Photochemically Sensitive Glass. Many glasses can be made photosensitive by doping with metals and appropriate sensitizers that

Table 3 Selected glass-bonded composite materials

Glass composition type	Crystal components	Application	Ref
Lead borosilicate	Pd/Au, Pt/Au	Conductor	2
	Pd/Au	Conductor	3
Proprietary glass	Pd/Ag	Conductor	23
CdBi lead borosilicate	Ba/Nd titanate	Capacitor	24
B_2O_3, lead borate	Ba titanate	Capacitor	25
B_2O_3	Ba ferrite	Magnets	26
Lead borate	Ag flake	Die attach	27
	Ag powder	Conductor (bus bars)	
Proprietary glass	AlN	Substrate	28

Table 4 Orientation of silicon films on selected glass substrates

Deposition process(a)	Substrate	Orientation	Ref
Plasma-enhanced CVD, 700 °C (1290 °F)	Fused quartz	(100), (110)	32
Low-pressure CVD, 700 °C (1290 °F)	Fused quartz	Random	33
Photo CVD	Fused quartz	(110)(b) (100)(c)	34
Reactive ion beam	Fused quartz, borosilicate	Random(d) (220)(e)	31
Plasma-enhanced CVD	Hoya NA-40	Amorphous(f)	30
Electron beam evaporation	Fused quartz	Amorphous(g)	35
Evaporation	Borosilicate, soda-lime	Amorphous(h)	36

(a) CVD, chemical vapor deposition. (b) <560 °C (1040 °F). (c) >575 °C (1065 °F). (d) 250 °C (480 °F). (e) 550–700 °C (1020–1290 °F). (f) 425 °C (795 °F). (g) <525 °C (975 °F). (h) <480 °C (895 °F)

Table 5   Properties of glass-polymer printed-circuit boards

Glass-polymer composition type	Glass transition temperature, T_g		Thermal expansion coefficient at $<T_g$				Dielectric constant, 1 MHz	Dissipation factor, 1 MHz	Surface resistivity, $\Omega/\square$	Water absorption, 24 h at RT(a)
			x-y		z					
	°C	°F	10^{-7}/K	10^{-7}/°F	10^{-7}/K	10^{-7}/°F				
FR-4 epoxy-glass	120–130	250–265	160–200	90–110	500–700	280–390	4–5.5	0.02–0.03	10^{14}	0.3
Polyimide-glass	210–220	410–430	140	80	400	220	4–5	0.01–0.015	10^{14}	0.15–0.4

(a) RT, room temperature. Source: Ref 16

will retain a latent image. With suitable heat treatment, the image can be nucleated and crystallized within the glass. The image or pattern can then be preferentially etched away. Corning's Fotoform process uses a lithium silicate glass to produce controlled through-holes, patterns, internal voids and channels, and so on with good dimensional control (Table 6). Fotoform opal and Fotoceram are derived from Fotoform glass by applying different heat treatments to yield a 28% and a 50% crystalline product, respectively. The minimum thickness available is 100 μm meters (0.004 in.). The UV transmission for Fotoform glass is shown in Fig 1 (curve E). Selected properties are shown in Table 2. Applications of photochemically sensitive glass in electronics include substrates, tooling and fixtures, memory disk headpads and sliders, and spacers for special vacuum tubes (Ref 66).

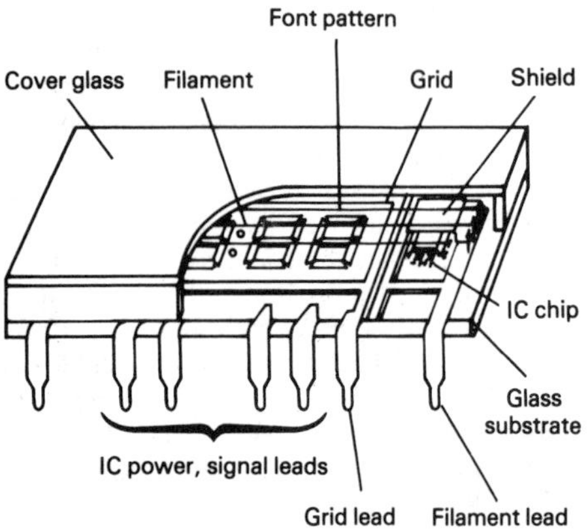

Data Storage Recording Heads. Fotoceram is used for constructing ferrite recording heads in magnetic disk storage systems. Its properties are shown in Table 2. It can be polished to <0.0025 μm (<0.1 μin.) roughness, is compatible with ferrite sealing glasses, can be worked to a flatness of 0.0001 cm/cm (0.0001 in./in.), has holes and slots as

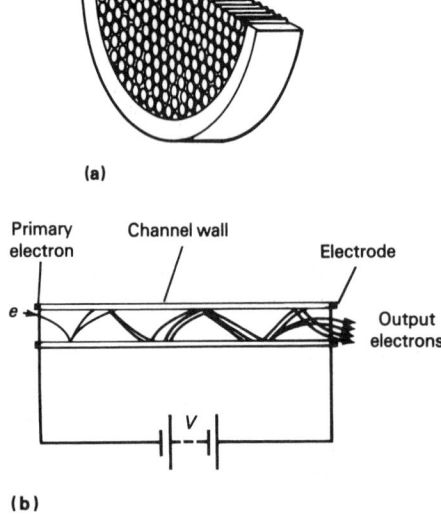

(a)

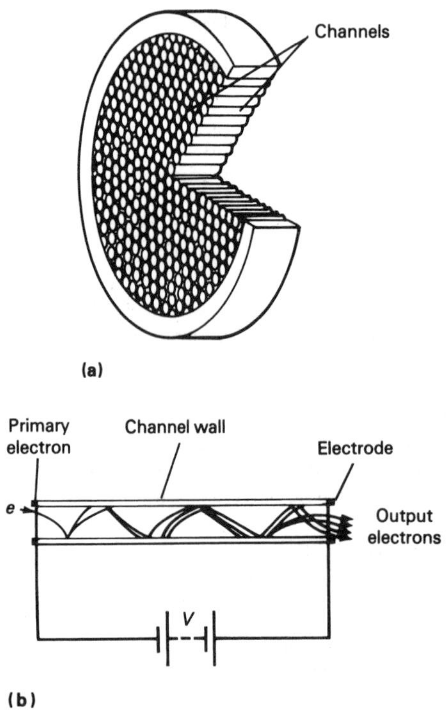

Fig 3 Schematic of a microchannel plate. (a) Single-disk construction showing device contains an array of parallel open glass microtubes. (b) Cross-section of an individual channel glass microtube

small as 38 μm (0.0015 in.), and finished parts are available down to 75 μm (0.003 in.) thick (Ref 67). Other commercial glasses are listed in Table 2. Still other glasses can be used for ferrite heads, as discussed in Ref 3, for a design that required high-temperature glass for the core gap and a low-temperature glass for bonding the head to the slider. A variety of simultaneous requirements were met by a barium borosilicate gap glass and a lead borosilicate bonding glass (Table 2).

Resistor Glasses. Discrete resistors are often made by depositing resistive metal or oxide films onto glass substrates, which may be glass sheet, rod, or tubing. The resistance value is obtained by cutting the film to an appropriate geometry (for example, helical grooves on a rod). Low-alkali glasses are used because of the performance sensitivity to alkali poisoning. The substrate generally controls the quality of the thin conductive layer and must be smooth, nonporous, stable, and have favorable electrical and compatible thermal properties with the film. References 58 and 68 discuss the overall requirements of resistor glasses. Resistance variations due to humidity, temperature, and load are kept within close specifications with the aluminosilicates (Table 2), borosilicates, and high-silica glasses used. Aluminosilicate and borosilicate sheets as thin as 50 to 84 μm (0.002 to 0.0034 in.) are available (Ref 69) as well as fused silica sheets as thin as 25 μm (100 μin.) (Ref 70). Resistor substrates sintered from glass powders have also been found useful and much less expensive (Ref 71).

Glass Capacitors. Discrete glass-dielectric capacitors are made with aluminum electrodes and lead glass. This provides a glass-sealed monolithic axial capacitor with good mechanical stability toward thermal cycling

Fig 2 Cutaway view showing the internal components of a chip-in-glass fluorescent indicator panel. Source: Ref 59

Table 6   Tolerance specifications for selected photochemically sensitive glasses

Capability	Glass composition type(a)	Hole and slot size tolerance		Centerline tolerance		Edge to vertical angle of etch, degrees	Minimum feature size		Maximum ratio of feature size to thickness	Thickness control		Flatness		Minimum thickness	
		±μm	±in.	±μm/cm	±in./in.		μm	in.		±μm	±in.	±μm/cm	±in./in.	μm	in.
Standard process	Fotoform glass	1	0.001	10	0.001	2–3	63	0.0025	8:1	50	0.002	10	0.001	500	0.020
	Fotoform opal	1	0.001	15	0.0015	2–3	75	0.003	8:1	50	0.002	10	0.001	500	0.020
	Fotoceram	1	0.001	15	0.0015	2–3	75	0.003	8:1	50	0.002	10	0.001	500	0.020
Ultimate precision	Fotoform glass	8	0.0003	1	0.0001	1–2	25	0.001	40:1	25	0.001	1(b)	0.0001(b)	100(b)	0.004(b)
	Fotoform opal	10.4	0.0004	2	0.0002	1–2	38	0.0015	40:1	13	0.0005	1(b)	0.0001(b)	75(b)	0.003(b)
	Fotoceram	10.4	0.0004	2	0.0002	1–2	38	0.0015	40:1	13	0.0005	1(b)	0.0001(b)	75(b)	0.003(b)

(a) Corning Glass Works Fotoform process product. (b) Dependent on size. Source: Ref 12

and vibration. These capacitors are used at high voltages and frequencies up to 250 MHz. Glass capacitors also show good radiation hardness. Reference 58 reviews the performance of glass capacitors under a wide range of conditions.

Tape-Reel Flanges. Solid glass tape-reel flanges solve a degradation problem encountered when using magnetic tape, in which the flange holes in metal reels induce tape packing defects. Glass reels show superior performance to metal reels, offer better dimensional stability in the 265 to 405 mm (10.5 to 16 in.) diameters, and allow easy monitoring of tape usage and packing. The reel flanges are ion-exchanged (Corning's Chemcor process) for greater strength and to induce a relatively safe granular fracture pattern (Ref 72). Alkali aluminosilicate glass compositions are typically used.

Ultrasonic Delay Lines. Glass is widely used as a delay-line component in electronics systems, particularly in color television (CTV) applications. An electrical signal is converted to an acoustic wave in glass, providing about 1 μs delay for every 2.5 mm (0.1 in.) of path length (Ref 58). High-quality fused quartz is often used, but near-zero temperature coefficients of delay are available with a tailored composition (Ref 73).

ACKNOWLEDGMENT

The writer gratefully acknowledges Vince Marcotte, Karl Puttlitz, and Carol Hand of IBM, Jonathan Rowe and Roger Whitney of Corning, Bruce Feller of Galileo Electrooptic Corp., and Richard Marlor of GTE Sylvania, for their helpfulness.

REFERENCES

1. G. Geiger, Glass in Electronic Packaging Applications, *Ceram. Bull.*, Vol 69 (No. 7), 1990, p 1131–1134, 1136
2. A.K. Varshneya and S.C. Cherukuri, Materials and Processing for Microelectronic Systems, *Ceramic Transactions*, Vol 15, K.M. Nair, R. Pohanka, and R.C. Buchanan, Ed., 1990, p 217–243
3. R.R. Tummala and R.R. Shaw, Glasses in Microelectronics in the Information-Processing Industry, *Advances in Ceramics*, Vol 18, *Commercial Glasses*, D.C. Boyd and J.F. MacDowell, Ed., American Ceramic Society, 1986, p 87–102
4. P.P. Bihuniak, High Silica Glass, *Advances in Ceramics*, Vol 18, *Commercial Glasses*, D.C. Boyd and J.F. MacDowell, Ed., American Ceramic Society, 1986, p 105–114
5. "Fused Quartz," GTE Sylvania, 1990
6. R. Clasen, Preparation of High-Purity Silica Glass Tubes by Centrifugal Casting of Colloidal Gels, *J. Mater. Sci. Lett.*, Vol 7, 1988, p 477–478
7. T. Mori, M. Toke, M. Ikejiri, M. Takei, M. Aoki, S. Uchiyama, and S. Kanbe, Silica Tubes by New Sol-Gel Method, *J. Non-Cryst. Sol.*, Vol 100, 1988, p 523–525
8. P.K. Bachmann, P. Gietner, E. Krafczyk, H. Lydtin, and G. Romanowski, Shape Forming of Synthetic Silica Tubes by Layerwise Centrifugal Particle Deposition, *Ceram. Bull.*, Vol 68 (No. 10), 1989, p 1826–1831
9. D.F. Horne, *Photomasks, Scales, and Gratings*, Adam Hilger Ltd, Bristol, 1983
10. W.H. Dumbaugh and P.S. Danielson, Aluminosilicate Glasses, *Advances in Ceramics*, Vol 18, *Commercial Glasses*, D.C. Boyd and J.F. MacDowell, Ed., American Ceramic Society, 1986, p 115–132
11. "Code 9741 UV Transmitting Glass," Corning Inc., 1981
12. "FOTOFORM," Corning Inc., 1989
13. "Code 0213 Solar Cell Glass," Corning Inc., 1990
14. "Code 7070 Lithium Potash Borosilicate Glass," Corning Inc., 1990
15. D.P. Seraphim, D.E. Barr, W.T. Chen, G.P. Schmitt, and R.R. Tummala, Chapt. 12, Printed-Circuit Board Packaging, *Microelectronics Packaging Handbook*, R.R. Tummala and E.J. Rymaszewski, Ed., Van Nostrand Reinhold, 1989
16. P.F. Aubourg and W.W. Wolf, Glass Fibers, *Advances in Ceramics*, Vol 18, *Commercial Glasses*, D.C. Boyd and J.F. MacDowell, Ed., American Ceramic Society, 1986, p 51–63
17. "Code 8161 Microchannel Plate Glass," Corning Inc., 1978
18. "Code 0120 Mounting Rim Glass," Corning Inc., 1978
19. Low-Noise, Long-Life High-Gain Microchannel Plate Glass, *NASA Tech. Briefs*, Aug 1990, p 47
20. "Magnetic Ferrite Head Glasses LS-0700 and FH-11," Nippon Electric Glass, 1990
21. "Code 1724 Aluminosilicate Glass," Corning Inc., 1990
22. "FOTOFORM Glass Substrate," Corning Product Information, 1989
23. S.J. Stein, C. Huang, and P. Bless, Pd/Ag Conductors Show Promise, *Electron. Man. Test*, 1986, p 29–31
24. D. Xie, Intermediate Firing Temperature Ceramic Dielectric for MLC, *39th Electronic Component Conference*, 1989, p 217–221
25. S.K. Sarkar and M.L. Sharma, Liquid Phase Sintering of $BaTiO_3$ by Boric Oxide and Lead Borate Glasses and Its Effect on Dielectric Strength and Dielectric Constant, *Mater. Res. Bull.*, Vol 24 (No. 7), 1989, p 773–779
26. S. Ram, Crystallization of $BaFe_{12}O_{19}$ Hexagonal Ferrite with an Aid of B_2O_3 and the Effects on Microstructure and Magnetic Properties Useful for Permanent Magnets and Magnetic Recording Devices, *J. Mag. Mater.*, Vol 82 (No. 1), 1989, p 129–150
27. N.M. Davey and F.W. Weise, Silver-Glass Die Attachment-Adhesion Mechanisms with Gold and Chromium/Gold Backed Die, *Hybrid Circ.*, Vol 13, 1987, p 25–29
28. L.M. Sheppard, Aluminum Nitride: A Versatile but Challenging Material, *Ceram. Bull.*, Vol 69 (No.11), 1990, p 1810–1812
29. L.F. Miller, Controlled Collapse Reflow Chip Joining, *IBM J. Res. Dev.*, 1969, p 239–250
30. A.C. Ipri and G. Kaganowitz, Low-Cost Process for Fabricating Polycrystalline Transistors, *IEEE Trans. El. Dev.*, Vol 37 (No. 7), 1990, p 1771–1772
31. H. Yamada and Y. Torii, Low-Temperature Polycrystalline Si Film Growth on Amorphous Insulators by Reactive Ion Beam Deposition, *J. Appl. Phys.*, Vol 65 (No. 3), 1989, p 1106–1111
32. S. Hasegawa, M. Morita, and Y. Kurata, Structural and Electrical Properties of P and B Doped Polycrystalline Silicon by Plasma Enhanced CVD and 700 Degrees C, *Jpn. J. Appl. Phys. 2, Lett.*, Vol 28 (No. 4), 1989, p 522–524
33. S. Hasegawa, S. Yamamoto, and Y. Kurata, Control of Preferential Orientation by In Situ Plasma Supply During Growth of Polycrystalline Films, *Appl. Phys. Lett.*, Vol 55 (No. 2), 1989, p 142–144
34. M. Okuyama, N. Fujiki, Y. Nakatani, K. Inoue, and Y. Hamakawa, Growth Behavior of Highly Oriented Si Films on Fused Quartz by Photo CVD, *Appl. Surf. Sci.*, Vol 41–42, 1989, p 41–42
35. N.K. Annamalai, N. Meyyappan, and A.N. Khondker, Properties of Polycrystalline Silicon Grown on Insulating Substrates by Electron Beam Evaporation, *Thin Sol. Films*, Vol 155 (No. 1), 1987, p 97–113
36. W. Schmolla, J. Diefenbach, G. Blang, B. Ocker, and W. Senske, Thin-Film Transistors from Very Low Temperature Polycrystalline Si on Glass Substrates, *Sol. St. Elec.*, Vol 32 (No. 5), 1989, p 391–396
37. S. Katayama, T. Tsuno, K. Enjoji, N. Ishii, and K. Sono, Magnetic Properties and Read-Write Characteristics of Multilayer Films on a Glass Substrate, *IEEE Trans. Magn. Jpn.*, Vol 24 (No. 6), 1988, p 2982–2984
38. M. Emmelius, G. Pawlowski, and H.W. Vollman, Materials for Optical Data Storage, *Angew. Chemie, Int. Ed.* (in English), Vol 28 (No. 11), 1989, p 1445–1471
39. K. Yokoyama, T. Tanaka, N. Koshizuka, and T. Okumura, Amorphous Tb-Fe-Co/Iron Garnet Composite Films on Glass Substrates, *IEEE Transl. J. Magn. Jpn.*, Vol TJMJ-2 (No. 11), 1987
40. D.A. Baert, J. Roggen, J.F. Nijs, and R.P. Mertens, Amorphous-Silicon Solar

Cells with Screen-Printed Metallization, *IEEE Trans. El. Dev.*, Vol 37 (No. 3), 1990, p 702–707

41. K. Kato and I. Kato, Fabrication of a-Si:H Thin Films Using a Microwave Discharge Under a Magnetic Field of Electron Cyclotron Resonance, *Jpn. J. Appl. Phys., 2*, Vol 28 (No. 3), 1989, p 343–345

42. M. Bhushan and J.M. Pawlikowski, Thin-Film $ZnSe-Zn_3P_2$ Heterojunctions for Solar Cells, *Electr. Technol.*, Vol 20 (No. 3–4), 1989, p 3–15

43. A.A. Zahab, M. Abd-Lefdil, and M. Cadene, Rectifying and Photovoltaic Parameters of Indium Telluride (p)-Cadmium Sulfide (n) Thin Film Heterojunctions, *Phys. St. Sol. A,* Vol 119 (No. 1), 1990, p K35-K39

44. S.A. Ringel, R. Sudharsanan, A. Rohgati, and W.B. Carter, Study of Polycrystalline Cd(Zn,Mn)Te/CdS Films and Interfaces, *J. Elec. Mater.*, Vol 19 (No. 3), 1990, p 259–263

45. J.T. Moon, K.C. Park, and H.B. Im, Photovoltaic Properties of CdS/CdTe Solar Cells Sintered with $CdCl_2$, *Sol. Energy Mater.*, Vol 18 (No. 1–2), 1988, p 53–60

46. G. Smestad, A. DaSilva, H. Tributsch, S. Fiechter, M. Kunst, N. Meziani, and M. Birkholz, Formation of Semiconducting Iron Pyrite by Spray Pyrolysis, *Sol. Energy Mater.*, Vol 18 (No. 5), 1989, p 299–313

47. S.S. Yeo and B.I. Ho, Microstructure and Optical Properties of Cd(1 − x)ZnxS Films Sintered with Cadmium Chloride, *J. Am. Ceram. Soc.*, Vol 72 (No. 11), 1989, p 2131–2135

48. N.A. Al-Rawi and A.A. Mohammad, Heat Distribution in Bi-Glass Encapsulation Silicon Solar Cell, *Sol. Cells*, Vol 25 (No. 1), 1988, p 39–51

49. A. Kimura, Cover Glass for Solar Cells in Space, *New Glass*, Vol 4, 1987, p 11–14 (Japanese)

50. F. Garnich and E. Sailer, $CuSiO_2$/Cu Cermet Selective Absorbers for Solar Photothermal Conversion, *Sol. Energy Mater.*, Vol 20 (No. 1–2), 1990, p 81–89

51. D. Davazoglou and A. Donnadieu, Electrochromism in Polycrystalline WO_3 Thin Films Prepared by Chemical Vapor Deposition at High Temperature, *Thin Sol. Films*, Vol 164, 1988, p 369–374

52. H.G. Tompkins and C. Tracey, Desorption from Spin-On Glass, *J. Electrochem. Soc.*, Vol 136 (No. 8), 1989, p 2331–2335

53. F.A. Leon, Numerical Modeling of Glass Flow and Spin-On Planarization, *IEEE Trans. Comp.-Aided Des.*, Vol 7 (No. 2), 1988, p 168–173

54. L.K. White, "Modelling Spin-On Film Planarization Properties," Sixth International VLSI Multilevel Interconn. Conference, 1989

55. F.O. Sequeda, Thin Film Deposition Techniques in Micro-Electronics, *J. Metall.*, Vol 38 (No. 2), 1986, p 55–65

56. K. Law, J. Wong, C. Leung, J. Olsen, and D. Wang, Plasma-Enhanced Deposition of Borophosphosilicate Glass Using TEOS and Silane Sources, *Sol. St. Technol.*, 1989, p 60–62

57. K. Yamanaka, Degradation Caused by Substrate Glass in WO_3 Electrochromic Devices, *J. Appl. Phys.*, Vol 54 (No. 2), 1983, p 1128–1132

58. G.W. McLellan and E.B. Shand, Chapt. 18, Electronic Components, *Glass Engineering Handbook*, 3rd ed., 1984

59. W.H. Kohl, *Materials and Techniques for Electron Tubes*, 1960

60. Y. Murayama, K. Tsujikawa, H. Nagai, M. Hiroi, J. Yamamoto, T. Uchimura, M. Satonaka, and K. Tofuku, Chip-in-Glass Fluorescent Indicator Panel, *NEC Res. Dev.*, Vol 93, 1989, p 47–56

61. H. Kume, K. Koyama, K. Nakatsugawa, S. Suzuki, and D. Fatlowitz, Ultrafast Microchannel Plate Photomultipliers, *Appl. Opt.*, Vol 27 (No. 6), 1988, p 1170–1178

62. Private communication, W.B. Feller, Galileo Electro-Optics Corp., 1990

63. "Microchannel Plate Detectors," Galileo Electro-Optics Corp., 1988

64. G.W. Fraser, Microchannel Plates, Chapt. 3, *X-Ray Detectors in Astronomy*, Cambridge University Press, 1989

65. "Code 8870 and 7061 Diode Enclosure Glass," Corning Inc., 1987

66. R.K. Whitney, "Precision Photochemical Machining of a Glass without Photoresist," Proc. Tech. Pr. NEPCON West '89, p 495–506

67. "Code 8603 FOTOCERAM Magnetic Recording Head Material," Corning Inc., 1989

68. J.E. Hughes, A Review of Thin Film Resistors and Their Assembly Problems, *Hybrid Circ.*, Vol 13, 1987, p 42–45

69. "Code 0211 Microsheet Glass," Corning Inc., 1990

70. "Polished Fused Silica Substrates," Valley Design Corp., 1990

71. B. Herod and D. Lusniak-Wojcicka, Inexpensive Substrates from Moulded Glass Powder for Thin Film Potentiometers, *Hybird Circ.*, Vol 13, 1987, p 21–24

72. "Data Shield Precision Glass Tape Reels," Corning Inc., 1989

73. "Code 8875 OTC Delay Glass," Corning Inc., 1990

Glazes and Enamels

Richard A. Eppler, Eppler Associates and Laurence D. Gill, Mobay Corporation

VITREOUS CERAMIC COATINGS are applied over substrates for a number of reasons (Ref 1). These coatings may be applied to a substrate surface to render the surface:

- Chemically more inert
- Impervious to liquids and gases
- More readily cleanable
- Smoother and more resistant to abrasion and scratching
- Mechanically stronger
- Decorative
- Aesthetically pleasing

Vitreous coatings are thin layers of glass fused onto the surface of the substrate. When the substrate is a ceramic, the coating is called a glaze. When the substrate is a metal, the coating is called a porcelain enamel. When the substrate is a glass, the coating is called a glass enamel.

There are two general types of properties that must be present in a ceramic coating (Ref 2). The first requirement derives from the fact that the coating must be applied to and bond with the substrate. The composition must fuse to a homogeneous viscous glass at a temperature that is either coincident with the temperature at which the body matures or at a temperature sufficiently lower to prevent distortion of the substrate during glost firing.

During and after fusion of the coating materials, they must react with the substrate to form an intermediate bonding layer of proper thickness. If the bonding layer or interface is too thin, the coating will flake off after application and subsequent firing. If the bonding layer is too thick, the composition of the body or the coating may be degraded. The coating must also have a coefficient of thermal expansion that coincides or fits the substrate (Ref 3). When the fired ware is cooled, the coated substrate contracts. If the coefficients of thermal expansion of the coating and the substrate are not matched, stresses that lead to spalling or crazing of the coating will be introduced. The coating materials should have a low surface tension to minimize the crawling of the coating away from the edges or any holes present during firing.

The second group of properties are those associated with the use of the product, such as appearance, smoothness, porosity, and corrosion resistance to various liquids and gases. Almost all vitreous coatings are expected to be homogeneous, smooth, and hard and also to resist abrasion and scratching. Such a surface is also more apt to be impervious to liquids and gases and hence more readily cleanable. The sole exception to the desire for a smooth surface is the textured coating in which a pattern is applied for aesthetic purposes.

In many applications, chemical durability in severe service conditions is a principal reason for the selection of a ceramic coating (Ref 4). Vitreous coatings are formulated to be resistant to a variety of reagents ranging from acids, to hot water, to alkalis, to essentially all organic media. The only important exception is hydrofluoric acid, which readily attacks all silicate glasses.

For some applications, the finished ware is to be subjected to elevated temperatures while in service. This is a prime reason for the selection of ceramic coatings for cookware applications and for industrial and military applications.

The optical and appearance properties of any surface coating material are major considerations in determining which coating will be applied. Various possibilities can be called for to meet the requirements of a particular application. Because vitreous coatings can be transparent or opaque; high gloss, satin, or matte; smooth, patterned, or textured; and monochrome or multicolored, the combination of requirements that meets each particular application is extensive.

Glazes

A ceramic glaze is a vitreous coating applied to a ceramic substrate (usually a whiteware). A great variety of formulations are used as glazes (Ref 1). Ceramic ware is fired over a wide range of temperatures from 800 to >1400 °C (1470 to >2550 °F). No single glaze composition would be satisfactory over such a wide range of temperatures. A glaze that melts at low temperature will run off the substrate, react with the substrate, or volatilize if fired at a high temperature. A given glaze composition is generally useful over a temperature range of only 30 °C (55 °F). Typical firing temperatures for glazes based on application can be summarized as follows (Ref 5):

- Electronic substrates, ≈600 to 900 °C (≈1110 to 1650 °F)
- Artware glazes, ≈900 to 1050 °C (≈1650 to 1920 °F)
- Dinnerware and tile glazes, ≈1000 to 1150 °C (≈1830 to 2100 °F)
- Structural clay and sanitaryware glazes, ≈1180 to 1250 °C (≈2155 to 2280 °F)
- Porcelain glazes, ≥1300 °C (≥2370 °F)

Because of the toxicity of lead, glazes are often classified on the basis of the presence or the absence of litharge (PbO). Safe working practices require unfired glazes containing litharge to be carefully handled and to have an acceptable level of acid resistance in raw powdered form in the work environment. Finished tableware coated with a lead glaze must also possess an acceptable level of resistance to chemical attack.

Another way to classify a glaze is according to the way it is constituted. Raw glazes are prepared from mineral powders (usually oxides or carbonates) and are chemically limited to elements that can be obtained in materials having low water solubility. Frits are prepared to allow the use of water-soluble materials, which are first melted to an insoluble glass, and then quenched and broken up. For a partially fritted glaze, only the soluble materials and sufficient network former to make an insoluble glass are fritted. In an all fritted glaze, all components are melted into the frit except for necessary suspending agents.

Markets for Glazed Ceramics

The total whitewares market in the United States is reported to be $3.459 billion in 1989 (Ref 6). A breakdown of the U.S. market for products using glazes is shown in the chart of Fig 1.

A ceramic coating typically makes up 10 to 15% of the total manufacturing cost of a ceramic product. The value of the properties provided by the coating usually far outweighs this cost. The protective, functional, and decorative surface that is obtained often serves as a primary selling feature of the end product.

Ceramic glazes find their way into a wide range of applications ranging from coffee mugs to automotive sparkplugs. The major markets for ceramic coatings have different requirements, but one common theme is corrosion resistance and cleanability.

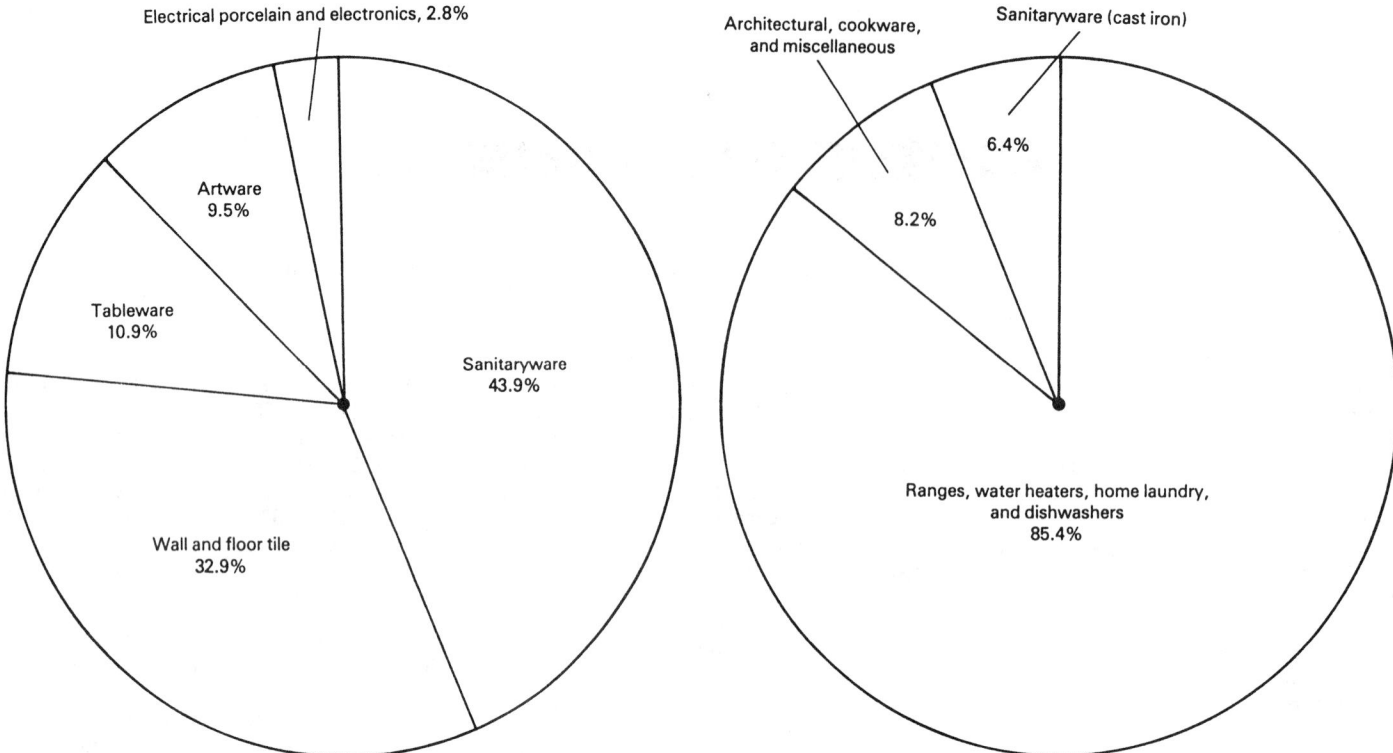

Fig 1 Chart showing breakdown of key segments of the $3.46 billion glaze market in the United States

Fig 2 Chart showing breakdown of key segments of the $5.486 billion porcelain enamel market in the United States

Rarely sold in a ready-to-use form, ceramic glazes are usually formulated and processed for application by the end user.

Role of Specific Oxides in Glazes

The commonly used oxides in glazes are SiO_2, ZrO_2, Al_2O_3, B_2O_3, CaO, SrO, BaO, MgO, ZnO, PbO, Li_2O, Na_2O, and K_2O (Ref 7). Small amounts of fluorine are sometimes used as a partial substitution for oxygen.

Silicon Dioxide. Most glazes contain more silica than all the other constituents combined. Silica promotes low expansion, high durability, and abrasion resistance. Its only serious deficiency is its high melting point, T_m, of 1723 °C (3133 °F).

Alkali Oxides. The foremost reason for adding other oxides is to reduce the maturing temperature. The alkalis are powerful fluxes at all maturing temperatures, and their use is limited by the high thermal expansion they impart to a glaze.

Alkaline Earths. The alkaline earths and magnesia are effective fluxes only at >1100 °C (>2010 °F).

Zinc oxide is effective at >1000 °C (>1830 °F) and in modest amounts it improves the effectiveness of other fluxes.

Lead monoxide is a powerful flux from the lowest temperatures to ≈1150 °C (≈2100 °F), above which volatilization becomes excessive.

Boric oxide is effective at all temperatures. As a network former, it can be used with other oxides to get a high fluxing level. Above 15% B_2O_3 concentration, however, it seriously degrades the durability of the glaze.

Alumina contributes to the working properties of a glaze, improves acid durability, and suppresses phase separation and crystallization of the glaze.

Zirconia is added to a glaze to improve the alkaline durability, and, in higher concentrations, as an opacifier.

Leadless Glazes

Tables 1 and 2 give the formulas of a number of commercial ceramic coatings. The first eight of these are leadless glazes. In these glazes, the alkali and alkaline earth oxides, together with MgO, ZnO, and B_2O_3, are used to provide the fluxing action.

Hard Porcelain Glaze. Glaze 1 is a feldspathic glaze suitable for use only on the highest firing hard paste porcelains (Ref 3).

Soft Porcelain Glaze. For porcelains fired at lower temperatures, such as soft paste porcelain or hard stoneware, glaze 2 would be satisfactory (Ref 8). This glaze is typical of that used on medieval Chinese porcelains.

Sanitaryware Glaze. Glaze 3 is a sanitaryware glaze (Ref 5). It is derived from the soft paste porcelain glaze by the addition of ZnO in large quantities.

Bristol glaze (glaze 4) is used to produce an opaque white coating on stoneware and other dark-colored bodies.

Fast-fire wall tile glazes are derived from the Bristol glaze by increasing the amount of

fluxes to increase the melting rate (Ref 9). Glaze 5 is a typical example.

Semivitreous Dinnerware Glaze. The development of glazes for dinnerware is more difficult because of the lower expansion of the bodies. Glaze 6 is an example of a glaze for semivitreous dinnerware (Ref 10). Several alkaline earths are used to improve the melting and surface properties.

Hotel China Glaze. Leadless glazes for vitreous hotel china (such as glaze 7) are a very recent development (Ref 11).

Low Expansion Glaze. Zircon and cordierite bodies are even lower in expansion, requiring a semicrystalline glaze, such as glaze 8 (Ref 12).

Lead-Containing Glazes

Litharge is used in glazes for several reasons (Ref 20). The strong fluxing action of PbO allows the formulation of glazes that mature at temperatures lower than their leadless counterparts, thus leading to greater flexibility in the formulation of the glaze to obtain low expansion, smooth surface, and maturing of the glaze over a wider firing range. Litharge imparts low surface tension for a smooth surface and a high index of refraction that results in a brilliant appearance. Glazes containing litharge heal over defects in the glaze surface more readily and are usually more corrosion resistant. This combination of desirable properties is difficult to achieve in leadless glazes on a production scale.

Table 1 Compositions of selected glazes and enamels based on mole ratio (Seger formula)

	Glazes						
Oxide	1 Hard porcelain	2 Soft porcelain	3 Sanitaryware	4 Bristol	5 Wall tile	6 Semivitreous dinnerware	7 Vitreous dinnerware
Li_2O	...	...	...	...	...	0.047	...
Na_2O	...	0.142	0.100	0.100	0.270	0.081	0.070
K_2O	0.30	0.135	0.100	0.100	0.040	0.115	0.069
MgO	...	0.043	...	0.200	0.010	0.066	0.037
CaO	0.70	0.680	0.600	0.400	0.350	0.580	0.391
ZnO	...	...	0.200	0.200	0.320	...	0.322
SrO	...	...	...	...	...	0.110	0.071
BaO	...	...	...	...	...	...	0.039
Al_2O_3	1.00	0.556	0.550	0.400	0.260	0.367	0.173
B_2O_3	...	...	...	...	0.050	0.171	0.188
SiO_2	10.00	4.570	3.000	3.500	2.650	2.721	2.224
ZrO_2	...	...	...	...	...	...	0.011
Ref	3	8	2	2	9	10	11

	Glazes (continued)						
Oxide	8 Low expansion	9 Cone 4 dinnerware	10 Cone 06 artware	11 Electronic	12 Opacified	13 Zinc matte	14 Lime matte
Li_2O	0.850	...	...	...	...	...	...
Na_2O	...	0.179	0.157	...	0.111	0.087	0.040
K_2O	0.150	0.066	...	...	0.083	0.052	0.059
MgO	...	...	...	...	...	0.097	...
CaO	...	0.494	0.218	...	0.407	0.152	0.524
ZnO	...	...	...	...	0.371	0.364	...
BaO	...	...	...	...	0.028	...	...
PbO	...	0.261	0.625	1.000	...	0.247	0.377
Al_2O_3	0.740	0.340	0.273	...	0.406	0.443	0.262
B_2O_3	0.200	0.314	0.507	...	0.143	0.142	0.176
SiO_2	2.510	3.369	2.792	0.500	2.019	1.566	1.746
ZrO_2	...	...	0.023	...	0.248	0.210	0.211
Ref	12	13	14	15	16	2	2

	Enamels							
Oxide	15 Ground coat	16 Home laundry	17 Hot-water tank	18 Continuous clean coating	19 Opaque cover coat	20 Semiopaque cover coat	21 Clear cover coat	22 Architectural cover coat
Li_2O	0.065	0.085	0.135	0.051	0.127	0.129	0.198	...
Na_2O	0.470	0.638	0.679	0.347	0.646	0.486	0.665	0.220
K_2O	0.054	0.052	...	0.046	0.277	0.341	0.137	0.680
MgO	...	0.014	...	...	...	...	...	...
CaO	0.244	0.157	0.110	0.034	...	...	...	0.007
ZnO	...	0.010	0.047	...	...	0.045	...	0.165
BaO	0.105	0.015	0.011	...	...	...	...	...
CoO	0.014	0.015	0.019	0.001	...	...	...	...
NiO	0.038	0.013	...	0.001	...	...	...	...
CuO	0.008	...	...	0.518	...	...	...	...
Al_2O_3	0.138	0.354	0.060	1.195	0.094	0.046	0.090	0.044
B_2O_3	0.489	0.721	0.330	0.050	0.986	0.833	0.344	0.402
Cr_2O_3	...	...	...	0.024	...	...	...	...
Sb_2O_3	...	...	...	0.003	...	...	...	0.008
SiO_2	1.623	2.171	2.820	1.186	2.902	2.730	3.313	2.026
ZrO_2	...	0.162	0.286	0.173	...	...	0.215	0.070
TiO_2	...	0.100	...	0.001	1.117	0.582	0.151	0.807
MnO_2	0.005	0.024	0.063	0.001	...	...	...	...
P_2O_5	0.011	0.010	...	...	0.039	...	...	0.011
Nb_2O_3	...	...	...	...	0.001	...	...	...
WO_3	...	...	...	...	0.001	...	...	...
MoO_3	...	...	...	...	...	...	0.011	...
F	0.316	0.381	0.349	0.112	0.710	0.724	0.417	0.082
Ref	2	17	17	18	19	19	2	2

Cone 4 Dinnerware Glaze. Glaze 9 in Tables 1 and 2 is an example of a lead-containing dinnerware glaze for cone 4 (1100 °C, or 2010 °F) (Ref 13).

Cone 06 Artware Glaze. Glaze 10 is an example of a clear glaze suitable for use on artware and hobbyware bodies at cone 06 (1000 °C, or 1830 °F) (Ref 14).

Electronic Glaze. The glazes used on alumina packages for integrated circuits (ICs) to seal the package represent the lowest firing lead-containing glazes. Glaze 11 is an example of such a glaze, which can be fired as low as 550 °C (1020 °F) (Ref 15).

Heavy-Metal Release

If lead-containing glazes are not properly formulated, they may be less resistant to acid attack, which results in the release of lead. If such glazes are used in contact with food or drink, lead poisoning of the user may result. Cadmium oxide, which is considerably more toxic than lead oxide (Ref 21), is only used in ceramic coatings in connection with the use of cadmium sulfoselenide pigments, which already contain large amounts of cadmium.

To control this problem under production conditions, standard tests have been developed for determining the lead and cadmium content released from glazed surfaces (Table 3). A sample is exposed to a 4% acetic acid solution for 24 h at room temperature while covered. The concentration in ppm of lead and cadmium in the solution is then determined.

FDA Guidelines. The current United States Food and Drug Administration (FDA) standard limits for lead released are 7 ppm average for six samples of flatware, 5 ppm maximum of six samples of small holloware, and 2.5 ppm maximum of six samples of large holloware (Ref 24). The standard limits for Cd are $1/2$ ppm average for six samples of flatware, $1/2$ ppm maximum of six samples of small holloware, and $1/4$ ppm maximum of six samples of large holloware. These limits are currently under review, and lower levels may possibly be applicable by the time this article is published (Ref 25). Because of statistical fluctuations in the measurements, operating standards must be less than half these guideline values (Ref 26).

In the FDA test, lead-containing commercial dinnerware glazes usually release $1/2$ to 2 ppm (Ref 20). Properly designed commercial artware will be somewhat higher. In contrast, ware implicated in health cases have all released in excess of 50 ppm Pb.

Most glazes have no added cadmium oxide. Those that do contain cadmium will release 0.1 to 0.2 ppm even when properly formulated (Ref 21). For this reason, it is best if CdS-Se colors are not used as surface decorations on glass surfaces that will come in contact with food or drink.

The issue of lead and cadmium release from glassware decorated with glass enamels has

However, litharge also has disadvantages as a glaze constituent. Lead glazed ware must be fired in a strongly oxidizing atmosphere because lead is readily reduced. Lead oxide cannot be used above 1150 °C (2100 °F) due to volatility. Most importantly, litharge is highly toxic. Moreover, lead poisoning is very difficult to diagnose because its symptoms are similar to other ailments. Therefore, every possible precaution must be taken when preparing lead glazes to avoid poisoning, and ware intended for food contact should be tested as noted below to ensure consumer protection.

Table 2 Compositions of selected glazes and enamels based on weight percent

	Glazes						
Oxide	1 Hard porcelain	2 Soft porcelain	3 Sanitaryware	4 Bristol	5 Wall tile	6 Semivitreous dinnerware	7 Vitreous dinnerware
Li_2O	...	...	...	...	...	0.51	...
Na_2O	...	2.24	2.05	1.98	6.54	1.81	1.81
K_2O	3.67	3.24	3.12	3.01	1.47	3.92	2.71
MgO	...	0.44	...	2.57	0.16	0.96	0.62
CaO	5.10	9.71	11.15	7.16	7.67	11.76	9.16
ZnO	...	...	5.39	5.19	10.18	...	10.94
SrO	...	...	...	...	...	4.12	3.07
BaO	...	...	...	...	...	...	2.50
Al_2O_3	13.24	14.44	18.58	13.01	10.36	13.53	7.37
B_2O_3	...	...	...	...	1.36	4.30	5.47
SiO_2	78.00	69.90	59.71	67.09	62.25	59.09	55.79
ZrO_2	...	...	...	...	...	...	0.57

	Glazes (continued)						
Oxide	8 Low expansion	9 Cone 4 dinnerware	10 Cone 06 artware	11 Electronic	12 Opacified	13 Zinc matte	14 Lime matte
Li_2O	9.08	...	...	...	...	...	...
Na_2O	...	3.06	2.46	...	2.50	1.91	0.85
K_2O	5.05	1.72	...	...	2.84	1.73	1.91
MgO	...	...	...	...	...	1.38	...
CaO	...	7.65	3.09	...	8.29	3.02	10.08
ZnO	...	...	...	...	10.97	10.48	...
BaO	...	...	...	...	1.56	...	...
PbO	...	16.08	35.30	88.14	...	19.52	28.87
Al_2O_3	26.98	9.57	7.04	...	15.04	15.99	9.17
B_2O_3	4.98	6.04	8.93	...	3.62	3.50	4.20
SiO_2	53.91	55.88	42.45	11.86	44.07	33.31	35.99
ZrO_2	...	...	0.72	...	11.10	9.16	8.92

	Enamels							
Oxide	15 Ground coat	16 Home laundry	17 Hot-water tank	18 Continuous clean coating	19 Opaque cover coat	20 Semiopaque cover coat	21 Clear cover coat	22 Architectural cover coat
Li_2O	0.88	0.81	1.33	0.52	0.89	1.10	1.76	...
Na_2O	13.15	12.60	13.92	7.30	9.41	8.58	12.23	4.21
K_2O	2.30	1.56	...	1.47	6.13	9.15	3.83	19.79
MgO	...	0.18	...	...	...	...	...	...
CaO	6.18	2.80	2.04	0.65	...	...	...	0.12
ZnO	...	0.26	1.27	...	...	1.04	...	4.15
BaO	7.27	0.73	0.56	...	...	...	...	...
CoO	0.47	0.36	0.47	0.03	...	...	...	...
NiO	0.29	0.31	...	0.03	...	...	...	...
CuO	0.20		...	13.99	...	...	...	...
Al_2O_3	6.35	11.50	2.02	41.38	2.25	1.34	2.72	1.39
B_2O_3	15.37	15.99	7.60	1.18	16.13	16.53	7.11	8.65
Cr_2O_3	...	...	...	1.24	...	...	...	...
Sb_2O_3	...	...	...	0.30	...	...	...	0.72
SiO_2	44.01	41.55	56.05	24.20	40.97	46.74	59.07	37.61
ZrO_2	...	6.36	11.66	7.24	...	...	7.86	2.67
TiO_2	...	2.55	...	0.03	20.97	13.25	3.58	19.93
MnO_2	0.20	0.66	1.81	0.03	...	...	...	...
P_2O_5	0.70	0.45	...	...	1.30	...	...	0.48
Nb_2O_3	...	...	...	...	0.06	...	...	...
WO_3	...	...	...	...	0.05	...	...	...
MoO_3	...	...	...	...	...	...	0.47	...
F	2.71	2.31	2.19	0.72	3.17	3.92	2.35	0.48

- Total glaze composition
- Thermal history during processing
- Glaze application techniques
- Glaze-body solution at the interface during firing
- Atmospheric conditions that exist during firing

Of all these parameters, the most important is the glaze composition.

Heavy-Metal Release Performance Rating. A figure of merit (FM) has been developed to predict the heavy-metal release or acid resistance of a glaze from its composition (Ref 14). Silica, alumina, zirconia, and similar ions such as titania and tin oxide are effective in lowering the lead release of a glaze:

$$\text{Good} = 2[Al_2O_3] + [SiO_2] + [TiO_2]$$
$$+ [ZrO_2] + [SnO_2] \qquad \text{(Eq 1)}$$

The concentrations given in Eq 1 are expressed in molar ratio. The factor 2 arises from the fact that there are 2 equivalents of aluminum ions per equivalent of Al_2O_3.

It has also been shown that alkalis, alkaline earths, B_2O_3, fluoride, phosphate, ZnO, CdO, and PbO are all more or less detrimental to the lead release in a glaze:

$$\text{Bad} = 2([Li_2O] + [Na_2O] + [K_2O]$$
$$+ [B_2O_3] + [P_2O_5]) + [MgO] + [CaO] \qquad \text{(Eq 2)}$$
$$+ [SrO] + [BaO] + [F] + [ZnO]$$
$$+ [PbO]$$

Combining these terms gives the figure of merit:

$$\text{FM} = \text{Good}/(\text{Bad})^{1/2} \qquad \text{(Eq 3)}$$

When the figure of merit is >2.05, the lead release is below the standard. When it is <1.80, some measurements are always greater than the standard. This figure of merit applies to all single-phase glazes. In a glass-crystalline system, the formulation of the least durable component must be used in the calculation.

Table 3 Standard test methods for determining the lead and cadmium content released from glazed and enamel surfaces used for food preparation when exposed to attack by acetic acid

ASTM standard	Test method	Ref
C 738	Lead and cadmium extracted from glazed ceramic surfaces	22
C 872	Lead and cadmium release from porcelain enamel surfaces	23
C 895	Lead and cadmium extracted from glazed ceramic tile	22
C 927	Lead and cadmium extracted from the lip and rim area of glass tumblers externally decorated with ceramic glass enamels	22
C 1034	Lead and cadmium extracted from glazed ceramic cookware	22

also been addressed through a voluntary quality-control program adopted by the ceramic industry (Ref 27). The lip and rim area (that is, the top 20 mm, or $^{25}/_{32}$ in.) of the decorated glass must not leach greater than 50 ppm Pb or 3.5 ppm Cd when analyzed by a mod-

ification of the existing ceramic ware test (see Table 3). Failure of one glass in a sample of six is cause for rejection.

Numerous factors must be considered in formulating and processing a glaze to achieve low lead and cadmium release:

Opaque Glazes

In opaque glazes, the transmittance of the glaze has been reduced to hide the body. Opacity is introduced into ceramic coatings by the addition of a substance that will disperse in the coating as discrete particles to scatter and reflect the incident light (Ref 5). This dispersed substance must have a refractive index that differs from that of the clear ceramic coating.

The refractive index, n_D, of most glasses is 1.5 to 1.6. Typical opacifiers include SnO_2 (n_D = 2.04), ZrO_2 (n_D = 2.40), $ZrSiO_4$ (n_D = 1.85), and TiO_2 (n_D = 2.5 for anatase and 2.7 for rutile). Opacified glaze (glaze 12) in Tables 1 and 2 is an example of a glaze fired at >1000 °C (>1830 °F) where zircon is the opacifier of choice (Ref 16).

In coatings fired at <1000 °C (<1830 °F), titania in the anatase phase is the best opacifier because of its high refractive index. The temperature limit is the point at which anatase inverts to rutile. Rutile crystals are unsuitable because they grow too large in glass for opacification and then have a yellow color.

Satin and Matte Glazes

Satin and matte effects are also due to dispersed crystals in the glaze (Ref 5). The crystals must be very small and evenly dispersed if the glaze is to have a smooth velvet appearance. Matte glazes are always somewhat opaque. The amount of opacity depends on the difference in refractive index between glaze and crystal and can be fairly small. Glaze 13 in Tables 1 and 2 is an example of a zinc matte glaze where the crystal is willemite (Zn_2SiO_4) (Ref 2). Glaze 14 is a lime matte glaze with a wollastonite ($CaSiO_3$) crystal.

Porcelain Enamels

The total market for porcelain enameled products in the United States was reported to be 5.486 billion in 1989 (Ref 28). The U.S. market breakdown is shown in Fig 2.

Similar to ceramic glazes, porcelain enamels provide an easy-to-clean corrosion-resistant surface for many consumer products. While the coating is primarily functional, the use of color provides an added sales feature, particularly in the architectural market. Overall corrosion resistance is important; alkali resistance and resistance to hot water corrosion are primary concerns for laundry, sanitaryware, and hot-water applications, while acid resistance comes to the forefront in coatings intended for ranges, architectural panels, and cookware.

Porcelain enamel coatings are rarely sold ready-to-use. Conventional porcelain enamel coatings are prepared in an aqueous system by ball milling and applied to the substrate by spray, dip, or flow coating. The coating is allowed to dry before firing. Newer technology involves dry application of powdered porcelain enamel by electrostatic spraying. This dry application eliminates the waste disposal problems associated with wet ball milling, subsequent application, and the washing and cleaning steps.

Adherence

A porcelain enamel must not only act as a protective and aesthetically pleasing surface, but it must also bond to the metal substrate. For proper adherence of the enamel to the metal, it is necessary to develop a continuous electronic structure or chemical bond across the interface (Ref 29). This continuity requires that both the enamel coating and the substrate metal be saturated at the interface with an oxide of the metal (Ref 30). This oxide, which for iron and steel substrates is ferrous oxide (FeO), must not be reduced by the metal when it is in solution in the glass.

Surface roughness also generally improves adherence by creating a greater contact area, but it is of little value if the chemical bond is weak.

To develop a continuity of atomic and electronic structure between an enamel and a metal substrate, a compatible transition layer is needed that is in equilibrium with both the metal substrate and the glass coating at the interface (Ref 29). This zone must include at least a monomolecular layer of an oxide of the metal and is stable only when both the metal and the glass at the interface are saturated with this metal oxide.

Certain transition metal oxides (for example, CoO, NiO, and CuO) can be added to an enamel formulation to improve the adherence between the metal and the substrate. These oxides increase the rate of the reactions in the adherence process and stabilize the saturation of the interface with the critical oxide. Ground coat enamels contain adherence oxides while cover coat enamels lack adherence oxides.

Ground Coat Enamels

Enamel 15 in Tables 1 and 2 is a typical example of a general-purpose ground coat enamel. These enamels are alkali borosilicates that contain small amounts of the adherence oxides to promote the bonding process. In addition to adherence, these coatings provide a chemically protective layer that minimizes surface defects.

Ground coat frits can also be formulated for particular end-use purposes.

Home Laundry Enamel. Enamel 16 in Tables 1 and 2 is a home laundry enamel that has been formulated for outstanding alkali resistance through the addition of large quantities of ZrO_2 (Ref 17).

Hot-Water Tank Enamel. An application with very stringent thermal and corrosion resistance requirements is a hot-water tank. Enamel 17 in Tables 1 and 2 is a hot-water tank enamel. The higher concentration of SiO_2 and the lower concentration of B_2O_3 reflect the higher firing temperature necessary to obtain the requisite thermal and chemical resistance properties.

Continuous Clean Coating. Enamel 18 in Tables 1 and 2 is a continuous clean coating that has been developed to provide a means of volatilizing and removing food soils from the internal surfaces of ovens during normal operation (Ref 18).

Cover Coat Enamels

Cover coat porcelain enamels are formulated to provide specific color and appearance characteristics, abrasion resistance, surface hardness, resistance to corrosion, and heat and thermal shock resistance. These enamels can be clear, semiopaque, or opaque depending on the application:

- Opaque enamels for white and pastel coatings
- Semiopaque enamels for most medium-strength colors
- Clear enamels to produce strong bright colors

Opaque cover coat enamels (for example, enamel 19 in Tables 1 and 2) are opacified with TiO_2 (Ref 19, 31, 32). Usually, all the TiO_2 is melted into a clear frit. During the firing process, this frit partially crystallizes to provide the required opacification. The enamels have excellent acid resistance and fairly good alkali resistance with reflectances ranging from 78 to 88%.

Semiopaque Cover Coat Enamel. Enamel 20 in Tables 1 and 2 is an example of a semiopaque cover coat enamel. These materials do not differ qualitatively from fully opaque enamels. The concentration of TiO_2 is reduced to the level at which the system is compatible with the use of pigments.

Clear cover coat porcelain enamels such as Enamel 21 in Tables 1 and 2 are used in conjunction with appropriate pigments for the production of strong and medium-strength colors and are similar to ground coat formulations without the adherence oxides.

Architectural Cover Coat Enamels. The architectural siding industry requires cover coat enamels of low gloss and good weatherability/durability. Enamel 22 in Tables 1 and 2 is an example of an enamel for this application. The high zinc oxide concentration is the source of the low gloss finish imparted to the coating.

Glass Enamels

Glass enamels are applied to glass for decorative purposes and not to improve chemical durability or cleanability. Because these coatings are matured at temperatures below the deformation point of the glass substrate (540 to 650 °C, or 1000 to 1200 °F) and require larger quantities of fluxing elements, corrosion resistance can be difficult to obtain.

Glass enamels are produced in ready-to-use form (for example, paste, thermoplastics, spray media, and ultraviolet curable media) by a few select manufacturers. These enamels are rarely compounded by the end user and represent a specialty product that is more akin to organic paints than to other ceramic coatings.

The markets for this specialty product are categorized as tableware, glass containers, architectural, lighting, and automotive. There are no published figures on the value of the specific portion of the glass market that is decorated with glass enamels.

As supplied to the user, glass enamels are mechanical mixtures of pigments (see the section "Ceramic Decoration" in this article), fluxes, and organic suspending media. The requirement for low maturing temperatures necessitates the use of very high lead-oxide containing borosilicates for the flux. The industry relies on in-house control tests to ensure acceptable levels of lead release from decorated areas that might come in contact with food or drink (see the section on "Heavy-Metal Release" in this article).

The organic suspending media for glass enamels are similar to materials used to make organic paints.

Ceramic Decoration

Techniques for Coloring Vitreous Coatings. There are a number of ways to obtain color in a ceramic coating (Ref 33). In one method, certain transition metal ions can be melted into a glass when it is made. While suitable for bulk glass, this method is rarely used for coatings because adequate tinting strength and the purity of color cannot be obtained by this process.

A second method to obtain color is to induce the precipitation of a colored crystal in a transparent matrix. Certain materials dissolve to some extent in a vitreous material at high temperature. When the temperature is reduced, the solubility is also reduced and precipitation occurs. This method is used for opacification (that is, the production of an opaque white color). Normally, some or all of the opacifier added to the coating dissolves during the firing process and recrystallizes upon cooling. For oxide colors other than white, however, this method lacks the necessary control for reproducible results and is seldom used.

The third method to obtain color in a vitreous matrix is to disperse in that matrix one or more insoluble crystals that are colored. The color of the crystal is then imparted to the transparent matrix. This method is the one most commonly used to introduce color to vitreous coatings.

Pigment Systems. To be suitable as a ceramic pigment, a material must have a high tinting strength, a high refractive index, and be free of grayness. It must also possess stability under the high temperatures and corrosive environments encountered in the firing of glazes (Ref 32). Most ceramic pigments are complex oxides (see Table 4) with the lone exception being the cadmium sulfoselenide red pigments.

Table 4 Inorganic pigments to impart colors to ceramic coatings (Ref 33, 34)

Pigment system	CAS registry number(a)	Chemical formula	DCMA number(b)
Pink and purple			
Chrome-alumina-pink spinel	68201-65-0	$Zn(Al,Cr)_2O_4$	13-32-5
Chrome-alumina-pink corundum	68187-27-9	$(Al,Cr)_2O_3$	3-03-5
Manganese-alumina-pink corundum	68186-99-2	$(Al,Mn)_2O_3$	3-04-5
Zirconium-iron pink zircon	68187-13-3	$(Zr,Fe)SiO_4$	14-44-5
Chrome-tin orchid cassiterite	68187-53-1	$(Sn,Cr)O_2$	11-23-5
Chrome-tin pink sphene	68187-12-2	$CaOSnO_2SiO_2:Cr$	12-25-5
Brown			
Zinc-iron-chromite brown spinel	68186-88-9	$(Zn,Fe)(Fe,Cr)_2O_4$	13-37-7
Iron-chromite brown spinel	68187-09-7	$Fe(Fe,Cr)_2O_4$	13-33-7
Iron-titanium brown spinel	68187-02-0	Fe_2TiO_4	13-34-7
Nickel-ferrite brown spinel	68187-10-0	$NiFe_2O_4$	13-35-7
Zinc-ferrite brown spinel	68187-51-9	$(Zn,Fe)Fe_2O_4$	13-36-7
Iron brown hematite	68187-35-9	Fe_2O_3	3-06-7
Chrome-iron-manganese brown spinel	68555-06-6	$(Fe,Mn)(Fe,Cr,Mn)_2O_4$	13-48-7
Chromium-manganese-zinc brown spinel	71750-83-9	$(Zn,Mn)Cr_2O_4$	13-51-7
Yellow			
Zirconium-vanadium yellow baddeleyite	68187-01-9	$(Zr,V)O_2$	1-01-4
Tin-vanadium yellow cassiterite	68186-93-6	$(Sn,V)O_2$	11-22-4
Zirconium-praseodymium yellow zircon	68187-15-5	$(Zr,Pr)SiO_4$	14-43-4
Lead-antimonate yellow pyrochlore	68187-20-2	$Pb_2Sb_2O_7$	10-14-4
Nickel-antimony-titanium yellow rutile	71077-18-4	$(Ti,Ni,Sb)O_2$	11-15-4
Nickel-niobium-titanium yellow rutile	68611-43-8	$(Ti,Ni,Nb)O_2$	11-16-4
Chrome-antimony-titanium buff rutile	68186-90-3	$(Ti,Cr,Sb)O_2$	11-17-6
Chrome-niobium-titanium buff rutile	68611-42-7	$(Ti,Cr,Nb)O_2$	11-18-6
Chrome-tungsten-titanium buff rutile	68186-92-5	$(Ti,Cr,W)O_2$	11-19-6
Manganese-antimony-titanium buff rutile	68412-38-4	$(Ti,Mn,Sb)O_2$	11-20-6
Green			
Chromium-green hematite	68909-79-5	$(Cr,Fe)_2O_3$	3-05-3
Cobalt-chromite blue-green spinel	68187-11-1	$Co(Al,Cr)_2O_4$	13-29-2
Cobalt-chromite green spinel	68187-49-5	$CoCr_2O_4$	13-30-3
Cobalt-titanate green spinel	68186-85-6	Co_2TiO_4	13-31-3
Victoria green garnet	68553-01-5	$3CaOCr_2O_33SiO_2$	4-07-3
Nickel-silicate green olivine	68515-84-4	Ni_2SiO_4	5-45-3
Blue			
Cobalt-aluminate blue spinel	68186-86-7	$CoAl_2O_4$	13-26-2
Cobalt-zinc-aluminate blue spinel	68186-87-8	$(Co,Zn)Al_2O_4$	13-28-2
Cobalt-silicate blue olivine	68187-40-6	Co_2SiO_4	5-08-2
Cobalt-zinc-silicate blue phenacite	68412-74-8	$(Co,Zn)_2SiO_4$	7-10-2
Cobalt-tin blue-gray spinel	68187-05-3	Co_2SnO_2	13-27-2
Cobalt-tin-alumina blue spinel	68608-09-3	$CoAl_2O_4/Co_2SnO_4$	13-49-2
Zirconium-vanadium blue zircon	68186-95-8	$(Zr,V)SiO_4$	14-42-2
Black			
Iron-cobalt black spinel	68187-50-8	$(Fe,Co)Fe_2O_4$	13-39-9
Iron-cobalt-chromite black spinel	68186-97-0	$(Co,Fe)(Fe,Cr)_2O_4$	13-40-9
Manganese-ferrite black spinel	68186-94-7	$(Fe,Mn)(Fe,Mn)_2O_4$	13-41-9
Copper-chromite black spinel	68186-91-4	$CuCr_2O_4$	13-38-9
Chromium black hematite	68909-79-5	$(Cr,Fe)_2O_3$	3-05-3
Chromium-iron-nickel black spinel	71631-15-7	$(Ni,Fe)(Cr,Fe)_2O_4$	13-50-9
Gray			
Cobalt-nickel gray periclase	68186-89-0	$(Co,Ni)O$	6-09-8
Titanium-vanadium-antimony gray rutile	68187-00-8	$(Ti,V,Sb)O_2$	11-21-8
Tin-antimony gray cassiterite	68187-54-2	$(Sn,Sb)O_2$	11-24-8

(a) CAS, Chemical Abstract Service. (b) DCMA, Dry Color Manufacturers' Association

Red Pigments. Orange, red, and dark red colors are obtained only by the use of the cadmium sulfoselenide pigments (Ref 21, 35). Because cadmium compounds are highly toxic (Ref 36), the cadmium sulfoselenide pigments should not be used in applications that will come in contact with food and drink. Moreover, these pigments must also be han-dled with great care to avoid the possibility of ingestion.

Pink and Purple Pigments. Chrome alumina pinks (Ref 37) impart pink shades in glazes suitably formulated. The manganese alumina pink is a very pure clean pink, but it is difficult to manufacture. The most stable pink pigment is the iron-doped zircon system

(Ref 38) in which the shades of color extend from pink to coral. The chrome-tin system is the only family that can produce purple and maroon shades as well as a pink shade (Ref 39).

Brown Pigments. The zinc-iron-chromite spinels (Ref 40) produce a wide palette of tan and brown shades. The other browns are variations of this system for specific applications.

Yellow Pigments. Zirconia vanadia yellows are economical pigments for use with coatings fired at >1000 °C (>1830 °F) when high tinting strength is not required (Ref 41). Tin vanadium yellows are a strong yellow color but are very costly to produce (Ref 42). The praseodymium zircon pigments have excellent tinting strength in coatings fired to as high as cone 10 (Ref 38). For lower-temperature applications, the tinting strength of the lead antimonate pigment is unsurpassed with the exception of the cadmium sulfoselenides. The rutile pigments yield yellow, orange yellow, or maple shades useful in porcelain enamels and glass enamels.

Green Pigments. Green Cr_2O_3 may be used in a few applications (Ref 43). The zinc-alumina-chromite bluegreen pigments give shades from blue to blue-green when used in strong masstones. The Victoria green gives a transparent bright green color. Because of difficulties inherent in the use of chromium-containing pigments, many green ceramic glazes are now made with zircon pigments (Ref 38). Stable greens are obtained with a blend of about 2 parts of the praseodymium zircon yellow to 1 part of the vanadium zircon blue.

Blue Pigments. Cobalt blues, both the spinel ($CoAl_2O_4$) and the silicate (Co_2SiO_4) forms, are the highest tinting strength colors used in ceramics (Ref 44). At the higher firing temperatures used for ceramic glazes, they often cause a bleeding defect. At these specific temperatures, the vanadium-doped zircon blue is used (Ref 38).

Black Pigments. Black ceramic pigments are formed by mixtures of several oxides to form the spinel structure (Ref 45). The one exception is the chromium black hematite, an inexpensive pigment suitable for use in zinc-free coatings (Ref 46).

Gray Pigments. The cobalt-nickel gray periclase uses zirconia or zircon as a carrier for various ingredients of blacks such as cobalt, nickel, iron, and chromium oxides. The titanium-vanadium rutile can be used in porcelain enamels.

Pigment Application in Coatings. When selecting pigments for a specific coating application, consideration must be given to the following parameters (Ref 47):

- Processing stability requirements
- Pigment uniformity and reproducibility
- Particle size distribution
- Compatibility of all materials to be used

An engobe or body stain must be stable to the bisque fire. An underglaze color or a colored glaze must be stable to the glost fire and to corrosion by the molten glaze ingredients. An overglaze or glass enamel must only be stable to the decorating fire and to corrosion by the molten flux used in the application.

For most ceramic pigments, uniform and reproducible manufacture requires great care. To avoid specking in a blend, no component should be <10% of the mix.

Most calcined ceramic pigments are in the 1 to 10 μm (40 to 400 μin.) range in mean particle size, with no residue on a 325 mesh (44 μm) screen. The optimum particle size for an application is the largest size that gives uniform dispersion and adequate strength in opacified coatings.

A ceramic pigment must function as a component in a glaze or porcelain enamel system. Thus, it must be compatible with the other components (that is, the glaze itself, the opacifier(s), and the other additives).

Cost Factors and Product Availability

Owing to the limited market and the variety and complexity of the products, ceramic pigments are manufactured by specialty firms and not by the users. The principal producers are Ceramic Color and Chemical Corporation, Drakenfeld Colors Division of Ciba-Geigy Corporation, Englehard Corporation, Ferro Corporation, General Color and Chemical Corporation, O. Hommel Company, Mason Color and Chemical Corporation, and Mobay Corporation. The cost of ceramic pigments ranges from $9 to $50/kg ($4 to $20/lb) or even higher, depending on the elemental composition and the processing required.

REFERENCES

1. R.A. Eppler, Chapt. 4, Glazes and Enamels, *Glass Science & Technology*, Vol 1, Academic Press, 1983, p 301–337
2. R.A. Eppler, Glazes and Enamels, *Commercial Glasses*, J.F. MacDowell and D. Boyd, Ed., Vol 18, *Advances in Ceramics*, American Ceramic Society, 1986
3. P. Rado, *Introduction to the Technology of Pottery*, 2nd ed., Pergamon Press, Oxford, 1988
4. R.A. Eppler, Chapt. 12, Corrosion of Glazes and Enamels, *Corrosion of Glass, Ceramics, and Ceramic Superconductors*, D.E. Clark and B.K. Zoitos, Ed., Noyes Publications, 1991, to be published
5. F. Singer and W.L. German, *Ceramic Glazes*, Borax Consolidated, 1964
6. *Ceram. Ind.*, Aug 1990, p 36
7. D. Rhodes, *Clay and Glazes for the Potter*, Rev. ed., Chilton, 1973
8. R. Tichane, *Ching-th-Chen; Views of a Porcelain City*, N.Y. State Institute Glaze Research, 1983
9. W.H. Orth, Effect of Firing Rate on Physical Properties of Wall Tile, *Am. Ceram. Soc. Bull.*, Vol 46 (No. 9), 1967, p 841–844
10. J.E. Marquis and R.A. Eppler, Leadless Glazes for Dinnerware, *Am. Ceram. Soc. Bull.*, Vol 53 (No. 5), 1974, p 443–445, 449; Vol 53 (No. 6), 1974, p 472
11. E.F. O'Conor, L.D. Gill, and R.A. Eppler, Recent Developments in Leadless Glazes, *Ceram. Eng. Sci. Proc.*, Vol 5 (No. 11), 1984, p 923–932
12. E.F. O'Conor and R.A. Eppler, Semicrystalline Glazes for Low Expansion Whiteware Bodies, *Am. Ceram. Soc. Bull.*, Vol 52 (No. 2), 1973, p 180–184
13. J.E. Marquis, Lead in Glazes—Benefits and Safety Precautions, *Am. Ceram. Soc. Bull.*, Vol 50 (No. 11), 1971, p 921–923
14. R.A. Eppler, Formulation of Glazes for Low Pb Release, *Am. Ceram. Soc. Bull.*, Vol 54 (No. 5), 1975, p 496–499
15. R.R. Tummala and R.R. Shaw, Glasses in Microelectronics in the Information-Processing Industry, *Commercial Glasses*, J.F. MacDowell and D. Boyd, Ed., Vol 18, *Advances in Ceramics*, American Ceramic Society, 1986
16. F.T. Booth and G.N. Peel, Principles of Glaze Opacification with Zirconium Silicate, *Trans. Brit. Ceram. Soc.*, Vol 58 (No. 9), 1952, p 532–564
17. R.A. Eppler, R.L. Hyde, and H.F. Smalley, Resistance of Porcelain Enamels to Attack by Aqueous Media: I Tests for Enamel Resistance and Experimental Results Obtained, *Am. Ceram. Soc. Bull.*, Vol 56 (No. 12), 1977, p 1064–1067
18. P.G. Monteith, O.C. Linhart, and J.S. Slaga, Performance Tests for Properties of Low Temperature Thermal Cleaning Oven Coatings, *Proc. Porcelain Enamel Institute Tech. Forum*, Vol 32, 1970, p 73–79
19. R.D. Shannon and A.L. Friedberg, Titania-Opacified Porcelain Enamels, *Ill Univ. Eng. Exp. Sta. Bull.*, Vol 456, 1960
20. R.A. Eppler, Chapt. 10, Formulation and Processing of Ceramic Glazes for Low Lead Release, *Proceedings of the International Conference on Ceramic Foodware Safety*, J.F. Smith and M.H. McLaren, Ed., Lead Industries Association, 1976
21. R.A. Eppler and D.S. Carr, Cadmium in Glazes and Glasses, *Cadmium 81, Proceedings of the 3rd International Cadmium Conference*, International Lead and Zinc Research Organization, 1982, p 31–33
22. Glass, Ceramic Whitewares, part 15.02, *Annual Book of Standards*, American Society for Testing and Materials
23. Metallic and Inorganic Coatings, part 2.05, *Annual Book of Standards*, American Society for Testing and Materials
24. Cadmium 7117.06, and Lead 7117.07, U.S. Food and Drug Administration Compliance Policy Guides

25. J.A. Calderwood, *Soc. Glass Ceram. Decorators 1988/89 Seminar Proceedings*, 1989, p 12–15

26. C.F. Moore, *Trans. J. Brit. Ceram. Soc.*, Vol 76, 1977, p 52–57

27. *Federal Register*, Vol 43 (No. 242), 15 Dec 1978

28. *Ceram. Ind.*, Aug 1990, p 49

29. J.A. Pask, Chemical Reaction and Adherence at Glass-Metal Interfaces, *Proceedings of the Porcelain Enamel Institute Tech. Forum*, Vol 22, 1971, p 1–16

30. B.W. King, H.P. Tripp, and W.H. Duckworth, Nature of Adherence of Porcelain Enamels to Metals, *J. Am. Ceram. Soc.*, Vol 42 (No. 11), 1959, p 504–525

31. R.A. Eppler, Crystallization and Phase Transformation in TiO_2-Opacified Porcelain Enamels: II Comparison of Theory with Experiment, *J. Am. Ceram. Soc.*, Vol 52 (No. 2), 1969, p 94–99

32. R.F. Patrick, Some Factors Affecting the Opacity, Color, and Color Stability of Titania-Opacified Enamels, *J. Am. Ceram. Soc.*, Vol 34 (No. 3), 1951, p 96–102

33. A. Burgyan and R.A. Eppler, Classification of Mixed-Metal-Oxide Inorganic Pigments, *Am. Ceram. Soc. Bull.*, Vol 62 (No. 9), 1983, p 1001–1003

34. *DCMA Classification and Chemical Description of the Mixed Metal Oxide Inorganic Colored Pigments*, 2nd ed., Metal Oxides and Ceramic Colors Subcommittee, Dry Color Manufacturers' Association, 1982

35. R.A. Eppler, Ceramic Colorants, *Ullmann's Encyclopedia of Industrial Chemistry*, Vol A5, VCH Publishers, Weinheim, West Germany, 1986, p 545–556

36. N.I. Sax and R.J. Lewis, Sr., *Hazardous Chemicals Desk Reference*, Van Nostrand Reinhold, 1987

37. R.L. Hawks, Chrome-Alumina Pink at Various Temperatures, *Am. Ceram. Soc. Bull.*, Vol 40 (No. 1), 1961, p 7–8

38. R.A. Eppler, Zirconia Based Colors for Ceramic Glazes, *Am. Ceram. Soc. Bull.*, Vol 56 (No. 2), 1977, p 213–215, 218, 224

39. R.A. Eppler, Lattice Parameters of Tin Sphene, *J. Am. Ceram. Soc.*, Vol 59 (No. 9–10), 1976, p 455

40. S.H. Murdock and R.A. Eppler, Zinc-Iron-Chromite Pigments, *J. Am. Ceram. Soc.*, Vol 71 (No. 4), 1988, p C212-C214

41. F.T. Booth and G.N. Peel, Preparation and Properties of Some Zirconium Stains, *Trans. J. Brit. Ceram. Soc.*, Vol 61 (No. 7), 1962, p 359–400

42. E.H. Ray, T.D. Carnahan, and R.M. Sullivan, Tin-Vanadium Yellows and Praseodymium Yellows, *Am. Ceram. Soc. Bull.*, Vol 40 (No. 1), 1961, p 13–16

43. P. Henry, Ceramic Green Colors for Whiteware Glazes, *Am. Ceram. Soc. Bull.*, Vol 40 (No. 1), 1961, p 9–10

44. R.K. Mason, Use of Cobalt Colors in Glazes, *Am. Ceram. Soc. Bull.*, Vol 40 (No. 1), 1961, p 5–6

45. R.A. Eppler, Cobalt-Free Black Pigments, *Am. Ceram. Soc. Bull.*, Vol 60 (No. 5), 1981, p 562–565

46. S.H. Murdock and R.A. Eppler, The Interaction of Ceramic Pigments with Glazes, *Am. Ceram. Soc. Bull.*, Vol 68 (No. 1), 1989, p 77–78

47. R.A. Eppler, Selecting Ceramic Pigments, *Am. Ceram. Soc. Bull.*, Vol 66 (No. 11), 1987, p 1600–1604

Sealing Glasses

Carl J. Hudecek, Consultant, OI-NEG Technical Products, Inc.

SEALING GLASSES are used to mechanically (and, usually, hermetically) join glasses, metals, ceramics, or composites. High-lead (PbO) sealing glasses that can be used at relatively low temperatures are called solder glasses because they were first used to join two higher-temperature glasses together. This article focuses on solder glasses, whereas the higher-temperature sealing glasses and glass-to-metal seals are described extensively in the section "Joining" in this Volume.

Solder glass is chosen to seal glass, ceramic, or metal parts where:

- The parts being sealed or the article being encapsulated cannot be subjected to high temperatures that will cause component damage
- One or both of the parts being sealed are glass, and distortion or disannealing cannot be tolerated. Disannealing occurs when a part is heated above the annealing point, and, upon cooling, residual stresses are left in the glass member(s)
- The seals are geometrically complex and require a flowable hermetic sealant, which can easily be applied in paste or slurry form, and then fired to the "liquid" state
- Cost is a factor and mass production techniques, such as screen printing or extrusion, can utilize powdered solder glass and vehicle-binder systems
- Hermetic seals with good electrical insulating characteristics are required
- Salvage or "unsealing" of the parts is a desired option

Solder glass is preferred for sealing television bulbs and display panels. It is also used to seal high-performance ceramic microelectronic packages.

The two physical properties of greatest significance in a glass used for sealing are its thermal contraction coefficient and its sealing temperature. The thermal contraction coefficient is defined as the average rate of contraction between the setting temperature and the temperature of ultimate use. In the case of vitreous solder glass, the setting temperature corresponds approximately to its annealing point. In the case of crystallizing solder glass, the setting temperature corresponds to the sealing, or holding, temperature during the devitrifying period.

Sealing temperature is the peak temperature required to achieve a suitable seal for the application. A suitable seal has proper flow, thickness, filleting, and adhesion (bonding) to the parts being sealed.

Solder Glass Types

There are two basic types of low-temperature solder glasses: vitreous and crystallizing (devitrifying). In vitreous solder glasses, seal solidification after firing is the result of the normal viscosity-temperature characteristic of vitreous glass. In crystallizing solder glasses, a controlled crystallization during the firing produces a fine-structured interlocking crystal matrix. Both types of solder glass seals are generally made in an oxidizing (air) environment.

Vitreous solder glasses that are commercially available are identified in Table 1.

The advantages of vitreous solder glasses are:

- They can be reworked repeatedly, with little or no change in properties
- Stresses can be "adjusted" in the finished seal, through selection of a specific setting temperature (adjustments are made by changing the contraction coefficient of the sealing glass by annealing it at temperature slightly above or below its annealing point
- Low-expansion, insoluble filler additives can be incorporated with linear changes in expansion (thermal contraction) properties
- Seals can be made quickly, especially at higher seal temperatures

The disadvantages of these glasses are:

- The thermal contraction curve after sealing and upon cooling has a rather sharp hook, that is, it is nonlinear, resulting in high temporary stresses during cooling, and lower-temperature thermal shock resistance during thermal cycling
- Chemical durability is usually less than that of crystallizing solder glasses
- Seals may fail from surface defects (typical glass failure)

Crystallizing-type solder glasses that are commercially available are listed in Table 2.

Their advantages are:

- The fine-grained interlocking crystals produce a bulk crystalline body that has a more linear thermal expansion/contraction curve than that of vitreous solder glasses. Because the seal is usually formed at temperatures well below the annealing point of the members being sealed, so the divergence between the sealing and sealed members is kept to a minimum, resulting in seals with good thermal shock resistance
- Chemical durability is usually better
- Crystallized seals are less susceptible to failure from surface flaws
- Crystallized seals can be reheated well above the sealing temperature (75 °C, or 135 °F, above, in most cases) without appreciably changing the crystalline microstructure or seal fillet geometry
- The breaking strength of crystallized seals is typically 50% higher than it is in vitreous systems

Their disadvantages are:

- The sealing time-temperature profile is more critical than it is for vitreous seals, and it is usually not possible to quickly develop a proper crystal structure
- Certain filler materials may cause undesirable early nucleation and premature crystal growth

Physical Forms and Application Methods

Solder glasses are usually supplied in powder form. The powder, commonly called "frit," has an average particle size of 20 μm (0.8 mil) and a maximum particle size of 146 μm (6 mils), which is representative of 100 mesh. The individual particles are typically wedge-shaped and are the result of impact fractionation in a ball mill (Fig 1).

The user typically mixes the solder glass powder with a two-component vehicle. One of the components provides fluidity to the mix, and the other component acts as a binder and provides green strength. One common vehicle is 98.8% amyl or butyl acetate/1.2% nitrocellulose. This vehicle is used in extrusion applications, such as television bulb sealing, where the solder glass-vehicle mixture is extruded upon one of the surfaces to

Table 1 Vitreous solder glasses

Contraction range, 10^{-7}/K	Firing temperature		Annealing point		Material to which seal is applied	Solder glass designations(a)
	°C	°F	°C	°F		
45–65	625	1155	470	880	Kovar, 42SS(b), moly, 8337, 8487, K-650(c), EN-1(c), N-51A(c)	LS-0110, LS-0113, LS-0111M, LS-0810, CF-5, CF-6, CF-8, SG-7
60–95	600–650	1110–1200	420	790	Alumina, kovar, 44SS(b), 46SS(b), TM-5(c), KG-12(c), R-6(c), 8250, 8436, EN-1(c), 7052(d), titanium	LS-2001B, LS-0803, LS-2010, LS-3001, CT-410, VSG-300, T-18, T-191, XS1175M1, SG 401, SG 387
90–110	420–450	790–840	300–365	570–690	Forsterite, steatite, 446SS(e), 49SS(b), lead, soda-lime, TV glasses	SG-67, SG-100, LS-7105, 7570
100–125	410–440	770–824	290–350	555–660	No. 4 metal, 430SS(e), 304SS, 52SS(b), color TV glasses, copper	SG-68, 9776

(a) LS- and CF- designate NEG; SG- designates OI-NEG; T- designates Iwaki. (b) Nickel-iron binary sealing steel, where number designates percentage nickel (SS designation is customary in sealing trade). (c) Kimble technical glasses. (d) Corning technical glass. (e) Stainless steel sealing alloy

Table 2 Crystallizing solder glasses

Contraction range, 10^{-7}/K	Firing temperature		Firing time, min	Material to which seal is applied	Solder glass designation(a)
	°C	°F			
45–65	675	1245	60	Molybdenum, tungsten, N-51A(b), KG-34(b), EE-2(b), 8337, 8487	C-7574, CV-285, CV-635
50–75	640	1185	60	8409, EN-1(b), 7052(c), 2954, types 42SS and 44SS(d)	C-7578, CV-111
65–95	400–465	750–870	15–60	Alumina, kovar, 46SS(d), 8250, 8436, EN-1(b), 7052(c), platinum, titanium	LS-7105, CV-111, CQ-4794A, CT-480, 7583BF
90–110	450	840	30–60	Forsterite, steatite, 446SS(e), 49SS(d), lead, soda-lime, TV glasses	CV-97, CV-101, CV-137, 7583BF, C-7580, C-7581, C-7583, C-7588
100–140	440	825	30–60	No. 4 metal, iron, 430SS(e), 304SS, 52SS(d), 304SS, KG-1(b), 4210, color TV glasses	CV-685, CV-1610, CV-1608, ASF-1307, ASF-1307B, C-7590, C-7598, 7590BF

(a) C- designates Corning; CV- designates OI-NEG; LS- designates NEG; CQ- designates Sem-Com; ASF- designates Asahi. (b) Kimble technical glasses. (c) Corning technical glass. (d) Nickel-iron binary sealing steels. (e) Stainless steel sealing alloy

be joined. The extruded ribbon then dries in a semicylindrical configuration. Drying is achieved through evaporation of the amyl acetate. The nonvolatile nitrocellulose remains in the ribbon to provide green strength, but is later burned off in the early part of the firing cycle.

Another common vehicle is 80% terpineol (terpine alcohol)/20% styrene resin, which is used for screen printing of solder glass paste, as well as for spraying application. Screen printing is commonly used in the preglazing step associated with alumina microelectronic packages.

The simplest, most effective, and most frequently used application methods are extrusion (the extrudable material can also be applied with a brush or spatula) based on the amyl or butyl acetate/nitrocellulose vehicle, and screen printing or spraying methods using various vehicles. While in paste form, the solder flows to the approximate shape of the finished seal. Because the solder glass is available near its ultimate final location, the diffusion at the seal interface is enhanced.

Vitreous solder glasses can also be used in preformed ring shapes. This is a common approach in situations where wires are being sealed into round holes.

There has also been success using hot-dipping processes. Vitreous solder glass is melted in a platinum crucible, and parts that have been preheated to about 450 °C (840 °F) are dipped into the melted glass.

Firing Cycle Stages

After the solder glass has been applied in extruded, screen printed, or paste form, it undergoes a firing cycle, during which a series of physio-chemical reactions takes place. Understanding the sequence of events that occur during firing enhances the effectiveness of designing and monitoring the solder glass seal cycle.

Vitreous Sealing Cycle. To exemplify this cycle, consider the case of joining two pieces of G-12 type glass using a vitreous solder glass. The G-12 type is a lead-containing silicate glass that is used for sealing to Dumet wire, No. 4 stainless, and platinum. A common application is in the end seals of fluorescent lamps. The typical annealing point for G-12 glass is 440 °C (825 °F), and, for solder glass, 315 °C (600 °F), whereas the thermal contraction from 315 °C (600 °F) to room temperature is 100×10^{-7}/K for G-12 glass and 94×10^{-7}/K for solder glass. The firing cycle is 15 min at 400 °C (750 °F). An ideal sealing temperature profile is shown in Fig 2.

An eight-point analysis of the profile shown in Fig 2 reveals that:

1. The heating rate of 10 °C/min (18 °F/min) is selected for two reasons: (a) It is moderate enough to prevent breakage of the G-12 glass parts from thermal shock, even in thicknesses as great as 20 to 30 mm (0.8 to 1.2 in.), and (b) it is slow enough to permit burnout (oxidation) of the binder. The binder must be oxidized to prevent discoloration or lead reduction in the seal
2. At point 2 (about 350 °C, or 660 °F), burnout of the binder is essentially complete, and interparticulate fusion has begun
3. At 400 °C (750 °F), fusion is virtually complete. Because of the moderate heating rate, air bubbles between the particles have had time to escape
4. The solder glass now flows freely; its flow kinetics are limited by its own surface ten-

Fig 1 Typical wedge-shaped particle of solder glass powder (−270, +400 mesh fraction)

50 µm

sion and by interfacial tension with the surfaces being wet

5. During the latter part of the hold cycle, interfacial diffusion of elements between the solder glass and G-12 glass takes place (wetting)

6. This is the first cooling stage. The cooling rate of 5 °C/min (9 °F/min) is less than the heating rate of 10 °C/min (18 °F/min), because the exposed glass surfaces are in temporary tension during cooling, and these surfaces are likely to contain defects such as bruises or checks. The slow cooling rate prevents breakage due to thermal shock

7. The 15-min "plateau" at 325 °C (615 °F) allows the solder glass to anneal and establishes a new "setting point" with respect to the sealed surfaces. The lower setting point, in essence, lowers the thermal contraction coefficient of the solder glass. This annealing plateau is not used in crystallizing glasses and is not necessary in many situations that employ vitreous solder glasses

8. After the solder glass annealing plateau, cooling at the 5 °C/min (9 °F/min) rate is resumed down to 50 to 100 °C (120 to 212 °F), after which the sealed part can be exposed to ambient conditions as soon as practical

Crystallizing Sealing Cycle. To exemplify this cycle, consider the sealing of color television parts (cathode-ray tube) using a crystallizing solder glass. The typical annealing point for television glass is 505 °C (940 °F). Thermal contraction, from 440 °C (825 °F) to room temperature, is 109×10^{-7}/K for television glass and 106×10^{-7}/K for solder glass.

The firing cycle is 45 min at 440 °C (825 °F). The entire sealing profile is shown in Fig 3.

A seven-point analysis of this profile shows that:

1. Again, a heating rate of a moderate 10 °C/min (18 °F/min) is appropriate to avoid glass breakage due to thermal shock and to permit full oxidation of the binder

2. Burnout of binder is essentially complete, and interparticulate fusion has commenced in the solder glass

3. At 400 °C (750 °F), fusion is nearly complete. Wetting and flow are underway. The slower heating rate above 375 °C (705 °F) is selected so that the flow kinetics occur in a controlled fashion around the entire perimeter. The "approach" heating rate is typically 5 °C/min (9 °F/min)

4. About 5 or 10 min into the hold, the flow kinetics are essentially complete. Crystal growth begins around formed nuclei. Wetting of the sealing surfaces through interfacial diffusion continues, but the rate slows down due to diffusion mechanics and the formation of solder glass crystals

5. About 20 min into the hold cycle, the rate of crystal growth reaches a maximum. The crystals are nearly full size

6. In the last 15 to 20 min of the hold cycle, crystallization reaches a completion level. The glass is 92 to 98% crystallized, with a thin, glassy matrix remaining between crystals to provide hermeticity and adhesion

7. The cooling cycle, down to 50 to 100 °C (120 to 212 °F), is slow enough to prevent breakage of the parts

Common Problems

Figure 4 shows cross sections of a good seal (Fig 4a) and some poor seals (Fig 4b-d). The common problems associated with poor seals, and their possible causes, are:

- Drooping or sagging solder glass fillets—heating rate above 375 °C (705 °F) too rapid; hold temperature too high
- Thick seals, small fillets having re-entrant appearance—heating rate above 375 °C (705 °F) too slow; hold temperature too low
- Discolored seals; bumpy or rough seal fillets—contaminated seal edges (usually organic contamination)
- Bubbles within solder glass—usually due to inadequate drying of vehicle, and/or glazing over of surface. Additional drying

prior to firing, and then heating from "bottom up" could alleviate the problem. Note that some bubbles, 2 to 6% by cross-sectional area, are normal for solder glass applied as a powder-vehicle system

- Breakage of seals during cooling cycle—mismatched thermal contraction rates of parts or of frit. Frit contraction should be 2 to 8% below that of the parts being sealed. Another cause can be excessive cooling rate, which can be checked out by controlled cooling at 5 °C/min (9 °F/min), which most articles will withstand
- Poor wetting (stripping) at solder glass-part interface—contaminated seal edge or too low of a sealing temperature was used, which suppressed diffusion at interface
- Runny solder glass-vehicle paste during application—too much vehicle, or insufficient solids content in vehicle. Can also be caused by high shear during paste application. Solder glass paste systems are thixotropic, and after a thorough initial mix, should be mixed only gently. After application, wet paste should not be subjected to vibration
- Bubbles in mixed paste—mixture too thick or mixing time was insufficient (rolling in jars at 5 to 10 rev/min is an effective postmix method of eliminating bubbles and improving rheology)

Solder Glass Technology

Base Glasses. The basic compositions for vitreous solder glasses are primarily within the lead-borate binary system. Lead borate has a eutectic point at 87.5 wt% PbO and 12.5 wt% B_2O_3. Minor ingredients, such as SiO_2, BaO, ZnO, Al_2O_3, and CuO, are often added to stabilize the binary system, extend its glassy range, and lower its expansion coefficient (Fig 5).

The basic compositions for crystallizing solder glasses are primarily within the lead-zinc-borate ternary system. Minor ingredients, such as BaO, SiO_2, and Al_2O_3, are usually added to suppress the onset of crystallization and extend the glassy range.

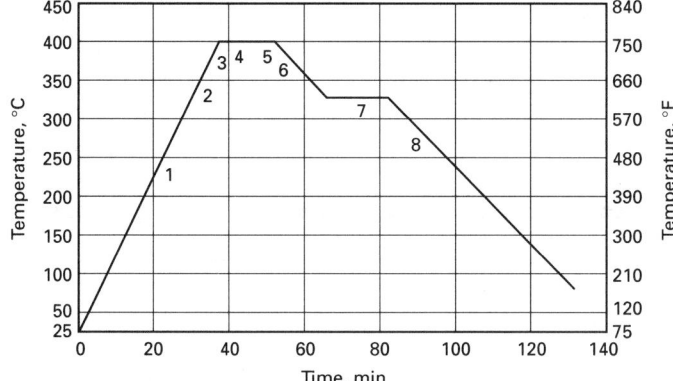

Fig 2 Vitreous solder glass sealing cycle. See text for explanation of numbering sequence.

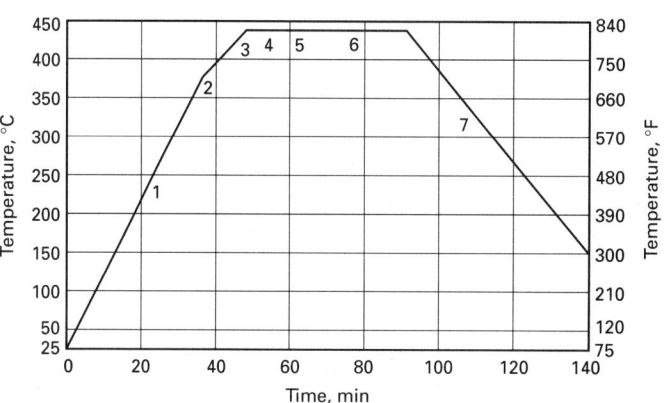

Fig 3 Crystallizing solder glass sealing cycle. See text for explanation of numbering sequence.

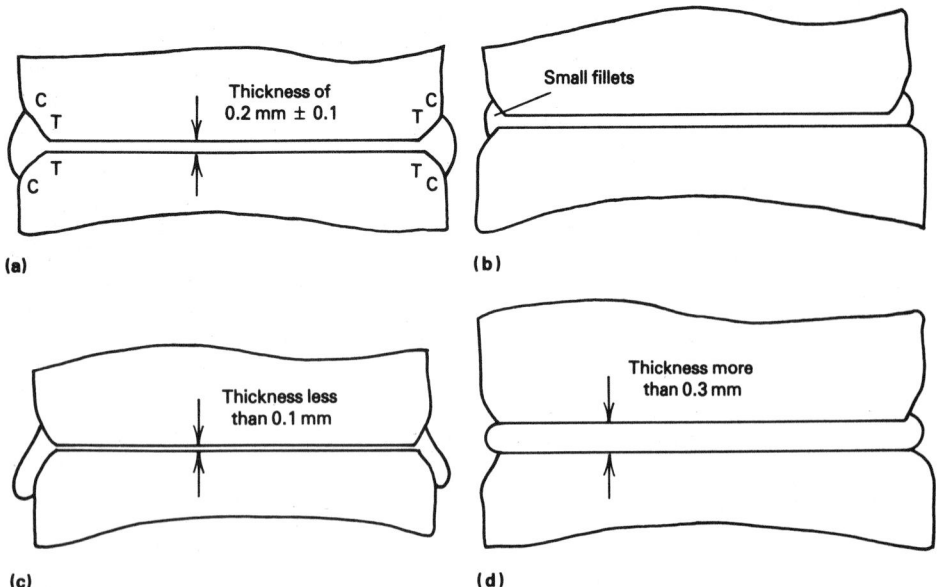

Fig 4 Seal geometry and stresses. (a) Ideal geometry and stress. (b) Insufficient solder glass. (c) Too-rapid heating rate in range from 375 to 440 °C (705 to 825 °F), resulting in thin seals and bead droop. (d) Too-slow heating rate above 375 °C (705 °F), resulting in thick seals

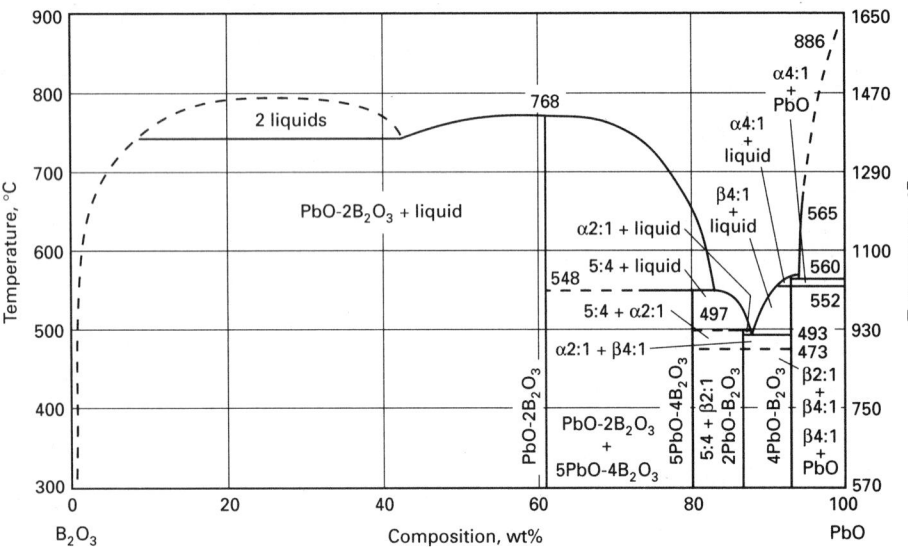

Fig 5 Binary system PbO-B₂O₃. Source: Ref 1

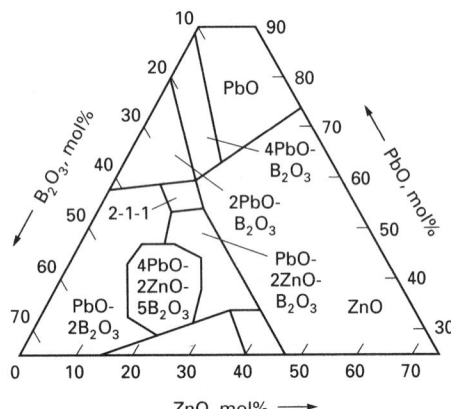

Fig 6 Ternary system PbO-ZnO-B₂O₃

Refractory fillers are commonly blended with powdered solder glass to increase the strength of the finished seal, and/or to lower the thermal contraction coefficient of the composite. Refractory particles also slow down or stop the propagation of cracks in vitreous solder glasses, making them less susceptible to failure from surface defects. As noted earlier, certain fillers can cause premature nucleation of the crystallizable solder glasses.

Some common refractory fillers are alumina, zirconium silicate, beta eucryptite, cordierite, tin oxide, and lead titanate. Table 3 identifies some filled vitreous solder glasses and their properties.

Health and Safety Issues

High-lead solder glasses are considered to be hazardous materials. The primary exposure route is through inhalation or ingestion. When working with these materials, general ventilation is satisfactory, unless active dust or fume generation occurs, in which case, local ventilation is recommended. If necessary, respirators approved by the National Institute of Occupational Safety and Health/Mine Safety Health Association (NIOSH/MSHA) for dust, mist, or fumes (as defined in the lead standard) are recommended. Eye, face, and skin protection is also recommended.

Spills should be immediately cleaned, using a high-efficiency filter vacuum. Disposal should be as hazardous waste, using the Environmental Protection Agency Waste ID number D008.

Suppliers

Suppliers of sealing/solder glasses are identified in Table 4.

Table 4   Sealing glass suppliers

Company	Location	Type
Kimble Glass Co.	Vineland, NJ	G
Sem-Com, Inc.	Toledo, OH	SG, G
OI-NEG Technical Products	Toledo, OH	TV, PS
Corning Asahi Video Products	Erwin, NY	TV, SG
Corning Incorporated	Corning, NY	G, SG
Iwaki Glass Co., Ltd.	Tokyo, Japan	TV, PS
Asahi Glass Co., Ltd.	Tokyo, Japan	TV, SG
Nippon Electric Glass Co., Ltd.	Osaka, Japan	PS, SG, G
Schott Glass Technologies, Inc.	Duryea, PA	G

TV, television sealing glass; PS, package sealant solder glasses; SG, general sealing solder glasses; G, general sealing hard glasses

The crystals that develop during devitrification are primarily 2-1-1 lead-zinc-borate, and 1-2-1 lead-zinc-borate. By blending certain refractory particles, such as zirconium silicate or zinc zirconium silicate, into the glass powder, the rate of crystallization and the type of crystal phase can be controlled (Fig 6).

Table 3   Filled vitreous solder glasses used in microelectronic package sealing

Contraction range, 10^{-7}/K	Firing temperature		Annealing point		Type of filler
	°C	°F	°C	°F	
60–65	400	750	300	570	Beta eucryptite or zirconium silicate
65–75	410–450	770–840	310	590	Beta eucryptite or cordierite plus tin oxide
70–80	420–450	790–840	315	600	Zinc silicate plus tin oxide or alumina titanate

REFERENCE

1. E.M. Levin *et al.*, *Phase Diagrams for Ceramists*, American Ceramic Society, 1964

SELECTED REFERENCES

- S. Claypoole, U.S. patent 2,889,952, June 1959

- R.E. Hogan, Solder Glasses, *Chem. Tech.*, Jan 1971
- K.G. Lusher, "Review of the Solder Glass Situation," Ninth Symposium on the Art of Glassblowing, American Scientific Glass Blowers Society, 1964
- N.B. Nofziger, U.S. patent 4,006,028, Feb 1977
- E.M. Rabinovich, Crystallization and Thermal Expansion of Solder Glass in the PbO-B$_2$O$_3$-ZnO System with Admixtures, *Bull. Am. Ceram. Soc.*, Vol 58 (No. 6), 1979
- W.T. Roubal, Use of Solder Glass by Unskilled Laboratory Workers, *J. Chem. Ed.*, Vol 47 (No. 2), Feb 1970
- W.M. Simpson, Solder Glass Technology, *Am. Glass Rev.*, Oct 1976

Ophthalmic and Optical Glasses

Emil W. Deeg, AMP Inc.

OPTICAL AND OPHTHALMIC GLASS-ES are primarily used as lenses (that is, as refractive elements). As such, they are usually components of image forming systems (for example, the combination of a spectacle lens and the lens of the human eye, a microscope, or a camera "lens"). For optically less demanding applications, lenses made of transparent organic polymers can be beneficial based on mechanical property and cost factors.

Lenses, windows, and so on for ultraviolet or infrared optical systems require special glasses (for example, "quartz glass," chalcide glasses, and calcium aluminate glasses). Depending on the spectral range, crystalline materials (calcium fluoride, alkali halides, silicon, germanium, and so on) may suffice as replacements for special glasses in selected applications.

Ophthalmic Glass Products

Correction of vision, protection against eye hazards, providing improved vision comfort, and utilization of the aesthetic appeal of glass are the main applications of ophthalmic glasses.

Products for correction of vision include single vision, one-piece multifocal, fused multifocal lenses, and extreme prescription lenses for low-vision aids. Glasses used can be colorless, tinted, or photochromic, but the lenses must meet certain impact resistance requirements. Guidelines for their expected optical and mechanical performance are given in ANSI Z80.1 (Ref 1). Low-vision aids can be headborne, hand-held, or stand-supported purely optical devices for persons with low vision that may be telescopic, microscopic, or telemicroscopic and assembled in binocular (two separate mutually aligned systems) or in biocular (single-element centered system) arrangements. Optical specification parameters include equivalent focal length and dioptric power, magnification, image distance, prism power, and transmittance. Focal length, magnification, image location, prism power, and transmittance are measured quantities while equivalent dioptric power is calculated from the equivalent focal length. For single-element systems with spherical surfaces, the power may also be calculated via the thick lens formula. Because of the highly individualized nature of each low-vision device, general performance standards are not possible. Terminology and some recommendations for low-vision aids are given in ANSI Z80.9 (Ref 2).

Glass products for protection against eye hazards include safety lenses (plano or prescription) in spectacles or eyecup goggles, plates in cover goggles, welding helmets, lift front helmets, and in hand-held shields. Depending on their design, such devices protect against impact, heat, chemicals, dust, and bright radiation within the near ultraviolet (UV), visible, and near infrared (IR) spectral regions. Glass elements incorporated in these protective devices can be laminated with other devices equipped with wavelength selective absorbing, transmitting, and antireflective coatings. The glasses used may be colorless or tinted. To reduce glare, polarizing foils of organic polymers may be sandwiched between two glass elements. Expected performance of eye and face protecting devices is covered by ANSI Z87.1 (Ref 3), which, however, does not apply to hazards in sports or other injuries caused by high-energy particle and wave (x-rays, γ-rays) radiation, microwave radiation, rf-frequency radiation, and laser radiation. Tinted ophthalmic glasses are exceptionally suited for protection against the monochromatic coherent radiation emitted by lasers. They are distortion-free, scratch resistant, chemically durable, and are free of the dye saturation and bleaching problems that are associated with plastics. At least 15 glass products are available at present to match specific laser wavelength and attenuation requirements.

Improved vision comfort and the aesthetic appeal of glass are utilized in sunglasses for casual wear, fashion eyewear, and recreational purposes. Such eyewear is equipped with substantially plano power lenses intended for attenuation of light in the entire optical spectrum (gray or neutral density lenses) or in selected ranges of the UV, visible, and IR spectral regions. The lenses may be single-piece or laminated, polarizing or nonpolarizing with uniform or gradient optical density. In addition, these lenses may also be photosensitive (that is, photo-chromic). Gradient optical density lenses are usually produced by the application of dielectric or metallic thin films. Photochromic gradient lenses can be produced by exposing unnucleated blanks to a temperature gradient. One line of photochromic lenses, available in gray or brown, blocks at least 94% of UV radiation. All sunglass lenses must be impact resistant under tests specified in ANSI Z80.3 (Ref 4). The optical performance of a sunglass lens is characterized by several wavelength-averaged transmittance values and by chromaticity coordinates for different standard illuminants (A, C, and D65). Of particular interest are the luminous transmittance in the visible spectral region (380 to 780 nm), and the mean transmittance in the near UV spectral region (310 to 380 nm), in the erythemal UV zone (290 to 315 nm), and the near IR zone (780 to 1400 nm).

The traffic signal transmittance must also be considered. Definitions, test methods, computing procedures, and performance requirements are covered in ANSI Z80.3 (Ref 4).

Of particular interest is a relatively new product providing eye comfort over a wide range of conditions of illumination. This lens is made of photochromic glass with high absorption in the UV and blue spectrum. This combination provides the advantages of lenses that adapt their transmittance to a given level of illumination as well as glare reduction desired by patients with hypersensitive eyes. Combinations of this lens and low-vision aid prescription lenses are possible by laminating the two components.

Optical Glass Products

Plano or curved windows and refractive and reflective elements for precision instruments and devices are made from optical glasses. Most products are silicates designed for maximal transmittance in the visible spectrum. Some glasses are prepared from extremely pure raw materials under special conditions to yield high UV transmission. Among IR transmitting glasses the chalcide glass As_2S_3, fused silica (waterfree, mostly prepared from natural quartz), and calcium-aluminate glasses are most common. Fused silica and As_2S_3 can be made in large pieces of very good quality for refractive elements. IR transmitting windows are made from all of them. Fluoride (with the exception of oxyfluoride) glasses offer high IR transmittance but are of limited utility due to their poor chemical durability.

Refractive optical elements include single lenses, prisms, and aberration corrected lens systems for a wide variety of consumer, industrial, medical, and military instruments. Precursors (blanks) for such elements and for optical windows are formed thermally by gobbing, pressing, extrusion, casting, or drawing from the melt. The desired shape of the surface is usually generated by a milling process with diamond studded tools. Subsequent loose abrasive grinding, polishing, and in some cases figuring by hand or computer-controlled machines provide the final shape and smoothness. Fire-polishing of generated surfaces is possible by short exposure (duration of a few seconds) of preheated blanks to high intensity UV (plasma or excimer laser) radiation. The physical and chemical properties of glasses permit surface trueness to fractions of a wavelength of light. The long viscosity versus temperature curve of many optical and ophthalmic glasses makes it also possible to slump (sag) glass blanks to final shape. Nonspherical surfaces can be produced (for example, by vacuum assisted sagging of spherical glass shells in porous ceramic molds). Condenser-quality lenses can be formed by precision pressing in highly polished pyrolytic graphite or special metal (for example, beryllium-copper) molds.

Reflective elements utilize the amorphous nature and homogeneity of optical glasses to provide grain-free surfaces on mirror, beam splitter, and interference filter substrates. The very low thermal and mechanical aftereffects of well annealed glasses, the high chemical durability, and the low thermal expansion of some optical glasses (particularly of optical-grade fused silica), offer additional and essential advantages. The substrates need to be coated with metallic or dielectric layers to produce the desired results. However, the effect of total internal reflection can make a glass surface itself serve as a reflective element. This is utilized, for example, in prisms for binoculars and single-lens reflex (SLR) cameras. Light propagation in step index fibers is also based on internal reflection at the interface of core and cladding.

High-index optical glasses are ideally suited for Cerenkov counters because they also meet the two additional requirements of transparency and nonconductivity. Dense flints with high contents of lead oxide find occasional use as magnetooptical switches and shutters because of their high Verdet constants. The Verdet constant is the proportionality constant in the equation of the Faraday effect. It is equal to the angle of rotation of plane-polarized light in a substance placed in a magnetic field divided by the product of the length of the light path in the substance and the magnetic field strength. In the blue spectrum and at room temperature, dense flints can reach about 0.2 but decrease rapidly with increasing wavelength to less than 0.1 [minute of angle/(Gauss · cm)] in the red. Special glasses for Faraday rotators cover a range from 0.5 in the blue to 0.25 [minute of angle/(Gauss · cm)] in the red. However, because of a broad absorption band in the blue, they have limited other uses.

Glass properties can be modified locally by selective ion-exchange at elevated temperature. This effect is utilized not only for toughening of optical windows and of ophthalmic lenses (chemtempering) but also for preparation of refractive elements with index gradients. The continuous local index change results in a continuous bending of a light ray. The most common gradient index lens is a rod with a radially changing refractive index, $n(r)$:

$$n(r) = n_0 \cdot [1 - K/(r^2)] \qquad \text{(Eq 1)}$$

where n_0 is the index in the axis; $\mathbf{r}$ is the radius vector in a plane normal to the axis; and K is an empirical constant. Both n_0 and K are wavelength dependent. In such a device, the rays follow peculiar paths. A screw ray, for example, spirals around the axis of the rod. Estimates of chromatic aberration of gradient index lenses require knowledge of the wavelength dependence of n_0 and K. Planar optical waveguides embedded in the surface of a glass substrate can be prepared by a combination of photolithography and subsequent ion-exchange. By choosing a suitable geometry a variety of passive optical circuits (for example, splitters, couplers, multiplexers, optical power taps) can be formed. Spatial optical solitons have been observed only recently in nonlinear glass waveguides prepared by ion-exchange. Still other experimental approaches for preparing optical waveguides embedded in glasses utilize the effect of permanent index changes by electron-beam radiation. Optimization of glass compositions and processes for embedding waveguides in glass are still ongoing.

Glass fibers for telecommunications and signal transmission are formed from special precursors prepared by precipitation of suitable compositions from the gas or vapor phase and subsequent heat treatment of the initial product. Resulting compositions are silica or fluorine-containing silica for the

Table 1   Nonoptical properties of single-component and multicomponent oxide glasses

	Physical properties				Mechanical properties					
			Poisson's ratio		Young's modulus				Microhardness, HK	
	Density		20 °C (70 °F)	900 °C (1650 °F)	20 °C (70 °F)		900 °C (1650 °F)			
Material	g/cm³	lb/in.³			GPa	10⁶ psi	GPa	10⁶ psi	MPa	ksi
Multicomponent oxide glass	2.3–6.3	0.083–0.23	0.19–0.31(b)		(b)		(b)		250–650	36–94
Single-component oxide glass (fused silica)(a)	2.201–2.203	0.07952–0.7959	0.17	0.19	70	10	81	12	900	130

| | Thermal properties | | | | | | | | | | | |
| | Linear thermal expansion coefficient, ppm/°C | | Transformation point, T_g | | Softening point (viscosity of $10^{6.6}$ Pa · s or $10^{7.6}$ P) | | Specific heat | | Thermal conductivity | | Strain point | |
Material	(−30)–70 °C [(−20)–160 °F]	0–100 °C (30–212 °F)	°C	°F	°C	°F	J/kg · °C	Btu/lb · °F	W/m · K	Btu · in./ ft² · h · °F	°C	°F
Multicomponent oxide glass	3.7–14.6	...	340–770	640–1420	471–825	880–1520	310–890	0.074–0.21	0.51–1.28	3.5–8.9	...	...
Single-component oxide glass (fused silica)(a)	...	0.51	...	...	1597–1727	2907–3141	750	0.18	1.38	9.57	1027–1077	1880–1971

(a) Ranges cover glass melted from quartz as well as synthetic material prepared by pyrolysis of silicon-tetrachloride or gaseous silicon-organic compounds. The low thermal expansion and the high thermal conductivity offer advantages for large mirror blanks. Caution: Prolonged exposure to about 1000 °C (1830 °F) can cause devitrification. (b) Young's modulus for multicomponent oxide glass is 40–129 GPa (6–19 psi × 10⁶); no temperature range specified.

Table 2 Letter designations for wavelengths (Fraunhofer lines) of optical and opthalmic glasses

Spectral range	Alpha symbol	Numeric value nm	Light source gas discharge in
Ultraviolet	i	365.0146	Mercury vapor
Violet	h	404.6561	Mercury vapor
Blue	g	435.8343	Mercury vapor
	F'	479.9914	Cadmium vapor
	F	486.1327	Hydrogen
Green	e	546.0740	Mercury vapor
Yellow	d	587.5618	Helium
	D	589.2938	Sodium vapor
Red	C'	643.8469	Cadmium vapor
	C	656.2725	Hydrogen
	r	706.5188	Helium
Infrared	s	852.1101	Cesium vapor
	t	1013.98	Mercury vapor

cladding and silica doped with index raising oxides (for example, GeO_2 and P_2O_5) for the core. The refractive index of the core glass of multimode fiber preforms depends on the radius in a way similar to that mentioned for the gradient index lens.

Fiberoptic faceplates are prepared by repeated drawing and pressure fusion of bundles of rods, each primary rod consisting of a high index core (lead-silicate), a low index cladding (borosilicate or soda-lime silicate) and usually a black, extramural absorbing glass surrounding the clad. The faceplates are cut from the final mosaic rod and their endfaces polished.

Microchannel Plates. A similar process is used to produce *microchannel plates* for image intensifiers. Here, an acid soluble glass is substituted for the high index core glass. After cutting the plates, the cores are dissolved, usually by dilute nitric acid. To sensitize the inner surface of the channels, a heat treatment under reducing conditions (hydrogen or forming gas) follows.

Flexible Fiber Bundles. Acid-soluble glasses are also used to prepare *flexible fiber bundles* for image transmitting industrial and medical fiberscopes. In this case, the acid-soluble glass replaces the extramural absorbing glass in a mosaic rod. After the mosaic rod is built up to contain the desired number of fibers, the acid soluble glass is leached out, except at both ends where the positions of the fibers relative to each other are preserved.

Filter Glasses. Optical-grade colored glasses are used as filters for isolation of broad regions in the UV, visible, and spectral regions. Ultraviolet transmitting filter glasses absorb the visible light and appear black (black light sources). Infrared absorbing glasses that transmit visible radiation serve as heat screen filters (for example, in slide projectors and occupational safety devices).

Illumination Optics. Colored glasses of compositions similar to those of optical filter glasses are used in illumination optics (for example, traffic and airport signals). These pressed optical elements have to meet spe-

cific chromaticity requirements but less stringent homogeneity and index specifications.

Laser glasses are prepared from exceptionally pure raw materials. Their base glasses were originally silicates but shifted to phosphates for reasons given below. The most frequently used active element is neodymium, which causes emission at 1060 nm. Several other rare earth ions (for example, erbium and terbium) either as single dopants or in combination with others have been shown to produce laser action in glasses. Presently, optical fibers with erbium-doped cores are of interest because of their potential as optical amplifiers. The doping level for such fibers is about 50 to 1000 ppm. The established neodymium-doped glasses for power lasers contain the active ion in concentrations of ~1% or more. Optimization of base glass and dopant concentration for fiber amplifiers is still in progress. The main advantages of glass as active laser elements is their high homogeneity and the capability to produce the glass in large pieces, and to form the glasses mechanically as well as thermally (for example, drawing into rods or fibers, slumping).

Materials that Compete with Glass

For less demanding instruments and devices, organic polymers are also used. Principal optical plastics are the thermoset polyallyldiglycolcarbonates (used in products such as ophthalmic lenses) and the thermoplastics polymethylmethacrylate (used in products such as windows, magnifiers, illumination optics, and short distance lightguides), polycarbonate (used in applications such as high impact resistant windows and safety eyewear), polystyrene, acrylate-styrene copolymers, and occasionally also polymethylpentene (TPX).

Refractive indexes for optical plastics range from about 1.465 for TPX to 1.586 for polycarbonate and are generally reported to the third decimal point. Advantages of optical plastics are their lower density and, for mass-produced items, lower cost per piece.

Infrared transmitting windows and optical elements are made from crystalline materials (for example, NaF, LiF, CaF_2, NaCl, CsBr, AgCl, MgO, Si, Ge, and diamond) whose selection depends on the desired wavelength

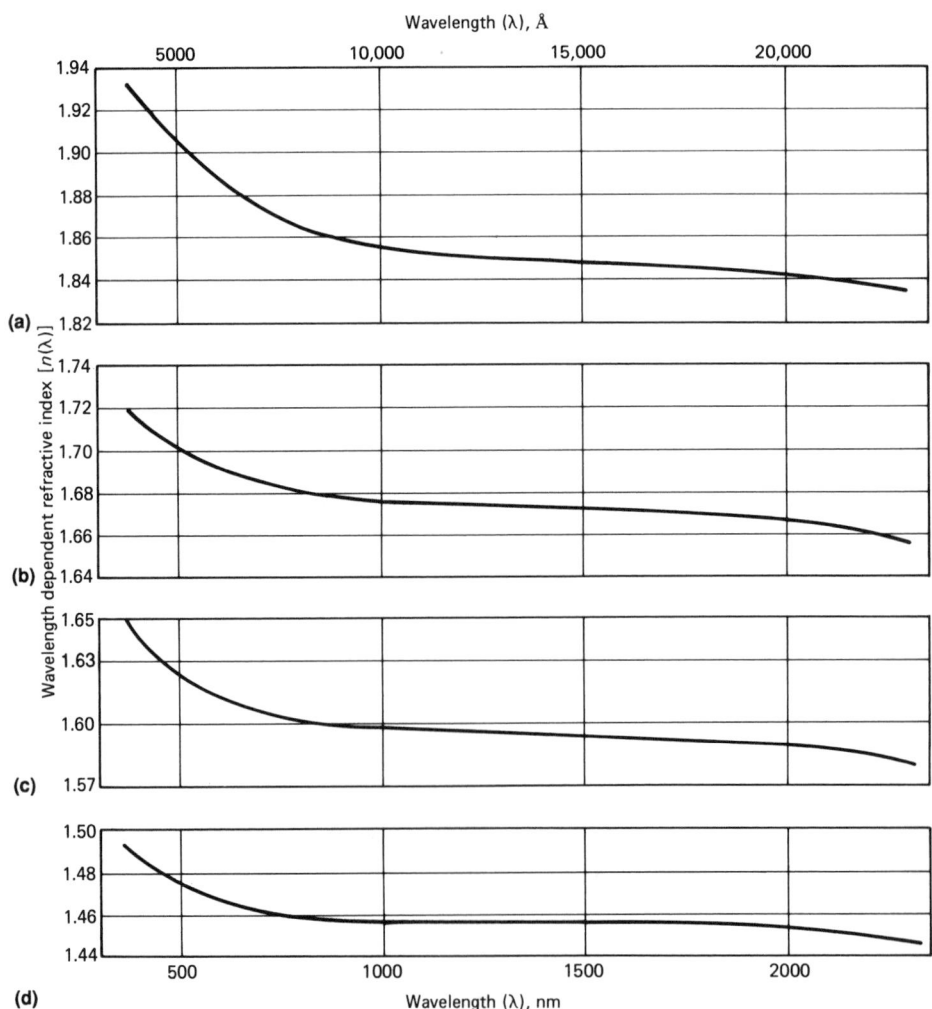

Fig 1 Wavelength dependence of refractive indexes of selected optical glasses. (a) 881 410. (b) 691 547. (c) 613 443. (d) 471 673

range and the environment to which the elements will be exposed. High refractive indexes of the chemically most durable crystalline materials (Ge, $n = 4.0$; Si, $n = 3.2$; and diamond, $n = 2.5$) require application of multilayer antireflective coatings.

Lightweight mirror substrates can be made of beryllium. Because beryllium is a polycrystalline material for use in high resolution instruments, an amorphous coating is applied before deposition of the reflective layer. Low-expansion, translucent glass-ceramics for mirror blanks are produced by optical glass manufacturers.

Properties of Optical and Ophthalmic Glasses

Optical glasses, which are inorganic amorphous materials that are identified by a set of properties at tolerance levels accepted by the precision optical industry, can be manufactured reproducibly with well-defined, low levels of defects (for example, striae and inclusions) and birefringence and are available in blanks (preforms) large enough for preparing the desired optical elements.

Their inorganic nature offers mechanical, thermal, and chemical stability. Because they are amorphous, their bulk properties are isotropic, which is an essential property for

refractive elements. The surface of optical glasses are grain-free, which is a necessary property for high resolution reflective optics. Because raw materials, composition, and processes are closely controlled to meet the stringent optical tolerance requirements, the nonoptical properties are also well defined. Ranges of some nonoptical properties of multicomponent oxide glasses and fused silica, a one-component oxide glass, are listed in Table 1.

Corrosion Resistance. The most important set of nonoptical properties is generally referred to as chemical durability, which includes resistance against attack by bases and acids and stability under temperature-humidity cycling (climate resistance). There is no set of laboratory tests that simulates completely all chemical environments that optical glasses are exposed to during manufacturing and as-finished optical elements. Most users of optical glasses rely solely on the data supplied by glass manufacturers, while some users supplement vendor information with their own process and product specific tests. A brief review of chemical durability testing is given in Ref 5.

Thermal Properties. The high glass transition temperatures of optical glasses are indicative of their thermal stability. Combined with a low rate of devitrification, this

permits extensive annealing for index adjustment (±0.000002) and for drastic reduction of birefringence (normal quality 10 to 25 nm/cm, special annealing 4 to 10 nm/cm) of pressings, blocks, ribbons, and other preforms.

Ophthalmic Glasses. Most of today's colorless (that is, clear) *ophthalmic glasses* are related to optical glass compositions. In fine-tuning glass compositions, special emphasis was placed on high chemical and mechanical durability and, at least in the United States, on their ability to permit chemical (ion-exchange) or thermal toughening of finished lenses. Ophthalmic glasses for production of fused multifocal lenses must be compatible in terms of their viscosity versus temperature curves, thermal expansion and contraction properties, and their surface tension. The glass formed in the fusion zone of two glasses must be stable against devitrification and reboiling to avoid hazy interfaces. Because of the different compositions of the crown glasses used for the distant vision portion (upper) and the flint glasses used for the near vision portion (segment), the development of optimal pairs and triplets of glasses presented a formidable challenge. These specific ophthalmic requirements do not apply to optical glasses used for instruments. However, they must be considered, when selecting combinations of core, cladding, acid soluble, and extramural absorbing glasses for optical waveguides used in medical and industrial fiberscopes, in fiberoptic faceplates, and in microchannel plates.

Refractive Index and Dispersion

Refractive Index. A key property of optical and ophthalmic glasses is the wavelength-dependent refractive index, $n(\lambda)$. It is defined experimentally through Snell's law at the interface of glass against air as

$$n(\lambda) = n_g/n_a = (\sin \theta_i)/(\sin \theta_r) \qquad (Eq\ 2)$$

where n_a is the refractive index of air; n_g is the refractive index of glass against vacuum; θ_i is the angle of incidence; θ_r is the angle of refraction; and λ is the vacuum wavelength of light. Wavelengths are identified by letters (Fraunhofer lines) instead of their numerical value (Table 2). The ~250 available optical glasses cover an approximate range of refractive indexes from 1.437 to 2.022 with deviation from catalog values less than ±0.002 for regular, ±0.000005 for precision quality. Of particular importance are the test certificates accompanying glass shipments by the leading manufacturers. They contain measurements of index values with ±0.00003 accuracy for regular products, ±0.00001 for precision-quality products, and corresponding dispersion values accurate to ±0.00002 for regular-quality products and ±0.000003 for precision-quality products.

Index values and quantities derived from them are listed in glass catalogs and certifi-

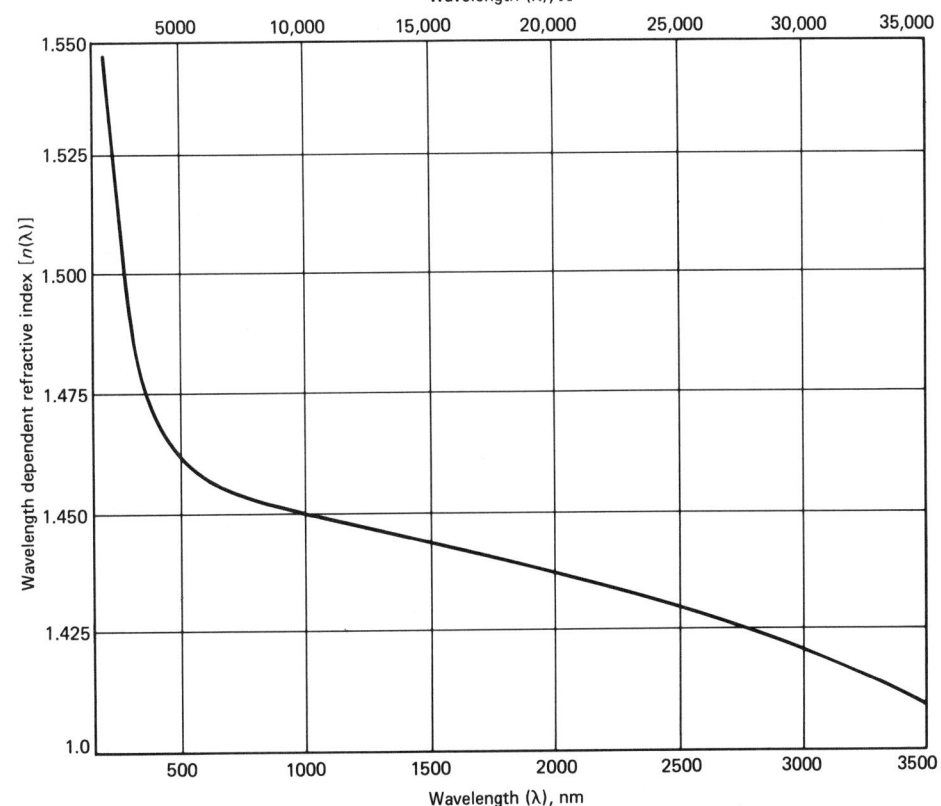

Fig 2 Wavelength dependence of the refractive index of fused silica for the wavelength range of 200 to 3500 nm

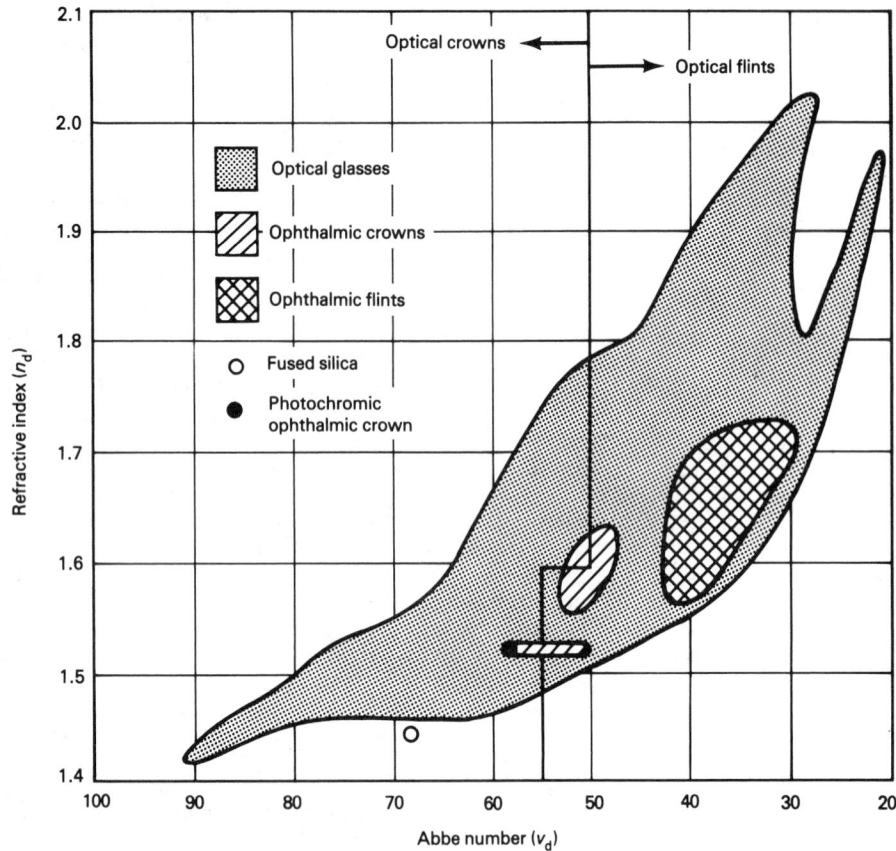

Fig 3 Map of optical and ophthalmic glasses in the $\{n_d, v_d\}$-plane. The solid line $\{2.1, 50\}$ to $\{1.6, 50\}$ to $\{1.6, 55\}$ to $\{1.4, 55\}$ separates optical crowns from flints. Ophthalmic crowns and flints do not follow the same distinction. The relationship of fused silica and photochromic ophthalmic crown glasses to these regions is also shown by individual points.

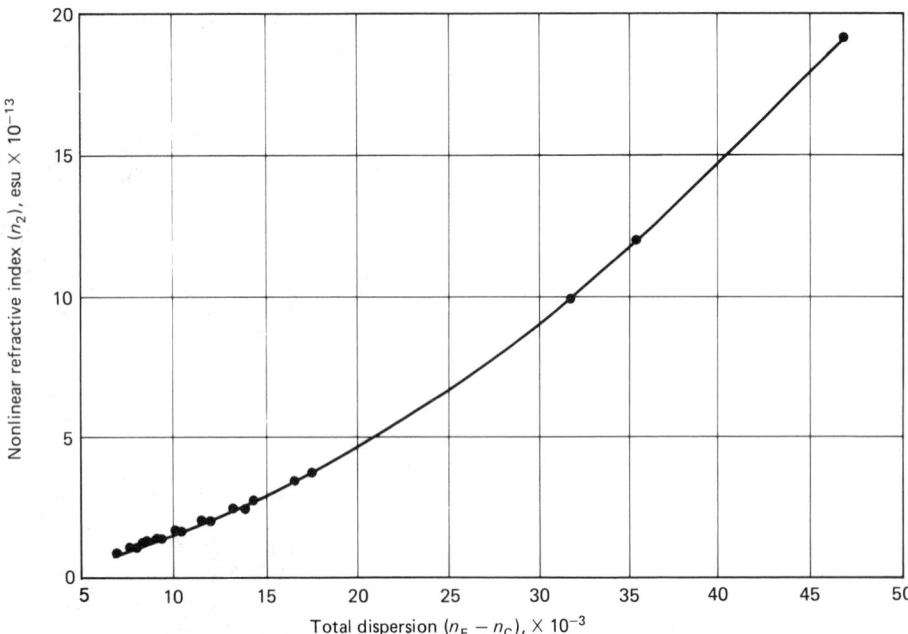

Fig 4 Nonlinear refractive index, n_2, of optical glasses as function of total dispersion $T = n_F - n_C$. A third order polynomial describes the curve very well (coefficient of determination is 100%): $n_2 = (-2.19 \times 10^{-4} + 0.12T + 6.19\,T^2 + 2.03\,T^3) \times 10^{-10}$, where $T = n_F - n_C$. Source: Ref 6

cates for a given temperature T and normal atmospheric conditions. Because n_a and n_g depend on temperature:

$$dn/dT = (1/n_a) \cdot (dn_g/dT) - (n_g/n_a^2) \cdot (dn_a/dT)$$

$$(Eq\ 3)$$

or because $n_a \sim 1$:

$$dn/dT \approx dn_g/dT - n_g \cdot (dn_a/dT) \qquad (Eq\ 4)$$

with dn/dT termed relative thermal coefficient and dn_g/dT absolute thermal coefficient of the refractive index. Both are wavelength dependent and are listed in catalogs for many glasses. The optical path length is the product of index and geometric path length. Its temperature dependence is determined by the coefficient of thermal expansion and the thermal coefficient of the refractive index. Some glasses have a negative thermal index coefficient that partially compensates for the temperature-induced change of the geometric path length resulting in a very low change of the optical path length. Leading manufacturers offer such thermooptically stable glasses. Corresponding relative thermooptical constants can range from -0.2 to 5.3 ppm/C.

The refractive index of ophthalmic crown glasses used for most single-vision prescription, safety, and one-piece multifocal lenses is held at 1.523. Two high index glasses with refractive indexes of 1.60 and 1.70 are available for preparation of lenses with extreme prescriptions. Both are compositionally related to optical flints. Lenses made from these glasses are less voluminous and more appealing (thinner edges for minus, thinner center for plus prescriptions) than the same lenses made from 1.523 crown glass or plastics. Ophthalmic flints are available in a range of refractive indexes ($1.56 \leq n \leq 1.72$), and are specified up to the fourth decimal place to satisfy design and manufacturing requirements for fused multifocals.

Because $dn/d\lambda \neq 0$, the power, ϕ, of a refractive element is also wavelength dependent. The thin lens with spherical surfaces in a vacuum may serve as an example:

$$\phi = 1/f = (n - 1) \cdot (1/R_1 - 1/R_2) \qquad (Eq\ 5)$$

where f is the focal length and R_1 and R_2 are the radii of curvature of the lens surfaces. Logarithmic differentiation of Eq 5 yields

$$d\phi/\phi = -df/f = dn/(n - 1) \neq 0 \qquad (Eq\ 6)$$

Equation 6 indicates that the focal lengths of the lens are different for different wavelengths. Examples of $n(\lambda)$ curves for some optical glasses are given in Fig 1 and a $n(\lambda)$ curve for fused silica is given in Fig 2.

For a given glass, the ratio $dn/(n - 1)$ is independent of the lens geometry but is a function of the wavelength. For practical reasons, the differential dn in Eq 6 is replaced by a difference $\Delta n_{x,y} = n_x - n_y$ of refractive indexes measured at wavelengths x and y. This difference is the dispersion over the wavelength range x to y. The terms $(n_F - n_C)$ and $(n'_F - n'_C)$ are total dispersions of a glass

Table 3 Approximate ranges of selected optical properties of optical and ophthalmic glasses

Property	Range
Refractive indexes	
h line (violet)	1.445 to 2.087
d line (yellow)	1.437 to 2.022
r line (red)	1.435 to 2.006
Principal dispersion (F-C)	0.005 to 0.05
Abbe number, v_d	20.4 to 90.7
Dispersion, P	
(blue)	0.53 to 0.63
(red)	0.44 to 0.57
Change in dispersion, ΔP	
(blue)	−0.01 to 0.04
(red)	−0.06 to 0.02
Verdet constant,	
(minimum angle/Gauss·cm)	≤0.5
Stress-optical coefficient (brewster)	−1.5 to 3.6
Relative thermooptical constant, ppm/°C	−0.2 to 5.9
Absolute thermooptical constant, ppm/°C	−0.8 to 4.4
Nonlinear index (n_2), esu	(1 to 19) × 10^{-13}
Ophthalmic glasses	
Refractive index, n_d	1.523 to 1.72
Abbe number, v_d	20 to 59

between the blue and the red end of the visible spectral region. The term $(n_F - n_C)$ is known as the principal dispersion.

Introducing a $\Delta n_{x,y}$ in Eq 6 yields a second material constant:

$$v_z = (n_z - 1)/(n_x - n_y) \qquad \text{(Eq 7)}$$

known as v-value, ν-value, Abbe value, Abbe number, constringency, constringence value, reciprocal dispersion, reciprocal dispersive power. Common combinations of wavelengths for v_z-values are:

$$\{x,y,z\} = \{F,C,d\} \qquad \text{(Eq 8a)}$$

and

$$\{x,y,z\} = \{F',C',e\} \qquad \text{(Eq 8b)}$$

In the early literature and for less demanding applications today, Abbe's original definition:

$$v_D = (n_D - 1)/(n_F - n_C) \qquad \text{(Eq 9)}$$

is found. Optical glasses cover a v_d-range from about 20 to 90, ophthalmic crowns from 51 to 59, and ophthalmic flints from 29 to 43. The set of $\{n_d, v_d\}$ pairs provides a mean of classifying or coding optical glasses. Initially the center of the two yellow sodium spectral lines (that is, $z = D$) was chosen as the basis for the classification. When more precise light sources became available, $\{n_d, v_d\}$ and $\{n_e, v_e\}$ replaced the original pair. Imaged onto a $\{n_z, v_z\}$-plane, such pairs form the map of optical glasses (Fig 3) that gives an overview of glasses offered by a manufacturer and permits a first selection of glasses for lens design. Traditionally, individual optical glasses were characterized by numerous alphanumeric codes such as FK 5, PK 51, K 3, BSC 510644. The letters identify a glass family and are abbreviations of chemical or optical, descriptive terms [for example, FK (fluor crown), BK or BSC (borosilicate crown), PK (phosphate crown), K (crown), HC (hard crown)]. The

numbers identify individual members within a family. Gradually a six-digit code, based on $\{n_d, v_d\}$ emerged. For example, given the glass BK 7 with $\{n_d, v_d\} = \{1.51680, 64.17\}$, certain digits are dropped:

$$\left.\begin{array}{l} n_2 = 1.51680 \rightarrow 1.517 \rightarrow 517 \\ v_2 = 64.17 \rightarrow 64.2 \rightarrow 642 \end{array}\right\}$$
$$\rightarrow \text{Identifier 517 642}$$

Both designations are merged and the above glass would be coded as BK 7–517 642 or BK7–517642.

Dispersion. Knowledge of n_z- and v_z-values is important for selecting glass combinations for achromatic lenses (common focus for two colors). For higher order correction of chromatic aberration (common focus for three colors), the relative partial dispersion $P_{x,y}$ of the glasses is required. It is the ratio of a dispersion ($n_x - n_y$) to a total dispersion. For example:

$$P_{g,F} = (n_g - n_F)/(n_F - n_C) \qquad \text{(Eq 10a)}$$

$$P'_{i,h} = (n_i - n_h)/(n'_F - n'_C) \qquad \text{(Eq 10b)}$$

The ranges (x to y) are smaller than (C to F) or (C' to F') and may lay outside of any of these two. For optical glasses it is $0.53 \leq P \leq 0.63$ in the blue and $0.44 \leq P \leq 0.57$ in the red.

The relationship between P- and v-values is important for optimal correction of chromatic aberrations. For normal glasses it is:

$$P = a + b \cdot v \qquad \text{(Eq 11)}$$

with empirical wavelength-dependent constants a and b. Some glasses deviate from this linear behavior and the degree of deviation is identified by the difference:

$$\Delta P_{x,y} = (P_{x,y})_{actual} - (P_{x,y})_{linear} \qquad \text{(Eq 12)}$$

P-values are listed and usually represented graphically in glass catalogs. The ranges of P-values, together with other optical properties, are summarized in Table 3.

The refractive index determines not only the refraction of a light beam but at optical interfaces also the reflectivity (Fresnel reflection). For $\theta_i \leq 50°$ and unpolarized light, the reflectivity r at a glass/air interface is given with good approximation by:

$$r = [(n - 1)/(n + 1)]^2 \qquad \text{(Eq 13)}$$

For precise calculation, particularly for polarized light, the complete Fresnel formulas must be used. Reflectivity at a single surface against air is about 3% for ophthalmic crowns but can reach about 11% for high index glasses ($n = 2.0$).

Nonlinear Refractive Index

For glasses exposed to high intensity optical radiation (for example, elements in laser systems or as laser glasses), the intensity dependence of the refractive index:

$$n(I) = n(E^2) \qquad \text{(Eq 14)}$$

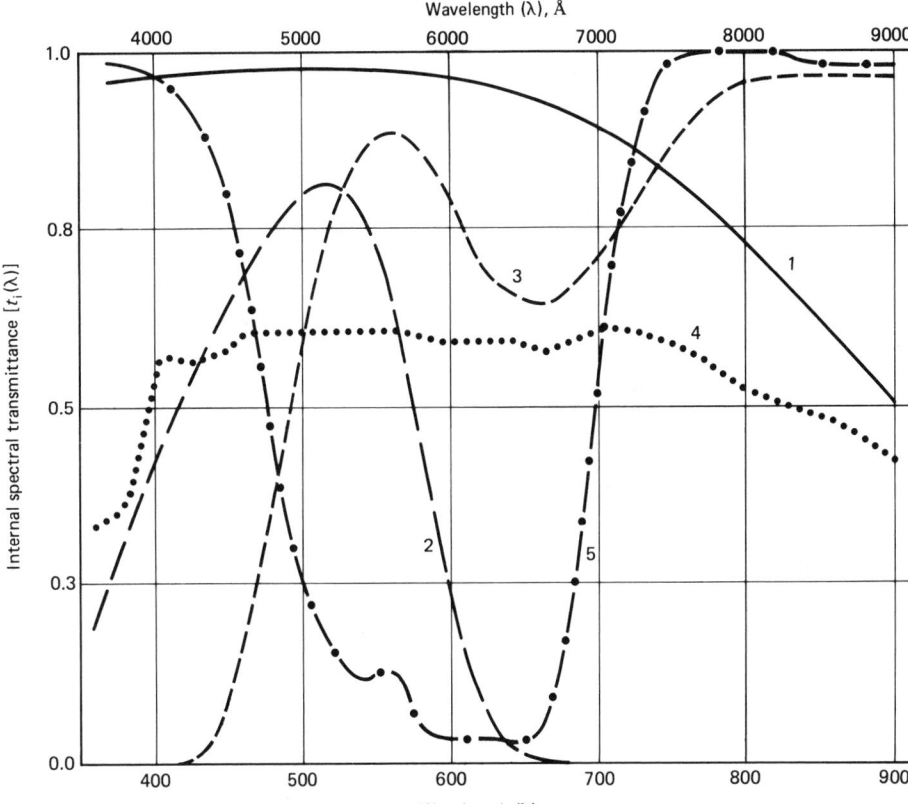

Fig 5 Internal spectral transmittance of selected optical filter glasses with 2 mm (0.08 in.) thickness. 1, Short-pass (heat screen); 2, band-pass (blue); 3, long-pass (yellow); 4, neutral (gray); 5, conversion filter

became important about twenty years ago. A series expansion of $n(E^2)$ yields:

$$n(E^2) \approx n_0 + n_2 \cdot E^2 \qquad \text{(Eq 15)}$$

where n_0 is the index for low intensity; n_2 is the nonlinear refractive index; and E is the electric field strength, usually given in electrostatic units (esu). The nonlinearity causes self-focusing of a laser beam that can lead to local destruction of the glass. For such systems, glasses with low values of n_2 are selected. Glasses with high n_2-values could be important for optical switching. From the compositional viewpoint, potassium-baria-phosphate glasses seem to have the lowest n_2 values while lithium-alumina-silicate glasses the highest n_2 values. Consequently, the more recently developed glasses for high-power lasers are phosphate-based. Several theoretical and empirical formulas were suggested relating n_2 to other glass properties. Since it was introduced in 1976, this author has found the correlation given in Fig 4 the most accurate and useful of any such equation.

Absorption and Color

In the absence of diffuse scattering and fluorescence, the intensity (that is, flux), I_t, transmitted through a planoparallel plate of geometric thickness, h, is with good approximation given by:

$$I_t = I_0 \cdot (1 - r)^2 \cdot 10^{-k \cdot x} \qquad \text{(Eq 16)}$$

where I_0 is the incident intensity; $k(\lambda)$ is the absorption coefficient; and x is the path length in the plate ($x = h$ for normal incidence). This expression neglects multiple internal reflections. Of particular importance for colored filter glasses and sunglasses is the dependence of x on the angle of incidence, θ_i, which, for a glass/air interface is:

$$x = (n \cdot h)/\sqrt{n^2 - \sin^2\theta_i} \qquad \text{(Eq 17)}$$

Two material constants are derived from Eq 16, the transmittance:

$$t = I_t/I_0 \qquad \text{(Eq 18)}$$

and the internal transmittance:

$$t_i = t/(1 - r)^2 \qquad \text{(Eq 19)}$$

Equation 18 includes the index-determined reflection losses, Eq 19 includes only absorption losses inside the glass. It is common to list t_i as a function of wavelength in leading optical and filter glass catalogs for given path lengths (thickness), usually 5 and 25 mm (0.20 and 1.0 in.) for optical glasses, 1 or 2 mm (0.04 or 0.08 in.) for filter glasses. The distinction between t and t_i becomes important if antireflective coatings are applied to lenses for optimized window selection and for systems with cemented lenses. Occasionally, the optical density, D, is expressed as follows:

$$D = -k \cdot h = -\log(t_i) \qquad \text{(Eq 20)}$$

and appears in filter glass catalogs. A quantity formally related to D but not excluding reflection losses is the transmission loss, L, expressed in dB/(unit length)

$$L = 10 \cdot \log(I_t/I_0) = 10 \cdot \log(t) \qquad \text{(Eq 21)}$$

Equation 21 is mostly used to describe losses in optical waveguides. Filter glasses are also characterized by the spectral diabatie value, Θ, given by:

$$\Theta = 1 - \log[\log(1/t_i)] \qquad \text{(Eq 22)}$$

Plotted against the wavelength, diabatie curves present more accurately the high transmission values than do $t_i(\lambda)$ curves because of the extended scale above 80%. Combining Eq 20 and Eq 22 shows an additional advantage: a linear translation of $\Theta(\lambda)$ parallel to the λ-axis permits quick extrapolation to different thicknesses. For many applications and for describing quantitatively the temperature dependence of the transmittance, the cutoff wavelength of optical and filter glasses which is the wavelength at which the internal transmittance reaches 50%, is important.

A key property of the ~100 optical filter glasses available is their spectral transmittance. Examples are given in Fig 5 for a neutral (gray), a color-temperature conversion, a bandpass, a long-wave pass, and a short-wave pass filter. In a wider sense, even clear optical glasses can function as long-wave pass filters (Fig 6).

Filter glasses are produced by adding colorants to a relatively small number of base glasses. The colorants are either metal ions (for example, copper, iron, rare earth metals, cobalt, chromium, and manganese) or colloidal particles of metallic (for example, copper, silver, and gold) or semiconductive nature (for example, cadmium selenides or sulfides).

Colloidally colored glasses develop their color during a well controlled heat treatment. For this reason they are also known by the glass engineering terms temperature colored or striking (struck) glasses. The most widely known among them is the gold-ruby glass leading to terms such as silver, copper, or cadmium-sulfide-selenide ruby although not all of them are ruby red. Silver-ruby glasses are normally yellow. A colorless alkali-lime-silicate glass with about 0.5% CdS and S added to the batch and melted under reducing conditions can yield a yellow, orange, or deep red glass depending on subsequent heat treatment. The term semiconductor doped glasses is used for a subset of the colloidally colored

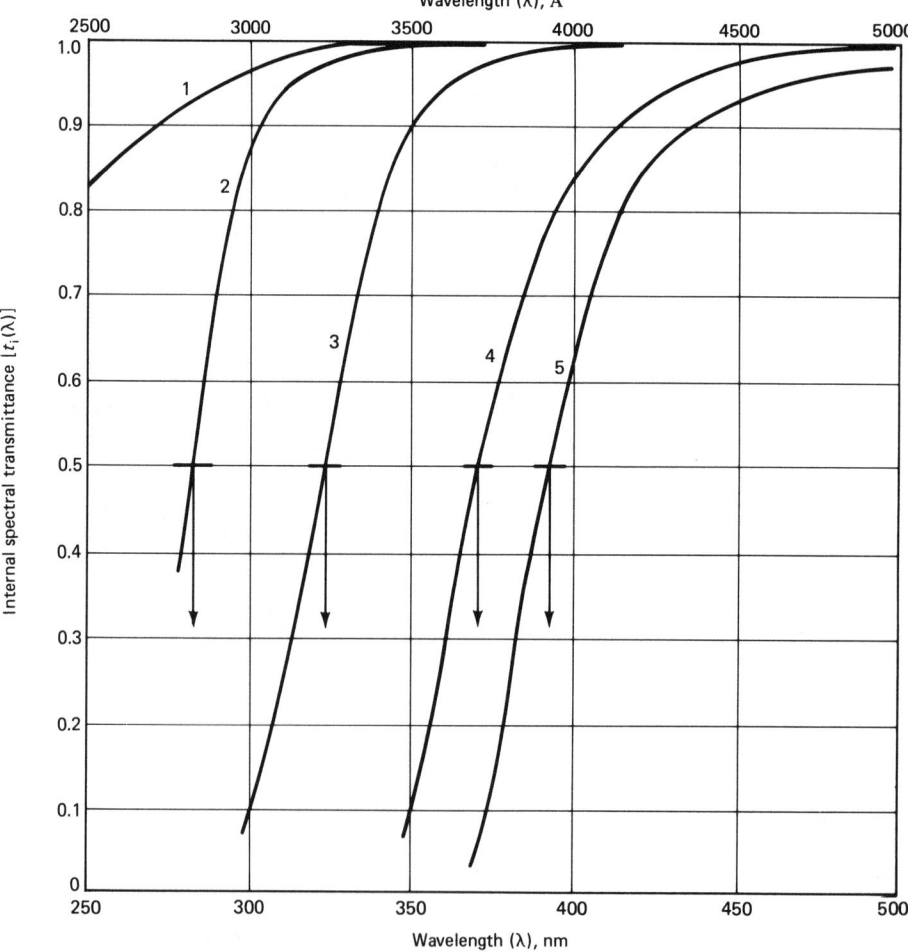

Fig 6 Short wave internal transmittance cutoff of some optical glasses of 5 mm (0.2 in.) thickness. 1, 548 743 (a UV transmitting glass that only recently became commercially available); 2, 487 704; 3, 486 818; 4, 807 316; 5, 953 204. In general, the higher the refractive index the farther the cutoff is shifted towards the red. Cutoff wavelengths are indicated by the arrows.

glasses in papers dealing with their recently rediscovered potential for optical switches (Ref 7, 8) of some of them. Photochromic glasses are related to the ruby glasses insofar as the silver-halide particles responsible for the photochromic effect are developed by heat treatment of pressings or plates. The nature of colloidally colored glasses does not allow extensive heating (required, for example, for slumping) but offers the advantage of producing articles with color gradients or gradient photochromic behavior. The spectral transmittance of the colloidally colored glasses can be understood by applying Mie's scattering theory that includes not only the wavelength dependent and the complex refractive index of the particles relative to the index of the matrix glass but also the shape and size distribution of the particles. Because of the dependence of the spectral transmittance on these parameters, it is possible to produce lenses with various tints from regular photochromic glasses by heating them in a reducing atmosphere. During this process, the silver-halide particles in the surface are reduced to metallic silver and also grow in size. One example is the formation of a thin layer of yellow silver ruby in the surface. Other colors (for example, the violet-purple tints known from aqueous colloidal silver solutions) are possible.

REFERENCES

1. "Prescription ophthalmic lenses, recommendations," ANSI Z80.1, American Standard for Ophthalmics, 1987
2. "Low vision aids," ANSI Z80.9, American National Standard for Ophthalmics, 1986
3. "American National Standard Practice for Occupational and Educational Eye and Face Protection," ANSI Z87.1, 1989
4. "Nonprescription Sunglasses and Fashion Eyewear—Requirements," ANSI Z80.3, American National Standard for Ophthalmics, 1986
5. E.W. Deeg, Optical Glasses, *Commercial Glasses*, Vol 18, *Advances in Ceramics*, D.C. Boyd and J.F. MacDowell, Ed., American Ceramic Society, 1986, p 9–34
6. H. Hack and N. Neuroth, Resistance of Optical and Colored Glasses to 3-nsec Pulses, *Appl. Optics*, Vol 21, 1982, p 3239–3248
7. G. Bret and F. Gires, Giant-Pulse Laser and Light Amplifier using Variable Transmission Coefficient Glasses as Light Switches, *Appl. Phys. Lett.*, Vol 4, 1964, p 175–176
8. E. Deeg, Ueber die Ursache der Abhaengigkeit der Lichttransmission gewisser Anlaufglaeser von der Lichtintensitaet, *Glastechn. Ber.*, Vol 40, 1967, p 177–181

Glass Containers

Peter J. Vergano, Department of Food Science, College of Agricultural Sciences, Clemson University

GLASS CONTAINERS are used to package and store a wide variety of foods, beverages, chemicals, pharmaceuticals, and cosmetics. In addition to providing information on these application areas, this article also reviews the types of glasses used for containers, design and strength requirements, surface treatments, and quality and regulatory requirements.

Compositions and Colors Used for Glass Containers

Base Compositions. The two base compositions used for glass containers are soda-lime glass and borosilicate glass. Soda-lime glass is used for all but a few specialty types of containers, primarily in pharmaceutical applications.

Typical compositions and properties of soda-lime and borosilicate container glasses are listed in Table 1. Due to the higher melting temperatures required for borosilicate glass, it is much more expensive than soda-lime glass. Borosilicate glass is used primarily for containers requiring its greater chemical durability.

Colors. Mass-produced glass containers are available in a limited range of colors, primarily because of manufacturing limitations imposed by the scale of facilities needed to keep costs low and competitive.

The colors in which glass containers are generally available are flint (clear), amber (brown), and emerald green (for example, 7-Up bottles). Raw materials selected for flint glass production have specifications to minimize concentrations of colorants such as transition metal oxides. Iron oxide levels, in particular, are kept low to minimize undesirable green or blue tints in flint glass. Amber glass is colored with higher iron and sulfur concentrations. The amber color is designed to meet and exceed the United States Pharmacopeia (USP) requirements for ultraviolet light absorption. Amber is the color of choice for pharmaceutical products and foods that are sensitive to ultraviolet light. Green glass is produced with iron oxide and chromium oxide colorants. Green glass is generally chosen for aesthetic reasons, for example, with yellow or clear liquids.

Glass containers in large numbers are produced in a variety of colors around the world. Light blue bottles for mineral water, for example, and a variety of green-brown colors are often produced for beer and wine bottles overseas.

Limited quantities of bottles in colors such as Georgia green (such as refillable returnable Coca-Cola bottles) and "dead leaf" green (wine bottles) are often produced by a process that stirs colored frit into molten glass as the glass is delivered from a furnace to a bottle forming machine. Still more limited quantities of specialty colors including opal (white) are produced for cosmetic ware for aesthetic effects.

Design/Shape of Containers

Parts of the Container. Illustrations of common glass containers are shown in Fig 1 with the parts of the containers identified. The top of the container to which the closure is fastened is the "finish" of the container. Other parts of the container include the neck, shoulder, body, heel, and bottom. The upper and lower contact points of the container shown in Fig 1(a) are also indicated. The contact points are the points at which two bottles of the same design will be in contact when placed side-by-side, for example, when placed on filling lines.

Design Considerations. The design of a glass container is based on a number of functional requirements including:

- Manufacturability
- Handling, filling, and processing conditions
- Aesthetics and tradition

The design drawing for a syrup bottle is presented in Fig 2. The design of this bottle is suitable for manufacture because it has proportions within the design range of manufacturing equipment. The more closely a bottle approximates a lightbulb shape, the more suitable it is for manufacture. The lower its center of gravity, the greater its resistance to tipping over, a handling problem in manufacture. This syrup bottle does not have the same diameter at its upper and lower contact points. This design feature would be detrimental to handling on high-speed filling lines.

This syrup bottle has a wide finish suitable for rapid filling. It is round and will rotate in a filling line to minimize binding against other containers. The bottle is designed to withstand the thermal shock of hot-filling syrup, but it is not optimized for pressure or impact strength.

The overall shape of the container has been set by tradition and aesthetics. It has a shape which is familiar to consumers as a liquid food bottle. A design change away from this shape would risk a loss in market share.

Containers whose design to a greater degree involves strength considerations are discussed below in the discussions on pressure ware and retorted ware.

Finish Designs. Detailed design and adherence to design dimensions is most critical in the finish of a glass container. Finishes are designed to accommodate a specific closure, ranging from the simplest crown finish for crimped metal caps to lug finishes suitable for tamper-evident button closures. The two most important objectives in the design of a finish and its closure are (1) maintenance of seal integrity through all process and distribution steps after capping, and (2) ease of removal by the user. Ease of removal is measured by determining the torque needed to rotate a cap off a finish.

Figure 3 presents several common finish designs. Figure 4 is a generalized finish design drawing which indicates the letters assigned to common finish dimensions. For example, the diameter of the finish measured between the outside of the threads is the "T" dimension.

Strength and Surface Treatments

Strength Determined by Surface Condition. As with all glass products, the strength of defect-free containers is determined by the condition of the glass surface. A pristine or acid-etched surface will exhibit a tensile fracture strength as high as 7 GPa (1×10^6 psi). However, glass subjected to normal handling will exhibit a tensile fracture strength on the order of 70 MPa (10 ksi). Normal handling brings glass containers in contact with metal,

Table 1 Glass container compositions and properties

	Soda-lime	Borosilicate
Composition, wt%		
SiO_2	70.0–74.0	80.5
$Na_2O + K_2O$	13.0–16.0	3.8
		0.4
$CaO + MgO$	10.0–13.0	…
		…
BaO	0–0.5	…
B_2O_3	…	12.9
Al_2O_3	1.5–2.5	2.2
Properties		
Annealing point, °C (°F)	548 (1018)	565 (1050)
Coefficient of thermal expansion from 0 to 300 °C	9×10^{-6}/°C	3×10^{-6}/°C

Source: Ref 1

other materials and each other following forming. Each contact of a glass surface with another surface, or even with dust, is believed to result in surface abrasions which lower glass strength.

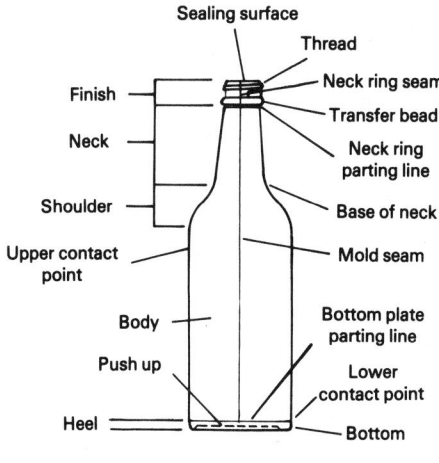

(a)

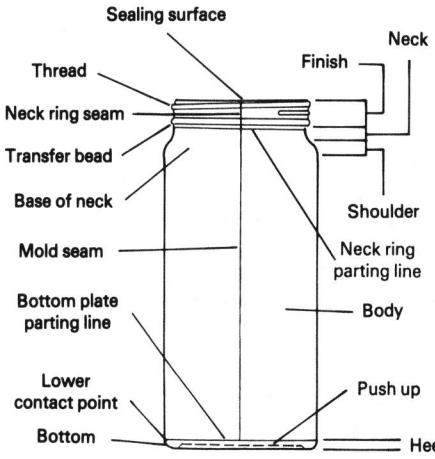

(b)

Fig 1 Bottle parts identification for (a) narrow neck and (b) wide mouth ware

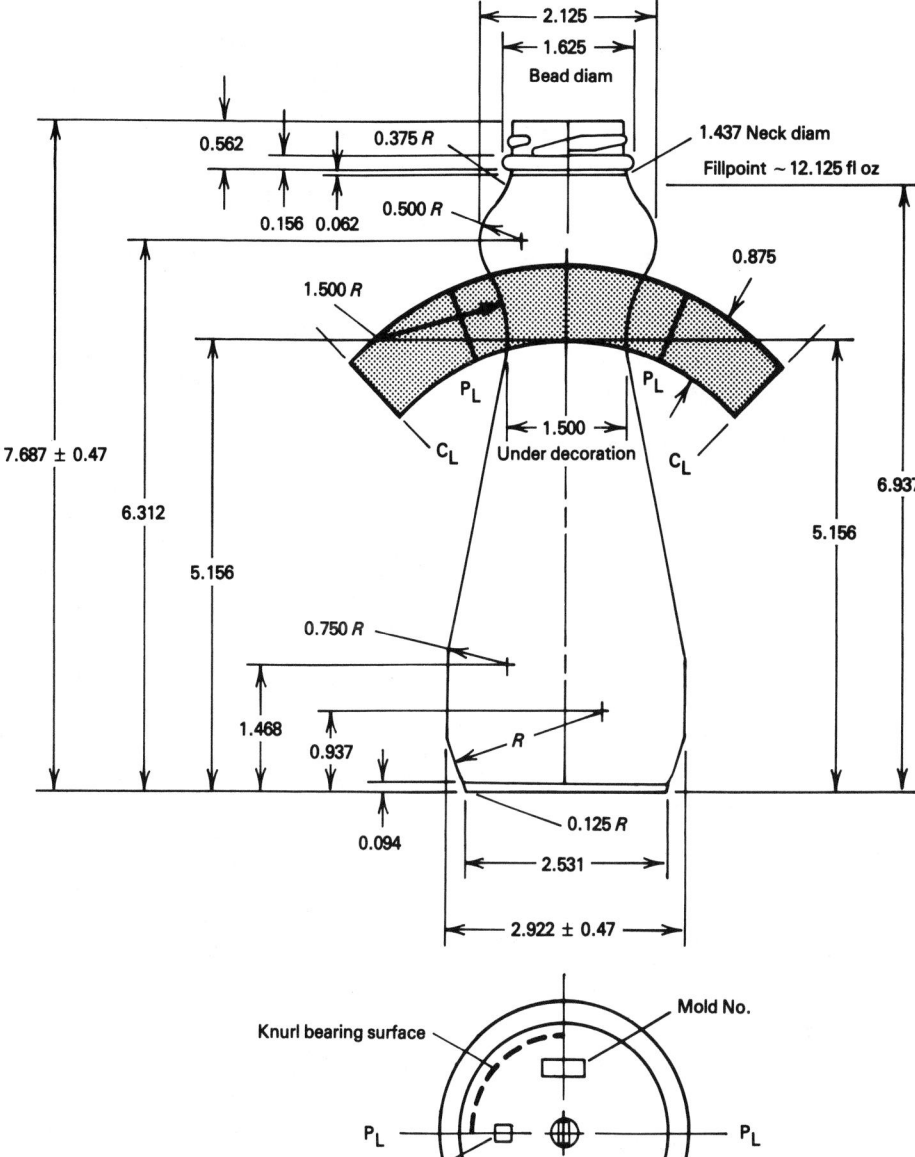

Fig 2 Engineering design drawing for a syrup bottle. The lower part of the figure shows the manufacturer's identification markings located at the bottom or heel of the container.

The use of lubricous coatings on glass containers has been found to preserve glass strength. Without lubricous coatings, glass containers tend to "bind" and scratch one another as they proceed down a filling line. With coatings, the containers do not bind and they rotate freely. Without lubricous coatings, glass containers could not be filled at rates of more than one thousand bottles per minute.

The choice of surface treatment for use on a particular glass container depends primarily on two considerations. These include (1) the filling line rate and conditions, and (2) the adhesive and label style. The most commonly used surface coatings are a combination of tin oxide and polyethylene and a combination of tin oxide and oleic acid. Polyethylene provides greater lubricity for high-speed filling lines. Oleic acid provides sufficient lubricity for slower filling lines and allows the use of less expensive adhesives in labeling. The quality of surface coatings is often gaged by rubbing two bottles together by hand. Improperly coated bottles will not slide freely over each other. Quantitative tests of surface

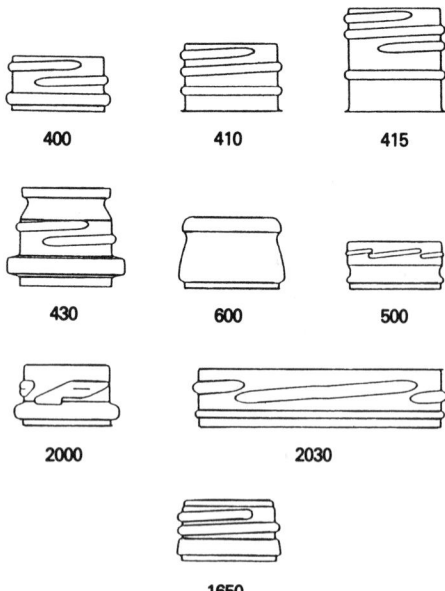

Fig 3 Common finish designs for glass containers. Numbers represent finish identifier.

coating quality utilize measurements of slip angle or scratch resistance. The slip angle determination utilizes two bottles lying on their sides and a third bottle resting on them. The three bottles are held in a fixture that can be tilted around a horizontal axis parallel to the bottom of the bottles. By raising the fixture, the angle at which the top bottle will slip off the bottom two bottles can be measured. This is a slip angle. Scratch resistance is measured with a device that applies a load to one bottle held with its sidewall resting on the sidewall of a second bottle. The load needed to cause scratching when the bottles are rubbed together is the scratch resistance of the coated bottles.

Surface coatings are applied to essentially all glass containers. The only important exceptions are returnable refillable beer and soft drink containers. As part of the cleaning process prior to refilling, glass containers are washed in caustic solution. The caustic attacks tin oxide, causing an undesirable pearlescent or "oily" appearance.

Glass container strength is routinely monitored only in the production of pressure ware (such as bottles for carbonated beverages).

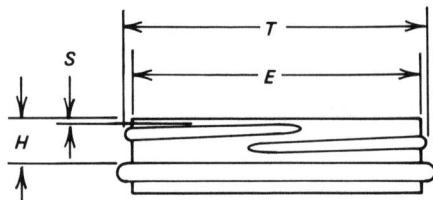

Fig 4 Finish design drawing showing the dimensions of the finish. *T*, diameter of thread; *E*, wall diameter; *H*, intersection of *T* with bead or shoulder; and *S*, start of thread

Designing for pressure strength is described in the section on pressure ware below.

Strengthening of Glass Containers. The two methods of strengthening glass that are utilized extensively in flat glass are thermal tempering and ion exchange ("chemical tempering"). These processes, however, are not generally used to strengthen glass containers. The relatively thin walls (as low as 0.75 mm, or 0.030 in., as described in the section on Voluntary Products Standards) and the complex geometry of containers limits the effectiveness of thermal tempering. Some thermal tempering strength enhancement does occur in most containers since they are not generally annealed to the point that all internal stresses are relieved. However, this strengthening is incidental in normal container manufacture.

Ion-exchange processes have been developed for soda-lime containers. The ion-exchange process for containers produces containers with a surface layer which is in compression. This compressive stress must be overcome by the application of tensile stress before the glass surface is subjected to a tensile stress. Glass containers with fracture strengths double or triple the "rule of thumb" practical fracture strength of 70 MPa (10 ksi) are produced by this process.

Glass containers strengthened by ion-exchange are not in general production because containers with sufficient strength are produced without this process and because localized defects that are the cause of failure in most weak containers produce some of the same weaknesses in ion-exchanged glass containers.

Applications

Pressure Ware. More than 50% of glass container production is devoted to containers for beer and soft drinks. Both of these products are carbonated. They subject their containers to internal pressures of up to 5 atm (500 kPa, or 75 psi) at room temperature and higher pressures in storage at higher temperatures. Containers for pressure ware normally carry a specification for a minimum internal pressure strength of 1380 kPa (200 psi) or greater.

Because beer and soft drink containers are high-volume production items, considerable attention has been focused on the development of design methods for them. A method of calculating a theoretical pressure strength for a container based on the geometry of the container has been developed. The key parameters in this theory are stress intensity factors. The stress intensity factor for each part of a glass container is determined primarily by the glass thickness and the radius of curvature of that part. Finite element analysis methods have also been developed for pressure ware design use.

Retort Ware. Glass containers utilized for products which are sterilized after filling and

sealing (generally low-acid foods) are subjected to pressure and temperature change conditions which limit design flexibility.

The sterilizing, or retorting, equipment and time-temperature schedule is selected to avoid exposing the glass container and its closure to high differentials in pressure and to avoid exposing the glass container to rapid changes in temperature. High differential pressures, particularly at high temperatures (for example 120 °C, or 250 °F) can produce product leakage through the closure seal. Rapid temperature changes, particularly in cooling, can cause thermal shock breakage of the glass. Glass containers for retort application generally specify an instantaneous maximum temperature drop of 40 °C (70 °F). Temperature drops of this magnitude are not unusual in retorting foods, where the process efficiencies are improved and residual viable microorganism growth is minimized by rapid cooling after retorting.

The familiar "spaghetti jar" shape for retorted foods, including many baby foods, is chosen to meet these requirements.

Liquor and Wine. Glass containers for noncarbonated wine and for liquor are generally designed for aesthetics and ease of dispensing rather than processing conditions.

White goods, that is, liquors such as gin and vodka, require a special internal treatment of their soda-lime containers to avoid the formation of $CaSO_4$ spicules. Although glass containers are accurately considered to be essentially free of product interactions, with these liquors this is not the case. The alcohol in these products reacts with the glass surface to produce crystals that are noticeable only if the liquor is agitated to raise the spicules from the container bottom.

In order to avoid this problem, liquor bottles are internally treated with a fluorine-containing gas at elevated temperatures (for example, 540 °C, or 1000 °F) to leach from the glass surface atoms susceptible to interaction with the product.

Hot Fill/Cold Fill Products. Glass containers used for products filled at temperatures at or below 100 °C (212 °F) and that require no retorting allow greater flexibility in design. The typical salad dressing jar, for example, is designed for ease in dispensing and to provide maximum area for an attractive label.

Pharmaceutical Ware. Glass containers used for pharmaceutical products are required to meet standards described in the USP. Glass containers are designated Type I, Type II, Type III, or Type NP, ranked in descending order of their chemical durability. The procedures for determining the chemical durability (the resistance to water attack) of glass containers are specified in the USP, Section XXI (661), Containers.

The durability test limits for the four types of glass are listed in Table 2. Type I requirements generally are met by the use of borosilicate glass containers. Preparations for use

Table 2   Glass types and test limits

Type	General description(a)	Type of test	Limits Size, mL(b)	mL of 0.020 N acid
I	Highly resistant, borosilicate glass	Powdered glass	All	1.0
			≤100	0.7
II	Treated soda-lime glass	Water attack	>100	0.2
III	Soda-lime glass	Powdered glass	All	8.5
NP	General-purpose soda-lime glass	Powdered glass	All	15.0

(a) The description applies to containers of this type of glass usually available. (b) Size indicates the overflow capacity of the container.

by parenteral administration utilize Type I glass containers. Type II glass, generally treated soda-lime glass, can also be used for acidic and neutral parenteral preparation and alkaline parenteral preparations for which their stability has been demonstrated.

Type III chemical durability requirements are generally met by soda-lime glass containers produced for the packaging of food and beverages. The chemical durability requirement for Type NP glass containers is easily met by essentially all mass-produced soda-lime-silicate glass containers.

Pharmaceutical ware which is offered as light-resistant ware must meet USP specifications for light transmission. The limits for light transmission are presented in Table 3.

Cosmetic. Glass containers for cosmetics, exemplified by bottles for perfume, emphasize aesthetics. Often these containers are designed by product designers unfamiliar with glass container manufacturing methods. Acute corner angles, for example, are often requested. The design of the container actually manufactured is a compromise between the designer and the bottlemaker.

A high premium is often placed on smooth surfaces on perfume bottles. Highly polished molds, sometimes coated by nickel plating, are used to produce smooth surfaces. An alternative process flame polishes the glass surfaces after forming.

Chemical/Acid. Glass containers used for packaging chemicals are often stock designs. That is, they are designs produced in large

Table 3   Limits for glass Types I, II, and III containers

Nominal size, mL	Maximum percentage of light transmission at any wavelength between 290 and 450 nm Flame-sealed containers	Closure-sealed containers
1	50	25
2	45	20
5	40	15
10	35	13
20	30	12
50	15	10

Note: Any container of a size intermediate to those listed above exhibits a transmission not greater than that of the next larger size container listed in the table. For containers larger than 50 mL, the limits for 50 mL apply.

quantity by container manufacturers but sold to many low-volume chemical manufacturers through distributors.

Bottles for acids and other hazardous chemicals are also available with a coating of ionomer polymer which is a fragment retentive coating.

Regulatory Requirements

Food and Drug Administration (FDA). Although the phrase "approved by the FDA" is still commonly applied to food and drug packaging materials, often it refers to an outdated concept. In order to meet FDA requirements for the introduction of a new packaging material for a food product, the user is required to run tests, for example, to measure the extent of contamination of the product by the material. These test results do not need to be submitted to the FDA. The material manufacturer generally does not have a letter on file from the FDA indicating approval. The material manufacturer, however, will provide users of their product with a letter indicating the product "meets FDA requirements" or a similar statement if the user's application is appropriate for the material.

For drug packaging, FDA approval is required.

Voluntary Product Standards (VPS). In 1977, the Glass Packaging Institute, Inc., and the National Soft Drink Association issued standards for manufacturing requirements for carbonated soft-drink glass containers. The standards were developed under procedures specified by the U.S. Department of Commerce. The standards were revised in 1989. They are published under the title "Voluntary Product Standards PS 73–89 Glass Bottles for Carbonated Soft Drinks." The standard is available from the Superintendent of Documents, U.S. Government Printing Office, Washington, DC, 20402.

The VPS is intended for soda-lime-silicate containers with capacities up to 1065 mL (36 fl oz) for soft drinks carbonated up to a maximum of five volumes. The VPS is applicable for nonrefillable bottles (both labeled and prelabeled by the bottle manufacturer) and refillable bottles.

The VPS provides manufacturing requirements for temper number, thermal shock resistance, detection of visual defects, wall thickness, dimensional tolerances for height

and maximum outside diameter, tolerances for capacity and mass (weight), internal pressure strength, impact resistance, abrasion resistance, perpendicularity, bottom characteristics, and bottle identification.

Most of the standards do not distinguish among nonrefillable, prelabeled nonrefillable, and refillable ware. The specification for minimum wall thickness, however, does distinguish among these three categories of bottles. The prelabeled nonrefillable bottle specification calls for a minimum thickness of 0.75 mm (0.030 in.). For bottles which are nonrefillable but not prelabeled, the minimum thickness is 1.1 mm (0.045 in.).

Quality Considerations

Glass container manufacturers lead the rigid container industry in the sophistication of electro-optical equipment used for on-line inspection of their product. There are measures of quality for glass containers of which the user should be aware. These are briefly described below.

Annealing Grade. The annealing grade of glass containers is a measure of residual stresses in the glass caused by differential cooling during container forming. The annealing grade is measured using equipment and procedures described in ASTM C 148 (see "Standard Methods of Polariscopic Examination of Glass Containers" in Volume 15.02).

Cord is the term applied to inhomogeneous, off-composition regions in the glass. In flat glass, this same characteristic is known as ream. The presence of cord does not render a glass container unsuitable for use. However, if present in large quantities or if the cord is far off in terms of composition, cord can produce stresses which weaken the container.

Seeds and Blisters. Seeds are fine bubbles in the glass. They may be filled with gases or vacuum. Generally, seeds are an aesthetic defect, although seeds located in critical areas of the finish can be a serious problem. Blisters, bubbles larger than 0.8 mm ($^1/_{32}$ in.) in diameter, can be stress concentrators which weaken the container.

Thin-walled blisters on the inside surface of the container can result in pieces of glass in the product.

Stones are crystalline inclusions in the glass. The severity of a stone defect depends on its size and the difference in thermal expansion between the stone and the glass. Although some types of stones are not detrimental to glass strength, others initiate fractures in the glass.

Checks and splits are small cracks in the glass. Generally they are characterized by location in the container (such as a heel check or a split finish). Any check or split in a container, no matter how small, is cause for rejection.

Geometric Deficiencies. Deficiencies in the geometry or shape of parts of glass containers is a quality category which covers a wide range of defects. Often, these defects concern the finish or the seams of the container. Finish defects can result in poor sealing of the container. Seam defects can be weak points or fracture origins.

Various surface marks appear on glass containers as a result of contacts between the hot or cold glass with other materials. Often, these marks are of aesthetic concern only. Checks or splits associated with marks are serious defects.

Defect Levels. Defects in glass containers are generally categorized as critical, serious, or appearance defects. Critical defects generally are those that result in pieces of glass in the product in the container. Glass protrusions inside the container which can be broken off in filling, for example, are critical

defects. Serious defects are those which often produce glass breakage in filling and handling, resulting in production downtime. Appearance defects are those which do not affect the performance, the strength, or functioning of the glass container.

Manufacturers Identification

The trade association of glass container manufacturers, the Glass Packaging Institute, has adopted a standard system for identification of a glass container's manufacturer, the plant in which it was made, and its mold number. The system involves symbols and code numbers located on the bottom or around the heel of the container (see Fig 2).

SELECTED REFERENCES

- *Glass Bottles for Carbonated Soft Drinks*, Voluntary Product Standard PS 73–89, U.S. Department of Commerce, National Institute of Standards and Technology, 1989
- M.J. Renaud, Ed., *Brockway Glass School Participants Manual*, Brockway Glass Containers, 1987
- R.J. Ryder and J.P. Poole, Container Glass, *Commercial Glasses*, D.C. Boyd and J.F. MacDowell, Ed., American Ceramic Society, 1986, p 35–41
- G.W. McLellan and E.B. Shand, *Glass Engineering Handbook*, 3rd ed., McGraw-Hill, 1984
- *The United States Pharmacopeia, 21st Revision*, 1985

Glasses for Laboratory and Process Applications

Pronob Bardhan, Robert M. Morena, and David L. Tennent, Corning Incorporated

GLASS REACTION VESSELS, containers, and components such as joints, supports, or tubing are commonly used in the laboratory because of their robust thermal, chemical, and mechanical properties. The chemistry and the properties of the glass containers, microscope slides/cover slips, and strengthened glass labware are the focus of the first section of this article. Subsequent sections emphasize the chemistry and applications of high surface area glasses (for example, controlled pore glass) in chromatography, the nature of ion selective glass electrodes, and details on the use of glass systems in process scale equipment.

Laboratory Glassware

Glassware in the laboratory is required to have good resistance to both chemical attack and thermal shock. Because of these dual demands, glasses for labware are typically grouped into several distinct classifications depending on coefficient of thermal expansion and corrosion resistance. The American Society for Testing and Materials divides laboratory glassware into three classes (Ref 1):

- Type I, class A: a low-expansion borosilicate glass (0 to 300 °C, or 32 to 570 °F expansion of 32 to 33 $\times$ 10^{-7}/°C, or 18 $\times$ 10^{-7}/°F) with excellent chemical durability (measured on powdered samples in very pure water under autoclave conditions)
- Type I, class B: an aluminoborosilicate glass with moderately low expansion (48 to 56 $\times$ 10^{-7}/°C, or 27 to 31 $\times$ 10^{-7}/°F) and excellent chemical durability
- Type II, classes A or B: a soda-lime-silica glass with moderate expansion (90 to 93 $\times$ 10^{-7}/°C, or 52 $\times$ 10^{-7}/°F) and moderate chemical durability

The low coefficient of expansion of the borosilicate glasses results in a high resistance to thermal breakage, which is frequently an important time-saving factor in laboratory testing because the glassware can be heated and cooled quickly. The aluminoborosilicate glass has a higher thermal expansion than the borosilicate glass but is comparable in chemical durability. Soda-lime-silica glasses (which formed the great bulk of laboratory glassware at the start of the 20th century) are used in applications in which some chemical contamination (such as alkali leaching) can be tolerated or where rapid heat-up or cool-down is not required. Nonglass competitive materials include a range of polymers and metals such as stainless steel. However, neither of these material groups combine the properties of transparency, relatively-high continued-use temperature (about 450 °C, or 840 °F), chemical durability, and low density that glass possesses. The composition and properties of some commercial glasses are shown in Table 1 (Ref 2).

Microscopic Slides and Cover Glasses. Glass for microscope slides and cover glasses must be a low-cost expenditure yet possess good chemical durability as well as the capability to be formed to close dimensional tolerances without an undue number of defects. These requirements are especially stringent for microscope cover glasses because cover glasses must be free of undue optical distortion as they form part of the optical path during specimen viewing. Both microscope slides and cover glasses must also be water-white (that is, clear and colorless) so that no interfering color effects are present. Glass for microscope slides is typically soda-lime-silica. However, in applications where corrosion may be a problem, borosilicate glass is used. Soda-lime-silica glass is generally not suitable for microscope cover glasses because of the difficulty in obtaining both exceptional thinness and uniformity during the drawing process. Cover glass compositions must resist weathering or corrosion by atmospheric gases such as water vapor and carbon dioxide. Because of the extreme dimensional flatness of cover glasses, these products are frequently coated with a very thin layer of a nonreactive agent such as ammonium chloride to minimize sticking together of the cover glasses. The composition and the properties of microscope slides and cover glasses are shown in Table 1.

Strengthened Glassware. Certain labware is available as strengthened glassware. Articles available in strengthened form are those which are either subjected to high stresses in service (centrifuge tubes), or, as in the case of pipettes, are prone to impact failures because of their long shape. Because of the geometry of these particular articles (small mouths and long tubes), strengthening is generally accomplished by chemical ion-exchange processes (see the article "Ion Exchange" in this Volume) rather than by thermal tempering. One glass manufactured for strengthened labware applications is an alkali-aluminosilicate that contains Li_2O. Strengthening in this instance involves treatment in a $NaNO_3$-Na_2SO_4 bath for several hours at a temperature just below the strain point of the glass. The resulting sodium for lithium replacement in the glass produces strengthening by creating a compressive surface stress in the glass workpiece. The glass composition is approximately 66 wt% SiO_2, 20 wt% Al_2O_3, 9 wt% Na_2O, and 5 wt% Li_2O. The final properties of the glass are a function of the chemical tempering process and thus are not listed in Table 1.

The mechanical strength of labware tempered in this fashion is several times that of comparable standard (nonstrengthened) borosilicate glassware. The additional process steps involved in the strengthening treatment result in a two- to three-fold increase in cost over that of the comparable borosilicate ware. Because of the surface compressive stress, strengthened borosilicate glassware is more resistant to scratching and resulting loss of optical clarity than is standard borosilicate glass. The alkali-aluminosilicate glass mentioned previously in this section also possesses better resistance to alkali and acidic environments than does borosilicate glass. Because the strengthening mechanism in strengthened labware depends on a residual stress state, these materials must not be used at temperatures anywhere near their strain point. The ion-exchange strengthened aluminosilicate glass has an upper use temperature of 300 °C (570 °F).

High Silica Glass. Labware for particularly high-temperature applications or thermal shock applications are formed by a special type of borosilicate. The starting Na_2O-B_2O_3-SiO_2 glass is subjected to a heat treatment that results in phase separation in the glass (see the section "Controlled Pore Glass" in this Article) to two glassy phases: one phase rich in Na_2O and B_2O_3 and another phase rich in SiO_2. The former has extremely poor dura-

Table 1 Composition and properties of selected glasses used for laboratory glassware applications

Material	Composition, wt%									Density		Young's modulus	
	SiO$_2$	B$_2$O$_3$	Al$_2$O$_3$	Na$_2$O	K$_2$O	CaO	MgO	ZnO	TiO$_2$	g/cm^3	lb/in.^{3}	GPa	10^6 psi
Borosilicate glass:													
Low-expansion	81	13	2	4	...	...	...	...	...	2.23	0.0806	63	9.1
Alumina	72	11	6	7	1	1	...	...	...	2.36	0.0853	...	...
Soda-lime glass	73	...	2	14	...	7	4	...	...	2.40	0.0867	...	...
Zinc-titania cover glass	65	9	2	7	7	...	...	7	3	2.57	0.0929	74	10.7
High-silica glass(a)	96.5	3	0.5	...	...	...	...	...	...	2.18	0.0788	73	10.6

Material	Coefficient of thermal expansion				Strain point		Annealing point		Softening point		Working point		Refractive index
	0–300 °C (32–570 °F)		25 °C (75 °F) to setting point										
	10^{-7}/°C	10^{-7}/°F	10^{-7}/°C	10^{-7}/°F	°C	°F	°C	°F	°C	°F	°C	°F	
Borosilicate glass:													
Low-expansion	32.5	18.1	35	19	510	950	570	1060	821	1510	1252	2286	1.474
Alumina	50	28	53	29	533	991	576	1070	795	1465	1189	2172	1.491
Soda-lime glass	89	49	...	...	511	952	545	1015	724	1335	...	...	1.515
Zinc-titania cover glass	74	41	84	47	508	946	550	1020	720	1330	1008	1846	1.52
High-silica glass(a)	7.5	4.2	...	...	890	1635	1020	1870	1530	2785	...	...	1.458

(a) Vycor

bility and can be extracted by a chemical dissolution treatment. The residual glass is then consolidated, and all micropores are eliminated (bodies such as beakers, crucibles, dishes, flasks, ball-and-socket joints, process and water still heaters, tubing, and rods are made by this process). The composition and properties of the material (Vycor) are shown in Table 1.

Silica-Based Supports for Chromatography Applications

Silica used in chromatographic applications is prepared by a number of techniques resulting in several shapes and sizes. Irregular silicas are prepared from diatomaceous earth or by the precipitation of silicic acid. Spherical silicas are prepared by the agglomeration and consolidation of smaller particles into larger particles. Both irregular and spherical-shaped particles are available in 30 to 70 µm (0.0012 to 0.0021 in.) particle size. These materials can either be surface coated (pellicular) with a stationary phase or have a porous silica layer deposited over a solid core. Porous particles of this size are avoided because the volume of pores deep within a large particle represents a stagnant mobile phase. This diffusion-limited accessible volume results in poor separations. However, small porous particles (<20 µm, or 800 µin.) are used extensively because of the higher relative stationary phase per unit volume when compared to pellicular particles. This results in higher loadings and better detection sensitivity. All types of silica are available with a variety of silane treatments to create the reverse phase environment needed for a given separation (Ref 3). In commercially available silicas, the pore size ranges from 4 to 110 nm (40 to 1100 Å) with most materials having an average pore diameter of 6 to 10 nm (60 to 100 Å).

Controlled pore glass (CPG) is prepared by a heat treatment to phase-separated borosilicate glass (see the article "Porous and Reconstructed Glasses" in this Volume). The glass is leached with an acid to remove the soluble boric-acid-rich regions (Ref 4). This creates a porous structure with pore diameters ranging from 3 to 6 nm (30 to 60 Å). Larger pores of up to 400 nm (4000 Å) can be created by a mild caustic treatment (Ref 5). The pore size created is a function of the time and temperature of the heat treatment used to cause the phase separation and subsequent coarsening of the phase-separated microstructure. Thus, the coarser the sodium-borate-rich phase, the coarser is the resulting pore size.

This high-purity material has several advantages for chromatography:

- It has very low levels of transition metal ions to irreversibly bind biomolecules. As a result, adsorption occurs only through the silanols or the Lewis acid sites (Ref 6)
- Because CPG material can withstand temperatures to 500 °C (930 °F), any organic residue can be easily removed. Obviously, CPG material can also be sterilized without deterioration
- It can withstand high pressures without compression. The compression of organic resins leads to high back pressures in high-pressure liquid chromatography (HPLC) and in large preparative columns. In a related area, CPG material does not absorb organic solvents and, as a result, does not swell. In the presence of several solvents, polymeric resins can swell leading to compression
- The CPG surface can be modified with coupling agents to create the desired surface properties. The surface can be modified with the proper silane to create environments that are strongly hydrophobic, hydrophilic, reactive, or which will bind metal ions
- Processing can control the pore size so that a reasonably narrow distribution of pore diameters is obtained (Ref 7)

One concern regarding CPG material in chromatographic applications is its resistance to caustic. The solubility of controlled pore glass is a function of time, temperature, pH, solution content, and surface area. While the nominal durability of CPG material is essentially that of silica and is very good, the high surface area results in a higher than normal solubility. While CPG material is most durable in the acidic to neutral pH, the durability decreases at higher pH. However, the addition of hydrophilic silane coupling agents dramatically improves durability (Ref 8).

Applications of chromatography with controlled pore glass have been demonstrated in several areas including thin layer and liquid chromatography. One of the earliest demonstrations for controlled pore glass in chromatography was by MacDonnell. In this application, a thin sheet of CPG material was used like a thin-layer chromatography plate to characterize water-soluble inks (Ref 6).

In the area of liquid chromatography, the surface chemistry plays an important role. At times, it is important to deactivate the Lewis acid sites created by the residual boron that is concentrated on the surface of the pores. These sites will strongly attract polar molecules and can result in poor separations. The boron can be removed by further leaching of the glass.

As mentioned above, the CPG surface can be modified with silanes. Lynn and Filbert (Ref 7, 8) have described the application of several silanes to CPG material to achieve a variety of functional surfaces. These chromatographic supports were stable up to 300 °C (570 °F). Filbert and Weetall (Ref 5) describe, in detail, the chemistry required to use CPG material treated with an aminopropyl silane in affinity chromatography applications.

Glass Electrode and Sensors

Glass has played an important active role in analytical chemistry as an electrochemical device. Glass electrodes, in the form of pH electrodes or sodium-ion selective electrodes, have been used extensively since pH meters were marketed by Arnold Beckman in the 1930s. This is because glass electrodes are simple in construction, robust, portable, and relatively inexpensive. Glass stability and inertness allow electrodes to be used in nonaqueous media even in the presence of organic solvents and the response is insensitive to anions. However, glass electrodes do have some drawbacks:

- Specificity and selectivity of ions can be compromised due to other cations (at certain concentrations)
- Susceptibility to electrical noise, and, in flow-through applications, the sensor response may be sluggish (Ref 9)
- Glass electrodes have nonlinear temperature dependence and need temperature compensation where high reliability is desired (Ref 10)

Today, optical sensors using silica-based glass fibers are beginning to be used for measuring chemical and physical parameters. However, their variety and early stage of application research precludes them from consideration in this brief review. The basis and application of glass electrodes only will be discussed in this section.

Glass electrodes are constructed so that a thin-wall bulb made of an ion-sensitive glass composition is sealed at the end of a high-resistivity, inert, nonresponsive glass tube. The solution whose response is to be monitored thus affects only the glass membrane tip. The bulb itself contains an appropriate electrolyte. Shielding and lead wires are introduced through the high-resistivity glass tube, rod, or stem for connecting to the necessary electronics. Because of their wide applicability, commercial electrodes come in appropriate sizes and shapes, including microelectrodes, robust pressure-suited electrodes, and high-temperature electrodes.

The fundamental functional response for a glass electrode is the development of a potential dependent on ionic activity, described by the Nernst equation for charged ions as:

$$E = E° + \frac{RT}{F} \ln a_i \qquad \text{(Eq 1)}$$

where E is the potential measured; $E°$ is the standard potential; R is the gas constant; T is the absolute temperature; F is Faraday con-

stant; and a_i is the activity of the ion. In situations where there is a mixture of two univalent ions (for example, H^+ and Na^+ mixture) the potential is described by

$$\bar{E} = E° + RT \ln (a_{H^+} + Ka_{Na^+}) \qquad \text{(Eq 2)}$$

A similar expression applies for a mixture of alkali ions of a constant pH.

The origin of the potential difference is not completely understood but appears to involve different cation activity of the solution-glass electrode interface. At this interface, the glass electrode surface has a hydrated region (that is, the glass wall may be considered to be a sandwich with an external hydrated gel layer), the unmodified or bulk glass wall, an internal hydrated layer (Ref 11). While the bulk glass is ≈50 μm (0.002 in.) thick, the gel layer is only 5 to 100 nm (50 to 1000 Å) thick. An ion-exchange process occurs in the gel layer resulting in a buildup of cation activity preferentially for one (ideally) cation resulting in observed membrane potential for a particular ion-exchange equilibrium.

The ion selectivity of the electrode is a function of the composition and resultant glass structure. The two composition classes of glass electrodes that will be discussed are: the hydrogen ion (pH) sensitive type and the sodium-selective type. The pH glass started with a 72 wt% SiO_2, 22 wt% Na_2O, 6 wt% CaO glass (sometimes referred to as NCaS22–6 or Corning 015 glass) and has a Nernstian behavior to a pH of ≈12 (Ref 12). At higher pH values, the glass electrode begins to select more alkali ions and an alkaline error can result. To overcome and virtually suppress the alkaline error, there are similar glasses where Li_2O replaces Na_2O and there is barium oxide in place of calcium oxide. It is worth noting that ceramic electrodes are said to be free of alkaline error (Ref 13). In order to preferentially select alkali metal ions, glass compositions need cations such as aluminum present. The trivalent ion results in AlO_4 groups with excess negative charge to which the alkali ion with appropriate charge-to-size ratio is drawn (for example, glasses such as 71 wt% SiO_2, 11 wt% Na_2O, and 18 wt% Al_2O_3 are particularly sodium selective relative to other alkali ions).

In addition to the pH or ion-sensitive glass, high-resistivity glasses are used to form the inert stem. Borosilicate and aluminosilicate glasses, with (dry) room-temperature resistivities in excess of 10^{15} $\Omega \cdot$ cm are both excellent insulating bulk glasses.

In addition, specialized glass electrodes have also been claimed. Thus, chalcogenide glassy alloys for bromine-selective electrodes, ion-doped chalcogenides for cupric ions, heavy-metal fluoride glasses for fluoride glass electrodes, microporous 96 wt% silica (Vycor) junction materials for reference electrodes, and silver-magnesium phosphate glasses for ammonia have been proposed or reported in the literature (Ref 13).

Glass Sensors. Ion-selective electrodes have found a large number of applications and these are briefly enumerated below. The pH sensor is used routinely in analytical chemistry laboratories, in industrial process for monitoring and control, in the assessment of boiler feed water, and more recently, in environmental studies such as natural water acidification including the use of microelectrodes for soil and water analysis. In clinical applications, in physiological and biomedical laboratories, glass capillary microelectrode sensors are also common. Glass ion-selective sensors are also used for the determination of sodium in clinical, nutritional, industrial, and agricultural applications.

Process Systems

Glass, particularly low-expansion borosilicate glass (for example, Pyrex), finds numerous applications as a material of construction in the chemical, food, pharmaceutical, petrochemical, and electronics industries. The principal attributes of interest in these applications are its chemical durability, thermal stability, resistance to thermal shock, ease of cleaning, transparency and economy; requirements that are identical to that in the laboratory but now scaled up to reflect process capacity. Modular glass components may be used to make systems such as distillation columns, boilers, scrubbers, and so on. Glass piping systems, glass-lined steel reactor vessels, and glass heat exchangers are some examples where corrosion or product purity needs dictate its use. Increasingly, glass reaction vessels are also being used in the treatment of hazardous (and, again, corrosive) wastes. Some of these are explained briefly.

Glass-lined steel vessels are used as reaction vessels where the additional mechanical integrity of the outer steel shell is a safety imperative. An acid recovery or concentration unit is a good example. Thus, a sulfuric acid concentrator would consist of a glass-lined steel container with a horizontal tantalum boiler. Waste sulfuric acid streams can be concentrated to 90 wt% H_2SO_4. The recovered concentrate may still contain trace amounts of organics. Vapor that arises during the process is condensed in yet another glass heat exchanger and then goes into a glass-phase separator where the organic and the water are separated for reuse. The glass linings or vessel walls are completely inert to sulfuric acid in all concentrations so that the purity of the product is dependent only on the quality of the feed and make-up fluids. There are two additional points:

- Abrasion resistance is not an issue with glass linings but can be a concern when alternate fluorocarbon linings are considered
- Glass lining can be damaged through internal mechanical impact or abuse; in such cases, the repair is performed with tantalum patches and so on

Table 2 Comparison of heat transfer coefficients of borosilicate glass to that of selected metals

| Material | Tube wall thickness, x | | Thermal conductivity, k $(Btu/h \cdot ft^2 \cdot °F)/ft$ | Coefficients | | | | | | | |
| | | | | Outer film, h_0 | | Inner film, h_i | | Wall heat transfer, h_w(a) | | Overall heat transfer, U(b) | |
	m	ft		$W/m^2 \cdot K$	$Btu/ft^2 \cdot h \cdot °F$	$W/m^2 \cdot K$	$Btu/ft^2 \cdot h \cdot °F$	$W/m^2 \cdot K$	$Btu/ft^2 \cdot h \cdot °F$	$W/m^2 \cdot K$	$Btu/ft^2 \cdot h \cdot °F$
Borosilicate glass	0.0012	0.0039	0.66	64.7	11.4	53	9.4	960	169	28.4	5.00
Type 316 stainless	0.00049	0.0016	9.4	64.7	11.4	53	9.4	3.339×10^4	5875	29.2	5.15
Aluminum	0.00049	0.0016	128	64.7	11.4	53	9.4	4.5×10^5	8.0×10^4	29.3	5.16
Copper	0.00049	0.0016	225	64.7	11.4	53	9.4	7.98469×10^5	1.40625×10^5	29.3	5.16

(a) $h_w = k/x$. (b) $U = [1/h_0 + 1/h_i + 1/h_w]^{-1}$

Modular glass components are employed in solvent recovery where the use of glass again ensures product purity.

Glass heat exchangers are a counter-intuitive example of the application of glass in process systems. Glass has poor thermal conductivity, but, as will be shown below, the heat transfer properties of borosilicate glass compare well with other construction materials. Recovering waste heat and cutting fuel consumption to reduce operational costs are important in any process. Unfortunately, recirculating hot exhausts, whether to process air inlet or make-up air inlet, is generally not possible because of the presence of contaminants. Exhaust streams need to be isolated and the heat extracted so that contaminants are left behind. This may be accomplished by drawing process intake air through banks of glass tubing (borosilicate, for example, for its low expansion and other desirable properties) positioned in the flowing hot exhaust gases. The physical separation avoids contamination. Table 2 shows a comparison of the heat transfer coefficient for the borosilicate glass (Pyrex), stainless steel, aluminum, and copper. The thermal conductivity of the tube material has a negligible role in the overall efficiency of the heat recovery unit. The critical factors in determining the heat transfer are the inner and outer gas film coefficients, which are nearly equal regardless of the tube material, resulting in an overall heat transfer coefficient for glass that is within 3% of that of copper. The following advantages accrue with glass:

- Cooling below the dew point has no associated risk of the condensate causing corrosion so that the latent heat is recovered in addition to the sensible heat
- Glass surface does not foul readily nor does it have particulate buildup problems
- Glass tubes are cleaned more easily with built-in spray systems
- Glass tubing assemblies provide faster payback because they are less expensive than the metal units

REFERENCES

1. E 438, *Annual Book of ASTM Standards*, American Society for Testing and Materials, 1990
2. D.C. Boyd and D.A. Thompson, Glass, *Encyclopedia of Chemical Technology*, Vol 11, 1980, p 807
3. M.E. Nordberg, *J. Am. Ceram. Soc.*, Vol 27, 1944, p 299–305
4. M.L. Hair and A.M. Filbert, *Research and Development*, 1969, p 34
5. A.M. Filbert and H.H. Weetall, *Methods in Enzymology*, Vol 34, 1974, p 59–72
6. H.L. MacDonnell, J.M. Noonan, and J.P. Williams, *Anal. Chem.*, Vol 33, 1961, p 1552
7. A.M. Filbert, Joint CIC-ACS Meeting, Toronto, 1970
8. M. Lynn and A.M. Filbert, *Bonded Stationary Phase*, E. Gruska, Ed., Ann Arbor Science, 1974
9. R.L. Solsky, *Anal. Chem.*, Vol 60, 1988, p 106R
10. D. Midgley, *Analyst*, Vol 112, 1987, p 581
11. H.H. Willard, L.L. Merritt, and J.A. Dean, Chapt. 20, *Instrumental Methods of Analysis*, Van Nostrand, 1974
12. J. Koryta, *Ann. Rev. Mater. Sci.*, Vol 16, 1986, p 13
13. M.A. Arnold and M.E. Meyerhoff, *Anal. Chem.*, Vol 56, 1984, p 20R
14. CorTherm™ product literature, Corning Incorporated

SELECTED REFERENCES

- D.L. Eaton, *Lab. Manag.*, 1967
- H.P. Hood and M.E. Nordberg, Borosilicate Glass, U.S. patent 2,221,709, 1940
- L.R. Snyder and J.J. Kirkland, *Introduction to Modern Liquid Chromatography*, 2nd ed., Wiley Interscience, 1979

Dental Applications

Arthur E. Clark and Kenneth J. Anusavice,
Department of Dental Biomaterials, College of Dentistry, University of Florida

GLASSES AND GLASS-CERAMICS play an important role in dentistry. Applications currently utilizing these materials include resin-composite restorative materials, cementation agents, crown and bridge materials, and implant materials. Each is discussed in the following sections in conjunction with a brief description of the dental application.

Dental Glass in Composite Resin Materials

Resin-composites have been used to repair and rebuild teeth at an ever-expanding rate since their introduction to dentistry in the early 1960s based on the original patent by Bowen (Ref 1). These materials consist of an organic polymer matrix, an inorganic filler phase, and a coupling agent which serves as a filler-matrix binder. Typically, a composite resin is available as either a paste-paste system (which when mixed, undergoes a chemically activated polymerization) or as a single-paste system (in which polymerization is activated by either an ultraviolet or visible light source). Most dentists prefer the light-activated systems as they provide more working time. To ensure longevity, several important requirements should be met. To rebuild or replace tooth structure, a restoration must have sufficient resistance to material loss whether from stresses generated by loading, wear, or other factors. Ideally, the margin or interface between the tooth and the restorative material should be sealed to prevent microleakage and the resultant caries that will occur. The restoration should be aesthetically acceptable to the patient. From the prospective of the dentist, the material should be easy to manipulate and insensitive to technique. Finally, the material must be compatible with biological tissues. Because this article describes materials already in use, biocompatibility is not stressed. However, in the development of any new material, biocompatibility is a critical factor which must be addressed. For any dental material, assessment of biocompatibility falls under guidelines detailed in the American National Standard/American Dental Association Document No. 41 for Recom-

mended Standard Practices for the Biological Evaluation of Dental Materials. This document has been approved by the Council on Dental Materials, Instruments and Equipment of the American Dental Association.

Dental composites have gained a wide acceptance in large part due to their superior aesthetic properties. Traditionally, the most extensively used material to restore teeth has been dental amalgam, a silver-tin-copper-mercury alloy. Amalgam restorations have demonstrated excellent durability over long periods of time in millions of patients. However, the dark silver-colored appearance associated with amalgam restorations is objectionable to many patients. An additional complicating factor is a concern over potential health effects associated with the mercury in amalgam alloys. While the documented evidence to date does not support the removal of serviceable amalgam restorations for the purpose of removing a toxic substance from the body (Ref 2–5), nevertheless there is increased pressure to develop a dependable cost-equivalent alternative to amalgam. Dental composites currently are potential replacement materials.

The polymer phase most commonly used in dental composites is an aromatic dimethacrylate monomer, Bisphenol A-glycidyl methacrylate (BIS-GMA). A few systems employ diurethane dimethacrylates. In general, the properties of an unfilled resin are inadequate for a restorative material. Properties which affect clinical performance include polymerization shrinkage, thermal expansion, water sorption, modulus of elasticity, tensile strength, polymerization shrinkage, and wear resistance.

Filler particles are incorporated into the resin matrix phase to improve properties and enhance the clinical longevity of restorative resins. The properties of the different filler materials, the particle size, shape, and filler fraction all affect the properties of the composite.

Typically, the filler phase particles are made of quartz, colloidal silica, or silicate glasses containing strontium or barium. The coupling agent which is coated onto the filler particles is a silane. A key factor in the performance of filled resins is the effective long-term

bonding of these particles to the matrix via the coupling agent.

From the mid 1960s to the late 1970s, quartz was the primary filler used in dental composites. Quartz is chemically inert, improves the hardness, and reduces the thermal expansion of an unfilled restorative resin. However, its hardness also contributes to difficulty in producing fine particles and in polishing to a smooth surface. Rough surfaces on restorations facilitate plaque accumulation which can lead to pulpal and periodontal problems. An additional problem associated with rough surfaces is an increased tendency to discolor via retention of food stains. Discoloration is a primary indication for replacement. Because of the lack of x-ray opacity of quartz, diagnoses of recurrent caries and identification of improper contours or voids in a restoration are compromised. An unexpected problem identified with quartz-filled materials was excessive wear noted in posterior restorations subjected to the forces of occlusion (Ref 6). The particles of the early materials were in the range of 30 to 50 μm, and they had irregular shapes produced by grinding. Because quartz is considerably harder than the resin matrix, stresses are concentrated at localized sites where the particles are embedded in the matrix. Cracking develops in the matrix and particle loss occurs. The abrasive particles thus contribute to the wear process.

To address these problems, two main types of filler materials have been developed: silicate glasses and colloidal silica. The silicate glasses typically contain (in mole %) 30 to 70% SiO_2, 0 to 30% B_2O_3, 0 to 15% Al_2O_3, 0 to 20% BaO, 0 to 23% ZnO, 0 to 30% SrO. Because these glasses are softer than quartz, powder production is facilitated. Particle sizes of 1 to 5 μm are used, and the number of filler particles per unit volume is significantly increased. Compared with the quartz-containing materials, there is a decrease in the stress around the particles. Furthermore, the softer glass particles absorb more of the stresses encountered during chewing and exhibit increased wear. The net effect is an improvement in clinical wear resistance. Polishing to a smoother finish is possible, enhancing appearance and reducing plaque accumulation. The radiopacity associated with the barium or strontium filler particles also enhances radiographic interpretation of caries around the restoration as well as inappropriate contours or bulk defects within the restoration (Ref 7).

As was mentioned earlier, a key to the enhanced properties of composite resins is associated with effective coupling or bonding of the filler to the matrix. Silane treatment is most often done with gamma-methacryloxy-propyltrimethoxy silane from an aqueous solution or by dry blending. In an aqueous solution, the silane coupling agent is hydrolyzed to form a triol with the fourth silicon bond linked to a reactive organic group. Upon drying, a condensation of the silanol groups occurs with bonding between silane molecules as well as hydrolyzed silicon atoms in the glass particles (Ref 7).

The second major alternative to large quartz filler particles is colloidal amorphous silica particles. So called pyrogenic silica is synthesized by burning silicon tetrachloride in a mixture of hydrogen and oxygen gas. Sol-gel processing can also be employed to manufacture these particles. The resultant filler particles are less than 0.05 μm in diameter and cannot be detected visually. These microfilled composites, as they are termed, can be polished to a smoother finish than composites containing the 1 to 5 μm glass particles. Thus, the tendency for plaque accumulation and food staining is reduced. In applications where aesthetics is of paramount importance, the microfilled materials currently are used almost exclusively.

In general, many mechanical properties of microfilled composites are inferior to composites with the 1 to 5 μm size particles. For example, the thermal expansion, polymerization shrinkage, and water absorption are greater for the microfilled materials. Also, the modulus of elasticity, tensile strength, and hardness are lower for the microfilled materials. These differences can be accounted for when one considers the amount of filler incorporated into the polymer. Typical composites with 1 to 5 μm size particles have a 75 to 85 wt% filler content. The microfilled composites have a 35 to 50 wt% filler content. The extremely small silica particles have such a high surface area that when they are added to the unpolymerized resin matrix, the viscosity rapidly increases producing a consistency which is very difficult to manipulate. One technique which is utilized to increase the filler content involves loading of the resin with as much colloidal silica as possible despite the increased viscosity. The mixture is polymerized and then ground to produce filled polymerized composite particles in the 20 μm size range. The prepolymerized particles are then added to a resin matrix which also contains colloidal silica. The resulting mixture can be adequately manipulated; however, the filler content still is only in the 50 wt% range.

Regardless of the filler type, filler shape, or filler content, resin composites have not demonstrated the longevity of amalgams for posterior restorations (Ref 8). Failure models associated with posterior composites include wear, recurrent caries, pain, and bulk fracture. The failures have been attributed to limitations in properties and to the degree of difficulty in manipulating composite materials during their placement. Two critical factors in placing any restorative material are the achievement of intimate interfacial contact with the remaining tooth structure and the reestablishment of any contact which had existed with adjacent teeth. Both of these goals are much more difficult to achieve with composites compared to amalgams. Furthermore, the shrinkage associated with polymerization as well as expansion and contraction associated with thermal cycling can compromise the interfacial integrity and lead to caries around the restoration.

The failures associated with wear and/or bulk fractures may be due in large part to a breakdown in the bond between the filler particles and the matrix (Ref 9). It has been documented that water exposure can leach out filler elements and induce filler failures and filler-matrix debonding (Ref 10). With the increasing demand for a more durable, easier-to-manipulate composite resin, emphasis is being placed on improving the filler-matrix bond strength as well as its resistance to breakdown.

Glasses in Dental Cements

Applications of dental cements include:

- Cementation of crowns, fixed partial dentures (bridges), or orthodontic bands
- Bases underneath other restorative materials
- Restorations themselves

Thus, the term cement is somewhat misleading. Furthermore, when actually used as cementing agent, mechanical retention rather than true adhesion is usually achieved. One of the few cements with documented chemical bonding to tooth structure is a glass-ionomer cement which is described later in this article. Retention is determined by a number of factors such as the strength and stiffness of the cement, the surface roughness of the preparation, the degree of taper of the preparation, and the proper manipulation of the material. Properties important to the performance of any cement include:

- Adequate strength and stiffness
- Resistance to dissolution in the oral environment
- Biocompatibility with pulpal tissues
- Ability to set with a low (25 to 50 μm) film thickness
- Bonding mechanism to tooth structure and to restorative materials (ideally chemical but usually mechanical)
- Ability to release fluoride to prevent caries

There currently are four cement systems that are used for luting or cementing applications, each of which is differentiated by their matrix phase: phosphate, phenolate (eugenolzte), polycarboxylate, and polymethacrylate. The two with the longest history of use are the zinc phosphate and zinc oxide and eugenol (ZOE) systems. Zinc oxide is the main powder constituent in both, and the liquid component is either phosphoric acid or eugenol (oil of cloves). Zinc phosphate typically is used for permanent cementation, and ZOE is used for temporary cementation. There are modified ZOE cements available for long-term cementation. These systems rely on retention and not adhesion. Both of these cement systems can also be used as a base. Bases are used in situations where a significant amount of tooth structure (dentin) has been removed and the pulp tissues are approximated (0.5 to 2 mm). Rather than placing a metallic restorative material (amalgam or a casting) with a relatively high thermal conductivity near the pulp, an insulating base is first placed to protect the pulp.

A cement system which has zinc oxide as the main powder component and a solution of polyacrylic acid as the liquid component was developed in the late 1960s (Ref 11); it is called a polycarboxylate. With this cement, bonding to clean enamel (Ref 12) and dentin can occur via chelation of carboxyl groups to calcium ions. It should be noted that bonding to dentin as compared to enamel is often limited by debris or contamination. Bond strength to dentin can be improved by treatment of the dentin with a remineralizing solution (Ref 13). The contribution of retention provided by the adhesion is still unknown, but carboxylate cements are used both as luting agents and as bases.

A modified class of carboxylate cement was developed by replacing the zinc oxide powder component with silicate glass particles (Ref 14, 15, 16). Typical compositional ranges (in wt%) of the glass particles are: 20 to 36% SiO_2, 15 to 40% Al_2O_3, 0 to 40% CaO, 0 to 35% CaF_2, 0 to 5% Na_3AlF_6, 0 to 6% AlF_3, 0 to 10% AlP. When the glass particles are mixed with an aqueous solution of polyacrylic acid, a series of reaction steps occur. Hydrogen ions are removed from the carboxyl groups of the acid solution. Simultaneously the glass particles undergo a solution reaction in which Al^{3+}, Ca^{2+}, Na^+, F^-, and PO_4^{3-} ions are released into solution to form a silica-rich gel on the particle surfaces. Polyacrylate salts of calcium and aluminum undergo precipitation and gelation. The calcium salts form first, followed by the aluminum salts. The set cement consists of a matrix phase of aluminum and calcium polyacrylates interspersed with glass particles with a silica gel surface layer. As was noted earlier, the fluoride content can be quite high. Fluoride ions perform two important functions. For any dental cement, the mixed powder and liquid should remain in a smooth consistency for the entire working time and then go through a rapid set. Initially, the fluoride complexes with the calcium and aluminum ions as they are removed from the glass, temporarily preventing these ions from bonding to the polycarboxylate chains; later, a rather sharp setting reaction occurs. Once the cement has set, there is still a significant amount of residual fluoride which can be released to exhibit a potential cariostatic effect.

The glass ionomer cements demonstrate adhesion or bonding to enamel, dentin, and metals via the same mechanism as the zinc polycarboxylate cements. They are used as cementing or luting agents and as bases. In addition, due to the translucency imparted by the glass particles, glass ionomer cement is used as a restorative material. Most commonly, they are placed in what are termed class V lesions which typically occur on the facial surface of teeth near or below the gum line. Their appearance and release of fluoride are primary indicators for this specific application. Glass ionomer cements used for cementation typically contain particle sizes of less than 25 μm, while the materials used for restorative applications contain particle sizes of 40 μm.

Because the complete setting reaction extends over several hours and the cement is subject to breakdown on exposure to water prior to complete setting, it is recommended that any exposed material be coated with a varnish which will provide temporary protection.

The fourth cement system based on polymethacrylates is the most recent to be developed and does not have the history of use of the previous three. One class based on the dimethacrylates (BIS-GMA) contains borosilicate or silicate glass powder as a filler (Ref 17). Currently, these materials are used to cement cast restorations which have been etched to produce an increased irregular surface to a similarly etched enamel surface. Thus, mechanical interlocking is the bonding mechanism. A problem with these cements is the potential for microleakage at the margin or tooth interface created by the polymerization shrinkage during setting.

For additional information on dental cements including information on comparative physical properties, readers are referred to the detailed reviews of Smith (Ref 18) and O'Brien (Ref 19).

Glass and Glass-Ceramic Crown and Bridge Materials

Approximately 80% of all fixed prostheses which were placed in the United States were fabricated from metal-ceramic systems (Ref 20). These prostheses are also referred to as porcelain-fused-to-metal (PFM) restorations. They generally consist of a cast metal substrate, a metal oxide (which promotes adhesion to porcelain), and several layers of porcelain (which bond to the metal oxide, mask out the color of the metal, provide translucency and color, and are thermally compatible with the metal). Noble metal alloys which have a majority of either gold or palladium contain several percent of oxidizable elements such as indium or tin. Alloys based on Ni-Cr, Ni-Cr-Be, or Co-Cr systems are also available for PFM restorations.

Recently, a variety of new dental ceramics has been introduced which can produce excellent aesthetics and optimum translucency compared with metal-ceramic crowns and bridges (fixed partial dentures). However, these products are not yet considered suitable for bridges because of the unpredictable levels of tensile strain which develop under a wide range of intraoral loading rates, magnitudes, and orientations. Long-term clinical data are not available to judge their durability under relatively severe stress states. Several *in vitro* studies have indicated that the strength of all-ceramic crowns is considerably less than that of PFM crowns. Nevertheless, their excellent aesthetic capability will serve as a principal advantage over PFM restorations in the years ahead.

Metal-Ceramic Prostheses

Adhesion of Porcelain to Metal Substrates. Ceramic-to-metal adhesion is primarily controlled by the structure and properties of the metal oxides which are formed during fabrication of prosthetic units. Mackert *et al.* (Ref 21) have shown for one gold alloy, four Ni-Cr alloys, and two Co-Cr alloys that the magnitude of the tensile bond strength of metal oxide to its metal substrate showed a strong positive correlation ($r^2 = 0.947$) to the area fraction of porcelain which was retained to the metallic substrate after testing in a "constant-strain flexure apparatus." No attempt was made to determine the influence of thermal incompatibility stress in porcelain on the measured bond strength values. Coffey *et al.* (Ref 22) reported that bond test specimens failed at lower loads when the thermal expansion coefficient of the metal was less than that of the bonded porcelain layer. Mackert *et al.* (Ref 23) observed that five of these alloys exhibited oxide-to-metal bond strengths which were greater than the reported tensile strength of opaque porcelain. One Co-Cr alloy with or without the use of its recommended bonding agent had an oxide adhesive strength less than that of opaque porcelain, while another Co-Cr alloy and a Ni-Cr alloy had significantly lower oxide adhesive strength values than opaque porcelain.

Pask and Tomsia (Ref 24) suggested that the bond strength of a metal-ceramic structure based on a substrate of an 80Ni-20Cr alloy is dependent on the strength of the individual oxide layers. Pask and Tomsia (Ref 24) recommended that the alloy composition should be adjusted so that the oxide formed during the preoxidation process is a monolayer of Cr_2O_3 which will ensure good oxide adherence and overall strength. Baran (Ref 25) has shown that the principal oxides which are formed on Ni-Cr alloys are strongly influenced by the firing temperature. Baran (Ref 26) also showed that a discrete and unaffected oxide layer remains after fusion of porcelain to Ni-Cr alloy.

Thermal compatibility of a metal-ceramic system is achieved when:

- Transient stresses in clinical prostheses of a wide range of shapes and sizes that develop during cooling are insufficient to cause immediate crack formation in the ceramic
- Residual stresses that remain in the ceramic after they have cooled to room temperature are insufficient to induce crack formation via a stress-corrosion mechanism or bulk fracture of the ceramic under the combined influence of the residual stress and masticatory stress

To determine the characteristics of compatible and incompatible systems, measurements must be made on clinically successful and unsuccessful metal-ceramic systems.

The analysis of thermal compatibility is confounded by the tendency of some dental porcelains to undergo a microstructural change and a corresponding increase or decrease in thermal contraction coefficient with each successive firing cycle (Ref 27–30) and the lack of a viscoelastic model to analyze thermally induced stresses in ceramic layers for clinical-like shapes, although such models are in various stages of development (Ref 31, 32). Mackert *et al.* (Ref 21) hypothesized that thermally induced decreases in porcelain expansion may result from one of three mechanisms: decoupling of leucite from the glass matrix, conversion of leucite to sanidine, and retention of cubic leucite during quenching. Three mechanisms were also proposed to explain an increase in thermal expansion after multiple firing cycles: recoupling of leucite to the glass matrix, crystallization of additional leucite, and volume relaxation of the glass matrix. Other problems which are associated with the analysis of thermal stresses include the poor correlation between dilatometry data determined at initial cooling rates much slower (3 to 5 °C/min, or 5 to 9 °F/min) than those experienced in commercial dental laboratories (200 to 500 °C/min, or 360 to 900 °F/min) and the upward shift in the thermal contraction coefficient and set point (temperature at which incompatibility stresses first develop in porcelain) with an increase in cooling rate. To establish a definition of porcelain-metal compatibility, Cascone (Ref 33) proposed that: "accurate contraction data for both porcelains and metals must be developed. This information combined with a three-dimensional thermal and stress analysis of several simple geometric designs, may be adequate to define the actual compatibility for the metal-ceramic system." Bertolotti (Ref 34) stated that: "There is also disagreement as to what constitutes thermal expansion matching of porcelain and metal. It is now clear that thermal expansion/ contraction values alone are not sufficient to predict thermal expansion compatibility; equally important are thermal history, geometry, and processing variables."

Longevity of Metal-Ceramic Restorations. In a Belgian study, Coornaert et al. (Ref 35) reported that only 2.4% of 2181 metal-ceramic units with a gold alloy substrate had failed after a period of 7 years. The prepared tooth forms were based on beveled shoulders, although preparations with a chamfer configuration were employed for optimum aesthetics. The minimal metal thickness was 0.2 mm (0.008 in.) and the thicknesses of opaque and body porcelains were 0.35 mm and 0.65 mm (0.014 in. and 0.025 in.), respectively.

In a U.S. study, Morris (Ref 36) reported excellent clinical results over a five-year period for five metal-ceramic systems which were used for metal-ceramic restorations in over 600 patients. Morris and his team of Veterans Administration clinicians observed a failure rate of only 1 to 2% (for the 450 remaining patients) over the five-year period for the following systems: Olympia (a Au-Pd alloy) and Ceramco porcelain, Ceramalloy II (a Ni-Cr alloy which requires a bonding agent) and Ceramco porcelain, Ticon (a Ni-Cr-Be alloy) and Ceramco porcelain, Microbond NP2 and Ceramco porcelain, and Will-Ceram W-1 (a Pd-Ag alloy) and Will-Ceram porcelain.

Despite the numerous pitfalls which can contribute to the ultimate clinical failure of metal-ceramic restorations, the longevity of these aesthetic crowns, pontics, and bridges appears to be comparable to that for the high-gold-content, cast-metal restorations except for rare situations in which impact forces or extremely high localized loads on the ceramic surface may occur.

Glass-Ceramic Prosthetics Formed by a Casting Process

The lost-wax method of casting glass objects of art was developed in the 1930s by Frederick Carder of the Steuben Division of Corning Glass Works. The transition from weaker glass to stronger castable ceramics occurred in 1957 with the development of glass-ceramics by S.D. Stookey of Corning Glass Works. This group of materials employs a nucleating agent to initiate the controlled growth of crystals within the amorphous glass matrix. The properties of glass-ceramics are affected by the type of crystals formed and the amount of growth within the glass matrix. The microstructures produced are nonporous and the grain size ranges from several hundred angstroms to a few microns.

Compared with the conventional porcelains used for all-ceramic dental restorations, castable ceramic systems offer great promise for dental applications because of their ease of fabrication, low processing shrinkage, high

strength, translucency control, insensitivity to abrasion damage, thermal shock resistance, chemical durability, and polishability. McMillan (Ref 37) reported that of all glass-ceramics available, the greatest flexure strength has been achieved with Li_2O-SiO_2 materials nucleated with P_2O_5. Several other glass-ceramics have been introduced for dental applications which must be externally colored with shading porcelain to achieve acceptable aesthetics.

One of the major goals of dental treatment is to replace, restore, and modify teeth which are missing, decayed unsightly, malformed, and malpositioned. Dental ceramic restorations which are used for these purposes should ideally be:

- Biocompatible
- Inert in the oral environment and resistant to the release of elements, ions, compounds, fragments or particles during wear or accidental fracture which are potentially harmful
- Capable of developing aesthetic characteristics similar to those of natural teeth
- Resistant to localized or generalized fracture due to stresses which are induced during placement on prepared teeth, during mastication, or during minor traumatic incidents
- Easily fabricated and finished at a reasonable cost
- Capable of being processed accurately to meet clinically acceptable standards of fit and morphology
- Resistant to distortion which may occur during processing
- Readily polishable yet resistant to abrasion
- Similar to tooth structure in thermal coefficient of expansion

The potential of using glass-ceramics for dental applications was first recognized by MacCulloch (Ref 38). Because of the excessive shrinkage, loss of contour, and porosity which occurred during processing of aluminous porcelain, he attempted to fabricate denture teeth from a Li_2O-ZnO-SiO_2 system.

The first demonstrated use of a castable glass-ceramic system for the fabrication of dental inlays, onlays, crowns, dentures, and veneers was reported by Hench et al. in 1971 (Ref 39). This system was based on a composition of 20 wt% Li_2O and 80% SiO_2. Although a nucleating agent was not used for this system, lithium ions are believed to induce phase separation in this system (Ref 40). Its diametral tensile strength is 124 MPa (18 ksi) which is approximately twice that of conventional feldspathic porcelain (61 MPa, or 8.85 ksi). The crystalline phase formed in this system was lithium disilicate ($Li_2O \cdot 2SiO_2$). Coloring elements and oxides were also added to a frit of the parent glass, and the mixture was melted, cast, and annealed. The resulting glass-ceramic produced a range of colors which could be varied further with control of the time of ceramming. One of the main

problems with this system was the presence of microcracks around the crystals which presumably formed because of a large volume change during crystallization.

Abe and Fukui (Ref 41) reported the use of a CaO-P_2O_5 glass-ceramic system for use in the production of dental crowns with the use of the lost-wax technique. Small concentrations of Al_2O_3 (1.5 wt%) were added to CaO to achieve an atomic percent ratio of 0.5 to 1.7. The frits were melted at 1300 °C (2370 °F) for one hour and after reheating to 635 °C (1175 °F), the glass was transformed to the glass-ceramic. A molar ratio, CaO/P_2O_5, of 55/45 (1.22) was found to produce the most favorable results. The strength of this material is reported as 117 MPa (17 ksi) compared with 61 MPa (8.85 ksi) for feldspathic dental porcelain. Linear shrinkage resulting from conversion of glass to glass-ceramic was less than 1%. The hardness (370 to 380 HV) and expansion coefficient (11.3 ppm/°C) of this glass-ceramic are similar to those of tooth structure. Coloration is produced by an external layer of shading porcelain.

Improvements in the castability, chemical durability, and resistance to microcracks of the lithia-silica glass-ceramics were reported by Barrett (Ref 42) and Barrett, Clark, and Hench (Ref 43) for lithium-silicate glass-ceramics through the addition of CaO in amounts of 1.0 to 11.0 wt% and Al_2O_3 from 1.0 to 10.0 wt% to a glass frit containing 25 to 33 mole% Li_2O and 53 to 73.5 mole% SiO_2. It was claimed that these additions substantially improved both the castability of the glass and the chemical durability of the glass-ceramic. Mechanical properties were further improved by combining platinum metal and Nb_2O_5 as nucleating agents. Through the use of these nucleating agents, the crystallization process for these lithia-alumina-calcia-silica (LACS) glass-ceramics resulted in a fine-grained crystal phase uniformly dispersed within a vitreous matrix. Flexural strengths of 214 MPa (31 ksi), which are 31 to 42% higher than the corresponding values reported for Dicor glass-ceramic, indicate that the LACS glass-ceramic may be more resistant to posterior failures when used for crowns and onlays. After exposure to a test bath of H_2O at 100 °C (212 °F) over a period of 6 h, these compositions released 17.2 to 22.0 ppm Li^+ for 0.01 wt% Pt and crystallization times of 0.5 h and 4.0 h, respectively, compared with 169 ppm Li^+ for the original Li_2O-SiO_2 glass-ceramic.

Additional improvements in properties of lithia-alumina-calcia-silica glass-ceramics were claimed in the patent of Wu, Cannon, and Panzera (Ref 44). They used P_2O_5 as a nucleating agent, which was found to produce greater flexure strengths (275 to 360 MPa, or 40 to 52 ksi) and higher softening temperatures.

Kihara et al. (Ref 45) introduced a castable calcium phosphate glass-ceramic composed of 30.1 wt% CaO, 68.4 wt% P_2O_5, and 1.5

wt% Al_2O_3. The glass can be melted and cast into cristobalite investment preheated to 600 °C (1110 °F). Thermal processing at 645 °C (1195 °F) for 12 h is required for crystal growth, and this heat treatment yields a glass-ceramic with a hardness of 380 HV and a flexure strength (modulus of rupture) of 116 MPa (16.8 ksi). Thus, this material appears to be too weak for crown and bridge applications, especially in posterior sites.

Hobo and Iwata (Ref 46–48) and Hobo (Ref 49) have reported the development of a castable apatite ceramic, CeraPearl, which is composed of 45 wt% CaO, 15% P_2O_5, 5% MgO, and 34% SiO_2. This material is melted at 1460 °C (2660 °F), heated to 750 °C (1380 °F) for 15 min, and crystallized at 870 °C (1600 °F) for 1 h to form $Ca_{10}(PO_4)_6O$. This structure is unstable and reacts with H_2O to form hydroxyapatite, $Ca_{10}(PO_4)_6(OH)_2$. The tensile strength of CeraPearl (150 MPa, or 21.8 ksi) is more than twice the diametral tensile strength of feldspathic porcelain. It has an elastic modulus of 103 GPa (15×10^6 psi), a Knoop hardness of 350, a compressive strength of 590 MPa (85.5 ksi), and an expansion coefficient of 11.0 ppm/°C. CeraPearl crowns are colored externally by applying stains formulated from K_2O-Al_2O_3-B_2O_3-SiO_2 with traces of various metal oxides. The stains are fired at 790 °C (1455 °F). As is true for most other ceramics and glass-ceramics, crowns of hydroxyapatite can be mechanically and chemically treated to promote adhesion of the interior surface of the crown to glass ionomer cements.

The use of a machinable glass-ceramic which can be melted and cast into the form of dental inlays, onlays, and crowns was introduced by Adair (Ref 50) and Grossman (Ref 40). This glass-ceramic system is based on the growth of fluorine-containing, tetrasilicic mica crystals which was first reported in a patent by Grossman in 1973. Compositions enriched in K_2O and SiO_2 can be melted from 1350 to 1400 °C (2460 to 2550 °F) and subsequently be cerammed to yield a tetrasilicic fluormica glass-ceramic (K_2O-MgF_2-MgO-SiO_2). Addition of ZrO_2, in amounts up to 7 wt%, is believed to improve chemical durability and enhance translucency. Glass-ceramics are formed by a fine-scale phase separation that occurs at 625 °C (1160 °F), followed by crystallization of spherical mica grains approximately 400 μm in diameter (Ref 51). Heating to 940 °C (1725 °F) causes recrystallization into block-shaped crystals. The mica platelets grow by a dissolution and reprecipitation mechanism at 1075 °C (1970 °F). A holding time of 6 h is considered optimum for strength and other desirable properties. Like all other castable glass-ceramics, this product, which is marketed under the commercial name Dicor (Dentsply International, York, PA), must be stained on the external surface with shading porcelain to achieve acceptable aesthetics. If too thin a layer is applied, the entire colorant layer may be lost during routine dental prophylaxis because of the abrasiveness of the prophy paste. In addition, the daily use of acidulated fluoride gels may cause the dissolution of the shading porcelain. This situation would require costly remakes, loss of wages, and additional trauma to the patient. One of the potential uses of Dicor or other machinable glass-ceramics is as a material to be used for the production of machined inlays, onlays, and crowns by means of computer-aided design and manufacturing systems such as that described by Mörman *et al.* (Ref 52).

Properties of Dicor Glass-Ceramic. When comparing the properties of different materials such as metals, ceramics, and polymers for restorative applications, performance cannot be determined from a single property. For example, ceramics most frequently fail under applied tensile stresses which may vary widely in magnitude depending on the loading rate, load orientation, and the concentration and size of surface flaws. Their compressive strengths are much higher because of a low resistance to crack initiation and propagation in the presence of surface flaws. One cannot use compressive strength values solely to make a prediction of clinical performance to applied intraoral forces. Ceramic engineers rely on data generated from bending tests which can be used to calculate modulus of rupture (MOR) values. These values essentially represent the levels of bending stress at which failure occurs. The modulus of rupture for a glass ceramic is twice that for feldspathic (low fusing) porcelains. However, the clinical failure rates of PFM crowns in molar sites are much lower (2 to 5% over a 7- to 10-year period, Ref 53) than that of Dicor crowns (5 to 30% over a 3-year period, Ref 54). These data were obtained for zinc phosphate cement used as a luting agent. Recent studies suggest that fewer failures of Dicor crowns occur when the resin-bonding technique is used.

The thermal expansion coefficient of the cast ceramic is 7.2×10^{-6}/° C (4×10^{-6}/° F), which is closer to that of enamel than composite resin or amalgam. Because the thermal conductivity and diffusivity are lower than those of metallic restorations, one would expect glass-ceramics like Dicor to exhibit relatively little expansion or contraction during thermal cycling and serve as excellent thermal insulators.

The density and hardness of Dicor are similar to the values for enamel. Thus, this material would not be expected to exhibit excessive wear of opposing enamel surfaces. In two-year clinical studies, no gross wear has been observed for Dicor crowns or opposing tooth surfaces (Ref 54). The wear of unstained Dicor glass-ceramic specimens is 3 to 4 times that of the stained ceramic because of the higher wear resistance of the shading porcelain. The results of a study by Grossman and Walters (Ref 55) indicate that the chemical stability of Dicor crowns is comparable to that of dental porcelains over a wide pH range. This study was conducted at elevated temperatures to accelerate the potential effects of corrosion in water or acidic environments.

Shrinkage-Free Core Ceramic

Products of this type are used to fabricate the core portion of porcelain jacket crowns. These materials are provided in the form of disks, which are heated and injected into an evacuated gypsum mold to form the desired prosthetic shape. After devesting the green coping from the investment, a heat-treatment process is used to strengthen the core by formation of an approximately 90 wt% crystalline material. The remaining 10% is an interstitial glass phase of an alkaline-earth aluminosilicate glass. The crystalline material is approximately 40 wt% α-alumina, 40 wt% magnesium aluminum spinel, and 20 wt% barium osumulite. The density of the core ceramic is 2.8 g/cm³ with a porosity content of about 0.2%. The thermal expansion coefficient is about 6.3 ppm/°C. The original product of this type, Cerestore, has been discontinued and reintroduced with a higher strength core ceramic under the name Alceram. The flexure strength reported for the low-strength core material (Cerestore) was 118 MPa (17 ksi). The higher strength core material (Alceram) has a flexure strength of 162 MPa (23.5 ksi). In combination with the controlled expansion of an epoxy resin die, the heat-treatment process also permits an expansion of the coping to compensate for any sintering shrinkage. Thus, this product is referred to as a "shrinkage-free ceramic." The remaining shaded porcelain layers are built up using veneering porcelains and firing is accomplished using procedures similar to those used for aluminous porcelain jacket crowns. The incidence of clinical failure of anterior crowns ranges from 0.8 to 4.3% (Ref 56), which is comparable to that of the cast ceramic crowns.

High-Strength Feldspathic Porcelain

OPTEC HSP, manufactured by Rx Jeneric, is a high-strength feldspathic porcelain, which like Dicor castable glass-ceramic, does not require an opaque core material. Only body and incisal porcelain are used because the opacity provided by the leucite crystals does not require the use of a core porcelain. The flexural strength reported by the manufacturer is approximately 172 MPa (25 ksi). Compared with conventional feldspathic porcelain, the increased strength of OPTEC HSP is derived from the nucleation and growth of small leucite ($KAlSi_2O_6$) crystals. No long-term clinical data are available on this material. It has been used primarily for the fabrication of inlays, onlays, crowns, and veneers.

CAD-CAM Ceramic Inlays

One commercial system, CEREC, has been recently introduced for the computer-aided design (CAD) and computer-aided manufacture (CAM) of inlays from pigmented blocks of either two materials, Vitablock and Dicor. This computer-driven system provides a means of preparing ceramic inlays within a 1-h period without the need for a dental impression.

Longevity of All-Ceramic Restorations

There are few published reports on the long-term clinical performance of porcelain jacket crowns (PJC). McLean (Ref 57) reported the results of a comprehensive analysis of the clinical performance of porcelain jacket crowns constructed with an aluminous-core that has an inner layer of a 25 μm thick platinum foil electroplated with a 0.2 μm thick layer of tin which was subsequently oxidized. Although the theory of this approach was that bonding of core porcelain to the plated and oxidized tin film on platinum foil would eliminate open surface defects from which tensile failure may originate, McLean (Ref 57) indicated that the cumulative failure rate for these crowns over a 7-year period was 15.2% for molar crowns, 6.4% for premolar crowns, 2.1% for incisor crowns, and 1.3% for incisors. The failure rates for premolar and molar crowns were unacceptably high and suggest that extreme caution should be exercised before using PJC restorations of posterior teeth on a routine basis.

The high failure rates of aluminous core jacket crowns in molar sites is consistent with the 4-year failure levels of 18.5% for Cerestore molar crowns and 3% for anterior crowns reported by Linkowski (Ref 56). However, Wohlwend *et al.* (Ref 58) indicate that the manufacturers of Cerestore "shrink-free" porcelain have introduced a new core material with a flexure strength of 162 MPa (23.5 ksi), which is nearly twice as high as that (88 MPa, or 12.8 ksi) of the conventional material. A new material, In-Ceram, exhibits a flexure strength of ≈460 MPa (67 ksi). This strength is achieved by infiltration of the 98% Al_2O_3 core with glass at an elevated temperature.

For Dicor, a tetrasilicic fluormica glass-ceramic, Moffa (Ref 54) reported a failure rate for molar, premolar, and anterior crowns of 35.3%, 11.8%, and 3.5%, respectively, during the first three years. Graser *et al.* (Ref 59) reported an 80% failure rate within the first year for Dicor three-unit fixed partial dentures. These prostheses were formed in two pieces with male and female connectors and were joined together with a composite resin. It should be noted that if miscasts or failures occur with castable ceramics, it is impossible to use the units again for the remake procedures.

Glass Implant Materials

For many years it was considered that the normal interfacial tissue response to implant materials was the formation of a fibrous capsule. This viewpoint came into question in the early 1970s with the discovery that a certain compositional range of calcium phosphate silicate-based glasses did not produce a fibrous capsule when implanted in bone, but instead produced a strongly adherent bonded interface (Ref 60), where the strength of the interface was equal to or exceeded that of the bone with which it was bonded (Ref 61). These glasses contain SiO_2, Na_2O, CaO, and P_2O_5 in specific proportions. Three key compositional features distinguish these glasses from traditional soda-lime-silica glasses that do not bond to tissues containing: less than 60 mole% of SiO_2, high Na_2O and CaO content (up to 30 mole%), and a high Ca/P_2O_5 ratio. These features make the surface highly reactive when exposed to an aqueous medium and result in rapid formation of hydroxycarbonate apatite on the surface (Ref 62). The tradename Bioglass was coined to describe this compositional range of bioactive glasses. Subsequently the term *bioactive* has been used to describe implant materials which form a bond with living tissues. Other examples of bioactive-glass containing systems include:

- The SiO_2-CaO-MgO-P_2O_5 system which contains apatite and wollastonite as crystalline phases and are referred to as AW glass-ceramics (Ref 63)
- The machinable glass-ceramics based on the Na_2O/K_2O-MgO-Al_2O_3-SiO_2-F system containing phlogopite and apatite crystals (Ref 64)
- The Na_2O-CaO-P_2O_5-B_2O_3-Al_2O_3-SiO_2 system (Ref 65)

Currently, the only example of a bioactive glass product available in the United States is based on a Bioglass composition 45 wt% SiO_2, 24.5 wt% CaO, 24.5 wt% Na_2O, 6 wt% P_2O_5. The product is called an ERMI (endosseous ridge maintenance implant). These implants, which are solid glass cones of various sizes, are designed to be placed into tooth sockets immediately after extraction of teeth. Their purpose is to maintain the bone level and contours which under normal circumstances resorbs after tooth extraction. The ERMIs are totally buried and not subjected to direct occlusal loading. They are indicated for patients scheduled to receive complete dentures after the loss of all their teeth as well as for individuals who have lost one or more teeth and will be restored with fixed bridgework or removable partial dentures.

While this is the only currently available bioactive glass product in the United States, there has been a significant amount of research and clinical trials devoted to these materials on a worldwide basis. A recent handbook has been published describing the basic science investigations and potential clinical applications of these implant materials (Ref 66).

Health and Safety Regulations

On May 28, 1976, Congress passed the Medical Device Amendments to the Federal Food, Drug and Cosmetic Act. As a result of this legislation, the Food and Drug Administration (FDA) received authority to assess and evaluate the biological effects of all materials and devices used in human beings. Included in this category are most dental products. The primary purpose of this legislation was to ensure that devices are safe and effective before they are placed on the market. An excellent review of this legislation and a discussion of some of the limitations were published by Kessler *et al.* in 1987 (Ref 67). A synopsis of key points follows. For a more complete discussion, the reader is referred to the original article.

FDA Categories and Approval Procedures. Devices were divided in two ways. In one system, devices are designated as Class I, II, or III with a Class III device being the most potentially hazardous. For a Class I device, the manufacturer is required to comply with regulations regarding registration, premarketing notification, record keeping, labeling, reporting of adverse experiences and good manufacturing practices (GMP). Collectively, these requirements are referred to as general controls. These requirements apply to all three classes. For a Class II device, the manufacturer must meet federally defined performance standards. For a Class III device, the manufacturer must demonstrate the safety and efficacy to the satisfaction of the FDA. Class III devices undergo the most extensive regulation.

In the second classification system, devices are placed in one of seven categories:

- Preamendment devices, which are devices on the market before the amendments were enacted in 1976
- Postamendment devices, which are devices put on the market after 1976
- Substantially equivalent devices, which are postamendment devices that are substantially equivalent to preamendment devices
- Implanted devices, which are devices that are inserted into a surgically formed or natural body cavity and intended to remain there for 30 days or more
- Custom devices that are not generally available to other licensed practitioners and not available in finished form
- Investigational devices, which are unapproved devices undergoing clinical investigation under the authority of an Investigational Device Exemption (IDE)
- Transitional devices that were regulated as drugs before enactment of the amendments but are now defined as devices

The class or category into which a material or device is placed is of concern to the manufacturer as there is direct correlation between the cost and the amount of documentation required by the FDA. Preamendment devices are presumed to be placed in Class I unless FDA determines that their safety and efficacy cannot be ensured without the stricter regulation of Class II or III requirements. Postamendment devices are automatically placed in Class III. However, the manufacturer may petition FDA for reclassification. Substantially equivalent devices are placed in the same class as their preamendment predecessor. Implanted devices are placed in Class III, unless the manufacturer can provide documentation that a less-regulated class will ensure safety and efficacy. Custom devices are exempt from premarket testing and performance standards but are subject to general controls. Investigational devices are exempt from the 3 classes based on their Investigational Device Exemption (IDE). Transitional devices are assigned as Class III, but again the manufacturer may petition for reclassification to a less regulated class.

There are two avenues that a manufacturer can take to secure FDA approval to market a new product. If substantial equivalence to a preexisting product can be documented, premarket clearance is sufficient. Under section 510(k) of the amendments, a manufacturer of a substantially equivalent device must notify FDA of its intention to place the device on the market, and if there is no objection by the FDA within 90 days, the device may be marketed. The second route requires the manufacturer to secure Pre-Market Approval (PMA) from the FDA.

Before granting PMA, the manufacturer must provide documentation that the device is both safe and effective. It should be noted that safety is evaluated by weighing the probable benefits to health against the probable risks of injury. Furthermore, it is not necessary for the manufacturer to prove that the product or device will never cause an injury or will always be effective. For many devices whether via a PMA or a 510(k) application, the FDA requires documentation supporting evidence in the form of clinical trials. In the past, the requirements for a 510(k) were not as demanding as for a PMA. Recently, the FDA has been recommending more stringent guidelines for a 510(k) submission essentially similar to that required for a PMA. Requirements of clinical trials include the minimum number of patients, the minimum length of evaluation of each patient, the design of the trial, and the number of independent trials.

ADA Evaluation. In addition to the regulation of the Food and Drug Administration, the American Dental Association (ADA) is involved in the evaluation and indirectly in the regulation of devices through the Council on Dental Materials, Instruments, and Equipment, and the Accredited Standards Committee MD 156 Dental Materials, Instruments and Equipment. The ADA in conjunction with the American National Standards Institute (ANSI) has specifications and acceptance program guidelines for a variety of materials and equipment. Currently approved specifications include ANSI/ADA Specification No. 8 for Dental Zinc Phosphate Cement, No. 27 for Direct Filling Resins, No. 30 for Zinc

Oxide-Eugenol and Non-Eugenol Cements, No. 61 for Zinc Polycarboxylate Cement, No. 66 for Glass Ionomer Cements, and No. 38 for Ceramic-Metal Systems, which are subject areas referred to in this article. In addition, there is a specification in preparation for Dental Ceramics (No. 69).

Typically, ADA specifications require documentation of physical properties. For example, specification No. 8 for Zinc Phosphate Cements stipulates minimum and maximum setting times, minimum compressive strength at 24 h, maximum allowable film thickness, and maximum solubility and disintegration at 24 h. Other areas addressed are composition and biocompatibility. As mentioned earlier, appropriate tests to verify biocompatibility for all types of dental materials are included in ANSI/ADA Document No. 41 for Recommended Standard Practices for the Biological Evaluation of Dental Materials. It should be noted that these tests are *in vitro* lab tests, tissue culture tests, and animal experiments. A complete list of all specifications and copies of individual specifications can be obtained from the Council on Dental Materials, Instruments and Equipment, American Dental Association, 211 E. Chicago Avenue, Chicago, IL 60611.

To obtain certification by the ADA, a manufacturer must submit documentation that their product meets or exceeds the requirements of the appropriate specification. At times, the FDA uses ADA specifications as required performance standards for approval to market.

In situations where there is no appropriate specification, the ADA has an Acceptance Program. The manufacturer submits an application to the Council on Dental Materials, Instruments and Equipment. The submission should include the following items:

- Name and use of item
- Numbers of any patents relating to the product
- Composition, physical, and chemical properties of dental materials, if applicable
- Materials used in construction and method of operation of a dental instrument or equipment, if applicable
- Evidence of premarket clearance by the Food and Drug Administration
- Evidence of safety and usefulness of product based upon *in vitro*, biological and clinical evaluations when applicable (see section below)
- A description of quality control processes routinely performed on the subject product
- Names of owners, officers of the firm, or other individuals who are authorized to furnish information to the Council and represent the firm to the Council
- Names and qualifications of scientific personnel responsible for formulation and testing of item or product
- All other information required by the

Guidelines for a specific Acceptance Program

Evidence of safety and usefulness of the product may be in the form of published reports or unpublished information obtained from appropriate scientific studies employing *in vitro*, biological, and clinical observations. Biological evaluations should be in compliance with American Dental Association specification No. 41 for Biological Evaluation of Dental Materials.

Evidence should be both sufficient in quantity and adequate in quality to permit sound conclusions. This requirement is especially important since most clinical studies involve subjective interpretations on the part of both the observer and the patient. All clinical reports must be signed by the principal investigator.

To provide a sufficient quantity of data, not only should the number of clinical cases available for observation in a single study be adequate, but reports of additional investigations by independent groups are usually required.

Evidence of adequate quality generally necessitates the employment of controls or comparison products evaluated under the same conditions of use as the test products. Double blind studies are desirable whenever possible.

The Council may approve the material as Acceptable, or Provisionally Acceptable which indicates a lack of sufficient evidence to justify classification as Acceptable, but for which there is reasonable evidence of safety and usefulness including clinical feasibility. A third option is disapproval.

ACKNOWLEDGMENT

The authors gratefully acknowledge support of NIDR/NIH grant P50 DEO9307.

REFERENCES

1. R.L. Bowen, "Dental Filling Material Comprising Vinyl Silane Treated Fused Silica and a Binder Consisting of the Reaction Product of Bisphenol and Glycidyl Acrylate," U.S. patent 3,066,112, Nov 1962

2. T. Axell, K. Nilner, and B. Nilsson, Clinical Evaluation of Patients Referred with Symptoms Related to Oral Galvanism, *Swedish Dent. J.*, Vol 7, 1983, p 169–178

3. B. Johansson, E. Steniman, and M. Bergman, Clinical Study of Patients Referred for Investigation Regarding So-Called Oral Galvanism, *Scand. J. Dent. Res.*, Vol 92, 1984, p 469–475

4. M. Alqwist, C. Bentsson, B. Furnunes, L. Hollender, and L. Lapidus, Number of Amalgam Tooth Fillings in Relation to Subjectively Expressed Symptoms in a Study of Swedish Women, *Commun. Dent. Oral Epidemiol.*, Vol 16, 1988, p 227–231

5. T. Kallus, Incidence of Adverse Reac-

tions from Dental Materials, *J. Dent. Res.*, Vol 64 (Abst. No. 21), 1985, p 758

6. K.F. Leinfelder, Current Developments in Posterior Composite Resins, *Adv. Dent. Res.*, Vol 2, 1988, p 115–121

7. K.J. Söderholm, Filler Systems and Resin Interface, *Posterior Composite Resin Dental Restorative Materials*, G. Vanherle and D.C. Smith, Ed., International Symposium, the Netherlands, Peter Szulc Publishing, 1985, p 139–160

8. J.P. Moffa, Comparative Performance of Amalgam and Composite Resin Restorations and Criteria for their Use, *Quality Evaluation of Dental Restorations, Criteria for Placement and Replacement*, K.J. Anusavice, Ed., Quintessence, 1989, p 125–138

9. K.J. Söderholm and M.J. Roberts, Influence of Water Exposure on the Tensile Strength of Composites, *J. Dent. Res.*, Vol 69, 1990, p 1812–1816

10. K.J. Söderholm, M. Zigan, M. Regan, W. Fischlchewiger, and M. Bergman, Hydrolytic Degradation of Dental Composites, *J. Dent. Res.*, Vol 63, 1984, p 1248–1254

11. D.C. Smith, A Review of the Zinc Polycarboxylate Cements, *J. Can. Dent. Assoc.*, Vol 37, 1971, p 22–29

12. A.D. Wilson and H.J. Prosser, A Survey of Organic Polyelectrolyte Cements, *Br. Dent. J.*, Vol 157, 1984, p 449–484

13. D.R. Beech, R. Soloman, and R. Bermer, Bond Strength of Polycarboxylate Cements to Treated Dentin, *Dent. Mater.*, Vol 1, 1985, p 154–157

14. S. Crisp and A.D. Wilson, Reactions in Glass Ionomer Cements—I Decomposition of the Powder, *J. Dent. Res.*, Vol 53, 1974, p 1408–1413

15. S. Crisp and A.D. Wilson, Reactions in Glass Ionomer Cements—II An Infrared Spectroscopic Study, *J. Dent. Res.*, Vol 53, 1974, p 1414–1419

16. S. Crisp and A.D. Wilson, Reactions in Glass Ionomer Cements—III The Precipitation Reaction, *J. Dent. Res.*, Vol 53, 1974, p 1420–1424

17. R. Simonsen, V.P. Thompson, and G. Barrack, *Etched Cast Restorations: Clinical and Laboratory Techniques*, Quintessence Publishing, 1983

18. D.C. Smith, Dental Cements: Current Status and Future Prospects, *Restorative Dental Materials an Overview*, Vol 1, J. Reese and T.M. Valega, Ed., Quintessence Publishing, London, 1985, p 439–465

19. W.J. O'Brien, Ed., *Dental Materials: Properties and Selection*, Quintessence Publishing, 1989

20. Alloy Direct Mail Research Report, Rand Research, Inc., 1981

21. J.R. Mackert, Jr., R.D. Ringle, E.E. Parry, A.L. Evans, and C.W. Fairhurst, The Relationship between Oxide Adherence and Porcelain-Metal Bonding, *J.*

22. J.P. Coffey, K.J. Anusavice, P.H. DeHoff, R.B. Lee, and B. Hojjstie, Influence of Contraction Mismatch and Cooling Rate on Flexural Failure of PFM Systems, *J. Dent. Res.*, Vol 67, 1988, p 61–65

23. J.R. Mackert, Jr., E.E. Parry, D.T. Hashinger, and C.W. Fairhurst, Measurement of Oxide Adherence to PFM Alloys, *J. Dent. Res.*, Vol 63, 1984, p 1335–1340

24. J.A. Pask and A.P. Tomsia, Oxidation and Ceramic Coatings on 80Ni-20Cr Alloys, *J. Dent. Res.*, Vol 67, 1988, p 1164–1171

25. G.R. Baran, Oxidation Kinetics of Some Ni-Cr Alloys, *J. Dent. Res.*, Vol 62, 1983, p 51–55

26. G.R. Baran, Dissolution of Oxides on Ni-Cr Alloys during Simulated Porcelain Fusion, *J. Dent. Res.*, Vol 66 (Abst. No. 1122), 1987, p 247

27. C.W. Fairhurst, K.J. Anusavice, D.T. Hashinger, R.D. Ringle, and S.W. Twiggs, Thermal Expansion of Dental Alloys and Porcelains, *J. Biomed. Mater. Res.*, Vol 14, 1980, p 435–466

28. R.P. Whitlock, J.A. Tesk, G.E.O. Widera, A. Holmes, and E.E. Parry, Consideration of Some Factors Influencing Compatibility of Dental Porcelains and Alloys: Part 1, Thermophysical Properties, *Proceedings of the 4th Annual Meeting*, International Precious Metal Institute, Pergamon Press, Toronto, 1980

29. P. Dorsch, Stresses in Metal-Ceramic Systems as a Function of Thermal History, *Ber. Dt. Keram. Ges.*, Vol 58, 1981, p 1–7

30. P. Dorsch, Thermal Compatibility of Materials for Porcelain-Fused-to-Metal Restorations, Ceramic Forum International, *Ber. Dt. Keram. Ges.*, Vol 59, 1982, p 1–5

31. K. Asaoka and J.A. Tesk, Transient and Residual Stress in a Porcelain-Metal Strip, *J. Dent. Res.*, Vol 69, 1990, p 463–469

32. P.H. DeHoff and K.J. Anusavice, Effect of Visco-Elastic Behavior on Stress Development in a Metal-Ceramic System, *J. Dent. Res.*, Vol 68, 1989, p 1223–1230

33. P. Cascone, Effect of Thermal Properties on Porcelain-to-Metal Compatibility, *IADR Prog. Abst.*, Vol 58 (Abst. No. 683), 1979, p 263

34. R.L. Bertolotti, Porcelain-to-Metal Bonding and Compatibility, *Dental Ceramics: Proceedings of the First International Symposium on Ceramics*, J.W. McLean, Ed., Quintessence Publishing, 1983, p 414–431

35. J. Coornaert, P. Adriaens, and J. DeBoever, Long-Term Study of Porcelain-Fused-to-Gold Restorations, *J. Prosthet. Dent.*, Vol 51, 1984, p 338–342

36. H.F. Morris, Clinical Performance of Metal-Ceramic Restorations—VA Cooperative Studies Project No. 147/242: A Five-Year Progress Report, *Quality Evaluation of Dental Restorations: Criteria for Placement and Replacement*, K.J. Anusavice, Ed., Quintessence Publishing, 1989, p 325–342

37. P.W. McMillan, The Properties of Glass-Ceramics, *Glass-Ceramics*, 2nd ed., 1979

38. W.T. MacCulloch, Advances in Dental Ceramics, *Br. Dent. J.*, Vol 124, 1968, p 361–365

39. L.L. Hench, R.E. Going, F.A. Peyton, B.H. Bell, N. Ingersoll, and G. Kluft, Glass-Ceramic Dental Restorations, *IADR Prog. Abst.*, Vol 50 (No. 322), 1971

40. D.G. Grossman, Cast Glass Ceramics, *The Dental Clinics of North America*, W.J. O'Brien, Ed., W.B. Saunders, 1985, p 725–739

41. Y. Abe and H. Fukui, Studies of Calcium Phosphate Glass-Ceramics-Development of Dental Materials (Part I), Shika Rikogaku Zasshi, *J. Jpn. Soc. Dent. Appl. Mater.*, Vol 15, 1975, p 196–202

42. J.M. Barrett, "Chemical and Physical Properties of Multicomponent Glasses and Glass-Ceramics," M.S. thesis, University of Florida, 1978

43. J.M. Barrett, D.E. Clark, and L.L. Hench, "Glass-Ceramic Dental Restorations," U.S. patent 4,189,325, Feb 1980

44. J.M. Wu, W.R. Cannon, and C. Panzera, "Castable Glass-Ceramic Composition Useful as a Dental Restorative," U.S. patent 4,515,634, May 1985

45. S. Kihara, A. Watanabe, and Y. Abe, Calcium Phosphate Glass-Ceramic Crown Prepared by Lost-Wax Technique, *Commun. Am. Ceram. Soc.*, Vol 67, 1984, p C100-C101

46. S. Hobo and T. Iwata, Castable Apatite Ceramics as a New Biocompatible Restorative Material. I. Theoretical Considerations, *Quint. Int.*, Vol 16 (No. 2), 1985, p 135–141

47. S. Hobo and T. Iwata, Castable Apatite Ceramics as a New Biocompatible Restorative Material. II. Fabrication of the Restoration, *Quint. Int.*, Vol 16 (No. 3), 1985, p 207–216

48. S. Hobo and T. Iwata, A New Laminate Veneer Technique Using a Castable Apatite Ceramic Material. I. Theoretical Considerations, *Quint. Int.*, Vol 16 (No. 7), 1985, p 451–458

49. S. Hobo, Castable Hydroxyapatite Ceramic Restorations, *Dental Ceramics, Proceedings of the Fourth International Symposium on Ceramics*, J.D. Preston, Ed., Quintessence Publishing, 1988, p 135–152

50. P.J. Adair, "Glass-Ceramic Dental Products," U.S. patent 4,431,420, Feb 1984

51. D.G. Grossman, The Science of Castable Glass Ceramics, *Perspectives, in*

Dental Ceramics, Proceedings of the Fourth International Symposium on Ceramics, J.D. Preston, Ed., Quintessence Publishing, 1988, p 117–133

52. W. Mörman, M. Brandestini, and F. Lutz, Das Cerec System: Computergestützte Herstellung direkter Keramikinlays in Einer Sitzung, *Quintessenz.*, Vol 38, 1987, p 457

53. K.J. Anusavice, Criteria for Selection of Restorative Materials: Properties Versus Technique Sensitivity, *Quality Evaluation of Dental Restorations: Criteria for Placement and Replacement*, K.J. Anusavice, Ed., Quintessence Publishing, 1989, p 15–60

54. J.P. Moffa, "Clinical Evaluation of Dental Restorative Materials—Final Report," Interagency Agreement No. 1Y01-DE40001–05, Letterman Army Institute of Research, 1988

55. D.G. Grossman and H.V. Walters, The Chemical Durability of Dental Ceramics, *J. Dent. Res.*, Vol 63 (Abst. No. 574), 1984, p 234

56. G. Linkowski, "Langzeituntersuchung bei Einem Vollkeramik-Cronensystem (Cerestore)," medical dissertation, Zurich, preparation in 1989

57. J. McLean, The Future for Dental Porcelain, *Dental Ceramics, Proceedings of the First International Symposium on Ceramics*, J.W. McLean, Ed., Quintessence Publishing, 1983, p 14–40

58. A. Wohlwend, J.R. Strub, and P. Schärer, Metal Ceramic and All-Porcelain Restorations: Current Considerations, *Int. J. Prosthodontics*, Vol 2 (No. 1), 1989, p 13–26

59. G.N. Graser, M.L. Meyers, D.G. Grossman, and V.T. Cammarato, Preliminary Clinical Evaluation of Cast Ceramic Fixed Partial Dentures, *J. Dent. Res.*, Vol 64 (Abst. No. 1688), 1985, p 362

60. L.L. Hench, R.J. Splinter, T.K. Greenlee, and W.C. Allen, Bonding Mechanism at the Interface of Ceramic Prosthetic Materials, *J. Biomed. Mater. Res.*, Vol 2, 1971, p 117–141

61. G. Pitrowski, L.L. Hench, W.C. Allen, and G.J. Miller, Mechanical Studies of the Bone Bioglass® Interfacial Bond, *J. Biomed. Mater. Res.*, Vol 6, 1975, p 47–61

62. A.E. Clark, L.L. Hench, and H.A. Paschall, The Influence of Surface Chemistry on Implant Interface Histology: a Theoretical Basis for Implant Materials Selection, *J. Biomed. Mater. Res.*, Vol 10, 1976, p 161–174

63. T. Kukubo, S. Ito, S. Saka, and T. Yamamuro, Formation of a High-Strength Bioactive Glass-Ceramic in the System MgO-CaO-SiO$_2$-P$_2$O$_5$, *J. Mater. Sci.*, Vol 21, 1986, p 536–540

64. W. Vogel, W. Holand, E. Naumann, and J. Gummel, Development of Machinable Bioactive Glass Ceramics for Medical Uses, *J. Non-Cryst. Solids*, Vol 80, 1986, p 34–51

65. O.H. Andersson, K.H. Karlsson, K. Kangasniemi, and A. Yli-Urpe, Models for Physical Properties and Bioactivity of Phosphate Opal Glasses, *Glastech. Ber.*, Vol 61, 1988, p 400–405

66. T. Yamamuro, L.L. Hench, and J. Wilson, Ed., *Handbook of Bioactive Ceramics Volume I Bioactive Glasses and Glass-Ceramics*, CRC Press, 1990

67. D.A. Kessler, M.P. Stuart, and D.N. Sundwall, The Federal Regulation of Medical Devices, *N. Eng. J. Res.*, Vol 317, 1987, p 357–366

Consumer Houseware Applications

G.J. Fine, Corning Incorporated

THE UTILITY OF GLASS for the mass production of a variety of consumer housewares was probably first recognized by the ancient Romans. By the end of the first century A.D., inexpensive tableware and drinkware made of glass were common throughout the empire. The glass compositions used by the Romans were relatively unsophisticated relative to the wide range of glass compositions now accessible to commercial glass manufacturers, but the Romans clearly admired glass, citing its appearance, adaptability, and low cost (Ref 1). These features, when combined with the durability and resistance to thermal breakage offered by more modern compositions, continue to create demand for a variety of consumer housewares fabricated from glass.

Glass housewares fall into four general categories:

- Tableware (including dinnerware, cups, and mugs)
- Drinkware
- Bakeware (or ovenware)
- Top-of-stove

While specific estimates are difficult to obtain, the market for glass housewares is undoubtedly on the order of $1 billion in the United States. Depending upon the specific category, a variety of other materials vie for market share, including ceramics, metals, and plastics.

Material Requirements

The environments that glass housewares are exposed to can be extremely harsh and place constraints on the range of usable compositions. A simple example is the hot, highly basic environment of the home dishwasher, which can readily dissolve the surface of many glasses. In general, these environments are well-known and characterized. The definition of material requirements for glass housewares is highly subjective, however, and is often the result of a delicate and often confusing balancing act between consumer needs and expectations, cost, and safety. For example, the dishwasher durability of tableware decorations varies considerably depending upon the manufacturer. While it is tempting to assume that high cost tableware is also more durable, this is often not the case. Despite this ambiguity, it is possible to summarize the general material requirements for each segment of the consumer housewares market and the tradeoffs implicit in selection of a glass composition for any housewares application.

Durability is probably the most important prerequisite. Glass and decoration must be durable enough to withstand chemical attack by hot fluids and solids over a wide range of pH. Typical commercial dishwasher detergents have a pH on the order of 10 to 11, while some foods can be quite high in acetic or citric acid content. Temperatures of exposure vary depending upon the application; top-of-stove ware can exceed 500 °C (930 °F) when used improperly. A variety of tests are used to quantify glass durability. For example, detergent durability is often evaluated using a 0.3% solution of SuperSoilax detergent held at 95 °C (200 °F). Other solutions used include hydrochloric acid, citric acid, and acetic acid. After set periods of time, weight loss and appearance are used to form a qualitative assessment of glass durability. Quantitative results are often not meaningful, because weight loss and appearance after exposure do not vary proportionally and it is difficult to assign a relative importance to either factor. Decoration durability after exposure is typically evaluated on the basis of four criteria: color, gloss, staining tendency, and adherence to the glass surface.

Housewares must also be strong enough to withstand household or commercial use. A variety of strength tests are commonly used, including rim impact, drop strength and flexural strength (modulus of rupture; MOR). The latter is geometry independent and is probably the most critical for materials evaluation. In general, modulus of rupture is usually measured on ware abraded to simulate the effects of use. Glasses usually have MOR of approximately 70 MPa (10 ksi) after abrasion; it can be doubled by tempering. Glass-ceramics, discussed below, have a range of flexural strengths.

Requirements for thermal shock resistance and stability vary with application. Obviously, they are more stringent for top-of-stove than tableware. Thermal downshock in excess of 500 °C (900 °F) is known to occur during top-of-stove use, while tableware downshock is on the order of 100 °C (180 °F). Thermal shock resistance is a function of coefficient of thermal expansion (CTE) over the temperature range of interest and ware geometry, especially thickness. Tempering more than doubles the thermal shock resistance of glass.

Modern cooking practices dictate that glass housewares should not heat excessively in microwave ovens. Direct measurement of the heat-up temperature of a particular geometry is the most reliable measure. There is a reasonable correlation between the loss tangent of glass or glass-ceramic and microwave heating. The loss tangents of a variety of commonly used materials are available in the literature (Ref 2).

Safety and Health

As with all consumer goods, the safety and health of the consumer is a primary concern. Release of toxic elements and materials into food has received special attention of late. In the United States, the Food and Drug Administration (FDA) has the power to issue a recall of any glassware that significantly adulterates food or drink. The FDA sets no absolute limit, but in conjunction with a variety of trade organizations has formulated a set of industry standards, above which the FDA will take action, including product recall.

Lead and cadmium are specific concerns. Lead has been used in a variety of commercial glazes, while cadmium is often used to create red colors for decorations. To test for lead and cadmium release, glassware is immersed in a 4% solution of acetic acid for 24 h and the concentrations of lead and cadmium dissolved in the solution are then measured. Maximum permissible limits are under discussion at the time of publication. In general, the glass industry is moving toward the elimination of lead and cadmium from all its products.

Tableware

Typical compositions for glass tableware are given in Table 1. Four general classifications of glass tableware are currently available in the U.S.: soda-lime, opal, laminate, and glass-ceramic.

Soda-Lime. Transparent soda-lime dinnerware is probably the least expensive currently sold in the United States. The advantages of soda-lime dinnerware are twofold: batch costs are low and forming utilizing conventional pressing or spinning techniques is straightforward. Soda-lime glass can also be readily colored; for instance, a line of black soda-lime dinnerware colored by adding a combination of MnO_2 and Cr_2O_3 is sold.

As a tableware material, soda-lime glass has disadvantages. First, while the durability is adequate, it is not exceptional. Possible reactions may take many forms, but two are important. In acidic solutions, exchange between H^+ ions exposed to the surface of the glass, and alkali ions in the glass can readily occur. This ion exchange often results in the formation of an iridescent, soluble layer on the glass surface. Conversely, exposure to basic solutions can cause disruption of the glass structure by the introduction of OH^- ions and can eventually result in complete dissolution of the glass. The formation of a white film on glassware after prolonged exposure to dishwasher detergent is a well-known example of the latter process.

The high coefficient of thermal expansion of soda-lime glass results in minimal thermal shock resistance. A $15 \times 15 \times 6.4$ mm (0.6 $\times$ 0.6 $\times$ 0.25 in.) piece of annealed soda-lime glass can survive thermal downshock of only 50 °C (90 °F). The abraded flexural strength is about 70 MPa (10 ksi), typical for

glass (Ref 2). To improve both strength and thermal shock resistance, most soda-lime dinnerware is tempered.

Opals. Inexpensive opaque tableware is usually made from phase-separated, or opal, glass. Opal glasses usually fall into two general categories: crystalline opals and liquid-liquid opals. In the former, crystals spontaneously form during manufacturing either upon cooling or subsequent heat treatment. The crystalline phase formed in dinnerware is usually either NaF or CaF_2, although many other crystalline opals have been discovered. Liquid-liquid opals are formed by separation of the glass into two discrete glassy phases while the glass is still in its softened state. In both types of opals, the opacity of the glass is caused by light scattering and is thus controlled by three factors: the difference in refractive index between the two phases in the glass, the degree of phase separation, and the size and distribution of the lesser phase. Opal glasses can vary from being translucent to nearly opaque, depending on these variables, and are usually creamy white (Ref 3).

Low batch cost and ease of forming combined with acceptable appearance are primary advantages of opal glasses. The disadvantages are the same as those of soda-lime dinnerware: mediocre durability, mechanical strength, and resistance to thermal breakage. Durability problems are caused by a number of different phenomena. In the fluorine-containing opals, durability is compromised either by the mobility of alkali ions in the glass or by the mobility of fluorine. In liquid-liquid opals, one phase is usually less durable than desirable. Depending upon the continuity of the phase, poor durability can result. Durability problems are most common in the more opaque, or dense, opals, which usually contain more alkali or fluorine. Opal glasses, like

soda-lime glasses, are usually tempered. Ware is also relatively thick to maintain high mechanical strength.

Laminates. A novel attempt to rectify the deficiencies of opal glasses was developed in the 1960s and resulted in the development of Corelle (Ref 4). Corelle is a laminate of two glasses: the core (or body) of a Corelle plate is a CaF_2 opal with a coefficient of thermal expansion (CTE) of 71×10^{-7}/°C; the cladding (or skin) is an alkaline-earth aluminosilicate with a CTE of 48×10^{-7}/°C. A Corelle laminate is formed by melting the two glasses in separate furnaces and then delivering the glass through a common orifice to form a hot, laminated sheet. During delivery, the CaF_2 core glass forms a spontaneous opal. The sheet is then cut into the desired shape, decorated, and tempered.

The advantages of a Corelle laminate are threefold: First, the alkaline-earth aluminosilicate cladding glass is exceptionally durable relative to typical dense opal glasses. The opal glass in the Corelle system is not exposed, but the aesthetic appearance of a dense opal can be created without the usual poor durability. Second, the differential thermal expansion of the core and skin glasses create compressive stresses in the glass surface upon cooling that are well in excess of those normally created by tempering alone. After tempering, abraded MOR of a Corelle laminate exceeds 250 MPa (36 ksi). Resistance to rim impact and bruise checks are also notably improved. Third, the high strength of the laminate allows thinner ware, resulting in a lighter weight product.

Glass-Ceramics. Another unique approach to tableware takes advantage of the properties of glass-ceramics. Glass-ceramics constitute a bridge between glasses and crystalline ceramics. Glass-ceramics are best defined as

Table 1 Common tableware compositions and properties

Compound/property	Composition, wt%(a)							
	1	2	3	4	5	6	7	8
SiO_2	71.5	75.3	58.9	71.1	58.0	64.0	43.3	67.1
Al_2O_3	5.6	3.2	10.5	1.6	15.0	6.0	29.8	1.8
B_2O_3	...	13.5	1.4	...	6.0	5.0	5.5	0.3
Li_2O	...	...	...	...	...	...	...	0.8
Na_2O	13.0	6.6	8.4	14.2	...	3.0	14.0	3.0
K_2O	...	...	...	0.5	...	3.0	...	4.8
MgO	...	...	...	1.1	6.0	1.0	...	14.3
CaO	...	0.7	5.8	9.8	15.0	15.0	...	4.7
BaO	6.3	...	...	...	...	...	...	...
ZnO	...	...	8.7	...	...	...	...	...
TiO_2	...	...	...	...	...	...	6.5	...
P_2O_5	...	...	...	...	...	...	...	1.0
As_2O_3	...	...	...	...	...	...	0.9	...
Sb_2O_3	...	...	...	...	...	...	...	0.2
Fluorine	3.7	0.8	4.2	...	3.0	...	...	3.5
Softening point, °C	715	764	792	...	890	...	...	...
CTE ($\times 10^{-7}$/°C) at 25–300 °C	75	44	80	...	48	71	95	118
Type	NaF opal	Liquid-liquid opal	CaF_2 opal	Soda-lime	Borosilicate cladding	CaF_2 opal core	Nepheline ceramic	K-richterite glass-ceramic

(a) Numbers refer to the following products: 1, Durand Table; 2, Durand Arcopal; 3, Corning Code 6720; 4, Durand Soda-Lime; 5, Corning Corelle Core Glass; 6, Corning Corelle Skin Glass; 7, Corning Pyroceram Code 9609; 8, Corning Pyroceram Code 0308

microcrystalline solids produced by the controlled devitrification of glass. Glasses are melted and formed using standard techniques and then heat treated to produce uniformly grained crystalline materials. Glass-ceramics are distinguished from opal glasses by the degree of crystallinity; glass-ceramics generally contain >50% crystals by volume.

The specific properties of any glass-ceramic are controlled by both the physical properties of the individual crystals and by the textural relationship between the crystals and the residual glass. As a result, glass-ceramics can offer a wide variety of properties not attainable in conventional glasses, including durability, strength, and exceptional thermal shock resistance (Ref 5).

An example of this phenomenon is Pyroceram tableware. Two compositions have been sold under this trademark. The oldest crystallizes nepheline and celsian upon heat treatment. Titania is added to the glass to enhance internal nucleation. This material has a relatively high CTE of $95 \times 10^{-7}/°C$. This permits application of a glaze with considerably lower CTE, resulting in compression strengthening of the ware. The net result is a high gloss, durable, high-strength dinnerware (Ref 6).

The second composition crystallizes potassium fluorrichterite and cristobalite, rather than nepheline and celsian, as the primary phases. Potassium fluorrichterite is a chain silicate with a rod-like morphology; glass-ceramics consisting of chain silicates have unique microstructures best described as acicular (Fig 1). In an acicular microstructure, the interlocking rods effectively deflect crack propagation and strengthen and toughen the material (Ref 7). The abraded MOR of a fluorrichterite glass-ceramics is about 150 MPa (22 ksi). Since these glass-ceramics have CTEs on the order of $115 \times 10^{-7}/°C$, the application of lower CTE glazes further enhances ware strength via surface compression.

The primary disadvantage of glass-ceramic tableware is its relatively high cost. There are three reasons: First, unlike soda-lime or opal dinnerware, glass-ceramics must be given a secondary heat treatment to form crystals. Second, a glaze is necessary for glossy surface. Glaze application adds cost. Third, during heat treatment, glass viscosity is often low enough that deformation results. The ware must thus be supported by ceramic formers, which greatly complicates the forming process and adds additional expense.

Alternate Materials. The primary competition for glass and glass-ceramic tableware comes from ceramics. There are a variety of types of ceramic tableware; none are reviewed here. It is worth noting that all must be glazed with glass to produce a glossy surface. Four typical glaze compositions are given in Table 2 (Ref 8). Glazes must have lower CTE than the ware body for strength and must be fired at a low enough temperature to preclude ware distortion during firing.

Drinkware

Material requirements for drinkware are similar to those for tableware, although thermal shock resistance is less important and durability is emphasized. Commercial drinkware compositions are given in Table 3. Three general classifications are prevalent.

The least expensive drinkware is usually made of tempered soda-lime glass; compositions do not differ significantly from those used for dinnerware and the properties are approximately the same.

The most expensive and probably best known compositions are made out of lead glass, also known as lead crystal. In Europe, the term "lead crystal" can only be used to describe compositions containing greater than 24 wt% PbO. In the United States, regulations are less specific. Lead glass compositions have high refractive indices relative to most glasses, a feature which creates an aesthetically pleasing brilliance in lead crystal. To maintain this brilliance, special care must also be taken to use pure batch materials during melting; contaminants such as iron oxide create undesirable colors in the glass that detract from the brilliance.

Finally, a number of manufacturers use a glass composition best characterized as alkaline-earth aluminosilicate. These glasses have refractive indices approaching those of lead crystal, but are apparently easier to manufacture and are slightly more durable.

In general, studies have found little functional difference between the three types of drinkware. Strength is a function of tempering and geometry; little difference exists between the three compositions. All have similar coefficients of thermal expansion (in the 80 to $100 \times 10^{-7}/°C$ range), which minimizes thermal shock resistance, but makes tempering straightforward. The durability of all the compositions is comparable, but none are exceptional relative to other glasses.

Ovenware

Material parameters are the same for ovenware, although obviously thermal shock resistance is much more important. In gen-

Table 2 Common glaze compositions for tableware

Compound	Composition, wt%(a)			
	1	2	3	4
SiO$_2$	59.1	55.8	55.9	42.5
Al$_2$O$_3$	13.5	7.4	9.6	7.0
B$_2$O$_3$	4.3	5.5	6.0	8.9
LiO$_2$	0.5	...	...	...
Na$_2$O	1.8	1.8	3.1	2.5
K$_2$O	3.9	2.7	1.7	...
MgO	1.0	0.6	...	...
CaO	11.8	9.2	7.7	3.1
BaO	...	2.5	...	...
SrO	4.1	3.1	...	...
ZnO	...	10.9	...	...
ZrO$_2$	...	0.6	...	0.7
PbO	...	...	16.1	35.3

(a) Numbers refer to the following products: 1, Semivitreous lead-free dinnerware glaze; 2, Vitreous lead-free dinnerware glaze; 3, Leaded dinnerware glaze; 4, Leaded low-temperature dinnerware glaze. Source: Ref 8

Fig 1 Microstructure of a potassium fluorrichterite glass-ceramic. The long acicular blades yield a material with unusual strength and toughness. 30,000×

Table 3 Drinkware compositions and properties

Compound/property	Composition, wt%(a)				
	1	2	3	4	5
SiO$_2$	59.9	72.6	68.4	69.4	69.4
Al$_2$O$_3$	...	1.3	0.2	0.6	1.4
Na$_2$O	3.3	13.5	10.6	10.7	9.8
K$_2$O	10.3	...	5.5	7.2	5.6
MgO	...	4.3	...	1.2	...
CaO	1.3	6.8	10.0	4.9	7.6
BaO	...	...	4.5	6.0	5.7
PbO	24.4	...	...	...	...
B$_2$O$_3$	...	...	0.8	...	0.7
Softening point, °C	647	736	695	682	704
CTE ($\times 10^{-7}/°C$) at 25–300 °C	89	86	98	99	89
Type	Lead crystal	Soda-lime	Alkaline earth silicate	Alkaline earth silicate	Alkaline earth silicate

(a) Numbers refer to the following producers: 1, Durand; 2, Libbey; 3, Schott; 4, Rosenthal; 5, Bormioli

Table 4 Ovenware compositions and properties

Compound/property	Composition, wt%(a)					
	1	2	3	4	5	6
SiO_2	76.4	74.0	82.0	69.7	68.8	66.1
Al_2O_3	1.8	2.0	2.0	17.8	19.2	20.2
B_2O_3	14.7	0.5	12.0	...	...	...
Li_2O	...	...	...	2.8	2.7	3.9
Na_2O	5.4	13.5	4.0	...	0.2	0.4
K_2O	...	...	...	...	0.1	...
MgO	...	...	...	2.6	1.8	1.0
CaO	0.7	10.0	...	...	...	...
BaO	...	...	...	...	0.8	1.8
ZnO	...	...	...	1.0	1.0	1.0
TiO_2	...	...	...	4.7	2.7	3.0
ZrO_2	...	...	...	...	1.8	1.7
As_2O_3	0.4	...	...	0.6	0.8	1.0
Sb_2O_3	...	0.2	...	...	...	...
CTE ($\times 10^{-7}$/°C) at 25 to 300 °C	...	86	37	4–20	5–7	1
Type	Tempered borosilicate	Tempered soda-lime	Tempered borosilicate	β-spodumene glass-ceramic	β-quartz glass-ceramic	β-quartz glass-ceramic

(a) Numbers refer to the following products/producers: 1, Anchor Hocking; 2, Corning Code 0281; 3, Corning Code 7251; 4, Corning Corning Ware; 5, Corning Visions; 6, Durand Arcoflam Clear-Line

eral, ovenware should be able to withstand thermal downshock of 150 °C (270 °F) without breakage. Glass and glass-ceramic compositions typically used for ovenware are given in Table 4.

Glass. Pyrex is probably the best known example of glass ovenware. Two separate glass compositions are sold as Pyrex, a tempered soda-lime glass and a borosilicate.

The soda-lime glass is not significantly different from those described previously in this article. Borosilicate glasses offer a number of advantages over soda-lime. The effect of the addition of B_2O_3 to the glass compositions is twofold. First, B_2O_3 lowers the CTE by up to 50%, thereby increasing the resistance of the ware to thermal shock. Second, the durability of the glass is dramatically enhanced. The causes for enhanced chemical durability in borosilicates are unclear. A prevalent theory for the enhanced durability is that as the glass cools, it separates into two immiscible phases, a discontinuous phase rich in Na_2O and B_2O_3, and a continuous SiO_2-rich phase. Because the SiO_2-rich phase is dominant, the durability of the glass is enhanced without loss of the ability to form the glass at reasonable temperatures (Ref 9). Since the diameter of the discontinuous phase is well below the wavelength of visible light, the transparency of the material is maintained.

Despite the low thermal expansion of the borosilicate composition sold as Pyrex, the composition is easily tempered and is sold in tempered form.

Glass-ceramic ovenware is generally made by the heat treatment of Li_2O-Al_2O_3-SiO_3 glass containing small amounts of TiO_2 or a combination of both TiO_2 and ZrO_2. Upon heat treatment, small crystals of rutile or zirconium titanate precipitate from the glass. These small crystals then provide nucleation sites for the growth of lithium aluminosilicate crystals from the glass. Depending upon the temperature of heat treatment and the composition of the glass, the lithium aluminosilicate phase is either β-spodumene or β-quartz. β-spodumene glass-ceramics typically have CTEs on the order of 12×10^{-7}/°C to 15×10^{-7}/°C, while β-quartz compositions can have lower, or even negative, CTEs. Given the low thermal expansions, these glass-ceramics can withstand the sudden temperature changes associated with oven usage (Ref 6).

Compositions which crystallize β-spodumene are best exemplified by Corning Ware, a durable, opaque cooking vessel. In certain β-quartz composition, the size of each of the crystals is significantly below the wavelength of light and the refractive index of the crystals can be made to match that of residual glass in the ware. As a result, light scattering does not occur and the materials are transparent.

Top-of-Stove

The thermal requirements for top-of-stove ware are extremely stringent. Probably as a consequence, only two general classifications of glassware have been sold for top-of-stove use. β-quartz glass-ceramics have the low CTE and thermal stability required for top-of-stove use and have found widespread acceptance. Some glassware has also been sold for top-of-stove use. Aluminosilicate glass compositions are most suitable, with relatively low CTEs and softening points high enough to prevent deformation during use. A typical aluminosilicate composition has a softening point of 915 °C (1680 °F) and a CTE of 42×10^{-7}/°C. When tempered, ware has sufficient thermal stability for top-of-stove use. Tempered aluminosilicates are also used for thermal servers such as percolator pots, although borosilicates are also used.

ACKNOWLEDGMENT

Corelle, Pyrex, Corning Ware, Pyroceram, and Visions are registered trademarks of Corning Inc.

REFERENCES

1. D.F. Grose, Innovation and Change in Ancient Technologies: The Anomalous Case of the Roman Glass Industry, *High Technology Ceramics: Past, Present and Future*, W.D. Kingery, Ed., American Ceramic Society, 1986, p 65–80
2. D.C. Boyd and D.A. Thompson, Glass, *Kirk-Othmer: Encyclopedia of Chemical Technology*, Vol 11, John Wiley & Sons, 1980, p 807–880
3. J.E. Flannery and D.R. Wexell, Opal Glasses, *Commercial Glasses*, D.C. Boyd and J.F. MacDowell, Ed., American Ceramic Society, 1986, p 141–150
4. W.H. Dumbaugh, J.E. Flannery, and J.E. Megles, Strong Composite Glasses, *J. Non-Cryst. Solids*, Vol 38 and 39, 1980, p 469–474
5. B.H.W.S. DeJong, J.W. Adams, B.G. Aitken, J.E. Dickinson, and G.J. Fine, Glass Ceramics, *Ulmann's Encyclopedia of Industrial Chemistry*, Vol A12, VCH Publishers, 1990, p 433–448
6. G.H. Beall, Glass-Ceramics, *Commercial Glasses*, D.C. Boyd and J.F. MacDowell, Ed., American Ceramic Society, 1986, p 157–173
7. G.H. Beall, Design of Glass-Ceramics, *Rev. Solid State Sci.*, Vol 3, 1989, p 333–354
8. R.A. Eppler, Glazes and Enamels, *Commercial Glasses*, D.C. Boyd and J.F. MacDowell, Ed., American Ceramic Society, 1986, p 141–150
9. R.H. Doremus, *Glass Science*, John Wiley & Sons, 1973

Electrical/Electronic Applications for Advanced Ceramics

Chairman: R.C. Buchanan, Department of Materials Science and Engineering, University of Illinois at Urbana-Champaign

Introduction

CERAMIC MATERIALS serve important insulative, capacitive, conductive, resistive, sensor, electrooptic, and magnetic functions in a wide variety of electrical and electronic circuitry.

Traditional voltage insulative uses have involved mainly dielectric isolation of conductors and various active devices. Glass insulators, ceramic heater substrates for microelectronics packaging, are all primarily used in this mode. Often, however, the ceramic material must exhibit other important characteristics, including temperature, corrosion and environmental stability, high mechanical strength, and good thermal shock resistance. As substrates for packaging applications, more demanding requirements may include the promotion of good bonding and adhesion to various materials, stability to reduced ambient conditions, and compatibility with solder reflow processes. In high density packaging, accompanying requirements are for (1) thermal-expansion match with silicon in order to reduce thermal stresses, (2) low dielectric constant to enhance signal processing, and (3) high thermal conductivity for heat dissipation from the hard driven integrated circuit devices. To meet these sometime conflicting requirements, different substrate formulations (Al_2O_3, BeO, mullite [$3Al_2O_3 \cdot 2SiO_2$], AlN, Si_3N_4, and SiC) have evolved which exhibit many, but rarely all, of the desired properties. Alumina (Al_2O_3-96 to 99 wt%) is currently the most widely used ceramic substrate material, since it satisfies most criteria. BeO, because of its high thermal conductivity, is utilized where heat dissipation is critical. AlN combines high thermal conductivity with an expansion coefficient relatively close to that of silicon and, in consequence, is being rapidly developed for substrate use.

Thin-film insulators, including SiO_2 and other oxides, glasses, as well as Si_3N_4, have been developed as interlayer dielectric and as thin-film passivation for integrated circuit devices. Also undergoing rapid development are diamond insulating films as well as ferroelectric thin films for nonvolatile memory (RAM, SRAM) applications.

Use of ceramic materials in the highly complex and multi-infacial regime of a microelectronic package requires close attention to processing details and control of precursor, synthesis, and impurity levels. These characteristics are also important for dielectric capacitance uses, where metal electrode interfaces and thin layer dielectrics also place high demand on the ceramic. $BaTiO_3$-based compounds, which exhibit high dielectric constants and which are formulated to satisfy specified capacitive and temperature characteristics (NPO, Z5U, X7R, and so on) are the most widely used materials for disc, multilayer, and thick-film capacitors. However, relaxor-based formulations are now being developed. Typically these consist of lead magnesium niobate (PMN) [or $Pb(Mg_{1-3}Nb_{2-3})O_3$] modified with (Zn, Fe, or Ni) as divalent cations and (Ti, Nb, Ta, or W) as higher valence cations. Key characteristics of these relaxors are their high dielectric content, broad dielectric maxima, and low firing temperature, which allow use of less expensive silver palladium electrodes and cofiring in air ambient.

Current trends in capacitor technology are increasingly towards integrated device packages where the capacitor is processed in thin- or thick-film form. Development of ferroelectric thin films for memory applications should accelerate this trend. Primarily these films are based on lead zirconate-titanate (PZT) formulations, either vapor deposited or sol-gel derived, but other formulations (PT, PLZT, and relaxor compositions) are also being developed. In bulk form these materials are all strongly piezoelectric and rapid strides have been made in their development and use as actuator motors, ultrasonic transducers, and for electromechanical sensing. Piezoelectrics (PLZT) also elicit wide interest as electrooptic materials for such applications as high-speed shutters, switches, and light modulators and displays.

Besides piezoelectrics, a wide range of ceramic materials have been developed as gas, oxygen, temperature, voltage, and humidity sensors. Typically, these are based on semiconducting oxides such as TiO_2, ZnO, modified $BaTiO_3$, SnO_2, and $MgCr_2O_4$. Since many of these sensors can provide a feedback signal for process control (smart sensors), these materials represent an expanding area of ceramic exploration and development. For the oxygen sensors, solid electrolyte conductors such as yttria-stabilized ZrO_2 are the most widely used particularly for automotive applications, but TiO_2 is also being developed as an oxygen sensitive resistive sensor. Other categories of ceramic materials include ferrites and ceramic (high T_c) superconductors. Recent developments in ferrites have been related primarily to their use as recording heads, but research in the area of microwave ferrites is also active.

Ferrites are ceramic oxide materials which exhibit ferrimagnetic properties by virtue of the opposed (A-B coupling) and variable cation orientation within the material. Categories of ferrite include hard, soft, and microwave materials. All are based primarily on Fe_2O_3. The hard ferrites are of composition (MeO, $6Fe_2O_3$, where Me = Sr and Ba). This class of materials is used as permanent magnets and for motor applications. Soft ferrites (MeO, Fe_2O_3) are based on the cubic spinel structure (Me = Ni, Co, Cu, Zn, and Mn) and are readily switchable and modified. These are used in computer memory cores, deflection yokes, E-cores, and for other applications. Microwave ferrites are based on the garnet structure and also contain iron-oxide ($3Y_2O_3$, $5Fe_2O_3$). They feature high-resistivity and high-frequency capability.

Recent developments with ceramic superconductors have targeted a number of important potential applications for these materials. Projected areas of major impact include thin-film devices, microwave applications, energy storage, and microelectronics. The structure of the ceramic superconductors is based mainly on an oxygen-deficient, layered perovskite structure. To date, formulations in the (Y-Ba-Cu-O) system (123) have been most successfully developed, but many other compositions have been explored. Processing of superconductors to achieve useful shapes, controlled microstructures, and high-current density materials are major challenges.

This brief review has highlighted the broad range of ceramic materials that are used for electrical and electronic applications. Typically, each category of materials exhibits unique and desirable property characteristics which directly reflect their composition, processing, and microstructure. The following articles will further develop and expand on these aspects of the materials.

Ceramic Substrates

David G. Wirth, Coors Ceramics Company

FOR HIGH-RELIABILITY electronic circuitry, ceramics historically have been the choice for both substrates and semiconductor packages. Traditionally, alumina ceramics dominate this area of application for their outstanding thermal and corrosion stability and their ability to provide hermeticity. However, new materials and processes may be needed to meet more demanding requirements such as denser line spacings, faster response, and better heat dissipation.

Increased circuit processing speed requires low-resistance conductors such as gold, silver, and copper. This allows the decrease of line widths and thicknesses. At the same time, ceramic insulators with lower dielectric constants would be required to minimize the delay of signal propagation. Materials such as glass-ceramics appear to be excellent candidates for these requirements (Table 1). The disadvantages of these materials are their low mechanical strength and poor thermal dissipation.

As silicon devices become larger, a closer match of expansion coefficients is needed for the supporting structure, substrate, or package. Materials such as mullite or aluminum nitride appear to be excellent choices for this need. In addition, as device requirements go to higher power, the need for improved heat removal becomes critical. This has traditionally been provided by the use of beryllia ceramics with their high thermal conductivity. More recently, the development of AlN with high thermal conductivity is very attractive for these applications, because it does not have the inherent toxicity problems associated with BeO processing.

These new materials and their added capabilities to provide solutions to future requirements appear very promising. Considerable development of metallization systems and material processing is, however, required before the same reliability of alumina substrate and package systems is realized.

Alumina Substrates

Depending on the subsequent processing, three types of alumina substrates are used to generate the required circuitry:

- Thick-film substrates
- Thin-film substrates
- Cofired alumina packages

Thick Film

The standard for thick-film substrates is the 96% alumina formulation, which has been used widely in the fabrication of hybrid microelectronic circuits. This is due in large part to their technical qualities, availability in large quantity at low cost, and their high quality. The substrates have excellent electrical insulating properties, good thermal conductivity, mechanical strength, chemical durability, and dimensional stability.

Composition. Thick-film alumina substrates having a nominal composition of 96% alumina are available from a number of qualified suppliers. Each supplier has its own proprietary formulation and manufacturing process, which may account for minor differences. In general, all formulations contain an alkaline-earth/aluminosilicate-glass phase that bonds the α-alumina crystals together. The glass phase also provides a bonding mechanism for adherence of the thick film with both conductor and resistor materials. The alkali content of the alumina and other additives must be kept at a very low level to prevent interaction of the substrate with the glass phase of the conductor and resistor materials, thus causing erratic behavior relative to absolute resistor values obtained as well as variability in the thermal coefficient of resistance. Typical thick-film alumina compositions fall in the ranges indicated in Table 2.

The fired microstructures obtained from such formulations in general consist of the α-alumina crystals and a glassy bond phase. Depending on thermal history during processing and on specific composition, the glass phase may often times partially recrystallize into binary and tertiary silicate and aluminate phases, which may cause undesirable properties based on thermal expansion incompatibility with both alumina and the residual gas phase. This may evidence itself as a network of small microcracks and may also result in inferior chemical durability during subsequent processing. In addition, upon further circuit generation, difficulty may be noted in conductor adhesion, conductor solderability, and anomalous resistance values.

Processing. Thick-film alumina substrates are fabricated by dry pressing or by a tape process, powder roll compaction, or doctor blade casting. The size of the desired parts usually dictates the fabrication process. In general, thickness requirements of less than 1.5 mm (0.060 in.) are usually met by a tape casting process, whereas dry-press fabrication is used for greater thicknesses. For large area parts, greater than 50 × 50 mm (2 × 2 in.), the tape process is favored with sizes up to 125 × 230 mm (5 × 9 in.) generally available.

Parts can be fabricated in a variety of geometrics with or without the presence of holes and slots (Fig 1). In addition, specific sizes and geometries may be fabricated by laser machining the fired substrate blanks. Table 3 lists typical dimensional specifications which can be readily obtained.

Properties and Functions. Table 4 lists the typical properties of importance for thick-film alumina. In addition to maintaining the proper purity and phase content during processing, the physical condition of the surface is fairly important. The achievement of a 3.75 g/cm^3 density along with a surface finish of approximately 0.63 μm (25 μin.) center line average (CLA) maximum is important in subsequent thick-film processing. Excessive surface porosity may lead to "flow out" of the glass binders used to obtain conductor adherence or resistor uniformity. The proper grain size and resulting surface finish is of prime importance in conductor adhesion and solderability particularly with respect to glass-bonded conductors. Glass-bonded conductors adhere to the substrate by reacting with the surface composition. If the substrate surface contains excess amounts of sodium or calcium, the softening point of the glass in the conductor will be lowered; consequently, its viscosity at temperature will also be reduced. The glass flows out from under the conductor causing a stained color and reduced adhesion. This may particularly affect aged adhesion results.

Conductor solderability is also enhanced by uniform surface composition. It has also been shown that variable surface composition can affect resistance values, the coefficient of variation, and the temperature coefficient of resistance, particularly with small-geometry, high-resistance formulations.

Thin Film

The standard for thin-film alumina substrates is the tape cast 99.5% Al_2O_3 formu-

Table 1 Physical properties of substrate materials

	96% Al₂O₃	99.5% Al₂O₃	BeO	AlN	Mullite	Various glass-ceramic materials
Density, g/cm³	3.75	3.90	2.85	3.25	2.82	2.5–2.8
Flexural strength, MPa (ksi)	400 (58)	552 (80)	207 (30)	345 (50)	186 (27)	138 (20)
Thermal expansion from 25–500 °C (77–930 °F), 10⁻⁶/°C	7.4	7.5	7.5	4.4	3.7	3.0–4.5
Thermal conductivity at 20 °C (68 °F), W/m·°C	26	35	260	140–220	4	4–5
Dielectric constant at 1 MHz	9.5	9.9	6.7	8.8	5.4	4–8
Dielectric loss at 1 MHz (tan δ)	0.0004	0.0002	0.0003	0.001–0.0002	0.003	>0.002

Table 2 Typical chemical composition of thick-film alumina substrates

Substrate	Composition, wt%
Al₂O₃	96
SiO₂	2.5–3.0
MgO	0.75–1.0
CaO	0.10–0.25
Na₂O	0.05–0.10 max
Fe₂O₃	0.03–0.05
ZrO₂	0–0.5

Table 3 Standard as-fired dimensional tolerances of thick-film alumina substrates

Dimension	Tolerance
Length, width, and hole centers	Not less than ±1% or ±0.08 mm (0.003 in.)
Thickness	Not less than ±10% or ±0.38 mm (0.0015 in.)
Camber	0.003 mm/mm (0.003 in./in.)
Hole diameters of:	
0.38–0.74 mm (0.015–0.029 in.)	±0.05 mm (0.002 in.)
0.76–2.54 mm (0.030–0.1 in.)	±0.08 mm (0.003 in.)
>2.54 mm (0.1 in.)	Not less than ±1%, or ±0.13 mm (0.005 in.)
Perpendicularity (squareness)	≤0.004 mm/mm (0.004 in./in.)
Edge straightness	≤0.003 mm/mm (0.003 in./in.)

lation which has been widely used for sputtered, evaporated, and chemically vapor deposited metals for circuit generation. The requirements of both surface finish and degree of surface perfection (freedom from defects) dictated by the thin-film metal deposition geometries also necessitates the use of very fine and reactive alumina raw materials to result in a small grain size of fired surfaces.

Composition. Thin-film alumina substrates are fabricated from alumina raw materials which are in the micron size range. The average particle size may vary from 0.5 to 1.5 μm for standard thin-film substrates to ultrasmooth fired finishes. In addition, the raw materials are of very high purity, low-alkali content, and very low in iron impurities. The only additives used in the formulations are a source of MgO, usually MgCO₃, which is added in small amounts to control the grain growth during densification. In some cases, a small amount of silica may also be added to assist in densification. In the standard thin-film alumina substrates, a fired average grain size of 2 to 2¹/₂ μm is obtained with the resulting surface finish of <0.15 μm (<6 μin.)

CLA. In ultrasmooth thin-film substrates, an average grain size of approximately 1 μm is obtained with a surface finish of <0.075 μm (<3 μin.) CLA.

Processing. The slurry of alumina and additives are normally ground in a ball mill to completely deagglomerate the alumina and to thoroughly mix the additional inorganic additives. An organic binder and plasticizer is then added and thoroughly mixed into the slurry during the latter stages of milling. The slurry is then de-aired and fed to a continuous caster where it is deposited onto a moving plastic carrier. A doctor blade with a constant-level slurry reservoir is used to uniformly spread the slurry onto the plastic carrier and control the wetsheet thickness at this point. The tape then is passed through a temperature and humidity dryer and then separated from the plastic carrier and rolled into spools.

Parts are punched from the spools of tape into various geometries (Fig 2). These parts are loaded onto refractory setters and fired in continuous kiln for densification. Parts may be refired, if necessary, on ground setters to improve their flatness.

Table 5 lists the surface quality attributes for the "A" surface for two grades of thin-film alumina substrates. The concentration and size of surface defects must be controlled at very low levels to give satisfactory yields during circuit fabrication.

Properties. Table 6 lists the typical material characteristics for both grades of alumina thin-film substrates. The 99.5% alumina materials are manufactured with a controlled range of density which is an important factor in controlling variations in surface finish, dielectric constant, and the resulting sheet resistivity. Tight density control ensures uni-

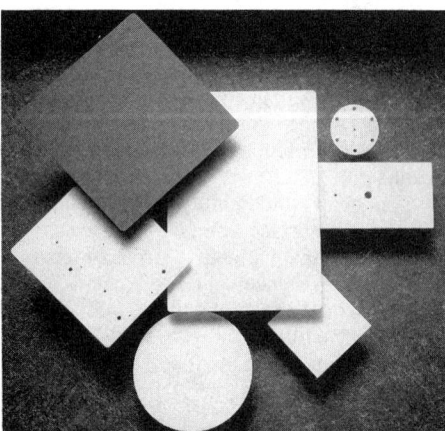

Fig 1 Various thick-film alumina substrates

Table 4 Typical properties of thick-film alumina (96% Al₂O₃)

Property	Conditions	Test	Property value
Surface finish (CLA), μm (μin.)	As-fired	Profilometer 0.762 mm (0.030 in.) cutoff	<0.6 (25)
Flexural strength, MPa (ksi)	21 °C (70 °F)	ASTM F 417	400 (58)
Thermal conductivity, W/m·°C	20 °C (68 °F)	ASTM C 408	26
	100 °C (212 °F)		20
	400 °C (750 °F)		12
Thermal coefficient of expansion, 10⁻⁶/°C	25–200 °C (77–390 °F)	ASTM C 372	6.3
	25–500 °C (77–930 °F)		7.1
	25–800 °C (77–1470 °F)		7.6
	25–1000 °C (77–1830 °F)		8.0
Volume resistivity, Ω·cm	25 °C (77 °F)	ASTM D 1829	>10¹⁴
	300 °C (570 °F)		5.0 × 10¹⁰
	700 °C (1290 °F)		4.0 × 10⁷
Temperature, °C (°F), at which resistivity is 1 MΩ·cm		ASTM D 1829	1000 (1830)
Dielectric strength (60 cycle average), kV$_{rms}$/mm (V$_{rms}$/mil)	0.635 mm (0.025 in.) thick specimens	ASTM D 116	23.6 (600)
Dielectric constant (relative permittivity) at 25 °C (77 °F)	1 kHz	ASTM D 150	9.5
	1 MHz		9.5
	100 MHz		9.5
Loss tangent (dissipation factor)	1 kHz	ASTM D 150	0.0010
	1 MHz		0.0004
	100 MHz		0.0004
Loss index	1 kHz	ASTM D 150	0.009
	1 MHz	ASTM D 150	0.004
	100 MHz	ASTM D 2520	0.004

Fig 2 Various thin-film alumina substrates

Table 6 Typical material characteristics of thin-film alumina substrates

Property	99.5% Al_2O_3	99.6% Al_2O_3	Test method
Bulk density, g/cm³			ASTM C 373
Typical	3.89	3.88	
Range	3.86–3.92	3.86–3.90	
Hardness, HR45N	87	87	ASTM E 18
Surface finish (as-fired center line			Profilometer: cut-off: 0.75 (0.030 in.),
average), μm (μin.)	≤0.15 (6)	≤0.075 (3)	stylus diam: 0.01 mm (0.0004 in.)
Average grain size, μm (mil)	≤2.2 (0.087)	≤1.2 (0.047)	Intercept method
Water absorption, %	Nil	Nil	ASTM C 373
Flexural strength, MPa (ksi)	570 (83)	570 (83)	ASTM F 394
Modulus of elasticity, GPa (10⁶ psi)	370 (54)	370 (54)	ASTM C 623
Poisson's ratio	0.20	0.20	ASTM C 623
Coefficient of linear thermal expansion,			
10^{-6}/°C, from 25 °C (77 °F) to:			
300 °C (570 °F)	6.7	6.7	ASTM C 372
600 °C (1110 °F)	7.5	7.5	
800 °C (1470 °F)	8.0	8.0	
1000 °C (1830 °F)	8.3	8.3	
Thermal conductivity, W/m·°C, at:			
20 °C (68 °F)	34.7	34.7	Various
100 °C (212 °F)	25.5	25.5	
400 °C (750 °F)	12.6	12.6	
Average rms dielectric strength (60			
cycle ac), kV/mm (V/mil):			
0.635 mm (0.025 in.) thick specimen	670	670	ASTM D 116
1.3 mm (0.050 in.) thick specimen	450	450	
Dielectric constant (relative			
permittivity) at:			
1 KHz	9.9 (±2%)	9.9 (±1%)	ASTM D 150
1 MHz	9.9 (±2%)	9.9 (±1%)	
Dissipation factor (loss tangent) at:			
1 KHz	0.005	0.005	ASTM D 150
1 MHz	0.002	0.002	
Loss index (loss factor) at:			
1 KHz	0.005	0.005	ASTM D 150
1 MHz	0.002	0.002	
Volume resistivity, Ω·cm, at:			
25 °C (77 °F)	>10^{14}	>10^{14}	ASTM D 1829
300 °C (570 °F)	>10^{12}	>10^{12}	
500 °C (950 °F)	>10^{9}	>10^{8}	
700 °C (1290 °F)	>10^{8}	>10^{8}	

formity and reproducibility leading to consistently higher yields during circuit generation.

The 99.6% alumina formulation has a finer surface finish and grain structure which is very uniform, thus permitting very fine line width and spacings to be used. This leads to an extremely low level of surface defects which results in even higher yields than the standard 99.5% alumina material.

Multilayer Packages

Integrated circuit packages from high-alumina ceramics is one of the most important applications of ceramics in the electronics industry. Ceramic multilayer co-fired packages may take different forms, such as dual-in-line packages, chip carriers, or pin-grid arrays; all, however, allow the housing of the semiconductor devices in a strong, thermally stable, and hermetic environment.

Composition. Formulations for alumina multilayer packages have generally fallen in the range of 90 to 94% alumina. The balance of the material is SiO_2 with additives to form a glass phase such as MgO and CaO. In addi-

tion, some formulations are dark in color and may contain Cr_2O_3, MnO_2, TiO_2, or Fe_2O_3 additions to affect the opaque character to the body. In some cases, the nonalumina additives may be melted to form a glass and then added to the alumina in that prereacted form.

Whatever the formulations, they must be compatible with the printed metallization,

usually either tungsten or molybdenum, which is co-fired along with the green ceramic into a dense, monolithic structure containing the internal conductor paths.

The processing of the alumina body formulation for multilayer packages is very similar to that of thin-film substrate fabrication up through the doctor blade casting to form the green tape. From that point, the green tape is punched, via-hole coated and metallized by screen printing, and then laminated into a multiple layered structure (Fig 3). The laminated (both ceramic and refractory metal) structure is then co-fired to temperatures in excess of 1600 °C (2900 °F) in a protective atmosphere of either H_2 or a mixture of H_2 and N_2. Figure 3 shows a process flow diagram for processing these materials. After sintering, the multilayer structure is then nickel plated to cover the exposed molybdenum or tungsten metallization. Metal lead frames then are brazed to the plated lead tabs, and the package then is usually gold plated.

Properties and Functions. The primary function of the multilayer package is to house the silicon device and provide the electrical interconnection system to the next level of circuitry. The silicon chip can be attached by several methods depending on package design. Common methods include gold-silicon eutectic die attachment, epoxy die attachment, and

Table 5 Quality attributes for two grades of thin-film alumina substrates

Visual attribute	Precision-resistor-grade acceptance level	Conductor-grade acceptance level
Burrs	>0.013 mm (0.0005 in.) high, none > 0.13 mm (0.005 in.) diam	>0.025 mm (0.001 in.) high, none >0.25 mm (0.010 in.) diam
Pits, holes, and pocks	<0.075 mm (0.003 in.) diam NIF, none >0.13 mm (0.005 in.) diam(a)	<0.075 mm (0.003 in.) diam NIF, None >0.25 mm (0.010 in.) diam(a)
Stains, spots, and contamination	None	None
Blisters	None	None
Scratches	None > 0.005 mm (0.0002 in.) deep and > 0.635 mm (0.025 in.) length	>0.18 mm (0.007 in.) deep and >0.635 mm (0.250 in.) length
Fins and ridges	None	None
Chips (width only to be considered)	³/₄% of substrate length that is not less than 0.13 mm (0.005 in.)	1% of substrate length that is not less than 0.25 mm (0.010 in.)
Cracks	None	None
Allowable density of defects	1/in.², noncumulative	1/in.², noncumulative
Inspection level	100%	100%
Camber	≤0.002 mm/mm (in./in.)	≤0.003 mm/mm (in./in.)
Surface finish, μm (μin.)		
99.5% alumina	≤0.15 (6)	≤0.15 (6)
99.6% alumina	≤0.075 (3)	≤0.075 (3)

(a) NIF, not inspected for

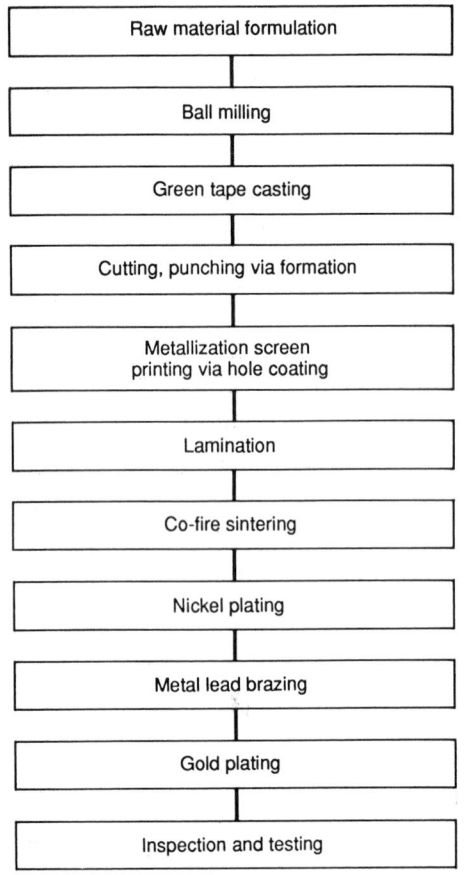

| Raw material formulation |
| Ball milling |
| Green tape casting |
| Cutting, punching via formation |
| Metallization screen printing via hole coating |
| Lamination |
| Co-fire sintering |
| Nickel plating |
| Metal lead brazing |
| Gold plating |
| Inspection and testing |

Fig 3 Process flow diagram for multilayer alumina co-fired packages

the use of high-lead solders as practiced in the controlled-collapse, chip-connection method. Lead attachment is done commonly by wire bonding by thermocompression, ultrasonic, and thermosonic bonding techniques. In addition, tape-automated-bonding (TAB) may be used. This involves thermocompression bonding of the silicon chip to a patterned metal on a polymer tape followed by encapsulation. The TAB process does not involve an alumina package.

The mounted silicon chip is enclosed in the package by making a hermetic seal which may take the form of a glass sealed lid, a welded metal lid, or a brazed or soldered metal lid, depending on package type.

Typical properties of a nominal 92% alumina opaque ceramic used for multilayer packages are listed in Table 7.

Beryllia Substrates

Beryllium oxide is a ceramic material normally associated with applications that require high levels of heat dissipation. Its thermal conductivity at room temperature is approximately ten times that of aluminum oxide. It is typically used in high-power electronic assemblies where heat removal is critical.

Composition. The most widely used composition of beryllia is the 99.5% BeO mate-

Table 7 Typical properties of opaque alumina (92% Al$_2$O$_3$)

Density, g/cm^3	3.72
Flexural strength, MPa (ksi)	324 (47)
Thermal expansion, 10^{-6}/°C from 25–500 °C (77–930 °F)	7.2
Thermal conductivity (at 20 °C, or 77 °F), W/m·°C)	15
Dielectric constant at 1 MHz	9.6
Dielectric loss at 1 MHz	0.002

rial. The only additives to this material are small amounts of MgO and SiO$_2$ to promote densification. Typical properties of a 99.5% BeO body are shown in Table 8.

Processing. Most beryllia bodies are fabricated by either dry pressing or isostatic pressing. Extrusion processes are used to make shapes with cylindrical geometry whereas tape casting is used for substrate geometries. Particular care must be taken when processing beryllia due to its toxicity. The tendency for beryllia to combine with water vapor at high temperature and form a surface layer usually precludes the use of natural gas in its sintering. Electric kilns employing dry air have been very successful in providing acceptable beryllia ceramics for a variety of uses.

Beryllia substrates are normally metallized by techniques using molybdenum in a process very similar to the refractory metal processing of alumina. The subsequent molybdenum layer is nickel plated and then brazed to metal leads and assemblies, the final step usually being a final gold plate. Thick-film metal mixtures are also available and used for beryllia.

Aluminum Nitride

Aluminum nitride has received much attention in recent years, particularly in the electronics area, because of its high thermal conductivity, high strength, and lower thermal expansion than alumina. In addition, it represents none of the toxicity problems associated with beryllia. The major obstacle in further commercialization of aluminum nitride has been its cost; the powders cost four to ten times that of electronic-grade alumina.

Composition. Because of the covalent bonding nature of aluminum nitride, high temperatures and high pressures or the presence of sintering aids are necessary in obtain-

Table 8 Typical properties of 99.5% beryllia

Density, g/cm^3	2.85
Flexural strength, MPa (ksi)	241 (35)
Thermal expansion, 10^{-6}/°C, from 25 °C (77 °F) to:	
200 °C (390 °F)	6.4
500 °C (930 °F)	7.2
Thermal conductivity, W/m·°C, at:	
20 °C (68 °F)	248
100 °C (212 °F)	188
Dielectric constant at 1 MHz	6.5
Dielectric loss tangent at 1 MHz	0.0004

ing high densities. Sintering aids have generally included CaO, Y$_2$O$_3$, or CaO with Al$_2$O$_3$ to achieve densification. In addition, carbon additives have been used to minimize the amount of oxide on the surface of the aluminum nitride powders.

Processing. The methods used in the fabrication of alumina substrates are also used for aluminum nitride. Dry pressing is commonly employed for small shapes with the body prepared by spray drying. Care must be taken in both the ball milling and slurrying of the aluminum nitride powder as well as in the spray drying to minimize the hydration or oxidation of the fine powder. Tape casting employing a nonaqueous vehicle has been used for large-area substrates. Sintering is normally conducted in nitrogen atmospheres at temperatures in excess of 1800 °C (3275 °F). In addition, hot pressing has been employed to fabricate high-density AlN packages which contain buried refractory metal conductors.

Properties and Functions. Table 9 gives typical properties for aluminum nitride. Because many different materials are available or being developed at this time from different suppliers, a range of property values are given.

Aluminum nitride substrates have been metallized by various techniques with varying degrees of success. Thin-film metallization formed by evaporation and sputtering has shown reasonable success. Thin films of Ni-Pd-Au have given good adhesion strength with good reliability and stability. In addition, Ti-Pd-Au thin films have also performed satisfactorily.

The use of thick-film metallization pastes developed for alumina have had limited success with application on aluminum nitride. Some success has been obtained with gold, silver-palladium, and copper thick-film conductors under the same firing conditions as used for alumina. Compatible thick-film resistor materials have had only limited success and require further development. Direct-bonded copper has had limited success.

Reasonable success has been obtained with tungsten co-firing of aluminum nitride. Standard molybdenum-manganese type of metallization systems have met with only limited success.

Glass-Ceramic Materials

Glass-ceramic materials normally refer to the combination where a ceramic or crystal-

Table 9 Typical properties of aluminum nitride

Density, g/cm^3	3.25–3.30
Flexural strength, MPa (ksi)	207–310 (30–45)
Thermal expansion, 10^{-6}/°C, from 25–500 °C (77–930 °F)	4.3–4.6
Thermal conductivity at 25 °C (77 °F), W/m·°C	130–200
Dielectric constant at 1 MHz	8.6–9.0
Dielectric loss at 1 MHz	0.001

Table 10   Typical properties of various glass-ceramics

Thermal expansion, 10^{-6}/°C, from 25–500 °C (77–930 °F)	3.0–7.0
Thermal conductivity at 25 °C (77 °F), W/m·°C	1–5
Dielectric constant at 1 MHz	4–8
Dielectric loss at 1 MHz	>0.002
Flexural strength, MPa (ksi)	140 (20)

line phase is formed by recrystallization from a glassy phase. For the majority of glass-ceramic materials as used for substrates, the resultant structure is formed by adding the two components, glass and crystalline ceramic, together.

These materials represent potentially one of the best combinations of properties to make them suited for substrates. Their low dielectric constant (coupled with low thermal expansion and the ability to use high-conductivity metals such as gold and copper in a co-fire process) make them particularly attractive for ceramic packages. The only drawbacks to these materials are their low strength and thermal conductivity relative to other materials. Table 10 lists typical properties for various glass-ceramic materials.

Composition. Typically, the glasses that have been employed in glass-ceramic substrates and packages have been borosilicate and alkaline-earth aluminosilicate compositions. The fillers, or crystalline material added to the glasses, are alumina, mullite, cordierite, and silica. A variety of properties are obtained depending on glass type, filler type, and amount.

Processing. In general, glass-ceramic materials are fabricated into substrates and packages by doctor blade tape casting. These materials may be screen printed with gold or silver-palladium thick-film materials, laminated, and co-fired to form multilayer packages. Copper also may be used, but this requires protective atmospheres in sintering to prevent excessive oxidation.

SELECTED REFERENCES

- L.M. Sheppard, Aluminum Nitride: A Versatile but Challenging Material, *Ceram. Bull.*, Vol 69 (No. 11), 1990, p 1801
- "Thick Film Substrates," Design Standard 49, Coors Ceramics Co., 1988
- "Thin Film Substrates," Design Standard 14, Coors Ceramics Co., 1989
- R.R. Tummala and E.J. Rymaszewski, *Microelectronics Packaging Handbook*, Van Nostrand Reinhold, 1989
- Y. Kurokawa, *et al.*, AlN Substrates with High Thermal Conductivity, *IEEE Trans. Comp. Hybrids Manuf. Technol.*, Vol 8 (No. 2), June 1985, p 248

Ceramic Capacitor Dielectrics

Ian Burn, Du Pont Electronics

CAPACITORS are essential components of most electronic circuits; not only can they store electrical energy, but these devices can be used for filtering out electronic noise and for high-frequency tuning as well as many other purposes. A ceramic disk capacitor is one of the simplest forms of capacitor: it consists essentially of a pair of parallel metal plates separated by an electrically insulating material, a dielectric. Green (unfired) disks are made by pressing powder or by punching from ceramic tape, and then they are fired at about 1250 to 1350 °C (2280 to 2460 °F) to produce a dense ceramic. Metal electrodes are applied to both surfaces of the disk, usually by screen-printing a silver paste, which is fired at about 750 °C (1380 °F) to sinter the metal and bond it to the ceramic. Leads are soldered to the electrodes, and then the disks are coated with an epoxy or with wax-impregnated resin to encapsulate them. Although the capacitance of disk capacitors can be increased by making them very thin (for example, 0.2 mm, or 0.008 in.) they become fragile. A more efficient way to increase capacitance is to make a multilayer structure which is co-fired with the electrodes. The thickness of the dielectric can then be decreased to about 0.015 mm (0.0006 in.) or less. Also, by connecting alternate layers in parallel, the capacitance increases in proportion to the number of layers, because the area of each plate is added together. The internal electrode is usually palladium, or a palladium-silver alloy with a melting point higher than the sintering temperature of the ceramic. After firing, the internal electrodes are connected at each side of the capacitor by applying a termination paste, usually containing silver or silver with added palladium. Multilayer ceramic capacitors (MLCs) can be used leaded and encapsulated, as for disks, but most are now used as "chips" soldered directly to a circuit board without any leads or encapsulation. Capacitors with a ceramic dielectric now account for about 90% of the total number of capacitors used, and most of these are MLCs, about 70 billion of which were manufactured worldwide in 1990. The rapid growth in the use of MLCs has stemmed from their unique combination of properties: small size, low cost, good high-frequency performance, and suitability for surface mounting to circuit boards. As will be described in this article, a variety of dielectric compositions have been developed to provide useful electrical properties for a wide range of applications, not only in MLCs but also in other types of ceramic capacitors.

Barium Titanate-Based Dielectrics

Barium titanate ($BaTiO_3$) is used widely in capacitors because of its unusually high dielectric constant (κ) compared with polymeric dielectrics and most other ceramics (Table 1). $BaTiO_3$ powder can be made by reacting (calcining) an intimate mixture of barium carbonate and titanium dioxide at about 1100 °C (2010 °F), but liquid-mix techniques are now being used increasingly for enhanced capacitor performance and better reproducibility. These wet chemical methods, such as the oxalate, alkoxide, or hydrothermal processes, are more expensive than the traditional mixed powder process but can provide powder with high purity, submicron particle size, and accurately controlled barium/titanium ratio.

In the oxalate process, aqueous solutions of barium chloride and titanium tetrachloride are reacted with oxalic acid to precipitate barium titanium oxalate [$BaTiO(C_2O_4)_2 \cdot 4H_2O$] which is then thermally decomposed into $BaTiO_3$ by calcination at about 900 to 1000 °C (1650 to 1830 °F). Another approach is to hydrolyze a mixture of titanium alkoxide and either barium alkoxide or barium hydroxide solution to form a precipitate of fine $BaTiO_3$ powder. Also, hydrous titanium oxide can be used with barium hydroxide solution to precipitate $BaTiO_3$ with or without hydrothermal conditions. These and other methods for making $BaTiO_3$ have been reviewed recently by Phule and Risbud (Ref 1).

When $BaTiO_3$ powder is sintered into a dense ceramic material, the dielectric constant, κ, depends strongly on the sintered grain size and on the temperature of measurement. There is a sharp peak in the dielectric constant at 130 °C (265 °F), and smaller peaks near 0 °C (32 °F) and −75 °C (−103 °F) where cubic-tetragonal, tetragonal-orthorhombic, and orthorhombic-rhombohedral phase transitions take place, respectively. Below 130 °C (265 °F), the Curie temperature, the material is ferroelectric and retains an electrical charge after the application and removal of an applied voltage. Dielectric constant near room-temperature is an optimum when the sintered grain size is close to

0.8 μm (32 μin.), as illustrated in Fig 1, so it is necessary to start with powder close to this size to obtain the highest value of κ. However, $BaTiO_3$ is rarely used without chemical modification because of the difficulty of controlling grain size during sintering, and because of the need to adjust the electrical characteristics to meet certain standards of capacitor performance (Table 2).

Barium Titanate Modifications

Several modifications are possible to modify the properties of $BaTiO_3$ capacitor materials. These include isovalent substitution and the use of donor and acceptor dopants.

Isovalent Substitutions. Ions with size and charge similar to those of Ba^{2+} or Ti^{4+} can substitute in the $BaTiO_3$ perovskite crystal lattice quite readily. For example, strontium or lead can replace barium in a continuous series of solid solutions. Strontium incorporation reduces the Curie temperature of $BaTiO_3$ linearly with concentration, whereas lead raises it. Calcium, on the other hand, appears to have less than a 20 at.% solubility and has only a small influence on the Curie temperature; it is frequently included in capacitor dielectric compositions to broaden the peak in dielectric constant at the Curie temperature. Similarly, titanium can be replaced by ions such as Zr^{4+}, Hf^{4+}, or Sn^{4+}, all of which lower the Curie temperature. Recent work on $BaTiO_3$-$BaZrO_3$ solid solutions made by coprecipitation (Ref 3) suggests that zirconium can also raise the peak value of the dielectric constant to 50,000 or more. Figure 2, taken from the excellent review by Jaffe et al. (Ref 4), summarizes the effect on the phase transition temperatures of substituting isovalent ions in $BaTiO_3$. Compositions which have a very high dielectric constant at room temperature and meet the appropriate requirements for dependence of κ on temperature can be formulated by combining additives which depress the Curie peak with those which broaden it. For example, $CaZrO_3$ or $CaSnO_3$ have been used frequently as additives to $BaTiO_3$ for this purpose. However, it should be noted that if modifiers are added to premade $BaTiO_3$ rather than during the manufacturing process, some grain growth during sintering is usually necessary to achieve solid solution.

Acceptor Dopants. Ions which substitute in the $BaTiO_3$ and have a lower charge than the ion they replace behave as electron accep-

Table 1 Dielectric constants at 25 °C (75 °F)

Material	Dielectric constant, κ
Teflon	2.1
Silica glass	3.8
Polyvinylidene fluoride	8.4
Alumina	10
Magnesium oxide	20
Barium tetratitanate	40
Titanium dioxide	100
Calcium titanate	160
Strontium titanate	320
Barium titanate	1000–5000
Barium zirconium titanate	20,000
Lead magnesium niobate	20,000

Table 2 Commercial capacitor requirements

EIA(a)	Temperature °C (°F)	Capacitance deviation(b)	Maximum dissipation factor (DF), %	Dielectric constant, κ, for present technology
COG	−55 to 125 (−65 to 255)	30 ppm/°C	0.1	75–100
X7R	−55 to 125 (−65 to 255)	±15%	2.5	3000–4000
Z5U	10 to 85 (50 to 185)	+22, −56%	4.0	8000–10,000
Y5V	−30 to 85 (−20 to 185)	+22, −82%	4.0	12,000–20,000

(a) Electronic Industries Association. (b) 25 °C (75 °F) reference.

tors. Most of the time these will be *B* site ions (substituting for Ti) and the solubility is usually limited to a few tenths of a percent. The possibility of substitution of monovalent ions (such as sodium, potassium, or silver) for barium is not well documented. Acceptor dopants such as $Mn^{2+,3+}$, $Co^{2+,3+}$, $Fe^{2+,3+}$, Ni^{2+}, and Zn^{2+}, or impurities such as Al^{3+} or Mg^{2+} incorporated during the manufacture of sintering of the $BaTiO_3$, induce vacancies in the oxygen sublattice in order to balance charge. These oxygen vacancies promote the potential for electrolytic migration of oxygen ions when a strong direct current (dc) bias voltage is applied, leading to a degradation of the electrical resistance of the material and possible capacitor failure. On the other hand, acceptor dopants generally produce a decrease in alternating current (ac) voltage loss (dissipation factor, or DF) of $BaTiO_3$ but increase the aging rate, that is, the tendency of the dielectric constant to become lower with time after firing, a characteristic of ferroelectric materials (Ref 5).

Donor Dopants. Certain rare earth ions, such as Nd^{3+}, can substitute for barium up to several atomic percent and similar amounts of ions such as Nb^{5+}, Ta^{5+}, or W^{6+} can replace titanium, the excess charge normally being accommodated by the creation of vacancies on titanium sites. These donor dopants have several strong effects. A few tenths of one percent can neutralize acceptor impurities and remove impurity-related oxygen vacancies. This amount can also increase the tendency of the ceramic to become semiconducting after firing, particularly if there is appreciable grain growth. When donor levels above about 0.5 at.% are added to premade $BaTiO_3$ powder, grain growth during sintering is strongly suppressed; this approach is frequently used to produce a ceramic with a fine, dense microstructure. When smaller amounts of acceptor dopants are also added, the temperature stability of the dielectric constant can be adjusted to achieve Electronic Industries Association (EIA) X7R capacitor (Table 2) characteristics. In such cases, a heterogeneous core-shell type of microstructure results, with the $BaTiO_3$ grains having a reacted surface layer contain-

ing the dopants. This has been confirmed by transmission electron microscopy in a number of studies, in which a ferroelectric domain structure has been observed only in the cores of the grains (Ref 6).

Donor and acceptor dopants have increased solubility when added together in charge-compensating amounts with appropriate stoichiometry adjustment and can produce large shifts in the Curie peak, which remains sharp. A typical example is the solid solution of $BaTiO_3$ and $BaZn_{1/3}Nb_{2/3}O_3$. Barium titanate based compositions meeting EIA Y5V capacitor requirements (Table 2) can be designed with κ of 20,000 using this approach.

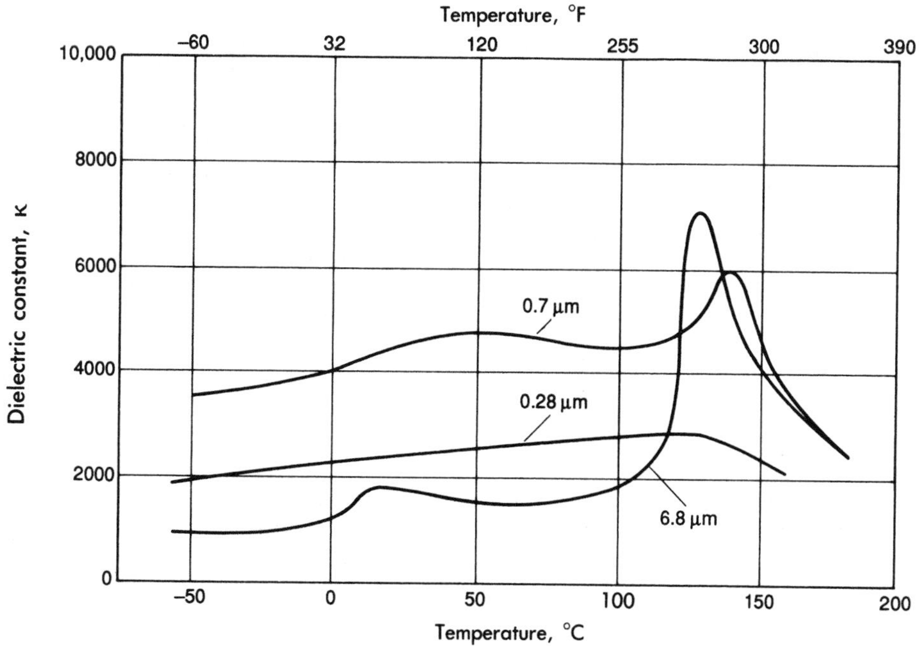

Low Sintering Temperatures

Ceramic compositions based on $BaTiO_3$ are usually fired at temperatures between 1250 and 1350 °C (2280 and 2460 °F), but lower sintering temperatures are needed when the ceramic is co-fired with certain metals, such as palladium-silver alloys, that have a melting temperature below this range. Addition of low-melting glasses or fluxes to $BaTiO_3$ often results in an appreciable decrease in dielectric constant. However, low-firing (1100 °C, or 2010 °F) Z5U or Y5V dielectrics with high κ can be designed by considering the composition of the glass, or flux, and the ceramic as a whole and then adjusting the cation stoichiometry (A/B ratio) and donor/acceptor balance to permit a small amount of uniform grain growth during sintering (Ref 7). Low-fire compositions with EIA X7R temperature dependence can be obtained by adjusting the donor/acceptor balance and/or the A/B stoichiometry to suppress grain growth with the flux added.

Reduction-Resistant Compositions

To avoid the cost of using precious metals as electrode materials, dielectrics have been designed that are compatible with electrodes of nickel or copper. Atmospheres of low oxygen content are needed during the firing process to prevent oxidation of the metal, but exposure of the ceramic to such atmospheres can often lead to degradation of electrical properties. This is particularly true for $BaTiO_3$ or other titanate-based dielectrics, for which oxygen loss during firing can cause an appreciable decrease in insulation resistance by

Fig 1 Dielectric constant of barium titanate versus temperature for various grain sizes. Source: Ref 2

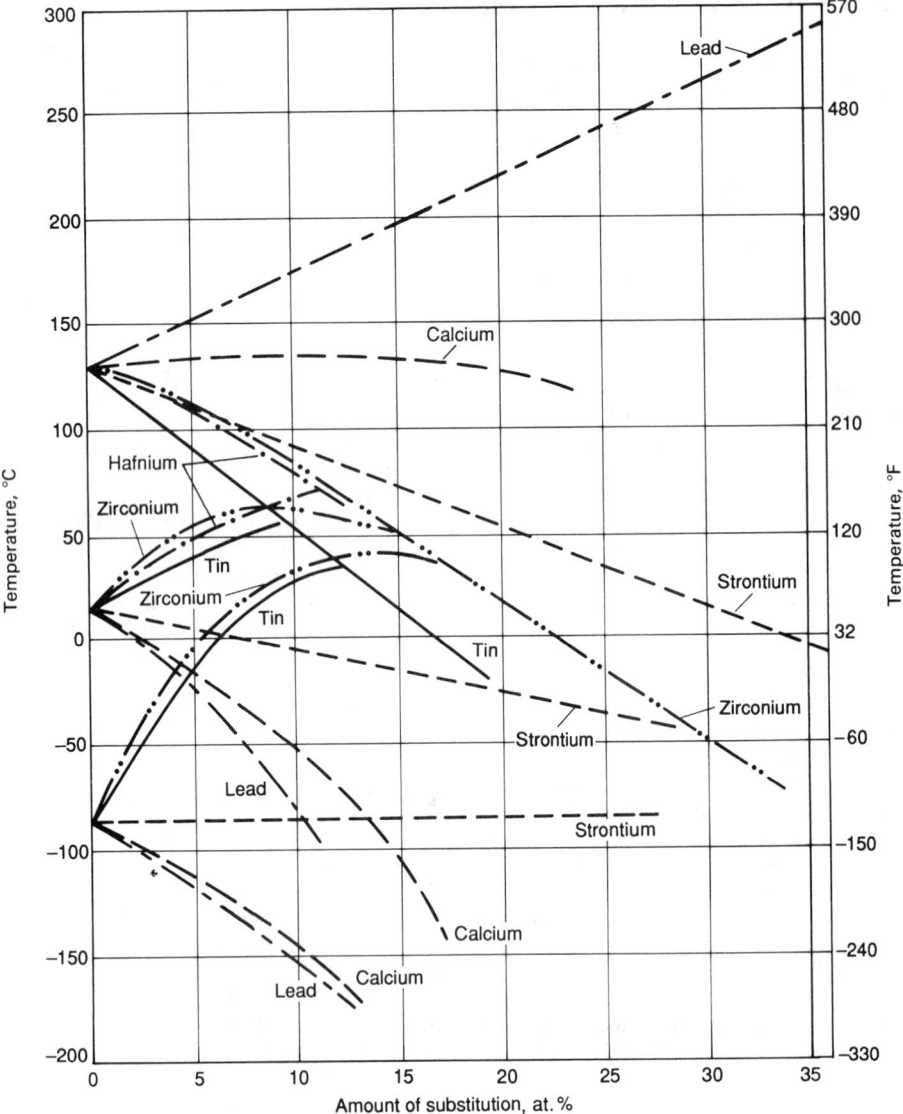

Fig 2 Effect of isovalent substitutions on the transition temperatures of barium titanate. Source: Ref 4

making the material semiconducting. The effect is magnified if any higher titanate phases are present as a result of the A/B cation ratio being less than 1. Addition of a small amount of acceptor dopant to barium titanate-calcium zirconate solid solutions can inhibit semiconduction in high-κ Z5U or Y5V dielectrics, but dc bias reliability can be degraded because of the acceptor additions.

A solution to this problem is to include a small amount of donor additive to prevent the formation of impurity-related oxygen vacancies and to add dopants with more than one valence state, for example, manganese, cobalt, chromium, or cerium, which act as oxidizers during the firing process. Also, the benefit of including some CaO in the dielectric to suppress semiconduction has been documented by Sakabe (Ref 8). This could be a result of an improved stability toward oxygen loss or because Ca^{2+} appears to be capable of acting as an acceptor by substituting on some tita-

nium sites. Development of nickel-compatible X7R dielectrics with κ up to 3300 has been described by Fujikawa (Ref 9) for BaTiO$_3$ with additions of 1 mol% CaZrO$_3$, Y$_2$O$_3$ (a donor), and MnO, and by Kishi *et al*. (Ref 10) for a Ba$_{0.95}$Sr$_{0.03}$Ca$_{0.02}$TiO$_3$ ceramic with 3% alkaline-earth lithium silicate glass fired at 1200 °C (2190 °F).

Reduction-resistant Y5V compositions based on BaTiO$_3$ which sinter below the melting point of copper (1083 °C, or 1981 °F) to allow co-firing with copper electrodes have been reported with a κ value of about 10,000 (Ref 11). BaTiO$_3$ was modified with zirconium, zinc, niobium, and manganese, and a zinc borate flux was added.

Other Titanate-Base Dielectrics

Strontium titanate (SrTiO$_3$) has a dielectric constant of about 320 at room tem-

perature and is cubic and paraelectric. The κ increases to near 20,000 at very low temperatures (near 0 K) with no evidence of a ferroelectric transition. A solid solution of 30 mol% SrTiO$_3$ in BaTiO$_3$ will produce a dielectric with $\kappa > 5000$ at room temperature, but the Curie peak is normally too sharp for capacitor applications. However, donor-doped strontium titanate is useful for barrier layer capacitors (Ref 12), which are described in the section "Surface Barrier-Layer Capacitors" in this article. Donor dopants are used to achieve high conductivity in the ceramic grains after firing. Firing is normally done in an atmosphere with low oxygen content, but firing in air is also possible if the stoichiometry of the SrTiO$_3$ is adjusted to allow for compensation of the donor by strontium vacancies (Ref 13). Typically, 0.5 to 1.0 mol% of niobium, yttrium, or neodymium is used.

Compositions based on SrTiO$_3$ have also been used for high-voltage capacitors. For example, Nishigaki *et al*. (Ref 14) have described materials with κ near 2000 that are very stable with applied voltages up to 5 kV/mm (125 kV/in.). These were based on the system $(1 - x)(Sr_{0.5}Pb_{0.25}Ca_{0.25})TiO_3 + x(Bi_2O_3 \cdot 3TiO_2)$ with the best properties being obtained for $x = 0.043$.

Calcium titanate has similar properties to SrTiO$_3$ except that the dielectric constant is about 50% lower (Table 1). It is used mainly for temperature-compensating capacitors for which a linear dependence of the dielectric constant on temperature is needed. The calcium titanate is usually blended with other additives such as oxides of lanthanum or bismuth.

Magnesium titanate has a different crystal structure (ilmenite) than the perovskite titanates discussed above and does not form solid solutions with them. The material has very low dielectric loss and a κ of 20, which increases slightly with temperature. This slightly positive dependence can be adjusted to meet EIA code COG requirements (Table 2) by blending the magnesium titanate with a small amount of calcium titanate (about 6 mol%). Low-firing compositions can be produced by adding fluxes, as for BaTiO$_3$ ceramics. While the low loss of magnesium titanate makes it attractive for use in capacitors for high-frequency applications, performance at frequencies over 1 MHz is strongly influenced by the conductivity of the electrodes in the capacitor and by the capacitor's inductance. A magnesium titanate-based dielectric with copper electrodes is an ideal combination for MLCs made for use at high frequencies (Ref 15).

Barium-Neodymium Titanate. Compositions with EIA COG temperature characteristics and with dielectric constant as high as 65 to 90 can be formulated from the barium oxide-neodymium oxide-titanium oxide system. This unique combination of small temperature dependence and high dielectric constant was discovered at the University of

Illinois in the late 1960s by R.L. Bolton and W.J. Muhlstadt, who blended rare-earth oxides with barium tetratitanate ($BaTi_4O_9$), a material with a κ of 40 and a temperature coefficient of -65 ppm/°C. Barium neodymium titanate compositions are now used worldwide; a typical composition might contain 12 to 20 mol% BaO, 12 to 20 mol% Nd_2O_3, and 60 to 70 mol% TiO_2, although the composition of the primary phase in this system has not been clearly established (Ref 16). Often the neodymium oxide is partially substituted with other rare earth oxides. Partial substitution of BaO with PbO can raise the dielectric constant close to 100, as can additions of bismuth oxide or bismuth titanate.

Dielectric Compositions with High Lead Content

Lead titanate is difficult to sinter into a dense ceramic. It is ferroelectric, with a Curie point at 490 °C (915 °F) and a room-temperature κ of about 350. High dielectric constant at room temperature (9000) can be obtained by lowering the Curie point by substituting equal molar amounts of W^{6+} and Mg^{2+} for Ti^{4+} and by partial substitution of lead with strontium. Sintering temperature is low (near 1050 °C, or 1920 °F) without the addition of fluxes. There is an extensive amount of patent literature related to dielectrics with high lead content. References for modified lead titanate compositions and other systems to be described below are summarized in Table 3, some of which is taken from the recent excellent review by Shrout and Dougherty (Ref 17). Abbreviated chemical nomenclature is used, as is customary with these materials.

Solid solutions of lead titanate with lead zirconate (known as PZT) are widely used for piezoelectric applications. Modification of PZT with lanthanum oxide (PLZT) provides unique electrooptic properties, and some compositions have attractive properties for capacitor applications, particularly because the dielectric constant remains high under the application of high dc bias voltage (Ref 18). A typical composition might be $Pb_{0.88}La_{0.12}Zr_{0.70}Ti_{0.30}O_3$, with a small excess of lead oxide to help sintering. Also, small additions of silver oxide, barium oxide, and bismuth oxide can be added to adjust the temperature dependence to meet EIA X7R requirements with a κ of 2000 and a sintering temperature of about 1100 °C (2010 °F).

Lead magnesium niobate (PMN) is the classic high-κ relaxor dielectric, so called because its dielectric maximum and loss factor (DF) change with the measuring frequency (Ref 17). A peak in the dielectric constant of about 20,000 normally occurs at 0 to -10 °C (32 to -15 °F) for measurement frequencies near 1 kHz, but the value of κ depends strongly on the purity of the raw materials and method of preparation. Even with high-purity materials, κ can be depressed if the fired ceramic contains some low κ py-

rochlore phase instead of the $PbMg_{1/3}Nb_{2/3}O_3$ perovskite phase. This can be minimized by prereacting the magnesium oxide and niobium oxide to form columbite and/or by using an excess of magnesium oxide. Sintering temperatures near 1000 °C (1830 °F) are possible by adding a small excess of lead oxide, or firing in a closed container in which a PbO vapor pressure is maintained. The use of lithium-containing fluxes has also been suggested to improve sinterability. Dielectric constant can be increased by adding 7.5 to 10.0 mol% lead titanate (PT) to move the peak in κ closer to room temperature.

Combinations of PMN-PT with $Pb(Ni_{1/3}Nb_{2/3})O_3$ (PNN) have been described by Ochi et al. (Ref 19). PNN is a relaxor itself, with a Curie peak at about -120 °C (-184 °F). Powder of the composition 0.2 PMN, 0.2 PT, 0.6 PNN was made by chemical coprecipitation using the alkoxide process. MLCs could be made with this powder at layer thicknesses as low as 6 μm (240 μin.) and sintered at 950 °C (1740 °F). Dielectric constant was close to 20,000.

Lead zinc niobate (PZN) has a Curie peak at 140 °C (285 °F) with very high values of κ (50,000) in single crystals, but a low-κ pyrochlore phase is usually obtained for polycrystalline ceramics. However, substitution of small amounts of barium, strontium, or calcium for lead can stabilize the perovskite phase. Compositions based on PZN with additions of PMN and/or lead titanate have very high resistivity, good dc bias characteristics, and κ as high as 12,000. The properties of dielectrics in the system $PZN-BaTiO_3$ have been described by Halliyal et al. (Ref 20). Also, dielectrics with EIA X7R characteristics and κ close to 5000 have been described for a sintered mixture of PZN with barium titanate (Ref 21).

Lead Iron Niobate-Lead Iron Tungstate (PFN-PFW). Low-firing (900 °C, or 1650 °F) ceramics with κ about 20,000 can be obtained by reacting $Pb(Fe_{1/2}Nb_{1/2})O_3$ with 30 to 35 mol% $Pb(Fe_{2/3}W_{1/3})O_3$ (Ref 22). PFN is a relaxor with Curie peak near 110 °C (230 °F), whereas the relaxor PFW has a Curie peak near -95 °C (-140 °F). Inclusion of a small amount of manganese in place of iron ($<1\%$) is necessary to decrease dielectric losses to an acceptable level. Mechanical strength of the fired ceramic is relatively low, as is the case for most lead-based dielectrics; however, a slight niobium excess is reported to increase the strength somewhat. Dielectric constants over 30,000 have been obtained when small amounts (2%) of barium-copper tungstate (BCuW) are added to the PFN-PFW system. Dielectrics based on the ternary system PFN-PFW-PZN have also been developed.

Other Relaxor Dielectrics. Various combinations of the above systems have been investigated with the objective of combining high dielectric constant, low firing temperature, improved mechanical strength, and high reliability (Table 3). While most of these are aimed at meeting EIA Y5V or Y5U temperature stability, some efforts have also been made to develop high-κ compositions meeting X7R specifications. For example, flattened temperature dependence can be obtained by blending various PFN-PFW solid solutions and firing with small additions of rare-earth oxide to suppress grain growth.

Also worthy of note are the relaxor compositions developed by Kato et al., which are resistant to reduction in atmospheres of low oxygen content and can be co-fired with copper electrodes (Ref 23). These compositions are based on 0.8PMN-0.125PT-0.075PNW, with a small (1.0 to 2.5%) substitution of lead with calcium. Reduction-resistant relaxors

Table 3 Lead-based dielectrics

Material designation	U.S. patent number and year	Inventor
PT-ST-PMW	4,582,814 (1986)	G.L. Thomas (Du Pont)
PT-PMW-PNN	4,450,240 (1984)	H. Miyamoto, *et al.* (NEC)
PLZT-Ag$_2$O-Bi$_2$O$_3$-BaO	4,324,750 (1982)	G.H. Maher (Sprague)
PMN-PT	4,265,668 (1981)	S. Fujiwara (TDK)
PMN-PT-PNN	4,712,156 (1987)	P. Bardhan (Corning)
PMN-PT-PNW (+Ca)	4,751,209 (1988)	Y. Yokotani, *et al.* (Mutsushita)
PMN-PT-PZN	4,735,905 (1988)	G. Nishioka, *et al.* (Murata)
PMN-PT-PZN (+Ca)	4,772,985 (1988)	T. Yasumoto, *et al.* (Toshiba)
PMN-PT-PFW (+Ba, Sr)	4,897,373 (1990)	Y. Inoue, *et al.* (MMCC)
PMN-PZN-PNN	4,749,668 (1988)	M. Fujino, *et al.* (Murata)
PMN-PZN-PFW	4,542,107 (1985)	J. Kato, *et al.* (Matsushita)
	4,555,494 (1985)	N. Nishida, *et al.* (Matsushita)
PMN-PZN	4,637,989 (1987)	H.C. Ling, *et al.* (AT&T)
PZN-PT (+Ca, Ba, Sr)	4,767,732 (1988)	O. Furukawa, *et al.* (Toshiba)
PFN-PFW	4,885,267 (1989)	H. Takahara, *et al.* (NEC)
	4,078,938 (1978)	M. Yonezawa, *et al.* (NEC)
PFN-PFW-BCuW	4,544,644 (1985)	Y. Yamashita, *et al.* (Toshiba)
PFN-PFW-PZN	4,236,928 (1980)	M. Yonezawa, *et al.* (NEC)
PFN-BCuW	4,772,985 (1988)	T. Yasumoto, *et al.* (Toshiba)
PFN-PFW-PNN	4,379,319 (1983)	J.M. Wilson (Ferro)
PFN-PMN	4,550,088 (1985)	H.D. Park, *et al.* (Union Carbide)
	4,216,102 (1980)	K. Furukawa, *et al.* (TDK)
PFW-PT	4,308,571 (1981)	H. Tanei, *et al.* (Hitachi)
PFW-PT-PNN	4,624,935 (1986)	Y. Sakabe, *et al.* (Murata)
PFW-PZ	4,235,635 (1980)	O. Iizawa, *et al.* (TDK)

Some data from Ref 17

based on the system PMN-PT-PZN have also been described, again with partial replacement of lead by calcium (Table 3).

Processing of Disks and Tubulars

Standard Disks. As mentioned earlier, green ceramic disks are formed from pressed powder or from ceramic tape. In the case of pressed powder, the ceramic ingredients are first milled in water with a small amount of binder (for example, polyvinyl alcohol) and then spray-dried to obtain a free-flowing powder which will pack uniformly into a die cavity. The main problems with this approach are the fragility and nonuniformity of density for disks with large diameter-to-thickness ratios. In the case of ceramic tape, the ceramic powder is blended with binder to a high viscosity and then extruded by pressure through a narrow slit. The tape is relatively tough, so that thin (<0.1 to 0.2 mm, or 0.004 to 0.008 in., thick) disks can be punched from it. The disks are usually fired in stacks, each disk being lightly dusted with a refractory powder to prevent them from sticking to one another.

Although silver is the popular choice of electrode material, base metals such as copper or nickel can be used, but the electrode paste must be fired in nitrogen instead of air. Also, the ceramic must be specially formulated to tolerate this processing without any degradation in performance as discussed earlier. Alternatively, an electrode can be formed by plating with copper and/or nickel, but the electrode area must be pretreated to prevent plating over the edges of the disk or the edges must be ground free of metal after plating.

Surface barrier-layer capacitors are processed initially in the same way as described above; that is, they are formed the same way and then fired in air. The dielectric is usually an EIA X7R formulation based on $BaTiO_3$ and often contains niobium and/or neodymium oxide to provide a fired ceramic with a small grain size of 1 to 2 μm (40 to 80 $\mu in.$). The donor dopants also help to produce a highly conducting ceramic when the disks undergo the next stage of processing, which consists of a heat treatment at 900 to 1000 °C (1650 to 1830 °F) in a gas mixture of hydrogen and nitrogen. This reduction treatment turns the disks from yellow or light green to black by removing a small amount of oxygen from the ceramic. In the next processing step, the conducting disks are printed with silver paste which, in addition to silver powder and glass for bonding, can also contain oxidizing agents such as manganese oxide and/or copper oxide. When the silver paste is fired (at 850 to 900 °C, or 1560 to 1650 °F in air) a thin layer of ceramic under the electrode is oxidized and becomes insulating again. This produces two capacitor layers in series, one on each major surface. These are connected by the ceramic, which is still conducting in the interior of the disk.

Grain boundary barrier-layer capacitors are usually based on $BaTiO_3$ or $SrTiO_3$ and are formulated so that large conducting grains of 50 to 100 μm (2000 to 4000 $\mu in.$) are produced when the disks are fired (Ref 24). If the ceramic is cooled slowly, or if oxygen is added to the furnace during cooling, oxygen will diffuse into the ceramic along the grain boundaries and reoxidize a surface layer on each conducting grain. This can lead to very high apparent dielectric constants (50,000 to 100,000) because of the thin surface layers of a few microns and the high effective surface area of the grain boundaries. However, this method of reoxidation usually results in dielectric layers with high dissipation factor and poor breakdown strength. Improved dielectric properties can be achieved by coating the surface of the fired disk with low-melting oxidizing agents, which penetrate the grain boundaries by liquid-phase reaction during a second firing, usually at a lower temperature (900 to 1000 °C, or 1650 to 1830 °F). The oxidizing agents can be mixtures of bismuth and copper oxides, sometimes with manganese oxide added, and can be incorporated into the electrode paste, so that the grain boundaries can be oxidized when the electrode paste is fired on (at 900 to 925 °C, or 1650 to 1700 °F, for silver paste).

Instead of making disks, the ceramic may be extruded into thin, spaghetti-like tubes. These long tubes are cut into small units and then processed like disks, except that the inside and outside of each tube are coated with silver to form electrodes. Metal end caps with leads are fitted on the tubes, and then the (tubular) capacitor is encapsulated. This type of configuration, known as an axial, is more suitable for packaging in reels ready for automatic insertion into wiring boards than disk capacitors, with their radial lead configuration. Tubulars also fit closer to the board.

Processing of MLCs

Multilayer ceramic capacitors are manufactured by two quite different processes: the dry method and the wet method. These processes are shown schematically in Fig 3, and details can be found in the recent review by Kahn *et al.* (Ref 5). In the dry method, ceramic tape is made by casting a slurry of the dielectric powder onto a carrier of polymer film or stainless steel. Often the slurry is a suspension of the powder in water, with a latex binder to give the dried tape adequate handling strength. A water-based system is preferable to one using an organic solvent for environmental considerations, but control of dispersion and rheology of some powders is technically more difficult in a water-based system. Ways of casting and handling thinner and thinner tape are being devised so that fired thicknesses below 10 μm (400 $\mu in.$) can be achieved, at least in the laboratory. However, lamination of the stack without introducing

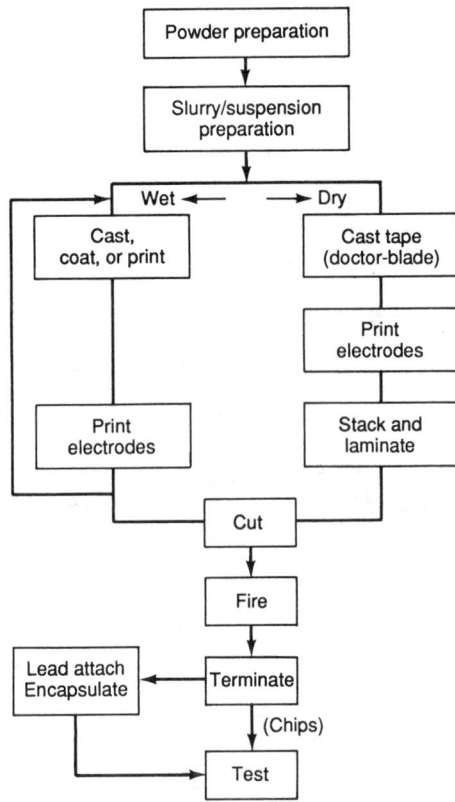

Fig 3 Manufacturing procedures for multilayer ceramic capacitors

problems related to nonuniform green density (such as mechanical flaws after firing) becomes more difficult as the ratio of electrode print thickness to dielectric thickness becomes larger.

The wet method avoids the problem of handling thin layers of tape by depositing layers of wet slurry onto a temporary substrate. This can be accomplished by doctor-blade, curtain-coating, or screen-printing processes. Layers of ceramic are built up sequentially by repeatedly casting, drying, and then screen-printing electrode paste onto the dried ceramic. This process is thought to permit fired layer thickness down to 5 μm (200 $\mu in.$) to be obtained, but very smooth electrode prints and absence of pinholes and airborne particles are required to avoid defects. Use of slurries with Newtonian rheology permits planarization of each dielectric layer over the electrode print (Ref 25). This adjusts for the extra thickness in the center of the capacitor which could be caused by the electrode prints. The finished build is then cut (diced) into individual green capacitors, which are released from the temporary substrate and then fired.

Firing involves first removing the organic binder from the ceramic by heating slowly over a period of many hours up to a temperature of about 750 °C (1380 °F). Heating too rapidly during this phase can lead to cracking or splitting (delamination) of the capacitor. The final sintering temperature is usually between 1000 and 1350 °C (1830 and 2460 °F) depending on the nature of the dielectric and the melting point of the electrodes. The capacitor chips are usually fired in air on zirconia setters or buried in zirconia sand. However, if the MLC has internal nickel or copper electrodes, then an atmosphere must be used that protects the electrodes from oxidation during firing. Various gas mixtures can be used to give a controlled, low level of oxygen in the furnace. Atmospheres of nitrogen (N_2), $H_2 + N_2$, or $CO + CO_2 + N_2$ have been used, and a variety of other atmospheres, such as $H_2 + H_2O + N_2$ or $CO_2 + H_2 + N_2$, can also be employed. The appropriate low level of oxygen partial pressure can be obtained by regulating the gas mixture according to gas-equilibrium thermodynamics and can be monitored with oxygen sensors in the furnace.

External connection of the alternate layers of the sintered capacitor, called termination, is accomplished by applying and firing a metal paste on the ends of the capacitor where the electrodes are exposed. The fired chips are usually tumbled in milling jars containing abrasive powder to provide good electrode exposure, and to round the corners of the capacitor for more uniform coverage of the termination paste. The termination is normally a silver-palladium alloy for good resistance to leaching during soldering to circuit boards. Silver terminations can be used if they are plated with a thin barrier layer of nickel to increase leach resistance. A thin coating of tin on the nickel maintains solderability.

Finally, a quite different type of MLC process, known as the lead injection process (Ref 26) should be mentioned. In this approach, a multilayer is made by either the wet or dry process but with electrode prints containing a combustible material such as carbon powder instead of metal. On firing, the carbonaceous material is removed, leaving thin cavities instead of electrodes. The cavities are then impregnated with molten lead or tin-lead alloy under pressure, through a porous termination previously applied to the chip. This procedure avoids the use of precious metals without the need for firing in special atmospheres.

Processing of Thick Film and Thin Film Capacitors

Thick film capacitors are prepared by screen-printing the dielectric onto a ceramic substrate, which is normally made from alumina. The substrate is used to support electronic circuitry consisting of printed and fired conductor lines and resistors, together with insulating layers of glass-ceramic where conductor lines cross over. Screen-printed capacitors can be included in the printed circuit instead of using MLC chips when discrete components are subsequently mounted onto the substrate.

Screen-printing pastes are high-solids dispersions of metal or ceramic powder with additions of glass and/or low-melting oxides to help bond the material to the substrate. A high molecular weight polymer, such as ethyl cellulose dissolved in a slow-drying solvent such as terpineol, is used in the dispersion to give the appropriate printing characteristics and adequate dried (green) strength. Each printed component of the circuit is usually fired separately, on a belt furnace, at a temperature of 850 to 925 °C (1560 to 1700 °F).

Generally, the procedure for thick film capacitors is to print, dry, and fire a palladium-silver bottom conductor and electrode area, then print, dry, and fire two layers of dielectric paste, followed by a top conductor and electrode pad. Even with fluxes added, the dielectric may still have open porosity after firing, because shrinkage is limited to the vertical direction only. Accordingly, thick film capacitors are normally coated with a fired top layer of glass encapsulant, or a cured organic encapsulant (silicone) can be used. Matching of suitable conductors with the dielectric is important, because the bonding agents in the conductors can react with the dielectric, adversely affecting capacitor performance.

Currently, thick film capacitors based on $BaTiO_3$ have dielectric constants of about 1500, with EIA X7R characteristics, at a dielectric thickness of 35 to 40 μm (1400 to 1600 μin.). However, Z5U compositions based on relaxor dielectrics have been reported with much higher dielectric constants, for example by using a PFN-BCuW system (Table 3) with silver electrodes, or PMN-PT-PZN (with calcium added) for firing with copper electrodes in nitrogen (Ref 27). Copper-compatible thick film relaxor compositions have also been reported for the PMN-PT-PNW system (Table 3).

Thin film capacitors are normally made using the chemical vapor deposition (CVD) or sputtering techniques employed for semiconductor device technology. Dielectric and conductor thicknesses are in the range 0.1 to 1.0 μm (4 to 40 μin.). Thin film capacitors with excellent high-frequency (100 MHz to 1 GHz) performance can be made in the form of chips of similar dimensions to MLCs, so they can be surface-mounted on hybrid circuits. Low-κ dielectrics such as SiO_2 or Si_3N_4 are deposited on a ceramic substrate with copper or aluminum conductors to provide capacitance up to ≈140 pF.

Higher κ dielectrics based on $BaTiO_3$ or lead-based materials such as PZT can be prepared in thin films by sputtering, but the films are usually amorphous and must be thermally annealed to produce a crystalline film with the optimum κ. Thin films of PZT, $BaTiO_3$, and other dielectrics have also been prepared by sol-gel techniques in which the dielectric is prepared using wet chemistry. The sol-gel, or a colloidal suspension, is usually applied to the substrate by spin-coating or dipping,

and then thermally decomposed and crystallized by heating to about 600 °C (1110 °F; Ref 28).

While thin film dielectrics are useful for capacitor applications, there is currently a very strong interest in using thin films of ferroelectric materials (particularly PZT) as memory devices for semiconductor circuits. Ferroelectric random access memories (FRAMS) retain information when power is disrupted and are faster and longer lasting than other types of nonvolatile memory devices (Ref 29).

Integrated Capacitors

To provide more available space on the surface of circuit boards, methods are being developed to build passive components, such as capacitors and resistors, into the substrate itself. This work has been pioneered in Japan, where a monolithic multicomponent ceramic (MMC) substrate has been developed using ceramic tape technology (Ref 30). Layers of high-κ dielectric based on PFN-PFW (κ = 7000) were sandwiched, with palladium-silver or gold electrodes and appropriate interconnection, between layers of glass-ceramic substrate material. The composite was then co-fired at 900 °C (1650 °F). Capacitor values of 10 pF to 3 μF were obtained.

A glass-ceramic substrate with integrated $BaTiO_3$ capacitors and copper conductors has also been reported (Ref 31). In this case the high-κ dielectric was screen-printed onto glass-ceramic green-tape, which was then laminated and fired at 950 °C (1740 °F).

Technically, it is very difficult to successfully co-fire materials with different shrinkages and expansion coefficients. In addition, chemical interaction between the high-κ dielectric and the glass-ceramic must be minimized. Consequently, this type of work is still in an evolutionary state.

REFERENCES

1. P.P. Phule and S.H. Risbud, Low-Temperature Synthesis and Processing of Electronic Materials in the BaO-TiO$_2$ System, *J. Mater. Sci.*, Vol 25 (No. 2B), 1990, p 1169–1183
2. G. Arlt, D. Hennings, and G. de With, Dielectric Properties of Fine-Grained Barium Titanate Ceramics, *J. Appl. Phys.*, Vol 58 (No. 4), 1985, p 1619–1625
3. R. McSweeney, K. Zuk, and D. Williamson, Square Loop Ba(Ti,Zr)O$_3$ Capacitors Based on Alkoxide Derived Powder, Proceedings of the First International Conference on Ceramic Powder Processing Science, Orlando, FL, 1987, *Ceramic Transactions: Ceramic Powder Science IIB*, 1988, p 1156–1166
4. B. Jaffe, W.R. Cook, Jr., and H. Jaffe, *Piezoelectric Ceramics*, TechBooks, 1971 (1990 reprint)

5. M. Kahn, D.P. Burks, I. Burn, and W.A. Schulze, Ceramic Capacitor Technology, *Electronic Ceramics*, L.M. Levinson, Ed., Marcel Dekker, 1988, p 191–274

6. T.R. Armstrong and R.C. Buchanan, Influence of Core-Shell Grains on the Internal Stress State and Permittivity Response of Zirconia-Modified Barium Titanate, *J. Am. Ceram. Soc.*, Vol 73 (No. 5), 1990, p 1268–1273

7. I. Burn, M.T. Raad, and K. Sasaki, New High-Performance, Low-Fire MLC Dielectrics for SMT Applications, *Ceram. Trans.*, Vol 8, 1990, p 20–34

8. Y. Sakabe, Dielectric Materials for Base-Metal Multilayer Ceramic Capacitors, *Am. Ceram. Soc. Bull.*, Vol 66 (No. 9), 1987, p 1338–1341

9. N. Fujikawa, T. Shintome, H. Utaki, and N. Yokoe, Temperature-Stable X7R Multilayer Ceramic Capacitor with Base Metal Electrodes, *IMC Proceedings*, Kobe, 1986, p 202–205

10. H. Kishi, S. Murai, H. Chazono, M. Oshio, and N. Yamaoka, Properties of Glass-Added Barium Titanate Based Ceramics Fired in a Reducing Atmosphere, *Jpn. J. Appl. Phys.*, Vol 26 (Suppl. 26–2), 1987, p 31–33

11. I. Burn, "High K Dielectric Composition for Use in Multilayer Ceramic Capacitors Having Copper Electrodes," U.S. patent 4,855,266, 1989

12. N. Yamaoka and T. Matsui, Properties of $SrTiO_3$-Based Boundary Layer Capacitors, *Advances in Ceramics*, Vol 1, American Ceramic Society, 1981, p 232–241

13. S. Neirman and I. Burn, Dielectric Properties of Donor-Doped Polycrystalline $SrTiO_3$, *J. Mater. Sci.*, Vol 17, 1982, p 3510–3524

14. S. Nishigaki, K. Murano, and A. Ohkoshi, Dielectric Properties of Ceramics in the System $(Sr_{0.50}Pb_{0.25}Ca_{0.25})TiO_2$-$Bi_2O_3 \cdot 3TiO_2$ and Their Application in a High Voltage Capacitor, *J. Am. Ceram. Soc.*, Vol 65 (No. 11), 1982, p 554–560

15. I. Burn and W.C. Porter, MLC's with Copper Electrodes for High Frequency Applications, *Proceedings of the 40th Electronic Comp. and Technology Conference*, Las Vegas, 1990, p 277–283

16. T. Jaakola, A. Uusimaki, R. Rautioaho, and S. Leppavuori, Matrix Phase in Ceramics with Composition near $BaO \cdot Nd_2O_3 \cdot 5TiO_2$, *J. Am. Ceram. Soc.*, Vol 69 (No. 10), 1986, p 234–235

17. T.R. Shrout and J.P. Dougherty, Lead Based $Pb(B_1B_2)O_3$ Relaxors vs. $BaTiO_3$ Dielectrics for Multilayer Capacitors, *Ceram. Trans.*, Vol 8, 1990, p 3–19

18. G.H. Haertling, Electro-optic Ceramics and Devices, *Electronic Ceramics*, L.M. Levinson, Ed., Marcel Dekker, 1988, p 371–492

19. A. Ochi, K. Utsumi, T. Mori, M. Yonezawa, J. Morishita, and T. Yoshimoto, A New Dielectric Ceramic Material for Capacitors with High Specific Capacitance, *Ceram. Trans.*, Vol 8, 1990, p 45–53

20. A. Halliyal, U. Kumar, R.E. Newnham, and L.E. Cross, Stabilization of the Perovskite Phase and Dielectric Properties of Ceramics in the $Pb(Zn_{1/3}Nb_{2/3})O_3$-$BaTiO_3$ System, *Am. Ceram. Soc. Bull.*, Vol 66 (No. 4), 1987, p 671–676

21. Y. Yamashita, O. Furukawa, H. Kanai, M. Imai, and M. Harata, A Low-Firing Dielectric for MLC's Based on Relaxor/BT Ceramic Composite, *Ceram. Trans.*, Vol 8, 1990, p 35–44

22. M. Yonezawa, Low-Firing Multilayer Capacitor Materials, *Am. Ceram. Soc. Bull.*, Vol 62 (No. 12), 1983, p 1375–1383

23. J. Kato, Y. Yokotani, H. Kagata, S. Nakatani, and K. Kugimiya, Dielectric Material in Lead-Based Perovskite and Fabrication Process for Multilayer Ceramic Capacitor with Copper Internal Electrode, *Ceram. Trans.*, Vol 8, 1990, p 54–68

24. G. Goodman, Capacitors Based on Ceramic Grain Boundary Barrier Layers—A Review, *Advances in Ceramics*, Vol 1, American Ceramic Society, 1981, p 215–231

25. I. Burn, "Newtonian Ceramic Slip and Process For Using," U.S. patent 4,510,175, 1985

26. T.C. Rutt and J.A. Stynes, Fabrication of Multilayer Ceramic Capacitors by Metal Impregnation, *IEEE Trans.*, Vol PHP-9 (No. 3), 1973, p 144

27. T. Yasumoto, N. Iwase, M. Harata, and M. Segawa, High Dielectric Constant Thick-Film Capacitors Fireable in Air and Nitrogen Atmosphere, *Int. J. Hybrid Microelectron.*, Vol 12 (No. 3), 1989, p 156–161

28. D.A. Payne, Thin-Layer Dielectrics by Sol-Gel Methods, *Bull. Am. Phys. Soc.*, Vol 34 (No. 3), 1989, p 991

29. J.F. Scott and C.A. Paz De Araujo, Ferroelectric Memories, *Science*, Vol 246, 1989, p 1400–1405

30. K. Utsumi, Y. Shimada, T. Ikeda, and H. Takamizawa, Monolithic Multicomponents Ceramic (MMC) Substrate, *Ferroelectrics*, Vol 68, 1986, p 157–179

31. K. Hoshi, S. Tosaka, and N. Yamaoka, Properties of Co-fired $BaTiO_3$ Based Capacitor with Low Temperature Fired Multilayer Ceramic Substrates, *Jpn. J. Appl. Phys.*, Vol 28 (Suppl. 28–2), 1989, p 50–51

Piezoelectric Ceramics

Kenji Uchino, Department of Physics, Sophia University, Japan

PIEZOELECTRIC CERAMICS are dielectric materials that are often used in both sensors and actuators. In addition to conventional piezoelectric vibrators, pressure and acceleration sensors are now commercially available, as well. Precision positioners and pulse drive linear motors that utilize ceramic piezoelectric elements have already been installed in precision lathe machines, semiconductor manufacturing apparatus, and OA equipment. Ultrasonic motors, which are high-power piezoelectric resonators, are the subject of recent developments. However, the focus of this article is newly developed piezoelectric and electrostrictive ceramic materials and their applications, particularly in actuators (Ref 1, 2).

Piezoelectric and Electrostrictive Materials

In the early 1950s, barium titanate ($BaTiO_3$) was used in Langevin-type piezoelectric vibrators. Later, PZT [$Pb(Zr,Ti)O_3$] was discovered to exhibit piezoelectric constants twice as large as those of $BaTiO_3$ around the morphotropic phase boundary between the rhombohedral-tetragonal boundary (Fig 1) (Ref 3).

Successively modified PZTs (by doping and using ternary solid solutions with a different perovskite) have been investigated intensively. Examples include:

$Pb(Mg_{1/3}Nb_{2/3})O_3$, $Pb(Mn_{1/3}Sb_{2/3})O_3$,
$Pb(Co_{1/3}Nb_{2/3})O_3$, $Pb(Mn_{1/3}Nb_{2/3})O_3$,
$Pb(Ni_{1/3}Mo_{2/3})O_3$, $Pb(Sb_{1/2}Sn_{1/2})O_3$,
$Pb(Co_{1/2}W_{1/2})O_3$, $Pb(Mg_{1/2}W_{1/2})O_3$.

The electromechanical parameters associated with PZT ceramics are summarized in Table 1.

In recent actuator applications, large-magnitude electric fields (1 kV/mm, or 25 kV/in.) are applied to the material, thereby generating large stresses (9.8 MPa, or 1400 psi) and strains ($\Delta L/L = 10^{-3}$). Therefore, electrical insulation strength and mechanical toughness are necessary material characteristics, in addition to an adequate electrostrictive response.

Lead zirconate titanate (PZT) based ceramics are currently the primary materials used for piezoelectric applications (Ref 4). Strain curves for one such composition are shown in Fig 2(a). When the applied electric field is small, the induced strain is nearly proportional to the electric field. However, as the field becomes larger (greater than about 0.1 kV/mm, or 2.5 kV/in.), the strain curve deviates from this linear trend and a significant hysteresis is exhibited due to polarization reversal. This limits the use of the material in actuator applications that require a linear, nonhysteresis response.

Previously, electrostriction, a second-order phenomenon of electromechanical coupling, was considered to be a negligible effect, and was therefore not studied from a practical point of view. However, recent research and development on lead magnesium niobate $Pb(Mg_{1/3}Nb_{2/3})O_3$ (PMN)-based ceramics (Ref 5) have kindled new interest in the use of such electrostrictive materials. PMN-based ceramics exhibit significant strains up to 0.1% (that is, a 10 mm, or 0.4 in. sample can elongate by as much as 10 μm, or 40 μin.). Another attractive feature of these materials is the near-absence of hysteresis (Fig 2b).

Piezoelectricity is also found in polymers (Ref 6). Piezoelectric polymers have the following characteristics:

- Small piezoelectric d constants (for actuators) and large g constants (for sensors)
- Light weight and soft elasticity, leading to good acoustic impedance matching to water or human body
- Low mechanical quality factor, giving a wide frequency band resonance

Recent developments in composites of piezoelectric ceramics and polymers are remarkable (Ref 7). Superior piezoelectricity has been achieved, while maintaining the mechanical flexibility of polymers.

Piezoelectric Ceramic Applications

Transformers. One very basic application of piezoelectric ceramics is a gas igniter. Very high voltage generated in a piezoelectric ceramic under an applied mechanical stress can cause sparking and gas ignition.

When input and output terminals are fabricated on a piezoelectric device, and input/output voltage is changed through vibration energy transfer, the resulting product is a piezoelectric transformer. Since the 1957 proposal by Rosen (Ref 8), there have been a variety of investigations. A fundamental structure, in which two differently poled parts coexist in one piezoelectric plate, is shown in Fig 3. A standing wave with a wavelength equal to the sample length is excited, where a half wavelength exists on the input (L_1) and output (L_2) parts, respectively.

Piezoelectric Vibrators. When using mechanical vibrations with filters or oscillators, the size and shape of the device is very important. The vibration mode and the ceramic material must also be considered. Devices with a rather low resonance frequency are used for speakers or buzzers that are audible to human ears. Examples are a bimorph, comprising two bonded piezoelectric ceramic plates and a metal plate. A piezoelectric buzzer, shown in Fig 4, has merits such as high electric power efficiency, compact size, and long life.

Ultrasonic waves are now used in various fields. The sound source is made of piezoelectric ceramic, as well as magnetrostrictive materials. A liquid medium is usually used for sound energy transfer. Ultrasonic washers and microphones for short-distance remote control are widely used in factories, and ultrasonic scanning detectors are useful in medical electronics.

Surface Acoustic Wave. SAW filters have been considered for application to intermediate frequency image transfer signal in color TVs, because of their excellent time delay characteristics. The fundamental structure of a SAW filter is illustrated in Fig 5, where a pair of interdigital electrodes is fabricated on the piezoelectric crystal. A surface wave generated at the input-side electrode is transferred and picked up as an electric signal at the output side. This is very useful for a high-frequency filter.

Four materials are presently used for SAW devices: PZT-based ceramics, ZnO thin films, and $LiNbO_3$ and $LiTaO_3$ single crystals, the characteristics of which are summarized in Table 2. When a polycrystalline material is used, poreless and homogeneous samples with good reproducibility must be produced. Advanced technology for the fine ceramic preparation is actually required.

Piezoelectric Actuators. The need for new displacement elements, particularly in such fields as optics, precision machinery, and small

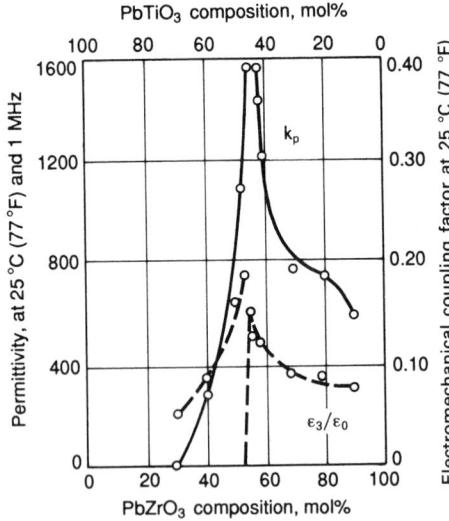

Fig 1 Permittivity, ϵ, and the planar coupling factor, k_p, in PbZrO$_3$-PbTiO$_3$ ceramics

Table 1 Electromechanical parameters in lead zirconate titanate and lead magnesium niobate ceramics

Parameters	PCM-5A(a)	PCM-33	PCM-4B	PCM-67	PCM-80
Electromechanical coupling factors					
k_p	0.65	0.65	0.70	0.32	0.58
k_{31}	0.38	0.39	0.43	0.19	0.35
k_{33}	0.71	0.74	0.72	0.48	0.69
k_{18}	0.70		0.63	0.37	0.64
k_t	0.50	0.50	0.50	0.40	0.47
Piezoelectric constants					
d_{21}, m/V	-186×10^{-12}	-263×10^{-12}	-247×10^{-12}	-42×10^{-12}	-122×10^{-12}
d_{33}, m/V	357×10^{-12}	575×10^{-12}	490×10^{-12}	109×10^{-12}	273×10^{-12}
d_{15}, m/V	579×10^{-12}	...	599×10^{-12}	131×10^{-12}	412×10^{-12}
g_{31}, V·m/N	-12.3×10^{-3}	-8.4×10^{-3}	-11.8×10^{-3}	-7.7×10^{-3}	-11.3×10^{-3}
g_{33}, V·m/N	25.7×10^{-3}	18.4×10^{-3}	20.6×10^{-3}	19.8×10^{-3}	25.5×10^{-3}
g_{15}, V·m/N	33.9×10^{-3}	...	29.7×10^{-3}	24.7×10^{-3}	31.9×10^{-3}
Permittivity					
$\epsilon_{33}^T/\epsilon_0$	1710	3530	2380	620	1210
$\epsilon_{11}^T/\epsilon_0$	1830	2740	2280	600	1460
Loss tangent, %	1.50	2.30	1.40	0.64	0.64
Elastic constants, GPa (10^6 psi)					
$1/s_{11}^E$	62 (9.0)	69 (10.0)	63 (9.1)	110 (16.0)	88 (12.8)
$1/s_{33}^E$	55 (8.0)	52 (7.5)	51 (7.4)	108 (15.7)	68 (9.9)
$1/s_{44}^E$	20 (2.9)	...	22 (3.2)	42 (6.1)	31 (4.5)
Density, g/cm^3	7.7	7.7	7.8	7.7	7.9
Mechanical quality factor	60	50	64	3130	2080
Curie temperature, °C (°F)	326 (619)	182 (360)	264 (507)	340 (644)	283 (541)

(a) PCM's are commercial products manufactured by Matsushita Panasonic.

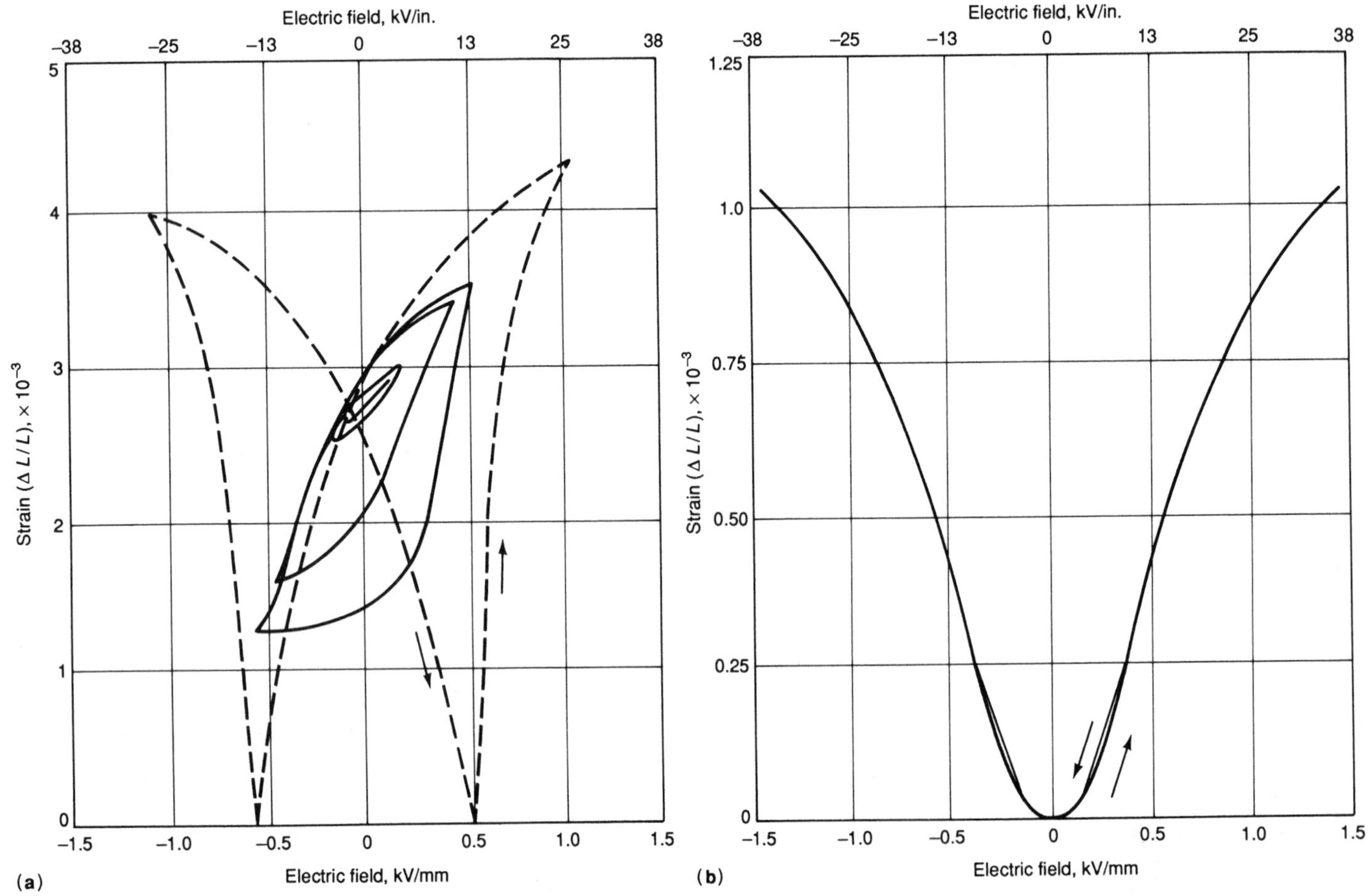

Fig 2 Field-induced strain in ceramics. (a) Piezoelectric material (Pb,La)(Zr,Ti)O$_3$(7/62/38). (b) Electrostrictive material 0.9Pb(Mg$_{1/3}$Nb$_{2/3}$)O$_3$-0.1PbTiO$_3$

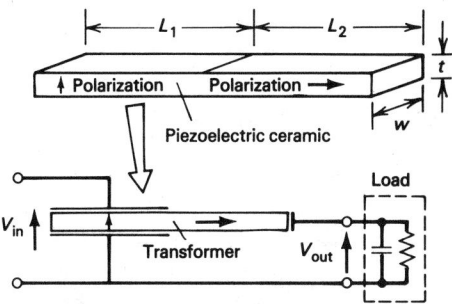

Fig 3 Piezoelectric transformer

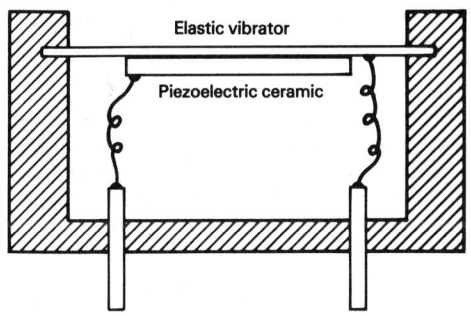

Fig 4 Piezoelectric buzzer

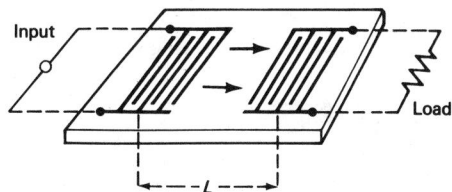

Fig 5 Fundamental structure of a surface acoustic wave filter

Table 2 Characteristics of surface acoustic wave filter substrate materials

Parameters	LiTaO$_3$	LiNbO$_3$	PZT(a)	ZnO
SAW velocity, m/s (10^3 ft/s)	3295 (10.8)	3960 (13.0)	2430 (8.0)	3150 (10.3)
Coupling factor (k^2), %	0.7	6.0	2.9	0.6
Temperature coefficient of frequency, 10^{-6}/K	−31	−78	−17	−15
Permittivity, ϵ_s	47.9	67.2	350	8.84
Curie point (T$_C$), °C (°F)	618 (1145)	1210 (2210)	300 (570)	1200 (2190)

(a) Pb(Mn$_{1/3}$Nb$_{2/3}$)O$_3$–PbZrO$_3$–PbTiO$_3$

signal, and pulse-driven motors that operate in a simple on/off switching mode (Ref 9).

Because the material requirements for these devices are somewhat different, certain compounds will be better suited to particular applications. For example, the ultrasonic motor usually requires the conventional hard-type piezoelectric material with a high mechanical quality factor. The servo displacement transducer suffers most from strain hysteresis, and therefore requires a PMN electrostrictor. The pulse-driven motor requires a low-permittivity material suitable for quick response, rather than a small hysteresis, which means that soft-PZT piezoelectrics are most appropriate for this application.

Described below are three typical product applications.

Deformable mirrors have been proposed to control the phase of the incident light wave in the field of optical information processing. The deformable mirror can be made either more convex or concave, as necessary. This type of mirror, used as an accessory device on observatory telescopes, effectively corrects for image distortions that result from fluctuating airflow.

An example of a deformable mirror is a multilayered two-dimensional multimorph, shown in Fig 7 (Ref 10). When three layers of thin, electrostrictive PMN plates are bonded to the elastic plate of a glass mirror, the mirror surface is deformed in various ways corresponding to the strain induced in the PMN layer. The nature of the deformation is determined by the electrode configuration and the distribution of the applied electric field. Trial units have been designed in which the first layer, with a uniform electrode pattern, produces a spherical deformation (that is, refo-

motors, has been rising rapidly. The processing accuracy of optical devices such as lasers and cameras, along with the positioning accuracy required in the processing of semiconductor chips, are now typically on the order of micron and submicron levels. The need for a reliable microscale positioner has motivated a surge of activity in the development of ceramic actuators that operate on the principle of electric-field-induced strain.

There are two categories of piezoelectric/electrostrictive actuators, based on the type of driving field applied to the device and the nature of the strain induced by that field (Fig 6). Rigid displacement devices, for which the strain is induced unidirectionally by an applied dc field, represent one category, while resonating strain devices, for which the mechanical resonance is excited by an ac field, represent the other category. Rigid displacement devices can be subclassified as servo displacement transducers controlled by a feedback system through a position detection

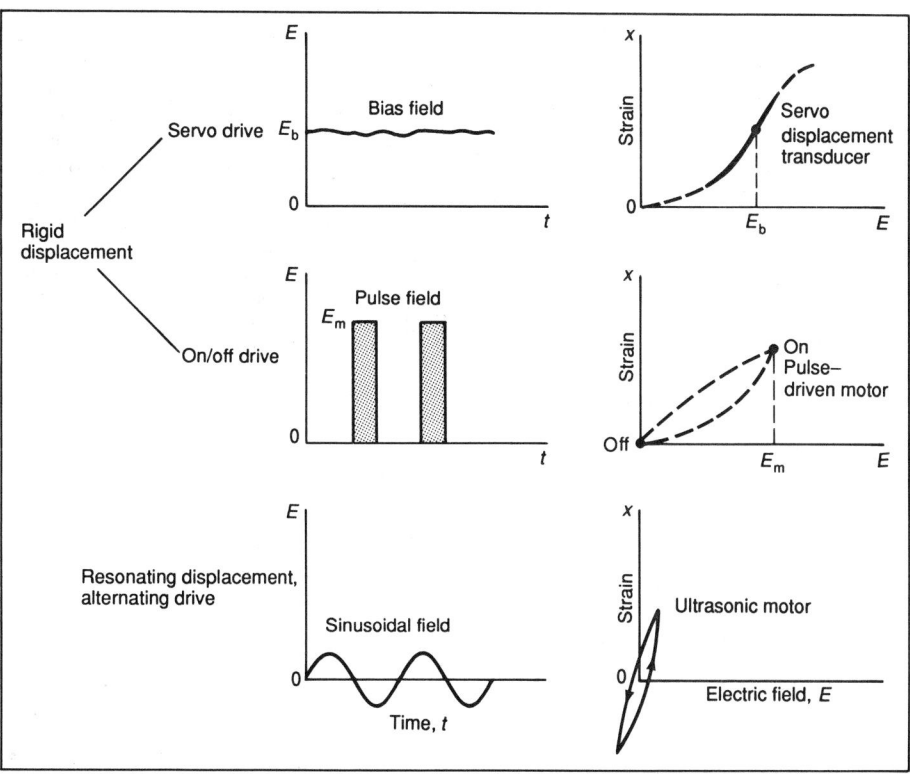

Fig 6 Categories of piezoelectric/electrostrictive actuators

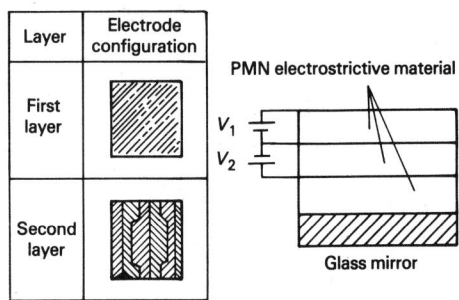

Fig 7 (a) Structure of multilayer bimorph deformable mirror and (b) actual control of wavefront

cusing), while the second layer, with an electrode pattern of six minute divisions, corrects for coma aberration.

Impact dot-matrix printers are common among the various printing devices currently in use. Each character formed by such a printer is initially composed of a 9 × 9 dot matrix. A printing ribbon is subsequently impacted by a multiwire array. The printer head is shown in Fig 8(a).

The basic actuator assumes a multilayer configuration in which roughly 100 thin piezoelectric ceramic sheets are stacked. This

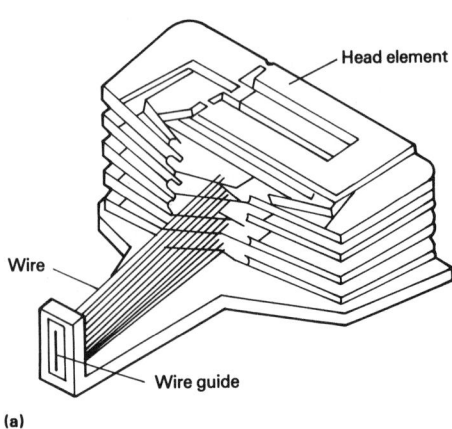

(a)

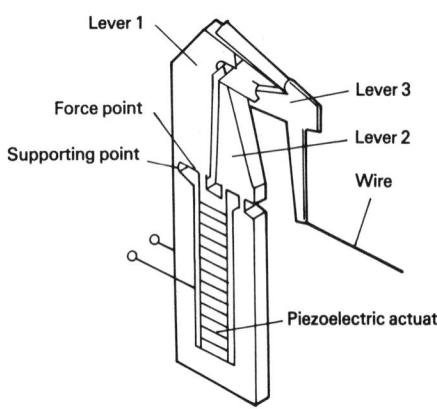

(b)

Fig 8 Structure of (a) printer head and (b) differential type head element

actuator is installed in a specially designed displacement magnification unit (Fig 8b) to drive the top printer pins. This unique magnification unit is based on a monolithic hinge lever with a magnification of 30. It realizes an amplified displacement of 300 μm (12 mils) and an energy transfer efficiency greater than 50%.

The merits of the piezoelectric impact printer, compared to the conventional electromagnetic type, are: (1) higher printing speed by an order of magnitude, (2) lower energy consumption by an order of magnitude, and (3) reduced printing noise, due to a complete sound shield that can be employed with the heatless drive.

Ultrasonic Motors. Efforts have been made to develop high-power ultrasonic vibrators as replacements for conventional electromagnetic motors. Two actuators in particular are currently being investigated for this application: a vibratory-coupler type and a surface-wave type (Ref 12).

The basic design of the coupler type is shown in Fig 9. The Langevin-type piezoelectric vibrator generates a flat, elliptical movement at the tip of the vibratory piece. When contacting a rotor at a slight angle, the vibratory piece generates a rotational torque.

This simple design does have several drawbacks. Its two most serious problems are defacement of the rotor caused by mechanical friction at the contact point and lack of control in both the clockwise and counter-clockwise directions. The friction problem is alleviated by securely pressure-fitting the vibratory reed and rotor together in order to restrict sliding during operation, as much as possible. The model in Fig 10 is one such modified unit (Ref 13). At a rotational speed of 600 rev/min, this motor has performance characteristics that surpass normal electromagnetic motors with a rotational torque of 1.3 N · m and an energy conversion efficiency of 80%.

The operating principle of the motor type that utilizes surface wave vibrations is illustrated in Fig 11. By means of the traveling elastic wave, a slider in contact with the "rippled" surface of the elastic body is driven in

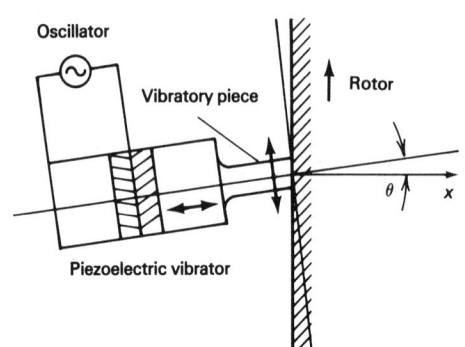

Fig 9 Structure of vibratory coupler type ultrasonic motor

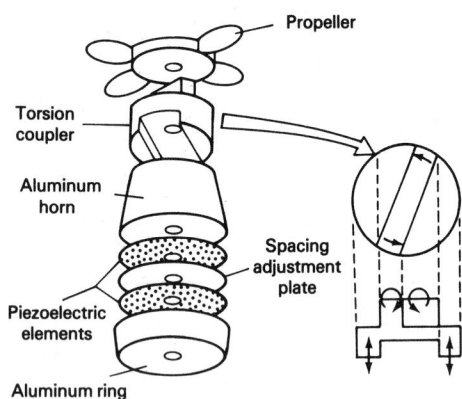

Fig 10 Ultrasonic motor using a torsion coupler

the direction indicated. Both linear and rotational-type motors of this type are possible. The structure of a surface-wave rotational-type motor is shown in Fig 12 (Ref 14). Although the energy transformation efficiency (30%) and rotational torque (50 N·m) of the surface-wave device is rather low compared to the vibratory-coupler type, its merits are its ability to rotate in both directions and its thin

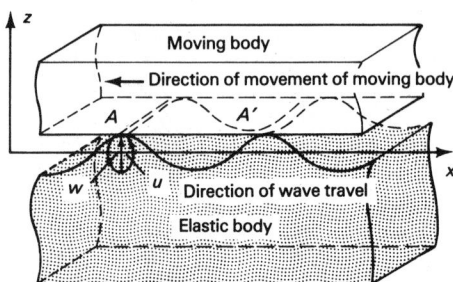

Fig 11 Operating principle of surface wave linear motor (transverse propagating wave being excited to an elastic body)

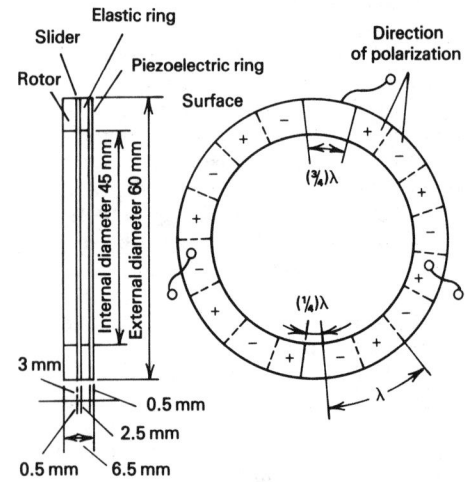

Fig 12 Structural example of surface wave rotational device. λ = wavelength

design, which makes it suitable for installation in video or movie cameras as an automatic focusing device.

REFERENCES

1. K. Uchino, *Piezoelectric/Electrostrictive Actuators*, Morikita Publishing, Tokyo, 1986
2. K. Uchino, J. Kuwata, S. Nomura, L.E. Cross, and R.E. Newnham, *Jpn. J. Appl. Phys.*, Vol 20, Suppl. 20–4, 1981, p 171
3. B. Jaffe, R.S. Roth, and S. Marzullo, *J. Res. Natl. Bur. Stand.*, Vol 55, 1955, p 239
4. K. Furuta and K. Uchino, *Adv. Ceram. Mater.*, Vol 1, 1986, p 61
5. L.E. Cross, S.J. Jang, R.E. Newnham, S. Nomura, and K. Uchino, *Ferroelectrics*, Vol 23, 1980, p 187
6. H. Kawai, *Jpn. J. Appl. Phys.*, Vol 8, 1969, p 975
7. K.A. Klicker, J.V. Biggers, and R.E. Newnham, *J. Am. Ceram. Soc.*, Vol 64, 1981, p 5
8. C.A. Rosen, Proceedings of the Electronic Component Symposium, 1957, p 205
9. K. Uchino, *Bull. Am. Ceram. Soc.*, Vol 65, 1986, p 65
10. T. Sato, H. Ishikawa, O. Ikeda, S. Nomura, and K. Uchino, *Appl. Optics*, Vol 21, 1982, p 3669
11. T. Yano, I. Fukui, E. Sato, O. Inui, and Y. Miyazaki, Electr. & Commun. Soc., Proc. Annual Mtg., Spring, 1984, p 1–156
12. Y. Akiyama, Ed., *Ultrasonic Motors/Actuators*, Triceps, Tokyo, 1986
13. A. Kumada, *Jpn. J. Appl. Phys.*, Vol 24, Suppl. 24–2, 1985, p 739
14. *Nikkei Mechanical*, 28 Feb 1983, p 44

Electrooptic Ceramics and Devices

Gene H. Haertling, Clemson University

CERAMICS have long been known for their desirable structural, electrical, and electro-mechanical (piezoelectric) properties; how-ever, it has only been within the last two decades that they have also found application in the field of electrooptics. Basically, elec-trooptic ceramics are polycrystalline, ferro-electric (FE) materials which, in addition to their many other characteristics, possess both high optical transparency and voltage-vari-able electrooptic (E/O) behavior. Taken together, these two properties have been the key to the successful utilization of ceramics in such electrooptic devices as shutters, mod-ulators, displays, and image storage devices (Ref 1–10).

Ceramics versus Single Crystals

Ceramics, in the narrow sense of the term, are inorganic, non-metallic solids which are polycrystalline in nature, and as such, are composed of an assemblage of microscopic crystallites (grains) intimately bonded together but randomly oriented with respect to each other, thus forming a structure where grain sizes and grain boundaries play a significant role in their behavior. Single crystals, on the other hand, are crystals of a much larger, macroscopic size with a high degree of inter-nal order and few or no grain boundaries. This fundamental difference between ceramics and single crystals is important to note and often leads to significant differences in properties and phenomena associated with their physical nature, and consequently, their application. This is especially true of optical effects such as light scattering, birefringence, and depo-larization which are of special importance in electrooptic materials.

In comparison to single crystals, electroop-tic ceramics have several specific character-istics which make them well suited for a variety of electrooptic applications. These include:

- Small (on the order of the grain size, that is, 1 to 10 μm) local areas of the ferro-electric ceramic can be electrically switched independently of other adjacent areas
- The light transmission characteristics of the switched areas depend on the thor-oughness and direction of switching or poling
- The switched areas are stable with time in memory materials or stable with applied electric field in non-memory materials
- The switched areas exhibit either electri-cally variable light scattering behavior or electrically variable optical birefringence
- Light is preferentially scattered along the switching (polar) direction
- Ferroelectric memory materials with large grain sizes (5 to 10 μm) scatter light more efficiently than those with small (1 to 2 μm) grain sizes; hence, the optical retar-dation (birefringence) effect is signifi-cantly more prominent at small grain sizes, whereas, scattering effects dominate at the larger grain sizes
- Ceramics can be hot pressed or sintered in a wide variety of sizes and shapes with a high degree of optical uniformity on a macroscopic scale

While some of the effects mentioned here are responsible for making possible novel electrooptic devices from ceramics which cannot be duplicated with single crystals, it should be remembered that ceramics are gen-erally less desirable than single crystals in regard to overall transparency and optical birefringence on a microscopic scale.

Electrooptic Ceramics as Ferroelectrics

A ferroelectric crystal is characterized by a net spontaneous polarization which can be reversed or reoriented along certain crystal-lographic directions of the crystal when the polar direction is changed, usually with an electric field. The spontaneous polarization has its origin in a noncentrosymmetric arrangement of the ions in the unit cell which produces an electric dipole moment associ-ated with the cell. The acentric symmetry also produces a distortion of the unit cell, usually in the form of an elongation along one of the crystal axes (the polar axis) and contractions along the other two axes. Adjacent unit cells tend to deform in the same direction to form a region of homogeneous polarization called a domain, and the interface between domains is referred to as a domain wall. Except for very small crystals, multiple domains and domain walls are usually found to exist in as-grown ferroelectrics.

Although there are several basic structures into which ferroelectrics commonly crystal-lize (perovskite, tungsten bronze, pyro-chlore, and bismuth layer), the one which best typifies most of the commercially important materials is the ABO_3 perovskite unit cell which forms the basic building block of the internal structure. As shown in Fig 1, it con-sists of a linked network of oxygen octahedra with the B ions (Ti^{4+}, Zr^{4+}, Nb^{5+}) occupying the sites within the oxygen octahedral cage (B sites) and the A ions (Pb^{2+}, Ba^{2+}, La^{3+}) situated in the interstices (A sites) created by the linked octahedra. When off-valent ionic substitutions are made into this structure (for example, Nb^{5+} for Ti^{4+} or La^{3+} for Pb^{2+}), electrical neutrality is automatically main-tained by the creation of A-site or B-site vacancies. The creation of these defects (vacancies) has a profound effect upon the properties of the material and can often lead to unusual phenomena which may have a positive or negative impact on their applica-tion (Ref 11, 12).

In the case of an electrooptic ceramic which is composed of an aggregate of small, ran-domly oriented crystallites, each possessing ferroelectric properties, there is no net elec-trooptic activity until all of the crystallites are constrained to act in concert with each other. This condition is achieved through a process known as poling, that is, the application of an electric field which is high enough to obtain a reorientation of the domains within the material. Once poled, the ceramic acts much like a ferroelectric single crystal but with somewhat less optimum properties. The extent to which it acts as a single crystal is depen-dent upon its specific crystal structure. For example, in the tetragonal system, the expected net polarization of a poled ceramic compared to a single crystal of the same symmetry is 83.1%, for rhombohedral symmetry this value rises to 86.6% and for orthorhombic symmetry it reaches a maximum of 91.2% (Ref 13).

Thus, it can be concluded that essentially all electrooptic ceramics are ferroelectric since their spontaneous polarization must be reo-riented in order for them to act as a single

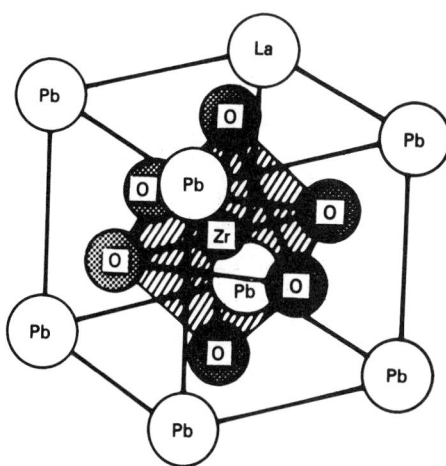

Fig 1 The ABO_3 perovskite structure showing the location of the ions and the octahedral cage (shaded). Example shown: PLZT

entity with cooperative optical and electrooptic behavior.

Materials

Although a number of polycrystalline, optically transparent ceramic materials have been developed over the years for applications dealing with components such as infrared (IR) windows, lamp envelopes, and missile radomes, the concept of utilizing transparent FE ceramics as an E/O medium has only existed since the mid-1960s. At that time, research surrounding the study of the properties of fully dense, chemically modified, lead zirconate titanate (PZT) resulted in noticeable improvements in their optical translucency and the discovery of the existence of significant electrooptic activity (Ref 14, 15). By 1969, this effort culminated in the development of the fully transparent, lanthanum-modified PLZT compositions which are still in use today (Ref 1).

Other materials which have subsequently been investigated for electrooptic applications include:

(Pb,La)(Hf,Ti)O₃   Pb(Sc,Nb)O₃
(Pb,Ba)(Zr,Ti)O₃   (Pb,Sr)(Zr,Ti)O₃
(Pb,Ba,La)Nb₂O₆   K(Ta,Nb)O₃
(Ba,La)(Ti,Nb)O₃₀

(Ref 10). Specific compositions within these systems have been reported to possess enhanced optical transparency and high electrooptic coefficients; however, in ceramic form, none of these have yet been found to exceed those of the PLZT materials.

PLZT Compositions

The room-temperature phase diagram of the PLZT compositional solid solution system is given in Fig 2. It can be seen that extensive solid solution of La₂O₃ in PbZrO₃ and PbTiO₃ occurs throughout the phase diagram, leading to large areas of homogeneous single phases. The actual extent of solubility of lanthanum in either PbZrO₃ or PbTiO₃ depends on the

ratio of the two compounds in solid solution with each other. Excess lanthanum results in an undesirable condition from the standpoint of optical transparency, that is, mixed phases (double cross-hatched region in Fig 2) and opacity. Within the regions of homogeneous solid solubility, lanthanum additions reduce the Curie point (ferroelectric-to-cubic transition temperature) in a linearly decreasing manner to room temperature and below. In fact, it is this very ability of lanthanum to significantly lower the Curie point (~36 °C, or 65 °F/at.% La), which is largely responsible for the enhanced optical transparency. Lowering the Curie point has the effect of reducing the ABO_3 unit cell distortion which results in less optical anisotropy and less internal refraction at grain boundaries and domain walls which scatter light. Another important but lesser effect of the lanthanum is its ability to produce a significant number of vacancies on the A and B sites of the unit cell which results in an enhancement of the densification process.

Also to be noted from Fig 2 is the boundary region between the FE and cubic phases denoted by the single cross-hatched area. Compositions within this region are broadly categorized as relaxors, possessing metastable FE properties, that is, they are macroscopically non-FE and undergo a phase change to a polar FE state when subjected to an electric field. These materials, while not possessing permanent spontaneous polarization like the FEs, mimic them in almost every other way and can even duplicate them when subjected to an electrical biasing field (Ref 16).

The five shaded and lettered areas of Fig 2 denote compositional regions of materials which are of specific current industrial interest. Region A includes phase boundary compositions possessing maximum piezoelectric effects (sonar, ultrasonics, filters, speakers), region B involves materials with easy ferroelectric domain switching (memories, image storage), region C covers the major relaxor compositions (shutters, actuators), region D denotes antiferroelectric, voltage-stable materials (capacitors), and region E identifies some special thin-film electrooptic compositions for integrated optics (switches, modulators).

A general formula describing all of the compositions in the PLZT system is given by:

$$Pb_{1-x}La_x(Zr_yTi_{1-y})_{1-x/4}O_3$$

where La^{3+} ions replace Pb^{2+} ions (ions of similar size) on the A sites of the unit cell and charge balance is maintained by the creation of lattice site vacancies on the B sites. In reality, both A and B sites possess vacancies (roughly, the A/B-site vacancy ratio is the same value as the zirconium/titanium ratio in the composition) when stoichiometric balance is achieved, usually resulting in the expulsion of some free PbO during sintering. The free PbO has been credited with playing a key role in the complete densification of the material via liquid-phase sintering.

Powders and Processes

Ceramics are traditionally prepared from powders formulated from the individual oxides; however, early attempts to produce the PLZT powders by this method proved to

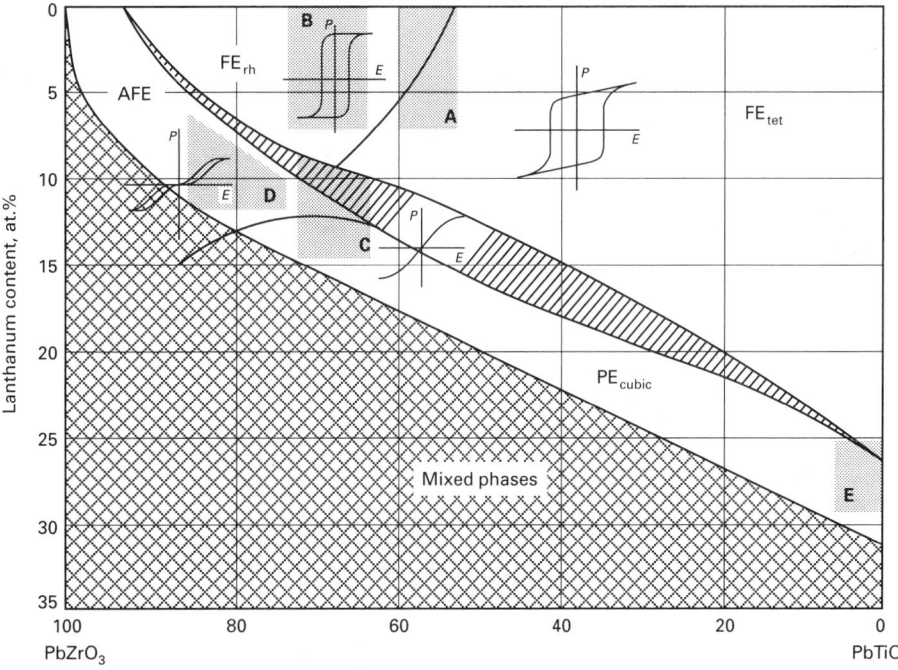

Fig 2 Room-temperature phase diagram of the PLZT system illustrating hysteresis loops of the various regions and application areas (shaded) of commercial importance

be inadequate from the standpoint of chemical and optical uniformity. As a result, various chemical coprecipitation (CP) processes, which utilized liquid precursors (in part or in whole) as the starting materials, were developed. This was done in an effort to achieve (1) more intimate chemical mixing before forming the powders, (2) finer powder particle sizes via controlled precipitation, and (3) purer powders since the precursors could be chemically purified before precipitation.

The earliest chemical coprecipitation technique (alkoxides) was developed in order to achieve high levels of chemical and optical uniformity but at the expense of higher material costs. Chemical methods developed later (nitrates, oxalates) stressed lower material costs, supposedly without sacrificing optical quality. The most economical method is the mixed oxide (MO) process and there have been continuing attempts to perfect this method for the PLZT materials. Considerable progress has been made with the MO process in the last few years, and excellent transparency and optical homogeneity can usually be achieved on a regular basis.

A flowsheet describing the essential steps for both the mixed oxide and coprecipitation processes is given in Fig 3. The essential difference between the two processes occurs in the powder forming state. For the MO technique, this very simply consists of wet milling the individual oxides, drying and calcining (a high-temperature, solid-state chemical reaction) at 800 to 900 °C (1470 to 1650 °F). In the CP process, the starting materials are usually solutions which are mutually soluble in each other, thus producing an atomically homogeneous precursor solution which is precipitated with one or more other solutions to produce a very fine particle size material that is subsequently filtered and dried. Because the particle size of the CP powders are usually much finer than the MO powders (0.03 to 0.1 μm versus 1 μm), they are more reactive and are calcined at a lower temperature of approximately 500 °C (930 °F). After calcining, the powder is wet ball milled, dried, and stored for further processing.

Fabrication

In addition to composition and powder preparation, consolidation of the powder into a fully dense slug is the third area which is critical to achieving optical transparency. The method of densification which has been found to be the most reliable is that of hot pressing. Using this technique, external pressure is applied to the cold-compacted powder during sintering at an elevated temperature in an oxygen-enriched environment. However, this operation, like that of powder preparation, is undergoing a change to more economical methods of production involving conventional sintering. These two alternatives are shown in the flowsheet of Fig 3.

Over the years, several variations of both processes have been developed in an effort to

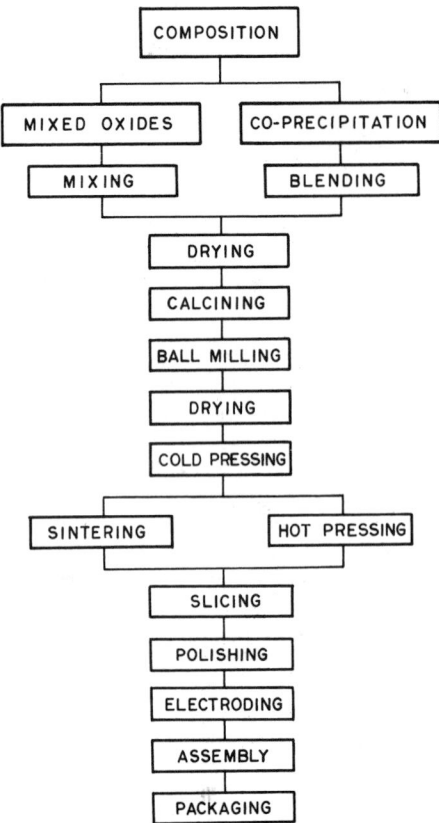

Fig 3 A process flow chart for the preparation of bulk PLZT materials and devices

increase the ease of fabrication, reduce fabrication costs, facilitate volume production, and maximize material quality. Although all of them have been reasonably successful in producing transparent material, not all of them are easy to implement, especially at affordable capital costs. The two basic methods that will most likely prevail in the future are vacuum/oxygen hot pressing for a high-quality, premium product, and vacuum/oxygen sintering for a somewhat lesser quality, more economical, higher volume product.

Properties

All materials which are used for electrooptic devices must undergo careful inspection and evaluation in order to determine their properties which pertain to the particular application. These include:

- Physical properties—microstructure, porosity, and grain size
- Electrical properties—dielectric constant, FE hysteresis loop characteristics, and piezoelectric coupling constants
- Optical properties—transmittance and haze
- Electrooptic properties—linear r_c coefficient, quadratic R coefficient, and half-wave voltage

In addition, several other properties such as overall stress/strain, mottling (index of refraction variations), and stars (small stress

centers) are evaluated under polarized light since such defects are very noticeable under crossed polarizers when in operation as part of the device.

Properties which are specific to given compositions within the PLZT system are described in detail under separate sections. The properties which are essentially the same for all of the PLZT compositions are:

Property	Value
Flexural strength	103 MPa (14.7 ksi)
Linear thermal expansion	5.4×10^{-6}/°C
Heat capacity	420 J/kg·°C
Thermal conductivity	1.5 W/m·K

Source: Ref 10, 17

Optical Properties. The most outstanding feature of the PLZT materials is their high optical transparency. Transparency is both a function of the concentration of lanthanum and the zirconium/titanium ratio of the composition. Maximum transparency is possessed by compositions along the FE-cubic phase boundary in Fig 2. For example, the 65/35 ratio compositions are most transparent when the lanthanum content ranges from 8 to 12 at.%.

A typical transmission curve for a 9/65/35 (La/Zr/Ti) material is given in Fig 4. As noted, the material is highly absorbing in the ultraviolet (UV) below 370 nm which is the commonly accepted value of the absorption edge for bulk PLZT. Maximum transmission extends throughout the visible region and out into the IR to approximately 6.5 μm (see inset). At 12 μm, the material is, once again, fully absorbing. The high surface reflection losses (31%) are a function of the index of refraction ($n = 2.5$ at 633 nm) of the PLZT. Also illustrated in Fig 4 are examples of the sizes and transparency of the PLZT materials.

Compositions throughout the solid-solution portion of the PLZT system characteristically exhibit a highly uniform microstructure consisting of randomly oriented, equiaxed grains. An example of such a microstructure is shown in the inset of Fig 4. Depending upon the fabrication conditions, the average grain size of any given material may vary from about 2 to 15 μm with a typical size being approximately 8 μm diameter.

Electrical Properties. Although the unique, high optical transparency of many compositions in the PLZT system makes them particularly attractive for electrooptic applications, the electrical properties of these materials are also of interest. These materials are, in general, characterized by high dielectric constants (K = 500 to 5000), low-to-moderate dielectric losses (tan δ = 0.1 to 10%), high specific electrical resistivity (10^{13} Ω · cm), moderate dielectric breakdown strengths (E_c > 250 kV/cm), and variable hysteresis loop properties. A tabulation of some of the common properties for specific PLZT composi-

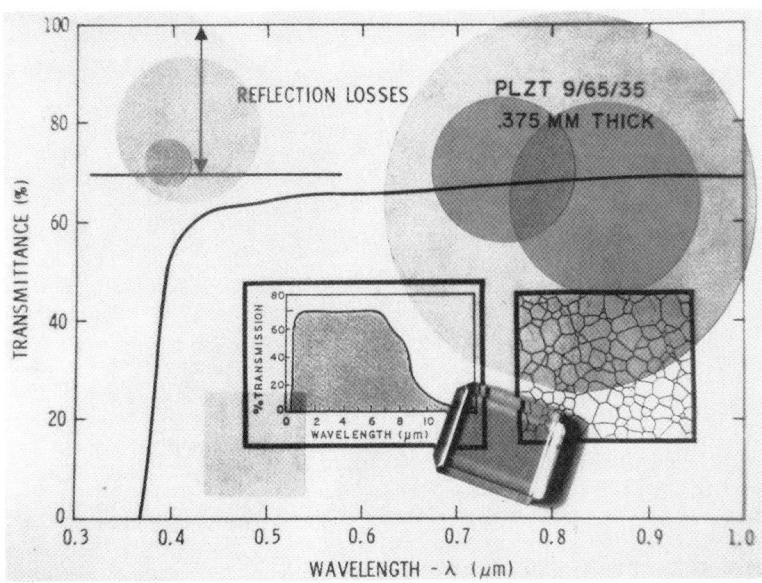

Fig 4 Spectral response of a quadratic, 9/65/35, PLZT material in the visible and IR (left inset). Also shown are an etched microstructure (right inset) and typical transparencies of 7/65/35 (upper left), 9/65/35 (right), and 8/70/30 (lower left).

tions is given in Table 1. Dielectric (K and tan δ), hysteresis loop (P_R and E_c), piezoelectric (d_{33}, g_{33}, and k_p) and electrostrictive (Q_{11} and Q_{12}) properties are listed in descending order of the zirconium/titanium ratio. Since the temperature of the dielectric constant maximum, T_m (Curie point) and the temperature at which there is a loss of permanent polarization (T_t) often do not agree in the slim loop ferroelectric (SFE) relaxor materials, both T_m and T_t are included in Table 1.

The hysteresis loop, that is, the polarization (charge) versus the electric field (voltage), is the single most important electrical measurement which can be made on an FE ceramic. Typical loops obtained from various PLZT compositions are shown in the phase diagram of Fig 2 (see lettered regions). These include: A, FE, hard, linear; B, FE, soft, memory; C, SFE, slim loop FE, quadratic; D, AFE, antiferroelectric; and E, SFE, slim loop FE, linear/quadratic.

The memory materials possess remanent polarization (P_R) when switched at electric fields higher than the coercive field, E_c. Electrical and optical information is stored in the ceramic by the incremental switching of polarization on traversing the hysteresis loop. Desirable characteristics for the memory materials are low E_c, moderate to high P_R, and a square loop.

The linear materials, like the memory ones, also possess remanent polarization; however, in this case the material is permanently poled to saturation remanence (P_R) and left to remain in that state. Optical information is extracted from the ceramic by the action of the positive or negative electrical field which causes slight changes (linear) in the optical birefringence of the ceramic, but polarization reversal should not occur. Desirable properties of linear materials are high E_c and high electrooptic sensitivity.

The SFE, quadratic materials do not possess permanent polarization but do exhibit substantial, induced polarization when subjected to an electrical field. Optical effects in this material are momentary, and noticeable only while the electric field is applied; removal of the field allows the material to relax back to its isotropic, nonactive state. Desirable characteristics in this material are complete non-memory ($P_R = 0$) in the unactivated state and a large induced optical birefringence in the activated state.

The AFE materials are essentially nonpolar, non-FE materials in their natural state and will revert to an FE state when subjected to a sufficiently high electric field. Like the SFE materials, electrooptic effects in these ceramics are only noticeable while the electric field is applied. Outwardly, they differ from the SFE materials in that they usually possess lower dielectric constants, higher electric fields are usually required to induce the FE state, and the onset of the FE state and the return of the AFE state are usually more abrupt, thus giving the loop an appearance of two sub-loops which are positive and negative biased.

The electrooptic properties of the PLZT ceramics are intimately related to their FE properties. Consequently, varying the polarization with an electric field, as in a hysteresis loop, produces a change in the optical activity of the ceramic. Moreover, the magnitude of the observed effect is dependent on both the strength and direction of the electric field.

Ferroelectric ceramics display optically uniaxial properties on a microscopic scale and also on a macroscopic scale when polarized with an electric field. In uniaxial crystals there is one unique symmetry axis, the optic axis

Table 1 Selected properties of PLZT compositions

PLZT type(a)	Density, g/cm³	T_t(b), °C	T_m(c), °C	K(d)	tan δ(e), %	P_R(f), μC/cm²	E_c(g), kV/cm	k_p(h)	d_{33}(i), C/N × 10¹²	g_{33}(j), V·m/N × 10³	Q_{11}(k), m⁴/C²	Q_{12}(l), m⁴/C²	$s_{11}E$(m), m²/N × 10¹²
8/90/10	7.83	...	...	303	0.4	0	0	0	0	0	...	...	...
7.4/70/30	7.84	52	...	4500	4.8	29	3.7	0.61	...	...	...	...	...
7.6/70/30	7.84	40	100	4940	5.4	28	3.6	0.65	...	...	...	...	...
7.8/70/30	7.83	29	...	5545	5.6	23	2.5	0.59	520	...	...	...	...
8/70/30	7.83	20	85	5100	4.7	0	0	0	0	0	0.010	−0.008	...
2/65/35	7.98	320	325	650	2.5	40	13.7	0.45	150	23	...	...	...
7/65/35	7.84	95	150	1850	1.8	34	5.3	0.62	400	22	0.022	−0.012	13.5
8/65/35	7.82	55	110	3400	3.0	30	3.6	0.65	682	20	0.018	−0.008	12.4
9/65/35	7.81	5	80	5700	6.0	0	0	0	0	0	0.020	−0.010	...
9.5/65/35	7.79	−10	75	5500	5.5	0	0	0	0	0	0.021	−0.009	...
10/65/35	7.78	−25	70	5100	5.4	0	0	0	0	0	...	...	...
12/65/35	7.74	...	60	2200	4.6	0	0	0	0	0	...	...	...
8/40/60	7.84	240	245	980	1.2	28	17.7	0.34	...	...	...	...	...
12/40/60	7.74	140	145	1300	1.3	25	12.5	0.47	235	12	...	...	7.5
8/10/90	7.84	...	355	360	0.4	29	37.5	0.21	...	...	...	...	...
30/0/100	7.41	10	...	2140	0.1	0	0	0	0	0	...	...	...

(a) La/Zr/Ti compositions with lanthanum given in at.%. (b) T_t is the temperature at which there is a loss of permanent polarization. (c) T_m is the temperature at the maximum dielectric constant. (d) K is the dielectric constant. (e) tan δ is the dielectric dissipation or the deviation from an ideal capacitor. (f) P_R is the remanent polarization. (g) E_c is the coercive field. (h) k_p is the planar coupling coefficient. (i) d_{33} is the longitudinal coupling coefficient. (j) g_{33} is the longitudinal open circuit voltage. (k) Q_{11} is the longitudinal electrostrictive coefficient. (l) Q_{12} is the lateral electrostrictive coefficient. (m) $s_{11}E$ is the compliance at constant electric field. Source: Ref 10, 18–20

(colinear with the FE polar vector), which possesses different optical properties than the other two orthogonal axes. That is, light traveling along the optic axis and vibrating in a direction perpendicular to it has a different index of refraction (n_0) than light traveling in a direction 90° to the optic axis and vibrating in a direction parallel to it (n_e). The absolute difference between the two indices is defined as the birefringence ($n_e - n_0 = \Delta n$). In ceramic materials where a statistical array of randomly-oriented crystallites is being considered, the effective birefringence is designated by $\Delta\bar{n}$. On a macroscopic scale, $\Delta\bar{n}$ is equal to zero before poling and has some finite value only after application of an electric field. The $\Delta\bar{n}$ value is a meaningful quantity in that it is related to the observed optical activity of the material.

A typical setup for detecting the voltage-variable change in optical birefringence in terms of light amplitude is shown in Fig 5. Linearly polarized light, on entering the activated ceramic, is resolved into two perpendicular components whose vibration directions are defined by the axes of the crystallites as they are influenced by the electric field. In the uniaxially negative PLZT materials, the electric field vector determines the overall optic axis of the ceramic; thus polarized light vibrating in the electric field direction also determines n_e. Because of the different refractive indices, n_e and n_0, the propagation velocity of the two components will be different within the material and will result in a phase shift called retardation. The total retardation, Γ, is a function of both Δn and the

optical path length t (generally, t = plate thickness) according to the relation, $\Gamma = \Delta nt$. When the proper voltage is applied to the ceramic, a Γ of $\lambda/2$ is achieved for one component relative to the other. The net result is one of rotating the vibration direction of the linearly polarized light by 90°, thus allowing it to be transmitted by the second polarizer in the ON condition. Switching of the ceramic from a state of $\Gamma = 0$ (no voltage) to $\Gamma = \lambda/2$ will create an ON/OFF light shutter.

In general, after application of an electric field, electrooptic ceramics behave in three distinct ways:

- As quadratic (Kerr cell) electrooptics having a squared relationship between Δn and E (the applied electric field)
- As linear (Pockel cell) electrooptics possessing a linear relationship between Δn and E
- As memory materials with switchable remanent electrooptic states

The general relationship that describes the ratio of the transmitted light (I_t) to the incident light (I_i) intensity in the configuration of Fig 5 is:

$$I_t/I_i = \sin^2 \pi\Gamma/\lambda = \sin^2 \pi\Delta nt/\lambda$$

where λ is the wavelength of the light, Γ is the total optical retardation, and t is the optical thickness.

A convenient method of characterizing electrooptic materials is by their respective electrooptic coefficients, R and r_c (Ref 1). These may be determined from the following relationships:

$$R = -2\Gamma/n^3tE^2$$
$$= -2\Delta n/n^3E^2 \text{ (quadratic)}$$
$$r_c = -2\Gamma/n^3tE = -2\Delta n/n^3E \text{ (linear)}$$

A comparison of the electrooptic coefficients of some of the PLZT ceramic materials with several of the more common electrooptic single crystals are given in Table 2. As can be seen, the ceramics compare favorably with most of the single crystals, and given their availability in sizes up to 150 mm (6 in.) diam, are a good choice for large-area shutters, modulators, filters, and displays.

Special Effects. The PLZT materials are known to possess a number of special photosensitive phenomena which are directly linked to their chemical, structural, electronic, and optical properties. These effects include photoconductivity, photovoltaic properties, photo-assisted domain switching, ion implantation enhanced photosensitivity, photochromic effects, photomechanical (photostrictive) behavior, photorefraction, and photoexcited space charge phenomena (Ref 6, 10, 11, 20, 25). While materials with such a multitude of properties and special effects hold promise for many new applications for the future, it should also be remembered that these same effects can, and often do, limit their application.

Applications

Modes of Operation. Figure 5 also illustrates the two basic modes of operation used in electrooptic devices, that is, the transverse and longitudinal modes. In the transverse mode, the electric field is applied in a direction normal to the light propagation direction while in the longitudinal mode, the field is applied along the light direction. In general, the transverse mode is most effective for variable birefringence devices (Fig 5a and 5b), while the longitudinal mode is better suited for variable light scattering devices (Fig 5c). Also, color effects can be produced with variable birefringence whereas they cannot with scattering. Variable birefringent devices always require the use of polarized light; however, scattering devices do not. It should be recognized that in order to produce polarized light from an incandescent white light source there is a substantial loss in light intensity. In the case of an ideal, linear polarizer, this loss amounts to 50% of the incident light; and furthermore, this loss increases to a total of about 70% with the use of the more economical plastic sheet polarizers.

In the transverse mode devices such as shutters or variable density filters, the electric field is generally applied by means of suitable electrode patterns on one or both major surfaces of a polished plate of material. Since viewing is accomplished through the gap between the positive and negative electrodes, it follows that the activating voltage can be reduced, for a given overall viewing area, by

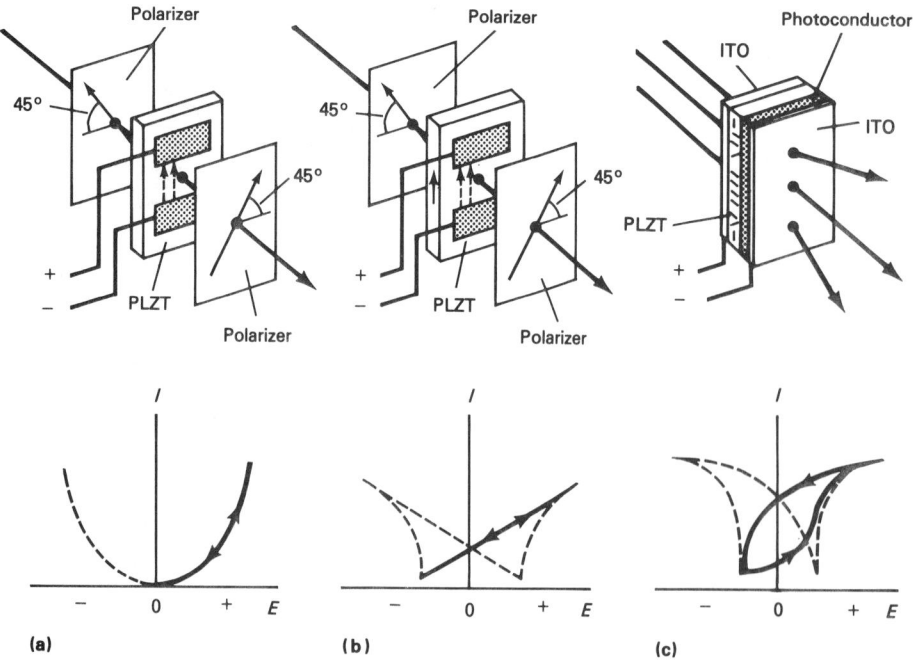

Fig 5 Operational configurations and light output responses of (a) quadratic, (b) linear, and (c) memory PLZT materials. The heavy accented portions of the response curves indicate the usable range. Note that in (b) the plate is prepoled and in (c) no polarizers are used. ITO designates indium-tin oxide transparent conductor typically consisting of 91 mol% In_2O_3 and 9% tin oxide.

Table 2   Electrooptic properties of ceramic and single-crystal materials

Material	K(a)	n(b)	r_c(c), $\times 10^{10}$ m/V	R(d) $\times 10^{16}$ m²/V²
Ceramic				
PLZT 8.5/65/35	5000	2.50	. . .	38.6
PLZT 9/65/35	5700	2.50	. . .	3.8
PLZT 9.5/65/35	5500	2.50	. . .	1.5
PLZT 8/70/30	5400	2.48	. . .	11.7
PLZT 8/40/60	980	2.57	1.02	. . .
PLZT 12/40/60	1300	2.57	1.20	. . .
PLZT 14/30/70	1025	2.59	1.12	. . .
Single-crystal				
LiNbO₃ (r_{33})	37	2.20	0.32	. . .
LiNbO₃ (r_{13})	37	2.29	0.10	. . .
BaTiO₃ (r_{33})	373	2.36	0.28	. . .
BaTiO₃ (r_{51})	372	2.38	8.20	. . .
KNbO₃ (r_{33})	30	2.17	0.64	. . .
KNbO₃ (r_{42})	137	2.25	3.80	. . .
SBN (T = 560 K)(e)	119	2.22	0.56	. . .
SBN (T = 300 K)(e)	3400	2.30	13.40	. . .
Ba₂NaNb₅O₁₅	86	2.22	0.56	. . .

(a) K is the dielectric constant. (b) n is the index of refraction at 633 nm. (c) r_c is the linear electrooptic coefficient. (d) R is the quadratic electrooptic coefficient. (e) SBN, strontium barium niobate. Source: Ref 1, 2, 10, 20–24

reducing the gap width and increasing the total number of gaps. This results in a number of narrow, interdigital electrodes on a given plate. By placing the device out of the focal plane of the optical system, the fine electrodes are virtually invisible, and image quality through the device is usually excellent. In contrast to the longitudinal mode, the transverse mode produces larger electrooptic effects. In the activated or ON state, the material is optically clear with essentially no scattering. Devices utilizing this mode may or may not exhibit memory, depending on the composition.

In the longitudinal mode, voltage is applied through the thickness of plate, necessitating the use of transparent electrodes such as tin oxide or indium-tin oxide (ITO). Since this mode generally aligns the macroscopic optic axis of the material parallel to the direction of viewing, optical birefringent effects are minimal. In this mode, the operating voltage is dependent on the thickness of the plate but is independent of the viewing area.

Device Considerations. Thin, polished plates of PLZT, when used with polarized light, make excellent wide aperture electronic shutters. Their advantages include fast response (0.01 to 50 μs), light weight, thin profile, wide viewing angle, and wide operating temperature range (−40 to 80 °C, or −40 to 175 °F). Their disadvantages are low ON state transmission of about 15 to 20% and high operating voltages (100 V or more) required to reach the full ON state. Despite the low ON state transmission, contrast ratios in excess of 2000 to 1 are easy to achieve.

In the actual design of a device, several factors must be taken into consideration. These include:

- Material composition (non-memory quadratic, memory hard, memory soft)
- E/O effect to be employed (birefringence, scattering)
- Optical address (transmission, reflection)
- Read address (direct viewing, projection)
- Write address (transverse, longitudinal)
- Address scheme (direct, matrix, electron beam)
- Construction details such as electrodes, polarizers, antireflection coatings (if needed) and mounting schemes

A typical device consists of a polished and electroded PLZT wafer sandwiched between two crossed polarizers and held in place mechanically or suspended in a very pliable potting gel which serves to locate and hold the part without imparting stress to the PLZT, eliminate dielectric breakdown over the surface of the PLZT across the electrode gap, and provide some measure of index matching between the polarizers and the PLZT.

The actual operating voltage of a typical transverse-mode, Kerr-type shutter device can be determined from the relationship:

$$V = g(2\Gamma/n^3 tR)^{1/2}$$

where g is the gap between the electrodes, Γ is the total phase retardation for the shutter (for example, 265 nm for white light), n is the index of refraction (2.5), t is the plate thickness, and R is the quadratic electrooptic coefficient (see Table 2). For example, assuming a quadratic-type, non-memory (9/65/35 material), crossed polarizer shutter with double-sided electrodes and a gap of 0.5 mm (20 mils) on a 0.25 mm (10 mils) thick plate, this calculation results in an operating voltage of 300 V. This voltage can be reduced by narrowing the gap between the electrodes provided at least a 1:1 gap-to-thickness is maintained.

This shutter can also be used as a color filter or modulator by simply increasing the voltage. By doing this, a series of colors are produced in the ascending order of yellow, red, blue, green, and so on into repeating color cycles or "orders." Again, using the above formula, the voltage required to achieve these colors can be calculated in the same manner as that for the full ON clear white state. One merely needs to substitute in the proper retardation (Γ) required to produce that color, remembering that these colors are of the subtractive interference type, that is, in order to obtain red (magenta), the full wave retardation for green (Γ = 530 nm) must be used. This calculation results in a voltage of 423 V in order to produce a first order red filter.

Devices. Like most materials, the successful application of the E/O ceramics is highly dependent on the relative ease with which they can be adapted to useful and reliable devices. The key words are simplicity, cost, and reliability. Successful devices which have been developed, to date, are in the area of optical shutters and include eye protective devices for the military and linear light gate arrays for optical recording. Applications involving spatial light modulators, color filters, and displays are still being developed as an outgrowth of the optical shutter technology. Continuing work on image storage devices also promises to open up new areas such as electronic photography.

Thin Films

By far, the most development work and the largest number of applications in ferroelectrics are associated with bulk materials, but a trend toward thin films has recently developed and is steadily increasing. Aside from the obvious advantages such as smaller size, less weight, and easier integration to silicon logic circuitry, FE thin films offer additional benefits including lower operating voltage, higher speed, and the ability to fabricate unique nano-level structures. Equally important, but not so obvious, is the fact that many materials which are difficult, if not impossible to fabricate into a dense form as a bulk material, are relatively easy to produce as thin films. In addition, the sintering temperatures of the films are usually hundreds of degrees centigrade lower than that of the bulk, and this can often be the deciding factor in a successful design and application.

The methods utilized in successfully fabricating FE thin films can be classified into two general categories—vacuum deposition and chemical processing techniques. In both cases, it is highly desirable to produce epitaxial growth onto a carefully selected substrate of the proper crystalline orientation, but this is only really necessary in the most demanding of applications such as optical waveguides. Recent results show that epitaxial film growth is more readily achieved with radio frequency magnetron sputtering (Ref 24), whereas wet chemical processing methods (sol-gel, metal-organic decomposition, and chemical vapor deposition) have proved to be more popular for most applications because of their ease of fabricating the films and the lower capital equipment costs (Ref 26–30). Laser ablation has also recently been reported to be a viable technique for film preparation and in the future may prove to be one of the preferred processes.

Although FE films have been investigated for such applications as capacitors, buffer layers, sensors, transducers, spatial light modulators, and displays, at the present time most of the interest centers around random access memories (RAMs) and integrated optical components (total internal reflection switches, modulators, and couplers) for data processing and communications. Working prototypes of both of these applications have been reported and offer promise for the future (Ref 31–33).

REFERENCES

1. G.H. Haertling and C.E. Land, Hot Pressed (Pb,La)(Zr,Ti)O$_3$ Ferroelectric Ceramics for Electrooptic Applications, *J. Am. Ceram. Soc.*, Vol 54, (No. 1–11), 1971

2. J.T. Cutchen, J.O. Harris, and G.R. Laguna, PLZT Electrooptic Shutters: Applications, *Appl. Opt.*, Vol 14, 1975, p 1866–1873

3. J.R. Maldanado, D.B. Fraser, and A.H. Meitzler, Display Applications of PLZT Ceramics, *Advances in Image Pickup and Display*, Vol 2, B. Kazan, Ed., Academic Press, 1975, p 65–168

4. K. Hardtl, Ferroelectric Displays, *Nonemissive Electrooptic Displays*, A. Kmetz and F. von Willisen, Ed., Plenum Publishing, 1976, p 241–259

5. A. Kumada, K. Suzuki, and G. Toda, Display Applications of Field-Enforced Phase Transitions in PLZT Ceramics, *Ferroelectrics*, Vol 10, 1976, p 25–28

6. C.E. Land, Optical Information Storage and Spatial Light Modulation in PLZT Ceramics, *Opt. Eng.*, Vol 17, 1978, p 317–326

7. H.E. Samek and M.A. Munoz, Light Transmission Characteristics of a High Density Electrooptical Light Gate Array Based on PLZT Ceramics, *Proc. SPIE*, Vol 307, 1981

8. G. Wessel, Electro-optical Ceramic (PLZT) as Light-Gate Array for Non-Impact Printer, *Proc. SPIE*, Vol 396, 1983, p 53–57

9. G.H. Haertling, PLZT Reflective Displays, *Ferroelectrics*, Vol 50, 1983, p 63–72

10. G.H. Haertling, Electro-optic Ceramics and Devices, *Electronic Ceramics*, Lionel Levinson, Ed., Marcel Dekker, 1988, p 371–492

11. A. Krumins, Specific Solid State Features of Transparent Ferroelectric Ceramics, *Ferroelectrics*, Vol 69, 1986, p 1–16

12. A. Sternberg, Transparent Ferroelectric Ceramics: Properties and Applications, *Ferroelectrics*, Vol 91, 1989, p 53–67

13. B. Jaffe, W.R. Cook, Jr., and H. Jaffe, *Piezoelectric Ceramics*, Academic Press, 1971

14. P.D. Thacher and C.E. Land, Ferroelectric Electrooptic Ceramics with Reduced Scattering, *IEEE Trans. Elect. Dev.*, Vol ED-16, 1969, p 515–521

15. G.H. Haertling, Hot-Pressed Ferroelectric Lead Zirconate Titanate Ceramics for Electro-Optical Applications, *Bull. Am. Ceram. Soc.*, Vol 49, 1970, p 564–567

16. L.E. Cross, Relaxor Ferroelectrics, *Ferroelectrics*, Vol 76, 1987, p 241–267

17. H. Jaffe and D.A. Berlincourt, Piezoelectric Transducer Materials, *Proc. IEEE*, Vol 53, 1965, p 1372–1386

18. S.T. Liu, S.Y. Pai, and J. Kyonka, A Study of Piezoelectric Properties of PLZT Ferroelectric Ceramics, *Ferroelectrics*, Vol 22, 1978, p 689–690

19. C.E. Land, P.D. Thacher and G.H. Haertling, Electrooptic Ceramics, *Applied Solid State Science, Advances in Materials and Device Research*, Vol 4, R. Wolfe, Ed., Academic Press, 1974

20. G.H. Haertling, PLZT Electrooptic Materials and Applications: A Review, *Ferroelectrics*, Vol 75, 1987, p 25–55

21. B.A. Tuttle, Electronic Ceramic Thin Films: Trends in Research and Development, *Bull. Mater. Res. Soc.*, Vol XII, 1987, p 40–45

22. R.L. Holman, L.M. Johnson, and D.P. Skinner, The Desirability of Electrooptic Ferroelectric Materials for Guided-Wave Optics, *Proceedings of 6th International Symposium on Applications of Ferroelectrics*, 8–11 June 1986, Lehigh University, p 32–41

23. P.D. Thacher, Refractive Index and Surface Layers of Ceramic (Pb,La) (Zr,Ti)O$_3$ Compounds, *Appl. Opt.*, Vol 16, 1977, p 3210–3213

24. K.L. Bye, High Birefringence PLZT Materials for Low Voltage Displays, *Ferroelectrics*, Vol 10, 1976, p 29–33

25. G.H. Haertling, Photoelectronic Effects in PLZT Ceramics, *Ceramic Transactions: Electro-optics and Nonlinear Optic Materials*, A. Bhalla, E. Vogel, and K. Nair, Ed., The American Ceramic Society, 1990, p 21–50

26. H. Adachi, T. Kawaguchi, M. Kitabatake, and K. Wasa, Dielectric Properties of PLZT Epitaxial Thin Films, *Jpn. J. Appl. Phys.*, Vol 22, Suppl. 22–2, 1983, p 11–13

27. K.D. Budd, S.K. Dey, and D.A. Payne, Sol-Gel Processing of PbTiO$_3$, PbZrO$_3$, PZT, and PLZT Thin Films, *Proc. Brit. Ceram. Soc.*, Vol 36, 1985, p 107–121

28. R.W. Vest, Metallo-organic Decomposition (MOD) Processing of Ferroelectric and Electrooptic Films: A Review, *Ferroelectrics*, Vol 102, 1990, p 53–68

29. C.J. Brierly, C. Trundle, L. Considine, R.W. Whatmore, and F.W. Ainger, The Growth of Ferroelectric Oxides by MOCVD, *Ferroelectrics*, Vol 91, 1989, p 181–192

30. G.H. Haertling, PLZT Thin Films Prepared from Acetate Precursors, *Ferroelectrics*, Vol 116, 1991, p 51–63

31. J.T. Evans and R. Womack, An Experimental 512-Bit Nonvolatile Memory with Ferroelectric Storage Cell, *IEEE J. Sol. State Cir.*, Vol 23, 1988, p 1171–1175

32. D. Bondurant and F. Gnadinger, Ferroelectrics for Nonvolatile RAMs, *IEEE Spectrum*, Vol 26, 1989, p 30–33

33. H. Higashino, T. Kawaguchi, H. Adachi, T. Makino, and O. Yamazaki, High Speed Optical TIR Switches Using PLZT Thin-Film Waveguides on Sapphire, *Jpn. J. Appl. Phys.*, Vol 24, Suppl. 24–2, 1985, p 284–286

Oxygen Sensors

Meilin Liu and Ashok V. Joshi, Ceramatec, Inc.

THE OXYGEN SENSOR is a key element in the monitoring and control of many processes. Liquid-electrolyte oxygen sensors are typically operated in polarographic, voltammetric, or galvanic mode (Ref 1). Due to the volatility of the solvents involved, these sensors can be used only at ambient temperatures. Solid-state sensors based on ionic conductors or semiconductors, on the other hand, have found wide applications in high-temperature processes, including control of air-to-fuel ratio in combustion (vehicles, boilers, coal gasification, and so forth), monitoring of oxygen in molten metals, control of atmosphere in materials processing or heat-treatment furnaces, and thermodynamic study of oxide or mixed-oxide systems.

This article emphasizes various aspects of solid-state oxygen sensors, including fundamental principles of transducing processes, basic characteristics of sensor response, operational and environmental effects on sensor performance, materials processing and device fabrication techniques, applications, and challenges for future developments. For detailed background information and engineering design of sensor devices, however, interested readers are referred to several reviews (Ref 2–5) and a monograph (Ref 6).

Principle of Operation

The oxygen sensor is an element that transduces oxygen partial pressure or activity into an electrical signal, which can be further electronically processed as desired to perform control activities. According to the key material used in the device, ceramic oxygen sensors can be classified into two categories: solid-electrolyte sensors and semiconductor sensors. The former is based on solid ionic conductors incorporating electrochemical phenomena, whereas the latter is based on oxide semiconductors incorporating electronic transport phenomena modulated by surface reactions with environment. Because of their simplicity, the most well-developed and widely used sensors in many applications are potentiometric sensors based on stabilized zirconia and conductimetric sensors based on TiO_2 semiconductors.

Solid Electrolyte Sensors

Sensors based on solid electrolytes are further divided into a number of operation modes in terms of the electrochemical phenomena associated with the transducing process such as potentiometry, amperometry, voltammetry, coulometry, and electrochemical pumping. Table 1 lists terms typically encountered in articles on oxygen sensor technology.

Potentiometric sensors are based on Nernst potential originated from the activity or concentration difference of chemical species. An electrochemical cell is a potentiometric sensor if it is constructed in such a manner that the potential of the indicating electrode with respect to the reference electrode, and hence the electromotive force (emf) of the cell, is a measure of the concentration of the chemical species to be detected. In general, the emf of the cell can be expressed as (Ref 7):

$$E = E^\circ + \left(\frac{RT}{nF}\right) \sum_j (v_j \ln a_j) \qquad \text{(Eq 1)}$$

where E° is the standard equilibrium potential, R is the gas constant, T is the absolute temperature, F is the Faraday constant, n is the number of electrons exchanged in the electrode reaction, v_j is the stoichiometric coefficient of species j, and a_j is the activity of species j involved in the overall reaction.

A potentiometric automobile sensor based on solid electrolyte (Ref 8) is schematically shown in Fig 1(a). One side of the electrolyte is exposed to a reference atmosphere (air) while the other side is exposed to a sample gas (exhaust), that is,

$$P'_{O_2}, M_1 \mid solid\ electrolyte \mid M_2, P''_{O_2} \qquad \text{(Eq 2)}$$

where p'_{O_2} and p''_{O_2} are partial pressures of oxygen at the sensing and the reference electrodes, M_1 and M_2, respectively. The emf of the cell is given by (Ref 9, 10):

$$E = \left(\frac{RT}{4F}\right) \int_{p'_{O_2}}^{p''_{O_2}} t_{O^=} d(\ln p_{O_2}) \qquad \text{(Eq 3)}$$

where $t_{O^=}$ is the transport number of oxygen anions in the solid electrolyte. Under the assumption that the electrolyte is predominantly ionically conductive, the emf of the cell can be approximated by

$$E = \left(\frac{RT}{4F}\right) \ln\left(\frac{P''_{O_2}}{P'_{O_2}}\right) = k_0 - k_1 \ln\left(P'_{O_2}\right) \qquad \text{(Eq 4)}$$

where k_0 is a constant determined by the reference atmosphere, k_1 is the sensitivity of the sensor, and p'_{O_2} is the partial pressure of oxygen in the sample gas.

If air is used as reference environment, the temperature dependence of cell emf can be expressed as (Ref 4)

$$\frac{dE}{dT} = \left(\frac{R}{4F}\right) \ln\left(\frac{P''_{O_2}}{P'_{O_2}}\right) \qquad \text{(Eq 5)}$$

thus, the temperature coefficient of the cell emf is close to 0.05 mV · K^{-1} per decade ratio p''_{O_2}/p'_{O_2}. If a metal/metal-oxide system is used as reference, however, the reference partial pressure of oxygen depends on temperature as well. Accordingly, the temperature dependence of sensor response is more complicated. Potentiometric sensors based on stabilized zirconia are widely used because of their simplicity.

Amperometric sensors are based on the dependence of diffusion-limited current on bulk concentration of an electroactive species. An electrochemical (electrolytic or galvanic) cell forms an amperometric sensor if it is constructed and operated under the condition that the observed current, or the rate of the overall process, is completely controlled by the diffusion of the species to be detected through a diffusion barrier. In addition, the observed current should be due merely (or predominantly) to the electrolysis of the species of interest. Electrolysis of any other species present in the sample gas will complicate the measurement. The diffusion barrier can be formed either by depositing a porous layer on the cathode or by a small cavity with an aperture (or a pin hole) near the cathode compartment (Fig 2). The diffusion limited currents, i, flowing through the cell can be expressed as a function of the concentration or partial pressure of oxygen (C_{O_2}) in the sample gas as (Ref 11, 12):

$$i = 4FAD_{O_2}\left[\frac{C'_{O_2}}{\delta}\right] = \left[\frac{4FAD_{O_2}}{RT\,\delta}\right] P'_{O_2} = k_2 P'_{O_2} \qquad \text{(Eq 6)}$$

where D_{O_2} is the diffusion coefficient of oxygen through the diffusion barrier whereas A and δ are the area and thickness of the diffusion barrier, respectively. Sensor constant, k_2, which can be determined by calibration, depends on the temperature and the geometric characteristics of the diffusion barrier (Ref 16).

Table 1 Oxygen sensor nomenclature

Symbol	Definition	Units
A	Area of a diffusion barrier	cm^2
a_j	Activity of species j	$\cdots$
C_j^b	Concentration of species j in a bulk phase	$mol \cdot cm^{-3}$
C_{O_2}	Concentration of oxygen	$mol \cdot cm^{-3}$
$\bar{d}$	Mean diameter of pores in a porous diffusion barrier	cm
D_{O_2}	Diffusion coefficient of oxygen	$cm^2 \cdot s^{-1}$
E	emf of a potentiometric sensor	V
E_a	Activation energy for conduction in a semiconductor	eV or J
$E°$	Standard potential of an electrode	V
E_s	Nernst potential across the gage cell in a double-cell sensor	V
e	Elementary charge	1.602×10^{-19} C
e', h	Symbol for electrons and electron holes	$\cdots$
F	Faraday's constant	96487 $C \cdot equiv^{-1}$
i	Current	A
I_0	Applied current in a coulometric sensor	A
k_0	A constant determined by the reference atmosphere	V
k_1	Sensitivity of a potentiometric sensor, $k_1 = \partial E/\partial(\ln P'_{O_2}) = (RT)/(4F)$	$\cdots$
k_2	Sensitivity of an amperometric sensor, $k_2 = \partial i/\partial(P'_{O_2}) = (4 FA D_{O_2})/(RT\delta)$	$\cdots$
k_3	Sensitivity of a coulometric sensor, $k_3 = \partial q/\partial(P'_{O_2}) = (4 FV_1)/(RT)$	$\cdots$
k_4	Sensitivity of an amperometric sensor with double-cell, $k_4 = \partial i_s/\partial(P'_{O_2}) = (4 FAD_{O_2})[1 - \exp(4 FE_s/RT)]/(RT \delta)$	$\cdots$
k_B	Boltzman constant	1.38×10^{-23} J/K
M	Molecular weight of oxygen	$\cdots$
M_1	Sensing electrode	$\cdots$
M_2	Reference electrode	$\cdots$
n	Number of electrons exchanged in an electrode reaction or excess electron density in a semiconductor	$number \cdot cm^3$
N	Number of moles of oxygen in the sample gas of volume	V_1
$\dot{N}$	Feeding rate of species to be detected	$mol \cdot s^{-1}$
O^{2-}, $O^=$	Oxygen anions	$\cdots$
p	Excess electron holes density in a semiconductor	$number \cdot cm^3$
p'_{O_2}	Partial pressure of oxygen in a sample gas	$\cdots$
p''_{O_2}	Partial pressure of oxyen in a reference environment	$\cdots$
P	Absolute pressure	atm
q	Cumulative charge	C
R	Universal gas constant	8.314 $J \cdot mol^{-1} \cdot K^{-1}$
t_1	Time needed for complete electrolysis in a coulometric sensor	s
t_{O^-}	Transference number of oxygen anions in a solid electrolyte	$\cdots$
$t_{O^=}$, $t_{O^{2-}}$	Transference number of oxygen anion	$\cdots$
T	Absolute temperature	K
v	Volume flow rate of sample gas	$cm^3 \cdot s^{-1}$
V_1	Volume of sample gas	cm^3
$V_{\ddot{o}}$	Symbol for oxygen vacancies	$\cdots$
μ_e	Electron mobility	$cm^2 \cdot s^{-1} \cdot V^{-1}$
δ	Thickness of a diffusion barrier	cm
Δ_n	Excess electron density	$\cdots$
α	Constant with value between 1.5 and 2	$\cdots$
β_j	Constant	$\cdots$
v_j	Stoichiometric coefficient of species j in an electrode reaction	$\cdots$
σ	Conductivity	$\Omega^{-1} \cdot cm^{-1}$
θ	Surface coverage of adsorbed oxygen molecules or atoms	$\cdots$

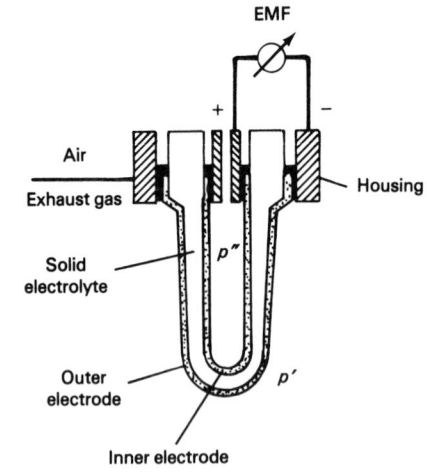

(a)

(b)

Fig 1 Examples of potentiometric sensors. (a) Automobile sensor. (b) Heated thin-film sensor. Partial oxygen pressures at the sensing and reference electrodes are shown as p' and p'', respectively. Source: Ref 48, 49

3. If the pore size or the dimension of the aperture is on the order of the mean free path of the gas molecules, both bulk and Knudsen diffusion will occur. Thus the temperature dependence can be tailored by the microstructure of the diffusion barrier

Amperometric sensors have higher sensitivity within a narrow concentration range than potentiometric sensors. The structure, fabrication, and operation of amperometric sensors, however, are more complicated than potentiometric sensors.

Coulometric sensors are based on Faraday's law of electrochemical equivalence. An electrolytic cell establishes a coulometric sensor if the cell operates at 100% current efficiency of the electrolysis of the species to be detected and if the end-reaction occurs at a potential sufficiently different from the electrolysis reaction. The requirements of 100% current efficiency and detectable end-point response are crucial to a practical coulometric sensor. For an electrolysis cell with enclosed volume of V_1 (Fig 3a), the total charge, q, due to the electrolysis of O_2 in

Effect of temperature on amperometric sensors depends strongly on the microstructure of the diffusion barrier:

1. If the pore size of the porous layer and the dimension of the aperture is much smaller than the mean free path of the gas molecules, Knudsen diffusion occurs in the diffusion barrier and the diffusivity can be expressed as (Ref 13):

$$D_{O_2} = \frac{1}{3}\left(\frac{8k_B T}{\pi M}\right)^{1/2} \bar{d} \propto T^{1/2} \qquad (Eq\ 7)$$

where k_B is the Boltzmann constant, M is the molecular weight of oxygen, and $\bar{d}$ is the mean diameter of the pores in the diffusion barrier

2. If the pore size of the porous layer or the dimension of the aperture is much greater than the mean free path of the gas molecules, bulk diffusion occurs in the diffusion barrier and the diffusivity can be expressed as:

$$D_{O_2} \propto \frac{T^\alpha}{P} \qquad (Eq\ 8)$$

where P is the absolute pressure and α is a constant with a value between 1.5 and 2

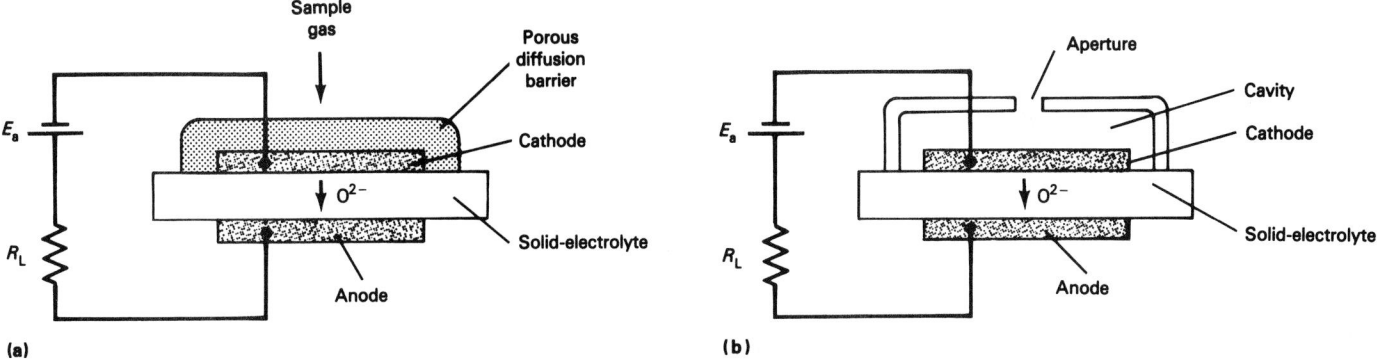

Fig 2 Schematic pictures of amperometric sensors with two types of diffusion barriers: (a) porous layer and (b) aperture. Source: Ref 16

sample gas of volume V_1 can be expressed as (Ref 14):

$$q \approx nFN = 4F\left(\frac{V_1 p'_{O_2}}{RT}\right) = k_3 p'_{O_2} \qquad \text{(Eq 9)}$$

where N is the mole number of O_2 in the sample gas of volume V_1 and n, F, R, and T are as defined earlier. If a constant current, I_0, is applied, the O_2 concentration in the sample gas can be determined as

$$p'_{O_2} \approx \left(\frac{RT}{4F}\right)\left(\frac{I_0 t_1}{V_1}\right) = k'_3 t_1 \qquad \text{(Eq 10)}$$

Thus, the partial pressure of oxygen is directly proportional to the measured value t_1, the time needed to complete the electrolysis. The sensor constant, k'_3, depends on temperature, the applied current, and the enclosed volume.

For a flow-through system (Ref 15), as shown in Fig 3(b), the sample gas is fed at a constant rate into an electrolysis vessel. Under the condition of complete electrolysis, the observed current is then controlled by the feeding rate and is given by the Faraday's law,

$$i = n F N = 4F v C'_{O_2} = \frac{4Fv}{RT}P'_{O_2} = k''_3 P'_{O_2} \qquad \text{(Eq 11)}$$

where $\dot{N}$ is the feeding rate of the chemical species to be detected (moles/s) and v is the volume flow rate of the sample gas (cm^3/s).

Sensors incorporating oxygen pumping (Fig 4) consist of two cells, a pumping cell and a sensing cell (Ref 16). This arrangement allows several modes of operation: coulometric, amperometric, and potentiometric (Ref 17).

Coulometric. A double-cell coulometric sensor (Ref 18) is schematically shown in Fig 4(a). The operation of this sensor involves two steps: 1, electrochemically pump the oxygen in the enclosed volume V_1 out of the chamber until the gage emf reaches a relatively high value, indicating that p'' is very low and O_2 left in the chamber is negligible. 2, reverse the pumping direction and monitor the cumulative charge passed, q, until $p'' = p'$ or the gage emf is zero ($E_s = 0$). The partial pressure of oxygen in the sample gas can be calculated from the observed charge using Eq 5. If a constant current is applied, the oxygen partial pressure is proportional to the time of pumping (t_1) and the absolute temperature as described by Eq 10. In addition, this device

can be alternatively operated in an oscillatory mode. Interested readers are referred to the original papers (Ref 19, 20) for detailed information.

Clearly, a double-cell coulometric sensor is better than a single-cell coulometric sensor in selectivity and accuracy. The disadvantages, however, include sophisticated structure, operation, and electronics as well as slow response. The operation is intermittent and the response time depends on the cycle time of the device.

Amperometric. Figure 4(b) shows a double-cell amperometric sensor (Ref 21). When the pumping current is automatically and continuously adjusted by an operational amplifier to keep the gage emf at a constant value, E_s, the steady-state current can be expressed as:

$$i_s = 4FAD_{O_2}\left[\frac{P'_{O_2} - P''_{O_2}}{RT\,\delta}\right]$$

$$= \left(\frac{4FAD_{O_2}}{RT\,\delta}\right)\left[1 - \exp\left(\frac{4FE_s}{RT}\right)\right]P'_{O_2}$$

$$= k_4 P'_{O_2} \qquad \text{(Eq 12)}$$

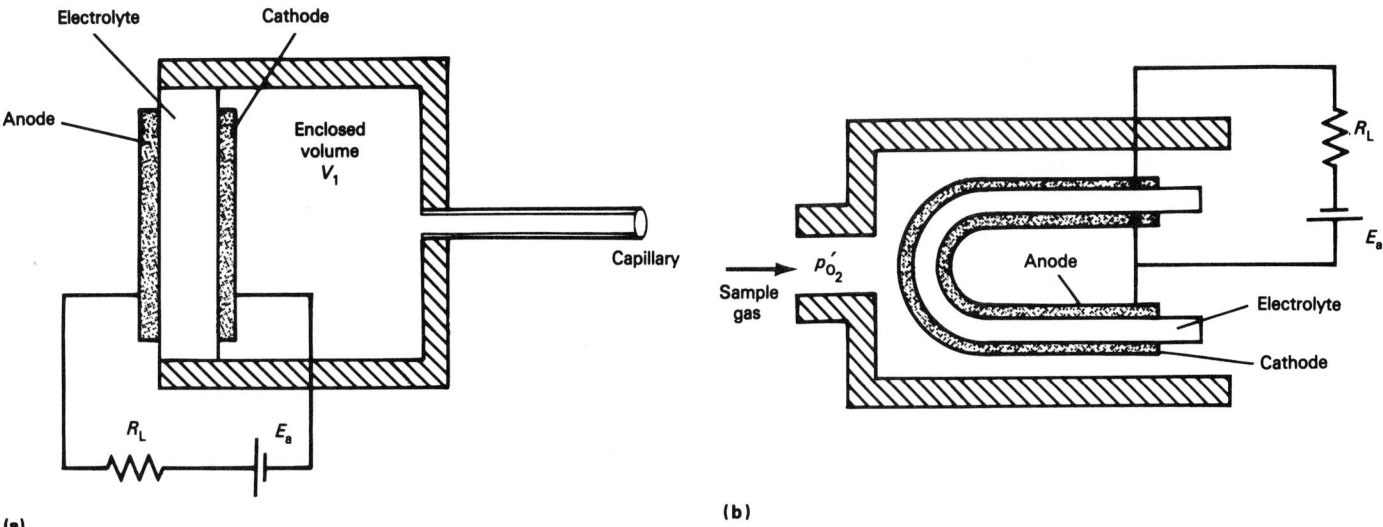

Fig 3 Coulometric sensor with (a) an enclosed volume and (b) a flow-through sensor. Source: Ref 14, 15

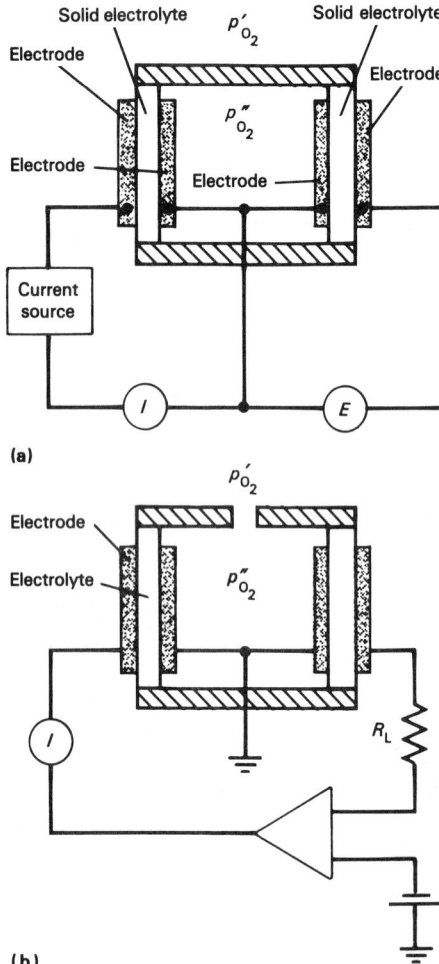

Fig 4 Sensors incorporating oxygen pumps operated at two modes: (a) coulometric and (b) amperometric. Source: Ref 4

where E_s is the Nernst potential across the gage cell and is given by Eq 3. The other parameters are as defined earlier. Thus, the steady-state current is directly proportional to the oxygen partial pressure in the sample gas for a given gage emf E_s at a given temperature. Sensor constant k_4 depends not only on temperature and geometric characteristics of the diffusion barrier but also on the gage emf.

Unlike a single-cell amperometric sensor, a double-cell amperometric sensor does not have to be operated under the condition where diffusion completely controls the overall process. In fact, a double-cell sensor can be operated at any level of current or polarization; the only requirement is that the pumping voltage is sufficient to provide the current necessary to keep E_s constant. Thus, a double-cell amperometric sensor effectively eliminates all the problems associated with the application of appreciable current or potential to a single cell.

Other types of solid-electrolyte sensors include impedance-based sensors and thick-film sensors. Impedance-based sensors (Ref 4) are based on the measurement of the impedance of electrode-electrolyte interfaces exposed to a sample gas. Thick-film sensors (Ref 22) or voltammetric sensors are based on the current-voltage characteristics of a flow-through cell.

Semiconductor Sensors

Oxide semiconductor sensors are based on the change in conductivity of semiconductors due to surface adsorption or interaction with the environment; thus, they are also called conductimetric sensors.

When a semiconductor is exposed to a sample gas, the molecules in the gas can react with the surface of the semiconductor and are adsorbed. The bulk concentration of the species to be detected in the sample gas is related to the surface concentration of the adsorbed species by a suitable isotherm. A number of reaction mechanisms have been proposed to describe surface adsorption processes (Ref 23) on various semiconductors. Langmuir isotherm (Ref 24), for instance, is based on the fact that the surface or interface adsorption sites are limited and hence predicts

$$\frac{\theta}{1-\theta} = \beta_j C_j^b \qquad \text{(Eq 13)}$$

where θ is the surface coverage, β_j is a constant, and C_j^b is the concentration of species j in the gas phase.

If the interaction between the adsorbed molecules and the semiconductor is refined to the boundary between the two phases, and if the conductivity of the space-charge region near the interface is sensitive to the sample gas, the sensor is called a boundary conductivity sensor. The main advantage of this sensor is the flexibility to be operated at low temperatures, where bulk diffusion process is prohibited. On the other hand, if the surface absorbents can further interact and exchange electrons with the interior of the semiconductor so that the conductivity of the bulk material is controlled by the sample gas, the sensor is then called bulk conductivity sensor. These sensors are often operated at elevated temperatures, where bulk diffusion of lattice defects such as oxygen vacancies or metal interstitials is sufficiently fast for the system to reach an equilibrium with the environment.

In both cases, it is the dramatic change in conductivity of semiconductor due to a slight change in surface concentration of the adsorbed donors or acceptors which constitutes the basis for semiconductor sensors. In an n-type semiconductor, such as TiO_2 or SnO_2, the majority charge carriers are electrons and the change in conductivity can be expressed as (Ref 6):

$$\Delta\sigma = e\mu_e \Delta n \qquad \text{(Eq 14)}$$

where e is the electron charge, μ_e is the electron mobility, and Δn is the excess electron density.

Clearly, the conductivity of an n-type semiconductor will be enhanced by surface adsorption of electron donors (such as hydrogen) and reduced by surface adsorption of acceptors (such as oxygen). Similarly, the conductivity of a p-type semiconductor will be increased by surface adsorption of acceptors while reduced by surface adsorption of donors.

Ceramic semiconductors, such as suitably doped oxides of transition metals (Ref 3), have been widely used for various gas sensors. For instance, n-type semiconductors (such as ZnO, SnO_2, Fe_2O_3, TiO_2, MnO_2, WO_3, ThO_2, and CdO, and so forth) have been used for sensing donor-type gases whereas p-type semiconductors (such as CoO, Cu_2O, NiO, Cr_2O_3, and so forth) have been used for sensing acceptor-type reducible gases (Ref 25).

For conductimetric oxygen sensors based on metal oxides or perovskites, the interaction between the oxygen molecules in a sample gas and the semiconductor can be described as (Ref 26):

$$O_O^X = \tfrac{1}{2}O_2(g) + V_O^{\cdot\cdot} + 2e' \qquad \text{(Eq 15)}$$

at low oxygen pressures, where oxygen is being released from the lattice, and

$$\tfrac{1}{2}O_2(g) + V_O^{\cdot\cdot} = O_O^X + 2h \qquad \text{(Eq 16)}$$

at high oxygen pressures, where oxygen is being incorporated into the lattice. $V_O^{\cdot\cdot}$ represents oxygen vacancies. Accordingly, the law of mass action predicts that at a given temperature, the concentrations of excess electrons (e) or electron holes (h) in a pure oxide satisfy

$$n \propto p_{O_2}^{-1/6} \quad \text{and} \quad p \propto p_{O_2}^{1/6} \qquad \text{(Eq 17)}$$

since electrical neutrality requires that $2[V_O^{\cdot\cdot}] = n$. For a doped oxide, however, the concentration of oxygen vacancies is predominantly determined by the level of dopant and is essentially independent of the external oxygen partial pressure (Ref 27). Thus, the relationship

$$n \propto p_{O_2}^{-1/4} \quad \text{and} \quad p \propto p_{O_2}^{1/4} \qquad \text{(Eq 18)}$$

prevails for a doped oxide at a given temperature.

Experimental observations indicate that the dependence of conductivity of an oxide semiconductor on temperature and oxygen partial pressure (p_{O_2}) is given by (Ref 28–32):

$$\sigma = k_5 \exp\left(\frac{-E_a}{k_B T}\right) p_{O_2}^{\gamma} \qquad \text{(Eq 19)}$$

where k_5 is a constant, E_a is the activation energy for conduction, and the sign and the value of the exponent γ depends on the nature of the lattice defects in the oxide semiconductor.

Because the conductivity of a semiconductor depends exponentially on temperature, a thermistor or a heater is necessary to compensate for the temperature variations (Fig 5). Common drawbacks of semiconductor sensors include poor selectivity, considerable drift, and deterioration in sensitivity. Thin-

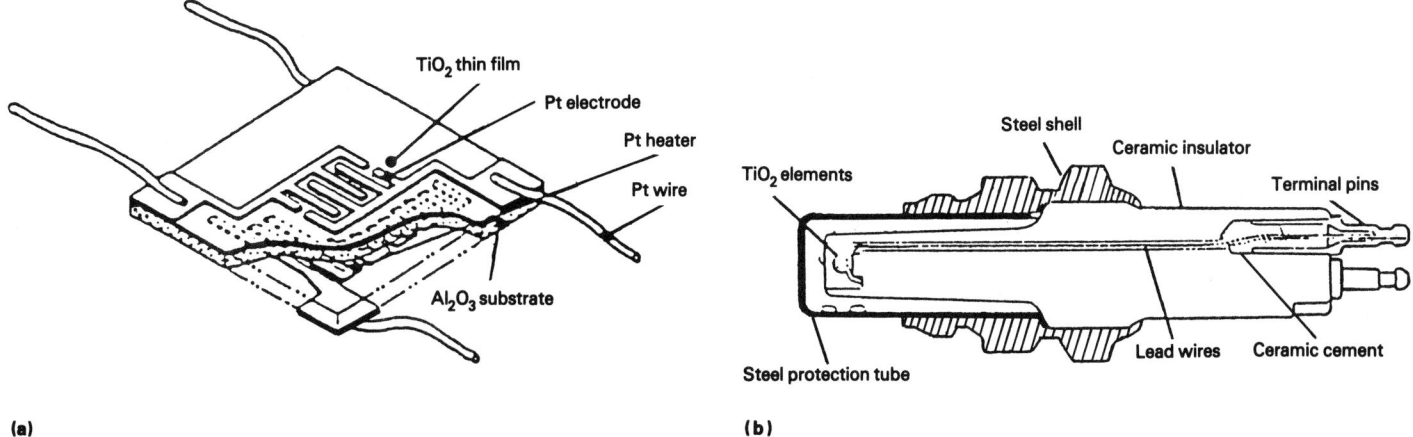

Fig 5 Conductimetric (semiconductor) oxygen sensors based on (a) thin-film oxide semiconductor and (b) thick-film oxide semiconductor. Source: Ref 29, 33

film technology, however, promises improvement in sensor performance.

Characteristics of Sensor Response

The performance of a sensor is typically evaluated in terms of

- Selectivity
- Sensitivity
- Range of operation
- Reproducibility and stability
- Speed of response
- Durability and lifetime

In general, the requirements regarding performance or the relative importance of these characteristics depend on application. For automobile exhaust sensors in stoichiometric combustion, for instance, the speed of response and the amplitude of potential generated are vital whereas the accuracy and sensitivity are relatively unimportant. Sensors for lean combustion, however, require high sensitivity within a narrow concentration range. Furthermore, for sensors used in thermodynamic study, accuracy and reliability are critical while speed of response is marginal.

Selectivity is the ability of a sensor to respond selectively to a specific species while inactive to other species in the sample gas. It is the selectivity which makes a sensor specific to a chemical species of interest. Clearly, a sensor without selectivity has limited value.

For a semiconductor sensor, selectivity depends merely on the specific adsorption of the semiconductor surface. For a solid-electrolyte sensor, however, selectivity is determined not only by the adsorption properties of the electrode or electrode-electrolyte interface but also by the transport properties of the electrolyte. In addition, for amperometric or coulometric sensors, reducible gases (such as H_2O or CO_2) present in a sample gas could interfere with the reduction of oxygen if a potential higher than the decomposition potentials of these species is applied to a sin-

gle-cell sensor. In this case, the selectivity can be improved by using double-cell sensors.

Sensitivity is usually defined as the slope of sensor response versus concentration of the chemical species to be detected, such as $\partial E/\partial(\ln p_{O_2})$ for potentiometric sensors, $\partial i_1/\partial p_{O_2}$ for amperometric sensors, and $\partial \sigma/\partial p_{O_2}$ for conductimetric sensors. Thus, sensitivity is a measure of the ability of a sensor to respond to small amount of change in concentration.

Sensitivity is largely determined by the principle of operation and materials properties, but it also is influenced strongly by sensor design and operating conditions. For example, amperometric sensors have high sensitivity to O_2 in lean-combustion region, while the sensitivity of a potentiometric sensor is inadequate for this application.

Range of Operation. Linear dynamic range is typically defined as the concentration range within which sensitivity is greater than zero and the sensor response is linearly related to the concentration of the species to be analyzed.

The range of operation of a sensor is preferably within or overlapped with the linear dynamic range, although it is possible to extend to the concentration range over which the sensor response obeys a theoretical or empirical equation. The range of operation depends on the principle of operation, the materials employed, sensor design, as well as operating conditions. Linear dynamic range can be easily determined by calibration.

Potentiometric sensors have a wide range of operation, up to 20 orders of magnitude, because of the logarithmic dependence of cell emf on concentration. Amperometric sensors, however, have much narrower range of operation, because the current passing through an interface usually varies only in a few orders of magnitude. Conductimetric sensors cannot be operated in a wide concentration range either, because the internal impedance of the sensor could be extremely high if the resistivity changes in several orders of magnitude.

Reproducibility and Stability. Reproducibility is the dispersion of the sensor output signal and typically characterized by standard deviation. Stability is the percent of change of the offset and/or sensitivity in time. Reproducible and stable sensor response is necessary for reliable monitoring and control.

Reproducibility and stability are vital characteristics for a sensor. These parameters depend critically on the stability of materials and reproducibility of the chemical, physical, or electrochemical phenomena on which the sensor operation is based. In addition, environmental contamination also plays an important role. Clearly, degradation of electrolyte, electrode, and electrolyte-electrode interface will inherently cause problems with reproducibility and stability (Ref 34).

Speed of Response. The speed of response refers to the kinetics to reach equilibrium. Fast response is crucially important when the sensor is used to initiate a control function. For instance, the fast response of a combustion-control sensor can significantly enhance energy efficiency while reducing pollutant emission in combustion processes. If the control system spends a large amount of time in either the lean or the rich region during switching, NO_x and CO emission will be high because more time is spent in the far rich or far lean region, even though the average air to fuel ratio might be close to stoichiometry region.

It is a common observation (Ref 35) that when a sensor, initially at equilibrium, is suddenly exposed to an environment containing different gas composition, the response of the sensor is not instantaneous. Obviously, the kinetics of sensor response varies with the principle of operation and sensor configuration.

For potentiometric sensors, the speed of response is primarily affected by the following factors (Ref 4, 36):

- Hydrodynamics in the gas phase, which determines the kinetics of the oxygen partial pressure in the bulk gas transporting to the sensor interface (that is, the gas diffusion through porous protective coating and platinum electrode)

- Electrode kinetics for oxygen reduction or evolution, which determines the rate at which the electrolyte-electrode interface reaches equilibrium with the gas phase
- The double layer charging process, which is a direct consequence of the gradient of chemical potential across the electrode-electrolyte interface
- The changes in stoichiometry of the electrolyte in the double layer to accommodate the double layer charging/discharging processes
- The changes in stoichiometry throughout the electrolyte and related charge transport processes

The overall process, however, has not been well understood, and the mathematical formulation has therefore not been well established, although several models have been developed based on experimental data (Ref 37).

Potentiometric and amperometric sensors typically respond faster than coulometric sensors or the sensors based on electrochemical pumping because these sensors typically involve a whole operation cycle.

For conductimetric sensors based on oxide semiconductors, the speed of response is strongly influenced by three factors (Ref 28):

- The dynamics of gas transport through the porous semiconductor materials
- The kinetics of the interaction between surface absorbents and lattice defects
- The diffusion of charge carriers such as oxygen vacancies and electrons in the bulk of semiconductors

Durability and Lifetime. This is an engineering or economic parameter. The factors influencing the durability and lifetime of a sensor include: thermodynamic and kinetic stability of sensor materials, thermal stability or thermal-shock resistance of the device to thermal cycle or temperature gradient, fabrication processes, as well as operating conditions.

Operational and Environmental Effects on Sensor Performance

Temperature. The impact of temperature on sensor performance depends on the principle of operation. When a sensor response is a strong function of temperature, it becomes necessary to incorporate a thermistor or a heater into the sensor device in order to compensate for temperature change. Mass and charge transport processes associated with sensor response are typically thermally activated processes. Accordingly, as temperature increases, the kinetics of sensor response is generally accelerated. However, the aging or

deterioration of sensor materials is also accelerated at high temperatures for the same reason (Ref 30). Therefore, sensors are preferably operated at the lowest possible temperature, where the performance is satisfactory. On the other hand, if the operating temperature is too low, non-equilibrium effect will severely interfere with sensor response. In this case, an embedded heater can be introduced to the sensor device to modulate the operating temperature and hence to improve sensor performance.

Sampling includes flow rate, pressure, and volume of sample gas. This is an engineering problem which depends critically on sensor design. For instance, potentiometric sensors are typically sensitive to the pressure and flow rate of sample gas. Flow-rate sensitivity can be minimized or eliminated by appropriate sensor design (Ref 38).

Combustibles in Sample Gas. Because of the high operating temperature (above 350 °C, or 660 °F) of these sensors, any combustibles present in the sample gas could burn at the surface of the electrode or the electrode-electrolyte interface and hence alter the concentration of oxygen in the sample gas (Ref 39). This effect is true for all types of sensors. Possible solutions include: lowering the operating temperature to where combustion is prohibited, poisoning the electrode for combustion, and employing high-selectivity electrode for oxygen. Efforts are underway worldwide to eliminate this problem by developing suitable electrode-electrolyte systems which can be operated at temperatures (below 250 °C, or 480 °F) where combustion is negligible (Ref 40).

Oxygen-Containing Reducibles in Sample Gas. Because the amperometric or coulometric sensors rely on the ionic current or the charge due to the reduction of oxygen in the sample gas, reducible species present in the sample gas may contribute to the sensor current when a potential higher than the decomposition potentials of the reducible species (such as H_2O and CO_2) is applied to a single cell. Double-cell sensors, however, can effectively reduce the interference with reducible species.

Materials Considerations and Processing Techniques

Solid-Electrolyte Sensors

Requirements for Electrolytes. Although the specific requirements for electrolytes depend on application, it is generally true that a suitable electrolyte should have:

- Sufficient chemical stability in contact with both the reference material and the sample (gas or molten metal) to be analyzed
- Satisfactory ionic conductivity to enhance the speed of response and minimize errors due to ohmic effect

- Minimum electronic conductivity to minimize errors due to electronic transport
- Acceptable mechanical strength for structural integrity and handling
- Adequate thermal stability to withstand thermal cycling or temperature gradients

Various oxygen anion conductors, including ZrO_2, ThO_2, CeO_2, and Bi_2O_3-based electrolytes, have been investigated for oxygen sensors. Stabilized zirconia, however, is by far the most commonly used electrolyte material for commercial sensors. A number of stabilizing oxides (such as Yb_2O_3, Y_2O_3, CaO, and MgO) and other additives (such as Al_2O_3 and SiO_2) have been added to zirconia in order to improve the electrical, mechanical, and thermal properties. Conductivity varies with the type of stabilizer in the order of $Yb_2O_3 > Y_2O_3 > CaO > MgO$. The amount of dopant is typically kept at the minimum level necessary to achieve a fully stabilized cubic phase, which has a high ionic conductivity. However, partially stabilized materials with tetragonal precipitates demonstrate high fracture toughness and good thermal shock resistance yet sufficient conductivity.

For molten metal sensors, additional requirements include adequate thermal-shock resistance, wetting with molten metal, and intimate contact with the reference materials (Ref 41, 42). Partially stabilized, multiphase, magnesia-zirconia ceramic is widely used as the electrolyte for molten-metal sensors due to its excellent thermal-shock resistance and good mechanical strength. In partially stabilized zirconia with 3 wt% of MgO, the volume contraction due to the phase transformation from low-temperature monoclinic phase of ZrO_2 to high-temperature tetragonal phase at 1200 °C (2200 °F) compensates for the thermal expansion of the cubic phase. In addition, the dispersed tetragonal phase further inhibits crack growth and hence increases the toughness of the material.

Electronic conductivity of an electrolyte typically increases with decreasing p_{O_2} and thus sets a limit on the lowest oxygen partial pressure which can be measured by oxygen-ion conductors. The low limit is a function of temperature and can be estimated from thermodynamic data (Ref 43, 44), although sluggish kinetics may extend the actual limit further.

In order to lower the operating temperature and accelerate the speed of response, electrolytes having high ionic conductivity, such as stabilized CeO_2 (Ref 45) and Bi_2O_3 (Ref 46), have been explored. Stabilizers studied include CaO for ceria and MoO_3 for bismuth oxide. To date, however, no significant improvement has been achieved. This, in fact, is not surprising because the electrolyte resistance is not the limiting factor to sensor response.

Preparation of ZrO_2-Based Electrolytes. An ideal starting material should have uniform chemical composition, suitable particle size, narrow particle size distribution, and adequate reactivity (high specific surface area). Stabilizing additives should be uniformly dispersed in zirconia powder on an atomic scale. The powder should be coarse enough to handle and press with minimum agglomeration, whereas the specific surface area (m^2/gram) should be sufficiently high to allow sintering at the lowest possible temperature (Ref 47).

Solid-state calcination often results in a decrease in activity or sinterability of the powder, while controlled co-precipitation of hydroxides produces completely homogeneous distribution of additives in zirconia. Investigations of various wet chemical methods for the precipitation of the hydroxide (Ref 48) concluded that the gel-route, where the agglomerates were weakly bound, produced the most active powders suitable for sintering at low temperatures. This process led to good compaction characteristics and high fired densities (>95% dense at 1180 °C, or 2150 °F, with a grain size of 0.3 μm). In addition, chemical vapor deposition (CVD) and hydrothermal oxidation methods were also investigated for preparation of fine active powders of zirconia (Ref 49–51).

Commercially available, active, sinterable powder of yttria-stabilized zirconia is typically produced by hydrolysis of a mixture of $ZrOCl_2$ and YCl_3 to precipitate the mixed hydroxide. After azeotropic distillation and drying, the resultant powders are then calcined at 850 to 950 °C (1560 to 1740 °F) to control the crystallite size of the powder. Ball milling is conducted to disperse the powder, which is then loosely agglomerated with or without binders during the spray drying process (Ref 47).

Electrodes. Electrode materials and the process of applying an electrode material to an electrolyte have significant impact on sensor performance (Ref 30). For instance, it is electrode kinetics, not electrolyte resistivity, that determines the lowest temperature at which a potentiometric sensor can be operated satisfactorily (Ref 52).

An appropriate electrode material should have adequate chemical stability, sufficient electronic conductivity, satisfactory bonding with electrolyte materials, and high catalytic activity for oxygen adsorption (or desorption) and reduction (or evolution).

Electrode processes at the cathode and the anode are of oxygen reduction and evolution, respectively. Clearly, these reactions occur only at the triple points, electrolyte-electrode-gas boundaries, where e′, O$^=$, $\ddot{V}_O$, and O$_2$ are simultaneously available.

Numerous electrode materials have been reported for sensor applications, including precious metals (Pt, Au, Ag, and so on), alloys (Ag-Pd, for example), and perovskites (Ref 53) such as LSM, LSCo, LSCoNi, LSCoFe. Platinum, however, is by far the most widely used material for electrode in commercial oxygen sensors. Processing techniques commonly employed to apply electrode to elec-

trolyte include brush painting or screen printing of slurry, sputtering, and electroless plating (Ref 54).

The critical issues which require special attention include 1, sheet resistance of electrode as prepared; 2, contact or bonding of electrode-electrolyte interface; 3, catalytic activity of the interface for oxygen reduction and evolution; 4, long-term stability of electrode performance; and 5, reproducibility of electrode processing.

Reference Oxygen Materials for Potentiometric Sensors. For potentiometric sensors in the control of combustion processes, the most commonly used reference is air. In general, however, any material having a fixed oxygen activity could serve as reference electrode for potentiometric sensors. In fact, many metal/metal-oxide systems have been investigated for this application, including Pd/PdO, Ir/IrO$_2$, Os/OsO$_2$, Ru/RuO$_2$, Cu/Cu$_2$O, Bi/Bi$_2$O$_3$, Pb/PbO, Ni/NiO, Re/ReO$_2$, Co/CoO, Fe/FeO, Zn/ZnO, Cr/Cr$_2$O$_3$, Ta/Ta$_2$O$_5$, and so forth (Ref 55–57). For molten metal sensors, in particular, Cr/Cr$_2$O$_3$ has received considerable attention as reference material.

Protective Coating. Materials used for protective coating include various oxides, such as alumina and spinel. The porous oxide layer, typically thermally sprayed on a sensing electrode, protects the electrode from erosion and contamination. The microstructure of this layer is important because it affects the oxygen equilibrium conditions and hence the speed of response.

Heated ZrO$_2$ Sensors. In order to minimize the dependence of sensor response on the temperature of sample gas, heated ZrO$_2$ sensors have been developed. Sensors of this type (Fig 1b) typically have a thin-layer sensing element and consist of three layers of tape-cast electrolyte (Ref 58). The first electrolyte layer is screen-printed with platinum sensing and reference electrodes, the second layer has an air duct to lead the air to the reference electrode, and the third layer carries a screen-printed resistive heater. The three green sheets are laminated together and sintered to form a single chip.

Semiconductor Sensors

Requirements for Semiconductors. An ideal semiconductor for oxygen sensors should have:

- Adequate chemical and thermal stability
- Rapid interaction with oxygen (fast adsorption/desorption)
- Dramatic change in conductivity with small change in oxygen concentration

Various metal oxides (Ref 24) and perovskite ceramics (Ref 59) have been investigated for oxygen sensors. The TiO$_2$ semiconductor is the most commonly used material, although Nb$_2$O$_5$ has been suggested as a substitute (Ref 60).

In order to minimize the effect of bulk diffusion on sensor response, the sensing element is typically made highly porous and is composed of finely divided TiO$_2$ particles. In addition, to accelerate the rate of interaction between oxygen and the TiO$_2$ semiconductor (particularly at low temperatures), the TiO$_2$ materials are further impregnated with platinum or other catalytic metals to assist oxygen adsorption-desorption processes at semiconductor surface.

Due to the high sensitivity to temperature, a TiO$_2$ sensor requires either a heater to modulate the operating temperature or a thermistor to compensate for temperature variations (Ref 27). Thermistors typically consist of dense TiO$_2$ ceramics.

Preparation of TiO$_2$ Semiconductors. Titania (TiO$_2$) ceramic material of high porosity and high mechanical strength can be obtained from powders produced by a chemical precipitation method (Ref 61). An aqueous solution of titanyl chloride is made by slowly adding TiCl$_4$ to ice water. The solution is neutralized by NH$_4$OH to precipitate TiO(OH)$_2$. After washing several times to remove the soluble NH$_4$Cl and followed by filtration, the precipitate is frozen in liquid nitrogen and dried under vacuum at 75 °C (165 °F).

It was observed that the mechanical strength of the sintered material is a strong function of the drying procedure of the gel. The air-dried gel forms a hard, nonfriable powder which is difficult to press and sinter to shape and the sintered material has low mechanical strength. The freeze drying process, however, crystallizes the water in the gel and prevents the formation of a large rigid structure, resulting in powders that are friable and easily pressed into a pellet shape. The sintered material has excellent mechanical strength. Precipitated titanium hydroxides have been reported to be either polymeric or amorphous hydrates (Ref 62) with loosely bound capillary water, which assists in achieving high porosity.

The dried powder has anatase structure and typically has specific surface area of 220 m^2/g. The sintered material has over 50% porosity after the phase transformation from anatase to rutile at 830 °C (1525 °F).

Processing of Thin or Thick Films. Sensing elements of TiO$_2$ are in the form of porous disk or thick film. Films of semiconductor can be deposited either on an insulating substrate (Al$_2$O$_3$) or on a porous precious metal electrode (Pt) and subsequently contacted with a second electrode (Ref 25, 27, 63).

The TiO$_2$ films are typically prepared by CVD, slurry painting, flame-spraying, plasma-spraying, and sputtering. Film thickness usually ranges from 10 to 100 μm (Ref 25).

Electrodes. Platinum electrodes for porous disk-type sensors are typically applied by thermal cladding (Ref 26). A platinum wire is inserted into each hole drilled in a pressed pellet and subsequently heated to 920 °C

(1690 °F) for one hour. About 24% shrinkage of TiO$_2$ securely anchors the wires in the pellet. Electrodes for film-type sensors can be prepared by slurry painting, spraying, and sputtering.

Catalyst Impregnation. Impregnation of porous TiO$_2$ pellet with a noble metal catalyst is typically accomplished by dipping the pellet in a solution of either chloroplatinic or bromoplatinic acid. Solution impregnation under vacuum displaces the trapped air in the pores with solution to achieve complete filtration. The impregnated pellet is then dried and heated to 500 °C (930 °F) in air to decompose the halogenated platinum.

Challenges for Future Developments

Continuous exploration of novel materials (electrolytes, semiconductors, and electrodes), fabrication process, and device design is vital to achieve higher selectivity, greater sensitivity, lower operating temperature, faster response, longer lifetime, and lower cost.

Fundamental Research. Critical understanding of the physical, chemical, and electrochemical processes associated with the transducing process is critical to advance the technology. Investigation of electrode kinetics and transport processes at electrode-electrolyte interfaces (Ref 64) and adsorption-desorption processes at semiconductor surfaces will provide valuable information for further development. It is also expected that exploration of high-conductivity electrolytes, high-catalytic-activity electrodes, and inherently bonded electrolyte-electrode interfaces can open new opportunities for high-performance sensors. In addition, fundamental investigations of the mechanisms governing the sensor response process can furnish guidance to facilitate sensor response effectively.

Sensors for Stoichiometric Combustion. The largest market for oxygen sensors today is, perhaps, for stoichiometric air-to-fuel-ratio control in combustion processes. Potentiometric ZrO$_2$ sensors essentially dominate the market because of their fast response and weak temperature dependence. In addition, the matured technology can produce millions of sensors each year with reproducible properties and adequate (5 year) durability (Ref 65). Nevertheless, improvements in performance, lifetime, and cost are expected to be pursued continuously. Conductimetric TiO$_2$ sensors, on the other hand, offer the advantages of compactness, simplicity, and low cost. The strong dependence of sensor characteristics on the properties of TiO$_2$ material, however, makes it difficult to manufacture TiO$_2$ sensors with reproducible properties. It is hoped that the introduction of thick-film technology will revolutionize the fabrication processes for high-performance TiO$_2$ sensors.

In addition, it has been observed that, at low exhaust gas temperatures (<350 °C, or 660 °F), departure from thermodynamic equilibrium results in nonideal behavior, such as shift and change in characteristics. Accordingly, the development of ZrO$_2$ and TiO$_2$ sensors incorporating a heater can improve sensor performance significantly at low exhaust temperatures.

It has recently been recognized that the monitoring of oxygen concentration in exhaust alone is inadequate to control precisely the combustion stoichiometry because the gas composition in the exhaust of a combustion process is typically away from the equilibrium composition at the temperature of the exhaust. It is therefore necessary to develop multifunction sensors which can monitor not only oxygen but also other species (CO or NO$_x$) and hydrocarbons at the same time.

Sensors for Nonstoichiometric Combustion. Recent investigations indicate that combustion under a slightly lean-fuel (excess air) condition has higher energy efficiency and less NO$_x$ emission. Potentiometric sensors and conductimetric TiO$_2$ sensors cannot be used for this application, because the former have inadequate sensitivity and the latter have high temperature sensitivity. Appropriate modification of these sensors can extend the range of operation from stoichiometry (air/fuel ratio, A/F = 14.5) to A/F = 18 to 20 (Ref 66), which is still inadequate for this application. Amperometric sensors or sensors incorporating oxygen pumping, in principle, can be operated from far-lean to far-rich combustion region. Preliminary results with these sensors are very promising. The successful development of a commercial device, however, requires substantial efforts.

Another challenge in this area is the development of sensors for reciprocating engines and diesel engines, which are typically operated at far-lean region (A/F from 18 to 60).

Molten Metal Sensors. The processing of high-quality metal or alloys, such as steel, requires extremely low oxygen concentration to minimize the formation of oxide inclusions during solidification. Potentiometric sensors based on solid electrolytes are widely used in metal or alloy processing, where a precise control of oxygen concentration is important. It is estimated that more than two million sensors were consumed in 1983 in steelmaking processes (Ref 67). These sensors were used to monitor O$_2$ concentration in molten steel and to determine the amount of aluminum or titanium that must be added to the melt in order to remove the excess oxygen. The critical issues to be addressed in this application include substantial electronic conductivity of magnesia-stabilized zirconia and poor thermal-shock resistance of yttria-stabilized zirconia.

Magnesia-stabilized zirconia (with 3 wt% MgO), currently used in the steelmaking industry, exhibits significant electronic conductivity at temperatures of molten steel under oxygen partial pressure less than 20 ppm. This sets the limit of detection of 20 ppm of O$_2$ in molten steel. For high-quality steel, however, a low concentration of oxygen is critical. Electrolytes based on ThO$_2$ have higher thermodynamic stability and less electronic conductivity under similar conditions than ZrO$_2$-based electrolytes and thus are preferred for this application. Unfortunately, ThO$_2$-based materials present some environmental problems. Yttria-stabilized zirconia has low electronic conductivity but has poor thermal-shock resistance, which has been improved recently (Ref 68). Another alternative is to develop a composite electrolyte with high thermal-shock resistance.

Oxygen concentration analyzers have been widely used for process monitoring. The challenge in this application is to improve the insensitivity to combustible species in a sample gas, either by lowering the operating temperature or by poisoning the combustion process at the sensing electrode.

Miniaturization. Microelectronic fabrication techniques can produce low-power, multisensor, single-chip, integrated devices, which offer potentially significant reductions in production cost and improvements in sensor performance. Challenges in miniaturization include:

- Development of micro-amperometric and micro-coulometric sensors to improve selectivity, sensitivity, and speed of response
- Exploration of multi-sensor array on a single chip incorporating materials deposited by CVD and doped by ion implantation

ACKNOWLEDGMENT

The authors gratefully acknowledge Professor Anil V. Virkar and Professor Jiri Janata for their careful review of the manuscript and valuable comments and suggestions.

REFERENCES

1. R.P. Buck, *Chemical Sensors*, PV 87-9, D.R. Turner, Ed., Electrochemical Society, 1987, p 344
2. D.E. Williams and P. McGeehin, *Electrochem.*, Vol 9, 1984, p 246
3. H. Dietz, W. Haecher, and H. Jahnke, *Adv. Electrochem. Electrochem. Eng.*, Vol 10, 1977, p 1
4. W.C. Maskell, *J. Phys. E.: Sci. Instrum.*, Vol 20, 1987, p 1156
5. W.C. Maskell and B.C.H. Steele, *J. Appl. Electrochem.*, Vol 16, 1986, p 475–489
6. J. Janata, *Principles of Chemical Sensors*, Plenum, 1989
7. J. O'Bockris and A.K.N. Reddy, *Modern Electrochemistry*, Plenum, 1970
8. H. Dueker, K.H. Friese, and W.D.

Haecker, Technical paper 750223, ASE Congress, 1975

9. C. Wagner, *Advances in Electrochemistry and Electrochemical Engineering*, Vol IV, Interscience, 1966, p 1

10. L. Hyene, *Solid Electrolytes*, S. Geller, Ed., Springer-Verlag, 1977, p 169

11. A. Bard and L.R. Faulkner, *Electrochemical Methods, Fundamentals and Applications*, Wiley, 1980

12. J.S. Newman, *Electrochemical Systems*, Prentice-Hall, 2nd ed., 1991

13. F. Reif, *Statistical and Thermal Physics*, McGraw-Hill, 1975

14. L. Hyene, *Measurement of Oxygen*, H. Degn, *et al.*, Ed., Elsevier, Amsterdam, 1976, p 65

15. J. Tenygl, *Electroanalytical Methods in Chemical and Environmental Analysis*, R. Kalvoda, Ed., Plenum, 1987, p 128

16. E.M. Logothetis and R.E. Hetrick, *Fundamentals and Applications of Chemical Sensors*, ACS, 1986

17. W.C. Maskell, H. Kaneko, and B.C.H. Steele, *Proceedings of the 2nd International Meeting on Chemical Sensors*, J.L. Aucouturier, Ed., Bordeaux Chemical Sensors, University of Bordeaux, 1986, p 302

18. D.M. Haaland, *Anal. Chem.*, Vol 49, 1977, p 1813

19. R.E. Hedrick, W.A. Fate, and W.C. Vassell, *IEEE Trans. Elect. Dev.*, Vol ED-29, 1982, p 129

20. H.L. De Jong, U.S. patent 4,384,935, 1983

21. R.E. Hedrick, W.A. Fate, and W.C. Vassell, *Appl. Phys. Lett.*, Vol 38, 1981, p 390

22. S. Kimura, S. Ishitani, and H. Takao, *Fundamentals and Applications of Chemical Sensors*, ACS, 1986

23. I. Lundstrom and C. Svensson, *Solid State Chemical Sensors*, J. Janata and R.J. Huber, Ed., Academic Press, 1985, p 1

24. A.W. Adamson, *Physical Chemistry of Surfaces*, Wiley, 1982

25. J. Fexa, *Electroanalytical Methods in Chemical and Environmental Analysis*, R. Kalvoda, Ed., Plenum, 1987, p 109

26. C.B. Choudhary, H.S. Maiti, and E.C. Subbarao, *Solid Electrolytes and Their Applications*, E.C. Subbarao, Ed., 1980, p 1

27. K. Kiukkola and C. Wagner, *J. Electrochem. Soc.*, Vol 104, 1957, p 379

28. J.L. Pfeifer, T.A. Libsch, and H.P. Wertheimer, Technical paper 840142, Society of Automotive Engineers, 1984, p 69

29. M.J. Esper, E.M. Logothetis, and J.C. Chu, Technical paper 790140, Society of Automotive Engineers, 1979, p 19

30. M.J. Kaiser and E.M. Logothetis, Technical paper 830167, Society of Automotive Engineers, 1983, p 61

31. J.A. Cook, *et al.*, Technical paper 830985, Society of Automotive Engineers, 1983, p 1

32. E.F. Gibbons, *et al.*, Technical paper 750224, Society of Automotive Engineers, 1979, p 1

33. T. Takeuchi, H. Takahashi, K. Saji, H. Kondo, and I. Igarashi, Technical paper 830929, ASE Congress, 1983, p 627

34. M. Liu, B. Handerson, and A. Joshi, *Ext. Abstr. Electrochem. Soc.*, Vol 90–2, 1990, p 1088

35. C.T. Young, Technical paper 810380, Society of Automotive Engineers, 1981, p 29

36. J.C. Myland and K.B. Oldham, *J. Electrochem. Soc.*, Vol 131, 1984, p 1815

37. L. Hyene and D. de Engelsen, *J. Electrochem. Soc.*, Vol 124, 1977, p 727

38. A.V. Joshi, Patent disclosure, 1990

39. K. Saji, K. Kondo, H. Takahashi, T. Takeuchi, and I. Igarashi, *Proceedings of the International Conference on Sensors and Actuators*, Philadelphia, IEEE, 1985, p 336

40. H. Arai, K. Eguchi, and T. Inoue, *Proceedings of the Symposium on Chemical Sensors*, ACS, 1988, p 122

41. W.L. Worrell, *Electrochemistry and Solid-State Science Education*, W.R. Smyrl and F. McLarnon, Ed., Electrochemical Society, 1986

42. W.L. Worrell, *Metal-Slag-Gas Reactions and Processes*, Z.A. Foroulis and W.W. Smertzer, Ed., Electrochemical Society, 1975, p 822

43. T.A. Ramanarayanan, *J. Electrochem. Soc.*, Vol 128, 1981, p 2487

44. H.L. Tuller and A.S. Nowick, *J. Electrochem. Soc.*, Vol 122, 1975, p 255

45. H. Aria, K. Eguchi, and H. Yahiro, *Proceedings of 2nd International Meeting on Chemical Sensors*, Bordeaux, 1986, p 335

46. T. Susuki, K. Kaku, S. Ukawa, and Y. Dansui, *Solid State Ionics*, Vol 13, 1984, p 237

47. R. Stevens, *Zirconia and Zirconia Ceramics*, Magnesium Elektron, 1986

48. M.A.C.G. Van de Graaf and A.J. Berg-graaf, *Advances in Ceramics*, Vol 12, American Ceramic Society, 1983, p 744

49. S. Hori, M. Yoshimura, and S. Somiya, *Advances in Ceramics*, Vol 12, American Ceramic Society, 1983, p 794

50. H. Toraya, M. Yoshimura, and S. Somiya, *Advances in Ceramics*, Vol 12, American Ceramic Society, 1983, p 806

51. S. Somiya, M. Yoshimura, and S. Kikugawa, *Emergent Process Methods For High Technology Ceramics*, Vol 17, R.F. Davis, *et al.*, Ed., p 155

52. S.P.S. Badwal, M.J. Bannister, and W.G. Garrett, *Solid State Ionics*, 1984

53. H. Arai, K. Eguchi, and T. Inoue, *Chemical Sensors*, Electrochemical Society, 1987, p 122

54. T. Watanabe and S. Wada, U.S. patent 4,170,530, 1979

55. O. Kubaschewski and C.B. Alcock, *Metallurgical Thermochemistry*, 5th ed., Pergamon, Oxford, 1979

56. R.A. Rapp and D.A. Shores, *Physicochemical Measurements in Metals Research*, Part II, R.A. Rapp, Ed., Wiley-Interscience, 1970, p 123

57. S. Seetharaman and K.P. Abraham, *Solid Electrolytes and Their Applications*, Plenum, 1980, p 127

58. N. Higuchi, S. Mase, A. Lino, and N. Kato, *Sensors and Actuators*, 1985, p 93

59. Y. Shimizu, *et al.*, *Fundamentals and Applications of Chemical Sensors*, ASC, 1986, p 83

60. H. Kondo, H. Takahashi, T. Takeuchi, and I. Igarashi, *Proceedings of the 3rd Sensor Symposium in Japan*, 1983, p 185

61. A.L. Micheli, *Ceram. Bull.*, Vol 63, 1984, p 694

62. T. Bardsdale, *Titanium, Its Occurrence, Chemistry and Technology*, Ronald, 1966, p 71

63. D.S. Howarth and A.L. Micheli, Paper 840142, SAE Congress, Detroit, Society of Automotive Engineers, 1984

64. M. Liu and A. Khandkar, *Ext. Abstr. Electrochem. Soc.*, Vol 90–2, 1990, p 1062

65. E.M. Logothetis, *Chemical Sensors*, ACS, 1987, p 142

66. H.M. Wiedenmann, L. Raff, and R. Noack, Paper 840141, SAE Congress, Detroit, Society of Automotive Engineers, 1984

67. K. Nagata and K.S. Goto, *Solid State Ionics*, Vol 9/10, 1983, p 1249

68. W.L. Worrell and Q. Liu, *Chemical Sensors*, ACS, 1987, p 113

Thick Film Circuits

Charles C. Y. Kuo, CTS Corporation

THICK FILM CIRCUITS most commonly involve the screen printing process of forming a functional composition through a patterned mesh screen onto glasses or ceramics. The printed film is dried and fired, typically in the range of 500 to 900 °C (930 to 1650 °F) to sinter the film and adhere it to the substrate with the desired characteristics for circuit applications. The functional compositions in general include conductors, resistors, and dielectrics. The dielectrics may be used as overglazes, crossovers, capacitors, or multilayers.

The term thick film originated from the term thin film. Thin films are deposited by vacuum technology, either by evaporation or sputtering. In the early 1960s, the Bell Telephone Laboratories of AT&T decided to use sputtered tantalum nitride films as resistors and capacitors in miniaturized circuits to match the advent of small-sized transistors (Ref 1). By that time, Du Pont had introduced a palladium oxide-silver thick film resistor system for the same applications (Ref 2). The tantalum nitride films initially were deposited on glass due to the smooth surface, and then later on fine 99.5% (Al_2O_3) substrates. The 99.5% Al_2O_3 substrates improve thermal conductivity and mechanical strength and are smooth enough for thin film applications. However, thick films can be applied on various substrates including glass, porcelainized metals and different kinds of ceramics. In addition, one of the main constituents in the thick film composition itself is glass frit. Therefore, ceramics and glass are vital to thick film technology. Thick films are products of ceramics and glass. In recent years, a variety of new ceramics, such as aluminum nitride and low-temperature co-fired multilayer substrates, have been used in thick film circuits.

Originally, the difference between thin films and thick films was defined by the film thickness. The usable film thickness range from vacuum deposition is between 10 and 500 nm (100 and 5000 Å). On the other hand, the thickness of thick film is in the range of 10 to 25 μm (400 to 1000 μin.). Therefore, in general, the thickness of a thick film is about two orders of magnitude greater than that of a thin film.

In addition to the film thickness, one of the main differences between thin and thick films is processing. Usually, thin films are deposited by vacuum technology over the entire substrate. Then they are photo-etched or otherwise processed to become a desired pattern for circuit applications. In a broad sense, any film deposited on glass or ceramic and fired at a high temperature to yield the required characteristics may be called a thick film.

The advantages of thick films over other processes include simple operation, easy automation, high production rate, and relatively low capital and operating costs. The thick film processes are simple and fast and can be used to make circuits on large (125 × 175 mm, or 5 × 7 in.), or small substrates. Thick film can be printed into multilayers up to several layers. Relatively new low-temperature co-fired tape multilayer substrates make thick film even more attractive to microcircuits.

General Compositions for Thick Films

Thick film compositions, in general, consist of three main ingredients: a vehicle or screening agent, a glass frit, and functional materials.

A vehicle is a homogeneous liquid with a certain viscosity which will be well-mixed with a glass frit and functional materials (mostly in the solid form) to form an ink with the required rheology for easy application to the substrate.

The vehicle generally consists of a solvent, a binder, and a wetting or dispersing agent. The solvent is used for adjusting the ink to the right viscosity or rheology for easy application, usually for screen printing. The solvent will not evaporate at room temperature during the printing period but rather is dried at an elevated temperature (usually 100 to 150 °C, or 210 to 300 °F) for a short time (less than 20 min) to yield a uniform dense film. For practical applications, the solvent must meet boiling point, evaporation rate, surface tension, solubility, compatibility, viscosity, flowability, order, hazardous nature, availability, and cost requirements. The solubility parameter, hydrogen bonding effects, and fractional polarity are used as guidelines to adjust for the required rheology (Ref 3). The common solvents used for thick film inks include pine oil, terpineol, Texanol (2,2,4 trimethyl-1,3-pentanediol monoisobutyrate), castor oil polymers, butyl carbitol acetate, hexyl carbitol, and numerous others. The solution theory can be used to predict the solubility of polymers in organic solvents in a three-dimensional approach.

The binder is usually a resin which can be dissolved in solvents and wetting agents to be used as a screening agent. The purpose of the binder is to temporarily bond the functional materials (mostly in powder form) and glass frit together on the substrate after printing and drying, with enough mechanical strength for handling. During high-temperature firing, resins must burn off completely. With the advent of non-noble metal systems such as copper conductors, and base metal resistors and dielectrics, it is even more difficult to select a suitable resin for such applications, because most of the binders for non-oxygen firing leave carbon residues. The carbon residues in resistors will affect sheet resistance and yield other unqualified properties. For copper conductors, carbon residue even in the range of 50 to 100 ppm will greatly reduce film adhesion. For dielectrics, carbon residues will decrease the breakdown voltage, produce blistering bubbles and other unqualified properties, and even short the circuits.

Binders for nitrogen firing include pyrolytic materials such as long chain depolymerizable compounds which can be decomposed into short chain volatile monomers that will escape through the exhaust chamber. Binders which leave negligible carbon residue in nitrogen firing include thermoplastic acrylic ester resins, nitrocellulose, copolymers of ethylene and vinyl acetate, and polybutene polymer resins. Most binder materials used in thick films to be fired in air include ethyl cellulose of different viscosities. The nitrogen-firing binders can also be used for air firing.

The glass frit is one of the main constituents in the thick film inks. Its purpose is to bind all the solid materials together. Adhesion to the substrate for conductors is more important than it is for resistors and dielectrics. Currently, certain functional oxides are incorporated in the glass during glass formation. For improving adhesion, especially aged adhesion, copper and/or bismuth oxides are added to or melted with the ingredients to create a homogeneous glass. For conductor applications, the high-lead oxide (PbO) and/ or bismuth oxide (Bi_2O_3) glasses are normally used. These particular glasses incorporated into the thick film inks also play an important role in the improvement of the conductor, resistor, and dielectric characteristics.

The processing involved in making glass frit includes: thoroughly mixing the fine ingredient powders; melting; quenching; and ball-milling to a particle size of less than 5 μm (200 μin.) with a tight particle size distribution. Density and surface area will depend on the glass composition. Important thermal properties of glass include softening point, annealing point, melting point, viscosity, and thermal coefficient of expansion. Chemical properties include moisture sensitivity, cation migration, environmental hazards, and acid-base properties.

Alkali-metal containing glasses are avoided because ion migration may occur which will affect the characteristics of resistors, conductors, and dielectrics. Many types of oxides may also be admixed, or melted, with glass as a flux to make soldering easier or to improve the adhesion of conductors. The glass can be vitreous or devitreous. In general, the melting point of the glass frit is preferably 200 °C (360 °F) lower than the ink firing temperature. The glass frits used for nitrogen firing are more difficult to choose than those used for air firing. Easily reducible oxides with a free energy more positive than −80 kcal/mole should be carefully used (Ref 4).

Functional Materials and Additives. The important physical properties of powders generally include particle size and particle size distribution, morphology, tap density, and surface area. For thick film inks, the particle size of the metal powders is preferably less than 2 μm (80 μin.) with a tight size distribution. The particle shape is of less importance, but most of the metal powders have spherical shapes. Flake-type powders, or mixtures of flake and spherical shapes, are used for improving conductivity by providing more contact areas. If the particle size becomes too fine, as with gold, the film may blister and bubble during firing.

In order to yield a dense fired metal film, the powder density should be high and the surface area should be low. Some highly catalytic metals, such as palladium or platinum, have surface areas that are usually higher than others. Therefore, the properties of the surface area must be carefully considered. For binary or ternary metal compositions, a coprecipitated or alloy powder may be better than an admixture.

Powder purity, oxygen content, and coating materials have to be considered, particularly for nitrogen firings. Purity can be determined by thermal gravimetric analysis (TGA) under air, nitrogen, and hydrogen conditions (Ref 5). For conductor applications, the metal powder should be easily roll-milled, well-dispersed in the vehicle, and not metallized on the rolls during ink preparation. Most powder properties will affect ink rheology, which will eventually affect not only printing quality, but also electrical and mechanical properties of the resulting thick film.

Other Thick Film Materials

Metallo-organic deposition (MOD; Ref 6, 7) or metal resinates have been used for decoration on glass and ceramic tableware for more than 100 years. MOD is a thermal decomposition process used to deposit thin films of metals or their compounds. By the MOD technique, films of specular precious metals and alloys can be obtained by firing in air, while films of base metals and their alloys can be obtained by firing under a neutral or reducing atmosphere. Films of ternary oxides with perovskite and pyrochlore structures and amorphous glass can also be obtained by using mixed metallo-organic compounds. Fired films can be used for conductors, resistors, dielectrics, insulators, magnetics, transparent conductors, photomasks, and recently for superconductors. Although the film thickness deposited by the MOD process is in the range of vacuum processing (about 100 to 200 nm, or 1000 to 2000 Å), the high-temperature firing process allows MOD to be classified as a thick film also. In recent years, patterned films generated by techniques such as writing with Micropen (Ref 8) or by laser instead of screen printing may also be called thick films because these films have to be fired at a high temperature to generate the desired characteristic.

Thick Film Tape. Thick film compositions, such as conductors, resistors, and dielectrics, can be made into a flexible tape in a thickness of from 0.025 to 0.075 mm (0.001 to 0.003 in.) (Ref 9). These tapes can be laser- or die-cut to a desired configuration, then placed on a substrate. The processes of firing and the characteristics of the fired films from tape are about the same as those from screen printing films. The advantages of the tape process include uniform film thicknesses to improve the yield, ability to conform to a variety of surface contours, and fast turn-around time for making circuits because no artwork and screenmaking are involved.

Compositions and Processes for Thick Film Conductors

Requirements for Conductors. In the early stages, thick film conductor applications were limited to replacing individual wires for the simple interconnection of electrical components. Solderability was one of the most important factors. Later, thick film conductor properties were extended so they could be used as terminations for resistors and capacitors, semiconductor attachments, internal electrodes for multilayer structures such as capacitors and packages, and for through-hole connections. Different applications have different requirements.

The requirements for thick film conductors include conductivity, adhesion, wettability or solderability, bondability, fine line capabil-ity, and chemical, thermal, and physical compatibility with related materials. For general conductor applications, conductivity, adhesion, and solderability are the most important requirements. For certain applications, aged adhesion is as important as initial adhesion. For resistor and capacitor terminations, material compatibility after firing is also important. For internal electrodes, the conductors must be compatible with the dielectric materials, but solderability may be unnecessary. For through-hole applications, the thermal expansion of the conductors must match that of the substrate or dielectric materials. Again, no solderability is required. Therefore, the compositions and processes have to be adjusted to meet different requirements. If soldering is required, solder leach in the solder pot must be kept to a minimum. Minimizing cation migration during high-humidity operating conditions is another important factor for a high-quality conductor.

Materials for conductors include powders of metals, alloys, and glasses; additives; and a screening agent. Each one plays an important role in meeting the broad requirements of different applications. Noble metals can be categorized as silver-bearing materials, which include silver, silver-palladium, silver-palladium-platinum, silver-platinum; and nonsilver-bearing materials, which include gold, gold-palladium, gold-platinum, and gold-palladium-platinum.

Silver is one of the most useful precious metals because of its excellent conductivity, solderability, and low cost. However, silver alone is not usually used for conductors because of solder leach and silver migration concerns. Palladium or platinum is usually added to minimize these concerns. The ratio of palladium to silver has to be optimized to meet different applications. Too much palladium in the silver will greatly decrease solderability and conductivity and increase material cost. For hybrid applications, the silver to palladium ratio is usually between 6:1 and 3:1.

The cost of platinum is usually higher than that of palladium. However, silver-platinum soldering alloys require much less platinum than silver-palladium alloys to produce the same results (Ref 10). In addition to the cost savings, the conductivity of silver-platinum for these applications is better than that of silver-palladium conductors.

For internal electrode applications, a high melting point and low cost are most important. To increase the firing temperature of multilayer Al_2O_3 capacitors to 1200 °C (2190 °F) and above, the palladium to silver ratio is usually 7:3 or higher. The cost savings gained by using palladium-silver, as compared with gold-palladium and gold-palladium-platinum, which have been used for a long time, is obvious. The gold conductors are generally used as pads to attach semiconductors to the circuits through gold or aluminum wire by using wire-bonding techniques.

For base metal conductors, the most usable metals include refractory metals such as molybdenum, molybdenum-manganese, and tungsten. They have been successfully used as internal electrodes for high-temperature firings, such as with Al_2O_3 multilayer substrates. These refractory metals usually are fired at 1500 °C (2730 °F) or higher under a hydrogen atmosphere. The thermal coefficient of expansion of these metals (5 to 6 × 10^{-6}/°C, or 2.8 to 3.3 × 10^{-6}/°F) matches more closely with Al_2O_3 than other metals. However, these metals cannot be directly soldered and must be electrolytically plated to become solderable.

Copper conductors provide high conductivity, high leaching resistance, low migration, and low cost. They can replace gold conductors for certain applications. Compositions of copper conductors have been well documented (Ref 11). Recently, copper conductors have been widely used for multilayers to interconnect integrated circuits. Copper conductors also offer better solderability than gold conductors for multilayer applications. Low-temperature co-fired tape multilayer substrates have been developed with copper to replace more than 5 layers of thick film processes. Nickel conductors have not been widely used because, so far, the adhesion and solderability are not as good as copper conductors.

Base metals can be fired in air if oxidation can be prevented during firing. One method for air firing is to add a reducing agent such as carbon, silicon, or boron. Boron is the most effective element for copper, nickel, and cobalt (Ref 12). Due to a layer of boron oxide film on the metal surface, the air-fired base metal conductors have to be burnished or chemically treated to be solderable. To avoid the burnish process for soldering, boron can be applied at the top of the copper conductor to be fired and then can be easily washed away (Ref 13). A mixture of aluminum powder with low-melting frit can be fired in air to yield a highly conductive conductor without boron added (Ref 14). It is expected the aluminum conductors are not easily soldered with ordinary solders.

Adhesion and Failure Mechanism. With the early thick film technology, only low-melting vitreous glass was used as a bonding agent. The bonding force is due to the interfacial energy between glass powder and substrate. This force is a mechanical or Van der Waals force, which is adequate for most applications. However, the thermal expansion of the glass may not match that of the substrate. After a few thermal shocks from −55 to 125 °C (−65 to 225 °F), which may occur during an application, the mismatch expansion of the glass and substrate will greatly decrease the adhesion.

Because most substrates used are Al_2O_3, some oxides can react with Al_2O_3 at the firing temperature to become a new compound which can enhance the adhesion. This approach may be called reactive or molecular bonding. From phase diagram and test results, oxides of copper, bismuth, vanadium, and lead react with Al_2O_3 substrates below 950 °C (1740 °F) and improve adhesion, especially aged adhesion (Ref 15).

When both glass frit and reactive materials are used, it is called mixed bonding. Three different bonding mechanisms can be used for the conductor compositions depending on the application. Mixed bonding may also be used on aluminum nitride (AlN) substrates, but only after the substrate surface is oxidized. Solders containing tin can react with silver and copper through grain-boundary diffusion to become intermetallic compounds, such as Cu_3Sn, Cu_4Sn, Cu_6Sn_5, $Cu_{31}Sn_8$, $Cu_{41}Sn_{11}$, and Ag_3Sn. Different intermetallic compounds of Pb-Sn are also found. These intermetallics are brittle, mismatch the substrate thermal coefficient of expansion, and will decrease adhesion during aging.

Composition and Processes for Thick Film Resistors

Formulating resistor compositions is challenging work because many stringent specifications must be met. These specifications include a sheet resistance ranging from 0.1 to 10^7 ohms per square ($\Omega/\square$) with a temperature coefficient of resistance (TCR) less than ±100 ppm/°C. These resistor inks can be blended to the desired values. Other requirements include minimum changes in voltage coefficient of resistance (VCR), short time overload (STOL), electrostatic discharge (ESD), temperature cycle, thermal shock, current noise (Ref 16), contact resistance (Ref 17), and processing variables. Long-term stability requirements during thermal, direct current (dc) or alternating current (ac) load, and humidity conditions also must be met. For surge protection applications, resistors must withstand lightning strike and ac voltage surge conditions, plus meet Underwriters' Laboratories (UL) requirements.

Resistor Materials. Before the 1960s, only polymer carbon and polymer silver were used for printed resistors and conductors. The materials formed on plastic substrates became known as printed circuit boards. The early palladium oxide-silver resistors yielded characteristics better than those of carbon resistors. At that time, it was said if there were more than three resistors in a circuit, it would be better to use the thick film approach. A second generation of ruthenium-base resistors has been developed to cover a wider sheet resistance range, offer tighter resistance and less sensitivity to processing conditions, and yield better performance characteristics than the first original system. CTS Corporation has developed and shipped ruthenium-base system resistance networks since the early 1960s. The ruthenium-base system is still the most widely used resistor system in the thick film field. Recently, emerging base metal resistors fired under nitrogen have begun proliferating for different circuit applications.

Ruthenium-base resistor materials include ruthenium oxide (RuO_2) and its pyrochlore and perovskite structures. Oxides of rhodium and iridium and related compounds can also be used for resistors of high quality (Ref 18), but the cost of rhodium and iridium is much higher than that of ruthenium. Thallium oxide (Tl_2O_3) and Tl_2O_3/RuO_2 resistors were introduced in the 1970s. Due to the toxicity of thallium, Tl_2O_3-base resistors have never been popularly used.

Several approaches have been suggested to solve the incompatibility of RuO_2 resistors with inexpensive copper conductors. One approach is to use certain types of ruthenates which can be fired under a nitrogen atmosphere (Ref 19). Du Pont introduced the MIDAS system to fire ruthenium-base resistors in air with the regular process and then post-fire the copper film at a low temperature of 600 °C (1110 °F) under a nitrogen atmosphere. Another approach is to fire a copper composition in air at a high temperature (850 to 950 °C, or 1560 to 1740 °F) to oxidize the copper and create adhesion to the substrate. After ruthenium-base resistors are applied and fired in air by the conventional process, the entire unit is then reduced in a hydrogen-nitrogen gas mixture (4 to 7% H_2, 96 to 93% N_2) at a temperature of 260 to 400 °C (500 to 750 °F) (Ref 20).

Indium oxide (In_2O_3) and tin oxide (SnO_2) are n-type semiconductors with negative temperature coefficient of resistance, and both have been used for resistors. Indium-base resistors can be fired in air or under nitrogen (Ref 21, 22). To adjust the temperature coefficient of resistance of SnO_2 to positive values, oxides of iron, cobalt, nickel, tantalum, niobium, zirconium, and hafnium are added (Ref 23).

Both In_2O_3 and SnO_2 resistors can be used in the $k\Omega/\square$ and above resistance range. For lower resistances, the covalent intermetallic compounds of certain conductive carbides (Ref 24), borides, nitrides and silicides (Ref 25) have been suggested. Borides of lanthanum and zirconium, tantalum nitride, and titanium silicide (Ref 26–29) and combinations thereof have been used as base metal resistors. Molybdenum trioxide (MoO_3) has also been suggested for use in the $k\Omega/\square$ range.

It is a challenge to extend the resistance ranges of both the high- and low-end resistor materials. For very low-resistance compositions (in the range of 0.1 to 10 $\Omega/\square$ with a temperature coefficient of ±50 ppm/°C capability), special alloys have to be chosen. It is well known that alloys exhibit a high resistivity due to a reduction in the length of the mean free path and a low temperature coefficient of resistance compared to their individual metals. The lowest resistance values are normally obtained with alloy compositions near the center of a binary phase diagram but usually not at 50/50. Alloys of palladium-silver, copper-nickel, nickel-chromium, and others with excellent resistor characteristics and high performance have been used for thick films (Ref 30, 31).

In general, the conductive phases in resistors of low value have metallic properties and those with high values have semiconductive properties. The weight ratios of the metallic conductives to glass for low-end members are <1, usually about 30/70. On the other hand, the weight ratios of semiconductives to glass in high-end members are >1, usually 70/30. These ratios hold true for both SnO_2 and RuO_2, regardless of the similarity in molecular weight of these materials.

The conduction mechanism may explain why the weight percentage between the glass and conductives is different for the high- and low-end members. In addition to the early theory of particle-particle contact, thick film conduction can be affected by microstructure development (Ref 32), thermal expansion, tunnel effect, and the conductive solubility in the glass. Both SnO_2 and RuO_2 are slightly soluble in glasses which can be used for color purpose; but in resistors, they may show different effects. If tin oxide is dissolved in the glass, the conductivity may be decreased. On the other hand, if RuO_2 is dissolved in the glass, ruthenium-base glass may develop a microstructure to extend and enhance the conductivity. RuO_2 shows metallic properties with a very positive temperature coefficient of resistance, but metal ruthenates may be amphoteric with unique conductor properties.

Adjusting Properties. In order to adjust resistance to within ±100 ppm/°C, valence doping, oxygen deficiency, and other methods are normally used. The Ta^{5+} in Ta_2O_5 is the most useful adjustment for high-resistance value compounds, such as semiconductive SnO_2, metallic RuO_2, and also for low-

value covalent intermetallic compounds, such as lanthanum and zirconium borides. In addition to Ta_2O_5, the sheet temperature coefficient of resistance and temperature coefficient of resistance of ruthenium-base resistors can be adjusted by the addition of oxides of titanium, rhodium, and manganese, and some inert materials such as Al_2O_3, SiO_2, and ZrO_2. These materials can be added in any form including admixture, which combines the conductive material and/or glass into a new compound.

Figure 1 shows that the various materials described have been used for thick film resistors. The horizontal axis represents relative concentration of conductives. As can be seen, it is difficult to use the same material to cover the entire range of resistance values from 0.1 to 10^7 $\Omega/\square$ with desired requirements. Several materials must be chosen. The ruthenium-base system approaches the ideal condition, which covers the entire range of resistance in a linear fashion with a slope close to unity. But ruthenium-base resistors still lack the very low sheet resistance capability. Gold, silver, and copper cannot be used as resistors because it is difficult to control the resistance ranges at the nearly 90° slope, and these metals also exhibit very positive temperature coefficients of resistance.

Compositions and Processes for Dielectrics

Thick film dielectrics can be used as overglazes, crossovers, capacitors, and multilayers. These dielectrics have to meet the requirements of dielectric constant, insulation resistance, dissipation factor, dielectric strength, and dielectric loss factor. The thermal conductivity, thermal expansion, thermal shock resistance, and mechanical properties

of dielectrics are also of importance (Ref 33). Overglazes are usually applied over already fired conductive and resistive films. They provide for the encapsulation of packages, hermetic sealing, and protection against chemical and environmental hazards. Changes in properties of the films underneath should be kept to a minimum. The glass composition for an overglaze should be chosen so that the firing temperature of the overglaze is lower than those of the films underneath to ensure negligible interaction. Most low-melting point glasses, such as lead and/or zinc borosilicates, can be used.

Crossover and multilayer dielectrics are preferably made with high softening point glasses which are not suitable for conductors and resistors. Multilayer dielectrics can be used for crossovers. The most important properties for crossovers and multilayers are that they do not flow and interact with additional layers of materials during subsequent firings. This control is accomplished with either of two approaches. One approach is to use devitrifying or crystallizing glass, which becomes soft and then crystallizes during firing so the film more closely resembles a ceramic than a glass. This allows layers of conductors, resistors, and additional dielectrics to be fired on top without affecting the already fired materials. The devitreous glasses can be made by the addition of devitrifying agents such as oxides of titanium, zinc, or zirconium. The crystallized temperature can be determined by the phase transition.

The second approach is to mix a vitreous glass with refractory oxides such as Al_2O_3 or SiO_2. When the composition is fired, the glass softens and dissolves a portion of the oxides. During this process, the softening point of the glass is raised so the glass will be less affected by subsequent firings. The choice of glasses and oxides depends on thermal expansion, softening point, viscosity, and the electrical properties desired such as dielectric constant and applied voltage. The physical properties of glass frits and fillers such as particle size and distribution, surface area, and density are also important to control final characteristics.

When a large substrate is used, bowing becomes a major concern. Matching the thermal coefficient of expansion of Al_2O_3 is one of the important factors in controlling bowing. Multilayers can be built with thick film dielectrics up to five layers with good properties and yields. Above this, bowing becomes a problem. For high dielectric constant capacitors, the ferroelectric compounds such as barium titanate ($BaTiO_3$), a Curie point modifier, and a glass frit can be used. Various capacitor compositions have been formulated for different applications.

Recent Developments. Recently, low-temperature co-fired multilayer ceramic (LTCC) substrates have been developed (Ref 34–36). Instead of the 1500 to 1700 °C (2730 to 3090 °F) firing temperature required for Al_2O_3 multilayer substrates, temperatures

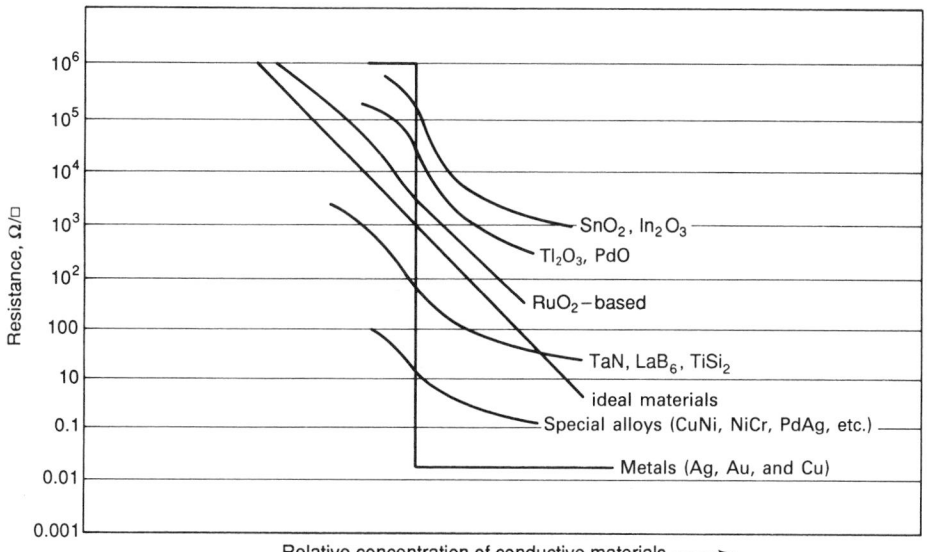

Fig 1 Resistance of metallic-base thick film resistor materials as a function of the concentration of conductive substances in the resistor materials

below 1000 °C (1830 °F) can be used to fire the LTCC packages. This allows the use of high-conductivity, low-melting point metals such as gold, silver, and copper as internal electrodes.

The materials for LTCC are usually formulated by the addition of glass to the ceramic or devitrifying glass to make a flexible tape. The tape is printed with internal conductors such as gold or copper with designed patterns and through holes to interconnect the layers. All the printed layers are laminated in a press at about 3000 psi (21 MPa) and 70 to 90° C (160 to 195 °F) and then cut to a desired dimension. The laminated substrates, which may have as many as 30 to 50 layers, are fired at about 850 to 900 °C (1560 to 1650 °F). These fired substrates have a controllable dielectric constant, thermal expansion that may be designed to match silicon, and gold or copper metallization for high power distribution and high signal speed. The LTCC offers high density with good thermal, mechanical, electrical, and dimensional control properties. Aluminum nitride substrates offer high thermal conductivity and low thermal expansion properties comparable to that of silicon. A recent innovation is the incorporation of aluminum nitride substrates for thick film circuits used for power hybrid applications.

Applications

In the early 1960s, the IBM 360 solid logic technology (SLT) (Ref 37) and RCA micromodule (Ref 38) were based on the first-generation palladium-base resistor system. Since then, thick film circuits have been widely used worldwide in all segments of electronics. During the past 30 years, the idea of combining the active and passive components into a small silicon chip has never been realized because of difficult processes and insufficient characteristics and performance. Therefore, the hybrid approach continues to be the most economical and effective way to fabricate microcircuits. The high-power integrated circuits also need high thermal management and a high-density substrate to complement their active component properties.

In the past, thick film technology has been successfully used in packages of several stacked Al_2O_3 and beryllia (BeO) substrates. They have been used for amplitude regulators, sonar transmitters, active filters, amplifiers, transducers, and many other applications. They can also be made for analog, digital, and microwave applications (Ref 39). Recent advances in new thick film materials offer improvements in the characteristics and processes used for chip resistors and resistor networks (Ref 40). The low-temperature, co-fired multilayer substrates make thick film processing more challenging, allowing 30 to 50 layers on a single substrate containing in excess of 100 chips. These substrates are made with low dielectric constants and the high conductivity of copper, gold, or silver for high power distribution and high signal and system speed. Thick film technology can be used for applications ranging from advanced high-speed computers to sophisticated medical, military, aerospace, and instrumentation devices, to consumer electronics.

REFERENCES

1. R. Berry *et al.*, *Thin Film Technology*, Van Nostrand, 1968
2. L. Hoffman, *Ceram. Bull.*, Vol 42 (No. 9), 1963, p 490
3. A. Barton, *Handbook of Solubility Parameters*, CRC Press, 1983
4. C. Kuo, *Proceedings of the International Society of Hybrid Microelect.*, 1989, p 150–166
5. C. Kuo and T. Martin, *Proceedings of the 4th International SAMPE Conference*, 1990, p 576–588
6. R.W. Vest, MOD, Metallo-Organic Materials, Purdue University Final Report, 1980
7. C. Kuo, *Solid State Technol.*, Feb 1974, p 49–55
8. W. Mathias, Micropen, *Proc. ISHM 1990*, p 31–35
9. C. Kuo, *Hybrid Circuit Technol.*, Aug 1987, p 29–33
10. Sayers, Kuo, and Cheng, *25th Elect. Components Conf. 1972*, p 159–167
11. V. Suita, U.S. patents 4,521,329, 4,514,321, 4,874,550, 4,540,604, and 4,687,597; A. Probhu and K. Hang, U.S. patents 4,863,517, 4,619,836, and 4,377,642; J. Steinberg, U.S. patent 4,868,034; C. Kuo and T. Bloom, U.S. patent 4,623,482
12. C. Kuo, U.S. patent 4,122,232; J. Provance, U.S. patents 4,322,316 and 4,317,750
13. C. Kuo, U.S. patent 4,409,261
14. R. Loasby *et al.*, *Solid State Technol.*, May 1972
15. C. Kuo, *Int. J. Hybrid Microelect.*, Vol II (No. 1), 1988, p 17–22
16. C. Kuo, *Proc. ISHM 1969*, p 153–160
17. C. Kuo, *Proc. ISHM 1969*, p 263–270
18. C. Kuo, U.S. patents 3,769,382 and 3,786,560
19. J. Steinberg, U.S. patent 4,818,107; H. Hankey, U.S. patent 4,536,328
20. C. Kuo, U.S. patent 4,316,942
21. M. Block and A. Mones, *Proc. ECC Elect. Components Conf. 1966*, p 191–196
22. A. Prabhu *et al.*, U.S. patents 4,467,009, 4,380,750, and 4,379,195
23. J. Hormadaly, U.S. patents 4,548,742, 4,539,223, and 4,707,346; C. Kuo, U.S. patents 4,711,803, 4,720,418, 4,698,265, and 4,655,965; R. Wahler, *et al.*, U.S. patents 4,215,020 and 4,065,743; J. Mochel, U.S. patents 3,564,707 and 2,490,825
24. Rao and Murry, U.S. patents 3,180,841, 3,277,020, and 3,788,997
25. C. Huang, U.S. patents 3,394,087, 4,256,796, and 3,506,801
26. P. Donohue, U.S. patents 4,585,580, 4,225,468, and 3,503,801
27. H. Watanabe, *et al.*, *IMC 1988 Proc.*, Tokyo, p 37–41
28. Shapiro, U.S. patent 4,205,298
29. C. Kuo, U.S. patent 4,639,391
30. C. Kuo and T. Martin, *Proc. ISHM 1990*, p 431–437, p 438–444
31. R. Howell, U.S. patent 3,744,518
32. R. Vest, "Conduction Mechanism in Thick Film Microcircuits," *Purdue University Final Report*, 1975
33. R. Buchanan, *Ceramic Materials for Electronics*, Marcel Dekker, 1986
34. J. Steinberg *et al.*, *Solid State Technol.*, Jan 1986
35. R. Tummala *et al.*, *Microelectronic Packaging Handbook*, Van Nostrand, 1989. J. Licari *et al.*, *Hybrid Microcircuit Handbook*, Noyes, 1988
36. T. Provo, *Hybrid Circuit Technol.*, Nov 1990, p 41–43
37. E. Davis *et al.*, *IBM J. Res. Dev.*, Vol 8, 1964, p 102
38. B. Schwartz, *Am. Ceram. Soc. Special Publ. No. 3*, 1969, p 12–25
39. C. Harper, *Handbook of Thick Film Microelectronics*, McGraw-Hill, 1974
40. L. Slack *et al.*, *Proc. ISHM*, 1990, p 423

Thermistors and Related Sensors

Bernard M. Kulwicki, Texas Instruments Incorporated

SEMICONDUCTING CERAMICS are designed and used to sense and control environmental variables such as temperature, relative humidity, or the presence of undesirable, toxic, or combustible gas concentrations in the air. All semiconductors exhibit sensitivity to these variables. It is therefore essential that the sensor exhibit maximum response, be selective, detecting the variable of prime interest, excluding others insofar as possible, and that its response characteristic is stable in the sensing environment. The choice of a suitable material, with a particular, desirable complement of microchemical and microstructural features and electrical response, depends on which variable is to be sensed as well as on the conditions of use.

Negative temperature coefficient (NTC) thermistors are typically formed from oxides with the spinel crystal structure, commonly based on the chemical formula $NiMn_2O_4$. They are bulk-effect devices possessing very stable temperature-dependent electrical resistivity. Principal applications include precise temperature measurement and control and temperature compensation of other nonlinear components and circuits. All the other ceramic sensors are barrier layer devices whose usefulness derives from the existence of resistive electrical junctions at the grain boundaries.

Positive temperature coefficient (PTC) resistors are produced from donor-doped solid solutions with the perovskite lattice structure, based on the formula $BaTiO_3$. The PTC corresponds to a substantial increase in electrical resistivity with temperature, near and immediately above the ferroelectric/paraelectric phase transition. The PTC region typically spans 3 to 7 orders of magnitude of resistivity over a temperature range of 50 to 150 °C (120 to 300 °F). Principal applications include self-regulating heaters, current surge devices, overcurrent protectors, and temperature sensors.

Gas (including humidity) sensors are produced from a variety of oxides including SnO_2, ZnO, $MgCr_2O_4$, and $ZnCr_2O_4$. Like PTC devices, their resistivity is controlled by electrical junctions at grain boundaries. These ceramics are porous, allowing free access to the air, and the resistance is modulated by selective adsorption of atmospheric constituents. Humidity sensors operate at ambient temperature, gas sensors at elevated temperatures. Applications include cooking controls for microwave ovens, low-cost humidistatic

controls, and alarms for dangerous buildup of carbon monoxide or combustible gases.

Several excellent reviews have been published describing specific sensor materials, processing technologies, and detailed behavior in a variety of applications. NTC thermistors are described in Ref 1 and 2; PTC resistors in Ref 1, 3 to 5; gas sensors in Ref 6 to 8; and humidity sensors in Ref 9 to 11. General papers that address these and related technologies (Ref 2, 12, 13) provide additional perspective. The original technical literature has been adequately covered in these publications, and in the interest of conciseness will not be reproduced in detail here.

Some common features link oxide thermistors and gas sensors:

- Ceramic, polycrystalline microstructure
- Semiconductivity
- Point defects
- Electrical contacts
- Electrical junctions at grain interfaces
- Complex impedance
- Voltage sensitivity
- Stability of electrical response

Contacts and barrier layers are topics of sufficiently general interest to merit preliminary discussion.

Electrical Contacts

A common feature of all electrical sensors is the requirement for adherent, stable, low-resistance, ohmic contacts. In some cases the contacts must be solderable. In other cases they must possess a low sheet resistance to permit uniform passage of substantial electric current without localized overheating. In still others they must be porous or possess a certain catalytic activity. In all cases they must be inexpensive.

For n-type semiconductors without surface states, achievement of ohmic contact is relatively straightforward. A metal with a low work function will produce an ohmic contact, and one with a high work function a rectifying contact or Schottky barrier. The reverse is true for a p-type semiconductor, but the choice is more limited; fewer metals with high work functions are available. In both cases, however, contact resistance can be overcome by introducing a high carrier density near the contact surface to thin the barrier. This is frequently required for p-type semiconductors.

Oxide semiconductors readily acquire adsorbed layers and usually possess surface states. Thus, an additional requirement is that such a surface layer should be disrupted. This can be accomplished either chemically or mechanically. For n-type oxides, the metal should possess a high oxygen affinity, or an ingredient of the electrode system should possess a high oxygen affinity.

In the case of n-type semiconducting titanates, utilized in PTC resistors, ohmic contacts are produced rather easily with aluminum and zinc, not so readily with nickel and silver, and hardly at all with copper, platinum, and gold. Ohmic contact with nickel and silver may be facilitated by doping, for example, Ni:B, Ni:P, Ag:Ga, and Ag:Zn. Clearly, the requirement for a high oxygen affinity is at least as important as the need for a low work function.

Silver is a common contact for NTC thermistors (p-type), and ruthenium oxide (RuO_2) is used for humidity sensors based on the chromite spinels (also p-type).

Although SnO_2 gas sensors are n-type, the contacts are usually platinum or iridium/palladium. Noble metals are required because of their stability in contact with the semiconductor at the operating temperature (frequently as high as 400 °C, or 750 °F).

A final consideration regarding the selection of a contact metal pertains to stability. Contact corrosion can be a major contributor to device failure. More subtle instability can arise from gradual chemical reaction between the metal and the semiconductor, accompanied by diffusion of point defects, possible formation of a resistive or capacitive interface, and subsequent drift in the electrical response of the sensor. These issues must be considered in design and selection of a sensor for a particular application.

Barrier Layers

The oxide ceramics used for sensor applications are generally sintered in an oxidizing atmosphere, frequently in air. Oxygen can adsorb during cooling from the sintering temperature, and an oxidized monolayer can form on the interfacial surfaces of the sintered grains. It is important to control the cooling rate and, sometimes, the oxygen partial pressure in the sintering atmosphere.

Since oxygen is an electron acceptor, free carriers originating within n-type grains can be trapped at the grain boundary, forming depletion layers on both sides of the interface. The potential energy of these "surface" traps can be influenced by minor compositional modifications. For example, small concentrations of transition metal ions (like manganese or iron in the case of PTC resistors) can significantly affect the surface potential and maximize the barrier strength. This is desirable for PTC resistors, which are often used at line voltage, and must operate without danger of thermal runaway (Ref 4).

In PTC resistors, the junction potential varies with temperature because the host semiconductor possesses nonlinear dielectric properties. Below the Curie temperature, ferroelectric polarization partially compensates the trapped electron charge, and the resistivity is low. Above the Curie temperature, the polarization vanishes and the barrier potential increases in response to falling permittivity:

$$\varphi_0 = \frac{en_s^2}{8\epsilon\epsilon_0 N_d} \qquad (Eq\ 1)$$

where e is the electron charge, n_s is the filled surface trap concentration, N_d is carrier (donor) concentration, ϵ is the dielectric constant, and $\epsilon\epsilon_0$ is permittivity. Since ϕ_0 varies much more rapidly with temperature than kT, the conductivity σ decreases exponentially:

$$\sigma = \sigma_0 \exp\left(\frac{-e\varphi_0}{kT}\right) \qquad (Eq\ 2)$$

Eventually barrier growth pulls the surface states up to the Fermi level, releases the trapped carriers, and the resistivity traverses a maximum.

It is important that stable barriers be formed that are reasonably unreactive in the presence of deoxidizing gases. Otherwise the PTC property will degrade, in some cases resulting in device failure (Ref 4). This possibility must be recognized during device design and precautions taken if indicated. For example, a lower Curie temperature might be selected to retard the rate of degradation by gasoline vapor in an automotive application; or a substitute packaging material for a self-regulating heater might replace one that could release degradant during operation (for example, ammonia from phenolic resin).

Gas Sensors. Precisely the opposite requirement exists for gas sensors, where the aim is to achieve a maximum response to the presence of deoxidizing gas molecules in the air. Gas sensors typically operate at sufficiently high temperature that both the oxidation and reduction reactions occur reversibly.

The role of barrier layers in the function of humidity sensors is somewhat more complex. In this case the host semiconductor is p-type, and donor molecules such as hydroxyl groups become attached at Cr^{3+} sites by chemisorption of water molecules at elevated temperature. Since the majority carriers are holes, potential barriers form and provide a

stable, high base resistivity at low humidity. The device operates at room temperature. Further absorption of water molecules occurs in stages. Initially, H_3O^+ ions are produced, assisted by high electric charge density in the neighborhood of the hydroxyl sites, and protonic conduction to adjacent sites becomes possible. At high humidity, condensation of water in the capillary-like pores leads to a liquid-like layer, and electrolytic conduction ensues.

At elevated temperature, the chromite spinels can also function as gas sensors (Ref 11). However, since they are p-type semiconductors, their response characteristic is such that their resistivity increases during the adsorption of reducing gases and vice versa.

NTC Thermistors. Grain-boundary barriers do not play a significant role in the conduction mechanism of NTC thermistors. The latter are p-type semiconductors, so adsorption of oxygen at the interfaces between grains would be expected to lead to accumulation layers (that is, regions enriched in majority carriers), simply improving grain to grain electrical contact. The microstructure does not appear favorable for the formation of hydroxyl-induced barriers as the ceramics are quite dense, and the charge density at the octahedral ion sites is not nearly as intense as in the case of the chromites.

NTC Thermistors

Most NTC thermistors are solid solutions in the oxide system $(Mn,Ni,Fe,Co,Cu)_3O_4$. Materials in this system possess high-temperature coefficients ($B = E_a/k = 3000$ to 4500 K) and excellent stability. Other, more specialized NTC materials include titanium-doped Fe_2O_3 and lithium-doped $(Ni,Co)O$.

The conduction mechanism requires that ions of different valence exist on the octahedral B-site ($A^{2+}B^{3+}_2O_4$). Alternate distributions are possible, even for fixed chemistry (Ref 1):

$$(Ni^{2+}_{1-x}Mn^{2+}_x)(Ni^{2+}_x Mn^{3+}_{2-2x}Mn^{4+}_x)O_4 \qquad (Eq\ 3)$$

and

$$(Mn^{2+}_{1-x}Mn^{3+}_x)(Ni^{2+}Mn^{3+}_x Mn^{4+}_{1-x})O_4 \qquad (Eq\ 4)$$

In both cases charge transport can occur, for ions with different valence are present on the closely spaced B-sites. The loosely bound electrons ($Mn^{4+} + e^- \leftrightarrow Mn^{3+}$) become mobile by virtue of the availability of nearby, unoccupied sites. The carrier density N_d is:

$$N_d = \frac{16P_d}{a^2c} \qquad (Eq\ 5)$$

where P_d is the probability that a B-cation is an electron donor, and a and c are the lattice parameters. Mobility μ is related to the probability that lattice vibrations will induce hopping, and is thermally activated:

$$\mu = \left(\frac{ea^2\nu}{8kT}\right)P_a \exp\left(\frac{-E_a}{kT}\right) \qquad (Eq\ 6)$$

where P_a is the probability that an adjacent B-site contains an acceptor cation, and ν is a frequency factor. Thus, the conductivity is:

$$\sigma = e\mu N_d = \left(\frac{2e^2\nu}{ckT}\right)(P_a P_d)\exp\left(\frac{-E_a}{kT}\right) \qquad (Eq\ 7)$$

The high activation energy for hopping gives rise to a correspondingly high conductivity dependence on temperature, making such materials especially suitable for application as temperature sensors.

Figure 1 illustrates how the resistivity can be adjusted by varying the composition in the manganese-nickel-iron spinel system. The resulting NTC slopes are typically -3.4 to $-5.0\%/°C$ at room temperature and -1.3 to $-2.0\%/°C$ at 200 °C.

For the intended applications, stability is an essential requirement. Potential sources of instability include:

- Changes in composition due to oxidation/reduction reactions, point defect migration, or diffusion of ions from the contacts
- Ionic rearrangements due to more favorable free energy at the temperature of use in comparison with the temperature of preparation
- Possible changes in surface states at the grain boundaries (for example, by adsorption of electron donor molecules)

The first source can be mitigated to some extent by proper choice of composition, such that the operating range corresponds to the saturation region between extrinsically and intrinsically activated conduction (Ref 2, 12). The second can sometimes be minimized by annealing or preaging the device. The last may require encapsulation.

Highly stable thermistors are commercially available. However, there always exists a trade-off between stability and cost that must be considered in the selection of a sensor for a given application.

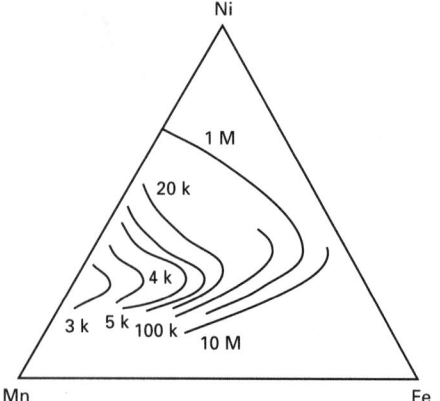

Fig 1 Room-temperature resistivities ($\Omega \cdot$ cm) for various compositions in the ternary spinel system Mn-Ni-Fe. Source: Ref 2

PTC Resistors

Ceramic PTC resistors are manufactured from donor-doped solid solutions of barium titanate $[(Ba,Sr,Pb)_{1-x}Ln_xTiO_3]$ where a trivalent lanthanide ion (in this example) resides on the perovskite A-site ($A^{2+}B^{4+}O_3$). Conduction electrons are produced by the resulting transformation of an equivalent number of titanium ions from valence four to three. Base resistivities commonly range between 20 and 20,000 $\Omega \cdot$ cm. The rise in resistivity with temperature is associated with the Curie transition: 130 °C (265 °F) for barium titanate. The transition temperature is decreased by substitution of strontium for barium and increased by substitution of lead for barium. Devices with Curie temperatures from 0 to 300 °C (32 to 570 °F) are commercially available. Typical resistivity versus temperature data for a variety of resistors having a wide range of Curie temperatures are displayed in Fig 2. Resistivity ratios R_{max}/R_{min} can exceed six decades. Maximum slopes are typically $\approx 20\%/$°C but can exceed $100\%/$°C for pure $Ba_{1-x}Ln_xTiO_3$.

Ceramic PTC resistors are normally produced in the form of pellets, discs, or bars, usually less than 50 mm (2 in.) in diameter (length) and 10 mm (0.4 in.) in thickness. Multicellular, extruded honeycombs and internally metallized honeycombs are also manufactured. The former are used in line-voltage air heaters and the latter in low-voltage automotive air-fuel heaters.

Properly manufactured, PTC resistors exhibit excellent long-term stability, both with high-temperature exposure and with electrical cycling. Certain instabilities exist that must be considered in device design and may require control, but they tend not to seriously impede the development of any but the most demanding applications (Ref 4). Examples include:

● *Aging.* Changes in resistivity as a function of time
● *Cycle aging.* Changes in resistivity as a function of electrical cycling
● *Current failure.* Cracking due to rapid self-heating and thermal expansion stress
● *Voltage failure.* Damage due to thermal runaway
● *Degradation.* Changes in resistivity due to prolonged exposure in a hostile environment
● *Banding.* Formation of a stable heated region (band) in the center of a resistor that limits power output

The incidence of degradation with exposure to vacuum or chemical reducing agents suggests that gas sorption and diffusion along grain boundaries occur rather easily. The problem of chemical degradation must be addressed in the product design stage. In some hostile environments (for example, spacecraft), the application of PTC resistors may not be feasible without special precautions. In most applications, however, degradation

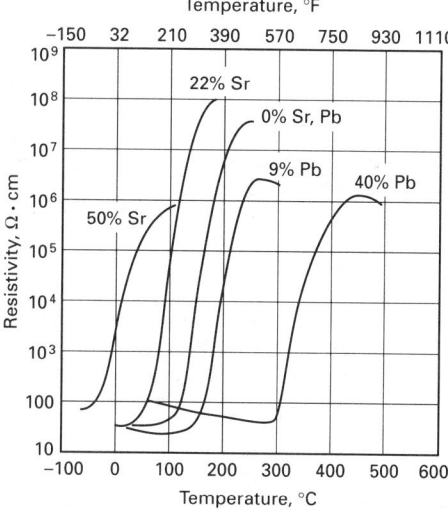

Fig 2 PTC resistors with Curie temperatures from −25 to 300 °C (−15 to 570 °F)

rates are sufficiently low that product reliability can be assured with good design practice and accelerated testing. For the few cases where reason for serious concern exists, lengthy, elaborate testing may be required because of the complexity of the reactions and present lack of detailed theoretical base from which predictive models can be applied.

Devices are capable of exhibiting excellent reliability (Ref 4, 5). Stability during electrical cycling is generally superb. Many devices must withstand 10,000 to 100,000 electrical cycles with an inrush current of 10 to 20 A and steady-state electrothermal power dissipation well above the Curie temperature, without a substantial change in resistance or in the PTC curve. Automotive honeycombs, for example, must survive 20,000 cycles at 14 V, plus an additional 2500 cycles at 24 V, with a total resistance change less than 10%. Additional specifications pertain to constant energization (500 h), gasoline exposure, contact adhesion, cold strength, and cell breakage strength.

PTC resistors are susceptible to similar instabilities as other ferroelectric ceramics, though the form is usually somewhat different. The more severe failure modes are the best understood and most readily controlled. The details of internal stress development and equilibrium domain patterning associated with ferroelectric aging are not as well understood, but the related instabilities tend to be relatively minor and have been minimized so as not to deter the proliferation of applications. Degradation by vacuum and reducing atmospheres is the least tractable because of its complexity both in terms of the range of potential degradants and the mechanism by which they react with the titanate and diffuse in. Rates must be determined in each case through experimentation, but can be minimized by reducing Curie temperature (reaction rate) and open porosity (diffusion rate).

PTC resistors can be produced with a nearly continuous range of transition temperatures. They withstand high transient overloads, but careful design is required to avoid failure. The minimum resistivity is limited to approximately 20 $\Omega \cdot$ cm, which limits current application to relatively small components and to inrush current levels below 50 A.

Gas Sensors

Gas and humidity sensors use another feature of the ceramic microstructure, porosity, to achieve properties that are otherwise difficult to realize. Porosity is normally an undesirable microstructural attribute, since it detracts from other useful properties such as strength, conductivity, and chemical stability. In some cases, however, porosity may be desirable, for example, to reduce the elastic modulus or maximize access speed of minor atmospheric constituents to the large internal surface area of the semiconductor. Control of the size and distribution of micropores is important in achieving reproducible, optimal sensor response.

The first oxide semiconductor to receive significant attention for gas sensing, and still the most widely used, is tin oxide (SnO_2), with additives to enhance sensitivity to particular gases. Sensors with strong response to hydrocarbons, carbon monoxide, polar molecules found in smoke, or alcohol vapor are available. In addition, n-type sensors based on zinc oxide (ZnO) and p-type sensors based on $MgCr_2O_4$ exist.

The response of a SnO_2 sensor to various gas concentrations is shown in Fig 3. The adsorption of electron acceptors such as O_2 decreases the electrical conductivity, and the

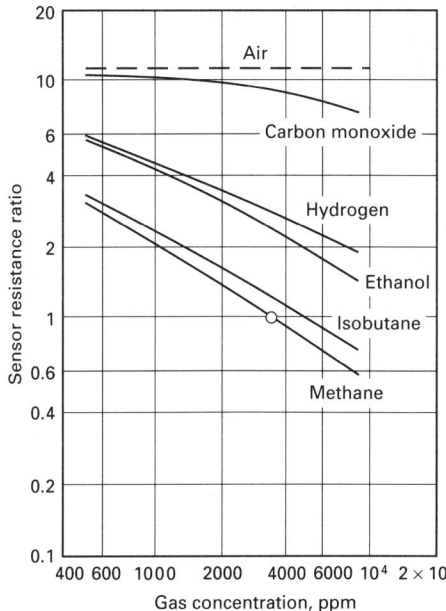

Fig 3 Response of a commercial SnO_2 gas sensor to various reducing gas concentrations in air. Courtesy Figaro Engineering Inc., Figaro USA

adsorption of donors increases it. The response of p-type sensors such as $MgCr_2O_4$ is just the opposite, with the oxidizing gases causing the conductivity to rise (Ref 11).

The SnO_2 sensing element is frequently prepared in the form of a porous ceramic pellet with an embedded heating element. A low carrier concentration is desired in order that reasonably strong barriers are generated at the grain boundaries by oxygen adsorption and electron trapping. High porosity facilitates access of the ambient atmospheres to the grain boundaries and intergranular necks.

When the sensor is heated to near 400 °C (750 °F), reducing gases present in the atmosphere can adsorb, react with the chemisorbed oxygen, and release the trapped electrons as the reaction product desorbs, resulting in an increase in conductivity. The mechanism has been confirmed by using a mass spectrometer to monitor the carbon dioxide (CO_2) released after exposure of a sensor to a gas containing carbon monoxide (CO).

The need to operate the sensor at elevated temperature arises from its sensitivity to water vapor. It is likely that dissociative chemisorption of H_2O molecules occurs on the surface of the oxide, and that other gases such as O_2 and CO adsorb on the resulting hydroxyl sites. Further, physical adsorption of water at low temperature can alter the surface electrical fields in such a way that oxygen desorption is promoted, restoring the trapped electron to the conduction band and thereby increasing the conductivity. However, if the temperature is sufficiently high, physical adsorption of water does not occur, and stable sensor operation is ensured.

Additional problems associated with the application of gas sensors concern their relatively poor selectivity (although improvements continue to advance). The difficulty is that any adsorbate that tends to react with chemisorbed oxygen will donate electrons to the conduction band, and vice versa. One approach is to search for dopants that catalyze certain reactions, or provide favorable conditions for the selective adsorption of certain reactants. Another strategy involves balancing the operating temperature to maximize sensitivity to a particular reactant molecule. Carbon monoxide sensors typically operate at intermediate temperature (150 °C, or 300 °F) to avoid potential interferences due to alcohols and hydrocarbons. But special precautions involving cyclic heating are needed to avoid confounding of the response by water vapor reactions.

Humidity Sensors

Ceramic humidity sensors can be fabricated from a wide variety of semiconducting oxides. Commercial devices based on porous, anodic alumina (Al_2O_3) have been available for a number of years. The latter are capacitive devices that require relatively complex circuitry and frequent recalibration, especially after exposure to wet ambients. The capacitance is generally in the range 50 to 500 picofarads (pF), and may increase by 25% over the humidity span of 10 to 100 %RH.

A "low-cost" resistive sensor with a more stable resistivity-humidity characteristic was originally developed for use in microwave ovens (Ref 14). The sensing element is a small, porous, rectangular wafer composed of a $MgCr_2O_4$-TiO_2 spinel solid solution (approximately 30 mol% TiO_2), with porous ruthenium oxide (RuO_2) electrodes and a coil heater for self-cleaning. The sensor is installed in the ventilation area of the microwave oven. It detects the rapid rise in humidity corresponding to the onset of cooking. Prior to each use the sensing element is cleaned by energizing the heater until the sensor reaches 500 °C (930 °F). During sensing, the element operates at room temperature.

More recently, resistive sensors based on $ZnCr_2O_4$-$LiZnVO_4$, SiO_2-ZnO, and zirconia (ZrO_2) have been developed. Comparative response characteristics of some ceramic humidity sensors are displayed in Fig 4. Typical room-temperature response of a resistive sensor at 1 kHz corresponds to a decrease in resistance of several decades between 10 to 100 %RH, for example, from 10^6 to 10^3 Ω.

Sensor construction can be as simple as employment of a ceramic pill or chip with a porous, ohmic contact metallization on both faces, connected to the control circuit using bonded lead wires, or just mechanically held by a spring. Alternatively, the oxide layer could be deposited over an inert electrode with interdigitated fingers. The sensing signal is usually pure alternating current (ac) since even a small direct current (dc) component could lead to polarization, electrolysis, corrosion of the electrical contacts and terminations, and rapid failure.

A comprehensive examination of the characteristics of a wide range of different humidity sensor materials, salt-doped and undoped, resistive and capacitive, concluded that all operate by means of the same physical mech-

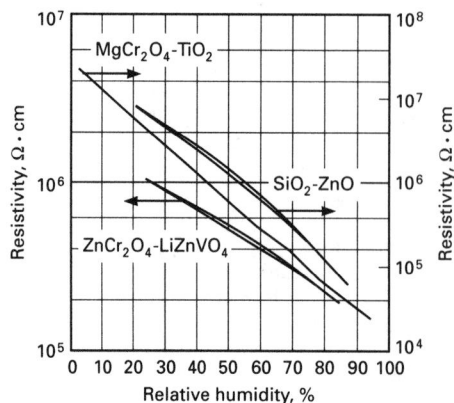

Fig 4 Electrical response of three ceramic humidity sensors at room temperature and 1 kHz

anisms (Ref 9). At low humidity, small ions (such as Li^+ or Cr^{3+}) present in the surface layer of the grains possess high local charge density and strong electrostatic field, and represent very good sites for the chemisorption of H_2O molecules. Upon exposure to the atmosphere, strongly bound water molecules quickly occupy the available sites. This layer, once formed, is not further affected by exposure to humidity, but it can be thermally desorbed.

Once the first layer has been formed, subsequent layers of water molecules are physically adsorbed. The physisorbed water dissociates due to the high electrostatic fields in the chemisorbed layer, yielding H_3O^+ and OH^- ions.

It is estimated that the fraction dissociated is on the order of 1%, or a factor of 10^6 greater than that in liquid water. Charge transport occurs when the hydronium ion releases a proton to a neighboring water molecule, which accepts it while releasing another proton, and so forth. This is known as the Grotthuss chain reaction. It is thought to represent the conduction mechanism in liquid water as well as in the surface layers of humidity-sensitive oxides. At high relative humidity (>40%) liquid water condenses in the pores as already described, and electrolytic conduction takes place in addition to the protonic transport in the adsorbed layers. Similar mechanisms are believed to be responsible for the properties of capacitive sensors, the principal distinction being the different substrate and pore geometries.

Aging mechanisms in humidity sensors include:

- Adsorption of contaminants preferentially on the cation sites
- Loss of surface cations due to vaporization, solubility, and diffusion, or annealing to a less reactive structure
- Migration of cations away from the surface due to thermal diffusion

Generally, the more sensitive a material is to humidity, the more susceptible it tends to be toward aging. This is a significant problem in practical devices, primarily because the sensing mechanisms depend on changes that take place at surfaces where water molecules adsorb. Different surface reactions, with altered kinetics, can occur when atmospheric contaminants are present. Various strategies have been devised to deal with this problem.

The operating range of the sensor could be restricted. This is a common practice for Al_2O_3 sensors, which are generally recommended for use only in low-humidity environments. Exposure to conditions near saturation leads to systematic aging and the need for periodic recalibration. This occurs when the dewpoint rises to within 10 °C (18 °F) of the ambient temperature, corresponding to a relative humidity greater than approximately 50%.

A heater can be employed to regenerate the

sensor material prior to measuring its resistance. This approach was utilized with the $MgCr_2O_4$-TiO_2 sensor in microwave oven control systems. The heater is periodically energized, driving the sensor to 500 °C (930 °F) and burning out residual adsorbed gases, thereby recovering the electrical response of the virgin device. This strategy is effective but is relatively complex and too costly for many applications.

More recently, efforts have led to the development of ceramic sensors without heaters that are capable of stable operation near saturation. One means to achieve this end has been to introduce a second phase which forms a thin coating on the grains, effectively isolating the active sensor material from condensate at high humidity and from contaminants. This procedure has been at least partially successful for the $ZnCr_2O_4$-$LiZnVO_4$ sensor (Ref 15).

Therefore, if exposure to significant chemical concentrations is likely in a particular application, life testing under reasonable conditions of exposure would be prudent. The ceramic sensor (without heater) might not always be the best choice where significant contaminant levels are present. In reasonably clean environments, particularly where low cost is important, this type of device could be perfectly suitable.

REFERENCES

1. E.D. Macklen, Thermistors, Electrochemical Publications Ltd., Ayr, Scotland, 1979
2. D.C. Hill and H.L. Tuller, Ceramic Sensors: Theory and Practice, *Ceramic Materials for Electronics*, R.C. Buchanan, Ed., Marcel Dekker, 1986
3. J.M. Herbert, *Ferroelectric Transducers and Sensors*, Gordon and Breach Science Publishers, 1982
4. B.M. Kulwicki, Instabilities in PTC Resistors, *Proceedings of the 6th International Symposium on Applied Ferroelectrics*, Lehigh University, Bethlehem, PA, 8–11 June 1986, p 656–664
5. B.M. Kulwicki, Trends in PTC Resistor Technology, *Electronic Materials and Processes*, N.H. Kordsmeier, C.A. Harper, and S.M. Lee, Ed., SAMPE, 1987, p 441–451; also *SAMPE J.*, Vol 23, 1987, p 34–38
6. S.R. Morrison, Semiconductor Gas Sensors, *Sens. Actuators*, Vol 2, 1982, p 329–341
7. J. Watson, The Tin Oxide Gas Sensor and Its Applications, *Sens. Actuators*, Vol 5, 1984, p 29–42
8. D.E. Williams and P. McGeehin, Solid-State Gas Sensors and Monitors, *Electrochemistry*, Vol 9, 1984, p 246–290
9. W.J. Fleming, "A Physical Understanding of Solid State Humidity Sensors," *Proceedings of the International Automotive Meeting*, Detroit, Paper No. 810432, Society of Automotive Engineers, 1981, p 51–62
10. T. Seiyama, N. Yamazoe, and H. Arai, Ceramic Humidity Sensors, *Sens. Actuators*, Vol 4, 1983, p 85–96
11. T. Nitta, Development and Application of Ceramic Humidity Sensors, *Chem. Sens. Technol.*, Vol 1, 1988, p 57–78
12. H.L. Tuller, Review of Electrical Properties of Metal Oxides as Applied to Temperature and Chemical Sensing, *Sens. Actuators*, Vol 4, 1983, p 679–688
13. B.M. Kulwicki, Ceramic Sensors and Transducers, *J. Phys. Chem. Sol.*, Vol 45, 1984, p 1015–1031
14. T. Nitta, Z. Terada, and S. Hayakawa, Humidity-Sensitive Electrical Conduction of $MgCr_2O_4$-TiO_2 Porous Ceramics, *J. Am. Ceram. Soc.*, Vol 63, 1980, p 386–391
15. Y. Yokomizo, S. Uno, M. Harata, and H. Hiraki, Microstructure and Humidity-Sensitive Properties of $ZnCr_2O_4$-$LiZnVO_4$ Ceramic Sensors, *Sens. Actuators*, Vol 4, 1983, p 599–606

Varistors

Tapan K. Gupta, Alcoa Electronic Packaging, Inc., Alcoa Technical Center

VOLTAGE-DEPENDENT RESISTORS OR VARISTORS have been in use, in one form or another, for many years to regulate transient voltage surges of unwanted magnitude. The oldest of these devices are based on selenium rectifiers, used extensively in older telephone systems and other surge-suppressing applications. With the advent of new technology, they have been largely replaced by the single-crystal silicon devices (avalanche or Zener diode) for low-voltage (signal) applications and by polycrystalline ceramic nonlinear devices of silicon carbide (SiC) and zinc oxide (ZnO) for high-voltage and power-related applications. Lately, as a result of vast improvement in performance, ZnO devices are systematically replacing the SiC devices in most areas of power applications. Also, as a result of size reduction, their applications in low-voltage automotive, telephone, and microelectronic industries are continuously increasing. Because of their increasing importance, this article will discuss ZnO varistors only.

Zinc oxide varistors (Ref 1, 2) are electronic ceramic devices whose primary function is to sense and limit (clamp) transient voltage surges and to do so repeatedly without being destroyed. Their current-voltage (I-V) characteristic is nonlinear, similar to that of a Zener diode. But unlike a diode, varistors can limit overvoltage equally in both polarities, thus giving rise to an I-V characteristic which is analogous to two back-to-back diodes. Varistors can be used in alternating current (ac) or direct current (dc) fields over a wide range of voltages, from a few volts to tens of kilovolts, and a wide range of currents, from microamperes to kiloamperes. Varistors have the additional property of high energy absorption capability, ranging from a few joules to thousands of joules. Their versatility has made varistors useful in both the power industry as well as in the semiconductor industry. Advantages of ZnO varistors include:

- High nonlinear coefficient
- Sharp breakdown voltage with no hysteresis
- High energy absorption capability
- Low power loss
- Fast response to sharp wave front
- No power follow current
- Gapless design
- Small size and optional geometries
- Long life under hostile environment
- Competitive material cost

Table 1 lists the electrical properties of ZnO varistors.

Electrical Characteristics of ZnO Varistors

The most important property of the ZnO varistor is its nonlinear I-V characteristic, as illustrated in the log-log plot of current and voltage in Fig 1. The curve has three characteristic regions. In the pre-breakdown region, the current increases linearly with voltage up to a current density of $\approx 10^{-4}$ A/cm^2 (6.5×10^{-4} A/in.^{2}). In this region, the device acts as an ohmic resistor. Then, as the electrical field strength is incrementally raised, the device acts increasingly as a non-ohmic resistor and conducts a large quantity of current for a negligible increase in voltage up to a current density of ≈ 100 A/cm^2 (645 A/in.^{2}). This is known as the nonlinear region and is the very heart of the ZnO device. In this region, the I-V characteristic can no longer be described by Ohm's law and is generally expressed by the relation: $I = (V/C)^{\alpha}$, where C is a constant and α is the nonlinear coefficient. This nonlinear region may extend over 6 to 7 orders of magnitude of current. It is this large nonlinearity over a wide range of current densities that makes the ZnO varistor distinctly different from any other nonlinear resistors discovered to date (compare the schematically drawn I-V curve for SiC) and thus makes it useful for a variety of applications. The degree of nonlinearity is determined by the flatness of the nonlinear region; the flatter the I-V curve in this region, the better the device. Finally there is an upturn region in the I-V curve at a current density of ≈ 1000 A/cm^2 (6450 A/in.^{2}). Similar to the trend in the low-current region, the current again increases linearly with voltage in this high-current region. Figure 1 also shows the ac and dc characteristics in the pre-breakdown region, with dc current considerably smaller than the ac current. In the breakdown region, they are identical.

The effect of temperature on the I-V curve is pronounced only in the pre-breakdown region. As the temperature is increased, the I-V curve in this region is shifted to the right (Fig 2). This implies that for a given applied voltage, the leakage current is increased with the ambient temperature. Since the leakage current induces a power loss in the device, the control of device temperature is, therefore, very important for the reliability of the device.

To characterize a ZnO device, it is desirable to determine the I-V curve for all three regions of the varistor. However, due to the wide range of currents involved, it is not possible to use the same measurement techniques for all regions. Usually, I-V characteristics below 100 mA/cm^2 (645 mA/in.^{2}) are measured by dc or 60 Hz ac, and those above 1 A/cm^2 (6.45 A/in.^{2}) are measured by impulse current with a typical waveshape of 8 μs rise time to peak value and 20 μs decay time to half the peak value (known as 8×20 μs waveshape). With shorter dwell time at high currents, the devices are restricted from heating during tests.

Microstructures of ZnO Varistors

The ZnO varistor is a multiphase polycrystalline ceramic (Ref 3–8) made by conventional sintering of a mixture of oxides such as those of bismuth, antimony, cobalt, manganese, and chromium, the major component being the ZnO powder. Upon sintering, four basic compounds are formed in the ZnO varistor: ZnO, spinel, pyrochlore, and several bismuth-rich phases. The typical grain size in a commercial varistor ranges from 15 to 20 μm (600 to 800 μin.) and the grains are always accompanied by twins. The spinel and pyrochlore phases retard grain growth, while compounds like BaO and TiO$_2$ accelerate grain growth.

Microstructural-Electrical Model. From the accumulated knowledge of microstructure and chemistry, combined with the electrical characteristics of the varistor, a microstructural-electrical model emerges for the ZnO varistor (Ref 8). It is now recognized that the basic building block of the ZnO varistor is

Table 1 Electrical properties of ZnO varistors

Property	Value
Grain-boundary resistance	$\approx 10^{12}\ \Omega \cdot cm$
Grain resistance	$\approx 1{-}10\ \Omega \cdot cm$
Donor concentration	$\approx 10^{17}/cm^3$
Trap density	$10^{14}/cm^3$
Voltage drop at grain boundary	$\approx 2{-}4\ V$
Nonlinear voltage	$\approx 1{-}10\ kV/cm$
Nonlinear coefficient	$\approx 15{-}100$
Grain-boundary capacitance	$\approx 0.2\ \mu F/cm^2$
Apparent relative dielectric constant	$\approx 10^3{-}10^4$
Depletion layer thickness	$\approx 50{-}100\ nm$ ($\approx 500{-}1000$ Å)
Energy absorption capability	$\approx 100{-}300\ J/cm^3$ ($\approx 0.5{-}1.5\ kW \cdot h/in.^3$)
Response time	$\approx 5{-}10\ ns$
Power loss	$\approx 10{-}100\ mW/cm^3$
Life	$>30\ years$

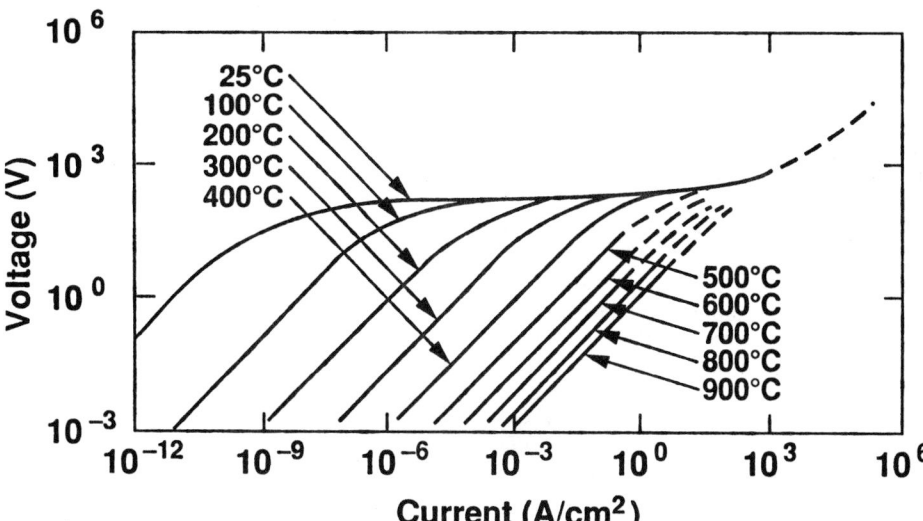

Fig 2 Temperature dependence of the current-voltage characteristic of a ZnO varistor containing five additives. Source: Ref 1

the ZnO grain formed as a result of sintering. During this process, various chemical elements are distributed in such a way in the microstructure that the near-grain-boundary region becomes highly resistive ($\approx 10^{12}\ \Omega \cdot$ cm) and the grain interior becomes highly conductive (≈ 1 to $10\ \Omega \cdot$ cm). These resistivities can be readily estimated from the slopes of the I-V curve shown in Fig 1. Figure 3(a) shows a schematic diagram of the apparent grain and grain-boundary resistivities. A sharp drop in resistivity from grain boundary to grain (Fig 3b) occurs within a distance of ≈ 50 to 100 nm (≈ 500 to 1000 Å). This area is known as the depletion layer. Thus, at each grain boundary, there exists a depletion layer on both sides of the grain boundary extending into the adjacent grains. The varistor action arises as a result of the presence of this depletion layer within the grains, and not as a result

of the existence of an intergranular insulating layer that has been observed in the ZnO microstructure. This depletion layer provides a barrier to the transport of electrons across the grain boundary. The theoretical and experimental studies of this barrier conform to a classical Schottky barrier model with positive changes residing in the depletion layer and the compensating negative charges at the grain-boundary interface as traps (Ref 9–20). Furthermore, these studies suggest that the conduction through the barrier takes place either by thermoionic emission or by tunneling.

The presence of two depletion layers on both sides of the grain boundary makes the ZnO varistor insensitive to polarity changes. In this respect, the varistor appears as a back-to-back diode. Furthermore, since the region

near the grain boundary is depleted of electrons, a voltage drop appears across the grain boundary upon application of an external voltage. This is known as the barrier voltage (V_{gb}) and typically is of the order of ≈ 2 to 4 V/grain boundary. This voltage is comprised of a resistive (R) and a capacitive (C) component, as shown schematically in the equivalent electrical circuit of Fig 3(c). Thus, when a voltage stress is applied to a ZnO varistor in the pre-breakdown region, the leakage cur-

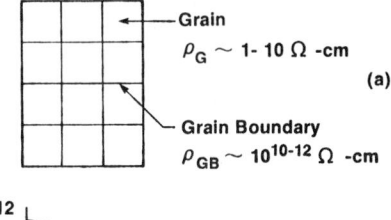

(a)

(b)

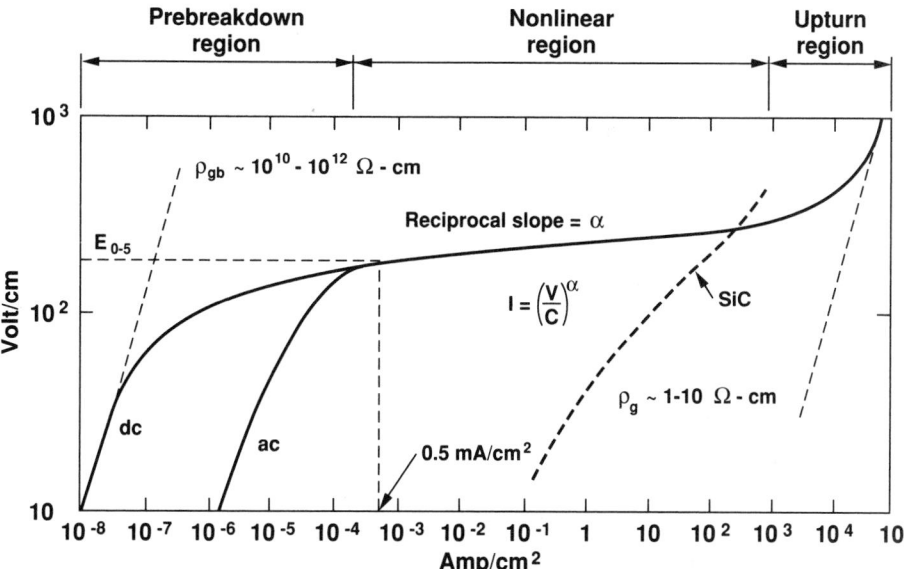

Fig 1 Typical current-voltage (I-V) characteristic of a ZnO varistor. The I-V characteristic of SiC is shown for comparison. Source: Ref 2

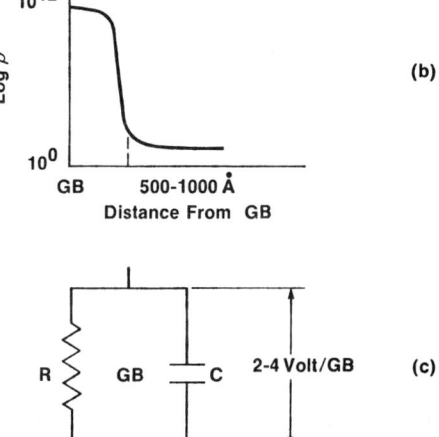

(c)

Fig 3 Schematic of microstructure and electrical characteristics. (a) Grain versus grain-boundary resistivity. (b) Resistivity profile at the depletion layer. (c) Equivalent electrical circuit at the grain boundary. Source: Ref 2

rent that flows through the device is entirely of grain-boundary origin, and, in the ac mode, this current consists of a resistive component and a capacitive component. The resistive component is typically lower than the capacitive component by a factor of about 5. In the breakdown region, the current is mostly resistive.

Referring to Fig 1, the following correlations between microstructure and electrical characteristics (Ref 8) are cited: the low-current pre-breakdown linear region has been identified to be controlled by the grain-boundary resistance and capacitance, and, on the other end of the I-V curve, the high-current linear region has been identified to be controlled by the impedance of the grain. The intermediate nonlinear region—the region of major importance for a variety of applications—is indirectly controlled by the resistivity differential between the grain boundary and the grain. This connection of the I-V curve with the microstructure of the ZnO varistor provides the materials scientists with an important tool to adjust the electrical property of the grain boundary as well as that of the grain to fit the requirements of a given application.

Application of ZnO Varistors: Critical Parameters

Varistor applications (Ref 2) take advantage of all of the regions of the I-V curve shown in Fig 1. The low-current linear region determines the watt loss and, hence, the operating voltage during the steady application of an external voltage. The nonlinear region determines the clamping voltage upon application of a transient surge. The high-current region presents the limiting condition for protection from high-current surges such as those found in lightning. Devices where the upturn occurs at an increasingly higher current density are, therefore, most desirable for applications involving high magnitudes of currents, since the voltage rise can be minimized with such devices.

There are several critical application parameters (Ref 2) associated with the various regions of the I-V curve and which serve various functions in the design and operation of a surge protector. These parameters are summarized in Table 2 with their functions. The most desirable device should have a high value of nonlinear coefficient, an acceptable rating of nonlinear voltage, a low value of leakage current, a long varistor life, and a high energy absorption capability. Most often, however, a tradeoff is required wherein various parameters are optimized for a given application.

The two most important application parameters are the nonlinear coefficient α and the energy density parameter E, measured in joules/cm³. As stated earlier, the higher the

Table 2 Parameters for device application

Parameter	Function	Equation(a)
Nonlinear coefficient	Protective level	$I = \left(\dfrac{V}{C}\right)^{\alpha}$
Nonlinear voltage	Voltage rating	$C = \left(\dfrac{V}{I^{1/\alpha}}\right) V$ at 1 mA
Leakage current	Watt loss/ operating voltage	$I_R = \dfrac{V_{ss}}{R_{gb}}$
Life	Stability	$P_G < P_D$
Energy absorption	Survival	$J = IVt$

(a) I is current, V is voltage, C is a constant, I_R is resistive current, V_{ss} is steady-state voltage, R_{gb} is grain-boundary resistance, P_G is power generated, P_D is power dissipated, J is joules, and t is time.

value of α, the better the device. In this respect, the ZnO varistor outranks other varistors in α value. While the reported α values for a ZnO varistor have been claimed from 25 to 100, that for a pure resistor is ≈ 1, for SiC ≈ 5, for selenium ≈ 8, and for a silicon diode ≈ 35.

Since the nonlinear coefficient α is defined by $\alpha = d \ln I / d \ln V$, it is clear from Fig 1 that the nonlinear coefficient α varies with current density. The α value increases in the pre-breakdown region, attains a maximum value in the breakdown region, then falls off sharply in the upturn region (Ref 21). For practical applications, α is typically calculated over a region of interest of currents and voltages such that $\alpha = [\log(I_2/I_1)/\log(V_2/V_1)]$ where I_2 and I_1 are the currents at V_2 and V_1, with $V_2 > V_1$. The voltage at which the onset of nonlinearity occurs has been conveniently defined as the voltage at 0.5 mA/cm² (3.2 mA/in.²), and is designated as $E_{0.5}$ in Fig 1. This nonlinear voltage is a measure of the voltage rating of the device at which the clamping action occurs. During device operation, the steady-state voltage is set at a certain percentage of the nonlinear voltage. For power application, this is typically 70 to 80% of $E_{0.5}$.

Next to nonlinearity, the energy absorption capability, E, measured in J/cm³, is the second most important property of a ZnO varistor and can be defined in terms of materials and electrical parameters as follows (Ref 2):

$$E = \rho C_p \Delta T = \rho C_p (T_2 - T_1) \qquad \text{(Eq 1)}$$

where ρ is density (g/cm³), C_p is the specific heat (in J/g · °C), and ΔT (or $T_2 - T_1$, where $T_2 > T_1$) is the temperature rise due to energy absorption. I is given in A/cm², V in V/cm, and t is the time in seconds. It is obvious from the above definition that the device size can be greatly reduced as the energy density of the varistor is increased. The energy levels of the present day varistor may be stated as between 200 and 250 J/cm³ (9×10^{-4} to 1.1×10^{-3} kW · h/in.³). This energy level will increase the varistor temperature < 100 °C (<180 °F) depending on the density of the

varistor. In recent years, varistors with energy levels of 1000 J/cm³ (5×10^{-3} kW · h/in.³) have been reported. The expected temperature rise in the varistor will be in the range of ≈ 225 to 275 °C (405 to 495 °F).

The importance of the energy absorption can be understood from the fact that a varistor can often be subjected to various switching surges which differ in peak value and duration in the system. The distinction between these surges is one of time variation, which ranges from microseconds to milliseconds. Against this background, the varistor must absorb the transient surges of varying times without causing excessive rise in temperature so that thermal runaway and the destruction of the device can be prevented (Ref 22).

Reliability of ZnO Varistors

Zinc oxide varistors are known to exhibit a degradation (Ref 23) in the current-voltage (I-V) characteristic when subjected to a continuously applied electric field. However, the varistor application requires that the I-V characteristic of a reliable varistor must remain stable to its original value during usage and after experiencing the electrical stresses as a result of absorbing the oncoming surges. Varistor life depends on the prevention of varistor degradation (Ref 24). One measure of a reliable varistor is that the value of $E_{0.5}$ (Fig 1) or the resistive component of leakage current (I_R) at a given steady-state voltage must remain constant with time. If, upon application of a steady voltage stress, the value of $E_{0.5}$ decreases or I_R increases as shown in Fig 4, the varistors are considered unstable and will degrade in properties.

The degradation phenomenon (Ref 25–31) has been studied under ac, dc, and pulsed electric fields, and phenomenological rate equations have been derived from these studies. Several mechanisms have been proposed to explain the observed degradation. The mechanisms suggested to account for this degradation are electron trapping, dipole ori-

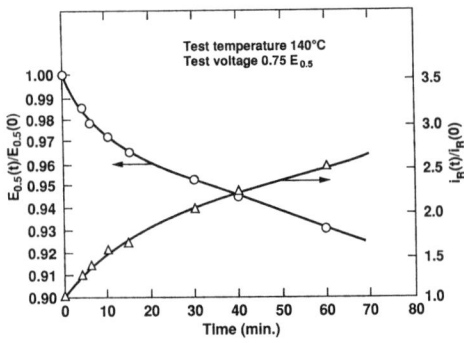

Fig 4 Decrease in $E_{0.5}$ and corresponding increase in I_R with time when the varistor is energized at temperature. The data are normalized with regard to values at time zero. Source: Ref 2

entation, ion migration, and oxygen desorption. Among these, ion migration has found strong support on the basis of experimental evidence. The experimental results to date indicate that:

- Degradation is a grain-boundary phenomenon
- Degradation is the result of ion migration
- Migrating ion is predominantly interstitial zinc
- Interstitial zinc is a result of the nonstoichiometric nature of zinc oxide

The model that agrees with the experimental data on degradation assumes that the Schottky barrier has a metastable component formed as a result of "frozen-in" zinc interstitials that are "trapped" in the depletion layer during cooling from varistor fabrication temperature. When an external stress is applied, the degradation arises as a result of field-assisted diffusion of zinc interstitial in the depletion layer followed by chemical interaction with the grain-boundary defects—a process that leads to a decreased barrier height and an increased leakage current. Based on this concept, a grain-boundary defect model (Ref 32) for the varistor has been developed which satisfactorily describes not only the degradation phenomenon in ZnO varistors but also the means to prevent degradation. The key to preventing varistor degradation is to eliminate the metastable barrier once it is formed, or to minimize the formation of the metastable barrier during fabrication of the varistor. The metastable component, once formed, can be removed by thermal anneal or prevented from forming by chemical means.

Annealing. It is well known in the manufacture of ZnO varistors that in order to make a stable varistor the device must be subjected to a post-sintering heat treatment (annealing) in an oxidizing atmosphere. Recent studies (Ref 33) have also confirmed that the optimum temperature of annealing is between 600 and 700 °C (1110 and 1290 °F), at which

temperature the varistor becomes highly stable. It has been shown that during annealing the zinc interstitial in the depletion layer is permanently diffused to the grain boundary and reacts with oxygen to form ZnO lattice at the grain-boundary interface. Additionally, it has been shown that a phase transformation occurs in the intergranular Bi_2O_3 phase (Ref 34). The effect of heat treatment on I_R as a function of time is shown in Fig 5. The 600 °C (1110 °F) annealed samples show no rise in I_R with time compared with other samples. A similar response of I_R with time is also observed when the varistor is chemically doped with a small amount of sodium or potassium during sintering (Ref 35). Here, the varistors are prevented from forming a metastable barrier.

Fabrication of ZnO Varistors

ZnO varistors are formed by mixing the semiconducting ZnO powder of other oxides such as those of bismuth, antimony, cobalt, manganese, and chromium. The ZnO is the major constituent, and the amount of ZnO can range from 85 to 98%. The exact composition is largely proprietary. The powders are mixed, pressed into desired shape and sintered in air for 2 to 10 h at temperatures ranging from 1100 to 1350 °C (2010 to 2460 °F). In this respect, the processing of a ZnO varistor is no different from that of any other multicomponent ceramic.

Effect of Sintering on Microstructure. During sintering, densification, grain growth, and chemical reactions occur that result in the

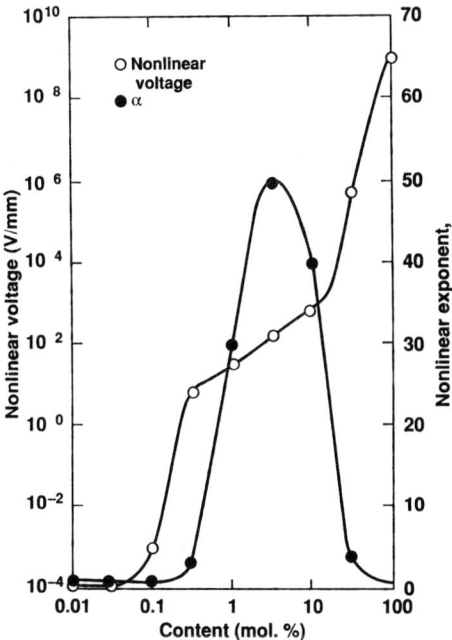

Fig 6 Effect of content of five additives on the nonlinear coefficient and nonlinear voltage of a ZnO varistor. Source: Ref 1

unique microstructure of the ZnO varistor. The varistor properties are developed during cooling; the grain boundaries become highly resistive, while the grain interior remains relatively conductive (see Fig 1). Various phases, as mentioned in the section on microstructure, are also formed during heating and cooling cycles. Inada (Ref 6) has extensively studied the formation of chemical phases in the varistor system, and has shown that the

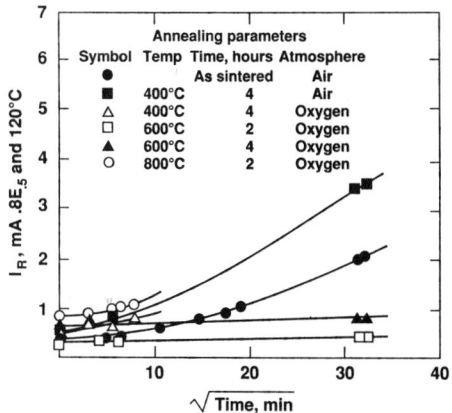

Fig 5 Effect of annealing on varistor stability as shown by resistive current I_R versus square root of time t. Note that I_R curve at 600 °C (1110 °F) is flat with time. Source: Ref 2

Table 3 Electrical properties of ZnO varistors as a function of additives.
Concentration of each additive was 0.5 mol%

Additive(s)	Sintering temperature		Nonlinear voltage		Nonlinear coefficient, α	Average grain size	
	°C	°F	V/mm	V/in.		μm	μin.
Bi_2O_3	1150	2100	10	250	4.0	20	800
Sb_2O_3	1150	2100	65	1650	3.1	3	120
Bi_2O_3 Sb_2O_3	1250	2280	30	750	13.0	25	1000
Bi_2O_3 MnO	1350	2460	50	1250	18.0	30	1200
Bi_2O_3 CoO MnO	1350	30 2460	30	750	22.0	30	1200
Bi_2O_3 CoO MnO Cr_2O_3	1250	2280	48	1220	21.0	20	800
Bi_2O_3 CoO MnO Cr_2O_3 Sb_2O_3 (a)	1350	2460	135	3430	50.0	10	400

(a) Amount of additive, 1.0 mol% Source: Ref 1

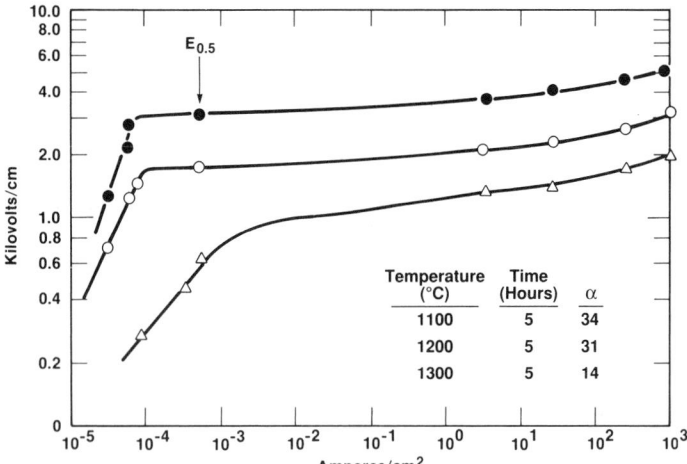

Fig 7 Effect of sintering temperature on the *I-V* characteristic of a ZnO varistor. Note the change in α value with temperature at the constant sintering time of 5 h. Source: Ref 2

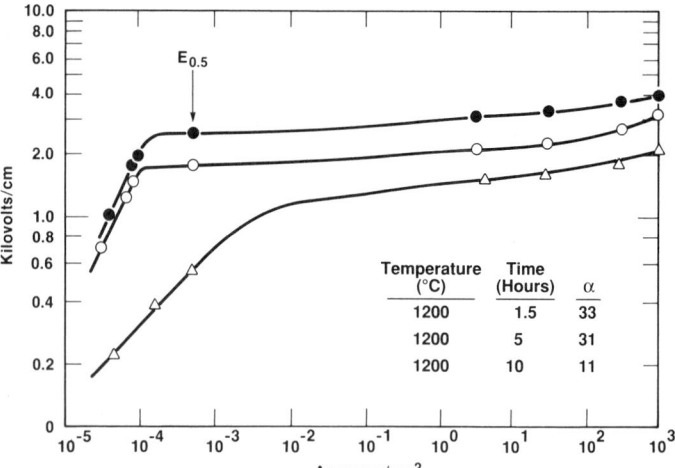

Fig 8 Effect of sintering time on the *I-V* characteristic of a ZnO varistor. Note the change in α value with time at the constant sintering temperature of 1200 °C (2190 °F). Source: Ref 2

formation of ZnO ceramics is affected both during heating and cooling. A complex series of reactions occur between several phases such as pyrochlore, $Zn_2Bi_3Sb_3O_{14}$, spinel $Zn_7Sb_2O_{12}$, and liquid Bi_2O_3. Other studies indicate that densification is aided by a liquid phase formed between ZnO and Bi_2O_3 and retarded by the formation of pyrochlore phase. The formation temperature of liquid phase depends on antimony/bismuth (Sb/Bi) ratio: about 750 °C (1380 °F) for Sb/Bi < 1, and about 1000 °C (1830 °F) for Sb/Bi > 1.

Additives. A number of studies (Ref 1, 6) exist to determine the roles of various additives. From these studies, it is possible to divide these additive oxides into three general categories. The first is the varistor former. These consist of oxides that must be added to ZnO to form a varistor and without which the varistor action cannot be obtained. The oxides in this group are those of bismuth, antimony, barium, praseodymium, and so forth. The second group of oxides can be labeled as performance enhancers. Their function is to improve the characteristics of the *I-V* curve, such as improving the nonlinearity of the curve or enhancing the energy absorption capability. The oxides in this group are those of cobalt, manganese, nickel, chromium, and so forth. A third group can be labeled as performance highlighters. Their function is to heighten the performance of a specific property, such as the stability of the varistor. The oxides in this group are those of sodium, potassium, aluminum, gallium, and so forth. The amount of these oxides needed to form a commercial varistor also follows a general pattern. The varistor formers are generally present in the amount of several volume percent, whereas performance highlighters are required only at parts per million (ppm) levels. The amount of performance enhancers is intermediate between the two extremes.

Table 3 shows the effects of various additives on varistor electrical properties. Table 3 indicates that the more additives there are in the composition, the better the nonlinear properties and the higher the voltages. Depending on the properties desired, the commercial varistor may contain as many as ten oxides.

The effects of total amount of additive (Ref 1) on nonlinear coefficient and nonlinear voltage are shown in Fig 6. The α goes through a maximum, whereas the voltage keeps increasing with the additive content reflecting the inhibiting effect of additive on grain growth (see Table 3). The effects of sintering temperature and time on *I-V* curves (Ref 2) are shown in Fig 7 and 8. The increasing temperature and time not only cause a lowering of $E_{0.5}$ (as expected from an increase in grain size) but also a change in the shape of the *I-V* curves, which are shifted to the right in the pre-breakdown region. This has the effect of increasing the leakage current at a given voltage, which can be detrimental to device stability. The nonlinear region is also changed considerably by a change in temperature and time. Both higher temperature and longer time cause a reduction in the value of α from 33 or 34 to 11 to 14. The α values are calculated for the current ranges of 0.5 to 250 A/cm² (3.2 × 10⁻⁶ to 1.61 kA/in.²).

The energy absorption capability of the varistor may also vary depending on the fabrication procedure, and may not increase with the increase in nonlinearity. As a result, when an application requires a high value of α combined with a high value of energy absorption, there may indeed be a conflict in the processing schedule. It is, therefore, imperative that the end-use of a varistor be incorporated into the processing technology to achieve the desired *I-V* characteristic, combined with energy absorption capability and stability of the device.

REFERENCES

1. M. Matsuoka, Non-ohmic Properties of Zinc Oxide Ceramics, *Jpn. J. Appl. Phys.*, Vol 10, 1971, p 736–746
2. T.K. Gupta, Application of Zinc Oxide Varistors, *J. Am. Ceram. Soc.*, Vol 73 (No. 7), 1990, p 1817–1840
3. D.R. Clarke, The Microstructural Location of the Intergranular Metal Oxide Phase in a Zinc Oxide Varistor, *J. Appl. Phys.*, Vol 49, 1978, p 2407
4. A.T. Santhanam, T.K. Gupta, and W.G. Carlson, Microstructural Evaluation of Multicomponent ZnO Ceramics, *J. Appl. Phys.*, Vol 50 (No. 2), 1979, p 852–859
5. J. Wong, Nature of Intergranular Phase in Non-ohmic ZnO Ceramics Containing 0.5 mol% Bi_2O_3, *J. Am. Ceram. Soc.*, Vol 57, 1974, p 357–359
6. M. Inada, Crystal Phases of Non-ohmic Zinc Oxide Ceramics, *Jpn. J. Appl. Phys.*, Vol 17 (No. 1), 1978, p 1–10
7. E. Olsson, I.K.L. Falk, G.E. Dunlap, and R. Österlund, The Microstructure of a ZnO Varistor Material, *J. Mater. Sci.*, Vol 20, 1985, p 4091–4098
8. T.K. Gupta, Influence of Microstructure and Chemistry on the Electrical Characteristic of ZnO Varistors, *Tailoring Multiphase and Composite Ceramics*, R.E. Tressler, G.L. Messing, C.G. Pantino, and R.E. Newnham, Ed., Plenum, 1986, p 493–507
9. W.G. Morris, Physical Properties of Electrical Barriers in Varistors, *J. Vac. Sci. Technol.*, Vol 13 (No. 4), 1976, p 926–931
10. J.O. Levine, Theory of Varistor Electrical Properties, *CRC Crit. Rev. Solid State Sci.*, Vol 5, 1975, p 597–608
11. L.M. Levinson and H.R. Philipp, The Physics of Metal Oxide Varistors, *J. Appl.*

Phys., Vol 46, 1975, p 1332–1341

12. P.R. Emtage, The Physics of Zinc Oxide Varistors, *J. Appl. Phys.*, Vol 48, 1977, p 4372–4384

13. K. Eda, Conduction Mechanism of Non-ohmic Zinc Oxide Ceramics, *J. Appl. Phys.*, Vol 49, 1978, p 2964–2972

14. P.L. Hower and T.K. Gupta, A Barrier Model for ZnO Varistors, *J. Appl. Phys.*, Vol 50, 1979, p 4847–4855

15. G.D. Mahan, L.M. Levinson, and H.R. Philipp, Theory of Conduction in ZnO Varistors, *J. Appl. Phys.*, Vol 50, 1979, p 2799–2812

16. R. Einzinger, Metal Oxide Varistor Action—A Homojunction Breakdown Mechanism, *Appl. Surf. Sci.*, Vol 1, 1978, p 329–341

17. J. Bernsconi, S. Strässler, B. Knecht, H.P. Klein, and A. Menth, Zinc Oxide-Based Varistors: A Possible Mechanism, *Solid State Commun.*, Vol 21, 1977, p 867–870

18. G.E. Pike, Electronic Properties of ZnO Varistors: A New Model, *Mater. Res. Soc.*, Vol 15, 1982, p 369–379

19. G. Blatter and F. Greuter, Carrier Transport through Grain Boundaries in Semiconductors, *Phys. Rev. B*, Vol 33, 1986, p 3952–3966

20. Y. Suzuoki, A. Ohki, T. Mizutani, and T. Ieda, Electrical Properties of ZnO-Bi_2O_3 Thin Film Varistors, *J. Phys. D.*, Vol 20, 1987, p 511–517

21. H.R. Philipp and L.M. Levinson, High-Temperature Behavior of ZnO-Based Ceramic Varistors, *J. Appl. Phys.*, Vol 50 (No. 1), 1979, p 383–389

22. K. Eda, Destruction Mechanism of ZnO Varistors Due to High Currents, *J. Appl. Phys.*, Vol 56, 1984, p 2948–2955

23. T.K. Gupta and W.G. Carlson, Barrier Voltage and Its Effect on Stability of ZnO Varistor, *J. Appl. Phys.*, Vol 53 (No. 11), 1982, p 7401–7409

24. T.K. Gupta, Effect of Material and Design Parameters on the Life and Operating Voltage of a ZnO Varistor, *J. Mater. Res.*, Vol 2 (No. 2), 1987, p 231–238

25. K. Eda, A. Iga, and M. Matsuoka, Degradation Mechanism of Non-ohmic Zinc Oxide Ceramics, *J. Appl. Phys.*, Vol 51 (No. 5), 1980, p 2678–2684

26. C.G. Shirley and W.M. Paulson, The Pulse-Degradation Characteristic of ZnO Varistors, *J. Appl. Phys.*, Vol 50 (No. 9), 1979, p 5782–5789

27. S. Tominaga, Y. Shibuya, Y. Fujiwara, M. Imataki, and T. Nitta, "Stability and Long-Term Degradation of Metal Oxide Surge Arrester," presented at 1979 IEEE Summer Power Meeting, Vancouver, BC, Canada

28. W. Moldenhauer, K.H. Bäther, W. Brückner, D. Hizn, and D. Bühling, Degradation Phenomena in ZnO Varistors, *Phys. Status Solidi A*, Vol 67 (No. 2), 1981, p 533–542

29. T.K. Gupta, W.G. Carlson, and P.L. Hower, Current Instability Phenomena in ZnO Varistor Under a Continuous AC Stress, *J. Appl. Phys.*, Vol 52 (No. 6), 1981, p 4104–4111

30. K. Takahashi, T. Miyoshi, K. Keada, and T. Yamazaki, Degradation of Zinc Oxide Varistors, presented at the Annual Meeting of the Materials Research Society, Boston, 16–19 Nov 1981

31. K. Sato and Y. Takada, A Mechanism of Degradation in Leakage Currents through ZnO Varistors, *J. Appl. Phys.*, Vol 53, 1982, p 8819

32. T.K. Gupta and W.G. Carlson, A Grain-Boundary Defect Model for Instability/Stability of ZnO Varistor, *J. Mater. Sci.*, Vol 20, 1985, p 3487

33. T.K. Gupta and W.D. Straub, Effect of Annealing on the ac Leakage Current of the ZnO Varistor. I, Resistive Current, *J. Appl. Phys.*, Vol 68 (No. 2), 1990, p 845–850; II, Capacitive Current, p 851–855

34. A. Iga, M. Matsuoka, and T. Masuyama, Effect of Phase Transformation of Intergranular Bi_2O_3 Layer in Non-ohmic ZnO Ceramics, *Jpn. J. Appl. Phys.*, Vol 15 (No. 6), 1976, p 1161–1162

35. T.K. Gupta and A.C. Miller, Improved Stability of ZnO Varistor via Donor and Acceptor Doping at the Grain Boundary, *J. Mater. Res.*, Vol 3 (No. 4), 1988, p 745–754

High-Temperature Superconductors

R.B. Poeppel, M.T. Lanagan, U. Balachandran, S.E. Dorris, J.P. Singh, and K.C. Goretta,
Materials and Components Technology Division, Argonne National Laboratory

APPLICATIONS OF HIGH-TEMPERATURE SUPERCONDUCTORS depend on the ability to fabricate these materials into useful shapes. In addition, suitable electrical and mechanical properties must be maintained in wires and tapes over long lengths. High critical current densities found in single crystals and epitaxial thin films have approached 10^7 A/cm^2; however, critical current is limited at the grain boundaries of polycrystalline materials. Ceramic processing is focused on the enhancement of critical current density by alignment of anisotropic grains and control of reaction kinetics for powder synthesis and sintering. Composite wires and tapes are necessary for many applications and are fabricated by co-processing superconductors with insulators and normal conductors.

This article briefly reviews both the processing and applications of ceramic superconducting materials. More detailed information can be found in the references that accompany this article as well as in Volume 2, 10th Edition, of *Metals Handbook* (see the Section entitled "Superconducting Materials" on pages 1026–1089).

Critical Parameters of Superconductivity

Three important parameters that describe the utility of a bulk superconductor are critical current density (J_c), critical magnetic field (H_c), and the transition temperature (T_c) (Ref 1). These interdependent variables describe the nature of the normal conducting/superconducting transition of a material. Before the development of high-temperature superconductors, materials with T_c values below 25 K were observed (Ref 2). The superconducting phase $YBa_2Cu_3O_{7-x}$ (YBCO) was the first superconducting material with a T_c greater than the boiling point of liquid nitrogen (Ref 3). Since then, other compounds in the Bi-Sr-Ca-Cu-O (BSCCO) and Tl-Ca-Ba-Cu-O (TCBCO) systems have been found to have T_c values exceeding 120 K (Ref 4, 5). These breakthroughs offered the possibility of cooling with liquid nitrogen instead of liquid helium. Advantages for potential applications include lower cooling costs and more efficient devices.

The critical current density, J_c, is defined as the amount of supercurrent that can be passed through a defined area of conductor. The J_c values for polycrystalline specimens are generally below 5×10^3 A/cm^2 and are dependent on the magnetic field (Fig 1). Many applications require J_c values of 10^4 A/cm^2 in a field of 1 T (1 T = 10^4 G). High critical current densities found in single crystals and epitaxial thin films have approached 10^7 A/cm^2. The contrast in transport properties between single crystals and polycrystalline ceramic samples has been attributed to weak-link behavior, which can be described as a normal conducting boundary between adjacent superconducting regions (Ref 6).

In the case of polycrystalline material, the superconducting grains are bounded by nonsuperconducting grain boundaries. Chemical inhomogeneity, microcracking, and incoherent grain boundaries are potential barriers to current flow. Grain-boundary composition has been studied by energy-dispersive x-ray analysis with a scanning transmission electron microscope (Ref 7). In many sections of the grain boundary, areas of copper excess and oxygen depletion relative to the grain interior were observed. Liquid phases are formed during the sintering process and solidify at the grain boundaries. This problem is common for compositions that have an excess of copper and barium (Ref 8). In addition, J_c degradation has been correlated with grain-boundary orientation (Ref 9).

Powder Synthesis

All fabrication processes originate with a homogeneous powder of suitable particle size. The YBCO system is presented as a specific example, and the Bi- and Tl-based superconductors are processed in a similar manner. YBCO is a line compound with a Y:Ba:Cu cation ratio of 1:2:3; however, superconductivity in BSCCO and TCBCO compounds occurs over a range of cation stoichiometries.

Powder is made by the calcining of oxide and carbonate precursors (Y_2O_3, $BaCO_3$, and CuO). The decomposition of $BaCO_3$ is essential for complete reaction and is influenced by CO_2 partial pressure. The CO_2 released during decomposition can react with YBCO to form $BaCO_3$, Y_2O_3, CuO, and $Y_2Cu_2O_5$, depending on temperature and pressure (Ref 10, 11). The kinetics of CO_2 removal become important for large batch sizes, because of the large volume of CO_2 produced during the reaction. Vacuum calcination has been found to enhance the kinetics of CO_2 removal and has been utilized to form phase-pure powders in yttrium- and bismuth-based systems. Oxygen pressures between 2.6×10^2 and 1.3×10^3 Pa have been used in the calcination; however, lower pressures may cause decomposition of the YBCO compound (Ref 12). An increased reaction rate under vacuum may result from the enhanced reactivity of copper oxide (Ref 13). Vacuum calcination allows for the production of a phase-pure powder and can easily be scaled up to large batch sizes.

Scanning electron microscopy reveals the particles to be platelike with large aspect ratios (Fig 2). The crystallographic c-direction is along the small dimension of the plate. The powder is milled to particle sizes between 1 and 10 μm for subsequent processing. BSCCO and TCBCO compounds also have micaceous morphologies.

Plastic Extrusion

Wires and tubes can be formed by plastic extrusion. The high-T_c powder is first combined with a set of organics and mixed in a blender. A solvent provides the basic vehicle into which the oxide powder and other organics are placed. Care must be taken in selecting a solvent that is compatible with the YBCO powder and the other organic constituents. Typical organic solvents include methyl ethyl ketone, methanol, and xylene. Dispersants are used to deflocculate the inorganic particles in the solvent and to assist in obtaining higher unfired densities. Binders impart strength to the unfired body, and plasticizers promote flexibility.

The extrusion process consists of placing a high pressure (~20 MPa, or 3 ksi) on the plastic mass and forcing it though a small aperture. Wires and tubes with radii between 0.1 and 10 mm (0.004 and 0.04 in.) have been manufactured in lengths of well over 15 m

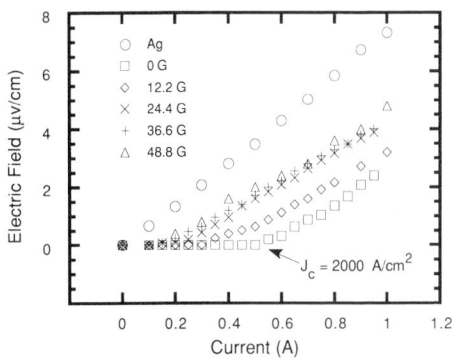

Fig 1 Critical current density of a polycrystalline ceramic specimen compared with a silver substrate as a function of applied magnetic field

Fig 3 Heat-treatment schedule for YBCO coils. Stage I is for removal of organics. Stages II and III are for sintering and oxidation, respectively.

(50 ft). The wire has great flexibility in the unfired state, and coils are made by wrapping the wire around a mandrel.

The extruded wire must be heated for powder consolidation. The heat-treatment schedule for fabricated shapes is divided into three parts (Fig 3). Initially, a slow increase in temperature is required to remove organics. If the organics are removed rapidly, the final product will have large voids and a bloated appearance. Densification occurs at temperatures between 870 and 1000 °C (1600 and 1830 °F) and low-melting liquid phases must be controlled for sintering at high tempera-

tures. Steps I and II are carried out at pressures of 1.3×10^3 Pa. Reduced oxygen pressures have been found to enhance binder removal and densification kinetics (Ref 14). The final step is an annealing procedure to incorporate oxygen into the YBCO lattice to form the superconducting phase. Step III is carried out at oxygen pressures of 1×10^5 Pa. The relationship between oxygen content and phase transition to the superconducting orthorhombic phase has been studied extensively (Ref 15).

Critical current density measurements carried out on the coils revealed values in the range of 120–225 A/cm². While the J_c values were not exceptional, it should be pointed

out that the coils contain long lengths of continuous superconductor (Fig 4). The J_c measurements were made for wire lengths up to 12.2 m (40 ft) for a five-layer coil.

Melt-Solidification Processing

Melt-solidification processing has been used by many researchers to achieve a high degree of crystallographic alignment and high critical current density. Figure 5 is an example of the microstructure that results from a zone-melt process. Crystal growth in the *a-b* plane is faster than that of the *c*-axis, and a lamellar morphology results. Several advantages are apparent from this microstructure. Microcracks due to anisotropic thermal expansion are minimized. The highest J_c values in single crystals are along the *a-b* plane, and these planes are aligned in melt-solidification processing. Initial work focused on melting the entire specimen and subsequent slow cooling through the liquid/solid transition. Critical current densities of 19,000 A/cm² are possible with this technique (Ref 16, 17).

More recently, a process has been developed in which a YBCO sample is heated to high temperature and quenched. The quenched material is subsequently heated to a temperature in which the 211 compound, Y_2BaCuO_5, and liquid are in equilibrium and is then cooled slowly through the liquid/solid transition. The melt-quench-growth method results in samples with excellent magnetic properties and it

Fig 2 Scanning electron micrograph of YBCO particles

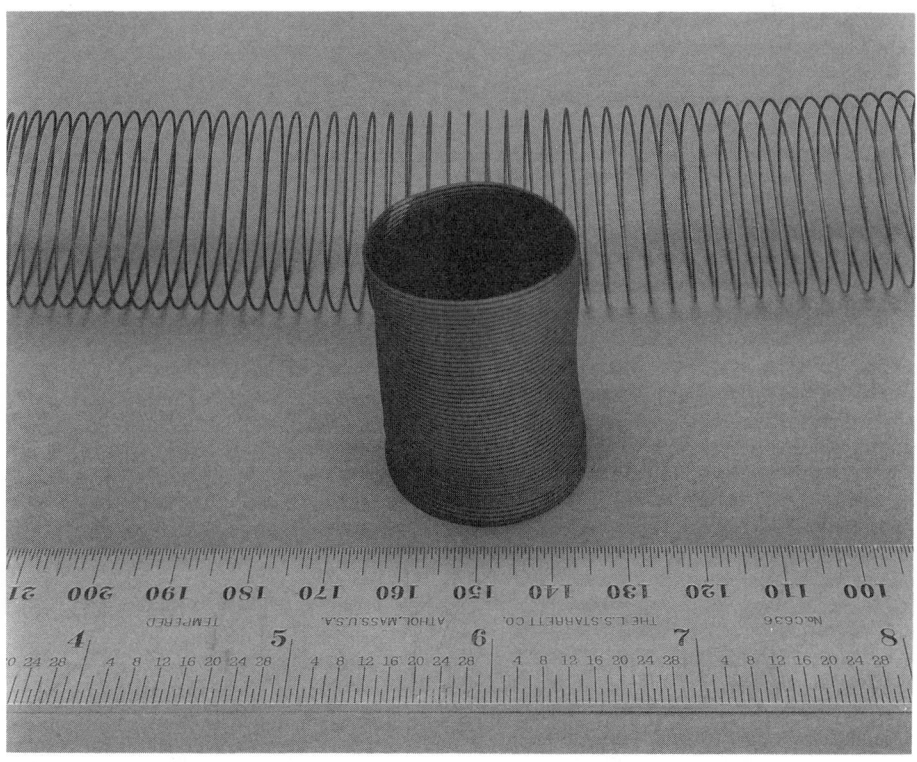

Fig 4 YBCO coils after heat treatment. Precipitates of Y_2BaCuO_5 are dispersed in a matrix of $YBa_2Cu_3O_x$.

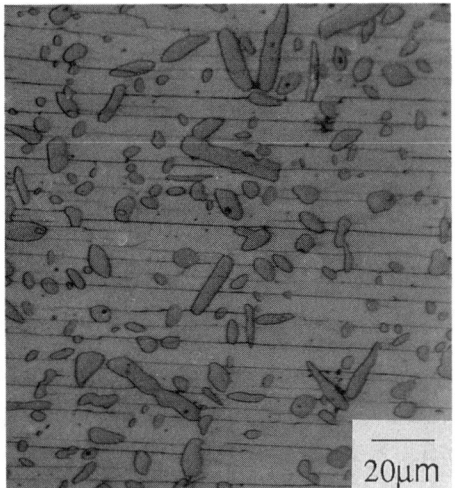

Fig 5 Scanning electron micrograph of zone-melted YBCO

has been suggested that the 211 precipitates that result from this process act as magnetic flux-pinning sites (Ref 18).

Many schemes have been attempted for heating the specimens above the melting point and for providing a larger thermal gradient at the liquid/solid interface. The zone-melt process consists of heating a section of the bulk specimen. A gradient is maintained across the liquid/solid interface by a localized heat source. The molten zone is traversed across the specimen and highly oriented crystals are left in its wake. The molten zone must travel some distance to achieve steady-state composition in the melt, which is a condition where compositions of the feed rod and the quenched material are the same.

Lasers provide a well-defined heat source and have been used successfully on monolithic and thick-film BSCCO specimens. Excellent texture and electrical properties have been achieved by this process (Ref 19, 20). Highly textured YBCO specimens have been fabricated by creating a large thermal gradient in a furnace and slowly moving the specimen through the furnace. Large critical current densities have been measured for specimens made by this technique (Ref 21).

At present, melt-textured specimens are limited to short lengths, which restricts immediate application of this process. Both YBCO and BSCCO melt peritectically, and directional solidification of these compounds is governed by the thermal gradient across the liquid/solid interface and the interface velocity (Ref 22). Specimens are produced at rates of less than 10 mm/h (0.4 in/h), and the amount of nonsuperconducting phase will increase as the speed of the solidification front increases.

Composite Structures

Silver has been incorporated into YBCO as a means of improving mechanical properties (Ref 23). With a silver content of 20 vol%, flexural strength doubled without degradation in J_c. Stress relaxation and crack pinning have been suggested as possible mechanisms for mechanical property improvement. In addition, silver also provides a useful substrate material because of its chemical stability and high thermal expansion coefficient. Highly textured BSCCO and YBCO thick films have been deposited on silver substrates (Ref 24).

For many applications, the high-T_c superconductor must be bonded to a normal conductor and an insulator. If the superconductor should quench, the current will shunt across the normal conductor. The generated thermal energy can also be removed from the superconductor by a good thermal conductor to prevent quenching. High-power applications will also require that the superconductor be in contact with a thermal conductor such as a metal. Insulators are necessary in coils for current isolation if the superconductor should quench. Current will be directed along the normal conducting path and not short across the windings.

The prototype multilayer coil shown in Fig 6 is fabricated by laminating $YBa_2Cu_3O_{7-x}$, Y_2BaCuO_5, and silver layers. Y_2BaCuO_5 can be used as the insulator because it has an acceptably high resistivity (1000 $\Omega \cdot$ cm) and is chemically compatible with YBCO during firing (Ref 25).

Long lengths of composite wire can be fabricated by powder-in-tube processing. Superconducting powder is loaded in a silver tube and the assembly is compacted by swaging, drawing, or rolling. The silver sheath provides a barrier that protects the superconductor from damage and environmental degradation. The reactivity of high-T_c superconductors with moisture is well documented (Ref 26).

Mechanical deformation has been used to align platelike particles of YBCO and BSCCO. The *a-b* crystallographic plane is oriented perpendicularly to the pressing direction. A general trend is that critical current density as wire thickness is reduced. Pole figure and neutron diffraction data on YBCO/silver composites suggest that increased J_c is related to degree of crystallographic orientation (Ref 27, 28).

High J_c values (1.5×10^4 A/cm^2) in magnetic fields up to 26 T have been reported by Heine *et al.* on BSCCO-silver composites (Ref 29). The best specimens have a high degree of texture, which is obtained by partial melting after mechanical deformation. Long hold times (100 h) at 840 °C (1545 °F) are required after melting to transform the melted material to the superconducting $Bi_2Sr_2CaCu_2O_x$ phase. The properties of these powder-in-tube specimens made from BSCCO have surpassed those of conventional superconductors in high magnetic fields at 4.2 K in small specimens. Longer specimens (280 mm, or 11 in.) made recently have values of 1.0×10^4 A/cm^2 at 4.2 K and 10 T (Ref 30).

Applications

Potential uses for high-temperature superconductor components include power generation and storage, medical diagnostics, transportation, large magnets, and microelectronics (Ref 31, 32). Motors, generators, and transmission lines with superconducting components will be more efficient. Generators and motors will require operating J_c values of 10^4 to 10^5 A/cm^2 in magnetic fields of 2 to 3 T. Iron cores in conventional motors and generators are currently needed in order to contain high magnetic flux densities and are a source of electromagnetic loss. Omission of the core will reduce weight and increase network stability under fault conditions.

Large magnetic fields produced with superconducting windings are needed for superconducting magnetic energy storage (SMES) units, magnetically levitated (maglev) trains, and magnetic resonance imaging devices (MRI). Conventional superconductors such as niobium-titanium alloys are presently used for these devices.

Prototype devices made from ceramic high-temperature superconductors have been fabricated. An electrical connector for a SMES was fabricated to transport 2000 A of current from a copper conductor operated at or below 77 K to a load in liquid helium at 4.2 K (Ref 33). Because the thermal conductivity of YBCO is lower than that of copper, less helium will be lost if a high-T_c superconducting connector is used in place of a copper bus bar extended directly into liquid helium. The reduced lower loss due to the zero resistance of YBCO should also minimize loss of liquid helium.

Superconductors have a finite resistance for alternating electric current, but this resistance may be lower than that of metallic conductors. Microelectronics and other high-frequency applications include narrow-band filters, stabilized oscillators, and directive antennas. Applications of bulk high-temperature superconductors are presently limited to frequencies below 10 GHz and low incident-power levels (Ref 34).

Electromagnetic shielding materials for biomedical applications may be an early application for high-T_c materials. It has been proposed that high-temperature superconductors are useful for isolating sensitive equipment from extraneous magnetic fields (Ref 35). Biomagnetic response analysis in humans requires the measurement of very small magnetic fields. Large shields must be made to isolate the patient from magnetic fields produced by outside sources. Bulk materials can be fabricated in large sections and may provide the necessary shielding required for this application.

ACKNOWLEDGMENT

The photograph (Fig 5) supplied by Dr. P. McGinn of Notre Dame University is greatly

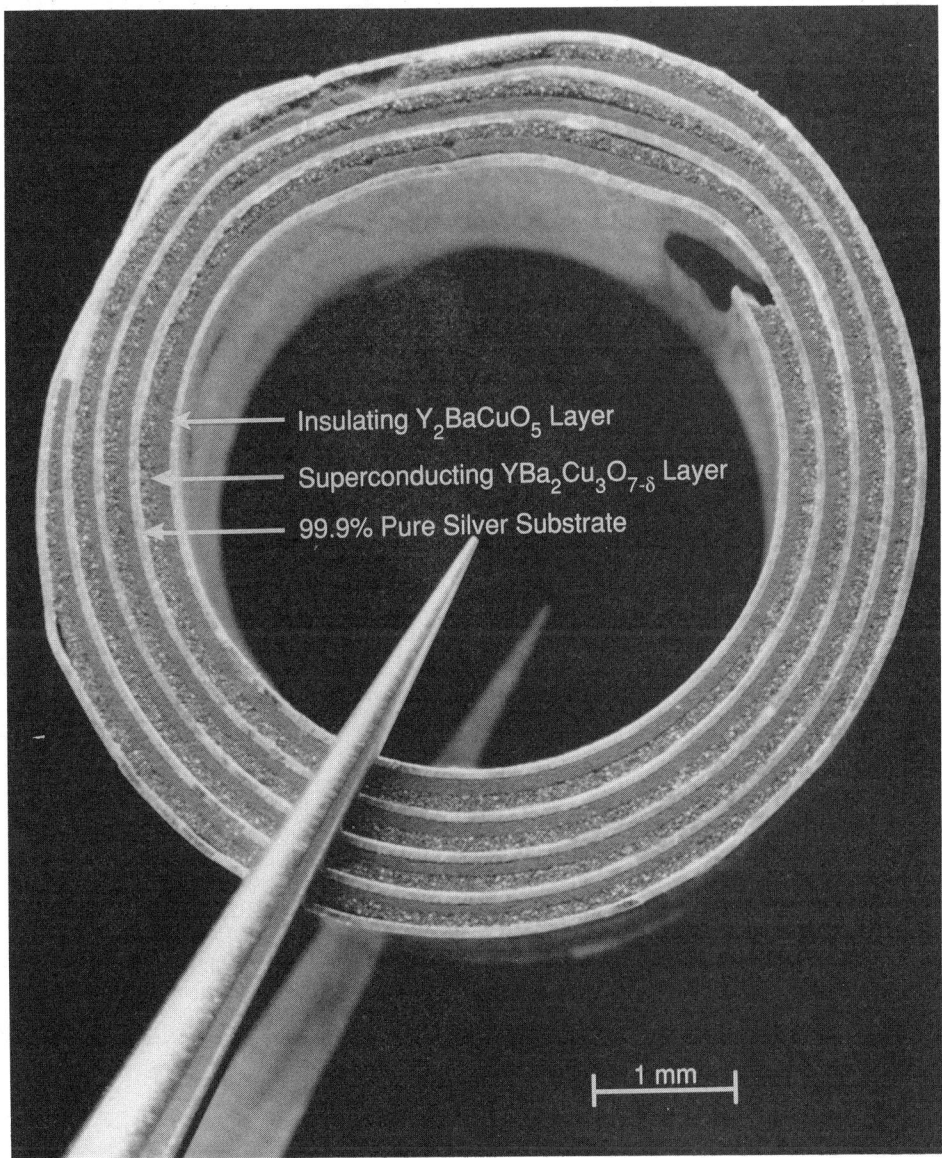

Fig 6 Prototype multilayer composite coil

appreciated (Ref 5). This work was supported by the U.S. Department of Energy (DOE), Conservation and Renewable Energy, as part of a DOE program to develop electric power technology, under Contract W-31–109-Eng-38.

REFERENCES

1. C. Kittel, Chapt. 12, *Introduction to Solid State Physics*, 5th ed., John Wiley & Sons, 1976

2. J.G. Bednorz and K.A. Muller, Possible High-T_c Superconductivity in the Ba-La-Cu-O System, *Z. Phys. B*, Vol 64, 1986, p 189–193

3. M.K. Wu, J.R. Ashburn, C.J. Torng, P.H. Hor, R.L. Meng, L. Gao, Z.J. Huang, Y.Q. Wang, and C.W. Chu, Superconductivity at 93 K in a New Mixed-Phase Y-Ba-Cu-O Compound System at Ambient Pressure, *Phys. Rev. Lett.*, Vol 58, 1987, p 908–910

4. H. Maeda, Y. Tanaka, M. Fukutomi, and T. Asano, A New High-T_c Oxide Superconductor Without a Rare Earth Element, *Jpn. J. Appl. Phys.*, Vol 27, 1988, p L209–210

5. Z.Z. Sheng and A.M. Herman, Superconductivity in the Rare-Earth-Free Tl-Ba-Cu-O System Above Liquid Nitrogen Temperature, *Nature*, Vol 332, 1988, p 55–58

6. R.L. Peterson and J.W. Ekin, Josephson-Junction Model of Critical Current in Granular YBa$_2$Cu$_3$O$_{7-\delta}$ Superconductors, *Phys. Rev. B.*, Vol 37, 1988, p 9848–9851

7. S.E. Babcock and D.C. Larbalestier, Evidence for Local Composition Variations within YBa$_2$Cu$_3$O$_{7-\delta}$ Grain Boundaries, *Appl. Phys. Lett.*, Vol 55, 1989, p 393–395

8. T. Aslege and K. Keefer, Liquidus Relations in Y-Ba-Cu Oxides, *J. Mater. Res.*, Vol 3, 1988, p 1279–1291

9. D. Dimos, P. Chaudhari, J. Mannhart, and F.K. LeGoues, Orientation Dependence of Grain-Boundary Critical Currents in YBa$_2$Cu$_3$O$_{7-\delta}$ Bicrystals, *Phys. Rev. Lett.*, Vol 61, 1988, p 219–222

10. Y. Gao, Y. Li, K.L. Merkle, J.N. Mundy, C. Zhang, U. Balachandran, and R.B. Poeppel, J_c Degradation of YBa$_2$Cu$_3$O$_{7-x}$ Superconductors Sintered in Co$_2$/O$_2$, *Mater. Lett.*, Vol 9, 1990, p 347–352

11. U. Balachandran, R.B. Poeppel, J.E. Emerson, S.A. Johnson, M.T. Lanagan, C.A. Youngdahl, D. Shi, K.C. Goretta, and E.G. Eror, "Synthesis of Phase-Pure Orthorhombic YBa$_2$Cu$_3$O$_{7-x}$ under Low-Oxygen Partial Pressure, *Mater. Lett.*, Vol 8, 1989, p 454–456

12. P. Meuffels, R. Naeven, and H. Wenzl, Pressure-Composition Isotherms for the Oxygen Solution in YBa$_2$Cu$_3$O$_{6+x}$, *Physica C*, Vol 161, 1989, p 539–548

13. G.S. Grader, P.K. Gallagher, and D.A. Fleming, Effect of Starting Particle Size and Vacuum Processing on YBa$_2$Cu$_3$O$_x$ Phase Formation, *Chem. Mater.*, Vol 1, 1989, p 665–668

14. N.C. Chen, D. Shi, and K.C. Goretta, Influence of Oxygen Concentration on Processing YBa$_2$Cu$_3$O$_{7-x}$, *J. Appl. Phys.*, Vol 66, 1989, p 2485–2488

15. R. Bormann and J. Nolting, Stability Limits of the Perovskite Structure in the Y-Ba-Cu-O System, *Appl. Phys. Lett.*, Vol 54, 1989, p 2148–2150

16. S. Jin, T.H. Teifel, R.C. Sherwood, R.B. van Dover, M.E. Davis, G.W. Kammlot, and R.A. Fastnacht, Melt-Textured Growth of Polycrystalline YBa$_2$Cu$_3$O$_{7-\delta}$ with High Transport J_c at 77 K, *Phys. Rev.*, Vol B37, 1988, p 7850–7853

17. K. Salama, V. Selvanickam, L. Gao, and K. Sun, High Current Density in Bulk YBa$_2$Cu$_3$O$_x$ Superconductor, *Appl. Phys. Lett.*, Vol 54, 1989, p 2352–2354

18. M. Murakami, M. Morita, K. Doi, and K. Miyamoto, A New Process with the Promise of High J_c in Oxide Superconductors, *Jpn. J. Appl. Phys.*, Vol 28, 1989, p 1189–1194

19. M.J. Cima, X.P. Jiang, H.M. Chow, J.S. Haggerty, M.C. Flemings, H.D. Brody, R.A. Laudise, and D.W. Johnson, Influence of Growth Parameters on the Microstructure of Directionally Solidified Bi$_2$Sr$_2$CaCu$_2$O$_y$, *J. Mater. Res.*, Vol 5, 1990, p 1834–1849

20. M. Levinson, S.S.P. Shah, and D.Y. Wang, Laser Zone-Melted Bi-Sr-Ca-O Thick Films, *Appl. Phys. Lett.*, Vol 55, 1989, p 1683–1685

21. P. McGinn, W. Chen, N. Zhu, M. Lan-

agan, and U. Balachandran, Microstructure and Critical Current Density of Zone-Melt Textured $YBa_2Cu_3O_{7+x}$, *Appl. Phys. Lett.*, Vol 57, 1990, p 1455–1457

22. M.C. Flemings, Chapt. 3, 4, *Solidification Processing*, McGraw-Hill, 1974

23. J.P. Singh, H.J. Leu, R.B. Poeppel, E. van Voorhees, G.T. Goudey, K. Winsley, and D. Shi, Effect of Silver and Silver Oxide Additions on the Mechanical and Superconducting Properties of $YBa_2Cu_3O_{7-x}$ Superconductors, *J. Appl. Phys.*, Vol 66, 1989, p 3154–3159

24. C.L. Bohn, J.R. Delayen, U. Balachandran, and M.T. Lanagan, Radio Frequency Surface Resistance of Large-Area Bi-Sr-Ca-Cu-O Thick Films on Ag Plates, *Appl. Phys. Lett.*, Vol 55, 1989, p 304–306

25. S.E. Dorris, M.T. Lanagan, D.M. Moffatt, H.J. Leu, C.A. Youngdahl, U. Balachandran, A. Cazzato, D.E. Bloomberg, and K.C. Goretta, Y_2BaCuO_5 as a Substrate for $YBa_2Cu_3O_x$, *Jpn. J. Appl. Phys.*, Vol 28, 1989, p L1415–L1416

26. N.P. Bansal and A.L. Sandkuhl, Chemical Durability of High Temperature Superconductor $YBa_2Cu_3O_{7-x}$ in Aqueous Environments, *Appl. Phys. Lett.*, Vol 52, 1988, p 323–325

27. K. Osamura, T. Takayama, and S. Ochiai, Effect of Cold-Working on the Critical Current Density of Ag-Sheathed $YBa_2YCu_3O_{6+x}$ Tapes, *Supercond. Sci. Technol.*, Vol 2, 1989, p 111–114

28. M. Okada, A. Okayama, T. Matsumoto, K. Aihara, S. Matsuda, K. Ozawa, Y. Morii, and S. Funahashi, Neutron Diffraction Study on Preferred Orientation of Ag-Sheathed Y-Ba-Cu-O Superconductor Tape with $J_c = 1000$–3000 A/cm^2, *Jpn. J. Appl. Phys.*, Vol 27, 1988, p L1715–1717

29. J. Tenbrink, M. Wilhelm, K. Heine, and H. Krauth, "Development of High-T_c Superconductor Wires for Magnet Applications," Proceedings of Appl. Superconductivity Conference, Snowmass, CO, 24–28 Sept 1990

30. K. Heine, J. Tenbrink, and M. Thoner, High-Field Critical Current Densities in $Bi_2Sr_2CaCu_2O_{8+x}$/Ag Wires, *Appl. Phys. Lett.*, Vol 55, 1989, p 2441–2443

31. D.J. Scalapino, D.R. Clarke, J. Clarke, R.E. Schwall, A.F. Clark, and D.F. Finnemore, New Research Opportunities in Superconductivity, *Cryogenics*, Vol 28, 1988, p 711–723

32. A.M. Wolsky, R.F. Giese, and E.J. Daniels, The New Superconductors: Prospects for Applications, *Sci. Am.*, Vol 260 (No. 2), Feb 1989, p 60–69

33. J.R. Hull, High Temperature Superconducting Current Leads for Cryogenic Apparatus, *Cryogenics*, Vol 29, 1989, p 1116–1123

34. K. Shigematsu, H. Ohta, K. Hoshino, H. Takayama, O. Yagashita, S. Kamazaki, H. Takahara, and M. Aono, Magnetic Shield of High-T_c Oxide Superconductors at 77 K, *Jpn. J. Appl. Phys.*, Vol 28, 1989, p L813-L815

35. C. Zahopoulos, W.L. Kennedy, and S. Sridhar, Performance of a Fully Superconducting Microwave Cavity Made of the High-T_c Superconductor $YBa_2Cu_3O_y$, *Appl. Phys. Lett.*, Vol 52, 1988, p 2168–2170

Magnetic Ceramics (Ferrites)

Alex Goldman, Ferrite Technology Worldwide Inc.

MAGNETIC CERAMICS are iron oxide compounds that contain an additional metallic ion. Magnetic iron oxides (Fe_3O_4 and Fe_2O_3) are usually categorized as ferrites. The Fe^{3+} ions can be replaced by any of the transition metals, such as Ni^{3+}, Mn^{3+}, Cr^{3+}, or Co^{3+}, as well as Al^{3+} and Ga^{3+}. Although magnetite, a naturally occurring ferrite, has been recognized since ancient times, its poor properties preclude its use in modern applications. Therefore, synthetic ferrites have been formulated to obtain specific magnetic properties in a wide variety of applications. Compared to their metal counterparts, many ferrite compounds are relatively new.

Rapid growth in ferrite usage was occasioned by the trend toward higher frequencies of operation in telecommunications and power electronic applications. Major applications of ferrites include magnetic cores for inductors and transformers (for spinels, up to 200 MHz; for planar hexaferrites, from 200 to 800 MHz), permanent magnets (for uniaxial hexaferrites), microwave devices (spinels, garnets), magneto-optic devices (garnets), and electromechanical transducers (spinels). A high resistivity is desired for most of these applications in order to minimize eddy currents or dielectric loss.

Properties

Ferrites exhibit some properties that are vastly different from those of magnetic metals. Oxygen ions in the ferrite structure are responsible for the high resistivity of the crystal lattice. Unfortunately, the dilution of the magnetic ion portion of the lattice volume by the presence of the large, nonmagnetic oxygen ions leads to both a lowering of magnetic moment and a decrease in the corresponding bulk property, that is, saturation magnetization. Contributing further to the decrease in saturation is the fact that the net moment in ferrites is the result of two anti-parallel spin moments on different lattice sites (see Table 1). This should be compared to the ferromagnetism of metals, where the individual moments are parallel.

In Fig 1, the induction, B, is plotted against the magnetizing current, H, in what is known as a hysteresis loop. The induction is defined as $B = H + 4\pi M_s$. Shown below the loop in

Fig 1 is the sine current wave shape that produced the loop. One current cycle produces one traversal of the loop. In soft materials, the saturation, or peak induction (B_s), is practically equivalent to $4\pi M_s$.

One difference between metals and ferrites is that ferrites have lower saturation inductions and higher resistivities, compared to those of metals (see Table 2). For permanent magnetic ferrites, the quadrant of the hysteresis loop that is important is the second quadrant, commonly called the demagnetization curve (see Fig 1). The lower moment in ferrites results in a remanent induction, B_r (the residual induction after the magnetizing field is removed) that is lower than the corresponding value for alnico metal magnets. However, the coercive force, H_c (the reverse field required to reduce the induction to zero), is higher.

Another difference between metals and ceramics is the greater importance in ferrites of microstructural effects, such as grain boundaries, porosity, and grain-size dependence. These effects tend to limit the domain wall movements, compared to relatively clean alloy structures such as Permalloy (80% Ni-Fe). Therefore, many soft ferrite materials have permeabilities of only about 2000 (although 15,000 has been obtained). These are low compared to the values for Permalloy (see Table 2). Apparently the domain walls can move easily across the grain boundaries in the Permalloy while experiencing retardation in the ceramics.

The ferrite grain-size effect is also more important in both high-permeability and high-frequency applications. The net result of both the lattice resistivity and the grain boundary resistivity allows lower losses and operation at much higher frequencies (see Tables 2 and 3). The advantage of using higher frequencies can be shown by:

$$E = 4.44\, BNAf \times 10^{-8} \text{ (sine wave)}$$

where E is the induced voltage (in volts), B is the maximum induction (in tesla), N is the number of turns, A is the cross-sectional area (in cm^2), and f is the operating frequency (in hertz).

Theoretically, if the frequency can be raised by a factor of 10^4 or higher to obtain the same induced voltage, the cross-sectional area can

be reduced by a corresponding factor. This can also reduce the length of winding per turn, lowering the winding losses as well. The advantage of the frequency increase using ferrites is much greater than the opposing effect of lower inductions in ferrites. At higher frequencies, losses that are called eddy current losses occur. These are dissipated in the form of heat. Unchecked, these losses can damage the component catastrophically as the frequency is increased. The eddy current losses are given by:

$$P_e = \frac{KB^2f^2d^2}{\rho}$$

where P_e is the eddy current loss, K is a constant depending on the sample shape, B is the maximum induction, f is the frequency (in hertz), d is the smallest dimension transverse to the flux, and ρ is the resistivity (in $\Omega \cdot$ cm).

In metals, the eddy current losses may be lowered by reducing d, which is the thickness, or gage, of the strip. At very thin gages, this rolling operation is costly, and there is a practical limit of reduction.

In ferrites, the same effect is accomplished by increasing the resistivity, as shown in Table 2. Therefore, by switching from 50 to 60 Hz in a silicon-iron transformer to 100 kHz in a ferrite core, the transformer efficiency can be raised from a 30 to 40% level to an 80 to 90% level. This leads to lower operating costs.

One limitation arising in the use of ferrites in large transformers is their low thermal conductivity (also caused by the oxide nature of ferrites), which makes it difficult to remove heat from the inside of the core. Amorphous metal alloys have somewhat higher resistivities than the crystalline metals and can be made thin without costly rolling. Therefore, these alloys may compete with ferrites at intermediate high frequencies.

In the microwave region, ferrites may be the only bulk magnetic material available. Ferrites can be made either with round loops for low losses or with square loops for recording or logic applications. They also can be made either magnetically hard for permanent magnet use or magnetically soft for transformer or inductor applications. At microwave frequencies, the unique interaction of electromagnetic radiation and the pre-

Table 1   Metal ion distribution in ferrites

Type of ferrite	Tetrahedral sites (A)		Octahedral sites (B)		Resultant moment
	Ions	Moments	Ions	Moments	
Zinc ferrite ($ZnFe_2O_4$), normal spinel	Zn^{2+}	...	Fe^{2+}	↑ ↓	0
Nickel ferrite ($NiFe_2O_4$), inverse spinel	Fe^{3+}	↓	Fe^{3+}	↑	↑
			Ni^{2+}	↑	
Nickel-zinc	Fe^{3+}	↓	Fe^{3+}	↑	↑
$Ni_{0.5}Zn_{0.5}Fe_2O_4$	Zn^{2+}		Ni^{2+}	↑	↑

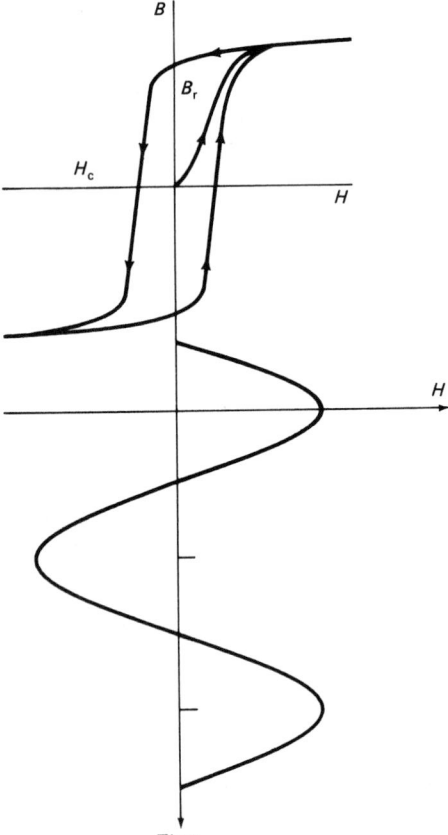

Fig 1 Magnetization curve and hysteresis loop of a ferrite showing how a current sine wave cycle produces the alternating field, *H*, that produces one traversal of the hysteresis loop

cessing electron spin array of a biased ferrite component allow important functions to be performed on the microwave beam.

In common with most other ceramic materials, ferrites have some mechanical and ther-

mal limitations when compared to metallic magnetic materials. Ultimate tensile strength values are low, ranging from 20 to 60 MPa (3 to 9 ksi), although ultimate compressive strength is high, ranging from 200 to 700 MPa (30 to 102 ksi). Young's modulus values range from 80 to 150 MPa (12 to 22 ksi). Hardness varies between 600 and 900 on the Vickers scale; the upper values are necessary for wear resistance in magnetic head applications. Specific heat values range from 700 to 800 J/kg · K (165 to 190 Btu/lb · °F), whereas thermal conductivity, as mentioned earlier, is low, ranging from 3.5 to 5 J/mm/s/K.

Ferrite Compositions

Magnetic ceramic structures can be classified into three categories of chemical composition and crystal structure.

A spinel has the general formula of either $MO \cdot Fe_2O_3$ or MFe_2O_4, where *M* can be a divalent ion such as Mn, Ni, Co, Zn, Fe, Cu, Mg, or the Fe^{3+}-compensated monovalent Li ion. It can also be the defect structure γ-Fe_2O_3, in which the divalent ions are substituted by a combination of cation vacancies and Fe^{3+} ions. If more than 50 mol% Fe_2O_3 is used in the initial mix, the surplus iron usually appears as divalent ferrous ion, thus producing some magnetite.

These ferrites crystallize in the cubic spinel structure with eight formula units per unit cell. The metal ions are coordinated by either four oxygen ions (tetrahedral, or A, sites) or by six oxygen ions (octahedral, or B, sites). There are 8 tetrahedrally coordinated metal ions and 16 octahedrally coordinated metal ions per unit cell.

The moments on the tetrahedral sites are oriented antiparallel (antiferromagnetically) to the octahedral sites. The net moment is the vector sum of the combined moments. Var-

ious ions have different site preferences, and skillful material design can use these site preferences to good advantage. For example, the nonmagnetic zinc ion is often substituted for part of the divalent ion content to disproportionate the Fe^{3+} ions on the two different sites and thus create large moments.

In addition to the major elements, additions of CaO, TiO_2, V_2O_5, SnO_2, and Al_2O_3 may be made to improve the properties. As an important minor impurity, SiO_2 often has its content controlled.

Magnetoplumbite structures have the general chemical formula of either $MO \cdot 6Fe_2O_3$ or $MFe_{12}O_{19}$, where *M* is usually Ba, Sr, or Pb. The crystal structure is hexagonal with a unique "*c*" axis. It can also be visualized as the combination of three rhombohedral unit cell structures joined at a common point to form a structure with a hexagonal cross section. The two radial axes are perpendicular to the *c* axis. More complex structures in this group can be constructed by layering magnetoplumbite and spinel layers in various proportions (see Table 4). In the basic magnetoplumbite structure, the *c* axis is the easy direction of magnetization. However, in some of the more complex substituted structures, the *c* axis becomes the difficult direction of magnetization; the easy direction is in the plane perpendicular to the *c* axis, with little particular preference of the direction in the plane. This type of material is called "Ferroxplana."

Magnetic garnet structures, sometimes called the "rare earth garnets," have the general formula $3M_2O_3 \cdot 5Fe_2O_3$ or $M_3Fe_5O_{12}$, where *M* is either yttrium or a rare earth ion. Here, all the metal ions are trivalent in the simple unsubstituted garnets. Various substitutions of Al, Ca, V, Bi, or other ions can be used to modify the properties.

Magnetic garnets crystallize in the 12-sided dodecahedral structure similar to the mineral

Table 3   Permeabilities and frequency ranges of ferrites

Ferrite	Permeability	Frequency range, MHz
Mn-Zn	750	0.01–2
Mn-Zn	2,000	0.01–0.5
Mn-Zn	5,000	0.01–0.2
Mn-Zn	10,000	0.01–0.1
Ni-Zn	40	10–50
Ni-Zn	100	2–30
Ni-Zn	800	1.5–5

Table 2   Saturation flux densities, resistivities, and permeabilities of magnetic materials

Material	Saturation flux density		Resistivity, $\Omega \cdot cm$	Permeability
	T	gauss		
Iron (unpurified)	2.15	21,500	10×10^{-6}	150
Silicon-iron (oriented)	2.00	20,000	50×10^{-6}	1800
80% Nickel-iron	0.80	8,000	55×10^{-6}	100,000
Mn-Zn ferrite	0.40–0.50	4,000–5,000	$10–10^3$	750–15,000
Ni-Zn ferrite	0.30–0.40	3,000–4,000	$10^5–10^{10}$	10–1000
Yttrium-iron garnet	0.18	1,750	$10^{10}–10^{12}$	...

Table 4   Designation and composition of several hexagonal ferrites

Ferrite designation	Chemical composition
M (Magnetoplumbite)	$BaO \cdot 6Fe_2O_3$
W	$BaO \cdot 2MeO \cdot 8Fe_2O_3$
S (spinel)	$MeO \cdot Fe_2O_3$
Z	$3BaO \cdot 2MeO \cdot 12Fe_2O_3$
Y	$2BaO \cdot 2MeO \cdot 6Fe_2O_3$

Note: W, Z, and Y are common designations for hexagonal ferrite types.

garnet. In addition to the tetrahedral and octahedral sites found in spinels, there is one more site, called the dodecahedral, or C, site. The spin moment is due to the disproportionation of the Fe^{3+} ions, but there may also be an orbital moment, that is not found in ferrites, involving transition metal magnetic ions.

Influence of Crystal Structure on Magnetic Properties

The choice of crystal structure and the magnetic ions contained in it will greatly affect the magnetic properties of ferrites and their ultimate application.

For soft ferrites, the cubic spinel is often chosen, in which there is little preference for the magnetic moment to lie in any specific crystallographic direction (low magnetocrystalline anisotropy energy). This is one requirement for increasing permeability. Optimization of the chemistry (particularly iron content) produces a low magnetostriction, which is another requirement for increasing permeability. On the other hand, the hexagonal ferrites have a preferred uniaxial crystallographic direction (high magnetocrystalline anisotropy energy) and therefore retain the magnetization in that direction, making them suitable as permanent magnets.

By choosing the proper divalent ions, the magnetic moment of a spinel can be increased, thus leading to the higher saturation magnetizations that are needed for high-power applications. Thus, crystal structure can be tailored for desired properties and applications.

Processing of Ferrites

Most ferrites are prepared by conventional ceramic techniques performed in this order:

1. Blending of raw materials
2. Calcining or presintering
3. Milling
4. Granulation or spray drying
5. Pressing to shape
6. Firing
7. Tumbling or grinding to finished shape

In low-cost materials such as those used for consumer electronic applications, the calcining and milling processes may be omitted. Blending (sometimes combined with a milling action) can be performed as a wet, semiwet, or dry process. Pressing is usually performed dry; the addition of a fugitive binder aids in compaction. In the case of some permanent anisotropic magnets, the ferrite particles may be oriented by a magnetic field during pressing (either wet or dry) to obtain better properties.

In the case of the Mn-Zn ferrites, the final firing must be done with great regard to the oxygen partial pressure upon the cores, especially during the cooling portion of the firing cycle. This provision is necessary to ensure

the proper Fe^{2+} content in the lattice, which in turn affects the magnetic properties. Finally, the cores are either tumbled to remove sharp edges, in the case of toroids, or flat ground to ensure good mating surfaces, in the case of E-cores and other power shapes. In inductor cores, such as pot cores, a gap may be ground in the center leg.

Nonconventional methods of making ferrites also exist. The most promising is known as co-spray roasting, in which a solution of mixed Mn and Fe chlorides is sprayed into a large roaster to recover the HCl; by oxidative hydrolysis, a mixture of Mn and Fe oxides is formed. The zinc oxide is added and the process is finished conventionally. Other methods of noncommercial, nonconventional processing are coprecipitation, decomposition of organic precursors, freeze drying, fused salt synthesis, and sol-gel processing.

Single crystals have also been grown by the modified Bridgeman and the Czochralski methods, and by using garnets from a melt of PbO and PbF_2. Ferrite and garnet films have been made by chemical vapor deposition (CVD), liquid phase epitaxy (LPE), sputtering, vapor deposition, and spin plating.

Functions and Applications

The distinct functions for which ferrites can be used are:

- As a permanent magnet, the function of which is merely to create a magnetic field
- As a low-level transformer that converts voltages, currents, or impedances of very low signal levels, such as those encountered in telecommunications
- As a high-level or power transformer to change voltage levels at high frequencies and high drive levels
- As a low-level inductor providing the inductance in an inductive-capacitive (L-C) electronic circuit that is used to maintain the frequency in a resonant circuit
- As a high-level inductor to remove some of the high-frequency ac components in a direct current
- As a component in a microwave circuit to interact with the electromagnetic wave, either switching it into the desired channel, or permitting passage of the wave in one direction while attenuating the reflected wave in the opposite direction
- As a bistable digital storage device that can participate in binary logic operations
- As a medium on which digital or analog signals can be recorded and read
- As the gapped core to provide the alternating field to impress and read electrical signals on the recording media described above
- As an electrical-to-mechanical transducer
- As a means of directing an electron beam in both horizontal and vertical directions at high frequencies

These ferrite functions can be used in a great variety of applications.

Permanent Magnets. In addition to the common mechanical uses of a magnet to lift, separate metal, hold objects, or seal refrigerator gaskets, for example, there are many electrical and electronic applications that use magnets. Speaker magnets and microphones are two applications, but by far the largest tonnage requirement for ferrite magnets is as field segments in small motors commonly used in automobiles and cordless electrical appliances. There are nonoriented and oriented hard ferrites; the oriented type has the best properties. Table 5 compares some commercially available hard ferrites.

Telecommunications. One of the earliest uses for ferrites was as channel filters in telephony. In one current application, the ferrite is used as the inductor part of an *L-C* channel filter, which serves to separate the many telephone conversations that are sent over a common line. Stability of inductance with temperature, time, and drive level are important in ferrites used in telecommunications applications. Compositions are chosen to give the best stability. Figure 2 shows the variation of some Mn-Zn ferrites with temperature. Although still in use in the early 1990s, their applicability diminished because of competition from solid-state filters and fiber optics.

Ferrites are also used as inductors in the tone-generating circuitry of touch-tone telephones. In rural areas outside the United States, they are used as load coils in transmission lines to provide the inductance that prevents signal loss over long distances.

As transformers, ferrites are used in television and radio applications to transform voltages, currents, and impedances. They are also used as tuning slugs for television channels.

The largest use of soft ferrites, in terms of tonnage, is in the manufacture of deflection yokes that fit over the neck of a television tube. When the proper signal is sent through the windings on the yoke, the electron beam is swept across the television screen both horizontally and vertically to form the transmitted image. Another part of the horizontal sweep uses a large fly-back transformer, which stores the magnetic energy while the horizontal electron beam is made to rapidly return (fly back) to the start of the next line down on the raster.

Another low-level, or telecommunications, use of ferrites is as a wide-band transformer. In this case, it is desirable to pass a signal over a rather large frequency band, but attenuate it at frequencies above and below this band. This application often requires very high permeability ferrites.

Power Transformers and Inductors. Whereas ferrite use in the telecommunications market has declined since 1980, the market for ferrites used in high power applications has increased dramatically. The dc-

Table 5 Properties of commercial ferrite permanent magnet materials

Grade	B_r		H_c		H_{ci}		
	T	gauss	10^6 A/m	Oersted	10^6 A/m	Oersted	BH_{max}(a)
Isotropic	0.27	2,700	0.18	2,300	0.26	3,250	1.05
High B_r	0.39	3,850	0.16	2,000	0.16	2,020	3.5
High H_{ci}	0.37	3,700	0.28	3,550	0.40	4,950	3.2
Premium high BH_{max}	0.44	4,400	0.23	2,850	0.23	2,880	4.5
Alnico 5	1.28	12,800	0.05	640	...		5.5

(a) Values should be multiplied by 10^6 gauss-Oersteds.

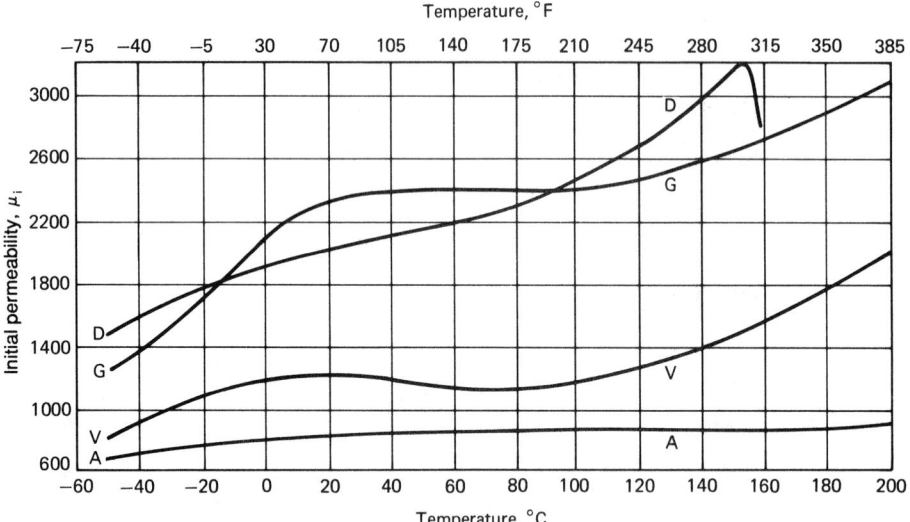

Fig 2 Initial permeability versus temperature curve for several different MnZn ferrites of varying permeabilities: A = 800, V = 1200, D = 2000, and G = 2300

to-dc converter (the magnetic heart of the switching power supply) often takes a dc (or rectified ac) signal and inverts it to a high frequency through a transistor switch. Then, a ferrite power transformer is used to convert the high-frequency voltage to a desired level with a required amount of regulation. Ferrites are also used in the power supply as input and output filters and in the switching regulator.

Memory and Recording. A magnetic hysteresis loop (see Fig 1) has two states of saturation. If the loop can be made rather square, the two stable states of remanence can be regarded as the "1" and "0" of binary logic. The first ferrite-based memories were in the form of very small toroids threaded with magnetizing wire. A digital signal could be impressed on each core and then read back. Nearly all of these toroidal memory cores have been replaced by semiconductor memories for computer central processing units. In the case of external memory storage, the cores have been supplanted by reel-to-reel tape, tape cassettes, and floppy disks, as well as hard disks. The original toroids have been replaced with very small areas of acicular particulate magnetic media having square loop properties. These are magnetized by the field in the

gap of a recording head (also often made of ferrite), and behave as binary bits, like the toroids.

Analog recording for audio tapes is another application. A form of ferrite (γ-Fe_2O_3) that has the spinel structure, but with cation vacancies instead of divalent ions, is the most common form of magnetic media. A cobalt oxide-coated form of this medium improves properties greatly, even though the cobalt oxide film is extremely thin. The media is mixed in an organic suspension and coated onto the tape or disk.

Another magnetic recording application that uses ferrites is known as magnetic bubble technology. This technology involves films that are mainly garnets in which the bubbles (cylindrical reverse domains in a magnetically biased single domain matrix) can be generated, moved, and stored in special locations. The presence or absence of a bubble in a location represents the two states of binary logic.

Microwave applications employ yet another feature of ferrites. At the high frequencies (over 10^9 Hz) of microwaves, the domain structure of ferrites breaks down. Rather than transmitting the electrical energy in wires using

conventional means of distributing voltages and currents, the energy is transmitted in the form of an electromagnetic wave contained in waveguides that can be cylindrical or rectangular. The size of the waveguide is determined by the wavelength of the microwaves being transmitted. Ferrites offer gyromagnetic properties, which involve a phenomenon in which the magnetic moments of the ferrite precess around an external magnetic field at a precessional frequency determined by the ferrite and the strength of the external dc field. When plane-polarized microwave excitation propagating down the waveguide near the precessional frequency interacts with the ferrite, energy is absorbed. The plane of polarization is rotated a certain angle (45°) in a particular direction determined by the field direction. When a wave propagating from the opposite direction interacts with the ferrite, it is rotated the same amount in the same absolute direction as the original wave. This makes the second (reflected) wave 90° out of phase with the original wave. This is a means of isolating transmitted and reflected waves using microwave Faraday rotation.

Other devices, such as a microwave circulator, can also be constructed. Here, a wave entering one of the ports is circulated and leaves at the next port, rotating clockwise (see Fig 3).

Another microwave ferrite application is radar-absorbing paint, which is used to make planes such as the Stealth bomber invisible to radar.

In magnetostrictive transducers, high-frequency electrical energy is converted to magnetic energy. A ferrite with a special composition converts the magnetic energy into mechanical energy so that the core vibrates ultrasonically by expanding and contracting. This function can be used to make ultrasonic cutting tools and cleaning baths.

Another application that uses the same principle is sonar, in which an ultrasonic burst is transmitted (below water, for example). The time elapsed before the reflected wave is detected denotes the distance of the object that reflected the sonar, such as a submarine or a school of fish.

Other applications that use ferrites include temperature or proximity sensors, xerography powders, electrodes, ferrofluids, and magnetic ink.

Future Ferrite Technology

As mentioned earlier, ferrites are one of the newest magnetic materials, offering some distinct advantages that make their future appear promising. There is a wide variety of compositions that are still untested with respect to their constituents, additives, processing variables, and shapes. In addition, the raw materials are available at a relatively low cost. This may be enhanced once ferric oxide, representing about 70 wt% of a ferrite, becomes

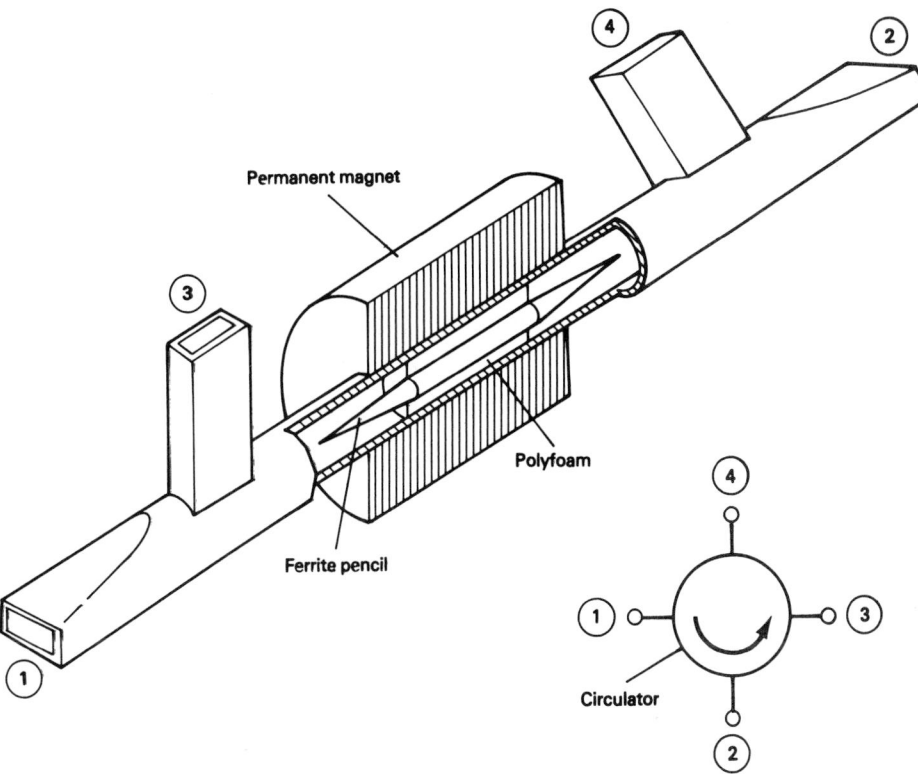

Fig 3 A Faraday rotator microwave circulator. Order of entry and exit of a wave is shown in the rotation in schematic at right. Source: Ref 1

phous metal alloys are limited mostly to toroid configurations, whereas ferrites can be molded to almost any shape.

REFERENCE

1. A.G. Fox, S.E. Miller, and M.T. Weiss, *Bell System Telephone J.*, Vol 34 (No. 5), 1955

SELECTED REFERENCES

- A. Goldman, *Modern Ferrite Technology*, Van Nostrand Reinhold, 1990
- A. Goldman, Magnetic Ceramics, *Electronic Ceramics*, L. Levinson, Ed., Marcel Dekker, 1988
- C. Heck, *Magnetic Materials and Their Applications*, Crane Russack, 1974
- E.C. Snelling, *Soft Ferrites, Properties and Applications*, 2nd ed., Butterworths, 1988
- M. Srwastava and J.J. Patni, Advances in Ferrites, *Proceedings of the Fifth International Conference*, Vol 1, Oxford and IBH Publishing, 1989
- M. Srwastava and J.J. Patni, Advances in Ferrites, *Proceedings of the Fifth International Conference*, Vol 2, Trans-Tech Publications
- F.F.Y. Wang, *Advances in Ceramics*, Vol 15, Part 1, Fourth International Conference on Ferrites, American Ceramic Society, 1985
- F.F.Y. Wang, *Advances in Ceramics*, Vol 16, Part 2, Fourth International Conference on Ferrites, American Ceramic Society, 1985

more available as a by-product of pickle liquor from steel mills, which are gradually converting to hydrochloric acid pickling for environmental reasons.

The amorphous metal alloys pose some competition to ferrite, but it must be shown that very thin gages of the FeBSi ferrite variety can be made reproducibly and economically. Another disadvantage of this amorphous material compared to ferrites is the difficulty in punching laminations or slitting strip because of their hardness. Therefore, amor-

Metric Conversion Guide

This Section is intended as a guide for expressing weights and measures in the Système International d'Unités (SI). The purpose of SI units, developed and maintained by the General Conference of Weights and Measures, is to provide a basis for world-wide standardization of units and measure. For more information on metric conversions, the reader should consult the following references:

- "Standard for Metric Practice," E 380, *Annual Book of ASTM Standards*, Vol 14.02, 1988, American Society for Testing and Materials, 1916 Race Street, Philadelphia, PA 19103
- "Metric Practice," ANSI/IEEE 268–1982, American National Standards Institute, 1430 Broadway, New York, NY 10018

- *Metric Practice Guide—Units and Conversion Factors for the Steel Industry,* 1978, American Iron and Steel Institute, 1133 15th Street NW, Suite 300, Washington, DC 20005
- *The International System of Units,* SP 330, 1986, National Bureau of Standards, Order from Superintendent of Documents, U.S. Government Printing Office, Washington, DC 20402-9325
- *Metric Editorial Guide,* 4th ed. (revised), 1985, American National Metric Council, 1010 Vermont Avenue NW, Suite 320, Washington, DC 20005-4960
- *ASME Orientation and Guide for Use of SI (Metric) Units,* ASME Guide SI 1, 9th ed., 1982, The American Society of Mechanical Engineers, 345 East 47th Street, New York, NY 10017

Base, supplementary, and derived SI units

Measure	Unit	Symbol	Measure	Unit	Symbol
Base units			Entropy	joule per kelvin	J/K
			Force	newton	N
Amount of substance	mole	mol	Frequency	hertz	Hz
Electric current	ampere	A	Heat capacity	joule per kelvin	J/K
Length	meter	m	Heat flux density	watt per square meter	W/m²
Luminous intensity	candela	cd	Illuminance	lux	lx
Mass	kilogram	kg	Inductance	henry	H
Thermodynamic temperature	kelvin	K	Irradiance	watt per square meter	W/m²
Time	second	s	Luminance	candela per square meter	cd/m²
			Luminous flux	lumen	lm
Supplementary units			Magnetic field strength	ampere per meter	A/m
			Magnetic flux	weber	Wb
Plane angle	radian	rad	Magnetic flux density	tesla	T
Solid angle	steradian	sr	Molar energy	joule per mole	J/mol
			Molar entropy	joule per mole kelvin	J/mol·K
Derived units			Molar heat capacity	joule per mole kelvin	J/mol·K
Absorbed dose	gray	Gy	Moment of force	newton meter	N·m
Acceleration	meter per second squared	m/s²	Permeability	henry per meter	H/m
Activity (of radionuclides)	becquerel	Bq	Permittivity	farad per meter	F/m
Angular acceleration	radian per second squared	rad/s²	Power, radiant flux	watt	W
Angular velocity	radian per second	rad/s	Pressure, stress	pascal	Pa
Area	square meter	m²	Quantity of electricity, electric		
Capacitance	farad	F	charge	coulomb	C
Concentration (of amount of			Radiance	watt per square meter steradian	W/m²·sr
substance)	mole per cubic meter	mol/m³	Radiant intensity	watt per steradian	W/sr
Conductance	siemens	S	Specific heat capacity	joule per kilogram kelvin	J/kg·K
Current density	ampere per square meter	A/m²	Specific energy	joule per kilogram	J/kg
Density, mass	kilogram per cubic meter	kg/m³	Specific entropy	joule per kilogram kelvin	J/kg·K
Electric charge density	coulomb per cubic meter	C/m³	Specific volume	cubic meter per kilogram	m³/kg
Electric field strength	volt per meter	V/m	Surface tension	newton per meter	N/m
Electric flux density	coulomb per square meter	C/m²	Thermal conductivity	watt per meter kelvin	W/m·K
Electric potential, potential difference,			Velocity	meter per second	m/s
electromotive force	volt	V	Viscosity, dynamic	pascal second	Pa·s
Electric resistance	ohm	Ω	Viscosity, kinematic	square meter per second	m²/s
Energy, work, quantity of heat	joule	J	Volume	cubic meter	m³
Energy density	joule per cubic meter	J/m³	Wave number	1 per meter	1/m

Conversion factors

To convert from	to	multiply by
Area		
in.^{2}	mm^2	6.451 600 E + 02
in.^{2}	cm^2	6.451 600 E + 00
in.^{2}	m^2	6.451 600 E − 04
ft^2	m^2	9.290 304 E − 02
Bending moment or torque		
lbf · in.	N · m	1.129 848 E − 01
lbf · ft	N · m	1.355 818 E + 00
kgf · m	N · m	9.806 650 E + 00
ozf · in.	N · m	7.061 552 E − 03
Bending moment or torque per unit length		
lbf · in./in.	N · m/m	4.448 222 E + 00
lbf · ft/in.	N · m/m	5.337 866 E + 01
Current density		
A/in.^{2}	A/cm^2	1.550 003 E − 01
A/in.^{2}	A/mm^2	1.550 003 E − 03
A/ft^2	A/m^2	1.076 400 E + 01
Electric field strength		
V/mil	kV/m	3.937 008 E + 01
Electricity and magnetism		
gauss	T	1.000 000 E − 04
maxwell	μWb	1.000 000 E − 02
mho	S	1.000 000 E + 00
oersted	A/m	7.957 700 E + 01
Ω · cm	Ω · m	1.000 000 E − 02
Ω · circular-mil/ft	μΩ · m	1.662 426 E − 03
Energy (impact, other)		
ft · lbf	J	1.355 818 E + 00
Btu(a)	J	1.054 350 E + 03
Btu(b)	J	1.055 056 E + 03
cal(a)	J	4.184 000 E + 00
cal(b)	J	4.186 800 E + 00
kW · h	J	3.600 000 E + 06
W · h	J	3.600 000 E + 03
Flow rate		
ft^3/h	L/min	4.719 475 E − 01
ft^3/min	L/min	2.831 000 E + 01
gal/h	L/min	6.309 020 E − 02
gal/min	L/min	3.785 412 E + 00
Force		
lbf	N	4.448 222 E + 00
kip (1000 lbf)	N	4.448 222 E + 03
tonf	kN	8.896 443 E + 00
kgf	N	9.806 650 E + 00
Force per unit length		
lbf/ft	N/m	1.459 390 E + 01
lbf/in.	N/m	1.751 268 E + 02
kip/in.	N/m	1.751 268 E + 05
Fracture toughness		
ksi $\sqrt{\text{in.}}$	MPa $\sqrt{\text{m}}$	1.098 800 E + 00
Heat content		
Btu/ft^3 (volume)	kJ/m^3	3.725 895 E + 01
Btu/lb (mass)	kJ/kg	2.326 000 E + 00
cal/g	kJ/kg	4.186 800 E + 00
Heat flow intensity		
Btu/ft^2 · h	W/m^2	3.154 591 E + 00
cal/cm^2 · h	W/m^2	1.163 000 E + 01
Heat input		
J/in.	J/m	3.937 008 E + 01
kJ/in.	kJ/m	3.937 008 E + 01

To convert from	to	multiply by
Impact strength		
ft · lbf/ft	J/m	4.448 222 E + 00
ft · lbf/ft^2	J/m^2	1.459 002 E + 01
ft · lbf/in.	J/m	5.337 866 E + 01
ft · lbf/in.^{2}	J/m^2	2.102 043 E + 03
Length		
A	nm	1.000 000 E − 01
μin.	μm	2.540 000 E − 02
mil	μm	2.540 000 E + 01
in.	mm	2.540 000 E + 01
in.	cm	2.540 000 E + 00
ft	m	3.048 000 E − 01
yd	m	9.144 000 E − 01
mile	km	1.609 300 E + 00
Length per unit mass		
in./lb	m/kg	5.599 740 E − 02
yd/lb	m/kg	2.015 907 E + 00
Mass		
oz	kg	2.834 952 E − 02
lb	kg	4.535 924 E − 01
ton (short, 2000 lb)	kg	9.071 847 E + 02
ton (short, 2000 lb)	kg × 10^3(c)	9.071 847 E − 01
ton (long, 2240 lb)	kg	1.016 047 E + 03
Mass per unit area		
oz/in.^{2}	kg/m^2	4.395 000 E + 01
oz/ft^2	kg/m^2	3.051 517 E − 01
oz/yd^2	kg/m^2	3.390 575 E − 02
lb/ft^2	kg/m^2	4.882 428 E + 00
Mass per unit length		
lb/ft	kg/m	1.488 164 E + 00
lb/in.	kg/m	1.785 797 E + 01
Mass per unit time		
lb/h	kg/s	1.259 979 E − 04
lb/min	kg/s	7.559 873 E − 03
lb/s	kg/s	4.535 924 E − 01
Mass per unit volume (includes density)		
lb/ft^3	g/cm^3	1.601 846 E − 02
lb/ft^3	kg/m^3	1.601 846 E + 00
lb/in.^{3}	g/cm^3	2.767 990 E + 01
lb/in.^{3}	kg/m^3	2.767 990 E + 04
oz/in.^{3}	kg/m^3	1.729 994 E + 03
Power		
Btu/s	W	1.055 056 E + 03
Btu/min	W	1.758 426 E + 01
Btu/h	W	2.930 711 E − 01
erg/s	W	1.000 000 E − 07
ft · lbf/s	W	1.355 818 E + 00
ft · lbf/min	W	2.259 697 E − 02
ft · lbf/h	W	3.766 161 E − 04
hp (550 ft · lbf/s)	kW	7.456 999 E − 01
hp (electric)	kW	7.460 000 E − 01
Power density		
W/in.^{2}	W/m^2	1.550 003 E + 03
Pressure (fluid)		
atm (standard)	Pa	1.013 250 E + 05
bar	Pa	1.000 000 E + 05
in. Hg (32 °F)	Pa	3.386 380 E + 03
in. Hg (60 °F)	Pa	3.376 850 E + 03
lbf/in.^{2} (psi)	Pa	6.894 757 E + 03
torr (mm Hg, 0 °C)	Pa	1.333 220 E + 02

To convert from	to	multiply by
Specific area		
ft^2/lb	m^2/kg	2.048 161 E − 01
Specific energy		
cal/g	J/g	4.186 800 E + 00
Btu/lb	kJ/kg	2.326 000 E + 00
Specific heat capacity		
Btu/lb · °F	J/kg · K	4.186 800 E + 03
cal/g · °C	J/kg · K	4.186 800 E + 03
Stress (force per unit area)		
tonf/in.^{2} (tsi)	MPa	1.378 951 E + 01
kgf/mm^2	MPa	9.806 650 E + 00
ksi	MPa	6.894 757 E + 00
lbf/in.^{2} (psi)	MPa	6.894 757 E − 03
MN/m^2	MPa	1.000 000 E + 00
Temperature		
°F	°C	5/9 · (°F − 32)
°R	K	5/9
°F	K	5/9 · (°F + 459.67)
°C	K	°C + 273.15
Temperature interval		
°F	°C	5/9
Thermal conductivity		
Btu · in./s · ft^2 · °F	W/m · K	5.192 204 E + 02
Btu/ft · h · °F	W/m · K	1.730 735 E + 00
Btu · in./h · ft^2 · °F	W/m · K	1.442 279 E − 01
cal/cm · s · °C	W/m · K	4,184 000 E + 02
Thermal expansion		
μin./in. · °C	10^{-6}/K	1.000 000 E + 00
μin./in. · °F	10^{-6}/K	1.800 000 E + 00
Velocity		
ft/h	m/s	8.466 667 E − 05
ft/min	m/s	5.080 000 E − 03
ft/s	m/s	3.048 000 E − 01
in./s	m/s	2.540 000 E − 02
km/h	m/s	2.777 778 E − 01
mph	km/h	1.609 344 E + 00
Viscosity (dynamic and kinematic)		
poise (P)	Pa · s	1.000 000 E − 01
cP	Pa · s	1.000 000 E − 03
lbf · s/in.^{2}	Pa · s	6.894 757 E + 03
ft^2/s	m^2/s	9.290 304 E − 02
ft^2/h	mm^2/s	2.580 064 E + 01
in.^{2}/s	mm^2/s	6.451 600 E + 02
Volume		
in.^{3}	m^3	1.638 706 E − 05
ft^3	m^3	2.831 685 E − 02
fluid oz	m^3	2.957 353 E − 05
gal (U.S. liquid)	m^3	3.785 412 E − 03
Volume per unit time		
ft^3/min	m^3/s	4.719 474 E − 04
ft^3/s	m^3/s	2.831 685 E − 02
in.^{3}/min	m^3/s	2.731 177 E − 07
Wavelength		
Å	nm	1.000 000 E − 01

(a) Thermochemical. (b) International Table. (c) kg × 10^3 = 1 metric ton, or 1 megagram (Mg)

SI prefixes—names and symbols

Exponential expression	Multiplication factor	Prefix	Symbol
10^{18}	1 000 000 000 000 000 000	exa	E
10^{15}	1 000 000 000 000 000	peta	P
10^{12}	1 000 000 000 000	tera	T
10^{9}	1 000 000 000	giga	G
10^{6}	1 000 000	mega	M
10^{3}	1 000	kilo	k
10^{2}	100	hecto(a)	h
10^{1}	10	deka(a)	da
10^{0}	1	BASE UNIT	
10^{-1}	0.1	deci(a)	d
10^{-2}	0.01	centi(a)	c
10^{-3}	0.001	milli	m
10^{-6}	0.000 001	micro	μ
10^{-9}	0.000 000 001	nano	n
10^{-12}	0.000 000 000 001	pico	p
10^{-15}	0.000 000 000 000 001	femto	f
10^{-18}	0.000 000 000 000 000 001	atto	a

(a) Nonpreferred. Prefixes should be selected in steps of 10^3 so that the resultant number before the prefix is between 0.1 and 1000. These prefixes should not be used for units of linear measurement, but may be used for higher order units. For example, the linear measurement, decimeter, is nonpreferred, but square decimeter is acceptable.

Abbreviations, Symbols, and Tradenames

Abbreviations and Symbols

a empirical wavelength dependent constant; thermal diffusivity of a product

a_i ion activity

A   ampere; area

A_1, A_2 constants associated with the mechanics of compaction

Å   angstrom

A area

AAS   atomic absorption spectroscopy

ac   alternating current

A-D   Anderson-Darling

AEM   analytical electron microscope

AES   arc emission spectroscopy; Auger electron spectroscopy

$A(g)$ geometrical constant

AM   active matrix

AMLCD   active matrix light-emitting diode

AMMRC   Army Materials and Mechanics Research Center

AMRAAM   advanced medium-range air-to-air missile

APS   air plasma spraying

AQL   acceptable quality level

AR   antireflection

AS   aluminum silicate

a-Si   amorphous silica

atm   atmospheres (pressure)

at.%   atomic percent

ATTAP   Advanced Turbine Technology Applications Project

AWJ   abrasive waterjet machining

AZS   glass contact refractory of alumina and zirconia crystals in an aluminosilicate glass matrix

b   Burgers vector

b empirical wavelength dependent constant

B collection of material and geometric constants; magnetic flux density; magnetic saturation

B_0 collection of constants that depend on surface energies, atomic size, atomic vibration frequencies, and system geometry

bal   balance

bcc   body-centered cubic

bct   body-centered tetragonal

BET   Brunauer-Emmett-Teller

$B(g)$ geometrical constant

BOS   basic oxygen steelmaking

BPSG   borophosphosilicate

B&W   black and white

C   coulomb; heat capacity

C capacitance of junction; proportionality constant

C_f grain solubility in liquid

C_l solubility of solute

C_p heat capacity at constant pressure; specific heat of a product

C_0 bulk impurity concentration

CAC   controlled atmosphere chamber

CAM   computer-aided manufacturing

CARES   Ceramic Analysis and Reliability Evaluation of Structures

CAS   chemical abstract system

CBN   cubic boron nitride

cd   candela

CEC   cation exchange capacity

$C(g)$ geometrical constant

CG   ceramic grade

CIE   Commission Internationale de l'Éclairage

CIP   cold isostatic pressing

cm   centimeter

CMC   ceramic matrix composite

CNB   Chevron notch bend

CNC   computer numerical control

C_4P tetracalcium phosphate ($Ca_4P_2O_9$)

CPG   controlled pore glass

CRT   cathode ray tube

CS   combustion synthesis

cSt   centiStokes

CT   computed tomography

CTE   coefficient of thermal expansion

CTP   climbing temperature program

CTV   color television

CVD   chemical vapor deposition

CVI   chemical vapor infiltration (impregnation)

CUPE   Cranfield Unit Precision Engineering

d initial particulate size; interplanar spacing; powder size diameter

D diffusion coefficient; lamella diameter; particle diameter

D_b grain boundary diffusion constant of a solute; grain boundary diffusivity

D_b^* boundary diffusion coefficient

D_f fill density of powder; relative final density

D_g compact density at applied pressure; green density

D_G vapor diffusion coefficient

D_L lattice diffusion coefficient

D_s surface diffusivity

D_v volume diffusivity

D_0 relative green density

dB   decibel

dBa   adjusted decibel

dc   direct current

DCB   double-cantilever beam

DCCA   drying control chemical additive

DCMA   Dry Color Manufacturers' Association

DCP   direct current plasma

dhcp   double hexagonal close-packed

Di   didymium (mixture of the rare earth elements praseodymium and neodymium)

diam   diameter

$d\ell/dt$ average rate of grain growth

dn/dT relative thermal coefficient of the refractive index

dn_g/dT absolute thermal coefficient of the refractive index

$d\mu/dx$ chemical potential gradient per ion across an interface

D.S.   driving stress

DSC   differential scanning calorimetry

DSR   differential surface refractometer

DT   double torsion

DTA   differential thermal analysis

DV   double vision

e natural log base, 2.71828; charge of an electron

E cell emf; elastic modulus; electric field strength; maximum thermodynamic efficiency; measured potential; modulus of elasticity; total x-ray energy

E   activation energy; applied voltage; Young's modulus

$E°$ standard potential

EAF   electric arc furnace

EDM   electrical discharge machining

EDS   energy-dispersive x-ray spectrometry

EGA   evolved gas analysis

EIA   Electronics Industries Association

EL   electroluminescent

emf   electromotive force

EPMA   electron probe microanalysis

EPROM   erasable programmable read-only memory

Eq   equation

ESCA   electron spectroscopy for chemical analysis

ESD   electrostatic discharge

ESR   electron spin resonance

esu   electrostatic units

et al.   and others

eV   electron volt

f focal length

f_c weight fraction of component element

f_g fractional sintered density

f_i fractional density at the beginning of the second stage of sintering

f_n normal load

f_{oxide} weight fraction of component oxide

f_s fractional green density; sintered fractional density

F Faraday constant; ratio of relative green density to final density

F_b average driving force for boundary motion

F_b^{max} maximum driving force for grain growth

F_b' driving force for boundary motion

F_p attachment force

fcc face-centered cubic

fct face-centered tetragonal

FDA Food and Drug Administration

FEA finite-element analysis

FGM functionally graded material

Fig figure

FIP fluorescent indicator panel

FM figure of merit

$f(n)$ function that depends on the order of reaction n

FOD foreign object damage

FPI fluorescent penetrant inspection

FRAM ferroelectric random access memory

fss ferrite solid solution

ft foot

FTIR Fourier transform infrared

g gram

G gauss

G grinding ratio; mean grain size; shear modulus

G_0 initial grain size

gal gallon

GC/MS gas chromatograph/mass spectrometer

GPa gigapascal

gr grain

GRIN gradient refractive index

h hour

h thickness of planoparallel plate

H henry

H change in height (thickness); degree of homogenization; enthalpy; hardness; magnetic field

HA hydroxylapatite

HB Brinell hardness

HCA hydroxyl-carbonate apatite

HCFS hypersonic combustion flame spray

hcp hexagonal close-packed

HDTV high-definition television

HIP hot isostatic pressing

HK Knoop hardness

hp horsepower

HPCS high-pressure self-combustion sintering

HPLC high-pressure liquid chromatography

HPR high-penetration resistant

HPSN hot-pressed silicon nitride

HR Rockwell hardness (requires scale designation, such as HRC for Rockwell C hardness)

H_2/N_2 forming gas

HV Vickers hardness

HVR high volume resistivity

Hz hertz

i current (measure of number of electrons)

I current; emergent intensity

I_t intensity transmitted through a planoparallel plate

I_0 incident intensity

IC integrated circuit

ICF inertial confinement fusion

ICI International Commission on Illumination

I.C.I. Illuminant C (the common or standard light source under average daylight conditions)

ICP inductively coupled plasma

IF indentation crack length/fracture

in. inch

IPS inert plasma spraying

IPTS International Practical Temperature Scale

IR infrared

IS indentation strength

ISA Instrument Society of America

ISO International Organization for Standardization

ISS ion scattering spectroscopy

IT isothermal technique

ITE irreversible dilatometry

ITO indium-doped tin oxide

ITS International Temperature Scale

J joule

JIS Japanese Industrial Standard

k karat

k Boltzmann constant; sum of the absorption factors for coloring oxides; thermal conductivity; wave number

k_0 preexponential constant

k_1, k_2 constants associated with the mechanics of compaction

K Kelvin

K empirical constant; interface reaction; mean integrated thermal conductivity; thermally activated parameter

K_{Ic} plane-strain fracture toughness

K_{ISCC} threshold stress intensity to produce stress-corrosion cracking

K_p mean integrated thermal conductivity for particulates

K_α continuous spectral radiation

K_β spectral radiation

kg kilogram

km kilometer

kN kilonewton

kPa kilopascal

K-S Kolmogorov-Smirnov

ksi kips (1000 lbf) per square inch

kV kilovolt

kW kilowatt

$k(\lambda)$ absorption coefficient

l mean free path

L combustion or plasma flame length; load; transmission loss in dB/(unit length)

L_0 initial length

LAS Li_2O-Al_2O_3-SiO_2 system

lb pound

lbf pound force

LBM laser beam machining

LCD liquid crystal display

L/D length-to-diameter ratio

LED light-emitting diode

LEFM linear elastic fracture mechanics

LFM liquid-film migration

LLNR Lawrence Livermore National Laboratory

ln natural logarithm (base e)

log common logarithm (base 10)

LPS liquid-phase sintering

LTCC low-temperature co-fired multilayer ceramic

LTI low-temperature isotropic

LVR low-volume resistivity

m meter

m compaction constant dependent on packing and deformability of powders; ion mass; Weibull modulus

M molecular weight

M_b average grain boundary mobility

M_p particle mobility

M_0 mobility of an ion crossing a boundary

mA milliampere

MAS MgO-Al_2O_3-SiO_2 system

MAS magnesium-aluminosilicate

MBI methyl blue index

MCP microchannel plate

MeV megaelectronvolt

mg milligram

Mg megagram (metric tonne, or kg $\times 10^3$)

MIM metal-insulator-metal

min minimum; minute

mL milliliter

MIT Massachusetts Institute of Technology

MLC multilayer ceramic capacitor

MLE maximum likelihood estimator

mm millimeter

MMC metal-matrix composite; monolithic multicomponent ceramic

MOD metallographic deposition

MOE modulus of elasticity

MOR modulus of resilience; modulus of rupture

mPa millipascal

MPa megapascal

MPCD minimum perceptible color difference

mpg miles per gallon

mph miles per hour

mR/h milliroentgens per hour

MRR material removal rate

MRR' unit-width material removal rate

ms millisecond

MS megasiemens

MSDS material safety data sheet

mT millitesla

mV millivolt

MV megavolt

n growth exponent

n integral number; number of equivalents transferred; order of reaction

n_a refractive index of air

n_g refractive index of glass against vacuum

n_0 index for low intensity; index in the axis

n_2 nonlinear refractive index

N newton

N available sites or activated atoms; density of particles at the grain boundary

N_K Knudsen number
N_0 total atoms
NASA   National Aeronautics and Space Administration
NBS   National Bureau of Standards
NIOH/MSHA   National Institute of Occupational Safety and Health/ Mine Safety Health Association
NIST   National Institute of Standards and Technology
nm   nanometer
NMR   nuclear magnetic resonance
No.   number
$n(r)$ radially changing refractive index
NRCA   National Roofing Contractors Association
NTC   negative temperature coefficient
$n(\lambda)$ wavelength dependent refractive index
Oe   oersted
OES   optical emission spectroscopy
ORNL   Oak Ridge National Laboratory
oz   ounce
p pressure
P power; vapor pressure
P_a applied pressure
P_C joining pressure
P_{gas} gas pressure
P_H upper punch hold-down pressure
$(P_{O_2})_{reference}$ oxygen partial pressure at the reference electrode
$(P_{O_2})_{unknown}$ oxygen partial pressure at the unknown electrode
$P_{x,y}$ relative partial dispersion
P_y apparent granule yield pressure
p   page
Pa   pascal
PA   prealloyed
PACVD   plasma-assisted chemical vapor deposition
PC   programmable controller
PCB   printed circuit board
PCS   photocorrelation spectroscopy
PDP   plasma display panels
PE-CVD   plasma-enhanced chemical vapor deposition
PFN-PFW   lean iron niobate-lead iron tungstate
pH   negative logarithm of hydrogen-ion activity
PID   proportional integral differential
PLC   programmable logic controller
PLZT   lead lanthanum zirconate titanate
P/M   powder metallurgy
PMMA   polymethyl methacrylate
PMN   lead magnesium niobate
PNN   lead nickel niobate
ppb   parts per billion
ppba   parts per billion atomic
ppm   parts per million
ppt   parts per trillion
PSG   phosphosilicate
psi   pounds per square inch
psig   gage pressure (pressure relative to ambient pressure) in pounds per square inch
PSZ   partially stabilized zirconia
PTC   positive temperature coefficient

PVA   polyvinyl alcohol
PVB   polyvinyl butyral
PVC   polyvinyl chloride
PVD   physical vapor deposition
PZN   lead zinc niobate
PZT   lead zirconium titanate
Q activation energy; heat generated in the reaction; partition coefficient for the impurity between the boundary and bulk crystal
Q' specific grinding rate
Q_L particulate heat content per unit volume of the liquid at the melting point
QC   quality control
QSA   quantitative spectrographic analysis
r particle radius; radius of curvature; reflection loss of 4% for a glass having refractive index of 1.5; reflectivity; spherical particle radius
r   radius vector in a plane normal to the axis
r_p pore radius
r_s radius of solid particle
R   roentgen
R average particle radius; gas constant; resistance
R_a surface roughness in terms of arithmetic average
R_q, R_{rms} root-mean-square roughness
R_t total roughness peak-to-valley
R_y maximum peak-to-valley roughness height
R_1, R_2 radii of curvature of lens surfaces
rad   absorbed radiation dose
RBSC   reaction-bonded silicon carbide
RBSN   reaction-bonded silicon nitride
RCF   rolling contact fatigue
RE   rare earth
Ref   reference
rem   remainder; roentgen equivalent man
rf   radio frequency
RH   relative humidity
RHM   refractory heavy mineral
RHP   reaction hot pressing
RIM   reaction injection molding
rms   root mean square
RS   Raman spectroscopy
RT   room temperature
RUM   rotary ultrasonic machining
s second
S siemens
S distance travelled
SAM   scanning Auger microscopy
SANS   small-angle neutron scattering
SCG   slow crack growth
SENB   single-edge notch beam
SEM   scanning electron microscopy
SEP   Société Europénne de Propulsion
SEPB   single-edge precracked beam
scfm   standard cubic foot per minute
sfm   surface feet per minute
SG   standard grade
SHS   self-propagating high-temperature synthesis
SI   Système International d'Unités
SIMS   secondary ion mass spectroscopy
sLm   standard liter per minute
SLR   single-lens reflex
SLT   solid logic technology

SOG   spin-on glasses
S.P.   strain point
SPC   statistical process control
SPIE   Society of Photo-Optical Instrumentation Engineers
SSS   solid-state sintering
std   standard
STN   supertwisted nematic (high-performance version of intrinsically addressed LCDs)
STOL   short time overload
Sv   sievert
t isothermal sintering time; thickness; time
t_i time corresponding to the onset of the intermediate stage
$t_i(\lambda)$ internal spectral transmittance
T   tesla
T absolute temperature; temperature; transmittance; total dispersion
T_{ad} adiabatic temperature
T_c steam outlet temperature
T_H steam inlet temperature
T_m homologous temperature; melting temperature
T_s particulate surface temperature
T_0 initial temperature
T_∞ uniform temperature
T^* value of temperature below the adiabatic temperature
TA   transferred arc
TBC   thermal barrier coating
TCP   tricalcium phosphate $[Ca_3(PO_4)_2]$
TE   reversible dilatometry
TEC   thermal expansion coefficient
TEM   transmission electron spectroscopy; transverse electromagnetic
TEOS   tetraethylorthosilicate
TFEL   thin-film electroluminescent
TFT   thin-film transistor
TG   thermogravimetry
TGA   thermogravimetric analysis
TIR   total indicator reading; total indicator runout
TM   thermomagnetometry
TMAH   tetramethylammonium hydroxide
TMOS   tetramethylorthosilicate
TPX   polymethylpentene
tsi   tons per square inch
TTZ   transformation-toughened zirconia
u velocity of the combustion front
U_∞ flow velocity
UAM   ultrasonic abrasive machining
UKAEA   United Kingdom Atomic Energy Authority
ULTI   ultralow-temperature isotropic
UNS   Unified Numbering System
USBM   United States Bureau of Mines
USM   ultrasonic machining
UTS   ultimate tensile strength
UV   ultraviolet
v workpiece velocity
v_b boundary velocity
v_d droplet impact velocity
v_f film velocity
v_p particulate velocity
v_z reciprocal dispersive power
V   volt
V accelerating voltage; compact volume at

pressure; potential drop; velocity; volume

V_g compact volume

V_{gb} barrier voltage

V_l volume fraction of liquid

V_p volume fraction of polymer; volume fraction of pores

V_s volume fraction of solid; volume fraction of solid powder

V_0 initial volume of compact

V^* fractional volume compaction

V_∞ volume of compact at theoretical density

VCR   voltage coefficient of resistance

VCS   viscous composite

VFD   vacuum fluorescent display

VGS   viscous glass sintering

VLS   vapor feed gases-liquid catalyst-solid crystalline whisker growth

vol   volume

vol%   volume percent

VPS   vacuum plasma spraying

W   watt

W total radiation

WORM   write once read many

wt%   weight percent

WXRFS   wavelength-dispersion x-ray fluorescence spectroscopy

x axial distance; path length in a planoparallel plate

X neck diameter

XPS   x-ray photoelectron spectroscopy

XRD   x-ray powder diffraction

XRFS   x-ray fluorescence spectroscopy

XRS   x-ray spectroscopy

Y scale of microstructural segregation

z ion charge

Z atomic number; impedance

ZTA   zirconia-toughened alumina

°   angular measure; degree

°C   degree Celsius (centigrade)

°F   degree Fahrenheit

⇌   direction of reaction

÷   divided by

=   equals

≅   approximately equals

≠   not equal to

≡   identical with

>   greater than

≫   much greater than

≥   greater than or equal to

∞   infinity

∝   is proportional to; varies as

∫   integral of

<   less than

≪   much less than

≤   less than or equal to

±   maximum deviation

−   minus; negative ion charge

×   diameters (magnification); multiplied by

·   multiplied by

/   per

%   percent

+   plus; positive ion charge

√   square root of

~   approximately; similar to

∂   partial derivative

$\partial\eta/\partial t$ kinetic function

□   cation vacancies

α   criterion for delineation of combustion modes; powder fluidity index

$\bar\alpha$ isobaric coefficient of expansion

γ   surface energy; surface tension

γ_b interfacial energy

γ_s surface energy

γ_{sv}, γ_{ss}, γ_{sl}, γ_{lv} interface energies (subscripts s, l, and v represent solid, liquid, and vapor, respectively)

Γ   exponentially temperature-dependent reaction constant

δ   grain boundary width; thickness of liquid boundary

δ_b interface thickness

$\delta_f(\ell)$ film thickness

δ_I width of a zone over which impurities interact with the boundary

Δ   change in quantity; an increment; a range

ΔA_{sv}, ΔA_{ss}, ΔA_{sl}, ΔA_{lv} changes in various interfacial areas (subscripts s, l, and v represent solid, liquid, and vapor, respectively)

ΔG change in free energy

$\Delta G°$ standard free energy of formation

ΔH change in enthalpy

ΔL change in compact length from the initial dimension

ΔT temperature difference between sample and reference material

ε   strain

$\dot\epsilon$ strain rate

η   viscosity of liquid

η^* fraction reacted <1

θ   angle of incidence; geometrical constant

θ_i angle of incidence

θ_r angle of refraction

κ   boundary curvature; dielectric constant

λ   vacuum wavelength of light

λ_{min} minimum wavelength

μ   friction coefficient; hole mobility; magnetic permeability; mean viscosity of plasma; powder die-wall coefficient of friction; viscosity of particulate; x-ray absorption coefficient

μB   Bohr magneton

μin.   microinch

μm   micron (micrometer)

μs   microsecond

ν   Poisson's ratio; velocity

ν_f volume fraction

π   pi (3.141592)

ρ   mean density; relative density; theoretical density

ρ_C resistivity of graphite

ρ_g green density of a molded blend of polymer and ceramic powders

ρ_p density of polymer

ρ_s density of solid powder

σ   green tensile strength; Stefan-Boltzmann constant; strength; stress; total collision cross section of the plasma particles; ultimate tensile strength of material

σ_a applied stress

σ_f applied axial load

σ_m heat transfer coefficient required to temper glass

σ_p vapor pressure inside pore

σ_s transient stress limit

Σ   summation of

τ   shear stress

τ_f shear stress at failure (or flow)

φ   angle of internal friction; dihedral angle; porosity; power

Φ   reaction rate (kinetic function); volume concentration of impurity ions

ψ   geometrical constant

ω_c mass absorption coefficient of each component element

$\omega_{mixture}$ mass absorption coefficient of a mixture

ω_{oxide} mass absorption of the oxide

Ω   atomic volume; electrical conductivity; molecular volume of solute; ohm; volume of matter transported along with rate limiting ion

ℓ average grain size

ℓ_0 initial grain size

ℓ^+ limiting matrix grain size

ℓ^* size of the stable six-sided grain

ℓ' grain boundary

$\mathscr{F}$ Faraday constant

Greek Alphabet

A, α	alpha
B, β	beta
Γ, γ	gamma
Δ, δ	delta
E, ε	epsilon
Z, ζ	zeta
H, η	eta
Θ, θ	theta
I, ι	iota
K, κ	kappa
Λ, λ	lambda
M, μ	mu
N, ν	nu
Ξ, ξ	xi
O, o	omicron
Π, π	pi
P, ρ	rho
Σ, σ	sigma
T, τ	tau
Υ, υ	upsilon
Φ, φ	phi
X, χ	chi
Ψ, ψ	psi
Ω, ω	omega

Tradenames

Al$_2$O$_3$ Fiber FP, Ludox, and **Teflon** are registered tradenames of E.I. DuPont de Nemour & Company, Inc.

Bioglass is a registered tradename of the University of Florida

Calcitite is a registered tradename of Calcitek

Ceravital and **Palavital** are registered tradenames of E. Leitz, Inc.

CMC is a registered tradename of Aqualon Company

Corelle, Fotoform, Macor, Pyrex, Pyroceram, Vycor, 2403, 7971 ULE, 9068, 9600, and **9608** are registered tradenames of Corning Incorporated

Cusil-ABA and **Cusin-1-ABA** are registered tradenames of GTE Products Corporation

Dicor is a registered tradename of Dentsply sply

Dimox and **NX-3400** are registered tradenames of Lanxide Corporation

Eccospheres is a registered tradename of Emerson-Cuming Corporation

G999 is a registered tradename of Radio Corporation of America

HPZ is a registered tradename of Dow Corning Corporation

Lucalox is a registered tradename of General Electric Company

Methocel is a registered tradename of Dow Chemical Company

Microgrit is a registered tradename of Micro Abrasives Corporation

Nextel is a registered tradename of 3M Corporation

Nicalocoat and **Nicalon** are registered tradenames of Nippon Carbon Company

PT-28C is a registered tradename of Nippon Electric Glass

Q-32 and **V-44** are registered tradenames of GTE Valenite Corporation

Thixo is a registered tradename of Eimco Process Equipment Company

TL-28 is a registered tradename of Owens Illinois

Tyranno is a registered tradename of Ube Industries

Zerodur and **8051** are registered tradenames of Schott Glass Technologies, Inc.

5008 is a registered tradename of Asahi

9075 is a registered tradename of Samsung Corning

Index